ENCYCLOPEDIA OF
WORLD PROBLEMS AND
HUMAN POTENTIAL

Union of International Associations, Brussels

München·NewYork·London·Paris

Selected Publications of UIA

Yearbook of International Organizations
27th edition, 3 volumes, 1990/1991, ISSN 0084-3814.

    Vol.1 Organization Descriptions and Index
        27th edition, 1990/91, 1776 pages + Appendices (14). ISBN 3-598-22205-X.

    Vol.2 International Organization Participation: Country Directory of Secretariats and Membership (Geographic Volume)
        8th edition, 1990/91, 1760 pages. ISBN 3-598-22206-1.

    Vol.3 Global Action Networks : Classified Directory by Subject and Region (Subject Volume)
        8th edition, 1990/91, 1684 pages. ISBN 3-598-22203-3.

International Congress Calendar
31st edition, 1991, quarterly. ISSN 0538-6349.

Encyclopaedia of World Problems and Human Potential
3rd edition, 2 volumes, 1991. ISBN 3-598-10842-7.

    Vol. 1 World Problems

    Vol. 2 Human Potential

International Association Statutes Series
1st edition, 1988, 600 pages. ISSN 0933-2588.0. ISBN 3-598-21671-8

Who's Who in International Organizations
1st edition, 1991. ISBN 3-598-10908-3.

ENCYCLOPEDIA OF WORLD PROBLEMS AND HUMAN POTENTIAL

Volume 2: Human potential

Edited by

Union of International Associations

3rd edition

K·G·Saur München·New York·London·Paris

ENCYCLOPEDIA OF WORLD PROBLEMS AND HUMAN POTENTIAL

The following people worked on the preparation of the current edition in different capacities and for different periods of time.

Editorial Staff
Marie Aeles *(Bibliographies)*
Nancy Carfrae
Anne Degimbe
Kristof Elst
Carine Faveere
Martine Gosse
Jon Jenkins *(World Problems)*
Maureen Jenkins
Jacqueline Nebel *(Human Development)*
Tarja Ryynänen *(World Problems)*
Rosemary Staniforth
Cecile Vanden Bloock

Computer support
Elisabeth Gale
Bernhard Knutsen
Paul Montgmery
Colin Mainoo
Stewart Woung

The programme through which this publication is produced is orchestrated by Anthony Judge.

Published Jan 1991 by
K.G. Saur Verlag KG
Ortlerstrasse 8
D-8000 München 70
Federal Republic of Germany

Information collected and edited by
Union of International Associations
40 rue Washington
B-1050 Bruxelles, Belgium

Computer typeset by
Computaprint Limited
39A Bowling Green Lane
London EC1R ONE, United Kingdom

Cover design by
Tim Casswell

Printed and bounded in Federal Republic of Germany

Deutsche Bibliothek Cataloguing-in-Publication Data

Encyclopedia of world problems and human potential / ed. by Union of International Associations. - München; New York; London; Paris: Saur.

ISBN 3-598-10842-7

NE: Union of International Associations; World problems and human potential

Vol. 2. Human potential. - 3. ed. - 1991

Copyright 1991 by Union of International Associations. All rights reserved. No part of this work may be reproduced or copied in any form or by any means - graphic, electronic or mechanical, including photocopying, recording, taping, or information and retrieval systems - without written permission of the Secretary General, **Union of International Associations.**

ISSN 0304-0089 ISBN 3-598-10842-7

UAI Publication 299

Contents: Volume 1

INTRODUCTION

User guide	12
Overview	15
Content	29
Method	35
Assessment	43
Integrative insights	55

WORLD PROBLEMS

- **Section P: World Problems**

Section PA: Abstract fundamental problems	77
Section PB: Basic universal problems	109
Section PC: Cross-sectoral problems	137
Section PD: Detailed problems	235
Section PE: Emanations of other problems	425
Section PF: Exceptional problems	627
Section PG: Very specific problems	843
Section PJ: Problems under consideration	845

- **Section PX: VOLUME INDEX (for Volume 1)** — 847

- **Section PY: Bibliography** — 1085

- **Section PZ: Notes** — 1141

Contents: Volume 2

INTRODUCTION

    User guide   12
    Overview   15

HUMAN POTENTIAL

- **Section H: Human Development**

Section HH: Human development concepts	27	Section HX: Index to concept sets	315
Section HM: Modes of awareness	139	Section HY: Bibliography	335
		Section HZ: Notes	361

- **Section K: Integrative Knowledge**

Section KC: Integrative concepts	397	Section KX: Index	465
Section KD: Embodying discontinuity	433	Section KY Bibliography	473
Section KP: Patterning disagreement	457	Section KZ: Notes	497

- **Section M: Metaphors and Patterns**

Section MM: Metaphors	519	Section MX: Index	573
Section MP: Patterns of concepts	533	Section MX: Bibliography	581
Section MS: Symbols	561	Section MZ: Notes	587

- **Section T: Transformative Approaches**

Section TC: Transformative conferencing	641	Section TX: Index	677
Section TP: Transformative policy cycles	667	Section TZ: Notes	685

- **Section V: Values**

Section VC: Constructive values	717	Section VX: Index	805
Section VD: Destructive values	743	Section VY: Bibliography	817
Section VP: Value polarities	769	Section VZ: Notes	825
Section VT: Value types	801		

VOLUME INDEX

- **Section X: Index to Volume 2**   841

APPENDICES (Section Z)

    Statistics   935
    Computers   939
    Union of International Associations: Profile   949

Introduction

User guide

    Access   12
    Warning   13
    Errata   14

Overview

    Contextual challenge   15
    Existential challenge   17
    Strategic assumptions   19
    Objectives   21
    Significance   23

Content   *See Volume 1*

    General structure
    Sections and sub-sections
    Modifications, improvements and omissions

Method   *See Volume 1*

    Logistical challenge
    Procedures
    Classification policy
    Language-determined distictions
    Response to diversity

Assessment   *See Volume 1*

    International organizations as a source
    Biases
    Strengths and weaknesses
    Criticism
    Global modelling perspective
    Future possibilities
    Implications
    Processing system

Insights   *See Volume 1*

    Section interrelationship
    Comprehension of sustainable integration
    Problem perception and deception
    Incommunicability of insights
    Problem perception and level of awareness
    Phases of human development through challenging problems
    Integration of perceived problems
    Barriers to transcendent insight and social transformation
    Interrelating possible viewpoints
    Human impotence and potential
    A new global organizational order?

***   This introduction repeats information appearing in the Introduction to Volume 1 which is more extensive.

User guide: access

- **Volumes**

This Encyclopedia is divided into two volumes:

-- Volume 1: World problems
-- Volume 2: Human potential

- **Sections**

Each volume is divided into sections and sub-sections. Each section is denoted by one code letter (e.g. P= World problems; V= Human values). Each sub-section is denoted by two code letters (e.g. HH= Human development concepts; VC= Constructive values). Sections and sub-sections all appear in alphabetic order by code letter. The code letters also have some mnemonic significance. All sections and sub-sections are listed on the contents page.

- **Entries**

Each sub-section is composed of a series of entries. Each entry is numbered using the code letters of the sub-section (e.g. PE2370 = Abuse of tax havens). The entries appear in numeric order within the sub-section.

- **Volume indexes**

The easiest way to find an entry on a specific topic is by consulting the Volume Index, where names of all entries are listed together in alphabetic order by keyword. The index gives the sub-section and number where the entry is to be found. There is a separate Volume Index for Volume 1 (see Section PX) and for Volume 2 (see Section X).

- **Section indexes**

An alternative way to find an entry in Volume 2 is by using the mini-indexes located near the end of individual sections. They are more convenient for scanning the range of entries in a section. These provide an overview of entries within a sub-section. Section VX is the index for Section V, for example.

- **Explanations**

A brief introduction and commentary is provided at the beginning of each section and of each sub-section. More detailed comments are provided in the Notes at the end of each section. For example, the Notes for Section H are at HZ, the Notes for Section K are at KZ.

- **Cross-references**

Cross-references between entries are explained in the sub-section introductions where appropriate.

- **Bibliographical references**

Several sections have bibliographies. These are located near the end of each section. For example, the Bibliography for Section H is at HY, the Bibliography for Section K is at KY.

- **Classified index**

A classified index by subject (3000 categories) is provided to the World Problems section (Section P) in a companion series: *Yearbook of International Organizations* (Vol. 3). This also includes international organizations and treaties dealing with the same subject.

User guide: warning

1. Inconsistencies
The information collected in the Encyclopedia, and especially in the world problems section, is derived from a very wide range of sources. These reflect many levels of insight and expertise, as well as many cultures, ideologies, beliefs, priorities and biases. No attempt has been made to eliminate inconsistencies, although incompatible items have been treated as separate entries where appropriate. For example, both "capitalism" and "communism" are treated as world problems.

2. Juxtaposition
This Encyclopedia is deliberately organized in such a way as to juxtapose bodies of information which are normally kept apart. The hard reality of the "world problems" section is counterbalanced by various sections highlighting human values and development. Within the world problems section itself, for example, "counterarguments" are given (where such information is available) questioning or denying the facts presented in the problem description.

3. Perceptions
Wherever possible the information is compiled using extracts from documents of international bodies, whether governmental or non-governmental, formal or informal. In this sense the information may be viewed as factual. Given the different interpretations of these facts however, the information presented, especially in the case of world problems, can best be viewed as a collection of perceptions with which significant international constituencies identify strongly in advocating (or resisting) any social change. The Encyclopedia provides an overview of the world's hopes and worries, whether real or imaginary.

4. Editorial intervention
In honouring the biases active in the international community in this way, the editors have limited themselves to ensuring that the texts in the main sections, especially on world problems, make their point strongly and in as clear and concise a manner as the available material permits. In this period of imminent crisis, the editors have however accepted the need for a higher level of risk in exploring innovative possibilities. Some of the smaller sections are therefore the result of deliberate editorial experiments in gleaning and presenting information to highlight such possibilities, despite the risks of inadequacy and error.

5. Editorial bias
The basic bias of the editors is against limitation of information to reflect only a single viewpoint or paradigm, whether ideological, cultural, scientific or religious. Within any such paradigm, the information here also reflects different levels of ignorance, rather than attempting only to reflect a consensus prevailing amongst an elite group of authoritative experts (whose views may be poorly received outside their own circle). The bias is therefore to include information from some constituencies which may well be judged qualitatively inferior, misleading, irresponsible, or irrelevant by some other constituency.

6. Significance
The amount of information given on any problem, for example, does not reflect an editorial evaluation of its importance. Problems commonly accepted as important may be documented only briefly. This may be because of resource limitations, because of the profusion of relatively diffuse material available on them, or because they can be more effectively documented through their sub-problems. Little-known problems may be given relatively extensive coverage precisely because their existence is not well-recognized. Inclusion of information in this publication implies only that the editors considered the source from which it derived sensitive to and capable of reflecting the views of an international constituency, and therefore as being of significance to a wider audience.

7. Naivety
Information on phenomena such as world problems, values or modes of human development is widely assumed to be relevant to the design of any new broad-based initiatives in response to the global problematique. The editors have accepted the need for a certain naivety to break through the conceptual frameworks determining the general indifference of academic and governmental authorities to any questions concerning the actual number and variety of such phenomena. In identifying such phenomena within an open framework, some entries (on which whole libraries of books have been written) must necessarily appear naive. But despite the availability of such a wealth of detailed information, to the point of overload, there is a poverty of information on how to connect together this fragmented pattern. It is to this condition that this project responds by indicating possibilities, even if at times the result appears superficial or naive.

8. Pragmatism
The production of this book, within the constraints of modest resources, has been feasible only because of an extremely pragmatic approach to the collection and processing of information. Within these constraints the editors have deliberately set out to "open up", or highlight, neglected categories of information, fleshing out the content to the extent possible. Where there has been conflict between ability to locate and process adequate information (within a reasonable time period) and the elaboration of the pattern of categories, the latter has been given priority. The intention has been to provide as broad a coverage as was feasible. Hopefully, even where the information supplied is inadequate, readers will be oriented to new features of the global system which others view as meriting their attention.

9. Non-completion
This Encyclopedia (in its third edition) is the product of an ongoing project to explore ways of identifying and presenting categories of information relevant to the development process as perceived by international organizations. Major refinements will therefore continue to be made to many of the sections, and to the pattern of cross-references, especially in response to feedback on inadequacies. In this sense the book cannot be regarded as an unfinished product.

10. Solutions
This book in no way attempts to present an editorial view of "the answer" to the world's problems. Some sections in the past editions indicated the various kinds of answer, or bases for an answer, which are favoured within the international community. The editors have however endeavoured to respond to the challenge of how to interrelate inherently incompatible answers. The concern has been to respond to the possibilities of formulating an appropriate meta-answer of practical significance in such paradoxical circumstances.

User guide: errata

A publication of this scope, based on a multiplicity of sources of information, necessarily contains errors of the following kinds:

- Errors of content, due to the sources of information used;

- Errors of interpretation, due to the manner of selection and representation of the information used by the editors;

- Errors of typography and form;

- Errors arising from the process of selecting and registering cross-references;

- Errors arising from circumstances unforeseen in the design of the many computer programmes through which the data has been processed.

Considerable editorial effort has been made to reduce the number of trivial formal errors, but it has not been considered feasible to eliminate all of them within the resources and time available.

With regard to substantive errors, many of the entries on world problems, for example, contain information from one international group which some other international group would consider erroneous. In this sense this book documents the fallacies which are active in the international community by juxtaposing incompatible perceptions.

Through each successive edition, the editors have attempted to respond to error in the spirit advocated by Donald Michael:

"Changing towards long-range social planning requires that, instead of avoiding exposure to and acknowledgement of error, it is necessary to expect it, to seek out its manifestations, and to use information derived from the failure as the basis for learning through future societal experiment. More bluntly, future-responsive societal learning makes it necessary for individuals and organizations to embrace error. It is the only way to ensure a shared self-consciousness about limited theory as to the nature of social dynamics, about limited data for testing theory, and hence about our limited ability to control our situation well enough to expect to be successful more often than not." ("On the requirement for embracing error"; In: On Learning to Plan and Planning to Learn. 1973).

Introduction \ Overview

Overview: contextual challenge

Much has been written on the challenge of our times. It could be argued that further statements on the dimensions of the crisis are both repetitive and counter-productive. Specific problems are the topic of frequent media coverage and of reports by bodies of the highest authority. There are also many positive indications that nurture hope that major crises may be averted. The events in Eastern Europe are an example -- although it is not clear whether the excitement at such breakthroughs, and the new possibilities they offer, do not also serve to obscure other emerging crises to which we prefer not to give attention.

In such a context, what then is the value of a new edition of an Encyclopedia of this kind ? Especially when information overload has itself become more than a minor problem, do we really need yet another book on the problems of the world ?

The programme through which this Encyclopedia is produced is based on the assumption that our difficulties in responding to the challenge of the times lie as much in how we process information with a view to action as in the process of implementing solutions. There seems to be a prevailing confidence in the methods of the international community in response to the problems of the times. This is less than warranted by the very partial successes of the strategies implemented -- and the dimensions of the many problems that continue to grow. This confidence is sustained in part by the methods of the academic community, to the extent that their theoretical preoccupations are brought to bear on issues faced by society.

It would appear that a number of unquestioned assumptions are made in responding to the problems of society. The assumptions are implicit in difficulties such as the following:

1. Pseudo-objectivity of analyses of problems
The vast majority of descriptions of problems recognized by the international community are produced using methods which depend upon authoritative interpretations of the significance of data, whether quantitative or not. The manner by which the data is selected varies, as does the basis for the interpretation of any such data by different international organizations (or other constituencies). The importance, and even the existence, of many problems thus becomes questionable in the debates between constituencies. "Over-population" is the most striking example. And yet reports continue to be produced claiming objectivity in exhorting use of particular strategies, whilst implicitly or explicitly suggesting that other interpretations are suspect. The dynamic between such opposing perceptions and priorities is not captured. As a result any remedial programmes are undermined by the dynamics inherent in the relationship with any opposing perception. This can then be used as a convenient scapegoat in the event of failure.

2. Withholding of relevant information
The reports produced are those based on readily available, acceptable information. In the case of official reports, it is conveniently forgotten that standard procedures require that information embarrassing to particular governments or interests should be omitted or toned down. Much information is available only on a restricted basis, if at all. Information is only classified because of its importance, which suggests the conclusion that much that is made available for use in public reports is of little real consequence. Data significant to understanding of problems may simply be withheld, especially in the case of embarrassing incidents in which a cover-up policy is implemented. Issues relating to the incidence of leukaemia in people exposed to nuclear tests in the 1950s are an example. The non-disclosure of comparative international statistics on crime is another. This dimension is seldom reflected in authoritative reports.

3. Issue avoidance
Within a pattern of institutions mandated to deal with recognized problems, any indication of the emergence of problems that are inadequately handled is perceived as a threat to those whose budgets and careers depend upon positive evaluations of their incumbency. The tasks of organizations are complicated enough as they now stand. Further complication is therefore resisted. Reporting procedures, basic to budget and career assessments, therefore tend to avoid mention of programmatic weaknesses or the emergence of new problems (unspecified within the unit's mandate). The emphasis is on "upbeat" reporting in order to conceal deficiencies. Bad news is unwelcome at any level of an institutional hierarchy. The bearer, as was traditionally the case, may be severely penalized. Evaluations of institutional responses to problems tend to fail to reflect this dimension.

4. Misrepresentation of information
Information made available tends to be presented in such a way as to encourage the most favourable interpretations. Thus, aside from the process of issue avoidance, active steps may be taken to cast a positive light upon events or to support favoured arguments. Information may be "adjusted", especially in the case of statistics or the results of monitoring exercises. If necessary various forms of deception may be practised, even by official bodies. Disinformation is one such practice. The production of some reports can be seen as an effort to dissuade and to distract rather than to provide a basis for more appropriate action.

5. Biased expertise
Experts of any discipline survive through the fulfilment of contracts, as is increasingly the case for industry-funded university departments. Their continuing survival depends on their ability to meet the requirements of funding bodies. The work of eminent specialists may even be undertaken through organizations "fronting" for vested interests. Experts must therefore be sensitive to the kinds of conclusions that are acceptable. In such circumstances experts can, if necessary, be found to support any position -- if only that there is "no proven link" calling for politically sensitive action. The conclusions of some highly authoritative reports may therefore be pre-determined by ensuring the presence of appropriate experts on the investigatory body (a procedure known as "stacking" a committee). Any subsequent effort to question the report from other perspectives can be disparaged as quibbling by the unqualified serving other interests. The question of the degree to which some of the major reports have been biased in this way has not been explored.

6. Corruption
Many programmes have been carefully designed in response to problems and yet have failed or under-performed. Whatever the official explanation, there is much evidence to indicate that an important factor in such failure is the activity of involved individuals in attempting to profit to an unforeseen degree from the resources and influence that they control during the execution of the programme. Much of the evidence is anecdotal, although well-known to those with any experience in the field. Much is reported on a daily basis in the quality press. It often touches those at ministerial level. Such corruption, although possibly occurring in many forms, is not confined to any particular group of countries, as some in the industrialized countries would like to claim, although it does tend to be more discreetly organized there. Such semi-formal subversion of programmes is tacitly accepted, even at the highest level within international organizations. No international study of corruption has ever been made. Its potential for undermining new strategies is never officially mentioned when they are advocated.

7. Harassment
Individuals aware of any of the issues noted above are not free to report on them in written form -- or rather they do so at their own risk. Typically most official bodies require that employees sign non-disclosure agreements which may well apply after termination of their employment. Any attempt to provide hard evidence can severely affect career opportunities (such as through "blacklisting") in the case of official actions. In the case of unofficial actions, as with various systems of corruption, it can lead to severe peer group pressure from those who do not wish "the boat to be rocked". Harassment can also take physical form, especially against external activists and "whistleblowers", where cases of assassination have

been documented. The action of the French government against the Rainbow Warrior in New Zealand is an example. Naturally official reports tend to be discreet where such threats may be brought to bear.

8. Short-termism
In producing official surveys on the challenge of the times, the priority of many is naturally to ensure their survival through to the next budgetary cycle. In a political environment it is short-term issues which are the guarantee of survival. Longer-term issues can safely be given lower priority, even when they are exacerbated by short-term decision-making. A focus on short-term action creates the impression of effective action even though it may be counter-productive in the longer-term. This cannot be effectively addressed in reports addressed to bodies governed by short-term priorities.

9. Ambiguity
Consensus under the above constraints can be most easily achieved through use of ambiguity, each constituency projecting onto an agreement its own interpretation. "Development" is the most tragic example. According to a definition favoured by commercial interests, any degradation of the environment can be interpreted as a development achievement (as typified by "clearing" the land). "Sustainable development" can thus be widely approved through being understood as "sustainable competitive advantage". Clearly international reports run the risk of being shelved if they do not permit such ambiguity of interpretation. This effectively undermines what many are led to assume is the purpose of such strategies.

10. Unaccountability
The accountability of institutions and those responsible for them is limited. Many institutions cannot be effectively held accountable for their abuse of the social or natural environments. Thus the World Bank has been able to resist any sensitivity to environmental issues for at least a decade after these formally became a concern to the United Nations. Senior management can seldom be held accountable for the unethical actions of their employees, even when they are responsible for the pressure giving rise to such actions. The whole framework of "plausible deniability" runs throughout organizations. The the chief executive is thus well insulated and able to deny knowledge of any unethical action by the body for which he is responsible.

11. Violation of commitments
Commitments in response to problems are violated, whether they take the form of breaches of electoral promises, neglect of resolutions of international meetings, or failure to conform to the agreed provisions of intergovernmental treaties and conventions. Given the development of contract law and the regulation of advertising claims, it might be asked why similar standards are not applied to the promises and claims made by those seeking political power.

12. Loss of integrity
In the context defined by the above issues personal integrity is easily compromised or readily sacrificed. In any competition for resources, it becomes a luxury. In any particular case, it is unclear how compromised an eminent authority may be in supporting some position. This is also true of countries whose declared support for positive initiatives is often totally compromised by previous or parallel commitments to programmes having the reverse effect (typified by military aid to repressive regimes). In the case of international organizations, the most striking symbol of loss of integrity has been the Waldheim incident and the silence of the great powers aware of the facts (unless it is to be assumed that their intelligence agencies are unbelievably incompetent). It is extremely difficult to raise such questions in relation to new initiatives.

13. Loss of credibility
Much has been learnt as a result of the limited success of programmes over the past decades. People are increasingly aware of the issues indicated above. As a result there is a widespread erosion of the credibility of institutions and official expertise. Increasingly this extends to any organized activity. There is awareness of the ways in which the media are used to manipulate opinion and to spread disinformation. Much of this loss of credibility may indeed be unjustified and even paranoid. It nevertheless affects the ways in which reports and new programme initiatives are received.

14. Disparagement of complementary initiatives
The great majority of views on the problems of society, social directions, and possible alternatives are posited with little or no reference to any other views, past or present. When such reference is made, it tends to be made disparagingly or with condescension. Competition for scarce resources obliges organizations, and departments within organizations, to define an approach which establishes the irrelevance of other initiatives whose activities might under other circumstances have been considered complementary.

15. Dubious standards of proof
Within these constraints, the tragedy is that any truth about the challenge of the times has become something movable, an illusion to be marketed for the benefit of the few. There is no standard by which the pronouncements of collectivities can be assessed with any degree of confidence. The greater the resources controlled, the greater the pressure to deceive unless constrained to do otherwise. Standards of proof developed by science or judicial systems have been shown to lend themselves to abusive manipulation even in the most respected democracies. It has not been possible to prove, in the scientific sense of the term, that such abuse is exceptional rather than systemic. There appears to be no way that powerful institutions can prove their integrity or that of their representatives. To use the favoured phrase, there is "no proven link" between statements emanating from such collectivities and the reality with which the world is faced. It is clearer to state that their reports have higher or lower degrees of correspondence with that reality, according to the pressures to which they are subject.

It would be both naive and presumptuous to assume that any body could escape the above constraints. However, rather than seeking coherence in the presentation of information from a selected group of authorities -- the approach of many reports on the condition of the world -- a different approach is possible. This can open the door to more radical insights into the dilemma of the times precisely because it explicitly recognizes difficulties such as those indicated above.

Overview: existential challenge

Corresponding to the contextual challenge in responding to their environments, individuals ar faced with an existential challenge in redefining their self-image and the mind-set with which they respond to the world. The following are some of the features of this challenge.

1. Proliferation of explanations and injunctions
The hyper-development of the ability to explain and to label has fostered the pervasive illusion that this necessarily ensures that an environment so treated is somehow under control. Much effort is devoted to this process, whether by researchers, educators, legislators, administrators or managers. There is a resemblance to the enthusiastic reliance on pesticides by the agri-business. This process does appear to freeze portions of the environment, since readily comprehensible explanations tend to be in static terms. Not only does this render invisible any dynamic relationship to other aspects of the environment, but it also defines the explainer at the same reduced level of complexity -- at least in the relationship to what has been so explained.

2. Simplistic responses
Action on the environment, perceiving and responding to problems, is then viewed primarily as a question of reordering explained categories into a more appropriate pattern -- "sustainability" being the latest criterion. "Profitability" is a competing criterion. Irrespective of the criterion, there is a resemblance to the procedures by which radioactive products are handled in laboratories through "glove-box" manipulators. The person controlling the manipulators is of a much higher order of complexity than that aspect which is manifest through the possible movements of the manipulators. And yet problems are perceived and acted on at the level of complexity of the manipulators. The glove-box delimits the reality to which society is prepared to respond and constrains the manner of that response. But above all it protects the users of the glove-box from exposure to the invisible challenge of the products therein.

3. Paradigmatic entrapment
Action on problems thus becomes a matter of shuffling categories and institutional elements, combining and recombining them in an effort to increase the effectiveness of response. New categories and institutions are invented within the same pattern. Blame for problems is reallocated in a similar manner. In this way much change is apparent, together with many explanations as to why such change is sufficient to the challenge of the times. And yet this perception tends to remain unchallenged.

4. Failure to act on knowledge
Explanations do not respond to present (or future) suffering, although they may reduce anxiety about it. A physician, fully informed of the dangers of smoking, may continue to do so, irrespective of the recognized effects on his own health or the indirect consequences for others. Similarly a factory may continue to discharge pollutants, despite the manager being fully informed of the consequences for the environment. A walker may point with complaint to a piece of rubbish in a forest but not feel called to remove it. Such examples are indicative of the protection offered by the existential glove-box. It permits those using it to feel uninvolved. The pattern of explanations and injunctions has a numbing effect by which individuals are protected from any challenge to their own pattern of behaviour.

5. Unaccountability
The professionalism of international responses to the challenges of the times also protects individuals from any need to be personally concerned whether a programme succeeds or fails (provided explanations can be found to deflect any negative consequences for career advancement). But how to distinguish between the necessary detachment of a surgeon whose skills are unable to save a patient, and the indifference of a surgeon whose inappropriate action is aggravating the condition of the patient?

6. Disempowerment
The quantity of explanations and injunctions, and the eminence of those offering them, disempowers non-specialists. Those who are not mandated to provide authorized explanations are forced into a position of dependency for the construction of the reality in which they live. Imagination is crushed by the weight of explanations and by those who are empowered to impose them. Imagination itself is only acknowledged in those who have proven their commercial worth. As such it has become a product for consumption. In the glove-box, images are generated which trap the unwary into belief in their reality.

7. Unchallenged self-image
There has been much concern voiced about the need for a new paradigm and for non-linear approaches to the complexities of the environment. This seems to address the simplistic, even mechanistic, pattern of category shuffling -- a recognition that the glove-box only permits a restricted pattern of movements. But much of that discussion still seems to be calling for what amounts to a more complex set of manipulators for the same glove-box. The relationship of the user of the glove-box to the manipulations therein is not called into question. The dualistic relationship is not challenged. However rich the paradigm, to what extent will it call for a new self-image on the part of the user of the glove-box implying a new involvement in action? To what extent do perceived problems become existential challenges rather than merely a flavour-of-the-month?

8. Unchallenged relationship to the environment
The fashion for "holographic" paradigms, with the implication that everything is reflected in everything else, is an intellectual challenge calling for a broader and richer understanding. The Gaia model is of this kind. But whilst these call for a higher degree of responsibility and accountability on the part of individuals, they fail to render explicit the challenge to the articulation of the individual's relationship to the environment. A non-dualistic framework continually questions both what it means to be and act as an individual and what is the nature of the environment in which action is taken.

9. Need for a different mode of thought
It is indeed possible to avoid this challenge. It is possible to assume that one does not need to change and that the paradigm of "development decades" and "international organizations" is sufficient for the times. It may be assumed that "win-win" solutions are possible and that there need be no losers -- gain without pain. But there is increasing recognition that unless the life-style in industrialized countries is radically changed, the current system will become increasingly unsustainable. This calls for a different mode of thought.

10. Existential sacrifice
The prevailing mode of thought makes no explicit call for existential sacrifice, since sacrificing a category within the glove-box is not an existential operation. And yet in the larger reality people are indeed sacrificed through the imposition of austerity programmes -- structural adjustment with an "inhuman" face. A non-dualistic approach sees winning and losing as complementary phases in necessary learning cycles. Like inspiration and expiration, they are both necessary processes in a growing organism. To what sacrifice do administrators of programmes expose themselves in order to comprehend more subtle approaches to the environment?

11. Change vs pseudo-change
It is ironic that it is only those who are least appreciated who consciously expose themselves to being sacrificed in contemporary society. The dramatic examples are terrorists on suicide missions, self-immolating protesters, and soldiers in a jihad. However society also requires human sacrifices before legislative changes are considered necessary: children have to die before dangerous foodstuffs are prohibited by law, and demonstrators have to be willing to suffer, or lose their lives, before their cause receives attention. Real change is accomplished when people expose themselves to an existential challenge, thus becoming agents of change. Pseudo-change occurs when the initiators engage in manipulations within the glove-box which leave them totally untouched (other than through any loss of status or pride, as in losing a wager).

12. Distorted value of life
Again it is ironic that there is less and less in modern society that people are prepared to die for, or to allow others to die for. Whole societies can now be held to ransom for a single known hostage. Millions can be spent to maintain a comatose, brain-damaged patient on life-support for decades. Euthanasia is illegal, no matter what the desire of the person concerned. Exposure to risk is progressively designed out of society, to be replaced by vicarious experiences through videos or with the protection of required safety devices. The paradox is that unknown numbers are however sacrificed through carcinogenic products, abortion, structural violence, massacres, gang murders, cult rituals, "snuff" movies and associated perversions, or a failure of food and medical supplies.

13. Value sacrifice
The attitude to life has become as immature as that to death. Millions are spent on efforts to maintain youthfulness, whether through cosmetics, cosmetic surgery or attempts to reverse the ageing process. Every other value is sacrificed to save lives in industrialized societies, whilst allowing others to die elsewhere. Individuals in industrialized societies are prosecuted for life-endangering neglect. But these same societies fail to apply the same standards in their policies towards other societies. Reproduction is tacitly encouraged without any provision for the resulting population growth or the effects on the environment. Society evokes problems to provide solutions for its own irresponsibility -- a control mechanism for the immature lacking the insight for a healthy relationship to cycles.

14. Personal transformation
The challenge of the times would seem to involve a call for personal transformation through which social and conceptual frameworks can be viewed anew. Willingness to sacrifice inherited perspectives is an indication of the dimension of the challenge -- most dramatically illustrated by willingness to risk death. However physical death is not the issue, and may easily be a simplistic, deluded impulse lending itself to manipulation. Destruction of frameworks valued by others is equally suspect. Such dramatics provide rewards within the very frameworks whose nature the individual needs to question, but by which he may need to choose to be constrained.

15. Existential discipline
What are the existential disciplines by which an individual can progressively redefine what he or she is in relationship to the environment -- at present and in response to the emerging future? What does this imply for the organizations through which individuals may work or for the conceptual frameworks appropriate to that work? How does this understanding affect the individual's relationship to the piece of rubbish at his feet? The many spiritual traditions provide clues for further exploration. But their advocates are often dangerously enthusiastic about their own insights and disparaging of others. Insight is buried in dogma. The letter obscures the spirit and denies the awareness by which they may be distinguished. Instant faith is demanded to avoid the long-term challenge of disciplined acquisition of insight. This suggests that any such disciplines must also be applied in response to the purveyors of insights and to their products. But where there is no exposure to risk and the possibility of error, there is no learning or possibility of meaningful change.

16. Transformation of perspective
There is an irony in the call to ensure humanity's continued existence on the planet. It is a challenge to our existence, but it has not yet been recognized as an existential challenge. As with suppositions about the life hereafter, it is assumed that no change is implied for our understanding of ourselves. And yet it may be that such transformation of perspective is the key which the nature of the crisis will force us to recognize. As the Sufis suggest, the trick of insight required may be to remove the point from which we currently view. There is an illusion of who we are that needs to be sacrificed to give birth to a sustainable future.

17. Persistent personal egotism
There are many admirable advocates of change, whether involving social transformation or personal transformation. The most eminent, whether in the world of science, art, politics, or religion, have made striking contributions to this process and will continue to do so. And yet the level of egotism among those with most insight remains a major challenge, as illustrated by the following comment made by Richard Gardner and quoted by Michael Marien (*Societal Directions and Alternatives*, 1976): *"We are afflicted not only by national but by personal egoism. That is what could eventually destroy us.. Many of these eminent people have such big egos that their principal preoccupation in life is to establish a piece of intellectual turf and preserve it against all comers, whatever the consequences. They're prepared to sacrifice truth -- prehaps not consciously, but subconsciously -- to the pursuit of idelgoy and the pursuit of ego."*

Overview: strategic assumptions

Over the past 25 years, from the first International Development Decade, international groups and organizations have implemented or advocated every conceivable strategy offering some promise of counteracting the emergence of a crisis of crises. Whatever the successes, it is widely acknowledged that the basic trend has not been significantly affected. This recognition has itself been voiced so frequently through the Secretary-General of the United Nations that it has itself become an outworn generalization associated in the minds of many with the of loss of credibility of existing institutions, of democratic political processes and of academic research, all of which have proven incapable of more than token response to the global problematique. The series of special international commissions (Brandt, Palmer, Brundtland, South, etc) convened over the years to report on particular aspects of the emerging crisis have proven to be as much a symptom of collective impotence as capable of offering a foundation for new initiatives.

The same 25 years have seen the emergence of a widespread counter-culture which has offered the hope of alternative approaches. These have borne fruit in the form of new communities, personal growth movements, political activism, volunteer programmes, alternative technology, computer-supported networking and the green movement. These developments have been sustained in part by exciting breakthroughs in comprehension of the nature of self-organization, paradigm change, holism, implicate order, and the relationship between physics and consciousness. Nevertheless whilst these continue to offer the possibility of significant impact on the global problematique, this has not been forthcoming. And to a large extent such alternative approaches have appeared as luxuries irrelevant to the priorities of developing countries.

In envisaging the design of a project to respond to the challenges noted above, the following strategic constraints have been assumed:

1. Maintenance orientation
It would appear that collective ability to respond to the crisis of crises has been effectively paralyzed. The 1980s have seen the emergence of a sense of apathy, defeatism and despair in the international development community and in grass roots movements. This is largely disguised by public information programmes and media events designed to maintain confidence in projects and campaigns which do indeed have some measure of success. But as the food crisis in Ethiopia has demonstrated, although a magnificent one-time attempt can be made to remedy short-term problems in the spotlight of media coverage, the solutions to the underlying longer-term problems are not in sight. At this point in time programmes are deemed a success if they can slow the trend toward major crisis. An acceptable criterion is maintenance of the status quo, provided it lends itself to being described as innovation. Significant social innovation is seldom sought, however eloquently it is advocated.

2. Solution production
Many "answers", whether explanations, programmes, strategies, ideologies, paradigms or belief systems, are put forward in response to the current crisis, however it is perceived. The proponents of each such answer naturally attach special importance to their own as being of crucial relevance at this time, whether in the short-term for tactical reasons, or in the long-term as being the only appropriate basis for a viable world society in the future. This widespread focus on "answer production", is a vital moving force in society. However it obscures both the significance of the lack of fruitful integration between existing answers and the manner in which such answers undermine each other's significance. This mind-set also fails to recognize the positive significance of the continuing disruptive emergence of new "alternative" answers.

3. Questionable truths
Amongst this multitude of answers, explanations put forward as factual by scientific or government authorities are increasingly questionable because of peer group, religious, political, military, security and commercial pressures guiding objective evaluation and reporting. Recent examples include dubious evaluations by authorities of nuclear reactor and toxic waste hazards, official denial of the impact of acid rain on forests, and reassessment of the world population problem as non-critical. The situation has been epitomized in NASA, the model of western high-tech management, by the top executive pressures on engineers to withhold information on the gravity of problems associated with low-temperature effects on space shuttle launchings. Middle management in any bureaucracy is under considerable pressure to report positive achievement in the light of pre-defined policy objectives, rather than to indicate the dimensions of problems detected in the process. There is no assurance that such pressures do not affect the reporting of many other facts of social significance. Self-censorship is increasingly practised as in biology textbooks (to meet creationist objections) and in encyclopedias (to avoid raising unwelcome political questions concerning such social realities as corruption and institutionalized torture). Even in courts of justice, an expensive (astute) lawyer considerably increases the probability of a judgement favourable to his client. The truth of facts has become a question of interpretation, leaving authorities free to deny politically unacceptable conclusions by selecting experts prepared to declare that "there is no proven causal link" between the problems in question (even though such a link may be accepted by equivalent bodies in other countries).

4. Gladiatorial arena
Policy integration initiatives at this time are themselves fragmented and mutually hostile, to a degree usefully interpreted in terms of the metaphor of a "gladiatorial arena". The survival of any integrative answer must be bought at the price of the elimination of all other competitors. There is considerable confusion about the nature of integration and it is difficult to imagine that integrative processes favoured by one group would be considered to be of much significance by another. This phenomenon cannot be disguised by simply opting for consensual procedures, "networking" processes or by viewing it as a "healthy" feature of academic or political debate.

5. Irrelevance of alternatives
The most characteristic response to this confusion is to simplify the situation by establishing or affirming, explicitly or implicitly, the fundamental irrelevance of any other answers and perspectives that are viewed as incompatible, if their existence is recognized at all. The preoccupations of the other constituencies are thus defined as dangerously misguided or agonizingly irrelevant. As a consequence there is always a perfectly valid reason for not instigating any advocated course of action or for not considering any alternative perspective.

6. Projection of blame
Many would reject any such recognition of paralysis. But the basis for their rejection is that, "if only" some other portions of society would cease to block effective change then this would release the resources that would demonstrate the collective paralysis to be only momentary. Unfortunately it is precisely the number and variety of such "if only theys", which has ensured the spread of this paralysis and which guarantees that it will prevail for some time to come.

7. Assumption of innocence
Corresponding to this projection of blame onto other groups, as suitable scapegoats, is a widespread assumption of the unquestionable innocence of one's own group. This may well be perceived as making an untarnished significant contribution to the well-being of society. Whether it be academic disciplines or their corresponding professions, national or international organizations, public or private bodies, benevolent or alternative groups, each acts as though its contributions to society constituted an unmitigated good. However valuable these may be, the suspect consequences of these contributions can only be questioned at the risk of ridicule. Sanctions may be applied against those voicing such criticism, from within or without, whether in the case of the United Nations or of alternative groups. A perfect disguise is therefore provided for every possible systematic abuse.

8. Reinforcement of fragmentation
One major characteristic of the plethora of material documenting the ills of the global community is that it tends to reinforce the plaintive or angry plea, noted above, that "if only" some other group would act in some other way all could be well. Each such report focuses on one part of the network of problems, explicitly or implicitly denying the relevance of some other part with which others identify. It is understandable that any such other group would not be strongly motivated to respond to the concluding pleas of such a report. Furthermore it will probably associate itself with some other report denying, explicitly or implicitly, the relevance of the priorities laid out in the first. This process can be observed between the Specialized Agencies of the United Nations or their equivalents at the national level. It can be seen in the failure of the Brundtland Report to build on the Brandt Report, followed by the failre of the South Report to build on the Brundtland Report. It is however far from being limited to governmental bodies.

9. Narrow information base
The consequence of this process is that no group is motivated to recognize or document the full range of perceptions of the ills and opportunities of the world. Such information exists but has to be culled from documents in different locations, which very few are inclined or able to do. If such perceptions are not interrelated the chances of reducing the level of paralysis are handicapped. The argument is therefore that recognizing the full range of ills and opportunities by which groups are touched, and with which they identify, is a minimum requirement for exploring the ways in which they can be collectively empowered to release their contribution to the paralysis.

10. Single-focus dependence
Such fragmentation encourages, and is further reinforced by, dependence on single-factor explanations and single-policy initiatives. Each such initiative may necessarily formulated be in terms of a limited information base. This is usually discipline-oriented in the case of the academic community, but ideological, action-preference, "priority" and other filters may also be used. The integration of the approach is thus achieved artificially by deliberately avoiding the encouragement of a variety of complementary approaches capable of counteracting each other's weaknesses. When the opponents of such a unified approach can demonstrate its weaknesses, they then move to implement another simplistic approach of a countervailing nature in order to remedy them. Society thus moves spastically from policy to policy without any ability to acknowledge the merit of an ecology of policies and of alternation through a cycle of policies.

11. Initiative obsolescence
Single-focus dependence leads directly to the repetition of initiatives of a form which has failed in the past or whose success has been only marginal. Questioning strategies based on thinking of this kind, especially when they are defined with politically acceptable trigger words (population, energy, environment, food, health, education), may be considered tantamount to questioning the merits of motherhood. In the Club of Rome's terms, many such initiatives are maintenance-oriented and are incapable of innovative breakthroughs. The need to break through to new forms of initiative is not accepted by the international community. Even the eloquent pleas for a new order are made on the assumption that well-tried conceptual, policy, programme, organization and conference forms are appropriate to its conception and implementation, with perhaps some minor adaptation.

12. Disagreement phobia
Society has been unable to design any framework, whether conceptual or organizational, in which disagreement is an accepted, permanent integral feature. The frameworks now used are based on the assumption that consensus is the keystone on which any viable organization must depend. As a consequence, disagreement can never be accepted as an integral feature of society, except through structures or processes designed to eliminate it (conflict resolution, mediation, arbitration). These include competition and violent conflict, in which victory is sought, through the downfall of the opponent, Although disagreement is a daily and often creative reality, the fear of situations in which disagreement prevails is such that they are shunned, whether unconsciously or by well-rationalized processes. When they cannot be avoided, much effort is devoted to amplifying the significance of whatever minor items can be discovered on which agreement has been achieved. Agreement then becomes an essentially superficial pretence of little operational significance. Conceptual, organizational or legal structures based on such agreement are consequently totally inadequate to the innovative requirements of any dynamic development process in which disagreement is inherent. Stressing consensus as a key to development and social transformation comes dangerously close to destroying the basis of its dynamism. Development can only occur if there is disagreement with those maintaining the status quo.

13. Self-reflective paradox
Any attempt to reflect the widest possible range of perspectives on the ills and opportunities of the world is bedevilled by an interesting paradox. Given the prevalence of disagreement, whatever method is employed must necessarily engender disagreement. It cannot be expected to result in some ideal, objective approach that would engender universal consensus. Indeed the very attempt to reflect the fullest range of perspectives must naturally remain suspect to those with the vested interests necessary for any specific form of action. Any breakthrough into a more fruitful mode must therefore endeavour to give explicit recognition to this paradox and to the dynamics associated with it. In this light it would be unproductive to attempt to produce yet another "answer" to the condition of the world, however inadequate it might be.

14. Embodiment of discord
The widespread tendency to produce incompatible answers is a symptom of the underlying paralysis noted above. Any endeavour to break out of this paralysis must respond to this dynamic, if it is to be of any relevance to the current conditions. Under the prevailing linear approach, a particular position is taken up and defended, as required by the militaristic conventions of academic, religious, political or ideological debate. This could be contrasted with a complementary non-linear response, whereby such positions are identified both as perceived by those who hold them and by those who consider them nonexistent, irrelevant, misleading or downright evil.

A valid response is therefore to attempt to design a framework to internalize or embody discord, contradictions and logical discontinuity. The status within the framework of the perspective that the attempt itself represents must necessarily remain a paradox. A further step is therefore called for within such a framework to explore the adequacy of conceptual language to contain such incommensurable perspectives so characteristic of the dynamics of global society. The ultimate question is therefore how to interrelate inherently incompatible answers without producing yet another answer to compete with them in a process which has proved unable to transcend itself.

Overview: objectives

1. Organizational context
One of the original objectives, in initiating this project in 1972, was to endeavour to document how the network of international bodies focused on the network of world problems. Clearly some key problems attract much attention and many others attract very little, if any. But of greater interest is whether the organizations focusing on the same problem are in communication with each other, or whether the organizations dealing with one problem that closely affects another are in contact with the bodies dealing with the latter problem. How do problems escape the net of organizations? How does the network of organizations fail to encompass and contain the network of problems? What kinds of information would then be required to enhance transnational initiatives?

In addition to the major role that international organizations perform in identifying problems, many of them perform an important function in relation to human values. In fact the two functions are often intimately linked, as in the case of human rights issues. A significant group of organizations is also concerned with human development in its less material sense, as is the case of bodies concerned with religion and personal development. Again the question may be asked with which of the specific values is an organization associated, given that many of them carefully identify values in their statutes? Clearly there are very "popular" values, such as peace and justice, but are there values with which few, if any, bodies are associated? And to what extent are such values vital to the functioning of society? Is there also a mismatch between the network of organizations and the network of values?

2. Core concerns

(a) Clarification of "fuzzy" domains: The core concerns of the international community, whether problems, values or human development, remain conceptually "fuzzy". They are a continuing challenge to both scholars and practitioners. It is from this fuzziness that dilemmas, contradictions and paradoxes emerge. In considering the role of integrative approaches, and even the power of metaphor, in responding to this "fuzziness", it is apparent that here too there is considerable ambiguity and confusion. For this reason, despite the vital importance of these concerns, they are especially difficult to handle within information systems. Few information systems attempt to do so, preferring to deal with harder data. Experimenting with computer-based procedures to do so therefore constitutes a valid preoccupation.

(b) Recognition of "vectors of concern": The many international constituencies tend to disparage each others concerns, if they recognize the existence of concerns other than their own. Mapping the "vectors of concern" in relationship to one another provides a means of defining the nature and dimensions of the communication space within which the dynamics of the international community operate. Recognizing such vectors determines with what concerns constituencies identify, thus clarifying what moves them to act as well as the nature of the social reality within which they perceive themselves as functioning.

(c) Bridging between incommensurable domains: A major underlying concern is to create a framework within which it is possible to register links between specific world problems, human values and modes of human development. The intimate relationships between these domains call for more effective ways of processing information on them as a complex system.

(d) Anchoring transient insights: Insightful perceptions of subtle challenges and opportunities, of potentially major significance, appear in the literature. Because of their nature, and the categories and perspectives that they call into question, there are few places at which this kind of information can be collected. A suitable context is required to hold and "anchor" that information in relation to more conventional perspectives.

(e) Transformation enablement: There is a need for information in a
form which enables social transformation in the light of more appropriate values. This suggests as a valid objective the extension of information system design to incorporate the fuzzy conceptual dimensions which catalyze and motivate such transformation.

3. Objectives
The objective of the project through which this Encyclopedia is produced is threefold and may be described as follows:

(a) Collection and presentation of information
At this level the intention is to demonstrate the feasibility and value of assembling information reflecting the perspectives of a very wide range of international constituencies. In contrast to normal practice, this information should not be filtered by some particular criteria of "truth" or "importance". Every effort should be made to present it in terms of what is held to be true by the constituencies from which it originates, even if the information totally contradicts that from some other international constituency. It is a basic assumption of this project that it is the dynamics inherent in the interaction of such conflicting biases which reflect the reality of global society, as much as the fundamental insights emerging from any particular analysis of the global system in the light of criteria carefully selected by leading experts.

In organizing the information into the sections in this Encyclopedia, the intention has been to group material into classes corresponding to the terms conventionally used to describe and order any response to the global problematique and the possibilities of human development. Each of these tends in some way to be of fundamental concern to any international constituency, whatever the differences about the appropriate content of such classes.

In designing a framework "hospitable" to such a wide range of perspectives, whether mutually indifferent or inherently incompatible, a secondary objective has been to seek ways to juxtapose such perspectives in order to highlight the variety of relationships between them. The framework therefore contains the variety of incompatible perspectives by fragmenting the information into a very large number of descriptive entries. This deliberate disorganization is counter-balanced by a very extensive network of cross-references between such entries. When appropriate information has been obtained from appropriate sources, some form of counterargument is included in many entries, illustrating the limited or misleading nature of the perspective presented.

Metaphor: star mapping This objective can be usefully described in terms of the metaphor of an astronomical telescope. Whereas a limited number of astronomical objects are visible to the naked eye, their visibility from Earth is determined both by their intrinsic brightness and by their distance from the observer. The major problems cited by any international constituency are equivalent to the brightest of those objects. Others may be barely visible to them. By the use of a telescope the number of visible astronomical objects, whether stars or galaxies, increases enormously, depending on the resolving power of the telescope. The brightness of some of them, to an observer located elsewhere, may be very much greater than those visible from Earth. So for some other international constituency, a different, but possibly overlapping, set of problems appears to be of major importance. The challenge of this project is conceived as one of designing a telescope of sufficient resolving power to collect information from distant sources on the phenomena which are highly visible to them. This is achieved by using the whole array of international bodies as collectors, thus constituting (as with a radio telescope with a long base line) a much more powerful telescope than that based on dependence upon any one of them or upon any small group of them. As with recent discoveries concerning the dangers of exposure to low-level radiation, this may also help to demonstrate that long-term exposure to less visible problems can be as dangerous as short-term exposure to the more visible problems.

It is hoped that the collection and presentation of information in this reference book form will meet the information needs of many users.

(b) Clarification of conceptual challenge
At this level the intention is to clarify the challenge of interrelating perceived patterns of information with which people and constituencies can identify and by which they are empowered. In one sense this project is an endeavour to document the perceptions active in global society. For the resulting quantity of information to begin to become meaningful as a whole, this calls for new approaches to communication, with an emphasis on patterns of concepts. The perceptions documented are those with which different people identify and by which they are motivated. For such motivations to reinforce each other to achieve the required political will to change, greater understanding is required of how patterns of concepts may be nested together without doing violence to the particular perceptions with which people identify. For such social change to be fruitful, there is the even more challenging requirement of ensuring a comprehensible relationship between mutually incompatible patterns of concern that can correct each others' inadequacies and excesses.

Metaphor: electricity generation This second objective can to some extent be described in terms of the metaphor of electricity generation. The electrical current produced by some form of generator depends upon the degree to which opposite charges can be simultaneously generated within the same framework and conducted together (but insulated from each other) to the point where the difference between the charges can be used to do work. This project endeavours to accumulate and juxtapose within the same framework both extremely negatively charged information on world problems, and extremely positively charged information on human potential in various forms (values, subtler states of awareness, etc), rather as in the design of a battery. The hypothesis is that it is through an appropriate juxtaposition of the "bad news" and the "good news" that the generation of the will to change is effectively generated. This is in strong contrast to many other initiatives which endeavour to focus only on positive initiatives (solutions, values, etc), only on negative doom-mongering, or on a mixture from which the opposing charges cannot be effectively separated so as to empower people to act. In the light of this metaphor the latter efforts are as likely to succeed as attempts to design monopolar batteries or an electrical circuit with a single wire. When they do succeed in mobilizing people, their subsequent failures could be usefully compared to the dangerous discharges resulting from the generation of static electricity.

It is hoped that the information presented here will stimulate some users to contribute further to the clarification of this challenge.

(c) Enablement of paradigm alternation
At this level the intention is to explore indications of ways of moving beyond the sterile relationships between the existing paradigms within which the perceptions documented in this Encyclopedia are generated. For although the strengths and weaknesses of such paradigms continue to be demonstrated in many studies, the purpose of such studies tends to be that of proving the merits of some existing or alternative paradigm. The challenge then is to explore ways of moving beyond prevailing conceptual fragmentation whilst avoiding the opposite danger of simplistic holism under the guise of sterile relativism. The challenge is made more dramatic by the irresponsibility of experts. Whilst these may be qualified to justify some particular position, they are totally unable to offer any guidance to voters and decision-makers as to the manner by which their position can be reconciled with some totally contradictory position justified on other grounds.

In an isolated local context, or a simpler world, this difficulty may be avoided by establishing certain perceptions as true and others as false, misleading or totally irrelevant. Some people are then empowered by the acceptance of such a coherent pattern of truths and the challenge of articulating them. Others are empowered by the process of denying the corresponding falsehoods.

In the more complex modern world of interacting contexts, decision-makers are forced to recognize pragmatically that contradictory positions may both be true, possibly under different conditions, even though there is no coherent framework within which they may be reconciled. Some are even empowered by the opportunity this provides to "divide and rule" by "playing one side against the other". But there is also the recognition by others that neither position need be true, and they are then empowered by the process of rejecting the system constituted by both together.

Metaphor: fusion reactor This third objective can also be described in terms of the metaphor of the current technological challenge of designing a suitable magnetic container for plasma to enable nuclear fusion to take place. In order to generate energy in a fusion reactor, the problem is to discover the particular configuration of magnetic fields, values of plasma parameters, and means of protecting the plasma from contact with any material surface which would quench it. This can be achieved by "bouncing" the plasma around within the configuration of a magnetic cavity (or "bottle"). As in the case of plasma, any comprehensive understanding of the human condition (encompassing both the global problematique and the associated opportunities for human development) is "quenched" by any efforts to contain it within a particular conceptual framework. And as with plasma, transcending this difficulty seems to require the design of a container which ensures that such understanding can only emerge, exist and develop if it is continually "bounced" or alternated between an appropriate configuration of different conceptual perspectives. Although there are indications as to the possible design of such a container, the multi-perspective containers that have so far been designed reflect the lowest common denominator of the participating perspectives, rather than the highest common insight by which appropriate action in response to the global problematique could be empowered.

As this metaphor illustrates, this project is in many ways about the adequacy of the language used amongst international constituencies. To what extent are the challenges of society and the possibility of innovative response determined by the distinctions and connections permitted or forbidden by the language of the international community (and its various jargons)? Can the many distinct problems, values and strategies engendered by that language be meaningfully distinguished? Is it in some way fundamentally inadequate as a means of formulating distinctions and relationships that are required to respond appropriately to the global problematique?

It is hoped that this Encyclopedia may to some degree be used to explore the nature of the art of alternating between paradigms, languages or viewpoints as a way of enabling individuals and societies to be appropriately empowered in response to the conditions of the moment. The challenge appears to be to discover a comprehensible conceptual dynamic of sufficient complexity to permit an appropriate conscious alternation between the different combinations of acceptance and denial. This has been admirably illustrated in drawings by the artist Escher, especially as analyzed by Douglas Hofstadter in *Gödel, Escher, Bach*. As in the relationship between male and female, or between parent and child, it is the collective equivalent of the art of saying "yes" or "no" under changing conditions. This is at its most frustrating and enchanting as it explores the excluded middle ground forbidden by the boundaries of Aristotelian logic, however vital the latter may be in particular circumstances.

Introduction \ Overview

Overview: significance

The significance of each of the sections of this publication is treated separately in the introduction at the beginning of each section and in the Notes on each of them.

The significance of this publication as a whole can best be briefly illustrated by the following quotations which, taken together, indicate the importance of exploring the kind of approach attempted here.

1. Antiquated concepts and attitudes
"It is unforgivable that so many problems from the past are still with us, absorbing vast energies and resources desperately needed for nobler purposes: a horrid and futile armaments race instead of world development; remnants of colonialism, racism and violations of human rights instead of freedom and brotherhood; dreams of power and domination instead of fraternal coexistence; exclusion of great human communities from world co-operation instead of universality; extension of ideological domains instead of mutual enrichment in the art of governing men to make the world safe for diversity; local conflicts instead of neighbourly co-operation. While these antiquated concepts and attitudes persist, the rapid pace of change around us breeds new problems which cry for the world's collective attention and care: the increasing discrepancy between rich and poor nations; the scientific and technological gap; the population explosion; the deterioration of the environment; the urban proliferation; the drug problem; the alienation of youth; the excessive consumption of resources by insatiable societies and institutions. The very survival of a civilized and humane society seems to be at stake. The world is bursting out of its narrow political vestments. The behaviour of many nations is certainly inadequate to meet the new challenges of our small and rapidly changing planet. International co-operation is lagging considerably." (U Thant, Secretary-General of the United Nations on the occasion of United Nations Day, 1970).

2. Inadequacy of fractional approaches
"Many of the most serious conflicts facing mankind result from the interaction of social, economic, technological, political and psychological forces and can no longer be solved by fractional approaches from individual disciplines...Complexity and the large scale of problems are forcing decisions to be made at levels where individual participation of those affected is increasingly remote, producing a crisis in political and social development which threatens our whole future." (Bellagio, Declaration on Planning, 1968)

3. Crisis of crises
"What finally makes all of our crises still more dangerous is that they are now coming on top of each other. Most administrations...are not prepared to deal with...multiple crises, a crisis of crises all at one time... Every problem may escalate because those involved no longer have time to think straight." (John R Platt. What we must do Science, 1969)

4. Interwoven networks
"Society is not a crowd or cluster or clump of human beings; it is a set of networks of relations between human beings. Every human being is linked with others in a number of networks which are not mutually exclusive and are also not coextensive with each other." (Arnold Toynbee. Aspects of Psycho-history. Main Currents in Modern Thought, 1972)

5. Mismatch between organizations and problems
"The map of organizations or agencies that make up the society is, as it were, a sort of clear overlay against a page underneath it which represents the reality of the society. And the overlay is always out of phase in relation to what's underneath: at any given time there's always a mismatch between the organisational map and the reality of the problems that people think are worth solving... There's basically no social problem such that one can identify and control within a single system all the elements required in order to attack that problem. The result is that one is thrown back on the knitting together of elements in networks which are not controlled and where the network functions and the network roles become critical." (Donald Schon. What can we know about social change?, BBC Listener, 1970)

6. Entrapment by problems
"When anything becomes a problem we are caught in the solution of it, and then the problem becomes a cage, a barrier to further exploration and understanding." (J Krishnamurti. The Urgency of Change, 1971)

7. Arrogance of the disciplines
"...how is a practitioner of any one discipline to know in a particular case if another discipline is better equipped to handle the problem than is his? It would be rare indeed if a representative of any one of these disciplines did not feel that his approach to a particular organizational problem would be very fruitful, if not the most fruitful..." (R L Ackoff. Systems, organizations, and interdisciplinary research, General Systems, 1960)

8. Helplessness in the face of complexity
"Because our strength is derived from the fragmented mode of our knowledge and our action, we are relatively helpless when we try to deal intelligently with such unities as a city, an estuary's ecology, or the quality of life." (Editorial. Fortune, 1970)

9. Dependence on information
"Today, as we have seen, information is not primarily the triumphant standard of progress. It is the only means of maintaining sufficient control of evolution in order that humanity, strengthened by its knowledge and experiences and making appropriate use of all available information, can always maintain itself ahead of any threat which may lead to catastrophe." (Helmut Arntz, President, International Federation for Documentation, 1975)

10. Information overload
"The problem is that in most, if not all spheres of inquiry and choice, quantities of raw information overwhelm in magnitude the few comprehensive and trusted bodies or systems of knowledge that have been perceived and elaborated by man... Where, for example, does the novice urban mayor turn to comprehend the dynamic interrelationships between transportation, employment, technology, pollution, private investment, and the public budget; between housing, nutrition, health, and individual motivation and drive? Where does the concerned citizen or Congressman interested in educational change go for the best available understanding of the relationship between communications, including new technology, and learning?" (McGeorge Bundy. Managing knowledge to save the environment, US House of Representatives, 1970)

11. Category obsolescence
"The most probable assumption is that every single one of the old demarcations, disciplines, and faculties is going to become obsolete and a barrier to learning as well as to understanding. The fact that we are shifting from a Cartesian view of the universe, in which the accent has been on parts and elements, to a configuration view, with the emphasis on wholes and patterns, challenges every single dividing line between areas of study and knowledge." (P F Drucker, The Age of Discontinuity; guidelines to our changing society. 1968)

12. Evocation of fragmentation
"...the penalty for any principle which fails to express the whole is the necessity to co-exist with its opposite." (Lancelot Law Whyte. The Next Development in Man, 1950)

13. Challenge of synthesis
"...in face of the growing specialization of thought and action brought about by diversification in research and the division of labour, Unesco has a duty to promote interdisciplinary activities and contacts and to encourage broad views, in short, to emphasize the vital importance of the spirit of synthesis for the health of our civilization. I say vital advisedly since man - and I mean his essence, which is to say his judgment and his freedom of choice - is just as likely to be smothered by his knowledge as paralysed by the lack of it. Similarly, he is quite as likely to lose his identity in the confusion of competing social pressures as to atrophy in the condition known as under-developed."

(René Maheu, Director-General of Unesco, Address to a symposium on science and synthesis, 1967)

14. Integration of elements of thought
"*The synthesis we need involves a better integration of the elements of our thought, policies and institutions in order to solve national problems effectively, via the efficient achievement of national goals through the use, where appropriate, of such means as science and technology.*" (Robert W Lamson, Office of Exploratory Research and Problem Assessment, National Science Foundation, USA as read into the Congressional Record 93rd Congress, 2nd Session, 1974 by Charles S Gubser)

15. Facilitation of new forms of knowledge
"*Interdisciplinary knowledge can only develop through interdisciplinary education; it is a question of facilitating the emergence of a new form of knowledge... Whilst operating according to the norms of his specific dimension, the researcher must be able to encompass a mental space vaster than the epistemological cell within which his research runs the risk of confining him... The new understanding must be based on an affirmation of the functional unity of the human being as a focal point for all research intentions in the different domains of knowledge... This new understanding must be embodied in a new pedagogy, oriented to compensating for the deficiencies of specialization by stressing the combined unity of all domains of knowledge.*" (Georges Gusdorf. Interdisciplinaire (connaissance). In: Encyclopedia Universalis)

16. Need for unified personalities
"*The development of a world culture concerns mankind at large and each individual human being. Every community and society, every association and organization, has a part to play in this transformation; and no domain of life will be unaffected by it. This effort grows naturally out of the crisis of our time: the need to redress the dangerous overdevelopment of technical organization and physical energies by social and moral agencies equally far-reaching and even more commanding. In that sense, the rise of world culture comes as a measure to secure human survival. But the process would lose no small part of its meaning were it not also an effort to bring forth a more complete kind of man than history has yet disclosed. That we need leadership and participation by unified personalities is clear; but the human transformation would remain desirable and valid, even if the need were not so imperative. The kind of person called for by the present situation is one capable of breaking through the boundaries of culture and history, which have so far limited human growth. A person not indelibly marked by the tattooings of his tribe or restricted by the taboos of his totem: not sewed up for life in the stiff clothes of his caste and calling or encased in vocational armour he cannot remove even when it endangers his life. A person not kept by his religious dietary restrictions from sharing spiritual food that other men have found nourishing; and finally, not prevented by his ideological spectacles from ever getting more than a glimpse of the world as it shows itself to men with other ideological spectacles, or as it discloses itself to those who may, with increasing frequency, be able without glasses to achieve normal vision. The immediate object of world culture is to break through the premature closures, the corrosive conflicts, and the cyclical frustrations of history. This breakthrough would enable modern man to take advantage of the peculiar circumstances today that favor a universalism that earlier periods could only dream about.*" (Lewis Mumford. The Transformations of Man. 1956)

17. Mental defences
"*...That since wars begin in the minds of men, it is in the minds of men that the defences of peace must be constructed...a peace based exclusively upon the political and economic arrangements of governments would not be a peace which could secure the unanimous, lasting and sincere support of the peoples of the world...*" (Constitution of the United Nations Educational, Scientific and Cultural Organization)

18. Fostering self-knowledge
"*The relations between world culture and the unified self are reciprocal. The very possibility of achieving a world order by other means than totalitarian enslavement and automatism rests on the plentiful creation of unified personalities, at home with every part of themselves, and so equally at home with the whole family of man, in all its magnificent diversity... Without fostering such self-knowledge, balance, and creativity, a world culture might easily become a compulsive nightmare.*" (Lewis Mumford. The Transformations of Man, 1956)

19. Concealed neurotic processes
"*The fact which confronts us is that cultural change is limited by the restrictions imposed on change in individual human nature by concealed neurotic processes. At the same time there is continuous cybernetic interplay between culture and the individual, ie between the intra-psychic processes which make for fluidity or rigidity within the individual and the external processes which make for fluidity or rigidity in a culture. It would be naive to expect political and ideological liberty to give internal liberty to the individual citizen unless he had already won freedom from the internal tyranny of his own neurotic mechanisms... Therefore, insofar as man himself is neurotogenically restricted, he will restrict the freedom to change of the society in which he lives. This interplay is sometimes clearly evident, sometimes subtly concealed; but it is the heart of the solution of the problem of human progress.*" (Lawrence S Kubie. The nature of psychological change and its relation to cultural change. In: Ben Rothblatt (Ed) Changing Perspectives on Man, 1968)

20. New vision of selfhood
"*Man's principal task today is to create a new self, adequate to command the forces that now operate so aimlessly and yet so compulsively. This self will necessarily take as its province the entire world, known and knowable, and will seek, not to impose a mechanical uniformity, but to bring about an organic unity, based upon the fullest utilization of all the varied resources that both nature and history have revealed to modern man. Such a culture must be nourished, not only by a new vision of the whole, but a new vision of a self capable of understanding and co-operating with the whole. In short, the moment for another great historic transformation has come. If we shrink from that effort we tacitly elect the post-historic substitute. The political unification of mankind cannot be realistically conceived except as part of this effort at self-transformation; without that aim we might produce uneasy balances of power with a temporary easing of tensions, but no fullness of development.*" (Lewis Mumford. The Transformations of Man, 1956)

21. Choice
"*We can either involve ourselves in the recreative self and societal discovery of an image of humankind appropriate for our future, with attendant societal and personal consequences, or we can choose not to make any choice and, instead, adapt to whatever fate, and the choices of others, bring along.*" (Center for the Study of Social Policy of the Stanford Research Institute. Changing Images of Man, 1974)

In summary: It seems appropriate to attempt to bring together and interrelate within one framework information on: the problems with which humanity perceives itself to be faced; the organizational, human, and intellectual resources it believes it has at its disposal; the values by which it is believed any change should be guided; and the concepts of human development considered to be either the means or the end of any such social transformation.

Problems, organizations, concepts and human development are usually considered as though they were unrelated. But it is necessary to have a progressively more integrated conceptual structure in society before the interrelationships between the newer problems can be perceived. Both are needed before an attempt can be made to interrelate organizational units to handle the interlinked problems. Individual ability to tolerate and comprehend the complexity and dynamism of these interrelationships is directly related to the individual's own degree of personal development. Furthermore, a general increase in integration in any of these four domains will tend to increase integration in the other three. Equally, progressive fragmentation in any of the domains will provoke disintegrative tendencies in the others.

Even if the constraints make it impossible to achieve a satisfactory result through this particular exercise, it is to be hoped that through the process outlined here it will be possible to learn more about how information from very diverse sources can be concentrated and structured to the critical level required to provide the kind of integrative overview necessary for all to develop a sufficiently complex and strategically sound response to the world problem complex as it is now emerging.

Human Development   H

Scope

The purpose of this section is to describe briefly the complete range of concepts of human development with which people identify, consider meaningful or reject in their search for growth and fulfillment in life. The scope of this section has been deliberately extended beyond the unrelated concepts accepted with great caution by intergovernmental agencies: the job-fulfillment orientation of ILO, the health-oriented concepts of WHO and the education-oriented concepts of UNESCO. It includes concepts legitimated by the psychological and psychoanalytical establishments as well as those promoted by the various contemporary growth movements. It also includes concepts from religions and from belief systems of different cultures. Entries are included on explicit concepts of human development and on therapies, activities or experiences in which a particular understanding of human development is implicit.

Sub-sections

The section contains 4,051 entries. It is divided into two parts: Section HH and Section HM. Section HH describes 1,292 concepts of human development. Section HM endeavours to describe 2,759 modes of awareness, namely the experiential states associated with different stages in the process of human development as perceived by different groups (and preferably using wording with which such groups would identify).

The entries have been interlinked by 15,027 cross-references. These either indicate relationships between more general or more specific concepts, or, especially in Section HM, the relationship between succeeding modes of awareness in some process of human development (whether linear or cyclical).

Method

The procedures used in preparing this section are discussed in detail in Section HZ.

Overview

Detailed comments are provided in Section HZ. This section indicates ways in which people struggle within themselves for fulfillment and the experiences associated with that struggle which they find meaningful (whether or not such experiences are considered totally deluded or inappropriate by different scientific or religious establishments). That many of these experiences cannot be effectively "put into words" is indicated by the use of metaphors or symbols in naming them. These appear as strange to Western eyes as do others to Eastern cultures.

Index

A keyword index to entries is provided in Section HX. The keywords are also incorporated into the index for Volume 2 (Section X)

Bibliography

Bibliographical references, by author, are given in Section HY.

Context

The contents of this section may be considered as complementing the other sections in ways such as the following:

Integrative knowledge: By the manner in which advances in the integration of knowledge are paralleled by integration of the individual and of society and require such integration in order to be meaningful.

World problems: By the manner in which human development is frustrated and impeded by world problems, and through the world problems engendered by the unbalanced pursuit of particular forms of human development (or by the conflict between different forms of human development).

Metaphors and patterns: By the manner in which human development options are communicated, and through the evolution of new forms of communication following efforts to communicate new insights into human development possibilities.

Transformative approaches: By the importance of taking into consideration different modes of human development in designing techniques to be of relevance to people with different needs and modes of response.

Human values: By the manner in which values acquire their significance through the pursuit of different modes of human development and through the association of many specific modes of awareness with the experience of particular values.

Concepts

Rationale

The concept of human development is central to the preoccupations of educators and individuals concerned with their own maturation and fulfillment. It is also central to the declared concerns of policy-makers and those concerned with social development and learning. There is however no agreement on what is meant by the term and little recognition of the variety of meanings associated with the term (and its synonyms) in different sectors of society. Each group implicitly rejects concepts of human development favoured by others, usually without any understanding of what is implied. As a consequence reports on human development tend to stress relatively narrow ranges of meanings thus failing to recognize the variety of concepts with which people identify, consider meaningful or reject in their search for growth and fulfillment in life, however misguided their understanding from others' point of view.

Contents

The section contains brief descriptions of 1,292 concepts of human development. Entries are included on explicit concepts of human development and on therapies, activities or experiences in which a particular understanding of human development is implicit. The entries do not include subjective experiences described in the following section.

Method

The information used was obtained from a wide range of specialized reference books as well as from reports published by intergovernmental organizations. A detailed description of the method is given in Section HZ.

Index

A keyword index to entries is provided in Section HX. The keywords are also incorporated into the index for Volume 2 (Section X)

Bibliography

Bibliographical references, by author, are given in Section HY.

Comment

Detailed comments are given in Section HZ.

Reservations

Many of the concepts of human development do not lend themselves to brief verbal description, particularly where their proponents sharply distinguish them from other concepts that could well be described using the same words. In editing the descriptions no attempt has yet been made to clarify such difficulties.

Possible future improvements

In addition to refining entries and extending the range, consideration could again be given to the possibility of including specific bibliographic references as was done for the previous edition.

ENTRY CONTENT AND ORGANIZATION

Ordering of entries
Entries are in **numeric order**. Entry numbers have been **allocated randomly**; they have no significance other than as a permanent point of reference to facilitate indexing, cross- referencing, and updating between editions.

Index access to entries
The location of an entry in this sub-section may be determined from:
- the **Volume Index** (Section X) on the basis of keywords in the name of the entry or its alternate names
- the **Section Index** (Section HX), if the entry forms part of a numbered set

Structure of entries
Entries may be composed of the following descriptive elements:

(a) **Entry number** This number has **no significance**, except as a convenient method of identifying the entry (particularly for indexing purposes), of filing information on it, and as an identifier to which cross-references from other entries (possibly in other sections) may refer in this and future editions. The first letter of the entry number refers to the section of this volume in which the sub-section, denoted by the second letter, is located.

(b) **Entry name** This is printed in bold characters. It may be followed by alternative names when appropriate. In certain cases these may include non-English terms when these are recognized as being more precise than the English translation.

(c) **Description** Brief description of the concept or approach.

Bibliographical references
Entries may be followed by some bibliographical references. Further bibliographical details to the publications cited are given, by author, in the sction bibliography (Section HY)

Cross-referencing of entries
At the end of any entry, there may be cross-references to other entries. These indicate the number and name of the cross- referenced entry, whether within this Section or in other Sections. There are 5 possible types of cross-references:
       Broader = Broader or more general concept
       Narrower = Narrower or more specific concept
       Related = Related or associated concept
       Preceded by = Concepts that precede the entry in some developmental sequence
       Followed by = Concepts that follow the entry in some developmental sequence

♦ **HH0002 Psychosynthesis**

Description Psychosynthesis is a comprehensive psychological and educational approach to the development of the whole person. With roots in East and West, it is a synthesis of many traditions. While most Eastern approaches have tended to emphasize the spiritual side of being, and Western approaches have usually focused on the personality level, psychosynthesis attempts to view man as a whole and to accord to each aspect its due importance. It postulates in the individual a transpersonal essence, and at the same time holds that the individual's purpose in life is to manifest his essence, or self, as fully as possible in the world of everyday personal living. Everyone is in a constant state of growth, ethically, aesthetically and in religious terms, and the will is seen as integral in the formation of choices and decisions which affect future development.

The concept and experience of a higher unconscious, or superconscious, is a central aspect to psychosynthesis. Like traditional psychoanalysis, psychosynthesis recognizes a primitive or lower unconscious (source of our atavistic and biological drives,) but it also posits a superconscious, an autonomous realm from which originate man's more highly evolved impulses: altruistic forms of love and will, humanitarian action, artistic and scientific inspiration, philosophic and spiritual insight, and the drive for purpose and meaning in life. Psychosynthesis maintains that man suffers not merely from repression of his basic biological drives, but that he can be equally crippled by repression of the sublime, by failure to accept his highest nature. It is therefore concerned both with integration of material from the lower unconscious, and with releasing and actualizing the content of the superconscious (that which is beyond our ordinary level of awareness). To this end, it has developed a wide range of techniques for contacting this realm, and establishing a bridge with that part of our being where true wisdom is to be found. This realm is accessible, in varying degrees, to the seeker, and can provide a great source of energy, inspiration, and direction.

One of the central tenets in psychosynthesis is the existence of the self as an entity supraordinate to the various aspects of the personality (body, feelings, and mind). This self is viewed as a center of awareness and purpose, around which integration of personality takes place. A distinction is made between the personal self (the I or center of individual consciousness) and the transpersonal Self, which is a deeper and inclusive center of identity where individuality and universality blend. The two central functions of the personal self are consciousness and will. The consciousness aspect of the self enables one to be clearly aware of what is going on within and around him, to perceive without distortion or defensiveness. To the extent that an individual is able to achieve this vantage point, the claims of the personality and its tendency to self–justification no longer stand in the way of clear vision. There are a variety of techniques in psychosynthesis to help people gain access to this vantage point, from which the most effective work on oneself can be done.

A major difficulty for individuals in learning to act from center is the larger number of false identifications they make with partial aspects of themselves. Much basic work in psychosynthesis is aimed at helping people to recognize and harmonize subpersonalities, so that they are no longer helplessly controlled by them but can learn to bring them increasingly under conscious direction of the personal self. This involves the central processes of dis–identification from all that is not the self, and self–identification, or the experience of our true identity as a center of awareness and purpose.

The overall psychosynthetic process consists of two consecutive stages (though there usually is a considerable degree of overlap): personal psychosynthesis in which integration of the personality takes place around the personal self and in which the person attains a level of functioning in terms of his work and relationships that would generally be considered optimally healthy by current standards of mental health; and transpersonal psychosynthesis, in which the person learns to achieve alignment with, and to transmit the energies of the transpersonal Self; thus manifesting social responsibilities, a global perspective, altruistic love, and transpersonal purpose. There are a wide variety of methods employed in psychosynthesis to meet the diversity of needs presented by different situations and different people. Some of the methods more commonly used include guided imagery, movement, gestalt, self–identification, creativity, meditation, training of the will, symbolic art work, journal–keeping, ideal models, and development of intuition. The emphasis is on fostering an on–going process of growth that can gain momentum and bring about a more joyful, balanced actualization of one's life.

Through the will of the personal self individuals gain freedom of choice, the power of decision for their actions and the ability to actively regulate and direct the many personality functions. In this way they can free themselves from helpless reaction to unwanted inner impulses, and to the expectation of others, and (being truly centered,) become able to choose a path in accordance with what is best within themselves. This implies development of a sense of values and a healthy functioning of the will, two basic aims of psychosynthesis.

The transpersonal Self, through the mediating function of the superconscious, acts as a source of wisdom, and of love. Through an increasing contact with the Self, individuals liberate the synthesizing energies that organize and integrate the personality, and become able to function in the world more serenely and effectively, in a spirit of cooperation and goodwill.

Refs Assagioli, Roberto *Psychosynthesis* (1965); Assagioli, Roberto *Dynamic Psychology and Psychosynthesis* (1958); Assagioli, Roberto *A New Method of Healing* (1927); Crawford, John D *Psychosynthesis* (1916); Psychosynthesis Institute *Psychosynthesis*.

Related Paideia (#HH0010) Accepting the shadow (#HH0204)
Spiritual development (#HH0017) Higher self (Psychosynthesis, #HM2970)
Conscious self (Psychosynthesis, #HM2123) Lower unconscious (Psychosynthesis, #HM2790)
Higher unconscious (Psychosynthesis, #HM2057) Middle unconscious (Psychosynthesis, #HM2306)
Collective unconscious (Psychosynthesis, #HM2811)
Ways to spiritual realization (Esotericism, #HH2665)
Personal consciousness field (Psychosynthesis, #HM2383).

♦ **HH0003 Psychotherapy**

Description The term (meaning literally treatment of the mind or psyche,) refers to any of a variety of psychological means used to modify mental, emotional, and behavioural disorders, including the relief of distress and disability. Drugs may be used as aids in the process, but the process is primarily based on the use of words and actions that are believed by the individual, the therapist and the group to which they belong, to have healing powers and that create an emotionally charged relationship between them. This therapeutic relationship assumes that the parties to the relationship share certain values mutually and are convinced of the value of self–examination to make conscious the unconscious attitudes, fears and conflicts that determine behaviour.

The components common to the many varieties of psychotherapy are: one or more persons (patients) with some awareness of neglected or mishandled life problems; one or more persons (therapists) with relative lack of disturbance who perceive the distress of the patients and believe themselves capable of helping the patients to reduce distress; a positive regard of patients for therapists and vice versa; understanding and empathy of therapist for patient (the therapist enters into the patient's suffering and often suffers vicariously); perception by patient of the positive regard for and empathic understanding of him by the therapist; provision by the therapist of more correct information for the patient regarding the realities of his environment; help that the patient may achieve a better self–evaluation; emotional catharsis; a gradually increasing number of tasks for the patient to perform between therapy sessions in applying new information about himself and his environment; and a gradual process whereby the patient learns to become independent of the therapist.

The many varieties of psychotherapy may be distinguished as focusing primarily on one of the following: conceptual restructuring (attitude change or insight); changing emotional states or emotional responsiveness and sensitivity; modifying behaviour. Methods are supportive (using the patient's own resources to restore balance); re–educative (aimed at the individual achieving greater insight into conscious conflicts); and reconstructive (where emphasis is on insight into unconscious conflicts). All forms of psychotherapy are basically a discussion between two persons. The simplest form is verbal reassurance to an anxious person. The term *general psychotherapy* is used to denote all influences assisting a patient in fighting disease; *specific psychotherapy* refers to specific treatment of patients with borderline forms of neuropsychic disorder.

Psychotherapy has had a significant impact on Western culture and continues to influence attitudes to education, law enforcement and many other areas of life. It is a major tool for research into the intricacies of human nature.

Refs Carpenter, J T *Meditation, Esoteric Traditions*; Ellis, A *Humanistic Psychotherapy* (1973); Frank, J D *Persuasion and Healing* (1961); Reynolds, David K *Naikan Psychology* (1983).

Narrower Taiken (#HM2858) Rebirthing (#HH1330) Group therapy (#HH0851)
Dream therapy (#HH2116) Naikan therapy (#HH1324) Morita therapy (#HH1121)
Shadan therapy (#HH0483) Family therapy (#HH0362) Primal therapy (#HH2087)
Gedatsu therapy (#HH0972) Radical therapy (#HH2675)
Psychobiologic therapy (#HH1236).

Related Shaman (#HH0973) Psychoanalysis (#HH0951) Hasidism (Judaism, #HH0597)
Physical therapy (#HH0379) Analytical psychology (#HH1420)
Soul care (Christianity, #HH1199) Structural integration (#HH0082)
Spiritual reductionism (#HH1981) Psychotherapeutic self examination (#HH0922)
Psychospiritual therapy (Christianity, #HH4201).

♦ **HH0004 I Ching** (Taoism)

Book of Changes

Description The Book of Changes embodies a special system which represents the results of three thousand years of Chinese reflection on the nature of change. In Chinese thought it is only in constant change and growth that life can be grasped at all; the opposite of change is neither rest nor standstill, for these are aspects of change. Rather the opposite of change is held to be regression, or growth of what should decay. Change is therefore natural movement, namely development which can only reverse itself by going against nature. The special importance and use of the system in relation to human development lies in its application of this notion to change within individuals and social groups as well, clarifying the manner in which man's growth and development takes place in typical forms and through interlinked transformations governed by laws of change.

The concept of change is here conceived not as an external, normative principle that imprints itself upon phenomena, but rather as an inner tendency according to which development takes place naturally and spontaneously and with which it is wise to be consciously in harmony. The occidental notion of progress, incorporated in the idea of cyclic movement by the image of the spiral, is alien to this system as being an attempt to exalt the new over the old in a manner which does not accord with change in nature. Here the concept of a cycle of transformations, which eventually return to any initial condition, is basic.

The ultimate frame of reference for the changes identified in the system is therefore the unchanging; the world is held to be a system of homogeneous relationships. As a foundation for the emergence of these relationships, two interacting primordial principles (of determination, and of dependence,) are distinguished. They are united by a relation based on homogeneity; they do not combat, but complement each other in their complex interplay. Their difference in level creates a potential, by virtue of which movement and living expression of energy become possible.

The system contains the patterns and scope of all possible (qualitative) transformations. These patterns are also present in the mind of man, the insights into the cosmic process being related directly to the human psyche and to social relations. So that everything that undergoes transformation must obey the laws prescribed by the mind of man, who is therefore free to act to his own advantage in the light of these laws governing the perceived reality of space–time conditions. The system is so widely applicable because it contains only those relationships which are so abstract that they can find expression within every framework of reality.

The special figurative notation used in the system provides representations of the essence of all earthly phenomena. In the manner by which they interact and are transformed into one another within the system, they provide a representation of the interrelationships of all events and conditions, both in the macrocosm and in the microcosm. The mathematically perfect structure of this notation and the absolutely logical construction of the system as a whole thus yield a strict norm which underlies individual, changing situations and at the same time provides the frame for life in all its comprehensiveness.

With the help of an understanding of the system, it is possible for an individual to arrive at a complete realization of his innate capacities. This unfolding rests on the fact that man has innate capacities resembling the fundamental principles of the cosmos, of which he is a microcosm. Since the laws governing the relationships between these principles are reproduced within the system, the individual is provided with the means of shaping his own nature so that his inborn potentialities can be completely realized.

Aside from its descriptive function, which serves as a meditation aid to further intuitive understanding of conditions in the world, the system has also been extensively used for prediction. The possible change relationships between events or conditions, represented by the manner in which the figures transform into one another, provide a comprehensive view of the laws relating any sequence of events, thus permitting prediction of a pattern of future conditions as a guide to action. An individual can therefore select courses of action to ensure conditions most facilitative for his development, before a sequence of events has actually begun. For the questioner, an ordered framework within a fully rounded system is therefore available, within which a point to be established would give his momentary situation and what it implied for his own further growth. For the individual, this map of possible changes provides him with the opportunity of shaping his life harmoniously, so that life in its turn becomes a reproduction of the laws of change. By embracing, through the system, the essential meaning of the various situations of life, an individual is in a position to shape and develop his life meaningfully, acting in accordance with order and sequence, and responding in each case according to the requirements of any situation. The individual's actions are thus set in order, to the satisfaction of his mind; for when the microcosmic conditions with which he may be faced (within the possible pattern of changes) are meditated upon, he intuitively perceives their interrelationships in the macrocosm. Thus the major value of the use of the system, for prediction in everyday life, lies in its ability to shift the questioner's point of view and to lift him out of personality–bound perception. It enables him to look at events from the point of view of ever transforming cycles of change, whilst at all times impressing upon him the need for self–knowledge.

How the system works is not understood, particularly in the West, since it is based on principles alien to the occidental mind. (The Chinese themselves have always held that the many levels of meaning contained within the framework of the system can only be assimilated through prolonged reflection and meditation). It is suggested, however, that in part it simply provides a functional framework for the intuition of the user, particularly when used as a predictive device. The view of the cosmos embodied in the system is in a way comparable to that of the modern physicist whose model of the world is a decidedly psycho–physical structure. The microphysical event includes the observer just as much as the reality underlying this system's structure comprises subjective (namely psychic) conditions in the totality of the momentary situation. Just as the physicist's principle of causality describes the sequence of events, so the principle of synchronicity of the Chinese mind deals with the coincidence of events, the starting point for organizing observation is the unity of the universe, as experienced by the individual at any one moment, namely as an underlying, all-inclusive flux. The principle of synchronicity takes the coincidence of events in space and time as meaning something more than mere chance, namely a peculiar interdependence of objective events amongst themselves as well as with the subjective (psychic) states of the observer (s).

Refs Lee, Jung Young *The I Ching and Modern Science* (1971); Whilhelm, Richard *The I Ching or Book of Changes* (1967).

Broader Maps of the mind (#HH8903).

Related Change (#HH1116) Totality (#HH0534) Divination (#HH0545) Synchronicity (#HH4269) Feng-shui (Chinese, #HH2901) Human development (Taoism, #HH0689) Balancing yin and yang (Taoism, #HH4997).

♦ HH0007 Colour therapy

Chromotherapy

Description Universal interpretations of specific colours associated with the environment of primitive man (for example: yellow - sun - day - activity; dark blue - night - ceasing activity) determine that variations in colours can cause physiological responses. Since colour affects the physical and mental condition, it may be used to facilitate therapeutic personality change. The Luscher colour test - a method of determining physical and psychological information through an individual's choice or rejection of particular colours - has been used in a wide range of therapeutic and counselling situations.

Refs Birren, Faber *Color Psychology and Color Therapy* (1961).

Broader Therapy (#HH0678).

♦ HH0010 Paideia

Lifelong educational transformation of the human personality

Description The Greek term paideia refers to education looked upon as a lifelong transformation of the human personality, in which every aspect of life plays a part. Unlike education in the traditional sense, paideia does not limit itself to the conscious learning processes, or to inducting the young into the social heritage of the community. By considering character formation and the appreciation of values of equal importance as the acquiring of knowledge, it becomes the task of giving form to the act of living itself: treating every occasion of life as a means of self-fabrication, and as part of a larger process of converting facts into values, processes into purposes, hopes and plans into consummations and realizations. It is therefore not merely learning: it is a making and a shaping; and man himself is the work of art that paideia seeks to form. Some leisure activities, particularly those with an element of risk or uncertainty, may be considered a necessary part of paideia as they allow for creative expression and may lead to flow experience and ego-transcendence.

Related Psychosynthesis (#HH0002) Spiritual development (#HH0017).

♦ HH0011 Fasting

Abstinence

Description Fasting (complete or partial abstention from food of any kind) or abstinence (absention from particular kinds of food, in particular meat or meat products), may be undertaken as an exercise in restraint, as a religious practice - penitence, propitiation, purification rites, mourning, for example - or in accordance with custom. It may also be used as a means of obtaining vivid dreams or visions. The purpose of the fast may be as an ascetic practice or as an exercise in self control, and may be accompanied by other austerities. Such practice is a recognised as a valuable means to self-renunciation and self-discipline inherent in the way of devotion. Abstention from foods but not liquid can be undertaken by the normal person for from 20 to 40 days without starvation, although very much shorter periods are far more usual. Absolute fasts - when even liquids are not taken - are also referred to, but naturally these can only occur for short times (up to three days).

Regular weekly fasts were practised in the early Christian Church; and the abstaining from certain foods was also established very early in the Church's history. Periods of the year during which fasting or abstinence are observed are common in many traditions: - the Christian *lent*, observed not only during the 40 days preceding Easter but sometimes also, in the Eastern Church, the Lent of the Holy Apostles (June 16-18), Mary's Lent (August 1-14) and preceding Christmas (15 Nov - 24 Dec); the Islamic Ramadan when, for the month commemorating the receiving of the Qu'ran, no food or drink is consumed between the hours of sunrise and sunset. In many traditions, times of fast are accompanied by prayer, worship and other religious practices, in particular absolute fasting before partaking of sacrificial food.

Fasting is associated with *spiritual detachment* releasing the spirit to *creative inventionality*, many find it a necessary accompanying means to prayer. It causes the individual to acknowledge contingency and come to terms with the reality of death. The physical experience of hunger, tedium and lassitude contrasts with experiences of intense lucidity, euphoria and engagement; there may be a feeling of ecstasy. Undesirable qualities which control the individual - anger, pride, fear, and so on - which might otherwise remain undetected, often surface during a fast and can therefore be faced up to and dealt with. Human cravings and desires can be controlled, the individual thus being liberated from slavery to appetites not only for food.

Refs Foster, Richard J *Celebration of Discipline* (1984).

Related Feasting (#HH0288) Saum (Islam, #HH4216) Asceticism (#HH0556) Temperance (#HH0600) Continence (#HH0070) Mortification (#HH0780) Spiritual poverty (#HH0190) Devotion (Christianity, #HH0349).

♦ HH0017 Spiritual development

Spiritual perfection

Description A principle aim of monasticism and related religious endeavours is the spiritual development or spiritual perfection of the individual, in the imperfect state perceived as mistakenly identifying himself with an ego which is not in fact his true self. Discovery of the true self can only be accomplished following a process by which various imperfections, impediments and false identifications are overcome. Discipline, control, chastisement, self-mortification and other psychophysical procedures are used, accompanied by meditation and contemplation on some set of spiritual concepts together with other practices such as prayer, incantation, reading religious texts, worship, propitiation, pilgrimage, fasting and various forms of self-abasement or self-inflation.

Buddhism aspires to spiritual perfection in one lifetime, complete fulfilment and maturity expressed as an arhat, worthy to achieve nirvana, with perfect insight into relation between life, suffering, cause and effect.

In the Catholic Church, spiritual perfection is considered in two senses, the first referring to the completeness or wholeness possessed when participating, through grace, in the supernatural life of God; and the second referring to the attainment, through charity, of union with God. Some Christians would look on perfection as the ultimate in holiness and sanctification, claiming that while the two latter may be achieved, perfection may not; but perfection is usually understood as ethical and spiritual completeness resulting from a religious life of faith and discipline. St Paul, in his epistle to the Ephesians, speaks of being "imitators" of God and "walking in love"; and in St John's first epistle God's love is referred to as "being perfected in us". This perfection is generally considered the destruction of sin but not of all human limitations (lack of knowledge, for example). The notion of being perfected in love is central to Methodist belief, but still with the possibility of involuntary transgressions. The Anglican Church looks to Christ's incarnation "without spot of sin" as being able to "make us clean from all sin". Lutheran and Calvinist Churches, on the contrary, look to absolute perfection only achieved after this life, as did St Augustine, who maintained that God retains some sin in the most perfect of lives, this being essential for true humility. Mystical Christianity has much in common with Buddhism on this point, seeking perfection in union with the uncreated essence when there is no longer distinction between the creator and the created.

Over the centuries numerous sects have arisen which claim perfection exclusively for themselves, assuming they are in a state beyond moral law. However, few established religions, particularly in the West, emphasize spiritual development to any great extent. Rather, the emphasis is on moral and ethical development and on worship or spiritual exercises. The experience of the spiritual, or spirituality, has until recently been looked on as a private concern of the individual. Since Vatican II there has been some redressing of the balance towards the spiritual as the experience of the whole of life being shot through with faith.

Broader Human perfectibility (#HH0212).

Narrower Sufi path to perfection (Sufism, #HH1747).

Related Paideia (#HH0010) Completeness (#HM3070) Spirituality (#HH5009) Perfectionism (#HH1466) Arhat (Buddhism, #HH0233) Psychosynthesis (#HH0002) Moral perfection (#HH5098) Perfect man (Hinduism, #HM2183) Ontological perfection (#HH0794) Spirituality (Christianity, #HH0792) Perfection of the soul (Christianity, #HH7343).

♦ HH0018 Anthroposophical system

Anthroposophy

Description Literally *wisdom about man*, anthroposophy is a system developed by Rudolf Steiner as a means of arriving at true knowledge and liberation from enslavement to the material world. Derived but distinguished from *theosophy*, anthroposophy postulates two selves in each person, the lower self which knows and the higher ego which is known; and a previous mode of consciousness, from which man's present intellectual capacity has evolved, which brought direct experience of the *transcendental*, the clairvoyance of ancient times. It teaches that Christ is the one divine manifestation or *avatar*, making possible a "resurrection" of consciousness which will enable the clarity and objectivity of the present day intellect to be carried into new modes of spiritual perception - *Imagination*, *Inspiration* and *Intuition*. As a path of knowledge, anthroposophy aims to guide the spiritual in the human being to the spiritual in the universe; this knowledge being communicated in a spiritual way. "At the very frontier where the knowledge derived from self-perception ceases, there is opened through the human soul itself the further outlook into the spiritual world". Methods for growth are prescribed at four levels of apprehension corresponding to four levels in the self.

Refs Steiner, Rudolf *Esoteric Development* (1982).

Related Theosophy (#HH0660) Inspired consciousness (#HM3125) Spiritual intuitive cognition (#HM2829) Imaginative consciousness (Anthroposophy, #HM2901).

♦ HH0019 Archetype

Primordial image

Description An archetype is the symbolised or pictorial representation in the unconscious of an idea or mode of thought derived, according to C G Jung, from the life experience of all the progenitors of the person considered. Thus the collective experience of humanity, its typical ideas, modes of throught and patterns of reaction, are present in each individual psyche. They are by definition unconscious; their presence can only be intuited in the powerful motifs and symbols that give definite form to psychic contents. Archetypes are actualized both externally and in inner consciousness; they provide a system from which to act while at the same time comprise the images and emotions. They may be expressed in the profoundest human experience and arise in religious thought and in great literature; they are also apparent in psychotic thinking when the ego system is overwhelmed by the archetypes of the collective unconscious. Archetypes give rise to images in primitive tribal lore, in myths and fairy tales, and in the contemporary media. Evidence of archetypal images common to all individuals is demonstrated in the appearance of the same archetypes in totally different cultures and in the "mythicization" of historic personages. Myth can be said to be the "acting out" of archetypal images and gods the metaphors of archetypal behaviour. Particularly similar archetypes are associated with basic and universal experiences - birth, marriage, death. Because of the predominance of archetypal imagery in dreams and psychic experience in general, these have a numinous quality.

Context The term derives from the Greek and can mean the mould common to all, whether making the creation as a copy of reality or making man in the image of God. In traditional cultures, archetypes have a supernatural or transcendent origin as a sacred reality revealed to mankind; they provide models for social institutions and norms for behaviour categories. In modern terms the archetype is the pattern or paradigm that determines human experience.

Refs Stevens, Anthony *Archetype* (1982).

Narrower Androgyny (#HM0212).

Related Duality (#HH0872) Symbols (#HH0690) Rites of passage (#HH0021) Accepting the shadow (#HH0204) Schizophrenic fantasy (#HH0939) Collective unconscious (Jung, #HM0085) Emblem archetypes in the psyche (Tarot, #HM2201) Collective unconscious (Psychosynthesis, #HM2811).

♦ HH0021 Rites of passage

Life-crisis rites — Rites at adolescence — Maturity rituals

Description Virtually every culture provides ceremonies, rituals, or *rites of passage*, which assist the individual to make the transitions necessary for development, transitions which may be seen as death to the old circumstances and rebirth in the new. This may be marked by the initiated passing into a stylized embryonic state. A celebration invariably consists of three stages (although not every stage may receive the same emphasis) - *separation*, *transition* and *incorporation* - which can be thought of as corresponding to the three stages of the mystical life: purification, illumination and union; or the three stages of realization: experience, reflection and understanding. In many cases, the details of the rite are kept secret from the uninitiated, thus increasing its mystery.

The special circumstances surrounding a rite of passage raise an individual to a transitory *sacred* state and enable him and society to cope with the transition. Once incorporated into the new circumstances (mature status, marriage, warrior, motherhood), the public recognition of the change assists the individual on his to return to a mundane state and supports him in integrating the previously unencountered archetypal processes into his psyche.

Primitive culture divides life into a series of well defined positions with a definite rite of passage from one position to the next. The incidence and belief in the efficacy of such rites decreases with loss in belief in magic and with increasing secularization of society. It has been postulated that the absence of rites of passage in modern Western society and the decline in ceremony and ritual make it more difficult for the individual to overcome regressive and immature attitudes, leading to psychological problems, aggressiveness and immature behaviour.

Broader Rites (#HH0423).

Narrower Resurrection (#HH4553).

Related Archetype (#HH0019) Initiation (#HH0230)
Dreaming experience (#HM3908) Schizophrenic fantasy (#HH0939)
Psychic inertia (Physical sciences, #HM3421).

♦ HH0024 Arica training
Description Literally "open door" (language of the Bolivian Indians), Arica is the name of a system/school for spiritual development and the attainment of full enlightenment and freedom by means of nine levels of training which lead, through a combination of physical work and meditation, to permanently altered states of consciousness. The aim of Arica is the achievement of a metasociety where human relationships are based on unity rather than competition.

♦ HH0025 Induction technique
Description A number of physical techniques have been described for opening the awareness to the mystical or transcendent experience. These normally involve intense concentration on one activity which is said to achieve a heightened awareness through damping of all other inputs and consequent inhibition of the cerebral cortex. Systems may involve: the control of breathing; repetitive chanting or internal repetition of a mantra; ritual dance or gestures, including sequences of martial arts techniques; concentration on a source of light; or self-monitoring through biofeedback techniques.
Related Chanting (#HH1257) Biofeedback training (#HH0765)
Pranayama (Hinduism, Yoga, #HH0213) Dance-induced experience (#HM3213)
Light-induced experience (#HM3559) Mantra-induced experience (Yoga, #HM2464)
Concentration on the flow of breath (#HH0687).

♦ HH0027 Conscientization
Description This is the process by which the individual learns to perceive social, political and economic contradictions and to take action against the oppressive elements of reality. It involves seeing the distinction between the inevitability of nature and the changeableness of cultures, thereby seeing through arguments that blur the distinction and exploring available alternatives. Such a process acts as critical vehicle for challenging the social, political and religious status quo.
Refs Freire, Paulo *Pedagogy of the Oppressed* (1970).
Related Consciousness raising (#HH0427).
Followed by Involvement (#HH0030) Dehumanization (#HM3030).

♦ HH0029 Artistic education
Aesthetic education — Aesthetic development
Description The inculcating of an interest in beauty and the capacity to perceive beauty, and making it part of the individual's make-up. This should lead to the awakening of creative enthusiasm and a perception of the world from a higher level of existence, thus communicating with and understanding the natural and social environment.
Broader Art (#HH0570) Education (#HH0945).
Related Wabi (Japanese, #HM1365) Sabi (Japanese, #HM4350)
Ushin (Japanese, #HM2762) Furyu (Japanese, #HM3387)
Shibumi (Japanese, #HM3597) Child creativity (#HH0500)
Aesthetic LSD experience (#HM2542).

♦ HH0030 Involvement
Related Compassion (#HM2919).
Preceded by Conscientization (#HH0027).

♦ HH0031 Ke hare kegare cycle (Japanese)
Matsuri
Description In Japanese thought there is no dichotomy between the sacred and the profane. Thus the sacred festivals or matsuri consist of a cycle. They commence with self purification through removal of pollution, fasting and abstinence leading to a sacred state of *hare*. There is then the ceremony of prayers, praise and offerings to the kami or sacred beings, but these are followed with celebrations of a more mundane nature including dancing, theatrical performances, wrestling matches and feasts. Here the matsuri returns to the *ke* of ordinary life. When the energy of hare which has been channelled into the ke is exhausted there arises a state of kegare, or impure sacredness, defilement; the functioning of ordinary life is impaired and again a period of self purification, the hare or pure sacred events is enacted and the cycle recommenced. The sacredness of kegare is marked by the carrying out of normally impure acts; for example, shoes may well be worn indoors at funerals.
Related Kami hotoke (Japanese, #HH0408).

♦ HH0034 Potential directiveness
Goal-directedness
Description This defines the manner in which a particular individual is likely to behave under certain conditions; in other words, under an impulse which says "hunger", "thirst", and so on, and in a particular state of alertness, what a particular individual would do. The mental state of a person may thus be defined in terms of the goal-directed behaviour such a state would lead to under specific conditions.

♦ HH0035 Baptism
Description This is a symbolic representation of self-purification (from the Greek meaning plunge, dip), and of death to the old and rebirth to the new life which is being entered upon. Invariably it involves either complete immersion in water or washing or sprinkling with water and sometimes with blood or saliva. Although most familiar through the Christian ceremony (particularly baptism of babies, when it is combined with name-giving) it has its place in many cultures and the combination with the giving of names is not uncommon; after the ceremony, the initiate is a recognized member of the group. Infant baptism is also linked with the ritual cleansing of mother and child after parturition in order to remove ceremonial uncleanness and protect them from evil. Adult baptism may be of a similar passive nature, or may involve active reaffirmation of faith.
Broader Rites (#HH0423).
Related Initiation (#HH0230) Confirmation (#HH0433) Purification (#HH0401)
Reconciliation (Christianity, #HH0164) Human development (Christianity, #HH2198)
Spiritual rebirth (Christianity, #HH4098).

♦ HH0037 Integral yoga (Yoga)
Purna yoga — Perfect yoga — Satyeswarananda
Description Integral yoga is primarily concerned with developing the downflow of creative, inspirational energies from the supramental centres of an individual's being, with the goal of complete self-integration. It is a modern form of yoga derived from many of the traditional systems practised in India, incorporating the truth in each while maintaining a balance between the attributes emphasised. Rather than freedom from nature and society, integral yoga looks for emancipation by transforming the bonds of attachment to become instruments of spiritual development. It sees no separation between body and spirit, but rather discovers psychic integration through intelligent fulfilment of instinctual desires. In this way the individual finds a source of psychic energy available to him.
Refs Chaudhuri, Haridas *Integral Yoga* (1965).
Broader Yoga (Yoga, #HH0661).
Related Integral meditation (#HH0579).

♦ HH0039 Blessing
Description In addition to the general use of blessings or curses, traditional times for blessing are the ceremonial blessing of children by their parents, of the Church for the married (the marriage itself being simply a personal arrangement until quite late in the history of Christianity), and on the sale of goods or property. Two aspects of blessing are the calling down of God's bounty upon a person, a group or upon mankind; and the returning of thanksgiving to God. This may be in the context of a sacrament, as at the end of Eucharist, as a sign of the sanctifying of the congregation and the receiving of spiritual and temporal favour and as the sacrifice or surrender of these gifts to God; or in everyday life, as in greetings (Arabic "Salaam"; English "Goodbye" - or "God be with you"; French "Adieu", etc). The magical efficacy of the words, whether spoken or written, of a blessing (or a curse) has at times been held to exist whatever the intention of the person instrumental in the act.
An act of blessing may not be, or not only be, through words but also through physical contact, such as in the laying on of hands. Physical contact may indicate the passing of power to the blessed, a bestowal of *mana*; the touch acting as a means of conducting power. Blessing (or cursing) is generally thought more efficacious when carried out by someone in contact with the supernatural, and also by someone who is dying. The blessing of someone who has nothing else to give (that is, the very poor) or lacking in power (women, servants) is also traditionally more efficacious.
Refs Fox, Matthew *Original Blessing* (1983).
Related Cursing (#HH0594) Sacrifice (#HH0577) Blessedness (#HM2741).

♦ HH0042 Care
Love of neighbour
Description This is the selfless concern for others from which flow profound human deeds; the benefits arising from the active participation in such care being not only material improvements but the inspiration such sacrificial action produces.
Broader Love (#HH0258).
Related Compassion (#HM2919) Corporateness (#HH0080)
Love of God (Christianity, #HH5232).

♦ HH0043 Needs
Description Individual and social development are said to come together in the striving for satisfaction of human needs. However, there is some disagreement on the definition of such needs, some claiming that true needs are universal and objective, while others consider them to be subjective and dependent on society and the environment. These latter needs may perhaps be defined as desires, the desire for a particular means of satisfaction of a need which is universal: for example, the need for mobility may be translated into the desire for a car. The basic needs which require satisfaction may be considered under the headings *welfare*, *security*, *identity* and *freedom*. Material needs are usually better satisfied in developed countries while non-material needs in less developed countries, although this situation is changing.
Refs Alderfer, Clayton P *Existence, Relatedness and Growth* (1972); Aronoff, Joel *Psychological Needs and Cultural Systems* (1967); Byrd, B *Cognitive Needs and Human Motivation.*
Related Needs hierarchy (#HH0913).

♦ HH0045 Celibacy
Sexual abstinence
Description Religious celibacy is most clear in systems arising from dualistic tradition, which regard physical matter as the source of evil and self-realization as conditional on the spiritual, with the extinction of impulses arising from the physical and the environment. Some Buddhist orders and also priests and medicine men of indigenous American peoples practised strict celibacy; and the eightfold path of yoga as set down by Patanjali also demands restraint from any sexual behaviour. Celibacy, particularly female celibacy or *virginity*, before marriage is considered a virtue in most traditions. Temporary celibacy as a form of corporal purification has also been and is used in many traditions, although it is mainly in the the Christian tradition that totally abstaining from marriage is considered commendable. The Roman Catholic and Orthodox Churches exclude ordained clerics from marriage; and the Roman Catholic Church only allows ordination of married men under special conditions. Such tradition considers the freely undertaken abstinence from marriage for the purpose of practising chastity and of dedicating one's life wholly to the service of God to be an efficacious incentive to charitable behaviour to all mankind.
Narrower Virginity (#HH0272).
Related Continence (#HH0070) Brahmacarya (#HH0978) Purification (#HH0401).

♦ HH0046 Social development
Socio-economic development
Description Social development comprises improvement in living standards for the widest social masses; education, housing, health care and related concerns, increase in employment opportunities and improvement in the distribution of income and of social opportunities. Such preoccupations are not only justified social aims, they also improve the physical capability, the capacities and qualifications of man, all of which are of the utmost importance whenever economic progress is concerned. Similarly, the greatest possible involvement of all those able to work, in productive activities or in any other socially useful action, is not only a social objective, it can also have significant economic effects and help to raise man's abilities.
Individual development and that of society are intimately linked; however personal self-transformation may seem, it nevertheless resonates in society as a whole. Indeed, personal development may be seen as extending the self in harmony with society and with nature. Integration of the individual with society has been defined as extending the boundary of the self. In this context, society is seen as a network of extended selves. A *human-centred society* favours the individual development of its members, ensuring that such development is equally available to all, that the integrity of other societies is respected and that, while not pursuing a course that will threaten similar opportunities for future generations, those alive at present are not jeopardised for some uncertain benefit yet to come. In such a society economic development is looked on primarily as a means of human enhancement, and human development as a synthesis of individual and societal developments.
The importance of the individual in social development may be demonstrated by the fact that economic efficiency depends primarily on human qualities, a capacity for rational action, innovative ability, energy, etc. Not only is the contribution of each individual of concern, but also the way it expresses itself in the social structure, where it acquires a new dimension and different values. Because of the importance of the human factor in increasing the economic efficiency of each society, the transformation of man (of his behaviour and of his socio-productive features) becomes an important point of concentration in any development strategy. In this way, it becomes possible to harmonise social and economic development objectives. A development strategy oriented towards the transformation of man cannot be implemented simply by increasing social outlays and calling for change. It must assume the form of an integrated programme covering the

whole field and all factors affecting man.
Refs Levine, Louis *Personal and Social Development* (1963).

♦ HH0047 Vocational training

Human resource training
Description This is the process of training for a specific occupation which is given in special institutes or settings where the working conditions are closely similar to those existing on the production line. Economically, such training is designed to provide industry with the skilled human resources needed to keep abreast of progress and economic expansion. Socially, it is important not to debar major sectors of the population of working age from all the benefits of modern life as well as ensuring equality of opportunity.

♦ HH0053 Charisma

Bakarah (Islam)
Description Charisma, and the gifts it bestows, or *charismata*, may refer to any superhuman power received from contact with supernatural beings, which may include revelation and the power to mediate spiritual grace to others. This is a special inspiration - not, for example, the carrying out of routine priestly functions. Such supernatural powers or *siddhis* were said to have been possessed by the Yogins as a result of extreme ascetic practices. Similarly, in Islam, the term *bakarah*, or blessing, connotes the power to work wonders as a special gift from God. Originally denoting an extraordinary "gift of grace", for example the power of healing, in theological terms described as a gift of the *Holy Spirit*, the term charisma has been extended, in sociological terms, to denote an extraordinary quality possessed by persons or objects, in particular to evoke an immediate positive response from others. Charisma may be intrinsic to an individual or be derived from the sacred nature of his office; but in either case is frequently the means for change in otherwise rigid or hidebound, authoritarian situations, which authority is challenged by a *charismatic leader*.
Related Hero worship (#HH0653) — Mass hypnosis (#HH2134) — Siddhis (Yoga, Buddhism, #HH0380) — Charismatic renewal (Christianity, #HH3124).

♦ HH0055 Motivational development

Description Motivation is a concept that refers to some condition or force within the human organism that compels it to respond to stimulation or to perform some action. A human being has physiological needs, which in turn generate drives. Once these primary physiological needs are satisfied, they give rise to secondary drives. The ideational activities associated with such secondary drives undergo regular changes with development. Such motivational activities (including wishing, striving, hoping) may be distinguished from primary drives. They represent what the individual desires to experience or possess. Such activities are cognitive and have no necessary relation to overt action.
At any given stage of development some incentives (activating motivational states) are more potent than others and tend to dominate the child's ideational activities. Children differ in the ease with which an incentive can evoke a particular motivation.
Early motivational development is associated with efforts toward: gratification, reduction of uncertainty and anxiety, affiliation, genital stimulation, instrumental help, affection, effectance and hostility. No universal consensus has been reached on the list of human needs on the psychological level. They may include such needs as: achievement, deference, autonomy, exhibition, affiliation, dominance, aggression. They may also include: belongingness, love, esteem, and self-actualization. Some theories suggest that the motives of mankind are essentially the same from birth until death, whilst others hold that in the course of childhood development it is important first of all that the basic drives be gratified so that the child may later be freed to adopt less self-centered (growth) motives. Thus a child which has already known basic drive-gratification and security can in later life tolerate a frustration of these same drives more readily than a person whose whole personality is permanently pivoted on needs that were never adequately gratified. The latter theories allow for the extensive transformation in motives from infancy to maturity, or for the extreme diversity of motives found in adulthood. More recent theories tend also to allow for competence, self-actualization and ego autonomy as equally basic features of human motivation.
Refs Arnold, W J and Jones, M R (Eds) *Nebraska Symposium on Motivation*.
Broader Motive (#HH0181).

♦ HH0056 Spirit possession

Description A part of many religious systems and cults, the possession by a spirit may be a privilege (when the person possessed may become capable of spiritual healing, divination, or the detecting of witchcraft or the cause of some misfortune); or an affliction (see the numerous accounts in the Bible of the casting out of evil spirits). Spirit possession is initially a disorder and is not the same as being a medium, although it may lead to becoming one. The individual is impelled to carry out actions alien to his or her normal personality which are not remembered after the event. It may give rise to distressing physical symptoms, from faintness and exhaustion to paroxysms, trance, to the devouring of inedible and dangerous objects (knives, hot metals) and regurgitating them without apparent injury. Resistance to the frenzy may cause insanity or death. Such possession is particularly characteristic of religious traditions giving prominence to ancestral spirits, where the spirits possessing a *shaman* assist him in foretelling the future and in carrying out priestly functions.
Broader Communication with supernatural beings (#HH1306) — Anomalies in integrity of consciousness (#HM4521).
Related Voodoo (#HH0061) — Shaman (#HH0973) — Spirit (#HH0770) — Trance (#HM3236) — Frenzy (#HM3594) — Obsession (#HM2809) — Invocation (#HH1876) — Channelling (#HH0878) — Spiritualism (#HH0479) — Spiritual healing (#HH0458) — Possession (Psychism, #HM3111).

♦ HH0058 Community

Commune — Religious community
Description A group of people living together in fellowship in order to share their common customs and interests, and frequently participating in common ownership. As members of a community, individuals subordinate their own concerns to the concerns of a common cause, learning to depend on each other and serve the good of the whole.
Related Friendship (#HH0156) — Corporateness (#HH0080).

♦ HH0061 Voodoo

Description An African-based religion particularly prevalent in Haiti, with a complex system of duties, voodoo has came to be associated with any African-derived American ritual involving magic and spirit possession.
Related Spirit possession (#HH0056).

♦ HH0064 Personal consecration

Description This may signify the total dedication of a person to God, a setting apart, thus separating him or her from ordinary human ties and becoming exclusively the servant of God. It can also be taken to mean not only belonging to God but the free choice to accept and live in accordance with particular vows or responsibilities, whether those fundamental to baptism or these taken in addition, for example a vow of virginity. Such consecration is often accompanied by a rite or ceremony symbolizing the type of consecration involved. Once consecrated, the rest of life requires perpetual vigilance as the tension between divine absolutes and temporary realities are acted out. In many orthodox and devotional traditions, such as Jewish and Hindu, this implies the carrying out of even the most simple daily acts in accordance with divine precepts.
Related Consecration (#HH0238).

♦ HH0070 Continence

Abstinence — Self restraint
Description Continence is normally accepted to be the control or moderation of any *appetite*, whether for food, drink, sexual satisfaction or whatever. Greek philosophy considers such moderation necessary to avoid pleasure passing beyond the bounds of reason. The Christian tradition normally associates continence with control of sexual appetite, voluntary abstention from conjugal acts being a characteristic of many traditions at specific times, for example before or during a hunting trip or a battle, or a religious festival. Abstinence from particular kinds of food - whether meat in general (vegetarianism), pork (Jewish and Muslim faiths), beef (Hindu) or alcohol (a number of protestant sects), may becoming the distinguishing mark of a particular group and serve to underline the individual's membership of such a group. In any case, the disposition of *will* to exercise self-restraint, holding firm against the disorderly impulses inherent in sensual appetite, has the function of training the individual to face any circumstance without violent or disorderly reactions. Such self-control inherently gives responsibility in action and is compatible with freedom and self-understanding. Continence, a virtue of the will, is contrasted with temperance, when the desires themselves are modified.
Related Will (#HH0920) — Fasting (#HH0011) — Celibacy (#HH0045) — Watchings (#HH1008) — Temperance (#HH0600) — Self control (#HH0778) — Self responsibility (#HH0187) — Chastity (Christianity, #HH0583).

♦ HH0072 Mental imagery therapy

Directed fantasy — Psycholytic therapy — Hypnotic imagogy — Autogenic imagogy
Description Various forms of mental imagery, using impressions derived through one or several of the senses, and including dreams and waking fantasies, may be used by the individual himself or with the aid of a therapist to facilitate personality change.
Broader Therapy (#HH0678).
Related Fantasy (#HM0892) — Guided fantasy (#HH0627) — Pseudo-hallucination (#HM4336) — Hypnotic states of consciousness (#HM2133).

♦ HH0075 Psychedelic drugs

Psychotomimetic drugs — Hallucinogenic drugs — Mind-expanding drugs — Drug-induced experience
Description Psychedelic drugs can, under certain conditions, provide one of the most direct and effective ways to self-discovery and personal growth and have a long history of use in the cultures of India, Iran and of the American Indians, although adequately controlled clinical studies of psychedelic therapy are not yet complete. Under the influence of such agents, the normal inhibitors to the *superconscious* seem themselves to be inhibited, so that areas of consciousness beyond normal ego-consciousness are available. Under these conditions, the ego changes its usual context of perception and may even seem so under attack that a feeling of impending death is experienced.
Psychedelic drugs do not, however, automatically and necessarily facilitate changes in behavioural and personality patterns. When and if such changes take place, they are functions of: the basic intentions of the individual using the drug, the skill of the therapist (or other mentor), the environment and cultural milieu within which the drug is taken, and other factors. Such drugs are subject to abuse and may, in the case of individuals, have dangerous consequences for either physical or mental health. The apparent abuser of drugs may unwittingly be actually treating himself for some distressing psychological problem as much as for the achievement of euphoria or relief from dysphoria, but continuing use of the drug leads to diminishing mood-elevating effects and the need to increase the quantity taken even to remain what was previously regarded as normal. The dangers of this situation are aggravated by the fact that the drugs are distributed and sold illegally in most countries and are in consequence frequently adulterated by the substitution of either inert materials or active materials such as smaller quantities of more dangerous drugs, or even toxic substances.
Use of these drugs may give rise to any of a very large range of experiences. From a therapeutic point of view, an individual may develop increased ability to examine non-defensively his habitual actions and responses, to reinterpret experiences of the past, and to accept new insights (effectively from himself) regarding such matters. Symbols having universal or integrative meanings may be projected onto external objects or perceived within, thus leading to an integrative change of self-image, personality characteristics and life pattern. A form of higher awareness resembling the characteristics of mystic experience may also be experienced. This may include:
1. An awareness of inner undifferentiated unity through a fading away of the multitude of external and internal sense impressions (including time and space) and the usual sense of identity or individuality, leaving only pure consciousness, possibly without empirical content or distinctions.
2. An awareness of external unity or underlying oneness behind the empirical multiplicity of external sense impressions; the essences of external objects are experienced intuitively at the deepest level.
3. A loss of the usual sense of space and time (and possibly past and future) providing a transcendent experience of eternity and infinity.
4. A deeply positive mood defined by a sense of joy, blessedness, peace, and possibly love.
5. A sense of sacredness, namely a nonrational, intuitive, response of awe and wonder, and possibly a sense of reverence.
6. A sense of insightful knowledge or illumination felt at an intuitive, non-rational level and gained by direct experience. This is reinforced by a sense of the authoritative nature of the objectivity and reality of the experience. This may relate to being and existence in general or to the individual himself.
7. A paradoxical quality to the experiences which render them mutually contradictory according to conventional logic.
Such essentially transient experiences tend to lead to: increased integration of the personality, a stengthening of inner authority and dynamism of life, creativity and efficiency, optimism, happiness, joy, and peace; increased tolerance, sensitivity, love, and authenticity; improved philosophy of life, values, meaning and purpose, commitment, reverence for life, and appreciation of life.
Chemically the drugs are classified under indalkaloids (derivates of lysergine, dimethyltrptamine, bufotenin, psilocybin, ibogaine, harmine), derivates of phenylethylamine (mescalin), derivates of piperidine (belladonna-alkaloids and other anticholinergic substances, phencyclidine), and tetracannabinnols. The pattern of action of these drugs is similar but the time required to reach peak effect and the duration of that effect vary between the drugs.
Refs Aaronson, Bernard and Humphrey, Osmund *Psychedelics* (1971); Clark, Walter Houston *Chemical Ecstasy* (1969); Masters, R E L and Houston, Jean *The Varieties of Psychedelic Experience* (1967).

Narrower Bad trip (#HM0818) Aesthetic LSD experience (#HM2542)
Psychodynamic LSD experience (#HM3271)
Drug use for transcendent experience (#HH1123).
Related Sacred drugs (#HH0788) Psychopharmacology (#HH0394)
Alternative pursuits (#HH5110) Transcendental experience (#HM2712)
Modes of awareness associated with use of hallucinogens (#HM0801)
Modes of awareness associated with psychoactive substances (#HM0584).

♦ HH0077 Conversion

Description The change or transformation in a person's way of acting or thinking with reference to a particular object, which presupposes a prior attitude of opposition, hostility or indifference. This is usually considered in terms of conversion to a particular way or religion and implies a change in moral or spiritual direction, the experience of which will depend on the psychological constitution of each individual. It may be very sudden (as with St Paul's conversion on the road to Damascus) or happen in a gradual process; it may be predominately in accordance with intellectual motives (such as the need for truth), with a desire for a moral ideal of purity or goodness, or in response to an emotional appeal (as in mass conversions).
Conversion may be considered in the context of: (1) the traditional social and cultural background of the individual; (2) the transformation or process of the change taking place (conversion is often preceded by a period of anguish, despair and other difficulties), leading to richer self realization; and (3) transcendence, the encounter with the sacred that many believe is both the source and the goal of conversion. Although conversion is apparently an individual process, it has been argued that it is interactive, with consequences for the community, and thus part of an overall evolving process. The church recognizes: *moral* conversion – in response to conscience and the fundamental knowledge of good, evil and natural law; *religious* conversion – based on a consciousness of one's existence as a creature and dependence on a supreme being; and *confessional* conversion – turning from one religious life to another.
However conversion takes place, the individual is fully aware that a radical change has occurred which results in a deliberate turning of the will and commitment to the new way. That person is fully convinced of a calling and of an obligation to respond with faith. He is assisted if his intellectual, emotional and practical needs are met by the new way of life and if its belief system and way of life can be modified to fit his individual needs and aspirations. The new way of life to which the individual is converted is often spoken of in terms of rebirth, regeneration, these terms indicating the fundamental change that has taken place.
Related Change (#HH1116) Penance (#HH0790) Contrition (#HH4112)
Backsliding (#HH2321) Religious experience (#HM3445)
Metanoia (Christianity, #HH0888) Spiritual rebirth (Christianity, #HH4098).

♦ HH0079 Avatar

Description Belief in the physical incarnation of God in one man is central to the Christian religion but also to many others. The Hindu religion, for example, believes that in every age God is incarnate to draw men to himself. The Meher Baba, in this century, is considered by himself and his disciples to be the avatar of this age, the Kali Yuga (or iron age) which ends the cycle (Gold, Silver, Bronze, Iron) leading to ultimate dissolution; to be followed by regeneration in a new Golden Age. The role of the avatar is to pass his teaching to a few followers who in turn teach others the spiritual path of return to God.
Related Kali yuga (#HH0239) Theosophy (#HH0660) Spiritual way (#HH1867)
Body of glory (Psychism, #HM4185) Human development (Hinduism, #HH0330).

♦ HH0080 Corporateness

Description Acting together as a community or as a group with a common cause, people achieve greater results than each individual member of the group would or could achieve acting on his or her own. Individual care is forged into a totality of common purpose resulting in increased power, fellowship and obedience.
Related Care (#HH0042) Community (#HH0058).

♦ HH0081 Altruism

Description A theory of conduct, inherent in atheistic humanism, which regards the good of others as the end of moral action. In opposition to divisive social tendencies and exclusive interests, altruism is evidenced in all acts of philanthropy and personal service made in a spirit of unselfish devotion. It has been argued that such behaviour is natural, since cooperation is more successful than competition in evolutionary terms – and also that all altruism is, at bottom, self–regarding, since it results in praise from others and a feeling of pleasure in having done the right thing. However, although purely altruistic action is probably rare, altruism is usually contrasted with egoism which leads to predominantly self–interested actions, and religion teaches an altruistic ethic. Both altruism and egoism require the recognition of individual, as opposed to group or tribal, responsibility for actions and a certain self–regard.
Refs Berkowitz, L and Macauley, J (Eds) *Altruism and Helping Behavior* (1970); Krebs, D *Altruism* (1970); Sorokin, Pitirim A *Explorations in Altruistic Love and Behavior* (1950).
Related Egoism (#HH0130) Philanthropy (#HH1206).

♦ HH0082 Structural integration

Rolfing
Description This is a method of deep massage or manipulation aimed at facilitating the integration of the individual by focusing primarily on the physical body. The method leads to the release of chronically contracted or relatively inelastic muscles, allowing the body to move back toward its natural symmetry, with the weight redistributed about a vertical axis, thus unblocking the physical basis of experience. Breathing is freed and becomes delicately responsive to the immediate situation and the metabolism of the muscles and of the entire body is re–energized. The person is then able to reclaim his right to be balanced in space, with the force of gravity working through his structure rather than pulling it down.
The body expresses emotional states which in turn give rise to physical attitudes. Similarly physical (structural) states give rise to emotional attitudes. The method integrates the physical and emotional states leading to awareness of self, awareness of new strength and of new possibilities of movement.
During a standard series of 10 one–hour sessions, the therapist (rolfer) applies pressure through finger tips, knuckles, the elbow and the forearm in order to stretch muscles and loosen obstructive tissue so that parts of the body can return to their proper positions. The musculature is conceived as containing emotional memories and memories of physical pain which are released as the therapist works. The individual may experience considerable physical and sometimes emotional pain during this process.
Refs Greenwald, J A *Structural Integration and Gestalt therapy* (1969); Rolf, Ida *Structural Integration* (1963).
Broader Physical therapy (#HH0379) Human potential movement (#HH0398).
Related Psychotherapy (#HH0003) Neuromuscular therapy (#HH4332).

♦ HH0084 Discursive power

Description The ability to understand a system as a series of related points when the whole is too complex to comprehend in a single act.

♦ HH0085 Martial arts

Martial way — Budo
Description Practice of martial arts techniques allows the individual to profit from the learning situations of self–discovery and confrontation with death inherent in combat. Such confrontation is typical of all spiritual systems: the ego, and the fear with which the ego confronts the reality of mortality, are revealed and can be overcome. Constant living with the awareness of death makes available the power to use time to the utmost and eliminate pettiness and self indulgent behaviour.
In time, a proponent of a technique such as *aikido* is able to counter an attack without physical contact with his opponent; the aim is to harm no–one, not even an attacker. The ultimate opponent is seen to be one's self. The state of inner conflict and separation which encourages an individual to demonstrate physical superiority to others, to be aggressive and arrogant, is seen to be the hindrance to spiritual development and it is that which has to be attacked.
The martial arts are more accurately now termed the martial *way*, and previous combat techniques are now used for the development of skills through physical exercise, establishment of objective standards with opportunities for competition, and many of the qualities of a sport. Practices such as *kendo*, *karate*, *judo*, *aikido*, are used to develop intentionality and strengthen the will, making it sensitive and clear in response. At the highest level, the intention is expressed as universal love, with the body, spirit (breath) and mind working in unison. The inner energy of the universe is brought into order, protecting the peace of the world and not simply preserving but moulding everything in nature into its right form. The love of Kami (the deity begetting, preserving and nurturing everything in nature) is strengthened within body and soul.
Refs Payne, Peter *Martial Arts* (1981); Ueshiba, Kisshomaru *The Spirit of Aikido* (1984).
Narrower Aikido (#HH0252) Grounding (#HH0318) Ki energy (#HH0620)
No–mind (Zen, #HM2163) Motionless meditation (#HH5256)
Controlled spontaneity (#HH0418).
Related Acheta (Japanese, #HM4390) Realization of nen (#HM0220)
Sport–induced modes of perception (#HM0062).

♦ HH0086 Latihan

Subud
Description Latihan is a spiritual exercise used in the movement known as Subud (which derives its name from a contraction of Sanskrit words *susila*, *buddhi* and *dharma*, meaning "right living in accordance with the will of God; the inner force residing in the nature of man himself; and surrender and submission to the power of God"). The exercise is generally conducted twice a week by a group of people (from a few to several hundred) gathering together in a large room or hall, where they remain for half an hour. During that time, their effort as individuals is to receive and submit to the power of God. To do so no specific rules are laid down. Some people make movements of the head, the body, the arms or the legs. Others walk, dance, run, lie down, make noises (including chanting, shrieking and animal sounds), or simply relax and do nothing. Individuals are not supposed to feel or do anything in particular, or to relate their actions or experiences to any system of ideas. The latihan is said to be a process of purification in which the individual is gradually freed from the dominance of animal lower energies and dependence on the intellect by opening him or herself to subtle and potentially more powerful divine energies. The latihan exercise provides a setting within which individuals can be injected with such higher energies, which are held to be transmittable integrally and without diminution from one human being to another. All that is required, if it is present in the group, is that the individual should be willing to receive it. By opening himself to a higher energy, the whole individual is gradually brought into a new harmony, with negative aspects of the personality departing, leading to emotional (and physical) healing. However, the process is not clearly understood by the lower nature which is acted upon during the exercise and so brought into harmony.
Although individuals have to submit completely to the action of the latihan, the submission must be absolutely voluntary and can be turned off at will (unlike a trance or hypnotic state). Partial submission brings energy into contact only with that part that submits, fractioning the individual and supplying him or her with a quality of energy lower than that received during total submission.
Refs Bapak Muhammad Subuh *Subud in the World* (1965); Bissing, Hurbert *Songs of Submission* (1982).
Broader Human potential movement (#HH0398).

♦ HH0088 Ahimsa

Non–violence
Description One of the two central concepts of Hindu ethics, the other being *satya* (truth), ahimsa is the inspiration of soul force and passive resistance and was said by Gandhi to be common to all religions. Patanjali considered the practice of ahimsa one of the prerequisites of the development of ethical consciousness, itself necessary before embarking on the practice of yoga. Ahimsa is also one of three key elements of Jainism, where non–injury is extended to all living beings and elaborate means may be practised to avoid injuring the smallest of creatures.
Ahimsa is a positive quality of universal love, not simply a negative harmlessness. It gives the individual an inward harmony with living things and recognition of the underlying unity of life; others are affected by the love which he generates, his compassion and his service to others. Adoption of ahimsa, a non–violent attitude to life, is a natural result of self–discovery and self–knowledge, where such discovery and knowledge lead to an understanding of the essential oneness of existence. Human life is seen as an integral part of nature, and each individual's life as an essential part of human life. Each individual must therefore be respected, and the conception one person may have of the truth may not be imposed on another. The attitude of ahimsa is developed through clear discrimination, *buddhi*, and results in right action whatever the cost and whatever the circumstances.
Thus the adoption of ahimsa leads to a belief in the mutual coexistence of all living things, and many who adopt non–violence also adopt vegetarianism and protection of animals as logical consequences of their beliefs. Those who do not become vegetarians nevertheless respect the life which was taken to supply food (and thus life) for them; this respect is typical in many meat–eating cultures, for example among Eskimos. Respect for life is also demonstrated in the saying of grace before meals, the sacrificial slaughter of animals and, ultimately, in the eating of the mystical body and blood of Christ. Practice of ahimsa may also lead to conscientious objection to military service which may be extended to refusal to pay taxes which would be used for armaments. In Jainism, it involves rejecting the whole mechanism of aggression, possession and consumption, and choosing a profession minimizing injury to others or destruction of living beings.
Context One of the five moral qualities of yama (restraint).
Refs Altman, Nathaniel *The Nonviolent Revolution* (1988); Indra, M A *Ahimsa Yoga*.
Broader Restraint (#HH6876).
Related Soul force (#HH0138) Passive resistance (#HH0759)
Conscientious objection (#HH1219) Non–harmfulness (Buddhism, #HM2608)
Listening with the heart (#HH5124) Human development (Jainism, #HH0622)
Human development (Hinduism, #HH0330).

♦ **HH0089 Spiritual unfoldment**
Description A system of spiritual philosophy, involving the use of meditation (the particular practice being chosen to suit the individual's needs) and psychological purification and healing, as the person learns to overcome his own weaknesses and problems.
Refs Abhedananda, Swami *Spiritual Unfoldment* (1978).

♦ **HH0090 Cultural science psychology**
Social science psychology
Description The creation in the early 20th century in Germany of an objective, logical foundation for the cultural sciences, apart from the natural sciences, led to the use of the term *Geisteswissenschaftliche psychologie* to describe psychological trends which concentrate on the investigation of meaningful individual experiences, value factors and purposes in the mental or spiritual sphere. As such experiences never occur as isolated phenomena but are part of a complex psychic structure, the concepts of ganzheit (wholeness) and gestalt (configuration) were developed to understand and describe feelings, aspirations, motives and character. Geisteswissenschaftliche psychologie uses of understanding, re-experiencing, description and interpretation in studying the confrontation of the ego with the environment.

♦ **HH0091 Active therapy**
Description By being forced to repeatedly face situations which he fears, an individual gains insight into such situations and learns that there are no insurmountable difficulties. He or she is not called upon to exert willpower or not to feel afraid, this only increases the problem. Simple repetition of an act (or even a thought) which is feared can succeed in eliminating the fear.
Broader Therapy (#HH0678).

♦ **HH0093 Dharma**
Dhamma — Sanatana dharma — Sadharana dharma
Description Literally meaning the right, something held fast or kept, a law or custom, dharma is – in Hinduism – the universal moral law, the basis of religion, maintaining and sustaining the world, human society and the individual. *Sanatana dharma* refers to the eternal law to which everything in the universe conforms through laws applying to its own particular nature; *sadhatana dharma* is more a code of ethics to be followed in this life, the virtue of adhering to one's duty. Following the meaning of sanatana dharma, for Brahmanism dharma signifies cosmic law; and in Jainism it refers to movement and the universal rule of non-violence. In Buddhism it has a variety of interpretations relating to the appearances, characteristics or reality culminating in *nirvana*, the object of Buddhist teaching; it is one of the three members of a trinity of jewels – Buddha, the law and the priesthood. The meaning can be extended to include deeds of merit which propitiate deities and lead to the attainment of heavenly bliss. In Zen, it refers to self nature, the support or basis of everything else, including body and mind.
In Hinduism, dharma, as righteousness, is one of the four great aims of life, the others being artha (wealth), kama (culture and art) and moksha (spiritual freedom). When these are in harmony there is no discord between the natural and spiritual life. The Bhagavad Gita is a story used to encourage fortitude in the cause of truth and justice and the rule of law; one is encouraged to follow one's own dharma, or duty, however badly rather than attempt to follow another's, however well. Being one's self as fully as one can one acknowledges each human consciousness as a unique and precious opportunity for finding the way to and discovering the essential truth.
Buddhism divides dharmas in a system describing the objective foundation of a bodhisattva's activity. Fist there are wholesome, unwholesome and indeterminate dharmas. The unwholesome are to be avoided, the indeterminate ignored. The wholesome is further divided into worldly (found in ordinary people, with outflows), which should be shunned by the saints as not being an antidote to seizing on the self, and supramundane (without outflows), which should be accepted. Again, the supramundane may be conditioned (relating to the empirical, conventional world, included in the triple universe, depending on causes and conditions) and unconditioned (relating to ultimate reality, not included in the triple universe, not depending on conditions). The unconditioned dharmas are further subdivided into common (manifesting in the spiritual stream of all the saints) and uncommon, such as the ten powers (manifesting only in Buddha).
Narrower Preaching the dharma (Buddhism, #HH1022)
Seventy-five dharmas (Buddhism, #HH1108)
Listening to the dharma (Buddhism, #HH1183).
Related Kama (Hinduism, #HH0754) Artha (Hinduism, #HH0353)
Moksha (Hinduism, #HM2196) Nirvana (Buddhism, #HM2330)
Human development (Taoism, #HH0689) Human development (Hinduism, #HH0330)
Religious experience as the realization of duty (#HM4327).

♦ **HH0095 Client-centred therapy**
Non-directive counselling — Relationship therapy
Description Client-centred therapy is a process of disorganization and reorganization of the self. The new organization contains more accurate symbolization of a much wider range of sensory and visceral experience, a reconceptualized system of values based on the person's own feelings and experiences (in place of the old, largely borrowed, second-hand values). The change in organization is made by the individual because of the attitude of acceptance of the therapist to both the old and the new, enabling the person to handle the new and difficult reality perceptions necessary for the reorganization.
The required characteristics of the therapist therefore include: a strong, consistent effort to understand the client's situation; an effort to communicate this understanding to the client; occasional presentation of a synthesis of expressed feelings; refusal to offer interpretations other than to summarize what the client is feeling; refusal to promote insight directly, or to give advice, praise, blame, or to teach or suggest programmes of activities, or to ask questions or suggest areas of exploration.
The permissive attitude is based on the view that the client has basic potentialities within him for growth and development. The main function of the therapist is to provide the atmosphere within which the person feels free to explore himself, to acquire deeper understanding of himself, and gradually to reorganize his perceptions of himself and the world about him by mobilizing his potentialities in the solution of his own problems. Diagnosis is held to be not only unnecessary, but unwise and detrimental. The job of the therapist is to communicate his sincere feeling that the client is a person of unconditional self-worth, of value regardless of his attitudes, ideas, and behaviour, and to reflect what the client is saying in such a way as to clarify his thoughts and make it clear that his feelings are fully understood. All responsibility for the course and direction of therapy is left to the client, including the speed with which the client faces certain problems in his life and the follow-up on apparently significant areas.
Application of this form of therapy has been extended beyond individual psychotherapy to marriage guidance, education, play therapy, group therapy, industrial and business administration, various aspects of religious work, in the training of counsellors and therapists, etc.
Refs Rogers, Carl R and Dymond, R F (Ed) *Psychotherapy and Personality Change* (1954).
Broader Therapy (#HH0678).
Narrower Re-evaluation counselling (#HH0342).

♦ **HH0096 Manipulation**
Description Change in individual behavioural patterns may be achieved by manipulation. This is commonly defined as the management and direction of human beings by a clever use of their desires and qualities in order to control them for scientific, social or political ends not of their own choosing. This may be accomplished by intensive repeated information (propaganda, advertisements), and exposure to stress situations (sensory over-stimulation or sensory deprivation). Inherited traits may also be changed by genetic manipulation. Manipulation may also be used to denote the inadequately legitimated management and direction of a population by an elite, although more generally any interaction between groups of individuals attempting to advance a particular viewpoint may be interpreted as manipulative. This may even be the case when such change is advocated and promoted in the interests of individuals other than those engaged in the advocacy process.
All therapeutic processes (including the use of drugs or surgery) which aim to correct defects of body or mind may be interpreted as containing an element of manipulation since they interfere with the nature of another person even though this may be done in the interests of healing such defects.
Refs Shostrom, E *Man the Manipulator* (1968).

♦ **HH0099 Ecumenicism** (Christianity)
Description Although originally referring to the whole of the inhabited world or mankind as a whole, ecumenicism is generally used in the sense of the undivided Christian Church. The *Ecumenical Movement* refers to present day attempts to unite the church through such organizations as the *World Council of Churches (WCC)*; or to informal and radical Christian witness and service not related to specific denominations.
Related Syncretism (#HH3214) Conciliarity (#HH3211) Religious education (#HH0709)
Inter-religious dialogue (#HH2434).

♦ **HH0100 Psychic abnormality**
Description It appears there is no hard and fast rule on the contribution of psychic abnormality for or against an individual's development. Although it is clear that pathological conditions have to be treated and that psychologically disturbed personalities may attribute false religious or spiritual meanings to their experiences, there seems to be a connection between some illnesses and the manifesting of an individual's gifts. Mental activity seems to be stimulated by some abnormalities so that great artists, writers and so an may produce their greatest works when ill. Similarly, a period of psychological turmoil, anguish, despair, conflict and other problems may precede religious conversion or other abrupt and radical change.
Related Spiritual life (#HH0101).

♦ **HH0101 Spiritual life**
Description The spiritual life is seen as a constant striving for higher values. Although at the level of the cosmic whole all may be said to be one, there are endless means by which such cosmic awareness may be approached; whether through a personal God, an impersonal truth, the spirit of nature, equality, freedom and so on. Whatever the means, the spiritual life is marked by progressive change in a person's behaviour and, as Christian teaching puts it, his or her conduct demonstrates more and more clearly the spiritual relationship to God. Christian teaching also emphasizes the effect of physical and mental health on progress in the spiritual life, such progress being impeded by over-zealous physical austerities, and this avoidance of extreme asceticism is echoed in the Vedas. However, most religions warn against the idea that the spiritual life is easy: "the good is one, the pleasant another; both command the soul. Who follows the good, attains sanctity; who follows the pleasant, drops out of the race. (Katha Upanishad)". "I am come to set a man at variance against his father, and the daughter against her mother. He that taketh not his cross and followeth after me is not worthy of me" (St Matthew's Gospel). The apparent contradiction is overridden by the fact that, however hard the going may be, the spiritual life brings peace of mind, "that peace which the world cannot give". This "spiritual wellbeing" has been shown to be more widespread among individuals having a set of spiritual beliefs than among those who do not.
Spiritual life involves loss of self, sacrifice, inner transformation and change. The change in attitude effects every element from the most trivial and earthy to the highest. It is not enough to take part in spiritual exercises such as yoga or meditation – these may be useful but they may simply bolster up the ego or false self, and bring no real change. The letting go and surrender have to be constant, every moment of the day, not just in those parts of life one may label spiritual. Nevertheless, spiritual exercises of stillness, meditation and prayer are necessary. Detachment, withdrawal into the soul's ground, awareness of nothingness before God may eventually be experienced throughout life but will probably start through spiritual exercise. This has to be balanced with committed activity, not swamped by it.
Refs Lovelace, R *Dynamics of Spiritual Life* (1979); Mork, Wulstan *A Synthesis of the Spiritual Life* (1962).
Related Health (#HH0509) Sharia (Islam, #HH0738) Purification (#HH0401)
Mental hygiene (#HH0296) Psychic abnormality (#HH0100)
Human psychological development (#HH0447) Psychic health and spirituality (#HH0109)
Spiritual aspects of psychic health (#HH0108).

♦ **HH0102 Stages of spiritual life**
Paths of wisdom
Description Most religious and philosophical traditions see spiritual progress as a number of discrete steps as the individual grows in awareness and spiritual maturity. Such stages may not be followed in sequence – there may be times when a "beginner" has insight associated with a higher stage and times when an "experienced follower" falls back – but, in general, these stages are seen as a clear progression. Sometimes the stages may be described as a series of concentric circles (different schools of Zen) or as progress along a path. Death itself may be seen as the gateway to a path; the Upanishads see four paths for the soul after death – direct transference to its new home, return to the universe, the path of the fathers (to the moon, the place of the dead), the path of the gods to the regions of light.
A path may be the process by which inner strength grows as physical strength weakens in some period of self-denial, as in a vision quest where the hero is seen on the journey back to the cosmos, to unity, away from chaos and increasing entropy. In many systems the initiate is seen as following the footsteps of a sacred hero, and the journey may actually be re-inacted physically, as in pilgrimages to Makkah. The mediaeval cathedrals are built to enact a sacred journey – passing by monsters at the entrance, the font (signifying baptism), moving through the phases of the mass (a representation of the death, descent and resurrection of Christ) ending at the altar from where the individual reenters the world as a renewed being.
Narrower Vision quest (#HH0897) Seven stages of life (#HH0460)
Eightfold way (Buddhism, #HM2339) Paths of wisdom (Judaism, #HM2509)
Paths of effect (Tibetan, #HM3249) Fountains of light (Sufism, #HM3039)
Paths of calming (Buddhism, #HM2987) Eightfold path of yoga (Yoga, #HH0779)
Harmonies with enlightenment (Buddhism, #HH0603)
Upanishadic stages of awareness (Hinduism, #HM2392)
Awareness of relative reality (Hinduism, Yoga, #HM2032)
Consciousness ascension stages game (Buddhism, Tibetan, #HH4000)

Sutra paths of accumulation and application in the ascension stages game (Buddhism, Tibetan, #HM2962)
Tantric paths of accumulation and application in the ascension–stages–game (Buddhism, Tibetan, #HM2848).
Related Maps of the mind (#HH8903) Human perfectibility (#HH0212)
Circles of enlightenment (Zen, #HH0350) Stages of personality development (#HH0285).

♦ HH0103 Atman (Hinduism)
Paramatman — Self
Description The aim of Hinduism is to see all the objective things of the world – all that is composed of the five physical elements, the intelligences (manas and buddhi) and the ahamkara (ego sense) – as the not–self. These eight elements constitute the body and mind of man. The self, or Atman, is subject, never object and therefore cannot be comprehended by the mind. It is the first vital principle and all other objects exist only through the Atman. The eight elements must be transcended for enlightenment. Then, when the Atman is known, everything else is also known and the Atman is realized to be identical with the Absolute, Brahman. Atman as the supreme Self is sometimes referred to as *Paramatman* for the purposes of distinguishing that Self from the individual soul or Atman.
Related Self (#HH0706) Brahman (#HH1226) Anatma (Buddhism, #HH0241)
Samsara (Hinduism, #HM1006) Advaita vedanta (Hinduism, #HH0518)
Human development (Hinduism, #HH0330).

♦ HH0105 Secular humanism
Description As a secular alternative to religion, the term secular humanism generally covers any philosophy which rejects religious beliefs and instead defends the intrinsic dignity and ideals of human nature in its own right. Man is the centre of the intellectual universe and the interpretation of experience is the primary aim of philosophy which is, in itself, capable of carrying out such interpretation. There is no transcendent truth or reality – what is truth and reality for man is attainable by man and sufficient for man.
Broader Humanism (#HH0674).
Related Secular quest (#HH4994).

♦ HH0106 Dianetics
Scientology
Description Dianetics is concerned with theory and therapy related to mental health and is developed within the religio–scientific movement called Scientology. The human mind is held to be engaged upon perceiving and relating data, composing or computing conclusions and posing and resolving problems. This activity and the healthy survival of man are inhibited by recollections of incidents (engrams) that contain physical pain or painful emotion and force the individual to respond in an unconscious, automatic, stimulus–response manner. The therapy uses a monitoring technique (with what is essentially a simplified polygraph) to guide the direction of sessions in which the individual is constantly brought into confrontation with such incidents until the analytical mind can handle them without detectable emotional reaction. By the use of this technique, referred to as *auditing*, on individuals in a normal state of mental health, the process creates a condition of positive mental health and the development of individuals of high worth to society.
Refs Hubbard, L Ron *Dianetics and Scientology Technical Dictionary* (1975); Hubbard, L Ron *Dianetics*.
Narrower Clearing (#HH1076).

♦ HH0107 Spiritual integration
Description Vocational commitment, such as is inherent in the religious vows of a Roman Catholic priest, involves renunciation of and abstention from activities, behaviour and sources of love normally open to people in their everyday lives. Such commitment is intended to channel an individual's energies towards higher, universal purposes. Spiritually mature individuals are able to sublimate fundamental instinctual forces to higher objectives and values in the service of mankind and of God. Less mature people find the discipline of such vows has a negative effect and, rather than integrating their theological life, it causes mental and spiritual suffering and internal conflict with concommitent disillusionment and personality problems.

♦ HH0108 Spiritual aspects of psychic health
Description Individual human development may be said to depend on three factors or forces – the psychic forces imposed by nature; the moral forces based on the values accepted by the individual; and the spiritual forces offered by God for acceptance by the individual. It is clear that the intrinsic psychic forces may, where lacking or unbalanced, make manifestation of the sanctity or spirituality of the individual impossible or unclear. Such lack of manifestation is not the fault of the individual and every aid must be given to strengthen and improve mental health for the full flowering of individual sanctification.
Related Spiritual life (#HH0101).

♦ HH0109 Psychic health and spirituality
Description Psychic problems and infantile mentality make observable progress in mature spiritual conduct more difficult. Opinions differ as to whether mental illness is the result of sin, although it is usually considered that the psychic rather than the moral personality is the basis of such illness and medical treatment is recommended for psychic problems. Neither achievement of moral perfection nor good mental health is found more often among fervently religious people than otherwise – since it is the effort towards exercising virtue that is important rather than the evident results.
Refs Gandolfo, Joseph B *Spiritual Psychic Healing* (1986).
Related Spirituality (#HH5009) Spiritual life (#HH0101) Spiritual healing (#HH0458).

♦ HH0110 Declaration
Witness
Description An external manifestation of internal belief and certainty, declaration involves a person in bearing witness to that for which he stands. This inherent desire to "stand up and be counted", to manifest in deed what is truly believed, results in power to change – for example, passive resistance in India before independence.
Related Martyr (#HH0838) Soul force (#HH0138).

♦ HH0111 Puja yoga (Yoga)
Description By directing the will from self–concerns and towards religious and social ideals, taking part in ceremonial worship and prayer, the individual, according to Master Da Free John, approaches stable maturity in the first three stages of life – the vital–physical, emotional–sexual and mental–volitional. Puja yoga, in combination with karma yoga, is at the level of the self–actualizing individual or hero who embodies the ideal of the "good person".
Broader Yoga (Yoga, #HH0661).
Related Prayer (#HH0185) Worship (#HH0298) Karma yoga (Yoga, #HH0372)
Self–actualization (#HH0412) Vital–physical stage of life (#HM3320)
Emotional–sexual stage of life (#HM2954) Mental–volitional stage of life (#HM3464).

♦ HH0112 Mandalas
Yantras
Description A mandala is a diagram designed to hold and focus the attention during meditation and support an individual's effort to maintain concentration while practicing a particular technique. Mandalas are a circular form of yantra which may be a geometric design, often interlaced triangles super–imposed on flower–type structures with varying numbers of petals, usually framed by a square with doors or gates at the four sides. The main structural features of a mandala are: a center, symbolizing eternal potential, the ultimate source of energy, the eternal now, and the centre of being; symmetry, symbolizing balance and integration of polarities and modes of being; and cardinal points, symbolizing the major psychic and cosmic forces. The whole represents supreme enlightenment, a symbolic representation of satori.
A mandala consists of a series of concentric forms which symbolize and suggest passage between different levels or dimensions of awareness, both in the macrocosm and the microcosm, which it thus interlinks. Through the concept and structure of the mandala, the individual may therefore be projected into the universe and the universe into the individual. The meditator experiences his essential relatedness to cosmic rhythms and intuits himself as both an organic set of on–going interrelationships of structures and systems, and as within a greater frame of reference. The mandala becomes a chart for a gradually unfolding series of visualization exercises, functioning as a map of a series of stages by which an experience of wholeness and unity is reached through knowledge and understanding of the processes of creation and dissolution of the forms and images in the meditator's consciousness.
The mandala may be used as a therapeutic device or as a ritual, meditative technique, but in both the aim is a higher level of integration. It is suggested that a mandala is an archetype of psychic integration. By projecting his own complexes upon the balanced complexity of the mandala, the person helps to liberate himself from his obsessions. Construction of a mandala is therefore a self–integrating ritual (which should be undertaken with care and attention). A true mandala should be a spontaneous creation which originates at a deep layer of personality, when, like all true symbols, it has an innate revealing power, and so can be used as an object of contemplation (in the same way as a crucifix or ikon).
Complementary to its use as a tool of transformation and integration, developed to transmute sensory experience from a binding to a liberating role, is the use of the mandala as a universal art form. As such it symbolizes integration, harmony, and transformation, giving form to a primordial intuition of the nature of reality. In addition, it functions as a key to systems of symbols which may be structured and represented in the form of a mandala defining the processes of nature as a set of interrelationships unified into a coherent whole. Since it is the mind of man that realizes and integrates the various parts of any such system, the mandala also constitutes a map of consciousness.
Refs Arguelles, José and Miriam, T *Mandala* (1972); Jung, C G *Concerning Mandala Symbolism* (1959); Jung, C G *Mandala Symbolism* (1972).
Related Yantras (#HH2662) Yantra yoga (Yoga, #HH1224) Feng–shui (Chinese, #HH2901)
Tantra (Buddhism, Yoga, #HH0306) Tantric visualization (Yoga, #HM1690).

♦ HH0115 est
Erhard seminars training
Description This is a practical philosophical training developed by Werner Erhard following an experience of transformation in his own life. Standard training is a 60–hour (4–day) course (two consecutive weekends) covering: epistemiology, and the realization of one's self as the source of experience; ontology, and the discovery of what it means to be; ethics, personal integrity and responsibility as the implications of what one is and knows. The results of such training include: a more satisfied approach to life; an ability to deal with barriers to satisfaction; acceptance and expansion of life's experiences.

♦ HH0117 Personal career development
Personal self–development
Description Large organizations are tending to change from stable hierarchical structures to more flexible, continuously evolving forms better able to respond to the rapidly changing demands of society. As a result the relationship between employee and organization has loosened, with employees making greater demands for involvement in their own career development, improvement of professional competence and choice of goals. Career development therefore becomes primarily self–development with the individual's obligation to develop himself becoming greater than his loyalty to his employer.
Refs Bolles, Richard N *What Color is Your Parachute* (1987).

♦ HH0120 Abstraction
Description The state of normality for an adult is characterized by the ability to develop concepts. Such a process of abstraction may be: generalization (separation of essential characteristics of a group of things, ideas, etc, from the apparently inessential); isolation (consideration of one specific aspect); and idealization (the construction of an ideal model following generalization and isolation).
Related Pratyahara (Yoga, #HH0829).

♦ HH0126 Physique and temperament
Sheldon's types
Description The three physical types – *endomorph* or "oval", *ectomorph* or "angular", and *mesomorph* or "muscular" – are said to correspond to three temperaments – *viscerotonic* (with a passion for eating), *cerebrotonic* (with a tendency to worry mentally over everything), and *somatotonic* (with the emphasis on action and risk–taking). According to W.H. Sheldon, one can use this correspondance to quantify temperament on a scale indicated by the body's physical characteristics.
Refs Otto, H A and Mann, J (Eds) *Ways of Growth* (1971).
Related Temperament (#HH0831).

♦ HH0129 Personal integrity
Description An individual is said (John Kleinig) to acquire a specific personality or identity through learning, constructing a unity through deliberation, experimentation, and accommodation of inherent capacities which are too many and various for all to be developed fully or even developed at all. This building is a life's work for which each individual is responsible, partially dependent on formative training in early years and partly on the coherence and cohesiveness which the individual manages to develop himself. He may or may not succeed to his own satisfaction, as a lack in maturity of purpose may lead to careless and foolish departure from his own more permanent and central projects and commitments.
This approach to personal integrity differs from that which assumes that everyone is born with a particular fixed pattern of development, with which no–one else has the right to interfere, and which will lead naturally to the most propitious results.
Related Integrity (#HH0234) Paternalism (#HH0735).

♦ **HH0130 Egoism**
Hedonism
Description Egoism is the attitude, in opposition to *altruism*, which holds that each individual should seek his or her own good, only considering another's good when it is to his or her advantage. Egoism includes *hedonism*, living for one's own pleasure; will–to–power, or the aim of achieving dominance over others; and perfectionism, which looks on self development as the only reason for existence. Theoretically, it is opposed to natural law and theocentric systems; but historically its main proponents have modified egoistic doctrines. Epicureans emphasised kindliness and friendship; and 17th and 18th century materialists, seeing good as, by definition, the object of man's desire and war as the natural state of endeavouring to destroy or subdue those who desire the same thing (Thomas Hobbes), derived a less harsh system (Jeremy Bentham) which declared benevolence to be the main source of egoistic satisfaction. Most moral philosophies are the antithesis of egoism in that they admit the existence of a supreme creator in whose service men best achieve happiness.
Related Altruism (#HH0081) Pleasure (#HH0194) Solipsism (#HH0191) Self–love (#HM0478) Egoism (Yoga, Zen, #HM3477).

♦ **HH0134 Elevation of man** (Christianity)
Description In the Christian tradition, the elevation of man is the granting by God to man of a supernatural destiny (made in the image of God), together with the means of attaining that end. The original elevation of man is seen, in Adam, to have been destroyed by sin and restored through Christ. This second elevation (through Christ) has in common with the first elevation a state of divine grace, and an indwelling of the Holy Spirit and of the theological and moral virtues; in addition, the second elevation is achieved through the sacraments of Baptism, Confirmation and Eucharist.

♦ **HH0136 Unity of personality**
Description In psychology, this term may be used to mean or include the following:
1. Self–consciousness, namely a subjective point of reference by virtue of which the individual feels that there is coherence between his memories of the past and his plans for the future, and a centre of an orderly psychological universe.
2. Imagination, namely the unifying capacity by which the individual plans his life and pursues long–term goals.
3. Self–esteem or egoism which focuses the emotional life of the individual and enhances the unity of his personality.
4. Intention, namely an ego–ideal governing the life pattern of an individual, may constitute the basis for unity of personality.
5. Homeostasis of the endocrines provides a biological basis for unity, preserving functional integrity in the course of growth
6. Temperament, particularly when determined by heredity, contributes to the stabilization of the course of personality development and its internal consistency
7. Temporary convergence occurs when psychophysical energies are mobilized in one maximally integrated activity in response to a particular environmental situation
8. Interdependence of traits provides unity through the pattern of their overlapping
9. Unity of personality occurs when the individual acquires a coherent self–image, namely when all perceptions of himself in relation to others are accepted into the organized conscious concept of self
10. Personality may be viewed as a Gestalt that has greater or lesser unity, depending upon its individual nature, upon the condition of the individual and the environment within which it is behaving.

In philosophy, the term may be used to mean or include the following:
1. Everything that an individual does or says presupposes his unity as a personality
2. The self is a passive guarantor of personal unity throughout life
3. The self is an agent that wills, directs and selects conduct and thus actively forges unity of personality
4. Man strives for a higher degree of perfection and ultimate unity and, insofar as he pursues this possibility, by reason and by choice, he achieves a factual unification of his personality
5. Unity lies in the essential systemic nature of personality
6. Unity lies in a tendency of the personality to stability
7. Unity lies in the tendency of human beings to grow into their perfected form
8. Unity emerges through the process of striving towards some goal which integrates all aspects of the personality toward the achievement of that goal; ideas whose inconsistency is recognized as the personality develops in terms of that goal are not accepted.

Refs Allport, Gordon W *The Unity of Personality*.
Narrower Unity of personality (Judaism, Christianity, #HH8206).
Related Strength of personality (#HH0198).

♦ **HH0138 Soul force**
Satyagraha
Description According to Gandhi, soul force, or the use of the permanent and superior nature of the soul, is infinitely superior to physical force when attempting to redress a wrong. The Sanskrit term literally implies insistence on truth. Use of soul force requires great courage and the recognition of the existence of the soul as separate from and superior to the body, together with the ability to count the body as nothing in comparison with the soul.
Related Soul (#HH0501) Ahimsa (#HH0088) Declaration (#HH0110) Passive resistance (#HH0759).

♦ **HH0141 Avadhoot**
Description Having shaken off attachments, including attachment to social convention, the avadhoot lives in a state of *divine madness*, the state of mind being unlimited by conventional, linear notions. Because of the expanded perception there is apparent madness since the avadhoot has no obligations to the conditional self. Teaching from such an adept will seem wilful but, founded in wisdom and love, it is also skilful and intended to assist in the pupil's ultimate enlightenment.
Related Guru (#HH0805).

♦ **HH0142 Soul–making**
Re–visioning
Description According to James Hillman, the purpose of life may be to make psyche from the psychological: to find connections between life and the soul, that which is there when everything else – including consciousness and ego – is not there. He defines the perspective of the soul as: the deepening of events into experiences; the significance derived from relationship with death; the recognition that all realities are symbolic and that therefore imagination – reflection, dreams, fantasy and so on – is also recognized as real. Archetypes (and thus primitive images, totems and idols) have real significance, as do the personifications of such metaphorical ideas as time, fame, friendship, etc; this is referred to as *en–souling* such ideas. En–souling and giving of names to things other than one's self helps to define the self (as other than those things named).
By seeing the many personalities and roles one possesses as also "other" one frees one's self from self–tyranny and learns to trust the imagination. Psychological reflection in soul–making is work on the anima through images, personifying the multiple voices which are seen as enacted by the (depersonified or impersonal) "me" that is projected by the soul. Depth psychology may thus be redefined as simply the making of the soul.
Related Soul (#HH0501).

♦ **HH0145 Empathy**
Sympathy — Alienation
Description *Empathy* is a psychological term, also used in aesthetics, to describe the transferring of feelings evoked by an object to that object itself. Thus a room which evoked sad feelings in an individual would be thought of by him or her as a "sad" room. The term is also used to describe the apprehension, whether intellectual or imaginative, and whether intentionally or not, of another person's state of mind. Thus *to empathize* is to experience vicariously the thoughts, feelings and actions of another person, thereby understanding (and, to some extent, being able to predict) his behaviour and his capacity to deal with a given situation. The knowledge that one is thus understood is liberating and stimulates the growth of a relationship.
The term *sympathy* refers to the existence of feelings similar to those experienced by someone else, to put one's self in the place of another; it may result in being overwhelmed by those feelings. This is contrasted with *identification* which refers to similar feelings based on some unconscious quality held in common, even where there is no emotional attachment. In contrast, *alienation* is a reaction to, and rejection of, the experience of empathy.
It has been said that sympathy with others and awareness of them as other selves is a step in the knowledge of one's self as a person. Feelings aroused may be in line with those already experienced; where they are at odds with those of the individual he may well overcome this sypathetic response (bravery in the face of cowardice, for example). Adam Smith referred to sympathy as the fundamental fact of moral consciousness.
In contrast to views, expressed in the past (Piaget, for example) that empathy with others could not be felt until development of cognitive abilities allowing the child to see things from another's perspective, at the age of about 7 or 8, recent psychological research has shown empathetic response in children as young as 9 months old, who will cry in response to another infant's cry. Even newborn babies cry more loudly when they hear another crying than when they hear computer simulation of such cries. And by the time a child can walk it will take active measures to assist another in distress (fetching a comforter, for example).
Refs Comfort, Alex *Reality and Empathy* (1984); Eisenberg, Nancy and Strayer, Janet (Eds) *Empathy and Its Development* (1987); Goldstein, Arnold P and Michaels, Gerald Y *Empathy* (1985); Morrison, Karl F *I am You* (1988).
Related Spiritual renaissance (#HH1676) Attentiveness to nature (Zen, #HH0463).

♦ **HH0146 Cognitive growth and development**
Intellectual growth and development — Cognitive evolution — Mental growth and development
Description Cognition comprises the processes by which an individual obtains knowledge of an object or of its environment. It includes: perception, discovery, recognition, imagining, judging, memorizing, learning, thinking, and frequently speech. These processes develop through a series of stages from birth onwards giving rise to a progressive increase in the ability to construct and express fundamental physical concepts (such as objects and space) and logical concepts (such as coherence and classification). Mental growth may therefore be defined as the progressive expansion of an individual's ability to deal effectively with encountered environmental situations. As such it is to be clearly distinguished from brain growth and neurological maturation, although a close relationship exists between them.
In the course of development from stage to stage, the child's interactions increasingly shift from (a) motions made in response to environmental stimulation to (b) actions upon and thoughts about the environment and himself as an object in the environment to (c) thoughts about his own thoughts and about possibilities as well as actualities. Ten such stages have been identified and may be classified as sensorimotor and symbolic operation (occasionally split into concrete operations and formal operations):
Sensorimotor stages (0 to 15 months) These correspond to the stages of sensorimotor development described separately. They comprise stages of: radical egocentrism, anticipating and generalizing, static coordinating, mobile coordinating and signalling, experimenting, and symbolizing.
Symbolic operational stages These comprise the stages of: preconceptual gestural verbal acts (1.5 to 4 years), intuitive quasi–verbal acts (4 to 7 years), concrete verbal acts (7 to 10 years), formal operations (10 to 15 years).
Refs Eliot, John (Ed) *Human Development and Cognitive Processes* (1971).
Related Evolution (#HH0425) Atman project (#HM2153) Primary growth (#HH6669) Religious growth (#HH1321) Moral development (#HH0565) Cognitive dissonance (#HH8204) Sensorimotor development (#HH0195).

♦ **HH0147 Human resources** (United Nations)
Description Human resources, as distinct from material resources, comprise the skills, knowledge and capacities of all the human beings actually or potentially available for economic and social development in a community. They are not limited to the resources of the working population but include also the actual, potential and prospective contribution to economic and social development of other persons. In this way, the concept of human resources comprises both men and women, whether technically they belong to the labour force or not, by virtue of the goods, services and care they provide, actually and potentially.
Refs United Nations *Development and Utilization of Human Resources* (1967).
Related Human resources development (United Nations, #HH0745).

♦ **HH0149 Expressive therapy**
Suppressive therapy
Description By reversing the normal defensive or "covering up" process and expressing (whether verbally, emotionally or by acting out) his ideas and feelings, a person is encouraged to understand the emotional roots of his symptoms or illness. The aim is to shift the roots of emotional and mental illness from the unconscious to the conscious mind. Although this self revelation may be painful and disillusioning, it aims towards self realization and a clearer view of *reality*. In contrast, *suppressive therapy* seeks to maintain a comfortable and peaceful *illusion* and to strengthen the repressive and defensive forces of a personality.
Broader Therapy (#HH0678) Psychoanalysis (#HH0951).

♦ **HH0151 Flagellation**
Description Flagellation was and is practised in rituals of sects and religions including Christianity. It appears to arouse biological and psychic excitation and may result in *paranormal consciousness*, religious fervour and psychopathological mystic states. When used as a form of voluntary penance, flagellation aims to promote conquest of self, expiation of an individual's sin or that of others, and the receiving of divine grace. As an ascetic exercise or "discipline" of the monastic life it was considered second only to martyrdom in bringing the individual close to Christ and his passion, and is still considered a useful corporal penance if undertaken under spiritual direction. Public processions of flagellants, particularly common during the middle ages and especially

during times of plague, were a form of group penance and expiation.
Broader Mortification (#HH0780) Penitential practices (#HH4653).

♦ **HH0155 Samkhya** (Hinduism)
Description A metaphysical orientation within Hinduism, which posits two eternal principles – *purusha*, the transcendental self and *prakriti*, the eternal force of conditioned nature. Followers distinguish between self and nature, renouncing that which pertains to nature. Samkhya is part of the scientific base behind yoga but differs in that it is atheistic as a doctrine whereas yoga (for example, in the sutras of Patanjali) accepts the existence of Ishvara or God.
Related I–transcendent stage of life (#HM3060).

♦ **HH0156 Friendship**
Description Friendship is associated with a need to know the self and to correlate emotional experiences with those of another person. It involves a reciprocal relationship based on affection or sympathy but is usually taken to exclude, or at least to occur in spite of, or separate from, erotic attachment. In contrast with erotic attachment, it frequently depends on a number of shared conditions, including activities, goals or membership of a common group. Whether complementarity or similarity is the main cause of friendship depends on the nature of the particular bond. Aristotle distinguishes the former (complementarity) as an attraction of opposites, based on utility, and the latter (similarity) as based either on pleasure or on virtue. Of the three he prefers that based on virtues as being the most perfect, when friends are drawn to each other because of what they are as opposed to what they can give. Cicero also agreed that the best friendship is based on virtue and its comparative rarity is a function of the rarity of men of virtue. Friendship is characterized by dynamism, in seeking out the other's company and sharing his or her thoughts; reciprocal affection, deepening and growing in psychological resonance; a spiritual or intellectual union; disinterested concern for the other; and in perfect friendship, a union where everything is held in common and the idea of "you" and "I" disappears.
Related Love (#HH0258) Community (#HH0058).

♦ **HH0157 Final integration of the personality**
Total integration of the personality
Description This degree of integration is characteristic of persons with only one dominant philosophy of life from which every attitude, trait, and individual act must proceed. It is the final, and possibly ideal, stage in the integration of the personality.
The term has also been used to denote the goal of personal development, through access to new states of transcendent awareness, and the development of a new self–concept and world–view.
Refs Arasteh, A Reza *Final Integration in the Adult Personality* (1965).

♦ **HH0158 Spiritual reading**
Divine reading (Christianity) — Lectio divina
Description In contrast to meditative reading or study, spiritual reading may be said to involve a receptive rather than an active approach but one which disposes the reader towards contemplation. It assists in progressive knowledge of spiritual concerns; strengthens against the negative impulses of everyday life; encourages personal striving for perfection; and disposes towards meditation and of contemplation of the Almighty.
The scripture is read so that it is really listened to. Repeating the words with one's lips the body enters the process. Gradually the capacity to listen at ever increasing depth of attention is cultivated. This is reflection or meditation. The response to this reflection is affective prayer, spontaneously speaking with God about what has been read. Eventually practice leads to resting in the presence of God, contemplation.
Related Meditative reading (#HH5550) Centering prayer (Christianity, #HH1994).

♦ **HH0160 Nada yoga** (Yoga)
Natha yoga
Description Related to *mantra yoga*, in that consciousness is transformed through the vehicle of sound. Closely related to *shabd yoga*, which is the Sikh version.
Broader Yoga (Yoga, #HH0661).
Narrower Shabda yoga (#HH0539).
Related Laya yoga (Yoga, #HH0782) Mantra yoga (Yoga, #HH0931) Mystical stage of life (#HM3191) Mantra–induced experience (Yoga, #HM2464).

♦ **HH0161 Personal growth facilitation**
Description The distinction between personal growth and therapy is blurred in normal usage. Both therapists and growth centres claim to achieve greater effectiveness, heightened awareness, increased self–understanding, and better ways of relating to others. The medical approach is however characterized by a rational–intellectual approach and the view that the individual's problems are in some way pathological. The growth–oriented approach is characterized by an emotional–intuitive approach and a concern for the individual as a fellow human being. Both approaches may be adopted by therapists however, so that growth may take place in therapy sessions.
A more useful distinction is to view the growth facilitation process as one into which the facilitator enters without a preconceived image of human nature in terms of which he wishes to shape the growing individual. The growth facilitator aims to create a phenomenological space for the person within which he can engage in an exploration of ever–deeper levels within himself. In the medical approach, the therapist perceives his task as one of getting the individual to conform to a predetermined image of the ideal human. Genuine growth is not just change, or change for the better; it is a kind of change which involves the emergence and integration of material from the person's inner depths.
Refs Fadiman, J and Frager, R *Personality and Personal Growth* (1976); Fadiman, James and Katz, Richard *Transformation*; McCarroll, Tolbert *Exploring the Inner World* (1974).

♦ **HH0162 Encounter group**
Description The term encounter group covers a wide variety of planned intensive group experience. The basic encounter group tends to emphasize personal growth and development and improvement of interpersonal communications and relationships through an experiential process. The aim is to assist in three needs inherent in the concept of self: confident *inclusion* in social dealings with others; balanced attitudes towards *control* of a situation; and solution of problems of *affection* and acceptance.
Associated group activities (described separately) include: sensitivity training group, T–group, sensory awareness group, organizational development group, team building group, task–oriented group. All these forms tend to have some common characteristics. The group is usually small (from 8 to 18 members), relatively unstructured, choosing its own goals and personal directions. The experience frequently, although not necessarily, involves some cognitive input presented to the group. Usually the leader's responsibility is primarily the facilitation of the expression of both feelings and thoughts on the part of group members. There is a focus on the process and dynamics of immediate personal interactions. Such groups may meet intensively over several days or weeks or regularly once or twice a week over an extended period. In one form, a group may meet continuously for 24 hours or more and is then known as an encounter marathon (described separately).
Carl Rogers has formulated a number of practical hypotheses which tend to be held in common by all such groups:
1. A facilitator can develop, in a group which meets intensively, a psychological climate of safety in which freedom of expression and reduction of defensiveness gradually occur.
2. In such a psychological climate many of the immediate feeling reactions of each member toward others, and of each member toward himself, tend to be expressed.
3. A climate of mutual trust develops out of this mutual freedom to express real feelings, positive and negative. Each member moves toward greater acceptance of his total being – emotional, intellectual, and physical – as it is, including its potential.
4. With individuals less inhibited by defensive rigidity, the possibility of change in personal attitudes and behaviour, in professional methods, in administrative procedures and in relationships becomes less threatening.
5. With the reduction of defensive rigidity, individuals can hear each other, can learn from each other, to a greater extent.
6. There is a development of feedback from one person to another such that each individual learns how he appears to others and what impact he has in interpersonal relationships.
7. With this greater freedom and improved communication, new ideas, new concepts, new directions emerge. Innovation can become a desirable rather than a threatening possibility.
8. These learnings in the group experience tend to carry over, temporarily or more permanently, into the relationships with spouse, children, students, subordinates, peers, and even superiors following the group experience.
The encounter group technique has been used in a variety of settings including: universities, industry, church, government agencies, penitentiaries, and educational institutions. The participants therefore include every variety of person but particularly those combinations of persons that can benefit from better interpersonal communication and sensitivity in their daily activity.
In such a group, the individual comes to know himself and each of the others more completely than is possible in the usual working or social relationships. He becomes more deeply acquainted with his own inner self which tends normally to be hidden behind masks and facades; and discovers that his real feelings are quite acceptable to members of the group. The degree to which he is accepted by the others increases, as his ability to communicate genuinely with them increases. Participants experience a closeness and intimacy which they may not even have experienced with members of their own family. A sense of confidence and trust is therefore built up enabling him to relate better to others, both in the group and outside it.
Refs Burton, Arthur (Ed) *Encounter Groups (Behavioral Sciences Service)* (1969); Solomon, L and Barzon, B (Ed) *New Perspectives on Encounter Groups* (1972).
Broader Human potential movement (#HH0398).
Narrower Encounter marathon (#HH0817).
Related Self responsibility (#HH0187) Self–realization therapy (#HH0784).

♦ **HH0163 Discipline**
Description Discipline involves the organization of behaviour and attitudes in accordance with particular teaching. The term discipline implies discipleship, or the acceptance of a teacher or guide, so that the disciple organizes his behaviour and attitude in accordance with one who has greater knowledge and experience, and accepts the authority of that guide: "whoever travels without a guide needs two hundred years for a journey of two days" (Rumi). Undertaking a discipline does not imply hardship or punishment so much as freedom from lesser instincts and desires in order to follow that which is understood to be true. One is said to be most free, most what one truly is and can be, when one is most disciplined, every discipline implying a corresponding freedom. However, disciplinary proceedings are usually taken against members of an organization who have fallen into error, whether to reclaim them from error or to protect the organization from such error.
Refs Foster, Richard J *Celebration of Discipline* (1984).
Narrower Self–discipline (#HH0877) Ethical discipline (#HH1086) Ecstatic discipline (#HH5282) Spiritual discipline (#HH1021).
Related Discipleship (#HH0376) Vinaya (Buddhism, #HH1376) Self–discipline (Christianity, #HH2121).

♦ **HH0164 Reconciliation** (Christianity)
Description In Christian teaching, this is the belief that man's state of enmity with God is replaced by friendship, through God's act of reconciliation through Christ (redemption) and the individual's acceptance of that reconciliation by Baptism and Penance. Reconciliation, whether between God and humanity or between individual human beings, expresses the result of a restored relationship. Practice of reconciliation demonstrates the religious and ethical meaning of forgiveness.
Refs Shaw, Franklin J *Reconciliation* (1966).
Broader Forgiveness (Christianity, #HH0899).
Related Baptism (#HH0035) Penance (#HH0790) Redemption (Christianity, #HH0167) Justification (Christianity, #HH0341).

♦ **HH0166 Concept progress**
Description The median number of added concepts per teaching step occurring in the basal text along a teaching path in a teaching programme.
Related Concept development (#HH1222).

♦ **HH0167 Redemption** (Christianity)
Description In Christian teaching, redemption refers to the delivery of mankind from the slavery of sin and restoring man to a state of grace as sons "by adoption" through the life and death of the incarnate Word of God, the "only begotten son" Jesus Christ. This act of redemption is seen as one of the great mysteries spanning the whole history of mankind, an act for once and for all in which mankind is in the position of recipient, each individual having the free choice to respond or not. Man cannot, by himself, remove the guilt of rejection established by original sin; but divine love freely and permanently asserted over this rejection overcomes guilt and makes forgiveness possible.
Objective redemption is seen as an event preceding justification and subjective redemption or *sanctification*. The event which makes objective redemption possible is the cross of Jesus Christ, making mankind always and everywhere open to God's forgiving love, as a whole and despite their culpable destiny. As Christ's suffering accomplished redemption, so it is on the path of suffering that the Church meets man. In Christ "every man becomes the way for the Church".
Refs Werblowsky, R J Zwi and Bleeker, C J (Eds) *Types of Redemption* (1970).
Related Suffering (#HM0471) Sanctification (#HH0428) Jesus as saviour (#HH2877) Justification (Christianity, #HH0341) Reconciliation (Christianity, #HH0164).

♦ **HH0168 Reputation**
Description The positive qualities thought to be possessed by an individual constitute his or her reputation; and the dealings that individual has with others are judged in the light of this reputation. To deprive someone of his or her good name means that that person is no longer accepted unconditionally. It is considered an infringement of human rights to deliberately and unnecessarily reveal facts which might damage a person's reputation unless failure to do so could in some way make that person a danger to society.

♦ HH0169 Grace
Description Psychologically, there are three "graces" beyond the individual, personal, self-conscious self – animal, human and spiritual grace. The first comes with physical harmony in accord with one's biological nature, and manifests as well-being, an awareness of the intrinsic goodness of life. The second comes from human relationships and human teachers; from membership of a family, group, community or nation; and from the external projection of one's own wishes, hopes and imaginings. Even response of prayer to the saints or to a deity by a devoutly religious person may be seen in this light. Lastly, spiritual grace can only be accepted when the preoccupation with "I", "me" and "mine" is forgotten, if only for a moment. This moment may be one of a totally altered state, when there is an awakening to a whole new dimension of knowledge and an awareness of the unity of creation. It has been described as the "wholly other".
Such cosmic consciousness or gratuitous grace is seen in Catholic teaching as divine influence, or the love of God operating in mankind as a free and unmerited favour, which calls for a free response through faith, hope and charity to the invitation to increase personal efforts towards inner and external knowledge. Gratuitous grace brings great joy, but the invitation to further efforts may be lost by dwelling on the experience of grace itself. Continuous receiving of spiritual grace comes to those who no longer act from their individual selves but from God in them. This spiritual grace is said to be true grace when the whole personality is changed and the individual devotes himself to his true end in God.
Grace may be seen from three standpoints: (1) As forgiveness, under the doctrine of justification by faith. Since humans are unable to fulfil the requirements of perfect love and obedience, grace as forgiveness transcends ethical categories and restores a morally right relationship between humans and God and between humans themselves, despite the violation of divine law by sin. (2) As the power to lead a moral life. Despite individual freedom and moral responsibility, many would argue that only through empowerment from outside is the individual free to become what he desires to become. (3) As a means of ensuring meaningful existence and wholeness of life. Ethical principles and action are insufficient to ensure wholeness of life despite ambiguity and failure in human moral existence.
In contemporary Christian terms (Paul Tillich) sin is regarded as the state of separation from oneself, others and God; this separation is simultaneously one's fate and one's guilt. To be struck by grace is neither belief in something nor moral progress; both can lead to pride or despair. It is to recognize and acknowledge one's state of separation, to hear that one is accepted in spite of the separation and to accept one's acceptance. Only by God's grace can the individual be again united with God. As John Ruusbroec says, without the mediation of God's grace and our free-willed and loving conversion, no-one can be saved.
Narrower Actual grace (Christianity, #HH6548) Sanctifying grace (Christianity, #HH5116).
Related Salvation (#HH0173) Prasada (Hinduism, #HH4330)
Gratitude (Christianity, #HH4202) Spiritual self-abandonment (#HH0868)
State of grace (Christianity, #HM6445).
Followed by Freedom (Christianity, #HH4020).

♦ HH0170 Determining mental factors (Buddhism)
Viniyata — Yul-nges
Context One of the six classes of mental factors which make up the knowing (as opposed to the observing) part of consciousness in Tibetan Buddhism.
Broader Secondary mental factors (Buddhism, Tibetan, #HH1348).
Narrower Aspiration (#HM2578) Mindfulness (#HM2847)
Stabilization (#HM2440) Mental wisdom (#HM2655)
Belief (Buddhism, Yoga, #HM2933).
Related Five powers (Buddhism, #HM3238) Five forces (Buddhism, #HM2948).

♦ HH0171 Propensity to become
Description This term refers to the likelihood and not just the innate possibility that a particular feature of human potential may be realized. It may be considered in stages, as such propensity will be greater or lesser depending on different conditions – if "a" happens then "b" is more likely to occur; if "b" occurs, then "c" is less likely; and so on. The problem is to consider *development to* rather than *development from*; the latter will in some ways affect the former but progress in the desired direction can be made by supplying additional conditions. This is one of three concepts suggested by Israel Scheffler to be useful in defining *human potential*.
Preceded by Capability to become (#HH0902).

♦ HH0172 Body movement therapy
Psychomotor therapy
Description The kinesthetic sense can be awakened and developed in using any and all kinds of body movement. In contrast to physical education, psychomotor therapy aims to establish a subjective connection to movement which renders it a conscious action. The individual is thereby enabled to experience his own personal identity anew through each movement and its special quality.
Refs Pesso, Albert *Movement in Psychotherapy* (1969).
Broader Therapy (#HH0678).

♦ HH0173 Salvation
Description The following of a spiritual path which will lead from a human condition of insecurity and disease to one of safety and wholeness may be described as salvation; and such paths are offered by "salvation religions". Some religions view salvation as an act of God, other systems look on it as deriving from human religious activity. Taoism emphasizes the elimination of disharmony; Mahayana Buddhism looks towards an intellectual vision parallelling Buddha's enlightenment; theism centres on a right relationship with the creator; Hinduism sees salvation as union with God; esotericism looks to knowledge of the self to remain on the path and be free of the cycle of birth and death. In Christian teaching, mankind has fallen from a state of right order towards God by the act of original sin; salvation is the generic term used to describe the action of God in restoring man to his original state. This hope of salvation from sin by the action of a deity or heroic figure is common to many religions.
Specifically, Christianity states that salvation cannot be won by human efforts, neither for the individual nor for the human race. By Christ's sacrifice of himself God receives the homage mankind had refused to render and the way is open for salvation. Mankind has only to respond in order to receive the grace which will free them from the slavery of sin. Currently, the definition of salvation in the Church has been widened to include struggle not only of hope against despair in personal life but also for economic justice instead of exploitation, for human dignity instead of political oppression and for solidarity instead of alienation.
Refs Kulshreshtha, Saroj *The Concept of Salvation in Vedanta* (1986); Rohde, S *Deliver Us from Evil* (1946).
Narrower Personal salvation (Christianity, #HM6335).
Related Grace (#HH0169) Liberation (#HH0388) Soteriology (#HH1166)
Moksha (Hinduism, #HM2196) Human development (Christianity, #HH2198).

♦ HH0175 Growth games
Transformative games
Description Growth games are aids to self-awareness developed within the content of the human potential movement. Such games may involve several persons or may be engaged in by one person alone. They are designed for a variety of purposes such as: sharpening the senses, expanding consciousness, enlivening the body, letting oneself go and releasing tensions, self study and being oneself, building interpersonal warmth and trust. Participants carry out techniques to remove physical tensions and inhibitions; this is often followed by the disappearance of the cause of the problem. Games may be used during the course of group therapy and encounter groups sessions.
Refs Lewis, Howard R and Streitfeld, Harold S *Growth Games* (1970).
Broader Games (#HH0705) Human potential movement (#HH0398).
Narrower Transformation game (#HH5634).

♦ HH0176 Human relations
Description This term designates an area of management concern in ensuring the smooth functioning of a corporation or enterprise. Productivity, satisfaction at work, and health are all intimately interwoven. Human relations skills are required to balance the various perceptions and perspectives of individuals performing different, but interdependent functions within an enterprise in order to avoid conflict, ill-will and ill-health. The focus is therefore on interpersonal relations and the aim is to increase productivity by improving the working atmosphere.

♦ HH0178 Guidance
Educational guidance — Vocational guidance
Description Based on an authoritarian relationship, guidance aims to advise and instruct an individual in the path he should follow to achieve a particular goal or avoid particular conflicts or anxieties. Specific forms of guidance are: (1) *Educational guidance*; (2) *Vocational guidance*; (3) *Spiritual guidance*.
Narrower Vocational guidance (#HH8005) Spiritual direction (#HH0442)
Educational guidance (#HH5786).

♦ HH0179 Pedagogy
Description In the narrowest sense of the word, pedagogy is the science or profession of *teaching*, but the meaning of the term can be extended to include the complete and systematic upbringing, education or moulding of a human being, the shaping of the *personality* by goal-orientated activities, and the forming of a fully developed person within the family and societal environment. The science of pedagogy covers both teaching (the transmission of knowledge, skills and so on) and supervision (the monitoring of a student's work); it is one of the basic social sciences and covers a variety of methods and disciplines depending on the philosophical/religious system of a society under consideration. From the 17th century onwards, with the development of the philosophy of enlightenment, the importance of instructing the individual to make him more able to strive for his or her personal utilitarian wellbeing and, subsequently, to strive for his or her personal spiritual wellbeing came to the fore.
Related Education (#HH0945) Philosophy of enlightenment (#HH0591).

♦ HH0180 Learning
Description This term covers broadly the acquiring of knowledge, abilities and habits, in fact all that may be considered a person's experience. Basically, *obligatory learning* is the acquisition of skills necessary to every individual and mainly takes place by *imprinting*; *facultative learning* refers to the particular changes in behaviour due to the specific experiences of an individual and occurs mainly through *habituation* and *imitation*. Although learning is usually defined as adapting to specific conditions, much interest has been focused on the transferring of what is learned under one set of conditions to situations with radically different conditions, on *latent learning* and on the formation of internal controls on behaviour. Learning is an open-ended process; working on partial information, there is a process of experimentation, *learning by doing*, under the guidance of higher modes. This yields a sense of direction for continuing the process rather than a finalized structure.
Refs Jantsch, Erich and Waddington, Conrad H (Eds) *Evolution and Consciousness* (1976).
Narrower Virtual learning (#HM1217) Conscious learning (#HM0168)
Innovative learning (#HH0812) Functional learning (#HM0196)
Accelerated learning (#HH4511) Superconscious learning (#HM1244).
Related Education (#HH0945) Educational mobilization (#HH0924)
State of learning (Buddhism, #HM3662).

♦ HH0181 Motive
Motivation
Description All human activity is prompted by motives, whether instinctive, emotional, or dependent on attitude or ideals. In biology, instinctive acts (hereditary) or conditioned reflexes (acts reinforced by experience) are prompted by active states of brain structures or systematically organized stimuli of the central nervous system, and these states and stimuli are referred to as motivations. *Individual* motivations refer to the most basic needs (adequate food, warmth, avoidance of pain); *group* motivations include caring for children, maintenance of society, and jockeying for position within that society; and motivations involved in play and exploratory behaviour are referred to as *cognitive* motivations. The presence of same need – motivational excitation – is necessary before activation of the emotions can take place. Behavioural science defines a motive or *motivational variable* as any stimulus which induces or activates behaviour. Depth psychology refers to biological instincts and affinities as *principal* motives, these being partially suppressed under different societal conditions. Thus a primary motive is that object which satisfies a simple material need. Subsequent motives include motivating ideas and conscious goals. Of the several motives prompting any activity the leading motive will affect the quality of that activity and its meaning to the person carrying it out. As a person learns to evaluate the complex and often contradictory motivations which caused activity (rationalization) and establishes a definite hierarchy of motives, then principal motives are subordinated to social or spiritual motives and activities are carried out with reference to ideals rather than material requirements.
Refs McClelland, D C, et al *The Achievement Motive* (1953).
Narrower Metamotivation (#HH0847) Achievement motivation (#HH0456)
Motivational development (#HH0055) Educational mobilization (#HH0924).

♦ HH0182 Experience
Description Experience refers to the body of knowledge possessed by an individual, based on the sum of his sensory impressions. There is philosophical disagreement as to the significance of experience to total knowledge. Idealist empiricists hold that experience is based not on objective reality but on subjective sensations and impressions (Berkeley, Hume); material empiricists assume that the source of all experience is the material world. In contrast, rationalists (Descartes, Leibniz) hold that experience provides confused knowledge and cannot be the basis of logical throught; and that the truth can be directly arrived at by reason without empirical or sensory cognition. Teilhard de Chardin points out that great changes are so slow as to be imperceptible in the experience of the individual, or even of mankind.

♦ HH0183 Holiness
Holiness code
Description Holiness may be said to be that which defines religion; there may be no conception

of a divinity but there must be a distinction between the holy and the profane. Under this definition, some forms of magic as well as Buddhism and other higher forms of salvation not involving belief in God can be considered religions. However, holiness normally refers to being of divine origin, transcendent and set apart, and therefore in Christian teaching, for example, only God can be holy in the full sense of the word. However, in that God sanctifies or mediates through a person, that person may be referred to as holy by derivation.

Although that which is holy is generally looked upon as being set apart, as opposed to the "normal" or profane (so that it is beyond the ordinary and cannot be used in the ordinary way), spiritual religion would abolish this external distinction and attempt to bring the whole of life under holiness. This is contrasted with the mysterious holiness or *mana* of the supernatural in many religions. It is interesting that the setting apart, the danger or taboo, of the holy makes it in some ways pure but in others unclean – too close contact with the holy for the ordinary person means, in many instances, the need to be purified before continuing a normal life. This is seen in specific rites associated with birth, marriage, death, royalty, priesthood and so on. Their close contact with the holy makes different standards expected of the priesthood as compared with the ordinary person.

With emphasis on the importance of the individual as opposed to the group, holiness and purity become personal qualities of deity and man rather than substances within things. Holiness is a divine influence – God makes man holy but this obliges man to strive for his own perfection. In this sense holiness implies ethical goodness and purity. The *Law of Holiness* or *Holiness Code* refers to those parts of the Book of Leviticus in the Bible which describe regulations on moral and ritual sanctity and the observance of festivals and sacrifice. The ritual and moral purity achieved by observance of this code is a means of mirroring the holiness of God.

In the Church, an individual is referred to as holy not because of moral perfection but because of the God-given quality, something divine or supernatural. A saint is made a saint because of veneration by the religious community and, to some extent, because of miracles performed by that person before or after his death. In the Methodist tradition holiness (or sanctification) and freedom from sin are integral parts of justification.

Again, in Christian terms, it is the new covenant which God made with man in the person of Jesus Christ, who sanctified Himself to His holy Father, that allows His followers to be made holy. A Christian, made holy in Christ, becomes part of a holy nation. This brings a responsibility to be holy in conduct. Holiness, initiated by God, is distinct from all that is not of God and implies dedication to God's purpose and participation in God's righteousness. There is no longer a question of being sacred at one time and place and profane at another. Holiness is an inner transformation received as a gift from God requiring a complete break with the former sinful state and a horror of being again defiled. This being a member of God's elect is mysterious when considered with respect to God's desire to save everyone.

François Fénelon says of holiness that it depends on being in the particular state that God requires one to be. It may, on appropriate occasions, require contemplation, it may require other forms of prayer, it may require the normal tasks expected of a good neighbour or a good citizen. The law involving the nature of holiness requires that it adapt to every situation.

Refs Goldbrunner, Josef *Holiness is Wholeness and Other Essays* (1964).
  Narrower Human holiness (Christianity, #HM0446).
  Related Mana (#HM2686)   Purification (#HH0401)   Sanctification (#HH0428)
  State of holiness (Christianity, #HM1723)
  Awareness of the sacred (Christianity, #HM0876).

♦ **HH0185 Prayer**
Saying of prayers — Praying
Description A basic distinction is made between petitionary prayers (saying prayers) and praying. The former usually takes place as ritual repetitions or requests made at specific times and places and which may be automatic. Prayer restricted to such conventional, ceremonial worship is the mark of the spiritually immature. Nevertheless, some feeling of respite and awe is evident, as the petitioner pours out frustration and misery or expresses joy and emotion. It may be a means of expressing relief and thanks at the experience of some good, or of pain and concern for the petitioner's loved ones, or for his own future; prayers for a good harvest are an example.

Ritual prayer or incantation, where the words themselves are believed to have magical powers to produce desired results, is part of many traditional rites and folklores. In some cases, the perfect repetition of particular forms is held essential, with dire results following a mistake. Such prayer approximates closely to the repeating of charms or spells; often the overriding feeling towards the deity or spirit involved is more of fear than of reverence.

True praying, however, demands the presence and attention of the whole person. The objective of praying as a form of spiritual exercise, may be described as an effort to raise the whole psycho-physical complex which is man as a biological entity, to the level of the life of the spirit (pneuma) and achieved by participation in the ever-flowing life of the Creator-Spirit, Love.

Prayer arises from the tension between the realization of present imperfection and incompleteness, and the longing for perfection and wholeness. It is a movement of mind and soul into the source of all being, the That which is called God. There a person opens himself by breaking through the narrow confines of his egohood to become one with the All and the Infinite. This experience is then translated into a response to the everyday world, a realization of the infinite in the finite, of the eternal in the temporal, and a translation of spirituality into selfless action and an all-embracing charity towards everyone and everything.

Aspects of prayer which have been distinguished include: prayer as a way of union with God and the realization of the self; the psychological aspect of prayer whereby the division between the conscious and the unconscious is overcome; prayer as a direction of the heart towards completeness, perfection and unity with the whole; prayer as a movement into the realm of the mystical; prayer as a source of power; prayer as an exploration of the reality; prayer as a means of enriching and fructifying daily life.

Five degrees of prayer have been distinguished: vocal prayer primarily in the form of adoration, thanksgiving, confession and petition; meditation at the level of the intellect, namely methodical and discursive prayer which includes considerations, arguments, and resolutions; affective prayer, in which affection and aspiration play the predominant part rather than petition; prayer of quiet simplicity or interior silence in which spiritual intuition replaces the intellect; and prayer of contemplation in which the self transcends the stages of symbol and of silence and enthusiastically energizes those levels which are dark to the intellect but radiant to the heart.

Refs Brame, Grace *Receptive Prayer* (1985); Happold, F C *Prayer and Meditation* (1971).
  Narrower Hesychasm (Christianity, #HH0205)   Centering prayer (Christianity, #HH1994)
  Liturgical prayer (Christianity, #HH4209)   Discipline of prayer (Christianity, #HH1193)
  Contemplative intuitive meditation (#HH0816).
  Related Puja yoga (Yoga, #HH0111)   Hope (Christianity, #HH0199)
  Practice of silence (#HH0983)   Devotion (Christianity, #HH0349).

♦ **HH0187 Self responsibility**
Description With reliance on the self and the consequent reduction in the role of authority, an individual takes responsibility for his or her own behaviour. This self reliance is reflected in systems of medicine (holistic) and education where self discipline rather than discipline from outside is the keynote.
  Related Continence (#HH0070)   Self-learning (#HH0417)   Encounter group (#HH0162)
  Holistic medicine (#HH0581)   Self-determination (#HH0317).

♦ **HH0188 Homiletics**
Description This term refers to the body of concepts, principles and practice which together govern effective preaching and determine the role of preaching in missionary work. It may also be considered to include the supernatural efficacy of preaching and its role in salvation.

♦ **HH0190 Spiritual poverty**
Self-simplification — Poor in spirit — Poverty of the spirit
Description This is the intensification of interior discipline, such as fasting, promoting spiritual detachment and the external indication of commitment to a specific way. Such poverty encompasses *affective* poverty – indifference to money and what it can buy. Poverty of spirit as an aspect of self-denial is part of the Islamic faith, abstinence from material goods and sensual pleasures having developed from abstinence from sin. The Sufis consider poverty as one of the *six stations of spirituality*, a means of achieving liberation from that which distracts one from God; poverty of spirit acknowledges need for God. To St John of the Cross, spiritual poverty provides the detachment for quiet and repose, with none of the fatigue inherent in covetousness. Concern is then only for the essentials of devotional life and with using things which assist in devotion. Not coveting material or spiritual things, the only coveting is for being right with and pleasing God. Fixing the eyes on the reality of inner perfection there is no interest in self-gratification.
Refs Javad, Nurbakhsh and Lewishon, Leonard *Spiritual Poverty in Sufism* (1984).
  Related Fasting (#HH0011)   Faqr (Sufism, #HM3427)   Religious poverty (#HH0991)
  Devotion (Christianity, #HH0349)   Awareness of spiritual poverty (ICA, #HM2286).

♦ **HH0191 Solipsism**
Description Believing the self to be the only thing that truly exists, or can be known to exist, the solipsist may became an egoist, a hedonist, etc.
  Related Egoism (#HH0130).

♦ **HH0192 Honour**
Description This refers to a quality worthy of respect, esteem or praise for its own sake and not for its utilitarian or pleasurable result. It also refers to the response paid to such a quality in someone else and the cultivation of a sense of when such a response is due. When associated with gallantry and behaviour in armed combat it has come to be associated with codes of chivalry and, derived from the knightly virtues, the correct behaviour of a gentleman. To the Japanese, honour implies "glory of the name"; honour involves respect for the family name and thus for the family or clan, shame to one's self implies loss of honour to ancestors and descendants, upholding one's name is related to the upholding of personal honour. In religious terms, honour has been considered, with fear, as a component of reverence – honour should be paid to God and, by reflection, to all that is placed in a position of honour by God; the corollary is the implied dignity of God's creatures rather than that of the roles they play. The desire for honour may motivate virtuous conduct and thus assist in human development, with the proviso that hypocrisy is avoided.
  Related Chivalry (#HH0329)   Nobility (#HH0499)   Reverence (#HM3023).

♦ **HH0194 Pleasure**
Description Pleasure is said to be either sensible (that is, resulting from gratification of the senses) or intellectual (the spiritual delight from possession of truth). Either or both may accompany uninterrupted activity which is directed towards an end. Aristotle indicates that one tends to do best what one most enjoys doing, the pleasure arising in a given activity possibly an indication that one was morally wise in adopting that activity. Although no value judgement is put on the experience of pleasure (goodness or badness depending more on what a person finds pleasurable than on the pleasure itself), most religions agree that the pursuit of pleasure as an end in itself is not morally defensible. For example, in the Katha Upanishad, death says, "The good is one, the pleasant another; both command the soul. Who follows the good, attains sanctity; who follows the pleasant, drops out of the race. Every man faces both. The mind of the wise man draws him to the good, the flesh of the fool drives him to the pleasant". Christian teaching indicates that in order to be honourable one must sometimes forgo pleasure; nevertheless the honourable course brings with it its own pleasure. In contrast, hedonism looks on pleasure as the ultimate goal of mankind, to be sought everywhere, while utilitarianism requires moral and religious tolerance in a framework where each person has increasing overall satisfaction, or has his or her individual pursuit of happiness free from interference by those who would menace the kind of life which would tend towards such happiness.
  Related Unlust (#HM3094)   Egoism (#HH0130)   Pleasure (#HM2883)
  Life instinct (#HH0750).

♦ **HH0195 Sensorimotor development**
Description The intelligence of the young child is taken to be sensorimotor, when thought and action are virtually inseparable. Six stages of sensorimotor development may be distinguished.
1. The first stage (0–1 month) is characterized by stabilization of the reflexes. Performance is improved and a generalizing and differentiating tendency is evident.
2. The second stage (2–6 months) is characterized by the stabilization of acquired (as opposed to unborn) behaviour patterns, indicating the influence of ontogenetic experience. Reflexes are coordinated, resulting in new and more complex behaviour patterns.
3. The third stage (5–8 months) is characterized by acquisition of new patterns of action that are more than the extensions of the schemata of the previous stage. There is a shift in interest to the external effects of actions and a search for actions that have interesting external effects.
4. The fourth stage (8–12 months) is characterized by the intentional application in new situations of familiar action patterns as means toward the attainment of desired ends and goals.
5. The fifth stage (12–16 months) is characterized by the child's ability to invent new behavioural actions as a result of a trial-and-error, goal-directed search. He capitalizes on accidental discoveries and uses these as a means towards desired goals.
6. The sixth stage (16 months onwards) is characterized by the ability to form mental representations (namely the capacity to imagine the external world and its possible transformations in the absence of perceptual cues) and the ability to form mental inventions (namely the capacity to think out actions before executing them).
  Related Primary growth (#HH6669)   Cognitive growth and development (#HH0146).

♦ **HH0196 Maturation**
Description Maturation is an autonomous process of somatic, psychological and mental differentiation and integration. The process is spread over stages concerning which there are no universally agreed precise boundary limits; new-born baby, baby, infant, child, juvenile or adolescent, young adult, mature adult, old adult. It is through the interaction and combination of the different phases of this process that the individual's growth is completed and consolidated, permitting him to adapt to life. Maturation is therefore a process of unfolding of the potential of the human organism, a ripening of the physical equipment coupled with a change in the organism's capacity to perform. There is considerable difficulty in ascribing distinct definitions to concepts such as growth, development, maturation and learning, since they are all closely intertwined. Distinction is usually made however between those changes that take place as a consequence of learning and those produced by growth and maturation.

HH0196

In psychological usage, maturation describes that part of development that takes place in the absence of learning (including specific experience or practice), whether the process is viewed as a ripening of the individual's capacity or the growth of some neuroanatomical structure permitting a performance impossible before that time. In the first case the nervous system as a whole ripens in its capacity for the retention of experience and for intelligent adaptation to new situations. Similarly, certain intellectual capacities (talents for art, music, mathematics, etc) undergo this kind of maturation. In the second case, there is the unlearned ripening of motor coordinations and capacities such as creeping, walking, climbing and vocalization. Some psychologists hold that maturation involves much more than simple physical readiness and that development in such areas as language and the acquisition of syntax proceeds through predetermined, unlearned, emergent and developmental sequences.

Maturation contributes to the development of personality by bringing out every inherited feature, including physical structure, temperament, talent, capacity for intelligent modification of behaviour, peculiarities of physical growth and decay, latent sexual functions, and numerous specific locomotor and vocal patterns. Maturation presents the individual with new internal situations to which he must adjust but, except at a rudimentary motor level, it does not provide him with ready-made instruments with which to do so.

In education or training, and the management of the learning process, the idea of maturation has as its counterpart the notion of readiness to perform. This is seen as the point in time of human development when the presentation of adequate stimuli will result in a reasonable gain in capacity to perform In order to learn a new response, the child must be physically and neurologically mature enough to attend, discriminate, and respond to relevant cues in his environment.

The most usual assumption of educators is that maturation is primarily a genetic process which triggers off in due time a capacity of the child to profit from the educational experience. Maturation is therefore most often judged by the chronological age of the child.

The process of maturation and that of learning or education should blend harmoniously. Maturation which is too rapid or too slow can be the cause of dysharmony in physical, mental and spiritual development as well as affecting the processes of social adaption. On the other hand, periods of dependence and dysharmony may be favourable to psychophysical humanization, language acquisition and socialization.

Although physical maturity is being reached earlier than previously, emotional development or maturity is said to be reached more slowly than before. This is ascribed, at least in part, to the decrease both in the number of adults with whom a child or young person has contact, and in the time spent with those adults. It is suggested that higher education should take this factor into consideration.

Refs Bühler, Charlotte *Maturation and Motivation* (1951).

◆ **HH0197 Discursive meditation** (Christianity)
Description By focusing the intellect, the imagination, the emotions and the will on events in the life of Christ, reflecting on the significance and the lessons to be drawn and turning to God in love and trust, the meditator follows a formal method of mental prayer of which there are several systems.

◆ **HH0198 Strength of personality**
Refs Allport, Gordon W *Pattern and Growth in Personality* (1961); Fadiman, J and Frager, R *Personality and Personal Growth* (1976); Hayakawa, S I *The Fully Functioning Personality* (1956); Jourard, Sidney M *Healthy Personality and Self-Disclosure* (1959); Kluckhohn, Clyde K and Murray, Henry A (Eds) *Personality in Nature, Society and Culture* (1953); Myers, F W H *Human Personality and its Survival* (1920); Sontag, L W; Baker, C T and Nelson, V L *Mental Growth and Personality Development* (1958).
   Related Unity of personality (#HH0136)   Personality development (#HH0281)
   Integration of the personality (#HH0561).

◆ **HH0199 Hope** (Christianity)
Description An as yet unfulfilled desire which nevertheless has the expectation of being fulfilled, hope is listed by St Paul, together with faith and charity, as one of the three great abiding virtues. From the eschatological viewpoint, hope is the clear and intuitive vision of the object of belief, that is, the eternal good of a future life in the presence of God as he is in himself. Hope for temporal and perishable goods is of secondary importance, although the "theology of hope" movement looks to social and political change based on hope for a new and coming future in this world. Hopefulness here-and-now also counteracts the despair which can cripple moral activity while fostering the courage which is a prerequisite for effective moral agency. Nevertheless, not all hope is positive. Discrimination is required to distinguish between hopes which are fostered by the Christian faith and are basic to human fulfilment and moral existence, and hopes which are empty in moral terms and possibly demonic. The Lord's prayer is said to be a perfect expression of hope but all prayer is a manifestation and interpretation of hope in some sense.
Context One of the three supernatural or abiding virtues.
   Broader Virtue (#HH0712).
   Related Prayer (#HH0185)   Despair (#HM3405)
   State of hope (Christianity, #HM7656).

◆ **HH0200 Habits**
Description Repetition of an activity establishes a habitual response in the subconscious so that similar situations to those in which the activity occurred will provoke habitual physical and emotional reaction.
Refs Teune, Henry *The Learning of Integrative Habits* (1964).
   Related Habit (#HH3665).

◆ **HH0203 Humanness**
Humanity
Description Humanness is generally taken to be that quality which is proper to human beings and which differentiates them from other entities and species. In particular, it describes the moral and ethical attitude including love of one's fellow men and human understanding which, as opposed to use of force and coercion, is thus the aim of civilization and the development of moral and political life.
Refs Korzybski, A *Manhood of Humanity* (1971).
   Related Humanism (#HH0674)   Transparency (#HH0673)
   Immediacy of fulfilled living (#HH0223).

◆ **HH0204 Accepting the shadow**
Integrating the shadow
Description The archetypal scapegoat or shadow present in everyone is that part of the psyche normally the focus of blame or attack when the individual feels it necessary to vindicate himself or justify his own behaviour. It is not normally recognized as part of the self and thus the blame or attack is usually received by someone else who has sparked off the disquieting view of the shadow. *Psychosynthesis* aims, by *psychoanalysis*, to assist in acknowledgement of this lower aspect of man's nature and thus create inner harmony. Such acknowledgement takes courage but results in integrating the shadow in a constructive manner, leading first to humility and humanness and eventually to new insight and expanded horizons.

It is postulated that the inability to accept that the "enemy" is in fact one's own lower nature is the cause of all bias, discrimination and conflict. Acknowledgement of the collective shadow might well prevent nationalistic or racialistic over-reactions to atrocities and barbarism which effectively are merely responding in kind. By accepting that everyone, as a human being, holds in himself collective responsibility for every development may well be the key to the next stage in human evolution.
   Related Evil (#HH1034)   Archetype (#HH0019)   Wholeness (#HM2725)
   Psychosynthesis (#HH0002)   Consciousness spectrum (#HM2530).

◆ **HH0205 Hesychasm** (Christianity)
Jesus prayer
Description The term derives from Hesychius, who founded a sect in 5th century Jerusalem. An inner tranquillity is achieved through combining a simple spoken prayer, *Kyrie eleison* – in full, "Lord Jesus Christ, Son of God, have mercy upon me, a sinner" – with the rhythm of the heartbeat or of breathing. The prayer may be shortened, sometimes to the simple repetition of the name Jesus. The power of this name is indicated in the New Testament where it castes out devils, heals the sick and justifies people. Jesus told his followers that whatever they asked in his name would be granted.

The intent is to move from verbal repetition of the prayer to mental prayer to inward prayer (prayer of the heart), so it emerges as a prayer of the spirit. The person does not so much offer a prayer as become a prayer. Practice of the prayer in time with breathing makes it a response of the whole person and eventually breathing and praying become synonymous, prayer arising without any conscious effort to pray. The repetition is not simply mechanical but an expression of the attitude of the heart. Thus although tranquillity and concentration may arise they are not the purpose, behind this is faith, faith in who Jesus is and what he is doing for the individual person.

In the tradition of hesychasm there is the one truly Christian initiation to the mysteries, although there is no specific rite of initiation. The repetition of the Jesus prayer as an exercice is warned against unless one has a spiritual master. If such a master is not to be found, then one must rely on books and caste one's self on the mercy of Christ as the source of instruction. This is a similar process to the Sufi *dhikr* or the Hindu *japa yoga*. Behaviour in the presence of the master or *geront* must be as though in the presence of Christ – but the geront is not initiator. Only Christ is the initiator and only His sacraments support the initiatic (as well as the exoteric) path. The ultimate purpose of spiritual endeavour is *deification*, or the highest possibilities salvation can comprise.
Refs Pallis, Marco *The Veil of the Temple* (1986).
   Broader Prayer (#HH0185).
   Related Mantras (#HH0386)   Mantra yoga (Yoga, #HH0931)
   Heart-awakening stage of life (#HM3518).

◆ **HH0206 Astral body**
Etheric Body — Second body — Etheric pathology
Description According to occult theory, every individual has a *second body* which is non-material and comprises the emotional part of the person and the higher intellectual faculties. Just as the physical body is liable to disorder, so is the astral body; and since it controls the instincts, senses, brain consciousness, passions, pleasure and pain, many diseases (mental and physical) and allergies are considered due to disorders of the etheric system or to lack of harmony between it and the physical system. Associated with belief in the second body are theories which explain ghosts, apparitions, zombies, werewolves, vampires, and traditional stories of transformation of men into wild beasts and so on, in terms of etherosis.
   Related Psychic atavism (Psychism, #HM0401).

◆ **HH0208 Hyper-personalisation**
Description Teilhard de Chardin's concept of unity as harmonised complexity entails each component of any grouping, whether, for example, a member of a society, a part of a body, or an element of a spiritual synthesis, in being perfected and fulfilled individually in order to create a harmonious whole. In particular, the summation of consciousness is not seen as a pantheist merging of indifuality where each individual "grain of salt" is dissolved in a uniform ocean of the great All; but as an accentuation or "hyper-personalisation" of each individual "self". This is in direct conflict with materialisation and the standardisation of humanity into well-ordered workers in a factory, soldiers on a paradeground and so on, with the resultant mechanization and statistical average, approaches which neglect the due place of the individual person. Teilhard de Chardin maintains that individual consciousnesses are immiscible but warns that individuality is not to be confused with "personality" when a particular element seeks tobecome separate from all other elements. Cf St Paul's 1st epistle to the Corinthians XII 14–17,20: "For the body is not one member, but many. If the foot shall say, because I am not the hand, I am not of the body; is it therefore not of the body? If the whole body were an eye, where were the hearing?.... But now are they many members, yet but one body".

◆ **HH0210 Loss of soul**
Description This expression is used by Jung to describe severance of relations with the individual psychic life. Often occurring at midlife, there is a lack of energy, of a sense of purpose, of personal responsibility; it is marked by the preponderance of affect and *abaissement du niveau mental*.
   Related Dasein (#HM2632)   Emotion (#HH0819)
   Abaissement du niveau mental (Jung, #HM3615).

◆ **HH0211 Sentics**
Description Sentics is concerned with the quality of experience and the way in which specific emotions have their own specific neurological pattern, by considering the correlation between brain-waves and the emotions. With the use of a "sentic cycle cassette" an individual can experience a complete cycle of the emotions, enabling a greater understanding of the emotions and the self. This is said to reduce anxiety and relieve psychosomatic problems.

◆ **HH0212 Human perfectibility**
Human perfection
Description Perfection may be seen as the actualizing of the highest human potential. Religious tradition sees perfectibility of human beings as realized in identification or union with the perfect in ultimate reality.

A number of different kinds of human perfection may be distinguished as follows:

Technical perfection by which man becomes perfect by improving his ability to perform a task and through his perfect performance of that task. This raises the question as to whether it is sufficient to perfect one's self as a human being merely by perfecting one's ability in a task and performing it perfectly. The answer is normally that this needs to be accompanied by some other form of perfection.

Obedientiary perfection by which man becomes perfect by absolute obedience to the will of some superior authority (possibly God or some elite) who is perfect and therefore guarantees the perfection of man through that obedience. The perfection of man's conduct when he obeys the authority then lies, however, not in his obedience as such but in the fact that his conduct reflects

the perfection of that authority.
Teleological perfection by which man becomes perfect through the attainment of some natural end in which it is his nature to find final satisfaction (which may be identified with happiness or well-being). Whereas technical perfection entirely depends on talent and skill, man can achieve this form of perfection as a gift, or by luck, rather than as a result of effort.
Moral perfection by which man becomes perfect through becoming, in some absolute sense, a better person. This implies that all human potentialities actualized in the person by this process are in fact potentials for good. It has been suggested that criminality, for example, is not a potentiality capable of being actualized but rather a defect or imperfect actualization of a potentiality.
Metaphysical perfection by which man becomes perfect or complete through reducing and eliminating his dependence for existence on anything else, since such dependence is an indication of his incompleteness. This requires that man stand aside from life and not allow himself to be influenced by anything which happens to him; that he become less finite and less temporal by freeing himself from all concern and uniting himself with some superior being. This removes final perfection beyond human aspiration and implies that no finite being could be absolutely perfect, although some exceptional individuals may be identified as such, particularly in certain religions.
Aesthetic perfection by which man becomes perfect through the full and harmonious, orderly development of all his faculties, physical, emotional, intellectual and moral. Aesthetic perfection is related to technical perfection but implies the perfect performance of tasks in a flawless social context, with the latter taking on the attributes of metaphysical perfection, namely unity, immutability and self-sufficiency. A society which values unity, harmony and stability above all else may do so only at the cost of suppressing human freedom and creative experiment.
Exemplary perfection by which man becomes perfect by conforming to an ideally perfect model of a human being or alternatively by becoming godlike (deiform perfection).
The Greeks established an ideal of metaphysical perfection and the notion that human beings could share some features of the perfection of a supreme being. They demonstrated that perfection involved knowledge, including a rational understanding of the universe, and that this knowledge could only be achieved by withdrawing from the world, although it was possibly achievable in an ideal society. They also attempted to set up a lower level of morality as a preliminary step on the ordinary citizen's path towards perfection.
The vedanta see human beings as imperfect in that they do not know their identity with the Absolute, Brahman, which is perfect. Because of this inability to perceive reality as it is, human actions are also imperfect. Such action or karma leads to the cycle of existence and reincarnation. Perfection is achieved with perfect realization that the Atman, individual soul, is one with the Brahman. Both Hindu and Buddhist traditions set down certain paths for perfection – the Hindu through *karma marga*, or unattached action, through *jnana marga*, wisdom through meditation, or through *bhakti marga*, devotion; and the Buddhist through the the three stages of *sila* (ethical conduct), *samadhi* (concentration) and *prajna* (wisdom). The first five perfections, or *paramitas*, of Buddhism and Zen, which are of the mind, are said to lead to the quest of the Arhat for final perfection beyond human perfection.
The major sects of Christianity have been opposed to the view that man can in this life lead a flawless existence, promising perfection only in some future life to some or all, and maintaining that in this life only progress towards perfection could be achieved. Others, however, have sought complete perfection, whether by renunciation of the world, or by direct union with God, by an overwhelming conversion, by placing themselves entirely in God's hands, or by an exercise of will. According to St Augustine, full perfection can be achieved only through grace – the lowly estate of man compared with God, and original sin which makes it imposssible for man even to will that perfection of which he is capable, make it impossible for man to achieve perfection on his own. Thomas Aquinas held that envangelical perfection could nevertheless be achieved by removal of all mortal sin and cultivation of the love of God. This quest for perfection was that of Christian mystics, with severe asceticism often used to subdue fleshly desires; perfection would be achieved through freedom from sin and union with God. John Wesley also held that perfection was possible, by grace and through faith which must be actively sought through following God's commandments.
In secular terms, following the Renaissance, hope for perfection was associated with the improvement of man's relationship with man rather than with God, in moral rather than metaphysical terms, and particularly in the notion of doing the maximum of good. Man could then be improved to any degree by the use of appropriate mechanisms such as education or behaviouralist procedures. The concern was whether such social action would be employed in the interest of freedom rather than in the interest of absolute authority.
History, as understood by Hegel, showed the gradual coming to self-consciousness of an absolute idea, sometimes identified with God, by means of dialectical processes. Marx showed that this dialectic worked through class-struggles which would finally result in a class-less, state-less, society. His view was that, under capitalism, man is alienated from his fellow-men and in consequence from his human essence. Private property alienates him, preventing him from working out of joy in his creative powers and his control over nature, thus denying himself in his work, developing no free physical and mental energy, mortifying his flesh and ruining his mind. Communism as the complete and conscious restoration of man to himself would free him from this alienation. As such it was the true resolution of the conflict between existence and essence, objectification and self-affirmation, freedom and necessity, individual and species.
Communism argues that following social revolution, the possibility of improvement is endless, provided that the bourgeois society is completely overthrown to eliminate the fundamental flaw in the civilized world due to which progress has dehumanized rather than humanized man. It is held that some form of antagonism or dialectic is essential to man's development, although not to his final condition.
Theories of natural evolution, following Darwin, suggested that the future perfection of man would continue as a result of the continuing process of the selection for survival of the fittest. In opposition, it is suggested that man is unique, particularly in his freedom and his ability to cooperate with the processes of evolution. Self-consciousness is an emergent quality, which suggests that some new and improved lifeform, some superhuman being, might yet evolve, with unforeseeable powers.
Teilhard de Chardin suggests that evolution proceeds towards the perfection of man in new forms of social organization which then make possible the ultimate union of man with God. The conflict between perfection within the religious framework and as a result of natural development is therefore eliminated.
Refs Passmore, John A *The Perfectability of Man* (1971).
Narrower Paramitas (Zen, #HH0219)
Moral perfection (#HH5098)
Spiritual development (#HH0017).
Related Self-perfection (#HH0519)
Perfect man (Hinduism, #HM2183)
Ontological perfection (#HH0794)
Stages of spiritual life (#HH0102).

♦ HH0213 Pranayama (Hinduism, Yoga)

Regulation of vital force
Description One of the two principal steps of *hatha yoga*, the other being *asana*, *pranayama* literally means control of breath. However, the breath referred to here is the five vital forces, or prana, which maintain the activity of the physical body. The aim is to master the vital forces, and to release the latent Kundalini energy. When this energy is able to flow, the individual starts on the path of reintegration with the ultimate ground of existence.
The practice of regulation of prana leads to an inner harmony, with one pointed attention and a dwindling of the five defects which cause all negative mental activity: passion or rage (raga); delusion (moha); attachment (sneha); desire (kama); and anger (krodha).
Context The fourth component of the eightfold path of yoga.
Refs Lysebeth, André Van *Pranayama* (1979).
Broader Hatha yoga (Yoga, #HH0862)
Eightfold path of yoga (Yoga, #HH0779).
Related Kriya yoga (Yoga, #HH0969)
Prana-loka (Leela, #HM4399)
Induction technique (#HH0025)
Asana (Hinduism, Yoga, #HH0669)
Concentration on the flow of breath (#HH0687).

♦ HH0214 Multiple intelligences

Description The idea of multiple intelligences put forward by Howard Gardner proposes that, rather than considering intelligence as one function with many facets (all of which may be assessed together as one *intelligence quotient* or *IQ*), there are many separate intelligences independent of each other and which therefore need to be considered separately. These separate intelligences are postulated as those which can be shown to occur (or to be missing) selectively. Studies of individuals with brain damage or highly gifted in some way, and of cultures with particular attributes, have demonstrated the independence of the following:
(1) *Linguistic intelligence*. This is demonstrated by a sensitivity to sounds, rhythms, inflections and meter, a special clarity of awareness of the core operation of language.
(2) *Musical intelligence*. Such intelligence has as its centre the relating of emotional and motivational factors to the perceptual ones; music is a way of capturing and communicating feelings and knowledge about feelings.
(3) *Logical/mathematical intelligence*. This is developed first from the ability to recognize classes or sets of physical objects; and later by conceptualizing classes or sets of objects or ideas in the mind and understanding logical connections among them.
(4) *Spatial intelligence*. An accurate perception of the physical world, an ability to transform or modify these perceptions, and the recreating of certain aspects of visual experience without relevant physical stimuli – these are all part of spatial ability.
(5) *Bodily-kinesthetic intelligence*. Skill in controlling bodily movements and in the ability to manipulate objects combine in this intelligence, which has been valued in many cultures as the harmony between mind and body – the mind trained to use the body properly and the body to respond to the mind.
(6) *Personal intelligences*. These are centred on the concept of the individual self and may be considered as: (a) *Access to one's own feeling life* – this is the development of the internal aspects of a person and the ability to detect and symbolize complex and highly differentiated sets of feelings. (b) *Ability to notice and make distinctions among individuals* – to read even the hidden intentions and desires of others and to use this knowledge to influence their behaviour.
All these foregoing intelligences are the basic competences on which diverse capacities are built up, depending on an individual's environment. It may be through acquiring skills in a non-literate society; through acquiring literacy in a traditional or religious-based school; or through scientific, secular education. Each of these systems will develop some aspects but none develops them all. The first emphasizes bodily/spatial/interpersonal forms; the second de-emphasizes these in favour of linguistic forms; and the last looks to logical/mathematical and intrapersonal approaches.
Refs Gardner, Howard *Frames of Mind* (1984).
Narrower Musical intelligence (#HH0961)
Spatial intelligence (#HH1075)
Personal intelligences (#HH1274)
Linguistic intelligence (#HH0843)
Bodily-kinesthetic intelligence (#HH0662)
Logical-mathematical intelligence (#HH1456).
Related Intelligence (#HH0431).

♦ HH0215 Enlightenment humanism

Description A form of humanism developed by philosophers of enlightenment in the 18th Century, emphasizing the transforming of man into full humanness through education.
Broader Humanism (#HH0674).
Related Philosophy of enlightenment (#HH0591).

♦ HH0216 Healing

Description Healing was the term used by Jung to express the wholeness of the individual which, rather than *cure*, was the aim of analysis. Such healing implied the compassion and empathy of the therapist and a therapeutic relationship between the therapist and the person under analysis. This implies an "inner wound" within the analyst – he is not all-powerful and strong, bringing a cure to the passive and dependent patient; his experience of wounded-ness enables him to be aware of the inner world and is projected into the patient, who himself is cut off from inner health if he thinks of himself, or is thought of, simply as "ill".
Refs Frank, J D *Persuasion and Healing* (1961); Sanford, Agnes *The Healing Gifts of the Spirit*.
Related Cure (#HH1264)
Spiritual healing (#HH0458).

♦ HH0217 Ego strength

Stability of the ego
Description Individuals differ considerably in the forms and effectiveness of the functioning of their egos and this gives rise to the concept of ego strength. An individual of strong ego tends to have the following characteristics: objectivity in apprehension of the external world; objectivity in self-knowledge or insight; organization of activity over longer time spans; maintenance of schedules and plans; ability to conceive and act on a self-selected course of action; ability to choose between alternatives; direction of drives into socially useful channels; and resistance to immediate environmental and social pressures. The related concept of *ego stability* refers to the normal as opposed to the neurotic, and to an emphasis on real situations as opposed to imagination and wishful thinking. The ego-weak individual is less capable of productive activity because his energy is drained into the protection of warped and unrealistic self-concepts due to a distorted perception of reality and self.
Related Identity-strength (#HM1907).

♦ HH0218 Rhythmic sensory bombardment therapy

Description Periods of exposure to sonic, photic or tactile stimulation applied intermittently or rhythmically are said to produce favourable responses in some mental disturbances.
Broader Therapy (#HH0678).
Related Rhythmic sensory bombardment (#HM6533).

♦ HH0219 Paramitas (Zen)

Perfections of zen
Description The first five perfections of mind, heart and will are possessed by the mature or perfect mind ready then to achieve *Prajna* or great perfection, *maha paramita*. Each of these is related to freedom from separate self-hood.
Broader Human perfectibility (#HH0212).

Narrower Shila (#HH0429) Virya (#HH0677) Kshanti (#HH0813) Prajna paramita (#HH0550) Dana (Buddhism, Zen, #HH1234) Dhyana (Hinduism, Buddhism, #HM0137).

♦ HH0221 Alchemy (Esotericism)

Hermetic philosophy

Description Whilst many alchemical works may be considered solely as a prelude to chemistry, there are a number of works in which alchemy is looked upon as both an operational and spiritual technique. The language of chemistry employed by the genuine alchemists is understood to be a metaphor for the work, the opus, of psychic transmutation. The transmutation of base metals into gold is the transmutation of psychophysical elements within man from an impure, obstructed state to a fine state of responsiveness to high-frequency energy.

The acknowledged pursuit of the alchemists was the elixir vitae and the philosopher's stone; that is, the conquest of immortality and achievement of absolute freedom. The process of individuation must be regarded as prefiguration of the *opus alchymicum* or, more accurately, an unconscious imitation (for the use of all beings) of the extremely difficult initiation process. This process undertaken by the unconscious without the permission of the conscious, and mostly against its will, leads man towards his own centre, the Self.

The unconscious undergoes processes which express themselves in alchemical symbolism, tending towards psychic results which correspond to the results of hermetic operations. Dreams or waking dreams containing alchemical symbols constitute a series, the development of which accompanies a process of individuation. Such products of the unconscious are therefore neither anarchic nor gratuitous. They pursue a precise goal, individuation, which may be considered to represent the supreme ideal of every human being, namely the discovery of the possession of one's own Self.

The alchemist's endeavour to conjure out of matter the "philosophical gold", or the "panacea", or the "wonderful stone", may then be considered as only an illusion in part. For the rest it corresponded to certain psychic concerns that are of great importance in the psychology of the unconscious. In this sense the alchemist projected an understanding of the process of individuation into the phenomena of chemical change. The principal, or underlying, intent of alchemy is then understood as one of making of the body a spirit and of the spirit a body through transmuting the bodily consciousness into spirit and through fixing the spirit in the body. This assumes the possibility of changing the "state of aggregation" of a body in conformity with the metallurgical symbolism that is used to describe the process.

Alchemy is a sacramental science in which material phenomena are not considered autonomous, but represent only the "condensation" of psychic and spiritual realities. Through its processes, when the spontaneity and mystery of nature are penetrated, nature becomes transparent. It is transfigured under the lightning flashes of divine energies, whilst incorporating and symbolizing those angelic states of awareness which ordinarily can only be briefly experienced, such as when listening to music or when contemplating a human face. This transfiguration of nature takes effect in the heart of man. It is the eye of the heart that transforms lead into gold as the metal whose luminous density is thought best to express the divine presence in the mineral realm. As such alchemy is an immense effort to awaken man to the divine omnipresence.

The mode of operation of nature in the universe of form is perceived as a continuous rhythm of coagulations and dissolutions. Form is impressed on matter and matter dissolves it in order to offer itself to another form, in a continuing pattern of alternation and transformation. In this interplay of perpetually interacting tensions, these neutralize each other at one moment by their very opposition and then destroy each other only to arise again in a new guise. The logic of alchemy is two-fold, involving a reintegration of manifestation with its principle and appearance with reality, as well as a dialectic of the living tension of all complementarities.

As in other traditions, two paths are recognized in alchemy:

(a) Through the first path, or "humid way", natural energy is drawn into itself in order to transform it into fervour. This path involves the phases of "blackening" (nigredo, melanosis), "whitening" (albedo, leucosis), and "reddening". In other, more detailed, explanations this path is understood to be made of a progressive alternation between critical periods ("putrefactions") and stages of further maturation or integration ("coniunctio"). Each maturational stage is thus preceded by a critical period involving a process of profound psychobiological transformation during which the individual has to make fundamental psychic as well as physiological adjustments. It is during these periods that unconscious symbols, such as those emerging in dreams, will conform to the alchemical patterns of mortification and putrefaction, subsequently to change into symbols of death and rebirth. Following the progressive individuation, through which a child matures to adulthood, this cycle of regressive individuation parallels the potential processes of maturation through to death. The first crisis of adulthood, symbolized by the "prima materia", is transcended through the first (Earth) "coniunctio", leading to a second crisis in middle age, symbolized by the "black putrefaction" (nigredo). This is transcended through the second (Moon) "coniunctio", leading to a third crisis in late middle age, symbolized by the "yellow putrefaction". This is transcended through the third (Sun) "coniunctio", leading to a third crisis in old age, symbolized by the "red putrefaction" (rubedo). This in turn is transcended through the fourth (Heaven) "coniunctio", symbolized by death.

(b) Through the second path, or "dry way", considered more rapid and more dangerous, there is a direct and unmediated path from the ego to the inner man, which slowly takes into itself the Soul of the World. It is initiated by a still more radical "descent into hell". This path bears some similarity to the yoga of knowledge and to the direct path of tantrism.

As with tantrism, identification with the world in its positive aspect is seen as the necessary first step in the process of liberation. Its methods are more concerned with processes of detachment than of renunciation, and with purified participation in the world rather than escape from it. The major practice of alchemy is based on the use of imagination (as distinct from fantasy) to cultivate the golden instants which occasionally illuminate ordinary life. Through such imagination the continuing creation of the world is reproduced within the individual. Alchemy enobles the soul's need to dream, expanding the dream beyond the limitations of individuality to encompass the cosmos – finally to raise the question "Who dreams". Alchemists thus come to see with the eyes of the spirit, as God dreams the phenomena of nature in sustaining its processes. The alchemist reverses cosmogony, dissolving material appearances in pure life. He makes in himself, by meditating on natural beauty and on the "sympathy" which holds all things together, the unity of the Soul of the World, until through his own heart he causes the solar fire of the Spirit to rise. Through this process the Spirit is experienced as embracing and transforming both matter and the body of man.

Jung saw the symbolic aspects of alchemy as a precursor of modern study of the unconscious and the transformation of personality; and its goals as metaphors for psychological growth and development. The combination of two opposite elements resulting in a new entity – seen symbolically as the conjunction of male and female producing an infant – are particularly clear symbols of the intrapsychic processes and the development of the individual personality.

Context In most traditions, alchemy is none other than the science through which the sacred nature of rhythm was expressed and the link to eternity established. It acquired a special significance in the monotheistic traditions, especially Christianity. Through its more general expression in the form of hermeticism, it was the only cosmological doctrine to survive in the Christian world, complementing Christianity up to the appearance of Gothic art. Despite the insistence of historians of science, alchemy was never, except in its degenerate aspects, a primitive form of chemistry. The importance of alchemy lies in its emphasis on the divine omnipresence in the depths of material heaviness, where it appears least likely to become apparent. The most detailed overview of alchemical processes and traditional imagery is that by a modern author writing under the traditional pseudonym of Johannes Fabricius.

Refs Aniane, Maurice *Notes on Alchemy* (1976); Burckhardt, Titus *Insight into Alchemy* (1987); Burckhardt, Titus *Alchemy* (1967); Canseliet, Eugène *Alchimie*; Eliade, Mircea *The Forge and the Crucible* (1971); Fabricius, Johannes *Alchemy; the medieval alchemists and their royal art* (1989); Franz, Marie-Louise von *Alchemical Active Imagination* (1979); Franz, Marie-Louise von *Alchemy* (1980); Jung, C G *Psychology and Alchemy* (1968); Jung, C G *Alchemical Studies* (1968).

Broader Maps of the mind (#HH8903).

Narrower Alchemical whitening (#HM6328) Alchemical reddening (#HM3631) Alchemical blackening (#HM5345) Alchemical imagination (#HM1409).

Related Coniunctio (#HH1125) Grail quest (#HH1179) Alchemy (Taoism, #HH5887) Mystical marriage (#HM6321) Tantra (Buddhism, Yoga, #HH0306).

♦ HH0223 Immediacy of fulfilled living

Description This refers to the living in the present that fulfilled living entails. In such a state the individual responds immediately to any situation which prevents itself. Being fully awake, the intellect and the emotions which might otherwise dwell on the past or the future react appropriately and urgently to what is required. Thus one cries at what is sad, laughs at what is happy, responds humanely to a particular need, independent of what has gone before or what is to follow.

Related Humanness (#HH0203) Effulgence (#HM2521).

♦ HH0225 Discipline of submission (Christianity)

Description Self-will implies the heavy burden of requiring to get one's own way and of suffering if thwarted. The discipline of submission implies giving way to another and this leads to an ability to discriminate matters of intrinsic importance from those important only to self-will. It is important to distinguish mere outward obedience disguising inner rebellion from a spiritual submission and self denial which truly and unconditionally gives up individual ideas of rights and personal success in order to serve others. Interestingly, the act of self-denial, rather than submerging the individual in loss of identity, leads to a fuller awareness of identity and self worth. Voluntary submission to another, even in conditions where such submission would be coerced if not accepted voluntarily, implies the acceptance of moral responsibility and the freedom to make the decision.

The corollary of this approach is the self-denial which compels the dominant partner in a situation not to take advantage of this dominance but to exercise consideration and restraint. This mutual submission in, for example, a master-slave relationship, sets both parties free to be more truly themselves and to appreciate each other. A further corollary is that, where it is necessary to refuse to submit for moral or religious reasons, such refusal can be submissive, with the implication that the consequences of such a refusal will be accepted without complaint.

Seven acts of submission are enumerated, first to God, second to scripture, and then in widening order, family, neighbours, community, the despised and oppressed, the world at large.

Refs Foster, Richard J *Celebration of Discipline* (1984).

Broader Spiritual discipline (#HH1021).

Related Self-denial (#HH0964).

♦ HH0227 Group counseling

Description In group counseling one person (the counselor) attempts to help the members of a group to focus on the emotional content of their interpersonal and intrapersonal relationships and their effects on the relationship of group members to the outside world. Frequently the focus is on what is currently occuring in the group. Problem solving focuses on the understanding of emotions and the resolving of emotional conflicts. Three-way interaction prevails with emphasis on group-to-member and member-to-member responses. Forms of interactions are: clarification; reflection; summarization; syntheses; and nonverbal, supportive gestures.

Mutual support increases the possibility of discussing the meaningful problems of the members. There is concentrated and careful listening not only to the content but even more significantly to its revelation of the speaker. In such a climate the ability and readiness to share oneself with others broadens and deepens. Creative differences in values, ideas, and feelings are expected, accepted and encouraged. Members are encouraged as they see that others are maturing.

Broader Human potential movement (#HH0398).

♦ HH0229 Detached self-observation meditation

Description Passive self-observation activates the unconscious mind, allowing previously hidden or forgotten thoughts, memories, desires and impulses to surface and be faced calmly and boldly. As they come under the focus of consciousness, the individual is able to come to understand the fullness of his nature. Such detached observation finally transcends beyond mental flux, when the meditator is aware of himself observing the flow in the mind, and finally beyond the ego, when observer and observed are one.

This process leads to authentic self-existence and the permanent search for higher values, when the frontiers of existence are always enlarging in a process of dynamic growth. It also leads to improved understanding of others and better relationships with them.

Related Self-observation (#HH0586).

♦ HH0230 Initiation

Rite of passage — Sacrament

Description This is a ceremony marking the transition from one state to another, usually indicating the end of childhood and acceptance into adult society, or the end of a period of novitiate and entry into a particular group. Initiation is very often accompanied by rites or ordeals which frequently symbolize death (to the old life) and rebirth (to the new status). They may include symbolic burial or forgetting of the past, and then reentry into the womb, sometimes referred to as being "devoured" and "regurgitated" from a monster, usually in the form of a building where novitiates are isolated. The newly initiated may receive a new name, may speak a new language. Some rites and ordeals may require courage and endurance. They may involve physical mutilation (male or female circumcision, tattooing, ear piercing, ritual incisions and scarring) and the withstanding of torture and terrifying experiences. The initiate may be expected to perform deeds of valour; perhaps he would wear the skin of an animal and symbolically metamorphose as that animal, or he might have to kill a wolf, boar or lion singlehanded. There may be a re-enactment, possibly in dance, of death, burial and resurrection. Having passed through these rituals, the initiate is considered worthy to receive the privileges of adulthood/his or her new status. Initiation is usually accompanied by instruction and the divulging of secret knowledge or mysteries; it is followed by a renewal of relations with the ordinary world, when there may be symbolic relearning of the things connected with that life – walking, eating, speaking. The previous *candidate* (or uninitiated) is *introduced* (initiated) into the community.

Initiation may be obligatory for all members of a particular society as they effect the transition to adulthood; it may be for comparatively small groups only, usually of one sex, as they enter a secret society; or it may be in connection with a mystic vocation, for instance as medicine man or shaman, when it is usually accompanied by an ecstatic element. Initiation may be for a number of persons of a similar age, thus forging a strong relationship among them; or it may be for an individual, when the relationship is with his inner self; or it may involve both. As well as the puberty rites common to many societies, specific examples are: baptism; bar mitzvah; conferring of knighthood; masonic

rituals. Other less formal rites, many of which are replacing traditional rites, are entering military service, first pregnancy and owning an automobile.
In keeping with the latin word for initiation, from which the term *sacrament* derives, the new mode of existence may be considered as born in spirit. This spiritual person – the real person – is not the automatic result of a natural process but made according to models revealed by the supernatural. There is an ontological change which is reflected in the new status, and the rite may be thought of as safeguarding the person from disintegration while the change, involving temporary loss of ego, takes place. Coupled with initiation are an expansion of awareness or consciousness and the ability (temporary or permanent) to transcend illusions and limitations. The stretching of the mind towards universal understanding invariably has some permanent effect and it does not return to its original "dimensions".
The different religions have different "initiatic methods" in order to foster spiritual realization. In Tibetan Buddhism, every detail of spiritual activity has an initiatic purpose. In particular, there are a number of special initiations, or wang-kur, each giving access to one particular form of meditation based on a mandala. There is a clear distinction between those who receive initiation for mixed motives and those who follow the path with full intent, keeping the end in view. In Zen, the spiritual training combining stringent discipline and koans (conundrums) constitutes an initiatic process.
The ritual process of initiation can be considered to include ceremonies to mark rites of passage between childhood and maturity, admission into secret societies, as well as spiritual initiations into some new mode of awareness. There are strong resemblances and overlaps between these three forms. Rites of passage may involve entry into a secret society of adults. Admission into secret societies (or into higher grades in those societies) may involve rites of passage. Both may, or may not, be associated with significant changes of awareness. Some rituals of initiation may symbolize changes of awareness without there being any expectation that the initiant should actually experience such a change, whether immediately or after the experience has been digested.
Initiation cannot simply be defined as admission to groups, whether secret or not. Secrecy is not an invariable accompaniment of initiation rites, althouth this is often the case. Oaths an affirmations are commonly found as are tests and ordeals of varying severity. Initiation defines boundaries between members of a group and outsiders, between different statuses and between contrasted ideas. The rites often involve ideas of hierarchical order, through which the initiates are not only transformed but also gain status. Each stage reveals the ignorance of the earlier one, changing and deepening the initiates understanding of their world and its control by ritual. The knowledge acquired is often equated with power, based on the control of mystical rather than material resources. Although such spiritual power conflates power defined as a coercive force (as the result of physical or economic pressure) and as authority (understood as the right to command). The legitimacy conferred on ritual officiants is that of traditional knowledge: the information, understanding and experience needed to ensure correct performance.
Initiation rituals present offices to the individual as the creation and possession of society into which he is incorporated through the office. They mobilize incontrovertible authority behind the granting of office and status, guaranteeing legitimacy and imposing accountability for the proper exercise of an office. Rituals are purposive in that the participants believe that they are accomplishing their aim through what they do. Participants belief that rituals effect change. They are not simply the expression of ideas and social meaning. They are seen as effective action, whether transforming individuals or the material world. The performance is not an end in itself, but a means to achieve other ends.
The transformation of individuals, by the ritual which transfers them from one state to another, is a demonstration of the power of ritual knowledge, experienced by the initiates and other participants alike, reaffirming the positions of all involved. Initiation is a process through which individuals cross a boundary between two states. The contrast between the two states is symbolized in the ritual. The ritual thus provides a symbolic mould through which many fundamental oppositions can be contrasted: child/adult, human/spirit, wild/cultivated, and others. The full meaning of a ritual, like that of a play, relies on a set of shared conventions and assumptions that may be quite difficult for an outsider, first to elicit from those involved, and then to understand. Some of these refer to the symbols, which condense many layers of meaning, drawn from both tradition and from daily life.
Refs Eliade, Mircea *Rites and Symbols of Initiation* (1958); Eliot, John *Rites and Symbols of Initiation* (1965); Pallis, Marco *The Veil of the Temple* (1986).
Broader Rites (#HH0423).
Narrower Initiation (Magic, #HH3432) Thinking (Buddhism, #HM3966)
Peer group initiation (#HH4502) Spiritual initiation (Esotericism, #HH3390)
Initiation into secret societies (#HH9807) Initiation as spiritual rebirth (Yoga, #HH3465).
Related Magic (#HH0720) Baptism (#HH0035) Ordeals (#HH0377)
Twice-born (#HH0715) Discipleship (#HH0376) Vision quest (#HH0897)
Rites of passage (#HH0021) Mysteries and religion (#HH4032)
Initiation (Christianity, #HH6211) Discipline of the secret (#HH0959).

♦ HH0232 Service
Discipline of service (Christianity)
Description Service has been defined as the spontaneous result from the soul having such definite and fixed contact that it pours through the instrument of the body without conscious action on the part of the individual. It cannot therefore be learned but is evidence of the soul manifesting itself at the physical level. Development of service requires deliberate effort, conscious wisdom and the ability to work without results. As these are achieved, the inherent desire to serve will be satisfied as the individual draws on all the sources of spiritual strength.
A distinction is drawn between self-righteous service consciously engaged in and defined, responding to a need as defined by those carrying out the service and normally expecting both recognition and reward; and the lifestyle of true service which is unpretentious and based on the needs of those requiring the service, whoever they are. The former leads to glorifying of the individual at the expense of community, while the latter to drawing the community of individuals together. If any and all are served, whether they respond with thanks or by taking advantage of the service, then there can be no manipulation and great freedom arises.
Broader Spiritual discipline (#HH1021).
Related Humility (Christianity, #HH0861).

♦ HH0233 Arhat (Buddhism)
Arahant
Description One ready for, deserving of, great perfection beyond human perfection – *prajna paramita*. According to Buddha, an arhat has overcome the first five "fetters", or character defects which obstruct realization, and is "ready" (the literal meaning of "arhat"), with perfect or mature mind, to face and overcome the next five. In the hinayana tradition, this is one who retires from the world to strive for his own enlightenment.
Related Mythic yoga (Yoga, #HH3405) Prajna paramita (Zen, #HH0350)
Bodhisattva (Buddhism, #HH1380) Spiritual development (#HH0017)
Character defects (Buddhism, #HH0470) Hinayana Buddhism (Buddhism, #HH0845).

♦ HH0234 Integrity
Description The church teaches that integrity is a gift of grace which enables a person to subject the body and the appetites of the senses to the soul and the power of reason. In paradise, the gift of integrity (or completeness) was freely given and implied freedom from sickness, suffering and death, as well as from negative concupiscence. The church further teaches that, although original sin destroyed immunity to concupiscence and that integrity can only be fully restored at the resurrection of the body, nevertheless by increase in grace and the virtues during life, an individual can gain more and more control over the disordered powers, thus reducing their harmful effects. They no longer have power over him, he can be free and act in that freedom for the good of mankind as a whole.
Perfect integrity and perfect humility are virtually coincidental. They require being one's self and not attempting to be someone else. In the Bhagavad Gita, we are told that it is better to do one's own task, however badly, than another's task, however well. Lack of integrity means that one may waste years of one's life attempting to be some other person, to possess someone else's spirituality. There may be an attempt at self-magnification by imitating the popular instead of thinking things out for one's self. The truly humble person is unlike everybody else in that he is himself and all individuals are really unique. It is not necessary to try to be different, on the surface of everyday life people may appear to be alike, with the same tastes, opinions and so on. The difference lies deep in the soul.
An *integral life* is lived in response to the actuality of existence, positively accepting with gratitude all that life brings, disinterested but never uninterested. Moral integrity implies psychological and physiological completeness, with none of the parts predominating under the dominance of a particular impulse; the whole human being is integrated into the morally good fundamental option of love.
Refs Aldrich, V C *Theory and the Integrity of Experience* (1946); Drakeford, John W *Integrity Therapy* (1967); Storr, A *The Integrity of the Personality* (1960).
Related Honesty (#HH5304) Paternalism (#HH0735) Sanctification (#HH0428)
Personal integrity (#HH0129) Humility (Christianity, #HH0861).

♦ HH0235 Counter-transference
Description A necessary part of psychoanalytic therapy, counter transference involves the unconscious needs and conflicts of the *analyst*, and the relationship which a particular patient and patient's treatment produces with some object from the analyst's past. These (usually partial and short lived) identifications enable the analyst to comprehend confusing productions of the patient. Long-lasting and deep-rooted counter transference due to personality disturbance of the analyst, however, are counterproductive to treating a patient's problems.
Broader Analytical therapy (#HH0507).
Related Transference (#HH0668).

♦ HH0237 Kundalini yoga (Yoga)
Kundali yoga
Description This form of yoga includes those practices which are specifically directed to the arousal of the kundalini serpent power, the central psycho-physical force in the human body, and to the purification of the elements of the body which takes place upon that event. This awakening is achieved by will power, with the aid of the most rigorously controlled practices of bodily purification as each of the psychic forces exalts the physical consciousness through the ascending planes, leading finally to ultimate union with the supreme self, and consequent enlightenment and transcendent bliss. Seven centres, also called "lotus" or chakras, are defined; these psychological centres are distributed throughout the body and may be activated by a rising spiritual power, by "awakening the coiled serpent" which otherwise sleeps at the lowest of the seven centres, leaving the other centres unactivated.
This form of yoga is part of the tantric system and has much in common with what is termed tantric yoga. It is also considered to be an aspect of raja yoga.
Refs Amma *Dhyan-Yoga and Kundalini Yoga* (1969); Avalon, A *The Serpent Power* (1974); Hans-Ulrich, Rieker and Becherer, Elsy *The Yoga of Light* (1974); Mookerjee, A *Kundalini* (1982); Sannella, L *Kundalini* (1976); White, John *Kundalini, Evolution and Enlightenment* (1979).
Broader Yoga (Yoga, #HH0661) Raja yoga (Yoga, #HH0755).
Narrower Ajna (#HM2144) Anahata (#HM3242) Manipura (#HM3063)
Muladhara (#HM2893) Vishuddha (#HM3461) Sahasrara (#HM3398)
Soma chakra (#HM0176) Svadhishthana (#HM3174).
Related Maithuna (Yoga, #HH2034) Mystical stage of life (#HM3191)
Kundalini conscious energy (Yoga, #HM2871) Chakra centres of consciousness (#HM2219).

♦ HH0238 Consecration
Dedication
Description A consecrated object is considered a link with higher reality and often ritually delimited from the everyday world. The act of consecration is thus a deliberate attempt to establish a link with the divine. This link may be in the form of a consecrated building which may contain the relic or material traces of a holy person who has left this world – this is true of relics of the Buddha in stupas and pagodas, and in relics of Christ or the saints in Christian churches. Where a physical relic may not be present, a church is nevertheless consecrated in the name of a saint. The link may be the image of a deity, which will be symbolically transformed so that it represents the deity during a ceremony and then the life of the deity made symbolically to leave the image after the ceremony is over. Consecration considered as permanent may be taken to exemplify the absoluteness of the divine; as temporary, the limitedness of human effort and of the material world.
During the act of consecration, the consecrator himself may need to be consecrated. This is true of the priest at mass, who represents Christ, and who is consecrated for life; or of the individual Hindu consecrating an image for a ceremony, who must first take on the aspects of the divine through a preliminary self-consecration. This giving of divine authority is further exemplified in the consecration or anointing of kings or rulers. The outward ritual giving power to consecrate is related to the inner ritual of contemplation, and so related to spiritual discipline; in fact, the outward ritual may not always be necessary. However, the power to consecrate, whether for the good of others (as in Catholicism) or for the benefit of the recipient (as in Buddhism), generally comes through some recognised tradition or source and is passed through established channels.
A specific example of consecration over which much controversy has arisen is the consecration of bread and wine at communion. Whether the bread and wine are actually transubstantiated on consecration although maintaining their outward appearance, or whether they are viewed as symbolically the body and blood of Christ, the centrality of such consecration to the Christian religion is undeniable. In fact, in many cultures, the ingesting of food is a potent means for transmitting psychic substance, although in Hindu tradition, for example, communion with the deity through consecrated food is without mystery. Similarly, the laying on of hands, as in ordination, expresses the continuity of saving grace from a senior to a junior.
Broader Rites (#HH0423).
Related Personal consecration (#HH0064).

♦ **HH0239 Kali yuga**
Iron age
Description According to some teachings, the cycles of existence (or *mahayuga*) are each divided into four ages or yuga, the gold, silver, bronze and iron ages, after which creation is drawn in to itself to re-emerge in a new golden age. The present age, kali yuga, the lowest and last of the cycle, is doomed to end in catastrophe and entails more suffering than the other ages. Suffering can thus be seen as understandable and bearable, the result of accumulated wrongs in previous ages. It is also said that the kali yuga is nevertheless that overcome by love; and that the avatar or reincarnation of the absolute in this age is to teach return to God through response to God's love.
Related Avatar (#HH0079) Twice-born (#HH0715) Primitivism (#HH1778) Cyclic history (#HH0933).

♦ **HH0241 Anatma** (Buddhism)
Anatta — Not-self — Impersonality — Egolessness
Description The Buddhist doctrine of the nature of man as not conceivable by the human mind is also accepted by Zen. The mind can only know objects. What is referred to as "myself", object, cannot be one's self; it is simply the five *skandhas*, or tendencies - form, emotions, perceptive faculties, character habits or tendencies, and mental ability or discrimination. This false self reincarnates, always subject to change, until replaced by truth, when only the not-self or truth would be as "suchness". This should not be confused with the Hindu anatman, referring to the objective things of the world, including ahankara, buddhi and manas; in these terms the atman is the true Self identical with the Absolute, Brahman.
Context One of the three characteristics of the phenomenal world, the others being anicca (transitoriness) and dukkha (suffering).
Related Self (#HH0706) Atman (Hinduism, #HH0103) Anicca (Buddhism, #HM9021) Duhkha (Buddhism, Hinduism, #HM2574).

♦ **HH0244 Language**
Description The structure of language can be seen to hold truths which day-to-day usage has forgotten. Thus the concept of duality equating with evil, non-unity, comes into 'dubious', 'dishonourable', 'two-timing'. Conceit and concept have the same root, as do wise and wizard. Thus the value of words is not just of themselves but the reality they manifest. This is specifically true for 'Logos', the 'Word' (as in the First Chapter of St John's gospel). However, words have their shortcomings. Their very discreteness leads to the perception as the things they describe as discrete entities and not part of a continuum, and this is the reason that experience is diminished when described verbally. Although verbal symbols are more precise and expressive than other communication means, they can confuse. It is for this reason that, for example, deliberate extravagances are used in Zen Buddhism - to break through 'reasonable' language to intuitive intellect.
Refs Mead, G H *Language and the Development of the Self* (1952).

♦ **HH0246 Evolution of the human mind**
Description An individual's mind may be defined as the integrated totality of the conscious and unconscious processes involved in acquiring, storing, and utilizing information in his interactions with his environment. Through the physical evolution of animals, leading to the emergence of man, mental processes have also evolved and themselves have adaptive significance. Their evolution has paralleled that of the nervous system, particularly the progressive increase in the size and complexity of the brain.
It is argued (and speculated) on the basis of available evidence that the future of man is an evolving psychosocial condition, with closer communication and cooperation between all men guided by what has been termed the noosphere, namely the sphere of collective mind or thought (by analogy with the biosphere). Other views focus primarily on the ability of mankind to control its own evolution; or are more concerned with the possibility of a quantum jump in individual mental capacities to a permanent higher state of creative consciousness, better adapted to handle the complexity of the emergent planetary society.
Refs Munn, N L *The Evolution of the Human Mind* (1971).
Broader Evolution (#HH0425).

♦ **HH0247 Human development strategy** (United Nations)
Description Within the framework of the United Nations International Development Strategy and the United Nations Development Decade (1970-1980), a strategy for human development was specifically identified as follows:
"Those developing countries which consider that their rate of population growth hampers their development will adopt measures which they deem necessary in accordance with their concept of development. Developed countries, consistent with their national policies, will provide support on request, through the supply of means for family planning and further research. International organizations concerned will continue to provide, when appropriate, the assistance that may be requested by interested governments. Such support or assistance will not be a substitute for other forms of development assistance.
Developing countries will make vigorous efforts to improve labour force statistics in order to be able to formulate realistic quantitative targets for employment. They will scrutinize their fiscal, monetary, trade and other policies with a view to promoting both employment and growth. Moreover, for achieving these objectives they will expand their investment through a fuller mobilization of domestic resources and an increased flow of assistance from abroad. Wherever a choice of technology is available, developing countries will seek to raise the level of employment by ensuring that capital-intensive technology is confined to uses in which it is clearly cheaper in real terms and more efficient. Developed countries will assist in this process by adopting measures to bring about appropriate changes in the structures of international trade. As part of their employment strategy, developing countries will put as much emphasis as possible on rural employment, and will also consider undertaking public works that harness manpower which would otherwise remain unutilized. These countries will also strengthen institutions able to contribute to constructive industrial relations policies and appropriate labour standards. Developed countries and international organizations will assist developing countries in attaining their employment objectives.
Developing countries will formulate and implement educational programmes taking into account their development needs. Educational and training programmes will be so designed as to increase productivity substantially in the short run and to reduce waste. Particular emphasis will be placed on teacher-training programmes and on the development of curriculum materials to be used by teachers. As appropiate, curricula will be revised and new approaches initiated in order to ensure expansion of skills at all levels in line with the rising tempo of activities and the accelerating transformations brought about by technological progress. Increasing use will be made of modern equipment, mass media and new teaching methods to improve the efficiency of education. Particular attention will be devoted to technical training, vocational training and retraining. Necessary facilities will be provided for improving the literacy and technical competence of groups that are already productively engaged, as well as for adult education. Developed countries and international institutions will assist in the task of extending and improving the systems of education of developing countries, especially by making available some of the educational inputs in short supply in many of these countries and by providing assistance to facilitate the flow of pedagogic resources among them.
Developing countries will establish at least a minimum programme of health facilities comprising an infrastructure of institutions, including those for medical training and research, to bring basic medical services within the reach of a specified proportion of their population by the end of the Decade. These will include basic health services for the prevention and treatment of diseases and for the promotion of health. Each developing country will endeavour to provide an adequate supply of potable water to a specified proportion of its population, both urban and rural, with a view to reaching a minimum target by the end of the Decade. Efforts of the developing countries to raise their levels of health will be supported to the maximum feasible extent by developed countries, particularly through assistance for the planning of health promotion strategy and the implementation of some of its segments, including research, training of personnel at all levels and supply of equipment and medicines. A concerted international effort will be made to mount a world-wide campaign to eradicate by the end of the Decade, from as many countries as possible, one or more diseases that still seriously afflict people in many lands. Developed countries and international organizations will assist the developing countries in their health planning and in the establishment of health institutions.
Developing countries will adopt policies consistent with their agricultural and health programmes in an effort towards meeting their nutritional requirements. These will include development and production of high-protein foods and development and wider use of new forms of edible protein. Financial and technical assistance, including assistance for genetic research, will be extended to them by developed countries and international institutions.
Developing countries will adopt suitable national policies for involving children and youth in the development process and for ensuring that their needs are met in an integrated manner.
Developing countries will take steps to provide improved housing and related community facilities in both urban and rural areas, especially for low-income groups. They will also seek to remedy the ills of unplanned urbanization and to undertake necessary town planning. Particular effort will be made to expand low-cost housing through both public and private programmes and on a self-help basis, including through co-operatives, utilizing as much as possible local raw materials and labour-intensive techniques. Appropriate international assistance will be provided for this purpose.
Governments will intensify national and international efforts to arrest the deterioration of the human environment and to take measures towards its improvement, and to promote activities that will help to maintain the ecological balance on which human survival depends."
A report of the Integration Group A of the "Goals, purposes and indicators development project" of the United Nations University (June 1983) suggests that human development must start at the micro level and questions the premise that the local cannot control the global. The "participation" and "perceptual" gaps between the individual person and global processes, and the concommitant cycle of apathy and myth of incompetence, can be bridged through motivating to learn and participate.
A false image of detachment from the world as a whole can be superseded by demonstrating the nearness of the people involved in "world scale" activities and decisions. Local people could cooperate in participatory learning with people of similar interests in their locality (chambers of commerce, churches, unions, and so on). The need for local experts in global problems would reduce the "brain drain" of such experts to the major centres.
Such an organic growth of communities could be independent of and across national borders, with regional interests having priority over national interests; local people would define their own communities in terms of their own needs and interests. Border areas would be junctions rather than barriers betwen countries. In this case, national and international policies would stem from rather than be imposed on "grass roots" organizations.
To enable the effective participation of individuals and local units in global affairs which influence their lives, a new set of indicators must be developed. These would offer insight into everyday affairs and not those concerning only an elite set of politicians and planners. Having developed a conceptual basis for indicators at the local level, these concepts could be adopted to improve indicators at the national and international levels, which are seen as still retaining importance.
Refs United Nations *Human Development*.
Related Human development (United Nations, #HH3290).

♦ **HH0248 Folk healing**
Description In response to increased disease among tribal peoples due to interaction with western society, a number of tribal movements offering spiritual healing have developed. Methods include physical treatment and psychological help (reconciliation, community support, confession) but are based on faith, prayer, exorcism, laying on of hands, and so on. Since other medical services are often inadequate or non-existent, these faith healers provide a valuable service.
Broader Spiritual healing (#HH0458).
Related Primal healing (#HH2314).

♦ **HH0249 Life space**
Body space — Personal space
Description *Life space* may be defined as the physical "field" within which an individual acts and which is, to a greater or lesser extent, affected by the behaviour of the individual. This *territory of the self* (Ewing Goffman) is defended against actions by others which threaten the boundaries of that territory. A particular example is the *body space* or *personal space* surrounding an individual, which may be observed when people are crowded together (on a beach, at a party and so on). The entry of another person into the physical space an individual considers to be his own will cause him to feel threatened, and social rituals have been evolved to relate to the intruder in this situation. Non-spatial territories (a person's body, possessions) are also protected by the individual according to socially accepted conditions, depending on the circumstances.

♦ **HH0250 Independence**
Refs Linden, W *The Relation Between the Practicing of Meditation by School Children and Their Levels of Field dependence-independence, test anxiety and reading achievement* (1972).

♦ **HH0251 Social character**
Description Necessary for the understanding of social processes and the channelling of group energy into a productive force, social character is that central nucleus of the character of each individual in a group which has developed through experiences similar to, or in common with, the other members of the group; and formed by the dynamic adaptation of needs to the specific mode of existence of that group.
Since social character changes in response to changed conditions, the collective energies of that society will respond with new ideas and needs, this determining the actions of the individuals in that society, which again changes social character. Individual character and behaviour are thus inextricably linked with the society in which the individual lives; in this context, socially unacceptable behaviour may be looked upon as the natural response to a particular environment.
Broader Character (#HH0895).

♦ **HH0252 Aikido**
Description Aikido, literally "way of harmony of spirit or energy" or "way of harmony with *ki*", is one of the martial arts which has made the transition from being simply a traditional form of physical self–defence to that of being a spiritual martial art and as such seeks to abolish conflict. Holistic self–discipline is stressed through mind–body awareness. Into the techniques and movements of Aikido are woven elements of philosophy, psychology and dynamics. These lead to a training of the mind, improvement of health and the development of an unbreakable self–confidence. The aim is therefore not strength but justice, not victory over the enemy but victory over Self, through understanding of correct principles.
Hard physical discipline, realizing universal love through rigorous training of the body, takes place together with mental development and spiritual growth. At the very heart of aikido is the realization of *nen*, or one–pointedness of concentration. Harmony, blending and centering are developed as a meditative attitude, rather than the use of strength against strength. There is a mutual echoing of the resonance of the body with that of the universe, and from this arises heat, light and power, united in a fuly realized spirit. The mind and body are fused at the centre of gravity of the body, the mind directing the body. Responses are more subtle and refined than simple instinct and the body is physically conditioned through a series of exercises and movements – *katas* – so that each act is an integrated response of body and mind. The mind is free from fears, desires or thought of self, and in such a state of harmony, or *no–mind* an opponent can be thrown, sometimes even without physical contact.
A true martial art or *budo*, aikido is the working of love in the universe. It protects living things, is the means by which everything is given life in its respective place. Aikido is the creative source, not only of the true martial art but of everything, nurturing growth and development. It manifests ultimate reality – the flowing and spontaneous movement of nature full of the power of *ki*. The aim is to unify body and mind in the formation of the ideal human self and to attain dynamic life both in activity and in stillness. While the fundamental principle is spiritual, the execution is rational.
Refs Heckler, Richard S (Ed) *Aikido and the New Warrior* (1985); Tohei, Koichi *Aikido in Everyday Life* (1966); Ueshiba, Kisshomaru *The Spirit of Aikido* (1984); Westbrook, A and Ratti, O *Aikido and the Dynamic Sphere* (1970).
Broader Martial arts (#HH0085).
Related Ki energy (#HH0620) Hara (Japanese, #HM2662) Realization of nen (#HM0220) Dance–induced experience (#HM3213).

♦ **HH0256 Aversion therapy**
Description Originally deriving from behaviour therapy, aversion therapy aims to produce a negative association between an undesirable behaviour pattern and some unpleasant stimulation or to make the unpleasant stimulation a consequence of the undesirable behaviour. This tends to lead to a reduction in the undesirable behaviour and the substitution of alternative behaviour patterns fostered by the therapeutic process. Aversion therapy is mainly used for the treatment of such disorders as alcoholism and sexual deviation. The unpleasant (averse) stimuli used in this therapy include electric shock, nausea–producing chemical drugs, drugs which induce disorders as alcoholism and sexual deviation.
Aversion therapy is based on the Pavlovian conditioning process, in the same way as people are and have been trained out of undesirable traits through corporal punishment, taboos, feelings of guilt, and so on. New forms of this therapy are increasingly based on cognitive manipulations. For example, *covert sensitization* attempts to build up, entirely at the level of imagination, an association between the undesirable activity and an unpleasant effect.
Broader Therapy (#HH0678).
Related Implosive therapy (#HH1107).

♦ **HH0257 Healthy personality**
Description This is usually defined as the personality of an individual who actively masters his environment, shows a certain unity of personality, and is able to perceive the world and himself correctly.
Refs Jourard, Sidney M *Healthy Personality and Self–Disclosure* (1959); Jourard, Sidney M *Healthy Personality* (1958); Maslow, Abraham and Hung–Min Chiang *Healthy Personality*; Senn, M J E (Ed) *Symposium on the Healthy Personality* (1950).

♦ **HH0258 Love**
Charity — Affection — Eros — Agape — Friendship — Self–transformation through love
Description Various descriptions of love refer to it as: the greatest of the virtues (I Corinthians 13); the expansion of the self or the identity to include those closest to one; the fusion of selfishness and unselfishness to become both giver and receiver. It can be considered four–fold (affection, friendship, eros and charity); or as need and gift; or as ontological, sociological and practical. It may also be considered from the viewpoint of the objects of love – brotherly love, motherly love, erotic love, self–love and love of God. Contrasts have been drawn between sensible love (which requires the satisfaction of animal needs) and rational love (objective response to something of worth); between concupiscent and benevolent love (both aspects of rational love, the former referring to enhancement of the beloved, the latter to the beloved for whom enhancement is desired); and eros and agape, eros referring to the fulfilment of the lover, agape to the fulfilment of others. Christian agape is impossible to mankind except for God's grace. Plato's eros is the quest of the individual for his own highest spiritual good. The different types of love may be distinguished by their objects (love of mother, father, child, of one's brother, erotic love, love of one's self).
The benefit of romantic love is more to the lover than to the beloved. The need behind falling in love may be that of requiring change, and this change may be more important than the love which triggered it. There is a separation, a disengagement from old commitments (whether to parents, friends, previous lover) and union or re–engagement to the beloved. It has been said that the basis of love is the desire for unity and the misery at being separate. The paradox of development through self–assertion and being separate and yet not being alone is solved. The changes arising from love to produce this new, expanded self may be more permanent than the love which provoked them. Although arising from need and imagination, love is real and asserts this reality by the intensity with which it is felt and by the permanent changes it effects. The self is expanded and enriched in this creative achievement which love is, synthesizing the gratification of wishes and desires at all developmental levels. Love as it affirms and values another person as he or she actually is, rather than an ideal which one would like or a projection of one's own mind, allows one to value the other person in total as an individual, accepting negative and positive qualities. Accepting the other's totality is accepting the shadow.
Philosophically, love is not a feeling (which can come and go) but a permanent virtue manifested in self–surrender to the needs of one's neighbour, action on behalf of one's brother. Its sublime nature transforms the most mundane situation. As such, love is not something which one is capable of without effort. It is an art which has to be learned, and requires knowledge; is expressed through joyful giving, caring, responsibility and respect.
The ultimate love, of and for God, is said to be a nearness of approach to God and acceptance of his love. This is distinguished from nearness meaning similar, as when a person is most God–like he is not necessarily nearer in love. In fact, love when most like God is said to be most demonic as it is then that it can wrongly be mistaken for God. A clear distinction is to be drawn between "God is love" and "love is God".
Many have argued that love is the single most potent force in the universe. As Teilhard de Chardin put it, "Love alone is capable of uniting living beings in such a way as to complete and fulfil them, for it alone takes them and joins them by what is deepest in themselves". It is thus the binding power between human groupings, whether family, tribe, nation; and the motive power behind religious, cultural and social systems. Charitable love has been considered a moral and religious obligation in most societies; the poor and the stranger are often considered as under the direct protection of God or gods.
Context One of the three supernatural or abiding virtues, love is the foundation, source or principle of all the virtues (St Augustine).
Refs Fromm, Erich *The Art of Loving* (1956); Harper, Ralph *Human Love* (1926); Lewis, C S *The Four Loves* (1960); May, Rollo *Love and Will* (1969); Nygren, Anders *Agape and Eros* (1953).
Broader Virtue (#HH0712) Creative existence (#HH0513).
Narrower Care (#HH0042) Eros (#HH7231) Love (Islam, #HH5712) Bhakti marga (Hinduism, #HH0628) Degrees of love (Christianity, #HH3012).
Related Justice (#HH2117) Affection (#HH0422) Friendship (#HH0156) Compassion (#HM2919) Philanthropy (#HH1206) Romantic love (#HM5087) Apokatastatis (#HH6071) Affinity (Islam, #HM5304) Existential love (#HM0334) Dana (Buddhism, Zen, #HH1234) Love consciousness (#HM2000) Aspiration (Buddhism, #HM2578) Love of God (Christianity, #HH5232) Forgiveness (Christianity, #HH0899) Spiritual love (Christianity, #HM8977).

♦ **HH0260 Affective transformation**
Description Common to many disturbances, particularly schizophrenia, affective transformation is the tendency to increase the dimensions of emotionally toned and affectively emphasized values.

♦ **HH0263 Koan** (Zen)
Description These meditation topics are drawn from ancient enigmatic sayings of Chinese masters and used in Zen monasteries to focus and subsequently baffle thought. When at last the intellect lets go and spontaneous response arises, then a state of transrational insight is achieved and the meditator has made progress in the search for illumination.
Refs Miura, Isshu and Sasaki, Ruth Fuller *The Zen Koan*.
Broader Human development (Zen, #HH1003).
Related Rinzai Zen (Zen, #HH6129) Insight–provoking tales (#HH0370) Transrational insight (Zen, #HM2084) Inward zazen meditation (Zen, #HH1134).

♦ **HH0265 Teamwork mentality**
Togetherness
Description This system of working requires all individuals in a concern to put the interests of the concern before self–interest. No individual desks or offices are allowed and every individual is part of a team at his particular level. Promotion is based largely on length of service (on the time taken to *assimilate* the company's ideals) and moral and social behaviour are emphasized as much as business acumen. Financial reward is not high, but staff and their families are provided with living accommodation, and medical, social, educational and recreational facilities. Although ambitious individuals may not find the system sufficiently flexible and may therefore leave, those who are prepared to allow the company to dominate their lives in this way are devoted to it, and the method is successful both in its effect on individuals and its ability to run a profitable business.

♦ **HH0266 Conditioning**
Conditioned reflex — Conditioned response
Description Reflexes are behavioural acts without a specific genetic or physiological background and rely entirely on a particular sensory event. One method of accomplishing control, and of describing the data of such modification of behaviour on conceptual and physical levels, is conditioning. Conditioning is an experimental procedure for bringing natural reflexes within experimental manipulation. Instrumental conditioning occurs when the reinforcement, upon which the conditioned bond depends, occurs after the response has been made. The response is thus understood to be instrumental in obtaining the reinforcement. In the case of operant conditioning, behaviour which produces satisfactory consequences is strengthened, whilst behaviour which produces unsatisfactory consequences is weakened. Response can be varied by arranging that reinforcement conform to schedules dependent upon fixed or changeable intervals of time, or permanent or temporary ratios of action. By combining these possibilities into sequential or concurrent assemblies, behaviour of a specific kind can be produced and maintained.
Refs Honig, W K (Ed) *Operant Behavior* (1966).

♦ **HH0267 Atonement**
Propitiation
Description Literally "at–one–ment", this refers to the reconciling, bringing together or "setting at one" of two previously estranged parties. The sense of expiation is also included as, in the sense normally used, it is one party who has in some way wronged the other to cause the estrangement. Atonement is thus an act or way of life that expiates wrongs previously committed, either by the person carrying out the act of atonement or by another for whom that person takes the role of mediator. In Christian teaching, although Christ is said to atone for the sins of the whole world, mankind must personally cooperate in this atonement by working under the influence of Christ's grace. According to Patanjali, when the soul and objective forms have reached equal purity then there is total atonement (at–one–ment); that which previously hindered expression of full divinity having become a means of expression and service.
Refs Clark, Gordon H *The Atonement* (1987).

♦ **HH0269 Thetan**
Spiritual being
Description According to the tenets of the *Church of Scientology*, thought – otherwise *prana*, *logos*, *entelechy*, *elan vital* – is spiritual energy. It is not part of the physical universe but the motionless, static ground of all being. It is the play of this static against the kinetic of physical matter which produces energy. The *Thetan* – the individual soul or *Atman* – is the real person, a unit of "awareness of awareness"; and this, rather than body, name or possessions, is what constitutes the identity of the being.
The perception of changes in physical matter is what constitutes *time*. Man is essentially good but is encumbered in his quest for survival by painful past experiences and harmful acts against others. By joining with other Thetans in *lila*, the game of life, through right action and reverence for life, the aim of the Scientologist is to understand first the self and then proceed outward through seven further dynamics, progressively leaving behind materiality and discord, and growing in love.
Related Lila (#HM2278) Engrams (#HH0514).

♦ **HH0270 Root afflictions** (Buddhism)
Mulaklesha — Rtsa–nyon
Description According to Tibetan Buddhism, these afflictions or *knowers* are the root of all others, and the cause of suffering the cycle of birth and rebirth.

Context The afflictions, together with the secondary mental factors, are comparable with the mental activities or formations of Hinayana Buddhism.
Broader Afflictions and hindrances (Buddhism, #HH5007).
Narrower Doubt (#HM4467) Desire (#HM2433) State of anger (#HM2959)
Afflicted views (#HM2996) Pride (Christianity, #HM2823)
Ignorance (Buddhism, Yoga, Zen, #HM3196).
Related Four noble truths (Buddhism, #HH0523) Sense consciousness (Buddhism, #HM2664)
Secondary afflictions (Buddhism, Tibetan, #HH0781)
Awareness as mental-formation group of conscious existence (Buddhism, #HM2050).

♦ HH0272 Virginity
Description This state of innocence as regards sexual intercourse is extended in Biblical terms to refer to the proper state of one about to be married - a moral quality of sexual purity as opposed to the physical state of virgo intacta. It implies that body and soul are such that their forces may be dedicated in obedience to God directly through "taking the veil", or reserved for fulfilment in marriage.
Broader Celibacy (#HH0045).
Related Life of chastity (ICA, #HM2506) Chastity (Christianity, #HH0583).

♦ HH0273 Psychodiagnostics
Description This term covers the many methods for discerning what lies within the hidden motivations and unconscious mind of the individual. External appearance and behaviour are studied through a variety of more or less accepted techniques including: *graphology* (the study of handwriting); *palmistry* (markings on the hands); *stoleomancy* (the clothes a person wears and his manner of wearing them); *phrenology* (the shape of the head); *laliognomy* (speech analysis); *pateomancy* (gait and manner of walking). *Xenoanalysis* studies unguarded speech and behaviour under some influence (as drugs, alcohol, shock, pain) and particularly in dreams - *oneiromancy*. Particular tests involve word-association, response to images (as in the Rorschach ink-blots test), free association; and emotional maturity may be demonstrated in an individual's response in intelligence tests. Group therapy and relations with family and friends, and also reactions to emotive subjects, provide further material from which to understand the "inner man".

♦ HH0276 Human engineering
Human-factors engineering — Ergonomics
Description This is the application of information about human characteristics, capacities and limitations to the design of machines, machine systems and environments so that people can live and work safely, comfortably and effectively. Major aims of human engineering also include: increasing the efficiency with which machines can be operated, increasing productivity in industrial and systems operations, decreasing the amount of human effort required to operate machines, and increasing human comfort in man-machine systems.

♦ HH0277 Naqsbandi tradition (Sufism)
Description A system of Sufi discipline involving a number of meditative exercises. These have been enumerated as: (1) Yad kard: Meditation on the inward and held breath, remembering the name of God orally and in the mind until the heart is conscious of the truth and inattention vanishes. (2) Baz gast: On the same breath, remembering there is no god but God and Mohammed is his prophet; "it is for Thou, Lord, that I aspire". (3) Negah dast: One the same breath, aware of consciousness, to make a profession of faith with no swerving of spirit. (4) Yad dast: To remember the presence of the Truth, a flavour which consciousness never lacks. (5) Hus dar dam: Breathing out, to continue full attention. (6) Safar dar vatan: On the spiritual journey to replace human with angelic qualities. (7) Nazar bar qadam: To be aware of the steps on the journey - never allowing thought to stray from whence one has come, from where one is and to where one is going. (8) Xalvat dar anjoman: Allowing the two states - of remaining externally in the company of men and retreating internally close to the truth - to complement each other. (9) Voquf-e zamani: Taking the time necessary to consider one's activities, to give thanks for the good and to ask pardon for the bad. (10) Voquf-e 'adadi: To enumerate in one's heart what one has learned, taking wandering thoughts into consideration. (11) Voquf-e qalbi: Ensuring that the heart has no aim but God by engraving the name of God on the heart.
Broader Human development (Sufism, #HH0436).
Related Fountains of light (Sufism, #HM3039).

♦ HH0278 Observing moral precepts (Buddhism)
Sila — Panca sila
Description Undertaking the set of precepts or rules of training - in particular the *panca sila* or five precepts - may be a formal undertaking following the taking of refuge in the three jewels of Buddha, the law (dharma) and the priesthood. This amounts to the acceptance of Buddhist teaching and practising it in day-to-day life. The panca sila refer to taking life, taking what has not been given, wrong use of sense pleasures, untruthful speech and use of intoxicants such as alcoholic drink.
Context The second of the ten meritorious deeds of South Asian Buddhist thought.
Broader Ten meritorious deeds (Buddhism, #HH0446).
Related Human development (Buddhism, #HH0650).
Followed by Bhavana (Buddhism, #HH0551).
Preceded by Dana (Buddhism, Zen, #HH1234).

♦ HH0281 Personality development
Depersonalization
Description Personality is the relatively stable organization of a person's motivational dispositions, arising from the interaction between biological drives and the social and physical environment. The term implies both cognitive and physical attributes, but usually refers chiefly to the affective-connative traits, sentiments, attitudes, complexes and unconscious mechanisms, interests and ideals, which determine man's characteristic or distinctive behaviour and thought. Many different definitions of personality exist and a number of divergent theories of personality are currently in use.
Personality development refers to a description and theoretical understanding of the establishment of those stable response dispositions that differentiate adult humans. Although it was believed previously that all personality traits were formed by the age of five years, this is now disputed. The current belief is that personality continues to develop during childhood and adolescence and maybe into young adulthood. There is also controversy on the relative importance of heredity and environment on personality development.
Depersonalization is equated with loss of identity, when the individual no longer feels familiar even with himself. Such feelings may be symptoms of mental disease, or a reaction to conditions of extreme sensory deprivation, torture, pain, emotional stress or to circumstances involving grief, ecstasy or unexpected escape from death.
Refs Anderson, H H *Creativity as Personality Development* (1959); Jung, C G *The Development of Personality* (1954); Sontag, L W; Baker, C T and Nelson, V L *Mental Growth and Personality Development* (1958).
Narrower Stages of personality development (#HH0285).
Related Depersonalization (#HM1248) Cultural personality (#HH0724)
Complete personality (#HH3998) Disruption of lifestyle (#HH3356)
Strength of personality (#HH0198).

♦ HH0282 T'ai Chi Ch'uan
Description This is a Chinese system of exercises or movement schemas in which the accuracy, perfection and balance of the forms of physical movement and the inner meditative experiences or mental "motions" by which the physical actions are controlled are equally important. Movements thus controlled are essentially natural and are the basis of an ancient martial art, fighting with the fists, which bears the same name. In fact, the name T'ai Chi Ch'uan has been translated as "Grand terminus fist". A person learns to move with a focus on the subtleties of his energy flows, particularly equilibrium and harmony.
Five qualities are developed: slowness (with the inherent poise and patience); lightness (allowing continual, almost effortless, motion); clarity (so that the mind controls action which is accurate and pure); balance (leading to action without strain); and calm (providing the concentration to sustain the complete movement from one phase to the next).
The essence is continuity of action, where each movement evolves from and grows out of the preceding movement, and links onto and activates the succeeding movement. This leads to physical and mental coordination, the power to control the self, and immunity from destructive external forces and poor health. With the technique of T'ai Chi Ch'uan, true energy can be controlled. Strength, balance and vitality are increased by exercising the body in such a way as not to strain the muscles nor to over-activate the heart, and not to exert the body excessively. The basic philosophy is that in order to prolong the life of the body, to stabilize the life of the emotions, and to intensify the life of the mind, conscious cooperation of the mind with activity is a fundamental necessity.
Refs Cheng, Man-ch'ing and Smith, Robert W *T'ai Chi* (1967); Feng, Gia-Fu and Kirk, Jerome *Tai-chi* (1970).
Related Quietism (#HH2006) Hara (Japanese, #HM2662).

♦ HH0284 Massage
Description As a therapeutic method, massage has been used since ancient times, both to treat physical disorders (injuries of the motor and support apparatus; disturbances of the metabolism; diseases of the nervous, circulatory and respiratory systems); and to promote general physical, emotional and mental wellbeing, by maintaining athletic form, combatting fatigue and restoring physical strength.
Broader Physical therapy (#HH0379).

♦ HH0285 Stages of personality development
Sequence of psychological stages in the life cycle
Description A number of efforts have been made to chart the stages of personality development that are universal. In one view, for example, there are eight such stages: oral, anal, phallic, latency, adolescence, genital, adulthood and senescence. These stages are the result of the epigenetic unfolding of the ground plan of personality that is genetically transmitted. The way in which they manifest will depend on the environmental experiences which occur at each stage. According to Jung, the psychological transition at mid-life is crucial, with the first and second half of life equivalent to the two phases of the individuation process; but eight major crises may be associated with the stages which the individual faces in the course of his lifetime. According to Erik H Erikson, the eight distinct stages, each with its own psychological conflict and resolution, contributing to a major aspect of personality, are: infancy; early childhood; play age; school age; adolescence; early adulthood; adulthood; old age. These may be said to represent: trust: autonomy; initiative; industry; identity; intimacy; generativity; and integrity. The person must adequately resolve each crisis in order to progress to the next stage of his development in an adaptive and healthy fashion. In fact, each stage may be seen as a potentiality for change, when new possibilities open for mutual interaction with his surroundings.
Refs Bühler, Charlotte and Massarik, Fred (Eds) *The Course of Human Life* (1968); Erikson, E *Identity and the Life Cycle* (1959); Erikson, E H *Growth and Crises of the 'Healthy Personality'.* (1950).
Broader Personality development (#HH0281).
Related Life cycle (#HH0511) Generative man (#HH2764) Stages of faith (#HH2097)
Midlife transition (#HH1366) Stages of spiritual life (#HH0102).

♦ HH0286 Mudras
Description Mudras are special ritual gestures of the hands that channel energy flow through the body. To the observer they symbolize different attitudes and states of feeling; to the practitioner they express these attitudes and feeling states. Mudras represent one of the tantric tools of transformation and integration developed to transmute sensory experience from a binding to a liberating role, and are also used to induce altered states of consciousness in *hatha yoga*.
Related Dance (#HH0445) Hatha yoga (Yoga, #HH0862)
Tantra (Buddhism, Yoga, #HH0306) Dance-induced experience (#HM3213).

♦ HH0288 Feasting
Description Feasting is the complement of fasting in that, in order to feast one must break one's fast; every culture has the experience of fasting, which can be considered an acknowledgement of death, normally succeeded by a feast as a celebration of life. Traditionally, at least since the 4th century, the Church requires a certain period of fasting before Eucharist, which was at first celebrated as a communal meal or love feast, *agape*.
Related Fasting (#HH0011).

♦ HH0289 Megavitamin therapy
Orthomolecular psychiatry
Description Massive doses of specific vitamins and minerals and, in some cases, a high protein, low carbohydrate diet, are used in the treatment of schizophrenia, depression, anxiety, alcoholism and drug addiction; and of hyperactivity, autism and other problems in children. Such therapy reduces or replaces other treatments and the need for hospitalization.
Broader Therapy (#HH0678).

♦ HH0295 Genius
Giftedness
Description Although giftedness is a term used to denote the highest talent in a particular field, genius has connotations of creativity and discovery which go beyond normal ability. It is applied to persons of extraordinary intellectual power, either as measured by some test of intelligence or as demonstrated by some actual achievement involving creative ability of an exceptionally high order. It is associated with creativeness, originality and the ability to think and work in subject areas not previously explored and thus to give the world something of value it would not otherwise possess; and has been described as a unique combination of intellect, zeal and power of working.
Psychologically, it has not proved possible to define a particular personality type to which genius may be said to belong. It does not even seem to be related to "cleverness" or high intelligence in general. Although gifted people tend to come from gifted families, lead lives above average in

health and happiness, and be emotionally stable, many persons classed as "genius" lead unhappy and deprived lives, often suffering depression from early loss of a parent. They tend to be emotionally unstable (sometimes verging on madness), and many have become insane, although again it is not possible to generalize. Certainly fantasy and playing games with reality have a large part to play. Political and social judgement among acknowledged geniuses has been shown to be remarkably lacking. Nevertheless, the creative process manifested by a genius is not different in kind from that inherent in any gifted person. This has not prevented some suggestions that people of genius belong to a separate psychobiological species, differing as much from ordinary man as man does from the ape; or that they have developed their ability over many lifetimes.
Refs Kretschmer, E *The Psychology of Men of Genius* (1931); Storr, Anthony *The School of Genius*; Terman, Lewis B *Genetic Studies of Genius* (1947).
Related Creativity (#HH0481) Precocious talent (#HH0611).

♦ HH0296 Mental hygiene
Mental health
Description Mental hygiene includes all measures taken to promote and to preserve mental health, or the condition of the individual relative to his capacities and to his socio–environmental contact. It represents a variety of human aspirations: rehabilitation of the mentally disturbed, prevention of mental disorder, reduction of tension in a conflict–laden world, and attainment of a state of well–being consistent with the individual's mental and physical potential.
In practice, mental hygiene is primarily concerned with care for the mentally and emotionally sick, improvement in the treatment and care of the emotionally sick and feebleminded, and clarification of the part played by psychological and mental disturbance in child–rearing, employment and criminology. Institutional care concentrates primarily on these activities. A major factor is also the effect of physical ill–health or handicap on mental health.
However, to the extent that institutionalized mental hygiene can focus on the improvement of mental health (as opposed to the care of those defined as sick), its aims may include: the development of the power of self–discovery through experiences of self and self–knowledge; development of the struggle for a true self–affirmation; the capacity to give other people the same value as the individual claims for himself; unhampered development of a power to love, unhampered performance of normal functions; and development of the power to make appropriate, unprejudiced judgments.
Efforts have been made to elaborate a concept of positive mental health (separately described).
Refs Ahrenfeldt, R H and Soddy, K *Mental Health in a Changing World* (1965).
Related Spiritual life (#HH0101) Positive mental health (#HH0675).

♦ HH0297 Polytheism concept
Description The idea of worshipping several gods although presupposing some supraordinate principle was used by Jung to contrast the multiplicity of archetypes with the supraordinate self.

♦ HH0298 Worship
Discipline of worship (Christianity)
Description To worship is to honour (by some rite or ritual) supernatural power, God, or some person or thing representing such power or considered *worthy* of respect. Different religions, and sects within religions, have developed liturgical systems and laid down forms of worship, usually with a congregation praying, praising and listening to readings from scripture, mainly under the direction of a cleric. The form of worship may vary depending upon language, culture and diversity of expression, but there are common factors: adoration; penitence; intercession; petition; thanksgiving.
In Christian terms at least, the purpose is: to be cleansed and forgiven; to rededicate one's self and what one has in response to the gospel and the sacraments; and to receive again the commission to the world. It is the meeting between sacred and secular and affirms the continuing call to holy living. The response of man to the promise of God is a life of work and worship in which work becomes worship and worship becomes part of the work to be performed.
Whatever religion or sect, the general pattern appears to be the arising of a moral or mystical teacher whose original ceremonies, songs and prayers are ritualized according to some socially recognized pattern. This pattern is corrupted through syncretism, whereupon a subsequent prophet or reform may arise who purifies the system and assists it to revert to its original form. Thus, although worship may be carried out by all members of a society its basis invariably goes back to the inspiration and experience of one individual. Most participants would agree that the form of worship must originally have been and must continue to be the result of divine inspiration although, particularly in the Church, contemporary concern is to change the style of the liturgy in line with current life–styles.
Although worship is frequently a group activity, its effect may vary from one individual to another. Examples of heightened states of awareness arising during worship vary from reverence, mysticism, love and sanctification to possession and shamanic trance.
Refs Cowan, Michael A, et al *Alternative Futures for Worship* (1987).
Narrower Salat (Islam, #HH0536) Corporate worship (Christianity, #HH2768).
Related Rites (#HH0423) Puja yoga (Yoga, #HH0111)
Devotion (Christianity, #HH0349)
Discipline of worship (Christianity, Judaism, #HH3776)
Religious experience as personal devotion and worship (#HM0561).

♦ HH0299 Mind
Mind–body dualism
Description Inquiry into the function of the mind as distinct from the body, and its substantiality, spirituality and mortality, has developed into modern psychology. Despite the close connection between the mind and the bodily functions, for example the influence of mind on bodily sickness and health, their qualities are sufficiently different to posit the hypothesis that the mind could exist without the body and could survive physical death. The mind is also distinguished from the personality and from the soul in philosophic thought, and the continuance of soul after death of body/mind is another proposition.
Refs Gardner, Howard *Frames of Mind* (1984); Hofstadter, Douglas R and Dennett, Daniel C *The Mind's I* (1981); Karlins, Marvin and Andrews, Lewis M *Biofeedback* (1973).
Related Soul (#HH0501) Psyche (#HH0979)
Manas awareness (Hinduism, #HM2902).

♦ HH0300 Giri (Japanese)
Social obligations
Description A traditional Japanese concept, whereby social obligations are implied to those with whom one has social relations, whether in terms of superior/inferior or in being beholden for some favour. Such social norms are expected to be upheld even where this conflicts with *ninjo* or natural feelings. A related concept, on, is used to refer to the social and psychological debt felt upon receiving a favour or gift of major proportions. As an inordinate debt, its payment is binding in expectation, although its profound significance implies that an individual that an individual can never be finally cleared from it.
Related Ninjo (Japanese, #HH0912).

♦ HH0303 Mystery
Sacred mysteries
Description Originally, mystery referred to sacred rites which were revealed only to the initiated and thus cultivated an awareness of the numinous. These rites might include the acting out of a mysterious event, such as myths explaining the death and rebirth of crops from winter to summer, the seasons of the year, and so on. In the Christian tradition, *spirituality*, the living of Christian life with intensity, manifests as an interpersonal relationship with Christ and a sharing in the sacred events or *mysteries* of his life. Awareness of the mystery in every relationship brings a freedom and commitment that leads to transparency, knowledge of self, manifesting as certitude, peace, joy and immortality, in the midst of negativism, problems, tragedy and contingency.
Related Transparency (#HH0673) Sense of mystery (#HM3608)
Mystical cognition (#HM2272) Spirituality (Christianity, #HH0792)
Psychic development (Psychism, #HH0855).

♦ HH0304 Rebirth
Psychic transformation
Description According to Jung, tales of death and rebirth abound and are so compelling because they parallel the psychological process of transformation or transcendence experienced as a reality, whether during a sacred rite; connected with altered states of consciousness perhaps induced by some magical procedure involving incantation, mesmerism or drugs; or during the process of individuation, with the feeling of rebirth into a larger personality. The rebirth experience is said to be the result of an encounter with the archetype of transformation.
Refs Tatz, Mark and Kent, Jody *Rebirth* (1977).
Related Transcendence (#HH0841) Transformation (#HH1039).

♦ HH0305 Courtesy
Description Literally to "adopt court manners", courtesy implies responding to the highest and most laudable aspects of a person or of the role he or she is playing. This in turn will tend to elicit a response from those laudable attributes within the person.

♦ HH0306 Tantra (Buddhism, Yoga)
Tantric yoga — Maithuna
Description Tantra is a psycho–religious movement connected with the worship of Sakti, the archetypal feminine aspect of God, seen as creative energy or the divine mother. In Tantric understanding there is total acceptance and affirmation of life. The world is a manifestation of the dynamic aspect of divinity, not to be devalued and renounced as suffering but to be celebrated and enjoyed with insight and understanding. Tantra advocates a practical, experimental approach to self–realization, rather than a philosophical one, and can better be described as a complete life–style with its own attitudes and values rather than a philosophy or a religion. Buddhist Tantrism is based on the mahayana system, the Hindu on advaita–vedanta. Tantra emphasizes a total involvement with life, expressing a deep and healthy respect for the human body as the basic and essential tool for salvation. The harmony and perfection of the physical body reflect the harmony and perfection of the universe, a microcosmic image of the macrocosm.
Absolute reality is a union of the opposing forces of pure consciousness (Siva) and pure energy (Sakti). At the time of creation their primordial unity is broken and the world of duality brought to birth. The bondage of duality may be escaped through realizing the union of opposites within the human body and mind. This cannot arise in an uninitiated, unguided and natural state. Inner divine union arises only when mind and body have been purified, controlled and perfected. Depending on the psycho–physical nature of the disciple the guru prescribes different techniques - orthodox, external worship for the tamasic (mind steeped in inertia) sadhaka, intense and experimental discipline for the rajasic (energetic) sadhaka. This latter, after a long and arduous sequence of spiritual discipline, undertakes the ritual of five essential elements, the methodical and ceremonial use of wine, meat, fish, cereal aphrodisiacs and sexual intercourse. Throughout the ritual there is strict emphasis on the use of Yogic bandhas and mudras. By proper control the sadhaka heightens psychic tension so that it has power to arrest allmental processes. This cessation of the modifications of the mind is the basis of spiritual ecstasy.
Countering the ascetic tradition, the role of the body and of sense experience are glorified as, although under most circumstances they are obstacles to self–realization (maya–Sakti), they may be transformed into vehicles of liberation and the attainment of higher consciousness (cit–Sakti). Since the human race is in the midst of the kali yuga or dark age, human wickedness is such that man is incapable of approaching divinity using the older techniques, practical tricks and devices at this stage being permissible. Several tantric tools have been developed to transmute sensory experience from a binding to a liberating role. These tools include: mandalas (or yantras), geometrical designs used for the transformation of visual sense experience; mantras, for the transformation of auditory sense experience; and mudras, for the transformation of bodily experience through the channelling of energies by means of special postures and gestures. Of these methods, use of mantras is pre–eminent.
Inner development is the progressive extrication from the profane life. In Hindu Tantrism, this progression is the standard way of the householder to householder–ascetic and finally mendicant. Kulanarva Tantra defines seven stages of initiation or degrees of spiritual maturity, in two groups, the first four being the path of worldly activity, *pravritti–marga*: veda acara, vaisnava acara, saiva acara, daksina acara; the last three being the path of cessation, *nivritti–marga*: vama acara, siddhanta acara, kaula acara. The first stage of initiation on the nivritti marga is a ritual known as *panca tattva* or *panca makara*, There are two paths, the right–hand path which treats the ritual metaphorically and the left–hand path which is literal. The last of the five components of the ceremony is *maithuna* or ritualized coition. This practice embodies the union of the male and female aspects of primal reality polarized in man, a union basic to Tantrism which sees the phenomenal world of *samsara* and the transcendence of *nirvana* as two poles of one unity, whose essential identity may be realized when human consciousness is cleansed of all misconceptions. Here there is no following of a path to reach a goal but a resting, fulfilled, in the all pervasive original mind.
Refs Allen, Marcus *Tantra for the West* (1981); Avalon, A *Tantra of the Great Liberation* (1972); Blofeld, J *The Tantric Mysticism of Tibet* (1970); Manoranjan Basu *Fundamentals of the Philosophy of Tantras* (1986); Mehta, Rohit *The Secret of Self–Transformation* (1987); Mookerjee, Ajit *Tantra Asana* (1971).
Broader Yoga (Yoga, #HH0661) Maps of the mind (#HH8903).
Related Mudras (#HH0286) Mantras (#HH0386) Yantras (#HH2662)
Mandalas (#HH0112) Maithuna (Yoga, #HH2034) Sacramental sex (#HM0843)
Nirvana (Buddhism, #HM2330) Sarnsara (Hinduism, #HM1006) Sexual awareness (#HM2374)
Sexual yoga (Taoism, #HH9112) Alchemy (Esotericism, #HH0221)
Mystical stage of life (#HM3191) Tantric visualization (Yoga, #HM1690)
Spirituality in tantric yoga (#HH6210) Mantra–induced experience (Yoga, #HM2464)
Tantric formless path (Buddhism, Yoga, #HM4603)
Tantra in meditation on emptiness (Buddhism, Tibetan, #HM2864).

♦ HH0307 Moral re–armament
MRA
Description A modern, non–denominational revivalistic approach which seeks to deepen the moral and spiritual life. Programmes include: development of theatre plays and films emphasizing

cooperation, honesty and mutual respect between opposing groups; and spiritual houseparties, similar to religious retreats. Individuals are encouraged to confess their deficiencies in group sessions.

♦ **HH0309 Vajrayana Buddhism** (Buddhism)
Diamond vehicle Buddhism
Description One of three branches of Buddhism, combining the realism of Hinayana and the idealism of Mahayana together with Tantric and Taoist mysticism, and including specific rituals. Developing later than the other two branches (around the 8th Century), it may roughly be equated with the northern Buddhism of Tibet, Mongolia, the Himalayas, China and the USSR.
Context According to Master Da Free John, characteristic of the Fourth Stage of development.
Refs Price, A F and Wong Mou-Lam (Trans) *The Diamond Sutra and the Sutra of Hui-Neng* (1969).
Broader Human development (Buddhism, #HH0650).
Related Hinayana Buddhism (Buddhism, #HH0845)
Mahayana Buddhism (Buddhism, #HH0900)
Way of the vajra masters (Buddhism, Tibetan, #HM3090).

♦ **HH0310 Narcotherapy**
Description Sodium amytal or sodium pentothal is slowly injected intravenously to produce complete relaxation and a feeling of wellbeing, under which conditions previously repressed memories and conflicts may be expressed. The therapist can them guide the patient through mental, emotional and behavioural rehabilitation.

♦ **HH0311 Unitary life**
Related Unitary consciousness (#HM2702).

♦ **HH0312 Self-knowledge**
Description This is the individual's knowledge of his ego, of his self, of his abilities, dispositions, errors and weaknesses, and response patterns. It is considered of considerable importance in some philosophical systems. In psychology, the potential for self-deception led to its replacement by operationally definable concepts such as: self-image, self-concept, and self-assessment.
In Christian terms, self-knowledge is knowledge of one's self as a person towards whom God has love; and awareness of those aspects of one's self which aid or hinder progress in the spiritual life and the realization of divinely given potential. Growth in response to grace is not achieved, however, by self-knowledge sought at the expense of other duties or regardless of one's neighbour.
According to Zen, because one's nature cannot be expressed in terms of things seen or known to the mind, becoming aware of one's own nature is referred to as no-seeing. True awareness of self is of the form known by small babies before the awakening of mental operations, *adhyasa*, which ascribe properties on the basis of having been aware of them elsewhere. Self-knowing is the supreme insight, the I looking directly at the I, *prajna paramita*.
To the Sufi, the knower of the Self is the lover of God. Distinctions between what is and what is "mine" disappear in transcendent and unitive experience of the truth (al-Haqq).
In order to understand and change society it is said to be necessary to understand and change one's self, since politics and society originate in the depths of the human heart. What one is unconsciously trying to achieve may be the very opposite of what one thinks one is aiming for at the conscious level. However good the action, if it is instigated by an unworthy motive, however unconscious, it will in the end result in harm rather than good. This is why self-knowledge is necessary.
Refs Mahadevan, T M P *Self-knowledge* (1975); Needleman, Jacob *On the Way to Self-Knowledge* (1976); Walker, Jeremy *Self-Knowledge as the Way to God* (1984).
Narrower Ming (Zen, #HM0765).
Related Self-knowing (#HM3599)
Self-examination (#HH0314)
Self-understanding (#HH0693)
Ways of knowing (Christianity, #HH1616)
Psychotherapeutic self examination (#HH0922).

♦ **HH0314 Self-examination**
Self-assessment
Description The conscience is aided in its development by critical comparison between one's conduct and the standards one accepts. By basing such assessment on a system, such as the ten commandments, a framework may be given to the examination.
Related Self-knowledge (#HH0312)
Self-understanding (#HH0693)
Psychotherapeutic self examination (#HH0922).

♦ **HH0315 Action learning**
Description This is a management system pioneered and developed by R. Revans, which stresses the need to learn from everyday tasks and from each other. In particular, it encourages the use of non-experts (sometimes reversing roles, for example temporarily putting a bank manager in charge of a soap factory and vice versa) so that a new, unbiased approach may be brought to bear, and to force communication and interdependence at all levels.

♦ **HH0317 Self-determination**
Description The process by which a group of people, usually possessing a certain degree of collective or national consciousness, form their own state and choose their own government.
Related Self responsibility (#HH0187).

♦ **HH0318 Grounding**
Description An expression used in martial arts to denote firm contact with the earth and a strong sense of the physical body. Both physically and spiritually, strength is said to arise from the solid ground under one's feet - a comparison is drawn with the tree whose roots planted in the ground provide stability and nourishment. Attention is centred at the centre of the body, where the legs meet the spine, an inner awareness of this part of the body favouring the establishment of physical and spiritual integration.
Broader Martial arts (#HH0085).

♦ **HH0320 Omnipresent mental factors** (Buddhism, Tibetan)
Sarvatraga — Kun-'gro — Ever-recurring mental factors
Context One of the six classes of mental factors which make up the knowing (as opposed to the observing) part of consciousness in Tibetan Buddhism.
Broader Secondary mental factors (Buddhism, Tibetan, #HH1348).
Narrower Contact (Buddhism, #HM2708)
Feeling (Buddhism, #HM2270)
Intention (Buddhism, #HM2589)
Discrimination (Buddhism, #HM2399)
Mental engagement (Buddhism, #HM3237).
Related Sense consciousness (Buddhism, #HM2664).

♦ **HH0321 Syntone**
Cyclothyme
Description This term describes a person who is in harmony, particularly emotional harmony, with the environment. An exaggerated emotional reaction to the environment beyond what is considered normal is exhibited by a *cyclothyme* personality.

♦ **HH0322 Myth of incompetence**
Description Individuals who are not in a position to participate directly in the world affairs which affect their daily lives tend to view such affairs as distant and originating from some national or international centre unrelated to their own experience. They fail to see their own neighbourhood as contributing to global issues such as unemployment, inflation, and so on. In particular, "foreign" affairs are controlled by a small elite, and local individuals or groups do not participate. These gaps in perception and participation lead to a *cycle of apathy*: lack of opportunity to participate means there is little motivation to learn about international affairs; and lack of knowledge reduces still further the opportunity to participate. This leads to the myth that foreign policy issues cannot be determined by the people themselves as the ordinary citizen is, by definition, incompetent to deal with such matters. As a result foreign policy is usually determined without prior consultation at the local level, the people virtually being "told" what is the right approach.
Related Apathy (#HH0467).

♦ **HH0324 Transcendent function**
Description The tendencies of the conscious and the unconscious together make up the transcendent function, thus making the transition from one attitude to another organically possible without loss of the unconscious. The transcendent function manifests itself as a quality of conjoined opposites. So long as these are kept apart (in order to avoid conflict), they do not function and remain inert. In whatever form the opposites appear in the individual, it is basically always a matter of consciousness lost, and obstinately sticking to one-sideness when confronted with the image of instinctiveness, wholeness and freedom.
In therapy, the therapist mediates the transcendent function for the individual by helping him to bring conscious and unconscious together and so arrive at a new attitude. The transcendent function may be produced through access to unconscious material (dreams, spontaneous fantasies).
Transcendent functioning is a level of behaviour rather than a type of behaviour, which goes beyond or transcends average behaviour. It is more creative, more efficient, more productive, and qualitatively superior to habitual behaviour.
Refs Watkins, Mary Maria *Creative Imagination and the Transcendent Function* (1972).
Related Transcendence (#HH0841).

♦ **HH0325 Addiction group therapy**
Verbal attack therapy
Description This form of group therapy is a means of aiding the addictive personality, primarily conceived for use without standard professional principles or therapists. It is administered by a group of peers who have similar problems and is used, for example, in cases of addictive persons (drug addicts, alcoholics) who tend to need help urgently and are unlikely to persist with any indirect therapeutic approach. Great emphasis is placed upon unrestrained directness in communication aimed at by-passing defences, rationalizations, evasions and politeness. Hostile confrontation and verbal brutality between participants is encouraged in what are called games, which offer immediate catharsis of aggressive feelings. This form of therapy is associated with the first example of its kind known as *Synanon*, which was a community where the participants lived together and where the emotional growth of each member was an ongoing concern for each.
Refs Yablonsky, Lewis *Synanon* (1965).
Broader Therapy (#HH0678)
Human potential movement (#HH0398).

♦ **HH0326 Open learning systems**
Description Characteristic of such systems is that the process of learning as it evolves affects the way in which learning continues, rather than being based on a fixed, predetermined step or series of steps. This exchange with the environment is typical of natural processes and living systems, and of human development systems with implied positive feedback as an intrinsic characteristic.
Related Process thinking (#HH1643)
General system theory of human development (#HH1127).

♦ **HH0327 Orgone therapy**
Description Orgone energy is the primordial cosmic energy, universally present and demonstrable visually, thermically, electroscopically and by means of Geiger-Mueller counters. In the living organism it is recognized as bioenergy or life energy. Orgastic potency, the goal of orgone therapy, is essentially the capacity for complete surrender to the involuntary convulsion of the orgasm and complete discharge of the excitation of the acme of genital embrace. It is always lacking in neurotic individuals but it is usually not distinguished from erective or ejaculative potency, both of which are only prequisities of orgastic potency. It presupposes the absence of pathological character armor (the sum total of typical character attitudes which an individual develops as a blocking against his emotional excitations) and muscular armor (the sum total of the muscular attitudes which an individual develops as a block against the breakthrough of emotions and organ sensations).
Two types of orgone therapy are distinguished. Physical orgone therapy is the application of physical orgone energy concentrated in an orgone energy accumulator to increase the natural bioenergetic resistance of the organism against disease. Psychiatric orgone therapy is the mobilization of the orgone energy in the organism, namely the liberation of biophysical emotions from muscular and character armorings with the goal of establishing, if possible, orgastic potency.
Refs Reich, Wilhelm *The Function of the Orgasm* (1967).
Broader Therapy (#HH0678)
Human potential movement (#HH0398).

♦ **HH0328 Autonomous discipline**
Description Where spiritual development takes place without resort to an external teacher or system, then the guidance for development arises deep within the seeker's personal being. This may be said to have been the case with the Buddha, for the set of practices and beliefs developed during the enlightenment and subsequent life of the Buddha arose, according to the traditional stories of his life, through insight gained by himself. This knowledge and insight were beyond mind and were such that the mind could not shake them.
Broader Spiritual discipline (#HH1021).

♦ **HH0329 Chivalry**
Related Honour (#HH0192).

♦ **HH0330 Human development** (Hinduism)
Description Rather than a separate religion, Hinduism is more an evolving tradition of such diversity that some major movements within it could be considered separate religions in their own right.
The Vedic texts, dating back several thousand years, include details of rituals both for priests and for the householder. Rituals of the former include heightened spiritual awareness through drinks

containing the sacred plant *soma*. Different stages are marked by use of rituals to placate the gods or as a means in their own right of ensuring the welfare of man and the universe. This is the *Vedic* tradition, culminating in the internal sacrifice of asceticism and meditation and the mystical experience of the one self, the identity of the individual Atman with Brahman. Such realization brings release from *samsara*, the endless cycle of birth and death, in which the conditions of each life are dependent on the acts (karma) of previous lives. This condition is inherently one of suffering and spiritual activity is directed at release from it. Sannyasa, or renunciation, is one means of release. The practitioner abandons all aspects of the world and performs the austerities and spiritual practices of some form of yoga; he seeks to become liberated while still alive – *jivanmukta*. For the *grihastha*, or householder, life is lived in the world; but liberation is achieved by a final sannyasa stage in life after family responsibilities have been fulfilled.

In contrast, classical Hinduism places more emphasis on life in this world. Society is divided into 4 castes, each with specific duties, rights and responsibilities; these are the priests and teachers (brahmans), the kings and warriors (kshatriyas), merchants and farmers (vaishyas), and menial workers (shudras). The dharma, or moral code, details behaviour for each caste. The first three castes, the *twice born*, have full religious rights. The shudras are more in relation to the other castes as dependents or children and have lesser rights. It should be noted that these four castes or varnas are not the same as the caste system of *jati* currently in operation in India, although the one may have derived from the other. Nevertheless, varnashrama dharma is taken to be obedience to the laws of one's jati. The regulations involve rituals for pollution and purification, particularly in relation to birth and death. They also cover whom the individual may marry, the vocation he may follow in life, what he may eat and who may prepare it, and so on. The way in which an individual complies with dharma affects his store of merit for this life and thus the life to which he will be born next time.

The four great aims of life are dharma (righteousness), artha (wealth), kama (culture and art) and moksha (spiritual freedom – this fourth aim was not always included). When these are in harmony there is no discord between the natural and spiritual life. Three ways to salvation are defined in the *Bhagavad Gita*: jnana marga, the way of knowledge or enlightenment; bhakti marga, the way of devotion to the Lord; karma marga, the way of action and religious rites. Although bhakti marga is the most emphasized overall, the influential *advaita vedanta* advocates the way of knowledge. This holds that nothing except Brahman, even the gods or devas, is real; all else is an illusion. On realization of this, enlightenment arises and the human spirit merges with Brahman. A development towards a more theistic approach where the soul, although of the essence of God, still maintains an eternal relationship with God, is the *Vishishtadvaita* or differentiated non–duality; this system was the basis of a number of schools, all emphasizing bhakti or devotion which has gradually come to imply not so much a way of salvation but salvation itself.

The Hindu religion accepts the existence of avatars, divinity incarnated as man, who mediate between God and men. These avatars may be worshipped, as may gods who embody specific attributes of the one God. Ultimate reality is thought of as both immanent and transcendent. Whether impersonal – nirguna brahma, beyond the gunas, or personal – saguna brahma, mystic experience of transcendent reality is sought in samadhi through devotion to the gods in worship, through ascetic practice and yoga, through any of the ways indicated above, or through communal festivals, pilgrimages, ritual image worship, reading or singing of holy scripture and songs, or pilgrimage to a sacred city.

Refs Daniélou, Alain *The Gods of India*; Dasgupta, S N *Hindu Mysticism* (1927); Sen, Kshiti Mohan *Hinduism* (1961).

Broader Human development through religion (#HH1198).
Narrower Ashramas (#HH2563) Jnana marga (#HH0495) Karma marga (#HH1211) Bhakti marga (#HH0628) Advaita vedanta (#HH0518) Vishishtadvaita (#HH3772).
Related Karma (#HH0567) Dharma (#HH0093) Avatar (#HH0079) Ahimsa (#HH0088) Brahman (#HH1226) Twice–born (#HH0715) Sacred drugs (#HH0788) Sannyasa (Yoga, #HH4210) Kama (Hinduism, #HH0754) Atman (Hinduism, #HH0103) Artha (Hinduism, #HH0353) Moksha (Hinduism, #HM2196) Samsara (Hinduism, #HM1006) Spirituality (Hinduism, #HH3000) Samadhi (Hinduism, Yoga, #HM2226) Human development (Sikhism, #HH6292) Indian approaches to consciousness (Hinduism, #HM2446).

♦ HH0331 Sainthood
Canonization

Description Particular individuals may be recognised within a religion or sect as having lived lives of such holiness that, after their death, their personal charisma is evident to those who did not know them in life. Their spritual perfection may have been evident in a number of ways, whether as prophet, ascetic, ruler or as a simple but pious person. Exemplifying the highest values of their faith they act as models for others to follow; devotees may call upon the supernatural powers inherent in the holiness of the saint to assist them in their own spiritual journey, whether as intercessor, worker of miracles or helper. The saint may exemplify the ways of action, wisdom or love; their love for God overflowing into love for humanity and healing and redeeming actions. Although some religions and sects to not recognise saints as such, nonetheless an individual may be venerated by his followers in a manner similar to that reserved for saints – this is true of Judaism and, to some extent, of Protestant Christianity which does not generally accept the idea of an intercessor, God's grace being the sole means of redemption. There is a close connection between *beatification*, when the person concerned is venerated at least locally after his death, and canonization. In the Roman Catholic Church, canonization is a formal matter and accepted only when there is evidence of miracles having taken place as well as of public veneration. This is true also of the Orthodox Churches, where incorruption of the saint's body is also taken as evidence. Canonization implies recognition of the saint having been accepted into heaven; prayers are no longer made for saint's soul, instead his intercession is invoked.

Related Saintliness (#HH1188) Beatification (#HH0608) Sanctity (Christianity, #HM0560).

♦ HH0333 Solitude
Discipline of solitude (Christianity)

Description The truest solitude is not something outside and not the absence of people or sound; it is an abyss opening up at the centre of the soul. The abyss of interior solitude is created by a hunger that cannot be satisfied. It is found by hunger, sorrow, thirst, poverty and desire. The soul which has found this solitude is empty, as if emptied by death. This solitude is everywhere but there are mechanisms for finding it.

All Christian religious communities have recourse to use means for finding solitude within their systems, possibly through *retreat* and/or periods of *silence*. This deliberate separation from the company of other people can help lead to deep inner silence and aid *recollection* and *prayer* without distractions. Solitude and silence are closely related states, the former giving the spiritual stillness to maintain the latter, and to speak only when words are really required. In many traditions a life of solitude as a *hermit*, whether totally alone or grouped around a common place of worship, has been used to achieve solitude of the soul. Frequently such retreat is to a place of harsh physical conditions (desert, mountain, forest); nevertheless, simply taking periods of solitude during the day whenever they may be experienced may be very efficacious.

The retreat is not to escape for its own sake or from others or the world because it is unpleasant. Peace will not be found nor will solitude. To seek solitude because it is what one prefers will never result in escape from the world and its selfishness. The interior freedom required to be truly alone is not present. Solitude is to be alone in the healing silence of recollection and the untroubled presence of God. The state of solitude results in a deeper understanding and love for one's fellows.

Thomas Merton says that solitude is not separation. If one goes into the desert it must not be to escape from others but to find them in God. There is a real need for solitude at a time when love and conformity are equated. True solitude is that of a person constituted by a uniquely subsisting capacity to love, it is not the refuge of an individualist. Physical solitude is contrasted with the escape into a crowd. Lost in a crowd, the person does not know he is alone but neither does he function as a person in a community. Not facing the risks or responsibilities of true solitude, his other responsibilities removed from his shoulders by the multitude, he is burdened by diffuse anxiety, nameless fears, the petty lusts and pervading hostilities which fill mass society. To remain human one must have true communion and dialogue with others – not simply living with others and sharing only the common noise and general distraction. True solitude is interior solitude, which is possible only for those accepting their right place in relation to others.

Refs Fritz, Mary *Take Nothing for the Journey* (1985); Storr, Anthony *Solitude* (1988).

Broader Spiritual discipline (#HH1021).
Related Retreat (#HH0420) Recluse (#HH0958) Loneliness (#HM1260) Pain of solitude (#HM1539) Practice of silence (#HH0983) Christian recollection (Christianity, #HM2077) Dark night of the soul (Christianity, #HM3941).

♦ HH0334 Noetic superstructure

Description A term in strata theory to describe individual thought processes which, together with voluntary activities constitute the *personal superstructure* and which take place above the *endothymic* basis. Noetic science studies physical and mental variations in different modes of awareness.

♦ HH0336 Sociotherapy
Milieu therapy — Therapeutic community

Description Sociotherapy covers all therapy emphasizing adjustment related to the socioenvironment and interpersonal factors. In particular, it refers to the setting up of a therapeutic community.

Broader Therapy (#HH0678).

♦ HH0337 Bhakti yoga (Yoga)

Description This yoga is one of the forms of Laya yoga and is primarily intended for those of emotional or devotional disposition. Devotion is centred on God, on a great saint, or on some great task. It is the path of devotional submission to the divine and to one's guru or spiritual master. As the way of devotion or of love, the aim of this yoga is to develop the individual's love of God and his surrender to this love, which is a spiritual transformation of all the types of love felt for friends, family and spouse. God is sensed as a distinct personality, however exalted his attributes of wisdom, compassion and grace. The sense of the otherness of God leads the individual to adore him with every element of his being. The aim of the individual then becomes to love God only (and others through him), for no ulterior motive (not even the desire for liberation), but for love's sake alone. Insofar as he succeeds, he will experience pure joy, and to the extent that he strengthens his affections for God, he will weaken the grip of the material world upon his attention.

Mystical disciplines of this kind yield a trance–like state (bhava samadhi) in which, after concentration on the personal God and his vision, a state of complete merging, quiet or silence ultimately supervenes. This last stage involves a complete relaxation of thoughts and emotions, and a relinquishment of the ego focus.

Refs Vivekananda, Swami *Bhakti–Yoga* (1964); Vivekananda, Swami *Bhakti Yoga*.

Broader Yoga (Yoga, #HH0661).
Related Mantra yoga (Yoga, #HH0931) Bhakti marga (Hinduism, #HH0628) Krishna consciousness (#HM2300) Heart-awakening stage of life (#HM3518) Resignation to the will of God (#HH1016).

♦ HH0338 Man of knowledge
Man of power

Description The ultimate goal of the system of *Carlos Castaneda* is to become a *man of knowledge* although, due to man's impermanence, this is never achieved. By following the teaching of a Shaman (Don Juan) and through supernatural intervention through an *ally* (hallucinogenic drugs), one is show the right way to live. The drugs induce a state of *non–ordinary reality* and such experience leads to new ways of behaviour. The pursuit of knowledge involves enormous effort; one is said to be a *warrior* on the path of such pursuit, living a frugal life and strengthening qualities of respect, fear, awakened intent and self control. This leads to *stopping the world*, when commonsense views are seen as only an interpretation and not reality as such.

Related Stopping the world (#HH0489).

♦ HH0340 Numerology

Description This is a method of studying the use of numbers as symbols for states of consciousness. It is held that numbers, like geometrical figures, can be taken to indicate different states in the process of an individual's inner growth and integration.

♦ HH0341 Justification (Christianity)

Description The act of being made worthy of salvation, the term used in Christian teaching being "justification by faith", which precedes sanctification. Here an offence against God is set right by an act of God, expressing the fact of restored relations between God and humanity.

Broader Forgiveness (Christianity, #HH0899).
Related Redemption (Christianity, #HH0167) Reconciliation (Christianity, #HH0164).

♦ HH0342 Re–evaluation counselling
Co–counselling — Counselling

Description A derivative of *client–centred therapy*, re–evaluation counselling enables counsel to be given and received by peers. Groups meet at regular intervals in pairs, each member of the pair giving full attention one to the other for about one hour each. Encouraged by the acceptance of the other member of the pair, the individual uses the experience as an outlet for accumulated frustrations, tensions and blocked emotions. The technique has been extended in marriage counselling, where the couple may become a "pair", each co–counselling the other, listening with attention and without interruption to what the other has to say.

Broader Counselling (#HH0842) Client–centred therapy (#HH0095).

♦ HH0343 Naturalness (Zen)
Shizen — Nature of being

Description In *shizen* is implied a mode of being, that which is as it is of itself; nature is the power of spontaneous self development and all that arises from that power. In Zen, the individual accepts his own nature, his own body, cooperating with nature in natural balance and harmony. The self–nature, or spirit, is considered as much of the natural world as body and mind; it is wise to be obedient to it and to accept it, which leads to joy. Such naturalness implies using wisdom,

judgement and restraint, as these too are natural.
Related Contemplation of nature (#HH0580).

♦ **HH0345 Compulsion**
Description Compulsions are said to occur from repeated and longterm *repression* of feelings considered in some way wrong. This act of repression reaches the stage that an *obsession* with the idea of carrying out an unreasonable and senseless act can no longer be resisted, resulting in irresistible compulsion. Common compulsions such as frequent handwashing, masturbation, and the counting or touching of particular objects, are characteristic of these particular neuroses; the will has no power to prevent the individual carrying out such acts, any attempt to do so causing even more tension.
Related Obsession (#HM2809).

♦ **HH0347 Healthy human growth and development**
Description Human development embraces every aspect of the maturation process, including its physical, biological, psychological, and social aspects. To bring about healthy development and to realize human potential, it is necessary to draw upon many areas of scientific knowledge and many components of the health service. Such areas as nutrition, communicable diseases, human reproduction, mental health, handicaps, and many others, together with the corresponding services, are related to human development. Many of these services have their greatest impact on development when they are employed early in the individual's life.
The principal requirement for healthy growth and development is good nutrition. Closely related to this is the requirement for measures to prevent infection; to be effective, such measures must be taken early in life and pursued in later years. Human development is favoured at the outset by the careful management of pregnancy and by the practice of family planning. For those problems of mental and social health that have their origins in childhood, early preventive and educational measures help to forestall or obviate problems in adulthood. The early detection and treatment of handicapping conditions will improve the long-term prospects for almost all children with handicaps.
The promotion of healthy development cannot be achieved by measures that derive from any single health discipline, nor can health measures be considered independently of the broader educational, social, economic, and administrative factors that are crucial to human development. Health measures aimed at promoting biological development (especially the control of infection, the improvement of nutrition, the management of pregnancy, and family planning) can improve the health of people; higher levels of health in turn contribute to socio-economic development. It was at one time assumed that if a child survives he must be well, but studies of the health of survivors now show that even when mortality rates are falling, many health problems remain unresolved. In fact, there is evidence that many children do not attain their full developmental potential because of adverse environmental factors before and after birth.
Environment. The environment (natural or man-made, physical, chemical, biological, and social) has a significant effect, whether direct or indirect, on human development. While scientific knowledge of the environment is growing, there is still much to be learned about its immediate and long-term effects on human development. Furthermore, current knowledge is not always applied, or not used in the best way to improve the quality of human life. There still exist numerous environmental hazards that could be at least partially controlled.
If the environment of the community is grossly unhygienic, the child is exposed to a host of infectious agents that he is physically unable to withstand. Such exposure, with modifications, continues throughout life, but children who survive the intitial encounter with infection and infestation acquire their own defences of active immunity. For millions of people in poor communities, the hygiene of the physical environment remains extremely unsatisfactory and dangerous. The absence of adequate facilities for the disposal of excreta accounts for the high prevalence of bacterial, viral, protozoal, and helminthic diseases. The lack of a plentiful, convenient, and safe water supply complicates efforts to maintain a sanitary environment, to keep food clean, and to practice good personal hygiene. Overcrowding and lack of ventilation predispose children to a host of airborne infections. Despite notable advances in control, insect vectors continue to be major sources of infection in many parts of the world.
Economic and social factors. Human growth and development are influenced at every stage by the customs and beliefs of the community. Where rapid technological changes are occurring, adjustments of social organization and of educational patterns are required. The type and degree of success of these adjustments affect the maturing individual profoundly. Efforts to improve health will not be successful unless they take into account the social and cultural characteristics of the communities in which they are made. Since economic resources, in the broadest sense, constitute our material and physcial environment and are underlying determinants of many cultural, social, and behavioural patterns, they are dealt with in this section. A few significant examples of the social aspects of the human environment are discussed below.
1. *Economic resources:* The abolition of poverty remains a prerequisite for healthy development, even though affluence, if misdirected, may bring new problems in its wake. The importance to health of the economic factor extends well beyond the ability to pay for health services. Family income influences the kind and amount of food eaten, the quality of housing, the type and duration of education and the whole range of social and economic factors bearing on human development. The family's economic resources may be especially strained in the early years of marriage when the children are young and their needs are greatest.
2. *Education:* If mankind is to maintain or, if possible, improve its level of development in a changing world, it must be helped to adapt to its cultural milieu and to its biological and physical environment. Education, which engenders change and fosters adaptation, helps to make this adjustment and contributes significantly to mankind's level of development. At present, one of the major problems in all countries is the education of children to live in a world that changes rapidly, constantly, and unpredictably, rather than in a social organization that is stable.
3. *Cultural factors:* Cultural factors have an enormous impact on health and human development. Attitudes towards life and death determine the value placed on health by a community, and may thereby decide the demand for health services and the use made of them. The status of women affects pregnancy, parturition, and lactation: whether women are expected to do heavy physical work even during the perinatal period, whether their diet is restricted by food taboos or by the custom of serving women last, or whether breastfeeding is customary, will establish the perinatal environment of the child. The status of children, preferences for boys or girls, and child-rearing practices all influence the course of human development.
4. *Changes in the human setting.* Human development is influenced by the rapid changes that are taking place in the human setting. Modern technology, industrialization, the expansion of cities, migration, and population growth result in many ecological, social and biological changes. These changes are not confined to the developed countries; in fact they are relatively far more rapid and occur in far more difficult situations in developing countries. The demographic transition is the most dramatic example. Rapidly falling death rates and persisting high natality rates influence the family and the community profoundly, and directly affect plans for the provision of resources for optimum development.
Some changes are beneficial to human development: the improved production and distribution of food have helped to control famine; the application of scientific knowledge has helped to control disease and extend life expectancy. But other changes have created new problems and exacerbated old ones. Unwelcome are: the degradation of natural ecosystems under the impact of human settlement, the deterioration of cities as a result of uncontrolled migration from rural to already overcrowded urban areas, the extensive pollution of air and water with the unregulated expansion of industry, and many other problems.
5. *Genetic factors.* Growth and development are the result of interactions between the genetic information contained in the zygote and the environmental variables encountered in the course of time, from fertilization through birth to adulthood. Mutations occur, from time to time, usually for unknown reasons but occasionally from known environmental causes such as radiation. Slow modification of gene frequencies is occurring all the time under the influence of natural selection.
Perinatal factors. Perinatal factors include: the duration of gestation, birth weight, and other factors that operate through the mother during pregnancy and parturition. The origin of much faulty development is to befound in various abnormalities of pregnancy. The birth process itself may be hazardous, and the establishment of respiration may be traumatic.
Family planning. Family planning is its broadest sense includes the planning of pregnancies so that they occur at the desired time, the spacing of births for the optimum health of all family members, and the prevention of further births when the family has reached the total size desired. Family planning as a means of promoting human development rests on a number of associations between health (as measured by maternal, perinatal, infant, and child mortality and morbidity rates and by indices of physical growth and mental development) and reproduction (as expressed in the timing, number, and spacing of pregnancies). For example, there is a poor prognosis for the infant born of a mother who began childbearing at a very young age and has already had many closely spaced pregnancies. Family planning will have a positive effect on health when it is used to postpone the first pregnancy until the woman has completed her own growth, thus avoiding the double burden of growth and reproduction; when it is used to space births and extend the interval between pregnancies, so that the woman has time to recuperate after each birth; and when it is used to limit the total number of births, since a poor outcome of pregnancy is associated with both high parity and advanced maternal age.
Nutrition. Nutrition is fundamental to human development and influences it throughout the life span; it has far-reaching effects on physical, intellectual, emotional and psychomotor development and indirectly affects social development. Malnutrition and undernutrition constitute major health problems not only in developing countries, but in many of the developed parts of the world as well. The important role played by malnutrition in the high mortality and morbidity attributed to infectious diseases is also recognized. The highest toll is exacted during the critical stages of development, especially during early childhood.
Infection. One of the major hazards of severe or prolonged infection is its potentially deleterious effect on growth and development, especially in the presence of malnutrition. Any significant infectious or parasitic disease affects the biology of living tissue or organ by producing increased catabolism, which may result in weight loss. The effect on growth and development is much the same as that of dietary deprivation. The clinical literature indicates that severe, prolonged infection causes growth retardation in muscles and bones and that it culminates in stunting, especially when it occurs very early in life. The combination of infection and malnutrition may also have an untoward effect on mental development and neurological function. Infection also tends to impair nutrition in a number of ways. Besides its general catabolic effect, infection can alter absorption, metabolism, and the excretion of specific nutrients. It can also diminish food intake – either directly through anorexia, or indirectly through the widespread customs of withdrawing solid food, using starvation therapy, and occasionally using purgatives in the treatment of febrile illnesses. The results can be serious, especially in infants and young children whose previous level of nutrition was marginal.
Refs WHO *Human Development and Public Health* (1972).

♦ **HH0348 Physiognomonic thinking**
Syncretic thinking — Concrete thinking — Abstract thinking — Categoried thinking
Description The first stage in a child's development of thought, when the ego is projected into inanimate objects, such as playing with a stick and calling it a horse, is referred to by Kasanin as *physiognomonic* thinking and by Piaget as *syncretic* thinking. The next stage, where words are used to describe a specific object, but not generalized, is referred to as *concrete* thinking. Here the word "table" refers to a particular table but not to all tables in general. Following some degree of education and experience, after adolescence, comes *abstract* or *categoried* thinking, when abstract and generalized thought is said to appear.

♦ **HH0349 Devotion** (Christianity)
Description According to St Thomas Aquinas, devotion is the virtue which directs an individual to render to God that worship which is his due. It is one of the eleven acts said to constitute the most perfect worship that mankind is capable of. In contrast to feelings and emotions, devotion is an act of will, although by that act of will emotions will be aroused towards the object of devotion.
An inner, intimate, essential side of worship, devotion thus implies self-dedication to the divine will and a movement in intention towards unity with God. This approach of the soul towards the divine is marked by a particular quality of prayer, quietness and union "emptied of all forms, species and images", (Mme Guyon). There is also an act of praise and homage when divine love overwhelms the soul's sense of guilt in mercy and grace. Such devotion is inseparable from adoration and thanksgiving. The practice of devotion also includes *meditation* or *contemplation*, preceded by *recollection*. This raises the whole personality to a higher level and releases otherwise hidden spiritual force. Another essential aspect of the life of devotion is *self-discipline* leading to *self-surrender* and a sincere and utter detachment from earthly things. This is not so much outward poverty as *self-simplification* and may involve almsgiving and fasting. Such practices may lead to the experience of spiritual rapture and ecstasy.
For William Law, prayer is only a particular instance of devotion. Devotion implies a life devoted to God, when the person no longer lives to his own will or to the way of the world, but solely to God. God is considered and served in everything. Every part of common life is done in the name of God and under rules conforming to His glory. Any activity that is not carried out strictly according to the will of God is as absurd as prayers not according to His will. This is made clear in the scriptures, where the religion or devotion covering ordinary actions is continually emphasized whereas there is little insistence on public worship. It is in everyday life that humility, self-denial, renunciation of the world, poverty of spirit and heavenly affection are to be demonstrated if one is to live as a Christian.
Refs Stanwood, P G (Ed) *William Law* (1979).
Related Prayer (#HH0185) Worship (#HH0298) Fasting (#HH0011)
Spiritual poverty (#HH0190) Bhakti marga (Hinduism, #HH0628)
Contemplative intuitive meditation (#HH0816) Christian recollection (Christianity, #HM2077)
Religious experience as personal devotion and worship (#HM0561).

♦ **HH0350 Circles of enlightenment** (Zen)
Description The use of the circle to indicate the contents of enlightenment is common in Zen teaching. In particular, the five stages of a disciples way can be represented by: (1) A black disc with a white section at the bottom, indicating the lord looking down on the servant; (2) A white disc with a black section at the bottom, indicating the disciple himself as the willing servant; (3) A white disc with a black dot at the centre, where the disciple realizes he is peripheral to the

lord; (4) A totally white disc, where he realizes he is within the lord; and (5) An all black disc, where the servant has become lord, they are all one and in complete unity. The stages are realized in each disciple's life but cannot be known until they are experienced, when they are recognized.
Related Stages of spiritual life (#HH0102).

♦ HH0351 Image of man
Concept of man
Description This may be considered to be the set of assumptions held about the human being's origin, nature, abilities and characteristcs, relationships with others, and place in the universe. Such a coherent image may be held by any individual or group, a political system, a religious group, a culture, or a civilization. It may consist of beliefs as to whether individuals are basically good or evil, whether the individual will is free or determined by external forces, whether he is cooperative or competitive, whether individuals are essentially physical or spiritual in nature, and whether men and women are essentially equal. It includes both what man is currently conceived to be and what it is considered that he ought to be. It is therefore a Gestalt perception of humankind, both individual and collective, in relation to the self, others and the cosmos. Such images are held with varying degrees of awareness by persons and societies, although in most cases such assumptions are held subconsciously.
A particular image of man may be appropriate for one phase in the development of a society, but once that stage is completed, the use of the image as a continuing guide to action may well create more problems than it solves. Man is now believed to stand at the end of the industrial era, and that the images of man that dominated the last two centuries will be inadequate to the post–industrial period. In response to this situation, a new image (or possibly images) of man is believed to be emerging. This emerging image reinstates the transcendental, spiritual side of man, which has received little attention from modern science.
From the nature of contemporary societal problems, studies of plausible alternative futures, and the role of any dominant image of man in society, characteristics that a new image must possess if it is to become dominant and effective in the future may be specified. One such set of characteristics requires that the image would need to: convey a holistic sense of perspective or understanding of life; entail an ecological ethic, emphasizing the total community of life–nature and the oneness of the human race; entail a self–realization ethic, placing the highest value on development of selfhood and declaring that an appropriate function of all social institutions is the fostering of human development; be multi–faceted, and integrative, accomodating various culture and personality types; involve the balancing and coordination of satisfactions along many dimensions rather than the maximizing of concerns along one narrowly defined dimension; and be experimental, open–ended and evolutionary.
Refs Center for the Study of Social Policy *Changing Images of Man* (1974).
Related Imitation (Christianity, #HH0810).

♦ HH0352 Alexander technique
Description The Alexander technique is a system by which conscious control of body movements and posture is acquired. The student becomes aware of poor habits of posture and carriage and learns, by conscious and reasoned control, to inhibit such habits and to replace them with improved habits. Through this technique the person develops a means of becoming conscious of his physical self and of his everyday movements. Once such attention to the physical self is developed, improvements and elimination of strain arising from poor habits become possible.
Derived from this technique is a system of correcting unbalanced bodily movements to correct the mental and emotional unbalance that such movements have developed. As the natural reflexes which enable the body to support itself against the downward pull of gravity are developed, the individual gains *poise*. This is used for treatment of many muscular, neurological, mental and psychosomatic disorders.
Refs Rosenfeld, E and Rubenfeld, Ilana *The Alexander Technique*.

♦ HH0353 Artha (Hinduism)
Wealth — Knowledge
Description One of the four great aims of life according to Hindu philosophy, the pursuit of wealth is seen as quite legitimate as long as it is done in harmony with dharma (righteousness), when it may lead to moksha (liberation). The desire for power and property, the economic and political aspects of human life, all these are seen as natural and not fundamentally at war with the spiritual life. However, the harmonious balance within and between political and spiritual lives is to be achieved in the regulations of dharma. Artha may also refer to spiritual wealth, to love, faith, devotion, goodness, beauty, peace and joy within. In this context, artha is also seen as one of the three categories of knowledge, the true knowledge about an object or its meaning which, according to Patanjali, is normally confused with *shabda*, knowledge based only on words and *jnana*, knowledge based on the perception of the sense organs and reasoning of the mind. The three components are separated and their correct relation seen in *savitarka samadhi*.
Context According to Patanjali, one of the three forms of knowledge which need to be distinguished from each other and separated if confusion *samkirna* is to be avoided.
Broader Knowledge (Buddhism, #HH1971).
Related Dharma (#HH0093) Kama (Hinduism, #HH0754) Moksha (Hinduism, #HM2196) Sabda (Yoga, Buddhism, #HH0538) Savitarka samadhi (Hinduism, #HM3526) Human development (Hinduism, #HH0330) Jnana (Yoga, Buddhism, Hinduism, #HH0610).

♦ HH0355 Metapsychiatry
Description This designates an area of study at the interface between psychiatry and mysticism. Metapsychiatry encompasses not only parapsychology but also all other suprasensory, suprarational and so–called supernatural manifestations of consciousness that are in any way relevant to the practice of psychiatry.

♦ HH0356 Psychological reform
Description Some models of mental disorder tend to reject the concept of mental "illness". Clinical psychology has therefore to cease approaching the problem as one that needs remedial treatment by removing focal infection.

♦ HH0357 Human genetic improvement
Eugenics — Dysgenics — Genetic counselling — Genetic selection — Genetic engineering
Description Although natural selection operating on spontaneously occurring variations would possibly improve the human condition in time, by suitable selection and manipulation of the genetic make–up, both the physical and psychological aspects of human beings can be intentionally improved. The term eugenics is applied to theories and practices designed to improve the human condition from a genetic point of view and particularly by eliminating hereditary defects and diseases; conversely, dysgenics applies to prevention of the decline of the human race. Genetic counselling aims to improve mankind as a species by discouraging reproduction among those likely to perpetuate genes that lead to physical or behavioural defects. Genetic selection may be accomplished by withholding job and other opportunities from groups whose genes are considered undesirable. Such techniques have been used by governments as an instrument of racial advancement and purification.
Thus, negative eugenics, or dysgenics, is the use of techniques to eliminate undesirable characteristics; positive eugenics aims to increase the frequency of superior hereditary endowments. Use of such techniques is considered necessary (even urgent) as a means of counterbalancing both the unchecked increase in mankind's genetic load (namely the deviation of observed fitness from the optimum), and the probable increase in mutation rates due to increased exposure to radiation.
All aspects of this approach are matters of controversy, whether due to lack of consensus on the characteristics to be eliminated or encouraged, or to the means of preventing undesirable reproduction or encouraging desirable reproduction, or to the human rights issues associated with these issues, or to the urgency of the question, if any. Particular controversy has been raised over the possibilities of *cloning* exceptionally gifted people, the use of host mothers, and the abortion of defective foetuses.
Many claim that nature has its own dysgenic methods and that non–interference in the laws of natural selection (apart from the relief of pain) would produce the desired result – the infirm (particularly the very old and the very young) would die naturally, unfit foetuses would not be brought to term, and biologically infertile couples would not produce babies. It is also claimed that active interference in natural selection might cause as yet unimagined genetic problems.
Related Euphenics (#HH0747) Evolution (#HH0425) Lamarckism (#HH1673) Gene therapy (#HH1334).

♦ HH0358 Identification
Description This term is used by Jung to describe the unconscious projection of one's own personality onto that of another who is able to provide a way of or reason for being. This may be a person, place or thing, and is part of normal development although its extreme may lead to *identity*, when the individual shares the life of another to the extent of having no existence of his own, or to *inflation*.
Related Inflation (#HM1288).

♦ HH0360 Socialist humanism
Description Sharing with all humanism a belief in the unity of humanity and in the possibility of human perfectability, socialist humanism stresses the need for a change in economic organization and the ameliorating of the dehumanizing effect of both impoverished and affluent alienation.
Refs Fromm, Erich Ed *Socialist Humanism* (1965).
Broader Humanism (#HH0674).

♦ HH0361 Asteya
Abstaining from misappropriation
Description This is not simply restraint from stealing or misappropriation of physical property, but also involves intangibles such as credit for something not done or undeserved privileges.
Context One of the five moral qualities of yama (restraint).
Broader Restraint (#HH6876).

♦ HH0362 Family therapy
Family group therapy — Family psychotherapy — Marriage therapy — Structural family therapy — Conjoint family therapy
Description A method of psychotherapy in which the family, instead of the individual, is considered as the unit in which emotional and psychological problems occur, particularly since it is this group to which the individual needs to learn to relate. Each member of the family may be perceived as a threat by each other member, thus provoking an exaggeration of the unsatisfactory existing dynamic as a defence against the threatening features perceived in any relationship which would be more balanced and fulfilling for all concerned. The family as a whole therefore requires education in more effective interpersonal relations.
In addition, every family has a large family potential which can be identified and used for its more productive functioning. Once a family begins consciously to seek out and develop such potentials, this process itself leads to a strengthening of the family and a consequent reduction in the significance of the difficulties previously experienced.
Structural family therapy involves the therapist in counselling on the functioning of the family and the changes necessary to recreate a healthy family structure. *Conjoint* therapy focuses on the problem of an individual member and his relationships with other members of the family. This type of therapy is particularly relevant in *marriage therapy*, which is concerned primarily with the marital relationship of husband and wife.
Refs Satir, Virginia *Conjoint Family Therapy* (1967).
Broader Therapy (#HH0678) Psychotherapy (#HH0003).
Related Child guidance (#HH0850).

♦ HH0364 Treasures of the godly
Qualities of the godly
Description Chapter XVI of the Bhagavad Gita lists the 26 characteristics of the divine nature, the treasures of the godly said to lead to liberation. These are: overcoming fear; purity of heart; steadfastness in knowledge and yoga; readiness to give totally without claim; rising above the senses; sacrifice and service; enquiring deeply into the self; austerity; uprightness; abstaining from violence; truthfulness; controlling anger; renouncing the things of this world; peacefulness; refraining from criticism; compassion for all beings; overcoming desires; gentleness; modesty; control of agitation and unnecessary activity; illumination; forbearance; purity and cleanliness; controlling malice and hatred; absence of self esteem. These are contrasted with demonic or ungodly tendencies: lack of knowledge of what should or should not be done; lack of purity; lack of truth; lack of understanding both of the scriptures and of the universe itself.
Related Heroism (#HH0929) Liberation (#HH0388) Fearlessness (#HM3553) Divine nature (#HH0700) Moksha (Hinduism, #HM2196) Non–anger (Hinduism, #HM6898) Eightfold path of yoga (Yoga, #HH0779).

♦ HH0366 Ontogenesis
Ontogeny
Description The development through successive transformations of an individual organism in a lifetime is referred to as ontogeny. It consists of growth and differentiation morphologically, physiologically and biochemically. During this process the hereditary information contained in each cell of the organism is realized. The interaction of the many influencing factors (including environmental conditions and reactions between cells and tissues) determines the complex ontogenetic patterns which constitute the program of ontogeny, the three main processes being growth, development and aging. Some sources quote the perinatal phase being experienced as a complete lifecycle before birth.
Related Life cycle (#HH0511) Ontogenetic model of human consciousness (#HM1105).

♦ HH0367 Not–doing
Description The objective of not–doing is to produce an awareness for the "gaps" in between the elements of ordinary perception, allowing the underlying phenomenon of perceiving to become conscious of itself. In loosening the grip of the tonal, that is, stilling mental and emotional activity while "gapping" (perceiving non–perception within given perceptions), aspects of underlying nagual may become aware. Not–doing, although specifically outlined in works as C

Castaneda's and the Tao Te King of Laotse, seems to be a general technique for stopping the activity of mind during activity which has acquired a number of different names and procedural directions.
Related Dreams (#HM2950) Tonal state of awareness (Amerindian, #HM2066) Nagual state of awareness (Amerindian, #HM2036).

♦ **HH0368 Fusion**
Description The two primal instincts, life and death, are said to be separate at birth but to fuse later. De-fusion of instincts may occur in psychiatric conditions where a threat (for example, to the ego) causes the repression of some impulse, which regresses and may then fuse with some lower-level instinct.

♦ **HH0370 Insight-provoking tales**
Fables - Parables
Description Meaningful stories and fables can be used to reach knowledge through the deep intuition which such stories reveal in the hearer. In Sufi teaching, the hearer is asked to choose stories which particularly appeal to him and to turn them over in his mind; this can effect a breakthrough into higher wisdom.
Related Koan (Zen, #HH0263).

♦ **HH0371 Ego development**
Description The ego is held to develop from the id and to this extent a child develops an identity that is progressively more coherent and independent. The ego develops as an adaptive unit which mediates the adjustment of his instincts to the demands made upon him by external reality and to the painful stimuli to which some action gives rise. In this way the ego progressively removes the child's instincts from the relatively direct contact they had with the external world in his early life. The learned separation of stimulation and response allows the interposition of more complex intellectual activities such as thinking, imagining and planning.
Suggested stages of ego development are; differental of self from nonself; exercise of will in response to impulses; opportunistic manipulation and exploitation of interpersonal relations; internalization of rules, and obedience to them, because conformity confers acceptance; individual morality given precedence over rules, and obligations assessed by inner standards; recognition of individual autonomy and inevitable mutual interdependence; integration. At this stage (which few are believed to reach, because not many realize their full potential), the person proceeds beyond coping with conflict to reconciliation of conflicting demands, renunciation of the unattainable, cherishing individual differences, and achievement of a sense of integrated identity. The major causes of ego development are held to be hereditary rational instinct, nonrational instinctual drives and the external environment (which delays gratification, restricts desired action and gives rise to behavioural habits in accordance with parental training).
Refs Loevinger, J *Ego Development* (1976); Loevinger, J *The Meaning and Measurement of Ego Development* (1966).
Related Atman project (#HM2153).

♦ **HH0372 Karma yoga** (Yoga)
Yoga of action
Description This form of yoga is based on the performance of right action under any circumstances. As such it is a part of all yoga practices. It may however also be viewed as a special method in its own right, distinct from the other major forms of yoga and based primarily upon renunciation of the fruits of all actions. As such it is an aspect of Raja Yoga.
Karma yoga is used by persons of action-oriented disposition or, in the system of Master Da Free John, one who has reached stable maturity in the first three stages of life and is set on the path of self-transcending service to others. In this form the individual transcends himself through application to work, whether approached philosophically (under the jnana or knowledge mode), or with an attitude of love (under the bhakti or devotion mode). Work becomes a vehicle for self-transcendence for the individual, in the first case by identifying himself with the impersonal Absolute at the core of his being or, in the second case, by shifting the centre of interest and affection to a personal God, experienced as distinct from himself. In both cases, every act done without thought of self diminishes self-centredness until finally no barrier separates from the identification of consciousness with ultimate reality, however conceived.
Whether in the philosophical or devotional mode, both approaches effectively starve the finite personality by deflecting it from the consequences of the actions on which it feeds. Pain, loss, and shame become accepted with equanimity because they are seen to touch only the superficial levels of the individual, whereas his identification is progressively transferred (through the work experience) to the underlying immutable self whose limitless joy and serenity are not disturbed by the ephemeral consequences and possibilities of the world of action.
Refs Vivekananda, Swami *Karma Yoga*.
Broader Yoga (Yoga, #HH0661) Raja yoga (Yoga, #HH0755).
Related Karma (#HH0567) Puja yoga (Yoga, #HH0111) Kriya yoga (Yoga, #HH0969) Karma marga (Hinduism, #HH1211) Resignation to the will of God (#HH1016).

♦ **HH0374 Samyama**
Description The central process of yoga - *dharana* (concentration), *dhyana* (meditation) and *samadhi* (contemplation) - eventually leading to liberation beyond object, power and knowledge. The process is a continuous sinking into the centre of one's consciousness and mastery of this process leads to knowledge and to super-physical powers or *siddhis*. By performing samyama or total concentration on a subject, total knowledge of that subject is revealed - for example, performing samyama on the sun brings knowledge of the solar system; performing samyama of a sound brings knowledge of its meaning (whatever language, even of animals). Finally, samyama on the process of time leads to *vivekajam jnanam*, awareness of ultimate reality.
Broader Illumination (#HH0804).
Narrower Dharana (Hinduism, #HM2566) Samadhi (Hinduism, Yoga, #HM2226) Dhyana (Hinduism, Buddhism, #HM0137).
Related Siddhis (Yoga, Buddhism, #HH0380) Eightfold path of yoga (Yoga, #HH0779) Jnana (Yoga, Buddhism, Hinduism, #HH0610) Awareness of the ultimate reality (#HM1456).

♦ **HH0375 Status**
Description Status denotes the rights, duties and respect accorded by society or social grouping to the function a person carries out in such a grouping. This is distinct from the role such a person plays, which refers to what he is expected to do. Since a person may have several functions (depending on his or her occupation, position in the family, position in leisure-time occupation, etc) the status he or she is accorded will depend on the function relevant at that time. In addition, status may be accorded through birth into a particular family or through meritorious achievements.

♦ **HH0376 Discipleship**
Description Discipleship - membership of a group following the way of a particular teacher - requires the disciplines of: personal decentralization; relinquishing personal preferences; and service to the group. This forgetting of the personal self can be an uncomfortable and painful experience, as illusions are transcended and consciousness widened; but as consciousness is thus widened, the disciple is initiated deeper into understanding the group and reality as a whole.
Refs Bailey, Alice A *Discipleship in the New Age* (1986).
Related Initiation (#HH0230) Discipline (#HH0163) Iradat (Sufism, #HM8901) Mujahadat (Sufism, #HM5215) Group consciousness (#HM3109).

♦ **HH0377 Ordeals**
Description An ordeal is a test of authenticity of an oath or of a person's innocence, when divine or supernatural judgement is invoked in some test involving the ability to survive pain, fire, water, poison or some form of combat. The accused may be required to grasp a heated metal bar or plunge his hand into boiling oil; or the ordeal may not, in itself, be dangerous but becomes so if the accused swears innocence when in fact guilty. Although not always recognized under law, trial by ordeal has been popular in many cultures, based on the instinctive belief that if a person is in the right then that person will receive divine approbation. Conversely, guilt is thought to weaken the wrong-doer and make him or her susceptible to either natural or magical injury.
Related Rites (#HH0423) Penance (#HH0790) Initiation (#HH0230) Divination (#HH0545).

♦ **HH0379 Physical therapy**
Kinesiotherapy — Physical process therapy — Physiotherapy — Somatherapy
Description Physical therapy is a branch of medicine involving the therapeutic and prophylactic used of physical treatment (exercise, massage, manipulation, application of heat, light, cold, electricity) to correct some bodily disfunction. Kinesiotherapy uses similar methods, but the aim here is not simply to correct or treat physical problems but, by correcting misuse of the body, to train the individual to use the body so as to improve the mental faculties, reduce stress and avoid fatigue and tension. Also possible to include under this heading are a variety of physical process techniques, such as the external use of herbs, poultices, baths (hot, cold, steam), fumigation, fasting, acupuncture and even the treatment of mental illness by diet and rest.
Broader Therapy (#HH0678).
Narrower Massage (#HH0284) Structural integration (#HH0082).
Related Psychotherapy (#HH0003).

♦ **HH0380 Siddhis** (Yoga, Buddhism)
Description Physical austerities leading to power to heal and to work miracles, including the ability to become invisible, levitation and occult powers. According to Patanjali, the performing of *samyama* leads to these extraordinary powers but the occult powers are of trivial importance compared with the higher states of consciousness also included in the meaning of siddhis. The former can distract the mind from its true goal of transcending the illusory side of life, since they are part of the illusion. They may appeal to negative traits which have been dropped through incessant practice of inner unification, such as pride, hedonism and extroversion - this temptation pandering to the appetite for power and domination. Therefore the student of yoga should remain indifferent to them until he has completely conquered his lower nature, when such powers may be used to help others. The adept will have supreme wisdom which makes misuse of these powers impossible, the importance of siddhis being the means for finding the ultimate truth through self-realization.
Examples of the powers of siddhis are: the ability to withdraw into one's causal body and thus obtain direct knowledge of one's previous incarnations; through knowledge of mental images to establish clairvoyant knowledge of the minds of others; by performing samyama on form, prevent light from the body reaching the eye of another and thereby be rendered invisible; by performing samyama on the nature of change to understand the nature of time itself. There is some discussion as to whether siddhis refer to the physical level or whether they are subtle powers which do not manifest physically.
Narrower Levitation (#HM8634).
Related Samyama (#HH0374) Charisma (#HH0053) Yogic feats (#HH4398) Parinama (Yoga, #HH4561) Tapas (Hinduism, #HH1248) Mystical stage of life (#HM3191) Magico-religious powers (#HH0497) Supernormal powers (Buddhism, #HH5652) Seventh chakra: plane of reality (Leela, #HM0754) Magical-forces awareness (Buddhism, Tibetan, #HM2679).

♦ **HH0382 Tarot** (Tarot)
Book of Thoth
Description The Tarot is an attempt to express the archetypes of the many phases of psychic transformation in direct, visual form. The form is designed to resonate in the mind and feelings of the perceiver without the culture-bound intermediary of languages or codes associated with the linguisitic and literary media of a particular period or school of thought. It does not expound any definite spiritual doctrine, but rather has the purpose of expanding the abilities of a person, particularly those associated with the mental faculties. It is, in effect, a philosophical machine which keeps the attention from wandering while preserving the intitiative and liberty of the mind. As such it is a guide to creative thinking, to the development of concentrated thinking and to the direction of thoughts and feelings into certain channels, as an approach to the ultimate mystery of all-encompassing unity.
Built into the Tarot are certain ideas and conceptions of what a human being is, his functions in the cosmos, the possibilities for development open to him, and how these possibilities may be realized. It provides the possibility of exercises in the practical use of mental equations provided by theosophical operations with the visual forms, leading to stabilization of the mental processes in the individual's mind and to the evolution of relationships of the macrocosm and microcosm.
The instructional aspect is associated with a prediction technique (without which it is suggested that it would not have survived the vagaries of history). The classical Tarot embraces four sub-divisions of occultism which are expressed as alchemy, astrology, kabbalah and magic.
Refs Mound Sadhu *The Tarot* (1962).
Broader Maps of the mind (#HH8903).
Narrower Tarot arcana of conscious states (#HM2097).
Related Kabbalah (#HH0921).

♦ **HH0383 Attending to the needs of superiors** (Buddhism)
Context The fifth of the ten meritorious deeds of South Asian Buddhist thought.
Broader Ten meritorious deeds (Buddhism, #HH0446).
Followed by Transferring merit (Buddhism, #HH1266).
Preceded by Respect for superiors (Buddhism, #HH0604).

♦ **HH0384 Illumined thinking**
Description As an individual's personality is clarified and transformed he is said to become capable of consciously working to lift and enlighten the sum total of human ignorance, fear and greed. This illumined thinking has great energy and is said to be wiser used as a group that by an individual alone, when any insufficiency in selflessness, self discipline or love of humanity might produce dangerous personal results.

♦ HH0385 Individuation

Description Individuation is a process (not a realized goal) in which the ego becomes increasingly aware of its origin from, and dependence upon, an archetypal psyche. The realization of the reality of the psyche, and the new world-view which it makes possible, can only be achieved by the individual alone working laboriously on his own personal development, since individuation implies division from all other beings. Each new level of integration must submit to further transformation if development is to proceed. The individuation urge promotes a state in which the ego is related to the self without being identified with it. Out of this state there emerges a more or less continuous dialogue between the conscious ego and the unconscious, and also between outer and inner experience. A two-fold split is healed to the extent that individuation is achieved, namely the split between conscious and unconscious (which began at the birth of consciousness), and the split between subject and object. The dichotomy between outer and inner reality is replaced by a sense of unitary reality.

Four principle stages are identified: the archetypes of the shadow, the soul-image, the "old wise man" in men or the "magna mater" in women, and that of the self. The process of individuation is symbolized by the trinity archetype, while the quaternity symbolizes its goal or completed state when, through the *transcendent function*, integration is complete. Three is the number for egohood; four is the number for wholeness, the self. But since individuation never truly completes integration, each temporary state of completion or wholeness must be submitted once again to the dialectic of the trinity in order for life to go on.

Refs Adler, Gerhard *Living Symbol* (1961); Goldbrunner, Josef *Individuation* (1955); Mahler, M S *Thoughts about Development and Individuation* (1963).

Narrower Adaptation (#HH1233).
Related Cure (#HH1264) Inflation (#HM1288) Analytical psychology (#HH1420) Ego transcendence (Jung, #HM2230).

♦ HH0386 Mantras

Description Mantras are particular sounds or groups of sounds which are used for tuning and centering auditory perception in such a way that the precise intoning of the selected sounds can induce a definite state of consciousness. In more complex meditation, the sounds (and even the physical shape of the letters representing the sounds in Sanskrit characters) are used as anchoring devices for the movements of consciousness during meditative practice. The use of the mantra is based on the belief that the spiritual teacher is able to determine what particular vibratory pattern is most appropriate for the future development of the person to whom it is given. Concentration of awareness on the mantra causes the meditator's consciousness to be modulated and extended in the desired manner. Particular mantras become associated, during a course of training, with particular features and processes of the meditation experience, thus effectively playing a role in the mapping of consciousness for the person. Mantras represent one of the tantric tools of transformation and integration developed to transmute sensory experience from a binding to a liberating role.

Mantras are said to have originated in the practice of magic and their use, in addition to identification, where consciousness coincides with the reality denoted by the mantra, is cited in the warding off of evil spirits or illness (propitiation) and in the acquiring of magic powers (acquisition).

Although the mantra is extensively used, and occupies an important place, in the spiritual exercises of Hinduism and Buddhism, it is also known in the West as the Prayer of Jesus used within the Eastern Orthodox Church (Hesychasm). It has also been suggested that many spiritually-minded Catholics, particularly in religious orders, make use of the rosary as a form of mantra, a means of keeping the mind in a state of recollection.

Related Affirmations (#HH1453) Mantra yoga (Yoga, #HH0931) Tantra (Buddhism, Yoga, #HH0306) Hesychasm (Christianity, #HH0205) Mantra-induced experience (Yoga, #HM2464).

♦ HH0388 Liberation

Salvation — Collective liberation

Description In his "Perennial Philosophy", Aldous Huxley defines liberation as "the process of waking up out of the nonsense, nightmares and illusory pleasures of what is ordinarily called real life into the awareness of eternity". He equates such an experience with *enlightenment, deliverance, felicity* and *beatitude*. Virtually all spiritual paths have such liberation as their goal, which is said to be realized when the identity of the individual with the universal spirit is perceived. There are and have been many methods of approach - physical austerity and the mortification of the flesh; renouncing preoccupation with the individual self through intellectual mortification; right action; spiritual simplicity. However, one is warned against the illusions of idolatry and superstition - although rites and rituals are said to have their place on the path to liberation.

Different religions and spiritual movements posit different causes for the requirement of salvation. Thus many Indian systems blame *avidya* (ignorance), while the Christian system has the doctrine of original sin. However, individual personal salvation in isolation is not seen as sufficient or even possible. Since human beings are interrelated and interdependent, true liberation requires the collective salvation of all mankind. Both Christianity and Buddhism affirm that their founders rejected personal salvation in order to liberate humanity. This is mirrored in the need for guidance on the path to salvation, whether from guru, shaman or priest for the individual's day to day development; and on some great spiritual leader both for knowledge of the way and for intercession with God.

Refs Avalon, A *Tantra of the Great Liberation* (1972); Evans-Wentz, W Y (Ed) *The Tibetan Book of the Great Liberation, or the Method of Realizing Nirvana through Knowing the Mind*; Huxley, Aldous *The Perennial Philosophy* (1980).

Narrower Personal salvation (Christianity, #HM6335).
Related Mukti (#HH0613) Salvation (#HH0173) Soteriology (#HH1166) Enlightenment (#HM2029) Moksha (Hinduism, #HM2196) Treasures of the godly (#HH0364).

♦ HH0389 Presence

Description Development on the community or historical level can occur when individuals use their very presence to initiate change, acting as a catalyst so that their visible example is followed by others. This is seen in individual actions bearing testimony to human rights, or setting the example in a community project. The individuals concerned are aware, consciously initiating the action, although those following their example may not be so aware.

Refs Beardsworth, Timothy *A Sense of Presence* (1977); Sampson, Tom *Cultivating the Presence* (1977).

♦ HH0390 Five kleshas (Yoga, Zen)

Description The series of five human troubles as given by Patanjali and referred to in Zen literature. These are ignorance, egoism, attachment, revulsion and possessiveness; they are given up in sequence, starting with possessiveness and finally giving up egoism.

Narrower Egoism (#HM3477) Revulsion (#HM3620) Attachment (#HM3119) Possessiveness (Yoga, #HM2911) Ignorance (Buddhism, Yoga, Zen, #HM3196).
Related Abhinvesa (Yoga, #HM3871) Afflictions (Yoga, Hinduism, #HH1047).

♦ HH0392 Kinesics

Body language

Description Different bodily movements, gestures and stances are developed in different social environments to communicate non-verbally. Such body language is a mix of inherited, imitated and learned behaviour; it is frequently more direct than the spoken word, which it may complement or interpret.

♦ HH0393 Scepticism

Description Scepticism is said to be an attitude of permanent doubt which does not allow for certitude but believes that definitive answers cannot be given to major philosophical problems. *Universal scepticism* refers to doubting all knowledge; because of the limits to human knowledge, reason cannot penetrate the transcendental; transcendental knowledge may exist, but beyond sensibility and understanding (Kant). *Partial scepticism* doubts knowledge from a particular source.

Related Doubt (#HM2490) Agnosticism (#HH0904).

♦ HH0394 Psychopharmacology

Description The administration of psychotropic drugs may give relief in cases of mental depression or in other forms of mental illness. Drugs which are or have been used include monoamine oxidase inhibitors (MAOI), tricyclics, lithium salts, and sympathomimetic amines (tranquillizers); bromides, calmatives and nervines (sedatives); amphetamines (stimulants); and hallucinogens.

Related Psychedelic drugs (#HH0075).

♦ HH0396 Self-manipulation

Description Individuals may adopt personal strategies to manage and direct their lives and change their own personalities and behaviour patterns. This may be done to further the person's social, career or intellectual ends. Such self-manipulation may also be undertaken by an individual through self-administration of drugs either to avoid suffering or distress, to improve bodily or mental performance, or to experience altered states of consciousness, including forms of ecstasy.

♦ HH0397 Doctrine of three ages (Christianity)

Description This view of development, from Old to New Testament revelation and culminating in the Age of the Spirit when outward authority would be replaced by inner spirituality, was taught by Montanus and further elaborated by Joachim.

Narrower Age of the Son (#HM3087) Age of the Father (#HM2895) Age of the Holy Spirit (#HM3499).

♦ HH0398 Human potential movement

Description From the 19th century onwards (Gurdjeff), many groups and movements have originated which aim to promote self realization and enlightenment. Many use techniques and practices derived from ancient philosophies and religions (examples: meditation, yoga), while others use totally new techniques (biofeedback, est, re-birthing). Such systems have proliferated since the 1960s and, although some have been attacked as harmful, many of the techniques developed (in particular *encounter groups*) are now used by more traditional institutions.

Refs Drury, Nevill *The Elements of Human Potential* (1989); Leonard, George *The Human Potential* (1968); Scheffler, Israel *Of Human Potential* (1985).

Narrower Latihan (#HH0086) Growth games (#HH0175) Group therapy (#HH0851) Orgone therapy (#HH0327) Social renewal (#HH0542) Encounter group (#HH0162) Gestalt therapy (#HH0751) Group counseling (#HH0227) Nude group therapy (#HH0482) Encounter marathon (#HH0817) Biofeedback training (#HH0765) Experiential therapy (#HH0885) Group-centred therapy (#HH0607) Consciousness raising (#HH0427) Structural integration (#HH0082) Addiction group therapy (#HH0325) Creativity group workshop (#HH0917).
Related New age movement (#HH3172) Humanistic psychology (#HH0793) Existential unity of being (#HH0602).

♦ HH0399 Afterlife

Description A belief in some from of afterlife is common to many religions, most of these considering that conformity or not to some ethical standard during this life will affect the type of afterlife an individual experiences. Some visions of this afterlife describe it in terms of bliss or torment (heaven or hell) in physical terms, others as a dim reflection of earthly existence (Sumerian, Ancient Greek). Related to the concept of afterlife are those of *soul*, generally considered the immortal or indestructible component (Amerindian "free soul", Egyptian "Ba"), and *spirit* (Amerindian "life" or "breath" soul, Egyptian "Ka"). The spirit is frequently seen as remaining for a while on earth near the remains of the physical body, or as returning to earth. This is the basis of funeral and memorial ceremonies in which food and household objects are provided for the use of the departed. Some Christians look on resurrection of the body in physical terms others as a metaphor for some kind of survival of the personality; according to Catholic teaching most souls judged worthy of heaven have to undergo prior purification in purgatory. Hindus and Buddhists look for many reincarnations in this life, but with periods of time in heavens or hells, all depending on the actions in particular embodiments.

Related Soul (#HH0501) Spirit (#HH0770) Soteriology (#HH1166).

♦ HH0401 Purification

Ritual cleansing — Lustration

Description A ceremony of ritual cleansing either from spiritual impurity (sin) or physical impurity is common to most religions and may involve sacrifice, washing, purging, anointing, sprinkling, burning or cutting. Such cleansing may take place subsequent to exposure to some source of physical or ritual uncleanness such as bloodshed (whether shedding the blood of another in battle or otherwise, hunting and killing an animal, giving birth, or menstruating), death, birth, or contact with a place or person considered unclean; or prior to some special rite. Infringement of a taboo, whether voluntarily or involuntarily, may render an individual dangerous to himself and to others until he has been ritually cleansed of its effect. The danger of remaining unclean may include being driven mad, pollution of food, illness or death. Sometimes more than one person, for example a whole village, may be affected, as in the death of a headman, when complex rituals may be required of the whole community.

Means of purification may be: cleansing the body with water; anointing it with oil; burning incense; jumping through smoke and/or over fire. Blood, although also a source of uncleanness, is used in many cultures for purifying the unclean; and sacrifice, whether of animals or (sometimes) humans is a widespread phenomenon. The transferring of guilt to a sacrificial victim or scape goat (as in the Old Testment) is also common. The prime example of this is the crucifixion of Christ, re-enacted in the taking of Holy Communion: "This is my blood of the New Testament, which is shed for you and for many for the remission of sins"; "our souls washed through his most precious blood". *Lustration*, originally a particular ceremony of purification and driving-out of evil conducted every five years in ancient Rome, is the name now appled to any ritual purification involving a slow, solemn religious procession. This is also an example of ritual purification on the basis of the calendar, other examples being ceremonial cleansing at New Year in many present-day cultures

(Peru, Mexico, China) and previously in ancient Babylon and by North American Indians.
Refs Buddhaghosa, Bhadantacariya *The Path of Purification* (1980); Buddhaghosa, Bhadantacariya *Visuddhimagga* (1976).
Broader Rites (#HH0423).
Narrower Purification of the self (Sufism, #HM1495)
Purification of the self (Sufism, #HH5092).
Related Taboo (#HH1101) Purity (#HH0719) Baptism (#HH0035)
Celibacy (#HH0045) Holiness (#HH0183) Spiritual life (#HH0101)
Triple way (Christianity, #HH0631) Forgiveness (Christianity, #HH0899)
Via purgativa (Christianity, #HH4090).

♦ **HH0405 Capacity to become**
Description Human potential may be understood as the capacity to become what at the present time it is not, or to manifest particular attributes or qualities which at the present time it does not. Capacity does not take into account the probability of a particular feature being realized. Such particular features may or may not be welcome, and may be encouraged or discouraged by action before that feature is realized. Such capacity is dependent on time and context, and its attributes and projected changes reflect the assumptions of those defining it. A systematic study would seek to determine factors which prevent the acquiring of conditions which render possible desirable potential attributes, and further, ways in which such factors may themselves be prevented. This is one of three concepts suggested by Israel Scheffler to be useful in defining *human potential*.
Followed by Capability to become (#HH0902).

♦ **HH0406 Agni yoga** (Yoga)
Actualism
Description In this form of yoga, the individual learns to carry the light of his own inner fire into his daily activity as a growing creative expression. Energy is concentrated where thought is focused and thus the inner fire is used to bring light into darkened areas of consciousness. This burns out obstructions to the free flow of energy from inner sources. The actual design of personality systems is in this way illuminated and the distortions and obstructions produced by conditioning eliminated.
Actualism is a teaching derived from agni yoga, with the goal of *unity mergence*; dynamic and magnetic energies are harmonized and balanced and inner wholeness is perceived. This leads to full self–expression and the ability to communicate at every level of consciousness.
Refs Serner, H W *Agni Yoga and Physics* (1973).
Broader Yoga (Yoga, #HH0661).

♦ **HH0407 Maturity as genitality**
Psychosexual development
Description Genetic maturity refers to the ability to reproduce. The psychoanalytic concept of maturity developed from the concept of genital primacy. Genitality is the potential capacity to develop orgastic potency in relation to a loved partner of the opposite sex. The principal characteristic of one concept of maturity is therefore held to be the ability of an individual to attain complete orgastic genital gratification with the opposite sex, since this is the best measure of his ability to surmount the repressive forces of society as well as the press of infantile sexual fixations. Mature sexuality is thus a synthesis of psychobiological activity based in the human body as an organism and an activity as a person grounded in consciousness and culture. Although adult motivation is in this way largely identified with the sex drive, it may be argued that if such a drive is handled in a mature way, it may well harmonize with, and reinforce, general maturity. It is however stressed that it is not in virtue of the fact that genital relationships are satisfactory that other kinds of cooperative relationships by the differentiated individual with differentiated objects are also satisfactory; rather it is in virtue of the fact that such satisfactory mature object–relationships have been established by the individual that true genital sexuality is attained.
Related Sexual maturity (#HH1325).

♦ **HH0408 Kami hotoke** (Japanese)
Sacred
Description Kami refers to the sacred in Shinto beliefs; this may be the beings or gods, the numinous, spirits, personal and impersonal, and also the individual himself when he has achieved a pure, honest and truthful heart. Hotoke are the Buddhist deities, a Buddha who has achieved dharma; so that truth incarnated to enlighten mankind are also hotoke. This concept is extended to ancestors revered for their qualities in this life. Thus the combined concept of kami–hotoke covers the overall concept of the sacred in Japanese thought.
Related Ke hare kegare cycle (Japanese, #HH0031).

♦ **HH0409 Working–through**
Working–over
Description In order to profit from therapy or any system of self discovery, there seem to be periods of inactivity when no apparent progress takes place. Referred to as "working–through", these periods are of assimilation and coming to terms with what has been experienced before new progress can be made. Such periods also occur after crises such as bereavement. *Working–over* occurs when excitations in the psyche are rearranged internally so as to reduce or prevent the harmful effect of being unable to express them outwardly.

♦ **HH0410 Recreation**
Leisure
Description Leisure may be defined as that part of a person's life devoted neither to gainful employment nor to basic nonproductive activities such as travelling to and from work, sleeping, eating and so on. It covers those activities free from the demands of work or duty.
Recreation – literally "re–creation" – can mean simply the use of leisure time in a manner which will renew energies for more efficient working. In this context, the type of activities considered recreational will vary depending on the habitual work carried out by an individual – passive recreation to replenish the body's energy resources; or physical activity, sport, amateur dramatics and so on, depending on the physical, intellectual, emotional work load. Whether social or individual in nature, such recreation would promote harmonious development.
However, both leisure and recreation can be considered on a higher plane. Leisure activities may demand more energy and commitment than a job; they nevertheless are generally accepted to be those activities which constitute a person's happiness. This is because it is during leisure time that those activities connected with a man's highest powers are pursued, activities which transcend the workaday world and have the purpose of achieving self–realization. Re–creation and self–renewal are part of holistic development, perhaps precipitated by crisis, periods of transition in which to reorientate for the future. Holistic recreation centres and retreats offer opportunities for replenishing and reconnecting resources and reviewing alternative new modes, particularly for those facing transitions in job, family, growing–up or dying.
Related Games (#HH0705) Flow experience (#HM2344).

♦ **HH0412 Self–actualization**
Self–fulfilment — Self–realization
Description Self–actualization is the basic tendency of individuals to persist in attempting to develop and manifest their latent potentialities – physical, mental, emotional and spiritual. Rather than being a seeker of homeostatic equilibrium and tension reduction, man is seen as restlessly and creatively evolving higher levels of being within himself, with the material world, with people close to him, and with the human race. A self–actualizing person is characterised by the following:
Capacity for acceptance. This acceptance extends beyond himself, to others and to the world as a whole. The wide–ranging acceptance of the world is possible because he has accepted himself with all his known limitations.
Efficient perception of reality. His essential acceptance enables him to see reality more clearly and have more comfortable relations with it. He sees human nature as it is and not as he would want it to be. He negates the distortion implicit in immediate sensory responses to the world and in the exaggerated inferences derived from such immediate responses.
Spontaneity. He is simple, direct and spontaneous in his responses in most situations he encounters, without the need to engage in tortuous rationalizations. He grows through his mistakes but without anxiety and an oppressive awareness of the opinions of others, or the need for token gestures that restrict meaningful involvement.
Transcendence of self–concern. He centers his attention on non–personal issues and problems that cannot be grasped at the level of egoistic encounters, giving himself the opportunity to extend his mental horizon and recreate his image of the world. Through his concern with wider needs in interpersonal relations and society, he can move freely between larger and more limited perspectives, thereby attaining a clearer perception of what is essential.
Detachment. He places his valuation of being human in a fundamental ground of being that goes beyond the levels at which he interacts with others. His friendships and attachments to family are not of the clinging, intrusive, possessive variety, cringingly dependent on the need to interact with others to the point of psychic exhaustion. He appreciates the need for self–examination and knows that in order to meet this need he requires solitude, privacy and quiet reflection. He is thus able to stand back and view his activities free of ego–centricism.
Independence of culture and environment. He gives existential authenticity to the abstract notion of individual autonomy as an agent, a knower and an actor in society. He has a sharp sense of his own individuality and of the boundaries of himself.
Transcendence of environment. The sense of inner space enables him to recognize more alternatives than appear on the surface and to feel himself capable of choosing meaningfully among them. This awareness takes the form of a freshness that he brings to bear on his appreciation of persons and situations and of particular moments. This freshness is accompanied by a sense of self–expansion and peak experiences, in which he is immersed in the vastness and richness of the world.
Social feeling and compassion. He has a basic ability to relate to many different types of people, to identify with their problem situation and be affectionately supportive.
Deep but selective social relationships. He is capable of unusually close personal attachments whilst handling less deep relationships, outside his chosen network, smoothly and with little friction.
Tolerance and respect. He demonstrates respect for any human being as an individual and irrespective of any special grouping to which he may belong.
Ethical certainty. He is unlikely to confuse means with ends or be unsure about what is right or wrong in daily living.
Unhostile sense of humour. He tends to exhibit thoughtful, philosophical humour, which is intrinsic to the situation rather than added to it, and which is spontaneous rather than planned.
Creativeness. His creativity will enable him to recognize opportunities for growth where other men see only limitations. His style of living has a certain strength and individuality in whatever he is engaged.
It is suggested that consciously or unconsciously every person is seeking some form of self–realization or to become a self–actualizing person, fully expressing his own innate potentialities as an individual, and in full recognition of his own uniqueness as a personality. It is believed that there is a diversity of paths and processes (of possibly unknown extent) by which self–actualization emerges, and that this diversity should itself be protected.
Refs Craig, R *Characteristics of Creativeness and Self–Actualization*; Fisher, Gary *Self–Actualization of Paranormals* (1971); Roberts, T *Beyond Self–Actualization* (1978).
Related Actualization (#HH0764) Perfectionism (#HM0213)
Puja yoga (Yoga, #HH0111) Self–expression (#HH0791)
Self–transcendence (#HH0526).

♦ **HH0413 Three gunas**
Sattva — Rajas — Tamas
Description These are the three fundamental properties underlying the manifest world. Rajas, or activity, and tamas, or stability and inertia, are balanced by sattva, cognition and harmony, the aspect of purity and neutrality, literally "being–ness". The action of the gunas is said to have to produced the entire cosmos: disturbance of the balance of the gunas in prakriti first produces buddhi, with a predominance of sattva; from this arises ahamkara under the impact of rajas; then a prevalence of tamas starts the branching off, with development of the mind and senses on the one hand and of the five potentials, *tanmatra*, and their material elements on the other. The whole creation is also said to be powered by a combination of these gunas, all being present all of the time but in different circumstances different gunas predominating. This is the nature of the phenomenal world, providing purusha, through the organs of sense and the elements, with experience and liberation. The aim of yoga (and of many paths of self realization) is to reduce the rajasic and tamasic elements of the citta, or heart, by increasing the sattvic element and allowing the citta to reflect purusha or consciousness most clearly.
Related Purusha (Yoga, #HM2396) Parinama (Yoga, #HH4561)
Kaivalya (Yoga, #HM3869) Saguna brahman (#HM0576)
Nirguna brahman (#HM4331) Stages of the gunas (Yoga, #HM2805)
Asamprajnata samadhi (Hinduism, #HM3041).

♦ **HH0414 Meekness**
Description The meek person surrenders his immediate interests, not only counting it "better to suffer than to do wrong" (Plato) but by not resisting evil, by refraining from self–assertion. No longer locked in his own ego, he is at one with the Almighty. According to Christian ethics, moral perfection is achieved through humblemindedness, gentleness, consideration for others, reverence, humility, and self respect without vanity, all of which together may be defined as meekness. Usually compared with anger (one of the seven deadly sins) or, as Aristotle put it, in the mean position with respect to anger, meekness is extolled in the Beatitudes – "Blessed are the meek, for they shall inherit the earth". This is not simply passivity, since righteous anger is not condemned, but the absence of the sin of anger, misdirected anger.
Related Passivity (#HH3229) State of anger (Buddhism, #HM2959)
Spiritual meekness (Christianity, #HM6226).

♦ HH0416 Logotherapy
Existential analysis
Description A method of psychotherapy aimed at the activation of meaning in the individual's life. It is suggested that when the individual does not have a meaningful aim in life an existential vacuum is created which makes him prone to neurosis. Tension is perceived as a necessary and healthy result of man's quest for meaning. Anticipatory anxiety which precipitates what is most feared and prevents what is most desired can be confronted and refocused by this therapy; and use of *paradoxical intention*, which is a technique of intending what is feared, helps restore a sense of proportion and of humour.
Refs Frankl, Viktor E *Man's Search for Meaning* (1963).
Related Abulia (#HM0672) Dasein analysis (#HH0695).

♦ HH0417 Self–learning
Self–education
Description Self–learning is a process which, with few exceptions, does not arise from the individual's spontaneous development. Learning to learn denotes a specific pedagogic approach that teachers must themselves master in order to be able to pass it on to others. It involves the acquisition of work habits and the awakening of motivations which are shaped in childhood and adolescence by the programmes and methods in schools and universities; and implies independent study of some field of interest to the learner.
An individual's aspirations to self–learning are realized by providing him with the means and incentives for making his personal studies into a fruitful activity. The diversification of education paths, the increasing facilities available to people who wish to teach themselves, all combine to disseminate the practice and affirm the principle of self–learning.
Refs Edgar Faure (Ed) *Learning to Be* (1972).
Broader Education (#HH0945).
Related Self responsibility (#HH0187).

♦ HH0418 Controlled spontaneity
Description A technique used in martial arts, where the natural movements produced through unity of body and mind are so perfected through long discipline that precise and fully appropriate actions occur spontaneously without conscious thought.
Broader Martial arts (#HH0085).

♦ HH0419 Metatalk
Description By using clichés and careless phrases a speaker can give the impression of meaning something different and more weighty than was actually intended, particularly where social or cultural barriers are being bridged. The intentional use of metatalk to create such impressions can be a powerful tool in influencing a listener or in changing the direction of a conversation.

♦ HH0420 Retreat
Description By retiring to a place of seclusion and solitude, groups or individuals are sustained and renewed in a particular vocation. During the time of a retreat, profound corporate reflections are possible on relatively common situations through a comparison of personal responses. Systematization of retreats is evident in many means of spiritual development, notably in the Spiritual Exercises of St Ignatius of Loyola for the Society of Jesus (Jesuits). Entry into the Society was preceded by approximately one month's retreat divided into four meditations, the first concerning the Purgative Way, the second and third the Illuminative Way and the fourth the Unitive Way, to bring the soul in closer union with God. Regular week–long and other retreats were also conducted. In other disciplines, short periods (even hours or minutes) of retreat from the bustle of life – simply sitting quietly alone – are recommended.
Narrower Spiritual retreat (Sufism, #HH1875).
Related Recluse (#HH0958) Solitude (#HH0333) Practice of silence (#HH0983)
Flight from the world (#HH3442).

♦ HH0421 Management of conflict
Description The management of conflict by maintaining means for containing and resolving it is essential for the maintenance and development of any human system. One method is to restrict the extent to which parties to a conflict have opportunity to communicate information on the subject, whether such "communication" is via force and violence or calculated falsification.
Constraints and alterations which conflict imposes on human relationships are almost more important than the conflict itself, conflict of loyalties being one of the most commonly recognized. Failure to manage conflict results in polarization and possible breaking into two or more systems (as, for example, in civil war or wars of secession). Such breakdown is particularly destructive with regards individual and social development.

♦ HH0422 Affection
Description The humblest and least distinct form of love, said to be the closest to the animal sense of need/gift love inherent in the relationship between a mother and her young, affection contains no cognition or volition. It is very wide in scope and can occur between people of totally different qualities and backgrounds. Such an emotion, in that it requires no specific qualities, can be twisted on these grounds to demand acceptance of undesirable behaviour, or to require evidence of being needed where no need exists. It therefore needs to be balanced by reason and commonsense.
Related Love (#HH0258) Kindness (#HH0457) Affection (#HM5409)
Feeling consciousness (Psychism, #HM3590).

♦ HH0423 Rites
Rituals — Cults — Ceremonies — Sacraments
Description Rituals and ceremonies are part of every formal social or religious activity; repeated, structured practices, whether simply a prescribed form of dress to be worn on such–and–such an occasion or an elaborate form of words and actions built up over centuries of custom. Every society builds its own set of rituals, which activities are closely connected with mythical/spiritual/intellectual beliefs, with religious practice and with behaviour of that society. In traditional societies, virtually every aspect of life has its own religious aspect and associated ritual, with numerous ceremonies and sacrifices, in particular with respect to rites of passage or the transition from one position in society to another.
The importance of such rituals to the individual and to the society of which he or she is a part is greater than might be immediately apparent. They are, as the Book of Common Prayer says of sacraments, "...an outward and visible sign of an inward and spiritual grace... and a pledge to assure us thereof." In other words, they act as a reminder to the participant of the true nature of things, symbolizing very vividly if less accurately facts which might less easily be understood in words. Invariably a society's rituals will stem (or be held to stem) from its gods, heroes or mythical ancestors. This is the divine model or archetype of the ritual, and by these means a society lives out its beliefs – the creation of the universe, the marriage of heaven and earth – whether as a continuation of the cycle of existence or a step on the way to the end of existence altogether. Failure to carry out a ritual or breaking of a taboo may thus be seen as breaking of the moral law and require some ritual act of contrition. Similarly, ritual objects may be the personification of a deity or a country – the latter being true, for example, of the ghost horn of the Kimbu (Tanzania).
The constant repetition of ritual forms is said to produce a lasting effect, that effect which induces a sense of awe in any visitor to a place constantly used for worship, whether or not that visitor is generally psychically "aware". It is also noticeable that neglect of devoted repetition of such rituals diminishes and finally removes completely the numinous presence previously associated with such a place.
Although rituals are very important, they may, in the end, become a stumbling block to those on the spiritual path. This occurs when the repetition of ritual acts becomes an end in itself, producing emotional satisfaction but not spirituality or deliverance, or when the rite is repeated in ignorance, with incomplete or distorted understanding of its significance. It is interesting that such "vain repetition" is condemned by religious writers of many denominations. It can also be a channel for receiving psychic power for its own sake – an argument frequently used against occult practices. Nevertheless there are many seekers of truth who find participation in ceremonies and sacramental rites an effective means of reaching that truth of which the ritual is a symbol.
Refs Bettelheim, Bruno *Symbolic Wounds* (1962); Eliot, John *Rites and Symbols of Initiation* (1965).
Narrower Taboo (#HH1101) Ordeal (#HH1141) Baptism (#HH0035)
Sacrifice (#HH0577) Initiation (#HH0230) Invocation (#HH1876)
Consecration (#HH0238) Purification (#HH0401) Rites of passage (#HH0021)
Ritual pattern–making (Magic, #HH5120).
Related Mana (#HM2686) Magic (#HH0720) Ordeals (#HH0377)
Worship (#HH0298) Symbols (#HH0690) Penance (#HH0790)
Chanting (#HH1257) Obsession (#HM2809) Tradition (#HH0440)
Ill–wishing (#HH3873) Sacred drugs (#HH0788) Vision quest (#HH0897)
Self adornment (#HH0555) Discipline of the secret (#HH0959).

♦ HH0424 The Path
The Way
Description In general, the path or road is used to describe the journey back to perfection from which mankind started, the particular path being different for each individual and depending on previous lifetimes and present goals; there is a sense of peace and confidence when one is on the "right track". The term has come to be specifically connected with the theosophical journey towards a state of supreme bliss and unity with divine consciousness. Having become a disciple of a "perfected man", the individual has four tasks: to know what is real; to reject what is unreal; to acquire some degree of control over thought, control over action, tolerance, endurance, faith and balance; to desire to be one with God. He learns meditation and works for the good of his fellows.
There are four initiations, one at each stage of the path: (1) Casting aside the illusion of self; clearing doubt with knowledge; and clearing superstition with discovery of truth. (2) Consciousness of the shortness of earthly life. (3) Freedom from desire and aversion; clear knowledge deep in the heart of things. (4) Freedom from desire for life and from the sense of individual difference from others.
Broader Spiritual way (#HH1867).
Related Theosophy (#HH0660).

♦ HH0425 Evolution
Description The concept of biological evolution is most commonly associated with Charles Darwin's *The Origin of Species by Means of Natural Selection*, subtitled *or the Preservation of Favoured Races in the Struggle for Life.* Thus evolution is contrasted with the fixity of specially created species and presupposes development. Darwin indicated a mechanism for what had already been postulated, that the variety of life is the outcome of a gradual process of descent from one or more very simple forms.
This apparent upset to established religious ideas, based on the first chapter of Genesis, lead to the presumption of a divine "orderer", development thus following as a consequence of fundamental regularities or laws in the universe. There is no rational interpretation of the laws of natural selection resulting in the survival of the fittest as therefore implying the survival of the best, unless the best is simply defined as the fittest to survive; "Scientific statements of facts and relations... cannot produce ethical directives" (Albert Einstein). Nevertheless, the development of more complex and, arguably, better forms does arise from the simple. Long before "The Origin of Species" Aristotle argued that life arose from metamorphosis of inorganic matter; plants arose first, with powers of nourishment and reproduction; then zoophytes, then animals with powers of sensibility and, to some extent, thought; then man, with the ability for abstract thought. Viewed this way, nature is an evolution from the lower to the higher, a struggle towards perfection. Aristotle himself was unable to resolve the question as to whether the perfecting principle, or *efficient cause*, was an original impulse which thereafter remained outside of nature, or whether it is continuously at work.
In scientific terms, it is clear that mankind has now theoretically the power to decide which species shall survive and, within our own species, to determine genetically how mankind itself will survive. Thus there is an element of choice in the matter of how development will continue; but this choice still has to be based on ethical arguments, the facts of themselves do not determine how the choice should be made.
Refs Jantsch, Erich and Waddington, Conrad H (Eds) *Evolution and Consciousness* (1976); McWaters, Barry *Conscious Evolution* (1981).
Narrower Lamarckism (#HH1673) Social evolution (#HH2653)
Cultural evolution (#HH0926) Evolution by prosthesis (#HH3662)
Evolution of consciousness (#HM2140) Evolution of the human mind (#HH0246)
Technological implications for human evolution (#HH0615).
Related Change (#HH1116) Euphenics (#HH0747)
Human genetic improvement (#HH0357) Cognitive growth and development (#HH0146).

♦ HH0427 Consciousness raising
Description Consciousness raising is a term used to denote the experiential process by which individuals change their whole perception of reality or world view. In this sense, consciousness is not simply a set of opinions, information or values, but a total configuration in any given individual. Included in this idea of consciousness is a person's background, education politics, insight, values, emotions and philosophy, but consciousness is more than these or the sum of them; it is the whole human being and his way of life; it is that by which he creates his own life and thus creates the society in which he lives.
As a mass phenomenon, consciousness is formed by the underlying economic and social conditions. Conscientization philosophy involves recognition of the inevitability of nature compared with the changeability of culture, and the dangers of confusing the two (in particular in the hands of an oppressor). It explores possible alternatives in the current epoch for liberation, assuming a teleological view of human nature and emphasizing the role of consciousness in overcoming "limit situations".
When human civilization changes slowly, the existing consciousness is likely to be in substantial accord with underlying material realities. In a rapidly evolving society, the consciousness of many begins to lag increasingly far behind reality or to lose touch with a portion of reality altogether. Individuals may maintain a level of consciousness appropriate to an earlier society of small towns, face–to–face relationships and individual economic enterprise. They may have a consciousness formed by organized technological and corporate society, but far removed from the realities of

human needs. Such combinations of anachronistic consciousness (characterized by myth) and in human consciousness (dominated by the machine reality of corporate state) may, through the process of consciousness raising, come to be perceived as utterly unable to manage, guide or control the immense apparatus of technology and organization. A new consciousness emerges appropriate to the new social, economic and technological realities.
The present inadequate level of consciousness is characterized by: subordination of man's nature to his role in the economic system (either on the basis of economic individualism or on the basis of participation in organization); domination of the environment by technology; subordination of man to the State (either by the theory of the unseen hand, or by the doctrine of the public interest); conception of man as basically antagonistic to his fellow man; and definition of man's existence and progess in material terms.
The emerging level of consciousness is characterized by: restoration of the non-material elements of man's existence; transcendence of science and technology (restoring them to their role as tools rather than determinants); assertion that the individual should have the power to choose a way of life. It does not describe a society in terms of its structures and institutions but in terms of the way of life, for the locus of any new social order is not in politics or economics but in how and for what ends the individual lives. A fundamental object of the consciousness raising process is transcendence or personal liberation. It is a liberation that is both personal and communal, an escape from the limits fixed by custom and society, in pursuit of something better epitomized by choice of a life-style and the idea that the individual need not accept the pattern that society has formed for him, but may make his own choice.
Consciousness raising groups meet in order that, through the group process, individuals may become more aware of their relation to their social context, political situation, economic situation, sex roles, civil and legal rights, and of the possibilities for change.
Refs Reich, Charles *The Greening of America* (1970).
Broader Human potential movement (#HH0398).
Related Radical therapy (#HH2675) Conscientization (#HH0027) Psychotechnology (#HH4221).

♦ HH0428 Sanctification
Subjective redemption
Description Rather than the realization of moral perfection, sanctification is said to be the free consent by an individual to an act of God. It is based on the idea of setting apart for ritual purposes, of consecrating in the sacrificial sense, and is as open to the psychically tormented or limited personality as to the individual whose balanced personality allows the manifestation of the results of such sanctity. Psychic, moral and religious forces are all called into play. What is involved is the personal consecration of an individual when he or she takes as a personal destiny the obligation inherent in responsibility. This requires total giving and total dedication in each individual situation. Behind this total dedication is a continuous caring for the whole world.
In specifically Christian terms, sanctification is the deliverance of men from guilt and the power of sin, and consecration to love and serve God, through the action of the Holy Spirit. Again, this is the subjective side of salvation, an act of grace, although dependent upon the prior striving and self-dedication of the believer. In these terms sanctification is a continuous process rather than a unique act, intended to be finally complete and perfect whether in this life or the next; but once the process has begun the person can be referred to as sanctified. Thus, sanctification can be contrasted with sanctity or holiness, which are states rather than processes, and with the moral quality of purity.
François Fénelon indicates that even souls perfected in love and the subjects of sanctification do not cease to grow in grace. He indicates two degrees of sanctification, *holy resignation* and *holy indifference*.
Narrower Holy resignation (Christianity, #HM8022) Holy indifference (Christianity, #HM5776).
Related Holiness (#HH0183) Integrity (#HH0234) Prajna paramita (Zen, #HH0550) Imitation (Christianity, #HH0810) Redemption (Christianity, #HH0167) Spirituality (Christianity, #HH0792).

♦ HH0429 Shila (Zen)
Harmony — Good conduct — Morality
Context The second of five perfections or *paramitas* of Zen which, when achieved, render the mind ready for the sixth - *maha paramita* or great perfection, *prajna*.
Broader Paramitas (Zen, #HH0219).
Followed by Kshanti (Zen, #HH0813).
Preceded by Dana (Buddhism, Zen, #HH1234).

♦ HH0431 Intelligence
Description Although the term intelligence has come to imply mental capacity, it was previously used (Thomas Aquinas) to express the spiritual function of the soul as participating in the divine nature, the means by which the soul acquires knowledge of the universal. It would thus be a direct and immortal creation of God. It is contrasted with *oestimativa* or *practical judgement*, the lower form of abstraction found also in animals which distinguishes instinctively and spontaneously the significance of objects for the needs of physical life. The latter implies no knowledge of causes or of relation of means to ends. *Cognitiva*, or understanding, arises during man's natural development, and this includes reasoning, inference from one particular to another or from a series of particulars to a different case of a similar kind. The final step to actual intelligence is made with the conception of general principles, with intuition of the universal and necessary. It is at this third stage that true knowledge and science are said to begin.
Current definitions of intelligence still relate intelligence with knowledge, rationality and the employment of the mind; and tests of intelligence - to reveal an individual's *intelligence quotient* or *IQ* - attempt to measure in one overall figure the varying capacities and potential capacities of the individual to perform different mental tasks. There is considerable discussion as to whether intelligence can be considered as a single, overall concept, or whether such measures as IQ should be broken down to measure ability or capacity in a number of discrete functions. Clearly, some capacities are generally available to the majority of people (it is rare for someone to be incapable of learning speech) whereas others show widely differing individual achievement (for example, music). It is still not fully clear whether each individual has general powers capable of being put to numerous uses or whether a specific individual has a greater inate proclivity to carry out certain intellectual operations whilst being incapable of developing the ability to execute others. These differing abilities have been linked with the functioning of particular areas of the brain; and biochemical, genetic and neurophysiological research aims at giving a scientific basis for the nature of human intellectual competence. For example, the question as to whether different parts of the nervous system are able to carry out a variety of operations or are restricted to particular intellectual operations is being addressed from the level of the functioning of the individual cell to that of the two halves of the brain.
With regard to human development, there is also the question as to how intellectual capacity and potential may vary within an individual under differing circumstances. Current opinion seems to indicate a factor of approximately 0.6 as the heriditary component of intelligence, presumably implying that a factor of 0.4 depends on environmental factors after birth. If intelligence is significantly variable then appropriate intervention at crucial times may yield an individual with greater or lesser range and depth of mental capacities and limitations. This point is important in making education effective in allowing the individual to achieve his full intellectual potential.
Refs Krishnamurti, J *The Awakening of Intelligence*.
Narrower Artificial intelligence (#HH2189).
Related Multiple intelligences (#HH0214) Instrumental enrichment (#HH6131).

♦ HH0432 Social innovation
Description There are enormous possibilities in the technological power available to mankind. These may lead to tyrannical and corrupt systems based on power over nature, or to a society where sources of human misery have been eliminated and each individual may live free from the oppression of war, poverty and disease.
Social innovation is necessary in a society where the transition from civilization to the postcivilization age is taking place. Basic changes are occurring in family and social structure, child rearing, mobility, average age of population and population expansion, religious beliefs and ideology. Not only is there a danger that the technological transitions taking place so differently in communist and capitalist systems may lead to conflict, war, and no stable social system in either case; but also, if there is in fact no conflict, that a single drab uniformity may result. Such uniformity, whether of culture or of race, is contrary to change and development. Hybrid strains need pure strains to be maintained if they are to remain healthy.
The necessity for social innovation is highlighted by the imbalance between the goals of decision-making and influential groups such as trade unions, professional bodies and so on (normally the exploitation of technology to improve material well-being) and the goals of individuals within such a group (where the emphasis may be on ethical, aesthetic and spiritual considerations). The result is a rejection by the individual of the social consensus of the group of which he is a part, while at the same time the determination to profit to the maximum from the material advantages offered by membership of such a group.
Attempts at harmonizing such imbalance have focussed on:
- The definition of a *social added value* similar to *economic added value*, where instead of considering the controlled energy applied in materials handling, the controlled energy applied by society as a whole (social and human activity, use of collective equipment) would be applied to education, health, cultural and civic activities.
- The financing of projects on dissemination of social innovation, the protection of legal and moral rights of social innovators and the determination of a means for evaluating an innovation and its market possibilities.

♦ HH0433 Confirmation
Description The individual act of confirmation derives from the laying on of hands or other action signifying the conferring of the gift of the Holy Spirit. It is closely connected with baptism and the two in many instances form a single rite.
Related Baptism (#HH0035).

♦ HH0434 Interpersonal therapy
Description A neurosis affecting more than one person - for example, several members of one family - may be treated by a therapist having a sequence of individual sessions with those involved before subsequently returning to his original patient, the cycle being repeated if necessary.
Broader Therapy (#HH0678).

♦ HH0436 Human development (Sufism)
Erfan — Tasawwuf
Description The main aim of Sufism is *self cognition*, or inward cognition of truth, based on the teachings of the Prophet Mohammad, "Whoever knows himself truly knows his God". There is awareness of the paradox that God is both infinitely transcendent and also closer to man than his own body. Awe and fear at the tremendous mystery of God is combined with a sense of this divine closeness and desire for greater intimacy with the source of inexhaustible mercy and loving concern. Seeking oneness with the Divine, the Sufi is able to discover the truth. This search or mystical pilgrimage begins with the stripping off of all earthly desires, progresses through love and knowledge, and a sense of amazement, to ultimate annihilation in uniting with God. This unity is the law of Sufism, which sees no opposition between mind and matter, or between physics and metaphysics. Mind, body and spirit are one complex unity.
The Sufi aims to be devoted to God, preferring God to everything and preferred, loved by Him. Pleased with all God does, God is pleased with all he does. He becomes dead to his own attributes, the Beloved's attributes manifesting through him. His inner and outer self are both totally surrendered. Love for the Absolute and unity as of a lover with the beloved are central, as is the path by which the Sufi is transformed spiritually. Perfection is not achieved through the orthodoxy of religion or belief in God from a ritual standpoint. These are not really true. Through Erfan, or the reality of religion, each individual is introduced to his inherent values and true personality. He is trained to develop all his creative abilities so as best to benefit from the resources provided by nature and live in peace, knowledge and justice. Emphasis is on the need for a spiritual guide - "whoever travels without a guide needs two hundred years for a journey of two days". There are numbers of mystical "schools" which describe journeying on this path and the stations or maqamat on the way; but all describe the states of *fana*, annihilation in the divine, and of *baqa*, of permanence or abiding.
The Sufi way of life relates to the inward transformation of man. Realization of moral ideals and higher values is from the pursuit of inward purity rather than from theoretical principles and doctrinal knowledge. Truth - al-Haqq - is understood from experience and from living with it. Emphasis is on discipline of the soul and personal experience - love of God and of his creation leads automatically to following of a moral path and loving one's fellow men. Training is through control of the desires of the lower self and discipline to subdue worldly temptations. Rather than burdening others, the Sufi takes on others' burdens. He treats others as he would wish to be treated, with unfailing kindness and with good manners. The heart is purified from recurrence of inborn weakness, natural characteristics are left behind, attributes of human-ness are extinguished and the Sufi holds aloof from temptations of the senses.
From the ethical standpoint, the Sufi system of progress towards unification with God considers moral attitude as the most important condition for the attainment of spiritual perfection. Moral development implies the purification of the soul necessary for realization of divine attributes. From the psychological standpoint, saintly progression towards God is through the experience of mujahada, mortification of the self, passing through several psychic states to attain the purity of heart to reach higher spiritual states and eventually leading to spiritual perfection. From the philosophical standpoint, the outward qualities of the ego are differentiated from inward manifestation, so that a true relationship is experienced with God, leading to annihilation of human qualities and consciousness of Godly attributes with subsistence in the creative truth.
Refs Arasteh, A Reza *Growth to Selfhood* (1980); Arberry, A J *Sufism* (1972); Bhatnagar, R S *Dimensions of Classical Sufi Thought* (1984); Cassim, K M P Mohamed *Sufism*; Stoddart, William *Sufism*; Trimingham, Spencer J *The Sufi Orders in Islam* (1971).
Broader Human development (Islam, #HH1799) Human development through religion (#HH1198).
Narrower Fana (#HM1270) Sabr (#HM3418) Ma'rifa (#HM2254) Tasawwuf (#HM0575) Divine path (#HM3371) Love of God (#HM4273) Inward purity (#HM3602) Spiritual master (#HH2080) Self-mortification (#HH0464) Fountains of light (#HM3039) Naqsbandi tradition (#HH0277)

Spiritual experiences (#HM3542) Sufi infused awareness (#HM2476)
Sufi path to perfection (#HH1747) Sufi contracted–consciousness (#HM2800)
Stations of consciousness – Ansari (#HM2317) Stations of consciousness – Simnami (#HM2341)
Ultimate Sufi state of consciousness (#HM2318)
Stations of consciousness – ibn–Abi'l–Khayr (#HM2424).
Related Spirituality (Islam, #HH5902) Creative imagination (#HM0786)
Self–cognition (Sufism, #HM3604) Unification with God (Sufism, #HM3864).

♦ HH0437 Economic and social development
Development
Description Economic and social development is generally accepted as meaning first and foremost economic development. It implies an effort on the part of each country, where necessary with outside assistance, to take stock of its natural resources and to develop them to their fullest extent. With the birth of the *new international economic order*, this responsibility is extended beyond national boundaries.
Economic and social aspects of development are considered to be so closely interwoven that frequently it is not considered useful to make any distinction between them. The ultimate object of development is to bring about sustained improvement in the well–being of the individual and bestow benefits on all.
The concept of development is in some cases considered to be broader, deeper and more varied, going beyond the purely economic aspects of improving man's living condition. Man is considered to be both the means and the end of development, not a one–dimensional abstraction of homo economicus, but a living reality, a human person, in the infinite variety of his needs, his potentialities and his aspirations. Development is meaningful only if man, who is both instrument and beneficiary, is also its justification and its end. It must be integrated and harmonized; in other words, it must permit the full development of the human being on the spiritual, moral and material level, thus ensuring the dignity of man in society through respect of individual human rights. Development is not development unless it is total, and cultural development is considered by some to be a part of total development.
Refs Fathi El–Rashidi *Human Aspects of Development* (1971).

♦ HH0439 Sociopsychosis
Socioneurosis — Ethnopathology — Sociopathology
Description This may be described as the communication of unhealthy attitudes to the individual by a sick society, due to the reciprocal relationship between an individual and his surroundings. Symptoms are a general increase in crime, drug abuse, violence, sexual promiscuity, and the breakdown of family relationships. This decline in moral standards is shown historically to be related to an increase in material prosperity which confronts the individual with a multitude of different values and standards. The loss of faith in spiritual reality causes internal confusion for which no solution is in evidence.

♦ HH0440 Tradition
Description Tradition may be said to encompass symbols, myths, rituals, outlooks and behaviour which structure the social and cultural circumstances in which an individual lives and ensure his connection with the past. As such it is a manifestation of human creativity and a force in shaping and renewing individuals and in determining the way they interact. Ceremonies and rites of passage weave together traditional culture; changes in traditional symbols and methods demonstrate the effect of religious change and the cultural impact of new orientations.
Refs Smith, Huston *Forgotten Truth* (1976).
Related Rites (#HH0423) Transformation (#HH1039).

♦ HH0441 Repentance
Description Literally "to change direction", repentance has came to mean facing up to and feeling remorse for some negative action/emotion and making a new start. It is seen as a necessary pre–requisite for spiritual and emotional health; the soul breaks away from the past so as to set out on ethical reformation. Thus it is man's nature as a moral being and his power of self–judgement that makes him capable of repentance. Socrates said that only by being convinced of one's own ignorance could one gain knowledge; and Plato that the potential faculty within everyone to distinguish the lesser from the greater good made it possible to renounce the lesser for the greater.
In Judaism, Islam and Christianity, repentance is not just a first step but a permanent condition for spiritual achievement. Indeed, Luther maintained that the whole of life should be a penitential act – sorrow for sin and faith in Christ; while repentance was a true change of mind. Repentance is thus the answer to the question of past sin, as to how the individual can be freed from the burden of wrongdoing done of free will and the creation of free spiritual activity. Repentance for past wrongdoing shows that the sinner has risen above his previous personality; what he condemns in himself cannot be his true self and this leaves the true self free to repent or, as in Islam is the meaning of the word "tawbah", to "turn to God". In Sufism, this turning to God is a favour granted by God. Repentance from sin leads to contrition which supplies the moral strength needed on the way to God.
True repentance brings new insight and illumination; not only the anxiety to make atonement for the person wronged or the spiritual order violated, but also a readiness for any task. There is an insight into duty and a new energy and inspiration of will.
Related Penance (#HH0790) Tawba (Sufism, #HM4062) Spiritual healing (#HH0458)
Metanoia (Christianity, #HH0888).

♦ HH0442 Spiritual direction
Spiritual guidance — Spiritual guide — Spiritual preceptor — Spiritual mentor — Educative spiritual direction
Description In order to guide a soul on the path of perfection, most teachings demand direction by someone further along that path. This will be a person not only having greater knowledge and insight but also greater holiness. Guidance may be through counselling, through confession (as in the Catholic church), and through spiritual direction. The guide is the archetypal intermediary between the material and the spiritual; passing the understanding of the teaching and spiritual discipline on to successors, both as written scriptures and as an oral tradition outside the written scriptures, is recognized as a vital and precarious process.
Although much has been written down for the direction of others, the written word does not replace the personal guidance of another. "The Cloud of Unknowing", for example, written as guidance in prayer and contemplation, is clear about the discrimination required in judging how useful its contents are at any given time and assumes that the reader also has regular personal meetings with a spiritual director. Although written direction may be systematic and ordered, it lacks the immediacy, the give and take, of personal meetings.
Many systems agree that there are very rarely those who have their teacher within themselves; disagreement arises as to whether this can be sufficient. In the Ch'an Buddhist tradition, it is said that a self–awakened person cannot rely on teachers from outside; but others suggest that the inner guru leads the seeker to an outer guru, and the outer guru reveals the inner guru. According to Sufism, when a seeker has no guide (or *shaykh*) then Satan becomes his shaykh. In the Catholic tradition, although the 14th and 15th centuries were times when spiritual guidance was practised extensively, following the Council of Trent (1545–1563) the practice narrowed to a role of guarding orthodoxy and avoiding both heresy and questionable mysticism.
The vocation of spiritual guidance has been considered one not to be undertaken lightly – mere age does not qualify, but spiritual discrimination and experience. In addition, having found a teacher the testing of a follower's resolve may involve him in seeming abuse or rejection before being accepted, particularly as the guide may be considered to become accountable before God for his disciples as well as for himself. While a master is not bound to accept an individual as disciple, once he has been accepted the disciple must not have concealed from him the spiritual chain represented by the master nor his spiritual predecessors.
Refs Coelho, Mary C and Neufelder, Jerome N (Eds) *Writings on Spiritual Direction by Great Christian Writers* (1982); Cullingan, Kevin (Intro) *Spiritual Direction* (1983); Kelsey, Morton T *Companions on the Inner Way* (1983).
Broader Guidance (#HH0178).
Narrower Practice of the confidence (Christianity, #HH3007).
Related Guru (#HH0805) Mentor (#HH3219) Channelling (#HH0878)
Spiritual way (#HH1867) Hasidism (Judaism, #HH0597)
Soul care (Christianity, #HH1199) Heteronomous discipline (#HH0925)
Mystical journey (Christianity, #HM0402) Discipline of guidance (Christianity, #HH1244)
Meditation way of the bodhisattvas (Buddhism, #HM2769).

♦ HH0444 Reality
Description Pluralistic philosophy assumes no unity in reality, and exalts the many above the one. By contrast, monistic belief in the eternal, non–temporal, as the ultimate reality puts present life either as a preparation for union with that eternal or as a continuous cycle of suffering life and death. It has been said that by viewing reality through three essential aspects – the pure non–temporal, the dynamic universal and the unique individual – the monistic and pluralistic approaches are transcended and *being* is experienced beyond strictly human categorical schemes.
Refs Berger, Peter L and Luckmann, Thomas *The Social Construction of Reality* (1972); Muses, C and Young, A M (Eds) *Consciousness and Reality* (1974); Watson, Thomas *The Doctrine of Repentance* (1987).
Narrower Social reality (#HH0714).

♦ HH0445 Dance
Dance therapy
Description As part of a ritual or recreative activity, dance may produce an inner calmness or an altered state of consciousness. Dance therapy is said to promote emotional and physical integration and may be used in two ways: (a) It may be used to encourage discovery of form and pattern through movement. The individual is encouraged to move with the music according to his own spontaneous inspiration, thus developing self–awareness. (b) Alternatively, the therapist may require the individual to perform a series of precisely defined movements until some degree of competence is achieved to the satisfaction of both the individual and the therapist. In both cases the individual becomes more aware of the body, learns to communicate through the body and ultimately improves his or her feeling of self worth.
This therapy has also been used as a means of encouraging individuals to take a renewed interest in their surroundings when they run the risk of sinking into a state of apathy (as in psychiatric institutions, convalescent homes and old people's homes).
Broader Therapy (#HH0678).
Related Mudras (#HH0286) Dance–induced experience (#HM3213)
Bodily–kinesthetic intelligence (#HH0662).

♦ HH0446 Ten meritorious deeds (Buddhism)
Dasakusalakamma
Description According to South Asia Buddhist teaching, merit may be achieved through: (1) Giving (dana); (2) Observing moral precepts (sila); (3) Meditation (bhavana); (4) Showing respect to superiors (apacayana); (5) Attending to the needs of superiors (veyyavacca); (6) Transferring merit (pattidana); (7) Rejoicing at the merit of others (pattanumodana); (8) Listening to the dharma (dhammasavana); (9) Preaching the dharma (dharmadesana); (10) Having right beliefs (ditthijukamma). Acquiring merit is an activity both of lay persons and of monks.
Broader Merit (#HH0859).
Narrower Bhavana (#HH0551) Right beliefs (#HH0974)
Dana (Buddhism, Zen, #HH1234) Transferring merit (#HH1266)
Preaching the dharma (#HH1022) Respect for superiors (#HH0604)
Listening to the dharma (#HH1183) Observing moral precepts (#HH0278)
Rejoicing at the merit of others (#HH0932) Attending to the needs of superiors (#HH0383).
Related Formation of merit (Buddhism, #HH5122)
Formation of demerit (Buddhism, #HH3226)
Formation of the imperturbable (Buddhism, #HH5543).

♦ HH0447 Human psychological development
Personality development — Human behaviour development
Description Human development is a continuous dynamic process through which an individual's capacities emerge, expand and are integrated, to enable him to cope with the external world. Results of the process are evident in changes in the form, structure and functioning of the individual. Internal genetic factors which stimulate the growth process and set its limits include both physical and mental capacity; and external forces impinging on the process include the physical environment (starting from the prenatal period and continuing throughout life) and the social environment, represented by the individual's interactions with other people.
Controversies concerning the nature of this process exist in the following areas:
1. The degree to which the child is viewed as active or passive in response to his environment.
2. The degree of continuity to be expected in psychological development (whether by gradual accretions, in sudden dramatic shifts, or segmented into stages).
3. How much development proceeds towards an idealized goal, rather than being an open–ended programme with no special terminal state defining the goal of growth.
Development in this context can be endowed with wide connotations or it can be given a limited meaning within a highly restrictive framework. The precision of definition within any particular theory often depends on whether the writer is more interested in describing the achievement of broad stages or plateaux of behaviour, or in the mechanisms which apparently govern the transitions between stages. How development is defined subsequently restricts what is then observed.
Refs Tanner, J M and Inhelder, B (Ed) *Discussion on Child Development* (1956).
Related Behaviourism (#HH1150) Spiritual life (#HH0101).

♦ HH0449 Surrender
Spiritual surrender
Description According to Vedic law, surrender of the will is part of the first *commandment* of all. In Christian teaching, no good or perfect work can be done without resignation or surrender of the individual will to the will of God; while surrender is the actual meaning of the word Islam. In psychology, similarly, the will to surrender his or her neurosis is the first sign that a mental patient is willing to be cured. However, schizophrenic surrender is a term used to describe passive

repression without any attempt by the schizophrenic to regain the objective world.
Related Sacrifice (#HH0577) Self–oblation (#HH0968)
Human development (Islam, #HH1799).

♦ **HH0452 Taraka yoga** (Yoga)
Description A version of yoga in which the ascent of attention is achieved through light phenomena.
Broader Yoga (Yoga, #HH0661).
Related Mystical stage of life (#HM3191) Light–induced experience (#HM3559).

♦ **HH0455 Togetherness**
Description A term in psychology to describe the earliest stage in the development of the personality when the child has little or no recognition of the self but is in a state of primary psychic union with its mother.

♦ **HH0456 Achievement motivation**
Success motivation
Description Achievement motivation may be characterized by the tendency of individuals to maintain and increase proficiency in all areas in which a standard of quality is taken as binding. The strength of such motivation varies from individual to individual, but remains relatively constant for a given individual. This seems to be dependent on early childhood experiences, particularly the way in which a child learns independence. Success motivation is also closely correlated, with failure–motivated individuals having unrealistically high or low levels of aspiration and willingness to take risks. Achievement motivation has been shown to be significant for the process of economic development; and programmes have been developed to alter motivation in this connection.
Refs McClelland, D C, et al *The Achievement Motive* (1953).
Broader Motive (#HH0181).
Related Success (#HH3188) Ambition (#HH6089).

♦ **HH0457 Kindness**
Description The term kindness originally indicated kinship or near relationship. It has come to refer to a complex emotional state, one of the dispositions constituting love, and is particularly used with respect to the love of a mother for her child. It has been postulated that as mankind progresses morally, then the circle within which kindness manifests also widens from the home ultimately to include all humanity and all creatures. Kindness does not necessarily entail softness or sentimentality, as is inherent in the aphorism "being cruel to be kind"; but is closely associated with pity and moral indignation. It may sometimes be used as a synonym for *affection*.
Related Affection (#HH0422) Love consciousness (#HM2000).

♦ **HH0458 Spiritual healing**
Faith healing — Psychic healing — Trance healing — Mental healing — Affirmative healing — Psychic healer — Spiritual therapy
Description In general this refers to healing of disease through psychic powers or supernatural agencies. The practitioner (whether medicine man, witch doctor, shaman, faith healer, or one who works through the laying on of hands or through prayer) may be a medium who goes into a deep trance and through whom the healing spirit is said to work.
Traditional healing has evolved through working out of cultural patterns over centuries and may use the motivation behind the cause of the illness as the means of cure. Its tools may be; surgery; herbal or pharmacological; prayer or incantation. Where disease is interpreted as of religious or magic origin, then clearly the method of healing will also involve religion or magic. This may require the propitiation of a dead person, exorcism of a supernatural agency, or confession and penance for wrongdoing. In traditional society, the healer will be one who knows the hidden powers of nature and can apply them, such ability being handed on from one healer to another, often within a family. It may be that such powers are divinely derived, perhaps from shaman–like visitation to other worlds. While in a trance state, a shaman may challenge or willingly be taken over by the spirit causing the disease and, in overcoming this spirit, cure the diseased person – usually of a mental or psychosomatic disorder.
In the Church, spiritual healing is primarily and originally intended as relief from suffering caused by emotional or internal hurts and the saving of the soul, but is also concerned with the healing of the physical body. Depending on the particular problem, different methods are used; but prayer in conjunction with the laying on of hands, and possibly anointing the forehead with oil, is typical. When an emotional factor is involved, such as resentment or bitterness, then some form of repentance precedes the ceremony. Simple physical cure is said to be incomplete as healing goes, the question being whether the previously sick person has been helped to a recovery which allows growth in love and service both of God and of his fellow men. The exercise of the spiritual gift of healing does not imply a guaranteed cure for all who benefit from the ministry – simply trust in God's care whatever the result.
The gift of spiritual healing is distinguished from faith healing – spiritual healing does not seem to depend on particularly strong faith on the part of the sick person, whereas faith healing and healing by miracles depend on belief of the person practising the cure or on the faith of the patient in the practitioner. This is particularly true in the case of so–called *affirmative* or *mental healing*, when the individual is healed through fixing attention on positive virtues and thus excluding negative thoughts which inhibit capacity for good health, perhaps through repeating daily a group of words signifying good health. This latter technique may work by inculcating a change in beliefs which bring conscious and subconscious in line with bodily health, by auto-suggestion, or by consciously telling the body cells of the area that requires their healing. The power to heal may thus be within the sufferer, having to be activated by the action of the healer. *Mental self healing* uses the mental healing technique to cure the person exercising the technique. Many individuals within and outside the Church have the ability to effect cures by "faith, piety and prayer". Some religious people put forward the caveat that physical healing which does not change the spiritual state of the sufferer may even in the long run be harmful and may be the result of evil forces. Spiritualism asserts that healing ability is dependent on acceptance of the knowledge that all are fragments of the universal whole and can be controlled in accordance with divine will.
Refs Bletzer, June G *The Donning International Encyclopedic Psychic Dictionary* (1987); Eddy, Mary Baker *Science and Health with Key to the Scriptures* (1875); Goldsmith, Joel S *Realization of Oneness* (1974); Heijkoop, H L *Faith Healing and Speaking in Tongues*; Inayat, Khan *The Development of Spiritual Healing* (1985); Jeter, Hugh P *By His Stripes* (1977); O'Neill, Andy *Power of Charismatic Healing* (1985); Peel, Robert *Spiritual Healing in a Scientific Age* (1987).
Broader Holistic medicine (#HH0581).
Narrower Folk healing (#HH0248) Spirit healing (#HH2997)
Magnetic healing (#HH2552).
Related Trance (#HM3236) Healing (#HH0216) Repentance (#HH0441)
Absolution (#HH0472) Spiritualism (#HH0479) Primal healing (#HH2314)
Spirit possession (#HH0056) Psychic health and spirituality (#HH0109)
Charismatic renewal (Christianity, #HH3124)
Communication with supernatural beings (#HH1306).

♦ **HH0460 Seven stages of life**
Description According to Master Da Free John, although to the fully enlightened adept there is no need for a spiritual path of seeking, since the seeker is always already in the condition to which he aspires, there appears to the unenlightened ego to be a series of stages in which realization is achieved. In fact, these are said to be the progressive falling away of illusions or superimpositions on reality. Most individuals are arrested at the first three stages, physical, emotional and mental. Only when maturity at these three stages is reached, with the will directed away from self–centredness towards the social and religious, does further self–development take place.
Refs Da Free, John *The Basket of Tolerance* (1989).
Broader Stages of spiritual life (#HH0102).
Narrower Mystical stage of life (#HM3191) Vital–physical stage of life (#HM3320)
I–transcendent stage of life (#HM3060) Heart–awakening stage of life (#HM3518)
Emotional–sexual stage of life (#HM2954) Mental–volitional stage of life (#HM3464)
Unconditional nirvikalpa samadhi (#HM3619).

♦ **HH0461 Human potential**
Description This is the potential of individuals for self–fulfillment, self–realization or self–actualization. Self–actualizing individuals are characterized by: capacity for acceptance, efficient perception of reality, spontaneity, transcendence of self–concern, detachment, independence of culture and environment, transcendence of environment, social feeling and compassion, deep but selective social relationships, tolerance and respect, ethical certainty, unhostile sense of humour, creativeness.
Realization of human potential is also characterized by expression and defence of values such as: truth, goodness, beauty, wholeness, aliveness, uniqueness, perfection, completion of growth and development, justice and order, simplicity, richness and variety, effortlessness and unstrained action, playful amusement, and self–sufficiency.
Achievement of full human potential is also characterized by higher or transcendent states of consciousness giving: an awareness of undifferentiated inner unity, a sense of underlying oneness behind the empirical multiplicity of external sense impressions, a sense of transcendence of space and time, a deeply felt positive mood of joy and peace, a sense of sacredness, a heightened sense of objectivity and insight into reality and a transcendence of self.
Every human being, in order to maintain a state of health at all levels, requires to mitigate the effects of low potential in specific areas and to augment those areas where his potentiality is high. These requirements have been defined as basic human needs. Various definitions or divisions of potential into specific areas have been postulated, with every potential having its corresponding need. The favourable resolution of a potential/needs polarity results the development of a particular strength or virtue. Explorations of the nature of human potential may occur: as research, through humanistic and transpersonal psychology; as therapy, through some schools of therapy; or through a variety of personal growth techniques.
Refs Drury, Nevill *The Elements of Human Potential* (1989); Leonard, George *The Human Potential* (1968); Scheffler, Israel *Of Human Potential* (1985).

♦ **HH0463 Attentiveness to nature** (Zen)
Description This quality is emphasized in Zen and is similar to being in sympathy with or empathy with nature; it includes awareness of another's point of view and the ability to put one's self in his place.
Related Empathy (#HH0145) Communion with nature (#HM0915)
Contemplation of nature (#HH0580).

♦ **HH0464 Self–mortification** (Sufism)
Mujahada — Subjugation of the nafs
Description This is recommended by the Sufis for the training of the soul. It assists in realization of perfect resignation from the empirical self. The carnal self (nafs) is subdued by living in seclusion, practising abstinence from food and, through stilling of external activity and of internal appetites and emotions, living a life of quietism. Dwelling in a state of purity, he may achieve a higher stage of self–mortification when spiritual perfection is gained through complete surrender to God. Negation of nafs, emptying one's self of self, brings the discovery of the true self.
Broader Human development (Sufism, #HH0436).
Related Quietism (#HH2006) Mortification (#HH0780) Mujahadat (Sufism, #HM5215)
Purification of the self (Sufism, #HH5092).

♦ **HH0465 Spiritual marriage** (Christianity)
Spiritual union
Description The goal of Christian mysticism, spiritual marriage is seen as a permanent transformation of an individual's whole being through the love of God. This union occurs with complete detachment and perfect resignation from everything but God, and cannot be perfect unless the soul itself is perfectly purified and clean.
Related Triple way (Christianity, #HH0631) Mystic union with God (Christianity, #HM3889).

♦ **HH0466 Conscience**
Description Conscience as a term has a range of related meanings which include: awareness of the nature and origins of rules of conduct, and consequent ability of the person to evaluate his own actions and intentions; avoidance of proscribed behaviour and tendency to act in meritorious ways; and feelings of obligation, guilt and remorse. It has even been argued that conscience is what distinguishes mankind from artificial or computer intelligences (although conscience has been postulated as theoretical design features for them too).
Different schools of thought differ on whether conscience and moral values are innate human qualities which are always present (possibly in response to mankind's relationship with God); or whether they derive primarily from childhood restrictions and prohibitions, conscience developing in adolescence from "must not" to "should not", and taking on the feeling of obligation. Conscience is generally understood to give intuitively authoritative judgments regarding the moral quality of a particular action.
One interpretation is that conscience is the awareness of responsibility arising in the experience of freedom, and it has been equated with the heart of the individual. Action in accordance with conscience then becomes action from an inner conviction, the establishment of which implies giving depth and precision to reflection on reality.
Refs Aronfreed, J *Conduct and Conscience* (1968).
Related Freedom of conscience (#HH2344).

♦ **HH0467 Apathy**
Description In situations where an individual has little control over circumstances and is unable to participate in affairs which affect him he has no motivation to learn about such affairs. This lack of education leads to even less ability to participate, and thus a cycle of apathy is set up.
Related Patience (#HH1304) Myth of incompetence (#HH0322).

♦ **HH0470 Character defects** (Buddhism)
Description Buddha listed ten fetters or character defects which obstruct essential realization. When the first five, representing the following of old, unphilosophical mental habits, are overcome, the seeker is an *Arhat*, or ready to face the next five.

Narrower Spiritual pride (#HM2852)
Emotional aversions (#HM3504)
Dependence on ceremonies (#HM0509)
Delusion of the personal self (#HM3152)
Doubt of the efficacy of the good life (#HM1294)
Ignorance of the nature of the essential self (#HM3493).
Emotional desires (#HM3301)
Desire for life in form (#HM3583)
Desire for formless life (#HM1261)
Notion of one's self as entity (#HM3612)
Related Arhat (Buddhism, #HH0233).

♦ HH0471 Self-concept
Self-identity — Ego-concept
Description According to one view, the concepts of self and of ego are interchangeable terms; other views differentiate them in many ways. The former view holds that self (or ego) is not a unitary structure appearing fully-formed in the infant. It develops first as a perceptual system based on differentiation of the individual's body from his surroundings, and becomes a progressively more complex conceptual system. It is therefore the totality of attitudes, judgements and values of an individual, relating to his behaviour, abilities and qualities, including the awareness of these variables and their evaluation. Self concerns become involved in the operation of motivational urges and the regulation of many psychological processes such as perceiving, learning, evaluating and remembering.
The unique formation that takes place as the self or ego of the human person therefore consists of a set of attitudes that define his stabilized bearings in relation to the physical and social surroundings. These attitudes include the individual's cherished commitments, stand on particular issues, acceptances, rejections, expectations in interpersonal relations, identifications with persons or values, and personal goals for the future. All these set him apart as a person with a unique sense of identity.
Where a distinction is made between self and ego, self may be used to refer to the individual as known to himself and ego to mean the group of the individual's activities concerned with the enhancement and defence of the self. Another view is that the self is made up primarily of perceptual components and ego consists of these and of affectively changed conceptual components. Again, self is said to be the real individual, while ego is a superimposition isolating the individual and confusing illusion with reality. Other variations have been proposed.
Refs Hartley, Margaret *Changes in the self-concept during psychotherapy* (1951); Zuger, B *Growth of the Individual's Concept of Self* (1952).
Related Superego (#HH0493).

♦ HH0472 Absolution
Description When an offender has repented of sin, confessed it and is willing to make reparation, then he may be granted absolution. Such absolution is often associated with, and may be a prerequisite for, healing and exorcism.
Related Penance (#HH0790) Spiritual healing (#HH0458) Conviction of sin (#HM1160)
Personal absolution (ICA, #HM2370).

♦ HH0474 Life themes
Description Life themes are the major preoccupations or concerns an individual develops which then affect his behaviour in other spheres. They demonstrate the way in which the person develops psychologically following infancy and help to explain differences in personality and social adjustment. They may also be used to show how each stage of development has affected the following stages and why the characteristics of an adult have developed one way rather than another, so that resulting perceptions of the same situation may be widely different. This is therefore a way of demonstrating how each individual has a different personality; similar personalities would be expected to have similar life themes. It is considered preferable to classifying people by personality "type", "traits" or development history.

♦ HH0475 Biological rhythms
Biorhythm
Description Rhythm underlies most of the processes by which man is surrounded. There are rhythms of: gravity, electromagnetic fields, light waves, air pressure, and sound. The rotation and revolution of the earth and the revolution of the moon lead to rhythms such as alternation of light and darkness, tides and seasons. Plants, animals and man respond to these various rhythms by rhythmic alterations in behaviour. Individuals fluctuate in energy, mood, well-being, and performance each day. There are longer, more subtle behavioural alterations each week, month, season and year.
Study of such biological rhythms provides new understanding of many aspects of human variability in: symptoms of illness, stress, (psychosomatic and emotional disease), sense of time and timing, learning, response to medical treatment (including drug toxicity), and job performance. It is suggested that there may be in a man a "combination lock" to his activity and rest, his moods, his illnesses, and his productiveness. By cultivating his understanding and awareness of these rhythms, he may learn to utilize his subjective sense of time.
Refs Gaer Luce, Gay *Biological Rhythms in Human and Animal Physiology* (1971); Reinberg, A and Ghata, J *Biological Rhythms* (1965); Sollberger, A *Biological Rhythm Research* (1965); Weraliz, Hans J *Biorhythm* (1961).
Narrower Aesthetic perception (Psychism, #HM0218).
Related Daily variations in consciousness (#HM1397).

♦ HH0476 Constructive personality change
Constructive psychotherapeutic change
Description This has been defined as an alteration of the personality structure of the individual towards greater maturity, integration and energy utilizable for effective living. It is suggested that the necessary and sufficient conditions for such change to occur are as follows: *Existence of a relationship* Significant positive personality change does not occur except in a relationship which requires psychological contact, namely awareness of the presence of the other. *Incongruence* A difference between the actual experience of the individual and his self picture, leading to some degree of anxiety. *Congruently functioning therapist* In a therapeutic relationship, the therapist must be freely, deeply, genuinely himself, with his actual experience accurately represented by his awareness of himself. *Acceptance by therapist* In a therapeutic relationship, the therapist must experience a warm and unconditional acceptance of each aspect of the person's experience and have a positive regard for him as a person. *Empathy by therapist* In a therapeutic relationship, the therapist must experience an empathic understanding of the person's awareness of his own experience and must try to communicate this understanding to the person. *Communication of therapist's empathy and regard* In a therapeutic relationship, the person must perceive that the therapist empathizes and accepts.
A further necessary condition suggested is that the therapist should have adequate knowledge and be adequately trained.
Related Depersonalization (#HM1248).

♦ HH0477 Alternative lifestyles
Alternative ways of life (AWL) — Dominant ways of life (DWL)
Description In any society there is a dominant way of life, made up of the different patterns of life of the people in that society. This is a complex structure in which the life pattern of each group affects, and is affected by, all the other groups (whether groups are considered by sex, age, education, occupation, or whatever).
The existence of this dominant way of life makes it very difficult to develop alternative ways of life within that society, although individual concern for human development means that such attempts are made. These attempts necessarily require structural change in order for them to be implemented. They are frequently means of countering mechanisms of exploitation and oppression inherent in the dominant way of life, and based on the vision of a more desirable society. Although they operate on a microsocial level, with limited possibility of influencing the macrosocial, they can and do provoke change and demonstrate alternative structural systems in action.
Related Inappropriate patterns of development (#HH0984).

♦ HH0479 Spiritualism
Description In general, spiritualism may be said to cover any belief in the existence of a reality not perceived by the senses and may include belief in: God; the immortality of the soul; cosmic forces; universal mind; the immateriality of will and intelligence. Many philosophical and all religious teachings are compatible with this definition in that they assume an independent reality superior to the material.
A less wide but no less popular definition of spiritualism involves a belief in life after death and the possibility of communicating with departed spirits. Such a definition includes the manifesting of phenomena such as clairvoyance and miracle cures. Several systems exist within this more narrow concept, generally including: belief in God; life after death (sometimes reincarnation), and the possibility of communication with this world after death; the need for living a life of love and service; responsibility for one's actions and retribution for evil deeds; and assistance on the path of perfection by spirit guides.
Spiritualism may be practised on several levels, its merit depends on the level of approach. It includes all efforts to determine the true nature of death and what the hereafter will bring and thus requires the services of a psychic medium able to contact subjective worlds. These contacts are usually at the astral level though they may be raised to mental levels and (rarely) intuitive or buddhic planes.
Related Spirit (#HH0770) Seance (#HH3487) Psychism (#HH1775)
Spirit possession (#HH0056) Spiritual healing (#HH0458)
Communication with supernatural beings (#HH1306).

♦ HH0481 Creativity
Inventiveness — Problem-solving ability
Description An ability to see new relationships, to make or otherwise bring into existence something new (possibly of aesthetic value), to produce new ideas and new solutions to problems, and to deviate from traditional patterns of thinking. When the innovation refers to the world of values of an entire culture, original and unique in terms of society and history and therefore unrepeatable, then it is distinguished as of such a standard as to be social creativity. When it refers to the world of individual experience, and is new as far as the individual is concerned, then it is termed individual creativity. This does not contribute directly to the progress of the culture, but is essential for the development of the individual.
Creative thinking may be divided into a number of stages. The preparation stage includes description of the problem and collection of information on it, which thus constitutes the raw material on which the individual works to reach the scientific solution, the artistic creation or any creative product. Predominant qualities at this stage are: sensitivity to new stimuli, discrimination, deliberate naiveté, deferral of judgment and categorization, use of analogies. The incubation stage takes place in the unconscious and involves weighing of the problem in the light of the acquired information and maturation of an idea. This stage is characterized by: restless moods, doubt, perplexity, and inferiority feelings requiring recognition by the person of his own value and considerable tolerance of unfamiliar situations. This is followed by the insight stage which is an involuntary sudden moment in which the material becomes clearly conscious and meaningful, accompanied by extremely powerful and often uncritical emotions. The qualities of the individual at this stage are: tolerance of situations of conflict, the knowledge that he cannot be accepted by everyone, a sort of sense of destiny, and ability to integrate contradictions. The final stage is that of verification and communication in which the subjective recognition is moulded into symbolic objective forms, usually accompanied by impatience and tension. The qualities required of the individual at this stage are: preference for complexity, independent judgment, search for possible implications, perseverence and audacity.
From the above, creativity may therefore be defined as a personal quality which enables the individual to withstand the presence of the emotions accompanying the creative process, to tolerate the paradoxical situations involved in the process, namely: involvement with maintenance of distance, freedom of creation and discipline of execution, integration and dispersal, restriction to the known from which confidence is drawn and a drive to explore the new and unknown, focus on a detail and on the whole, and to maintenance of a simultaneously active and passive attitude. For some, the need to be innovative may become compulsive and dominate their lives.
The growth of a creative process is facilitated in an atmosphere of freedom and security. When the individual feels secure he can leave the hot-house of the closed world for the new, the unknown, the open, accepting the stimuli of the external world. When he feels free, he can be himself, make use of the abilities he has in him (and not only those which he is supposed to have in him) and can dare to react to external stimuli by means of the talents in his internal world. A mildly elated mood in which self-judgement is temporarily suspended facilitates creativity, although manic elation overwhelms it. Some researchers have correlated the existence of creative and of manic depressive persons in the same family, suggesting the same genetic tendency nurtures both, the former simply having smaller swings of mood than the latter. Creativity requires a degree of uncertainty and freedom of choice. Some societies and environments therefore have greater potential than others for the encouragement of creativity – being watched while working or having to produce creative ideas to demand are both said to stifle the trait. An index of creativity may be used to evaluate inventiveness and to identify factors which stifle or encourage it.
Highly creative people tend to exist in a state of tension between the establishment and maintenance of environmental constancies and the interruption of achieved equilibria in the interest of new possibilities of experience. Such people are usually highly intelligent (although their abilities may not show up in conventional tests which do not always allow for original and nonconformist thinking); but the converse is not necessarily true. Whilst occasionally giving an impression of psychological imbalance or extremism, such persons may have exceptionally deep, broad and flexible awareness of themselves. It is suggested that the unconventionality may be in part a resistance to acculturation and the surrender of the person's unique and fundamental nature.
Refs Anderson, H H *Creativity as Personality Development* (1959); Barron, Frank *Creativity and Psychological Health* (1963); Barron, Frank *Creativity and Personal Freedom* (1968); Karagulla, Shafica *Breakthrough to Creativity* (1967).
Related Genius (#HH0295) Teleology (#HH0601) Uncertainty (#HM3051)
Freedom of choice (#HH0789).

♦ HH0482 Nude group therapy
Description Group nudity and its beneficial effects have been long accepted in traditions such as Nordic sauna. In some forms of group therapy nudity may be considered normal or else may

be encouraged whenever participants desire it. Such nudity is held to increase interpersonal transparency, remove inhibitions in the area of physical contact, decrease the sense of personal isolation and estrangement, and culminate in a feeling of freedom and belongingness. Nudity strips the individual of his ego pretensions and sometimes, of his ego defences. It helps to break down the walls of isolated cells of psychological privacy. Nudity in a group that encourages skin contact is held to be therapeutic in itself.
Refs Ruitenbeek, H H *The Nude Group Therapies.*
Broader Therapy (#HH0678) Human potential movement (#HH0398).

♦ **HH0483 Shadan therapy**
Broader Therapy (#HH0678) Psychotherapy (#HH0003).

♦ **HH0484 Buddhi yoga** (Yoga)
Description A "realist" yoga practised in Hinayana or Thervada Buddhism. According to Master Da Free John, this "non-dualistic" yoga is a tradition of the sixth or I-transcendent stage of life.
Broader Yoga (Yoga, #HH0661).
Related Buddhi awareness (Hinduism, #HM2099)
I-transcendent stage of life (#HM3060).

♦ **HH0485 Language development**
Description This is the process whereby a child acquires the ability to name things as a basis for comprehending, organizing and experiencing the world. Language also serves as the most effective means of communication of socialization demands and pressures. In modern culture, language becomes, as an individual grows up, a factor of increasing importance to the conduct of life. This importance means that inadequate or mistaken understanding of what society means by particular expressions may cause severe psychological disturbance. The importance is refected in the term *time-binding*, the means by which, using language, man comprehends past, present and future.
Refs Eliot, John (Ed) *Language.*
Related General semantics (#HH0585).

♦ **HH0486 Recognition of ignorance**
Description This recognition has been cited as a necessary prerequisite to real development, since it is based on the assumption not that future events are totally predictable but that they are unexpected; successful systems will have the resilience to deal with the unexpected and must be designed to absorb and accommodate any eventuality. This may be through generating detectable but controllable failures in order to prevent a too rigid or stable situation. It may be defined as a safe-fail rather than a fail-safe strategy.
Related Ignorance (Buddhism, Yoga, Zen, #HM3196).

♦ **HH0487 Transactional analysis**
Description A system of group psychotherapy based on the assumption that there are three positions from which one individual communicates with another (child, adult, parent) and three kinds of growth (stimulus-hunger, recognition-hunger and structure hunger) resulting in typical sequences in behaviour or games. Transactional analysis diagnoses the ego states in any transaction and "crossed transactions" when the ego states of the two parties to a transaction are different, with consequent breakdown in communication. Typical sequences of behaviour, or *games*, are identified, together with the intensity, tenacity and flexibility with which they are "played". The aim is to build up the "adult" state by filtering out the "child" and "parent" states, allowing the adult to work out a new system of values, and enabling the individual to relearn the capacity for awareness, spontaneity and intimacy, which results in autonomy.
Refs James, M and Jongeward, D *Born to Win* (1971).
Related Group therapy (#HH0851).

♦ **HH0488 Beauty**
Description Love of beauty is equated, by Socrates, by Buddha and by Plotinus, with love of good. Plotinus states that to make the soul good and beautiful is to make it like God, because God is beauty; and Plato, that beholding true beauty with the eye of the mind generates and nourishes virtue and makes the beholder a friend of God.

♦ **HH0489 Stopping the world**
Separate reality
Description This philosophy of don Juan/Carlos Castaneda was developed through the use of hallucinogenic drugs. By transcending ordinary reality one acquires wisdom of the right way to live. Ordinary reality, or the world view, is simply one interpretation of what really is. In order to stop the world "one has to learn the new description in a total sense, for the purpose of pitting against it the old one, and in that way break the dogmatic certainty, which we all share, that the validity of our perceptions, or our reality of the world, is not to be questioned".
Refs Castaneda, Carlos *Journey to Ixtlan* (1972); Castaneda, Carlos *The Teachings of Don Juan* (1968); Castaneda, Carlos *A Separate Reality* (1971).
Related Man of knowledge (#HH0338) Way of the warrior (Amerindian, #HH8219)
Tonal state of awareness (Amerindian, #HM2066)
Nagual state of awareness (Amerindian, #HM2036).

♦ **HH0491 Quietness**
Moral tranquillity — Soma
Description Moral tranquillity - the freeing from alternate subjection to love and hate, joy and sorrow - is said to prepare the soul for the "inflowing and teaching of the Holy Ghost", (St John of the Cross). Brooding over anything which disturbs the soul creates an interior disorder which hinders moral good and prevents spiritual progress. In addition to its positive effect on the spiritual life, quietness aids in relieving those pains and troubles of the mind which exacerbate the adversities of life. It is for this reason that schools of wisdom and the great religions have taught the establishment of *inner tranquillity*. Every individual is said to know the value of such tranquillity in which the fulfilment of life is experienced; but external circumstances and personality both affect the awareness of such knowledge. Systems of discipline such as the *Japanese cult of stillness* seek to cultivate freedom from external distraction and restlessness, and promote inner harmony. The individual learns to react to external situations so that, instead of being subject to them, the mind uses them to evoke a characteristic calmness. Systems of yoga also teach restraint of the external sense organs and the overcoming of the mind's restlessness.
Related Practice of silence (#HH0983) State of tranquillity (Buddhism, #HM3492)
Christian recollection (Christianity, #HM2077).

♦ **HH0492 Activity group therapy**
Situational therapy
Description Using a group activity (such as a trip, picnic, etc), or the physical environment in which the group is placed, has beneficial effects not only on the social behaviour but also on the personalities of the individuals taking part in such group psychotherapy.
Broader Therapy (#HH0678).
Narrower Group therapy (#HH0851).

♦ **HH0493 Superego**
Description According to Freud the superego is that aspect of personality which, among other functions, develops ideals, self-observation and moral conscience. Its role is comparable with that of judge or censor of the ego; and it has been postulated that it is formed during the pre-Oedipal stage.
Related Ego (#HH0636) Self-concept (#HH0471).

♦ **HH0495 Jnana marga** (Hinduism)
Jnana kanda — Way of knowledge — Way of enlightenment
Description Ignorance, or *avidya*, prevents the seeker from seeing things as they really are and is based on *maya*, illusion. The whole material world is the result of this illusion. It is ignorance which causes *samsara*, the cycle of birth, death and rebirth; therefore by destroying ignorance one can eventually obtain release from *samsara*. This does not imply salvation through knowledge such as learning the scriptures, the law and so on, but true knowledge of the Atman, the Self, as Brahman, the Self-existent Absolute. This is the knowledge of the *Vedanta* as taught in the *Upanishads* and has come to be known as *advaita*, non-duality.
Other approaches to jnana marga - the Sankhya school - put emphasis on the deliverance from samsara through knowledge of *purusa* (soul or spirit) as utterly distinct from *prakriti* (matter); the connection between the two is only apparent. Prakriti would remain unconscious if not acted upon by purusa; and the sufferings of prakriti are no longer experienced when not illuminated by purusa. The yoga system of Patanjali is closely connected with this school, although the final aim is union of the individual soul with God, here referred to as *Isvara*.
Context One of the three ways of salvation of orthodox Hinduism, the others being *karma marga* and *bhakti marga*. No one is said to be complete without some influence of the other two.
Broader Human development (Hinduism, #HH0330).
Related Jnana yoga (Yoga, #HH0927) Karma marga (Hinduism, #HH1211)
Bhakti marga (Hinduism, #HH0628) Advaita vedanta (Hinduism, #HH0518)
Religious experience as meditational insight (#HM1593).

♦ **HH0496 Music therapy**
Chants — Music
Description Every known society has some form of musical expression; and a significant use of music is in communication with the supernatural, in particular repetitious vocal music or chanting. Song may disclose beliefs, attitudes and values not otherwise communicable. It is suggested that by the use of music, deeper levels of the personality can be reached and influenced than is possible with the spoken word. This was certainly believed in previous generations, when physical disease as well as psychological problems were treated with very specific musical cures. Warning was (and is) given of the deleterious effects of some music, which by working directly on a person's mood can agitate, excite and make furious. Some pop music is said to be very harmful (even to plants).
Music therapy is a method of psychotherapy which attempts to obtain therapeutic effects in individuals through exposing them to various kinds of music. The music acts upon the mood of the individual and also upon his physiological functions. The method may be used either in individual treatment or in group therapy. Individuals may listen passively or they may actively practice on simple instruments. The latter demands much greater commitment of the whole personality.
Broader Therapy (#HH0678).
Related Sound health (#HH2543) Human development through music (#HH3003).

♦ **HH0497 Magico-religious powers**
Description Magical powers may be defined as the ability to cause a desired effect by means which mystify the beholder while being understood by the individual mastering the power; whereas religious powers also mystify the operator, perhaps the medium of some higher being who is presumed to understand the cause. Three main types of power are classified (Mircea Eliade): ecstatic, shamanic powers in which the individual attempts to leave his physical body; yogic powers, internally directed or enstatic, with the attempt to withdraw from the physical world; power of a liberated while living *jivanmukti*.
According to tantric explanations every being has contact, albeit unconscious, with the pure and subtle energy of the cosmos. Esctatic or contemplative techniques may be used to abandon identification with the gross matter of mind and body and allow a subtle form to emerge that identifies directly with this subtle energy. The yogin is then no longer subject to the laws of the material world. This is mirrored in the images of dimemberment and subsequent phenomena of the shaman. Both yogin and shaman may use their powers, if they are sufficiently directed by compassion for others, to assist their communities and the world.
The following are two well elaborated systems of superknowledge and magical powers:
(1) the Buddhist *Abhidharma Pitaka* classifies six superknowledges - magical powers or *riddhi* deriving from concentration of the will and making possible thought transference, teleportation and the ability to see things from a distance; clairaudience; telepathy; memory of past lives; liberation, the cessation of mental defilement.
(2) The sutras of Patanjali indicate magical powers or *siddhis* achieved through yogic practices and the withdrawal of subtle energy from the gross physical world. They may be classed as (a) Extrasensory perception and mental powers, including the first five Buddhist superknowledges, plus heightened senses, knowledge of the cosmos, of the internal workings of the body and of the meaning of language and animal sounds. (b) Physical powers including powers of the body (to shrink to the microworld, expand to the macroworld, levitate, increase in weight, travel through the universe, control one's physiology and others' thoughts and actions; plus ability to transmit psychic energy, generate inner bliss, raise the dead, create mass illusions). (c) Wisdom and transcendent powers, including knowledge of reality, the absolute, subtle and causal realms, law, oneness, bliss, and so on. (d) Ecstasy arising from union with the divine.
Related Power (#HH0848) Magic (#HH0720)
Siddhis (Yoga, Buddhism, #HH0380)
Magical-forces awareness (Buddhism, Tibetan, #HM2679).

♦ **HH0498 Homeopathy**
Description A system of treating disease with small doses of substances which cause symptoms similar to the disease being treated. A related system of tissue salt therapy has been developed based on salts and trace elements necessary for the healthy functioning of bodily tissues. One of a group of holistic methods of therapy, homeopathy involves the cooperation of the individual in effecting a cure and correcting the lack of harmony deep in the psyche which is manifesting physically as illness.
Broader Holistic medicine (#HH0581).

♦ **HH0499 Nobility**
Description Originally referring to hereditary ownership of land, titles and privileges, to be a member of the nobility distinguished a person from other groups by status, education and way of

life. Although conferring many advantages (in particular a right to rent and labour from an estate, nobility confers responsibility as well, as the expression *noblesse oblige* indicates. The word nobility now frequently refers to a person who demonstrates the qualities expected of one highly born (generosity, excellence, dignity), whatever his or her birth.
Related Honour (#HH0192).

♦ HH0500 Child creativity
Description By exercising and developing creativity, a child develops intellectually and emotionally; this experience leads to development of motor skills, practise with various instruments, criteria for determining position in life, and the benefit of group interaction. Such creativity is the potential for the development of society in general and may be developed through technical (machine) and artistic (drawing, literary, etc) improvisation.
Related Artistic education (#HH0029).

♦ HH0501 Soul
Description According to Plato the individual soul is that part of the individual person which mirrors the one Universal soul and which exists before the birth of the physical body and continues to exist in itself alone after the death of the body. It will then be reborn in another body. The physical body is in fact a hindrance to the philospher in that its requirements for food, clothing and physical pleasure prevent attention on the acquirement of knowledge and also it is an inaccurate observer through the senses. The soul, then, when it has as little to do with the body as possible is most near to having true existence revealed. It is the observer of the absolute qualities of justice, beauty and good. Lovers of wisdom should not then fear death as, unlike lovers of the body, of money or of power, it is after death that they will more nearly approach what they love.
A consequence of the fact that the soul is immortal is the need to care for matters which concern the soul in this life. Depending on how this life is lived a soul will be more or less able to deal with the next world, and will be reborn in fitting circumstances in this world.
Aristotle considered the soul to be the realization of the body, the means behind its purposeful activity, and he distinguished three types: nutritive or vegetative; sensitive or animal (capable of perceiving through the senses); rational or human. Christian teaching further develops both these concepts in conformity with immortality, individuality and personality. Descartes viewed soul as spirit manifesting in various states of consciousness, the body being material and extensional. Later views limit the soul/body interaction to a matter of feelings, desires and mental phenomena, excluding metaphysics; for example, Hume doubted the reality of soul since it cannot be derived from an empirical description of mental life. To some extent the concept of the soul has been replaced by that of the psyche or the ego. However, Jung used the expression "soul" to indicate the impenetrability of the psyche, when contrasting its depth, variety and plurality with any discernible pattern, order or meaning; and later analytical psychologists have used it to indicate a perspective which concentrates on depth imagery and the conversion by the psyche of events into experiences.
Refs Hofstadter, Douglas R and Dennett, Daniel C *The Mind's I* (1981); Jung, C G *Modern Man in Search of a Soul* (1936); Yogeshwaranand, Swami *Science of Soul* (1972).
Related Ego (#HH0636) Mind (#HH0299) Psyche (#HH0979)
Spirit (#HH0770) Oversoul (#HH0504) Afterlife (#HH0399)
Soul force (#HH0138) Soul-making (#HH0142) Pre-existence (#HH1339)
Perennial philosophy (#HH0665).

♦ HH0504 Oversoul
Ultimate reality
Description A term used by Emerson to express the absolute spiritual reality of the universe, related in idea to *universal mind* or *world soul*, the unity of all individual souls and of all that could be referred to as soul in the entire universe. There is no barrier between the oversoul and the individual soul; by silencing all that is private or exclusive to the individual self and "listening greatly", the individual allows the oversoul to flow into his soul, increasing wisdom and goodness beyond the narrow, human range. Developed from the writings of Plato and Plotinus, there is said to be one supreme mind at the centre of nature and of each individual soul.
Related Soul (#HH0501) Ultra-consciousness (#HM2156)
Cosmic consciousness (#HM2291) Oversoul awareness (Psychism, #HM0827).

♦ HH0506 Autogenic training
Description This method of psychotherapy combines both psychophysiological exercises and, once the physiological exercises are being satisfactorily performed, meditation.
The six standard exercises are based on the key phrases: my right arm is heavy; (leading to muscular relaxation); my right arm is warm (leading to dilation of the blood vessels); heart beat calm and regular (leading to cardiac regulation); it breathes me (leading to completely relaxed, natural breathing); my solar plexus is warm (leading to a calming effect on central nervous activity); my forehead is cool (leading to drowsiness and sedation). The meditation exercises focus on: spontaneous phenomena; experience of colours; imaginary objects; abstract ideas; emotions and feelings; interrogation of the unconscious.
Refs Luthe, W *Autogenic Training* (1970); Schultz, J H and Luthe, W *Autogenic Training* (1959).

♦ HH0507 Analytical therapy
Analytic group therapy
Description The purpose of both individual and group analytic therapy is to consciously re-establish the normally unconscious compensation mechanism which has been disturbed by over-strong contrast. The psychoanalytic viewpoint is applied to individuals or, in techniques of group psychotherapy, in groups of four or more participants, some of whom may also be receiving individual analysis. The ultimate aim, in either case, is the facilitation of the fullest possible communication of unconscious material; and transference relationships, catharsis, insight, reality testing, etc. are features of the dynamics.
In group therapy, the task of the therapist may be eased by the possibility of demonstrating and interpreting the unreality of any transference reactions which occur in the group situation between participants; although analysing and interpreting the network of these transferences may be much more difficult.
Refs Slavson, S R *Analytic Group Psychotherapy* (1950).
Broader Therapy (#HH0678).
Narrower Counter-transference (#HH0235).

♦ HH0508 Free self-inquiry meditation
Description As a method of meditation, free self-inquiry considers the nature of the self in a series of questions on the theme "who am I". Identification of "I" with the body is followed by deeper insight as the consciousness of "I" dissociates itself from the body, the senses, the mind, the intellect, the spirit, until there is pure unobjective consciousness. Meditation on "Who am I" may thus lead to cosmic or super consciousness where the subject-object relationship no longer exists.
Broader Meditation (#HH0761).
Related Cosmic consciousness (#HM2291).

♦ HH0509 Health
Description Full health encompasses not only freedom from disease and infirmity but, as stated by the World Health Organization, "a state of complete physical, mental and social well-being". This state is nonetheless not easy to define objectively and it is evident that an individual's "degree of health" is dependent on a number of factors not quantitatively measurable, although Jung equated *wholeness* with health. Considerable progress has been made in the reduction and eradication of disease; but the prevention of ill health and the interplay of spiritual, mental and physical variables in sickness and in health are still imperfectly understood. A society which considers good health dependent on not arousing the hostility of a sorcerer, or of someone hiring a sorcerer, may see healthy people as good or as socially adept. Where good health is seen as a matter of luck, then health may not be seen as related to other personal qualities. Current Western attitudes are that bad health may, in part, be due to not assimilating and accepting everyday experience without stress. However, it is clear that spiritual development can and does continue into old age when physical (and sometimes mental) faculties progressively deteriorate. Conversely, individuals experiencing a high degree of physical health are not necessarily mentally or spritually well developed. It has been stated, however, that a person on a path of spiritual development should aim at maintaining physical health. Hatha yoga, for example, is a system where mastery over the body is treated as a means of mastery over the mind and one of its results is good physical health and longevity.
Most religions deplore over-asceticism as prejudicial to spiritual development. In the Chandogya Upanishad, Shwetaketu is told by his father Uddlaka "mind comes from food" and that abstaining from food inhibits memory. "Moderation in all things" would appear to be universally agreed, the consequent good health being conducive to providing the strength required on the spiritual path: John Locke (quoting Juvenal - "Mens sana in corpore sano") says, "A sound mind in a sound body, is a short but full description of a happy state in this world". But unavoidable ill-health is not seen as detrimental to an individual's progress, when the value of each effort is not measured by observable results. It has even been held that ill health may in some cases be a requisite for spiritual development.
Narrower Curative education (#HH1208).
Related Wholeness (#HM2725) Spiritual life (#HH0101) Fitness for work (#HH0758)
Physical perfection (#HH0722).

♦ HH0510 Common sense
Description The spontaneous attitudes brought about in a person by his experience of every day life, based on practical activity and commonly accepted morality, may be said to comprise common sense. Because it does not rise beyond this immediate practical relationship with the world, common sense will then vary with circumstances and has been variously held: to prove the existence of God; to prove that God cannot exist; to be equivalent to personal utility or advantage in a given situation. However, a definition of common sense as the inborn principle in the human spirit which makes experience possible, leads to the belief that it is through common sense that man derives his belief in God and in the existence of reality.
Refs Ferguson, Marilyn *The New Common Sense* (1991).
Related Understanding (#HH0658).

♦ HH0511 Life cycle
Social phases of life
Description The phases through which an organism passes as it develops, matures and becomes capable of giving rise to another generation determines the length of its life cycle. In human terms, these phases may be seen as stages in personality development. They are generally thought of as extending from birth to death, although during the period from conception to birth the individual may be considered as a social entity and most religions posit life after death. But, by analogy with the life-cycle of any organism, the human life-cycle may be seen as the social path from birth to adulthood, possibly followed in old age by loss of qualities associated with the mature, social adult. Thus, in social terms, stages of growth may be marked by an ability and readiness to participate in social roles and institutions. Individuals mature, gain adult qualities, as they pass through the socialization experiences which enable them to learn how to conduct themselves at each particular stage. Thus the individual progresses through infancy, childhood (7-13 years), youth (13 to 25 years), maturity (26-60 years) and old age (61 years onwards). In some societies, such as Japan, social recognition continues after death with a number of celebrations and regular rituals. The social framework varies with time and, although in past centuries change was very slow, twentieth century change in society's schedule has in many cultures been dramatic.
Refs Lidz, Theodore *The Person* (1968).
Related Ontogenesis (#HH0366) Stages of faith (#HH2097)
Stages of personality development (#HH0285).

♦ HH0512 Psychodrama
Sociodrama — Physiodrama — Axiodrama — Hypnodrama — Psychomusic — Psychodance — Therapeutic motion picture
Description Psychodrama is a form of group therapy in which one or more persons act out situations in a dramatic setting which enable them to gain some insight into the emotional problems deliberately drawn into the scenario. Ultimately, the intention is to encounter the self by transcending everyday life, and to form a deeper relationship with existence.
The individuals may either act out their own real life role in the scenario, or else take up a counter-role through which they gain understanding not easily available otherwise. The therapist may also take up roles, and he and his assistants may also find themselves obtaining insights into some of their own problems, patterns of behaviour or motives. For this reason psychodrama need not be confined to groups of people whose problems are pathological; it is basically a training in acting out past and present problems, both realistically and symbolically, both alone and with others, and develops a spontaneity and awareness from which perfectly normal people can also benefit. In fact, it has been said that games in which children play roles (mothering a doll, shooting toy pistols, and so on), a natural part of growing into adulthood, are everyday forms of psychodrama.
An important justification for this technique is that individuals may well have difficulty in giving verbal expression to their emotional problems, particularly when these are intense. Acting out a situation may prove both easier and more meaningful, given that this takes place in a setting which approximates most closely to the problem-producing situation of life. Psychodramatists contend that irrational and compulsive patterns are more readily seen and treated in the situation which involves action rather than just conversation.
Refs Moreno, J L *Psychodrama*.
Related Theatre games (#HH0582).

♦ HH0513 Creative existence
Description The fourfold aspects of creative existence are said to be *aspiration* (the initial decision to search for, and reorganize one's life on the basis of, spiritual values and the experience of God), *action* (which itself has three aspects - self expression, self poise and self-donation), *meditation* and *love*. These four will lead to *existential integration* and a harmonious and fruitful life.

Narrower Love (#HH0258) Meditation (#HH0761)
Aspiration (Buddhism, #HM2578).
Related Existential integration (#HH0832).

♦ HH0514 Engrams

Description The term *engram* is used in Scientology to denote mental images of pain and unconsciousness taken into the memory but neither analyzed nor integrated. Such images are carried in the "reactive" mind from one lifetime to another and are superimposed over the "analytical" mind or reason in times of stress. Although such reactions have their uses they are also the cause of *psychosomatic illness*, misery and conflict. A part of Scientology religious practice is *auditing* which has the aim of reversing this pattern.
Related Thetan (#HH0269).

♦ HH0515 Superstition

Description The observance of same rite or practice in response to some fear or scruple, without belief in the basis or morality of the religion, philosophy or tradition which developed such a practice.

♦ HH0516 Leadership development

Description Leadership is a quality that is developed. Personal qualities which are traditionally associated with the requirements of leadership can be developed as the person himself develops; these include: eloquence, decisiveness, physical energy, chairpersonship, drive, clarity of thought, maturity, good memory, diplomacy, personality, experience, sincerity and knowledge.
Leadership may be described as: the ability to motivate self, and secondly, to motivate others; the ability to see a problem, recognize it, plan a solution, and execute that solution without having to be prompted by someone else; the ability to lift man's vision to higher planes, to raise man's performance to a higher standard, to build man's personality beyond its normal limitations; the ability to make people want to do things they would not ordinarily think of doing; the ability to make people want to accept as their own the object of the social group of which they are a part. According to a recent survey among acknowledged leaders, they themselves considered the most important quality in a leader was the ability to make a decision.
It is also imperative that a leader possess a conscious awareness of his position and what it implies. The aspects of personal growth which contribute to development of leadership potential may be distinguished as: physical development (leading to physical energy, drive), mental development (decisiveness, eloquence, maturity and knowledge), financial development, social development, family life development, spiritual development.
Related Leadership intelligence (#HM5451) Organizational development group (#HH0617).

♦ HH0517 Human synergy

Description Mind may be viewed as the total of the continually expanding imagery shared by mankind, where images are mental representations of anything presented through sensory experiences or created by feeling and thought. The unique characteristic of human mind to acquire and store knowledge beyond the capability of the individual central nervous system enables individuals to contribute to the communal imagery of mankind as cells do to the body. This process, envisaged as culminating in a noosphere resulting from the eventual coalescence of all individual minds requires synergy to make such a coalescence more than a melting pot. However, whether or not this view is correct, the evolution of imagery initially based on biological evolution has now become a synergic process in its own right. The characteristics of synergy in this context are:
Synergy involves all components of a situation and brings about positive interaction even between ostensibly diametrically opposed aspects, making these cooperate in an unexpected manner.
Synergy, with the involvement of man in his environment, leads to the concept that man is directing his own evolution.
Some writers correlate synergy with love, by defining love as the expansion of the identity to include others, desiring to share experience with them or to bear unpleasant experiences for them.
Related Transcendence (#HM4104) Planetary consciousness (#HM2006).

♦ HH0518 Advaita vedanta (Hinduism)

Non–duality
Description Literally the following of a philosophy of non–duality, practitioners follow the Upanishad tradition through the path of wisdom, *jnana yoga*, as founded by Badarayana and expounded by Sankara. They allow no duality between creator and created – all are reflections or manifestations of the *one*, which is not and cannot be an object of sense ("not this, not this") but which is the underlying reality or consciousness, the subject of which all else is the object ("thou art that"). By the removal of ignorance – *avidya* – the Self, the vital principle, the *Atman*, comes to be seen as identical with the first principle, the all–pervading power, the *Brahman*.
Non–duality is not meant to imply simply one–ness; the distinction between what is and what is not is to be found in the permanence or changeableness of what is being considered. That which observes does not change with what is observed, the ultimate being perception or consciousness itself which is unchanging. Rationally it is clear that there can be no being beyond consciousness and that consciousness and real existence are inseparable. Consciousness and "is–ness" are Brahman. It is the deluded sense of separation from Brahman, of separate individuality, which is the cause of pleasure and pain; identification with Brahman is bliss.
Some advaita traditions expect the following of an ascetic life but in others the ways is also open to the householder. Again, some systems demand simply the practice of listening, *sravana*, to the point of really hearing, holding that no other work or practice is necessary or even useful; whereas others require practical self–discipline in addition, and subsequent profound enquiry into the Self which becomes deep meditation, *dhyana*. Necessary preparations before listening can be effective include development of the powers of discrimination, renunciation, equanimity and a longing for truth.
Refs Deutsch, Eliot *Advaita Vedanta* (1969); Misra, Ram Shankas *The Integral Advaitism of Sri Aurobindo* (1957); Satprakashananda, Swami *Methods of Knowledge According to Advaita Vedanta* (1975).
Broader Human development (Hinduism, #HH0330).
Related Brahman (#HH1226) Saccidananda (#HM0592) Maya (Hinduism, #HH1004)
Atman (Hinduism, #HH0103) Jnana yoga (Yoga, #HH0927)
Jnana marga (Hinduism, #HH0495) Unitary consciousness (#HM2702)
Vishishtadvaita (Hinduism, #HH3772) I–transcendent stage of life (#HM3060).

♦ HH0519 Self–perfection

Description The systematic improving of qualities and habits of behaviour, whether moral, physical or ethical, constitutes the activity of self–perfection. This will normally be in order to correspond with some determined ideal, depending on the social and historical conditions under which a person lives. Once an individual has reached a sufficiently high level of consciousness and self–knowledge he or she has the capacity both for self analysis and observation, and for understanding the actions of others. Motivated by the wish for acceptance and authority in a peer group, an individual develops personal qualities by self–perfecting. On a different level, the aim of an individual on a spiritual path such as yoga is also self–perfection.
Refs Chimnoy, Sri *The Inner Promise, Paths to Self–Perfection* (1970).
Related Human perfectibility (#HH0212).

♦ HH0520 Progress

Description Philosophically, progress supposes development of some system or part of some system from a lower or less perfect state to a higher or more perfect state. Classical thought was divided between no–progress (cyclical) existence or negative progress (decline from a golden age). Although Christian thinking allows for progress outside the scope of history, the social development of the 17th century brought with it the the belief that the *reign of reason*, lying in the future, was a goal to which humanity was progressing historically (Philosophy of Enlightenment). Later, Hegel defined history as progress in the consciousness of reason, the new always being founded on the old, while Marxism says that societies evaluate history subjectively from their individual positions, always seeing themselves as the culmination of preceding stages, and that objective assessment of progress must be based on the material progress of a society. However, believers in democracy emphasize the individual and the ability to dominate material things with the mind, measuring progress in the "capacity to become a free and responsible person in a world of free and responsible nations". Currently, the progressive spirit of cooperation and understanding among major religions is held by some to indicate progress towards redemption by God of all human life.
Refs Hocart, A *The Progress of Man* (1933).
Related Progress (Christianity, #HH6434) Philosophy of enlightenment (#HH0591).

♦ HH0521 Advanced training

Description This term denotes continuing specialized education for improving the skills of workers and providing new theoretical knowledge, whether during or outside working hours and whether at the place work or at a specialized institute.
Related Educational self–development (#HH0955).

♦ HH0522 Instruction

Description This is a dual process involving teacher and student in the transmitting and acquiring of knowledge. Humanity's experience is transmitted and personal qualities and cognitive abilities developed in the student. Different methods of instruction emphasize theoretical learning applied in practice, or practical training from which theoretical understanding may be developed. The use of instruction simply to train in a particular skill reduces instruction to the level of animal–type training, whereas *cognitive theory* regards learning as the formation of cognitive structures, a strictly human activity. Although the nature of instruction and degree of student participation may vary, the assimilation of knowledge is considered to be the result of the students own cognitive activity.
Refs Da Free John *The Paradox of Instruction* (1977).

♦ HH0523 Four noble truths (Buddhism)

Ariya — Suffering — Dukkha (Pali) — Duhkha
Description The four truths revealed to the Buddha in mystic illumination: that the cycle of birth and rebirth is basically suffering; that this suffering is caused by *ignorance* and the other *afflictions*, and by actions taking place under their influence; that transcendence of suffering is *nirvana*, the cessation of sufferings and their origins such that they will never return; and that there are means of attaining nirvana, the *true paths* or consciousnesses.
1. *Suffering* (dukkha). This is suffering in the sense of transience, of being in the created world and subject to change and loss. Even in happiness there is sorrow because there is realization that the happiness of this world cannot last. The pleasures of the world, being transient, are not satisfying as spiritual pleasures are, and this lack of satisfaction and harmony motivates the individual to spiritual effort leading to mental awakening, in particular through the shocks of birth, sickness, old age and death. The first truth is to be fully comprehended.
2. *Arising of suffering* (paticca samuppada). Suffering is caused through thirsting for pleasures of the senses, *tanha*, through desiring that things should be different from the way they are. Life as it happens now depends on the result of previous acts carried out in ignorance of the divine law, such distorted activity necessarily producing distorted results. This is the effect of clinging to the transient as though the transient were substantial. Forgetting that life is a moving pattern, there is a clinging to life and thus a clinging to death. Change appears as suffering because of the desire to possess or selfishly enjoy something. It is only through attachment to things that change in them becomes distressing. The forcing of experience through cravings creates a future prison which may, however, be escaped from through understanding the nature of the process. The second truth is to be abandoned.
3. *Cessation of suffering* (nirvana). The cessation of suffering is supreme bliss and final liberation, with the ending of ignorance and craving experienced like health after illness as the cause of suffering is removed. Although it is easier to speak of nirvana in terms of what it is not rather than what it is, this does not necessarily imply that nirvana is annihilation, since there is bliss. It lies beyond positive and negative. The third truth is to be made visible.
4. *The way to cessation of suffering* (eightfold ariya path). The ordinary (lokiya) and transcendent (lokuttara) paths are not so much a set of steps as a naturally flowing progression of states of mind, the transcendent path leading to harmonious balance with transcending of ordinary understanding and the acquisition of direct knowledge of unconditioned truth. This is the middle way, *majjhima patipada*, avoiding extremes of indulgence or rigorous self–mortification. The fourth truth is to be brought into being, *bhavana*, meditation.
Context One of the paths of calming according to Buddhist teaching.
Refs Bowker, John *Problems of Suffering in the Religions of the World* (1970).
Broader Paths of calming (Buddhism, #HM2987).
Narrower Nirvana (#HM2330) Eightfold way (#HM2339)
Duhkha (Buddhism, Hinduism, #HM2574)
Causally continuous doctrine of being (#HH1099).
Related Suffering (#HM0471) Doubt (Buddhism, #HM4467) Bhavana (Buddhism, #HH0551)
Right acts (Buddhism, #HM0198) Right speech (Buddhism, #HM1157)
Right outlook (Buddhism, #HM1280) Right resolves (Buddhism, #HM1142)
Right endeavour (Buddhism, #HM1295) Root afflictions (Buddhism, #HH0270)
Right livelihood (Buddhism, #HM1341) Human development (Buddhism, #HH0650)
Right mindfulness (Buddhism, #HM0704) Ignorance (Buddhism, Yoga, Zen, #HM3196)
Cessation of suffering (Buddhism, #HH2119) Steps to enlightenment (Buddhism, #HH4019)
Two–fold knowledge of truths (Buddhism, #HM3527)
Right rapture of concentration (Buddhism, #HM0931)
Secondary afflictions (Buddhism, Tibetan, #HH0781)
Meditation way of the four truths (Buddhism, #HM2245)
Religious experience as meditational insight (#HM1593)
Understanding as knowledge of suffering (Buddhism, #HM6870)
Tibetan meditative states of form concentrations (Buddhism, #HM2693)
Understanding as knowledge of the origin of suffering (Buddhism, #HM7290)
Understanding as knowledge of the cessation of suffering (Buddhism, #HM8163)
Understanding as knowledge of the way leading to cessation suffering (Buddhism, #HM7645).

♦ HH0524 Sexual synergism

Description According to Freud, any significant physical process of the body, whether pleasant or unpleasant, contributes to the arising of sexual excitement.

♦ **HH0525 Rejuvenation**
Description The myth of rejuvenation has been described as the human response to fear of death, and to be founded in the experience of changing seasons, where the withering and decay of winter is always followed by the miraculous new growth in spring. Rituals and therapies aimed at reversing the ravages of time have emphasized a return to the source of life, to the womb, and subsequent rebirth. From observation of nature have arisen myths of: rejuvenating sleep; shedding of the skin of old age to emerge once again young. Other means of restoring youth have been special fruits, elixirs and fountains (the latter thought by some to be related to veneration of water as the female element). Such means often invoke the divine, as for example, the fruit of the tree of life in the Garden of Eden. In religious thought, however, length of life is usually thought of in spiritual terms, with either the hope of heaven after death or liberation from earthly existence. Liberation is more from the ravages of sin than of old age although rituals echo myths of rejuvenation in this life.
Secular society, with no recourse to spiritual or divine aid for length of life in this world or the next, has emphasized strictly practical and physical methods for rejuvenation and resistance to the effects of old age.
These range from consuming the sexual organs of wild animals, as in ancient India and China, through transplantation of animal sex organs, shunamism, to the hormone and vitamin therapy of the present day. Inspired by the hope that future medical science will effect rejuvenating and resuscitation techniques, some individuals have directed that, after their death, their bodies be stored indefinitely at freezing temperatures for possible subsequent restoration to life. This may perhaps be compared with embalming techniques practised in Ancient Egypt.
Refs Rechelbacher, Horst *Rejuvenation* (1987).
Related Shunamism (#HH3452) Immortality (#HH0800) Reincarnation (#HH0686).

♦ **HH0526 Self-transcendence**
Description It is suggested that the theories of man, which are circumscribed by the individual himself (whether upon the reduction of his tension, as in homeostasis theory; or in the fulfilment of the greatest number of immanent possibilities, as in self-actualization), are inadequate. An adequate view of man can however be formulated when it goes beyond homeostasis and beyond self-actualization to that transcendent sphere of human existence in which man chooses what he will do and what he will be in the midst of an objective world of meanings and values. In this sense, self-transcendence can be considered the fulfilment of self-actualization.
To speak of the world as merely a design produced by the cognitive subject is to do injustice to the full phenomenon of the cognitive act which is the self-transcendence of existence toward the world as an objective reality. The more cognition actually becomes mere self-expression and a projection of the knowing subject's own structure, the more it becomes involved in error. Cognition is true cognition only to the extent in which it is contrary of mere self-expression and to which it involves self-transcendence.
Refs Frankl, Viktor E *Self Transcendence as a Human Phenomenon* (1966); Hammond, G B *The Power of Self-Transcendence* (1966).
Related Transcendence (#HH0841) Self-actualization (#HH0412)
Journeying within transcendence (Christianity, #HH6505).

♦ **HH0527 Centering**
Description Centering exercises are used as an introduction to meditation and as a means of aiding relaxation and facilitating concentration, leading to a sense of inner equilibrium. The process may begin by focusing attention on the physical centre of gravity of the body and feeling the relationship of the body to the earth and the space surrounding it. Experiencing the sense of balance and support provided by the floor or a chair helps to focus attention on physical sensations. Attention may be directed to the surface of the skin, body boundaries, and the flow of energy associated with the circulatory system. Focusing attention on the centre of physical energy (associated with the solar plexus according to Eastern traditions) while noticing the movements of breathing may also be used.
Centering may also be used as a general term for any mental or imaginative exercises designed to facilitate access to a higher state of consciousness.
Refs Hendricks, Gay and Wells, Russel *The Centering Book* (1975); Richards, Mary C *Centering* (1969); Richards, Mary C *Centering in Pottery, Poetry and the Person* (1962).
Related Centre (Taoism, #HM1233) Centering prayer (Christianity, #HH1994)
Disruptive thoughts (Christianity, #HH4001).

♦ **HH0528 Racial memory**
Description The archaic heritage of fragments of phylogenetic origin with which each individual is endowed at birth, racial memory is a blue print enabling development and reactions to follow a particular model. Some authorities claim that memories of the experiences of previous generations are also included, when they were significant enough or repeated sufficiently often. Such memories become conscious when the remembered event occurs again, and may invoke obsessive reactions.

♦ **HH0529 Materialism**
Dialectical materialism
Description Basically materialism considers the physical body and the physical world to be the overriding reality, as rather than thought, attention, art and law which, according to Plato and in idealistic philosophy, are prior to the physical body and are responsible for the great and primal works and actions. As opposed to agnosticism, which holds that it is impossible to characterize reality objectively, materialism recognizes the world as cognizable and considers mankind capable of reaching objective truth.
Dialectical materialism involves the logical testing of ideas via thesis, antithesis and synthesis, stressing universal relationships and progressive change. Consciousness is held to be a function of the brain which, in turn, is said to be the most highly organized form of matter's motion, matter and motion being eternal and infinite. Knowledge, however, is relative; human thought exists only as the thought of past, present and future generations and truth emerges historically as the expression of this cognitive process. Based on scientific and social progress, dialectical materialism itself is the basis of Marxist-Leninist teaching.
Refs Buckarin, N *Historical Materialism* (1925).
Narrower Medical materialism (#HH3376) Spiritual materialism (#HH0828).
Related Idealism (#HH0893) Agnosticism (#HH0904).

♦ **HH0530 Thought processes**
Thinking — Thought
Description Thinking is the establishment of order in the world as it is apprehended, and in the relations between objects and between representations of objects. Each thought process is a result of autistic (self absorbed) and environmental influences and any kind of thinking relates intrinsic activity with external stimulation. Opinion has differed as to the verbal and non-verbal nature of thinking; at one time all thought was supposed to be linguistic but it has been shown (Vygotski, Piaget) that children are capable of reasoning by the assembling and combining of non-verbal acts. Thought processes are generally considered to be the comparison and association of similar or contiguous ideas. In a sense, thought is the interface between the external and internal world.
Narrower Linear thinking (#HH2287) Process thinking (#HH1643)
Positive thinking (#HH1088) Maternal thinking (#HH4646)
Associative thinking (#HH0644) Pre-conscious thinking (#HM1084).

♦ **HH0534 Totality**
Description An awareness of the totality of existence relates each individual's experience with the experience of everyone else. Each is dependent on everyone so that every action is of total significance. As Einstein said "The falling of the tiniest leaf affects the farthest star".
Refs Chang, G C C *The Buddhist Teaching of Totality* (1971); Johnsen, Carsten *Man the Indivisible* (1971).
Related Completeness (#HM3070) I Ching (Taoism, #HH0004)
Unitary reality (#HH0726).

♦ **HH0536 Salat** (Islam)
Namaz — Worship
Description This worship consisting of a sequence of utterances and actions is repeated five times a day. Although exact performance may vary between different schools, proper observance of salat is required duty and the basic essentials are the same throughout Islam. It symbolizes submission or surrender to God and is the essence of Islam.
Context One of the five pillars of Islam.
Refs Nizami, Ashraf *Namaz*.
Broader Worship (#HH0298) Human development (Islam, #HH1799).
Related Religious experience as personal devotion and worship (#HM0561).

♦ **HH0538 Sabda** (Yoga, Buddhism)
Shabda — Knowledge
Description Sabda is described as that knowledge based on theory, on words, and not connected with the object under consideration itself.
Context According to Patanjali, one of the three forms of knowledge which need to be distinguished from each other and separated if confusion *samkirna* is to be avoided.
Broader Knowledge (Buddhism, #HH1971).
Related Artha (Hinduism, #HH0353) Shabda yoga (Yoga, #HH0539)
Jnana (Yoga, Buddhism, Hinduism, #HH0610).

♦ **HH0539 Shabda yoga** (Yoga)
Shabd yoga — Nam simran
Description A version of *nada yoga*, using sound as the means for transforming consciousness, as practised in the Sikh religion. It includes the practice of *simran*, internal repetition, of "five names", the names of the five internal spiritual regions of ascent in mystical experience: Jyoti Niranjan; Onkar; Rarang; Sohang; Sat Nam. Names of God or yogic mantras may also be used, all with the aim of focusing the attention inwards and upwards and ultimately transcendence of body and mind in *nirvikalpa samadhi*.
The difference between shabda and mantra yoga is that here one does not start with a mantram but sets out to discover the inner sound and to identify one's self with the universal sound current. This may at first be perceived by the inner ear in a variety of forms - bells and other instruments, animal and human voices, water, thunder. It may be in systematic sequence and refer to the energy centres in the body where the sounds seem to occur as practice proceeds. Some sounds are not of this world and bring ineffable states, some represent the action of vast cosmic voices into which the small self is absorbed. It is an exploration of the states known to Jewish and Muslim mystics but more particularly in the aural mode.
Refs Godwin, Joscelyn *Harmonies of Heaven and Earth* (1987).
Broader Yoga (Yoga, #HH0661) Nada yoga (Yoga, #HH0160).
Related Sabda (Yoga, Buddhism, #HH0538) Mystical stage of life (#HM3191)
Human development (Sikhism, #HH6292) Nirvikalpa samadhi (Hinduism, #HM2061)
Human development through music (#HH3003).

♦ **HH0540 Thought-pressure**
Pressure of ideas
Description In some mental disorders - schizophrenia, manic depressive and some psychoneurotic states - ideas seem to be forced into the mind, sometimes apparently due to pressure brought to bear on the individual by others (schizophrenia), or as an irresistible force in his or her own mind.

♦ **HH0542 Social renewal**
Description The *Wrekin Trust* founded by Sir George Trevelyan, bases a series of conferences and courses on the premise that social renewal is possible though an inner, willed change in *personal consciousness*. Changes at material, psychic and spiritual level are not independent; and in this materialistic age the search through the innermost feelings for a meaning and purpose to life is expressed at the social level in the present widespread dissatisfaction, social unrest, terror, war and the breakdown of institutions. Courses and studies concerned with the evolution of consciousness and the deepest truths of all systems of philosophy and religion are designed to pass on ancient knowledge to those capable of using it, in order to attune to the creative intelligence of the universe and encourage the present revival in such spiritual knowledge.
Broader Human potential movement (#HH0398).

♦ **HH0543 Psychic isolation**
Description According to Jung, psychic isolation occurs when an individual, for no apparent reason, becomes consciously aware of some material in the collective unconscious which would normally be hidden. This experience is so overwhelming and uncanny that the individual feels estranged from others, partly from a fear of being considered insane; the psychic atmosphere is animated and a "momentous alteration of the personality" is signified.

♦ **HH0544 Abhyasa**
Practice — Effort — Application
Description Any practice or system in which effort is directed towards attainment of a state of uninterrupted awareness of reality can be referred to as abhyasa. Such a practice or system may be traditional or newly developed, followed by many or adapted to the specific temperament and requirements of one individual. Application towards realization of transcendence is the essence of yoga. It requires discrimination and discernment between that which is real and wholesome and that which is transient and unworthy of being the focus of motivation.

♦ **HH0545 Divination**
Mantic arts
Description The use of magical or supernatural means to discover information, often about the future or about things not otherwise open to normal perception, has been and is common in most civilizations. Commonly it is used to discover the means of curing a particular ailment, the best course for future action, or events taking place some distance away. It may either be part of the established religion (for example, the Delphic oracle, and many African religions) or parallel with or in opposition to religion, Christianity being in general opposed to such methods. The diviner may

enter a special state or trance and pass on information, or may interpret enigmatic data (as, for example, examining animal entrails to foretell the future). A common method is for a diviner, or for any individual requiring treatment or advice, to sleep in a particularly hallowed place, his or her dreams being subsequently interpreted. A specific form of divination is the use of ordeals to determine the innocence or guilt of a suspect, when some test is imposed when the guilty will suffer but the innocent escape injury (trial by ordeal).
Related to positive actions taken to determine information is the interpretation of naturally occurring phenomena as omens of the future, whether these are seen in reality or in dreams. For example, sneezing or stumbling may be taken as bad omens, as may chancing on a particular animal or bird. The mantic arts use this belief - that every occurrence is pre-revealed symbolically and the predictable result of causes which are discernible by divination - as an aid in solving problems from personal to political and as a guide to future conduct.
Related Magic (#HH0720) Shaman (#HH0973) Ordeals (#HH0377)
I Ching (Taoism, #HH0004).

♦ HH0546 Programmed learning
Description Although there is no generally agreed definition of the exact nature of programmed learning, methods of teaching usually considered under this heading generally require the existence of feedback on the student's progress at each step, so that the teacher can alter instruction to fit each individual's pace. The process depends on computer technology and the dividing of the instruction into small, discrete steps which favour such a system and optimise control of the acquiring of knowledge by the student.
Broader Education (#HH0945).

♦ HH0547 Sex education
Description Sex education starts at a very early age, when the child of 2 or 3 years old starts to ask questions such as "where did I come from ?" and to notice the physical differences between boys and girls; it continues with the aim of encouraging the harmonious development of young people and the understanding of responsible childbearing. The changes in social morality and outlook brought about with the decline in religion in the XXth Century have contributed to marked increases in abortion, unwanted pregnancies, single parent families and sexually transmitted diseases; sex education is aimed at protecting young people from this trend and increasing understanding and a responsible attitude.
Broader Education (#HH0945).

♦ HH0550 Prajna paramita (Zen)
Great perfection — Maha paramita — Prajna
Description This is the perfection of the arhat who has succeeded in perfecting heart, mind and will and is deserving of that perfection beyond human perfection. Even books of scripture dealing with this perfection in transcendent wisdom are treated with reverence in Zen practice.
Context The sixth, or culmination of the first five perfections or *paramitas* of Zen.
Broader Paramitas (Zen, #HH0219).
Related Virya (Zen, #HH0677) Kshanti (Zen, #HH0813) No-thought (Zen, #HM2500)
Arhat (Buddhism, #HH0233) Sanctification (#HH0428) Prajna (Hinduism, #HM3455)
Essential wisdom (#HM0107) Dana (Buddhism, Zen, #HH1234)
Prajnaparamita (Buddhism, #HM1344) Dhyana (Hinduism, Buddhism, #HM0137).

♦ HH0551 Bhavana (Buddhism)
Buddhist meditation
Description This system of meditation is based on stillness of mind, *samatha*, and experiential insight, *vipassana*, which are eventually harmonized in the achieving of *lokuttara consciousness*. The term bhavana literally means "bringing into being", in this case the bringing into being of the *eightfold path*. Different approaches give greater emphasis to vipassana, sometimes omitting later stages of samatha which, similar to dhyana yoga, is seen as a danger in being too pleasant in itself to encourage further development. Others emphasise samatha, considering premature development of higher insight as a disadvantage. There is a creative tension between the two techniques, the one emphasizing cognitive skills and the other emotional attitude. Samatha is seen more as an ecstatic experience, "devouring" the universe while vipassana is enstatic, "excreting" the universe.
Context The third of the ten meritorious deeds of South Asian Buddhism.
Broader Meditation (#HH0761) Ten meritorious deeds (Buddhism, #HH0446).
Narrower Citta-bhavana (#HH5973) Samadhi-bhavana (#HH8231)
Samatha-bhavana (#HH0710) Vipassana-bhavana (Buddhism, Pali, #HH0680).
Related Mark of calm (Buddhism, #HM8673) Eightfold way (Buddhism, #HM2339)
Four noble truths (Buddhism, #HH0523) Lokuttara consciousness (Buddhism, #HM2120).
Followed by Respect for superiors (Buddhism, #HH0604).
Preceded by Observing moral precepts (Buddhism, #HH0278).

♦ HH0552 Astrology
Synastry
Description Astrological symbolism is said to have been designed to help man in his evolutionary development, and to help him to liberate his higher will by making clear some of the factors opposing it. Astrological factors may be worked with, assimilated and transmitted in the process toward the goal of individuation. Astrologers tend to regard features of a horoscope as metaphors for psychological processes. The meaning is derived in part from traditional sources, in part from experience and intuition and in part from astronomy.
Astrological horoscope casting functions in one respect as a framework for intuitive perception. A horoscope constitutes, for an astrologer, a person-centered symbol carrying a formulation of the person's personality and destiny. It suggests how the individual can best actualize the innate potentialities of his particular and unique selfhood. Astrology is therefore a language of symbols, and as such implies a process of unfoldment and of changing relationships. It is based on the assumption that features of a horoscope can contain a description of: all the essential functions implied in the existence of an organized field of activity, and especially of a human being; all archetypal classes of experience necessary for the development of a natural individual; and the basic modes of energy, or archetypal qualities of being which essentially colour any functional activity operating in their field. The experiences of the person as a whole are limited and periodical. There are only a certain number of basic meanings to be gathered by a human being in his lifetime, and these can be seen in terms of structural and cyclic sequence.
Astrological horoscopes first became available to the general public, as opposed simply to the rich or influential, with the introduction of the Julian calendar, when the practice became less difficult and expensive. The sun entered a new sign of the zodiac at about the same date each month and each day of the week was named after a planet. Some astrologers denounce the majority practice of mass-produced printed horoscopes dealing with every aspect of life and suggest that only application of individually calculated astrological techniques is adequate, accurate and useful to a specific individual.
Synastry is the branch of astrology concerned with techniques of comparing the birth charts of two or more persons, especially to understand the mututally constructive aspects of any relationship. It is much used in assessing potential marital relationships.
In the past, criticism of astrology centred on the conflict between a predetermined future and the role of free will. Nowadays, many claim that modern scientific techniques and powerful telescopes, which show astrology to have been based on an over simplified model, have disproved completely its occult implications. Others have put forward a number of models to attempt to supply a basis for astrology. These include:
Physical model (Other than the symbolic perspective noted above) Some evidence exists for sun-planet interaction in which the planets, through their interacting gravitational and magnetic fields, modulate the basic radiation cancelled by the sun. The different frequencies channelled in this way are linked to various (cyclic) biological, psychological, and social processes on earth and to changes in the glandular activity of particular personality types.
Planetary heredity. It is suggested that an inherited genetic factor, responsive to planetary action, stimulates the onset of labour and hence largely determines the time of birth of a person. The tendency to be born at the time of a particular planetary configuration is therefore inherited, and the configuration therefore indicates the presence of a genetic factor governing a particular personality disposition.
Synchronicity. This suggested principle takes the coincidence of events in space and time (such as a birth at the time of a particular planetary configuration) as meaning something more than mere chance, namely a peculiar interdependence of objective events among themselves as well as with the subjective (psychic) make-up of the individual (s) involved.
Refs Ganquelin, F and Ganquelin, M *A Possible Hereditary Effect on Time of Birth in Relation to the Diurnal Movement of the Moon and the Nearest Planets* (1967); Meyer, Michael R *The Astrology of Relationship* (1976).
Related Maps of the mind (#HH8903)
Zodiacal forms of awareness (Astrology, #HM2713).

♦ HH0554 Satya
Truth
Description Since discrimination or intuition, *buddhi*, is clouded by any untruthfulness, utter truthfulness is necessary for making clear the spiritual path. Truthfulness also prevents the complications arising as more and greater lies have to be told to cover up a first, even minor, lie. These complications cause strain in the mind and are the basis for emotional disturbance. Any quest for reality necessitates a straightforward life and harmony with the law of truth to provide the necessary inner tranquillity, when the mind mirrors the divine reality.
Context One of the five moral qualities of yama (restraint).
Broader Restraint (#HH6876).

♦ HH0555 Self adornment
Dress — Uniform — Fashion consciousness
Description Dress is the most obvious status symbol, and to be dressed attractively and in fashion gives a sense of well-being and confidence. The human-being uses self adornment (dress, cosmetics, tattoos for example) to express physically his or her interior being and personality. The clothes a person wears and the way in which he or she wears them, the colours he or she chooses, express a fundamental and basic reality; and traditionally specific societal groupings have adopted particular forms of dress. The scriptures refer to the removal of earthly garments and being clad in heavenly garments; Christ refers to the guest being excluded from the wedding becomes he has not put on the appropriate clothing.
In present-day society, uniform is seen as having no links with individual personality, since it irons out all individual and social distinctions; the desire to express individuality is cited as one reason for the habit of ordained priests to wear lay clothes and of military men to prefer to wear civilian dress. On the other hand, uniform is a powerful symbol in that it demonstrates visibly the role an individual plays. Some "uniforms" are highly symbolic in content. For example, the habit of a Zen patriarch demonstrates, through its round hat a knowledge of the heavens, through its square shoes a knowledge of the earth, and through its bells an understanding of universal harmony. The wearing of white (for baptism, confirmation, and as a bride) are traditional demonstrations of inner purity. The Chinese nobleman demonstrated in his clothing: harmony (12 bands representing the 12 months of the year), rounded sleeves (graceful bearing), straight-cut back (rectitude), and the horizontal lower edging (a heart at peace). The muraqq'a or patched frock granted to a novice by his shaykh at the end of mystical training in the Sufi system may indicate: by its collar, patience or annihilation of intercourse with others; by its sleeves, fear and hope or observance and continence; by its two gussets, contraction and dilation or poverty and purity; by its belt, self-abnegation or persistence in contemplation; by its hem, soundness in faith or tranquillity in the presence of God; by its fringe, sincerity or settlement in the abode of union.
The tendency not to wear uniform or formal clothing therefore demonstrates a diminishing sense of symbolism and consecration and, in a way, a rejection of the society or grouping that the uniform represents. It may also represent rapid change and mingling of societies - clerical dress, once the sign of service, may come to signify male domination; or what indicated virginity in one culture (the wearing of white) may mean mourning in another.
As well as dress, other more permanent forms of adornment - tattoos, filing teeth, binding feet - also may have deep ritual significance and be used as demonstrating rites of passage. And the changing of personal appearance through plastic surgery has deep psychological effects.
Related Rites (#HH0423) Physical appearance (#HH0993).

♦ HH0556 Asceticism
Description A voluntary renunciation of all sensuous gratification, sometimes in an attempt to train the will to endure enforced deprivation of pleasure. Asceticism is usually associated with the fulfilment of the ideal way of life postulated by some religious and even political systems; or as the preparation for some initiation. Christian asceticism, for instance, is concerned with distancing the soul from sin and wrong inclinations, to advance in the cultivation of virtues and to rest in loving union with God, spiritual perfection only being possible with the renunciation of the things of this life. Such renunciation is likened to the strict training of an athlete, as it is also in Greek philosophy where asceticism is synonymous with training of the mind, mental culture and ethics. It may also be seen as a preparation for death, either as a confrontation with the mortal condition by dying to the world or as death to one's self preparatory to experience of the divine.
True asceticism is said to occur naturally, since one's desires have to be curbed in order to dedicate one's behaviour to the needs of others; it should not be used as a defence against unacceptable feelings or as a therapy. The dangers of over-asceticism are mentioned in many religions, the chief danger being that of pride in one's ability to do without. Other criticisms demonstrate the duality of the approach which attempts to free man from the "defilement" of the body and its desires in order to progress spiritually, although this is hardly an argument against asceticism as such. Marxism rejects asceticism as contrary to the total and harmonious development of man.
Refs Colliander, Tito *The Way of the Ascetics.*
Related Fasting (#HH0011) Mortification (#HH0780) Mythic yoga (Yoga, #HH3405)
Via purgativa (Christianity, #HH4090) Ascetical theology (Christianity, #HH3652).

♦ HH0558 Commitment
Refs Wieman, H N *Man's Ultimate Commitment* (1958).
Related Ideological commitment (#HH0571).

♦ HH0559 Prophecy
Prophesy
Description Prophecy is considered theologically to be a gift to chosen individuals who are empowered to communicate the word of God and act as his spokesman. This may be a on a particular occasion as a gift of the Holy Spirit in specific circumstances; or a person may have the vocation of prophet on a permanent basis. The scriptures refer to both possibilities. In contemporary renewal movements the former case, congregational prophecy, is what is being experienced. The prophet may consciously or unconsciously be able to foresee the result of activities already set in motion. Certainly, in the sense of telling openly or proclaiming the word of God, prophecy may illuminate the past, the present or the future. The intent is to encourage or exhort, strengthen or console. It is not the same as inspired preaching or teaching, however, which are spiritual gifts there all the time. Prophecy only arises when there is inspiration to prophesy. This inspiration is referred to as *being in the spirit*; the prophet may not himself know the import of what he is saying.
The individual may be unsure as to whether or not he is called to make a prophetic statement. Very often a thought arises on what to say which persists and nudges, but even so it should be considered, prayed over and then the right occasion waited for and not forced. When the appropriate moment arrives, the statement should be made with confidence. It is as important to know when to stop as when to start - simplicity is necessary and once inspiration fades elaboration should not be forced.
Related Premonition (Psychism, #HM3606)
Communication with supernatural beings (#HH1306).

♦ HH0560 Catharsis
Description The essence of aesthetic experience, said by Plato to consist of the freeing of the soul from the body and from passions and pleasures. Whether this term should refer to the elimination or the harmonizing of particular emotions is a matter of controversy; but its original sense involved purification or purging - the freeing from undesirable emotions through vicariously experiencing them represented on the stage; it has been used to describe aesthetic, ethical and medical phenomena.
In psychology, catharsis represents the release of repressed emotions which the patient has previously been struggling not to express. This is one of the basic aims of psychotherapy.

♦ HH0561 Integration of the personality
Disintegration
Description This is the process by which all the systems of response that represent an individual's characteristic adjustments to his various environments are formed or coordinated into a consistent and coherent totality. It is the consistency with which diverse individual processes or actions, and in particular the individual's decisions, prove effective in functional dependence on each other.
Integration of the personality may also mean the process of formation of the whole personality through individual effort at personal development, possibly with psychotherapeutic assistance. In this sense it is analogous to mental health or maturity and the necessary grounding for individuation.
According to one definition used by Jung, integration may imply a diagnosis of the psychological situation of an individual, including examination of conscious-unconscious interaction, masculine and feminine components of personality, the various opposites, relation of the ego to the shadow and movement between function and attitudes of consciousness.
Disintegration is the disorganization of the psychic processes leading to inconsistencies and a general weakening of the effect of the volition over thinking, feeling and acting.
Related Dissociation (#HH1294) Strength of personality (#HH0198).

♦ HH0562 Growth centres
Description Growth centres are places to which individuals can go to learn or experience personal development. They vary greatly in importance and the range of techniques used. Some use only one technique, others use a very wide range.

♦ HH0563 Predestination (Christianity)
Description This term may be used in a wide sense to refer to the will of God in revealing Himself though creation and providence. It is a philosophical conception fundamental to religious belief, implying prescience (foreknowledge) and prevenience (acting upon that knowledge). In a narrower sense, predestination may refer to the predetermination of whatever occurs, in particular the destiny of good and evil, including eternal damnation of the wicked. There is some inconsistency here between human freedom of choice and moral responsibility for action, and foreknowledge of such action on the part of God. Such discrepancy is bound up with the concept of time and eternity and in the necessary limitation on human knowledge imposed by time. However, it has been argued that freedom of choice is not simply self-will but the growth in the faculty of cooperation with divine purpose; there is the choice to close one's self against assimilation of the proffered wisdom and life - "Our wills are ours to make them Thine".
According to Christian ethics, mankind is the end of physical creation, and history the progression by which means spiritual man evolves. It is the nature of God to actualize Himself in humanity and the human spirit recognizes itself as being divine in principle. In Christ there is perfect consciousness of this divinity and his incarnation demonstrates the unity of the human and the divine, revealing mankind's divine nature to mankind and stimulating the human spirit to new life. This revelation is an act of grace not of mankind's growth to consciousness of divinity. Christ's life, death and resurrection reveals the will of God for the salvation of mankind.
Refs Ockham, William of *Predestination, God's Foreknowledge and Future Contingents* (1983).
Related Freedom of choice (#HH0789).

♦ HH0565 Moral development
Moral education — Moral growth — Development of moral values — Moral delinquency — Degeneracy
Description Moral development occurs as one result of an increasing ability to perceive objective reality, to organize and integrate experience, and to form finer discriminations and more comprehensive generalization. One necessary (but not sufficient) condition for principled morality is therefore general intelligence. The main experiential determinants of moral development are held to be the amount and variety of social experience, and the opportunity to take a number of roles and to encounter other perspectives. Developmental change can be fostered in a classroom situation with children exposed to a moral education curriculum of a series of increasingly complex moral dilemmas. Stimulating such development depends upon identification of the child's current stage and presenting him with dilemmas that cannot be easily solved with the modes of thinking characteristic of that stage. The child then becomes uneasy with his own judgments and finds superior principles which provide more equilibrium and logical satisfaction.
Moral and religious education are distinguished from each other since moral duty may be taken as duty towards one's fellow men, whereas religious duties are to one's God. It has been objected that morality is dependent on religion, but this is true of any subject in a religious environment and does not preclude their treatment as separate subjects. Moral education has in some cases replaced religious education in current Western secular society and is part of the normal school curriculum in many countries.
To some extent moral education must depend upon the society in which it takes place; but a morality taught simply on the basis of obedience to the present regime (whether parental, scholastic or social) is held to be non-progressive. A more enlightened aim is to communicate a sense of personal, social, civic and international responsibility. Using *justice* as the organizing principle for moral education is held to meet satisfactorily all the necessary criteria: guarantee of freedom of belief; use of a philosophically justifiable concept of morality; and the basis of the psychological facts of human development. Individuals are believed to acquire and refine the sense of justice through a sequence of invariant developmental stages which are:
1. Orientation to punishment and reward, and to physical and material power.
2. Hedonistic orientation with an instrumental view of human relations with notions of reciprocity based mainly on exchange of favours, not on loyalty, gratitude or justice.
3. Conformity to stereotypical images of majority or "natural" behaviours, seeking to maintain expectations and win approval of the immediate group morality.
4. Orientation toward authority, law, the maintenance of a fixed order (whether social or religious) which is assumed as a primary value.
5. A social-contract orientation, generally with legalistic and utilization overtones, emphasizing equality and mutual obligations within a democratically established order.
6. Orientation toward the decisions of conscience and toward self-chosen ethical principles appealing to logical comprehensiveness, universality and equality of human rights, respect for the dignity of human beings as individual persons.
A more recent (1984) approach of L. Kohlberg to these stages lists 7, each describing the distinctive way in which people at that stage reason out a decision in the face of conflict between competing claims. They are in many ways analagous to the above 6: 1. Heteronomous morality. 2. Individualistic, instrumental morality. 3. Interpersonally normative morality. 4. Social system morality. 5. Human rights and social welfare morality. 6. Morality of universalizable, reversible and prescriptive general ethical principles. 7. A hypothetical stage based, not on reason, but on the more metaphysical or religious question of why be moral at all.
Other approaches consider moral judgement and reasoning too narrow a focus for the discussion of moral development; and attempt to include other factors, such as intention, perception, outlook, disposition, habit and values, among others. This would focus more on development of personal responsibility in relation to character and relate this intentionality to those moral intentionalities that transcend the life of the individual and, indeed, social life. An 8-stage epigenetic cycle relating ego strength development to basic human virtues (Erik Erikson) goes some way to formulating such a system.
Moral delinquency or degeneracy, previously equated with physical characteristics from general ugliness to flat feet, is now said to derive from continual exposure to stimulations which dull the senses (in particular, to over-loud pop music) and to mass media which titillate the weak-minded to emulate the scenes of brutality and pornography they witness. Higher degenerates are said to be those with high verbal ability but no moral stamina, who gravitate towards cults, politics and fads which demand much talk but little disciplined action. They simultaneously hold that freedom is everything, while vehemently denying it to those whose opinions do not agree with theirs.
Refs Kohlberg, L *Essays on Moral Development* (1981); Kohlberg, L and Kramer, R B *Continuities and Discontinuities in Childhood and Adult Moral Development Revisited* (1973); Kohlberg, Lawrence *The Development of Moral Character and Ideology* (1964); Kohlberg, Lawrence *Stages of Moral Development as a Basis for Moral Education* (1971).
Related Virtue (#HH0712) Maturity (#HH0971) Atman project (#HM2153)
Religious growth (#HH1321) Moral perfection (#HH5098) Maternal thinking (#HH4646)
Cognitive growth and development (#HH0146).

♦ HH0566 Convulsive therapy
Shock therapy
Description The convulsions of the epileptic state may be artificially induced by the injection of certain drugs or by the use of electric shock treatment. The violence of the convulsions may be reduced with other drugs and the pain largely eliminated by combining these with intravenous anaesthesia. The shock to the nervous system caused by these epileptic convulsions favourably influences the course of mental disease and is held to be antagonistic to the schizophrenic state.
Broader Therapy (#HH0678).

♦ HH0567 Karma
Kamma — Action
Description According to Panini, karma is the aim of any action, that which the person instigating the action desires as the result of that action (karturipsitatamang karma). Since actions are rarely done for their own sake but for some other result (which is, by definition, a further action) karma will therefore lead to further action and has come to mean the action itself, and also the sum of the effects of past actions. This in turn leads to karma as the cycle of birth, death and rebirth, and to the conditions into which an individual is born for this incarnation. According to the vedas, only by performing actions for their own sake (akarmaka) can the individual free himself (moksha) from the cycle of birth and death; actions done with good intent lead to a propitious rebirth whereas those done with evil intent lead to a rebirth under unfavourable conditions.
Thus karma determines the conditions of existence of an individual, and his or her ultimate progress in terms of the main goal of existence - according to Hinduism, Buddhism and Jainism - freedom from the laws of cause and effect and from the ties of mortal existence. The present is thus reaping the fruits of past actions while sowing the seeds for the future. One is born with one's karma for this life - 'ripe karma' - but as one goes through life one pays off old karma and makes new karma. Buddhism, however, rather than rebirth looks to rebecoming, with no continuous substantiality or ego, although the new entity originates in dependence upon the preceding existence. Through true knowledge, the traces left by activities - samskaras - are eradicated and the karmic cycle suspended. This requires profound inner reorientation, metanoia or spiritual conversion, as exemplified on the *noble eightfold way*. Theosophy points to the gradual building up of strength of character, directing thought constantly to the positive attributes that are lacking and eventually, over many lifetimes, eradicating the negative attributes caused by that lack.
Some hold the view that, to escape the cycle of birth and death, it is necessary only to do actions which pay off old karma and do no action to create new karma. This is all that is necessary for enlightenment or liberation. There is even a view that, after reaching liberation, a sage remains alive until all old karma is paid off.
Refs Neufeldt, Ronald W (Ed) *Karma and Rebirth* (1986); Rama Rao Pappu, S S (Ed) *The Dimensions of Karma* (1987).
Related Fate (#HH1362) Merit (#HH0859) Purpose (#HH0691)
Pre-existence (#HH1339) Karma yoga (Yoga, #HH0372) Samsara (Hinduism, #HM1006)
Eightfold way (Buddhism, #HM2339) Karma awareness (Buddhism, #HM2048)
Human development (Hinduism, #HH0330).

♦ HH0568 Goodness
Description According to Mencius the nature of mankind is naturally disposed to love, right, courtesy and wisdom. Every person has within him or herself the seeds of these qualities which

HH0568

he or she can encourage to sprout and grow – not for some selfish motive but simply because it is good to do so.

♦ HH0570 Art
Description A work of art is at the same time a material structure, whether experienced through sound (music or words), vision (colour or form) or whichever of the physical senses, and a spiritual image of reality. It acts as a means of communication among people and of enriching their knowledge of the world and of themselves. Meaning may be at the personal level, symbolizing the artist's own experience; specific to an individual culture; or universally understood independent of culture.

The primary function of a work of art may be said to consist in its capacity to extend vision, purify perception, elevate consciousness, intensify sensibilities and illuminate experience in time with timeless reality. Even when art is considered as simply a representation or imitation of what is commonly perceived as reality it has value in that, together with its aesthetic enjoyment, it instils a specific system of values and arouses emotions similar to "real" experiences. It may express in ritual what is considered dangerous or taboo if expressed openly. On a higher level, it may be thought of as the highest form of language, that of communicating perception. This is witnessed by the struggle some artists experience in the effort to express exactly what they (perhaps unconsciously) intend; and by the ability demonstrated by "non–artists" to produce one particular work in response to an inner need for expression, this ability disappearing when the need is fulfilled. The capacity to appreciate the work of art of another demonstrates a kinship with the artist, some spark of his genius existing in every individual.

For the artist, to produce a work of art is consciously to experience his inner perceptions and images, the work itself being an autonomous self–representation of perception. Nature (as the artist) may in a sense be said to experience nature (as the perception) – subject and object thus becoming one.

The importance of the individual in transmitting cultural skills for the purpose of preserving traditional art forms is recognized in Japan by the designation of people particularly gifted in this sense as living national treasures.
Refs Dewey, John *Art as Experience* (1959); Wade, Nicholas *The Art and Science of Visual Illusions* (1983).
Narrower Artistic education (#HH0029).

♦ HH0571 Ideological commitment
Description This refers to the commitment to a coherent totality of ideas and ideals which goes beyond knowledge of such ideas and the passive acceptance of certain principles, and which implies consistent loyalty at all levels – theoretical and practical. Ideological commitment will cause an individual to see and act beyond his or her own narrow or local sphere of interests and to assess any new ideas or circumstances in the light of this ideology. It acts as a source of energy, moral courage and enthusiasm, and leads to acts of heroism.
Related Commitment (#HH0558).

♦ HH0574 Death–urge
Thanatos — Life–urge — Eros
Description It has been said that no–one can be truly aware that he is alive unless he is aware of his own death as a physical fact; this is the basis of tragedy and the authenticity of the human struggle. Life–urge and death–urge are therefore inseparably part of life, and the one cannot be understood without the other.
Related Eros (#HH7231).

♦ HH0575 Physical education
Description This includes physical exercise and conditioning, calisthenics, gymnastics and athletics. It is designed to train the human body in order to improve its function, and to provide an outlet for the natural need for action. This relates to one definition of health which is the capacity of all body organs and systems for high–level function, in contrast to the narrowest definition of health, namely the absence of disease. Individuals may be particularly concerned to develop their musculature in order to increase their strength and to improve their physical appearance. Physical education programmes have been considered by governments as an important component of national (or racial) advancement.
Refs Paplauskas–Ramunas, Anthony *Development of the Whole Man through Physical Education* (1968).
Broader Education (#HH0945).
Related Physical culture (#HH0730)   Mysteries and religion (#HH4032).

♦ HH0576 Born–again (Christianity)
Spiritual renewal
Description A contemporary, fundamentalist Christian movement in the USA, *Born–again Christians* base their spiritual rebirth on Christ's admonition (St John III 3), "Except a man be born again, he cannot see the Kingdom of God". Through repentance and faith, the individual makes a personal commitment to Christ and receives an experience of assurance imparted by the Holy Spirit. The spiritual renewal obtained signifies a new beginning so radical it can be thought of as a rebirth.
Broader Spiritual rebirth (Christianity, #HH4098).
Related Twice–born (#HH0715)   Second birth (ICA, #HM3175)   Initiation (Christianity, #HH6211)
Spiritual rebirth of the ego (#HH0698)   Charismatic renewal (Christianity, #HH3124)
Initiation as spiritual rebirth (Yoga, #HH3465).

♦ HH0577 Sacrifice
Description Essentially that part of a religious ritual in which gifts are offered to a deity or spirit, sacrifice ranges variously from the dedication of an action or a song of praise, to the shedding of blood of animals or humans to propitiate some god. Typically, a person authorized to represent a group of worshippers dedicates a material offering by changing it in such a way as to withdraw it from profane use and bring it within the sacred sphere. Sacrifice and belief in the magical power of that which is sacrificed are closely linked, primitive belief being that the gods fed on what was sacrificed to them. Although this may no longer be believed in the sense of physically devouring food, the spiritual aspect of a god feeding on worship offered with devotion is not so far-fetched. The ritual of offering food may be said to demonstrate at the physical level an action which is taking place spiritually. Many such rituals include a communion in which participants join in consuming the food ritually offered. According to some systems of thought, every action is an offering to some "god", depending on the intention behind the action.

In line with a common use of the word sacrifice, a sacrificial act normally includes an aspect of renouncing or forgoing as well as making sacred. What is offered is ritually forfeited, and may be destroyed, in order to establish relations with a source of spiritual strength for the benefit of one in need of such strength. It is an expression of adoring self–surrender and a sign of God's willingness to enter communion with man. The act of sacrifice is thus, according to Jung, one of renunciation to a principle supraordinate to present consciousness. At some point every individual has to give up some psychological attitude, whether or not neurotic. In order to go on to a more meaningful or significant ego position a lesser one has to be renounced. This transition occurs when conscious contents present themselves and there is a conflict of opposites; and sacrifice is the price paid for consciousness. Again, just as religious and ritual sacrifice means the giving of something as though it were to be destroyed, so the transition requires giving up part of one's personality and self esteem, the price paid for being human.
Broader Rites (#HH0423).
Narrower Self sacrifice (#HH1241).
Related Blessing (#HH0039)   Surrender (#HH0449).

♦ HH0578 Virtuous mental factors (Buddhism)
Kushula — Dge–ba
Description This group of eleven mental factors are the natural virtues defined in Tibetan Buddhism as being virtuous independent of other factors such as motivation. They may be attained at birth, through application, through intention, through assisting others, in overall behaviour, in overcoming the unfavourable and objects of abandonment, by pacification, or by means of a favourable power or clairvoyance.
Broader Secondary mental factors (Buddhism, Tibetan, #HH1348).
Narrower Effort (#HM2389)   Pliancy (#HM3162)   Non-hatred (#HM2744)
Equanimity (#HM7769)   Basic faith (#HM3209)   Non–ignorance (#HM2695)
Embarrassment (#HM2210)   Non–attachment (#HM2128)   Sense of shame (#HM2881)
Non–harmfulness (#HM2608)   Conscientiousness (#HM3220).
Related Sense consciousness (Buddhism, #HM2664).

♦ HH0579 Integral meditation
Description As part of the practice of integral yoga, integral meditation involves the following sequence: (i) active or dynamic offering of the self to the cosmic reality; (ii) in the state of self-offering or self–observation, to explore the self (psychic observation); (iii) this observation eventually activates the energy of the unconscious psyche (sometimes called kundalini, serpent power); (iv) this power has to be judiciously channelled and controlled in purity of heart and involves critical evaluation; (v) direct insight, or existential experience, of the ultimate ground of existence, which brings the experience of unity referred to in various systems as cosmic consciousness, bodhi, satori, samadhi.
Broader Meditation (#HH0761).
Related Integral yoga (Yoga, #HH0037).

♦ HH0580 Contemplation of nature
Description This is employed in a practice of meditation where one particular aspect or object of nature is contemplated so that the qualities of that which is contemplated become one with the consciousness of the person practising meditation, leading first to a oneness with nature and then with the infinite.
Related Meditation (#HH0761)   Deep ecology (#HH2315)   Naturalness (Zen, #HH0343)
Communion with nature (#HM0915)   Attentiveness to nature (Zen, #HH0463).

♦ HH0581 Holistic medicine
Holistic health care movement
Description A system of medicine, encompassing a variety of new and not-so-new approaches, which has as its major tenets: health for the whole person; change in attitude conducive to healthy body chemistry and internal harmony; self–care and practitioner cooperation with the patient/client, rather than an authoritarian approach; careful diet and use of therapeutic plants rather than orthodox medicine; stimulation of the body's own healing processes rather than imposing external cures on it; discovering and removing the cause of ill health rather than treating the effects; the social and family context of health. Techniques used include homeopathy, acupuncture, spiritual healing, esoteric healing, bioenergetics and gestalt therapy. Self–reliance and self–responsibility are emphasized, and the mind and body are treated as a single whole, taking cognizance of lifestyle, environment and spiritual development.
Broader Holistic approach (#HH0649).
Narrower Homeopathy (#HH0498)   Reiki therapy (#HH0731)
Spiritual healing (#HH0458)   Bioenergetic analysis (#HH0652).
Related Primal healing (#HH2314)   Gestalt therapy (#HH0751)
Self responsibility (#HH0187)   Humanistic psychology (#HH0793).

♦ HH0582 Theatre games
Description A theatre game is based on an outline scenario designed around a specific problem to be solved through the development of the drama. The solution is not written into the scenario but must emerge from the player–participants' improvised response to the situation and to the evolution of each other's roles. Provided the environment permits, each player–participant may learn whatever he chooses to learn. The game experience therefore increases the individual's capacity for experiencing and for surfacing unforeseen personality potentials. Through spontaneity in such a setting, individuals re–form their images of themselves and of each other. Spontaneity then provides a moment of personal freedom for the individual to explore reality, how he perceives it, and how the aspects of his personality function in relation to it, as an organic whole.
Refs Spolin, Viola *Improvisation for the Theater* (1963).
Broader Games (#HH0705).
Related Psychodrama (#HH0512).

♦ HH0583 Chastity (Christianity)
Description In Christian terms, chastity is said to be primarily an aspect of love in expressing respect for other human beings and of one's own nature, both physical and psychic, as creations of God and redeemed in Christ. It implies sexual purity – virginity for the unmarried, loyalty to one's partner for the married and continence in the widowed. The state of remaining physically chaste is mirrored spiritually in devoting the powers of body and soul to God; Christ's teaching against adultery includes the desire as well as the physical act. Other traditions similarly value chastity, in particular the virginity of a woman when she marries and restraint from intercourse at times of religious or ritual ceremonies.
Refs Felix, M *New Look at Chastity* (1974).
Related Virginity (#HH0272)   Continence (#HH0070)   Brahmacarya (#HH0978).

♦ HH0584 Self–accusation
Description A feature of *gedatsu therapy* when the individual (usually a woman) blames herself for her suffering and for the suffering of others, even where it appears that it is another who is in the wrong (when the ego must apologize for allowing the other to do wrong). Interpersonal conflict is treated by demonstrating the fault in the other person to be a reflection of the individual's own disposition. Suffering is therefore said to be rooted in the mind of, and the fault of, the sufferer.
Broader Gedatsu therapy (#HH0972).
Followed by Allocentric attribution (#HH1249).

♦ HH0585 General semantics
Linguistics — Time–binding
Description General semantics is a discipline whose aims are to explain to individuals how to use their nervous systems most efficiently and to provide training toward that end. It involves the

formulation of a non–aristotelian system of orientation. The methods used focus on psychosomatic factors which help to balance and integrate the functions of the nervous system. They eliminate or alleviate: different semantogenic blockages; many emotional disturbances (including some neuroses and psychoses), various learning, reading, or speech difficulties; and general maladjustments in professional and personal life. The structure and form of language may be said to some extent to determine the perception of reality of the speaker. Neurotic behaviour is held to stem from a lack of clear understanding in the use of words and their meanings. Therapy consists primarily in teaching individuals correct word habits to replace previously acquired faulty orientation. The individual is then able to use symbols efficiently, to think critically about his values and lifegoals, and to substitute reality oriented ends and means for self–defeating patterns.
The misuse of *neuro–linguistic* and *neuro–semantic* mechanisms is said to be a major reason for deterioration in human values. In general semantics, the main emphasis is placed on non–identification, or the elimination of erroneous identifications in relating to and defining the objects of perception. Man is conceived as a *time–binder*, stratifying reality into many orders of abstraction and comprehending the chain of past-present-future. This results in the ability and the desire to transmit ideas from generation to generation through the study and refinement of language. Realization of such stratification of reality is held to eliminate identification, leading to consciousness of the abstracting process and enlargement of the field of consciousness in general. Training in consciousness of abstracting provides a simple psychophysiological means by which to integrate the functioning of the human nervous system and for the consequent elimination of many human difficulties.
Refs Murray, Elwood et al *Integrative Speech* (1953).
Related Language development (#HH0485).

♦ HH0586 Self–observation
Description Self–observation is the process of studying one's intellectual, emotional and other functions separately from each other, and the manner in which one is involved and identified with them. It is always directed at some definite function, whether observation of thoughts, movements, emotions or sensations. There is always a definite object to be observed within oneself.
Self–observation leads on to self–remembering (described separately) in which the person is not himself divided by the observation process but experiences himself as a whole.
Related Introspection (#HH0824) Self–remembering (#HM2486)
Self–observation therapy (#HH2276)
Detached self-observation meditation (#HH0229).

♦ HH0587 Non–temporal dimension of being
Siva
Description There are said to be three essential aspects of *reality*– the pure non–temporal, the dynamic universal and the unique individual. The non–temporal is the sustaining dimension of the cosmic whole, the ground of existence beyond subject and object or mind and matter. The inner dichotomy which produces hate in the name of love, war in the name of peace and falsehood in the name of truth is dissolved in realization of the non–temporal which goes beyond the quasi–permanence and security of fixed forms and ideas. In realization of the non–temporal the separate consciousness or ego is dissolved to be spiritually reborn, blissfully aware of its oneness with God.

♦ HH0588 Mass communication
Mass media
Description By systematically disseminating information from one source to a massive audience (via radio, television, the press, and so on), the intellectual and moral values of numbers of individuals can be affected. The media are therefore very powerful in political and ideological terms.

♦ HH0589 Ch'an (Zen)
Hsin tsung — Mind doctrine — Dhyana
Description This is the doctrine of ultimate or Buddha mind which, on reaching Japan from China, was altered to become Zen. The essence is meditation, a conscious, prolonged and persistent effort towards direct awareness of the source of that inner consciousness that is the source of striving to live and live more fully. The movement started as individuals in isolation leading a life of harmony and meditating to attain tranquillity and intuitiveness. Although some such individuals acquired followers, and monasteries and temples were set up, the emphasis was still on the individual and also on physical work as well as meditation.
Related Dhyana yoga (Yoga, #HH0827) Human development (Zen, #HH1003)
Dhyana (Hinduism, Buddhism, #HM0137).

♦ HH0591 Philosophy of enlightenment
Description A movement arising in the 17th and 18th centuries, the philosophy of enlightenment can be considered the emergence of the individual from intellectual dependence. Human reason, rational self–consciousness, was held to be capable of discovering the truth about humanity, the creation and the divine. Supernatural revelation was either denied or considered unnecessary for such discovery. The motto of the movement was *sapere aude* - have the courage to think - and the movement instigated the "Age of Reason" or "Age of Philosophers", believing that society could be rebuilt on rational foundations with the downfall of religious dogmatism and the triumph of science over the mediaeval, traditional scholasticism of the Church. According to Kant, man's state of dependence and inability to use his intellect without guidance was his own fault, and required decision and courage to change. The movement called for freedom from controls on individual independence and liberty. The emphasis on what man could do with his own powers led to scientific and mathematical analysis, experiment and discovery; to humanistic and atheistic tendencies; to scepticism and destructive criticism; and precipitated the French Revolution. Theologically, there seems a discrepancy between the struggle throughout history for freedom and the perfection of human endeavour through the intercession of God.
Related Pedagogy (#HH0179) Progress (#HH0520) Humanism (#HH0674)
Agnosticism (#HH0904) Illumination (#HH0804)
Enlightenment humanism (#HH0215).

♦ HH0592 Mode of life
Lifestyle
Description This is a sociological term which covers distinctive features of all spheres of human activity and categorizes a particular socioeconomic group, whether defined by class, working conditions, education or domicile (urban or rural, for example). Less restricted than *standard of living*, the expression "mode of life" includes qualitative characteristics such as means of expression, rules of behaviour and use of leisure activities. The expression "lifestyle" has similar connotations but refers not to a group, but to the individual's chosen pattern of existence.
Related Life–style (#HH0941) Consciousness shift (#HM1793)
Disruption of lifestyle (#HH3356).

♦ HH0593 Sacred places
Sanctuary — Holy places
Description Holy places are areas particularly associated with the presence of a spiritual force or god. This may be a building such as a temple, built in a place where a god may have manifested his presence; or simply a place such as a spring, tree or mountain, often associated with particular rites (eg. fertility). Areas like this come to be taboo and persons entering them have to fulfil certain criteria and perform certain rituals.
Related Sense of sacredness (#HM2302).

♦ HH0594 Cursing
Official curse
Description The verbal repudiation, for example, of a person or even of God (cf Job... "Curse God and die"). A belief is implied in the power of the spoken word as opposed to simple wishing; this power then may be sufficient to carry out the curse (or blessing) spoken into existence, or it may be an appeal to a higher (probably supernatural) power to carry it out. Cursing in a ritual or official way by someone in authority induces terror in the victim who may feel unable to resist the power of such a curse. Although the supernatural quality of words may no longer be theologically admitted, the Book of Common Prayer of Edward VI contains "A Commination or Denouncing of God's Anger and Judgements Against Sinners" to be used as a "warning" for the cursing of various misdoers.
A curse may involve some other activity as well as the speaking of certain words; the curse may empower some object of action, such as a knife or a poison, and may be accompanied by appropriate gestures. It may also be passed on from the object of hostility to his relations and descendants. Depending upon the spiritual power of the person effecting the curse it may be more or less efficacious; in psychic terms, there is focusing on a thought form which is moulded into an "elemental" that is then invoked in the curse.
Related Blessing (#HH0039) Ill–wishing (#HH3873).

♦ HH0595 Mime
Description The techniques of mime can help to integrate mind and body. Harmony of mind is developed through the use of imagination and by becoming aware of the movements, actions, and reactions required by the individual to present himself

♦ HH0596 Occupational therapy
Diversional therapy
Description Individuals who are inactive or confined to bed usually do not recuperate as well as those who are in some way occupied, particularly if they are also psychologically disturbed. Occupational therapy involves such persons in tasks appropriate to their capabilities and disabilities. It is suggested that the method can assist individuals to gain some insight into their psychological condition, leading to improved sociability and self–confidence.
Diversional therapy comprises those activities in occupational therapy designed simply to occupy the patient and divert his or her mind from the surroundings. Other activities involved in occupational therapy may also have the goal of overcoming specific disabilities or handicaps, improving motor coordination, or of rendering the patient physically capable of rehabilitation into useful work. The social aspect of group working has a beneficial recuperative effect, but occupational therapy in the patient's home or at other time's when the patient is alone is also of value.
Broader Therapy (#HH0678).

♦ HH0597 Hasidism (Judaism)
Description This 18th and 19th century movement of non–exclusive, non–esoteric Judaic mysticism rejected formal ritual in favour of personal relationship between the individual human being or ego and the Thou of faith. The spiritual leader, or *zaddik*, forms a relationship with his disciple and acts as his spiritual guide simply by being himself. Their respective roles have much in common with the psychotherapist and his patient. The emphasis is not on what to do but on how to do it; and on relationships with other people as the nearest approach to relationship with God, tempered with periods of solitude and silence. There is an inate reverence for life and joy in the everyday things of life.
Refs Hoffman, Edward and Schachter, Zalman M *Sparks of Light* (1983).
Related Psychotherapy (#HH0003) Mystical cognition (#HM2272)
Spiritual direction (#HH0442) Fundamental dialogue (#HM3225)
Soul care (Christianity, #HH1199).

♦ HH0599 Cosmic identification
Description In psychological terms, this refers to the failure on the part of an individual (usually schizophrenic) to differentiate between him or herself and the outside world.
Related Herd instinct (#HM2559).

♦ HH0600 Temperance
Abstinence — Sobriety — Self control
Description The virtue of moderation, controlling the sense of appetite with respect to food and drink (abstinence) or intoxicating drink (temperance), as an exercise of self restraint, is considered a training in *will power* and in encouraging the ability to act in accordance with *reason*. Thomas Aquinas defined temperance as the order of reason with regard to the passions when these draw us to something irrational.
Context Also referred to as self control, temperance is one of the four cardinal or principal virtues recognized by Plato in the Republic and featuring prominently in mediaeval Christianity.
Broader Virtue (#HH0712).
Related Fasting (#HH0011) Justice (#HH2117) Prudence (#HH7902)
Fortitude (#HH5640) Continence (#HH0070).

♦ HH0601 Teleology
Description The concept that behind every change there is a directional force, design or purpose.
It has been said (A K Sinha) that it is the telic nature of human personality which produces its internal order and gives rise to the ability to make free and rational decisions, to plan and to seek adventure. Purposive behaviour is closely related to creativity. As the individual matures it is said that interest shifts from biological creativity (procreation) to utilitarian, aesthetic and intellectual creativity and appreciation, the requisite subjective or psychic order for the latter arising through intuitive comprehension and creative imagination. Such development arises by reflection in solitude, and is fostered by: individual freedom and opportunity; creative education; value–consciousness which encourages ethical, juridical and cultural conscience and works against egoism.
In these terms, society is seen as a system of telic persons organized to realize the same ends, interpersonal harmony being achieved through critical appraisal, formal approval or emotional involvement, these all being expressions of mutual appreciation. Such a society is culturally dynamic and progressive. In turn, a dynamic and well integrated sociocultural system is a factor in stable world order, since such stability is only possible where each component society is dynamic but with minimal disorder. Mutual tolerance and appreciation between societies is

therefore an extension of such tolerance and appreciation among telic persons and lasting peace may be achieved through telic unity and creative advance.
Refs Rosenblueth, A; Wiener, N and Bigelow, J *Purpose and Teleology* (1943).
Related Creativity (#HH0481).

♦ HH0602 Existential unity of being
Description An individual normally has a sense of his or her presence in the world as a real, live, whole, and in a temporal sense, continuous person. As such, he or she can live out into the world and meet others: a world and others experienced as equally real, alive, whole and continuous. In ordinary circumstances, the individual then experiences his own being as differentiated from the rest of the world so clearly that his identity and autonomy are never in question, being seen as having an inner consistency, substantiality, genuineness and worth and as being spatially coextensive with the body.
Such a basically ontologically secure person will encounter all the hazards of life (social, ethical, spiritual, biological), from a firm central sense of his own and other people's reality and identity. It is often difficult for a person with such a sense of his integral selfhood and personal identity (of the permanency of things, of the reliability of natural processes, of the substantiality of others) to transpose himself into the world of an individual whose experiences may be utterly lacking in any unquestionable self-validating certainties such as an over-riding sense of personal consistency or cohesiveness.
Refs Laing, Ronald D *Divided Self* (1960).
Related Completeness (#HM3070) Divine indwelling (#HH0814)
Human potential movement (#HH0398).

♦ HH0603 Harmonies with enlightenment (Buddhism)
Yogic paths
Description According to Buddhist teaching, these are divided into seven sections. All 37 harmonies are attained when the state of *foe destroyer* is actualized.
Broader Stages of spiritual life (#HH0102).
Narrower Suchness (#HM2435) Cessations (#HM3310)
Five powers (#HM3238) Five forces (#HM2948)
Eightfold way (#HM2339) Thorough abandonings (#HM3057)
Forms of enlightenment (#HM2886) Establishments in mindfulness (#HM2997)
Prerequisites of manifestation (#HM3170).

♦ HH0604 Respect for superiors (Buddhism)
Apacayana
Context The fourth of the ten meritorious deeds of South Asian Buddhist thought.
Broader Ten meritorious deeds (Buddhism, #HH0446).
Followed by Attending to the needs of superiors (Buddhism, #HH0383).
Preceded by Bhavana (Buddhism, #HH0551).

♦ HH0606 Self-therapy
Description A series of techniques aimed at facilitating personal development and intended for use by an individual without the aid of a therapist.
Refs Schiffman, M *Gestalt Self Therapy and Further Techniques for Personal Growth* (1971).
Broader Therapy (#HH0678).

♦ HH0607 Group-centred therapy
Description Group-centred therapy is the application of client-centred therapy in groups of about six people (together with the therapist), all usually having similar backgrounds and problems. Such groups tend to meet on a regular basis, usually twice a week for one hour sessions. Participants bring up any topics they consider significant, the therapeutic role often falling to participants other than the therapist; the main role of the therapist being acceptance, reflection, and clarification of attitudes and feelings presented by other members of the group. The group setting provides many opportunities for interpersonal interactions which constitute the testing ground and support for the most important changes in personality.
Refs Hobbs, Nicholas *Group-Centered Psychotherapy*.
Broader Therapy (#HH0678) Human potential movement (#HH0398).

♦ HH0608 Beatification
Description An individual may have lived a life of such goodness that, after death, that person is publically venerated. The Roman Catholic Church permits the formal title of Beatus or Beata, 'Blessed' and allows veneration by exhibition of relics or pictures and saying of the particular office either in the region from which the person came or in the order of which he or she was a member. This limited recognition is widened to include the whole Church if the person involved is *canonized*.
Related Sainthood (#HH0331).

♦ HH0609 Spontaneity training
Description Through learning to deal with new situations and roles in a graduated series of constructed situations, the individual learns how to integrate these roles into his personality; he is then better able to act instantaneously in a new situation without loss of spontaneity.

♦ HH0610 Jnana (Yoga, Buddhism, Hinduism)
Knowledge
Description There are a number of variations in the exact meaning of the term, but in general jnana is the knowledge which is based on perception by means of the senses and their interpretation by the mind, in other words knowledge about something. Such cognition may or may not be accurate depending on misperceptions, errors, opinions and so on. Correct interpretation such that the knowledge is truly knowledge is referred to as *prama*. The term *wisdom* takes an evaluation of such knowledge into account, and implies the use of discrimination, *buddhi*. Beyond this is *prajna*, perfect or essential wisdom. Some commentators have indicated that jnana is all knowledge which may be perceived by the individual. It does not belong to that individual but is all said to be present, more or less hidden by ignorance, *avidya*.
Context According to Patanjali, one of the three forms of knowledge which need to be distinguished from each other and separated if confusion *samkirna* is to be avoided, the others being shabda and artha.
Refs Mishra, Rammurti S *The Textbook of Yoga Psychology*; Puligandla, Ramakrishna *Jnana-Yoga* (1985); Vivekananda, Swami *Jnana Yoga*.
Broader Knowledge (Buddhism, #HH1971).
Related Samyama (#HH0374) Ma'rifa (Sufism, #HM2254) Artha (Hinduism, #HH0353)
Jnana yoga (Yoga, #HH0927) Essential wisdom (#HM0107)
Sabda (Yoga, Buddhism, #HH0538) Ignorance (Buddhism, Yoga, Zen, #HM3196).

♦ HH0611 Precocious talent
Exceptionally gifted children
Description Two approaches have been used in developing the talents of exceptionally gifted children. One involves accelerated education and the other enrichment of the standard syllabus. Results are not conclusive on the preferability of the systems; much seems to depend on vocational interest, intention, and on sex differences.
Refs Greenacre, Phyllis *The Early Years of the Gifted Child* (1962); Hallahan, Daniel P and Kauffman, James M *Exceptional Children* (1988); Khatena, Joe *The Creatively Gifted Child*; Patton, James R, et al *Exceptional Children in Focus* (1987); Perino, Sheila and Perino, Joseph *Parenting the Gifted* (1981).
Related Genius (#HH0295).

♦ HH0613 Mukti
Salvation — Liberation
Description Liberation, or release from the cycle of birth and death *samsara*, is normally seen in Hindu tradition as only occurring at death after a life of devotion to God. However, some traditions see the possibility of such detachment from all that is liable to change and of realization of the immutable pure consciousness of the essential nature of being that liberation may be achieved while living in this lifetime – *jivanmukti*. This discrimination of the permanent from the transient frees the individual from suffering so that there is a state of bliss, *ananda*. After death, the sequence of causes for further action, *karman*, ceases and the individual is not reborn but achieves union with Brahman.
In the Advaita school this implies seeing all plurality, even the plurality of the individual *atman*, as an illusion; everything in reality is Brahman, and awareness of this is liberation. Another approach (Samkhya) sees mukti as being achieved on perceiving the spirit (purusa or atman) as entirely other than nature (prakriti), as simply witnessing the suffering arising from lack of discrimination. This perception leads to *kaivalya*; the individual in a state of serene wisdom lives on until his karman is exhausted, although it is no longer added to, and after death he is not reborn. Tantric sects look on mukti as the purifying of the impure body into pure substance; the actual body of such a *siddha* is considered by some to be immortal. For the Buddhist, mukti implies a period of sentient existence freed from suffering; this entails detachment from objects and persons – neither attachment nor aversion, as both are considered desires, and freedom is freedom from desire. This period in heaven eventually culminates in death; permanent release is achieved only by an *arhat*. Although Zen puts little emphasis on desires and their results, it makes clear that the enlightenment does in fact bring automatic freedom from desire and the cycle of birth and death.
Broader Emancipation of the self (#HH0907).
Narrower Cetovimutti (Pali, #HM0125) Videha-mukti (Yoga, #HM4489)
Pannavimutti (Pali, #HM8125) Jivanmukti (Hinduism, #HM3890)
Ubhatobhagavimutti (Pali, #HM7983).
Related Liberation (#HH0388) Soteriology (#HH1166) Moksha (Hinduism, #HM2196)
Samsara (Hinduism, #HM1006) Bhakti marga (Hinduism, #HH0628)
Duhkha (Buddhism, Hinduism, #HM2574) Ananda (Hinduism, Buddhism, Yoga, #HM3227).

♦ HH0614 Learning theory therapy
Description This therapy is based on an understanding of why an individual does not learn to adjust behaviour automatically. It is based on the *pleasure principle*, that one seeks to avoid situations which carry painful reactions; and emphasizes new, pleasurable situations in which to build up new, well-adjusted habits. Maladaptive behaviour patterns are unlearned by counterconditioning the anxieties that motivate them.
Broader Therapy (#HH0678).

♦ HH0615 Technological implications for human evolution
Description The rate at which computer technology has evolved has given some pointers as to the expected path for other, slower evolution. Assuming human evolution to follow a similar path, it may be expected that: continuing refinement of the human body will incorporate light into its substance, leading to faster, wider and more precise thinking; a higher proportion of the body will be devoted to storage of memory (which will be considerably expanded at all levels), including "memories" of muscles, skin, and so on, to which there will be access at will. This memory storage also includes the conscious use of the emotions for storage of thought. The speed at which comparisons may occur will lead to finer discrimination and judgement. Since direct communication will occur between persons, memory will be selective, as there will be instant access to other's memories for specialized knowledge. The senses will be used for communication of emotion rather than facts, as factual information will be retrieved via minds rather than through audio-visual stimuli. The mind will suppress sense signals which are unnecessary, the senses thus becoming more sensitive. Physical life will be longer, with "single-purpose" organs being replaced by consciously controlled multi-purpose tissue.
Broader Evolution (#HH0425).
Related Evolution by prosthesis (#HH3662).

♦ HH0616 Biotechnology
Description Cells and other constitutents of the body have physical properties such as semiconduction, capacitance and microwave transmission. These electronic solid state properties permit the use of devices to provide an interface between man's psychoenergetic existence and his technological creations. Biotechnology is therefore a general term covering such devices as: acupuncture, biofeedback devices, biotelemetry, brain-implanted stimulation receivers, pacemakers and Kirlian photography.
Users of such devices may develop a degree of sensitivity and control that permits them to achieve the same effects without the need for the technological support.
Refs Manganelli, Louise A *Biomagnetism* (1972).

♦ HH0617 Organizational development group
Leadership development group
Description This is a form of intensive group interaction aimed at improving leadership skills.
Related Leadership development (#HH0516).

♦ HH0618 Regeneration (Christianity)
Description Regeneration may be defined as a spiritual and psychological change occurring on reconciliation with God. It is a spiritual renewal or revival. The defilement, disorder, desolation and chaos of sin has made man unfit to live with God. Regeneration means man has been cleansed through a new creation and, by constant renewal, kept steadfast. The radical and thorough change that occurs when those previously steeped in sin enter on the Christian life can only be expressed figuratively. Scriptures refer to it as repentance (literally the turning away from the old life to the new); as becoming like a little child, depending on the Father's love; as a seed being quickened by the divine Word; of rebirth into a new world, a new atmosphere, a new environment; of becoming a new creature; of putting off old garments and putting on new; of passing from death to life; of burial and resurrection.
The change is new, it is a new person that is created. It is not just a strewing of flowers and perfume on a dead body but a vital principle is put into the heart. Another and higher will sustains one's will, subdues one's nature to itself and natural life becomes spiritual life. A new personality arises, the multitude of selves comprising the person are unified into one self. The change cannot be worked by the individual on himself, he needs the inward power of the Holy Spirit, who stands against the possession of evil spirits, the new power on earth.
Related Spiritual rebirth (Christianity, #HH4098).

♦ **HH0619 Vocation** (Christianity)
Calling
Description Christians receive a call or a calling to share in redemptive blessings to which God has appointed them. Since this call comes from God, it is referred to in scriptures as a "high calling" or a "heavenly calling". While there is the moral freedom to respond or not to the call, God works with the Christian both in his will, to respond to the call, and in the working of that response. The Christian is also said to be called to earn his living so that the work which he does is pleasing to God and is done to the best of his ability. The term "vocation" is therefore not simply a calling to the Christian life but implies a certain quality in the daily work of the Christian.
Related Primal vocation (ICA, #HM2621) Realized vocation (ICA, #HM2441) Religious vocation (ICA, #HM2293) Historical vocation (ICA, #HM2616).

♦ **HH0620 Ki energy**
Kiai — Ch'i
Description Although it has a multitude of meanings, ki or ch'i has been described as the breath, spirit or nature of things. In neo–Confucianism the concept was expanded and precised to indicate that which causes the physical manifestation (variable) of the metaphysical (always good). A flawed ki thus indicates a flawed moral nature; purification of ki leads to enlightenment, with consciousness and action in perfect unity.
Ki energy is both personal and impersonal, concrete and universal. Transcending time and space, it is the basic creative energy or force in life. As a psychophysical energy important in the martial arts, *ki* energy is developed by breathing exercises and implemented by concentration of the will. *Ki* is twofold – the unity of the individual–universe and the free and spontaneous expression of breath–power. The former, inheriting the ideas of ancient Chinese thinkers, is realized through unifying *ki*, mind and body in, for example, aikido training. It acts as defence against physical attack and may be used at a distance. The *kiai* shout focuses all the bodily and spiritual energy into one sound, unifying the proponent's powers and disconcerting the opponent.
Harmony of *ki*, or *ai–ki*, is manifest when mind and body are unified, its subtle working is the maternal source that affects changes in breath. As delicate changes in breath power occur spontaneously, proper technique flows freely.
Awareness of delicate changes may show fierce and potent or slow and stolid movements of *ki* in the void, leading to discernment of the degree of concentration or unification of mind and body. *Ki* can be taken to mean mind, spirit or heart, and thus to have *ki* out of order in some way means a diminished state of consciousness. This may be insanity, nervousness, depression; it may be an emotional condition (quick tempered, shy). It is the *ki* and not the individual that is referred to in these cases.
Refs Koichi, Tohei *Book of Ki*; Reed, William *Ki*; Siu, R G H *Ch'i* (1974).
Broader Martial arts (#HH0085).
Related Ri (#HH3544) Aikido (#HH0252) Quietism (#HH2006) Acupuncture (#HH0728) Realization of nen (#HM0220) Spirituality (Taoism, #HH5117) Qi gong (#HH3862).

♦ **HH0621 Outward zazen meditation** (Zen)
Soto zazen meditation
Description The preferred method of meditation of the Soto school, also called serene reflection (mo–chao) or the quietness of no–thought (wu–nien). The aim is to observe the mind in tranquillity.
Broader Soto Zen (Zen, #HH0928) Zazen meditation (Zen, #HH0882).
Related Inward zazen meditation (Zen, #HH1134).

♦ **HH0622 Human development** (Jainism)
Description This religious tradition pre–dates Buddha but nevertheless rejects the idea of God although affirming the existence of liberated souls. All *jivas* – sentient, living things – are eternal and discrete entities which have always existed and always with a material component, *ajiva*, due to karma. The true nature of the jiva is realized when it loses its material, karmic ajiva which obscures, although does not destroy, the jiva. Freedom from ajiva implies no further birth and the jiva abides for ever in perfect knowledge and self–containment in the siddha–loka, or abode of liberated souls. This is the highest level of the universe, lower levels being urdhva–loka (the celestial world), madhya–loka (the terrestrial world) and adha–loka, the underworld.
Jains are divided into two groups, depending on monastic practice: the Shvetambaras who wear white clothing and who admit women to full vows; and the Digambaras who take renunciation to include that of clothing and who consider women must await reincarnation as men before being admitted to full vows. All Jains practice *ahimsa*, harmlessness, as part of the path to liberation; this includes not only abstinence from killing or injuring living beings (and therefore strict vegetarianism) but also absence of aggression and possessiveness. The three "jewels" or ways are right faith, right knowledge and right conduct. Philosophically, Jainism rejects *ekantavada* or onesidedness, holding that this must inevitably fail to take in all equally important dimensions of existence; rather, it supports *anekantavada*, an approach which synthesizes conflicting approaches to reality and includes both the inherent unchangeability of the soul and its ability to alter qualitatively.
Following a rigorous discipline of renunciation under the guidance of teachings of 24 great teachers or *jinas*, the individual can, by his own efforts and without supernatural aid, achieve release from the bondage of physical existence and the cycle of birth and death. The jinas or *tirthankaras* (crossing makers) show the way to recovery of the true jiva nature. This is a process which takes numerous lifetimes. There are fourteen stages, *gunasthanas*, from total bondage to karma to total freedom. A life of piety may be lead at the fourth stage; but the sixth stage must be reached before commencing the life of a monk. Then the Jain makes five mahavratas or vows – observance of ahimsa and refraining from lying, stealing, sexual intercourse and possessions – and lives a life of increasing asceticism which may culminate in old age with a ritual fasting to death, this latter to avoid negation of progress through clinging to material existence. The monastic pattern is mirrored in lay practice, when eleven stages or *pratimas* bring the lay person to a condition of renunciation similar to that achieved by those in orders: right views; taking of vows; meditation leading to equanimity; holy day fasts; pure food; refraining from sexual intercourse during the day; refraining from all sexual intercourse; abandoning of household activity; abandoning possessions; abandoning concern for the life of the householder; renouncing all family ties. This is propitious for rebirth into a life where the monastic path to liberation is possible. Full enlightenment, *kevalajnana*, is said to have been achieved by some practitioners of Jainism, who transcended the separation of eternal transcendent reality from manifestation.
Refs Jaini, Padmanabh S *The Jaina Path of Purification* .
Broader Human development through religion (#HH1198).
Related Ahimsa (#HH0088) Adha–loka (Jainism, #HM3697) Siddha–loka (Jainism, #HM0569) Urdhva–loka (Jainism, #HM0131) Madhya–loka (Jainism, #HM1357) Kevalajnana (Jainism, #HM3948) Spirituality (Jainism, #HH8712) I–transcendent stage of life (#HM3060).

♦ **HH0623 Wisdom**
Description Wisdom implies not simply knowledge, but that type of knowing which is experienced and gives the capacity to judge; it has been defined as the correct and timely use of knowledge, knowledge itself arising from the practice of awareness. Such wisdom is inherent in the word *philosophy*, love of wisdom. Every culture and religion has prized wisdom, whether in the practical sense (knowledge of right living and the ability to pass that knowledge to others); or as an intellectual attainment: "...the price of wisdom is above rubies". In the practical sense, wisdom is demonstrated in the proverbs of China, and in the rational and practical good sense of the Old Testament; whereas intellectual wisdom gave rise to the Indian *Vedic* and classical Greek literature. The Socratic and Platonic traditional approach implies that wisdom brings knowledge of the *self* and of one's own ignorance, and also an awareness of orderly conduct and an appreciation of the beautiful. In Zen literature, when something is known or valued for its relation to life, that is said to be wisdom or *buddhi*; perfect wisdom, *prajna* is what is known by the 'Buddha–mind'. Some traditions suggest that the assimilation of wisdom is permanent and does not need to be re–acquired in subsequent lifetimes.
Wisdom as the perfection of man suggests, whether directly or indirectly, a knowledge of God. "Fear of the Lord, that is wisdom". According to St Thomas Aquinas there are three *realizations* of wisdom: *metaphysics*, based on reason; *theology*, based on faith and reason; and the gift of the Holy Spirit, *mystical wisdom*.
Refs Brunton, Paul *The Wisdom of the overself* (1970); Watts, A W *The Wisdom of Insecurity* (1968).
Narrower Prudence (#HH7902).
Related Essential wisdom (#HM0107) Knowledge (Buddhism, #HH1971) Buddhi awareness (Hinduism, #HM2099) Dhyana (Hinduism, Buddhism, #HM0137).

♦ **HH0624 Indicators of development**
Description Present development, based on the macro–scale and neither considering nor motivating the individual, has to be replaced by a system where individual people are brought to the centre of development processes. This might be done by creating a framework of development indicators which would show how present goals deviate from previously defined proper development.

♦ **HH0625 Total fitness**
Description This programme sets out three levels of physical training, depending on whether the body is to receive at least the minimum exercise necessary for fitness, whether it is required to go through a normal day without fatigue, or whether it needs to be prepared for particular exertion. Pulse rate and diet are monitored to determine the amount of exercise required for health, over exertion and too–strict discipline being considered as unnecessary and possibly harmful.

♦ **HH0627 Guided fantasy**
Guided imagery
Description A process in which individuals, often in therapy, use their imagination to create a new experience for themselves. The purpose is to create an experience which, at least in part if not in its entirety, has not previously been encountered. Guided fantasy is therefore most appropriately used when an individual's representation of his environment is too impoverished to offer an adequate number of choices for coping with that environment. At the same time it provides the therapist with an experience which can be used to challenge the individual's presently impoverished model of reality. Guided fantasy also serves to create a reference structure in terms of which the individual can be encouraged to order his experience. Guided fantasies often take the form of metaphors, rather than a direct representation of the problem with which the individual claims to be faced.
This is a method of facilitating concentration and directing attention. Directed fantasy sessions are useful for learning specific content, while open–ended fantasies evoke creativity and aid self–discovery.
Most of the education in modern society emphasizes verbal knowledge and reasoning and is considered to be primarily a function of the left–hand hemisphere of the brain which is predominantly sequential and analytic in its processing of clearly defined symbols. The right hemisphere is spatially oriented, however, and functions in terms of pictures, perceives patterns as a whole, and operates in an intuitive, emotional, and receptive mode. Guided fantasy offers the possibility of engaging the right half of the brain in the learning process. For example, in guided imagery, members of the group are guided, by suggestion, to visualize or fantasize pictures as an aid to developing their imagination.
Related Fantasy (#HM0892) Guiding images (#HH3020) Mental imagery therapy (#HH0072) Guided visualization (Magic, #HH6053) Neuro–linguistic programming (NLP, #HH4872).

♦ **HH0628 Bhakti marga** (Hinduism)
Way of loving faith
Description The term *bhakti* implies devotion arising as a result of faith – faith which may have arisen from knowledge or from good works with pure intent and whose results are surrendered to God. In religious terms it thus implies recognition of a supreme deity, and *bhakti marga* the way of devotion to that deity; it is therefore the preferred way of theistic traditions. Bhakti may be said to arise in five stages – resignation, obedience, friendship, tender fondness and passionate love; and bhakti marga requires constant awareness of its Lord through meditation, prayer and repetition of his name. This way implies that salvation cannot be achieved through efforts of man alone but requires God's grace.
Context One of the three ways of salvation of orthodox Hinduism, the others being *karma marga* and *jnana marga*. No one is said to be complete without some influence of the other two.
Refs Chhaganlal, Lala *Bhakti in Religions of the World* .
Broader Love (#HH0258) Human development (Hinduism, #HH0330).
Related Mukti (#HH0613) Moksha (Hinduism, #HM2196) Bhakti yoga (Yoga, #HH0337) Prasada (Hinduism, #HH4330) Jnana marga (Hinduism, #HH0495) Karma marga (Hinduism, #HH1211) Devotion (Christianity, #HH0349) Religious experience as personal devotion and worship (#HM0561).

♦ **HH0629 Shakti**
Cosmic creativity — Dynamic universal
Description Shakti is understood as the underlying, undifferentiated creative principle comprising the totality of all past, present and future phenomena in every part and level of creation. An awareness of the evolutionary nature of creation is said to be essential for an understanding of the human position and of the diversity, individuality and changeable nature of things.
Refs Narayanananda, Swami *The Primal Power in Man, or the Kundalini Shakti*.
Related Shakti yoga (Yoga, #HH3912) Kriya sakti (Hinduism, #HH4309).

♦ **HH0630 Peri–natal psychology**
Psychological development of infants
Description The experiences received in the womb mean that the neonate is not entirely "sensorily bereft". When a child is born it already has psychophysiologically stabilities. Studies of sensory range and responses indicate very young babies to be active rather than reactive, and to discriminate sufficiently to solve such problems as how to suck when bottle rather than breast–fed.
In addition, the emotional state of the other both before and after the birth of the child evokes

matching responses in the child, which in turn amplify the feelings of the mother, with resultant self-reinforcing and cumulative effects. Neurotic, hyperactive and hypoactive traits can therefore be alleviated, with the proviso that an already anxious woman is not made more anxious by knowing that this very anxiety is likely to have ill-effects on her child.

♦ **HH0631 Triple way** (Christianity)
Three spiritual ways
Description Rather than three successive stages, the triple way refers to the three separate but parallel courses which, followed together, lead to internal order and mystical union with God. According to the Franciscan school, these three exercises of hierarchical actions are said to be *purgative*, *illuminative* and *unitive*; and are said to be achieved through organized meditation, through prayer and through contemplation. However Pseudo-Dionysius and others also class the three ways in a hierarchy so that "beginners" are those attempting to purify themselves of sin, "proficients" are seeking illumination, and the "perfect" act in union with God.
Broader Spiritual way (#HH1867).
Narrower Via unitiva (#HH4100) Via purgativa (#HH4090) Via illuminativa (#HH6030).
Related Illumination (#HH0804) Purification (#HH0401) Spiritual marriage (Christianity, #HH0465).

♦ **HH0632 Free-thinking meditation**
Independent reflection
Description Free-thinking meditation consists of unbiased and unprejudiced investigation of the thoughts of one's inner self on a particular subject of spiritual or intellectual importance. The meditator writes down the ideas that come to mind when he or she concentrates on the chosen subject intelligently and with an open mind. This is followed by logical inquiry into the possible truth of these ideas when they are referred to the inner self. Thus the meditator learns his or her own inner reactions to ideas, unrelated to what has been heard or read by an authority on the subject. This brings insight and flexibility of mind and an ability to discriminate the truth from habitual ways of thinking and attractive ideas.
Broader Meditation (#HH0761).
Related Reflection (#HH1204).

♦ **HH0633 Play therapy**
Ludo therapy
Description A child expresses himself and reveals unconscious material more usefully in play situations than by verbalizing. The technique is used with children in circumstances where an adult would receive psychoanalysis; it is useful in assisting children to face their feelings of insecurity, hostility and anxiety.
Broader Therapy (#HH0678).
Related Games (#HH0705).

♦ **HH0634 Self-esteem**
Self-regard — Self worth — Ego level
Description The way in which an individual views his own abilities and worth will have a significant effect both on the goals he or she chooses and on the probability of achieving those goals. It also determines the front which the individual adopts to deceive both himself and the rest of the world into believing him different from what he actually is. This persona is then used to rationalize behaviour or attitudes which the individual feels he needs to justify. The result of exposing this pretence for what it is is said to "shatter a person's ego", and is the prime cause of being "put to shame".
Refs Branden, Nathaniel *The Psychology of Self-Esteem* (1971); Cook, Patrck *Self Esteem* (1983).
Related Self-respect (#HH1384).

♦ **HH0636 Ego**
Self
Description The terms ego and self are variously viewed as equivalent or distinct, although in some cases one or both are held to be meaningless terms.
A new-born child is said to have no self sense and therefore to perceive no separation between the body and the environment, nor to view events as occurring objectively. This state of non-differentiation is gradually left behind as the child develops, and life may be thought of as a move from the sub-conscious to the self-conscious state, as the ego develops.
Ego is viewed as a source or objective of motivation, as the organizing centre for behaviour and experience, and as the totality of experiential content and behaviour patterns relating to the individual's own person. It is also viewed as a part of the psychic structure (contrasting with the id and the super-ego) permitting adaptation of reality and as the decisive component of the psyche. It gives continuity and consistency to behaviour by providing a personal point of reference. The term can be used to describe that part of the growing human personality that tends, under suitable conditions, to become integrated into a unit. It can also be described as an individually characteristic centering of the personality.
Beyond this view is the second stage in the life cycle when, after the development of the self-conscious ego, a super-conscious state of transpersonal unity is achieved.
Refs Fingarette, H *The Ego and Mystic Selflessness* (1958); Loevinger, J *Ego Development* (1976); Spitz, R *A Genetic Field Theory of Ego Formation* (1959); Vaughan, Frances and Walsh, Roger (Eds) *Beyond Ego* (1980).
Narrower Altered ego states (#HM4120).
Related Soul (#HH0501) Self (#HH0706) Psyche (#HH0979) Superego (#HH0493).

♦ **HH0637 Pilgrimage**
Description A pilgrimage is usually a journey to a sacred place or shrine (possibly with sacred relics) undertaken by individuals for a variety of religious and spiritual motives. The significance for the individual may lie in the journey itself and the effect of its experiences on his own spiritual development, or possibly in the special insight he believes he can receive through visiting the sacred place. Visiting the tomb of a saint provides a physical link with those now with God. Following prescribed rituals or penances purges away past mistakes and sins. Although current pilgrimage may not result in the miraculous casting aside of sickness common in the past as the psychodrama of the saint's life was re-enacted, pilgrims often find peace and acceptance. Non-believers may also have special places which have become "sacred" to them - perhaps linked with childhood memories - which provide renewal when visited. Such "secular" pilgrimages may also be to places of historical, cultural or political significance. Returning to one's roots may enhance the sense of continuity with the past, reinforcing cultural identity, helping to make sense of the world.
The concept of pilgrimage may be extended to cover a journey not in space but in time, when important moments of revelation or vision associated with a particular place are called to mind in order to seek purification and renewal. Disciplines aimed at enlightenment and the experience of sacred presence in the depth of one's being are often referred to as "the Way", a reference to the discipline of pilgrimage.
Refs Jha, Mhakan *Dimensions of Pilgrimage* (1985).
Related Hajj (Islam, #HH1265).

♦ **HH0638 Habit training**
Description This is the acquisition of specific behaviour patterns involved in eating, dressing and toilet training, which the child learns by conditioning. In psychoanalysis, habit training is seen as the clinging to particular habits as a defence mechanism against unconscious wishes or desires.
Related Habit (#HH3665).

♦ **HH0639 Ethnocentrism**
Ethnocentricity
Description An element of personality. The individual views his own group as the centre and refers all else to it, leading to an exaggeration of similarities within the group and differences between it and what is outside the group; this leads to a strengthening of the group.

♦ **HH0640 Abreaction**
Description A technique in psychoanalysis when the individual relives a traumatic experience through hypnosis or in the waking state, "depotentiating" its affecting activity. The technique on its own was eventually considered insufficient or even harmful by both Jung and Freud, the former holding that integration of dissociation should be the aim, the re-experience enabling relationship with the positive side of the neurosis and hence bringing the affect under control.
Broader Psychoanalysis (#HH0951).
Related Trauma (#HM3724).

♦ **HH0641 Sarvodaya concept**
Description This movement, which is a descendent of the programme of construction in India started by Gandhi, seeks to develop the individual as a participant in an all-embracing development of his community which takes place through the willing sharing of human and material resources in that community. The keynote is human development rather than economic development. By means of harmonious, local, decentralized structures which allow the individual to participate in decisions which affect him, community consciousness is reinforced through autonomous, self-reliant development, with simple technology appropriate to the situation.

♦ **HH0642 Natural law**
Description Whatever source is studied, all philosophic and religious teachings seem to indicate a common agreement of right and wrong, a natural law for individual behaviour which goes beyond social customs and conventions. These have been formulated as: the law of general beneficience (goodwill and generous behaviour to mankind in general); the law of special beneficience (specific goodwill and action to individuals in particular); duties to parents, elders, ancestors; duties to children and posterity; the law of justice (honesty, fidelity, justice before the law); the law of good faith and veracity; the law of mercy; the law of magnanimity.

♦ **HH0643 Hostile feelings**
Hostility
Description The emotional reaction to some threat, frustration or feelings of hatred or anger, hostility is an enduring emotion directed to the offending object, whether or not the person experiencing the emotion is aware of it. Particularly when hostile feelings are repressed or displaced towards a less threatening object, such hostility breeds hostile attitudes in others - for example, children pick up hostile attitudes from their parents.
Related Antipathy (#HM2481).

♦ **HH0644 Associative thinking**
Description As opposed to *free association*, which can be considered regressive and deals with problems originating in infancy, associative thinking is lateral, an associative *verbal catharsis*, enabling a person's immediate, everyday problems to be dealt with.
Broader Thought processes (#HH0530).

♦ **HH0645 Psychic ontogeny**
Description This refers to the development of the mind, particularly in relating inner needs with environmental demands. It is said to include object relationship development (autoerotic - narcissistic - homoerotic - heteroerotic - alloerotic); libidinal phases corresponding to vicissitudes of the drives (pre-superego sexuality: oral; anal; phallic - latency - genitality); and the development of mechanisms to achieve these.

♦ **HH0647 Development of the quality of human life**
Description Underlying the concern for human development in both industrially developed and developing societies is the question of quality of life. It has been customary to use crude economic indices like the gross national product in order to measure social progress and human welfare. The concern for quality of life represents an attempt to find complementary elements or correctives which include such (national and international) criteria as: distributive justice, participation in decision-making, community life, and spiritual welfare. However, the measurement of these is exceedingly difficult. A more fully human perspective is required to integrate these values into national and international planning.
Quality of life requires that every person have the right to live a meaningful life in which he or she can develop his or her own personality as a socially responsible human being. Each person has the right to a minimum of material requirements necessary to such a life. Social structures necessary to achieve quality of life differ widely in different countries and are only capable of comparison in terms of patterns of relationships between resources, hopes, possibilities and actual achievements. However, both the quantitative and qualitative components of the concept convey intricate relationships between the basic human needs of persons and the human aspirations of persons to reach their full development.
Beyond meeting the basic human needs which all human beings require and which can be quantified, the criteria for a higher quality of life can only be stated in general terms:
1. New forms of community life which will replace the present compartmentalized form of technological industrialization, emphasizing, on the one hand, new patterns of home, work, school, church, citizenship, recreation, and on the other, a new appreciation of the traditional restrictions of rural life. These new patterns of community should provide for close relationships among generations, a place for the unmarried and the widowed, the widening of affection and emotional support, care of the young, the weak and the old, conservation of resources and energy, equal participation of women and members of different racial and ethnic groups, sharing of burdens and responsibility of each for each and each for all.
2. Structures which permit children to develop the capacity to participate in societies as whole human beings.
3. Productive processes that are resource and energy conserving, and as closely tied as economically feasible to local resource, local use and local skill.
4. Opportunities for creativity, innovation and the realization of potentialities, beginning at the local community level and extending to the global community.

5. The fullest attainable participation of all members of society in appropriate levels of decision-making.
6. Recognition of different cultural and linguistic values and structural provisions for freedom of choice among them, consonant with the well-being of the wider community.
7. Social structures which, while encouraging peoples of each nation to devote themselves to the preservation of their land, their people, their cultural traditions and their descendants, help at the same time to increase awareness of the interdependence of all countries.
Related Standard of living (#HH0768).

♦ HH0649 Holistic approach

Description Rather than seeing society as a structure in which some parts are causal and the rest simply effects of those causes, the holistic approach considers society as an interdependent network of relations in which every individual has a role to play. Such an approach rejects the idea that society is merely an aggregation of individuals and also the idea that the individual exists only as a reflection of the society he or she lives in. These are both seen as incompatible with a human-centred view.
Refs Boerstler, Richard W *Letting Go* (1982); Cantore, Enrico *Holistic Development of Man in the Scientific-Technological Age* (1973); The Theosophy Science Study Group *Holistic Science and Human Values*.
Narrower Holistic medicine (#HH0581).

♦ HH0650 Human development (Buddhism)

Description The keynote of Buddhism is enlightenment. Buddhism denies reality to a supreme, conscious self, either in man or in the universe; but it does not deny the apparently real necessity of attaining truth through use of the mind. Much of the teaching concerns the mind - how it works, how it experiences and remembers. For precision, this implies the double negative - the more one knows what the mind is, the more one knows what it is not. The more one knows what it is not, the more one knows what it is. Topics include the awareness of consciousness, the way in which things exist, the transmission of instinct for further cognition from lifetime to lifetime. This path of learning brings understanding of how the omniscient mind of Buddha knows everything and ultimately the development of buddha mind in the self. Ultimately the aim is nirvana or nibbana, the end of suffering and attainment of perfect happiness. In that "Buddha" means one who has awakened or who knows the dhamma (the true nature of things), then all who achieve the goal of the Buddhist path are buddhas. Their lives exemplify the dhamma and follow a pattern laid down by dhamma, passing through numerous incarnations, the last of which is divine.
The spirit of Buddhist development is inner self-control over the psychological life, meditation, search of soul and forbearance. There is the vision of invisible solidarity among all living beings in universal life, of all minds in the eternal spirit. This has to be practised and promoted in inner and outer loves. Inner strength must be cultivated first, only then do compassion and loving-kindness to others become possible.
Although there are many schools of Buddhism, all somewhat influenced by other religions sharing the same geographical area, all agree on the basic concern of spiritual development of the individual through meditation and training, leading to right insight which transforms him so that he reaches liberation. Peripheral religious activity, so long as it does not deflect from the main aim, is not discouraged, which makes Buddhism very flexible as far as traditional religious and cultural practices are concerned. Three main schools, *hinayana*, *mahayana* and *vajrajana* are roughly equivalent to the three main geographical areas of South Asia, East Asia (where it is closely related with Taoism, Confucianism, Shinto and folk religions) and Northern Asia, notably Tibet (directly derived from classical Indian Buddhism). All three ultimately derive from the Theravada (elders' teaching) of the Buddhist sangha or community.
The original teachings of Buddha were intentionally limited to that necessary for spiritual development and avoided all extremes, whether religious or materialist, ascetic or self-indulgent. They are probably related to Brahmanism and also influenced Hinduism. Subsequent teachings emphasized concern for practical and moral welfare and for compassion; but *dana* or giving, which is a foundation of Buddhist practice, relates to sacrifice of one's belongings as a religious act in the prescribed manner to a person of religious commitment and brings great rewards in this and subsequent lives. Dana, together with *sila*, following of precepts, form the first part of Buddha's step-by-step discourse. These are reinforced by knowledge of the operation of *kamma* or law of action, when every act of the will is a seed from which future conditions and activity grow; this liberates from fatalism and makes one responsible for one's actions and condition of life. The first part of the discourse thus concerns the way life is lived outwardly. Inner purity is dealt with in the second part, which covers the defects of sensuality and the positive gain in freedom from it. This assists development of those following the spiritual way and leads to mental transformation. More profound teaching, the four noble truths, is intended for those thus transformed. It covers the cause of suffering and its transcendence in *nirvana*.
Refs Galtung, Johan *Buddhism* (1988); Kornfield, J *Living Buddhist Masters* (1977); La Vallée Poussin, Louis de *The Way to Nirvana* (1917); Nyanatiloka, Mahathera *Buddhist Dictionary* (1972); Robinson, Richard H and Johnson, Willard L *The Buddhist Religion* (1982); Stcherbatsky, T *Buddhist Logic* (1962).
Broader Human development through religion (#HH1198).
Narrower Hinayana Buddhism (#HH0845) Mahayana Buddhism (#HH0900) Vajrayana Buddhism (#HH0309) Human development (Zen, #HH1003) Steps to enlightenment (#HH4019) Nichiren shoshu buddhism (#HH3443) Consciousness in Buddhism (#HM2200).
Related Nirvana (Buddhism, #HM2330) Dana (Buddhism, Zen, #HH1234) Spirituality (Buddhism, #HH3907) Karma awareness (Buddhism, #HM2048) Four noble truths (Buddhism, #HH0523) Observing moral precepts (Buddhism, #HH0278) Religious experience as meditational insight (#HM1593).

♦ HH0651 Self-development

Description Constructive self-development usually involves some ethical discipline such as a philosophical system or yoga. The striving towards self-development, self-awareness and a whole-hearted approach to life requires mutual support and encouragement. Self-development is not an automatic process, but requires individual will and effort. Starting with curiosity or the "desire to know", the individual is impelled to understand, systematize, organize and analyse more and more, the more he or she learns, both in depth and breadth. By doing so the person becomes something more than previously.
Self-development carries with it the need for self-expression to be creative in the sense that children are creative, in contrast to the predictable responses acquired with enculturation. Ultimately it leads to spontaneous behaviour without motivation and is experienced as the mystic awe and delight associated with the term *moksha*. At this level the ego-centred self is enlarged and extended to include and integrate with others, and is expressed as a responsible, giving love based on knowledge. The ultimate aim can never be said to have been reached, as the more satisfying the pursuit of self-development the more one is impelled to pursue it.
Related Self (#HH0706) Enculturation (#HH0954) Moksha (Hinduism, #HM2196) Existential integration (#HH0832) Self-culture (Christianity, #HH3710) Christian self-love (Christianity, #HH6732).

♦ HH0652 Bioenergetic analysis

Description Depersonalization and alienation of the individual are said to have resulted in diverting the body from its true role and exaggerating the role of the ego, thus putting body and ego in conflict. Bioenergetic analysis aims to redress the effect of repressing body feelings and provide an adequate body image. In this form of therapy the body is considered to have one fundamental energy, bioenergy, manifested by emotions and muscles alike. Every neurotic problem is held to have its counterpart in the structure and functioning of the body. Muscular tension is considered to be a basic factor in emotional disturbances and the latter may be dispelled by integrating analytical work aimed at intellectual understanding with physical exercises and postures.
The body is involved as follows: individuals wear tights or other minimal clothing so that bodily tensions may be visible in group sessions; its energy processes are mobilized through breathing and movement in order to facilitate expression of feeling; a setting is provided for the release of repressed feelings through bodily action (e.g. hitting, kicking, etc); direct physical contact occurs between the individual and the therapist.
The capacity for emotional expression is to the degree of muscular coordination. In the emotionally disturbed person, portions of the body may be too rigid and he cannot move gracefully. As individuals learn to relax or let go in postures or in such activities as striking a couch or kicking their legs, their general muscular coordination increases spontaneously. Feelings are expressed through movement, so every muscular tension is a barrier that blocks or limits the expression of that feeling. By eliminating the tension, the feeling can be released; conversely, by expressing the feeling the tension can be reduced.
Refs Lowen, Alexander *The Betrayal of the Body* (1967); Lowen, Alexander *Bioenergetics* (1975).
Broader Holistic medicine (#HH0581).

♦ HH0653 Hero worship

Description Reverence for and emulation of a hero figure and his admirable qualities may, as in Japan, result in the virtual deification of a now dead hero with charismatic qualities. Shrines may be dedicated and cults surrounding such a hero used to encourage patriotism, for example.
Related Charisma (#HH0053) Revered hero (ICA, #HM2654).

♦ HH0654 Risk-taking

Description Risks can be taken in the sense of gambling on a result which is of a chance nature; or of making a decision where insufficient information is available for the result to be certain. In the latter case, the risk of making a decision based on insufficient information is weighed against the gravity of the decision and the cost and difficulty of obtaining further information. Western society values risk-taking more highly than caution, and individuals take risks in order to be equal to more risk-orientated peers or as part of a group with a high-risk-taking leader.
Related Danger (#HM1363) Uncertainty (#HM3051) Freedom of choice (#HH0789).

♦ HH0655 Perceptual development

Description This is the process whereby a child learns to acquire precepts that are a progressively more satisfactory and comprehensive representation of reality. Of particular interest is the development, up to adolescence, of perception of: distance; size and distance; form; and apparent movement. The developmental relationship of perceptual judgments, sensorimotor acts and contemplative operations are such that with increasing age the first two become increasingly subordinate to the higher mental functions of contemplation. Consequently, it is highly unlikely that pure perceptual judgments exist in normal adult behaviour independent of contemplative judgement.
Refs Jenkin, N and Pollack, R H (Eds) *Perceptual Development* (1925); Murphy, G and Hochberg, J *Perceptual Development* (1951); Wapner, S and Werner, H *Perceptual Development* (1957).

♦ HH0656 Aspiration level

Description The level of aspiration of an individual or group is the possible goal which the individual or group sets for performance. Its formation is influenced by the interaction of momentary achievement, longstanding achievement confidence, momentary achievement impulse, seriousness of the situation and the type of objective. Those confident of success set an aspiration level somewhat above previous actual achievement, whereas those preoccupied by the possibility of failure prefer an aspiration level far above or far below their actual capacity.

♦ HH0658 Understanding

Description It is generally accepted that understanding is the power that forms concepts, thus ordering the data received by the senses. It is contrasted with *reason*, which is said to deal with the ordering of ideas, and with *knowledge* of causes in the scientific sense. It is that insight into the human mind which enables sympathetic "understanding" of another person's situation. In this sense, *mutual understanding* between individuals in different groupings (by sex, race, nation, religion and so on), is frequently put forward as a necessary precursor to the solving of conflicts and to true tolerance and freedom.
One exposition (B J F Lonergan) puts understanding in the pivotal position between experience and judgement. It is the component that says, with wonder, on receiving an impression from the senses, "What is that ?"; and which is then subjected to the judgement, "Is it that ?". The most basic type of understanding would be of the *common sense* type; there are also scientific or abstract understanding; that which anticipates or discovers; and that which infers results from activities.
Refs Lonergan, B *Insight, A Study of Human Understanding* (1970).
Related Wonder (#HM3197) Symbols (#HH0690) Common sense (#HH0510).

♦ HH0659 Initiation (Freemasonry)

Masonic degrees

Description Freemasonry, as the world's largest secret society, has a well-defined procedure for admitting candidates and advancing members through its organizational structure. There are many different masonic groups using different structures of varying complexity. During its history, different schools have incorporated a number of different "mysteries", so that there is considerable variation both in rite and legend. Common to most is the sequence of three craft degrees (entered apprentice, fellow craft, and master mason). In the Scottish Rite, for example, this is however followed by a sequence of a further thirty degrees into which the member may progressively be initiated. At each stage it is implied that the individual obtains access to greater insight, notably through understanding of symbols of greater power.
The degrees of *freemasonry*, commonly accepted to be thirty three in all, may be considered as an ascending scale of development of the individual, although few freemasons ever proceed beyond the third degree or *master mason*. Each ascent from one level to the next is marked by a specific ritual relevant to the stage which has been reached, although some orders of freemasonry dispense with a number of rituals between the 18th and the 31st degrees. Although the higher echelons of masonry are cloaked in secrecy, they are invariably men who have achieved very high distinction in their chosen profession and/or in public life.
Refs Beha, Ernest *A Comprehensive Dictionary of Freemansonry* (1962); Carlile, Richard *Manual of Freemansonry* (1845); Lawrence, John *Freemansonry* (1982); Mackey, Albert G *Encyclopedia*

of Freemansonry (1946).
Broader Initiation into secret societies (#HH9807).

♦ HH0660 Theosophy

Description In general, the term refers to mystical knowledge of God and the study of divinity based on subjective mystic experience. The phenomenal universe is seen as derived from the play of forces within the divine nature. Inner yearning concerned with the imperfection of this world and with religious needs lead to the search for experience of actualization on a higher plane of being. Although more in line with Hindu movements, the theosophical approach is also typified by Meister Eckhart and Jakob Böhme. In the 20th century, theosophy has come to be associated with specific teachings of H P Blavatsky and of the Theosophical Society founded by Blavatsky and H S Olcott. These teachings involve the uniting of many creeds through an understanding of their religious symbols; and the creation of a new religion based on the teaching of *avatars* or inspired teachers through the ages. Through occult knowledge and supernatural powers, universal brotherhood would be achieved, together with an understanding of previously unexplained natural laws and of mankind's hidden forces.
Refs The Theosophy Science Study Group *Holistic Science and Human Values.*
Related Avatar (#HH0079) The Path (#HH0424)
Anthroposophical system (#HH0018).

♦ HH0661 Yoga (Yoga)

Description Literally meaning "union" and "control", yoga is a system of methods of physical, mental, moral and spiritual development. Its premise is that through self-estrangement, alienation from existence, man has lost contact with the infinite ground of existence. Restless search for truth and happiness outside himself occurs because he has forgotten that these are only to be found at the inmost centre of his own being. The purpose of all forms of yoga is to unite finite man in full consciousness with the infinite, by whatever name this ultimate reality is known. The term is used to designate any ascetic technique and any method of meditation (Indian thought), although classical (systemic), popular (nonsystemic), and non-Brahmanic (Buddhist, Jainist) yogas are distinguished.
The fundamental goal of union with the spiritual world presupposes a preliminary detachment from the material world. Yoga therefore places emphasis on the self-discipline by virtue of which the individual can obtain the necessary concentration of spirit prior to any experience of true union. The exercises directed toward this end are extremely practical and guard against side-tracking into fanciful contemplation and dilettante exploration of profound ideas.
In addition to being a system of methods, yoga is also a system of philosophy (although not in the Western sense) lived out through these methods. It is a system of coherent affirmations, coextensive with human experience (which it attempts to interpret in its entirety), with the aim of liberating man from ignorance and allowing him to experience the *supra-conscious* component of his personality. As such it is one of the six orthodox Indian systems of philosophy.
Yoga methods (as practiced in India) are not supposed to be learnt by the individual working in isolation. The guidance of a master (guru) is considered necessary. The individual begins by giving up the profane world with which he has been associated and, guided by his guru, applies himself to passing successively beyond the behaviour patterns and values proper to the human condition. Through this process the individual may achieve (if successful) a form of death and rebirth into a transcendent mode of liberated being which enables him to maintain the experience of spiritual union whilst retaining a relationship to his ongoing, limited, earthly experience.
Within the general system of yoga, four more specialized systems are distinguished, although few gurus would make use of one yoga to the point of excluding use of the other three in his practices (to the point that each guru effectively creates his own yoga combining elements from each of the others). These four correspond to the developmental priorities and endowments of four basic personality types: reflective, emotional, action-oriented, and empirical or experimental. All forms of yoga require the individual to cultivate practices such as: truthfulness, self-control, cleanliness, harmlessness. All of them usually require some practice of the first form of yoga, namely Hatha Yoga, which gives the body the necessary health and strength (particularly through breathing control) to endure the hardships of the more advanced stages of training. The other three forms are: Laya Yoga, leading to control over the mind and emotions (defined to include mastery over will, love, energy, sound and form); Raja Yoga, leading to control over consciousness (defined to include discrimination, intellect, action, psychic-nerve energy, and ecstasy); and Dhyana Yoga which is concerned with meditation, and as such is common to all the yogas, since it is an essential practice in each of them.
Each of the first three yogas therefore leads to the perfection of one aspect of man's nature, respectively the physical man, the emotional/mental man, and the spiritual man. Taken together, yoga is therefore conceived to be a method of bringing about man's complete development by means of mental powers and spiritual forces, experienced in and through the human organism.
Refs Asrani, U A *Yoga Unveiled* (1977); Ayyangar, T R S *The Yoga Upanisads* (1952); Chaudhuri, Haridas *Integral Yoga* (1965); Choisy, M *La métaphysique des yogas* (1948); Dassgupta, Surendranath *Yoga as Philosophy and Religion* (1986); Eliade, M *Patanjali and Yoga* (1969); Feuerstein, G *Textbook of Yoga* (1975); Feuerstein, Georg *The Essence of Yoga* (1974); Ghosh, J *A Study of Yoga* (1933); Ghosh, Shyam *Original Yoga* (1980); Pavitrananda, Swami *Common Sense About Yoga*; Rai, Ram Kumar *Encyclopedia of Yoga*; Sivananda, Swami *Science of Yoga*; Taimni, I K *The Science of Yoga* (1961).
Narrower Maithuna (#HH2034) Agni yoga (#HH0406) Raja yoga (#HH0755)
Atma yoga (#HH0774) Laya yoga (#HH0782) Puja yoga (#HH0111)
Nada yoga (#HH0160) Karma yoga (#HH0372) Hatha yoga (#HH0862)
Jnana yoga (#HH0927) Kriya yoga (#HH0969) Bhakti yoga (#HH0337)
Dhyana yoga (#HH0827) Mantra yoga (#HH0931) Buddhi yoga (#HH0484)
Taraka yoga (#HH0452) Shabda yoga (#HH0539) Mythic yoga (#HH3405)
Shakti yoga (#HH3912) Yantra yoga (#HH1224) Samadhi yoga (#HH4321)
Asparsa yoga (#HH3904) Integral yoga (#HH0037) Kundalini yoga (#HH0237)
Breath control (#HH2763) Animistic yoga (#HH7219) Six-member yoga (#HH2077)
Sexual yoga (Taoism, #HH9112) Asana (Hinduism, Yoga, #HH0669)
Tantra (Buddhism, Yoga, #HH0306) Eightfold path of yoga (#HH0779)
Initiation as spiritual rebirth (#HH3465)
Tantra in meditation on emptiness (Buddhism, Tibetan, #HM2864).
Related Sannyasa (Yoga, #HH4210) Concentration (#HM4226)
Yoga sastra (Yoga, #HH1784) Universal religion (Yoga, #HH0746)
Control of involuntary functions (#HH1988)
Transformation of consciousness (Yoga, #HM4423).

♦ HH0662 Bodily-kinesthetic intelligence

Description Skill in controlling bodily movements and in the ability to manipulate objects combine in this intelligence, which has been valued in many cultures as the harmony between mind and body – the mind trained to use the body properly and the body to respond to the mind. It reaches its height in dance, which has supernatural connotations in some cultures, and in other performing roles. Low bodily-kinesthetic intelligence is equated, in India for example, with immaturity.
Context One of the separate or multiple intelligences put forward by Howard Gardner.
Broader Multiple intelligences (#HH0214).
Related Dance (#HH0445).

♦ HH0663 Modification

Environment
Description Although individuals may have virtually identical hereditary characteristics, environmental influences cause modifications in behaviour patterns. Work on pairs of twins who have been split up since early childhood has revealed striking differences (but also striking similarities) in behaviour.

♦ HH0665 Perennial philosophy

Description According to A Huxley, perennial philosophy comprises the highest common factor of all theologies, that which recognizes divine reality in the mind and in what is substantial to the senses; and which is found to be similar to, or one with, the soul. It is what was and is known to comtemplatives and mystics of every religion, the "eternally complete consciousness", the basis of the material world actualized in particular individuals in every age of mankind through annihilation of the individual ego and awareness of the unity of all, within and without.
Refs Huxley, Aldous *The Perennial Philosophy* (1980).
Related Soul (#HH0501) Syncretism (#HH3214).

♦ HH0666 Socialization

Description Psychological development is immersed in the family, the culture, and significant other individuals, who will most greatly shape the adult form that the individual assumes. Socialization is the process whereby the individual, whether child or adult, develops the qualities essential to his functioning effectively in those sectors of the society within which he moves. Socialization therefore includes the process of learning: how he ought to feel and behave; what he ought to value; how to act in ways acceptable to others; appropriate ways to think about the real world.
In early childhood, when the socialization process commences, children must develop cognitive skills that will enable them to deal intelligently with the complex and changing requirements of everyday social life. They must become skilled in the intricate social behaviours that facilitate reciprocally satisfying relationships. They must become proficient in verbal communication in order that they can both influence and be influenced in daily life. They must also come to learn to value social approval and other symbolic rewards which make them amenable to social influence.
The primary agents of socialization (namely people usually with social roles and functions embedded firmly in the social or personal institutions of the culture) exert their influence early in the psychological and social development of the child. Socialization is however a process that continues throughout life. As with the child, the agents of socialization that affect adults use a combination of calculated deprivations and gratifications to achieve their ends. Adult socialization is essentially the process of recruitment of developing persons into full participation in the multitude of roles society has available, such as universities, clubs, trade unions, business organizations, sports clubs and social circles. Social, legal, educational, and religious organizations, mass media, and a multitude of extra-familial agents contribute in varying degrees to the types of value and response patterns instilled in group members. The process is further complicated by the fact that these multiple sources of influence frequently act in conflicting directions.
Effective socialization involves some fundamental alteration of the image an individual has of himself. For adults exposed to new situations that require new roles and new patterns of behaviour, sometimes only partial socialization occurs and the adult learns to display the minimum necessary conformity in appearance and behaviour without becoming attached to or identified with the social institutions within which he is enmeshed.
The ultimate aim of the socialization process is the substitution of internal controls for external sanctions in any given setting. Once the individual has learnt to internalize the controls on his behaviour and develop a self-regulatory system, his self-demands and self-reactions to his own behaviour serve as his main guides and deterrents. Adherence to societal norms occurs in the absence of external pressures and social surveillance.
The process of socialization can be interrupted almost from the beginning of life. Physical malformation can set immediate limits on a child's capacity to respond to the efforts of adults to socialize him. Destructive psychological happenings during early childhood will be equally effective in distorting the socialization of the child. The normally expected ages and stages of social growth may then fail to appear according to the accustomed schedule.
Refs Kohlberg, Lawrence *Stage and Sequence* (1969).
Related Enculturation (#HH0954).

♦ HH0668 Transference

Psychological rapport
Description This is a specific form of generalization when an individual perceives or interprets a present situation in terms of similar situations or experiences in the past. In psychoanalytic terms, transference involves the unconscious transfer of childhood feelings, previously held towards those with whom the individual had close emotional links (parents, siblings), to the analyst conducting the treatment. These feelings, whether positive or hostile, are contrasted with feelings which would normally arise in reaction to events and behaviour during the course of the relationship. Types of transference are generally classed as *libidinal*, *aggressive*, *libidinal-defensive*, *aggressive-defensive*, *anxious defensive*.
Related Transference love (#HM7045) Counter-transference (#HH0235).

♦ HH0669 Asana (Hinduism, Yoga)

Description Practice of these physical exercices and postures may bring marked changes in the body and assist in making the body perfectly healthy and fit. Hatha yoga incorporates the formal practice of asanas, also raja yoga uses the technique as a means of eliminating disturbances in the physical body so that the yogi can forget it and place his full attention on eliminating disturbances of mind and achieving altered states of consciousness. Although there may be initial difficulty in maintaining the posture, with practice it may be maintained for considerable periods of time; an asana is said to be mastered when it can be maintained for three hours without strain. Of the theoretical 8,400,000 asanas there are 84 which are considered best and 32 said to be the most conducive to good health.
The particular posture usually associated with yoga is intended to make the body an instrument of self-elevation. The posture is assumed with an inner disposition that includes awareness and peace. The members of the body are united in an homogenous whole, awakening the individual to the intrinsic aliveness of the body, the first stage of directing consciousness inwards. In this posture, the spine is symbolically assimilated to the world axis.
Context The third component of the eightfold path of yoga.
Refs Mookerjee, Ajit *Tantra Asana* (1971).
Broader Yoga (Yoga, #HH0661) Eightfold path of yoga (Yoga, #HH0779).
Related Raja yoga (Yoga, #HH0755) Hatha yoga (Yoga, #HH0862)
Pranayama (Hinduism, Yoga, #HH0213) Dance-induced experience (#HM3213).

♦ HH0672 Sensitivity training

Training laboratories — T-Groups — Human relations groups
Description Sensitivity training is a method used in group therapy sessions to assist the

individuals who participate to increase their effectiveness in self-fulfilment and in relating to others. Personal experience in the group is used to make individuals more aware of themselves and of the manner in which they affect others and are in turn affected by them. It is thus possible to reduce fixed reactions toward other people and to achieve social sensitivity.
The main feature of the method is the expression by each individual of his perception of others present, thus revealing and sharing personal concerns, emotional conflicts, and commonalities of experience in relation to others. The trainer in the group refrains from taking on any traditional role as a group leader or lecturer and attempts instead to clarify the group processes using incidents as examples to clarify general points or provide feedback.
Sensitivity training is in many respects similar to group psychotherapy but tends to focus only on matters which can be dealt with reasonably within the relatively brief period which is usually available. It does not inquire into historic roots of behavioural patterns, into socially taboo areas such as sexual mores, or into the realm of the truly unconscious impulses and defences. Its aim is more re-educative than re-constructive, depending more upon insight and corrective emotional or behavioural experiences than those of genuine therapy. The method is widely used in management training sessions in business and industry.
Refs Back, W *Beyond Words* (1972).
Related Self-realization therapy (#HH0784).

♦ HH0673 Transparency
Description When one is so totally committed that each action is absolutely free, then one is transparent in every situation. One develops insight into true humanity or *humanness* and loses the sense of being lost or that things are not quite right. Transparency brings a sense of certitude despite the apparent absolute nothingness, peace and freedom from problems, joy in tragedy and a knowledge of eternity in the midst of contingency. This transparency arises with the awareness of the mystery beyond apparent reality, relating to that mystery.
Related Mystery (#HH0303) Humanness (#HH0203).

♦ HH0674 Humanism
Description Humanism is opposed to absolutism and to transcendence of truth and reality. Absolute truth and reality can only be defined as unattainable by man and absolutist doctrine cannot really establish a relationship between its use of the terms absolute truth and reality and the human use of theses terms; but the truth and reality for man are attainable by man and are sufficient for man. Claims of metaphysical systems to absolute and exclusive truth are refuted by the multiplicity of differing systems making such claims. Humanism does not deny the right to set up a metaphysical system; it does deny the right to claim infallibility and to force others to accept what is essentially a philosopher's personal system. Humanism thus substitutes for belief in an absolute and immutable truth a common truth which rests on social agreement and grows as human knowledge grows. It implies toleration, not persecution, and respect for common sense rather than dielectical subtleties.
Modern humanism starts with a feeling of wonder at the unfathomable mystery of life, which the individual can explore using his own natural (and not infallible) faculties. Within the limit of his own abilities, he is capable of achieving an existence worthy of humanity, not in slavery to inefficacious instincts. He makes use of all his natural faculties - the *concept of reality* as received through his impressions, the *passions*, *instincts*, and the *sense of value* or power of *discernment*. Through *humanist counselling*, the individual is aided in appraising his own experiences by self-examination and by projection, leading to the fullest possible maturity and actualization.
Refs Fromm, Erich Ed *Socialist Humanism* (1965); Lerner, Max *Education and a Radical Humanism* (1962); Maritain, Jacques *Integral Humanism* (1968); Reiser, Oliver *Cosmic Humanism* (1966).
Narrower Secular humanism (#HH0105) Socialist humanism (#HH0360)
Ecological humanism (#HH4965) Humanistic capitalism (#HH6209)
Enlightenment humanism (#HH0215).
Related Humanness (#HH0203) Agnosticism (#HH0904)
Philosophy of enlightenment (#HH0591).

♦ HH0675 Positive mental health
Description Mental health and well-being are usually taken for granted so that the main focus of attention in the mental health field has been on various forms of mental illness. However, mental health is not only a question of treating individuals, nor of implementing measures which are preventive and orientated towards the individual. Modern mental health considers man in his interplay with his contemporary milieu (from family and working environment to the wider economic and ecological milieu) at any given moment.
An important task for mental health is therefore to organize the environment in such a way that man can thrive. Social, ecological or physical purification of the environment and other mental hygienic interventions thus clearly come under this heading. This is why mental health has become a political issue, and why reconsidering concrete tasks for mental health also requires consideration of the concept of mental health in relation to mankind's current situation.
For instance, rapidly increasing urban slums are areas where the low quality of life causes a variety of mental distress, this being excaberated by the numbers of individuals living alone or in very small family groups. Community mental health projects can promote mutual support and self-help. Raising social consciousness in people who previously accepted resentfully their deprived situation can simultaneously enable them to improve their own present mental health by learning to work together and to cooperate in problem solving; and improve the situation for the future by dealing with behaviour problems and psychosocial ills in the community. Similarly, migration (particularly selective migration of workers who leave their families behind) can cause a variety of mental disturbances. Legislation can increase these problems by denying rights to migrant workers and their families; but such legislation - and legislation in every field affecting human relations and the life of the individual - could and should be directed towards positive aspects of mental health. Laws which now consider only the physical wellbeing of claimants for welfare should be altered to take their mental wellbeing into account also.
Another area in positive mental health which deserves attention is the strengthening of the family and the way it functions through clear-sighted social policies, better community organization and special intervention in risk situations. Again the advantages are two-fold - improvement in the mental climate within the family; and better care in cases where treatment is necessary by using the family and community to assist in that treatment, thus "re-humanizing" the health services and diverting much needed resources to other areas of need. This improvement in family and community attitudes has also to be followed in the training of health workers at all levels in a more humane and understanding approach to mental disability.
Increasingly, therefore, the various kinds of mental disease are considered as given, with specification of positive mental health and its conditions as the problem to be solved.
Refs Davis, James A *The Meaning of Positive Mental Health*; Jahoda, Marie *Current Conceptions of Positive Mental Health* (1958).
Related Mental hygiene (#HH0296) Psychoneuroimmunology (PNI, #HH1401).

♦ HH0676 Attitude therapy
Description A method of re-educational psychotherapy which helps an individual substitute attitudes that foster harmonious relationships for previously held maladaptive attitudes.
Broader Therapy (#HH0678).
Related Attitude change (#HH0856).

♦ HH0677 Virya (Zen)
Bravery — Vigour
Context The fourth of five perfections or *paramitas* of Zen which, when achieved, render the mind ready for the sixth - *maha paramita* or great perfection, *prajna*.
Broader Paramitas (Zen, #HH0219).
Related Effort (Buddhism, #HM2389) Prajna paramita (Zen, #HH0550).
Followed by Dhyana (Hinduism, Buddhism, #HM0137).
Preceded by Kshanti (Zen, #HH0813).

♦ HH0678 Therapy
Description Modern therapeutic methods, whether for the prevention or curing of disease, have eradicated some diseases completely (smallpox) or dramatically reduced their effects (vaccines against poliomyelitis, typhoid fever; antibiotics for tuberculosis; insulin against diabetes); this in turn has significantly lengthened the average life span. Modern therapeutic practice has also alleviated unpleasant physical symptoms through physical methods. These techniques complement rather than supplant traditional methods; in fact ancient techniques such as acupuncture and homeopathy are becoming more popular.
Narrower Dance (#HH0445) Qi gong (#HH3862) Art therapy (#HH0775)
Reflexology (#HH0987) Sociotherapy (#HH0336) Self-therapy (#HH0606)
Play therapy (#HH0633) Will therapy (#HH0883) Hypnotherapy (#HH0962)
Gene therapy (#HH1334) Aromatherapy (#HH5080) Group therapy (#HH0851)
Shock therapy (#HH0975) Reiki therapy (#HH0731) Music therapy (#HH0496)
Colour therapy (#HH0007) Active therapy (#HH0091) Orgone therapy (#HH0327)
Family therapy (#HH0362) Shadan therapy (#HH0483) Rehabilitation (#HH0874)
Morita therapy (#HH1121) Naikan therapy (#HH1324) Gestalt therapy (#HH0751)
Gedatsu therapy (#HH0972) Aversion therapy (#HH0256) Physical therapy (#HH0379)
Attitude therapy (#HH0676) Multiple therapy (#HH0846) Breathing therapy (#HH0752)
Religious therapy (#HH0762) Behaviour therapy (#HH0795) Implosive therapy (#HH1107)
Expressive therapy (#HH0149) Nude group therapy (#HH0482) Analytical therapy (#HH0507)
Convulsive therapy (#HH0566) Persuasion therapy (#HH0696) Suggestion therapy (#HH0796)
Assignment therapy (#HH0802) Megavitamin therapy (#HH0289)
Occupational therapy (#HH0596) Electroshock therapy (#HH0866)
Experiential therapy (#HH0885) Group-centred therapy (#HH0607)
Body movement therapy (#HH0172) Interpersonal therapy (#HH0434)
Neuromuscular therapy (#HH4332) Client-centred therapy (#HH0095)
Mental imagery therapy (#HH0072) Activity group therapy (#HH0492)
Child guidance therapy (#HH0894) Psychobiologic therapy (#HH1236)
Addiction group therapy (#HH0325) Learning theory therapy (#HH0614)
Self-realization therapy (#HH0784) Conditioned reflex therapy (#HH0985)
Assertion-structured therapy (#HH0685)
Rhythmic sensory bombardment therapy (#HH0218).

♦ HH0679 Creative intelligence
Description Creative intelligence corresponds to the level of the logical self and is said to be the source of consciousness or the reservoir of all possibilities. It combines scientific and creative theories of consciousness and is approached practically in transcendental meditation.
Related Pre-conscious pure intelligence (#HM2788).

♦ HH0680 Vipassana-bhavana (Buddhism, Pali)
Cultivation of experiential insight — Nana-dassana
Description By practising constant awareness, conscious of the body, feelings, state of mind and the mental processes, an individual achieves insight into reality. The system involves repeated meditation on major items of Buddhist doctrine until they are completely internalized and the individual cognitive and perceptual systems operates only in these terms - the old cognitive and perceptual habit patterns are removed. This is done under the supervision of individuals who have already achieved insight and is usually preceded by *samatha* meditation to bring a state of calm (although nowadays this step is sometimes omitted as being distracting).
The technique teaches the individual to know and be conscious of something new about the way things are. It is seen as solving the basic problem of ignorance, believing that inaccurate understanding leads to misinterpretation and misunderstanding both of the human experience and of the nature of the world. In contrast, knowledge brings control so that the individual has power over his own destiny and is no longer subject to repeated rebirths and deaths.
With practice, the person meditating becomes aware of impermanence (anicca), unsatisfactoriness (dukkha) and insubstantiality (anatha). If the ensuing detachment is not so satisfying as to deter from continuing progress, further insight on the Buddhist path follows, culminating in *lokuttara*, transcendent consciousness.
Context One of the two systems of Buddhist meditation, the other being samadhi-bhavana (development of concentration).
Broader Meditation (#HH0761) Bhavana (Buddhism, #HH0551).
Related Enstasy (#HM3169) Samatha-bhavana (#HH0710) Intuition (Buddhism, #HM1632)
Samadhi-bhavana (Buddhism, #HH8231) Eightfold path of yoga (Yoga, #HH0779)
Lokuttara consciousness (Buddhism, #HM2120) Imperfection of insight (Buddhism, #HM9722)
Yatha-bhuta-jnana-darsana (Buddhism, #HM7549).

♦ HH0681 Psycho-cybernetics
Description This is a method of achieving a satisfying life by improving the self-image. It is based on a series of practice exercises designed to change thinking habits and acquire the habit of happiness. The aim is to redirect goal-striving subconscious mind towards success-oriented goals.

♦ HH0682 Transcendental meditation (TM)
Description A technique which, by turning the attention inwards, allows the conscious mind to experience increasingly more subtle states of thought until the mind transcends the experience of the subtlest thought-state and arrives at the source of thought. This expands the conscious mind and brings it in control with the unlimited reservoir of energy and creative intelligence. The practice enables the individual to use his full potential in all fields of thought and action. For this increase in capacity and fulfilment to take place, it is necessary for the person to be in contact with transcendental pure consciousness, his own essential nature. Without having to observe or struggle with himself, quite spontaneously the individual's natural inclinations then begin to come into harmony with the natural laws of the evolution of life. His desires become increasingly life supporting and, simultaneously, increasingly fulfilled.
The premise of transcendental meditation is that the attention of the mind has as its special characteristic the tendency to scan the field of its experience in search for a greater intensity of happiness than it is at the moment experiencing. The attention is governed and directed by this natural tendency of the mind, to search for greater satisfaction, greater joy, greater fulfilment than it is at the moment experiencing. Transcendental meditation is thus understood simply as a means of opening the wandering mind to the riches of happiness which it has erroneously been seeking in external objects and events. As the mind moves in the direction of the absolute

bliss of the transcendental being, it finds increasing charm at every stage, whether or not the person is emotionally or intellectually developed.

In transcendental meditation each person practices in solitude with a sound as a focus for meditation. Each person is given the sound most suited to his temperament. As the mind transcends during meditation, natural physiological changes occur in the body corresponding to changes in mental activity and the level of experience. The mind and body achieve a state of restful alertness which, on the psychical level, corresponds to the state of pure consciousness. The physiological processes, seemingly more efficiently than during sleep, relieve the stresses which accumulate in the nervous system.

The reported physiological changes during practice of this (and other) forms of meditation include: reduction of the metabolic rate by up to 25–30 per cent; reduction of the total oxygen consumption by up to 20 per cent; reduction of the breathing rate to 4–6 (from 12–14) per minute; an increase of the amount of alpha and theta waves of the brain; a reduction of the blood pressure by 20 per cent in hypertensive patients; decrease in the cardial output (heart blood flow) by 25 per cent. Regular practice of TM is said to improve both the physical and the psychological quality of life, allowing deep rest, improved health and the regaining of mental stability. It is a very effective way of removing stress and works towards the achievement of full human potential. Meditators have been shown to be happier and less dependent on their surroundings than non-meditators; they experience more enjoyment, develop deeper personal relationships and have a deeper sense of purpose.

Refs Aron, Arthur and Aron, Elaine *The Maharishi Effect* (1986); Nidich S, Seeman W and Dreskin T *Influence of Transcendental Meditation* (1973); Russell, Peter *The TM Technique* (1976); Wallace, Keith *Physiological Effects of Transcendental Meditation*.
Narrower Technique of no technique (#HH1092).
Related Levitation (#HM8634)   Transcendence (#HH0841)   Maharishi effect (#HH3764)
Restful alertness (#HM0996)   Unity consciousness (#HM3193)
Path of effortlessness (#HH2188)   Transcendental consciousness (#HM2020)
Refined cosmic consciousness (#HM3222)
Transcendental cosmic consciousness (#HM2405)
Swapna state of dream consciousness (Yoga, #HM2781)
Jagrat state of waking consciousness (Yoga, #HM2141)
Susupti state of unconsciousness (Hinduism, Zen, #HM2957).

♦ **HH0683 Nembutsu** (Japanese)
Contemplative invocation
Description In the hope of rebirth into the pure land of the Buddha Amida, the devotee makes the invocation "I take my refuge in the Buddha Amida" (namu amida butsu). This current practice is based on a traditional varied practice of a number of sects, including contemplation (where the characteristics of a Buddha were envisioned) and meditation on the spiritual qualities of a Buddha. Nembutsu was also directed to Buddhas other than Amida and towards goals other than pure land rebirth. Nembutsu-zammai, for example, had the goal of enlightenment – samadhi.
Related Pure-lands awareness (Buddhism, Tibetan, #HM2168).

♦ **HH0685 Assertion-structured therapy**
Description Behavioural possibilities are selected by the perceiving-acting person to meet the situations confronted. They are not a function of the "depths" of the person, namely the unconscious mind, but rather of the choices made. The person therefore lives by assertions about the situations confronted, which have varying probabilities of confirmation or disconfirmation. The neurotic is essentially a person who is constantly betting on a set of assertions or assumptions that have a high probability of disconfirmation. Certain patterns of behaviour tend to be repeated in a circular self-defeating way, leading to redundancy and inflexibility in the organism. The task of the psychotherapist is to reduce the redundant condition of disconfirmation and the resulting tension, so that the individual can engage in effective problem-solving and thus be more firmly oriented in reality.
Broader Therapy (#HH0678).

♦ **HH0686 Reincarnation**
Immortality — Rebirth — Metempsychosis — Recurrence — Transmigration — Palingenesis — Re-embodiment — Past life regression
Description Reincarnation is a term loosely applied to cover a cluster of meanings which are themselves sometimes distinguished by certain authors or else used interchangeably by others. It may refer to the idea that, since perfection cannot be achieved in one lifetime but many lives are required, the soul moves from one human body to another until it perfects itself. This process may envisage the soul taking forms other than human. Conversely, the idea of rebirth may itself be used to combat the idea that there exists such a thing as the soul. The idea of recurrence envisages the endless repetition of the same life in all its details until consciousness or remembrance is attained and freedom genuinely desired. Spiritual progress may be thought of as occurring through a series of incarnations, the more progress being made in this life, the more propitious the next life; or else incarnation itself may be seen as a sign that significant progress has not yet been made. In either case, there is no finality – even heaven is temporary, as are the gods, and all will be dissolved at the end of this creation. The only escape from reincarnation in this case is through self realisation and unity with the absolute. Meanwhile, one is always given another chance, another lifetime in which to achieve the goal. The doctrine of reincarnation is accepted in many religions, including the early Christian Church; and also by many philosophers, from Pythagoras and Socrates to Schopenhauer.

The concept of *immortality* – defined by Kant as "the infinitely prolonged existence and personality of one and the same rational being" – may include reincarnation on this earth or the doctrine of eternal life in some other world (heaven or hell) depending on the way in which the present life has been led. Some form of these beliefs are known to have been held in virtually all civilisations, with elaborate funeral and burial customs designed to benefit the soul in its next life. The widespread belief in rebirth (which may be considered as an antidote to the apparent arbitrariness of life) is one of the primordial affirmations of mankind. The psychic events underlying such affirmations are therefore valid subject matter for psychology, whether or not their metaphysical and philosophical bases are considered significant.

Refs Badham, Paul and Badham, Linda *Death and Immortality in the Religions of the World* (1987); Clow, B *Eye of the Centaur* (1986); Head, J and Cranston, S L *Reincarnation in World Thought* (1967); Head, Joseph and Cranson, S L (Eds) *Reincarnation* (1968); McClain, F *A Practical Guide to Past Life Regression* (1986); Murray, David C *Reincarnation* (1988).
Related Embodiment (#HM3409)   Immortality (#HH0800)
Rejuvenation (#HH0525)   Pre-existence (#HH1339)
Psychological growth in a dying person (#HM2553).

♦ **HH0687 Concentration on the flow of breath**
Description As a method of meditation, the passive and detached observation of the flow of breath precedes awareness of the mystery of life. A mantram is slowly repeated while holding the breath; on inhaling the individual focuses attention on the filling of the system with love, light, purity, etc, while during exhalation attention is on the elimination of impurities and disease.
Broader Meditation (#HH0761).
Related Hatha yoga (Yoga, #HH0862)   Induction technique (#HH0025)
Pranayama (Hinduism, Yoga, #HH0213).

♦ **HH0689 Human development** (Taoism)
Tao — Michi
Description Antedating all formulations, Taoism is associated with a recondite realm of the mind where the customary divisions of thought do not exist. Taoism recognizes the potential of humanity to acquire extraordinary powers and modes of perception. The dilemma concerning the individual and collective benefit or harm deriving from the exercise of such possibilities leads some practitioners to subordinate everything they do to the quest for permanent stabilization of consciousness, namely the purification and deautomatization of the mind so as to make it sensitive to reality. To this end a practice of central importance is that of "balancing yin and yang", through which unconditioned energy is gathered. One method of doing so, in the case of the individual, is through the practice of alchemical procedures. A closely related method, traditionally used in the governance of any collectivity, is that associated with the I Ching. In such cases, practitioners commonly practice quiet, meditative sitting as an aid to development of their understanding. This may involve meditative exercises such as the "waterwheel exercise". Some traditions also practice a form of sexual yoga (the twin cultivation of yin and yang), considered as complementary to meditation. Taoism recognizes that no practice, include the stillness produced by concentration, is necessarily sufficient to break through the boundaries of psychological conditioning and may only hold it in temporary abeyance. In particular altered states produced during such exercises are viewed as "phantom elixir" in that they vanish with time. The art that is cultivated is that of blending and transcending the elements of duality, notably stillness and movement, in order to unite the celestial and mundane within humanity.

Tao is literally the *path* or *way*, in Japanese *michi*, and in many ways equivalent to the Buddhist *dharma*. Tao is also ultimate reality, the unity underlying plurality. It is the motive of all movements, the mother of all substance. Its nature or virtue is referred to as *teh*, harmony or strength, whose presence indicates health and strength of the body and harmonious relations with all. With *Tao* as the source and natural way of things at all levels, from the physical to the spiritual, Taoism sees all misfortune as deriving from separation from this source and deviation from this way. Returning to the source means the cultivation of simplicity, and becoming child-like by mortification of the will, by recollection and meditation. Only by obedience and study can spontaneous, "unstudied" action truly arise. Although deliberate right choice and obedience to ethical codes have merit, they are very much "second best". Unitive knowledge of Tao – the ground and the logos – is true charity, achieved at the spiritual level by means of morality and kindness. Basic to the principle of Taoism is the relativity of attributes – there is no ultimate attribute, a thing is only long in comparison with something which is shorter and which has been taken as standard. Such a standard cannot be absolute or objective although it may be thought of as such. In Tao, opposites are blended and contrasts harmonized.

Despite this description, Tao is in fact said to be the wordless doctrine, being beyond words, which are irrelevant to the deep experience Tao brings. The Tao that can be expressed is not the Tao. One does not worship the Tao, one harmonizes with it. All forms and names, all external characterizations are simply mental gymnastics.

Context Taoism uses many media of expression, of which many are recognized has having degenerated over time, obscuring any underlying link which may have existed between them. These forms of expression, doctrine and praxis continue to evolve and adapt in response to contemporary needs. Central to taoist understanding is the statement: A path that can be verbalized is not a permanent path; terminology that can be designated is not a permanent terminology. Some authors identify 3,600 practices in Taoism. Others repudiate all such formal practices as "sidetracks" which merely hint at the metaphysical nature of Tao and confirm the contention that truth is formless. Most didactic devices employed by Taoism can therefore be interpreted in different ways. No standard scheme is universally applied. This apparent confusion follows directly from the fact that Taoism endeavours to deal with the process of formalization as it assists and restricts human development.

Refs Chang, Po-tuan *The Inner Teachings of Taoism* (1986); Chang, Po-tuan *Understanding Reality* (1987).
Broader Human development through religion (#HH1198).
Narrower Alchemy (#HH5887)   Tso-wang (#HM0346)   Sexual yoga (#HH9112)
Balancing yin and yang (#HH4997)   Ultimate accomplishment (#HM1342)
Circulation of the light (#HH6452)   Merging with the ordinary world (#HM0486).
Related Dharma (#HH0093)   Quietism (#HH2006)   Yu (Taoism, #HM1771)
Non-action (#HH1073)   I Ching (Taoism, #HH0004)   Feng-shui (Chinese, #HH2901)
Spirituality (Taoism, #HH5117)   Mystical stage of life (#HM3191)
Refining the gold pill (Taoism, #HH4880)   Release from the matrix (Taoism, #HM0635).

♦ **HH0690 Symbols**
Myths
Description Symbols and myths are traditional means for explaining and understanding things which in themselves are or seem beyond understanding, or as a tangible expression or reminder of the intangible. For example, a sacrament is described in the Catechism is "an outward and visible sign of an inward and spiritual grace".

For non-materialist, archaic man, physical objects and acts are only real and only have value by their participation in a transcendent reality. The act of eating renews communion, consecration makes sacred by uniting an object with God. Every action is seen as re-enacting what has been done many times before by others, and originally by something higher than mankind, a *celestial archetype*. Building a temple to a divine pattern means physically representing something which exists already at a higher plane. It is said, for example, that temples and palaces are representatives of the holy mountain at the *centre of the world* where heaven and earth meet.

Although problems may arise when different religious or cultures use different symbols to represent the same thing, and dissent occurs when unfamiliar rituals or expressions are encountered, it is interesting that many basic symbols are common to all traditions.

Symbolism is inherent in the relating of things because of their apparently superficial similarity, as is typical in *archaic thought*.

The world speaks through symbols, reveals itself through them. Symbols are not replicas of objective reality. They reveal something deeper and more fundamental. Symbols are capable of revealing a modality of the real or a dynamic of the world which not available on the plane of immediate human experience. They can operate, not in the first instance, on the level of rational cognition but rather on the level of apprehension by the active consciousness prior to reflection. Symbols which touch on patterns reveal a deeper life, more mysterious than what is grasped by everyday experience. Symbols are always religious, in that they always point to something real or to a world pattern; and the real is the powerful, the significant, the living and therefore the sacred. They imply an ontology which is a judgement of the world and human existence and which can not always be translated into concepts. Another characteristic of symbols is that they are multivalent; they express simultaneously several meanings the unity of which is not obvious on the plane of immediate human experience nor by critical reflection. This unity is the result of a mode of viewing the world. As a result of its multivalent capacity it also reveals a perspective in which diverse realities can be fitted together or integrated into a system of thought. Also, they have the capacity to reveal paradoxical aspect of a single reality. For example, the verbal symbol fire points to comfort and pain, a tool and a source of destruction and to a warm and cozy evening before the fireplace and a towering conflagration and the smell of burning flesh resulting from fire bombing. Finally, symbols have an existential value. They can reveal that the modalities of the spirit are manifestations of life. They bring meaning to mundane human living. A person who

understands a symbol not only opens himself up to the objective world but is connected to the universal.
Refs Journal of Transpersonal Psychology *Symbols of Transpersonal Experiences* (1969).
Related Rites (#HH0423)   Archetype (#HH0019)   Understanding (#HH0658)
Guiding images (#HH3020)   Archaic-paralogical thinking (#HH0748).

♦ HH0691 Purpose
Description A sense of purpose, or coherent end, for behaviour integrates actions into a sequence or system. Originally "purpose" denoted "reason for existence", with the teleological implication of divine purpose behind creation and development. Rationalism treats purpose as human activity, associated to reason. Some philosophies include the juxtapositioning of causality with purpose (indeterminism); and purpose as the reflection of objective needs, a plan of action determining the nature of various operations and their place in the overall system (Marxism).
Related Karma (#HH0567).

♦ HH0692 Organized games
Generalized other
Description Within the structured activity of organized games the individual learns to take the role of all the others involved in the game. The organization of these roles determines the actions of the individual who experiences these roles as the *generalized other*.
Broader Games (#HH0705).
Related Physical culture (#HH0730)   Being all the other (ICA, #HM2983).

♦ HH0693 Self-understanding
Description Understanding one's self is similar to understanding any other person and requires the same capacity to compare and give coherence to experiences. It requires a continuous appraisal of the meaning behind personal experience and the search for personal metaphors to highlight past experiences and unite them with present activities and hopes for the future.
Refs Malamud, D and Machover, S *Toward Self-Understanding* (1965).
Related Self-knowledge (#HH0312)   Self-examination (#HH0314).

♦ HH0694 Faith
Description Faith is generally considered belief and acceptance of something as true when such a thing is not or cannot be known as a fact. In religious terms, it is a belief and confidence in God as revealed through scripture and teaching, often with reference to the future. Religion may be said to presuppose faith. In Zen, where the duality implied in dependence on God as some other being is contrary to basic tenets, faith is nevertheless considered fundamental with reference to the existence of the Buddha mind and in meditation as the means of achieving it; it is the beginning of the Buddha mind where the self-image made by the ordinary mind is not yet dislodged. As such it may be compared with *divine discontent*, where there is no satisfaction in even the best that can be produced by the ordinary mind.
Although not based on rational deduction, faith is nevertheless considered an act of the intellect. For example, that which was experienced as knowledge on a particular occasion can be accepted by faith when the experience of knowledge is no longer present. Christianity teaches that faith is a gift of God, through grace, to which an individual may or may not give personal assent. Religious writers are unanimous in considering faith as a clearing ground for right action, a means of setting aside all hindrances to the task in hand.
Nothing that can be seen can either be God or represent Him to us. Nothing that can be heard is God. Similarly, God cannot be imagined. To find God we must enter darkness, silence, obscurity without images or the likeness of anything created. He can be understood only by Himself. To know Him means being transformed so as to know Him as he knows Himself, and faith is the first step in that transformation. The simple act of submission to the authority of the Church through believing some article of faith brings the gift of a pure and simple inner light which perfects the intellect. Feeling the weakness and instability of the spirit in the presence of God's mystery brings a subjective sense of helplessness, so that there are still doubts in a sense. In fact, such doubt may grow as faith grows. The obscurity of faith appears as darkness to the mind because it transcends the mind's weakness. Just as the darkness is at its deepest, when the mind is truly liberated from weak and created lights, then one is filled with the infinite light of faith which is pure darkness to reason. In perfection of faith, when God becomes the light of the darkened soul, faith becomes understanding.
Faith is thus acceptance of God and the beginning of communion. It is intensive and yet it affects everything one does. It gives simplicity and depth to apprehensions and experiences, so that life is a mystery of which only a small part is what one can rationalize and the unknown is incorporated into everyday life. Not only does it bring us into contact with the authority of God and teach us truths about God, it reveals the unknown in the self in as much as it lives in God. Faith is the only way of opening up the depths of reality, even one's own reality. It integrates the whole unconscious – that above as well as that below the conscious – with the rest of life. It enables one to accept one's animal nature and to govern it according to divine will; and it subjects reason to hidden spiritual forces that are above it. Thus the whole man is subject to the "unknown" above him.
Context One of the three supernatural or abiding virtues.
Refs Chirban, J T *Human Growth and Faith* (1981); Fowler, J W *Stages of Faith* (1981); Haught, John F *Religion and Self-Acceptance* (1980); Merton, Thomas *Seeds of Contemplation* (1972).
Broader Virtue (#HH0712).
Narrower Stages of faith (#HH2097)   Passive way of faith (#HH3412).
Related Basic faith (Buddhism, #HM3209)   Belief (Buddhism, Yoga, #HM2933)
Passive state (Christianity, #HM8123).

♦ HH0695 Dasein analysis
Existence analysis
Description This is a form of psychotherapy with philosophical foundations from phenomenology. The individual is not seen as separate from but part of the environment, the *dasein* defining the existence of a being whose essence is "being in the world". Mental disease is seen as a modification of the fundamental structure or modes of transcendence, the product of one's own "world design".
Refs Boss, Medard *Psychoanalysis and Daseinsanalysis* (1963).
Related Logotherapy (#HH0416).

♦ HH0696 Persuasion therapy
Description A form of psychotherapy in which an attempt is made to influence the patient by rational argument to reach a reasonable judgement of the illness.
Broader Therapy (#HH0678).

♦ HH0697 Deviance
Description Although non-conformity to social norms may indicate mental illness, it can also indicate conformity to a sub-group rather than to the dominant social group, and may be considered the action of a saint or martyr by societies existing elsewhere or during another period.

♦ HH0698 Spiritual rebirth of the ego
Description Integral philosophy maintains that when the separate, individual ego is dissolved in the One of the eternal, it is reborn in the enlightened, spiritual sense to a cosmocentric existence in profound joy and harmony. Such teaching has echoes in Christianity, with the image of the resurrection of the dead "... born a natural body, raised a spiritual body" and the need for the seed to "die" before it can grow and bring forth fruit – "Except a corn of wheat fall into the ground and die, it abideth alone; but if it die, it bringeth forth much fruit".
Related Self-denial (#HH0964)   Born-again (Christianity, #HH0576)
Spiritual rebirth (Christianity, #HH4098)   Initiation as spiritual rebirth (Yoga, #HH3465).

♦ HH0700 Divine nature
Description The divine nature of mankind is said to be the perfect, pure and eternal nature of every human being, however dark, imperfect and apparently impermanent the outer characteristics of that person may be. By strengthening the true nature, the divinity behind the appearance of mortality is revealed.
Related Treasures of the godly (#HH0364).

♦ HH0702 Naturalism
Description Seeing all phenomena as determined by and explicable by the laws of nature, naturalism considers all human behaviour as subject to deterministic laws not requiring the assumption of human purpose or freedom of choice. This may be considered the opposite of *Idealism*.
Refs Congresso Nazionale di Filosofia *L'Unificazione del Sapere* (1967).
Related Idealism (#HH0893).

♦ HH0703 Creative thinking
Productive thinking
Description New or original ideas and thoughts, those leading to something new for the individual or for humanity in general, are termed creative or productive thinking.
Refs Davis, G A and Scott, J A (Ed) *Training Creative Thinking* (1971); de Bono, Edward *Masterthinker's Handbook* (1985).
Related Innovative learning (#HH0812).

♦ HH0704 Deutero learning
Secondary learning
Description Learning which results incidentally as a result of learning something else rather than as the result of a conscious effort. In particular, this is one of the aspects of *enculturation*, when values, norms and styles of learning are absorbed without being taught and may be crucial in determining a person's future behaviour and learning patterns.
Related Enculturation (#HH0954).

♦ HH0705 Games
Play
Description Games are commonly accepted as not only a means of socializing but also of learning useful skills and of inculcating desirable roles. Three types of games are identified – competitive (leading to a sense of personal merit), chance (giving the impression of being favoured or otherwise by fate), and strategy (political integration). The less complicated activity of playing is also useful in taking on the role of another and thus experiencing the related perceptions; in addition it may release from habitual inhibitions.
Refs Caillois, Roger *Man, Play and Games* (1961).
Narrower Lila (#HM2278)   Growth games (#HH0175)   Theatre games (#HH0582)
Organized games (#HH0692)
Consciousness ascension stages game (Buddhism, Tibetan, #HH4000).
Related Recreation (#HH0410)   Play therapy (#HH0633).

♦ HH0706 Self
Description The individual self is the subject of successive states of consciousness, but although each individual has an acute awareness of self, it is difficult for him to determine exactly what he is aware of; some thoughts and acts are more self-relevant than others. It has been said that the true self can only be defined by what it is not, every denial "this is not I" more clearly highlighting what is.
Distinctions may be made between self and ego in many different ways, or they may be treated as equivalent. For example, self as known to the individual is known as the "me" or empirical ego (see Self-concept). As known, it is the "I" or pure ego.
To Jung, the self is the unity of the whole personality, the archetypal image of man's fullest potential; it is the centre of psychological life and also the whole circumference, the unifying principle, the totality comprised of conscious and unconscious elements; and as such extends beyond the ego both in scope and intensity, the ego being the centre of the conscious mind. Again, to Jung, the process of relating ego to self is unending, the self demanding realisation but being beyond the range of human consciousness. Positive response to interventions of the self, rather than simple submission to or ignoring of the archetype, is the discriminatory function of consciousness. This definition of self as archetypal urge to deal with the tension of opposites makes it the means for confronting good and evil, human and divine, an interaction in which freedom of choice is exercised towards the inconsistent demands of life. Archetypal symbols have a numinous, transcendent, god-like quality.
Followers of Jung have extended the concept of self development to refer to the expression during life of innate, archetypal potential existing at the outset; the process implies *deintegration* of emerging potentials and their *reintegration* through reactive response and internalisation of the response. Others see the mother's self acting as a mirror for the child, whose ego emerges with gradual separation from the mother and whose attitude to the self is therefore related to his relationship with his mother.
The Vedas refer to the true self or *Atman* and the false self or *Ahamkara*, the latter being identified with the ego-sense. In the individuation (or coming-to-be of the self), the true self emerges as the goal of the whole personality, and this is quite distinct from the coming of the ego into consciousness. In Buddhism, *anatma* or not-self is described as man's true nature, nothing that can be referred to by the individual as "myself" being, in fact, the self.
Refs Hofstadter, Douglas R and Dennett, Daniel C *The Mind's I* (1981).
Related Ego (#HH0636)   Atman (Hinduism, #HH0103)   Anatma (Buddhism, #HH0241)
Self-development (#HH0651)   Self-preservation (#HH0711)
Ego awareness (Hinduism, #HM2059)
Psychological approach to transcendence (#HH1763).

♦ HH0707 Spiritual exercises
Spiritual disciplines
Description Intellectual or physical actions undertaken to attain spiritual goals. Such actions may include meditative devotional prayer, examination of conscience, meditation on the life of a holy person, reading of sacred texts, fasting, or the use of specific bodily postures.
The main discipline of the Christian life is the presence at Holy Communion or Eucharist. This sacrament is a profound means of grace, each time bringing fresh entry into God's kingdom. It

brings the individual to practice many of the disciplines - preparation through fasting, meditation and confession; taking part in prayer, praise, reading the scriptures, silence, fellowship, praise, making offerings, reciting the Lord's prayer and singing hymns. Eventually prayer and fasting become a natural part of life. The gift of these disciplines is received from the Church; they may not be mastered or understood immediately but are basic to growing in grace and to continuing communion with God.
Refs Campbell, Joseph (Ed) *Spiritual Disciplines* (1985).
Narrower Saviors of God (Christianity, #HM3420)
Spiritual exercises of St Ignatius (Christianity, #HH9760).

♦ HH0708 Tragic suffering
Inner contradictions — Tragedy — Basic conflict
Description The concept of the tragic is a consequence of accepting the individual as an autonomous being and of the insoluble social and historical conflict arising from the individual's free actions under self-determination. The suffering of the hero manifests his dignity and nobility as he is not reconciled to fate even in defeat. Aristotle demonstrated the tragic aspects of human life - its changing nature where sorrow follows joy - as the necessary result of the *nous*, eternal and self-contained, surrendering to its other being and becoming temporal and subject to necessity. As this is recognized by the mind, the contrast between the present state of guilt, crime and gloom and the earlier, beatific innocence leads to tragic pathos; but this recognition is then followed by cathartic retribution when the passions are purged and the mind's equilibrium restored. This is similar to the interpretation (Schelling) of the tragic as the struggle and defeat by fate when the hero voluntarily atones for his predetermined guilt through acceptance of punishment, this acceptance constituting freedom.
Suffering and death are always necessary if there is to be rebirth, whether in romance, in mysticism or in religion. The pain arises in giving birth to the divine world in one's life. This is the price for transformation, inevitably arising on the way to consciousness. Conscious, voluntary acceptance of suffering, living through the death of the ego, one ceases to seek for the divine world in another and finds one's own inner life. This is a psychological and a religious act.
Broader Suffering (#HM0471).
Related Will (#HH0920) Freedom of choice (#HH0789)
Suffering servant (ICA, #HM3126) Duhkha (Buddhism, Hinduism, #HM2574).

♦ HH0709 Religious education
Description Education is a means whereby traditional and cultural knowledge is passed on to the next generation; continuous monitoring is necessary to ensure that prejudice and misunderstanding are identified and removed from what is taught. This is particularly true in the case of religious education seen in the light of ecumenicism. Christian education which extends beyond the Church and involves both social action and spirituality is seen as a process which frees rather than limits. This is particularly true of less formal educational methods emphasizing justice, freedom, human rights and dignity and which meet the need for change.
Since the Christian believes that God wills the salvation of every person, then each person should be assisted in growing to his full stature in Christ and to fully develop his gifts for the service of others. Education should therefore be open to all, whatever race, religion, sex or social group; and education without religious belief as a basis is lacking in an essential element.
A classroom approach to teaching of religion threatens that religion with subversion; no religion could survive on classroom teaching alone. The scriptures of particular religions tend not to refer to teaching of religion as such but to teaching of the way of life, and such education is the job of the community rather than the school. Despite the similarities among religions there can be no "standard", homogenous, religious education in which all the various religions are included. This is quite different from the teaching of other subjects, where emphasis is on clarity, rationality and a dispassionate approach to evidence.
Refs Moran, Gabriel *Religious Education Development* (1983); Nipkow, Karl Ernst *Grundfragen der Religionspädogogik (1975–1982)*; Taylor, Marvin J (Ed) *Changing Patterns of Religious Education* (1984).
Broader Education (#HH0945).
Related Ecumenicism (Christianity, #HH0099) Human development (Christianity, #HH2198).

♦ HH0710 Samatha-bhavana
Shamatha — Cultivation of tranquillity
Description By training in alertness and joyful contentment, and disciplining the activities of mind and body (in particular, overcoming anger, sloth, excitement/guilt, doubt and the desire for sense objects), the individual achieves inner tranquillity. Dhyana (meditation) achieves higher states of consciousness with related psychic and clairvoyant sensitivity. The technique aims at enstasy and is designed to reduce the contents of consciousness, focusing awareness on a single point and eventually halting mental activity. Samatha, with its emphasis on attitude, usually precedes the practice of *vipassana*, which emphasises cognition, when the transitory effects of samatha are rendered more permanent.
Broader Bhavana (Buddhism, #HH0551).
Related Meditation (#HH0761) Samadhi-bhavana (Buddhism, #HH8231)
Dhyana (Hinduism, Buddhism, #HM0137) Eightfold path of yoga (Yoga, #HH0779)
Lokuttara consciousness (Buddhism, #HM2120) Vipassana-bhavana (Buddhism, Pali, #HH0680).

♦ HH0711 Self-preservation
Description Three paths of morality may be distinguished, each respectively treating self-preservation as of no, some or considerable importance.
1. Some oriental mystical pantheism or nihilism requiring the overcoming of the individual will, its desires, and the body as bearer of these desires, together with western pessimism with the negation of the will to live, treat self-preservation as of no importance or as a positive hindrance to progress.
2. Implicit in most ethical systems, however, is the duty of preserving the physical self, of maintaining the body in health, of avoiding unnecessary risk and of self-defence against violence. Included here is the general moral and religious argument against suicide.
3. A third system, the individualistic morality which looks on morality as a product of individual selves, puts self-preservation as the highest or supreme principle. Here: (a) all so-called good actions are said, in fact, to be aimed at self-preservation; (b) the social system implies a struggle between individuals, the supreme good being, for each individual, to survive this struggle and the supreme duty to acquire the power to do so; (c) the genetic or hereditary aspect of how social morality and restraints developed are emphasized.
For this third system, the good is that which attracts the individual (whether an appetite or a desire) and evil is that which repels or causes displeasure. Since this results in an endless struggle for survival and power, there would be permanent war of all against all if reason did not demand a renunciation of the natural right to wage such an offensive so a common power may preserve order and make peace possible. If self defence is a duty then "to seek peace and follow it" is a law of nature. However, where the state seeks to take away the individual's life, the individual's duty for self-preservation implies his right to resist the state whether by legal or illegal means (Thomas Hobbes).
This philosophy of each seeking his own advantage is also inherent in the system of Spinoza, where virtue is self-expansion and attainment of positive good. Self-assertion is useful to all since a person can only be useful to another when he has something positive to offer. Again, good is relative, something is good because it is desired. The more clear and rational the self the more adequately it expresses the nature of God, since the truly conceived self is one with the nature of God. Knowing all things as modes of God and thus absolutely determined, there is no subjection to passion but joy arises in the intellectual love of God and man's highest good. Freedom from passion implies freedom from contention with others if they also recognize the natural order of things.
The biological theory of evolution lead to the contention that morality is the latest phase of life and moral law a development of general law. This is the condition of self-preservation as developed in the struggle for existence of individuals and groups. Longevity, or quantity, is of greater good than breadth, or diversity, of life; however, removal of pleasure removes value from life (Herbert Spencer). The self here is solely the product of impersonal forces beyond it; the universe determines moral conduct for man in preserving itself. For Nietzsche, the essence of each creature is the will to power, the value of life being the glory and excitement of the struggle for survival and power. Conventional morality and religion is a slave morality encouraging the weak and inferior. The impulsive self has the goal of power for power's sake.
Related Self (#HH0706) Life instinct (#HH0750) Self sacrifice (#HH1241).

♦ HH0712 Virtue
Cardinal Virtues
Description Many attempts have been made to define the cardinal virtues, or those virtues most central to excellence in humanity, or even to define what is meant by a virtue. Plato's Republic lists four cardinal or principal virtues: prudence (or practical wisdom); justice; temperance (or self control); fortitude (or courage). They are linked to the three parts of the soul (reason, spirit and desire) and to the three parts of government (philosopher-kings, guardians and artisans). These four cardinal virtues feature prominently in mediaeval Christianity.
It is generally agreed that virtue as such is more that attitude which coordinates the virtues, that which makes the virtues virtuous - specific virtues being unified in virtue. For example, St Augustine defines the four virtues as "forms" of love with God as their object. He says that virtues which do not refer to God and which are sought with no reference to God are not real virtues. Thomas Aquinas also stated that charity was the "form" of the virtues, that natural or acquired virtues of prudence, justice, temperance and fortitude needed to be "formed" by the supernatural virtues of faith, hope and charity. In addition, others have indicated that particular virtues are to be associated with different stages in life - obedience in childhood, wisdom in old age.
There is much controversy over the extent to which virtues can be possessed by those who are not good or who are not in union with God. This is partly because virtues seem intermediate between the interior disposition and intention and action. Thus Aristotle emphasizes the acquiring of virtue through action while St Augustine emphasizes orientation to God. Again, do the cardinal virtues form a schema for classifying good character dispositions or are they distinct causal powers effective in subordinate virtues ?
Narrower Love (#HH0258) Faith (#HH0694) Justice (#HH2117)
Prudence (#HH7902) Fortitude (#HH5640) Temperance (#HH0600)
Hope (Christianity, #HH0199).
Related Moral development (#HH0565).

♦ HH0713 Sublimation
Description Although commonly considered a defence mechanism (the diversion of sexual instinct to a non-sexual aim which is socially more acceptable), sublimation is considerably more than simply a defence mechanism and is, in fact, postulated by Freud to be the means by which a development and transformation of infantile drives achieves satisfaction in artistic, ethical or cultural activity.

♦ HH0714 Social reality
Description Maintenance of the individual's stable view of the existence of phenomena is validated by consensus of the group; loss of faith in the nature of reality as expressed through the culture of a society results in the individual being cut off from meaningful association with others.
Broader Reality (#HH0444).

♦ HH0715 Twice-born
Dvija (Hinduism)
Description The three upper castes of the Hindu system (that is, all but the artisan or labouring class) have a sacrament of initiation signifying a second or spiritual birth. The individual leaves behind the "womb" of guide-lines fostering him in immaturity and is born again as a competent adult functioning rationally and leaving childhood behind. He may then, like the deities, wear the sacred thread. The practice may perhaps be related to the concept of the twice-born deities of fertility and light, dying at the summer solstice and reborn at the winter solstice - and further, to the idea of rebirth in Christ; see also St Paul, I Corinthians 13 "when I became a man, I put away childish things".
Refs Begbie, T *Twice Born Men* (1909).
Broader Spiritual rebirth (Christianity, #HH4098).
Related Kali yuga (#HH0239) Initiation (#HH0230) Cyclic history (#HH0933)
Born-again (Christianity, #HH0576) Human development (Hinduism, #HH0330).

♦ HH0716 Task-oriented group experience
Description This is a form of intensive group interaction widely used in business and industry. The focus is placed on the group and the means of improving its accomplishment through better interpersonal interaction amongst the group members.

♦ HH0717 Team building group
Description This is a form of intensive group interaction designed to develop more closely-knit and effective working teams. It is primarily used in business, industry and military sectors.

♦ HH0718 Being-psychology
Onto-psychology — Transcendental psychology — Psychology of perfection — Psychology of ends
Description This covers experiences and states which are intrinsically important and not means to an end, and similarly treats individuals as they are, requiring no justification. The states of having arrived, experiencing pure joy, gratifying of all needs and also of total hopelessness and despair; and states perceived as whole or perfect, all are considered here. The possibility of defining a quantitative scale of humanness based on extrapolation from ideal goals and producing a definition of perfection of humanity is included. The study would be a means of approaching ultimate unity and the resolution of dualities.

♦ HH0719 Purity
Tahara — Satya samshuddhi
Description Although a distinction can be drawn between spiritual and physical purity, the use of physical rituals of purification to demonstrate spiritual cleansing means that much emphasis is

laid in most religions on being physically purified. Islam, for example, lists sexual intercourse, menstruation and childbirth as religious impurities, and certain objects and discharges as actual impurities; all of these have physical rituals of purification usually involving pure, running water. Other examples of religious significance attached to physical impurity are the complex regulations in Judaic law involved with the preparation of food, the "untouchables" of the Indian caste system, and the taboos related to leprosy in many systems. The emphasis on physical impurity and taboo is characteristic of animist cultures where the widespread belief in supernatural power attached to material (or natural) things and persons involves a danger which, if it cannot be avoided, can only be mitigated by ritual purification. In the Baghavad Gita, among the treasures of the godly are listed purity of heart (satya samshuddhi) and purity or cleanness (shaiycha), the latter relating to the vedic law "Be clean: reflect the absolute; do not steal". In the Christian tradition, sin or spiritual impurity "builds up" on original sin, creating spiritual disorder; order can only be restored by suffering, purification or "nights of the soul" and "nights of the senses".

Purity connotes the singleness or simplicity of a nature which finds satisfaction in it desires in God. The opposite of purity is uncontrolled or misdirected desire. It forms a part of that self–control which includes sexual purity but extends to include renunciation of the world and mortification of the flesh. It is not only the abstinence of illicit pleasure but the positive integrity of a will dedicated to God in perfect simplicity of purpose. It is not simply the sacrifice of innocent desires but the consecration of them, in striving after goodness in the widest sense. Purity of intention consists in seeking to please God in all things and to make His glory the object of every act and word. The pure heart constantly seeks God, holding fast to that purpose amid the multiplicity of calls and duties, responsibilities and claims.

The reward of purity is vision, insight and illumination. It is a certain freedom from intellectual illusion and error. The pure heart seeks not God's gifts but Himself. It knows that He is, not what He gives is the source of the true life of humans.
   Related Sauca (Yoga, #HH1391)   Purification (#HH0401) Inward purity (Sufism, #HM3602).

♦ **HH0720 Magic**
Magical development — Magical training — High magic — Low magic
Description Magic is a set of methods for arranging awareness according to patterns; it is not a truth or a religion. Nor is it even a philosophy, in the strict sense of the word, although there are echoes of profound philosophy in most magical traditions. It is basically an artistic science in which the practitioner controls and develops imagination to cause changes in the outer world. The serious application of magical methods leads to transformation and it is the transformation which is of value and not the methods themselves. All magic derives from controlled work with the imagination. Magic does not work because its propositions are essentially real or true; it works because practitioners become imaginatively involved in these propositions. Thus for controlled periods of time under non–habitual circumstances, they behave as if they were true. It is not a question of becoming habituated to falsehood but rather of the magician growing through the patterns, whether true or not, and emerging beyond them into a clarity of awareness that was not possible before the experience of transition and transformation.

In early magical training there is an extended period of confusion in which personal weaknesses and problems (especially self–inflation), become highly amplified before they are destroyed and the energies in question are absorbed into a balanced inner pattern.

From the perspective of a magician, the propensity of people for engaging daily in activities which they know are fruitless or harmful, sustained by a pattern of values and habits, achieves its apparent coherence through a form of fantasy–sharing that holds the illusion together collectively and individually. This same propensity is used by magic to motivate inner transformation rather than outer identifications. When the awareness of values changes (in contrast to changes of values) the externally perceived world may be transformed by magical means. This possibility is facilitated when the symbols used are those of the culture with which the practitioners are familiar. Once the perception of the external world can be transformed by such means, magic then enables changes within the individual through which further methods applicable to the transformed consciousness may be inwardly apprehended. Magic thus attempts to relate human consciousness to divine consciousness through patterns inherent in each. This is otherwise known as the Great Work.

A major premise of magic is that access may be obtained to many worlds or worldviews. The transformations which occur within the magician enable access to such innerworlds of consciousness in ways which transcend the limitations of purely intellectual endeavour or the inspirations of folklore. Images are deliberately evoked and cultivated as part of this process. Initially magic alters the focus or area of attention, drawing the vital; energies together with the discipline of a tradition and its restricting vessel or matrix. In a second stage the energies are redirected and gradually amplified through attuning to richer, more complex and more encompassing patterns. These integrative patterns have a resonant effect on the psyche. They may take the form of simple symbols, or may be imaginatively recreated as complex scenes, beings or other patterns. As such they may be used to focus and direct a wide spectrum of personal and group energies on many levels of awareness. In a third stage, the awareness having been attuned to various patterns normally inaccessible to everyday consciousness, begins to operate in other worlds or dimensions through the effect of the magical patterns and key symbols. Finally the practitioner is projected into the alternative worlds of experience, often with considerable energy.

The increasing ability to change worldviews follows from a reassembly and redirection of the practitioner's energies. Such changes enable the practitioner to gain a more accurate understanding of the shared world. The value of such transitions to other world realities is that they contribute to the overall liberation from the particular illusion of the coagulated consensual worldview. They also ensure fruitful exchanges between such distinct realities and the entities that inhabit them. The intent is therefore not to escape this world but rather to transform it. The transformation begins within new directions of awareness sought in early training. It finally permeates the practitioner through to the physical body. Whereas religions seeks to save the world, the magical disciplines affirm a particularly subtle aspect of this insight, namely the possibility of transforming all worlds.

There are five fundamental magical arts: concentration, meditation, visualization, ritual pattern making, mediation. Although each of these disciplines of consciousness may be developed separately from the others, they are in fact harmoniously interwoven in any well balanced magical work. These all lead consciousness to change its direction, moving inwards rather than fixating outwards as it does in daily habitual life.

Through the practice of these arts during magical development, the individual progressively learns to balance the reality–worlds within individual consciousness through ritual and planned activity by which life becomes attuned and rhythmic rather than random and chaotic. At the same time the individual endeavours to energize the imaginative constructs and the contacts established through transformative rituals and powerful mediation. The spiritual power of the practitioner is directed outwards towards material ends, flowing through the psychic body complex, transforming the awareness of the practitioner before it reaches any other defined goal. These two processes may be integrated in one harmonious living pattern, a magical life of enlightenment, in which the practitioner seeks a continual interaction between the individual and the worlds occupied by his awareness.

Some sources distinguish between *high magic*, a form of magic in which the magician depends only on psychic power without recourse to external aids or rites, and *low magic*, when various external aids – talismans, ritual chants or movements – are used by the magician to stimulate his psychic energy and enable him to perform magic feats.
Context The term magic is frequently abused and separated from a spiritual foundation. In any historical period, as with religion, magical arts are taken up in fashionable and often bizarre forms, by various groups and movements as continues to occur at this time. The enduring magical tradition is derived from perennial philosophy, sustained by myth, legend, visionary cosmology and poetic insight. In some cultures many perverted forms of magic continue to be practised for ignorant or selfish ends. Trivial, resource-consuming, or ultimately sinister practices are degraded forms of the enduring tradition that can lead to dangerous forms of imbalance.

Magic is frequently associated with the occult as the preoccupation of secret cults in pursuit of secret powers in order to manipulate others. As with other disciplines, it can attract self-centred individuals of extremely dubious motivation. Through their efforts to draw attention to themselves, wider understanding of magic as a discipline is distorted. The potent powers to which magic offers access are the common energies and properties of humankind and are not the monopoly of any conspiracies that may endeavour to exploit them.

Magic has frequently been considered evil, especially by organized religion and as a result of the actions of those who exploit the gullible. As a neutral set of artistic and scientific techniques for controlling the imagination, magic (as with any set of methods), may indeed be employed by those who are imbalanced to enhance their own image of themselves. Evil may then be considered as associated with that imbalance, but not with the principles, however they are abused. Many modern religions, especially Christianity, make use of magical practices identical in principle to those of the pagan religions they displaced. Such religions also exhibit special concern at the evocation of gods and goddesses as being a completely regressive spiritual tendency. However this reservation should now be seen in the light of the insights of archetypal psychology in which the imaginative value of such symbols for the psyche is recognized as one way of facilitating individuation. Just as some religions make specific use of icons and other images as an aid to prayer, magical traditions use specific images of deities to gain specific results with the imagination and its effects upon the outer world.

Magic relies very strongly on tradition, namely on the collective stream of information, methods and ethics, none of which are fully written down even in literate cultures. Modern magical arts are a clarification or restating of ancient enduring traditions. By comparison the magical practices of the nineteenth century were replete with mystification, ignorance and quasi–religious posturing reflecting the intellectual ambience of the times.

Because its means are mysterious, magic whether "black" or "white" evokes feelings of fear and insecurity – it threatens confidence in the three-dimensional, rational view of things, making it in some way inadequate. The widespread belief in astrology, superstition and symbolic representation has been used as evidence that magic is intrinsic to humanity. The very violence and intensity of witch hunts and resistance to the occult may be used as evidence for the validity of magical and occult practices, despite the fact that rationalists and sociologists have chosen to ignore folk beliefs in magic and concentrated on religion.

In fact it has been argued that the emergence of individual awareness and personal ego commenced in primitive societies by magical, symbolic means. The dasein (being-there-ness) must be defended against loss, just as the physical body has to be defended against sickness and death. Otherwise the individual is totally overwhelmed by the environment and ceases to be able to support him or herself (fascinans). This defence is symbolically enacted for a whole group by a *shaman* who first allows himself to be possessed by a spirit and then fights off the spirit, regaining his "self-possession" and, in doing so, that of the group.

With the rise of religious and moral structures, the individual self has a stronger backing and the need for such defence lessens. Nevertheless, the self can feel threatened when an individual knows he has gone against moral authority. It has been postulated that the feeling of guilt produced when refusing to assist a dependent and destitute relative, excaberated by that relative cursing one's hard–heartedness, could produce "magical" illness in the same way as spirit possession, such illness only being exorcised by punishing the cause of the guilt – in this case, the unfortunate dependent or *witch*.

Magic may be viewed as expressive symbolism, cathartic in relieving anxiety; and magic rituals, spells and incantations may act as symbolic representation of cherished values and beliefs. In societies where magic is a part of social organization the magician plays a role including religious and medical specialist. Where religion and science fail to solve pressing problems then magic may act as a substitute adjustment, a means of dealing with issues beyond the capacity of scientific and technical knowledge. It may also be used to motivate the individual away from antisocial behaviour, whether to avoid being accused of the practice of magic or to avoid its being practised against one.

Common to most religions, but particularly clearly defined in Islam, is the concept of the True God, worshipped by true believers whose perception is broadened and deepened by revelation; and the untrue or illusory vision of the supernatural portrayed by magicians through "sihr" or "glamour", when the true nature or form of something is transformed to the unreal or mere appearance. Knowledge of such magic is said to be divisive, producing duality, and to harm those who possess it, so that they forfeit their place in the next world.
Refs Bailey, Alice A *A Treatise on White Magic* (1934); Drury, N *Inner Visions* (1979); Drury, Nevill *Don Juan, Mescalito and Modern Magic* (1978); O'Keefe, Daniel *Stolen Lightning* (1982); Regardie, I *Foundations of Practical Magic* (1979); Stewart, R J *Living Magical Arts* (1987); Stewart, R J *Advanced Magical Arts* (1988).
   Narrower Witchcraft (#HH1909)   White magic (#HH3556)
Mediation (Magic, #HH4521)   Magical meditation (#HH5547)
Guided visualization (Magic, #HH6053)   Ritual pattern-making (Magic, #HH5120).
   Related Mana (#HM2686)   Rites (#HH0423)   Shaman (#HH0973)
Divination (#HH0545)   Initiation (#HH0230)   Initiation (Magic, #HH3432)
Magical thinking (#HM0359)   Bewitched (Psychism, #HM0819)
Magico–religious powers (#HH0497)   Neuro-linguistic programming (NLP, #HH4872).

♦ **HH0721 Utopia**
Description Utopia is an expression derived from Thomas More's book of the same name (literally "nowhere"), to describe an ideal society or social system, usually accepted as unrealizable in fact. The chief argument levelled against endeavours to achieve "Utopian" society is that the totalitarian nature of such a system is against human nature and threatens individuality. Fears of the manipulation of personal behaviour and consciousness have resulted in such literary works as Aldous Huxley's "Brave New World" and George Orwell's "1984".

♦ **HH0722 Physical perfection**
Description Physical perfection is that state of the physical body in which harmonious development and fitness render it most efficient for achieving good health and meeting the demands made on it by day-to-day work and other activities. Communist state systems put physical perfection as one of the three aims of the upbringing of the *new man*, the others being *intellectual wealth* and *moral purity*.
   Related Health (#HH0509)   Physical culture (#HH0730)   Fitness for work (#HH0758)
Physical appearance (#HH0993)   Physical development (#HH0836).

♦ **HH0723 Sports psychology**
Description Emotional pressure has been shown to have a deleterious effect on physical achievement in sport; this system was developed to come to terms with such pressure and thus maximize alertness, concentration, strategy and good judgement when playing team sports. It also (by improving total awareness) helps to minimize negative psychological tactics of the opposing players. The programme, which includes a *sports emotional reaction profile* (SERP), has been extended to general activities, improving concentration and relaxation, and enabling the participant to be responsible for enjoying life to the full.
Related Sport-induced modes of perception (#HM0062).

♦ **HH0724 Cultural personality**
Description If personality is taken to refer to the behaviour and way of thinking and feeling of the individual, then the life-style, ideas and values prevalent in a particular society will affect the personality of an individual inhabiting that society. It has been said (Ruth Benedict) that by the internalization of the same cultural ethos then individuals will come to share some of the same basic psychological structures.
Related Personality development (#HH0281).

♦ **HH0725 Ethos**
Description The dominant or essential emotional aspects of consciousness which give rise to the distinctive character and guiding beliefs of an individual or a group. In the past referred to as the soul or genius of a nation or race.
Related Enculturation (#HH0954).

♦ **HH0726 Unitary reality**
Description A sense of unitary reality is experienced by an individual when the dichotomy between inner and outer reality has been transcended. The original unconscious wholeness and oneness with life from which man emerged is recovered in part on a conscious level.
Related Totality (#HH0534).

♦ **HH0727 Complete man**
Universal man
Description The physical, intellectual, emotional and ethical integration of the individual into a complete man is a broad definition of the fundamental aim for education. This pedagogic ideal is to be found throughout history, in almost all countries, among philosophers and moralists, and among most theoreticians and visionaries of education. It has been one of the fundamental themes for humanist thought in all times. It may have been applied imperfectly, but it has been fruitful and helped to inspire many of the noblest educational enterprises.
Contemporary science has shown that man is biologically unfinished. In effect he may be said never to become an adult and that his existence is an unending process of completion and learning. It is essentially his incompleteness that sets him apart from other living things, the fact that he must draw from his surroundings the techniques for living which nature and instinct fail to give him. He is obliged to learn unceasingly in order to survive and evolve.
Refs Edgar Faure (Ed) *Learning to Be* (1972).
Related Perfectionism (#HH1466).

♦ **HH0728 Acupuncture**
Description Acupuncture is based on the theory that all disease is due to some deficiency or congestion of energy flow along a system of meridian lines on the body. Different organs or parts of the body may be treated by stimulation and unblocking of the appropriate acupuncture points on these meridians. Using the same knowledge that practitioners of martial arts use in their combat, acupuncture and similar techniques draw on ki energy to stimulate or calm the energy in one's own or another's body.
Related Ki energy (#HH0620).

♦ **HH0729 Endurance**
Perseverance — Staying power — Fortitude
Description Spiritual development is marked by periods of progress interspersed with times when no progress is made. Enduring periods of discouragement leads to strength and, in the long-term, contributes to progress as well. In contrast, the Church indicates that without persevering, and in allowing one's self to fall away from grace, one may eventually fall into perdition.
Related Heroism (#HH0929) Fortitude (#HH5640) Frustration (#HM3084)
Backsliding (#HH2321) Discouragement (#HM2498)
Dark night of the soul (Christianity, #HM3941).

♦ **HH0730 Physical culture**
Description Physical cultural activities such as sports and athletics, whether part of the educational system or as leisure activities, not only strengthen a person's physical capabilities and improve health but also provide opportunities for friendly competition and for relationships on the individual, group, national and international levels.
Related Organized games (#HH0692) Physical education (#HH0575)
Physical perfection (#HH0722) Physical development (#HH0836).

♦ **HH0731 Reiki therapy**
Universal life energy
Description A system of holistic healing in which universal life energy from the hands of the Reiki therapist accelerates the body's ability to heal itself physically, balancing body, mind and spirit and opening the mind and spirit to the cause of disease and pain. There is a change in attitude and feeling of peace; and a realization of the need to take responsibility for one's life.
Broader Therapy (#HH0678) Holistic medicine (#HH0581).

♦ **HH0734 Psychic energy**
Description The energy which provides the force behind any mental activity, with its source in the id. In the context of seeking pleasure and avoiding displeasure, accumulation of psychic energy results in a striving for discharge to reduce the tension, and thus impels behaviour based on instinctual drive. Small increases in psychic energy may create signal anxiety or bring an unconscious idea into consciousness.

♦ **HH0735 Paternalism**
Patriarchialism — Mentor relationships
Description This is descriptive of behaviour which assumes that the person behaving in a "paternalistic" manner knows better than the individual he is advising or remonstrating with. To some, such behaviour is seen as offensive, as it indicates incompetence, incapacity and a lower standing (such as may be assumed in a father/child relationship). However, it may be argued that benevolent interference in a person's conduct or choice, when it is clear that his behaviour deviates from his own preferred desires and dispositions, does not constitute a violation of his personal integrity. Such an argument is based on the knowledge that an individual acquires a specific personality or identity through learning, constructing a unity through deliberation, experimentation, and accommodation of inherent capacities which are too many and various for all to be fully developed or even developed at all. It is in fact a positive assistance in the development of personal integrity to receive support in resisting the tendency to lapse or deviate from one's personal permanent and central commitments, in particular when such a lapse could put such significant aims or preferences in jeopardy.
It is further argued that it is insufficient to stand by while another makes a mess of his life, simply extending sympathy after it has happened. Although it involves taking a moral risk, positive interference to prevent catastrophe is necessary in a world in which ideal models of individual excellence are frequently aimed for but rarely achieved.
Related Mentor (#HH3219) Integrity (#HH0234) Individuality (#HH0776)
Maternal thinking (#HH4646) Personal integrity (#HH0129).

♦ **HH0736 Reinforcement**
Description This is the phenomenon by which learning is aided by an external positive impulse when a correct response is given during the learning situation. Such an impulse may be simply a smile or the offering of a reward in terms of food, etc. In other words, an individual is more likely to respond in the manner desired if by doing so he or she will receive something in return.

♦ **HH0737 Cultural development**
Description Culture has been described as the sum total of distinctive spiritual, material, intellectual and emotional features characterizing a given society or social group. Cultural development is thus both a means to, and an end of, general development. Action to promote cultural development comprises:
- Encouragement of artistic creation. Creativity is linked to each particular civilization and is, of all human activities, the one that best assembles, sums up and expresses a civilization.
- Education of the professional artist in order that he may be capable of responding to human needs in a rapidly-changing society and that he is at the same time in touch with the many and ever-evolving art forms.
- Control of the cultural environment. If culture is to help improve the quality of life, cultural problems must be taken into account in urban planning, which should not be based exclusively on economic considerations. Man is obliged either to manage his spatial environment or endure it as a form of alienation.
- Dissemination of cultural products by a wide variety of recording and distribution media. The accelerated development of new techniques, as well as their increasing variety, radically transforms access to culture and its products. Preservation of the cultural heritage of mankind including historic monuments, sites, art objects and other forms of cultural property. Each individual should thus have the opportunity to develop his personality through direct participation in the creation of human values, and thus to control his own situation, whether locally or internationally.
Refs Hawkes, J, et al *History of Mankind, Cultural and Scientific Development* (1964).
Related Cultural identity (#HH1929).

♦ **HH0738 Sharia** (Islam)
Spiritual path — Law of religion
Description Sometimes referred to as the word of the Prophet, the Sharia comprises two main groups of Islamic prescriptions for life in all its aspects - spiritual, political, social, domestic, public, private - according to infallible doctrine set down in the Qu'ran. The two groups refer to religious ritual and worship, and to juridicial and political regulations, although the borderline between the groups is not clear-cut. As is generally the case with a state religion, there is not always complete harmony between laws as laid down by the Islamic religion and laws enforced by Islamic states; and the emphasis an individual regulations varies. Nevertheless, a devout follower of Islam will attempt to follow the spiritual path in unquestioning obedience to the Sharia system. Sharia is also the first stage on the threefold journey to God of Sufism, the others being *tariqa* and *haqiqa*.
Broader Spiritual way (#HH1867).
Related Nasut (Sufism, #HM3579) Haqiqa (Sufism, #HM0124)
Ma'rifa (Sufism, #HM2254) Spiritual life (#HH0101)
Spirituality (Islam, #HH5902) Divine path (Sufism, #HM3371)
Mysteries and religion (#HH4032).

♦ **HH0739 Causative factor for altered states**
Trigger factor
Description Whether an altered mode of awareness or mystic experience is achieved by deliberate practice of a technique or whether it occurs spontaneously, there seems invariably to be a sudden occurrence which triggers the change taking place or a set of circumstances which activate the mechanism for sudden illumination. Often this may be music, or beautiful or awe-inspiring surroundings. Other factors have been variously described: childbirth; grief; illness; nearness of death; sensual experience; flight; sports; achievement, and so on.

♦ **HH0740 Interactive discipline**
Description Such discipline consists of the internal authority of the seeker centering in an attentive wholeness on the external, natural world with its traditions and ethos. The interactive experience allows the seeker to experience a oneness with the rhythmic flow of life; this is particularly clear in aesthetic experience, the musician at one with his instrument, the sculptor with a piece of marble, and is exemplified in the Japanese haiku tradition - short poems composed in moments of sublime understanding of the world.
Broader Spiritual discipline (#HH1021).

♦ **HH0741 Derwish** (Islam)
Dervish — Meditation in motion
Description A member of an Islamic sect following a particular ritual, usually emphasising the emotional part of religious life based on the repetion of *dhikr* or religious formulae; and the achieving of a state of consciousness characterised by dreaminess and reinforced by dancing, drumming and so on, during which a highly complex sequence of movements may be performed without conscious thought and in a state of total absorption. There is an experience of spiritual elevation and unity.
Related Kirtan (Hinduism, #HH2390) Dance-induced experience (#HM3213)
Recollection (Islam, Sufism, #HM2351).

♦ **HH0742 Labour training**
Description Depending on the ideological system of a particular society, individuals in that society will be educated to have a different approach to work and workmanship. In some societies such training starts in childhood, children being taught to work in society and expected to make their own contribution and perform useful tasks (for example, in the Soviet Union). In other societies, labour training is more a matter of training in attitudes to work, with the emphasis on discipline, conscientiousness and attention to detail. Whatever the system, such training is generally considered essential both for the harmonious, free development of the individual and for the attainment of his or her useful place in the community.

♦ **HH0743 Self-existence**
Description One result of the practice of passive, detached self-observation, authentic self existence becomes clear as the mind is gradually emptied of personal likes and dislikes, bias and confusion.

♦ **HH0745 Human resources development** (United Nations)
Manpower resource development — Manpower training — Mobilizing human resources
Description The development of human resources has three essential and interdependent components: the building of appropriate incentives, the promotion of effective training of employees, and the rational development of formal education.
Building incentives. This is to encourage individuals to prepare for and engage in the kinds of productive activity which are needed for accelerated growth. Large outlays for education are, for example, unlikely to produce the kind of high-level manpower needed if the proper incentives are lacking.
Training of employees. Training involves the development of specific skills which are needed to perform a particular job or series of jobs. It is a continuous process of human resource development rather than a simple pre-employment indoctrination.
Development of formal education. Traditionally, human resource development has focused primarily on education and secondarily on health. Human resource development is frequently equated with investments in formal education. The essential function of formal education is to prepare people for training rather than to train people for particular occupations. The main purpose of primary education is to make people literate and to make them more effective citizens in the modernizing society. The major mission of secondary education is to give students firm grounding in verbal and written communication skills, mathematics, foreign languages, history, social studies and science, aiming for breadth rather than specialization at an early stage, and especially if a large part of the occupational skill development is to be left to later training in employment or to post-secondary educational institutions. The mission of higher education is to provide liberally educated persons for positions of leadership in the modernizing society and to develop high-level technical manpower.
Vocational training and management development are operational approaches to human resources development. An emphasis on school and university education has recently shifted from national economic and technical requirements towards the socialization function of schooling, examination of the screening hypothesis and studying labour market segmentation. Effective educational planning is now seen to be based on realistic assessment of the operations of the labour market which are in a continuous state of flux, in particular in terms of employment patterns. It is difficult to avoiding dealing with what was required yesterday rather than what is required today, as old skills are outdated and new skills required. The vocational and job-specific education, previously considered economically necessary, is now superseded by the demand for more flexibility. Instead of trying to forecast the impossible, the system must be capable of responding rapidly and effectively to changes in the economy and in society. This means emphasizing recurrent education and training and retraining of adults.
Three components are indicated:
- *Creating human resources*, focusing primarily on education and training.
- *Deploying human resources*, emphasizing entrepreneurial and managerial abilities, research, technology, general skill formation, participation in decision-making. The possibility of deploying the human resources which have been invested in over a wide range of social, economic and cultural activities must be assured. Since the population is widely educated it must, necessarily, be included in decision-making.
- *Incentive structures to realize desirable deployment*, through action to: (a) adapt the income structure (monetary and non-monetary) so that occupations and education are chosen in line with future development of society; (b) adapt decision-making structures to be horizontal instead of hierarchical, and to provide a maximum of participation together with speed and efficiency; (c) redirect educational spending with priority for primary and lower secondary education, selected research institutes and middle level management. Additionally, changes are necessary in educational and research structures and facilities.
Finally there is the question of innate ability, motivation and achievement. This varies globally, so that some areas in the world are clearly more successful than others and the people are psychologically orientated towards success. The human factor may be the limiting factor in economic and social development.
The ultimate objective of economic development should be human development. However, there is no commonly accepted definition as to what in practice constitutes human resources development. According to a report of the United Nations Development Programme, human resources development should be broadly defined, because of its intersectoral links, as "the maximization of the human potential as well as the promotion of its fullest utilization for economic and social progress". This definition "requires that people be given the opportunity to apply the full range of their skills and abilities, fulfil their desires and ambitions and make their contributions to the improvement of their lives and their society. Thus, human resources development depends upon a political and social utilization of human potential. It thus depends on the very nature of society and its economic history and culture". Regional, subregional and national differences must be taken into account; an operational human resource development strategy can be articulated only within the context of each country's special conditions.
Human resources development is a crucial requirement not only to build up technical knowledge and capabilities, but also to allow people to better use the resources they command, to stretch them further and to create new values to help individuals and nations cope with rapidly changing social, environmental and development realities. Knowledge shared globally would assure greater mutual understanding and create greater willingness to share global resources equitably. This calls for broadening education (improving skills necessary for pursuing sustainable development, providing insights on the interaction between natural and human resources, between development and environment, raising global awareness) and devising new methods for environmental education and new incentives for engaging into sustainable development paths.
Refs Cole, S *Human Resources Development and Long-Term Forecasting* (1989); Committee for Development Planning *Human Resources Development* (1988); Emmerij, Louis *The Human Factor in Development*; Haq Kadija and Huner Kirdar (Eds) *Human Development* (1986); Progoff, Ira *Depth Psychology and Modern Man* (1959); Sachs, Céline *Exploring the Human Dimensions of Development* (1989); UNESCO *UNESCO and the Development of Human Resources* (1988); United Nations *Development and Utilization of Human Resources* (1967); United Nations *Human Resources Development* (1989); United Nations Development Programme *Human Resources Development* (1986).
Related Human-centred development (#HH3607)
Human resources (United Nations, #HH0147)
Enhancement of human capabilities (#HH2806).

♦ **HH0746 Universal religion** (Yoga)
Universal spirituality
Description All the great religions have been said (Swami Omkarananda) to have the same essentials - love, truth, goodness, faith, devotion, knowledge. This writer and others indicate that when a sincere man, whatever his creed, reaches the goal of his spiritual path he achieves a personal realization of truth which takes him beyond his particular religion to an expanded consciousness of direct insight into reality. This true wisdom is the goal of *yoga*, which may be practised by any sincere individual whatever his beliefs or non-beliefs and may ultimately lead to cosmic integration.
Related Yoga (Yoga, #HH0661) Spirituality (#HH5009)
God experience (#HM4100) Spiritual renaissance (#HH1676)
Human development through religion (#HH1198).

♦ **HH0747 Euphenics**
Description The management of the developmental patterns of individuals as an environmental support for eugenical measures leading to improvements in genetic endowments. This may include the treatment of genetic disease by medical and surgical means - *therapeutic medical genetics*. However, treatment of the symptom but not the cause may lead to the survival of an individual who would probably otherwise have died in infancy. Since there is a likelihood of any offspring of the individual so treated also suffering from the same malfunction and therefore also requiring treatment, euphenics often sometimes argued as contributing to genetic deterioration of the gene pool. There is therefore the ethical dilemma of whether individuals who might be helped should be refused treatment on eugenic grounds.
Related Evolution (#HH0425) Lamarckism (#HH1673) Gene therapy (#HH1334)
Human genetic improvement (#HH0357).

♦ **HH0748 Archaic-paralogical thinking**
Archaic thought — Primordial thinking — Primitive thinking
Description This type of thinking, commonly classified with *prearchaic* and *paralogical-logical* thinking, demonstrates a general impairment in the ability to generalise or think in abstract terms, and is said to be typical of primitive man and also to occur in schizophrenic disorders. Reasoning and the ability to differentiate are diminished; and feeling, perception and seeing in concrete terms dominate. There is a lack of differentiation between the self and objects and between objects and a fusion of sense modalities.
Related Symbols (#HH0690) Magical thinking (#HM0359)
Pre-conscious thinking (#HM1084).

♦ **HH0749 Excellence**
Description Excellence may be measured by comparing the achievements either of different people carrying out a similar task or of the same person carrying out similar tasks on several occasions. The former gives an idea of the comparative excellence of individuals in a particular field, the latter a yardstick for measuring how well an individual is performing within his or her ability. Different societies and different times have honoured excellence in different fields. It is argued that, in order for individuals and society to develop, it is essential that aiming for excellence should be encouraged, with the emphasis on the aiming rather than the achievement. Such arguments go back as far as Plato who, in the Republic, makes provision for special privileges, training and responsibilities for the most able. This argument is counterbalanced by the dangers inherent in the *superman* doctrine of Nietsche as realized in fascist and racist movements. However, the over-egalitarian society my stifle excellence, resulting in a mediocre lowest common denominator.
Refs Gardner, John W *Excellence* (1961).

♦ **HH0750 Life instinct**
Psychic energy
Description According to Freud, life instinct is the energy behind all human behaviour, the drive behind wishes, desires and striving for pleasure.
Related Pleasure (#HH0194) Instinct (#HM3517) Self-preservation (#HH0711)
Libido development (#HH0986).

♦ **HH0751 Gestalt therapy**
Description The immediate aim of Gestalt therapy is the restoration of awareness, with the ultimate goal of restoring the functions of the organism and the personality, within a holistic frame of reference, in order to make an individual whole and to release his potentialities. It assumes that awareness itself is sufficient to bring about development and change, through *organismic self regulation*. In modern culture, however, the awareness of the average person of the integrity of his thinking, feeling and acting is fragmented. Gestalt therapy is the effort to heal individuals of their dualism of being and to redevelop the unitary outlook.
Gestalt (approximated in English to imply configuration, meaningfully organized whole, structural relationship, theme) is the perception of a meaningfully organized whole of a figure and its background. The healthy personality has a permanent, meaningful emerging and receding figure and ground. Attention, interest, excitement, grace, concentration, and concern are characteristic of such a person's figure/ground formation as compared to the case of a neurotic for which such elasticity is replaced by either rigidity (fixation) or absence (repression) of figure formation, resulting in confusion, boredom, compulsions, anxiety and self-consciousness. The task of the therapist is to help the person complete the Gestalt by reintegrating attention and awareness.
A basic procedure in Gestalt therapy is to orient the person towards experiencing a continuum of awareness and to return constantly to any splits in attention and awareness which indicate that focused organismic attention is developing outside of awareness (for example, the person may be talking about one problem whilst sensing and acting in a number of other unrelated ways). A second technique is that of dramatization which allows the therapist to lead the person to identify and become aware of his unconscious alienated activity, particularly by taking up any alienated role, merging with his actions and feelings in that role, and having them express what they wish.
The artificial nature of the boundary between the self and the not-self, understood to be an illusory separateness from the stream of life, is held to lie at the root of all inner conflicts. Individuals are held to be living in only a fragment of themselves, holding on to a pre-established self-image and rejecting as not-self all that is conflicting with it or that is expected to be painful. Gestalt therapy suggests that the self-image be regarded as the figure in the figure-ground relationship that is involved in all perception. The therapeutic approach encourages the individual to reverse the figure-ground relationship involved in this self-perception and start experiencing himself as the background, namely not as the person who is unfulfilled in some way, but rather as that which causes this unfulfillment. Only by sensing how he does this reversal can a person cease to do it and to waste his energy. In this way different personality functions can be brought into relationship and integrated in a process of intra-psychic encounter. The integration of these fragments of the person's being, through the person's full acceptance of how he is (rather than of how he should be), leads to the possibility of change.
Refs Fagan, Joel and Shepherd, I L (Eds) *Gestalt Therapy Now* (1972); Kogan, J *Gestalt Therapy Resources* (1971); Naranjo, Claudio I *I and Thou* (1967); Perls, F S *Gestalt Therapy and Human Potentialities* (1966).
Broader Therapy (#HH0678) Human potential movement (#HH0398).
Related Holistic medicine (#HH0581) Living in the present (#HH2345).

♦ **HH0752 Breathing therapy**
Description Breathing is used as a means for opening sensory consciousness throughout the organism. Awareness of the breathing process as a sensation is used by the person to

re-establish contact with his physical organism. Since the sensory life of the organism is the basic building block of all forms of experience, extension of such awareness immediately relates to all feelings, images, movements and thoughts. Breathing is thus used to reclaim and expand the sensory life of the self by combining certain breathing techniques with subtle motions to cultivate perception. The technique lends itself to self-exploration by concentration in an introverted attitude upon oneself.
Refs Proskauer, Magda *Breathing Therapy* (1968).
Broader Therapy (#HH0678).

♦ HH0754 Kama (Hinduism)
Artistic life — Cultural life
Description Hindu code of practise sees four great aims in human life underlying the unity which draws together the instinct to procreate, love of power and wealth, desire for the common good and longing for spiritual communion. Kama, mankind's artistic or cultural life, or the satisfaction of desires, is one of these four aims.
Related Dharma (#HH0093) Artha (Hinduism, #HH0353) Moksha (Hinduism, #HM2196) Human development (Hinduism, #HH0330).

♦ HH0755 Raja yoga (Yoga)
Description This is one of the main forms of yoga and is primarily intended for those with an empirical or experimental disposition. The individual practices certain mental exercises (preferably under the guidance of a guru) in an attempt to achieve complete cessation of mental function so as to observe the light of the spirit within. These exercises are preceded by training in *yama* (moral discipline) and *niyama* (religious observances) which are the first two steps on the the *eightfold path of yoga*. The practice of an asana, or physical posture, assists in eliminating disturbances arising from the physical body and allows the body to remain relaxed and completely at rest for long periods of time.
The series of practices assists in the control of the mind by the will and leads the individual to direct personal experience of progressively deeper levels of his own being, by focusing all the mental energies on one particular and relevant idea or truth. The intent is to drive man's psychic energy into the deepest part of his being. If the procedures are correctly carried out, the personality is able to absorb and integrate the new forces that are tapped and is able to experience new levels of self-awareness and self-control. Finally, the mental level is transcended; the individual is aware of his true self, not just a mirror image of that self – he is in a state of *nirvikalpa samadhi*.
According to Master Da Free John, Raja yoga is a way practised at the fifth, or mystical, stage of life. Aspects of this form of yoga include: Jnana (or Gnani) yoga, Karma yoga, Kundalini yoga and Samadhi yoga.
Refs Vivekananda, Swami *Raja-Yoga* (1970).
Broader Yoga (Yoga, #HH0661).
Narrower Jnana yoga (#HH0927) Karma yoga (#HH0372) Samadhi yoga (#HH4321) Kundalini yoga (#HH0237).
Related Kaivalya (Yoga, #HM3869) Asana (Hinduism, Yoga, #HH0669) Mystical stage of life (#HM3191) Dhyana (Hinduism, Buddhism, #HM0137) Eightfold path of yoga (Yoga, #HH0779).

♦ HH0756 Attention
Awareness
Description Voluntary, conscious and deliberate attention is an act of will, and said to be a characteristic specific to man. The capacity for persistent attention, its range, distribution and arousal, are significant factors in the understanding of the psyche. According to William James, it is where one directs one's attention that determines the merit of one's life. It is the different aspects of life to which people direct their attention that determines their different experiences.
The practice of attention to present experience is used in several traditions of spiritual discipline. It focuses on full awareness of the here and now and avoids any escape into thoughts about the past or the future (considered to be rarely objects of wise reflection but rather objects of day-dreaming and vain imaginings, which waste a considerable amount of mental and emotional energy).
In the Gestalt point of view, for example, the healthy individual is constantly attending to matters of importance to his maintenance or survival. Such matters of importance are organism-environment transactions that keep or restore equilibrium or smooth functioning. Attending here refers to a behavioural focusing of parts of the organism towards relevant parts of the environment, and not to a conscious state.
Active, selective attention implying spontaneous reaction to experience is contrasted with passive acceptance of experience (empiricism); and with idealist subordination of experience to evolution of the idea.
Refs Se, Anima *Attention and Distraction* (1983).
Narrower Wise attention (Buddhism, #HM4309).
Related Attention (#HM1817).

♦ HH0757 Tenko
Ideological conversion
Description The individual, under some degree of coercion, formally rejects ideological commitment or admits to a political change of direction. This is a means of dealing with "thought criminals" by inducing the "criminal" to recant, often by instilling a sense of guilt and obligation, which frequently results in some genuine internal change. Specifically, Tenko refers to recanting of individuals from the Japanese Communist Party and acknowledgement of the authority of the Emperor.
Related Thought reform (Brainwashing, #HH0865).

♦ HH0758 Fitness for work
Description Individuals vary in their capacity to perform general types of work depending on their physical fitness. Where hazardous or particularly severe conditions are involved then regular physical check-ups are particularly necessary. Individuals of limited fitness for normal work, or who are temporarily or permanently disabled from working, are usually registered as such under national legislation.
Related Health (#HH0509) Physical perfection (#HH0722).

♦ HH0759 Passive resistance
Civil disobedience
Description Those who wish to change or reform society against the resistance of a ruling elite, but who are unwilling on moral or other grounds to do so by violent means, have the option of "passive resistance". This system, developed particularly by Mahatma Gandhi, usually involves: deliberate flouting of laws to which there are objections and the obstruction of law enforcement agencies attempting to rectify the position, often by mass sitting or lying down in front of traffic; boycotting of particular services; hunger strikes; and even suicide. The essence of passive resistance is that the persons involved are prepared to suffer the consequences of their actions, even if this involves imprisonment, injury or death, in order to draw attention to their cause. Such action can only be successful when the ruling body is sufficiently humanitarian, or sensitive to public opinion, to respond in the required manner.
Related Ahimsa (#HH0088) Soul force (#HH0138) Conscientious objection (#HH1219).

♦ HH0760 Epigenesis
Description The concept that totally new properties and phenomena may emerge as systems develop, which did not exist previously even in undeveloped form.

♦ HH0761 Meditation
Description The term has been used to designate a seemingly wide variety of practices, some of which require concentration on mental images, whilst others discourage attention to imagery. Some require action and movement whereas others stress inaction. The practice normally involves narrowing the focus of consciousness to one object while remaining cognitively and intellectually aware. Irrespective of the medium or form used, whether images, physical experiences, verbal utterances, etc, the essential objective of meditation is the same, namely the development of a presence, a modality of being, which may be expressed or developed in whatever situation the individual may be involved.
The practice of meditation consists of a persistent effort to detect and become free of all conditioning, compulsive functioning of mind and body, and of habitual emotional responses that may contaminate the utterly simple situation required by the individual. Long-term results include a capacity for serenity and the ability to carry out tasks single-mindedly and competently. It presents a new view of reality leading to self awareness and therefore to increased calm in relationships.
According to Patanjali, there are six stages in the process of meditation: aspiration; concentration; meditation; contemplation; illumination; and inspiration. All meditation involves a conscious exercise of awareness. This leads to a spontaneous flow of experience to which the person becomes a receptive onlooker. In its fullest development, the feeling of a separate self is lost and a degree of union with the object of meditation is achieved. This state may be called contemplation, when the individual becomes aware of his absolute relation with the absolute. As such it is said to be the highest form of prayer.
When deliberately carried out as a spiritual exercise, meditation commonly contains the following elements: achievement of physical relaxation (often by adoption of some special posture); quieting of the emotions (often by the use of some breathing technique); elimination of intruding thoughts; concentration of the mind.
Importance may be attached to the choice of the object of meditation. Objects with a symbolic significance serve the double function of a target of attention and a reminder of the correct attitude which is both the path and goal of meditation. Characteristics of the most universal objects of meditation, whether visual, verbal or accoustical are: centrality, namely a centre of balance, source or end around which actions flow, or a centre of radiation or emanation (e.g. cross, sun, heart, lotus, seed); order, regularity and lawfulness, particularly as conciliation of opposites or a representation of unity in multiplicity; the religious quality. Object-centred meditation is a dwelling of the individual upon his deepest identity, upon the reflection of himself in the mirror of symbolism, which functions as a reminder of the central core of the individual's being and its relationship to the world.
Forms of meditation exist which do not require the individual to place himself under the influence of a symbol. The person then allows himself to be guided by the experiences provided by his own deeper nature, which are progressively refined as a result of the attention accorded to them. Another approach is to focus on the rejection of all objects of meditation, whether derived from without or within. By so excluding the irrelevant, the individual is open to the relevant and to a conceptual awareness of his true nature.
Refs Davidson R J, Goleman D and Schwartz G E *Attentional and Affective Concomitants of Meditation* (1976); Davidson, J M *The Physiology of Meditation and Mystical States of Consciousness* (1976); Fish, Sharon and McCormick, Thomas *Meditation* (1983); Fox, Douglas *Meditation and Reality* (1986); Goleman, D *Meditation and Consciousness*; Griffiths, Bede *Return to the Center* (1976); Hanh, Thich N *The Miracle of Mindfulness* (1988); Jarrell, Howard R *International Meditation Bibliography, 1950–1982* (1985); Johnson, Williard *Riding the Ox Home* (1987); Johnston, William *Silent Music* (1979); Jyotir Maya Nanda, Swami *Concentration and Meditation* (1971); Norbu, Namkhai *The Crystal and the Way of Light* (1986); Rajneesh, Bhagwan Shree *The Orange Book* (1983); Reyes, Benito F *Meditation* (1978); Reynolds, David K *Naikan Psychology* (1983); Russell, Peter *The TM Technique* (1976); Schuman, Marjorie *The Psychophysiological Model of Meditation and Altered States of Consciousness* (1980); Shafii, Mohammd *Freedom from the Self* (1988); Shapiro, Deane *Meditation* (1980); Shapiro, Deane H *Meditation* (1980); Shapiro, Deane H and Walsh, Roger N (Eds) *Meditation* (1984); Syed, Abdullah *Meditation*.
Broader Creative existence (#HH0513).
Narrower Cave meditation (#HH1932) Dharana (Hinduism, #HM2566) Bhavana (Buddhism, #HH0551) Guided meditation (#HH2304) Active meditation (#HH4008) Burmese meditation (#HH3783) Pyramid meditation (#HH2213) Process meditation (#HH2054) Magical meditation (#HH5547) Integral meditation (#HH0579) Kriya sakti (Hinduism, #HH4309) Mahamudra meditation (#HH8677) Motionless meditation (#HH5256) Structured meditation (#HH3287) Kinerhythm meditation (#HH2031) Hypergnostic meditation (#HH1562) Free-thinking meditation (#HH0632) Jewish meditation (Judaism, #HH1097) Dhyana (Hinduism, Buddhism, #HM0137) Kasina meditation (Buddhism, #HH3246) Free self-inquiry meditation (#HH0508) Dynamic self-opening meditation (#HH0787) Subjects for meditation (Buddhism, #HH3987) Vipassana-bhavana (Buddhism, Pali, #HH0680) Christian meditation (Christianity, #HH5023) Concentration on the flow of breath (#HH0687) Discipline of meditation (Christianity, #HH1688).
Related Phosphenism (#HH6321) Self-oblation (#HH0968) Contemplation (#HM2952) Samatha-bhavana (#HH0710) Seed meditation (#HM3692) Witness meditation (#HM3911) Silva mind control (#HH3635) Practice of silence (#HH0983) Biofeedback training (#HH0765) Contemplation of nature (#HH0580) Religious experience as meditational insight (#HM1593).

♦ HH0762 Religious therapy
Description Lack of religious motivation and repression of the religious instinct are said to be the cause of much mental and physical ill health. Religious therapy (including counselling, conversion and a return to prayer and worship) leads to a change in direction – *metanoia* or *re-pentence* – and personal reconciliation with the transcendent. This brings about wholeness and healing.
Broader Therapy (#HH0678).
Related Metanoia (Christianity, #HH0888).

♦ HH0763 Multiple personality
Dissociative states — Alter personality — Multiple personality disorder — MPD — Co-presence — Co-consciousness
Description Multiple personality disorder may affect individuals who have suffered childhood trauma, usually of repeated emotional, physical and sexual abuse. It is perceived as the co-existence simultaneously in a single body of several (usually between 8 and 13) totally distinct and complex personalities. The control of the body "switches" from one personality to another,

often without the individual being aware of the previous personality.

The alter personalities may have different physiological characteristics (brainwave patterns, immune status, left and right-handedness), behaviour patterns, and skills (including ability to speak foreign languages). An awareness that one alter personality has of another is referred to as co-consciousness; and the ability of one to influence the behaviour of another as co-presence. Common to many sufferers of MPD is an *inner self helper*, apparently an inner spirit guide, which is not an alter personality as such but which knows and understands the individual's past, predicts future behaviour and advises (accurately) on therapy.

Conflicting attributes and subpersonalities are evidently present in everyone. Different situations call into activity endless alterations and recombinations of the elements of the self; and depending on conditions the various "selves" are fixed and perpetuated. Study of individuals in which *dissociative states* are so extreme is hoped to shed light in general on the inner workings of the mind. There are links between dissociative states, hypnotic trance and epilepsy. The alter personalities appear to be related to different levels or states of consciousness produced by different emotional states, this being related to the state when particular behaviour patterns were learned.

Refs Bliss, Eugene L *Multiple Personality, Allied Disorders and Hypnosis* (1986).
Broader Anomalies in integrity of consciousness (#HM4521).
Related Fugue state (#HM7131) Sub-personalities (#HM0463) Double mindedness (#HM3362).

♦ HH0764 Actualization
Description The contents of memory become conscious and actual according to conscious or unconscious association mechanisms.
Related Self-actualization (#HH0412).

♦ HH0765 Biofeedback training
Biotelemetry

Description An instrumental technique for human self-monitoring and control of physiological processes and psychological states, using technological devices to present information to a person externally about the state of internal processes. Biofeedback training is a procedure whereby an individual can tune his involuntary bodily processes and, eventually, to some extent control them. It is thought that there is improved coordination between various parts of the brain due to global cortical damping. The technique has been described as a "new mode of communication between the physical self and the mental self". It can therefore open the door to learning (or in the case of neurosis, to unlearning) physiological and psychological data previously beyond the range of unassisted awareness. Conscious control can be developed over the theta and alpha brain waves (electrical rhythms of 4-7 and 8-12 cycles per second respectively, as measured by an electroencephalograph), heartbeat, blood pressure, body temperature and gastrointestinal activity. This has the effect of control of attitudes, thoughts and emotions; and of awareness of inner and external processes and perceptions.

Of special interest is the relationship between meditation and biofeedback. Adept practioners show almost continuous alpha waves (normally associated with a state of relaxed alertness) during meditation. Alpha waves are inhibited by sensory stimuli and intellectual activity, when they are replaced by beta waves (13-30 cycles per second) of a lower amplitude. Their frequency drops with diminution of the level of central excitation (such as in sleep).

Although it is not suggested that the state of consciousness of adept practitioners of meditation can be equalled by that of someone using biofeedback therapy to produce almost continuous alpha rhythm, the possibility exists that one of the consequences of the practice of meditation for many years is the production of a high alpha state. Consequently, if this state can be achieved by an individual in a psychophysiological laboratory with just a few hours of training, such an individual might thus eliminate the need for much preliminary work in attempting to achieve the same state of consciousness as an adept practitioner. Certainly, after initial training in the art of meditation, biofeedback techniques have been shown to produce a hierarchy of states which may be distinguished physiologically as well as experienced. So far six levels have been documented; stage 5 is said to be just above the fourth stage of traditional meditation and to be preceded by 'sudden excursions into other realms'. However, questions are raised as to the quality of an experience resulting from attempts to produce physiological changes compared with one in which physiological changes occur more as a by-product unrelated to intention.

Refs Biofeedback Research Society *Biofeedback and Self-Regulation* (1973); Danskin, D G and Walters, E D *Biofeedback Training as Counseling* (1975); Karlins, Marvin and Andrews, Lewis M *Biofeedback* (1973).
Broader Human potential movement (#HH0398).
Narrower Internalization awareness (#HM1044) Autocontrol of consciousness (#HH1235) States of awareness in biofeedback training (#HM3744).
Related Meditation (#HH0761) Brain waves (#HH3129) Induction technique (#HH0025) Electronic meditation (#HH1100) Alpha wave consciousness (#HM2345) Autogenic relaxation sequence (#HH2338).

♦ HH0766 Autosuggestion
Coéism

Description A method of psychotherapy in which the person frequently repeats to himself some such phrase as: "Every day, and in every way, I am feeling better and better". Sometimes referred to as a form of self-hypnosis, the method is held to have effected organic change in people.

♦ HH0767 Mutational phases of life
Description It has been postulated (Timothy Leary) that human life on this planet is only at an embryonic stage, and that the first four mutational phases are about to be completed. The circuits of the nervous system at present manifest - ventral-dorsal; vertical; 3-dimensional; and protective-incorporative - are those necessary for "Newtonian", gravitational life on earth. Four "post-Newtonian" or "Einsteinian", gravity-selective phases are predicted, when survival in post-terrestrial space wil be achieved through four Einsteinian circuits of the nervous system involving: mastery of the body as a time-vehicle; employing the nervous system as a self-directed bio-electronic computer; using the genetic code as a molecular intelligence system; and mastery of meta-physiological quantum mechanical force fields.

♦ HH0768 Standard of living
Level of living

Description The level of living of an individual, group or country refers to some qualitative/quantitative assessment of the conditions in which that person or those people are living, in terms of budget available for goods and services, the costs of those goods and services, and the proportion of income spent on basic commodities (food, housing, clothing, fuel, lighting). This is termed *objective* standard of living. The *subjective* standard of living is seen as the conditions to which a person or people aspire.
Related Development of the quality of human life (#HH0647).

♦ HH0769 Inter-personality
Description Right from baby-hood, an individual's development depends very heavily on his or her relationships with others. One learns to refer one's behaviour and achievements to the responses they elicit from other people and to modify them to produce desired-for responses. The first interpersonal experiences are to do with an awareness of separateness from the world in general and mother in particular; then there is learning to talk and also the development of rationality. All these are aided or hampered by the responses of other people. Particularly in close relationships, there is interpersonal exchange of feelings, when one person "picks up" feelings of cheerfulness, misery, guilt, energy, listlessness and so on, from another.
Related Transpersonal psychology (#HH0916).

♦ HH0770 Spirit
Description Spirit is the immaterial principle as opposed to the material principle of existence. Philosophy disagrees on the relative importance of the two principles; materialism stating that spirit is the highest creation of matter, whereas spiritualism and idealism consider spirit prior to and in itself higher than matter, the vital essence or animating force in living creatures. According to Jung, spirit can be neither described or defined, nor is it subject to human expectations nor the demands of will; it is linked to the intuitive force or purpose connecting different events and endeavours. Religion considers spirit divine in origin, in the Bible it is the creative or animating power of God. The idea of spirit as the essence or the meaning is carried over into such expressions as the "spirit of the law". In the Bhagavad Gita, Krishna says "That which is not, shall never be; that which is, shall never cease to be". Spirit is interwoven with creation but is beyond destruction, leaving the physical body like a worn-out garment and going on to one that is new. The Upanishads refer to spirit as "the good of all" and austerity, self-control and meditation as the foundations of spiritual knowledge. In essence, spirit is that which transforms inanimate matter into life and which returns it to matter after death.

Refs Davis, Charles *Body as Spirit* (1976).
Related Soul (#HH0501) Psyche (#HH0979) Afterlife (#HH0399) Spiritualism (#HH0479) Spirit possession (#HH0056).

♦ HH0771 Behaviour modification
Description The techniques of behaviour therapy may be applied to individuals who are not suffering from behaviour disorders as defined in terms of strictly medical considerations. Individuals may be defined as social deviants for political or ideological reasons, and subjected to techniques such as conditioning to encourage them to conform to behaviour patterns consistent with the views of a dominant group. A clear distinction has not been made between beneficial behaviour therapy techniques and abuse of such techniques to serve interests other than those of the individual subjected to them.

♦ HH0772 Brain building
Brainpower enhancement

Description A number of techniques are claimed to increase the brain power of the individual. These include devices to induce meditative trance, such as synchroenergizers; and low voltage electrical devices which are said to enhance relaxation, improve memory, lessen drug use, increase intelligence and allow expansion of brain cells and tissues.

♦ HH0773 Patterning
Description By imposing intense physical activity on brain damaged individuals, first by volunteers moving the limbs, then by the individual being encouraged to carry out a programme of violent exercise, it is said that new neurological pathways are forged in the brain, bypassing the injured parts. Persons barely able to function can be assisted to lead approximately normal lives.

♦ HH0774 Atma yoga (Yoga)
Description Form of yoga in which the automatism of *citta* is controlled and a state achieved where neither pain not pleasure affects the mind or influences the actions; the yogi absorbs and knows the bliss of Brahman.
Broader Yoga (Yoga, #HH0661).
Related Chitta-dependent consciousness (Yoga, #HM3011).

♦ HH0775 Art therapy
Description Art therapy views any artistic product as a means towards promoting emotional growth. It does not matter what a person produces, but rather how the person adjusts to, uses, and organizes the realities of the artistic media. The content often expresses unusual or even pathological psychological experiences. The ability to compose, to bring about the aesthetic balance of an art work, requires control over the material presented from the unconscious or preconscious.

The material represented and the methods of its representation of strong or morbid emotions are similar in the fine arts and in that of a patient undergoing therapy. The difference in their creative processes is that the artist is able to evoke emotions from inner depths and be their master while the patient and emotionally disturbed person is dominated by them and carries out their orders, unconsciously choosing the subject, the technique, the colour and the style. The symbolism in the material represented often remains hidden from the patient, while the symbols are actually manipulated by the artist to establish communication with the spectator at a different level than that of a pure description. This conscious manipulation of the symbolic makes it difficult for the artist to have sufficiently free association in the execution of his work to make art therapy successful.
Broader Therapy (#HH0678).

♦ HH0776 Individuality
Description This is the unique nature of the personality of a human being, whether understood as being the unique qualitative characteristic or as the unique quantitative characteristic (in the sense of a uniqueness of the individual configuration of factor values in the relevant ability and personality factors).

Genuine individuality, embodying both discrete particularity and individual uniqueness, is a characteristic of human life. Although the human baby is born more helpless than the young of other species, less "finished", he has within himself the ability and responsibility to complete himself, each stage being an individual "choice". The capacity for self-transcendence is the basis of discrete individuality, for this self-consciousness involves the sharp distinction between the self and the totality of the world. Self-knowledge is thus the basis of discrete individuality.

Such individuality, rather than isolating the individual, enables him to rise above the limitations of his immediate group and to develop a sense of belonging to the whole human race. This gives clarity of approach to human conflicts and inner strength to stand against hypocrisy and falsehood.

Refs Moustakas, Clark E *Individuality and Encounter* (1968).
Related Paternalism (#HH0735) Individualism (#HH1162).

♦ HH0777 Self-made person
Description A person having succeeded by his own efforts, especially after beginning life without money, education, or influence.

♦ **HH0778 Self control**
Self harmony
Description Under the harmony of self control the mind remains of fixed intent, not swayed by selfish desire, lust or pleasure. Such control of the will leads to responsible action in which the individual is free and understands himself. Self-control, and its associated responsible but free and self-understanding activity, is contrasted with neurotic abstention, as for example abstinence arising from fear or hatred of the sexual impulse.
Refs Barber, T X, et al (Ed) *Biofeedback and Self-Control* (1970).
Related Continence (#HH0070) Ethical discipline (#HH1086).

♦ **HH0779 Eightfold path of yoga** (Yoga)
Astanga yoga
Description According to Patanjali, yoga should not be embarked upon without a preliminary practice of ethical discipline. This involves abstention from evil doing (yama), observing the noble principles (niyama), truthfulness (satya), non-violence (ahimsa), refraining from stealing (asteya), lack of self-indulgence (brahmacarya), and absence of greed (aparigraha). One is then in a state of sufficiently strong ethical consciousness to embark on the psycho-physical exercises and spiritual training that lead to self-realization or integration with the ultimate ground of existence.
Another interpretation of Patanjali's eightfold path implies the eight levels of yama (restraint), niyama (discipline), asana (bodily posture), pranayama (breath control), pratyahara (withdrawal of the senses), dharana (concentration), dhyana (meditation) and samadhi (identification). These are the components of classical yoga and are not so much approached in order as complementary to each other, so that yama and niyama are clearly practised together, and also if the components are considered on a circular plan, opposite pairs are functionally related - yama with pratyahara etc.
Broader Yoga (Yoga, #HH0661) Stages of spiritual life (#HH0102).
Narrower Restraint (#HH6876) Pratyahara (#HH0829) Niyama (Hinduism, #HM3280) Dharana (Hinduism, #HM2566) Asana (Hinduism, Yoga, #HH0669) Samadhi (Hinduism, Yoga, #HM2226) Pranayama (Hinduism, Yoga, #HH0213) Dhyana (Hinduism, Buddhism, #HM0137).
Related Samyama (#HH0374) Raja yoga (Yoga, #HH0755) Samatha-bhavana (#HH0710) Six-member yoga (Yoga, #HH2077) Eightfold way (Buddhism, #HM2339) Treasures of the godly (#HH0364) Vipassana-bhavana (Buddhism, Pali, #HH0680) Religious experience as meditational insight (#HM1593).

♦ **HH0780 Mortification**
Description Mortification of the flesh, the deliberate restraining and disciplining of natural impulses in obedience to illuminated reason, is characteristic of the spiritual life. In addition to the active avoidance of the temptations of "the world, the flesh and the devil", most religions demand some ritual fasts and abstinence (for example, Islamic "Ramadan", Christian "Lent"). The further along the spiritual path an individual progresses the more he or she will require such disciplines. These disciplines are subject to the approval of an acknowledged wise director as they can other wise lead to problems of pride and self-deceit.
Mortification means, literally, putting to death, so here the natural or selfish desires are put to death. The individual practices of self denial which come under the heading of mortification, although useful, are probably less helpful than the willing acceptance of stresses and difficulties arising from following one's vocation, when all other desires are submerged in the desire to follow the way demanded by God.
Broader Penitential practices (#HH4653).
Narrower Flagellation (#HH0151).
Related Fasting (#HH0011) Asceticism (#HH0556) Self-denial (#HH0964) Self-mortification (Sufism, #HH0464).

♦ **HH0781 Secondary afflictions** (Buddhism, Tibetan)
Upaklesha — Nye-nyon
Description A group of twenty afflictions or *knowers* which, in Tibetan Buddhism, are said to be part of or close to one of the six *primary afflictions*.
Context The afflictions, together with the secondary mental factors, are comparable with the mental activities or formations of Hinayana Buddhism.
Broader Afflictions and hindrances (Buddhism, #HH5007).
Narrower Spite (Buddhism, #HM2778) Deceit (Buddhism, #HM3246) Jealousy (Buddhism, #HM2638) Lethargy (Buddhism, #HM2926) Laziness (Buddhism, #HM3163) Non-shame (Buddhism, #HM2986) Non-faith (Buddhism, #HM2461) Resentment (Buddhism, #HM2134) Excitement (Buddhism, #HM2534) Concealment (Buddhism, #HM2605) Miserliness (Buddhism, #HM2964) Haughtiness (Buddhism, #HM2528) Harmfulness (Buddhism, #HM3210) Distraction (Buddhism, #HM3154) Belligerence (Buddhism, #HM2264) Dissimulation (Buddhism, #HM3093) Forgetfulness (Buddhism, #HM3053) Non-embarrassment (Buddhism, #HM3203) Non-introspection (Buddhism, #HM2063) Non-conscientiousness (Buddhism, #HM2314).
Related Root afflictions (Buddhism, #HH0270) Four noble truths (Buddhism, #HH0523) Sense consciousness (Buddhism, #HM2664) Awareness as mental-formation group of conscious existence (Buddhism, #HM2050).

♦ **HH0782 Laya yoga** (Yoga)
Description This is one of the main forms of yoga and is concerned primarily with the methods of acquiring control over the mind and more particularly the will-power.
Specific yogas which depend upon the conscious exercise of the controlled power of the mind include; Bhakti yoga, namely a devotional approach to the power of divine love; Shakti yoga, namely an energy-focused approach to divine power; Mantra yoga and Yantra yoga, namely the use of specially selected sounds and specially designed diagrams to focus communication with psychic powers (which are invoked by the individual to assist him in his development).
Laya yoga tends to be taught after the individual has been somewhat prepared by the practice of Hatha yoga and as a preliminary to Dhyana yoga and Raja yoga. The emphasis is on concentration on sound, or superphysical *nada* sound, which can be heard at certain points within the body. It is a necessary grounding for the performance and understanding of classical Indian music.
Refs Goswami, Shyam Sundar *Layayoga* (1980).
Broader Yoga (Yoga, #HH0661).
Narrower Shakti yoga (#HH3912) Yantra yoga (#HH1224).
Related Nada yoga (Yoga, #HH0160).

♦ **HH0783 Assimilation**
Cultural assimilation — Acculturation
Description New facts and ideas presented to an individual are modified or distorted depending on that person's expectations based on previous experience. Such previous experience has been *assimilated* into the self due to prolonged contact with a particular environment. *Cultural assimilation* or *acculturation* describes the process of absorbing/becoming like the *social* environment and adopting the attitudes of that society. Data subjected to "assimilation" will be then passed on to others in its modified/distorted form. According to *assimilation-contrast theory*, information received by an individual is accepted and assimilated, or rejected, depending on whether it falls within the tolerance of his or her opinion. When rejected it may have the affect of producing more extreme views to counter such information.

♦ **HH0784 Self-realization therapy**
Description A number of techniques may be used whose aim is to produce, with group support, sufficiently strong emotional reaction to break the self out of some constricting mould and allow intense experience or change. The experiences produced are of varying intensity. At one level they may be simply recreational, as teaching adjuncts, as means for improving industrial efficiency or for training political action groups. Or they may precede religious experience and mystical revelations of great intensity, although, with no ultimate religious aim, this is usually a secularization of the transcendental experience. Virtually all point to a discovery of the individual's maximum potential or real self.
Common to all techniques is the use of feedback, of group interaction, as events within the group and the effects of taking part are described, analyzed and evaluated by those involved. Although regular meetings may be held, it is more usual for the groups to take several days away from habitual surroundings and thus make the experience more intensive Also common to such groups is that participants do not use the group as medical treatment but to enhance their experience or self-expression. In fact, use of such groups for medical therapy may have an unfortunate letdown effect after the group session is over and the caring, supportive environment withdrawn. Another limitation is that the aim may become egoistic with their emphasis on the individual and not on humanity as a whole.
Refs Chung-yuan Chang *Self Realization and the Inner Process of Peace* (1956).
Broader Therapy (#HH0678).
Related Group therapy (#HH0851) Encounter group (#HH0162) Sensitivity training (#HH0672).

♦ **HH0785 Bombu Zen** (Zen)
Meditation
Description The term Bombu Zen refers to meditation for whatever purpose - physical, mental or spiritual - as an exercise for the mind within the context of *mushinjo*.
Broader Human development (Zen, #HH1003).
Related Mushinjo (Zen, #HM2974).

♦ **HH0786 Integration**
Description Psychoanalytically, *primary* integration is the process of bringing together the separate parts of the mind which takes place in the first five years of life, so that the child starts to see itself as a whole and distinct from the environment. *Secondary* integration is the coordination of individual components into unified and socialized action.
In general terms, integration is the process by which disparate units of behaviour are formed or coordinated into larger and more inclusive patterns, through conditioning, generalization of habit, and all associational processes.
1. Simple reflexes in which integration takes the form of a limited number of nerve cells functioning together because of some synaptic affinity between them. A hierarchy of levels (between the cell and the total personality) may be distinguished as produced by integration.
2. Conditioned reflexes, in which integration takes the form of substitution of associated stimuli for congenitally effective stimuli with the result that the individual performs innate acts to altered stimulus conditions.
3. Habits, namely integrated systems of conditioned responses, involving altered responses as well as an extended range of effective conditioning, leading to fairly stereotyped forms of response in the face of recurrent situations of a similar type.
4. Traits, resulting (at least in part) from the integration of specific habits expressing characteristic modes of adaptation of the individual to his surroundings.
5. Selves, integration of systems of traits that are coherent among themselves, although likely to vary in different situations.
6. Personality, the progressive final integration of all the systems of response that represent an individual's characteristic adjustment to his various environments.
7. Final perfect personality integration through which a completely unified personality emerges (this is considered to be a theoretical possibility).

♦ **HH0787 Dynamic self-opening meditation**
Description The attention is sequentially focused on the physical, emotional and intellectual components of the individual, his aesthetic, social and moral behaviour. By opening his being to the higher divine consciousness he is physically strengthened and freed from illness and unwanted habits; emotionally purged of selfishness, destructive impulses and inner conflicts, becoming a channel for the universal life force; and mentally enlightened, broadened and freed from fixed ideas so as to articulate the supreme truth. His whole being vibrates in harmony with the cosmic dance. Upward and inward self-opening transforms the personality to a centre of the divine.
Broader Meditation (#HH0761).

♦ **HH0788 Sacred drugs**
Soma
Description Mind-expanding drugs are often used for religious and ritual purposes, in particular to open the awareness to other worlds. Not only are sacred drugs used extensively in "primitive" and magical cultures, they also seem to have an important role in the main religious traditions. For example: soma in the vedic religion; hashish in various Islamic, Sufi and Hindu (Shaivism) traditions; alcohol (wine) in Christianity and Sufism. In Sufism, "intoxication with the divine spirit" is a common phrase. The Sufi is "drunk with love", lost in ecstasy.
Related Rites (#HH0423) Psychedelic drugs (#HH0075) Human development (Hinduism, #HH0330) Modes of awareness associated with use of cannabis (#HM3691).

♦ **HH0789 Freedom of choice**
Description Such freedom requires implicitly that there should be a number of alternatives and that their outcome is not completely predictable. This is the only situation in which creative expression is possible. Totally unpredictable existence, or one in which every action is the result of free choice would, however, engender panic or anomie; therefore a certain stability of basic parameters is desirable. Freedom without stress is thus most effectively exercised in a limited stimulus field.
The mere ability to make choices between good and evil is the lowest limit of freedom because the only aspect that is free is that good can be chosen. When evil can be chosen freedom is lost. An evil choice always destroys freedom. Total spiritual freedom is the complete inability to choose evil. When everything desired is truly good and every choice aspires to that good and attains it, there is freedom because everything desired is done and every act of will ends in perfect fulfilment. Freedom is not the equal balance between good and evil but in the perfect love and acceptance of what is good and the hatred and rejection of what is evil so that everything done is good and results in happiness. It is in refusing and denying and ignoring every possibility that might lead to unhappiness, self-deception and grief. Perfect freedom can only be found in perfect union and submission to the will of God. Perfect freedom is the ability to do God's will. To be able to resist

God's will is not to be free but to be enslaved.
Related Will (#HH0920) Creativity (#HH0481) Uncertainty (#HM3051) Risk-taking (#HH0654) Flow experience (#HM2344) Tragic suffering (#HH0708) Evolutionary choice (#HH4712) Freedom (Christianity, #HH4020) Predestination (Christianity, #HH0563).

♦ HH0790 Penance
Description This term is generally used to describe an external act signifying internal repentance, consciousness of previous sin/wrong-doing, and conversion. Forms include public confession, baptism, fasting, prayer, wearing of uncomfortable clothing (hair shirts), physical disfigurement, sacrifice and acts of charity. Penance may be individual, group, or by an individual acting symbolically for a group (for example a priest or a king). In the Christian tradition penitential acts and abstinence are also used to emulate the life of Christ and master human inclinations.
As a sacrament of the Church, penance may be equated with the sacrament of reconciliation. In order to rid him or herself of sin committed since baptism, the penitent makes full and sincere confession, expressing genuine sorrow for sins committed. The priest then, having counselled the penitent, prescribes the act of penance and the penitent receives absolution.
Related Rites (#HH0423) Ordeals (#HH0377) Repentance (#HH0441) Conversion (#HH0077) Absolution (#HH0472) Reconciliation (Christianity, #HH0164).

♦ HH0791 Self-expression
Description The performance of some activity, whether artistic or mundane, in order to express one's self independent of another's appreciation of the action, may be referred to as self-expression. This seems to be the spontaneous response to an inner necessity; it may take the form of play, as with children, or to some extension of the concept of play, where a common factor seems be the satisfaction of using the faculties employed to the point of complete exhaustion. This is true of artistic self expression, when the act of creation employs all the energies while the state is of being most one's self and yet beyond one's self; but the concept can be extended to the experience of work and to the religious life. Religion has been regarded as "the spontaneous act of the soul in response to its most intimate sense of absolute worth". Often the process of self-expression may be arduous, through "cries and groanings of the spirit"; but when the individual may be said fully to express himself there is the experience of *self-realization*.
Related Peak experiences (#HM2080) Self-actualization (#HH0412).

♦ HH0792 Spirituality (Christianity)
Description Human beings are created such that they are incomplete when encapsulated in the self. There is a deep yearning for self-transcendence and surrender. Spirituality is the response to this yearning for surrender to something greater than one's self. When experienced, this surrender gives the feeling of having found one's place. Someone who is evidently spiritual is not necessarily religious but is directed from the inner person and the unseen, inner worlds are of great importance. When the focus of self-transcendence is seen as a power or being, and a relationship is made with this power or being, then this is a development from natural spirituality to religious spirituality.
The Christian develops this process of probings and response to spiritual longing in the context of Christian faith and community. He or she lives in the Spirit of God, and awareness of being in the presence of God increases and deepens. Spirituality thus begins and grows through the interior life, relating to God in the depths of the self. It is here that spiritual transformation takes place, a spiritual growth starting in the heart and working outwards.
The leading of a spiritual life while remaining in the world is, to the Christian, a question of living integrally in the life of Christ, resulting finally in the soul's union with God. It may be attempted in many ways: through austerities (so long as the result of such deprivations is indifference to the inessential and liberation from worldly things); it may be sought in the monastic life, away from the direct influence of the world; through contemplation of creation, mysticism, devotion.
By whatever means it is approached, the spiritual life requires some intensity of purpose, and response to God's love by loving one's fellow men. It is the intellectual and affective realization of God through direct experience, and a personal response to God; a narrow path between acceptance of the means of achieving spirituality as ends in themselves, and rejection of the spiritual life altogether.
Refs Bouyer, L *Introduction to Spirituality* (1961); Bouyer, Louis *Orthodox Spirituality and Protestant and Anglican Spirituality* (1982); Cousins, Ewert, et al (Eds) *Christian Spirituality* (1987); Cousins, Ewert, et al (Eds) *Christian Spirituality* (1987); Cousins, Ewert, et al (Eds) *Christian Spirituality*; Cox, Michael *Handbook of Christian Spirituality*; Faricy, Robert *The Lord's Dealing* (1988); Kaschmitter, William A *The Spirituality of the Catholic Church* (1982); Monk of the Eastern Church *Orthodox Spirituality* (1968); Spidlik, Tomas *The Spirituality of the Christian East* (1986); Underhill, Evelyn (Ed) *The Cloud of Unknowing* (1970).
Broader Spirituality (#HH5009).
Related Mystery (#HH0303) Reverence (#HM3023) Sanctification (#HH0428) Spiritual renaissance (#HH1676) Spiritual development (#HH0017) Psychic development (Psychism, #HH0855) Christian mysticism (Christianity, #HH5100).

♦ HH0793 Humanistic psychology
Description Humanistic psychology is a comprehensive orientation, an affirmation of a positive psychology for the future of mankind. Given the growing, changing, varied nature of the orientation, it is not considered possible or desirable by its proponents for it to be delimited by particular methods (experimental or statistical). Humanistic psychology attempts to deal with whole, living persons in the process of becoming, rather than isolated aspects of human behaviour. It is therefore especially concerned with topics such as authenticity, encounter, self-actualization, search for meaning, creativity, intentionality, psychological health, being motivation, values, love, identity and commitment.
As a result of the work of C. J. Jung, who formulated a psychology of positive ends and purposes, the main consideration has become the essential spirit of man, and ways and means of achieving it, rather than the pathology of man and its reduction. This approach is characterized by the writings of Goldstein, Fromm, Horney, Rogers, Maslow, Allport, Angyl, Buhler, etc, as well as by certain aspects of the writings of Jung, Adler, and the psychoanalytic ego-psychologists.
Refs Bugental, J F T *Challenges of Humanistic Psychology* (1967); Cohen, John *Humanistic Psychology* (1962); Jourard, Sidney M *Disclosing Man to Himself* (1968).
Related Holistic medicine (#HH0581) Human potential movement (#HH0398).

♦ HH0794 Ontological perfection
Description The *Perfect Prayer* in the Isha Upanishad reads:
"That is perfect. This is perfect. Perfect comes from perfect. Take perfect from perfect, the remainder is perfect".
The word "perfect" is derived from the latin "per facere" and implies thoroughly made, a state of completion or totality. That which is perfect can be said by definition to lack nothing, so no change can improve it, nor is it subject to decay. Absolute perfection is therefore eternal and immortal and, if conceived of as having an existence beyond an ideal in the mind, identical with the concept of God. On the other hand, "relative" perfection, perfection within certain limits, may be used to refer to the greatest perfection which something intrinsically removed from absolute perfection is capable of achieving. The process of perfection, or "making perfect", meaningless in absolute terms, can have meaning in terms of relative perfection; for example, in relation to human development. A thing is said to be perfect in this sense when it achieves final fulfilment in attaining its end or destiny.
Greek thought (and many subsequent philosophies) considered that the state of becoming or change was illusory or imperfect, only being is perfect. Different interpretations on the "oneness" of being and the illusion of plurality or multiplicity include Plato's argument that multiple, created things, although imperfect in themselves, are perfect in that they participate in the single, infinite whole. This also seems implicit in the above quotation from the Isha Upanishad. Opinions have differed as to the possibility, nonetheless, of being aware of the perfect being, as for instance Kant, who said that knowledge is simply of appearances, of that which the senses make available to consciousness, and that the perfect being is unknowable. This interpretation, however, denies the validity of knowledge in belief or inner experience, the comprehensive awareness of the nature of existence; and it is towards this knowledge that those on different spiritual paths are said to strive.
Related Human perfectibility (#HH0212) Spiritual development (#HH0017).

♦ HH0795 Behaviour therapy
Description Behaviour therapy includes a number of therapeutic methods sharing certain theoretical conceptions, although the methods themselves may vary from desensitisation to aversion treatment. Practitioners hold that neurotic behaviour and other types of disorder are acquired in a manner subject to the established laws of learning. Such laws encompass both the acquisition of new behaviour patterns, and the reduction or elimination of existing (undesirable) behaviour patterns. *Continuous positive reinforcement* may be used to encourage the acquisition of desired behaviour and discourage the undesirable.
Behaviour therapy has been successfully used in the treatment of a variety of neurotic conditions including phobias, sexual disorders, obsessional disorders and children's speech disorders.
Refs Chesser, E and Meyer, V *Behaviour Therapy in Clinical Psychiatry* (1970).
Broader Therapy (#HH0678).

♦ HH0796 Suggestion therapy
Description A method of psychotherapy based on the communication to the individual of positive suggestions appealing to the emotions and frequently without rational foundation. It is closely related to magic and folk healing methods.
Broader Therapy (#HH0678).

♦ HH0797 Strength of character
Broader Character (#HH0895).

♦ HH0799 Internal marriage
Description According to Jung, since masculine and feminine factors exist in each individual, then the personality may be viewed as the balance of these factors in an internal 'marriage'.
Related Betrothal initiation archetypal image (Tarot, #HM3028).

♦ HH0800 Immortality
Description Immortality implies the continued existence in some form of consciousness of the rational unity of an individual personality for an unlimited period of time; this in turn implies belief in the personality's survival of death of the body, a belief which is and has been held in most societies and civilizations. Belief that terrestrial life is not the only life profoundly affects the way in which terrestrial life is lived. For example, materialism and concern for events in this life are diminished in the wider view of immortality, although fulfilment of duties and enjoyment of benefits in this life are nonetheless considered by many to be consistent with such belief. Lack of concern for events in this life has lead to great acts of altruism. The transcendent experience of the soul as a more exalted function than materiality is taken as evidence of immortality. Ultimately, most religions posit salvation, paradise or heaven (which are not identical in all systems) as the result of accepting God, and hell as the result of rejecting him. Again, some systems would declare that hell as immortal suffering while others look on it as temporal or as annihilation.
Refs Badham, Paul and Badham, Linda *Death and Immortality in the Religions of the World* (1987); Case, Charles J *Beyond Time* (1985); Lifton, Robert J *The Future of Immortality* (1987); Lu Kuan Yü *Taoist Yoga* (1970).
Related Rejuvenation (#HH0525) Reincarnation (#HH0686).

♦ HH0801 Social group-work
Description An educational process in which individuals are developed through the group experience while at the same time the group performs a social function.

♦ HH0802 Assignment therapy
Description By placing an individual in a group whose social, sexual, racial and leadership structure best match his position in the community, he is better enabled to adjust to the group and ascend to a position in the group to express his spontaneous abilities.
Broader Therapy (#HH0678).

♦ HH0804 Illumination
Illuminism
Description The first act in creation as described in Chapter I of Genesis, when "darkness was upon the face of the deep" and "the Spirit of God moved upon the face of the waters", was when God said, "Let there be light: and there was light". In the Kena Upanishad, Light is given supremacy because it was the first to recognise spirit; and in the Chandogya Upanishad, the first created by mere being was Light. Chapter I of St John's Gospel says "In him was life; and the life was the light of men. And the light shineth in darkness; and the darkness comprehended it not... That was the true Light, which lighteth every man that cometh into the world". Clearly, the act of illuminating or being illuminated has enormous philosophical and spiritual significance. According to St Augustine, the individual soul is like an unlit lamp, in darkness until illuminated by the true light. Divine illumination lights up the intellect and will, so that mankind receives true knowledge. This idea is carried over in the common use of the terms "bright" and "brilliant" to describe someone who is very intelligent. According to Patanjali, illumination is the result of sanyama.
Two different teachings on illumination of the intellect come under the heading *illuminism* – that which considers illumination of the mind to come from a higher source, direct revelation to the chosen few (for example, gnosticism, Rosicrucians); and that characteristic of the *Philosophy of Enlightenment*, which claimed that pure, enlightened reason was possible within each individual, dispensing with the need for traditional culture, religion or government and emphasising naturalism with respect to rights, society and religion.
Narrower Samyama (#HH0374).
Related Enlightenment (#HM2029) Mystical cognition (#HM2272) Cosmic consciousness (#HM2291) Triple way (Christianity, #HH0631) Philosophy of enlightenment (#HH0591) Via illuminativa (Christianity, #HH6030) Mystical theology (Christianity, #HH5217).

♦ **HH0805 Guru**
Spiritual mentor — Adept — Staretz — Geront
Description Literally "heavy" or "weighty" - and thence "great" or "respected" - a guru, or *teacher of spiritual truths* is one who possesses sacred knowledge and who passes that knowledge on to others less far on the spiritual path. Such spiritual knowledge, received by direct experience, is deeper and more fundamental than intellectual knowledge gained second-hand by study; and is taught in a practical way so the pupil may acquire similar experience or *realisation* of truth. An alternative interpretation of the term guru is "obstructing of darkness".
In particular, a perception of the unity of the individual with the universal will draw others with a glimpse of such a vision towards the teacher. They see God through such a teacher and see the teacher as God. It is for this reason that the Hesychast disciple behaves in the presence of his teacher as in the presence of Christ, that the Tibetan looks upon his lama as the Buddha himself, that Hindu teaching identifies the guru with God, the direct representative of the sad-guru. This is in line with the belief that the wholly enlightened being is literally identical with God, beyond ego-identity. Such an adept will not have an individual teaching method but will respond spontaneously to any situation, totally in the present. He will use any method to dislodge the self-inflicted limitations and complacency of his disciples' egos and allow them also to direct their attention away from personal concerns and to glimpse the truth of their own inherent enlightenment, the transcendental "I am".
Such direct experience of a guru's teaching allows progress on the difficult path to freeing attention from attachment and reaction to externals to be achieved considerable more quickly than the individual could achieve by his own self effort. This is because self effort is inherently based in the ego which the individual is trying to transcend. Spiritual identification with the samadhi of an adept is the practice of *satsang*, the depth of response achieved naturally depending upon the stage of spiritual life of the devotee and of the adept. Although direct experience such as described is independent of time and environment, a guru's intepretation of such experience will depend on the local language, symbols and social and religious structure. This accounts for apparent differences of perception as "Buddhist", "Christian", "Islamic", "Hindu", and so on. Nevertheless, the vision transcends such differences, and is essentially one. A spiritual master with be aware of both the relativity of forms and their necessity, so his knowledge must transcend form. Although he may not have practical familiarity with other traditions he will not be limited by his own. No true master puts himself outside a tradition or religion, he knows its divine origin.
Related Adept (#HH1552) Mentor (#HH3219) Avadhoot (#HH0141)
Spiritual way (#HH1867) Spiritual direction (#HH0442) Spiritual discipline (#HH1021)
Satsang (Hinduism, Yoga, #HM3329) Spiritual master (Sufism, #HH2080)
Human development (Sikhism, #HH6292) Discipline of guidance (Christianity, #HH1244).

♦ **HH0806 Humanistic biology**
Description Humanistic biology suggests that, rather than studying "statistically average" samples of the population, it is more useful to study the ablest individuals in the particular attribute being considered. This is far more likely to indicate the best that the human race is capable of achieving. To the criticism that such judgements as "best" are not capable of being made objectively, it replies that such judgements are in fact made all the time in research, whether or not the researcher is aware of it; and demonstrates that it is possible to discover the intrinsic values of human beings. Rather than the value-free approach to those sciences which involve human problems, humanistic biology starts from the premise that values exist in reality, that there is such a thing as a "good" society (that which fosters the fullest degree of humanness and the fullest development of human potential); and that, for example, self-actualizing people are better able than others to distinguish right from wrong.
It is postulated that since the human race is now capable of being responsible for its own evolution, a taoist (loving) objectivity is more likely to bring the best decisions than classical (non-caring) objectivity. Given a person's intrinsic goodness or badness, society could be organized to favour the good behaviour naturally manifesting. Emphasis would be on improving the means by which a desirable end is reached rather than the end itself. A society where basic-need satisfactions are lost will produce more physically as well as mentally sick individuals and is less likely to achieve the aims it sets itself.
Refs Maslow, Abraham H *Towards a Humanistic Biology* (1970).

♦ **HH0807 Monasticism**
Description Monasticism is a term encompassing religious institutions, ritual and belief systems whose agents, members or participants attempt to practice religious works that are above and beyond those required by the religious teachings of their society or of exceptional individual religious and spiritual leaders in their society. They therefore make a radical interpretation of the tenets that apply to all believers or to the whole society. Monasticism may involve such practices as celibacy, and asceticism; its practitioners usually isolate themselves physically from the surrounding society.
The purposes of monasticism include: discovery of the true self, emancipation of the self, and the improvement of society. Discovery of the true self is accomplished by overcoming imperfections of the body and mind (through discipline, control, chastisement, self-mortification, or other psychophysical procedures) and by spiritual perfection (using processes of intense contemplation, possibly accompanied by prayer, worship, incantation, propitiation, self-abasement, or self-inflation). Emancipation of the self is accomplished by salvation from bondage, by redemption as deliverance from the spiritual effect of past transfressions, and by liberation from cycles of birth and death.
Refs Leclercq, Jean *The Love of Learning and the Desire for God* (1961).

♦ **HH0808 Cosmic integration**
Description Concommitant with the integration of the psyche is the awareness of its relationship to nature and to society, the fundamental reality of the psyche-cosmos continuum. As John Donne put it "No man is an island, entire of itself". One cannot exist without being related to everyone and everything else; and psychic growth cannot be healthy when there is no integral relationship with the total environment. This may express itself in: reverence for life; social action; and subordinating self-interest to the interests of family, friends, community, nation and, ultimately, humanity and the cosmos as a whole. Such cosmic integration is not achieved through reason; but only through stepping into supra-cosmic realization.
Related Psychic integration (#HH0823) Existential integration (#HH0832)
Conscious self-relatedness (Yoga, #HM0467).

♦ **HH0809 Gracefulness**
Description Gracefulness is said to be most purely expressed by a body which either has no self-reflective consciousness at all, or when consciousness has returned following cognition has just passed through an "infinite" and there is infinite self-reflective consciousness.

♦ **HH0810 Imitation** (Christianity)
Imitation of Christ
Description The words "imitation" and "image" have the same root, so that imitation of something or someone involves the reflecting of some or all of their attributes. Christians are encouraged to imitate Christ as part of the process of *sanctification*, so that their lives are a reflection or image of the life of Christ in his sonship of God and in his virtues, behaviour and intention. As is recorded in the first chapter of Genesis, "God created man in his own image". Imitation involves not simply the focus of spiritual and moral attention on the example of Christ, although this is important; it also implies the conforming or moulding of the life of the Christian, through grace, to the likeness of Christ.
Related Image of man (#HH0351) Sanctification (#HH0428).

♦ **HH0812 Innovative learning**
Description As opposed to *maintenance learning* (which, although necessary, simply reinforces and enhances already existing outlooks, methods, rules and ways of life), *innovative learning* operates in open situations and open systems, and derives from dissonance among contexts. Through problem formulating and clustering it achieves integration, synthesis, and broadening of horizons. Where appropriate contexts and analogies do not already exist, innovative learning encourages the development of a capacity to construct suitable alternative mental frameworks, and enlarge the range of options available for sound decision-making. By this means, human capacity to act in new situations and deal with unfamiliar events is enhanced.
Broader Learning (#HH0180).
Related Creative thinking (#HH0703).

♦ **HH0813 Kshanti** (Zen)
Patience
Context The third of five perfections or *paramitas* of Zen which, when achieved, render the mind ready for the sixth - *maha paramita* or great perfection, *prajna*.
Broader Paramitas (Zen, #HH0219).
Related Patience (#HH1304) Prajna paramita (Zen, #HH0550)
Mahayana receptivity-application awareness (Buddhism, Tibetan, #HM2247).
Followed by Virya (Zen, #HH0677).
Preceded by Shila (Zen, #HH0429).

♦ **HH0814 Divine indwelling**
Description As a person progresses on the spiritual path, the inner presence of God becomes apparent. Doctrines differ on the precise meaning of this "indwelling", whether it is different in kind from God's presence in all things (as a special gift of grace), or whether it is the same form of indwelling as occurs in the whole creation but, as the individual is perfected, the way is cleared for it to show. One of the deepest mysteries of the Christian faith, it brings Christian doctrine close to the doctrine of *unity* or *oneness* with God common to eastern religious.
Related Transcendence (#HH0841) Unitary consciousness (#HM2702)
Existential unity of being (#HH0602).

♦ **HH0815 Time binding**
Description The ability of the human being to profit from previous generations and project his influence into the future through social planning, etc, means that the present contains both the past and the future. Creative growth thus occurs in time, with habits and disposition deriving from the past, and ideals, hopes and aspirations extending into the future.

♦ **HH0816 Contemplative intuitive meditation**
Contemplative prayer — Mystical prayer — Simple contemplation — Imageless prayer (Christianity) — Prayer in the spirit
Description Contemplation may be conceived as a descent into the depth-consciousness in which is stored the collective wisdom and insight of the human race and where the inner deity is enthroned. This form of contemplation precedes that of the advanced contemplative. Two types of perception, the intellectual-rational and contemplative-intuitive are used alternately and intermingled. The individual strives to adapt his mode of perception to enter through contemplation into an awareness of oneness rather than multiplicity, in order to apprehend the world and himself as they really are. This leads to a gradual transformation of the personality or sense of selfhood and a new vision of the nature of reality. The psychical man slowly and laboriously grows into the stature of the spiritual man through the realization in direct experience of the presence within him (other than the phenomenal ego) of an inner deity which he may apprehend as Spirit, God, Self, the Christ (or Buddha) within, or Atman, namely his essential Self. This realization may take a personalistic or non-personalistic form.
In Christian terms, the Cloud of Unknowing describes the signs for transition from "intellectual" to simple, passive, contemplative prayer. The transition indicates a more complete surrender to God in all aspects of life and inviting God increasingly to take the lead. A background of reading, reflecting and praying will have led to a longing for a life of contemplation and prayer. This longing is more persistent than any other; if it disappears it does so only to return with greater joy. Contemplative prayer is loving rather than knowing, even the most holy thoughts are hidden in the cloud of forgetting. A single word arises, encompassing all the love and longing that is felt. Discursive thought and imagination are a hindrance. Similarly, Meister Eckhart speaks of imageless prayer but, while rejecting discursive reasoning or thought as a means of attaining God, he indicates that purest love is heightened, purified, intensified in intellect, a higher knowing or spiritual knowledge. Then there is not only an aspiring to the goal but a uniting with it.
Meister Eckhart has practical advice for those who are troubled by distractions in the mind. Like other authors, he warns against trying to forcibly eliminate such distractions, which is ineffective and tiring. His advice is firmly and resolutely to be detached from them, not to cling to or identify with such images, which will always be there. They must simply be gently put aside. This unencumbered, virginal state is not the highest, however. There is a spark or dart of longing love, as is also indicated in the "Cloud of Unknowing", which is likened to a wife, receiving God into the self and giving birth to Him. Here is the birth of God in the soul's ground.
The essence of contemplative prayer is the way of pure faith. It does not have to be felt but it does have to be practised. A process of interior transformation, this is a conversation initiated by God. If the person consents it leads to divine union. In the process, the person's way of seeing reality changes. There is a restructuring of consciousness empowering the person to perceive, relate and respond with increasing sensitivity to the divine. There is detachment from rather than absence of thought as heart, body and emotions are opened to God.
The practice of contemplative prayer is an education imparted by the Spirit; the person's participation is self denial, denial of his innermost self. This implies detachment from habitual functioning of the intellect and will. It may demand the letting go of the most devout reflections and aspirations as well as ordinary thoughts, if these has been treated as indispensable as a means for going to God. Thought becomes presence, an act of attention rather than understanding. Attention is given to the presence of Jesus without adverting to any particular detail.
As energies from the unconscious are released, two states arise: there is the experience of personal development as spiritual consolation, charismatic gifts or psychic powers; and there is humiliating self-knowledge as human weakness is experienced. The one balances the other, so there is no extreme of pride or despair.
Refs Keating, Thomas *Open Mind, Open Heart* (1986); Underhill, Evelyn (Ed) *The Cloud of Unknowing* (1970).
Broader Prayer (#HH0185).
Related Contemplation (#HM2952) Deautomatization (#HH2331)

Meditative prayer (#HH4089) Communion with deity (#HM2675)
Devotion (Christianity, #HH0349) Spiritual knowledge (Christianity, #HM7554)
Birth of God in the soul's ground (Christianity, #HM6522).
Preceded by Centering prayer (Christianity, #HH1994).

♦ HH0817 Encounter marathon

Marathon group

Description This is a special form of encounter group (see separate description) in which the group meets continuously over a period of 24 hours or more. In this form of intensive group interaction, all the emphasis is placed on immediate experience. It aims to simulate the world of emotionally significant others. The ways in which the member relates to this world reflect the core pattern of his being. The group members' reactions give cues as to the effect his behaviour patterns have on the world. The individual is offered the option of coming to terms in the immediate experience without postponing or evading such a confrontation. Members are expected to show themselves as they are in the world outside.
Nonverbal techniques may be used in the marathon experience, especially in dealing with openly expressed anger or hostility. Physical contact may occur and often helps to break down intellectual defences. The emotional strain on a participant may be balanced by an offer of physical comfort. In general a group will not intrude upon a participant who is undergoing any such deeply felt sense of strain, desolation and often loneliness, but will accept his feelings with warm silence.
The marathon group departs radically in form and content from the ordinary group–therapy group and from traditional approaches to psychotherapy.
Broader Encounter group (#HH0162) Human potential movement (#HH0398).

♦ HH0819 Emotion

Feeling — Affect

Description Emotion is a term variously used to describe either transient feelings, or deeper qualities or ideals which have some permanence even in day–to–day life. Thomas Aquinas listed eleven such: love, desire, joy, hate, aversion, sorrow, hope, despair, courage, fear and anger; although many other lists have been attempted, with little agreement between them. In psychological terms, these latter "emotions" are usually referred to as affects, the energies which generate feelings. The transient feelings are considered: (1) to be generated to suit particular goals and support the carrying out of an intention (Adler); (2) the physical response (facial or bodily) to perception of some object; (3) the feeling aroused by such a physical response in relation to an exciting object. According to Jung, an affect is an emotion or feeling of sufficient intensity to cause nervous agitation and to only with difficulty be commanded by the will. Thus the affect roused when a psychic "wound" is touched is a means of identifying a complex. Certainly the emotions are acknowledged as important for morality and psychologically in shaping behaviour; neglect or over–emphasis of the emotions can result in damage to the personality. An inappropriately high level of emotion may inhibit necessary action; similarly, too low a level leads to indifference and failure to trigger the necessary response.
Although the path to self realization and knowledge of unity is said to be beyond the emotions, an emotional religious response may be a possible first step on the way. Religious ritual seems important in patterning emotional experience, although in religion as elsewhere the power of emotional response may be manipulated to achieve the ends of the manipulator.
Refs Clyne, Manfred *Sentics* (1989).
Narrower Affects (#HM7132) Dissociation of affect (#HM6087).
Related Mood (#HM1748) Sentiment (#HM3497) Loss of soul (#HH0210)
Feeling consciousness (Psychism, #HM3590).

♦ HH0820 Saivism

Description This Kashmiri tradition is based on the philosophy of cosmic existence emanating in direct hierarchical sequence from the divine. The conditional self is really the formless divine; the conditional self is really the emanated divine, both are really Siva. Four ways or stages of self–realization are suggested: (1) The individual way, or way of absorption in the object. (2) The energic way, or way of absorption in energy. (3) The divine or superior way, or way of absorption in the void. (4) The null way, or superior way of absorption in bliss. Although the path may be followed progressively through the four stages, it may also happen that stages are pursued simultaneously or a later stage may arise without the preparation of an earlier stage.

♦ HH0823 Psychic integration

Description Psychic integration comprises the harmonious and balanced growth of the personality as opposed to the extreme cases where one aspect is cultivated to the exclusion of others and the detriment of the whole. Thus, psychic integration avoids hedonism (search for instinctual pleasure by following every passing impulse and restraining reason), which in the end can lead only to a chaotic or aimless life lacking fulfilment. It also avoids over austerity or puritanism, when the suppression of the unconscious desires by the super ego can lead to eccentric behaviour and psychic disturbance. Finally, it avoids spiritual development at the expense of contact with basics, or "down–to–earthness". By reconciling impulse and reason, the basic unity of existence is perceived, and one is able to come to terms with the self and society.
Related Cosmic integration (#HH0808) Existential integration (#HH0832).

♦ HH0824 Introspection

Description Introspection is the act of observing the workings of one's own mind and behaviour. External and internal events may be considered as simply a stream of experiences, the former having no awareness of themselves, the latter representing some form of knowledge; but the turning of the mind's attention to the state in which these events are observed is an act of awareness, of introspection. Rather than acting as a passive spectator, the mind thus controls the tendency for thoughts to follow their own course and the will to express passing impulses. The observer therefore shows partial independence of his own environment. Such observation is rarely, if ever, totally objective, and various systems have been devised by psychologists to enable it to be more systematic and controlled. It may be contrasted with the passive awareness that a person may have of his or her own mental state. The dichotomy involved in separating the "observed" self from the "observing" self has been a problem faced by philosophy and psychology since their inception. As it says in the Upanishad "You cannot know the knower of the known".
Analytical introspection in the system of Wundt and Tidnirer, involves a frame of reference which excludes all but perceptions, ideas and feelings and their qualities; whereas Ryle considered introspection by the self in the present to be the objective perception of the past self (this is open to the criticism that one can only recall mental events of which one was aware at the time). However, although somewhat lacking in objectivity, introspection is the only direct means of psychological investigation (it is the vital difference between human and veterinary medicine). From Aristotle onwards the introspective method has been essential in acquiring psychic facts and solving psychological problems, albeit on a subjective basis.
Related Self–observation (#HH0586) Self–remembering (#HM2486)
Self–cognition (Sufism, #HM3604) Self–conscious consciousness (#HM2610).

♦ HH0825 Discipline of confession (Christianity)

Description The very act of confessing one's guilt and hearing words of forgiveness has the effect of lifting the weight of guilt feelings and of the moral paralysis that sometimes accompanies them. Further, because of Christ's sacrifice of himself on the cross, confession is not merely psychologically therapeutic, it also heals and transforms the inner spirit. There is a subjective change in the individual and an objective change in his relationship with God. Although this is received as a demonstration of God's grace, it is a requirement, and a corporate requirement – confession of sins is intended to be made in the hearing of others, or at least one other, as well as privately without human mediation. There is a transforming of humanity when sins are confessed openly, a freeing from fear and pride as there is recognition that all are sinners and, through the forgiving words spoken out loud, a sense of the reality of God's presence and forgiveness. This freedom spreads, as the confession of sins by one person to another gives the other the freedom to confess his own sins or come out with something within his life that troubles him.
Three things are said (St Alphonsus Lugiori) to be necessary for confession: examination of conscience; sorrow; determination to avoid sin. Inward healing is ignited by confession of specific sins rather than generalizations. These sins must be taken seriously. And there must be a desire and will to be delivered from sin. Once there is awareness of the horror of one's own sin, one can hardly be horrified by the sins of another. There is an end to pretence. Just as honesty leads to confession, confession leads to change.
Broader Spiritual discipline (#HH1021).
Related Guilt (#HM3391) Admission of sin (#HH4012) Conviction of sin (#HM1160)
Soul care (Christianity, #HH1199).

♦ HH0827 Dhyana yoga (Yoga)

Path of meditative absorption

Description One of the four main types of yoga, this form of yoga is primarily concerned with meditation, leading to control of the mind and intuitive wisdom. Since some form of mind–control is an aim of all forms of yoga, dhyana yoga practices tend to be an essential aspect of the practice of all forms of yoga.
Refs Dewana, Mohan Singh Uberoi *Dhyana Yoga*.
Broader Yoga (Yoga, #HH0661).
Related Ch'an (Zen, #HH0589) Mystical stage of life (#HM3191)
Dhyana (Hinduism, Buddhism, #HM0137) Tantric visualization (Yoga, #HM1690).

♦ HH0828 Spiritual materialism

Description A fundamental distortion caused when spiritual techniques serve only to strengthen egocentricity although the individual believes himself to be following a spiritual path. It is caused by intellectual rationalization of knowledge and experience; and the inability to surrender hopes, expectations and fears in order to open one's self to reality.
Refs Trungpa, Chogyam *Cutting Through Spiritual Materialism* (1975).
Broader Materialism (#HH0529).

♦ HH0829 Pratyahara (Yoga)

Abstraction — Withdrawal of the senses

Description The senses lead the individual to project consciousness outwards rather than bringing it to rest within and guiding it back to the transcendental source. Unless a tight rein is kept upon the senses, then, there is the beginning of a chain of action described in the Bhagavad Gita, through contact, desire, anger, bewilderment, disorder of memory, destruction of the faculty for wisdom and the man is lost. The initial stages of yoga reverse the tendency for senses to flow outwards to sense objects. This is only possible if there is an inner distance from mundane things. Although the senses appear to be controlled by the mind, in fact the mind withdraws into itself with no external object of attraction. It may be achieved by force of will (raja yoga) or by the attraction of an internal object of devotion (bhakti yoga).
Context Interpreted by some as the fifth component of Patanjali's eightfold path of yoga.
Broader Eightfold path of yoga (Yoga, #HH0779).
Related Abstraction (#HH0120).

♦ HH0830 Eurhythmics

Description By translating sound into physical bodily movement, with different movements to represent the notes, the Delcroze method not only assists an understanding of harmony and music but develops the power of concentration and improves reactions.

♦ HH0831 Temperament

Refs Plomin, Robert and Dunn, Judy *The Study of Temperament* (1986).
Related Physique and temperament (#HH0126).

♦ HH0832 Existential integration

Description This is seen as the final step in integral self development when total harmony, beyond reason, is achieved with timeless being, such true harmony being impossible within the constraints of time. Nevertheless, even this eternal of which existential experience is aware and which is the transcendent dimension of the mystic, is not the whole story. The fullness of being integrates the eternal with the temporal.
Related Lila (#HM2278) Self–development (#HH0651) Cosmic integration (#HH0808)
Creative existence (#HH0513) Psychic integration (#HH0823).

♦ HH0834 Magnanimity

Description On the path of perfection, the magnanimous individual is inclined towards generous and valiant deeds and achieves meritorious actions, not for the glory and praise such deeds attract but in the pursuit of excellence. It is the quality of magnanimity which makes a person aware of his or her capacity to accomplish difficult and dangerous tasks (as such it may said to be part of fortitude), while at the same time it is compatible with humility, because the accomplishment of such tasks is seen as due to divinely–given ability and not primarily praiseworthy in the person carrying them out. Good will, self sacrifice, courage and moral uprightness are embodied in such a person, who is free from meanness, selfishness and vindictiveness.
Related Humility (Christianity, #HH0861).

♦ HH0835 Growth spirals

Description Human consciousness is said to evolve from relationships between the mind and its environment. According to D. Johnson, each stage or spiral is followed by a phase of transition to the next spiral when the contradictions and limitations which have arisen cause expansion or evolution by the synthesis of a new spiral, accompanied by expanded awareness.
Transitions through infancy, childhood and adolescence to social adulthood are largely dependent on external factors, although they may be hindered by lack of incentive to evolve, with consequent immature behaviour with respect to social norms for individuals at that particular age. But further development through individual self–actualization to true freedom and thence to the vision of a *seer* is dependent on the individual him or herself. Failure to develop beyond the "social adult" stage results in stagnation, but many individuals do break through to being their "own person". Development may be hindered by psychological "assistance" as this attempts to re–establish the "social adult" norm; or it may be assisted by a guru and the incorporation of compassion for others

into every action. Beyond the seer it is not possible to describe in words, although evidence of this mystic state shows that it is the ultimate in human potential and a stage of absolute enlightenment.
Refs Johnson, Dwight *Spirals of Growth* (1983).

♦ HH0836 Physical development
Physical growth
Description This is the process by which the individual increases in body size and weight with increase in age. Growth is rhythmic, depending on seasonal and even 24-hour cycles, and may be increased or retarded by external conditions. The speed at which physical maturity is reached is largely genetically programmed but is subject to environmental influences including nutritional standards and social conditions, material and non-material wealth, culture, physical exercise (which may be used to develop particular muscles selectively) and upbringing. Physical development is one major indicator of public health.
Refs Paplauskas-Ramunas, Anthony *Development of the Whole Man through Physical Education* (1968).
Related Physical culture (#HH0730) Physical perfection (#HH0722).

♦ HH0837 Cultural revolution
Description Cultural revolution may be distinguished from political revolution in that the political revolutionary aims to improve existing institutions (in terms of their productivity and the quality and distribution of their products). The cultural revolutionary believes that these institutions are based on consumption habits which have radically distorted the view of what human beings can have and want. He questions the reality that others take for granted and risks the future on the educability of man.
Cultural revolution involves a fundamental change in the relationship between society and education. Education becomes woven into the social, political and economic fabric, covering both the family unit and civic life. Every citizen then has the means of learning, training and cultivating himself, under all circumstances.

♦ HH0838 Martyr
Witness
Description The bearing witness or testimony to unpopular beliefs has historically met with such violent opposition that the originally almost juridical sense of the Greek word martyr has come to mean someone who voluntarily suffers death rather than retract.
Related Declaration (#HH0110).

♦ HH0839 Personal demoralization
Description When a person becomes unsatisfied with, and rejects, his or her life style, without being able to reorganize and develop a satisfactory new lifestyle, then demoralization sets in. It is particularly characteristic of periods of rapid breakdown in social systems when new social systems are unable to keep pace.

♦ HH0840 Feeling of I
Sense of I
Related Identity (#HH0875) Self-respect (#HH1384) Ego consciousness (#HM0570) Ego awareness (Hinduism, #HM2059).

♦ HH0841 Transcendence
Description The freedom of humans to transcend their individual, social, environmental and natural conditions is said to be their peculiar characteristic and is fundamental to ethics and religion. Marx emphasized the ability of humanity as a whole to transcend socioeconomic alienation; while Christianity looks not only to transcending the natural self in concerns and activities with non self-centred goals, but also to the ultimate, supernatural, transcendent reality which makes human transcendence possible at all. In fact, transcendence may be defined as an encounter with the holy which may constitute the source and goal of conversion. It is the domain of the sacred.
According to Kant, although transcendent objects are beyond theoretical knowledge they can be apprehended by faith and supported by practical reason. According to Jung, transcendence may be experienced in connection with sacred rites of renewal or the seeing of visions; but the experience is said to effect no permanent change or *transformation*.
For the Christian it is divine transcendence which differentiates the creator from the created and which, because it is the basis of human transcendence, cannot be separated from divine immanence by which the Holy Spirit is present in and acts through the creature. Arguments which have been put forward to rebuff the apparent implication that therefore divine transcendence inhibits human freedom include: the liberating rather than oppressive conception of the future as a paradigm for transcendence (Moltmann); the liberating immanence of divine incarnation and presence.
Refs Dowdy, Edwin (Ed) *Ways of Transcendence* (1982); Kulandran *Concept of Transcendence* (1982); Moltmann, J *The Future of Creation* (1979).
Related Rebirth (#HH0304) Immanence (#HM3558) Transcendence (#HM4104) Transformation (#HH1039) Divine indwelling (#HH0814) Self-transcendence (#HH0526) Transcendent function (#HH0324) Transcendental experience (#HM2712) Transcendental meditation (TM, #HH0682) Way of transcendence (Christianity, #HH6566) Journeying within transcendence (Christianity, #HH6505).

♦ HH0842 Counselling
Description Counselling with another, as a means to achieving self understanding and the solution of internal conflict or of some external problem, is an ancient technique although not always referred to under this name. Simply by listening to the other's problem, encouraging him and giving time for a train of thought to be considered, the counsellor acts as a mirror to reflect the other person, who will be able to see himself from previously unconsidered angles. As the counselling process develops, the focus tends to become less and less self-centred, this change in perception being followed by changes in choice and action so that behaviour becomes more reasoned. The "image" mirrored by the counsellor which the client perceives as superior to previous pictures is eventually enacted and a more objective approach permanently adopted.
Counselling is the basis of *client-centred therapy* and of *re-evaluation counselling*, the former normally being directed by a therapist with the aim of curing some abnormality and the latter as a formal extension of counselling for some specific problem.
Narrower Re-evaluation counselling (#HH0342).
Related Soul care (Christianity, #HH1199).

♦ HH0843 Linguistic intelligence
Description This is demonstrated by a sensitivity to sounds, rhythms, inflections and meter, a special clarity of awareness of the core operation of language. Such gifts are particularly characteristic of poets; but are said to be universally relevant in order: to use rhetoric in order to convince others; to remember information mnemonically; to explain something clearly to others (even when what is being explained is mathematical, logical or whatever); and to understand language itself. This intelligence is shown to be rooted in the left-hemisphere of the brain; and although the right-hemisphere may be used to learn both to read and to speak, such ability will be somewhat restricted.
Context One of the separate or multiple intelligences put forward by Howard Gardner.
Broader Multiple intelligences (#HH0214).

♦ HH0844 Mercy
Description Mercy is that quality, allied to compassion and charity, which, rather than identifying with another's misfortune at the emotional level, unites with that person at the spiritual level and seeks to alleviate the problem as if it were one's own. It is that which, in religion, "hates the sin but loves the sinner". Nevertheless, it does imply inequality, whether natural or spiritual, between the giver and the receiver, although this does not signify condescension.
Mercy can thus be expressed as forgiveness, and in Christian terms believers are expected to emulate God's mercy and forgiveness in their own lives. As opposed to justice, which behaves towards another in accordance with right, mercy may be extended to another when the other has no claim on such mercy or in disregard of one's own claims against the other.

♦ HH0845 Hinayana Buddhism (Buddhism)
Theravada — Small vehicle Buddhism — Lesser vehicle Buddhism — Southern Buddhism
Description One of three branches of Buddhism, emphasizing the transcendence of experience in extreme realism, this is the early Buddhism taught by disciples of Gautama Buddha. The Hinayana goal of becoming an *arhat* is sometimes said to be less worthy and compassionate and more selfish than the *mahayana* path aiming at full Buddhahood. However, if the concern of the arhat is seen as the welfare and benefit of this world much of this criticism may be seen not to apply. In fact, many of the various hinayana traditions are derived from mahayana. Geographically, this branch of Buddhism may be related to Burma, Cambodia, Laos, Sri Lanka and Thailand, although some devotees are also found in Bangladesh, India and Viet Nam and as emigrants in the west from any of these countries.
Refs Buddhaghosa, Bhadantacariya *The Path of Purification* (1980); Buddhaghosa, Bhadantacariya *The Path of Purity* (1975).
Broader Human development (Buddhism, #HH0650).
Narrower World cycles (#HH2002) Ascetic practices (#HH4298).
Related Arhat (Buddhism, #HH0233) Ascetic states (Buddhism, #HM5354) Mahayana Buddhism (Buddhism, #HH0900) Vajrayana Buddhism (Buddhism, #HH0309).

♦ HH0846 Multiple therapy
Description In multiple therapy two counsellors work together with one or more individual, each therapist interacting with each individual and with each other. This method emerged from the recognition that an individual may have unresolved problems in his work with a therapist of one sex and may benefit from discussion with a therapist of the opposite sex (or vice versa). The two therapists in this method are therefore usually of opposite sex.
At least four dynamically distinct relationships are possible when two therapists work with one individual. These recapitulate the individual's relationships with his parents in their simplest form. It is in the addition of interrelationships between two people through the presence of a third or fourth person that an opportunity is gained for both conflict generation and resolution, which is not readily available in dyadic therapy. The method has been used particularly in dealing with married couples and their problems.
Broader Therapy (#HH0678).

♦ HH0847 Metamotivation
Description Self-actualizing individuals, namely those who are more mature and more fully human, are by definition already suitably gratified in their basic needs. Self-actualizing people do not (for any length of time) feel anxiety-ridden, insecure, unsafe, alone, ostracized, rootless, unlovable, rejected, etc. They are therefore no longer primarily motivated by such basic needs but rather by higher motives and needs, termed metaneeds and metamotivations respectively.
Metamotivations include: delight in bringing about justice, delight in stopping cruelty, fighting lies and untruths, love of virtue being rewarded, enjoyment of happy endings and completions, dislike of sin and evil being rewarded, enjoyment of doing good, avoidance of publicity and honours, enjoyment of peace and calm, response to challenge in a job, attraction to unsolved problems, enjoyment in bringing about law and order, enjoyment in bringing about happiness, enjoyment of responsibility, enjoyment of greater efficiency and elegant work operations, and enjoyment and acceptance of the world as it is.
Refs Maslow, Abraham H *A Theory of Metamotivation* (1967).
Broader Motive (#HH0181).

♦ HH0848 Power
Description Power is the energy or ability to cause a desired effect. As far as human development is concerned this may be regarded as an expression of the desire to understand, control or transform nature, whether this refers to inner understanding, control and transformation, or to external natural and social environments. When power is seen as deriving from God, irresponsible exercise of power results in slavery to that which was intended to be commanded. Nor is it possible simply to abdicate power, this also being irresponsible and slothful. Power bring obligation. Since it is through power that everyday life and society are organized and goods are distributed, not sharing in this power means failure to join in the life of the community.
In the Christian tradition there has been a development from the tendency to accept earthly powers almost uncritically to the present attitude which is more that it is a duty to use any power one has to alter political institutions for moral purposes. This is inherent in the concept of a responsible society, when power is controlled and exercised by legitimate authority.
Related Magico-religious powers (#HH0497).

♦ HH0850 Child guidance
Description By diagnosing and treating children's problems, whether behavioural, learning or psychosomatic, child guidance seeks to prevent and/or cure serious disorders. The system is based on teamwork among psychiatrist, psychologist and social worker based in special child guidance clinics, and includes parent counselling.
Related Family therapy (#HH0362) Child guidance therapy (#HH0894).

♦ HH0851 Group therapy
Group psychotherapy — Mass therapy
Description Group therapy is a term applied to any form of psychotherapy in which several persons (usually from 6 to 12, equally balanced by sex) are treated simultaneously in the presence of one or more therapist. Dynamic interaction among the members causes a holistic and synergistic concept to arise, and the resulting group morale is beneficial to all the individuals taking part. Practitioners of most individual therapy systems have experimented with group work, often combining several techniques in working with the same group. Group psychotherapy is designed to help those who suffer from minor reactions to situational stress and those who have severe neurotic disturbances and psychoses.
It is useful to distinguish between four major types of therapy group; the therapeutic social group through which members increase skill in social participation; groups oriented towards providing emotional support to members as a means of combatting the vicious circle of loss of self-

esteem, withdrawal, and further damage to self–esteem, associated with such habits as alcoholism, drug–taking, over–eating, etc; psychodrama by which members attempt to free blocked spontaneity; and analytically oriented group therapy representative of the mainstream of group therapy.
In the case of psychodrama, the aim is the integration of the participant's self against the uncontrolled forces around him through free and spontaneous interaction. In psychoanalytic group therapies the ultimate aim is the facilitation of the fullest possible communication of unconscious material.
In family group therapy the mental health of each member is improved by bringing about a more realistic, less repressed equilibrium in family interactions, which is only possible by treating the family as a whole rather than each individual member. In group–centered therapy feelings of acceptance are stressed in a setting which offers the participant an immediate opportunity to test the effectiveness of his ability to relate to people and to improve his skills in interpersonal relations.
In pedagogical or didactic group therapy the stress is placed on the communication of mentally healthful concepts to members of the group through lectures, symposia, readings and other educational means.
Situational therapy uses the therapeutic effect of the physical environment and the social relationships, which is also true of activity group therapy.
Therapists differ widely in their goals and procedures, some being interested in cohesiveness, smooth functioning and nondirective procedures, whilst others aim for as much emotional stress among participants as the group can handle without serious disruption. In all cases, however, therapists aim to free the participants' spontaneity and capacity for emotional growth so that they may become more comfortable, effective and emotionally mature.
In general the therapeutic value of participation in a group is the opportunity it offers the individual to; examine his own fears and distrust; confront himself as he is within a supportive atmosphere; confront his own needs to control and influence; confront dissonance in his own internal world as he moves towards new levels of creative integration; and obtain glimpses of what a person can become.
Refs Coffey, H S *Socio and Psyche Group Process* (1952); Mowrer, O Hobart *New Group Therapy* (1964).
Broader Therapy (#HH0678)   Psychotherapy (#HH0003)
Activity group therapy (#HH0492)   Human potential movement (#HH0398).
Related Group consciousness (#HM3109)   Transactional analysis (#HH0487)
Self–realization therapy (#HH0784).

♦ **HH0855 Psychic development** (Psychism)
Spiritual development
Description This is the process whereby an individual develops ability in telepathy (perception of another person's on–going mental activities without the use of any sensory means of communication), in clairvoyance (ability to know directly without normal sensory means information or facts about events occurring in remote locations), in precognition (ability to know of events or happenings in the future without sensory or inferential means of knowing), and in psychokinesis (movement of matter by nonphysical means or direct mental influence over physical objects or systems). Such abilities may develop naturally or may be enhanced by various forms of training which may include prayer and meditation.
Advocates of psychic development frequently use it as being synonymous with spiritual development, although those concerned with transcendental forms of consciousness consider the phenomena associated with psychic development to be more of a hindrance to the achievement of higher states of consciousness.
Refs Bletzer, June G *The Donning International Encyclopedic Psychic Dictionary* (1987); UNESCO *The Parasciences* (1974).
Related Mystery (#HH0303)   Channelling (#HH0878)
Spirituality (Christianity, #HH0792).

♦ **HH0856 Attitude change**
Description Attitudes may be observed by inference from behaviour and are said to be based on social influences. They are very resistant to alteration, particularly when based on more than one function – for example, an ego–defence attitude may be supported by the utilitarian function of wishing to retain acceptance in a particular group. This may be because an individual's convictions often form a coherent pattern which is an expression of some deep–seated trend in his personality. Attitudes which might be termed prejudice are amenable to psychotherapy.
Refs Hovland, C J and Rosenber, M J *Attitude Organization and Change* (1960).
Related Attitude therapy (#HH0676).

♦ **HH0859 Merit**
Description In Christian terms, an individual's efforts to use the gifts received from God to serve God through good acts are deemed worthy of reward, although much controversy has arisen over the relative merits of good deeds and faith. An act thus worthy of reward is termed meritorious, and both Old and New Testaments encourage the notion of reward for good and punishment for evil behaviour. Similarly in Vedic philosophy, the Bhagavad Gita encourages right action, "...the fruit of a meritorious action is spotless and full of purity", (Chapter XIV). Hinayana Buddhism lists "ten meritorious deeds" and Buddhism has many ways of acquiring merit, including observance of moral precepts. In Hindu, Jainism and Buddhism the acquiring of merit (or demerit) affects the circumstances of one's rebirth in the next life. The Judaic tradition is that blessing and welfare in this life and for future generations arise from observing the law. Islam recognizes praiseworthy acts and also acts which are required of all who are able, such as the pilgrimage to Makkah; but it does not consider salvation can be "earned" by such acts, this being dependent only on God's grace and His knowledge of what might be expected of each individual. Some Christian writers emphasize the inherent lack of merit of man's actions as compared with God, in the extreme putting forward the doctrine of total depravity; others agree that Christ alone has merit but that, through Him, man may also achieve merit.
Narrower Formation of merit (Buddhism, #HH5122)
Formation of demerit (Buddhism, #HH3226)
Ten meritorious deeds (Buddhism, #HH0446)
Formation of the imperturbable (Buddhism, #HH5543).
Related Karma (#HH0567).

♦ **HH0861 Humility** (Christianity)
Description It is said that, claiming none of his good qualities as his own, the humble person attributes all good to God. Otherwise mortifications, penances and so on would simply be a source of self esteem and lead to fruitless comparison with others. Humility is a characteristic mark of charity, a necessary condition for love, while at the same time love makes possible true humility. True humility does not entail a false or pretended low opinion of one's self, nor self–loathing. What is felt is lowliness as unimportance with respect to God, a reverence for glory, and a self–emptying which is the opposite of pride and self–centredness.
St John of the Cross says that, while all who set out on the spiritual path will make mistakes, although some have more imperfections than others, those who are humble make progress and have their spiritual life built up. They are not self–satisfied with their own behaviour, while they respect the progress of others. They are not over concerned with what others are doing, certainly not judging them in comparison with themselves. They are only sad when they do not serve God as well as others do. They are always ready to assist others to serve God. Whatever their good deeds, they know that God deserves all and that there is actually very little they can do for Him. Simply desiring to please God, they have the wisdom of God's Spirit living within them. They have the grace to keep God's treasures in secret while expelling from themselves all that is evil. Even if they do fall into sin they set their hope in God and behave towards Him with humility, meekness and reverence.
Although humility is a quality that cannot be gained by seeking, it can arise from a spiritual shift occurring through service to accentuate the good of others. It is the best defence against despair, since all selfishness disappears and therefore all self–pity which is the basis of despair. It also destroys self–centredness, thus giving the possibility of joy. The humble receive praise with no embarrassment, since they know that the good in themselves comes from God. They are no longer concerned with themselves, but with God and with reality as it is and not how they would like it to be. This brings freedom from fear – fear of failure, fear of themselves. They are confident in God's power against which there can be no obstacle and compared with which no other power has meaning.
Humility involves accepting one's self for the person one is and not attempting to be someone else. This would mean saying that one knows better than God. One just has to be the person God intended one to be. This honesty of approach may lead to the charge of pride, and the anguish is to keep balanced, not getting tough about continuing to be one's self and not asserting the false self against the false self of others. Defending the false self brings loss of peace of heart. Taking one's self too seriously makes one the prisoner of one's own vanity, one sees others' sins and faults to bolster up one's own ideas of one's self. Humility brings freedom from all this, from attachment to one's own works. There is perfect joy in self–forgetting. Paying no more attention to one's own activities, reputation and excellence brings freedom to serve God in perfection for His own sake.
Refs Merton, Thomas *Seeds of Contemplation* (1972).
Related Service (#HH0232)   Integrity (#HH0234)   Magnanimity (#HH0834)
Pride (Christianity, #HM2823).

♦ **HH0862 Hatha yoga** (Yoga)
Ashtanga yoga
Description This form of yoga is concerned with mastery of the body by rigorous physiological purification as a prior condition for the achievement of any adequate purification of the psyche. Changes in consciousness arise through the setting in motion of kundalini and pranic forces in the body. The result of hatha yoga practice is great physical health and longevity, providing the physical fitness necessary to support the strain on the nervous system of the spiritual life.
The methods include some unusual and difficult physical postures and breathing exercises and for higher spiritual activities such as concentration and meditation. The postures (asanas) are of many kinds and few are immediately attainable. Months or even years are required for competence in a series of 15 to 20 postures, which should be undertaken daily or else progress is likely to be very slow. An asana is said to be mastered when it can be maintained with out strain for three hours. The postures are normally complemented and completed by several kinds of disposition and movement of hands and other bodily parts (such as the head, abdomen and anus), together with breathing exercises. When a certain degree of competence has been achieved, the breathing rhythm controls and paces an exercise sequence.
Recent research has shown that the special postures produce a rhythm in the neuro–muscular mechanism, thus improving muscles to which emotions and perceptions are related. Other favourable effects are the toning of veins and arteries and possible improvement in the function of ductless glands. Breathing exercises are also recognized as being able to decrease the domination of egotistic thought, and to give emotional stability.
Refs Iyangar, S *The Hathayogapradipika* (1972); Satchidananda, Swami *Integral Yoga Hatha* (1970).
Broader Yoga (Yoga, #HH0661).
Narrower Yogic feats (#HH4398)   Pranayama (Hinduism, Yoga, #HH0213).
Related Mudras (#HH0286)   Asana (Hinduism, Yoga, #HH0669)
Mystical stage of life (#HM3191)   Concentration on the flow of breath (#HH0687).

♦ **HH0865 Thought reform** (Brainwashing)
Brainwashing — Ideological reform — Menticide — Sensory deprivation
Description A system involving the subjection of an individual to conditions of physical and even psychic duress in order to persuade him to alter his viewpoint or convictions, or otherwise to convert him to a desired viewpoint or action. Thought reform consists of two basic elements: confession, or the exposure and renuciation of past and present "evil"; and re–education, namely the remaking of the individual in the desired new image. These elements are closely related and overlapping, since they both bring into play a series of pressures and appeals (intellectual, emotional, and physical) aimed at social control and individual change. The death and rebirth of the personality arouses a sense of participation as the deprivations of the initial part of the process are replaced by the joy of encouragement and acceptance into a "reformed" group.
Extreme forms of brainwashing involve total breakdown of the victim's will and power to resist, so that personal autonomy and identity are surrendered, followed by a period of mental and emotional blankness when a totally different personality may be grafted on through imprinting of behaviour patterns. Such a process is referred to as *depersonalization*. However, some forms of psychotherapy or even group pressure in everyday life may be considered less extreme forms of the same phenomenon. Logic and reason are bypassed, suggestions being accepted by the subconscious either following deliberate confusion or by constant repetition (advertising or political slogans, for example).
Refs Lifton, Robert J *Thought Reform and the Psychology of Totalism* (1961).
Narrower Self–betrayal (#HM2229)   Final confession (#HM3156)
Transitional limbo (#HM2934)   Desperate gratitude (#HM3017)
Ideological rebirth (#HM2843)   Recognition of guilt (#HM2923)
Channelling of guilt (#HM3204)   Logical dishonouring (#HM2564)
Compulsion to confess (#HM2745)   Deprivation of selfhood (#HM3002)
Breaking point of basic fear (#HM2799)   Breaking point of total conflict (#HM2354)
Sense of harmony through progress (#HM2149).
Related Tenko (#HH0757)   Depersonalization (#HM1248)   Sensory deprivation (#HH1478)
Ideological re–education (#HH0996)   Destructive manipulation (#HH1020).

♦ **HH0866 Electroshock therapy**
Electric convulsive therapy (ECT) — Electroconvulsive therapy — Sleep–electroshock therapy — Regressive electroshock therapy
Description Electrodes are attached to the scalp of the person and an electrical current is passed through to produce convulsions and unconsciousness. This form of therapy is held to be the most reliable and simple treatment with the least likelihood of unforeseen serious complications, particularly in the case of acute catatonia and severe depression. If the patient has developed a fear of such treatment, particularly in the case of brief stimulus therapy when lower electrical energy is required to produce the shock, it is carried out on the point of waking from drug–induced sleep. Regressive electroshock therapy, where the patient is subjected to several shocks a day for several days, produces a state of non–contact with surroundings, incontinence, inability to feed

one's self, and generally slow and uncertain behaviour.
Among the ethical problems raised by this treatment is the desirability of relieving depression as a symptom and then returning the person involved to the same environment in which the depression was generated.
Broader Therapy (#HH0678) Shock therapy (#HH0975).

♦ HH0868 Spiritual self-abandonment
Non-attachment
Description Self-abandonment is the surrender of the self to the divine will, said to be the total expression of love for God. In Christian terms, this self-abandonment is similar to the Vedic non-attachment to the fruits of actions. Renouncing preoccupation with "me" and "mine" is the liberating influence which eventually brings joy - since only by regarding nothing as one's own can one realize that everything in fact is ours for the enjoying (as it is everyone else's). As the Isha Upanishad puts it, "Claim nothing; enjoy, do not covet, his property.
Related Grace (#HH0169) Self-oblation (#HH0968)
Abandonment to God (Christianity, #HH1223).

♦ HH0871 Metaneeds
B-values
Description Self-actualizing people are said not to be motivated by basic needs, which in their case are already satisfied, but to be *metamotivated* by metaneeds or *B-values.* These metaneeds, being common to the individual and what is outside the individual, contribute to the breaking down of the distinction between the self and the not-self, and therefore to self-transcendence.
The intrinsic values to which a self-actualized person responds are said to be necessary not only to achieve fullest humanness but also to avoid illness, or "diminution of humanness". Examples are truth, goodness, justice, order. Their absence leads to such symptoms as alienation, apathy, despair, insecurity, and so on.

♦ HH0872 Duality
Description Creation and destruction may both be considered as the juxtaposition of opposing forces and this duality exists throughout nature. The whole universe is conceivable only in terms of polarities, these giving the coordinates of space and time (left and right, vertical and horizontal, past and future). This duality is reflected in the brain, the two hemispheres performing different functions and controlling opposite halves of the body. Duality implies conflict, in the case of the brain of the logical versus the intuitive. IQ has even been defined as "the ratio of left to right brain hemispheric capacity" (Fischer and Rhead). Duality is seen when separate researchers may produce opposing hypotheses each of which is apparently valid in experience.
The concept of the *archetype* also presupposes duality, in that it is, for example, conscious and unconscious, psychic and non-psychic. This "instinct's perception of itself" may be seen not in terms of opposites but of complementarity, and is the means of actualizing the archetype at both the objective (outer) level and in inner experience. Such duality precludes the necessity for considering mind and body as separate entities, and approaches such as *mentalism* (when mind and body are considered separate and equal), *vitalism* (when the mind is said to dominate the body), *epiphenomenalism* (the mind being dominated by the body) and *interactionism* (concerned with their mutual influence), are superseded.
Even the ultimate in non-duality, the awareness that this Atman (the essence of self or reality within) is that Brahman (the essence of reality without) can only be expressed in terms of complementary duality.
Refs Bakan, David *Duality of Human Existence* (1926).
Related Archetype (#HH0019).

♦ HH0873 Ways of knowing (Buddhism)
Lo-rig (Tibetan) — Tibetan studies
Description There is a uniform system of education used in the Buddhist monasteries of Tibet through which novices, who start learning at the age of eight years, receive their Geshe degree when they are at least 25 years old. The system aims to prepare the novices to think for themselves and to develop to their full potentials. Each monastery uses different books and provides different explanations as a spur to further debate and developing the novices to become great teachers, progressing on the path to enlightenment and helping others on the same path.
The first class is collected definitions (Dura) when, having learned definitions, students pair off and debate among themselves the implications, defence and possible refutations. Having learned the fundamentals of debating they enter the second class at the age of nine or ten; this is on ways of knowing (Lo-rig). The third class is on ways of reasoning (Tag-rig), and on completion of this class they approach the five primary subjects for the degree, having by this time memorized all the major texts. These are: (1) The perfection of discriminating awareness, *prajnaparamita.* (2) The middle way, *madhyamaka.* (3) Validity, *pramana.* (4) Laws, *vinaya.* (5) General knowledge, *abhidharma.*
The ways of knowing as studied in the second class and are memorized in such a way as to make clear the most important definitions and divisions for use in debate. They are divided into valid and invalid ways of knowing.
Refs A-kya Yong-dzin Yang-chan ga-wai lo-dro *A Compendium of Ways of Knowing* (1980).
Narrower Presumption (Buddhism, Tibetan, #HM4032)
Bare perception (Buddhism, Tibetan, #HM4651)
Distorted cognition (Buddhism, Tibetan, #HM2668)
Indecisive wavering (Buddhism, Tibetan, #HM3074)
Subsequent cognition (Buddhism, Tibetan, #HM3508)
Conceptual cognition (Buddhism, Tibetan, #HM0713)
Induced attentiveness (Buddhism, Tibetan, #HM3378)
Inattentive perception (Buddhism, Tibetan, #HM3348)
Seemingly bare perception (Buddhism, Tibetan, #HM3265).
Related Secondary mental factors (Buddhism, Tibetan, #HH1348).

♦ HH0874 Rehabilitation
Social therapy — Social treatment
Description Individuals suffering from mental or physical handicaps are doubly penalized in that, in addition to suffering the handicap itself, they are prevented from fully taking part in normal human activities and occupations. Rehabilitation programmes and agencies provide money and resources to train handicapped people to care for themselves and their families and to qualify for salaried employment. Such programmes may be for people who have always been handicapped or for those who have to be retrained to cope with everyday life after having become disabled (due to accident, ill health, drug addiction and so on). In this way their social dependency is reduced.
Social treatment or therapy covers rehabilitation therapy and any other treatment involving the social environment of the individual concerned.
Broader Therapy (#HH0678).

♦ HH0875 Identity
Ego identity — Personal identity
Description Ego identity denotes certain comprehensive gains which the individual, at the end of adolescence, must have derived from all of his pre-adult experience in order to be ready for the tasks of adulthood. Identity, in outbalancing at the conclusion of childhood the potentially malignant dominance of the infantile superego, permits the individual to forego excessive self-repudiation and the diffused repudiation of otherness. Such freedom provides a necessary condition for the ego's power to integrate matured sexuality, ripened capacities, and adult commitments.
The term identity points to an individual's link with the unique values, fostered by a unique history, of his people. Yet it also relates to the cornerstone of the individual's unique development. It is the identity of a feature of the individual's core with an essential aspect of a group's inner coherence. The term expresses a mutual relation in that it connotes both a persistent sameness within the individual and a persistent sharing of some kind of essential character with others. In different connotations it may refer to: a conscious sense of individual identity, an unconscious striving for a continuity of personal character, a process of ego synthesis, and a maintenance of inner solidarity with a group's ideals and identity.
Refs Erikson, E *Identity and the Life Cycle* (1959).
Related Feeling of I (#HH0840) Depersonalization (#HM1248)
Individual development (#HH1543)
Anomalies in awareness of one's self (#HM8186).

♦ HH0877 Self-discipline
Tapasya (Yoga)
Description One of the essential steps in *jnana yoga*, the way of knowledge, self discipline is said to consist of six elements: sama - calmness (overcoming of the mind's restlessness); dama - restraint (restraining of the senses, or organs of the mind); uparati - renunciation (freedom from emotional ties); titiksa - forbearance (endurance of inner conflict, indifference to praise and blame); sraddha - faith (to withstand times of discouragement and lack of vision); samadhana - self-settledness (single-minded devotion to the completion of the spiritual task).
Broader Discipline (#HH0163) Jnana yoga (Yoga, #HH0927).
Related Tapas (Hinduism, #HH1248) Niyama (Hinduism, #HM3280)
Self-discipline (Christianity, #HH2121).

♦ HH0878 Channelling
Spirit medium — Spiritual guidance
Description An individual may act as a "channel" for messages, teaching and guidance from one or more individuals or entities in another plane of being. Frequently these are, or claim to be, individuals who have previously been incarnate this world and who wish to help the world in its current state of development and promote individual well-being. They are said to exist, and to be aware of existing, as a spark of consciousness at many levels, that creates thought and reality. Unlike possession, when the possessed person is taken over by the personality of the individual who is communicating, the channeller simply acts as a means of communication in the manner of a receiver and transmitter, although this may occur in a trance-like state. Communication may be direct with an intelligence in the etheric world or through means of a spirit guide. To be a medium requires a special nervous system and bodily chemistry, which may be developed, combined with a willingness to act in this way. The more prepared the medium is, through meditation and regular practice, and the more worthy his or her morals and lifestyle, the higher the intelligence attracted and the more accurate the information forthcoming.
Broader Anomalies in integrity of consciousness (#HM4521).
Related Psychism (#HH1775) Mediation (Magic, #HH4521) Spirit possession (#HH0056)
Spiritual direction (#HH0442) Semi-trance (Psychism, #HM0887)
Trance channelling (Psychism, #HM3456) Psychic development (Psychism, #HH0855)
Psychic inspiration (Psychism, #HM1094) Self-induced trance (Psychism, #HM1964)
Communication with supernatural beings (#HH1306).

♦ HH0881 Sexual socialization
Description The adaptation of the individual to society and its norms of behaviour is rendered more difficult in this case by the sudden jump between the prohibitions, taboos and denials related to sexual activities at puberty and the expectations of behaving with fully skilled sexual behaviour immediately on marriage.

♦ HH0882 Zazen meditation (Zen)
Zen meditation
Description Zazen meditation involves a variety of training techniques designed to guide the individual to experience a turning point termed satori (separately described) which is a major shift in the method of experiencing oneself in relation to the world, and which is an important step on the way to enlightenment. Satori is considered to be a condition which is always present and is not the product of some particular technique, a sudden flash of true wisdom or intuition when discursive thought is left behind. The main aim of zazen is to help the individual to let go of mind and body; not to rest the mind but rather to establish a base, a foundation, or a centre for it.
This is a strenuous task of consciousness, whatever technique is adopted. The precise techniques used vary with the school and mentor. Generally the individual sets aside a portion of each day for sitting motionless and engaging in some concentration exercise. The object of concentration varies considerably and may be changed as the individual progresses. The aim is to suspend the ordinary flow of thoughts without falling into a stupor. The achievement of undistracted concentration is the first means of coming to grips with the purely conceptual mode of experience.
Importance may be attached to the precise posture adopted; fore example, the eyes are kept partly open to avoid sleep or stupor.) Concentration may be specially focused on breathing, particularly as a means of developing concentration. Since the bodily state is indicative of the condition of the whole person, the correct harmonization of body and breath lead to the right state of mind.
Various techniques are used for quieting the mind. The individuals may concentrate on a koan, a statement which is impossible to comprehend rationally but meaningful to a person who has experienced satori. The individual therefore attempts to penetrate to its meaning and to the state of mind which it expresses. This technique seems to deepen the intellectual crisis preceding satori and to produce a deeper and more vivid experience. The attempt to solve the koan becomes almost a surrogate for the individual's struggle to solve his own life.
Once the ability to concentrate has been developed a condition of relaxation and self-immersion is achieved. This is accompanied by the emergence of internal distractions which may arouse anxiety. This is followed by a stage in which nothing definite is thought, planned, striven for, desired or expected. The condition is focused in no particular direction and is accompanied by a sense of calm stillness, energy, vitality, invulnerability and potency. This state of mind is traditionally described by analogy to a mirror, which reflects many things, yet is itself unchanged by them. In this condition receptivity to previously excluded experience is increased, together with the ability to deal with it in a detached, non-anxious manner.
Various dangers and obstacles are associated with the practice of zazen. These include: a condition of depression or a kind of melancholy accompanied by sleepiness, possibly deepening into a non-conscious state; a condition of attention to the many thoughts and ideas running through the mind; indulgence in a sense of elation, ecstasy or quietism; unconscious projection of emerging material and loss of awareness of its subjective origin; or concern with paranormal psychic functions. Only after all these have been mastered does satori follow as joy emanating

from a mind which has transcended all relative joys and sorrows.
Refs Fromm, Erich; Suzuki, D T and de Martino, R *Zen Buddhism and Psychoanalysis* (1963); Lesh, T *Zen Meditation and the Development of Empathy in Counselors* (1970).
Broader Human development (Zen, #HH1003).
Narrower Saijo Zen (#HH1009) Inward zazen meditation (#HH1134) Outward zazen meditation (#HH0621) Oxherding pictures in Zen Buddhism (#HM2690).
Related Jiriki (Zen, #HM3215) Satori (Zen, #HM2326) Ji hokkai (Zen, #HM3168) Ri hokkai (Zen, #HM3352) Hara (Japanese, #HM2662) Ji-ji-mu-ge (Zen, #HM2034) Ji-ri-mu-je (Zen, #HM3164) Transrational insight (Zen, #HM2084).

♦ HH0883 Will therapy
Description Active encouragement to assert one's self, so as to strengthen and develop the will, is used in cases where neurosis due to birth trauma has caused conflicting desires to return to the womb (security) and to re-enact separation and become independent.
Broader Therapy (#HH0678).

♦ HH0885 Experiential therapy
Description In this form of psychotherapy the therapist and the individual enter into a special kind of relationship of which the most therapeutically significant is recognized as a form of joint fantasy experience in which each is responding maximally to the unconscious of the other. The individual deepens his symbolic involvement with the therapist, whilst the therapist sees the patient (emotionally) as the child-self of the therapist. Because the needs of the therapist are intra personal, this leads him to achieve a better integration of his own self through the individual as his child-image of himself. The therapist then feels an urgency and significance in the acceleration of the growth of the individual, for whom greater continuity is established through the stabilized joint fantasy, and an increased capacity for the realization of biological needs. The psychotherapeutic focus is concentrated on growth and maturity in the sense of integration of the biological effects of previous experiences on the organization of current experience, perceived as essentially emotional rather than analytical and logically causal (namely id processes rather than ego-level processes).
Refs Malone, T P and Whitaker, C A *The Roots of Psychotherapy* (1953).
Broader Therapy (#HH0678) Human potential movement (#HH0398).

♦ HH0886 Intersubjectivity
Description Subjective experience becomes intersubjectivity when several individuals experience and report the same thing. Such an association of subjectivities is necessary, according to one view, for an objective world to exist.
Refs Mensch, James R *Intersubjectivity and Transcendental Idealism* (1988).

♦ HH0887 Ageing
Revitalization
Description The individual inevitably degenerates physically, and in behaviour and attitude, with age. As the body structure changes, so the senses become less acute, the brain cells atrophy and one becomes more susceptible to mental illness. It has been shown that these processes can be slowed down considerably by realizing that the most important factor is the mind, as this is the way of understanding how the brain affects the functions of the body; and by then deciding which factors in the ageing process one would most like to change. Diet (particularly the eating of fresh, "living" foods) is also held to be of prime importance.

♦ HH0888 Metanoia (Christianity)
Description In Catholic teaching, this converting from sin to God is primarily involved with the repudiation of sin first through sorrow, then confession, amendment and satisfaction. It is a change of mind based not only on changed attitude but also on external demonstration of this change through confession, fasting, etc. Lutheran teaching disagrees that mankind through its works can satisfy for sin, as this would be blasphemous rejection of Christ's work; protestant thought therefore considers repentance or change of heart to be the keystone, and that although one can lead a new life this does not or cannot make satisfaction for the past.
Refs Pathrapankal, J *Metanoia, Faith, Covenant* (1971).
Related Shalom (#HM1606) Conversion (#HH0077) Repentance (#HH0441) Contrition (#HH4112) Religious therapy (#HH0762).

♦ HH0889 Swadeshi
Description The principle of concern for the immediate neighbourhood. Such concern leads to the ability to start where one is, whether geographically, spiritually or emotionally. Development cannot occur on a global level without each individual manifesting his basic value commitments at his actual location.

♦ HH0892 Spiritual world order
Unified world order
Description The predicted fulfilment of cosmic evolution, a world order subject to spiritual values, is the goal of religion and of spiritual paths. The Judaeo-Christian tradition visualizes man's liberation through the supernatural intervention of God, whereas Hindu-Buddhism sees it in terms of transcendence. In contrast, present-day political effort relies on socio-political reforms to bring a *new world order* of peace and progress based on materialistic goals. Integral yoga attempts to unite these approaches, de-emphasizing the distinction between nature and spirit and looking for spiritual regeneration, a true change of heart, but rooted in nature.

♦ HH0893 Idealism
Ethical idealism — Immaterialism — Absolute idealism
Description As opposed to materialism (which considers matter to be eternal and mind simply a highly ordered form of matter), and naturalism (which regards the human mind as a by-product of the operation of natural laws), idealism considers matter to be a creation or an illusion; and mind, spirit or God to be ultimate reality. For example, Berkeley would say that material objects exist only in that they are experienced as collections of ideas; the consistency with which they are experienced is because all exist as ideas in the mind of God (empiricism). Leibnitz came to the same conclusion of *immaterialism* but through argument, denying the validity of experience (rationalism). The *absolute idealism* of Hegel is not immaterialist but states that all philosophy is idealist; that the finite presupposes the infinite within which the finite is dependent; and that human knowledge progresses away from sense experience towards philosophical completeness.
Related Naturalism (#HH0702) Materialism (#HH0529).

♦ HH0894 Child guidance therapy
Description Potential and actual problems of the child, whether emotional, sociological or psychological, are treated by positively influencing the child's family and social environment and treating the neurotic anxieties of the parents, particularly the mother.
Broader Therapy (#HH0678).
Related Child guidance (#HH0850).

♦ HH0895 Character
Character development — Character education — Character building
Description Character is the indivisible, individual distinctiveness of a person or self, which is exhibited in certain modes of individual experience and experiencing. It is specific to an individual and distinguishes him from other individuals. The modes of experience are organized as wholes and subject to change, but they persist in essence. In this sense, character is the form of a person, and a stage in the formation or development of his personality.
Thus character may be considered as comprising the *fundamental* character, which remains more-or-less constant; and *empirical* character which is dependent on age, education, environment, and so on, and which may be said to be the creation and means of self expression of the fundamental character. Other opinions consider that character is developed entirely through education, in other words, simply a product of the society in which a person lives; or that it is a "predisposition" of the emotions, instincts and impulses, lying between natural dispositions and ethical values.
The development or education of character may be looked on as assisting an individual to learn moral and spiritual values, *character building*, or as a political act of inculcating those characteristics demanded by a particular society.
Refs Kohlberg, Lawrence *The Development of Moral Character and Ideology* (1964); Peck, R and Havighurst, R *The Psychology of Character Development* (1960).
Narrower Social character (#HH0251) Strength of character (#HH0797).

♦ HH0896 Synthesis
Creative synthesis — Distributive synthesis — Syllabic synthesis
Description The formation of an integrated whole from its component parts, synthesis refers to the integration of all the factors of which the personality comprises, to the intactness of the personality and to the ability of the psyche to keep its component parts from dissociating. It thus denotes a condition of health as opposed to neurosis. Synthesis may also refer to the ability to comprehend a situation as a whole. It covers the range of simplification, generalization and harmonization and, ultimately, of understanding.
Distributive synthesis involves the consideration of the factors necessary to achieve security, reviewed after each session of psychotherapy.
Creative synthesis is a principle by which the psychic structure represents more than the sum of its elements, conceived to include not only feelings, sensations and objective circumstances but also subjective values and goals.
Syllabic synthesis refers to the condensing of two or more words into one new word, such as occurs in dreams and schizophrenic fantasies.

♦ HH0897 Vision quest
Description A custom of the North American Indian religion, a vision quest or search for a vision of a supernatural being is undertaken in order to acquire supernatural power and usually as a rite of passage. After various ascetic practices, the individual goes to a remote retreat where, after prayer and further practices of asceticism, the vision of a guardian spirit may be granted. The guardian spirit, in the form of an animal, teaches various rituals and indicates whether the initiate might become a medicine man.
Broader Ecstatic discipline (#HH5282) Stages of spiritual life (#HH0102).
Related Rites (#HH0423) Shaman (#HH0973) Initiation (#HH0230) Resurrection (#HH4553).

♦ HH0899 Forgiveness (Christianity)
Description In religious terms, offending against the holiness of God by human offence leading to guilt and enmity between God and man. This enmity can only be set aside by God's forgiveness of the offence. The need for assurance that sin has been pardoned and right relations restored has led to rituals of actual or symbolic cleansing both at the individual and the collective level. The setting right of the relationship is what is termed *justification*, while the result of this fact, the actually restored relationship, is termed *reconciliation*. The two may be considered to form part of forgiveness, which is the divine and human practice both of the setting right and the resultant restored relationship. In Christian terms, the death and resurrection of Jesus Christ is the atonement (at-one-ment) between God and humanity. Humanity no longer needs to get right with God. The freedom this brings makes it possible to trust God, to trust neighbours, to trust enemies.
The relationship between forgiveness and justice and love is at the very centre of Christian ethics. It provides the context for human freedom and human responsibility in renewal and fulfilment. God's presence and grace are a sign of His love; His pardoning or doing away with of human disbelief and violation of this loving initiative is forgiveness; His presence underlying human aspirations and struggle to be human, experienced as the setting right the wrongs in human interactions, is justice. This practice of love through the means of forgiveness and justice is expressed in the individual through his behaviour towards God and towards his fellows, in the intention that God's will should be done and in the knowledge that it is done through God's power.
Narrower Justification (#HH0341) Reconciliation (#HH0164).
Related Love (#HH0258) Justice (#HH2117) Purification (#HH0401) Journeying within transcendence - from bondage to freedom (Christianity, #HH2661).

♦ HH0900 Mahayana Buddhism (Buddhism)
Great vehicle Buddhism
Description One of three branches of Buddhism, marked by idealism and devotional practices, this tradition may be broadly related with the geographical areas of China, Japan, Korea and Vietnam. It has become closely associated with various indigenous religions and many see no difficulty in adhering simultaneously to Buddhist and Confucian, Shinto or Taoist practices. Immediate personal enlightenment is seen as a lesser goal, the importance being altruism and compassion with the ultimate aim for all of becoming a *bodhisattva*, seen as superior to that of the Hinayana tradition of arhatship. Understanding of the world needs to be transformed until nothing created is seen to have real independent existence, or any permanence. Balance and harmony are achieved - devotional and visionary experience warm the otherwise cold, dry, emptiness experience, which in turn protects from the trap of marvellous visions.
Broader Human development (Buddhism, #HH0650).
Narrower Great-vehicle lower-path awareness (Buddhism, Tibetan, #HM2268) Great-vehicle higher-path awareness (Buddhism, Tibetan, #HM3048) Great-vehicle middle-path awareness (Buddhism, Tibetan, #HM2160).
Related Bodhisattva (Buddhism, #HH1380) Human development (Zen, #HH1003) Hinayana Buddhism (Buddhism, #HH0845) Vajrayana Buddhism (Buddhism, #HH0309).

♦ HH0902 Capability to become
Description In human potential terms, an individual showing capability to realize a particular outcome will be more likely to achieve success than another individual with the same environmental circumstances but lesser capability. The outcome may thus be predicted reasonably accurately if the capability is known, and is in part dependent on the choice of the agent carrying out the task (one could if one so chose, rather than one could if circumstances were different). Desirable results may therefore be achieved by motivating the person (s) with capability to carry out the

necessary actions. This is one of three concepts suggested by Israel Scheffler to be useful in defining *human potential*.
Followed by Propensity to become (#HH0171).
Preceded by Capacity to become (#HH0405).

♦ **HH0904 Agnosticism**
Description This term describes the belief that God (or indeed anything beyond experiential phenomena) is unknowable to the human mind. It does not necessarily preclude belief in God – many agnostics have professed such belief, although others prefer to make no judgement – but holds that the reason cannot under any circumstances prove such existence. Emphasis is therefore on the experience of phenomena, objective characterization of reality being impossible.
Related Humanism (#HH0674) Scepticism (#HH0393) Materialism (#HH0529) Philosophy of enlightenment (#HH0591).

♦ **HH0906 Self-analysis**
Description A form of psychoanalysis in which an attempt is made to free the individual from the ideal image he has of himself in order to enable him to become aware of his real self in the present, rather than hope to compete within himself with his imaginary ideal. The neurotic alienation of the individual from his true self, and the associated anxiety, is held to be fostered by the conflicting social values in modern society.

♦ **HH0907 Emancipation of the self**
Salvation — Redemption — Liberation of the self
Description The ultimate aim of monasticism and related religious endeavours is the attainment by the individual of a state free from bondage as understood within the belief system in question. The meanings attached to such terms as salvation, liberation and emancipation vary from culture to culture and are closely related to the society's concept of the individual, the mind, the soul, the spirit, and his place within the universe.
Various concepts may be grouped under the idea of emancipation, all referring to realization of the self. Examples are: *moksha*, liberation through love; *kaivalya*, liberation through yogic practice; *nirvana*, the result of practising the Buddhist eightfold way. It appears that none of these is identical to another although all imply surrender of the individual self to realize awareness of the eternal one Self. Some schools of thought see the final act of emancipation as irreversible and coinciding with death of the body and mind – *videha mukti*; this is the basis of mythic yoga. Liberation in this life – *jivan mukti* – is, however, deemed possible by some schools as it is in Buddhism – *nirvana* as opposed to *pari–nirvana*.
Broader Emancipation (#HH0953).
Narrower Mukti (#HH0613) Kaivalya (Yoga, #HM3869) Moksha (Hinduism, #HM2196) Nirvana (Buddhism, #HM2330).

♦ **HH0908 Upbringing**
Description As Wordsworth said, "The child is father to the man" and, although upbringing may continue in adulthood, it is true to say that the upbringing one receives as a child will shape one's life. This is a combined influence received from family, peers, school, community and society (the media and day–to–day contact). In this way social and spiritual values are preserved and developed and similarities and differences made permanent. Basic to all societies are rituals and games (sometimes differing for boys and girls) and the passing–on of experience and behaviour patterns from the older to the younger generation. In addition, work skills, religion, class distinction and formal education all have a part to play. Present–day state education systems, and national and global media, have increased the significance of the whole of society and commensurately diminished the influence of individuals on a person's upbringing, with a resultant tendency to stereotyping.
Related Education (#HH0945) Education for early achievement of excellence (#HH6018).

♦ **HH0910 Changeable mental factors** (Buddhism)
Aniyata — Gzhan–'gyur (Tibetan)
Description One of the six classes of mental factors which make up the knowing (as opposed to the observing) part of consciousness in Tibetan Buddhism. The individual factors are: sleep; contrition; investigation; and analysis. These may be virtuous, non-virtuous or neutral depending on circumstances.
Narrower Sleep (#HM7921) Analysis (#HM5089) Contrition (#HM7209) Investigation (#HM1177).
Related Sense consciousness (Buddhism, #HM2664).

♦ **HH0911 Pleroma**
Description Beyond the constraints of time and place, where tension between opposites is extinguished or resolved, is pleroma, the implicate or enfolded order of reality which lies within, behind and underneath reality as ordinarily perceived. It is not a state as such, but apperception of the pleroma is an understanding of the condition of wholeness and also of a number of mystical states. It is not the same as individuation or wholeness, since it is a "given" not an achievement.

♦ **HH0912 Ninjo** (Japanese)
Natural feelings
Description A Japanese expression covering those feelings which arise naturally towards others – love, pity, etc. It is compared and contrasted with *giri*, or social obligations, which there is some pressure to put first if the two are in conflict.
Related Giri (Japanese, #HH0300).

♦ **HH0913 Needs hierarchy**
Description According to Maslow there are five *stages* in human motivation: (1) *Physiological* needs which cover basic biological drives (sex, food and drink, sleep and avoidance of pain). This is the most primitive condition. (2) *Safety* needs, which predominate when physiological needs are met and the desire is for security, change is looked on as a threat, and thinking is in black–and–white terms. Authoritarianism is typical of this level. (3) *Love–belongingness* emphasizes membership of a group, warm human relations and the goodness of conviviality. (4) Need for *esteem*, when a person has a feeling of self–worth and values being looked up to for his or her ability more than simply belonging to a group. (5) *Self–actualization*, when a person's reference group has narrowed from people in general, then colleagues, to finally the person himself. Personal development and self–exploration become of the greatest importance.
Although no one person is totally at one stage, the lower stages have to be satisfied before the higher stages appear, and each person has one stage which predominates. Sudden loss at a lower stage will cause temporary reversion to that stage while the need is attended to.
Refs Roberts, T B *Maslow's Human Motivation Needs Hierarchy* (1973).
Related Needs (#HH0043) Concretization (#HH4302).

♦ **HH0916 Transpersonal psychology**
Description Transpersonal psychology is an approach to the study of man distinct from the three major approaches – behaviourism, psychoanalysis and humanistic psychology – which are viewed as limited in their ability to comprehend human capacities and potentialities. Transpersonal psychology offers a more inclusive vision of human potential, suggesting both a new image of man and a new world view. An underlying assumption of transpersonal psychology is that physical, emotional, intellectual, and spiritual growth are interrelated; and that optimal educational environment stimulates and nurtures the intuitive as well as the rational, the imaginative as well as the practical, and the creative as well as the receptive functions of each individual. It focuses attention on the human capacity for self–transcendence as well as self–realization, and is concerned with the optimum development of consciousness.
No specific course of action is recommended, there being many ways towards self–fulfillment; but the concept of a centre or essence, the core of a person's being (whatever term may be used to refer to this), is fundamental. Such a centre, the "Self", is seen as distinct from, and deeper than, the personality.
Most topics being investigated by transpersonal psychologists include aspects of at least one of the following: altered states of consciousness (including meditation, dreams, etc); self–realization and self–transcendence; impulses toward higher states (such as peak experiences); spiritual growth; parapsychology and psychic phenomena; voluntary control of internal states (biofeedback); and the sacralization of everyday life. It further assumes that these are biologically rooted and positive experiences, namely healthy and good. All of them are seen to be related to the ultimate development of man, both as an individual and as a species, and not simply as a return from unhealthiness to normality.
Refs Tart, C *Transpersonal Psychologies* (1983).
Related Inter–personality (#HH0769).

♦ **HH0917 Creativity group workshop**
Description This is a form of intensive group interaction in which the emphasis is placed on creative expression through various art media. The aim is the development of individual spontaneity and freedom of expression.
Broader Human potential movement (#HH0398).

♦ **HH0920 Will**
Good will — Free will
Description *Deterministic* philosophy regards will as being directed from outside (whether from physical, psychological, sociological or supernatural/divine causes) whereas *indeterminism* looks on will as autonomous and self–realizing – voluntarism considering it the primary basis of all human activity. Whether or not choice is *free* (and most religious writers assume free will a basic truth, and necessary for salvation to have any meaning) philosophic teachings are clear that good will – the aspiring to do great things – is of prime importance, the acts proceeding from that good will being of lesser importance. For William Law, heaven is when man's will works with God, and hell when it works against. According to Kant, the only thing which can be called "good" without qualification is good will, the so–called good qualities presupposing good will and no longer being unequivocally good in the absence of such will.
No path to liberation or realization can be followed without a strong will which, together with endurance, keeps the aspirant on the "journey".
Many philosophies consider ignorance to be basically the will not to know. Disturbance of the will – usually a lack of will, or changeability or stubborn fixity of will – is a symptom of schizophrenia.
Refs May, G *Will and Spirit* (1982); May, Rollo *Love and Will* (1969).
Related Continence (#HH0070) Tragic suffering (#HH0708) Freedom of choice (#HH0789).

♦ **HH0921 Kabbalah**
Qabalistic tree of life — Tree of sephiroth — Qabalah
Description The Kabbalah is based upon a diagram known as the Tree of Life. This diagram can be interpreted and used in many ways. Its chief use is as a map of stages and levels of consciousness. The Tree structure includes references to all parts of human being: thoughts, feelings, sensations, deeper levels of awareness and archetypes. The structure also indicates the various pathways that lead to increased personal, interpersonal and transpersonal clarity. Various techniques and exercises can be used to explore these pathways.
The Kabbalah may be described in many ways. But the truly universal and living Kabbalah is based on direct experience. From this perspective the Tree of Life is a living entity through which people may communicate individual experiences to others.
Ten levels or *sephiroth* are defined, by means of which man may make a mystical ascent. These ten are divided into the seven lower levels, representing levels of spiritual attainment which are symbolized in mythology but still within the sphere of the fallen; and into three higher levels, unaffected by the fall and the great abyss, representing a great trinity. Beyond and above these is the unknown and formless, limitless light, *Ain soph aur*, while below is an obverse, "black" tree of life with branches reaching far below the earth, representing the lower, less–evolved, animal aspects of personality, which have to some extent been overcome.
The Tree of Life is said to have three pillars – positive, negative and neutral. The manifestation of the universe begins in pure energy. Then, with the interplay between force and form, between male and female polarities, the world comes into being. The aim to universalize consciousness is achieved, having overcome the lower, material side of personality, in retracing these stages. The ten levels each have their own divine name and mantra – they can be viewed as metaphors for an orderly and gradual ascent by stages in the transformation of personality. Within the whole, each level to some extent reflects every other.
In the system of Tarot, there are 22 paths of wisdom or doorways connecting the ten levels. The magician ascends via these paths but is always aware that the reverse tree, transcendental evil, is lurking to trick the unwary on the pathway to transcendental good. The Hebrew Kabbalah identifies 32 paths of wisdom.
Context The Kabbalah is considered one of the foundations of the western esoteric tradition. It constitutes a great body of theoretical and practical philosophy and psychology interwoeven into the religious texts of Judaism and in a vast complex of alchemical, astrological, occult, Rosicrucian and Masonic symbolism, including the Tarot. The Hebrew word Kabbalah means both to "receive" and to "reveal". The Tree of Life has been described as a "mighty, all–embracing glyph of the Soul and the Universe." The sefiroth, the ten metaphysical numbers or numerations of the divine aspects, are the principal keys to the mysteries of the Torah of Judaism. They form a tenfold hierarchy and in their totality constitute the doctrinal basis of Jewish esotericism. They are to the Kabbalah, or mystical tradition of Judaism, what the Ten Commandments are to the Torah, as the exoteric law.
Refs Berg, Philip S *Kabbalah for the Layman*; Halevi, Z'ev ben Shimon *Kabbalah*; Halevi, Z'ev ben Shimon *Kabbalah and Psychology* (1986); Hoffman, Edward (Ed) *Path of the Kabbalah*; Kaplan, A *Meditation and Kabbalah* (1982); Parfitt, Will *The Living Qabalah* (1988); Schaya, Leo *The Universal Meaning of the Kabbalah* (1971); Scholem, Gershom G *On the Kabbalah and its Symbolism* (1965); Suarès, Carlo *The Sepher Yetsira* (1976).
Broader Maps of the mind (#HH8903).
Narrower Sphere of glory (Kabbalah, #HM2238) Sphere of beauty (Kabbalah, #HM3031)

Sphere of wisdom (Kabbalah, #HM2348) Sphere of victory (Kabbalah, #HM2362)
Sphere of judgement (Kabbalah, #HM2290) Sphere of knowledge (Kabbalah, #HM0126)
Sphere of the kingdom (Kabbalah, #HM2288) Sphere of understanding (Kabbalah, #HM2372)
Sphere of the foundation (Kabbalah, #HM2410)
Sphere of mercy and greatness (Kabbalah, #HM2420)
Sphere of the of the supreme crown (Kabbalah, #HM3132).
Related Tarot (Tarot, #HH0382) Pathworking (#HH8003)
Paths of wisdom (Judaism, #HM2509) Jewish mysticism (Judaism, #HH1232)
Shekinah awareness (Judaism, #HM2002) God as wholly other (Judaism, #HM4501)
Tarot arcana of conscious states (Tarot, #HM2097).

♦ HH0922 Psychotherapeutic self examination

Self evaluation
Description A technique used in psychoanalysis and psychotherapy, when the therapist assists the individual to make conscious those unconscious attitudes, fears and conflicts causing current behaviour. This may lead to resolution of problems, a desired change in behaviour and an aptitude for self perception leading to self knowledge.
Related Psychotherapy (#HH0003) Psychoanalysis (#HH0951)
Self-knowledge (#HH0312) Self-examination (#HH0314).

♦ HH0924 Educational mobilization

Educational motivation
Description It has been postulated (R A LeVine and M I White) that differences in academic achievement between countries, social classes and ethnic groups are not so much due to differences in intellectual potential but more dependent on motivation: of those teaching, of those learning, and of every individual in society. This implies that increasing investment of time, space, attention and money, both at the individual and the societal level, and increasing ideological commitment to coordinating educational activities at both these levels, would lead to increased educational performance.
Societies which traditionally value education and encourage (even by threats, punishments and beatings) a child's interest in education are demonstrated to have produced high achievers. And new systems have been demonstrably successful, for example in Japan, with educational mobilization of the whole nation (involving parental commitment and community interest); and in China, where the emphasis of educational mobilization is on moral commitment to education as a virtuous member of society, caring for and educating children to produce a morally sound and motivated population.
Problems arising in educationally mobilized societies can be due to the relative impossibility of achieving the ideal model established in a society, such *relative deprivation* producing feelings of hopelessness, failure and inferiority. There is also the *competitive pressure* which leads to excessive family expenditure of time, money and effort towards academic achievement. And there is a *narrowing of life course values*, as the focus on education and career reduces the individual's interest and care in relationships and in the wider meaning of life.
Broader Motive (#HH0181).
Related Learning (#HH0180).

♦ HH0925 Heteronomous discipline

Description In this approach on the spiritual path the seeker finds fulfilment and meaning in service to a master, whether his present teacher (as in Ch'an Buddhism) or some personal or impersonal authority whose service must be learned from another, as from a rabbi in Judaism, from a Buddhist arhat, etc. In Islam, service to the will of God is by following the guidance of Muhammed. To follow this guidance so as to have direct experience of God, particularly according to the Sufi practice and beliefs, one must first seek out a teacher to guide one through the stages of the path or spiritual journey. The spiritual duties performed during these stages gradually rid the seeker of all imperfections and he takes on the qualities of the divine.
Broader Spiritual discipline (#HH1021).
Related Spiritual direction (#HH0442).

♦ HH0926 Cultural evolution

Description This is the process by which man's cultural products (such as tools, art, language, belief systems, disciplines) evolve over time from their primitive origins. At the present time, it is evident in the emergence of new cultural products, some of which will survive for longer periods of time as more successful reflections of man's needs and of the evolution of the human mind. At the same time there is an awareness of the need to safeguard the achievements of every culture and make them accessible to all.
Refs Huxley, Julian *Cultural Process and Evolution* (1958); Steward, J *Cultural Evolution*.
Broader Evolution (#HH0425).
Related Culture (#HH1754) Cultural identity (#HH1929).

♦ HH0927 Jnana yoga (Yoga)

Gnani yoga — Sankya yoga
Description This form of yoga is one of the essential components of raja yoga. It is primarily intended for individuals with strong intellectual interests and for whom philosophical ideas and the search for truth are of first importance and the determining factor in their way of life.
A series of meditations and logical demonstrations is provided to convince the individual of the distinction between his true self and the veil of personality that is its present disguise. By turning his awareness inward, the individual is led to pierce and dissolve the numerous layers of the manifest personality until all strata are cut through. He then arrives at the anonymous and unconcerned actor/observer who stands beneath, at the core of his being, with whom he can subsequently identify. Seven steps are prescribed: discrimination, detachment, self-discipline, longing for freedom, hearing, reflection, meditation. The aim of this yoga is therefore to distinguish the larger self obscured by the foreground of daily habits and preoccupations. It leads to self-realization through yogically induced insight into the true nature of existence.
Refs Vivekananda, Swami *Jnana Yoga*.
Broader Yoga (Yoga, #HH0661) Raja yoga (Yoga, #HH0755).
Narrower Self-discipline (#HH0877).
Related Jnana marga (Hinduism, #HH0495) Advaita vedanta (Hinduism, #HH0518)
I-transcendent stage of life (#HM3060) Jnana (Yoga, Buddhism, Hinduism, #HH0610).

♦ HH0928 Soto Zen (Zen)

Ts'ao tung
Description This branch of Zen emphasizes gentle methods of realization, zazen meditation without the use of koans as practised in the Rinzai school. There are five stages or relations: (1) The individual accepts or discovers for himself that there is a reality, a superior self which, although unseen, overshadows the "little" self which only seems to exist. These are referred to as the prince and his minister or the host and his guest. At this stage the superior self is unseen or unknown, dark but looming over the known "smaller" self. (2) The individual recognizes that he is in fact the servant of the prince – the spiritual is his master and he begins to feel that he should act accordingly. Here the greater has more significance, the "little" self is seen as the dark or shadowy figure. Attachments die away and to outsiders the individual appears like a "withered log". (3) Spiritual living becomes positive, the servant sees the real and higher as actually within himself, the guest sees the host's interests as his own, the withered log blooms again. At this stage, "not-seeing" diminishes, the servant becomes luminous with spirit as the mind declines. (4) Prince and minister are one, there are no other motivations. The individual works only from the inner promptings and not from thinking or planning of the separate self. (5) Oneness with the prince is now realized as oneness with all, even with what before seemed erroneous – only the "seeming" was wrong.
Broader Human development (Zen, #HH1003).
Narrower Outward zazen meditation (#HH0621).

♦ HH0929 Heroism

Courage — Fortitude
Description Heroism is distinguished as *individual* when a person carries out an individual act of great courage; or as *of the masses* when not just the moral strength of one individual but the combined courage of whole communities lies behind single acts or continued day-to-day heroic existence. Traditionally, the "Age of Heroes" of mythology denotes a phase in human development when the qualities of courage, endurance, self-sacrifice and firmness were taken as those of exemplary behaviour. Such behaviour in the struggle against oppression has, for example, given rise to the titles "Hero of the Soviet Union" and "Hero of Socialist Labour" in the USSR.
Narrower Female heroism (#HH6432).
Related Fortitude (#HH5640) Endurance (#HH0729) Fearlessness (#HM3553)
Treasures of the godly (#HH0364) Human development (Existentialism, #HH3098).

♦ HH0931 Mantra yoga (Yoga)

Japam — Japa yoga
Description In the process of unification, the use of a mantra is referred to as mantra yoga, whether in meditation, ritual recitation or chanting (bhakti yoga). Repetition of mantra (in particular a single word or phrase given to an individual by his or her teacher at initiation) is referred to as *japa* or *mantra japa*. This repetition will be practised during meditation, but also throughout waking life, until it and its embodied truth become a permanent source of strength and inspiration. The syllables, words or sentences have a symbolic meaning, the knowledge or consideration of which may or may not be involved in the practice. The mantram may be accompanied by visualization or there may be concentration exclusively on sound. There may also be variation in the degree of repetition and the extent to which mantra are invested with cosmic or magical significance. These all depend on the school or individual to decide.
The mantram is not an end in itself but an aid on the path to meditation. Concentration on the sound replaces the continuous chatter in the mind, concentration on the symbol replaces the endless flicker of images. This brings a state of inner quiet when the mantram can be discarded as true meditation begins.
Refs Sivananda, Swami *Japa Yoga*.
Broader Yoga (Yoga, #HH0661).
Related Mantras (#HH0386) Sound health (#HH2543) Nada yoga (Yoga, #HH0160)
Bhakti yoga (Yoga, #HH0337) Hesychasm (Christianity, #HH0205)
Mystical stage of life (#HM3191) Heart-awakening stage of life (#HM3518)
Mantra-induced experience (Yoga, #HM2464) Human development through music (#HH3003)
Indian approaches to consciousness (Hinduism, #HM2446).

♦ HH0932 Rejoicing at the merit of others (Buddhism)

Pattanumodana (Pali)
Context The seventh of the ten meritorious deeds of South Asian Buddhist thought.
Broader Ten meritorious deeds (Buddhism, #HH0446).
Followed by Listening to the dharma (Buddhism, #HH1183).
Preceded by Transferring merit (Buddhism, #HH1266).

♦ HH0933 Cyclic history

Eternal return
Description The concept of the cyclic nature of time and the periodic regeneration of creation is basic to many cultures and religions. This concept of periodicity is inherent in belief in astrology and the governing of events by the rotation and repetitive movements of the stars and planets. In particular, periods of suffering, pestilence and misery are found easier to tolerate and come to terms with when related to (and effectively identified with) similar events in the past, the expiation of sin, and the progress towards another golden age. Historical people and events become identified, and sometimes confused, with mythical people and events.
Although Christian thought theoretically precludes such a concept of cyclicism, because of the "once and for all" nature of the fall of man and redemption by Christ, nevertheless theories of perfection of humanity through new creations do exist.
The historicism implicit in dialectical materialism also precludes the "cyclic history" hypothesis. Although it postulates the development of history towards some golden age it does not consider such a golden age to have occurred in the past; mankind has to make new, creative acts with no basis or precendent in the mythical past. Archaic or traditionalist response would be that such acts are limited to the few, whereas the traditional approach allows everyone to participate in the creative act as it is echoed in everyday life.
Related Kali yuga (#HH0239) Twice-born (#HH0715).

♦ HH0936 Emotional growth and development

Description Emotional development occurs as a consequence of maturation and learning. The infant at birth can express only the most rudimentary emotions. At a later stage he learns a wide range of emotions, and he learns when certain emotions are appropriate and when they are not. Almost any important emotional situation has the form of its expression determined by the culture that surrounds it.
With increasing age, the individual learns to increase the amount of control he has over his emotional response. An adult is considered immature if he fails to attain a level of development at which he is capable of control overt response to emotional stimuli until an appropriate time and place. The exact mechanism by which this evolution in response takes place remains obscure.
Emotion first manifests itself in infancy as undifferentiated excitement due to increase or change in stimulation, and may be expressed as delight or distress. Such expression is innate (for example, blind children will respond with the same facial expressions as sighted children). Insufficient variation in environmental stimulation in childhood results in lethargic behaviour and slower reactions as an adult. It may also affect biological resistance to stress (whether physical or psychological) and rate of physical growth. Current theoretical conceptions suggest that all emotions are derived originally from primitive physiological states of the organism. The physiological reaction of individuals to emotions and their degree of emotional maturity may therefore affect physical growth and development.
Refs Greenacre, Phyllis *Emotional Growth* (1971).

♦ HH0937 Human capital formation

Description The process of acquiring and increasing the number of persons who have the skills, education, and experience which are critical for the economic and political development of a country. Human capital formation is thus associated with investment in man and his development as a creative and productive resource. It includes investment by society in education, investment by employers in training, as well as investment by individuals of time and money in their own development. Such investments have both qualitative and quantitative dimensions. In other words, human capital formation includes not only expenditures for education and training, but also the development of attitudes toward productive activity.

♦ **HH0938 Atheistic education**
Description Education, as practised in the Soviet Union and elsewhere, which systematically propagates a scientific–materialistic world view and disseminates the origins and essence of religion as opposed to science and incompatible with communist ideology, may be referred to as atheistic education. This is intended to liberate people from the vestiges of religion and commences at the kindergarten stage, continuing throughout the school syllabus. It emphasizes the antiscientific and reactionary essence of religion and is linked with studies intended to satisfy cultural and leisure requirements. Such studies are intended to influence the intellect, feelings and frame–of–mind of the pupil, and are reinforced by propaganda through literature, the arts and the media and by introducing civil rites to replace religious ones and drive religion from daily life. When dealing with religious believers, such education draws attention to internal contradictions and inconsistencies in their views and persuades them to understand the dialectical, materialist view of reality.
Broader Education (#HH0945).

♦ **HH0939 Schizophrenic fantasy**
Description Schizophrenics have been shown to be part of a network of disturbed and disturbing patterns of communication. Their fantasies, previously thought of as unreal and needing to be interpreted for understanding of their individual problems, are real and meaningful experiences which can assist in the understanding of their inner selves and of the fear which attempts to divide the self from the body. The attitude to schizophrenic fantasy is held to be symptomatic of the general denial by society of the self and of experience. Acceptance of the fantasy aspect of experience would promote better understanding of the self and of others.
It is interesting that during the 'journey' that the schizophrenic makes into the psyche he encounters the archetypal, symbolic figures universal to traditional mythologies; and that, according to some sources, by successfully continuing the journey he may recover his balance through completion of the equivalent of the mythological spirit journey, having experienced chaotic and then harmonizing encounters which give new courage.
Related Fantasy (#HM0892)   Archetype (#HH0019)   Rites of passage (#HH0021)
Modes of awareness associated with schizophrenia (#HM2313)
Anomalies in experience of the self as distinct from the outside world (#HM4754).

♦ **HH0941 Life–style**
Description The relationship between material and spiritual aspects; social and individual aspects; the balance of work, pastimes and leisure; and the type of relationships a person has within and outside the family – all these go towards defining an individual's life–style, the life he or she leads outside the working environment. Such a life–style will depend heavily on the mode–of–life and standard–of–living of an individual's family or community. In life–styles of international communities or socialist countries where, respectively, national and class distinctions are diminished, individual needs and tastes are not necessarily brought to a common level and specific characteristics are retained.
Related Mode of life (#HH0592).

♦ **HH0945 Education**
Description The physical, intellectual, emotional and ethical integration of the individual into a complete man is a broad definition of the fundamental aim of education. Within this fundamental aim are the general, ultimate aims essentially laid down by society, and these aims dictate the specific ends. Every educational act is part of a process directed towards such an end. The objective reality of a given situation necessarily conditions present aims for education in each particular national context.
A free, self governing society would hope to be guided, strengthened and defended through the education of its citizens. However, a distinction has to be made between the initiation of the young into what the elders in a society feel to be true and what they feel to be useful, the latter being simply indoctrination. The alternative, giving no indication at all of what the educators feel, leaves the way open for an uncomprehending following of the latest (and not necessarily the best) philosophy, and possibly to a breakdown of that society.
Each society needs a certain number of educated citizens, more or less specifically qualified, at this or that level and with one or another prospect in view, including that of structural changes. Generally speaking, this need stems in the first place from the economy, but it may also be generated by a variety of other sources, including the State itself, which has to recruit administrative personnel and may also have manifold political motives for pushing educational development. The most positive among these is that of raising the people's cultural level and enhancing their consciousness, out of concern to create the conditions for greater mass participation in democratic processes.
Education is both a world in itself and a reflection of the world at large. It is subject to society, whilst contributing to its goals, and in particular it helps society to mobilize its productive energies by ensuring that required human resources are developed. Education also contributes to bringing about the objective conditions of its own transformation and progress. It has to prepare for changes, show people how to accept them and benefit from them, create a dynamic, non–conformist, non–conservative frame of mind. Concurrently it has to function as an antidote to the many distortions within man and society, and as such must be able to provide a remedy to frustration, to the depersonalization and anonymity in the modern world and, through lifelong education (described separately), reduce insecurity and enhance professional mobility.
Education can no longer be defined in relation to a fixed content which has to be assimilated, but must be conceived of as a process in the human being, who thereby learns to express himself, to communicate and to question the world, and increasingly to fulfil himself. It is not sufficient to balance technical skills (giving the individual sufficient factual knowledge to make a contribution to the material well–being of society) with emotional guidance (education in the humanities, to enrich the mind in order to appreciate the great visions of life). What is required is the nourishing of a critical, questioning spirit to discriminate in all fields of life, not simply to appreciate specific areas of knowledge and culture.
Refs Edgar Faure (Ed) *Learning to Be* (1972).
Narrower Sex education (#HH0547)   Self–learning (#HH0417)
Physical education (#HH0575)   Artistic education (#HH0029)
Curative education (#HH1208)   Atheistic education (#HH0938)
Programmed learning (#HH0546)   Religious education (#HH0709)
Ideological re–education (#HH0996)   Seishin education (Japanese, #HH1109)
Education for early achievement of excellence (#HH6018).
Related Learning (#HH0180)   Pedagogy (#HH0179)   Upbringing (#HH0908)
Educational self–development (#HH0955).

♦ **HH0947 Metapathology**
Description Deprivation of metaneed gratification or frustration of metamotivations leads to disturbances, illnesses, pathologies or diminution in full humanness or of the human potential. Such diminutions are termed metapathologies and include: alienation, anomie, anhedonia, loss of zest in life, meaninglessness, inability to enjoy, indifference, boredom, ennui, worthlessness of life, existential vacuum, noogenic neurosis, philosophical crisis, apathy, resignation, fatalism, valuelessness, desacralization of life, spiritual illnesses and crises, axiological depression, death wishes, sense of being useless or unwanted, hopelessness, despair, anguish, cynicism, and aimlessness.

♦ **HH0948 Expurgation**
Description Part of *gedatsu therapy* when the individual ritually purges away the obsession with "I". The ritual may also involve bodily purgatives – a right diet and a body free from impurities signifying freedom from spiritual impurities.
Broader Gedatsu therapy (#HH0972).
Preceded by Identity interchange (#HH1002).

♦ **HH0950 Vigilance**
Revolutionary vigilance
Description Unflagging alertness, attention to hostile forces, lack of complacency and a conviction that one is working for a just cause, are all hallmarks of vigilance. In socialist countries *revolutionary vigilance* – lack of trust in deceitful demagoguery and awareness of dangerous trends and subversive actions – has always been encouraged to protect against threats to the achievements of the socialist revolution at the ideological level.

♦ **HH0951 Psychoanalysis**
Repression
Description The analysis of the separate components of the psyche into the *id* (a non–discriminating reservoir of energy and instinctive desires), the *conscious ego* (as an impression of external reality on the id), and the *superego* (the source of moral attitudes based on education, etc, and manifesting as conscience). Guided by the demands of reality, the ego represses the irrational impulses of the id, these *repressions* acting as a defence against excessive tension if the unconscious desires were given expression.
Psychoanalysis as a psychotherapeutic method has the object of bringing to light the unconscious meaning of words, actions and mental images. Individuals are encouraged to talk freely about the nature and sources of their problems and thus (in some instances) overcome emotional disturbances. The interpretation of dreams and unconscious verbal slips has been developed as an important aid to the free association process used as an extension of this form of analysis.
Psychoanalysis is based on the fundamental hypothesis that all psychic events have causes which usually derive from the unconscious. When the psychic life is disturbed by conflicts between drives and defence mechanisms, or because a drive has become too strong, a direct link is established between the unconscious and reality. The sources of such disturbances are considered to be primarily of a sexual nature. The defence mechanisms include: repression, projection, rationalization, introjection, regression, turning against self, thought dissociation, isolation, reaction formation, and denial of reality.
When the individual is able to respond, and show by his dreams, his remarks and his behaviour that he is already on the point of understanding the meaning of what is being brought into the light of consciousness, then the analyst can make interpretations to guide the development of his understanding and acceptance of what has previously been suppressed. By this means the individual progressively uncovers the unconscious motivation of his behaviour, mitigates the intensity of any conflict, and enriches his personality by recovering access to energies whose existence has been hitherto denied.
Refs Jones, Ernest *The Life and Work of Sigmund Freud*; Moore, Burness E and Fine, Bernard D (Eds) *Psychoanalytic Terms and Concepts* (1990).
Narrower Abreaction (#HH0640)   Dream therapy (#HH2116)
Expressive therapy (#HH0149)   Integral psychoanalysis (#HH7123).
Related Dreams (#HM2950)   Dissociation (#HH1294)   Psychotherapy (#HH0003)
Structural change (#HH2178)   Analytical psychology (#HH1420)
Psychotherapeutic self examination (#HH0922).

♦ **HH0953 Emancipation**
Narrower Emancipation of the self (#HH0907).

♦ **HH0954 Enculturation**
Description The process of learning a culture. As opposed to acculturation (which refers to groups), enculturation is the acquisition by the individual, whether consciously or unconsciously, of the specific cultural standards and symbols of his environment. The enculturation process usually progresses in stages, six–year olds tending to be more enculturated than three–year olds, and teenagers having almost completed the process (or believing they have). Acculturation of groups refers to their tendency to adapt to beome almost indistinguishable from the dominant culture of their society.
Related Ethos (#HH0725)   Socialization (#HH0666)   Deutero learning (#HH0704)
Self–development (#HH0651).

♦ **HH0955 Educational self–development**
Development of personal cultural potential — Lifelong education — Continuing education
Description The lifelong concept covers all aspects of education, embracing everything in it, with the whole being more than the sum of its parts. Lifelong education is not an education system but the principle on which the over–all organization of a system is founded, and which should accordingly underlie the development of each of its component parts. In the light of such a principle, education is continued at all ages of man, according to each individual's needs and convenience. He therefore has to be oriented from the outset and from phase to phase, keeping the real purpose of education in mind: personal learning, self–teaching, and self–training.
Society is currently characterized by constant feats of creativity, invention, discovery and an increase in the importance of leisure, scholastic, para–scholastic and post–scholastic structures. These structures tend to enable a doubly privileged group (of those having received higher education and being the children of educated parents) to attain cultural self–fulfilment. This therefore constitutes a system of continuing education, the spontaneous existence of which draws attention to the inadequacy of school by itself and necessitates the radical overhaul of the school system in order that future secondary school pupils should have the greatest possibility of developing their individual cultural potential throughout life.
There are three requirements in training pupils for self–development: a common curriculum dictated by the requirements of society; optional subjects, which cater for individual differences; and free activities, selected by the individuals and groups themselves. The job of the educator is to direct pupils who want information to the various appropriate sources in the community.
Refs UNESCO *The School and Continuing Education* (1972).
Related Education (#HH0945)   Advanced training (#HH0521)
Education for early achievement of excellence (#HH6018).

♦ **HH0956 Discipline of celebration** (Christianity)
Jubilee of the spirit
Description The original Jubilee of the Old Testament was a freeing from possessions, a restructuring of social arrangements when slaves were released. It is also a metaphor for the coming of the Gospel or good news of Jesus Christ. Celebration brings joy and a carefree spirit into apathy and melancholy; it is central to all spiritual disciplines, so that instead of being dull and breathing death they become festive and carefree with a sense of thanksgiving. The genuine joy that is manifested in celebration arises only in obedience. This is the secret of a happy life, without it joy becomes artificial and hollow. The secret is not to bypass misery but to transform it. It is the spiritual disciplines that lead to the transformation which brings joy. Another necessity is to be without care or anxiety, trusting entirely in God to provide for one's needs, making one's requests known to God through prayer and thanksgiving.
Celebration is a discipline because it requires an act of will, of consciously choosing to dwell on the good things in life, of choosing the way one thinks and lives. It requires that one does not take one's self too seriously, and brings a sense of perspective. When celebrating the glory of God, no-one is higher or lower than another, there is no sense of judging another. Traditionally, group celebration has included music, dancing and lots of noise, although these are not necessary. All the rites of passage and traditional feasts are occasions for celebration, as are all the disciplines.
Broader Spiritual discipline (#HH1021).

♦ **HH0957 Study of sacred scriptures**
Svadhyaya (Hinduism)
Context One of the five components of *niyama*, self-discipline.
Broader Niyama (Hinduism, #HM3280).
Related Kriya yoga (Yoga, #HH0969) Discipline of study (Christianity, #HH1323).

♦ **HH0958 Recluse**
Hermit — Anchorite — Anchoress — Eremetical life
Description In order to achieve spiritual union with God through a life of uninterrupted prayer and penance, religious minded people may retire as completely as possible from human society, either by taking up habitation in a deserted or secluded region on by being "walled in" to a small room or cell, often attached to a church or monastery. Such practices were common in the early Christian church and in the middle ages, when perpetual or long-term inclusion involving severe penances and asceticism were often practised.
Related Retreat (#HH0420) Solitude (#HH0333).

♦ **HH0959 Discipline of the secret**
Secrecy
Description The rites and secrets of many religious, particularly pagan mystery religious, are commonly shared only by the initiated; this practice of secrecy become common in the early Church (3rd to 5th centuries). The principle is that the mysteries of the faith do not pertain to this world and are only divulged gradually as the individual is spiritually capable of understanding. Vestiges of this practice occur in the liturgy. The most commonly accepted secular use of such discipline is in the order of Free Masons but other organizations exact similar promises.
Related Rites (#HH0423) Initiation (#HH0230).

♦ **HH0960 Integrated type**
Description Although the orientation may be more inward or more ontward, the integrated type thinks holistically and is characterized by wide ranging attention and the ability to change.

♦ **HH0961 Musical intelligence**
Description Such intelligence has as its centre the relating of emotional and motivational factors to the perceptual ones; music is a way of capturing and communicating feelings and knowledge about feelings Musical ability is centred in the right-hemisphere of the brain and varies widely among individuals and cultures It seems to be used in exploring and interpreting other forms of intelligence
Context One of the separate or multiple intelligences put forward by Howard Gardner.
Broader Multiple intelligences (#HH0214).

♦ **HH0962 Hypnotherapy**
Self-hypnosis — Guided-hypnosis — Biomagnetic therapy — Heterohypnotherapy — Hypnotic suggestion
Description Hypnosis is used in several ways in therapy. Sleep may be induced where prolonged periods free from stress may be beneficial. It may be used to facilitate the emergence of memories and associations in classical psychoanalysis or to produce cathartic abreactions during treatment of post-traumatic neuroses. Hynosis may also be used as a means of communicating direct suggestions to the person when suggestibility is high, thus offering the possibility of relief of specific physical or mental symptoms without attempting to influence underlying causes. The person may also be helped to gain insight into the dynamic unconscious sources of his difficulties as a result of the therapist demonstrating to him the meanings of his symptoms in terms of repressed conflict by inducing experimental conflicts in the course of the hypnotic trance.
Self-hypnosis, when self-suggestion is used to induce the hypnotic state, can be used to discover the reason behind a person's problems and to condition or de-condition, adjust and strengthen the responses. *Guided-hypnosis*, when the subject is hypnotized by a therapist, is said to lead to a deeper level of consciousness; *heterohypnotherapy* is another term used to describe therapy under guidance, where suggestions or instructions assist in altering behaviour patterns and in revealing previous experiences which may have a subconscious inhibiting effect.
Early demonstrations of hypnosis or *biomagnetic therapy* inspired preliminary discovery and analysis of the subconscious mind.
Refs Gordon, J *Handbook of Clinical and Experimental Hypnosis* (1967).
Broader Therapy (#HH0678).
Narrower Mass hypnosis (#HH2134)
Mono-motivational hypnotic state of transcendence (#HM4401).
Related Hallucination (#HM4580) Hypnoidal state (#HM1971)
Deep state of hypnosis (#HM1226) Medium state of hypnosis (#HM5856)
Hypnotic states of consciousness (#HM2133).

♦ **HH0964 Self-denial**
Description By denying one's own personality, emptying one's personal self and refusing to gratify personal inclinations, one starts to achieve *unselfishness*. There is no denial or rejection of self-worth as would occur in self-contempt. Self-denial celebrates the intrinsic value of the self in each individual and reaches out to others from confident self-love. Particular exercises of self-denial are undertaken as a discipline in order to become closer to God and, in Christian terms, to help nullify the effects of original sin. The greater the sanctity of a person's life, the greater self-denial is necessary - although most teachings require the more extraordinary measures of mortification and denial to be carried out with the approval and advice of a spiritual guardian.
Related Mortification (#HH0780) Self-discipline (Christianity, #HH2121)
Spiritual rebirth of the ego (#HH0698) Spiritual self-denial (Christianity, #HH4522)
Discipline of submission (Christianity, #HH0225).

♦ **HH0966 Psychosocial isomorphism**
Description Psychosocial isomorphism exists whenever the relations within a personality are similar to those within a society, namely whenever a social structure was similar to a personal structure. For example, it is suggested that there are structural similarities between war propaganda and some times of mental illness. Both are characterized by inner conflicts or self-contradictions, by denial or repression of these conflicts, by projection of the denied material (usually dominance and aggression in war propaganda, but usually sex and aggression in mental illness) onto other nations or other persons, and by justification of aggression, for example, as an appropriate response to the aggression projected onto others. Such forms of mental illness and war propaganda are different in their content but show a similarity between their structures.
Refs Eckhardt, William *Psychosocial Isomorphism.*

♦ **HH0968 Self-oblation**
Self-surrender
Description In Christian terms, self-oblation has overtones of sacrifice, the offering of the self to God in union with Christ's sacrifice of himself on the cross, whether this be to physical persecution or by inner self-denial. As an act of devotion, such self surrender may be said to be perpetual if every act is seen as self-offering, done to the glory of God. As a method of *meditation*, self-offering gradually permeates the whole being and enters every action. The meditator starts by realizing "I do nothing at all", it is not I who meditate but God through me. Surrendering his nature to the divine will, he or she offers all that arises in the mind and rejects nothing, aware of all that gradually becomes exposed.
Related Surrender (#HH0449) Meditation (#HH0761)
Spiritual self-abandonment (#HH0868) Resignation to the will of God (#HH1016)
State of self-surrender (Christianity, #HM5467).

♦ **HH0969 Kriya yoga** (Yoga)
Description A preliminary to systematic concentration and meditation, kriya yoga denotes union with the Infinite through certain actions (kri) and rites. A particular example is the *pranayama* method, a tradition of the *Paramahansa Yogananda* schools, taught in combination with a combined Indian and Christian approach.
Refs Davis, Roy E *Science of Kriya Yoga* (1984); Davis, Roy Eugene *The Science of Kriya Yoga*; Gupta, Sailendra Bejoy Das *Kriya Yoga and Swami Sri Yukteshvar*, Kriyananda, Goswami *The Spiritual Science of Kriya Yoga.*
Broader Yoga (Yoga, #HH0661).
Related Tapas (Hinduism, #HH1248) Karma yoga (Yoga, #HH0372)
Soul vision (Psychism, #HM3050) Kriya sakti (Hinduism, #HH4309)
Pranayama (Hinduism, Yoga, #HH0213) Study of sacred scriptures (#HH0957)
Resignation to the will of God (#HH1016).

♦ **HH0971 Maturity**
Mature personality — Psychological maturity — Fully developed personality — Self-actualizing personality — Healthy personality
Description The characteristics of psychological maturity vary according to the authority.
For Marie Johoda a healthy personality is characterised by: ability to actively master the environment; demonstration of a certain unity of personality; and ability to perceive the world and one's self correctly.
For E H Erikson the characteristics emerge at different periods of the life cycle: a basic sense of trust (infancy); a sense of autonomy (early childhood); a sense of initiative (play age); industry and competence (school age); personal identity (adolescence); intimacy (young adult); generativity (adulthood); integrity and acceptance (mature age).
For A H Maslow, self-actualizing personalities are characterised by: more efficient perception of reality and more comfortable relations with it; acceptance of self, others, nature; spontaneity; problem centering; detachment; independence of culture and environment; continued freshness of appreciation; limitless horizons; social feeling; deep but selective social relationships; democratic character structure; ethical certainty; unhostile sense of humour; creativeness. They are in good psychological health, with their basic needs satisfied; their life is centred round a vocation, what they do is for intrinsically worthwhile principles - *metaneeds* or *B-Values.*
For G W Allport they are: an extension of the sense of self; a warm relationship of self to others; emotional security and self-acceptance; realistic perception and assessment of skills and assignments; self-objectification with insight and humour; and a unifying philosophy of life.
An extension of the sense of self requires that the individual participate in an authentic manner in some significant spheres of human endeavour. This activity should extend beyond simple task-involvement to ego-involvement. True participation gives direction to life. Maturity advances in proportion as lives are decentered from the clamorous immediacy of the body and egocenteredness. Self-expression requires of the individual the capacity to lose himself in the pursuit of objectives, not primarily referred to the self. Unless directed outward toward socialized, culturally compatible ends, unless absorbed in causes that outshine self-seeking and vanity, a person must necessarily remain immature.
A warm relationship of self to others may be of two kinds. By virtue of self-extension, an individual is capable of great intimacy in his capacity for love within the family or in friendship. But the person also has a certain detachment, taking the form of compassion, which makes him respectful and appreciative of the human condition of all men. By contrast the immature person feels that only he and his groups have the distinctively human experiences, and that all else is alien, dangerous, and to be excluded.
Emotional security and self-acceptance distinguish the person with emotional poise and include the ability to avoid overreaction to matters pertaining to segmental drives. The mature person tolerates frustration and is prepared to take blame if it is appropriate to do so. He has learned to live with his emotional states in such a way that they do not betray him into impulsive acts nor interfere with the well-being of others. He expresses his convictions and feelings with consideration for the convictions and feelings of others and does not feel threatened by his own emotional expression or by theirs. He possesses integrative values that control and measure the flow of emotional impulse.
Realistic perception and assessment of skills and assignments is required for solving objective problems. Mature people are problem oriented; something objective is worth doing and egoistic impulses of drive-satisfaction can be put aside for long periods. Not only are perceptions veridical (since maturity does not bend reality to fit the individual's needs and fantasies), and cognitive operations accurate and realistic, but appropriate skills are acquired to handle the problem-situations confronted.
Self-objectification, with insight and humour is characteristic of the mature person because he has the most complete sense of proportion concerning his own qualities and cherished values and is able to perceive their incongruities and absurdities in certain settings. Self-objectification gives the mature person detachment when he surveys his own pretensions in relation to his abilities, his present objectives in relation to possible objectives for himself, his own equipment in comparison with the equipment of others, and his opinion of himself in relation to the opinion

others hold of him.
A unifying philosophy of life or sense of directedness is more marked, more outwardly focused, than in the less mature. Such a philosophy is not necessarily articulated in words. It may take the form of a value-orientation of which religions may be the most comprehensive and integrative.
Refs Abrahamsen, David *The Road to Emotional Maturity* (1958).
Related Old age (#HH5223) Moral development (#HH0565).

♦ **HH0972 Gedatsu therapy**
Description Gedatsu is a religious cult based in Japan whose psychotherapeutic methods of self-reconstruction, based on traditional Japanese culture and moral values, emphasize the feminine approach – women being expected to be less assertive and autonomous, more passive and maternal. The technique involves self-accusation, allocentric attribution, identity interchange and expurgation, reinforcing the Japanese moral ideal of selflessness.
Broader Therapy (#HH0678) Psychotherapy (#HH0003).
Narrower Expurgation (#HH0948) Self-accusation (#HH0584) Identity interchange (#HH1002) Allocentric attribution (#HH1249).

♦ **HH0973 Shaman**
Shamanism — Witch doctor
Description Individuals who have achieved spiritual or supernatural powers as the result of contact with the spiritual world through a *vision quest* or some other means, shamans are concerned with diagnosis and cure of disease, communication with the dead, divination and magic. They act as intermediary between the world of spirits and the material world, travelling in a trance state through the world of spirits to find knowledge necessary for this world; and help they their people maintain the balance of communal life and survival. In the realm of illness, shamans may perform psychic surgery, when diseased tissue is removed by hand movements on the surface of the body. Many have a remarkable understanding of psychiatric symptoms.
A call to become a shaman may be through epilepsy, disease or some disorder; or it may be through a dream or simply superior knowledge or learning. There is often difficult, and sometimes dangerous, initiation and long study under guidance of someone who has already travelled the path. After initiation the shaman continues to observe special taboos and requirements. Each shaman is individual in the ceremonies and rites performed, as opposed to a priest, who will follow traditional rites. They and their initiates are powerful members of their communities, spiritually and politically, and are feared because of their power. Shamanism is particularly developed in Siberia, Manchuria and extended across through Lapland and the North American Indians; it is also prevalent in South America and Sumatra. Although sometimes exclusive, it usually coexists with other religions and forms of magic.
Refs Eliade, Mircea *Shamanism* (1964); Harner, Michael *The Way of the Shaman*; Larsen, S *The Shaman's Doorway* (1976); Walsh, Roger *The Spirit of Shamanism*.
Related Magic (#HH0720) Shamanism (#HM1189) Divination (#HH0545) Vision quest (#HH0897) Resurrection (#HH4553) Psychotherapy (#HH0003) Shamanic journey (#HM6120) Spirit possession (#HH0056) Communication with supernatural beings (#HH1306) Human development through primal religion (#HH1902).

♦ **HH0974 Right beliefs** (Buddhism)
Ditthijjukamma
Context The tenth of the ten meritorious deeds of South Asian Buddhist thought.
Broader Ten meritorious deeds (Buddhism, #HH0446).
Related Belief (Buddhism, Yoga, #HM2933).
Preceded by Preaching the dharma (Buddhism, #HH1022).

♦ **HH0975 Shock therapy**
Insulin coma — Convulsive therapy
Description Drug induced coma (using insulin) or convulsions followed by coma (using Metrazol, Cardiazol) have high success rates in treating schizophrenics and affective disorders as for manic depressives, and some success with psychosomatic conditions. Convulsive therapy is nowadays more commonly carried out using electroshock treatment (see separate entry) than with drugs.
Broader Therapy (#HH0678).
Narrower Electroshock therapy (#HH0866).

♦ **HH0976 Reflective rational meditation**
Thought meditation
Description Concentration of the mind upon a particular spiritual object or idea and, while holding it there, as far as possible not deviating from it and examining it in its various aspects. As such this is an intellectual exploration into the nature of the self and the cosmos. Depending upon the nature of the object or idea chosen, and on the depth of the exploration, the meditation may become more spiritualized, more intuitive in character, less dependent upon use of words and symbols, and closer to contemplative intuitive meditation.
The following types or foci of pre-contemplative meditation have been distinguished: on the nature of the self; on the vehicles of consciousness (physical, emotional and mental); on dispassion and detachment; on motive; on the nature of faith and knowledge; on the dogmas of faith.
Refs Happold, F C *Prayer and Meditation* (1971).

♦ **HH0977 Reflexivity**
Reflection
Description The process of thinking is raised to a second level, so that one is not simply thinking about something but thinking about that thinking, not simply self-conscious but conscious of that self-consciousness. This capacity is said to be characteristically human and to be that which enables the development of complex symbolic systems about symbol systems and also the possibility of laughing at one's self. This latter is due to the possibility of exchange between the self which perceives and that which reflects, to Aristotle (among many examples) the divinity within man. Extension of the concept of reflexivity to apply to groups rather than individuals has demonstrated (Victor Turner) that cultural representation in ritual, theatre, games, etc, may be considered plural reflexivity. It is further demonstrated by several authors that the "pattern of patterns" of plural reflexivity is the function of religion, the means by which man interprets the world to himself; and that religion uses systems which are intrinsically reflexive – meditation, contemplation, prayer, confession – to turn from the world to the self.
Refs Bellah, Robert *Beyond Belief* (1970); Hofstadter, Douglas R and Dennett, Daniel C *The Mind's I* (1981); Oliver, W D and Landfield, A W *Reflexivity* (1963).
Related Reflection (#HH1204) Reflexivity (#HM0087) Human development through religion (#HH1198).

♦ **HH0978 Brahmacarya**
Abstention from craving for sensual enjoyment — Chastity
Description Different authorities differ on the details of what is literally "moving in Brahman". Although some require total abstention from sexual intercourse, whether in practice or in thought or speech, others make some exceptions in the case of husband and wife. The fruits of brahmacarya are said to be increase in physical and mental vitality. The senses are restrained, enabling concentration on approaching the self and allowing freedom from sin.
Context One of the five moral qualities of yama (restraint).
Broader Restraint (#HH6876).
Related Celibacy (#HH0045) Ashramas (Hinduism, #HH2563) Chastity (Christianity, #HH0583) Brahmacharyashrama (Hinduism, #HH1987).

♦ **HH0979 Psyche**
Description The total conscious and unconscious personality of the individual is referred to as psyche (Greek: "soul"). In that duality is intimated in describing one individual as separate from others, this suggests duality. It is in the unconscious psyche that race memory, powerful symbols or archetypal images, is embedded, and it is for this reason that images of God the Father, Mother Nature, and so on have universal appeal. The unconscious psyche is also the source of energy to grow and develop, and of sexual energy. The so-called *kundalini serpent power*, dominant in the personality, becomes when activated a source of creative energy.
Refs Dourley, John P *Psyche as Sacrament* (1981); Edinger, Edward F *Ego and Archetype* (1972).
Related Ego (#HH0636) Mind (#HH0299) Soul (#HH0501) Spirit (#HH0770).

♦ **HH0980 Homosexuality**
Description Although it may be what remains from infantile sexuality, homosexuality is an inevitable and potentially psychologically valuable factor of the inner world. Jung saw the homosexual as having a great capacity for friendship, with tenderness between persons of the same sex and the opposite sex. In the male homosexual there is feminine good taste and aesthetic sense, the feminine insight and tact may make him a good teacher. He may cherish the value of the past and be endowed with greater religious feelings and spiritual receptivity.
Refs Licata, Salvatore J and Petersen, Robert P *Historical Perspectives on Homosexuality* (1982).

♦ **HH0981 Psychic research**
Parapsychology
Description Psychic research is concerned with extra-sensorimotor phenomena in general. This covers the following areas: *Extrasensory perception* (also called bioinformation) including telepathy, clairvoyance, precognition and retrocognition. *Psychokinesis* (also called bioenergetics) including teleportation, materialization, psychic healing, psychic photography, and out-of-body experience. *Survival phenomena* including apparitions, hauntings, mediumship, and poltergeists (where not treated as psychokinesis).
Parapsychology is not synonymous with psychic research, confining itself primarily to the laboratory study of extrasensory perception and psychokinesis. Paraphysics, defined as the study of those paranormal phenomena which can be viewed as extensions and generalizations of physical phenomena, is especially concerned with relationships between psychic research and physics, including such phenomena as dowsing, radiesthesia, physical mediumship and the physical aspects of parapsychological research and of paranormal healing.
The status of parapsychology remains controversial. Spontaneous phenomena continue to be reported frequently, but there has been little success in applying experimental control or in arriving at any real theoretical understanding of the matter. Such phenomena may be explicable within the framework of modern physics.
The potential importance of this field of study for human development lies in the possibility that understanding such phenomena may require a revision of the concept of individuals as psychically isolated. For if the conventional concept of human beings as separate egos is correct, there cannot be any actual expansion of the self. But if the self in some real sense encompasses its social and physical environment and extends beyond the borders of is organism, then science, technology and social policy cannot be expected to respond to its needs until the possibility of an extrasomatic self is recognized.
Refs Ashby, R *The Guide Book to the Study of Psychical Research* (1972); Rhine, J B and Pratt, J G *Parapsychology*.

♦ **HH0983 Practice of silence**
Description Silence as a form of worship is practised in most religions, the discipline of inner and external silence bringing a special depth of communication with God. Of the three forms of prayer – oral, prayer in the heart, and spiritual – Francisco de Osuna says of the third, or spiritual, "The greater is this love, the fewer are its words. Because true love works in silence... ". Silent prayer is experienced by many *mystics* as being made by God himself, and the practices of *meditation* and *adoration* are carried out in silence, during which a special quality of external quiet is achieved. The awareness of the *presence of God*, both in private and in public worship, occurs in and produces silence, as, for instance, during the *expectant silence* of the Quakers and at the sacramental presence of God in the Roman Catholic Mass. The practice of silence is said also to give great discrimination about when to speak, so that what is spoken is directed right to the point and has great power.
Related Muni (#HH5450) Prayer (#HH0185) Retreat (#HH0420) Silence (#HM3603) Solitude (#HH0333) Quietness (#HH0491) Adoration (#HM2412) Meditation (#HH0761) Voice of silence (#HM2227) Practice of the presence of God (#HH0992) Christian recollection (Christianity, #HM2077) Dark night of the soul (Christianity, #HM3941).

♦ **HH0984 Inappropriate patterns of development**
Description Dominant patterns of development which are subject to protest from a wide range of groupings may be considered inappropriate for the fulfilment of basic human urges and the augmenting of human dignity and human potential. Gross inequalities, exploitation and dehumanizing domination characterize not only relations between nations but also relations between individuals. Such alienation and lack of harmony leads individuals to question whether an alternative mode of development needs to be formulated, to consider how this may be achieved and how it might be implemented. Current thought tends towards bringing each individual to the centre of the development process, to be an active participant in shaping his or her own destiny.
Related Alternative lifestyles (#HH0477).

♦ **HH0985 Conditioned reflex therapy**
Description In this therapy, the Pavlovian conditioning process is applied to clinical procedures. On the basis that man's behaviour is inseparably rooted in animal nature, neurotic behaviour is rendered healthy by direct re-conditioning, which takes the form of strong positive suggestions. The emotionally ill person is held to have experienced, in one way or another, some inhibiting conditioning, and needs to be encouraged to develop the excitatory reflexes of which he is deprived in modern society. The healthy person is considered to be one who acts without thinking on the basis of spontaneous, outgoing feeling; whereas the emotionally ill think without acting.
Broader Therapy (#HH0678).

♦ HH0986 Libido development

Description Each stage of personality development may be conceived as the consequence of the investment of libidinal energy in a particular bodily zone. Libidinal energy (largely synonymous with the life instinct) is conceived as the drive toward gratification of wishes and motives, as the striving for pleasure, and as the energy behind all human behaviour. This instinctual energy activates the mode of functioning of the zone in question. The bodily zone and mode of functioning that mature during a given period of development determine the person's adaptive activity at that stage of his life. The nature of the zone's operation therefore constrains, or provides the limits, within which the person's identity will be formed and within which his social interaction may take place.

Libido development occurs in distinct stages:

1. *Oral stage* (1 year). The child's first maturational stage of psychosexuality is the result of libidinal investment in the oral zone of his body which serves the primary self-preservative functions of breathing, drinking and eating. This stage of functioning is therefore the basis of all human trust and social communality, with its associated sharing of a common cultural world. Associated with this stage is a crisis of trust or confidence.
2. *Anal stage* (1–3 years). The instinctual urges then shift to the anal and erogenous zones and their functions of retention and evacuation, which require the child to attempt to master his own impulses. Associated with this stage is the crisis in which the child either develops a feeling of autonomy or one of shame and doubtfulness about his actions and his capacity to behave in an independent and appropriate fashion.
3. *Phallic stage* (3–5 years). The next shift is from himself to another person, usually a member of the child's family. The child identifies with the same–sex parent in such a way that his instinctual urge is to usurp that parent's role with respect to the parent of the opposite sex. The consequences of this (oedipal) crisis are primary identification with mother or father, and relatively normal or abnormal personality development. Associated with resolution of this conflict between the child's initiative and his consequent guilt is his acquisition of a sense of moral responsibility. The child begins to understand, and to operate in accordance with, the rules and regulations of his social milieu.
4. *Latency stage* (6–11 years). In this subsequent stage there is a reduction of libidinal energy invested in the genital zone and the onset of a period of sexual retrogression or quiescence. The psychosocial crisis that arises during this stage of identity formation centers on whether the child will become adequately industrious, in his own and others' eyes, or will feel inferior and inadequate.
5. *Adolescent stage* (12–15 years). In this stage the person's sexual impulses shift to a new sexual object, namely a person of the opposite sex outside the family. The crisis is one of identity adoption and repudiation versus identity diffusion, with the danger that the individual will not further develop his ego autonomy or sense of integrity with a hierarchy of values that make certain things in life particularly meaningful to him.
6. *Genital stage* (16–18 years). The young adult's instinctual energy and sexual gratification then become fully centered upon the genital erogenous zone. This coordination under the control of the genital zone is directed by the aim of obtaining full sexual gratification in another person. The crisis is one of achieving and obtaining full gratification from intimacy and solidarity with others versus isolation and withdrawal from partnerships. Associated with this, the individual must develop skills that permit him to work competently and with a sense of responsibility and directed attention in the face of the routine of daily life.
7. *Adult stage*. This is characterized by the utilization of sexual pleasure for purposes of propagation, the establishment of a family and a circle of significant relations. The adult becomes part of and helps to construct the social order. The crisis is one of generativity, making happen and caring for versus self-absorption and a pervading sense of stagnation and interpersonal impoverishment. The adult who achieves gratification from helping others to grow and from creating is equipped with the personal integrity necessary to face the final crisis of life, namely his own disintegration and death.

Refs Erikson, E H *Childhood and Society* (1950).

Related Life instinct (#HH0750).

♦ HH0987 Reflexology

Zone therapy — Pressure therapy

Description Based on the same theory as acupuncture, reflexology attempts to bring relief to either specific or general physical problems by the application of pressure in particular zones of the body. Massage is concentrated on the reflex areas where the nerves are nearest to the surface, particularly on the feet. The method is also used instead of classic anaesthetics and for general revitalization of the whole body.

Broader Therapy (#HH0678).

♦ HH0989 Benevolence

Description Universal benevolence is rooted in the knowledge that all the problems of the world – past, present and future – are my problems. Identifying world suffering as my suffering and my suffering as world suffering, one becomes aware of one's self born of humanity and responding to humanity. Knowing the limitations of one's humanness means also knowing its extent. Benevolence realizes there is never an excuse not to act. Thus, caring is unlimited and, in a state of complete selflessness, action takes place in the present moment and no action has a significance which has been pre–determined.

Related Philanthropy (#HH1206).

♦ HH0990 Virtue of simplicity

Description In spiritual terms, simplicity indicates unity of purpose and heart, and a purity of approach which rejects intentions which, however worthy, distract from the love of God. This unity of purpose opposes duality and thus duplicity, deceit and double dealing. As Kierkegaard put it, "Purity of Heart is to will One Thing; and that is the Good".

Broader Spiritual discipline (#HH1021).

Related Prayer of simplicity (Christianity, #HM0238).

♦ HH0991 Religious poverty

Voluntary poverty

Description Poverty, or lack of possessions, means total insecurity as to the future and inability to provide one's self with what are considered necessities. The more complex and sophisticated a civilization, the more possessions are considered necessary. In the spiritual sense, total lack of possessions and total trust in God to provide what is required, is an aid both in understanding the greatness of God and the insignificance of one's self (and thus achieving true humility); and in enabling one to accept everything as a gift from God, abandoning one's self to the will of God instead of depending on human guarantees and assurances. Religious organizations in the Christian church frequently include the observance of poverty among their vows, members having "all things in common", although this does not imply material deprivation.

Related Spiritual poverty (#HH0190).

♦ HH0992 Practice of the presence of God

Sakina — Isvarapranidhanadva — Presence of the alone

Description The awareness of God's presence, whether purely spiritually or through some sign appreciated by the senses (fire; cloud; light; still small voice) is one experience common to all religions. As a spiritual exercise in the Christian church, the *practice of the presence of God* is at the same time an act of the will and of the intellect, and an affective act. The mind is turned to God, the will is strengthened and the desire to reject sin and serve God is awakened. The practice is particularly associated with the life of Brother Lawrence (1611–1691); it has been related to the sadhana of Isvarapranidhanadva mentioned in the Yogasutras of Patanjali. The term *Sakina* in Islam comes form the Hebrew *Shekina* – it refers both to the presence of God in, for example, the Ark of the Covenant, and as it occurs in an individual (the experience of spiritual peace and security), and again as it manifests in a person's character (calm, dignity). Such presence is described as radiance or glory of God in the midst of his people, in no way distant or aloof. In yoga, the presence of the alone is seen as the silent appreciation of the absolute good, a stilling of heart and mind and quietening of all other demands. It is this experience which is sometimes referred to by Christians as the presence of the Holy Ghost.

Refs Lawrence of the Resurrection, Brother *The Practice of the Presence of God*.

Related Presence of God (#HM2961) — Practice of silence (#HH0983) — Shekinah awareness (Judaism, #HM2002) — Practice of the presence of God (#HH0992) — Sense of the presence of God (Christianity, #HM6989) — Discipline of worship (Christianity, Judaism, #HH3776).

♦ HH0993 Physical appearance

Aesthetic surgery — Cosmetic surgery — Reconstructive surgery

Description Physical deformities have a great effect on the psychological state of the people suffering from them. Aesthetic surgery to alter the shape of the nose, remove unwanted deposits of fat, enlarge breasts, reduce double chins, can change the person's whole attitude, giving self confidence and an improved ability to deal with life. Similar techniques are used in cosmetic as in reconstructive surgery, when people suffering the after–effects of accidents (burns, for example) are treated surgically to heal and then disguise the effects of their injuries.

Related Self adornment (#HH0555) — Physical perfection (#HH0722).

♦ HH0994 Assertiveness training

Description Non–assertiveness brings discomfort in social situations, fear and sometimes illness. Developed from behaviour therapy, assertiveness training works on the premise that assertive behaviour builds self–esteem and gives emotional freedom. The setting and achievement of specific goals are assisted by exercises and assignments. In this way, habits can be changed and roles altered. It has been particularly useful in assisting women to change their (socially reinforced) non–assertive roles.

Related Self assertion (#HM3587).

♦ HH0995 Electrical stimulation of the brain

Description This is a technique for controlling, understanding, and possibly extending the field of consciousness. It is currently used either in experiments on the functioning of the brain, or else, through the implantation of electrodes in the brain, to control behaviour.

Refs Delgado, José *Physical Control of the Mind* (1969).

♦ HH0996 Ideological re–education

Description Educational methods can be used for ideological ends to accomplish shifts and alterations in the individual's sense of inner identity. Four general approaches may be used, singly or in combination: coercion, exhortation, therapy and realization.

Coercion. The message to the individual is that he must change and become what the authority is telling him to become or else suffer the consequences (varying from social ostracism, through emotional and physical pain to, in some instances, death). The aim is to produce a demoralized follower. It is directed at the most primitive of human emotions and stimulates the desire to submit completely.

Exhortation. The message to the individual is that he should change (if he is a moral person) and become what the authority is telling him to become because of the moral superiority of that state. The goal is to create converts and disciples, changed in accordance with the specific ideological convictions of the authority. It focuses on the individual's desire to improve himself and reject his current shameful and morally inferior state.

Therapy. The message to the individual is that he can change from his present unhealthy state, if he has a genuine desire to be healthy (in the manner defined by the authority), and if he is willing to follow the authority's method and guidance. Its goal is physical and emotional health. The appeal is directed to the individual as a reasonable, health–seeking, balanced person.

Realization. The message to the individual is that he can change in such a way as to express more fully his own potential, provided that he is willing to confront himself with ideas and approaches which challenge his present ways of knowing and acting. Its goal is to produce an individual who expresses his creative potential to the full and to the limit of his capability.

These approaches may be combined by some authority in subtle ways with the object of achieving some form of totalitarian thought reform. It may then be very difficult to distinguish between the desirable application of such techniques and applications which infringe upon human rights, particularly since the undesirable features may closely resemble normal processes of human change in such settings as:

Education. In the student's act of attaining knowledge (with every new concept or technique acquired), his previous patterns of identity as well as belief must be altered and rearranged, however slightly, in accordance with the educator's view or the viewpoints embodied in (or excluded from) the materials provided. Students may feel unable to question such presentations, or to inquire into alternative perspectives, without some form of sanction. In this way the individual's intellectual growth and his quest for realization are hampered. The positive alternative is a rejection of omnipotence on the part of the educator, a balance between a vigorous presentation of available knowledge, and encouragement of those aspects of the student's imagination which may eventually transcend that knowledge in new discovery.

Psychology. Psychoanalysis and psychotherapy both focus on re–educating an individual concerning his understanding of himself at the most fundamental levels of his being. But therapists may perceive their particular professional body as imbued with a near–mystical aura, demanding from them a degree of ideological purity, making them hesitant to critize its teaching and thus leading to adoption of a pattern of intellectual conformity. The therapist's notion of reality may also be coloured by his own ideological convictions about such matters as psychological health, social conformity, and maturity, and their relations to the problems of personal identity which trouble his client.

Religion. Organized religion may lead to exaggerated control and manipulation of the individual, the flooding of the social environment with a perspective of guilt and shame, and emphasis upon the individual's depravity and worthlessness and upon his need to submit abjectly to a vengeful deity. This may be done within the framework of an exclusive and closed system of ultimate truth. Alternatively, religious settings may stress, for example, the individual's worth and possibilities as well as his limitations, and his capacity to change as well as impediments to that change. This opens the way to emotional and intellectual growth and a broadened sensitivity.

Science. Similarly, political life may be subject to purges and inquisitions of varying degrees of seriousness. The scientific method may be deified and advocated as the sole basis for liberating man from social encumbrances and from all irrational features of society, by exposing these as unscientific and unnecessary in a truly scientific environment.
Broader Education (#HH0945).
Related Thought reform (Brainwashing, #HH0865).

♦ HH1002 Identity interchange
Description A component of *gedatsu therapy* when the problems of an individual are seen as due to vicariously suffering the result of unresolved suffering or wrong-doing of an ancestor. This may be resolved through redeeming the sin of the ancestor as a substitute for the sinner, the individual taking on the role of the sinner and carrying out prescribed rituals in the sinner's place.
Broader Gedatsu therapy (#HH0972).
Followed by Expurgation (#HH0948).
Preceded by Allocentric attribution (#HH1249).

♦ HH1003 Human development (Zen)
Description The Japanese school of Mahayana Buddhism, derived from *Ch'an* in China and therefore with Taoist influences, Zen is a philosophy of absolute negations which are at the same time absolute affirmations. It aims to transcend objective and subjective viewpoints in awareness beyond mind. This places Zen beyond verbal teaching, with attainment of Buddha-nature by direct transmission in a state of Alaya (the true home), empty of all material or mental objects. The aim is for realization of the illusory nature of the individual which has been fostered by past habits. This illusion creates a great burden from which realization brings great release, even partial realization for a short time; but finally the aim is for total destruction of belief in the ego personality and in personal attainment. Central to the practice of Zen is *zazen meditation*, centering or concentrating the mind so as to effect a major shift in the method of experiencing oneself in relation to the world, achieving *satori*, knowledge without thought. The mind may also be focused on a *koan* or enigmatic saying to focus and subsequently baffle thought, so that the intellect is forced to let go and a spontaneous response arises.
In Zen there is no interest in the shifting play of viewpoints, or the kinds of interest with which an 'object' may be looked at, with the 'subject' remaining always on one and the same level of daily experience. The intent is rather to think in terms of two totally different dimensions of consciousness, namely the interest is in a sudden, abrupt shift on the part of the perceiving subject from the daily consciousness to that of supra-consciousness.
Intellectual concepts and definitions are pushed aside as the person is pointed towards experiencing first-hand knowing, towards what the Buddha experienced. Self reliance is emphasized, with no leaning on authority, even the authority of Buddha. Nothing, however wonderful or wise, comes between the person and the experience of what is.
Two main sects are Rinzai (emphasizing sudden enlightenment, which it is said cannot be achieved in gradual stages, and the use of koans) and Soto (followers of Dogen, emphasizing discipline, self control and philosophical questioning); a smaller sect, Obaku, emphasizes study of the sutras. Artistic and cultural developments of Zen include the well known Japanese arts of ink painting, flower arranging, No drama, landscaping and the tea ceremony. Traceable to Zen also is the bushido ethic of the *samurai*.
Refs Cook, Francis Dojun *How to Raise an Ox*; Suzuki, D T *Essays in Zen Buddhism* (1970); Watts, Alan W *Way of Zen* (1957); Wood, Ernest *Zen Dictionary* (1977).
Broader Human development (Buddhism, #HH0650)
Human development through religion (#HH1198).
Narrower Koan (#HH0263) Ming (#HM0765) No-mind (#HM2163)
Mushinjo (#HM2974) Soto Zen (#HH0928) Bombu Zen (#HH0785)
No-thought (#HM2500) Rinzai Zen (#HH6129) Non-abiding (#HM3485)
Formlessness (#HM0910) Zazen meditation (#HH0882) Transcending mind (#HM3489)
Seeing into self-nature (#HM0917) Oxherding pictures in Zen Buddhism (#HM2690).
Related Ch'an (Zen, #HH0589) Satori (Zen, #HM2326)
Alaya-vijnana (Buddhism, #HM2730) Mahayana Buddhism (Buddhism, #HH0900).

♦ HH1004 Maya (Hinduism)
Sheaths masking reality — Illusion
Description Apparent reality as viewed by the human being is the result of five outer sheaths masking or veiling the truth. These sheaths - annamayakosha (physical body), pranamayakosha (breath or life energy), manomayakosha (mind), vijnanamayakosha (intelligence), anandamayakosha (awareness of bliss) - have to be progressively transcended in order to attain ultimate realization. It is not that what is perceived does not exist but that erroneous perception sees it for what it is not. There is attachment to the world of forms, for example, unless one retains sight of the void behind it.
Related Change (#HH1116) Illusion (#HM2510) Parinama (Yoga, #HH4561)
Advaita vedanta (Hinduism, #HH0518).

♦ HH1008 Watchings
Abstention from sleep
Description An individual may abstain from sleep in order to attend to prayer or other spiritual duties.
Related Continence (#HH0070).

♦ HH1009 Saijo Zen (Zen)
Description After satori has been experienced it is fulfilled by this last stage in zazen. With body relaxed and mind tranquil, the clarity and strength achieved in conscious awareness is taken deeper into the meditation. The exercise may be referred to as 'just sitting' in a state of great alertness *'harikitte'*, with no sense of hurry or striving *'nombiri'*, in complete steadiness of purpose *'dosshiri'* and totally withdrawn from the senses *'rinzen'*.
Broader Zazen meditation (Zen, #HH0882).

♦ HH1010 Five eyes (Buddhism)
Five levels of understanding — Madhyamaka buddhism — Middle way
Description Madhyamaka Buddhism recognizes five eyes or levels of understanding which must be purified by the bodhisattva so that the elements are not clung to.
1. Mamsacaksus (the eyes of flesh): These see only the immediate and are purified by the performance of moral deeds.
2. Divyacaksus (the Deva eyes): These see the karmic chain and its factors, such as good and bad, causal relations. Like the eyes of flesh, these eyes see in terms of poles and dualities. They are purified by dhyana (meditation).
3. Prajnacaksus (The eye of wisdom): This eye sees Nirvana, undifferentiated oneness, the universe through no categories. Different views disappear, activities of the mind return and enter dharmata, there is no other sphere for the mind to reach. All words cease and the world is seen in its true nature as no different from Nirvana. The wisdom by which this ultimate truth is seen is the eye of wisdom. Having once seen with this eye of wisdom the bodhisattva no longer clings to the elements of analysis, mundane, transmundane or Nirvana. This is contrasted with viewing only the mundane, which brings a false notion of existence, or the transmundane (incomposite), bringing a false notion of non-existence. The middle way (madhyamaka) abandons both these extremes. With unerring wisdom all elements of ignorance, whether general or particular, are put to an end.
4. Dharmacaksus (the eye of dharma): This eye reveals its wisdom from infinite compassion for all sentient beings, it sees all the different ways in which people conceptualize. The Bodhisattva sees that the only way to save all beings, as he has promised, is by uncovering the eye of dharma and seeing the conceptual networks of people. Then conceptual monsters can be seen and defined, shown to be unreal, relative constructs which are empty, mutually dependent poles. Thus others can be guided to Nirvana through the particular paths suited to their own circumstances.
5. Buddhacaksus (the eye of Buddha): This eye is not separate from the other four but it is not the same. The other eyes are consummated in this eye, it incorporates all the other levels into a basic integration.
Refs MacDowell, Mark *A Comparative Study of the Teachings of Don Juan and Madhyamaka Buddhism* (1986); Ramanan, Vetaka K *Nagarjuna's Philosophy as Presented in the Maha-Prajñaparamita-Sastra* (1971).
Related Madhyamaka (Buddhism, #HH1038).

♦ HH1016 Resignation to the will of God
Isvara pranidhana (Hinduism) — Devotion to the lord — Self surrender
Description This way to the realization of self in samadhi is through surrender of attachments and personal desires until the mind comes to rest. One path is through not simply a pious resignation of the will of the ego to that of God, but a real desire to become the instrument of God in the unfolding of the divine will and the scrupulous following of right action insofar as the individual is able to discriminate at this stage; there is consistent effort to remove consciousness from the ego to that of the Absolute being. Another path is through the emotions, with subordination of the will to that of the Absolute through love for the Absolute. The two paths are the equivalent of karma yoga and bhakti yoga respectively.
Context One of the five components of *niyama*, self-discipline.
Broader Niyama (Hinduism, #HM3280).
Related Renunciation (#HM3118) Self-oblation (#HH0968)
Karma yoga (Yoga, #HH0372) Kriya yoga (Yoga, #HH0969)
Bhakti yoga (Yoga, #HH0337) Human development (Islam, #HH1799).

♦ HH1020 Destructive manipulation
Brainwashing cults
Description Individuals are kept from making their own decisions, and ideas and suggestions implanted without their knowledge, through a number of techniques including: lack of sleep, lack of protein, group hypnosis, telepathic suggestion and sleep learning. A complete shift in consciousness may result so that the whole person works only for the organization by which he or she has been manipulated, functioning at the level of instinctual awareness. The subconscious mind may be irreversibly dulled; even if this does not occur, *deprogramming* is necessary to reverse the process.
Narrower Psychological imprisonment (#HM4539).
Related Floating (#HM3903) Blissing out (#HM1089) Deprogramming (#HH4102)
Ego destruction (#HM0655) Heavenly deceit (#HM3421)
Thought reform (Brainwashing, #HH0865).

♦ HH1021 Spiritual discipline
Description Discipline implies discipleship, being the follower of a teacher and being prepared to learn. The quest for spiritual development has shown that to achieve progress it is normally necessary to follow a teaching, a set of values, a guide or a teacher and to practice some specific set of rules. Among the numerous paths available one must be chosen and adhered to as the *way*, this term being used in many traditions (eg. Tao). As one's behaviour and attitudes are organized according to the discipline, the way will lead to a specific goal of development. Paradoxically, it is by strict adherence to such discipline that true freedom is experienced. Every discipline has its corresponding freedom.
Heteronomous discipline implies an external authority. The disciple may act in obedience to a teacher actually present and near at hand, for example a ch'an Buddhist master; or he may learn from the experience of another disciple who is further along the spiritual path, as from a Jewish rabbi, Buddhist arhat or Sufi master. In either case the disciple will follow duties appropriate to his spiritual position or progress on the path.
Autonomous discipline Here the teacher is not external to the seeker but deep within him, an inner wisdom that must be discovered. This is typified in the path followed by Siddhartha Gautama and his development as Buddha; he himself did not wish the role of master but recommended observance of his teachings - the eightfold path - for their own sake and not from loyalty to him.
Interactive discipline Here there is interaction between external traditions or structures and the seeker's internal authority which responds to and is given form by the external teaching. Such discipline is typical of the artist producing a work of art in accordance with the rules of the medium he is using but in response to an inner will which arises in the state of complete attention. In the Japanese tradition, interactive discipline is said to lead to satori which arises in carrying out the simplest everyday activity.
In practice, none of these three methods tends to be exclusive and each contains some aspects of the other two.
Disciplines may have a number of qualities. *Ecstatic discipline* tends to culminate in an an "out of body" experience in which the seeker is no longer limited by time and space, often as a result of ascetic practices. An example is the vision quest of North American Indians. *Constructive discipline* builds on desirable characteristics by faithful imitation of an ideal model. *Discipline of the body*, as in Christian monastic and Yogic ascetic traditions, emphasizes control of the physical body to free the mind of distractions of the senses. *Discipline of the mind* likewise frees the disciple from being carried away by thoughts, whether in the "not this, not this" of Vedic tradition or the "via negativa" of Christianity. *Discipline of the heart* is based upon the "bhakti" of the vedas, the charity of Christianity, the divine love exemplified in the Judaic Song of Songs. *Discipline of enduring relationships* is based on maintenance of proper relations and the social structure as a sacred duty, exemplified in the Jewish laws and in the caste system of India.
Self discipline is said to be best practised in conjunction with normal life and the transformation effected should be seen in ordinary relationships. Nor should it be enacted with a long face, as the freedom to be obtained is a cause for joy. A stumbling block arises when the disciple is to carry out discipline through the will alone, when success leads to worship of the will. There is also a danger of treating a discipline as law, with pride and fear as results. Spiritual discipline is only of value if it results in bringing a person to God and the granting of liberation. The inner change effected by continued discipline without self congratulation on results is then an act of grace.
Refs Foster, Richard J *Celebration of Discipline* (1984).
Broader Discipline (#HH0163).
Narrower Service (#HH0232) Solitude (#HH0333) Virtue of simplicity (#HH0990)
Autonomous discipline (#HH0328) Interactive discipline (#HH0740)
Heteronomous discipline (#HH0925) Discipline of study (Christianity, #HH1323)

Discipline of prayer (Christianity, #HH1193) Discipline of guidance (Christianity, #HH1244)
Discipline of confession (Christianity, #HH0825) Discipline of submission (Christianity, #HH0225)
Discipline of meditation (Christianity, #HH1688) Discipline of celebration (Christianity, #HH0956)
Discipline of worship (Christianity, Judaism, #HH3776).
Related Guru (#HH0805) Mentor (#HH3219) Ethical discipline (#HH1086).

♦ HH1022 Preaching the dharma (Buddhism)
Dhammadesana
Context The ninth of the ten meritorious deeds of South Asian Buddhist thought.
Broader Dharma (#HH0093) Ten meritorious deeds (Buddhism, #HH0446).
Followed by Right beliefs (Buddhism, #HH0974).
Preceded by Listening to the dharma (Buddhism, #HH1183).

♦ HH1024 Neurosis
Description According to Jung, a neurosis arises as a result of unbalanced or one-sided development, or a failure to adapt. Neurotic symptoms may be indicative of this one-sidedness but also of an attempt at self-healing, a drawing of the attention to the condition; they may be the result of a frustrated attempt to find meaning in life. Neurosis may be caused by any of a number of factors, including guilt, sexual repression and childhood trauma. Unlike psychosis, neurosis allows maintenance of contact with reality and only in extreme cases manifests as actual mental derangement.
Refs Horney, Karen *Neurosis and human growth* (1970); Maslow, Abraham H *Neurosis as a Failure of Personal Growth* (1967).
Related Adaptation (#HH1233).

♦ HH1028 Revelation
Description Since God is hidden and cannot be fully grasped by human thought or described in terms of concepts derived from experience of the world, knowledge of God can never be obtained from natural science and his attributes cannot be described directly but only through analogy. God may, however, "choose" to make himself known and revelation is direct communication to man from God. To the Christian, as to those of many religions, this is the revelation of God in history experienced as guidance in the individual's life. Man may seek for revelation in terms of his practical needs, desiring good relations with the provider of material blessings; it may be for understanding of mysteries inherent in this life, such as birth and death, and of how life should be lived; or it may be for understanding for its own sake. All may be desired and most religious would say that all could be revealed – if God is all powerful, then it is within his power to make himself known. All revelation thus depends ultimately on God's will, but man has a part to play in its effectiveness. He has to be prepared to accept revelation through open eyes and ears – in this sense revelation is in two parts, the giver and receiver. Primarily the experience is of God himself and of his purposes (as in Krishna's revelation to Arjuna in the Bhagavad Gita); although the promulgation of a code of laws may be part of what is received (Moses receiving the ten commandments). To the Christian the greatest revelations are of Christ on earth and of his church.
Revelation is distinguished from magic, in that the former implies power over the divine while the latter is freely given by the divinity. Inasmuch as they are manifestations of divine grace, mystic and gnostic experiences may be included under this heading. The veracity of revelation experienced cannot be determined objectively; however, those who have been considered the most sane and trustworthy have also been those claiming experiences of revelation. Revelation is known by its fruits. The form of revelation varies enormously – from oracles, dreams, beauty, visions, to the still, small voice. The glories of creation reveal the even greater glory of their creator. Sacred writings, as revealed truth, may be considered revelations to the reader. As well as for personal instruction or persuasion, the person receiving the revelation may be a shaman, prophet or mediator who is inspired to divine mission or to act as an oracle.
The term revelation points to five different and often competing understandings of God's interaction with man: 1) Revelation is divinely authoritative doctrine inerrantly proposed as God's word by the Bible or by official church teaching; 2) Revelation is the manifestation of God's saving power by his great deeds in history; 3) Revelation is the self-manifestation of God by his intimate presence in the depths of human spirit; 4) Revelation is God's address to those whom he encounters with his word in Scripture and Christian proclamation; 5) Revelation is a breakthrough to a higher level of consciousness as humanity is drawn to a fuller participation in the divine creativity.
Refs Balsys, Bodo *Revelation* (1983).
Narrower Private revelation (#HH3238).
Related Inspiration (#HH2139) Experience of revelation (#HM3538).

♦ HH1031 Humour
Description The quality said to make man human, humour ensures self-criticism and a sense of proportion that protects the individual from self-importance. It guards against fanaticism, not by bringing in equally fanatical alternatives but by weakening it and protecting against excess.
Related Sense of humour (#HM3596).

♦ HH1032 Self-transcendent systems
Description Natural systems (whether physical entities like atoms or trees or social systems, languages or any evolving process) are permanently in a condition of change. They are not only changed by their environment but they also change their environment which consequently alters the effect the environment has on them; they can therefore be referred to as self-transcendent. Human development systems based on self-transcendence – which are open to change while in progress – might be expected to be more effective and in tune with the natural order of things, even if less predictable.
Related Process thinking (#HH1643)
General system theory of human development (#HH1127).

♦ HH1034 Evil
Description Jung saw good and evil as judgement based on experience, a subjective response so that what appeared evil at one stage of development might in fact appear good at a higher stage. Finally, however, the individual has to deal with evil as the necessary opposition to good as shadow is to light. On the contrary, rather than seeing good and evil as equally balanced, religion sees evil as ultimately subordinate to good and while evil has always to be guarded against in this life it will ultimately be conquered. Some authorities would say evil to be derivative, that it exists only as a distortion of the good. It is the result of the free choice of a created will setting itself in opposition to God, a renunciation of good; and therefore of no independent reality without good. Theologically, evil thus commenced with the fall of man, when man set himself up as autonomous with respect to God and sought fulfilment in himself rather than in the creative and selfless love of the creator. Rather than disorder or disruption, although this may be the result, evil is then the attempt to be self-sufficient. For Kierkegaard, purity of heart is to will one thing, and that is good; he demonstrates that the good is always single, unity, and that evil is therefore fragmented, "legion", duality.
Refs Rohde, S *Deliver Us from Evil* (1946).
Narrower Awareness of angelic evil (#HM2212) Emblems of personified evil (Tarot, #HM2378).
Related Wickedness (#HM3789) Accepting the shadow (#HH0204).

♦ HH1038 Madhyamaka (Buddhism)
Middle way
Description This is study of the profound teachings of voidness and the ten perfections. In madhyamaka the dialectic is not the means to an end but the end in itself. It is implied that the real is overlaid with notions and views, which are ignorance (avidya) screening the real. The dialectic puts forward four logical alternatives: A is B; A is not B; A is both B and not B; A is neither B nor not B. All these four facets are used in philosophical discussion as the warp and weft of conceptual knowledge are exposed. Reality is simply experienced (prajnaparamita).
Context The second of the five primary subjects for the Geshe degree in Tibetan Buddhism.
Refs MacDowell, Mark *A Comparative Study of the Teachings of Don Juan and Madhyamaka Buddhism* (1986).
Related Five eyes (Buddhism, #HH1010) Emptiness (Buddhism, Zen, #HM2193).
Followed by Pramana (Buddhism, #HH4388).
Preceded by Prajnaparamita (Buddhism, #HM1344).

♦ HH1039 Transformation
Description Literally a radical or structural change, an experience of transformation is contrasted with one of transcendence in that the change to one's being is lasting. Such change may be psychopathological, as in identification or possession; it may arise through altered states of consciousness, be induced by drugs or arise in response to rituals and magical procedures such as initiation rites, or arise spontaneously during the process of individuation when the individual has the sense of being reborn as a larger personality. A cure may be regarded as the transformation from sickness to health. Transformation is central to the experience of conversion. It may commence in a state of turmoil and difficulty as thoughts, feelings and actions are all changed from traditional, habitual ways of behaviour and a new awareness of commitment takes over.
Refs Da Free John *Enlightenment and the Transformation of Man* (1983); Durckheim, K G von *The Way of Transformation* (1971).
Narrower Cure (#HH1264).
Related Change (#HH1116) Rebirth (#HH0304) Tradition (#HH0440)
Transcendence (#HH0841) Religious experience (#HM3445).

♦ HH1047 Afflictions (Yoga, Hinduism)
Klesas — Causes of misery
Description Unable, through deluded sense of egoism, to be aware of true reality, the individual is swayed by attraction and repulsion, attachment to possessions and to the present life. This is the cause of all misery and affliction. The practice of yoga is designed to free one's self from this delusion and achieve objective awareness of reality free from the cause of misery.
Narrower Desire (#HM2433) Asmita (Yoga, #HM2937) Abhinvesa (Yoga, #HM3871)
Revulsion (Yoga, Zen, #HM3620) Ignorance (Buddhism, Yoga, Zen, #HM3196).
Related Five kleshas (Yoga, Zen, #HH0390).

♦ HH1049 Deification
Description Where no hard and fast distinction is drawn between men and gods, extraordinary qualities of past or present heroes may be rewarded by honours and rituals reserved for deities.
Refs Atreya, B L *Deification of Man*.

♦ HH1066 Narcissism
Self-love
Description Primary narcissism, the love of one's self, is necessary before one is capable of relating to or loving others. This self-love is inhibited in its development if a child does not feel loved by its parents and results in lack of self esteem in later life. In secondary narcissism there is an inability to be aware of the world as apart from one's self; the omnipotence and grandiosity being the distorted image of the relationship the child felt he should have had with his parents. Healthy narcissistic development, the concentrating of time and effort on developing ideals, values and aims in life, is a continuing necessity for individuation.
Related Self-love (#HM0478) Narcissistic equilibrium (#HM2630).

♦ HH1073 Non-action
Wu-wei
Description Every action is according to nature, not the result of personal prejudice nor of desires. Non-action is not idleness or passivity but refraining from acting against the grain and from using force when perception sees that this must fail.
Related Satori (Zen, #HM2326) Inner peace (#HM3575)
Human development (Taoism, #HH0689).

♦ HH1075 Spatial intelligence
Description An accurate perception of the physical world, an ability to transform or modify these perceptions, and the recreating of certain aspects of visual experience without relevant physical stimuli – these are all part of spatial ability. Centred in the right-hemisphere of the brain, spatial skills are typical of cultures where tracking, hunting and visual recognition of the environment are paramount; but present-day Western culture requires it no less, whether for the architect or the mathematical topologist or the molecular biologist.
Context One of the separate or multiple intelligences put forward by Howard Gardner.
Broader Multiple intelligences (#HH0214).

♦ HH1076 Clearing
Auditing
Description Unpleasant events which have caused repression of associated emotions are examined and their associated emotions discharged so that the mind is cleared. This process is part of many therapeutic techniques.
Auditing is a monitoring process used in dianetics to detect and eliminate *engrams*, by using an "E-meter". This process clears the individual's mind of psychological and psychosomatic disorders, leaving him "clear" and free to develop his potential or *thetan*.
Broader Dianetics (#HH0106).
Related Bathing (Taoism, #HH3664).

♦ HH1086 Ethical discipline
Moral discipline
Description Despite an inner knowledge of what is the right or moral course, the individual habitually discovers that his desires prevent him following such a course. There is always the tendency to follow the easiest rather than the best. Aristotle says "the incontinent man's impulses run counter to his reason"; and this is echoed by Death speaking to Nachiketas in the Katha Upanishad "The good is one the pleasant another; both command the soul" and "the mind of the wise man draws him to the good, the flesh of the fool drives him to the pleasant". The will requires some form of chastisement to cooperate with reason and not give way to the irrational element; the desires need to be reformed to be consistent with right reason.
The first stage of ethical discipline is that received by a child from external authority and accepted

without an understanding of why. Some systems continue to demand unquestioning obedience but true morality is said to be achieved only through voluntary discipline of the self. This includes *discipline of the the intellect*, as only the disciplined and trained mind has the necessary control to systematize and unify incoming data – *apperception* – rather than functioning through simple association of ideas. There must also be *discipline of the will*, moral training giving the power to direct and control natural impulses and to free the individual from domination by the idea of the moment. This implies self control and attention to the likely result of conduct. *Discipline of the emotions* supplies the control for power over feelings and passions which would otherwise cloud moral decision–making.
Broader Discipline (#HH0163).
Related Self control (#HH0778) Apperception (#HM1961) Moral perfection (#HH5098) Spiritual discipline (#HH1021) Self-discipline (Christianity, #HH2121).

♦ **HH1088 Positive thinking**
Positive thought
Refs Meyer, Donald *The Positive Thinkers* (1988).
Broader Thought processes (#HH0530).
Related Mind cure (#HH1772) Super–immunity (#HH3755) Psychoneuroimmunology (PNI, #HH1401).

♦ **HH1092 Technique of no technique**
Description In transcendental meditation it is the initial starting point which counts. Once the correct conditions are applied there is no technique to follow, the meditation proceeds naturally. This is said to reflect experience of the physical world of mechanics when the minimum possible energy for an action determines the path of physical events. Transcendental meditation is therefore said to follow the path of not doing although this is distinguished from doing nothing.
Broader Transcendental meditation (TM, #HH0682).
Related Path of effortlessness (#HH2188).

♦ **HH1097 Jewish meditation** (Judaism)
Refs Kaplan, Aryeh *Jewish Meditation* (1985).
Broader Meditation (#HH0761).

♦ **HH1098 Crusade evangelism** (Christianity)
Mass evangelism — Television evangelism
Description This approach is usually quite separate from the Church in the community where the individual lives. Often it is focused on a particular communicator who uses modern marketing and advertising techniques to communicate a message. The method may have a well organized follow–up procedure but there is still some difficulty in maintaining a conversion originally taking place through hearing verbal proclamation only.
Broader Evangelism (Christianity, #HH3798).

♦ **HH1099 Causally continuous doctrine of being** (Buddhism)
Paticca samuppada — Noble system — Paccayakara
Description This logical sequence which argues that death, decay and suffering are absent only when birth ceases is a central doctrine of early *Buddhism*. It covers the sequence of birth – death – birth through to enlightenment, or "coming to be" and "ceasing". Ignorance in a previous life resulted in actions whose results are transmitted into the present life, where the consciousness sequence continues leading to future birth (whether in earth, a heaven, a hell), decay and dying and so on.
Context Sometimes referred to as the second of the four noble truths.
Broader Four noble truths (Buddhism, #HH0523).

♦ **HH1100 Electronic meditation**
Description By monitoring progress on the EEG instrument there is enhanced capability to achieve meditative states of awareness – high alpha rhythm.
Related Biofeedback training (#HH0765) States of awareness in biofeedback training (#HM3744).

♦ **HH1101 Taboo**
Tabu
Description This may refer to the nature of anything that is forbidden; but that which is taboo is usually prohibited because it is in some way sacred, mysterious or extraordinary. This may be particularly the case with anything connected with death. Rituals and ceremonies have grown up in many cultures and religions around what is taboo, to avoid danger associated with mana, or with positive holiness working as a power. Although breaking a taboo is thought of as morally evil, it is due to the fault of an individual and not as serious an offence as an antisocial action which involves an insult to the ancestors of the group, when guilt is considered collective and has to be expiated collectively.
Refs Freud, Sigmund *Totem and Taboo* (1952); Watts, Alan W *The Book – on the Taboo against Knowing Who You Are* (1967).
Broader Rites (#HH0423).
Related Mana (#HM2686) Purification (#HH0401) Sense of sacredness (#HM2302).

♦ **HH1107 Implosive therapy**
Description This therapy is based upon the assumption that conditioned aversive stimuli produce neurotic symptoms as an avoidance response. Rather than direct inhibition of of these symptoms, implosive therapy reproduces verbally or otherwise the maximum number of stimuli of the original conditioning, excluding the unconditioned aversive stimulus; particular emphasis is placed on the stimuli for which the individual can give no explanation. This diminishes the aversiveness if such stimuli are removed from the avoidance responses making up the neurotic behaviour.
Broader Therapy (#HH0678).
Related Aversion therapy (#HH0256).

♦ **HH1108 Seventy–five dharmas** (Buddhism)
Vaibhasika system
Description The 75 dharmas, classified in a number of ways, are intended to provide a classification of all possible types of existence. They may be considered in 5 different types: physical; mental; those related to mind; those unrelated to matter or mind; unconditioned dharmas. The properties associated with the five sense organs and their objects are physical dharmas, (taste, tangibility, smell, etc), as is unmanifest physical form; the mind is the one mental dharma; mentally related dharmas number 46 categories in all, covering all mental events and each distinct from the others. These include attitudes and emotions, formation of categories, consciousness. Through learning to define and understand the dharmas, each experience can be analysed into its component parts, each transitory mental event from which the continuity of everyday experience is constructed being labelled and identified.
Broader Dharma (#HH0093).

♦ **HH1109 Seishin education** (Japanese)
Spiritual education
Description This training provided for employees of Japanese companies emphasizes the concept of morality as being loyalty and hard work for the company. Training is residential and takes place over several months. An appreciation of the burdens of leadership and of the need for cooperation is built up as the employees take responsibility for every aspect of their life together, from cleaning and kitchen duties to scheduling of programmes. The seishin aspect of every activity, inside or outside the training, starts to become evident. Emphasis is on harmonious teamwork, teams being the basis for all activities whether or not competitive; and on the quality of persistence. The conformity and acceptance required to work in a group are considered difficult and therefore useful in self development. There is concern for team spirit and the composure of the individual so that mind and body work harmoniously together for effective action.
The key element for the individual taking part in seishin training is the spiritual struggle to combat emotional wavering and complete whatever test is underway. Competition is within the self more than to achieve better than others and success is completion of the test, each completed test making completion of the next test less difficult. Personal difficulties are overcome by identifying and dealing with incorrect attitudes so that there is ready acceptance of unpleasant or difficult tasks as being a necessary part of life. Problems are regarded as useful opportunities to test and acquire spiritual strength. The selfish self is defeated, resulting in contentment and freedom from confusion and frustration.
Broader Education (#HH0945).

♦ **HH1116 Change**
Mental transformation
Description Change represents a dynamic as opposed to a static state and is characteristic of the universe and all that is in it. An awareness of appearance as *maya*, or the veils covering reality, demonstrates this – that the apparently solid and stable is actually a state of flux and nothing can be said to be identically the same from one instant to the next. It is maya which makes the static state appear real. The process of change or of transformation is manifested through alterations in an individual's thoughts, feelings and actions. In religious terms this may imply a conversion to a new faith or stronger commitment to a previous faith. The process of change may be very uncomfortable and is the subject of study of psychology.
Related Evolution (#HH0425) Conversion (#HH0077) Parinama (Yoga, #HH4561) Maya (Hinduism, #HH1004) I Ching (Taoism, #HH0004) Transformation (#HH1039) Consciousness shift (#HM1793).

♦ **HH1121 Morita therapy**
Broader Therapy (#HH0678) Psychotherapy (#HH0003).

♦ **HH1123 Drug use for transcendent experience**
Description The use of drugs to alter consciousness chemically as an aid to achieve mystic experience is common in religious and other traditions. It is one of several techniques used by mystics, said (Aldous Huxley) to be no more artificial in itself than some ascetic practices taken to extremes to achieve the same purpose (lack of food, flagellation, breathing exercises and continual chanting all having marked effects on body chemistry). However, others assert that the following of ascetic practice is necessary to prepare the person for religious or mystic experience which then arises spontaneously rather than to order and, some believe, has more long–lasting effects. Independent tests show that, in a religious setting, drugs may enhance experience, possibly but not necessarily triggering what then requires considerable effort to integrate into everyday life.
Refs Clark, Walter Houston *Chemical Ecstasy* (1969); Huxley, Aldous *The Doors of Perception* (1954); Watts, Alan *The Joyous Cosmology* (1962); Watzlawick Paul, Weakland John and Fisch Richard *Change* (1974).
Broader Psychedelic drugs (#HH0075).
Related Aesthetic LSD experience (#HM2542) Psychodynamic LSD experience (#HM3271) Modes of awareness associated with use of hallucinogens (#HM0801) Modes of awareness associated with psychoactive substances (#HM0584).

♦ **HH1125 Coniunctio**
Description In alchemical terms this refers to the union of two opposites to produce offspring, the birth of a new element combining in a greater wholeness attributes of both. For Jung, this is the archetype of psychic functioning, demonstrating symbolically the relationship between unconscious factors. The resultant symbolic projections are a useful source of insight in analysis. They may represent: the relationship formed between analyst and subject (patient); the relation between the unconscious and consciousness of the subject or of the analyst; the merging of the spiritual and the physical.
Related Alchemy (Esotericism, #HH0221).

♦ **HH1126 Self discovery**
Refs Beesing, Maria et al *The Enneagram* (1984).

♦ **HH1127 General system theory of human development**
Description Rather than a process whose outcome is fixed at inception, human development is seen as evolving interactively, new situations giving rise to new structures, likened to standing–wave patterns, these structures defining new processes which give rise to new structures and so on. This complementarity is the basis of a dynamic general system theory encompassing, among other factors: limitations on systems taken beyond practical complexity; interaction between system and environment (in general, complementarity of space–time structure and function); conditions giving rise to temporary structural stability at discrete levels of complexity; evolving rather than absolute nature of time as a property of the system.
Refs Jantsch, Erich and Waddington, Conrad H (Eds) *Evolution and Consciousness* (1976).
Related Process thinking (#HH1643) Open learning systems (#HH0326) Self–transcendent systems (#HH1032).

♦ **HH1134 Inward zazen meditation** (Zen)
Rinzai zazen meditation
Description The preferred method of meditation of the Rinzai school. The aim of this meditation is to realize one's own Buddha mind. The student is given a hard task for the mind, perhaps the treatment of a koan problem or even by being physically startled by a shout or a blow, so that the mind–process is knocked off its customary position.
Broader Rinzai Zen (Zen, #HH6129) Zazen meditation (Zen, #HH0882).
Related Koan (Zen, #HH0263) Outward zazen meditation (Zen, #HH0621).

♦ **HH1141 Ordeal**
Trial by ordeal
Description In many cultures, trial by ordeal has been used as a test of innocence or guilt. The person or persons involved perform some act which would normally result in injury (handling red hot metal, plunging an arm into boiling water); if there is no injury or if injury is limited to a predetermined extent then the person is innocent. A conviction of innocence on the part of the

person taking part is said to have actually resulted in succeeding in the ordeal.
Broader Rites (#HH0423).

♦ HH1150 Behaviourism

Description The behaviourist approach maintains that any reference to a psychological or mental state is ultimately reducible to, or should in fact be replaced by, some expression about the disposition of an organism to behave in a particular way. This conflicts with the Cartesian view that inner states are not behavioural dispositions themselves and are capable of causing such dispositions.
Related Human psychological development (#HH0447).

♦ HH1162 Individualism

Refs Jacobi, Jolande *The Way of Individualism* (1967).
Related Individuality (#HH0776).

♦ HH1166 Soteriology

Way of salvation
Description The concept of salvation presupposes a problem in the human condition which it is possible to ameliorate, whether through the individual's own efforts or as a result of a saving God. Different religions have defined this problem in different ways – for example ignorance or avidya in Hindu and Buddhist thought, original sin in the Judaic and Christian traditions, giving rise to the endless cycle of birth and death in the former and separation from God in the latter cases. The character of salvation is also different in different schools of thought. It may be that salvation is achieved in this life (for example, jivanmukti) or after death, or having worked through a series of situations (heavens and hells, as in Buddhism; purgatory, as in Catholicism). A utopian blessed state for all in this life is the hope not only of some religions but of secular movements as well, when practical goals such as a classless socio–economic system (Marxism) may emphasize community as opposed to individual salvation.
Related Mukti (#HH0613) Salvation (#HH0173) Afterlife (#HH0399) Liberation (#HH0388) Moksha (Hinduism, #HM2196) Jivanmukti (Hinduism, #HM3890).

♦ HH1167 Names of Christ (Christianity)

Description Fray Luis de León clearly states that language about God is inadequate. God can only be known and named when finally standing in His presence. Until that time, the various names which Christ has been given each reveal or provide the means of exploring some facet of God's revelation of Himself. The greatness, perfection, functions, benefits of Christ are beyond the ability of a soul to encompass them, let alone a single name describe them. Each name is like a single drop of water poured into a glass, each adding to understanding because one is incapable of receiving more at any given time. This gradual self–revelation of God through the Spirit is different and appropriate for each individual.
Refs Byrne, Lavinia (Ed) *Traditions of Spiritual Guidance* (1990).
Related Ninety–nine names of Allah (Sufism, #HH2561).

♦ HH1171 Via negativa

Description Because God transcends all qualities it is impossible to describe him in terms of such qualities, or in terms of what he is, but rather in terms of what he is not. Names and definitions of God are misleading because they limit what is essentially indefinable. What is above knowledge cannot be an object of knowledge. The via negativa seeks to have knowledge of God through union with him, transcending reason and avoiding definitions; instead there is self–purification and worship both in prayer and praise and in silence until, in pure devotion, God's presence is experienced. This state of "unknowing" is not the absence of knowledge but the surrender of the mind which then becomes the instrument in union and contemplation. This Christian approach is analagous to the "neti, neti" (not this, not this) of Vedic teaching, to the indescribable nature of Tao, to the Nirvana of Buddhism which cannot be described or analyzed. All these traditions emphasize meditation, contemplation and reflection as a means of approach. The via negativa does not imply that the positive way is actually wrong but that there is a danger of becoming lost in a system of concepts without real experience of God.
Related Ineffability (#HM3589) Via positiva (#HH6222) Via purgativa (Christianity, #HH4090) Via illuminativa (Christianity, #HH6030).

♦ HH1173 Traditionalism (Christianity)

Description This view that truth cannot be known through reason, but only through authoritatively attested divine revelation, is rejected by the Catholic Church since it overlooks the fact that tradition and revelation are addressed to human reason, which is capable of responsible judgement.

♦ HH1179 Grail quest

Refs Jung, Emma and von Franz, M L *The Grail Legend* (1971).
Related Alchemy (Esotericism, #HH0221).

♦ HH1183 Listening to the dharma (Buddhism)

Context The eighth of the ten meritorious deeds of South Asian Buddhist thought.
Broader Dharma (#HH0093) Ten meritorious deeds (Buddhism, #HH0446).
Followed by Preaching the dharma (Buddhism, #HH1022).
Preceded by Rejoicing at the merit of others (Buddhism, #HH0932).

♦ HH1187 Purification by overcoming doubt (Buddhism)

Purity of transcending doubt — Kankhavitarana–visuddhi–niddesa (Pali) — Lesser stream–enterer — Junior stream–winner
Description Here knowledge is established, as doubt about the three divisions of time is overcome or transcended through discerning the causes of mentality–materiality or name and form, 'nama' and 'rupa'. They are seen to be neither created by a creator or causeless, but occurring due to conditions. Further, these conditions may be seen as both general and particular. Another may put away doubt through seing the condition of mentality–materiality through their dependent origins in reverse order, ending with ignorance, or in direct order, starting with ignorance. Another means is by the consideration of karma and karma result. All this – whether thought of as correct knowledge, right seeing or overcoming doubt – arises in one practising insight; he is called a *lesser stream enterer*.
Context One of the five purifications or purities which constitute the trunk or body of understanding according to Hinayana Buddhism, the others being: purification or purity of view; purification by knowledge and vision of what is the path and what is not the path or purity of knowledge and discernment of the right path and the wrong path; purification by knowledge and vision of the way or purity of knowledge and discernment of the middle way; purification by knowledge and vision or purity of knowledge and discernment.
Broader Purifications (Buddhism, #HH3875).
Related Full understanding as the known (Buddhism, #HM3490).
Followed by Purification by knowledge and vision of what is the path and what is not the path (Buddhism, #HH4007).
Preceded by Purification of view (Buddhism, #HH2718).

♦ HH1188 Saintliness

Description Despite the great diversity of saints and their admirable qualities, it is possible to see a similarity among the saintly. It is not so much the individual virtues but the overall way the virtues are displayed, exemplified by a maturity of personality in which the natural character is perfected by grace.
Related Sainthood (#HH0331).

♦ HH1193 Discipline of prayer (Christianity)

Description Although the life of prayer has been derided as lacking in activity, the most active of religious leaders have systematically devoted much time to prayer, considering it the main business of their lives. Examples are Martin Luther and John Wesley, who both spent several hours a day in prayer. Like all disciplines, prayer must be learned and practised. The disciples asked Jesus to teach them how to pray. There is a realization that the universe is not so fixed as to be unchangeable and a conviction that that which is prayed for is the will of God. Learning takes place in practice until what started as verbal request becomes silent listening. Eventually prayer may accompany and infuse all activity. Although answered petitions may be dismissed as coincidence, it has been said (William Temple) that coincidences occur more frequently when one prays.
Broader Prayer (#HH0185) Spiritual discipline (#HH1021).

♦ HH1198 Human development through religion

Description Some aspects of how this subject has been approached are touched on here.
1. Religion reveals the ultimate, eternal reality in whose unity and completeness there is perfection. Humanity is imperfect. Religion provides the means of overcoming the separation between the perfect and the imperfect, whether by human initiative or divine grace, or more usually by both, the one arising in response to the other.
2. The individual is not an absolute entity – "No man is an island" – everyone is essentially associated with something beyond himself. There is a feeling of psychic incompleteness in one's self while at the same time a consciousness of a relationship with something beyond the self which may be obscured by everyday life and current opinions.
3. It may be argued that it is perfectly possible to lead a happy and successful life without a religion; but the question "why do I need religion ?" is countered by another question "what is life, existence, for ?". It is this question which sets a seed of doubt into complacency and takes life to a level where everything else is meaningless except the search for a meaning, the ordinary mode of being is broken through; and it is here that religion becomes a necessity. Often it requires a serious event – coming face–to–face with death, for example, or the death of someone much loved – which makes the questioning arise. This questioning is specific for each individual. But as questioning continues, the underlying unity of all things in reality – "to see the world in a grain of sand" – demonstrates the unreality of separation; in reality subject and object are indissolubly part of the greater unity.
4. Although things in the external world are thought of as real they are rarely experienced as real but simply observed, with the relation of observed to observer as external to internal, with a division between the two. Even internal thoughts and feelings are rarely experienced; these also are usually observed, with separation between the self observing and the feeling being observed. This division results in a mechanistic view of the world, each ego entirely separate and surrounded by lifeless matter. There is a need to return to see life as a whole, lived through all living things with psychic sympathy between all living things.
5. The search for meaning and the answers arrived at depend upon the individual's level of development and his cultural and religious experience; different faiths have different practices; but ultimately all imply a theory of reality, and "a man's religion, if it is sincere, is that consciousness in which he takes up a definite attitude to the world and gathers to a focus all the meaning of his life" (E Caird).
6. The supreme reality, God, is that mighty being whose consciousness transcends the manifested universe. Jung considers God a mystery knowable only to God. All man can do is to speak in terms of images; the God–image in each person is the self (distinguished from the ego). Religion is then an attitude of consciousness when changed by experiencing the numinous.
7. Despite widely diverse concepts of God, there is a characteristic conviction of a personal relationship with some external, transcendent power, this power being intimately involved in the person's needs, behaviour and interests at every level. Different religions and cultures will typically have different deities or aspects of the deity to deal with the same functions. Moreover, these concepts of God will satisfy emotional and spiritual needs and supply role models in individual development.
8. As a reflexive cultural system, religion provides models of and for self and society, the pattern behind all patterns. It is an interpretation of existence which may itself be interpreted. Myths, rituals and sacred symbols re–enact comprehensive ideas of order and paradoxical and ambiguous related disorder. Reflexive religious practices enable withdrawal from the world and a turning inward towards the self.
9. Rather than an outburst of irrational spirit, religion enlightens rationality. Purification through faith in the fundamentals of one religion and endeavour to achieve the goal set by that religion will be transformed into universal faith, since differences between faiths are superficial. The one religion of which all faiths are facets is based on love.
10. Different aspects of different religions may co–exist and complement each other. This is particularly true in China, where Confucianism provides a set of ethics for public life and the rites of passage within it; Taoism shapes attitudes to the natural world and the related festivals, and deals with healing; and Buddhism deals with the cares of the world, with death and its rituals, and with salvation in the afterlife. In total, there is joy and meaning in life and death which can be shared in a social context.
11. The mentality of a particular country or region may have a marked effect on the way religion is lived. Teilhard de Chardin indicates three spiritual modalities in the Far East: In *India* the attitude is of a mysticism of God. There is emphasis on the One and the divine, with an awareness of the unreality of all phenomena and of the invisible as more real than the visible. These leads to a pantheistic or theistic attitude even within Buddhism and to a conception of unity within Hindu theism. In *China* the mysticism is of the individual faced with and coming to terms with the world; emphasis is on harmony and equilibrium of an established order. There is a naturalist or humanist attitude, the tangible is supreme in both Taoism and Confucianism; and this is reflected in Chinese Buddhism with Boddhisattva Amida being substituted for Nirvana. *Japan* considers the mysticism of the social, and the humanistic attitude centres not so much on the individual as on the group, on movement, on conquest.
Refs Bellah, Robert *Beyond Belief* (1970); Bowker, John *Problems of Suffering in the Religions of the World* (1970); Chirban, J T *Human Growth and Faith* (1981); Dowdy, Edwin (Ed) *Ways of Transcendence* (1982); Hinnells, John R (Ed) *A Handbook of Living Religions* (1984); Hinnells, John R Ed *The Penguin Dictionary of Religions* (1984); Nishitani, Keiji *Religion and Nothingness* (1982); Yogananda, Paramahansa *The Science of Religion* .
Narrower Syncretism (#HH3214) Human development (Zen, #HH1003) Human development (Islam, #HH1799) Human development (Taoism, #HH0689) Human development (Sufism, #HH0436) Human development (Judaism, #HH3029) Human development (Sikhism, #HH6292) Human development (Jainism, #HH0622) Human development (Buddhism, #HH0650) Human development (Hinduism, #HH0330)

Human development (Baha'ism, #HH2435) Human development (Christianity, #HH2198)
Human development (Zoroastrianism, #HH1903)
Human development through pantheism (#HH5190)
Human development through panentheism (#HH1908)
Human development through primal religion (#HH1902)
Human development through new religious movements (#HH1523).
Related Reflexivity (#HH0977) Spiritual way (#HH1867) Supernaturalism (#HH4213)
Religious growth (#HH1321) Religious experience (#HM3445)
Mysteries and religion (#HH4032) Universal religion (Yoga, #HH0746)
Inter-religious dialogue (#HH2434) Human development through theism (#HH4663).

♦ HH1199 Soul care (Christianity)

Spiritual friendship — Soul friendship — Spiritual guidance

Description It has been advocated that every Christian should have a spiritual guide with whom he or she share her spiritual experiences. The caring implied here derives from the idea of a spiritual director or mentor but without the authoritarian implications, with the aim of aiding spiritual growth. Hunger for a deeper spiritual life and awareness of the limitations involved in trying to carry on alone lead to agreeing with a spiritually mature and respected person to meet on a regular basis within defined expectations and roles, although the structure of the relationship allows considerable choice and depends on the nature of the two people concerned. Although usually a relationship between a more and a less spiritually mature person, two people may act as spiritual guides for each other or they may be group guidance. Whatever the structure, of central importance is the discerning of the leadership of the Holy Spirit and a nurturing in Christian spirituality and growth. Some also stress the importance of the goal of a growth in prayer and of life in the Spirit, a "dying" to the sinful impediments to union with God and experience of His forgiveness.

The main aims are soul cure, through remedy for sin, and soul care as spiritual development through the various stages from depravity to holiness. Overall, in line with Christ's teaching, what is sought is a total reorientation of the person's life. It is clear from Christ's teaching that he considered the individual person to be of immense worth. There is an obvious connection between soul care and both psychotherapy and pastoral counselling. Where soul care is said to differ is that, rather than seeking to alleviate anxiety, or using anxiety as a means of growing and of understanding what God is saying, the anxiety is not the focus but rather the longing to give one's self to God more fully and know more deeply His presence in one's life. The main aim is spiritual rather than psychological growth, although both may proceed together, and psychological spirituality is not confused with Christian spirituality.

Soul care is referred to in ancient times. Socrates referred to himself as a healer of souls (a term from which "psychiatrist" derives). The Old Testament refers to wise men who counselled and whose function is now performed by the rabbi. And the role of the good shepherd is described in both Old and New Testament. The tradition of soul friend dates from earliest Christian periods, particularly in monastic orders. The rules of St Columbanus, for example, state that everything must be done with counsel, this bringing safety of conscience and exoneration of the soul. The counsel was not necessarily from a priest but should be from someone knowledgeable of the scriptures. Children when they were confirmed would have a soul friend with whom they would read psalms, hymns and the rules of the church. Clearly to be a soul-friend in monastic terms could seem an onerous task. Advice is given on the necessary qualities, in particular the need for holiness and wisdom. Correction should be offered without sharpness or reproof although the seeker of a soul friend was recommended to seek out one who would spare him the least. If a penitent did not follow the soul friend's advice then he could refuse to continue as soul friend. A soul friend was confessor in the broadest sense as well as director of souls. The desert fathers also acted as guides on the spiritual road; again, theirs was no authoritarian leadship and teaching was primarily by example. Other examples are the *startsy* of the Russian Church, who took Christ the good shepherd as their example and who suffered with and for their "sheep".

However, although the 14th and 15th centuries were times when spiritual guidance was practised extensively, following the Council of Trent (1545-1563) the practice narrowed to a role of guarding orthodoxy and avoiding both heresy and questionable mysticism. In the protestant tradition, as well, mutual admonition and guidance came to replace spiritual guidance of a personal and individual nature and also the direct action of God in the individual life was stressed.

Refs Benner, David G *Psychotherapy and the Spiritual Quest* (1988); Byrne, Lavinia (Ed) *Traditions of Spiritual Guidance* (1990); Ives, Kenneth H *Nurturing Spiritual Development* (1982).
Narrower Practice of the confidence (#HH3007).
Related Counselling (#HH0842) Psychotherapy (#HH0003) Hasidism (Judaism, #HH0597)
Spiritual direction (#HH0442) Discipline of guidance (Christianity, #HH1244)
Psychospiritual therapy (Christianity, #HH4201) Discipline of confession (Christianity, #HH0825).

♦ HH1204 Reflection

Description The process of reflection is a turning inwards of the mind on itself so that a deliberative step intervenes between a stimulus and reaction to it. At the instinctive level this is a compulsive act but is transformed, when raised to conscious awareness, to be purposive and individually oriented introspection. This consciousness, being recognized as more than just knowledge, makes possible the balancing of opposites and seeing within. To Jung, reflection is the culture instinct demonstrating and maintaining its position as superior to nature, a capacity demonstrated by the animus which, in confrontation and integration with the anima giving relatedness, is manifest creatively in transformation of their relationship.
Related Reflexivity (#HH0977) Quiet reflection (#HM0798)
Free-thinking meditation (#HH0632).

♦ HH1206 Philanthropy

Charity

Description Literally the "love of mankind", the showing of generosity to others, whether dependants, friends or strangers, is emphasized in all major religions. Both Christianity and Buddhism regard good works as "laying up treasure in heaven" and as beneficial to the giver as well as the receiver; Islam extols the giving of alms above the obligatory duties of taxpaying; Judaism demands the payment of one tenth of one's income for philanthropic purposes, the remission of debts in sabbatical and jubilee years, and the rights of the landless and poor to glean fields after harvest. Philanthropy has been the basis for the foundation of universities, schools, hospitals, alms houses and religious institutions. It is particularly characteristic of individualistic societies emphasizing personal freedom where activities associated with philanthropy are not automatically handed over to state or organized religion.
Related Love (#HH0258) Altruism (#HH0081) Benevolence (#HH0989)
Dana (Buddhism, Zen, #HH1234).

♦ HH1208s Curative education

Description Any study which encompasses the cause and treatment of disease. Particularly used to cover alternative or unorthodox medicine, natural means for the body cells to normalize and heal themselves, and attitudes (possibly in previous incarnations) which may have lead to disease.
Broader Health (#HH0509) Education (#HH0945).

♦ HH1211 Karma marga (Hinduism)

Way of action

Description This is traditionally the way of the householder, when family, social and religious duties are faithfully carried out for their own sake and not for the sake of the fruits of such activity. This is also, then, a way of renunciation. Karma marga will not itself lead to liberation, only to auspicious rebirth; but it may be preceded by many years of study and be followed by years, free of family duties, of meditation or wandering from village to village following the way of knowledge, jnana marga. In this way salvation may be achieved.
Context One of the three ways of salvation of orthodox Hinduism, the others being *bhakti marga* and *jnana marga*. No one is said to be complete without some influence of the other two.
Broader Human development (Hinduism, #HH0330).
Related Karma yoga (Yoga, #HH0372) Jnana marga (Hinduism, #HH0495)
Bhakti marga (Hinduism, #HH0628)
Religious experience as the realization of duty (#HM4327).

♦ HH1219 Conscientious objection

Description This refusal to participate in military service on moral or religious grounds may arise from a variety of reasons. The individual may object to taking part in any form of bloodshed (universal objection) or because of belief that a particular war is unjust (selective objection). Although many states recognize the right of conscientious objection, most require some alternative form of service (often for a longer period of time); this is intended, among other reasons, to test the individual's sincerity. An extension of the principle includes objection to other civil duties, such as oath-taking or tax-paying; and the carrying out of illegal activity essential to particular beliefs, such as polygamy or rituals involving prohibited drugs. There is an argument that conscientious objection harms nobody and that it is a private affair between the individual and the state; it may be said that the paternalism of the state should yield to the conscience of the individual.
Related Ahimsa (#HH0088) Passive resistance (#HH0759).

♦ HH1222 Concept development

Concept formation

Description Concept development during individual conceptualization is the process of abstraction of qualities or properties. Piaget's concept development scheme and Vygotsky's process of concept formation both map the process of development through phases from childhood to adolescence, all (particularly the first two phases) depending upon the use of words: the grouping of different objects under a common name on the basis of their external relationship; the use of objective and connective thinking in the formation of potential concepts, in uniting and generalizing single objects and in picking out individual common attributes; the formation of genuine concepts by considering, detaching, abstracting and isolating individual elements outside the actually existing bond.
Related Concept progress (#HH0166).

♦ HH1223 Abandonment to God (Christianity)

Self-renunciation

Description It is not that everything is renounced, so that the soul becomes brutish; everything is renounced except God's will. The soul does not only forsake outward things; it also forsakes itself. Self-abandonment or self-renunciation is not abandoning faith or love, for example, only selfishness. There will be temptation, specifically on those points to which renunciation relates. All inward supports are removed and one is made to live only by faith. These trials are an inward purgatory but suffering is for no longer than is necessary. One must beware of assuming that inward crucifixion is complete, there may still be reservations, despite the sincerity of the act of consecration. One is tried by God and exposes one's self to illusion and injury if one concludes prematurely that one is wholly given to the Lord.

Once having given one's self to God there is no taking back – what is given away is no longer at one's disposal. Abandonment is in great faith, not listening to reason or reflection, allowing one's self to be entirely guided by God. The outward and the inward self are in the hands of God, so one forgets one's self and thinks only of God. Then the heart is free and contented. There is a continual loss of one's own will in God's will. All natural inclinations, however good they may appear, are renounced so one can choose as God chooses. The past is forgotten, the future left to providence, the present given to God.

Refs Fénelon, François *The Maxims of the Saints*; Guyon, J M B de la Mothe *A Short Method of Prayer and Spiritual Torrents* (1875).
Related Spiritual self-abandonment (#HH0868).

♦ HH1224 Yantra yoga (Yoga)

Description This form of laya yoga emphasizes concentration on yantras or mandalas and mastery of form. The symbolism of the yantra is rendered intelligible as the practitioner centres on it and thus on himself.
Refs Pott, P H *Yoga and Yantra* (1966).
Broader Yoga (Yoga, #HH0661) Laya yoga (Yoga, #HH0782).
Related Mandalas (#HH0112).

♦ HH1226 Brahman

Description The all-pervading power or consciousness itself which is the first principle, the eternal; and with which, according to *advaita vedanta*, the Self or *Atman* is identical. Mundane experience which appears to deny the fact that there is nothing but Brahman is clouded by ignorance which sees illusion as reality. True liberation is achieved when this illusion is seen for what it is and the soul is no longer tied to the cycle of existence but experiences itself as one with Brahman. This is ultimate enlightenment and absolute freedom.
Narrower Saguna brahman (#HM0576) Nirguna brahman (#HM4331).
Related Atman (Hinduism, #HH0103) Samsara (Hinduism, #HM1006)
Advaita vedanta (Hinduism, #HH0518) Human development (Hinduism, #HH0330).

♦ HH1232 Jewish mysticism (Judaism)

Refs Hoffman, E *The Way of Splendor* (1981); Matt, Daniel Chanan (Trans) *Zohar*, Scholem, Gershom G *Major Trends in Jewish Mysticism* (1954); Scholom, Gershom *Jewish Gnosticism, Merkabah Mysticism and Talmudic Tradition* (1960); Sharot, Stephen *Messianism, Mysticism, and Magic* (1987).
Broader Mystical cognition (#HM2272).
Related Kabbalah (#HH0921) God as wholly other (Judaism, #HM4501).

♦ HH1233 Adaptation

Description An essential aspect of individuation, adaptation refers to the relating of external and internal factors and the different demands they may make upon the individual. Failure to come to terms with or balance the needs of internal and external worlds may be considered as neurosis; similarly there may be excessive dependence on one particular mode of adaptation, or a neurotic concentration on satisfying the demands of the external or the internal world. A certain breakdown in the adaptation such an individual may appear to have achieved is part of the psychoanalytical process, since such adaptation may be spurious – it can then be replaced by a

mode which better balances the individual's particular needs on the personal–collective spectrum, although perfect adaptation is seldom, if ever, achieved.
Refs Lifton, Robert J *Adaptation and Value Development* (1968).
Broader Individuation (#HH0385).
Related Neurosis (#HH1024).

♦ HH1234 Dana (Buddhism, Zen)
Giving — Charity — Love
Description The act of giving as a religious act directed towards a person of religious commitment is said to be of great potency, even akin to magic. The power of sacrifice of personal possessions is said to transform the mind, reducing possessiveness and selfishness and leading to sensitivity to others' needs. It leads naturally to the next stage of commitment to control of external activity.
Context The first of five perfections or *paramitas* of Zen which, when achieved, render the mind ready for the sixth – *maha paramita* or great perfection, *prajna*. Also the first of the ten meritorious deeds of South Asian Buddhist thought.
Broader Paramitas (Zen, #HH0219)   Ten meritorious deeds (Buddhism, #HH0446).
Related Love (#HH0258)   Zakat (Islam, #HH5643)   Philanthropy (#HH1206)
Prajna paramita (Zen, #HH0550)   Human development (Buddhism, #HH0650).
Followed by Shila (Zen, #HH0429)   Observing moral precepts (Buddhism, #HH0278).

♦ HH1235 Autocontrol of consciousness
Description A technique in biofeedback training for changing anxiety levels.
Broader Biofeedback training (#HH0765).

♦ HH1236 Psychobiologic therapy
Description This technique, developed by Adolph Meyer, emphasizes integration of psychotherapy with biological and medical approaches to treatment, in particular working on complete medical and social histories of the individuals concerned and encouraging expression of their own views on the problem. There is conscious effort by the therapist to avoid allowing his own biases to influence the patient The system of diagnosis is based on *reaction types*, categorized by symptoms; character difficulties of which the patient has become a prisoner are held to arise from childhood experiences. Treatment techniques are flexible, directive or non-directive, suggestive or passive, being designed so as to best accomplish collaboration with each individual treated.
Broader Therapy (#HH0678)   Psychotherapy (#HH0003).

♦ HH1239 Studying sutras
Description Literally a "thread" or "stitch", a sutra binds together the essential aspects of a subject, wisdom distilled into a short statement or verse which may be memorized and reflected upon. Reflection may lead to discovery of the hidden thread of the argument under what appears to be a collection of unrelated ideas. In fact, the sutra contains everything needed to be known for a proper understanding of the subject, although considerable effort may need to be made to fully elucidate what it has to say. This effort is as valuable as the resultant knowledge and is the reason for the apparent obscurity of the sutra. Knowledge is thoroughly assimilated and the powers and faculties of mind developed. Each word may imply a whole pattern of thought; and study in the original language is thus more illuminating than study of a translation, when it is virtually impossible to carry over the full meaning.

♦ HH1241 Self sacrifice
Description This has been defined as that which surrenders the pleasures of the moment for nobler ends, a voluntary endurance of what would normally be avoided or doing without what which is preferred, not with the aim of a greater good for one's self but for purely altruistic reasons. There is no ethical value in a grudging sacrifice of one's own desires. While developing the individual, then, self sacrifice improves the life of others – in fact family, community and society demand that each sacrifices his own needs for the good of others or anarchy prevails. In religious terms, self sacrifice is considered necessary at the most mundane level to overcome selfishness and too much attachment to worldly goods and at the highest level, in laying down one's own life for others as the greatest act of charity.
Broader Sacrifice (#HH0577).
Related Self-preservation (#HH0711).

♦ HH1244 Discipline of guidance (Christianity)
Description The teaching of Christ suggests that divine guidance is most likely to be received when a group of people fasts, prays and worships together. The individual may be guided to speak in guidance to such a group which then, under divine guidance, comes to an agreement. Examples of such guidance are cited in the Acts of the Apostles, in the lives of the saints, and in the Church today, when an individual decision (on a call to the ministry, on the advisability of a marriage) may be prayerfully deliberated and advice given. A specific example of calling on divine guidance is dependence on the advice of a *spiritual director*, as is currently practised within the Catholic monastic system.
François Fénelon warns against trusting too literally to dreams, visions and revelations. These may as easily be due to a disordered state of the physical system as to divine guidance and do not constitute holiness. Above all, these visions and remarkable states, which may be from imperfect experience, should not be taken as a guide for life separate from and above God's written law. The soul's guide should be faith in God, His Word and His Providence, the divine Word being interpreted by the holy heart. Giving one's self wholly to God allows the Holy Spirit to dwell in the heart and He guides into all the truth that will be necessary. Thus a truly holy soul, who continually looks to God for understanding of His word, may trust in the confidence that He will guide him aright. Such a holy soul may deduce views from the Word of God but cannot add anything to it. There is only one legitimate Originator in the whole universe, the business of man is to concur.
Refs Fénelon, François *The Maxims of the Saints*.
Broader Spiritual discipline (#HH1021).
Related Guru (#HH0805)   Spiritual direction (#HH0442)
Soul care (Christianity, #HH1199).

♦ HH1248 Tapas (Hinduism)
Austerities
Description By practice of fasting, physical austerities, observance of vows, the will is strengthened and association with the physical body weakened. The lower nature is purified and brought under control in a way analagous with the refining of gold by fire. Tapas is said also to lead to occult powers and siddhis. Not all austerities and siddhis are considered worthy – some vows may be foolish or demonic and the misuse of siddhis by the morally or spiritually undeveloped brings trouble in this life and the next.
Context One of the five components of *niyama*, self–discipline.
Broader Niyama (Hinduism, #HM3280).
Related Kriya yoga (Yoga, #HH0969)   Self-discipline (#HH0877)
Siddhis (Yoga, Buddhism, #HH0380).

♦ HH1249 Allocentric attribution
Description A component of *gedatsu therapy* when the individual's experience is evaluated from another's viewpoint, thus sensitizing the ego and attributing achievements to numerous benefactors both living and dead. The individual is not only responsible for her own suffering, but also for the suffering of these benefactors and of her descendants, as long as her suffering lasts.
Broader Gedatsu therapy (#HH0972).
Followed by Identity interchange (#HH1002).
Preceded by Self-accusation (#HH0584).

♦ HH1257 Chanting
Incantation
Description Chanting is practised in a number of systems as a technique for inducing mystic or transcendent experience. This includes some Sufi schools and Zoroastrianism, and may be one factor in experiences described by some Christian mystics. The harmony of chanting is said to aid meditation and to be a factor in healing disease and disunity between people. The chanting of some sacred formulae, ritual chants or incantations may be used to conjure spirits, to keep someone under a spell, in healing ceremonies, to conjure up spirits.
Related Rites (#HH0423)   Invocation (#HH1876)   Sound health (#HH2543)
Induction technique (#HH0025).

♦ HH1264 Cure
Cure of souls
Description In that cure signifies a transformation from illness to health, Jung considered an objective, once–for–all, cure under analysis unlikely. Nevertheless, there may be an acceptance of the illness, for it is the very illness which may provide the opportunity for development through reflection on inappropriate forms of ego adaptation. This may give rise to awareness of attitudes which are more adequate and the possibility of adjustment. The adjustment may not be long-lasting; but as the problem reasserts itself the experience may be integrated, leading to individuation.
Broader Transformation (#HH1039).
Related Healing (#HH0216)   Individuation (#HH0385).

♦ HH1265 Hajj (Islam)
Pilgrimage
Description The great pilgrimage to the sacred monuments of Islam in and around Makkah is required of all Muslims once in their lifetime, supposing they are physically and financially able to make the journey.
Context One of the five pillars of Islam.
Broader Human development (Islam, #HH1799).
Related Pilgrimage (#HH0637).

♦ HH1266 Transferring merit (Buddhism)
Pattanumodana (Pali)
Context The sixth of the ten meritorious deeds of South Asian Buddhist thought.
Broader Ten meritorious deeds (Buddhism, #HH0446).
Followed by Rejoicing at the merit of others (Buddhism, #HH0932).
Preceded by Attending to the needs of superiors (Buddhism, #HH0383).

♦ HH1267 Bio–globalism
Bio–revolution — Ecological awareness — Ecological planetary consciousness
Refs Anderson, Walter Truett *To Govern Evolution* (1987).
Related Deep ecology (#HH2315)   Ecological humanism (#HH4965)
Planetary consciousness (#HM2006)   Environmental mysticism (#HM0800).

♦ HH1268 Aparigraha
Non–possessiveness
Description The tendency to amass wealth and possessions diverts time and energy in the acquisition and maintenance of what one owns, and mental energy in concern over losing it, either temporarily in this world or permanently on death. Although it is possible to have possessions without attachment to them, the aspiring yogi is advised to reduce worldly possessions to the minimum. He is thus protected from vanity and greed, from wasting limited internal resources and from disturbance of mind.
Context One of the five moral qualities of yama (restraint).
Broader Restraint (#HH6876).

♦ HH1274 Personal intelligences
Description These are centred on the concept of the individual self and may be considered as: (a) *Access to one's own feeling life* – this is the development of the internal aspects of a person and the ability to detect and symbolize complex and highly differentiated sets of feelings (b) *Ability to notice and make distinctions among individuals* – to read even the hidden intentions and desires of others and to use this knowledge to influence their behaviour Development of these intelligences leads to self-maturity and to personal knowledge of one's self as a unique individual
Context One of the separate or multiple intelligences put forward by Howard Gardner.
Broader Multiple intelligences (#HH0214).

♦ HH1276 Positivism
Logical positivism
Description This philosophy holds that only facts of experience are worth knowing and scientific achievement is of the greatest worth. Knowledge is said to have past through three stages: (1) theological, or the childhood of mankind, when all was explained by the existence of spirit behind phenomena, every occurrence being the action of some god; (2) metaphysical, when fancy and speculation result in explanations based on principles, abstractions and metaphysical theories; (3) positive or scientific, when facts are explained by causes or necessary relations discovered through observation, experiment, induction and generalization. *Logical positivism* rejects personal experience as the basis for knowledge, accepting only experimental verification.

♦ HH1281 Endeavour
Struggle
Description Spiritual progress is not a simple growth. It requires permanent effort to direct attention towards the goal and not to be diverted by old habits and impulses arousing other desires, whether these are triggered by material objects, by other people, by considering one's own past, or by disturbing thoughts and ideas. It is will that is important, not opinion.

♦ HH1294 Dissociation
Description This expression was used by Jung to indicate a disunion within the self and a breakdown in the potential to embody wholeness, an aspect of neurosis, when there is a discrepancy between conscious attitudes and the trends of the unconscious. Jung further described analysis as a healing of dissociations, although in some psychoses dissociation was

to great for this to be achieved.
The term dissociation is also used to describe conscious fragmenting in order for analysis to take place, where an all-embracing approach might, in fact, be preferable.
Related Psychoanalysis (#HH0951) Dissociation of affect (#HM6087) Integration of the personality (#HH0561).

♦ HH1300 Regression
Description To Freud regression was indicative of failure and needed overcoming; but to Jung this temporary surrender of ego-ic, adult behaviour provides the background of regeneration for subsequent progress to take place and is compared with the death necessary before rebirth is possible.
Refs McNeill, J T *A History of the Cure of Souls* (1951); Prince, R and Savage, C *Mystical States and the Concept of Regression* (1966).
Related Incest (#HH1308) Ego death (#HH6921) Biosocial consciousness (#HM0044).

♦ HH1301 Dreaming (Australian)
Aboriginal dream-time
Description The dreaming is a mystical way of life for the Australian Aborigine offering a place of metaphysical repose. The dreaming may be understood as a pure asceticism of nature, an attainable condition within any individual provided certain insights are appropriately cultivated. When the dreaming is understood as a spiritual condition (rather than simply as a term denoting the creation-time of the Aborigines), it can then provide an unchanging (and on-going) metaphysical referent to order the relationship between an individual and the natural environment, with the spirit of the individual acting as nature's consort. The dreaming is thus a quality of spiritual perception offering an ability to live in the shade of primordial creation as if it were eternally recurring.
The Australian Aborigine accepts the land on which he lives as an extension of himself, not as a separate entity that should be simply exploited for material gain. The land and the people on it are understood to be manifestations of a continuing and recurrent creative process, metaphysical in nature, outside space and time. When the chthonic rhythms of nature are respected, humans are able to live in harmony with nature. When they are disrupted, by wounding the land through exploitation, natural forces (such as hurricanes and earthquakes) are engendered in response. The Aborigine tradition maintains practices which ensure a deep and reverential contact with the power of the land. For them, true imagination is the power to see subtle processes of nature and their angelic prototypes in the form of spirits of the dreaming. The significance of this bond is focused for the individual through a particular totem that offers him an alternative perspective from which to relate to the environment. They are thus able to reproduce within themselves the cosmogenic unfolding, the permanent creation of the world, in the sense in which all creation is finally only divine imagination.
In terms of this understanding, land which remains wild and unpacified retains a powerful mysterious force that can function as a regenerative force among men. Within the landform itself a life-essence (kurunba) is present as a cultural layer that has been inspired by mythological contact with the dreaming. The landform thus becomes iconic in essence embodying metaphysical significations. The culture of the landscape, namely the sacred or primordial history attributed to it, thus becomes a reality of other-wordly dimension and a gift to those who live upon it. The topography becomes a mnemonic device through which the events of the dreaming can be recalled and relived. Landscape becomes an active participant in the creation of culture. It thus serves as a manifold mytho-poetic edifice which the individual can enter into in times of personal or cultural renewal. As such any rearrangement of it does violence to the bond between humans and the metaphysical reality.
Refs Cowan, James *Mysteries of the Dream-Time* (1989).
Related Feng-shui (Chinese, #HH2901).

♦ HH1304 Patience
Description Mental discipline which creates indifference to pleasure or pain, and may be considered apathetic, is contrasted with patient acceptance of misfortune which nonetheless maintains hope in the ultimate good. This submission which still cares is said to be particularly characteristic of the Christian. Suffering with patience leads to perfection.
Related Apathy (#HH0467) Kshanti (Zen, #HH0813).

♦ HH1306 Communication with supernatural beings
Nubuwah (Islam) — Messengers of God
Description This aspect of the spiritual or religious life is often through means of intermediaries. An individual may feel called to speak for some supernatural force or God, as with the Old Testament prophets, or be commissioned to speak, as with oracles and shamans, these latter sometimes producing intelligible messages, sometimes mysterious or unintelligible. In Islam, *nubuwah* is the continuous communication of God with humanity through a chain of message-bearers, from Adam to Muhammad, all communicating God's desire that man should surrender to his will. Each of these is related to his predecessor although many have specific messages as well. Muhammad, as the last in the line, has all the qualities of those who precede him and receives the revealed truth of God in its perfect form.
There are common links among all messengers: all are chosen by God rather than choosing or wanting the task; they lead exemplary lives themselves with unity of purpose allowing nothing else in their lives to share in God's power; the message is accepted by some but rejected by others (who prefer duality and polytheism), who may physically harm them; not only is the joy of salvation held out to those who accept the message, but those refusing the message are warned of the punishment awaiting evildoers; the obedience of followers towards God and towards his prophet are not distinguishable, so that a community of believers based on the revelation of God may be set up.
Narrower Spirit possession (#HH0056) Trance channelling (Psychism, #HM3456).
Related Shaman (#HH0973) Prophecy (#HH0559) Channelling (#HH0878) Spiritualism (#HH0479) Spiritual healing (#HH0458) Human development (Islam, #HH1799).

♦ HH1308 Incest
Description According to Jung, the incestuous feelings or fantasies experienced by a child may be an unconscious attempt to enrich his or her personality through deep and meaningful emotional contact with a parent. While this enrichment occurs, with an emotional bond assisting in the maturation process, the incest taboo makes the child normally safe from having these fantasies physically acted out; it also makes the focus of the sexual impulse a definite individual and assists in viewing the marriage partner also as an individual when this stage is reached.
An adult may regress in an incestuous manner, although without necessarily a focus on one person. This is an attempt at spiritual and psychological regeneration, a mystic or creative reverie from which may spring the artistic process.
Related Regression (#HH1300).

♦ HH1321 Religious growth
Moral growth
Description In the attainment of moral and religious maturity, a child passes through a number of natural changes, based partly upon genetic factors and partly upon experience. Such growth is contrasted with the development which occurs through manifestation of instincts and impulses. Growth is the specific and particular direction given to tendencies through the individual's own experience in a particular environment. In this sense the cultural progress of the child is not simply development, it is growth. The scheme given below is based on a Western, Christian environment at the turn of the century, but is said to be adaptable and applicable to other environments and religions. It combines three points of view: phenomena arising from increase of knowledge or of physical or mental powers; relative prominence at different ages of instincts or unlearned tendencies; changing contacts with social environment.
– *The pre-individual stage, first year*: Here there is little difference in reaction to persons or things, and little realization of one's individual selfhood or the thoughts and feelings of others. Association with persons can, however, lead to genuinely social habits if such association is connected with pleasure. This is a foundation for quickening intelligence and the beginnings of moral and spiritual interpretations.
– *Preliminary socialization by imitation, ages 1–2*: Although there is as yet no steady distinction of one's self from others, increased abilities in walking, talking and control of hand and arm movements increase participation with others and in the common consciousness of the family. Antipathies and likings exhibited by others in his environment are acquired by the infant as well, so that attitudes and habits may be formed which are difficult to dislodge at a later stage.
– *Preliminary individualization, ages 3–5*: There is discovery of one's self as an individual, leading to experimentation with self as opposed to things and other persons. There may be the appearance of contrariness, and refusal, apparently unmotivated, to conform with social expectation – these being further experimentation. This individualization is a necessary foundation for character and offers new opportunities for moral and religious growth. Although self-assertion implies no moral fault, the child starts to learn that certain behaviour brings mutual pleasure while other brings mutual pain. He may, in a family where religious faith is expressed, assimilate religion from this early stage.
– *Socialization through regulation and competition, ages 6–11*: Both at school and at home the child is now expected to obey certain rules and comply with its social environment. Games with other children are also in conformity with rules which channel increasing strength, initiative and resourcefulness in non-destructive ways. These games develop tendencies of mastery and submission, approving and scornful behaviour, emulation and rivalry. Games extend from seeking individual success, through that of success within a team, to that of the team for which the child is playing. Thus, as socializing instincts begin to bloom, individualistic and socializing tendencies are in competition. Girls spontaneously form sets, and boys form gangs.
Although the adult world may chiefly be perceived as an inhibitor to freedom, the family environment will determine whether freedom is perceived within obedience to social conditions or whether arbitrariness of response intensifies individualistic self assertion. The child may become aware of the family, the Church and society in terms of fellowship, as a means of interpreting the values of his own competitive systems and to reinforce the developing social instinct. Although progress in character is more on the basis of laws, rights and penalties and the force of social opinion than of "grace", nevertheless children of both sexes and from a very early age show parental instincts, towards babies and smaller children and also to animals, dolls and even parents. This can be developed as a socializing force. They also respond with real devotion towards affection and there is a genuine family loyalty and pride.
Providing the child is in a religious environment, he will conform to it as he conforms socially in other respects. Religion becomes an extension of social experience at other levels. Although there is no intellectual or moral depth, religious experience will include consequences of right and wrong, admiration and condemnation, and the inclusion of God in the family life with loyalty to Jesus as one's leader.
– *Early adolescence, ages 12–14*: Changes focus on the approach and attainment of puberty, thus arising some years earlier for girls than boys (about one year at the beginning of adolescence to 3 or 4 years at its completion). Simultaneous with a self-assertive or independent attitude towards the family and to social authority in general, there is a deepening social attachment and loyalty to peer groupings. In the extreme this may display itself as juvenile crime, commonly carried out in gangs or triggered and supported by gang activity and feelings. The new individualization and socialization are a preparation for ending passive attachment to arbitrary or chance rules and social groupings and commitment to personal convictions and profounder loyalties. Parents and teachers are now in a position to assist the release from childhood constraints and promote free devotion to worth-while social activities. Religious and moral activities tend to be based on social organizations for groups of young people which also emphasize heroic attributes. Many young people are drawn to church membership at this age.
– *Middle adolescence, ages 15–17*: Moral growth is again determined by instinctive development as the previous apparent repulsion between the sexes is replaced by attraction. Although sexual powers may be misused there is an instinctive preparation for family life. The capacity for feeling and the requirement to look upon one's self as an adult and no longer a child indicate that growth to self-conscious individuality is reaching a climax. This is the age where religious conversion is most likely to occur, not so much because of an awakening capacity for religion (religious confirmation in nearly all societies focuses on an earlier age, about 14) but because of responsiveness to an emotional appeal. Despite the fact that many youngsters in this age group are working, moral interests clearly demonstrate a need for education to continue, allowing freedom, play, idealization and experiment.
– *Later adolescence, ages 18–24*: At this stage there are great changes in social relationships and functions. A full member of society and the state, the individual achieves legal responsibility, starts on a career and probably enters into marriage. Emotional ferment tends to be checked and moral and religious growth is characterized by independent reflection and solidifying of will through responsibility. Responsibility for one's own thoughts leads to doubts on commonly accepted views of religion or politics but, unlike in earlier adolescence, dissent is less from arbitrary impulse than from a steady questioning into the basis of things. It is at this stage that mental and moral life take a stance and fix on interests that will determine fundamental motivations of maturity, although naturally these continue to increase and differentiate.
Naturally, these stages are actually part of a continuous whole, with no sharp distinction between them. Throughout all periods there is a single, central principle of moral and religious growth, the reciprocal individuation and socialization of consciousness through social participation – it is primarily through dealings with others that reflective self-consciousness is attained. There is no "pre-religious" or "pre-moral" period. Even the youngest child in a religious family adjusts to the concept of a heavenly father. Even unfortunate hereditary tendencies can be ameliorated by fortunate social and moral contacts in the formative years, while the most auspicious heredity can be turned to vice and crime through perverted growth. Arrested growth, when a mode of functioning persists beyond its normal period, and perverted, disproportionate growth of natural factors in any period thus reflect on the provision society makes for the moral life of its young people, and entrenched social wrongs depend on specific social experiences of the young. In religion there is the broadest and deepest sense of social connectedness together with profound realization of personal freedom. It can be said that only religious faith, hope and loyalty gives total expression to the social principle of moral growth.

Refs Kirkpatrick, E A *The Individual in the Making*.
  Related Moral perfection (#HH5098)   Moral development (#HH0565)
  Cognitive growth and development (#HH0146)   Human development through religion (#HH1198)
  Psychospiritual growth (Christianity, #HH2909).

◆ **HH1323 Discipline of study** (Christianity)
Description This is a means of changing the inner spirit through application of the mind. True study involves understanding of what is studied, interpretation of what is meant and evaluation of the meaning. Four stages are defined: (1) *Repetition*. Since thought processes take on the order of what is studied, repeated and attentive study will form ingrained habits of thought. (2) *Concentration*. With singleness of purpose the attention is centred on the object of study, enhancing the natural ability of the brain. (3) *Comprehension*. A basis for true perception of reality is provided through insight and discernment. (4) *Reflection*. Pondering on the meaning of what has been studied demonstrates its significance and leads to understanding of self.
  Broader Spiritual discipline (#HH1021).
  Related Study of sacred scriptures (#HH0957).

◆ **HH1324 Naikan therapy**
Description By guided pondering on his past, the individual reviews what he has received in this life, how he has repaid it and what trouble he has caused others. Rancour and arrogance are replaced by blaming the self and by gratitude and humility. Self–image is radically re–evaluated and guilt is expiated by working to repay good received and for the good of society.
  Broader Therapy (#HH0678)   Psychotherapy (#HH0003).

◆ **HH1325 Sexual maturity**
Description Sexuality is intrinsically and inextricably bound up with the learning processes for moral development; and sexual development may be considered central to that line of moral development which implies the capacity for valuing another as a person. Thus sexual behaviour which intentionally depersonalizes one's self or another may be considered sexual immaturity. Behaviour which demonstrates a mature capacity for empathy, seeing one's self as separate and respecting the different perspective of another, indicates the high degree of emotional and intellectual development which is the distinguishing mark of both moral and sexual maturity. This includes the ability to consider alternative courses of action and act in accordance with one's deliberations, to act with knowledge and self–consciousness, to know and predict the consequences of one's actions and to distinguish activity in the world from fantasy. When such action involves another person then there is an awareness born of experience which relates the action to feeling and to knowing what the other is feeling. The sexually mature will have developed a personal and coherent system of moral values which encompasses their own desires and abilities and which allows for a complete sexual act in which bodily desire and mutual relationship with another are essential.
  Related Maturity as genitality (#HH0407).

◆ **HH1330 Rebirthing**
Description Negative feelings on the birth process are worked through with the aid of a physical enactment of the prebirth position. Under guidance from a therapist and using breathing apparatus, the individual is submerged at body temperature and in the foetal position. Subconscious impressions of the birth trauma may be replaced by those of pleasure.
  Broader Psychotherapy (#HH0003).
  Related Spiritual rebirth (Christianity, #HH4098).

◆ **HH1333 Absent healing**
Description The healer, in a physically relaxed state and with conscious mind passive, heals an individual situated elsewhere.

◆ **HH1334 Gene therapy**
  Broader Therapy (#HH0678).
  Related Euphenics (#HH0747)   Human genetic improvement (#HH0357).

◆ **HH1339 Pre–existence**
Prior existence
Description The belief that the soul existed prior to union with the physical body is common to many religions and philosophies, although details and interpretations differ. Many look on life in the physical body as a punishment or the inevitable result of deeds in a previous life (Hindu, Buddhist). For Plato, the embodied soul has three elements: the cognitive soul controlling the body but debased by it; the appetitive soul (the senses); the courageous soul, linking the first two. Generally only the first is thought of as pre–existing. A common view is the essential nature of the soul, as the breath of life permeating the body or as life itself clothed with the body, which is subordinate.
  Related Soul (#HH0501)   Karma (#HH0567)   Reincarnation (#HH0686).

◆ **HH1348 Secondary mental factors** (Buddhism, Tibetan)
Description These are the means by which what is perceived by the means of sense consciousness are interpreted by the mind. There are 51: 5 ever–recurring or omnipresent; 5 object–attentive or determining; 11 virtuous; 6 root deluded; 20 auxiliary deluded; 4 changeable. Every act of cognition involves some of these factors – those always present (neither beneficial nor detrimental) and others beneficial or not. Cognition may become valid and virtuous through learning discernment.
Context The secondary mental factors, together with the afflictions, are comparable with the mental activities or formations of Hinayana Buddhism.
  Narrower Omnipresent mental factors (#HH0320)
  Virtuous mental factors (Buddhism, #HH0578)
  Determining mental factors (Buddhism, #HH0170).
  Related Ways of knowing (Buddhism, #HH0873)   Sense consciousness (Buddhism, #HM2664)
  Awareness as mental–formation group of conscious existence (Buddhism, #HM2050).

◆ **HH1354 Journeying within transcendence – from blindness to sight**
Description Confession of ignorance is followed with confession of faith, while certainty leads to ignorance. Abstract vision sees impersonally and is unable to go beyond basic assumptions or to treat people as subjects rather than objects. Everyday vision desires no changes and keeps its distance from new situations. Vision that looks on any situation as a means to gain is controlled by power and by fear; anything threatening power threatens existence and must be eliminated; this is ignorance of truth and leads to blindness. In this case, habitual vision, built up as a defence and enabling growth and development, becomes blind persistence in routine; awareness that it is there is the first step to returning to sight. Self–conscious (as opposed to self–aware) sight is like the critical judgement of a third party on one's actions; love is tempered by fear which holds it back, preventing spontaneity. There is the vision arising in moments of stillness; the ordinary way of functioning disappears and there is apparent blindness which develops into a new kind of seeing, in the present and with faith as connections are formed with what is deep down within the self. There is the vision of the contemplative, of God, accepting things as they are, disorienting other vision so it can be reoriented. This is the vision of the self manifest in myths, symbols and dreams, revealing the blindness, drawing closer to wholeness. Finally is the vision which wants to learn and be changed, that is prepared to risk transformation, when it is seen that all the types of vision are within one's self.
Context The eleventh section of St John's Gospel, Chapter IX 1–41, is related to a stage in the spiritual journey of the individual.
  Broader Journeying within transcendence (Christianity, #HH6505).
  Followed by Journeying within transcendence – from outside to inside (#HH3215).
  Preceded by Journeying within transcendence – from bondage to freedom (Christianity, #HH2661).

◆ **HH1357 Authenticity**
Description In existential terms this indicates that a person has become fully him or herself, through intelligence, conscience and integrity refusing to follow unthinkingly conventional standards and judgement on morality. Each situation becomes unique, governed neither by rules or general principles. The argument that such an approach could result in anarchy, racism or tyranny is countered by the assertion that authenticity requires to take into account affirmative relations with others. In Christian terms, concern for interests of the community and love are necessary requirements for true authenticity.
  Refs Bugental, J F T *The Search for Authenticity* (1965).

◆ **HH1362 Fate**
Destiny — Kismet
Description The view that each individual's life is determined by forces outside his own control. This attitude is common to religious and spiritual beliefs of most civilizations and is only superseded with the onset of a belief in individual freedom with the obligation but not the necessity to follow a particular moral path. It may be, nonetheless, that refusal to follow a moral path will bring retribution – the distinction is whether the individual is actually free to choose such a path or whether his apparent free will decision is actually forced upon him by fate. Destiny, the fate of a particular person, may in a sense imply a goal towards which that person must endeavour or it may be used to refer to fate as applied to some role he is obliged by fate to fulfil.
Particular examples of fatalistic belief are those based on astrology or omen, when future occurrences may be predicted and planned for or, in some cases, averted. In China, natural law implied an impersonal and ordered nature to which mankind was subject. In Hindu and Buddhist thought, fate is pictured as a ceaseless cycle of action and the results of action, although, through self–realization, the individual may escape this eternal treadmill. Judaism and Islam both have an element of fatalism, as does Christianity. The individual's life is seen as governed more by external events than by his own plans; this is demonstrated particularly in the inevitability of death. Again, fate can be transcended through free action from a position of total dependence on God; paradoxically, it is from the realization of obligation within freedom that freedom from destiny arises.
  Refs Cahn, Steven M *Fate, Logic and Time* (1986).
  Related Karma (#HH0567)   Kismet (#HH3773).

◆ **HH1366 Midlife transition**
Stages of life
Description According to Jung, the psychological achievements of the first phase of life – separation from the mother, achievement of a strong ego, giving up of infant and childhood status and acquiring of an adult identity – lead to a life based on social position and relationships such as marriage and parenthood. However, midlife may present a crisis period if the individual has failed to anticipate and adapt to the demands of the second phase of life. At this stage, consciousness shifts from the external, interpersonal to the internal dimension, a conscious relationship with the inter–psychic process. Dependence upon the ego must be replaced by a relationship with the Self, with concern for meaning and spiritual values modifying dedication to outer success. The emphasis is on a sense of purpose involving self–acceptance, the satisfaction of a life lived in accordance with one's potential and the reality of approaching death.
  Related Self–acceptance (Jung, #HM3912)   Stages of personality development (#HH0285).

◆ **HH1376 Vinaya** (Buddhism)
Discipline
Description The vinaya is the first "basket" or collection of teachings of Buddha and describes the course of action for spiritual progress towards nirvana. The code delineates monastic and secular disciplines of behaviour (the monk from the way of the householder); and the mental discipline for avoiding unhealthy states of mind. The code consists of over 200 rules, the majority related to the *pancasila*. Fulfilment of vinaya requires the mental clarity and wisdom associated with a life of meditation and the accumulation of knowledge as well as moral concern, the former being essential for the restraint needed to carry out actions under the latter.
  Related Pancasila (#HH4536)   Discipline (#HH0163)   Vinaya (Buddhism, #HH3398).

◆ **HH1380 Bodhisattva** (Buddhism)
Description An enlightened being in the mahayana tradition, who defers his own enlightenment in order to serve the enlightenment of others as a spiritual practitioner. This is contrasted with the Arhat who retires from the world to work for his own enlightenment. Ascending orders of bodhisattva commence with the streamwinner and end in full Buddhahood, 20 different kinds of saint or bodhisattva being enumerated.
  Refs Katz, Nathan *Buddhist Images of Human Perfection* (1982).
  Related Arhat (Buddhism, #HH0233)   Mahayana Buddhism (Buddhism, #HH0900)
  State of bodhisattva (Buddhism, #HM1225).

◆ **HH1384 Self–respect**
Self–feeling
Description Regard for one's own worth and dignity as a human being includes concern for personal integrity and awareness of self and may be an inner guarantee of objectively moral conduct. Although a sense of duty will ensure conduct to a level of what is required, self–respect goes beyond this. Morality here is not what instinctive or customary reaction will produce but an inner and voluntary act of self determination.
True self–respect is contrasted with that induced through pride in one's worldly position, fear of not coming up to some moral norm because of what others may think, or desire to impress others. Rather, self–respect is responsive to external impulse and to sympathy and duties towards others, respecting them as persons also having self–respect and worthy of it. It requires pursuit of common ends which are beyond the personal; and leads to respect for social institutions and to a reform of those institutions so as to allow self–respect to the under–privileged within them.
  Related Self–love (#HM0478)   Self–esteem (#HH0634)   Feeling of I (#HH0840)
  Self–remembering (#HM2486).

◆ **HH1391 Sauca** (Yoga)
Purity
Description The state of impurity and sin arising from many lifetimes of progressive secession from the absolute source of all being is combatted through practice of inner and bodily purity. The practitioner of yoga is on his guard against the body; and desire for inner purity loosens

HH1391

attachment to the body and false identification with it. He works in seclusion, avoiding the polluting influence of others.
Context One of the five components of *niyama*, self-discipline.
Broader Niyama (Hinduism, #HM3280).
Related Purity (#HH0719).

♦ **HH1401 Psychoneuroimmunology** (PNI)
Description It has been shown that the brain may influence the response of the immune system, so that mental states may inhibit the body's resistance to disease; and that there is equally an influence from the immune system to the brain, so that hormones can effect the state of mind. PNI has demonstrated that there is a shared consciousness in mind and body; medical and social sciences combine to treat the two together, emphasizing that not only can mental states cause disease but can also be effective in producing positive health – a serene mental state can act therapeutically on disease. For example, it has been demonstrated that psychological states can alter the concentration of the cells used by the body to fight viruses and tumours.
Related Super-immunity (#HH3755) Positive thinking (#HH1088)
Positive mental health (#HH0675).

♦ **HH1407 Talkin**
Instructing the dying
Description This Indonesian custom of whispered instructions to the dying, also repeated at the end of the funeral, concentrates on making the dying experience a positive one. It assists in the transition from life to death and reminds the dying or dead person of the proper answers to questions from the angels of death.
Related Bardo consciousness (Buddhism, Tibetan, #HM0698).

♦ **HH1409 Mansions of the soul** (Christianity)
Interior castle of St Teresa
Description The progress of the soul in the mystical life is viewed as a progress through seven mansions as it is transformed from a creature of sin to the bride of spiritual marriage.
Refs Teresa of Avila *Interior Castle* (1988).
Narrower Mansions of humility (#HM3382) Mansions of exemplary life (#HM2969)
Mansions of incipient union (#HM4110) Mansions of spiritual marriage (#HM3971)
Mansions of the practice of prayer (#HM3218) Mansions of spiritual consolations (#HM3443)
Mansions of favours and afflictions (#HM2470).

♦ **HH1420 Analytical psychology**
Depth psychology
Description In general, this refers to psychological methods which emphasize the reduction of the subject matter to its elements. The term is particularly associated with Jung, who referred to it as embracing the psychoanalytic method of Freud as well as the individual psychology of Adler. All experiences are retained in the unconscious, which is complementary to consciousness and communicates with it through universal images or *archetypes*. A person's motivation is determined by revealing this underlying imagery and exploring its significance for him. The methods of *psychotherapy* in practice aim towards *individuation* of the self and are contrasted with the "cures" for neuroses which explain away symbols and have the result of diminishing the person's vitality. Analytical psychology has been the basis for the study of religion, of paranormal phenomena and of alchemy, among many others.
Related Gnosis (#HM0413) Psychotherapy (#HH0003) Individuation (#HH0385)
Psychoanalysis (#HH0951) Depth consciousness (Jung, #HM2173)
Emblem archetypes in the psyche (Tarot, #HM2201).

♦ **HH1432 Mystical transformation**
Interiorization
Description The first phase concerns outward objects, the sensory world. This is drained of importance and emotional centrality for the individual. There is a reordering of behaviour and of thought processes, they are reoriented to the primary objects of the religious world. In mediaeval times, the totality of sacraments and scripture of the Church served as an object for this transition. This is religious consciousness, when the physical, sensory world is derealized and a virtual world erected along religious dimensions which becomes a new reality permeating everyday life.
The next phase is a derealization of religious objects in their outward form, even of religious myths and history. They are internalized into the structure and rhythms of inner spiritual life. For the Christian, for example, the Trinity loses its significance as the theological scheme of the macrocosm; of more importance is the trinitarian organization of the soul. In Judaism, the predominance of the "tent tabernacle" is displaced by the "inner tent" which is the dwelling place of divinity in man. So the virtual religious kingdom, established in the first phase, is transposed into the wider, inner, spiritual domain.
In the final phase there is a relinquishment of even the religious dimensions which have been transported into the interior. External aids are now hindrances and all spiritual practices are given up. This is the way of annihilation. All remnants of the personal self, all images which have assisted the movement to the interior, all forms and names of the Godhead are renounced. There is a radical interiority but no detectable content. Embracing this is the mystic union, coming home to the One, a foretaste of eternity.
In this interiorization, all explicitly Christian forms, for example, are dissolved in the final emptying.
Refs Valle, Ronald S and Eckartsberg, Rolf von (Eds) *The Metaphors of Consciousness* (1981).
Related Religious consciousness (#HM3008).

♦ **HH1433 Failure**
Loss
Related Success (#HH3188).

♦ **HH1449 Anthropomorphism**
Description The use of anthropomorphic analogy, while remaining aware of its analagous nature, is a means of penetrating mystery and assists knowledge of the nature of God.

♦ **HH1453 Affirmations**
Description The daily repetition of groupings of words of truth is said to arouse inner perfection and change outward manifestation. The conscious repetition gradually affects the unconscious which in turn improves specific aspects of life, such as inner harmony, health, education.
Related Mantras (#HH0386).

♦ **HH1456 Logical–mathematical intelligence**
Description This is developed first from the ability to recognize classes or sets of physical objects; and later by conceptualizing classes or sets of objects or ideas in the mind and understanding logical connections among them. Central features are: the ability to identify and then solve significant problems; memory for repetitive patterns and the ability to compare and operate upon such patterns mentally; and an intuitive feel for logical relationships.
Context One of the separate or multiple intelligences put forward by Howard Gardner.
Broader Multiple intelligences (#HH0214).

♦ **HH1466 Perfectionism**
Description In the sense of completeness or wholeness, perfection may refer to the full development of the individual's human capacities – physical, cognitive, aesthetic, moral, religious. While originally referring to the perfection of the individual, as in the doctrine of Plato and Aristotle, the notion was developed by Hegel to refer to the individual as part of the world spirit, holding that moral perfection is achieved only as part of the social whole.
Refs Warfield, Benjamin B *Studies in Perfectionism* (1958).
Related Complete man (#HH0727) Perfectionism (#HM0213)
Spiritual development (#HH0017).

♦ **HH1471 Uncertainty** (Christianity)
Description Uncertainty may be described as a genuine expression of faith when security is repudiated and life is lived in full openness and questioning, based on the certainties of God, the gospel and the Church.
Related Uncertainty (#HM3051).

♦ **HH1478 Sensory deprivation**
Isolation initiation
Description The effect of sensory deprivation depends on the level of stimulation the person is normally comfortable with. Deprived of such stimulation, there may be attempts to seek stimulation by moving about or making noises in order for consciousness to have its usual stimulus. When subjected to almost zero sensory input one suffers mental confusion which may lead to hallucinations, temporal disorientation, visions. There is anxiety, restlessness, irritability and boredom. the feeling of disorientation may be beneficial. The result may be the therapeutic release of pent-up emotion. An example is the Tibetan isolation initiation. On spending days in an underground tomb entirely cut off from sensory input, the person communes with his inner self and is purified of karma. This is an initiation process for priesthood and psychic healing. Another example of sensory deprivation is that employed to achieve altered modes of awareness through floating in an isolation tank of concentrated salt solution at body temperature. The feeling of weightlessness and independence from heat and cold is accompanied by deprivation from other sensory input (sight, sound, taste, smell). The experience is conducive to meditation and surfacing of repressed thoughts. Experiments in an isolation tank under the influence of measured doses of LSD have resulted in awareness of birth and expansion of the universe, and of the micro dimensions down through body cells to fundamental particles.
Refs Zubek, J P (Ed) *Sensory Deprivation* (1969).
Related Primal therapy (#HH2087) Thought reform (Brainwashing, #HH0865).

♦ **HH1523 Human development through new religious movements**
Cargo cults
Description The impact of the major religions (Christianity, Islam) on local, tribal religions in primal societies has resulted in the development of new movements, and this has been accelerated by rapid change in previously traditional cultural and social behaviour. They arise in answer to a search for spiritual meaning and spiritual power, both eroded when traditional belief and practice as well as the new, imported religion fail to respond to the new cultural circumstances. Very often the basis is a mystical experience or revelation received by the movement's leader detailing new rituals and codes of morality together with the receiving of a gift of spiritual healing. Local spirits and divinities become subservient to one supreme and personal god, who may replace a traditional, remote single deity. Often the new religion transcends single tribal groupings to cover wider areas; examples are Godian and Aladura movements in Nigeria; although, on the contrary, some movements may be local and short-lived, as for cargo cults in Melanesia.
Despite a complete break with tradition, in which symbols of the old ways are destroyed and in which the leader may be young and female in a society previously dominated by male elders, many old beliefs are incorporated into the new movement; and worship may consist of a mix of old and new forms. Faithful observance of the new rituals and forms of worship – which, although sometimes permissive, more often include ascetic practices and disciplines more rigorous than those of the religions from which they are derived – promises the immediate benefit of mental and physical healing, revelations and protection from evil powers, together with future prosperity and possibly the coming of a new order with paradise here on earth. The supernatural help may be from God, Jesus or from ancestors, heroes or spirit beings. It results in the infusion of dignity, self-confidence and self-respect to people whose morale has been lowered by the effects of invasion or colonialism. Many new movements have been recognized by the Church as genuine, grass-roots Christianity.
Broader Human development through religion (#HH1198).
Related Syncretism (#HH3214) Human development (Christianity, #HH2198)
Human development through primal religion (#HH1902).

♦ **HH1542 Human being**
Description Opinion is very divided on what actually constitutes a human being. Some definitions include: the biological, chemical and mechanical components comprising the physical body; a being in the image of God, potentially having all the attributes of God; an organism capable of receiving input from numerous forms of energy and of altering its own frequency through thought and action; a receiver for every nature of atmosphere, thought, consciousness and energy fields; a person (man, woman or child), as distinguished from an animal or a supernatural being; a three-dimensional mass of energy with intelligence capable of making decisions and with the desire to evolve; spirit and soul – physical – perisprit (Allan Kardec); an internal and external balance of forces, a "human gyroscope"; an intersynergistic higher order of mental and psychic patterns (Beal).
Refs Byrne, Edmund F and Maziarz, E A *Human Being and Being Human* (1969); Rasey, Marie (Ed) *Nature of Being Human.*

♦ **HH1543 Individual development**
Description Normal individual development in modern society may be examined in three stages:
(1) *The small child*, who establishes a natural identity, gradually distinguishing his own body from the environment. He has needs which must be met; and these he learns to negotiate through claims and responses which work, first with his mother and later with peers and siblings as well. These are elaborated by role testing in different situations. Internal behaviour controls and role responsibilities are built up through growth in the family environment and experience of the Oedipal crisis.
(2) *The school child*: there is rapid cognitive, emotional, moral, physical and interactive development during which the child achieves an identity defined by role. Conflict and depression mark a coming to terms with the development process and release from internalized mother- and father-concepts as the young person is prepared for departure from the parental family. Intermediate between attachment of the libido to previous and new objects there is a narcissistic stage with characteristic moral instrumentalism. The individual learns to distinguish between norms and the principles which legitimate them, discriminating those which seem worthy and attempting, despite

conflicting role expectations and systems, to consistently embody some principle.
(3) *The young adult*, where there is integration into the whole society, a balancing between different areas of life and and overall unifying interpretation of life–history. There is also interactive participation where there is attempted exploitation of the situation for his own wealth and prestige, but within the bounds of socially acceptable rules.
These three stages are referred to by L Kohlberg as *pre–conventional* (obedience to avoid punishment and for personal gratification), *conventional* (respect for authority and maintenance of social order) and *post–conventional*, within which several levels are defined culminating in orientation to general principles such as sanctity of life and furthering the development of mankind. A fourth stage is postulated, where logic and reason no longer seem sufficient and there is post–conventional religious orientation, where the morally mature question and seek answers to the meaning of life in a non–egoistic and non–dualistic approach.
Refs Döbert, R et al *Entwicklung des Ichs* (1977); Kohlberg, L and Kramer, R B *Continuities and Discontinuities in Childhood and Adult Moral Development Revisited* (1973); Riegel, K F and Meacham, J A *Developing Individual in a Changing World* (1976); Winnicott, D W *The Family and Individual Development* (1965).
Related Identity (#HH0875) Stages of faith (#HH2097).

♦ HH1552s Adept
Description As well as referring to an illuminated individual in this world, the term may be used in the psychic sense to describe a soul–mind in the etheric world working to assist the advance of mankind.
Related Guru (#HH0805).

♦ HH1562 Hypergnostic meditation
Refs Ichazo, Oscar *Hypergnostic Meditation Training Manual* (1986).
Broader Meditation (#HH0761).

♦ HH1579 Social intelligence
Intelligence revolution
Description Human development depends on man's intellectual capability, the capacity to create, acquire, distribute and use various types of knowledge. This capability has been amplified, imitated and advanced through artificial means. Currently there is radical expansion of intellectual capacity in industrialized countries, both of individuals and of their social structures which could be called an *intelligence revolution*, a revolution in cognitive power. If intelligence is defined as the ability of a system to adjust appropriately to a changing world, then computer, communications and information technology are radically changing humanity's social intelligence. There is a growth of intelligence that could be referred to as an intelligence revolution. To keep abreast of this revolution, countries need computer literacy campaigns just as campaigns for reading and writing literacy were necessary in the past.

♦ HH1616 Ways of knowing (Christianity)
Description Meister Eckhart distinguishes two ways of knowing. The first, "lower knowing" or discursive reasoning, is common to all and is neither spiritual nor leads to the attainment of God. "Higher knowing" or spiritual knowledge is consistent with imageless prayer. A spark is kindled in the soul, a fire which leaps up and strives towards heaven.
Refs Smith, Cyprian *The Way of Paradox* (1987).
Narrower Spiritual knowledge (#HM7554) Discursive reasoning (#HM6997).
Related Self–knowledge (#HH0312).

♦ HH1643 Process thinking
Description This concept refers to thinking in an evolving situation. Rather than the paradigm of a fixed structure in a condition of equilibrium, whose internal and external relations are static or changing in a concrete, step–wise process, human systems are regarded in terms of natural, open learning systems constantly changing with their environment. The former entails a predetermined "good" end goal, treating the system as a machine to be "engineered" from outside. The latter implies positive feedback as an intrinsic characteristic.
Refs Jantsch, Erich and Waddington, Conrad H (Eds) *Evolution and Consciousness* (1976).
Broader Thought processes (#HH0530).
Related Open learning systems (#HH0326) Self–transcendent systems (#HH1032)
Ontogenetic model of human consciousness (#HM1105)
General system theory of human development (#HH1127).

♦ HH1672s Old soul
Description An individual who seems endowed from an early age with unusual maturity, confidence, intelligence, "other–worldliness" and, despite an attractive personality, independence from social norms, may have achieved these qualities through many, or well–lived, previous lives.

♦ HH1673 Lamarckism
Organic evolution
Description According to Lamarck, evolution of a species occurs in response to a need developing due to change in habits of that species; use of a particular organ or aptitude thus enlarges, strengthens or encourages hereditary development of that organ or aptitude, while disuse diminishes it. Thus by changing habits evolutionary changes may be induced through inheritance of acquired characteristics – parents who spent much time and effort practising piano playing would have children with a natural aptitude for playing the piano. This theory has been disproved practically through tests in breeding animals and theoretically through an understanding of genetically inherited characteristics.
Broader Evolution (#HH0425).
Related Euphenics (#HH0747) Human genetic improvement (#HH0357).

♦ HH1676 Spiritual renaissance
Global spirituality
Description Despite an apparent general spiritual impoverishment and disregard for highest values there is, paradoxically, a new spirituality emerging from this secularization. There is a spiritual impulse acting in all spheres of life as people become aware of a holistic vision that is personal yet transcends the individual and there seems to be a general participation in the externalization of essential spirituality. The search to experience higher states of consciousness is combined with a commitment to work together so that the resulting perceptions may be manifest in human affairs. Inspiration is sought from all cultures and times, the heroes of each revealing some aspect of corporate inner destiny. The theme of unity and global cooperation is arising from numerous directions, indicating that the corporate soul or divine presence is at work in the world. Self–interest is being sacrificed and concern is emerging for the improvement of all life in a spiritual renaissance.
An example of this concern is the increasing prevalence of the empathetic approach as opposed to the objective, reductionist approach in study. Progress is based on intuitive understanding and a recognition of the causal nature of consciousness. There is a desire to explore inner as well as outer space.
Rather than participation in intense private devotions or "escapist" religious life, this new spirituality looks to secular as well as religious institutions and relates to everyday life. It is a dynamic process of transformation integral to human development and permeating all human activity. Its practice demands the use of the imagination as a source of insight. In the contemporary cross–cultural environment, this spirituality implies a pluralistic perspective in time as well as space.
Refs Cousins, Ewert, et al (Eds) *World Spirituality*; Harman, Willis *Global Mind Change* (1988); Harman, Willis W and Rheingold, Howard *Higher Creativity* (1984); King, Ursula *The Spirit of One Earth* (1989).
Related Empathy (#HH0145) Spirituality (#HH5009)
Universal religion (Yoga, #HH0746) Spirituality (Christianity, #HH0792).

♦ HH1688 Discipline of meditation (Christianity)
Description In Christian terms, this discipline of being open to communion with God, away from noise, hurry and crowds, is detachment from confusion leading to greater attachment to God and to our fellow men. It leads to an inner wholeness and to spiritual perception; and is a means of redirecting life to deal with its problems. Rather than requiring a mediator with God it provides direct contact. Methods include recollection through surrendering concerns to God and centering on some aspect of the creation; or there may be meditation upon some portion of scripture or some aspect of current affairs whose significance needs elucidating. Regular practice leads to a constant learning process.
Broader Meditation (#HH0761) Spiritual discipline (#HH1021).
Related Christian recollection (Christianity, #HM2077).

♦ HH1747 Sufi path to perfection (Sufism)
Description This path is likened to a journey during which the seeker passes through seven valleys:
1. *Quest*. By refusal to be distracted from the goal whatever the situation, the seeker gives up all his possessions and devotes himself to the quest. Nothing matters except pursuit of this aim; dogma and belief or its absence no longer exist.
2. *Love*. In this valley the seeker experiences agitation and distress until he reaches the object of his desire, the perception of whose value results in true love. Love allows insight to manifest and provides the strength to identify with the object of desire. Good and evil no longer exist.
3. *Understanding*. Insight leads to this valley which has neither beginning nor end. The knowledge gained here is not permanent but, depending on his receptiveness, the seeker receives some understanding of truth; he learns to understand the relatedness of things and realizes his potentiality. Desire to become what he should be enables him to cross the valley as greater understanding is sought.
4. *Independence*. Increase of understanding brings detachment both from the desire to possess and from the desire to discover. In the awareness that the world is an insignificant speck in the universe there can be no action or inaction, simply contemplation of the source of existence.
5. *Unity*. As conventional meaning is lost there is a feeling of fragmentation until the discovery that the universe and growth all derive from the same source, that the many arose from the one. Then the past and future no longer exist, in the discovery of the invisible source the visible becomes nothing.
6. *Astonishment*. The individual is depressed and despondent as sorrow and eagerness arise, yet if one has achieved unity all is forgotten, even one's self. Certain of nothing, understanding nothing, unaware of himself, full of love with no knowledge as to its object, the heart is both full and empty of love.
7. *Poverty and nothingness*. Deprivation takes the place of bewilderment in a state of mute emptiness, when enlightenment arises. Then the secret of creation is revealed and existence is no longer separate.
Refs Arasteh, A Reza *Growth to Selfhood* (1980); Chittick, William C *The Sufi Path of Knowledge* (1989); Chittick, William C *The Sufi Path of Love* (1984).
Broader Spiritual development (#HH0017) Human development (Sufism, #HH0436).

♦ HH1754 Culture
Description In human development terms, culture may be defined as that which holds the individuals in a community together, accentuating what they have in common and creating continuity from the heritage of the past through the language, lifestyle, artistic expression of the present to the handing on of this heritage to future generations. It is the similarities within a culture which form the differences between cultures and enrich human experience, although these differences can become the cause of strife. Cultural differences are a manifestation of the plurality of creation although, in theological terms, these differences are transcended in the creator of all cultures.
Refs Rossi, I (Ed) *The Unconscious In Culture* (1974).
Related Cultural evolution (#HH0926).

♦ HH1763 Psychological approach to transcendence
Description Psychological description of transcendence implies technological control by the individual of himself and his environment during the experience, this further implying information about threats and ways in which to meet them. Three possible dynamics are described, depending upon whether society, the self or the sensory environment assume precedence:
1. *Bothersome child – Biosocial consciousness, consciousness one*. Social psychology would suggest a dynamic of psychological regression to characterize the experience, man being determined by society. The "real" world is that taken to be such by social consensus. Deviation from this reality is viewed as childlike fantasy, a regression to childish ways of dealing with reality, with divine powers seen as projections of the meaning of relationships with parents in the first years of life. Transcendent experience of the regressive type is seen as requiring therapy emphasizing re–socialization.
2. *Wise adult – existential consciousness, consciousness two*. From the viewpoint of ego psychology, the dynamic of transcendent experience is psychological progression, with man able to fulfil his potential as a self–actualizing person despite societal constraints. For example, Jung considered the inner world of fantasy as powerful and necessary as physical and social realities. Fantasy must be reckoned with, as must dying, death and rebirth. Transcendent experience of the progressive type brings shattering realization of the frailty and impermanence of physical existence and may lead to spiritual or religious experience independent of the individual's cultural or religious background. The goal of fulfilment of human potential is seen as a dying and rebirth process; transcendence is a peak experience which may permanently affect the person. The world is seen as good, beautiful and worthwhile, despite pain and suffering.
3. *Data processor – transpersonal awareness, consciousness three*. Dependence on the sensory environment assumes a dynamic of psychological alteration, as in perception and cognition psychology. Human sensory processing is likened to management of input to a computer. The senses are data reduction systems, screening out all but a particular range of data and habituating what is recovered. Psychological alteration (change in state of consciousness) arises during changes from the habitual sensory input to which it is not necessary to respond. When a practical system of habituated attention or perception is set aside or broken down, the result is transcendent experience of breaking down automatic response to new situations. The ego, simply an extension of the body, is discontinuous, an illusion; and during the moment of de–automatization there is

transpersonal experience beyond the normal boundaries of ego, time and space – pure awareness.
Refs Bowker, John *The Religious Imagination and the Sense of God* (1978); Dowdy, Edwin (Ed) *Ways of Transcendence* (1982); Progoff, Ira *The Symbolic and the Real* (1963).
Related Self (#HH0706)   Peak experiences (#HM2080)   Ego consciousness (#HM0570)
Ego awareness (Hinduism, #HM2059)   Biosocial consciousness (#HM0044)
Transpersonal awareness (#HM0768)   Existential consciousness (#HM3336).

♦ **HH1765 Abhidharma** (Buddhism)
Abhidhamma (Pali) — General knowledge — Metaphysics
Description This is the study of metaphysics and cosmology.
Context The fifth of the five primary subjects for the Geshe degree in Tibetan Buddhism.
Refs Narada, M *A Manual of Abhidhamma* (1968).
Preceded by Vinaya (Buddhism, #HH3398).

♦ **HH1772 Mind cure**
Description A system developed in the 19th Century by Phineas Quimby, mind cure is based on the premise that disease is an error of mind which, through wrong thinking, has shut us off from from our divine nature in which we take part in the life of God who alone fills the universe. This shutting off can be reversed by filling the mind with good, loving and curative thoughts to drive out badness and ill-health. The emphasis is on relaxation and allowing the goodness of God to take over.
Related Super-immunity (#HH3755)   Positive thinking (#HH1088).

♦ **HH1773 Journeying within transcendence – from son to father**
Description The concern is that the disciples may be protected from the world and have a true unity of relationship, a unity of differentiation, reflecting God's glory. Initial growth into consciousness is through manifestation of the self in the ego; but establishment and strengthening of the ego involves apparent loss of self. Further growth involves meeting the shadow, freeing the self to be the self while the ego becomes relative. The self bears life, it is conscious of God's love. Through suffering it transcends, through differentiating it unites, through glorifying it reveals.
Context The sixteenth section of St John's Gospel, Chapter XVII 1-24, is related to a stage in the spiritual journey of the individual.
Broader Journeying within transcendence (Christianity, #HH6505).
Followed by Journeying within transcendence – from humiliation to exaltation (#HH2329).
Preceded by Journeying within transcendence – from me to you (#HH5856).

♦ **HH1775 Psychism**
Mental psychism — Physical psychism
Description This includes the practice and study of influences of and communication with intelligences other than incarnate human, including channelling and the activities of spirit mediums. This is seen as personal communication at any level: five minds are distinguished – conscious, subconscious, superconscious, universal and subliminal. Physical psychism uses only the subconscious mind (personal physical psychism) or intelligences from other worlds (physical mediumship). Mental psychism uses any or all the levels of mind, whether mental personal psychism or mental mediumship, but whereas the former may be carried out intentionally and improved with practice, the latter, although under the control of the psychic, uses the energy of the other intelligence.
Related Channelling (#HH0878)   Panpsychism (#HH3876)
Spiritualism (#HH0479).

♦ **HH1778 Primitivism**
Fall of man — Cultural primitivism — Original sin — Ecosophy
Description In contrast to theories of human development which see the human race as ascending from a lower to a higher plane, much of philosophy and religion sees the decline of humanity from some Golden Age, a decline which each individual life is an attempt to redress. Some have seen this as the need to return to the natural, uncontrived state currently superseded whether due to historical development, to technological development, to social laws and customs, or to development of mental activity.
In classical thought, the Golden Age is distinguished as an age when mankind was naturally just and required no legal system. Cultural primitivism as lived by Aristhenes rejected luxury, property and social and moral rules. Similar (though more ascetic) life marked Judaic and early Christian primitivists, comparing such simplicity with the state of Adam and Eve in the Garden of Eden before the fall. In Hindu teaching, the human race is currently in the final and lowest of four ages, the Iron Age or Kali Yuga; by practice of discipline and realization of the self the individual can overcome external circumstances and live in the Golden Age. The cultural primitivists of the middle ages, however, rejected historical primitivism, seeing history as an ascent or development from the Ages of God the Father and God the Son to a third age, that of God the Holy Spirit, when men will be full of grace and spiritually, as opposed to intellectually, advanced. Later, the simple, uncorrupt life was stressed both by Luther and Calvin, and by the romanticists of the 18th century, notably Rousseau; while 19th and 20th centuries are marked by numbers of proponents of the simple, childlike or primitive existence.
Present day primitivism is demonstrated by those who currently hail the Age of Ecology as the development of humanity to a maturity beyond individual ideologies, when the boundaries between humanity and nature cease to exist. Consciousness includes the whole of nature, even of the cosmos. This is seen as a vindication of pagan and pantheist harmonious relationship with nature and as the culmination of mystical teachings of both East and West. The exclusiveness of "civilized" religions and their attempts to impose their authority on others are held to be disastrous.
Refs Williams, Norman P *The Ideas of the Fall and of Original Sin*.
Related Kali yuga (#HH0239)   Fall of man (#HH3567)
Cosmic consciousness (#HM2291)   Planetary consciousness (#HM2006)
Human development through pantheism (#HH5190).

♦ **HH1781s Noogenesis**
Description As superconscious learning adapts man to his cultural environment, it is central to the genesis of mankind's cultural and mental world.
Related Superconscious learning (#HM1244)
Ontogenetic model of human consciousness (#HM1105).

♦ **HH1784 Yoga sastra** (Yoga)
Description A system of spiritual exercises.
Related Yoga (Yoga, #HH0661).

♦ **HH1799 Human development** (Islam)
Surrender to the will of God
Description Notable for its prophetic tradition, Islam is surrender to God's will in obedience to the teachings of Muhammad, the last and greatest of the prophet messengers of God who covers the teachings of all those who precede him, including Abraham and Jesus, and perfectly reveals God's truth. The revelations of Muhammad are set down in the Qur'an, which is the word of God.

Behaviour is also modelled on the Hadiths which report the acts and sayings of Muhammad. The surrender of physical and spiritual activity to God enables transcendence from simple physical existence to becoming fully human as a worshipper of the divine; and ultimately to complete transcendence of human existence in eternal life. In particular, the way of inner religious discipline of the mystic, with its aim of self-transcendence and ultimate union with the Absolute as of the lover with his beloved, is the hallmark of *Sufism*.
In Islam there is no distinction between the sacred and the secular, it is a total way of life with prescribed order and behaviour for all occasions, so that individuals and all social groupings reflect the will of God. Man is formed potentially both good and evil, human life is a testing ground for which good is rewarded and evil is punished at the last judgement; in this life, the individual's sex, happiness, food and lifespan are determined before birth. There are five basic practices, the *Pillars of Islam*: profession of faith, *shahada*; worship, *salat*; giving of alms, *zakat*; fasting, *saum*; pilgrimage, *hajj*. A sixth practice, holy war or *jihad* is sometimes added to these five.
The Qur'an contains many injunctions to alleviate suffering and injustice. It promotes the organization of a society in which the justice and compassion of God can be extended and implemented. There are explicit directions in the Qur'an on family and social matters and certain things are totally prohibited (drinking alcohol, gambling, eating pork, for example). From the view of religion, men and women are equal, with the same religious and moral duties and the same rights to education, especially religious education. A man should be respectful and protective towards women; he is responsible for his wives and children and therefore inherits a higher percentage of wealth. Within this framework are many groupings and culturally different practices allowing different experience and celebration of religion.
Refs Cousins, Ewert, et al (Eds) *Islamic Spirituality*; Cousins, Ewert, et al (Eds) *Islamic Spirituality* (1987); Nasr, Seyyed Hossein *Islam and the Plight of Modern Man* (1975); Nicholson, R A *Studies in Islamic Mysticism* (1929).
Broader Human development through religion (#HH1198).
Narrower Saum (#HH4216)   Hajj (#HH1265)   Salat (#HH0536)
Zakat (#HH5643)   Jihad (#HH3687)   Shahada (#HH2341)
Human development (Sufism, #HH0436).
Related Surrender (#HH0449)   Spirituality (Islam, #HH5902)
Human development (Sikhism, #HH6292)   Human development (Baha'ism, #HH2435)
Resignation to the will of God (#HH1016)
Communication with supernatural beings (#HH1306).

♦ **HH1801 Family life** (Christianity)
Description The family is both the place where the individual learns how to conform to social requirements and where he or she receives support in social relationships. It is critical for the physical and psychological welfare of its members. Each member has his or her unique role in the family.
Because of its personal and institutional aspects the family is always subject to tensions, particularly in a time of swift social change. The secular position that marriage is a contract governed by law, entered into freely by both parties and dissolvable by law means that religious aspects are held to be a private matter. Conversely, Christianity looks on family life as a vocation in parallel with the single state in the service of God. It is not always clear, even in the New Testament, when the secular and when the religious aspect is being referred to.
Despite being based on romantic love, many marriages are in fact contractual, as romantic love in the long-term proves an inadequate basis. Another approach is that of natural union based on biological and social foundations; this emphasis on sexuality comes into conflict in Christianity where sex without procreation has always been an ethical difficulty. Then there is marriage as the command of God, where roles are explicitly laid down following religious or cultural norms. The metaphor of covenant or vocation, based on the view of the relation between God and the Church or mankind, again emphasizes role models but here there is a contract between unequals which may mean a demeaning status for women and children. However, within this latter, emphasis on the covenantal or sacramental side prevents the family from being too private.
Christian ethics have their place in the understanding of human sexuality, in the transformation of sexual desire into conjugal love, in emphasizing the covenantal and vocational nature of marriage and the family, in preparing for marriage, parenthood and upbringing of children, in daily life and in moments of crisis. Not just as a set of guidelines but as pastoral care and counselling the aim is to enable every Christian to embody his or her faith in the life of the family.
Related Parenthood (#HH3543).

♦ **HH1809 Extraordinary knowledge** (Buddhism)
Abhijna — Abhinna (Pali) — Mngon par shes pa (Tibetan)
Description The six types of extraordinary knowledge are: magical physical powers; the ability to hear sounds and voices from all over the universe; the ability to know the thoughts of others; memory of past lives; the ability to see all creatures in the world; the extinction of harmful mental states.
Related Steps to enlightenment (Buddhism, #HH4019).

♦ **HH1865 Purification of virtue** (Buddhism)
Purity of morality
Description *Virtue as volition* is present in one who, for example, abstains from killing living things, and who fulfils his duties. *Virtue as consciousness concomitant or mental properties* is the abstinence in one who follows the law. *Virtue as restraint* is that exercised through: the rules of the community (patimokkha); mindfulness; knowledge; patience; energy. *Virtue as non-transgression* means not to transgress in body or speech the precepts which one has undertaken. Purification of virtue is the purified four-fold virtue.
Context One of the two purifications or purities which constitute the roots of understanding according to Hinayana Buddhism, the other being the purification of consciousness or purity of mind.
Broader Purifications (Buddhism, #HH3875).
Followed by Purification of consciousness (Buddhism, #HH2332).

♦ **HH1867 Spiritual way**
Spiritual path
Description Any spiritual method is said to require two elements: discrimination or discernment between the real and the unreal; concentration on the real. Although the former does not require a particular religious form (although it does require metaphysical understanding), the latter can only arise through a tradition or religion. It is taught by a spiritual master who, once he has accepted a person as his disciple, passes on to him the doctrine and is the link in the chain going back to the founder of the religion (avatar). Perpetual concentration on the real does not arise through human means alone. A person's will is not strong enough to adhere to the Absolute. It requires integration of the whole being into a non- or supra-individual form.
There is said to be no spiritual path outside Judaism, Christianity, Islam, Buddhism, Hinduism and Tao. Within these traditions, the master transmits: the spiritual influence from God derived through the founder; the means or keys to understanding the method or meditation; the sacred supports required for concentrating on the real. Individual experimentation brings ruin.
Refs Bobgan, D and Bobgan, M *The Psychological Way/The Spiritual Way* (1979); Chittick, William C *The Sufi Path of Love* (1984); Foster, Richard J *Celebration of Discipline* (1984); Smith,

Margaret *The Sufi Path of Love* (1954).
Narrower The Path (#HH0424) Sharia (Islam, #HH0738)
Triple way (Christianity, #HH0631).
Related Guru (#HH0805) Avatar (#HH0079) Spiritual direction (#HH0442)
Spiritual master (Sufism, #HH2080) Human development through religion (#HH1198).

♦ **HH1875 Spiritual retreat** (Sufism)
Khalwat — Spiritual solitude
Broader Retreat (#HH0420).
Related Inabat (Sufism, #HM3484).

♦ **HH1876 Invocation**
Description This is a process where a god or spirit is invoked through ritual chanting or other ceremony, perhaps to enter the body of a magician who will then be enabled to carry out otherwise impossible feats for the benefit of those conducting the ceremony. The term is also used for the prayer used at the commencement of a church service.
Broader Rites (#HH0423).
Related Chanting (#HH1257) Spirit possession (#HH0056).

♦ **HH1890 Spiritual affections**
Religious affections — Gracious affections
Description According to Jonathan Edwards, although natural and spiritual affections appear to have much in common, there is about spiritual affection something which is totally absent from natural affection. There is a difference not only in degree and circumstances but also in its whole nature, While still a "natural" man and yet denying natural affections and leading a religious life is against nature. If corrupt nature is not mortified, sooner or later nature will prevail. Nature has to be mortified and a new and heavenly nature infused. Then man will walk in a new and heavenly nature.
Refs Edwards, Jonathan *Treatise Concerning Religious Affections* (1959).

♦ **HH1900 Journeying within transcendence - prologue**
Description The hymn of the Word revealed in flesh shows not simply creation generally proceeding from God but the individual's specific emergence from the creator, the mystery of existence. The emergence from is also an attraction to, providence is directing the journey of life towards meaning, intelligence and love. The individual self has within its subjectivity the numinous which must be reverenced, and so has all creation. Despite the existence of internal forces still inclined to remain in darkness, the answer to the demand for acceptance of God's presence in the self is "yes".
Context The first section of St John's Gospel, Chapter I 1-18, is related to a stage in the spiritual journey of the individual.
Broader Journeying within transcendence (Christianity, #HH6505).
Followed by Journeying within transcendence - from call to mission (Christianity, #HH3352).

♦ **HH1902 Human development through primal religion**
Totemic religion — Totemism
Description Tribal societies tend to have a close sense of unity with their surroundings which has been lost in so-called civilization. This oneness excludes sharp distinction between spiritual and physical and implies both spiritual and physical dependence on the natural scene, with its cycles and seasons. The precariousness of human existence is clearly felt, with consequent humility regarding creation and the spirits and powers with which it is peopled. There is contact for humans with the spirits, benevolent beings protecting humanity from evil forces. Behaviour in this life is modelled on the requirements of these spiritual beings. The whole of life is permeated with rituals so that religion is a permanent and living reality from day to day, although the world of the spirit may be only unclearly experienced. Focus for such experience may be through activities of shamans or medicine men, many of whom combine such activity with normal living. Death is seen as a transition, not the end of all, although beliefs vary as to the afterlife, whether remaining with the living as ancestral spirit, or being reborn in another incarnation, or attainment of eternal life in heaven.
Typical of Australian aboriginals, totemic religion involves a special relationship between a person (or group) and a species, object or phenomenon in his environment. It affirms the bond between man and nature, and each totemic group is responsible for rituals ensuring supply of or good relations with the totem (which may be an animal, such as the kangaroo, or natural phenomenon such as rain). Much of the ritual is secret and its publication to women or to those not initiated constitutes breaking of taboo, however unwittingly, and presages disaster. The same life essence exists in the individual and the land in which he lives, the spirit of each person continuing in that land after death.
Broader Human development through religion (#HH1198).
Related Shaman (#HH0973) Totemic awareness (#HM4341)
Dreaming experience (#HM3908)
Human development through new religious movements (#HH1523).

♦ **HH1903 Human development** (Zoroastrianism)
Iranis — Parsis
Description Zoroastrian religion is equated with high ethical standards and an altruistic philosophy of life. The one eternal God, Ahura Mazda, is wise, good and just but resisted by his twin, Angra Mainyu, the evil spirit. Through the seven holy immortals, Ahura Mazda made the world as the battle ground between good and evil. Through the greatest of these immortals, Spenta Mainyu, Ahura Mazda can become immanent in mankind. The instinct of creation is to strengthen good and combat evil; but only man can do so by deliberate choice. Each individual is judged at his death, so that if his good outweighs evil he ascends to heaven while if not he falls into hell. The aim of life is a virtuous striving for salvation of the world. The last days will be marked by great calamities and by the coming of a saviour born of a virgin, when, after a great battle between the lesser immortals and evil spirits and between good and evil men, good will be victorious. At the last judgement, all those who have died will be resurrected, the good having bodies as immortal as their souls and the evil being destroyed.
The Zoroastrian way of life emphasizes purity and the purifying quality of fire; there are special rites to heal the sick and restore lost purity; also dancing and singing, incense and prayers please the divine beings. Emphasis is on furthering the good of creation through deeds of benefit to others, and for society at large. Qualities include love of God and of fellow creatures (including animals), courage, integrity and resisting of evil.
Broader Human development through religion (#HH1198).

♦ **HH1908 Human development through panentheism**
Description As opposed to pantheism, which identifies the world with God, panentheism sees the creation as being within God, an interior modification and manifestation which exists only through God but into which God is not absorbed. In other words, all things are not the sum total of God; but God is in all things. This is a belief common to many religions, including Hinduism and (some interpretations of) Christianity. As St Thomas Aquinas put it, the Holy Trinity is in the whole creation, in every part of it.
Religions which see God as totally separate from his creation and ruling it from without are said to lead to a loss of sense of the sacred in all created things, and ultimately to the abuse of nature and the ecological problems which threaten to destroy the world. In this sense, the semitic religions emphasize the masculine aspects of God. Panentheism leads to a recognition of the feminine aspect, immanent in creation. A view of the earth as a living being, mothering and nourishing mankind, may lead to a greater respect for nature and the end to abuse of the environment. A return to panentheism may also lead to the recognition of God in each soul and of each soul in God rather than of the soul separated from God by sin so that there is a great gulf between God and humanity. This common experience of God in the depths of each individual being, beyond differences of body and soul, leads to a sense of common humanity, of community, so that individuals no longer see themselves as totally separate but as members one of another.
Refs Griffiths, Bede *Christianity in the Light of the East.*
Broader Human development through religion (#HH1198).
Related Planetary consciousness (#HM2006)
Human development through pantheism (#HH5190).

♦ **HH1909 Witchcraft**
Way of the goddess — Craft work — Wicca craft
Description The remains of a goddess cult, witchcraft works within specific and narrow limits to generate cathartic magical liberation. Rituals represent aspects of the cultural history of human consciousness. Witches claim to stand in a long tradition as adherents of an ancient religion which is counter to those artificially created and which springs from an inherent human capability to relate to, understand and use nature and natural forces, although these are unperceived by those who cannot exercise their latent powers in this way. Thus witches and their assemblies, which may include those wishing to develop such abilities, are the custodians at once of an ethic and philosophy of nature, and a knowledge of such things as natural medicine and healing. Renewed interest in the craft is characteristic of a time in which man and woman have recovered awareness of nature and each other.
Witches or traditional healers flourish in developing countries, especially in Africa. Whereas modern doctors are respected for relieving symptoms, many Africans believe that diseases have spiritual roots and that a thorough cure requires a healer in touch with such dimensions. They are therefore often useful in the treatment of psychosomatic ills. Traditional herbal remedies often anticipate developments in pharmacology.
Refs Farrar, Janet and Farrar, Stewart *The Witches' Goddess*; Scott, Gini G *Cult and Countercult* (1980); Stewart, R J *Living Magical Arts* (1987).
Broader Magic (#HH0720).
Related Feminine spirituality (#HH4003) Psychological development of women (#HH1976)
Acceptance of the female principle (#HH3663).

♦ **HH1929 Cultural identity**
Description Although cultural identity is continually under change from without and within, every people requires to have a solid link with its own history. Where external forces (colonization, for example) have prevented continuity, there is a great impetus to reforge the links and remedy the sense of loss such deprivation arouses.
Related Cultural evolution (#HH0926) Cultural development (#HH0737).

♦ **HH1932 Cave meditation**
Description Meditation is (over)practised in order to escape from problems rather than as an end in itself.
Broader Meditation (#HH0761).

♦ **HH1971 Knowledge** (Buddhism)
Description Knowledge refers to an awareness of things as they are, the disquieting and unpleasant as well as the inspiring and pleasant. According to Buddha, since birth, death, suffering, woe, lamentation and despair exist, one should follow a teaching which leads to the destruction of these things, not first demanding to know the answer to unprofitable questions such as whether or not the world is eternal.
Narrower Artha (Hinduism, #HH0353) Sabda (Yoga, Buddhism, #HH0538)
Jnana (Yoga, Buddhism, Hinduism, #HH0610).
Related Wisdom (#HH0623) True knowledge (Christianity, #HH5767).

♦ **HH1973 Hearing the music of the spheres**
Description What is called music in everyday language can be understood as merely a subset or miniature of that music or harmony of the whole music which is working behind everything, and which is the source and origin of nature. There is an age-old myth describing human development as a process of ascent through the spheres of the imaginal world. Experiences of this ascent are often described in terms of the music associated with the process and the domains encountered. The spheres may be associated symbolically with the planetary spheres, each associated with a particular tone such that their changing relationships produced an ever-changing musical harmony comprehensible as a whole. The movement of the individual through the spheres is then understood as a purification process. Thus for Jung, the ascent through the planetary spheres therefore meant something like a shedding of the characterological qualities indicated by the horoscope, namely a retrogressive liberation from the character originally imprinted by the rulers of those spheres. In this sense the ascent is like the overcoming of a series of psychic obstacles beyond which the harmony of the spheres is eventually experienced. The ascent is thus a journey, or a series of initiations in the imaginal world. The music that is heard there and whose supernatural beauty is always remarked upon is none other than the knowledge gained in those initiations by those who have attained the requisite stage of psychic growth. It is therefore in music the seer can see the picture of the whole universe. From this perspective the cosmic system works by and through the laws of music.
Refs Godwin, Joscelyn *Harmonies of Heaven and Earth* (1987).
Related Human development through music (#HH3003).

♦ **HH1976 Psychological development of women**
Context The development of women's studies and feminist scholarship has highlighted the similarities and differences between male and female psychological development.
Refs Belenky, M, et al *Women's Ways of Knowing* (1986); Bernard, Jessie *The Female World* (1981); Boulding, Elise *The Underside of History* (1976); Castilejo, Claremont De *Knowing Woman* (1973); Gilligan, Carol *In a Different Voice* (1981); Pearson, Carol *The Hero Within* (1986); Schaef, Anne Wilson *Women's Reality* (1981).
Related Witchcraft (#HH1909) Female heroism (#HH6432) Feminine spirituality (#HH4003)
Feminine consciousness (#HM2768).

♦ **HH1981 Spiritual reductionism**
Description Non-physiological medical problems are said to be of the spirit rather than the mind. Psychotherapy, in that it addresses the mind, misses the real point of the problem. Not only this, in treating the "old nature" which the scriptures say should be crucified, it compounds the problem and is an enemy of the spirit. Rather than strengthening the person, one should become weaker

so that God is one's strength. In these terms, psychotherapy is said to set itself up as a substitute for the work of the Holy Spirit. Spiritual reductionism denies that psychotherapy can help in identifying and crucifying the false self and in seeing who one really is. Non-physiological problems are rooted in personal sin and should be addressed by spiritual counselling in order to trust and follow the spiritual principles of the Bible.
Refs Benner, David G *Psychotherapy and the Spiritual Quest* (1988).
Related Psychotherapy (#HH0003) Psychospiritual therapy (Christianity, #HH4201).

♦ **HH1987 Brahmacharyashrama** (Hinduism)
Celibate stage of life
Description This period of about 12 years training follows initiation at about 6 to 8 years old. The young boy becomes the full-time disciple of a religious master and is initiated into knowledge of the scriptures. During this time he observes celibacy, and knowledge is taken in with a spirit of surrender and worship. The initiation is referred to as the *upanayana* ceremony when the boy is brought physically and spiritually near his guru. The ceremony includes the shaving of the head indicating removal of desires, and the wearing of the sacred thread, tied to indicate determination to unite the finite personality with the infinite self. The thread is also a symbol of the three states of mind or *gunas* and the three states of consciousness (waking, dreaming and sleeping) to be transcended in turiya, realization.
Context This is the first of the four ashramas or stages of life according to Indian tradition, leading eventually to self-realization.
Broader Ashramas (Hinduism, #HH2563).
Related Brahmacarya (#HH0978).
Followed by Grahastashrama (Hinduism, #HH2343).

♦ **HH1988s Control of involuntary functions**
Description Various systems of self-development give the ability to control bodily functions normally not controllable at will, such as blood flow, temperature regulation.
Related Yoga (Yoga, #HH0661).

♦ **HH1994 Centering prayer** (Christianity)
Description Centering prayer is a method of self surrender, teaching one not to be possessive but to let go. Having peace without thinking about having it means one has learned how to do it.
The practice is designed to prepare for the gift of contemplation or contemplative prayer, by reducing the obstacles to such contemplation. The intuitive faculties are refined through regular practice. All thoughts, any perception that appears on the inner screen of consciousness, is disregarded for the source of all thought, and a deeper kind of attention emerges. The method implies a relatively comfortable physical position in a quiet place at a time of day when one is most awake and alert. A sacred word is gently introduced to one's attention. Each time a thought intrudes, the sacred word, which has been chosen previously, is put in its place to direct the attention towards God. The word is not chosen for its meaning but as a pointer to reaffirm attention. It centres one's attention on God's presence within. After the practice there should be some minutes of quiet thought or vocal prayer before opening the eyes.
Refs Keating, Thomas *Open Mind, Open Heart* (1986).
Broader Prayer (#HH0185).
Related Centering (#HH0527) Spiritual reading (#HH0158)
Resting in God (Christianity, #HM9213).
Followed by Contemplative intuitive meditation (#HH0816).

♦ **HH2002 World cycles** (Buddhism)
Aeons of dissolution and evolution
Description Buddhism refers to millions of world cycles of dissolution and evolution or contraction and expansion with periods superseding contraction or expansion. Thus each cycle is in four parts or aeons. The world cycle may be destroyed by fire, water or air, when all is demolished up to a particular Brahma world. These are the limits on contraction or dissolution. In breadth one Buddha-field is demolished (in the field of his birth, ten thousand world spheres or systems; in the field of his authority, one hundred thousand million world spheres or systems; in the field of his sphere, to an infinite extent, as far as he wishes).
The world's destruction is because of unprofitable or immoral roots. When greed or lust is most conspicuous the world is destroyed by fire. When hate is superabundant the world is destroyed by water (some sources reverse these two). If delusion is more conspicuous then destruction is by wind. There are seven destructions by fire then an eighth by water until the 64th aeon which is destroyed by wind. Fire destroys to the Abhassara (streaming radiance) world, water to the Subhakinha (refulgent glory), and wind to the Vehapphala (great fruit) world.
Knowing the approaching end of an aeon, deities prophesy on earth. They entreat development of good qualities such as lovingkindness, compassion, gladness and equanimity leading to rebirth in the divine world. Those acquiring jhana in the divine world are reborn in the Brahma world.
Broader Hinayana Buddhism (Buddhism, #HH0845).
Related Knowledge of recollection of previous existence (Buddhism, #HM4297).

♦ **HH2003 Systematics**
Gradations of knowledge
Description Progressive increases in comprehension of complexity can be viewed through the number of terms by which knowledge at any stage is organized and through which it is experienced. Comprehension of such systems proceeds in a definite sequence, given the order of their emergence into awareness and the minimum number of terms required to exemplify the attributes of that knowledge at any stage. At each stage there is a field of experience coterminous with all experience. But a particular system never exhausts the possibility of description and comprehension for, whatever number of terms is reached, some degree of abstraction remains and additional terms must be admitted in order to move towards greater concreteness. The first three gradations of knowledge can be considered subjective because the object of knowledge is placed outside the act of knowing. Growth in understanding requires recognition of the representational power of successive systems and a deepening appreciation of their significance.
Context A classification of a progression of twelve modes of knowledge, developed by J G Bennett, inspired by G Gurdjieff.
Refs Bennett, J G *The Dramatic Universe* (1956); Kuchinsky, Saul *Systematics* (1985).
Narrower Two-fold knowledge (Systematics, #HM1234)
Six-fold knowledge (Systematics, #HM5187)
Ten-fold knowledge (Systematics, #HM1804)
Four-fold knowledge (Systematics, #HM1991)
Five-fold knowledge (Systematics, #HM8080)
Nine-fold knowledge (Systematics, #HM1408)
Three-fold knowledge (Systematics, #HM0811)
Seven-fold knowledge (Systematics, #HM0776)
Eight-fold knowledge (Systematics, #HM0344)
Eleven-fold knowledge (Systematics, #HM0065)
Twelve-fold knowledge (Systematics, #HM0707)
Non-discriminative knowledge (Systematics, #HM4002).

♦ **HH2006 Quietism**
Description This philosophy is typical of many schools but specifically of several schools in China in the third and fourth centuries before Christ. By stilling of external activity and of internal appetites and emotions one prepares one's self for self-perfection. Through working back through layers of consciousness until there is no more the things which are perceived but that which perceives, no more what is known but that by which one knows, through arriving at pure consciousness, there is stillness. This is the state of absolute joy and of power, a state which cannot be described but which can only be experienced by trying it out. In Quietism, the mind has perfect poise which manifests in perfect equilibrium of action through unity with the medium in which the activity takes place.
Related Passivity (#HH3229) Ki energy (#HH0620) T'ai Chi Ch'uan (#HH0282)
Human development (Taoism, #HH0689) Self-mortification (Sufism, #HH0464).

♦ **HH2007 Hell**
Description The idea of hell as a place of everlasting suffering where the wicked are punished for their faults is said to derive from ancient mythology, many cultures picturing the dead as inhabiting an underground realm. This underground realm became equated, in Christian terms, with descriptions of the end of the world in apocalyptic teachings. These literal portrayals, more-or-less universally accepted in the middle ages, have been dismissed by liberal reaction as simply a means to frighten people into moral behaviour. The argument is that an eternal state of evil conflicts with God's purpose for the world. This view, again, is challenged by Jungian philosophy, which demonstrates the shadow behind every virtue and that human beings are demonically bent upon destruction. Other commentators indicate that there can be no spiritual conflict (which evidently exists) without the existence of an antagonist. Perdition and hell are seen in this case as being very present realities.
Related Heaven (#HH3564).

♦ **HH2008 Psychological reductionism**
Description This attitude maintains that, although religion provided a prescientific perspective, modern psychology presents a fuller and more accurate picture of the personality. All non-physiological medical problems are psychological or physical. Mysticism can be explained as a regressive reactivation of the lack of ego boundaries of the infant or as obscure self-perception of the realm outside the ego, of the id. To remain scientific one must treat man as an advanced animal but nothing else. Psychoanalysis can study the human reality behind religion. Living in love and thinking truth are the question, not the symbol system used. Religion is only useful if it helps people to live more fully. Without dogma or religious symbols, psychotherapy can accomplish the same results as religion.
Refs Benner, David G *Psychotherapy and the Spiritual Quest* (1988); Fromm, Erich *Psychoanalysis and Religion*.
Related Psychospiritual therapy (Christianity, #HH4201).

♦ **HH2009 Non-deterministic psychology**
Refs Epstein, Gerald N (Ed) *Studies in Non-Deterministic Psychology* (1980).
Related Waking dreams (#HM1376).

♦ **HH2023 Laying on of hands** (Christianity)
Description This is used in the Christian faith as a symbolic transfer of the power of the Holy Spirit, whether in healing ceremonies or as a transfer in spiritual power (as in the ordination of Church leaders), or in the service of confirmation on receiving the gift of the Holy Spirit as a confirmed member of the Church.
Related Magnetic healing (#HH2552) Human development (Christianity, #HH2198).

♦ **HH2031 Kinerhythm meditation**
Refs Ichazo, Oscar *Kinerhythm Meditation* (1978).
Broader Meditation (#HH0761).

♦ **HH2034 Maithuna** (Yoga)
Yoga of sex — Tantric yoga
Description The principle of the Tantras is that orthodox spiritual techniques fail if, in stressing asceticism, they do not take the tremendous power of the human sexual drive into account. Rather than allowing suppressed sexuality to fester in the subconscious mind from where it may one day explode and do damage to the human psyche, this powerful drive is brought into the open so that it is thoroughly understood. Understanding means that the dangerous quality is lost and the drive can be easily controlled. The disciple rises by the very things which can cause his downfall.
Maithuna, or ritualized sexual union, plays an important part in some tantric activity, in particular as the final component of the *panca tattva* or *panca makara* ritual. This has led (particularly as a result of abuse) to considerable misunderstanding and condemnation. However, maithuna aims at the experience of complete union between the partners, dissolving of polarities and the mental rather than physical appreciation of the universe. In the Buddhist tradition seminal emission is totally avoided but Hindu Tantras normally include emission in the vedic idea of sacrifice, *yajna*. Certain aspects of maithuna have been used in the West as means of treating sexual problems.
Refs Van Vliet, C J *The Coiled Serpent*.
Broader Yoga (Yoga, #HH0661).
Related Sacramental sex (#HM0843) Sexual awareness (#HM2374)
Sexual yoga (Taoism, #HH9112) Kundalini yoga (Yoga, #HH0237)
Tantra (Buddhism, Yoga, #HH0306).

♦ **HH2054 Process meditation**
Refs Progoff, Ira *The Practice of Process Meditation* (1980).
Broader Meditation (#HH0761).

♦ **HH2056 Sanyasashrama** (Hinduism)
Mendicant stage of life
Description As a homeless beggar-saint, the sanyasi renounces the world and becomes fully attuned to the consciousness within. His fire-coloured ochre robe indicates the burning away of the body and of claims on society. Wanting nothing from the world, he is dedicated to giving, service and sacrifice for the bettering of humanity.
Context This is the fourth of the four ashramas or stages of life according to Indian tradition, leading eventually to self-realization.
Broader Ashramas (Hinduism, #HH2563).
Preceded by Vanaprasthashrama (Hinduism, #HH2782).

♦ **HH2076s Constructive imagination**
Description A controlled or deliberate calling to mind of images and pictures may result in an imaginative solution to problems. Repeated imagining of a desired happening is said to effect that happening if executed sufficiently often and with sufficient emotion.
Related Imaginative consciousness (Anthroposophy, #HM2901).

♦ **HH2077 Six-member yoga** (Yoga)
Sadanga yoga
Description Apparently developing under the influence of Buddhism and referred to in the Maitrayaniya Upanishad, this ancient system of yoga does not include as separate elements the first three components of classical or eightfold yoga, yama (restraint), niyama (discipline) or asana (posture); included instead is *tarka* or *anusmriti* (recollection), referring to the ultimate realization of yoga.
Broader Yoga (Yoga, #HH0661).
Related Eightfold path of yoga (Yoga, #HH0779).

♦ **HH2080 Spiritual master** (Sufism)
Shaykh — Pir
Description When the traveller on the spiritual path reaches the level of *murid* he comes under the guidance of a spiritual director from whom he receives mystical training. The director disciplines his soul through a mystical process during which spiritual progress is carefully monitored; and becomes the means for the making his disciple love God and for making him lovable to God.
Broader Human development (Sufism, #HH0436).
Related Guru (#HH0805) Murid (Sufism, #HM3887) Spiritual way (#HH1867) Identification with spiritual master (Sufism, #HM3944).

♦ **HH2087 Primal therapy**
Primal scream
Description After a period of isolation to weaken the defence mechanisms, the individual is lead by the therapist to relive the experience of a key scene in his life. This taps the repressed pain of not being understood and loved by one's parents and can be an experience of intense aloneness when the individual cries out in a *primal scream*. Feelings are de-anaesthetized so that it becomes possible to experience the full range of human emotions. The therapy assists in freeing the individual from mental or behavioural disorders.
Refs Janov, Arthur *The Primal Scream* (1970).
Broader Psychotherapy (#HH0003).
Related Sensory deprivation (#HH1478).

♦ **HH2097 Stages of faith**
Description James Fowler has developed a path of faith development which parallels the stages of individual human development described by Jean Piaget, Lawrence Kohlberg and Erik Eriksen. It includes six progressive stages, preceded by the undifferentiated faith of the baby, which is called a pre-stage.
Refs Fowler, J W *Stages of Faith* (1981).
Broader Faith (#HH0694).
Narrower Conjunctive faith (#HM3359) Mythic-literal faith (#HM1525) Universalizing faith (#HM1465) Undifferentiated faith (#HM8672) Intuitive-projective faith (#HM1425) Synthetic-conventional faith (#HM1814) Individuative-reflective faith (#HM1665).
Related Life cycle (#HH0511) Individual development (#HH1543) Stages of personality development (#HH0285).

♦ **HH2098s Affinities**
Description According to Rudolph Steiner, individuals may experience many lifetimes together, leading to special affinity in friendship, marriage or work.

♦ **HH2099 Licence**
Antinomianism
Description In the Christian faith, and in others, there has been held the view that faith abolishes the law so that the faithful are no longer subject to law. This view has been misinterpreted by various sects and has led to the countenancing and even encouraging of licentious conduct which goes against the basic principles of the faith concerned.

♦ **HH2116 Dream therapy**
Description Analysis of dreams may bring to light hidden talents or repressed experiences normally not available consciously.
Refs Rossi, Ernest L *Growth, Change, and Transformation in Dreams* (1971).
Broader Psychotherapy (#HH0003) Psychoanalysis (#HH0951).
Related Dreams (#HM2950).

♦ **HH2117 Justice**
Description Thomas Aquinas defined justice as the order of reason with regard to actions affecting others and present in the will. It involves commitment to putting right what is seen not to be right in relationships when openness and trust have been denied and enmity intensified.
Justice demands that the will should habitually be set towards impartiality. It requires the other three virtues to be realized effectively: discernment of the right means for the good to which justice disposes to arise (prudence); ordering of the passions so that the good may be pursued single-mindedly (temperance); and steadfastness to pursue the good even when the self is threatened (fortitude). As with all the virtues, justice should be shaped and directed by charity.
Justice is, however, sometimes contrasted with charity or love. Love is said to apply for relations between an individual and his neighbour, in sacrificial regard for his wellbeing (personal bonds), while justice applies in relations with third parties (impersonal bonds), although these are not exclusive and judgement takes a secondary role in the first case as love does in the second.
Justice is called into play in finite, sinful conditions where self-interested motives and perspectives dominate, love where conflict and inordinate self-assertion are absent. Using God's love as a paradigm, justice as it simply considers merit is in opposition to love for one's neighbour regardless of merit, unconditional and gracious. Love may demand more but never demands less than justice.
Another concept of justice (John Rawls) is of the rights and duties, and of benefits of social cooperation, of each individual as a free and equal participant in political society and following a life plan in accordance with a particular conception of the good (egalitarianism). It must apply in circumstances where there may be a shortage of resources and disagreement as to what kind of life constitutes human fulfilment. Principles of justice have then to be agreed with respect to political and economic institutions by persons who agree together as free and equal, having abstracted from their particular life plans. This procedure allows an egalitarian conception of justice to emerge, guaranteeing equality of liberty and opportunity and allowing social and economic inequalities to work for the benefit of the least advantaged.
In contrast to this is the utilitarian concept of justice, where maximum average desire-satisfaction or happiness is sought in society as opposed to respect for each individual's separate life plan. Other approaches include libertarianism, where priority is given to protecting freely consented (and thus uncoerced) arrangements; and socialism, where social and material equality are emphasized. Different systems are criticized as: interference with private freedom of choice (socialism); undermining of freedom by lack of attention to human needs (libertarianism); dominating the individual through accumulated political and economic power in private hands (libertarianism). These criticisms have also been in some sense levelled at egalitarianism.
Problems that arise in any system are the compromises necessary in balancing the individual's life prospects against maximizing happiness overall; in determining how much governmental intrusion is warranted into private engagements to ensure state powers in providing its citizens with basic human requirements, and how much human needs have a special priority so as to permit public intervention when private initiatives fail; and in deciding the balance between human needs and human preferences.
Problems also arise for the individual when social and political justice are contrasted with justice as a virtue; when individual behaviour is seen in the light of God's justice and of God's love; when theological and non-theological concepts of social justice are compared.
Context One of the four cardinal or principal virtues recognized by Plato in the Republic and featuring prominently in mediaeval Christianity.
Refs Rawls, John *A Theory of Justice* (1971).
Broader Virtue (#HH0712).
Related Love (#HH0258) Prudence (#HH7902) Fortitude (#HH5640) Temperance (#HH0600) Gratitude (Christianity, #HH4202) Measuring emblems (Tarot, #HM2178) Forgiveness (Christianity, #HH0899).

♦ **HH2119 Cessation of suffering** (Buddhism)
Cessation of ill
Description Suffering is said only to cease when its cause ceases; the cause of suffering is craving and the subsequent rebirth it entails, repeatedly "becoming again" as one craves for things of the senses, for becoming again, for not becoming again. The craving has to be rooted out, not through self-mortification, which addresses only the effects, but through dispassion or "fading away"; this is *nirvana*.
Related Tanha (Pali, #HM3707) Nirvana (Buddhism, #HM2330) Five aggregates (Buddhism, #HH3321) Duhkha (Buddhism, Hinduism, #HM2574) Four noble truths (Buddhism, #HH0523) Understanding as knowledge of the cessation of suffering (Buddhism, #HM8163) Understanding as knowledge of the way leading to cessation suffering (Buddhism, #HM7645).

♦ **HH2121 Self-discipline** (Christianity)
Self-denial
Description By disciplining himself in exercises of restraint, such as fasting or abstinence, the individual strengthens his will to be able to follow a spiritual path. This discipline is, in Christian terms, a response to self-love and it results in freedom, that of the spirit to dominate the personality.
Self-discipline commences with an effort of will towards self-purification, "putting off the old man", as St Paul puts it, through conversion and repentance, followed by the "putting on of Christ", the new self. It continues through fasting, almsgiving and prayer, the three duties of asceticism in the three areas of duty - to one's self, to man and to God. This is a continuous process, requiring endurance, readiness to meet emergencies, vigilance, sobriety and courage. By exercise of these qualities, the new personality or self attains full development and arrives at true spiritual liberty.
Although self-discipline gives the individual power to do great things, to endure and achieve much, it is through the mundane, obvious or trivial duties that control of appetite and strengthening of will are fostered. For spiritual freedom it is necessary not to be the slave of one's appetites. Permanently gratifying one's appetites puts them in a position of tyrants. The impression that one is real is a delusion, compulsions have actually reduced one to a shadow of a genuine person. The contemplative life, for example, requires learning to survive without habit-forming luxuries.
Related Discipline (#HH0163) Self-denial (#HH0964) Self-discipline (#HH0877) Ethical discipline (#HH1086) Via purgativa (Christianity, #HH4090) Christian self-love (Christianity, #HH6732) Spiritual self-denial (Christianity, #HH4522).

♦ **HH2134 Mass hypnosis**
Description Groups of people may be hypnotized to a greater or lesser extent en masse, whether deliberately or unintentionally, by a variety of means adopted by a speaker and/or by the playing of loud, rhythmic music. An audience can thus be lead to concentrate more deeply on and to accept unconditionally the content of what is being said. Many charismatic leaders are said to have a hypnotic effect on their audience. The beneficial effects of some religious or healing meetings may depend to some extent on the hypnotic state of the participants.
Broader Hypnotherapy (#HH0962).
Related Charisma (#HH0053).

♦ **HH2139 Inspiration**
Description This is said to arise from charismatic divine influence which may instigate and guide in the production of writings and works of art. For example, the writers of the books of both Old and New Testaments of the Bible are said to be truly the writers of these books (as opposed to by divine revelation) yet the books are still taken to be the infallible word of God.
Related Revelation (#HH1028) Inspired consciousness (#HM3125) Psychic inspiration (Psychism, #HM1094).

♦ **HH2143 Superman**
Description Visions of a superman with supra-normal powers, whether in mythology, fiction, history or philosophy, describe an individual whose qualities are those of ordinary human beings but occurring at an exceptional level. This may be physical prowess, will, mental acumen and force; or it may be spiritual perfection (Swami Omkarananda). Nietsche saw this as man's urge to surpass himself as a will for power, dominating the environment and those about him and contemptuous of those who fail to pursue power.
Refs Ettinger, R C W *Man into Superman* (1972).

♦ **HH2145 Contemplative life** (Christianity)
Description According to John Ruusbroec, the contemplative life may be entered by a fervent lover of God who: possesses God in blissful rest; possesses himself in dedicated and active love; possesses his entire life in virtues and righteousness. Such a person is chosen by God and raised to a state of super-essential contemplation in the divine light.
Refs Dupré, Louis and Wiseman, James A (Eds) *Light from Light* (1988); Wiseman, James A (Trans) *John Ruusbroec* (1985).
Broader Contemplation (#HM2952).
Related Mystical contemplation (#HM2710) Life of contemplation (ICA, #HM2109) Superessential contemplation (Christianity, #HM4340).

♦ **HH2161 Life reading** (Psychism)
Description By psychic means, information on previous lives is made available for use in the present life. This may be by means of a medium who, in a deep trance, makes known personality and experience from previous incarnations and answers questions on their relevance to this life; or it may be through parapsychology, when the scribbles of a child give insight into the subconscious mind and thence to past lives, and are used to foretell the child's future.

♦ **HH2178 Structural change**
Description These changes may occur as the result of psychoanalysis. There is a reduction in the conflict among the functions of the superego, the ego and the id so that relations among them are more harmonious.
Related Psychoanalysis (#HH0951).

♦ **HH2187 Ideological development**
Ideology
Description Religions and ideologies coexist, each exerting an influence on the other. The development of ideologies, particularly through liberation struggles, increases awareness and criticism of one's basic beliefs and desires, and of the interaction between one's self and God, whether or not one is actively actually involved in the liberation struggle. The acceptance of an ideology, although not the same as acceptance of a religion, has profound religious implications; and each individual has a moral as well as a social duty to assist in elaborating a coherent ideology in line with his or her beliefs. Ideology can, therefore, be the means for understanding of the self as well as being the dynamic factor in social change.

♦ **HH2188 Path of effortlessness**
Description Despite the apparent hardships and efforts required to follow paths of self-realization, many of the original teachers of such methods stress their ease. Christ said "my yoke is easy and my burden light". The hardships may arise because the person practising the method wants the personal self to be in control, as in meditation techniques where the injunction "the mind must become quiet" may be interpreted as "the mind must be made quiet". In fact, the mind cannot make the mind become still, because the "making" is a mental activity. The process of transcendental meditation has been described as the path of effortlessness because it involves a passive letting go rather than an active putting things away. By itself and without effort the mind is in a state of silence.
Related Technique of no technique (#HH1092) Transcendental meditation (TM, #HH0682).

♦ **HH2189 Artificial intelligence**
Simulation — Three world concept
Description Attempts to construct machines or write programs simulating what is generally recognized as intelligent behaviour have lead to further understanding of the interaction of the individual with the world outside him. An example is Karl Popper's *three world* concept. The conscious self or mind is referred to as World 2, primary reality. Secondary reality, the world of physical objects and states, is referred to as World 1. The brain is the part of World 1 which relates with the mind through a part which is termed the *liaison brain*. Also included in World 1 are the rest of the brain and the physical body, the outside world and World 3, objective knowledge or human culture as encoded in the physical world, increasingly through computerization. Books, for example, come within World 3, coding and decoding being through reading and writing. If computer programs are looked upon as means for telling a computer what type of machine to be to carry out a particular required activity, then that process simulates the intelligence process - although it is not necessarily organized in the same way as the human brain is organized.
Broader Intelligence (#HH0431).
Related Maps of the mind (#HH8903).

♦ **HH2198 Human development** (Christianity)
Description Tha main and essential feature of Christianity is the belief that Christ is THE incarnate word. Christians are those who perceive the birth, death and resurrection of Jesus Christ as the revelation of God and as the means of salvation to all who believe in Him. Despite an insistence on separateness, with God the creator of but not identical with the creation, many Christian mystics speak of mystical union with God and of an inherent unity. In this sense, man's separation from God is because of original sin, the determination of the ego to be separate; surrender to God then brings union.
The Christian way of life is basically that of a shared community of which Jesus Christ is head and the Church is the "body" of many members, with their own necessary and specific functions, the whole directed by God. From the first there was emphasis on morality, on worship and on close and caring relationships with others. This has lead to movements aimed at redressing social injustice and stressing the responsibility of Christians to the whole of society. Since Christ was incarnate, so must be his Church. Part of its mission is support for the world's poor and oppressed, and for social and economic reform to realize their revolutionary hopes. Christian society is marked by giving and established means of assisting the poor and the sick.
The Western Church is based on three confessions: (1) The Catholic Church, with its emphasis on the apostolic succession whose combined knowledge is the catholic or universal truth; and on the seven sacraments (baptism, marriage, confirmation, Eucharist, penance, anointing of the dying, ordination). (2) The Protestant Church which has three central affirmations: that salvation is by grace on divine initiative and not from mechanical following of particular rules; salvation is though faith and not through merit, although many preach that faith without good deeds is meaningless; knowledge is from scripture only. (3) The Anabaptist reformers who historically emphasized personal faith, free from established links with the state, with the corollary of freedom of conscience. Protestant and Anabaptist Churches are closely linked with much overlap. The Eastern Church (Orthodox Churches), places more emphasis on asceticism and mysticism, although mystic and contemplative traditions have also been accepted or tolerated by other branches of the Church. The mystical tradition teaches inner stillness and total renunciation when the self is lost in contemplation of a light so light it is darkness, approachable only by unknowing; whereas more activist traditions emphasize good works with dependence on Christ as intercessor. The XXth Century has seen rapid expansion of the Church in the Southern continents, particularly Africa, where local religious traditions have been assimilated, with dreams and mythology playing an important part, as do spirit mediumship and spirit healing.
Entry to a Church, whether as a child or as an active profession of faith by an adolescent or adult, is through baptism, symbolizing cleansing from sin, repentance and transformation to a new life; baptism with the Holy Spirit (in some Churches through a separate service of confirmation when the person indicates commitment, having previously been baptised as a baby) signifies God's initiative in salvation.
Refs Chirban, J *Developmental Stages in Eastern Orthodox Christianity*; Pope Paul VI *Encyclical Letter of His Holiness Paul VI to the Bishops, Priests, Religious, the Faithful and to All Men of Good Will* (1967); Van der Bent, Ans J *Vital Ecumenical Concerns* (1986).
Broader Human development through religion (#HH1198).
Narrower Evangelism (#HH3798) State of quies (#HM3771)
Charismatic renewal (#HH3124) Fund of the individual (#HM4497).
Related Baptism (#HH0035) Salvation (#HH0173) Religious education (#HH0709)
Initiation (Christianity, #HH6211) Glossolalia (Christianity, #HM1608)
Laying on of hands (Christianity, #HH2023) Christian mysticism (Christianity, #HH5100)
Mystical Christ-consciousness (Christianity, #HM2880)
Human development through new religious movements (#HH1523).

♦ **HH2207 Suffering** (Christianity)
Description Jean-Pierre de Caussade indicates that physical illness and feebleness, unless it absolutely requires lying in bed, should be ignored. Battling with the duties of daily life when not feeling healthy is an exercice in faith and submission, and improves strength of mind which is weakened by indulgence. The body may be worked to death through following what is ordained by God. Once human ingenuity realizes its own feebleness then God's purpose is manifest, leading souls past mortal perils and carrying them to heaven.
Refs Bowker, John *Problems of Suffering in the Religions of the World* (1970); Caussade, Pierre de *The Sacrament of the Present Moment* (1981).
Related Suffering (#HM0471).

♦ **HH2213s Pyramid meditation**
Description Facing east, the individual meditates under a suspended pyramid shape or within a large pyramid. The pyramid is said to be a source of energy if designed according to the proportions of the great pyramid and aligned with one face to the north.
Broader Meditation (#HH0761).

♦ **HH2267 Human aura**
Description The human aura is an energy field, or sequence of energy fields, normally only visible clairvoyantly, said to extend beyond the physical body and to vary in shape, quality and intensity depending upon the physical, emotional and spiritual condition of the individual and upon his lifestyle. Seven energy fields are described: electric; magnetic; infrared; sound and infrasound; ultraviolet; chemical; psychic. As intelligence, character and attitudes change so does the vibrational frequency of the aura, the frequency increasing in response to positive development.

♦ **HH2276 Self-observation therapy**
Description A therapy using altered modes of awareness to identify thoughts which trigger a particular reaction.
Related Self-observation (#HH0586).

♦ **HH2287 Linear thinking**
Description Thought is limited to what is physically seen.
Broader Thought processes (#HH0530).

♦ **HH2304 Guided meditation**
Description Attention is focused on a meditation leader who, through directing a system of, for example, chanting or relaxation exercices, brings both mental and physical quiet to the group.
Broader Meditation (#HH0761).

♦ **HH2314 Primal healing**
Description The physical symptoms of the patient are seen in the broader context not just of interpersonal relationships, but also of harmonious interaction with the whole cosmos. Primal healers are concerned with the whole man in his whole environment, with the mental and spiritual outlook within the community which is the context in which healing must take place.
Related Folk healing (#HH0248) Spiritual healing (#HH0458)
Holistic medicine (#HH0581).

♦ **HH2315 Deep ecology**
Ecology movement — Green movement — Biocentric equality — Ecological wisdom — Ecosophy
Description Conventional ecological lobbies work through the existing political system to alleviate or mitigate the negative effects of man on his environment. Although such lobbies are useful and necessary, they are criticized for resulting in short-term orientation and in compromise. In contrast, deep ecology focuses on the individual, on each person looking at himself and becoming more real. More and more deeply, the individual questions himself, the assumptions on which the dominant world-view of our culture is based, and on the meaning and truth of our reality. Different techniques are used to develop an attitude of open attention, of contemplation, replacing the strained kind of attention achieved by effort of will. This leads to a slowing down, a liberation from the frenzy of busy-ness, so that life is lived on a more even tempo and the individual can progress in working out his own salvation. This is referred to as real work. By cultivation of *ecological consciousness* there is a facing up to facts, life is lived deliberately and a new, ethically-centred conscience is developed.
A political strategy is developing which is based on personal ecological consciousness as well as addressing public policy issues. Deep ecology goes beyond factual, scientific issues to the level of self and earth wisdom; new outlooks on political and social policy are generated, which go beyond simple scientific consideration of ecological matters. It addresses the paradox of maintenance and increase in uniqueness of the individual self which is also an inseparable part of the whole system; and it rejects a system based on dominance, whether of mankind over non-human nature or of one section of humanity over another. The individual and the human race are considered as part of the organic whole and this is compared with similar questioning in the context of the different spiritual traditions. Self realization is seen as the awareness of the individual self as part of the Self of organic wholeness. This leads intuitively to the concept of *biocentric equality*, which sees all the organisms making up the whole as of equal intrinsic worth. Although there is mutual dependence in the sense of one species using another for food, harmony is achieved in living with minimum rather than maximum impact on other species - and this is extended to include the whole of life, whole ecosystems and the component natural features such as rivers and landscapes. Economic growth as the basis of society is not seen as central or even necessary to fulfilment of basic human needs. Simplicity of material living is emphasized, allowing more attention to be focused on the real work of spiritual growth.
In particular, children need to spend time in natural landscapes, with opportunities for watching, touching and moving in response to natural phenomena. Adolescents need the opportunity for risk-taking, solitude and group activity through such activities as mountaineering and sailing. Later in life there should be time to help the human community and the wider community of non-human species. Finally, the body dies, and the components of this temporary organism may become parts of unknown future beings.
Ultimately, there may be a need to reduce the human population of the planet to proportions which would allow a viable existence in a more natural state. Currently, the aim is to encourage development of more advanced stages of psychological and emotional maturity, which would be enhanced by small-scale communities which inhabit but do not command their region and minimize their impact on it (no litter, for example) and for leisure activities which respect and incorporate the environment in which they are carried out.
Refs Canadian Ecophilosophy Network *The Trumpeter* (1989); Devall, Bill and Sessions, George *Deep Ecology* (1985); McGurk, H (Ed) *Ecological Factors in Human Development* (1978).
Related Bio-globalism (#HH1267) Ecological humanism (#HH4965)
Communion with nature (#HM0915) Planetary consciousness (#HM2006)
Environmental mysticism (#HM0800) Contemplation of nature (#HH0580)
Christian stewardship (Christianity, #HH3121) Acceptance of the female principle (#HH3663)
Human development through pantheism (#HH5190).

♦ **HH2321 Backsliding**
Description A moral or religious lapse; may refer to the falling from grace after following a particular religion, reverting to old ways after conversion or to rejoining of a less than worthy organization or cult after apparently successful "deprogramming". Backsliding may be contrasted with endurance, the former showing a falling away in time of trial, adversity or lack of progress, the latter demonstrating an ability to keep going despite setbacks. Particularly distressing is the re-emergence of old habits which had apparently been overcome. However, although backsliding is common to those who have experienced a conversion, so also is the process of renewal after backsliding. The whole process is part of the natural rhythm of rise and fall of religious feeling. Whether change in attitude is accompanied by backsliding or endurance, the process of development may thus continue, such problems being inherent in the process.
Related Endurance (#HH0729) Conversion (#HH0077) Deprogramming (#HH4102).

♦ **HH2329 Journeying within transcendence - from humiliation to exaltation**
Description Despite his apparent freedom, Pilate is compelled by circumstance to condemn Jesus to death. Despite his apparent lack of freedom, Jesus is in control and, although in irony, is crowned king and lifted up in glory on the cross where a notice proclaims his kingship in all the languages of the known world. Just at the moment of death, when everything seems to fall apart, everything in fact holds together. All the facets of the personality are presented: love of money (Judas); the tendency to go along with the crowd; the impulsive violence and denial to follow the leadership of the emerging self (St Peter); the childish wishing to avoid responsibility (Pilate). Yet there is also the Jesus accepting the cross of freedom and his hour of suffering and glory. The opposites are in creative tension by something larger at the depth of being. The cross with its vertical masculine pole of consciousness; its vertical, feminine pole of unconscious grounding being; and its horizontal bar embracing the whole of life, is held in the circle of the hour and centres in joy. Each person's life can have this crowning in freedom.
Context The seventeenth section of St John's Gospel is related to a stage in the spiritual journey of the individual.
Broader Journeying within transcendence (Christianity, #HH6505).
Followed by Journeying within transcendence - from extraordinary to ordinary (Christianity, #HH4110).
Preceded by Journeying within transcendence - from son to father (#HH1773).

♦ **HH2331 Deautomatization**
Description The process of automatization, as a result of which an activity is carried out unconsciously, can be reversed through a resumption of the conscious mode of carrying out this activity, either to overcome some disability or the effects of an intoxicant or to improve performance or perception. Contemplative meditation is said to manipulate attention in this way, so that the perception of what is perceived is attended to, not the cognition of that perception.
When the process of deautomatization or depersonalization is intentional, the aim is thus to dispel blindness and reveal a new perception, variously referred to as enlightenment or illumination. Activities and perceptions, previously carried out automatically, are now invested with attention as they were when the automatic behaviour pattern was first built up. This "shaking up" may lead to an advance or a decrease in the level of organization.
Deautomatization has been related to the mystic techniques of renunciation and contemplation. For example, in contemplative meditation, the the technique is to manipulate attention in a way which is that required to produce deautomatization. Unusual perceptions of meditation subjects are then described (Arthur J Deikman) as sensory translation, reality translation and perceptual expansion. Imagery and thought are just such as observed in children and in primitive cultures - more sensuous, full of detail, with colour and vivacity of image. There is a decrease in distinction between the self and the object of observation.
Again, renunciation requires the attempt to banish from awareness the objects of the world and the desires directed towards those objects. This results in "starving" the perceptual and cognitive structures of nutriment which would tend to produce unusual experience. Decrease in responsiveness to distracting stimuli as measured by disappearance in alpha rhythm is noted in these cases. Long-term deprivation or decreased variability of stimulation would again produce the effects typical of deautomatization. There may be temporary stimulus barriers, such as have been postulated as operating in schizophrenia, which would produce a functional state of sensory isolation. Contemplative meditation and renunciation combined therefore have a powerful effect, the mystic becoming more and more committed to his goal as, the world having been abandoned, there is no other sustenance.
Refs Ornstein, Robert *The Psychology of Consciousness* (1986).
Related Renunciation (#HM3118) Contemplation (#HM2952)
Automatization (#HM3906) Mystical cognition (#HM2272)
Religious experience (#HM3445) Breakdown of automatization (#HM8105)
Contemplative intuitive meditation (#HH0816)
Deautomatization and the mystic experience (#HM4398).

♦ **HH2332 Purification of consciousness** (Buddhism)
Purity of mind
Context One of the two purifications or purities which constitute the roots of understanding according to Hinayana Buddhism, the other being the purification of virtue or purity of morality.
Broader Purifications (Buddhism, #HH3875).
Followed by Purification of view (Buddhism, #HH2718).
Preceded by Purification of virtue (Buddhism, #HH1865).

♦ **HH2334s Psychic ability**
Description This ability, which is either inborn or arises with practice if truly desired, implies the use of subtle forces not covered by the senses. Information arises and is interpreted by non-rational means, perhaps due to the mind functioning beyond the three dimensions. Those with psychic ability may be designated mediums, diviners, magicians, witches.

♦ **HH2338 Autogenic relaxation sequence**
Description By following a sequence of audible instructions the individual relaxes and quietens the mind, resulting in an alpha state similar to that achieved in biofeedback training.
Related Biofeedback training (#HH0765).

♦ **HH2341 Shahada** (Islam)
Profession of faith
Description With sincere intent to recite the Islamic confession of faith, "There is no god but God" and "Muhammad is the Messenger of God", this is the essential element of being a Muslim.
Context One of the five pillars of Islam.
Broader Human development (Islam, #HH1799).

♦ **HH2343 Grahastashrama** (Hinduism)
Householder stage of life
Description While leading a married life, living with his wife and children and performing obligatory duties, the householder practices discipline and is never self-indulgent. He fulfils sacramental rights in submission to the spiritual authority of the brahmins. The noble qualities of love, sacrifice and service are developed within the context of marriage.
Context This is the second of the four ashramas or stages of life according to Indian tradition, leading eventually to self-realization.
Broader Ashramas (Hinduism, #HH2563).
Followed by Vanaprasthashrama (Hinduism, #HH2782).
Preceded by Brahmacharyashrama (Hinduism, #HH1987).

♦ **HH2344 Freedom of conscience**
Description This can be understood in several different senses: freedom of the will to recognize or ignore the demands of conscience; freedom to obey conscience independent of external influence; freedom to live in society according to one's own conscience. None of these senses is dependent upon the adequate or erroneous recognition of the truth by the conscience. In Christian terms, conscience is distorted and undeveloped through man's fallen nature and can only be perfected through divine revelation. Nevertheless, decisions made in the situation in which the individual finds himself may be referred to the grace of God, which liberates his freedom.
Related Conscience (#HH0466).

♦ **HH2345 Living in the present**
Present-centredness
Description To live totally in the "here and now" is the aim of Gestalt. Avoidance of delay, reacting instantaneously to a good impulse is said (Swami Omkarananda) to awaken further latent goodness, strengthen good tendencies and create further opportunity for doing good. Delay, on the other hand, leads to further delay and puts all activity "off course".
Related Kairos (#HM2749) Gestalt therapy (#HH0751).

♦ **HH2378 Suggestopedia**
Description This wholistic learning system, developed by Dr Georgi Lazanov, substantially increases the speed of learning and improves recall by using the 90 percent of the brain not normally used. Information goes straight into the long-term memory system and the "forgetting curve" of traditional learning is dispelled.

♦ **HH2387 Adhicitta sikkha** (Buddhism)
Training of higher mind
Description Adhicitta or higher mind is the mind in the state of samadhi. It is the result of the system of mental training, the steady ascent through stages of development until it has raised itself above and remains above the condition where it is a slave to sensory impulses and to emotions.
Related Concentration (Buddhism, #HM6663).

♦ **HH2390 Kirtan** (Hinduism)
Description Dance whose movements are such as to alter the state of consciousness of the dancer and aid spiritual enlightenment.
Related Derwish (Islam, #HH0741).

♦ **HH2434 Inter-religious dialogue**
Living-in-dialogue
Description True dialogue involves the opening of heart and mind to another person, it presupposes that each side desires to learn something from the other. This implies recognizing common humanity, taking a risk and being sensitive to differences, to the variety of human life. Dialogue between religions does not necessarily imply syncretism, the conscious or unconscious selection and fusing of parts of more than one faith to create a new religion. Rather such dialogue can be the means of enriching and strengthening an individual's faith in his own religion; and a realization that others' views may be closer to the basic tenets of his religion than are his own, and more appropriate to present-day needs. It can be a creative interaction resulting in spiritual freedom from narrow and inward looking attitudes inherited along with, but not truly part of, his faith; often such attitudes may be of social rather than religious origin. In fact, there is some question as to the validity of a faith which is unable to appreciate or which is ignorant of other beliefs.
Faith is thus tested, refined and sharpened by deeper knowledge of other faiths, a knowledge which may be increased not only in formal discussion but by communal living with people of other religions. The notion of dialogue may thus be extended to all levels - social, intellectual and political as well as personal. So long as there is no dilution of conviction or mere intellectual sophistry, there may be all round enrichment, a discovery of a new dimension of truth. The result is mutual trust and peaceful negotiation where negative attitudes and conflict might otherwise have prevailed.
Related Syncretism (#HH3214) Conciliarity (#HH3211)
Ecumenicism (Christianity, #HH0099) Human development through religion (#HH1198).

♦ **HH2435 Human development** (Baha'ism)
Description The aim is peace and unity in the world and in the human institutions which make it up - a new world order - through transformation of each individual to become his true self on following teachings based on the revelations of Baha' Allah and their interpretations by Abd al-Baha and Shoghi Effendi. The process of transformation leads to the development of capacities for love (both the giving and attraction of love) and knowledge (including learning and teaching); and for their use in the service of mankind. Further, love for God opens a channel for God's love. The matrix for transformation is the Baha'ai community, commencing with tolerance for the diversity of members, which grows into understanding, love and appreciation. Anxiety and tension are not avoided but used in this process of development; and prejudice (knowledge and love in conflict) faced up to and thus eradicated.
Broader Human development through religion (#HH1198).
Related Human development (Islam, #HH1799).

♦ **HH2498 Journeying within transcendence - from isolation to imagination**
Description Living is seen not as isolated and abandoned in an uncaring environment, nor as subject to impersonal legislation, but as a dynamic, interpersonal existence in the community, with mutual dependence, despite isolation, loneliness and death. The inner battle for control, the loneliness, the demanding voices, are overwhelmed by a centering on inner stillness and quiet, a different kind of aloneness at the centre where a voice offers healing.
Context The seventh section of St John's Gospel, Chapter V 1 to VI 1, is related to a stage in the spiritual journey of the individual.
Broader Journeying within transcendence (Christianity, #HH6505).
Followed by Journeying within transcendence - from object to subject (#HH4188).
Preceded by Journeying within transcendence - from forgetfulness to memory (#HH5219).

♦ **HH2543 Sound health**
Vibratory maintenance — Right vibratoryhood
Description Creation myths have often been expressed in metaphors of sound. To the extent that form comes about through sound, then it is appropriate to explore the ways in which sound,

HH2543

especially vocal sound, facilitates the return journey. Whereas the creation may be seen as a cascade of increasingly impure sounds as the material comes to dominate awareness, the return journey then requires a progressive purification of these sounds. This involves the progressive purification of these sounds, going back through each of the energy centres (chakras) with which they are associated, mantrically chanting and purifiying them. In the language of Buddhism, this is the process of untying knots in the reverse order to that in which they were tied, undoing the original creation. This is achieved by involving the distractable mind in tasks which free the individual for more profound levels of awareness of emptiness, of clarity and ultimate enlightenment.
Sound health is a concept which is deeeply rooted in many languages (eg sound in mind and body, sound ideas). The body is an instrument which needs to be kept in tune. Its different parts constitute a symphony of vibrations (eg rhythms of walking and breathing, heartbeat, pulsation of organs and cells). It is the individual vibratory trueness and the relationship between these vibrations which maintains a healthy state. Sound may be used for healing and for the transformation of consciousness.
Context Many traditions recognize the vibratory nature of the world, expressing this understanding through their own cultural metaphors, and recognizing that it was necessary to participate consciously in some way in this vibratory ordering of nature. Failure to sustain such vibratory maintenance was even seen as endangering the world.
Refs Gerber, Richard *Vibrational Medicine* (1988); Purce, Jill *The Sound of Enlightenment.*
Related Chanting (#HH1257) Music therapy (#HH0496) Mantra yoga (Yoga, #HH0931)
Sound consciousness (#HM8232) Mantra-induced experience (Yoga, #HM2464)
Vibrational health (#HH4908).

♦ HH2552 Magnetic healing

Laying on of hands — Contact healing — Touch healing — Cheirothesy — Auric healing
Description Having diagnosed the area of congestion, the healer makes magnetic passes over that part of the body until there is a sense that the congestion has eased, when the body is better able to heal itself. There may be actual touching of the diseased area of the body, or simulated touching a little away from the body - the patient's auric field. The effect may be instantaneous or may take up to several days to be noticeable. It is said to be perhaps due to controlled transfer of human energy in the form of "magnetic fluid" from the healer to the patient, although the healer has no knowledge of the process. Sometimes, under hypnosis, the sufferer diagnoses his own complaint and its treatment.
Broader Spiritual healing (#HH0458).
Related Laying on of hands (Christianity, #HH2023).

♦ HH2561 Ninety-nine names of Allah (Sufism)

Most beautiful names — Names of God — Divine names — Divine attributes
Description Although religious scholars indicate that Allah has three thousand names - one thousand known only by angels, one thousand known only by prophets, 300 in the Torah, 300 in Zabar (Psalms of David), 300 in the New Testament and 99 in the Quran together with one known only to Allah but hidden in the Quran - the 99 names in the Quran are said to be Allah's alone. To the mystic, the Absolute reveals itself in the form of one of the divine names through divine presences, which are states or stations. His very presence is said to efface the name, so that God is present in any name at the same moment as he is absent.
Jami indicates that the perfection of the divine names can be realised in two ways - outward expression or *jala*, the visible manifestation, and inward display or *istijla*, the invisible Self, consciousness of God Himself. The calligraphic form of the divine names, as of any name in Arabic script, can be understood as a visual embodiment of divine revelation, the balance and rhythm of form and non-form, evoking a particular timeless quality and thus speaking to the mystic state it represents. The structure of any such calligraphic form, composed of a set of horizontal and vertical strokes, interweaves horizontal and vertical dimensions corresponding to the active and passive qualities in all things, evoking and reinforcing a particular mystical awareness, especially as a result of continual and uninterrupted repetition of invocation, thus writing his own inner complementary opposites.
The divine names give understanding of God unmanifest in form. With creation, the names are revealed as attributes, since so-called created things are divine attributes which, before their appearance, were in the form of the divine names. Teaching the names of things implies making a man conscious of the essence of those things, having full knowledge. This, in turn, implies the essence of the name becoming part of the being. Repeating the name "Allah" or some of His infinite attributes as indicated in the 99 names allows the believer to remember and come near Him. Anyone who learns, understands and enumerates the 99 names enters Paradise and achieves eternal salvation. Because the names are powerful, repetition must be with respect, care and good intention, or injury could result. One should also observe the proper ritual.
Refs Al-Halveti, Sheikh Tosun Bayrak al-Jerrahi (Comp) *The Most Beautiful Names* (1985); Childe, C *Social Evolution* (1951); Friedlander, Shems *Ninety-Nine Names of Allah* (1978); Poddar, H P *The Divine Name and Its Practice* (1965).
Broader Maps of the mind (#HH8903).
Narrower Allah (#HM4500)
Al-'Adl (#HM3807)
Al-Qawi (#HM4988)
Al-Ahad (#HM6117)
Al-Mani (#HM7220)
Al-Badi (#HM4299)
Al-Malik (#HM4061)
Al-Bari' (#HM3636)
Al-Basit (#HM0115)
Al-Hakam (#HM5014)
Al-'Azim (#HM5198)
Al-Muqit (#HM3171)
Al-Karim (#HM5289)
Al-Wasi' (#HM4169)
Al-Majid (#HM5446)
Al-Hamid (#HM0392)
Al-Mu'id (#HM4093)
Al-Wajid (#HM0743)
As-Samad (#HM4980)
Al-Akhir (#HM7112)
Al-'Afuw (#HM5543)
Al-Ghani (#HM3822)
Ar-Rahman (#HM3539)
Al-Jabbar (#HM0215)
Al-Wahhab (#HM0069)
Al-Khafid (#HM3644)
Al-Ghafur (#HM6255)
Al-Tawwab (#HM4334)
Al-Warith (#HM0113)
Al-Mudhill (#HM4510)
Al-Muhaymin (#HM0526)
Al-Muqaddim (#HM5331)
Al-Mutakabbir (#HM3850)
An-Nur (#HM1354)
Al-'Ali (#HM4162)
Al-Wali (#HM1633)
Al-Wali (#HM0525)
Ad-Darr (#HM4542)
Al-Baqi (#HM6786)
As-Salam (#HM1221)
Al-'Alim (#HM4010)
Al-Sami' (#HM0419)
Al-Latif (#HM0308)
Al-Kabir (#HM4437)
Al-Hasib (#HM1467)
Al-Raqib (#HM6098)
Al-Hakim (#HM4799)
Al-Wakil (#HM1382)
Al-Muhsi (#HM4544)
Al-Muhyi (#HM1007)
Al-Majid (#HM4526)
Al-Qadir (#HM1500)
Az-Zahir (#HM4042)
Ar-Ra'uf (#HM0206)
An-Nafi' (#HM6350)
Al-Quddus (#HM0235)
Al-Khaliq (#HM0034)
Al-Razzaq (#HM7021)
Al-Mu'izz (#HM1116)
Al-Ba'ith (#HM4017)
Al-Muqsit (#HM8001)
Ar-Rashid (#HM5234)
Ash-Shakur (#HM1934)
Al-Musawwir (#HM5786)
Al-Muta'ali (#HM8019)
Al-Mu'akhkhir (#HM4052)
Al-Rafi (#HM4040)
Al-Haqq (#HM3945)
Al-Hayy (#HM4236)
Al-Barr (#HM7022)
Al-Hadi (#HM1287)
Al-Rahim (#HM4043)
Al-'Aziz (#HM4562)
Al-Qabid (#HM2767)
Al-Basir (#HM6512)
Al-Halim (#HM4519)
Al-Hafiz (#HM7099)
Al-Jalil (#HM6771)
Al-Mujib (#HM0504)
Al-Wadud (#HM7712)
Al-Matin (#HM5633)
Al-Mubdi (#HM0622)
Al-Mumit (#HM3868)
Al-Wahid (#HM5506)
Al-Awwal (#HM6139)
Al-Batin (#HM1209)
Al-Jami' (#HM0009)
Al-Sabur (#HM4495)
Al-Mu'min (#HM1245)
Al-Qahhar (#HM1318)
Al-Fattah (#HM0002)
Al-Khabir (#HM4400)
Al-Qayyum (#HM5006)
Al-Mughni (#HM3633)
Al-Ghaffar (#HM1257)
Ash-Shahid (#HM0375)
Al-Muqtadir (#HM3576)
Al-Muntaqim (#HM4513)
Malik-ul-Mulk (#HM5902)
Dhul-Jalali wal-Ikram (#HM6719).
Related Non-faith (Buddhism, #HM2461) Recollection (Islam, Sufism, #HM2351)
Names of Christ (Christianity, #HH1167).

♦ HH2563 Ashramas (Hinduism)

Four stages of life
Description According to Hindu tradition, man's life proceeds through four stages or ashramas. These are *brahmacarya* or celibacy, *grahasta* or household, *vanaprastha* (recluse) and *sanyasa* (mendicancy). Although everyone does not need to go physically through all four stages for full development, the four stages nonetheless demonstrate pictorially the progress to full self realization.
Broader Human development (Hinduism, #HH0330).
Narrower Sanyasashrama (#HH2056) Grahastashrama (#HH2343)
Vanaprasthashrama (#HH2782) Brahmacharyashrama (#HH1987).
Related Brahmacarya (#HH0978).

♦ HH2633 Upper middle grade (Taoism)

Description Methods include formal religious practices including transmission of initiation and precepts, readings, recitations, preaching. Practices include stargazing and bowing to the stars; the practitioner may keep silence, do hard labour; he maintains outward virtues.
Context In the nine grades of practices which are side tracks and auxiliary methods in Taoism, this is the highest of the three middle grades.
Broader Sidetracks and auxiliary methods in Taoism (Taoism, #HH7004).
Followed by Lower upper grade (Taoism, #HH4711).
Preceded by Middle middle grade (Taoism, #HH5207).

♦ HH2651 Human scale development

Description Development is scaled: to allow satisfaction of fundamental human needs; to generate increasing levels of self-reliance; and to construct coherent and consistent relations of balanced interdependence of people with nature and technology, of global with local processes, of personal with social, of planning with autonomy, and of civil society with the state. All this requires creating conditions in which people take the lead in their own future. Their areas of activity must be respected both in their diversity and their autonomy. The individual as object cannot be transformed to the individual as subject in enormous, hierarchical systems where decisions flow from the top downwards.
Refs Max-Neef, Manfred et al *Human Scale Development* (1990).

♦ HH2653 Social evolution

Description According to Teilhard de Chardin, there is a natural tendency towards perfect unity and complexity. Physical evolution has perfected man physically and the next stage, currently occurring, is social evolution which will perfect man through cultural convergence of economic, political and thought processes.
Broader Evolution (#HH0425).

♦ HH2661 Journeying within transcendence - from bondage to freedom (Christianity)

Description A deep sense of shame brought about by inner accusations results in depression and despair. This is not a true reconciliation of conscience with Christ but of the ego and superego, an identification of the self with the act of wrongdoing, not leading to true forgiveness and therefore leaving one bound to the past in unhealthy guilt. Real, healthy guilt depends on the power of love, not fear of the loss of love, providing motivation to face the future. Distinguishing conscience from superego is necessary for inner wholeness, acceptance of past mistakes and errors and of the imperfection of the present as part of development which, through patience, can result in transformation.
Context The tenth section of St John's Gospel, Chapter VIII 1-12, is related to a stage in the spiritual journey of the individual.
Broader Journeying within transcendence (Christianity, #HH6505).
Related Forgiveness (Christianity, #HH0899).
Followed by Journeying within transcendence - from blindness to sight (#HH1354).
Preceded by Journeying within transcendence - from discussion to decision (#HH3366).

♦ HH2662 Yantras

Description A geometric figure usually drawn on some specially consecrated material but also simply drawn in the sand or even a three-dimensional structure, a yantra has intricate symbolism which becomes intelligible in meditative absorption. As well as representing the multiple strata of the universe in time and space, the cosmogram, it also acts as a psychocosmogram, the emanation and reabsorption of the universe being reflected within each individual being. A mandala is a complex, pictorial variant of a yantra.
Refs Madhu Khanna *Yantra* (1979).
Related Mandalas (#HH0112) Tantra (Buddhism, Yoga, #HH0306)
Tantric visualization (Yoga, #HM1690).

♦ HH2665 Ways to spiritual realization (Esotericism)

Seven rays
Description In Western esotericism there is a tradition of seven distinct ways to spiritual realization. Each person is understood to be pursuing one or more of these ways over any particular period and in so doing the ways manifest in the person as characteristic different aspects of their being. Most people are considered to be influenced by the way with which their personality is associated and by that with which their soul is associated. The ways or rays are also associated with numbers: (1) Will or power; (2) Love-wisdom; (3) Active intelligence; (4) Beauty or harmony; (5) Knowledge or science; (6) Devotion or idealism; (7) Ritual or organization.
Refs Assagioli, Roberto *Psychosynthesis typology* (1983); Bailey, Alice A *A Treatise on the Seven Rays.*
Narrower Way of power (#HM0454) Way of harmony (#HM0763)
Way of idealism (#HM6554) Way of knowledge (#HM1201)
Way of love-wisdom (#HM1507) Way of organization (#HM1605)
Way of active intelligence (#HM1997).
Related Psychosynthesis (#HH0002).

♦ HH2675 Radical therapy

Radical psychiatry
Description It has been argued that psychotherapy may be a means of oppression in that its aim is to assist the individual in adjusting to the status quo when in fact it is the status quo that should change. Radical therapy uses psychiatry as a force for liberation, taking the position that alienation is the result of oppression and that by discovering the reason for the alienation the individual will grow in self-knowledge.
Broader Psychotherapy (#HH0003).
Related Consciousness raising (#HH0427).

♦ **HH2718 Purification of view** (Buddhism)
Purity of views — Ditthi-visuddhi-niddesa (Pali) — Defining of mentality-materiality — Delimitation of formations
Description Correct seeing of mentality materiality or discernment of name and form is the purification or purity of view. In calm and serenity the person emerges from a jhana of the fine-material or immaterial spheres (except from neither-perception-nor-non-perception), discerns the jhana factors and the states associated with them. This is mentality or name. Scrutiny of the mentality or name reveals from whence it proceeds or by what it is supported. This is revealed as the heart basis or matter of the heart. The heart's support is revealed as primary elements. The remaining (derived) kinds of materiality or form have these elements as their support. All that defined as mentality or name - 'nama' - has the characteristic of bending upon the object. All that defined as materiality or form - 'rupa' - has the characteristic of being molested or changing (by the action of cold, for example). It is then possible to further discern materiality and mentality by different aspects, and this is the purification of view.
Context One of the five purifications or purities which constitute the trunk or body of understanding according to Hinayana Buddhism, the others being: purification or purity of overcoming doubt; purification by knowledge and vision of what is the path and what is not the path or purity of knowledge and discernment of the right path and the wrong path; purification by knowledge and vision of the way or purity of knowledge and discernment of the middle way; purification by knowledge and vision or purity of knowledge and discernment.
Broader Purifications (Buddhism, #HH3875).
Followed by Purification by overcoming doubt (Buddhism, #HH1187).
Preceded by Purification of consciousness (Buddhism, #HH2332).

♦ **HH2763s Breath control**
Description This technique is said to allow the body to remain alive without breathing for extended periods of time; instead, the vagus nerves supply the organs as necessary.
Broader Yoga (Yoga, #HH0661).

♦ **HH2764 Generative man**
Good man
Description Generative man is described by Erik Erikson. This is the good man who is dedicated to the maintenance and ecological strength of the human race. Fulfilment is in the confirming recognition of his children and the children of others. His hope is that each child born into the world is wanted and cared for. He realizes that growth of the human race must be severely limited if the species is to be maintained. Generative man is, in the wider sense, also committed not only to the physical preservation of the race but to its wider ecological integrity. This implies curtailing the uncontrolled economic and technical expansion which pollutes the natural environment and weakens the fabric of social existence.
The so-called lower man - the animal, primitive and child - contains important regulatory and organizing powers which are essential for maintenance and restoration of the ecological integrity of man; but it does not function with the precision and specificity of instinct in the animal world and has constantly to be restated in the context of man's more progressive capacities for autonomy, conscious reflection, responsibility and purposive activity. Generative man restores what is most truly animal and most truly human, childlike and adult, primitive and civilized. As well as prudent control of the excesses of modernity there is wise use of its opportunities. Structural differentiation, rapid social change and the multiplication of options in modern life have broken down the meaningful, shared rituals that give integrity and coherence to a civilization. When stable yet flexible patterns of mutuality do not exist then neither society nor the individual has the capacity to submit to meaningful change in the context of commonly acknowledged continuities. Appreciating the necessity for innovation and for tradition, generative man is a creative ritualizer. Modern techniques and technology have to be balanced - they may remove toil and disease from life but they must not expand unchecked in ecologically unsound gigantism. All human activity is judged from the perspective of its contribution to the generative task, the establishment and maintenance of succeeding generations so there can be no completely autonomous human enterprise - art, knowledge, pleasure cannot exist as ends in themselves. Because generative man has free access to his own childhood depths he is able to enter creatively into dialogue with his own and other children.
Refs Browning, Don S *Generative Man* (1975).
Related Stages of personality development (#HH0285).

♦ **HH2768 Corporate worship** (Christianity)
Description The act of corporate worship is both honouring God and at the same time enacting the self-understanding of each individual participating and of a specific people. Being a self is to honour, owe allegiance to and give one's life for some reality, some god which involves participating in some set of communal symbols through which one is enabled to become the self one understands oneself to be.
Christian worship is the portrayal of and dramatization for those gathered as the forgiven ones, the thankful ones and the dedicated ones as they must grasp themselves when God the Father, Son and Holy Spirit becomes their God. The implications of this are 1) participants in worship are not spectators but actors, 2) religious feeling may be the result of participation in worship but they are not the focus because it is the total person that is involved and, therefore, all feelings, ideas and actions, and 3) the god worshipped determines the structure of worship.
While there are private prayer and personal rituals, worship is necessarily corporate because of the self-understanding inherent in Christian faith. The person of faith recognizes that he must hear over and over again the word of forgiveness. And to hear it, it necessarily comes from another. Christians come together to worship to hear the word and to speak it to another.
Within the great variety of Christian worship the structure is basically the same. The first act has to do with confession and pardon; the second with praise and witness; and the third with offering and dedication. Some traditions place these together in the same ritual and others divide into separate ceremonies but all three are present in the life of the church. Neither the nature nor the order are arbitrary because these three acts in this sequence tell the story of a human who stands before the God in Christ. Confronted with the fact of one's guilt in failing God and man one is driven to acknowledge one's sin. This act of acknowledgement enables one to hear God's pardon which then throws one into a state of praise and thanksgiving. Freed from the bonds of sin one turns to the world dedicated to serve God and neighbour. Worship for the Christian is not a duty but a rehearsal of life lived in its fullness.
Broader Worship (#HH0298).
Related Liturgical prayer (Christianity, #HH4209).

♦ **HH2772 Abhyantara vritti yoga** (Yoga)
Description The breath is held - used in breathing exercises for higher attainment of soul mind.
Related Inward-flowing consciousness (Yoga, #HM2738).

♦ **HH2782 Vanaprasthashrama** (Hinduism)
Recluse stage of life
Description While maintaining relations with his family, the married man develops further mental abstinence and detachment, gaining in spiritual maturity. Living among his possessions he nevertheless develops the spirit of dispossession and maintains a sense of detachment in all his activities. When his sons come of age he may renounce possessions and family, retiring into solitude for contemplation and meditation.
Context This is the third of the four ashramas or stages of life according to Indian tradition, leading eventually to self-realization.
Broader Ashramas (Hinduism, #HH2563).
Followed by Sanyasashrama (Hinduism, #HH2056).
Preceded by Grahastashrama (Hinduism, #HH2343).

♦ **HH2806 Enhancement of human capabilities**
Description As opposed to the basic needs approach, which is a goods- oriented view of development, this approach puts people first. Human resources development has been defined as synonymous with enhancement of human capabilities.
Related Human resources development (United Nations, #HH0745).

♦ **HH2877 Jesus as saviour**
Joshua — I am — Yahweh
Description The Hebrew name Joshua or Jesus implies that Yahweh is salvation, through God's mercy. In that the name Yahweh itself implies "I am", salvation is through the one self, the subject of which the whole creation is object. Again, the word saviour implies redeemer, redemption of the world by the merciful act of its creator.
Related Redemption (Christianity, #HH0167).

♦ **HH2901 Feng-shui** (Chinese)
Geomancy
Description Feng-shui is an ancient Chinese practice concerned with the art of living in harmony with the land, and deriving the greatest benefit, peace and prosperity from being in the right place at the right time. It is also understood as the art of adapting the residences of the living (and the dead) so as to remain in harmony with the local currents of the cosmic breath. Each location, whether geographical or in-house, has topographical features which modify the local influence of various energies (ch'i) of nature. Of special concern is the form of hills and the directions of watercourses, but the heights and placement of buildings are of equal concern in an urban environment. Feng-shui recognizes, in addition to wind and water, other types of energy which permeate the earth and atmosphere and animate the forms of nature. Feng-shui may therefore be considered a code of practice whereby humans can govern their relationships to the environment. One well- recognized consequence is the great beauty of the siting of many farms, houses and villages throughout China. Harmony and balance (of yin and yang) are crucial factors in feng-shui. The practice may be understood as a way in which the lives of people are sanctified, attuning them to the rhythms of nature, and providing them with a sense of security and continuity. One tool of practitioners, the geomancer's compass (luopan) through which the appropriateness of the orientation of buildings is determined, takes the form of a complex mandala that permits a holistic balancing of environmental features.
Context Feng shui is a mystical combination of Chinese philosophical, religious, astrological, cosmological, mathematical and geographical concepts. There are many different traditions of feng-shui, some of which draw on insights from India, and Tibet. In China it is closely linked to Confucian and Taoist traditions.
Refs Eitel, Ernest J *Feng-Shui* (1984); Lip, Evelyn *Feng Shui* (1987); Rossbach, Sarah *Interior Design with Feng Shui* (1987).
Related Mandalas (#HH0112) I Ching (Taoism, #HH0004)
Dreaming (Australian, #HH1301) Human development (Taoism, #HH0689).

♦ **HH2909 Psychospiritual growth** (Christianity)
Description Psychospiritual growth may defined as the structural development of psychological growth, together with or followed by the direction of spiritual growth. The sense of self is developed and differentiated from the false self which has been generated as a defence against the anxiety of being a true self. The false self tends to be defended; only the true self can be transcended, and it is in its transcending that it is ultimately discovered. During this process of development of self there may be spiritual response to deeper spiritual realities, but hearing and response will be hindered by psychological conflict or problems. Again, psychological growth is not always followed by spiritual growth as the individual may be too satisfied with the self-understanding achieved to see that this is not an end in itself but a by-product on the way to self-transcendence and surrender.
The self-encapsulation and self-preoccupation resulting from sin and from psychopathology block spiritual growth. Only when the false selves are seen for what they are can they be given up and the true self seen, when the need for surrender to God is understood. Both Christian spirituality and psychotherapy involve the crucifixion of the false self or self-centredness. Then God rather than the self can become the lord over life.
In the model of psychospiritual development put forward by David Benner, a number of *structural milestones* (psychological development) and *directional milestones* (spiritual development) are included.
Structural milestones:
1. Symbiotic dependency: undifferentiated psychological fusion with the mother.
2. Differentiation of self from the mother.
3. Relatedness and attachment to others in non-symbiotic ways, recognizing the separateness of one's self from others and preserving ego boundaries.
4. Individuation, as the conscious and unconscious parts of the self are allied, the true self only being discovered as hidden aspects of the self that one did not with to accept are acknowledged.
5. Self-transcendence and self surrender to some higher purpose.
6. Integration of personality.
Directional milestones:
1. Development of basic trust, choosing to open one's self to others and the world.
2. Awareness of call to self transcendence, of deep inner strivings together with a longing to surrender or to find one's place, knowing that this cannot be found in the material or the temporal.
3. Recognition of call as from God and that longings are for God.
4. Awareness of insufficiency of self (sinfulness) in comparison with God's holiness and the standards of His law, and understanding that to approach God a divine intervention of grace is needed.
These 4 are described as preparation.
5. Receipt of divine forgiveness through grace.
This is described as justification.
6. Progressive freedom from sin, involving spiritual warfare with the forces of darkness.
7. Progressive evidence of the fruit of the spirit: love, joy, peace, patience, kindness, generosity, fidelity, tolerance and self-control.
8. Deepening intimacy with God, mystical unity which is the basis for spiritual unity and communion among all believers.
These 3 are described as sanctification.
Growth in neither sphere is linear, and growth in one sphere is dependent upon what is happening

HH2909

in the other. Growing spiritually presupposes a certain psychological maturity, and psychospiritual maturity is characterized by the integration of the personality in a context both of significant interpersonal relationships and of surrender to God, in which latter the true self is discovered. This integrated self is both an achievement and a gift, the self that God has given and what it was always intended one should be.
Refs Benner, David G *Psychotherapy and the Spiritual Quest* (1988).
    Related Religious growth (#HH1321).

♦ HH2918 Tenfold powers (Buddhism)
Meditating on the kasinas
Description There are ten kasinas (universals) or devices used as subjects for meditation in the fine-material sphere. These are: earth kasina; water kasina; fire kasina; air kasina; blue kasina; yellow kasina; red kasina; white kasina; light kasina; limited or separated space kasina. Each kasina is associated with special powers.
- Earth kasina: having been one, becoming many; by creating earth, walking, standing or sitting on space or water.
- Water kasina: diving in and out of earth; producing rain, rivers or oceans; making earth, mountains, palaces shake.
- Fire kasina: producing smoke, flames, sparks; destruction of fire with fire; burning whatever one wishes; making light to see physical objects with the divine eye; on attaining nibbana, to consume one's own body with fire.
- Air kasina: moving with the speed of wind; causing storms.
- Blue kasina: creating black forms; producing darkness; mastery through fair and ugly appearance; liberation through beauty.
- Yellow kasina: creating yellow forms; resolving that something should become gold; mastery through the way stated; liberation through beauty.
- Red kasina: creating red forms; mastery through the way stated; liberation through beauty.
- White kasina: creating white forms; putting away sloth and torpor; dispelling darkness; making light to see visible objects with the divine eye.
- Light kasina: creating luminous forms; putting away sloth and torpor; dispelling darkness; making light to see visible objects with the divine eye.
- Limited or separated space kasina: uncovering the hidden; creating space within earth and rock and maintaining postures there; travelling through walls.
By extension, the kasinas are perceived upward to the sky, downward to the earth, around as far as is desired.
    Related Kasina meditation (Buddhism, #HH3246)
    Supernormal powers (Buddhism, #HH5652)
    Subjects for meditation (Buddhism, #HH3987).

♦ HH2987 Journeying within transcendence – from death to life
Description Salvation is not at some vague future time, it is now, in the present. In faith and acceptance of a relationship there is a meeting with him who has the power to give life normally associated with the last days and the end of all. Contrasted with the pervading sense of death in current culture, which invites suicide as an escape, is the impossibility of imagining one's own death. Perhaps birth and death are simply two aspects of the same event.
Context The thirteenth section of St John's Gospel, Chapter XI 1 to XII 8, is related to a stage in the spiritual journey of the individual.
    Broader Journeying within transcendence (Christianity, #HH6505).
    Followed by Journeying within transcendence – from served to servant (Christianity, #HH6012).
    Preceded by Journeying within transcendence – from outside to inside (#HH3215).

♦ HH2992 Contingency
Description The creation is inherently contingent since, even if it were possible to prove that every occurrence is totally rational and necessary for the functioning of the whole, the existence of the whole can never be rationally explained. Acceptance of free will must necessarily include contingency. Thought is said to be the continuous combination of the antitheses of the rational with pantheism as its ultimate, and irrationality with the consequence of polytheism. Life in the world rationally demands a fixed and universal law together with irrational vitality, multiplicity and freedom. This leads to inevitable contradictions and incoherences despite traditions, such as Judaeo-Christian theism, which take cognizance of both sides at once.

♦ HH2997s Spirit healing
Healing through breath
Description The rhythmic breathing of a group in physical contact with each other, or of a single practitioner, focuses healing energy on the diseased part allowing it to heal itself.
    Broader Spiritual healing (#HH0458).

♦ HH3000 Spirituality (Hinduism)
Refs Cousins, Ewert, et al (Eds) *Early Hindu and Jain Spirituality*; Cousins, Ewert, et al (Eds) *Post-Classical Hindu and Sikh Spirituality*.
    Broader Spirituality (#HH5009).
    Related Human development (Hinduism, #HH0330).

♦ HH3003 Human development through music
Psycho-social transformation through music
Description The idea of music as an actual instrument of the well-being, and even the transformation, of society has a long history. It has frequently been recognized that changes to musical styles affect, and even pre-figure, changes in the more important laws and systems of organization in society. At its simplest, music has been held to have a civilizing function as a means of more creatively channelling energies. It has been used to give humanity a glimpse of alternative ways of being. It is argued that the great collective movements of the human soul have always been anticipated by innovations in music, namely through the embodiments of new patterns of order in forms which transcend habitual categories and attract a wider audience.
Each person may be perceived as a music which can be heard by the intuitive faculty. For some individuals there are dreams in which music acts a kind of leader of the soul into the life after death. Such music is felt to be of indescribable beauty, leaving behind a feeling of consolation and of certainty of the existence of timeless forces existing beyond death and transcending human experience.
In Hinduism, music is used as an aid to the attainment of higher states. All religious traditions that acknowledge the existent of angelic beings concur in giving them musical attributes that may also be understood as tonal qualities. Higher states of awareness may be comprehended and experienced in musical terms.
Refs Attali, Jacques *Noise* (1985); Berendt, Joachim-Ernst *Nada Brahma* (1988); Bouny, Helen and Savary, Louis *Music and Your Mind* (1973); Crandall, Joanne *Self-Transformation through Music* (1986).
    Related Music therapy (#HH0496)   Mantra yoga (Yoga, #HH0931)
    Shabda yoga (Yoga, #HH0539)   Musical inspiration (#HM4805)
    Hearing the music of the spheres (#HH1973).

♦ HH3006 Thelema (Islam)
Will
Description This is the position of some Islamic theologians, the later Ash'arites, for whom God's love is equated with his will. In fact, God has has neither anthropomorphic nor anthropopathic attributes, he does not experience the pleasure of satisfied love. What has been called God's love is actually his willing good to (some) men. There is no purpose behind his willing good to some and evil to others. The pleasure man feels at the beatific vision is not because love is satisfied but an independent, though simultaneously created, pleasure.
Refs Bell, Joseph Norment *Love Theory in Later Hanbalite Islam* (1979).
    Related Eros (#HH7231)   Nomos (Islam, #HH6501).

♦ HH3007 Practice of the confidence (Christianity)
Manifestation of conscience
Description On a regular basis the person has open, sincere and detailed conversation about his personal life with his spiritual director. This brings complete openness and understanding. Although the practice was previously an important part of the life of a religious order, generally taking place yearly or half-yearly (but currently in Opus Dei once a week), it is said to be open to abuse and was banned by the Catholic Church in 1890.
    Broader Spiritual direction (#HH0442)   Soul care (Christianity, #HH1199).

♦ HH3008 Interior life (Christianity)
Description John Ruusbruec describes the person as a double mirror. In the higher part of his being he receives the image of God and all His gifts. In the lower part of his being he receives corporeal images through the senses. Turning inward he can practice righteousness unhindered, but through inconstancy he turns outward. He then gets caught up in the activities of the senses and falls into daily faults. However, for the righteous person constantly moving inward, in love, these faults are merely a drop of water in a red-hot furnace.
Refs Dupré, Louis and Wiseman, James A (Eds) *Light from Light* (1988); Wiseman, James A (Trans) *John Ruusbroec* (1985).
    Related Meeting God (Christianity, #HM0541).

♦ HH3012 Degrees of love (Christianity)
Description In response to the first commandment, "Thou shalt love the Lord thy God", Bernard of Clairvaux sets out a number of degrees of love as the individual progresses in the spiritual life:
1. *The first degree of love: when man loves himself for his own sake*. What is natural should be at the service of the Lord of nature, but service of nature comes first. In bodily love, man loves himself for his own sake. This is innate, but may get out of control when it must be halted by loving one's neighbour as one's self. Loving one's neighbour with purity can only be done in God. This can only happen if one loves God. This love for God is brought about through God. Tribulation arises so that no rational being should claim the gifts of the Creator for himself – man honours God as He come to his aid. So, because man loves himself, he comes to love God for his own benefit.
2. *The second degree of love: when man loves God for his own good*. Loving God for his own sake and not for God's, man avoids hurting God by recognizing what he can do only with God's help. Frequent tribulations, bringing frequent turning to God and experiencing of God's liberation and generosity, soften man's heart and bring him to love God for Himself and not only for the benefit He bestows.
3. *The third degree of love: when man loves God for God's sake*. Frequent needs bring man often to call upon God. This frequent contact brings discovery of how sweet the Lord is. Truly loving God he also loves what is God's. This chaste love means that keeping the commandments is not burdensome. It is love given freely in truth and action, giving back what it has received. Love for God is not for the good that has been received but because God is good.
4: *The fourth degree of love: when man loves himself for the sake of God*. Man's love for himself is only for God's sake. This experience is very rare in this life, it is to become like God, when human affections dissolve and are poured into the will of God. This is the state to be desired. There is no more self-will, nothing of man remains in man. It can only arise when one loves God with all one's heart, soul and strength, the power of the soul is free of ties and strengthened by the power of God. In a spiritual and immortal body, the care of the weak, physical body no longer requires attention. The soul is caught up in a love that cannot be obtained by human effort but arises through God's power.
Refs Dupré, Louis and Wiseman, James A (Eds) *Light from Light* (1988); Evans, G R (Trans) *Bernard of Clairvaux* (1987).
    Broader Love (#HH0258)   Love consciousness (#HM2000).

♦ HH3013 False paths (Taoism)
Lowest paths
Description These are paths of confusion, misleading "mud and water alchemy". Their methods give sexual and quasi-sexual connotations to technical alchemical terms and include 72 schools of sexual play. Some look upon women as the alchemical cauldron, for example valuing menstrual fluid as medicine or the ultimate treasure, or using menstrual fluid and semen as the bases of the great elixir, there being among over 300 such practices.
Context In the nine grades of practices which are side tracks and auxiliary methods in Taoism, this is the lowest of the three lower grades.
    Broader Sidetracks and auxiliary methods in Taoism (Taoism, #HH7004).
    Followed by Outside paths (Taoism, #HH3667).

♦ HH3020 Guiding images
Guiding metaphors — Guiding myths — Guiding fantasies — Guiding archetypes
Description Metaphors, myths, fantasies and archetypes provide images through which individuals are able to grow in the development of self. Although their function can be very limited, such images may also, to different degrees, be embodied by a person and used as a way of structuring the reality to which he responds. Human thought processes can be considered as largely metaphorical, so that the human conceptual system can be considered as metaphorically structured and defined. Use of images can therefore powerfully affect the opportunities and style of human development open to an individual. However, when the images guiding habitual behaviour are not recognized and named, the individual becomes hostage to them and can do little but live out the plots they imply to their end. When they are named, the individual has a choice in responding to them. He can extricate himself from undesirable myths and/or learn to respect the archetypal pattern that is exerting control and learn the lessons it offers.
Refs Campbell, Joseph *Myths to Live By* (1985); Feinstein, David and Krippner, Stanley *Personal Mythology* (1988); Judge, A J N *Metaphoric Revolution* (1988); Latroff, George and Johnson, Mark *Metaphors to Live By* (1980); Pearson, Carol *The Hero Within* (1986).
    Related Symbols (#HH0690)   Golden age (#HH4572)   Guided fantasy (#HH0627)
    Archetypal psychology (#HH5119)   Neuro-linguistic programming (NLP, #HH4872).

♦ HH3025 Purification by knowledge and vision (Buddhism)
Purity of knowledge and discernment — Nanadassana-visuddhi-niddesa (Pali)
Description This is knowledge of the four paths. Having passed through the 9 knowledges of

purification by knowledge and vision of the way, there comes knowledge of change of lineage. It is intermediate between purification by knowledge and vision of the way and purification by knowledge and vision. Activated by knowledge of change of lineage, knowledge of the path of stream entry follows immediately, with nirvana as object. Its result is the immediate arising of two or three fruition consciousnesses. Here the stream enterer is called the second noble person. He reviews the path he has come by, the blessings he has obtained, the defilements abandoned, the defilements still remaining, and deathless nirvana as the state entered as object. The once-returner and non-returner do the same, except that the non-returner has no defilements still remaining. Thus further reviewing brings knowledge of the second path (the path of once-return), third (the path of non-return) and fourth path (the path of arahantship), with second, third and fourth fruition. The end of the first path is the first noble person, of first fruition the second noble person, of second path the third noble person and so on until after fourth fruition is the eighth noble person. This last is one of the great ones, all cankers destroyed, bearing his last body, his burdens laid down. He has reached his goal and destroyed the fetters of becoming, liberated with right or final knowledge.
Context One of the five purifications or purities which constitute the trunk or body of understanding according to Hinayana Buddhism, the others being: purification or purity of view; purification or purity of overcoming doubt; purification by knowledge and vision of what is the path and what is not the path or purity of knowledge and discernment of the right path and the wrong path; purification by knowledge and vision of the way or purity of knowledge and discernment of the middle way.
Broader Nanadassana (Buddhism, #HM6502) Purifications (Buddhism, #HH3875).
Related Knowledge of change of lineage (Buddhism, #HM4637)
Knowledge of the path of non-return (Buddhism, #HM6920)
Knowledge of the path of once-return (Buddhism, #HM7563)
Knowledge of the path of arahantship (Buddhism, #HM7055)
Knowledge of the path of stream entry (Buddhism, #HM1088)
Profitable consciousness in the supramundane plane (Buddhism, #HM4930)
Indeterminate consciousness in the supramundane plane - resultant (Buddhism, #HM5129).
Preceded by Purification by knowledge and vision of the way (Buddhism, #HH3550).

♦ HH3029 Human development (Judaism)

Description God works His redemption through history, with final redemption for all in the dawning of the messianic age. Israel is a special people with whom God has a covenant and as such the Jew must affirm God's unity and follow the commandments of God as appear in the Bible and as taught by the rabbis. Traditional Judaism centres around the halakhic rituals which determine the individual's approach to God, to his fellow men and to the world, and which centre on the family. Detailed rules apply to all secular activity, for example the preparation of food, as a sign that God must be served in the most basic activities of life. Prophecy is the means by which God communicates with man, and Moses is the greatest of the prophets. Good and evil acts receive their rewards; one is free to make one's own choices but must accept the consequences of that choice. After death there is resurrection which is a stage prior to the disembodied bliss of the soul.
Because unmitigated experience of the oneness of God can only result in physical death, all experience of God in this life is said to be veiled, so that no experience of God can be complete. Only the very wise and learned can approach the ecstatic revelation of God through mystical experience without madness or agonizing death. Even then, the *Kiss of God* of direct revelation ends life, but sweetly and softly. This is God as "Wholly Other". Conversely, transcendent experience of God as "Wholly the Same" arises in the normality of everyday life. The mind is fully engaged and God is seen as one with his creation. Although, again, utter oneness can only be achieved at the end of all things, the task is to work towards relating everything directly to God so that he is truly one. The material world is then a physical manifestation of the hidden or spiritual Torah. Revelation of the Torah reverses the process of Adam's rebellion and original sin; and matter is transmuted to spirit.
This balancing of wholly other and wholy the same is illustrated in prayer which is said standing in reverence at God's awesome presence but also silently because God is nearer than man's very heart. The *torah*, God's teaching, is both the explict, harmonious, transcendent structure of creation, whose practice represents creation as it was before the fall, and also the inner transformative process within all things, including the heart of man. Actualizing the torah enacts the will of God. Love and justice are actualized through a mysticism of the will requiring personal identity and discriminating consciousness. It is the very Name of God which is expanded through spiritual generation into the hidden and thence to the revealed torah. Separation of the material from the spiritual is the continuation of Adam's fall and the cause of darkness, opaqueness and evil. The task of the mystic is to transmute the physical to the spiritual so that the torah can shine through, breaking the casing of darkness that surrounds it.
Broader Human development through religion (#HH1198).
Narrower En-sof (#HM0934).
Related Shekinah awareness (Judaism, #HM2002)
God as wholly other (Judaism, #HM4501)
God as wholly the same (Judaism, #HM4291).

♦ HH3098 Human development (Existentialism)

Existential imagination
Description Existentialism is a philosophy that responds to an individual's sense of disorientation in an alienating society and the need to find new ways to come to terms with his existence. A sense of absurdness is experienced in the implicit antagonism between the individual mind and the collective world, in which both strain against each other and without the possibility of either satisfactory embracement or resolution. In the contemporary context in which traditional frameworks are viewed with scepticism and despair, the individual must make his own decisions. It is thus only the individual and his consciousness which are of consequence, and any attempt to come to terms with this must deal with this existential dimension and experience. The individual is thus obliged to seek within in unfamiliar territory and take absolute responsibility for that search. Existence has also been treated as the specifically religious response of an individual faced with decisions in the face of God, and as such with questions of ethics and salvation as opposed to abstract speculation. For the individual it represents a fierce and solitary encounter with nothingness and with a sense of nullification of self through which his salvation may then become possible. It is the acceptance of this initial position of maturity and responsibility which represents the key step in human development from this perspective. Existentialism can be viewed as a philosophy of resistance and liberation. As such it is an attempt to set free the individual's authentic self from the cage-like existence of his inauthentic self. The individual's very being rests on the freedom he gains when he chooses to make himself what he his. It serves therefore to awaken the individual from his apathy, obliging him to face his true self, no matter how unpleasant that confrontation may be. In stripping away the illusions by which he lives, and in confronting the very self that he tries to disguise, he may find a tentative peace as a result of a response from with himself. Existentialism therefore gives the individual a way of discovering what makes him unique, affording him a means of comprehending his situation. The existential imagination demonstrates that even at bay the individual still has the means to seek his personal identity and to find happiness even in failure. The individual then transcends his pettiness and becomes a "hero" worthy of his existence. His worthiness derives from his confrontation with his situation, no matter how disenchanting, difficult or frustrating.
Refs Barrett, William *What is Existentialism?* (1964); Karl, Frederick R and Hamalian, Leo (Eds) *The Existential Imagination* (1963).
Related Heroism (#HH0929).

♦ HH3117 Psychic growth

Development of psychosomatic power
Description Growth may be defined as the development of the psychosomatic power and aptitudes of the person, and of the display of new, more complex possibilities for fruitful interchange between the person and his or her environment. Intelligence grows from a stage of undifferentiated perception to that of unfolding abstract, logical-conceptual operations. While it develops, new, more complex and more powerful tools are acquired for knowing reality and adequately exchanging with it. These add to or integrate previous stages of the capacity for adaptation, so that each stage brings a perfection and complication of the cognitive structure which corresponds to the previous stage. The person matures, develops, increases his affectivity and perception of himself, his grasp and the structuring of social and natural reality. Growth in adaptation to the environment and possibility for interacting with the world is from a primary, affective symbiosis with the mother, through incipient structuring of the self leading to primary egocentrism to gradual integration with one's peers.
Refs Mallmann, Carlos A and Nudler, Oscar (Eds) *Human Development in its Social Context* (1986).
Narrower Primary growth (#HH6669) Maturity growth (#HH5434).

♦ HH3120s Cultural transformation

Description A collective evolutionary change is reflected in individual awareness of change and seeking out of a new modus vivendi. This in turn snowballs and effects cultural change, which may be inhibited by other individuals seeking to maintain the lifestyle of the past.

♦ HH3121 Christian stewardship (Christianity)

Description Teachings of both Christianity and Judaism are basically ecological and supportive of sustained habitation of the Earth. The philosophy of Christian stewardship towards life and the Earth is therefore very old, although recent centuries have seen little emphasis on it. The current environmental debate has brought it back into the limelight as Christians see creation being degraded and they move back to embrace the Creator and to speak out for the Creator's works. There is recognition that belief in God as Redeemer requires belief in God as Creator. There are three sources of Christian stewardship: a deep and reflective study of the scriptures; learning from the cosmos; nurturing a life of spirituality. Scriptures show that humanity often behaves contrary to harmony and order, and that stewardship will not arise from simply being human. When arrogance, ignorance and greed prevail there is death as the integrity of creation is degraded. Behaviour in harmony with the cosmos and the Creator leads to life. Scriptures indicate that mankind does not own the earth but that all creatures have intrinsic values. Every creature and all creation requires a sabbath of fulfilment so that mankind should engage in loving keeping of the earth and its creatures. Christian stewardship is a work of Redemption, restoration and service.
There is much to be learned from creation, as the environment acts on living beings, society and human culture and they act on it in the integrated fabric of the biosphere. Understanding of the beauty, harmony and greatness of the cosmos and its creatures elicits awe and humility, the heart touched and transformed at this reflection of the creator. A world view embracing Christian stewardship elicits active striving to preserve and restore creation's integrity.
Refs Canadian Ecophilosophy Network *The Trumpeter* (1989); Carmody, John *Ecology and Religion* (1983); DeWitt, Calvin B *A Sustainable Earth* (1987).
Related Deep ecology (#HH2315) Ecological humanism (#HH4965)
Planetary consciousness (#HM2006) Environmental mysticism (#HM0800).

♦ HH3124 Charismatic renewal (Christianity)

Pentecostal movement — Spiritual renewal
Description Personal baptism in the Holy Spirit is the focus of a current charismatic renewal in the Church, bringing spiritual power into people's lives and resulting in more authoritative prayer, worship, service and evangelization. There is real repentance and inner healing, and the movement is not confined to one denomination but seems to be arising spontaneously throughout the Church. Relationship with Jesus Christ becomes more personal and there is a powerful sense of union with other Christians in Christ. Spiritual growth becomes of immediate concern. Particular gifts are spiritual healing and speaking in tongues. Spiritual renewal is at the very heart of Christ's teaching; a new beginning so radical it can be thought of as a rebirth.
Within local congregations, individuals begin to respond at greater depth to the life of the Spirit. They are more aware of belonging to the body of Christ, and this description of the Church takes on more meaning. The ministry becomes more dynamic as mature members of the congregation share both administrative and pastoral responsibility. Large congregations are subdivided into smaller house groups. Individual ministries of administration, prophesy, healing and so on are recognized, as the ideal is approached of having each individual a functional member. Healing and counselling are an important component of the congregation's programme; and training and teaching become more central, not only by the priest but by other gifted individuals. Planned giving is accepted. Worship which is joyful and spontaneous and the flowing of gifts of the Spirit are included in the regular pattern. The congregation shares all kinds of united mission with other local churches. Worship involves individuals feeling more, and expressing this feeling in swaying of the body, raising hands and looking at each other during worship. The framework may be traditional but spontaneous and local additions are incorporated as the Spirit leads.
Distinctive features of charismatic renewal are: (1) Belief that the Holy Spirit is available to all who have repented of their sins and accept Jesus Christ, through baptism in the Spirit. (2) Belief that the gifts of the Spirit revealed in the New Testament are available to the Church and to Christians in every age. (3) Expectancy that the guidance of Christ will be experienced as every part of the individual's life is submitted to Him. (4) Reassessment and response to the word of God as revealed in the Bible. (5) An emphasis on joyful praise and thanksgiving in worship. (6) Experience of oneness in Jesus Christ beyond denominational and theological differences. (7) Recognition that every believer has spiritual gifts to be used in the ministry of the Church. (8) Awareness of evil as an objective power and of an inner battle being waged against its forces.
Refs Gunstone, John *Baptised in the Spirit* (1989).
Broader Human development (Christianity, #HH2198).
Related Charisma (#HH0053) Spiritual healing (#HH0458)
Born-again (Christianity, #HH0576) Glossolalia (Christianity, #HM1608)
Spiritual repose (Christianity, #HM6552) Personal salvation (Christianity, #HM6335).

♦ HH3129 Brain waves

Description Four major kinds of brain waves have been discovered. In terms of cycles per second, the delta waves (1 to 4 cycles) are slowest, followed by theta (4-8), alpha (8-13) and beta (13-26). Different wave types are correlated with different types of mental activity. Sharply focused attention on a mental activity leads to brain waves of all frequencies (desynchronized activity); non-focused attention leads to a predominance of alpha waves, whereas deep sleep involves low frequency, delta-activity. In meditation there is a predominance of rather slow alpha waves with unusually wide amplitude; although normally associated with the back of the head, in

HH3129

meditation alpha waves occur across the whole head and are synchronized with each other. They first build up on the left and then move to the right, indicating a transfer form active analytical to receptive synthetic; this is itself reduced. One researcher (William Condon) refers to six different brain wave frequencies, which include the highly active beta II (perhaps of to 40 cycles per second), and a second delta wave that may be the basic or background rhythm to human behaviour. Condon's work shows how the brain waves correlate to speech and behaviour and how individual waves may interlink and synchronize with other's individual and group waves.
Narrower Beta wave consciousness (#HM3476) Alpha wave consciousness (#HM2345) Theta wave consciousness (#HM2321) Delta wave consciousness (#HM1785).
Related Biofeedback training (#HH0765).

♦ **HH3172 New age movement**
Description All existence is said to be a manifestation of supreme consciousness the essence of which is love. The purpose of existence is to bring love fully into manifestation; all religions are the expression of this truth. Life perceived by the senses is an outer veil of otherwise invisible, inner and causal reality. The temporary human personality lasts only one life but the multidimensional inner being (soul, higher self) is eternal whose purpose through reincarnation is to unfold to become perfect love. Spiritual leaders are thus those who have become liberated and unconditional love. Life is interconnected energy and each individual bears joint responsibility for the state of himself, of all selves, of the environment and of life.
The evolution of humanity and of the planet is said to have reached a point where there is fundamental spiritual change in individual and mass consciousness, resulting from increasingly successful incarnation of cosmic love and of what is known in the west as the Cosmic Christ but is referred to by other names in other cultures. The resultant consciousness is demonstrated in an instinctive knowledge of the sacredness and interconnectedness of all existence. This new consciousness and new understanding of the dynamic interdependence of life mean that there is currently a process of evolving a new planetary culture attempting to work with and be open to the best of the old (old religions, science) and of the new (new psychotherapies, science). The aim is to be aware of the complete newness of these times while still retaining a sensitive an active awareness of responsibilities.
The new age movement is characterized by individuals who consciously (as with shamans) broaden perception, assimilate the experience within their cultural context and, by sharing with others, assist in broadening the range of available experiential knowledge. Others (as with mediums) depend on the authority or permission of other sources, surrendering personal discrimination and decision-making to various types of divination or to a dominant spiritual leader.
Related Human potential movement (#HH0398).

♦ **HH3179 Dignity**
Description Dignity has been defined (Swami Omkarananda) as the capacity to hold permanently in control the moods of the moment and to express cheerfulness.
Refs Gotesky, Rubin and Laszlo, Ervin (Eds) *Human Dignity, This Century and the Next* (1970); Pico della Mirandola, Giovanni *Oration on the Dignity of Man.*

♦ **HH3181 Sustainable development**
Description Sustainable development is that development which meets the needs of the present without compromising the ability of future generations to meet their own needs. It is a process of change in which the exploitation of resources, the direction of investment, the orientation of technological development and institutional change are all in harmony and enhance both current and future potential to meet human needs and aspirations. It must rest on political will. There are three key concepts: needs; limitations on the environment's ability to meet present and future needs; and maximum sustainable yield, or carrying capacity, which must be defined after accounting for system-wide effects of exploitation.
Related Human-centred development (#HH3607).

♦ **HH3188 Success**
Achievement
Description A specific aim of human development is to be perceived as a success by others and to feel that one has acted successfully and has a personal sense of being successful. Much ambition is focused on the process of achieving such success. Although every success is unique (Robert Heller) there are key supporting factors which successes have in common. Individuals need leadership, challenge, decisiveness, speed, clarity, mastery of the basics, firm objectives and acceptance of change. Above all, there is the inbuilt incentive to achieve. Each individual has a certain potential for success which may be accentuated or underexploited. However, failure to achieve more than relatively little of his imagined potential tends to make the individual deny his ambitions and reduces achievement still further. An environment of failure makes personal success difficult. A number of factors have been shown to improve individual performance:
1. Clear recognition and belief in talent. The individual builds on and develops the talents he has rather than spending time and resources on trying to create ability that does not exist.
2. Studying what has already achieved success. Rather than agonizing over what went wrong in the case of failure, the emphasis is on what went right in the case of success.
3. Recognition of excellence. The individual brings out the best in others when he recognizes excellent performance.
4. Moving from strength. Consolidation and developing strong points means that weaknesses can be managed.
5. Building relationships. The individual who encourages relationships with others finds they are more willing to work with or for him.
6. Matching of expectations with potential. People do not achieve if nothing is expected of them; conversely, repeated requests to achieve something for which the person asked has no response is destructive.
7. Decision-making. Delegation of decision-making to a person close to the action gives a sense of ownership and an impetus to do well.
8. Clear organization. Constant reorganization creates misunderstanding, while a clear concept of how things are done gives clarity of action.
9. Objective measurement of achievement. When the individual knows how well he is doing, his performance improves.
10. Teamwork. Success is based on maximizing the individual's talents within the overall team effort.
Refs Heller, Robert *Unique Success Proposition* (1989).
Related Failure (#HH1433) Ambition (#HH6089) Achievement motivation (#HH0456).

♦ **HH3198 Immaterial states as meditation subjects** (Buddhism)
Aruppa-niddesa (Pali) — Formless realms
Description The four immaterial states or formless realms are: sphere of boundless or unlimited space; sphere of boundless or unlimited consciousness; sphere of nothingness; sphere of neither perception nor non-perception. The meditator has already surmounted gross physical matter by entering the fourth jhana in any one of the kasinas except the limited-space kasina. However, because the kasina has materiality as its object (this materiality being viewed with dispassion or disgust), and because joy is its near enemy, he now approaches the immaterial states and the four jhanas associated with them.
Having spread out the kasina, attention is given to the space touched by it and the kasina is removed. With practice, mastery is achieved. Boundless space is then seen as being to close to fine-material jhana and boundless consciousness is seen as more peaceful. Attention is then on boundless consciousness pervading space and having boundless space as its object. Striking it with applied and sustained thought, first the mind becomes concentrated in access then, with further practice, in absorption. Having obtained mastery in the five ways again, the meditator goes on to the sphere of nothingness and finally to the sphere of neither perception nor non-perception.
Broader Subjects for meditation (Buddhism, #HH3987).
Narrower First absorption in the immaterial sphere (#HM2110) Third absorption in the immaterial sphere (#HM2027) Second absorption in the immaterial sphere (#HM3043) Fourth absorption in the immaterial sphere (#HM2051).
Related Jhana (Buddhism, Pali, #HM7193) Immaterial-sphere concentration (Buddhism, #HM0696) Formless-realm awareness (Buddhism, Tibetan, #HM3144) Profitable consciousness in the immaterial sphere (Buddhism, #HM4701).

♦ **HH3209s Self-metaprogramming**
Description This is a term used to describe the process of assimilation by the subconscious of experience-related beliefs and emotions, these being brought into effect by future experiences.

♦ **HH3211 Conciliarity**
Description The concept of actual unity of the Church although apparently separated by location or culture or time, with the goal of achieving fellowship in action.
Related Ecumenicism (Christianity, #HH0099) Inter-religious dialogue (#HH2434).

♦ **HH3214 Syncretism**
Syncretic religion
Description Attempts to fuse the characteristics of different religions are common and natural where cultures meet. The search for a common truth behind different systems has lead to the widening of worship of particular deities to include similar deities in other cultures; to the combination of Christianity with native elements in many parts of the world; to Hindu-Buddhist, Hindu-Islam (for example, Sikh), Hindu-Christian combinations; to combinations of Buddhism with Confucianism or Shinto; and to theosophical and anthroposophical movements.
Refs Burger, Hanry G *Syncretism, an Acculturative Accelerator* (1966).
Broader Human development through religion (#HH1198).
Related Perennial philosophy (#HH0665) Ecumenicism (Christianity, #HH0099) Inter-religious dialogue (#HH2434) Human development (Sikhism, #HH6292) Human development through new religious movements (#HH1523).

♦ **HH3215 Journeying within transcendence - from outside to inside**
Description The pen in which the sheep are locked is likened to the defensive structures built up in self protection; it may once have helped in adaptation to life but eventually stands in the way of further growth. The wolf is the appetites, demanding more and more. The doorway to freedom, opening on the threshold of the future, bridging conscious to unconscious, is approached through the dreams which are the break in the defence mechanism, to where one can know and be known.
Context The twelfth section of St John's Gospel, Chapter X 1-41, is related to a stage in the spiritual journey of the individual.
Broader Journeying within transcendence (Christianity, #HH6505).
Followed by Journeying within transcendence - from death to life (#HH2987).
Preceded by Journeying within transcendence - from blindness to sight (#HH1354).

♦ **HH3219 Mentor**
Description Seeking advice or simply unburdening problems to a respected mentor has the effect of clarifying the individual's position and enabling decision-making. Because the relationship is built up over a period of time there is a familiar, two-way, exchange resulting in an experience of confidence in conclusions reached.
Related Guru (#HH0805) Paternalism (#HH0735) Spiritual direction (#HH0442) Spiritual discipline (#HH1021).

♦ **HH3220 Steps to cosmic consciousness**
Description Swami Omkarananda enumerates a number of steps: (1) Inner purity and transformation which help manifest higher powers of consciousness; (2) Training of physical, mental and emotional capacities to yield higher experiences, a wider and larger field; (3) Illumination of the mind with spiritual truths; (4) Purifying of nature so that higher powers function unimpeded; (5) Generating a constant sense of God's presence; (6) Conditioning one's life to disciplines which have helped great philosophers; (7) Silencing the mind and senses; (8) Reflection on the nature of God as presented by those who have experience of God; (9) Looking at one's self in relation to the infinite; (10) Constant generation of the feeling that God is always with us; (11) Remaining in objectless consciousness.
Related Cosmic consciousness (#HM2291).

♦ **HH3221 Dehumanizing process**
Description As a means of therapy, under controlled conditions, the individual is isolated without the stimulus of sound, light or objects. He or she may react so as to discharge suppressed emotions. Dehumanizing may also be used destructively in brainwashing techniques.
Related Dehumanization (#HM3030).

♦ **HH3226 Formation of demerit** (Buddhism)
Description In Hinayana Buddhism, the unprofitable or immoral volitions of the sense sphere are conditions, as karma condition or decisive-support condition, for unprofitable resultant consciousnesses of the different kinds of rebirth linking and in the course of an existence in the sense sphere and the fine-material sphere. The Path of Purification details exactly how each unprofitable consciousness results in the appropriate resultant consciousness, both for unhappy and for happy destinies. Even Brahmas see, hear etc undesirable sights, sounds, etc, in the sense sphere although there are no such undesirable data in the Brahma world itself.
Refs Buddhaghosa, Bhadantacariya *The Path of Purity* (1975); Buddhaghosa, Bhadantacariya *The Path of Purification* (1980).
Broader Merit (#HH0859).
Related Formation of merit (Buddhism, #HH5122) Ten meritorious deeds (Buddhism, #HH0446) Formation of the imperturbable (Buddhism, #HH5543) Unprofitable consciousness in the sense sphere (Buddhism, #HM8375).

♦ **HH3228s Amping**
Description This is a general term covering systems used to tune the subconscious mind to receive psychic information which involve the electrical or nervous system of the body.

♦ **HH3229 Passivity**
Description Although each individual has to work out his own salvation, the Christian doctrine of redemption nevertheless assumes a passive acceptance and dependence upon God. Such passivity presupposes a receptiveness of spirit, an openness to take what life brings and a meekness to accept without complaint. Certain religions, notably Christianity and Buddhism, strongly advocate passivity, passive obedience and submissiveness, and passive endurance in the face of the pressure of the hostility and hatred of the world. This passivity does not imply inaction, in particular activity to redress wrongs suffered to others is considered obligatory.
Related Meekness (#HH0414) Quietism (#HH2006).

♦ **HH3238 Private revelation**
Description Revelation to the individual does not require faith from the church nor is it necessarily preserved or expounded by the church. It is for the guidance and salvation of the individual to whom it is revealed, although care must be taken not to confuse it with subjective fantasy or sudden manifestations of the subconscious. Again, although the revelation is genuine it may be distorted or misrepresented by the recipient. However, such a revelation may give a prophetic mission, providing the impetus for action according to the changing circumstances of the church in fidelity with the unchanging gospel.
Broader Revelation (#HH1028).

♦ **HH3246 Kasina meditation** (Buddhism)
Colour kasina — Nature kasina — Earth kasina — Sound kasina — Space kasina — Water kasina
Description Meditation commences with attention to a particular representation of the subject chosen - for instance, earth. With practice the subject is called to mind without seeing its physical representation - although any distortion or imperfection in the original representation is still visualized. Finally the essence of the subject is brought to mind in its perfect state. Some sources indicate that this form of meditation focuses attention the mind on a particular emotion or mood, or on what gives rise to this emotion or mood.
Broader Meditation (#HH0761) Subjects for meditation (Buddhism, #HH3987).
Related Tenfold powers (Buddhism, #HH2918).

♦ **HH3287 Structured meditation**
Description A form of group meditation which is guided by one person who may speak words specifically designed to assist achievement of the meditative state.
Broader Meditation (#HH0761).

♦ **HH3290 Human development** (United Nations)
Description The United Nations Development Programme (UNDP) defines human development as a process of enlarging people's choices. Although these can be infinite and change with time, the three essential ones are for a long and happy life, to acquire knowledge, and to have access to resources necessary for a decent standard of living. Other highly valued choices range from political, economic and social freedom to opportunities for being creative and productive and to enjoying personal self-respect and guaranteed human rights. One important choice is for income, but this is not the only choice. The focus is people themselves and not just expansion of income and wealth. In fact there is no direct correlation between income and human development levels on a country-by-country basis.
There are two sides to human development: the formation of human capabilities (for example, improved health, knowledge and skills); and the use people make of their capabilities (for leisure, productive purposes, or activity in cultural, social and political affairs). Development is thus both a process of widening people's choices and the level of well-being they achieve.
Human development may possibly be measured through focusing on longevity (life expectancy), knowledge (literacy is a crude measure of this but a possible starting point), decent living standards (perhaps using per capita income adjusted to include real purchasing power). These may be incorporated into a *Human Development Index*.
Refs United Nations Development Programme *Human Development Report 1990.*
Narrower Human development index (#HH5101).
Related Human development strategy (United Nations, #HH0247).

♦ **HH3296 Lower middle grade** (Taoism)
Description Practices involve eating and drinking in small or large quantities, abstaining from particular foods, eating filth, abstaining from cooked food, avoiding flavourings, enduring cold, lying on ice with an exposed back, performing strange feats.
Context In the nine grades of practices which are side tracks and auxiliary methods in Taoism, this is the lowest of the three middle grades.
Broader Sidetracks and auxiliary methods in Taoism (Taoism, #HH7004).
Followed by Middle middle grade (Taoism, #HH5207).
Preceded by Outside paths (Taoism, #HH4289).

♦ **HH3321 Five aggregates** (Buddhism)
Khandha (Pali) — Skandha — Aggregates of clinging — Khandhas — Awareness of inter-dependency of conscious existence phenomena
Description The five aggregates are those of: materiality (corporeality of form); feeling; perception; mental formations; consciousness. Although all except that of materiality can in some sense be free from cankers and not subject to them, and also not subject to clinging, all five have a sense in which they are subject to cankers, etc. Thus any kind of matter, whether past, present or future, internal or external, subjective or objective, gross or refined, superior or inferior, far or near, and which is subject to cankers and liable to clinging is referred to as the materiality aggregate of clinging. The same is true of feeling, perception, mental-formations and consciousness. They can give rise to the view "this is mine, it is I, it is myself". They have been described as: the hospital or sick room (materiality, where the "sick man" dwells); the disease (feeling, which is painful); that which provokes the disease (perception, giving rise to feelings); the root-cause of the disease (mental-formations, the store of unprofitable karma having caused birth in the first place); the sick man himself (consciousness, never free from the sickness of feeling).
The five clinging aggregates are thus the enemy on the road to freedom from birth and death, but they can be overcome. Seeing personal materiality as foul, no longer seeing beauty in the foul, no longer bound by desires of the senses, frees from clinging to sense-desires. Seeing feeling as painful (never free from suffering), no longer seeing pleasure in the painful, brings freedom from the canker of becoming and from attachment to rites and rituals. Seeing perception and mental-formations as other than the self and uncontrollable, no longer seeing the self in not-self, crossing the flood of wrong-views, cuts the conviction that "this" is the truth and is not attached to the theory of self. Seeing consciousness as lacking in permanence, always rising and falling, no longer seeing permanence in the impermanent, brings freedom from the bond of ignorance.
The five aggregates are said, in the Path of Purification, to be part of the soil in which understanding grows. In order to perfect understanding one should learn and question these things.
Conscious existence thus depends on the five groups of interacting physical and mental phenomena. Their interaction gives rise to the illusion of personality, individuality or ego, which has no existence of its own.
Context In Southern Buddhist Pali texts the Khandhas are listed in the series of dhammas, which are all the phenomena in existence. The phenomena are also called *sankhara*: dhammas = "things"; sankhara = "formations". The corresponding Sanskrit terms are skandhas and dharmas; and samskara, which is comparable to maya or lila. Nibbana (nirvana) is the state in which the "individual's" khandhas cease to "exist", stopping the process of karmic reincarnation and rebirth. "Nibbana" means "extinction".
Refs Buddhaghosa, Bhadantacariya *The Path of Purity* (1975); Buddhaghosa, Bhadantacariya *The Path of Purification* (1980).
Broader Phenomena awareness (Buddhism, #HM2551)
Ignorance in dependent origination formula (Buddhism, #HM3035).
Narrower Feeling aggregate (#HM4983) Perception aggregate (#HM4143)
Dispositions of consciousness (#HM2098)
Awareness of corporeality-group of conscious existence (#HM2108)
Awareness as mental-formation group of conscious existence (#HM2050)
Awareness of consciousness-group of conscious existence - senses and mind (#HM2556).
Related Clinging (Hinduism, #HM6153) Understanding (Buddhism, #HM4523)
Cessation of suffering (Buddhism, #HH2119)
Worlds of conscious existence (Buddhism, Pali, #HM2072).
Followed by State of universal cessation awareness (Buddhism, #HM2596).

♦ **HH3322 Way of the warrior**
Refs Croucher, Michael and Reid, Howard *The Way of the Warrior* (1987); Heckler, Richard S (Ed) *Aikido and the New Warrior* (1985); Millman, Dan *Way of the Peaceful Warrior* (1984); Robert Aubroy; Laing, Ronald and R Pflughaupt, Knut *The Way of the Warrior* (1982).
Narrower Way of the warrior (Amerindian, #HH8219).

♦ **HH3324 Veganism**
Description A way of living which seeks to exclude all forms of exploitation of animals for food, clothing or any other purpose. In dietary terms it entails dispensing with all animal produce (including meat, fish, poultry, eggs, non-human animal milks, honey and their derivatives). The organic agriculture that this necessitates works with nature, requiring less land.
Related Human development through diet (#HH7324).

♦ **HH3329 Anthropocentrism**
Description Although anthropocentrism may be seen as the imprisoning of the individual in a false autonomy which denies the love of God, there is a sense in which the centering on the self is the necessary prerequisite for departure from the self. An action is moral in that it is carried out by a subject grounded in God and through his own personal transcendence. Anthropocentrism may be seen as implying the Christocentrism of God.

♦ **HH3335 Way of paradox** (Christianity)
Description Meister Eckhart indicates that grasping the reality of God, the perception of all-embracing unity, is arrived at only through the tension and clash of opposites. To find new life in God, one must die to the life one is living. Not only must the will be crucified, but also thought and speech. In daily life, this is the practice of detachment. In thinking and talking, it involves paradox. There is a rhythm between opposites which, if not resisted, gradually kindles divine knowledge and opens the eye of the heart. The deepest truth can only be grasped through the perpetual alternation of opposites. A statement is both true and untrue at the same time. The highest truth transcends the principle of contradiction, beyond the truth or untruth of a statement. This paradox makes the normal intellect aware of its limitations, opening the possibility of a higher way of knowing. The contraries are all contained in an all-embracing unity. Neither this nor that, God cannot be grasped, imagined, or understood. Viewing and seeking to live the Christian revelation this way opens many mysteries and energizes spiritual life.
Following the way of transcendence we find that, although seeking God from outside means that He retreats within, if we seek Him within He affirms Himself without. Again, in pouring Himself out or melting God utters Himself totally, the whole mystery is uttered. The expression of Himself is Himself. God as speaker is the father. God as spoken is the Son. The going out and yet remaining within is the bond between the two, making the unity, the Holy Spirit. This going out and remaining within is the key to all spiritual life. One truly lives only inasmuch as one is caught up and lives in the Trinity. Pouring one's self out in the world and in relationships one remains inwardly detached, in peace, tranquillity and contemplation and yet untiring, accumulating rather than losing energy.
To be a person implies aiming perpetually at transcendence, oneness melting out as multiplicity, multiplicity melting back as oneness. It implies going out yet remaining within, simultaneously exerting power and restraint, transcending yet remaining one's self, in movement yet in repose.
One has first to enter the formless abyss in God and in the depths of one's self. Then one is reborn into the life of communion (represented by the Trinity). One is led to the abyss in the first place by the Word, the Son, who took human form as Jesus.
Seeing the world from in God, the universe is an echo or reflection of God. There is an emptiness at the heart of things which arises from the fact that the universe is not God. Creation has no being in itself but only as it is in the presence of God. Thus it, paradoxically, is and is not. It has duality. As God contemplates the universe He sees only His own reflection, hears only His own word. Seeing the world thus from within God creates the attitude of detachment, of lucid, compassionate awareness. Standing within the Word brings a proper love for the world which is the World's echo, a joyful understanding of the rhythm of birth and death, light and dark, breathing in and breathing out. In God the two movements are simultaneous, in creation they are successive and this arouses disillusionment and sorrow. But standing within the Word the meaningless flux comes together, the underlying harmony becomes audible. This is the key to the law of the universe, that of rhythm and dance.
Again, paradoxically, if one wants to change the world it is one's self that must be changed. It one wants to gain the world one must let it go and surrender to God. Although surrender to God and entry into the inner kingdom opens the outer kingdom, surrender must not be with this in mind. One gains only what one has fully let go.
Refs Smith, Cyprian *The Way of Paradox* (1987).
Related Melting (Christianity, #HM7384) Universal human nature (#HH4389)
Detachment (Christianity, #HM1534) Way of transcendence (Christianity, #HH6566)
Mystic union with God (Christianity, #HM3889)
Birth of God in the soul's ground (Christianity, #HM6522).

♦ **HH3352 Journeying within transcendence - from call to mission** (Christianity)
Description One is called to witness to Christ, to proclaim him and serve him. Even before one sees one must proclaim. The task is then to be attentive so that when Christ is revealed he is recognized. The moments of contact with the transrational draw one beyond the limits of the ego so that the deeper self is discovered to which one wishes to surrender one's whole being, all that

one has, one's limitations, restrictions, followers. This is an invitation, a gift of grace.
Context The second section of St John's Gospel, Chapter I 19 to II 1, is related to a stage in the spiritual journey of the individual.
  Broader Journeying within transcendence (Christianity, #HH6505).
  Followed by Journeying within transcendence – from water to wine (#HH4219).
  Preceded by Journeying within transcendence – prologue (#HH1900).

◆ **HH3356 Disruption of lifestyle**
Description Sudden changes in lifestyle brought about by some external event may set off internal changes so that personality is transformed, there is a new sense of direction and an increased feeling of self worth.
  Related Mode of life (#HH0592)   Personality development (#HH0281).

◆ **HH3366 Journeying within transcendence – from discussion to decision**
Description Between full rejection or full acceptance of Jesus there are a number of intermediate positions. The process of decision-making may be avoidance, a refusal to face up to what one really feels, thinks or wants, or an unwillingness to trust or risk one's way or truth. Or it may be that, after ridding one's self of deception and delusion, one looks into the darkness and decides in the light. The very process of making a choice, despite the distress it causes, is then the chance to be blessed.
Context The ninth section of St John's Gospel, Chapter VII 1 to VIII 58, is related to a stage in the spiritual journey of the individual.
  Broader Journeying within transcendence (Christianity, #HH6505).
  Followed by Journeying within transcendence – from bondage to freedom (Christianity, #HH2661).
  Preceded by Journeying within transcendence – from object to subject (#HH4188).

◆ **HH3376 Medical materialism**
Description Whatever the spiritual, altruistic or creative experience, this may be reduced to the malfunctioning of the body or mind, so there is little distinction to be drawn between the outpourings of a genius and his state of physical health or obsession. Much modern art can thus be interpreted as the result of scotomic vision during an attach of migraine. A Freudian analyst might interpret the struggle for beauty and for spiritual values as attempting to overcome premature ejaculation. Such a theory does not explain why physical disability does not always result in spiritual or artistic expression; and ignores the fact that, although an individual's physiology may predispose him to certain means of expression, it is only a conditioning or modulating not a creating or directing influence.
  Broader Materialism (#HH0529).

◆ **HH3390 Spiritual initiation** (Esotericism)
Description Spiritual initiation can be considered as a progressive sequence of directed energy impacts. Each initiation is a process of energy transmission from a higher centre of energy to a lower, enhancing the sensitivity of the initiate and displacing lower patterns of energy. This sensitivity acts as an attractive force draws to the individual those forms of being which he can instruct and aid. These energy impacts are characterized by points of tension and lead inevitably to points of crisis. Each initiation is in reality a crisis, a climaxing event which is only truly brought about when the individual has learnt patience, endurance and sagacity in emerging from the many preceding and less important crises. It is a culminating episode, made possible because of the self-inspired discipline to which the individual has forced himself to conform. Initiation admits a person into some area or level of divine consciousness, into a state of being hitherto regarded as sealed and closed.
Refs Bailey, Alice A *The Rays and the Initiations* (1960); Bailey, Alice A *Initiation, Human and Solar* (1978).
  Broader Initiation (#HH0230).
  Narrower Initiation of birth (#HM1337)   Initiation of baptism (#HM1267)
  Initiation of refusal (#HM1020)   Initiation of decision (#HM0322)
  Initiation of transition (#HM1258)   Initiation of revelation (#HM0181)
  Initiation of renunciation (#HM0210)   Initiation of resurrection (#HM1153)
  Initiation of transfiguration (#HM0428).
  Related Group spiritual initiation (Esotericism, #HM1417).

◆ **HH3398 Vinaya** (Buddhism)
Laws
Description This is the study of disciplinary laws and of the laws of cause and effect.
Context The fourth of the five primary subjects for the Geshe degree in Tibetan Buddhism.
  Related Vinaya (Buddhism, #HH1376).
  Followed by Abhidharma (Buddhism, #HH1765).
  Preceded by Pramana (Buddhism, #HH4388).

◆ **HH3405 Mythic yoga** (Yoga)
Path of renunciation
Description This type of yoga is centred on reaching further and further into the inner depths accompanied by total avoidance of and insulation from the external world. Systematic withdrawal from external reality is accompanied by viewing the body as the source of all evil and the female sex as particularly responsible for temptation and spiritual degradation. The yogin looks to permanent liberation from the cycle of birth and death through entering the Absolute with full enlightenment awareness. It is the way of the arhat, the sannyasin, the ascetic, the muni. The final and irreversible act of emancipation and liberation, *kaivalya*, is also the end of temporal body and mind in death.
  Broader Yoga (Yoga, #HH0661).
  Related Muni (#HH5450)   Asceticism (#HH0556)   Sannyasa (Yoga, #HH4210)
  Arhat (Buddhism, #HH0233)   Videha-mukti (Yoga, #HM4489)
  Non-attachment (Buddhism, #HM2128).

◆ **HH3412 Passive way of faith**
Description Here the creature is taken out of his own capacity, receiving an infinite capacity in God. Thus the soul loses the human so it can lose itself in the divine which, mystically, becomes its being and subsistence. Having lost its old life it receives a new life in God, no longer living and working of itself but God lives, acts and operates in it. It possesses nothing but is possessed. It is indifferent to all, for all is equally God. The insensibility is not that of death and decay but that of an elevation above all sentiments, tastes, views and opinions.
Refs Guyon, J M B de la Mothe *A Short Method of Prayer and Spiritual Torrents* (1875).
  Broader Faith (#HH0694).
  Related Passive state (Christianity, #HM8123).

◆ **HH3421 Heavenly deceit**
Misuse of scripture
Description Selected passages from the Bible, chosen and interpreted idiosyncratically, are emphasized and repeated to reinforce the message of a particular organization and assist in changing an individual's belief system.
Context Used in destructive manipulation by some cults.
  Related Destructive manipulation (#HH1020).

◆ **HH3424s Adore**
Description A personal belonging or its replica which contains special attributes of a deity is used as a psychic channel to receive revelation or phenomena of the deity.
  Related Adoration (#HM2412).

◆ **HH3432 Initiation** (Magic)
Description The cumulative effect of discipline restricts the dissipative habitual flow of the practitioner's life- energies and attention. It is the discipline, through imposing a willed set of controls in the place of illusory limitations, that brings initiation and subsequent liberation. The disciplinary methods employed in magic direct or bind the flow of these energies into a harmonic matrix of consciousness/energy. This can be a magical tradition or a coherent set of symbols for inner transformation. This builds higher levels of energy. During initiation, a window of opportunity is offered, both within the individual psyche and in imaginal worlds, through which these pressurized or shaped energies of the initiate are channelled. The initial experience is gained through dissolution of the habitual conditioned personality. Following this breakdown, a balancing power restructures experience into a new configuration. The newly liberated energies take on a pattern of simplicity, of health or of harmony. From this perspective the magical transformation does not involve the acquisition of new powers, rather it opens up a new worldview in which properties of consciousness/energy are released to act more appropriately than was possible when constrained by habitual patterns. During an initiation, experienced practitioners ensure the appropriate conditions, integrating through ritual the vital pattern of innerworld contacts. However the initiate must make the transition by which the ritual pattern is engendered. The initiation is not conferred. The first initiation is the beginning of a traditional process or journey whereby, through a series of encounters, the initiate transits through inner worlds. It is this apparent succession of encounters which can be understood as a sequence of initiations, each a harmonic of the first. Although explanations of magical arts tend to be divided and graded due to the nature of literary and verbal communication, experience occurs in highly-energized and concentrated bursts which may take years to decode into serial outer awareness.
Refs Maruyama, Magoroh *Paradigmatology and its application to cross-disciplinary, cross-professional and cross-cultural communication* (1974).
  Broader Initiation (#HH0230).
  Related Magic (#HH0720).

◆ **HH3435 Eccentricity**
Description Despite apparently odd behaviour, eccentrics suffer less than the general population from mental illness and stress. At the opposite extreme from neurotics, who may lack sense of humour and take themselves too seriously, the deviant behaviour of the eccentric may be healthy and life-enhancing. Eccentrics are a liberating influence from inhibiting convention.
Refs Ward, Kate *Eccentrics*.

◆ **HH3442 Flight from the world**
Description Although earthly values may be positive, practical expression of one's readiness to accept self-communication of God's love may require one to flee from them in preparation for sharing the death which is the fate of the world and as an expression of the will to grace (given by God) beyond earthly sense and meaning. In a broader sense, flight from the world implies conscious and positive withdrawal from all who have closed themselves to God, and from the relationships they have created where these imply temptation to further sinful rejection of God, with the intention of saving the world.
  Related Retreat (#HH0420).

◆ **HH3443 Nichiren shoshu buddhism** (Buddhism)
Nam-myoho-renge-kyo — Lotus sutra
Description This branch of buddhism aims to overcome the human problems of daily life and manifest the latent qualities in the individual so that he is able to develop himself and have a positive effect on an ever increasing sphere. Emphasis is on wisdom, courage, compassion and life-force. A happy life is said to be based on the attitude towards problems and suffering; the individual develops through his suffering, he grows as a person by chanting on his suffering. Based on the teachings of Nichiren Daishonin (1222-1282), said to be the Buddha for the present age, the system is based on the practice of reciting chapters of the lotus sutra and of chanting every morning and evening the mantra "nam-myoho-renge-kyo", the invocation of the lotus sutra, the sutra for the present evil age. Hope is for a future in which the human race as a whole develops so as to overcome its problems. Although written and theoretical proof of a system are important, the most important is said to be that it works in practice.
Refs Causton, Richard *Nichiren Shoshu Buddhism* (1988).
  Broader Human development (Buddhism, #HH0650).
  Narrower Six paths (#HM1914)   Ten worlds (#HM2657)   State of hell (#HM4282)
  State of anger (#HM2959)   State of hunger (#HM0150)   Four evil paths (#HM1252)
  Three evil paths (#HM0923)   State of rapture (#HM1973)   Four noble paths (#HM4026)
  State of learning (#HM3662)   State of animality (#HM0847)
  State of buddhahood (#HM1873)   State of realization (#HM0450)
  State of bodhisattva (#HM1225)   State of tranquillity (#HM3492).

◆ **HH3452 Shunamism**
Shunamitism — Psychic vampirism — Parasitism
Description In shunamism, flagging physical vigour is said to be restored by drawing health from the heat, breath, bodily odours of and physical contact with young people. Examples where strength is received from a chaste relationship of an elderly man with a young girl are numerous, ranging from King David in the Old Testament to Gandhi. In contrast, psychic vampirism or parasitism involves the drawing of psychic or emotional energy from another. This may be deliberate, as in the case of black magic or of some spiritualists or clairvoyants; or it may be unsuspecting, the psychic "vampire" unknowingly gravitating towards those most able to give psychic energy, the latter often feeling drained by the experience. In these cases, the giver and receiver are born with these capacities. Spiritual healers are notable for their innate ability to donate psychic energy, and they too may be drained after the practice of healing. Deliberate parasitism of the old on the young has been recommended as a similar practice to shunamism. The disadvantage is that although the intention is for the young person to benefit the old, the flow may in this case be in the other direction, the vitality of the older person actually being drained by the young. It is a fairly frequently observed phenomenon that adults often associating with teenagers are easily fatigued.
  Related Rejuvenation (#HH0525).

◆ **HH3456 Spirit transformation in organizations**
Description Transformation is the organizational search for a better way to be when old modes of action are no longer appropriate. Although the results of transformation appear with the emergence of new organizational form, the essence of transformation lies in the odyssey or passage of the human spirit as it moves from one formal manifestation to another. The process is a movement through form, whether the organization links two individuals or a complex institution. Development needs to be distinguished from transformation both in effect and in function. Transformation consists of making an organization different because the environment

is so unstable and/or radically altered that the earlier form is no longer viable. Development is a matter of making an existing organization better and presumes a reasonably stable environment. Organizations in transformation are confusing in that forms appear and disappear. This is characterized, as experienced intuitively, by the very complex and often turbulent energy flow, namely the flow of the spirit as it moves from one manifestation ot another.
Refs Harrison, Owen *Spirit* (1987).

♦ HH3465 Initiation as spiritual rebirth (Yoga)
Description In yogic practice, an adept initiates his disciple in a ceremony symbolizing spiritual rebirth. Mundane life is renounced and the novice is reborn with teaching as father and teacher as mother. Both must be shown respect. Because the teacher has followed the path already, he knows it and can show the way.
Broader Yoga (Yoga, #HH0661) Initiation (#HH0230).
Related Born-again (Christianity, #HH0576) Spiritual rebirth of the ego (#HH0698) Spiritual rebirth (Christianity, #HH4098).

♦ HH3466 Entrainment
Description A process in both the physical and the organic world typified by the manner in which a group of fireflies have a tendency towards blinking in unison. In this sense one nervous system drives another, or two drive each other, to the highest frequency of any oscillatory behaviour which they are both exhibiting. This tendency makes interactional synchrony possible.
Related Interactional synchrony (#HM1210).

♦ HH3487 Seance
Home circle
Description A group of individuals, in harmony with each other and including at least one medium, meets together to produce physical phenomena and/or to receive counsel and knowledge on psychism and etheric realms. Participants sit in a circle to ensure concentration of psychic energy and to keep out negative vibrations; the seance may be preceded by prayer, meditation or other activity conducive to building up necessary vibrations.
Related Spiritualism (#HH0479).

♦ HH3498 Human bonding
Affectional bonding — Parent–infant bonding — Pair–bonding — Social bonding — Sexual bonding
Related Affectionate bonding (#HM0973).

♦ HH3534 Divine abidings as meditation subjects (Buddhism)
Brahmavihara–niddesa — Divine states
Description These are four - lovingkindness, pity or compassion, gladness or sympathy and equanimity or even–mindedness.
Broader Subjects for meditation (Buddhism, #HH3987).
Narrower Pity (#HM0513) Gladness (#HM5224) Equanimity (#HM7769) Lovingkindness (#HM7607).

♦ HH3543 Parenthood
Family life
Description Although primarily a relationship through procreation, one can also be a parent through adoption, through marriage to a partner already having children or, in the widest sense, by assuming the relationship of parent to anyone younger (whether physically or spiritually) than one's self.
Parenting implies both care and nurture on the one hand, and discipline and the setting of limits on the other. Thus it contributes to the psychosocial and to the spiritual development of the child. It fosters a sense of individual self–worth in the framework of inter–personal responsibility. Where parenthood is not undertaken in the right way then the result may be physical or psychological abuse, failure to set or enforce limits (a mistaken idea of what it means to love), or rigid division of love from authority (perhaps on lines of gender).
Parents provide a role model for children and educate them in social skills. Over–identification here may lead to parents trying to live through the lives of their children or demanding from them an unrealistic perfection. Parents also, in their role of developing a family to act as a social unit, develop a feeling of personal allegiance in their children and assign tasks to the children in the family context. As a bridge between the family and the rest of the world they develop social responsibility and foster explicit recognition of the family's interdependence with other families and communities. It is the parents who provide love and support in times of failure. Conversely, bad parenting may include neglected responsibilities, the encouragement of isolationism (thus abusing loyalty), and refusal to admit to their own or their children's errors.
Particularly in a multicultural society or one which rejects religion, conflicts and confusion arise as to what constitutes good family life. It may eventually be seen as an arbitrary matter or one for individual decision. A number of major influences are at work here: the theories of experts (which may be conflicting and which certainly change with time); diversity of religious tradition; demands of popular culture; the upbringing received by the parents themselves; the attitudes of parents' and children's peer groups. These issues have to be addressed in the light of belief and experience.
Related Family life (Christianity, #HH1801).

♦ HH3544 Ri
Description In Confucianism, this is the law or principle which is the source of and which underlies all existence and the rules (both natural and social) which govern it. In some schools of thought in Japan, the meaning of ri is restricted to natural and social law, or simply to natural laws; other schools give it a more metaphysical sense as the principle which, in conjunction with *ki* – vital breath – is the source of all.
Related Ki energy (#HH0620) Ri hokkai (Zen, #HM3352) Ji–ri–mu–je (Zen, #HM3164).

♦ HH3546 Gnosticism
Description The predominant view is that of a radically dualistic religious movement positing the existence of two equal and contrary forces in the universe: a good God (transmundane, unknowable) and the evil Demiurge (responsible for the creation of the world and all the calamities therein). Gnosticism sees man as a misplaced spark of the divine light, engulfed in darkness through no fault of his own. Man must struggle to free himself from the mortal encasement in order to return to the empyrean realm from which he came.
Context Gnosticism flourished before, during and after the rise of Christianity. Gnosticism can be considered a purely historical phenomenon as distinct from gnosis which is phenomenological.
Refs Avens, Roberts *The New Gnosis* (1984); Jonas, Hans *The Gnostic Religion* (1963); Needleman, Jacob (Ed) *The Sword of Gnosis* (1986); Schoedel, William *Gnostic Monism and the Gospel of Truth* (1980).
Related Gnosis (#HM0413).

♦ HH3550 Purification by knowledge and vision of the way (Buddhism)
Purity of knowledge and discernment of the middle way — Patipada–nanadassana–niddesa (Pali)
Description The culmination of insight is reached in the eight knowledges plus a ninth, *knowledge in conformity with truth*. These are now free of the ten imperfections and the three characteristics may be observed.
Context One of the five purifications or purities which constitute the trunk or body of understanding according to Hinayana Buddhism, the others being: purification or purity of view; purification or purity of overcoming doubt; purification by knowledge and vision of what is the path and what is not the path or purity of knowledge and discernment of the right path and the wrong path; purification by knowledge and vision or purity of knowledge and discernment.
Broader Nanadassana (Buddhism, #HM6502) Purifications (Buddhism, #HH3875).
Related Imperfection of insight (Buddhism, #HM9722)
Knowledge of change of lineage (Buddhism, #HM4637)
Knowledge of appearance as terror (Buddhism, #HM7114)
Knowledge in conformity with truth (Buddhism, #HM5403)
Knowledge of desire for deliverance (Buddhism, #HM0766)
Knowledge of contemplation of danger (Buddhism, #HM7297)
Knowledge of contemplation of dispassion (Buddhism, #HM5364)
Knowledge of contemplation of reflection (Buddhism, #HM0169)
Knowledge of equanimity about formations (Buddhism, #HM0424)
Knowledge of contemplation of dissolution (Buddhism, #HM3385)
Knowledge of contemplation of rise and fall (Buddhism, #HM3723).
Followed by Purification by knowledge and vision (Buddhism, #HH3025).
Preceded by Purification by knowledge and vision of what is the path and what is not the path (Buddhism, #HH4007).

♦ HH3556 White magic
Refs Bailey, Alice A *A Treatise on White Magic* (1934); Knight, G *A History of White Magic* (1978).
Broader Magic (#HH0720).

♦ HH3562 Open mindedness
Description The relative openness or closedness of a mind cuts across specific content. It is not uniquely restricted to any one particular ideology, religion, philosophy, or scientific viewpoint. Open mindedness is more easily defined in contrast with close mindedness and with dogmatic attitudes involving intolerance, authoritarianism and righteousness. Open mindedness is concerned with various forms of acceptance, whether of ideas, people or authority, or of change in its many forms. It has been argued that individuals and societies survive and grow through alternating between periods of opening and closing to information.

♦ HH3564 Heaven
Description The concept of heaven has been used as an incentive for moral behaviour in this world. Achieving the highest good here leads to reward or consummation of that good in the hereafter. This concept has been attacked by both secular and religious sources, the "modern" view even denying that morality has a basis in the eternal order of things. Others look on the other–worldliness of heaven as simply an escapist dream from the hard facts of "real" life. The complexity of motives inherent in any activity has led to questioning of a providential moral order at all.
Against this negative attitude must be balanced the survival of traditional beliefs in another world and their tying–in with revelations arising in dreams, in parapsychology, in self–transcendent experiences of ecstatics. All these indicate that individual human beings are not entirely "self–enclosed".
In Christian terms, discipline and asceticism in this life are not only valuable in their character–strengthening properties but also provide the basis for life in eternity. Heaven is thus the goal of ascetic practice. Denying gratification of the appetites, so long as this is balanced with reason and done with positive purpose, is part of the ascent to heaven. Loving sacrifice emulates the life of Christ and reflects the Holy Spirit's dwelling within. Heaven is then the mystical union and eternal bliss promised in the future when virtuous conduct now is both initiated and perfected by God.
Related Hell (#HH2007) Heaven (#HM4302).

♦ HH3567 Fall of man
Original sin
Description If sin may be defined as "separation from God" (Paul Tillich), then the fall of man as described in Genesis is the separation of mankind from God due to disobedience or to exercise of free will. This separation is inherent in man from birth and divides him from God, from other men and from himself. In these terms, human development may be seen as the attempt by man to return to unity with God, to overcome the split in his own personality part of which seems to lead him to self destruction. To the Christian, this split may be overcome by the acceptance of God's grace.
Refs Williams, Norman P *The Ideas of the Fall and of Original Sin.*
Related Primitivism (#HH1778) Admission of sin (#HH4012).

♦ HH3607 Human–centred development
Description This seeks to enhance the full capacities and capabilities of human beings. It is often considered synonymous with human resources development and is increasingly regarded as a right to which all people are entitled. Current concepts of human development encompass the multiple dimension of the development process as well as the need to seek endogenous roots rather than rely on mimetic approaches. It is a development process in which the human element is not only the means but the end, and may be considered both as analytically complex and, at the same time, holistically.
Analytically, the subprocesses of development – economic, political, social and so on – may each individually be treated from the human–centred perspective. This implies a sharp change of focus. With economic development, for example, a human–centred economic development implies the internalizing of so–called externalities, so that ecological, social, cultural and psychological preconditions and consequences of economic activity are considered. Quality of work is considered in terms of its impact on workers' well–being and creativity, scale of production processes, decentralization of economic wealth and power, limits to growth, as well as tangible output. The challenge to such an approach is to make it truly operative – examples are Vanek's work on self–management and work inspired by Gandhi's view of economic progress in conditions of mass poverty.
Going on to political aspects, one of the main problems is to link global or macro views with local or micro. Macro views emphasize economic and political restraints arising from the structure of the world system. Micro views stress consciousness–raising, participation, direct democracy and community building. The difficulty is to bridge the two levels. Local people need to learn, through new kinds of local education and organization, how to confront powerful global forces in their local communities. Both economic and political approaches require an increased openness and sensitivity to the complexities of the real world and a refusal to remain in narrowly defined limits.

HH3607

In social development, the human–centred approach does not equate development simply with satisfaction of basic needs. Although this is important, the satisfaction of other needs must not be postponed or left to chance. It may be that the process of satisfying basic needs may be the means of satisfying other needs – the basic need of housing being fulfilled with active participation of those concerned in the design and building of the houses, for example. The human–centred approach goes beyond even the satisfaction of human needs at all, since this looks on the human being as a product or reflection of his social structure. Although influenced by this structure, each individual has the potential for being an autonomous centre of experience and freedom. This implies the notion of personal growth through enriching and expanding the realm of inner experience, not in an individualist sense but as a requirement of development planning so that rigid, homogenizing formulae are avoided.

Holistically, the atomistic approach to various dimensions of human development must also be avoided. Although each dimension may be unique and not reducible to another, none is really independent. In fact, the complex interrelatedness of the different dimensions and levels of the developmental process must be taken into account by any adequate theory. For example, the cultural context is the most powerful source of day–to–day living, its holistic qualities permeating and providing meaning to any activity taking place in a social setting.

A conceptual framework for relationships between development process and indigenous culture is still lacking, but one step on the path adopts the organic/inorganic distinction. In organic development (possible equatable with developed countries), development is induced internally and economic, social and cultural evolution are closely connected. In inorganic or nonorganic development (possibly equatable with underdeveloped countries), development is externally induced and connections between economic, social and cultural evolution are loose. However, although development in the west has been promoted internally and in that sense has been organic, the model of development which has been followed has been, to a significant extent, nonorganic, so that in fact there may be said to be currently two nonorganic forms of development. One, called underdevelopment, is mainly generated from outside. Penetration from the ever-expanding modern system produces a cultural shock with a number of disruptive effects, traditional equilibrium being destroyed without being replaced by another. The second, possibly labelled overdevelopment, is mainly generated from inside through forces linked to the development model and leading to violence over nature, conviviality and inner balance. The two forms of nonorganic development are said to be the basis of the current world crisis in its many manifestations.

A human–centred perspective of development would therefore have to take into account the complex interrelations between different aspects and levels of social life and would need to be internally motivated, rooted in people's culture, values and collective will. This would therefore be organic and not mechanical. This model is no more a utopia than the present mechanical model was in the past; and the present model, because it shows signs of being deeply affected by the overall crisis, is open to change. Just as the present model exists only because of the collective view of the world, so this may change with a change in collective world view, and human–centred development is a practical response to the current crisis. A crucial component is self–reliance which must not imply abandoning of high technology – or backwardness and increased dependency may result. Historically it would seem that there must be a mix between openness and closedness. If there is full mobilization of both human and non–human resources and full reliance on them, then foreign advanced technology, selectively incorporated, might reinforce rather than weaken national autonomy. However, at a national level, physical or social limitations prevent sufficient internal integration to avoid external dependency, and collective rather than national self–reliance seems more likely to succeed.

A human–centered society should satisfy the following criteria:

(1) *Social equity*, with human development equally possible to all members of society.

(2) *Interregional and international equity*, so that, while promoting the human development of its own members, a society should respect the integrity of other societies and not indulge in any activity which might prevent members of those societies from achieving human development.

(3) *Living presence of the future*, so that present generations, in pursuing their own human development, do not endanger that of future generations, whether environmentally or in terms of respecting historical achievements and values that go towards defining cultural identity.

(4) *Sensitivity to the present*, so that (conversely) the development of future generations is not at the cost of imposed deprivation or oppression of the current generation.

(5) *Participation and meaning*, so that society provides not only equity but also a meaning frame for human existence, common feelings and goals being shared and each person having an opportunity to contribute to their realization without loss of personal freedom.

Refs Haq Kadija and Huner Kirdar (Eds) *Human Development* (1986); Miles, Ian *Social Indicators for Human Development* (1985); Streeten, P *Mobilizing Human Potential* (1989); United Nations University *Goals, Processes and Indicators of Development Project* (1983).

Related Sustainable development (#HH3181) Expression of human entitlements (#HH6115) Human resources development (United Nations, #HH0745) Structural adjustment with a human face (United Nations, #HH4009).

♦ HH3635 Silva mind control

Refs McKnight, Harry F *Silva Mind Control Through Psychorientology.*

Related Meditation (#HH0761).

♦ HH3652 Ascetical theology (Christianity)

Perfection in spiritual life — Spiritual theology

Description The emphasis here is on progress in spiritual life through ascetic practices and leading a physically blameless life. Progress is through various stages as the individual develops a full Christian life in perfect charity. Although extremes of ascetic practice and arguments over whether salvation is achieved through good works or through faith have at times made ascetical theology the object of abuse, it has a solid basis in New Testament writings where both Jesus and St Paul ascribe great importance to spiritual effort and exercise. It is the abuse of pious and penitential practices which were the subject of Reformation reaction, and good works are clearly part of Protestant as well as Roman Catholic tradition.

Context This branch of theology, with its emphasis on religious practice as opposed to mysticism, was, after the Reformation, looked upon as the first stage in religious life, with ascetic practices and practice of Christian virtues seen as a necessary first step before spiritual perfection was achieved through the stages of mystical theology. Previously, mystical theology had covered all three stages of the spiritual life (purgation, illumination, contemplation). Currently, *spiritual theology* is seen as a totality comprising the two branches in which neither has superiority.

Related Asceticism (#HH0556) Via purgativa (Christianity, #HH4090) Mystical theology (Christianity, #HH5217).

♦ HH3662 Evolution by prosthesis

Technological development

Description It has been postulated (Dr John McHale) that technological development is part of man's evolution and has, up to now, been as little under his control as other aspects of the evolutionary process. Such development includes means to sense, monitor and control the environment, the evolution of man's *external metabolic system*. It could be said to have commenced with the use of hand–tools to extend the power of the hand, this having extended to the manufacturing process of the automated factory (where the control systems may be considered extra brains). Protective enclosures acting as extended functions of the skin could be interpreted as including clothing, housing, cars etc. Means of transport are not simply extensions of the skin giving protection against environmental extremes but also as extended capabilities of the legs. The development of aeroplanes is "easier" in environmental terms than developing physical wings on the body. The senses have also been extended. Vision has been amplified through microscope, telescope and camera; mankind can now "see" in frequencies beyond the visible spectrum – ultraviolet, infrared, x–ray. Hearing is extended to radio frequencies. Instruments can "feel" more sensitively than the most sensitive skin. Meanwhile, the *internal metabolic system* has evolved to the extent of increasing life expectancy through biophysical developments, from heart valves to transplanted organs.

External and internal development combine in the extension to sensing, monitoring and control of the whole planet – as man's affairs extend he develops the tools necessary, both conceptually and physically, to deal with this expansion. It is postulated that the next stage in evolution is a radical change in social, ethical and economic ways in which society operates. Scientific and technological developments are such that evolution is no longer beyond man's control but a matter of conscious choice as regards socio–ethical decisions, even as to whether the the social and physical environments remain habitable.

Broader Evolution (#HH0425).

Related Technological implications for human evolution (#HH0615).

♦ HH3663 Acceptance of the female principle

Maternal conception of social responsibility

Description This view is held by the branch of the women's movement which does not so much try to expunge all intrinsic differences between men and women as to re–emphasize and revalue feminine qualities. Paternalistic attitudes, preoccupied with the individual and with a view of the world as a succession of things, are superseded by a more socially oriented outlook and one which supports a holistic view more usually associated with women than with men. This leads to a liberating effect on men as they are "released from their emotional cage", as well as liberating women from their fate as second class citizens. Ultimately, the female imperative to protect could save the earth from exploitation, in line with current green and ecological movements.

Related Witchcraft (#HH1909) Deep ecology (#HH2315) Feminine consciousness (#HM2768).

♦ HH3664 Bathing (Taoism)

Washing — Cleaning

Description A relaxation of intensive effort in spiritual practice. This process prevents excess of striving in meditation from resulting in counterproductive impulses such as eagerness or unbalanced force. This process may be related to, or distinguished from, that of "washing the mind". This is the process of cleaning concentration of mental contaminants that may be retained or even exaggerated in the effort to focus the mind on something else. This process may be a stage in any cyclic meditation process.

Refs Chang, Po–tuan *Understanding Reality* (1987).

Broader Alchemy (Taoism, #HH5887).

Related Clearing (#HH1076) Circulation of the light (Taoism, #HH6452).

♦ HH3665 Habit

Description The dulling of perception inherent in habitual activity may be considered a hindrance to development in that it implies lack of consciousness and attention. Even sudden and inhabitual behaviour triggered, for example, by conversion or being "born again", can with time become habitual; or it may in time be erased and the old habits recur. Nevertheless, replacing bad habits with good habits may be the first step to replacing all spiritual habit with full attention and conscious action.

Related Habits (#HH0200) Habit training (#HH0638).

♦ HH3667 Outside paths (Taoism)

Middle–grade low paths

Description There are over 300 fallacious techniques in the 84 schools of sexual intercourse, with 36 methods of culling the female principle. As in the previous grade, technical terms are given biological associations and practices include ingesting sexual material such as placenta and one's own seminal fluid. The outside paths are deviations from the way.

Context In the nine grades of practices which are side tracks and auxiliary methods in Taoism, this is the middle of the three lower grades.

Broader Sidetracks and auxiliary methods in Taoism (Taoism, #HH7004).

Followed by Outside paths (Taoism, #HH4289).

Preceded by False paths (Taoism, #HH3013).

♦ HH3687 Jihad (Islam)

Holy war

Description Revivalist movements within Islam, concerned at social and moral degradation, call for a return to original Islam with rejection of later additions to the religion and promotion of reform through armed force. This holy war or jihad is referred to by some as the sixth pillar of Islam.

Broader Human development (Islam, #HH1799).

♦ HH3698 Developmental psychology

Refs Datan, Nancy et al (Eds) *Life–Span Developmental Psychology* (1986).

♦ HH3710 Self–culture (Christianity)

Self–development

Description The self–love which it is proper for a Christian to feel, to love one's self as one is loved by God, implies care for body, mind and spirit and thus self culture as a moral duty. One is thus a suitable instrument for service of God and of mankind.

Physically, this implies; temperance; proper balance between work and recreation; cultivation of endurance.

Intellectually and imaginatively one should open the mind to humanity and its highest interests. One–sidedness in the spiritual nature should be corrected and the faculty of judgement encouraged, with a wide knowledge of many subjects and of the different methods employed balanced by realistic awareness of one's own limitations. The gifts of nature and civilization, of culture and the arts are all for mankind's use but the individual should be used for spiritual ends under the guidance of the Holy Spirit. In particular, since emotions and conduct are so influenced by imagination, the imaginative faculty should be cultivated with care.

The will should be harnessed by self control so that character may be complete, growing by acts of moral decision. By dedicating the will to goodness, character is restrained and inspired by this one aim, becoming fixed and stable.

The singleness of purpose which is the aim of self–culture results in body, soul and spirit all acting in obedience to the love of God.

Related Self–development (#HH0651) Christian self–love (Christianity, #HH6732).

♦ **HH3755 Super–immunity**
Description Control of thoughts and emotions and "thinking well" has been shown capable of controlling the body's immune system to achieve protection from disease, assistance in healing if disease or injury arises and a high degree of good health which is more than simply absence of disease. Hormone–linked ways of functioning may be too "hot" or too "cold" and lead to a tendency to disease, which can be ameliorated by balancing the functioning at an optimum level. Disease is treated as a challenge from which it possible to learn, although conventional treatment and medical science are not rejected.
Refs Pearsall, Paul *Superimmunity* (1987).
Related Mind cure (#HH1772) Positive thinking (#HH1088)
Psychoneuroimmunology (PNI, #HH1401).

♦ **HH3764 Maharishi effect**
Theory of collective consciousness
Enhancing coherence of collective consciousness ???? **Description** When a required number of individuals participate in the Maharishi Technology of the Unified Field, then there is increased positive change in society and world events. It is postulated that there is a relationship between individual consciousness, collective consciousness and the unified field of natural law. It is also said that there is a relationship between violence and stress in collective consciousness and that world peace can be achieved through creation of coherence in collective consciousness at all levels of society. Research indicates that if one percent of a population participates in the transcendental meditation programme or if the square root of one percent of a population participates together in group practice of the TM–Sidhi programme then there is a measurable and holistic influence of harmony and integration of the entire population. Slightly more than 7,000 persons collectively practising the Maharishi technology of the unified field would raise the level of coherence in collective consciousness so as to provide a stable basis for lasting world peace. This is because, in systems governed by wavelike interactions, the strength of elements interacting coherently is proportional to the square of their number while of those acting incoherently only to their number, so that the coherent influence outweighs the incoherent.
Refs Maharishi European Research University *Maharishi's Programme to Create World Peace* (1987); Orme–Johnson, David W and Dillbeck, Michael C *Maharishi's Program to Create World Peace* (1987).
Related Levitation (#HM8634) Planetary noogenesis (#HH4621)
Transcendental meditation (TM, #HH0682).

♦ **HH3772 Vishishtadvaita** (Hinduism)
Differentiated non–duality — Vaishnavism
Description Distinguished from *advaita*, or utter non–duality, this system of Ramanjua teaches that although the soul, Atman, and God, Brahman, are of the same essence, the individual soul will always maintain its self–consciousness and therefore be able to exist eternally in relationship with God. This is the basis of *Vaishnavism.*
Broader Human development (Hinduism, #HH0330).
Related Advaita vedanta (Hinduism, #HH0518).

♦ **HH3773 Kismet**
Description This concept assumes that divine will, providence or chance predestines one's existence; fate and destiny are outside the hands of the individual. The result could be a negative effect on development as the attitude predisposes to indifference and inactivity.
Related Fate (#HH1362).

♦ **HH3774 Spirituality** (Sikhism)
Refs Cousins, Ewert, et al (Eds) *Post–Classical Hindu and Sikh Spirituality.*
Broader Spirituality (#HH5009).
Related Sikh mysticism (Sikhism, #HH6199) Human development (Sikhism, #HH6292).

♦ **HH3776 Discipline of worship** (Christianity, Judaism)
Description Worship of God is the first commandment in both Judaic and Christian religions. It is therefore of prime importance and comes before service to others. The discipline of worship requires an inner stillness and openness to God at all times. Although possibly accompanied by outward show of ritual and liturgy, true worship is said to be contact between the individual spirit and the spirit of God, such contact making the form of worship unimportant. This is *Shekinah*, the presence of God in everyday experience, permeating the whole of life and making acts of corporate worship more meaningful when they occur. This worship may be total silence, it may be singing, it may be bodily movements such as kneeling or raising the hands, and it may be dancing or shouting with joy. True worship involves the whole being, physical, mental, emotional and spiritual; the person worshipping must be prepared to change and to accept the call to service consequent on worship.
Broader Spiritual discipline (#HH1021).
Related Worship (#HH0298) Presence of God (#HM2961)
Practice of the presence of God (#HH0992)
Religious experience as personal devotion and worship (#HM0561).

♦ **HH3778 Nominalism**
Description This theory holds that thought and speech operate through use of linguistic or other symbols and that there is no difference in kind between thought and speech; thinking is therefore simply the ability to use symbols correctly and does not involve conceptualization or generalization. Universals are held not to exist.

♦ **HH3783 Burmese meditation**
Description A non–ritual means of working through individuality to the universal whole.
Broader Meditation (#HH0761).

♦ **HH3796 Journeying within transcendence - from secular to sacred**
Description Just as Jerusalem represents the sacred and the profane, one is not whole until one comes to experience the opposites within one's self and to accept them, the polarities preserving their separateness in their unity. Then the whole of life can be embraced and not be polarized in one part. The task is to continue in practice of spiritual disciplines while awaiting the message from the deeper self. Then one becomes in one's self a temple of prayer.
Context The fourth section of St John's Gospel, Chapter II 12 to III 1, is related to a stage in the spiritual journey of the individual.
Broader Journeying within transcendence (Christianity, #HH6505).
Followed by Journeying within transcendence - from head to heart (#HH4974).
Preceded by Journeying within transcendence - from water to wine (#HH4219).

♦ **HH3798 Evangelism** (Christianity)
Mission
Description Those outside the Church are the focus of its mission, as they also belong to the people of God and it is the Church's job to be the "sacrament of unity of the whole human race". There is an urgency in a technological age to replace the worship of creature comforts rather than their creator by the truth of God. Men are seen as alienated from God with a need for familiarization with the purpose of life. Christians are bound by their faith to "go out into the world". Through the inspiration of the Holy Spirit there is a need to return to the spirit of early Christians who "went everywhere preaching the word". Knowledge of languages, of cultures, of social structures, of other religions are all required in the Church for this outreach. Caution is necessary in personal interpretation, as human elements may distort the whole truth of the gospel.
Evangelism may be construed as the proclamation of the gospel, with the declaring of the word and the mighty acts of God in salvation and liberation; it may be construed as growth of the Church, the establishment of local churches far and wide; it may be construed as "revival meeting" conversion of the individual; it may be construed as a sharing of faith on a one–to–one basis; it may be construed as not only bringing the individual to the decision to follow Christ but also following up with a programme of instruction so that the individual may then evangelize others. All five methods of evangelism are important and have their place. To avoid conflict between the different methods, a proposed definition is as a set of activities with the goal of primary initiation into the kingdom of God. This primary initiation is the beginning of a complex process. Although the first response may be non–cognitive (stimulated emotions through music and other techniques), leading to some kind of religious decision or experience, the mind must eventually be engaged in cognitive initiation. The issues raised are far–ranging and complicated and have to be worked through as ability and opportunity allow. Since they can never be fully worked through by any individual, however able or long–lived, they have to be based on the basic creed or beliefs of the Church, which provides an intellectual map to help the individual on the way.
Refs Abraham, William J *The Logic of Evangelism* (1989).
Broader Human development (Christianity, #HH2198).
Narrower Crusade evangelism (#HH1098).
Related Initiation (Christianity, #HH6211) Personal salvation (Christianity, #HM6335).

♦ **HH3862 Qi gong**
Qigon therapy — Qigong
Description A mystical traditional art in China, based on breathing exercises, which assists in relaxation. The art has been extended to include a number of more dramatic effects. Practitioners claim to heal the sick, to transport themselves instantaneously around the country, to be unharmed by bullets, to predict the future.
Refs Takahashi, Masaru and Brown, Stephen *Qigong for Health*; Zhang, Mingwu and Sun, Xingyuan (Eds) *Chinese Qigon Therapy.*
Broader Therapy (#HH0678).
Related Ki energy (#HH0620).

♦ **HH3873 Ill–wishing**
Hexing
Description Harm is brought to another by use of magic rites, combined with concentrating the mind on the misfortune to be produced. The effect is similar to cursing, and both may be described as "putting a hex" on someone. The effects may be counteracted by a "hex doctor" or by use of special symbols or amulets.
Related Rites (#HH0423) Cursing (#HH0594).

♦ **HH3875 Purifications** (Buddhism)
Purities — Visuddhimagga (Pali) — Path of purification — Path of purity
Description In Hinayana Buddhism, the roots and trunk or body of understanding are said to be the seven purifications or purities. One studies the aggregates, bases (sense–organs), elements, faculties, truths, dependent origination, which are the soil of understanding. One perfects or fulfils the roots which are the purifications of virtue and consciousness, and develops the "trunk" which is the five purifications of view, of overcoming doubt, of knowledge and vision of what is and what is not the path, of knowledge and vision of the way, and of knowledge and vision. The last purification, that of knowledge and vision, is that of the arahant. Having passed through that path he becomes one of the great ones, all cankers destroyed, bearing his last body, his burdens laid down. He has reached his goal and destroyed the fetters of becoming, liberated with right or final knowledge.
Refs Buddhaghosa, Bhadantacariya *The Path of Purity* (1975); Buddhaghosa, Bhadantacariya *The Path of Purification* (1980).
Narrower Purification of view (#HH2718) Purification of virtue (#HH1865)
Purification of consciousness (#HH2332) Purification by overcoming doubt (#HH1187)
Purification by knowledge and vision (#HH3025)
Purification by knowledge and vision of the way (#HH3550)
Purification by knowledge and vision of what is the path and what is not the path (#HH4007).
Related Understanding (Buddhism, #HM4523).

♦ **HH3876 Panpsychism**
Description Everything that exists is seen a distinct and individual soul or spirit, more complex entities consisting of agglomerates of more simple entities, all aware of the other entities and of the possibility of change and of striving for or against direction of such change for expansion or development.
Related Psychism (#HH1775).

♦ **HH3900 Spiritual discernment** (Christianity)
Discretion
Description Particular practices, religious or otherwise, may be valuable at one time and not at another. This includes praiseworthy practices like fasting, silence and solitude. Discernment should be employed to be sure which practice is valuable at a given time. This discernment is taught through contemplation; the more one gives one's self to loving God in contemplation, the more surely is one aware of which activities and ways of life are in harmony with this love. Discernment is also learned by consulting with a spiritual director. Eventually one will increasingly come to know when to speak and when to keep silent, for example.
Spiritual discernment is nurtured by exercising spiritual disciplines and leads to right use of the gifts of the Holy Spirit. It has to be practised with respect to all experiences arising on the spiritual path. Emphasis may wrongly be given to superficial emotions at the expense of true religious experience. Faults in external behaviour can be traced to inner attitudes. False paths may be followed when evil appears disguised as good. In every case the religious experience may be evaluated by the resulting way in which the person lives; the main signs to look for are those of charity (love).
Refs Ducharme, A *Spiritual Discernment and Community Deliberation* (1974).
Related Discernment (Buddhism, #HM1143) Consolation (Christianity, #HM3502).

♦ **HH3904 Asparsa yoga** (Yoga)
Refs Cole, Colin A *Asparsa Yoga* (1982).
Broader Yoga (Yoga, #HH0661).

♦ **HH3907 Spirituality** (Buddhism)
Refs Cousins, Ewert, et al (Eds) *Buddhist Spirituality*; Cousins, Ewert, et al (Eds) *Buddhist Spirituality.*
Broader Spirituality (#HH5009).
Related Human development (Buddhism, #HH0650).

♦ **HH3908 Personification**
Description A common process of growth and of cultural preservation is that of becoming like one's father, teacher, master or guide through identification with him.
Related Identification with spiritual master (Sufism, #HM3944).

♦ **HH3912 Shakti yoga** (Yoga)
Description This branch of laya yoga emphasizes mastering of energy and the energizing forces of nature.
Broader Yoga (Yoga, #HH0661) Laya yoga (Yoga, #HH0782).
Related Shakti (#HH0629).

♦ **HH3945 Enneagram patterning** (Sufism)
Enneagram of personality types
Description The enneagram is a unique graphic representation of the interrelationship between nine points. In one application, each point is used to represent a personality type or feature, thus providing a cognitive/emotional map of personality as a useful guide to understanding and transforming personalities. Where, for a given individual, one of these features is the central axis around which the delusional aspects of his personality revolve, recognizing it within the context of the enneagram offers the person a major insight into the way in which his life is defective as well as offering a general outline of the ways to work in changing it.
The enneagram is not a map of a static system. The interconnecting lines between the points indicate the potential for dynamic movement. Although individuals may discover themselves to be especially linked to a particular point, the diagram suggests the possibility of a versatility of movement between the points.
The enneagram goes beyond preoccupations with ordinary life and makes it possible to explore the existential and spiritual virtues that could be developed if an individual can recapture the essential life energy that is normally displaced into pathological defences against his real nature. It assists people to appreciate the predisposition that each personality type has for higher human capacities such as empathy, omniscience and love. As such the enneagram can be used to describe the levels of humanity's possible evolution beyond personality. Its power derives from the recognition it offers of how the limited, and seemingly neurotic, normal habits of heart and mind may be used as potential access points into higher states of awareness. It is thus not limited to the pathological perspective.
The nine types mapped by the enneagram are: perfectionist, giver, performer, tragic romantic, observer, devil's advocate, epicure, boss, and mediator. With each type are associated typical defence mechanisms, intuitive styles, pathologies, passions, preoccupations, virtues.
In addition to its use in the understanding personality, the enneagram may also be used as an instrument through which to learn to think triadically rather than linearly and sequentially. Specifically it may be used to analyze patterns of human activity and projects in order to determine what is possible and impossible in human undertakings. In this sense it may be used to determine whether developmental processes are sustainable, or what is required to make them so.
Context The term enneagram was introduced by G I Gurdjieff as a description of a traditional Sufi technique. Its uses were developed by one of his students J G Bennett. Further attention was subsequently drawn to it by Oscar Ichazo. It is a tool of an oral teaching tradition that views personality preoccupations as indicators of latent abilities that unfold during the development of higher consciousness.
Refs Beesing, Maria et al *The Enneagram* (1984); Bennett, J G *Enneagram Studies* (1983); Palmer, Helen *The Enneagram* (1988); Popoff, Irmis B *The Enneagram of the Man of Unity* (1978).
Related Nine-fold knowledge (Systematics, #HM1408).

♦ **HH3987 Subjects for meditation** (Buddhism)
Description Hinayana Buddhism enumerates forty subjects as suitable for meditation. These are:
Ten kasinas (universals) or devices: earth kasina; water kasina; fire kasina; air kasina; blue kasina; yellow kasina; red kasina; white kasina; light kasina; limited or separated space kasina.
Ten kinds of foulness: swollen or bloated; discoloured or livid; festering; fissured or cut-up; gnawed or mangled; dismembered or scattered; hacked and dismembered or scattered; bleeding; worm-infested; skeleton.
Ten recollections: of the Buddha (the enlightened one); of the law (dhamma); of the order or community (sangha); of morality or virtue; of generosity or liberality; of the deities (deva); mindfulness of death; mindfulness of the body; mindfulness of breathing or respiration; of calm or peace.
Four divine states or abidings: lovingkindness; pity or compassion; sympathy or gladness; equanimity.
The four immaterial states or formless realms: sphere of boundless or unlimited space; sphere of boundless or unlimited consciousness; sphere of nothingness; sphere of neither perception nor non-perception.
One perception: that of the abominableness or repulsiveness of food.
One definition or specification: that of the four elements.
They cannot be developed by those hindered by karma, those of wrong view, those hindered by karma-result (birth through no or two moral causes), those lacking faith, zeal or understanding.
Some of these subjects bring "access" only, in other words unification of mind but not absorption. These are the recollections, except for those of body or breathing, and the one perception and one definition. The others bring absorption or ecstatic concentration, giving rise to the four jhanas as follows: the ten kasinas and mindfulness of breathing bring all four jhanas; the ten kinds of foulness and mindfulness of body bring the first jhana; the first three divine abidings bring three jhanas; the fourth divine abiding and the four immaterial states bring the fourth jhana.
There are two ways of transcending or surmounting: of factors and of objects. Those subjects bringing three or four jhanas imply transcending factors, as the second jhana has to be reached in the same object by transcending the previous jhana factors (applied thought and sustained thought), the third by transcending the second, and so on. Similarly in the fourth divine abiding, as this is reached by surmounting the other abidings in the same object. Transcending of objects is as in the case of the four immaterial states, where boundless space is reached by transcending one of the first nine kasinas as object, this is then transcended as object by boundless consciousness, and so on. There is no transcending with the rest.
Another development is the extension or not of the subject of meditation. This refers only to the ten kasinas, when hearing with the divine ear, seeing with the divine eve, knowing the minds of other beings with the mind depends upon the amount of space on which one is intent. In the case of the other subjects, it may be possible to extend the subject (perceiving a large skeleton as opposed to a small) but, as there is no advantage, this is not recommended. Other subjects (formless objects, for example) cannot, of their nature, be extended.
The subjects of meditation may or may not have "counterpart signs" as object, or individual essence as object. There maybe moving objects in the early stage (although the counterpart is not moving). Again, some of the subjects do not occur among deities, in the Brahma world or in formless existence, although all may arise among human beings – but the divine abiding of equanimity and the four formless states are not for beginners. The means of apprehension in the early stages may be sight, touch or hearsay, or some combination of these. A meditation subject may be conditional on another, for example: the kasinas (except the space kasina) are conditions for the immaterial state; three divine abidings are conditional for the fourth, and so on. All cause blissful life, insight and fortunate rebirth.
Particular meditation subjects are more suited to particular temperaments. The ten kinds of foulness and mindfulness of body are suited to the lustful or greedy person. The divine abidings and the four colour kasinas are suitable for one who hates. Mindfulness of breathing is suited to the deluded or the speculative. The first six recollections are suited to the faithful. Mindfulness of death, recollection of peace, specification of the four elements, perception of repulsiveness in food are suitable for the intelligent. Additionally, the kasinas should be limited in the case of the speculative and extended for the deluded. Again, development of foulness assists in suppressing lust and greed, of lovingkindness in abandoning ill will, mindfulness of breath in cutting off applied thought, impermanence in eliminating pride in "I am".
Broader Meditation (#HH0761).
Narrower Kasina meditation (#HH3246) Recollections as meditation subjects (#HH6221)
Divine abidings as meditation subjects (#HH3534)
Immaterial states as meditation subjects (#HH3198).
Related Tenfold powers (Buddhism, #HH2918) Dhyana (Hinduism, Buddhism, #HM0137)
Access concentration (Buddhism, #HM4999) Absorption concentration (Buddhism, #HM0311)
Trances and mental absorptions (Buddhism, #HM2122)
First absorption in the immaterial sphere (Buddhism, #HM2110)
Third absorption in the immaterial sphere (Buddhism, #HM2027)
Second absorption in the immaterial sphere (Buddhism, #HM3043)
Fourth absorption in the immaterial sphere (Buddhism, #HM2051).

♦ **HH3997 State socialist human development**
Description Like liberal capitalism, state-socialism is externalist. At the core of the model is the importance of material achievement, in particularly in terms of applying technology. Unlike liberal capitalism, however, it rejects individualism in favour of strengthening and growth of the state and achievement of a communist society, intervening in personal areas such as choice of work-place, residence and studies. As in liberal capitalist societies there is a growing dominance of secondary (rational, bureaucratic) organizations, but here power is concentrated in the Communist Party. State socialism is thus a model where externalism and bureaucratic collectivism are blended.
Despite official condemnation of the competitiveness, crude materialism and lack of social solidarity of liberal capitalism, these are also features of life in state socialist countries. Again, like liberal capitalism, state socialism has a mechanical, "outside-in" view of the human being. Also, bureaucracy and authoritarianism inherent in the system severely restrict full expression and realization of human potential.
Context One of four current models of human development, that described as collective external.
Refs Mallmann, Carlos A and Nudler, Oscar (Eds) *Human Development in its Social Context* (1986).
Related Communalist human development (#HH5522)
Liberal humanist human development (#HH4033)
Liberal capitalist human development (#HH6122).

♦ **HH3998 Complete personality**
Description Rooted in the heart rather than the intellect, the complete personality is the acceptance by the individual's free decision that he is a person, that is: that life is a dialogue ordered to mystery; that he accepts freedom, duty, responsibility, sinfulness, the individuality of his neighbour, pain and death.
Related Wholeness (#HM2725) Personality development (#HH0281).

♦ **HH4000 Consciousness ascension stages game** (Buddhism, Tibetan)
Description The mind is led to different domains of awareness by words and pictures, by programmed action (ritual) and guided behaviour. Games have the characteristic of both programmed action (or ritual) and guided behaviour. Some games can symbolize the phenomena of life, referring also to differing kinds of knowledge and levels of perception of reality. The Tibetan Consciousness Ascension Stages games are played on variously designed boards, some with 72 squares, others with many more. The games can be used to describe one's past, present and future existences, or to teach a view of the universe in which the interrelatedness and transitory nature of phenomena are put forth, and the way to escape (nirvana) from the samsara (wheel of rebirth).
Refs Tatz, Mark and Kent, Jody *Rebirth* (1977).
Broader Games (#HH0705) Maps of the mind (#HH8903)
Stages of spiritual life (#HH0102).
Narrower Way of the vajra masters (#HM3090)
Emanation-body awareness (Buddhism, #HM2340)
Form-realm consciousness (Buddhism, #HM2257)
Desire-realm consciousness (Buddhism, #HM2733)
Meditation way of the bodhisattvas (Buddhism, #HM2769)
Supreme heavens in the ascension stages game (#HM3214)
Masters of wisdom in the ascension stages game (#HM3066)
Religious traditions in the ascension stages game (#HM3341)
Sutra paths of accumulation and application in the ascension stages game (#HM2962)
Tantric paths of accumulation and application in the ascension-stages-game (#HM2848).
Related Lila (#HM2278) Buddha-consciousness (Buddhism, Tibetan, #HM2735).

♦ **HH4001 Disruptive thoughts** (Christianity)
Description Thoughts inevitably arise when practising a discipline such as centering. They appear disruptive but are an integral part of the healing and growing process initiated by God. Rather than looking on them as painful distractions, one should see the broad perspective which includes interior silence and thoughts, the latter just as valuable in the process of purification as are moments of profound tranquillity. Thomas Keating defines five types of thought which come down the stream of consciousness when one starts to quieten the mind.
– *Wool gathering*: These are superficial thoughts brought up by the imagination because of its natural inclination to be in perpetual motion. They should simply be accepted and not given attention so that the attention may be as undivided as circumstances allow.
– *Emotionally attractive thought*: Here something in the thought excites the interest or curiosity. To return to loving attention on God some action is necessary. Here the method of *centering* implies resting the attention on the sacred word without annoyance at or involvement in the thoughts.
– *Insights and breakthroughs*: These are spiritually very tempting as the thought arises to attend to the insight so it will not be forgotten. That thought if dwelled upon breaks the refreshment of interior silence. The discipline or self denial in allowing the insight to be experienced without thinking about it is an asceticism attacking the roots of attachment to the false self.
– *Self-reflection*: Settling into deep peace, free from particular thoughts, the desire arises to relect upon what is happening. The choice is to reflect upon what is going on or to let the experience go. If the latter is chosen, then one goes deeper into the interior silence. If the former, then one comes out of the inner stillness and has to start again. this is because reflection is one step back from experience – reflect on the experience and the experience itself is lost.

– *Interior purification*: Meditation or prayer which transcends thinking starts the process of dynamic interior purification. Deep-rooted tensions are released in the form of emotionally charged thoughts. If one returns to the sacred word then the undigested psychological material of one's lifetime is gradually evacuated, early childhood programmes for instinctual happiness are dismantled and the false self gives way to the true self.
Refs Keating, Thomas *Open Mind, Open Heart* (1986).
Related Centering (#HH0527).

♦ **HH4003 Feminine spirituality**
Refs Carmody, Denise L *Seizing the Apple* (1984); Conn, Joan W (Ed) *Women's Spirituality* (1986); Faricy, Robert *The Lord's Dealing* (1988); Giles, Mary E *The Feminist Mystic and Other Essays on Women and Spirituality* (1982); O'Brien, Theresa K (Ed) *The Spiral Path* (1988); Ochs, Carol *Women and Spirituality* (1983); Shorter, Bani *An Image Darkly Forming* (1987); Spretnak, Charlene *Politics of Women Spirituality* (1982); Starhawk *The Spiral Dance* (1979); Starhawk *Dreaming the Dark* (1982).
Broader Spirituality (#HH5009).
Related Witchcraft (#HH1909) Psychological development of women (#HH1976).

♦ **HH4006d Practice of presence of mindfulness** (Buddhism)
Satipatthana (Pali)
Description This path of observational analysis is compared with (and opposed to) that of following enstatic practices. The ultimate aim is attainment of final deliverance from suffering. The method implies: direct confrontation with actuality; merging of everyday life with meditative practice; transcending conceptual thought by direct observation and introspection; emphasis on the here and now. The presence of mindfulness is cultivated through four foundations: body contemplation on the body; feeling contemplation on feelings; mind contemplation on mind; mind object contemplation on mind objects. This is made possible through giving up: fondness for activity (busy-ness), talking, sleeping and company; lack of sense control; immoderate eating. The middle teaching of dependent origination transcends all concepts of monism, pluralism and dualism. Once the disciple has received information on the way and the goal he has two rules for successful spiritual practice – begin and continue (endure). Eventually life should become one with the spiritual practice and practice become full-blooded life. The little things of life become teachers of great wisdom, gradually revealing their own immense dimension of depth.
Refs Engler, J *Buddhist Satipatthana-Vipassana Meditation and An Object Relations Model of Therapeutic Developmental Change* (1983); Nyanaponika, Thera *The Heart of Buddhist Meditation* (1972); Nyanaponika, Thera *The Power of Mindfulness* (1968).

♦ **HH4007 Purification by knowledge and vision of what is the path and what is not the path** (Buddhism)
Purity of knowledge and discernment of the right path and the wrong path — Maggamagga-nanadassana-visuddhi-niddesa (Pali)
Description It is first necessary to apply one's self to the inductive insight from contemplation or comprehension of groups, where states are differentiated into past, future and present (overcoming doubt). Having begun insight, knowledge of what is and what is not the path arises during full understanding as investigation.
Imperfection is avoided when the skilful meditator defines and examines illumination or whatever has arisen, understanding that it is impermanent, formed, subject to destruction, fall, fading away, and so on. Or he may see that it is not self being taken as self. Seeing the illumination or whatever has arisen as not mine, not I, not myself, there is no wavering. Seeing them as states that arise on the way but not the path itself, the meditator keeps to the course of insight knowledge, free of imperfections, that is the path.
Context One of the five purifications or purities which constitute the trunk or body of understanding according to Hinayana Buddhism, the others being: purification or purity of view; purification or purity of overcoming doubt; purification by knowledge and vision of the way or purity of knowledge and discernment of the middle way; purification by knowledge and vision or purity of knowledge and discernment.
Broader Nanadassana (Buddhism, #HM6502) Purifications (Buddhism, #HH3875).
Related Principal insights (Buddhism, #HM3630) Imperfection of insight (Buddhism, #HM9722) Full understanding as investigation (Buddhism, #HM4552).
Followed by Purification by knowledge and vision of the way (Buddhism, #HH3550).
Preceded by Purification by overcoming doubt (Buddhism, #HH1187).

♦ **HH4008 Active meditation**
Refs Japikse, Carl and Leichtman, Robert R *Active Meditation* (1983).
Broader Meditation (#HH0761).

♦ **HH4009 Structural adjustment with a human face** (United Nations)
Description The accent on restoring macro-economic equilibria has shifted discussion from long-term goals of human development to short-term concerns in which structural adjustment programmes are accompanied by austerity measures which have resulted in neglect of human resources development and human development, and in increased poverty and inequality. Adjustment with a human face as promoted by UNICEF combines the promotion of economic growth with the protection of the vulnerable and macro-economic adjustment.
Why focus on adjustment rather than development ? Adjustment policy is the dominating economic preoccupation for setting the frame and constraints within which all other economic and development issues have to be considered. The issue should not be adjustment or growth, but adjustment for growth. The need for more growth-oriented adjustment policies has been widely recognized. The UNICEF study recognizes that the primary cause of the downward economic pressures on the human situation in most of the countries affected is the overall economic situation, globally and nationally, not adjustment policy as such. Indeed, without some form of adjustment, the situation would often be far worse. However, many past adjustment policies have been inadequate. There is a recognized need for broader approach to adjustment – satisfactory adjustment involves restructuring the economy so that major imbalances are eliminated at a desirable level of output, investment and human needs protection, while keeping the economy in shape for future growth and sustained development.
Refs Cornia, Giovanni, et al (Eds) *Adjustment with a Human Face* (1987).
Related Human-centred development (#HH3607).

♦ **HH4012 Admission of sin**
Mortal sin
Description Sin can only be committed as a free act. In mortal sin there is rejection of the creator's will for his creation and of God's will to communicate himself to the creature in grace. Thus the sinner contradicts his own nature and the purpose of his freedom which is to love God. The sinner then hides the fact of sin even from himself – *repression*. The admission of sin is then the first effect of the redemptive revelation and grace of God.
Related Fall of man (#HH3567) Discipline of confession (Christianity, #HH0825).

♦ **HH4019 Steps to enlightenment** (Buddhism)
Bodhi — Wu (Zen) — Byan-chub (Tibetan) — Satori
Description Enlightenment is freedom from suffering caused by selfishly trying to have reality the way one wants it rather than the way it is.
In order to understand the world, fully and compassionately, without disfiguring it with one's own desires, expectations or habits, one follows the progress of the Gautama under the Bodhi tree. First there is the struggle with the demon Mara to control entrapping emotions – when the temptations, desires and fears which normally delimit one's identity are confronted and defeated. Secondly there is entering the four levels of dhyana or jhana, until there is mastery of all stages of meditative concentration and ease in moving from one level to another. Through the subsequent series of insights there comes perception of the six types of extraordinary knowledge, abhijna or abhinna. This perception enables understanding of the nature of suffering. Thirdly comes meditative analysis of one's life, to understand how the total of past actions determine the present. There is understanding that one is one's self responsible for one's own personality. Gautama is said to have remembered all his previous lives in order and seen how they led to the present one. Fourthly there is the ability to understand other people's idiosyncracies and predicaments in the same way as one understands one's own. They create their own problems although they may not be aware of the fact, and one can respond fully and compassionately to them. In the divine vision of the Gautama, he saw all former lives of all beings. Finally, having destroyed the source of asrava or asava, the poison of the mind, one comes to know the four noble truths. Having gained this insight, Gautama saw that suffering comes from desiring things the way one wants them rather than the way they are. Nirvana is the "blowing out" of this desire. Having reached such enlightenment, Gautama became Buddha.
Theravada recognizes three levels of enlightenment: savakabodhi, paccekabodhi and samma-sambodhi. It also relates enlightenment to the extinction of desire in cessation, *nirodha*, the third noble truth. There is freedom from ignorance (avidya, avijja) and craving (trisna, tanha). Other related states are cessation of passion, hatred and illusion (ragaksaya, dosaksaya, mohaksaya) and unconditioned existence (asamskrita).
Mahayana describes enlightenment as seeing everything as it is, the "suchness" of all. This may be to see that they are empty of essential or substantial being. Being what they are and nothing more or less, they cannot be characterized. Awareness of all things and perfect enlightenment – *sambodhi* – is wisdom which, if cultivated, leads to understanding the emptiness of the world.
Broader Human development (Buddhism, #HH0650).
Narrower Sambodhi (#HM7195) Savakabodhi (#HM0680) Paccekabodhi (#HM3955) Samma-sambodhi (#HM7091).
Related Satori (Zen, #HM2326) Enlightenment (#HM2029) Nirvana (Buddhism, #HM2330) Suchness (Buddhism, #HM2435) Thatness (Buddhism, #HM8003) Ragaksaya (Buddhism, #HM6093) Dosaksaya (Buddhism, #HM5676) Mohaksaya (Buddhism, #HM6124) Asamskrita (Buddhism, #HM6504) Dhyana (Hinduism, Buddhism, #HM0137) Four noble truths (Buddhism, #HH0523) Path of enlightenment (Buddhism, #HH5645) Extraordinary knowledge (Buddhism, #HH1809).

♦ **HH4020 Freedom** (Christianity)
Description Christian freedom is responsible living in a the world as it actually is, being sensitive to the deeps of human life, daring to act when necessary, and living a life given on behalf of all men.
The free person is free to be lucid about the horror and inseparable glory of the world; it is tragic, transitory, the only one there is and the one of choice. This free person is free to be lucid about the neuroses, biases, illusions imbedded in his own consciousness and about his own greatness. He is lucid about those around him; they too are neurotic, biased and filled with illusions. They too are filled with unique greatness and possibility. He is lucid about his unconditional acceptance in Christ and assumes responsibility for being accepted.
The free person is free to be sensitive to every dimension of life, passionate and disinterested, engaged and detached. The free person is free to be sensitive to all of creation, listen to history, witness pain and suffering and participating in the chaos of his time. He is free to be sensitive to the specific situation in which he finds himself, free to affirm and create life in that situation, as it is. He is sensitive to the mystery, depth and greatness of the situation, its past and its potential. He is detached in his sensitivity. He is nonchalant about his own significance. He is an actor playing the role required to create a more human present and future. He is a lover giving all he has to this moment. He is a conqueror, requiring everything from himself to achieve victory without having to have the fruits of winning.
The free person is free to be exposed to history and neighbour through the deeds he performs. The free person abandons the need for security and risks making decisions. He dares to change, to act out his care. He freely accepts the radical ambiguity of decisions. He knows there is no certainty on his part about the outcome of his decisions. At the same time he is free to use his critical intelligence to make choices. He is free to bring all the wisdom he can to a situation. He builds models using data coming from the situation, the past, his colleagues and his own intuitions and he acts. He freely accepts the consequences of his deeds. He knows there is no ultimate justification for his actions. They are the basis of the next decision.
The free person is free to be disciplined. The free person is disciplined in being what he is, lucid, sensitive and exposed. He freely imposes on himself the decision to live a life given to history in this freedom. He freely surrounds himself with the mechanism to remind himself of his decision to be free. He arranges his life so that others of freedom are around him to remind him of his decision; he is of the church. He exposes himself to the symbols of the church and others which enable him to be this freedom. He exposes himself to the means of grace that he might never forget his acceptance, his freedom and his obligation. Finally, he is free to be judged by history, he renders his deeds up to history for what they are, his deeds. He is accountable to God until death.
Related Freedom of choice (#HH0789).
Preceded by Grace (#HH0169).

♦ **HH4022 Quaker meditation** (Christianity)
Refs Steere, Douglas V *Quaker Spirituality*.
Broader Christian meditation (Christianity, #HH5023).

♦ **HH4032 Mysteries and religion**
Esoteric and exoteric practice
Description The exoteric has been defined (René Guénon) as the individual human interest, while the esoteric is the reaching beyond the individual to embrace superior states of being, aspiring even to the supreme state, transcendence beyond all possibility of comparison. Whatever the formal framework of a religion, no authentic and integral tradition can exist on solely the exoteric and collective. Nor can it exist on esoterism. The one is religion without a heart, the other a wholly subjective and quasi-abstract spiritual life, all heart and no body. Because of the inclusiveness of Christianity, its formal framework is lacking in the esoteric. Similarly, some neo-Vedantist and other movements lack the exoteric.
Islam defines the two areas clearly – the law (shariah) being the exoteric and spiritual vision, the tasawwuf of the Sufi, being esoteric. Judaism divides the two symbolically by the veil in the temple – on the one side is the religion as practised by all, on the other the empty holy of holiness, entered

by the naked priest. This veil is rent by Christianity, it was never there in Buddhism. In Christianity this has resulted in a paucity of esoteric expression, in Buddhism a regret for the necessity of the exoteric. Nevertheless, in Christianity and in Buddhism (Tibetan, Zen, Jodo) initiatic ceremonies can be taken at both exoteric and esoteric levels. In Buddhism, particularly, this is clearly delineated.
Refs Pallis, Marco *The Veil of the Temple* (1986); Steiner, Rudolf *Esoteric Development* (1982).
   Related Initiation (#HH0230)   Sharia (Islam, #HH0738)   Physical education (#HH0575)
Esoteric development (#HH6902)   Initiation (Christianity, #HH6211)
Human development through religion (#HH1198).

♦ **HH4033 Liberal humanist human development**
Description In contrast with the two dominant systems, liberal capitalism and state socialism, liberal humanism is internally orientated. Emphasis on external achievement in human development is replaced by the ability (a) to experience the world in non-stereotyped, intense ways and (b) to be in contact with, recognize and deal non-defensively with deep feelings and drives. The former ability allows the world to be perceived without rubricizing it as occurs under liberal capitalism and state socialism. The latter ability may be looked on as the denied or largely underestimated ability to develop a holistic understanding of reality, without which the unity of the human being is destroyed, being dissociated into body and mind, present and future, reason and emotion. Humanism look on the ability to grasp meaningful wholes without analysing component parts as not only acceptable at the pre-scientific level but also needing to be integrated with analytical capabilities to allow the attainment of an extended awareness. Human development occurs in stages based on previous stages and emerging from them, self actualization only being expressed and satisfied after more basic needs are first satisfied.
Like liberal capitalism, liberal humanism emphasizes the individual as its central feature. The individual/internal approach partly explains why no viable human development model for a real complex society has come out of the system – humanists are interested in the micro-psychological level and do not relate it to the macro-social, they do not question the liberal social system as a whole.
However, some humanists really care about developing a comprehensive humanist social alternative, gradually disseminating from small groups and communities based on humanist ethics, although it is hard to see how this could succeed in bringing about social change without a simultaneous transformation of the power structures which penetrate everyday life-styles. A workable humanist alternative might emerge if liberal humanism were to be integrated with actual social forces like the green movement (socialist humanism).
Context One of four current models of human development, that described as individual internal.
Refs Mallmann, Carlos A and Nudler, Oscar (Eds) *Human Development in its Social Context* (1986).
   Related Communalist human development (#HH5522)
State socialist human development (#HH3997)
Liberal capitalist human development (#HH6122).

♦ **HH4066 Ondinnonk**
Innermost benevolent desires of the soul
Description The belief of the Iroquois Indians that the soul makes known its natural desires through dreams parallels modern psychology. Here the impulses of the soul are not the aggrandizement of desires but benevolent, life centred and health-giving, in other words a good kind of *id*. If these desires are thwarted then the soul revolts against the body and makes it sick. If the good, altruistic impulses of human nature are listened to, then development occurs in unexpected ways.

♦ **HH4087 Dream yoga** (Buddhism)
Lucid dreams
Description Just as the shaman directs his journey through lower, middle or upper worlds, the lucid dreamer directs his dream and visualizes travelling through other worlds or this world, meeting other beings, exploring, questioning and learning. The technique is used in Tibetan dream yoga where there is a journey to various heavenly realms to seek teachings from the Buddhas.
   Related Dreams (#HM2950)   Lucid dreaming (#HM3618)   Shamanic journey (#HM6120).

♦ **HH4089 Meditative prayer**
Description As opposed to contemplation, this form of prayer proceeds according to discursive reason. Having previously delimited the subject matter of the meditation, and made a preparatory prayer, the subject is represented in the imagination by application of the senses, the whole man is turned to God in unrestrained interior prayer. For much of the individual's religious development, meditation is considered indispensable for really interior and personal prayer.
   Related Christian meditation (Christianity, #HH5023)
Contemplative intuitive meditation (#HH0816).

♦ **HH4090 Via purgativa** (Christianity)
Description Although in its extreme form, this way of Pseudo-Dionysius is criticized as nurturing obsessively negative attitudes to the flesh or physical side of life, and to dualism, where the flesh is sinful of itself and at war with the spirit, the prolonged and strenuous self-discipline of this first stage of the threefold path (purgation, illumination, mystical contemplation) is an intrinsic part of the process of spiritual development. Obstacles to divine grace are removed and an inner freedom is achieved which allows the Christian to respond lovingly to Christ's call.
Broader Triple way (Christianity, #HH0631).
Related Asceticism (#HH0556)   Via negativa (#HH1171)   Via positiva (#HH6222)
Purification (#HH0401)   Self-discipline (Christianity, #HH2121)
Mystical theology (Christianity, #HH5217)   Ascetical theology (Christianity, #HH3652).
Followed by Via illuminativa (Christianity, #HH6030).

♦ **HH4098 Spiritual rebirth** (Christianity)
Second birth
Description This term has been used to describe the process of conversion, when the convert is born into a spiritual life. It is what is meant by being born not of the flesh but of the spirit. When Christianity was largely a faith passed on to adults, who were converted and baptised, the rite of baptism was a symbol of the process of dying to the old life and being born again, purified, as one came up out of the water. Infant baptism and the sprinkling with water rather than total immersion have changed this, so that there is no distinct single initiation; the sacrament of confirmation symbolizes the baptism with tongues of fire that the coming of the Holy Spirit is described as being in the Acts of the Apostles. Spiritual rebirth is a personal experience of conversion symbolized by these rites, the encounter with God that brings forgiveness of sins and a new life. A key text is from St Paul's epistle to Titus III 5: "according to his mercy he saved us, by the washing of regeneration, and renewing of the Holy Ghost".
Narrower Born-again (#HH0576)   Twice-born (#HH0715)
Related Baptism (#HH0035)   Rebirthing (#HH1330)   Conversion (#HH0077)
Regeneration (Christianity, #HH0618)   Spiritual rebirth of the ego (#HH0698)
Ideological rebirth (Brainwashing, #HM2843)   Initiation as spiritual rebirth (Yoga, #HH3465).

♦ **HH4099 Creative psychology**
Description A synthesis combining physiology and psychology, eliminating the artificial barrier between them which has been created by dualistic theories which separate mind and matter.
Refs De Ropp, Robert S *The Master Game* (1969).

♦ **HH4100 Via unitiva** (Christianity)
Description Contemplation, the *via illuminativa*, is a step on the path to full union, the *via unitiva* of Christian mystical teaching.
Broader Triple way (Christianity, #HH0631).
Related Mystical contemplation (#HM2710).
Preceded by Via illuminativa (Christianity, #HH6030).

♦ **HH4102 Deprogramming**
Icebox effect
Description An individual who has been brainwashed, for example by some cult, is deprogrammed through a system of interrogation. Questions he or she has not been programmed to answer force the mind, after initial confusion and frustration, to make its own decisions. Thoughts and questions suppressed by brainwashing now surface and are addressed, so that instead of permanently centering on cult needs the individual faces up to conflicts normal to his or her development. This latter is likened to "defrosting" a mind "frozen" by indoctrination.
Related Floating (#HM3903)   Backsliding (#HH2321)
Destructive manipulation (#HH1020).

♦ **HH4110 Journeying within transcendence – from extraordinary to ordinary** (Christianity)
Description After his resurrection, Christ appears to his disciples in ordinary situations, when they receive a healing mission which will also heal them, in Thomas' case suspicion replaced by faith. Spiritual integration occurs as the division of the split soul is overcome, resulting in (Tad Dunne) a capacity to move meaningfully among the realms of common sense, theory, method, religious transcendence and story.
Context The eighteenth section of St John's Gospel, Chapter XX 1–20, is related to a stage in the spiritual journey of the individual.
Broader Journeying within transcendence (Christianity, #HH6505).
Followed by Journeying within transcendence – from agape to philo (#HH5908).
Preceded by Journeying within transcendence – from humiliation to exaltation (#HH2329).

♦ **HH4111 Intoxication**
Refs Siegel, Ronald *Intoxication* (1989).

♦ **HH4112 Contrition**
Attrition
Description Whether imperfect, as attrition, when the individual is moved by God's justice to turn from sin, or perfect, as true contrition, when he is moved by love of God to turn from sin, contrition is the conversion of a sinner to God. By his nature, man gives his heart to either the true God or to something he has deified. Rejection of the latter for the former is true conversion, with contrition or sorrow for the moral worthlessness both of past deeds and for the attitude which provoked them whether or not the purpose or result was good.
Related Conversion (#HH0077)   Metanoia (Christianity, #HH0888).

♦ **HH4188 Journeying within transcendence – from object to subject**
Description Physical hunger and its satisfying exemplify a spiritual hunger. There is development from a quest for some magical answer to problems of existence, which is sought by the powerful "I", to the presence of another "I", the awareness of which grows through intuition and wonder but which cannot be known, and which draws one onward in the journey. There is no choice as to how this greater "I" is accepted. Failure to accept is the negative, hostile way of the power complex, seeing the partial as the totality; acceptance means trusting intuition, remaining in the question "are you God ?".
Context The eighth section of St John's Gospel, Chapter VI 1 to VII 1, is related to a stage in the spiritual journey of the individual.
Broader Journeying within transcendence (Christianity, #HH6505).
Followed by Journeying within transcendence – from discussion to decision (#HH3366).
Preceded by Journeying within transcendence – from isolation to imagination (#HH2498).

♦ **HH4201 Psychospiritual therapy** (Christianity)
Christotherapy
Description Although psychotherapy and spiritual guidance are usually thought of as separate although related activities, the functions of the two may be combined. Both deal with spiritual and psychological aspects of the person (neither can be artificially excluded when dealing with the other) although emphasis may be different. Such a combination is contrasted with psychospiritual dualism, psychological reductionism and spiritual reductionism which seek to separate the parameters or ignore one of them.
One specific example of the combination is in *christotherapy* developed by Bernard Tyrrell. The argument is that Christ is the true healer, not the therapist, and that His healing is intended for the whole person. Christ is present in all healing and growth, whether or not His presence is recognized. Christotherapy actively acknowledges the presence of Christ in the healing process and uses the healing made available through His life, death and resurrection.
The system integrates healing and growth as principles present in Christian revelation with the Spiritual exercises of St Ignatius and the techniques of secular psychotherapy. The first goal is *reformation*, with awareness of one's position before God, the reality of rebellion and awareness of the need for the grace of Christ's redemption. This is followed by *conforming*, the turning to Christ which follows turning from sin. Conforming the self to the mind of Christ brings a new disposition of heart and mind, allowing growth and deepening in God's love. Then there is *confirmation*, or the affirming of death to sin and life as a new creation in Christ. The initial turning from sin is confirmed, the individual seeks to be one with Christ in His suffering and self-sacrifice through service to others. Finally, *transformation* moves from identifying with Christ in His death to contemplating Him in His glory. The Holy Spirit empowers a fuller turning to Christ and transformation in His image.
The therapist uses the methods of: (1) *Existential loving*, which affirms the good and immense value of the other person beyond neurosis and sin, and attempts to see the other person as he or she is seen by God. This is a loving of the whole person and must involve liking the person. (2) *Existential diagnosis*, which is a discovery of the existential meaning of whatever difficulties are affecting the person seeking help. Here insight is sought into the central beliefs, values and assumptions of the person's life, these generating psychospiritual and physical illness or wholeness. With the guidance of the Holy Spirit, the aim is to identify the basic errors about what fulfilment is and how it is realized. (3) *Existential appreciation*, which, so as not to focus only on

problems, involves identifying and reinforcing the qualities in the person that enhance life. (4) *Existential clarification* which, through prayer, meditation, confrontation, encouragement, etc, attempts to provide the best conditions for God's gift of existential understanding leading to replacement of erroneous ways of approaching life with truthful ways and to enhancement of those qualities already reflecting truth and righteousness. (5) *Mind fasting*, which is used throughout the therapy and is practised by both therapist and subject of therapy with the aim of becoming a permanent habit. There is focusing on a problem or negative experience and prayer for discernment, leading to insight and commitment to actions associated with this insight. (6) *Spiritual feasting*, which is like mind fasting but with the focus on positive experience.
Refs Benner, David G *Psychotherapy and the Spiritual Quest* (1988); Tyrrell, Bernard J *Christotherapy II* (1982).
Related Psychotherapy (#HH0003) Soul care (Christianity, #HH1199) Spiritual reductionism (#HH1981) Psychospiritual dualism (#HH4455) Psychological reductionism (#HH2008).

♦ **HH4202 Gratitude** (Christianity)
Description Christians believe that all they have and are come from the grace of God, that they live through God's grace. The appropriate response to the gift of grace is gratitude (both words come from the same Greek root). Gratitude for past benefits is seen as a more worthy motive for moral life now than future punishment or reward. It is a virtue linked to justice. Although some may consider it an "optional" virtue, which has to arrive spontaneously and cannot be forced, others look on gratitude as a moral obligation from recipient to donor. This moral obligation is significant in the personal relationship it implies between recipient and donor and also in the larger cultural setting of obligation as a practice of social morality.
Because a gift, although given, still implies links with the giver and the intention in giving the gift. It must be followed not only by grateful conduct towards the giver but also by grateful use of the gift in accordance with the donor's intentions. These duties are not in a position to be demanded by the donor and the recipient may choose how and whether to fulfil implied obligations.
The enrichment of life by persons or powers outside the individual logically implies generosity to others on the part of the recipient. This enrichment arises when the gift is given as expressing generosity and benevolence. As an instrument of domination it elicits resentment.
Related Grace (#HH0169) Justice (#HH2117).

♦ **HH4209 Liturgical prayer** (Christianity)
Description The liturgy takes the individual as the sinner he is, seeking the mercy of God. Although the words spoken out loud may not necessarily express the feeling at that moment in the heart, personal sentiments are put aside in order to be united with the thoughts and desires of the group as they are expressed in the prayer. These thoughts and desires then become the individual's own. He or she is thus raised above the individual level to that of the mystical Christ.
Refs Merton, Thomas *Spiritual Direction and Meditation* (1975).
Broader Prayer (#HH0185).
Related Mental prayer (Christianity, #HH8672) Corporate worship (Christianity, #HH2768).

♦ **HH4210 Sannyasa** (Yoga)
Renunciation
Description Through abandoning of home, social position and the bondage of the world, and by practising severe austerities through the following of some form of yoga, the individual achieves *jivanmukti* – liberation while still alive.
Related Yoga (Yoga, #HH0661) Renunciation (#HM3118) Mythic yoga (Yoga, #HH3405) Jivanmukti (Hinduism, #HM3890) Human development (Hinduism, #HH0330).

♦ **HH4213s Supernaturalism**
Description This may be defined as the belief that there is something, whether magical or divine, which is outside nature and in contemplation of which a sense of awe arises. It implies a power, whether evil or good, and those dealing with the power (magicians, priests) are also invested with this sense of awe, which is channelled and satisfied in traditional activity. This belief in otherworldly reality is intrinsic to all religion, whether the division between nature and supernature is considered as a broad gulf, or whether there is no real distinction between the two, the supernatural pervading all the natural.
Related Mana (#HM2686) Sense of sacredness (#HM2302) Human development through religion (#HH1198).

♦ **HH4216 Saum** (Islam)
Fasting
Description During the month of Ramadan, all men and all women not bleeding from menstruation or childbirth keep a fast which, between the hours of sunrise and sunset, requires no eating or drinking, no intentional vomiting, and sexual abstinence. Talk should not be indecent or slandrous and there should be no gossip. Despite the regulatory nature of the fast it is in fact an act of personal piety.
Context One of the five pillars of Islam.
Broader Human development (Islam, #HH1799).
Related Fasting (#HH0011).

♦ **HH4219 Journeying within transcendence – from water to wine**
Description This is the stage of transformative change. What changes are occurring in one's life and how does one respond to them ? There may be signs, but while signs reveal God's glory to the believer they hide it from those who do not believe. So do these changes simply touch the surface, or do they indicate a process of individuation, a purposive passage of time where there is conscious coming to knowledge and possession of the deepest self ? There is attraction in the power of change which one wants for one's self, the black magician in the psyche battling for control and entrapping love. The miracle of change is the transforming of power into love.
Context The third section of St John's Gospel, Chapter II 1–12, is related to a stage in the spiritual journey of the individual.
Broader Journeying within transcendence (Christianity, #HH6505).
Followed by Journeying within transcendence – from secular to sacred (#HH3796).
Preceded by Journeying within transcendence – from call to mission (Christianity, #HH3352).

♦ **HH4221s Psychotechnology**
Change in awareness
Description Many systems, both ancient and modern, are aimed at deliberately producing changed consciousness. These techniques often involve some form of meditation and may be based on the use of scientific instruments (such as biofeedback methods), bodily movements (flow experience, dance, derwishes), sounds (chanting, mantras) and light (candle flame). Other techniques involve breathing exercises, sensory deprivation or overload, guided or mutual help as in many group therapies, mental aids such as koans.
Related Consciousness raising (#HH0427).

♦ **HH4223 Human development through atheism**
Description Life for the practising atheist is lived without significant influence of any possible theoretical recognition of the existence of God. Militant atheism, in fact, considers that the doctrine of atheism should be preached for the good of mankind and that there is no infinite and absolute reality distinct from man. Religion is thus a destructive aberration.
Related Human development through theism (#HH4663).

♦ **HH4233 Mutual trust**
Mutual confidence
Description In this state, barriers preventing communication and dividing families, institutions and peoples are broken down, allowing free human contact and exchange.

♦ **HH4269 Synchronicity**
Description Links are formed between apparently unrelated events in the phenomenal world, so that a need is fulfilled by some agency outside the individual's conscious control. This relationship is exploited in the throw of the die in *Lila*, the game of knowledge, and in looking for advice to the I Ching.
Refs Bolen J S *The Tao of Psychology* (1982); Bolen, Jean S *The Tao of Psychology* (1979); Peat, David F *Synchronicity* (1987).
Related Lila (#HM2278) I Ching (Taoism, #HH0004).

♦ **HH4289 Outside paths** (Taoism)
Upper-grade low paths
Description These paths include 400 recipes for material alchemy to make potions to be ingested. They are outside paths, deviations from the way.
Context In the nine grades of practices which are side tracks and auxiliary methods in Taoism, this is the highest of the three lower grades.
Broader Sidetracks and auxiliary methods in Taoism (Taoism, #HH7004).
Followed by Lower middle grade (Taoism, #HH3296).
Preceded by Outside paths (Taoism, #HH3667).

♦ **HH4298 Ascetic practices** (Buddhism)
Description The pursuit of virtue in *Hinayana Buddhism* is accompanied not only by meditative practices but by ascetic practices. These purify virtue and lead to special qualities such as fewness of wishes, contentment, effacement, solitude, loss of sin, energy and modest requirements. Thirteen kinds of ascetic practice are allowed, each having three grades of severity depending on the vow undertaken. The practices are: the wearing of refuse rags; the wearing of a triple robe; the eating of alms food; house–to–house seeking of alms; the eating of food at one session; the eating of food from one bowl; the refusal of later food; dwelling in the forest; dwelling at the root of a tree; dwelling in the open air; dwelling on a ground where dead bodies are burnt; accepting any bed; remaining sitting as opposed to lying down while sleeping. Detailed instructions are given for these practices, including advantages in carrying out such a vow and conditions under which the vow is broken.
The practices may be carried out in combination (except, of course, where one includes the other or where they are mutually exclusive). Again, some are unsuitable for women, some more appropriate for different types of person. In general, asceticism is said to be good for individuals of greedy disposition, in that with progress greed subsides, and for the deluded in that delusions are got rid of by the diligent in effacement. Dwelling in the forest or at the root of a tree are said to be good for someone with the disposition to hate.
No ascetic practice is seen as unprofitable – all are profitable or indeterminate. The ascetic states that go with volition of ascetic practices bring non–greed, so that greed for forbidden things and indulgence in pleasure of those things which are allowed are both shaken off; they also bring absence of delusion, so that the dangers of forbidden things are no longer hidden and there is no indulgence of self–mortification through excessive self–effacement in ascetic practice.
In that the practices are seen as an aid to meditation, they are recommended only in that they assist in progress or do not decrease progress; where an ascetic practice makes the meditation subject deteriorate then it should not be cultivated.
Broader Hinayana Buddhism (Buddhism, #HH0845).
Narrower Ascetic states (#HM5354).

♦ **HH4301 Spirituality** (Confucianism)
Refs Cousins, Ewert, et al (Eds) *Confusian Spirituality*.
Broader Spirituality (#HH5009).

♦ **HH4302 Concretization**
Description A way of life that denies the spirit, with concrete rewards being the goal, has been postulated as normal where basic needs are not met. However, just as meditational states and some forms of asceticism may become a psychologically defensive denial of the suffering in real life, spirit denying consumerism and the obsessive, compulsive concept of monetized work may be a neurotic defence mechanism against recognition of the decadence and degradation which arise with privilege. If the hierarchy of needs implies that freedom can be an ego–defence and cannot be a substitute for "bread", the converse is also true, "bread" can become an ego–defence and is not a real substitute for freedom.
Refs Mallmann, Carlos A and Nudler, Oscar (Eds) *Human Development in its Social Context* (1986).
Related Needs hierarchy (#HH0913).

♦ **HH4309 Kriya sakti** (Hinduism)
Description A meditation method used in healing.
Broader Meditation (#HH0761).
Related Shakti (#HH0629) Kriya yoga (Yoga, #HH0969).

♦ **HH4312 Projections**
Imaginary relationships
Description The blurred perception of another person that arises because the perceiver sees the person as an aspect of himself rather than as an other. It is neither an abnormal nor a disturbed condition, except when carried to extreme. Each person creates a series of more or less imaginary relationships based essentially on projections of this kind. Three levels of projection have been distinguished: (a) parallel projection, the most superficial form, in which a person tends to see others as themselves; (b) unconscious projection, involving the repression by the person of painful or unacceptable feelings and then perceiving such feelings in others (following Freud); (c) mythic projection, involving the projection of powerful archetypal images onto others (following Jung).
Refs Halpern, James and Ilsa *Projections* (1983).

♦ **HH4321 Samadhi yoga** (Yoga)
Description Under the general heading of raja yoga, samadhi yoga emphasized mastery of self and leads to powers of ecstasy.
Refs Sivananda, Swami *Samadhi Yoga*.

HH4321

Broader Yoga (Yoga, #HH0661) Raja yoga (Yoga, #HH0755).
Related Samadhi (Hinduism, Yoga, #HM2226).

♦ **HH4330 Prasada** (Hinduism)
Grace
Description In theistic Hindu traditions, this is the act of the deity in granting salvation, when the devotee on the *bhakti marga*, or way of love, is absorbed into a state of enstasy.
Related Grace (#HH0169) Enstasy (#HM3169)
Bhakti marga (Hinduism, #HH0628).

♦ **HH4332 Neuromuscular therapy**
Description A natural therapy using a combination of massage techniques to exert pressure at specific parts of the body, thus releasing the stresses which caused the body to malfunction and the emotions which caused them, and reducing nervous activity.
Broader Therapy (#HH0678).
Related Structural integration (#HH0082).

♦ **HH4388 Pramana** (Buddhism)
Validities
Description This is the study of logic, of the mind and of the theory of learning.
Context The third of the five primary subjects for the Geshe degree in Tibetan Buddhism.
Followed by Vinaya (Buddhism, #HH3398).
Preceded by Madhyamaka (Buddhism, #HH1038).

♦ **HH4389 Universal human nature**
Description Meister Eckhart uses this term to describe the loftiest and truest things that humanity has in common, that which unites us at the highest level. Essentially capable of union with God, we are most human and most one with each other when we are detached from self and from creatures, totally humble and receptive, open to God. In the soul's ground we are what we were created for. If it is not seen so it is because we live our lives at the subhuman level – what we see as normal is actually abnormality and sickness.
Refs Smith, Cyprian *The Way of Paradox* (1987).
Related Way of paradox (Christianity, #HH3335) Way of transcendence (Christianity, #HH6566)
Birth of God in the soul's ground (Christianity, #HM6522).

♦ **HH4398 Yogic feats**
Description The control of that practice of hatha yoga produces over bodily functions is said to allow feats such as: remaining on one foot for months; holding the arms in the air for extended periods; remaining waist deep in cold water for days on end.
Broader Hatha yoga (Yoga, #HH0862).
Related Siddhis (Yoga, Buddhism, #HH0380).

♦ **HH4455 Psychospiritual dualism**
Description This attitude considers the spirit as a separate part of the personality, dealing particularly with relations with God, while psychological aspects of personality are those involved with normal life and relationships. Spiritual problems, under this definition, become problems involving relations with God and are assumed to be the result of sin. In the extreme form, psychological aspects are said to be irrelevant to spirituality.
Refs Benner, David G *Psychotherapy and the Spiritual Quest* (1988).
Related Psychospiritual therapy (Christianity, #HH4201).

♦ **HH4477 Quietism** (Christianity)
Selfless love
Description François de Lamothe–Fénelon reinterprets spiritual life by distinguishing between the two loves: conscupiscent, mercenary or selfish love of God with hope for future benefits, which may be assisted by discursive meditation; and totally selfless love, expressed only in acts of perfect simplicity. This perfect love, expressed through total indifference to the vicissitudes of life, leads to union. There is a gradual surrender of active control of the spiritual life as one develops towards pure love. Eventually the soul acquires a holy passivity which leaves all initiative to the divine impulses of grace.
Refs Dupré, Louis and Wiseman, James A (Eds) *Light from Light* (1988).
Related Love of God (Christianity, #HH5232).

♦ **HH4502 Peer group initiation**
Work group initiation
Description In contrast to traditional initiation practices, whether rites of passage or involving membership of secret societies, this contemporary form is practised by many groups in modern society. Most typically it is associated with the acceptance of a new member of a student body, a new military recruit, a new member of a work force, or a new member of gang. In each case some form of "initiation" may be considered traditional and appropriate. The process may be limited to challenging the newcomer with some ordeal, possibly designed to humiliate, or some form of ritual may also be involved.
Broader Initiation (#HH0230).

♦ **HH4508 Archaic spirituality**
Ancient spirituality
Refs Cousins, Ewert, et al (Eds) *Asian Archaic Spirituality*; Cousins, Ewert, et al (Eds) *European Archaic Spirituality*; Cousins, Ewert, et al (Eds) *Ancient Near Eastern Spirituality*.
Broader Spirituality (#HH5009).

♦ **HH4509 Nihilism**
Description Strictly this term implies the philosophy that nothing exists or that, if it does exist, nothing can be known about it. Nevertheless, it has come to be used to describe any theory which implies undermining of the truth, denial of the possibility of metaphysics or failure to agree to accepted moral standards or convictions. It was used in this sense by Nietzsche, who stated that for radically new moral notions to be set up the old, traditional ones must be destroyed.
Related Nihilism (#HM2830).

♦ **HH4510 Social imagination**
Social imaging of the future — Inventing the future — Imagining utopias — Imagining alternative futures
Description Social imagination is a necessary ingredient in social intelligence. Imagining the future involves individuals in new ways of using their minds to grasp complexities that they have not yet understood. It enables people to construct new social realities, whether locally or globally. This is not undertaken merely for academic curiosity, but by integrating feelings and actions it enables the individual to remake himself and the world. This approach does not belittle the cognitive/analytic mode, rather it enhances it and sets the individual free to use more of his potential in social action. The social imagination can be thought of as a problem–solving faculty, continually reworking human experience by means of image formation. The obstacles to such imaging lie partly in social institutions, including schools, which discourage imaging because it leads to visualizing alternatives which challenge existing social arrangements. Other obstacles lie in the minds of older generations that have lost the capacity to use such faculties. These may however be re–engaged by techniques similar to those used in meditation. Imaging, like meditation, requires an emptying the mind. Imaging then puts experience together in new ways, possibly as a scenario. It has an experiential quality. The mind is actively engaged in the work of processing experience during imaging. In focused social imaging a person may work with a group of colleagues in terms of some normatively defined goals in order to envision a social conditions in which specific current problems have been resolved. From this activity a working group can emerge with shared imagery in terms of which to design concrete projects to bring about elements of their vision of the future.
Refs Boulding, Elise *Building a Global Civic Culture* (1988); Zdenek, Marilee *Inventing the Future* (1987); Ziegler, W *Mindbook for Imaging/Inventing a World Without Weapons* (1987).
Related Social innovation (#HH8963).

♦ **HH4511 Accelerated learning**
Suggestology
Refs Lozanov, Georgi *Suggestology and Outlines of Suggestopedy* (1978); Rose, Colin *Accelerated Learning* (1987).
Broader Learning (#HH0180).

♦ **HH4521 Mediation** (Magic)
Description In mediation during a magical operation, the practitioner reaches into states of consciousness or inner worlds where archetypal and transhuman entities are met. These may be perceived as aspects of divinity, god–forms or entities who have transcended human expression. Alternatively they may be experienced as a specific power, consciousness or energy. These two approaches are interchangeable, but as a rule they are used individually because of the effort required to maintain the necessary images to enable either to function. In mediation, the practitioner acts as a focus or gateway for the consciousness of such entities. Mediation is not however a matter of passive reception (as is the case of mediumship) but rather of refining and clarifying personal modes of awareness engendered by the resonance between these entities and their reflection within the practitioner.
As one of the most ancient and enduring inner disciplines, mediation is often undertaken as part of higher visualization and meditation, in which the transformed consciousness, guided by imagery, assumes or takes on the energies and qualities of a chosen god–form. Through this process the mediator acts as a channel for a specific power to flow out into a group or to the wider world. Practitioners can become deluded by the forces that they seek to mediate, especially when they are vulnerable to personal inflation. It is then only through experience that the mediation experience, however total, is recognized as temporary. Mediation is then accompanied by a disturbing echo of undeniably deep personal insight, frequently indicating areas of weakness requiring attention.
The principal aim of effective mediation is defined by the pattern of any ritual or visualization in which assembled energies are given inhabitual form and direction. Such work endeavours to realign energies that have become out of phase with higher patterns or with some form of spiritual awareness.
Context A basic distinction is made between mediumship and mediation. A medium is passive and generally unaware of what is communicated (to the point of being unconscious), while a mediator is active and in a state of heightened consciousness without any loss of individuality. There is also claimed to be a significant difference in the quality, power and type of entity contacted.
Refs Stewart, R J *Advanced Magical Arts* (1988).
Broader Magic (#HH0720).
Related Channelling (#HH0878) Trance channelling (Psychism, #HM3456)
Psychic inspiration (Psychism, #HM1094).

♦ **HH4522 Spiritual self–denial** (Christianity)
Spiritual self restraint
Description Spiritual self–denial (St John of the Cross) implies not only discipline with respect to physical gluttony but to the seeking of pleasure in spiritual exercises. The aim is not to seek spiritual delights and blessings from God, which is childish and self–indulgent. Such behaviour makes travelling the road of the cross very difficult. Eventually, through temptation, spiritual dryness and other trials the person is led to spiritual self–restraint, self–denial, reverence and submission. Even spiritual pleasures are perfect only in being able to refuse them.
Related Self–denial (#HH0964) Self–discipline (Christianity, #HH2121)
Spiritual gluttony (Christianity, #HM0507).

♦ **HH4536 Pancasila**
Fivefold moral law
Description This overarching set of five moral maxims dating from the time of Buddha is the basis of a number of religious groupings. It calls for abstention from: killing (especially taking of human life); sexual misconduct; taking what is not given; making false spiritual claims; taking of intoxicants.
Related Vinaya (Buddhism, #HH1376).

♦ **HH4552 Liberation theology** (Christianity)
Description Emphasis is on the social consequences inherent in God's love for the poor, the insignificant and the oppressed. There is opposition to the form of capitalism exhibited in third world countries and support for human aspirations for justice.

♦ **HH4553 Resurrection**
Description During the initiatory process, the would–be shaman experiences an illness and is close to death. He heals himself and is thus empowered to heal and give psychic counsel to his tribe.
Context A practice of North American indians.
Broader Rites of passage (#HH0021).
Related Shaman (#HH0973) Vision quest (#HH0897).

♦ **HH4561 Parinama** (Yoga)
Mental transformation — Change
Description In yogic philosophy, all transformation occurs because of change in the balance of the three *gunas*; mastery of this balance follows the same laws at all levels, so that the ability to control mental and spiritual change is said to extend to control of the physical and thus ability to perform *siddhis*. The parinamas or modes of mental transformation of yogic tradition arise when the yogi is able to pass into the samadhi state as he wishes.
Narrower Samadhi parinama (#HM3207) Nirodha parinama (#HM3437)
Ekagrata parinama (#HM3337).
Related Change (#HH1116) Three gunas (#HH0413) Maya (Hinduism, #HH1004)
Siddhis (Yoga, Buddhism, #HH0380).

♦ **HH4572 Golden age**
Paradise
Refs Heinberg, Richard *Memories and Visions of Paradise* (1989).
Related Guiding images (#HH3020).

♦ **HH4621 Planetary noogenesis**
Planetary consciousness
Description The processes of evolution may be leading to the emergence of consciousness at the planetary level, whereby the planet achieves its own equivalent of consciousness as a form of "global brain". It has been suggested that the majority of humans currently alive may experience an evolutionary shift from ego-centered awareness to a unified field of shared awareness. This would constitute a completely new level of evolution, as different from consciousness as consciousness is from life, as as life is from matter. This shift is catalyzed by environmental crises, technological breakthroughs, developments in information systems, and the rapid spread of consciousness-expanding techniques. Indeed the hypothesis is that popular attainment of higher states of awareness are a prerequisite to such planetary noogenesis.
Refs Russell, Peter *The Global Brain* (1983).
Related Maharishi effect (#HH3764) Planetary consciousness (#HM2006).

♦ **HH4646 Maternal thinking**
Mothering — Maternalism
Description Motherhood, like all activity, brings its own distinctive perception, conduct and perspectives. The process of mothering, whether of one's biological children or of any child - even normally accepted mothering activity when carried out by a man, brings its own work and thought forms. Mothering therefore not only impinges on the child being mothered but also on the person doing the mothering, developing virtues, insights and intelligence which, if not common to all mothers, most mothers would accept as what they aspire to. This process of shaping heart and mind of both mother and child may be referred to as a *moral education*. Both the physical care of the child and contact with its strong and contradictory passions, its nature and its will has the effect of sharpening or moulding qualities such as respect for nature, ability to accept change and a tolerance for emotional ambivalence.
Refs Ruddick, Sara *Maternal Thinking*.
Broader Thought processes (#HH0530).
Related Paternalism (#HH0735) Moral development (#HH0565)
Feminine consciousness (#HM2768).

♦ **HH4653 Penitential practices**
Penitential routines
Description With the intention of subjugating the physical body and the desires, a number of pentitential practices, whether personal or prescribed to members of a group, have been and are employed. Mortifications include the practice of silence and sleeping on the ground or on bare boards. Other practices which are more violent have been dropped by some orders as they are considered psychologically unhealthy. They are still practised by others, however. Opus Dei, for example, prescribes the use of a cilice, or spiked bracelet, around the thigh for 2 hours a day, and self-flagellation once a week.
Narrower Flagellation (#HH0151) Mortification (#HH0780).

♦ **HH4663 Human development through theism**
Description Formed as a view of God common to all religions, as a sovereign being acting personally and continuously present and active in the world, theism was designed to hinder the spread of atheism. There is a fundamental belief that God is essentially unknown so that no statement can be made about his being, but that nonetheless he is in sympathetic solidarity with mankind.
Related Human development through atheism (#HH4223)
Human development through religion (#HH1198).

♦ **HH4711 Lower upper grade** (Taoism)
Description There are a number of practices including mirror gazing, meditative breathing, massage and physical exercices, extended pronunciation of certain sounds, mentally gazing at the top of the head or keeping the attention on the navel, keeping the mind on the elixir fields of the torso and head, staring at the nose, swallowing saliva.
Context In the nine grades of practices which are side tracks and auxiliary methods in Taoism, this is the lowest of the three upper grades.
Broader Sidetracks and auxiliary methods in Taoism (Taoism, #HH7004).
Followed by Middle upper grade (Taoism, #HH6304).
Preceded by Upper middle grade (Taoism, #HH2633).

♦ **HH4712 Evolutionary choice**
Description Hanna Newcombe indicates that the design for a better world should be dynamic rather than static, dealing ideally not with a final state but with a process. It would not be a single design - at various points there will be several alternatives or a continuous range of alternatives. Problems may often be solved in a number of almost equally good ways. Choice opens the way to future negotiations among decision-makers. The decision does not have to be made once and for all and for all time. Since human institutions are only the creation of human beings they should not be endowed with quasi-divine powers and worshipped as sanctified by tradition. They should be changed as often as is desired to suit evolving needs.
Refs Newcombe, Hanna *Design for a Better World*.
Related Freedom of choice (#HH0789).

♦ **HH4742 Highest vehicle** (Taoism)
Description This is the ineffable way of supreme ultimate reality. The most developed people can practice the subtlety of this supreme vehicle and, when accomplishment is fulfilled, character is well developed and the practitioner transcends all at once to completion. He is physically and spiritually sublimated and merges with the Tao in reality. The cauldron is cosmic space, the furnace is the absolute. The foundation of the elixir pill is clear serenity, its matrix nondoing. Lead is essence, mercury life. Concentration is water, insight is fire, the mixing of the two is controlling desire and anger. Combining of metal and wood is the unification of essence and sense, cleaning the mind is bathing. Maintenance of sincerity and settling the will is stabilization. The three essentials are discipline, concentration and insight, the mysterious pass is the centre. Clarifying the mind is miraculous experience, seeing the essence of mind is crystallization. The spiritual embryo is the merging of the three bases in one, completion of the elixir pill is unification of essence and life. Release from the matrix is having a body outside the body. Breaking through space is perfect attainment.
Refs Cleary, Thomas (Trans) *The Book of Balance and Harmony* (1989).
Related Refining the gold pill (Taoism, #HH4880)
Release from the matrix (Taoism, #HM0635).
Preceded by Three vehicles of gradual method (Taoism, #HH5342).

♦ **HH4746 Spirituality** (Amerindian)
Refs Cousins, Ewert, et al (Eds) *North American Indian Spirituality*.
Broader Spirituality (#HH5009).

♦ **HH4776 Spirituality** (Judaism)
Refs Blumenthal, David R *God at the Center* (1988); Cousins, Ewert, et al (Eds) *Jewish Spirituality* (1988); Cousins, Ewert, et al (Eds) *Jewish Spirituality* (1987).
Broader Spirituality (#HH5009).

♦ **HH4777 Skill in absorption** (Buddhism)
Skill in ecstasy
Description There are ten skills enumerated in Hinayana Buddhism which are recommended as aids for those in whom absorption does not arise swiftly:
1. Clean physical basis: The body should be neat and clean (internal basis); the clothes and lodging should also be kept clean (external basis). Then the consciousness and its concomitants that arise are clean and purified and the subject of meditation grows and develops.
2. Maintenance of balanced faculties: Faith, energy, mindfulness, concentration and understanding faculties should all be balanced. If one is too strong then the others cannot perform their specific functions. It is particularly necessary to balance faith with understanding and concentration with energy. But mindfulness should always be strong.
3. Being skilful in the sign: once the sign of the subject of meditation is produced and developed, it should be protected.
4. Exerting or upholding the mind when this should be done: If the mind is slack it should be exerted by developing enlightenment factors of investigation of states or doctrine, energy, and happiness or rapture, rather than those of tranquillity, serenity and equanimity. Wise attention to these factors acts as fuel to exerting the mind.
5. Restraining the mind when this should be done: If the mind is agitated it should be restrained by developing enlightenment factors of tranquillity, serenity and equanimity, rather than those of investigation of states or doctrine, energy, and happiness or rapture. Wise attention to these factors acts as fuel to restraining the mind.
6. Encouraging the mind when this should be done: If the mind is listless and dissatisfied it should be stimulated by reviewing the eight reasons for urgency - birth, ageing, sickness, death, suffering of the states of woe or loss; suffering through births in the past, suffering through births in the future, suffering in the present due to searching for food. Confidence or satisfaction arises on recalling the qualities of the Buddha, of dhamma (law) and of sangha (the order).
7. Viewing the mind with equanimity when this should be done: If the mind of one who practices is serene, if it is not slack or excited, if it rests evenly on its object, neither idle nor agitated, then it should be viewed with equanimity.
8. Avoiding persons who are not concentrated: People who have never followed the path of renunciation, who are busy with affairs and whose hearts are distracted should be avoided.
9. Cultivating persons who are concentrated: People who have followed the path of renunciation and obtained concentration should be sought out from time to time.
10. Resoluteness or intentness on that: Concentration should be given its importance, there should be tending, inclination and leaning towards it.
Related Equanimity (Buddhism, #HM7769) Wise attention (Buddhism, #HM4309)
Dhyana (Hinduism, Buddhism, #HM0137) Enlightenment factors (Buddhism, #HM6336)
Absorption concentration (Buddhism, #HM0311)
Appearance of absorption (Buddhism, #HM1618).

♦ **HH4864 Lower vehicle path** (Taoism)
Description This is a method of comfort and bliss and includes over 100 operations. It will nurture life if the practitioner can forget feelings. In method this is similar to the three upper grades of practices but in application it is different. Terminology includes body as alchemical cauldron, mind as furnace, vitality and energy as medicinal ingredients, heart as fire, genitals water, five internal organs as five forces, liver as dragon, lungs as tiger, semen as true seed. Year, month date and hour are significant in the firing process. Swallowing saliva is bathing (irrigation of the digestive system). The three essentials are mouth and nostrils, the mysterious pass is the space in front of the kidneys and behind the navel. Completion of the elixir pill is through merging of the five forces.
Context The lowest of the three vehicles of the gradual method, superior to the nine grades of practices which are sidetracks and auxiliary methods.
Broader Three vehicles of gradual method (Taoism, #HH5342).
Followed by Middle vehicle path (Taoism, #HH6345).

♦ **HH4865 Greek spirituality**
Refs Cousins, Ewert, et al (Eds) *Classical Mediterranean Spirituality*.
Broader Spirituality (#HH5009).

♦ **HH4872 Neuro-linguistic programming** (NLP)
Description NLP is the study of the structure of subjective experience. NLP makes explicit patterns of behaviour and change that have previously been only understandable intuitively. It is the basis for a particular approach to therapy, although, in terms of the NLP perspective, each school of psychotherapy is a metaphor designed to help and expand the limitations of a client's personal metaphors. And just as the form and organization of an individual's map of existence deeply affect his experience so too do the different treatments metaphors offer limitations as ways to learn and grow. But when a metaphor of personality becomes so wordy and entrancing that the graceful art of change is buried by concepts and analysis, then a reorientation, such as that facilitated by NLP, is called for. Continuing to embrace a particular metaphor in preference to another inhibits the process of human development. Presuppositions reinforced by a particular metaphor limit the range of personal and professional choices to which a person is exposed.
Whereas many therapies facilitate a degree of change in clients, this change tends to be limited to assisting the individual in coping with currently experienced difficulties. NLP aims to enable individuals to create systematically a reference structure, or set of experiences, that permits them to change their coping patterns in response to new difficulties as they emerge. This involves the systematic demystification of normally out-of-awareness aspects of communication that give the person a heightened sense of control over himself and his environment.
Human beings live in a real world. They do not however operate directly or immediately upon that world. Rather their relationships to it are mediated by neurological filters. Because sensory organs vary greatly between people, each perceives the world differently through different models or maps of reality. A series of such maps is used to guide behaviour. These maps, or representational systems, necessarily differ from the territory which they model. They have built-in errors. These are due to three processes characteristic of human modelling: generalization (the assumption of conformity to a general pattern), deletion (failure to attend to significant details), and distortion (alteration of perception of sensory input). The limitation that people experience are typically in their representation of the world, not in the world itself.
NLP focuses on the human language as the best understood of the representational system of maps and on transformational grammar as the best model of it. This provides a meta-model, a representation of the structure of human language, which is itself a representation of the world of experience. NLP has adapted this for therapeutic purposes. Using this appropriate grammar for

therapy, people can be assisted in expanding the portions of their representations of the world which impoverish and limit them.
Context Originally created by behavioural modellers John Grinder and Richard Bandler, NLP intersects with the theoretical material of several fields including cybernetics, linguistics, psychotherapy and personality theory.
Refs Bandler, Richard and Grinder, John *The Structure of Magic* (1975–76); Gordon, David *Therapeutic Metaphors* (1976); Lankton, Steve *Practical Magic* (1980); Lewis, B A and Pucelik F *Magic Demystified* (1982).
Related Magic (#HH0720) Guided fantasy (#HH0627) Guiding images (#HH3020) Therapeutic double bind (#HH6112).

♦ HH4880 Refining the gold pill (Taoism)
Arriving at reality
Description Cauldron and furnace are action and stillness, water and fire are vitality and energy, the evolutionary mechanism is body and mind, the medicinal ingredients are essence and sense. Keeping centred on mindfulness of the celestial one finds the mysterious pass, gathers vital energy of the sense of essence at the appropriate time, and then withdraws into watchful passivity in the proper manner. Essence, sense, spirit, vitality and will are unified. Firmness and flexibility, creativity and receptivity, movement and stillness are kept in balanced proportion. Combining will and an inner sense of true essence one returns to the fundamental, goes back to the basis, reverts to the root, returns to life. With the work complete and the spirit prepared the ordinary is shed and one becomes immortal. This is the completion of the elixir pill.
The mysterious pass or "centre" is not a place but is when thoughts do not arise. It is the point of movement that arises when quietude reaches its consummation, and appears spontaneously when attention is applied to the point where the mind and active thought are roused. "Gathering medicine" is gathering the true sense of the essence of consciousness within one's self, by first quietening the mind so that impulses of arbitrary feelings are stilled. In perfection of this stillness there is movement of unconditioned energy, energy of true sense whose first movement arising from stillness is called return of yang. By fostering this, sense and essence, energy and spirit are united. Then there is withdrawal into watchful passivity – intensive concentration after the point of sufficiency means the work is wasted. The cycle of work is movement to stillness to movement to stillness. After persevering for a long time comes a gradual crystallization, stabilization of real consciousness. Nonsubstance produces substance, the spiritual embryo. As mind and body unite, spirit and energy merge, sense and essence conjoin, the elixir pill is completed.
Thus is the elixir completed.
Refs Cleary, Thomas (Trans) *The Book of Balance and Harmony* (1989).
Related Highest vehicle (Taoism, #HH4742) Human development (Taoism, #HH0689) Complete awareness (Buddhism, #HM0634).

♦ HH4908 Vibrational health
Crystal consciousness — Meditation with crystals — Crystal healing
Refs Baer, R and Baer, V *The Crystal Connection* (1986); Bonewitz, R *Cosmic Crystals* (1983); Bonewitz, R *The Cosmic Crystal Spiral* (1986); Gerber, Richard *Vibrational Medicine* (1988); Norbu, Namkhai *The Crystal and the Way of Light* (1986); Raphaell, K *Crystal Enlightenment* (1985).
Related Sound health (#HH2543).

♦ HH4965 Ecological humanism
Description Basically, ecological humanism maintains that humankind is the universe as it is in the process of being made. One strives for meaning through one's own existential efforts, gives meaning to the universe through acquired humanity. Aesthetic sensitivity is acquired as part of the evolutionary process. Mind and its cognitive capacities are acquired through the strivings of mankind and of evolution. Spirituality is acquired as the result of evolutionary unfolding. Godliness is acquired as, at the end of the evolutionary journey, we make gods of ourselves. Although this cannot be proved by reason, it can be incorporated into the structure of one's life. In some cases, whether consciously or (more often) subconsciously it has been incorporated into existing ways of life. Rather than defying reason, ecological humanism is an expression of reason seen in evolutionary unfolding. Long–term biological and environmental survival depends on humankind's capacity to remake the world from within, transcending present conditions, adapting to the frugality which is a precondition of inner beauty. If humankind does not reach beyond it will be swept away from where it is.
Refs Skolimowski, Henryk *Eco–Philosophy* (1981).
Broader Humanism (#HH0674).
Related Deep ecology (#HH2315) Bio–globalism (#HH1267) Environmental mysticism (#HM0800) Christian stewardship (Christianity, #HH3121).

♦ HH4974 Journeying within transcendence - from head to heart
Description Established tradition meets the new and the free. Seeing with the mind is to know; seeing with the heart is to be known. There is uneasiness as a suspicion arises that one's own efforts and accomplishments are insufficient, there is a need to be born again. Traditional, reasoned knowledge takes one further and further from the mythic world into which one was born; the knowledge of the immediate in Jesus connects with the inner world and demonstrates that other knowledge is in fact weakness. Wisdom and understanding come with awareness of the good, of value, as well as simply the true.
Context The fifth section of St John's Gospel, Chapter II 1 to IV 1, is related to a stage in the spiritual journey of the individual.
Broader Journeying within transcendence (Christianity, #HH6505).
Followed by Journeying within transcendence - from forgetfulness to memory (#HH5219).
Preceded by Journeying within transcendence - from secular to sacred (#HH3796).

♦ HH4994 Secular quest
Refs Cousins, Ewert, et al (Eds) *Spirituality and the Secular Quest.*
Broader Spirituality (#HH5009).
Related Secular humanism (#HH0105) Secular transcendence (#HM2804).

♦ HH4997 Balancing yin and yang (Taoism)
Blending celestial consciousness with earthly consciousness
Description Basic to the taoist approach to human development is the process of balancing yin and yang or earth and heaven. Here heaven may be interpreted as a world– transcending higher consciousness, beyond the levels of ordinary thought and emotion, whereas earth may be understood as the everyday mundane experience of an individual in the world. The challenge for human development, to achieve the complete or "real" human being, is to balance appropriately the combination of these two levels of experience. In practice this means maintaining contact with the higher form of consciousness whilst living effectively in the earthly domain. Celestial consciousness thus guides earthly consciousness. Given the predominant influence of mundane awareness, much of the practice of Taoism is concerned with "repelling yin" and "fostering yang". This often involves practices of standing aloof from mundane awareness in order to increase awareness of the celestial or primordial mind. Emptiness and stillness are cultivated by quieting the mental activity (of the "wandering mind" or "human mind") which sustains acquired world views and habitual involvement therein. The object of such practice is not stillness itself but rather, through stillness, to become conscious of the underlying awareness ("the shining mind" or "mind of Tao") which is normally obscured by habitual entanglements. The purpose is not to suppress mundane conditioning but rather to enlighten it from the complementary celestial perspective. By so doing the individual acquires autonomy from particular conditions which can then be responded to according to the needs of the moment. The individual then transcends yin and yang, reaching an undefinable state of awareness in which he "does nothing, yet does anything", participating in mundane activities or withdrawing from them as necessary means to an ultimate balance and completeness. The correct balance of yin and yang may also be considered as a balance of flexibility and firmness (or movement and stillness) in action, whether in the social or the spiritual life. The challenge in practice is then one of discerning the exact quality of experience to determine the degree to which it is governed by the human mind or by the primordial mind of Tao transcending dualistic distinctions.
The many stages and conditions of balancing yin and yang are symbolized by the 64 hexagrams of the I Ching.
Refs Chang, Po–tuan *Understanding Reality* (1987).
Broader Human development (Taoism, #HH0689).
Related I Ching (Taoism, #HH0004).

♦ HH5007 Afflictions and hindrances (Buddhism)
Description In various Buddhist traditions, considerable importance is attached to fundamental afflictions as the cause of suffering (and as responsible for maintaining the cycle of rebirth). All other problems are seen as engendered by them. In the Visuddhimagga of Buddhaghosa, prepared in the 5th century AD, the following detailed checklist is given (followed there by indications of which forms of knowledge ensure release from them in each case). The seeming duplication is due to the emphasis on the different ways a limited set of "problems acts, as indicated by the often metaphoric categories:
– Fetters: greed for material benefits, greed for non–material benefits, conceit/pride, excitement/agitation, ignorance, delusion of selfhood (false view of individuality), doubt, susceptibility to rites and rituals, greed for sense desires, and resentment.
– Corruptions/Defilements: greed, hatred, delusion, conceit/pride, false view, uncertainty, mental sloth, excitement/agitation, consciencelessness, shamelessness.
– Wrongnesses: wrong view, wrong thinking, wrong speech (falsehood), wrong action, wrong livelihood, wrong effort, wrong mindfulness, wrong concentration, possibly together with wrong understanding of deliverance and wrong knowledge.
– Worldly conditions (despondency/servitude to states): gain, loss, fame, disgrace, pleasure, pain, blame, praise.
– Meannesses (kinds of avarice): avarice about dwellings, families, gain, dhamma, praise.
– Perversions (reversals): perversion of perception, of consciousness, and of view (whereby, in each case, the inappropriate is misapprehended as the appropriate).
– Ties: covetousness, ill–will, susceptibility to rites and rituals, dogmatic misinterpretation of truth.
– Tendencies to inappropriate action: partiality (desire/zeal), hatred, delusion, fear.
– Bonds (cankers/yokes/floods): sensuous lust, lust for rebirth, wrong views, uncontrolled sensuousness, being swept into becoming, difficulty of overcoming.
– Hindrances: sensuous desire, ill–will, sloth/torpor, distraction (agitation/worry), doubt.
– Misapprehension/Wrong views: ignoring essentials in favour of non–essentials.
– Graspings/Clingings: clinging to views, susceptibility to rites and rituals, clinging to selfhood, desire.
– Inherent tendencies/Biases: sensuous passion, resentment, conceit/pride, false view, doubt, craving for existence, ignorance.
– Stains/Taints: greed, hatred, delusion.
– Courses of immoral action: life–taking, theft, sexual misconduct, lying, slanderous speech, harsh speech, gossip, covetousness, ill–will, wrong view.
– Immoral states of consciousness: eight rooted in greed, two rooted in hate, two rooted in delusion.
Refs Buddhaghosa, Bhadantacariya *The Path of Purification* (1980); Buddhaghosa, Bhadantacariya *The Path of Purity* (1975).
Narrower Root afflictions (#HH0270) Secondary afflictions (Buddhism, Tibetan, #HH0781).
Related Lower fetters (Buddhism, #HM6342) Five hindrances (Buddhism, #HH6773).

♦ HH5009 Spirituality
Spiritual growth
Refs Carreiro, Mary E *The Psychology of Spiritual Growth* (1987); Cousins, Ewert, et al (Eds) *World Spirituality*; Cousins, Ewert, et al (Eds) *Encounters of Spiritualities*; Cousins, Ewert, et al (Eds) *Encounters of Spiritualities*; Cousins, Ewert, et al (Eds) *Dictionary of World Spirituality*; Cully, Iris V *Education for Spiritual Growth* (1984); Fox, Matthew (Ed) *Western Spirituality* (1981); Jones, Cheslyn, et al *The Study of Spirituality* (1986); Steiner, Rudolf *The Bridge Between Universal Spirituality and the Physical Constitution of Man* (1979); Thompson, Helen *Journey Toward Wholeness* (1982).
Narrower Secular quest (#HH4994) Greek spirituality (#HH4865) Spirituality (Taoism, #HH5117) Spirituality (Judaism, #HH4776) Spirituality (Sikhism, #HH3774) African spirituality (#HH5224) Spirituality (Buddhism, #HH3907) Egyptian spirituality (#HH7234) Spirituality (Amerindian, #HH4746) Spirituality (Esotericism, #HH5234) Spirituality (Confucianism, #HH4301) Spirituality (Islam, #HH5902) Roman spirituality (#HH6291) Native spirituality (#HH6121) Spirituality (Jainism, #HH8712) Archaic spirituality (#HH4508) Spirituality (Hinduism, #HH3000) Feminine spirituality (#HH4003) Formative spirituality (#HH7403) Environmental mysticism (#HM0800) Spirituality (Christianity, #HH0792) Spirituality in tantric yoga (#HH6210).
Related Spiritual renaissance (#HH1676) Universal religion (Yoga, #HH0746) Spiritual development (#HH0017) Psychic health and spirituality (#HH0109).

♦ HH5022 We-psychology (Christianity)
Religious psychology
Description This psychology developed by Fritz Kunkel looks on development and operation of egocentricity as the major obstacle to surrender to God. Here the child is originally in a state of interpersonal connectedness when experience with others is integrated as part of the self, the self not being limited to one's own self but including the "we" experience, the own self with others. The inherent experience of interconnectedness is lost as a child comes in contact with the less then complete (because egocentric) love of his parents. This leads to egocentricity on the part of the child, the ego becoming the (limited) centre of existence instead of the (wider) self. This egocentric life results in alienation, being cut off from creativity and energy which arise from deep relations with others. With the ego forming a protective shell around the personality the resources for truly living are left outside and this leads to crisis. The cure is to lose what appears to be life (the system comprising mistaken ideas and values which forms the ego) in order to gain real life through commitment and engagement involving others. Initially there is an emptiness and indifference. There is knowledge of powerlessness in the individual but of power and responsibility

as part of a larger unit. This "we" is seen as created, sent, supported, endowed and used by God, in whom one lives and moves and has one's being. There is then the awareness of being a tool in the hand of God, commissioned with a concrete task. Spirituality is thus self-transcendence and self-surrender.
Refs Benner, David G *Psychotherapy and the Spiritual Quest* (1988).

♦ **HH5023 Christian meditation** (Christianity)
Description As in many meditative techniques, Christian meditation often centres on the repetition of a single word or phrase – a mantra. This is true of the system recommended in the mediaeval mystic writing of "The Cloud of Unknowing" and of systems recommended at the present time. The importance is to ignore anything that occurs during meditation, and simply continue to repeat the mantra. This is the reverse of the search for experience for its own sake which, by itself, could lead to spiritual anarchy. Faithful practice leads to the experience of awakening, of spiritual vision. Experience of the Kingdom of God permeates every dimension of solitary and relational living. Rather than "experience of God", with its implication of duality, there is the experience of being taken into the self-knowledge of God, of being one with God.
Practice of meditation, then, leads to freedom from the ever-diminishing, grey kingdom of the ego which is spiritual death. Meditation is always approached without demands or expectations. What is received is a gift of spiritual knowledge, the grounding realization that we are. This consciousness that we are fills with joy; consciousness that being is joy transforms experience. Every practice of meditation leads back to this consciousness of being. Perception of the mystery of life becomes more firmly rooted and it becomes possible to communicate this perception to others with joy. The being of God fills our being, purifying the heart and leading deeper into the vision of God which is his own self-knowledge. This is the state of living no longer but of Christ living in us.
For Madame Guyon, meditation and meditative reading are two means of being led into the higher forms of prayer. Meditation should be engaged in at a chosen time, not at the time of meditative reading. By an act of faith one brings one's self into the presence of God. The attention is fixed by reading something substantial. The faith of the presence of God within the heart leads to entering within one's self and to drawing near to God, with thoughts collected and not wandering, distracting thoughts got rid of and outward things lost sight of. God dwells and is found in the secret place of the heart. Buried in one's self, penetrated with the presence of God within, senses drawn to the centre, the soul gathered within itself and occupied with feeding upon the truth read (not reasoning on it, exciting the will by affection rather than the understanding by consideration), the affection reposes sweetly and at peace. In this loving repose, full of respect and confidence, what has been tasted and masticated is swallowed.
Thomas Merton warns that it may appear that meditation has failed. One may feel helpless to know God while desiring more and more to see and know Him. This tension between desire and failure generates a painful longing for God which nothing satisfies. This is, in fact, bringing one close to God where one is forced to reach out in blind faith, hope and love. Meditation is a spiritual work of love and desire. It requires effort, at least at first, the sincerity, humility and perseverance of effort depending on the desire, which is a gift of grace. One must first pray for the desire and grace to meditate. The idea is to awaken the interior self and be inwardly attuned to the Holy Spirit so as to respond to His grace. Years of mental prayer will have refined and purified interior perceptivity. One must be attuned to unexpected movements of grace, ready to cooperate with humiliating as well as consoling graces, lights which blast self-complacency as well as those which exalt. Meditation is always associated with abandoning to the will and action of God, with self-renunciation and obedience to the Holy Spirit. If it does not attempt to bring the whole being into conformity with God's will it will be sterile and abstract. In contrast, sincere interior prayer is always rewarded by grace and acts as a sanctifying force in one's life. St Teresa of Avila believed that no-one could lose his soul if he was faithful in the practice of meditation.
Refs Amaldas, Swami *Christian Yogic Meditation*; Brame, Grace A *Receptive Prayer* (1985); Guyon, J M B de la Mothe *A Short Method of Prayer and Spiritual Torrents* (1875); Helleberg, Marilyn M *Beyond T M* (1981); Helleberg, Marilyn M *A Guide to Christian Meditation* (1985); Himalayan International Institute (Eds) *Meditation in Christianity*; Hulme, William E *Celebrating God's Presence* (1988); Main, John *The Present Christ* (1986); Merton, Thomas *Seeds of Contemplation* (1972); Merton, Thomas *Spiritual Direction and Meditation* (1975); Underhill, Evelyn (Ed) *The Cloud of Unknowing* (1970).
Broader Meditation (#HH0761).
Narrower Quaker meditation (#HH4022).
Related Meditative prayer (#HH4089) Meditative reading (#HH5550)
Mental prayer (Christianity, #HH8672)
Sense of the presence of God (Christianity, #HM6989).

♦ **HH5080 Aromatherapy**
Refs Jackson, Judith *Aromatherapy* (1987).
Broader Therapy (#HH0678).

♦ **HH5092 Purification of the self** (Sufism)
Tadhkiya-i nafs
Description The *nafs* is the very essence of man, and may be variously interpreted as the soul, the ego (when falsely imagined as separate) or the seat of passion and lust, the sensual self. In the course of *suluk*, advancement in spiritual life, the *nafs* is cleansed. In its natural state it is attributed with animal qualities and naturally tends towards evil, commanding the person to do evil, and known as imperious (nafs-i ammara). As it is purified and starts to avoid evil it reproaches itself (nafs-i lawwama). Finally, attaining the love of God, it is no longer the source of evil but acquires angelic attributes and has the faculty for doing right. It is no longer self-accusatory but becomes tranquil (nafs-i mutma'inna).
In its impure state the nafs is the greatest obstacle to spiritual progress, its desires acting as veils separating from union with God. The process of purification requires practice of austerities to: overcome slavery to carnal desires; replace hypocrisy with truth and sincerity; remove ostentation and dissimulation (which is false worship); renounce the claims of the ego to divinity (in its pride, it sets itself up as God, whereas only God has the attributes of existence); relinquish attachment to worldly possessions. Negation of nafs brings the discovery of the true self.
Context The first of four contemplative disciplines the salik (seeker after God) must pass through to obtain ma'rifa. Various Sufi orders may differ in approach but all require the same spiritual concentration; manifestation in its many forms may be dealt with in many ways but the realization of spiritual reality behind the forms is the same.
Broader Purification (#HH0401) Awareness of the mystic journey (#HM2900).
Related Subtle faculties (Sufism, #HH6282) Self-mortification (Sufism, #HH0464).
Followed by Cleansing of the heart (Sufism, #HM6932).

♦ **HH5098 Moral perfection**
Moral perfectibility
Description This implies that the individual becomes perfect through becoming, in some absolute sense, a better person. This implies that all human potentialities actualized in the person by this process are in fact potentials for good. It has been suggested that criminality, for example, is not a potentiality capable of being actualized but rather a defect or imperfect actualization of a potentiality.
Broader Human perfectibility (#HH0212).
Related Religious growth (#HH1321) Moral development (#HH0565)
Ethical discipline (#HH1086) Spiritual development (#HH0017).

♦ **HH5101 Human development index** (United Nations)
Description In order to index human progress, the *United Nations Development Programme (UNDP)* has focused attention on: (1) deprivation of life expectancy; (2) literacy; (3) income for a decent living standard. The first two are the commonly used concepts; the third is an adjusted figure for per capita income based on purchasing power of gross domestic product (GDP) figures. The index scale is based upon minimum and maximum national values of the indicators in 1987 (the GDP figure being represented by its logarithm), the three maxima and minima being numbered zero and one. The average measure on the three scales gives a country's average human deprivation index.
Refs United Nations Development Programme *Human Development Report 1990*.
Broader Human development (United Nations, #HH3290).

♦ **HH5110 Alternative pursuits**
Alternatives to drug abuse
Description This approach is based on the belief that drug abuse is linked to psycho-social malaise and the difficulty the individual in the developed world has in finding purpose and meaning in life. Alternative means are sought for experiencing such meaning, with the assumption that achievement of a drugless "high" would help turn the user away from drugs (direct alternatives). These include meditation, aesthetic or religious experience, mind trips, sensory awareness training, biofeedback, bioenergetics, group work, diet and yoga. Indirect alternatives designed to fulfil emotional needs that turn the individual to drug abuse cover a wider field – from sport to vocational training, from social confidence training to creative artistic experience. Informal centres are set up to provide such input, frequently run by ex-drug users and usually catering specifically for young people.
Related Psychedelic drugs (#HH0075).

♦ **HH5116 Sanctifying grace** (Christianity)
Description According to Catholic doctrine, this is a supernatural quality dwelling in the human soul and the means by which the person shares in the life of God. Its effects are to make the person holy and pleasing to God, to make him an adopted child of God, to make him a temple of the Holy Spirit and to give him a right to heaven. Because it is the supernatural life, sanctifying grace is necessary for attainment of the supernatural happiness of heaven.
Broader Grace (#HH0169).
Related Actual grace (Christianity, #HH6548) State of grace (Christianity, #HM6445).

♦ **HH5117 Spirituality** (Taoism)
Refs Cousins, Ewert, et al (Eds) *Taoist Spirituality*.
Broader Spirituality (#HH5009).
Related Ki energy (#HH0620) Human development (Taoism, #HH0689).

♦ **HH5119 Archetypal psychology**
Imaginal psychology — Re-souling the world
Description Archetypal psychology (pioneered by James Hillman) is a post-Jungian way of looking at human experience. It stresses the uses of metaphor, symbol and imagination in contrast to a literal or material view. Archetypal psychology is not however a psychology of archetypes. Its primary activity is not that of matching themes in mythology and art to similar themes in life. The aim is rather to enable the individual to see every fragment of life, and every dream, as myth and poetry. This presupposes a poetic basis to mind through which consciousness is freed from its thin, hard crust of literalism to reveal the depth of experience. Experiencing through the soul turns events into experiences and effectively re-souls the world. Images are therefore sought in events that give rise to meaningfulness, value and the full range of experience. There is consequently a drive for depth, resonance and texture in all that is experienced. Archetypal psychology uses the penetrating vision of the imagination to perceive those fundamental fantasies (or archetypes) that animate all of life. In this sense archetypal means fundamentally imaginal. Psychoanalytic concepts and ideas have therefore to be heard as expressions of imagination and read as metaphors. Imagination is given absolute priority over ego understandings and applications. Psychology can only tend to the soul when soul is perceived properly through image. The process of human development can therefore be seen as one of healing, or making whole, through imagery. The language of the soul is image. Soul is the unknown component that makes meaning possible. It "works" through the metaphor of deepening, namely deepening events into experiences, and is thus the medium through which individuals are able to reflect on their existence. Soul is thus the imaginative potential within the individual, the ability to experience through reflective speculation, dream image and fantasy. Through its special relationship with the mythical underworld and with death, it gives life and death meaning and purpose.
Refs Avens, Roberts *Imagination is Reality* (1980); Bleakley, Alan *Earth's Embrace* (1989); Bolen, Jean S *Gods in Everyman* (1989); Bolen, Jean S *Goddesses in Everywoman* (1985); Hillman, James *Archetypal Psychology* (1983); Hillman, James *The Myth of Analysis* (1978); Hillman, James *Loose Ends* (1975); Hillman, James *Re-Visioning Psychology* (1975); Hillman, James *Peaks and Vales*; Hillman, James *Healing Fiction* (1983); Hillman, James *The Essential James Hillman* (1989); Hillman, James, et al *Facing the Gods* (1980).
Related Guiding images (#HH3020).

♦ **HH5120 Ritual pattern-making** (Magic)
Ceremonial magic
Description A ritual of any kind sets up specific conditions (or a specific context) in both the operator and the "real" world as it is intended that it should be perceived. The main function of ritual in the magical tradition is to set up some particular state of emotion or awareness. Pattern-making through ritual is one of the five magical arts. The pattern acts as a matrix for energies arising within the consciousness of participants. Under specific conditions it can involve the bio-electrical energies of the body and psyche. The consciousness which merges with and consists of such energies is both individual and collective. It is expressed as a sequence of integrative insights shared by the group within its imagination. One interpretation is that traditional rituals of speech, movements, consecration conjure spirits and by means of their services bring about magical results which may be beneficial – exorcism, healing, knowledge, prosperity.
Contrary to a widespread assumption, powerful rituals may be quite simple in form and language, even though they have complex effects and relationships upon awareness. Mystique, romanticism and pseudo-learning are unnecessary, especially when deliberately designed to obscure and impress in lengthy, repetitive rituals. But curious words, chants, vocal tones and other verbal symbols may be used when these have significance for all participants. Magical operations generally employ a combination of expressed modes of communication: words, music, dance, formal movement, scents, colours, sounds, objective symbols and implements. These are only of value when they complement each other so as to enhance a pattern which captures the imagination. Hours of complex ritual may often be more effectively replaced by a simple ceremony or a basic meditation.

Refs Stewart, R J *Advanced Magical Arts* (1988); Tyson, Donald *The New Magus* (1988).
Broader Magic (#HH0720) Rites (#HH0423).

♦ **HH5122 Formation of merit** (Buddhism)
Description In Hinayana Buddhism, the profitable or moral volitions of the sense sphere and the immaterial sphere are conditions, as karma condition or decisive–support condition, for profitable resultant consciousnesses of the different kinds of rebirth linking and in the course of an existence in the sense sphere and the fine–material sphere. The Path of Purification details exactly how each profitable consciousness results in the appropriate resultant consciousness, both for happy and for unhappy destinies. Even in hell a desirable object may be encountered.
Refs Buddhaghosa, Bhadantacariya *The Path of Purity* (1975); Buddhaghosa, Bhadantacariya *The Path of Purification* (1980).
Broader Merit (#HH0859).
Related Formation of demerit (Buddhism, #HH3226)
Ten meritorious deeds (Buddhism, #HH0446)
Formation of the imperturbable (Buddhism, #HH5543)
Profitable consciousness in the sense sphere (Buddhism, #HM4447)
Profitable consciousness in the fine–material sphere (Buddhism, #HM5338).

♦ **HH5124 Listening with the heart**
Negotiating from openness
Description Particularly in conflict situations, negotiation usually takes place from a position of self–defence and attempts to manipulate the other party, from attempts to convince the other party that one is right and they are wrong. If, in contrast, mind and heart are opened to the other person so that one is willing to see if one can learn from their point of view, fear and pride are moved through; and the other person is given the psychological space to receive one's own viewpoint. Being willing to work from this position of vulnerability and openness releases the energy normally used for self–defence and manipulation so it can be used to find creative and permanent solutions to problems which will not then arise again at some future date.
Related Ahimsa (#HH0088).

♦ **HH5190 Human development through pantheism**
Description Far eastern and, in the West, pre–Christian pantheism see divinity in every natural phenomenon, a multitude of gods. Activities such as cultivation of the soil become religious rites. Although not polytheistic, modern pantheism still emphasizes experience of nature or *ecological consciousness*. Contact with nature "in the wild" leads to spiritual catharsis and peak experience. Intellectually, there is an acceptance of evolution and scientific discovery as a means of spiritual enrichment – knowledge of how a rainbow is formed, for example, enriches without removing the mystery. Worship consists of reverencing something for what it is rather than as a symbol of something else. Nature ceases to be viewed as an external resource to be exploited – the forces and workings of nature are God, the sacred exists within the natural world and may be experienced with joy, the impulse to use nature for one's own ends is checked. This is monism, the universe being one sole substance, God and nature. This holistic view sees no division between the individual person and his physical surroundings. God is not personified, is neither masculine nor feminine; anthropomorphism is rejected as too limiting, reality is more diverse and abundant.
Particularly characteristic of Hindu religion, but approached by mystics of many religions, is the belief that God is the totality of existence. *Pancosmism* sees God as only the physical creation, with nothing existing outside the creation, a view some have taken to be atheistic; whereas *acosmism* sees the creation as an illusion in that it appears to change and God the only reality, unchanging, the view of Hinduism and Brahmanism. These are viewed by some authorities as the two extremes of pantheism, whereas the latter is sometimes referred to as *panentheism*.
Broader Human development through religion (#HH1198).
Related Primitivism (#HH1778) Deep ecology (#HH2315)
Planetary consciousness (#HM2006) Environmental mysticism (#HM0800)
Human development through panentheism (#HH1908).

♦ **HH5207 Middle middle grade** (Taoism)
Description Here there are all sorts of artificial visualizations, swallowing fog, culling the light of the sun and the moon and the energies of water and fire, drinking the lights of the stars, taking in the energies of the five directions.
Context In the nine grades of practices which are side tracks and auxiliary methods in Taoism, this is the middle of the three middle grades.
Broader Sidetracks and auxiliary methods in Taoism (Taoism, #HH7004).
Followed by Upper middle grade (Taoism, #HH2633).
Preceded by Lower middle grade (Taoism, #HH3296).

♦ **HH5213 Li**
Sense of propriety — Confucianism
Description The tradition of Confucianism stresses the religious significance of acting in accordance with an informed sense of propriety, *li*, with emphasis on a carefully structured system of human relations.

♦ **HH5217 Mystical theology** (Christianity)
Spiritual theology
Description The framework for the life of the spirit enunciated by Pseudo–Dionysius identifies a progression of three stages: purgation; illumination; mystical contemplation, or complete union with the divine. These three stages were later equated with the three stages of growth of supernatural charity in the soul: beginners; those making progress; perfection. Progress is not necessarily a move from one to another, the stages may be mutually present in one individual. The first stage is distancing one's self from sin and from inclination to do wrong; the second is to cultivate virtue; the third is loving union with God.
Although thus once referring to the whole of spiritual life, mysticism gradually came to refer only to the highest peaks of such life, *ascetical theology* being the term equated with the preparatory stages. Currently ascetical theology and mystical theology are together component parts of *spiritual theology*, the one referring to the practical side of Christian life and including asceticism, the other to mystical experience, including *infused contemplation*.
Related Illumination (#HH0804) Religious experience (#HM3445)
Mystical contemplation (#HM2710) Via purgativa (Christianity, #HH4090)
Ascetical theology (Christianity, #HH3652).

♦ **HH5219 Journeying within transcendence – from forgetfulness to memory**
Description An unfocused life, forgetful of the past and drifting into the future, is replaced by prayerful remembering. This leads to a deeper remembering of who one is before God. Being in apparent control is replaced by an awakening response, to the desire for more than just pleasure but a deeper meaning to life, As one begins to accept one's self, one's vision expands to see the truth.
Context The sixth section of St John's Gospel, Chapter IV 1–41, is related to a stage in the spiritual journey of the individual.
Broader Journeying within transcendence (Christianity, #HH6505).
Followed by Journeying within transcendence – from isolation to imagination (#HH2498).
Preceded by Journeying within transcendence – from head to heart (#HH4974).

♦ **HH5221 Infallibility**
Description Despite the fallibility of all human beings, including the Pope, in personal conduct of personal views, the Church as a tangible entity remains in the truth. This could not be the case if its own statements of belief were wrong. There are historical limitations attached to dogma, since every human statement is exposed to misunderstanding, is capable of interpretation, is in need of development; but realization of the truth of statements and propositions brings realization of the basic and original truth of God's revelation of himself to man.

♦ **HH5223 Old age**
Spiritual maturity — Rejuvenescence
Description Despite the negative response usually given to growing old, in fact there are many advantages in reaching mature years. Some cultures have valued this period of life above all others and in most cultures old age is treated with respect and deference to the acquired wisdom that experience brings. Depending on the mental attitude of the individual concerned, old age may bring the opportunity to develop those qualities which have been shelved due to the necessity of dealing with pressing everyday matters.
Related Maturity (#HH0971).

♦ **HH5224 African spirituality**
Refs Cousins, Ewert, et al (Eds) *African and Oceanic Spirituality*.
Broader Spirituality (#HH5009).

♦ **HH5232 Love of God** (Christianity)
Description According to St John of the Cross, spiritual friendship with another person leads to love which encourages love of God. Remembering love of God brings greater desire for God. Adding goodness to goodness makes goodness grow. Conversely, friendship for another person based on lust causes love of God to diminish. Remorse occurs as the person becomes aware that the soul is forgetting God in favour of the friend. Then the growing love for God can make the lustful love diminish and become forgotten.
According to François Fénelon, there are three kinds of love of God:
Mercenary or selfish love, which is love for God originating in desire for one's own happiness. If this is the only love one feels for God, then God is only a means to an end and it is sacrilegious and impious, seeking only self–gratification.
Mixed love, where a regard for one's own happiness, although there, is subordinate to a regard to the glory of God. This is not necessarily wrong, as loving God has He ought to be loved and one's self no more than one ought is both unselfish and right.
Pure love is mixed love carried to its true result. This result implies that the motive of God's glory so fills the mind that the motive of one's own happiness is practically annihilated. God becomes the centre of the soul to whom all affections tend, the sun from whom all light and warmth proceed. Ones own happiness and all that regards one's self is lost sight of. It is not that it is wrong to desire one's own good, simply that when God is in the soul who can think of himself ? God alone is loved, all other things in and for God.
Refs Fénelon, François *The Maxims of the Saints*.
Related Love (#HH0258) Care (#HH0042) Love to God (Islam, #HM5116)
Love of God (Sufism, #HM4273) Quietism (Christianity, #HH4477)
State of perfect love (Christianity, #HM0074).

♦ **HH5234 Spirituality** (Esotericism)
Refs Cousins, Ewert, et al (Eds) *Modern Esoteric Spirituality*.
Broader Spirituality (#HH5009).
Related Esoteric development (#HH6902).

♦ **HH5256s Motionless meditation**
Description A technique practised regularly during sessions of karate to clear the mind.
Broader Meditation (#HH0761) Martial arts (#HH0085).

♦ **HH5282 Ecstatic discipline**
Description This tends to culminate in an an "out of body" experience in which the seeker is no longer limited by time and space, often as a result of ascetic practices. An example is the vision quest of North American Indians. An enduring out–of–body experience is that moved through by the soul in the period between death and rebirth of Tibetan Buddhism.
Broader Discipline (#HH0163).
Narrower Vision quest (#HH0897)
Bardo consciousness (Buddhism, Tibetan, #HM0698).
Related Ecstasy (#HM2046) Out–of–body experience (Psychism, #HM5534).

♦ **HH5304 Honesty**
Description In thought, word and deed the individual behaves in a way which does not conflict with his basic beliefs. He demonstrates integrity and his conduct is upright and straightforward.
Related Integrity (#HH0234).

♦ **HH5328 Becoming Christ**
God frequency
Description Christ is the second member of the godhead. He is the aspect which creates life, born of the Father issuing forth spirit into matter. Christ, as the force of life, is potential in all aspects of creation, inherent in the soul–centre of every person. This potential is dormant, but can be awakened so that the qualities underlying the essence of life – light and love, and also wisdom and power in combination – become conscious. This requires transformation to another level, that of pure spirit, through raising the vibrational frequency of the energy which the person expresses. The spiritual level, experienced in the past by mystics and sages, is the next stage in evolution from the present, in which the world is at its most dense. Not just worshipping those who have reached this level in whatever religion, of whom Master Jesus is the fullest expression, each individual has now to touch and hold the spiritual levels which are potentially available.
The first stage is awakening. A response in the human heart is stimulated by ritual and worship, although the pull of matter is still strong. Interpretations and dogma may be an obstacle to progress, as the individual worships something external, inhibiting awakening to the internal light.
In order to become conscious of the internal Christ light, the personality must be light and warm, vibrations being of a higher frequency than before. When a certain overall level is reached light and love shine forth. This already happens during certain peak experiences. The aim is to raise the overall vibrational frequency so that the individual not only touches but lives in *God frequency*. The key is to *'be still and know that I am God'*, through the process of meditation. Eventually living and meditating become synonymous, but before then regular practice of meditation, especially in groups, is the way to throw off the blocks that prevent rise in frequency. Other tools may also

be useful – psychotherapy and other techniques of personal growth – but they are all simply means to this end.
The next step is initiation, contact with the *'still, small voice'* that is the inner voice of the soul. This gradually develops with practice, at first only present for short times but eventually always there to be heard and obeyed. Constantly living in the presence of God means that frequencies are raised and the voice is not separate from normal consciousness.
Ultimately the person is totally identified with higher spiritual frequencies – Christ frequencies. Not listening or responding to the Christ consciousness within, the person is Christ in the same sense that *'I and my Father are one'*.
There are no short cuts in this way. Attempting to achieve this state, through drugs for example, could bring disorientation or insanity. There must be a steady progress through all levels for the light to shine through. At the end of the Piscean age, entering the age of Aquarius, the light attracts those who are already light. The stages are: (1) birth of the Christ in the heart; (2) connection with the inner light; (3) preparation and testing (overcoming the temptations of the soul); (4) crucifixion or death of the ego, as the spirit overcomes matter; (5) resurrection, or wakening to the reality of the risen Christ. In a sense these are sequential, although progress may be made across several stages and the next stage is always visible from the previous one.
Related Christ consciousness (Christianity, #HM2013)
Mystical Christ–consciousness (Christianity, #HM2880).

♦ **HH5342 Three vehicles of gradual method** (Taoism)
Description The firing process is refining thought by the mind. The fire is nurtured through ceasing thought. Stabilization is through keeping brilliance to one's self. In the "battle in the field" inner demons are conquered. The three essentials are body, mind and will. The mysterious pass is the heart of heaven.
Refs Cleary, Thomas (Trans) *The Book of Balance and Harmony* (1989).
Narrower Lower vehicle path (#HH4864) Middle vehicle path (#HH6345)
Higher vehicle path (#HH7386).
Followed by Highest vehicle (Taoism, #HH4742).
Preceded by Sidetracks and auxiliary methods in Taoism (Taoism, #HH7004).

♦ **HH5343 Human development through humour**
Refs Keller, Daniel *Humor and Therapy* (1984); Kuhlman, Thomas L *Humor and Psychotherapy* (1984); McGhee, Paul E *Humor and Children's Development* (1989).

♦ **HH5434 Maturity growth**
Psychic growth
Description Unlike primary growth, which proceeds through infancy, childhood and adolescence and may be said to cease on reaching adulthood, maturity growth does not recognize time limits. It may potentially continue throughout life, declining perhaps in senility or ceasing only in death. This type of growth is slower and less spectacular than primary growth. It is not determined in intensity or duration by the person's genetic code, nor expressed through the individuals physical size or acquisition of sensorimotor or intellectual capabilities. It develops or is restricted by interaction with a given environment and is a certain type or expression of psychological growth. It is expressed through strengthening and enriching the personality, involving a process of constructing and consolidating the self, of developing certain capacities and powers. Provided there is a favourable socio–cultural context, these are encountered in a state of active tension or latency in the individual once he has begun to pass through the final stages of primary growth.
The following facets are those through which maturity growth is revealed: profound emotivity; capacity for communication; rationality, imaginative capacity and intuition; constructing the self and consolidating personal identity; sensitivity and experiential openness; creative and expressive capacity; psychosomatic integration; adult realism; capacity for constructive work; impulse toward active social transcendency.
Refs Mallmann, Carlos A and Nudler, Oscar (Eds) *Human Development in its Social Context* (1986).
Broader Psychic growth (#HH3117).
Related Primary growth (#HH6669).

♦ **HH5450 Muni**
Way of silence
Description This is the path of the vedic ascetic who renounces the world and practices silence as a sacred act, conserving life force and gaining power over the numinous.
Related Mythic yoga (Yoga, #HH3405) Practice of silence (#HH0983).

♦ **HH5487 Niskama karma** (Hinduism, Yoga)
Action without desire
Description Under the influence of *kama* (personal desire), ordinary actions are performed which result in the building up of a personal karma. This is because such actions are performed while indentifying with the ego, seeking fulfilment of desires, and result in both pleasurable and painful experience. However, when the person completely dissociates himself from the ego and performs actions while identifying completely with the Supreme Spirit working through the ego, then such actions without desire, *niskama*, produce no more karma, despite possibly being engaged in the affairs of the world. Consciously identifying with the divine, with no taint of personal motive, no more karma being formed, it remains to work off karma or sanskaras previously accumulated.
Broader Karma awareness (Buddhism, #HM2048).

♦ **HH5522 Communalist human development**
Description Encounter groups and alternative communities in search of a humanist way of living are, in the first case, parallel to everyday social life and in the second, small scale and devoid of weight in present–day society.
Context One of four current models of human development, that described as collective internal.
Refs Mallmann, Carlos A and Nudler, Oscar (Eds) *Human Development in its Social Context* (1986).
Related State socialist human development (#HH3997)
Liberal humanist human development (#HH4033)
Liberal capitalist human development (#HH6122).

♦ **HH5543 Formation of the imperturbable** (Buddhism)
Preparation for stationariness
Description In Hinayana Buddhism, this is the condition for the resultant consciousnesses of the immaterial or formless sphere, whether in the course of existence or at rebirth linking, in immaterial becoming.
Refs Buddhaghosa, Bhadantacariya *The Path of Purity* (1975); Buddhaghosa, Bhadantacariya *The Path of Purification* (1980).
Broader Merit (#HH0859).
Related Formation of merit (Buddhism, #HH5122)
Formation of demerit (Buddhism, #HH3226)
Ten meritorious deeds (Buddhism, #HH0446)
Indeterminate consciousness in the immaterial sphere – resultant (Buddhism, #HM4982).

♦ **HH5547 Magical meditation**
Description Meditation is used as one of the five magical arts to collect and re–focus awareness towards the core of ritual or ceremony. As such it is a means to an end rather than an end in itself. In magical work there is a balance between inner and outer action. The outer action is defined by words, movements and physical symbols. The inner action is established in meditation which therefore resembles the controlled consciousness–in–movement to be found in the genuine martial arts. Inner and outer awareness fuse together to give a perfect balance. It is not a question of withdrawing consciousness from the outer world as in other mediation techniques.
At a deeper level of meditation, the practitioner (or group) attunes to specific energies, filling the imagination and even the total awareness with the energy in question. Magical meditation does not necessarily pass on into a contemplative stage of formless awareness. In practice in a group different members may take different responsibilities, such that some may act in contemplation as poles of transcendent awareness that may be directed by fellow members using ritual and imagery. Higher consciousness is thus brought into the time– bound world in a highly concentrated form and channelled through mediating patterns towards its goal.
Broader Magic (#HH0720) Meditation (#HH0761).

♦ **HH5550 Meditative reading**
Description According to Madame Guyon, this is one means of being led into the highest forms of prayer. The other is meditation. In meditative reading, a doctrinal or practical truth is read, two or three lines at a time, remaining on each short passage and seeking to enter the full meaning of the words for as long one finds satisfaction in it, before going on to the next passage.
Refs Guyon, J M B de la Mothe *A Short Method of Prayer and Spiritual Torrents* (1875).
Related Spiritual reading (#HH0158) Christian meditation (Christianity, #HH5023)
Sense of the presence of God (Christianity, #HM6989).

♦ **HH5634 Transformation game**
Description This board game is a joyful way of understanding and transforming the way people play their lives. It focuses people on the kinds of experiences that they can create, allowing them to observe how they respond, and how they can use the insights for their own growth. The game enables people to become more aware of their own strengths and to learn lessons that can deepend their understanding of how to operate on the physical, emotional, mental and spiritual levels. It can also be used as a powerful interactive tool, by both individuals and groups, for solving problems and achieving desired goals. By providing new perspectives on current life issues, the game helps people clarify old beliefs and attitudes and transform reaction patterns. It offers the possibility of discovering new dimensions and allowing changes to occur with greater understanding and ease. The game is a multi–purpose tool which can be played for fun or easily adapted by therapists, counselors and professional facilitators to promote more in–depth experiences to facilitate decision–making processes.
Context The game was developed by Joy Drake and Kathy Tyler through the Findhorn Foundation. It exists in initial and in advanced versions which may be conducted over several days by trained guides.
Refs InnerLinks *The Transformation Game* (1987).
Broader Growth games (#HH0175).

♦ **HH5640 Fortitude**
Courage
Description Courage is one of the virtues necessary for all moral conduct, a mean between rashness and cowardice. It is the capacity of human character to resist pain, danger or adversity by enduring what cannot be changed and changing what cannot be endured in the love of God and neighbour. It has been defined as the emotion involved in practising fortitude. Christianity has regarded fortitude as a passive form of courage, exemplified by the experiences of the martyrs. Thomas Aquinas defined fortitude as the order of reason with regard to the passions when these draw us away from a reasonable course of action.
Courage, according to Aristotle, is that which engenders respect when, from a higher or noble motive, danger is despised.
Context One of the four cardinal or principal virtues recognized by Plato in the Republic and featuring prominently in mediaeval Christianity.
Broader Virtue (#HH0712).
Related Heroism (#HH0929) Justice (#HH2117) Prudence (#HH7902)
Endurance (#HH0729) Temperance (#HH0600).

♦ **HH5643 Zakat** (Islam)
Almsgiving
Description A pious act of giving the surplus to one's own requirements for the benefit of the needy.
Context One of the five pillars of Islam.
Broader Human development (Islam, #HH1799).
Related Dana (Buddhism, Zen, #HH1234).

♦ **HH5645 Path of enlightenment** (Buddhism)
Bodhi–marga
Related Steps to enlightenment (Buddhism, #HH4019).

♦ **HH5652 Supernormal powers** (Buddhism)
Iddhi (Pali) — Psychic powers — Direct knowledge — Higher knowledge
Description Five kinds of direct knowledge are enumerated: knowledge of supernormal powers such as one becoming many; deva–hearing (divine ear element); penetration of minds (knowledge of others' thoughts); recollection of previous existences; knowledge of passing away and rebirth of beings.
Only those having vast previous endeavour are endowed with such knowledge. For others wishing achieve it, the Path of Purification gives details as to how such knowledge is attained. First, the meditator must achieve the eight attainments (jhana) in each of the first eight kasinas, ending with the white kasina. He has complete control of his mind so that he may subdue it in fourteen ways:
1. He repeatedly attains jhana in each kasina in order, starting with earth and ending with white.
2. He repeatedly attains jhana in each kasina in reverse order, starting with white and ending with earth.
3. He repeatedly attains jhana in each kasina in order and in reverse order, from earth to white and then from white to earth.
4. He repeatedly attains from the first jhana to the base of neither perception nor non–perception in the order of the jhanas.
5. He repeatedly attains from the the base of neither perception nor non–perception to the first

HH5652

jhana in the reverse order of the jhanas.
6. He repeatedly attains from the first jhana to the base of neither perception nor non–perception and from the base of neither perception nor non–perception to the first jhana in the order and then the reverse order of the jhanas.
7. Following the order of the kasinas from earth to white, he skips alternate jhanas in each kasina (first, third jhanas, base of boundless space, base of nothingness).
8. Following the order of the jhanas he skips alternate kasinas, so that the first jhana is attained in the earth kasina, then in the fire, then in the blue, then in the red.
9. Skipping jhanas and kasinas he goes from the first jhana in the earth kasina to the third jhana in the fire kasina, then boundless space after removing the blue kasina, then nothingness from the red kasina.
10. Transposing factors he goes from the first jhana in the earth kasina to the other jhanas in the same kasina.
11. Transposing the object he goes from the first jhana in the earth kasina to the first in the water kasina and so on.
12. From the first jhana in the earth kasina he goes to the second jhana in the water kasina up to the base of neither perception nor non–perception from the white kasina – transposing factor and object.
13. Definition or fixing of factors – the first jhana has five factors, the second three, up to the base consisting of neither perception nor non–perception.
14. Definition or fixing of only the object – this has earth kasina as object, etc.
Having reached this stage, which is very difficult and achieved by very few, the meditator may begin to accomplish transformation by supernormal or psychic power. The mind is purified and bright, unblemished and devoid of evil, it has become supple and is ready to act, it is firm and steady and thus imperturbable. When consciousness possesses these eight factors it may be directed to realization through direct or psychic knowledge of states which can be realized by such knowledge. Numerous stories are told of those who, achieving this state, work miracles.
For direct knowledge, further to the practice described above, the four planes or stages, the four bases or roads, the eight steps, the sixteen roots of supernormal power must be accomplished. The planes are the four jhanas: first, seclusion; second, happiness and bliss; third, equanimity and bliss; fourth, neither pain nor pleasure. The roads are: concentration due to purpose, zeal, right effort (will to strive); concentration due to energy and right effort (will to strive); concentration due to natural purity of consciousness and right effort (will to strive); inquiry and right effort (will to strive).
The steps on the road are: obtaining unification of mind or concentration supported by zeal; obtaining unification of mind or concentration supported by energy; obtaining unification of mind or concentration supported by purity of consciousness; obtaining unification of mind or concentration supported by inquiry.
The roots are of imperturbality of mind are: undejected consciousness; unelated consciousness; unattracted consciousness; unrepelled consciousness; independent consciousness; untrammelled consciousness; liberated consciousness; unassociated consciousness; consciousness rid of barriers; unified consciousness; consciousness reinforced by faith; consciousness reinforced by energy; consciousness reinforced by mindfulness; consciousness reinforced by concentration; consciousness reinforced by understanding; illuminated consciousness.
Having accomplished all these things, the meditator attains jhana as basis for direct knowledge and emerges from it and repeats the process depending on what he resolves, for example, resolving to become many. Then, when he emerges from the jhana and resolves to become many he appears as, say, 100 persons. Similarly the visible may be made invisible and the invisible visible. Or there may be control of the elements through their respective kasinas, so that earth may become as water and be dived into and out of, or earth may be made in space so the meditator travels in space as he does on earth. Similarly what is far can be made near, what is little can be made much, what is much can be made little. There may be transformation into another body or creation of another body. Numerous examples of mind–made changes are enumerated.
Narrower Knowledge of supernormal powers (#HM7672).
Related Siddhis (Yoga, Buddhism, #HH0380) Tenfold powers (Buddhism, #HH2918)
Unified consciousness (Buddhism, #HM7843) Unelated consciousness (Buddhism, #HM5623)
Liberated consciousness (Buddhism, #HM8241)
Undejected consciousness (Buddhism, #HM6521)
Unrepelled consciousness (Buddhism, #HM5908)
Unattracted consciousness (Buddhism, #HM5335)
Independent consciousness (Buddhism, #HM6243)
Illuminated consciousness (Buddhism, #HM6208)
Untrammelled consciousness (Buddhism, #HM6791)
Unassociated consciousness (Buddhism, #HM8092)
Consciousness rid of barriers (Buddhism, #HM8621)
Consciousness reinforced by faith (Buddhism, #HM7902)
Magical–forces awareness (Buddhism, Tibetan, #HM2679)
Consciousness reinforced by energy (Buddhism, #HM4396)
Consciousness reinforced by mindfulness (Buddhism, #HM5499)
Consciousness reinforced by concentration (Buddhism, #HM5588)
Consciousness reinforced by understanding (Buddhism, #HM5901).

♦ HH5712 Love (Islam)
Mahabba
Description Many kinds of love are enumerated within Islam. One (threefold) division is love to God, love in God and love with God. A further division is between beneficial and harmful. The beneficial are love to God and love in God plus love for whatever assists in obedience to God. The harmful are love which diminishes or cuts short love to God, love of what is hated by God and love with God which is idolatry. One of the main functions of faith is to enable the person to distinguish between loves which are harmful and those which are beneficial. Love for one's lawful wife or concubine is normally beneficial, protecting from sin, whereas passion for forbidden objects for their own sake, such as that for small boys, is harmful.
The stages by which both sacred and profane loves develop are also described by a number of teachers. Even the most profane have their mystical interpretations.
Refs Bell, Joseph Norment *Love Theory in Later Hanbalite Islam* (1979).
Broader Love (#HH0258).
Narrower Love to God (#HM5116) Love in God (#HM4712)
Love with God (#HM5580) Passionate love (#HM3656).
Related Mahabba (Islam, #HM6523).

♦ HH5767 True knowledge (Christianity)
Description In a state of simple faith, souls are taught by the indwelling Spirit of God. The knowledge so received could never be imparted by the wisdom of the world, yet they remain respectful of religious teachers, docile to Church instructions and conforming to scriptural precepts.
Related Knowledge (Buddhism, #HH1971).

♦ HH5786 Educational guidance
Description Aid is given in finding training commensurate with a person's abilities and preferences. This may be direct guidance to the individual involved or, particularly, where the education of a child is considered, advice to parents and educators on the physical, intellectual and moral development of that person. The aim here is to promote independence and self discipline. Special educational guidance is important in the case of handicapped people and for people with personality defects.
Refs Hopke, William E (Ed) *Encyclopedia of Career and Vocational Guidance* (1987).
Broader Guidance (#HH0178).

♦ HH5856 Journeying within transcendence - from me to you
Description Although the disciples have misunderstood Jesus they still follow him in faith, unlike the Jews who seek his death. His death will break and shatter them at many levels, but there is still hope. They will receive the strengthening and power of the Holy Spirit which will transform the negative to the positive and act as another advocate for them with God. The Spirit witnesses in the heart so that the disciple may witness externally to others. At the centre of the self there is an openness to God and humanity; and it is from there that the energy, breath, life–giving force acts to lead one out of one's self, beyond the limits, there is the revelation of who "I am". From here there is a moving out to others, a refusal to move out diminishes the spirit within.
Context The fifteenth section of St John's Gospel, Chapter XIII 31 to XVI 31, is related to a stage in the spiritual journey of the individual.
Broader Journeying within transcendence (Christianity, #HH6505).
Followed by Journeying within transcendence – from son to father (#HH1773).
Preceded by Journeying within transcendence – from served to servant (Christianity, #HH6012).

♦ HH5887 Alchemy (Taoism)
Description Taoist alchemy may be described in terms of processes whereby real knowledge (symbolized by water) is retrieved from the overlay of artificial conditioning. The real knowledge is then used to replace the mundanity infecting conscious knowledge (symbolized by fire), thereby restoring the basic completeness of the primordial celestial mind. Expressed differently, the challenge of human development is that consciousness is normally volatile, given to imagination and wandering thought. Real knowledge then tends to become submerged in the unconsciousness, sinking into oblivion. There is no appropriate integration of the two forms of knowledge which act separately from one another. Through the alchemical process, these two forms of knowledge are forced to interact. Real knowledge (water) stabilizes consciousness (fire) and removes its volatility, while consciousness brings real knowledge into action in life.
The task of alchemy is therefore twofold, to "empty the mind" and to "fill the belly". The first is that of cultivating essence, displacing the mundane preoccupations of the human mind. The second is that of cultivating life. When the "belly is full", sane energy arises through accumulation of right action, and the energy of mundane conditioning dissolves of itself. An alternative representation of the task is that of discovering the flexibility within strength and the strength within flexibility. Another is that of seeking sense through essence and returning essence to sense, meaning that essence and sense unite.
The "firing process" is a metaphor employed in alchemical texts for the order of practical spiritual work, namely the order of application of effort in the cultivation of reality. Associated with this process is the notion of a "crucible" which is subjected to the firing and within which transmutation takes place. It is through the firing process that the encrustations of the faculties are burnt away to expose the awareness of the original spirit. This requires an appropriate combination over time of inward discipline, deflection of externals, application of effort, gentle nurturing, and use and withdrawal of energy.
Under the normal conditions, natural processes result in a life cycle of seven states: the generative state of the womb; the state of birth and infancy; the state of childhood; the recognition of dichotomies following the division of yin and yang; the separation of the five elements, imbalancing one another and encrusting the senses; the predominance of acquired conditioning, governing emotions and desires; and finally, the complete domination of mundanity, with the extinction of positive energy, leading to death.
This natural cycle is reversed by the alchemical firing process in seven corresponding stages through which immortality is achieved. Through this process reality is cultivated, restoring the self, which otherwise gradually dies through the natural process described above. Real celestial positivity returns in the midst of total mundanity.
Refs Chang, Po–tuan *Understanding Reality* (1987); Chang, Po–tuan *The Inner Teachings of Taoism* (1986); Cleary, Thomas (Trans) *The Book of Balance and Harmony* (1989); Wilhelm, Richard (Trans) *The Secret of the Golden Flower* (1962).
Broader Human development (Taoism, #HH0689).
Narrower Bathing (#HH3664) Refining the self (#HM4007)
Unification of energy (#HM4762) Transcending the world (#HM5265)
Merging of yin and yang (#HM1330) Recognition of the true mind (#HM6633)
Assembling the five elements (#HM4338)
Restoration of celestial awareness within the mundane (#HM6534).
Related Alchemy (Esotericism, #HH0221) Circulation of the light (Taoism, #HH6452).

♦ HH5902 Spirituality (Islam)
Refs Cousins, Ewert, et al (Eds) *Islamic Spirituality* (1987); Cousins, Ewert, et al (Eds) *Islamic Spirituality*.
Broader Spirituality (#HH5009).
Related Sharia (Islam, #HH0738) Human development (Islam, #HH1799)
Human development (Sufism, #HH0436).

♦ HH5908 Journeying within transcendence - from agape to philo
Description Even now there is a narrowness and limitedness to the individual who is called by the deeper self to develop beyond psychic infancy, to come out of the confines of limited thinking, to love and care for the fragmented parts of existence. The way to something larger is pointed and one is invited to follow in faith the larger vision, trusting and valuing one's own experience and calling, while respecting those of others. Transformation is not yet complete, the journey with the transcendent will last a lifetime.
Context The nineteenth section of St John's Gospel, Chapter XXI 1–24, is related to a stage in the spiritual journey of the individual.
Broader Journeying within transcendence (Christianity, #HH6505).
Preceded by Journeying within transcendence – from extraordinary to ordinary (Christianity, #HH4110).

♦ HH5973 Citta–bhavana (Buddhism)
Development of mind
Description This practice of samadhi implies both mental and physical training. Systematic meditation trains the mind until, with the full development of samadhi, mental power is concentrated and the mind achieves self–mastery. It can resist the current of pleasant or painful feeling and is unshaken by sense stimuli, even enduring deadly pain.
Broader Bhavana (Buddhism, #HH0551).

♦ HH6012 Journeying within transcendence - from served to servant (Christianity)
Description As a symbol of utter humility and sacrifice in death, Christ washes the disciples' feet, not playing the role of servant but being who he is. This is compared with Judas who, although

outwardly a disciple, has only been playing a role. He cannot accept the invitation to integrate the role with his life but refuses self-knowledge. Life can be in service of others needs, or a selfish attempt to protect one's self from death. Fidelity to the ego may be infidelity to the self. The self that I am is a gift to be accepted or rejected.
Context The fourteenth section of St John's Gospel, Chapter XIII 1–31, is related to a stage in the spiritual journey of the individual.
  Broader Journeying within transcendence (Christianity, #HH6505).
  Followed by Journeying within transcendence – from me to you (#HH5856).
  Preceded by Journeying within transcendence – from death to life (#HH2987).

♦ **HH6018 Education for early achievement of excellence**
Description Young (pre-school) children, given the opportunity to absorb facts on every subject and to excel physically, in a warm and loving atmosphere, are shown to excel socially as well. Research has shown that, having received such an impetus before attending normal school, children maintain their level of excellence, profiting from school when they start to attend despite being ahead of their peers, particularly when given challenges to match their level of attainment. The children show self-confidence and independence and seem to act as role models for the others.
  Refs Harvey, Neil *The Renaissance Children* (1988).
  Broader Education (#HH0945).
  Related Upbringing (#HH0908)   Educational self-development (#HH0955).

♦ **HH6019 Following Jesus** (Christianity)
Description This requires of the believer the readiness in faith to yield full power over himself to the Kingdom of God present in Jesus, to deny himself and to take up the cross of Christ.

♦ **HH6030 Via illuminativa** (Christianity)
Description As God's light is imparted to Christians, so they must allow their good works to shine like a light before others so that God is glorified.
Contemplation, the *via illuminativa*, is a step on the path to full union, the *via unitiva* of Christian mystical teaching.
  Broader Triple way (Christianity, #HH0631).
  Related Via negativa (#HH1171)   Via positiva (#HH6222)
  Illumination (#HH0804)   Mystical contemplation (#HM2710).
  Followed by Via unitiva (Christianity, #HH4100).
  Preceded by Via purgativa (Christianity, #HH4090).

♦ **HH6053 Guided visualization** (Magic)
Description As one of the five basic arts of magic, visualization is used to contact and develop subtler levels of consciousness. This discipline should be distinguished from recent initiatives in mental therapy to use relaxing guided fantasies in some forms of therapy. For more challenging experiences, several conditions should be fulfilled: the symbolism needs to be coherent and related to a specific tradition; no attempt should be made to complete the visualization or render it all-inclusive since this inhibits imaginative participation; the sequence of symbols should include challenging and even disturbing phases, and not be simply supportive and comforting; opportunities should be made for silent meditation to explore any insights that are triggered by the sequence; traditional symbols are more effective than those from popular culture; the visualization should bear some structural relationship to magical pattern-making. Visualizations should be characterized by intellectual, psychological, topological and cosmological clarity through which related realms of consciousness merge, dissolve and re-emerge in a master pattern. A complex visualization moves through several levels of consciousness or magical worlds.
  Refs Stewart, R J *Advanced Magical Arts* (1988).
  Broader Magic (#HH0720).
  Related Pathworking (#HH8003)   Guided fantasy (#HH0627)
  Active imagination (#HM0867)   Alchemical imagination (Esotericism, #HM1409).

♦ **HH6071 Apokatastatis**
Description The restitution of all things to their essential perfection. This possibility is held to be latent in man and all creation. Love is considered to be the prime agent of apokatastasis, notably in Christian and pre-Christian traditions.
  Related Love (#HH0258)   Poetic imagination (#HM1538).

♦ **HH6089 Ambition**
  Related Success (#HH3188)   Achievement motivation (#HH0456).

♦ **HH6109 Ethological interpretation of behaviour**
Description A pure survival value is placed on all human behaviour. Whatever one feels to the contrary, a baby smiles at its mother to create a feeling of pleasure in the mother and thus foster maternal feeling which ensures the baby's survival. This interpretation assumes innate response must necessarily be divorced from conscious experience and ignores unselfish love or attachment.

♦ **HH6112 Therapeutic double bind**
Description A situation imposed upon a client by a therapist, in which any response by the client will be an experience, or reference structure, which lies outside the client's current model of the world. Such double binds thus implicitly challenge the client's model by forcing him into an experience which contradicts the impoverishing limitations of his own model. The experience then serves as a reference structure which expands the client's model of the world.
  Related Neuro-linguistic programming (NLP, #HH4872).

♦ **HH6115 Expression of human entitlements**
Description Each person in society can command a set of alternative commodity bundles using the totality of rights and opportunities that he or she faces. On the basis of this entitlement, a person can acquire some capabilities and fail to acquire others. The process of economic development can be seen as a process of expanding the capabilities of people. Given the functional relationship between entitlements of persons over goods and their capabilities, a useful (though derivative) characterization of economic development is in terms of expansion of entitlements. For most of humanity, about the only commodity a person has to sell is labour power, so that the person's entitlements depend crucially on his or her capability to find a job, the wage rate for that job and the price of commodities that he or she wishes to buy.
When the concern is for such notions as the well-being of a person, or standard-of-living, or freedom in the positive sense, there is a need for the concept of capabilities. The concern is with what a person can do. This is not the same thing as how much pleasure or desire fulfilment he gets from these activities ("utility"), nor what commodity bundles he can command ("entitlements"). Ultimately one has to go not merely beyond the calculus of the national product and aggregate real income but also beyond the calculus of entitlements over commodity bundles viewed on their own. The focus on capabilities differs also from concentration on the mental metric of utilities, and this contrast is similar to the general one between pleasure on the one hand and positive freedom on the other. The particular role of entitlements is through its effects on capabilities. It is a role that has substantial and far-reaching importance but remains derivative on capabilities.
  Refs Sen, A *Poverty and Famines* (1981); Sen, A *Resources, Values and Development* (1984); Sen, A *Hunger and Entitlements, Research for Action* (1987).
  Related Human-centred development (#HH3607).

♦ **HH6121 Native spirituality**
  Refs Cousins, Ewert, et al (Eds) *South and Meso-American Native Spirituality*.
  Broader Spirituality (#HH5009).

♦ **HH6122 Liberal capitalist human development**
Description Western thinking has always believed in the radical separateness of individual consciousness, even to philosophical doubt about the existence of the physical world and other minds. Economics puts the individual on one side, society on the other; and the problem is to devise a compromise sufficiently fair to both. Liberal capitalism, in common with other western thinking, emphasizes the individual and individual rights. Personal development is a private affair. The paradox is that general wellbeing can only be attained if people work with enthusiasm for their own self-interest.
In internal/external terms, liberal capitalism emphasizes achievement in the external world and equates development with success, whether due to knowledge, courage or intuitive ability, although it does not claim that success always accompanies achievement. This attitude is based on protestant ethics and on a behaviourist approach, the latter taking externalism to such an extreme that it virtually denies the existence of an inner life. (More sophisticated approaches such as cognitive psychology and artificial intelligence research still view man as a machine).
Externalism and individualism collide because the latter affirms the irreducible originality and non-interchangeability of human beings, so that this model of human development has an unresolved tension between freedom and determinism, inner autonomy and external manipulation. A new model, that of liberal humanism, is developing.
Context One of four current models of human development, that described as individual external.
  Refs Mallmann, Carlos A and Nudler, Oscar (Eds) *Human Development in its Social Context* (1986).
  Related Communalist human development (#HH5522)
  State socialist human development (#HH3997)
  Liberal humanist human development (#HH4033).

♦ **HH6129 Rinzai Zen** (Zen)
Lin chi
Description To prevent simply dreaming or using explanations which are just mental activity, the Rinzai school uses direct shock which may be physical violence or the use of a verbal conundrum – koan – to aid spiritual enlightenment.
  Refs Miura, Isshu and Sasaki, Ruth Fuller *The Zen Koan*.
  Broader Human development (Zen, #HH1003).
  Narrower Inward zazen meditation (#HH1134).
  Related Koan (Zen, #HH0263).

♦ **HH6131 Instrumental enrichment**
Intelligence enhancement
Description This system is based on the belief that apparent low intelligence, or intellectual shortcomings in otherwise gifted children, may be due to the fact that some basic thinking skills which would normally have been passed on by parents or grandparents have, for some reason, not been transmitted. Individuals particularly helped are children whose intelligence has been impaired by cultural breakdown, but experiments have also demonstrated in prison systems where adult offenders analyse their own situations and learn the intellectual control to prevent them committing impulsive criminal acts. The theory is that intelligence is the ability to learn from experience. Skills such as comparison of objects, focusing of attention for reasonable lengths of time, understanding the nature of cause and effect and of space and time are taught sympathetically using instrumental enrichment exercises developed by Professor Reuven Feuerstein. This leads to dramatic increases in intelligence scores. Perceptive faculties are sharpened among children who have never learned to analyse visual and auditory impressions, and who have no strategy or rules to interpret events.
  Related Intelligence (#HH0431).

♦ **HH6199 Sikh mysticism** (Sikhism)
  Refs Mohan, Singh Diwana *Sikh Mysticism* (1968).
  Broader Mystical cognition (#HM2272).
  Related Spirituality (Sikhism, #HH3774)   Human development (Sikhism, #HH6292).

♦ **HH6209 Humanistic capitalism**
Description Current capitalist society has failed to give each individual the opportunity for full and valued participation with the feeling of belonging and being useful. There is no synergism of decisions at the personal and organizational level leading to satisfactory macro-decisions at the level of society as a whole. And it has failed to achieve a satisfactory redistribution of power and wealth. Willis Harman suggests that these failures can only be resolved if growth and consumption are replaced by ecological and self-realization ethics. The first would foster a sense of total community of man, in oneness with the human race and in partnership with nature; and the second would counter the current alienation and anomie, placing the highest value on the development of selfhood. Under such ethics, institutions would serve the needs of those whose lives are touched by them; and the incentive structure of society would mean that the interest of society as a whole was promoted by the individual pursuing his own self-interest.
The self-realization ethic follows as a natural consequence of man's experience of his dual (physical and spiritual) nature. In business, also, it appears that the values necessary for the putting together and operating of current highly complex social-technological tasks are similar to those for supporting quality of life and the continuation of the earth as a habitable place. The distinction between "work" and "non-work" becomes blurred, as does the division of individuals into managers, owners, workers, consumers and the public. Where what is good for business conflicts with what is good socially, then the cultural system and the institutions serving it should be – and are – modified. Opportunities for full and valued participation in society are made available to all who require them. Economic efficiency becomes less dominant than the actualization of human potential, achieving community and being socially responsible.
  Refs Harman, Willis W *Humanistic Capitalism*.
  Broader Humanism (#HH0674).

♦ **HH6210 Spirituality in tantric yoga**
Description Spirituality is the awakening of divinity in consciousness, when consciousness is freed from the trap of mind and body. Freedom arises as the sense consciousness (mind) is gradually transformed. Initially sense consciousness perceives the world and produces uncontrolled thoughts. It desires, it senses pleasure and pain, it thinks, wills and seeks pleasure, in the latter aspect sometimes committing excesses. Transforming sense consciousness brings free-

dom from the slavery of mind, lust, greed, uncontrolled thoughts and internal dialogue. The other aspect of being can be experienced, when mind is detached from the sensory world, no longer thinking, desiring or willing. "I"–ness moves away from the cycle of birth and death and merges into supreme consciousness.
Refs Johari, Harish *Chakras* (1987).
Broader Spirituality (#HH5009).
Related Tantra (Buddhism, Yoga, #HH0306).

♦ **HH6211 Initiation** (Christianity)
Becoming a Christian
Description When the process of evangelization has occurred, bringing the new Christian into the body of Christ, then the process of initiation may be said to take place. This is very complex. The individual is said to be "born again", to be fully justified or acquitted before God, to be raised from death to life or awakened from a deep sleep. Converted from darkness to light, adopted into the family of God, a member of the body of Christ, he or she has entered the new covenant, incorporated into the kingdom of God, sanctified and set aside for the service of God in the world. Another description is as bond–slave of Jesus Christ. Having repented of sin the individual is enlightened and convicted by the Holy Spirit. Trust is not in his or her own merits but in the mercy and grace of God. Saved from sin, reconciled to God through the death of Jesus Christ, he or she has the internal witness of the Holy Spirit. Set free to love God, sent to love one's neighbour as one's self, equipped to stand up against the forces of evil which would hold bondage, the individual has been baptized and filled with the Holy Spirit, baptized with water, tested the powers of the age to come.
A metaphor for the process of initiation is the tearing of the veil of the temple "from top to bottom". The barrier between the area where all are admitted and the holy of holies where, naked, man enters the mysteries, is torn down by Christ's death on the cross. There is a complete and irrevocable change, so that the exoteric and the esoteric or religious and mysterious are no longer separated, the mysteries are open to all. Because the exoteric and esoteric are, to appearances, merged, no formal expression of their separation is possible although each is real in its own order. Only the context of the central rites, not their content, would indicate the esoteric or exoteric nature of what was occurring.
This explains why Christian spirituality tends not to include esoteric ways. It is for this reason that gnosticism is condemned by the Church. However, although formal esoteric traditions may be absent, some esoteric element is necessary for the tradition to have a "heart" as well as a "body". This part of the faith has not received emphasis, the interior part of spirituality being minimized while the exterior, peripheral and collective are the focus of attention. The liturgy (in the Orthodox Church) and the sacraments are mysteries. Thus baptism, while an exoteric experience for the person having no inkling as to its significance, may be an initiatic experience to the person aware of what is involved. Having accepted Christ, been baptized, one is filled with the Holy Spirit in confirmation or Chrismation, another initiation for the aware. Similarly the other sacraments, though treated exoterically by many, are inward to those who treat them that way, above all this being true of the Eucharist where the partaker receives the "holy mysteries" which are the body and blood of Christ.
The Christian tradition has thus conserved the virtuality of inner life, but following an esoteric way in Christianity is difficult under today's conditions, particularly because qualified spiritual instruction is lacking and because the contemplative life in monastic institutions is not contemplative in the sense that the term is understood in other religions.
Refs Abraham, William J *The Logic of Evangelism* (1989); Eliot, John *Rites and Symbols of Initiation* (1965); Pallis, Marco *The Veil of the Temple* (1986).
Related Initiation (#HH0230) Mysteries and religion (#HH4032) Evangelism (Christianity, #HH3798) Born–again (Christianity, #HH0576) Human development (Christianity, #HH2198).

♦ **HH6221 Recollections as meditation subjects** (Buddhism)
Mindfulness
Description Recollection is the mindfulness that arises at the appropriate time in a wellborn person who has entered the religious life through faith. Ten recollections are enumerated: of the Buddha (the enlightened one); of the law (dhamma); of the order or community (sangha); of morality or virtue; of generosity or liberality; of the deities (deva); mindfulness of death; mindfulness of the body; mindfulness of breathing or respiration; of calm or peace.
Broader Subjects for meditation (Buddhism, #HH3987).

♦ **HH6222 Via positiva**
Related Via negativa (#HH1171) Via purgativa (Christianity, #HH4090) Via illuminativa (Christianity, #HH6030).

♦ **HH6282 Subtle faculties** (Sufism)
Latifas — Alam–i–amr — Alam–i–khalq
Description Of the ten subtle faculties of man, five are concerned with the world of command - *Alam–i–amr* - and five with the world of creation - *alam–i–khalq*. The faculties of the world of command are qalb, ruh, sirr, khafi and akhfa. Those of the world of creation are the four elements (earth, water, air and fire) and the nafs. The alam–i–amr and the nafs are connected with the inner life of the individual and are located at particular areas of the body. The *dhikr of dhikrs* of the shaykhs of the Naqshbandiyya–Mujaddidiyya order involves remembering God (dhikr) at the different positions in order, ending with latifa–i–nafs (which is the essence in reality of the four elements), and then with the whole body, so that the movement of the dhikr is felt in all and is it is heard with the heart. At this stage of perfection the voice of all created beings is heard praising God.
Related Latifa–i–qalb (Sufism, #HM1905) Cleansing the sirr (Sufism, #HM5353) Cleansing of the heart (Sufism, #HM6932) Purification of the self (Sufism, #HH5092) Illumination of the spirit (Sufism, #HM6162).

♦ **HH6291 Roman spirituality**
Refs Cousins, Ewert, et al (Eds) *Classical Mediterranean Spirituality*.
Broader Spirituality (#HH5009).

♦ **HH6292 Human development** (Sikhism)
Description Sikhism may be seen as a movement for reconciliation of Hindu and Islam faiths in the worship of one, formless, God. To be a Sikh implies being a follower of the founder, Guru Nanak, who taught from the basis that God is neither Hindu or Muslim and that the path to be followed is God's. He stressed an egalitarian approach in society with emphasis on human rights and greater equality for women. Birth and family no longer determine caste, all Sikhs being considered kshatriyas - warriors or leaders. The seeker of liberation should seek out a guru, not one who is born a Brahman but one who is directly commissioned by God; basically, it is God himself who manifests in the teacher or teachings. In practice, this means learning from the teachings of the first leaders of the Sikh religion who were called Gurus, the tenth Guru naming the Adi Granth or teachings as his successor, the Guru Granth Sahib. Although God is not manifest in the form of an avatar he is manifest in speech; and spiritual liberation is achieved by obedient living to the inner voice of God. By divine grace, the effects of karma will be worked off so that the spirit, after death, lives in the divine presence. Sometimes called *shabda yoga*, the yoga of sound, this is based upon meditation on the name of God, *nam simran*. The whole being is permeated with an awareness of God, of divine unity, overcoming the illusion of duality.
Despite the emphasis on the individual, there is also place for corporate worship; nor is the way of asceticism encouraged, rather the following of a normal family and business life, serving one's fellow men while uncontaminated by lust, greed, attachment, anger or pride. The duties of a Sikh are to constantly remember the name of God, to earn an honest living and to give to charity.
Refs Cousins, Ewert, et al (Eds) *Post–Classical Hindu and Sikh Spirituality*; Mohan, Singh Diwana *Sikh Mysticism* (1968).
Broader Human development through religion (#HH1198).
Related Guru (#HH0805) Syncretism (#HH3214) Shabda yoga (Yoga, #HH0539) Spirituality (Sikhism, #HH3774) Sikh mysticism (Sikhism, #HH6199) Human development (Islam, #HH1799) Human development (Hinduism, #HH0330).

♦ **HH6304 Middle upper grade** (Taoism)
Description Practices involve holding the breath and circulating psychosomatic energy. There are exercices of bending and stretching, massage, focusing the mind on the forehead or the umbelical region, twisting the spine, exercising the eyes. The point between the eyes may be considered the mysterious pass. The practitioner may chatter the teeth (gate of heaven), imagine the basic spirit going in and out through the top of the head. Silent court may be paid to the supreme god. Oblivion may be considered entry into a trance. The firing process may be counting of breaths. Some imagine the merging of the black and white energies of the heart and the genitals.
Context In the nine grades of practices which are side tracks and auxiliary methods in Taoism, this is the middle of the three upper grades.
Broader Sidetracks and auxiliary methods in Taoism (Taoism, #HH7004).
Followed by Higher upper grade (Taoism, #HH7734).
Preceded by Lower upper grade (Taoism, #HH4711).

♦ **HH6321 Phosphenism**
Phosphenic mixing
Description Phosphenism refers to a body of experiential knowledge derived from image retention by the human retina. Closing the eyes after a sharp and sustained stimulation of white light produces the appearance of a "phosphene", an acute image of the original light source. The phosphene progressively dissipates but it is asserted that its intensity, shape and travel across the retina can be controlled by the involved self. Dr Lefebure asserts that phosphenism (largely his own creation) permits direct contact with the subconscious mind and that it induces a transactional rapport beneficial to total well being. Phosphenism may be deemed to belong to the vibrational group of healing techniques.
Phosphenic mixing aims to develop memory and intelligence. The technique involves alternately gazing at a bright light and then its after image or phosphene. Once the phosphene is present, one forms a mental image of an object, for example of someone one holds dear, bringing to memory their appearance and gestures, preferably miniaturized in the phosphene. Here there is an intensified feeling towards the person imagined and a better understanding of them. Mixing using a written foreign word as object brings improved memory both of that word and of others learned at the same time; using a geometric problem brings understanding of a solution. It can also be used to promote and intensify spiritual experiences. In general, memory, intelligence, ideational activity are all stimulated, then interest, intellectual curiosity, spirit of initiative, sociability and intuition.
Refs Lefébure, Francis *Le Mixage Phosphénique* (1970); Lefébure, Francis *Du Moulin à Prière à la Dynamo Spirituelle* (1984).
Related Meditation (#HH0761) Phosphenes (#HM0794) Light–induced experience (#HM3559).

♦ **HH6345 Middle vehicle path** (Taoism)
Description Much like the lower vehicle, the middle vehicle includes dozens of operations. It is the method for nurturing life and, practised diligently, can prolong life. Terms include the trigrams of heaven and earth as cauldron and furnace, the trigrams of water and fire as water and fire, sun and moon as medicinal ingredients. Vitality, spirit, higher soul, lower soul and will are the five forces. Tiger is body, dragon is mind, energy the true seed. Cold and hot seasons of the year are the firing process, showering with holy water bathing. Stabilization is inward states not going out nor external objects getting in. The three essentials are the head, solar plexus and pubis. The mysterious pass is the top centre of the brain. The elixir pill is the merging of vitality and spirit.
Context The middle of the three vehicles of the gradual method, superior to the nine grades of practices which are sidetracks and auxiliary methods.
Broader Three vehicles of gradual method (Taoism, #HH5342).
Followed by Higher vehicle path (Taoism, #HH7386).
Preceded by Lower vehicle path (Taoism, #HH4864).

♦ **HH6432 Female heroism**
Description The patterns of female and male heroism are quite similar on the archetypal level. However those of women differ profoundly in tone, detail, and meaning from analogous stories about men. The journey of the female hero tends to be more optimistic, more democratic, and more egalitarian than that of the male.
Refs Campbell, Joseph *The Hero with a Thousand Faces* (1970); Pearson and Pope *The Female Hero in American and British Literature* (1981).
Broader Heroism (#HH0929).
Related Psychological development of women (#HH1976).

♦ **HH6434 Progress** (Christianity)
Description François Fénelon indicates that, while progressing in the Christian life, the early stages may be marked by motives of personal happiness. Appeals to the fear of death, judgement of God, terrors of hell and joys of heaven are recognized in holy scriptures and render powerful assistance to beginners in repressing passions and strengthening practical virtue. The grace of God is manifest in this "inferior'" form of religion, and one is intended to follow God's grace. Progress to the higher state of pure or perfect love is made step by step, carefully watching God's providence, receiving increased grace as one improves the grace one has.
Refs Fénelon, François *The Maxims of the Saints*.
Related Progress (#HH0520).

♦ **HH6451 Education for international understanding**
Description According to the constitution of Unesco, since wars begin in the minds of men, it is in the minds of men that the defences of peace should be constructed. Strengthening international understanding ensures that the processes of détente become irreversible. Education and information make a significant contribution to this task, since they help to form knowledge, beliefs and ideas concerning the contempraory world and to shape the principles which condition individual and collective behaviour. Although it is equally evident that education and information can be made to serve antithetical purposes and have often done so with disastrous effect.
Refs Boulding, Elise *Building a Global Civic Culture* (1988); Heater, Derek *World Perspectives*

(1986); International Bureau of Education *Education for International Education* (1968).
Related Planetary consciousness (#HM2006) International consciousness (Yoga, #HM0099).

♦ HH6452 Circulation of the light (Taoism)
Waterwheel exercise
Description The life–energies of an individual are understood within Taoism as having a natural tendency to flow "downward", namely into the outer world. From a Jungian perspective this may be understood as allowing victory of the anima (p'o soul) over the animus (hun soul). Through appropriate practices the individual may lead the life–energy through a "backward–flowing" process of six stages through which it is conserved and made to "rise" rather than dissipate. The animus thus becomes victorious and there is persistence of the ego after death when it becomes a spirit (shen) or god. By maintaining this practice over an extensive period of time, the individual may succeed in developing an immortal spirit body (after developing a "spiritual embryo"), symbolized as the stage of the "golden flower". The ego is then freed from the conflict of opposites and becomes part of the undivided Tao.
Context The practice of circulation of the light is closely related to practices of Taoist alchemy, especially to the waterwheel exercise. The practice is the subject of an extensive commentary by C G Jung.
Refs Wilhelm, Richard (Trans) *The Secret of the Golden Flower* (1962).
Broader Human development (Taoism, #HH0689).
Related Alchemy (Taoism, #HH5887) Bathing (Taoism, #HH3664)
Sexual yoga (Taoism, #HH9112).

♦ HH6501 Nomos (Islam)
Description As opposed to eros, which is the life urge of a soul which is individual and exists prior to the body, nomos assumes the soul to be breathed into the body after the body is created (in a foetus of about 4 months). It is essentially corporeal; and it is as a whole man, including the body, that resurrection occurs. Man is saved by following the law. The community, not the individual, is emphasized.
The nomos, or orthodox, pattern is opposed to thelema, where God and man are considered so unlike that neither feels love for the other, God simply willing good to some men for no purpose. In the nomos pattern, all the attributes traditionally ascribed to God are true, although inexplicable. God truly loves and experiences joy. God's love for man follows naturally from his self–love, and is different from the creative aspect of his will. The acts of God have a wise purpose. Man loves God, truly, in and for himself. The consummation of man's love to God is the pleasure of the beatific vision.
Refs Bell, Joseph Norment *Love Theory in Later Hanbalite Islam* (1979); Nygren, Anders *Agape and Eros* (1953).
Related Eros (#HH7231) Thelema (Islam, #HH3006).

♦ HH6505 Journeying within transcendence (Christianity)
Description In his book on the Gospel according to St John through a Jungian perspective, Diarmuid McGann traces the spiritual journey of the individual as he contemplates the mystery of life and wonders at his own being.
Refs McGann, Diarmuid *Journeying within Transcendence* (1989).
Narrower Journeying within transcendence – prologue (#HH1900)
Journeying within transcendence – from me to you (#HH5856)
Journeying within transcendence – from water to wine (#HH4219)
Journeying within transcendence – from head to heart (#HH4974)
Journeying within transcendence – from death to life (#HH2987)
Journeying within transcendence – from son to father (#HH1773)
Journeying within transcendence – from agape to philo (#HH5908)
Journeying within transcendence – from call to mission (#HH3352)
Journeying within transcendence – from secular to sacred (#HH3796)
Journeying within transcendence – from object to subject (#HH4188)
Journeying within transcendence – from outside to inside (#HH3215)
Journeying within transcendence – from served to servant (#HH6012)
Journeying within transcendence – from bondage to freedom (#HH2661)
Journeying within transcendence – from blindness to sight (#HH1354)
Journeying within transcendence – from discussion to decision (#HH3366)
Journeying within transcendence – from forgetfulness to memory (#HH5219)
Journeying within transcendence – from isolation to imagination (#HH2498)
Journeying within transcendence – from humiliation to exaltation (#HH2329)
Journeying within transcendence – from extraordinary to ordinary (#HH4110).
Related Transcendence (#HH0841) Self–transcendence (#HH0526).

♦ HH6509 Self-surrender in love
Description There is an inherent paradox in seeking total union with another if, in that fusion, the other ceases to exist, thus destroying that which was loved and with which one sought union. In the brief moments of self–surrender in passionate love, separate selves are mingled and ego boundaries transcended so that the self is enlarged and enhanced. Self–surrender is here an empowering act. Even more extensive self–surrender may bring self–worth in devotion to the beloved or a finding of meaning in connection with the beloved. But extreme forms lead to impoverishment of the self, becoming simply an appendage of the beloved.
Refs Person, Ethel Spector *Love and Fateful Encounters* (1990).
Related Ego death (#HH6921).

♦ HH6548 Actual grace (Christianity)
Description According to Catholic doctrine, this is supernatural help from God by means of which one's mind is enlightened and one's will strengthened to do good and avoid evil. Without actual grace, those who have obtained the use of reason cannot for long resist the power of temptation or perform actions which merit a reward in heaven. However, because of free will, one is not forced by God to accept His grace.
Broader Grace (#HH0169).
Related State of grace (Christianity, #HM6445) Sanctifying grace (Christianity, #HH5116).

♦ HH6566 Way of transcendence (Christianity)
Description Stripping away the projections and fantasies one has of God there is revealed, not an empty nothingness but a deeper truth. Union with God is not mediated through an image, nor is there a sense of distinction or separateness. God is no longer the purely external. Meister Eckhart indicates two veils which must be stripped away in order to reveal the truth.
(1) The material world. Not only is God not the limited, material things such as money, security or status, He is not the liturgy which one prefers, or a particular way of devotion, or even a particular way of life such as that of a monk or nun. There must be detachment towards all these things.
(2) The mental, imaginative and emotional life. This refers especially to the religious area of life, to prayer, worship, theological and spiritual study. The images and symbols associated with this life are perfectly valid. But if one is called to do so, one should also use inward prayer not involving images or concepts. That one consciously needs this kind of prayer is a sign of aptitude for the spiritual way. One needs to be shocked into an awareness of the inadequacy of one's own idea of God. God is not even good or wise in our own, limited, understanding of these terms. Even the personal images of Trinity, of Father, Son and Holy Spirit, even of God Himself, have to go. Ultimate reality has no name, cannot be described through images, cannot be ascribed any qualities. Sceptics and unbelievers may also be on the path if their rejection of religion is of limited symbols as they hinder perception.
The depth of God beyond God is reflected in the depth of the self. Emphasis is on what in the self is most like God and how that likeness can be heightened and transcended until it becomes unity. Not only does God call us to this unity, but there is within the soul an element of affinity with God. Being in the image and likeness of God is not only with the persons of the Trinity but there is also a self within the self that has an affinity with the God beyond God. This ground of the soul is the true image of God within man, it is that in man which is neither this nor that. This true self is unrelated to roles and functions, to thoughts and emotions, it is the 'I' which watches them all, detached and serene. This *permanent self* is a refuge from the tyranny of thoughts and emotions. Acting from this permanent self or ground of the soul is the only true action, all other "action" is actually reaction in response to external stimuli. No longer caring about past or future, one is aware only of the present moment as it impinges on eternity.
Thus, what is normally regarded as the self must be stripped away in order to find the true self within. Without goal or ambition, not doing anything or being anything, one has to learn simply to be. Without even the goal of finding the true self, the true self is found. The journey into the depths of the self is the same experience as the journey into the depths of God. Paradoxically, it is having discovered the fundamental reality that one melts and flows out of one's self to be a person. As a person, one participates in the mystery, the secret, of communion with the Three–in–One, the Trinity. Following the way of transcendence we find that, although seeking God from outside means that He retreats within, if we seek Him within He affirms Himself without. As opposed to standing within the world, when God appears wholly transcendent and other, one stands within God and the world is other, a pale, imperfect reflection of that truth in which one lives and makes one's home.
Refs Smith, Cyprian *The Way of Paradox* (1987).
Related Transcendence (#HM4104) Transcendence (#HH0841)
Melting (Christianity, #HM7384) Universal human nature (#HH4389)
Way of paradox (Christianity, #HH3335)
Birth of God in the soul's ground (Christianity, #HM6522).

♦ HH6632 Optimism
Description It is claimed that human beings are able to improve themselves individually and the human condition in general; this improvement may arise under the guidance of an ultimate reality which is good. The religious believer holds that God, in His goodness, wisdom and power, has created everything to be the best it can possibly be. The unbeliever holds that human beings of themselves can enrich human life, for example by applying scientific knowledge. Whether human beings are basically good, or capable of improvement and achieving wellbeing and happiness, or good through God's guidance and grace, performing good works and achieving saintliness, the future will be better than the past. Teleologically, this implies that the end to which history is moving is both rational and good.
Arguments levelled against optimism point to: the evil co–existing with good in individual lives, with despair often triumphing over hope; the existence of original sin which makes each human being a sinner, who may be forgiven but is always capable of sinful and cruel acts; the capacity of mankind for mass destruction. The Christian answer to this is that Christ has "overcome the world", that the victory is already won, that there are indeed grounds for hope and joy.

♦ HH6669 Primary growth
Description Human life is generally defined according to biological limits and considered as a process in three stages – growth (childhood, adolescence), permanence (adulthood), decadence (old–age, senility). It is thus defined in terms of time. Primary growth is that which covers the process of development of a child from infancy through adolescence to the final mutation as an adult, when primary growth ceases. At the same time as corporal development and organic maturation there is rapid psychic development. It constitutes, at least in part, a biological demand on the organism and in general may be said to respond to internal laws of the organism, taking elements from wherever is available to accomplish an inner developmental logic. Primary growth is contrasted with maturity growth which does not recognize time limits.
Refs Mallmann, Carlos A and Nudler, Oscar (Eds) *Human Development in its Social Context* (1986).
Broader Psychic growth (#HH3117).
Related Maturity growth (#HH5434) Sensorimotor development (#HH0195)
Cognitive growth and development (#HH0146).

♦ HH6732 Christian self–love (Christianity)
Description Although self–love in the purely selfish sense is to be avoided, there is a self–love which is considered the duty of all Christians. This is both in that each self is a reflection of Christ as the archetype of mankind and in that man is enjoined to love his neighbour as himself. It is part of, and coincides with, love of God.
If man is to love himself as he is loved by God, this implies awareness of what he is, what God would have him be and what he is in the process of becoming. The personality should be cared for, even reverenced, at all levels – body, mind and spirit – through self–discipline and self–culture; until, through a gradual process of self–surrender, the self is fully realized.
Related Self–development (#HH0651) Self–culture (Christianity, #HH3710)
Self–discipline (Christianity, #HH2121).

♦ HH6773 Five hindrances (Buddhism)
Gogai (Zen) — Five covers
Description These are the mental and moral hindrances of sense desire, anger (ill–will) drowsiness (sloth and torpor), excitability (restlessness and worry) and doubt.
Related Afflictions and hindrances (Buddhism, #HH5007).

♦ HH6876 Restraint
Poise — Yama (Yoga)
Description Through practice of restraint the mind is purified and there is freedom from thoughtless and compulsive activity. Failure to practice such restraint results in slavery to the cycle of birth and death, with rebirth into the lowest of states. As *yama*, the rules comprising restraint are the basic requirements of ethics and are basic to all religions. They regulate social behaviour and are: ahimsa (harmlessness, non–hurting); truthfulness; asteya (not stealing); chastity; freedom from greed. Another source quotes ten yamas, the additional five being fortitude, kindness, straight–forwardness, moderation in diet and purity (bodily cleansing).
Context The first component of the eightfold path of yoga.
Broader Eightfold path of yoga (Yoga, #HH0779).
Narrower Satya (#HH0554) Ahimsa (#HH0088) Asteya (#HH0361)
Aparigraha (#HH1268) Brahmacarya (#HH0978).
Related Lord–of–death awareness (Buddhism, Tibetan, #HM2088).

♦ HH6902 Esoteric development
Refs Steiner, Rudolf *Esoteric Development* (1982).
Related Mysteries and religion (#HH4032) Spirituality (Esotericism, #HH5234).

♦ **HH6921 Ego death**
Description Dissolution or death of the ego in its old form may occur in *regression* if this is seen as an attempt to refuel the personality through merger with a parental or god–image. The ego potential may then reconstitute in a more adequate or conscious form. This loss of ego control, even if temporary, is dangerous; and it is only after emergence of an enriched personality that the "death" of the ego may be seen to have been the precursor of transformation.
Taking this change and death as a threat, the ego may resist and feel impending disaster. But as the ego sacrifices its old world, point–of–view and attitude, a new set of values and a new synthesis become possible.
Related Regression (#HH1300) Ego destruction (#HM0655)
Self–surrender in love (#HH6509) I–transcendent stage of life (#HM3060).

♦ **HH6973 Striving against the soul** (Islam)
Description Since the soul innately loves passions (hawa) which need to be controlled, there is a striving against the soul when one tries to harness the passions. The resistance is *sabr* (patient endurance). Conflict between reason and passion occurs in the heart. As the five senses bring useful information to the heart they also bring what is harmful. The paths bringing worldly preoccupations have to be blocked.
Refs Bell, Joseph Norment *Love Theory in Later Hanbalite Islam* (1979).
Related Hawa (Islam, #HM0129) Sabr (Sufism, #HM3418)
Striving against the soul (Islam, #HM4766).

♦ **HH7004 Sidetracks and auxiliary methods in Taoism** (Taoism)
Description There are nine grades of practice. The three lower grades comprise over 1,000 items and are practised by lustful and greedy people. The three middle grades gradually approach the Tao and the three upper grades, also over 1,000 practices, are performed by mediocre practitioners who can use them to ward of sickness. Beyond these nine grades are the three vehicles of the gradual method.
Refs Cleary, Thomas (Trans) *The Book of Balance and Harmony* (1989).
Narrower False paths (#HH3013) Outside paths (#HH3667)
Outside paths (#HH4289) Lower upper grade (#HH4711)
Lower middle grade (#HH3296) Upper middle grade (#HH2633)
Middle upper grade (#HH6304) Higher upper grade (#HH7734)
Middle middle grade (#HH5207).
Followed by Three vehicles of gradual method (Taoism, #HH5342).

♦ **HH7123 Integral psychoanalysis**
Analytical trilogy
Refs Pacheco, Dr Cláudia Bernhardt *The ABC of Analytical Trilogy* (1988).
Broader Psychoanalysis (#HH0951).

♦ **HH7219 Animistic yoga** (Yoga)
Refs Woodward, Kimm *Animistic Yoga.*
Broader Yoga (Yoga, #HH0661).

♦ **HH7231 Eros**
Life–urge
Description To Plato, eros is the quest of the individual for his own highest spiritual good. Man is a prisoner in the material world. The soul exists previously in some higher existence; of its nature, it differs from the body and only the soul and not the body can achieve salvation.
Refs Nygren, Anders *Agape and Eros* (1953).
Broader Love (#HH0258).
Related Death–urge (#HH0574) Nomos (Islam, #HH6501) Thelema (Islam, #HH3006).

♦ **HH7234 Egyptian spirituality**
Refs Cousins, Ewert, et al (Eds) *Classical Mediterranean Spirituality.*
Broader Spirituality (#HH5009).

♦ **HH7324 Human development through diet**
Shojin–ryori
Description We are what we eat; and spiritual and intellectual progress may be helped or hindered by what food is eaten and when. Homeopathic and holistic methods emphasize the importance of diet. The Tantric ritual of five elements uses food and sex to gain inner knowledge. Traditional Chinese methods divide food into cold or warm and prescribe them in treatment of persons who are predominantly warm or cold. Buddhism and Hinduism look to the harmonizing the five tastes – sweet, salty, vinegary, bitter, hot. Shojin–ryori, Japanese vegetable cooking in the Zen tradition, emphasizes the five tastes plus "soft" taste, such as tofu. It aims to restore connections with the natural environment which has been distanced through modern, centrally heated and air-conditioned living. In this tradition, even the tableware relates to the season.
Refs Dogen Zen Master and Uchiyama Kosho *Refining Your Life*; Santa, Maria Jack *Anna Yoga.*
Related Veganism (#HH3324).

♦ **HH7343 Perfection of the soul** (Christianity)
Description According to William Law, the soul can only reach perfection through change and exaltation of its first properties. In nature all things are born and die, unrelated to reason. No good can come to the soul apart from the entrance of the Deity into the properties of its life. Nature has to be set right, its properties enter the process of new birth, and work to the production of light before the spirit of love can be born in it. Love is delight and delight can only arise in a creature if its nature is in a delightful state or possessed by that in which it must rejoice. God must become man because birth of the Deity must be found in the soul, giving to nature all that it lacks, or the soul will never be in a delightful state working with the spirit of love. In its natural life, the soul in a state such as nature without God is in. The properties of nature work in blindness, the soul governed and tormented by restless and contrary passions, in a hell of hunger, anguish, anxiety, contrariety and self–torment. The natural (or "old man") cannot change of itself. Despite rules and moral behaviour, nature is not changed and all that happens is that man learns to conceal his inward evil and appear not to be under its power. Only when Deity becomes man, is born in fallen nature, is united to it and becomes its life can nature be overcome. The work of morality is the doctrine of the cross, to resist and deny nature so that supernatural power of divine goodness possesses it and brings it new light.
Refs Stanwood, P G (Ed) *William Law* (1979).
Related Spiritual development (#HH0017).

♦ **HH7386 Higher vehicle path** (Taoism)
Description This path ressembles that of the middle vehicle although its application is different. It involves about a dozen items and is the path of extending life. A superior practitioner, carrying it out consistently from beginning to end, can ralize the way of immortals. Heaven is the cauldron, earth the furnace, sun the fire, moon is water, the mechanism of evolution is yin and yang. The five elements are lead, mercury, silver, sand and earth, essence the dragon, sense the tiger, thought the true seed. The firing process is refining thought with mind; the fire is nurtured through ceasing thought, stabilization is keeping brilliance to one's self. The battle in the field is conquering inner demons. The three essentials are body, mind and will. The mysterious pass is the heart of heaven. Completion of the elixir pill is sense coming back to essence. Bathing is being suffused with harmonious energy.
Context The highest of the three vehicles of the gradual method, superior to the nine grades of practices which are sidetracks and auxiliary methods.
Broader Three vehicles of gradual method (Taoism, #HH5342).
Preceded by Middle vehicle path (Taoism, #HH6345).

♦ **HH7403 Formative spirituality**
Refs Van Kaam, Adrian L *Formative Spirituality.*
Broader Spirituality (#HH5009).

♦ **HH7634 Compromise** (Christianity)
Description The Christian is required to make efforts always to improve the status quo, whether in personal life, in the corporate life of the Church and in the world at large. Attempts are impeded by the carrying over of sin from the past and by the need for agreement with others which involves compromise with one's own wishes. There is always tension between what one wishes and what appears possible. Whether at the personal or collective level the individual has to respond in decisions of his own conscience. Despite individual fallibility due to sin, the possibility of error, and failure through sin to implement a decision once made, there is still a duty attempt both to discover the right course of action and to carry it through. This is compromise between the radical nature of the gospel and the immediate possibilities open to one.

♦ **HH7637 Mythical control of human development**
Psychic control of human development — Spiritual control systems
Description In periods of social crisis, people develop a strong desire for contact with superior minds and beings capable of offering guidance. Human life is to a large degree ruled by imagination and myth. It has been suggested that such levels of human belief are vulnerable to being controlled and conditioned, namely that there is a level of control in society which is a regulator of human development. In this sense mythology rules at a level of social reality over which normal political and intellectual systems have no real power. For example, reports of unidentified flying objects and related encounters can be understood as corresponding to reinforcement schedules in some conditioning process from that level. If UFOs act at the mythic and spiritual level, it is to be expected that they would be almost impossible to act by conventional methods. Such devices can then be understood as a focus for psychic phenomena, evoking a deep emotional reaction in those experiencing them, possibly leading to reinforcement of various salvation myths.
Refs Vallee, Jacques *Dimensions* (1988).
Related Alien consciousness (#HM2329).

♦ **HH7734 Higher upper grade** (Taoism)
Description The practitioner exercises vitality and energy, tunes the internal organs, visualizes pure lands, concentrates fixedly on the elixir fields, circulates psychophysical energy through the three elixir fields, swallows noon sunlight. The vitality aroused through sexual intercourse or inner concentration may be rerouted so that it travels up the spine to boost the brain. Another exercice is inward gazing.
Context In the nine grades of practices which are side tracks and auxiliary methods in Taoism, this is the highest of the three upper grades.
Broader Sidetracks and auxiliary methods in Taoism (Taoism, #HH7004).
Preceded by Middle upper grade (Taoism, #HH6304).

♦ **HH7902 Prudence**
Practical wisdom
Description Thomas Aquinas defined prudence as the good of reason present in judgement.
Context One of the four cardinal or principal virtues recognized by Plato in the Republic and featuring prominently in mediaeval Christianity.
Broader Wisdom (#HH0623) Virtue (#HH0712).
Related Justice (#HH2117) Fortitude (#HH5640) Temperance (#HH0600).

♦ **HH8003 Pathworking**
Description Pathworking is an exercise through which symbols associated with particular states of consciousness are visualized as uplifting the awareness of the practitioner to higher states. The states are those very specifically defined by the Kabbalah as being connected by a network of paths forming the Tree of Life. Pathworking requires very demanding visualization to stabilize understanding of each state and the connecting paths in the network.
Related Kabbalah (#HH0921) Guided visualization (Magic, #HH6053).

♦ **HH8005 Vocational guidance**
Description Help is given in deciding on a likely future career, on the training necessary for that career and in finding suitable employment.
Broader Guidance (#HH0178).

♦ **HH8007 Bandhas** (Yoga)
Description These are physical devices which lock areas temporality containing energy so that the energy may be directed as the yogi desires.

♦ **HH8204 Cognitive dissonance**
Description Knowledge of one's own and other people's attitudes, feelings and behaviour are the cognitive elements which are normally held in harmony by the individual. When there is incompatibility between the elements, cognitive dissonance arises, leading to psychological tension. The individual then takes steps to reduce the tension by decreasing dissonance. This may be by seeking social support for a view which is causing tension.
Related Over–valued ideas (#HM1466) Secondary delusions (#HM2795)
Cognitive growth and development (#HH0146).

♦ **HH8206 Unity of personality** (Judaism, Christianity)
Description The individual person is looked upon as a totality and it is this total "I" which confronts God, not the body or the mind or the spirit as separate entities. The whole person sins and the whole person repents for being that kind of total person which originates and commits sinful acts. It cannot be said that one "has" a body or "has" a spirit – one is an embodied spirit. Although the psyche (soul, life constituted in man) and pneuma (spirit, life having origin in God) are two separate terms, they can be used interchangeably. Both describe the psychospirituality which is the inner core of human personality.
Broader Unity of personality (#HH0136).

♦ **HH8219 Way of the warrior** (Amerindian)
Description In Amerindian terms, if the brave becomes a warrior, and lives like a warrior, then tonal and nagual can be unified. This is a life in which whatever the task may be it is done in the

best possible way, not clinging and with the highest excellence. It is comparable with mindfulness and one-pointedness in the life of a Bodhisattva.
The system is described by Carlos Castaneda from his meetings with don Juan. Once the teacher is aware that the pupil is ready he embarks him on the life of the hunter. In order for the hunted not to turn the tables on the hunter, the hunter learns to dispense with all habits. He learns non-attachment, "letting go, but not letting go", so that total energy is concentrated on the task and yet there is complete control over the situation.
This is the exoteric situation. In the esoteric situation, the hunter trains to become the warrior, what is hunted is power - the perpetual and mysterious flow of force underlying appearance. He learns to peep behind appearance without concentrating on making it yet another appearance. When he has learned to face this power with abandon and control he is ready to be a warrior, although there is no strict demarcation between the two. When he has reached the warrior stage he can face any task on his own and without attachment to views, fears, enlightenment or bravado. The impeccable warrior trusts his personal power, whether small (in the case of the young or inexperienced) or enormous (in the case of the older and relatively more experienced, having accumulated much merit). In Buddhist terms, the warrior trusts to the positive karma he has so far accumulated. There is no more a feeling of helplessness in dealing with the world. There is freedom to act, all results follow regardless of the attitude of the actor. The warrior takes responsibility for the act and lives with the consequences - he is not attached to the consequences either. He is thus free to use the power of any facet of karma to its best advantage.
An aid to the denial of all attachments is using death as advisor. Impending death is a reminder that there is no time to waste in perfecting himself. All his action is the action of a man who knows he is going to die.
The four natural enemies of man are fear, clarity, power and old age. The first is combatted as death itself is used to combat fear of death. With fear vanquished, the warrior looks at the contradictions of the world with great clarity. But clarity is only part of possible experience - it is used impeccably when it is advantageous to do so but is overcome when it is realized that power lies outside language. Then power has to be defied deliberately - the power seemingly conquered is never the warrior's. Without control over himself, handling carefully and faithfully all he has learned, clarity and power are worse than mistakes. With this control he will reach the point where everything is held in check and the third enemy is defeated. Handling power unclingingly the warrior becomes a man of knowledge and fends off the invincible enemy, old age. Finally, because the warrior cannot be cold, lonely and without feelings, he has a great love, that of the earth, which he loves with unbending passion and is so released from sadness. It is the equivalent of the compassion of Buddhahood.
Refs Castaneda, Carlos *Journey to Ixtlan* (1972); MacDowell, Mark *A Comparative Study of the Teachings of Don Juan and Madhyamaka Buddhism* (1986).
Broader Way of the warrior (#HH3322).
Related Stopping the world (#HH0489) Mindfulness (Buddhism, #HM2847)
Concentration (Buddhism, #HM6663)
Tonal state of awareness (Amerindian, #HM2066)
Nagual state of awareness (Amerindian, #HM2036).

♦ **HH8231 Samadhi-bhavana** (Buddhism)
Cultivation of concentration
Description This is the development or cultivation of concentration, through methods and practices of mental development. Since concentration is a result attained by mental discipline it is still acting upon the surface level of consciousness and cannot itself cope with residual dispositions (asavas) of the mind, or dispel ignorance and uproot the causes of the miseries of existence.
Context One of the two systems of Buddhist meditation, the other being vipassana-bhavana (development of insight).
Broader Bhavana (Buddhism, #HH0551).
Related Samatha-bhavana (#HH0710) Vipassana-bhavana (Buddhism, Pali, #HH0680).

♦ **HH8487 Synergic enlightenment**
Description This is a means of getting in touch with one's self, experiencing one's own power and potential, and visualizing the cooperative potential of humankind. It implies:
(1) Experiencing one's self both emotionally and intellectually as responsive and self-directed. One exists inexorably in a state of interdependence with other people.
(2) Experiencing and understanding how through both action and inaction one partially or wholly causes what happens to one's self and to others and that one is therefore responsible for these effects that one has caused.
(3) Visualizing how one can work with others in the design and building of a caring community which is fit for fully-evolved human beings.
Refs Marguerite, Craig and James, H *Synergic Power* (1979).

♦ **HH8672 Mental prayer** (Christianity)
Description This type of prayer is personal and individual; it has the aim of awakening the Holy Spirit within the individual and bringing the heart into harmony with Him, so that it is the Holy Spirit praying through the individual's voice and affections. The individual is then, as far as possible, conscious of this prayer in the heart. The difficulty is that, while nothing should be said in mental prayer that is not sincerely meant, it is sometimes difficult to fix the attention on God, the mind is full of distractions which possess the heart, and prayer is started from routine rather than from desire to pray. Here the need is to admit the fact, to be humble and to recognize the need for effort. There may then be the arising of compunction, which can transform prayer from a cold formality, with the emphasis on one's self, to a living act, bringing one's self into a real, spiritual and personal relationship with God face to face, the I-Thou relationship.
Refs Alcantara, Peter S *A Golden Treatise of Mental Prayer* (1978); Lehodey, Dom W *The Ways of Mental Prayer* (1982); Merton, Thomas *Spiritual Direction and Meditation* (1975).
Related Fundamental dialogue (#HM3225) Compunction (Christianity, #HM4251)
Liturgical prayer (Christianity, #HH4209) Christian meditation (Christianity, #HH5023).

♦ **HH8677 Mahamudra meditation**
Refs Brown, D P *Mahamudra Meditation* (1981).
Broader Meditation (#HH0761).

♦ **HH8712 Spirituality** (Jainism)
Refs Cousins, Ewert, et al (Eds) *Early Hindu and Jain Spirituality*.
Broader Spirituality (#HH5009).
Related Human development (Jainism, #HH0622).

♦ **HH8903 Maps of the mind**
Mental maps — Maps of consciousness
Description Mental maps are produced as, more or less, diagrammatic representations of the structure of the human mind and/or of the different modes of awareness open to an individual. There are many such maps with little consensus on how they should be combined. They tend to be generated by disciplines such as psychology, psychoanalysis, philosophy, biology, the neurosciences, and cybernetics. Maps with a similar function have also been generated within many of the spiritual traditions of the world, notably Buddhism and Taoism. Where these address the question of transcending duality, the maps make extensive use of metaphor and symbol, particularly by referring to sets of domains, realms, heavens, hells, and other worlds, which individuals may inhabit under certain conditions of awareness.
The process of psychological transformation may be considered as a journey or a path. The deep experiences of consciousness can be "mapped" as a guide for this journey. Fixed mental, emotional, perceptual and behavioural patterns are conditioned into each level of consciousness. There exist both traditional and newly developed maps of consciousness and its evolutionary development. They imply a model of man oriented toward some form of personal development and often include coded instructions concerning the practice of certain psychological techniques. Such maps are best used with guidance by others who have already had practical experience of them and can help the individual in the selection of the most useful version for him. However, it is characteristic of such maps and systems that they are written on many different levels, and individuals derive from them what information they can assimilate according to their level of understanding at the time.
Drug-induced experience has been used as a guide in mapping consciousness. Individuals have been guided in a pattern of descent (Masters and Houston), when four levels of experience were noted and hypothesized to represent four major levels of the psyche: sensory; recollective-analytic; symbolic; integral. The journey inwards focuses energy more and more on areas of experience alien to the ego.
Refs Clark, John H *A Map of Mental States* (1983); Collingwood, R G *Speculum Mentis* (1963); Eckartsberg, Rolf von *Maps of the Mind*; Fischer, R *Cartography in Inner Space*; Fischer, R A *A Cartography of the Ecstatic and Meditative States* (1971); Hampden-Turner, Charles *Maps of the Mind* (1981); Masters, R E L and Houston, Jean *The Varieties of Psychedelic Experience* (1967); Metzner, Ralph *Maps of Consciousness* (1971); Ring, K *Mapping the Regions of Consciousness* (1976).
Narrower Lila (#HM2278) Kabbalah (#HH0921) Tarot (Tarot, #HH0382)
I Ching (Taoism, #HH0004) Alchemy (Esotericism, #HH0221)
Tantra (Buddhism, Yoga, #HH0306) New religious modes (ICA, #HM3004)
Ninety-nine names of Allah (Sufism, #HH2561)
Zodiacal forms of awareness (Astrology, #HM2713)
Other world in the midst of this world (ICA, #HM2614)
Consciousness ascension stages game (Buddhism, Tibetan, #HH4000).
Related Astrology (#HH0552) Artificial intelligence (#HH2189)
Stages of spiritual life (#HH0102)
Modes of awareness associated with use of hallucinogens (#HM0801).

♦ **HH8963 Social innovation**
Social invention
Description A social invention is a new or imaginative way of responding to a social problem or improving the quality of life. Unlike a technological invention, it tends to be a new service, rather than a product or patentable process. It does not necessarily generate funds. It may take the form of a new law, organization or procedure that changes the way in which people relate to themselves or to each other, whether individually or collectively. Each social system ultimately combines a series of such social inventions. Some, such as education, may be relatively developed, while others, such as intergroup relations, may have so few methods to rely on that the system is more a constellation of problems than a cluster of solutions. Human development may be viewed in terms of the manner in which social innovation is facilitated.
Refs Albery, Nicholas and Yule, Valerie (Eds) *Encyclopaedia of Social Inventions* (1990).
Related Social imagination (#HH4510).

♦ **HH9112 Sexual yoga** (Taoism)
Description Sexual intercourse (whether physical or mental) may be used to cultivate energy, bliss and health, especially in the case of practitioners of advanced age. The basic aim of sexual yoga in Taoism is to gain control and conscious use of sexuality, since neither celibacy nor indulgence balance psychic energies in a manner which leads to their transcendence. Whether or not physical intercourse is involved, the process of sexual intercourse is recognized as a fundamental metaphor, or set of imagery, through which the art of fruitfully balancing duality can be explored. In one form of sexual yoga inner heat is generated by sexual arousal. This is then circulated by concentration through the active and passive psychic channels, up the spine, through the head, and down the centre line of the front of the torso. In another form, practised in conjunction with sexual intercourse, ejaculation by the male practitioner is suppressed (either by muscular contraction or external pressure) creating an extremely intense and prolonged orgasm, the heat of which is then conducted by concentration up the spine by the active channel into the brain, where it "burns" away mundane thoughts and feelings through bliss. It is recognized, as with breathing exercises and psychosomatic mediation practices, that sexual techniques may be hazardous if improperly performed, namely if the practitioner controls his energies or is controlled by them. Stress is placed on the complementary role placed by meditation, both in ensuring the appropriately serene attitude to undertake such practices effectively and, subsequent to them, in achieving unconditioned transcendence.
Refs Chang, Po-tuan *Understanding Reality* (1987).
Broader Yoga (Yoga, #HH0661) Human development (Taoism, #HH0689).
Related Maithuna (Yoga, #HH2034) Tantra (Buddhism, Yoga, #HH0306)
Circulation of the light (Taoism, #HH6452).

♦ **HH9219 Eastern mysticism**
Refs De Riencourt, A *The Eye of Shiva* (1981).
Broader Mystical cognition (#HM2272).

♦ **HH9760 Spiritual exercises of St Ignatius** (Christianity)
Description Three methods of prayer are used. For the first week there is discursive meditation according to memory, intellect and will. Memory recalls the point previously chosen as the subject of discursive meditation. The intellect rests on the lessons to be drawn from that point. The will makes and keeps resolutions based on the point to put the lessons into practice, thus leading to reform of life.
The second week one gazes on a subject of contemplation in the imagination - seeing persons in the Gospel as though they were there, hearing what they are saying and relating and responding in words and actions. This is the development of affective prayer.
The third week one successively applies in spirit the five senses to the subject of meditation, and one is introduced to contemplation as it is traditionally understood.
Refs Mottola, Anthony (Trans) *Spiritual Exercises of St Ignatius* (1964); Sheenan, John F *On Becoming Whole in Christ* (1978).
Broader Spiritual exercises (#HH0707).

♦ **HH9764 Superhuman qualities** (Buddhism)
Uttarimanussa
Description Ten superhuman qualities are described in the Sutta Vibhanga as being special attainments of insight above that of ordinary men. They are necessary for members of the sangha.

HH9764

Narrower Vimoksa (#HM5977)
Vinivaranata (#HM0927)
Concentration (#HM6663)
Jhana (Buddhism, Pali, #HM7193).
Samapatti (#HM0159)
Kilesappahana (#HM7345)
Phala sacchikiriya (#HM0932)
Nanadassana (#HM6502)
Magga-bhavana (#HM0796)
Sunnagare abhirati (#HM5134)

♦ **HH9807 Initiation into secret societies**
Description Rituals concerned with groups, such as initiation into secret societies and into cults, are normally considered separately from those concerned with the status of individuals. The organization of secret societies is usually hierarchical. Initiation places the new member at the foot of a ladder of position, each rung of which is characterized by its own secrets and initiation rituals. The elaboration of this internal structure may be extremely complex. The final stages in all such rituals is one in which the initiate receives new powers. The ritual itself is a display of esoteric lore, which is only partly explained to the initiated. Much of it remains mysterious, indicating deeper levels of understanding, associated with higher rank in the organization.
Broader Initiation (#HH0230).
Narrower Initiation (Freemasonry, #HH0659).

Modes of awareness

Rationale

In contrast to the previous section on concepts of human development, this section focuses on distinct subjective experiences during the process of human development. The variety of such states of consciousness or modes of awareness, essential to any adequate comprehension of human development, is a fundamental characteristic of some approaches to human development. This is especially true of "alternative" approaches, traditional approaches and those originating in non-Western cultures. Recognition of the existence of such subjective states is not acknowledged in official studies of human development, except through externally defined terms such as fulfillment, satisfaction, well-being, and happiness or their opposites. They are frequently only considered because of their economic significance in ensuring a productive labour force and consumer satisfaction. Since fulfilling or challenging modes of awareness are key indicators for a person in assessing his or her own development, a clarification of their variety is therefore a way of establishing a bridge to active constituencies and to traditional frameworks in which human development processes have a well-established meaning with which people identify. Of great importance to many is that in a number of cases these modes are grouped into sets, cycles or series which enable people to order their psychic life and understand how they may aspire and progress to greater fulfillment, or deprive themselves of it.

Contents

The 2,759 entries in this section each endeavour to describe a mode of awareness that is named in the literature of some particular approach to human development. The descriptions are provided as far as possible in the terms of those to whom the mode is meaningful, selecting and presenting the information in a way that may help others to find it meaningful as well. Where such modes have metaphorical names these have been retained, however odd they may seem to people from other cultures or belief systems. Where the distinction between modes is more precisely defined in another language (e.g. Sanskrit), the transliterated name in that language is also given. Where the mode is part of a set or a progression, this is indicated by a contextual paragraph and cross-references.

Method

The information used was obtained from a wide range of specialized reference books. The method is reviewed in detail in Section HZ.

Index

A keyword index to entries is provided in Section HX. The keywords are also incorporated into the index for Volume 2 (Section X)

Bibliography

Bibliographical references, by author, are given in Section HX.

Comments

Detailed comments are given in Section HZ.

Reservations

Verbal descriptions of subjective experiences can only be inappropriate, if not insensitive or totally inadequate. Nevertheless such descriptions are attempted by those drawing attention to the importance of particular modes of awareness. Attempting to group such descriptions into one framework obscures implicit subtle distinctions. It is for this reason that no attempt has been made to combine entries on different approaches to what might be considered the same mode of awareness.

Possible future improvements

In addition to refining entries, it would be possible to extend the range to offer better coverage of modes of awareness recognized in non-Western cultures. A pattern of cross-references to specific symbols and to specific values could also be developed.

ENTRY CONTENT AND ORGANIZATION

Ordering of entries
Entries are in **numeric order**. Entry numbers have been **allocated randomly**; they have no significance other than as a permanent point of reference to facilitate indexing, cross- referencing, and updating between editions.

Index access to entries
The location of an entry in this sub-section may be determined from:
- the **Volume Index** (Section X) on the basis of keywords in the name of the entry or its alternate names
- the **Section Index** (Section HX), if the entry forms part of a numbered set.

Structure of entries
Entries may be composed of the following descriptive elements:

(a) **Entry number** This number has **no significance**, except as a convenient method of identifying the entry (particularly for indexing purposes), of filing information on it, and as an identifier to which cross-references from other entries (possibly in other sections) may refer in this and future editions. The first letter of the entry number refers to the section of this volume in which the sub-section, denoted by the second letter, is located.

(b) **Qualifier** following entry number. As a tentative indication of different levels of awareness, the modes are classified on a scale "a" to "g". The basis for designating the qualifier is a combination of number of other entries cited in the entry; number of narrower entries in a chain; number of broader entries in a chain; and a series of other criteria (see Section HZ)

(c) **Entry name** This is printed in bold characters. It may be followed by alternative names when appropriate. In certain cases these may include non-English terms where these are recognized as being more precise than the English translation.

(d) **Description** Brief description of the mode of awareness.

(e) **Context** Indication of the context within which the mode of awareness is recognized.

Bibliographical references
Entries may be followed by some bibliographical references. Further bibliographical details to the publications cited are given, by author, in the section bibliography (Section HY)

Cross-referencing of entries
At the end of any entry, there may be cross-references to other entries. These indicate the number and name of the cross- referenced entry, whether within this Section or in other Sections. There are 5 types of cross-references:

> **Broader** = Broader or more comprehensive mode of awareness
> **Narrower** = Narrower or more specific mode of awareness
> **Related** = Related or associated mode of awareness
> **Preceded by** = Preceding or prior mode of awareness (in a succession or cycle)
> **Followed by** = Following or subsequent mode of awareness (in a succession or cycle)

♦ **HM0002d Al-Fattah** (Sufism)
Description The believer is aware of Allah as the opener, solving all problems, eliminating obstacles. Knock on the door of al-Fattah, Allah will open to you. But one must also open the door of generosity and mercy to others.
Context A mode of mystical awareness in the Sufi tradition associated with the eighteenth of the ninety-nine names of Allah.
Broader Ninety-nine names of Allah (Sufism, #HH2561).
Followed by Al-'Alim (Sufism, #HM4010).
Preceded by Al-Razzaq (Sufism, #HM7021).

♦ **HM0003g Paraesthesia**
Paresthesia
Description Malfunctioning or distortion of sensory perception results in misinterpretation, heightened awareness or total lack of awareness.
Narrower Anaesthesia (#HM3898) Synesthesia (#HM1241)
Hyperaesthesia (#HM1479).

♦ **HM0008d Plane of fragrance** (Leela)
Gandha-loka
Description Transmutation of the sense of small brings experience of divine fragrances while meditating. The body itself no longer produces bad smells but instead a fragrance of sandalwood or lotus flowers.
Context The 33rd state or square on the board of *Leela*, the game of knowledge, appearing in the fourth row.
Broader Fourth chakra: attaining balance (Leela, #HM3291).

♦ **HM0009d Al-Jami'** (Sufism)
Al-Jame'
Description The believer is aware of Allah as the gatherer of whatever, whenever, wherever he wishes. He gathers together the cells that constitute our bodies and can disperse them again at will. Flesh, heart, mind, soul, all the little "I"s, "me"s and "mine"s, all are gathered in one person. The only companion is a man's deeds. The heedless do not see this and are preoccupied with the things of the world, enslaved by the flesh and the ego. However, all will be gathered together at the last judgement, His friends in paradise, His enemies in hell.
Context A mode of mystical awareness in the Sufi tradition associated with the eighty-seventh of the ninety-nine names of Allah.
Broader Ninety-nine names of Allah (Sufism, #HH2561).
Followed by Al-Ghani (Sufism, #HM3822).
Preceded by Al-Muqsit (Sufism, #HM8001).

♦ **HM0019e Asomatic experience** (Psychism)
Description Consciousness is mentally projected to another location, but without awareness of the physical body there.

♦ **HM0020e Xenophrenia** (Psychism)
Description This is any state other than the normal waking state; it may be triggered by some external circumstance and may or may not be remembered on return to normality. The state has been described as the etheric body coming out of alignment with the physical body.

♦ **HM0021d Ataraxia**
Non-agitation
Description This state of inner calm, unmoved by triumph or adversity, is typical of Stoic philosophy.

♦ **HM0034d Al-Khaliq** (Sufism)
Description The believer is aware of Allah as the creator. Man is the supreme creation, all was created for him, man was created for God. One should find and use the beneficence and wisdom in creation and the order which follows, feel the blessing of being part of creation, which reflects the creator.
Context A mode of mystical awareness in the Sufi tradition associated with the eleventh of the ninety-nine names of Allah.
Broader Ninety-nine names of Allah (Sufism, #HH2561).
Followed by Al-Bari' (Sufism, #HM3636).
Preceded by Al-Mutakabbir (Sufism, #HM3850).

♦ **HM0035c Immediate boundary awareness**
Alert immediacy
Description Immediate awareness is a state of mind that a person lives but does not make. It is not experienced as produced by something, nor is it a representation. It is our alertness as we live, and as such is particularly at issue in sickness or health. It is an individual's open sense for the limits of who he is and for the limits of selfhood. This reduces the threat otherwise experienced when encountering the boundaries with other vastly different kinds of awareness. Recognizing the boundaries in mind, has the effect of freeing the person for boundary awareness, for experience of radical otherness, and for all the contingencies in being. Such immediate boundary awareness is always intuitive, as distinct from conceptual objectivity, and is often affective. Such states may be approached by following feelings, remaining uncritically alert with affect, or giving focus to emotions. These states find expression in directions of all sorts: innuendo, irony, style, images, fantasies, depth feelings, desires, etc.
Immediate awareness is that region of awareness that is neither moral nor immoral, that is not fixed by any overarching law, but that is nevertheless alert and alive as the occurrence of many relations of regions and things variously and finitely constituted. A person opens to immediacy of awareness by allowing his thinking to be centered in openness, disclosure, or transience. Each moment then dissolves as it happens. There is then a freedom in transience, a translucence and lightness in things in contrast to normal material heaviness. This awareness is lost when the person imposes desires on the occurrence of things, allowing insistence and holding to dominate as the mediation of desires that provide connections between things through which they may be constructed. Environments are thus constructed as mediated realities which may absorb attention to the point that the existence of the unmediated regions is forgotten. These regions can be understood as distinct from conscious processes. They can be viewed as not susceptible to description or else precede observation and thus are already past when seen or spoken of. They may be considered as nonconceptual and fundamentally changed when grasped conceptually.
Refs Scott, Charles E *Boundaries in Mind* (1982).
Related Reverie (#HM1576) Acceptance (#HM1875) Transcendence (#HM4104)
Fundamental dialogue (#HM3225).

♦ **HM0036f Erotic frenzy**
Erotomania
Description Nymphomania in women and satyriasis in men are said to be the result of excessive and persistent sexual desire and may give rise to erotic frenzy. Because the language used to describe ecstatic religious experiences of mystical union with God is so similar to that used in erotic language, some psychologists have attributed all such spiritual states to overactive eroticism.
Broader Frenzy (#HM3594).

♦ **HM0043d Mythical imagination**
Mythical consciousness
Description Primitive mentality is said (Cassirer) not to invent myths but to experience them. They are original and involuntary revelations of the pre-conscious psyche and promote a consciousness of solidarity of all life, a unity of feeling among individuals and a sense of harmony with nature and life as a whole. This primitive, mythical world embodies a unitary energy of the spirit, manifesting as creating gradually an intermediate reality, a realm of the image and of pure imagination which exists on the borderline between wholly subjective (internal) and wholly objective (external) reality. The mythical image is not a representation of something, but replaces that something, it is that something, so that phenomenal and real are fused, every phenomenon becomes an incarnate pure expression not a representation. Reality and appearance cannot be contrasted, reality is fully present in its appearance. If something affects mind, feeling or will then it is a fully objective and undoubted reality, so that being effective is the same as being.
It follows that the mythical mode is not so much a personalization of natural forces and events as objectivizations of intense if fleeting impressions occupying primitive consciousness. There is only a fluid boundary between personal and impersonal, it and thou, everything being interconnected and this very interconnectedness being strangely impersonal. The sense of "I" only emerges gradually and mythical consciousness is never fully lost, so that the original potency of the myth in unifying internal and external, seeing the universe in the particular, continues and asserts itself in the entire field of consciousness. The human spirit has no absolute past. What has passed it gathers into itself and preserves as present.
Refs Avens, Roberts *Imagination is Reality* (1980); Cassirer, E J *The Philosophy of Symbolic Forms* (1953).
Related Mythical consciousness (#HM2078).

♦ **HM0044d Biosocial consciousness**
Consciousness as ego conflict — Regression
Description Consciousness is derived from conflict of the ego with the real objects which make up the world, or from interaction with them. The world thus presents a threat; and control is a matter of engaging this threat, accepting that the ego is changed as a result of the engagement. The ego is therefore also a product of the external world.
Context One of three possible modes by which the relationship of consciousness with ego may be regarded, each implying a particular strategy of control or of seeing the way through limitations, what may be termed a form of transcendence.
Broader Ego consciousness (#HM0570).
Related Regression (#HH1300) Transpersonal awareness (#HM0768)
Existential consciousness (#HM3336)
Psychological approach to transcendence (#HH1763).

♦ **HM0046e Laziness** (Yoga)
Description Due to lack of will, the individual is unable to work sufficiently hard to achieve what he truly aspires to achieve, and there is a lack of correspondence between what he knows and what he does.
Context One of the nine obstacles to soul cognition enumerated by Patanjali.
Broader Obstacles to soul cognition (Yoga, #HM3182).
Related Sloth (#HM3116) Accidie (#HM3259).
Preceded by Carelessness (Yoga, #HM0361).

♦ **HM0047c Faculty of recollection** (Buddhism)
Satindriya (Pali)
Broader Faculty of recollection (Buddhism, #HM4248).

♦ **HM0050c Itmianan** (Sufism)
State of securing the self
Context The eighth psychic state listed by A Reza Aresteh as progress on the inner self through divine attraction, *kedesh-jazba*, the outer self already having been purified through conscious effort, *kushesh*.
Broader Psychic states (Sufism, #HM4311).
Followed by Yaqin (Sufism, #HM4426).
Preceded by Mushahada (Sufism, #HM3521).

♦ **HM0051d Stasis of thought** (Buddhism)
Cittassa thiti (Pali)
Broader Sukha (Buddhism, #HM2866).

♦ **HM0052c Emptiness of what is free of permanence and annihilation** (Buddhism, Tibetan)
Atyantashunyata
Context One of the eighteen emptinesses comprising the paths of view in Tibetan Buddhism.
Broader Emptinesses on the paths of the view in Buddhism (Buddhism, Tibetan, #HM2944).

♦ **HM0059c The cry** (Christianity)
Description Free from the mind, the heart and the temptation of the heart, suddenly a cry for help tears through one's heart. It is one's duty every moment, day and night, in sorrow or joy, in the midst of mundane activity to discern this cry rending the whole heart of humankind. One discerns it with vehemence or restraint, according to one's nature, with laughter or weeping, in action or thought, and strives to find out who is imperilled.
Context The call to join the march in the spiritual exercises of Nikos Kazantzakis.
Broader Saviors of God: the march (Christianity, #HM3439).
Followed by First step: the ego (Christianity, #HM3748).

♦ **HM0062b Sport-induced modes of perception**
Spiritually evocative sports — Mystical sensations in sport — altered perceptions in sport
Refs Murphy, Michael and White, Rhea A *The Psychic Side of Sports* (1978).
Related Martial arts (#HH0085) Flow experience (#HM2344) Sports psychology (#HH0723).

♦ **HM0063b Exhilaration**
Related Depression (#HM2563) Cyclothymia (#HM1616).

♦ **HM0065c Eleven-fold knowledge** (Systematics)
Experience of domination
Description Domination is the power that reconciles order and disorder through the agency of creativity. It is the highest form of relatedness and requires eleven independent terms to be experienced. It provides the conditions for mutual completion of structures of different kinds.
Context The eleventh in a sequence of twelve modes of knowledge, identified by J G Bennett,

inspired by G Gurdjieff.
Broader Systematics (#HH2003).
Followed by Twelve-fold knowledge (Systematics, #HM0707).
Preceded by Ten-fold knowledge (Systematics, #HM1804).

♦ **HM0066c Faculty of faith** (Buddhism)
Saddhindriya (Pali)
Narrower Assurance (#HM1687) Confidence (#HM1998) Having faith (#HM3158) Power of faith (#HM4404) Belief (Buddhism, Yoga, #HM2933).

♦ **HM0069d Al-Wahhab** (Sufism)
Description The believer is aware of Allah as the bestower. As thanks are due to those through whom the gifts of Allah are received, so much more so are they due to Allah who initiates such giving.
Context A mode of mystical awareness in the Sufi tradition associated with the sixteenth of the ninety-nine names of Allah.
Broader Ninety-nine names of Allah (Sufism, #HH2561).
Followed by Al-Razzaq (Sufism, #HM7021).
Preceded by Al-Qahhar (Sufism, #HM1318).

♦ **HM0074c State of perfect love** (Christianity)
Pure love — State of transformation
Description François Fénelon indicates that in all who feel pure love for God, the motive of God's glory so fills the mind that the motive of one's own happiness is practically annihilated. God becomes the centre of the soul to whom all affections tend, the sun from whom all light and warmth proceed. Ones own happiness and all that regards one's self is lost sight of. God alone is loved, all other things in and for God. In this state of perfect love the individual has all moral and Christian virtues in himself, since love is the foundation, source or principle of all the virtues (St Augustine).
This highest religious state has also been referred to as a state of transformation. The soul has no preferences of itself, it will move in any direction God gives it.
Refs Fénelon, François *The Maxims of the Saints.*
Broader Love consciousness (#HM2000).
Related Love of God (Christianity, #HH5232).

♦ **HM0077d Vacillation** (Buddhism)
Thambhatattam (Pali)
Broader Perplexity (Buddhism, #HM0812).

♦ **HM0082c Wisdom** (Buddhism)
Panna (Pali)
Broader Faculty of wisdom (Buddhism, #HM3233).

♦ **HM0085g Collective unconscious** (Jung)
Description For Jung, reborn in each individual brain structure there is a powerful spiritual inheritance of human development which is distinct from the personal unconscious and contains the archetypes, which are forms universally valid for everyone and into which the personal element of the individual enters as content.
Refs Adler G, Fordham M and Read H (Eds) *The Collected Works of C G Jung*; Robertson, Robin *CG Jung and the Archetypes of the Collective Unconscious* (1987); Samuels, Andrew et al *A Critical Dictionary of Jungian Analysis* (1986).
Related Archetype (#HH0019) Collective unconscious (Psychosynthesis, #HM2811).

♦ **HM0087c Reflexivity**
Awareness of coding
Description Awareness of paradigmatic codes as codes, namely the deliberate awareness of constructing and using a code, and the having of that awareness as part of the code. From this perspective, any spiritual tradition or psychological discipline may be considered heuristically valuable as a tool, without requiring that they be "true". This avoids the tendency of a person to map any one of these codes onto his entire ontology. It does not involve making everything conscious, rather it includes the possibility that the code of which the person is aware is fed by sources that lend themselves only to indirect awareness.
Refs Atlan, Henri *Le Cristal et la Fumé* (1979); Attali, Jacques *Noise* (1985); Bateson, Gregory *Where Angels Fear.*
Related Reflexivity (#HH0977) Composition (#HM1406) Self-remembering (#HM2486) Ontological love (#HM1211).

♦ **HM0092f Visual hallucination**
Micropsia
Description There may be flashes, flickering, lifelike appearances (perhaps overlarge or diminished in size, the latter being referred to as micropsia), or whole scenes. Usually the hallucination is three-dimensional, casts a shadow and appears solid.
Broader Hallucination (#HM4580).
Related Doppelgänger (#HM7342) Clairvoyance (Psychism, #HM3498) Negative hallucination (#HM5904).

♦ **HM0098d Conceit** (Leela)
Mada
Description Intoxicated with vanity or pride in possessions and achievements, the player creates bad karma. His desires lead him to seek out bad company which supports him in this tendency. Seeking good company can halt the process.
Context The seventh state or square on the board of *Leela*, the game of knowledge, appearing in the first row.
Broader First chakra: fundamentals of being (Leela, #HM4103).
Preceded by Bad company (Leela, #HM0764).

♦ **HM0099d International consciousness** (Yoga)
Context A step on the path to cosmic consciousness, when the individual has not yet achieved conscious self-relatedness and lacks the courage to go against mankind as a group.
Related Conscious self-relatedness (Yoga, #HM0467) Education for international understanding (#HH6451).
Followed by Cosmic consciousness (#HM2291).
Preceded by National consciousness (Yoga, #HM1370).

♦ **HM0101d Yoga of the intermediary state** (Buddhism)
Bar-do (Tibetan)
Context The fifth branch of the Tibetan Tantric path of form according to the Naropa school.
Broader Tantric visualization (Yoga, #HM1690).
Followed by Yoga of consciousness transference (Buddhism, #HM5122).
Preceded by Yoga of the clear light (Buddhism, #HM3690).

♦ **HM0102c Beginner in spiritual life** (Christianity)
Spiritual immaturity
Description After individuals have been converted to God's service and have begun a contemplative life, they are at first nurtured by God's grace. They experience a taste of spiritual depth, largely without effort on their part. They have great delight in spiritual exercises; they find joy in prayer, there is pleasure in penances and fasting. The sacraments bring them consolation. Spiritual people use these activities carefully and find them useful and some discover they are weak and imperfect in the spirit. At the same time they have certain faults because they have not matured through struggling with goodness. The spiritually immature often fall into one or more of the seven deadly sins.
Context This is the first stage in the Dark Night by St John of the Cross.
Broader The dark night (Christianity, #HM1714).
Narrower Spiritual envy (#HM3818) Spiritual lust (#HM4180) Spiritual sloth (#HM0491) Spiritual anger (#HM0936) Spiritual pride (#HM4533) Spiritual avarice (#HM0642) Spiritual gluttony (#HM0507).
Followed by Dark night of the senses (Christianity, #HM1727).

♦ **HM0103c Sixth order perceptions - control of relationships**
Description Relationships, the way things go together, are something which is learned, so that one makes appropriate responses in a given situation. It is the awareness that a café is somewhere that one eats. One may recognize a configuration (for example, one's mother), be aware of transition (she is walking towards one), be capable of physical response, know that physical response is required and yet not know what that response is (it is appropriate to kiss her, to say hello). That response would be a sixth order perception.
Context The sixth of ten orders in the perceptual system described by William Glasser.
Broader Orders of perception (#HM0988).
Followed by Seventh order perceptions - programme control (#HM1001).
Preceded by Fifth order perceptions - control of sequence (#HM0772).

♦ **HM0107b Essential wisdom**
Prajna (Zen, Yoga) — Transcendent wisdom — Superconsciousness — Superconscious knowledge
Description (1) This is the experience an individual has when he feels the infinite totality of things in its most fundamental sense. In psychological terms, it is when the finite ego, breaking its hard crust, refers itself to the infinite which envelops everything that is finite, limited and transitory. The experience is somewhat similar to a totalistic intuition of something that transcends all particularized, specified experience. Superconscious knowledge can thus be said to be based on the direct perception of truth.
(2) In Zen writings, prajna or transcendent wisdom is that which transcends the knowledge of things and of the mind, of *sanjna* and *vijna* which are the awareness of material things and their use and value. It is beyond the duality of subject and object. Such wisdom is equated to Buddha mind: subject, predicate and object are all the same. This special process of knowing, *abruptly seeing* or *seeing at once*, does not follow the general laws of logic or result from reasoning. When reasoning is abandoned as futile, use of the will-power finished, then one suddenly finds one's self facing *sunyata* or emptiness. Seeking for such knowledge implies the desire to know, and therefore to have an underlying interest in, life itself. It is the basis of love and the knowledge of this is wisdom - as is implicit in the word *philosophy*, love of wisdom. It is said there is no real meditation without prajna, and without meditation no prajna. The working of prajna is the basis of religious knowledge, when contradictions, absurdities, paradoxes and impossibilities are accepted as revealed truth. What is normally seen is turned upside down, like the turning of a brocade cloth. What was seen on the surface may have been bewildering; abruptly turning it over, interrupting the course of the eyesight, reveals the whole scheme. The Zen experience is the seeing into the working of prajna, which is where the ordiary world of contradictions starts.
(3) According to Patanjali, prajna stands for all the states of consciousness in *samadhi*, from *vitarka* to *asmita*. When pure awareness of reality takes place then prajna comes to an end in *vivekajam jnanam*, awareness of ultimate reality.
Related Wisdom (#HH0623) Sanjna (Yoga, #HM3160) Vijna (Buddhism, #HM3617) Prajna (Hinduism, #HM3455) Sambodhi (Buddhism, #HM7195) Prajna paramita (Zen, #HH0550) Sarvajnata (Buddhism, #HM5334) Understanding (Buddhism, #HM4523) Mental wisdom (Buddhism, #HM2655) Samadhi (Hinduism, Yoga, #HM2226) Perfect wisdom (Buddhism, #HM7844) Emptiness (Buddhism, Zen, #HM2193) Jnana (Yoga, Buddhism, Hinduism, #HH0610).

♦ **HM0110e Lack of dispassion** (Yoga)
Addiction to objects
Description An inability to remain detached from sense objects and a desire for material and sensuous things.
Context One of the nine obstacles to soul cognition enumerated by Patanjali.
Broader Obstacles to soul cognition (Yoga, #HM3182).
Followed by Erroneous perception (Yoga, #HM1077).
Preceded by Laziness (#HH0963).

♦ **HM0113d Al-Warith** (Sufism)
Description The believer is aware of Allah as the ultimate inheritor, that all he has is only lent to him and will revert to Allah. Knowing this, he follows neither the desire of the flesh nor the command of the ego but does all he does for the sake of Allah. He thus becomes one with Allah, eternal and everlasting.
Context A mode of mystical awareness in the Sufi tradition associated with the ninety-seventh of the ninety-nine names of Allah.
Broader Ninety-nine names of Allah (Sufism, #HH2561).
Followed by Ar-Rashid (Sufism, #HM5234).
Preceded by Al-Baqi (Sufism, #HM6786).

♦ **HM0115d Al-Basit** (Sufism)
Description The believer is aware of Allah as the expander and releaser. Opening His hand he releases abundance, joy, relief, ease. These times must not be periods of forgetfulness and belief that one is the cause of one's own success, but of thankfulness. Knowledge that good and bad all come from Allah provides a balanced state.
Context A mode of mystical awareness in the Sufi tradition associated with the twenty-first of the ninety-nine names of Allah.
Broader Ninety-nine names of Allah (Sufism, #HH2561).
Followed by Al-Khafid (Sufism, #HM3644).
Preceded by Al-Qabid (Sufism, #HM2767).

♦ **HM0116c Malaqut** (Sufism)
Description This is the spiritual state of angelic nature, when the follower enters the spiritual path in the real sense.
Context The second of four stages on the Sufi spiritual path.

Broader Divine path (Sufism, #HM3371).
Followed by Jabarut (Sufism, #HM4392).
Preceded by Nasut (Sufism, #HM3579).

♦ **HM0123e Akshara consciousness** (Psychism)
Description At this level of consciousness it is possible to tune into unmanifest intelligence.

♦ **HM0124c Haqiqa** (Sufism)
Truth
Description The mystical journey to God has been described in three main stages in which the life of the Prophet is imitated and his sayings observed. If shari'a is the word and tariqa the action, haqiqa is the inward state. The soul is illuminated and, by God's grace, the seeker sees the vision of God and experiences knowledge of God.
Related Sharia (Islam, #HH0738) Ma'rifa (Sufism, #HM2254) Divine path (Sufism, #HM3371).

♦ **HM0125d Cetovimutti** (Pali)
Liberation of the mind
Description This liberation is achieved through perfect mastery of enstatic practices and leads to nibbana. It is also described as freedom from passion.
Broader Mukti (#HH0613).
Related Pannavimutti (Pali, #HM8125) Ubhatobhagavimutti (Pali, #HM7983).
Followed by Nirvana (Buddhism, #HM2330).

♦ **HM0126c Sphere of knowledge** (Kabbalah)
Daath–sephira
Description This sphere of awareness is associated with the development of knowledge. It corresponds to the recognition of meaning in experience. Associated with the Tree of Life, it signifies the experience of knowledge without understanding. It represents both no dimensions and all dimensions of awareness. It is also understood as a centre of generation and regeneration and is related to the expression of spiritual knowledge. It is symbolized by the positive and negative archetypes of transformation, and by the planet Pluto.
Context Known as the sphere without a number, associated with the ten spheres described in the Kabbalistic system of sephiroth.
Broader Kabbalah (#HH0921).
Related Vishuddha (Yoga, #HM3461).

♦ **HM0128d 'Ibadat** (Sufism)
Worship of God
Description With body seeking to be of service, heart exhuberant with love and head in roaring quest of contemplation, the devotee reads the Quran and invokes the name of God during the day and in the night remains standing.
Context According to Shaykh Abu Sa'id ibn Abi'l–Khayr, this is the 15th of 40 stations or maqamat the Sufi must possess for his journey on the path of Sufism to be acceptable.
Broader Stations of consciousness – ibn–Abi'l–Khayr (Sufism, #HM2424).
Followed by Wara (Sufism, #HM0286).
Preceded by Zuhd (Sufism, #HM4450).

♦ **HM0129c Hawa** (Islam)
Passion — Concupiscible appetite — Desire
Description This is the appetite for what is necessary to the individual for his survival. It is that which encourages him to eat, drink or mate, for example. Unfortunately it is not usual to stop at the limits of what is beneficial, and the tendency is for temporal pleasure to be sought without concern for the outcome. Becoming addicted to desires, the individual invariably derives less pleasure from them without, however, giving them up. Restraining the passions, or striving against the soul, is essential for progress.
Refs Bell, Joseph Norment *Love Theory in Later Hanbalite Islam* (1979).
Narrower Nazar (#HM6003) Passionate love (#HM3656).
Related Striving against the soul (Islam, #HH6973).

♦ **HM0130c Mundane resultant consciousness** (Buddhism)
Karma–result consciousness
Description The Path of Purification (Hinayana Buddhism) describes this consciousness as being of 32 kinds, 10 from the five senses (each being profitable karma resultant or unprofitable karma resultant) and 22 from mind consciousness. Each is the result of stored up karma. In the sense sphere, 16 arise with the sense sphere formation of merit as condition (profitable result) and 7 with sense sphere formation of demerit as condition (unprofitable result). In the fine–material sphere, 5 arise with fine–material sphere formation of merit as condition, leading to dwelling in the first to the fifth jhanas. In the immaterial sphere there are 4 kinds which are formed from good karma having been stored up in connection with the formless realm, leading to surmounting perception of material form, abiding in rapt meditation accompanied by perception of boundless space, boundless consciousness, nothingness and finally neither–perception–nor–non–perception.
Resultant consciousness arises during a particular existence or at rebirth–linking.
In the course of an existence, this depends on profitable or unprofitable result. The five sense consciousnesses, profitable or unprofitable depending upon whether the object is desirable, desirable–neutral, undesirable–neutral or undesirable, are followed by receiving, investigating and registering through profitable or unprofitable resultant mind–element and mind–consciousness elements. The mind–consciousness elements are also active in the function of life–continuum after rebirth linking and in death at the end of the course of existrence. Consciousnesses in the fine–material and immaterial spheres accomplish the functions of life–continuum and death.
Broader Dispositions of consciousness (Buddhism, #HM2098)
Indeterminate consciousness in the sense sphere – resultant (Buddhism, #HM5721)
Indeterminate consciousness in the immaterial sphere – resultant (Buddhism, #HM4982)
Indeterminate consciousness in the fine–material sphere – resultant (Buddhism, #HM0594).
Narrower Eye consciousness (#HM2074) Ear consciousness (#HM2169)
Mind consciousness (#HM3323) Nose consciousness (#HM2364)
Body consciousness (#HM2562) Tongue consciousness (#HM2263)
Mental consciousness (#HM2838) Mind–consciousness element (#HM6173).
Related Rebirth–linking (Buddhism, #HM8266) Mundane concentration (Buddhism, #HM7234)
Mundane understanding (Buddhism, #HM7628).

♦ **HM0131c Urdhva–loka** (Jainism)
Description According to the Jain system, this is the second level of the universe, above Mount Meru and the abode of the celestials, with many levels depending on the celestial beings inhabiting them.
Related Adha–loka (Jainism, #HM3697) Siddha–loka (Jainism, #HM0569)
Madhya–loka (Jainism, #HM1357) Human development (Jainism, #HH0622).

♦ **HM0134f Modes of awareness associated with alcohol consumption**
Hangover
Description In small doses, alcohol is consumed (usually in social circumstances) to produce good cheer and fellowship. Inhibitions are reduced and the individual is more vivacious and gregarious. Heavy drinkers may also desire to reduce anxiety and tension, although anxiety in fact increases measurably with alcohol consumption. Of all psychoactive substances, alcohol produces the most aggression. After–effects of excessive consumption, popularly referred to as *hangover*, include headaches, nausea, loss of memory and general debility. Alcohol is addictive and high intake leads to physical dependence, as well as to irreversible brain and liver damage. Cessation may lead to hallucination, usually unpleasant, whether visual or auditory. The latter may be hissing, buzzing or (more usually) voices, often discussing the individual in the third person. Hallucinations may last for weeks or even months.
Broader Modes of awareness associated with psychoactive substances (#HM0584).
Narrower Delirium tremens (#HM7805).
Related Formication (#HM4094).

♦ **HM0135c Suppression** (Yoga)
Niruddha
Description This state is devoid of thought. Through constant practice of cessation of thought, the world of names and forms is seen as a product of the mind. When the mind ceases to exist in the practical sense then everything else dissolves.
Context One of five states of mind classification identified in yoga.
Related Nirodha parinama (Yoga, #HM3437).
Preceded by One–pointedness (Yoga, #HM5734).

♦ **HM0137c Dhyana** (Hinduism, Buddhism)
Jhana (Pali) — Meditation — Contemplation — Absorption
Description This is the practice of intensive concentration of the mind, when distractions through intruding thoughts have been eliminated. It is the stage beyond concentration known as dharana when distractions still appear. It is a refinement of perception which leads to a smooth flow of awareness when the duality of perceptor and perceived is merged. In Book VII of the Chandogya Upanishad, dhyana in placed between *citta* or mind's mother substance and *vidya* or wisdom, as worthy of worship. The fifth of five perfections or *paramitas* of Zen which, when achieved, render the mind ready for the sixth – *maha paramita* or great perfection, *prajna*. According to Patanjali, this is the seventh stage of the eightfold path of *raja yoga*, leading to *samadhi*.
In Buddhism there are four degrees of dhyanas or jhanas, with preliminary practices of the first five constituents of yoga as preparatory to the first. At this stage one is accustomed to a concentration of the mind in which the senses are suppressed and the state of ecstasy is achieved which may be equivalent to dhyana. The first dhyana is described as joy and gladness at being separated from sensuality and sin and as being reflective and investigative. The second is deep tranquillity with the suppression of reflection and investigation; thought is tranquillized and intuition predominates. In the third there is tranquil serenity and patience when passion is destroyed, an awareness in the body of joyful and conscious delight. The fourth is without sorrow or joy, pure equanimity and recollection where previous gladness and grief are destroyed. Attainment of these four dhyanas determines one's position in the rupaloka (world with form) heavens. The first three dhyanas belong to the first seven grades of holy paths, while the final is that of an Arhat.
Further developments in Mahayana Buddhism distinguish many more kinds of dhyana, being the way of bodhisattvas but not an Arhat. While Zen – the very name of which derives from the Sanskrit "dhyana" through the Chinese Ch'an – makes dhyana central to the entire teaching.
Context The seventh component of the eightfold path of yoga.
Refs Goldstein, Joseph *The Experience of Insight* (1987).
Broader Samyama (#HH0374) Meditation (#HH0761) Paramitas (Zen, #HH0219)
Eightfold path of yoga (Yoga, #HH0779).
Narrower Dwelling in the first jhana (Buddhism, #HM4298)
Dwelling in the third jhana (Buddhism, #HM5643)
Dwelling in the second jhana (Buddhism, #HM7121)
Dwelling in the fourth jhana (Buddhism, #HM8087).
Related Wisdom (#HH0623) Ch'an (Zen, #HH0589) Raja yoga (Yoga, #HH0755)
Samatha–bhavana (#HH0710) Dhyana yoga (Yoga, #HH0827) Dharana (Hinduism, #HM2566)
Prajna paramita (Zen, #HH0550) Jhana (Buddhism, Pali, #HM7193)
Concentration (Buddhism, #HM6663) Skill in absorption (Buddhism, #HH4777)
Rupaloka consciousness (Buddhism, #HM2536) Steps to enlightenment (Buddhism, #HH4019)
Subjects for meditation (Buddhism, #HH3987)
Form–realm consciousness (Buddhism, #HM2257)
Absorption concentration (Buddhism, #HM0311)
Chitta–dependent consciousness (Yoga, #HM3011)
Religious experience as meditational insight (#HM1593)
First trance of the fine–material sphere (Buddhism, #HM2450)
Third trance of the fine–material sphere (Buddhism, #HM2062)
Second trance of the fine–material sphere (Buddhism, #HM2038)
Fourth trance of the fine–material sphere (Buddhism, #HM2586).
Preceded by Virya (Zen, #HH0677).

♦ **HM0140d Uprisedness of thought** (Buddhism)
Attamanata cittassa (Pali)
Broader Fivefold happiness (Buddhism, #HM0747).

♦ **HM0144d Koinonia** (Christianity)
Description One of the three ways in which the mission of the Church is manifest, this refers to a state of participative fellowship in the act of God in Christ, arising through preaching.
Related Kerygma (Christianity, #HM4256) Diakonia (Christianity, #HM0810).

♦ **HM0145d Apana–loka** (Leela)
Description The player learns to maintain in harmony the airs in his body and realizes the importance of apana, which is responsible for the elimination of energy from the body. Yogic practice and proper diet lead to fusion of prana with apana. The player is rejuvenated, with the vitality, stamina and power of a sixteen year old.
Context The 39th state or square on the board of *Leela*, the game of knowledge, appearing in the fifth row.
Broader Fifth chakra: man becomes himself (Leela, #HM0933).

♦ **HM0147g High dream** (Psychism)
Description This form of dream, characterized by a feeling of heightened emotion or of being accompanied by a higher being, or of transcendent consciousness, has a lasting effect on the dreamer – whose attitude or personality may be changed – and also leaves a clear and lasting impression.
Broader Dreams (#HM2950).
Related Emotional high (#HM1400).

♦ **HM0148g Jada samadhi** (Yoga)
Unconscious enstasy — Jadya
Description This psychosomatic state similar to hibernation is a pseudo enstatic state characterized by unconsciousness - jadya - rather than luminosity of consciousness as arises in real enstasy.
Related Enstasy (#HM3169).

♦ **HM0150g State of hunger** (Buddhism)
Greed
Description Here one is ruled by the desires, which are the main driving force for life. Failure to control one's desires makes one their slave, never satisfied. Indeed, one may reach a state of being dissatisfied immediately a desire is fulfilled. Desire can be so strong as to distort the perspective of reality, extreme hunger leading to the impossibility of exercising judgement. Nichiren Shoshu Buddhism teaches that this desire can be eradicated without eradicating life itself. Desires have to be sublimated and reorientated towards creative and valuable ends.
Context One of the ten worlds described in Nichiren Soshu Buddhism.
Broader Six paths (Buddhism, #HM1914) Ten worlds (Buddhism, #HM2657)
Four evil paths (Buddhism, #HM1252) Three evil paths (Buddhism, #HM0923)
Nichiren shoshu buddhism (Buddhism, #HH3443).
Related Greed (Buddhism, #HM3283).
Followed by State of animality (Buddhism, #HM0847).
Preceded by State of hell (Buddhism, #HM4282).

♦ **HM0158c Understanding** (Buddhism)
Pajanana (Pali)
Broader Faculty of wisdom (Buddhism, #HM3233).

♦ **HM0159c Samapatti** (Buddhism)
Attainment
Context One of ten superhuman qualities described in the Sutta Vibhanga as being special attainments of insight above that of ordinary men.
Broader Superhuman qualities (Buddhism, #HH9764).

♦ **HM0160c Matins** (Christianity)
Hours of matins
Description The canonical hours of matins traditionally take place between 12 and 3 am. They dramatize waiting for God in the midst of the darkest moments of life. The mood is watchful waiting.
Broader Canonical hours (Christianity, #HM1167).
Followed by Lauds (Christianity, #HM0894).
Preceded by Compline (Christianity, #HM4321).

♦ **HM0168d Conscious learning**
Description Characteristic of self–reflective consciousness or apperception and the development of language, this is the third mode of learning posited by Jantsch and Waddington in their ontogenetic model of human consciousness. It seems to be the primary vehicle for epigenetic development for conscious and creative interaction of the individual with the environment, and central to sociogenesis. This advanced human stage is equivalent to personal consciousness, commencing in an awareness of free space at birth, leaving foetal consciousness behind, and culminating in death of ego consciousness and transcendence of personal boundaries; it is characterized by conscious action. On a human scale, transition from conscious to superconscious learning is equivalent to the struggle between cultures, life styles and world views resulting in a sense of the wholeness of the process of mankind. It is at this transition that inner and outer reality are in perspective so as to guide thinking within the wider existential context.
Context One of four learning processes distinguished (Lazslo, 1972) by the type of consciousness brought into play.
Broader Learning (#HH0180) Ontogenetic model of human consciousness (#HM1105).
Followed by Superconscious learning (#HM1244).
Preceded by Functional learning (#HM0196).

♦ **HM0169d Knowledge of contemplation of reflection** (Buddhism)
Knowledge of analysis
Related Purification by knowledge and vision of the way (Buddhism, #HH3550).
Followed by Knowledge of equanimity about formations (Buddhism, #HM0424).
Preceded by Knowledge of desire for deliverance (Buddhism, #HM0766).

♦ **HM0172d Liquid plane** (Leela)
Jala–loka
Description The heated energy of violence is cooled in pure water - which dissolves identification with form. As water takes the form of the vessel in which it is contained, the player takes the form of whatever confronts the self.
Context The 53rd state or square on the board of *Leela*, the game of knowledge, appearing in the sixth row.
Broader Sixth chakra: time of penance (Leela, #HM4412).

♦ **HM0176c Soma chakra** (Yoga)
Twelve–petalled lotus — Sixteen–petalled lotus
Description A minor chakra in the seventh chakra, this is located above the third eye, in the centre of the forehead. Meditating on this chakra stops the downward flow of amrita or nectar through performance of khechari mudra. This brings immortality in the physical body so that the process of aging is stopped. There is youth, vitality, stamina, and victory over disease, decay and death. Through the union of Shiva and Shakti one enjoys eternal bliss.
Broader Sahasrara (Yoga, #HM3398) Kundalini yoga (Yoga, #HH0237)
Chakra centres of consciousness (#HM2219).

♦ **HM0177d Doubting** (Buddhism)
Kankhayana (Pali)
Broader Perplexity (Buddhism, #HM0812).

♦ **HM0180c Correct recollection** (Buddhism)
Sammasati (Pali)
Broader Faculty of recollection (Buddhism, #HM4248).

♦ **HM0181d Initiation of revelation** (Esotericism)
Description The liberating experience associated with this initiation is that of freedom from blindness, permitting the initiate to see a new vision concerning the reality lying beyond any reality hitherto sensed or known. It is therefore an emergence from a tomb of darkness and an entrance into a light of an entirely different nature. Life in form is revealed as being death, so that form dies for the initiate so that he knows a new expansion of life and undergoes a new understanding of living. The significance which becomes apparent opens new configurations of loyalty and goals of a higher level of inspiration. A synthesis is revealed of the overall pattern of energies and forces upon which he can call and the nature of the opportunity it implies. The challenge of developing the will–to–good (spiritual will) becomes apparent, both in its comprehension and its use in world service. The individual is called to formulate his intention. Intuition, regulated by creative imagination, develops to further group activity in this light.
Context The fifth initiation in the tradition of western esotericism.
Broader Spiritual initiation (Esotericism, #HH3390).
Related Muladhara (Yoga, #HM2893) Way of power (Esotericism, #HM0454).

♦ **HM0183e Astral consciousness** (Psychism)
Description A mental projection or dream state when there is a sensation of lightness and of seeing with the whole body.

♦ **HM0184c Delusion** (Buddhism)
Moha (Pali)
Narrower Delusion (#HM0918) Nescience (#HM1394) Stupidity (#HM1857)
Not–seeing (#HM1510) Unawakened (#HM1362) Unawareness (#HM1725)
Childishness (#HM1864) Non–grasping (#HM0451) Utter delusion (#HM1371)
Non–penetration (#HM1637) Un–comprehension (#HM4090) Being unobservant (#HM1965)
Non–consideration (#HM1902) Complete delusion (#HM1956) Not understanding (#HM0403)
Flood of ignorance (#HM3145) Snares of ignorance (#HM1013) Fetters of ignorance (#HM4470)
Inability to compare (#HM0427) Inability to demonstrate (#HM1746)
Obsession with ignorance (#HM3829) Tendency towards ignorance (#HM1375)
Delusion an unwholesome root (#HM1439) Ignorance (Buddhism, Yoga, Zen, #HM3196).

♦ **HM0188c Unremorsefulness** (Buddhism)
Anottappa (Pali)
Related Power of unremorsefulness (Buddhism, #HM1915).

♦ **HM0191d Mildness** (Buddhism)
Maddavata (Pali)
Broader Flexibility of body (Buddhism, #HM3370)
Flexibility of thought (Buddhism, #HM1805).

♦ **HM0192c Splendour of wisdom** (Buddhism)
Pannapajjota (Pali)
Broader Faculty of wisdom (Buddhism, #HM3233).

♦ **HM0194a Dharma megha samadhi** (Yoga)
Description This highest state of samadhi is irreversible and passing through it leads to *kaivalya* or liberation. Free from the world of dharmas which cloud reality, this is the culmination of the mutually reinforcing practices of viveka khyati and para vairagya. Descriptions of this blissful state fall so far short of the experience that Patanjali, for example, does not attempt it.
Broader Samadhi (Hinduism, Yoga, #HM2226).
Related Viveka khyati (Yoga, #HM4493).

♦ **HM0196d Functional learning**
Description Characteristic of reflective consciousness or simple perception, with the testing of primarily metabolic functions in terms of relations with the environment, this is the second mode of learning posited by Jantsch and Waddington in their ontogenetic model of human consciousness. This is the mode characteristic of the whole bioorganismic world, closely linked to genetic communication and evolution. It is equivalent to perinatal consciousness, commencing in an "oceanic" awareness in union with the mother and culminating in leaving the womb with the development of personal consciousness; it is characterized by behaviour. On a human scale, there is a dualistic awareness of the world and God, resulting in tension, doubt and criticism; transition from the functional to the conscious learning level is equivalent to the struggle between social systems resulting in the growth of cultures.
Context One of four learning processes distinguished (Lazslo, 1972) by the type of consciousness brought into play.
Broader Learning (#HH0180) Ontogenetic model of human consciousness (#HM1105).
Followed by Conscious learning (#HM0168).
Preceded by Virtual learning (#HM1217).

♦ **HM0198c Right acts** (Buddhism)
Samyak–karmanta — Right behaviour — Right aims of action — Right action
Description Right outlook and right speech bring right acts. This involves moral conduct, in particular abstaining from killing, stealing and lechery. The characteristic is originating or producing; the function is abstaining; manifestation is in abandoning wrong action.
Context One of the characteristics of the Eightfold Way of Buddhism attained with the path of bhavana (meditation) and which, together with right speech and right livelihood, constitutes ethical conduct (sila).
Broader Eightfold way (Buddhism, #HM2339).
Related Four noble truths (Buddhism, #HH0523).
Followed by Right livelihood (Buddhism, #HM1341).
Preceded by Right speech (Buddhism, #HM1157).

♦ **HM0201d Life** (Buddhism)
Description Life here refers to immaterial states or things. It is by means of life that associated states live.
Context One of the formations aggregate (mental coefficients) of Hinayana Buddhism, being listed among the constant states which appear in their true nature, and as general primary (always present in any consciousness).
Broader Awareness as mental–formation group of conscious existence (Buddhism, #HM2050).

♦ **HM0203c Egoic musical inspiration**
Description This is the third of the three levels of musical improvisation. It is the inspiration which derives from the creator's own ego and from the models he sees around him in the world and from his subconscious (not his superconscious) mind. This is the level of self–expression of the talented artist who may achieve unusual levels of virtuosity without being truly inspired. It does not involve the constant contemplation of the musical models awakened within.
For the listener, only when listening is concentrated unbrokenly on the music does he enter into an experiential phase comparable to this third degree of creative inspiration. He then shares in the personality of the composer by means of responses which may be visceral, emotional or mental. Visceral, body–music is best experienced by actual participation in movement and gesture, reaching a peak of subtlety when empathetic feelings dance with the music. Heart– music grips the emotions, which at its best refines and ennobles them by a process of entrainment. Head–music draws a response from the critical intellect enhancing the capacity to understand and evaluate musical forms. Listening with a combination of bodily, emotional and thinking responses provides a rich and rewarding experience.
Broader Musical inspiration (#HM4805).

♦ **HM0205d Aesthetic emotion** (Hinduism)
Rasa
Description A rasa is the involuntary state produced in the spectator originally of drama, but the theory was later extended to the other arts. The state might be romantic, comic, sorrowful, violent, heroic, terrifying, repulsive or marvellous (plus, in non–dramatic literature, peaceful). The determinant is what is occurring in the drama, the state is what arises from the determinant (a ghostly figure produces apprehensiveness, shock and fear) and the consequent is the physical response – bulging eyes, trembling, cry of alarm. There are thirty three transitory or subsidiary states and eight or nine permanent or predominant states, each associated with a particular rasa – love, mirth, grief, fury, resoluteness, fear, revulsion, wonder (and peace). The rasa is not experienced as its permanent state is experienced in real life. The experience is direct and vivid just as if it were one's own response to real circumstances yet with complete detachment. For example, one experiences grief neither of one's self in a particular situation nor of any other person distinct from one's self – it is grief generalized. It is the function of literature to generalize emotion so that it can be tasted like this. The response arises only inasmuch as the evoked emotion is within one's experience – it is a re–experience of one's own emotion, a calm, unthreatening, recreative ordering of what is already within.
Distinct from conation (kriti) and cognition (jnapti), rasa is the nature of enjoyment or actual aesthetic experience – it is the aesthetic enjoyment which is an emotional experience. Different commentators have described this state, which is experienced by persons of taste as sattva or purity predominates and the other gunas are suppressed. Rajas would produce restlessness and tamas unconsciousness, whereas here there is the manifestation of the experience through sattva. The emotion is entire and indivisible, self–luminous, made up up cognition and bliss, free from the cognition of other objects and related to realization of the Absolute, of the essence of transcendental wonder. During the experience the individual psychic components cannot be distinguished, it is a mass of feelings, emotions and sentiments. Not just an affective experience but also cognitive, it is composed of both cognition and joy. Being both self–luminous and self–aware, it is experienced by itself and not through some other mental mode. It contains a self–conscious element of cognition and is self–aware. The experience is free from the touch of cognitions of other objects – lost in his own aesthetic enjoyment, the person is unconscious of all other objects. He experiences aesthetic enjoyment as a yogin experiences Brahman, direct and immediate, lost in the enjoyment as the yogin is lost in Brahman, although the experience is inferior to the intuitive experience of Brahman. In this ecstasy of joy, distinction of subject and object is lost. The essence of the enjoyment is transcendental wonder, transcendental in that it is felt as an appreciative spectator would identify himself with a person whose emotion is expressed by an actor on a stage. This wonder is of the nature of expansion of the mind and constitutes the core of the aesthetic enjoyment.
Aesthetic enjoyment or relish is essentially joyful in nature, arising from the bliss of the self as the meaning of poetry or drama is appreciated. Skilful acting, for example, excites the permanent emotional dispositions of the spectator, who forgets the distinction between himself and others and experiences the intense joy of the self evoked in him. This extraordinary emotion is free from consciousness of self and not–self, friend and foe, and devoid of all distinctions of space and time, free from all obstacles. Unlike flashes of intuition, perception, recollection and other kinds of knowledge, or the intuitive experience of Brahman, it is associated with permanent emotional dispositions of love, mirth and so on.
There can be no aesthetic enjoyment without subconscious impressions or emotional dispositions like those of love or anger. These impressions or vasanas are innate or acquired and are necessary conditions for such enjoyment. Thus philosophers, devoid of innate emotional dispositions, or some affectionate people, devoid of acquired emotional dispositions, are dead to all aesthetic enjoyment since an emotional disposition is an indispensable prerequisite. Also essential is taste, persons devoid of taste cannot experience the ecstatic joy of aesthetic enjoyment. Rasa springs from the bliss of the self as it realizes the meaning of poetry, for example, and enjoys it. It cannot be made known to others since it exists only as it is experienced, and similarly it cannot be proved – its only proof is the experience, it cannot be proved by some other kind of knowledge. The emotion is identical with the actual enjoyment, aesthetic enjoyment being the only proof of aesthetic emotion and being identical with it. It is not a mental structure but a mental function, a concrete actualized emotion felt by a person of taste as a concrete emotion.
The various conditions which produce an aesthetic emotion do so through activity – vyapara – known as sympathetic identification. This sympathy is with emotions experienced long ago by the personages represented by actors on the stage, for example and enables the person to identify with the persons represented. Without this sympathetic rapport there would be no aesthetic emotion – there must be an illusory sense of identity, a feeling of at–one–ment, projection and identification. There is a peculiar sense of make–believe as the emotion is experienced as the spectator's own and yet not his own.
The aesthetic enjoyment (carvana) is identical with aesthetic emotion (rasa) but is not consistently there with it – it appears and disappears and is only experienced occasionally. An aesthetic emotion is not an effect although its appearance and disappearance may lead it to appear so. It is extraordinary since it is proved by its own enjoyment and is incompatible with the ordinary processes of knowledge. Whereas common emotion is interested, aesthetic emotion is disinterested. Common emotion is immediately personal, aesthetic emotion is impersonal – it is generic, common to all trained spectators. It contains elements of transcendental wonder, the nature of the expansion of the mind awakened by the marvellous. The experience is unique and underived although evoked by a variety of conditions.
Different aesthetic emotions have different effects on consciousness. Erotic and ludicrous emotions bring blooming of consciousness; heroic emotion and the emotion of wonder bring consciousness expansion; horror and fear produce agitation of consciousness; fury and pathos bring about obstruction of consciousness. Although they are produced by different causes they have the same effect on consciousness.
Refs Sinha, Jadunath *Indian Psychology* (1986).
Narrower Comic sentiment (#HM0645) Erotic sentiment (#HM1780) Heroic sentiment (#HM4085) Odious sentiment (#HM0966) Furious sentiment (#HM8762) Pathetic sentiment (#HM0399) Terrible sentiment (#HM1644) Marvellous sentiment (#HM0291).
Related Vasana (Yoga, #HM3231) Plane of taste (Leela, #HM0878) Aesthetic enjoyment (Hinduism, #HM7219).

♦ **HM0206d Ar–Ra'uf** (Sufism)
Description The believer is aware of Allah as benignly compassionate. In infinite mercy and clemency is shown to all of creation, but in particular to mankind, believer and unbeliever alike. Reflecting on this clemency, the believer tries to serve Allah with all he has.
Context A mode of mystical awareness in the Sufi tradition associated with the eighty–third of the ninety–nine names of Allah.
Broader Ninety–nine names of Allah (Sufism, #HH2561).
Followed by Malik–ul–Mulk (Sufism, #HM5902).
Preceded by Al–'Afuw (Sufism, #HM5543).

♦ **HM0210d Initiation of renunciation** (Esotericism)
Initiation of detachment — Initiation of crucifixion
Description The liberating experience associated with this initiation is that of freedom from all self–interest, and the renouncing of the personal life in the interest of the larger whole. Even soul–consciousness ceases to be of importance and a more universal awareness, and one closer to the divine mind, takes its place. Part of the crisis of this experience is the major sense of abandonment that is suffered until the lower dependence is renounced. The sacrifice made is one through which the individual is made holy and set apart for spiritual development and service. The individual dies to all that is material and physical. Prior to renunciation conflict is always present. It is replaced by crises of decision, based on perceptions of the most appropriate form of action, and leading to right discrimination. The individual is presented with a battleground and a field of experience in which he makes great experimental choices through which right orientation is learnt. The cleavage between desirable spiritual values and undesirable material values becomes clear. The individual begins to act entirely out of intuition, enabling him to live in the light of direct knowledge, expressing wisdom itself in all affairs. A rapport is thus established between intuitive understanding and the physical world. He learns to define the wholeness which is his divine right and prerogative.
Context The fourth initiation in the tradition of western esotericism.
Broader Spiritual initiation (Esotericism, #HH3390).
Related Anahata (Yoga, #HM3242) Way of harmony (Esotericism, #HM0763).
Preceded by Initiation of transfiguration (Esotericism, #HM0428).

♦ **HM0212c Androgyny**
Soul as androgyne — Androgynous love
Description Understood in its broadest sense, androgyny signifies the One which contains the Two, namely both the male and the female. It is an archetype inherent in the human psyche. Androgyny continually represents itself in myths and symbols, which have the capacity (if recognized and invoked) to energize the creative potency of men and women in ways that are not usually imagined. As possibly the oldest archetype of which individuals have any experience, it appears to individuals as an innate sense of a primordial cosmic unity, having existed in oneness or wholeness before any separation was made. The concept of a divine androgyny is a consequence of the concept that divine being consists of a unity–totality within which are seen to exist all the conjoined pairs of opposites at all levels of potentiality. Androgyny refers to a specific way of joining the "masculine" and "feminine" aspects of a single human being. As a state of consciousness, androgyny is far from ordinary and threatens an individual's state of equilibrium as identitified with a particular gender. Through this state there is recognition a dynamic inner oscillation of forces between masculine and feminine poles of being. Conscious awareness of these forces within, of their continuing separation and reunion, is an essential part of the inner development of the androgyne. The quality of androgyny is experienced as much in the body as in consciousness. The bodily experience can find its ultimate expression through sexual intercourse, when this is experienced in the spirit of evolutionary consciousness as a mystical marriage and a blending of gender identity. Androgyny is the experience of the flow between opposites.
Refs Critchlow, Keith *The Soul as Sphere and Androgyne* (1980); Singer, June *Androgyny* (1976).
Broader Archetype (#HH0019).
Related Sexual experience (#HM0829).

♦ **HM0213c Perfectionism**
Description This may refer to a state of perfect wholeness and regeneration when attitude and motive are without sin, even though conduct may, through lack of knowledge, be at fault. Philosophically, such wholeness would be the full development of the individual's capacity at all levels, although Hegel considered moral perfection achievable only as a social whole.
Related Completeness (#HM3070) Perfectionism (#HH1466) Self–actualization (#HH0412).

♦ **HM0215d Al–Jabbar** (Sufism)
Description The believer is aware of Allah as the compeller, repairing the broken, completing the lacking. Despite freedom of choice, submission to this forcefulness is necessary. Before receiving punishment for disobedience and revolt, if one takes refuge in the mercy of Allah one feels the reflection of al–Jabbar.
Context A mode of mystical awareness in the Sufi tradition associated with the ninth of the ninety–nine names of Allah.
Broader Ninety–nine names of Allah (Sufism, #HH2561).
Related Latifa–i–qalb (Sufism, #HM1905).
Followed by Al–Mutakabbir (Sufism, #HM3850).
Preceded by Al–'Aziz (Sufism, #HM4562).

♦ **HM0216f Archaic ego states**
Description In Freudian psychology, these states are described as a carrying over into adult life of feelings normally associated with early stages of ego development. Examples are *oceanic feeling*, where there is a limitless experience of the self and the outside world; and experience of the *uncanny*, said to derive from an early narcissistic phase of animism acceptance.
Narrower Omnipotence (#HM2750).
Related Altered ego states (#HM4120).

♦ **HM0218g Aesthetic perception** (Psychism)
Description In psychic terms, a cycle of 43 days during one half of which there are days of low or stagnant emotion and the other half when one is emotional very easily.
Broader Biological rhythms (#HH0475).
Related Aesthetic state (Psychism, #HM3416).

♦ **HM0220c Realization of nen**
Single hearted concentration — One–pointedness in martial arts training
Description Realization of *nen* is the key to aikido, for example. It is not concerned with winning or losing and grows through being properly connected with the *ki* of the universe. It seeks the unity of the order in the universe and becomes the source of the subtle working of *ki*. It is the line connecting the individual *ki* through mind and body to the universal ki.
Refs Ueshiba, Kisshomaru *The Spirit of Aikido* (1984).
Related Aikido (#HH0252) Ki energy (#HH0620) Martial arts (#HH0085) One–pointedness (Yoga, #HM5734).

♦ **HM0223d Mental perturbation of thought** (Buddhism)
Cittassa manovilekha (Pali)
Broader Perplexity (Buddhism, #HM0812).

♦ **HM0226d Tranquillity of consciousness** (Buddhism)
Calming of mind — Serenity of mind — Repose of citta (Pali)
Description This refers to the tranquillizing of the aggregate of consciousness. The characteristic is pacifying suffering and quieting disturbance of the citta, the function is to crush its suffering and disturbance. It is manifest as unwavering, inactive and coolness. Consciousness is the proximate cause and it is the opponent of corruption or defilement causing disturbance or lack of peace in the citta due to distraction, agitation and so on. Tranquillity of mental body and tranquillity of

consciousness are considered together.
Context One of the formations aggregate (mental coefficients) of Hinayana Buddhism, being listed among the constant states which appear in their true nature, and as profitable primary (always present in any profitable or profitable–resultant consciousness).
  Broader Awareness as mental-formation group of conscious existence (Buddhism, #HM2050).
  Related Tranquillity of mental body (Buddhism, #HM4704).

♦ **HM0227d Avarice** (Buddhism, Pali)
Meanness
Description In Hinayana Buddhism, avarice is the state of being mean. It is looked on as mental ugliness or disfigurement. It has the characteristic of concealing one's own current or coming success. Its function is not bearing to share one's property with others. It manifests as shrinking from sharing or as niggardliness. The proximate cause is one's own property or success.
Context One of the formations aggregate (mental coefficients) of Hinayana Buddhism, being listed among the inconstant states which are immutable by nature, and as unprofitable secondary (sometimes present in any unprofitable or unprofitable–resultant consciousness).
  Broader Awareness as mental-formation group of conscious existence (Buddhism, #HM2050).
  Related Avarice (Leela, #HM0987)   Miserliness (Buddhism, #HM2964)
  Spiritual avarice (Christianity, #HM0642).

♦ **HM0229d Endeavour** (Buddhism)
Vayama (Pali)
  Broader Faculty of energy (Buddhism, #HM1470).
  Narrower Wrong endeavour (#HM1585)   Correct endeavour (#HM1657).

♦ **HM0234e Psychical unease born of contact with element of mind–consciousness** (Buddhism)
Manovinnanadhatusampassa jam cetasikam asatam (Pali)
  Broader Feeling (Buddhism, #HM2270).
  Related Mind-consciousness element (Buddhism, #HM6173).

♦ **HM0235d Al-Quddus** (Sufism)
Description The believer is aware of Allah as holy, free from defect. The believer, knowing this, wishes to praise Allah for His perfection.
Context A mode of mystical awareness in the Sufi tradition associated with the fourth of the ninety-nine names of Allah.
  Broader Ninety-nine names of Allah (Sufism, #HH2561).
  Followed by As-Salam (Sufism, #HM1221).
  Preceded by Al-Malik (Sufism, #HM4061).

♦ **HM0238c Prayer of simplicity** (Christianity)
Second degree of prayer — Prayer of repose — Prayer of silence
Description Having become aware of a facility in recognizing the presence of God, the soul collects itself more easily, prayer is natural and pleasing and the soul knows it leads to God. Bringing one's self into the presence of God by an act of faith, one remains silent. There may be a taste of the presence of God. If this state of peace leaves, it may be brought back through exciting the will through some tender affection. This going to God is to please Him and do His will. Since it is not to obtain things from Him, one will not be disappointed by times of aridity, not by His apparent indifference or repulses.
  Refs Guyon, J M B de la Mothe *A Short Method of Prayer and Spiritual Torrents* (1875).
  Related Virtue of simplicity (#HH0990)   Spiritual repose (Christianity, #HM6552).
  Preceded by Sense of the presence of God (Christianity, #HM6989).

♦ **HM0239c Nonecstatic perception** (Hinduism)
Viyukta perception
Description Valid perception of subtle, hidden and remote objects through four-fold contact of the sense organs with these objects, sense organs with manas, manas with self. It is experienced by yogis who have fallen out of ecstasy, though supernatural power arising from special merit due to meditation.
Context One of two kinds of yogic perception distinguished by Prasastapada.
  Broader Yogic perception (Hinduism, #HM5005).
  Related Ecstatic perception (Hinduism, #HM7100).

♦ **HM0248d Escapist daydreaming**
Description Unlike free-floating fantasy, this is not pure play. It has the intention of removing the individual from immediate reality and creates alternative and highly personalized realities. These may range from creative to destructive and may or may not be beneficial to the dreamer. The beneficial effect depends on the feedback system between the dream world and the real world.
  Refs Boulding, Elise *Building a Global Civic Culture* (1988).
  Broader Daydream (#HM2138).
  Related Lucid dreaming (#HM3618)   Free-floating fantasy (#HM1011).

♦ **HM0249c Differentiation** (Buddhism)
Paccupalakkhana (Pali)
  Broader Faculty of wisdom (Buddhism, #HM3233).

♦ **HM0251d Mental** (Buddhism)
Manasa (Pali)
  Related Manas awareness (Hinduism, #HM2902)
  Cognition born of contact with element of mind-consciousness (Buddhism, #HM1501).

♦ **HM0252d Leaving off, abstaining, totally abstaining and refraining from the wrong livelihood** (Buddhism)
Miccha ajiva arati virati pativirati veramani (Pali)
  Broader Correct livelihood (Buddhism, #HM0549).

♦ **HM0253c Straightness of thought** (Buddhism)
Cittujukata (Pali)
  Narrower Rectitude (#HM4250)   Untwistedness (#HM1008)   Uncrookedness (#HM4037)
  Undeflectedness (#HM1054)   Straightness of the aggregate of consciousness (#HM1338).

♦ **HM0261c Faculty of life** (Buddhism)
Jivitindriya (Pali)
  Broader Faculty of living (Buddhism, #HM3404).

♦ **HM0264g Torpor** (Buddhism)
Middha (Pali)
Description In Hinayana Buddhism, some sources list stiffness and torpor as a single formation. There is general paralysis due to lack both of urgency and vigour. In particular, torpor has the characteristic of unwieldiness. Its function is to smother or bind associated states; it manifests as laziness or tardiness in grasping an object or as drowsiness, blinking the eyes, nodding and sleep. As with stiffness, the proximate cause is unwise attention to boredom, discontent, sloth, etc.
Context One of the formations aggregate (mental coefficients) of Hinayana Buddhism, being listed among the constant states which appear in their true nature, and as unprofitable secondary (sometimes present in any unprofitable or unprofitable–resultant consciousness).
  Broader Awareness as mental-formation group of conscious existence (Buddhism, #HM2050).
  Related Torpor (#HM1483)   Sleep (Buddhism, #HM7921)
  Stiffness (Buddhism, #HM5667).

♦ **HM0265g Pressure vertigo**
Alternobaric vertigo
Description A sensation of dizziness or spinning which can occur on failure to "clear the ears" properly during changes of pressure. Particularly experienced by divers and pilots.

♦ **HM0267e Focus twelve** (Psychism)
Description There is no awareness of having a physical body.
Context A state occurring during astral projection.
  Broader Astral projection (Psychism, #HM1887).

♦ **HM0282c Indeterminate consciousness in the immaterial sphere – functional** (Buddhism)
Indeterminate consciousness in the formless realm – inoperative
Description As with profitable consciousness, a total of four kinds of consciousness may arise here. These are associated with the four immaterial states or jhanas. The first is based on the jhana of boundless or unlimited space; the second with the jhana of boundless or unlimited consciousness; the third with the jhana of the sphere of nothingness and the fourth with the jhana of the sphere of neither perception nor non-perception. But in contrast to the profitable, in this case the consciousness only occurs in arahants.
All four give rise to materiality and to postures but not to intimation.
Context In Hinayana Buddhism, 89 consciousnesses are enumerated in aggregate (khanda). Of these, 21 are profitable or moral, 12 are unprofitable or immoral and 56 are indeterminate (resultant or functional). The unprofitable all arise in the sphere of sense and desire, whereas profitable and indeterminate arise in sense, fine-material, immaterial and supramundane spheres.
  Broader Dispositions of consciousness (Buddhism, #HM2098).
  Related Impulsion (Buddhism, #HM7268)
  Profitable consciousness in the immaterial sphere (Buddhism, #HM4701).
  Indeterminate consciousness in the immaterial sphere – resultant (Buddhism, #HM4982).
  Preceded by Indeterminate consciousness in the fine-material sphere – functional (Buddhism, #HM4761).

♦ **HM0284d Experience of miracles** (Christianity)
Sense of wonder
Description A miraculous event is experienced as one defying rational explanation. It communicates with and transcends the empirical world of man, leading to perceptivity for permanent proximity of God beyond the world of experience. This perceptivity must be constantly renewed under the dulling and stifling of worldly influence. The ordinary effect of such a sudden confrontation with what appears a bizarre fact is rejection by the mind, which refuses to consider it. One knows at one level that something is impossible and could not have happened and yet one is convinced at another level that it has. Faith and devotion to God preserve the sense of wonder so that, having determined that the miraculous event is not explicable in natural terms, it is accepted as an event directly intended and caused by God with the aim of an historical dialogue with God.
Miracles demonstrate the multifaceted nature of reality, each part depending on the others which together form a whole greater than the sum of the individual parts. If the function of whole of material nature is accepted as to express the will of God, God's miraculous intervention in history is seen as the bringing into play of a new mode of that function. Nature is at that point released from the confines of scientific law into a higher realm of natural law, as part of God's self-communication in free grace to man, providing a testimony of the work of God's saving will. The decisive miracle is the resurrection of Jesus Christ, anticipating the final destiny of man and his promised perfection.
Some would claim that miracles can be performed by great exponents of evil as well as of good, a capacity of the mind which can be used by either. They are said to obey laws at a higher plane, supernatural in that they are above the nature one knows. This is claimed by those who look on such experiences as a break-through from another plane of being to which most are strangers, but natural to, for instance, Jesus. Aldous Huxley discounts the miraculous as clearly existing but being of relatively little importance.
  Refs Foundation for Inner Peace *A Course in Miracles* (1975); Lewis, C S *Miracles* (1974).
  Related Wonder (#HM3197).

♦ **HM0285d Yoin** (Japanese)
Resonance
Description A moving experience causes profound emotional vibrations so an eternal moment is re-experienced. The memory of some past experience is triggered but, unlike nostalgia, the consciousness is of resonance in the present rather than of recapturing the past, there is resonance with something that is as well as something that was.
  Related Nostalgia (#HM1503).

♦ **HM0286c Wara** (Sufism)
Parhiz — Abstinence from wrongdoing
Description The traveller may refrain not only from what is forbidden but from what is questionable in the law or shari'a; he may use his own moral discrimination to determine from what he should abstain; he may refrain from anything which separates him from God. Ultimately the state is of one who looks for detachment from everything except God.
Context The second stage in a systematic account of various stations of the Sufi spiritual path. According to Shaykh Abu Sa'id ibn Abi'l-Khayr, this is the 16th of 40 stations or maqamat the Sufi must possess for his journey on the path of Sufism to be acceptable.
  Broader Mystic stations (Sufism, #HM3415)
  Stations of consciousness – ibn-Abi'l-Khayr (Sufism, #HM2424).
  Followed by Zuhd (Sufism, #HM4450)   Ikhlas (Sufism, #HM4663).
  Preceded by Tawba (Sufism, #HM4062)   'Ibadat (Sufism, #HM0128).

♦ **HM0291c Marvellous sentiment** (Hinduism)
Aesthetic emotion of wonder
Context One of eight kinds of aesthetic sentiment or rasas in Indian psychology.
  Broader Aesthetic emotion (Hinduism, #HM0205).

♦ **HM0294d Selfless service** (Leela)
Parmarth
Description Rather than arising in temporary activities, this is a permanent mode of being in which the player, living in harmony and in the present moment, gives up his individual self for a higher cause, in selfless disregard for the outcome of his actions. This totally dutiful behaviour lifts the player to the fifth or human plane.
Context The 27th state or square on the board of *Leela*, the game of knowledge, appearing in the third row.
Broader Third chakra: theatre of karma (Leela, #HM0717).
Followed by Human plane (Leela, #HM5155).

♦ **HM0297c Concentration of the first jhana of five** (Buddhism)
Description Having suppressed hindrances, five factors remain in the first jhana: applied thought; sustained thought; rapture or happiness; ease or bliss; concentration.
Context According to Hinayana Buddhism, concentration is considered as of one kind (monad), of two kinds (dyads), of three kinds (triads), of four kinds (tetrads) or of five kinds (pentad). In the pentad, fivefold according to the five jhana factors, this is the first concentration. It is equivalent to the first concentration in the third tetrad.
Broader Concentration (Buddhism, #HM6663).
Related Dwelling in the first jhana (Buddhism, #HM4298)
Concentration of the first jhana of four (Buddhism, #HM4456).
Followed by Concentration of the second jhana of five (Buddhism, #HM4575).

♦ **HM0298e Inability to achieve concentration** (Yoga)
Description In order to be renewed, concentration – the focusing and control of the mind – has to be achieved. This is hindered by the failure to deal with wrong conditions.
Context One of the nine obstacles to soul cognition enumerated by Patanjali.
Broader Obstacles to soul cognition (Yoga, #HM3182).
Related Mind drift (#HM4006).
Followed by Failure to hold meditative attitude (Yoga, #HM0610).
Preceded by Erroneous perception (Yoga, #HM1077).

♦ **HM0303d Kut**
Spiritual dance performance
Description This is a Korean expression for a shamanic healing trance when the shaman puts on and removes different articles of clothing as the different spirits enter and leave. The term has been extended in the west to describe the trance like state in which different articles of clothing are tried on in a "shopping trip" taken to heal the person of a disappointment in love or in preparation for going on a date.
Broader Shamanism (#HM1189).

♦ **HM0308d Al–Latif** (Sufism)
Description The believer is aware of Allah as the subtle one, the all beautiful. Even the finest details and most hidden beauty are known to Him, especially the knowledge of Allah in the heart of the believer. Al–Latif may manifest as quietness amidst worldly activity, as blessing within punishment.
Context A mode of mystical awareness in the Sufi tradition associated with the thirtieth of the ninety–nine names of Allah.
Broader Ninety–nine names of Allah (Sufism, #HH2561).
Related Al–Qahhar (Sufism, #HM1318).
Followed by Al–Khabir (Sufism, #HM4400).
Preceded by Al–'Adl (Sufism, #HM3807).

♦ **HM0309d Sensation of ease born of contact with the psychical** (Buddhism)
Cetosamphassaja sata vedana (Pali)

♦ **HM0310d Ignorance** (Leela)
Avidya
Description Forgetting the illusory nature of existence, attachment to emotional states and sensory perceptions draws the player down to the sensual plane of the first chakra. This ignorance can only arise where wisdom is also possible; eventually right knowledge will prevail and carry the player beyond the chakras.
Context The 44th state or square on the board of *Leela*, the game of knowledge, appearing in the fifth row.
Broader Fifth chakra: man becomes himself (Leela, #HM0933).
Related Ignorance (Buddhism, Yoga, Zen, #HM3196).
Followed by Sensual plane (Leela, #HM3627).

♦ **HM0311d Absorption concentration** (Buddhism)
Collectedness of mind — Ecstatic concentration
Description Absorption concentration is more controlled and prolonged than access concentration. Unlike access concentration, where the jhana factors are not strong, here they are strong. Having interrupted the flow of bhavanga (life–continuum), the mind may continue in profitable impulsion. Collectedness of mind in the subjects of meditation not covered under access concentration brings to the saint who has entered jhana a state of living in bliss; to probationers and average persons it causes insight.
Context According to Hinayana Buddhism, concentration is considered as of one kind (monad), of of two kinds (dyads), of three kinds (triads), of four kinds (tetrads) or of five kinds (pentad). In the first dyad, this is the unification that follows access concentration.
Broader Concentration (Buddhism, #HM6663)
Appearance of absorption (Buddhism, #HM1618).
Related Dhyana (Hinduism, Buddhism, #HM0137)
Skill in absorption (Buddhism, #HH4777)
Subjects for meditation (Buddhism, #HH3987).
Preceded by Access concentration (Buddhism, #HM4999).

♦ **HM0312c Third order perceptions – configurations**
Description Here there is perception of a form as unique, for example the word "dog" in whatever accent it is spoken, the form "chair" whether high, low, simple or a throne. This order of perception makes the world easier to comprehend, things have a name and a shape, can be distinguished, are grouped and classified.
Context The third of ten orders in the perceptual system described by William Glasser.
Broader Orders of perception (#HM0988).
Followed by Fourth order perceptions – control of transitions (#HM1516).
Preceded by Second order perceptions – sensation (#HM1932).

♦ **HM0317d Unknowing** (Taoism)
Darkness
Description When subject to the conditioning of mundane circumstances, the awareness of the human mind is that of external objects and not of inner essence, of which it is ignorant. The situation is reversed from the perspective of the celestial awareness of Tao, in that there is no fixation on externals but rather a sense of detachment. Darkness can thus symbolize unknowing in the sense of ignorance or in the sense of detachment.

♦ **HM0318e Clairscent** (Psychism)
Description The awareness of a smell with no apparent physical cause can be taken as an omen of a future event. There is often a simultaneous sense of taste.
Related Clairvoyance (Psychism, #HM3498) Clairaudience (Psychism, #HM4333)
Clairgustance (Psychism, #HM0412) Olfactory hallucination (#HM4559)
Extrasensory perception (ESP, #HM2262).

♦ **HM0322d Initiation of decision** (Esotericism)
Description The liberating experience associated with this initiation is that of freedom of choice, unconstrained by the individual's past. The revelation accorded to the initiate gives him a complete picture of the proceses that have brought him to this creative moment of decision. This if followed by a revelation of what he may be. The individual commits himself to the use of a particular form of energy through which to work and which determines the future pattern of his service. This energy is used to impart, strengthen and enlighten existing initiatives which are already well coordinated and approriately inspired in order for them to express more effectively the all–encompassing whole. This involves a will to synthesis. The initiate becomes a conscious aspect of that of which he forms an integral part. He reaches an understanding of the nature of the creative process, of the forms to which it gives rise, and of his involvement in that process.
Context The sixth initiation in the tradition of western esotericism.
Broader Spiritual initiation (Esotericism, #HH3390).
Related Ajna (Yoga, #HM2144) Way of knowledge (Esotericism, #HM1201).

♦ **HM0330a Baqa** (Sufism)
Subsistence in God — Abiding
Description Although not distinguished by some from *fana*, annihilation of human attributes, others see baqa as transformation into the cosmic self, when human attributes are replaced by the divine attributes of knowledge, justice and gratitude. Living and moving in the Godhead, *perfect man* abides permanently in the world of divinity and becomes a pattern for his fellow men.
Context The fourteenth stage in a systematic account of various stations of the Sufi spiritual path. However, according to Shaykh Abu Sa'id ibn Abi'l–Khayr, this is the 22nd of 40 stations or maqamat the Sufi must possess for his journey on the path of Sufism to be acceptable.
Broader Mystic stations (Sufism, #HM3415)
Stations of consciousness – ibn–Abi'l–Khayr (Sufism, #HM2424).
Followed by 'Ilm al–yaqin (Sufism, #HM0930).
Preceded by Fana (Sufism, #HM1270).

♦ **HM0332d Spontaneous out–of–body experience**
Description Although the experience is similar to that of a shamanic journey, the person concerned may never have heard of the existence of such a phenomenon. Such experiences may, in fact, have been the inspiration for intentionally induced journeys.
Related Lucid dreaming (#HM3618) Shamanic journey (#HM6120)
Near death experience (NDE, #HM0777) Out–of–body experience (Psychism, #HM5534).

♦ **HM0334 Existential love**
Description A form of love that comes with full maturity and old age. It celebrates the present in poignant awareness that life is the more precious because it is fleeting. It is the source of the unique patience so often seen in grandparents.
Refs Weg, R (Ed) *Sexuality in the Later Years* (1983).
Related Love (#HH0258).

♦ **HM0335c Khalifa** (Sufism)
Description The name of a traveller on the *tariqa* or Sufi spiritual path at the level of a vice regent appointed by God over mankind.
Context Spiritual progress on the spiritual path is marked by the attainment of specific *maqamat* or station.
Broader Mystic stations (Sufism, #HM3415).

♦ **HM0344c Eight–fold knowledge** (Systematics)
Revealed knowledge — Experience of individuality
Description Individuality (whether actual or potential) is the source of initiative residing in organized structures and can only be experienced through a minimum of eight independent terms. The selfhood associated with individuality cannot be readily communicated. It is the nature of being a free agent, both as a unique centre of subjective experience and as a source of initiative. This form of knowledge is only reached when all ordinary knowledge is laid aside and consciousness lies open to the light of revelation. The essential feature of such knowledge (which is not confined to religious experience) is that it cannot be ascribed to any functional activity within the whole by which it is received. A barrier of separateness is broken down giving such knowledge the character of totality and unity. This eight–fold knowledge is able to represent organized structures and historical processes ranging in scale from unity to totality. Its value is classificatory, interpretative, heuristic and predictive.
Context The eighth in a sequence of twelve modes of knowledge, identified by J G Bennett, inspired by G Gurdjieff.
Broader Systematics (#HH2003).
Preceded by Seven–fold knowledge (Systematics, #HM0776).

♦ **HM0345c Scientific consciousness**
Description The science developed by the intuitive mind and the holistic mode of consciousness reveals aspects of nature which are necessarily invisible to the verbal–intellectual mind and the analytical mode of consciousness. The ability to work with this intuitive approach, to see what it reveals, not only requires a transformation of scientific method but of the scientist himself. The result of such a transformation is a radical change in awareness of the relationship between man and nature. Knowledge of a phenomenon is intimately related to the phenomenon itself since the state of "being known" is understood as a further stage of the phenomenon itself, namely the stage which the phenomenon reaches in human consciousness. The knower is thus not an onlooker but a participant in nature's processes, which now act in consciousness to produce the phenomenon consciously as they act externally to produce it materially. In Goethe's words: "through the contemplation of an ever creating nature, we should make ourselves worthy of spiritual participation in her production". When consciousness is properly prepared it becomes the medium in which the phenomenon itself comes into presence. The act of knowing is an evolutionary development of the phenomenon and not just a subjective activity of man. The scientist himself becomes the apparatus in which the phenomenon appears. The difficulty in in comprehending this comes from the fact that a way of seeing cannot be grasped like an object, to appear as a content of perception. What is encountered in the way of seeing is the organization or unity of the world. It can only be understood through participation. This is not however purely subjective. When it is recognized that what is experienced as a way of seeing is the unity of the phenomenon, then it is an original event of participation in which people can learn to participate, instead of repeating

HM0345

something which once happened and now has gone. A way of seeing thus has the temporal quality of belonging to the present instead of the past.
Context Goethe elaborated a scientific way of seeing that focused on qualitative dimensions in contrast to the quantitative focus of mainstream science. It offers a new way of doing science and of seeing nature whole. This is to be understood as complementary to norma analytical modes of science.
Refs Bortoft, Henri *Goethe's Scientific Consciousness* (1986).

♦ **HM0346c Tso-wang** (Taoism)
Description Literally "sitting with no thoughts", this state has been compared with transcendental consciousness.
    Broader Human development (Taoism, #HH0689).
    Related Transcendental consciousness (#HM2020).

♦ **HM0348g Stage-three sleep**
Description This is a stage when between 20 and 50 percent of EEG activity is delta wave. There are no eye movements and muscle tension is usually low.
    Broader Sleep (#HM2980).
    Related Delta wave consciousness (#HM1785).
    Followed by Deep sleep (#HM6307).
    Preceded by Sleep spindles (#HM4806).

♦ **HM0349d Good quality of the aggregate of sensation, of the aggregate of cognition, and of the aggregate of synergies** (Buddhism)
Vedanakkhandhassa sannakkhandhassa sankharakkhandhassa (Pali)
    Broader Fitness of mental body (Buddhism, #HM1455).

♦ **HM0351g Oculoagravic illusion**
Elevator illusion — Oculogyral illusion
Description An illusion engendered by horizontal linear acceleration which alters the effective direction of the gravitational force. It causes an upright passenger to feel tilted and to see the visual world as tilted. Apparent movement occurs when the accelerative force changes, and objects appear to swing into a new orientation. When such linear acceleration occurs in a vertical direction, a variant occurs which is also known as the elevator illusion. The effective force increases on the passenger when the lift accelerates on the way up or decelerates on the way down. With increasing force, the framework of the life appears to rise for a moment, while under decreasing force it appears to sink. An oculogyral version of the illusion is engendered with angular acceleration.
    Broader Illusion (#HM2510).

♦ **HM0352c Power of shame** (Buddhism)
Hiribala (Pali)
    Narrower Being ashamed of what one ought to be ashamed (#HM3828)
    Being ashamed of acquisition of sinful and unwholesome dharmas (#HM0552).

♦ **HM0357b States of prayer** (Christianity)
Visions of the Trinity
Description Marie of the Incarnation, in her account of the states of prayer, describes three intellectual visions of the Trinity and concludes with a description of the permanent state of union which she experienced.
First Vision affecting the understanding): Light and insight way beyond the powers of expression flood intelligence. The soul is seen as created in the image of God. Memory relates it to the eternal Father; understanding relates it to the Son; and will relates it to the Holy Spirit. Just as the Trinity is threefold in person but one in essence, so the soul is threefold in its powers but one in substance.
Second Vision affecting the will: Engulfed in the presence of Father, Son and Holy Spirit, and acknowledging the lowliness of the soul, there is awareness of Christ as true spouse of the faithful soul. The soul experiences the presence of this Being which has taken possession of her in spiritual marriage, inflaming her and consuming her with fire in so agreeable and pleasant a manner it is impossible to describe. In this spiritual marriage the soul changes state, no longer tending towards and expecting a grace seen from afar, but actually experiencing possession of Him whom she loves and therefore having no more desires.
Third Vision: Loving Christ, the soul becomes the abode of the three Divine Persons of the Trinity, They possess the soul entirely. Possessed by them, she also possesses them. The Father becomes her father, the Word her spouse; and the Holy Spirit is the one by whom the divine impressions are received. Even though the self is seen as nothingness, nevertheless it belongs entirely to God.
Union: The Word Incarnate is love itself, united to the spirit, the spirit united to His. Intimately united in Him, the soul is also united with the Father and the Holy Spirit. This is no imaginative image but a real experience, with constant communication in delicate, simple and deep manner. Communication in the language of the spirit is through His impulse. Rather than an action it is an atmosphere in the centre of the soul, impossible to describe in its simplicity. Sometimes there is an experience of being totally consumed by love which, if its splendour were not immediately tempered by a second impression, would be too great to bear in this present life. The second impression refers to the Word as Divine Spouse. The effect of this state are an annihilation, a knowledge of one's nothingness to keep one humble, a fear without anxiety of being deceived in the ways of the spirit (useful for abnegation and the spirit of compunction), and patience with one's crosses and inclination towards peace and benignity with everyone from one's whole soul. Sufferings are accepted in a spirit of love and union with the Word. The state brings love for the vocation to which God calls the soul, love for everything practised in the Church of God, and an urge to allow one's self to be guided by those holding the place of God and submitting one's judgement to them.
Refs Hall, Darl M *The Management of Human Systems* (1971).

♦ **HM0359d Magical thinking**
Description A primitive or archaic type of thinking, pre-logical and (in the schizophrenic) a symbolic equivalent of action. Such thinking may be a symptom of fatigue or neurosis, and also occurs in early childhood and amongst people of primitive cultures. The individual allows for relationships between things which are not observable in fact and, by mystical participation or simply wishing, attempts to alter the course of events.
    Related Magic (#HH0720)   Pre-conscious thinking (#HM1084)
    Archaic-paralogical thinking (#HH0748).

♦ **HM0361e Carelessness** (Yoga)
Light-mindedness
Description The mental attitude which leads to the formation of thoughts and the scattering of attention, thus hindering the achievement of one-pointedness.
Context One of the nine obstacles to soul cognition enumerated by Patanjali.
    Broader Obstacles to soul cognition (Yoga, #HM3182).
    Followed by Laziness (Yoga, #HM0046).
    Preceded by Wrong questioning (Yoga, #HM1080).

♦ **HM0362d Angst**
Description In existential terms, this state is marked by general dread and despair at the human condition. It may arise when all the advantages of civilization and sophistication seem only to suffocate and alienate and there seems no purpose to life.
    Related Anxiety (#HM2465)   Melancholy (#HM3141)   Weltschmerz (#HM0900).

♦ **HM0375d Ash-Shahid** (Sufism)
Description The believer is aware of Allah as the witness of all that appears. He will be the witness for every action of every man on the day of judgement.
Context A mode of mystical awareness in the Sufi tradition associated with the fiftieth of the ninety-nine names of Allah.
    Broader Ninety-nine names of Allah (Sufism, #HH2561).
    Followed by Al-Haqq (Sufism, #HM3945).
    Preceded by Al-Ba'ith (Sufism, #HM4017).

♦ **HM0376d Creature consciousness**
Description This is the awareness of the smallness and insignificance of one's individual self faced with an encounter with the numinous or mysterium tremendum.
    Related Numinosum (Jung, #HM3811).

♦ **HM0377a Devekuth** (Judaism)
Description This is the Jewish equivalent of the Christian mystical union with God and the Buddhist nirvana.
Refs Katz, Steven *Mysticism and Religious Traditions* (1983); Katz, Steven T (Ed) *Mysticism and Philosophical Analysis* (1978).
    Related Mystic union with God (Christianity, #HM3889).

♦ **HM0392d Al-Hamid** (Sufism)
Description The believer is aware of Allah as the most praiseworthy. He is to be honoured with respect and thankfulness through speech, through action, through one's very existence, as the source of all gifts and perfection. Performing one's duty when it is due every hour of one's life brings material benefit and spiritual joy and wisdom, it is manifest praise. The opposite is denial and leads to being cursed by all creation.
Context A mode of mystical awareness in the Sufi tradition associated with the fifty-sixth of the ninety-nine names of Allah.
    Broader Ninety-nine names of Allah (Sufism, #HH2561).
    Followed by Al-Muhsi (Sufism, #HM4544).
    Preceded by Al-Wali (Sufism, #HM1633).

♦ **HM0399c Pathetic sentiment** (Hinduism)
Aesthetic emotion of grief — Sorrowful rasa
Context One of eight kinds of aesthetic sentiment or rasas in Indian psychology.
    Broader Aesthetic emotion (Hinduism, #HM0205).

♦ **HM0400d Turmoil of thought** (Buddhism)
Bhantattam cittassa (Pali)
    Broader Excitement (Buddhism, #HM1469).

♦ **HM0401e Psychic atavism** (Psychism)
Description This term may refer to regression to primitive or pre-civilized behaviour, possibly under hypnosis; or to change in physical appearance (were-wolf).
    Related Astral body (#HH0206)   Metamorphosis (Psychism, #HM0551)
    Shape shifting (Psychism, #HM3202).

♦ **HM0402b Mystical journey** (Christianity)
Description Those on the mystic journey live as though nothing exists but God and the obligation to follow the path of duty. Even participation in God's mysterious purpose outside themselves, although voluntary and tangible, is at the same time innate and mystical. What they might otherwise have to achieve through their own endeavour is brought to pass for them by God. Activity is confined to what is necessary for fulfilling the duty of the present moment, obligations are scrupulously fulfilled, but there is passivity in everything else. This may draw criticism as particular acts of piety may be expected from them by others. However, although such exhortations to pious acts may be ignored by those on the way to divine union and for whom such acts are unnecessary, they still seek spiritual guidance from one in whom they are led by God to confide. In fact to reach divine union requires a spiritual adviser.
    Related Kairos (#HM2749)   Spiritual direction (#HH0442)
    Waiting on God (Christianity, #HM6129).

♦ **HM0403d Not understanding** (Buddhism)
Ananubodha (Pali)
    Broader Delusion (Buddhism, #HM0184).

♦ **HM0404d Volition born of contact with element of mind consciousness** (Buddhism)
Manovinnanadhatu samphassaja cetana (Pali)
    Related Mind-consciousness element (Buddhism, #HM6173)
    Element of mind-consciousness (Buddhism, #HM1834).

♦ **HM0405c Remembering** (Buddhism)
Anussati (Pali)
    Broader Faculty of recollection (Buddhism, #HM4248).

♦ **HM0409d Straightness of the aggregate of sensation, of the aggregate of cognition, and of the aggregate of synergies** (Buddhism)
Vedanakkhandhassa sannakkhandhassa sankharakkhandhassa ujukata (Pali)
    Broader Straightness of body (Buddhism, #HM1424).

♦ **HM0411d Tribal consciousness** (Yoga)
Context A step on the path to cosmic consciousness, when the individual has not yet achieved conscious self-relatedness and lacks the courage to go against his immediate group.
    Related Conscious self-relatedness (Yoga, #HM0467).
    Followed by Racial consciousness (Yoga, #HM1070).

♦ **HM0412e Clairgustance** (Psychism)
Description There is a sensation of a particular taste when the substance or food is not in the mouth. Often occurring at the same time as *clairscent*, such experience is often taken as an omen or warning.

Related Clairscent (Psychism, #HM0318) Clairaudience (Psychism, #HM4333) Extrasensory perception (ESP, #HM2262). Clairvoyance (Psychism, #HM3498) Gustatory hallucination (#HM6995)

♦ **HM0413b Gnosis**
Gnostic knowledge — Salvational knowledge
Description Gnostic knowledge is the knowledge of the soul. Its aim is not to prove or to explain the soul but to transform it. It is based on an insight into the strict correspondence between knowledge and being, through which knowledge or thinking is inseparable from being and vice versa. Between believing and knowing there is a third mediating function associated with inner vision. Cosmologically this is effectively an intermediate and mediating world (forgotten by official philosophy and theology), namely the imaginal world, the world of the soul or psyche. Gnosis is then the cultivation of the soul as a source of knowledge. It is not a question of distinguishing between faith and reason but rather between consciousness of one's own states (whether derived from reason or emotion) and the unconscious reactions (whether of thought or emotion). The focus is therefore on self-attention. Gnosis is a salvational, redemptive knowledge, because it has the virtue of bringing about the inner transformation of man. In contrast to theoretical learning, it is knowledge that changes and transforms the knowing subject. The intensity of the search for knowledge itself becomes an ontologically transforming force. From this perspective salvation results in nothing for there is not something to be acquired as a possession, rather it is an event within the soul. However that soul is also the soul of the world. Salvational knowledge is thus equally concerned with resouling the world. It is a recollection, a remembering of a worldly soul and of an ensouled world. It is a process of allowing the world to be.
Context Historically, gnosis constitutes the esoteric element in the official or exoteric religious traditions of the world. As such it should be distinguished from gnosticism. Gnosticism is a purely historical phenomenon, whereas gnosis is phenomenological. Gnosis, as a lived practice, can exist without gnosticism.
Refs Avens, Roberts *The New Gnosis* (1984); Needleman, Jacob (Ed) *The Sword of Gnosis* (1986).
Related Gnosticism (#HH3546) Himma (Sufism, #HM8667) Ma'rifa (Sufism, #HM2254) Somatic experience (#HM1766) Negative capability (#HM6133) Analytical psychology (#HH1420).

♦ **HM0419d Al-Sami'** (Sufism)
Description The believer is aware of Allah as the one who hears all, He is absolute perfection. Humanity hears only imperfectly, but that hearing is to lead to perfection when he know the perfect attributes of Allah.
Context A mode of mystical awareness in the Sufi tradition associated with the twenty-sixth of the ninety-nine names of Allah.
Broader Ninety-nine names of Allah (Sufism, #HH2561).
Followed by Al-Basir (Sufism, #HM6512).
Preceded by Al-Mudhill (Sufism, #HM4510).

♦ **HM0424d Knowledge of equanimity about formations** (Buddhism)
Knowledge of indifference to complexes
Related Purification by knowledge and vision of the way (Buddhism, #HH3550).
Preceded by Knowledge of contemplation of reflection (Buddhism, #HM0169).

♦ **HM0425d Elation** (Buddhism)
Odagya (Pali)
Broader Fivefold happiness (Buddhism, #HM0747).
Related Elation (#HM4899).

♦ **HM0427d Inability to compare** (Buddhism)
Apariyogahana (Pali)
Broader Delusion (Buddhism, #HM0184).

♦ **HM0428d Initiation of transfiguration** (Esotericism)
Description The liberating experience associated with this initiation is that of freedom from the ancient authority of the threshold personality. It is a recognition of complete spiritual control of the personality which is thereby transfigured. The new awareness produces its main effects upon the mind. It enables the initiate to use the mind as his major instrument in the work to be done. The individual becomes a transmitter of energy which is stepped down into the activities of groups, whether or not he is conscious of the process. His thought life therefore becomes the field of his major effort as the arena in which the direction of energy is determined. He endeavours to concentrate within himself so that he may eventually come to control the flow of such energy. The individual's task thus becomes greater personality integration, so that he becomes an increasingly soul-infused personality, as well as greater integration with his environment in a service role. He is no longer troubled by any sense of separatness and division. He feels and knows something of the essential unity of all manifested life.
Context The third initiation in the tradition of western esotericism.
Broader Spiritual initiation (Esotericism, #HH3390).
Related Ajna (Yoga, #HM2144) Way of knowledge (Esotericism, #HM1201).
Followed by Initiation of renunciation (Esotericism, #HM0210).
Preceded by Initiation of baptism (Esotericism, #HM1267).

♦ **HM0436d Agitation** (Buddhism)
Avupasama (Pali)
Broader Excitement (Buddhism, #HM1469).

♦ **HM0439c Correct action** (Buddhism)
Sammakammanta (Pali)
Narrower Unaffected (#HM3375) Correct action (#HM0687) Not incurring guilt (#HM3413) Leaving undone (#HM3429) Destroying cause (#HM1098) Not trespassing limit (#HM3435) Leaving off, abstaining, totally abstaining and refraining from the three deviations of body (#HM1133).

♦ **HM0442e Active trance** (Psychism)
Description In this *hypnotic* state the individual is able, when directed, to create works of art as though by a particular deceased artist and with an ability far exceeding that of which he is capable in normal life.
Broader Trance (#HM3236) Hypnotic states of consciousness (#HM2133).

♦ **HM0446c Human holiness** (Christianity)
Description In Roman Catholic doctrine, man may participate in the holiness of God through justification by sanctifying grace and through total surrender to God. Holiness may develop through God's grace until there is attainment of the sanctity of the saints.
Broader Holiness (#HH0183).
Related Sanctity (Christianity, #HM0560).

♦ **HM0449d Unperturbedness of thought** (Buddhism)
Avisahara (Pali)
Broader Sukha (Buddhism, #HM2866).

♦ **HM0450c State of realization** (Buddhism)
Engaku (Japanese) — Absorption — Two vehicles
Description This is the state in which, through one's own efforts, some understanding of life is achieved. Together with the state of learning the state of realization comprises the two vehicles and has the goal of self-betterment. It corresponds to wisdom or insight, where understanding comes from one's own experiences and reflections. and reflections. Learning may be thought of as trying to understand, realization as actually understanding. Both worlds have limitations. They are self-centred, self-betterment for one's own sake, which may lead back to the world of anger, looking down on others. It may also lead back to self-absorption so that all else, including the implications of what one is doing, is obliterated from view. The other limitation is a difficulty in accepting that conclusions one has come to one's self are not totally correct. These limitations are dangerous when it is realized that people in these worlds are leaders and experts. Based on a desire to improve the human condition they may lead to beneficial results; based on a desire for profit for profit's sake they may lead to degradation of the environment and the destruction of life.
Context One of the ten worlds described in Nichiren Soshu Buddhism.
Broader Ten worlds (Buddhism, #HM2657) Four noble paths (Buddhism, #HM4026) Nichiren shoshu buddhism (Buddhism, #HH3443).
Followed by State of bodhisattva (Buddhism, #HM1225).
Preceded by State of learning (Buddhism, #HM3662).

♦ **HM0451d Non-grasping** (Buddhism)
Asagahana (Pali)
Broader Delusion (Buddhism, #HM0184).

♦ **HM0453d Istihsan** (Islam)
Admiration
Description The object of one's gaze is found to be beautiful.
Context The second cause of profane love occurring in the beholder according to Ibn al-Qayyim.
Refs Bell, Joseph Norment *Love Theory in Later Hanbalite Islam* (1979).
Followed by Al-fikr fi 'l-manzur (Islam, #HM1264).
Preceded by Nazar (Islam, #HM6003).

♦ **HM0454c Way of power** (Esotericism)
Way of will
Description Characteristic of people who are wilful, one-pointed, quick to act and decisive. On this way they are challenged by the task of tempering will with love, becoming more understanding and cooperative, and being able to build as well as to destroy.
Context The first of seven ways to spiritual realization characteristic of Western esotericism.
Broader Ways to spiritual realization (Esotericism, #HH2665).
Related Sphere of judgement (Kabbalah, #HM2290) Initiation of revelation (Esotericism, #HM0181).

♦ **HM0455d Clarity of consciousness** (Leela)
Swatch
Description Purified in purgatory from the opacity of bad karma and having worked his way through doubts and irreligiousity, the player has clear, unclouded understanding beyond the simply intellectual reasoning of the third chakra. Sanctified and with good tendencies he is enters the upward flow of energy lifting him to the fifth chakra, that of Man.
Context The 36th state or square on the board of *Leela*, the game of knowledge, appearing in the fourth row.
Broader Fourth chakra: attaining balance (Leela, #HM3291).

♦ **HM0460d Fewness of wishes** (Buddhism)
Description This is a state of non-greed.
Context In Hinayana Buddhism, one of the five ascetic states.
Broader Ascetic states (Buddhism, #HM5354) State of not feeling greed (Buddhism, #HM1825).

♦ **HM0461e Altered states of consciousness induction device** (ASCID)
Description This device, developed by Jean Houston and Robert Masters, induces trance-like states through a swinging motion on a subject who is supported physically and blindfold. The effect is similar to that achieved by psychedelic drugs or hypnosis; however, the individual has more autonomy although being open to suggestion from outside. The effect may be profound, described as experiencing the source of creation, death and rebirth, the ground of being. The results are considered important in discovering the capacities available to man but not normally used; and in control of perception and self-healing.
Refs Masters, R E L and Houston, Jean *The Varieties of Psychedelic Experience* (1967).

♦ **HM0463d Sub-personalities**
Description The existence of a sub-personality is recognized when a person finds himself, in a particular situation, acting in ways which he does not like or which appears to go against his interests, without being able to change this by an act of will or conscious decision for the period in which the situation persists (whether minutes or hours). As such a sub-personality is a semi-permanent and semi-autonomous region of the personality capable of acting as a person. It goes beyond the notion of an ego state as a coherent system of behaviours with more or less permeable boundaries, or of patterns of feelings, thoughts and perceptions coalescing in response to particular situations. It is a unique configuration or system of psychological structures to which the sense of "I" is given.
Refs Rowan, John *Subpersonalities* (1990).
Related Fugue state (#HM7131) Double mindedness (#HM3362) Multiple personality (#HH0763).

♦ **HM0464f Delusional perception**
Description Although objects and events are appropriately perceived, they are endowed with vague significance or new meaning. Often they are uncanny, eery, mystifying. The new meaning may refer to the individual himself.
Broader Primary delusions (#HM4604).

♦ **HM0466c None** (Christianity)
Hours of none
Description The canonical hours of none traditionally take place between 3 and 6 pm. They dramatize the fact the day is nearly over and the task is simply keeping going in one's work. The mood is steadfast perseverance.

Broader Canonical hours (Christianity, #HM1167).
Followed by Vespers (Christianity, #HM1468).
Preceded by Sext (Christianity, #HM3725).

♦ **HM0467c Conscious self-relatedness** (Yoga)
Self-consciousness
Description This is the mature selfhood which allows critical evaluation and comprehensive vision. When a person has discovered his own individuality he is able to rise above tribal, racial, national and international consciousness to full cosmic consciousness. He has the inward strength and freedom to rise above the limitations of the group, allowing the critical evaluation and comprehensive vision to take a stand for justice and peace and against the conflicts arising between races and nations. Gradually he discovers his relatedness to the eternal.
Refs Chaudhuri, Haridas *Integral Yoga* (1965).
Related Cosmic integration (#HH0808) Cosmic consciousness (#HM2291)
Self-awareness (Psychism, #HM2436) Tribal consciousness (Yoga, #HM0411)
Racial consciousness (Yoga, #HM1070) National consciousness (Yoga, #HM1370)
International consciousness (Yoga, #HM0099).

♦ **HM0471d Suffering**
Description Suffering is essential to the nature of man; it seems to belong to man's transcendence, one of those points in which man is destined to go beyond himself. In fact, suffering is almost inseparable from man's earthly existence; it evokes compassion and respect, it intimidates. Pope John Paul II indicates that the need of the heart commands us to overcome fear and the imperative of faith allows us to dare to touch the intangible mystery of man in his suffering. Suffering is wider than the sickness treatable by medicine, physical pain, or moral suffering, pain of the soul, although it accompanies both physical and moral pain. Man suffers whenever he experiences evil - the "sins" of Christianity are the "afflictions" of Buddhism; but it is said that the hurt of suffering is caused by aversion, and that acceptance can bring joy.
Buddhist and Hindu thought relates suffering to attachment and the release from suffering to non-attachment. Other thought suggests that suffering is the inevitable result of the process of change, and therefore necessary for development. Again, Pope John Paul II says that man can be said to suffer because of a good in which he does not share, either because he is cut off from it or because he has deprived himself of it. Every individual, through suffering, becomes part of the world of suffering in which all who suffer are brought together. Such suffering can become a source of strength. St Paul rejoiced in his suffering for others when the meaning of his suffering was discovered and shared; and redemption for all was achieved through the sufferings of Christ.
Meister Eckhart distinguishes two kinds of suffering. One arises in selfishness, the clinging to creatures. It is hard to bear and crushes. The other arises from trying to detach one's self from creatures and cling to God. It strengthens and is, in a way, easy to bear, both because it can be accepted as a necessary step on the road to freedom and also because God, in effect, carries it in one's place. When this suffering touches us it touches God first. There is even a kind of joy in suffering which is part of learning detachment and brings increased communion with God.
Refs Bowker, John *Problems of Suffering in the Religions of the World* (1970); Pope John Paul II *Salvifici Doloris* (1984).
Narrower Tragic suffering (#HH0708).
Related Pain (#HM4031) State of hell (Buddhism, #HM4282)
Suffering (Christianity, #HH2207) Redemption (Christianity, #HH0167)
Duhkha (Buddhism, Hinduism, #HM2574) Four noble truths (Buddhism, #HH0523).

♦ **HM0472b Spiritual hunger**
Description There is an innate hunger for knowledge and perfection without which there is no rest and no true satisfaction. It is this hunger that stimulates the search for the beautiful, the good, the truth and the longing for peace and joy. (Swami Omkarananda).

♦ **HM0476d 'Ilm al-basir** (Sufism)
All-pervading knowledge
Description Prolonged practice of remembering reality on the continuous motion in the body means that consciousness of this knowledge may be acquired without first being aware of it in the physical heart. Attention is elevated to the whole body, at first for short periods of time, then for longer, and finally all the time.
Context This is the sixth state arising in the method of dhikr described by Shaykh Kalimallah.
Broader Remembrance of God (Sufism, #HM6562).
Related Ghaflah (Sufism, #HM7383).
Followed by Continuous all-pervading knowledge (Sufism, #HM4788).
Preceded by Total annihilation (Sufism, #HM5775).

♦ **HM0478c Self-love**
Description Self-love may be referred to as that love of self which puts a person's individual good or desires above the general or divine good. In this sense it implies seeking one's own advantage regardless of anyone else's and is particularly frowned on in Christian teaching where mortification, humility, obedience and charity have variously been recommended to counteract such tendencies. However, in the sense in which the Christian is enjoined to love his neighbour as himself, lack of self-love is not taken to imply masochism; can one put one's neighbour's interests first without, through self-love, knowing by experience what these interests are likely to be ? In a sense, it could be said that all activity proceeds from self-love in response to the desire for happiness with the consequent reflection on how to act in conformity with such desire. The affirmation of one's own life, happiness, growth, and freedom is rooted in one's capacity to love. One can only really love productively if one can love one's self. If one is only able to love others one cannot real love at all. It follows that selfishness and self-love are not the same, they are opposites. It is loving one's self too little, self-hatred, that causes the concern to compensate by snatching from life the satisfaction that one blocks one's self from attaining.
A part of a person may love itself with particular intensity and such love may ensure the part's separation from the whole. Such exaggerated affection for itself is the part's denial of its transcendence and of its mind which is its relatedness in the whole of the world. Such destructive partiality is seen both when a disrelated part of the person attends primarily to itself but also when there is insistence that the personal reflection dominate in all experiences. The self-love of the part, instead of enjoying its own being in the whole, attempts to suffuse the whole with its own limited way of being.
Refs Fromm, Erich *The Art of Loving* (1956); Zwieg, Paul *The Heresy of Self Love* (1980).
Related Egoism (#HH0130) Narcissism (#HH1066) Self-respect (#HH1384)
Self-aggrandizing love (#HM4671).

♦ **HM0481d Plane of dharma** (Leela)
Dharma-loka
Description Dwelling in and at one with reality, the player carries out conscious acts in harmony with the needs of the moment and according to the rules regulating his existence. He recognizes the virtue of doing good for others and of avoiding their harm. Deviating from dharma causes a downward flow in energy which will carry the player back to lower regions.
Context The 22nd state or square on the board of *Leela*, the game of knowledge, appearing in the third row.
Broader Third chakra: theatre of karma (Leela, #HM0717).
Followed by Positive intellect (Leela, #HM1789).

♦ **HM0485c Four discriminations** (Buddhism)
Four analyses
Description The four kinds of understanding as knowledge - knowledge about meaning, knowledge about law, knowledge about language, knowledge about kinds of knowledge or perspicuity, grouped together as the four discriminations or analyses can be subdivided in several ways: - two planes or spheres: as of probationers or trainers; as of non-probationers or trainers.
- five aspects: attainment or achievement of arahantship; study or mastery of the scriptures, the word of Buddha; hearing and learning the dhamma through careful attention; questioning through discussion difficult passages and their interpretation in texts and commentaries; previous application or effort, so that there is constant devotion to the subject of meditation even while walking on the alms round.
- another division goes: previous application; knowledge of a science or art; knowledge of language; study of scripture; questioning (even if only one verse); achievement of arahantship or entering the stream; living close to learned teachers; friendship with such teachers.
Context On the Path of Purification of Hinayana Buddhism; panna (understanding) is considered as of one kind (monad), of two kinds (dyads), of three kinds (triads) or of four kinds (tetrads). There are five dyads, four triads and two tetrads. All have the characteristic of penetrating the individual essences or true nature of states (monad). The four kinds of understanding in the second tetrad are together referred to as the four discriminations or analyses.
Broader Understanding (Buddhism, #HM4523).
Narrower Understanding as knowledge about law (#HM4726)
Understanding as knowledge about meaning (#HM1663)
Understanding as knowledge about language (#HM5026)
Understanding as knowledge about kinds of knowledge (#HM4958).

♦ **HM0486b Merging with the ordinary world** (Taoism)
Harmonizing illumination
Description From the perspective of Taoism, the challenge for those who have achieved illumination is to merge with the ordinary world in order to channel its processes more appropriately. For the practitioner, this involves the art of mixing with people of varying degrees of enlightenment without revealing his inner state. It is a question of being in the world and yet transcending it, responding to people and yet remaining autonomous. The phenomena of the world are used to practice the principles of Tao.
Broader Human development (Taoism, #HH0689).
Preceded by Transcending the world (Taoism, #HM5265).

♦ **HM0487d Focusing** (Buddhism)
Vyappana (Pali)
Related Fixation (Buddhism, #HM1617).

♦ **HM0491d Spiritual sloth** (Christianity)
Description The beginner on the road to the human deeps finds that most spiritual things are tedious and do not give any tangible pleasure and thus fall into spiritual sloth by passing them up. They feel that all spiritual things should be delightful and so quickly become bored with things which give them no pleasure. If contemplation, prayer or meditation give no pleasure then they give them up. They avoid the road to perfection, i.e. the way of saying no to their own will and pleasure for God's sake. They seek the road of pleasure and the satisfaction of their own wills. Rather than bring their own will into line with the will of God they attempt to bring God's will into line with their will. They are like those brought up in luxury and who run from the thought of any thing hard. They are offended by spirit and physical suffering. They in their sloth are too weak to face the trials of the spirit.
Context This is an imperfection of beginners in the Dark Night by St John of the Cross.
Broader Beginner in spiritual life (Christianity, #HM0102).
Related Spiritual envy (Christianity, #HM3818) Spiritual lust (Christianity, #HM4180)
Spiritual anger (Christianity, #HM0936) Spiritual pride (Christianity, #HM4533)
Spiritual avarice (Christianity, #HM0642) Spiritual gluttony (Christianity, #HM0507).

♦ **HM0496d Measureless concentration** (Buddhism)
Infinite concentration
Description This is unification associated with the noble paths.
Context According to Hinayana Buddhism, concentration is considered as of one kind (monad), of two kinds (dyads), of three kinds (triads), of four kinds (tetrads) or of five kinds (pentad). In the fourth triad, this is the third concentration.
Broader Concentration (Buddhism, #HM6663).
Related Eightfold way (Buddhism, #HM2339).
Preceded by Exalted concentration (Buddhism, #HM4036).

♦ **HM0502d Time-gap experience**
Description Particularly in western culture, life is clearly structured by time requirements, so that one is habitually aware of time as measured by the clock and the passing of time. However, the experience of having no awareness of a period of time is common, when time passes without the events habitually taken as time-markers being registered. The awareness is not of the time which is being "lost" but of "waking up" after the time-gap experience and realizing that time has past without one being aware of it. The experience may arise when one is carrying out a complex series of activities, such as driving a car, but these activities become habitual and not requiring conscious attention, for example in quiet traffic conditions, when the task has relatively unchanging demands. Strictly speaking, then, the gap is not in time but in alertness or conscious attention.
Refs Reed, Graham *The Psychology of Anomalous Experience* (1988).
Broader Time consciousness (#HM2601).

♦ **HM0504d Al-Mujib** (Sufism)
Description The believer is aware of Allah as responsive to every need or prayer, so close He knows our needs before we do ourselves, satisfying them before they arise. Mankind must therefore be responsive and attentive to Allah, praising and praying, following the duties He lays down and responding to the needs of creation.
Context A mode of mystical awareness in the Sufi tradition associated with the forty-fourth of the ninety-nine names of Allah.
Broader Ninety-nine names of Allah (Sufism, #HH2561).
Followed by Al-Wasi' (Sufism, #HM4169).
Preceded by Al-Raqib (Sufism, #HM6098).

♦ **HM0507d Spiritual gluttony** (Christianity)
Description Nearly all beginners on the path of self-consciousness fall at one time or another into one or more of the many dimensions of spiritual gluttony. The journey of the spirit has many

delights and pleasures found in doing spirit exercises and the spiritual glutton begins doing them for the pleasure they bring and not for the growth they promote. Rather than seek purity of the soul and devotion to God they desire pleasures of the spirit. This leads them to going to extremes. They may even kill themselves or do deep physical or psychological damage to themselves doing penances or fasting. They avoid advice about these things and may do just the opposite of what they are told. They feel self-inflected, physical punishment is preferable to submitting themselves to advice from experienced guides. Rather than taking the penance of obedience to someone wiser they become proud and spiritual gluttons. Any form of obedience to another is so distasteful that they modify any suggestion made to them. They may reach the point where they lose the desire to do any exercise suggested to them and only do what they want to do. They feel the only way they can serve God is to do as they want, as though pleasing and satisfying themselves was pleasing and satisfying God. Rather than lovingly fearing God they forget their own spiritual immaturity and incompleteness. They attend services and participant in religious exercises far more than they are advised to do. They spend so much energy trying to receive some kind of delight that they believe that if they experience nothing they have received nothing. Even though the greatest benefits from a spiritual life are not tangible. God may hold back tangible blessings when we pray so that we can more closely focus our attention on his will. When this happens these people may stop praying altogether. They read books, and try one kind of meditation now and then another for the sole purpose to experience the delights of the spirit. They oppose denial of the self as a road to spiritual growth.
Context This is an imperfection of beginners in the Dark Night by St John of the Cross.
Broader Beginner in spiritual life (Christianity, #HM0102).
Related Spiritual envy (Christianity, #HM3818) Spiritual lust (Christianity, #HM4180)
Spiritual sloth (Christianity, #HM0491) Spiritual anger (Christianity, #HM0936)
Spiritual pride (Christianity, #HM4533) Spiritual avarice (Christianity, #HM0642)
Spiritual self-denial (Christianity, #HH4522).

♦ **HM0509d Dependence on ceremonies** (Buddhism)
Context The third fetter or character defect referred to by Buddha. The first five fetters are marked by automatic response due to the following of old, unphilosophical mental habits. When these are overcome, the disciples is ready (arhat) to face the final five.
Broader Character defects (Buddhism, #HH0470).
Followed by Emotional desires (Buddhism, #HM3301).
Preceded by Doubt of the efficacy of the good life (Buddhism, #HM1294).

♦ **HM0513d Pity** (Buddhism)
Compassion
Description Having reflected on the evils of lack of compassion and the blessings of compassion, the meditator does not at first direct compassion towards those to whom he is antipathetic, to very dear friends, to a neutral person, to an enemy or hostile person or to a member of the opposite sex or to a dead person. He becomes versatile in the unspecified pervasion of compassion (five ways), specified pervasion (seven ways) and directional pervasion (ten ways); and eleven advantages arise.
Context One of the four divine abidings or states described as subjects for meditation in Hinayan Buddhism. As experienced in the sense sphere, one of the formations aggregate (mental coefficients) of Hinayana Buddhism, being listed among the inconstant states, and as general secondary (sometimes present in any profitable or profitable-resultant consciousness).
Broader Divine abidings as meditation subjects (Buddhism, #HH3534)
Awareness as mental-formation group of conscious existence (Buddhism, #HM2050).
Related Compassion (#HM2919) Gladness (Buddhism, #HM5224)
Equanimity (Buddhism, #HM7769) Compassion (Buddhism, #HM5634)
Lovingkindness (Buddhism, #HM7607).

♦ **HM0514f Hypomania**
Description In this (pathological) state, the individual experiences inappropriately inflated self esteem and high spirits, hyperactivity and a craving for new experiences and stimuli. As opposed to mania, there is a relatively healthy ego which moderates behaviour; the sense of reality is not compromised and critical self-awareness is not completely lost.
Broader Anomalies in the flexibility of associations (#HM4312).
Related Mania (#HM2787) Elation (#HM4899).

♦ **HM0521b Pure consciousness** (Psychism)
Description This has been described as simply being, a perfect state not comprehensible to or describable by those who have not accomplished it.
Related Prajna (Hinduism, #HM3455).

♦ **HM0525d Al-Wali** (Sufism)
Description The believer is aware of Allah as the governor, managing the whole of creation. All is planned and predestined. Man has will, whether to enjoy and benefit from what is destined for him; or to revolt, so that what would have happened still happens but, resentful and out of harmony, one is unaware that it is happening. With eyes open one sees the creation of Allah manifesting His beautiful names. With eyes closed, one does not see them but they are still there. One has, and should use, the will to see that one is part of this divine order.
Context A mode of mystical awareness in the Sufi tradition associated with the seventy-seventh of the ninety-nine names of Allah.
Broader Ninety-nine names of Allah (Sufism, #HH2561).
Followed by Al-Muta'ali (Sufism, #HM8019).
Preceded by Al-Batin (Sufism, #HM1209).

♦ **HM0526d Al-Muhaymin** (Sufism)
Description The believer is aware of Allah as the protector. Consciousness and control of one's own thoughts, words and actions reflects al-Muhaymin in one's self.
Context A mode of mystical awareness in the Sufi tradition associated with the seventh of the ninety-nine names of Allah.
Broader Ninety-nine names of Allah (Sufism, #HH2561).
Followed by Al-'Aziz (Sufism, #HM4562).
Preceded by Al-Mu'min (Sufism, #HM1245).

♦ **HM0529d Perplexity** (Buddhism)
Vicikiccha (Pali)
Broader Perplexity (Buddhism, #HM0812).

♦ **HM0530c Warid** (Sufism)
Mystical experience
Description In this state, the mystic is aware that soul and body are internal and external aspects of one absolute unity and that this unity with God exists and has always existed.

♦ **HM0541b Meeting God** (Christianity)
Description John Ruusbroec refers to a number of ways in which Christ, coming from above, like a mighty lord and benefactor, and from within outward, meets with us, coming from below like poor servants, and from outside inwards.
1. Natural union, no intermediary: All persons, both good and bad, naturally possess a nobility in the essential unity of spirit which, in its bare nature and in the highest part of its being, is naturally united with God.
2. Meeting God with intermediary: The spirit exists in unity in its activity, in itself, as its created, personal mode of being. In this unity it is either like God (by means of grace and virtue) or unlike God (because of mortal sin). Human beings are made in the likeness of God, in the grace of God. Losing this likeness is damnation. Whenever we turn to God, so that God finds in us a capacity for receiving His grace, He gives us life and makes us like Himself by means of His gifts. Christ enters us with gifts (with intermediary). We enter Christ with our virtues (with intermediary). There is constant renewal, with the giving of new gifts and the spirit turning back to God in accordance with how it has been called and gifted, and in this encounter receiving new gifts. Christ imprints His own image and likeness on us, delivers us from our sins, sets us free and makes us like Himself.
3. Meeting God without intermediary: Similarly, Christ enters us above all gifts (without intermediary); we enter Christ above all virtues (without intermediary). This is the most interior way of life. Incomprehensible light transforms and pervades the spirit's inclination to blissful enjoyment in a way devoid of particular form. It can only be known through itself. God calls in an overflow of essential resplendence which, enveloping in love, makes one lose one's self and flow into the darkness of the Godhead. One with the Spirit of God, one meets God with God to possess eternal blessedness with Him and in Him.
There are three modes in which the interior way of life is practised: emptiness; active desire; resting and working in accordance with righteousness. These are described in separate entries.
Refs Dupré, Louis and Wiseman, James A (Eds) *Light from Light* (1988); Wiseman, James A (Trans) *John Ruusbroec* (1985).
Narrower Meeting God with emptiness (#HM8772)
Meeting God with active desire (#HM6204)
Meeting God both resting and working in accordance with righteousness (#HM5309).
Related Interior life (Christianity, #HH3008) Essential unity of being (Christianity, #HM6111)
Superessential contemplation (Christianity, #HM4340).

♦ **HM0543c First order perceptions - intensity**
Description One perceives that something is there or that something is there which differs from what had been there. One is not normally conscious of these perceptions, despite the fact that they are the only direct contact with the outside world. These are intensity of feeling, for example, as of the buttocks on a chair, although one is only conscious of this feeling when the chair becomes uncomfortable and one registers the discomfort. All one feels is the energy of the signal or disturbance. First order perceptions do not register cause.
Context The first of ten orders in the perceptual system described by William Glasser.
Broader Orders of perception (#HM0988).
Followed by Second order perceptions - sensation (#HM1932).

♦ **HM0549c Correct livelihood** (Buddhism)
Samma ajiva (Pali)
Narrower Unaffected (#HM3375) Leaving undone (#HM3429)
Correct livelihood (#HM1763) Not incurring guilt (#HM3413)
Not trespassing limit (#HM3435)
Leaving off, abstaining, totally abstaining and refraining from the wrong livelihood (#HM0252).

♦ **HM0551e Metamorphosis** (Psychism)
Description This is the reported ability of a being (either living or dead) to transform the body at will and appear in, for example, animal form. The effect is usually temporary.
Refs Schactel, E *Metamorphosis* (1959); Wachsmuth, G *Reincarnation as a Phenomenon of Metamorphosis*.
Related Shape shifting (Psychism, #HM3202) Psychic atavism (Psychism, #HM0401).

♦ **HM0552d Being ashamed of acquisition of sinful and unwholesome dharmas** (Buddhism)
Hiriyati papakanam akusalanam dhammanam samapattiya (Pali)
Broader Power of shame (Buddhism, #HM0352).

♦ **HM0554d Jealousy** (Leela)
Dwesh
Description Over-indulgence in fantasy leads to confusion, to loss of contact with reality and an inflated ego. The behaviour of others is not in conformity with the player's own self image - he is plagued by doubts and lack of self-confidence, is suspicious and fears rivals. Growing self-hatred is projected as hatred of others, and the player must return to the first chakra to regain his sense of security.
Context The 16th state or square on the board of *Leela*, the game of knowledge, appearing in the second row.
Broader Second chakra: realm of fantasy (Leela, #HM3651).
Followed by Greed (Leela, #HM1931).

♦ **HM0560b Sanctity** (Christianity)
Sainthood
Description According to Thomas Merton, sanctity consists of discovering one's true self, making the choice to be really that self; and offering to God the worship of that true self in imitation of God, actively participating in one's own life by choosing truth. In this state it is not being admirable to others which counts, but finding it possible to admire everyone else, to have the compassion which sees good in everyone. One is aware that one is a sinner among sinners and that all need the mercy of God.
Refs Merton, Thomas *Seeds of Contemplation* (1972).
Related Sainthood (#HH0331) Sanctity (Sufism, #HM4041)
Human holiness (Christianity, #HM0446).

♦ **HM0561c Religious experience as personal devotion and worship**
Description Liberation through devotion, bhakti marga, is said to be open to all. The experience of divine presence may be shattering and awe-inspiring, witness the transfiguration of Christ, and the appearance of Krishna in his fullness in the Bhagavad Gita. The Children of Israel could not look at the face of Moses after his encounter with God. The experience of the holy, of the numinous, may be inwardly transforming and liberating. It contains experience beyond rational expression which is the source of bliss. There is a uniquely religious feeling; and within this deeply felt and transforming sense of the divine is deep intimacy and awareness of grace so that the most simple act of worship or devotion, if wholehearted, may become the medium of salvation.
Refs Batson, Daniel and Ventis, W Larry *The Religious Experience* (1982); Smart, Ninian *The Religious Experience of Mankind* (1971); Unger, Johan *On Religious Experience* (1976).
Broader Religious experience (#HM3445).

Related Worship (#HH0298) Salat (Islam, #HH0536) Khidmat (Sufism, #HM8776)
Bhakti marga (Hinduism, #HH0628) Devotion (Christianity, #HH0349)
Discipline of worship (Christianity, Judaism, #HH3776).

♦ **HM0563a Divine consciousness**
God consciousness — Kingdom of heaven awareness
Description Spiritual practices, such as meditation or unceasing prayer of the heart, may lead to the cessation of physical reaction to external circumstances and of limitations of the heart and feelings. This state has been described as divine or God-consciousness, living in the Kingdom of Heaven, and corresponds to removal of ignorance, avidya or forgetfulness of self.
Related God-consciousness (#HM2166) Ignorance (Buddhism, Yoga, Zen, #HM3196)
Esoteric divine consciousness (Psychism, #HM4024).

♦ **HM0567d Right knowledge** (Leela)
Suvidya
Description A combination of awareness with experience, right knowledge takes the player beyond the chakras to the plane of cosmic good. Consciousness is nourished, the inner voice is strengthened, knowing and that which is known become one.
Context The 45th state or square on the board of *Leela*, the game of knowledge, appearing in the fifth row.
Broader Fifth chakra: man becomes himself (Leela, #HM0933).
Followed by Plane of cosmic good (Leela, #HM4343).

♦ **HM0568d Fayd** (Sufism)
Description This is a form of enlightenment when the believer receives and achieves truth in a manner beyond the normal intellect.
Broader Enlightenment (#HM2029).
Related Al-Wadud (Sufism, #HM7712).

♦ **HM0569b Siddha-loka** (Jainism)
Description According to the Jain system, this is the uppermost part of the universe, the abode of liberated souls, the jiva finally free from their ajiva component and from karma.
Related Adha-loka (Jainism, #HM3697) Urdhva-loka (Jainism, #HM0131)
Madhya-loka (Jainism, #HM1357) Human development (Jainism, #HH0622).

♦ **HM0570c Ego consciousness**
Description Consciousness has been defined as undirected attentional energy devoid of object but which may be linked to objects of consciousness, which rise and fall from its field with time. The ego is one object of consciousness, as are sensory objects, personal impulses, needs, etc, and also the relationships, roles, patterns of exchange which make up society. It is what, among these objects, is referred to as "me". Three possible modalities for the coupling of ego with consciousness, each implying a particular control strategy, are described.
Narrower Biosocial consciousness (#HM0044) Transpersonal awareness (#HM0768)
Existential consciousness (#HM3336).
Related Feeling of I (#HH0840) Ego awareness (Hinduism, #HM2059)
Psychological approach to transcendence (#HH1763).

♦ **HM0573e Aia** (Psychism)
Description In the psychic sense, an individual who no longer has to incarnate on earth, having earned the right to live in higher realms.

♦ **HM0574d Mass hysteria**
Description All forms of this shared consciousness seem to be manifestations of the prevailing sociopsychosis. There may be a great upsurge of enthusiasm, fear, panic or other emotion which, rather than affecting everyone at once, seems to start on a small scale and spread rapidly to all in the vicinity, engulfing large numbers of people in highly charged emotion. Individuals seem deprived of their senses, behave and speak irrationally and manifest symptoms resembling those of possession; even those simply intending to watch what is happening may be caught up. Mass hysteria has manifested itself in religious wars, witch hunting and flagellation. In more secular terms, it is demonstrated at "pop" festivals, and football matches.
Related Hysteria (#HM3010) Herd instinct (#HM2559).

♦ **HM0575c Tasawwuf** (Sufism)
Desirelessness — Purity — Spiritual vision — Sufism
Description This is the old name given to the way of the Sufi mystics, particularly characteristic of the zuhd (austerity) and faqr (poverty) stages in development. Tasawwuf indicates not only renunciation of but also lack of desire for wealth and the superficial things of the world. There is no showiness or desire for recognition but perfect spiritual apathy, free from ties, when intellectual riches are laid aside, worldly riches are given away, there is no turning back. Not only does the Sufi possess nothing, he is possessed by nothing. Having left behind the unreal, he lives in the reality of complete dependence on God. Through purification there is attachment neither to outward nor inner activity; this leads to vision of God and a feeling of affinity with him. Inner purity leads to realization of moral ideals and values, not through rules and regulations or acquisition by instruction or exertion, but as is learned through experience.
The true Sufi is purified of all desire, is inwardly purified from wretchedness. His speech is never inadvertent, thoughtless or calumnious. His is radiant of mind, with eyes turned from the things of the world and he is fully instructed with truth.
Context According to Shaykh Abu Sa'id ibn Abi'l-Khayr, this is the last of 40 stations or maqamat the Sufi must possess for his journey on the path of Sufism to be acceptable.
Broader Human development (Sufism, #HH0436)
Stations of consciousness - ibn-Abi'l-Khayr (Sufism, #HM2424).
Related Zuhd (Sufism, #HM4450) Faqr (Sufism, #HM3427)
Inward purity (Sufism, #HM3602) Desirelessness (Hinduism, #HM4406)
Non-desire (Christianity, #HM9330).
Preceded by Nihayat (Sufism, #HM3377).

♦ **HM0576c Saguna brahman**
Description Reality as grasped through name and form, Brahman in terms of the three gunas.
Broader Brahman (#HH1226).
Related Three gunas (#HH0413).

♦ **HM0581d Ku** (Buddhism)
State of latency — Neither existence nor non-existence
Description This is the state of not manifesting, such as anger which disappears when its manifestation disappears but is always ready to arise again, or music on a cassette which is not being played, or flowers on a tree in winter, or memories that are not this instant being recalled.
Related Ten worlds (Buddhism, #HM2657).

♦ **HM0584f Modes of awareness associated with psychoactive substances**
Description In general, psychoactive substances are used because, however temporarily, they general produce a pleasurable sense of feeling better. This may be relief from feelings of dysphoria (perhaps induced by some psychological disorder) or to aid feelings of euphoria. It may simply be to relax tensions or allay anxiety. Some drugs (in particular, alcohol) may be used to induce conviviality and fellowship. Stimulants are used for their properties of elevating mood and (in the case of hallucinogens) for intensifying of sensory awareness, in particular visual awareness, and for the inducing of transcendent experience. Depressants lead to relaxation, tranquillizing and withdrawal - they may even end in stupor or coma; and anaesthetic drugs are used to produce a high, as are solvents and other substances. Transcendent experience may arise through use of drugs other than those usually defined as hallucinogens, for instance alcohol, laudanum (opium) and nitrous oxide.
However, not all drugs invariably produce the desired effect and they may sometimes produce the reverse of the response aimed for. There is also an opposite effect as the effect of the drug wears off, with feelings of dysphoria or worse. The after effects of alcohol consumption are well known, another example is post-amphetamine depression. Over time, steadily increasing doses may be required to produce the desired effect. Eventually, neuro-physiological changes take place which diminish the pleasurable feeling zone so that higher doses are needed with reducing response and ordinary pleasurable activities are hardly registered.
The mystical state induced by some substances has been and is used in some religious rituals for direct experience of God or a superlative state of human awareness. The existence of such states has called into question the traditional division of states into "sane" and "insane". Current definitions would include a continuum, starting with the disordered and disintegrated states of insanity, then sanity, and finally a superlative state which might be labelled *unsanity* and which may characteristically slip over into the insane.
Refs Clark, Walter Houston *Chemical Ecstasy* (1969); Huxley, Aldous *The Doors of Perception* (1954); Watts, Alan *The Joyous Cosmology* (1962).
Narrower Bad trip (#HM0818) Aesthetic LSD experience (#HM2542)
Psychodynamic LSD experience (#HM3271)
Modes of awareness associated with use of khat (#HM4277)
Modes of awareness associated with use of cocaine (#HM1594)
Modes of awareness associated with use of caffeine (#HM1368)
Modes of awareness associated with use of cannabis (#HM3691)
Modes of awareness associated with use of nicotine (#HM1881)
Modes of awareness associated with use of inhalants (#HM3743)
Modes of awareness associated with use of sedatives (#HM4546)
Modes of awareness associated with alcohol consumption (#HM0134)
Modes of awareness associated with use of hallucinogens (#HM0801)
Modes of awareness associated with use of opium and similar drugs (#HM4060)
Modes of awareness associated with use of amphetamines or similar drugs (#HM0803)
Modes of awareness associated with use of phencyclidine and similar drugs (#HM4254).
Related Psychedelic drugs (#HH0075)
Drug use for transcendent experience (#HH1123).

♦ **HM0592a Saccidananda**
Sat-cit-ananda
Description The bliss (ananda) of being, existence, truth (sat) and of knowing that one is that being, existence, truth (cit). The being and the awareness of being are indivisible, there is no duality (advaita). Although a sanskrit term used in vedic philosophy (Hindu), Abhishiktananda (Henri Le Saux) demonstrates the universality of the term and it's relationship to Christianity, equating "sat" with the Father, "cit" with the Son and "ananda" with the Holy Spirit. Saccidananda symbolizes the innermost mystery of God and also the divine presence in the innermost being of man. It is said to be indescribable but capable of being experienced in profound silence of spirit. It is compared with the silence following pronunciation of the sacred syllable "Om", when all sound is left behind.
Refs Abhishiktananda *Saccidananda* (1984).
Related Bliss (#HM4048) Advaita vedanta (Hinduism, #HH0518)
Ananda (Hinduism, Buddhism, Yoga, #HM3227)
Chitta-dependent consciousness (Yoga, #HM3011).

♦ **HM0594c Indeterminate consciousness in the fine-material sphere - resultant** (Buddhism)
Indeterminate consciousness in the form-realm - resultant
Description As in the profitable sphere, these are associated with the jhana factors, that is: (i) with applied thought or inception of thought; sustained thought; happiness, joy or zest; bliss or ease; and concentration. (ii) with sustained thought; happiness, joy or zest; bliss or ease; and concentration. (iii) with happiness, joy or zest; bliss or ease; and concentration. (iv) with bliss or ease and concentration, happiness having faded away. (v) with concentration and equanimity. But whereas with the profitable consciousnesses these occurred in a cognitive series through right attainment, here it is by birth, through rebirth-linking (reconception), life-continuum and death.
All five give rise to materiality but not to postures or to intimation.
Context In Hinayana Buddhism, 89 consciousnesses are enumerated in aggregate (khanda). Of these, 21 are profitable or moral, 12 are unprofitable or immoral and 56 are indeterminate (resultant or functional). The unprofitable all arise in the sphere of sense and desire, whereas profitable and indeterminate arise in sense, fine-material, immaterial and supramundane spheres.
Broader Dispositions of consciousness (Buddhism, #HM2098).
Narrower Mundane resultant consciousness (#HM0130).
Related Rebirth-linking (Buddhism, #HM8266) Mind consciousness (Buddhism, #HM3323)
Profitable consciousness in the fine-material sphere (Buddhism, #HM5338)
Indeterminate consciousness in the fine-material sphere - functional (Buddhism, #HM4761).
Followed by Indeterminate consciousness in the immaterial sphere - resultant (Buddhism, #HM4982).
Preceded by Indeterminate consciousness in the sense sphere - resultant (Buddhism, #HM5721).

♦ **HM0600d Yoga of the dream state** (Buddhism)
Rmi-lam (Tibetan)
Context The third branch of the Tibetan Tantric path of form according to the Naropa school.
Broader Tantric visualization (Yoga, #HM1690).
Followed by Yoga of the clear light (Buddhism, #HM3690).
Preceded by Yoga of the illusory body (Buddhism, #HM1222).

♦ **HM0606c Nabi** (Sufism)
Description The name of a traveller on the *tariqa* or Sufi spiritual path at the level of a divinely inspired prophet.
Context Spiritual progress on the spiritual path is marked by the attainment of specific *maqamat* or station.
Broader Mystic stations (Sufism, #HM3415).

♦ **HM0607d Hatred** (Buddhism)
Dosa (Pali)
Broader Hatred (Buddhism, #HM4502).

♦ **HM0610e Failure to hold meditative attitude** (Yoga)
Description The result of failure to deal with wrong conditions.
Context One of the nine obstacles to soul cognition enumerated by Patanjali.
Broader Obstacles to soul cognition (Yoga, #HM3182).
Preceded by Inability to achieve concentration (Yoga, #HM0298).

♦ **HM0622d Al-Mubdi** (Sufism)
Description The believer is aware of Allah as the originator of all. Before creation only Allah existed. He created creation, building within it its own means of continuity. The believer must contemplate his own origins, his own marvellous existence, and confirm Allah as the originator of it all.
Context A mode of mystical awareness in the Sufi tradition associated with the fifty-eighth of the ninety-nine names of Allah.
Broader Ninety-nine names of Allah (Sufism, #HH2561).
Followed by Al-Mu'id (Sufism, #HM4093).
Preceded by Al-Muhsi (Sufism, #HM4544).

♦ **HM0629e Focus one** (Psychism)
Context The first state occurring during astral projection.
Broader Astral projection (Psychism, #HM1887).

♦ **HM0633c Wisdom as a goad** (Buddhism)
Patoda (Pali)
Broader Faculty of wisdom (Buddhism, #HM3233).

♦ **HM0634b Complete awareness** (Buddhism)
Gold pill (Taoism) — Absolute (Confucianism) — Calm stability
Description The absolute is movement and stillness without beginning. The infinite absolute is the limit of the unlimited. Calm and stable, as yet unaffected by things, the human mind is merged in the celestial design - the subtlety of the absolute. Once affected by things, there is partiality, change of the absolute. In the calm and stable state and careful of attention the celestial design is always clear, there is unobscured open awareness, autonomy in action and one can deal with all that arises. The practice of calm stability, when mature, gives rise to the spontaneous true restoration of the infinite. The subtle responsive function of the absolute is clear, the design of the universe and of all things is complete in one's self.
Refs Cleary, Thomas (Trans) *The Book of Balance and Harmony* (1989).
Related Refining the gold pill (Taoism, #HH4880)
Merging of yin and yang (Taoism, #HM1330).

♦ **HM0635b Release from the matrix** (Taoism)
Description This is the Taoist equivalent of nirvana, when the matrix of mundanity is shed. One has a body "outside the body". Vitality is refined into energy, energy into spirit, spirit into emptiness and then the fundamental is embraced and one returns to openness, which is the same as the Buddhist ultimate emptiness.
Related Nirvana (Buddhism, #HM2330) Highest vehicle (Taoism, #HH4742)
Human development (Taoism, #HH0689).
Followed by Spontaneity (Taoism, #HM1865) Transcendent liberation (Taoism, #HM1833).

♦ **HM0642d Spiritual avarice** (Christianity)
Spiritual greed
Description For those at the beginning of the journey of the spirit who become dissatisfied with the spirituality they have been given, do not find comfort they desire in the spiritual things they have and are unhappy and complaining, are experiencing spiritual avarice. They constantly listen to the advice of others, go to spiritual counsellors, attend retreats, spend their time buying and reading books on spirituality. They thus avoid killing off their own human desires and growing spiritually. They gather around themselves spiritual artifacts, especially those of value. They collect, barter and trade sacred icons, medals, medallions, scrolls, relics and beads. They collect valuable things related to the spirit in direct opposition to the devotional life.
Context This is an imperfection of beginners in the Dark Night by St John of the Cross.
Broader Beginner in spiritual life (Christianity, #HM0102).
Related Avarice (Buddhism, Pali, #HM0227) Spiritual envy (Christianity, #HM3818)
Spiritual lust (Christianity, #HM4180) Spiritual sloth (Christianity, #HM0491)
Spiritual anger (Christianity, #HM0936) Spiritual pride (Christianity, #HM4533)
Spiritual gluttony (Christianity, #HM0507).

♦ **HM0645c Comic sentiment** (Hinduism)
Aesthetic emotion of mirth
Context One of eight kinds of aesthetic sentiment or rasas in Indian psychology.
Broader Aesthetic emotion (Hinduism, #HM0205).

♦ **HM0649d Shamelessness** (Buddhism)
Anuttassa (Pali) — Fearlessness of blame
Description In Hinayana Buddhism, this is the formation that is neither ashamed about misconduct nor anxious or agitated about evil; its characteristic is absence of fear or dread of evil and its proximate cause is lack of respect for others.
Context One of the formations aggregate (mental coefficients) of Hinayana Buddhism, being listed among the constant states which appear in their true nature, and as unprofitable primary (always present in any unprofitable or unprofitable-resultant consciousness). Although shamelessness and consciencelessness are considered closely related they are listed as separate formations.
Broader Awareness as mental-formation group of conscious existence (Buddhism, #HM2050).
Related Shame (Buddhism, #HM8112) Non-shame (Buddhism, #HM2986)
Non-embarrassment (Buddhism, #HM3203) Consciencelessness (Buddhism, Pali, #HM4394).

♦ **HM0650g Foundation awareness**
Description Some sources quote this inattentive perception of objects, the basis of cognition, with the 6 sense consciousnesses and with delusion consciousness as one of the 8 types of consciousness.
Related Sense consciousness (Buddhism, #HM2664)
Delusion consciousness (Buddhism, #HM1119).

♦ **HM0652c Samvriti** (Buddhism)
Mundane awareness
Description This is the Madhyamaka Buddhism equivalent of tonal, it is all that can be described in speech. Here lies all duality, including that which is implied by samvriti and paramartha.
Related Paramartha (Buddhism, #HM1067) Discursive reasoning (Christianity, #HM6997)
Tonal state of awareness (Amerindian, #HM2066).

♦ **HM0655g Ego destruction**
Description An awareness of one's self as an individual is lost or diminished, existence is only as part of the group. There is no inward motivation to advance or grow.
Related Ego death (#HH6921) Destructive manipulation (#HH1020)
I-transcendent stage of life (#HM3060).

♦ **HM0665e After death dream state** (Psychism)
Description A trance condition after (often a sudden) death when the physical world is still seen and, not realizing it is dead, the soul-mind wonders at not being able to communicate with those alive.
Related Dream-state bardo awareness (Buddhism, Tibetan, #HM2371).

♦ **HM0666d Flow experience through thought**
Flow experience of the mind — Enjoying thought
Description Here the activity enjoyed through flow experience is symbolic, depending on natural language, mathematics or some other notational system, the action being a mental manipulation of concepts. For flow experience to arise there must be skill, rules, a goal and a way of obtaining feedback. The mental condition must be ordered so that one can concentrate and interact with opportunities at the level of one's skills.
Normally one is programmed in a series of activities where psychic energy is channelled by habit and thoughts are on "automatic pilot". If there is nothing to do, thoughts tend to take on random patterns until attention is attracted to whatever is most problematic. This is neither useful nor enjoyable and the usual response is escape, probably into television where the attention is structured. An alternative is to acquire habits of control over the mental processes instead of having them controlled from outside. this requires practice, with goals and rules intrinsic to flow experience.
All mental flow depends on memory and feats of memory bring the satisfaction of flow experience. Exercise of memory through recitation of tribal knowledge and the setting and solving of complex riddles were both common in ancient cultures. Rote learning, far from being incompatible with creativity, gives the mind stable content and makes it richer. Learning complex patterns of information brings control of consciousness. Again, the facts of memory must fit into patterns, with likenesses and regularities, in order for what takes place in the mind to take shape. Names and concepts are necessary - in the Bible, God speaks the universe into existence by naming it's constituents, man names the animals. In the vedanta, things exist in name and form only. Systematic rules combine this information and thinking becomes a game in the mind, the exhilerating playing with ideas, whether philosophy or science or the playing with words of the writer.
Extrinsically applied education ideally opens the mind for intrinsically motivated education. The mind is controlled by the individual himself, free from externally applied opinions of the media and those he meets.
Refs Csikszentmihalyi, Mihaly *Flow* (1990).
Broader Flow experience (#HM2344).

♦ **HM0670d Irreligiousity** (Leela)
Adharma
Description Adharma arises wherever life is lived contrary to the laws of nature when the three gunas are no longer in balance. This may be through the player denying, rejecting or praising his own self. Blind faith without regard for the laws of nature or his own dharma, not in tune with his inner vibrations, leads the player to delusion and the first chakra level.
Context The 29th state or square on the board of *Leela*, the game of knowledge, appearing in the fourth row.
Broader Fourth chakra: attaining balance (Leela, #HM3291).
Followed by Delusion (Leela, #HM0697).

♦ **HM0672f Abulia**
Aboulia — Emptiness — Spiritual emptiness
Description This apathetic indifference marked by absence of interest either in one's self or in others arises in a number of disturbed mental conditions. The associated feelings of helplessness and lack of faith or hope distinguish the state from *accidie*, which is more indicative of lack of concern, and from *anomie*, where lack of adjustment to changed circumstances causes alienation. In abulia, every effort is meaningless for there is no sense of direction; the result may be uninhibited or unconsidered activity - *hyperabulia* - or extreme hesitancy. Emptiness is characterized by a feeling of difference from others, an inability to love or care about others or to respond to affection and attention. It may appear to be the entirety of experience when all other feeling is precluded. Spiritual emptiness even under conditions of success and prosperity may be due to a breakdown in social standards and values. *Logotherapy* is one system intended to counteract this tendency. The term abulia is also extended to include a level of consciousness in hypnosis when the conscious mind becomes passive and the will is relinquished.
Broader Hypnotic states of consciousness (#HM2133).
Related Anomie (#HM2947) Accidie (#HM3259) Logotherapy (#HH0416).

♦ **HM0673d Fear of God** (Christianity)
Holy fear
Description According to St Francis de Sales, one fears God out of love rather than loving him out of fear. This holy fear is an element of adoration and of the recognition of the individual's dependence and sinfulness. It also includes an element of fear for his own salvation in the light of God's justice which may lead to repentance, this being morally justifiable if it acts as preparation for justification, when it may be transformed and integrated by the love of God for his own sake. The fear then becomes loving reverence.
Broader Fear (Christianity, #HM3311).
Related Khawf (Sufism, #HM1047).

♦ **HM0679e Auric clairvoyance** (Psychism)
Description The mind in a clear state (alpha rhythm or after meditation) either deliberately or spontaneously sees the electromagnetic energy field or aura surrounding another.
Broader Extrasensory perception (ESP, #HM2262).

♦ **HM0680d Savakabodhi** (Buddhism)
Description This is the enlightenment gained through hearing the lessons of the Buddha.
Context One of three types of enlightenment recognized in Theravada Buddhism.
Broader Steps to enlightenment (Buddhism, #HH4019).

♦ **HM0684d Not being covetous** (Buddhism)
Anabhijjha (Pali)
Broader Non-attachment (Buddhism, #HM2128).

♦ **HM0687d Correct action** (Buddhism)
Sammakammanta (Pali)
Broader Correct action (Buddhism, #HM0439).

♦ HM0696d Immaterial-sphere concentration (Buddhism)
Formless-realm concentration
Description This is profitable unification of mind or collectedness of moral thought associated with the immaterial sphere or jhana.
Context According to Hinayana Buddhism, concentration is considered as of one kind (monad), of two kinds (dyads), of three kinds (triads), of four kinds (tetrads) or of five kinds (pentad). In the fifth tetrad, this is the third concentration.
Broader Concentration (Buddhism, #HM6663).
Related Unincluded concentration (Buddhism, #HM5768)
Sense-sphere concentration (Buddhism, #HM1097)
Formless-realm consciousness (Buddhism, #HM2281)
Fine-material-sphere concentration (Buddhism, #HM4265)
Immaterial states as meditation subjects (Buddhism, #HH3198).

♦ HM0697d Delusion (Leela)
Attachment — Moha
Description One of the obstacles to spiritual growth, delusion implies that the player sees the phenomenal world as the only possible way in which reality manifests. It may arise through irreligiousity (living out of harmony with universal law) and the mind cannot perceive truth. It is the first square on which the player lands when he enters the game, and the one to which he is destined to return as long as he sees the world as closed to change.
Context The sixth state or square on the board of *Leela*, the game of knowledge, appearing in the first row.
Broader First chakra: fundamentals of being (Leela, #HM4103).
Preceded by Irreligiousity (Leela, #HM0670).

♦ HM0698b Bardo consciousness (Buddhism, Tibetan)
Description These are the states which are experienced between death and rebirth. Various deities are encountered in these states; it is said that liberation is achieved when these are recognized as luminosity and emptiness of one's own mind. The dying or dead person is guided through the intermediate stages of the departed soul to ensure a profitable rebirth or freedom from the cycle of rebirth. This is done through complex ritual involving reading from the Tibetan Book of the Dead.
In broader terms, *bardo* can refer to any situation of radical transformation where the psyche is in a confused and buffeted state between fear and longing. The recognition that this is a state of transition helps deal with the confusion and one can traverse it without fear.
Broader Ecstatic discipline (#HH5282).
Narrower Dream-state bardo awareness (#HM2371)
Meditation-state bardo awareness (#HM2311)
Life-state bardo awareness (Buddhism, #HM2347)
Death-bardo clear-light awareness (Buddhism, #HM2335)
Death-bardo heaven-reality awareness (Buddhism, #HM2407)
Death-bardo rebirth-seeking awareness (Buddhism, #HM2431).
Related Talkin (#HH1407)   Near death experience (NDE, #HM0777).

♦ HM0700d Grasping of view (Buddhism)
Gaha (Pali)
Broader Wrong view (Buddhism, #HM1710).

♦ HM0704c Right mindfulness (Buddhism)
Samyak-smriti
Description Exerting right effort, and with right outlook, somatic and mental processes, such as breathing, normally semi- or unconscious, are made conscious. Realizing what the body, feelings, heart and mental states are, the individual is free from the wants and discontent attendant on these things and is ardent, alert and mindful. The characteristic is establishing and attending; the function is not confusing or forgetting; manifestation is in abandoning wrong mindfulness.
Context One of the characteristics of the Eightfold Way of Buddhism attained with the path of bhavana (meditation) and which, together with right endeavour and right rapture of concentration constitutes mental discipline (samadhi).
Broader Eightfold way (Buddhism, #HM2339).
Related Smrityupasthana (Pali, #HM6754)   Four noble truths (Buddhism, #HH0523).
Followed by Right rapture of concentration (Buddhism, #HM0931).
Preceded by Right endeavour (Buddhism, #HM1295).

♦ HM0705d Intention (Buddhism)
Sancetana (Pali)
Related Intention (Buddhism, #HM2589).

♦ HM0707c Twelve-fold knowledge (Systematics)
Autocracy
Description Autocracy is the primary affirmation by which all possible experience is brought into existence whether as the potential pattern or the actual pattern of the universe. It can only be experienced through a minimum of twelve independent terms. This twelve-fold knowledge is the first through which all the main elements of experience can be represented. It combines dynamism and diversity, or relativity and relatedness. It is the culmination of the transformations whereby the structure of existence is first disordered, then corrected, then redeemed, then finally perfected. Through the form of knowledge action is possible without domination, willing without reaction, to unify all possibilities.
Context The last in a sequence of twelve modes of knowledge, identified by J G Bennett, inspired by G Gurdjieff.
Broader Systematics (#HH2003).
Preceded by Eleven-fold knowledge (Systematics, #HM0065).

♦ HM0709c Ceto samatha (Buddhism)
Tranquillity of thoughts — Mental quiescence
Description The state of samadhi as it calms mental wavering and agitation, subdues trepidation and establishes and maintains inner serenity.
Broader Calm abiding (Buddhism, #HM2147)   Concentration (Buddhism, #HM6663).

♦ HM0713c Conceptual cognition (Buddhism, Tibetan)
Description This way of knowing takes metaphysical entities as its appearing objects. It is contrasted with *bare perception* of objective entities. Metaphysical entities are permanent, not products of cause or circumstance and with no ability to produce an effect. They cannot be said to exist objectively but can be known through conceptual cognition through a mental label – clear appearance of a metaphysical entity to consciousness is known through conceptual cognition.
Context One of the valid ways of knowing of Tibetan Buddhism.
Broader Ways of knowing (Buddhism, #HH0873).
Related Bare perception (Buddhism, Tibetan, #HM4651).

♦ HM0715d Unperturbed mindedness (Buddhism)
Avisahatamanasata (Pali)
Broader Sukha (Buddhism, #HM2866).

♦ HM0717c Third chakra: theatre of karma (Leela)
Celestial plane
Description The player is dominated by recognition of the ego, by the search for immortality of the body. This is the level of power and worldly achievement, negative if the will of the ego is imposed on others but positive if organizational skill is developed through altruism. Mastery of the chakra brings healing power, and knowledge of disease and sorrow and of different worlds.
Context The third row of the board of Leela, the game of knowledge, this is represented by the celestial, karma and dharma planes, and also groups the squares representing the states of charity, atonement, bad company, good company, sorrow and selfless service. Located at the root of the navel, connected with the source of sleep and thirst, this chakra is related to the element fire, to the sense of sight and to the colour red. It is compared to the octave comprising the ages 14 to 21 of temporal life. Charity, selfless service and the plane of dharma are the arrows at this stage, and bad company the snake.
Broader Lila (#HM2278).
Narrower Sorrow (#HM1782)   Charity (#HM4387)   Atonement (#HM3837)
Bad company (#HM0764)   Good company (#HM1355)   Plane of karma (#HM0948)
Plane of dharma (#HM0481)   Celestial plane (#HM1052)   Selfless service (#HM0294).
Related Manipura (Yoga, #HM3063).
Followed by Fourth chakra: attaining balance (Leela, #HM3291).
Preceded by Second chakra: realm of fantasy (Leela, #HM3651).

♦ HM0719d Hostility (Buddhism)
Pativirodha (Pali)
Broader Hatred (Buddhism, #HM4502).

♦ HM0724c Rasul (Sufism)
Description The name of a traveller on the *tariqa* or Sufi spiritual path at the level of an apostle giving the message of God to humanity.
Context Spiritual progress on the spiritual path is marked by the attainment of specific *maqamat* or station.
Broader Mystic stations (Sufism, #HM3415).

♦ HM0728d Fulfilment
Description This is a sense of having achieved one's full potential; the satisfaction of fully realizing one's capacities.
Related Effulgence (#HM2521).

♦ HM0730d Concentration without happiness (Buddhism)
Concentration without rapture
Description Here there is unification of the mind in the jhanas not associated with concentration with rapture, two of both the fourfold reckoning and the fivefold reckoning. This is one way of experiencing access concentration.
Context According to Hinayana Buddhism, concentration is considered as of one kind (monad), of two kinds (dyads), of three kinds (triads), of four kinds (tetrads) or of five kinds (pentad). In the third dyad, this contrasts with concentration with happiness.
Broader Concentration (Buddhism, #HM6663).
Related Access concentration (Buddhism, #HM4999)
Concentration with happiness (Buddhism, #HM5767).

♦ HM0739g Workability of body (Buddhism)
Kayakammannata (Pali)
Narrower Being workable (#HM3716)   Working ability (#HM0979)
Workability of the aggregate of sensation, of the aggregate of cognition, and of the aggregate of synergies (#HM1381).

♦ HM0741c Lahut (Sufism)
Al-haqq
Description Absorbed in the essence of God the seeker attains divine qualities and realizes the truth, al-haqq.
Context The last of four stages on the Sufi spiritual path.
Broader Divine path (Sufism, #HM3371).
Related Al-Haqq (Sufism, #HM3945).
Preceded by Jabarut (Sufism, #HM4392).

♦ HM0743d Al-Wajid (Sufism)
Description The believer is aware of Allah as He who finds and obtains His wishes. All are in His presence, no-one can hide from Him. Unlike human aids – doctors, lawyers – who are not always available, Allah is always there, we are always in His presence when we need his aid. Not only at the appointed times of prayer but at every moment we should present Him with our needs.
Context A mode of mystical awareness in the Sufi tradition associated with the sixty-fourth of the ninety-nine names of Allah.
Broader Ninety-nine names of Allah (Sufism, #HH2561).
Followed by Al-Majid (Sufism, #HM4526).
Preceded by Al-Qayyum (Sufism, #HM5006).

♦ HM0747c Fivefold happiness (Buddhism)
Piti (Pali) — Rapture — Bliss — Zest — Happiness — Joy
Description Piti arises in opposition to ill-will and brings about, by degrees, an expansion of interest in the object of contemplation. Five kinds of happiness are enumerated in the "Path of Purification". These are: *minor happiness*, sufficient only to raise the hairs on the body; *momentary happiness*, like flashes of lightening; showering or *flooding happiness*, breaking over the body like waves on the sea shore; uplifting or *transporting happiness*, sufficient to lift the body into the air and transport it; *pervading happiness*, when the body is totally suffused. This last is characteristic of the first jhana. When fivefold happiness is conceived and matured it perfects bodily and mental tranquillity, which, in turn, perfects bodily and mental bliss. Conceived and matured, this bliss perfects momentary, access and absorption concentration.
Context One of the formations aggregate (mental coefficients) of Hinayana Buddhism, being listed among the constant states which appear in their true nature and as general secondary (sometimes present in any consciousness). One of the five constituent factors of jhana.
Narrower Mirth (#HM1421)   Delight (#HM4065)   Elation (#HM0425)
Felicity (#HM1545)   Merriment (#HM1061)   Delightful (#HM1398)
Joyful feeling (#HM8737)   Delightfulness (#HM4246)
Uprisedness of thought (#HM0140).
Related Jhana (Buddhism, Pali, #HM7193)   Dwelling in the first jhana (Buddhism, #HM4298)
Second trance of the fine-material sphere (Buddhism, #HM2038).

♦ **HM0748d Knowledge of passing away and reappearance of beings** (Buddhism)
Knowledge of divine sight — Deva sight — Direct knowledge — Higher knowledge
Description Just as people normally watch with eyes of flesh, he who has developed divine sight watches with the divine, purified and superhuman eye. Such sight includes seeing objects far away, hidden behind walls. Although the moment of passing away or reappearance cannot be seen, with divine sight he sees passing away and reappearing the inferior or superior, the fair or ugly, the happy or unhappy; he understands how this arises depending on the deeds of these beings. He sees how the ill–conducted with wrong views are reborn with an unhappy destiny, perhaps in hell, while those who are well–conducted and have right views have a happy destiny in the heavenly world.
Development of divine sight is through making sure that the jhana with the fire, white or (preferably) light kasina as object is in every way susceptible to guidance. Resorting to the access jhana but not arousing absorption, the kasina is extended. Repeating preliminary work constantly extends the kasina light until all is visible. He then proceeds with the visible datum as object, following the process of appearance of absorption until the fourth or fine–material jhana is reached and the conascent knowledge of divine sight and appearance and reappearance of beings arises.
Context One of the five kinds of direct or higher knowledge developed in Hinayana Buddhism.
Related Appearance of absorption (Buddhism, #HM1618)
Knowledge of supernormal powers (Buddhism, #HM7672)
Knowledge of penetration of minds (Buddhism, #HM5232)
Knowledge of the divine ear element (Buddhism, #HM5982)
Knowledge of recollection of previous existence (Buddhism, #HM4297).

♦ **HM0750e Lucid awareness** (Psychism)
Fifth state of consciousness
Description A state typified by a specific pattern of brainwaves and, in particular, observed in metaphysical healers while healing; a state in the process of spiritual healing when the brainwaves of healer and patient are said to be similar.

♦ **HM0754c Seventh chakra: plane of reality** (Leela)
Description At this chakra, the player is beyond all pleasure. Full of supreme consciousness and bliss, he is master of all eight siddhis, and can create anything at will. He can still be overcome by egotism and become the prisoner of the siddhis, or be drawn to illusion through tamas.
Context The seventh row of the board of Leela, the game of knowledge, this is represented by the planes of reality, primal vibrations, and radiation, and by the gaseous plane, and groups squares representing the states of egotism, positive intellect, negative intellect, happiness and tamas. Located at the crown of the head, it is compared to the octave comprising the ages 42 to 49 of temporal life, that of settled searching for truth. Egotism, negative intellect and tamas are the snakes at this stage.
Broader Lila (#HM2278).
Narrower Tamas (#HM6238) Egotism (#HM1726) Happiness (#HM5099)
Gaseous plane (#HM6434) Plane of reality (#HM1293) Plane of radiation (#HM5028)
Positive intellect (#HM1789) Negative intellect (#HM3623)
Plane of primal vibrations (#HM3732).
Related Sahasrara (Yoga, #HM3398) Siddhis (Yoga, Buddhism, #HH0380).
Followed by Beyond the chakras: the gods themselves (Leela, #HM1141).
Preceded by Sixth chakra: time of penance (Leela, #HM4412).

♦ **HM0756c Understanding free from cankers** (Buddhism)
Description This is understanding unaffected by cankers, not the object of cankers.
Context On the Path of Purification of Hinayana Buddhism, panna (understanding) is considered as of one kind (monad), of two kinds (dyads), of three kinds (triads) or of four kinds (tetrads). There are five dyads, four triads and two tetrads. All have the characteristic of penetrating the individual essences or true nature of states (monad). In the second dyad, this is contrasted with understanding subject to cankers.
Broader Understanding (Buddhism, #HM4523).
Related Understanding subject to cankers (Buddhism, #HM4194).

♦ **HM0757d Irascibility**
Related Peevishness (#HM2472) Irritability (#HM1706).

♦ **HM0762c Showq** (Sufism)
Intense longing for divine unity — Longing — Shugh — Shawq
Description Experience of unity in the diversity of form leads to intense longing for greater intimacy. Al–Ansari describes this on three levels: longing for paradise; longing for God because of God's kindness, with attachment of the heart to the divine attributes; longing desire when earthly life is spoiled, no comfort is possible and which no consolation deters except meeting with the beloved.
Context The third psychic state listed by A Reza Aresteh as progress on the inner self through divine attraction, *kedesh–jazba*, the outer self already having been purified through conscious effort, *kushesh*.
Broader Psychic states (Sufism, #HM4311).
Related Uns (Sufism, #HM1957) Shawq (Islam, #HM0877)
Firaq (Sufism, #HM4112).
Followed by Mehr (Sufism, #HM1266).
Preceded by Nearness–awareness (Sufism, #HM2377).

♦ **HM0763c Way of harmony** (Esotericism)
Way of beauty
Description Characteristic of people who are creative and intuitive , often with a deep understanding of themselves and their environment. Such people cultivate beauty for its own sake. On this way they are challenged by the task of cultivating persistence and will, making real choices and asserting themselves effectively, handling more skillfully the influence of others, and of giving form to their insights.
Context The fourth of seven ways to spiritual realization characteristic of Western esotericism.
Broader Ways to spiritual realization (Esotericism, #HH2665).
Related Sphere of victory (Kabbalah, #HM2362) Initiation of renunciation (Esotericism, #HM0210).

♦ **HM0764d Bad company** (Leela)
Ku–sang–loka
Description Lacking the strength to fulfil desire for ego–identification, the player seeks the help of a group of others on the same path. Wrong vibrations may lead to a group not acting in accordance with dharma, whose (probably charismatic) teacher is not at a sufficiently high level and where wrong traits in the player's character are either ignored or encouraged. Personal problems are seen as the fault of others, the player is deluded and grows in selfishness and swollen ego. He is transported down a snake to conceit in the first chakra.
Context The 24th state or square on the board of *Leela*, the game of knowledge, appearing in the third row.
Broader Third chakra: theatre of karma (Leela, #HM0717).
Followed by Conceit (Leela, #HM0098).

♦ **HM0765c Ming** (Zen)
Self–knowledge
Description This state of self–knowledge has been equated with transcendental consciousness. It is the becoming aware of one's own nature and is referred to as no–seeing, the supreme insight, the I looking directly at the I.
Broader Self–knowledge (#HH0312) Human development (Zen, #HH1003).
Related Transcendental consciousness (#HM2020).

♦ **HM0766d Knowledge of desire for deliverance** (Buddhism)
Knowledge of desire for release
Related Purification by knowledge and vision of the way (Buddhism, #HH3550).
Followed by Knowledge of contemplation of reflection (Buddhism, #HM0169).
Preceded by Knowledge of contemplation of dispassion (Buddhism, #HM5364).

♦ **HM0767d Won** (Korean)
Unwillingness to let go of illusion
Description Even while understanding that there is a way out of the spiritual dilemma one is in, one refuses to let go of the illusion enslaving one to suffering. This unwillingness to let go of lost opportunities, regrets, resentments and negative experiences of life brings continual rebirth in the same old order and a lost opportunity to grasp a less illusory reality.

♦ **HM0768d Transpersonal awareness**
Consciousness three — Alteration
Description Consciousness is the primary reality, both ego and the external world being derived from it. The basis for control is through self image, and this underlies not only mind–cure movements but also Buddhist traditions; and is the psychological basis for prayer and hope theology in Christianity. Through non–attachment, suffering is accepted or even embraced; as are also threats to the ego. Control comes through inducing an event through mental imagery.
Context One of three possible modes by which the relationship of consciousness with ego may be regarded, each implying a particular strategy of control or of seeing the way through limitations, what may be termed a form of transcendence. Only in this third mode may the integrity of the unity of consciousness and ego be affirmed independent of considerations of manipulation of the environment.
Broader Ego consciousness (#HM0570).
Related Biosocial consciousness (#HM0044) Existential consciousness (#HM3336)
Psychological approach to transcendence (#HH1763).

♦ **HM0772c Fifth order perceptions - control of sequence**
Description The ability to control sequence is that which transforms recognition of transition (fourth order) into carrying out transition (walking or talking for example).
Context The fifth of ten orders in the perceptual system described by William Glasser.
Broader Orders of perception (#HM0988).
Followed by Sixth order perceptions - control of relationships (#HM0103).
Preceded by Fourth order perceptions - control of transitions (#HM1516).

♦ **HM0776c Seven–fold knowledge** (Systematics)
Structural knowledge — Experience of structure
Description A structure, as a self–regulating system capable of relatively independent existence, can only be experienced through seven independent elements. Such a system is no longer closed and changes in the environment accompany any changes in the structure. This seven–fold knowledge provides a transformational superstructure to reconcile the self–realization requirement of the well–defined entity (namely the acquisition of new properties that were previously neither potential nor possible) and the dissolution of identity required for integration as a part within a whole. Self–regulation is a property that can not be foreseen in the light of the first six gradations of knowledge. They are required but are transcended and transformed by this seven–fold knowledge. Such knowledge is directly self–verifying in that it penetrates into the very structure of reality. The dualism of freedom and necessity is harmonized.
Context The seventh in a sequence of twelve modes of knowledge, identified by J G Bennett, inspired by G Gurdjieff.
Broader Systematics (#HH2003).
Followed by Nine–fold knowledge (Systematics, #HM1408)
Eight–fold knowledge (Systematics, #HM0344).
Preceded by Six–fold knowledge (Systematics, #HM5187).

♦ **HM0777b Near death experience** (NDE)
Core–experience — Out–of–body experience
Description Comparison of well documented experiences of individuals who have been declared clinically dead, or who have been very near death, but who have subsequently recovered, show remarkable similarities. They usually commence with the certainty that they are dead, often with hearing themselves pronounced to be dead or that there is a grave medical complication. At this point there is no fear and no pain. Some then describe an *out–of–body experience* when they see themselves and the people round them from outside their own bodies. They describe hearing and seeing with particular clarity what is independently agreed to have occurred, with no feeling of having a body, simply being "I", and usually positioned somewhere near the ceiling.
The *transcendent* stages follow, of floating through a peaceful, dark or black, space or tunnel and emerging into brilliantly beautiful golden light. There is said to be a feeling of bliss beyond any previously known. Another world of beautiful landscapes and music is experienced. There may be a retrospective, panoramic view of past life, sometimes including encounter with a *presence* with whom their life is discussed. There is then a boundary beyond which no return is possible, and the choice of whether or not to return (for the sake of family, friends, or unfinished tasks). The choice to return is invariably with reluctance, requiring great effort of will - sometimes apparently against the individual's will.
Although there are (less frequently recorded) negative experiences which mirror this sequence, with predominant feelings of loneliness and desolation, the after–effects tend to be similar. Near death experience has a profound effect on those experiencing it. Not only do they no longer fear death, and are convinced that life continues after death, but personality changes include increased *self–confidence* and more *loving* relationships with other people. Orientation becomes less materialistic and the individual may be said to have experienced a spiritual rebirth and received a new purpose in living. The experience and its sequel are similarly experienced regardless of race, culture or religion. Being religious does not seem to affect the likelihood or depth of the experience. Interestingly, it seems more likely to arise in persons never having heard of such an experience than with those who know of its existence.
Since the understanding changes in accordance with the state of being of the experiencer, the unusual state of being inherent in the near death experience may be of assistance in rendering accessible a not otherwise achievable understanding of reality. Evidence suggests that although the brain and the thinking, conscious mind act closely together, they are not the same. By comparing near death experiences with cases of people who have predicted their own deaths and

subsequently died (sometimes for unknown causes and sometimes after apparently living with impossibly grave conditions under which they would have been expected to die long before), evidence also shows that the *will* (including the will to live or to die) may be part of the *transpersonal self* such as is manifest in meditative states and peak experiences. The "everyday I" which most would consider their real selves would in that case be simply a projection of the real self.

Not only have near death experiences brought renewed confidence and improved quality of life to those experiencing them, they may possibly provide a means for systematic study of the mystic experience, thus bringing scientific study of the physical world closer to incorporation into the mystic. It has been suggested that the increasing frequency with which such experiences are being recorded may indicate a possible individual and collective change in humanity to a new awareness and reinterpretation of the truth and an understanding of the unity underlying existence.

Narrower Near death fear (#HM3177) Near death hell (#HM3165)
Near death peace (#HM2494) Near death detachment (#HM3349)
Near death transition (#HM3294) Near death black void (#HM2213)
Near death evil force (#HM3422) Near death review of life (#HM2888)
Near death awareness of light (#HM3381) Near death entering the light (#HM3180)
Near death negative detachment (#HM3115)
Near death encounter with the higher self (#HM3059).
Related Shamanic journey (#HM6120) Unconscious stupor (#HM2473)
Death consciousness (Psychism, #HM1966) Out-of-body experience (Psychism, #HM5534)
Spontaneous out-of-body experience (#HM0332)
Bardo consciousness (Buddhism, Tibetan, #HM0698)
Psychological growth in a dying person (#HM2553).

♦ **HM0778c Christ consciousness as empty mind** (Christianity)
Description Consciousness is not bound by any particular pattern or form of perception or conception; the language and thought patterns of a particular culture no longer bind reality. The mind is flexible and open to other concepts. As Christ does not define himself in terms of place, any place is his home. He does not define himself in terms of a social group, he is open to any person independent of social status, he can challenge any social structure. He does not define himself in terms of one legal system. Respecting a system of laws as an expression of a mode of consciousness, he is nonetheless not bound by it. Respecting places, status of persons, law, he recognizes that they are not absolute. As Christ consciousness, the empty mind is also Christian consciousness.
Broader Christ consciousness (Christianity, #HM2013).
Related Christ consciousness as loving mind (Christianity, #HM0976)
Christ consciousness as dying-rising (Christianity, #HM1402)
Christ consciousness as forgiving mind (Christianity, #HM1728).

♦ **HM0786c Creative imagination**
Refs Corbin, Henry *Creative Imagination in the Sufism of Ibn 'Arabi* (1969).
Related Active imagination (#HM0867) Human development (Sufism, #HH0436).

♦ **HM0793c Muga** (Japanese)
Description This is pure doing, a state in which what is done is done with total attention and wholeheartedness, no hesitation or criticism, doubt or inhibition. Such spontaneity requires complete forgetfulness of self.
Related Satori (Zen, #HM2326) Innocent cognition (#HM0888).

♦ **HM0794g Phosphenes**
After-images
Description These are the images which persist after staring at bright light. They have no real distance or location, existing only as stimulation from the retina. They can be "projected" onto different surfaces, and their apparent size grows with the apparent distance of the surface.
Broader Illusion (#HM2510).
Related Phosphenism (#HH6321).

♦ **HM0796c Magga-bhavana** (Buddhism)
Description Cultivation of the 37 constituents of enlightenment (bodhipakkhiya).
Context One of ten superhuman qualities described in the Sutta Vibhanga as being special attainments of insight above that of ordinary men.
Broader Superhuman qualities (Buddhism, #HH9764).

♦ **HM0798b Quiet reflection**
Description A state arising during meditation when body, emotions and mind are at peace and the pure self is reflected in consciousness.
Related Reflection (#HH1204).

♦ **HM0800c Environmental mysticism**
Eco-spirituality — Green spirituality
Refs Berman, Morris *The Reenchantment of the World* (1981); Carmody, John *Ecology and Religion* (1983); Grandberg-Michaelson, Wesley *A Worldly Spirituality* (1984); Hart, John *The Spirit of the Earth* (1984); Phipps, John-Francis *The Politics of Inner Experience* (1990); Spretnak, Charlene *The Spiritual Dimension of Green Politics* (1986).
Broader Spirituality (#HH5009) Mystical cognition (#HM2272).
Related Deep ecology (#HH2315) Bio-globalism (#HH1267)
Ecological humanism (#HH4965) Planetary consciousness (#HM2006)
Christian stewardship (Christianity, #HH3121)
Human development through pantheism (#HH5190).

♦ **HM0801f Modes of awareness associated with use of hallucinogens**
Use of psychedelics — Psychedelic experience — Consciousness-expanding drugs
Description Hallucinogens are stimulants producing a state of sympathetic dominance. They have traditionally been used for transcendent experience and their ingestion, it has been hypothesized, may account for primitive conceptions of God. The use of LSD or peyote alters and expands human consciousness in a very potent manner. There are changes in perception of the senses, in particular a heightened sense of colour. Other changes are in experience of space and time, in rate and content of thought, in body image. There are: hallucinations and eidetic images (seen with the eyes closed); abrupt and frequent changes in mood and affect; heightened suggestibility; enhanced memory; depersonalization and dissolution of the ego; dual, multiple and fragmentized consciousness; apparent awareness of internal organs and body processes; upsurging of material from the subconscious; awareness of linguistic nuances; increased sensitivity to non-verbal clues, with a sense of capacity to communicate better non-verbally, even telepathically; feelings of empathy; regression and primitivization; apparent heightened concentration; magnification of character traits and psychodynamic processes. The psychodynamic processes appear "naked" so that interaction of ideation, emotion and perception with each other and with inferred unconscious processes become evident. There is concern with philosophical, cosmological and religious questions. The world is apprehended as having slipped from the constraints of normal, categorical ordering so that there is intensified interest in self and world. The range of responses moves from extremes of anxiety to extremes of pleasure.

Commonly the experience is said to be indescribable but the following perceptions may occur:
1. A dissolving of boundaries between what is within and outside "me", so there is oneness with the universe.
2. An awareness of greater reality than normal.
3. Time and space no longer exist.
4. A sense of sacredness, awe and significance.
5. Visual beauty or blazing white light.
6. A positive mood of joy, ecstasy, bliss, peace or love.
7. Resolution of paradoxes and reduction of opposites to components of the same thing.
8. Abrupt alteration of values and beliefs. This may be permanent or temporary. Temporary change leaves a feeling of guilt and despair, a feeling of having failed.

The difference between the above states occurring "naturally" or as the result of ingesting chemicals is not clear, although the former tend to be more valued, perhaps because the person experiencing such a state naturally has gone through rigorous preparation enabling him to profit more fully from it. Insights brought about by the use of psychedelics are rarely of lasting use or significance, probably because they arise too fast and lack either the preparation or the surrounding support necessary for a permanent effect.

The effects are very dependent on setting, dose, prior expectations and personality. They may continue for months or even years, although the individual is usually aware of their hallucinatory nature. On the negative side, there may be paranoid ideation, anxiety, depression, fear of going insane. Synaesthesia may occur, as may depersonalization and derealization. If a mood disorder pre-exists, then taking a hallucinogen to elevate the mood may result in more severe depression.
Refs Huxley, A *Moksha* (1977); Leary Timothy, Ralph Metzner and Richard Alpert *The Psychedelic Experience* (1971); Leary, Timothy *The Politics of Ecstasy* (1970); Solomon, D (Ed) *LSD* (1964); Stace, Walter T *Mysticism and Philosophy* (1960).
Broader Modes of awareness associated with psychoactive substances (#HM0584).
Narrower Aesthetic LSD experience (#HM2542) Psychodynamic LSD experience (#HM3271).
Related Synesthesia (#HM1241) Peak experiences (#HM2080)
Maps of the mind (#HH8903) Psychedelic drugs (#HH0075)
Religious experience (#HM3445) Cosmological mysticism (#HM1635)
Induced religious experience (#HM3221) Introvertive mystical experience (#HM0899)
Drug use for transcendent experience (#HH1123).

♦ **HM0802e Greed** (Buddhism)
Lobha (Pali)
Broader Greed (Buddhism, #HM3283).

♦ **HM0803f Modes of awareness associated with use of amphetamines or similar drugs**
Description The effect of taking amphetamines is like a long-acting cocaine, with mood elevation and energizing and a feeling of euphoria; the tired feel more awake, the sad feel brighter. There is a heightened sensitivity to weak stimuli and awareness is "de-automatized". However, following the *high* experienced on taking amphetamine there is a subsequent post-amphetamine depression, where the individual may be paranoid or suicidal, with delusions of persecution. There may also be delirium, accompanied by touch or smell hallucinations and violent or aggressive behaviour. Repeated use induces neuro-physiological changes which diminish the pleasurable feeling zone so that higher doses are needed for reducing response and ordinary pleasurable activities are hardly registered.
Broader Modes of awareness associated with psychoactive substances (#HM0584).

♦ **HM0805e Psychically unpleasant** (Buddhism)
Cetasikam dukkham (Pali)
Broader Feeling (Buddhism, #HM2270).
Related Psychically pleasant (Buddhism, #HM1862).

♦ **HM0807c Tawakkul** (Sufism)
Trust in God — Confidence — Self-surrender — Tawakul
Description This is a state of total trust in God and perfect faith based on the beginning of affinity with him, when the traveller depends entirely on God's mercy. It arises from serious pursuit of *riyadat* (mystical training) and *mujahada* (mortification of the self). Although this implies effort on the part of the Sufi, in the process of courageously emptying himself, it is through the grace of God that success is granted. Some believe it arises gradually from the practice of patience. Tawakkul strengthens separation from the world and the worldly mentality, and also strengthens the attainment of relatedness to divine unity. Trust in God also aids in getting away from the ego and surrendering the self. As self-surrender, all previous stages in the path are encompassed. In seeking to lose the phenomenal self and what the environment has created in him and utilizing the image of God as a motivation for attraction and divine unity the Sufi becomes more God-like.
Context The sixth stage in a systematic account of various stations of the Sufi spiritual path, including the path to sainthood of the Naqshbandiyya order. According to Shaykh Abu Sa'id ibn Abi'l-Khayr, this is the 13th of 40 stations or maqamat the Sufi must possess for his journey on the path of Sufism to be acceptable.
Refs Arasteh, A Reza *Growth to Selfhood* (1980).
Broader Mystic stations (Sufism, #HM3415)
Suluk of the Naqshbandiyya order (Sufism, #HM4356)
Stations of consciousness - ibn-Abi'l-Khayr (Sufism, #HM2424).
Related Al-Wakil (Sufism, #HM1382).
Followed by Rida (Sufism, #HM4190) Sabr (Sufism, #HM3418)
Zuhd (Sufism, #HM4450).
Preceded by Sabr (Sufism, #HM3418) Taslim (Sufism, #HM5290)
Contentment (Sufism, #HM7024).

♦ **HM0808d Succumbing to hesitation** (Buddhism)
Samsaya (Pali)
Broader Perplexity (Buddhism, #HM0812).

♦ **HM0809b Nirvicara samadhi** (Yoga)
Description There is ontic identification with the internal reality of the object contemplated, transcendental reflection but no cogitation.
Context The fourth level of samprajnata samadhi.
Broader Samprajnata samadhi (Yoga, #HM2896).
Preceded by Vicara samadhi (Yoga, #HM3880).

♦ **HM0810b Diakonia** (Christianity)
Diaconia
Description One of the three ways in which the mission of the Church is manifest, this refers to a state in which Christian love and compassion are extended in the service of man's needs, to the ministry of the Church in the world. In that material poverty and oppression may act as an obstacle to salvation, diaconia is the liberating means of salvation; but the term refers more to the intent and the quality of the loving and giving in the eucharistic sense.
Related Kerygma (Christianity, #HM4256) Koinonia (Christianity, #HM0144).

♦ **HM0811c Three-fold knowledge** (Systematics)
Relational knowledge — Experience of relatedness
Description All real relationships are experienced as reducible to a combination of three independent elements standing to one another as affirming, denying and reconciling influences. A relationship is not itself a whole, nor is it a property of the wholes which it relates. Hence no principle of relatedness can be reached merely by combining wholeness and polarity. Relatedness goes beyond logic. The appearance of relational knowledge coincides with the birth of understanding. It can only arise through experience, which enables one inner functional order to reconcile two others, whether outer or inner. Otherwise unrelated facts can then be welded together into a coherent system. The triadic relationship breaks through the polar barrier that separates the subjective and objective experiences of the individual.
Context The third in a sequence of twelve modes of knowledge, identified by J G Bennett, inspired by G Gurdjieff.
Broader Systematics (#HH2003).
Followed by Four-fold knowledge (Systematics, #HM1991).
Preceded by Two-fold knowledge (Systematics, #HM1234).

♦ **HM0812c Perplexity** (Buddhism)
Vicikiccha (Pali) — Uncertainty
Description Perplexity is that without the wish to cure (thought). It is regarded as obstructing theory and harming attainments. It has the characteristic of doubt and of shifting mind. Its function is to waver or tremble; it manifests as indecision or taking several sides. The proximate cause is unwise attention.
Context One of the formations aggregate (mental coefficients) of Hinayana Buddhism, being listed among the constant states which appear in their true nature, and as unprofitable secondary (sometimes present in any unprofitable or unprofitable-resultant consciousness).
Broader Awareness as mental-formation group of conscious existence (Buddhism, #HM2050).
Narrower Doubt (#HM4347) Dubiety (#HM3934) Evasion (#HM3394)
Doubting (#HM0177) Puzzlement (#HM4008) Perplexity (#HM0529)
Uncertainty (#HM0870) Vacillation (#HM0077) Indecisiveness (#HM3745)
Lack of real grasping (#HM4228) Standing at crossroads (#HM1286)
Succumbing to hesitation (#HM0808) Mental perturbation of thought (#HM0223)
Being in doubt before two alternatives (#HM1561).

♦ **HM0815c 'Abid** (Sufism)
Description The name of a traveller on the *tariqa* or Sufi spiritual path at the level of a worshipper of God.
Context Spiritual progress on the spiritual path is marked by the attainment of specific *maqamat* or station.
Broader Mystic stations (Sufism, #HM3415).

♦ **HM0818f Bad trip**
Description A distressing experience, the result of taking a hallucinogenic drug, when vision and emotion are distorted; there is a feeling of dissociation from reality, of dismemberment.
Broader Psychedelic drugs (#HH0075)
Modes of awareness associated with psychoactive substances (#HM0584).
Related Psychodynamic LSD experience (#HM3271).

♦ **HM0819e Bewitched** (Psychism)
Spellbound
Description The psychic influence of another transfixes the bewitched person who, often unaware that he is not carrying out actions of his own volition, experiences events not due to his own will.
Related Magic (#HH0720).

♦ **HM0820b Quickened consciousness** (Psychism)
Description An apparently inauspicious event gives rise to activity in order to ameliorate its effects, the generation of such energy stimulating a higher state of consciousness.

♦ **HM0822e Adi plane** (Psychism)
Description In *psychic* terms, the seventh or highest level of consciousness, that in which the world was first formed.

♦ **HM0824d Perceptual concentration**
Description This is the context of the mystic experience when the person becomes aware of intra-psychic processes not normally in the scope of awareness.
Related Deautomatization and the mystic experience (#HM4398).

♦ **HM0827e Oversoul awareness** (Psychism)
Description Not bound by narrow, human qualities, the individual operates from the wisdom and love within the oversoul.
Plato and Plotinus, and thence Emerson, refer to this unity in which all souls that are, of everything in the universe which can be referred to as soul, are united in the one, all-inclusive soul. Man and nature in its myriad forms are expressions of this oversoul.
Related Oversoul (#HH0504) Ultra-consciousness (#HM2156).

♦ **HM0829g Sexual experience**
Sexual orgasm — Sexual ecstasy — Sex as flow experience — Enjoying sex
Description Initial experiences of sex are pleasurable enough to keep the experiencer in a state of flow for weeks at a time as new challenges are faced. To maintain the flow experience of sex with the same partner, however, the relationship must become more complex, the partners must invest attention in each other and discover new potentialities both in each other and in themselves. They need to know the thoughts, feelings and dreams in their partner's mind. Sex needs to be taken control of and cultivated in the direction of greater complexity.
Refs Csikszentmihalyi, Mihaly *Flow* (1990).
Broader Flow experience (#HM2344).
Related Androgyny (#HM0212).

♦ **HM0831c Ceto vimutti** (Buddhism)
Mental release — Phala samadhi
Description Samadhi or concentration is developed to the culminating point on the path to Nirvana. It is associated with full knowledge and releases the mind from defilements, yielding the fruits of Arhatship. The highest and finest state of samadhi, or phala-samadhi, this is concentration conducive to the fruit of Arhatship. The term can also be used generally to denote the different stages of samadhi, but always refers to an advanced state.
Broader Concentration (Buddhism, #HM6663).

♦ **HM0832d Envy** (Leela)
Eirsha
Description Not yet ready to remain on the higher plane, which he has reached by chance, the player still has negative vibrations and is envious of those remaining steadily on higher planes. He lacks self-confidence and reverts to first chakra strategy, being brought down to the first level by a snake which, on the positive side, assists in purifying the thought process.
Context The 12th state or square on the board of *Leela*, the game of knowledge, appearing in the second row.
Broader Second chakra: realm of fantasy (Leela, #HM3651).
Followed by Avarice (Leela, #HM0987).

♦ **HM0836g Being contacted** (Buddhism)
Samphusitattam (Pali)
Broader Contact (Buddhism, #HM2708).

♦ **HM0837e Psychic attack** (Psychism)
Description Three forms of attack are defined:
(1) Negative thoughts and feelings are transferred deliberately and telepathically from one person to another, particularly if the two people are very close physically or emotionally. (2) An individual who is vulnerable from being low on the moral scale or recently mentally ill has a "leaking" aura which allows the entry of an etheric world entity to haunt him - and usually can only be removed by exorcism. (3) Negative thoughts such as rage, hate or resentment form an *elemental* which disrupts the individual's life and is intensified, weakened or dissipated as thoughts persist or diminish and with change in attitude.
Narrower Accidental psychic attack (#HM4033).
Related Psychic transfer (Psychism, #HM3571).

♦ **HM0843h Sacramental sex**
Refs Evola, Julius *The Metaphysics of Sex* ; Gallaghar, Charles A, et al *Embodied in Love* (1983).
Related Maithuna (Yoga, #HH2034) Tantra (Buddhism, Yoga, #HH0306).

♦ **HM0847g State of animality** (Buddhism)
Foolishness — Thoughtlessness — Mindlessness
Description This is the state of instinctual behaviour, when desires dominate. Here there is the means to adapt to and take advantage of the environment and ensure the survival of one's self and of the species. Here also, however, is the tendency to threaten the weak and fear the strong, which manifests particularly in hierarchical organizations. Animality also manifests in the law of the jungle, exercising power for selfish and partisan ends, supporting a group to the detriment of the wider community. The clearest characteristic is foolishness, when immersion in the present and acting according to instinct one forgets or ignores consequences for the future. this thoughtlessness or mindlessness leaves one open to manipulation by others.
Context One of the ten worlds described in Nichiren Soshu Buddhism.
Broader Six paths (Buddhism, #HM1914) Ten worlds (Buddhism, #HM2657)
Four evil paths (Buddhism, #HM1252) Three evil paths (Buddhism, #HM0923)
Nichiren shoshu buddhism (Buddhism, #HH3443).
Followed by State of anger (Buddhism, #HM2959).
Preceded by State of hunger (Buddhism, #HM0150).

♦ **HM0850e Crisis apparition** (Psychism)
Description There is clairvoyant prevision of some crisis (death, accident) by a close friend or relative of the person affected. This is said to be due to a desire of the emotional soul-mind to intimate what is to occur.
Related Extrasensory perception (ESP, #HM2262).

♦ **HM0853b Jihi** (Buddhism)
Description The spirit of universal and impartial compassion which gives happiness and removes suffering.

♦ **HM0856d Purgatory** (Leela)
Narka-loka
Description Here the player, who has attained freedom of action and is responsible for the fruits of his actions, bears the consequences of his acts. Having made his way through these levels of the narkas the player is then ready to pass on to heaven. Although painful, arising from bad karma or from violence, this is not a punishment but rather a means of purification. The player at this level recognizes that this is not a failure of the personal ego but sees the imperfections of his activities and the need to improve.
Context The 35th state or square on the board of *Leela*, the game of knowledge, appearing in the fourth row.
Broader Fourth chakra: attaining balance (Leela, #HM3291).
Preceded by Plane of violence (Leela, #HM1276).

♦ **HM0859d Plane of agnih** (Leela)
Plane of fire — Agnih-loka
Description This is the state of fire, of life itself. The body is only a vehicle for this life. In the light of fire, the eternal witness expressing man's innate nature, self deception is impossible.
Context The 42nd state or square on the board of *Leela*, the game of knowledge, appearing in the fifth row.
Broader Fifth chakra: man becomes himself (Leela, #HM0933).

♦ **HM0863d Wrong path** (Buddhism)
Micchapatha (Pali)
Broader Wrong view (Buddhism, #HM1710).

♦ **HM0866e Higher state of consciousness** (Psychism)
Description Personal consciousness is united with universal consciousness.
Broader Higher states of consciousness (#HM0935).
Related Superconscious (#HM2960) Ultra-consciousness (#HM2156).

♦ **HM0867c Active imagination**
Description A purposive turning to the transpersonal unconscious which is not only the basis of the conscious mind but also the subjective or inner aspect of nature. In active imagination the "otherness" and genuine autonomy of an objective impersonal psyche is recognized. The images appearing during this activity must however be allowed to speak for themselves, with the ego's task to be an onlooker whilst simultaneously being involved. Recognition of the nature of the drama leads to experience of a new level of being, a world in which the ego participates not as director but as the actively experiencing one.
Context A concept developed by Jung and distinguished from Freudian free association technique which is based on the premise that all significant imagining represent the fulfilment of certain infantile wishes. It is also distinguished from ordinary conscious imagining (fancy, daydreams, reveries), namely passive fantasy dominated by the ego.
Refs Avens, Roberts *The New Gnosis* (1984); Hull, R F C *Bibliographical Notes on Active Imagination in the Works of C J Jung* (1971).

Related Creative imagination (#HM0786) Guided visualization (Magic, #HH6053)
Imaginative consciousness (Anthroposophy, #HM2901).

◆ **HM0870d Uncertainty** (Buddhism)
Indecision — Anekamasaggaha (Pali)
  Broader Perplexity (Buddhism, #HM0812).

◆ **HM0872d Being of good quality** (Buddhism)
Pagunabhava (Pali)
  Broader Fitness of mental body (Buddhism, #HM1455)
  Fitness of consciousness (Buddhism, #HM1810).

◆ **HM0875c Poetic enthusiasm**
Description To the poet, surrendering to enthusiasm opens the experience of that magical and apparently spontaneous generation of poetic form which comes about when the inspiring spirit descends upon the poet. Such spontaneity is at once the expression of the energizing of imaginative thought, and the evidence of it.
  Related Poetic imagination (#HM1538) Religious enthusiasm (#HM2146).

◆ **HM0876c Awareness of the sacred** (Christianity)
Holy
Description Although the sacred is experienced in the profane world as absence, nonetheless it is necessary for sense to be made of existence. The sacred only shows its true nature at the high points of human life (birth, falling in love, death, for example) and then withdraws. This gives the sacred the character of an event, which is further seen in the holiness of God as power intervening to shape history. In Christian terms, the definitive and universal salvation given in Jesus means that the holy and sacred are no longer set apart from the world.
The failure of Christianity to keep the non-rational element alive in religious experience is said to have given an intellectualistic and rationalistic interpretation of God. Holiness does not arise from ethics. The experience of the holy without ethics is referred to as numinous.
  Refs Otto, Rudolf *The Idea of the Holy* (1950).
  Related Holiness (#HH0183) Numinosum (Jung, #HM3811) Mystical cognition (#HM2272)
  Religious experience (#HM3445).

◆ **HM0877d Shawq** (Islam)
Longing — Desire — Ishtiyaq
Description Man longs for God while his knowledge of God is incomplete. The more he knows God the greater his longing. Since there is no end to what may be known of God, the longing is also unending. Some would say that, even when knowledge is complete, desire can continue. Even in paradise the beatific vision only occurs twice a day – in between must arise longing. There is a level of shawq that can exist even on union and which increases with contemplation, just as, in terms of allowed human attachments, it does not cease on intercourse. The longing when the beloved is absent is different from the longing in the presence of the beloved. This second level of longing is termed ishtiyaq.
  Related Showq (Sufism, #HM0762).

◆ **HM0878d Plane of taste** (Leela)
Rasa-loka
Description At this level, taste is no longer simply a sensory perception but is purified. Not only are no flavourings added to food, so the player is aware of the true taste of what he eats; but aesthetically his good taste is appreciated by everyone and admirers seek to vibrate in the same frequencies.
Context The 34th state or square on the board of *Leela*, the game of knowledge, appearing in the fourth row.
  Broader Fourth chakra: attaining balance (Leela, #HM3291).
  Related Aesthetic emotion (Hinduism, #HM0205).

◆ **HM0881d Grasp** (Buddhism)
Paggaha (Pali)

◆ **HM0882g Empty-field myopia**
Description Under conditions in which there is nothing on which the eye can focus, the eye tends to focus at about 3 metres, with the result that any object in the distance will be out of focus and may not be noticed. The condition is experienced by pilots flying in cloud, by divers in cloudy water, or by skiers in a snow-storm.
  Broader Illusion (#HM2510).

◆ **HM0883d Dejection** (Hinduism)
Visada
Description This is depression of spirit expressed in gloomy appearance. There is languor of cognitive and motor sense-organs.
  Related Grief (Hinduism, #HM9392).

◆ **HM0885e Psychic field consciousness** (FC)
Description The individual's awareness of self is expanded to include and merge with the immediate environment.

◆ **HM0887e Semi-trance** (Psychism)
Description In this relaxed and dreamy state, the body may stand or sit and the individual is aware enough to respond to external stimuli such as answering questions and to know what is happening, but will not remember what has occurred. The state may be achieved by synchronization of a medium with an intelligence to transmit messages.
  Broader Trance (#HM3236).
  Related Channelling (#HH0878) Medium state of hypnosis (#HM5856).

◆ **HM0888c Innocent cognition**
Suchness — Sonomama (Japanese)
Description When something is viewed as the only thing there is, itself in its surroundings, the whole is accepted as such. It is related to that innocence before the fall when, not having eaten of the tree of knowledge of good and evil, there is no intellectualizing into separate categories. By achieving unitive consciousness it is said that this primal innocence may be recaptured.
  Related Muga (Japanese, #HM0793) Suchness (Buddhism, #HM2435)
  Innocence (Christianity, #HM3412).

◆ **HM0889d Sabsung** (Thai)
Slaking emotional or spiritual thirst — Being revitalized
Description One is aware that a hard-to-define but real need is being satisfied. Whether listening to beautiful music, seeing a work of art, having some revitalizing experience when the stresses and exhaustions of the world had previously threatened to overwhelm one, there is a sense of psychic and spiritual revitalization as one's spiritual thirst is slaked.

◆ **HM0891b Degrees of consciousness** (Hinduism)
Conditions of the self
Description In the Mandukya Upanishad the soul is said to have four conditions – the waking state or vaisvanara, the dream state or taijasa, deep sleep or prajna and the overarching state which completes the whole, turiya or superconsciousness, transcending both consciousness and unconsciousness. Other sources quote a fifth state beyond these four, turyatita, when the yogin is pervaded by the experience of non-difference between the individual self and the supreme or universal self – and approaches the experience of liberation.
  Narrower Prajna (#HM3455) Taijasa (#HM3040) Turyatita (#HM3139)
  Vaisvanara (#HM2336) Turiya awareness (#HM2395).

◆ **HM0892e Fantasy**
Description Fantasy is an expansive force in the life of an individual. It reaches and stretches beyond the immediate people environment or event which may otherwise contain him. Sometimes these fantasy extensions can gather such great force and poignancy that they achieve a presence which is more compelling than real-life situations. Fantasy is of prime importance in therapy, since it can ensure a renewal of energy. It often marks a new course in the individual's sense of self.
  Narrower Free-floating fantasy (#HM1011)
  Related Reverie (#HM1576) Waking dreams (#HM1376) Guided fantasy (#HH0627)
  Schizophrenic fantasy (#HH0909) Mental imagery therapy (#HH0072).

◆ **HM0894c Lauds** (Christianity)
Hours of lauds
Description The canonical hours of lauds traditionally take place between 3 and 6 am. They dramatize giving praise to God for having survived the darkest hours of life. The mood is ecstatic praise.
  Broader Canonical hours (Christianity, #HM1167).
  Followed by Prime (Christianity, #HM1904).
  Preceded by Matins (Christianity, #HM0160).

◆ **HM0899f Introvertive mystical experience**
Mysterium
Description This is a state occurring at the deepest phenomenological layer of the person's experience. There are particularly rich sensory level phenomena and may be some variety of cosmological mysticism leading towards a more profound mystical state or mysterium experienced as the source level of reality. Concepts such as "primordial essence" and "ultimate ground of being" suddenly take on a previously hitherto unknown immediacy and clarity; the semantic of theological discourse become visceral realities.
  Refs Masters, R E L and Houston, Jean *The Varieties of Psychedelic Experience* (1967).
  Related Cosmological mysticism (#HM1635)
  Modes of awareness associated with use of hallucinogens (#HM0801).

◆ **HM0900d Weltschmerz**
Description The individual, usually young and well-endowed, is overwhelmed with a languid sorrow, a death-loving romanticism which may result in suicide.
  Related Angst (#HM0362).

◆ **HM0901e Unease experienced as born of contact with the psychical** (Buddhism)
Cetosamphassajamasatam vedayitam (Pali)
  Broader Feeling (Buddhism, #HM2270).

◆ **HM0903b Vairagya** (Yoga)
Viraga — Dispassion
Description The beginning of dispassion is the decision to take on a disciplined course of action such as demanded of the traveller on the yogic path. Practice of discipline sharpens dispassion, and the two are interdependent. Finally, the adept reaches a state of thirstlessness, *vitrisna*, when the mundane life is abandoned and there is absence of all desire for the pleasures and values of the world.
  Related Vitrisna (Yoga, #HM1029).

◆ **HM0904d Horror**
Description State typical of an emotional response to viewing or hearing of acute danger or suffering in others. This might be a motor accident between others, whether narrowly averted or the sight of injuries when such an accident actually happens; witnessing cruelty to a child; hearing a sudden cry of pain.
  Refs Atreya, B L *Deification of Man*.
  Related Shock (#HM1962) Fear (Christianity, #HM3311).

◆ **HM0906e Mental inertia** (Yoga)
Description This is due to a lethargic, heavy mental rhythm which has to be replaced by a developing intellectual interest, and control by the mind as opposed to the feelings, before a mental body able to think clearly about the means of realization is achieved.
Context One of the nine obstacles to soul cognition enumerated by Patanjali.
  Broader Obstacles to soul cognition (Yoga, #HM3182).
  Followed by Wrong questioning (Yoga, #HM1080).

◆ **HM0910c Formlessness** (Zen)
Wu-hsing
Description One remains in form yet is detached from it.
Context One of the three principles of Zen of Hui-neng, together with wu-nien (no-thought) and wu-chu (non-abiding).
  Broader Human development (Zen, #HH1003).
  Related No-mind (Zen, #HM2163) No-thought (Zen, #HM2500)
  Non-abiding (Zen, #HM3485).

◆ **HM0915d Communion with nature**
Description The contemplation of nature, its beauty and grandeur, is a continual source of inspiration and raises the individual to a state of transcendent joy.
  Related Deep ecology (#HH2315) Contemplation of nature (#HH0580)
  Planetary consciousness (#HM2006) Attentiveness to nature (Zen, #HH0463).

◆ **HM0917c Seeing into self-nature** (Zen)
Chen-hsing
Description Although the Northern School of Dhyana, Shen-hsui's school, pays more attention to the body aspect of self-nature, with the aim of seeing body and mind as free from impurity and mediation as quiet contemplation, keeping watch over purity, the Southern School of Prajna,

Hui–neng's school, looks on this as binding one's self with a self–created rope. In the latter schools, seeing into one's own nature makes it clear that body is only use, that there is no body in self–nature. Use is the body seeing itself in itself.
Refs Suzuki, D T *The Zen Doctrine of No–Mind* (1970).
Broader Human development (Zen, #HH1003).

♦ **HM0918d Delusion** (Buddhism)
Moha (Pali)
Broader Delusion (Buddhism, #HM0184).

♦ **HM0921g Dawning consciousness** (Psychism)
Description The state in which an inspired thought appears in the conscious mind.

♦ **HM0922c Perfect attainment** (Taoism)
Description This is breaking through space, merging with cosmic space.
Preceded by Transcendent liberation (Taoism, #HM1833).

♦ **HM0923d Three evil paths** (Buddhism)
Description Three of the ten inner states of being or worlds described in Nichiren Soshu Buddhism, these are conditions of unhappiness.
Broader Six paths (Buddhism, #HM1914) Ten worlds (Buddhism, #HM2657)
Nichiren shoshu buddhism (Buddhism, #HH3443).
Narrower State of hell (#HM4282) State of hunger (#HM0150)
State of animality (#HM0847).
Related Four evil paths (Buddhism, #HM1252).

♦ **HM0927c Vinivaranata** (Buddhism)
Description The absence of hindrances of lust, hate and delusion.
Context One of ten superhuman qualities described in the Sutta Vibhanga as being special attainments of insight above that of ordinary men.
Broader Superhuman qualities (Buddhism, #HH9764).

♦ **HM0929e Hypertrance** (Psychism)
Description Describes the trance state when psychically very alert although physically, mentally and emotionally dormant.
Broader Trance (#HM3236).

♦ **HM0930d 'Ilm al–yaqin** (Sufism)
Certain knowledge of God — Science of certainty
Description Looking from this state, all from the highest heaven to the lowest level of earth is seen without a veil.
Context According to Shaykh Abu Sa'id ibn Abi'l–Khayr, this is the 23rd of 40 stations or maqamat the Sufi must possess for his journey on the path of Sufism to be acceptable.
Broader Stations of consciousness - ibn–Abi'l–Khayr (Sufism, #HM2424).
Followed by Haqq al–yaqin (Sufism, #HM4556).
Preceded by Baqa (Sufism, #HM0330).

♦ **HM0931c Right rapture of concentration** (Buddhism)
Samyak–samadhi — Right unification — Right meditative stabilization — Right concentration
Description Supreme mindfulness guards the mind. Unification or collectedness of the mind associated with right outlook destroys wrong concentration. There is progressive introversion and dismantling of consciousness, even of enstatic states. Stripped of lusts and wrong dispositions, the individual develops and abides in the zest and satisfaction of the first ecstasy. With the laying to rest of observation and reflection there is an inward serenity, a focusing of heart in the zest and satisfaction of the second ecstasy, born of concentration. From here he passes to the third and fourth ecstasies. The characteristic is not wavering, non–distraction; the function is concentrating, placing the mind well on its object; manifestation is in abandoning wrong concentration.
Context One of the characteristics of the Eightfold Way of Buddhism attained with the path of bhavana (meditation) and which, together with right endeavour and right mindfulness, constitutes mental discipline (samadhi).
Broader Eightfold way (Buddhism, #HM2339) Concentration (Buddhism, #HM6663).
Related Four noble truths (Buddhism, #HH0523) Correct concentration (Buddhism, #HM1735).
Preceded by Right mindfulness (Buddhism, #HM0704).

♦ **HM0932c Phala sacchikiriya** (Buddhism)
Description Realization of the fruit of the four stages of the path of the Arhat.
Context One of ten superhuman qualities described in the Sutta Vibhanga as being special attainments of insight above that of ordinary men.
Broader Superhuman qualities (Buddhism, #HH9764).

♦ **HM0933c Fifth chakra: man becomes himself** (Leela)
Human plane
Description The player has realized compassion, and desires to pass the fruits of his experience to others. Representing higher, not animal, nature, he is born as man and understands fire (agnih). Knowledge is demonstrated in knowing the scriptures without being formally instructed. Manifesting the virtues, he suffers neither sorrow nor disease. Although understanding and merciful to all, there may be the problem of authoritarianism and obsession with logic. The presence of one who has mastered the chakra brings knowledge of the Self and an understanding of nature and the presence of the divine in all that exists. Meditation leads to control of hunger and thirst and achievement of steadiness. The player is able to rejuvenate himself.
Context The fifth row of the board of Leela, the game of knowledge, this is represented by the human and agnih planes, and also groups the squares representing the states of knowledge (jnana or gyana), prana–loka, apan–loka, vyan–loka (the planes of the three life breaths), birth of man, ignorance and right knowledge. Located in the throat, this chakra is related to akash (ether) and to a smoky purple colour. It is compared to the octave comprising the ages 28 to 35 of temporal life, that of parent and teacher. Gyana and right knowledge are the arrows at this stage, and ignorance the snake.
Broader Lila (#HM2278).
Narrower Knowledge (#HM3846) Ignorance (#HM0310)
Prana–loka (#HM4399) Apana–loka (#HM0145)
Vyana–loka (#HM6002) Human plane (#HM5155)
Birth of man (#HM0978) Plane of agnih (#HM0859)
Right knowledge (#HM0567).
Related Vishuddha (Yoga, #HM3461).
Followed by Sixth chakra: time of penance (Leela, #HM4412).
Preceded by Fourth chakra: attaining balance (Leela, #HM3291).

♦ **HM0934c En–sof** (Judaism)
Awareness of infinity
Description This state, described in Judaism, has been compared with transcendental consciousness.
Broader Human development (Judaism, #HH3029).
Related Transcendental consciousness (#HM2020).

♦ **HM0935b Higher states of consciousness**
Superior consciousness — Wisdom of the heart
Description If the normal human condition, customary experience, is in two realms, sleeping and waking, then higher states correspond with reality which is finer and more subtle. The following of prescribed methods of a religious discipline, for example, may bring states characterized by enhanced faculties of attention, thought, feeling and sensation. Perception, awareness and experience all conform more fully and adequately to various levels of reality and truth as they exist in the universe.
These states cannot be considered the same as changes in mood or as other phenomena brought about through normal thought or feeling, nor can they be produced through refining of rational thought or the intense emotions, all these things serve only the egoistic aspect of human nature unless accompanied by spiritual development. Superior states reflect the arising of exceptional attention and awareness and generate new powers of the self. There are new feelings, sensitivities and cognitions, the development of "wisdom of the heart" with unmediated contacts with reality allowing comprehension and experience of life's meaning, value and purpose.
The journey from lower to higher states of consciousness is depicted in many philosophical and religious traditions. Plato refers to shackled prisoners in a cave, with limited vision of shadows on the wall and hearing only echoes, being freed to turn, to ascend, to hear and see true reality. The journey is arduous and confusing. Hindu, Buddhist, Sufi and Christian teachers all indicate stages on this journey, and in modern times psychologists have also looked at these levels.
William James refers to mystical states that are inaccessible to the rational mind but nonetheless impart exceptional meaning and understanding. Rational consciousness is only one part of what can be experienced, it is surrounded by entirely different potential forms. These different forms have four salient qualities: (1) There is a noetic or cognitive aspect, wisdom, a power of heightened intellectual discernment and relational understanding. Positioning, valuation and function are apprehended and apparently disparate facts properly ranked and organized. (2) The ineffability of the experience means that it cannot adequately be described in words. (3) The experience tends to be transient, and marked by an awareness of being in the present moment. (4) There is feeling of passivity, one's own will seems suspended and one is open to a superior or higher force. On is not one's self, another power, person or force seems to be operating through one.
Refs James, William *The Varieties of Religious Experience* (1961).
Narrower Higher state of consciousness (Psychism, #HM0866).
Related Mystical cognition (#HM2272) Altered states of consciousness (#HM2391).

♦ **HM0936d Spiritual anger** (Christianity)
Description Many beginners on the path of the spirit have a worldly desire to spirituality and when they have religious experiences they are often tinged with the sin of spiritual anger. When they come to the end of a spiritual experience they naturally become upset and take their displeasure out on themselves and others. They become upset over the smallest of things. This in itself is not the problems but when this leads to wallowing in disappointment it becomes a hindrance to growth in the spirit. Another form of this is people who become irritated at the faults of others. They begin to watch out for these errors and even continually point them out. They become arbiters of goodness. A third form of this is becoming annoyed when they become aware of their own imperfections; which is quite the opposite of spiritual meekness and humility. They think spiritual perfection is to be reached overnight. They aim to do a great deal and time and again make resolutions and promises. Because they are over confident and not humble they fail in their intentions and become even more angry. They do not have the patience to wait for the spirit to come as and when it will.
Context This is an imperfection of beginners in the Dark Night by St John of the Cross.
Broader Beginner in spiritual life (Christianity, #HM0102).
Related Peevishness (#HM2472) State of anger (Buddhism, #HM2959)
Spiritual envy (Christianity, #HM3818) Spiritual lust (Christianity, #HM4180)
Spiritual sloth (Christianity, #HM0491) Spiritual pride (Christianity, #HM4533)
Spiritual avarice (Christianity, #HM0642) Spiritual gluttony (Christianity, #HM0507)
Spiritual meekness (Christianity, #HM6226).

♦ **HM0945d Churlishness** (Buddhism)
Candikka (Pali)
Broader Hatred (Buddhism, #HM4502).

♦ **HM0946c Deification** (Sufism)
Identification with God
Description Rather than knowing the spiritual master as God personified, this is direct awareness of the creator. This may arise as a continuation of the journey started in personification, or it may arise directly through concentration on union with God as he reveals himself in the Qur'an. The process is: *tawaja* (intense concentration); assimilation of attributes of quality, will and desire of the object of desire; stabilization of this state so that infusion of wills occurs. It has been described as: *hulyl*, infusion of God's attributes; *fana*, passing away in God's existence; *itihad*, union.
Context The second category of experience related to the three types of object of desire.
Related Identification with God (Sufism, #HM3520).
Followed by Unification with life essence (Sufism, #HM3340).
Preceded by Identification with spiritual master (Sufism, #HM3944).

♦ **HM0948d Plane of karma** (Leela)
Karma–loka
Description Here the dominant concern is with power and with identifying the ego and extending its influence. Fantasy is replaced with the practical, with interaction, with the cycle of birth and rebirth. Manifest body and unmanifest being, both are integral parts of the universal whole.
Context The 19th state or square on the board of *Leela*, the game of knowledge, appearing in the third rcw.
Broader Third chakra: theatre of karma (Leela, #HM0717).

♦ **HM0953d Maya** (Leela)
Description Awareness of unity is lost as the player becomes fascinated with the play, setting up rules and committing himself to play it through to the end. The one reality becomes the many of illusion, of ignorance, of lack of knowledge. The phenomenal world evolves under the play of the three gunas - sattva, rajas and tamas - and cosmic consciousness becomes individual consciousness.
Context The second state or square on the board of *Leela*, the game of knowledge, appearing in the first row. The two represents delusion and duality.
Broader First chakra: fundamentals of being (Leela, #HM4103).
Preceded by Tamas (Leela, #HM6238).

♦ **HM0956d Understanding as interpreting the internal** (Buddhism)
Description This understanding is initiated when one has grasped the complex aggregate of elements of which one is made up.
Context On the Path of Purification of Hinayana Buddhism, panna (understanding) is considered as of one kind (monad), of two kinds (dyads), of three kinds (triads) or of four kinds (tetrads). There are five dyads, four triads and two tetrads. All have the characteristic of penetrating the individual essences or true nature of states (monad). In the fourth triad, this is compared with understanding as interpreting the external or as interpreting the internal and the external.
Broader Understanding (Buddhism, #HM4523).
Related Understanding as interpreting the external (Buddhism, #HM1128)
Understanding as interpreting the internal and the external (Buddhism, #HM4490).

♦ **HM0966c Odious sentiment** (Hinduism)
Aesthetic emotion of disgust — Repulsive sentiment — Rasa of revulsion
Context One of eight kinds of aesthetic sentiment or rasas in Indian psychology.
Broader Aesthetic emotion (Hinduism, #HM0205).

♦ **HM0967g Reverie** (Psychism)
Description In this state, with its high theta brain-wave content, unconscious mental processes are revealed as images, symbols and patterns of thought. A level used by psychics for producing psychic information.
Related Reverie (#HM1576) Theta wave consciousness (#HM2321)
Hypnagogic consciousness (Psychism, #HM3511).

♦ **HM0972d Neither psychical ease nor unease born of contact with element of eye-consciousness** (Buddhism)
Cakkhuvinnanadhatusamphassajam cetasikam neva satam nasatam (Pali)
Broader Feeling (Buddhism, #HM2270).
Related Eye consciousness (Buddhism, #HM2074).

♦ **HM0973c Affectionate bonding**
Mature love — Companionate love
Description Characterized by the development of deep, mutual caring which is understood to stand the test of time and difficult circumstances. Each has a realistic appraisal of the other, accepting both strengths and weaknesses and thus avoiding the delusions of the mutual idealization of passionate lovers. A relationship based on passion is recognized as a mere prelude, in healthy and optimal situations, to this mature form of love. A certain measure of emotional maturity is required to transform the passionate phase into a steady-state mature relationship. Affectionate bonding is thus based on mutuality, warmth, and mutual affirmation, but especially on trust and loyalty. The lovers recognize the optimal distance between them, allowing for union without subverting autonomy through domination or submission.
Refs Johnson, Robert A *The Psychology of Romantic Love* (1984); Person, Ethel Spector *Love and Fateful Encounters* (1990).
Broader Love consciousness (#HM2000).
Related Human bonding (#HH3498).
Preceded by Romantic love (#HM5087).

♦ **HM0974d Harmony**
Description There is an awareness of harmony in life when conscious and subconscious are at ease with decisions made and one's lifestyle is congruent with one's beliefs.
Refs Cleary, Thomas (Trans) *The Book of Balance and Harmony* (1989); Delza, Sophia *Body and Mind in Harmony* (1961); Panikkar, Raimundo *The Invisible Harmony* (1987).

♦ **HM0975d Excitement of thought** (Buddhism)
Cittassa uddhacca (Pali)
Broader Excitement (Buddhism, #HM1469).

♦ **HM0976c Christ consciousness as loving mind** (Christianity)
Description The mind which is empty is also loving. Love responds in the most appropriate way in relation to each person or situation met with. It is empty and open to exchanges, forms and structures which it can enter, shed or challenge as the situation requires. Unbound by place, status or custom, the loving and empty person is open to all and can respect and challenge all. There is a rhythm in this love, emptiness and fullness, dying and rising.
Broader Christ consciousness (Christianity, #HM2013).
Related Christ consciousness as empty mind (Christianity, #HM0778)
Christ consciousness as dying-rising (Christianity, #HM1402)
Christ consciousness as forgiving mind (Christianity, #HM1728).

♦ **HM0978d Birth of man** (Leela)
Manushya-janma
Description Living in harmony with the laws of truth the player ceases to be a mere creature in human form and takes birth truly as man, the son of God. He needs to please no-one, devotees are no longer of interest. Direct experience of truth makes truth the only aim.
Context The 43rd state or square on the board of *Leela*, the game of knowledge, appearing in the fifth row.
Broader Fifth chakra: man becomes himself (Leela, #HM0933).

♦ **HM0979g Working ability** (Buddhism)
Kammannattam (Pali)
Broader Workability of body (Buddhism, #HM0739)
Workability of thought (Buddhism, #HM1584).

♦ **HM0987d Avarice** (Leela)
Matsar — Matsarya
Description Deluded that he is a separate reality and certain he is better than others, the player feels he deserves what others have more than they do. Any means of achieving his desires seems justified. He dislikes the other players, is spiteful, wants their possessions. Greed combined with envy increases and he is plagued with problems.
Context The eighth state or square on the board of *Leela*, the game of knowledge, appearing in the first row.
Broader First chakra: fundamentals of being (Leela, #HM4103).
Related Avarice (Buddhism, Pali, #HM0227).
Preceded by Envy (Leela, #HM0832).

♦ **HM0988b Orders of perception**
Description All one knows of the external or "real" world is from energy which comes from this world and strikes sensory perceptors of the perceptual system. Everything else that one claims to be the real world is our own perception of that world, and this one constantly tries to change so as to coincide with the world in one's head. Right from the first cry after birth one only cares for what is going on in the real world as long as one cannot change it to be like the world in one's head. It is really only the world in the head that counts. The world around is experienced through an orderly hierarchy of perceptions from low or simple to high or complex. These orders of perception are what prevents one from leading an otherwise random, disorganized life as one struggles to gain order from a haphazard world that makes little sense. Although everything is initially perceived through the sensory perceptors it may go through as many as ten orders of perception in order to make sense from it. The input or awareness of the world is not always conscious, if consciousness is defined as perception that one senses as going on right now. There are three kinds of perception - controlled, uncontrolled and new-information. Normally one is consciously aware only of uncontrolled perceptions or perceptions associated with a controlled system where there is a substantial error.
Refs Glasser, William *Stations of the Mind* (1981).
Narrower First order perceptions - intensity (#HM0543)
Second order perceptions - sensation (#HM1932)
Third order perceptions - configurations (#HM0312)
Fifth order perceptions - control of sequence (#HM0772)
Seventh order perceptions - programme control (#HM1001)
Eighth order perceptions - control of principles (#HM1325)
Fourth order perceptions - control of transitions (#HM1516)
Sixth order perceptions - control of relationships (#HM0103)
Ninth order perceptions - control of systems concepts (#HM4000)
Tenth order perceptions - universal oneness - meditation (#HM1717).

♦ **HM0992g Moha** (Hinduism)
Slight unconsciousness — Intoxication
Description One is less unconscious here than in a swoon.
Related Swoon (Hinduism, #HM1159) Samnyasa (Hinduism, #HM1356).

♦ **HM0996d Restful alertness**
Description With body and emotions passive, the mind is alert although not conscious to external stimuli.
Context Said to be a state achieved in transcendental meditation.
Related Alert passivity (#HM3959) Pure alertness (Psychism, #HM1461)
Transcendental meditation (TM, #HH0682).

♦ **HM1001c Seventh order perceptions - programme control**
Description Here a complex programme can be controlled, using perceptions of all the previous levels, so that, for example, a journey can be planned and carried out within the plan and with response to the unexpected. Preparations for the journey may include filling the petrol tank of the car, planning a route; adaptations may include altering the route during the journey to suit unexpected road conditions. Before actually carrying out a programme, simulated programmes may be made in the head, compared, rejected, accepted, modified. There are not only logic and deduction but also superstition, perceptions of one's own making (not walking under a ladder, for example).
Context The seventh of ten orders in the perceptual system described by William Glasser.
Broader Orders of perception (#HM0988).
Followed by Eighth order perceptions - control of principles (#HM1325).
Preceded by Sixth order perceptions - control of relationships (#HM0103).

♦ **HM1006b Samsara** (Hinduism)
Description This is the endless cycle of births and deaths characterized by suffering, *duhkha*, to which the soul is subject through *karma* until it attains liberation, *moksha* or *mukti* in awareness of the identity of the individual self, *atman*, with *brahman*.
Related Karma (#HH0567) Mukti (#HH0613) Brahman (#HH1226)
Atman (Hinduism, #HH0103) Moksha (Hinduism, #HM2196)
Tantra (Buddhism, Yoga, #HH0306) Duhkha (Buddhism, Hinduism, #HM2574)
Human development (Hinduism, #HH0330)
Consciousness states in cyclic existence (Buddhism, #HM2177).

♦ **HM1007d Al-Muhyi** (Sufism)
Description The believer is aware of Allah as the giver of life. He has created life and death. He gave us our lives as a gift, lent to us, and we choose what we shall do with that gift, whether to believe or not, whether to obey or not. After death we shall be brought back to life whether to qualify for heaven or for hell. Here and now we should work with thanks to serve His creatures as though we were never to die, yet be continually aware of our own mortality and work towards salvation.
Context A mode of mystical awareness in the Sufi tradition associated with the sixtith of the ninety-nine names of Allah.
Broader Ninety-nine names of Allah (Sufism, #HH2561).
Followed by Al-Mumit (Sufism, #HM3868).
Preceded by Al-Mu'id (Sufism, #HM4093).

♦ **HM1008d Untwistedness** (Buddhism)
Akutilata (Pali)
Broader Straightness of body (Buddhism, #HM1424)
Straightness of thought (Buddhism, #HM0253).

♦ **HM1010d Concentration of easy progress and sluggish direct-knowledge** (Buddhism)
Concentration of easy progress and sluggish intuition
Description Cultivating what is suitable, carrying out preparatory tasks, not overwhelmed by craving, having practice in serenity (absorption concentration), having a mild lower nature, all these mean progress is easy. Cultivating what is unsuitable, lacking skill in absorption, overwhelmed by ignorance, lacking practice in insight, having dull faculties, all these mean intuition is sluggish.
Context According to Hinayana Buddhism, concentration is considered as of one kind (monad), of two kinds (dyads), of three kinds (triads), of four kinds (tetrads) or of five kinds (pentad). In the first tetrad, this is the third concentration.
Broader Concentration (Buddhism, #HM6663).
Related Concentration of easy progress and swift direct-knowledge (Buddhism, #HM4977)
Concentration of difficult progress and swift direct-knowledge (Buddhism, #HM7859)
Concentration of difficult progress and sluggish direct-knowledge (Buddhism, #HM5992).

♦ **HM1011d Free-floating fantasy**
Description With no other purpose, this is pure play which is complete in itself. Although it cannot be harnessed deliberately, it is to be encouraged as an end in itself, as it makes the person more human and more able to do human things.
Refs Boulding, Elise *Building a Global Civic Culture* (1988).
Broader Fantasy (#HM0892).
Related Lucid dreaming (#HM3618) Escapist daydreaming (#HM0248).

♦ **HM1013d Snares of ignorance** (Buddhism)
Avijjalanga (Pali)
Broader Delusion (Buddhism, #HM0184).

MODES OF AWARENESS   HM1081

♦ **HM1015d Restlessness** (Yoga)
Vyagra
Description This is the state characteristic of most spiritual devotees. The mind is sometimes calm, sometimes disturbed. In a state of temporary calm the mind may understand when it hears of the real nature of subtle principles and may contemplate them for extended periods. But, because the mind is restless, concentration does not last long and then distraction arises again, so liberation or salvation is not achieved. Only the mind free from distractions and one-pointed may achieve this.
Context One of five states of mind classification identified in yoga.
  Followed by One-pointedness (Yoga, #HM5734).
  Preceded by Stupefaction (Yoga, #HM1346).

♦ **HM1020d Initiation of refusal** (Esotericism)
Description The liberating experience associated with this initiation is that of freedom from all possible forms of enticement, particularly with reference to subtler realms of experience. The revelation at this initiation concerns the nature of being and existence and the divine purpose of the solar system. It involves a refusal to become involved in initiatives at an inappropriate level (despite the calls to do so) in the light of a recognition of the subtler levels at which the dramatic difficulties of the lower levels need to be resolved.
Context The ninth initiation in the tradition of western esotericism.
  Broader Spiritual initiation (Esotericism, #HH3390).
  Preceded by Initiation of transition (Esotericism, #HM1258).

♦ **HM1023d Not ashamed of ought to be ashamed** (Buddhism)
Na kiriyati hiriyitabbena (Pali)
  Narrower Not ashamed of acquisition of sinful and unwholesome dharmas (#HM3257).

♦ **HM1024d Physical plane** (Leela)
Bhu-loka
Description The player is trapped by the lower self and is mainly concerned with physical survival and material achievement. Recreation is physical (competitive sport) and may include violence. This plane is related to earth, to the mother, and represents the necessary ground from which to realize other planes.
Context The fifth state or square on the board of *Leela*, the game of knowledge, appearing in the first row. Five is the number of subtle elements, of sense organs, of organs of work. This is the first of eight lokas represented on the spine of the board and marking a stage in the development of the individual's consciousness. The first seven represent a psychic centre or chakra on the spine of the human body, a level of psychic evolution. An eighth loka lies beyond the chakras. Each loka is defined by the nature of the matter of which it is composed. The eight other states in the same row may be considered as special regions within the loka.
  Broader First chakra: fundamentals of being (Leela, #HM4103).

♦ **HM1026c Kedesh-jazba** (Sufism)
Divine attraction
Description Following purification of the achieved self, *kushesh*, progress on the inner self takes place through divine attraction and the individual passes from one state or maqam to another. These maqamat vary in number according to paths prescribed by different teachers. A. Reza Arestah speaks of 9 psychic states.
  Related Psychic states (Sufism, #HM4311).
  Preceded by Kushesh (Sufism, #HM3335).

♦ **HM1027f Theomania**
Furor divini
Description A state of frenzy and exaltation arising under an excess of devotional or religious fervour, when the person may seem "possessed" by a deity, theomania may arise from any activity where there is preoccupation with religion or which causes xenophrenic states – asceticism, chanting, prayer, drugs – and may manifest in unusual behaviour and phenomena, from convulsions to clairvoyance and glossolalia, from healing powers to immunity to pain. It is not always easy to make a distinction between divine frenzy (furor divini), where a divine presence is manifest, and ecstasy or possession by spirits or demons. In the extreme the person may believe himself to be God.
  Broader Frenzy (#HM3594).
  Related Mania (#HM2787)   Religious enthusiasm (#HM2146).

♦ **HM1028g Non-rigidity** (Buddhism)
Akakkhalata (Pali)
  Broader Flexibility of body (Buddhism, #HM3370)
  Flexibility of thought (Buddhism, #HM1805).

♦ **HM1029c Vitrisna** (Yoga)
Thirstlessness
Description By continuous practice and the cultivation of dispassion, *vairagya*, the traveller on the yogic path reaches a state in which there is no thirst for worldly pleasures, objects or values.
  Related Vairagya (Yoga, #HM0903).

♦ **HM1031e The unpleasant experienced as born of contact with the psychical** (Buddhism)
Cetosamphassajam dukkham vedayitam (Pali)
  Broader Feeling (Buddhism, #HM2270).

♦ **HM1044e Internalization awareness**
Description *Biofeedback* apparatus is utilized to determine the manner in which external stimuli are internalized.
  Broader Biofeedback training (#HH0765).

♦ **HM1045c 'Arif** (Sufism)
The knower — Gnostic
Description The name of a traveller on the *tariqa* or Sufi spiritual path at the level of a gnostic to whom spiritual knowledge is revealed because of the illumination of his soul. Spiritual practice has lead to inner transformation, he is pure and detached from worldly desires. His soul is joyful in the revelation of God in His divine glory, known with such clarity that it cannot be doubted. At this stage the individual is no longer in himself. He himself has no knowledge, hearing or sight. His existence is in God, his actions held by God, God is his ear through which he hears, his eyes through which he sees. He is conscious of no form of otherness. He is taken to the mystic stage of *fana*.
Context Spiritual progress on the spiritual path is marked by the attainment of specific *maqamat* or stations.
  Broader Mystic stations (Sufism, #HM3415).

  Related Ma'rifa (Sufism, #HM2254)
  Experience of mystical knowledge (Sufism, #HM4349).
  Followed by Fana (Sufism, #HM3799).

♦ **HM1047c Khawf** (Sufism)
Fear of God
Description Whether fear of God's possible vindictiveness, fear at the thought of separation from God, or (the most perfect) fear of the knowledge of God's majesty, this state arises because of the proximity of God. Seeing His justice, there is no hope in being obedient and the seeker melts in fear.
Context According to Shaykh Abu Sa'id ibn Abi'l-Khayr, this is the 19th of 40 stations or maqamat the Sufi must possess for his journey on the path of Sufism to be acceptable.
  Broader Nearness-awareness (Sufism, #HM2377)
  Stations of consciousness – ibn-Abi'l-Khayr (Sufism, #HM2424).
  Related Fear of God (Christianity, #HM0673).
  Followed by Raja (Sufism, #HM4393).
  Preceded by Sidq (Sufism, #HM6631).

♦ **HM1052d Celestial plane** (Leela)
Swarga-loka
Description Beyond material and sensual realms, the player begins to comprehend the nature of his own identity and look to immortality of the ego in a heaven of infinite pleasure beyond pain and suffering, the goal of all religion. He issues challenges and asserts himself. This is the domain of Indra, where organs of sense and activity are mastered, the home of saints and devotees. Things still exist on the physical plane but in harmony, without lower desires, attachments or greed.
Context The 23rd state or square on the board of *Leela*, the game of knowledge, appearing in the third row. This is the third of eight lokas represented on the spine of the board and marking a stage in the development of the individual's consciousness. The first seven represent a psychic centre or chakra on the spine of the human body, a level of psychic evolution. An eighth loka lies beyond the chakras. Each loka is defined by the nature of the matter of which it is composed. The eight other states in the same row may be considered as special regions within the loka.
  Broader Third chakra: theatre of karma (Leela, #HM0717).
  Preceded by Purification (Leela, #HM1773).

♦ **HM1054d Undeflectedness** (Buddhism)
Ajimhata (Pali)
  Broader Straightness of body (Buddhism, #HM1424)
  Straightness of thought (Buddhism, #HM0253).

♦ **HM1061d Merriment** (Buddhism)
Pahasa (Pali)
  Broader Fivefold happiness (Buddhism, #HM0747).

♦ **HM1066d Constancy of thought** (Buddhism)
Cittassa santhiti (Pali)
  Broader Sukha (Buddhism, #HM2866).

♦ **HM1067c Paramartha** (Buddhism)
Supramundane awareness
Description The Madhyamaka Buddhist equivalent of nagual.
  Related Samvriti (Buddhism, #HM0652).
  Nagual state of awareness (Amerindian, #HM2036).

♦ **HM1070d Racial consciousness** (Yoga)
Context A step on the path to cosmic consciousness, when the individual has not yet achieved conscious self-relatedness and lacks the courage to go against his racial group.
  Related Conscious self-relatedness (Yoga, #HM0467).
  Followed by National consciousness (Yoga, #HM1370).
  Preceded by Tribal consciousness (Yoga, #HM0411).

♦ **HM1073d Limited concentration** (Buddhism)
Description This is a state of unification on the plane of access.
Context According to Hinayana Buddhism, concentration is considered as of one kind (monad), of two kinds (dyads), of three kinds (triads), of four kinds (tetrads) or of five kinds (pentad). In the fourth triad, this is the first concentration.
  Broader Concentration (Buddhism, #HM6663).
  Related Access concentration (Buddhism, #HM4999).

♦ **HM1077e Erroneous perception** (Yoga)
Description The inability to see things as they really are, this is counteracted by study of all the means of perception – physical vision, etheric vision, clairvoyance, symbolic vision (this being the cause of wrong perception, leading to illusion and error); and pure vision (or pure knowledge), spiritual vision (true perception), cosmic sight (of nature inconceivable to man), superseding yet including the first four means and leading to true perception.
Context One of the nine obstacles to soul cognition enumerated by Patanjali.
  Broader Obstacles to soul cognition (Yoga, #HM3182).
  Followed by Inability to achieve concentration (Yoga, #HM0298).
  Preceded by Lack of dispassion (Yoga, #HM0110).

♦ **HM1078e Vampire** (Psychism)
Description In this state an entity, whether materializing from the dead or changing form to take astral flights, appears in animal shape to suck the blood of humans, sometimes leaving the victim in a hypnotized state.

♦ **HM1080e Wrong questioning** (Yoga)
Doubt
Description Such questioning is based on identification with the illusory mental body and leads to doubting the existence of eternal truths and searching for solutions to problems by means of the transitory and ephemeral. It is contrasted with intelligent enquiry, freed from all externally imposed dogma, which may lead to perception of the truth.
Context One of the nine obstacles to soul cognition enumerated by Patanjali.
  Broader Obstacles to soul cognition (Yoga, #HM3182).
  Followed by Carelessness (Yoga, #HM0361).
  Preceded by Mental inertia (Yoga, #HM0906).

♦ **HM1081d Strong grip of the burden** (Buddhism)
Dhurasampaggaha (Pali)
  Broader Faculty of energy (Buddhism, #HM1470).

♦ **HM1084d Pre–conscious thinking**
Description Normally present in small children but recurring in adulthood (in dreams, fatigue or as symptoms of hysteric, neurotic or psychotic individuals), pre–conscious thinking is a preverbal, prelogical mode of thought, with primitive, archaic, magical features and ruled by the emotions. Thus it is not in accordance with observed reality, it is tinged with wishes and fears, and the pictorial view of objects is related to objects of similar actual or symbolic appearance. With the advent of verbal ability, thinking becomes logical and organized and the conscious and unconscious are differentiated.
Broader Thought processes (#HH0530).
Related Magical thinking (#HM0359) Archaic–paralogical thinking (#HH0748).

♦ **HM1087c Fana** (Islam)
Passing away
Description As opposed to Sufism, which equates annihilation in God as the goal of the mystic journey, Islam in general frowns on the idea that man the created can become one with God the creator. Ibn al–Qayyim, for example, defined fana as the following three states: (1) Passing away from existence of that which is other than God. This is the mistaken ideal of the monists who attempt to free themselves from (as they see it) the illusion of God and man as duality. (2) Passing away from contemplation of that which is other than God, again an illusion, where the distinction between God and man and between obedience and disobedience is acknowledged but no longer perceived. (3) Passing away from worship or love of that which is other than God. This is legitimate and occurs simultaneously with perceiving the other, serving to increase the believer's love for God.
Related Fana (Sufism, #HM1270) Fana (Sufism, #HM3799).

♦ **HM1088d Knowledge of the path of stream entry** (Buddhism)
Description This follows uninterruptedly from knowledge of change–of–lineage. The masses of greed, hate and delusion are pierced and exploded for the first time. The suffering of the round of rebirths is dried up. All doors to the states of loss are closed. The seven noble treasures – faith, virtue, conscience, shame, learning, generosity and understanding – are actually experienced. The eightfold wrong path is abandoned. Enmity and fear are allayed. It leads to the state of the breast–born son of the perfect Buddha. This is the first noble person. Supramundane path moment stream entry consciousness is followed by two or three fruition moment stream entry consciousnesses. This is the second noble person.
Related Purification by knowledge and vision (Buddhism, #HH3025).
Followed by Knowledge of the path of once–return (Buddhism, #HM7563).
Preceded by Knowledge of change of lineage (Buddhism, #HM4637).

♦ **HM1089e Blissing out**
Description An ecstatic state induced through brainwashing techniques used by some cults; the feeling of joy is followed by one of confusion.
Related Destructive manipulation (#HH1020).

♦ **HM1090e Sensation of unease born of contact with the psychical** (Buddhism)
Cetosamphassaja asata vedana (Pali)
Broader Feeling (Buddhism, #HM2270).

♦ **HM1094e Psychic inspiration** (Psychism)
Inspired medium — Inspirational art — Inspirational speaking — Inspirational writing — Inspirational thought
Description Either consciously, or wholly or partially in a trance, the individual carries out some creative activity of which he or she is not normally capable or at a higher level than normal. This may be artistic, speaking or writing. The medium may plan the work which is then carried out while in a light trance under instruction from another intelligence; or the medium may be in a deep trance and not be conscious at all of what is happening, merely acting as an instrument. Sometimes the different styles make it appear that several different intelligences have been involved in the activity. In the case of speaking or writing, words may be jumbled, illegible or mis–spelled as information arises too fast to be interpreted. In the case of thought, mind impressions seem to bypass the belief system and information which could not have been known to the individual sometimes flows so fast that, again, it is too rapid for the brain to process coherently. Information gained is frequently philanthropic and intended for the benefit of mankind.
Refs Bletzer, June G *The Donning International Encyclopedic Psychic Dictionary* (1987).
Related Channelling (#HH0878) Inspiration (#HH2139)
Mediation (Magic, #HH4521) Inspired consciousness (#HM3125).

♦ **HM1097d Sense–sphere concentration** (Buddhism)
Description This is the concentration associated with the realm of sense; it is all kinds of access unification.
Context According to Hinayana Buddhism, concentration is considered as of one kind (monad), of two kinds (dyads), of three kinds (triads), of four kinds (tetrads) or of five kinds (pentad). In the fifth tetrad, this is the first concentration.
Broader Concentration (Buddhism, #HM6663).
Related Sense consciousness (Buddhism, #HM2664)
Access concentration (Buddhism, #HM4999)
Unincluded concentration (Buddhism, #HM5768)
Immaterial–sphere concentration (Buddhism, #HM0696)
Fine–material–sphere concentration (Buddhism, #HM4265).

♦ **HM1098d Destroying cause** (Buddhism)
Setughata (Pali)
Broader Correct speech (Buddhism, #HM1821) Correct action (Buddhism, #HM0439).

♦ **HM1099d Humaneness**
Refs Lorenz, Konrad *The Waning of Humaneness* (1988); Wynne–Tyson, Jon (Ed) *The Extended Circle* (1985).

♦ **HM1105b Ontogenetic model of human consciousness**
Description A theory developed by Erich Jantsch and Conrad Waddington, this models human consciousness on a double learning spiral through four learning modes – *virtual* or sperm–egg consciousness; *functional* or perinatal, foetal consciousness; *conscious* or personal consciousness; and *superconscious* or transpersonal consciousness. Transition from one state of consciousness to another is equivalent to a death–rebirth process.
At each level the evolutionary learning process is open–ended, based on experimentation which may be likened to a strategic exploration of options available when the choice made is subsequently vindicated. This is partially informed learning–by–doing guided by higher modes of learning, an aspect of self–transcendence. What is learned is a sense of direction, the overall process being regarded as evolutionary experimentation. The learning hierarchy requires a process of feedback between levels, so that there is an intra and intersystemic process. Transition from virtual to functional learning levels is equivalent to the struggle of the individual for survival resulting in development of social systems; from the functional to the conscious learning level is equivalent to the struggle between social systems resulting in the growth of cultures; and from conscious to superconscious learning is equivalent to the struggle between cultures, life styles and world views resulting in a sense of the wholeness of the process of mankind. The next transition is from superconscious to virtual again, going forward is also linking backwards to the origin (re–ligio).
Refs Jantsch, Erich *Consciousness Evolving* (1975); Jantsch, Erich and Waddington, Conrad H (Eds) *Evolution and Consciousness* (1976).
Narrower Virtual learning (#HM1217) Conscious learning (#HM0168)
Functional learning (#HM0196) Superconscious learning (#HM1244).
Related Noogenesis (#HH1781) Ontogenesis (#HH0366)
Process thinking (#HH1643).

♦ **HM1111d Maliciousness** (Buddhism)
Byapatti (Pali) — Byapada
Broader Hatred (Buddhism, #HM4502).

♦ **HM1114e Bridge of consciousness** (Psychism)
Description A state of transition between alert wakefulness and deep trance, described by mediums, when the conscious mind relinquishes control to the subconscious.

♦ **HM1116d Al–Mu'izz** (Sufism)
Description The believer is aware of Allah who honours and glorifies. The state of pride and dignity so received is not what the believer feels he deserves but respect paid to the honour he gives God.
Context A mode of mystical awareness in the Sufi tradition associated with the twenty–fourth of the ninety–nine names of Allah.
Broader Ninety–nine names of Allah (Sufism, #HH2561).
Followed by Al–Mudhill (Sufism, #HM4510).
Preceded by Al–Rafi (Sufism, #HM4040).

♦ **HM1118e Emotional clairvoyance** (Psychism)
Description There is a detailed vision of an event happening or about to happen either to the individual or to someone close to him.
Broader Clairvoyance (Psychism, #HM3498).

♦ **HM1119e Delusion consciousness** (Buddhism)
Description Some teachings quote this cognition which takes as permanent that which is perceived in foundation awareness, together with the 6 sense consciousnesses and with foundation awareness, as one of the 8 types of consciousness.
Related Foundation awareness (#HM0650) Sense consciousness (Buddhism, #HM2664).

♦ **HM1122b Anomalies in consistency of consciousness**
Description Inconsistencies include *déjà vu* and *jamais vu*, where there is a discrepancy between actual perception of a scene and the memory, or its lack, of what is being seen. They also include depersonalization, with a feeling of unreality even though one knows who one is. Normal self–monitoring is not related to actual performance, so that diminished ability may be accompanied by heightened emotion which makes one believe one is functioning at a high level of cognitive efficiency. The sense of well–being is likely to be accompanied by diminished self–criticism. In the search for consciousness broadening, 'highs' or psychedelic experiences there is a reduction in self–judgement and relief from responsibility. All involve a diffusion of ego individuality. While being convinced of heightened sensitivity and mental accuity, the individual's cognitive functioning is actually impaired. There may be intensified sensory input or diminished input (such as in meditation techniques). Inconsistency is even more apparent in the hypnotic and post–hypnotic states.
Narrower Déjà vu (#HM1240) Jamais vu (#HM3384) Depersonalization (#HM1248)
Rhythmic sensory bombardment (#HM6533).
Related Altered states of consciousness (#HM2391).

♦ **HM1128c Understanding as interpreting the external** (Buddhism)
Description This understanding is initiated when one has grasped the complex aggregate of elements of which another person or inanimate matter – external material not bound up with the faculties – is made up.
Context On the Path of Purification of Hinayana Buddhism, panna (understanding) is considered as of one kind (monad), of two kinds (dyads), of three kinds (triads) or of four kinds (tetrads). There are five dyads, four triads and two tetrads. All have the characteristic of penetrating the individual essences or true nature of states (monad). In the fourth triad, this is compared with understanding as interpreting the internal or as interpreting the internal and the external.
Broader Understanding (Buddhism, #HM4523).
Related Understanding as interpreting the internal (Buddhism, #HM0956)
Understanding as interpreting the internal and the external (Buddhism, #HM4490).

♦ **HM1133d Leaving off, abstaining, totally abstaining and refraining from the three deviations of body** (Buddhism)
Tihi kayaduccaritehi arati virati pativirati veramani (Pali)
Broader Correct action (Buddhism, #HM0439).

♦ **HM1134e Bodily disability** (Yoga)
Description This is said to be overcome by: (1) Eradicating present disease, refining and rebuilding the body, and protecting it from future attack, so as to render it immune to disease and indisposition. (2) Fine–tuning of the *etheric body*. (3) Awakening of the etheric bodily centres and centralizing its fires for union with the fire of the soul. This stage is not attempted until the first three stages of yoga have been developed. (4) Coordinating of the physical body and aligning it with the soul by means of the *sutratma*.
Context One of the nine obstacles to soul cognition enumerated by Patanjali.
Broader Obstacles to soul cognition (Yoga, #HM3182).

♦ **HM1136e Focus ten** (Psychism)
Description The mind is fully awake although the body seems asleep.
Context A state occurring during astral projection.
Broader Astral projection (Psychism, #HM1887).

♦ **HM1141c Beyond the chakras: the gods themselves** (Leela)
Plane of cosmic consciousness
Description Beyond the chakras are the nine God–forces. The player has to realize cosmic consciousness, or return to earth through tamoguna (responsible for evolution) and rejoin play until cosmic consciousness is reached.
Context The eighth and top row of the board of Leela, the game of knowledge, this is represented by the planes of cosmic consciousness, inner space, of bliss, of cosmic good, and the absolute plane, and groups squares representing the states of tamoguna, rajoguna, sattoguna and the phenomenal world. Tamoguna is the snake at this stage.

Broader Lila (#HM2278).
Narrower Satoguna (#HM4310) Rajoguna (#HM3281) Tamoguna (#HM5561)
Plane of bliss (#HM6901) Absolute plane (#HM4326) Phenomenal plane (#HM3907)
Plane of inner space (#HM6287) Plane of cosmic good (#HM4343)
Plane of cosmic consciousness (#HM1301).
Preceded by Seventh chakra: plane of reality (Leela, #HM0754).

♦ **HM1142c Right resolves** (Buddhism)
Samyag-samkalpa — Right realization — Right thought — Right thinking — Right aims
Description This is the resolve to renounce the world, to practice benevolence and to do no harm or hurt (ahimsa). It arises from right outlook. The characteristic is right directing of the mind; the function is application of the mind, bringing absorption of consciousness on nirvana as object; manifestation is through abandoning wrong thinking.
Context One of the characteristics of the Eightfold Way of Buddhism attained with the path of bhavana (meditation) and which, together with right outlook, constitutes wisdom (panna).
Broader Eightfold way (Buddhism, #HM2339).
Related Direct perception (#HM4469) Four noble truths (Buddhism, #HH0523).
Followed by Right speech (Buddhism, #HM1157).
Preceded by Right outlook (Buddhism, #HM1280).

♦ **HM1143c Discernment** (Buddhism)
Sallakkhana (Pali)
Broader Faculty of wisdom (Buddhism, #HM3233).
Related Spiritual discernment (Christianity, #HH3900).

♦ **HM1153d Initiation of resurrection** (Esotericism)
Description The liberating experience associated with this initiation is that of freedom from the hold of the phenomenal life of the seven planes of planetary life. As such it constitutes a lifting out of those constraints and, in a special sense, a return to an original state of being or originating source. It is a liberation into a consciousness of universal life. He is accorded a revelation of the quality of love–wisdom which must express itself through all creative forms. This revelation is divorced from all considerations of form and the initiate becomes a concentrated point of living light. He knows for the first time that life is all that is, and that it is this life and its real fullness which makes him part of that "life more abundant" which lies beyond planetary life — the greater whole of which planetay life is a part.
Context The seventh initiation in the tradition of western esotericism.
Broader Spiritual initiation (Esotericism, #HH3390).
Related Sahasrara (Yoga, #HM3398) Way of love–wisdom (Esotericism, #HM1507).
Followed by Initiation of transition (Esotericism, #HM1258).

♦ **HM1155d Malleability of consciousness** (Buddhism)
Pliancy of mind — Pliancy of citta (Pali)
Description This refers to the malleability or pliancy of the aggregate of consciousness. The characteristic is suppressing or quieting rigidity of the citta, the function is to crush its stiffening. It is manifest as non–resistance. Consciousness is the proximate cause and it is the opponent of defilements causing stiffness in the citta. Malleability of mental body and malleability of consciousness are considered together.
Context One of the formations aggregate (mental coefficients) of Hinayana Buddhism, being listed among the constant states which appear in their true nature, and as profitable primary (always present in any profitable or profitable–resultant consciousness).
Broader Awareness as mental–formation group of conscious existence (Buddhism, #HM2050).
Related Pliancy (Buddhism, #HM3162)
Malleability of mental body (Buddhism, #HM3696).

♦ **HM1157c Right speech** (Buddhism)
Samyag–vaca
Description With right outlook and right resolves, bad verbal conduct is abolished. This involves guarding one's language, with abstinence from lies, slander, abuse and gossip. The characteristic is embracing or seizing the listener or associated states; the function is abstaining from wrong speech; manifestation is in abandoning wrong speech.
Context One of the characteristics of the Eightfold Way of Buddhism attained with the path of bhavana (meditation) and which, together with right acts and right livelihood, constitutes ethical conduct (sila).
Broader Eightfold way (Buddhism, #HM2339).
Related Four noble truths (Buddhism, #HH0523).
Followed by Right acts (Buddhism, #HM0198).
Preceded by Right resolves (Buddhism, #HM1142).

♦ **HM1159g Swoon** (Hinduism)
Murccha
Description There is a cessation of determinate cognition due to some external factor such as a blow on the head. It is distinguished from waking and dream because there are no cognitions. It is not deep sleep as it manifests differently. Nor is it death, because one recovers and the functions of life go on. One is more unconscious than in intoxication (moha) and less than in apoplexy (samnyasa). It is thus a distinct condition.
Related Moha (Hinduism, #HM0992) Samnyasa (Hinduism, #HM1356).

♦ **HM1160d Conviction of sin**
Description This is a state of self–blame and contempt, with feelings of wretchedness and unworthiness together with fear of punishment for sin, which precedes conversion and is said to be the first stage on the way to salvation. Although this may be a valuable experience, attempts to deliberately induce conviction of sin in the hope of securing conversion are criticized for their manipulative nature and also for the confirmation of neurotic tendencies which may actually inhibit conversion.
Related Absolution (#HH0472) Discipline of confession (Christianity, #HH0825).

♦ **HM1166e Sense of presence**
Description Despite the lack of physical data corresponding to the feeling (no visual, audible or other cues) the individual has a sense of not being alone. This is most commonly experienced in eery or strange surroundings, which suggests heightened sensibility where some natural stimulus may open itself to misinterpretation without the individual realizing it. Silence is a frequent accompaniment to such sense, which may often be dispelled by familiar noises or even singing to one's self. This may be because such sound masks smaller, unidentifiable sounds or because silence offers and unstructured field in which diffuse fears become conscious as if projected on an empty screen.
In mentally disturbed patients the sense of presence becomes a full–scale hallucination, when the person may be aware of another person who he cannot see but who he may describe in detail or locate exactly as to position.
Refs Reed, Graham *The Psychology of Anomalous Experience* (1988).
Related Doppelgänger (#HM7342) Hallucination (#HM4580)
Psychic presence (Psychism, #HM3546).

♦ **HM1167b Canonical hours** (Christianity)
Description The Canonical hours are a set of 8 liturgies which rehearse in ritual form the journey of the spirit.
Narrower Sext (#HM3725) None (#HM0466) Lauds (#HM0894)
Prime (#HM1904) Terce (#HM2965) Matins (#HM0160)
Vespers (#HM1468) Compline (#HM4321).

♦ **HM1169d Cave phenomenon**
Jogging meditation — Zen of running
Description A feeling of pure sensation and beauty when intellectual thought seems arrested, achieved when running for long periods of time - it has been described after jogging for a considerable period, when thoughts may no longer be organized and perception changes. There is a mystical unity with the surroundings. Such experience is marked by ease in solving problems, the answers arising without conscious effort.
Related Flow experience (#HM2344).

♦ **HM1172c Joy**
Description Writers distinguish between joy as a pleasant state achieved on satisfaction of some aim or desire, and spiritual joy or bliss experienced in knowledge or union with God. This latter joy is independent of external circumstances and various writers speak of their joy even in misfortune, rejection and pain. There is joy in the acceptance of every experience in life as useful in development of the self. The Taittinya Upanishad speaks of joy (ananda) as freedom from fear; and Sadi as being joyful because God is the source of joy.
Refs Schutz, William *Joy* (1967).
Related Bliss (#HM4048) Joy (Hinduism, #HM8098)
Ananda (Hinduism, Buddhism, Yoga, #HM3227).

♦ **HM1175d Limited concentration with a limited object** (Buddhism)
Description Concentration is limited and thus unfit as a condition for a higher jhana; the object of concentration is unextended and therefore limited.
Context According to Hinayana Buddhism, concentration is considered as of one kind (monad), of two kinds (dyads), of three kinds (triads), of four kinds (tetrads) or of five kinds (pentad). In the second tetrad, this is the first concentration.
Broader Concentration (Buddhism, #HM6663).
Related Limited concentration with a measureless object (Buddhism, #HM5547)
Measureless concentration with a limited object (Buddhism, #HM4653)
Measureless concentration with a measureless object (Buddhism, #HM5214).

♦ **HM1177c Investigation** (Buddhism)
Vitarka — Rtog–pa (Tibetan) — Vitakka (Pali) — Applied thought — Reasoning
Description Vitakka is the right thinking which eliminates sloth and torpor and applies the mind and its concomitants to the object of concentration.
In Tibetan Buddhism, investigation refers to inquiry into the rough entity or the name of an object, and is virtuous, non–virtuous or neutral depending upon motivation and object of investigation. Virtuous generates pleasant effects and leads to abiding in contact with happiness. Non–virtuous generates unpleasant effects and leads to not abiding in contact with happiness.
In Hinayana Buddhism, applied thought is when the mind is directed onto the object like the striking of a bell. It is what first holds the object in mind for further perusal (sustained thought).
Context One of the four changeable factors referred to in Tibetan Buddhism. These may be virtuous, non–virtuous or neutral depending on circumstances. One of the formations aggregate (mental coefficients) of Hinayana Buddhism, being listed among the constant states which appear in their true nature, and as general secondary (sometimes present in any consciousness). One of the five constituent factors of jhana.
Broader Changeable mental factors (Buddhism, #HH0910)
Awareness as mental–formation group of conscious existence (Buddhism, #HM2050).
Related Analysis (Buddhism, #HM5089) Examining (Buddhism, #HM7324)
Jhana (Buddhism, Pali, #HM7193) Dwelling in the first jhana (Buddhism, #HM4298).

♦ **HM1182g Suspended animation**
Description All activities characteristic of life cease and the actually living body appears dead.
Related Absence of consciousness (#HM2670).

♦ **HM1189e Shamanism**
Shamanistic trance
Description In shamanistic rituals the neophyte becomes a "soft man being", namely an androgyne whose essential mission is to be an intermediary between the cosmological planes of earth and sky. The effort to incorporate this paradox involves the shaman in the constant practice of transformation, as if moving from one point of view to another provides the experimental ground of understanding, of wisdom, of true perspective.
Refs Avens, Roberts *The New Gnosis* (1984); Halifax, Joan *Shamanic Voices* (1979).
Narrower Kut (#HM0303).
Related Shaman (#HH0973) Shamanic journey (#HM6120).

♦ **HM1191d View** (Buddhism)
Ditthi (Pali)
Broader Wrong view (Buddhism, #HM1710).

♦ **HM1194d Fear** (Hinduism)
Bhaya
Description Fear is a response to the threat of danger and is said to be overcome by true knowledge. It is a painful emotion evoked through perceiving a cause of possible future pain.
Related Fear (Christianity, #HM3311).

♦ **HM1195d Atmanubhava** (Hinduism)
Being Atman
Description This is the integral experience of the Self.
Related Unconditional nirvikalpa samadhi (#HM3619).

♦ **HM1201c Way of knowledge** (Esotericism)
Way of science
Description Characteristic of those who use their minds, both to handle abstractions and in concrete applications, in the tireless pursuit of knowledge for its own sake. On this way they are challenged by the task of developing and appreciating feeling and sensitivity to others, as well as discovering more significant uses for their skills.
Context The fifth of seven ways to spiritual realization characteristic of Western esotericism.
Broader Ways to spiritual realization (Esotericism, #HH2665).
Related Sphere of glory (Kabbalah, #HM2238) Initiation of decision (Esotericism, #HM0322)
Initiation of transfiguration (Esotericism, #HM0428).

♦ **HM1208d Feeling remorse of acquisition of sinful and unwholesome dharmas** (Buddhism)
Ottappati papakanam akusalanam dhammanam samapattiya (Pali)
Broader Power of remorse (Buddhism, #HM2538).

♦ **HM1209d Al-Batin** (Sufism)
Description The believer is aware of Allah as the inward, the hidden. Even though all can see signs of existence, His essence is hidden. The knowledge, mind and understanding of the created are too limited to really comprehend the creator who is unlimited. Allah should only be contemplated in His attributes. Because all things are from Him, nothing can truly be known fully in its essence – attempts to do so finally end in awe.
Context A mode of mystical awareness in the Sufi tradition associated with the seventy-sixth of the ninety-nine names of Allah. Contemplation of this name in the Naqshbandiyya order follows the stages of major saintship, and the strengthening and widening of the internal spiritual state through contemplation of the name *az-Zahir*. It implies superior saintship or saintship of the angels.
Broader Ninety-nine names of Allah (Sufism, #HH2561).
Related Superior saintship (Sufism, #HM7089).
Followed by Al-Wali (Sufism, #HM0525).
Preceded by Az-Zahir (Sufism, #HM4042).

♦ **HM1210 Interactional synchrony**
Syncing — Being in sync
Description Individuals in a group have a tendency to synchronize their behaviour, especially body language, without necessarily being consciously aware of this process. An individual is then said to be "in sync." Although individuals experience some degree of discomfort when they are not synchronized with each other, this existence of interactional synchrony is most clearly demonstrated by very detailed analysis of films of interactional behaviour.
Refs Gatewood, J B and Rosenwein, R *Interactional Synchrony*; McDowell, Joseph J *Interactional Synchrony*.
Related Entrainment (#HH3466).

♦ **HM1211c Ontological love**
Description Experience of oneself, rather than the experience of God through some form of ecstatic mystical union. Whereas the mystic aspires to ascend to love of God or union with God, this form of enlightenment involves real knowledge and feeling for the person's own ontological dilemma. This experience is characterized by openness and vulnerability through a bodily presence in the world rather than an endeavour to escape from it. It is the art of living an ordinary life in an extraordinary manner.
Refs Needleman, Jacob *Lost Christianity* (1982).
Broader Love consciousness (#HM2000).
Related Reflexivity (#HM0087) Self-remembering (#HM2486).
Followed by Composition (#HM1406).

♦ **HM1212f Fragmentation of thinking**
Description This occurs due to disturbance of the thinking process caused, for example, by mental illness such as schizophrenia. The mind is so confused that logical pathways of thought association are no longer perceived and bizarre associations lacking concept or purpose dominate.

♦ **HM1217d Virtual learning**
Description Characteristic of nonreflective consciousness and physical development, this is the first mode of learning posited by Jantsch and Waddington in their ontogenetic model of human consciousness. It is equivalent to sperm-egg consciousness, commencing in an awareness of source and culminating in conception with the development of perinatal consciousness; it is characterized by response. Changes forecast are very short-range. On a human scale, there is naive participation in the surroundings with submission to things as they are; transition from virtual to functional learning levels is equivalent to the struggle of the individual for survival, resulting in the development of social systems.
Context One of four learning processes distinguished (Lazslo, 1972) by the type of consciousness brought into play.
Broader Learning (#HH0180) Ontogenetic model of human consciousness (#HM1105).
Followed by Functional learning (#HM0196).

♦ **HM1219b Lowered state of consciousness** (Psychism)
Description This term is used in some circles to describe the state of inner quiet and freedom from stress when perception is enhanced.

♦ **HM1221d As-Salam** (Sufism)
Description The believer is aware of Allah as the source of peace. The peace of As-Salam in the heart brings tranquillity, freedom from panic and saving from danger.
Context A mode of mystical awareness in the Sufi tradition associated with the fifth of the ninety-nine names of Allah.
Broader Ninety-nine names of Allah (Sufism, #HH2561).
Followed by Al-Mu'min (Sufism, #HM1245).
Preceded by Al-Quddus (Sufism, #HM0235).

♦ **HM1222d Yoga of the illusory body** (Buddhism)
Sgyu-lus (Tibetan)
Context The second branch of the Tibetan Tantric path of form according to the Naropa school.
Broader Tantric visualization (Yoga, #HM1690).
Followed by Yoga of the dream state (Buddhism, #HM0600).
Preceded by Yoga of the inner fire (Buddhism, #HM3863).

♦ **HM1225d State of bodhisattva** (Buddhism)
Description In this altruistic state there is joy in helping others. Normally compassion has a hierarchy, probably first for one's children, then one's spouse and immediate family, and so on. The bodhisattva nature extends to all mankind. In extending compassion in a particular way (through nursing, through the arts) one is embodying the qualities of one of the great Bodhisattvas. Acting as a bodhisattva also implies teaching the law to others so they can develop their inherent Buddhahood. In doing this one develops the qualities of the four bodhisattva leaders who are themselves manifestations of Buddhahood. This state is not perfection, however, and there may still be some superior or condescending manner The only wholly positive state is buddhahood itself.
Context One of the ten worlds described in Nichiren Soshu Buddhism.
Broader Ten worlds (Buddhism, #HM2657) Four noble paths (Buddhism, #HM4026)
Nichiren shoshu buddhism (Buddhism, #HH3443).
Related Contact (Buddhism, #HM1380) Bodhisattva (Buddhism, #HH1380).
Followed by State of buddhahood (Buddhism, #HM1873).
Preceded by State of realization (Buddhism, #HM0450).

♦ **HM1226f Deep state of hypnosis**
Somnambulistic hypnosis
Description The body becomes limp and consciousness is suspended, although the subconscious functions and is open to suggestion by the therapist; at the instigation of the therapist the individual will answer questions and describe the experience. Upon reverting to normal consciousness there is no memory of what has passed or even of how long the state lasted. Suggestions made during this state are accepted and this acceptance may be lasting, so that hypnosis may be used for therapeutic purposes to alter attitudes and habits.
Broader Hypnotic states of consciousness (#HM2133).
Related Hypnotherapy (#HH0962) Delta wave consciousness (#HM1785).

♦ **HM1233d Centre** (Taoism)
Description The state of mind before it is affected by feelings. It is the point of balanced intent at the hub of experience, where the celestial and the earthly, detachment and involvement, and firmness and flexibility, are all in dynamic equilibrium.
Related Centering (#HH0527).

♦ **HM1234c Two-fold knowledge** (Systematics)
Discriminative knowledge — Polar knowledge — Experience of polarity
Description Any pair of terms between which both connection and disjunction are experienced. Few pairs stand in more than weak opposition to one another or with more than insignificant connection. Through polarity, everything is experienced as being in a state of strain which polarity itself can do nothing to relieve. It gives rise to force which may be transformed into direction. It can neither show how oppositions arise nor how they may be resolved. Its closure is not that of completeness. The power to recognize differences of quality demands the separation of consciousness characteristic of polarity. Polarity is experienced as too uncomfortable to be endured, as such it is a source of disturbance compelling deeper penetration into what is experienced.
Context The second in a sequence of twelve modes of knowledge, identified by J G Bennett, inspired by G Gurdjieff.
Broader Systematics (#HH2003).
Followed by Three-fold knowledge (Systematics, #HM0811).
Preceded by Non-discriminative knowledge (Systematics, #HM4002).

♦ **HM1237g Perceptual narrowing**
Tunnel vision
Description Under conditions of narcosis, drunkenness, schizophrenia, anxiety and many other forms of stress, an apparent loss of peripheral vision may occur, accompanied by distortions of apparent size and distance. As a consequence there is a lack of attention to peripheral phenomena, or a redistribution of attention over the visual field. The term is also used metaphorically to describe the reduction in ability to take into account concerns which are considered peripheral to some central preoccupation which is a source of anxiety or stress.
Broader Illusion (#HM2510).
Related Group think (#HM6119).

♦ **HM1240d Déjà vu**
Déjà raconté — Déjà entendu — Déjà eprouvé — Déjà fait — Déjà pensé — Déjà voulu
Description In Freudian terms, there is a subjective falsification of experience leading to belief that current experience has occurred previously. Unreal and dreamlike feelings predominate, and the uncanny sense of predicting each element of the experience as it occurs; or of having been in the same place or situation before, although objectively this is known to be impossible.
Broader Altered ego states (#HM4120)
Anomalies in consistency of consciousness (#HM1122).
Related Jamais vu (#HM3384) Isakower phenomenon (#HM5124).

♦ **HM1241g Synesthesia**
Synaesthesia — Secondary sensation
Description These are sensory perceptions experienced as accompanying sensations of another modality. The senses are transposed, stimulation appropriate to one sense organ being sensed by another, for example: different colours associated with musical notes or with words; or different tastes or smells, etc, associated with other senses. In some cases the stimulus for one sense elicits vivid perceptions in several senses. There are well-attested cases of people "seeing" with their fingers. This may be due to vestigial photosensitive cells on the skin surface or to nerve damage. Certainly, the connecting of a nerve fibre from the tongue to one from the ear to the brain allows taste to be sensed as noise.
Refs Cytowic, Richard E *Synesthesia*.
Broader Paraesthesia (#HM0003).
Related Agnosia (#HM8932) Perception through the senses (#HM3764)
Modes of awareness associated with schizophrenia (#HM2313)
Modes of awareness associated with use of hallucinogens (#HM0801).

♦ **HM1244d Superconscious learning**
Description Characteristic of a complex, self-reflective consciousness mirrored both in surface and in multilevel, super-consciousness, this is the fourth mode of learning posited by Jantsch and Waddington in their ontogenetic model of human consciousness. It is equivalent to transpersonal consciousness, commencing in an awareness of evolution beyond history and culminating in the death of mankind consciousness leading to integral awareness of evolution. Experientially accessible, present in later stages of social organization through morality (learning of inner constraints) and archetypes, it is central to noogenesis and characterized by superconscious self-regulation of cultural and mankind processes. Changes forecast are very long range.
Context One of four learning processes distinguished (Lazslo, 1972) by the type of consciousness brought into play.
Broader Learning (#HH0180) Ontogenetic model of human consciousness (#HM1105).
Related Noogenesis (#HH1781).
Preceded by Conscious learning (#HM0168).

♦ **HM1245d Al-Mu'min** (Sufism)
Description The believer is aware of Allah as the guardian of faith. With faith in the heart, one takes refuge in al-Mu'min, not denying help to others taking refuge in Him.
Context A mode of mystical awareness in the Sufi tradition associated with the sixth of the ninety-nine names of Allah.
Broader Ninety-nine names of Allah (Sufism, #HH2561).
Followed by Al-Muhaymin (Sufism, #HM0526).
Preceded by As-Salam (Sufism, #HM1221).

♦ **HM1248f Depersonalization**
Derealization — Dereism
Description A feeling of loss of personal identity, and that one is strange or unreal, may occur

in a variety of mental disorders. Organizing and arranging thoughts becomes difficult, the brain feels numb, and occasionally the sufferer feels giddy and fears collapsing in public. There may be a belief that part of one's self has disappeared, or that one has ceased to exist – *'délire de négation'* – or that one's body has swollen to an enormous extent – *'délire d'énormité'*. Depersonalization may be accompanied by *derealization*, when the environment appears to have changed in such a way that reality, the whole world, has changed, with possible feelings of imminent catastrophe. It may be that everything appears flat and lacking in significance. It may also be accompanied by *dereism*, when mental activity no longer takes the facts of reality into consideration, but deviates from laws of logic and experience. Impressions include the feeling of having no body, of being a ghost, of seeing someone else when one looks in a mirror, of having another's voice. The sufferer is a stranger to himself. There may be a residual sense of "I" as an outside observer; but there is no sense of "me" or of involvement. There is an overall sense of loneliness and isolation.

Depersonalization may be triggered by a number of circumstances, including sensory deprivation, extreme pain, hysteria, emotional distress, or under the effects of psychedelic drugs, as well as arising in schizophrenia and other mental disorders and nervous diseases. It may also occur in response to sudden disaster or threats to security, such as to someone reprieved at the last moment from the death sentence. Sufferers from allochiry (lack of perception of one half of the body) may also suffer from total or partial depersonalization. Depersonalization is a common feature of psychiatric cases, particularly obsessional disorders and in depressive states. The experience may be descibed as being cut off from the grace of God and, in combination with dissociation of affect, with loss of capacity to love their families, isolation and unworthiness. All feelings of depersonalization are not necessarily negative. There may be a feeling of altered identity and unreality when falling in love or coming into large sums of money. The term might also be extended to cover the transcending of reality which occurs in religious rites, mystic rituals and the taking of drugs. Depersonalization also gives rise to insight – there is an awareness of personal identity and also of the unreal quality of the experience.
   Broader Altered ego states (#HM4120)
   Anomalies in consistency of consciousness (#HM1122)
   Anomalies in experience of the reality of one's self and the environment (#HM2548).
   Narrower Koro (#HM4803).
   Related Identity (#HH0875)   Derealization (#HM5128)
   Personality development (#HH0281)   Thought reform (Brainwashing, #HH0865)
   Constructive personality change (#HH0476)
   Anomalies in awareness of one's self (#HM8186).

♦ **HM1252d Four evil paths** (Buddhism)
Description Four of the ten inner states of being or worlds described in Nichiren Soshu Buddhism, the three evil paths plus anger, these are conditions of unhappiness.
   Broader Six paths (Buddhism, #HM1914)   Ten worlds (Buddhism, #HM2657)
   Nichiren shoshu buddhism (Buddhism, #HH3443).
   Narrower State of hell (#HM4282)   State of anger (#HM2959)
   State of hunger (#HM0150)   State of animality (#HM0847)
   Related Three evil paths (Buddhism, #HM0923).

♦ **HM1256c Intuition** (Hinduism)
Refs Sinha, Jadunath *Indian Psychology* (1986).
   Narrower Yogic perception (#HM5005)   Occult perception (#HM1285)
   Flash of intuition (#HM4875)   Intuition of sages (#HM9124)
   Released soul's perception (#HM6558).

♦ **HM1257d Al–Ghaffar** (Sufism)
Description The believer is aware of Allah as the forgiver. Guilty of disrupting harmony within and around himself, he repents and begs not to sin again. His sin may be transformed into a good deed.
Context A mode of mystical awareness in the Sufi tradition associated with the fourteenth of the ninety–nine names of Allah.
   Broader Ninety–nine names of Allah (Sufism, #HH2561).
   Followed by Al–Qahhar (Sufism, #HM1318).
   Preceded by Al–Musawwir (Sufism, #HM5786).

♦ **HM1258d Initiation of transition** (Esotericism)
Description The liberating experience associated with this initiation is that of freedom from the reaction of consciousness and a liberation into a state of awareness, a form of conscious recognition which has no relation to consciousness as normally understood. This may be understood as complete freedom from sensitivity cmbined with a full flowering of compassion. A sense of planetary purpose is revealed as well as well as that by which it is threatened. The nature and purpose of duality is revealed in a new light.
Context The eighth initiation in the tradition of western esotericism.
   Broader Spiritual initiation (Esotericism, #HH3390).
   Followed by Initiation of refusal (Esotericism, #HM1020).
   Preceded by Initiation of resurrection (Esotericism, #HM1153).

♦ **HM1260d Loneliness**
Description This is a feeling of desolation or sadness arising either from lack of any human company or from separation from a particular person or situation on which there had been some dependence. The cause may be an inability to relate to others, despite constantly trying, derived from a basic inability to relate to or approve of one's self.
Refs Lynch, J *The Broken Heart* (1977).
   Related Solitude (#HH0333)   Pain of solitude (#HM1539).

♦ **HM1261d Desire for formless life** (Buddhism)
Context The seventh fetter or character defect referred to by Buddha. This is in the second set of five fetters, when automatic responses are already overcome and the disciple is ready (arhat) to face and overcome the second set.
   Broader Character defects (Buddhism, #HH0470).
   Followed by Spiritual pride (Buddhism, #HM2852).
   Preceded by Desire for life in form (Buddhism, #HM3583).

♦ **HM1263d Physical inception of energy** (Buddhism)
Cetasiko viriyarambho (Pali)
   Broader Faculty of energy (Buddhism, #HM1470).

♦ **HM1264d Al–fikr fi 'l–manzur** (Islam)
Contemplation of the object
Description The object of admiration, gaving been gazed upon and found to be beautiful, is now contemplated.
Context The third cause of profane love occurring in the beholder according to Ibn al-Qayyim.
Refs Bell, Joseph Norment *Love Theory in Later Hanbalite Islam* (1979).
   Followed by Tama' (Islam, #HM4772).
   Preceded by Istihsan (Islam, #HM0453).

♦ **HM1265c Phenomenon of presence** (Christianity)
Description This occurs in the sensation of plural separate realities in space and time experienced in a certain aspect as unity. Theologically, there is: (1) The mutual presence in transcendental unity of the divine persons in God; God's presence in the created world; the presence of God in man by self–communication. (2) Categorical unity, as between persons united in knowledge and love. (3) Based on the unity of space, the presence of entities in space. (4) Sacramental presence, as of the body of Christ in the Eucharist.

♦ **HM1266c Mehr** (Sufism)
Unconditional love for all
Context The fourth psychic state listed by A Reza Aresteh as progress of the inner self through divine attraction, *kedesh–jazba*, the outer self already having been purified through conscious effort, *kushesh*.
   Broader Psychic states (Sufism, #HM4311).
   Followed by Omid (Sufism, #HM4477).
   Preceded by Showq (Sufism, #HM0762).

♦ **HM1267d Initiation of baptism** (Esotericism)
Description The liberating experience associated with this initiation is that of freedom from the control of the emotional nature and the selfish sensitivity of the lower self. The individual becomes aware of a vision of the possibility of transcendence which is communicated through the symbolism of purification. There is a vision of a higher focus and his place in the larger whole begins slowly to reveal itself. A new creativity and new goals become the immediate focus. There is a deep dissatisfaction with things as they are and a recognition of the failure of old modes of activity employed. Their continued use, for lack of greater insight, results in intense personal suffering. The person recognizes that his emotional nature, his lower psychic faculties are arrayed against him. In this way he is freed from glamour, illusion and distortion. The group begins to mean more to the individual than himself.
Context The second initiation in the tradition of western esotericism.
   Broader Spiritual initiation (Esotericism, #HH3390).
   Related Manipura (Yoga, #HM3063)   Way of idealism (Esotericism, #HM6554).
   Followed by Initiation of transfiguration (Esotericism, #HM0428).

♦ **HM1270a Fana** (Sufism)
Annihilation — Passing away
Description In this annihilation of consciousness of humanity, transformation into the divine, and disappearance in relation to human qualities, the Sufi is at the last stage of his journey. First there is annihilation of human qualities, then of the vision of God, and finally annihilation of annihilation, the perfect state of fana. Existence is only through existence of God. He passes away from the human qualities such as ignorance, injustice and ingratitude, although the essence of humanity itself is not destroyed but transformed by the divine light.
This state of mystical annihilation, of absorption in God, literally of "passing away", has been equated with transcendental consciousness.
Context The thirteenth stage in a systematic account of various stations of the Sufi spiritual path. However, according to Shaykh Abu Sa'id ibn Abi'l–Khayr, this is the 21st of 40 stations or maqamat the Sufi must possess for his journey on the path of Sufism to be acceptable.
   Broader Mystic stations (Sufism, #HM3415)   Human development (Sufism, #HH0436)
   Stations of consciousness – ibn-Abi'l–Khayr (Sufism, #HM2424).
   Related Fana (Islam, #HM1087)   Fana (Sufism, #HM3799)
   Fani (Sufism, #HM4064)   Tawhid (Sufism, #HM3438)
   Love of God (Sufism, #HM4273)   Transcendental consciousness (#HM2020)
   Identification with God (Sufism, #HM3520)   Consciousness of mystical poverty (#HM3241).
   Followed by Baqa (Sufism, #HM0330).
   Preceded by Raja (Sufism, #HM4393)   Ma'rifa (Sufism, #HM2254).

♦ **HM1271d Cognition born of contact with element of eye–consciousness** (Buddhism)
Cakkhuvinnanadhatusamphassaja sanna (Pali)
   Broader Cognition (Buddhism, #HM1389).
   Related Eye consciousness (Buddhism, #HM2074).

♦ **HM1272g Somnambulism**
Sleep walking
Description In a state of deep sleep when there are no rapid eye movements to indicate dream, the person appears to respond to his surroundings and may carry out difficult or even dangerous tasks while apparently being unconscious of what is occurring, or at least being unable to recall events afterwards. Activities may be impossible to reproduce in the waking state (walking along a high ledge, singing the whole of a piece of music only heard once). Somnambulism seems to occur in response to an abnormal response of the nervous system.

♦ **HM1273g Hypercognition**
Description The cognitive faculties work at unusually fast speeds – said to occur in conditions of high levels of alpha brainwaves.
   Related Alpha wave consciousness (#HM2345).

♦ **HM1274d Relaxed thinking**
Context Normal consciousness is said to be alert thinking (concentrated) interspersed at regular intervals by relaxed thinking (concentrated) or daydreaming (diffuse). This mirrors the regular pattern of paradoxical or rapid eye movement sleep which occurs about every 90 minutes during the night.
   Related Daydream (#HM2138)   Alert thinking (#HM4967).

♦ **HM1276d Plane of violence** (Leela)
Himsa–loka
Description Knowing that no threat comes from outside, and perfectly self–confident, the player uses violence not as self–defence but as the willed result of a claim to have the whole truth – to be God or his agent, and forcing others to recognize this, through their physical death if necessary. Lacking fluidity and spiritual devotion the player is forced back to purgatory for even harder penances before he can continue on the path.
Context The 52nd state or square on the board of *Leela*, the game of knowledge, appearing in the sixth row.
   Broader Sixth chakra: time of penance (Leela, #HM4412).
   Followed by Purgatory (Leela, #HM0856).

♦ **HM1279d Solar plane** (Leela)
Yamuna
Description In the plane of male energy, the witness–player who has passed through wisdom and

right knowledge is able nonetheless to balance the play of energies and avoid over concern with destruction, power and self identification. Solar and lunar energies do not simply entwine, they become one.
Context The 48th state or square on the board of *Leela*, the game of knowledge, appearing in the sixth row.
Broader Sixth chakra: time of penance (Leela, #HM4412).

♦ **HM1280c Right outlook** (Buddhism)
Samyag–drishti — Right understanding — Right view
Description This is to know suffering, its origin and cessation, and the path which leads to its cessation. The transience of conditioned existence is realized. Nirvana is the object of understanding, this eliminating the innate tendency to ignorance. The characteristic is seeing rightly; the function is setting forth or revealing the elements; manifestation is in dispelling the darkness of ignorance.
Context One of the characteristics of the Eightfold Way of Buddhism attained with the path of bhavana (meditation) and which, together with right resolves, constitutes wisdom (panna).
Broader Eightfold way (Buddhism, #HM2339).
Related Wrong view (Buddhism, Pali, #HM5324) Four noble truths (Buddhism, #HH0523).
Followed by Right resolves (Buddhism, #HM1142).

♦ **HM1285c Occult perception** (Hinduism)
Siddhadarsana
Description This is valid immediate knowledge of hidden and remote objects which are present through the sense organs. The power of the sense organs is enhanced by the occult power of medicine and incantations which strengthen and purify the sense organs but is basically due to the special merit of the occultist.
Broader Intuition (Hinduism, #HM1256).

♦ **HM1286d Standing at crossroads** (Buddhism)
Dvedhapatha (Pali)
Broader Perplexity (Buddhism, #HM0812).

♦ **HM1287d Al–Hadi** (Sufism)
Description The believer is aware of Allah as the guide, directing his servants to good, to the satisfaction of their own needs and to benefitting others, and thus developing faith. When one is lead astray the result is infidelity - but Allah only leads astray those who misuse their will. He who is well–guided knows the truth, living and dying for it.
Context A mode of mystical awareness in the Sufi tradition associated with the ninety–fourth of the ninety–nine names of Allah.
Broader Ninety–nine names of Allah (Sufism, #HH2561).
Followed by Al–Badi (Sufism, #HM4299).
Preceded by An–Nur (Sufism, #HM1354).

♦ **HM1288f Inflation**
Description There is either a feeling of immense power or of worthlessness, in either case the disorientation being due to identification with the collective psyche as consciousness is extended or invaded by unconscious archetypal contents and discrimination is lost. The ego is inflated until it may identify with the self so that there is no distinction between the individual person and the image of God and individuation becomes impossible.
Related Individuation (#HH0385) Identification (#HH0358).

♦ **HM1293d Plane of reality** (Leela)
Plane of truth — Satya–loka
Description In a state of samadhi, of pure bliss, the player becomes realized. He is sat–chit–ananda - knowledge, consciousness and bliss. There is harmony as the forces of the cosmos are balanced. Although there are still snakes ahead which can lead back into the game through doubt or laziness, right karma will mean attainment of the goal.
Context The 59th state or square on the board of *Leela*, the game of knowledge, appearing in the seventh row. This is the seventh of eight lokas represented on the spine of the board and marking a stage in the development of the individual's consciousness. It is the last of seven representing a psychic centre or chakra on the spine of the human body, a level of psychic evolution. An eighth loka lies beyond the chakras. Each loka is defined by the nature of the matter of which it is composed. The eight other states in the same row may be considered as special regions within the loka.
Broader Seventh chakra: plane of reality (Leela, #HM0754).

♦ **HM1294d Doubt of the efficacy of the good life** (Buddhism)
Context The second fetter or character defect referred to by Buddha. The first five fetters are marked by automatic response due to the following of old, unphilosophical mental habits. When these are overcome, the disciples is ready (arhat) to face the final five.
Broader Character defects (Buddhism, #HH0470).
Followed by Dependence on ceremonies (Buddhism, #HM0509).
Preceded by Delusion of the personal self (Buddhism, #HM3152).

♦ **HM1295c Right endeavour** (Buddhism)
Samyag–vyayama — Right effort
Description Established in right speech, right acts and right livelihood, and in association with right outlook, energy cuts off idleness. The individual makes every effort to prevent those bad or wrong qualities which have not yet arisen from arising, and to renounce those which have already arisen. He fosters good qualities which have not yet arisen and establishes, develops and perfects those which are already there. Particularly through guarding the senses, wholesome volition is cultivated and unwholesome mental activity prevented. The characteristic is exerting or upholding; the function is non–arousal of unprofitable or immoral states or qualities; manifestation is in abandoning wrong effort.
Context One of the characteristics of the Eightfold Way of Buddhism attained with the path of bhavana (meditation) and which, together with right mindfulness and right rapture of concentration, constitutes mental discipline (samadhi).
Broader Eightfold way (Buddhism, #HM2339).
Related Four noble truths (Buddhism, #HH0523) Correct endeavour (Buddhism, #HM1657).
Followed by Right mindfulness (Buddhism, #HM0704).
Preceded by Right livelihood (Buddhism, #HM1341).

♦ **HM1301d Plane of cosmic consciousness** (Leela)
Vaikuntha–loka
Description This is the primal element, the source of all elements. Achievement of this state is the aim of the game and the player's greatest desire, even though he may be sidetracked by lesser desires during the progress of the game. All paths have this as their aim, whether through steady progress in eightfold yoga or from the spiritual devotion of following dharma. There is the choice of whether to resume the game, to remain beyond it or to return and help others reach the goal.
Context The 68th state or square on the board of *Leela*, the game of knowledge, appearing in the eighth row. This is the last of eight lokas represented on the spine of the board and marking a stage in the development of the individual's consciousness. The first seven represent a psychic centre or chakra on the spine of the human body, a level of psychic evolution. This eighth loka lies beyond the chakras.
Broader Beyond the chakras: the gods themselves (Leela, #HM1141).
Preceded by Spiritual devotion (Leela, #HM1475).

♦ **HM1317e Audible thought**
Psychic hallucinations — Inner voices
Description The experience of hearing one's inner thoughts as if they were spoken from outside, although almost no sensory components are actually involved.
Related Hallucination (#HM4580) Clairaudience (Psychism, #HM4333)
Auditory hallucination (#HM6704)
Modes of awareness associated with schizophrenia (#HM2313).

♦ **HM1318d Al–Qahhar** (Sufism)
Description The believer is aware of Allah as the subduer, dominating creation with irresistible power which is balanced by al–Latif (loving finesse). The light of al–Latif shines on beautiful characteristics, the darkness of terror on negative qualities such as rebellion and denial. Refuge from Allah al–Qahhar is sought in Allah al–Latif.
Context A mode of mystical awareness in the Sufi tradition associated with the fifteenth of the ninety–nine names of Allah.
Broader Ninety–nine names of Allah (Sufism, #HH2561).
Related Al–Latif (Sufism, #HM0308).
Followed by Al–Wahhab (Sufism, #HM0069).
Preceded by Al–Ghaffar (Sufism, #HM1257).

♦ **HM1325c Eighth order perceptions - control of principles**
Description This is the order of morality, responsibility and values, where concepts such as good and bad, moral and immoral, just and unjust, are necessary if one's conflicting needs are to be satisfied. Rather than fixed behaviour in response to a given situation, with an inbuilt programme saying when to procreate, when to turn out the young on their own as an animal has, man can (although he does not always) step outside the programme in response to his own needs. This is the level of making value judgements, it is also the order at which conflicts arise. If each side in a conflicting decision is of equal value then there can be no satisfactory resolution.
Context The eighth of ten orders in the perceptual system described by William Glasser.
Broader Orders of perception (#HM0988).
Followed by Ninth order perceptions - control of systems concepts (#HM4000).
Preceded by Seventh order perceptions - programme control (#HM1001).

♦ **HM1330c Merging of yin and yang** (Taoism)
Formation of gold elixir — Spirit of open consciousness — Valley spirit
Description Following the assembly of the five elements, yin and yang may be merged into one (such that within yin there is yang, and within yang there is yin). This is equivalent to a recovery of child–like (in contrast with childish) awareness. Acquired conditioning is now strongly governed by primordial insight and no longer acts in a harmful manner. Although this merging process is symbolized by sexual intercourse (and by other combinations of polarized energies), the "gold elixir", as the primordial energy of life, cannot be formed other than through the crystallization of the energy of primordial nothingness.
This elixir is a metaphor of the essence of true consciousness, which is fundamentally complete and illumined. It is the unique energy of primordial nothingness and in the darkness of profound abstraction. Although neither matter nor emptiness, it is both material and empty at the same time. It is the border of the mind of Tao and the human mind, the root of real knowledge and of conscious knowledge, the place where distinctions are born. It is the innate knowledge and capacity fundamental to human life. It is a combination of polar energies, with firmness and flexibility in proper balance, engendering the vitality of real unity. Under temporal conditioning, this awareness recedes and distinctions predominate. When the valley spirit is stable, the mind of Tao is always there, and the human mind is guided by it. Real knowledge and conscious knowledge then unite. Innate knowledge and capacity integrate with the celestial design. The gold elixir is the energy of true unity, or primordial nothingness. In connection with he root of consciousness, when chaos first becomes differentiated, it acts as the generative energy which produces beings.
Context The fifth of seven stages of the Taoist alchemical firing process through which reality is cultivated and the self is restored.
Broader Alchemy (Taoism, #HH5887).
Related Complete awareness (Buddhism, #HM0634).
Followed by Unification of energy (Taoism, #HM4762).
Preceded by Assembling the five elements (Taoism, #HM4338).

♦ **HM1333d Energy** (Buddhism)
Viriya (Pali)
Description Supporting and consolidating coexistent or conascent states, viriya is the state of the energetic or vigorous which manifests as not giving way or collapsing. Agitation or a sense of urgency, or basic grounds for making energy, are the proximate cause.
Context One of the formations aggregate (mental coefficients) of Hinayana Buddhism, being listed among the constant states which appear in their true nature, and as general secondary (sometimes present in any consciousness).
Broader Faculty of energy (Buddhism, #HM1470)
Awareness as mental–formation group of conscious existence (Buddhism, #HM2050).

♦ **HM1336f Anomalies in experience of the self as recognized in personal performance**
Loss of personal attribution
Description Although appreciating himself as a discrete entity, the individual does not recognize thoughts, imagery and activities as his own, perhaps attributing some of them to outside agencies.
Broader Anomalies in awareness of one's self (#HM8186).
Narrower Thought broadcasting (#HM9384) Alientation of thought (#HM6508)
Thought insertion (Psychism, #HM5479).

♦ **HM1337d Initiation of birth** (Esotericism)
Description The liberating experience associated with this initiation is that of freedom from the control of the physical body and its appetites. It can be regarded as the goal and reward of the mystical experience. It produces a measure of order in the emotional processes of the initiate instituting a new attitude towards relationships. Individual consciousness is expanded into a growing group awareness. The person becomes able to recognize the kind of energy he has to offer and is able to establish certain service relationships.
Context The first initiation in the tradition of western esotericism.
Broader Spiritual initiation (Esotericism, #HH3390).
Related Svadhishthana (Yoga, #HM3174) Way of organization (Esotericism, #HM1605).

♦ **HM1338d Straightness of the aggregate of consciousness** (Buddhism)
Vinnanakkhandhassa ujukata (Pali)
  Broader Straightness of thought (Buddhism, #HM0253)
  Dispositions of consciousness (Buddhism, #HM2098).

♦ **HM1341c Right livelihood** (Buddhism)
Samyag-ajiva
Description Right speech and right acts are purified. In association with right outlook, the disciple supports himself without resorting to wrong means of livelihood in a profession which does others no harm. The characteristic is cleansing; the function is the mode or occurrence of proper livelihood; manifestation is in abandoning wrong livelihood.
Context One of the characteristics of the Eightfold Way of Buddhism attained with the path of bhavana (meditation) and which, together with right speech and right acts, constitutes ethical conduct (sila).
  Broader Eightfold way (Buddhism, #HM2339).
  Related Four noble truths (Buddhism, #HH0523)
  Abstinence from wrong livelihood (Buddhism, Pali, #HM7887).
  Followed by Right endeavour (Buddhism, #HM1295).
  Preceded by Right acts (Buddhism, #HM0198).

♦ **HM1342b Ultimate accomplishment** (Taoism)
Perfect attainment
Description Following attainment of Tao, which results in the creation of a spiritual body outside the physical body, the practitioner is physically and mentally sublimated. The task that then emerges is that of storing the spiritual body and secretly developing spiritual powers. The spiritual body is then becomes the vehicle through which transmutation spontaneously occurs, progressively increasing its degree of openness. Although nothing then exists, the spiritual body transmutes in countless ways. There is a complete cessation of effort. Using the metaphor of birth employed in the alchemical process, this state is symbolized by the child producing children and grandchildren, who produce in their turn.
  Broader Human development (Taoism, #HH0689).
  Preceded by Transcending the world (Taoism, #HM5265).

♦ **HM1344c Prajnaparamita** (Buddhism)
Intuitional insight — Perfection of discriminating awareness
Description Hidden meanings of voidness teachings are studied within the teachings of the enlightened motive of *bodhicitta*. Stages and paths to enlightenment are dealt with specifically.
Context The first of the five primary subjects for the Geshe degree in Tibetan Buddhism.
  Related Prajna paramita (Zen, #HH0550)
  Tantra climax-application awareness (Buddhism, Tibetan, #HM2220).
  Followed by Madhyamaka (Buddhism, #HH1038).

♦ **HM1346d Stupefaction** (Yoga)
Mudha
Description The mind is obsessed with matters concerned with the senses and one is unfit to think of subtle principles. The person may, for example, be obsessed with family or wealth to the point of infatuation.
Context One of five states of mind classification identified in yoga.
  Followed by Restlessness (Yoga, #HM1015).
  Preceded by Autism (Yoga, #HM4058).

♦ **HM1347d Internal awareness** (Psychism)
Description The individual is conscious of and to some extent is capable of controlling activity in the silent mind.

♦ **HM1351d Plane of balance** (Leela)
Maha-loka — Mahar-loka
Description This is the centre of balance of the whole game and the player speaks from the heart, which is filled with devotion. The hands express gestures balancing the flow of energy, the voice becomes soft and gentle, and he starts to attract followers who seek the same vibrational patterns.
Context The 32nd state or square on the board of *Leela*, the game of knowledge, appearing in the fourth row. This is the fourth of eight lokas represented on the spine of the board and marking a stage in the development of the individual's consciousness. The first seven represent a psychic centre or chakra on the spine of the human body, a level of psychic evolution. An eighth loka lies beyond the chakras. Each loka is defined by the nature of the matter of which it is composed. The eight other states in the same row may be considered as special regions within the loka.
  Broader Fourth chakra: attaining balance (Leela, #HM3291).
  Preceded by Charity (Leela, #HM4387).

♦ **HM1352d Neither suffering nor pleasure experienced born of contact with the psychical** (Buddhism)
Cetosamphassajam adukkhamasukham vedayitam (Pali)
  Broader Unperturbedness (Buddhism, #HM4572).

♦ **HM1353f Diminished range of awareness**
Diminished focus of awareness
Description Diminished consciousness may arise for a number of reasons, including intoxication, fear or shock. An example is *narrowing of object awareness* or *tunnel vision*, when only some of the data presenting itself, that of most relevance to a current concern, is assimilated. Other, objectively more important, features are not regarded. Another example is the domination of one idea or fear in *tunnel thinking*. Love causes the lover to be "blind" to imperfections in the beloved. Anger causes "blindness" to the other side of the argument. The extreme case of restricted awareness is a *twilight state*.
  Narrower Twilight consciousness (#HM2406).

♦ **HM1354d An-Nur** (Sufism)
Description The believer is aware of Allah as the light making apparent the whole creation. When the heart is illuminated by the light of faith, the devil and the ego will not enter. The gate to the heart, the mind, is illuminated by knowledge which extinguishes the evil of ignorance, hypocrisy, imagination and arrogance. Similarly, the soul needs the light of consciousness. The physically blind may be lead by the hand, but the blind heart is lost forever.
Context A mode of mystical awareness in the Sufi tradition associated with the ninety-third of the ninety-nine names of Allah.
  Broader Ninety-nine names of Allah (Sufism, #HH2561).
  Followed by Al-Hadi (Sufism, #HM1287).
  Preceded by An-Nafi' (Sufism, #HM6350).

♦ **HM1355d Good company** (Leela)
Su-sang-loka
Description In fellowship with others seeking realization and forming a group round a teacher whom they emulate, the player has his good and bad tendencies mirrored so he can act upon them. All aspects of the self are confronted and worked on, problems in the previous two chakras disappear and there are no further hindrances to reaching the fourth level of the game.
Context The 25th state or square on the board of *Leela*, the game of knowledge, appearing in the third row.
  Broader Third chakra: theatre of karma (Leela, #HM0717).

♦ **HM1356g Samnyasa** (Hinduism)
Apoplexy
Description Here there is complete unconsciousness.
  Related Moha (Hinduism, #HM0992)
  Swoon (Hinduism, #HM1159).

♦ **HM1357c Madhya-loka** (Jainism)
Description According to the Jain system, this is the third level of the universe, including Mount Meru and the terrestrial world, the abode of humans.
  Related Adha-loka (Jainism, #HM3697)
  Siddha-loka (Jainism, #HM0569)
  Urdhva-loka (Jainism, #HM0131)
  Human development (Jainism, #HH0622).

♦ **HM1359d Illogical thinking**
Description Such thinking, which has obvious internal contradictions or arrives at obviously erroneous conclusions, arises when a person is distracted or fatigued, as well as in mental disorders.
  Related Delusive consciousness (#HM2600)
  Logical and metalogical awareness (#HM2320).

♦ **HM1362d Unawakened** (Buddhism)
Asambodha (Pali)
  Broader Delusion (Buddhism, #HM0184).

♦ **HM1363d Danger**
Risk
Description Danger is an elusive experience, partly engendered by the environment and partly by the person experiencing it. The experience may be enjoyed by some because of the risks. These can cause a feeling of arousal and a sense of superiority that comes from acting in a riskier manner than others. Hardship and danger may then appear attractive in their own right.
  Related Risk-taking (#HH0654).

♦ **HM1364b Insight**
Description In contrast to *intuition*, insight may be thought of as the understanding of something which may be described in concrete terms. Such understanding may, in some sense, be interpreted as *wisdom* as it connects different perceptions and passes them on to another in the form of a skill or ability. Insight also refers to an apprehension of one's own situation or problems and may arise after gradual accretion of self-knowledge, with facilitation of psychic reorganization and development of new emotional freedom.
  Refs Goldstein, J *The Experience of Insight* (1976); Lonergan, B *Insight, A Study of Human Understanding* (1970).

♦ **HM1365d Wabi** (Japanese)
Aesthetic simplicity
Description An aesthetic and moral principle, emphasizing the experience of beauty as simple and austere through a serene and transcendental frame of mind. It originates in a notion of poverty and loneliness as a liberation from material and emotional worries, turning the absence of apparent beauty into a new and higher form of beauty. Richness is experienced through poverty and beauty through simplicity. In art it is associated with the beautiful, distinctive, aesthetic flaw that distinguishes the spirit of the moment in which an object was created from all other moments in eternity. It is the inspired limitation that gives elegance to the whole. Learning to perceive wabi in the environment is a process of aesthetic development.
  Related Sabi (Japanese, #HM4350)
  Furyu (Japanese, #HM3387)
  Artistic education (#HH0029).

♦ **HM1367c Love of the whole**
Description A form of love in which the whole occurs in immediate openness to itself in the occurrence of the parts, such as when a group of people act in a way that reflects their bonding commitments as they live or work together. This occurs when a person is open to the unruly parts of his nature, to the awesome transcendence of mind, and to the boring, negative aspects of his own character — and when these same aspects are also open to other people. The openness itself reflects the whole as the parts live out their destinies and preferences. The whole of mind finds this stage of love as various parts become free for their openness. Mental accord happens as differences resonate each other in hearing openness as an immediate awareness of the whole in differentiating activity. Love is not then an attitude of a subject, but an open accord of the parts at once. The whole comes repeatedly to itself through the way the parts happen.
  Refs Scott, Charles E *Boundaries in Mind* (1982).
  Related Transference love (#HM7045).

♦ **HM1368f Modes of awareness associated with use of caffeine**
Description Caffeine-containing beverages and plants are consumed for their stimulant effect and increased conviviality. They give a lift and assist clear thinking. Caffeine intoxication may also be unpleasant, causing restlessness and insomnia, for example. Overdose may induce rambling of thoughts and speech, and sensory disturbance or, in extreme cases, seizure and respiratory failure.
  Broader Modes of awareness associated with psychoactive substances (#HM0584).

♦ **HM1369c Faculty of mind** (Buddhism)
Manindriya (Pali)
  Related Manas awareness (Hinduism, #HM2902).

♦ **HM1370d National consciousness** (Yoga)
Context A step on the path to cosmic consciousness, when the individual has not yet achieved conscious self-relatedness and lacks the courage to go against his national group.
  Related Conscious self-relatedness (Yoga, #HM0467).
  Followed by International consciousness (Yoga, #HM0099).
  Preceded by Racial consciousness (Yoga, #HM1070).

♦ **HM1371d Utter delusion** (Buddhism)
Pamoha (Pali)
  Broader Delusion (Buddhism, #HM0184).

♦ **HM1375d Tendency towards ignorance** (Buddhism)
Avijjanusaya (Pali)
Broader Delusion (Buddhism, #HM0184).

♦ **HM1376g Waking dreams**
Refs Epstein, Gerald N *Waking Dream Therapy* (1980); Watkins, Mary M *Waking Dreams* (1977).
Broader Dreams (#HM2950).
Related Fantasy (#HM0892)   Lucid dreaming (#HM3618)
Non–deterministic psychology (#HH2009).

♦ **HM1377d Composedness of the aggregate of sensation, of the aggregate of cognition, and of the aggregate of synergies** (Buddhism)
Vedanakkhandhassa sannakkhandhassa sankharakkhandhassa passaddhi (Pali)
Broader Composedness of body (Buddhism, #HM1867).

♦ **HM1380g Contact** (Buddhism)
Samphusana (Pali)
Broader Contact (Buddhism, #HM2708).
Related State of bodhisattva (Buddhism, #HM1225).

♦ **HM1381g Workability of the aggregate of sensation, of the aggregate of cognition, and of the aggregate of synergies** (Buddhism)
Vedanakkhandhassa sannakkhandhassa sankharakkhandhassa kammannata (Pali)
Broader Workability of body (Buddhism, #HM0739).

♦ **HM1382d Al-Wakil** (Sufism)
Description The believer is aware of Allah as the trustee. Only He acts, completing His work and solving all problems. He is the only one who can be trusted. The believer, having done his best, leaves the outcome to Allah and is at peace. Trusting in Allah, *Tawakkul*, he strives for what he wishes, knowing that such effort is a prayer and not counting on his own efforts but on Allah who can be trusted.
Context A mode of mystical awareness in the Sufi tradition associated with the fifty–second of the ninety–nine names of Allah.
Broader Ninety–nine names of Allah (Sufism, #HH2561).
Related Tawakkul (Sufism, #HM0807).
Followed by Al-Qawi (Sufism, #HM4988).
Preceded by Al-Haqq (Sufism, #HM3945).

♦ **HM1389b Cognition** (Buddhism)
Sanna (Pali)
Narrower Cognition born of contact with element of eye–consciousness (#HM1271).

♦ **HM1394d Nescience** (Buddhism)
Annana (Pali)
Broader Delusion (Buddhism, #HM0184).

♦ **HM1397b Daily variations in consciousness**
Circadian rhythm
Description During the day consciousness changes many times, so there are states of sleep, dreams, wakefulness, daydreams and intermediate states. The basis underlying this is the circadian rhythm, which depends on factors such as light and dark, sun position, working hours. When this regular pattern is replaced by an environment without such clues, then the circadian rhythm goes from 24 to 25 hours. The rhythm is evident in body temperature, which has peaks, troughs and levelling off at different times. Each person has his or her own pattern, feeling most alert and capable when the body temperature is high.
Refs Ornstein, Robert *The Psychology of Consciousness* (1986); Strober, C F and Luce G *The Importance of Biological Clocks in Mental Health* (1968).
Related Biological rhythms (#HH0475).

♦ **HM1398d Delightful** (Buddhism)
Pamojja (Pali)
Broader Fivefold happiness (Buddhism, #HM0747).

♦ **HM1399d Non–hatred** (Buddhism)
Adosa (Pali)
Broader Non–hatred (Buddhism, #HM2744).

♦ **HM1400e Emotional high**
Emotional peak experience — Hypnotic ecstasy — Hysteria
Description Generally describes the feeling of joy and ecstasy and of oneness achieved either spontaneously or as the result of effort and triggered by any of: meditation; chanting; music; prayer; consciousness, religious or other group activity. In the latter cases, the repetitive rhythm or the exhausting effect of dance may result in a hypnotic or hysteric trance–like state.
Related Euphoria (#HM3763)   Peak experiences (#HM2080)
High dream (Psychism, #HM0147)   Aesthetic state (Psychism, #HM3416)
Extreme exaltation (Psychism, #HM4152).

♦ **HM1402c Christ consciousness as dying–rising** (Christianity)
Description The Christian has died in Christ and is risen with Christ. The total self–emptying of Christ's death, actively and intentionally making himself vulnerable and giving himself over to the power of another, is the act in which he is raised and filled. Jesus is not bound by any forms, even those of life and death. In Christian consciousness, to hold onto one's life boxes it in, confines it and closes it from possible new directions. To lose one's life is to allow it to flow, to be open, shared or empty. To die means to rise, to be empty is to be full.
Broader Christ consciousness (Christianity, #HM2013).
Related Christ consciousness as empty mind (Christianity, #HM0778)
Christ consciousness as loving mind (Christianity, #HM0976)
Christ consciousness as forgiving mind (Christianity, #HM1728).

♦ **HM1403d State of unabated endurance** (Buddhism)
Anikkhittadharata (Pali)
Broader Faculty of energy (Buddhism, #HM1470).

♦ **HM1404d Comoción** (Spanish)
Crowd emotion
Description There is a feeling of emotional resonance with a large group of people, whether a whole nation (united in grief at the death of a beloved personage, for instance) or a physical crowd such as at a football match or a political rally. The feeling is thrilling but dangerous, as it can be incited by an orator or rabble–rouser to bring together an otherwise diverse group and encourage them to behave in a way which the individual members might normally find repugnant.
Related Group unconsciousness (#HM2602).

♦ **HM1405c Contemplative indifference**
Anonymous accord
Description A state in which a person's presence with things is through lack of interest, indifference, and this presence through a form of open emptiness is an awareness quite distinct from awareness through and in activities. There is increased awareness of things in their eventfulness. Increasingly the non–interest of the awareness and the non–interest of things, of whatever kind, becomes the sensed region of kinship. The very happening of awareness–with–things gains luminosity or audibility in this way. A sense of anonymous accord occurs, and if appropriated into on's other sensibilities, may grow into pervasive quality in one's usual consciousness. In this form of contemplation the person discovers that one kind of serenity happens with fantasies, things and people. Awareness in this case happens as a non–interfering region of acceptance which is not under normal intentional control. Happening resonates happening, and this resonance without subject or object is the serenity of contemplation.
Refs Scott, Charles E *Boundaries in Mind* (1982).
Related Acceptance (#HM1875)   Contemplation (#HM2952)
Fundamental dialogue (#HM1762).

♦ **HM1406c Composition**
Self–conscious coding
Description A radical socially–oriented form of awareness in which the body is treated as capable not only of production and consumption, but also of autonomous pleasure. Through this the individual creates his own relation with the world, endeavouring to tie other people into the meaning thus created. The individual identifies with his code creating activity, leading to the emergence of a free act of self–transcendence and a pleasure in being rather than in having. Composition is in effect trusting in direct experience, taking pleasure in the somatic experience of the body and ensuring an exchange between bodies. The world is transformed into an art form and life into a shifting pleasure. There is no fixation on artificially trying to recreate old codes through which to order communications between people. There is an experiential commitment to the permanent frailty of meaning unconstrained by the trap of repetitive communication habits.
Refs Attali, Jacques *Noise* (1985).
Related Reflexivity (#HM0087).
Preceded by Ontological love (#HM1211)   Self–remembering (#HM2486).

♦ **HM1407d Distortion of views** (Buddhism)
Ditthivisukayika (Pali)
Broader Wrong view (Buddhism, #HM1710).

♦ **HM1408c Nine–fold knowledge** (Systematics)
Experience of pattern
Description Experience would lose all coherence if there were not always sources of order residing in the patterns of organized structures. All experience is pervaded by the influence of such active sources of order which require a minimum of nine independent terms to be experienced. The ideal completion of eight–fold knowledge does not take into account the uncertainty and hazard encountered in actual experience. This nine–fold knowledge permits the representation of everyday working structures (disturbed by environmental factors) in which harmony is established and maintained. The harmonization is dynamic and indeterminate.
Context The ninth in a sequence of twelve modes of knowledge, identified by J G Bennett, inspired by G Gurdjieff.
Broader Systematics (#HH2003).
Related Enneagram patterning (Sufism, #HH3945).
Followed by Ten–fold knowledge (Systematics, #HM1804).
Preceded by Seven–fold knowledge (Systematics, #HM0776).

♦ **HM1409c Alchemical imagination** (Esotericism)
Description The imagination (as distinct from fantasy) is the basis for the major practice of alchemy towards cultivating the golden instants which occasionally illuminate ordinary life. Through such imagination the continuing creation of the world is reproduced within the individual. The dream is expanded beyond the limitations of individuality to encompass the cosmos. Finally the question is raised "Who dreams". The practitioner comes to see with the eyes of the spirit, as God dreams the phenomena of nature in sustaining its processes.
Broader Alchemy (Esotericism, #HH0221).
Related Guided visualization (Magic, #HH6053).

♦ **HM1410d Autotelic experience**
Description This is the experience of carrying out an activity for its own sake and not for any end result. Even activities initially commenced as *exotelic*, or with an end in view, may become *autotelic* if they become intrinsically rewarding. It is a matter of where the attention is focused. It is the opposite of the feeling that time – whether work or leisure – is being wasted. Alienation, boredom and helplessness are replaced by involvement, enjoyment and a feeling of control. The experience of the present is rewarding in itself and one does not depend upon (hypothetical) future gain. The state is potentially addictive, and although it may arise in rewarding activities it also arises in less acceptable situations. For some the inflicting of pain or the carrying out of daring crimes may be difficult to give up. It has been postulated that juvenile delinquency may be motivated by the search for an autotelic or flow experience which does not arise in ordinary life. Different people have different ability to transform ordinary experience into flow experience. Some people find it easier to control consciousness than others, so that schizophrenics and those suffering from attentional disorders are unlikely to have autotelic experiences, they are unable to concentrate psychic energy, the attentional processes are fragmented. It is also a matter of controlling psychic energy so that it is not overly directed towards the self (not being too self–conscious or self–centred), with the result that attentional processes are excessively rigid. Social pathology hindering autotelic experience are those of anomie and alienation, which can be thought of as equivalent to attentional disorders and self–centredness.
Refs Csikszentmihalyi, Mihaly *Flow* (1990).
Related Flow experience (#HM2344)   Flow experience through work (#HM4877).

♦ **HM1412d Being intended** (Buddhism)
Cetayitattam (Pali)
Related Intention (Buddhism, #HM2589).

♦ **HM1417d Group spiritual initiation** (Esotericism)
Description Through group initiation several individuals experience the process of spiritual initiation together. This follows from a new approach to life conditions, producing a widespread tread towards group awareness. The process is based upon a uniform and united group will, consecrated towards the service of humanity and based upon loyalty, cooperation and interdependence.
Refs Bailey, Alice A *The Rays and the Initiations* (1960).

Related Group consciousness (#HM3109) Spiritual initiation (Esotericism, #HH3390).

♦ **HM1418c Stasis** (Buddhism)
Thiti (Pali)
Broader Faculty of living (Buddhism, #HM3404).

♦ **HM1421d Mirth** (Buddhism)
Hasa (Pali)
Broader Fivefold happiness (Buddhism, #HM0747).

♦ **HM1423c Mindfulness** (Buddhism)
Dharanata (Pali)
Refs Brown, D P and Engler, Jack *The Stages of Mindfulness Meditation* (1980); Deatherage, O G *The Clinical Use of Mindfulness Meditation Techniques in Short-Term Psychotherapy* (1975); Kornfield, J M *The Psychology of Mindfulness Meditation* (1976); Soma, Bhikku *The Way of Mindfulness* (1949).
Broader Faculty of recollection (Buddhism, #HM4248).
Related Mindfulness (#HM4543).

♦ **HM1424c Straightness of body** (Buddhism)
Kayujukata (Pali)
Narrower Rectitude (#HM4250) Untwistedness (#HM1008) Uncrookedness (#HM4037) Undeflectedness (#HM1054)
Straightness of the aggregate of sensation, of the aggregate of cognition, and of the aggregate of synergies (#HM0409).

♦ **HM1425c Intuitive–projective faith**
Stage one faith
Description This is a fantasy–filled, imitative phase. The child, typically aged 3 to 7, is powerfully and permanently influenced by the example, moods, actions and stories of the visible faith of adults related to the child. The stage is marked by relatively fluid thought–patterns, by continually encountering novelties for which no stable operations of knowing have been formed. There is no restraint or inhibition of underlying imaginative processes by logical thought. Imagination, acting with forms of knowing dominated by perception, produces longlasting images and feelings, both positive and negative. At the stage of first self–awareness, the child is egocentric as regards others' perspectives. There is the first awareness of death and of sex, together with the strong taboos by which both cultures and individual families insulate these areas.
Context Stage 1 in the system of faith development described by James Fowler.
Refs Fowler, J W *Stages of Faith* (1981).
Broader Stages of faith (#HH2097).
Followed by Mythic–literal faith (#HM1525).
Preceded by Undifferentiated faith (#HM8672).

♦ **HM1430d Mimesis**
Active emotional identification
Description Submission to the spell of a performer involving active emotional identification with a speaker or a chorus in a dramatic setting. Through this process knowledge is transmitted at a directly experiential level that may be both relaxing and erotic. Learning occurs at the level of the body without separate intellectual analysis, commenting on the world at setting it at a distance. The information presented may be in poetic form encouraging a form of autohypnosis through which it was memorized. Such performances have an absorbing, repetitive, trance–like aspect. Understanding is thus achieved through absorption and loss of psychic distance. Such participation or identification is highly sensuous in nature. It is a mode of knowing that cannot be intellectually refuted because of its immediate, visceral quality.
Refs Berman, Morris *Coming to our Senses* (1990).

♦ **HM1433d Anger** (Leela)
Krodh
Description The ego is confronted with an aspect of the self it has rejected as evil, and the existence of the illusory self with which it identifies is threatened. Although violent anger is destructive, anger expressed through non–violence creates moral strength. It is based on love of truth, is impersonal and assists spiritual growth.
Context The third state or square on the board of *Leela*, the game of knowledge, appearing in the first row. The 3 signifies fire and zeal.
Broader First chakra: fundamentals of being (Leela, #HM4103).
Related Anger (Hinduism, #HM8665).
Preceded by Egotism (Leela, #HM1726).

♦ **HM1434c Sophrosune**
Soundness of mind — Soundness of heart
Description A wholeness and fullness of life, or a sense of well–being. An established, appropriate balance among things, right accomplishment of character, a good control of oneself. That accomplishment is given meaning by the continuing presence of passions that tend to disorder and madness. It is the very possibility of such intemperate possibilities that gives temperance meaning in relation to ongoing events. It is a recognition of the relation of measure to the sensed immeasureableness of destiny and the non–rationality of events. Sophrosune names how madness happens by establishing an integrated relationship to it such that chaos and disorder do not prevail. There is recognition of continuously living in orders of disorder, and that the sound heart and mind is a region in which madness goes on without conquering the whole. And the whole happens without conquering the madness and foolishness occurring in the parts.
Context A concept elaborated in classical Greece.
Refs Scott, Charles E *Boundaries in Mind* (1982).

♦ **HM1435d Lightness of the aggregate of sensation, of the aggregate of cognition, and of the aggregate of synergies** (Buddhism)
Vedanakkhandhassa sannakkhandhassa sankharakkhandhassa labuta (Pali)
Broader Lightness of body (Buddhism, #HM4395).

♦ **HM1436d Strive** (Buddhism)
Uyyama (Pali)
Broader Faculty of energy (Buddhism, #HM1470).

♦ **HM1439d Delusion an unwholesome root** (Buddhism)
Moho akusalamulam (Pali)
Broader Delusion (Buddhism, #HM0184).

♦ **HM1440f Delusions of ill–health**
Description Depressive individuals may believe that they are incurably ill, with detailed (and sometimes bizarre) beliefs as to their physical condition. In the case of schizophrenics, these beliefs may be supported by somatic hallucinations.

Broader Delusive consciousness (#HM2600).

♦ **HM1442e Being greedy** (Buddhism)
Lubbhana (Pali)
Broader Greed (Buddhism, #HM3283).

♦ **HM1451d Correct construing** (Buddhism)
Sammasankappa (Pali)

♦ **HM1452d Kashf** (Sufism)
Unveiling of divine mysteries
Description There is no veil between God and the heart of the devotee and nothing is hidden from him.
Context According to Shaykh Abu Sa'id ibn Abi'l–Khayr, this is the 33rd of 40 stations or maqamat the Sufi must possess for his journey on the path of Sufism to be acceptable.
Broader Stations of consciousness - ibn–Abi'l–Khayr (Sufism, #HM2424).
Followed by Khidmat (Sufism, #HM8776).
Preceded by Wisal (Sufism, #HM7119).

♦ **HM1454d Concentration accompanied by bliss** (Buddhism)
Concentration accompanied by ease
Description Here, as with concentration accompanied by happiness, there is unification with the first two jhanas in the fourfold reckoning and the first three in the fivefold reckoning; but there is also unification with the third jhana in the fourfold reckoning and the fourth in the fivefold reckoning. It is one of the kinds of access concentration.
Context According to Hinayana Buddhism, concentration is considered as of one kind (monad), of two kinds (dyads), of three kinds (triads), of four kinds (tetrads) or of five kinds (pentad). In the third triad, this is the second concentration.
Broader Concentration (Buddhism, #HM6663).
Related Access concentration (Buddhism, #HM4999).
Followed by Concentration accompanied by equanimity (Buddhism, #HM5008).
Preceded by Concentration accompanied by happiness (Buddhism, #HM4433).

♦ **HM1455c Fitness of mental body** (Buddhism)
Proficiency of mental factors — Kaya–pagunnata (Pali)
Description This refers to the proficiency of the three aggregates of feeling, perception and formation or mental activities. The characteristic is healthiness of the mental body and freedom from illness, the function is to crush its unhealthiness. It is manifest as absence of disability and freedom from evil. Mental factors are the proximate cause. It is the opponent of corruptions like faithlessness or diffidence causing illness or unhealthiness in the mental body. Fitness of mental body and fitness of consciousness are considered together.
Context One of the formations aggregate (mental coefficients) of Hinayana Buddhism, being listed among the constant states which appear in their true nature, and as profitable primary (always present in any profitable or profitable–resultant consciousness).
Broader Awareness as mental–formation group of conscious existence (Buddhism, #HM2050).
Narrower Good quality (#HM3424) Being of good quality (#HM0872)
Good quality of the aggregate of sensation, of the aggregate of cognition, and of the aggregate of synergies (#HM0349).
Related Fitness of consciousness (Buddhism, #HM1810).

♦ **HM1456c Awareness of the ultimate reality**
Vivekajam jnanam (Yoga) — Transcendence of time
Description Performing samyama on the process of time leads to a knowledge born of the awareness of the ultimate reality, the discrimination and knowledge being so far beyond what are normally termed discrimination and knowledge that it is actual awareness. Although it can be thought of as passing from a less to a more real state of consciousness, which is normally a matter of discernment, here the change is tremendous. Even those things which are normally indistinguishable by ordinary methods, which can be distinguished by the time factor alone, may now be distinguished. This is the highest knowledge, the ultimate objective of yoga, the only knowledge which is not relative, when all objects and processes are cognized simultaneously: past, present and future and transcending the world process. Although the world of the real is seen, the world of the relative is not left behind, simply it is seen in its true nature and perspective. Thus the self–realized yogi lives and works in the relative world unaffected by the illusions created by Prakriti.
Related Samyama (#HH0374) Viveka (Yoga, #HM2998) Viveka khyati (Yoga, #HM4493).

♦ **HM1461e Pure alertness** (Psychism)
Description As opposed to alert passivity, when one is very aware of external stimuli, the mind in pure alertness is unattached to externals but awake to inner realizations. This is a meditative state, the body being asleep.
Related Alert passivity (#HM3959) Restful alertness (#HM0996).

♦ **HM1465c Universalizing faith**
Stage six faith
Description This stage arises very rarely, the person having generated faith composition in which the feeling of an ultimate environment includes all beings. These are the incarnators and actualizers of the spirit of an inclusive and fulfilled human community. For those around them there is liberation from social, political, economic and ideological constraints and an awareness of participating in a power that unifies and transforms the world. Because of this they may be experienced as subversive to the religious and other structures which sustain corporate survival, security and significance, they may even be killed by those they are trying to change and are usually more honoured after their death than during their lives. These persons have a special grace to appear more lucid and simple and yet more fully human and their community extends universally. The particular is valued because it is a vessel of the universal not from utilitarian considerations. They love life yet hold to it loosely. The person of universalizing faith is ready for fellowship with persons at any other stage and from any other tradition.
Context Stage 6 in the system of faith development described by James Fowler.
Refs Fowler, J W *Stages of Faith* (1981).
Broader Stages of faith (#HH2097).
Preceded by Conjunctive faith (#HM3359).

♦ **HM1466f Over–valued ideas**
Dominant ideas
Description An over–valued idea is one which may or may not be plausible but which is affectively loaded, which preoccupies an individual and which dominates his personality. It is not really a delusion, in that it is in line with the individual's life experience and personality; nor is it an obsession, because the individual himself thinks of it as important and desirable. The idea itself may be perfectly logical, or at least held in common with other people, but it dominates so that the individual defends the belief and tries to spread it, may even be prepared to die for it. The emotional intensity with which the idea is charged indicate that it is not based on rational analysis.

HM1466

The vigour with which an over-valued idea is held may be understood in terms of cognitive dissonance, the individual attempting to reduce psychological tension by gaining social acceptance for his idea.
Refs Reed, Graham *The Psychology of Anomalous Experience* (1988).
Related Obsession (#HM2809)   Cognitive dissonance (#HH8204)
Delusive consciousness (#HM2600).

♦ **HM1467d Al–Hasib** (Sufism)
Description The believer is aware of Allah as the reckoner, taking account of all that he does throughout his life. At the last judgement, on the day of reckoning, each will have to account for all the capital he has been lent, for how he has spent his time.
Context A mode of mystical awareness in the Sufi tradition associated with the fortieth of the ninety–nine names of Allah.
Broader Ninety–nine names of Allah (Sufism, #HH2561).
Followed by Al–Jalil (Sufism, #HM6771).
Preceded by Al–Muqit (Sufism, #HM3171).

♦ **HM1468c Vespers** (Christianity)
Hours of vespers
Description The canonical hours of vespers traditionally take place between 6 and 9 pm. They dramatize giving thanks to God for the day's work being completed. The mood is fervent thanksgiving.
Broader Canonical hours (Christianity, #HM1167).
Followed by Compline (Christianity, #HM4321).
Preceded by None (Christianity, #HM0466).

♦ **HM1469c Excitement** (Buddhism)
Uddhacca (Pali)
Narrower Agitation (#HM0436)   Turmoil of thought (#HM0400)   Psychical perplexity (#HM4425)
Excitement of thought (#HM0975).
Related Excitement (Buddhism, #HM2534).

♦ **HM1470c Faculty of energy** (Buddhism)
Viriyindriya (Pali)
Narrower Strive (#HM1436)   Ardour (#HM3357)   Energy (#HM1333)
Stamina (#HM1749)   Exercise (#HM1496)   Exertion (#HM3469)
Firmness (#HM1974)   Endeavour (#HM0229)   Power of energy (#HM1943)
Faculty of energy (#HM1829)   Correct endeavour (#HM1657)
Zeal (Buddhism, Pali, #HM4487)   State of unabated desire (#HM4131)
Strong grip of the burden (#HM1081)   State of unabated endurance (#HM1403)
Physical inception of energy (#HM1263)   State of unfaltering exertion (#HM3261).

♦ **HM1475d Spiritual devotion** (Leela)
Bhakti-loka
Description The most direct approach to cosmic consciousness, in this state the player is attracted by nothing but the beloved, the divine. When the pain of separation is replaced, through divine grace, with union, then there is awareness of non-duality. He accepts play – lila – as his basic nature. Each square is a manifestation of the divine and he is one with all. While the previous chakra produces the wise man, this produces the divine child held in his mother's lap and protected by his father.
Context The 54th state or square on the board of *Leela*, the game of knowledge, appearing in the sixth row.
Broader Sixth chakra: time of penance (Leela, #HM4412).
Followed by Plane of cosmic consciousness (Leela, #HM1301).

♦ **HM1478e Limbo**
Description This is a state of transition or intermediate state after death. According to Roman Catholicism it is the abode of unbaptized souls; in psychic terms it is one of seven planes of density or the level of the virtuous who died before the birth of Christ. According to some definitions it is an area of hell experienced as banishment from God.

♦ **HM1479g Hyperaesthesia**
Description In this state, one or more of the senses is hyperactive. There may be an ability to "read" written words by touch when blindfold, to read in the dark, to hear whispered words in another room, to respond to sights or smells imperceptible to others.
Broader Paraesthesia (#HM0003).

♦ **HM1483g Torpor**
Torpid consciousness
Description Torpor is sluggish mental reaction to stimuli, or reaction only to strong stimuli. There is drowsiness which is pathological, often resulting in falling into deep sleep. It is almost impossible to focus attention or thought; perception is difficult and fragmentary; there is no feeling of interest or motivation; affect is flattened.
Broader Diminished clarity of awareness (#HM6201).
Related Sopor (#HM8006)   Torpor (Buddhism, #HM0264)
Bettschwere (German, #HM1600)   Unconscious stupor (#HM2473)
Absence of consciousness (#HM2670).

♦ **HM1484c Remembrance** (Buddhism)
Saranata (Pali)
Broader Faculty of recollection (Buddhism, #HM4248).

♦ **HM1485d Plane of fantasy** (Leela)
Naga-loka
Description Imagination is unrestrained, nothing is impossible and the player may produce great works of art or invention. Carried away by fantasy, he may lose contact with reality and not see what lies ahead. This is the plane of the snake, of flexibility, of ability to change form.
Context The 15th state or square on the board of *Leela*, the game of knowledge, appearing in the second row.
Broader Second chakra: realm of fantasy (Leela, #HM3651).

♦ **HM1486c Jewel of wisdom** (Buddhism)
Pannaratana (Pali)
Broader Faculty of wisdom (Buddhism, #HM3233).

♦ **HM1490c Composedness of thought** (Buddhism)
Cittapassaddhi (Pali)
Narrower Composure (#HM1738)   Calming down (#HM1761)
Completely tranquillized (#HM1528)   Complete tranquillization (#HM3411)
Workability of the aggregate of consciousness (#HM4605).

♦ **HM1496d Exercise** (Buddhism)
Nikkama (Pali)
Broader Faculty of energy (Buddhism, #HM1470).

♦ **HM1498d Quietude** (Buddhism)
Samatha (Pali)
Broader Sukha (Buddhism, #HM2866).

♦ **HM1499c Erudition** (Buddhism)
Pandicca (Pali)
Broader Faculty of wisdom (Buddhism, #HM3233).

♦ **HM1500d Al-Qadir** (Sufism)
Description The believer is aware of Allah as the all powerful, able to do all He wills. Creation is a mirror reflecting His power. Contemplating the wonders of creation, how they are created at God's will, the believer prostrates himself in awe and respect.
Context A mode of mystical awareness in the Sufi tradition associated with the sixty-ninth of the ninety-nine names of Allah.
Broader Ninety-nine names of Allah (Sufism, #HH2561).
Followed by Al-Muqtadir (Sufism, #HM3576).
Preceded by As-Samad (Sufism, #HM4980).

♦ **HM1501d Cognition born of contact with element of mind-consciousness** (Buddhism)
Manovinnanadhatu samphassaja sanna (Pali)
Related Mental (Buddhism, #HM0251)
Element of mind-consciousness (Buddhism, #HM1834).

♦ **HM1502c Mourning**
Description Mourning runs through different stages and may, in the case of a death coming slowly (through illness, for example), commence before death occurs. If the person affected is unable to face up to the coming death there may be confusion and anger – they may even avoid the dying person which brings remorse after the death of that person. However, acknowledgement of the approaching death can bring reconciliation to any grievances that may have arisen during life. When the death actually takes place, there is emotional turmoil and the desire to do something which would protect or please the dead person. If this is not successful, if the funeral is not as he or she would have wanted it, for example, there may be panic and a feeling of being overwhelmed, perhaps incoherence. At the other extreme a state of dissociation may arise, self-protection through loss of recent memories. Then follows a phase of turning away from feelings through avoiding reminders of the death, although the dead person may seem alive in dreams. The mourner may be numb to all emotion, maybe abusing drugs or alcohol or taking part in frenzied activity, until the next phase where there is a mental reviewing of life with the deceased and the beginning of adjustment to the loss. At this time there may, however, be a reaction of recurring nightmares and overwhelming rage, despair, shame, guilt or fear. The next phase is intense yearning for the company of the dead person, again a denial of the death, until this yearning yields to emotional acceptance. Until this final stage is reached, not only is there an inability to work, to be caring or creative or to feel pleasant feelings, there may also be anxiety, depression, rage followed by shame and guilt. Until mourning is completed the mourner has no feeling of mastery over his or her life.
Related Grief (#HM2685)   Dissociation of affect (#HM6087).

♦ **HM1503d Nostalgia**
Description A state in which yearning is felt for something in the past which is irrecoverable.
Related Yoin (Japanese, #HM0285).

♦ **HM1507c Way of love-wisdom** (Esotericism)
Description Characteristic of people favouring inclusiveness, cooperation, brotherhood and group consciousness. On this way they are challenged by the task of refining their positive attitude, overcoming inertia, using their will, acting decisively, and taking hard and painful decisions.
Context The second of seven ways to spiritual realization characteristic of Western esotericism.
Broader Ways to spiritual realization (Esotericism, #HH2665).
Related Sphere of mercy and greatness (Kabbalah, #HM2420)
Initiation of resurrection (Esotericism, #HM1153).

♦ **HM1508c Conservation** (Buddhism)
Palana (Pali)
Broader Faculty of living (Buddhism, #HM3404).

♦ **HM1509f Phobia**
Description This is a state of probably irrational fear produced in response to a specific impulse which varies from person to person, for example agoraphobia (fear of open spaces).
Broader Fear (Christianity, #HM3311).
Related Mania (#HM2787).

♦ **HM1510d Not-seeing** (Buddhism)
Adassana (Pali)
Broader Delusion (Buddhism, #HM0184).

♦ **HM1511d Fetters of views** (Buddhism)
Ditthisannojana (Pali)
Broader Wrong view (Buddhism, #HM1710).

♦ **HM1513e Feeling greed** (Buddhism)
Lubbhitattam (Pali)
Broader Greed (Buddhism, #HM3283).

♦ **HM1516c Fourth order perceptions – control of transitions**
Description Here there is perception of motion as the same configuration presented in rapid sequence in slightly different positions. Music, for example, is the transition that changes sound into melody. In a ball game, perceiving the ball at the level of fourth order perception enables the body to respond automatically and hit the ball correctly. It will do so even more if one has previously watched a good player again at the level of the fourth order. Opening up higher stations decreases the control.
Context The fourth of ten orders in the perceptual system described by William Glasser.
Broader Orders of perception (#HM0988).
Followed by Fifth order perceptions – control of sequence (#HM0772).
Preceded by Third order perceptions – configurations (#HM0312).

♦ **HM1525c Mythic–literal faith**
Stage two faith
Description At the stage of the school child (although sometimes also dominant in adolescence and adulthood), stories, beliefs and observances, symbolizing the belonging to a particular community are appropriated with literal interpretations, as are moral rules and attitudes. Symbols are taken literally and one–dimensionally. The imaginative composing of the world in the previous stage is curbed and ordered in the rise of concrete operations. Whereas intuitive–projective faith was episodic, mythic literal faith has a more linear and narrative construction of coherence and meaning. Taking the perspective of other persons is marked by increased accuracy and a world is composed that is based on reciprocal fairness and an immanent justice founded on reciprocity. The actors in cosmic stories are anthropomorphic. The child is deeply affected by symbolic and dramatic materials and can describe in great detail what has occurred, although it does not step back to formulate reflective, conceptual meanings – the meaning is carried by but also trapped by the narrative. The rise of narrative and emerging story, drama and myth as ways of finding and giving coherence to experience is a new capacity or strength. The faith at this stage may be an overcontrolled or stilted perfectionism, "righteousness by works", or an abasing sense of badness held because of others' mistreatment, neglect or apparent disfavour. This is due to the limitations of literalness and over–reliance on reciprocity.
The clash or contradictions implied in stories may lead to reflection on meanings, and this initiates transition to Stage 3. Literalism is broken down and "cognitive conceit" leads to being disillusioned with old teachers and teachings. A mutual interpersonal perspective emerges, creating the need for more personal relations with the unifying power of the ultimate environment.
Context Stage 2 in the system of faith development described by James Fowler.
Refs Fowler, J W *Stages of Faith* (1981).
  Broader Stages of faith (#HH2097).
  Followed by Synthetic–conventional faith (#HM1814).
  Preceded by Intuitive–projective faith (#HM1425).

♦ **HM1528d Completely tranquillized** (Buddhism)
Patipassambhatattam (Pali)
  Broader Composedness of body (Buddhism, #HM1867)
  Composedness of thought (Buddhism, #HM1490).

♦ **HM1529g Birth consciousness**
Description The consciousness of the new-born child, said to be a subconscious level of awareness and thinking when logical reasoning is lacking; gradually superseded until around 7 years old when mental mind consciousness takes over.

♦ **HM1534c Detachment** (Christianity)
Abgeschiedenheit
Description Everything loved for its own sake, outside God alone, blinds the intellect and destroys judgement of moral values. It vitiates choices so that good from evil can not be distinguished and God's will can not be truly known. A person can have given up the pleasures and ambitions of the world but if he has acquired pleasures and ambitions of a spiritual nature, he is not detached. Spiritual detachment means giving up prayer, fasting, meditation, devotions, virtues and any other practice, idea or attitude that is valued for itself. Interior peace and a sense of the presence of God are as created as an automobile or a glass of beer and, as such, are imperfections preventing union with God. It is only when any attachment to all knowledge, all created wisdom, all pleasure, all prudence, all human joy, all striving, even after the love of God, and all human hope is defeated will a person be free. Because this detachment is a gift of God, it can not be bought by any act of will; it can not be acquired by spiritual exercises. At the same time a person can prepare himself to receive this gift by seriously undertaking to totally renounce all attachments. To rest in the beauty of God is natural, can be desired by nature and can be acquired by natural disciplines and can be a source of attachment; it also must be given up.
For Meister Eckhart, detachment is the immovability of the spirit despite joy or sorrow, honour or disgrace, because these are transient things, they are not God. This is how man is like God. Detachment leads man to purity, from purity to simplicity and from simplicity to understanding. By grace, man is drawn away from the temporal and purified from the transient. It is by the practice of detachment that one reaches the soul's ground and attains similarity and therefore union with God. Detachment does not imply indifference to the sufferings and failings of the world. They are real and so is one's duty to relieve them. But in attending to the needs of the world and alert to the demands of the moment one must not be self-seeking. There is obedience to God's will and indifference to one's own personal feelings.
Refs Merton, Thomas *Seeds of Contemplation* (1972); Smith, Cyprian *The Way of Paradox* (1987).
  Related Detachment (Hinduism, #HM5091)   Non–attachment (Buddhism, #HM2128)
  Way of paradox (Christianity, #HH3335).
  Birth of God in the soul's ground (Christianity, #HM6522).

♦ **HM1537d Alternate state of consciousness** (ASC)
Description Hilary Evans contrasts this state with the *usual* state of consciousness. Any state not experienced as usual is implied. It is characterized by significant changes in perception and behaviour and may arise spontaneously or be entered by choice. There is some loss of self-control and sense of identity, and an awareness of not being one's usual self.
Refs Evans, Hilary *Alternate States of Consciousness* (1989).
  Related Altered states of consciousness (#HM2391)
  Usual state of consciousness (USC, #HM1999).

♦ **HM1538c Poetic imagination**
Sacred analogy — Symbolic language of tradition — Language of divine analogy — Sacred imagination
Description A form of language which is proper to imaginative and creative discourse. It is through such discourse, at once universally and perennially the language of man's spiritual destiny, that the needs of the human spirit are met so as to allow man to reach beyond the confines of a merely quantitative order of things. Poetry is one such language of analogy whose terms establish relationships of a mental character, and within an immaterial order that does not exist for the positivist mentality. By means of analogical and symbolic thinking, by mediating images and the play of correspondences and unusual associations, and by virtue of a language through which is transmitted the very rhythm of being, the poet clothes himself in a surreality to which the scientific mode of thought cannot aspire.
Poetry may thus be experienced as a mode of thought, or kind of consciousness, in whose expression mythological themes, images, ideas, languages and music are indivisible. True poetry has the power of transforming consciousness itself by presenting icons, images of forms only partially and superficially realized in ordinary life. The beauty of their truths, rectifies and informs the formless reality (or unreality) of the everyday world. The recognition of such beauty is immediate and intuitive, but it is the response of a higher faculty than the discursive reason.
Knowledge of the symbolic language of tradition is essentially a kind of learning, but it is the learning of the imagination, not of the merely conceptual mind. It is the learning of poets and of symbolic art. From this perspective the created world is, at every level, a manifestation of anterior causes. Symbolic art is thus the natural language of such thought. Thus an image of apparent simplicity may contain a resonance that sets into vibration planes of reality and consciousness other than those of the sensible world. The language of symbolic analogy is only possible upon the assumption that these multiple planes exist. Those for whom the material world is the only plane of the real are unable to understand that the symbol (and poetry in the full sense of symbolic discourse by analogy) has as its primary purpose the evocation of one plane in terms of another.
All poets, and all readers of poetry who pass beyond the writing or reading of poetry for merely descriptive purposes, cross a frontier from the personal world into the world of those experiences which lie beyond the reach of everyday consciousness, but to which, in moments of greatest vision, of expanded consciousness, may be occasionally glimpsed. Those poets for who imagination is meaningful endeavour to embody insights from subtler planes into an imagery of perfect correspondence in which thought and image are one (simple), perfectly realized in the image (sensuous) and felt as living experience (passionate), and not merely conceptually comprehended. In contrast to the analytic distinctions of philosophy, such poetry brings together, creating wholes and harmonies in a synthesis that may be experienced. The poem is thus able to create in the reader a sense of the wholeness and harmony its symbols and its rhythmic unity both realize and affirm. It is impossible to experience such an interior and archetypal vision without at the same time experiencing it qualitatively, as an epiphany of knowledge for which such images are the vehicles.
Context In western thinking spiritual knowledge is embodied and transmitted principally within that tradition which descended from Orphism to Plato, to the neo–Platonists and the Gnostic sects, and their successors both within Christendom and without it. It is the language of alchemy and the kabbalah, and of allied ways of thought.
Refs Black, Michael *Poetic Drama* (1977); Raine, Kathleen *Defending Ancient Springs* (1985); Trevelyan, George *Magic Casements* (1980).
  Related Anamnesis (#HM1872)   Apokatastatis (#HH6071)   Poetic enthusiasm (#HM0875).

♦ **HM1539d Pain of solitude**
Loneliness
Description Being alone with nothing specific to do can bring an intolerable sense of emptiness, there seems to be a need for a sense of purpose, the exotelic experience of a goal to work for. Without external input, attention wanders, thoughts become chaotic. Without control of consciousness the mind relaxes and worries immediately creep in. Typically, the individual tries to escape from solitude through TV; or, more drastically, through consumption of alcohol or other drugs, so that the self is no longer responsible for the direction of psychic energy; or through obsessive habits. None of these develop attentional habits that might lead to a greater complexity of consciousness. On the contrary, the ultimate test of the ability to control the quality of experience is what a person does on his own. Filling free time with activities requiring concentration, increase skills and develop the self, while still coping with the threat of chaos, also leads to growth. It requires learning to use the time one has alone instead of simply escaping from it. This control of consciousness enables development of discipline and is particularly important for the young who need complex skills learned on their own in order to find well paid and satisfying jobs. They also need to develop discipline for later life, so that attention can be controlled in solitude when one will need to turn one's energies from mastery of the external world to a deeper exploration of inner reality.
Refs Csikszentmihalyi, Mihaly *Flow* (1990).
  Related Solitude (#HH0333)   Loneliness (#HM1260).

♦ **HM1545d Felicity** (Buddhism)
Vitti (Pali)
  Broader Fivefold happiness (Buddhism, #HM0747).

♦ **HM1547c Lustre of wisdom** (Buddhism)
Pannaabhasa (Pali)
  Broader Faculty of wisdom (Buddhism, #HM3233).

♦ **HM1554c Proficiency** (Buddhism)
Kosalla (Pali)
  Broader Faculty of wisdom (Buddhism, #HM3233).

♦ **HM1557d Non-hating** (Buddhism)
Adussana (Pali)
  Broader Non–hatred (Buddhism, #HM2744).

♦ **HM1561d Being in doubt before two alternatives** (Buddhism)
Dvelhaka (Pali)
  Broader Perplexity (Buddhism, #HM0812).

♦ **HM1563c Recollection** (Buddhism)
Sati (Pali) — Recolleection
Refs Bhikshu, Ven Sumedho *Handbook for the Practice of Dhamma*.
  Broader Faculty of recollection (Buddhism, #HM4248).
  Related Mindfulness (Buddhism, #HM2847).

♦ **HM1564e Afflatus**
Description In a relaxed state of consciousness, knowledge is transmitted from the superconscious to the conscious mind.

♦ **HM1576d Reverie**
Description A state of mind in which words come to presence bearing experiences and depth of meaning that originate outside a person's region of consciousness and identity. As control of language gives way to the words themselves, a state of mind develops in which mind occurs in awareness of itself as words come forth. A spaciousness emerges, a space of allowance, that nurtures words and relations of words. Fantasies grow and take on lives of their own, enabling the person to live through the sounds of a poem. There is an awareness without concepts and a deep unfolding of mind's engendering. The person may witness the mind's reflecting itself to itself in words of creation, learning through reverie to hear what is not spoken and appears initially incomprehensible. Heeding and disclosure of things outside the person's identity are the prominent characteristics.
Refs Bachelard, Gaston *The Poetics of Reverie* (1971).
  Related Fantasy (#HM0892)   Reverie (Psychism, #HM0967)
  Immediate boundary awareness (#HM0035).

♦ **HM1584c Workability of thought** (Buddhism)
Cittakammannata (Pali)
  Narrower Being workable (#HM3716)   Working ability (#HM0979)
  Workability of the aggregate of consciousness (#HM4605).

HM1585

♦ **HM1585d Wrong endeavour** (Buddhism)
Micchavayama (Pali)
Broader Endeavour (Buddhism, #HM0229).
Related Correct endeavour (Buddhism, #HM1657).

♦ **HM1586c Samatva** (Yoga)
Equanimity
Description Closely related with *samtosa*, or contentment, equanimity has several grades and levels: (a) Indifference to material things. (b) Indifference to sorrow and pleasure and to the fluctuations in life. (c) Sameness of vision, where what ever is looked upon reveals the One.
Related Samtosa (Yoga, #HM2898)   Equanimity (Buddhism, #HM7769).

♦ **HM1590d Conscience** (Buddhism)
Hiri (Pali) — Conscientiousness — Shame
Description In Hinayana Buddhism, this is the formation that has conscientious scruples and may also be referred to as modesty; its characteristic is disgust or abomination of evil and its proximate cause is respect for self.
Context One of the formations aggregate (mental coefficients) of Hinayana Buddhism, being listed among the constant states which appear in their true nature, and as profitable primary (always present in any profitable or profitable–resultant consciousness). Although shame and conscience are considered closely related they are listed as separate formations. Nevertheless there is some confusion, as, in Tibetan Buddhism, 'hri' is translated as 'shame'. Together they are referred to as the 'guardians of the world'.
Broader Awareness as mental-formation group of conscious existence (Buddhism, #HM2050).
Related Shame (Buddhism, #HM8112)   Conscientiousness (Buddhism, #HM3220)
Consciencelessness (Buddhism, Pali, #HM4394).

♦ **HM1591d Causing harm** (Buddhism)
Byapajjana (Pali)
Broader Hatred (Buddhism, #HM4502).

♦ **HM1593c Religious experience as meditational insight**
Description Experience shows that disciplines of study and meditation may lead to realization of identity of the individual self with the one Self, the ultimate reality. This realization is beyond intellectual or conceptual understanding but a total experience of unity, of identity with the ultimate. Many systems may be quoted from all religions and cultures and most involve the guidance of a teacher or sage who is further along the path of knowledge and can guide his disciple. Aids to attain this unitive insight include the numerous systems of meditation and yoga, in particular in Buddhism, such as the *eightfold path* and *four noble truths*; but similar meditative practices occur in the Confucian tradition, Taoism, Shintoism, and in movements within Islam (Sufism), Judaism and Christianity (as for example the contemplative hermits of the Orthodox Church).
Broader Religious experience (#HH3445).
Related Meditation (#HH0761)   Contemplation (#HM2952)
Jnana marga (Hinduism, #HH0495)   Dhyana (Hinduism, Buddhism, #HM0137)
Four noble truths (Buddhism, #HH0523)   Human development (Buddhism, #HH0650)
Eightfold path of yoga (Yoga, #HH0779).

♦ **HM1594f Modes of awareness associated with use of cocaine**
Use of crack
Description As a mood–elevating substance, cocaine produces very rapidly an intense euphoria which also dissipates very fast. There may be visual or tactile hallucinations. Intoxication may develop into delirium. The "crash" following the elevation of mood may include: dysphoria; craving for more cocaine; anxiety; irritability; depression; fatigue. These may continue long enough to be considered withdrawal symptoms. In this case there may be paranoid or suicidal feelings. Delusions may last over a year. The use of *crack* cocaine has even swifter effect than cocaine itself. The crack high reinforces feelings of power and aggression. It lasts a short time, and the high/crash cycle is typically only one hour.
Broader Modes of awareness associated with psychoactive substances (#HM0584).
Related Formication (#HM4094).

♦ **HM1596d Svabhava** (Buddhism)
Own–being — Inherent existence

♦ **HM1597d State of not feeling hatred** (Buddhism)
Adussitattam (Pali)
Broader Non-hatred (Buddhism, #HM2744).

♦ **HM1598d Hinekurata** (Japanese)
Warpedness — Bitterness
Description This is warped or twisted state of the embittered, begrudging and resentful person, a state arising from the frustration and consequent sulking when pretended indifference to one's needs brings envy towards others whose needs are apparently satisfied. It is contrasted with sunao–na, the uprightness and compliance which is one of the most valued personality traits in interpersonal relationships.
Related Sunao na (Japanese, #HM2917).

♦ **HM1600g Bettschwere** (German)
Decadent sleepiness
Description Despite no lack of sleep and no physical exhaustion from over–working, there is a feeling of languor or torpor so that the effort to leave bed is impossible and the individual remains semi–conscious or goes back to sleep.
Related Torpor (#HM1483).

♦ **HM1601d Conscience** (Leela)
Vivek
Description Inner wisdom differentiates the essence from the temporal form and the player is protected from reverting to attachment to objects of sense perception. Insight into the future arises in perception through the third eye; all the collective unconscious is at the player's disposal. Landing here the player is transported to the next chakra, to happiness.
Context The 46th state or square on the board of *Leela*, the game of knowledge, appearing in the sixth row.
Broader Sixth chakra: time of penance (Leela, #HM4412).
Followed by Happiness (Leela, #HM5099).

♦ **HM1602c Research** (Buddhism)
Pavicaya (Pali)
Broader Faculty of wisdom (Buddhism, #HM3233).

♦ **HM1605c Way of organization** (Esotericism)
Way of ritual
Description Characteristic of those skilled in organizing people and things, creating order out of chaos, restructuring the environment into new patterns, often for their own sake. On this way they are challenged by the task of transcending mechanistic and routine forms of organization and ensuring that the patterns with which they work are in harmony with the rhythms of nature.
Context The seventh of seven ways to spiritual realization characteristic of Western esotericism.
Broader Ways to spiritual realization (Esotericism, #HH2665).
Related Sphere of beauty (Kabbalah, #HM3031)   Initiation of birth (Esotericism, #HM1337).

♦ **HM1606b Shalom**
Kingdom of God
Description When the whole of life is in right relationship with God, with others and with the creation then that state has been described as shalom. This encompasses security for every part of creation, justice and love infusing every relationship. The result is not a static, non–evolving situation but one in which all situations are transformed, possessiveness becoming a willing sharing of resources, the power to dominate becoming the power to heal, estrangement becoming shared trust and confidence. The kingdom or realm of God, shalom, is said to be inherent to creation and available in all situations. The term *metanoia* is used to describe the transforming of a situation so that the realm of God is manifest, the aim being the restoration of integrity - wholeness - oneness.
Related Metanoia (Christianity, #HH0888).

♦ **HM1608d Glossolalia** (Christianity)
Speaking in tongues — Praying in tongues — Jubilatio — Xenoglossis
Description In this heightened state of consciousness, an individual or group (jubilatio) utters unintelligible but fluent sounds in praise or prayer to God. Although the individual has control of whether or not to speak, how fast or slowly, how loudly or quietly, the actual sounds are not of his volition, God praising God through the medium of the individual without him knowing what he is saying. Regarded as one of the nine gifts of the Holy Spirit, it is said to benefit the individual practising it and those for whom he prays, although an interpreter is normally necessary for individual public prayer in this manner. Interpretation is not translation but making the meaning clear, as in interpreting dreams. There are exceptions to the need for an interpreter when the person speaking in tongues utters intelligible sounds not of his volition but in a comprehensible language, perhaps a language unknown to the speaker. Speaking in tongues is said to be a clear channel between the speaker and God, since it is the Spirit praying and He knows what should be prayed for better than the speaker himself. It is a form of devotion which assists in building up the individual spiritually, a praying in the spirit rather than the mind.
Refs Gunstone, John *Baptised in the Spirit* (1989).
Related Human development (Christianity, #HH2198)
Charismatic renewal (Christianity, #HH3124).

♦ **HM1612b Purity of Heart**
Description Kierkegaard says "Purity of heart is to will one thing, and that is the good". Christ said that the pure in heart are blessed because they will see God. In this state there is deliverance from images and concepts, from the forms and shadows of things desired with the appetites. One is delivered from feeble and delusive analogies normally used to arrive at God in the grip of a deep and penetrating experience.
Refs Merton, Thomas *Seeds of Contemplation* (1972).

♦ **HM1613d Thicket of views** (Buddhism)
Ditthigahana (Pali)
Broader Wrong view (Buddhism, #HM1710).

♦ **HM1615e Akuhaiamio** (Psychism)
Description Psychic knowledge gives inner awareness of what to decide and to do.
Related Buddhi awareness (Hinduism, #HM2099).

♦ **HM1616f Cyclothymia**
Manic depression
Description Human life is seen as a cyclical process, with periods of exhilaration alternating with times of depression. In cyclothymia, these moods reach abnormal peaks and troughs which inhibit normal life and dealings with others.
Related Mood (#HM1748)   Mania (#HM2787)   Depression (#HM2563)
Exhilaration (#HM0063).

♦ **HM1617d Fixation** (Buddhism)
Appana (Pali)
Related Focusing (Buddhism, #HM0487)   Mark of calm (Buddhism, #HM8673).

♦ **HM1618c Appearance of absorption** (Buddhism)
Description Guiding the mind and confronting the subject of meditation (for example, the earth kasina), there is a knowledge that absorption will succeed. Mind-door adverting arises with the same kasina as object, cutting the life continuum or subconsciousness. Then follow four or five flashes of apperception, the last of the fine–material sphere or form–realm of the jhana to be entered, the others from the realm of sense. These latter have, nevertheless, stronger applied and sustained thought, happiness, bliss and unification of mind than usual. The first may be preliminary work, the second access, the third conformity and the third change of lineage, as it transcends the sphere of the senses; the fifth, in the fine material sphere, is absorption consciousness, when absorption is fixed. The first may be omitted, as is the case when there are four flashes.
Narrower Access concentration (#HM4999)   Absorption concentration (#HM0311).
Related Skill in absorption (Buddhism, #HH4777)
Dwelling in the first jhana (Buddhism, #HM4298)
Dwelling in the third jhana (Buddhism, #HM5643)
Dwelling in the second jhana (Buddhism, #HM7121)
Dwelling in the fourth jhana (Buddhism, #HM8087)
Dwelling in the fivefold jhana (Buddhism, #HM6553)
Knowledge of the divine ear element (Buddhism, #HM5982)
Knowledge of recollection of previous existence (Buddhism, #HM4297)
Knowledge of passing away and reappearance of beings (Buddhism, #HM0748).

♦ **HM1623c Breadth of wisdom** (Buddhism)
Bhuri (Pali)
Broader Faculty of wisdom (Buddhism, #HM3233).

♦ **HM1626c Al–khatim** (Sufism)
Description The name of a traveller on the *tariqa* or Sufi spiritual path - the seal who is the last prophet of God.
Context Spiritual progress on the spiritual path is marked by the attainment of specific *maqamat*

or station.
Broader Mystic stations (Sufism, #HM3415).

♦ **HM1632c Intuition** (Buddhism)
Vipassana (Pali) — Experiential insight
Description This is the insight when an object is thoroughly penetrated. It is particularly used to describe the full knowledge acquired by discerning transitoriness (anicca), suffering (dukkha) and non-self (anatta), the three characteristics of the phenomenal world.
Refs Vaughan, F *Awakening Intuition* (1979).
Broader Faculty of wisdom (Buddhism, #HM3233).
Related Vipassana-bhavana (Buddhism, Pali, #HH0680).

♦ **HM1633d Al-Wali** (Sufism)
Description The believer is aware of Allah as the protecting friend of His righteous servants. All their difficulties are eliminated by Him, they have guidance, peace and success now and hereafter. Their hearts are expanded, no longer attached to the present. They know the Lord as His friends. They learn from all they see and hear and are enlightened by divine light so that all who see them are reminded of Allah. They feel neither fear nor sadness, needing and expecting nothing from anyone except Allah. Being with them one learns to be like them.
Context A mode of mystical awareness in the Sufi tradition associated with the fifty-fifth of the ninety-nine names of Allah.
Broader Ninety-nine names of Allah (Sufism, #HH2561).
Followed by Al-Hamid (Sufism, #HM0392).
Preceded by Al-Matin (Sufism, #HM5633).

♦ **HM1635f Cosmological mysticism**
Description This may be an experience of reality illuminated from within. The "doors of perception are cleansed" so that "everything appears to man as it is, infinite". What has been the inspiration of poets and "nature mystics" to revel in the immanence in things, one step along the Mystic Path, also appears to the psychedelic subject as he gazes into the very heart of things.
Refs Masters, R E L and Houston, Jean *The Varieties of Psychedelic Experience* (1967).
Related Introvertive mystical experience (#HM0899)
Modes of awareness associated with use of hallucinogens (#HM0801).

♦ **HM1637d Non-penetration** (Buddhism)
Appativedha (Pali)
Broader Delusion (Buddhism, #HM0184).

♦ **HM1643d Sensation of neither suffering nor pleasure born of contact with the psychical** (Buddhism)
Cetosamphassaja addukkhamasukha vedana (Pali)
Broader Unperturbedness (Buddhism, #HM4572).

♦ **HM1644c Terrible sentiment** (Hinduism)
Aesthetic emotion of fear — Terrifying rasa
Context One of eight kinds of aesthetic sentiment or rasas in Indian psychology.
Broader Aesthetic emotion (Hinduism, #HM0205).

♦ **HM1645c Sword of wisdom** (Buddhism)
Pannasattha (Pali)
Broader Faculty of wisdom (Buddhism, #HM3233).

♦ **HM1646c Registering** (Buddhism)
Registration
Description This occurs after impulsion (apperception) in the case of a large or very vivid object (with a life of 16 conscious moments) at the "five-doors" (senses) or very clear object at the "mind-door". For beings in the sense-sphere the result is by means of a condition it has obtained previously, whether due to karma, impulsion consciousness or whatever. It occurs as one of the 8 indeterminate resultant consciousnesses with root cause or as one of the 3 indeterminate resultant mind-consciousness elements. Following the impulsion or apperception which has come into play, registration occurs with consciousness making the object of impulsion its object, although it is ready to make the object of life-continuum its object. When registration is complete, life continuum resumes.
Context In Hinayana Buddhism, this is the 13th mode of occurrence of consciousness in which the 89 kinds of consciousness proceed.
Broader Modes of occurrence of consciousness (Buddhism, #HM6720).
Related Indeterminate consciousness in the sense sphere - resultant (Buddhism, #HM5721).
Followed by Life-continuum (Buddhism, #HM6221).
Preceded by Impulsion (Buddhism, #HM7268).

♦ **HM1647d Approaching reflection** (Buddhism)
Upavicara (Pali)
Related Analysis (Buddhism, #HM5089).

♦ **HM1649d Faculty of mental gladness** (Buddhism)
Somanassindriya (Pali)
Related Gladness (Buddhism, #HM5224).

♦ **HM1655c Experience of An** (Sufism)
Experience of original moments — Creative moment
Description *An* is the precise instant, the real time, in transcendental and experiential Sufi psychology. Past and future exist only in the mind; *An* is the present moment, the moment of fulfilment, appearing when "temporal" time transcends itself and arrives at "eternal" time. *An* is the sharp sword which cuts the past from the future. "Not-being" in the form of successions of events appear one after the other and touch the shore of "being"; "I" becomes identical with "not-I"; the personal becomes universal. Time is no longer dualistic. This is the moment when the psyche has regained its total harmony and is free from internal conflict. An is manifest in "I am-ness", when knower and known become one and subject and object unite. It is through *An* that man transcends his inner core. *An* is instant illumination, through which experience real knowledge comes.
Each experience of *An* is a great potentiality, it is the mother idea giving birth to form. Each experience is one rung of a ladder up which one progresses, an arrival of one *An* following the fulfilment of another, each progressively more intense. To arrive intentionally at the experience of *An*, three illuminating techniques are necessary: (1) The illumination of names. This is the lowest category, where name symbolizes conceptual reality. One is aware of words in their constant relatedness to the objects which produce them. (2) The illumination of attributes, where an object is experienced as an aggregate of qualities. Here one is aware of the waves, light and original elements which are the source of shape, colour and sound. (3) Illumination of essence, unity, one sees an object for what it intrinsically is. This occurs only to the highly sensitive psyche, to the workers of miracles.
Refs Arasteh, A Reza *Growth to Selfhood* (1980).
Related Kairos (#HM2749).

♦ **HM1657d Correct endeavour** (Buddhism)
Sammavayama (Pali)
Broader Endeavour (Buddhism, #HM0229) Faculty of energy (Buddhism, #HM1470).
Related Right endeavour (Buddhism, #HM1295) Wrong endeavour (Buddhism, #HM1585).

♦ **HM1663d Understanding as knowledge about meaning** (Buddhism)
Attha-patisambhida (Pali)
Description The knowledge referred to is that of the result or "fruit" of a root condition or cause. In particular, it includes knowledge which arises on reviewing the meaning of: things produced by conditions or causes; nibbana; the sense or meaning of what is spoken; results (of kamma); action or functional consciousness. Examples of knowledge of meaning are: knowledge of suffering; knowledge about the cessation of suffering; knowledge about the result of a condition or root cause; knowledge about what has come to be, arisen, been born; knowledge about ageing and death; knowledge about the cessation of ageing and death; knowledge about the cessation of compound things or formations; knowledge of the meaning of the dhamma and the scriptures; knowledge about the result of moral or profitable states.
Context On the Path of Purification of Hinayana Buddhism, panna (understanding) is considered as of one kind (monad), of two kinds (dyads), of three kinds (triads) or of four kinds (tetrads). There are five dyads, four triads and two tetrads. All have the characteristic of penetrating the individual essences or true nature of states (monad). In the second tetrad, this is one of the four kinds of understanding as concerned with meaning, law, language and perspicuity. The four kinds of understanding are together referred to as the four discriminations or analyses.
Broader Understanding (Buddhism, #HM4523) Four discriminations (Buddhism, #HM0485).
Related Understanding as knowledge about law (Buddhism, #HM4726)
Understanding as knowledge about language (Buddhism, #HM5026)
Understanding as knowledge about kinds of knowledge (Buddhism, #HM4958).

♦ **HM1664c Aesthetic silence**
Description A region of awareness that lacks definiteness and location, that is different from particularity of any kind. This appears to occur with experienced events and to be one dimension of a person's happening. From this perspective total reliance on particular events or things, on familiarity, certainty, location, and identity is a kind of madness that is a tacit refusal of a dimension of the person's happening. The awareness is of a depth that underlies and accompanies surface events, a depth that is frequently silent in relation to usual sounds. It is a depth that changes the appearance of things as it is entered. The silence is itself immediate perception, as in a pause in music. The silence communicates the relevant. It is the significance of not-saying or the sense of absence in the presence of other events. It is a frequent characteristic of communication in therapy.
Refs Scott, Charles E *Boundaries in Mind* (1982).
Related Silence (#HM3603).

♦ **HM1665c Individuative-reflective faith**
Stage four faith
Description Transition to this stage is usually at young adulthood (although it may arise in the thirties or forties). It is when the late adolescent or adult takes seriously the responsibility for his or her own commitments, lifestyle, beliefs and attitudes. Unavoidable tensions will have to be faced: individuality as opposed to being defined by a group of which one is a member; subjectivity, and the power of strongly felt but unexamined feelings, as opposed to objectivity and its requirement for critical reflection; self fulfilment or actualization as opposed to service to and being for others; commitment to the relative as opposed to struggling with the possibility of an absolute. The self now claims an identity no longer defined by a composite of one's roles or meanings to others. This identity is sustained through composing a frame of meaning which is conscious of its own boundaries and inner connections, aware of itself as a "world view". The person's own self (identity) and outlook (world view) are differentiated from those of others and are acknowledged factors in reactions, interpretations and judgements made on his or her own actions or those of others. Its intuitions of coherence in an ultimate environment are expressed in terms of an explicit system of meanings. This is the stage of demythologizing, with minimal attention to unconscious factors influencing judgement and behaviour.
The strength of this stage is related to the capacity for critical reflection on identity, self and on outlook, ideology. The dangers are: excessive confidence in the conscious mind and in critical thought; a kind of narcissism when the self, now clearly bounded and reflective, over-assimilates reality and others' perspectives in its own world view.
Readiness for transition to Stage 5 arises when the person becomes restless with self-images and outlook. Anarchic and disturbing inner voices are attended to. There may be elements from the childish past, images and energies from the deeper self, a flatness and sterility in the meanings the person currently serves. The neatness of his previous faith is broken into by stories, symbols, myths and paradoxes from his own tradition or from others. The person is disillusioned with the compromises necessary to maintain the logic of clear distinctions and abstract concepts and recognizes that life is more complex than these can comprehend. There is pressure towards a more dialectical and multi-levelled approach to truth.
Context Stage 4 in the system of faith development described by James Fowler.
Refs Fowler, J W *Stages of Faith* (1981).
Broader Stages of faith (#HH2097).
Followed by Conjunctive faith (#HM3359).
Preceded by Synthetic-conventional faith (#HM1814).

♦ **HM1673d Not feeling remorse of what one ought to be remorseful** (Buddhism)
Na ottappati ottappitabbena (Pali)
Broader Power of unremorsefulness (Buddhism, #HM1915).

♦ **HM1683e Being infatuated** (Buddhism)
Sarajjana (Pali)
Broader Greed (Buddhism, #HM3283).

♦ **HM1687c Assurance** (Buddhism)
Abhippasada (Pali)
Broader Faculty of faith (Buddhism, #HM0066).

♦ **HM1688d Not infatuated** (Buddhism)
Asaraga (Pali)
Broader Non-attachment (Buddhism, #HM2128).

♦ **HM1689c Deep penetration by memory** (Buddhism)
Apilapanata (Pali)
Broader Faculty of recollection (Buddhism, #HM4248).

♦ **HM1690c Tantric visualization** (Yoga)
Path of form — Meditative absorption
Description The aim is to realize the whole through which the world becomes utterly transparent. According to a strict series of formulae, the content of the visualization is an iconographic type representing an enlightened entity with the corresponding level of being. Identification with the vividly imagined deity is the merging with that on which concentrated consciousness rests. This internal image is then dissolved so that it and the yogin's empirical identity are erased together and the yogin enters the great void. The path of form is contrasted with the formless method which does not produce the quantities of potential psychic energy. It may have numerous branches. One system, the Naropa school, distinguishes six.
Narrower Yoga of the inner fire (Buddhism, #HM3863)
Yoga of the dream state (Buddhism, #HM0600)
Yoga of the clear light (Buddhism, #HM3690)
Yoga of the illusory body (Buddhism, #HM1222)
Yoga of the intermediary state (Buddhism, #HM0101)
Yoga of consciousness transference (Buddhism, #HM5122).
Related Yantras (#HH2662) Mandalas (#HH0112) Dhyana yoga (Yoga, #HH0827)
Tantra (Buddhism, Yoga, #HH0306)
Tantric formless path (Buddhism, Yoga, #HM4603).

♦ **HM1691c Going on** (Buddhism)
Yapana (Pali)
Broader Faculty of living (Buddhism, #HM3404).

♦ **HM1700d Takuan** (Chinese)
Description In this state the individual understands that the world works in ways that seem strange and unfair and that things cannot always be understood. In a world dominated by ambition the individual can still accept obscurity, poverty and setbacks.

♦ **HM1705c Unforgetfulness** (Buddhism)
Asammusanata (Pali)
Broader Faculty of recollection (Buddhism, #HM4248).

♦ **HM1706d Irritability**
Related Peevishness (#HM2472) Irascibility (#HM0757).

♦ **HM1709d Ma** (Japanese)
Stillness — Interval
Description Designates an aesthetically placed interval in time or space. By the very absence of sound or colour, ma helps accentuate the overall rhythm or design. In drama, through maximizing stillness, ma provides the moment when emotional intensity is maximized and when the performer's heart may be revealed. In the experience of painting, ma is the meaningful void. In landscape gardening, open space at a strategic spot offers an experience which enhances the effect of the whole design.

♦ **HM1710c Wrong view** (Buddhism)
Micchaditthi (Pali)
Narrower View (#HM1191) Wrongness (#HM3356) Wrong path (#HM0863)
Erroneous way (#HM1912) Sectarian bias (#HM3685) Thicket of views (#HM1613)
Fetters of views (#HM1511) Scuffle of views (#HM2722) Grasping of view (#HM0700)
Sticking strongly (#HM3942) Distortion of views (#HM1407) Wilderness of views (#HM4548)
Inclination towards view (#HM3432) Grasping of inverted views (#HM1751)
Current views and opinions (#HM2054) Holding as paramount one's view (#HM1992).
Related Wrong view (Buddhism, Pali, #HM5324).

♦ **HM1714b The dark night** (Christianity)
Dark night of St John of the Cross
Description The spiritual road is described by St John of the Cross. On this road the soul will enter the dark night to be stripped of its imperfections and will be led to perfect union with God.
Refs Kavanaugh, Kieran (Ed) *John of the Cross* (1987); St John of the Cross *The Dark Night of the Soul* (1988).
Narrower Dark night of the soul (#HM3941) Dark night of the senses (#HM1727)
Beginner in spiritual life (#HM0102).

♦ **HM1715c Search of dharma** (Buddhism)
Dhammavicaya (Pali)
Broader Faculty of wisdom (Buddhism, #HM3233).

♦ **HM1717c Tenth order perceptions - universal oneness - meditation**
Description Here all perceptions of the world seem unified in one, it is the level of the mystic where one controls for nothing or, perhaps, for only one thing, successfully control for one idea (practice of Zen, for example). There is no error so that one is aware of the whole reorganization system as it idles along, since there is no error and no new information on which to act. The sense of oneness would be of the creative system in its pure, non-driven form.
Context The tenth of ten orders in the perceptual system described by William Glasser.
Broader Orders of perception (#HM0988).
Preceded by Ninth order perceptions - control of systems concepts (#HM4000).

♦ **HM1723c State of holiness** (Christianity)
Description Being holy enables the individual to rise above his merely human nature and to become a son of God. In this state there is no vengeance, the individual loves even his persecutors. The moral purity implied has a numinous quality, since holiness in human beings is a reflection of the holiness of God.
Related Holiness (#HH0183).

♦ **HM1724d Power of shamelessness** (Buddhism)
Ahirikabala (Pali)
Related Non-shame (Buddhism, #HM2986).

♦ **HM1725d Unawareness** (Buddhism)
Asampajanna (Pali)
Broader Delusion (Buddhism, #HM0184).

♦ **HM1726d Egotism** (Leela)
Ahamkara
Description Here the player directs all his attentions to the object of his desire. He becomes self-centred, adopting any means that will speed his journey with no regard to others. The ego resists its death in the unity of cosmic consciousness. This egotism is attached to the game so the player identifies with its object and responds to its ups and downs with elation and depression. Unable to merge upward in pure vibration, he fights the flow of sudharma and slides down into anger.
Context The 55th state or square on the board of *Leela*, the game of knowledge, appearing in the seventh row.
Broader Seventh chakra: plane of reality (Leela, #HM0754).
Followed by Anger (Leela, #HM1433).

♦ **HM1727c Dark night of the senses** (Christianity)
Purification of the senses — Night of correction
Description Beginners travel on the spiritual road to God in an inferior way and are wrapped up in their own selfish desires. They have lived for some time on the way of goodness, persevered at meditation and prayer. They find delight in having lost their love for the things of the world and found some spiritual strength in God. To some extent they have been able to curb human desires and have experienced some spiritual burdens without turning back. Just when they are experiencing the greatest of delight in spiritual exercises, God changes all of the light into darkness. He turns his face from them; the door to spiritual refreshment is closed and all is dark. Their mind becomes useless. They cannot meditate as they have done before. Not only is there no pleasure in doing spiritual exercises but these activities become insipid and bitter. They experience aridity.
Most often spiritual aridity comes from sins, imperfections, weakness, lukewarmness or one's physical or mental disposition rather than from this night. There are ways of telling which. 1) In all forms of aridity the things of God bring no pleasure but during the purification of the senses all other created things are experienced the same way. Nothing is attractive or desirable. During aridity resulting from sins, worldly things are pleasurable.
2) In the dark night of the senses the mind is often focused on God and one feels that one is not serving Him. The individual painfully believes he is backsliding. This is the difference between dryness and lukewarmness. Lukewarmness is characterized by weakness and neglect of the spirit and could not care less about serving God. Initially the soul finds only dryness and no pleasure in the pure spirit because it is not used to this. It finds itself striving harder and harder to please God. While the soul is gaining strength the person experiencing it is usually unaware. There is a strong desire to be quiet and alone without knowing why. It is at this point that the soul desires to actively pursue God and when it is best to be quiet if it only knew how.
3) The soul can no longer meditate the way it used to. God no longer communicates through the mind but rather through the spirit which the unspiritual part of the soul is unable to grasp. Meditation and recollection become useless. They move from meditation to contemplation.
This is a time of great trials, not because of the aridity but because the spiritual person feels he has lost the road to God. He tries to focus his mind. He tries to meditate. The soul needs a break from knowledge and thought, even though this seems to be a waste of time. This is when they should seek a guide. They should not give up but trust God and seek Him with a humble heart. This is the time to allow the soul to be quiet and restful and to refrain from all reasoning and meditation. Wait on God. Any striving will only hinder God's work and earn His displeasure.
There are four benefits produced in the soul by the night of the senses. 1) The joy of peace is found in the soul. 2) There is a constant memory of God. The first is self-knowledge and awareness of the miserable state of the soul. When everything was going well, the soul could not see its actual situation. It thought it was making progress on its own. It now realizes that it can do nothing, is worth nothing and has done nothing worthwhile. The soul learns to talk to God with greater respect and reverence because not only is it aware of its own miserable state, it sees the greatness and excellence of God. 3) The soul becomes clean and pure. 4) The soul practices new found virtues. The soul becomes spiritually humble, the opposite of spiritual pride. From this humility comes a love of neighbours, recognizing that it has no place in judging others. The greed for spiritual pleasures and exercises is gone, they no longer please and are nauseating. The soul no longer lusts for the spiritual. From spiritual gluttony it escapes. The soul no longer fines pleasure in sensual things whether from God or from the worldly. The soul experiences patience and long suffering. It has learned to be strong in the face of adversity because it is weak. It is no longer angry and upset because of its faults or of a neighbour's faults, nor is it displeased because God has not made the individual a saint overnight. Love toward others has replaced envy. What envy remains becomes a virtue in that it seeks to imitate others. Sometimes God will speak to the soul when it least expects it or fill it with spiritual love or delight. These spiritual blessing are of special importance because they are not felt by the senses. The soul no longer delights in blessings but only in God.
Context This is second stage on the road to perfect union with God described in the Dark Night by St John of the Cross.
Refs St John of the Cross *The Dark Night of the Soul* (1988).
Broader The dark night (Christianity, #HM1714).
Followed by Dark night of the soul (Christianity, #HM3941).
Preceded by Beginner in spiritual life (Christianity, #HM0102).

♦ **HM1728c Christ consciousness as forgiving mind** (Christianity)
Description The mind which is empty is also forgiving. The structure of values is not a logical returning in proportion to what has been received but a gratuitous, free and irrational bestowing of love where there is no objective deserving. Forgiveness implies flexibility, not holding past behaviour against a person and not restricted by accepted legal structures or social codes. The burden of past sins is lifted in the attitude and behaviour that is not bound by the brokenness of life as it is normally lived together. The behaviour attitude of forgiveness is love.
Broader Christ consciousness (Christianity, #HM2013).
Related Christ consciousness as empty mind (Christianity, #HM0778)
Christ consciousness as loving mind (Christianity, #HM0976)
Christ consciousness as dying-rising (Christianity, #HM1402).

♦ **HM1732d Plateau cognition**
Serene B-cognition
Description After an ecstatic or peak experience an individual cannot "unsee" what he has seen. Repetition of such experiences leads to an ease with them, a controlled serenity in the knowledge that such experiences arise.
Broader Being cognition (#HM2474).

♦ **HM1734d Unmaliciousness** (Buddhism)
Abyapada (Pali)
Broader Non-hatred (Buddhism, #HM2744).

♦ **HM1735d Correct concentration** (Buddhism)
Right concentration — Samma-samadhi (Pali) — Sammasamadhi
Description The true samadhi as opposed to a state leading to illusory conceptions.
Broader Sukha (Buddhism, #HM2866) Stabilization (Buddhism, #HM2440)
Concentration (Buddhism, #HM6663).
Related Ceto-samadhi (Buddhism, #HM5587) Wrong concentration (Buddhism, #HM1802)
Right rapture of concentration (Buddhism, #HM0931).

♦ **HM1736d Base of mind** (Buddhism)
Manayatana (Pali)
Related Manas awareness (Hinduism, #HM2902).

MODES OF AWARENESS

♦ **HM1738d Composure** (Buddhism)
Patipassaddhi (Pali)
   Broader Composedness of body (Buddhism, #HM1867)
   Composedness of thought (Buddhism, #HM1490).

♦ **HM1741d Good quality of the aggregate of consciouness** (Buddhism)
Vinnanakkhandhassa pagunata (Pali)
   Broader Fitness of consciousness (Buddhism, #HM1810).

♦ **HM1743d Mind** (Buddhism)
Mano (Pali)
   Related Manas awareness (Hinduism, #HM2902).

♦ **HM1746d Inability to demonstrate** (Buddhism)
Apacakkhakamma (Pali)
   Broader Delusion (Buddhism, #HM0184).

♦ **HM1748f Mood**
Affects
   Description A mood is a relatively stable and long-lasting affective state, a complex psychic state differing from the individual's normal condition. It typically lasts for some hours or days but may be much shorter. As well as a cognitive component (narrowing of mental content, alteration of some aspects of thinking) and a behavioural component, there is an affective component, in that the emotions are involved. By expressing a complex mixture of affects in a regulated way, a mood defends against potentially overwhelming discharge of these affects while allowing them some expression. While experiencing a mood there is a compromising of the ability to assess and deal with reality – impressions are coloured so that the individual may feel worthless and unappreciated or powerful and optimistic. A mood tends to be confirmed and extended because focus on internal and external impressions is restricted to those reinforcing the mood; similarly, behaviour while experiencing the mood affects others' responses in such a way as to reinforce the mood.
   Related Emotion (#HH0819)   Affects (#HM7132)   Cyclothymia (#HM1616).

♦ **HM1749d Stamina** (Buddhism)
Thama (Pali)
   Broader Faculty of energy (Buddhism, #HM1470).

♦ **HM1750c Sagacity** (Buddhism)
Medha (Pali)
   Broader Faculty of wisdom (Buddhism, #HM3233).

♦ **HM1751d Grasping of inverted views** (Buddhism)
Vipariyasaggaha (Pali)
   Broader Wrong view (Buddhism, #HM1710).

♦ **HM1756g Flow experience through the senses**
Enjoying the senses
   Description Since the body is the means through which one obtains information about the outside world it is like a probe full of sensitive devices, an instrument for getting in touch with the universe. The sensing devices produce a positive sensation and have thus a potential for flow experience as one develops skills and finds delight in what the body can do.
   Intense flow experiences may arise, not just with seeing great works of art, for example, but through mundane things such as the view from a train if one is prepared to invest psychic energy in the experience of seeing. Hearing is also a source of flow experience, in particular listening to music. Music helps to order the mind, reduces the disorder deriving from random input, and wards off boredom and anxiety. Again, it is not so much the hearing of music but really listening and paying attention which enables awareness of flow. There needs to be a strategy and a formulation of goals, having a challenge of greater complexity as one goes from the sensory experience through the analogic to the analytic mode of listening. Making music is even more rewarding, as the harmony of sound is seen to underlie universal harmony. Plato believed that by learning at an early age to pay attention to rhythm and harmony the whole consciousness would become ordered. Taste is also a source of flow experience if one takes control of the activity, approaching eating and cooking in a spirit of adventure and curiosity, but as in all cases of flow experience it must be for the experience itself and not for motives such as showing off one's expertise.
   Refs Csikszentmihalyi, Mihaly *Flow* (1990).
   Broader Flow experience (#HM2344).

♦ **HM1757d Feeling remorse of what one ought to be remorseful** (Buddhism)
Ottappati ottappitabbena (Pali)
   Broader Power of remorse (Buddhism, #HM2538).

♦ **HM1759d Power of concentration** (Buddhism)
Samadhibala (Pali)
   Broader Sukha (Buddhism, #HM2866).

♦ **HM1761d Calming down** (Buddhism)
Passambhana (Pali)
   Broader Composedness of body (Buddhism, #HM1867)
   Composedness of thought (Buddhism, #HM1490).

♦ **HM1762c Fundamental dialogue**
   Related Contemplative indifference (#HM1405).

♦ **HM1763d Correct livelihood** (Buddhism)
Samma ajiva (Pali)
   Broader Correct livelihood (Buddhism, #HM0549).

♦ **HM1766d Somatic experience**
Visceral experience — Body perception — Body image
   Description Immediate experience of the body understood as extending beyond the physical limits of the skin. This experience has a powerful influence on a person's response to every social situation, with the image changing depending on circumstances. The body image is highly plastic and completely inseparable from social interaction. Such a body image is not given once and for all, but is subject to a series of creations and de-creations. It may be extended to include products of the body. A person can act to extend his body image by incorporating other things into it. Influences between people can be sensed at this somatic level. When the mind is experienced as being part of the body for any length of time, the person is able to experience himself as a magic, self-sensing form. This leads to a corresponding shift in the physical environment. The elements of nature are then experienced as standing forth as living, communicative presences.
   Refs Abram, David *The Perceptual Implications of Gaia* (1985); Berman, Morris *Coming to our Senses* (1990); Schilder, Paul *The Image and Appearance of the Human Body* (1950).
   Related Gnosis (#HM0413).

♦ **HM1771d Yu** (Taoism)
Following the Tao
   Description The right way for following the tao (or path), *yu* has been variously described as wandering, walking without touching the ground, swimming, flying and flowing. Yu occurs when conscious mastery is transcended, when attention to the task in hand is developed over time until experience allows apparently spontaneous action to be sure and perfect.
   Related Flow experience (#HM2344)   Human development (Taoism, #HH0689)
   Flow experience through work (#HM4877).

♦ **HM1772f Virtual reality**
Cyberspace — Electronic high
   Description The individual experiences a world of electronic realities created by interaction between human beings and computers. The experience of reality transference is as powerful as that induced by psychedelic drugs, so that previously immutable reality becomes fluid with multiple references of reality valid at any given moment. Thoughts, no matter how sudden or indescribable, are recorded for later analysis, and one can experience being someone or something else, or combinations of other people and things. This experience permanently changes perception of the "real" world.
   Refs Heibrun, Adam and Stacks, Barbara *Virtual Reality*; McKenna, Terence *Virtual Reality and Electronic Highs* (1990).

♦ **HM1773d Purification** (Leela)
Tapah
   Description Feelings of emptiness and confusion are countered by altering the behaviour of sense organs, work organs and habitual daily existence. Various fasts and austerities lead to purification of the senses, increase the vibrational level of being and allow the player to transcend problems of the first and second chakras.
   Context The 10th state or square on the board of *Leela*, the game of knowledge, appearing in the second row.
   Broader Second chakra: realm of fantasy (Leela, #HM3651).
   Followed by Celestial plane (Leela, #HM1052).

♦ **HM1780c Erotic sentiment** (Hinduism)
Aesthetic emotion of love — Romantic rasa
   Context One of eight kinds of aesthetic sentiment or rasas in Indian psychology.
   Broader Aesthetic emotion (Hinduism, #HM0205).

♦ **HM1781d Leaving off, abstaining, totally abstaining and refraining from the four deviations of speech** (Buddhism)
Catuhi vaciduccaritehi arati virati pativirati veramani (Pali)
   Broader Correct speech (Buddhism, #HM1821).

♦ **HM1782d Sorrow** (Leela)
Dukh
   Description Refusing to confront an aspect of himself, to face the unacceptable and to lose his identity, the player suppresses it, creating depression, pain and introversion. It may be that he tries unsuccessfully to identify with his deity, senses the divine, but feels unworthy and unable to know it.
   Context The 26th state or square on the board of *Leela*, the game of knowledge, appearing in the third row.
   Broader Third chakra: theatre of karma (Leela, #HM0717).

♦ **HM1785d Delta wave consciousness**
   Description Four major kinds of brain waves have been discovered. In terms of cycles per second, the delta waves (1 to 4 cycles) are slowest. They are characteristic of the first 2 hours of normal night sleep and of the deep state of hypnosis. The mind in this condition is said to be most easily programmed by outside stimuli such as post hypnotic suggestion; and the psychic mechanism easily utilized. One researcher (William Condon) refers to six different brain wave frequencies, which include a second delta wave that may be the basic or background rhythm to human behaviour.
   Broader Brain waves (#HH3129).
   Related Deep sleep (#HM6307)   Sleep spindles (#HM4806)
   Stage-three sleep (#HM0348)   Deep state of hypnosis (#HM1226)
   Beta wave consciousness (#HM3476)   Alpha wave consciousness (#HM2345)
   Theta wave consciousness (#HM2321).

♦ **HM1789d Positive intellect** (Leela)
Subuddhi
   Description This is a state of non-dual consciousness when the divine is revealed in every phenomenon. Such right discrimination can arise through following the path of dharma.
   Context The 60th state or square on the board of *Leela*, the game of knowledge, appearing in the seventh row.
   Broader Seventh chakra: plane of reality (Leela, #HM0754).
   Preceded by Plane of dharma (Leela, #HM0481).

♦ **HM1792c Power of wisdom** (Buddhism)
Pannabala (Pali)
   Broader Faculty of wisdom (Buddhism, #HM3233).

♦ **HM1793b Consciousness shift**
   Description This is a spontaneous and lasting change in the mode of awareness which may occur instantaneously or over a period of time. The result is a change in lifestyle, with new friends, patterns of behaviour and even choice of foods.
   Related Change (#HH1116)   Mode of life (#HH0592).

♦ **HM1796d Flexibility of the aggregate of consciousness** (Buddhism)
Vinnanakkhandhassa muduta (Pali)
   Broader Flexibility of thought (Buddhism, #HM1805)
   Dispositions of consciousness (Buddhism, #HM2098).

♦ **HM1798d Feierabend** (German)
   Description At the end of a working day there is a feeling of warm, relaxed and comfortable euphoria when the pressure is off and the fun begins.

♦ **HM1799d Disinterestedness** (Buddhism)
Alobha (Pali)
Broader Non-attachment (Buddhism, #HM2128).

♦ **HM1802d Wrong concentration** (Buddhism)
Micchasamadhi (Pali)
Related Correct concentration (Buddhism, #HM1735).

♦ **HM1804c Ten-fold knowledge** (Systematics)
Experience of creativity
Description There is widespread evidence of a creative (pattern generating) activity that is not only the source of order but also the vehicle of disorder. This can only be experienced through a minimum of ten independent terms. Although it can be inferred elsewhere, an individual can only be directly aware of this power within his own consciousness. In creativity there is an authentic addition to the sum total of experience. At this ten-fold level of knowledge several sets of processes can be experienced as compensating for one another's defects to produce an overall harmony that reacts on, and sustains, the individual structures.
Context The tenth in a sequence of twelve modes of knowledge, identified by J G Bennett, inspired by G Gurdjieff.
Broader Systematics (#HH2003).
Followed by Eleven-fold knowledge (Systematics, #HM0065).
Preceded by Nine-fold knowledge (Systematics, #HM1408).

♦ **HM1805c Flexibility of thought** (Buddhism)
Cittamuduta (Pali)
Narrower Mildness (#HM0191)   Non-rigidity (#HM1028)   Non-stiffness (#HM3014)
Flexibility of the aggregate of consciousness (#HM1796).

♦ **HM1809d Limerence**
Being hooked — Romantic love
Description A condition of preoccupation, or total obsession, with the beloved. Whereas love is affection for the beloved, limerence is adoration of the beloved. There is no desire to do anything other than to be with the beloved, whatever the obstacles. It is the agony of such obstacles that contributes paradoxically to the exquisite pleasure of the experience, despite the anxiety, desperation and risk of losing one's self in the other. Lovers then love the state of being in love, which in a sense blinds them to each other.
Refs Tennov, Dorothy *Love and Limerence* (1979).
Broader Love consciousness (#HM2000).
Related Courtly love (#HM6122)   Romantic love (#HM5087).

♦ **HM1810c Fitness of consciousness** (Buddhism)
Fitness of thought — Proficiency of mind — Citta-pagunnata (Pali)
Description This refers to the proficiency of the aggregate of consciousness. The characteristic is healthiness of the citta and freedom from illness, the function is to crush its unhealthiness. It is manifest as absence of disability and freedom from evil. Consciousness is the proximate cause. It is the opponent of corruptions like faithlessness or diffidence causing illness or unhealthiness in the citta. Fitness of mental body and fitness of consciousness are considered together.
Context One of the formations aggregate (mental coefficients) of Hinayana Buddhism, being listed among the constant states which appear in their true nature, and as profitable primary (always present in any profitable or profitable-resultant consciousness).
Broader Awareness as mental-formation group of conscious existence (Buddhism, #HM2050).
Narrower Good quality (#HM3424)   Being of good quality (#HM0872)
Good quality of the aggregate of consciousness (#HM1741).
Related Fitness of mental body (Buddhism, #HM1455).

♦ **HM1814c Synthetic-conventional faith**
Stage three faith
Description This is the faith that rises in adolescence and may continue to be permanent in adult life. Such faith has to provide a coherent orientation in the midst of a complex and diverse range of involvements – family, school or work, peer group, street society, media and (perhaps) religion. It has to synthesize values and information and provide a basis for identity and outlook. The ultimate environment is structured in interpersonal terms, images of unifying value and power are derived from extending qualities experienced in personal relationships. Faith is tuned to others' expectations and judgements, not having a sure enough grasp of its own identity and judgement to construct and maintain an independent perspective. It is thus conformist, an ideology with a reasonably consistent clustering of values and beliefs which has not been objectified for examination. In a sense the person is unaware of having it. Although deeply felt, beliefs and values are probably held tacitly. If another's outlook is different, this is experienced as a different "kind" of person. Authority is located either in the incumbents of traditional authority roles (where these are seen as personally worthy) or else in the consensus of a "face-to-face" group that is valued.
What emerges is the capacity to form a personal myth of one's own becoming in identity and faith, which incorporates the past and anticipated future in an image of the ultimate environment which is unified by characteristics of personality. Possible dangers and deficiencies are: that expectations and evaluations of others are internalized so compellingly and so sacralized that subsequent autonomy of judgement and action is jeopardized; that interpersonal betrayal gives rise to nihilistic despair about a personal principle of ultimate being, or to a compensating intimacy with God which is unrelated to mundane relations.
Changes that give rise to a breakdown of this stage and transition to the next include emotionally or physically leaving home, which precipitates examination of self, background and life-guiding values. Other triggering events may be: clashes or contradictions between valued authority sources; marked changes by acknowledged leaders of practices or policies previously deemed sacred or unbreachable; encounter with experiences and perspectives leading to critical reflection on how one's faith and belief formed and changed and how much they depended upon one's particular group or background.
Context Stage 3 in the system of faith development described by James Fowler.
Refs Fowler, J W *Stages of Faith* (1981).
Broader Stages of faith (#HH2097).
Followed by Individuative-reflective faith (#HM1665).
Preceded by Mythic-literal faith (#HM1525).

♦ **HM1817b Attention**
Inward attention
Description Voluntary, conscious and deliberate attention is an act of will. It focuses on full awareness of the here and now and avoids any escape into thoughts about the past or the future. Inward attention involves a stilling of the mind to all external activity and is a state analagous to meditation or more easily achieved after meditation; practice and discipline render this state increasingly more easily achieved.
Refs Se, Anima *Attention and Distraction* (1983); Van Nuys, D *A Novel Technique for Studying Attention during Meditation* (1971).

Narrower Sated attention (#HM2580)   Wise attention (Buddhism, #HM4309)
Fatigued conscious awareness (#HM2423)   Spiritual attentiveness (Christianity, #HM7143).
Related Attention (#HH0756)   Quiet attentiveness (#HM4383)
Mental engagement (Buddhism, #HM3237).

♦ **HM1821c Correct speech** (Buddhism)
Sammavaca (Pali)
Narrower Unaffected (#HM3375)   Leaving undone (#HM3429)
Correct speech (#HM4608)   Destroying cause (#HM1098)
Not incurring guilt (#HM3413)   Not trespassing limit (#HM3435)
Leaving off, abstaining, totally abstaining and refraining from the four deviations of speech (#HM1781).

♦ **HM1825d State of not feeling greed** (Buddhism)
Alubbhitattam (Pali)
Description The ascetic states of fewness of wishes, contentment, effacement and seclusion that go with volition of ascetic practices bring non-greed, so that greed for forbidden things and indulgence in pleasure of those things which are allowed are both shaken off.
Broader Non-attachment (Buddhism, #HM2128)
Narrower Seclusion (#HM6312)   Effacement (#HM7703)
Contentment (#HM4563)   Fewness of wishes (#HM0460).

♦ **HM1829d Faculty of energy** (Buddhism)
Viriyindriya (Pali)
Broader Faculty of energy (Buddhism, #HM1470).

♦ **HM1830c Lightness of thought** (Buddhism)
Cittalahuta (Pali)
Narrower Non-inertness (#HM3374)   Capacity of easy transformation (#HM3808)
Lightness of the aggregate of consciousness (#HM3403).

♦ **HM1833c Transcendent liberation** (Taoism)
Description Returning to the root, returning to life, going back to the original beginning.
Followed by Perfect attainment (Taoism, #HM0922).
Preceded by Release from the matrix (Taoism, #HM0635).

♦ **HM1834d Element of mind-consciousness** (Buddhism)
Manovinnanadhatu (Pali)
Related Volition born of contact with element of mind consciousness (Buddhism, #HM0404)
Cognition born of contact with element of mind-consciousness (Buddhism, #HM1501)
Psychical ease born of contact with element of mind-consciousness (Buddhism, #HM1921).

♦ **HM1840c Faculty of wisdom** (Buddhism)
Pannindriya (Pali)
Broader Faculty of wisdom (Buddhism, #HM3233).

♦ **HM1846c Lamp of wisdom** (Buddhism)
Pannaaloka (Pali)
Broader Faculty of wisdom (Buddhism, #HM3233).

♦ **HM1857d Stupidity** (Buddhism)
Dummajjha (Pali)
Broader Delusion (Buddhism, #HM0184).

♦ **HM1858c Shining mind** (Taoism)
Unstirring mind — Mind of Tao
Description This mind is always calm even when active, always responding to change. It is subtle and hard to see but it is there, even in the ignoramus, even in the human mind and there is human mind even in the mind of Tao. The shining mind is constantly balanced in straightforwardness, discerning unity, never vacillating, always holding to the centre. When a student of immortality discerns and keeps to unity, holding to the centre, then the perilous becomes safe and the subtle obvious.
Refs Cleary, Thomas (Trans) *The Book of Balance and Harmony* (1989).
Related Wandering mind (Taoism, #HM5622).

♦ **HM1862d Psychically pleasant** (Buddhism)
Cetasikam sukham (Pali)
Related Pleasant feeling (Buddhism, #HM6722)   Psychically unpleasant (Buddhism, #HM0805).

♦ **HM1864d Childishness** (Buddhism)
Balya (Pali)
Broader Delusion (Buddhism, #HM0184).

♦ **HM1865b Spontaneity** (Taoism)
Description Even after release from the matrix there is evolution. It is not having a body outside the body which is a marvel, but revelation of complete reality when space is shattered. After release from the matrix one treads the ground of reality until one unites with space. This is the non-contrivance of Confucianism, the true emptiness of Buddhism. The fundamental is embraced, one returns to the origin and unites with cosmic space. This is the way unknowable to people with fixations, it is permeated with unity.
Preceded by Release from the matrix (Taoism, #HM0635).

♦ **HM1867c Composedness of body** (Buddhism)
Kayapassaddhi (Pali)
Narrower Composure (#HM1738)   Calming down (#HM1761)
Completely tranquillized (#HM1528)   Complete tranquillization (#HM3411)
Composedness of the aggregate of sensation, of the aggregate of cognition, and of the aggregate of synergies (#HM1377).

♦ **HM1872c Anamnesis**
Remembering
Description Poetry and the other arts exist in order to present images which have the power to awaken recollection of the paradise of all mythologies. That place, once and forever known, but lost, haunts people by offering a sense of inaccessible knowledge, and by a sense of estrangement from a place or state to which they feel they are native. This half forgotten paradise, whether felt to be a former state or one which we have never known, has, when brought to consciousness, a familiarity (as part of the psyche) of something recollected. That place is the ground of psyche itself and hence its deep familiarity.
Refs Raine, Kathleen *Defending Ancient Springs* (1985).
Related Poetic imagination (#HM1538).

MODES OF AWARENESS

♦ **HM1873c State of buddhahood** (Buddhism)
Description The result of acts in the state of bodhisattva, this is absolute happiness. Even birth, old age, illness and death are part of the joy of living. The whole universe is illuminated by the light of wisdom, man's innate nature is destroyed, the life-space of Buddha unites with the universe. As the self becomes the cosmos the life-flow encompasses all the past and the future. In the present moment, the fountain of energy which is the life-force of the cosmos flows outwards. Unlike rapture, this joy cannot be destroyed. It is said to be within the reach of all, but lies dormant until the person discovers how to manifest it, not as an extra state but as a quality which, when it predominates, enriches every moment.
Context One of the ten worlds described in Nichiren Soshu Buddhism.
    Broader Ten worlds (Buddhism, #HM2657)   Four noble paths (Buddhism, #HM4026)
    Nichiren shoshu buddhism (Buddhism, #HH3443).
    Related Sokushin jobutsu (Buddhism, #HM3514).
    Preceded by State of bodhisattva (Buddhism, #HM1225).

♦ **HM1874d Self-awareness** (Buddhism)
Sampajanna (Pali)
    Broader Self-remembering (#HM2486).

♦ **HM1875c Acceptance**
Description Through acceptance, without the qualifications of role or principle, a person experiences a different order of communication in comparison to role-governed or principle-governed communication. Thus when a dream is allowed without judgement or analysis, it often stays with a person initially with more power and with a greater tendency to remain. Whatever is present, not conceived, but known and allowed, gives its reality in the openness of the allowance and is in the allowance to some extend affirmed and loved. This unsentimental love appears to be regenerative, although seldom comforting, offering a form of release. It is an awareness that awakens a wide range of affections that are frequent agents for a person's changing his ideas and ways of acting.
Refs Scott, Charles E *Boundaries in Mind* (1982).
    Related Self-acceptance (Jung, #HM3912)   Contemplative indifference (#HM1405)
    Immediate boundary awareness (#HM0035).

♦ **HM1881f Modes of awareness associated with use of nicotine**
Tobacco smoking
Description Alterations of mood under the influence of nicotine are small if perceptible at all, and may be of relaxation or of stimulation. Some evidence suggests that when smoking cigarettes, short puffs stimulate while long puffs relax. However, it is very physiologically addictive; and withdrawal symptoms associated with discontinuing its use are severe, including anxiety, difficulty in concentrating, irritability and anger.
    Broader Modes of awareness associated with psychoactive substances (#HM0584).

♦ **HM1882d Baraka** (Islam)
Spiritual energy — Sense of blessedness
Description Rituals of prayer or dancing can generate spiritual energy which may be passed on directly, energizing those who receive it. Even the atmosphere surrounding a person touching on oneness is alive with this energy. Years of loving use of buildings or objects can impart baraka to them so that there is a sense of blessedness attached. The term may be extended to describe the "humanizing" effect that years of use can impart to a machine or instrument so that one has a rapport with it.

♦ **HM1886e Deep trance** (Psychism)
Full trance
Description A state self-induced by a medium through self-hypnosis in preparation for allowing another intelligence to enter the body. The eyes, limbs and voice are used by this intelligence, the medium subsequently having no memory of what has occurred.
    Broader Trance (#HM3236).

♦ **HM1887e Astral projection** (Psychism)
Astral flight
Description By an act of will, the soul-mind is said to leave the body and travel to another location where it may or may not be visible physically or clairvoyantly. The sense of time is distorted.
    Narrower Focus one (#HM0629)   Focus ten (#HM1136)
    Focus twelve (#HM0267)   Focus fifteen (#HM3975).
    Related Mental projection (#HM4529).

♦ **HM1889e Adrenergia** (Psychism)
Description In psychic terms, an excited inner state of mind rendering more successful the sending of mental messages.

♦ **HM1902d Non-consideration** (Buddhism)
Apaccavekkhana (Pali)
    Broader Delusion (Buddhism, #HM0184).

♦ **HM1904c Prime** (Christianity)
Hours of prime
Description The canonical hours of prime traditionally take place between 6 and 9 am. They dramatize turning to the day's work and dedicating these efforts to God. The mood is consecrated services.
    Broader Canonical hours (Christianity, #HM1167).
    Followed by Terce (Christianity, #HM2965).
    Preceded by Lauds (Christianity, #HM0894).

♦ **HM1905d Latifa-i-qalb** (Sufism)
Active with the remembrance of God
Description In a state of perfect purification and sitting in a quiet place, the seeker touches his palate with his tongue and thinks of his physical heart repeating "Allah, Allah". Knowing he is not listening to it but intending to try, he fully concentrates on listening to it. Through God's help there is a slight motion in the heart. Not being able to know if this is real or simply respiration or imagination, the seeker concentrates more and the motion becomes more audible so that he is certain of the heart palpitating and murmuring.
Context This is the first state arising in the method of dhikr described by Shaykh Kalimallah. It is said to be a manifestation of Allah as the compelling, *al-Jabbar*.
    Broader Remembrance of God (Sufism, #HM6562).
    Related Al-Jabbar (Sufism, #HM0215)   Subtle faculties (Sufism, #HH6282)
    Annihilation of the heart (Sufism, #HM7622).
    Followed by Khilwat dar anjuman (Sufism, #HM4987).

♦ **HM1907c Identity-strength**
Ego strength
Description The extent to which consciousness is exceptionally well-defended with respect to boundary situations, namely where the person has succeeded in making his own interests and energy, or sense of belonging, into a closed, habitual pattern of self-reference. This is associated with tight confidence, unambiguous clarity of view, a serious sense of rightness and, possibly, a feeling of narrow, if autonomous vitality and direction.
Refs Scott, Charles E *Boundaries in Mind* (1982).
    Related Ego strength (#HH0217).

♦ **HM1911d Philosophical awareness**
Description Being philosophically aware implies attention on questions and problems arising from the "fact" of existence and of attempting to establish relationships between particular manifestations of life implicit in questions such as: What is a living being ? What is the meaning of life ? Who am I ? Who are you ? What is knowledge ? Such questions are bedevilled by the limitations of language itself. Despite the seemingly insurmountable problems to enunciate answers, which, for example, resulted in Wittgenstein's "Philosophical Investigations" completely contradicting his previous work "Tractacus Logico-Philosophicus", perseverance is said to be rewarded (I Ching).

♦ **HM1912d Erroneous way** (Buddhism)
Kummagga (Pali)
    Broader Wrong view (Buddhism, #HM1710).

♦ **HM1914b Six paths** (Buddhism)
Description The first six of the ten worlds described in Nichiren Soshu Buddhism, these are not achieved by inner effort but are rather the spontaneous result of external factors. Ranging from hell to heaven, the individual bounded by them is trapped by the demands of his desires, slipping easily from one to the other. They are not the entirety of life, however. Looking beyond day-to-day exingencies one enters on the four noble paths and starts to develop true potential.
    Broader Ten worlds (Buddhism, #HM2657)
    Nichiren shoshu buddhism (Buddhism, #HH3443).
    Narrower State of hell (#HM4282)   State of anger (#HM2959)
    State of hunger (#HM0150)   Four evil paths (#HM1252)
    State of rapture (#HM1973)   Three evil paths (#HM0923)
    State of animality (#HM0847)   State of tranquillity (#HM3492).
    Related Four noble paths (Buddhism, #HM4026).

♦ **HM1915c Power of unremorsefulness** (Buddhism)
Anattappabala (Pali)
    Narrower Not feeling remorse of what one ought to be remorseful (#HM1673)
    Not feeling remorse of acquisition of sinful and unwholesome dharmas (#HM3353).
    Related Unremorsefulness (Buddhism, #HM0188).

♦ **HM1919e Psychic consciousness** (Psychism)
Psyche consciousness
Description This state may occur in dreams, hypnosis or meditation. The conscious, decision-making part of the mind makes way for the subconscious or soul-mind to improve itself. It is typical of the dreaming state, dreams being said to symbolize life up to that point and to extend help in decisions for the next day. It is said to be easy to achieve this state as the soul-mind is free to speak out and the state provides an awareness of self, of personal existence.
Refs Bletzer, June G *The Donning International Encyclopedic Psychic Dictionary* (1987).
    Related Dreams (#HM2950).

♦ **HM1921d Psychical ease born of contact with element of mind-consciousness** (Buddhism)
Manovinnanadhatu samphasajam cetasikam satam (Pali)
    Broader Feeling (Buddhism, #HM2270).
    Related Element of mind-consciousness (Buddhism, #HM1834).

♦ **HM1930d Stabilization of thought** (Buddhism)
Anupekkhanata (Pali)

♦ **HM1931d Greed** (Leela)
Lobh
Description Although his material needs are satisfied, the player feels unfulfilled. Since his skills are only those for maintaining physical existence, he forever seeks fulfilment in acquisition of material things and is forever disappointed. Greed arises from lack of belief in providence and misidentification with the self, bringing insecurity. Its positive aspect is manifest in greed for the spiritual, for knowledge and for love.
Context The fourth state or square on the board of *Leela*, the game of knowledge, appearing in the first row. The 4 represents impetus towards completion, manifested materially as greed.
    Broader First chakra: fundamentals of being (Leela, #HM4103).
    Related Greed (Buddhism, #HM3283).
    Preceded by Jealousy (Leela, #HM0554).

♦ **HM1932c Second order perceptions - sensation**
Description These perceptions are totally within the brain as first order perceptions are transformed into meaningful sensations. They are totally subjective, based on intensity signals but not intensity, a way in which one uniquely chooses to interpret simple-intensity perception. This is the level of emotions, where one becomes aware of painful feeling behaviours or pleasurable behaviours.
Context The second of ten orders in the perceptual system described by William Glasser.
    Broader Orders of perception (#HM0988).
    Followed by Third order perceptions - configurations (#HM0312).
    Preceded by First order perceptions - intensity (#HM0543).

♦ **HM1934d Ash-Shakur** (Sufism)
Description The believer is aware of Allah as the grateful, the appreciative one, who repays good deeds with greater rewards. Using Allah's bounty with thanks leads to further good fortune; denying his bounty brings destitution in the midst of abundance.
Context A mode of mystical awareness in the Sufi tradition associated with the thirty-fifth of the ninety-nine names of Allah.
    Broader Ninety-nine names of Allah (Sufism, #HH2561).
    Followed by Al-'Ali (Sufism, #HH4162).
    Preceded by Al-Ghafur (Sufism, #HM6255).

♦ **HM1943d Power of energy** (Buddhism)
Viriyabala (Pali)
    Broader Faculty of energy (Buddhism, #HM1470).

HM1954

♦ **HM1954e Greed an unwholesome root** (Buddhism)
Lobho akusalamulam (Pali)
 Broader Greed (Buddhism, #HM3283).

♦ **HM1955e Cataplexy**
Description A state of total immobility where the individual is conscious but unable to respond which may occur in response to an anaesthetic (when an operation may unwittingly be carried out on a conscious patient) or to a sudden physical and/or emotional shock. The effect is as if hypnotized by fright, and is the extreme case of the common being "paralyzed with fear". Unlike the *cataleptic* state, however, the individual remains conscious.
 Related Cataleptic trance (Psychism, #HM4319).

♦ **HM1956d Complete delusion** (Buddhism)
Sammoha (Pali)
 Broader Delusion (Buddhism, #HM0184).

♦ **HM1957c Uns** (Sufism)
Intimacy with God
Description In this state the seeker discovers of the unity of life and diversity of forms. Intimacy with the manifestation of God in nature makes life meaningful and God is closer than his own body. Thus, rather than seeking him outside himself, God is discovered within. Both the outer and the inner self are surrendered to God, all belongs to God and the devotee is entirely devoid of his own being. The ascendance of man to God becomes God's descending to man. While unity within diversity of form is experienced there is a great longing, *shugh*.
Context The sixth psychic state listed by A Reza Aresteh as progress on the inner self through divine attraction, *kedesh-jazba*, the outer self already having been purified through conscious effort, *kushesh*.
 Broader Psychic states (Sufism, #HM4311).
 Related Firaq (Sufism, #HM4112)   Showq (Sufism, #HM0762)
 Unification with God (Sufism, #HM3864).
 Followed by Mushahada (Sufism, #HM3521).
 Preceded by Omid (Sufism, #HM4477).

♦ **HM1959g Touching** (Buddhism)
Phusana (Pali)
 Broader Contact (Buddhism, #HM2708).

♦ **HM1961b Apperception**
Description Mental perception which allows understanding of previous experience, there is controlled and orderly reduction of data to unity and system; this is characteristic of the disciplined and trained mind and is necessary for clear judgement and fruitful, unprejudiced search for the truth.
 Related Ethical discipline (#HH1086)   Self-conscious consciousness (#HM2610).

♦ **HM1962e Shock**
 Related Horror (#HM0904).

♦ **HM1964e Self-induced trance** (Psychism)
Description This deep trance state is produced by a medium synchronizing with the intelligence with which he or she is in contact.
 Broader Trance (#HM3236).
 Related Channelling (#HH0878)   Cataleptic trance (Psychism, #HM4319).

♦ **HM1965d Being unobservant** (Buddhism)
Asamapekkhana (Pali)
 Broader Delusion (Buddhism, #HM0184).

♦ **HM1966d Death consciousness** (Psychism)
Description The state experienced during the process of dying, with an awareness of crossing a threshold.
 Refs Grof, S and Grof, C *Beyond Death* (1980).
 Related Near death experience (NDE, #HM0777).

♦ **HM1967d State of feeling hatred** (Buddhism)
Dussitattam (Pali)
 Broader Hatred (Buddhism, #HM4502).

♦ **HM1969c Duration of the formless dharmas** (Buddhism)
Arupinam dhammanam ayu (Pali)
 Broader Faculty of living (Buddhism, #HM3404).

♦ **HM1971f Hypnoidal state**
Description Intense concentration leads inadvertently to a mild state of self-hypnosis.
 Related Hypnotherapy (#HH0962).

♦ **HM1972c Progression** (Buddhism)
Iriyana (Pali)
 Broader Faculty of living (Buddhism, #HM3404).

♦ **HM1973c State of rapture** (Buddhism)
World of desires — World of form — World of formlessness — Heaven
Description This (temporary) state arises from the gratification of desire. The first level, *world of desires*, refers to gratifying desires of the six lower worlds (instinctive urges, social recognition, overcoming suffering). The second level, *world of form*, is rapture of the body, refreshed, wide awake, exercising one's natural talents. The third level, *world of formlessness*, is rapture of the spirit, the most satisfying and enduring of all, arising from creative life of expanding and richer experience. This latter may "spill over" into the worlds of form and of desire, so that living is more healthy and behaviour likely to cause suffering to others is eliminated. Suffering is contained and absorbed in a wider happiness. This is heaven and may be experienced here on earth.
Context One of the ten worlds described in Nichiren Soshu Buddhism.
 Broader Six paths (Buddhism, #HM1914)   Ten worlds (Buddhism, #HM2657)
 Nichiren shoshu buddhism (Buddhism, #HH3443).
 Followed by State of learning (Buddhism, #HM3662).
 Preceded by State of tranquillity (Buddhism, #HM3492).

♦ **HM1974d Firmness** (Buddhism)
Dhiti (Pali)
 Broader Faculty of energy (Buddhism, #HM1470).

♦ **HM1982d Application of reflection** (Buddhism)
Anuvicara (Pali)
 Related Analysis (Buddhism, #HM5089).

♦ **HM1991c Four-fold knowledge** (Systematics)
Value knowledge — Experience of subsistence
Description Subsistence is the limitation of existence within a framework that must be experienced through not less than four independent terms. It specifies and event. It is the form of all activities that lead to a change of order and as such is inherently inflexible. Its very nature is to be an activity of transformation. Its lack of central emphasis allows activity to be experienced as ordered diversity, but prevents the association of the activity with a particular entity. Indeed it does not allow for the existence of separate entities. This form of knowledge requires an active participation that is absent from the first three forms. Knowledge of self is significant only when it distinguishes between different levels of experience within the same being. Self-knowledge is thus the condition pre-requisite of value knowledge; for without it there can be no standards of comparison and therefore no discrimination of values. This knowledge conveys a feeling for values. All the individual can make a right assessment of experience, he remains closed to its hidden possibilities. The power of subsistential knowledge is to bring about an order within the functions by connecting intellectual and physical apprehension with an emotional attitude. Out of this connection comes the moment of knowledge by which the individual himself undergoes functional change.
Context The fourth in a sequence of twelve modes of knowledge, identified by J G Bennett, inspired by G Gurdjieff.
 Broader Systematics (#HH2003).
 Followed by Five-fold knowledge (Systematics, #HM8080).
 Preceded by Three-fold knowledge (Systematics, #HM0811).

♦ **HM1992d Holding as paramount one's view** (Buddhism)
Paramasa (Pali)
 Broader Wrong view (Buddhism, #HM1710).

♦ **HM1993c Onlooker consciousness**
Description This consciousness is based on an extreme separation between subject and object, consciousness and the world. This separation is a consequence of over-reliance on the intellectual mind and the analytical mode of consciousness with which it is associated. Thinking is equated with subjective experience which is identified with consciousness. Consciousness thus has the role of an onlooker to a world which is outside itself. The experiencer has the impression of being separate and independent from the world and detached from nature. The world may thus be treated as an object that can be operated on, manipulated and organized.
 Related Analytical consciousness (#HM2415).

♦ **HM1995c Imitative musical inspiration**
Description This is the second of three levels of musical inspiration. At this level the creative artist is inspired by traditional musical forms bequeathed by inspiration of the first level. The artist responds to them with musical forms according to his capacity, listening inwardly to a source of music in his soul as a form of mental improvisation. Creativity at this level may raise traditional forms to a transcendent level.
The listener at this level calls upon higher faculties (corresponding to the higher chakras) than those required for listening of the body–emotion–intellect (visceral–heart–head) variety.
With consciousness deliberately focused in the heart chakra, a higher octave of emotions is felt, namely the feeling qualities that underlie ordinary music of the heart. These are cosmic feeling-qualities beyond joy and sorrow. They are experienced as an ever-changing dilation and contraction, tension and release, to which none of the five external senses offer any parallel. These may be carried either by the harmony (Western music) or by a tonal centre of gravity (Oriental music).
With consciousness deliberately focused in the throat chakra, the larynx may actually respond as though singing the melody which tends to be exteriorized in an imaginary space of notes. This experience readily deteriorates into the normal experience of head music. If the observer avoids such observation of the melody and instead becomes identified with it, it becomes a vehicle for a journey through time. The listener is thus brought close to the creative source of the original melodic inspiration on which the composer drew. This is an experience of the nature of time.
With consciousness deliberately focused through the mid-eye chakra, provides a vision of music akin to insight. It is concentrated attention without selectivity, from which the listener may emerge to become identified with the music itself. Ego-bound consciousness is then supplanted by the state of the music itself. This total self-absorption in sound brings to being in the individual the primordial condition of the universe as a musical and temporal whole. The individual feels himself to be indistinguishable from the object of perception.
 Broader Musical inspiration (#HM4805).

♦ **HM1997c Way of active intelligence** (Esotericism)
Description Characteristic of the pragmatic and skilful, who are efficient and effective at getting things done, taking pleasure in action for its own sake. On this way they are challenged by the task of learning to be still, cultivating being as a complement to doing, developing an aesthetic appreciation of their environment, and discovering ways of acting to larger purpose.
Context The third of seven ways to spiritual realization characteristic of Western esotericism.
 Broader Ways to spiritual realization (Esotericism, #HH2665).
 Related Sphere of the kingdom (Kabbalah, #HM2288).

♦ **HM1998c Confidence** (Buddhism)
Okappana (Pali)
 Broader Faculty of faith (Buddhism, #HM0066).

♦ **HM1999d Usual state of consciousness** (USC)
Description The state is described by H Evans as the everyday condition, the equilibrium state to which the individual reverts after any alternate state experience. It is characterized by a degree of control over thoughts and actions and by a sense of identity. Surroundings are perceived apparently as they are perceived by others. Behaviour in the usual state may vary but the individual will be aware of what he is doing. The usual state varies between individuals, depending, for example, on differences in personality and cultural background. These differences will affect the likelihood of an individual experiencing an alternate state and how he reacts to experiencing such a state.
 Refs Evans, Hilary *Alternate States of Consciousness* (1989).
 Related Ordinary awareness (#HM2266)   Altered states of consciousness (#HM2391)
 Alternate state of consciousness (ASC, #HM1537).

♦ **HM2000b Love consciousness**
Conscious love
Description The affective aspect of love brings its object into the field of consciousness in a special way. What is loved is perceived as worthy of love and value may sometimes be imputed

to it with insufficient basis. What is loved also may become the focus of attention, and consciousness may be intensified until it has no other focus, as in passionate, erotic love, or as in mystical love which leads to divine union. This latter is variously referred to as *charity* or *naked intent* directed to God for God. Ultimately, love is said to be the knowledge God has of himself, which is one with himself.

Intense love of another or of an ideal can cause a state of close attention or scrutiny of the object and raise consciousness of it. Love of country may raise the consciousness of societal needs. Love of a single animal may make one more conscious of all animal life. Love of the condition of loving, or love of love, can cause one to investigate deeper conditions of love until it becomes a way of knowing reality.

Many kinds of love have been categorized, some of great depth some more limited, but none of the categories should be regarded as pure, in reality love is a combination of these categories, and one category can evolve into another (vanity–love into passionate love, romantic love to affectionate love, for example).

Context Love is an energy that can motivate many human endeavours. Conscious love is a state of awareness which can give purposiveness to actions, making a bridge between reason, which identifies the object of love, and will, which causes love to be expressed as action towards its object or objective. Love, therefore, is evoked in personal relations, in patriotism, in occupation or vocation, in artistic expression, in religious behaviour and attitudes, and in the contexts of human development and planetary consciousness, since its function is to promote wholeness.

Refs Johnson, Robert A *The Psychology of Romantic Love* (1984); Lewis, C S *The Four Loves* (1960); Person, Ethel Spector *Love and Fateful Encounters* (1990).

Narrower Limerence (#HM1809) Sexual love (#HM9406)
Infatuation (#HM5991) Courtly love (#HM6122)
Romantic love (#HM5087) Sympathy–love (#HM4905)
Neurotic love (#HM5702) Ontological love (#HM1211)
Transference love (#HM7045) Affectionate bonding (#HM0973)
Self–aggrandizing love (#HM4671) Degrees of love (Christianity, #HH3012)
State of perfect love (Christianity, #HM0074).
Related Love (#HH0258) Lila (#HM2278) Kindness (#HH0457)
Affinity (Islam, #HM5304) Hatred awareness (#HM4596)
Spiritual love (Christianity, #HM8977).

♦ HM2001c Total freedom from attachment (Hinduism)

Fifth plane of wisdom

Description As a result of previous practices there is total freedom from attachment and a conviction of the nature of truth.

Context The fifth of seven ascending planes of wisdom described in the Supreme Yoga, which require to be known so as not to be caught in delusion.

Broader Planes of wisdom (Hinduism, #HM3298).
Followed by Cessation of objectivity (Hinduism, #HM3354).
Preceded by Establishment in truth (Hinduism, #HM3006).

♦ HM2002a Shekinah awareness (Judaism)

Kavod

Description The awareness of the Shekinah, in the Jewish tradition, is said to be equivalent to the mystical way of the Tree of Life. That way is the ascent through the paths of wisdom to the stations or realms of the emanations or sephiroth through the four worlds. Shekinah can be found, it is said, in malkuth, the "kingdom", lowest of the sephiroth in the consciousness of this world; and it is sometimes identified only with this stage. It appears, however, that the Shekinah (dwelling or vehicle) of God can ascend and descend the Tree of Life. In its active or "masculine" role it has been considered an angel of liberation for human consciousness and given a position inferior only to God from the point of view of human salvation. Whether masculine or feminine in its personifications, the Shekinah is a mediating consciousness between the infinitude of God and limited human awareness. In Chasidic teachings the Shekinah is called *kavod*, the divine radiance or glory; and when considered as an angel it is related to the cherub on the throne of God.

Narrower World of creation (#HM3038) World of formation (#HM2360)
World of emanation (#HM2408) World of manifestation (#HM2312).
Related Kabbalah (#HH0921) Presence of God (#HM2961)
Paths of wisdom (Judaism, #HM2509) Human development (Judaism, #HH3029)
Practice of the presence of God (#HH0992).

♦ HM2005d Obedient son (ICA)

Description This is experienced when what is required is utterly impossible and violates all an individual's social and intellectual training and yet he knows it is exactly what he must do. It occurs when an individual decides to be obedient to the eternal context of his life and not to his own will, as when Jesus said "not my will, but thine, be done".

Context In the ICA New Religious Mode in the arena of acting out one's deed (the life of doing) the fourth formal aspect is becoming the symbol of the eternal context. At the third phenomenological level this occurs when the individual embraces his election to this life, this task.

Broader Symbolizing the eternal context (ICA, #HM3176).
Followed by Doing the mystery (ICA, #HM2143).
Preceded by Meaning creation (ICA, #HM2017).

♦ HM2006g Planetary consciousness

Ecological consciousness — Global consciousness — Evolutionary awareness — Planetization of consciousness

Description Ecological consciousness is said to arise from a pantheistic awareness of nature, an openness or receptivity so great that there is a spiritual catharsis when the personal self identifies with ultimate being - the "ecological self". Although such consciousness does not, itself, arise from intellectual activity, the concept may or may nor include any of the following components in a given instance: a general systems perspective of the integration of all social and natural system levels into a global system; more comprehensive (meta) models of reality; isomorphic relations between man the microcosm and the planet as the macrocosm (man as a whole creature woven into the collective fabric of humanity); a symbiotic and morphogenetic awareness of mankind; a sense of world community; a sense of man's role as a part of, and as potential controller of, the evolutionary process.

In general, it is held that modern technological and materialistic society has cut off the individual from his sense of connection with and intrinsic part of the surroundings. This has created and exacberated environmental problems such as fossil and nuclear fuel pollution and destruction of tropical forests. A return to the ecological consciousness typical of less developed peoples is advocated by environmentalist individuals and groups. It may be cultivated through work on the awareness of one's self, of becoming more real. There is a need to be open and receptive, to learn to listen and to appreciate silence and solitude. This process has been referred to (Theodore Roszak) as cultivation of *conscience*. Things move more slowly, more simply, there is liberation from haste and waste.

Refs Barron, Frank *Toward an Ecology of Consciousness* (1972); Devall, Bill and Sessions, George *Deep Ecology* (1985); Gandhi, Kishore *The Evolution of Consciousness* (1986); Russell, Peter *The Global Brain* (1983).

Related Primitivism (#HH1778) Deep ecology (#HH2315) Human synergy (#HH0517)
Bio–globalism (#HH1267) Planetary noogenesis (#HH4621)
Communion with nature (#HM0915) Environmental mysticism (#HM0800)
Christian stewardship (Christianity, #HH3121)
Human development through pantheism (#HH5190)
Human development through panentheism (#HH1908)
Education for international understanding (#HH6451).

♦ HM2007d Divine–animal–hell awareness (Buddhism, Tibetan)

Description This state of awareness is characterized by supernatural powers although beast-formed. It is represented as gardens of the gods where mythological beasts reside.

Context In Tibetan Sakya Buddhism this is one of the states in the "Ascension Stages Game". In some sets it is numbered 12 on the board.

Broader Desire–realm consciousness (Buddhism, #HM2733).
Followed by Asura–world awareness (Buddhism, Tibetan, #HM2579)
Animal–hell awareness (Buddhism, Tibetan, #HM2636)
Barbarian–state of awareness (Buddhism, Tibetan, #HM2091)
Hungry–ghosts–hell awareness (Buddhism, Tibetan, #HM2112)
Thirty–three–god–heaven awareness (Buddhism, Tibetan, #HM2606)
Four–great–kings–heaven awareness (Buddhism, Tibetan, #HM2082).
Preceded by Pure–lands awareness (Buddhism, Tibetan, #HM2168)
Animal–hell awareness (Buddhism, Tibetan, #HM2636)
Reviving–hell awareness (Buddhism, Tibetan, #HM2516)
Heaven–without–fighting consciousness (Buddhism, Tibetan, #HM2130)
Delightful–emanation–heaven consciousness (Buddhism, Tibetan, #HM2546).

♦ HM2008d Called to accountability (ICA)

Description The second formal aspect of the *life of meditation*, an arena in the ICA *New Religious Mode*, this represents an experience of being called to accountability by one's interior priors.

Broader Life of meditation (ICA, #HM3234).
Narrower Concerned judge (#HM2582) Universal father (#HM2418)
Heavenly advocate (#HM2784) Unfailing prompter (#HM2356).

♦ HM2009d Daring embracement (ICA)

Description This is experienced as an amazing dependency on life situations that are incomplete, imperfect and in some way less than desired. It occurs when all life gives is misfortune and stupidity and when good intentions come to naught. What is given is embraced, as when a community celebrates the disaster of monsoon rains destroying a bridge that months of their labour has built.

Context In the ICA New Religious Mode in the arena of acting out one's freedom (the life of prayer) the second formal aspect is affirmation of one's dependence or gratitude. At the first phenomenological level this occurs when one encounters the burden of one's own existence.

Broader Affirmation (ICA, #HM2199).
Followed by Splendid vices (ICA, #HM2165).

♦ HM2010d Heavenly–highway awareness (Buddhism, Tibetan)

Description This state of awareness is characterized as the present moment; the start of the stages of ascent or transformations of consciousness. It is represented by the choice of roads, one for each of the karmic destinations: to the four great kings, to the southern continent, to the asura–world, or to the hells.

Context In Tibetan Sakya Buddhism this is one of the states in the "Ascension Stages Game". In some sets it is numbered 24 on the board.

Broader Desire–realm consciousness (Buddhism, #HM2733).
Followed by Animal–hell awareness (Buddhism, Tibetan, #HM2636)
Asura–world awareness (Buddhism, Tibetan, #HM2579)
Reviving–hell awareness (Buddhism, Tibetan, #HM2516)
Southern–continent awareness of men (Buddhism, #HM2127)
Hungry–ghosts–hell awareness (Buddhism, Tibetan, #HM2112)
Four–great–kings–heaven awareness (Buddhism, Tibetan, #HM2082).

♦ HM2011c Path of imaginative intelligence (Judaism)

Description This state of awareness is of the laws by which creation develops from types, and types from archetypes, all things bearing the likeness of their models.

Context The twenty fourth of the spiritual paths expressed in the Jewish Kabbalah.

Broader Paths of wisdom (Judaism, #HM2509).
Followed by Path of intelligence of temptation (Judaism, #HM2503).
Preceded by Path of stable intelligence (Judaism, #HM2533).

♦ HM2012c Arupaloka consciousness (Buddhism, Pali)

Description In Pali texts, this is the second of the three higher states: that to which the mental absorptions (jhana) of human meditation rise and where the six higher spiritual powers start to appear.

Broader Worlds of conscious existence (Buddhism, Pali, #HM2072).
Narrower Third absorption in the immaterial sphere (Buddhism, #HM2027)
First absorption in the immaterial sphere (Buddhism, #HM2110)
Fourth absorption in the immaterial sphere (Buddhism, #HM2051)
Second absorption in the immaterial sphere (Buddhism, #HM3043).
Related Jhana (Buddhism, Pali, #HM7193) Phenomena awareness (Buddhism, #HM2551)
Rupaloka consciousness (Buddhism, #HM2536)
Formless–realm consciousness (Buddhism, #HM2281)
Tibetan meditative states of formless absorptions (Buddhism, #HM2669).
Preceded by Fourth trance of the fine–material sphere (Buddhism, #HM2586).

♦ HM2013b Christ consciousness (Christianity)

Presence of Christ — Christian consciousness

Description The presence of Christ within a person transforms that person such that he becomes a new creation. S E Fittipaldi describes how this is realised in different ways in different people, manifesting as: an empty (or un–bound) mind; a forgiving mind, stretching beyond the limits of what is normally accepted; a loving mind; and a dying–rising awareness of life–in–death and death–in–life. Christian consciousness is Christ consciousness, it is identifying with Christ that the person becomes human so that the process of Christification is that of humanization. This is the core of Christian mystical tradition.

Christ consciousness is comparable to the emptiness sought through Buddhist meditation. It is quoted in sayings of Jesus as arising when the lower vehicle of the senses is caste off, the emotions raised and united with reason, when two are one, the external pure so as to be united with the internal nature, male and female sex transcended, then the individual is one in Christ and attains Christ consciousness.

François Fénelon reminds us that no–one can gain victory over sin and be brought into union with God except through Christ. Every period of progress is with Christ as the way, life is derived from God through Him and for Him. The most advanced on the way are the most possessed with the thoughts and the presence of Christ.

Refs Fittipaldi, Silvio E *Human Consciousness and the Christian Mystic*; Valle, Ronald S and Eckartsberg, Rolf von (Eds) *The Metaphors of Consciousness* (1981).

Narrower Christ consciousness as empty mind (#HM0778)
Christ consciousness as loving mind (#HM0976)
Christ consciousness as dying–rising (#HM1402)
Christ consciousness as forgiving mind (#HM1728).
Related Becoming Christ (#HH5328) Ishvara–consciousness (Yoga, #HM2913)
Mystical Christ–consciousness (Christianity, #HM2880).

♦ **HM2014g Wakeful dream** (Hinduism)
Description The mind is fully awake to, and filled with, its own fancies.
Context The fourth of seven descending steps of ignorance described in the Supreme Yoga, which veil *self-knowledge*. Each has innumerable sub–divisions.
Broader Veils of delusion (Hinduism, #HM3592).
Followed by Dream (Hinduism, #HM3147).
Preceded by Great wakefulness (Hinduism, #HM3055).

♦ **HM2015d Uninterrupted and release paths** (Buddhism, Tibetan)
Anantaryamarga — Vimuktimarga
Description Mental contemplations that lead to sustained meditation and to release from the desire realm.
Context These paths or mental conditions are found among the *preparations for concentration* in Tibetan Gelugpa Buddhism.
Broader Special insight and preparations as states (Buddhism, #HM2623).
Related Final–training subtle contemplation (Buddhism, #HM2683).

♦ **HM2016a Selflessness** (Buddhism)
Self–emptiness
Description The awareness of true true selflessness, such as will lead to liberation from cyclic existence, entails analysis of the inherent existence of phenomena and the penetrating perception that objects do not exist as they appear to do. This is a far finer concept than simply not conceiving the seif of phenomena (which could be achieved through sleep or stupor). Such realization of the subtle emptiness of phenomena is the aim of *Tibetan meditation upon emptiness*, as it is of all Buddhist schools.
Narrower Calm abiding (#HM2147)
Only–a–beginner subtle contemplation (#HM2171)
Buddha–consciousness (Buddhism, Tibetan, #HM2735)
Special insight and preparations as states (#HM2623)
Tibetan meditative states of form concentrations (#HM2693)
Tantra in meditation on emptiness (Buddhism, Tibetan, #HM2864).

♦ **HM2017d Meaning creation** (ICA)
Description This is realizing that the significance or meaning of an event is what one decides it to be; when a person discovers that what he does is known in history but that the interpretation of an event is his to create. It might be that when Mahatma Gandhi was thrown off the train in South Africa, he decided that this was to be not just an unpleasant experience but one that would shape his future life.
Context In the ICA New Religious Mode in the arena of acting out one's deed (the life of doing) the fourth formal aspect is becoming the symbol of the eternal context. At the second phenomenological level this occurs when one recognizes one's life as sheer venture.
Broader Symbolizing the eternal context (ICA, #HM3176).
Followed by Obedient son (ICA, #HM2005).
Preceded by Final situation (ICA, #HM2119).

♦ **HM2019d Radiant guru** (ICA)
Description This is experienced as always having someone or something that is with one, reminding one that life is a great gift and that one can dare to live it no matter what happens. It occurs on accepting that there is a reality that is not going to let one escape from the life one has. It may be compared to Martin Luther King's relationship with Mahatma Gandhi and his decision to use the strategy of non–violence.
Context In the ICA New Religious Mode in the arena of knowing one's internal sociality (the life of meditation) the first formal aspect is one's encounter with the mediator who pronounces personal absolution. At the second phenomenological level, this occurs when a person finds his selfhood audited by that which commands his attention.
Broader Personal absolution (ICA, #HM2370).
Followed by Persistent friend (ICA, #HM2139).
Preceded by Word–bearing priest (ICA, #HM2236).

♦ **HM2020b Transcendental consciousness**
Fourth state (Yoga) — Wakeful hypometabolic state
Description Transcendental consciousness is said to be a fourth major state of consciousness after waking, dreaming and sleeping states. Its properties are described as empty, devoid of objects, yet wakeful, alert and conscious of self. Four states can be described as awake/asleep, with/without objects of consciousness. If waking is awake with objects of consciousness, dreaming is asleep with objects of consciousness and deep sleep is asleep with no objects of consciousness, then transcendental consciousness is awake with no objects of consciousness. In this latter case, although wide awake, mental activity ceases as normal thought is transcended. Where normally the object of experience is not the same as that which is experiencing (in other words, the subject), here subject and object are the same. The experience is that of bliss and attempts at its description have been made mystics and philosophers of many traditions, notably in the Upanishads, in the writings of Plotinus, in Buddhism, in mediaeval mysticism. It has been equated with states of quiet and enhanced awareness which include the *samadhi* of Hinduism, *fana* of Sufism, *ming* of Zen; and references also occur to it in poetry. Since it has none of the usual qualities of thought it nonetheless belies description. It is the state of the true self, pure and without attributes, which underlies all experience; the state of pure being, of oneness with the whole of creation.
The physiological characteristics are, among many others, alpha–wave production, lowering of blood pressure and skin resistance, and reduction in bodily consumption of energy. Breathing slows and may cease altogether for a minute or more, without effort.
Context Proposed originally by Maharishi Mahesh Yogi, transcendental consciousness is probably one of the scientifically best–documented of the so–called higher states of consciousness. Rooted in the Upanishad and Vedanta conception of consciousness (turiya), experimental observation strongly suggests similarities to states described in various other traditions. It is also claimed that this state is not a state of consciousness among others but rather an underlying phenomenon, accompanying all other conceivable states with more or less clarity.
Refs Wallace, R Keith *The Physiological Effects of Transcendental Meditation* (1970).
Broader Awareness of relative reality (Hinduism, Yoga, #HM2032).
Related Ming (Zen, #HM0765) Fana (Sufism, #HM1270) En–sof (Judaism, #HM0934)
Tso–wang (Taoism, #HM0346) Unity consciousness (#HM3193)
Samadhi (Hinduism, Yoga, #HM2226) Turiya awareness (Hinduism, #HM2395)
State of quies (Christianity, #HM3771) Transcendental meditation (TM, #HH0682)
Fund of the individual (Christianity, #HM4497).
Followed by Transcendental cosmic consciousness (#HM2405).
Preceded by Susupti state of unconsciousness (Hinduism, Zen, #HM2957).

♦ **HM2021c Path of intelligence of spiritual action** (Judaism)
Path of intelligence of the secret
Description This state of awareness encompasses the secret activities of the spiritual beings and the transmission of blessing.
Context The cognition of divine action is the nineteenth spiritual path expressed in the Jewish Kabbalah.
Broader Paths of wisdom (Judaism, #HM2509).
Followed by Path of intelligence of will (Judaism, #HM2629).
Preceded by Path of intelligence of influences (Judaism, #HM2045).

♦ **HM2022d Joyful–heaven awareness** (Buddhism, Tibetan)
Tushita
Description This state of awareness is described as the traditional realm of bodhisattvas before they take a final birth in human form.
Context In Tibetan Sakya Buddhism this is one of the states in the "Ascension Stages Game". In some sets it is numbered 30 on the board.
Broader Desire–realm consciousness (Buddhism, #HM2733).
Followed by Great–vehicle lower–path awareness (Buddhism, Tibetan, #HM2268)
Great–vehicle middle–path awareness (Buddhism, Tibetan, #HM2160)
Great–vehicle higher–path awareness (Buddhism, Tibetan, #HM3048)
Mahayana heat–application awareness (Buddhism, Tibetan, #HM2208)
Mahayana receptivity–application awareness (Buddhism, Tibetan, #HM2247)
Mahayana highest–teachings–application awareness (Buddhism, Tibetan, #HM2271).
Preceded by Form–realm awareness (Buddhism, Tibetan, #HM2142)
Non–emanating consciousness (Buddhism, Tibetan, #HM2070)
Great–vehicle lower–path awareness (Buddhism, Tibetan, #HM2268)
Heaven–without–fighting consciousness (Buddhism, Tibetan, #HM2130)
Pratyeka Buddha application awareness (Buddhism, Tibetan, #HM3020)
Delightful–emanation–heaven consciousness (Buddhism, Tibetan, #HM2546).

♦ **HM2023d Common earth** (ICA)
Description This is experienced as being a "comrade" of all of existence; when one can take or leave whatever situation, relationship, or moment one is given. It may be compared to the Zen story of two monks, an elder and a novice, who were travelling together. When they arrived at a deep, fast stream, a beautiful woman was standing there, unable to cross. The elder monk picked up the woman and carried her across. The two monks continued their journey, but the novice was obviously greatly troubled. Finally he asked the elder why, in light of their vow never to touch a woman, the elder had carried the woman across the stream. The elder said he had set her down at the edge of the stream; why was the novice still carrying her ?
Context In the ICA New Religious Mode in the arena of knowing one's disengagement (the life of poverty) the third formal aspect is being nonchalant about one's relations. At the fourth phenomenological level, this occurs when one is filled with the power of requiring nothing from a situation.
Broader Liberation from relationships (ICA, #HM3240).
Preceded by Abounding abasement (ICA, #HM2209).

♦ **HM2024c Life of obedience** (ICA)
Description An arena in the ICA *New Religious Mode* when one is aware of acting out one's engagement.
Broader New religious modes (ICA, #HM3004).
Narrower Embodying peace (#HM2876) Embodying equity (#HM2442)
Embodying charity (#HM3056) Embodying service (#HM3199).

♦ **HM2026b Subjective domains of consciousness** (Yoga)
Three worlds
Description Material forms produced by actions are the first domain of subjective consciousness; the subtle world of forms produced by emotions is the second domain; and the incorporeal world of forms produced by thoughts is is the third. The second and third domains are as subjectively real as the world around us. Experience of any of the three as the one and unique reality is due, according to yogic philosophy, to engagement of only the lower levels of mind (manas) and to the inaction of the higher faculties (buddhi, etc).
Related Indian approaches to consciousness (Hinduism, #HM2446).

♦ **HM2027d Third absorption in the immaterial sphere** (Buddhism)
Akimcanyayatana — Akincannaayatana (Pali) — Consciousness of nothingness absorption — Sphere of nothing at all — Third formless attainment (Tibetan)
Description *Hinayana Buddhism* At the perception "Nothing is there" the mind reaches the sphere of nothingness. The practitioner remains in equanimity, happiness and mindfulness, detached from joy.
Tibetan Gelugpa Buddhism The meditator in this state considers that there is nothing formed or formless to be apprehended.
Context *Hinayana Buddhism* This is the next to last of the jhanas.
Tibetan Gelugpa Buddhism This is the third of the four formless absorptions (arupayasamapatti).
Broader Jhana (Buddhism, Pali, #HM7193)
Arupaloka consciousness (Buddhism, Pali, #HM2012)
Trances and mental absorptions (Buddhism, #HM2122)
Immaterial states as meditation subjects (Buddhism, #HH3198)
Tibetan meditative states of formless absorptions (Buddhism, #HM2669).
Related Subjects for meditation (Buddhism, #HH3987)
Formless–realm consciousness (Buddhism, #HM2281).
Followed by Fourth absorption in the immaterial sphere (Buddhism, #HM2051).
Preceded by Second absorption in the immaterial sphere (Buddhism, #HM3043).

♦ **HM2028d Adamantine–hell awareness** (Buddhism)
Vajra–hell — Rdo–rje (Tibetan)
Description This state of awareness is characterized by fearful mental pain and is represented as a hell.
Context In Tibetan Sakya Buddhism this is one of the states in the "Ascension Stages Game". In some sets it is numbered 1 on the board.
Broader Desire–realm consciousness (Buddhism, #HM2733).
Followed by Lord–of–death awareness (Buddhism, Tibetan, #HM2088).
Preceded by Kriya–tantra awareness (Buddhism, Tibetan, #HM2558).

♦ **HM2029b Enlightenment**
Spiritual enlightenment — Illumination — Bodhi — Wisdom
Description Enlightenment can refer to Buddha's experience under the bodhi tree, in which the experiential and cognitive dimension is ontological, viz a mystical insight into the nature of reality; or enlightenment can refer to an age (as in 19th century Europe) characterized by a rational consciousness based on a materialist philosophy. A related term, illumination, can refer to mystical gnosis, and suggest an occult philosophy. All these cases of illumination or enlightenment are based on a fundamental and universal experience of humanity: when the unconscious or

subconscious releases or is caused to release a hold on a previously unknown and unexplored territory of the mind.
Enlightenment or illumination may be the consummation of *devotion* and can occur suddenly, as a personal revelation bursting forth in the mind and consciousness, or as a gradual attainment. In either case it is reflected as a harmony of love and wisdom at a different level and beyond the intellectual and emotional functions of mind. Such experiences can occur in an individual or a group. When it occurs in a group it is usually on indication of the advent of the numinous, an intersection of the spirit with the evolving line of human development, and it may signal the birth of a new age.
Refs Da Free John *Enlightenment and the Transformation of Man* (1983); White, John (Ed) *What is Enlightenment?* (1985).
Narrower Fayd (Sufism, #HM0568) Sambodhi (Buddhism, #HM7195).
Related Satori (Zen, #HM2326) Liberation (#HH0388) Illumination (#HH0804) Kaivalya (Yoga, #HM3869) Mystical cognition (#HM2272) Kevalajnana (Jainism, #HM3948) Samadhi (Hinduism, Yoga, #HM2226) Anuttara samyak sambodhi (#HM3263) Faculty of living (Buddhism, #HM3404) Breakdown of automatization (#HM8105) Steps to enlightenment (Buddhism, #HH4019).

♦ **HM2030d Ultimate reality** (ICA)
Description This state occurs when an individual experiences all he knew or counted upon in life as somehow taken away from him, as if he no longer had a place to stand, and a thick fog had descended around his world; all he had depended upon to inform and shape his anticipations of life are not there. For many people the assassination of President Kennedy was such a moment; other examples occur in Camus' work "The Stranger" and the movie "On the Beach", which aptly shows a situation where all the ground rules of life have been changed. It is like waking from a deep sleep in a strange room and not remembering where one is.
A dimension of this experience is a sense of awe, that is fear and fascination. A characteristic of such a time is profound fear that perhaps reality is not the way one had always thought it to be. A feeling that everything one does is wrong and one will never belong again is held in tension with a sense of being really alive, a sense of newness - "I've never been here before" - that seems to remove the old limits. One feels beckoned to create the new in the midst of this unknown world.
In this state one decides to trust what one knows about life and to keep moving, to live in the midst of this strange event rather than seek to escape. After such a time one thinks twice before acting and feels one's way. Being lost is not a problem and one develops a style for all situations.
Context This state is number 3 in the ICA *Other World in the midst of this World.*
Broader Aweful encounter (ICA, #HM2619).
Followed by Final limits (ICA, #HM2674) Second birth (ICA, #HM3175) Primordial wonder (ICA, #HM2186) Self transcendence (ICA, #HM2584) Temporal solidarity (ICA, #HM2363).
Preceded by Second birth (ICA, #HM3175) Absurd existence (ICA, #HM2331) Contentless word (ICA, #HM2373) Temporal solidarity (ICA, #HM2363) Transcendent immanence (ICA, #HM3034).

♦ **HM2031d Naga-world awareness** (Buddhism, Tibetan)
Description This state of awareness is characterized by generosity and represented as the world of half-human water serpents, mermen and mermaids.
Context In Tibetan Sakya Buddhism this is one of the states in the "Ascension Stages Game". In some sets it is numbered 13 on the board.
Broader Desire-realm consciousness (Buddhism, #HM2733).
Followed by Asura-world awareness (Buddhism, Tibetan, #HM2579) Animal-hell awareness (Buddhism, Tibetan, #HM2636) Hindu-states of awareness (Buddhism, Tibetan, #HM2115) Barbarian-state of awareness (Buddhism, Tibetan, #HM2091) Four-great-kings-heaven awareness (Buddhism, Tibetan, #HM2082) Thirty-three-god-heaven awareness (Buddhism, Tibetan, #HM2606).
Preceded by Animal-hell awareness (Buddhism, Tibetan, #HM2636) Howling-hells awareness (Buddhism, Tibetan, #HM2100) Crushing-hells awareness (Buddhism, Tibetan, #HM3142) Hindu-states of awareness (Buddhism, Tibetan, #HM2115) Pratyeka Buddha awareness (Buddhism, Tibetan, #HM2180) Hungry-ghosts-hell awareness (Buddhism, Tibetan, #HM2112) Wheel-turning-king awareness (Buddhism, Tibetan, #HM2058) Barbarian-state of awareness (Buddhism, Tibetan, #HM2091) Western-continent awareness of cattle (Buddhism, Tibetan, #HM2519) Eastern-continent awareness of noble figures (Buddhism, Tibetan, #HM2543).

♦ **HM2032b Awareness of relative reality** (Hinduism, Yoga)
Conditional consciousness states — Avasthas
Description In yoga there are three states (avasthas) of consciousness dominated by the subjective influence of the ego. They are the ordinary waking state (jagrat); the dreaming sleep state (swapna); and the deep, dreamless sleep state (susupti). The "fourth" state (turiya) is objective, pure knowledge which takes in its field the true self. Beyond this is a supra-conscious state (turyatita) of being in union with one's god.
Broader Stages of spiritual life (#HH0102).
Narrower Shushumna (Hinduism, #HM2053) Turyatita (Hinduism, #HM3139) Unity consciousness (#HM3193) Turiya awareness (Hinduism, #HM2395) Transcendental consciousness (#HM2020) Swapna state of dream consciousness (Yoga, #HM2781) Jagrat state of waking consciousness (Yoga, #HM2141) Susupti state of unconsciousness (Hinduism, Zen, #HM2957).
Related Crazy-monkey ego consciousness (Hinduism, #HM2590).
Preceded by Discipleship-vision awareness (Buddhism, Tibetan, #HM2240) Pratyeka Buddha arhat awareness (Buddhism, Tibetan, #HM2776) Pratyeka Buddha cultivation awareness (Buddhism, Tibetan, #HM2252).

♦ **HM2033c Direct non-conceptual cognition of emptiness** (Buddhism, Tibetan)
Tibetan definitions of awareness
Description Different levels of awareness are seen as stages on the path to direct non-conceptual cognition of emptiness, and are divided into two classes: *cognizing consciousness*, when there is valid cognition of the object of awareness by a valid cognizer (whether direct or inferential) and *non-cognizing consciousness*, where there is valid cognition by a non-valid cognizer.
The stages are referred to as: *wrong consciousness* (log-shes), *doubt not tending toward the fact, equal doubt, doubt tending toward the fact, correct assumption* - these five being of non-cognizing or non-valid consciousness; and *inference through the power of the fact* (dngos-stobs-rjes-dpag), *mental direct valid cognizer* (yid-kyi-mngon-sum-gyi-tshad-ma), *yogic direct valid cognizer* (rnal-'byor-mngon-sum-gyi-tshad-ma) - these three being of cognizing or valid consciousness. The path is seen as running, therefore, from a strong sense of the real existence of objects (wrong consciusness), through levels of doubt as the yogi hears and reflects on the meaning of emptiness and subsequently studies and meditates on what he has heard. He then reaches a conceptual understanding of emptiness which is, nevertheless, not inference although it is generated through the process of inference. When true inference occurs the yogi is realizing emptiness; he remains in this state as long as he can in order to develop the insight based on calm abiding and continue on the path to non-conceptual direct cognition of emptiness.

♦ **HM2034c Ji-ji-mu-ge** (Zen)
Description The experience of the world as it is, without hindrance of desire or fear, and the realization that this is *nirvana*. This is the aim of the *jiriki* approach described in Zen Buddhism.
Related Ji-ri-mu-je (Zen, #HM3164) Zazen meditation (Zen, #HH0882).
Preceded by Jiriki (Zen, #HM3215).

♦ **HM2035c Path of illuminating intelligence** (Judaism)
Description This state of awareness is of the secret doctrines concerning the stages of holiness.
Context The fourteenth among the spiritual paths expressed in the Jewish Kabbalah.
Broader Paths of wisdom (Judaism, #HM2509).
Followed by Path of constituting intelligence (Judaism, #HM2117).
Preceded by Path of uniting intelligence (Judaism, #HM2583).

♦ **HM2036c Nagual state of awareness** (Amerindian)
Description Nagual appears within the context of Central Amerindian tradition as relating to a variety of experiences linking to a reality beyond ordinary senses. As such it is often related to use of sacred drugs. In a more concrete sense, it signifies the indescribable, the unforseeable, the unexpected which often evokes fear, astonishment or grotesque feelings of humour or craziness when experienced. Originating from Amerindian experience, Nagual is generally associated with aspects of the other world, e.g. the spirit of the peyote-plant in peyote-sessions. C Castaneda uses the term to designate the true character of the world, of which we ordinarily only perceive a limited section (the island of Tonal) corresponding to the nature of our comprehension. If the latter shrinks through shock, or on purpose (dreaming, not-doing, sacred drugs) the vastness of existence may open up.
This is the state of consciousness in which all internal dialogue has ceased, that in which true "seeing" takes place. Thus, to really "see" a man (as opposed to looking at him), one views him (in as far as tonal language can describe the viewing) as luminous fibres forming a luminous egg - a parallel that also occurs in some Tibetan Buddhist writing.
In Amerindian terms, if the brave becomes a warrior, and lives like a warrior, then tonal and nagual can be unified. This is a life in which whatever the task may be it is done in the best possible way, not clinging and with the highest excellence.
Refs Castaneda, Carlos *Journey to Ixtlan* (1972); Castaneda, Carlos *The Teachings of Don Juan* (1968); Castaneda, Carlos *A Separate Reality* (1971); MacDowell, Mark *A Comparative Study of the Teachings of Don Juan and Madhyamaka Buddhism* (1986).
Broader Dreams (#HM2950).
Related Not-doing (#HH0367) Stopping the world (#HH0489) Paramartha (Buddhism, #HM1067) Dreaming experience (#HM3908) Way of the warrior (Amerindian, #HH8219) Tonal state of awareness (Amerindian, #HM2066).

♦ **HM2037d Spiritual creativity** (ICA)
Description The fourth formal aspect of the *life of chastity*, an arena in the ICA *New Religious Mode*, this is experienced as devoting one's entire being to guarding the spiritual dimension of existence.
Broader Life of chastity (ICA, #HM2506).
Narrower Divine captive (#HM3190) Eternal friends (#HM3095) Secondary integrity (#HM3086) Replication of Christ (#HM3140).

♦ **HM2038c Second trance of the fine-material sphere** (Buddhism)
Piti (Pali) — Dvitiyadhyana — Bsam-gtan-gnyis-pa — Second form-realm concentration
Description *Hinayana Buddhism* The goal of the second trance is to leave behind the mystical rapture (piti) which had been obtained in this stage of the jhanas. Only concentration (samadhi) and joy (sukha) are carried forward.
Tibetan Gelugpa Buddhism This is a more advanced stage of analytical meditation in which special insights are obtained.
Context This is one of the four form-concentrations *dhyana* found in Tibetan Gelugpa Buddhism.
Broader Rupaloka consciousness (Buddhism, #HM2536) Form-realm consciousness (Buddhism, #HM2257) Trances and mental absorptions (Buddhism, #HM2122) Tibetan meditative states of form concentrations (Buddhism, #HM2693).
Related Dhyana (Hinduism, Buddhism, #HM0137) Fivefold happiness (Buddhism, #HM0747) Dwelling in the second jhana (Buddhism, #HM7121).
Followed by Third trance of the fine-material sphere (Buddhism, #HM2062).
Preceded by First trance of the fine-material sphere (Buddhism, #HM2450).

♦ **HM2039c Buddha nature body** (Buddhism, Tibetan)
Buddha truth body
Description The Buddha nature body is the ultimate true cessation part of the *Buddha truth body*, and consists of the naturally pure nature body (the absence from beginningless time) and the adventitiously pure nature body (the absence of adventitious stains thanks to application of their antidotes) which allows for the spontaneity of Enjoyment and Emanation Bodies.
Broader Buddha-consciousness (Buddhism, Tibetan, #HM2735).
Related Buddha wisdom body (Buddhism, Tibetan, #HM2834) Buddha emanation body (Buddhism, Tibetan, #HM3019) Buddha enjoyment body (Buddhism, Tibetan, #HM3113).

♦ **HM2040d Cold-hells awareness** (Buddhism, Tibetan)
Description There are 8 cold-hell states of awareness represented. All the hells are for those whose karma is the outcome of hateful activities including heresy or other crimes against the Buddhist dharma. These states are characterized by punishment from cold.
Context In Tibetan Sakya Buddhism this is one of the states in the "Ascension Stages Game". In some sets it is numbered 7 on the board.
Broader Desire-realm consciousness (Buddhism, #HM2733).
Followed by Animal-hell awareness (Buddhism, Tibetan, #HM2636) Asura-world awareness (Buddhism, Tibetan, #HM2579) Reviving-hell awareness (Buddhism, Tibetan, #HM2516) Crushing-hells awareness (Buddhism, Tibetan, #HM3142) Hungry-ghosts-hell awareness (Buddhism, Tibetan, #HM2112) Western-continent awareness of cattle (Buddhism, Tibetan, #HM2519).
Preceded by Very-hot-hells awareness (Buddhism, Tibetan, #HM2576) Crushing-hells awareness (Buddhism, Tibetan, #HM3142) Temporary-hells awareness (Buddhism, Tibetan, #HM2454) Thirty-three-god-heaven awareness (Buddhism, Tibetan, #HM2606) Great-vehicle lower-path awareness (Buddhism, Tibetan, #HM2268).

♦ **HM2041b Brahma consciousness** (Hinduism)
Description Mystical realization of the god Brahma is said to be a state of awareness of the cosmic creative consciousness (the logos or demiurge of other systems). The name is from the root meaning to grow or expand; therefore Brahma the creator, Vishnu the "pervader" (and hence the universal sustainer) and Shiva the dancing god of disruption (who dances in order to produce change and development), taken together are the Trimurti, the triple aspect of reality.
Related Indian approaches to consciousness (Hinduism, #HM2446).

♦ **HM2042d Glorious mystery** (ICA)
Description The fourth formal aspect of the *life of knowing*, an arena in the ICA *New Religious Mode*, this represents an awareness of the final nothing of life, its glorious mystery.
Broader Life of knowing (ICA, #HM2801).
Narrower Eternal void (#HM2994) Immutable friend (#HM3124)
Everlasting enemy (#HM2821) Honouring the mystery (#HM3029).

♦ **HM2043b Sahaja samadhi** (Hinduism)
Description Sahaja samadhi is by far the highest level of samadhi. In this state one rests in highest consciousness but works in the gross physical world at the same time. The experience of Nirvikalpa Samadhi is upheld together with earthly activities. One has become Soul and uses the Body at the same time as perfect instrument. When reaching Sahaja Samadhi one behaves like an ordinary human being but the innermost heart is flooded with divine enlightenment. One has become lord and master of reality. Even reaching the highest level of spiritual perfection, one is very seldom blessed with Sahaja Samadhi. Very few spiritual masters have reached this level. Whoever has attained Sahaja Samadhi is manifesting God in every second consciously and perfectly.
Broader Samadhi (Hinduism, Yoga, #HM2226).
Related Indian approaches to consciousness (Hinduism, #HM2446).
Preceded by Nirvikalpa samadhi (Hinduism, #HM2061).

♦ **HM2044e Devachon awareness** (Psychism)
Description Consciousness on the mental plane when the soul functions in (and is limited by) the mental body. Although of a higher order than "ordinary" heaven, the soul is deprived of its astral body but is still within the lower world of form. Such awareness is said (Alice Bailey) to be the goal of the majority; but, for true realization, it must be transformed to that of *nirvana*.
Followed by Nirvana (Buddhism, #HM2330).
Preceded by Heaven awareness (Psychism, #HM2565).

♦ **HM2045c Path of intelligence of influences** (Judaism)
Description This state of awareness encompasses the *arcana of secret influences* and the hidden nature of things.
Context The eighteenth of the spiritual paths expressed in the Jewish kabbalah.
Broader Paths of wisdom (Judaism, #HM2509).
Followed by Path of intelligence of spiritual action (Judaism, #HM2021).
Preceded by Path of disposing intelligence (Judaism, #HM2569).

♦ **HM2046b Ecstasy**
Description Ecstasy is a state of consciousness characterized by one or more of the following: an experience of an inner vision of a supreme being; an experience of union with such a being; a state of rapture, often with a decrease of self–control and excessive movement; and illusions concerning the individual's relationship to the space–time framework. Traditionally four stages are involved: purgation of bodily desire, purification of the will, illumination of the mind, unification of being or will with the divine.
It is recognized that whilst ecstatic experience may lead to superior integration of the personality, it may equally result in a complete breakdown of all accepted values, in total indifference to good and evil, in madness and schizophrenia. Normal, creative, schizophrenic and mystical–ecstatic states may be conceived on a continuum of increasingly higher central nervous system arousal, accompanied by an increased rate of information processing to a peak when interpretive activity ceases in a final state of rapture. The three levels of mystic ecstasy are described as: suspension of the external senses, suspension of both external and internal senses, and direct contemplation of the divine.
Ecstasy may be experienced unsought and unaided, or as a result of the deliberate use of: drugs, alcohol, dancing, sexual orgies, sexual abstinence, self–inflicted torture and related means; or by meditation, contemplation and the spiritual concentration practised in yoga. These latter are not associated with the extreme frenzy that may arise as a result of the former group but are possibly accompanied by mystic trances. This state is sometimes referred to as enstasy to distinguish it from ecstasy, the former referring to a state or concentration when contacts with the external world are withdrawn, consciousness is empty of all content and wisdom is firm. Some forms of ecstasy may involve groups and even crowds of people.
One experience often classified as ecstatic is the *out–of–body* experience. Leaving the physical body behind the individual visits other places or times and may communicate with spirits from another world.
Refs Baer, Dobh and Jacobs, Louis *On Ecstasy* (1982); Greeley, A *Ecstasy* (1974); Laski, M *Ecstasy* (1961); Rama, Swami *Enlightenment Without God*.
Related Bliss (#HM4048) Enstasy (#HM3169) Euphoria (#HM3763)
Ecstatic discipline (#HH5282) Religious enthusiasm (#HM2146)
Out–of–body experience (Psychism, #HM5534).

♦ **HM2047c Path of renovating intelligence** (Judaism)
Path of renewing intelligence
Description This mode of awareness encompasses the phenomena of change and renewal and their causes.
Context The twenty sixth of the spiritual paths expressed in the Jewish Kabbalah.
Broader Paths of wisdom (Judaism, #HM2509).
Followed by Path of natural intelligence (Judaism, #HM2071).
Preceded by Path of intelligence of temptation (Judaism, #HM2503).

♦ **HM2048d Karma awareness** (Buddhism)
Kamma (Pali)
Description Volitional action (karma, kamma) causes rebirth and shapes the destiny of beings. Karma, in Buddhism, is the intention of the will that creates a disposition or directedness in the mental factors as traits or tendencies that behaviour inclines to follow. Karma is the cause of results of actions, not the results themselves. To the extent that volition can be conceived apart from action it is termed *cetana* but it is karma nonetheless, from a different point of view. All consciousness has to have volition, hence karma. The cessation of personal, ego consciousness causes the cessation of karma, and the cessation of karma causes the cessation of re–birth.
Narrower Niskama karma (Hinduism, Yoga, #HH5487)
Worlds of conscious existence (Buddhism, Pali, #HM2072).
Related Karma (#HH0567) Intention (Buddhism, #HM2589)
Human development (Buddhism, #HH0650) Phenomena awareness (Buddhism, #HM2551)
Indian approaches to consciousness (Hinduism, #HM2446)
Ignorance in dependent origination formula (Buddhism, #HM3035).

♦ **HM2050c Awareness as mental–formation group of conscious existence** (Buddhism)
Sankhara–khanda (Pali) — Formations aggregate — Aggregate of mental activities
Description One of the five interacting aggregates that produce the illusory ego, this group is unified by volition (cetana) which is inseparably bound–up with all conscious phenomena. Being volitional, it is therefore action and, as such, generative of karma, both good karma favourable to progression and bad karma, a spiritual hindrance.
In Hinayana Buddhism, mental–formation is considered as anything that is perceived as being mental–formation, whether past, present or future, internal or external, subjective or objective, gross or refined, superior or inferior, far or near.
The formations aggregate is the aggregate of all that has the characteristic of forming or composing and also of adding together to form an agglomerate or compound. The formations have the function of accumulating or combining. Their manifestation is intervening, being busy. As proximate cause are the other three immaterial or mental aggregates: consciousness, feeling and perception. As regards karma they are three–fold: when associated with profitable or moral consciousness they are profitable; associated with unprofitable or immoral consciousness they are unprofitable; associated with indeterminate consciousness they are indeterminate.
The group is analyzed, in Hinayana Buddhism, as consisting of 50 mental phenomena or concomitants of consciousness. Of these, 25 are karmically wholesome or neutral, 14 or 13 are karmically unwholesome, and 11 or 12 are general elements. Within these distinctions are primary (always present in that type of consciousness) and secondary (sometimes present).
The general elements include the 5 primary concomitants to all consciousness: contact, sensation or mental impression (phassa); volition (cetana); life or vitality (jivita); attention or advertence (bringing to mind – manasikara); and concentration (samadhi). Sometimes also distinguished is steadiness of consciousness, although others consider it a different grade of concentration and do not include it separately. These 5 or 6, along with the feeling and perception aggregates (khandas) (which are sometimes included in the list making it 52), are concomitants to any consciousness. The remaining 6 general elements of the mental formation group are not always present in all forms of consciousness: applied thinking; sustained thinking; resolve; energy; happiness (zest); zeal (desire).
The profitable elements include 19 primary concomitants to any profitable or indeterminate profitable–resultant consciousness: faith; mindfulness; conscience (sense of shame); shame (dread of blame); non–greed (absence of greed); non–hate (absence of hate); specific neutrality (equanimity); tranquillity or repose of mental body (mental factors); tranquillity or repose of consciousness (or mind); lightness or buoyancy of mental body (mental factors); lightness or buoyancy of consciousness (or mind); pliancy or malleability of mental body (mental factors); pliancy or malleability of consciousness (or mind); wieldiness of mental body (mental factors); wieldiness of consciousness (or mind); proficiency or fitness of mental body (mental factors); proficiency or fitness of consciousness (or mind); rectitude of mental body (mental factors); rectitude of consciousness (or mind). Secondary profitable elements, sometimes but not always present in profitable and indeterminate profitable–resultant consciousness are 6: Abstinence from bodily misconduct; abstinence from misconduct in speech; abstinence from wrong livelihood; compassion or pity; gladness or sympathy.
The unprofitable elements include 4 primary concomitants to any unprofitable or indeterminate unprofitable–resultant consciousness: Delusion; consciencelessness (having no conscientious scruples); shamelessness; agitation. Secondary unprofitable elements, sometimes but not always present in unprofitable and indeterminate unprofitable–resultant consciousness are 10: hate; envy; avarice; worry; greed; wrong view; pride (conceit); stiffness; torpor; uncertainty. Stiffness and torpor are sometimes considered together, making a total of 9.
These formations are variously associated with the the 89 kinds of consciousness comprising the vinnana–khanda, depending not only upon whether they are profitable, unprofitable or indeterminate but also upon the sphere (sense, fine–material, immaterial or supramundane).
For example, there are 36 elements associated with the sense sphere, profitable consciousness accompanied by joy, associated with knowledge, unprompted whether sometimes or always present. Of these, 27 are said to be constant (appearing in their true nature), 5 are inconstant (immutable by nature) and 4 described as "or–whatever–states". All these 36 (except for some exceptions in particular instances) are associated with all the states in the "profitable" consciousnesses of the consciousness aggregate. Exceptions include absence of happiness in states accompanied by equanimity, for example. Also exluded are the elements associated with jhana factors not included in a particular jhana in the fine–material, immaterial and supramundane spheres (the three abstinences, applied thought, sustained thought, happiness, compassion, gladness).
There are 17 elements associated with the sense sphere, first unprofitable consciousness rooted in greed, of which 13 are constant and 4 "or–whatever–states". There are slight differences for the other consciousnesses rooted in greed. There are 18 elements associated with the first kind of unprofitable consciousness rooted in hate, of which 11 are constant, 4 are "or–whatever–states" and 3 inconstant. There are slight differences for the other consciousness rooted in hate. There are 13 elements associated with the first kind of unprofitable consciousness rooted in delusion, of which 11 are constant and 2 are "or–whatever–states". There are slight differences for the other consciousness rooted in delusion.
Again, diiferent groups of elements are associated with indeterminate resultant and functional consciousnesses – for example, 4 constant (contact, volition, life and steadiness of consciousness) and one "or–whatever–state" (attention) are associated with eye, ear, nose, tongue and body consciousnesses.
Context The mental activities or formations are comparable with the secondary mental factors and the afflictions of Tibetan Buddhism.
Refs Buddhaghosa, Bhadantacariya *The Path of Purity* (1975); Buddhaghosa, Bhadantacariya *The Path of Purification* (1980).
Broader Five aggregates (Buddhism, #HH3321).
Narrower Life (#HM0201) Pity (#HM0513) Shame (#HM8112)
Greed (#HM3283) Worry (#HM9101) Energy (#HM1333)
Torpor (#HM0264) Hatred (#HM4502) Contact (#HM2708)
Analysis (#HM5089) Gladness (#HM5224) Jealousy (#HM2638)
Intention (#HM2589) Stiffness (#HM5667) Conscience (#HM1590)
Non–hatred (#HM2744) Perplexity (#HM0812) Mindfulness (#HM2847)
Distraction (#HM3154) Investigation (#HM1177) Stabilization (#HM2440)
Non–ignorance (#HM2695) Shamelessness (#HM0649) Non–attachment (#HM2128)
Mental engagement (#HM3237) Pride (Christianity, #HM2823)
Zeal (Buddhism, Pali, #HM4487) Specific neutrality (#HM7002)
Belief (Buddhism, Yoga, #HM2933) Avarice (Buddhism, Pali, #HM0227)
Fitness of mental body (#HM1455) Lightness of mental body (#HM7636)
Fitness of consciousness (#HM1810) Rectitude of mental body (#HM5402)
Resolution (Buddhism, Pali, #HM3800) Wrong view (Buddhism, Pali, #HM5324)
Wieldiness of mental body (#HM7969) Lightness of consciousness (#HM6033)
Rectitude of consciousness (#HM9001) Tranquillity of mental body (#HM4704)
Malleability of mental body (#HM3696) Wieldiness of consciousness (#HM6556)
Ignorance (Buddhism, Yoga, Zen, #HM3196) Tranquillity of consciousness (#HM0226)
Malleability of consciousness (#HM1155) Consciencelessness (Buddhism, Pali, #HM4394)
Abstinence from bodily misconduct (#HM5600)
Steadiness of consciousness (Buddhism, Pali, #HM6652)
Abstinence from wrong livelihood (Buddhism, Pali, #HM7887)
Abstinence from verbal misconduct (Buddhism, Pali, #HM7171).

Related Becoming (Buddhism, #HM5909) Root afflictions (Buddhism, #HH0270)
Feeling aggregate (Buddhism, #HM4983) Perception aggregate (Buddhism, #HM4143)
Conditions for consciousness (Buddhism, #HM7655)
Dispositions of consciousness (Buddhism, #HM2098)
Secondary afflictions (Buddhism, Tibetan, #HH0781)
Secondary mental factors (Buddhism, Tibetan, #HH1348)
Awareness of corporeality-group of conscious existence (Buddhism, #HM2108).

♦ **HM2051d Fourth absorption in the immaterial sphere** (Buddhism)
Naivasamjnanasamjnayatana — Nevasanna-n'asannayatana (Pali) — Bhavagra — Srid-rtse (Tibetan) — Peak of cyclic existence absorption — Sphere of neither cognition nor non-cognition — Sphere of neither perception nor non-perception — Fourth formless attainment
Description *Hinayana Buddhism* Overcoming the perception of nothingness the mind reaches the *sphere of neither-perception-nor-non-perception.* Happiness and sadness are both abandoned, joy and sorrow have disappeared. Equanimity, purity of mindfulness, arises.
Tibetan Gelugpa Buddhism This state is the highest in the samsara or wheel of existence and reincarnation. Beings born at this level have the longest lifespan in the three worlds of desire, form, and formlessness. However, if the results of all the concentrations and absorptions, even of the peak absorption, are not applied, the way out of cyclic existence will not be realized.
Context *Hinayana Buddhism* This is the last of the 4 jhanas of the immaterial sphere, the 8th jhana of all when meditation practice from commencing with the kasina is considered.
Tibetan Gelugpa Buddhism This is the last of the four formless absorptions (arupayasamapatti).
Broader Jhana (Buddhism, Pali, #HM7193)
Arupaloka consciousness (Buddhism, Pali, #HM2012)
Trances and mental absorptions (Buddhism, #HM2122)
Immaterial states as meditation subjects (Buddhism, #HH3198)
Tibetan meditative states of formless absorptions (Buddhism, #HM2669).
Related Subjects for meditation (Buddhism, #HH3987).
Followed by Nirvana (Buddhism, #HM2330) Attainment of cessation (Buddhism, #HM5438)
Meditation way of the four truths (Buddhism, #HM2245).
Preceded by Third absorption in the immaterial sphere (Buddhism, #HM2027).

♦ **HM2052d Interminable-hell awareness** (Buddhism, Tibetan)
Avici-hell
Description This state of awareness is characterized by punitive suffering for having committed crimes against society or against those closest to one, whether parent or religious teachers or others. It is represented as one of the 8 lower hells.
Context In Tibetan Sakya Buddhism this is one of the states in the "Ascension Stages Game". In some sets it is numbered 2 on the board.
Broader Desire-realm consciousness (Buddhism, #HM2733).
Followed by Very-hot-hells awareness (Buddhism, Tibetan, #HM2576)
Southern-continent awareness of men (Buddhism, #HM2127)
Hungry-ghosts-hell awareness (Buddhism, Tibetan, #HM2112).
Preceded by Very-hot-hells awareness (Buddhism, Tibetan, #HM2576)
Barbarian-state of awareness (Buddhism, Tibetan, #HM2091).

♦ **HM2053c Shushumna** (Hinduism)
Description Shushumna is the central energy channel in tantra and kundalini yoga which interconnects all chakras which each other. Physiological correspondence is the spinal nerve channel. The life-force at the base of the spine (kundalini) is drawn upwards by tantric ritual and kundalini yoga exercises, and merged with the supreme consciousness at the top of the head, resulting in samadhi. Psychic energy in this channel is the pervasive sustaining power of Self. Shushumna is analogous to the axis mundi, Jacobs's ladder, the stem of the tree of life in Kabbalah, the stem of the tree of the world in shaman drumming, the stem of the tree of Jesse, etc. Mastery of shushumna leads to integration of ida and pingala, a state in which there are neither days nor hours.
Broader Awareness of relative reality (Hinduism, Yoga, #HM2032).
Narrower Ida conscious energy (Yoga, #HM2087)
Pingala conscious energy (Yoga, #HM2890).
Related Kundalini conscious energy (Yoga, #HM2871)
Chakra centres of consciousness (#HM2219).

♦ **HM2054d Current views and opinions** (Buddhism)
Ditthigata (Pali)
Broader Wrong view (Buddhism, #HM1710).

♦ **HM2055d Demon-island awareness** (Buddhism, Tibetan)
Rakshasas
Description This state of awareness is characterized by night-presence and is represented as man-eating demons who haunt cemeteries.
Context In Tibetan Sakya Buddhism this is one of the states in the "Ascension Stages Game". In some sets it is numbered 14 on the board.
Broader Desire-realm consciousness (Buddhism, #HM2733).
Followed by Asura-world awareness (Buddhism, Tibetan, #HM2579)
Very-hot-hells awareness (Buddhism, Tibetan, #HM2576)
Crushing-hells awareness (Buddhism, Tibetan, #HM3142)
Tantra-beginner awareness (Buddhism, Tibetan, #HM2452)
Southern-continent awareness of men (Buddhism, #HM2127)
Great-path tantra awareness (Buddhism, Tibetan, #HM2656).
Preceded by Asura-world awareness (Buddhism, Tibetan, #HM2579)
Temporary-hells awareness (Buddhism, Tibetan, #HM2454)
Hungry-ghosts-hell awareness (Buddhism, Tibetan, #HM2112)
Bon-practitioner state of awareness (Buddhism, Tibetan, #HM2639).

♦ **HM2056d Seminal illumination** (ICA)
Description This is the dawning of sense, meaning and order to one's consciousness. Irrational at first, this dawning becomes a wild freedom to see all there is to see, know all there is to know. It is the radical "eureka" at the foundation of consciousness, when it suddenly all makes sense. It is like the child revealing the true nature of the make-believe clothes in the story of "The Emperor's New Clothes"; and Richard Leakey may have known this moment when he realized that what had been discovered in the soil of Kenya might be part of the missing history of early humanity.
A dimension of this experience is a sense of awe, that is fear and fascination. The fear is that one might forget and that it will all go away. One wonders how it can all make sense, how can this be. The fascination is that it does all fit together, it really does make sense; and the decision is to grasp the moment, to say "no" to any doubts one might have. Following such an experience, one sees everything from a new perspective and knows that, whatever appearances are, it does all fit together.
Context This state is number 49 in the ICA *Other World in the midst of this World.*
Broader Radical illumination (ICA, #HM2273).
Followed by Vital signs (ICA, #HM2875) Creative futility (ICA, #HM2493)
Ultimate awareness (ICA, #HM2388) Radical contingency (ICA, #HM2477)
Inclusive comprehension (ICA, #HM2256).
Preceded by Vital signs (ICA, #HM2875) Living death (ICA, #HM2808)
Ultimate awareness (ICA, #HM2388) Definitive effectivity (ICA, #HM2796)
Individual fatefulness (ICA, #HM2223).

♦ **HM2057b Higher unconscious** (Psychosynthesis)
Description According to R Assagioli the "higher" part of the mind whose contents are not accessible at will, is the "unconscious" super-conscious. It is the source of higher cognition and higher emotion, where intuition, inspiration, altruism, deeper love, genius and the higher psychic functions and spiritual energies are centred.
Related Superconscious (#HM2960) Psychosynthesis (#HH0002).
Followed by Collective unconscious (Psychosynthesis, #HM2811).

♦ **HM2058d Wheel-turning-king awareness** (Buddhism, Tibetan)
Cakravartin
Description This state of consciousness is characterized by the awareness of world order and is represented by the King of the World. This figure is conceived to be the monarch over all mankind. Peace-loving yet powerful, he is born like a Buddha, miraculously and already aware of his role as king-statesman for the nations.
Context In Tibetan Sakya Buddhism this is one of the states in the "Ascension Stages Game". In some sets it is numbered 26 on the board.
Broader Desire-realm consciousness (Buddhism, #HM2733).
Followed by Naga-world awareness (Buddhism, Tibetan, #HM2031)
Southern-continent awareness of men (Buddhism, #HM2127)
Thirty-three-god-heaven awareness (Buddhism, Tibetan, #HM2606)
Heaven-without-fighting consciousness (Buddhism, Tibetan, #HM2130)
Northern-continent awareness of community (Buddhism, Tibetan, #HM2067).
Preceded by Southern-continent awareness of men (Buddhism, #HM2127).

♦ **HM2059d Ego awareness** (Hinduism)
Ahankara — Ahamkara
Description The tendency in one's consciousness to conceptualize a proprietary unique selfhood is a function of the "I-maker" (ahamkara). The illusion of such a self nevertheless functions as a bridge between the spiritual principles in man (atma, buddhi, etc) and the person of corporeal form, senses, sense-data memory, and lower mind (manas), in order that a conscious ego may take part in the cosmic play (lila) yet have intimations of divine reality.
Another interpretation is that, from *Aham*, or awareness of "I am", comes *Ahankara* or "I am something", and then creation starts. Ahankara crystallises from the buddhi under the impact of rajas, the exciting and mobile guna. Then, under a prevalence of tamas evolve the eleven senses (including manas) and the five *tanmatra* or potentials. In this case it is manas that acts as the bridge or interpreter between the higher mind the five physical senses.
Related Self (#HH0706) Feeling of I (#HH0840) Ego consciousness (#HM0570)
Manas awareness (Hinduism, #HM2902)
Crazy-monkey ego consciousness (Hinduism, #HM2590)
Psychological approach to transcendence (#HH1763)
Indian approaches to consciousness (Hinduism, #HM2446).

♦ **HM2060c Kamaloka consciousness** (Buddhism, Pali)
Description States of consciousness exist as are appropriate to the beings in the sensous world (kamaloka). There are four sub-human realms of beings: hell, demon world, animal kingdom and ghost world. Then there is the human world; and above it the deva lokas of heavenly beings, of which there are six kinds in the sensuous sphere.
Context The mystical geography of southern Buddhism for this sphere is similar to that of the northern Buddhism of Tibet. The Pali has six levels; the Tibetan also has six, but the order is different; and the Pali uniquely has an infernal demon world while the Tibetan has a unique, superior demigod world. This means the human world is fifth in Pali but fourth in the Tibetan.
Broader Worlds of conscious existence (Buddhism, Pali, #HM2072).
Related Phenomena awareness (Buddhism, #HM2551)
Desire-realm consciousness (Buddhism, #HM2733).

♦ **HM2061a Nirvikalpa samadhi** (Hinduism)
Description Nirvikalpa samadhi is one of the highest samadhis. It is said to be equivalent to the yogic *asamprajnata samadhi.* Entering into it, one experiences the heart as being wider than the universe; infinite bliss; and immeasurable power exceeding any occult power. In acquiring nirvikalpa samadhi one generally does not want to enter back into the world. If one remains for 21 days in this state it is likely that one leaves the body. Beyond nirvikalpa samadhi only sahaja samadhi is known.
Another interpretation is that this is the culmination of the fifth stage, a yogic state of formless ecstasy when there is (temporary) absorption in divine reality and a loss of body sense. The individual does return from this state, with an accompanying sense of separate identity and the ego has been suspended but not transcended.
Broader Samadhi (Hinduism, Yoga, #HM2226).
Related Shabda yoga (Yoga, #HH0539) Unity consciousness (#HM3193)
Mystical stage of life (#HM3191) Asamprajnata samadhi (Hinduism, #HM3041)
Unconditional nirvikalpa samadhi (#HM3619)
Indian approaches to consciousness (Hinduism, #HM2446).
Followed by Sahaja samadhi (Hinduism, #HM2043).
Preceded by Savikalpa samadhi (Hinduism, #HM2650).

♦ **HM2062c Third trance of the fine-material sphere** (Buddhism)
Upekkha (Pali) — Tritiyadhyana — Bsam-gtan-gsum-pa (Tibetan) — Third form-realm concentration
Description *Hinayana Buddhism* The goal of the third trance is a state of equanimity (upekkha), attentiveness and clear consciousness. However, perceptions of joy or pleasure and of pain remain which must be given up to reach to next of the jhana trances.
Tibetan gelugpa Buddhism All the concentrations are obtained by means of the 7 preparations. Afflictions and hindrances are overcome by way of the 9 interrupted paths and the 9 paths of release; the ninth path of release from below is the third concentration. Each concentration has branches. The third has 5 branches, 2 types and 3 sub-types.
Context This is one of 4 form concentrations (dhyana) found in Tibetan Gelugpa Buddhism.
Broader Rupaloka consciousness (Buddhism, #HM2536)
Form-realm consciousness (Buddhism, #HM2257)
Trances and mental absorptions (Buddhism, #HM2122)
Tibetan meditative states of form concentrations (Buddhism, #HM2693).
Related Equanimity (Buddhism, #HM7769) Dhyana (Hinduism, Buddhism, #HM0137)
Dwelling in the third jhana (Buddhism, #HM5643).
Followed by Fourth trance of the fine-material sphere (Buddhism, #HM2586).
Preceded by Second trance of the fine-material sphere (Buddhism, #HM2038).

♦ **HM2063d Non-introspection** (Buddhism)
Asamprajanya — Shes-bzhin-ma-yin-pa (Tibetan)
Description The cause of infractions of ethical codes, due to unawareness of engaging in action, whether physical, mental or verbal.
Context One of the twenty *secondary afflictions* of Tibetan Buddhism.
Broader Secondary afflictions (Buddhism, Tibetan, #HH0781).

♦ **HM2064d Four doors of retention** (Buddhism)
Description These are: retention of *patience* (for lack of fear with respect to emptiness); retention of *secret speech* (for the ability to make pacifying spells to quieten the injurious); retention of *words* (not to forget names, thoughts and meanings); retention of *meaning* (not to forget the specific and general characteristics of phenomena).
Context One of the paths of special qualities according to Buddhist teaching.
Broader Paths of special qualities (Buddhism, #HM3079).

♦ **HM2066g Tonal state of awareness** (Amerindian)
Description A term used by C. Casteneda to describe that mode of awareness which is rational and appears in terms of an ordered and describable universe. It is said to complement *nagual*, the two modes being necessary for completeness; but tonal, which starts to develop soon after birth, tends to swamp the nagual which has been there from the beginning. Order in perception is the exclusive realm of the Tonal. In tonal, actions have a sequence - like stairways where one counts the steps. The tonal state of everyday awareness is one aspect of reality, the nagual another. Tonal corresponds to the verbal description of the world, everything that meets the eye, it ends at death.
Refs Castaneda, Carlos *Journey to Ixtlan* (1972); Castaneda, Carlos *The Teachings of Don Juan* (1968); Castaneda, Carlos *A Separate Reality* (1971); MacDowell, Mark *A Comparative Study of the Teachings of Don Juan and Madhyamaka Buddhism* (1986).
Broader Dreams (#HM2950).
Related Not-doing (#HH0367) Samvriti (Buddhism, #HM0652) Stopping the world (#HH0489)
Dreaming experience (#HM3908) Way of the warrior (Amerindian, #HH8219)
Nagual state of awareness (Amerindian, #HM2036).

♦ **HM2067d Northern-continent awareness of community** (Buddhism, Tibetan)
Kuru
Description This state of awareness is characterized by the felicity of communal living and is represented as the happiest of the four human islands. However, this state provides no motivation for escaping from the samsara of cyclic existence and leads to idyllic stultification.
Context In Tibetan Sakya Buddhism this is one of the states in the "Ascension Stages Game". In some sets it is numbered 20 on the board.
Broader Desire-realm consciousness (Buddhism, #HM2733).
Followed by Asura-world awareness (Buddhism, Tibetan, #HM2579)
Southern-continent awareness of men (Buddhism, #HM2127)
Four-great-kings-heaven awareness (Buddhism, Tibetan, #HM2082)
Thirty-three-god-heaven awareness (Buddhism, Tibetan, #HM2606)
Eastern-continent awareness of noble figures (Buddhism, Tibetan, #HM2543).
Preceded by Wheel-turning-king awareness (Buddhism, Tibetan, #HM2058)
Thirty-three-god-heaven awareness (Buddhism, Tibetan, #HM2606)
Pratyeka Buddha application awareness (Buddhism, Tibetan, #HM3020).

♦ **HM2069b Monadic awareness** (Amerindian)
Description Awareness is concerned with the formal, that is with the comprehensible, aspects of reality. It rules over a vast variety of different phenomena, all having order (of varying degree), that is to say a quantifiable relation to each other in common. As proposed by C Castaneda, the term awareness may be associated with all phenomena classically termed "spiritual", especially relating to the holy spirit and to the generally monistic outlook of experiences of this sort, owing to the systemic character of order.
Followed by Dyadic awareness (#HM2930).

♦ **HM2070d Non-emanating consciousness** (Buddhism, Tibetan)
Paranirmita-vasavartin
Description The field of consciousness at cosmic levels includes a state in which mental emanations are controlled to the extent that, in the realms of desire, whatever is wished for is created.
Context In Tibetan Sakya Buddhism this is one of the states in the "Ascension Stages Game". In some sets it is numbered 32 on the board.
Broader Desire-realm consciousness (Buddhism, #HM2733).
Followed by Form-realm awareness (Buddhism, Tibetan, #HM2142)
Joyful-heaven awareness (Buddhism, Tibetan, #HM2022)
Hungry-ghosts-hell awareness (Buddhism, Tibetan, #HM2112)
Thirty-three-god-heaven awareness (Buddhism, Tibetan, #HM2606)
Western-continent awareness of cattle (Buddhism, Tibetan, #HM2519).
Preceded by Pure-lands awareness (Buddhism, Tibetan, #HM2168)
Discipleship-vision awareness (Buddhism, Tibetan, #HM2240)
Delightful-emanation-heaven consciousness (Buddhism, Tibetan, #HM2546).

♦ **HM2071c Path of natural intelligence** (Judaism)
Description This state of awareness is said to be of the perfected nature of beings informed under the sun.
Context The twenty seventh of the spiritual paths expressed in the Jewish Kabbalah.
Broader Paths of wisdom (Judaism, #HM2509).
Followed by Path of active intelligence (Judaism, #HM2218).
Preceded by Path of renovating intelligence (Judaism, #HM2047).

♦ **HM2072c Worlds of conscious existence** (Buddhism, Pali)
Loka consciousness
Description There are 4 rounds (gati) of existence below the human level: the niraya or *downward path* of a hell for infernal beings, the asura-nikaya, or level of demons, the peta-loka or ghost level and the animal creation (tiracchana yoni). The human level of conscious existence (manussa-loka) is fifth in ascending order of rounds. Above the human are the divine gati or deva lokas of heavenly beings. These rounds of existence take place in a corresponding objectified lower world called *kama loka* or the sensuous-sphere. The conscious existences in this sphere exemplify a continuum of lower states of consciousness, therefore they may be considered together as the *sensuous-sphere consciousness*. Above this are three higher states: the *fine-material-sphere* (rupaloka) consciousness; the *immaterial-sphere* (arupaloka) consciousness; and the fourth state, the *supermundane* (lokuttara). The trances or mental absorptions (jhana) of huma meditations take place in the *fine-material-sphere*, and rise to the *immaterial sphere* where the six higher spiritual powers (abhinma) begin to be acquired. The sixth power takes the adept to the fourth state (lokuttara), or the *supermundane sphere*, and eventually to nirvana.
Context The devas or heavenly beings are classified as being of 6 grades in the sensuous-sphere, of 12 grades in the fine-material-sphere and of 4 grades in the immaterial-sphere. In the fourth state of consciousness of the supermundane (lokuttara) sphere the grades of gods or devas (divinity) are not given; but according to one reckoning it comprises nine conditions or sub-states of consciousness, and according to another, five. The total number of sub-classes of divine beings is 22 in the three lower spheres. Each of these is also said to have its own heaven. In the 2 lowest spheres are humans and the four lowest classes or rounds of beings with conscious existence. Thus six general classes of conscious beings may be counted; many more if sub-classes and gods are enumerated (perhaps 33 in all). Each being has a consciousness normative to its sphere. In the case of men, at least, each has also the possibility of the consciousness of the three higher spheres, depending on the disposition, or as a function, of the phenomena (dhammas) and particularly of the khandhas (the elements of mental interaction). The disposition of the elements in turn depends on karma and the dispensation of spiritual forces directed towards those disciples who have entered the stream (sotapan).
Broader Karma awareness (Buddhism, #HM2048)
Phenomena awareness (Buddhism, #HM2551)
Consciousness in Buddhism (Buddhism, #HM2200).
Narrower Kamaloka consciousness (#HM2060) Arupaloka consciousness (#HM2012)
Rupaloka consciousness (Buddhism, #HM2536).
Related Five aggregates (Buddhism, #HH3321)
Consciousness states in cyclic existence (Buddhism, #HM2177).

♦ **HM2073d Awareness of allegiance** (ICA)
Description The third formal aspect of the *life of doing* in the ICA *New Religious Mode*, this represents an allegiance to those who act on behalf of life as a whole.
Broader Life of doing (ICA, #HM3018).
Narrower People of God (#HM2484) Religious function (#HM2577)
Religious vocation (#HM2293) Primordial colloquy (#HM2352).

♦ **HM2074g Eye consciousness** (Buddhism)
Chakshurvinjana — Mig-gi-rnam-par-shes-pa (Tibetan) — Visual awareness
Description This is the consciousness that apprehends physical forms as colours and shapes. It is not the consciousness that interprets these forms as "a pot", "a horse", etc; that is the task of mental consciousness. It is twofold, that is: where eye consciousness is pleasant it is associated with indeterminate resultant consciousness with profitable result; where it is painful, it is associated with indeterminate resultant consciousness with unprofitable result.
Context One of the six consciousnesses defined in Buddhism as dependent on the individual senses, and with objects of sense as their focus.
Broader Sense consciousness (Buddhism, #HM2664)
Perception through the senses (#HM3764)
Mundane resultant consciousness (Buddhism, #HM0130)
Sense mode of consciousness occurrence (Buddhism, #HM4389)
Indeterminate consciousness in the sense sphere - resultant (Buddhism, #HM5721)
Awareness of consciousness-group of conscious existence - senses and mind (Buddhism, #HM2556).
Related Spatial awareness (#HM2176) Ear consciousness (Buddhism, #HM2169)
Nose consciousness (Buddhism, #HM2364) Body consciousness (Buddhism, #HM2562)
Tongue consciousness (Buddhism, #HM2263) Mental consciousness (Buddhism, #HM2838)
Mind-consciousness element (Buddhism, #HM6173)
Cognition born of contact with element of eye-consciousness (Buddhism, #HM1271)
Neither psychical ease nor unease born of contact with element of eye-consciousness (Buddhism, #HM0972).

♦ **HM2075g Psycho-physical faculties awareness** (Buddhism)
Indriya phenomena (Pali)
Description Each human body is the base for its sense organs (5) and for both feminine and masculine sex characteristics. To these 7 faculties can be added vitality which bridges between the bodily and mental natures. The remaining 14 phenomena are mental. They are: mind (mano, vinnana or citta) including the sub-conscious; the 5 feelings; the 5 mental powers (bala) of faith (saddha), energy (viriya), attentiveness (sati), concentration (samadhi) and wisdom (panna); and 3 supermundane faculties arising in the supermundane sphere.
Context These phenomena or faculties are also enumerated among the conditions (paccaya) that all things may take.
Broader Phase-conditions of consciousness (Buddhism, #HM2599).

♦ **HM2076d Passing awayness** (ICA)
Description This is the realization that not only is one contingent one's self; but that everything, including the universe itself, is passing out of existence. One sees death as ever present in one's life. Soren Kierkegaard referred to this when he wrote about the big fish eating the little fish, which in turn eat the smaller fish, and so on.
Context In the ICA New Religious Mode in the arena of articulating the word about life (the life of knowing) the second formal aspect is one's experience of the vanishing cosmos or of wonder at the world. At the first phenomenological level, this occurs when a person sees through the events of everyday life and his own subjectivity to discern existence clearly.
Broader Wonder at the world (ICA, #HM3062).
Followed by Stark givenness (ICA, #HM2101).

♦ **HM2077b Christian recollection** (Christianity)
Description In the Christian spiritual life recollection signifies a concentration of one's attention, or awareness, and other powers on God or on the expression of God in whatever closely relates to Him. It is the precursor of spiritual contemplation, the drawing together of the forces of inner life. The act of recollection may occur voluntarily and sporadically, as in moments of spiritual concentration, worship or service. It also may become habitual as an acquisition of the sense of the presence of God.
Recollection is essential to meaningful worship and takes the form of attention to prayer: its sounds, meanings, and object (the source of power which can grant fulfilment). In its higher stages, recollection during prayer or prayerful worship is no longer of the body of prayer (sound), or its soul (meaning or intention). It is rather an absorbed recollection of the spirit of prayer, that is, God. These degrees of recollection are acquired, or initiated by the religious. There is another degree, infused contemplation or infused recollection, which is said to operate under the grace or charismatic gift of the Holy Spirit since it cannot be cultivated by an individual's own efforts. This is considered the first degree of true, mystical prayer, in which the soul and all its faculties are gathered and concentrated on God. It is an absorption which, as St Theresa of Avila taught, leads to the transforming union or mystical marriage and the beatific vision of the divine world. In St Theresa's presentation, "The Interior Castle", there are three degrees of prayerful recollection. Two are preparatory, the third is true recollection when the faculties of the soul are gathered. The fourth state of prayer is the fruition of union, the mystical rapture itself.
Thomas Merton refers to recollection as necessary in order to meditate. The mind has to be drawn from all that prevents attention to God in the heart, and so the senses have to be recollected. This cannot be done at the moment of prayer if the senses have been uncontrolled all day, and so "moderate" recollection has to be preserved all the time, by living in an atmosphere of faith and with moments of prayer and attention to God. This is done through resisting the appeals that society makes to the senses, through the media for example, and through giving up luxuries until one has sufficient self control to use them without becoming their slaves.

Related Solitude (#HH0333) Quietness (#HH0491) Practice of silence (#HH0983)
Devotion (Christianity, #HH0349) Recollection (Islam, Sufism, #HM2351)
Discipline of meditation (Christianity, #HH1688)
Mystical Christ-consciousness (Christianity, #HM2880).

♦ HM2078c Mythical consciousness
Description This is considered a third stage in the evolution of consciousness, in which the soul becomes conscious of its internal reality. In dream and myth, mythical consciousness adds sun and heaven to earthbound *magical consciousness*, thereby creating a 2-dimensional structure of polarity. Conceptualized by Gebser, mythical consciousness is characterized by its imaginative pictoral awareness and its imbeddedness in natural time. Mythical consciousness originates with the development of seasonal ritual, astronomy and the calendar.
Broader Evolution of consciousness (#HM2140).
Related Mythical imagination (#HM0043).
Followed by Mental consciousness (#HM2319).
Preceded by Magical consciousness (#HM2090).

♦ HM2079e Double-awake body-asleep trance
Description In making use of the "sorcerers explanation of the world", intentional dreaming is an activity of tonal and nagual which activates the double or second life within a person. It is performed in a state of awareness during sleep which is capable of manipulating the results of dream activity to gain access to a separate reality. Dreaming with an intent to gain access to a different kind of reality has pervaded most magical cultures. As well as the use of sacred drugs, guiding symbolism and magical procedures have been used to gain control over the dream process.
Related Mind-awake body-asleep trance (Psychism, #HM2640).

♦ HM2080d Peak experiences
Core-religious experiences — Transcendent experiences
Description Peak experiences are, for an individual, spontaneous moments of greatest happiness, ecstasy, and rapture. They occur under a wide variety of circumstances, from being in love or giving birth to a child, to listening to music, creating a work of art, or climbing a mountain. They are also occasions of emotional growth, opening new horizons, enriching values, and releasing new potential.
According to Maslow, the complete attention characteristic of such experiences, experienced with the whole of being, is a cognitive experience undergone by self actualizing individuals. However, some evidence suggests that most people have, or can have, peak experiences. The content of such experiences is believed to be similar, although the situations which trigger off such occurrences vary from person to person. Although many people may have such experiences, it is not clear to what extent particular experiences differ in degree, quality or level of joy, ecstasy or transcendence. In particular it is not clear to what extent infrequently reported experiences such as mystic or similar illumination differ from more commonly encountered expressions of great joy or happiness. One distinction is that, although they are experienced in the same way, the deep mystic state of wakeful relaxation such as arises in transcendental meditation is a "planned" state, whereas a peak experience is sudden and unplanned, resulting from a flash of insight or a new and significant experience. Reported characteristics of peak experiences include:
1. Perception of the universe as an integrated whole.
2. Intense concentration of a kind that does not normally occur; the precept is exclusively and fully attended to.
3. The occurrence of a cognition of being that tends to perceive external objects, the world, and individual people as more detached from human concerns.
4. Perception is ego-transcending, self-forgetful, ego-less and unselfish; objects and people are more readily perceived as having independent reality of their own.
5. The experience is felt to be a self-validating, self-justifying moment which carries its own intrinsic value.
6. Such experiences are recognized as end-experiences rather than means experiences, namely they are part of the operational definition of such statements as: life is meaningful.
7. Disorientation in time and space, or even lack of consciousness of space/time, experienced as a sense of eternity and infinity.
8. The world is accepted as it is and experienced as beautiful, good, and worthwhile.
9. A sense of being god-like.
10. Overlapping or fusion between facts and values, and identification of the intrinsic values of being.
11. A form of cognition which is much more passive and receptive than normal.
12. Experience of emotions such as wonder, awe, reverence, humility, surrender, and worship before the immensity of the experience.
13. Transcendence or resolution of the dichotomies, polarities and conflicts of life.
14. Loss of anxiety, fear, inhibition, defence, confusion and restraint.
15. Immediate after effects upon the person ranging from the simply therapeutic to full religious conversion.
16. Recognition of the ever-present possibility of access to what is perceived as a personally-defined heaven.
17. Sense of self-determination, responsibility and creative ability as a free agent.
18. Resolution of the dichotomy between humility and pride.
19. Experience of unitive consciousness, namely a sense of the sacred perceived in and through the particular and the secular.
20. Greater ability to love and accept, leading to greater spontaneity, honesty, and innocence.
21. Greater sense of being a person rather than an object in the world.
22. Reduction in striving, needing and wishing
23. Experience of a sense of luckiness, fortune and grace.
Peak experiences tend not to occur in persons whose character structure forces them to try to be extremely rational, materialistic or mechanistic. However, precisely those individuals with the clearest and strongest sense of identity are most able to transcend the ego or self and to become selfless.
Refs Maslow, Abraham H *Religions, Values and Peak-experiences* (1964); Maslow, Abraham H *Music Education and Peak-Experiences* (1968); Maslow, Abraham H *Cognition of Being in the Peak Experiences* (1956); Wuthnow, Robert *Peak Experiences* (1976).
Related Focusing (#HM3003) Transcendence (#HM4104) Numinosum (Jung, #HM3811)
Emotional high (#HM1400) Flow experience (#HM2344) Self-expression (#HH0791)
Being cognition (#HM2474) Mystical cognition (#HM2272) Partial-mindedness (#HM2430)
Psychological approach to transcendence (#HH1763)
Modes of awareness associated with use of hallucinogens (#HM0801).

♦ HM2081c Path of faithful intelligence (Judaism)
Description This mode of awareness is characterized by an increase in all the spiritual virtues.
Context The twenty second of the spiritual paths expressed in the Jewish Kabbalah.
Broader Paths of wisdom (Judaism, #HM2509).
Followed by Path of stable intelligence (Judaism, #HM2533).
Preceded by Path of intelligence of reward (Judaism, #HM2105).

♦ HM2082d Four-great-kings-heaven awareness (Buddhism, Tibetan)
Description This state of human awareness is projected as the qualities ascribed to the deities in this heaven. They are able to regulate the mundane affairs of the world system, because they can control such things as the planetary influences, the winds of the four quarters and the four elements.
Context In Tibetan Sakya Buddhism this is one of the states in the "Ascension Stages Game". In some sets it is numbered 27 on the board.
Broader Desire-realm consciousness (Buddhism, #HM2733).
Followed by Reviving-hell awareness (Buddhism, Tibetan, #HM2516)
Southern-continent awareness of men (Buddhism, #HM2127)
Hungry-ghosts-hell awareness (Buddhism, Tibetan, #HM2112)
Thirty-three-god-heaven awareness (Buddhism, Tibetan, #HM2606)
Bon-practitioner state of awareness (Buddhism, Tibetan, #HM2639)
Western-continent awareness of cattle (Buddhism, Tibetan, #HM2519).
Preceded by Naga-world awareness (Buddhism, Tibetan, #HM2031)
Animal-hell awareness (Buddhism, Tibetan, #HM2636)
Temporary-hells awareness (Buddhism, Tibetan, #HM2454)
Heavenly-highway awareness (Buddhism, Tibetan, #HM2010)
Divine-animal-hell awareness (Buddhism, Tibetan, #HM2007)
Bon-practitioner state of awareness (Buddhism, Tibetan, #HM2639)
Western-continent awareness of cattle (Buddhism, Tibetan, #HM2519)
Pratyeka Buddha application awareness (Buddhism, Tibetan, #HM3020)
Northern-continent awareness of community (Buddhism, Tibetan, #HM2067).

♦ HM2083d Panic
Description This state occurs when the immediate task is seen as almost beyond the individual's capability to perform. Since it is not seen as totally impossible, there is an attempt to achieve success; but this attempt is based on impulsive and ill-considered rather than rational action, and in desperation rather than by choice.
Related Uncertainty (#HM3051) Flow experience (#HM2344).

♦ HM2084b Transrational insight (Zen)
Description A mental state which is the aim of zen meditation on *koans* (enigmatic and apparently absurd statements); the individual is released from following the intellect through logical propositions and achieves the ability to be spontaneous.
Related Koan (Zen, #HH0263) Zazen meditation (Zen, #HH0882).

♦ HM2085g Nightmares
Threatening dreams
Description Nightmares or threatening dreams occur when the higher mind is dormant or in abeyance, and are caused in part by the automatism of the lower mind and the emotional centre when the reasoning faculties are suspended. The past, and conversations and meetings with others, may be brought back; but also the future may be depicted, with warnings, scenes of judgement or deadly perils. There may be encounters with hideous, deformed beasts or demons which often appear to be crsuhing the breath out of the victim; and the whole is imbued with a sense of terror. Threatening dreams may arise out of guilt or anxiety feelings, or they may be true presentiments of events to come. Explanations vary, from the prosaic (indigestion or pressure on some part of the body causing disorder of internal organs) to the psychic. Dream night-journeys to awesome, unfamiliar surroundings, and terrifying dream night-beings encountered, may in fact belong to another reality which is accessible to the subtle body and its perceptive faculties. These experiences are conceived by some as "out-of-the-body", "astral" projections of the "double"; and by others as resulting from a telepathic link with beings, human or non-human, who are actually experiencing what the dreamer dreams, or whose experience is recorded in the akashic records being dream-read.
Broader Dreams (#HM2950).

♦ HM2087d Ida conscious energy (Yoga)
Description Ida is the name for the energy channel on the left side of shushumna in tantra and kundalini-yoga. Physiological correspondence lies in the parasympathetic nervous system. Psychological correspondence is the power of dissolution, symbolized by moon (night). Ida, pingala and shushumna are the three most important nadis or subtle energy channels through which prana or life-force flows. They interconnect and merge with each other in the chakras or centers of consciousness. In Indian and Buddhist approaches to consciousness they play an important role both in theory and practice.
Broader Shushumna (Hinduism, #HM2053).

♦ HM2088d Lord-of-death awareness (Buddhism, Tibetan)
Yama state
Description This state of awareness is represented as the gateway to hell where the mirror of past deeds and ripening karma must be looked into. The scales are here to try the soul. This state is characterized by judgement.
Context In Tibetan Sakya Buddhism this is one of the states in the "Ascension Stages Game". In some sets it is numbered 9 on the board.
Broader Desire-realm consciousness (Buddhism, #HM2733).
Related Restraint (#HH6876).
Followed by Great-black-lord awareness (Buddhism, Tibetan, #HM2118)
Great-path tantra awareness (Buddhism, Tibetan, #HM2656).
Preceded by Adamantine-hell awareness (Buddhism, #HM2028).

♦ HM2089c Emptiness of non-products (Buddhism, Tibetan)
Asamskrtashunyata
Context One of the eighteen emptinesses comprising the paths of view in Tibetan Buddhism.
Broader Emptinesses on the paths of the view in Buddhism (Buddhism, Tibetan, #HM2944).

♦ HM2090c Magical consciousness
Description The second stage within the evolution of consciousness, in which man is propelled from the O-dimensional identity with the universe, *being the world*, into 1-dimensional unity, *having the world.* The origin of will, magic becomes means of emancipation from nature. Conceptualized by Gebser, magical consciousness holds five characteristics: egolessness; pointlike, unitary world; time- and spacelessness; bondage to nature; magical reaction towards this bondage.
Broader Evolution of consciousness (#HM2140).
Followed by Mythical consciousness (#HM2078).
Preceded by Archaic consciousness (#HM2189).

♦ HM2091d Barbarian-state of awareness (Buddhism, Tibetan)
Description This state is characterized by remorseless carnivorousness and warfare, so that neither man not beast is spared. It is represented by the Central Asian Turkic tribes whose members were butchers in Tibet. Their religion is Islam which, however, is not considered independent, but as a link in Asian spiritual unfoldiment. Hence, should they enter the stream, all Islamic practitioners may rise to the spiritual state equivalent to "Hindu Wisdom-holder", according

to Sakya Buddhists, and to the higher states accessible to the dharma followers.
Context In Tibetan Sakya Buddhism this is one of the states in the "Ascension Stages Game". In some sets it is numbered 21 on the board.
Broader Religious traditions in the ascension stages game (Buddhism, Tibetan, #HM3341).
Followed by Desire-realm consciousness (Buddhism, #HM2733)
Naga-world awareness (Buddhism, Tibetan, #HM2031)
Animal-hell awareness (Buddhism, Tibetan, #HM2636)
Asura-world awareness (Buddhism, Tibetan, #HM2579)
Interminable-hell awareness (Buddhism, Tibetan, #HM2052)
Hungry-ghosts-hell awareness (Buddhism, Tibetan, #HM2112).
Preceded by Naga-world awareness (Buddhism, Tibetan, #HM2031)
Asura-world awareness (Buddhism, Tibetan, #HM2579)
Divine-animal-hell awareness (Buddhism, Tibetan, #HM2007)
Western-continent awareness of cattle (Buddhism, Tibetan, #HM2519).

♦ HM2092d Inventing humanness (ICA)
Description This is experienced as creating a life style that others can and should follow; when one chooses to be the human being one is, so that others may be released to be the human beings they are. It may be compared to MacMurphy in "One Flew Over the Cuckoo's Nest".
Context In the ICA New Religious Mode in the arena of knowing and doing intensified in one's life (life of being) the fourth formal aspect is the experience of transparent presence or effulgence. At the second phenomenological level this occurs when a person relates to the experience of nothingness.
Broader Transparent presence (ICA, #HM2462).
Followed by Human transformation (ICA, #HM2334).
Preceded by Being myself (ICA, #HM2154).

♦ HM2094f Co-consciousness
Subconsciousness — Unconsciousness
Description Those states of consciousness which are not in the focus of attention but are on the edge of the content of consciousness, co-conscious states coexist with but are dissociated from personal consciousness. They cover perception, thought, feeling, remembering, decision and action which are completely analagous to conscious functions although carried out unconsciously, and include pathologically split off states.

♦ HM2095c Path of administrative intelligence (Judaism)
Description This state of awareness is of the laws that keep the movements of the planets regulated.
Context The thirty second of the spiritual paths expressed in the Jewish Kabbalah.
Broader Paths of wisdom (Judaism, #HM2509).
Preceded by Path of perpetual intelligence (Judaism, #HM2643).

♦ HM2096b Awareness
Description Awareness is essentially an undefined term referring to a particular kind of immediate experience which may be distinguished from other states of consciousness. Awareness develops with, and is integrally part of, an organismic-environmental transaction. It includes thinking and feeling, but is always based on current perception of the current situation. It includes some intention and directionality of the self toward the world. In its pure form there is a momentary weakening of the self-other barrier and the object of awareness seems momentarily included in the self. A few people seem to experience this condition more or less continuously. The usual content of consciousness for many people, however, is a flow of fantasy-imagery and subvocal speech that is not deeply rooted in ongoing behaviour, but only tangentially related to it. Awareness is distinct from this unfocused reverie. In healthy life it is simply present, parelleling all behaviour. In therapy, however, when awareness develops where it has previously been blocked, it tends to be accompanied by a sense of release of tension and a feeling of increase in energy.
Awareness is distinct from introspection in which the self is split, with one part observing another part as an object, self-consciously. Awareness is the whole self, conscious of that to which the organism is attending. Introspection being relatively detached from ongoing total organismic concern, and being out of touch with the actual environment, can never discover anything very new, but only rearrange and rehash the remembered and hence unnourishing past. Awareness, being in contact with the current environment and organism, always includes something refreshingly new. Awareness is related to emotionally-rooted insight in that the latter is based on an expansion of awareness of an ongoing organism-environment relationship with its associated positive affect and sense of discovery, while intellectual insight lacks this crucial rootedness in the actual.
Awareness develops spontaneously where novelty and complexity of transaction are greatest, and the most possibilities (for good or ill) exist. It seems to facilitate the maximum efficiency by focusing the attention of the individual and concentrating his abilities on the most complex, possibility-loaded, situations. When awareness does not develop at this region of the organism-environment contact boundary, at which an especially important and complex transaction is occurring, something is going wrong. Therapy (of the Gestalt variety) consists of reintegration of attention and awareness.
Awareness is also used in the sense of task awareness or readiness when individuals are oriented to the performance of a task on receipt of some signal, without consciously registering that the instruction has been given.
Refs Bois, J S *Explorations in Awareness* (1957); Bois, Samuel *The Art of Awareness* (1966); Chaudhuri, Haridas *The Evolution of Integral Consciousness* (1977).

♦ HM2097b Tarot arcana of conscious states (Tarot)
Description The tarot picture cards, numbered 1 to 21 together with one unnumbered picture card, are referred to as the Greater Trumps or Arcana and are considered by specialists to represent in their symbolism the archetypal patterns present in all of creation. Just as the totality of all things can be represented by a circle, and the circle can be represented by the relationship of pi (3.142. 1), with a circumference between 21 and 22 and a diameter of 7; so the Greater Arcana of 22 symbols can be analyzed numerically. This analysis can be referred to states of consciousness using such image-ordering approaches as (2 x 11); (3 x 7) 1; and (4 x 5) 2; or by the permutations of 3 equals 7, and of 4 equals 15, so that the trumps are split into seven supreme Arcana, and fifteen Great Arcana of consciousness.
Context The cards of the Great Arcana are related with the archetypal emblems and with the spiritual journey through the Kabbalistic Tree of Life.
Broader Tarot (Tarot, #HH0382).
Narrower Death emblems (#HM3121)
Mutability emblems (#HM2250)
Emblems of disaster (#HM2350)
Emblems of supremacy (#HM2411)
Emblems of well-being (#HM2422)
Wisdom archetypal image (#HM2261)
Magician archetypal image (#HM2237)
Emblems of the mysterious (#HM2398)
Measuring emblems (#HM2178)
Emblems of renewal (#HM2315)
Emblems of totality (#HM2387)
Victimization emblems (#HM2190)
Natural forces emblems (#HM3127)
Pilgrim archetypal image (#HM2225)
Life-fluid emblems - star (#HM3137)
Hierophant archetypal image (#HM2260)
Charioteer archetypal image (#HM3045)
Old wise-man archetypal image (#HM2202)
Female sovereignty archetypal image (#HM2285)
Emblems of personified evil (#HM2378)
Life-fluid emblems - temperance (#HM3289)
Betrothal initiation archetypal image (#HM3028).
Related Kabbalah (#HH0921) Emblem archetypes in the psyche (Tarot, #HM2201).

♦ HM2098b Dispositions of consciousness (Buddhism)
Vinnana-khanda development (Pali) — Consciousness aggregate — Bare cognition — Awareness of consciousness-group of conscious existence — Vinnananakkhandha — Aggregate of consciousness
Description In Hinayana Buddhism, consciousness is considered as anything that is perceived as being consciousness, whether past, present or future, internal or external, subjective or objective, gross or refined, superior or inferior, far or near.
Taken together, everything which has the characteristic of cognizing comprises the consciousness aggregate. This is consciousness in the sense of bare cognition, apart from feeling, perception or formation (mental activity).
States of consciousness are disposed according to karma in a threefold manner - profitable (moral) associated with wholesome karma; unprofitable (immoral), associated with unwholesome karma; indeterminate (karmically neutral). There are four worlds or spheres (lokas) of conscious existence experienced by human beings. In the sensual desire sphere (the lowest) there are 54 dispositions of consciousness of which 8 are karmically wholesome and 12 are contaminated by greed, hate and delusion. A further 34 are indeterminate dispositions, karmically neutral. Of these, 23 are resultant and 11 functional or inoperative. In the three higher spheres unwholesome karma is not generated. The two middle spheres - fine-material and immaterial or form and formless (rupaloka and arupaloka) - are the worlds which meditation pierces. In the fine-material sphere, 5 dispositions arise as profitable or moral, producing wholesome karma, and 10 are karmically neutral or indeterminate, 5 of which are resultant and 5 functional or inoperative. There are 4 trances or absorptions classed as profitable in the immaterial sphere (arupaloka) and 8 karmically neutral or indeterminate, 4 of which are resultant and 4 functional or inoperative. Above these in the highest sphere - supramundane or transcendental (lokuttara) - there may be counted 4 conditions classed as profitable in the consciousness-group (vinnana khanda) of conscious existence, and 4 karmically neutral, resultant. These are 4 dispositions said to be peculiar to noble paths (profitable) and 4 to fruition moment in the noble paths (indeterminate resultant) in the supramundane sphere. Therefore there are 21 profitable and 12 unprofitable dispositions. Karmically neutral results or independent functions of consciousness (kiriya-citta) account for 56 of the 89 dispositions. All but two of the 20 indeterminate functional dispositions refer to ariya or arahants only.
Context In Hinayana Buddhism, 89 consciousnesses are enumerated in aggregate (khanda). Of these, 21 are profitable or moral, associated with good or wholesome karma; 12 are unprofitable or immoral, associated with bad or unwholesome karma; and 56 are indeterminate (resultant or functional), karmically neutral. The unprofitable all arise in the sphere of sense and desire, whereas profitable and indeterminate consciousnesses arise in sense, fine-material, immaterial and supramundane spheres.
Broader Five aggregates (Buddhism, #HH3321).
Narrower Trances and mental absorptions (#HM2122)
Mundane resultant consciousness (#HM0130)
Lightness of the aggregate of consciousness (#HM3403)
Profitable consciousness in the sense sphere (#HM4447)
Flexibility of the aggregate of consciousness (#HM1796)
Workability of the aggregate of consciousness (#HM4605)
Unprofitable consciousness in the sense sphere (#HM8375)
Straightness of the aggregate of consciousness (#HM1338)
Profitable consciousness in the immaterial sphere (#HM4701)
Profitable consciousness in the supramundane plane (#HM4930)
Profitable consciousness in the fine-material sphere (#HM5338)
Indeterminate consciousness in the sense sphere - resultant (#HM5721)
Indeterminate consciousness in the sense sphere - functional (#HM3852)
Indeterminate consciousness in the immaterial sphere - resultant (#HM4982)
Indeterminate consciousness in the immaterial sphere - functional (#HM0282)
Indeterminate consciousness in the supramundane plane - resultant (#HM5129)
Indeterminate consciousness in the fine-material sphere - resultant (#HM0594)
Indeterminate consciousness in the fine-material sphere - functional (#HM4761).
Related Vritti (Hinduism, #HM2692) Feeling aggregate (Buddhism, #HM4983)
Perception aggregate (Buddhism, #HM4143)
Consciousness-born materiality (Buddhism, #HM8273)
Phase-conditions of consciousness (Buddhism, #HM2599)
Modes of occurrence of consciousness (Buddhism, #HM6720)
Awareness of corporeality-group of conscious existence (Buddhism, #HM2108)
Awareness as mental-formation group of conscious existence (Buddhism, #HM2050)
Awareness of consciousness-group of conscious existence - senses and mind (Buddhism, #HM2556).

♦ HM2099b Buddhi awareness (Hinduism)
Discrimination — Understanding — Being consciousness — Direct intuitive knowledge
Description In Hindu psychology, buddhi (wisdom or clear discrimination) emerges when the distortions of the lower-mind are suppressed. Through buddhi, the divine self (atman) is known to be present. Buddhi is sometimes called *the witness* as the consciousness shifted to this level allows the individual to regard his mental phenomena from "outside", without becoming identified with them. This permits a new level of "I-ness" to evolve which is associated with the subtle bodies (koshas). Normally, buddhi functions through citta, and the mind is helped to know and understand objects, for the buddhi illuminates citta. However, while buddhi is functioning through the mind it is not possible to know pure consciousness. In samadhi, however, free from association of citta, it turns in on itself and illuminates consciousness, its own nature. This is sva-buddhi.
The understanding which is buddhi is of a thing as a whole and not as the sum of its parts. When all the data one knows about a given matter ccome together in integral knowledge there is a flash of perception and this is the action of buddhi. There is a sense of wonder and delight; even "I", although not absent, is forgotten. There is an understanding of what beauty is.
Some refer to buddhi as the generic being consciousness, being the first to evolve from the unmanifest core of the world and from which ahankara and the rest of development arises. This is *mahat*, the great, mediating between the unmanifest "world-ground" and the world of differentiated phenomena.
Related Wisdom (#HH0623) Buddhi yoga (Yoga, #HH0484)
Akuhaiamio (Psychism, #HM1615)
Atma awareness (Hinduism, #HM2103)
Manas awareness (Hinduism, #HM2902)
Citta (Hinduism, Buddhism, #HM3529)
Indian approaches to consciousness (Hinduism, #HM2446).

♦ HM2100d Howling-hells awareness (Buddhism, Tibetan)
Description These states have two degrees: howling-hell and great-howling-hell. They are the conditions of those who lie and of those who commit crimes while intoxicated; and are characterized by the presence and actions of tormenting, demonic beings.
Context In Tibetan Sakya Buddhism this is one of the states in the "Ascension Stages Game". In some sets it is numbered 4 on the board.
Broader Desire-realm consciousness (Buddhism, #HM2733).

Followed by Naga-world awareness (Buddhism, Tibetan, #HM2031)
Reviving-hell awareness (Buddhism, Tibetan, #HM2516)
Crushing-hells awareness (Buddhism, Tibetan, #HM3142)
Very-hot-hells awareness (Buddhism, Tibetan, #HM2576)
Temporary-hells awareness (Buddhism, Tibetan, #HM2454)
Hungry-ghosts-hell awareness (Buddhism, Tibetan, #HM2112).
Preceded by Asura-world awareness (Buddhism, Tibetan, #HM2579)
Reviving-hell awareness (Buddhism, Tibetan, #HM2516)
Crushing-hells awareness (Buddhism, Tibetan, #HM3142)
Formless-realm awareness (Buddhism, Tibetan, #HM3144)
Hungry-ghosts-hell awareness (Buddhism, Tibetan, #HM2112)
Bon-practitioner state of awareness (Buddhism, Tibetan, #HM2639).

♦ HM2101d Stark givenness (ICA)
Description This is experienced as the knowledge that the universe is indifferent to a one's existence - neither for nor against, just indifferent. The individual sees the universe as broken and hostile and yet full of wonder and glory; as in the movie "Three Days of the Condor", when the main character realizes that the situation is not going to change for him.
Context In the ICA New Religious Mode in the arena of articulating the word about life (the life of knowing) the second formal aspect is one's experience of the vanishing cosmos or the wonder at the world. At the second phenomenological level, this occurs when one is stripped of one's illusions and has to decide how to relate to one's new universe.
Broader Wonder at the world (ICA, #HM3062).
Followed by Sacramental universe (ICA, #HM2366).
Preceded by Passing awayness (ICA, #HM2076).

♦ HM2102b Growth of consciousness
Description This has been postulated as the development from the egocentricity experienced in childhood, to perspectivism as the child learns to distinguish between an event and his own and others' points-of-view related to that event. There is a mutually dependent relationship between the abstract implications developed through everyday experience and the learning of abstract ideas (as in science education at school) which is then used in concrete experience.
Related Consciousness expansion (#HM2126).

♦ HM2103b Atma awareness (Hinduism)
Description Atma is considered to be the ultimate self with which individual humans are endowed. It is the "thou" in the formula *Thou art That* "that" being god or brahma; therefore it is taught that awareness of the true self is awareness of God. The atman state is recognized through wisdom (buddhi).
Related Buddhi awareness (Hinduism, #HM2099)
Indian approaches to consciousness (Hinduism, #HM2446).

♦ HM2104d Human contingency (ICA)
Description This is the realization that no amount of self-sufficiency deals with the fact of one's eternal dependence; when one sees that no disengagement from things can deal with the question of detachment as a spirit discipline. It may be compared to the main character in Eric Newby's book "Love and War in the Appennines", when he escapes from a prisoner-of-war camp only to discover that he is absolutely dependent on the local people to hide and assist him.
Context In the ICA New Religious Mode in the arena of knowing one's disengagement (the life of poverty) the fourth formal aspect is the experience of spiritual denial or of the sacrificial offering. At the first phenomenological level, this occurs when one is no longer content with one's customary habits, but see in them a futility never before discerned.
Broader Spiritual denial (ICA, #HM2174).
Followed by Intentional self-negation (ICA, #HM2369).

♦ HM2105c Path of intelligence of reward (Judaism)
Path of intelligence of conciliation
Description This mode of awareness is said to be of the divine influence which flows into the consciousness.
Context The twenty first of the spiritual paths expressed in the Jewish Kabbalah.
Broader Paths of wisdom (Judaism, #HM2509).
Followed by Path of faithful intelligence (Judaism, #HM2081).
Preceded by Path of intelligence of will (Judaism, #HM2629).

♦ HM2106c Joyous awareness
Bast (Sufism)
Description The Sufis express mystical joy by the term *bast*. It is the awareness of the emotions engendered when God is revealed and expands (bast) the heart and hope of the mystic.
Broader Stations of consciousness - Simnami (Sufism, #HM2341).

♦ HM2107c Path of transparent intelligence (Judaism)
Path of intelligence of light
Description A mode of awareness represented as the image of magnificence, from whence visions and prophecies are attained.
Context This state of transparency is the twelfth of the spiritual paths expressed in the Jewish Kabbalah.
Broader Paths of wisdom (Judaism, #HM2509).
Followed by Path of uniting intelligence (Judaism, #HM2583).
Preceded by Path of fiery intelligence (Judaism, #HM2523).

♦ HM2108c Awareness of corporeality-group of conscious existence (Buddhism)
Rupa-khanda (Pali) — Materiality aggregate — Aggregate of matter
Description One of the five interacting aggregates that produce the illusory ego, this group includes the human body as well as other forms. Anything that is perceived as being material, whether past, present or future, internal or external, subjective or objective, gross or refined, superior or inferior, far or near, is included in this classification which corresponds to the Western idea of reality.
The corporeality-group is the outcome of the activities and combinations of the 4 root-elements, earth, water, fire and air or wind, conceived metaphysically. From the 4 root-elements are derived 24 secondary phenomena of sentience, which include the five senses and their organs, and the corresponding physical sense-objects. The latter four constitute the consciousness of form, sound, odour and taste along with other bodily consciousness as, for example, of pressure, pain, touch, etc. Among the 15 other derived corporeal elements is the seat of consciousness or physical base of mind (hadaya-vatthu). Figuratively, that base is said, by some commentators on the Pali tradition, to centre on the heart and is referred to as heart-basis.
Broader Five aggregates (Buddhism, #HH3321).
Related Feeling aggregate (Buddhism, #HM4983)
Perception aggregate (Buddhism, #HM4143)
Dispositions of consciousness (Buddhism, #HM2098)
Consciousness-born materiality (Buddhism, #HM8273)
Awareness as mental-formation group of conscious existence (Buddhism, #HM2050).

♦ HM2109c Life of contemplation (ICA)
Description An arena in the ICA *New Religious Mode* when one is aware of standing present to the mystery of being in life.
Broader New religious modes (ICA, #HM3004).
Narrower Awareness of depth (#HM2447)
Awareness of archaism (#HM3007)
Awareness of futurity (#HM2609)
Awareness of externality (#HM2637).
Related Contemplation (#HM2952)
Contemplative life (Christianity, #HH2145).

♦ HM2110d First absorption in the immaterial sphere (Buddhism)
Akasanantyayatana — Akasanancayatana (Pali) — Maulasamapatti — Dngos gzhii snyoms jug (Tibetan) — Limitless space absorption — Sphere of infinite space — First formless attainment
Description *Hinayana Buddhism* The mind reaches the *sphere of unbounded space* in the *first absorption*. Its consciousness is of limitless extension. This is the first *jhana*, when there are reasoning and deliberation. This state arises from separation, and joy and happiness are experienced.
Tibetan Gelugpa Buddhism In this stage of meditation the appearance of forms to the mind completely disappears. Space pervades everywhere.
Context *Hinayana Buddhism* The next three absorptions will be in the spheres of of *unbounded consciousness*, *nothingness*, and *neither-perception-nor-non-perception*.
Tibetan Gelugpa Buddhism This is the first of the 4 formless absorptions (arupayasamapatti).
Broader Jhana (Buddhism, Pali, #HM7193)
Arupaloka consciousness (Buddhism, Pali, #HM2012)
Trances and mental absorptions (Buddhism, #HM2122)
Immaterial states as meditation subjects (Buddhism, #HH3198)
Tibetan meditative states of formless absorptions (Buddhism, #HM2669).
Related Subjects for meditation (Buddhism, #HH3987)
Formless-realm consciousness (Buddhism, #HM2281).
Followed by Second absorption in the immaterial sphere (Buddhism, #HM3043).

♦ HM2111d Singular adoration (ICA)
Description This is experienced as a burning desire for the mystery; and with the awareness that one's whole life is one occasion after another of being overwhelmed by the mystery, one surrenders - utterly. The popular song "I don't know why I love you like I do", and the poetic line "And are we yet alive", seem to capture the significance of this state. There is an awareness that the object of affection is not one's conditional terms, but only that which has rendered one's terms impotent. With heartsick pain one declares that it is God alone one loves.
A dimension of this experience is a sense of awe, that is fear and fascination. The fear is of not knowing know what this will lead to; there is an estrangement from the world that leaves an individual wondering what could have been. And yet there is fascination with the fantastic nature of the event; he finds himself saying, "So this is what wonder and awe are really all about". A decision is made to abandon his own will to the mystery, which is followed by deep longing to maintain this state of being. He is alive in a new way to the possibilities and options before him.
Context This state is number 16 in the ICA *Other World in the midst of this World*.
Broader Infinite passion (ICA, #HM2234).
Followed by Total exposure (ICA, #HM2764)
Primordial wonder (ICA, #HM2186)
Ultimate awareness (ICA, #HM2388)
Definitive effectivity (ICA, #HM2796)
Destinal accountability (ICA, #HM2309).
Preceded by Total exposure (ICA, #HM2764)
Dynamic selfhood (ICA, #HM3091)
Contingent eternality (ICA, #HM2456)
Transcendent immanence (ICA, #HM3034)
Definitive effectivity (ICA, #HM2796).

♦ HM2112d Hungry-ghosts-hell awareness (Buddhism, Tibetan)
Pretas
Description This state of awareness is characterized by frustration of the appetites and represented as a hell for those who have been excessively greedy.
Context In Tibetan Sakya Buddhism this is one of the states in the "Ascension Stages Game". In some sets it is numbered 10 on the board.
Broader Desire-realm consciousness (Buddhism, #HM2733).
Followed by Naga-world awareness (Buddhism, Tibetan, #HM2031)
Animal-hell awareness (Buddhism, Tibetan, #HM2636)
Demon-island awareness (Buddhism, Tibetan, #HM2055)
Howling-hells awareness (Buddhism, Tibetan, #HM2100)
Temporary-hells awareness (Buddhism, Tibetan, #HM2454)
Eastern-continent awareness of noble figures (Buddhism, Tibetan, #HM2543).
Preceded by Cold-hells awareness (Buddhism, Tibetan, #HM2040)
Asura-world awareness (Buddhism, Tibetan, #HM2579)
Animal-hell awareness (Buddhism, Tibetan, #HM2636)
Howling-hells awareness (Buddhism, Tibetan, #HM2100)
Reviving-hell awareness (Buddhism, Tibetan, #HM2516)
Very-hot-hells awareness (Buddhism, Tibetan, #HM2576)
Crushing-hells awareness (Buddhism, Tibetan, #HM3142)
Heavenly-highway awareness (Buddhism, Tibetan, #HM2010)
Interminable-hell awareness (Buddhism, Tibetan, #HM2052)
Non-emanating consciousness (Buddhism, Tibetan, #HM2070)
Divine-animal-hell awareness (Buddhism, Tibetan, #HM2007)
Barbarian-state of awareness (Buddhism, Tibetan, #HM2091)
Four-great-kings-heaven awareness (Buddhism, Tibetan, #HM2082)
Heaven-without-fighting consciousness (Buddhism, Tibetan, #HM2130)
Western-continent awareness of cattle (Buddhism, Tibetan, #HM2519).

♦ HM2113c Indivisibility of essence (Sufism)
Description The plurality of modes are not included in the unity of Essence as is the part in the whole or the contained in the container, but as the cause englobes its concomitants; or, as with fractions of the whole, the fractions are not manifest without the whole.
Context The nineteenth illumination of Jami.
Broader Fountains of light (Sufism, #HM3039).
Followed by Unchanging reality of being (Sufism, #HM3316).
Preceded by Beyond individual distinctions (Sufism, #HM2681).

♦ HM2114d Existential guidance (ICA)
Description The third formal aspect of the *life of meditation*, an arena in the ICA *New Religious Mode*, this represents an internal encounter with the saints and receiving existential guidance.
Broader Life of meditation (ICA, #HM3234).
Narrower Revered hero (#HM2654)
Guardian angel (#HM3130)
Scorching avatar (#HM2921)
Ever-present brother (#HM2814).

♦ HM2115d Hindu-states of awareness (Buddhism, Tibetan)
Description This state of awareness is characterized by a cultured, full life, but with a superficial externalized and ritualized spiritual view. It is represented by the Brahmin culture of the Indo-Pakistan subcontinent. This state, despite high achievements, leads to subjectivity and meditative accomplishments that are not sufficiently stabilized or analytical and are thus bound to generate karma within cyclic existence.

Context In Tibetan Sakya Buddhism this is one of the states in the "Ascension Stages Game". In some sets it is numbered 22 on the board.
Broader Religious traditions in the ascension stages game (Buddhism, Tibetan, #HM3341).
Followed by Naga-world awareness (Buddhism, Tibetan, #HM2031)
Animal-hell awareness (Buddhism, Tibetan, #HM2636)
Asura-world awareness (Buddhism, Tibetan, #HM2579)
Very-hot-hells awareness (Buddhism, Tibetan, #HM2576)
Hindu-wisdom-holder awareness (Buddhism, Tibetan, #HM2723)
Great-vehicle lower-path awareness (Buddhism, Tibetan, #HM2268).
Preceded by Naga-world awareness (Buddhism, Tibetan, #HM2031)
Southern-continent awareness of men (Buddhism, #HM2127)
Thirty-three-god-heaven awareness (Buddhism, Tibetan, #HM2606)
Delightful-emanation-heaven consciousness (Buddhism, Tibetan, #HM2546).

♦ **HM2117c Path of constituting intelligence** (Judaism)
Description This mode of awareness is represented by the material of creation in a cloud of darkness.
Context The fifteenth among the spiritual paths expressed in the Jewish Kabbalah.
Broader Paths of wisdom (Judaism, #HM2509).
Followed by Path of triumphant intelligence (Judaism, #HM2593).
Preceded by Path of illuminating intelligence (Judaism, #HM2035).

♦ **HM2118c Great-black-lord awareness** (Buddhism, Tibetan)
Mahakala
Description This state of awareness is characterized by fierce determination and represented as a being of frightening appearance and attributes.
Context In Tibetan Sakya Buddhism this is one of the states in the "Ascension Stages Game". In some sets it is numbered 34 on the board.
Broader Tantric paths of accumulation and application in the ascension-stages-game (Buddhism, Tibetan, #HM2848).
Followed by Amoghasiddhi-karma awareness (Buddhism, Tibetan, #HM2783)
Fifth vajra-master awareness (Buddhism, Tibetan, #HM2763)
Guhya-samaja urgyan awareness (Buddhism, Tibetan, #HM2699).
Preceded by Lord-of-death awareness (Buddhism, Tibetan, #HM2088)
Amoghasiddhi-karma awareness (Buddhism, Tibetan, #HM2783)
Discipleship-vision awareness (Buddhism, Tibetan, #HM2240)
Rudra awareness of black freedom (Buddhism, Tibetan, #HM2603).

♦ **HM2119d Final situation** (ICA)
Description This is the awareness that each situation is of such significance that it could be lived as if it were the last moment of life. It occurs when one realizes that nothing can be treated as anything less than the most important moment of one's life, as when being saved from a fatal accident or like an opening night on Broadway.
Context In the ICA New Religious Mode in the arena of acting out one's deed (the life of doing) the fourth formal aspect is becoming the symbol of the eternal context. At the first phenomenological level this occurs when one is confronted by the fact that to live is to be of service.
Broader Symbolizing the eternal context (ICA, #HM3176).
Followed by Meaning creation (ICA, #HM2017).

♦ **HM2120b Lokuttara consciousness** (Buddhism)
Description This awareness is that experienced by a Buddha completely beyond this world. Persons and things are known not to be real; but the principles of the two kinds of void, of persons and things, are absolutely real (sanyata). According to Lokuttaravadins, the historical life and actions of Buddha are mental images and mere appearance.
Related Samatha-bhavana (#HH0710)
Bhavana (Buddhism, #HH0551)
Eightfold way (Buddhism, #HM2339)
Phenomena awareness (Buddhism, #HM2551)
Bodhi-pakkhiya dhamma (Buddhism, #HM2793)
Vipassana-bhavana (Buddhism, Pali, #HH0680)
Trances and mental absorptions (Buddhism, #HM2122)
Beyond-the-third-realm consciousness (Buddhism, Tibetan, #HM2653)
Tibetan meditative states of form concentrations (Buddhism, #HM2693).

♦ **HM2122d Trances and mental absorptions** (Buddhism)
Jhana (Pali)
Description Concentration of the mind allows it to enter the fine-material (rupaloka) sphere where it may progressively experience 4 trance states. Beyond this the trances may progress into the immaterial (arupaloka) sphere and 4 additional stages of absorption. The jhana lead to mundane and supermundane (lokuttara) insights or immediate intuitive knowledge (vipassana). Some 18 chief kinds of insight are enumerated.
Context In southern Buddhism, bhavana (yoga) with its mental trances and absorptions produces insight or wisdom (panna), tranquillity (samatha) and the gnosis of light (samadhi). The fourth trance of the five-material sphere is also the starting point for the development of the 6 kinds of *higher spiritual forces* (abhinna) and the *roads to power* (iddhi-pada).
Broader Dispositions of consciousness (Buddhism, #HM2098).
Narrower Third trance of the fine-material sphere (#HM2062)
First trance of the fine-material sphere (#HM2450)
Third absorption in the immaterial sphere (#HM2027)
Second trance of the fine-material sphere (#HM2038)
First absorption in the immaterial sphere (#HM2110)
Fourth trance of the fine-material sphere (#HM2586)
Fourth absorption in the immaterial sphere (#HM2051)
Second absorption in the immaterial sphere (#HM3043).
Related Jhana (Buddhism, Pali, #HM7193)
Concentration (Buddhism, #HM6663)
Rupaloka consciousness (Buddhism, #HM2536)
Lokuttara consciousness (Buddhism, #HM2120)
Subjects for meditation (Buddhism, #HH3987)
Tibetan meditative states of form concentrations (Buddhism, #HM2693)
Tibetan meditative states of formless absorptions (Buddhism, #HM2669).

♦ **HM2123d Conscious self** (Psychosynthesis)
Description According to R Assagioli, the truer self in experiential terms is the "I", that is the centre of consciousness, and not the contents of a particular state or formation that is identified as the personality. It is awareness in its pure potency. However, it is subject to sleep and to conditions like hypnosis, shock, narcosis and the like. The *conscious self* is able to reappear only because it depends from the *higher self*.
Related Psychosynthesis (#HH0002).
Followed by Higher self (Psychosynthesis, #HM2970).
Preceded by Collective unconscious (Psychosynthesis, #HM2811).

♦ **HM2124c Fiery awareness** (Astrology)
Triplicity of fire
Description An individual whose consciousness is influenced by the *triplicity* combining the zodiacal signs Aries, Leo and Sagittarius will demonstrate ardent and keen characteristics, having attributes associated with the element fire. Such attributes, although possibly destructive and consuming in the sense of insensitivity, over-vivaciousness and thoughtlessness, will create warmth and enthusiasm. The vitality and spontaneity are conducive to unconscious perception and intuition. The use of the senses to promote awareness of objects, and attention to detail, are both repressed in order to see the essence of a situation or the wider context. Practicalities may be beyond such an individual but he will have very clear perception of the deepest places in his spiritual life.
Context One of the four *triplicities* or *elements*, each of which combines three related signs of the zodiac.
Broader Zodiacal forms of awareness (Astrology, #HM2713).
Narrower Leo-consciousness (#HM2376)
Aries-consciousness (#HM2665)
Sagittarius-consciousness (#HM2726).
Related Airy awareness (Astrology, #HM2955)
Watery awareness (Astrology, #HM2384)
Earthy awareness (Astrology, #HM3235).

♦ **HM2125a Nirbija samadhi** (Hinduism)
Description Nirbija means seedless; nirbija samadhi is a condition in which there is nothing in the mind except consciousness and intense awareness of consciousness, consciousness of Purusa. The pure but partial knowledge of sabija samadhi is replaced by merging of the mind in the one reality. The mind exists only to radiate reality's effulgence.
Broader Samadhi (Hinduism, Yoga, #HM2226).
Related Purusha (Yoga, #HM2396)
Indian approaches to consciousness (Hinduism, #HM2446).

♦ **HM2126a Consciousness expansion**
Awareness expansion — Mind expansion
Description This is a general term used to denote a wide variety of processes and techniques of: self-actualization, self-realization, peak experience, personal awakening, psychedelic experience, altered states of consciousness, and meditation.
Consciousness expansion is aimed at awareness of the unity of all things and at seeing the contradictory nature of considering the universe as fragmented into individual, unrelated, separate bits, of which men (inhabited by their individual egos) should be the top, controlling species. It is characterized as follows: 1. Sensory level: subjective reports of alterations in space, time, body image and sensory impressions
2. Recollective-analytic level: novel ideas and thoughts concerning the individual's psychodynamics or conception of the world and his role in it.
3. Symbolic level: identification with historical or legendary personages, with evolutionary recapitulation, or with mythical symbols.
4. Integral level: religious or mystical experience in which an ultimate being is confronted or in which the individual dissolves into the energy field of the universe.
Expanded states of consciousness may occur spontaneously or may be induced through hypnosis or sensory bombardment. They are frequently brought about experimentally by the use of psychedelic drugs.
Related Growth of consciousness (#HM2102)
Anomalies in experience of the self as distinct from the outside world (#HM4754).

♦ **HM2127d Southern-continent awareness of men** (Buddhism)
Jambudvipa — Dzam-bu-gling (Tibetan)
Description This is the normative human state of awareness characterized by opportunity to escape from the cycle of death and rebirth by following the dharma. It is represented as a floating land south of Mount Meru.
Context In Tibetan Sakya Buddhism this is one of the states in the "Ascension Stages Game". In some sets it is numbered 17 on the board.
Broader Desire-realm consciousness (Buddhism, #HM2733).
Followed by Reviving-hell awareness (Buddhism, Tibetan, #HM2516)
Hindu-states of awareness (Buddhism, Tibetan, #HM2115)
Tantra-beginner awareness (Buddhism, Tibetan, #HM2452)
Discipleship-karma awareness (Buddhism, Tibetan, #HM2192)
Wheel-turning-king awareness (Buddhism, Tibetan, #HM2058)
Great-vehicle lower-path awareness (Buddhism, Tibetan, #HM2268).
Preceded by Form-realm awareness (Buddhism, Tibetan, #HM2142)
Animal-hell awareness (Buddhism, Tibetan, #HM2636)
Demon-island awareness (Buddhism, Tibetan, #HM2055)
Reviving-hell awareness (Buddhism, Tibetan, #HM2516)
Formless-realm awareness (Buddhism, Tibetan, #HM3144)
Heavenly-highway awareness (Buddhism, Tibetan, #HM2010)
Interminable-hell awareness (Buddhism, Tibetan, #HM2052)
Wheel-turning-king awareness (Buddhism, Tibetan, #HM2058)
Four-great-kings-heaven awareness (Buddhism, Tibetan, #HM2082)
Thirty-three-god-heaven awareness (Buddhism, Tibetan, #HM2606)
Heaven-without-fighting consciousness (Buddhism, Tibetan, #HM2130)
Northern-continent awareness of community (Buddhism, Tibetan, #HM2067).

♦ **HM2128c Non-attachment** (Buddhism)
Alobha (Pali) — Ma-chags-pa (Tibetan) — Non-greed — Disinterestedness — Detachment — Renunciation
Description As opposed to *mortification*, which implies the deliberate avoidance of sensual pleasure, detachment is a freedom from longing, a non-involvement with the objects of the senses, leading to clarity, efficient and effective action, calm and joy. It has been described as "thirstlessness", implying release from the duality implied in the element *water*.
Together with non-hatred and non-ignorance, non-attachment acts as an antidote to the three causes of misconduct - desire, hatred and ignorance. These three are related in both Tibetan and Hinayana Buddhism. It is related to all paths since, depending on the being's small, medium or great capacity, it leads: to seeking one's own welfare in future lives as opposed to the present life; to seeking release from all cyclic existence; or to seeking the non-abiding nirvana where, while remaining meditating on emptiness, one manifests to help others migrate from the cycle of birth and rebirth. According to Patanjali, non-attachment to aspiration after illumination and isolated unity brings awareness of spiritual knowledge; it appears as an overhanging cloud to be reached, used and penetrated.
In Hinayana Buddhism, absence of greed is the means for not being greedy, not being greedy in itself and merely not being greedy. Its characteristic is lack of desire, the mind is free from cupidity for an object of thought, it does not adhere. Its function is not laying hold of or appropriating.
Context One of the eleven *virtuous mental factors* referred to in Tibetan Buddhism. One of the formations aggregate (mental coefficients) of Hinayana Buddhism, being listed among the constant states which appear in their true nature, and as profitable primary (always present in any profitable or profitable-resultant consciousness).
Broader Virtuous mental factors (Buddhism, #HH0578)
Awareness as mental-formation group of conscious existence (Buddhism, #HM2050).
Narrower Not infatuated (#HM1688)
Disinterestedness (#HM1799)
Not being covetous (#HM0684)
State of not being greedy (#HM3947)
State of not feeling greed (#HM1825)
State of not being infatuated (#HM3183)
State of not feeling infatuation (#HM4081)
Disinterestedness as a wholesome root (#HM3752).
Related Greed (Buddhism, #HM3283)
Mythic yoga (Yoga, #HH3405)

Non–hatred (Buddhism, #HM2744) Equanimity (Buddhism, #HM7769)
Detachment (Hinduism, #HM5091) Non–ignorance (Buddhism, #HM2695)
Detachment (Christianity, #HM1534) Subtlety of mind (Hinduism, #HM3312).

♦ HM2129e Telepathic consciousness
Description Apparently related to *alien consciousness*, in that the telepath appears so different as to be hardly human, this type of consciousness is postulated as feeding vicariously on other people's innermost thoughts while at the same time the individual's own mind appears to diminish. Paradoxically, this ability cuts him off emotionally from others. In non–fictitious terms, that part of the self which intrudes on another's privacy diminishes the likelihood of the individual giving and receiving love.
Related Alien consciousness (#HM2329) Extrasensory perception (ESP, #HM2262).

♦ HM2130d Heaven–without–fighting conscic ness (Buddhism, Tibetan)
Yama–devas
Description This state of human awareness is projected as the qualities ascribed to the deities in this heaven. They are beyond strife, peace–loving and peace–keeping and have vanquished distraction, confusion and attraction to delightful experiences.
Context In Tibetan Sakya Buddhism this is one of the states in the "Ascension Stages Game". In some sets it is numbered 29 on the board.
Broader Desire–realm consciousness (Buddhism, #HM2733).
Followed by Joyful–heaven awareness (Buddhism, Tibetan, #HM2022)
Southern–continent awareness of men (Buddhism, #HM2127)
Hungry–ghosts–hell awareness (Buddhism, Tibetan, #HM2112)
Divine–animal–hell awareness (Buddhism, Tibetan, #HM2007)
Bon–practitioner state of awareness (Buddhism, Tibetan, #HM2639)
Delightful–emanation–heaven consciousness (Buddhism, Tibetan, #HM2546).
Preceded by Wheel–turning–king awareness (Buddhism, Tibetan, #HM2058)
Discipleship–vision awareness (Buddhism, Tibetan, #HM2240)
Thirty–three–god–heaven awareness (Buddhism, Tibetan, #HM2606)
Bon–practitioner state of awareness (Buddhism, Tibetan, #HM2639)
Pratyeka Buddha cultivation awareness (Buddhism, Tibetan, #HM2252).

♦ HM2131d Definitive predestination (ICA)
Description This is the realization that this is the only world one has. The cartoonist George Schultz portrays this very succinctly in his Peanuts comic strip when Lucy confronts Charlie Brown with the fact that this is the only world there is and then admonishes him to live in it; and the poet E.E. Cummins says it too, with tongue in cheek: "Listen: there's a hell of a good universe next door; let's go". It is like having been shot from a timeless cannon to a fixed target on the temporal plane. The singer Bruce Springsteen conveys the same experience in his song "Dancing in the Dark" when he sings: "There's something happening somewhere, Baby I just know that there is".
A dimension of this experience is a sense of awe, that is fear and fascination. While acknowledging the gift of the situation, one resents the limitations it imposes, and appears boxed in or imprisoned by limits of time, space and physiological constraints. Yet there is also a certain sense of victory at making it through this moment, somewhat akin to solving a very difficult problem or riddle. Realizing what one has on one's hands, one stops looking for an alternative to run to and decides to live the life one has, even if it is not what one ordered; one acknowledges that everyone else shares the same limitations. There is an inner sense of gratitude for the limits that life has imposed.
Context This state is number 34 in the ICA *Other World in the midst of this World*.
Broader Original gratitude (ICA, #HM3105).
Followed by Absurd existence (ICA, #HM2331) Temporal solidarity (ICA, #HM2363)
Universal compassion (ICA, #HM2734) Ancestral obligation (ICA, #HM3122)
Inclusive comprehension (ICA, #HM2256).
Preceded by Absurd existence (ICA, #HM2331) External relation (ICA, #HM2471)
Interior discipline (ICA, #HM2851) Ancestral obligation (ICA, #HM3122)
Individual fatefulness (ICA, #HM2223).

♦ HM2133b Hypnotic states of consciousness
Hypnosis
Description A mode of awareness with superficial resemblance to sleep or trance, which may be self–induced or induced by a hypnotist. It differs from dreaming in that the hypnotized person responds physically as directed, even to the extent of control not normally experienced in the normal waking state – altering body temperature, for example, or increasing pain thresholds. Also it is not a unitary or consistent state.
Experiments by Professor Charles Tart on deep, unguided hypnosis have shown this mode of awareness to be very flexible. As the subject goes deeper into a hypnotic trance, he experiences similar effects to those induced by psychedelic drugs. At a particular depth of experience there seems to be a consciousness threshold, beyond which the individual, personalized identity disappears and existence appears as potential identity; space and time are meaningless concepts; and there is no spontaneous thought. Studies on mutual hypnosis show that, at least at a depth similar to psychedelic drug hallucinations, experiences in hypnotic trance can be shared. The transcendental and transpersonal potentialities of hypnosis are yet to be fully explored.
Narrower Abulia (#HM0672) Active trance (Psychism, #HM0442)
Deep state of hypnosis (#HM1226) Light state of hypnosis (#HM4595)
Medium state of hypnosis (#HM5856).
Related Dreams (#HM2950) Trance (#HM3236) Hypnotherapy (#HH0962)
Mental imagery therapy (#HH0072).

♦ HM2134c Resentment (Buddhism)
Upanaha — 'Khon–'dzin (Tibetan)
Description The wish to harm or respond to harm, and the cause of *impatience*.
Context One of the twenty *secondary afflictions* of Tibetan Buddhism.
Broader Secondary afflictions (Buddhism, Tibetan, #HH0781).

♦ HM2135g Deep relaxation
Description This is a state extending beyond muscular relaxation into a state of detachment from everyday conflicts and problems. It is not a state of indifference, but rather an acceptance of a new perspective. This perspective allows the individual a new understanding of his inner and outer world; and may bring into this realm a state of balance, which may be expressed by the individual as a feeling of peace.

♦ HM2136b God–consciousness as buoyancy
Hermes awareness
Description Human development can lead to a state of consciousness that is unfettered by the body or the material world. In this state the mind is unanchored and buoyant, rising and moving in the currents of light. One flies, like a winged Hermes, flashing in crystalline form through all the realms of inner and outer space and all the realms of time.
Broader God–consciousness (#HM2166).

♦ HM2137a Unitive life
Jivana mukta state
Description This state is one of the aims and achievements of various yoga practices. Unitive life is the term used in Christian terminology. It is a state of waking consciousness in which thinking and action are not disjointed. When thinking is necessarily involved in action, concentration and attentiveness follow everything undertaken, including even such physiological activities as eating and drinking. But when thinking is not necessary for action, the mind is relaxed or disengaged, and lapses into a condition of harmony and equipoise. The man begins to live in the living present, and to be free of wishful thinking, brooding over the past or worrying about the future. It is a state characterized by self–confidence and self–reliance in action, but non–attachment and relaxation after work is over; most particularly, there is no lingering possessiveness about the results of one's work.
In addition, there is freedom from anxiety and other mental tensions – a freedom which extends even to a loss of philosophical inquisitiveness. Yet metaphysical problems seem to dissolve into a larger perspective, and the mind is not impelled to speculation. The relinquishing of ego values make the appreciation of other values almost spontaneous, and the fresh eye with which the world is viewed opens up whole ranges of aesthetic appreciation hitherto unknown.
This state therefore constitutes an inner environment in which the integration of the personality can take place. There results a natural, warm, loving and spontaneous attitude of openness to life and to others, without fear or condemnation.
Related Jivanmukti (Hinduism, #HM3890).

♦ HM2138g Daydream
Fantasying — Daydreaming
Description The mind is withdrawn from immediate physical surroundings and indulges in pleasing speculation or wanders through images pertaining to the individual's life. Apparently a time–filler, but possibly psychologically necessary, periods of relaxation or diffuse thinking tend to be interposed between more concentrated times on a regular basis. This may be an extension of the habit of "tuning out" habitual impulses, so that unusual events are noticed. When there are no unusual events, then daydreaming creates its own novelty.
Four types of daydream have been noted: (1) *Self recrimination*, with the theme of what one should have said or done on some previous occasion. (2) *Controlled and thoughtful*, when the day ahead is planned or some future event organized. (3) *Autistic*, when consciousness is interrupted by material usually associated with dreams at night. (4) *Neurotic or self–conscious*, including fantasies about the future.
The attraction of daydreaming and fantasy is that existential limitations are changed and one can act as if one had all the qualities one desired, being omnipotent, god–like and free of the limitations and pressures of actual events. The danger is that imagined possibilities may masquerade as realizable projects; and that not only may one fantasize upwards into the wish–worlds, but also downwards into dread and self–in–despair.
Context Normal consciousness is said to be alert thinking (concentrated) interspersed at regular intervals by relaxed thinking (concentrated) or daydreaming (diffuse). This mirrors the regular pattern of paradoxical or rapid eye movement sleep which occurs about every 90 minutes during the night.
Refs Ornstein, Robert *The Psychology of Consciousness* (1986).
Narrower Escapist daydreaming (#HM0248).
Related Alert thinking (#HM4967) Relaxed thinking (#HM1274)
Wish–dominated awareness (#HM2179).

♦ HM2139d Persistent friend (ICA)
Description This is experienced when an individual embraces the presence in his life which calls to a full life and makes that presence a companion on the journey. Then he engages that presence in a dialogue about his life and what he intends doing with it. An example is the movie "Patton", when the main character stands on the battlefield of Carthage and dialogues with the generals whose armies fought there several thousand years before.
Context In the ICA New Religious Mode, in the arena of knowing one's internal sociality (the life of meditation) the first formal aspect is one's encounter with the mediator who pronounces personal absolution. At the third phenomenological level, this occurs when one dialogues across time and space with those who also travel this spiritual journey.
Broader Personal absolution (ICA, #HM2370).
Followed by Eternal saviour (ICA, #HM2253).
Preceded by Radiant guru (ICA, #HM2019).

♦ HM2140b Evolution of consciousness
Description A cultural–anthropological scheme of development of global and individual consciousness from the early beginnings of archaic "paradisical" consciousness, via stone–age magical, iron–age mythical and modern mental, toward utopist integral consciousness, is proposed in the philosophy of Jean Gebser. Gebser's scheme includes scientific, religious, artistic, social and philosophic fact into its five–level description. All levels exist not only in their historic succession but also interrelated within the cultural structure of the present world–situation.
Broader Evolution (#HH0425).
Narrower Mental consciousness (#HM2319) Magical consciousness (#HM2090)
Archaic consciousness (#HM2189) Mythical consciousness (#HM2078)
Integral consciousness (#HM2152).

♦ HM2141c Jagrat state of waking consciousness (Yoga)
Description One of the three *avasthas* recognized in yoga as dominated by the subjective influence of the ego. The ordinary waking state, it is said by the Maharishi Mahesh Yogi to be the first in a series of seven states attainable to the human being, a state from which eventually *turiya*, or objective consciousness, may be reached. In the Mandukya Upanishad, Jagrat is said to be the field of Vaisvanara, the material condition.
Broader Awareness of relative reality (Hinduism, Yoga, #HM2032).
Related Vaisvanara (Hinduism, #HM2336) Turiya awareness (Hinduism, #HM2395)
Transcendental meditation (TM, #HH0682).
Followed by Swapna state of dream consciousness (Yoga, #HM2781).

♦ HM2142d Form–realm awareness (Buddhism, Tibetan)
Rupa–dhatu
Description This state of awareness is represented as occurring beyond the realms of the senses, high above Mount Meru. There are 4 trance states in this awareness. The first is represented as a 3–stage Brahma heaven where monotheistic conception dominates. The second trance states are represented as a 3–stage gods–of–light (abha) heaven. The triple third trance is represented as being among the gods–of–splendour (shubha). In the game of the Ascension Stages, gods represented as pertaining to the fourth trance are divided among the squares. Three are included in this station of awareness. They are called the unclouded, the merit–born, and the great fruit gods.
Context In Tibetan Sakya Buddhism this is one of the states in the "Ascension Stages Game". In some sets it is numbered 35 on the board.
Broader Form–realm consciousness (Buddhism, #HM2257).
Related Animal–hell awareness (Buddhism, Tibetan, #HM2636).

Followed by Pure-lands awareness (Buddhism, Tibetan, #HM2168)
Joyful-heaven awareness (Buddhism, Tibetan, #HM2022)
Formless-realm awareness (Buddhism, Tibetan, #HM3144)
Southern-continent awareness of men (Buddhism, #HM2127)
Great-vehicle lower-path awareness (Buddhism, Tibetan, #HM2268).
Preceded by Non-emanating consciousness (Buddhism, Tibetan, #HM2070)
Pratyeka Buddha cultivation awareness (Buddhism, Tibetan, #HM2252).

♦ HM2143d Doing the mystery (ICA)

Description This is experienced when a person lets go of all of his previous values and identities; the moment in which his life and the required deed are synonymous. It occurs when he is aware that all of creation depends on this moment and this situation, and that his act is this deed and this situation; for example, Thomas à Becket before his king or Jesus in Jerusalem.
Context In the ICA New Religious Mode in the arena of acting out one's deed (the life of doing) the fourth formal aspect is becoming the symbol of the eternal context. At the fourth phenomenological level this occurs when the unique greatness of one's expenditure is embodied.
Broader Symbolizing the eternal context (ICA, #HM3176).
Preceded by Obedient son (ICA, #HM2005).

♦ HM2144c Ajna (Yoga)

Awareness of authority — Sixth chakra
Description The mystic inward ear and eye are fully open, one is in heaven, the soul beholding its perfect object, God. All five elements are present in their rarified pure essence (tanmatra). Meditating on this chakra the person has all his sins and impurities eradicated. All those in his presence are calm and sensitive to the sound frequencies of AUM - the sound which generates from the person's body itself. He is beyond the desires that motivate life and impel movements in many directions. He is now one-pointed, a knower of past, present and future, beyond time in sushumna. As long as he is in the physical body there is no more backsliding, there is constant state of f nondual consciousness. Able to enter any body at will, he can comprehend the inner meaning of cosmic knowledge and generate scriptures. Revealing the divine within; he reflects divinity within others. Having evolved through the fourth and fifth levels (ananda - bliss and chit - cosmic consciousness) he becomes sat - truth and embodies sat-chit-ananda - "That I am; I am that". The realization "hamsa", "I am that", is the reverse of "soham" or "that I am" of the previous chakra. He is called paramhamsa.
Context The sixth "lotus" or chakra defined in kundalini yoga, said to be situated between and somewhat above the eyebrows.
Broader Kundalini yoga (Yoga, #HH0237) Chakra centres of consciousness (#HM2219).
Related Sphere of wisdom (Kabbalah, #HM2348)
Sphere of understanding (Kabbalah, #HM2372)
Initiation of decision (Esotericism, #HM0322)
Sixth chakra: time of penance (Leela, #HM4412)
Initiation of transfiguration (Esotericism, #HM0428).
Followed by Sahasrara (Yoga, #HM3398).
Preceded by Vishuddha (Yoga, #HM3461).

♦ HM2145b Meditative states of mental abidings (Buddhism)

Navakara chittasthiti — Sems gnas dgu (Tibetan)
Description To attain the higher meditative states of concentration and absorption one has to cultivate calm abiding (shamatha) and special insight (vipashyana). The former leads to an acquisition of requisite mental power, the latter to wisdom. During the cultivation of calm abiding a meditator passes through nine states of mind called mental abidings.
Broader Consciousness states in cyclic existence (Buddhism, #HM2177).
Narrower Calm abiding (#HM2147)
Only-a-beginner subtle contemplation (#HM2171)
Special insight and preparations as states (#HM2623)
Tibetan meditative states of form concentrations (#HM2693)
Tantra in meditation on emptiness (Buddhism, Tibetan, #HM2864).
Followed by Tibetan meditative states of formless absorptions (Buddhism, #HM2669).

♦ HM2146b Religious enthusiasm

Enthusiasm
Description Originally meaning "god within", enthusiasm referred to an altered or inspired state resulting from the in-dwelling of a deity in a human being. This is regarded in a positive sense, unlike possession which refers to in-dwelling of a noxious or evil power, although both may be referred to as *mania*. Union with a deity to produce an enthusiastic state has been represented as eating or drinking the deity or having sexual intercourse with the him or her. The soul may develop higher powers of vision and anticipate the future. To Plato, it is through enthusiasm that the philosopher has direct vision or intuition of deity. For Plotinus, there is union of the human soul with deity - but separated from the body, which implies ecstasy as opposed to enthusiasm.
Generally, enthusiasm is animated interest or preoccupation with something. Religious enthusiasm can include irrational behaviour and inexplicable psychological and psycho-somatic phenomena. These include insensitivity to pain, "exalted" states of consciousness, glossolalia (incomprehensible language), stigmata, "miraculous" healings, and genuine instances of extra-sensory perception. The term is also used to describe excessive display of piety, or the group fervour manifest by some Christian sects. In the negative sense, it implies that there is doubt as to the authenticity of the claimed source of inspiration. Although the inexplicable phenomena may be claimed as inspired, some regard them as emotionalism or as ways of working out inner conflicts in a manner acceptable to the group of which the individual is a member. Enthusiasts usually attack the status quo and have a very individualistic theology, including belief in the imminent end of the age, the return of Christ and the separation of the chosen elect from the rest.
Related Frenzy (#HM3594) Ecstasy (#HM2046) Theomania (#HM1027)
Poetic enthusiasm (#HM0875).

♦ HM2147c Calm abiding (Buddhism)

Shamatha — Samatha (Pali) — Zhi gnas (Tibetan) — Serenity
Description Calm abiding is stabilized meditation brought to fulfilment. The meditator acquires a mind of the next sphere, the *form realm*, and a stage of preparation for further states which is called the "not-unable" (anagamaya). The physical body is felt to be renewed and strengthened. The calm abider meditates in a favorable area, has few desires, knows satisfaction, has pure ethics, forsakes involvements and commotions, and lets go of sexual desires and other thoughts of gratification. His posture is stabilized in one of the recommended positions. With right motivation to benefit all sentient beings the meditation is calm and stabilized and the object - material, the meditators own mind, or the "body" of the Buddha (created in the imagination) - is engaged.
The second stage of *meditation on emptiness* is the cultivation of a "similitude" of calm abiding, stabilization resulting from meditation together with *pliancy* achieving a "similitude" of insight. Laziness is overcome by faith, aspiration, exertion and pliancy; forgetfulness of advice by mindfulness; laxity and excitement by introspection; non-application by application; and over-application by equanimity.
Context The calm abiding state in Tibetan Buddhism corresponds to *access concentration* (upachara) in the Pali tradition.
Refs Pal, Rai Sawindar *Samata Yoga*.

Broader Selflessness (Buddhism, #HM2016) Emptiness (Buddhism, Zen, #HM2193)
Meditative states of mental abidings (Buddhism, #HM2145).
Narrower Ceto samatha (#HM0709) Setting the mind (#HM2217)
Resetting the mind (#HM2265) Close setting the mind (#HM2157)
Continuous setting of mind (#HM2241) Mental abiding in equipoise (#HM2277)
Pacifying in mental abiding states (#HM2205) Disciplining in mental abiding states (#HM2181)
One-pointedness in mental abiding states (#HM2753)
Full pacification in mental abiding states (#HM2729).
Related Pliancy (Buddhism, #HM3162) Equanimity (Buddhism, #HM7769)
Access concentration (Buddhism, #HM4999).
Followed by Special insight and preparations as states (Buddhism, #HM2623).
Preceded by Only-a-beginner subtle contemplation (Buddhism, #HM2171).

♦ HM2149e Sense of harmony through progress (Brainwashing)

Description Nourishing of the new self is achieved through the emotional satisfaction of being accepted by the group in which the individual finds himself, as he gradually adapts to his situation. Recognition and adaptation are achieved only by progressive reform, which is far more real than the individual realizes, and he finds he is accepted despite the incompleteness of the reform at this stage.
Context A stage reached in thought reform, a system of organized, deliberate and total psychological training which effects individual change through two basic elements - *confession* (renouncing of past beliefs and attitudes) and *re-education* (remaking of the individual in the required image).
Broader Thought reform (Brainwashing, #HH0865).
Followed by Final confession (Brainwashing, #HM3156).
Preceded by Logical dishonouring (Brainwashing, #HM2564).

♦ HM2150b Angelic frame of awareness

Description Angels are a representation of invisible agents and operations considered to be involved in supernatural events and extra-sensory perception. However, sudden impulses of will, extraordinary imagination, and heightened cognition have been attributed to an angelic presence in man's inner world. Thus the characterization of angels ranges from unseen but sensed influences on mind and behaviour within normal limits, to such phenomena as mental communication and clairaudient perception (where the angelic agent is incorporeal) and to clairvoyant and sensible visions (where the angels have form and even substance, in as much as they are seen to conduct operations in the material world).
While the more popular view has always viewed angels as material beings of a higher order, philosophers (such as Philo) considered them to be incorporeal intelligences. On the other hand, psychologists, whether with Western scientific training or Eastern traditional backgrounds, have also considered that angels can represent a range of contents in the intermediate consciousness of man which lies between the personal and the transpersonal or universal. This may explain why, within the angelic frame of awareness, consciousness is so diversely characterized by the numerous classes of angels, and by "individual" beings within these, such as Michael, Gabriel, Lucifer and so many others. This frame of reference, therefore, is an implicit system of depth and height individual psychology. At the same time it expresses characteristics of social phenomena, since the world-wide diffusion of angels and related beings can be surveyed and studied in a variety of societies and cultures.
Context The greatest influences on Jewish angeology may have been from Babylonia and Persia; Parsee literature, for example, knows some 119 chief angels. But medieval Jewish, Islamic and Christian thinkers so developed the theory that individual angels were reckoned on a one-to-one-basis with the heavenly bodies, and also it would seem with the earthly, physical bodies of humans. Each soul could have its star and its angel by such perspectives.
The spheres of the star-angels reflected the psychic-projection of classifications for types of consciousness. Thus there were higher and lower, brighter and darker heavens and heavenly hosts and detailed classes that reflected human behaviour. Some classes reflected political power and organization, for example, in Christianity, *dominions* (or dominations) is a class, which is also called *lordships*. Dionysius describes this category as regulators of angels. *Princedoms* or principalities is another class, which, along with *thrones and powers* (or dynamis, potencies, potentates or authorities) bear names with a political sense. Another category belongs to this system which, while it is usually called *virtues*, corresponds mainly to the Hebrew *malakim*, the order whose name means kings.
In the Christian scheme of 9 orders of angels in the celestial hierarchy the 5 listed above are usually in the mid-range. The orders simply called *archangels* and *angels* come below, while the *seraphs* and *cherubs* maintain their Hebrew names and lead the lists. The ancient Hebrew names for most orders do not reflect political projections. Some of these are: fire-serpent (seraph), intercessor (cherub), zoa or divine animal (chayoh), wheel or chariot (ophan) and holy watchers (irin gaddisin). Moreover, individual angel names such as Raphael (the healer) and many such others indicate that "the names and presences were revealed to Israel on an experiential basis and were not elaborated at a single instant into a grand scheme".
The angelic hierarchies therefore reflect psychological, social and natural contexts whose integration was less successful than that of astrological dogma with which they have some relationship. The condition of the angelic system in the various faiths may be a psychological reflection on each. The greatest effort to codify and classify the heavenly hosts was spurred by the neo-Platonic revival in Islam, Judaism and Christianity during the Middle-Ages. Neo-Platonic philosophy provided the theory, while Aristotelianism provided the methodology and mysticism the authority. The Kabbalah, Sufism, and Dionysian-inspired Christian mysticism became the repository on the speculative side of hundreds of names of angels, and numerous classifications. The other side is the goetic, or medieval magic, whose practitioners were members of one or other of the three faiths. This had its own, very often debased, angelic literature which is still cultivated by modern occultists.
Invocations of angels, or prayers honouring angels or requesting their intercession, are still made in the Russian Orthodox, Roman Catholic and Anglican-Episcopalian branches of Christianity. Angels have been suppressed (along with saints, mysticism miracles and sacraments) in most Protestant and Judaic official worship; however, spiritually gifted authors such as Boehme, Swedenborg and Steiner among those with Protestant background, and Maimonides, Karo and Buber of Judaic faith, attest to dimensions of awareness in which the angels are encountered.
Narrower Awareness of angelic evil (#HM2212) Angelic awareness - Logos (#HM2248)
Angelic awareness - Uriel (#HM3123) Angelic awareness - Raphael (#HM2162)
Angelic awareness - Michael (#HM2198) Angelic awareness - Gabriel (#HM2222)
Angelic awareness - Zacharael (#HM3022) Angelic awareness - Zohorariel (#HM2282)
Orders of angelic awareness (Judaism, #HM2188)
Awareness of angelic-transformation (#HM3146).

♦ HM2151d Kalacakra-tantra shambhala awareness (Buddhism, Tibetan)

Description The time-wheel (kalacakra) tantra represents a state of awareness of timelessness as a place: an idyllic, never-changing Shangrilah (Shambhala). It is the yogi's goal as he loses the horizons of this world from his sight while crossing the high-land (Tibet) in his northward trek. This state of timelessness is epitomized in the representation of the "polar" palace-city, Kalopa, which is the heart of Shambhala. Below it (or "south" of it) the kalacakra tantra mandala turns, measuring out the aeons of the samsara. Shambhala is one of the magic terrestrial lands of attainment. Others are Potala Island, far to the south of Tibet, and the lotus-light palace of

Urgyan. Such places may also exist in the collective store–house (zelaya) consciousness of humanity. Their topography "constellates" psychic content.
Context In Tibetan Sakya Buddhism this is one of the states in the "Ascension Stages Game". In some sets it is numbered 59 on the board.
Broader Supreme heavens in the ascension stages game (Buddhism, Tibetan, #HM3214).
Followed by Potala–island awareness (Buddhism, Tibetan, #HM2175)
Great–path tantra awareness (Buddhism, Tibetan, #HM2656)
Tantra heat–application awareness (Buddhism, Tibetan, #HM2696)
Tantra climax–application awareness (Buddhism, Tibetan, #HM2220)
Mahayana heat–application awareness (Buddhism, Tibetan, #HM2208)
Mahayana receptivity–application awareness (Buddhism, Tibetan, #HM2247).
Preceded by Great–path tantra awareness (Buddhism, Tibetan, #HM2656)
Middle–path tantra awareness (Buddhism, Tibetan, #HM3026)
Great–vehicle higher–path awareness (Buddhism, Tibetan, #HM3048)
Tantra master in form–realm awareness (Buddhism, Tibetan, #HM2235).

♦ HM2152c Integral consciousness
Description The final stage in the evolution of consciousness in which abstraction of mental consciousness is concretized into a state of diaphanous transparence. The waking state of consciousness is not simply expanded (space–bound) but intensified. This intensification of consciousness leads to transparency of the presence. Conceptualized by Gebser, a likeness to Aurobindo's integral conception of yoga can be stated, with the characteristics of openness, freedom from time and space, aperspectivity, and inclusion of the four preceding steps in its capacity to concretize itself. Its origin lies within the presence and the present step within the evolution of consciousness of mankind.
Broader Evolution of consciousness (#HM2140).
Preceded by Mental consciousness (#HM2319).

♦ HM2153b Atman project
Description An ascent of consciousness described by K. Wilber, in which systems of cognitive, moral, ego and self development, etc, are compared. The conclusion is that, in the end, the unknown, unseen, unspoken Atman is realized to have been always there, ordinary and obvious. When this is remembered the wonder is that it was ever forgotten, or that the real was ever renounced for misery and nothingness.
Related Ego development (#HH0371) Moral development (#HH0565)
Cognitive growth and development (#HH0146).

♦ HM2154d Being myself (ICA)
Description This is experienced as being both at one with one's self, and unable to escape from one's self; when one chooses to be the human one really is. It may be compared to Mountain Rivera in the film "Requiem for a Heavyweight", when he comes to terms with the fact that his only choice is to become a wrestler.
Context In the ICA New Religious Mode, in the arena of knowing and doing intensified in one's life (life of being), the fourth formal aspect is the experience of transparent presence or of effulgence. At the first phenomenological level this occurs when a person experiences discontinuity at the centre of his life.
Broader Transparent presence (ICA, #HM2462).
Followed by Inventing humanness (ICA, #HM2092).

♦ HM2155d First scriptural bodhisattva awareness (Buddhism, Tibetan)
Bhumi — Pramudita
Description This state of awareness is characterized by freely giving out the truth of the dharma. The bodhisattva rejoices (pramudita) at this level which is represented as being in the beginning of the ten stages of his path.
Context In Tibetan Sakya Buddhism this is one of the states in the "Ascension Stages Game". In some sets it is numbered 71 on the board.
Broader Meditation way of the bodhisattvas (Buddhism, #HM2769).
Followed by Third vajra–master awareness (Buddhism, Tibetan, #HM2727)
Third scriptural bodhisattva awareness (Buddhism, Tibetan, #HM2215)
Second scriptural bodhisattva awareness (Buddhism, Tibetan, #HM2739).
Preceded by Hyper–bliss–realm awareness (Buddhism, Tibetan, #HM2337)
Amoghasiddhi–karma awareness (Buddhism, Tibetan, #HM2783)
Mahayana highest–teachings–application awareness (Buddhism, Tibetan, #HM2271).

♦ HM2156a Ultra–consciousness
Supra–consciousness — Cosmic consciousness — Universal consciousness
Description This term is used to designate the highest state of awareness to which individuals can have access. It corresponds to the state of cosmic or universal consciousness, and follows that of true self–awareness. Neither of these states occur automatically and both are comparatively rare experiences.
Related Oversoul (#HH0504) Cosmic consciousness (#HM2291)
Oversoul awareness (Psychism, #HM0827)
Higher state of consciousness (Psychism, #HM0866).

♦ HM2157d Close setting the mind (Buddhism)
Upasthapura — Nye bar jog pa (Tibetan)
Description Having abandoned distraction through recognizing it, the meditator in the state does not lose the object of meditation at all. The power of mindfulness associated with this and with the previous stage has matured.
Context This is the fourth of the nine states of mental abiding (navakara chittasthiti) in Gelugpa Tibetan Buddhism.
Broader Calm abiding (Buddhism, #HM2147).
Followed by Disciplining in mental abiding states (Buddhism, #HM2181).
Preceded by Resetting the mind (Buddhism, #HM2265).

♦ HM2158b Spiritual childhood (Christianity)
Description In Christianity, a state in which the individual is aware of God's fatherhood and his own dependence, "feeling and acting under the discipline of virtue as a child feels and acts by nature" (Pope Pius XI).

♦ HM2159d Seed state of wakefulness (Hinduism)
Description In pure consciousness, mind and jiva exist only in name.
Context The first of seven descending steps of ignorance described in the Supreme Yoga, which veil *self–knowledge*. Each has innumerable sub–divisions.
Broader Veils of delusion (Hinduism, #HM3592).
Followed by Wakefulness (Hinduism, #HM2567).

♦ HM2160c Great–vehicle middle–path awareness (Buddhism, Tibetan)
Mahayana path
Description This state of awareness is characterized by perseverance and preparation.
Context In Tibetan Sakya Buddhism this is one of the states in the "Ascension Stages Game". In some sets it is numbered 53 on the board.
Broader Mahayana Buddhism (Buddhism, #HH0900)
Sutra paths of accumulation and application in the ascension stages game (Buddhism, Tibetan, #HM2962).
Followed by Pure–lands awareness (Buddhism, Tibetan, #HM2168)
Asura–world awareness (Buddhism, Tibetan, #HM2579)
Temporary–hells awareness (Buddhism, Tibetan, #HM2454)
Discipleship–vision awareness (Buddhism, Tibetan, #HM2240)
Mahayana heat–application awareness (Buddhism, Tibetan, #HM2208)
Great–vehicle higher–path awareness (Buddhism, Tibetan, #HM3048).
Preceded by Joyful–heaven awareness (Buddhism, Tibetan, #HM2022)
Great–vehicle lower–path awareness (Buddhism, Tibetan, #HM2268).

♦ HM2161c Meditation way of the hearers (Buddhism)
Shravaka — Nyan thos (Tibetan)
Description This way applies the *absorptions* to deliverence from cyclic existence. It leads to the conscious states of the bodhisattvas.
Refs Wilber, Ken *The Atman Project* (1980).
Followed by Meditation way of the bodhisattvas (Buddhism, #HM2769)
Meditation way of the solitary realizers (Buddhism, #HM2709).
Preceded by Meditation way of the four truths (Buddhism, #HM2245).

♦ HM2162c Angelic awareness - Raphael
Description This mode of awareness may be considered the over–seeing eye of the inner nature that controls and sustains the autonomy of the psyche by an integrative and regenerative presence. Raphael has been represented as the great healing angel, and in post–biblical religion he is among the six or seven angels who most notably represent the deep structure of the psyche. Raphael is represented as bringing to those who are wise the great magical seal by which all the impulses of the lower nature may be subdued; and he is also depicted as the guardian of the Tree of Life.
Broader Angelic frame of awareness (#HM2150).

♦ HM2163c No–mind (Zen)
Wu–hsin — Mushin — Primordial awareness — Original mind
Description This expression is used in the martial arts to describe the tranquillity and freedom of the fundamental state of meditative practice, when the mind is totally open and free, with no distinction between the self and others. Mind and body, external and internal, are harmonized so there is no distortion of perception and no attachment.
Rather than the duality where pure mind is looked at and its pure light received, here there is no duality, simply pure seeing, when there is looking "as" reality as opposed to "at". In this looking there is no self, no attitude, no point–of–view. It is a state on non–ego, selflessness.
Context According to Bodhidharma, equivalent to wu–nien.
Refs Suzuki, D T *The Zen Doctrine of No–Mind* (1970).
Broader Martial arts (#HH0085) Human development (Zen, #HH1003).
Related No–thought (Zen, #HM2500) Non–abiding (Zen, #HM3485)
Mushin (Japanese, #HM3554) Formlessness (Zen, #HM0910)
Alaya–vijnana (Buddhism, #HM2730).

♦ HM2165d Splendid vices (ICA)
Description This is a passionate affirmation of one's own individuality. It occurs when one recognizes that even vices and neuroses are a part of life, and is like smiling at yourself when you catch yourself being incorrigibly yourself.
Context In the ICA New Religious Mode in the arena of acting out one's freedom (the life of prayer) the second formal aspect is affirmation of one's dependence or gratitude. At the second phenomenological level, this occurs on recognizing one's own full participation in one's fate.
Followed by Manifold blessings (ICA, #HM2269).
Preceded by Daring embracement (ICA, #HM2009).

♦ HM2166a God–consciousness
Description At the end of the individuation process (Jung) or as the culmination of self–actualization (Maslow) the creative person may gain experiences which give him the feelings of an omnipotent power, an ability to magically transform things, super–knowledge, and a perception that he empathizes with or pervades all things and witnesses all things. He has been possessed, as it were, by a god. To the extent that the divine attributes dwell in him his state can be described, for example, by the names Zeus, Apollo, Dionysos, and Hermes–Mercury. On a different level, the culmination of religion may be seen as the awareness of unity or one–ness with the focus of that religion whether Christ, Buddha, Krishna, Ishvara.
Narrower Krishna consciousness (#HM2300) Ishvara–consciousness (Yoga, #HM2913)
God–consciousness as power (#HM2760) God–consciousness as rapture (#HM2390)
God–consciousness as buoyancy (#HM2136) God–consciousness as adapability (#HM2393).
Related Unity consciousness (#HM3193) Divine consciousness (#HM0563).

♦ HM2167d Joy–land awareness (Buddhism, Tibetan)
Sukhavati
Description This state of awareness is characterized both by compassion, transmuted from passion, and discriminative understanding. It is represented as the pure land or Buddha–field of Amitabha and his boundless light. It is said to be that state attainable at the time of death, when the believer's aggregate of cosncious existence is drawn from the heart, in size about that of a small pearl, and dissolved in the heart of the Buddha's bliss and light. From there it will emanate to take birth from a lotus flower in Joy–land. Amitabha's subordinate budhisattva is said to be the Lord of Compassion, Avalokitesvara.
Context In Tibetan Sakya Buddhism this is one of the states in the "Ascension Stages Game". In some sets it is numbered 77 on the board.
Broader Supreme heavens in the ascension stages game (Buddhism, Tibetan, #HM3214).
Followed by Magical–forces awareness (Buddhism, Tibetan, #HM2679)
Third vajra–master awareness (Buddhism, Tibetan, #HM2727).
Preceded by Tantra receptivity–application awareness (Buddhism, Tibetan, #HM2756)
Mahayana receptivity–application awareness (Buddhism, Tibetan, #HM2247)
Mahayana highest–teachings–application awareness (Buddhism, Tibetan, #HM2271).

♦ HM2168d Pure–lands awareness (Buddhism, Tibetan)
Suddhavasa
Description This state of awareness is characterized by well–being and is represented as the highlands of the form–realm where five levels of gods exist, high about Mount Meru. Actually, the fifth state, the unsurpassed (akanistha), in the Game of Ascension, is not classed here but higher, as a "supreme heaven".
Context In Tibetan Sakya Buddhism this is one of the states in the "Ascension Stages Game". In some sets it is numbered 37 on the board.
Broader Form–realm consciousness (Buddhism, #HM2257).
Related Nembutsu (Japanese, #HH0683).
Followed by Non–emanating consciousness (Buddhism, Tibetan, #HM2070)
Divine–animal–hell awareness (Buddhism, Tibetan, #HM2007)

Great-vehicle lower-path awareness (Buddhism, Tibetan, #HM2268)
Great-vehicle higher-path awareness (Buddhism, Tibetan, #HM3048)
Mahayana highest-teachings-application awareness (Buddhism, Tibetan, #HM2271).
Preceded by Form-realm awareness (Buddhism, Tibetan, #HM2142)
Arhat sanctity awareness (Buddhism, Tibetan, #HM3024)
Discipleship-karma awareness (Buddhism, Tibetan, #HM2192)
Pratyeka Buddha arhat awareness (Buddhism, Tibetan, #HM2776)
Great-vehicle middle-path awareness (Buddhism, Tibetan, #HM2160)
Pratyeka Buddha cultivation awareness (Buddhism, Tibetan, #HM2252).

♦ HM2169g Ear consciousness (Buddhism)
Shotravijnana — Rna-ba'i-rnam-par-shes-pa (Tibetan) — Hearing awareness
Description This is the consciousness that apprehends objects of hearing, whether articulate or inarticulate, pleasant or unpleasant, and whether or not caused by elements conjoined with consciousness. It is twofold, that is: where ear consciousness is pleasant it is associated with indeterminate resultant consciousness with profitable result; where it is painful, it is associated with indeterminate resultant consciousness with unprofitable result.
Context One of the six consciousnesses defined in Buddhism as dependent on the individual senses, and with objects of sense as their focus.
Broader Sense consciousness (Buddhism, #HM2664)
Perception through the senses (#HM3764)
Mundane resultant consciousness (Buddhism, #HM0130)
Sense mode of consciousness occurrence (Buddhism, #HM4389)
Indeterminate consciousness in the sense sphere - resultant (Buddhism, #HM5721)
Awareness of consciousness-group of conscious existence - senses and mind (Buddhism, #HM2556).
Related Eye consciousness (Buddhism, #HM2074)
Nose consciousness (Buddhism, #HM2364)
Body consciousness (Buddhism, #HM2562)
Tongue consciousness (Buddhism, #HM2263)
Mental consciousness (Buddhism, #HM2838)
Mind-consciousness element (Buddhism, #HM6173).

♦ HM2170c Mountain of care (ICA)
Description An area in the ICA *Other World in the midst of this World* characterized by being of service, of the world and of agape.
Broader Other world in the midst of this world (ICA, #HM2614).
Narrower Singular mission (#HM3217) Transparent power (#HM2828)
Universal concern (#HM2774) Original gratitude (#HM3105).
Related Land of mystery (ICA, #HM2434) Sea of tranquillity (ICA, #HM3033)
River of consciousness (ICA, #HM2993).

♦ HM2171c Only-a-beginner subtle contemplation (Buddhism)
Las-dang-po-pa-tsam-kyi-yid-byed (Tibetan) — Manaskaradhikarmika
Description This constituent of the mental state preparatory to the first concentration is the same as the calm abiding state. It is regarded, however, not as a terminus of a meditation sequence, but as a transition phase, along with the other six; and with preparations (samantaka) and special insight (vipashyana), leading further. The yogi meditates on the selflessness of the person, watching for that "I" to appear which is fused with the idea of mind and body. The appearance of "I" is analyzed and seen to be non-existent.
Context This is the first of the preparations in Tibetan Gelugpa Buddhism, and the first stage of meditation on emptiness.
Broader Selflessness (Buddhism, #HM2016) Emptiness (Buddhism, Zen, #HM2193)
Meditative states of mental abidings (Buddhism, #HM2145).
Followed by Calm abiding (Buddhism, #HM2147).

♦ HM2172d Cessation-awareness (Buddhism, Tibetan)
Nirodha
Description This state is characterized as an escape from cyclic existence (samsara). However, it also implies the reality of cessation. The higher nirvana is dynamic so that bodhisattvas do not distinguish between the reality of nirvana and samsara. Hence arhats in cessation are represented as being roused by the boundless light of the Buddha Amitabha, so that they may advance into the mahayana and realize the bodhisattva ideal in the Great Vehicle.
Context In Tibetan Sakya Buddhism this is one of the states in the "Ascension Stages Game". In some sets it is numbered 48 on the board.
Broader Religious traditions in the ascension stages game (Buddhism, Tibetan, #HM3341).
Related Nirodha parinama (Yoga, #HM3437) Nirodhasamapatti (Buddhism, #HM6346)
Attainment of cessation (Buddhism, #HM5438).
Preceded by Arhat sanctity awareness (Buddhism, Tibetan, #HM3024)
Pratyeka Buddha arhat awareness (Buddhism, Tibetan, #HM2776).

♦ HM2173b Depth consciousness (Jung)
Description It is recognized that the psychology of Jung has penetrated even deeper into the unconscious than has that of Freud. Jung's psychology affirms the ego-modes of consciousness while also recognizing the limitations of such modes and the possibility of transcending them; and points to the reality of the creative tension inherent in the ego-self axis.
Refs Adler G, Fordham M and Read H (Eds) *The Collected Works of C G Jung*; Marlan, Stanton *Depth Consciousness*; Samuels, Andrew et al *A Critical Dictionary of Jungian Analysis* (1986).
Related Analytical psychology (#HH1420).

♦ HM2174d Spiritual denial (ICA)
Description The fourth formal aspect of the *life of poverty*, an arena in the ICA *New Religious Mode*, this is experienced as sacrificial offering.
Broader Life of poverty (ICA, #HM2299).
Narrower Human contingency (#HM2104) Defender of deeps (#HM2591)
Intentional self-negation (#HM2369) Awareness of spiritual poverty (#HM2286).

♦ HM2175d Potala-island awareness (Buddhism, Tibetan)
Description This state of awareness is characterized as a consciousness of ancestral and racial heritage and a search for origins. It is represented to Tibetans as an island south of the Indian sub-continent where the Tibetan progenitors, the compassion deities Avalokitesvara and Tara, dwell. By going to this island one recognizes the innate compassion buried deeply in human character. One ascends Potala Mountain to the lake of the "god-who-looks-down" (Avalokitesvara) and his stone image. Evoking the real forceful presence of the divinities, however, is said to be possible only by intense meditation. The Lhasa mountain palace has been named after Potala.
Context In Tibetan Sakya Buddhism this is one of the states in the "Ascension Stages Game". In some sets it is numbered 60 on the board.
Broader Supreme heavens in the ascension stages game (Buddhism, Tibetan, #HM3214).
Followed by Great-path tantra awareness (Buddhism, Tibetan, #HM2656)
Mahayana receptivity-application awareness (Buddhism, Tibetan, #HM2247)
Mahayana highest-teachings-application awareness (Buddhism, Tibetan, #HM2271).
Preceded by Middle-path tantra awareness (Buddhism, Tibetan, #HM3026)
Great-vehicle higher-path awareness (Buddhism, Tibetan, #HM3048)
Kalacakra-tantra shambhala awareness (Buddhism, Tibetan, #HM2151).

♦ HM2176g Spatial awareness
Visual consciousness
Description The human being has been described as inhabiting the space of the world, conscious of all that is around (and therefore not limited by) his physical body, in whatever sense modality or altered state of consciousness he may be aware.
Every form of consciousness may be described as *openness to the world*, but *visual consciousness* is of particular interest in that one may choose whether or not to be aware by simply opening or closing the eyes, by looking or turning away. The objects of visual consciousness do not therefore intrude on the body in the way that sounds, tastes, smells and touch do. The awareness of the distance of objects which may in the future be closer (as they are approached) or further away (as they are left behind) implies a looking into, or vision of, the future.
Broader Openness (#HM2770).
Related Eye consciousness (Buddhism, #HM2074).

♦ HM2177b Consciousness states in cyclic existence (Buddhism)
Samsara — Khor-ba — Tridhatu — Khams-gsum (Tibetan)
Description States of consciousness vary in the three realms and nine levels of conscious existence. The cycle of existence spans the Desire Realm, the Form Realm and the Formless Realm, all within the samsara. Beyond is the unchanging.
Context Tibetan Gelugpa Buddhism provides a full analysis of all the states of consciousness in the samsara.
Narrower Setting the mind (#HM2217) Form-realm consciousness (#HM2257)
Desire-realm consciousness (#HM2733) Formless-realm consciousness (#HM2281)
Meditative states of mental abidings (#HM2145)
Beyond-the-third-realm consciousness (Buddhism, Tibetan, #HM2653).
Related Samsara (Hinduism, #HM1006)
Worlds of conscious existence (Buddhism, Pali, #HM2072).

♦ HM2178c Measuring emblems (Tarot)
Justice
Description Among the archetypal emblems projected by the psyche are those that depict measuring, weighing, testing and analysis. One of the common elements to these operations is comparison against a standard, and traditional depictions of the *weighing of the soul* indicate one aspect of the emblems' meaning. In the Tarot system the emblem called "Justice" (number 8) has the conventional, seated female figure with scales in one hand, and the sword of discernment in the other. No further progress can be made on the path until retribution has been made for previous wrongdoing. The worldly aspects of self are overcome on the path towards integration with the higher, spiritual self.
Context In the Kabbalah, Justice is placed on the 22nd Path between and beauty (Tiphareth) and force or judgement (Geburah).
Broader Emblem archetypes in the psyche (Tarot, #HM2201)
Tarot arcana of conscious states (Tarot, #HM2097).
Related Justice (#HH2117) Sphere of beauty (Kabbalah, #HM3031)
Sphere of judgement (Kabbalah, #HM2290).

♦ HM2179d Wish-dominated awareness
Wishful thinking
Description In this mode of awareness a desired object or condition is visualized as being enjoyed, giving rise to pleasurable sensations and thus to reinforcement for prolonging the wish-gratifying fantasy. Children, the immature, schizoid personalities and those suffering from autistic behaviour may indulge in wishful thinking, particularly as a form of compensation for deprivation, impotence or inadequacy.
Related Daydream (#HM2138).

♦ HM2180d Pratyeka Buddha awareness (Buddhism, Tibetan)
Great awakening
Description This state of awareness, characterized as the great awakening, is represented as the karmic reward for individuals who follow the Dharma through ages when no Buddha teaches in the world. There are five stages: entering the solitary vehicle; path of application; path of vision; path of cultivation; and arhatship. These are surpassed by the states of Nirvana.
Context In Tibetan Sakya Buddhism this is one of the states in the "Ascension Stages Game". In some sets it is numbered 43 on the board.
Broader Religious traditions in the ascension stages game (Buddhism, Tibetan, #HM3341).
Followed by Naga-world awareness (Buddhism, Tibetan, #HM2031)
Reviving-hell awareness (Buddhism, Tibetan, #HM2516)
Discipleship-karma awareness (Buddhism, Tibetan, #HM2192)
Thirty-three-god-heaven awareness (Buddhism, Tibetan, #HM2606)
Great-vehicle lower-path awareness (Buddhism, Tibetan, #HM2268)
Pratyeka Buddha application awareness (Buddhism, Tibetan, #HM3020).
Preceded by Bon-wisdom awareness (Buddhism, Tibetan, #HM2663)
Discipleship-application awareness (Buddhism, Tibetan, #HM2716)
Delightful-emanation-heaven consciousness (Buddhism, Tibetan, #HM2546)
Eastern-continent awareness of noble figures (Buddhism, Tibetan, #HM2543).

♦ HM2181d Disciplining in mental abiding states (Buddhism)
Damana — Dul-bar-byed-pa (Tibetan)
Description In this stage the overly-quiescent mind is rescued from laxity and disciplined by strong introspection, the associated power. The meditator derives joy from clairvoyance and the ability to penetrate the meaning of difficult topics.
Context This is the fifth of the nine states of mental abiding (navakara chitta sthiti) in Gelugpa Tibetan Buddhism.
Broader Calm abiding (Buddhism, #HM2147).
Followed by Pacifying in mental abiding states (Buddhism, #HM2205).
Preceded by Close setting the mind (Buddhism, #HM2157).

♦ HM2182d Psychic helplessness
Description According to Freud, this is the state which produces anxiety in early infancy and which mirrors the physical reactions (high pulse rate, rise in blood pressure and so on) which were responses to the unpleasant sensations of birth. In contrast to anxiety in later life produced in response to some clear danger, psychic helplessness occurs when needs (food, warmth) which the infant is powerless to gratify remain unanswered. Psychic helplessness reduces as the ego matures, and persists only in certain stituations which may be related to a particular neurosis.
Related Anxiety (#HM2465).

♦ HM2183a Perfect man (Hinduism)
Bodily perfection
Description This state may be achieved when one realizes one's essential one-ness with God. It may be compared with Patanjali's description of *bodily perfection*, referring not to the physical

body but to the bodies on all planes: symmetry in form, beauty of colour, and the strength and compactedness of a diamond.
Related Human perfectibility (#HH0212) Spiritual development (#HH0017).

♦ **HM2184d Transparent lucidity** (ICA)
Description The second formal aspect of the *life of being*, an arena in the ICA *New Religious Mode*, this is the awareness of divination.
Broader Life of being (ICA, #HM3229).
Narrower Virgin birth (#HM2666) Heavenly secret (#HM2918)
Prophetic sight (#HM2846) Trust intuitions (#HM2761).

♦ **HM2185b Way of the Buddhas** (Buddhism, Tibetan)
Description Each buddha passes through the four stages of trance in the Form–realm. Then he rises through the four equalizations in the Realms of Formlessness. He achieves the balance and repose of cessation at the Peak of Existence. He then passes in reverse back through the eight stages. He moves upward and downwards in and out of nirvana and sangsara doing the work of buddhas or acting in their rest from work.
Related Enjoyment–body awareness (Buddhism, Tibetan, #HM2873)
Beyond–the–third–realm consciousness (Buddhism, Tibetan, #HM2653).
Preceded by Meditation way of the bodhisattvas (Buddhism, #HM2769).

♦ **HM2186d Primordial wonder** (ICA)
Description This state occurs when an individual experiences the wonder of his own existence. It is not something one conjures up, but is rather what is experienced whenever one opens one's eyes to the amazing reality of existence and of one's existence. People speak of being in "a cloud of awe" or of "being overcome by the wonder of it all." The photographs of the earthrise from the moon invoked such a feeling in many people. It may be compared with Moses before the burning bush or someone standing in a great temple: such moments of transfixing wonder are often experienced.
A dimension of this experience is a sense of awe, that is fear and fascination. The fear is of not knowing what is going to happen; one is afraid that one is the only person having such an experience and that it will never go away. Once having discovered the wonder of it all one is trapped forever over the unfathomable abyss. The fascination is that something wonderful is happening which is going to bring a great change to one's life. The decision is to walk with one's head in a cloud of awe, to trust what is happening and embrace everything that is going on. After such an experience one cannot erase it, one is alerted to everything as significant, and is permanently ready for such moments to recur again and again.
Context This state is number 4 in the ICA *Other World in the midst of this World.*
Broader Aweful encounter (ICA, #HM2619).
Followed by Total exposure (ICA, #HM2764) Incarnate living (ICA, #HM2394)
Dynamic selfhood (ICA, #HM3091) Perpetual becoming (ICA, #HM2717)
Sacramental universe (ICA, #HM2445).
Preceded by Ultimate reality (ICA, #HM2030) Dynamic selfhood (ICA, #HM3091)
Personal epiphany (ICA, #HM2595) Singular adoration (ICA, #HM2111)
Sacramental universe (ICA, #HM2445).

♦ **HM2187d First vajra–master stage** (Buddhism, Tibetan)
Vajracarya
Description This state of awareness is considered to be among the ten bodhisattva stages on the tantric path and is the first initiation of the ten said to confer visions of Buddha in the many buddha fields.
Context In Tibetan Sakya Buddhism this is one of the states in the "Ascension Stages Game". In some sets it is numbered 66 on the board.
Broader Way of the vajra masters (Buddhism, Tibetan, #HM3090).
Followed by Third vajra–master awareness (Buddhism, Tibetan, #HM2727)
Fourth vajra–master awareness (Buddhism, Tibetan, #HM2251)
Second vajra–master awareness (Buddhism, Tibetan, #HM2703).
Preceded by Tantra climax–application awareness (Buddhism, Tibetan, #HM2220)
Tantra receptivity–application awareness (Buddhism, Tibetan, #HM2756)
Tantra union–in–learning–application awareness (Buddhism, Tibetan, #HM2280).

♦ **HM2188b Orders of angelic awareness** (Judaism)
Seraphim
Description This mode of awareness, under the name of seraphim, is represented as placed before the divine throne of glory and characterized as the consciousness of the universal force of fiery–love and its light. Under the name of hayyoth these angels are said to be found in the seventh heaven, although a lower class of hayyoth are considered equivalent to the cherubim. The hayyoth are the camp or company of the shekinah who receive the holy effluence from above and disseminate it to the "movers of the cosmic wheels" below. The force of this level of consciousness is said to uphold the universe and support the throne of glory. The first order of angelic intelligence is characterized by the qualities ascribed variously to the personifications called: Michael, Satan (before his fall), Metatron, and sometimes, Uriel.
Broader Angelic frame of awareness (#HM2150).

♦ **HM2189c Archaic consciousness**
Description This consciousness is conceived of as zero–dimensional structure, the earliest state within the evolution of consciousness, characterized by dreamless time and complete indistiction between man and the universe. Conceptualized by the cultural anthropologist J Gebser, archaic consciousness may be localized within mythologies promoting the androgynous image of man.
Broader Evolution of consciousness (#HM2140).
Followed by Magical consciousness (#HM2090).

♦ **HM2190c Victimization emblems** (Tarot)
Hanged man
Description Among the images that the psyche may project on itself or in outward representation are those symbolizing personal misfortune of the must grievous, although not necessarily mortal, kind. The emblems of the archetypal victim are one example of this: the scapegoat, the sacrificial lamb or the crucifix; extending to all instruments of public execution: nooses, guillotines, and various chambers of death. In the Tarot System the emblem called the Hanged Man (number 12) is represented by a hanging man with a rope around his feet rather than his neck, as he is suspended upside down. One could interpret him also as the trapped man. But his head is illuminated as though it were a cup to be filled with higher, spiritual energies from Binah, the great mother. If this happens, he will reflect the transendent purity from beyond the abyss.
Context The hanged or hanging man is placed by Kabbalists in the path between glory and splendour (Hod) and strength or judgment (Geburah).
Broader Emblem archetypes in the psyche (Tarot, #HM2201)
Tarot arcana of conscious states (Tarot, #HM2097).
Related Sphere of glory (Kabbalah, #HM2238) Sphere of judgement (Kabbalah, #HM2290).

♦ **HM2191d Fourth scriptural bodhisattva awareness** (Buddhism, Tibetan)
Arcis–mati — Flaming enlightenment
Description This state of awareness is characterized by practice of the middle way between philosophies of destructive asceticism and those of self–indulgence, and between doctrines of non–reality and those of reality of phenomena. This is called the flaming (arcis–mati) enlightenment. The bodhisattva has passed the need for beautiful expression of the Dharma, and passed the wrangle of schools. No partisan, he welcomes all followers of the Buddha's path, all representations of truth.
Context In Tibetan Sakya Buddhism this is one of the states in the "Ascension Stages Game". In some sets it is numbered 78 on the board.
Broader Meditation way of the bodhisattvas (Buddhism, #HM2769).
Followed by Sixth scriptural bodhisattva awareness (Buddhism, Tibetan, #HM2385)
Fifth scriptural bodhisattva awareness (Buddhism, Tibetan, #HM2909).
Preceded by Jewelled–peaks–realm awareness (Buddhism, Tibetan, #HM2275)
Third scriptural bodhisattva awareness (Buddhism, Tibetan, #HM2215)
Second scriptural bodhisattva awareness (Buddhism, Tibetan, #HM2739).

♦ **HM2192d Discipleship–karma awareness** (Buddhism, Tibetan)
Sravaka–sambhara–marga
Description This state of awareness is characterized as a need to join the spiritual community, to worship regularly and do good works as a karmic preparation for higher practices.
Context In Tibetan Sakya Buddhism this is one of the states in the "Ascension Stages Game". In some sets it is numbered 38 on the board.
Broader Religious traditions in the ascension stages game (Buddhism, Tibetan, #HM3341).
Followed by Pure–lands awareness (Buddhism, Tibetan, #HM2168)
Animal–hell awareness (Buddhism, Tibetan, #HM2636)
Crushing–hells awareness (Buddhism, Tibetan, #HM3142)
Discipleship–vision awareness (Buddhism, Tibetan, #HM2240)
Discipleship–application awareness (Buddhism, Tibetan, #HM2716)
Great–vehicle lower–path awareness (Buddhism, Tibetan, #HM2268).
Preceded by Formless–realm awareness (Buddhism, Tibetan, #HM3144)
Pratyeka Buddha awareness (Buddhism, Tibetan, #HM2180)
Tantra–beginner awareness (Buddhism, Tibetan, #HM2452)
Southern–continent awareness of men (Buddhism, #HM2127)
Hindu–wisdom–holder awareness (Buddhism, Tibetan, #HM2723)
Great–vehicle lower–path awareness (Buddhism, Tibetan, #HM2268)
Western–continent awareness of cattle (Buddhism, Tibetan, #HM2519)
Eastern–continent awareness of noble figures (Buddhism, Tibetan, #HM2543).

♦ **HM2193c Emptiness** (Buddhism, Zen)
Shunyata — Sunyata — Voidness
Description This state is not attainable since, although it is always with one it vanishes as soon as one tries to hold it before one's eyes. It is beyond perception, beyond grasping, beyond being and non–being. If it were absolutely beyond all human attempts to take hold of it in any sense it would not have any value, nor come in the sphere of human interest. However, on the contrary, the fact that it is always in one and withone conditions all knowledge, all acts, life itself.
In Buddhism, the state called *Shunyata* can indicate an attitude towards, or a high stage of insight into, the emptiness of all phenomena. As an attitude, it frees the mind from attachment. In terms of its ontological dimension, it expresses reality as perceived in the nirvana or lokuttara states – the final mode of existence of all phenomena. It is said to be achieved through perfection of wisdom and omniscient liberation from cyclic existence.
The path of meditation on emptiness is based on the realization of emptiness in persons, based on the 20 false views of the real self: in viewing the form as self, forms as existing in self, self as inherently possessing form, as inherently existing in form; and similarly for feelings as self, discrimination as self, compositional factors as self and consciousness as self.
Refs Nishitani, Keiji *Religion and Nothingness* (1982).
Broader Consciousness in Buddhism (Buddhism, #HM2200).
Narrower Calm abiding (Buddhism, #HM2147)
Only–a–beginner subtle contemplation (Buddhism, #HM2171)
Special insight and preparations as states (Buddhism, #HM2623)
Tantra in meditation on emptiness (Buddhism, Tibetan, #HM2864)
Tibetan meditative states of form concentrations (Buddhism, #HM2693).
Related Nirvana (Buddhism, #HM2330) Essential wisdom (#HM0107)
Madhyamaka (Buddhism, #HH1038).

♦ **HM2195d Individual knowledge of the character** (Buddhism, Tibetan)
Subtle contemplation
Description This state, belonging to the preparations for each of the four concentrations on the Desire Realm (kamadhatu, dod khams) is an analytical contemplation of the preceeding concentrations, viewing them all as unsatisfactory to eliminate attachments.
Context This is the second of the preparations in Tibetan Gelugpa Buddhism.
Broader Special insight and preparations as states (Buddhism, #HM2623).

♦ **HM2196b Moksha** (Hinduism)
Liberation consciousness — Truth consciousness — Moksa
Description In ancient Hindu philosophy and religion, the ultimate goal of human development is frequently expressed as liberation (moksha). Unlike Kant, who denied that man could see things as they truly are, the teachings of yoga speak of a transcendental state of knowing beyond the intellect's habitual division into subject and object. Transcendental awareness or being extends and surpasses the bounds of ordinary awareness. Rather than knowing, this awareness is termed *realization.* Moksha is said to be realized through loving attachment to God, *bhakti*, or through unconditional offering of the self, *prapatti.* It is compared with liberation as *kaivalya*, the goal of Patanjali's yoga, which is consciousness of the self alone without realization of the "lord".
Although the intellect in its ordinary state is enslaved by desires, feelings and attitudes so that it cannot see true reality, not all methods to attain moksha are mystical, magical or anti–rational. In the nyaya darshana (method–philosophy), for example, dialectic is the key to salvation, through a rigorous system of logic. This philosophical development was parallelled and anticipated in Greece where the mystical liberation taught by Orpheus was supplemented by the dialectical training sponsored by the cult of Delphic Apollo; the goal being *truth.* In Christianity, the Bible also expresses the same concept of liberation, "You shall know the truth, and the truth shall make you free". Logic and dialectic rose to paramount places in the medieval Catholic church (Abelard, St Thomas, etc) alongside theology; and in another cultural parallel, they developed at the same time in Buddhism. The canon of Tibetan Buddhism, for example, contains extensive treatment of logic, codified at the same time that St Thomas Aquinas was writing his works. Liberation conceived as truth seems to be an innate idea of mankind that has been expressed variously at various times. The way of reason or dialectic is discovered by mystics intuitively, and moksha is perceived as the end–purpose of science.
Moksha is sometimes included with *dharma*, *artha* and *kama* as one of the great aims of life.
Refs Pramod Kumar *Moksa, The Ultimate Goal of Indian Philosophy.*
Broader Emancipation of the self (#HH0907).

Related Mukti (#HH0613) Dharma (#HH0093) Salvation (#HH0173)
Liberation (#HH0388) Soteriology (#HH1166) Kaivalya (Yoga, #HM3869)
Kama (Hinduism, #HH0754) Artha (Hinduism, #HH0353) Vimoksa (Buddhism, #HM5977)
Samsara (Hinduism, #HM1006) Self-development (#HH0651)
Bhakti marga (Hinduism, #HH0628) Treasures of the godly (#HH0364)
Human development (Hinduism, #HH0330)
Indian approaches to consciousness (Hinduism, #HM2446).

♦ **HM2198c Angelic awareness - Michael**
Description This mode of awareness may be considered the highest consciousness viewed in its dynamic mediating function between personal and transcendent being. The mind said to be "liberated by Michael" may rather release Michael from within itself as fire and light. As the "first" or foremost angel, this mode has been considered the expression of an utterly transcendent God (for example, the name Michael means, "God-like"), and thus corresponds to the ultimate threshold of personal consciousness. Michael has been regarded as the greatest of all angels, being considered chief of the Seraphim (the first order of angels), chief of the archangels, and prince of the Presence. He has been identified with Metatron and closely identified also with the Shekinah. The attributes assigned to Michael of active fire and light allow for his purging and destructive powers. Fire is alo the attribute of Metatron, the other personification of the supreme angel, who is also called "Moses' rod", a power giving both life and death.
Context Some Kabbalists affiliate Michael to tiphereth. In Islam, Mikal is somewhat subordinate to Djabrail (Gabriel) in role, although not hierarchially.
Broader Angelic frame of awareness (#HM2150).

♦ **HM2199d Affirmation** (ICA)
Description The second formal aspect of the *life of prayer*, an arena in the ICA *New Religious Mode*, this represents a positive statement of one's dependence and gratitude.
Broader Life of prayer (ICA, #HM2511).
Narrower Besetting sin (#HM2379) Heavenly sorrow (#HM2216)
Daring embracement (#HM2009) Personal violation (#HM2561).

♦ **HM2200b Consciousness in Buddhism** (Buddhism)
Self-consciousness
Description Buddhism denies reality to a supreme, conscious self in the universe; but it does not deny the apparently real necessity of attaining truth through use of the mind.
Consciousness is defined as apprehension of an object. In particular, mental phenomena apprehend special states of consciousness. In all cases consciousness is self-conscious - clearly, what is experienced is present to the mind. All mental phenomena must be conscious of their own existence. This feeling of own existence is referred to as direct or immediate cognition. When something external is perceived there is an internal emotion responding to it which is not the same as the object perceived. The internal emotion is a property of the self, not of the perceived object. This experience is knowledge; and it shows that we do experience our own knowledge. Self-consciousness is direct knowledge which makes our own self present to us.
Transcendental reality can be elicited. Contemplating this reality forces it into consciousness until the mind, which contains the image of the object contemplated, commences to achieve a condition of clarity. This is the first condition, when contemplation is in progress. The direct perception of the saint is the second condition when clarity is nearly complete, when progress towards it is still being made and when reality is veiled by the thinnest of clouds. Finally, there is direct knowledge of reality when it is perceived totally clearly.
Context There are distinctive features to the Buddhist path according to where it is practiced. In some countries the greater vehicle is taught, in others the lesser vehicle. There are the abrupt paths, the Tantra, Zen, Yogacarya, and the esoteric sects. All teach complex yogas and seek a nirvana which surmounts a number of states of consciousness.
Refs Stcherbatsky, T *Buddhist Logic* (1962).
Broader Human development (Buddhism, #HH0650).
Narrower Nirvana (#HM2330) Alaya-vijnana (#HM2730) Mental sensation (#HM6644)
Sense consciousness (#HM2664) Emptiness (Buddhism, Zen, #HM2193)
Ignorance in dependent origination formula (#HM3035)
Worlds of conscious existence (Buddhism, Pali, #HM2072).
Related Bodhi-pakkhiya dhamma (Buddhism, #HM2793)
Mystical Christ-consciousness (Christianity, #HM2880)
Indian approaches to consciousness (Hinduism, #HM2446).

♦ **HM2201b Emblem archetypes in the psyche** (Tarot)
Image awareness
Description The use of individual emblems in the Tarot system demonstrates one of many attempts to give direct, visual form to the archetypes representing phases in psychic transformation. Such emblems assist in concentrating the mind and holding the attention, directing the thoughts towards the mystery of unity and keeping them on the path towards such unity.
Narrower Death emblems (#HM3121) Measuring emblems (#HM2178)
Mutability emblems (#HM2250) Emblems of renewal (#HM2315)
Emblems of disaster (#HM2350) Emblems of totality (#HM2387)
Emblems of supremacy (#HM2411) Victimization emblems (#HM2190)
Emblems of well-being (#HM2422) Natural forces emblems (#HM3127)
Wisdom archetypal image (#HM2261) Pilgrim archetypal image (#HM2225)
Magician archetypal image (#HM2237) Life-fluid emblems - star (#HM3137)
Emblems of the mysterious (#HM2398) Hierophant archetypal image (#HM2260)
Charioteer archetypal image (#HM3045) Emblems of personified evil (#HM2378)
Old wise-man archetypal image (#HM2202) Life-fluid emblems - temperance (#HM3289)
Female sovereignty archetypal image (#HM2285) Betrothal initiation archetypal image (#HM3028).
Related Archetype (#HH0019) Analytical psychology (#HH1420)
Tarot arcana of conscious states (Tarot, #HM2097).

♦ **HM2202c Old wise-man archetypal image** (Tarot)
Hermit
Description One of the archetypal emblems projected by the psyche is the wise elder. This is frequently depicted as male, since a long beard can be attached to signify seniority. He is a patriarch whose splendour is concealed by a hooded cloak; nevertheless, this path (also called "Yod") is associated with Virgo and shows some balance of sexual polarities. In the Tarot, the emblem of the Hermit (number 9) represents, by the figure's hand-held lantern, the "midnight-oil-burning" sage or one who sees into obscure matters; it may also be seen as lighting his way as he journeys upwards towards the abyss.
Context In the Kabbalah, the Hermit is placed on the 20th path between kindness or mercy (Chesed) and beauty (Tiphareth).
Broader Emblem archetypes in the psyche (Tarot, #HM2201)
Tarot arcana of conscious states (Tarot, #HM2097).
Related Sphere of beauty (Kabbalah, #HM3031)
Sphere of mercy and greatness (Kabbalah, #HM2420).

♦ **HM2203b Objective awareness**
Objective reason — Objective consciousness
Description According to Gurdjieff, this is a meditative state when there is full awareness of external and internal phenomena. It is the maximum intensity of which man's consciousness is capable, only to be achieved by intense effort against the fantasy which is taken as reality in normal awareness.
Related Objective awareness (ICA, #HM2598).
Preceded by Self-remembering (#HM2486).

♦ **HM2204d Enchantment**
Disenchantment
Description When the world is seen as unsatisfactory and the present moment as untidy and incomplete, then the goal is always to strive for a future when external conditions are a fundamental improvement on those prevailing now. This disenchantment is said to be a by-product of rationalization and estrangement from nature. In contrast, a sense of purposeful belonging to a unified and spiritually encompassing creation is typical of more primitive societies, where individuals have a deeper and more complete understanding of and control over their surroundings than is possible for someone living in a complex, scientific and technologically developed environment. Enchantment with life can be recaptured in an activity such as mountaineering, when one creates the rules and understands fully each detail of the process, having chosen the environment rather than having had it imposed from outside.
Related Flow experience (#HM2344).

♦ **HM2205d Pacifying in mental abiding states** (Buddhism)
Shamana — Zhi-bar-byed-pa (Tibetan)
Description In this state the mind, revivified in the preceeding condition, is subtly excited. It is stilled with the aid of the power of introspection which is matured on this level, so that such subtle excitement is less frequent; there is little danger now that subtle laxity will arise.
Context This is the sixth of the nine states of mental abiding (navakara chittasthiti) in Gelugpa Tibetan Buddhism.
Broader Calm abiding (Buddhism, #HM2147).
Followed by Full pacification in mental abiding states (Buddhism, #HM2729).
Preceded by Disciplining in mental abiding states (Buddhism, #HM2181).

♦ **HM2206d Presence of Jesus Christ** (ICA)
Description This is the realization that everything contains the life-giving word and that, at the same time, one is always called to be that word. One has to reinvent what Jesus is, which gets the Christ word articulated to the situation in which one finds one's self.
Context In the ICA New Religious mode in the arena of articulating the word about life (the life of knowing) the second formal aspect is one's experience of the vanishing cosmos or the wonder of the world. At the fourth phenomenological level, this occurs when one recognizes the humiliation of standing before the final mystery of life in every moment.
Broader Wonder at the world (ICA, #HM3062).
Preceded by Sacramental universe (ICA, #HM2366).

♦ **HM2207d States of special insight** (Buddhism)
Vipashyana — Lhag-mthong (Tibetan)
Description Special insight is wisdom. Any of its sub-species, kinds of special insights, may facilitate calm abiding or arise as a result of calm abiding. Special insight realizing emptiness (shunyata) or the four truths (satya) may be included in the preparations for the first concentration.
Broader Special insight and preparations as states (Buddhism, #HM2623).

♦ **HM2208c Mahayana heat-application awareness** (Buddhism, Tibetan)
Dharmaloka-labdha-samadhi
Description This state of awareness is characterized by the empty nature of reality. It is represented as an experience of the mystical light of the Dharma generated from the heat of concentration and sustained compassion for all beings. At this stage the practitioner is said to develop the first of "four root-conditions of entering".
Context In Tibetan Sakya Buddhism this is one of the states in the "Ascension Stages Game". In some sets it is numbered 55 on the board.
Broader Sutra paths of accumulation and application in the ascension stages game (Buddhism, Tibetan, #HM2962).
Followed by Mahayana climax-application awareness (Buddhism, Tibetan, #HM2232)
Mahayana receptivity-application awareness (Buddhism, Tibetan, #HM2247).
Preceded by Joyful-heaven awareness (Buddhism, Tibetan, #HM2022)
Great-vehicle middle-path awareness (Buddhism, Tibetan, #HM2160)
Great-vehicle higher-path awareness (Buddhism, Tibetan, #HM3048)
Kalacakra-tantra shambhala awareness (Buddhism, Tibetan, #HM2151).

♦ **HM2209d Abounding abasement** (ICA)
Description This is experienced as continually raising the question "What does this other person need ?", in whatever situation and whoever the other person is; when one gives up the necessity of acting out one's own standards of behaviour and propriety. It may be compared to the situation of at the end of Charles Dickens' "Tale of Two Cities".
Context In the ICA New Religious Mode in the arena of knowing one's disengagement (the life of poverty) the third formal aspect is being nonchalant about one's relations. At the third phenomenological level, this occurs when one is able to so value the gift of life that nothing else is needed for one's fulfilment.
Broader Liberation from relationships (ICA, #HM3240).
Followed by Common earth (ICA, #HM2023).
Preceded by Serious sharing (ICA, #HM2323).

♦ **HM2210c Embarrassment** (Buddhism)
Apatrapya — Khrel-yod-pa (Tibetan)
Description Related to *shame*, embarrassment is a factor in preventing misconduct which would lead to displeasure of or despising by others.
Context One of the eleven *virtuous mental factors* referred to in Tibetan Buddhism.
Broader Virtuous mental factors (Buddhism, #HH0578).
Related Sense of shame (Buddhism, #HM2881).

♦ **HM2211d Tantra master-in-sense-realm awareness** (Buddhism, Tibetan)
Kamadera-vidyadhara
Description This state of awareness is represented by the Tibetan masters in western theosophic and occult literature, in which they are also associated with Shambhala. The vidyaharas (wisdom-holders) are especially connected with the kalachakra tantra and have been known to Buddhists and Hindus as the "Siddha gods", since their powers are god-like, equal to the divinities in the realm of sense-desire (kamadevas). According to Tibetan sources the masters antedate the Kalachakra and are the source of all tantra. The master's own ruler is sometimes termed the Emperor of the World, or the Wheel-turner King.
Context In Tibetan Sakya Buddhism this is one of the states in the "Ascension Stages Game". In some sets it is numbered 67 on the board.
Broader Masters of wisdom in the ascension stages game (Buddhism, Tibetan, #HM3066).

Followed by Great-path tantra awareness (Buddhism, Tibetan, #HM2656)
Wheel-turner-king awareness (Buddhism, Tibetan, #HM2759)
Middle-path tantra awareness (Buddhism, Tibetan, #HM3026)
Tantra master in form-realm awareness (Buddhism, Tibetan, #HM2235).
Preceded by Magical-forces awareness (Buddhism, Tibetan, #HM2679)
Middle-path tantra awareness (Buddhism, Tibetan, #HM3026).

♦ HM2212c Awareness of angelic evil

Azazel — Satan — Abaddon

Description This mode of awareness is focused on the destructive elements in the psyche. The Luciferian angeology projects a drama of fallen human nature in which misapplication and misdirection of libidinous energy is in the forefront. Hidden in the symbolism is perhaps a more important theme, one of dislocation, the suggestion being that some human faculty or organ is not in its proper place (as Plato wrote in the Timaeus). Conventional human psychology has required that human happiness depends on the projected "fallen angel of evil" being bound in a smoky, obscure abyss of repression. However, in this mode the binding energy itself is lost for constructive purposes. Therefore all legendary descents into Hades have had a liberating goal and it may be seen that the object was a release of energy from those regions. The risen Lazarus, still with his bindings, is an emblem of this type.
Broader Evil (#HH1034) Angelic frame of awareness (#HM2150).

♦ HM2213d Near death black void (NDE)

Description As opposed to the positive experience of seeing a light at the end of a tunnel, there is an experience of entering a black void.
Context The third stage in the (less frequently recorded) negative near death experience.
Broader Near death experience (NDE, #HM0777).
Related Near death transition (NDE, #HM3294).
Followed by Near death evil force (#HM3422).
Preceded by Near death negative detachment (NDE, #HM3115).

♦ HM2214c Identity with God (Sufism)

Description By letting go of other thoughts, beliefs, sensations, and destroying them, the identity of the individual with the One Self becomes so complete that even awareness of this identity disappears, leaving nothing but identity with God. Perfect poverty – not only unaware of himself but unaware of this unawareness, only the One remains.
Context The eighth illumination of Jami.
Broader Fountains of light (Sufism, #HM3039).
Followed by Annihilation of personal sensation (Sufism, #HM3358).
Preceded by Identifying with reality (Sufism, #HM2985).

♦ HM2215d Third scriptural bodhisattva awareness (Buddhism, Tibetan)

Prabha-kari — Illumination

Description This state of awareness is characterized by endurance. It is represented by the third stage bodhisattva who will bear anything for the sake of the dharma. He delights in the beauty that may be found by literary expression of the Buddha's way. His meditative powers are said to be phenomenal, attaining lordship over the desire-realm and our solar system in particular. His activity is illumination (prabha-kari).
Context In Tibetan Sakya Buddhism this is one of the states in the "Ascension Stages Game". In some sets it is numbered 79 on the board.
Broader Meditation way of the bodhisattvas (Buddhism, #HM2769).
Followed by Fifth scriptural bodhisattva awareness (Buddhism, Tibetan, #HM2909)
Fourth scriptural bodhisattva awareness (Buddhism, Tibetan, #HM2191).
Preceded by First scriptural bodhisattva awareness (Buddhism, Tibetan, #HM2155)
Second scriptural bodhisattva awareness (Buddhism, Tibetan, #HM2739).

♦ HM2216d Heavenly sorrow (ICA)

Description This is the moment when a person realizes that all existence is separation, when he sees that his entire life and every act is in the context of profound separation from himself, from others and from the mystery of life; as in Hamlet's soliloquy on being and not-being.
Context In the ICA New Religious Mode in the arena of acting out one's freedom (the life of prayer) the first formal aspect is one's struggle with repentance for one's sinfulness or confession. At the fourth phenomenological level this occurs when one knows that one's whole life is inescapably bound in this struggle.
Broader Affirmation (ICA, #HM2199) Struggle to confess (ICA, #HM2718).
Preceded by Besetting sin (ICA, #HM2379).

♦ HM2217d Setting the mind (Buddhism)

Chittasthapana — Sems-jogpa (Tibetan)

Description Strong faith and aspiration are applied to setting the mind on the object of meditation and stabilizing its condition. In this first stage, three kinds of mental torpor or sloth that permit distraction (vikshepa) are overcome. The three are: lack of impulse; attachment to habit; and anticipation of failure. The associated power in this state is hearing.
Context This is the first of the nine states of mental abiding (navakara chittasthiti) in Gelugpa Tibetan Buddhism.
Broader Calm abiding (Buddhism, #HM2147)
Consciousness states in cyclic existence (Buddhism, #HM2177).
Followed by Continuous setting of mind (Buddhism, #HM2241).

♦ HM2218c Path of active intelligence (Judaism)

Path of exciting intelligence

Description This state of awareness is said to be of the spirit and its proper motions and activities as applied to each being informed under the sun.
Context The twenty eighth of the spiritual paths expressed in the Jewish Kabbalah.
Broader Paths of wisdom (Judaism, #HM2509).
Followed by Path of corporeal intelligence (Judaism, #HM3149).
Preceded by Path of natural intelligence (Judaism, #HM2071).

♦ HM2219b Chakra centres of consciousness

Description The various mystical traditions of the world know a number of *chakras* or centres of consciousness in the body. Most emphasised are: the centre on the top of the head, the fontenelle *ajnakhya* or *sahasrara chakra*; the *visuddha chakra* between the front sinuses; the *anahata chakra* variously described as at the root of the nose or at the heart; the *mani-pura chakra* at the pit of the stomach; the *svadhishthana chakra* at the navel; and the *muladhara chakra* in the pubic region. Depending on the system in question, varying descriptions are used for up to twelve centres, some of them lying outside the body. Kundalini yoga recognizes seven chakras. The first five are connected with the five elements – earth, water, fire, air and ether, as are the planes of these five chakras. The final two chakras are beyond the elements. Through the practice of yoga one first reaches tapas loka (austerity). If self-realization is not complete one is born again as an ascetic, avatara, bodhisattva or prophet. However, it is still required to go beyond the gunas through the practice of awakening Kundalini which, after reaching the sahasrara chakra and uniting with Siva, descends and restores the powers of the chakras and the deities dwelling in them. From then on one lives in an extended state of consciousness – a changed person, exhausting karma, so that on leaving the body nirvana is achieved. "Lila", the game of life, also recognizes seven but with a plane outside the body said to be beyond the chakras. Each centre is ascribed a particular mode of perception of the world. The activated centres are the concomitants of corresponding qualities of subtle energy (prana).
Refs Johari, Harish *Chakras* (1987); Motoyama, Hiroshi *Theories of the Chakras*.
Narrower Ajna (Yoga, #HM2144) Anahata (Yoga, #HM3242)
Manipura (Yoga, #HM3063) Muladhara (Yoga, #HM2893)
Vishuddha (Yoga, #HM3461) Sahasrara (Yoga, #HM3398)
Soma chakra (Yoga, #HM0176) Svadhishthana (Yoga, #HM3174).
Related Shushumna (Hinduism, #HM2053) Kundalini yoga (Yoga, #HH0237)
Kundalini conscious energy (Yoga, #HM2871).

♦ HM2220c Tantra climax-application awareness (Buddhism, Tibetan)

Bodhicitta awareness — Conception

Description This state of awareness is characterized by the word *conception* and is represented as a mystical union or marriage of elements within the lama-yogi's own body. The state is described as corresponding to the secret initiation into the supreme tantra. The elements of the union are represented as blessed by the deities drawn into the body from the mandala. The climax is the union of the male white element (essential seed) with the female red element (the ovum and its life-support system), also viewed essentially. This is said to produce, conjointly, a mystic substance in the body called *bodhicitta*. The contributing male and female essences are also called *vajra* (technique or applied power or means) and *vidya* or *prajna* (wisdom). Instead of a ritual beverage, the initiates (male or female) may taste bodhicitta through their meditation activity in their throat-centre (cakra) and experience the joy of the spiritual void. Since this cakra is associated with spiritual love or agape it is said to be important to consider that the feminine component of boddhicitta is equally the power of compassion, and that the ethical excercise and development of compassion for all beings is the heart of the mystical, feminine matrix of life.
Similar descriptions may be found in other traditions. In the Kabbalah, the feminine *pillar of mercy* is paired with the masculine *pillar of justice or judgement* (the Tibetan discernment or means) as being essential to the ascent of awareness or consciousness. The Tibetan *Tara* is a feminine personification of compassion. She is their Eve and their father is the god of compassion, *Avalokitesvara*. The Jews have the Skekinah and the Christians the Holy Spirit, both with feminine characteristics, while the Roman Catholics have also denominated Mary as "Our Lady of Mercy". In modern analytical psychology, the integration of the psychic masculine and feminine are essential to individuation.
Bodhicitta-awareness is therefore a behavioural condition in which human development effort (masculine, working with progressively more subtle energies) is directed by a benevolent and sane (non-egoistical) disposition of the will and understanding (feminine, wisdom).
Context In Tibetan Sakya Buddhism this is one of the states in the "Ascension Stages Game". In some sets it is numbered 50 on the board.
Broader Tantric paths of accumulation and application in the ascension-stages-game (Buddhism, Tibetan, #HM2848).
Related Prajnaparamita (Buddhism, #HM1344).
Followed by First vajra-master stage (Buddhism, Tibetan, #HM2187)
Tantra receptivity-application awareness (Buddhism, Tibetan, #HM2756).
Preceded by Great-path tantra awareness (Buddhism, Tibetan, #HM2656)
Tantra heat-application awareness (Buddhism, Tibetan, #HM2696)
Kalacakra-tantra shambhala awareness (Buddhism, Tibetan, #HM2151).

♦ HM2222c Angelic awareness - Gabriel

Description This mode of awareness may be considered as a state of consciousness which perceives the ontological ground without loss of self-identity. Gabriel's annunciatory nature may, in addition, reflect prophetic consciousness and also creative aesthetic consciousness or inspiration. As Michael is identified as the Shekinah's male aspect, Gabriel may be the male, active aspect of the Holy Spirit.
Context Some Kabbalists affiliate Gabriel to Yesod. In Islam, Gabriel's (Djabrail's) functions are more often referred to than Michael's (Mikal).
Broader Angelic frame of awareness (#HM2150).

♦ HM2223d Individual fatefulness (ICA)

Description This is an awareness of the incredible fact that one made it into history. Of all the myriad chances of not showing up at all, one hit the right combination. There is a sense of "it's great to be alive " and one wakes up in the morning grateful to have made it through the night. The writer/palaeontologist Loren Eisley captures this well in his description of "the snout", the amphibious creature that made the mutation from being a fish to a reptile.
A dimension of this experience is a sense of awe, that is fear and fascination. In the midst of amazement at being here at all, an individual finds himself wondering if perhaps it is all an illusion. Surely there must be a price to pay for such an unbelievable gift as one's life? It can't go on forever and there must be some catch to it; it's like being awarded the supreme grand prize – the glory of consciousness – and he doesn't quite know what to do with it. As he marvels at the wonder of life, he finds himself developing reverence for the miraculous. He knows he won't ever be able to go back the self-deprecation he knew in the past; and yearns to remain in this state, fleeting as it is. All he can do is to live his fatedness and participate in the journey he is a part of.
Context This state is number 33 in the ICA *Other World in the midst of this World*.
Broader Original gratitude (ICA, #HM3105).
Followed by Primal sympathy (ICA, #HM2550) Global guardianship (ICA, #HM2817)
Radical contingency (ICA, #HM2477) Seminal illumination (ICA, #HM2056)
Definitive predestination (ICA, #HM2131).
Preceded by Ultimate awareness (ICA, #HM2388)
Global guardianship (ICA, #HM2817)
Radical contingency (ICA, #HM2477)
Diaphanous intuition (ICA, #HM2927)
Destinal accountability (ICA, #HM2309).

♦ HM2224d First duty (Christianity)

Description To see and accept the boundaries of the human mind without rebellion and to work within these limitations without ceasing is where the first duty lies. The mind can only perceive appearances and not the essences of things. It can not perceive all appearances but only those of matter. It can not even see the appearances of matter but only the relationships between them. These relationships are not real and independent of the human mind but its creation. These relationships are not the only one possible, but the most convenient for his needs.
Context The first duty in the preparation for the spiritual exercises of Nikos Kazantzakis.
Broader Saviors of God: the preparation (Christianity, #HM3264).
Followed by Second duty (Christianity, #HM2861).

♦ HM2225c Pilgrim archetypal image (Tarot)

Fool

Description The encounter of the psyche with a representation of a pilgrim or wayfarer is an archetypal image. Variations are the minstrel, the itinerant fool and the roving player or actor. This emblem is one of the profoundest of the archetypes of awareness. In the Kabbalah, the fool

has been associated with one of the two highest paths of wisdom on the sephirothic tree, between wisdom (the first emanation) and the crown (an aspect of revealed divinity). In the Tarot, he is a young wanderer represented as walking on the high–road with his kit bag of possessions and a staff. Before him lie many paths and his multicolored garments reflect the spectrum of possibilities. This is why, as the modern playing card the *joker*, he is free to assume different values; and one of the reasons why, in the Tarot, his card is numbered zero (by being unnumbered but placed in order before the first card). Within the Tarot system itself, the fool precipitates the cosmic process; his energies will descend to the chasm at the bottom of the tree of life, and what was previously manifest will come into being. There may be an indication that the psychic encounter with the pilgrim may be associated with the encounter with the chariot (or royal charioteer).
Context A path from the Crown (Kether) to the worlds below; that for the pilgrim is by means of Chokmah, wisdom, according to Tarot correspondences.
Broader Emblem archetypes in the psyche (Tarot, #HM2201)
Tarot arcana of conscious states (Tarot, #HM2097).
Related Sphere of wisdom (Kabbalah, #HM2348)
Sphere of the of the supreme crown (Kabbalah, #HM3132).

♦ **HM2226a Samadhi** (Hinduism, Yoga)
Savikalpa samadhi — Nirvikalpa samadhi — Enstasy
Description Samadhi is a state of consciousness for whose description words are considered to be totally inadequate. It is said to be the highest state of consciousness, associated with direct mystic experience of reality. Samadhi cannot be experienced until a condition of mindlessness has been created, through the deliberate elimination of the objects of thought from consciousness. The organs of sense perception are so controlled that they no longer pass to the mind their reactions to what is perceived. The mind loses its identity by absorption into a higher state which precludes any awareness of duality, although a form of unitary awareness of the conventional world is retained. Different levels of consciousness or degrees of samadhi are distinguishable. *Savikalpa samadhi* is subject to time and change; *nirvikalpa samadhi*, the higher state, is timeless. Samadhi is the end result and aim of the practice of yoga (separately explained). It is a term used in a different sense in Buddhism, where it is equivalent to concentration or one–pointedness of citta.
Context The eighth component of the eightfold path of yoga.
Refs Chakrabarty, Surath *Mysterious Samadhi* (1984); Sadhu, Mouni *Samadhi* (1962).
Broader Samyama (#HH0374) Eightfold path of yoga (Yoga, #HH0779).
Narrower Sahaja samadhi (Hinduism, #HM2043) Sabija samadhi (Hinduism, #HM2308)
Samprajnata samadhi (Yoga, #HM2896) Nirbija samadhi (Hinduism, #HM2125)
Dharma megha samadhi (Yoga, #HM0194) Savikalpa samadhi (Hinduism, #HM2650)
Savitarka samadhi (Hinduism, #HM3526) Nirvikalpa samadhi (Hinduism, #HM2061)
I–transcendent stage of life (#HM3060) Asamprajnata samadhi (Hinduism, #HM3041)
Unconditional nirvikalpa samadhi (#HM3619).
Related Enstasy (#HM3169) Satori (Zen, #HM2326) Enlightenment (#HM2029)
Nirvana (Buddhism, #HM2330) Essential wisdom (#HM0107)
Samadhi yoga (Yoga, #HH4321) Samadhi parinama (Yoga, #HM3207)
Concentration (Buddhism, #HM6663) Citta (Hinduism, Buddhism, #HM3529)
Human development (Hinduism, #HH0330) Transcendental consciousness (#HM2020).
Preceded by Super–contemplative state (Yoga, #HM2818).

♦ **HM2227c Voice of silence**
Related Silence (#HM3603) Practice of silence (#HH0983).

♦ **HM2228d Pratyeka Buddha vision–path awareness** (Buddhism, Tibetan)
Description This state is one of several described for independent (pratyeka) Buddhas in the Tibetan Ascension Stages game. The independent Buddha–path is followed with analytical meditation and purification of philosophical outlook. This stage is characterized by the beginning of the practice of the Noble Eightfold Path.
Context In Tibetan Sakya Buddhism this is one of the states in the "Ascension Stages Game". In some sets it is numbered 45 on the board.
Broader Religious traditions in the ascension stages game (Buddhism, Tibetan, #HM3341).
Preceded by Pratyeka Buddha application awareness (Buddhism, Tibetan, #HM3020).

♦ **HM2229e Self–betrayal** (Brainwashing)
Description Protestations of innocence are treated as proof of guilt and the individual is persuaded that for self–survival he may as well confess those things which are already known to his interrogators. In doing so he betrays his previous friends and colleagues, and his own most cherished principles. This produces real feelings of guilt and a sense of disloyalty and self–betrayal. He becomes more and more involved with his captors and with the doubts and antagonisms beneath the surface of his loyalties. Turning back becomes progressively more difficult.
Context A stage reached in thought reform, a system of organized, deliberate and total psychological training which effects individual change through two basic elements – *confession* (renouncing of past beliefs and attitudes) and *re–education* (remaking of the individual in the required image).
Broader Thought reform (Brainwashing, #HH0865).
Followed by Breaking point of basic fear (Brainwashing, #HM2799)
Breaking point of total conflict (Brainwashing, #HM2354).
Preceded by Recognition of guilt (Brainwashing, #HM2923).

♦ **HM2230b Ego transcendence** (Jung)
Description As the contents of the ego shift and change, the structure of the ego–complex expands and the previous "ego–image" – often mistakenly supposed to be the ego itself, and that which gives rise to the sense of "I" – is outgrown or transcended. This change in the ego constitutes, in Jungian terms, the *individuation* process and is said to have its equivalent in Eastern philosophies in the Buddhist "not–self" or "no soul".
Refs Adler G, Fordham M and Read H (Eds) *The Collected Works of C G Jung*; Samuels, Andrew et al *A Critical Dictionary of Jungian Analysis* (1986).
Related Individuation (#HH0385) Flow experience (#HM2344)
I–transcendent stage of life (#HM3060).

♦ **HM2232c Mahayana climax–application awareness** (Buddhism, Tibetan)
Aloka–vriddhi samadhi
Description This state of awareness is represented as a concentration in which the object of perception is annihilated. The formula of twelve–fold dependent origination is said to be fully understood as the Mahayana way. At this stage, one is characterized as embodying five cardinal virtues: "faith–works" or piety; potency; mindfulness; analytical concentration (samadhi); and wisdom (prajna or vidya). These virtues are all essential energies so that the mahayana adept is said to possess extraordinary powers to teach the dharma and reform the world.
Context In Tibetan Sakya Buddhism this is one of the states in the "Ascension Stages Game". In some sets it is numbered 56 on the board.
Broader Sutra paths of accumulation and application in the ascension stages game (Buddhism, Tibetan, #HM2962).
Followed by Mahayana receptivity–application awareness (Buddhism, Tibetan, #HM2247)
Mahayana highest–teachings–application awareness (Buddhism, Tibetan, #HM2271).
Preceded by Mahayana heat–application awareness (Buddhism, Tibetan, #HM2208).

♦ **HM2233g Concrete awareness**
Description This is the awareness of objects and people as concrete, bodily realities, identifiable by proper name, retained by and available to imaged memory.

♦ **HM2234d Infinite passion** (ICA)
Seduced by mystery
Description This is an experiential trek in the *Land of Mystery*, within the ICA *Other World in the midst of this World*, when the individual is aware of adoring being. The states are described in separate entries.
Broader Land of mystery (ICA, #HM2434).
Narrower Essential dubiety (#HM3135) Cryptic disclosure (#HM2824)
Singular adoration (#HM2111) Transcendent immanence (#HM3034).
Related Aweful encounter (ICA, #HM2619) Inescapable power (ICA, #HM3096)
Transformed state (ICA, #HM2386).

♦ **HM2235d Tantra master in form–realm awareness** (Buddhism, Tibetan)
Rupa–dhatu–vidyadhara
Description This state of awareness is represented by adepts who are mentally able to transverse the starry skies. They are said to inhabit a pure land in the Realm of Form and practice shabda–yoga, that is, mentally dwell in the sound of mantra without visualizing an object.
Context In Tibetan Sakya Buddhism this is one of the states in the "Ascension Stages Game". In some sets it is numbered 68 on the board.
Broader Masters of wisdom in the ascension stages game (Buddhism, Tibetan, #HM3066).
Followed by Great–path tantra awareness (Buddhism, Tibetan, #HM2656)
Middle–path tantra awareness (Buddhism, Tibetan, #HM3026)
Tantra heat–application awareness (Buddhism, Tibetan, #HM2696)
Kalacakra–tantra shambhala awareness (Buddhism, Tibetan, #HM2151).
Preceded by Tantra master–in–sense–realm awareness (Buddhism, Tibetan, #HM2211).

♦ **HM2236d Word–bearing priest** (ICA)
Description This is experienced on encountering people and events that remind one of one's capacity to live one's life, when one is conscious that one is being called to pick up one's own life and live it. An example is suddenly remembering something said to one a long time ago in another difficult situation, referring to the fact that one's life is one's own to create.
Context In the ICA New Religious Mode in the arena of knowing one's internal sociality (the life of meditation) the first formal aspect is one's encounter with the mediator who pronounces personal absolution. At the first phenomenological level, this occurs when that which is other intrudes upon one's consciousness.
Broader Personal absolution (ICA, #HM2370).
Followed by Radiant guru (ICA, #HM2019).

♦ **HM2237c Magician archetypal image** (Tarot)
Magus
Description The encounter of the psyche with the emblem or image of a magician is one of the most powerfully felt archetypes that express themselves in conscious awareness. The power or force of the magus is represented in the Tarot by the emblem of the Magician (number 1) depicted with the tokens and talismans of the extent of his powers and domains. These are the symbols of four elements of the material world. Above the abyss, the Magician is related with male but virginal purity, but having his being in a dimension seeking manifestation in nature. The world, or plane of manifestation, is represented by his table. Some tarot systems call him the Juggler. Since the only things depicted that he can juggle are the symbols of the four elements, his magic is transmutation, or transformation. On his table is a book which may lead him to the lady of books, wisdom, the High Priestess (Tarot emblem 2).
Context In the Tree of Sephiroth the path of the Magician is from the crown (Kether) to intelligence or understanding (Binah).
Broader Emblem archetypes in the psyche (Tarot, #HM2201)
Tarot arcana of conscious states (Tarot, #HM2097).
Related Sphere of understanding (Kabbalah, #HM2372)
Sphere of the of the supreme crown (Kabbalah, #HM3132).

♦ **HM2238c Sphere of glory** (Kabbalah)
Hod–sephira
Description This sphere of awareness is associated with passive action and with a calculating mentality and thus with the development of the ability to study. It corresponds to the developmental phase of childhood and learning. It is associated with the intellect and with rational thinking; also with the intellectual conquest of the animal instincts. It is symbolized by the positive and negative archetypes of the trickster, the crook and the planet Mercury, as well as by Aaron in the Jewish tradition. This state of consciousness corresponds to the kabbalistic World of Formation.
Context The eighth of ten spheres described in the Kabbalistic system of sephiroth.
Broader Kabbalah (#HH0921).
Related Manipura (Yoga, #HM3063) Emblems of renewal (Tarot, #HM2315)
Emblems of disaster (Tarot, #HM2350) World of formation (Judaism, #HM2360)
Emblems of well–being (Tarot, #HM2422) Victimization emblems (Tarot, #HM2190)
Way of knowledge (Esotericism, #HM1201) Emblems of personified evil (Tarot, #HM2378).
Followed by Sphere of beauty (Kabbalah, #HM3031)
Sphere of victory (Kabbalah, #HM2362)
Sphere of judgement (Kabbalah, #HM2290).
Preceded by Sphere of the kingdom (Kabbalah, #HM2288)
Sphere of the foundation (Kabbalah, #HM2410).

♦ **HM2239d Cancer–consciousness** (Astrology)
Individualized awareness
Description The Cancer mode acts as a stimulus to satisfy desires. It is the guardian of formative energy and growing life. Under optimal conditions it is associated with extreme alertness to ever–changing circumstances, in search of new opportunities that will permit response of a higher order to any challenge.
Context One of twelve conditions of being or streams of divine energy encompassed within the zodiac. It is said that all twelve conditions must be present and playing its proper part for any successful and fulfilling action to take place. This may be through a group of individuals with different endowments and soul qualities working together; or it may be through one individual who, by self–discipline, practice and meditation, has allowed all twelve conditions to arise within himself.
Broader Watery awareness (Astrology, #HM2384)
Cardinal awareness (Astrology, #HM2259)
Zodiacal forms of awareness (Astrology, #HM2713).
Followed by Leo–consciousness (Astrology, #HM2376).
Preceded by Gemini–consciousness (Astrology, #HM2924).

♦ **HM2240d Discipleship-vision awareness** (Buddhism, Tibetan)
Sravaka-darsana-marga — Paths of vision and cultivation
Description This state of awareness, called the Paths of Vision and Cultivation, is characterized by deep insight into the four noble truths. This insight destroys the idea of the reality of the self. In addition, through cultivation of analytical meditation, all attachment to pleasurable sensation is destroyed. One experiences the "adamantine trance" (vajropama samadhi).
Context In Tibetan Sakya Buddhism this is one of the states in the "Ascension Stages Game". In some sets it is numbered 40 on the board.
Broader Darsana (Buddhism, Tibetan, #HM2931)
Religious traditions in the ascension stages game (Buddhism, Tibetan, #HM3341).
Followed by Awareness of relative reality (Hinduism, Yoga, #HM2032)
Great-black-lord awareness (Buddhism, Tibetan, #HM2118)
Non-emanating consciousness (Buddhism, Tibetan, #HM2070)
Thirty-three-god-heaven awareness (Buddhism, Tibetan, #HM2606)
Heaven-without-fighting consciousness (Buddhism, Tibetan, #HM2130)
Delightful-emanation-heaven consciousness (Buddhism, Tibetan, #HM2546).
Preceded by Discipleship-karma awareness (Buddhism, Tibetan, #HM2192)
Discipleship-application awareness (Buddhism, Tibetan, #HM2716)
Great-vehicle middle-path awareness (Buddhism, Tibetan, #HM2160).

♦ **HM2241d Continuous setting of mind** (Buddhism)
Samsthapara — Rgyun-du-jog-pa (Tibetan)
Description The meditator in this stage is able to stay on the object longer than in the first stage, by forcibly returning attention again and again to the object of contemplation. The associated power in this state is thinking.
Context This is the second of the nine states of mental abiding (navakara chittasthiti) in Gelugpa Tibetan Buddhism.
Broader Calm abiding (Buddhism, #HM2147).
Followed by Resetting the mind (Buddhism, #HM2265).
Preceded by Setting the mind (Buddhism, #HM2217).

♦ **HM2242d Multifocal thought**
Asyndesis
Description The state in which an individual focuses thoughts on several levels and meanings simultaneously, together with their different objective situations.
Related Logical and metalogical awareness (#HM2320).

♦ **HM2243c Withdrawal** (Sufism)
Ghaiba
Description This level of awareness in Sufism is desirибed as a stage on the way to the mystic goal of fana (cessation). It is the withdrawal of all except Allah from the heart-consciousness. As man withdraws himself, so Allah may withdraw a man. He may be the Hidden Imam or Mahdi, the Hidden Pole, the Abdal, or al-Khadir (the "green man"); a secret spiritual master known only to a few as he is, but also possibly experienced as a spiritual force among his people.
Broader Stations of consciousness - Simnami (Sufism, #HM2341).

♦ **HM2245c Meditation way of the four truths** (Buddhism)
Chatvari satvani — Bden-pa-bzhi (Tibetan)
Description This is a supplemental excercise to the concentrations and absorptions, and consists of analytical meditations on the four truths and on the four aspects of each - for example, on the aspect of impermanence within the first noble truth of suffering existence (duhkhasatya). The sixteenth aspect discloses that the wisdom of realizing the non-existence of individual personality is the path to deliverance.
Context This way takes the practitioner beyond the absorptions (arupaya samapatti) according to Tibetan Gelugpa Buddhism.
Related Four noble truths (Buddhism, #HH0523).
Followed by Meditation way of the hearers (Buddhism, #HM2161)
Meditation way of the bodhisattvas (Buddhism, #HM2769).
Preceded by Fourth absorption in the immaterial sphere (Buddhism, #HM2051).

♦ **HM2246c Divine nature** (Sufism)
Jesus of one's being
Description This is the subtle organ which receives inspiration, that which announces the Name. The colour is luminous black.
Context The sixth of seven subtle stages relating to the seven major prophets of Semitic monotheism as described by Simnani.
Broader Stations of consciousness - Simnami (Sufism, #HM2341).
Followed by Divine essence (Sufism, #HM3513).
Preceded by World beyond form (Sufism, #HM4570).

♦ **HM2247c Mahayana receptivity-application awareness** (Buddhism, Tibetan)
Kshanti
Description This state of awareness is characterized as the dispersal of the subject in the object-subject illusory reality.
Context In Tibetan Sakya Buddhism this is one of the states in the "Ascension Stages Game". In some sets it is numbered 63 on the board.
Broader Sutra paths of accumulation and application in the ascension stages game (Buddhism, Tibetan, #HM2962).
Related Kshanti (Zen, #HH0813).
Followed by Joy-land awareness (Buddhism, Tibetan, #HM2167)
Hyper-bliss-realm awareness (Buddhism, Tibetan, #HM2337)
Great-path tantra awareness (Buddhism, Tibetan, #HM2656)
Mahayana highest-teachings-application awareness (Buddhism, Tibetan, #HM2271).
Preceded by Joyful-heaven awareness (Buddhism, Tibetan, #HM2022)
Potala-island awareness (Buddhism, Tibetan, #HM2175)
Mahayana heat-application awareness (Buddhism, Tibetan, #HM2208)
Great-vehicle higher-path awareness (Buddhism, Tibetan, #HM3048)
Kalacakra-tantra shambhala awareness (Buddhism, Tibetan, #HM2151)
Mahayana climax-application awareness (Buddhism, Tibetan, #HM2232).

♦ **HM2248c Angelic awareness - Logos**
Iesu — Messiah
Description This mode of awareness is of the Word, the reality said to lie behind forms and which creates them by its own super-sensible sound, motions or breaths. In terms of consciousness it is represented by the pulsations of mentatation: the mathematically expressible rhythms that evoke a personal cosmos out of an indifferentiated mental chaos. The Logos (word) was considered an angel by Philo, and by the rabbis as that angelic reality which is portrayed in the personification of Metatron. Medieval Kabbalists have also seen the Logos behind Michael, and have conceived that the Messiah also represents the Angel of the Logos. Early Christianity viewed the name *Iesu* as indicating an angelic nature and assigned this to the prince of all the angelic orders. Modern Catholicism sees the mother of Iesu, *Marianne*, as queen of angels who in the Kabbalah is the Shekinah.
Broader Angelic frame of awareness (#HM2150).

♦ **HM2249d Keeping conscience** (ICA)
Description This is having to make one's own decisions without being able to rely on rules, other people or precedents; and being willing to live with the consequences of those decisions; when one experiences the burden and gift of being truly free. It may be compared with the decision of President John Kennedy to assume responsibility for the Bay of Pigs invasion.
Context In the ICA New Religious Mode in the arena of knowing and doing intensified in one's life (life of being) the third formal aspect is the experience of transparent engagement or the experience of levitation. At the third phenomenological level this occurs when a person disciplines his life to be continually related to the experience of nothingness.
Broader Transparent engagement (ICA, #HM2736).
Followed by Cruciform exaltation (ICA, #HM2303).
Preceded by Impossible possibility (ICA, #HM2524).

♦ **HM2250c Mutability emblems** (Tarot)
Fortune-wheel
Description Among the archetypal emblems projected by the psyche are those that depict change. Change has many aspects. Sometimes it is beneficent, sometimes catastrophic. Sometimes it is represented as irreversible and final. In the Tarot, the emblem of the Wheel of Fortune (number 10) indicates that change may be regulated by periodic laws, as well as being cyclical; so that, for example, the seasons of life and their associated mental states may all come round in the circle of time. This path leads to Chesed, the region of the pure archetypes which reflect the trinity above the abyss, where the alternating polarities are fused as duality is overcome.
Context In the Kabbalah, the fortune-wheel is placed on path 21 between force or victory (Netzach) and mercy or kindness (Chesed).
Broader Emblem archetypes in the psyche (Tarot, #HM2201)
Tarot arcana of conscious states (Tarot, #HM2097).
Related Sphere of victory (Kabbalah, #HM2362)
Sphere of mercy and greatness (Kabbalah, #HM2420).

♦ **HM2251d Fourth vajra-master awareness** (Buddhism, Tibetan)
Description This state of awareness is characterized by the need to speak the dharma words of power. One is said to perform a Buddha's functions of instruction. The corresponding initiation object is the bell, symbolizing prajna or wisdom-power, which has as a handle the *dorje* or vajra thunderbolt, symbolic of understanding or means. These qualities are united in the symbolism of the object as they are said to be united in the adept.
Context In Tibetan Sakya Buddhism this is one of the states in the "Ascension Stages Game". In some sets it is numbered 75 on the board.
Broader Way of the vajra masters (Buddhism, Tibetan, #HM3090).
Followed by Sixth vajra-master awareness (Buddhism, Tibetan, #HM2287)
Fifth vajra-master awareness (Buddhism, Tibetan, #HM2763)
Seventh vajra-master awareness (Buddhism, Tibetan, #HM2789).
Preceded by First vajra-master stage (Buddhism, Tibetan, #HM2187)
Wheel-turner-king awareness (Buddhism, Tibetan, #HM2759)
Third vajra-master awareness (Buddhism, Tibetan, #HM2727)
Second vajra-master awareness (Buddhism, Tibetan, #HM2703).

♦ **HM2252d Pratyeka Buddha cultivation awareness** (Buddhism, Tibetan)
Description This state of awareness is represented as the fourth path of the independent Buddha and is characterized by detachment from everything except existence. This stage for a self-taught sage is represented as being associated with great powers and magical abilities.
Context In Tibetan Sakya Buddhism this is one of the states in the "Ascension Stages Game". In some sets it is numbered 46 on the board.
Broader Religious traditions in the ascension stages game (Buddhism, Tibetan, #HM3341).
Followed by Pure-lands awareness (Buddhism, Tibetan, #HM2168)
Form-realm awareness (Buddhism, Tibetan, #HM2142)
Arhat sanctity awareness (Buddhism, Tibetan, #HM3024)
Awareness of relative reality (Hinduism, Yoga, #HM2032)
Pratyeka Buddha arhat awareness (Buddhism, Tibetan, #HM2776)
Heaven-without-fighting consciousness (Buddhism, Tibetan, #HM2130).
Preceded by Pratyeka Buddha application awareness (Buddhism, Tibetan, #HM3020).

♦ **HM2253d Eternal saviour** (ICA)
Description This is the rare experience when a person's life is utterly transformed forever and he is never the same again. Such experiences may nurture and enable a person to deal with the great transformation occurring in his life. An example might be Helen Keller's reflection on the first time she linked words to experiences in her life; she said that she went back repeatedly to that moment in order to sustain herself in difficult times.
Context In the ICA New Religious Mode in the arena of knowing one's internal sociality (the life of meditation) the first formal aspect is one's encounter with the mediator who pronounces personal absolution. At the fourth phenomenological level, this occurs when a person is sustained by communion with people and events which serve as milestones for his consciousness.
Broader Personal absolution (ICA, #HM2370).
Preceded by Persistent friend (ICA, #HM2139).

♦ **HM2254b Ma'rifa** (Sufism)
Knowledge of God — Gnosis — Contemplative knowledge — Ma'rifat
Description This is the state in which divine knowledge is revealed by God, arising in pure love of God and for God, and as a reward for love of God. There is direct knowledge of God, knowledge which cannot be revealed by the intellect, which is an indirect process depending on limited human knowledge, nor through theological knowledge, nor through inspiration, which is subjective, nor through intuition, which is still veiled by finite human nature; but which can only be experienced directly. The barrier of humanity is transcended in contemplative knowledge. In this vision of God all veils which would hinder sight are removed, the inner light being far beyond the light of faith. God's divine glory is revealed in a manner which precludes doubt. Awareness of attributes of the self such as awe, fear, hope and love ceases. Only the attributes of God are known. There is a passing away from the self into God. No further ascent is possible. The true meaning of the soul's unification with God comes to be known. According to Ibn al-'Arabi, in ignorance there might be belief that the soul is part of the divine soul; in true knowledge comes awareness that the individual soul is the divine soul.
Although knowledge of God arises only by revelation there are three ways in which He may be known. Observance of the law, *shari'a*, when the seeker observes the law, results in knowledge being attained from God. Following the mystical path or *tariqa*, the soul experiences a number of states leading to knowledge with God. Revelation of the truth, *haqiqa*, when the soul is illuminated and by God's grace sees a vision of God, brings knowledge of God. All three are linked in the same universal truth, so that genuine knowledge is attained when following the mystic path in observance of the law.
Context The doctrines of *sabr*, *mahabbat* and *ma'rifa* were elaborated by the Sufi scholar Abu Talib al-Makki and followed and interpreted by many others. According to Shaykh Abu Sa'id ibn

Abi'l-Khayr, God is perceived through all creatures and all people. Kharraz indicates that when God has the desire to be united with His servant, He first opens the way to worship. If He takes delight in worship, He opens the way to proximity. Then he is raised to fellowship, then (tawhid) seated on the throne of unification. Then the veil is raised, God makes him enter into His own, unveils His glory. When the servant sees the glory and majesty he remains outside himself and, coming into the care of God, is freed for ever from the self. For Abu'l-Hasan Sumnun, love is the foundation and the principle of the way to God, all states and stations being stages of love. For Ahmad b Abu'l-Hasan al-Nuri, gnosis is a very high stage and the gnostic is one who speaks from this stage of annihilating the self in the ecstasy of love and knowledge of God. The knowledge of God is not rational knowledge, because there is then a veil between knower and known; it is revealed only in contemplation, *mushahada*, when the light of God's unicity is shed on reason. 'Amr b Uthman al-Makki regarded the inward experience of the real as absolutely personal and not possible to disclose.
Ma'rifa is the twelfth stage in a systematic account of various stations of the Sufi spiritual path. Also, according to Shaykh Abu Sa'id ibn Abi'l-Khayr, it is the 25th of 40 stations or maqamat the Sufi must possess for his journey on the path of Sufism to be acceptable.
Refs Bhatnagar, R S *Dimensions of Classical Sufi Thought* (1984).
Broader Mystic stations (Sufism, #HM3415) Human development (Sufism, #HH0436)
Stations of consciousness - ibn-Abi'l-Khayr (Sufism, #HM2424).
Related Gnosis (#HM0413) Sabr (Sufism, #HM3418) Sharia (Islam, #HH0738)
'Arif (Sufism, #HM1045) Haqiqa (Sufism, #HM0124) Jabarut (Sufism, #HM4392)
Mushahada (Sufism, #HM3521) Love of God (Sufism, #HM4273)
Divine path (Sufism, #HM3371) Jnana (Yoga, Buddhism, Hinduism, #HH0610).
Followed by Fana (Sufism, #HM1270) Jahd (Sufism, #HM6099).
Preceded by Wajd (Sufism, #HM3650) Haqq al-yaqin (Sufism, #HM4556).

♦ HM2256d Inclusive comprehension (ICA)
Description This is the experience of universalization of initial illumination, when the individual comprehends everything and knows what no one can tell him, the all-determining dynamic of being itself which directs the world of man and nature. It may be compared to first seeing the earth from the moon, it as was if we had a new vision of our planet and people began to speak of spaceship earth. It is like the Buddha after he has experienced illumination.
A dimension of this experience is a sense of awe, that is fear and fascination. The fear is in the sense of its fragility and that there is no reason to be assured; there is a sense of standing on the edge of the universe and of helping it unfold, and also a deeply fascinating self-assurance.
The decision is to risk one's self and be the wise one, to share this comprehension with the world around and not be shaken by what might be said or done to one as a result. There follows a radical self-confidence and knowledge of participating in giving a great gift to all of creation.
Context This state is number 50 in the ICA *Other World in the midst of this World.*
Broader Radical illumination (ICA, #HM2273).
Followed by Contentless word (ICA, #HM2373) Absurd existence (ICA, #HM2331)
External relation (ICA, #HM2471) Problemless living (ICA, #HM2747)
Spontaneous gratitude (ICA, #HM3025).
Preceded by External relation (ICA, #HM2471) Seminal illumination (ICA, #HM2056)
Spontaneous gratitude (ICA, #HM3025) Resurrectional existence (ICA, #HM2631)
Definitive predestination (ICA, #HM2131).

♦ HM2257c Form-realm consciousness (Buddhism)
Rupadhatu — Gzugs-khams (Tibetan)
Description There are seventeen classes of gods in the form realm. They exist in four regions which correspond to the four concentrations or *dhyana.* Any concentration attained by the meditator causes rebirth in one of the four related areas. The first concentration has three lands: Brahma-type, Brahma-front and Great Brahma. The second concentration also has three levels: Little Light, Limitless Light and Bright Light. The third concentration is divided into Little Bliss, Limitless Bliss and Vast Bliss. The fourth concentration is divided into eight lands where the meditator may be reborn in a corresponding state of consciousness. The first three are called: Cloudless, Born from Merit and Great Fruit Land. Beyond these are the five pure places: Not Great, Without Pain, Excellent Appearance, Great Perception and Not Low. All these state of being and consciousness are in the form-realm, therefore they may be embodied in forms of vast dimensions and enjoy aeons of existence.
Broader Consciousness states in cyclic existence (Buddhism, #HM2177)
Consciousness ascension stages game (Buddhism, Tibetan, #HH4000).
Narrower Form-realm awareness (Buddhism, Tibetan, #HM2142)
Pure-lands awareness (Buddhism, Tibetan, #HM2168)
First trance of the fine-material sphere (#HM2450)
Third trance of the fine-material sphere (#HM2062)
Second trance of the fine-material sphere (#HM2038)
Fourth trance of the fine-material sphere (#HM2586).
Related Dhyana (Hinduism, Buddhism, #HM0137)
Rupaloka consciousness (Buddhism, #HM2536)
Fine-material-sphere concentration (Buddhism, #HM4265).

♦ HM2258d Submissive obedience (ICA)
Description This is saying "yes" to a situation in order to be able to participate in it and bring about change; when an individual decides to be obedient to the present situation for the sake of bringing about a new situation. An example might be being called upon to play a number of roles in obedience to changing situations - as a mother or father, husband or wife, an employee or employer, a home owner, a car driver, etc.
Context In the ICA New Religious Mode in the arena of acting out one's engagement (the life of obedience) the first formal aspect is embodying peace as the establishment to enable the ordering of human affairs. At the second phenomenological level this occurs when one recognizes one's obligation to the rest of mankind.
Broader Embodying peace (ICA, #HM2876).
Followed by Radical incarnation (ICA, #HM2587).
Preceded by Missional engagement (ICA, #HM2343).

♦ HM2259c Cardinal awareness (Astrology)
Cross of transcendence
Description A person whose consciousness is influenced by the *cardinal quadruplicity* of Aries and Libra, Cancer and Capricorn will demonstrate a cosmic awareness, with an outward looking and active approach. This is the final cross to be surmounted when the cycle of existence is dominated and liberation is achieved. Under the influence of Aries the initiation into incarnation is commenced, to be lived through until the wheel is fixed and reversed; and under Capricorn it is culminated.
Context One of the three cardinal qualities, each uniting four signs of the zodiac in two pairs of opposites or polarities which are strongly linked and between which the influenced individual tends to oscillate, all four signs having much in common.
Broader Zodiacal forms of awareness (Astrology, #HM2713).
Narrower Aries-consciousness (#HM2665) Libra-consciousness (#HM2512)
Cancer-consciousness (#HM2239) Capricorn-consciousness (#HM2822).
Related Fixed awareness (Astrology, #HM2554) Mutable awareness (Astrology, #HM3092).

♦ HM2260c Hierophant archetypal image (Tarot)
High priest — Pope
Description The encounter of the psyche with the images of a patriarch, ecclesiastical supreme dignitary, or lord-mayor type with keys to the city; or with porters, janitors, custodians, sextons and similar "key" figures, as well as with other roles connected with the power of entry, expresses one of the archetypes that lies deepest within the psyche. Most versions of the Tarot express this as the fifth emblem, which is the Hierophant (also called the Pope). The hierophant is represented as having the power to sanctify the temporal world, and two figures seen genuflecting or kneeling before him may be taken as the King and Queen in the following Tarot emblems, or psychologically as the ego and its desires and driving power.
Context In the Kabbalistic tree of sephiroth this emblem is said to correspond to the path between Chesed (kindness or mercy) or the pillar of mercy and Chokmah (wisdom).
Broader Emblem archetypes in the psyche (Tarot, #HM2201)
Tarot arcana of conscious states (Tarot, #HM2097).
Related Sphere of wisdom (Kabbalah, #HM2348)
Sphere of mercy and greatness (Kabbalah, #HM2420).

♦ HM2261c Wisdom archetypal image (Tarot)
High priestess
Description The encounter of the psyche with the image of an august, imposing, feminine personage, "larger than life", is one of the most universal of the archetypes that humanity expresses in numerous ways. In the Western classical world she was embodied in many deities and so also in the East. In Hellenistic Judaism her archetypal presence was expressly indicated as being symbolic of wisdom. Being of Kether, she reaches above the abyss, pure and virginal. Among the Tarot emblems she is number 2; she precedes her representative on earth, the empress, and comes before the emperor and the Pope as well. However the latter emblem, called also, the Hierophant, may be encountered as an archetype in conjunction with that of wisdom, as the Tarot suggests.
Context Her Kabbalist path is represented as lying between the sephiroth of the crown (Kether) and beauty (Tiphareth).
Broader Emblem archetypes in the psyche (Tarot, #HM2201)
Tarot arcana of conscious states (Tarot, #HM2097).
Related Sphere of beauty (Kabbalah, #HM3031)
Sphere of the of the supreme crown (Kabbalah, #HM3132).

♦ HM2262e Extrasensory perception (ESP)
Clairvoyance — Bio-information — Psychotronics — Telepathy — Paranormal cognition
Description ESP is the acquisition of information from the external environment or from another mind other than through any known sensory channels. One hypothesis suggests this is perception of energy through some unexplained, spaceless function. Others point to biological harmony between sender and receiver (heartbeats, brainwaves); a relationship with electric and magnetic force fields; and cosmic static, or the interception of an intermediate third party. Some psychologists argue that, since sensory perception is not fully understood even in the case of the known channels, the term may overemphasize a difference between parapsychological phenomena and related events in orthodox areas of psychology.
Four basic forms are distinguished, although they may be closely interlinked: *telepathy*, or thought transference; *clairvoyance*, or seeing distant objects or events; *precognition*, or prophecy; and *psychokinesis*, or influencing physical events through mental operations.
The evidence for the existence of ESP is often persuasive, judged by scientific standards ordinarily applied in other areas of psychology. Reservations exist because: the phenomena are not as reproducible as in other disciplines; the phenomena require much firmer evidence than normal to justify such an implausible hypothesis; the effects claimed are ordinarily very slight; and statistical questions are raised concerning the manner in which the results of the large number of trials in each experiment are analyzed.
Refs Douglas, Alfred *ESP Powers* (1976); Gudas, Fabian *Extrasensory Perception* (1975).
Narrower Mayko (#HM3192) Clairempathy (Psychism, #HM4516)
Auric clairvoyance (Psychism, #HM0679).
Related Clairscent (Psychism, #HM0318) Clairvoyance (Psychism, #HM3498)
Passivity experiences (#HM7203) Clairaudience (Psychism, #HM4333)
Clairgustance (Psychism, #HM0412) Telepathic consciousness (#HM2129)
Crisis apparition (Psychism, #HM0850).

♦ HM2263g Tongue consciousness (Buddhism)
Jihvavijnana — Lce'i rnam par shes pa (Tibetan) — Taste awareness
Description This is the consciousness that apprehends objects of taste, whether sweet, sour, salt, pungent, bitter or astringent. It is twofold, that is: where tongue consciousness is pleasant it is associated with indeterminate resultant consciousness with profitable result; where it is painful, it is associated with indeterminate resultant consciousness with unprofitable result.
Context One of the six consciousnesses defined in Buddhism as dependent on the individual senses, and with objects of sense as their focus.
Broader Sense consciousness (Buddhism, #HM2664)
Perception through the senses (#HM3764)
Mundane resultant consciousness (Buddhism, #HM0130)
Sense mode of consciousness occurrence (Buddhism, #HM4389)
Indeterminate consciousness in the sense sphere - resultant (Buddhism, #HM5721)
Awareness of consciousness-group of conscious existence - senses and mind (Buddhism, #HM2556).
Related Eye consciousness (Buddhism, #HM2074)
Ear consciousness (Buddhism, #HM2169)
Nose consciousness (Buddhism, #HM2364)
Body consciousness (Buddhism, #HM2562)
Mental consciousness (Buddhism, #HM2838)
Mind-consciousness element (Buddhism, #HM6173).

♦ HM2264d Belligerence (Buddhism)
Krodha — Khro-ba (Tibetan)
Description The intention to harm another when one is in a situation of harmful intent.
Context One of the twenty *secondary afflictions* of Tibetan Buddhism.
Broader Secondary afflictions (Buddhism, Tibetan, #HH0781).
Related Anger (Hinduism, #HM8665).

♦ HM2265d Resetting the mind (Buddhism)
Avasthapana — Slan-te-jog-pa (Tibetan)
Description At this level of meditation the mind is able to return readily to its object through meditative stabilization, and any distraction is momentary. The associated power in this state is mindfulness.
Context This is the third of the nine states of mental abiding (navakara chittasthiti) in Gelugpa Tibetan Buddhism.
Broader Calm abiding (Buddhism, #HM2147).
Followed by Close setting the mind (Buddhism, #HM2157).
Preceded by Continuous setting of mind (Buddhism, #HM2241).

♦ **HM2266g Ordinary awareness**
Everyday thinking
Description Ordinary, every–day awareness of the external world is dominated by two conditions: dependency on sense data, and the necessary mental reconstruction of the discrete units of sense data into internally sensible wholes that can be recognized and identified. Crucial to this process, which results in "understanding", is memory wherein the mental wholes or patterns corresponding to the objective world's phenomena are stored. One peculiarity of everyday consciousness is the requirement that it should respond at enormous speeds to sense stimuli. In order to do this it takes a short cut to understanding by reducing recognition time. This is achieved by matching the reconstruction of discrete date to the memory at stages where such reconstruction is only partial. Normally, the correct match is made with the corresponding internal structure in the memory. Sometimes the match is premature and error results. However, such error is usually corrected with more sense–data input which may come in continuously from the stimulus–object. The real limitation and significant source of error in everyday consciousness lies in the structures, maps, schemata, types, concepts, forms, etc which are in the memory, and which, as an imposition on the understanding process, require sense–data constructs to conform to.
Related Usual state of consciousness (USC, #HM1999).

♦ **HM2267d Subtle contemplation of withdrawal or joy** (Buddhism, Tibetan)
Ratisamgrahakamanaskara
Description This state of contemplation continues the process of the mind's withdrawal from the desire realm.
Context This is the fifth of the preparations in Tibetan Gelugpa Buddhism.
Broader Special insight and preparations as states (Buddhism, #HM2623).

♦ **HM2268c Great–vehicle lower–path awareness** (Buddhism, Tibetan)
Mahayana–path
Description This state of awareness is characterized by study leading to wisdom, by moral living and by the accumulation of good karma.
Context In Tibetan Sakya Buddhism this is one of the states in the "Ascension Stages Game". In some sets it is numbered 52 on the board.
Broader Mahayana Buddhism (Buddhism, #HH0900)
Sutra paths of accumulation and application in the ascension stages game (Buddhism, Tibetan, #HM2962).
Followed by Cold–hells awareness (Buddhism, Tibetan, #HM2040)
Animal–hell awareness (Buddhism, Tibetan, #HM2636)
Joyful–heaven awareness (Buddhism, Tibetan, #HM2022)
Discipleship–karma awareness (Buddhism, Tibetan, #HM2192)
Great–vehicle higher–path awareness (Buddhism, Tibetan, #HM3048)
Great–vehicle middle–path awareness (Buddhism, Tibetan, #HM2160).
Preceded by Form–realm awareness (Buddhism, Tibetan, #HM2142)
Pure–lands awareness (Buddhism, Tibetan, #HM2168)
Bon–wisdom awareness (Buddhism, Tibetan, #HM2663)
Joyful–heaven awareness (Buddhism, Tibetan, #HM2022)
Arhat sanctity awareness (Buddhism, Tibetan, #HM3024)
Hindu–states of awareness (Buddhism, Tibetan, #HM2115)
Pratyeka Buddha awareness (Buddhism, Tibetan, #HM2180)
Southern–continent awareness of men (Buddhism, #HM2127)
Discipleship–karma awareness (Buddhism, Tibetan, #HM2192)
Hindu–wisdom–holder awareness (Buddhism, Tibetan, #HM2723)
Pratyeka Buddha arhat awareness (Buddhism, Tibetan, #HM2776)
Discipleship–application awareness (Buddhism, Tibetan, #HM2716)
Bon–practitioner state of awareness (Buddhism, Tibetan, #HM2639)
Delightful–emanation–heaven consciousness (Buddhism, Tibetan, #HM2546).

♦ **HM2269d Manifold blessings** (ICA)
Description This is a surprised acknowledgement that every gift and foolishness, every friend and enemy, every passion and passivity, makes the situation the rich thing that it is. It occurs when one accepts all factors as significant, as in being thankful for an enemy who has made things difficult, or the unexpected interventions that have made plans for action useless.
Context In the New Religious Mode in the arena of acting out one's freedom (the life of prayer) the second formal aspect is affirmation of one's dependence or gratitude. At the third phenomenological level this occurs when one acknowledges that which is beyond one's power to control.
Followed by Unspeakable joy (ICA, #HM2400).
Preceded by Splendid vices (ICA, #HM2165).

♦ **HM2270c Feeling** (Buddhism)
Vedana — Tshor–ba (Tibetan) — Sensation
Description The objects of feeling are pleasure, pain and neutrality; it is the individual experiencing of the fruit of virtuous and non–virtuous actions. Thus all pleasures arise from virtuous acts, all pains from non–virtuous acts. All three (pleasure, pain, neutrality) may be experienced physically or mentally, the former accompanying any of the five sense–consciousnesses and the latter accompanying mental consciousness. Again, feeling may be the base of attachment to the Desire Realm or deliverance from the Desire Realm; or materialistic (toward physical or mental aggregates) or non–materialistic (accompanying wisdom consciousness cognizing selflessness).
Context One of the five omnipresent mental factors defined in Tibetan Buddhism.
Broader Omnipresent mental factors (Buddhism, Tibetan, #HH0320).
Narrower Psychically unpleasant (#HM0805)
Sensation of unease born of contact with the psychical (#HM1090)
Unease experienced as born of contact with the psychical (#HM0901)
Sensation of unpleasant born of contact with the psychical (#HM4076)
The unpleasant experienced as born of contact with the psychical (#HM1031)
Psychical ease born of contact with element of mind–consciousness (#HM1921)
Psychical unease born of contact with element of mind–consciousness (#HM0234)
Neither psychical ease nor unease born of contact with element of eye–consciousness (#HM0972).
Related Feeling aggregate (Buddhism, #HM4983).

♦ **HM2271c Mahayana highest–teachings–application awareness** (Buddhism, Tibetan)
Laukikagra–dharma
Description This state is characterized by the understanding of the voidness of all phenomena. One is said to be prepared to be a bodhisattva ready for full awakening. The concentration of "non–hindered thought" (anantarya–citta–samadhi) is said to have been accomplished and the path to freedom in faith (adhimukti–carya), has been traversed.
Context In Tibetan Sakya Buddhism this is one of the states in the "Ascension Stages Game". In some sets it is numbered 64 on the board.
Broader Sutra paths of accumulation and application in the ascension stages game (Buddhism, Tibetan, #HM2962).
Followed by Joy–land awareness (Buddhism, Tibetan, #HM2167)
Tantra heat–application awareness (Buddhism, Tibetan, #HM2696)
First scriptural bodhisattva awareness (Buddhism, Tibetan, #HM2155).
Preceded by Pure–lands awareness (Buddhism, Tibetan, #HM2168)
Joyful–heaven awareness (Buddhism, Tibetan, #HM2022)
Potala–island awareness (Buddhism, Tibetan, #HM2175)
Mahayana climax–application awareness (Buddhism, Tibetan, #HM2232)
Mahayana receptivity–application awareness (Buddhism, Tibetan, #HM2247).

♦ **HM2272b Mystical cognition**
Mysticism — Mystic experience — Experiential wisdom — Unitive mysticism — Epithalamian mysticism
Description The mystical is a particular type of inner experience which may or may not be called religious or spiritual according to the meaning given to these words. Such experience may range from the intuitive to the mystical in its fullest sense. Although some argue that mysticism is the heart of true religion, others contend that it is only one aspect, on a level and to be contrasted with the prophetic and the devotional. It is rejected completely by Marxism–Leninism as being irreconcilable with the scientific, materialistic world view.
The mystical manifests itself at different levels and in different ways, but in all cases consciousness and cognition operate as though an inner, non–rational light is communicated, through which the mind is enabled to apprehend something which would otherwise be inaccessible. This has been called experiential wisdom. The mystical can therefore be defined as the complexion which reality assumes when perception and thought have moved into a particular dimension of consciousness which is at the intersection of rational and non–rational thought and perception. As such it is a response to an apprehension of a quality of Is–ness, whether of God or of the world. At the heart of mystical knowledge is awareness of undifferentiated unity or unity with the source. This may be interpreted as identical with God or the Absolute, or simply as knowledge of union. To the believer in God the content of the experience is godliness; to the non–believer it is goodness. It not only has an ethical content, it is primarily validated by being demonstrated in a life of godliness or goodness.
– William James systematized the mystical experience under four headings: ineffability; noetic quality; transiency; passivity, these being the effects that the experience has on the person experiencing it.
– Evelyn Underhill summarized mysticism as a specialized form of the search for reality and for a heightened and completed life which is a constant characteristic of human consciousness. The spiritual spark, which lies below the threshold in ordinary men, emerges in the mystic and becomes the dominant factor in his life. The powers of love and will become servant to it and are enhanced by it. It has the following characteristics: (1) It is the art and science of establishing conscious relation with the Absolute rather than just talking about it. (2) Having surrendered to the embrace of reality, and with transformed vision, the mystic sees a different world and lives a spiritual life. (3) Mysticism is the art of the arts, it is the source and end of artistic inspiration. The mystic attempts to communicate a vision of reality through symbol and image, although no representation can contain the full meaning of the mystical experience. (4) True mysticism is an active and practical process of life, neither passive nor theoretical. It is not a philosophy and has nothing to do with occult knowledge, nor is it concerned with manipulating the physical universe. Its aims are entirely transcendental and spiritual. (5) Although intellectual investigation and emotional longing must be present for union with the One, they are not enough. The mystic way involves an arduous physiological and spiritual process and the liberating of the latent state of consciousness which has been referred to as ecstasy or the unitive state. (6) In mysticism, the will and the emotions are united in the desire to be joined in love with the one, eternal and ultimate object of love perceived by the soul. (7) For the mystic, the one reality is an object of love drawing one homeward under guidance of the heart. Generous love in all aspects of life is the business and method of mysticism. It is a passion pursued only for the sake of love and never self–seeking. The great religious traditions restate the relevance of mysticism to contemporary life, each with a unique perception of the universal experience. Such restatements are also found in art and in the life of any human being who awakens to spiritual knowledge and ecstatic love.
– W T Stace distinguishes between extrovertive mysticism, when the whole of nature appears in harmony and at one, and introvertive mysticism, when everything is excluded from the mind except the self, and finally the self is absorbed in the void. Five characteristics are characteristic of both forms: blessedness, objectivity, holiness, paradoxicality and ineffability. The other two characteristics differ in the two forms: unity, which the extrovertive sees in the all while the introvertive in the one, void or pure consciousness; and apprehension which is of the inner life of all things in the extrovertive while in the introvertive is nonspatial and nontemporal. In fact, the extrovertive is seen to be the lower level of the introvertive, in which its tendencies are fulfilled. It is as though the multiplicity is half absorbed in unity in the first case and wholly obliterated in the second.
The mystical experience may be evident as an expansion of the range of perception and awareness permitting experiences which, though interpreted according to different religious philosophies, have the same character and presumably the same origin. Such experiences notably include permanent states of consciousness variously called *illumination*, *enlightenment*, *union with God*, *nirvana*, etc. The mystical may also take the form of a transitory experience, occurring perhaps only once or twice in a lifetime, having a unique quality and perception of reality which remains of great importance to the individual. Terms used to label such experiences include: peak experience, intuitive enlightenment, timeless moment, illumination. The mystical may also take the form of a special feeling or non–rational apprehension of the nature of reality. It may manifest itself as a contemplative insight culminating a process of rational and analytical thought.
For some, mystical experience does not necessarily imply religious experience. The experience seems to be the same independent of the religious tradition (or lack of it) of the experiencer. W T Stace, for example, sees God as an interpretation of the experience, not the experience itself. Others have taken mysticism to imply religious experience, an experiencing of God rather than intellectual belief. Meister Eckhart, in this sense, is referring to a way of life. It is not the transient experience such as arises on a drugs "trip", for example, but an abiding attitude which remains constant through all states of consciousness. This permanent state of letting go, of surrender, brings receptivity to whatever the present moment may bring. It is union with God through total surrender of the self.
There is a distinction to be drawn between mysticism as approached in western religion, the *epithalamian* tradition, with the goal of marriage between the soul and God; and the approach of eastern religious traditions, where fusion of the soul with God, unitive experience, is the aim. The former may be looked upon as *agape*, the latter *eros*. In unitive mysticism the soul reaches God, something which cannot be said to occur in epithalamian mysticism, where there always remains passion and longing.
While the more complete and permanent manifestations of the mystical are experienced by comparatively few, it is believed that it is possible for one who is in no way a contemplative, and who has never had any unique mystical experience, to think and apprehend mystically, and so to arrive at a mystical interpretation of the nature of the world, rather than assuming the intellect to be the ultimate manner of knowing. The intellect may even be considered a hindrance to mystical cognition, which may also oppose logic and philosophy.
Mystical experience is distinguished from visionary experience in that it is not described in sensory language and is marked by low levels of normal cognition and physiological activity.
Refs Cox, M *Mysticism* (1983); Coxhead, Nona *The Relevance of Bliss* (1985); Dupré, Louis and Wiseman, James A (Eds) *Light from Light* (1988); Ferguson, J *Encyclopedia of Mysticism* (1976); Katz, Steven *Mysticism and Religious Traditions* (1983); Katz, Steven T (Ed) *Mysticism and Philosophical Analysis* (1978); Krishna, Gopi *The True Nature of Mystical Experiences* (1978);

O'Brien, E *Varieties of Mystic Experience* (1965); Stace, Walter T *Mysticism and Philosophy* (1960); Underhill, Evelyn *Mysticism* (1977); Wainwright, W *Mysticism* (1981); Zaehner, R C *Mysticism, Sacred and Profane* (1961).

Narrower Eastern mysticism (#HH9219) Sikh mysticism (Sikhism, #HH6199) Environmental mysticism (#HM0800) Jewish mysticism (Judaism, #HH1232) Christian mysticism (Christianity, #HH5100).
Related Mystery (#HH0303) Illumination (#HH0804) Ineffability (#HM3589) Enlightenment (#HM2029) Numinosum (Jung, #HM3811) Hasidism (Judaism, #HH0597) Nirvana (Buddhism, #HM2330) Deautomatization (#HH2331) Peak experiences (#HM2080) Religious experience (#HM3445) Visionary experience (#HM9218) Higher states of consciousness (#HM0935) Awareness of the sacred (Christianity, #HM0876) Deautomatization and the mystic experience (#HM4398) Birth of God in the soul's ground (Christianity, #HM6522).

♦ **HM2273d Radical illumination** (ICA)
Certitude at the centre
Description This is an experiential trek in the *Sea of Tranquillity*, within the ICA *Other World in the midst of this World*, when the individual comes to know that life shines in the shadows of existence. The states are described in separate entries.
Broader Sea of tranquillity (ICA, #HM3033).
Narrower Contentless word (#HM2373) Personal epiphany (#HM2595) Seminal illumination (#HM2056) Inclusive comprehension (#HM2256).
Related Endless life (ICA, #HM2437) Unknowable peace (ICA, #HM3015) Contentment at the centre (ICA, #HM2529).

♦ **HM2275d Jewelled–peaks–realm awareness** (Buddhism, Tibetan)
Ratna–kuta
Description This state of awareness is characterized by understanding the nature of equality. It is the antithesis to mental justifications for egoism, for self–interest, or for hoarding of material goods. It also dissolves greed for immaterial and spiritual goods, such as secret knowledge or unusual powers. This state is represented as the Southern Pure Land of Ratnasambhava who created the Jewelled–peaks realm. Through his wish–granting gem the Buddha's power of providing means is afforded. Subordinate to him, it is said, is the bodhisattva called *space–nature* (akasa–garbha).
Context In Tibetan Sakya Buddhism this is one of the states in the "Ascension Stages Game". In some sets it is numbered 76 on the board.
Broader Supreme heavens in the ascension stages game (Buddhism, Tibetan, #HM3214).
Followed by Third vajra–master awareness (Buddhism, Tibetan, #HM2727) Second vajra–master awareness (Buddhism, Tibetan, #HM2703) Fourth scriptural bodhisattva awareness (Buddhism, Tibetan, #HM2191).
Preceded by Hyper–bliss–realm awareness (Buddhism, Tibetan, #HM2337).

♦ **HM2277d Mental abiding in equipoise** (Buddhism)
Samadhana — Mnyam–par–jog–pa (Tibetan)
Description In this stage the meditator is able to remain effortlessly in stabilized meditation. He has overcome the five faults: laziness, forgetting the precepts, laxity and excitement, non–application, and application–effort. A similitude of calm abiding is achieved.
Context This is the last of the nine states of mental abiding (navakara chitta sthiti) in Gelugpa Tibetan Buddhism, the second stage of meditation on emptiness, leading to calm abiding. The next stage is *special insight.*
Broader Calm abiding (Buddhism, #HM2147).
Preceded by One–pointedness in mental abiding states (Buddhism, #HM2753).

♦ **HM2278b Lila**
Leela — Game of life — Gyan chaupud — Game of knowledge — Cosmic love
Description The universal play of cosmic energy, or *lila*, in which each human being plays his or her part, has been formalized into a system reminiscent of "snakes and ladders" in which the player sees his progress through the play of life. As the individual plays he becomes identified with the persona, the role he is playing, and then his life is decided by karma. However, there are still moments when perception of non–duality breaks through. This perception may be strengthened and lengthened by practice. At a deep level of awareness the realization occurs that the world and God are a unity, the world being a diversified expression of God.
Perception of the positive non–duality of the timeless reality is referred to as *integral existential experience*, and is the basis of illumined creative living. This state is referred to in Hindu philosophy as reflecting cosmic love, in which the individual joyfully acts towards human evolution free from personal desires or attachments. Such an expression of cosmic love, created through the burning of the ego, is contrasted with ethical love which maintains egocentric individuality despite active involvement in social improvements.
The game "Leela" itself is played on a board whose squares are related numerologically to produce a perfect rectangle. The board comprises an octave of horizontal levels, equally representing stages in temporal human life and psychic centres or chakras in spiritual development. Each level is divided into nine squares, representing in all the 72 primary states of being. The pattern of an individual's existence becomes clearer as pathways in the game (progressing according to the throw of a die with leaps upwards by means of arrows or downwards along snakes) tend to repeat themselves. Each horizontal level represents particular energy vibration. A player who, by karmic chance, reaches a level with whose vibrations his own are not in accord will soon reach a snake and slide back down. Synchronicity determines the fall of the die. There is no death; and eventually there is harmony at all levels, so there are no "ups and downs". The game only comes to an end when the player attains total understanding of the game and arrives at the state of cosmic consciousness.
Refs Johari, Harish *Leela* (1980).
Broader Games (#HH0705) Maps of the mind (#HH8903).
Narrower Sixth chakra: time of penance (Leela, #HM4412) Third chakra: theatre of karma (Leela, #HM0717) Second chakra: realm of fantasy (Leela, #HM3651) Fourth chakra: attaining balance (Leela, #HM3291) Seventh chakra: plane of reality (Leela, #HM0754) Fifth chakra: man becomes himself (Leela, #HM0933) First chakra: fundamentals of being (Leela, #HM4103) Beyond the chakras: the gods themselves (Leela, #HM1141).
Related Thetan (#HH0269) Synchronicity (#HH4269) Love consciousness (#HM2000) Existential integration (#HH0832) Consciousness ascension stages game (Buddhism, Tibetan, #HH4000).

♦ **HM2280c Tantra union–in–learning–application awareness** (Buddhism, Tibetan)
Description This state of awareness is represented as the fourth and final Tantric initiation in which the initiate receives the "word" or teaching concerning the highest worldly dharma on the path of Tantra. This teaching explains that at the higher limit of the corresponding fourth meditative cultivation, the union of wisdom–knowledge (prajna–jnana), has given access to all the subtle–energy focal–points (cakras) in the body. As a result the body at the higher level of awareness with be one of the noble ones (adamantine or rainbow). In addition, perseverence in repeating the union of realization and emptiness leads to buddhahood, which is said to be the union surpassing union–in–learning.
Context In Tibetan Sakya Buddhism this is one of the states in the "Ascension Stages Game". In some sets it is numbered 58 on the board.
Broader Tantric paths of accumulation and application in the ascension–stages–game (Buddhism, Tibetan, #HM2848).
Followed by First vajra–master stage (Buddhism, Tibetan, #HM2187) Hyper–bliss–realm awareness (Buddhism, Tibetan, #HM2337) Third vajra–master awareness (Buddhism, Tibetan, #HM2727) Second vajra–master awareness (Buddhism, Tibetan, #HM2703).
Preceded by Tantra receptivity–application awareness (Buddhism, Tibetan, #HM2756).

♦ **HM2281c Formless–realm consciousness** (Buddhism)
Arupadhatu — Gzugs–sku (Tibetan)
Description There are four levels in this realm: Limitless Space, Limitless Consciousness, Nothingness, and the Peak of Cyclic Existence. The gods or beings of this realm are meditators who have been reborn here because they have attained the corresponding consciousness through practice of the absorptions.
Broader Consciousness states in cyclic existence (Buddhism, #HM2177).
Related Arupaloka consciousness (Buddhism, Pali, #HM2012) Immaterial–sphere concentration (Buddhism, #HM0696) Formless–realm awareness (Buddhism, Tibetan, #HM3144) First absorption in the immaterial sphere (Buddhism, #HM2110) Third absorption in the immaterial sphere (Buddhism, #HM2027) Second absorption in the immaterial sphere (Buddhism, #HM3043).

♦ **HM2282c Angelic awareness - Zohárariel**
Ariel — Jeu
Description This mode of awareness is represented as the theophanic consciousness, wherein the ego confronts its god, its archetypal forefather. In Jewish mysticism associated with ascension through the heavenly palaces (nekaloth) to reach god's throne (merkabah), the object of the quest is said to have been the vision of Zohárariel, a secret name of God in the aspect in which he discloses himself. Disclosed, as an object of awareness, he is an angel, that is, a subjective experience in the consciousness. Zohar–ariel means splendour–lion. To this name is added the four–letters JHVH to show its divinity. Ariel was also the name given to the prototypal Jerusalem, the heavenly city and its ruler, and seems to have been the older name for the lion–headed gnostic demiourgos Ialdalbaoth, also known as Iao and as Jeu (the name that is included in Jesu). He also may be identified with Abraxas, Brahma, and Janus, who like JHVH are fourfold or manifold divinities pointing to the indefiniteness of the Real in its experiential aspect.
Broader Angelic frame of awareness (#HM2150).

♦ **HM2283d Rational, calculative thinking**
Description As opposed to *intuitive, meditative thinking*, which is concerned with ontological reality, *rational, calculative thinking* is the calculative mode which predominates in modern, secular, technological society and is concerned with the ontic level of material being. This mode is said to be associated with a loss of awareness both of the true ground of being and of the sense of awe and wonder inherent in such awareness. Reality is seen as a duality of subject and object, as the objective world to be conquered by the will of man.
Related Intuitive, meditative thinking (#HM2785).

♦ **HM2284c Concentration of the third jhana of four** (Buddhism)
Description With the elimination or fading away of rapture or happiness, two factors remain in the third jhana: ease or bliss; concentration.
Context According to Hinayana Buddhism, concentration is considered as of one kind (monad), of two kinds (dyads), of three kinds (triads), of four kinds (tetrads) or of five kinds (pentad). In the third tetrad, fourfold according to the four jhana factors, this is the third concentration. It is equivalent to the fourth concentration of the pentad.
Broader Concentration (Buddhism, #HM6663).
Related Dwelling in the third jhana (Buddhism, #HM5643) Concentration of the fourth jhana of five (Buddhism, #HM4882).
Followed by Concentration of the fourth jhana of four (Buddhism, #HM7202).
Preceded by Concentration of the second jhana of four (Buddhism, #HM4380).

♦ **HM2285c Female sovereignty archetypal image** (Tarot)
Empress — Matrona
Description One of the archetypal emblems projected by the psyche is the matron, a mature woman of authority (mundane as opposed to spiritual). In the Tarot she is represented by the emblem of the Empress (image number 3). This is the realm of pure illumination, where the archetypal opposites - mother and father - are combined; on the Tree of Life, this is the highest form of balance. From the womb of the Empress flow all forms capable of manifest existence.
Context In the Kabbalah, the Empress is said to correspond to the 14th path from intelligence (Chokmah) to wisdom (Binah).
Broader Emblem archetypes in the psyche (Tarot, #HM2201) Tarot arcana of conscious states (Tarot, #HM2097).
Related Sphere of wisdom (Kabbalah, #HM2348) Sphere of understanding (Kabbalah, #HM2372).

♦ **HM2286d Awareness of spiritual poverty** (ICA)
Description This is the realization that one is incapable of doing with one's life what one senses one has been called upon to do. One sees that the task one has accepted is far beyond one's own capacity. St Francis of Assisi said that his body was like a donkey which continually did what it wished no matter how he beat it to get it to do his will.
Context In the ICA New Religious Mode in the arena of knowing one's disengagement (the life of poverty) the fourth formal aspect is the experience of spiritual denial or sacrificial offering. At the third phenomenological level, this occurs when one is able to so value the gift of life in God that nothing else is needed for one's fulfilment.
Broader Spiritual denial (ICA, #HM2174).
Related Spiritual poverty (#HH0190).
Followed by Defender of deeps (ICA, #HM2591).
Preceded by Intentional self–negation (ICA, #HM2369).

♦ **HM2287d Sixth vajra–master awareness** (Buddhism, Tibetan)
Description This state of awareness is characterized by innerness and is represented by the yogi's ability to travel through the system of his subtle body channels. He is said to reach his destination at the stop called the heart–knot. By threading through its maze, propelled by psychic winds he unravels its mystery.
Context In Tibetan Sakya Buddhism this is one of the states in the "Ascension Stages Game". In some sets it is numbered 82 on the board.
Broader Way of the vajra masters (Buddhism, Tibetan, #HM3090).
Related Vritti (Hinduism, #HM2692).

Followed by Ninth vajra-master awareness (Buddhism, Tibetan, #HM2325)
Eighth vajra-master awareness (Buddhism, Tibetan, #HM2301).
Preceded by Fourth vajra-master awareness (Buddhism, Tibetan, #HM2251).

◆ **HM2288c Sphere of the kingdom** (Kabbalah)
Malkuth sephira
Description This sphere of awareness is associated with the sub-conscious, the functions of the physical body, and thus with the development of physical or somatic awareness. It is the immediate consciousness, the materialization of ideas, and also the beginning or entrance to all spiritual journeys. It corresponds to the developmental stage of incarnation and birth. It is symbolized by the positive and negative archetypes of Earth and the body, as well as by David in the Jewish tradition. It is associated with the immediate environment, with the world of earth, crops, living things.
Context This sphere is the last or lowest of ten described in the Kabbalistic system of sephiroth.
Broader Kabbalah (#HH0921).
Related Muladhara (Yoga, #HM2893)   Emblems of renewal (Tarot, #HM2315)
Emblems of totality (Tarot, #HM2387)   World of manifestation (Judaism, #HM2312)
Emblems of the mysterious (Tarot, #HM2398)
Way of active intelligence (Esotericism, #HM1997).
Followed by Sphere of glory (Kabbalah, #HM2238)
Sphere of victory (Kabbalah, #HM2362)
Sphere of the foundation (Kabbalah, #HM2410).

◆ **HM2289c Jesus falls the third time under the cross** (Christianity)
Description The believer is aware of the weakness of Jesus which caused Him to fall a third time, and begs for strength to overcome the human traits and wicked passions which take the emphasis from and cause him to despise the friendship of Jesus. Loving Jesus more than himself, he repents with his whole heart and begs never again to separate himself from Him. He asks that he will love Jesus always and that Jesus will do with him as He wills.
Context The ninth station of the cross.
Broader Way of the cross (Christianity, #HM3516).
Followed by Jesus stripped of his garments (Christianity, #HM3166).
Preceded by Jesus speaks to the daughters of Jerusalem (Christianity, #HM2737).

◆ **HM2290c Sphere of judgement** (Kabbalah)
Geburah-din sephira
Description This sphere of awareness is associated with passive emotion and with a sense of discernment and thus with the development of discipline. It corresponds to the developmental phase of early middle age and the cultivation of determination. This state is associated with the universal laws or justice that measure and weigh all things, thus fixing their conditions and limitations in the whole. A purging and cleansing force, through which unwanted and unnecessary elements are destroyed with severity and justice. It is symbolized by the positive and negative archetypes of hero and inquisitor, and by the planet Mars, as well as by Isaac in the Jewish tradition.
Context This sphere is the fifth of ten described in the Kabbalistic system of sephiroth.
Broader Kabbalah (#HH0921).
Related Measuring emblems (Tarot, #HM2178)   Way of power (Esotericism, #HM0454)
World of creation (Judaism, #HM3038)   Victimization emblems (Tarot, #HM2190)
Natural forces emblems (Tarot, #HM3127)   Charioteer archetypal image (Tarot, #HM3045).
Followed by Sphere of understanding (Kabbalah, #HM2372)
Sphere of mercy and greatness (Kabbalah, #HM2420).
Preceded by Sphere of glory (Kabbalah, #HM2238)
Sphere of beauty (Kabbalah, #HM3031).

◆ **HM2291b Cosmic consciousness**
Illumination
Description This term has been applied to a level of consciousness and insight of which the prime characteristic is a consciousness of the cosmos and of the life and order of the universe as an experienced whole. It has been distinguished from mystical experiences and from spiritual or metaphysical awareness which imply that it is not also extremely concrete and physical although some of the features of such experiences may also be present in cosmic consciousness. Reported characteristics of such consciousness include: a sense of objective light, moral elevation, intellectual illumination, sense of immortality, loss of fear of death, loss of sense of sin, instantaneousness of the awakening, transfiguration of the person as seen by others. To the individual thus enlightened it appears as a vivid and overwhelming certainty that the universe, precisely as it is at this moment, as a whole and in every one of its parts, is so completely right as to need no explanation or justification beyond what it simply is. The universe may be seen, not as dead matter but as a living presence. The immediate now, whatever its nature, becomes the goal and fulfilment of living. An emotional ecstasy is associated with this core experience, a sense of intense relief, freedom, and lightness, and often of almost unbearable love for the world. Although the experience may last only a few seconds it may have a permanent effect on the life of the person experiencing it, a lasting knowledge that, whatever the appearances, this is how things really are.
In the 19th Century, William James defined four characteristics of the experience: (1) There is a sense of unity or oneness, so that experience is comprehensive, not fragmented, and one sees relationships between things that are normally separate. (2) There is an awareness that the experience of relations between things is more real, closer to the truth, than normal experience. (3) The experience is ineffable and cannot be communicated in words. (4) There is a vividness and richness, a freshness and clarity, not normally present.
One study of well-known cases suggests that cosmic consciousness appears in individuals (mostly men, who are otherwise highly developed physically, morally and intellectually,) between the age of 30 and 40. Intimations of this state may be experienced by many people for no apparent reason and possibly with the aid of psychedelic drugs.
Some believe that such consciousness is an emerging faculty which represents a direction of future human evolution; others equate it with the *illumination* achieved from the practice of yoga, for example, as described by Patanjali. The individual has achieved conscious self-relatedness and has the courage to go against his tribe, community or race in pursuit of international justice and peace.
Refs Bucke, Richard M *Cosmic Consciousness* (1972).
Related Oversoul (#HH0504)   Primitivism (#HH1778)   Illumination (#HH0804)
Ultra-consciousness (#HM2156)   Religious experience (#HM3445)
Free self-inquiry meditation (#HH0508)   Steps to cosmic consciousness (#HH3220)
Conscious self-relatedness (Yoga, #HM0467)
Transcendental cosmic consciousness (#HM2405).
Preceded by International consciousness (Yoga, #HM0099).

◆ **HM2292d Ninth scriptural bodhisattva awareness** (Buddhism, Tibetan)
Sadhu-mati — Appropriate intellect
Description This state of awareness is of the need to conform teaching to the student. It is represented in its developed form as the ability to speak in several languages and styles of delivery, with varying formulations adapted to hearers, all simultaneously – as practised by the ninth-stage bodhisattva expounding the dharma. This stage is called, appropriate intellect (sadhu mati).
Context In Tibetan Sakya Buddhism this is one of the states in the "Ascension Stages Game". In some sets it is numbered 95 on the board.
Broader Meditation way of the bodhisattvas (Buddhism, #HM2769).
Followed by Supreme-heaven awareness (Buddhism, Tibetan, #HM2813)
Tenth scriptural bodhisattva awareness (Buddhism, Tibetan, #HM2421)
Preceded by Fifth scriptural bodhisattva awareness (Buddhism, Tibetan, #HM2909)
Eighth scriptural bodhisattva awareness (Buddhism, Tibetan, #HM2816)
Seventh-scriptural bodhisattva awareness (Buddhism, Tibetan, #HM2361).

◆ **HM2293d Religious vocation** (ICA)
Description This state is experienced when a person acknowledges his covenant in the spirit with all who ever cared for creation, when he is called to display his commitment through a dedicated life in action. It is like Martin Luther King's "I have a dream" speech describing a covenant with all the suffering people of the world.
Context In the ICA New Religious Mode in the arena of acting out one's deed (the life of doing) the third formal aspect is one's awareness of one's allegiance to those who act on behalf of life as a whole. At the third phenomenological level this occurs when the individual embraces his election to this life, this task.
Broader Awareness of allegiance (ICA, #HM2073).
Related Vocation (Christianity, #HH0619).
Followed by Primordial colloquy (ICA, #HM2352).
Preceded by People of God (ICA, #HM2484).

◆ **HM2294c Emptiness of ultimate nirvana** (Buddhism, Tibetan)
Paramarthashunyata
Context One of the eighteen emptinesses comprising the paths of view in Tibetan Buddhism.
Broader Emptinesses on the paths of the view in Buddhism (Buddhism, Tibetan, #HM2944).

◆ **HM2295g Brain-based consciousness**
Description Different types of consciousness and thought processes are associated with major brain areas. The brain-evolution theories (Papez, MacLean) describe man as possessing three brains: the reptilian (top of brain stem); the old- or paleo-mammalian (limbic system or limbic cortex); and the new, neo-mammalian or neo-cortex. Anatomically, the stem is encircled by the limbic system, and this in turn is envelopped by the neo-cortex, thus constituting a pattern of brain-within-brain. This system can be described as the brain's vertical structure. The horizontal structure is the division of the cerebrum into two hemispheres: the left brain and the right brain which, unlike the vertical, is now well-integrated, joined by the great cerebral commissure, or corpus callosum, although several thousand years ago the hemispheres may have acted independently or bicamerally (J Jaynes). Another feature of the brain affecting consciousness is its processing of images and other inputs to compare them with stored memories or anticipations. This creates an interference pattern between the two components and sets up a holographic redundancy, or distribution of "learning-experience" to be stored in many areas by the brain (K Pribram).
Consciousness in its different forms can be affiliated to the anatomy of the brain: the triple brain and the left and right brain being, currently, the most prominent examples. Consciousness can also be related or correlated to brain and blood chemistry (taking hormones and other substances into account); and to electrical brain waves (alpha, beta, epsilon, theta, etc). Consciousness as a function of the brain is also influenced by genetic traits. Consciousness, however, may not be totally reducible to brain functioning as certain phenomena, viz creative genius and extra-sensory perception, are not fully understood.
Refs Annett, Marian *Left, Right, Hand and Brain* (1985); Goleman, D and Davidson, R J (Eds) *Consciousness* (1979).
Narrower Holistic consciousness (#HM2367)   Analytical consciousness (#HM2415)
Reptilian human brain consciousness (#HM2307)
Mammalian limbic human brain consciousness (#HM2355)
Neomammalian neocortex human brain consciousness (#HM2427).

◆ **HM2297b Tacit knowledge in consciousness**
Description In whatever direction awareness extends, it brings to bear an important amount of already stored experience that the mind associates as relevant. To a considerable extent, this tacit knowledge represents an incoherent assembly of valid and invalid information that includes theories as well as perception-processing techniques. According to M Polanyi, tacit knowledge situates new experience in a context that influences every level of judgement. It also is projected and fixed in the metaphors, myths and symbols in which mankind clothes its experience. Examination of metaphors in art and religion, for example, discloses the realm of tacit knowledge and its many forms and formulations.
Related Contextual consciousness (#HM2380).

◆ **HM2298d Discrete states of consciousness** (Physical sciences)
Description In atomic physics, certain energy states can occur only in a complete succeed or fail manner. There are no intermediate states composed of infinitesimal gradations. States of consciousness may be structured in an analogous (discrete or quantized) manner, and transitions may exhibit the equivalent of the quantum jump in high energy nuclear processes. Individual states themselves have properties that restrict their interactions and which provide them with character-istic structures.
Context There are at least five universal principles postulated in the physical sciences: duality; quantum discreteness; relativity; conservation; and least action. Their application to states of consciousness has been presented by Professor Charles Tart and others.
Refs Tart, Charles T *States of Consciousness and State-Specific Sciences* (1972).
Broader States of consciousness (Physical sciences, #HM2938).
Related Physical duality in conscious states (Physical sciences, #HM2381)
Physical relativity in conscious states (Physical sciences, #HM2322)
Physical conservation in conscious states (Physical sciences, #HM2357)
Least action principle in conscious states (Physical sciences, #HM2417).

◆ **HM2299c Life of poverty** (ICA)
Description An arena in the ICA *New Religious Mode* when one is aware of knowing one's disengagement.
Broader New religious modes (ICA, #HM3004).
Narrower Spiritual denial (#HM2174)   Disengagement from work (#HM2552)
Liberation from possessions (#HM3131)   Liberation from relationships (#HM3240).

◆ **HM2300b Krishna consciousness**
Description The final aim of devotees of the *International Society for Krishna Consciousness*, this is awareness of unity with Krishna as universal creator. By means of *bhakti yoga* (the yoga of devotion, followers learn first to know and to love Krishna and then approach self-realization.
Broader God-consciousness (#HM2166)
Related Bhakti yoga (Yoga, #HH0337).

♦ **HM2301d Eighth vajra–master awareness** (Buddhism, Tibetan)
Description This state of tantra awareness is characterized by projection and is represented by the master's ability to produce mentally an "apparent" vehicle for his subtle nature, in the Form–realm.
Context In Tibetan Sakya Buddhism this is one of the states in the "Ascension Stages Game". In some sets it is numbered 89 on the board.
Broader Way of the vajra masters (Buddhism, Tibetan, #HM3090).
Preceded by Sixth vajra–master awareness (Buddhism, Tibetan, #HM2287)
Fifth vajra–master awareness (Buddhism, Tibetan, #HM2763)
Guhya–samaja urgyan awareness (Buddhism, Tibetan, #HM2699).

♦ **HM2302b Sense of sacredness**
Consciousness of awe
Description The concept of sacredness arises from the unique and special value gives to encounters with the numinous or the holy, or with high reality, or with any intense experience evoking the deepest love, wonder, or awe. The cause of such a sense or consciousness of sacredness may lie in the range of natural or of supernatural perception, and may be objective or subjective. Regardless of cause, the outcome is that one feels that the experience is capable of being profaned in a number of ways. Therefore the encounter with awe or with the sacred may be kept secret and the individual experiencing it may believe himself to be privileged. Also, awe can be induced in people, causing them to suspend critical judgement and may lead to manipulation.
Related Awe (#HM2592) Taboo (#HH1101) Sacred places (#HH0593)
Supernaturalism (#HH4213).

♦ **HM2303d Cruciform exaltation** (ICA)
Description This is the joy of knowing that the death of your own illusions leads to new birth; that your own death means releasing others to freedom. It may be the decision to give up one's own life for the sake of others' lives, like the priest who offered to go to the gas chambers in another's place.
Context In the ICA New Religious Mode in the arena of knowing and doing intensified in one's life (life of being) the third formal aspect is the experience of transparent engagement or the experience of levitation. At the fourth phenomenological level this occurs when a person becomes one with the eternal.
Broader Transparent engagement (ICA, #HM2736).
Preceded by Keeping conscience (ICA, #HM2249).

♦ **HM2304d Mastery of stalking**
Description Mastery of *stalking* is one of three techniques described by Casteneda as leading to mastery of awareness. It involves systematic control of behaviour, which might be considered surreptitious in dealings with other people.
Related Mastery of intent (#HM3064) Mastery of dreaming (#HM3173).

♦ **HM2305c State–awareness** (Sufism)
Muraqaba — Muraqabat — — Watching — Observation of one's psychic stream — Constant attention
Description This state is defined as "the servant of God's constant awareness that the lord knows all his states". It is also said to be awareness of all one's states along with the awareness of God. The seeker is either in the *hal* of observing the action of God or the hal of being observed by God. The state of inner concentration is so great that external happenings are not responded to and the individual may not even breath perceptibly. The focus of this intent watching, concentration or self–observation is represented as being on the heart which, according to Rumi, is to be continually polished by the mind. It will become, he says, a mirror of the unseen, and images of angels and houris may dart in and out of its frame. It fully reflects the essence of the universe. Social and intellectual veils which diminish creative sensitivity are in this way removed.
Context The first psychic state listed by A Reza Aresteh as progress on the inner self through divine attraction, *kedesh–jazba*, the outer self already having been purified through conscious effort, *kushesh*. A preparation for this state is the actualization of *faqr*, divine poverty. According to Shaykh Abu Sa'id ibn Abi'l–Khayr, this is the sixth of 40 stations or maqamat the Sufi must possess for his journey on the path of Sufism to be acceptable.
Broader Psychic states (Sufism, #HM4311)
Higher states of consciousness (Sufism, #HM2365)
Stations of consciousness - ibn–Abi'l–Khayr (Sufism, #HM2424).
Related Cleansing the sirr (Sufism, #HM5353).
Followed by Sabr (Sufism, #HM3418) Nearness–awareness (Sufism, #HM2377).
Preceded by Faqr (Sufism, #HM3427) Mujahadat (Sufism, #HM5215).

♦ **HM2306g Middle unconscious** (Psychosynthesis)
Description According to R Assagioli, the second level of the unconscious is composed of elements similar to those in waking consciousness and thus it "speaks the language" of the everyday mind. Its accessibility shows that this is the inner region where daily experience is assimilated in a continuum with the consciousness, and to some extent, formulated as names, concepts, or imaginative products which are presented to the conscious mind.
Related Psychosynthesis (#HH0002).
Followed by Personal consciousness field (Psychosynthesis, #HM2383).
Preceded by Lower unconscious (Psychosynthesis, #HM2790).

♦ **HM2307g Reptilian human brain consciousness**
Description In man, the phytogenetically eldest brain functions are those performed by the primitive, reptilian–like, brain stem. This structure is akin to the brains found in lizards, alligators, snakes and the like, and is the centre of the instinctual life.
Broader Brain–based consciousness (#HM2295).
Followed by Mammalian limbic human brain consciousness (#HM2355).

♦ **HM2308b Sabija samadhi** (Hinduism)
Description This is a state of samadhi, in which there is a seed of thought left, a state of concentration which involves the suppression of intruding thoughts.
Broader Samadhi (Hinduism, Yoga, #HM2226).
Related Indian approaches to consciousness (Hinduism, #HM2446).

♦ **HM2309d Destinal accountability** (ICA)
Description This occurs when an individual gives up his old ties to this world, and enters into a marriage covenant with Being itself. He is henceforth responsible to God alone and every deed is rendered accountable to seeing life as just one deed - to invent the new creation that was the old self but is now transformed. Siddhartha in Hermann Hesse's book of the same name portrays the experience of this transformation.
A dimension of this experience is a sense of awe, that is fear and fascination. The fear is that maybe it doesn't matter, and this is accompanied by a kind of sorrow over all the violations that have been involved in the journey to this moment. The fascination is the sense that one's life is in touch with history; and there is an utter openness to the unknown in one's life and in history. The decision is to let history be the judge and to give one's self completely to it. This leaves a new sense of who or what one is obligated to; and that one's life is completed. One is full of peace as one grasps that that there is nothing else to do.
Context This state is number 32 in the ICA *Other World in the midst of this World*.
Broader Final accountability (ICA, #HM3206).
Followed by Perpetual becoming (ICA, #HM2717)
Archetypal humanness (ICA, #HM3112)
Contingent eternality (ICA, #HM2456)
Individual fatefulness (ICA, #HM2223)
Definitive effectivity (ICA, #HM2796).
Preceded by Primal vocation (ICA, #HM2621) Singular adoration (ICA, #HM2111)
Archetypal humanness (ICA, #HM3112) Contingent eternality (ICA, #HM2456)
Passionate disinterest (ICA, #HM2547).

♦ **HM2311c Meditation–state bardo awareness** (Buddhism, Tibetan)
Third antarabhava consciousness
Description This state of awareness is of the intermediate sphere of experience of meditation. Thus, achievement of samadhi and the clear light of the dharmakaya, or of any of the stages of educating awareness, have the profoundest effect on the after–life bardos.
Context This state is one of the six major bardo existences in Tibetan Buddhism.
Broader Bardo consciousness (Buddhism, Tibetan, #HM0698).
Followed by Death–bardo clear–light awareness (Buddhism, #HM2335).
Preceded by Dream–state bardo awareness (Buddhism, Tibetan, #HM2371).

♦ **HM2312c World of manifestation** (Judaism)
Olam–asiyah
Description This world, encompassing our own, is represented by the domain of sensory facts. It is associated with the five elements in man, the fifth of which is the corporeal quintessence (avir) or "ether". This world is the realm of the sphere (sephira) kingdom *Malkuth*, characterized as the substantial recipient of all the sephirotic emanations.
Context This is lowest of the Kabbalistic four worlds.
Broader Shekinah awareness (Judaism, #HM2002).
Related Paths of wisdom (Judaism, #HM2509) Sphere of the kingdom (Kabbalah, #HM2288).
Followed by World of creation (Judaism, #HM3038).

♦ **HM2313f Modes of awareness associated with schizophrenia**
Description (1) Delusions: Thoughts are disturbed so that there may be multiple, fragmented or bizarre delusions. Events, objects or other people may seem especially significant, usually in a negative way. There is a feeling that one's thoughts are broadcast so that others hear them, that thoughts are inserted into or withdrawn from one's head, that thoughts, feelings, impulses and actions are imposed by some external force.
(2) Disorder of thought form: Ideas shift from one thing to another with little or no connection so that communication becomes confused, apparently meaningless or incoherent.
(3) Hallucinations: There are major disturbances in perception, particularly the hearing of voices outside the head, which may be making derogatory remarks or commands which must be obeyed despite danger to the hearer or others. Hallucinations involving the other senses also occur, together with perceptual abnormalities such as a feeling of bodily change and synaesthesia.
(4) Affective disturbance: There may be flat affect, with lowered emotional response or no feelings at all; or there may be inappropriate affect, out of harmony with the person's ideas or speech.
(5) Ego–loss: The sense of unique self–hood is disturbed so that the person is confused about his or her identity and the meaning of existence.
(6) Volitional disturbance: interest and drive are diminished so that activity is not goal–directed and the individual cannot function in a role–playing manner.
(7) Withdrawal and detachment: The person is preoccupied with internal delusions, fantasies and ideas, these distorting or excluding the outside world and causing difficulties in personal relationships.
(8) Catatonic and psychomotor disturbances: There may be catatonic stupor with apparent unawareness of the environment, rigidity, extreme flexibility, bizarre movements and postures unrelated to external circumstances, resistance to being moved, odd mannerisms, facial grimaces.
Narrower Thought broadcasting (#HM9384) Alientation of thought (#HM6508)
Thought insertion (Psychism, #HM5479).
Related Synesthesia (#HM1241) Hallucination (#HM4580)
Audible thought (#HM1317) Schizophrenic fantasy (#HH0939)
Anomalies in awareness of one's self (#HM8186).

♦ **HM2314d Non–conscientiousness** (Buddhism)
Pramada — Bag–med–pa (Tibetan)
Description A laxity of mind which prevents the cultivating of virtuous phenomena.
Context One of the twenty *secondary afflictions* of Tibetan Buddhism.
Broader Secondary afflictions (Buddhism, Tibetan, #HH0781).

♦ **HM2315c Emblems of renewal** (Tarot)
Judgment
Description Among the archetypes projected by the psyche are emblems of self–renewal, of regeneration and renovation. A new personality is formed from the unharmonized and diverse aspects of the unenlightened man. As irrational, animal instincts are mastered the experiencer rises in triumph from the grave of ignorance. The emblems include the snake which sloughs off its skin, the caterpillar which turns into a butterfly and the self–renewing phoenix. All nature itself is a symbol of renewal, seen in the sprouting out of the ground, from fallen "dead" seeds, the green shoots of life. In the Tarot system it is the resurrection judgment (number 20) where Gabriel with his trumpet is depicted calling forth the dead from their graves.
Context Judgment is correlated by the Kabbalists to Path 31 between kingdom (Malkuth) and splendour, glory and intelligence. (Hod).
Broader Emblem archetypes in the psyche (Tarot, #HM2201)
Tarot arcana of conscious states (Tarot, #HM2097).
Related Sphere of glory (Kabbalah, #HM2238) Sphere of the kingdom (Kabbalah, #HM2288).

♦ **HM2316d Authentic relation** (ICA)
Freedom of awareness
Description This is an experiential trek in the *River of Consciousness* within the ICA *Other World in the midst of this World*, when the individual realizes "I am my consciousness". The states are described in separate entries.
Broader River of consciousness (ICA, #HM2993).
Narrower External relation (#HM2471) Ultimate awareness (#HM2388)
Self transcendence (#HM2584) Perpetual becoming (#HM2717).
Related Moral ground (ICA, #HM2721) Creative existence (ICA, #HM2894)
Final accountability (ICA, #HM3206).

♦ **HM2317b Stations of consciousness - Ansari** (Sufism)
Maqamat hundred
Description Among the Sufis, spiritual development is represented by progressive steps or stations (maqamat). One system indicates that there are 10 stations with 10 degrees in each. The first four stations are called: gateways; doors; conduct or behaviour; and character. The second cycle of stations is composed of the three decads called: principles; valleys; and mystical states. The third cycle contains two decads, sanctity and realities; while above this, like the top of a pyramid, is the one decad called: supreme goal.
Context This is the system of Ansari.
Broader Divine path (Sufism, #HM3371) Human development (Sufism, #HH0436).
Narrower Doors (#HM3451) Gateway (#HM3226) Conduct (#HM4648) Valleys (#HM2874) Nihayat (#HM3377) Sanctity (#HM4041) Character (#HM3759) Realities (#HM3755) Principles (#HM3069) Mystical states (#HM3607).
Related Ascension awareness (Sufism, #HM2327)
Higher states of consciousness (Sufism, #HM2365)
Stations of consciousness - Simnami (Sufism, #HM2341)
Ultimate Sufi state of consciousness (Sufism, #HM2318)
Stations of consciousness - ibn-Abi'l-Khayr (Sufism, #HM2424).

♦ **HM2318b Ultimate Sufi state of consciousness** (Sufism)
Gathering-separation — Jam-tafriqah — Eighth stage of progress
Description In some Sufi systems the eighth is the highest state of mystical consciousness, which has been described as total annihilation of the self and absorption into the infinite. As such it can be compared with the Buddhist *nirvana*. Among the Nakshbandi it is called "Alone in a crowd". In Ansari the eighth cycle is called "Sanctity" and described as the "gathering after separation", where the spiritual powers make their appearance (although Ansari gives "Realities" and "Mystical union" as above "Sanctity"). He describes the mystical station of gathering (jam) as world-renunciation, and the station of separation (tafriqah) the common view of one's individuality as being apart from the universe, ever among the pious. Thus the eighth is the station of "gathering", in which the mystic "beholds one moon plainly", conscious of nothing but unity. However a still higher state is the "gathering of the gathering" (jam al-jam) in which one beholds "three moons together": divine unity as essence, creator and creatures; or as essence, qualities and actions; or as the law, the way and the truth. Mystical stations are said to be fulfilled only through their opposites; so souls perfected by the gathering must be joined to forms which are in separation to effect unity of existence.
Broader Human development (Sufism, #HH0436).
Related Nirvana (Buddhism, #HM2330) Identification with God (Sufism, #HM3520)
Higher states of consciousness (Sufism, #HM2365)
Stations of consciousness - Ansari (Sufism, #HM2317).

♦ **HM2319c Mental consciousness**
Description This is the fourth stage in the evolution of consciousness, in which directed thought comes into operation. Discrimination and judgement in the waking state of consciousness determine the characteristics of reality on this level. Gebser, conceptualizing this level of consciousness, localizes its origin parallel to the establishment of patriarchy. Characteristics of mental consciousness are: abstraction; space-orientation; ratio (judgement); transformation of polarity into trinary synthesis.
Broader Evolution of consciousness (#HM2140).
Followed by Integral consciousness (#HM2152).
Preceded by Mythical consciousness (#HM2078).

♦ **HM2320d Logical and metalogical awareness**
Vertical and lateral thinking
Description Common thought processes have, as one nearly universal characteristic, the uncritical acceptance of situational premises and conditions. The given data are taken at face value. Subsequent, organized thinking based on unexamined premises passes as logical, but may be termed linear, vertical, or one-directional. Validation or examination of given or apparent facts creates a loop in the conscious mind and sets up conditions for multi-focal thought whose progression or movement may be lateral or tangental to the linear, single, logical approach. Logical, multilogical and metalogical thinking complement each other.
Related Multifocal thought (#HM2242) Illogical thinking (#HM1359).

♦ **HM2321g Theta wave consciousness**
Threshold awareness
Description There is some correlation between states of consciousness and brain-wave patterns as registered on an electroencephalograph (EEG) reading. Theta wave dominance is recorded when subjects are on the threshold of sleep and the conscious mind is passive. It might be tempting to call the theta wave reading the indicator of a threshold consciousness, but additional associations with the theta wave state need to be taken into account. One qualified investigator (E Green) has published experiments indicating there is a "theta power" associated with releasing psychic contents from the unconscious, and the authors of "Psychic Discoveries Behind the Iron Curtain" have reported research linking theta wave propagation with psychokinetic extra-sensory power, the ability to move small objects at a distance, "influence" the outcome of a roll of dice and other chance events, etc. More everyday, the threshold consciousness before sleep gives access to "visions", that is, images, impressions and communications from the inner world, a phenomenon particularly noted by creative people.
Broader Brain waves (#HH3129).
Related Reverie (Psychism, #HM0967) Beta wave consciousness (#HM3476)
Alpha wave consciousness (#HM2345) Delta wave consciousness (#HM1785)
Fringe of space-time (Psychism, #HM3770)
Theta-clear state of consciousness (Scientology, #HM2428).

♦ **HM2322b Physical relativity in conscious states** (Physical sciences)
Description Physical relativity theory situates any reality, matter or energy or a composite of both, in its own particular space-time, from which it acquires particular space-time dependent dualities. States of consciousness or awareness are also conditioned, in this case by such societal analogues to space-time as *zeitgeist*, class orientation, educational formation and family behaviour - in short, the environment of personal development.
Among other inferences that can be drawn from physical and psychological relativity is that truth and objectivity are unattainable within a single framework. Therefore diverse states of consciousness may complement one another in giving information about reality.
Context There are at least five universal principles postulated in the physical sciences: duality; quantum discreteness; relativity; conservation; and least action. Their application to states of consciousness has been presented by Professor Charles Tart and others.
Related Discrete states of consciousness (Physical sciences, #HM2298)
Physical duality in conscious states (Physical sciences, #HM2381)
Physical conservation in conscious states (Physical sciences, #HM2357)
Least action principle in conscious states (Physical sciences, #HM2417).

♦ **HM2323d Serious sharing** (ICA)
Description This is experienced as being somebody and being nobody at the same time; when a person loses the attachment to being someone he felt he needed to be in order to have a meaningful life. It may be compared to a prisoner-of-war who was an important officer before his capture using his status to help fellow prisoners-of-war without regard for his own life.
Context In the ICA New Religious Mode in the arena of knowing one's disengagement (the life of poverty) the third formal aspect is being nonchalant about one's relations. At the second phenomenological level, this occurs when a person takes charge of desires which formerly controlled his life.
Broader Liberation from relationships (ICA, #HM3240).
Followed by Abounding abasement (ICA, #HM2209).
Preceded by Eternal insecurity (ICA, #HM2517).

♦ **HM2325d Ninth vajra-master awareness** (Buddhism, Tibetan)
Description This states of awareness is characterized by the perception "mind is all". It is represented as a purification of the apparent body of form in which clear light, great bliss and divine play are all realized.
Context In Tibetan Sakya Buddhism this is one of the states in the "Ascension Stages Game". In some sets it is numbered 90 on the board.
Broader Way of the vajra masters (Buddhism, Tibetan, #HM3090).
Preceded by Sixth vajra-master awareness (Buddhism, Tibetan, #HM2287).

♦ **HM2326a Satori** (Zen)
Wu
Description Satori is the psychological result and aim of the practice of zazen meditation (described separately). This practice induces a special form of objective awareness of self associated with an experience of joy emanating from a mind which has transcended all relative joys and sorrows.
Experience is no longer mediated through concepts (which is the reason for the tendency to give illogical responses to requests for definitions of satori). In addition, the satori experience has a paradoxical quality, such as a feeling of oneness, which is inexpressible in a language posited on a subject-object dichotomy in a conventional space-time framework. The existence of a separate self is viewed as a fiction through the satori experience, which is an intuitive perception of the real self as the true author of the individual's behaviour and at the same time a part of the whole flux of the universe. Experience is felt to take place directly through the real self unmediated by conscious thought, and without consciousness of the process. The individual is content to permit his behaviour to bring out a self which cannot be fully conceptualized, trusting the self sufficiently to suspend conscious reflective control over it.
Satori may be experienced for shorter or longer times, depending on the length of training and the responsiveness of the individual to it. Brief experiences in exceptional situations may be developed so that the experience is achieved in a wider variety of conditions. Enlightenment, a nearly impossible ideal, is considered to be the constant experience of satori.
Related Non-action (#HH1073) Muga (Japanese, #HM0793) Enlightenment (#HM2029)
Vikalpa (Buddhism, #HM3306) Zazen meditation (Zen, #HH0882)
Kensho-godo (Japanese, #HM2700) Human development (Zen, #HH1003)
Seventh plane (Psychism, #HM4417) Samadhi (Hinduism, Yoga, #HM2226)
Steps to enlightenment (Buddhism, #HH4019).

♦ **HM2327b Ascension awareness** (Sufism)
Miradj
Description Muhammad the Prophet's ascension to heaven represents modes of awareness in a number of stages and contains diverse symbolical elements. The initial elements include: the preparation of Muhammad's body by the purification of the heart (performed by an angel); the place of ascent beginning at the center of the world (i.e. Mecca or Jerusalem) and the nocturnal journey there; and the mystic ladder (miradj) provided for the climb. However, a variant legend has the instrument of ascent to be not a ladder, but the beast, Burak (Lightening-flash). The power symbolized by this beast is also characterized as feminine, and since the beast itself is usually typified as a large ass, Burak becomes a winged mare in the traditions. She is represented also with a human face to signify that she is fact is an accessible force - her name connects phonetically with the spiritual power known as Baraka.
The additional symbolic elements are comprised in the ascent of Muhammad's awareness through hell, paradise and all the seven heavens. Eventually he converses with Allah, although perhaps through a veil. For mystics, the Miradj in a prototypal ascent.
Related Higher states of consciousness (Sufism, #HM2365)
Stations of consciousness - Ansari (Sufism, #HM2317)
Stations of consciousness - Simnami (Sufism, #HM2341).

♦ **HM2329g Alien consciousness**
Extra-terrestrial minds
Description The characteristics given to hypothetical extra-terrestrial intelligences correspond in part to alleged human psychic potentialities. In terms of powers, alien consciousness is typically telepathic, able to project holographic images, and psychokinetically empowered to manipulate objects or establish force fields. Less is said qualitatively about alien consciousness, but a typical characteristic imputed to it is that it is not unitary, there being a group or hive consciousness in each alien individual. Moreover, individuality as we know it may not exist. The aliens, for example, may be clones. Alien consciousness may be depicted as very-long lived, giving an experiential quality that may be termed "wisdom" if benevolent. On occasions the alien is a disembodied mind, sometimes associated with habitation in the further reaches of space. "Our" aliens always communicate with us, however, regardless of their life-form (once we encounter them) even if it is by acts of silent destructiveness. At present, aliens are what we project them to be. They are our inner selves.
Refs Vallee, Jacques *Dimensions* (1988).
Related Telepathic consciousness (#HM2129)
Mythical control of human development (#HH7637).

♦ **HM2330a Nirvana** (Buddhism)
Nibbana (Pali) — Pari-nirvana — Cessation — Annihilation
Description Derived from nir-va, to blow out, the term nirvana refers to a state of complete freedom, liberation and enlightenment, and in total peace and bliss; the goal of Buddhism. The path to Nirvana is the eightfold way, also referred to as the *noble path*: right view, right aim, right speech, right action, right living, right effort, right mindfulness, right contemplation. One who has achieved nirvana is called an Arhat, living in a condition where there is neither earth nor water nor fire nor air; neither infinite space nor infinite consciousness; perception or non-perception. The world is not so much overcome as removed. There is total emancipation from matter. Time ceases to exist.
Nirvana is the culmination of a "journey" through the four realms of form and formlessness. The sphere of neither conceptualization nor non-conceptualization is entirely transcended so that the practitioner remains in a state of cessation from both sensation and conceptualization. This state is characterized by utter bliss and liberation; there is cessation of all suffering - the third noble truth; ignorance and craving are dissolved in knowledge of unconditioned dharma. Activity is no

longer directed towards a definite goal.
Complete extinction, *pari-nirvana* or *nirvana without remainder* is that final and irreversible state of emancipation which may only arise on death of the body of the fully enlightened. Some equate this with total annihilation, self-identity ceasing when psycho-physical processes cease, although the orthodox view is that nothing can be known about such a state. Others identify attainment of cessation with nirvana, although there is some discussion as to whether nirvana as cessation is identical with pari-nirvana or whether it can arise in this life when it is as though physically dead. Nirvana with remainder is that state experienced by Buddha when meditating under the Bo tree. In this case there is sufficient remainder of past actions for the practitioner to continue to live and act.
When Buddha achieved this state he is reputed to have said: "I, Buddha, who wept with all my brother's tears, laugh and am glad, for there is liberty". It is notable that his first two word are "I" and his name "Buddha". As in the mystic Sufi doctrine of (fana) annihilation in God, what is existinguished or blown out is metaphorically heat (agitation). The light remains.
The characteristic of nirvana is peace; its function is not passing away and also comforting; its manifestation is as signlessness or as non-diversity. The *Path of Purification* gives insight into nirvana and its real existence. Although it cannot be apprehended by ordinary people, this does not mean it cannot be apprehended at all; it can certainly be reached by following the right way. Past and future aggregates are absent - although their absence does not necessitate nirvana; there is no clinging to the past nor arousing of future. Present aggregates remain, supporting the path moment as nirvana is entered. Again, defilements are absent, but their absence alone does not necessitate nirvana. If nirvana were simply destruction it would be temporary, would be formed, achieved independent of right effort. Being formed entails the fire of craving and therefore suffering, which cannot be nirvana. Nirvana is not created - it is reachable by the Noble Path but not created by it. Finally, it is not non-existent - there is that which is unborn, is not the result of becoming, is not made, is not conditioned or formed.
Context Sometimes referred to as the third of the four noble truths.
Refs Johansson, Rune E A *The Psychology of Nirvana* (1969); Welbon, Guy R *Buddhist Nirvana and Its Western Interpreters* (1968).
Broader Emancipation of the self (#HH0907) Four noble truths (Buddhism, #HH0523)
Consciousness in Buddhism (Buddhism, #HM2200).
Narrower Pari-nirvana (#HM4665) Sa-upadisesanibbana (Pali, #HM5023).
Related Dharma (#HH0093) Mystical cognition (#HM2272)
Tantra (Buddhism, Yoga, #HH0306) Eightfold way (Buddhism, #HM2339)
Seventh plane (Psychism, #HM4417) Samadhi (Hinduism, Yoga, #HM2226)
Emptiness (Buddhism, Zen, #HM2193) Human development (Buddhism, #HH0650)
Release from the matrix (Taoism, #HM0635) Cessation of suffering (Buddhism, #HH2119)
Steps to enlightenment (Buddhism, #HH4019) Attainment of cessation (Buddhism, #HM5438)
Ultimate Sufi state of consciousness (Sufism, #HM2318)
Anomalies in experience of the self as distinct from the outside world (#HM4754).
Preceded by Cetovimutti (Pali, #HM0125) Pannavimutti (Pali, #HM8125)
Ubhatobhagavimutti (Pali, #HM7983) Devachon awareness (Psychism, #HM2044)
Fourth absorption in the immaterial sphere (Buddhism, #HM2051).

♦ HM2331d Absurd existence (ICA)
Description This state occurs when an individual experiences the absurdity of life and in particular his own life. People frequently use such expressions as: "it's madness ", "but why would they do that?", and "nothing makes sense ". They all know moments when the absurdity of existence, of the eternal riddle, seems to overwhelm them, when they ask "why me?" and realize that finally there is no good reason for it. It is on occasions like these that the mystery of the absurdity of life overtakes them.
A dimension of this experience is a sense of awe, that is fear and fascination. One knows a deep fear of not being in charge of one's life, as if one's life was more than one can handle, things were out of control, that one had been cut lose from a critical link in this world. One asks "am I going crazy?" and yet, in the midst of such bewilderment, a fascination emerges as to what will come next. One is strangely willing to see what will follow. One abandons one's self to the situation with a sense of "it doesn't matter so let's go ahead."
With great courage, the individual decides go ahead without resolving the mystery; without knowing all the whys and wherefores he dares to participate in history, in life, in this absurd existence. Such intense moments generate a new sense of humour about life in general and especially about one's own existence, a new sensitivity to the mystery of the absurd to be found in every situation in every life.
Context This state is number 2 in the ICA *Other World in the midst of this World.*
Broader Aweful encounter (ICA, #HM2619).
Followed by Ultimate reality (ICA, #HM2030) External relation (ICA, #HM2471)
Ubiquitous otherness (ICA, #HM2570) Transformed existence (ICA, #HM2862)
Definitive predestination (ICA, #HM2131).
Preceded by Cryptic disclosure (ICA, #HM2824) Radical contingency (ICA, #HM2477)
Transformed existence (ICA, #HM2862) Inclusive comprehension (ICA, #HM2256)
Definitive predestination (ICA, #HM2131).

♦ HM2332g Experience of the unconscious
Description In modern psychology, the unconscious is referred to as underlying consciousness. A mass of psychological factors are buried there. They may surface in response to a conscious effort or quite unexpectedly and in a disguised form. Whatever mental contents or states that are not in awareness or recallable to it are also labelled *unconscious.* The term can cover morbid conditions such as coma, swoons, delirium, nightmares, dreamless sleep, somnambulism, hypnotic trances and drugged or toxic states as well as those experienced during paranormal mental activity such as prophetic or mystical trance or deep concentration. The entire contents of memory, various theorized psychical structures (such as archetypes), pathologies such as phobias and complexes, and significant features of the personality, may all be said to lie in the unconscious. Several governments sponsor research into its nature, since to control the unconscious is to control human behaviour.
Related No-thought (Zen, #HM2500).

♦ HM2333d Inclusive collegiality (ICA)
Description The second formal aspect of the *life of chastity*, an arena in the ICA *New Religious Mode*, this represents collegiality with symbols of profound being when the individual develops an ability to see through such symbols to their deepest significance.
Broader Life of chastity (ICA, #HM2506).
Narrower Raw reality (#HM3195) Symbol maker (#HM3103)
Transfigured man (#HM3161) Transparent existence (#HM3067).

♦ HM2334d Human transformation (ICA)
Description This is experienced as the joy and pain of creating a way of life for all of creation as God's co-worker; when, having surrendered the need to control history or to simply experience events, one accepts being directed by the eternal and the immediate at the same time. It may be compared to Jacob who, having wrestled with the angel all night, both received a blessing and was wounded for life.
Context In the ICA New Religious Mode in the arena of knowing and doing intensified in one's life (the life of being) the fourth formal aspect is the experience of transparent presence or the experience of effulgence. At the third phenomenological level this occurs when a person disciplines his life to be continually related to the experience of nothingness.
Broader Transparent presence (ICA, #HM2462).
Followed by Saving the mystery (ICA, #HM2413).
Preceded by Inventing humanness (ICA, #HM2092).

♦ HM2335s Death-bardo clear-light awareness (Buddhism)
Fourth antarabhava consciousness — Chikhai-bardo (Tibetan)
Description This state of awareness is represented by the first after-life bardo and is characterized by the perception of the dharmakaya. It is said to be an experience of the highest level of spiritual reality. If the deceased can sever all ties to the world he has left he will remain free in this state. During the first stage the consciousness principle is guided from the physical body; during the second, still under guidance, the thought forms of the death process are viewed.
Context This is one of the six major bardo conditions. It is experienced sometime in the first two days after death.
Broader Bardo consciousness (Buddhism, Tibetan, #HM0698).
Followed by Death-bardo heaven-reality awareness (Buddhism, #HM2407).
Preceded by Meditation-state bardo awareness (Buddhism, Tibetan, #HM2311).

♦ HM2336d Vaisvanara (Hinduism)
Material condition
Description The first "quarter" or condition of the self or *Atman* according to the Mandukya Upanishad, that which is common to all, enjoying material objects through the organs of sense, the motor organs, the pranas and the aspects of mind; its field the waking (outwardly cognitive) state. It is associated with the syllable "A" of the mystic word "AUM".
Broader Degrees of consciousness (Hinduism, #HM0891).
Related Turiya awareness (Hinduism, #HM2395)
Jagrat state of waking consciousness (Yoga, #HM2141).
Followed by Taijasa (Hinduism, #HM3040).

♦ HM2337d Hyper-bliss-realm awareness (Buddhism, Tibetan)
Abhirati
Description This state of awareness is characterized by imperturbable calm and represented as the eastern Pure Land of Buddha Aksobhya. His consort is said to be called the Eye-of-great-awakening.
Context In Tibetan Sakya Buddhism this is one of the states in the "Ascension Stages Game". In some sets it is numbered 85 on the board.
Broader Supreme heavens in the ascension stages game (Buddhism, Tibetan, #HM3214).
Followed by Second vajra-master awareness (Buddhism, Tibetan, #HM2703)
Jewelled-peaks-realm awareness (Buddhism, Tibetan, #HM2275)
First scriptural bodhisattva awareness (Buddhism, Tibetan, #HM2155).
Preceded by Mahayana receptivity-application awareness (Buddhism, Tibetan, #HM2247)
Tantra union-in-learning-application awareness (Buddhism, Tibetan, #HM2280).

♦ HM2338d Communion of saints (ICA)
Description This is the knowledge that all the people of the Way of the past, present and future are about the same task in history that one is about one's self, when a person suddenly realizes that he is not alone in his work, but surrounded by a host of colleagues. It is like St. Paul writing of his concern and responsibility for all the churches, past, present and future.
Context In the ICA New Religious Mode in the arena of acting out one's engagement (the life of obedience) the fourth formal aspect is embodying service to people's spirit life. At the fourth phenomenological level, this occurs when one lives on behalf of the evolution of human consciousness.
Broader Embodying service (ICA, #HM3199).
Preceded by Eternal identification (ICA, #HM2541).

♦ HM2339b Eightfold way (Buddhism)
Aryastangamarga — Noble eightfold path
Description These are the characteristics attained with the path of bhavana (meditation), the correct: views or understanding, realization or thought, speech, aims of action, livelihood, effort, mindfulness, meditative stabilization or concentration. These states of mind are not so much a series of steps as the harmonious flowing of all the teachings of Buddhism towards its goal. The first two constitute wisdom (panna), the next three faith or ethical conduct (sila), and the final three mental discipline (samadhi). The way is divided into two paths: lokiya (ordinary); lokuttara (transcendent), the latter leading to full, harmonious balance, transcendence of normal understanding and direct knowledge of the unconditioned truth.
Following the way leads to a gradual relinquishing of the individual's hold on the chain of causation. In effect, this is the reversing of the twelve links of conditioned genesis: cessation of ignorance leads to cessation of volition; this leads to cessation of consciousness, which leads to cessation of mental and physical phenomena, and so on, to cessation of repeated becoming. As a result of this, decay, death and suffering cease.
Context One of the sections of yogic paths or harmonies with enlightenment defined in Buddhist teaching. Sometimes referred to as the fourth of the four noble truths leading to the cessation of suffering.
Broader Stages of spiritual life (#HH0102) Four noble truths (Buddhism, #HH0523)
Harmonies with enlightenment (Buddhism, #HH0603).
Narrower Right acts (#HM0198) Right speech (#HM1157)
Right outlook (#HM1280) Right resolves (#HM1142)
Right endeavour (#HM1295) Right livelihood (#HM1341)
Right mindfulness (#HM0704) Right rapture of concentration (#HM0931).
Related Karma (#HH0567) Nirvana (Buddhism, #HM2330) Bhavana (Buddhism, #HH0551)
Eightfold path of yoga (Yoga, #HH0779) Lokuttara consciousness (Buddhism, #HM2120)
Measureless concentration (Buddhism, #HM0496)
Supramundane concentration (Buddhism, #HM4243)
Understanding as knowledge about law (Buddhism, #HM4726).

♦ HM2340c Emanation-body awareness (Buddhism)
Tulku (Tibetan) — Nirmana-kaya
Description This state of awareness is characterized by the simultaneous presence of mind in various stations and is represented as the Buddha's distribution of emanation-bodies in all the world systems to teach the dharma. Thus the world's saviours have a docetic or illusory bodily existence. One attaining this state enacts the role of such an incarnation in the nirmana-kaya, appearing to be born, experiencing a miraculous infancy and so on. Sakyamuni Buddha was such an emanated form or tuklu, and his life is a paradigm of buddhic development.
Context In Tibetan Sakya Buddhism this is one of the states in the "Ascension Stages Game". In some sets it is numbered 97 on the board.
Broader Consciousness ascension stages game (Buddhism, Tibetan, #HH4000).
Followed by Buddha-consciousness (Buddhism, Tibetan, #HM2735).
Preceded by Enjoyment-body awareness (Buddhism, Tibetan, #HM2873).

♦ **HM2341b Stations of consciousness - Simnami** (Sufism)
Maqamat seven–worlds
Description In Sufism, spiritual development is represented by progressive steps or stations (maqamat). One system (of Simnami) indicates that there are 7 maqamat in 7 worlds. The first two worlds are those of materiality (man in nature) and of form. Their corresponding stations of consciousness are those of the instincts and the vital senses. The next two worlds are those of subtle perception and of imagination. The respective station characteristics are the "spiritual, embryonic heart", the consciousness of one's real individual essence; and the "secret", the portal of the superconscious symbolized as the "Moses of one's self", of which the mystic colour is white. It is fourth of the seven stages, in the second pair of world stations. The next two worlds are those of formlessness, and of the immaterial divine nature. The corresponding station for the first called the *spirit*, is represented as the King David of one's self, coloured yellow and numbered fifth of the seven. The sixth station is called *inspiration* and represented as the Jesus of one's own being, the manifest Name of God. It's colour is luminous black. Unpaired on the fourth level so to speak, is the seventh station, the Prophet of one's self in the world of the divine essence. This stage is represented as green. There are variants on the alignments of personages with the stations. In the esoteric system of the Ismailiya, David is omitted, which makes Jesus fifth, and Muhammad sixth. The prophet to come, Ka'im, is seventh. Alternatively, perhaps according to the colour symbolism, one may associate the seventh level with al Khadir (also spelled Khidir, Kidar, Khihir) the "Green Man" (whose emblem is a fish and who lives at "the spring of life").
Refs Bakhtiar, Laleh *Sufi* (1987).
Broader Divine path (Sufism, #HM3371) Human development (Sufism, #HH0436).
Narrower Fana (#HM3799) Withdrawal (#HM2243) Divine nature (#HM2246)
World of forms (#HM3286) Divine essence (#HM3513) World of nature (#HM3425)
Joyous awareness (#HM2106) World beyond form (#HM4570)
World of imagination (#HM3616) Sufi infused awareness (#HM2476)
Perfect man consciousness (#HM2920) Sufi contracted–consciousness (#HM2800)
World of spiritual perception (#HM3974) Consciousness of human service (#HM3012)
Islamic transformation awareness (#HM2850) Awareness of the divine presences (#HM2416).
Related Ascension awareness (Sufism, #HM2327)
Awareness of the mystic journey (#HM2900)
Higher states of consciousness (Sufism, #HM2365)
Stations of consciousness - Ansari (Sufism, #HM2317)
Mystical Christ–consciousness (Christianity, #HM2880)
Stations of consciousness - ibn–Abi'l–Khayr (Sufism, #HM2424).

♦ **HM2343d Missional engagement** (ICA)
Description This is experienced as simply joining in covenants in society. Part of participating in these covenants is the conserving function or the maintaining of the covenant in being. It occurs when one joins in social covenants that involve a commitment to maintain and create order, as when everyone turns out to protect a small community faced with a flood.
Context In the ICA New Religious Mode in the arena of acting out one's engagement (the life of obedience) the first formal aspect is embodying peace as the establishment to enable the ordering of human affairs. At the first phenomenological level this occurs when a person becomes aware that he is not autonomous, but rather is bound unto death to other people.
Broader Embodying peace (ICA, #HM2876).
Followed by Submissive obedience (ICA, #HM2258).

♦ **HM2344d Flow experience**
Description A holistic awareness experienced when one acts with total involvement in a task, so that action succeeds action without conscious intervention by one's self. It is achieved in attempting to meet challenges which one has previously defined as difficult but possible: this may be in *competition* (whether sports, politics or whatever); in *chance* (challenging the unpredictable, as in gambling); in *mimicry* (transcending personal limits in theatre, dance, etc); or in *vertigo* (the altering of consciousness by intentionally placing one's self in uncertain conditions, as in mountaineering, skiing or sky–diving).
Such awareness is thus the result of freely choosing to express one's self creatively in conditions of uncertainty or risk. Movement becomes a continuous flow, a means of *self–communication* when action and awareness merge. Concentration is heightened and focused on the task in hand; there is acute awareness of the smallest sensation, little awareness of anything outside the present time and place. Individuality is transcended as consideration of the self becomes irrelevant, and there is no room for fear. There is an intense flow of energy used to maximum effect and a minimum of wasted effort.
The essence of the experience is that the task is not necessary but chosen for non–material ends. It is in the balance position between *boredom* and alienation (too little stimulus or challenge) and *panic* (when the task is seen as almost beyond one's capabilities). It is sought through a variety of engrossing, compulsive or addictive activities but, once experienced, may carry over into less challenging, more mundane events which are more rewarding as a result.
Refs Csikszentmihalyi, Mihaly *Flow* (1990); Mitchell, Richard J *Mountain Experience* (1983).
Narrower Sexual experience (#HM0829) Flow experience through work (#HM4877)
Flow experience through thought (#HM0666) Flow experience through the senses (#HM1756).
Related Panic (#HM2083) Yu (Taoism, #HM1771) Recreation (#HH0410)
Uncertainty (#HM3051) Enchantment (#HM2204) Dehumanization (#HM3030)
Cave phenomenon (#HM1169) Peak experiences (#HM2080) Freedom of choice (#HH0789)
Autotelic experience (#HM1410) Ego transcendence (Jung, #HM2230)
Sport–induced modes of perception (#HM0062).

♦ **HM2345g Alpha wave consciousness**
Description There is some correlation between states of consciousness and brain–wave patterns registered on an electroencephalograph (EEG) reading. Alpha wave dominance is recorded when there is a feeling of euphoria, when the mind and body are in a state of nerve relaxation, and also when there is a calm state of high awareness. The alpha wave state is therefore associated with altered and higher states of consciousness and various meditation states. Voluntary inducement of the alpha wave state is practised by many individuals using biofeedback techniques.
Broader Brain waves (#HH3129).
Related Drowsiness (#HM6231) Hypercognition (#HM1273)
Alert passivity (#HM3959) Biofeedback training (#HH0765)
Beta wave consciousness (#HM3476) Theta wave consciousness (#HM2321)
Delta wave consciousness (#HM1785) Fringe of consciousness (Psychism, #HM3667)
Theta–clear state of consciousness (Scientology, #HM2428).

♦ **HM2346d Surrender to one's calling** (ICA)
Description The first formal aspect of the *life of doing* in the ICA *New Religious Mode*, this represents an awareness of one's unique calling as a person.
Broader Life of doing (ICA, #HM3018).
Narrower Dying death (#HM3099) Suffering servant (#HM3126)
Actional existence (#HM3080) Unlimited commitment (#HM3230).

♦ **HM2347c Life–state bardo awareness** (Buddhism)
First antarabhava consciousness (Tibetan)
Description This state of awareness is of the intermediate sphere of existence from birth to death. It is illumined by meditation on the arising of all phenomena from mind, and on the dissolution of all phenomena into their source. The contents of the bardos of the dream–state, of the meditation–state, and of the life–state, constitute, in a large part, the determining characteristics of the bardos in the after–life.
Context This is one of the six major bardo existences in Tibetan Buddhism.
Broader Bardo consciousness (Buddhism, Tibetan, #HM0698).
Followed by Dream–state bardo awareness (Buddhism, Tibetan, #HM2371).

♦ **HM2348c Sphere of wisdom** (Kabbalah)
Chokmah sephira
Description This sphere of awareness is associated with active (as opposed to passive) intellect, with a sense of vision, and thus with the development of inspiration. It corresponds to the developmental phase of death and the realization of achievement. It is symbolized by the positive and negative archetypes of the animus and the wizard, and by the planet Uranus, as well as by Adam in the Jewish tradition. This state of consciousness is said to be the mystery by which the world exists; the potency whose seminal spark will conceive, in conjunction with the female vehicle, all the images of creation.
Context This sphere is the second of ten described in the Kabbalistic system of sephiroth.
Broader Kabbalah (#HH0921).
Related Ajna (Yoga, #HM2144) Paths of wisdom (Judaism, #HM2509)
Emblems of supremacy (Tarot, #HM2411) World of emanation (Judaism, #HM2408)
Pilgrim archetypal image (Tarot, #HM2225) Hierophant archetypal image (Tarot, #HM2260)
Female sovereignty archetypal image (Tarot, #HM2285).
Followed by Sphere of the of the supreme crown (Kabbalah, #HM3132).
Preceded by Sphere of beauty (Kabbalah, #HM3031)
Sphere of understanding (Kabbalah, #HM2372)
Sphere of mercy and greatness (Kabbalah, #HM2420).

♦ **HM2350c Emblems of disaster** (Tarot)
Tower — Ruin
Description Among the archetypal emblems projected by the psyche are those depicting calamity, destructive forces and desolation. These can include representations of conflagrations and floods, or ruins and wastelands. In the Tarot system the emblem called the Tower (number 16), is depicted as a lightning–struck building whose function is seen possibly to have served as a very high throne. This is indicated by the crowned occupants who have been knocked of the top of the tower and who appear to represent a king and a queen. Ignorance and limited, arrogant ideas are destroyed and cast down, like the Tower of Babel; but, if pride can be overcome, the tower potentially leads to Kether.
Context The Tower is placed by Kabbalists on the 27th path between splendour and glory (Hod) and force or victory (Netzach).
Broader Emblem archetypes in the psyche (Tarot, #HM2201)
Tarot arcana of conscious states (Tarot, #HM2097).
Related Sphere of glory (Kabbalah, #HM2238) Sphere of victory (Kabbalah, #HM2362).

♦ **HM2351c Recollection** (Islam, Sufism)
Dhikr — Zikr — Khikr — Kikr
Description A state of consciousness said to be required on the spiritual path is one in which the objective of the journey is continually kept in mind. Traditionally this state is achieved by repeating God's name, or a holy word or formula, either audibly or inaudibly; by repeating symbolic movements or adopting particular postures or by special placement of the hands; and and most difficult of all, by retaining in the mind itself a divine symbol, special condition of awareness, or "memory" of the goal. All these elements are what constitute, in their inner essence, the worship, the liturgy and the rituals of the world's faiths.
In Islam, the dhikr (remembrance) performed by glorifying Allah may be accompanied by appropriate movements and particular breathings. Among the dervishes, music and dancing are associated with dhikr as well. The Islamic fraternities each have their own special dhikr (remembrance ritual). The most usual verbal formula is "La ilaha illa Allah" - "There is no god but God". One peculiar quality of the special state of spiritual recollection, is that conditions are set up for an inner ritual that is taken over and performed "automatically" by the mind. This state is capable of being maintained by the practitioner while he on she is engaged in routine affairs, so that it provides a background of a continual spiritual presence to behaviour and to perception. Everything but the dhikr is seen as tribulation. Detached from the fears and sorrows of the world, the reflections of phenomenal forms veiling the heart are removed. There is no room in the heart for anything but love of God. Attachments are negated, regarded as false, while the love of God is affirmed until the heart is empty of love of things and becomes the essence of God's unity. The reality of the dhikr and the substance of the heart become one - an experience which has been referred to as the trans–substantiation of the heart, when the heart is so full with love for God that there is no room for any other thought. This experience is *tawhid*, the direct personal experience of reality. It is also related to the state of *qurb*, when there is experience of approaching one's goal, of nearness to God, the object of love.
Context According to Shaykh Abu Sa'id ibn Abi'l–Khayr, this is the eighth of 40 stations or maqamat the Sufi must possess for his journey on the path of Sufism to be acceptable.
Broader Stations of consciousness - ibn–Abi'l–Khayr (Sufism, #HM2424).
Narrower Numerical awareness (Sufism, #HM5118).
Related Derwish (Islam, #HH0741) Tawhid (Sufism, #HM3438)
Love of God (Sufism, #HM4273) Nearness–awareness (Sufism, #HM2377)
Remembrance of God (Sufism, #HM6562) Unification with God (Sufism, #HM3864)
Nakshbandi recollection (Islam, #HM2375) Ninety–nine names of Allah (Sufism, #HH2561)
Christian recollection (Christianity, #HM2077).
Followed by Rida (Sufism, #HM4190).
Preceded by Sabr (Sufism, #HM3418).

♦ **HM2352d Primordial colloquy** (ICA)
Description This is experienced when a person discovers that he is not just part of the league but he is the league, when he experiences dialogue with people from the past. There is a resonance with people across time and space who share a covenant, like meeting someone for the first time and feeling one has known them for years.
Context In the ICA New Religious Mode in the arena of acting out one's deed (the life of doing) the third formal aspect is one's awareness of one's allegiance to those who act on behalf of life as a whole. At the fourth phenomenological level this occurs when the unique greatness of one's expenditure is embodied.
Broader Awareness of allegiance (ICA, #HM2073).
Preceded by Religious vocation (ICA, #HM2293).

♦ **HM2354e Breaking point of total conflict** (Brainwashing)
Description The individual becomes confused as to the real truth of his past. He is aware that he disagrees with what is presently being asserted as fact, and is totally estranged from his environment. No longer able to communicate with or relate to any familiar exterior, his feelings of estrangement are excaberated by seeking to escape the situation through self–betrayal, and extend even to the self he previously was.
Context A stage reached in thought reform, a system of organized, deliberate and total psychological training which effects individual change through two basic elements - *confession*

HM2354

(renouncing of past beliefs and attitudes) and *re-education* (remaking of the individual in the required image).
Broader Thought reform (Brainwashing, #HH0865).
Related Breaking point of basic fear (Brainwashing, #HM2799).
Followed by Desperate gratitude (Brainwashing, #HM3017).
Preceded by Self-betrayal (Brainwashing, #HM2229).

♦ **HM2355g Mammalian limbic human brain consciousness**
Description The limbic system in man is where the functions of attention, learning, memories and emotions have been localized. It is the mediator between the human brain, the neomammalian, neocortex, and the information coming in from the senses. The limbic system is a partly successful organismic equilibrator or homeostatic regulator. It attempts to control such antithetical reactions as: rage/fear; fight/run; pleasure/pain; hope/reality-recognition; relaxation/tension; reward/punishment; socialization/ego-assertion; and affection-courtship/ antagonism-repulsion. The emotional nature of the limbic cortex is most pronounced, as well the failure of its design to be self-stabilizing, since, as a system, it is noted for its oscillations and run-away character. The autonomy of the limbic system appears to be physiologically based, there being no major links with the neocortex understood. It is conjectured that this system of independence works better than if fallible, so-called reason were to constantly over-rule the emotions.
Broader Brain-based consciousness (#HM2295).
Followed by Neomammalian neocortex human brain consciousness (#HM2427).
Preceded by Reptilian human brain consciousness (#HM2307).

♦ **HM2356d Unfailing prompter** (ICA)
Description This is the experience of persons or events in one's life that will not allow one to escape from dealing with the real situation. It occurs when one is reminded repeatedly of what is required of one's life; an example is Scrooge in Charles Dickens' story "A Christmas Carol", after he has been visited by the three ghosts. He starts to live a new life and the ghosts are seemingly always there to remind him of what he has decided to do.
Context In the ICA New Religious Mode in the arena of knowing one's internal sociality (the life of meditation) the second formal aspect is the experience of being called to accountability by one's interior priors. At the first phenomenological level, this occurs when that which is other intrudes upon one's consciousness.
Broader Called to accountability (ICA, #HM2008).
Followed by Concerned judge (ICA, #HM2582).

♦ **HM2357b Physical conservation in conscious states** (Physical sciences)
Description In chemical and physical processes, the sum of material mass and energy is conserved. While there may be transformation between mass and energy, nothing is lost. Consciousness depends on mental or psychic energy. Freud called this *libido*, Lao Tzu called it *ch'i*, and others have given it various names. If human energy is not free but latent in mass then, on analogous reasoning, consciousness is not developed. Conservation indicates that there is a potential for greater consciousness in the fact that some energy must lie in the psychic or somato-psychic analogues to material mass. These possibly include the crystallizations or structural formations within the ego, much of which constitute the personality. They may encompass such things as opinions (light mass), habits (medium mass), and instinctive feelings and behaviour (heavier mass). From another point of view, mass is food, energy is consciousness (and life) and - as indicated in the Upanishads - food becomes blood and genetic material. Thus conservation helps to explain the rationale of spiritual disciplines and the potential of human development.
Context There are at least five universal principles postulated in the physical sciences: duality; quantum discreteness; relativity; conservation; and least action. Their application to states of consciousness has been presented by Professor Charles Tart and others.
Related Discrete states of consciousness (Physical sciences, #HM2298).
Physical duality in conscious states (Physical sciences, #HM2381)
Physical relativity in conscious states (Physical sciences, #HM2322)
Least action principle in conscious states (Physical sciences, #HM2417).

♦ **HM2360c World of formation** (Judaism)
Olam-yetzirah
Description This world is said to correspond to the higher ego consciousness freed from illusion. It is the realm of the three spheres (sephiroth): glory, victory and foundation (hod, netzach and yesod), characterized as natural vital and psychic force, spiritual force, and will-directed conscious activity on the non-illusory level.
Context This is the third of the Kabbalistic four worlds.
Broader Shekinah awareness (Judaism, #HM2002).
Related Paths of wisdom (Judaism, #HM2509)   Sphere of glory (Kabbalah, #HM2238)
Sphere of victory (Kabbalah, #HM2362)   Sphere of the foundation (Kabbalah, #HM2410).
Followed by World of emanation (Judaism, #HM2408).
Preceded by World of creation (Judaism, #HM3038).

♦ **HM2361d Seventh-scriptural bodhisattva awareness** (Buddhism, Tibetan)
Duramgama
Description This state of awareness is characterized by the recognition of the need to work for the welfare of others and also by the spontaneous emergence of supreme artistic and literary skill. The stage is also called *far reaching* (duramgama). The bodhisattva has now outstripped the Arhats and Pratycka Buddhas in accomplishment, particularly by being in the state of "nirvana-in-motion". He remains in the samsara but transcends it, oscillating between the two conditions of experience: the fullness and the void. To develop others he disregards their path - great vehicle, lesser vehicle, Tantra, etc. He creates his own pure land for them to dwell in and cultivate their spiritual life, planting them there like wheat; and shines on them with far-reaching light until the ears of grain ripen.
Context In Tibetan Sakya Buddhism, this is one of the states in the "Ascension Stages Game". In some sets it is numbered 86 on the board.
Broader Meditation way of the bodhisattvas (Buddhism, #HM2769).
Followed by Ninth scriptural bodhisattva awareness (Buddhism, Tibetan #HM2292)
Eighth scriptural bodhisattva awareness (Buddhism, Tibetan, #HM2816).
Preceded by Amoghasiddhi-karma awareness (Buddhism, Tibetan, #HM2783)
Sixth scriptural bodhisattva awareness (Buddhism, Tibetan, #HM2385).

♦ **HM2362c Sphere of victory** (Kabbalah)
Netzach sephira
Description This sphere of awareness is associated with active (as opposed to passive) action, with a sense of initiative, and thus with the development of the use of skills in practice. It corresponds to the developmental phase of youth and self-assertion. Complementing the rational, this sphere is related to the arts, the emotions, the subjective; there is an emphasis on instinctual drive, love and spiritual passion. It is symbolized by the positive and negative archetypes of youth, maid and whore, and by the planet Venus, as well as by Moses in the Jewish tradition.
Context This sphere is the seventh of ten described in the Kabbalistic system of sephiroth.
Broader Kabbalah (#HH0921).
Related Manipura (Yoga, #HM3063)   Death emblems (Tarot, #HM3121)
Mutability emblems (Tarot, #HM2250)   Emblems of disaster (Tarot, #HM2350)
World of formation (Judaism, #HM2360)   Way of harmony (Esotericism, #HM0763)
Emblems of the mysterious (Tarot, #HM2398)   Life-fluid emblems - star (Tarot, #HM3137).
Followed by Sphere of beauty (Kabbalah, #HM3031)
Sphere of mercy and greatness (Kabbalah, #HM2420).
Preceded by Sphere of glory (Kabbalah, #HM2238)
Sphere of the kingdom (Kabbalah, #HM2288)
Sphere of the foundation (Kabbalah, #HM2410).

♦ **HM2363d Temporal solidarity** (ICA)
Description This state is an awareness of being related to all of creation, like having a rock for an uncle. The Australian aborigines embrace this reality in their religious understanding that they are inextricably linked to the land and to the forms of plant and animal life it contains. In their mythology of the Eternal Dreamtime, each person is linked to a particular totemic creature, which sustains him or her, and which in turn must be sustained. As spacemen have reported when they look back to earth, they experience a deep sense of kinship not with the various divisions it contains, but with the totality of life represented by this bright blue ball.
A dimension of this experience is a sense of awe, that is fear and fascination. As a an individual stands alongside the rest of creation, he develops an appreciation of even the most insignificant forms of life and honours the diversity of the planet, seeing it as a gift rather than a threat or a problem. There is an awesome feeling that he is are responsible for and accountable to everything and also an equal appreciation of the fragility of all things. He doubts whether he can live with only what he has been given. This knowledge of being related brings realization that one cannot separate one's self from anything nor dismiss anything as meaningless. Each moment is taken as one's big chance and one looks forward to engaging in the next with anticipation, experiencing a rapport with life that knows no bounds and a solidarity with the entire journey of consciousness.
Context This state is number 35 in the ICA *Other World in the midst of this World*.
Broader Original gratitude (ICA, #HM3105).
Followed by Contentless word (ICA, #HM2373)   Ultimate reality (ICA, #HM2030)
Sacrificial passion (ICA, #HM2641)   Sacramental universe (ICA, #HM2445)
Futuric responsibility (ICA, #HM3016).
Preceded by Ultimate reality (ICA, #HM2030)   Self transcendence (ICA, #HM2584)
Impactful profundity (ICA, #HM2607)   Futuric responsibility (ICA, #HM3016)
Definitive predestination (ICA, #HM2131).

♦ **HM2364g Nose consciousness** (Buddhism)
Ghranavijnana — Sna'i-rnam-par-shes-pa (Tibetan) — Fragrance awareness
Description This is the consciousness that apprehends objects of fragrance, whether pleasant or unpleasant, equal (non-pervasive) or unequal (pervasive). It is twofold, that is: where nose consciousness is pleasant it is associated with indeterminate resultant consciousness with profitable result, where it is painful, with indeterminate resultant consciousness with unprofitable result.
Context One of the six consciousnesses defined in Buddhism as dependent on the individual senses, and with objects of sense as their focus.
Broader Sense consciousness (Buddhism, #HM2664)
Perception through the senses (#HM3764)
Mundane resultant consciousness (Buddhism, #HM0130)
Sense mode of consciousness occurrence (Buddhism, #HM4389)
Indeterminate consciousness in the sense sphere - resultant (Buddhism, #HM5721)
Awareness of consciousness-group of conscious existence - senses and mind (Buddhism, #HM2556).
Related Eye consciousness (Buddhism, #HM2074)
Ear consciousness (Buddhism, #HM2169)
Body consciousness (Buddhism, #HM2562)
Tongue consciousness (Buddhism, #HM2263)
Mental consciousness (Buddhism, #HM2838)
Mind-consciousness element (Buddhism, #HM6173).

♦ **HM2365b Higher states of consciousness** (Sufism)
Ahwal — Hal
Description Islamic mystical literature describes a number of very high states of consciousness. One such listing of ten (in Sarraj) includes muraqaba (state-awareness), qurb (God-nearness awareness) and mphabba (love), and culminates with mushahada (contemplation) and yaqin (certainty).
Broader Divine path (Sufism, #HM3371).
Narrower Yaqin (#HM4426)   Mushahada (#HM3521)   Love of God (#HM4273)
State-awareness (#HM2305)   Nearness-awareness (#HM2377)
Divine-love awareness (#HM2401).
Related Ascension awareness (Sufism, #HM2327)
Stations of consciousness - Ansari (Sufism, #HM2317)
Stations of consciousness - Simnami (Sufism, #HM2341)
Ultimate Sufi state of consciousness (Sufism, #HM2318)
Stations of consciousness - ibn-Abi'l-Khayr (Sufism, #HM2424).

♦ **HM2366d Sacramental universe** (ICA)
Description This is experienced as the individual's decision to dare to live and act in this world which he did not create for himself; when he embraces the world as it is and dares to love it all. An example is in the movie "Norma Raye", when the husband of Norma Raye tells her he will see her through it all, no matter what life brings to them.
Context In the ICA New Religious Mode in the arena of articulating the word about life (the life of knowing) the second formal aspect is one's experience of the vanishing cosmos or the wonder of the world. At the third phenomenological level, this occurs when someone perceives in any experience its timeless authenticity.
Broader Wonder at the world (ICA, #HM3062).
Followed by Presence of Jesus Christ (ICA, #HM2206).
Preceded by Stark givenness (ICA, #HM2101).

♦ **HM2367g Holistic consciousness**
Right brain consciousness
Description A non-linear, simultaneous, intuitive mode of awareness, concerned more with relationships than with the discrete elements that are related. Relationships are experienced as real in themselves, whilst the elements related are experienced as of lesser significance. It is associated functionally with intuition, a-rational and holistic knowledge, perceptions of timeless-ness, simultaneity, and with visual, spatial structuring. Its aim is to synthesize new, whole patternings from the pieces of data at its disposal. Metaphorically, this kind of consciousness is associated with non-linear and lateral thinking; and with creativity in general. It is a way of seeing that can only be experienced in its own terms.
Context Associated with the right hemisphere of the brain and viewed as complementary to the verbal-analytical mode associated with the left hemisphere of the brain.
Refs Blakeslee, Thomas *The Right Brain* (1980).

Broader Brain–based consciousness (#HM2295).
Related Analytical consciousness (#HM2415).

♦ HM2368d Creative bisociated consciousness
Divergent thinking
Description Discovery may more often result when the mind is allowed to diverge from a concerted attack on a subject; when it is allowed to roam, associate, or brainstorm. Ideas are produced this way, coming out of differing perspectives, satisfying varying criteria, or having self–logical frames of reference. When these frames (or thought matrices) are allowed to converge and superimpose there is a synergy and dimensional depth to ideas, that would not have arisen in a convergent, linear approach.

♦ HM2369d Intentional self–negation (ICA)
Description This is experienced as saying no to one's own will; when one sees that one is required is to give up one's own peace, security and spiritual well–being for the sake of a larger task. It may be compared to Jean Valjean in "The Tale of Two Cities" when he reflects that it is a "far, far better thing" to negate than to affirm his own will.
Context In the ICA New Religious Mode in the arena of knowing one's disengagement (the life of Poverty) the fourth formal aspect is the experience of spiritual denial or sacrificial offering. At the second phenomenological level, this occurs when a person takes charge of desires which formerly controlled his life.
Broader Spiritual denial (ICA, #HM2174).
Followed by Awareness of spiritual poverty (ICA, #HM2286).
Preceded by Human contingency (ICA, #HM2104).

♦ HM2370d Personal absolution (ICA)
Description The first formal aspect of the *life of meditation*, an arena in the ICA *New Religious Mode*, this is an awareness of encountering a mediator who pronounces personal absolution.
Broader Life of meditation (ICA, #HM3234).
Narrower Radiant guru (#HM2019) Eternal saviour (#HM2253)
Persistent friend (#HM2139) Word–bearing priest (#HM2236).
Related Absolution (#HH0472).

♦ HM2371c Dream–state bardo awareness (Buddhism, Tibetan)
Second antarabhava consciousness
Description This state of awareness is of the intermediate sphere of sleep and dream consciousness. It is illumined by meditation and yogic practice. Deceptive dream–imagery is said to be transformed into forms of knowledge, and because experience in this state can be guided, it is an important preparation for the bardos in the after–life.
Context This state is one of the six major bardo existences in Tibetan Buddhism.
Broader Bardo consciousness (Buddhism, Tibetan, #HM0698).
Related After death dream state (Psychism, #HM0665).
Followed by Meditation–state bardo awareness (Buddhism, Tibetan, #HM2311).
Preceded by Life–state bardo awareness (Buddhism, #HM2347).

♦ HM2372c Sphere of understanding (Kabbalah)
Binah sephira
Description This sphere of awareness is associated with passive intellect, with the ability to reason, and thus with the development of contemplative ability. It corresponds to the developmental phase of old age and the cultivation of philosophy. This state is represented as the concealed world, or the mystery of the supreme world. It is characterized as the intelligence which applies wisdom (chokmah). Various names for it include, the Mother (Aima) and the Assembly (ekklesia) of Israel. It is the source of all images and forms, the womb of creation producing the archetypes in the manifest universe. It is symbolized by the positive and negative archetypes of the anima and the witch, and by the planet Saturn, as well as by Eve in the Jewish tradition.
Context This sphere is the third of ten described in the Kabbalistic system of sephiroth.
Broader Kabbalah (#HH0921).
Related Ajna (Yoga, #HM2144) World of emanation (Judaism, #HM2408)
Magician archetypal image (Tarot, #HM2237) Charioteer archetypal image (Tarot, #HM3045)
Path of resplendent intelligence (Judaism, #HM3426)
Female sovereignty archetypal image (Tarot, #HM2285)
Betrothal initiation archetypal image (Tarot, #HM3028).
Followed by Sphere of wisdom (Kabbalah, #HM2348)
Sphere of the of the supreme crown (Kabbalah, #HM3132).
Preceded by Sphere of beauty (Kabbalah, #HM3031)
Sphere of judgement (Kabbalah, #HM2290).

♦ HM2373d Contentless word (ICA)
Description This is experiencing that the only truth one will ever know is in the meaninglessness or undisclosed mystery that sits in the centre of every situation. With every illusion and every wish dream exposed, one stands on another shore, able to be what one is. D. H. Lawrence's poem "We Are Transmitters" speaks of this in the lines: "Even if it is a woman making an apple dumpling, or a man a stool. If life goes into the pudding, good is the pudding, good is the stool, content is the woman, with fresh life rippling in to her. Content is the man".
A dimension of this experience is a sense of awe, that is fear and fascination. The fear is that it will be more than one can handle and will overwhelm one's life. But one also feels a new kind of power and realizes that there is more to everything than one had ever thought possible. The decision is to be prepared – one can taking nothing for granted. This leads to fulfillment in that there is never any situation or act without significance.
Context This state is number 51 in the ICA *Other World in the midst of this World.*
Broader Radical illumination (ICA, #HM2273).
Followed by Blissful seizure (ICA, #HM3148) Ultimate reality (ICA, #HM2030)
Personal epiphany (ICA, #HM2595) Self transcendence (ICA, #HM2584)
Transcending hostility (ICA, #HM2658).
Preceded by Blissful seizure (ICA, #HM3148) Self transcendence (ICA, #HM2584)
Temporal solidarity (ICA, #HM2363) Everlasting community (ICA, #HM2777)
Inclusive comprehension (ICA, #HM2256).

♦ HM2374d Sexual awareness
Description In *tantric* philosophy, the fusion of male and female in sexual embrace is identified with the fusion of the many interpenetrating individual "bodies" which make up an individual human being; the bliss of such union at the physical level being compared with the bliss of self–realization at all levels, and re–enacted in the form of yogic *sadhana* couples. The unity of the person is also seen in the right (male) and left (female) sides of the body. This concept is common to many systems which emphasise integration through development of the less developed, weaker, female and unconscious aspects of the personality. The Hindu male deities are invariably accompanied by their female counterparts (the "forces" of the "forceful"), and there are also the familiar symbols such as yin and yang, lingam and yoni, Yab–Yum, which demonstrate the same concept.
Related Maithuna (Yoga, #HH2034) Tantra (Buddhism, Yoga, #HH0306).

♦ HM2375b Nakshbandi recollection (Islam)
Description The states of awareness and recollection (dhikr) cultivated by the Nakshbandi Order in Islam contains the following elements: oral and mental repetition of the Name of God (using the name received); pronunciation of the sacred formula "La ilaha illa'llah..."; regulation of breath; and fixing of attention; all this to eliminate distractions in the beginning. The fourth stage (yad dast) is termed *fixation of the presence of truth* (the Saviour) in the mind. The fifth stage (hus dar dan) is termed *breath–knowing*, in which the single object of consciousness is the vital wind. The sixth stage (safar dar vatan) is the *journey to the true homeland* in which one acquires the angel's powers in place of one's own. The seventh stage (nazar bar qadam) is *attention to direction or orientation* in which the pilgrim's consciousness is never distracted from the origin and goal of each footstep. The eighth stage (originally, in the Nakshbandi system, the last), is the mystery of the dual presence of the Self, apparently only going about in the affairs of men, but also internally being in a spiritual retreat attended by truth. As the comment on the first stage of this dhikr put it, "the heart is always conscious of truth". The eighth stage, for these reasons is called, "alone in a crowd" (xalvat dar anjoinan), a state in which a man can be a moral influence in his time, and with undistracted mind help others toward the goal.
Related Recollection (Islam, Sufism, #HM2351).

♦ HM2376d Leo–consciousness (Astrology)
Assertive self–consciousness
Description Leo denotes a sense of independence from the surroundings and from others that goes to the point of passionate opposition to them. Under optimal conditions it is associated with a total involvement, in which all faculties and resources are fully dedicated to the immediate challenge, and in which barriers between self and situation are dissolved in the interests of empathy and understanding.
Context One of the twelve conditions of being or streams of divine energy encompassed within the zodiac. It is said that all twelve conditions must be present and playing its proper part for any successful and fulfilling action to take place. This may be through a group of individuals with different endowments and soul qualities working together; or it may be through one individual who, by self–discipline, practice and meditation, has allowed all twelve conditions to arise within himself.
Broader Fiery awareness (Astrology, #HM2124) Fixed awareness (Astrology, #HM2554)
Zodiacal forms of awareness (Astrology, #HM2713).
Followed by Virgo–consciousness (Astrology, #HM2439).
Preceded by Cancer–consciousness (Astrology, #HM2239).

♦ HM2377c Nearness–awareness (Sufism)
Qurb — God–proximity — Closeness to God — Qurbat
Description This state is characterized as an awareness of one's nearness to God wherever one may be. It may arise in the traveller on the path: through devoted behaviour to God; through awareness of God's closeness to him; through closeness in which God does not make him aware of the attachment. Perfection is realized in affinity with God. The closeness results in *khawf* (fear) and *raja* (hope) and brings the seeker nearer to God, to fulfilment, to passage to unity. It is the key to transformation of the soul.
Realization of identity with God arises from following the path of love, either love of the pure, divine essence of God received as a gift of grace, or a love of God's attributes acquired through mystical efforts or purification of the soul when there is complete freedom from desires of the lower soul.
Context The eighth stage in a systematic account of various stations of the Sufi spiritual path. Also the second psychic state listed by A Reza Aresteh as progress on the inner self through divine attraction, *kedesh–jazba*, the outer self already having been purified through conscious effort, *kushesh*. According to Shaykh Abu Sa'id ibn Abi'l–Khayr, this is the 30th of 40 stations or maqamat the Sufi must possess for his journey on the path of Sufism to be acceptable.
Broader Psychic states (Sufism, #HM4311) Mystic stations (Sufism, #HM3415)
Higher states of consciousness (Sufism, #HM2365)
Stations of consciousness – ibn–Abi'l–Khayr (Sufism, #HM2424).
Narrower Raja (#HM4393) Khawf (#HM1047).
Related Recollection (Islam, Sufism, #HM2351).
Followed by Showq (Sufism, #HM0762) Tafakkur (Sufism, #HM6227)
Love of God (Sufism, #HM4273).
Preceded by Rida (Sufism, #HM4190) Wajd (Sufism, #HM3650)
State–awareness (Sufism, #HM2305).

♦ HM2378c Emblems of personified evil (Tarot)
Devil
Description One of the archetypal emblems projected by the psyche is that of the ugly monster. The monster personifies everything that is a hindrance or harm to human life and desires. Moral evils also take on emblematic form and they are all concentrated in the image of the prince of hell. In the Tarot system, the Devil (emblem number 15) has bat wings to indicate his nocturnal, cavern–dwelling nature; a long–eared, animal–like, horned head; and a hirsute, beastly lower torso to indicated his sub–human behaviour. At this stage one must consider one's weaknesses, in particular relating to worldly possessions and ideas of security. Illusion is revered rather than reality. Man is demonstrated in a sorry plight and a limited framework, a lesson on the search for unity.
Context In the Kabbalah, the Devil is placed on the path between splendour or glory (Hod) and beauty (Tiphareth).
Broader Evil (#HH1034) Emblem archetypes in the psyche (Tarot, #HM2201)
Tarot arcana of conscious states (Tarot, #HM2097).
Related Sphere of glory (Kabbalah, #HM2238) Sphere of beauty (Kabbalah, #HM3031).

♦ HM2379d Besetting sin (ICA)
Description This is the awareness that all of existence is nothing other than separation and desecration, and that one is at the centre of all disrelationships. It occurs when one acknowledges that every act, even the most creative, creates separation, every act is in the state of sin, as when Jesus says, "Let he who is without sin cast the first stone".
Context In the ICA New Religious Mode, in the arena of acting out one's freedom (the life of prayer) the first formal aspect is one's struggle with repentance for one's sinfulness or confession. At the third phenomenological level this occurs when one acknowledges that which is beyond one's power to control.
Broader Affirmation (ICA, #HM2199) Struggle to confess (ICA, #HM2718).
Followed by Heavenly sorrow (ICA, #HM2216).
Preceded by Personal violation (ICA, #HM2561).

♦ HM2380b Contextual consciousness
Existential dasein
Description According to existentialism (Heidegger, Jaspers, Sartre) individual consciousness has always the context of choices among a plurality of personally perceived "probable" realities. Consciousness is always delimited by man's presence in particular contexts, by being there (dasein). The context, that is *existence*, expresses itself as possibilities from which the consciousness must select, thus committing the perceiver to an orientation and to specific tendencies

toward, as well as to real courses of action in his engagement with the world. These viewpoints, the statement of Heidegger that "dasein is always its own possibility", and Sartre's remark that "the possible can come into the world only through a being which is its own possibility...human reality being its being in the form of an option on its being", show that where many traditional religious and idealist philosophies had situated freedom in the cosmic and psychic mind–stuffs (which by a process akin to imagination could create the variables of the universe), the existentialist conception, denying "mind", situates freedom in that aspect of consciousness called will or volition. The context of dasein allows for conscious free–choice of action in the existent world and denies any single, essential reality, or sets of inherent mental constructs, that predestine human behaviour.
Related Tacit knowledge in consciousness (#HM2297).

♦ HM2381b Physical duality in conscious states (Physical sciences)
Description Brain–based conscious states express the duality of the physical world in being either caused or affected by states of matter or energy. The duality of consciousness is the awareness of the multiplicity of the phenomenal world, balanced or counterpoised by the potential or realization of consciousness of the unity of all things. Ordinary consciousness is the analogue of matter, unitary consciousness is the analogue of energy.
Context There are at least five universal principles postulated in the physical sciences: duality; quantum discreteness; relativity; conservation; and least action. Their application to states of consciousness has been presented by Professor Charles Tart and others.
Related Discrete states of consciousness (Physical sciences, #HM2298)
Physical relativity in conscious states (Physical sciences, #HM2322)
Physical conservation in conscious states (Physical sciences, #HM2357)
Least action principle in conscious states (Physical sciences, #HM2417).

♦ HM2383g Personal consciousness field (Psychosynthesis)
Description According to R Assagioli, this term used in psychosynthesis refers to the field of mental contents: sensations, impulses, desires, feelings, imagery, thoughts and other subjective experience including memory and identity, which constitute the ordinary world of the personality and its frames of reference to common reality.
Related Psychosynthesis (#HH0002).
Preceded by Middle unconscious (Psychosynthesis, #HM2306).

♦ HM2384c Watery awareness (Astrology)
Triplicity of water
Description A person influenced by the element of water, by the combination of the zodiacal signs Cancer, Scorpio and Pisces, will be sensitive and emotional, outwardly calm but with hidden depths and, like the sea, capable of sudden and unexpected violent storms. Although in some contexts such awareness may appear damping or, like an uncontrolled torrent, overwhelming and destructive, watery awareness is notably compassionate. It demonstrates insight into the feelings of others combined with a sense of value for relationships. A subjective solicitude or "motherliness" may be suffocating, but the fluid nature of water makes the individual capable of taking the shape of and reflecting those with whom it comes into contact.
Context One of the four *triplicities* or *elements*, each of which combines three related signs of the zodiac.
Broader Zodiacal forms of awareness (Astrology, #HM2713).
Narrower Cancer–consciousness (#HM2239) Pisces–consciousness (#HM2856)
Scorpio–consciousness (#HM2615).
Related Airy awareness (Astrology, #HM2955) Fiery awareness (Astrology, #HM2124)
Earthy awareness (Astrology, #HM3235).

♦ HM2385d Sixth scriptural bodhisattva awareness (Buddhism, Tibetan)
Abhimukhi
Description This state of awareness is characterized by being face–to–face with reality which is neither samsara nor nirvana; it is represented by the sixth level bodhisattva who embodies wisdom, and compassion for those tormented by attachments and sufferings.
Context In Tibetan Sakya Buddhism this is one of the states in the "Ascension Stages Game". In some sets it is numbered 87 on the board.
Broader Meditation way of the bodhisattvas (Buddhism, #HM2769).
Followed by Eighth scriptural bodhisattva awareness (Buddhism, Tibetan, #HM2816)
Seventh–scriptural bodhisattva awareness (Buddhism, Tibetan, #HM2361).
Preceded by Fifth scriptural bodhisattva awareness (Buddhism, Tibetan, #HM2909)
Fourth scriptural bodhisattva awareness (Buddhism, Tibetan, #HM2191).

♦ HM2386d Transformed state (ICA)
Recreated by mystery
Description This is an experiential trek in the *Land of Mystery*, within the ICA *Other World in the midst of this World*, when the individual is aware that all things are new. The states are described in separate entries.
Broader Land of mystery (ICA, #HM2434).
Narrower Second birth (#HM3175) Vibrant powers (#HM2982)
Dynamic selfhood (#HM3091) Transformed existence (#HM2862).
Related Infinite passion (ICA, #HM2234) Aweful encounter (ICA, #HM2619)
Inescapable power (ICA, #HM3096).

♦ HM2387c Emblems of totality (Tarot)
World
Description Among the archetypes projected by the psyche are emblems of wholes, the universe of personal awareness, the inter–relatedness or network of all things. These emblems include the starry sky or vault of heaven, spheres, circles, the grand assembly of souls, and the zodiac. In the Tarot system, the related emblem is the World (number 21), which is depicted as a large wreath supported by the three animals and one human figure associated with the constellations in which the equinoxes and solstices occur. The spirit of the world is depicted in human form within the wreath; although adrogynous, representing male and female polarities, she appears as female. The representation is of the underworld of the subconscious reflecting the supreme father, Kether – the first step on the journey to cosmic consciousness, embodying spiritual rebirth.
Context The World is placed by Kabbalists in the 32nd path between kingdom or world (Malkuth) and foundation (Yesod).
Broader Emblem archetypes in the psyche (Tarot, #HM2201)
Tarot arcana of conscious states (Tarot, #HM2097).
Related Sphere of the kingdom (Kabbalah, #HM2288)
Sphere of the foundation (Kabbalah, #HM2410).

♦ HM2388d Ultimate awareness (ICA)
Description This is being jarred into awareness of everything going on in one's life and in the world around; as though one were frozen in a state of lucidity, standing outside of one's self observing everything going on inside. There is a sense of having one's eyelids permanently sewn open – no matter how hard one might try to rid one's self of this heightened awareness, it just won't go away. It may compared to a condemned man on his way to the gallows who finds it hard to fall asleep.
A dimension of this experience is a sense of awe, that is fear and fascination. One wonders how much more one can stand it and often longs to return to a life where things were simple and one was not so aware of what was happening; and also wonders if one's mind is playing tricks on one. At the same time there is the intriguing question of what it would mean to embrace this consciousness forever, and one experiences an exhilarating kind of liveliness that sustains in everything one does.
In the midst of this experience, people look straight at the real situation, yearning to communicate their lucidity with others as though they had a deep secret to share. They sense they know what truth is like – they have seen it, heard it, felt it, tasted it or touched it; and this creates a sense of freedom to pursue what has been revealed.
Context This state is number 17 in the ICA *Other World in the midst of this World.*
Broader Authentic relation (ICA, #HM2316).
Followed by Universal fate (ICA, #HM2687) Beyond morality (ICA, #HM3049)
External relation (ICA, #HM2471) Seminal illumination (ICA, #HM2056)
Individual fatefulness (ICA, #HM2223).
Preceded by Beyond morality (ICA, #HM3049) Singular adoration (ICA, #HM2111)
Original integrity (ICA, #HM2773) Radical contingency (ICA, #HM2477)
Seminal illumination (ICA, #HM2056).

♦ HM2389c Effort (Buddhism)
Virya — Brtson–'grus (Tibetan)
Description The quality on which all auspicious qualities depend, true effort involves fulfilling and accomplishing virtues; and is contrasted with striving to accomplish affairs in this lifetime, which is simply lazy attachment to bad action. It may be one of five types: *armouring*, or the prior willingness to engage in virtue; *application*, or the mental delight of carrying out that practice; *non–inferiority*, the delight in not questioning one's ability to do such actions; *irreversibility*, the delight which prevents circumstances from impeding the practice; and *non–satisfaction*, which spurs one on to greater virtues and is not satisfied with achieving the smaller.
Context One of the eleven *virtuous mental factors* referred to in Tibetan Buddhism.
Broader Virtuous mental factors (Buddhism, #HH0578).
Related Virya (Zen, #HH0677).

♦ HM2390b God–consciousness as rapture
Apollo awareness — Orpheus awareness
Description Human development can lead to a state of consciousness which expresses the attributes of the God, Apollo. This is, on the one hand, Apollo leader of the muses, the god of imperishable beauty; and on the other, it is Apollo the god, who represents mind and true being. In one figure Apollo brings together being, beauty and mind, and indicates that state of intellectual rapture induced by divine vision. This is the vision of beauty absolute (as in Plato's Symposium) or the sight of Beatrice or the Rose (in Dante). It is of such visions that Orpheus sang, with the promise of immortality.
Broader God–consciousness (#HM2166).

♦ HM2391b Altered states of consciousness
Altered states of awareness — Levels of awareness — States of non–ordinary reality
Description The normal state of consciousness is one in which any given individual spends the major portion of his waking time. It is universally assumed that the normal state of one individual is quite similar to that of all others, even though the evidence for this is questionable. It may be argued that man has functioned in a multitude of states of consciousness and that different cultures have very different views concerning the recognition, utilization and attitudes towards altered states.
An altered state of consciousness for a given individual is one in which he clearly feels a qualitative shift in his pattern of mental functioning. That is, he feels not just a quantitative shift (more or less alert, more or less visual imagery, sharper or duller, etc), but also that some quality or qualities of his mental processes are different. In such a case mental functions operate that do not operate ordinarily and perceptual qualities appear that have no normal counterparts. It is acknowledged that there are numerous borderline cases in which the individual cannot clearly distinguish precisely in what way his state of consciousness is different from normal, particularly where the quantitative changes are very marked. Altered states of consciousness may be positive or pathological (as in schizophrenia, for example); and they may involve inactivity of the senses or the mind.
If, in the normal state of consciousness, one can consider external input to be balanced with sampling of stored material, then in altered state there is a disturbance of this base line. This occurs in natural sleep due to reduced cortical activity; but in altered states there is previous inculcation of set to remain awake at some level. This may be due to the intention of the individual or to the instructions of a hypnotist. There is awareness of some classes of input while others are not processed. Altered states of consciousness may be characterized by:
(1) Alterations in thinking, including subjective disturbances in concentration, attention, memory and judgement.
(2) Disturbed time sense, including a subjective feeling of timelessness.
(3) Loss of control, a sense of helplessness and loss of grasp of reality.
(4) Change in emotional expression, including sudden and unexpected displays of more primitive and intense emotion.
(5) Body image change, including a sense of depersonalization.
(6) Perceptual distortions, including hallucinations, increased visual imagery and illusions of every variety, lack of response to external stimuli.
(7) Change in meaning or significance, including feelings of profound insight, illumination and truth.
(8) Sense of the ineffable and inability to communicate the nature or essence of the experience.
(9) Feelings of rejuvenation.
(10) Hypersuggestibility.
Examples of altered states of consciousness include: hypnagogic state (borderline between waking and sleeping); dream consciousness; meditative state; hypnotic trance; psychedelic drug–induced state.
Altered states of consciousness may be induced by various physiological, psychological, or pharmacological manoeuvres or agents. The methods may be classified as:
(1) Reduction of exteroceptive stimulation and/or motor activity. This includes mental states resulting primarily from the absolute reduction of sensory input, the change in patterning of sensory data, or constant exposure to repetitive, monotonous stimulation.
(2) Increase of exteroceptive stimulation and/or motor activity and/or emotion. This includes excitatory mental states resulting primarily from sensory overload or bombardment, which may or may not be accompanied by strenuous physical activity or exertion and profound emotional and mental fatigue.
(3) Increased alertness or mental involvement. This includes mental states which appear to result primarily from focused or selective hyperalertness with resultant peripheral hypoalertness over a sustained period of time.
(4) Decreased alertness or relaxation of critical faculties. This includes mental states which appear to occur mainly as a result of a passive state of mind in which goal–directed thinking is minimal.

It includes mystical, transcendental, or revelatory states (such as satori, samadhi, nirvana, cosmic consciousness) attained through meditation, as well as various trance states.
(5) Presence of somatological factors. This includes states arising primarily as a result of alterations to body chemistry or neurophysiology.
An altered state may be deliberately induced, as with taking psychedelic drugs, or may arise spontaneously.
Refs Tart, Charles T (Ed) *Altered States of Consciousness* (1971); Teyler, Timothy J *Altered States of Awareness* (1972).
Related Higher states of consciousness (#HM0935)
Usual state of consciousness (USC, #HM1999)
Alternate state of consciousness (ASC, #HM1537)
Anomalies in consistency of consciousness (#HM1122).

♦ **HM2392b Upanishadic stages of awareness** (Hinduism)
Description In the Chandogya Upanishad, it is said that a person must learn to worship Spirit in sixteen aspects. First as *nama* (name), then as *wac* (speech), as *manas* (mind), *sankalpa* (resolution), *citta* (mind's mother substance), *dhyana* (meditation), *vidya* (knowledge), *balam* (power), *anna* (food), *apa* (water), *tayjas* (light), *akasha* (ether), *smara* (memory), *asa* (hope), *prana* (life) and finally *satya* (truth). Having come to worship each of these as Spirit, the individual knows, speaks, thinks, is devoted to, acts and finds happiness in the truth beyond discussion, the unlimited.
Broader Stages of spiritual life (#HH0102).
Related Indian approaches to consciousness (Hinduism, #HM2446).

♦ **HM2393b God-consciousness as adapability**
Dionysos awareness — Phoenix awareness
Description Human development can lead to a state of consciousness which expresses the principle attribute of the God Dionysos; that is the power to change, metamorphose or adapt. Like the infant Dionysos, in order to respond to the challenge of survival and growth the human must change and renew his psychic formations. Also, like the phoenix which voluntarily immolates itself to rise anew from its own ashes, those who would perfect themselves must be aware of the regenerative power within them and willingly die psychologically in order to be reborn. In this "death" agony there may be an ecstatic madness, laughter, at the supreme joke and paradox, that we die and live alternately.
Broader God-consciousness (#HM2166).

♦ **HM2394d Incarnate living** (ICA)
Description This state occurs when an individual experiences being envelopped in what seems to be a cloud of mystery that invades every corner of life; as if, at every turn, he encounters the unknowable. The mundane activities of life are suddenly no longer commonplace but full of mystery which engenders wonder. It is like walking into a surrealist painting by Salvador Dali, and may be compared with the main character in the book "Journey to Ixtlan" when he experiences the world around him as being transformed by the Indian Shaman, Don Juan. The world has seemingly taken on an extra dimension that one experiences one's self as part of while still being part of the everyday world.
A dimension of this experience is a sense of awe, that is fear and fascination. The fear is primarily one of not knowing what to do and of inundation: "What will I do if this continues and there is no way to get away from it ?". The fascination is at the wonder of it all; there is a perpetual sense of surprise and of being able both to participate in events and to stand outside watching the participation.
The decision experienced in this state is to decide. One is driven to settle for nothing less than the whole of reality. Following such times one is able to see from many perspectives and has a new appreciation of ambiguity. There is also a new kind of restlessness in one's life.
Context This state is number 5 in the ICA *Other World in the midst of this World.*
Broader Inescapable power (ICA, #HM3096).
Followed by Vibrant powers (ICA, #HM2982) Universal fate (ICA, #HM2687)
Primal sympathy (ICA, #HM2550) Essential dubiety (ICA, #HM3135)
Ubiquitous otherness (ICA, #HM2570).
Preceded by Primal sympathy (ICA, #HM2550) Primordial wonder (ICA, #HM2186)
Essential dubiety (ICA, #HM3135) Creative futility (ICA, #HM2493)
Radical contingency (ICA, #HM2477).

♦ **HM2395b Turiya awareness** (Hinduism)
Transcendental awareness — Subramania
Description In yoga philosphy there are three states (avasthas) of ordinary consciousness (waking, dreaming, sleeping) and a fourth, superconscious or extraordinary meditation state (turiya), which unites the three and in which the true self may be discovered. According to the Mandukya Upanishad it is associated with the complete sound of the mystic syllable AUM, and completes the four "quarters" or conditions of the self or *Atman*; cognition is neither outward or inward, nor the presence or absence of these, but it is the indescribable essence of self-cognition common to all states; it the self to be realized. It is also the seventh of seven ascending planes of wisdom described in the Supreme Yoga, which require to be known so as not to be caught in delusion. In Hindu mythology the state is exemplified by the god Subramania or Kartikeya.
Broader Planes of wisdom (Hinduism, #HM3298)
Degrees of consciousness (Hinduism, #HM0891)
Awareness of relative reality (Hinduism, Yoga, #HM2032).
Related Prajna (Hinduism, #HM3455) Taijasa (Hinduism, #HM3040)
Vaisvanara (Hinduism, #HM2336) Transcendental consciousness (#HM2020)
Asamprajnata samadhi (Hinduism, #HM3041)
Swapna state of dream consciousness (Yoga, #HM2781)
Jagrat state of waking consciousness (Yoga, #HM2141)
Susupti state of unconsciousness (Hinduism, Zen, #HM2957).
Followed by Turyatita (Hinduism, #HM3139).
Preceded by Cessation of objectivity (Hinduism, #HM3354).

♦ **HM2396c Purusha** (Yoga)
Purusa — Witness-consciousness
Description In the Samkhya-yoga philosophy, the mythological, primordial father-mother appear as the metaphysical prakriti (the ground of matter; feminine), and purusha (the ground of consciousness; masculine). In man, purusha is the witness-consciousness that neither acts nor refrains from action. The universe of reality is perceived as material by purusha because the three qualities (gunas) of prakriti have become unbalanced. A man who controls the gunas in himself sees prakriti no more, but sees the true spirit.
Related Three gunas (#HH0413) Kaivalya (Yoga, #HM3869)
Nirbija samadhi (Hinduism, #HM2125)
Indian approaches to consciousness (Hinduism, #HM2446).

♦ **HM2397d Absolute-body awareness** (Buddhism, Tibetan)
Dharma-kaya
Description This state of awareness is characterized as the fullest *great awakening* and is represented as the innermost essence of Buddhahood. This body of the Buddha is said to be ineffable, neither existent nor non-existent. It is one yet manifold, since many attain it. This is the mystical body constituted by realized mankind and compares to the Pauline conception of Christ; to Zoroastrian and Aryan conceptions of the *grand man* (Anthropos); and to the mystic Adam among Judaeo-Christian, Mandaean, and Manchaean speculative philosophies.
Context In Tibetan Sakya Buddhism this is one of the states in the "Ascension Stages Game". In some sets it is numbered 93 on the board.
Broader Supreme heavens in the ascension stages game (Buddhism, Tibetan, #HM3214).
Followed by Enjoyment-body awareness (Buddhism, Tibetan, #HM2873).
Preceded by Supreme-heaven awareness (Buddhism, Tibetan, #HM2813)
Final vajra-master awareness (Buddhism, Tibetan, #HM2849)
Tenth scriptural bodhisattva awareness (Buddhism, Tibetan, #HM2421).

♦ **HM2398c Emblems of the mysterious** (Tarot)
Moon
Description Among the archetypes projected by the psyche are emblems indicating indeterminacy, the unknown, the forbidden and the hidden. These emblems include darkness, veils, masks, shadows and mysterious apparitions and shape-shifting forms. In the Tarot the emblem of the Moon (number 18) is shown with howling dogs and a deep pool with a bottom-crawling crayfish.
Context The Moon emblem is correlated by Kabbalists to the 29th path, between kingdom (Malkuth) and force or victory (Netzach).
Broader Emblem archetypes in the psyche (Tarot, #HM2201)
Tarot arcana of conscious states (Tarot, #HM2097).
Related Sphere of victory (Kabbalah, #HM2362) Sphere of the kingdom (Kabbalah, #HM2288).

♦ **HM2399c Discrimination** (Buddhism)
Samjna — 'Du-shes (Tibetan) — Upalakkhana
Description Two types of discrimination are noted: *non-conceptual*, which apprehends the non-common signs of an object which appears to the non-conceptual mind; and *conceptual*, which apprehends the uncommon signs of an object appearing to the conceptual mind.
Context One of the five omnipresent mental factors defined in Tibetan Buddhism.
Broader Faculty of wisdom (Buddhism, #HM3233)
Omnipresent mental factors (Buddhism, Tibetan, #HH0320).

♦ **HM2400d Unspeakable joy** (ICA)
Description This is an irrational wonder in the completeness and appropriateness of every aspect of every situation. It occurs when the possibility of working within a situation is not dependent upon your own intervention.
Context In the ICA New Religious Mode in the arena of acting out one's freedom (the life of prayer) the second formal aspect is affirmation of one's dependence or gratitude. At the fourth phenomenological level this occurs when one knows that one's whole life is inescapably bound in this struggle.
Preceded by Manifold blessings (ICA, #HM2269).

♦ **HM2401c Divine-love awareness** (Sufism)
Mahabba
Description This state is characterized as leading to the goal of mystical vision or union. It is said to be a mysterious degree of love which is put among the mystical, infused states since it cannot be acquired at will. Love is one, for outwardly and inwardly the devotee is with the One.
Context The tenth stage on the path to sainthood of the Naqshbandiyya order. According to Shaykh Abu Sa'id ibn Abi'l-Khayr, this is the 28th of 40 stations or maqamat the Sufi must possess for his journey on the path of Sufism to be acceptable.
Broader Higher states of consciousness (Sufism, #HM2365)
Suluk of the Naqshbandiyya order (Sufism, #HM4356)
Stations of consciousness - ibn-Abi'l-Khayr (Sufism, #HM2424).
Related Mahabba (Islam, #HM6523) Love in God (Islam, #HM4712)
Love to God (Islam, #HM5116) Love of God (Sufism, #HM4273)
Love with God (Islam, #HM5580).
Followed by Wajd (Sufism, #HM3650) Wilayat (Sufism, #HM3647).
Preceded by Rida (Sufism, #HM4190) Wilayat (Sufism, #HM3647).

♦ **HM2402c Objective fourth state of consciousness**
Description According to G I Gurdjieff, included among the states of consciousness, counting sleep and waking or clear consciousness, there are also flashes of self-remembering or self-consciousness of one's being. These are considered a third state, which is the precondition for a fourth, or objective, state of consciousness is which things are seen as they truly are. This latter is, therefore, enlightenment or truth, and is connected with higher mental functions. It is also connected with higher emotional functions and with conscience.

♦ **HM2403d Sadness**
Description The feeling of powerlessness or inability to cope with an unpleasant situation or the loss of some source of love and wellbeing produces the emotion of sadness, in particular in response to unfulfilled or conflicting desires. This is more marked in sensitive or melancholic individuals and is accompanied by reduction in bodily strength and an inability to act. As absorption in suffering increases, depression and sometimes stupor result.
Related Accidie (#HM3259).

♦ **HM2404d Solitary being** (ICA)
Description The first formal aspect of the *life of knowing*, an arena in the ICA *New Religious Mode*, this represents an awareness of being a self, a solitary individual.
Broader Life of knowing (ICA, #HM2801).
Narrower Horror of sin (#HM2720) Unveiled being (#HM2611)
Incarnate Christ (#HM2480) Representational existence (#HM2557).

♦ **HM2405b Transcendental cosmic consciousness**
Cosmic consciousness — Fifth state (Yoga)
Description Cosmic consciousness is said to emerge out of interaction of transcendental consciousness and everyday experience. Its characteristics are clarity and tranquillity of the mind and uninterrupted awareness of self during waking, dreaming and sleeping states, in a dual, non-interfering and unconnected manner. Scientific research suggests increased brainwave coherence.
Context Postulated by Maharishi Mahesh Yogi, cosmic consciousness has been studied along with transcendental consciousness in individual and societal context. Links between proposed ability of effortless functioning and brainwave studies provide objective data for validation. Cosmic consciousness seems to occur spontaneously and to disintegrate if not stabilized into God-consciousness or unity-consciousness.
Related Cosmic consciousness (#HM2291) Transcendental meditation (TM, #HH0682).
Followed by Refined cosmic consciousness (#HM3222).
Preceded by Transcendental consciousness (#HM2020).

♦ **HM2406g Twilight consciousness**
Amnesic behaviour
Description Unremembered behaviour, or short, absent periods of time may be due to epileptic seizure, to pathological alcoholism, or to schizophrenia. Acts and inner experiences occurring during absence of the conscious ego are a twilight zone between the "day" of full awareness and the "night" of permanent derangement or disability. this twilight consciousness may be characterized by the appearance of waking dreams in the case of schizophrenics, hallucinations in the case of alcoholics, and illusion or confusion in the case of epileptic seizures. Observation of mental patients shows periods of apparently (to others) normal behaviour, in which however, the patient was absent and was later unable to recall. The number of people who fulfil the requirements of everyday life (including working, marital relationships and leisure activities) and who are partly amnesic or in a state of twilight consciousness, is not known.
In the affective twilight state there is an extreme emotional reaction to a psychological trauma. Comprehension of the situation is restricted so that assessment is distorted. The individual may show irrational behaviour with lack of social restraint, and may be violent. The state may develop into an amnesic fugue. Twilight states with organic origin arise in temporal lobe epilepsy and punch drunkenness. Here activity may continue normally so the observer sees nothing wrong, while the sufferer feels confused and disoriented. After the attack there is total amnesia. There may also be clouding of consciousness.
Broader Diminished range of awareness (#HM1353).
Narrower Punch-drunk (#HM7174).
Related Fugue state (#HM7131) Clouding of consciousness (#HM5754).

♦ **HM2407c Death-bardo heaven-reality awareness** (Buddhism)
Fifth antarabhava consciousness — Chonyid bardo (Tibetan) — Chos-nid
Description This state of awareness is represented by the second after-life bardo and is characterized by the perception of the sambhogakaya. It is said to begin the descent from the perception of the clear light to that of a perception of a heaven in which the soul or awareness-principle experiences the fields of bliss on the plane of the bodhisattvas in view of Buddha Amitabha. The beautiful creatures encountered are images of the self the person has constructed. If attachment to them is not renounced they will turn into demonic monsters.
Context This state is one of the six major bardo conditions. It is experienced from the 4th through the 19th day after death.
Broader Bardo consciousness (Buddhism, Tibetan, #HM0698).
Followed by Death-bardo rebirth-seeking awareness (Buddhism, #HM2431).
Preceded by Death-bardo clear-light awareness (Buddhism, #HM2335).

♦ **HM2408c World of emanation** (Judaism)
Olam atziluth
Description This world is said to correspond to the divine essence or transcendental self of man. It is the realm of the three spheres (sephiroth): Kingdom, Wisdom, and Understanding (kether, chokmah and binah) characterized as essence, knowledge of God, and ability to discriminate between the real and unreal.
Context This is the highest of the Kabbalistic four worlds.
Broader Shekinah awareness (Judaism, #HM2002).
Related Paths of wisdom (Judaism, #HM2509) Sphere of wisdom (Kabbalah, #HM2348)
Sphere of understanding (Kabbalah, #HM2372)
Sphere of the of the supreme crown (Kabbalah, #HM3132).
Preceded by World of formation (Judaism, #HM2360).

♦ **HM2409b Happiness**
Satisfaction awareness — Feeling of well-being — Contentment
Description Degrees of satisfaction with one's actual experience compared to desired experience vary from acceptance of what is less-than-desirable, to contentment, joy and ecstatic bliss. Satisfaction involves a state of critical awareness of a standard or norm for experience. There is acceptance and a feeling of comfort with one's self. This experience may be personal and simple, on the physiological and sense levels, up through the emotional, rational, aesthetic, imaginative and other functions; or it may be personal and complex involving the ego in several dimensions, so that happiness as an ego state of awareness may take in unconscious and sub-conscious conditions as well as conscious. Experience may also involve awareness of the states of others, and of society. Dissatisfaction may result if the values or standards held as normative for others and for society are not met. In as much as there may be an innate sense of what is normative for others as there is for oneself (viz freedom from meaningless and unnecessary pain and suffering) human happiness in its fullest development may depend on conscious satisfaction not only with personal experience, but with interpersonal experience and the conditions of societal life as a whole.
Refs Al-Ghazzali *The Alchemy of Happiness.*
Narrower Self-satisfaction (#HM3288) Satisfaction (Hinduism, #HM8116).
Related Rida (Sufism, #HM4190) Contentment (#HM3729)
Sukha (Buddhism, #HM2866).

♦ **HM2410c Sphere of the foundation** (Kabbalah)
Yesod sephira
Description This sphere of awareness is associated with ego identity, with the psychological functions of the persona and its shadow, and thus with the development of a sense of acuity. It corresponds to the developmental phase of infancy and ego perception. It is the state related to the sexual instinct and to witchcraft. It is symbolized by the positive and negative archetypes of Diana, Hecate and the Moon, as well as by Joseph in the Jewish tradition.
Context This sphere is the ninth of ten described in the Kabbalistic system of sephiroth.
Broader Kabbalah (#HH0921).
Related Svadhishthana (Yoga, #HM3174) Emblems of totality (Tarot, #HM2387)
World of formation (Judaism, #HM2360) Emblems of well-being (Tarot, #HM2422)
Way of idealism (Esotericism, #HM6554) Life-fluid emblems - star (Tarot, #HM3137)
Life-fluid emblems - temperance (Tarot, #HM3289).
Followed by Sphere of glory (Kabbalah, #HM2238)
Sphere of beauty (Kabbalah, #HM3031)
Sphere of victory (Kabbalah, #HM2362).
Preceded by Sphere of the kingdom (Kabbalah, #HM2288).

♦ **HM2411c Emblems of supremacy** (Tarot)
Emperor
Description Among the archetypal images projected by the psyche are those representing paramount authority, supreme power and invincibility. In the Tarot system, the emblem called *the Emperor* (number 4) signifies power over others and over affairs. He faces the archetypal father beyond the abyss and draws on his energy; he is the father of the manifest universe through his union with the empress, and has the quality of mercy as well as of strength and vitality.
Context In the Kabbalah, the Emperor is placed on the 28th path between beauty (Tiphareth) and wisdom (Chokmah).
Broader Emblem archetypes in the psyche (Tarot, #HM2201)
Tarot arcana of conscious states (Tarot, #HM2097).
Related Sphere of beauty (Kabbalah, #HM3031) Sphere of wisdom (Kabbalah, #HM2348).

♦ **HM2412b Adoration**
Description The reverential attitude which acknowledges with a person's whole being the absolute immensity, holiness and glory of God, or, in secular terms, the greatness of that which one holds to be the final reality in life and one's entire subjection to it. An interior discipline of devotion and prayer which may be manifested physically by physical gesture/devout posture such as genuflection, prostration, kneeling.
Related Adore (#HH3424) Practice of silence (#HH0983).

♦ **HM2413d Saving the mystery** (ICA)
Description This is experienced as union with God; when one serves God and God's creation such that neither would be the same without that service. It may be compared to Moses leading the Israelites out of Egypt, across the Red Sea and through the desert to the promised land.
Context In the ICA New Religious Mode in the arena of knowing and doing intensified in one's life (the life of being) the fourth formal aspect is the experience of transparent presence or the experience of effulgence. At the fourth phenomenological level this occurs when a person becomes one with the eternal.
Broader Transparent presence (ICA, #HM2462).
Preceded by Human transformation (ICA, #HM2334).

♦ **HM2414c Emptiness of products** (Buddhism, Tibetan)
Samskrtashunyata
Context One of the eighteen emptinesses comprising the paths of view in Tibetan Buddhism.
Broader Emptinesses on the paths of the view in Buddhism (Buddhism, Tibetan, #HM2944).

♦ **HM2415g Analytical consciousness**
Left brain consciousness
Description The internalization of the experience of the closed boundaries around objects leads to a way of thinking which naturally emphasizes distinction and separation. Because of characteristic externality of such solid bodies, this way of thinking is necessarily analytical. It is also sequential and linear, proceeding from one element to another in a linear fashion. It is the elements that are related that stand out in this form of experience, with the relationships themselves but a shadowy abstraction. This mode is also associated with language, especially in terms of the subject-predicate features of grammar, which have the effect of dividing experience into separate elements. Because language is transparent in the act of disclosing the world, this encourages belief that the world is divided in the way provided in any given language. In emphasizing linearity, this mode is unable to acknowledge effectively the pattern of non-linear relationships between the elements of experience.
Context Associated with the left hemisphere of the brain and viewed as complementary to the holistic mode associated with the right hemisphere of the brain.
Broader Brain-based consciousness (#HM2295).
Related Holistic consciousness (#HM2367) Onlooker consciousness (#HM1993).

♦ **HM2416c Awareness of the divine presences** (Sufism)
Hadarat — Hadrat
Description In the Sufi *Ibn Arabi* there are five divine presences in the celestial worlds of conscious being. The same term (hadarat, hadrat) may be used for holy presences, as titles of respect: for Allah, saints, the Prophet, or any Koran-learned man. The presence (hadara, hudur) of Allah or the spiritual in the life of the pious is therefore expressed of different levels, also by the Dirvishes who call their Friday service *hadra*; and by Ibn Khaldun, who refers to a Plotinian word of divine emanations as hadarat.
Broader Stations of consciousness - Simnami (Sufism, #HM2341).

♦ **HM2417b Least action principle in conscious states** (Physical sciences)
Description In natural physical and chemical processes where there are alternative ways to effect a particular change, then the way which normally occurs is that requiring the least action or expenditure of energy. This principle seems to apply in biology and psychology, particularly in the areas of perception and comprehension, and in education. However, the ordinary human state does not evidence a true least-action pattern, but oscillates between action deficiency and excess.
Context There are at least five universal principles postulated in the physical sciences: duality; quantum discreteness; relativity; conservation; and least action. Their application to states of consciousness has been presented by Professor Charles Tart and others.
Related Discrete states of consciousness (Physical sciences, #HM2298)
Physical duality in conscious states (Physical sciences, #HM2381)
Physical relativity in conscious states (Physical sciences, #HM2322)
Physical conservation in conscious states (Physical sciences, #HM2357).

♦ **HM2418d Universal father** (ICA)
Description This refers to those moments when an interior dialogue develops in a person's life, when he thinks through what he is going to do about a situation which he has created with his actions. It occurs when a person finds his actions and the resulting situation being called into question by his own personal values, and he asks himself what he intends to do now. An example is the references in Nikos Kazantakis' book "Report to Greco" to having El Greco as that person who holds him accountable for living his life.
Context In the ICA New Religious Mode in the arena of knowing one's internal sociality (the life of meditation) the second formal aspect is the experience of being called to accountability by one's interior priors. At the third phenomenological level, this occurs when one dialogues across time and space with those who also travel this journey of the spirit.
Broader Called to accountability (ICA, #HM2008).
Followed by Heavenly advocate (ICA, #HM2784).
Preceded by Concerned judge (ICA, #HM2582).

♦ **HM2420c Sphere of mercy and greatness** (Kabbalah)
Chesed sephira
Description This sphere of awareness is associated with active (as opposed to passive) emotion, with expressive abilities, and thus with the development of the ability to love. It corresponds to the developmental phase of late middle age and the expression of generosity This state is represented as the place of revelations and identified with the four-letter name of God. It is associated with divine mercy and majesty, and with wisdom. Here is the source of all archetypal ideas, so that there can be no images beyond this level although images are used to describe the states beyond, the trinity beyond the abyss. It is symbolized by the positive and negative archetypes of the Great King and the wastrel, and by the planet Jupiter, as well as by Abraham in the Jewish tradition.
Context The fourth of ten spheres described in the Kabbalistic system of sephiroth.
Broader Kabbalah (#HH0921).
Related Mutability emblems (Tarot, #HM2250) World of creation (Judaism, #HM3038)
Natural forces emblems (Tarot, #HM3127) Way of love-wisdom (Esotericism, #HM1507)
Hierophant archetypal image (Tarot, #HM2260)
Old wise-man archetypal image (Tarot, #HM2202)
Path of perfect intelligence (Judaism, #HM3243).

Followed by Sphere of wisdom (Kabbalah, #HM2348).
Preceded by Sphere of beauty (Kabbalah, #HM3031)
Sphere of victory (Kabbalah, #HM2362)
Sphere of judgement (Kabbalah, #HM2290).

♦ **HM2421d Tenth scriptural bodhisattva awareness** (Buddhism, Tibetan)
Dharma–negha
Description This state of awareness is characterized by all–encompassing wisdom and power, and represented as the tenth bodhisattva stage leading into the dharma body of the Buddha.
Context In Tibetan Sakya Buddhism this is one of the states in the "Ascension Stages Game". In some sets it is numbered 94 on the board.
Broader Meditation way of the bodhisattvas (Buddhism, #HM2769).
Followed by Absolute–body awareness (Buddhism, Tibetan, #HM2397)
Supreme–heaven awareness (Buddhism, Tibetan, #HM2813).
Preceded by Ninth scriptural bodhisattva awareness (Buddhism, Tibetan, #HM2292)
Eighth scriptural bodhisattva awareness (Buddhism, Tibetan, #HM2816).

♦ **HM2422c Emblems of well–being** (Tarot)
Sun
Description Among the archetypes projected by the psyche are emblems of achievement, self–maintenance, success, happiness, fame and related states of awareness of well–being. Symbols can include a crown, a high seat, appearance before an audience, and being in the lime–light or a radiant place. In the Tarot, the related emblem is the Sun (number 19) shining on two children. There is a type of innocence and the combining of opposite polarities. At this stage, there is rigid self scrutiny as typified by the the dispersal of darkness by the angel of fire. The emblem suggests that well–being can never be a static situation, but one of growth and human development.
Context The Tarot emblem *the Sun* is placed by Kabbalists on the 30th path between foundation (Yesod) and splendour or glory (Hod).
Broader Emblem archetypes in the psyche (Tarot, #HM2201)
Tarot arcana of conscious states (Tarot, #HM2097).
Related Sphere of glory (Kabbalah, #HM2238)   Sphere of the foundation (Kabbalah, #HM2410).

♦ **HM2423g Fatigued conscious awareness**
Loss of attention — Boredom — Monotony
Description Awareness can be fatigued due to physiological or to psychological causes. Physiological causes include inadequate or disturbed sleep, nutritional deficiency, physical exhaustion, and various dysfunctionings due to pathological reasons. Psychological causes for fatigued mental states are: too heavy a demand on the attention, e.g. prolonged, intellectual analytical work; and too little demand on the attention, e.g. boredom or monotony resulting from routine. This fatigued state is a contraction of the field of cognition. Inconstant attention reduces sense stimuli, and apprehended stimuli are partly blocked by the fatigue condition from reaching the conscious comprehension stage. This state is responsible for fatal accidents, faulty product manufacturing, and insufficient dialogue in adversarial proceedings, to name only a few of the evils that results from the loss of attention.
Broader Attention (#HM1817).

♦ **HM2424b Stations of consciousness - ibn–Abi'l–Khayr** (Sufism)
Maqamat forty
Description In Sufism, spiritual development is expressed by progressive steps or stations (Maqamat). These stations are enumerated by different authors as being anywhere from 7 to 101. One description (Abu Sa'id ibn Abi'l–Khayr) comprises a 40 stage development. In this system, each stage belongs to a prophet among prophets, starting with Adam and finishing with Muhammad. Many of the stages are the same as stations described in other systems, others correspond to ahwal, or passing states granted by God, as described by other Sufis. The system is notable for its continuing beyond the state of *baqa'* which, as union with God, is usually considered the highest station. These higher stations may be considered journeying with God as opposed to journeying towards God. The whole journey goes from the state of being spiritually asleep to that of the true Sufi, the experience of the stations in between each being more intense than any worldly experience but all being open to anyone who will devote himself to the spiritual life; and approaches the experience not as an end in itself but as a step leading to the Absolute who is beyond all stations and states and yet is at the very core of his being, the origin of that which unites bodily, psychic and spiritual states with their common principle.
Broader Divine path (Sufism, #HM3371)   Human development (Sufism, #HH0436).
Narrower Sabr (#HM3418)   Rida (#HM4190)   Zuhd (#HM4450)
Sidq (#HM6631)   Raja (#HM4393)   Fana (#HM1270)
Baqa (#HM0330)   Jahd (#HM6099)   Wajd (#HM3650)
Wara (#HM0286)   Tawba (#HM4062)   Khawf (#HM1047)
Wisal (#HM7119)   Kashf (#HM1452)   Niyyat (#HM4901)
Inabat (#HM3484)   Iradat (#HM8901)   Taslim (#HM5290)
Ikhlas (#HM4663)   Tajrid (#HM7621)   Tafrid (#HM6762)
Tahqiq (#HM5563)   'Ibadat (#HM0128)   Ma'rifa (#HM2254)
Wilayat (#HM3647)   Khidmat (#HM8776)   Inbisat (#HM7098)
Nihayat (#HM3377)   Tawakkul (#HM0807)   Tafakkur (#HM6227)
Tasawwuf (#HM0575)   Mujahadat (#HM5215)   Muwafaqat (#HM7981)
'Ilm al–yaqin (#HM0930)   Haqq al–yaqin (#HM4556)   State–awareness (#HM2305)
Mukhalafat–i nafs (#HM7336)   Nearness–awareness (#HM2377)
Divine–love awareness (#HM2401)   Recollection (Islam, Sufism, #HM2351).
Related Higher states of consciousness (Sufism, #HM2365)
Stations of consciousness - Ansari (Sufism, #HM2317)
Stations of consciousness - Simnami (Sufism, #HM2341).

♦ **HM2426b Consciousness**
Description Consciousness has a multiplicity of meanings and degrees of significance depending on the school of thought within which the term is used. Meanings include: awareness or perception of an inward psychological or spiritual fact; inward awareness of an external state, object or fact; concerned awareness; mind, in the broadest and most inclusive sense; the totality of sensations, perceptions, ideas, attitudes, and feelings of which an individual (or group) is aware at any given time, or within a particular span of time; waking life (as distinct from sleep or unconsciousness); the part of mental life or psychic content that is immediately available to the ego; or a function of the brain in reflecting the external world when an individual assimilates historically elaborated forms of culture.
To the extent that consciousness is taken to mean mental alertness or wakefulness, different levels or degrees of consciousness exist between the completely unconscious state (which may be coma, faint, sleep or hypnotic trance), through normal states, to altered states of consciousness in which there is some heightened awareness, whether this manifests as a painful shock, brilliant alertness or calm awareness. It may also be said that persons lacking in one or more of the senses (blind, deaf, etc) are to this extent lacking in consciousness.
It is not well understood what are the neuro–physiological conditions for the functioning of normal cognitive capacities, nor at what level of life consciousness may be said to start being experienced. Opinions differ also on the origins of consciousness; Marxism holds that human consciousness developed from animal psychic activity through labour which enables the transformation of the environment. Other theories speak of conceptualized input; or of immanence proceeding from the depths of spirit.

♦ **HM2427g Neomammalian neocortex human brain consciousness**
Description In terms of brain evolution theory, the neocortex is specifically, in relative size, the human brain. The limbic system is only paleomammalian while the brain stem corresponds to the reptile stage of evolution. The neocortex specializes in learning new ways to adapt its functions for human survival. It is concerned with voluntary movements, whereas the lower brains affect the autonomous nervous system.
Broader Brain–based consciousness (#HM2295).
Preceded by Mammalian limbic human brain consciousness (#HM2355).

♦ **HM2428b Theta–clear state of consciousness** (Scientology)
Description According to scientology, there are eight states of progressively widening awareness: of self, of mate and family, of group or tribe, of aggregates of mankind (nations, races, all mankind), and of the material universe. In addition, there is a seventh state of awareness, that of the spiritual universe (termed Theta) and an eighth state, the awareness of the Supreme Being. Awareness is a function of the interaction of three minds in man: the reactive, instinct–dominated mind; the analytical, reasoning mind with its ability to store and recall data; and the ultimate mind, the spiritual being in man (termed the Thetan), which is pure awareness and also the mind which projects space, time and the conditions of the universe and its objects. The Thetan being is also considered to be static, the ground of existence which, interacting with the dynamics of the eight objects of awareness (from self to Supreme Being), creates all energy, mental and physical. Because the analytical mind is dominated by limiting experiences imposed by the reactive mind, in the form of conditioning controlled by complex–like "engrams" in the subconscious, the Thetan or inner spiritual being cannot sufficiently influence the analytical mind. As a consequence, progressive understanding of self, family, group, mankind and universe is blocked. If the engrams are eliminated or neutralized, a person may attain understanding up to the level of the spiritual universe, a Theta–clear state of consciousness, and this expansion would render him free to act with the new energy at his disposal. Under the influence as the Supreme Being, he would be truly able, wise and totally free. The developmental technique called *auditing* assists persons to attain the spiritual, Theta state.
Related Theta wave consciousness (#HM2321)   Alpha wave consciousness (#HM2345).

♦ **HM2429d Abject helplessness** (ICA)
Description This is experienced as the inability to achieve anything by one's self, sometimes even the most seemingly simple tasks. One discovers that one's achievements are not one's own. It occurs when one is shocked by one's complete ineptitude, such as getting huge supplies of food to Ethiopia and not being able to transport it to where it is needed.
Context In the ICA New Religious Mode in the arena of acting out one's freedom (the life of prayer) the third formal aspect is surrender to one's unlimited inadequacy, or petition. At the first phenomenological level, this occurs when one encounters the burden of one's own existence.
Broader Surrender to inadequacy (ICA, #HM2922).
Followed by Representational sign (ICA, #HM2532).

♦ **HM2430g Partial–mindedness**
Existential vacuum
Description A form of forgetfulness of peak experience and partial awareness described by Colin Wilson as the normal state of the individual. This partial–mindedness of boredom and "taking life as it comes", can be made complete–mindfulness through deliberate concentration beyond the point of fatigue, to trigger *peak experience*.
Related Focusing (#HM3003)   Preparedness (#HM2963)   Peak experiences (#HM2080).

♦ **HM2431c Death–bardo rebirth–seeking awareness** (Buddhism)
Sixth antarabhava consciousness — Sridpa bardo (Tibetan)
Description This state of awareness is represented by the third major after–life bardo and is characterized by the perception of the arising nirmanakaya, and the spiritual beings and Buddha–guides to the six worlds. This bardo is said to be the most important in the journey toward re–birth and is detailed in the Tibetan Book of the Dead. Fearing the terrors which the images of himself have become the person flees into new birth.
Context This state is the final bardo before rebirth. It is experienced from the 20th to the 49th day preceding rebirth.
Broader Bardo consciousness (Buddhism, Tibetan, #HM0698).
Preceded by Death–bardo heaven–reality awareness (Buddhism, #HM2407).

♦ **HM2432d Global brotherhood** (ICA)
Description This is experienced as the realization that one's actions, no matter where are performed, are related to all of mankind, that one is on a journey and that there are colleagues on this same journey who nurture and care for one. It is like the "crimson line " of which Nikos Kazantzakis wrote, the path of many people throughout history that we are called to follow.
Context In the ICA New Religious Mode in the arena of acting out one's engagement (the life of obedience) the fourth formal aspect is embodying service to people's spirit life. At the second phenomenological level, this occurs when one recognizes one's obligation to the rest of mankind.
Broader Embodying service (ICA, #HM3199).
Followed by Eternal identification (ICA, #HM2541).
Preceded by Ethical existence (ICA, #HM2661).

♦ **HM2433c Desire**
Wishes — Raga (Yoga) — 'Dod–chags (Buddhism, Tibetan)
Description Wishes, cravings and inclinations all form part of conscious desires; these may or may not be admitted, as some are not generally acceptable (death wish, incestuous desires). Similarly in the unconscious there are what may be considered biological wishes or "urges" which strive to be expressed and which may be more or less acceptable to the conscious personality. Some systems consider wishes to be synonymous with *needs* to the extent that when an individual says he "needs" something he frequently means that he desires it. Society has evolved in a manner which allows for an individual's conscious wishes to be gratified as long as this does not interfere with the wishes of others. According to most religious systems, however, a person develops when individual wishes and desires are controlled, and only the desire for truth remains.
One specific definition of desire is, for example, as one of the six *root afflictions* referred to in Buddhist teaching. Here it is perceived as generating *suffering*, in that it sees and seeks some contaminated phenomenon, whether internal or external, as pleasant. This may be what is generally considered unpleasant or perverse, or the most laudable and beautiful. It clings to such phenomena and is hard to separate from them. Thus it is attachment, whether to the Desire, Form or Formless Realm, and the cause of the cycle of birth and rebirth. Also one of the five afflictions or causes of misery in yoga.
Broader Root afflictions (Buddhism, #HH0270)   Afflictions (Yoga, Hinduism, #HH1047).

Narrower Attachment (Hinduism, #HM3914).
Related Greed (Buddhism, #HM3283) Clinging (Hinduism, #HM6153).

♦ **HM2434c Land of mystery** (ICA)
Description An area in the ICA *Other World in the midst of this World*, characterized by being in a state of humility, in wonder and in God.
Broader Other world in the midst of this world (ICA, #HM2614).
Narrower Aweful encounter (#HM2619) Infinite passion (#HM2234) Inescapable power (#HM3096) Transformed state (#HM2386).
Related Mountain of care (ICA, #HM2170) Sea of tranquillity (ICA, #HM3033) River of consciousness (ICA, #HM2993).

♦ **HM2435b Suchness** (Buddhism)
Tathata — Yathabhuta — Yan-dag-pa-ji-lta-ba-bzin-du (Tibetan)
Description A phenomenon of the pure class according to Buddhist teaching, this refers to the emptiness (which remains whether Buddhas appear or not); and to the "natural" nirvana of emptiness beyond inherent existence.
Broader Harmonies with enlightenment (Buddhism, #HH0603).
Related Innocent cognition (#HM0888) Steps to enlightenment (Buddhism, #HH4019).

♦ **HM2436b Self-awareness** (Psychism)
Description Self-awareness is one of the possible contents of any station of consciousness in the higher worlds. It is possible to be unaware of the self, but to be conscious. It is possible also to be aware of the true self in any of the worlds, and thus attain one of the highest degrees of consciousness. In psychic terms, there is acute discrimination between physical or mental and psychic impressions.
Related Self assertion (#HM3587) Self-remembering (#HM2486) Conscious self-relatedness (Yoga, #HM0467).

♦ **HM2437d Endless life** (ICA)
Everlastingness at the centre
Description This is an experiential trek in the *Sea of Tranquillity*, within the ICA *Other World in the midst of this World*, when the individual comes to question "Death, where is thy sting ?" The states are described in separate entries.
Broader Sea of tranquillity (ICA, #HM3033).
Narrower Living death (#HM2808) Everlasting community (#HM2777) Contingent eternality (#HM2456) Resurrectional existence (#HM2631).
Related Unknowable peace (ICA, #HM3015) Radical illumination (ICA, #HM2273) Contentment at the centre (ICA, #HM2529).

♦ **HM2438b Spiritual realization** (Yoga)
Description This is achieved after advanced yoga practice when the yogi, in a super-contemplative state, is fully aware of his spiritual identity. Having arrived at the heart of his being he is free from delusion and no longer subject to deception. He is unconcerned with the passing play of life, with pain or with pleasure, and is fully *self-realized*.
Related Awareness of spiritual identity (Yoga, #HM3232).

♦ **HM2439d Virgo-consciousness** (Astrology)
Intuitive consciousness
Description With intuition linked to conscience, Virgo represents a mode in which the individual understands his relationship with the oneness of the whole universe. This creates a new impulse to cooperate with others and serve them, transforming consciousness and involving a new sense of potentiality. Under optimal conditions it is associated with a mature and unprejudiced spirit of non-selfseeking service, in response to the intuited needs of the whole.
Context One of the twelve conditions of being or streams of divine energy encompassed within the zodiac. It is said that all twelve conditions must be present and playing its proper part for any successful and fulfilling action to take place. This may be through a group of individuals with different endowments and soul qualities working together; or it may be through one individual who, by self-discipline, practice and meditation, has allowed all twelve conditions to arise within himself.
Broader Earthy awareness (Astrology, #HM3235) Mutable awareness (Astrology, #HM3092) Zodiacal forms of awareness (Astrology, #HM2713).
Followed by Libra-consciousness (Astrology, #HM2512).
Preceded by Leo-consciousness (Astrology, #HM2376).

♦ **HM2440b Stabilization** (Buddhism)
Samadhi — Ting-nge-'dzin (Tibetan) — Concentration
Description In Tibetan Buddhism, stabilization is generated by mental attention to or consciousness of an internal object, in order to perceive such an object clearly, whether or not it is *real*.
In Hinayana Buddhism: The mind is steady, placed evenly and well on the object or is simply collected. The characteristic of concentration is not scattering or wandering, not wavering or distracted. It welds together coexistent or conascent states, it manifests as peace of mind. Bliss or ease is usually the proximate cause.
Context One of the five determining mental factors of Tibetan Buddhism. One of the formations aggregate (mental coefficients) of Hinayana Buddhism, being listed among the constant states which appear in their true nature, and as general primary (always present in any consciousness).
Broader Determining mental factors (Buddhism, #HH0170) Awareness as mental-formation group of conscious existence (Buddhism, #HM2050).
Narrower Correct concentration (#HM1735) Steadiness of consciousness (Buddhism, Pali, #HM6652).

♦ **HM2441d Realized vocation** (ICA)
Description This is experienced as a time when one cannot distinguish one's work from one's life; when one sees that only he who extends his engagement to all of society has a life-long vocation. It may be compared to Florence Nightingale's realization that just being a nurse was not enough, and working to create a world-wide movement to care for the sick.
Context In the ICA New Religious Mode in the arena of knowing one's disengagement (the life of poverty) the second formal aspect is being reliant upon the larger society or being disengaged from one's own work. At the fourth phenomenological level, this occurs when one is filled with the power of requiring nothing from a situation.
Broader Disengagement from work (ICA, #HM2552).
Related Vocation (Christianity, #HH0619).
Preceded by Social failure (ICA, #HM2728).

♦ **HM2442d Embodying equity** (ICA)
Description The second formal aspect of the *life of obedience*, an arena in the ICA *New Religious Mode*, this represents an awareness of embodying equity as the disestablishment to call forth human justice.
Broader Life of obedience (ICA, #HM2024).
Narrower Corporate duty (#HM2694) Loyal opposition (#HM2910) Individual rights (#HM2782) Perpetual revolutionary (#HM2839).

♦ **HM2443c Enquiry** (Hinduism)
Second plane of wisdom
Description This is characterized by the practice of direct observation.
Context The second of seven ascending planes of wisdom described in the Supreme Yoga, which require to be known so as not to be caught in delusion.
Broader Planes of wisdom (Hinduism, #HM3298).
Followed by Subtlety of mind (Hinduism, #HM3312).
Preceded by Pure intention (Hinduism, #HM3100).

♦ **HM2444c Path of occult intelligence** (Judaism)
Path of hidden intelligence
Description This state is represented as the splendour of all the virtues perceived in spirit and in faith.
Context The seventh among the spiritual paths expressed in the Jewish Kabbalah.
Broader Paths of wisdom (Judaism, #HM2509).
Followed by Path of perfect intelligence (Judaism, #HM3243).
Preceded by Path of intelligence of mediating influence (Judaism, #HM3089).

♦ **HM2445d Sacramental universe** (ICA)
Description In this state, an individual perceives the mystery of life within each and every thing. Examples are: seeing a patch of grass through the clouds from an aeroplane and noticing that it becomes a grave; or in Sartre's novel "Nausea", when he reflects on the meaninglessness of life as he describes a puddle of beer on a café table; or watching the flames of a fire and finding one's self recalling past events and dreaming of the future. D.H. Lawrence captures the mysteriousness of one tiny creature when he says: "The mosquito knows full well...he is a beast of prey...he only takes his bellyful, he doesn't put my blood into the bank."
A dimension of this experience is a sense of awe, that is fear and fascination. Even the most mundane thing explodes with meaning; one feels perpetually astonished with life. At the same time, one begins to ask one's self if there are any answers to anything. There is a sense of frustration at being wholly engulfed by the inexplicablity of life.
Standing in awe at life's unpredictiblity, the decision is to continue pursuing the unfathomable, daring to bleed the meaning out of every situation, knowing that one will finally come face to face with the mystery itself. Everything becomes a symbol of the sacred in life. Such a simple thing as a rock becomes a treasure as it continues to remind one of the mystery within everything.
Context This state is number 36 in the ICA *Other World in the midst of this World*.
Broader Original gratitude (ICA, #HM3105).
Followed by Primal sympathy (ICA, #HM2550) Invented history (ICA, #HM3245) Personal epiphany (ICA, #HM2595) Primordial wonder (ICA, #HM2186) Soteriological existence (ICA, #HM2904).
Preceded by Invented history (ICA, #HM3245) Primordial wonder (ICA, #HM2186) Perpetual becoming (ICA, #HM2717) Temporal solidarity (ICA, #HM2363) Definitive effectivity (ICA, #HM2796).

♦ **HM2446b Indian approaches to consciousness** (Hinduism)
Description In Hinduism, philosophic and experiential analysis of states of consciousness varies according to period from ancient Vedic and Upanishadic, to Tantric and modern syncretistic; and also according to schools of philosophic analysis ranging from materialist to idealist. Religious concepts of consciousness vary according to cults, for example, among those focused exclusively on Shiva, or on Vishnu, or on Krishma or Kali. The main characteristic of Hindu findings concerning consciousness (chitta or citta) is its creative power (maya, lila, sakti) and its scope (avasthas, lokas, Brahma), which give rise to techniques (yogas) ways (dharma) or paths (marga) of life that are world-denying and often psychologically and socially disruptive.
Refs Feuerstein, Georg *The Essence of Yoga* (1974).
Related Purusha (Yoga, #HM2396) Kaivalya (Yoga, #HM3869) Moksha (Hinduism, #HM2196) Mantra yoga (Yoga, #HH0931) Ego awareness (Hinduism, #HM2059) Sahaja samadhi (Hinduism, #HM2043) Atma awareness (Hinduism, #HM2103) Sabija samadhi (Hinduism, #HM2308) Karma awareness (Buddhism, #HM2048) Nirbija samadhi (Hinduism, #HM2125) Buddhi awareness (Hinduism, #HM2099) Human development (Hinduism, #HH0330) Savikalpa samadhi (Hinduism, #HM2650) Nirvikalpa samadhi (Hinduism, #HM2061) Brahma consciousness (Hinduism, #HM2041) Consciousness in Buddhism (Buddhism, #HM2200) Chitta-dependent consciousness (Yoga, #HM3011) Subjective domains of consciousness (Yoga, #HM2026) Upanishadic stages of awareness (Hinduism, #HM2392) Mystical Christ-consciousness (Christianity, #HM2880).

♦ **HM2447d Awareness of depth** (ICA)
Description The fourth formal aspect of the *life of contemplation*, an arena in the ICA *New Religious Mode*, this is an experience of dreadful "in-myself-ness".
Broader Life of contemplation (ICA, #HM2109).
Narrower All-being-in-myself (#HM3097) Appropriated passion (#HM3054) Unexplainable thereness (#HM3108) Irreplaceable uniqueness (#HM3155).

♦ **HM2448c Emptiness of the inherent existence of non-things** (Buddhism, Tibetan)
Abhavasvabhavashunyata
Context One of the eighteen emptinesses comprising the paths of view in Tibetan Buddhism.
Broader Emptinesses on the paths of the view in Buddhism (Buddhism, Tibetan, #HM2944).

♦ **HM2449d Dangerous intrusion** (ICA)
Description This is experienced as an event intruding into a person's life that appears as life threatening; when suddenly something totally unexpected and inexplicable is happening. An example is the main character in the book "Daniel Martin", when he receives a phone call in the middle of the night from a person he has not seen for years, who lives thousands of miles away, saying that he has cancer and asking Daniel to visit him.
Context In the ICA New Religious Mode in the arena of standing present to the mystery of being in life (life of contemplation) the first formal aspect is the experience of enigmatic not-me-ness or of externality. At the first phenomenological level this occurs when a person encounters an event in his life that is completely mysterious.
Broader Awareness of externality (ICA, #HM2637).
Followed by Everlasting inescapability (ICA, #HM2791).

♦ **HM2450c First trance of the fine-material sphere** (Buddhism)
Vitakka-vicara (Pali) — Prathamadhyana — Bsam-gtan-dang-po — First form-realm concentration
Description *Hinayana Buddhism* The mind detached from sensual objects by means of meditation may enter the first of the higher states of mental activity. One by one, these strip off the encumbrances of personality that mask reality. The object of the first trance is to lead to the

subsiding of discursive thinking (vicara) and of thoughts (vitaka). Detachment is achieved.
Tibetan Gelugpa Buddhism The meditator holds the mind inside – mind and mental factors are all equally operating on the object. The four elements have become balanced. Through the first concentration, on the basis of calm abiding, many meditative absorptions and extra-sensory cognitions are achieved.
Context This state begins the series of four form-concentrations (dhyana) found in Tibetan Gelugpa Buddhism.
Broader Rupaloka consciousness (Buddhism, #HM2536)
Form-realm consciousness (Buddhism, #HM2257)
Trances and mental absorptions (Buddhism, #HM2122)
Tibetan meditative states of form concentrations (Buddhism, #HM2693).
Related Dhyana (Hinduism, Buddhism, #HM0137)
Dwelling in the first jhana (Buddhism, #HM4298).
Followed by Second trance of the fine-material sphere (Buddhism, #HM2038).

♦ **HM2451d Worldly detachment** (ICA)
Description This occurs when an individual is reconciled with the fact that this world is not his home. He becomes detached from the world of relative relationships and things that used to give meaning to his life – family, nation, job or whatever – no longer wield power over him, so he is free to plunge into any activity with all his being.
A dimension of this experience is a sense of awe, that is fear and fascination. One has burnt all one's bridges and there is no going back. There is a sense of amazement that one could have really been so attached to all those things that previously claimed one's life and that one will no more allow one's self to be claimed in such a way. A new vitality comes with the realization that one could live the rest of one's life in this state of detachment.
There are no regrets, but a sense of single-mindedness, where even life and death decisions have no power. In the midst of this utter seriousness about his life, an individual is able to smile with a detached kind of humour that even his detachment is finally nothing.
Context This state is number 30 in the ICA *Other World in the midst of this World*.
Broader Final accountability (ICA, #HM3206).
Followed by External relation (ICA, #HM2471) Interior discipline (ICA, #HM2851)
Relational situation (ICA, #HM2978) Passionate disinterest (ICA, #HM2547)
Resurrectional existence (ICA, #HM2631).
Preceded by Original integrity (ICA, #HM2773) Cryptic disclosure (ICA, #HM2824)
Relational situation (ICA, #HM2978) Intentional conscience (ICA, #HM2892)
Resurrectional existence (ICA, #HM2631).

♦ **HM2452c Tantra-beginner awareness** (Buddhism, Tibetan)
Description This state of awareness is characterized by taking refuge in the Buddha, the Dharma, and the Sangha and by taking the Boddhisattva vow. It is particularly represented as the Tantric path, with its obligatory mantram repititions, mandala offerings and mystical communion with the saints or gurus of the personal esoteric transmission. This is the way of mantra-vidya (the gnosis of effective prayer) and the adamantine or diamond path of power (siddhi).
Context In Tibetan Sakya Buddhism this is one of the states in the "Ascension Stages Game". In some sets it is numbered 25 on the board.
Broader Tantric paths of accumulation and application in the ascension-stages-game (Buddhism, Tibetan, #HM2848).
Followed by Kriya-tantra awareness (Buddhism, Tibetan, #HM2558)
Magical-forces awareness (Buddhism, Tibetan, #HM2679)
Discipleship-karma awareness (Buddhism, Tibetan, #HM2192).
Preceded by Demon-island awareness (Buddhism, Tibetan, #HM2055)
Southern-continent awareness of men (Buddhism, #HM2127).

♦ **HM2454d Temporary-hells awareness** (Buddhism, Tibetan)
Description This state is characterized by isolation and is represented as a hell occurring in isolated places on this earth. It is represented as one of the higher hells as it is not created by the mass karma of all living beings like the lower hells.
Context In Tibetan Sakya Buddhism this is one of the states in the "Ascension Stages Game". In some sets it is numbered 8 on the board.
Broader Desire-realm consciousness (Buddhism, #HM2733).
Followed by Cold-hells awareness (Buddhism, Tibetan, #HM2040)
Animal-hell awareness (Buddhism, Tibetan, #HM2636)
Demon-island awareness (Buddhism, Tibetan, #HM2055)
Reviving-hell awareness (Buddhism, Tibetan, #HM2516)
Four-great-kings-heaven awareness (Buddhism, Tibetan, #HM2082)
Eastern-continent awareness of noble figures (Buddhism, Tibetan, #HM2543).
Preceded by Bon-wisdom awareness (Buddhism, Tibetan, #HM2663)
Howling-hells awareness (Buddhism, Tibetan, #HM2100)
Reviving-hell awareness (Buddhism, Tibetan, #HM2516)
Very-hot-hells awareness (Buddhism, Tibetan, #HM2576)
Hungry-ghosts-hell awareness (Buddhism, Tibetan, #HM2112)
Great-vehicle middle-path awareness (Buddhism, Tibetan, #HM2160).

♦ **HM2456d Contingent eternality** (ICA)
Description This is an awareness of life beyond the grave and before birth. It is not some mystical trick to avoid the unavoidable, but rather the state of existence in which an individual grasps, in the present moment, the utter reality of life's limits and the supreme worth of life within those limits. The grave is not victorious for no six-foot hole could contain the life he has experienced. It may be compared to Martin Luther King declaring he had " seen the promised land"; or to the end of the movie "2001 – A Space Odessey" when the old man and the foetus become one; or to St. Francis of Assisi's awareness of everything as awe-filled reality. It is the knowledge of living beyond the limits of temporal life.
A dimension of this experience is a sense of awe, that is fear and fascination. Everything is awe-filled; all events and things engender wonder and there is a foretaste of heaven. The person is sheer presence, he and Being are one reality and he knows himself to be God's messenger. He decides to say yes to the wonder of life and death – in a sense he is beyond decision; not passive or fatalistic but rather that the cares of this world that require "decisions" have been transcended. He knows that all history has conspired to create this moment and he is responsible for it.
Context This state is number 64 in the ICA *Other World in the midst of this World*.
Broader Endless life (ICA, #HM2437).
Followed by Personal epiphany (ICA, #HM2595) Singular adoration (ICA, #HM2111)
Radical contingency (ICA, #HM2477) Exclusive contradiction (ICA, #HM2951)
Destinal accountability (ICA, #HM2309).
Preceded by Final blessedness (ICA, #HM2958) Everlasting community (ICA, #HM2777)
Definitive effectivity (ICA, #HM2796) Exclusive contradiction (ICA, #HM2951)
Destinal accountability (ICA, #HM2309).

♦ **HM2458d Universal prior** (ICA)
Description This is experienced as the decision to be an organizing force in every social situation. It occurs when one realizes that no one else is going to play the role of creating order and that one has appeared as the organizing force, as when the presence alone of some people creates order out of chaos.
Context In the ICA New Religious Mode in the arena of acting out one's engagement (the life of obedience) the first formal aspect is embodying peace as the establishment to enable the ordering of human affairs. At the fourth phenomenological level, this occurs when one lives on behalf of the evolution of human consciousness.
Broader Embodying peace (ICA, #HM2876).
Preceded by Radical incarnation (ICA, #HM2587).

♦ **HM2460d Terrifying acceptance** (ICA)
Description This is experienced as grasping the reality of the word in one's own life; when one accepts one's acceptance and opens one's whole life in order to live it. In the movie "Little Big Man", the main character realizes that both the white man and the Indians are broken and failed human beings; and yet he decides to embrace them as part of his world and life.
Context In the ICA New Religious Mode in the arena of articulating the word about life (the life of knowing) the third formal aspect is one's experience of contentless transformation or the greatness of the word. At the second phenomenological level, this occurs when one is somehow stripped of one's illusions and must decide how to relate to one's new universe.
Broader Contentless transformation (ICA, #HM3106).
Followed by Classical story (ICA, #HM2757).
Preceded by Objective awareness (ICA, #HM2598).

♦ **HM2461d Non-faith** (Buddhism)
Ashraddhya — Ma-dad-pa (Tibetan)
Description The cause of *laziness* and itself caused by *ignorance*, this is a negative attitude to virtuous phenomena.
Context One of the twenty *secondary afflictions* of Tibetan Buddhism.
Broader Secondary afflictions (Buddhism, Tibetan, #HH0781).
Related Laziness (Buddhism, #HM3163) Ignorance (Buddhism, Yoga, Zen, #HM3196)
Ninety-nine names of Allah (Sufism, #HH2561).

♦ **HM2462d Transparent presence** (ICA)
Description The fourth formal aspect of the *life of being*, an arena in the ICA *New Religious Mode*, this is an experience of effulgence.
Broader Life of being (ICA, #HM3229).
Narrower Being myself (#HM2154) Saving the mystery (#HM2413)
Inventing humanness (#HM2092) Human transformation (#HM2334).

♦ **HM2463d Painful acknowledgement** (ICA)
Description This is experienced when an individual faces the reality of life without illusions; when the awareness of separation from himself and others, both created and simply fated, overwhelms his life. It may be compared to the experience of Oedipus Rex when he realises who his mother and father really are.
Context In the ICA New Religious Mode, in the arena of acting out one's freedom (the life of prayer), the first formal aspect is the struggle with repentance for one's sinfulness or confession. At the first phenomenological level this occurs when one encounters the burden of one's own existence.
Broader Struggle to confess (ICA, #HM2718).
Followed by Personal violation (ICA, #HM2561).

♦ **HM2464d Mantra-induced experience** (Yoga)
Mantra-consciousness
Description The use of a mantra or inner repetition of a word or phrase may induce an altered state of awareness or mystic experience. This is the technique used in some systems of meditation and yoga, notably Tantric yoga, in which it is said repetition of a mantra without developing mantra-consciousness is futile.
Related Mantras (#HH0386) Sound health (#HH2543) Nada yoga (Yoga, #HH0160)
Mantra yoga (Yoga, #HH0931) Induction technique (#HH0025)
Tantra (Buddhism, Yoga, #HH0306).

♦ **HM2465b Anxiety**
Station fear
Description Anxiety is generally recognized as a chronic condition akin to fear, present in most neuroses; and exhibited in response to certain conditions, in particular to uncertainty. Any of the normal disruptions of life may require facing unfamiliar situations or sudden change; and all may be sources of anxiety. The term *station fear*, originally descibing the anxiety experienced when starting or completing a journey or of missing the means of transport, has come to mean all anxiety faced at a time of transition.
It has been postulated that all needs are responses to anxiety and all efforts are towards reduction in such anxiety; but subsequent study has shown that curiosity and exploration are more hindered by anxiety than engendered by it.
Freud distinguished *objective anxiety*, or fear, as the response to some external danger, and *neurotic anxiety* as the response to an internal impulse. Although different conditions produce anxiety in different people, a single individual's anxious reponse to a given situation is predictable unless de-conditioning or counter-conditioning has taken place. Anxiety is characterized by muscular tension (giving rise to impaired coordination in movement), diminished mental concentration, reduced efficiency at work and impaired social and sexual behaviour.
Related Tapa (#HM3530) Angst (#HM0362) Unlust (#HM3094)
Fear (Christianity, #HM3311) Psychic helplessness (#HM2182).

♦ **HM2466d Action irrelevant** (ICA)
Description This is experiencing extremely effective activity as being nothing compared with the demands of history; when one acts without needing to justify the act by its effects and without the need to accomplish. It may be compared to Arjuna being told by Krishna that it is not in aiming to win the war but in the fighting itself that meaning is found.
Context In the ICA New Religious Mode in the arena of knowing and doing intensified in one's life (life of being) the third formal aspect is the experience of transparent engagement or the experience of levitation. At the first phenomenological level this occurs when a person experiences discontinuity at the centre of his life.
Broader Transparent engagement (ICA, #HM2736).
Followed by Impossible possibility (ICA, #HM2524).

♦ **HM2468c Continuance** (Buddhism)
Vattana (Pali)
Broader Faculty of living (Buddhism, #HM3404).

♦ **HM2469b Intentionality**
Description Consciousness has the structure of intentionality or may even be considered as intentionality. Since consciousness cannot be aware of itself, it must be consciousness of something other, an object or meaning for consciousness. Hence in cognitive perception there is an indissoluble unity between the conscious mind and the object of which it is conscious. The structural relationship between consciousness and what is observed is referred to as *intentionality*,

HM2469

such overcoming of the separation of subject and object being part of *phenomenology* theory. Whereas in the norma processes of cognition the dimension of mind is invisible, the intentionality of consciousness makes clear the difference between the meaning which is what is seen and the meaning of what is seen.

♦ **HM2470c Mansions of favours and afflictions** (Christianity)

Description In these mansions the soul may be granted many great favours, but it must also endure far greater trials. As the soul enter these mansions, it has been wounded with love for God and wants no other love.

While the trials of the seventh mansions are far greater, the trials here are great. First, people point out either how "holy" the person is getting or how deceptive she is by making others look sinful. Friends abandon the person and accuse her of being deluded by the devil. The worst of this trial is that it does not end. Second, she is well spoken of, which is even a greater trial. She knows that any good she may have is a gift of God and not her own. This becomes less intolerable for various reason. She learns that people speak equally easily of good and bad. Seeing that all good comes from God, she begins praising God when these things happen. Realizes that if people see her development they may gain from it. As the soul prizes honour and glory of God more than its own it is no longer tempted to think such praise will do it harm. Third, when publically and without reason praised she is greatly troubled. Fourth, she may experience great physical suffering. Fifth, a inexperienced or unspiritual confessor may attribute the whole thing to psychological problems or to the devil. Having gone to the confessor with fear of these things and knowing its own sinfulness it is greatly distressed. Even when God grants a favour which can come from no one else but God, it relaxes only to have this pass quickly and to return to torment. At this point the soul may come to believe that it has never know God and never will. The soul can only wait on the mercy of God, who at some unexpected moment with a single word lift the whole burden from the soul's shoulders. The soul realizes it is a powerless and miserable creature, though not devoid of grace because it endured these trials without even thinking of offending its lover. Comforts of this world can only torment it. Interior prayer is impossible. Vocal prayer is not understandable to it. Solitude is torment even though it is torture to be in the company of anyone. Despite all attempts to conceal it she becomes despondent and upset. The only thing to be done to enable her to endure this is to occupy oneself with works of charity and hope for God's mercy.

Many favours are granted in these mansions. God may fill the soul with fervent desire even when it is unprepared for such a thing and is not even thinking of God. The soul become highly aware that is called by God. The soul trembles and complains with words of love. It experiences itself as delectably wounded and would be glad if the wound never healed. It is aware of the presence of God without having the ability of enjoying this presence. This inability causes the soul a kind of sweet grief. The lover is calling. The soul desiring to respond knows not how. This pain does not last and none of the faculties can effect it. For years after this experience the soul is willing even desires to suffer for God's sake and consider itself well rewarded. Another way this happens is the soul seems to catch fire in some wonderful way or infused with the presence of God.

A second way the soul is awakened is through perceiving some word or phrase from God. This may actually be heard. It may be imagined. Or it may be imprinted in the deeps of the soul. Such locutions may come from God, the devil or one's own imagination regardless of how they are perceived. One test of them coming from God is the sense of power and authority they carry with them and in the actions which follow them. Second, a great peace dwells in the soul, its devotion is recalled and it is ready to sing praises to God. Third these words carry such power and result in such conviction they are from God that they do not leave the memory for a long time, if ever.

A third way the Lord speaks to the soul is a kind of intellectual vision. This vision occurs so deep within the soul, with such clarity and with such profundity that there is no doubt it comes from the Lord. Its effect are so wonderful that it could not come from neither the imagination nor the devil.

God may, in these mansions, grant raptures which carry the soul out of its senses. In one type of rapture the soul, though not actually engaged in prayer, is struck by some word, which it either remembers or hears spoken by God. The interior senses fully attuned and the soul has never before been so awake to the things of God or had such understanding. While in this state of suspension mysteries such as things of heaven are revealed. The soul sometimes loses its power of breathing. It cannot speak. Other times the hands and body become so cold that it seems dead. This total rapture does not last long and the body stirs, breathes and then returns to the rapture. The will can be so completely absorbed and the understanding so transported that the soul seems to be unable to grasp anything that does not awaken the will to love. Everything else it is asleep to. When the soul awakens from such a rapture it feels confusion and desires it be used for God in any way. It desires to do penance but what ever penance it does seems to count for little.

A second form of rapture is a kind of flight of the spirit. The soul becomes conscious of such rapid motion it is, especially at first, filled with fear. There is no certainty that this motion is caused by God. The soul seems to leave the body altogether, on the other hand it is clear that it is not dead. The soul feels that it has visited another world where in a single instant it was taught many wondrous things. The eyes of the soul see things and knows them well. Sometimes the soul just knows things without having seen them. Any resistance only makes matters worse, as it is no longer its own mistress. As the soul reflects on what God is doing to it and on what it is doing, it realizes how little it is doing to fulfil its obligations, how feeble is that which it does do and how full of faults and failures it is. If it does do good, it tries to forget it, to keep in mind only its failures and to throw itself on the mercy of God. Sometimes, the Crucified One comforts her by pointing out that she is receiving the pains and trials that He suffered so that they might be her own to offer God. Thereafter living on earth is a great affliction to her. Things that used to give her pleasure no longer do so. It is as though the soul is shown the road that it must suffer. At the same time it enjoys the greatness of God, the self–knowledge and humility at realizing its own baseness and the supreme contempt for earthly things save those which can be of service to God. Having won such favours, the soul now longs for death that it might be with God. The things of this world weary it. When it is alone in briefly finds relief only to have the distress return. Its love is so full of tenderness that any occasion which serves to increase this love causes the soul to take flight, even in public. It feels a great interior security and is in great distress from fear of the devil and from fear of offending its love. The road she is on is both dangerous and of great benefit. All it wants is to be left in the hands of God. It would like to flee other people and it would like to serve in the world in order to help one soul praise God more. It may experience a terrible sense of cowardliness, but this seems to be when God has left it to its own nature so that it realizes that any usefulness it may have is a gift of God. Tears may come that are comforting and tranquilizing when God pleases. If tears are not comforting or are the result of weak health there is danger of being lead away from prayer and the religious life.

Some times the soul finds itself in a strange kind of prayer and in jubilation. When this happens the soul wants to share its joy with all so that they might rejoice in God. The joy may be so great that the soul is conscious of nothing and unable to speak of nothing except praise of God.

It is at this point the soul sorrows even more for its sin, and for being ungrateful it has been. It is aghast at its boldness, its lack of reverence, for its foolish mistakes and for foresaking God for such base earthly things. Their greatest fear is not of Hell but of losing God.

Some souls seem to retain something of Divine favour from having reached a perfect state of contemplation after which they can no longer meditate on the Passion of Christ. This type of meditation is the road to Divine favour and having travelled it, it is no longer necessary. At the same time the mysteries of the are often in the mind and can be the source of recalling the need to return the gifts of the spirit. This type of practice is beneficial and those who say they cannot meditate in this way because they are continuously enjoying the gifts of the spirit should regard themself in danger.

Another type of favour granted in this mansion is an intellectual vision. When the soul is least expecting Christ to be granting a favour, it becomes aware that He is near, though He cannot be seen physically or by the soul. At first the person may be greatly perturbed because she cannot see anything. Then she realizes is Christ. The realization is so powerful that she has no doubts about its authenticity. This vision, unlike imaginary visions, may last for many days or even more than a year. She may even be comforted by hearing assurances from Christ. The benefits to the interior are great. She may, at times, be confused or have misgivings about why these blessings have come to her; she experiences great humility. She come to have a special knowledge and most tender love of God. She yearns to serve even more. The soul becomes sensitive to everything. When the Lord withdraws, it experiences a profound loneliness and any attempts to regain His companionship are of little avail. But the soul gains from this loss and even from attacks from the Devil if it does not become careless. It is wise to go to a learned and spiritual confessor explain what has happened and then drop the matter.

Imaginary visions, while granted from God and are the most beneficial of this mansion should not be desired because the Devil frequently uses them. These visions happen in a flash and the most glorious image of Christ's humanity is engraved upon the imagination. This image may be seen or heard but through a kind of protective veil. As a result this unbelievably majestic presence fills the soul with terror. This does not happen when it is strove for but when least expecting it. All of the faculties are thrown in fear and confusion. The whole interior world if filled with commotion and then, all at once, there is a blessed calm, and the soul is so completely instructed in such great truths that it is in need of no other master. For a while the soul enjoys great certainty that this vision comes from God. Later, especially when a confessor suggests that it is not possible, the soul begins to doubt such favours could be granted to such a sinner. Even so it is important to describe clearly and simply to a confessor the experience and do not scorn the image. Do not pray for such visions as this shows a lack of humility. This will lead to certain deception by the Devil or by the self. It is very presumptuous of the soul to choose one's path of the spirit on behalf of God. The trials of this path are very great. The thing that is desired will be lost. It is safest to only will God's will.

God also communicates in many ways through apparitions. Some of them come when one is about to experience great trials, so that the soul is comforted. Do not be distressed or uneasy because the Devil will use this to hinder the praise of God. The soul may be at prayer and in possession of all its senses when there comes a suspension in which the Lord communicates most secret things. And while this passes quickly, it leaves the soul greatly confused by showing how wrong it is in offending God. Give thanks for not being cast directly into Hell, endure all that is required and love those who have done us wrong, as God has forgiven those who have offended Him. Desire only to do His will. After years of receiving such favours, the soul is still signing and weeping and each new favour causes fresh pain because it sees how far it is from God and unable to enjoy union with Him, its desire only increases. The soul seems to be interiorly burning and a single word or remark may deal it a blow, wound to the depth of the soul so that everything that is earthly is reduced to powder. While this lasts nothing can be done that does not increase the pain. It is an enrapturing of the senses and faculties. The understanding wants to know why it feels absent from God. God aids it a knowledge so lively about Himself that the soul cries out loud from the pain deep within the soul. The person may be quite close to death, the pulse weak, the limbs disjointed, the body cools, and it feels no physical pain. The pain is so acute within that the person takes no notice of the body. A strange solitude fills the person so that no creature on earth or in heaven save God could be a companion. Earthly companionship is tormenting. The soul feels this affliction is so precious that it fully realizes it does not deserve it; none the less, suffers gladly and if so willed by God would suffer it all life long. This state only lasts three or four hours. These attacks cannot be resisted. When the torment is greatest, God may grant a deep rapture or some kind of vision which comforts her so she can wish to live as long as God wills it. The most wonderful effects are produced. The soul loses any fear of trials. The soul has great contempt for the world. The two dangers at this point are offending God and excessive rejoicing and delight.

Context The sixth of seven mansions of the soul's progress described by St Teresa of Avila.
Broader Mansions of the soul (Christianity, #HH1409).
Followed by Mansions of spiritual marriage (Christianity, #HM3971).
Preceded by Mansions of incipient union (Christianity, #HM4110).

♦ **HM2471d External relation** (ICA)

Description This occurs on realizing that there is absolutely nothing to fall back on or to cling to. What were thought to be solid foundations just crumble away beneath one, leaving one standing on nothing; the only thing that can be relied on is no–thing and that is utterly precarious. It may be compared to discovering that one's life–long best friend has just double–crossed one, and could be what Martin Luther knew when he proclaimed "Here I stand". It is the experience of a person who has never flown before, having to land an aeroplane because the pilot has just had a heart attack.

A dimension of this experience is a sense of awe, that is fear and fascination. One knows there is no second chance, it is do or die. The price for failure is one's own death; it is a case of winning big or losing big, no middle path, a simple roll of the dice can make all the difference. There is a distinct feeling that this is the moment one has been waiting for all one's life. The curtain has gone up and one is on stage.

In this state one decides to keep standing, not to collapse under the pressure and not to take no for an answer. There is a sudden dawning that it was really always like this; one is left with a sense of amazement, of "Wasn't that really something ?". One acknowledges that life has been taken out of one's hands and will never belong to one again.

Context This state is number 18 in the ICA *Other World in the midst of this World*.
Broader Authentic relation (ICA, #HM2316).
Followed by Self transcendence (ICA, #HM2584)
Relational situation (ICA, #HM2978)
Intentional conscience (ICA, #HM2892)
Inclusive comprehension (ICA, #HM2256)
Definitive predestination (ICA, #HM2131).
Preceded by Absurd existence (ICA, #HM2331)   Ultimate awareness (ICA, #HM2388)
Worldly detachment (ICA, #HM2451)   Intentional conscience (ICA, #HM2892)
Inclusive comprehension (ICA, #HM2256).

♦ **HM2472d Peevishness**

Chronic irritability — Fretfulness

Description An excess of sensibility makes the person over susceptible to every nervous stimulation. Even the most ordinary impressions of the senses exceed the limit of healthy or agreeable excitement. If the excess of sensibility is not restrained it becomes a tyrannical habit, manifesting as an irritable temperament. The irritation, rather than directed at the self, is blamed on others who may only be implicated incidentally. It impedes the growth of sympathy and encourages all kinds of malice.
Related Irritability (#HM1706)   Irascibility (#HM0757)
State of anger (Buddhism, #HM2959)   Spiritual anger (Christianity, #HM0936).

♦ **HM2473g Unconscious stupor**
Alternative mode of awareness while unconscious
Description In the schizophrenic condition known as catatonic stupor the patient is usually aware of the surroundings, their nature and details, although he may appear sensorially deadened. In the preliminary phase of death a true stupor manifests, although here too appearances can be deceptive, as those recovering from a clinically-pronounced death-state may report unbroken awareness. These two example of apparent stupor or death of the sensibility raise the possibility that awareness may have an alternative mechanism to that of the known senses.
Related Torpor (#HM1483) Absence of consciousness (#HM2670)
Near death experience (NDE, #HM0777).

♦ **HM2474a Being cognition**
B-cognition
Description A term used by Maslow to describe the perception of life as unitary reality. It may arise in the transient sense, for example in a peak experience, or serenely as one accepts and lives with the transcendent.
Narrower Plateau cognition (#HM1732).
Related Transcendence (#HM4104) Peak experiences (#HM2080)
Deficiency cognition (#HM3157).

♦ **HM2475b Omniscience** (Yoga)
Description According to Patanjali, omniscience is achieved when the yogi discriminates between soul and spirit, and achieves supremacy over all conditions. He has already achieved omnipresence, as he realizes his unity with all and the oneness of his soul with other souls. Now he is truly wise, and the next stage is *omnipotence* when he will achieve total power.
Related Released soul's perception (Hinduism, #HM6558).

♦ **HM2476b Sufi infused awareness** (Sufism)
Ilham
Description In Sufism, the sainted mystic is illumined or inspired by an angel-messenger who brings him a personal revelation (ilham).
Broader Human development (Sufism, #HH0436)
Stations of consciousness - Simnami (Sufism, #HM2341).

♦ **HM2477d Radical contingency** (ICA)
Description This state occurs when a human being experiences the mystery of death as the final reality of every life. Today's world is alive with images that communicate this reality. Just a few include: the nuclear threat; the daily news of famine, airline hi-jacking, civil wars, terrorist bombings; and advanced medical techniques that sometimes work and sometimes don't. Whatever occasions this state, everyone has those moments in which his own death becomes a reality to him.
A dimension of this experience is a sense of awe, that is fear and fascination, accompanied by a unique emotional tone. One experiences the shock and fear of realizing that death is the absolute, irreversible finality; the fear of not knowing how, when or where it will occur, just that it will. This is accompanied by a kind of amazement that one is, in that moment, still alive - that it is not me this time. This is the fascination which compels one to watch death occur at the same moment as one is tingling with one's own liveliness. During such a moment, one decides on present awareness of one's death; and more than that, to carry on living in and through such a moment. The residue of these events is a sense of not wanting to forget the reality of death that has been experienced, at the same time as a deeper appreciation for life, alive in a new way to the possibilities and options open to one.
Context This state is number 1 in the ICA *Other World in the midst of this World.*
Broader Aweful encounter (ICA, #HM2619).
Followed by Vibrant powers (ICA, #HM2982) Absurd existence (ICA, #HM2331)
Incarnate living (ICA, #HM2394) Ultimate awareness (ICA, #HM2388)
Individual fatefulness (ICA, #HM2223).
Preceded by Vibrant powers (ICA, #HM2982) Essential dubiety (ICA, #HM3135)
Seminal illumination (ICA, #HM2056) Contingent eternality (ICA, #HM2456)
Individual fatefulness (ICA, #HM2223).

♦ **HM2480d Incarnate Christ** (ICA)
Description This is the realization that in living authentically you are the truth about life. It occurs when you see others imitating you or you realize that there is no one other than your self. An example is the calling of Moses to lead the people of Israel out of Egypt.
Context In the ICA New Religious Mode in the arena of articulating the word about life (the life of knowing) the first formal aspect is one's experience of being a solitary individual or being a self. At the fourth phenomenological level, this occurs when one recognizes the humiliation of standing before the final mystery of life in every moment.
Broader Solitary being (ICA, #HM2404).
Preceded by Representational existence (ICA, #HM2557).

♦ **HM2481d Antipathy**
Description A fixed attitude when feelings of repugnance, dislike, and sometimes hostility, cause an individual to shrink from some person or object, whether rationally (based on some reason and feeling of which the individual experiencing antipathy is consciously aware) or irrationally.
The state is quite consistent with morality and may, indeed, be necessary at some stages of moral development. In fact, it is a mental or emotional state unrelated to ethical considerations.
Related Hostile feelings (#HH0643).

♦ **HM2483c Ultimate transparency of spirit** (Sufism)
Description Although apparently blinded (opaque) through his physical nature, the individual's spirit is transparent at the summit of vision. Man takes on the colour of his surroundings - so that hiding the opaque self from consciousness but manifesting ontological reality and understanding its true laws he appears as the wholeness of all that exists, a reflection of God; catching sight of himself, the individual realizes he is a manifestation of God and "I am the truth" becomes "he is the truth", revealed.
Context The sixth illumination of Jami.
Broader Fountains of light (Sufism, #HM3039).
Followed by Identifying with reality (Sufism, #HM2985).
Preceded by Perception of universality (Sufism, #HM2827).

♦ **HM2484d People of God** (ICA)
Description This is experienced when a person perceives his life as one component among many which comprise a movement that will in some way transform history, when he realizes that he is not alone in his care but rather part of a long chain of people who have given their lives to building a human world. It is like what Hermann Hesse in his book "Journey to the East" calls the *League.*
Context In the ICA New Religious Mode in the arena of acting out one's deed (the life of doing) the third formal aspect is awareness of one's allegiance to those who act on behalf of life as a whole. At the second phenomenological level this occurs when a person recognizes his life as sheer venture.
Broader Awareness of allegiance (ICA, #HM2073).
Followed by Religious vocation (ICA, #HM2293).
Preceded by Religious function (ICA, #HM2577).

♦ **HM2486b Self-remembering**
Self-awareness — Self-consciousness — Self-transcendence
Description The term self-remembering is used to signify the process of being aware of oneself, namely self-awareness or self-consciousness. Although Marxist thought considers self-consciousness to be the rational activity of evaluating one's own knowledge, outlook, behaviour and feelings as a totality and with reference to other people, in other systems it is not considered to be a function, a form of thinking, or a form of feeling, but rather a different state of consciousness that emerges from the mental process of trying to remember oneself, an attempt to create in oneself a state of consciousness without any relation to functions such as emotion and sensation.
Self-remembering or self-transcendence may be experienced as a result of some religious emotion, under the influence of a work of art, in the rapture of sexual love, or in situations of great danger and difficulty. It then constitutes a certain detachment of awareness from whatever an individual happens to be doing, thinking or feeling. There is an objective awareness of self, of being outside of, separated from the confines of the physical body, namely a state of non-identification with the observed world. It is an acute identity experience in which this greatest attainment of autonomy or selfhood is itself simultaneously a transcending of itself, beyond and above selfhood such that the person becomes relatively ego-less.
Self-remembering is considered to be distinct from self-observation which is always directed at some definite function, whether observation of thoughts, movements, emotions, or sensations. In self-observation there is always a definite object to be observed within oneself, whereas self-remembering does not divide the person in this way. It requires that the person remember the whole, experiencing the "I" of his own person. At a later stage, self-observation and self-remembering may occur simultaneously, but initially the process of self-observation leads to the understanding that the person does not remember himself except on rare occasions.
When a moment is experienced as part of eternity and the individual is aware of the unity in everything, this may be referred to as self-remembering. The mind's tendency to associate ideas and to think about what is being experienced is for that moment, checked. This may occur as the result of a shock or through disciplined self-observation and practice. According to Gurdjieff, self-remembering may be used to discourage the waste of fine energy by sifting the impressions which the human organism receives to remove attention from the negative and non-beneficial and focus attention on the beneficial.
Self-remembering is distinguished from the ordinary state of alertness of mind. It is a form of thinking or intellectual work which corresponds to awakening, and in this way it induces a moment of realization that the self-remembering state is as different from the normal state as the normal waking state is from the sleep state. When asleep an individual's world is limited by actual sensations, and on awakening he finds himself in an objective world which is much less limited. This is however only a half-awake state, which impedes awareness of a still richer world whose characteristics normally pass unnoticed; Gurdjieff refers to it as *waking consciousness.* Beyond self-remembering is a fourth state, *objective consciousness*, to which again one wakes from self-remembering as one wakes from sleep to waking consciousness.
Narrower Self-awareness (Buddhism, #HM1874).
Related Reflexivity (#HM0087) Self-respect (#HH1384) Introspection (#HH0824)
Transcendence (#HM4104) Self-observation (#HH0586) Ontological love (#HM1211)
Self-consciousness (#HM2571) Self-awareness (Psychism, #HM2436)
Anomalies in awareness of one's self (#HM8186).
Followed by Composition (#HM1406) Objective awareness (#HM2203).
Preceded by Waking consciousness (#HM3021).

♦ **HM2490d Doubt**
Description Doubt involves a suspension of judgement for or against a proposition, when serious evidence either way is lacking, *negative doubt*, or when evidence for and against is balanced, *positive doubt.* Ethics distinguishes further, between *speculative doubt* - objective morality, is a particular action intrinsically moral or not; and *practical doubt* - is the performance of this particular action in this particular situation moral or not ? According to Descartes, certitude can only be established by first subjecting every proposition to doubt, although there is discussion whether this is necessary if an assumption of the contradiction of a proposition is impossible to entertain - *fictitious doubt* - such as to the proposition "I exist". If doubt is accepted as the outcome of inquiry, this is referred to as *scepticism.*
In theological terms, doubt is distinguished from questioning, suggesting that doubt is the deliberate suspension of assent to knowledge whose import and basis are known. This doubting state may be due to an interior attitude which prevents complete recognition of revelation as the word of God or suspends previous judgement.
Refs Da Free, John *The Transmission of Doubt* (1984).
Related Scepticism (#HH0393).

♦ **HM2491c Niskama karma** (Hinduism, Yoga)
Action without desire
Description Under the influence of *kama* (personal desire), ordinary actions are performed which result in the building up of a personal karma. This is because such actions are performed while indentifying with the ego, seeking fulfilment of desires, and result in both pleasurable and painful experience. However, when the person completely dissociates himself from the ego and performs actions while identifying completely with the Supreme Spirit working through the ego, then such actions without desire, *niskama*, produce no more karma, despite possibly being engaged in the affairs of the world. Consciously identifying with the divine, with no taint of personal motive, no more karma being formed, it remains to work off karma or sanskaras previously accumulated.

♦ **HM2492c Ox-and-man-both-out-of-sight awareness** (Zen)
Description In the serenity prevailing as all confusion is set aside, there is no dualism, not even that implied by holiness.
Broader Oxherding pictures in Zen Buddhism (Zen, #HM2690).
Followed by Back-to-the-source awareness (Zen, #HM3251).
Preceded by Ox-forgotten-leaving-man-alone awareness (Zen, #HM2604).

♦ **HM2493d Creative futility** (ICA)
Description This occurs on discovering that any worldly hope or desire, any attachment to objects, ideas, relationships, or even to one's own sacrificial responsibility, is utterly futile; and at the same time that one's only hope and desire belong to that which gives and demands life - that which will never allow one to control the universe. It may be compared to the end of the book "The Ronin" when, after many years of work, the diggers break through in the wrong place and it will take many more years of digging to correct and make worthwhile the work already done. The older man turns his back on it and starts to walk out of the tunnel and, when the young man calls to him, responds: "To hell with it ". Or to the decision of the woman in a story of Nazi

concentration camps, who decides that while all are dying around her, she will continue to live.
A dimension of this experience is a sense of awe, that is fear and fascination. Hope is known to be the enemy. There is a kind of deep silence that comes from inside the person. There is nevertheless a fascination with standing on nothing, with Being itself; and a sense of victory that is not victory the way the world understands it. The decision is to keep on going, to trust the absence of hope; one's hope is in no hope. What remains is a sense that all is good, one has a different set of priorities and expectations from the rest of the world, one's hope is not an illusion but a tool to be used to create and release a more human life.
Context This state is number 53 in the ICA *Other World in the midst of this World.*
Broader Unknowable peace (ICA, #HM3015).
Followed by Vital signs (ICA, #HM2875) Living death (ICA, #HM2808) Universal fate (ICA, #HM2687) Incarnate living (ICA, #HM2394) Problemless living (ICA, #HM2747).
Preceded by Living death (ICA, #HM2808) Universal fate (ICA, #HM2687) Primal sympathy (ICA, #HM2550) Personal epiphany (ICA, #HM2595) Seminal illumination (ICA, #HM2056).

♦ **HM2494d Near death peace** (NDE)
Description There is a sense of peace, calm, joy, wellbeing and freedom from bodily pain.
Context The first stage in the process referred to as *near death experience.*
Broader Near death experience (NDE, #HM0777).
Related Near death fear (NDE, #HM3177).
Followed by Near death detachment (NDE, #HM3349).

♦ **HM2495d Understanding having a limited object** (Buddhism)
Description This understanding occurs contingent with the states of the realm of the senses or sense-sphere. Together with understanding having an exalted object it comprises mundane or worldly insight.
Context On the Path of Purification of Hinayana Buddhism, panna (understanding) is considered as of one kind (monad), of two kinds (dyads), of three kinds (triads) or of four kinds (tetrads). There are five dyads, four triads and two tetrads. All have the characteristic of penetrating the individual essences or true nature of states (monad). In the second triad, this is compared with understanding having an exalted or measureless object.
Broader Understanding (Buddhism, #HM4523).
Related Mundane understanding (Buddhism, #HM7628) Understanding having an exalted object (Buddhism, #HM3296) Understanding having a measureless object (Buddhism, #HM6681).

♦ **HM2498d Discouragement**
Description The taking up of a spiritual path which requires continual application of discipline requires fortitude and strength of character which even the strongest find hard. Periods of psychological growth are invariably followed by times of discouragement and often depression. These are said to be caused by the contrast between the purity and beauty of what is contemplated and the impurity and misery of the contemplating. Such discouragement may be viewed as evidence of development, and continuing on the chosen path despite aridity of spirit eventually leads to further growth.
Related Endurance (#HH0729).

♦ **HM2499g Embodied consciousness**
Description An individual's awareness of the surroundings is always with reference to the body and its sense organs. The differentiation of the person as phenomenologically open to the world through the body, but not at one with it, is a sophisticated, later development in mankind's development. Only when aware of this *embodied consciousness* can the personality distinguish either the world or itself.
Related Dasein (#HM2632).

♦ **HM2500c No-thought** (Zen)
Wu-nien — Unconscious — Suchness
Description This is closely related to wu-hsin, no-mind, which is said by some to designate the same experience. It can be referred to as the unconscious, since it is where conscious thoughts and feelings grow from. It implies having thoughts and yet not having them, untainted by conditions of life with which it is in contact, detached from objective conditions in one's own consciousness. It is to find the unconscious in consciousness. According to Shen-hui, this is suchness, not being attached to form, not thinking of being or non-being, of good or bad, of having limits or having no limits, of measurements or non-measurements, of enlightenment or being enlightened, of Nirvana or obtaining Nirvana, this is the unconscious which is prajnaparamita. Hui-neng says that no-thought-ness is seeing all things yet keeping the mind free from stain or attachment. This gives the perfect way through the world of multiplicities. Understanding the idea of no-thought-ness brings sight of the realm of all the Buddhas, attainment to the stage of Buddhahood.
Thus the intuition or apprehension of suchness occurs when the person is free of concepts, expectations and self-indulgent emotions. The realization of suchness may occur unexpectedly, for example in the beauty of a landscape or of a single flower. Keeping the mind in unison with suchness, the artist is aware of the same life animating him and the object. It is the object itself which works, the artist's brush, arms, fingers becoming servzants of the object which makes its own picture.
Context According to Hui-neng, this concept of the unconscious is the foundation of zen, together with formlessness, wu-hsing, and non-abiding, wu-chu.
Refs Suzuki, D T *The Zen Doctrine of No-Mind* (1970).
Broader Human development (Zen, #HH1003).
Related No-mind (Zen, #HM2163) Non-abiding (Zen, #HM3485) Formlessness (Zen, #HM0910) Prajna paramita (Zen, #HH0550) Experience of the unconscious (#HM2332).

♦ **HM2501c Emptiness of the unapprehendable** (Buddhism, Tibetan)
Anupalambhashunyata
Context One of the eighteen emptinesses comprising the paths of view in Tibetan Buddhism.
Broader Emptinesses on the paths of the view in Buddhism (Buddhism, Tibetan, #HM2944).

♦ **HM2502e Voodoo trance consciousness**
Description Voodoo trance states may be entered voluntarily or involuntarily through means of auto-suggestion, hypnosis or drugs. In one of the trance states of consciousness the devotee may become possessed by any one of a number of spirits (loa) which abound in the world. Haitian voodoo makes use of extensive rituals, and the powers of the priest (ongan) and priestess (manbo) are sufficient, through the stimulus of fear or dread alone, to induce ecstatic states or non-rational behaviour during the rites. One of the chief purposes of these rites is precisely to invoke the loa, using songs, prayers and vévé (magical diagrams).
Broader Trance (#HM3236).

♦ **HM2503c Path of intelligence of temptation** (Judaism)
Description This state of awareness is of those things that constitute the fundamental and primary obstacles for devotion to self-realization.
Context The twenty fifth of the spiritual paths expressed in the Jewish Kabbalah.
Broader Paths of wisdom (Judaism, #HM2509).
Followed by Path of renovating intelligence (Judaism, #HM2047).
Preceded by Path of imaginative intelligence (Judaism, #HM2011).

♦ **HM2506c Life of chastity** (ICA)
Description An arena in the ICA *New Religious Mode* when one is aware of being in the presence of transcendence.
Broader New religious modes (ICA, #HM3004).
Narrower Inventing essence (#HM3138) Spiritual creativity (#HM2037) Radical identification (#HM2758) Inclusive collegiality (#HM2333).
Related Virginity (#HH0272).

♦ **HM2507c Four immeasurables** (Buddhism)
Description These are equanimity, love, compassion and joy.
Context One of the paths of calming according to Buddhist teaching.
Broader Paths of calming (Buddhism, #HM2987).

♦ **HM2508c Palace of wisdom** (Buddhism)
Pannapasada (Pali)
Broader Faculty of wisdom (Buddhism, #HM3233).

♦ **HM2509b Paths of wisdom** (Judaism)
Description The Jewish Kabbalah draws on the Biblical sapiental tradition wherein God's wisdom is the instrument of creation. The theosophic mystics among medieval Jewry sought to trace the paths of consciousness or wisdom through the universe, upwards and downwards, from God to man. The system of 32 paths can be considered as a concise indication of stages in the mystical ascent in the four worlds. There are ten emanation-realms of the sephiroth, through whose qualities and virtues Wisdom creates the universe.
Broader Stages of spiritual life (#HH0102).
Narrower Path of fiery intelligence (#HM2523) Path of active intelligence (#HM2218) Path of stable intelligence (#HM2533) Path of occult intelligence (#HM2444) Path of natural intelligence (#HM2071) Path of intelligence of will (#HM2629) Path of uniting intelligence (#HM2583) Path of perfect intelligence (#HM3243) Path of radical intelligence (#HM3181) Path of faithful intelligence (#HM2081) Path of purified intelligence (#HM2943) Path of perpetual intelligence (#HM2643) Path of corporeal intelligence (#HM3149) Path of intelligence of reward (#HM2105) Path of disposing intelligence (#HM2569) Path of receiving intelligence (#HM3102) Path of collective intelligence (#HM3046) Path of renovating intelligence (#HM2047) Path of triumphant intelligence (#HM2593) Path of imaginative intelligence (#HM2011) Path of transparent intelligence (#HM2107) Path of resplendent intelligence (#HM3426) Path of sanctifying intelligence (#HM3567) Path of constituting intelligence (#HM2117) Path of illuminating intelligence (#HM2035) Path of intelligence of temptation (#HM2503) Path of intelligence of influences (#HM2045) Path of the admirable intelligence (#HM3013) Path of administrative intelligence (#HM2095) Path of the illuminating intelligence (#HM2612) Path of intelligence of spiritual action (#HM2021) Path of intelligence of mediating influence (#HM3089).
Related Kabbalah (#HH0921) World of creation (Judaism, #HM3038) Sphere of wisdom (Kabbalah, #HM2348) Shekinah awareness (Judaism, #HM2002) World of formation (Judaism, #HM2360) World of emanation (Judaism, #HM2408) World of manifestation (Judaism, #HM2312).

♦ **HM2510d Illusion**
Description Illusion, or faulty cognition, can result from deceptive sense stimuli, from defective sense mechanisms, from distortions in subconscious apprehension or from interferences with the intrinsic logic of comprehension. Perceived actuality in all these instances is altered to a grossly inaccurate comprehension state. What is perceived does not correspond to the objective physical properties. The misinterpretation which occurs may be based on a predisposition to a certain interpretation because of a preoccupation at the time, and because the situation or stimulus is itself ambiguous. Illusion may be due to misidentification, or (more correctly) to correct identification but misinterpretation.
Refs Leeuw, J J Van der *The Conquest of Illusion* (1968); Ross, Helen E *Behaviour and Perception in Strange Environments* (1974).
Narrower Pareidolia (#HM8135) Phosphenes (#HM0794) Coriolis illusion (#HM4664) Empty-field myopia (#HM0882) Perceptual narrowing (#HM1237) Audiogyral illusions (#HM7212) Oculoagravic illusion (#HM0351).
Related Maya (Hinduism, #HH1004) Hallucination (#HM4580) Disillusionment (#HM6174) Delusive consciousness (#HM2600).

♦ **HM2511c Life of prayer** (ICA)
Description An arena in the ICA *New Religious Mode* when one is aware of acting out one's freedom.
Broader New religious modes (ICA, #HM3004).
Narrower Affirmation (#HM2199) Struggle to confess (#HM2718) Surrender to inadequacy (#HM2922) Universal responsibility (#HM3110).

♦ **HM2512d Libra-consciousness** (Astrology)
Aspiration to unity
Description A mode of imaginative and intuitive balance among the faculties, based on a subtle sense of affinity with the environment. Under optimal conditions, Libra is associated with an integration of insights and perspectives, allowing appropriate values and priorities to emerge and following wherever these may lead.
Context One of the twelve conditions of being or streams of divine energy encompassed within the zodiac. It is said that all twelve conditions must be present and playing its proper part for any successful and fulfilling action to take place. This may be through a group of individuals with different endowments and soul qualities working together; or it may be through one individual who, by self-discipline, practice and meditation, has allowed all twelve conditions to arise within himself.
Broader Airy awareness (Astrology, #HM2955) Cardinal awareness (Astrology, #HM2259) Zodiacal forms of awareness (Astrology, #HM2713).
Followed by Scorpio-consciousness (Astrology, #HM2615).
Preceded by Virgo-consciousness (Astrology, #HM2439).

♦ **HM2516d Reviving-hell awareness** (Buddhism, Tibetan)
Description This condition is represented as the eight hot hells. It is the state for killers and also for wanton slaughterers of animals. Its character is that individual souls kill and are killed here. Those experiencing death are revived and the killing recommences until karma is exhausted.
Context In Tibetan Sakya Buddhism this is one of the states in the "Ascension Stages Game". In some sets it is numbered 6 on the board.

Broader Desire-realm consciousness (Buddhism, #HM2733).
Followed by Howling-hells awareness (Buddhism, Tibetan, #HM2100)
Crushing-hells awareness (Buddhism, Tibetan, #HM3142)
Temporary-hells awareness (Buddhism, Tibetan, #HM2454)
Southern-continent awareness of men (Buddhism, #HM2127)
Divine-animal-hell awareness (Buddhism, Tibetan, #HM2007)
Hungry-ghosts-hell awareness (Buddhism, Tibetan, #HM2112).
Preceded by Cold-hells awareness (Buddhism, Tibetan, #HM2040)
Howling-hells awareness (Buddhism, Tibetan, #HM2100)
Pratyeka Buddha awareness (Buddhism, Tibetan, #HM2180)
Temporary-hells awareness (Buddhism, Tibetan, #HM2454)
Southern-continent awareness of men (Buddhism, #HM2127)
Heavenly-highway awareness (Buddhism, Tibetan, #HM2010)
Four-great-kings-heaven awareness (Buddhism, Tibetan, #HM2082).

♦ HM2517d Eternal insecurity (ICA)
Description This is experienced when a person is unsure of what is required of him and what role he needs to play; when he is called upon to shift his understanding of who he is and what he is doing in the world. It may be compared with Dustin Hoffman's description of his experience in deciding to play in the movie "Tootsie."
Context In the ICA New Religious Mode in the arena of knowing one's disengagement (the life of poverty) the third formal aspect is being nonchalant about one's relations. At the first phenomenological level, this occurs when one is no longer content with one's customary habits, but sees in them a futility never before discerned.
Broader Liberation from relationships (ICA, #HM3240).
Followed by Serious sharing (ICA, #HM2323).

♦ HM2519d Western-continent awareness of cattle (Buddhism, Tibetan)
Apara-godaniya
Description This state of awareness is characterized by a heightened form of life and is represented as a floating island west of Mount Meru. Awareness here may include the crossing of the senses so that colours can be experienced by touch, taste, sound and smell, while sounds can be seen, and so on.
Context In Tibetan Sakya Buddhism this is one of the states in the "Ascension Stages Game". In some sets it is numbered 18 on the board.
Broader Desire-realm consciousness (Buddhism, #HM2733).
Followed by Naga-world awareness (Buddhism, Tibetan, #HM2031)
Animal-hell awareness (Buddhism, Tibetan, #HM2636)
Discipleship-karma awareness (Buddhism, Tibetan, #HM2192)
Barbarian-state of awareness (Buddhism, Tibetan, #HM2091)
Hungry-ghosts-hell awareness (Buddhism, Tibetan, #HM2112)
Four-great-kings-heaven awareness (Buddhism, Tibetan, #HM2082).
Preceded by Cold-hells awareness (Buddhism, Tibetan, #HM2040)
Non-emanating consciousness (Buddhism, Tibetan, #HM2070)
Four-great-kings-heaven awareness (Buddhism, Tibetan, #HM2082).

♦ HM2521b Effulgence
Fulfilment
Description Literally shining or radiant, effulgence is the joyful acceptance and fulfilment experienced during the carrying out of a task rather than in its completion; it is that situation in which an individual or a community engages in an action joyfully, in the fullness of life.
Related Fulfilment (#HM0728) Immediacy of fulfilled living (#HH0223).

♦ HM2522c Emptiness of the objects of sense and of mental consciousness (Buddhism, Tibetan)
Bahirdhashunyata
Context One of the eighteen emptinesses comprising the paths of view in Tibetan Buddhism.
Broader Emptinesses on the paths of the view in Buddhism (Buddhism, Tibetan, #HM2944).

♦ HM2523c Path of fiery intelligence (Judaism)
Description This state of awareness is represented by those that come before the veil to contemplate the secret display of the orders and dispositions of the principles (sephiroth) of cosmic genesis.
Context The condition of "fiery intelligence" is the eleventh of the spiritual paths expressed in the Jewish Kabbalah.
Broader Paths of wisdom (Judaism, #HM2509).
Followed by Path of transparent intelligence (Judaism, #HM2107).
Preceded by Path of resplendent intelligence (Judaism, #HM3426).

♦ HM2524d Impossible possibility (ICA)
Description This is acting toward a creative solution when nothing seems possible, or discovering that one has succeeded when at first it seemed that nothing could be done; when one chooses to act when neither action nor inaction seems relevant. It may be compared to the action of the assault team in the film "Guns of Navarone".
Context In the ICA New Religious Mode in the arena of knowing and doing intensified in one's life (life of being) the third formal aspect is the experience of transparent engagement or the experience of levitation. At the second phenomenological level this occurs when a person relates to the experience of nothingness.
Broader Transparent engagement (ICA, #HM2736).
Followed by Keeping conscience (ICA, #HM2249).
Preceded by Action irrelevant (ICA, #HM2466).

♦ HM2528d Haughtiness (Buddhism)
Mada — Rgyags-pa (Tibetan)
Description A "puffing-up" of the mind at one's own good qualities or fortune.
Context One of the twenty *secondary afflictions* of Tibetan Buddhism.
Broader Secondary afflictions (Buddhism, Tibetan, #HH0781).

♦ HM2529d Contentment at the centre (ICA)
Unspeakable joy
Description This is an experiential trek in the *Sea of Tranquillity*, within the ICA *Other World in the midst of this World*, when the individual comes to know that rapture walks with woe. The stages are described in separate entries.
Broader Sea of tranquillity (ICA, #HM3033).
Narrower Vital signs (#HM2875)
Final blessedness (#HM2958)
Blissful seizure (#HM3148)
Spontaneous gratitude (#HM3025).
Related Endless life (ICA, #HM2437)
Radical illumination (ICA, #HM2273).
Unknowable peace (ICA, #HM3015)

♦ HM2530b Consciousness spectrum
Transpersonal bands
Description According to K. Wilber, consciousness has evolved from an original, all-inclusive, timeless and non-dual awareness, through various levels of *maya* or illusion; each of these levels brings further separation and increased appearance of duality as the individual equates his own identity with the self (and everything else as not-self), as his body, as his ego, and finally as those facets of the ego he is prepared to accept. The spectrum of consciousness is a representation of these "identifications" of absolute subjectivity with one "set of objects" compared with all others; each set of bands being more narrow and exclusive than the last.
The separation inherent in maya can be considered as the setting of limits or boundaries, that within the boundary is one and that outside the boundary is another; in order to consider the one, the other is dismissed. This is the second or "existential" level of the spectrum, where the organism is felt and believed to be distinct from its environment. At this level the awareness of space and of time (the latter being an avoidance of death inherent in identification with the body - there is nothing before the birth of the body nor after its death) are generated. The anxiety caused by the fear of death leads to the next level, where the organism is further divided as ego and body, identification being with the ego which has, not is, the body. Between the first level (of mind) and the existential level are the *transpersonal bands*, the collective unconscious: astral projections, extrasensory perception and so on occurring at this level, where duality is not totally experienced. Communication among individuals may be at any of the consciousness levels described, or a combination of some or all of them. Misunderstanding of such communication leads to further duality, as the self-image built up by an individual as acceptable to himself rejects parts of the ego, which are retained as the "shadow".
In order to escape from the smaller and smaller view of the self that the individual creates one has to regain the sense of the timeless "now" - which is actually always present. This cannot be done by trying to retrieve what is already there but by removing what is not there: by active attention; by stopping the mental chatter and background "noise"; and by passive awareness - all leading to the end of the sense of "I", to "ego-death". Virtually all systems of self-realization and spiritual development have this aim in view.
Refs Wilber, K *Psychologia Perennis*; Wilber, Ken *The Spectrum of Consciousness* (1977).
Related Accepting the shadow (#HH0204).

♦ HM2531c Perception of hidden reality (Sufism)
Description The truth is always and everywhere present, seeing both the external appearance of things and their hidden reality. Since the truth is always gazing upon one, one should not look elsewhere; a split second's awareness of the truth is better than a lifetime beholding earthly beauty.
Context The third illumination of Jami.
Broader Fountains of light (Sufism, #HM3039).
Followed by Separation from the transient (Sufism, #HM2906).
Preceded by Single desiredness (Sufism, #HM2765).

♦ HM2532d Representational sign (ICA)
Description This is experienced as throwing oneself into one's own crippledness, when one decides to rejoice in one's own inadequacy and receive it as the gift that it is. This is like Beethoven continuing to write music when he could no longer hear.
Context In the ICA New Religious Mode in the arena of acting out one's freedom (the life of prayer) the third formal aspect is surrender to one's unlimited inadequacy, or petition. At the second phenomenological level, this occurs on recognizing one's own full participation in one's fate.
Broader Surrender to inadequacy (ICA, #HM2922).
Followed by Imploring succour (ICA, #HM2771).
Preceded by Abject helplessness (ICA, #HM2429).

♦ HM2533c Path of stable intelligence (Judaism)
Description This is awareness of the consistency of all numerations.
Context The twenty third of the spiritual paths expressed in the Jewish Kabbalah.
Broader Paths of wisdom (Judaism, #HM2509).
Followed by Path of imaginative intelligence (Judaism, #HM2011).
Preceded by Path of faithful intelligence (Judaism, #HM2081).

♦ HM2534d Excitement (Buddhism)
Auddhatya — Rgod-pa (Tibetan)
Description The scattering and non-peacefulness of mind which is caused by following desire for the pleasant.
Context One of the twenty *secondary afflictions* of Tibetan Buddhism.
Broader Secondary afflictions (Buddhism, Tibetan, #HH0781).
Related Excitement (Buddhism, #HM1469) Distraction (Buddhism, #HM3154).

♦ HM2535c Jesus condemned to death (Christianity)
Description The believer is aware that it is his own sin which led Jesus to be unjustly condemned to death. In the light of Jesus' journey to the cross he asks assistance for his soul in the journey to eternity. Loving Jesus more than himself, he repents with his whole heart and begs never again to separate himself from Him. He asks that he will love Jesus always and that Jesus will do with him as He wills.
Context The first station of the cross.
Broader Way of the cross (Christianity, #HM3516).
Followed by Jesus made to bear the cross (Christianity, #HM3500).

♦ HM2536c Rupaloka consciousness (Buddhism)
Jhana (Pali) — Fine-material sphere
Description The first of the three higher states in Pali texts, the *fine-material sphere* in which the *jhana* or mental absorptions of human meditation take place.
Broader Worlds of conscious existence (Buddhism, Pali, #HM2072).
Narrower Third trance of the fine-material sphere (#HM2062)
First trance of the fine-material sphere (#HM2450)
Second trance of the fine-material sphere (#HM2038)
Fourth trance of the fine-material sphere (#HM2586).
Related Dhyana (Hinduism, Buddhism, #HM0137)
Phenomena awareness (Buddhism, #HM2551)
Form-realm consciousness (Buddhism, #HM2257)
Arupaloka consciousness (Buddhism, Pali, #HM2012)
Trances and mental absorptions (Buddhism, #HM2122)
Tibetan meditative states of form concentrations (Buddhism, #HM2693).

♦ HM2538c Power of remorse (Buddhism)
Ottappabala (Pali)
Narrower Feeling remorse of what one ought to be remorseful (#HM1757)
Feeling remorse of acquisition of sinful and unwholesome dharmas (#HM1208).

♦ HM2541d Eternal identification (ICA)
Description This is the realization that one is responsible for, and at one with, the past and with the future, that all lives are both an extension of and a prelude to the present. It occurs when an individual becomes one with the peoples of all times who have trodden the path he has chosen.
Context In the ICA New Religious Mode in the arena of acting out one's engagement (the life

of obedience) the fourth formal aspect is embodying service to people's spirit life. At the third phenomenological level, this occurs when people willingly embrace their responsibility to society.
Broader Embodying service (ICA, #HM3199).
Followed by Communion of saints (ICA, #HM2338).
Preceded by Global brotherhood (ICA, #HM2432).

♦ **HM2542f Aesthetic LSD experience**
Description This state of awareness is induced by the use of psychedelic drugs and is apparently close to the siddhis achieved through the mystic training of yogis. The senses are altered and intensified; there may be sensations of physical transformation, lightness and clairvoyance, and a sense of unity.
Refs Grof, S *Varieties of Transpersonal Experiences* (1972).
Broader Psychedelic drugs (#HH0075)
Modes of awareness associated with use of hallucinogens (#HM0801)
Modes of awareness associated with psychoactive substances (#HM0584).
Related Artistic education (#HH0029) Transcendental experience (#HM2712)
Psychodynamic LSD experience (#HM3271)
Drug use for transcendent experience (#HH1123).

♦ **HM2543d Eastern-continent awareness of noble figures** (Buddhism, Tibetan)
Purva-videha
Description This state of awareness is characterized by mildness and tranquillity, and represented as beings of especially noble human form.
Context In Tibetan Sakya Buddhism this is one of the states in the "Ascension Stages Game". In some sets it is numbered 19 on the board.
Broader Desire-realm consciousness (Buddhism, #HM2733).
Followed by Naga-world awareness (Buddhism, Tibetan, #HM2031)
Asura-world awareness (Buddhism, Tibetan, #HM2579)
Animal-hell awareness (Buddhism, Tibetan, #HM2636)
Pratyeka Buddha awareness (Buddhism, Tibetan, #HM2180)
Discipleship-karma awareness (Buddhism, Tibetan, #HM2192).
Preceded by Temporary-hells awareness (Buddhism, Tibetan, #HM2454)
Hungry-ghosts-hell awareness (Buddhism, Tibetan, #HM2112)
Discipleship-application awareness (Buddhism, Tibetan, #HM2716)
Northern-continent awareness of community (Buddhism, Tibetan, #HM2067).

♦ **HM2546d Delightful-emanation-heaven consciousness** (Buddhism, Tibetan)
Nirvana-rati
Description This state of human awareness is projected as the qualities ascribed to the deities in this heaven. They are morally refined and intensely generous, and enjoy the emanations from their own minds more than material things.
Context In Tibetan Sakya Buddhism this is one of the states in the "Ascension Stages Game". In some sets it is numbered 31 on the board.
Broader Desire-realm consciousness (Buddhism, #HM2733).
Followed by Joyful-heaven awareness (Buddhism, Tibetan, #HM2022)
Pratyeka Buddha awareness (Buddhism, Tibetan, #HM2180)
Hindu-states of awareness (Buddhism, Tibetan, #HM2115)
Non-emanating consciousness (Buddhism, Tibetan, #HM2070)
Divine-animal-hell awareness (Buddhism, Tibetan, #HM2007)
Great-vehicle lower-path awareness (Buddhism, Tibetan, #HM2268).
Preceded by Discipleship-vision awareness (Buddhism, Tibetan, #HM2240)
Heaven-without-fighting consciousness (Buddhism, Tibetan, #HM2130).

♦ **HM2547d Passionate disinterest** (ICA)
Description This is the awareness of having totally committed one's life beyond all human relationships and covenants. Relationships in this world become nothing and one will never again answer to any of them. In the midst of passionate responsibility for this world one becomes passionately irresponsible to this world, not accountable to family or job, aware that that is not what life is about.
A dimension of this experience is a sense of awe, that is fear and fascination. The fear is that one will not succeed; a sense of having gone beyond all that bound one to this world leaves the fear that one will forever be bound with just one's self and history. The fascination is of of walking hand in hand with one's own death which one is now free to receive. There is a discovery that one has decided one is risking everything, a decisional objectivity about all that has been binding one to this world. The residue of such a time is a nonchalance that comes from knowing that one is in history for good or bad; it is no longer necessary to win - the act is it.
Context This state is number 31 in the ICA *Other World in the midst of this World.*
Broader Final accountability (ICA, #HM3206).
Followed by Self transcendence (ICA, #HM2584)
Impactful profundity (ICA, #HM2607)
Contextual world-view (ICA, #HM2842)
Everlasting community (ICA, #HM2777)
Destinal accountability (ICA, #HM2309).
Preceded by Cosmic sanctions (ICA, #HM2945) Worldly detachment (ICA, #HM2451)
Contextual world-view (ICA, #HM2842) Everlasting community (ICA, #HM2777)
Transcendent immanence (ICA, #HM3034).

♦ **HM2548f Anomalies in experience of the reality of one's self and the environment**
Loss of experience of reality
Description Very few people question the reality of themselves or of their environment; but a sense of unreality may arise, typically in experiences of severe discomfort or pain. There may be a feeling that it is not really one's self who is experiencing the discomfort at all, depersonalization. It is a common feature of psychiatric cases, particularly obsessional disorders and in depressive states. The experience may be described as being cut off from the grace of God and, in combination with dissociation of affect, with loss of capacity to love their families, isolation and unworthiness. All feelings of depersonalization are not necessarily negative. There may be a feeling of altered identity and unreality when falling in love or coming into large sums of money. The term might also be extended to cover the transcending of reality which occurs in religious rites, mystic rituals and the taking of drugs.
Derealization is the anomalous experience of the environment, when one's surroundings seem to lose reality although one may be convinced of the reality of one's self. Depersonalization and derealization often occur together.
Broader Anomalies in awareness of one's self (#HM8186).
Narrower Derealization (#HM5128) Depersonalization (#HM1248).
Related Breakdown of automatization (#HM8105).

♦ **HM2550d Primal sympathy** (ICA)
Description This is an awareness of caring about what's going on around one; not that one should care, one just does. It is what happens to someone who comes across a dead dog that has just been hit by a car. A sense of compassion wells up inside him, without even willing it. You don't set out to be your brother's keeper, you just wake up discovering yourself being it. It is like Walter Matthau in the movie "A New Leaf" when, after trying all his life to not do anything but spend his inherited fortune, he suddenly finds himself impelled to take charge of his newly acquired mansion and put it back into order.
A dimension of this experience is a sense of awe, that is fear and fascination. The realization of caring often comes as a surprise, like a package with one's name on it. One asks "Is this really me or have I mistaken myself for someone else ?". One finds one's self attracted to this capacity to care but knows that it always costs something; in fact, it can cost one's life. There is no end to it, it just keeps going on.
Experiencing himself as a compassionate human being, an individual finds he is responding to the needs of others, not out of any emotional sentimentality but out of a deeper sense of being bound one to another. He sees that all of life is accountable to all the rest of life, and knows he will never again ignore the demands of life. He resolves not to forget what he has experienced - it's like wearing a new set of glasses through which he can interpret and respond to what is going on around him.
Context This state is number 37 in the ICA *Other World in the midst of this World.*
Broader Universal concern (ICA, #HM2774).
Followed by Incarnate living (ICA, #HM2394) Creative futility (ICA, #HM2493)
Global guardianship (ICA, #HM2817) Universal compassion (ICA, #HM2734)
Diaphanous intuition (ICA, #HM2927).
Preceded by Universal fate (ICA, #HM2687) Incarnate living (ICA, #HM2394)
Sacramental universe (ICA, #HM2445) Diaphanous intuition (ICA, #HM2927)
Individual fatefulness (ICA, #HM2223).

♦ **HM2551c Phenomena awareness** (Buddhism)
Dhammas (Pali) — Sankhara — Dhatu
Description The term, dhamma (sanskrit, dharma), besides denoting the liberating Law discovered by Buddha and summed up in the four Noble Truths (ariya-sacca), means a thing or object that "bears" something. A stone bearing inscriptions of laws could be dhamma, but in the subtle and mental worlds, dhamma is an object of consciousness. Thus all mental (and material) phenomena could be comprised of elements called dhammas. More often dhamma is meant as phenomena generally, or mental phenomena. Physical and partly physical elements are called dhatu. All mental processes are based on 22 elements: the 4 producing corporeality, the 5 sense organs, the 5 corresponding objects, and the 5 specialized (corresponding) conformations of consciousness. In addition there is a mind-element (mano-dhatu) and its correspondin mind-object (dhamma-dhatu) and a mind-consciousness element (mano-vinnana-dhatu).
Narrower Five aggregates (#HH3321) Phase-conditions of consciousness (#HM2599)
Worlds of conscious existence (Buddhism, Pali, #HM2072).
Related Karma awareness (Buddhism, #HM2048)
Rupaloka consciousness (Buddhism, #HM2536)
Lokuttara consciousness (Buddhism, #HM2120)
Kamaloka consciousness (Buddhism, Pali, #HM2060)
Arupaloka consciousness (Buddhism, Pali, #HM2012).

♦ **HM2552d Disengagement from work** (ICA)
Description The second formal aspect of the *life of poverty*, an arena in the ICA *New Religious Mode*, this represents the experience of relying on the larger society and being disengaged from one's own work.
Broader Life of poverty (ICA, #HM2299).
Narrower Beyond success (#HM2941) Social failure (#HM2728)
Realized vocation (#HM2441) Historical vocation (#HM2616).

♦ **HM2553b Psychological growth in a dying person**
Death as an altered state of consciousness — Ars moriendi
Description An intimate confrontation with death, in addition to triggering deep biologically rooted anxiety related to the self-preservation instinct, represents a painful reminder of the ultimate limitations of man's efforts at control and mastery of his environment and his unconscious wish to be eternal. This shattering encounter constitutes an agonizing existential crisis for the individual who realizes that no matter what he does in his life he cannot escape the inevitable and will have to leave the world bereft of everything that he has accumulated, achieved, and become emotionally attached to. In this respect the process of dying resembles the process of birth. Exposure to the phenomenon of death tends to open up spiritual and religious dimensions that appear to be an intrinsic part of the human personality and are independent of his cultural and religious background.
Traditional practices for assisting the dying existed and exist in most civilizations. Life may be considered as an education in dying, and reminders of mortality were common in traditional victory and life-celebrating festivals. In modern Western society, a therapist working with healthy individuals may help to stimulate psychological growth by facilitating a confrontation for the individual with his fear of death in a supportive, educational, psychotherapeutic frame of reference.
Other forms of therapy may assist individuals in the process of dying to achieve psychological growth and altered states of consciousness. The general taboo and dread associated with even the idea of death in Western society is partly due to the lonely and depersonalized experience which modern medicine and surgery have made of death and of keeping people alive. This may be overcome by a renewed faith in interpersonal relationships and help in experiencing death through the emotional effects it has on the dying person and those closest to him. The dying individual is given the opportunity to face and deal with his mortality in a way that is worthy of his human dignity and which is therefore of potential value both to him and to his family. By transcending individual existence and interacting with others, and by accepting the changes which are occurring and the unknown future, death can become the culmination of life and the means of knowing the self. When mystical states of consciousness occur among the dying, they may impart a sense of new and profound insight into the meaning of life and a deeper sensitivity to values. They may provide the individual with an increased awareness of the significance of his existence and an enriched appreciation for the whole of creation.
Associated with the notion of death as an opportunity for psychological growth is the concept of death itself as an altered state of consciousness. Related to this is the belief in survival after death and the belief in reincarnation (separately described). In Christianity, for example, death is not simply the transition from one form of existence to another similar form but the transition from the temporal and incomplete to the eternal and complete. The whole of creation is seen as growing towards a definitive state through incarnate spiritual persons and their death, this growth from within continuing until the external intervention of God in the Day of Judgement.
Refs Kubler-Ross, Elisabeth (Ed) *Death* (1975); Ushabudha Arya *Meditation and the Art of Dying* (1979); Zinker, J C and Fink, S L *The Possibility for Psychological Growth in a Dying Person* (1966).
Related Reincarnation (#HH0686) Near death experience (NDE, #HM0777).

♦ **HM2554c Fixed awareness** (Astrology)
Cross of transmutation
Description The resistance to change implicit in the fixed cross of Taurus and Scorpio, Leo and Aquarius, brings a quality of stability and support to the consciousness of persons under the influence of this *quadruplicity*. They are characterized by the overcoming of desire and selfishness, to demonstrate selfless aspiration and service, the cross of discipleship being born by someone

who has already passed beyond the mutable cross of servitude. A systemic awareness develops.
Context One of the three cardinal qualities, each uniting four signs of the zodiac in two pairs of opposites or polarities which are strongly linked and between which the influenced individual tends to oscillate, all four signs having much in common.
Broader Zodiacal forms of awareness (Astrology, #HM2713).
Narrower Leo–consciousness (#HM2376) Taurus–consciousness (#HM2815) Scorpio–consciousness (#HM2615) Aquarius–consciousness (#HM2973).
Related Mutable awareness (Astrology, #HM3092) Cardinal awareness (Astrology, #HM2259).

♦ **HM2556d Awareness of consciousness–group of conscious existence - senses and mind** (Buddhism)
Vinnana–khanda (Pali)
Description The consciousness aggregate, vinnana–khanda, is one of the five interacting aggregates that produce the illusory ego. Here are at least six classes of sense consciousness: one for each sense–organ ("body" consciousness equals the "touch" organ) plus "mind" consciousness. The latter (bhavanga–mano) appears to be only the subconscious mind, or that part of the mind (mano–dhatu) that adverts to the sense–object.
Context The southern Buddhists, analyzing the 6 states of consciousness from the psychological view–point of volition and hence the moral view–point of Karma, find 89 states of consciousness to describe. Some northern Buddhists (Yogacarins in the Mahayana) add two more states in the vinnana–khanda: manovinnana (or mano–vijnana), and alaya–vinnana. The former receives data from the sixth state or function and conveys them to to eighth level, after investigation. The alaya–vinnana is considered a "repository" consciousness. In the Pali texts the seventh seems to appear as a karmically independent function and alaya may be synonymous with consciousness (citta) itself.
Broader Five aggregates (Buddhism, #HH3321) Phase–conditions of consciousness (Buddhism, #HM2599).
Narrower Eye consciousness (#HM2074) Ear consciousness (#HM2169) Body consciousness (#HM2562) Nose consciousness (#HM2364) Tongue consciousness (#HM2263) Mental consciousness (#HM2838).
Related Mental sensation (Buddhism, #HM6644) Dispositions of consciousness (Buddhism, #HM2098).

♦ **HM2557d Representational existence** (ICA)
Description This is the realization that history is riding on your back and that you live on behalf of all; you are aware that it is your life that is going on here and not some imitation. An example is what Jean–Paul Sartre wrote on the subject of knowing the world by daring to act and to change it.
Context In the ICA New Religious Mode in the arena of articulating the word about life (the life of knowing) the first formal aspect is one's experience of being a solitary individual or of being a self. At the third phenomenological level, this occurs when someone perceives in any experience its timeless authenticity.
Broader Solitary being (ICA, #HM2404).
Followed by Incarnate Christ (ICA, #HM2480).
Preceded by Horror of sin (ICA, #HM2720).

♦ **HM2558c Kriya–tantra awareness** (Buddhism, Tibetan)
Description This is the awareness is of the need to purify the elements of the practitioner's inner and outer world. In meditation, the ego is identified with the nature of a deity, verbalization with a mantram, and one's own locality with the corresponding pure realm of the god. It is an awakened state, yet one under guided initiation with prescribed ritual, chanted prayers and gift–offerings to the divinity and the teacher.
Context In Tibetan Sakya Buddhism this is one of the states in the "Ascension Stages Game". In some sets it is numbered 33 on the board.
Broader Tantric paths of accumulation and application in the ascension–stages–game (Buddhism, Tibetan, #HM2848).
Followed by Adamantine–hell awareness (Buddhism, #HM2028) Great–path tantra awareness (Buddhism, Tibetan, #HM2656) Middle–path tantra awareness (Buddhism, Tibetan, #HM3026) Rudra awareness of black freedom (Buddhism, Tibetan, #HM2603).
Preceded by Magical–forces awareness (Buddhism, Tibetan, #HM2679) Tantra–beginner awareness (Buddhism, Tibetan, #HM2452) Middle–path tantra awareness (Buddhism, Tibetan, #HM3026).

♦ **HM2559d Herd instinct**
Group feeling — Group formation
Description The desire to socialize with others is considered from a psychoanalytical viewpoint to be the result of childhood reaction to the hostility and jealousy felt towards siblings, and subsequent identification with them.
Related Instinct (#HM3517) Mass hysteria (#HM0574) Group unconsciousness (#HM2602) Cosmic identification (#HH0599).

♦ **HM2560c Herding–the–ox awareness** (Zen)
Description Keeping the mind on a tight rein prevents the confusion of endless trains of thought, the asserting of falsehood and self–deception.
Broader Oxherding pictures in Zen Buddhism (Zen, #HM2690).
Followed by Coming–home–on–the–ox's–back awareness (Zen, #HM3153).
Preceded by Catching–the–ox awareness (Zen, #HM2979).

♦ **HM2561d Personal violation** (ICA)
Description This is the horror of seeing that one is actively and passively involved in desecrating life; not only life in general but especially the people and situations one cares most about. It occurs on being forced to acknowledge that one is the cause of this disrelationship, as when someone who works for peace and disarmament encounters himself as the aggressor in family relationships.
Context In the ICA New Religious Mode in the arena of acting out one's freedom (the life of prayer) the first formal aspect is the struggle with repentance for one's sinfulness or confession. At the second phenomenological level this occurs on recognizing one's own full participation in one's fate.
Broader Affirmation (ICA, #HM2199) Struggle to confess (ICA, #HM2718).
Followed by Besetting sin (ICA, #HM2379).
Preceded by Painful acknowledgement (ICA, #HM2463).

♦ **HM2562g Body consciousness** (Buddhism)
Kayavijnana — Lus–kyi–rnam–par–shes–pa (Tibetan) — Touch awareness
Description This is the consciousness that apprehends objects of touch, whether the four elements (earth, water, fire or wind), or objects arising from those elements (smoothness, roughness, heaviness, lightness, cold, hunger, thirst). It is twofold, that is: where body consciousness is pleasant it is associated with indeterminate resultant consciousness with profitable result; where it is painful, it is associated with indeterminate resultant consciousness with unprofitable result.
Context One of the six consciousnesses defined in Buddhism as dependent on the individual senses, and with objects of sense as their focus.
Broader Sense consciousness (Buddhism, #HM2664) Perception through the senses (#HM3764) Mundane resultant consciousness (Buddhism, #HM0130) Sense mode of consciousness occurrence (Buddhism, #HM4389) Indeterminate consciousness in the sense sphere - resultant (Buddhism, #HM5721) Awareness of consciousness–group of conscious existence - senses and mind (Buddhism, #HM2556).
Related Contact (Buddhism, #HM2708) Body sense (Buddhism, #HM9123) Painful feeling (Buddhism, #HM8010) Pleasant feeling (Buddhism, #HM6722) Eye consciousness (Buddhism, #HM2074) Ear consciousness (Buddhism, #HM2169) Nose consciousness (Buddhism, #HM2364) Tongue consciousness (Buddhism, #HM2263) Mental consciousness (Buddhism, #HM2838) Mind–consciousness element (Buddhism, #HM6173).

♦ **HM2563f Depression**
Description Feelings of sadness, despondency, unresponsiveness and lack of drive are characteristic of depression, whether endogenous (hereditary), reactive (in response to stressful events), due to exhaustion (mental overstrain), or neurotic (arising from unresolved conflicts in the unconscious). There may be guilt superimposed on sadness. Together with a slowing of psychomotor activity and general lassitude there may be insomnia (or an increased need for sleep) and anorexia (or overeating). Depression may arise from a breakdown in self–esteem regulation due to loss of something or someone related to the individual's self–representation.
Related Cafard (#HM4344) Cyclothymia (#HM1616) Exhilaration (#HM0063).

♦ **HM2564e Logical dishonouring** (Brainwashing)
Description The first step in re–education, when the individual learns to see his whole previous life as a series of shameful acts not only in their opposition to the way of thought in which he is being trained but also to the very principles he then held dear. The negative parts of his identity which he previously de–emphasized are now given such emphasis that they swamp the positive aspects of the personality he once was. Profound shame and guilt are experienced as he confronts his human limitations. This existential guilt is used by his reformers to attack the basic meaning of his life, assisted by the intimate knowledge they have built up on him. His very vulnerability speeds the growth of a new personality.
Context A stage reached in thought reform, a system of organized, deliberate and total psychological training which effects individual change through two basic elements - *confession* (renouncing of past beliefs and attitudes) and *re–education* (remaking of the individual in the required image).
Broader Thought reform (Brainwashing, #HH0865).
Followed by Sense of harmony through progress (Brainwashing, #HM2149).
Preceded by Channelling of guilt (Brainwashing, #HM3204).

♦ **HM2565e Heaven awareness** (Psychism)
Description Consciousness on the astral plane based on forms of joy, following desire for rest, peace and happiness. Non–attachment has yet to be achieved and such awareness is enjoyed by the lower self.
Related Heaven (#HM4302).
Followed by Devachon awareness (Psychism, #HM2044).

♦ **HM2566b Dharana** (Hinduism)
Concentration
Description As *dharana*, concentration is the first stage leading to meditation and profound contemplation, when some object, symbol word or phrase of spiritual importance is the focus of attention. Concentration first on some concrete symbol is followed by concentration on the abstract which was symbolized and thence to inner illumination. Sustained dharana is *dhyana* - absorption or meditation.
Context The sixth component of the eightfold path of yoga. Together with dhyana and samadhi parinama, dharana is one of the three inner members of the eightfold path.
Refs Jyotir Maya Nanda, Swami *Concentration and Meditation* (1971).
Broader Samyama (#HH0374) Meditation (#HH0761) Eightfold path of yoga (Yoga, #HH0779).
Related Concentration (#HM4226) Samadhi parinama (Yoga, #HM3207) Concentration (Buddhism, #HM6663) Dhyana (Hinduism, Buddhism, #HM0137) Access concentration (Buddhism, #HM4999).

♦ **HM2567d Wakefulness** (Hinduism)
Description This is the state in which notions of "I" and "this" arise.
Context The second of seven descending steps of ignorance described in the Supreme Yoga, which veil *self–knowledge*. Each has innumerable sub–divisions.
Broader Veils of delusion (Hinduism, #HM3592).
Followed by Great wakefulness (Hinduism, #HM3055).
Preceded by Seed state of wakefulness (Hinduism, #HM2159).

♦ **HM2569c Path of disposing intelligence** (Judaism)
Description This state of awareness is represented as being clothed with the spirit and characterized as providing faith and a foundation for perseverance.
Context The seventeenth of the spiritual paths expressed in the Jewish Kabbalah.
Broader Paths of wisdom (Judaism, #HM2509).
Followed by Path of intelligence of influences (Judaism, #HM2045).
Preceded by Path of triumphant intelligence (Judaism, #HM2593).

♦ **HM2570d Ubiquitous otherness** (ICA)
Description This state occurs when an individual experiences that he is, for all time and in all places, bound to the inescapable power of the mystery that is present in his life; there is no where else to go; he is always on stage and can never be off stage. Movies such as "Three Days of the Condor" or "The Thirty–nine Steps" show situations in which people experience this inability to escape the situation. It may compared to G.M. Hopkins writing of the "hound of heaven" in one of his poems. There is an overwhelming awareness that the mystery is everywhere all the time, and there is nowhere to run and hide.
A dimension of this experience is a sense of awe, that is fear and fascination. The fear is that there will never be any let up; one always knows what is coming and is worn out by it. At the same time there is a fascination that it is never boring: one is intrigued by engaging with this awesome presence in one's mundane existence. Whether or not one continues trying to escape the presence of the mystery, one is aware of the futility of such a struggle. After such a time one knows one is never alone.
Context This state is number 6 in the ICA *Other World in the midst of this World.*
Broader Inescapable power (ICA, #HM3096).
Followed by Final limits (ICA, #HM2674) Cryptic disclosure (ICA, #HM2824) Relational situation (ICA, #HM2978) Universal compassion (ICA, #HM2734) Transformed existence (ICA, #HM2862).

Preceded by Incarnate living (ICA, #HM2394) Absurd existence (ICA, #HM2331)
Cryptic disclosure (ICA, #HM2824) Problemless living (ICA, #HM2747)
Universal compassion (ICA, #HM2734).

♦ **HM2571g Self-consciousness**
Consciousness of self
Description In the young child, the consciousness of self is a gradual and difficult achievement, although consciousness of a less personal order is no doubt present from birth. Until the child has a fairly definite conception of himself as an independent person he cannot conceptualize his relationship to the surrounding world, and hence lacks the subjective nucleus for the development of his own personality. Conditions responsible for the child's lack of self-consciousness include: deficiency in memory ability; ungraded and undifferentiated character of emotional responses; and deficiency in language.
Factors contributing to the growth of self-consciousness include: the fusion of sensory impressions (particularly around the kinesthetic sense of postural strain and position), possibly centering in the head; conscious recognition of recurring experiences; existence of anchorage points such as name, a peer group and characteristic clothing; and experiences of pain, frustration and social ridicule, which engender acute self-consciousness. More abstractly, it has been suggested that the locus of selfhood emerges as a recognition of that which lies between right and left, before and behind, above and below, and past and future.
Related Self-remembering (#HM2486).

♦ **HM2574d Duhkha** (Buddhism, Hinduism)
Dukkha (Pali) — Suffering — Unsatisfactoriness — Insufficiency
Description The normal awareness of creatures under delusion, *maya*, is of the transient as real; this results in suffering. Only the removal of this delusion can release the soul *jiva* from suffering and lead to liberation *mukti* and bliss *ananda*. Some Hindu traditions see release from suffering as possible in this lifetime, and with no rebirth after death. Buddhist tradition sees suffering, *dukkha*, as the inevitable result of indulging the senses, despite the apparent and transient delight or satisfaction such indulgence brings. Freedom from suffering entails rebirth into a heaven, achieved by stages of mental training, after which, despite a long sojourn there, the individual does eventually die. Final salvation in this case is to end life as an *arhat* or as a god in an immaterial heaven.
Three levels of dukkha are described: (1) *Dukkha-dukkha*, which is the fact of suffering inherent in the life process through birth, sickness, old age and death. (2) *Viparinama-dukkha*, which is the suffering of sentient creatures who are aware of the transitory nature of all things and of the gap between what may be desired and what may be obtained. (3) *Samkhara-dukkha*, which is the suffering inherent in human nature.
Context One of the three characteristics of the phenomenal world, the others being anicca (transitoriness) and anatta (non-self).
Refs Bowker, John *Problems of Suffering in the Religions of the World* (1970).
Broader Four noble truths (Buddhism, #HH0523).
Related Mukti (#HH0613) Suffering (#HM0471) Tanha (Pali, #HM3707)
Anatma (Buddhism, #HH0241) Anicca (Buddhism, #HM9021) Samsara (Hinduism, #HM1006)
Tragic suffering (#HH0708) Cessation of suffering (Buddhism, #HH2119)
Ananda (Hinduism, Buddhism, Yoga, #HM3227).

♦ **HM2576d Very-hot-hells awareness** (Buddhism, Tibetan)
Description These states of awareness are characterized by two degrees: ordinary hot-hell awareness and very-hot-hell awareness. They are experienced by offenders who misuse spiritual practices or hinder those engaged in them.
Context In Tibetan Sakya Buddhism this is one of the states in the "Ascension Stages Game". In some sets it is numbered 3 on the board.
Broader Desire-realm consciousness (Buddhism, #HM2733).
Followed by Cold-hells awareness (Buddhism, Tibetan, #HM2040)
Animal-hell awareness (Buddhism, Tibetan, #HM2636)
Crushing-hells awareness (Buddhism, Tibetan, #HM3142)
Temporary-hells awareness (Buddhism, Tibetan, #HM2454)
Interminable-hell awareness (Buddhism, Tibetan, #HM2052)
Hungry-ghosts-hell awareness (Buddhism, Tibetan, #HM2112).
Preceded by Demon-island awareness (Buddhism, Tibetan, #HM2055)
Howling-hells awareness (Buddhism, Tibetan, #HM2100)
Crushing-hells awareness (Buddhism, Tibetan, #HM3142)
Hindu-states of awareness (Buddhism, Tibetan, #HM2115)
Interminable-hell awareness (Buddhism, Tibetan, #HM2052).

♦ **HM2577d Religious function** (ICA)
Description This is the awareness that one is concerned with the spiritual well-being of society, that what one is doing with one's life is more than a job: there is dedication to the task. It occurs when one becomes aware of the effect one's life and work have on the task to which one has given one's self, as when deciding to work full time in the ecological movement or working to bring about the "green revolution" in Asia.
Context In the ICA New Religious Mode in the arena of acting out one's deed (the life of doing) the third formal aspect is one's awareness of one's allegiance to those who act on behalf of life as a whole. At the first phenomenological level this occurs when one is confronted by the fact that to live is to be of service.
Broader Awareness of allegiance (ICA, #HM2073).
Followed by People of God (ICA, #HM2484).

♦ **HM2578b Aspiration** (Buddhism)
Chhanda — 'Dun pa (Tibetan)
Description One of the five determining mental factors of Tibetan Buddhism, aspiration makes possible the generation of continuous effort and is one of the antidotes to laziness. Aspiration is said to have three types - the wish to meet, the wish not to separate and the wish to seek. It may be an emulation of the Buddha as he set out on his search for liberation from the cycle of birth and death.
As the first of the four fundamental principles of creative existence, aspiration is the decision, or evolution in the individual of the over-riding desire, to gain direct insight of and realize the eternal. In religion this is referred to as the love of God. It expands consciousness and results in selfless action.
Broader Creative existence (#HH0513)
Determining mental factors (Buddhism, #HH0170).
Related Love (#HH0258) Belief (Buddhism, Yoga, #HM2933).

♦ **HM2579d Asura-world awareness** (Buddhism, Tibetan)
Description This state of awareness is characterized by life in a city of light, and represented as a world of anti-gods who are jealous of the thirty-three upper divinities who cast them out of heaven.
Context In Tibetan Sakya Buddhism this is one of the states in the "Ascension Stages Game". In some sets it is numbered 15 on the board.
Broader Desire-realm consciousness (Buddhism, #HM2733).
Followed by Animal-hell awareness (Buddhism, Tibetan, #HM2636)
Demon-island awareness (Buddhism, Tibetan, #HM2055)
Howling-hells awareness (Buddhism, Tibetan, #HM2100)
Barbarian-state of awareness (Buddhism, Tibetan, #HM2091)
Hungry-ghosts-hell awareness (Buddhism, Tibetan, #HM2112)
Thirty-three-god-heaven awareness (Buddhism, Tibetan, #HM2606).
Preceded by Naga-world awareness (Buddhism, Tibetan, #HM2031)
Cold-hells awareness (Buddhism, Tibetan, #HM2040)
Bon-wisdom awareness (Buddhism, Tibetan, #HM2663)
Demon-island awareness (Buddhism, Tibetan, #HM2055)
Hindu-states of awareness (Buddhism, Tibetan, #HM2115)
Heavenly-highway awareness (Buddhism, Tibetan, #HM2010)
Barbarian-state of awareness (Buddhism, Tibetan, #HM2091)
Divine-animal-hell awareness (Buddhism, Tibetan, #HM2007)
Great-vehicle middle-path awareness (Buddhism, Tibetan, #HM2160)
Northern-continent awareness of community (Buddhism, Tibetan, #HM2067)
Eastern-continent awareness of noble figures (Buddhism, Tibetan, #HM2543).

♦ **HM2580d Sated attention**
Satiation
Description Attention may be diverted from an object due to an effect, or an emotional or negative value, that acts as an aversion. This can arise as a reaction against monotonous, repetitive sense-stimuli.
Broader Attention (#HM1817).

♦ **HM2581g Outward-flowing consciousness** (Yoga)
Bahir-vritti
Description Seeing personality and consciousness as constituted of the information passing through the senses to the brain, individuals with this type of awareness (by far the majority) derive their excitement from contact between the dense senses and even denser sense-objects. Being dependent on these exterior sense-objects, such individuals can never be free or self-dependent unless, through an act of will and perhaps through contact with an individual of *inward-flowing consciousness*, they reverse the outward flow of energy and realize their interior energies.
Related Inward-flowing consciousness (Yoga, #HM2738).

♦ **HM2582d Concerned judge** (ICA)
Description This is experienced as a voice within, which points out that one has really done an inhuman thing and asks what one is going to do about it. It occurs when a person is called to account for his actions and cannot escape the voice of accountability, as in Charles Dickens' story "A Christmas Carol", when Scrooge first encounters the ghosts and they point out to him how destructive all of his relationships have been.
Context In the ICA New Religious Mode in the arena of knowing one's internal socialty (the life of meditation) the second formal aspect is the experience of being called to accountability by one's interior priors. At the second phenomenological level, this occurs when a person finds his selfhood audited by that which commands his attention.
Broader Called to accountability (ICA, #HM2008).
Followed by Universal father (ICA, #HM2418).
Preceded by Unfailing prompter (ICA, #HM2356).

♦ **HM2583c Path of uniting intelligence** (Judaism)
Inductive intelligence of unity
Description This state of awareness is represented as the essence of glory, the consummation of truth of individual spiritual things.
Context The thirteenth of the spiritual paths expressed in the Jewish Kabbalah.
Broader Paths of wisdom (Judaism, #HM2509).
Followed by Path of illuminating intelligence (Judaism, #HM2035).
Preceded by Path of transparent intelligence (Judaism, #HM2107).

♦ **HM2584d Self transcendence** (ICA)
Description This state occurs whenever an individual realizes that he creates himself. There are no guidelines, no justifications, no excuses and no bounds to what he can create. From nothing, he invents who he is and what he can be. In the movie "Mask", the central character Rocky is a living embodiment of this reality. Born with a horribly disfigured face, he defies society's image of himself as an imbecile, a repugnant human being, and shows himself to be an intelligent, responsive and caring soul. Through his own decision to create who he is, he enables others to grasp hold of the possibility of doing the same with their lives.
A dimension of this experience is a sense of awe, that is fear and fascination. Shaping one's own life, one asks "Is this really me ?". There is a sense of marvel at what one is capable of becoming and yet had never quite glimpsed before. At the same time there is an incredible fear that it could all go up in smoke. The clay keeps shifting under one's feet and there are no guidelines or measures to tell one how to proceed. There is a feeling of being irreversibly changed, there is no going back to the person one was because one is forever becoming something new. The challenge is to appropriate the uniqueness that one is.
Context This state is number 19 in the ICA *Other World in the midst of this World.*
Refs Hammond, G B *The Power of Self-Transcendence* (1966).
Broader Authentic relation (ICA, #HM2316).
Followed by Cosmic sanctions (ICA, #HM2945) Contentless word (ICA, #HM2373)
Perpetual becoming (ICA, #HM2717) Temporal solidarity (ICA, #HM2363)
Contextual world-view (ICA, #HM2842).
Preceded by Cosmic sanctions (ICA, #HM2945) Ultimate reality (ICA, #HM2030)
Contentless word (ICA, #HM2373) External relation (ICA, #HM2471)
Passionate disinterest (ICA, #HM2547).

♦ **HM2585f Stress**
Description Referring to both physical and mental strain, stress occurs in response to some stimulus seen as endangering an individual's wellbeing or integrity, and in response to which protective action must be taken. Such stimuli may be short or long-term. The immediate response is an alarm reaction of shock and lowered resistance, followed by counter-shock, defensive mechanisms. On a long-term basis, adaptation produces optimal resistance which may finally be exhausted as energy reserves are depleted and energy available from other functions (such as digestion) is also reduced. Such long-term physiological stress reaction may lead to damaged tissue and disruption of adreno-cortical functioning; the sufferer has no behavioural escape mechanisms such as occur in neuroses.
Related Anomalies in experience of the unity of self (#HM9132).

♦ **HM2586d Fourth trance of the fine-material sphere** (Buddhism)
Appana samadhi (Pali) — Chaturthadhyana — Bsam gtan bzhi pa — Fourth form-realm concentration
Description *Hinayana Buddhism* This is the last trance in the fine-material sphere (rupaloka). It represents the mind's total overcoming of the lower and higher material realm perceptions of external objects by its absorption inward. Equanimity and concentration (samadhi) remain. The mind, in its first state of absorption, can now enter the immaterial world (arupaloka).

Tibetan Gelugpa Buddhism The ninth *path of release from below* is the fourth concentration; 7 stages of preparations have led to it. The fourth concentration eliminates the last of the 8 faults, those of inhalation, exalation and the feeling of bliss. In the 4 concentrations the main activity is that of analytical meditation. A feeling of sinking under the ground is generated by any of the concentrations. When the practioner passes beyond the fourth concentration and attains the formless absortions, feelings of bodily flying-off into space are generated.
Context This is the last of the 4 *form-concentrations* (dhyana) found in Tibetan Gelugpa Buddhism.
Broader Rupaloka consciousness (Buddhism, #HM2536)
Form-realm consciousness (Buddhism, #HM2257)
Trances and mental absorptions (Buddhism, #HM2122)
Tibetan meditative states of form concentrations (Buddhism, #HM2693).
Related Dhyana (Hinduism, Buddhism, #HM0137)
Dwelling in the fourth jhana (Buddhism, #HM8087).
Followed by Arupaloka consciousness (Buddhism, Pali, #HM2012).
Preceded by Third trance of the fine-material sphere (Buddhism, #HM2062).

♦ HM2587d Radical incarnation (ICA)
Description This is experienced as passionately doing one's job, as when a person discovers that throwing himself into his role in society can be an exciting adventure. It is like in the story "Good-bye, Mr. Chips", when the hero has to decide to be the headmaster of the school, even though this is not a role he has sought to play.
Context In the ICA New Religious Mode in the arena of acting out one's engagement (the life of obedience) the first formal aspect is embodying peace as the establishment to enable the ordering of human affairs. At the third phenomenological level, this occurs when people willingly embrace their responsibility to society.
Broader Embodying peace (ICA, #HM2876).
Followed by Universal prior (ICA, #HM2458).
Preceded by Submissive obedience (ICA, #HM2258).

♦ HM2588d Engagement to history (ICA)
Description The second formal aspect of the *life of doing* in the ICA *New Religious Mode*, this represents an extending of one's engagement ot include the whole of history.
Broader Life of doing (ICA, #HM3018).
Narrower Eternal moment (#HM2786)
Determining history (#HM2682)
Every situation (#HM2991)
Primordial sociality (#HM2877).

♦ HM2589b Intention (Buddhism)
Chetana — Cetana (Pali) — Sems-pa (Tibetan) — Volition
Description In Tibetan Buddhism this mental factor has the function of engaging the mind in its object, whether virtuous, non-virtuous or neutral, and thus unites and powers the minds and the mental factors. It may arise in respect to intended physical or verbal actions, or mental actions of intention. In Hinayana Buddhism it manifests as coordinating, is characterized by willing and has the function of accumulating. It drives associated states, causing them to be energetic in remembering urgent tasks, for example.
Context One of the five omnipresent mental factors defined in Tibetan Buddhism. One of the formations aggregate (mental coefficients) of Hinayana Buddhism, being listed among the constant states which appear in their true nature, and as general primary (always present in any consciousness).
Broader Omnipresent mental factors (Buddhism, Tibetan, #HH0320)
Awareness as mental-formation group of conscious existence (Buddhism, #HM2050).
Related Intention (Buddhism, #HM0705)
Karma awareness (Buddhism, #HM2048).
Being intended (Buddhism, #HM1412)

♦ HM2590d Crazy-monkey ego consciousness (Hinduism)
Lower manas and ahamkara
Description The consciousness of the lower mind, erratic and clever-animal-like, is a projection on the screen of manas (lower mind). It is composed of the perceptions of sense data and of memory, personalized as the experience of an ego by the action of ahamkara (I-am-this-ness).
Related Ego awareness (Hinduism, #HM2059)
Awareness of relative reality (Hinduism, Yoga, #HM2032).

♦ HM2591d Defender of deeps (ICA)
Description This is experienced as the decision to be a demonstration of profound living which will allow others to experience the profundity of their own lives; when a person surrenders his need for a rich spirit life in order to lead others. It may be compared to the guru who does not go to the profound centre of life because his task is to enable others to find it, as in the Buddha's return from enlightenment to service.
Context In the ICA New Religious Mode in the arena of knowing one's disengagement (the life of poverty) the fourth formal aspect is the experience of spiritual denial or sacrificial offering. At the fourth phenomenological level, this occurs when one is filled with the power of requiring nothing from any situation.
Broader Spiritual denial (ICA, #HM2174).
Preceded by Awareness of spiritual poverty (ICA, #HM2286).

♦ HM2592b Awe
Description Awe is the emotion felt by an individual in the presence of the numinous or mysterious. Such experiences of particular insight are common to those in the forefront of human knowledge, whether scientists, artists or mystics, and grow in depth and frequency with increasing understanding. Paradoxically, it is the very impossibility of explaining the mystery which is so evidently experienced that produces the sense of awe.
Related Wonder (#HM3197)
Reverence (#HM3023)
Sense of sacredness (#HM2302).

♦ HM2593c Path of triumphant intelligence (Judaism)
Path of eternal intelligence
Description This state of awareness is in the pleasure of divine glory. It is represented as a paradise prepared for the righteous.
Context The sixteenth of the spiritual paths expressed in the Jewish Kabbalah.
Broader Paths of wisdom (Judaism, #HM2509).
Followed by Path of disposing intelligence (Judaism, #HM2569).
Preceded by Path of constituting intelligence (Judaism, #HM2117).

♦ HM2595d Personal epiphany (ICA)
Description This is the experience or knowledge of being preposterously nominated to be the presence of God in the world. The decision is "I'm it. You want to know how life is? Look at me". It may be compared to Red McMurphy in "One Flew Over the Cuckoo's Nest" being the way for his fellows and taking upon himself the offense and persecution experienced by those who embody reality. It is the stance of the prophet Isaiah: "Here am I Lord, send me".
A dimension of this experience is a sense of awe, that is fear and fascination. The fear is not knowing where this could lead nor the price to be paid, which is sensed to be very high. The fascination is the chance to create one's life. There is a sense of power. The decision is to be both united with and open to the interior deeps of one's own life; to risk all one's being in living one's life. A style of audacious certainty develops with the knowledge of being a marked person.
Context This state is number 52 in the ICA *Other World in the midst of this World*.
Broader Radical illumination (ICA, #HM2273).
Followed by Creative futility (ICA, #HM2493)
Primordial wonder (ICA, #HM2186)
Exclusive contradiction (ICA, #HM2951).
Final blessedness (ICA, #HM2958)
Perpetual becoming (ICA, #HM2717)
Preceded by Contentless word (ICA, #HM2373)
Perpetual becoming (ICA, #HM2717)
Contingent eternality (ICA, #HM2456).
Final blessedness (ICA, #HM2958)
Sacramental universe (ICA, #HM2445)

♦ HM2596d State of universal cessation awareness (Buddhism)
Samvatta-kappa (Pali)
Description The world-dissolution period (pralaya, one of 4 aeons or kappas) culminates in the cessation of all consciousness except those of the Noble Ones (ariya, arahat).
Related Arhat sanctity awareness (Buddhism, Tibetan, #HM3024).
Preceded by Five aggregates (Buddhism, #HH3321).

♦ HM2598d Objective awareness (ICA)
Description This is experienced as seeing that the objective reality of life is that the word of life is in every situation; as being aware of the difference between saying that life is good and experiencing that life is good at every moment. An example is the father in the movie "The Hotel New Hampshire", who constantly picks up his life and lives it in spite of crushing blows.
Context In the ICA New Religious Mode in the arena of articulating the word about life (the life of knowing) the third formal aspect is one's experience of contentless transformation or the greatness of the word. At the first phenomenological level, this occurs when one sees through the events of everyday life and one's own subjectivity, to discern existence clearly.
Broader Contentless transformation (ICA, #HM3106).
Related Objective awareness (#HM2203).
Followed by Terrifying acceptance (ICA, #HM2460).

♦ HM2599c Phase-conditions of consciousness (Buddhism)
Paccaya — Vinnana-kicca (Pali)
Description The phenomena of consciousness are events in a network of preconditions or constraints, of concomitants and of resultant conditions. The paccaya show, in 24 phases, how any state of mental or physical phenomena is conditioned by these interdependent stages. They represent a key analytical element in Buddhist epistemology and are a logical paradigm or taxonomy applicable to classifying states and stages of consciousness.
Context While the paccaya may be an a priori excercise in classification, another system, also of an epistemological nature, seems to be experimentally or experientially based. The Functions of Consciousness (vinnana-kicca), unlike the paccaya network, appears to be strictly a time-successive, linear progression of 14 function activities involving the subconscious, the sense-consciousness, and those functions called, "mind-element", "mind-consciousness element", and "rootless-mind-consciousness element". This system is used to develop the 89 states of consciousness (vinnana-khanda).
Broader Phenomena awareness (Buddhism, #HM2551).
Narrower Psycho-physical faculties awareness (#HM2075)
Awareness of consciousness-group of conscious existence - senses and mind (#HM2556).
Related Dispositions of consciousness (Buddhism, #HM2098).

♦ HM2600f Delusive consciousness
Delusion
Description Every delusion is an incorrect understanding in its totality, although it may contain correct judgments in part. It implies belief based not on correct interpretation of wrong or partial information, but on incorrect interpretation of correct information due to erroneous reasoning. This may be from exaggerated scepticism, or from laziness of thought which refuses to consider or discuss certain subjects, or from mental illness. In the last case the false belief defies credibility and is not normally accepted by other members of the individual's culture or sub-culture. Then an individual holds, with complete conviction, a belief which is demonstrably false by the standards of his socio-cultural background. Typical examples are: belief that feelings, thoughts and impulses are being controlled from outside; exaggerated and grandiose sense of one's own importance; belief that the self or parts of the self do not exist; belief that the person or a group connected with him is being conspired against; belief that events or people in the vicinity are unusually significant.
True delusions remain firm in the face of any amount of persuasion, argument or even physical coercion and torture. Unlike normal beliefs, delusions are constantly ruminated upon and restated. New evidence is constantly sought. In the case of persecution delusions, the individual's life may be totalled geared to defensive measures or revenge.
Characteristically, delusions are wilful. They arise out of the desire for not knowing the truth oneself, or for keeping the truth from others. From the latter aspect, political history contains many instances of deluded citizenry, and of self-delusion among individual political leaders. Self-delusion is noted among leaders of all kinds. In the leader, the unconscious motive force for delusion is often ego-aggrandizement by satisfaction of power drives. In the follower, the unconscious motive for being deluded is ego-maintenance and the satisfaction of security needs. Leadership may be a role projected by followers on to someone who is invested with non-existent superior attributes.
Delusions are more likely to arise in people who have strong ego boundaries and who are rigidly detached from their environment. Delusion is contrasted with *illusion*, which is normally the result of inadequate or distorted sensory perception. It is only on refusal to accept rational explanation of erroneous sensory perception that illusion becomes delusion. Delusion and projection are often associated phenomena. Primary delusions are experiences of a change in significance and disorder. Secondary delusions, which may arise from primary delusion or from other disturbing phenomena, are an attempt to explain the primary experience.
The belief that delusion and projected (subjectively distorted) perceptions of reality are phenomena restricted to the mentally ill or the few, is itself a delusion. Although the delusions of the insane are sufficiently unreasonable to be more obvious, delusion in fact makes up a significant portion of the conscious understanding of everyday life. The state is typified by an overwillingness to accept unthinkingly some particular opinion and a refusal to change that opinion whatever facts or arguments are presented; and an unwillingness to focus at all on particular subjects considered too mysterious or holy.
Refs Reed, Graham *The Psychology of Anomalous Experience* (1988).
Narrower Delusions of love (#HM6276)
Delusions of guilt (#HM6036)
Delusions of poverty (#HM9701)
Delusions of grandeur (#HM9002)
Delusions of ill-health (#HM1440)
Primary delusions (#HM4604)
Secondary delusions (#HM2795)
Delusions of jealousy (#HM6993)
Passivity experiences (#HM7203)
Delusions of persecution (#HM7760)

Ignorance (Buddhism, Yoga, Zen, #HM3196).
Related Illusion (#HM2510) Hallucination (#HM4580) Psychotic state (#HM2887) Over-valued ideas (#HM1466) Illogical thinking (#HM1359).

♦ HM2601d Time consciousness

Description Consciousness of physical time is shown by the calendar-makings of all societies, based on the apparent motions of stars, sun, moon and planets. Other observations of physical time take in tidal movement, seasons and the growth and decay cycle of vegetation. Physical time may be experienced as an objective consciousness of time, of which one characteristic is the regularity of perception that the duration of an invariable event is always the same. However in subjective (personal or "psychological") consciousness of time, a day, for example, may appear to be longer or shorter.
Consciousness of biological time, in man, can be the awareness of his diurnal cycle, his biorhythms, and of other internal clocks. Objective physical or biological time may also be registered as a consciousness of its direction: irreversibly progressive or anomalously retrogressive or reversible. It may also be registered as a consciousness of its form: linear, non-linear; symmetrical, irregular; cyclical, helical; and also as a consciousness of its qualitative meaning or phase: constructive, destructive; upwards, downwards; evolute, involute; etc.
Time is, therefore, experienced quantitatively (duration of an interval, etc), qualitatively (phase, meaning etc), symbolically (form, "wave property") and logically (sequence, direction, causality). In all these modes the consciousness of time is subject to distortion. A possibility of non-distortion exists in the consciousness of sacred time, that is, the eternal. This state of time lies at the extreme end of the time arrow, in one direction towards total entropy, a universe of diffused energy where self-consciousness is only potential; and in the other direction to total negentropy, energy concentration and the consciousness of a god.
Narrower Time-gap experience (#HM0502) Consciousness of cyclical time (#HM2751)
Consciousness of the reversible direction of time (#HM2701)
Consciousness of the irreversible direction of time (#HM2651).
Related Kairos (#HM2749).

♦ HM2602d Group unconsciousness

Crowd consciousness — Collective mob mind
Description Unlike group meditation, brain-storming or other techniques such as dialogue, consciously practised by two or more of people to heighten their attention or awareness and develop their human potential, group activity without such an intent tends to allow for domination of the many by the few. In addition, such a group tends to be idea-poor, and to be guided by non-rational considerations that a skilful leader may manipulate. Groups can have the will of their individual members weakened and destroyed, starting with the initial surrender to authority, or to apparently innocent ideals such as mutual sharing and personal sacrifice to help others.
The fear of becoming a "groupie", an anomymous unit in a collective being, is real to many people. It is real to them in proportion to the weakness of their personalities. People who are described as "loners" and who particularly disdain participation in activities with others because of its group nature, simply show another kind of "unconsciousness". They constitute their own "group" characterized by individual isolation and resistance to worthwhile cooperation with others.
Refs Granmann, C F and Moscovici, S *Changing Conceptions of Crowd Mind and Behavior* (1985).
Related Herd instinct (#HM2559) Comoción (Spanish, #HM1404).

♦ HM2603d Rudra awareness of black freedom (Buddhism, Tibetan)

Mokshakala
Description This state is characterized as the power of yogism turned to selfish ends. It is represented as the lowest of the gods, Rudra, who was subjugated by a bodhisattva. Black freedom is criminally applied yogic power and abuse of Tantra.
Context In Tibetan Sakya Buddhism this is one of the states in the "Ascension Stages Game". In some sets it is numbered 16 on the board.
Broader Desire-realm consciousness (Buddhism, #HM2733).
Followed by Great-black-lord awareness (Buddhism, Tibetan, #HM2118).
Preceded by Kriya-tantra awareness (Buddhism, Tibetan, #HM2558).

♦ HM2604c Ox-forgotten-leaving-man-alone awareness (Zen)

Description The ox is seen only to be a symbol, the dharmas are one - it is not the means that are necessary but the end.
Broader Oxherding pictures in Zen Buddhism (Zen, #HM2690).
Followed by Ox-and-man-both-out-of-sight awareness (Zen, #HM2492).
Preceded by Coming-home-on-the-ox's-back awareness (Zen, #HM3153).

♦ HM2605c Concealment (Buddhism)

Mraksha — 'Chab-pa (Tibetan)
Description Through ignorance, the individual wishes to hide a fault from another.
Context One of the twenty *secondary afflictions* of Tibetan Buddhism.
Broader Secondary afflictions (Buddhism, Tibetan, #HH0781).
Related Ignorance (Buddhism, Yoga, Zen, #HM3196).

♦ HM2606d Thirty-three-god-heaven awareness (Buddhism, Tibetan)

Trayatrimsa
Description This state of human awareness is projected as the qualities ascribed to the deities in this heaven. They are powerful, endowed with supernatural sensory powers, yet are attached to pleasure. In this state one becomes a god, but only for a time before one is a dying god, destined for rebirth in a lower state when good karma is exhausted.
Context In Tibetan Sakya Buddhism this is one of the states in the "Ascension Stages Game". In some sets it is numbered 28 on the board.
Broader Desire-realm consciousness (Buddhism, #HM2733).
Followed by Cold-hells awareness (Buddhism, Tibetan, #HM2040)
Animal-hell awareness (Buddhism, Tibetan, #HM2636)
Hindu-states of awareness (Buddhism, Tibetan, #HM2115)
Southern-continent awareness of men (Buddhism, #HM2127)
Heaven-without-fighting consciousness (Buddhism, Tibetan, #HM2130)
Northern-continent awareness of community (Buddhism, Tibetan, #HM2067).
Preceded by Naga-world awareness (Buddhism, Tibetan, #HM2031)
Asura-world awareness (Buddhism, Tibetan, #HM2579)
Pratyeka Buddha awareness (Buddhism, Tibetan, #HM2180)
Non-emanating consciousness (Buddhism, Tibetan, #HM2070)
Divine-animal-hell awareness (Buddhism, Tibetan, #HM2007)
Wheel-turning-king awareness (Buddhism, Tibetan, #HM2058)
Discipleship-vision awareness (Buddhism, Tibetan, #HM2240)
Four-great-kings-heaven awareness (Buddhism, Tibetan, #HM2082)
Discipleship-application awareness (Buddhism, Tibetan, #HM2716)
Northern-continent awareness of community (Buddhism, Tibetan, #HM2067).

♦ HM2607d Impactful profundity (ICA)

Description This is discovering one has power to profoundly influence the lives of others by the clarity and passion with which one speaks. One's whole style of being emanates authority and authenticity. It may be compared to McMurphy in "One Flew Over the Cuckoo's Nest" as he breathes life into the dehumanizing environment of a mental institution, affecting both patients and staff with his presence. The Peter Sellers' movie "Being There" makes a spoof of this power as the childlike Chancy Gardener becomes the most respected voice in the United States, holding the nation spellbound by his supposed depth and lucidity.
A dimension of this experience is a sense of awe, that is fear and fascination. One experiences being forever on stage and utterly adequate for every moment. One has no power of one's self but life itself seems to be using one as an instrument for its purposes: whatever one says or does has a shocking accuracy about it. While discovering one has this charisma, one fears that one's own will might interfere; and has misgivings that people won't hear what one is saying and that one has set one's self up as a target.
Awareness of the power to transform another's life brings decision to live beyond one's own previously constructed boundaries, to dare to trust life's ability to flow through one and to speak with a sense of profound assurance. There is a firm belief that the "magic will work" and one is given courage to keep standing on centre stage.
Context This state is number 47 in the ICA *Other World in the midst of this World.*
Broader Transparent power (ICA, #HM2828).
Followed by Temporal solidarity (ICA, #HM2363)
Sacrificial passion (ICA, #HM2641)
Everlasting community (ICA, #HM2777)
Definitive effectivity (ICA, #HM2796)
Transcendent immanence (ICA, #HM3034).
Preceded by Interior discipline (ICA, #HM2851) Sacrificial passion (ICA, #HM2641)
Futuric responsibility (ICA, #HM3016) Passionate disinterest (ICA, #HM2547)
Transcendent immanence (ICA, #HM3034).

♦ HM2608c Non-harmfulness (Buddhism)

Avihimsa — Rnam-par-mi-'tshe-ba (Tibetan)
Description The essence of the teachings of Buddha, non-harmfulness wishes all sentient beings to be free from suffering and, as a component of *non-hatred*, is a patient lack of intention to injure.
Context One of the eleven *virtuous mental factors* referred to in Tibetan Buddhism.
Broader Virtuous mental factors (Buddhism, #HH0578).
Related Ahimsa (#HH0088) Non-hatred (Buddhism, #HM2744).

♦ HM2609d Awareness of futurity (ICA)

Description The third formal aspect of the *life of contemplation*, an arena in the ICA *New Religious Mode*, this is an experience of awesome "not-yet-ness".
Broader Life of contemplation (ICA, #HM2109).
Narrower Luminous change (#HM3065) Cut-off unknownness (#HM3167)
All-that's-yet-to-be (#HM3083) Frightful possibility (#HM3117).

♦ HM2610b Self-conscious consciousness

Apperceptive awareness
Description When the perceiver is aware that he is perceiving as well as being simultaneously aware of the object perceived, it is potentially a different state from perception without self-awareness since, by consciously willing attentiveness, perception comes be focussed and sharpened. An example of this is music playing in the background which is only subliminally heard, compared to the intent listening of a serious music lover. Self-conscious observation of an apperceptive nature can only be maintained with difficulty. Days, weeks and longer periods can go by in some people's lives without an experience of apperception. When it does occur it may last only for seconds. Contemplatives and yogis may train themselves for extended apperception, and some professions, such as acting and other performing arts, may require intermittent apperceptive states. Religious excercises and artistic expression are qualitatively raised by the self perceiving the self, so it is considered possible that more frequent recourse to this state by people generally would improve interpersonal relations and public life.
Related Apperception (#HM1961) Introspection (#HH0824).

♦ HM2611d Unveiled being (ICA)

Description This is experienced when one sees one's self in a mirror seemingly for the first time; when one experiences one's self doing something one simply can't believe one is doing. This revealing of a man to himself is described in John Fowles' novel "The Magus".
Context In the ICA New Religious Mode in the arena of articulating the word about life (the life of knowing) the first formal aspect is one's experience of being a solitary individual or of being a self. At the first phenomenological level, this occurs when a person sees through the events of everyday life and his own subjectivity to discern existence clearly.
Broader Solitary being (ICA, #HM2404).
Followed by Horror of sin (ICA, #HM2720).

♦ HM2612c Path of the illuminating intelligence (Judaism)

Description This state is represented as the crown of creation, exalted above every head and the splendour of unity.
Context The second among the spiritual paths expressed in the Jewish Kabbalah.
Broader Paths of wisdom (Judaism, #HM2509).
Related Sphere of the of the supreme crown (Kabbalah, #HM3132).
Followed by Path of sanctifying intelligence (Judaism, #HM3567).
Preceded by Path of the admirable intelligence (Judaism, #HM3013).

♦ HM2614b Other world in the midst of this world (ICA)

Description In an attempt to dilineate a system for spiritual and social renewal, a scheme of the "other world" of transcendence and mystery has been produced by the *Institute of Cultural Affairs (ICA)*, a group concerned at the present-day loss of the sense of human significance. Human life may be seen as a states of being indicated by the images of the *Land of Mystery*, the *River of Consciousness*, the *Mountain of Care* and the *Sea of Tranquillity*. The states are described in separate entries.
Refs Institute of Cultural Affairs (Eds) *The Other World* (1987).
Broader Maps of the mind (#HH8903).
Narrower Land of mystery (#HM2434) Mountain of care (#HM2170)
Sea of tranquillity (#HM3033) River of consciousness (#HM2993).
Related New religious modes (ICA, #HM3004).

♦ HM2615d Scorpio-consciousness (Astrology)

Self-sacrifice
Description This mode faces the transformative challenge of overcoming the attraction of self-centred desires, and leads to a sense of renewal and an appreciation of the whole. Under optimal conditions, such willing self-sacrifice of past attachments is associated with the emergence of natural spontaneity and awareness of higher goals and possibilities.
Context One of the twelve conditions of being or streams of divine energy encompassed within the zodiac. It is said that all twelve conditions must be present and playing its proper part for any successful and fulfilling action to take place. This may be through a group of individuals with different endowments and soul qualities working together; or it may be through one individual who,

by self-discipline, practice and meditation, has allowed all twelve conditions to arise within himself.
Broader Fixed awareness (Astrology, #HM2554)   Watery awareness (Astrology, #HM2384)
Zodiacal forms of awareness (Astrology, #HM2713).
Followed by Sagittarius-consciousness (Astrology, #HM2726).
Preceded by Libra-consciousness (Astrology, #HM2512).

♦ **HM2616d Historical vocation** (ICA)
Description This is experienced as the realization that no one work is going to define one's calling, rather the whole of one's life is encompassed in one's vocation; when one has to decide what one's life is all about. It may be compared with President John Kennedy of the U.S. saying "Ask not what your country can do for you, but what you can do for your country".
Context In the ICA New Religious Mode in the arena of knowing one's disengagement (the life of poverty) the second formal aspect is being reliant on the larger society or being disengaged from one's own work. At the second phenomenological level, this occurs when one takes charge of desires which formerly controlled one's life.
Broader Disengagement from work (ICA, #HM2552).
Related Vocation (Christianity, #HH0619).
Followed by Social failure (ICA, #HM2728).
Preceded by Beyond success (ICA, #HM2941).

♦ **HM2619d Aweful encounter** (ICA)
Impacted by mystery
Description This is an experiential trek in the *Land of Mystery*, within the ICA *Other World in the midst of this World*, when the individual is aware of being "up against" the ultimate. The individual states are described in separate entries.
Broader Land of mystery (ICA, #HM2434).
Narrower Absurd existence (#HM2331)   Ultimate reality (#HM2030)
Primordial wonder (#HM2186)   Radical contingency (#HM2477).
Related Infinite passion (ICA, #HM2234)   Transformed state (ICA, #HM2386)
Inescapable power (ICA, #HM3096).

♦ **HM2620d Boredom**
Enervated consciousness — Inactive awareness
Description There is a basic level of activity required in the field of conscious awareness to maintain mental health. A lack of stimulation, quantitative or qualitative can (as indicated by sensory deprivation experiments) produce severe, dissassociated and pathological mental states. Boredom, a milder form of deprivation, occurs when simple, undemanding, short-term tasks are repeated to the point of tedium. It may also be induced by rejection of stimuli based on a personal value system. In this state the individual is only partially aware of his surroundings and is barely conscious. There is loss of interest, wandering attention, low arousal; and working efficiency is impeded. A state in which it is not possible to stimulate the field of consciousness is pathological if involuntary. It may be achieved voluntarily, for example through meditation practices.
Related Uncertainty (#HM3051).

♦ **HM2621d Primal vocation** (ICA)
Description This is the experience of having one's name called by history and seeing one's life as having a sense of significant purpose or destiny. It is captured by Hermann Hesse's image of being a marked person and in Dag Hammarksjold's poetry: "Weep, Weep if you can, But don't complain, the Way has chosen you and you must be thankful". Perhaps General Patton experienced it as he stood on the plains of Carthage preparing to go into battle, sensing that he had been there before.
A dimension of this experience is a sense of awe, that is fear and fascination. One feels compelled to pursue one's quest, be it historical or personal. There is constant fear of losing one's nerve and giving up the task and one sometimes wonders if one is going psychotic. It becomes apparent that one is already a dead person and that ultimately no-one can hurt one. While being completely serious about honouring one's election, one also sees the final absurdity of whatever one does. One learns to trust history as one participates in creating it and feels in league with others who have gone before and trodden a similar path. There is a clear sense of the direction in which life is going and almost a sense of relief that one does not share others' ambivalence.
Context This state is number 28 in the ICA *Other World in the midst of this World*.
Broader Moral ground (ICA, #HM2721).
Related Vocation (Christianity, #HH0619).
Followed by Invented history (ICA, #HM3245)   Final blessedness (ICA, #HM2958)
Original integrity (ICA, #HM2773)   Perpetual becoming (ICA, #HM2717)
Destinal accountability (ICA, #HM2309).
Preceded by Cosmic sanctions (ICA, #HM2945)   Dynamic selfhood (ICA, #HM3091)
Final blessedness (ICA, #HM2958)   Perpetual becoming (ICA, #HM2717)
Archetypal humanness (ICA, #HM3112).

♦ **HM2623c Special insight and preparations as states** (Buddhism)
Vipashyana — Lhag mthong manaskara — Yid la byed pa — Samantaka — Nyer bsdogs (Tibetan)
Description This stage is based on the achievement of actual *calm abiding*; the concentrations and formless absorptions may be prepared for, and special insight (vipashyana) cultivated. Union of calm abiding and special insight with *emptiness* as object prepares the yogi for initial direct *cognition* of emptiness.
There are seven preparations (samantaka) or mental contemplations (manaskara) that are the means of attaining each of the four concentrations (dhyana) in the Form Realm. Clear conceptual perception of *suchness* is achieved during the *heat* stage of the preparation path; this is followed by a heightening of the perception in the *peak* period of the path, when one passes beyond sorrow at the instability or annihilation of roots of virtue. Then follows a period of *forbearance*. Here there is no contradistinction between the subject - mind of special insight, and object - emptiness of meditation. The yogi attains endurance and fearlessness with respect to emptiness. Finally there is the time of *supreme mundane qualities*, when the cognizing object can no longer be ascertained and the meditative stabilization so achieved precedes generation of the *path of seeing*.
Broader Selflessness (Buddhism, #HM2016)   Emptiness (Buddhism, Zen, #HM2193)
Meditative states of mental abidings (Buddhism, #HM2145).
Narrower States of special insight (#HM2207)   Belief-arising subtle contemplation (#HM2719)
Full-isolation subtle contemplation (#HM2743)   Final-training subtle contemplation (#HM2683)
Critical-analytical subtle contemplation (#HM2659)
Uninterrupted and release paths (Buddhism, Tibetan, #HM2015)
Individual knowledge of the character (Buddhism, Tibetan, #HM2195)
Subtle contemplation of withdrawal or joy (Buddhism, Tibetan, #HM2267).
Related Principal insights (Buddhism, #HM3630).
Followed by Tibetan meditative states of form concentrations (Buddhism, #HM2693).
Preceded by Calm abiding (Buddhism, #HM2147).

♦ **HM2624d Spontaneous religious experience**
Description An intense, individual flash of illumination which is overwhelmingly rapturous and unmistakenly enlightening. It is not confined to time, place, person or thing, and has been shown to happen to numbers of people. The experience seems to be no respecter of persons and may arise in someone who might be judged as unworthy and whose moral life is definitely not beyond reproach.
Refs Hardy, A *The Spiritual Nature of Man* (1979); Hardy, Sir Alister *Darwin and the Spirit of Man* (1984).
Broader Religious experience (#HM3445).
Related Induced religious experience (#HM3221).

♦ **HM2625d Utter awareness** (ICA)
Description This is an overwhelming awareness that one is obligated to all of creation and that one's every deed for better or worse affects all of life. It occurs when one understands that one's actions or inactions really do affect the lives and decisions of other people. To dare to act is to assume responsibility for history. It is like seeing famine victims from Africa on television and realizing they are part of one's world and that, if they are going to live, one has do something about it.
Context In the ICA New Religious Mode in the arena of acting out one's freedom (the life of prayer) the fourth formal aspect is one's experience of universal responsibility or intercession. At the first phenomenological level this occurs when one encounters the burden of one's own existence.
Broader Universal responsibility (ICA, #HM3110).
Followed by Particular concern (ICA, #HM2939).

♦ **HM2629c Path of intelligence of will** (Judaism)
Description This is awareness of the action of the will in spiritual unfoldment, with knowledge of primordial wisdom.
Context The twentieth of the spiritual paths expressed in the Jewish Kabbalah.
Broader Paths of wisdom (Judaism, #HM2509).
Followed by Path of intelligence of reward (Judaism, #HM2105).
Preceded by Path of intelligence of spiritual action (Judaism, #HM2021).

♦ **HM2630d Narcissistic equilibrium**
Description When ego and super-ego are in harmony, the ego not afraid of severe threats by the super-ego which does not demand more of the ego than it is capable of fulfilling, then a state of narcissistic equilibrium is experienced. This is likened to the harmony between loving parents and an obedient child.
Related Narcissism (#HH1066).

♦ **HM2631d Resurrectional existence** (ICA)
Description This is an experience of having been crucified, having been called from the comfort of the tomb to continue as a radically transformed human being. Such an individual carries with him the pain of all mankind; it is as though he walked through impenetrable barriers in this world for he lives another life, unblocked and unfettered by circumstances. It may be compared to Scrooge in Charles Dickens' story "A Christmas Carol" when, after he has been visited by the ghosts, he is transformed into a new man; or to Marilyn Monroe's comment in an interview following her marriage to Arthur Miller, that she would live her life exactly the same again if that was what it would take to bring her to this moment.
A dimension of this experience is a sense of awe, that is fear and fascination. Pain and joy are neither painful nor joyful but simply part of the unity of experience. There is fear of living in this world but one is driven to fulfil one's life, knowing that life and death are in God's hands. The fascination is that Being just never goes away. One has to decide to come back to this world or to embrace life after having given it up. There is a new sense of the vitality; one wants to bring others to this state.
Context This state is number 62 in the ICA *Other World in the midst of this World*.
Broader Endless life (ICA, #HM2437).
Followed by Problemless living (ICA, #HM2747)   Cryptic disclosure (ICA, #HM2824)
Worldly detachment (ICA, #HM2451)   Everlasting community (ICA, #HM2777)
Inclusive comprehension (ICA, #HM2256).
Preceded by Living death (ICA, #HM2808)   Problemless living (ICA, #HM2747)
Worldly detachment (ICA, #HM2451)   Interior discipline (ICA, #HM2851)
Spontaneous gratitude (ICA, #HM3025).

♦ **HM2632e Dasein**
Fascinans — Soul-loss — Being-in-the-world
Description The primitive realization, prior to true individuality and social acceptance of the individual soul, of nevertheless *being-in-the-world*. At this stage the ego is very fragile and may lapse into *fascinans*, or identification with some prepossessing external. Such a lapse may be referred to as *soul-loss*. It is said to be involved in all mental disturbances and to be resisted by anxiety. Magical powers, such as possessed by a *shaman*, will protect a community against soul-loss. This is the basis of ceremonies in which the shaman first allows himself to be overwhelmed by fascinans, identifying himself with the possible source of soul-loss; and then, through dasein, overwhelms the fascinans and passes this insight on to the community.
Dasein as transcendental imagination is not a subject but a trans- or pre-subjective self preceding the dichotomy of subject and object and rendering it possible. Such imagination is then an indeterminate unknown that lies between the subject and the object.
To be means to be open, unhidden, and Dasein means to be the place of this openness and unconcealment. Dasein is therefore primarily not man at all but the place of the presence and revelation of Being. Dasein is the truth of Being, namely Being in its openness. It is a basic misunderstanding to emphasize Being as a noun, suppressing its significance as a verb. Being is not a vacuous concept standing for something remote or abstract. Rather it is the most concrete and closest of presences. Ordinary human life moves within a preconceptual understanding of Being. Dasein's being-in-the-world is a unitary phenomenon, a primary datum within which the question of whether the world exists apart from the knowing subject simply does not arise.
Refs Avens, Roberts *The New Gnosis* (1984).
Related Loss of soul (#HH0210)   Embodied consciousness (#HM2499).

♦ **HM2635b Eight liberations** (Buddhism)
Description These are the eight means by which a being is liberated from the manifest activity of specific afflictions, although complete cessation (which includes freedom from the potential of the affliction) is not included in such liberation. These are: embodied looking at form; formless looking at form; beautiful form; infinite space; infinite consciousness; nothingness; peak of cyclic existence; equipoise of cessation.
Context One of the paths of calming according to Buddhist teaching.
Refs Hurvitz, Leon *The Eight Liberations* (1979).
Broader Paths of calming (Buddhism, #HM2987).

♦ **HM2636d Animal-hell awareness** (Buddhism, Tibetan)
Description This state of awareness is characterized by the appetites and dispositions of animals. It is represented as a hell where the soul becomes animal-formed: serpents come from anger,

lions from pride, dogs from arrogance, and the horse from unpaid debts. Other animals correspond to other faults.
Context In Tibetan Sakya Buddhism this is one of the states in the "Ascension Stages Game". In some sets it is numbered 11 on the board.
Broader Desire-realm consciousness (Buddhism, #HM2733).
Related Form-realm awareness (Buddhism, Tibetan, #HM2142).
Followed by Naga-world awareness (Buddhism, Tibetan, #HM2031)
Crushing-hells awareness (Buddhism, Tibetan, #HM3142)
Southern-continent awareness of men (Buddhism, #HM2127)
Divine-animal-hell awareness (Buddhism, Tibetan, #HM2007)
Hungry-ghosts-hell awareness (Buddhism, Tibetan, #HM2112)
Four-great-kings-heaven awareness (Buddhism, Tibetan, #HM2082).
Preceded by Cold-hells awareness (Buddhism, Tibetan, #HM2040)
Naga-world awareness (Buddhism, Tibetan, #HM2031)
Asura-world awareness (Buddhism, Tibetan, #HM2579)
Very-hot-hells awareness (Buddhism, Tibetan, #HM2576)
Crushing-hells awareness (Buddhism, Tibetan, #HM3142)
Formless-realm awareness (Buddhism, Tibetan, #HM3144)
Hindu-states of awareness (Buddhism, Tibetan, #HM2115)
Temporary-hells awareness (Buddhism, Tibetan, #HM2454)
Heavenly-highway awareness (Buddhism, Tibetan, #HM2010)
Hungry-ghosts-hell awareness (Buddhism, Tibetan, #HM2112)
Divine-animal-hell awareness (Buddhism, Tibetan, #HM2007)
Barbarian-state of awareness (Buddhism, Tibetan, #HM2091)
Discipleship-karma awareness (Buddhism, Tibetan, #HM2192)
Thirty-three-god-heaven awareness (Buddhism, Tibetan, #HM2606)
Great-vehicle lower-path awareness (Buddhism, Tibetan, #HM2268)
Western-continent awareness of cattle (Buddhism, Tibetan, #HM2519)
Eastern-continent awareness of noble figures (Buddhism, Tibetan, #HM2543).

♦ **HM2637d Awareness of externality** (ICA)
Description The first formal aspect of the *life of contemplation*, an arena in the ICA *New Religious Mode*, this represents the experience of enigmatic "not-me-ness".
Broader Life of contemplation (ICA, #HM2109).
Narrower Hallowed honour (#HM2680)
Being all the other (#HM2983)
Dangerous intrusion (#HM2449)
Everlasting inescapability (#HM2791).

♦ **HM2638d Jealousy** (Buddhism)
Irshya — Phrag-dog (Tibetan) — Envy (Pali)
Description In Tibetan Buddhism, jealousy is a deep disturbance of the mind involving hatred towards another for his good fortune, and stemming from an attachment to worldly goods. In Hinayana Buddhism, envying is that which envies. It is regarded as a fetter, as something which binds. It has the characteristic of being jealous of another's good fortune; its function is dissatisfaction and lack of delight in such good fortune. It manifests as turning away from and being adverse to the prosperity of others; the proximate cause is such success of another.
Context One of the twenty *secondary afflictions* of Tibetan Buddhism. One of the formations aggregate (mental coefficients) of Hinayana Buddhism, being listed among the inconstant states which are immutable by nature, and as unprofitable secondary (sometimes present in any unprofitable or unprofitable-resultant consciousness).
Broader Secondary afflictions (Buddhism, Tibetan, #HH0781)
Awareness as mental-formation group of conscious existence (Buddhism, #HM2050).
Related Envy (Christianity, #HM3098).

♦ **HM2639d Bon-practitioner state of awareness** (Buddhism, Tibetan)
Magical-shamanic state
Description This state of awareness is characterized by shamanic powers, magic, communion with the dead and flight through the air. It is represented by the pre-Buddhist Tibetan faith of Bön, considered to be inferior only to the Dharma. Its former blood-rites lead to the hells, its reformed practices lead to high states of consciousness.
Context In Tibetan Sakya Buddhism this is one of the states in the "Ascension Stages Game". In some sets it is numbered 23 on the board.
Broader Religious traditions in the ascension stages game (Buddhism, Tibetan, #HM3341).
Followed by Bon-wisdom awareness (Buddhism, Tibetan, #HM2663)
Demon-island awareness (Buddhism, Tibetan, #HM2055)
Howling-hells awareness (Buddhism, Tibetan, #HM2100)
Four-great-kings-heaven awareness (Buddhism, Tibetan, #HM2082)
Great-vehicle lower-path awareness (Buddhism, Tibetan, #HM2268)
Heaven-without-fighting consciousness (Buddhism, Tibetan, #HM2130).
Preceded by Four-great-kings-heaven awareness (Buddhism, Tibetan, #HM2082)
Heaven-without-fighting consciousness (Buddhism, Tibetan, #HM2130).

♦ **HM2640e Mind-awake body-asleep trance** (Psychism)
Unmani
Description In esoteric yoga practices it is claimed that the body can be put into a special condition of yogic sleep at any time. This may be akin to self-hypnotizing techniques developed in the West. One purpose of yogic sleep, however, is said to allow the adept of this practice to leave his or her body safely for limited periods of time. The American clairvoyant healer Edgar Cayce entered a kind of yogic sleep trance in order to diagnose illnesses.
Related Double-awake body-asleep trance (#HM2079).

♦ **HM2641d Sacrificial passion** (ICA)
Description This is being consumed by giving one's life to something new, to the point of feeling totally drained. Pouring one's heart out, one often finds one's self weeping with an indefinable sadness. In his book "Cry, the Beloved Country", Alan Paton depicts this state as he describes the pain of living under apartheid. In the movie "The Killing Fields", the American and the Cambodian reporters stand in this position as they decide to stay on in Cambodia after the fall of the Pnom Penh government. And it is reflected in the story of Jesus weeping over Jerusalem before his final entry into the city.
A dimension of this experience is a sense of awe, that is fear and fascination. Being consumed by care is like being on fire. One becomes one with the pain of innocent human suffering, not just with the particular concerns that are facing them. One is confronted by the temptation to succumb to the pain and become part of the problem instead of being part of the solution. Realizing the extent of the damage one could do, one nevertheless feels compelled to extend one's effort even more.
Caring with all one's heart brings an almost levitational quality in everything one does. Something beyond the normal sources of sustenance such as eating and sleeping keeps one going. One lets go of the props that supported one, the role one is playing and even the belief in one's own capacity to play it. There is a sense of being larger than life and more powerful than one had ever known one's self to be.
Context This state is number 39 in the ICA *Other World in the midst of this World.*
Broader Universal concern (ICA, #HM2774).
Followed by Final limits (ICA, #HM2674)
Futuric responsibility (ICA, #HM3016)
Soteriological existence (ICA, #HM2904).
Impactful profundity (ICA, #HM2607)
Transcending hostility (ICA, #HM2658)
Preceded by Final limits (ICA, #HM2674)
Universal compassion (ICA, #HM2734)
Contextual world-view (ICA, #HM2842).
Temporal solidarity (ICA, #HM2363)
Impactful profundity (ICA, #HM2607)

♦ **HM2642c Manifestation of the essence through veils** (Sufism)
Description The essence has, in itself, no name or attribute, no relationship or proportion. The Self, whose light is hidden in manifestation, is manifest by drawing veils in front of him. As the sun is seen more clearly through a veil of cloud, so the essence manifests through different veils. In these different manifestations, resemblances and relationships are multiplied.
Context The sixteenth illumination of Jami.
Broader Fountains of light (Sufism, #HM3039).
Followed by First individuation of essence (Sufism, #HM3614).
Preceded by Identity of attributes with their essence (Sufism, #HM3260).

♦ **HM2643c Path of perpetual intelligence** (Judaism)
Description This state of awareness is of the laws that keep the movements of the sun and moon regulated.
Context This awareness of the laws is the thirty first of the spiritual paths expressed in the Jewish Kabbalah.
Broader Paths of wisdom (Judaism, #HM2509).
Followed by Path of administrative intelligence (Judaism, #HM2095).
Preceded by Path of collective intelligence (Judaism, #HM3046).

♦ **HM2644d Universal Christ** (ICA)
Description This is experienced as the discovery that in every situation one is the transforming agent, that no matter what is happening, every situation contains the possibility for transformation. An example is when the main character in the movie "Cool Hand Luke" gets the chain gang to do a very hard job that should have taken hours in just a few minutes, and to have fun while they are doing it.
Context In the ICA New Religious Mode in the arena of articulating the word about life (the life of knowing) the third formal aspect is one's experience of contentless transformation or the greatness of the word. At the fourth phenomenological level, this occurs when one recognizes the humiliation of standing before the final mystery of life in every moment.
Broader Contentless transformation (ICA, #HM3106).
Preceded by Classical story (ICA, #HM2757).

♦ **HM2650b Savikalpa samadhi** (Hinduism)
Description In this state one loses all human consciousness for a short time, and perception of time and space is utterly different. One looks into a different world and sees that practically everything is already done. One is only an instrument. But everybody has to return from this samadhi back into ordinary human consciousness. There are various degrees of Savikalpa samadhi; different ideas and thoughts are experienced without disturbance of meditation. One step further, in Nirvikalpa samadhi, all thought ceases. Savikalpa samadhi is said to be synonymous with the samprajnata samadhi of yoga.
Broader Samadhi (Hinduism, Yoga, #HM2226).
Related Samprajnata samadhi (Yoga, #HM2896)
Indian approaches to consciousness (Hinduism, #HM2446).
Followed by Nirvikalpa samadhi (Hinduism, #HM2061).

♦ **HM2651g Consciousness of the irreversible direction of time**
Description People who have made errors in their past are accutely aware that the past cannot be relived. Those who are mature are conscious that their youth cannot be brought back. The entire world races against the clock to get done what it can before time passes. Physicists (based on Einstein) sometimes assert, while noting its implications for psychology, that time is irreversible at the fundamental, quantum level of elementary atomic processes and that this irreversible, progressive arrow of time is associated with the constant increase in entropy.
If entropy (that is, energy-release with loss of order and information) is universally progressive consciousness may be said (according to information theory) to suffer diminution or dilution of content. Thus, at various levels of consciousness, loss may be progressively initiated. For example, if progressive loss of information content begins at the level of biological time, specific instincts, biorhythms, and internal clocks (control activities) are lost. If this entropy of consciousnessness progresses to the subconscious level memory may be lost, along with control or master elements consisting of information structures (archetypes, patterns, clusters, etc), which also include language. At the conscious level, entropy loss would affect cognition and recognition. Sensation would lose its meaning and one would be unaware.
The emptying out of time and information from the instincts, the subconscious, and the consciousness seems to have the same effects as mystical concentration and yoga practices. These attempts at human development through attainment of altered states of consciousness and ego emptying seem to accelerate entropy and hasten time as they concurrently release energy.
Broader Time consciousness (#HM2601).

♦ **HM2653c Beyond-the-third-realm consciousness** (Buddhism, Tibetan)
Description This is the realm beyond cyclic existence. It is Buddhahood and beyond.
Broader Consciousness states in cyclic existence (Buddhism, #HM2177).
Related Lokuttara consciousness (Buddhism, #HM2120)
Way of the Buddhas (Buddhism, Tibetan, #HM2185).

♦ **HM2654d Revered hero** (ICA)
Description This is experienced as recalling the life of another person whom one feels one should follow, when a one becomes aware that one is being guided by someone else. It occurs when an individual encounters someone whom wishes to emulate, as in the book "Goodbye Mr Chips", when a former student recalls the life of a great teacher who shaped the course of his and many other students' lives.
Context In the ICA New Religious Mode in the arena of knowing one's internal sociality (the life of meditation) the third formal aspect is one's experience of being given existential guidance or encountering the Saints. At the first phenomenological level, this occurs when that which is other intrudes upon one's consciousness.
Broader Existential guidance (ICA, #HM2114).
Related Hero worship (#HH0653).
Followed by Scorching avatar (ICA, #HM2921).

♦ **HM2655b Mental wisdom** (Buddhism)
Prajna — Shes-rab (Tibetan) — Overcoming doubt — Knowledge
Description By differentiating both the faults and the virtues of what is observed, such wisdom generates certainty. This is one of the five *determining mental factors* of Tibetan Buddhism.
Broader Determining mental factors (Buddhism, #HH0170).
Related Essential wisdom (#HM0107)
Understanding (Buddhism, #HM4523).

♦ **HM2656c Great-path tantra awareness** (Buddhism, Tibetan)
Mahamudra
Description This state of awareness is characterized by identification with the Buddhas of the 5

tantric Buddha families. It is said to be achieved after the absolute truth of emptiness has been realized and the apparently personal mental elements (skandas) have been purified. The mind is represented in this stage as generating the divine mandala of the 5 high Buddhas who are addressed with the appropriate 5 gestures (mudras). At this stage there is no more regression into lower paths.
Context In Tibetan Sakya Buddhism this is one of the states in the "Ascension Stages Game". In some sets it is numbered 42 on the board.
Broader Tantric paths of accumulation and application in the ascension-stages-game (Buddhism, Tibetan, #HM2848).
Followed by Tantra heat-application awareness (Buddhism, Tibetan, #HM2696)
Tantra climax-application awareness (Buddhism, Tibetan, #HM2220)
Kalacakra-tantra shambhala awareness (Buddhism, Tibetan, #HM2151).
Preceded by Demon-island awareness (Buddhism, Tibetan, #HM2055)
Kriya-tantra awareness (Buddhism, Tibetan, #HM2558)
Lord-of-death awareness (Buddhism, Tibetan, #HM2088)
Potala-island awareness (Buddhism, Tibetan, #HM2175)
Middle-path tantra awareness (Buddhism, Tibetan, #HM3026)
Kalacakra-tantra shambhala awareness (Buddhism, Tibetan, #HM2151)
Tantra master in form-realm awareness (Buddhism, Tibetan, #HM2235)
Tantra master-in-sense-realm awareness (Buddhism, Tibetan, #HM2211)
Mahayana receptivity-application awareness (Buddhism, Tibetan, #HM2247).

♦ **HM2657b Ten worlds** (Buddhism)
Mutual possession of the ten worlds
Description These are the states of life described in Nichiren Soshu Buddhism. The first six arise spontaneously without any effort, but the four noble paths require inner effort. All ten worlds are possessed by everyone, but each individual tends to have one dominating, for one anger, another bodhisattva, and so on. Any activity under the influence of another world with tend to be coloured by the dominant world, for example compassionate acts by an anger-dominated person will tend to be accompanied by feelings of superiority. Thus each of the ten worlds may be said to possess the others. Mutual possession also explains why one can move from one state to another. The world of hell contains the world of rapture, for example, so that good news can still cheer one in a hellish state. All worlds not being experienced are nonetheless latent, in a state of *ku*.
Broader Nichiren shoshu buddhism (Buddhism, #HH3443).
Narrower Six paths (#HM1914) State of hell (#HM4282) State of anger (#HM2959)
State of hunger (#HM0150) Four evil paths (#HM1252) State of rapture (#HM1973)
Three evil paths (#HM0923) Four noble paths (#HM4026) State of learning (#HM3662)
State of animality (#HM0847) State of buddhahood (#HM1873)
State of realization (#HM0450) State of bodhisattva (#HM1225)
State of tranquillity (#HM3492).
Related Ku (Buddhism, #HM0581).

♦ **HM2658d Transcending hostility** (ICA)
Description This is the experience of having no earthly foes, nothing to hate, of ending one's private war with being, of being are eternally "in the other guy's shoes". Hatred becomes a luxury one can no longer afford. In the realization that God knows what is necessary, so as not to fall into passivity, there is memory that one alone is responsible for discerning and acting upon His will. This may be the experience of many, both Japanese and American, as they honour the anniversary of the atomic bomb blast in Hiroshima. It may be compared to eulogizing the deceased at a funeral.
A dimension of this experience is a sense of awe, that is fear and fascination. The fear is that hatred will return. The fascination is in the strange quality unlike anything previously experienced and that hate is simply no longer in one's being. There is a sense of having been cleansed. The decision is to trust in the awe which leads to detachment from relations and from hatred which comes from one's care. There is a new, near ruthless individuality to live one's own life.
Context This state is number 55 in the ICA *Other World in the midst of this World.*
Broader Unknowable peace (ICA, #HM3015).
Followed by Final limits (ICA, #HM2674) Blissful seizure (ICA, #HM3148)
Everlasting community (ICA, #HM2777) Contextual world-view (ICA, #HM2842)
Exclusive contradiction (ICA, #HM2951).
Preceded by Contentless word (ICA, #HM2373) Problemless living (ICA, #HM2747)
Sacrificial passion (ICA, #HM2641) Everlasting community (ICA, #HM2777)
Contextual world-view (ICA, #HM2842).

♦ **HM2659d Critical-analytical subtle contemplation** (Buddhism)
Mimamsamanaskara — Dpyod-pa-yid-byed (Tibetan)
Description This state of contemplation requires a critical awareness of remaining attachments in the mental aggregates, and is used to spur release from the Desire Realm.
Context This is the sixth of the preparations in Tibetan Gelugpa Buddhism.
Broader Special insight and preparations as states (Buddhism, #HM2623).
Related Critical awareness (#HM3609).

♦ **HM2660b States of understanding**
Comprehension
Description Understanding or comprehension is the end for which awareness, cognition and the entire structure and activity of consciousness exists. One may postulate conditions in which proportions could be developed between understanding and related conscious states. Among those in states of everyday waking consciousness, for example, different degrees of understanding of the same phenomena exist. Therefore higher states of consciousness, or heightened sense perception, or heightened perception (cognition or apprehension) by other faculties, including the alleged super-sensory, would exhibit variance in levels of correlated understanding as well. This suggests the explanation for ignorant yogis, saints and mystics (and some charismatic public figures) whose understanding has not benefited from "higher" or altered states of consciousness, yet who have gained the power to influence others. The key mental activity that correlates consciousness and comprehension with degrees of proportion is Apollonian reason (ratio). An altered state of consciousness without reason may be Dionysian, aimless frenzy.
Understanding, however, is only partly a product of a particular state of consciousness and a process of reason. At its deepest level it may also be influenced by the dispositions of the personality and the will, that is, the intentions to which the individual life-force has been focussed. Wisdom, the state of fullest understanding, may also be influenced by the degree of empathy attained with the wholes of which the personality is part: the true self; the selves of others; and the universe of life, "nature", consciousness and purpose of which mankind is a member.

♦ **HM2661d Ethical existence** (ICA)
Description This is experienced as the question of how all of one's life is going to be related to all of life. It occurs on seeing that one's concern is for the spirit dimension of life and that one is related to all of history, as when Eleanor Roosevelt was a delegate to the United Nations and worked to create the UN Declaration of Human Rights for all the people of the world.
Context In the ICA New Religious Mode in the arena of acting out one's engagement (the life of obedience) the fourth formal aspect is embodying service to people's spirit life. At the first phenomenological level, this occurs when a person becomes aware that he is bound unto death to other people.
Broader Embodying service (ICA, #HM3199).
Followed by Global brotherhood (ICA, #HM2432).

♦ **HM2662c Hara** (Japanese)
Description Hara is that state in which the individual has found his primal centre, and has proven himself by it. The concept is associated with the practice of Tai Chi Ch'uan, Aikido and Zen. It constitutes a state of perfect calm around which there may be great activity. When in this state, a person is able to move from a point of harmony to permit an opponent's own energy to defeat him, or to permit the person to carry out a delicate yet steady series of movements for a long time. The term "hara" is also used to denote the physical vital centre of the body, slightly below the navel.
Refs Durckheim, Karlfried *Hara* (1962); Yamaoka Haruo *Meditation Gut Enlightenment.*
Related Aikido (#HH0252) T'ai Chi Ch'uan (#HH0282)
Zazen meditation (Zen, #HH0882).

♦ **HM2663d Bon-wisdom awareness** (Buddhism, Tibetan)
Nyamsrtsal — Shen-siddhi
Description This state of awareness is repersented by the Bön adept. It is characterized in tantric terms as the union of male and female, the female being the wisdom (vidya) by which creation was originally emanated from mind, and which is possessed by the Bön practitioner. Among the non-Buddhist, Himalayan-region religions, Bön excels according to the Tibetan texts.
Context In Tibetan Sakya Buddhism this is one of the states in the "Ascension Stages Game". In some sets it is numbered 65 on the board.
Broader Masters of wisdom in the ascension stages game (Buddhism, Tibetan, #HM3066).
Followed by Asura-world awareness (Buddhism, Tibetan, #HM2579)
Pratyeka Buddha awareness (Buddhism, Tibetan, #HM2180)
Temporary-hells awareness (Buddhism, Tibetan, #HM2454)
Great-vehicle lower-path awareness (Buddhism, Tibetan, #HM2268).
Preceded by Bon-practitioner state of awareness (Buddhism, Tibetan, #HM2639).

♦ **HM2664b Sense consciousness** (Buddhism)
Sensual awareness
Description Six sense consciousnesses are enumerated in Buddhism: eye consciousness supported by the eye sense powers with visible forms as objects; ear consciousness supported by ear sense powers with sounds as objects; nose consciousness supported by nose powers with odours as objects; tongue consciousness supported by tongue sense powers with tastes as objects; body consciousness supported by the body sense powers with tangible objects as objects; and mental consciousness supported by mental sense powers with phenomena as objects. These give rise to six feelings (for example, from contact on the "aggregation of an ear sense", a sound and an ear consciousness).
That part of consciousness referred to as *mind* (chitta) is aware of an object as an entity; this knowledge is interpreted by the other part of consciousness, the *mental factor* (chaitta, sems-byung), which engages in it with respect to other features such as function. There are six types of mind corresponding to the six sense consciousnesses, and 51 mental factors classed in six groups: omnipresent factors (5); determining factors (5); virtuous factors (11); root afflictions (6); secondary afflictions (20); and changeable factors (4).
Broader Consciousness in Buddhism (Buddhism, #HM2200).
Narrower Eye consciousness (#HM2074) Ear consciousness (#HM2169)
Nose consciousness (#HM2364) Body consciousness (#HM2562)
Tongue consciousness (#HM2263) Mental consciousness (#HM2838).
Related Sensory awakening (#HM2972) Foundation awareness (#HM0650)
Root afflictions (Buddhism, #HH0270) Perception through the senses (#HM3764)
Delusion consciousness (Buddhism, #HM1119) Virtuous mental factors (Buddhism, #HH0578)
Changeable mental factors (Buddhism, #HH0910)
Sense-sphere concentration (Buddhism, #HM1097)
Secondary afflictions (Buddhism, Tibetan, #HH0781)
Secondary mental factors (Buddhism, Tibetan, #HH1348)
Omnipresent mental factors (Buddhism, Tibetan, #HH0320)
Sense mode of consciousness occurrence (Buddhism, #HM4389).

♦ **HM2665d Aries-consciousness** (Astrology)
Undifferentiated potentiality
Description A condition of willed thought associated with desire for creative action as a means of self expression. Under optimal conditions this will lead to a clear vision of goals, constraints and problem boundaries.
Context One of twelve conditions of being or streams of divine energy encompassed within the zodiac. It is said that all twelve conditions must be present and playing its proper part for any successful and fulfilling action to take place. This may be through a group of individuals with different endowments and soul qualities working together; or it may be through one individual who, by self-discipline, practice and meditation, has allowed all twelve conditions to arise within himself.
Broader Fiery awareness (Astrology, #HM2124) Cardinal awareness (Astrology, #HM2259)
Zodiacal forms of awareness (Astrology, #HM2713).
Followed by Taurus-consciousness (Astrology, #HM2815).
Preceded by Pisces-consciousness (Astrology, #HM2856).

♦ **HM2666d Virgin birth** (ICA)
Description This is both experiencing the depths of despair and knowing that one's coming into being was a perfect act of the Mystery. It occurs when one belongs to the nothing. It may be compared to the baptism of Jesus when the Heavens opened and a voice said: "Thou art my beloved son".
Context In the ICA New Religious Mode in the arena of knowing and doing intensified in one's life (life of being) the second formal aspect is the experience of transparent lucidity or the experience of divination. At the fourth phenomenological level this occurs when a person becomes one with the eternal.
Broader Transparent lucidity (ICA, #HM2184).
Preceded by Trust intuitions (ICA, #HM2761).

♦ **HM2668d Distorted cognition** (Buddhism, Tibetan)
Context One of the five invalid ways of knowing of Tibetan Buddhism.
Broader Ways of knowing (Buddhism, #HH0873).

♦ **HM2669b Tibetan meditative states of formless absorptions** (Buddhism)
Arupayasamapatti — Gzugs-med-kyi-snyoms-jug (Tibetan)
Description Concentration of the mind brings it to enter four states of formless absorption. These are the absorptions of limitless space, limitless consciousness, nothingness, and the "peak of cyclic existence" (samsara). This corresponds to the four higher trances (jhana) of southern Buddhism which have identical names for the first three stages. The Tibetan absorptions are in the Formless Realm which is the same as the arupaloka or immaterial sphere of the southern Buddhists.

HM2669

Context One of the paths of calming according to Buddhist teaching.
Broader Paths of calming (Buddhism, #HM2987).
Narrower First absorption in the immaterial sphere (#HM2110)
Third absorption in the immaterial sphere (#HM2027)
Second absorption in the immaterial sphere (#HM3043)
Fourth absorption in the immaterial sphere (#HM2051).
Related Arupaloka consciousness (Buddhism, Pali, #HM2012)
Trances and mental absorptions (Buddhism, #HM2122).
Preceded by Meditative states of mental abidings (Buddhism, #HM2145).

♦ **HM2670g Absence of consciousness**
Coma
Description In human beings the medical condition called coma is the only one known in which all consciousness is believed to be absent. In this state all voluntary mental activity is considered impossible. Even reflex functioning disappears and the individual cannot be wakened. The patient's life functions are maintained however, and recovery from coma and comatose conditions is common. On recovery there is complete amnesia about the period of coma.
Life without consciousness, the state of coma, raises interesting problems in that life functions must be regulated by something internally. To deny the internal controls or regulators a status of being a species of consciousness is to evade the issue in the philosophical mind-body problem. The state of coma and other conditions illustrate that the taxonomics for the properties and kinds of consciousness are far from being complete or coherent.
Broader Diminished clarity of awareness (#HM6201).
Related Torpor (#HM1483) Unconscious stupor (#HM2473)
Suspended animation (#HM1182).

♦ **HM2671c Emptiness of the five senses** (Buddhism, Tibetan)
Adhyatmashunyata
Context One of the eighteen emptinesses comprising the paths of view in Tibetan Buddhism.
Broader Emptinesses on the paths of the view in Buddhism (Buddhism, Tibetan, #HM2944).

♦ **HM2674d Final limits** (ICA)
Description This state occurs when an individual is conscious of the all-pervading mystery. He knows he is in a life and death struggle and that he will lose. Nothing he can ever do or know can compete with this omnipotence, the mystery always wins. It may be compared to the experience of Sophie in the film "Sophie's Choice", when she sees that there is no way for her to be happy whatever she decides about the two men in her life. Sartre's work "No Exit" dramatizes this sense of being overwhelmed by omnipotence.
A dimension of this experience is the sense of awe, that is fear and fascination. The fear is that one is placed in a state of utter impotence, for the all-powerful nature of God is not an abstract attribute but a concrete fact of existence; one knows one's self to be utterly inept and that one will always be that way. The fascination is that there is but one option and one chooses it: to end striving to escape this power and to "give in to God", to resign one's self to this presence in one's life. The residue of such times is a new awareness that one does not have to have it one's way. Indeed, one becomes "the Way" or, more exactly, "the Way" becomes one's way.
Context This state is number 7 in the ICA *Other World in the midst of this World.*
Broader Inescapable power (ICA, #HM3096).
Followed by Second birth (ICA, #HM3175) Total exposure (ICA, #HM2764)
Sacrificial passion (ICA, #HM2641) Contextual world-view (ICA, #HM2842)
Transcendent immanence (ICA, #HM3034).
Preceded by Ultimate reality (ICA, #HM2030) Sacrificial passion (ICA, #HM2641)
Ubiquitous otherness (ICA, #HM2570) Transcendent immanence (ICA, #HM3034)
Transcending hostility (ICA, #HM2658).

♦ **HM2675c Communion with deity**
Communion with God — Eucharist
Description To Meister Eckhart, taking of communion in the mystical eucharist is not something done by man but by God. Normally what one eats is transformed into one's self. Here it is God who transforms us into himself. Outwardly one is eating but inwardly one is being eaten, becoming food for God. Similarly, one does not pray but is prayed. In this prayer than can be no question of not being answered. One is not only united with God, one becomes God. Having become nothing, having let go of self and creatures, one is filled with the infinity of God.
Related Contemplative intuitive meditation (#HH0816).

♦ **HM2679d Magical-forces awareness** (Buddhism, Tibetan)
Eight siddhis
Description This state is characterized as the outcome of the exercise of the will, rather than that of the mind, whereby eight magical powers are obtained by strict meditative rituals (sadhara). These powers are represented as employing metallurgical and alchemical operations to fashion: an invincible sword; an eye salve producing clairvoyance; a foot balm for magical travel; pills to make one small; and a drink to reserve youthful vitality. There are also three powers of bodily form; shape-shifting; penetration of barriers; and sinking into the earth to control spirits and treasures.
Context In Tibetan Sakya Buddhism this is one of the states in the "Ascension Stages Game". In some sets it is numbered 72 on the board.
Broader Masters of wisdom in the ascension stages game (Buddhism, Tibetan, #HM3066).
Related Siddhis (Yoga, Buddhism, #HH0380) Magico-religious powers (#HH0497)
Supernormal powers (Buddhism, #HH5652).
Followed by Kriya-tantra awareness (Buddhism, Tibetan, #HM2558)
Middle-path tantra awareness (Buddhism, Tibetan, #HM3026)
Tantra master-in-sense-realm awareness (Buddhism, Tibetan, #HM2211).
Preceded by Joy-land awareness (Buddhism, Tibetan, #HM2167)
Tantra-beginner awareness (Buddhism, Tibetan, #HM2452).

♦ **HM2680d Hallowed honour** (ICA)
Description This is experienced as coming to peace with having the mystery ever-present in one's life; when one affirms the state of one's life as forever separated and yet bound. It may be compared to the young boy in the movie "Mask", when he has to come to terms with his illness and what it has done to his life.
Context In the ICA New Religious Mode in the arena of standing present to the mystery of being in life (life of contemplation) the first formal aspect is the experience of enigmatic not-me-ness or of externality. At the third phenomenological level a person experiences collegiality with the mystery of being that which he is bound to.
Broader Awareness of externality (ICA, #HM2637).
Followed by Being all the other (ICA, #HM2983).
Preceded by Everlasting inescapability (ICA, #HM2791).

♦ **HM2681c Beyond individual distinctions** (Sufism)
Description Passing beyond that which characterizes individuals, individual species, bodies, substance, intelligences, souls; passing beyond the distinction between substance and accident, all comes together in absolute being, the reality of existence itself. In the external world there is one, unique, ontological reality, although, clothed in modes and attributes, it gives the illusion of being multiple and numerous to those imprisoned in lower levels.
Context The eighteenth illumination of Jami.
Broader Fountains of light (Sufism, #HM3039).
Followed by Indivisibility of essence (Sufism, #HM2113).
Preceded by First individuation of essence (Sufism, #HM3614).

♦ **HM2682d Determining history** (ICA)
Description This is knowing that one's actions are part of history and that they shape the direction in which society is moving. It occurs when one realizes that one's every act is of significance relative to the future of one's life and to that of society, as when parents select a school to which to send their children, or the editor of a newspaper decides whether or not to publish a story.
Context In the ICA New Religious Mode in the arena of acting out one's deed (the life of doing) the second formal aspect is extending one's engagement to history. At the third phenomenological level this occurs when the individual embraces his election to this life, this task.
Broader Engagement to history (ICA, #HM2588).
Followed by Eternal moment (ICA, #HM2786).
Preceded by Every situation (ICA, #HM2991).

♦ **HM2683d Final-training subtle contemplation** (Buddhism)
Prayoganishthamanaskara — Sbyor-mthai-yid-byed (Tibetan)
Description The state leads, after developing into 9 "uninterrupted paths" and 9 "paths of release", to the first and to the successive concentrations. The series of seven subtle contemplations viewed as preparations for achieving each of the concentrations has mainly a function of detaching the mind from its identification with a particular state. The ninth path of release of the seventh preparation in the Desire Realm is the same as the first concentration in the Form Realm.
Context This is the seventh and last of the preparations in Tibetan Gelugpa Buddhism.
Broader Special insight and preparations as states (Buddhism, #HM2623).
Related Uninterrupted and release paths (Buddhism, Tibetan, #HM2015).

♦ **HM2684d Tariki** (Japanese)
Outside strength
Description An approach to enlightenment which relies on outside help in response to prayer.
Related Jiriki (Zen, #HM3215).

♦ **HM2685d Grief**
Sorrow
Description Feelings of grief may be accompanied by disbelief, shock, guilt, anger, blame and anguish. These may be mitigated by mourning rituals after the death of someone loved, for example.
Related Despair (#HM3405) Mourning (#HM1502) Grief (Hinduism, #HM9392)
Sad feeling (Buddhism, #HM7120).

♦ **HM2686e Mana**
Mana personalities
Description The power generated in a group by the re-enactment of a group custom. It is a physical awareness of the presence of society, which may manifest as a feeling of awe; an impersonal force which nevertheless attaches to persons (either directly or via spirits) giving them power which may be described as magical, social, religious, taboo, or even simply success in battle. The feeling of awe generated by group ceremony is said to be the start of religion; and the conceptualizing or giving a name to this feeling as the start of magic. Mana may be associated with objects which will then be regarded as sacred. It may be granted by the gods in return for ritual offerings and sacrifice.
Related Magic (#HH0720) Taboo (#HH1101) Rites (#HH0423)
Holiness (#HH0183) Supernaturalism (#HH4213).

♦ **HM2687d Universal fate** (ICA)
Description This state occurs when an individual encounters the sheer arbitrariness of his existence. Call it fortune or fate, the fact that each of us showed up at all was a one-in-a-million chance. What is more, everyone has his own particular set of unique characteristics that defines who he is. I can go on asking "why me ?" but the fact is this is just the way I arrived; like Luke in the movie "The Empire Strikes Back" when, having fought and defeated the figure of Darth Vadar, he pulls off the mask to find his own face there. It may be compared to coming to terms with being a homosexual, despite all attempts to hide it or run away from it; or what happens when a mother discovers she has given birth to a brain-damaged child.
A dimension of this experience is a sense of awe, that is fear and fascination. The acknowledgement of my fatedness is accompanied by the realization that no-one will ever understand the peculiarity that I am; and a sense of being trapped by existing. At the same time there is an awareness that if I don't embody this particular perspective that is me, nobody else will. I don't have to keep trying to become something else; just being who I am is the greatest chance I have had or ever will have to live an authentic and fulfilling life.
The decision is to take hold of the life one has been given and to live it. This is the one big chance and there is the gnawing knowledge that one could easily fail. The sense of mundaneness of one's life is matched by an equal sense of awe at the profundity of it. One is reminded of the songs of the African slaves working on the plantations of North America as they celebrated the pain and glory of their existence in the "New World."
Context This state is number 21 in the ICA *Other World in the midst of this World.*
Broader Creative existence (ICA, #HM2894).
Followed by Beyond morality (ICA, #HM3049) Primal sympathy (ICA, #HM2550)
Creative futility (ICA, #HM2493) Original integrity (ICA, #HM2773)
Relational situation (ICA, #HM2978).
Preceded by Incarnate living (ICA, #HM2394) Creative futility (ICA, #HM2493)
Perpetual becoming (ICA, #HM2717) Ultimate awareness (ICA, #HM2388)
Original integrity (ICA, #HM2773).

♦ **HM2690b Oxherding pictures in Zen Buddhism** (Zen)
Cow-herding simile
Description The path to Buddhahood as followed in the discipline of Zen has been represented by a sequence of oxherding pictures. As the ox is hunted, viewed, caught and tamed it may become progressively lighter in colour, ending in the disappearance of awareness both of the ox and the observer, return to source and Buddha-awareness. There are at least four varieties of such pictures. Individual entries describe those of Kaku-an Shi-en of the Rinzai school which depicts ten stages, although the Seikyo set is said to have been only six.
Refs Cook, Francis Dojun *How to Raise an Ox*; Johnson, Williard *Riding the Ox Home* (1987).
Broader Zazen meditation (Zen, #HH0882) Human development (Zen, #HH1003).
Narrower Seeing-the-ox awareness (#HM2755) Herding-the-ox awareness (#HM2560)
Catching-the-ox awareness (#HM2979) Searching-for-ox awareness (#HM3036)
Seeing-the-traces awareness (#HM3302) Back-to-the-source awareness (#HM3251)
Coming-home-on-the-ox's-back awareness (#HM3153)

Ox-and-man-both-out-of-sight awareness (#HM2492)
Ox-forgotten-leaving-man-alone awareness (#HM2604)
Entering-city-with-bliss-bestowing-hands awareness (#HM3068).

♦ **HM2692d Vritti** (Hinduism)
Skandha (Buddhism) — Khandha (Pali) — Aggregate — Individual tendencies of consciousness
Description In Hindu psychology, vritti is a term given to what depth psychologists call the structures or constellations in the sub-conscious. Positively, these may include guiding, developmental archetypes; negatively, complexes; and neutrally, dispositions and tendencies that delineate the personality. In Buddhism, the equivalent concept to vritti is that of skandhas, or personality tendencies that transmigrate. The vritti then are energizings of a particular content. From their dynamics they are waves or vortices in the ground-stuff of the mind (citta). Depending on their amplitude and direction they may obstruct higher consciousness.
Related Citta (Hinduism, Buddhism, #HM3529)
Dispositions of consciousness (Buddhism, #HM2098)
Sixth vajra-master awareness (Buddhism, Tibetan, #HM2287).

♦ **HM2693c Tibetan meditative states of form concentrations** (Buddhism)
Dhyana — Bsam gtan (Tibetan)
Description Meditation brings the mind to enter the Form Realm, above the Desire Realm of the gross material world. There is direct realization of emptiness, without the intervening medium of an image, when all appearance of subject and object is extinguished in *suchness* and the *path of seeing* is attained. The meditative equipoise of this path is divided into the uninterrupted path of eight forbearances and the eight knowledges of the path of release, as the artificial conceptions of inherent existence with respect to the *four noble truths* are abandoned.
There are 4 states of concentration of consciousness. The first 3 are subdivided into 3 levels each. The fourth has eight levels. The 4 states correspond to those described in southern Buddhism in the five-material sphere (rupaloka).
Context One of the paths of calming according to Buddhist teaching.
Broader Selflessness (Buddhism, #HM2016) Emptiness (Buddhism, Zen, #HM2193)
Paths of calming (Buddhism, #HM2987)
Meditative states of mental abidings (Buddhism, #HM2145).
Narrower First trance of the fine-material sphere (#HM2450)
Third trance of the fine-material sphere (#HM2062)
Second trance of the fine-material sphere (#HM2038)
Fourth trance of the fine-material sphere (#HM2586).
Related Mystical stage of life (#HM3191) Four noble truths (Buddhism, #HH0523)
Rupaloka consciousness (Buddhism, #HM2536) Lokuttara consciousness (Buddhism, #HM2120)
Trances and mental absorptions (Buddhism, #HM2122).
Followed by Tantra in meditation on emptiness (Buddhism, Tibetan, #HM2864).
Preceded by Special insight and preparations as states (Buddhism, #HM2623).

♦ **HM2694d Corporate duty** (ICA)
Description This is experienced when a person realizes that he has a duty to the community to see that justice is realized, when a personal complaint becomes a social contradiction within the community. It is like Dietrich Bonhoeffer deciding that his duty to German society was to organize an underground seminary during Hitler's rule.
Context In the ICA New Religious Mode in the arena of acting out one's engagement (the life of obedience) the second formal aspect is embodying equity as the disestablishment to call forth human justice. At the second phenomenological level, this occurs when one recognizes one's obligation to the rest of mankind.
Broader Embodying equity (ICA, #HM2442).
Followed by Loyal opposition (ICA, #HM2910).
Preceded by Individual rights (ICA, #HM2782).

♦ **HM2695c Non-ignorance** (Buddhism)
Amoha (Pali) — Gti-mug-med-pa (Tibetan) — Non-delusion — Absence of dullness — Absence of delusion
Description The ascetic states of effacement and seclusion that go with volition of ascetic practices bring absence of delusion, so that the dangers of forbidden things are no longer hidden; and there is no indulgence of self-mortification through excessive self-effacement in ascetic practice.
In Tibetan Buddhism it is said that application of hearing, thinking or meditating allows the knowledge of individual analysis to arise; or this may be present from birth as the fruition of actions of previous lives. Together with non-attachment and non-hatred, non-ignorance is the basis of all paths, of all virtuous practices and all means of ceasing wrong behaviour.
In Hinayana Buddhism, it is also related to non-attachment or non-greed and to non-hatred as the roots of all that is profitable or moral. Its characteristic is penetration of the intrinsic nature of things, of their individual essences. Its function is to illuminate the objective field and it manifests as non-perplexity, non-bewilderment.
Context One of the eleven *virtuous mental factors* referred to in Tibetan Buddhism. One of the formations aggregate (mental coefficients) of Hinayana Buddhism, being listed among the constant states which appear in their true nature, and as profitable secondary (sometimes present in any profitable or profitable-resultant consciousness).
Broader Faculty of wisdom (Buddhism, #HM3233)
Virtuous mental factors (Buddhism, #HH0578)
Awareness as mental-formation group of conscious existence (Buddhism, #HM2050).
Narrower Seclusion (#HM6312) Effacement (#HM7703).
Related Non-hatred (Buddhism, #HM2744) Equanimity (Buddhism, #HM7769)
Non-attachment (Buddhism, #HM2128) Ignorance (Buddhism, Yoga, Zen, #HM3196).

♦ **HM2696c Tantra heat-application awareness** (Buddhism, Tibetan)
Alchemical-flask
Description This state of awareness is characterized by empowerment and is represented as first great initiation into the Supreme (anuttara) Tantra. Three more initiations through the states of the tantric bodhisattva path lead toward four forms of Buddhahood. The first empowerment in this state is represented as a flask or alchemical vessel from the later contents of which will grow the form of Buddhahood called the *emanation-body* (nirmana-kaya, sprul-sku). In the initiation, an actual flask is filled with amrita, the sacred potion of some 25 ingredients compounded with water and spiritual power, which is drunk by the candidate. Thereafter he enters the mandala of the ritual which represents the consciousnesses that he is encountering.
Context In Tibetan Sakya Buddhism this is one of the states in the "Ascension Stages Game". In some sets it is numbered 49 on the board.
Broader Tantric paths of accumulation and application in the ascension-stages-game (Buddhism, Tibetan, #HM2848).
Followed by Tantra climax-application awareness (Buddhism, Tibetan, #HM2220)
Tantra receptivity-application awareness (Buddhism, Tibetan, #HM2756).
Preceded by Great-path tantra awareness (Buddhism, Tibetan, #HM2656)
Kalacakra-tantra shambhala awareness (Buddhism, Tibetan, #HM2151)
Tantra master in form-realm awareness (Buddhism, Tibetan, #HM2235)
Mahayana highest-teachings-application awareness (Buddhism, Tibetan, #HM2271).

♦ **HM2699d Guhya-samaja urgyan awareness** (Buddhism, Tibetan)
Description This state of awareness is characterized by a recognition that places may be imbued with spiritual force. It is represented by the Palace of Lotus-light at Lake Urgyan (northern Pakistan) as taught in the Secret Assembly (guhya-samaja) tantric texts. Special to this awareness is the recognition that ordinary places and ordinary acts of behaviour, even those involving strong passion, can be transmuted and informed with spiritual presence or significance. Secular life can be sacramentalized and ignorant persons may thus be drawn into the path of dharma.
Context In Tibetan Sakya Buddhism this is one of the states in the "Ascension Stages Game". In some sets it is numbered 61 on the board.
Broader Supreme heavens in the ascension stages game (Buddhism, Tibetan, #HM3214).
Followed by Supreme-heaven awareness (Buddhism, Tibetan, #HM2813)
Eighth vajra-master awareness (Buddhism, Tibetan, #HM2301)
Seventh vajra-master awareness (Buddhism, Tibetan, #HM2789).
Preceded by Great-black-lord awareness (Buddhism, Tibetan, #HM2118).

♦ **HM2700b Kensho-godo** (Japanese)
Description Looking directly into one's own nature one finds it to be the same as the ultimate nature of the universe. Attainment of this state is said to be a kind of *satori*, although there may be differences of clarity and depth.
Related Satori (Zen, #HM2326).

♦ **HM2701g Consciousness of the reversible direction of time**
Description The possibility that time could flow backwards has been advanced in high energy physics (Feynman) from experimental work with sub-atomic particles. Another approach resulting in the same hypothesis or postulate comes from experimental work with telepathy, in which signals were believed to travel faster than the speed of light. According to Einstein, anything travelling faster than light would travel backwards in time. Those with psychic gifts who claim to divine the past may have support in this theory. However, also according to Einstein, nothing within our universe can travel faster than the speed of light. Lord Kelvin remarked, "if living creatures could grow backward they would have conscious knowledge of the future but no memory of the past, and would become again unborn". In the theory of magick one produces effects because, violating the forward flow of time, one goes back to alter causes. A case can be made for the experience of reversed time in many parapsychological phenomena (hauntings, miraculous healings, precognition, etc).
Broader Time consciousness (#HM2601).

♦ **HM2702a Unitary consciousness**
Sense of oneness — Unity of being — Beyond number
Description The experience of undifferentiated unity is one of the important characteristics of mystical experience. Two aspects of such experience are distinguished, depending upon whether the subject-object dichotomy transcended is between the usual self and some inner world within the experiencer, or whether it is between the usual self and the external world of sense impressions outside the experiencer. Both forms of unity may be experienced successively and it is believed that the states of consciousness ultimately achieved in each case may be identical
Internal unity: This is characterized by loss of awareness of all normal sense impressions and the usual sense of individuality, although paradoxically a pure consciousness of what is being experienced remains and seems to expand as a vast inner world is encountered. In the most complete experience, this consciousness is a pure awareness beyond empirical content, with no external or internal distinctions. Although awareness of the empirical ego ceases, the individual does not become unconscious, but rather remains aware of a oneness or undifferentiated unity, associated with a sense of merging with a ground of being.
External unity: This is perceived outwardly with the physical senses through the external world. Awareness of one or more particular sense impressions grows in intensity until suddenly the object of perception and the empirical ego simultaneously seem to cease to exist as separate entities, while consciousness seems to transcend subject and object and become impregnated by a profound sense of unity, accompanied by the insight that ultimately all is one. The essences of external objects are experienced intuitively and felt to be the same at the deepest level and that all are a part of the same undifferentiated unity. In the most complete experience, the individual feels in a deep sense that he is a part of everything that exists. Despite this awareness the experiencer retains the knowledge that on another level, at the same time, he and the external objects may be considered separate.
Related Unitary life (#HH0311) Divine indwelling (#HH0814)
Unity consciousness (#HM3193) Advaita vedanta (Hinduism, #HH0518)
Anomalies in experience of the self as distinct from the outside world (#HM4754).

♦ **HM2703d Second vajra-master awareness** (Buddhism, Tibetan)
Description This state of awareness is represented as the initiation of the Crown. One is said to be prepared to become a universal ruler and at this stage acquires supernatural well-being.
Context In Tibetan Sakya Buddhism this is one of the states in the "Ascension Stages Game". In some sets it is numbered 73 on the board.
Broader Way of the vajra masters (Buddhism, Tibetan, #HM3090).
Followed by Wheel-turner-king awareness (Buddhism, Tibetan, #HM2759)
Fifth vajra-master awareness (Buddhism, Tibetan, #HM2763)
Fourth vajra-master awareness (Buddhism, Tibetan, #HM2251).
Preceded by First vajra-master stage (Buddhism, Tibetan, #HM2187)
Hyper-bliss-realm awareness (Buddhism, Tibetan, #HM2337)
Amoghasiddhi-karma awareness (Buddhism, Tibetan, #HM2783)
Jewelled-peaks-realm awareness (Buddhism, Tibetan, #HM2275)
Tantra receptivity-application awareness (Buddhism, Tibetan, #HM2756)
Tantra union-in-learning-application awareness (Buddhism, Tibetan, #HM2280).

♦ **HM2706d Levitational submission** (ICA)
Description This is experienced as voluntarily being driven by the direction which history is demanding, as when Teresa of Avila describes the Lord as "burning away" superficial cares and desires. It occurs when one's only desire is to participate unreservedly in life's creative process.
Context In the ICA New Religious Mode in the arena of acting out one's freedom (the life of prayer) the third formal aspect is surrender to one's unlimited inadequacy, or petition. At the fourth phenomenological level this occurs when one knows that one's whole life is inescapably bound in this struggle.
Broader Surrender to inadequacy (ICA, #HM2922).
Preceded by Imploring succour (ICA, #HM2771).

♦ **HM2708c Contact** (Buddhism)
Sparsha — Sparsa — Phassa (Pali) — Reg-pa (Tibetan) — Touch
Description In Tibetan Buddhism, this distinguishes the object as pleasant, unpleasant or neutral and is thus the basis of feeling and, thereby, of desire, hatred and ignorance. In Hinayana Buddhism, it is manifested in the sense sphere by the coinciding of the object, which is the physical basis, and consciousness. This does not refer only to touch but to any of the senses

where consciousness and object impinge.
Context One of the five omnipresent mental factors defined in Tibetan Buddhism. One of the formations aggregate (mental coefficients) of Hinayana Buddhism, being listed among the constant states which appear in their true nature, and as general primary (always present in any consciousness).
Broader Omnipresent mental factors (Buddhism, Tibetan, #HH0320)
Awareness as mental-formation group of conscious existence (Buddhism, #HM2050).
Narrower Touch (#HM4212) Contact (#HM1380) Touching (#HM1959)
Being contacted (#HM0836).
Related Body consciousness (Buddhism, #HM2562).

♦ HM2709c Meditation way of the solitary realizers (Buddhism)
Pratyckabuddha — Rang-sang-rgyas (Tibetan)
Description The is the state of the *great awakening* achieved by an arhat without having had the aid of a Buddha or without having had access to Buddhist teachings in his last lifetime.
Preceded by Meditation way of the hearers (Buddhism, #HM2161).

♦ HM2710b Mystical contemplation
Infused contemplation
Description Contemplation is the supreme manifestation of the indivisible power of knowing which lies at the root of all artistic and spiritual satisfaction. It is an act of the whole personality working under the stimulus of mystic love. It is not a simple state of consciousness, as is the case of meditation, governed by one set of psychic conditions. It is a general name for a large group of states, partly governed by the temperament of the individual, accompanied by states of feeling varying from extreme quietude to rapturous and active love in some cases combined with intellectual vision.
A distinction is made between acquired contemplation (which is the result of man's own efforts assisted by divine grace) and infused contemplation which is solely and entirely given by God. Psychologically, contemplation is an induced state, in which the field of consciousness is greatly contracted to focus the whole self upon the chosen object, allowing the reality of that object to penetrate consciousness and thus releasing new powers of perception and opening deeper layers of the personality. The whole personality, directed by love and will, transcends the sense world, rises to freedom, there to apprehend the supra-sensible by immediate contact.
In the true contemplative mystic, consciousness moves on to a higher level as the result of the emergence and deliberate cultivation of powers which in most people are latent or totally absent. There is a complete withdrawal of attention from the sensible world and a total dedication of action and mind towards a particular interior object. Consciousness is transformed and remade, resulting in a state of permanent illumination and the withering away of the sense of individual selfhood. Thus contemplation, the *via illuminativa*, is a step on the path to full union, the *via unitiva* of Christian mystical teaching. Throughout the Bible there are references to seeing God face-to-face as the aim of all contemplatives; this is said to make the contemplative's face shine with illumination.
Many attempts have been made to describe the stages of the ascent from meditation through contemplation to ecstasy; for example, St Teresa refers to four stages: the prayer of recollection, the prayer of quiet, the prayer of union and the prayer of ecstasy. According to Patanjali, in this final stage of meditation: physical, mental and emotional reactions are no longer regarded; all sense of separation, of a lower personal self, disappears; reality is revealed; a sense of unity with all beings is achieved; and the state of illumination, of samadhi, is reached.
Narrower Superessential contemplation (Christianity, #HM4340).
Related Contemplation (#HM2952) Via unitiva (Christianity, #HH4100)
Via illuminativa (Christianity, #HH6030) Mystical theology (Christianity, #HH5217)
Contemplative life (Christianity, #HH2145).

♦ HM2712b Transcendental experience
Description This form of consciousness is one characteristic of the mystic experience. The individual loses the usual sense of time and space and is not oriented in terms of any three-dimensional perception of his environment or any sense of past and future. He experiences a sense of timelessness and spacelessness which are felt to be related to the concept of eternity and infinity respectively.
The normal identity-anchored, space-time-bound experience is recognized by contemporary research to be historically and ontologically relative, as well as being relative from a cultural and socio-economic perspective. The consensual and interpersonal confirmation it offers does however provide a sense of ontological security whose validity is experienced by the individual as self-validating, despite the knowledge that historically, ontologically, socio-economically and culturally the apparent absolute validity is an illusion.
Related Transcendence (#HH0841) Psychedelic drugs (#HH0075)
Religious experience (#HM3445) Aesthetic LSD experience (#HM2542).

♦ HM2713b Zodiacal forms of awareness (Astrology)
Description The zodiac may be seen as a symbol of unity and wholeness encompassing twelve different conditions of being, each of which offers different strengths and sensitivities. These conditions may be looked upon as twelve streams of divine energy, the forces and influences of which are reflected in the body of man as a temple into which a divine being can extend. Two sequences of experience are involved: the progressive embodiment of spiritual consciousness; and the progressive liberation to that maturer and fuller consciousness of transformed spiritual individuality. It is said that every individual has the capacity for wholeness represented by the wheel of the zodiac.
The twelve different conditions may be considered as four *triplicities* corresponding to the four elements of *earth*, *water*, *fire* and *air*. Individuals influenced by a particular element will have that element's characteristics in common. The twelve may also be subdivided into the *cardinal*, *fixed* and *mutable quadruplicities*, also referred to as the *common cross*, the *fixed cross* and the *cardinal cross*; the qualities inherent in each of these quadruplicities will be evident in the nature of persons united in them. The individual is said to be born onto the common cross, to the wheel of incarnation; to come under the influence of the fixed cross as the wheel of incarnation is reversed, and to be liberated under the influence of the cardinal cross as the wheel is transcended. Again, the signs of the zodiac are alternately positive and negative (or masculine and feminine) - the fire and air signs coming under the former classification and demonstrating different qualities of the archetypal masculine; and the earth and water signs under the latter grouping and demonstrating qualities of the archetypal feminine.
All twelve conditions may be present and playing its proper part in an individual who, by self-discipline, practice and meditation, has allowed all twelve conditions to arise within himself. The particular signs and their groupings are treated more fully in separate entries.
Refs Bailey, Alice A *The Labours of Hercules* (1971).
Broader Maps of the mind (#HH8903).
Narrower Airy awareness (#HM2955) Fiery awareness (#HM2124)
Fixed awareness (#HM2554) Earthy awareness (#HM3235)
Watery awareness (#HM2384) Leo-consciousness (#HM2376)
Mutable awareness (#HM3092) Cardinal awareness (#HM2259)
Aries-consciousness (#HM2665) Libra-consciousness (#HM2512)
Virgo-consciousness (#HM2439) Cancer-consciousness (#HM2239)
Gemini-consciousness (#HM2924) Pisces-consciousness (#HM2856)
Taurus-consciousness (#HM2815) Scorpio-consciousness (#HM2615)
Aquarius-consciousness (#HM2973) Capricorn-consciousness (#HM2822)
Sagittarius-consciousness (#HM2726).
Related Astrology (#HH0552).

♦ HM2716d Discipleship-application awareness (Buddhism, Tibetan)
Sravaka prayoga-marga
Description This state of awareness is characterized by calm abiding and special insight. It is represented as achieving 5 mental powers: magical; clairaudience, telepathy, and past-life recollections of self and others. There is still another of the mental powers (abhijna) to come. This, the sixth, will eventuate when Nirvana is about to be realized.
Context In Tibetan Sakya Buddhism this is one of the states in the "Ascension Stages Game". In some sets it is numbered 39 on the board.
Broader Religious traditions in the ascension stages game (Buddhism, Tibetan, #HM3341).
Followed by Pratyeka Buddha awareness (Buddhism, Tibetan, #HM2180)
Discipleship-vision awareness (Buddhism, Tibetan, #HM2240)
Thirty-three-god-heaven awareness (Buddhism, Tibetan, #HM2606)
Great-vehicle lower-path awareness (Buddhism, Tibetan, #HM2268)
Eastern-continent awareness of noble figures (Buddhism, Tibetan, #HM2543).
Preceded by Discipleship-karma awareness (Buddhism, Tibetan, #HM2192)
Pratyeka Buddha application awareness (Buddhism, Tibetan, #HM3020).

♦ HM2717d Perpetual becoming (ICA)
Description This is the experience that the task of creating one's life is endless, without any hope of completion, like being unquenchably thirsty. It is reflected in the ancient Chinese wisdom that the essence of life is the void or emptiness which can never be filled up. Camus points to this reality in "The Myth of Sisyphus". Sisyphus pushes the rock up to the top of the hill, only to have it come tumbling down again. When he turns to go back down the hill he realizes this is the meaning of his life.
A dimension of this experience is a sense of awe, that is fear and fascination. In the process of perpetually inventing who one is, one asks "Can I make it through another round?", "Is there no rest?". There is fear of being overcome by the illusion that one has arrived at the endpoint, knowing that there is always more lying in wait. Yet the sheer fascination is the adventure that life has in store for one if one dares keep on creating one's self, the strong feeling that if one gives up the quest of inventing who one is one will just shrivel up and die.
The decision in this state of creation and re-creation is to stay in the race and not drop out; to plunge to new depths of one's being that one didn't know existed, like Alice discovering in Wonderland that there are endless doors. One keeps on opening them as one discovers the possibilities that life contains. Everything becomes radically relative and a sense of hope takes root.
Context This state is number 20 in the ICA *Other World in the midst of this World*.
Broader Authentic relation (ICA, #HM2316).
Followed by Universal fate (ICA, #HM2687) Primal vocation (ICA, #HM2621)
Personal epiphany (ICA, #HM2595) Archetypal humanness (ICA, #HM3112)
Sacramental universe (ICA, #HM2445).
Preceded by Primal vocation (ICA, #HM2621) Primordial wonder (ICA, #HM2186)
Personal epiphany (ICA, #HM2595) Self transcendence (ICA, #HM2584)
Destinal accountability (ICA, #HM2309).

♦ HM2718d Struggle to confess (ICA)
Description The first formal aspect of the *life of prayer* in the ICA *New Religious Mode*, this represents a battle to repent for one's sinfulness.
Broader Life of prayer (ICA, #HM2511).
Narrower Besetting sin (#HM2379) Heavenly sorrow (#HM2216)
Personal violation (#HM2561) Painful acknowledgement (#HM2463).

♦ HM2719d Belief-arising subtle contemplation (Buddhism)
Adhimokshikamanaskara — Mos-pa-las-byung-bai-yid-byed (Tibetan)
Description The mental disposition of faith is important in this state of contemplation.
Context This is the third of the preparations in Tibetan Gelugpa Buddhism.
Broader Special insight and preparations as states (Buddhism, #HM2623).
Followed by Full-isolation subtle contemplation (Buddhism, #HM2743).

♦ HM2720d Horror of sin (ICA)
Description This is the realization that everything one has done has been destructive of something; that there is nothing one can do to escape the profound separation that exists in all dimensions of one's life. St. Augustine experienced this when he was converted and was horrified at the realization of what his life had been up to that point, yet knew that even the new person he had become was finally disrelated from life at its most profound level.
Context In the ICA New Religious Mode in the arena of articulating the word about life (the life of knowing) the first formal aspect is one's experience of being a solitary individual or of being a self. At the second phenomenological level, this occurs when one is somehow stripped of one's illusions and must decide how to relate to one's new universe.
Broader Solitary being (ICA, #HM2404).
Followed by Representational existence (ICA, #HM2557).
Preceded by Unveiled being (ICA, #HM2611).

♦ HM2721d Moral ground (ICA)
Freedom of decision
Description This is an experiential trek in the *River of Consciousness*, within the ICA *Other World in the midst of this World*, when the individual realizes "I am my conscience". The states are described in separate entries.
Broader River of consciousness (ICA, #HM2993).
Narrower Beyond morality (#HM3049) Primal vocation (#HM2621)
Cosmic sanctions (#HM2945) Intentional conscience (#HM2892).
Related Authentic relation (ICA, #HM2316) Creative existence (ICA, #HM2894)
Final accountability (ICA, #HM3206).

♦ HM2722d Scuffle of views (Buddhism)
Ditthivipphandita (Pali)
Broader Wrong view (Buddhism, #HM1710).

♦ HM2723d Hindu-wisdom-holder awareness (Buddhism, Tibetan)
Vidyadhara
Description This state of awareness is represented as the highest for those following the noble Brahmanical tradition. The Hindu yogi rises to the Peak of Existence but is said to be unable to escape the sansara because he does not comprehend the unreality of subjective and objective phenomena.
Context In Tibetan Sakya Buddhism this is one of the states in the 'Ascension Stages Game'. In some sets it is numbered 62 on the board.

Broader Masters of wisdom in the ascension stages game (Buddhism, Tibetan, #HM3066).
Followed by Discipleship–karma awareness (Buddhism, Tibetan, #HM2192)
Great-vehicle lower–path awareness (Buddhism, Tibetan, #HM2268).
Preceded by Hindu–states of awareness (Buddhism, Tibetan, #HM2115).

♦ **HM2724c Reza** (Sufism)
Joyful satisfaction
Description The last stage in the conscious effort to achieve poverty and tranquillity, *kushesh*, when the Sufi is ready to start on the next stage of progress on the inner self, *keshesh–jazba*. There is simultaneously conscious satisfaction in one's self and total psychic commitment without consciousness.
Broader Kushesh (Sufism, #HM3335).

♦ **HM2725b Wholeness**
Completeness
Description In terms of the psychology of Jung, wholeness is the expression in the fullest possible way of all aspects of the personality in itself, in relation to other people and in relation to the environment. It can be equated with health. Fundamental wholeness is a state into which one is born, but this breaks down to re–form into something more differentiated. Subsequent achievement of conscious wholeness may be seen as the purpose of life. It may be helped or hindered by interaction with other people or one's environment, but cannot be actively followed as such. Being greedy for wholeness may actually be an escape from psychological conflict. Nevertheless, it can often be seen that wholeness is the hidden or secret end of life's experience. Again, Jung saw wholeness as linked to the coming together and synthesizing of two opposites, for example male and female elements (animus and anima) and good and evil elements (accepting the shadow).
Refs Hannah, Barbara *Striving toward Wholeness* (1971).
Related Health (#HH0509) Completeness (#HM3070)
Complete personality (#HH3998) Accepting the shadow (#HH0204).

♦ **HM2726d Sagittarius–consciousness** (Astrology)
Spiritual aspiration
Description This mode recognizes and identifies with the coordinating and synthesizing powers of the super–conscious. Under optimal conditions, Sagittarius is associated with the disciplined development of new levels of sensitivity and understanding, characterized by greater levels of significance.
Context One of the twelve conditions of being or streams of divine energy encompassed within the zodiac. It is said that all twelve conditions must be present and playing its proper part for any successful and fulfilling action to take place. This may be through a group of individuals with different endowments and soul qualities working together; or it may be through one individual who, by self–discipline, practice and meditation, has allowed all twelve conditions to arise within himself.
Broader Fiery awareness (Astrology, #HM2124) Mutable awareness (Astrology, #HM3092)
Zodiacal forms of awareness (Astrology, #HM2713).
Followed by Capricorn–consciousness (Astrology, #HM2822).
Preceded by Scorpio–consciousness (Astrology, #HM2615).

♦ **HM2727d Third vajra–master awareness** (Buddhism, Tibetan)
Description This state of awareness is characterized by mastery of tantric theory accomplished by non–discursive understanding that is Buddha–like.
Context In Tibetan Sakya Buddhism this is one of the states in the "Ascension Stages Game". In some sets it is numbered 74 on the board.
Broader Way of the vajra masters (Buddhism, Tibetan, #HM3090).
Followed by Fifth vajra–master awareness (Buddhism, Tibetan, #HM2763)
Fourth vajra–master awareness (Buddhism, Tibetan, #HM2251).
Preceded by Joy–land awareness (Buddhism, Tibetan, #HM2167)
First vajra–master stage (Buddhism, Tibetan, #HM2187)
Amoghasiddhi–karma awareness (Buddhism, Tibetan, #HM2783)
Jewelled–peaks–realm awareness (Buddhism, Tibetan, #HM2275)
First scriptural bodhisattva awareness (Buddhism, Tibetan, #HM2155)
Tantra union–in–learning–application awareness (Buddhism, Tibetan, #HM2280).

♦ **HM2728d Social failure** (ICA)
Description This is experienced as using one's own work in order to ensure the success of others in history; when one sees that one cannot focus on a particular situation but must move to encompass all that is going on. It may lead people to give up their professions or social status to take some social direction they consider important, like Dr. Albert Schweitzer.
Context In the ICA New Religious Mode in the arena of knowing one's disengagement (the life of poverty) the second formal aspect is being reliant upon the larger society or being disengaged from one's own work. At the third phenomenological level, this occurs when one is able to so value the gift of life that nothing else is needed for one's fulfilment.
Broader Disengagement from work (ICA, #HM2552).
Followed by Realized vocation (ICA, #HM2441).
Preceded by Historical vocation (ICA, #HM2616).

♦ **HM2729d Full pacification in mental abiding states** (Buddhism)
Vyvpashamana — Nye–bar–zhi–bar–byed–pa (Tibetan)
Description The meditator, through effort (the power associated with this stage) progresses in stabilizing his meditation, although subjective interruption is still possible.
Context This is the seventh of the nine states of mental abiding (navakara chittasthiti) in Gelugpa Tibetan Buddhism.
Broader Calm abiding (Buddhism, #HM2147).
Followed by One–pointedness in mental abiding states (Buddhism, #HM2753).
Preceded by Pacifying in mental abiding states (Buddhism, #HM2205).

♦ **HM2730c Alaya–vijnana** (Buddhism)
Consciousness storehouse — Citta — Stored consciousness — All–conserving mind
Description In Yogacara Buddhism, this is the term given for the state of consciousness in which the personal tendencies (skandhas) exist. It is more than a passive, unconscious store, however, as it takes from experience and constantly modifies itself. It is mind in its deepest and most comprehensive sense; while it manifests itself as individualized in empirical consciousness, it never loses its identity and eternality (Suzuki). It makes itself appear to be both the inner world of mental activity and the perceived external world, so that dreams and waking consciousness, by this standard, only differ by degree, not kind. The yogic goal is to attain cognizance of alaya, and place the light of meditation awareness upon it, so that the mind can be liberated from the alaya–reality illusion; after the pure emptiness of space there is awareness of universal light.
Broader Consciousness in Buddhism (Buddhism, #HM2200).
Related No–mind (Zen, #HM2163) Vijna (Buddhism, #HM3617)
Human development (Zen, #HH1003) Citta (Hinduism, Buddhism, #HM3529).

♦ **HM2731c Samachittata**
Equanimity
Related Equanimity (Buddhism, #HM7769).

♦ **HM2733c Desire–realm consciousness** (Buddhism)
Kamadhatu — Dod–khams (Tibetan)
Description States of consciousness exist as are appropriate to the beings in the Desire Realm. There are 6 types: hell–beings, hungry ghosts, animals, humans, demigods and gods. The hell–beings have 8 hot–hells, one on top of the other, and each of the 8, in each of its four corners, has 4 neighboring hells; altogether 108 neighboring hells. The hell–beings also have 8 cold hells. In addition to the major 16 hells there are special hells. The next type of conscious beings, the hungry ghosts are subdivided into 3; animals are divided into 2 types, one of which humans ordinarily never see. Humans are divided into 12 different types according to an invisible or archetypal geography of lands centered on Mount Meru. On the Mount's lowest places are 4 classes of demigods. Of gods there are the 4 Great Kings, and the 33 gods above. In the ether over the Mount are 3 classes of gods. All these beings are in the Desire Realm and correspond to states of consciousness.
Broader Consciousness states in cyclic existence (Buddhism, #HM2177)
Consciousness ascension stages game (Buddhism, Tibetan, #HH4000).
Narrower Adamantine–hell awareness (#HM2028)
Southern–continent awareness of men (#HM2127)
Cold–hells awareness (Buddhism, Tibetan, #HM2040)
Naga–world awareness (Buddhism, Tibetan, #HM2031)
Animal–hell awareness (Buddhism, Tibetan, #HM2636)
Asura–world awareness (Buddhism, Tibetan, #HM2579)
Demon–island awareness (Buddhism, Tibetan, #HM2055)
Reviving–hell awareness (Buddhism, Tibetan, #HM2516)
Howling–hells awareness (Buddhism, Tibetan, #HM2100)
Lord–of–death awareness (Buddhism, Tibetan, #HM2088)
Joyful–heaven awareness (Buddhism, Tibetan, #HM2022)
Crushing–hells awareness (Buddhism, Tibetan, #HM3142)
Very–hot–hells awareness (Buddhism, Tibetan, #HM2576)
Temporary–hells awareness (Buddhism, Tibetan, #HM2454)
Heavenly–highway awareness (Buddhism, Tibetan, #HM2010)
Interminable–hell awareness (Buddhism, Tibetan, #HM2052)
Non–emanating consciousness (Buddhism, Tibetan, #HM2070)
Hungry–ghosts–hell awareness (Buddhism, Tibetan, #HM2112)
Divine–animal–hell awareness (Buddhism, Tibetan, #HM2007)
Wheel–turning–king awareness (Buddhism, Tibetan, #HM2058)
Rudra awareness of black freedom (Buddhism, Tibetan, #HM2603)
Four–great–kings–heaven awareness (Buddhism, Tibetan, #HM2082)
Thirty–three–god–heaven awareness (Buddhism, Tibetan, #HM2606)
Western–continent awareness of cattle (Buddhism, Tibetan, #HM2519)
Heaven–without–fighting consciousness (Buddhism, Tibetan, #HM2130)
Northern–continent awareness of community (Buddhism, Tibetan, #HM2067)
Delightful–emanation–heaven consciousness (Buddhism, Tibetan, #HM2546)
Eastern–continent awareness of noble figures (Buddhism, Tibetan, #HM2543).
Related Kamaloka consciousness (Buddhism, Pali, #HM2060).
Preceded by Barbarian–state of awareness (Buddhism, Tibetan, #HM2091).

♦ **HM2734d Universal compassion** (ICA)
Description This is a realization that not only does one care, but one cares for the whole world. Nothing escapes one's care; in fact, what one's life is all about is acting out one's endless care. John Wesley captured this reality when he announced that: "All the world is my parish". Perhaps Martin Luther King knew this when he declared to a congregation of civil rights activists that even their most despised enemies were part of the movement because without them there would be no solution to the problem they were out to deal with.
A dimension of this experience is a sense of awe, that is fear and fascination. From this perspective, an individual asks himself whether his particular response to life's demands is really going to make any difference; he easily becomes overwhelmed by the complexity of life and yet find himself relentlessly driven to respond to it, sensing that no–one else is in any better position, and that ultimately nothing could be more adequate than his response. Whether it is helping to teach a small child how to tie a shoe lace or throwing one's self into major international efforts, one sees that it's worth one's life.
In caring for the whole world, a person is tempted by the possibility of abandoning everything, finding himself continually responding to whatever comes along as though nothing is unworthy of his care. The decision to care is known to have effected a permanent change as he decides to not look back.
Context This state is number 38 in the ICA *Other World in the midst of this World.*
Broader Universal concern (ICA, #HM2774).
Followed by Problemless living (ICA, #HM2747) Sacrificial passion (ICA, #HM2641)
Interior discipline (ICA, #HM2851) Ancestral obligation (ICA, #HM3122)
Ubiquitous otherness (ICA, #HM2570).
Preceded by Primal sympathy (ICA, #HM2550) Interior discipline (ICA, #HM2851)
Relational situation (ICA, #HM2978) Ubiquitous otherness (ICA, #HM2570)
Definitive predestination (ICA, #HM2131).

♦ **HM2735b Buddha–consciousness** (Buddhism, Tibetan)
Description The experience of total unity with the Buddha and a sharing of his experience of awakening and illumination. This experience is the basis of the aspiration to enlightenment for the sake of all sentient beings and, by whatever method prescribed, has the effect of attaining the Buddha's Form Body and the Buddha's Truth Body. The Form Body may in turn be considered as Enjoyment Body and Emanation Body, and the Truth Body as Nature Body and Wisdom Body. All mistaken dualistic appearances and obstructions to omniscience have been eradicated and the yogi is completely enlightened with respect to all phenomena. This is the last stage of the Tibetan meditation on emptiness; and is depicted in the final stages of the *Tibetan consciousness ascension stages game*, when the experiences are described as:
(1) *Setting forth* (pravrajita), when, after all attempts to prevent him have failed, the Buddha leaves the luxury of his life in the palace to seek either perpetual youth, health and immortality or that he may not be reborn.
(2) *Ascetic practices* (tapas). Rejecting painful forms of meditation since they do not lead to liberation, and the teaching of liberation of supreme self from the body (since the liberated soul is still subject to change and rebirth), the Buddha seeks, with five companions, to remove false views through asceticism. After six years of abiding in "space–pervading concentration", the Buddha sits motionless until his mother persuades him not to die and he perceives that this is not the way of liberation. He is abandoned by the five ascetics and, having eaten and washed, sits to meditate.
(3) *Conquering Mara*, chief of all gods in the Sense Desire Realm. Despite being assailed by all the fears and distractions that Mara personifies, the Buddha, seated beneath the Bodhi tree, is not moved; he calls on the Earth goddess to witness his fitness for enlightenment and at this the army of Mara flees.
(4) *Buddhahood* Following the defeat of Mara, the Buddha ascends and descends through four trances, seeing the suffering of the world as beings rise and fall on the karmic wheel of life. He

sees all his and others' past lives and seeks the cause and cure of suffering, formulating the *four noble truths*: rebirth is suffering, the cause of thirst and ignorance; nirvana is transcending rebirth and there is a means of attaining such transcendence. He understands the *twelve-fold chain of dependent origination* and is fully awakened. He then contemplates and travels through the world systems and seas. Although tempted to give up body and life and achieve nirvana, he decides to remain in silence and peace.
(5) *Turning the wheel of dharma.* At first reluctant to teach what will not be understood, the Buddha finally, out of compassion, leaves the forest to teach. Three times he turns the wheel of dharma, teaching first the middle-way, the *eightfold path*; then the *Madhyamika sutras* of essential emptiness; and finally the *Yogacara* doctrine of the absolute principle of emptiness. The two latter reveal the tantras.
(6) *Performing miracles.* In order to convert those who are not impressed by verbal teachings, the Buddha performs a number of miracles.
(7) *Nirvana* Leaving the physical body, having lived for three months by psychic powers and no longer by karmic forces, the Buddha rises through the four stages of trance, the Realm of Form and the four equalizations of Formless Realm to attain the equanimity of cessation. He descends through the eight stages and rises again through the four trances; he enters Nirvana from a karmically neutral position to pervade the universe with his emanations and be revered for the rest of the age.
Context In Tibetan Sakya Buddhism these are states in the "Ascension Stages Game". In some sets they are numbered 98 to 104 on the board.
Broader Selflessness (Buddhism, #HM2016).
Narrower Buddha wisdom body (#HM2834) Buddha nature body (#HM2039) Buddha emanation body (#HM3019) Buddha enjoyment body (#HM3113).
Related Consciousness ascension stages game (Buddhism, Tibetan, #HH4000).
Preceded by Emanation-body awareness (Buddhism, #HM2340) Tantra in meditation on emptiness (Buddhism, Tibetan, #HM2864).

♦ **HM2736d Transparent engagement** (ICA)
Description The third formal aspect of the *life of being*, an arena in the ICA *New Religious Mode*, this is the experience of levitation.
Broader Life of being (ICA, #HM3229).
Narrower Action irrelevant (#HM2466) Keeping conscience (#HM2249) Cruciform exaltation (#HM2303) Impossible possibility (#HM2524).

♦ **HM2737c Jesus speaks to the daughters of Jerusalem** (Christianity)
Description The believer is aware of Jesus, Himself burdened with sorrows, telling the women to weep for themselves and their children. Like them, he weeps for the offences he has committed, for the pains they deserved, for the displeasure they cause Jesus who loves so much. It is this love, rather than the fear of hell, which causes him to weep. Loving Jesus more than himself, he repents with his whole heart and begs never again to separate himself from Him. He asks that he will love Jesus always and that Jesus will do with him as He wills.
Context The eighth station of the cross.
Broader Way of the cross (Christianity, #HM3516).
Followed by Jesus falls the third time under the cross (Christianity, #HM2289).
Preceded by Jesus falls the second time under the cross (Christianity, #HM3824).

♦ **HM2738c Inward-flowing consciousness** (Yoga)
Antar-vritti
Description This consciousness is associated with individuals who live in the awareness of their cosmic connection and of the divine consciousness running through them. Their only action is to serve as a channel for the cosmic flow and, although many come to depend upon them, they depend on nobody.
Related Abhyantara vritti yoga (Yoga, #HH2772) Outward-flowing consciousness (Yoga, #HM2581).

♦ **HM2739d Second scriptural bodhisattva awareness** (Buddhism, Tibetan)
Vimala
Description This state of awareness is characterized by dispassionate action. It is represented by the second-stage bodhisattva who no longer produces karmic-stains by any action taken whatsoever in furtherance of the dharma. For this reason, his stage is called *immaculate* (vimala).
Context In Tibetan Sakya Buddhism this is one of the states in the "Ascension Stages Game". In some sets it is numbered 80 on the board.
Broader Meditation way of the bodhisattvas (Buddhism, #HM2769).
Followed by Third scriptural bodhisattva awareness (Buddhism, Tibetan, #HM2215) Fourth scriptural bodhisattva awareness (Buddhism, Tibetan, #HM2191).
Preceded by First scriptural bodhisattva awareness (Buddhism, Tibetan, #HM2155).

♦ **HM2741b Blessedness**
Blessed
Refs Fox, Matthew *Original Blessing* (1983).
Related Blessing (#HH0039).

♦ **HM2743d Full-isolation subtle contemplation** (Buddhism)
Pravivekyamanaskara — Rab-tu-dben-pai-yid-byed (Tibetan)
Description This state of contemplation requires physical, emotional and mental isolation.
Context This is the fourth of the preparations in Tibetan Gelugpa Buddhism.
Broader Special insight and preparations as states (Buddhism, #HM2623).
Preceded by Belief-arising subtle contemplation (Buddhism, #HM2719).

♦ **HM2744c Non-hatred** (Buddhism)
Advesha — Adhosa (Pali) — Zhe-sdang-med-pa (Tibetan) — Non-hate — Adosa
Description Related to non-attachment and non-ignorance in both Tibetan and Hinayana Buddhism. In Tibetan Buddhism, non-hatred includes an absence of the intent to harm and conquers the generation of hatred on observing of suffering, sources of suffering, or harmfulness in others. Hinayana Buddhism indicates that non-hatred is like a gentle or agreeable friend - the characteristic is lack of churlishness or savagery and freedom from resentment or opposition. The function is removing or dispelling annoyance and distress; it manifests as pleasing. With non-attachment and non-ignorance it is the root of all that is profitable or moral.
Context One of the eleven *virtuous mental factors* referred to in Tibetan Buddhism. One of the formations aggregate (mental coefficients) of Hinayana Buddhism, being listed among the constant states which appear in their true nature, and as profitable primary (always present in any profitable or profitable-resultant consciousness).
Broader Virtuous mental factors (Buddhism, #HH0578) Awareness as mental-formation group of conscious existence (Buddhism, #HM2050).
Narrower Harmless (#HM4565) Non-hatred (#HM1399) Non-hating (#HM1557) Unmaliciousness (#HM1734) State of not feeling hatred (#HM1597) Non-hatred as a wholesome root (#HM3560).
Related Hatred (Buddhism, #HM4502) Equanimity (Buddhism, #HM7769) Non-ignorance (Buddhism, #HM2695) Non-attachment (Buddhism, #HM2128) Non-harmfulness (Buddhism, #HM2608).

♦ **HM2745e Compulsion to confess** (Brainwashing)
Description In contrast to false confessions made when self-betrayal seems the only way to avoid further suffering, confessions at this stage are repentant and an attempt to win favour in a system the individual is beginning to accept and commit himself to. Confession leads to further confession as the individual is pressed to say more and is guided through his own inner fantasies. He starts, under supervision, to fill the space left by what he has rejected in himself.
Context A stage reached in thought reform, a system of organized, deliberate and total psychological training which effects individual change through two basic elements - *confession* (renouncing of past beliefs and attitudes) and *re-education* (remaking of the individual in the required image).
Broader Thought reform (Brainwashing, #HH0865).
Followed by Channelling of guilt (Brainwashing, #HM3204).
Preceded by Desperate gratitude (Brainwashing, #HM3017).

♦ **HM2747d Problemless living** (ICA)
Description This is experienced as the absence of worldly cares. In the never-ending responsibility for the world, the fear of death loses its power; and the individual trusts the fact of life, that significance is already given to his own most personal existence. This is security beyond security: perplexing questions are no longer blocks. It may be compared to the popular story of the Duke of Wellington dancing at a ball the night before the Battle of Waterloo; or to the old song, "Pack up your troubles".
lso A dimension of this experience is a sense of awe, that is fear and fascination. The fear is of being at life's mercy, the knowledge that there is nothing to hold on to, and at the edge of existence is death. It makes no difference but it is there nevertheless. At the same time there is fascination in noticing the insignificance of the things are that consume one's time and a sense of almost floating in the midst of life's cares; one gives one's self to this sense of detachment and the larger context. Such a time brings knowledge that one can go anywhere and do anything. Nothing of this world binds one.
Context This state is number 54 in the ICA *Other World in the midst of this World.*
Broader Unknowable peace (ICA, #HM3015).
Followed by Ubiquitous otherness (ICA, #HM2570) Relational situation (ICA, #HM2978) Spontaneous gratitude (ICA, #HM3025) Transcending hostility (ICA, #HM2658) Resurrectional existence (ICA, #HM2631).
Preceded by Creative futility (ICA, #HM2493) Universal compassion (ICA, #HM2734) Relational situation (ICA, #HM2978) Inclusive comprehension (ICA, #HM2256) Resurrectional existence (ICA, #HM2631).

♦ **HM2749c Kairos**
Auspicious moment
Description As opposed to *chronos*, or the linear time measured as past, present and future and experienced as the cycles of seasons, human existence, etc, *kairos* is a total living in the present moment, the biblical *now*, the period of historical decision of Greek philosophy, the time of salvation of the New Testament, the expression of God's dominion over time.
Related Time consciousness (#HM2601) Living in the present (#HH2345) Creative moment (Sufism, #HM4449) Experience of An (Sufism, #HM1655) Mystical journey (Christianity, #HM0402) State of self-surrender (Christianity, #HM5467).

♦ **HM2750f Omnipotence**
Oceanic feeling — Merging — Timelessness — Self-esteem
Description Before the conception of objects, when the outside world appears as part of and within the organism which is perceiving, there is a feeling of omnipotence sometimes referred to as *oceanic feeling*. The experience may include an awareness of extending beyond habitual limits of space (merging) or of time (timelessness). This occurs not only in very small children but also to adults suffering from delusions.
Broader Archaic ego states (#HM0216).

♦ **HM2751g Consciousness of cyclical time**
Description Observation that cycles of birth, growth, decay and death are universal in nature, that the year brings it circles of seasons, and that the stars move from their places and return after aeons of journeying, gives rise to the idea that time itself is cyclical. All human history, the affairs of nations and of men, have been considered to ebb and flow in predestined rhythm by the passage of time. The ancient Chinese had the Book of Changes to guide men through time's cycles. Other nations had other literature, as in Greece, with Hesiods "Works and Days"; and Rome, with Ovid's "Fasti". In India, time (kala) is also perceived as cyclical. The Vedas teach the theory of the Great Age (mahayuga) of 4.3 billion years, which is divided into four parts. The present age (Kaliyuga) is the last before cosmic dissolution (pralaya), which is followed by re-creation of the universe. The consciousness of cyclical time is embodied in ritual around the world; is a key element in mythology; and is expressed in the world's calendars. Notable secular Western cycles are the week, the year, the century and the millenium. The consciousness of cyclical time is closely connected with the concepts of reincarnation and the doctrines of astrology and occultism.
Broader Time consciousness (#HM2601).

♦ **HM2753d One-pointedness in mental abiding states** (Buddhism)
Ekotikarana — Rtse-geig-tu-byed-pa (Tibetan)
Description During the meditation session, the meditator has reached the ability to uninterruptedly engage his mind with perfect concentration on its object. Mind control and the controlling factor are equally balanced. Among the six powers, effort has matured.
Context This is the eighth of the nine states of mental abiding (navakara chittasthiti) in Gelugpa Tibetan Buddhism.
Broader Calm abiding (Buddhism, #HM2147).
Related Concentration (Buddhism, #HM6663).
Followed by Mental abiding in equipoise (Buddhism, #HM2277).
Preceded by Full pacification in mental abiding states (Buddhism, #HM2729).

♦ **HM2754d Sacrificial friendship** (ICA)
Description This is experienced as a realization that one's relationships are built on a shared concern for human suffering, that one's "friends" are those who have joined in acting out their care for humankind. It occurs when a person embodies his decision to care and finds that all his old relations have somehow lost their meaning and importance for them, as when the main character in the movie "Norma Raye" discovers that her colleagues in the struggle are the black people of the community with whom she had never associated.
Context In the ICA New Religious Mode in the arena of acting out one's engagement (the life of obedience) the third formal aspect is embodying charity as the transestablishment to act out social concern. At the fourth phenomenological level this occurs when one lives on behalf of

the evolution of human consciousness.
  Broader Embodying charity (ICA, #HM3056).
  Preceded by Disinterested collegiality (ICA, #HM2812).

♦ **HM2755c Seeing-the-ox awareness** (Zen)
Description The seeker finds the way through the harmonious ordering of the senses. When his eye is properly directed he will see that the "ox" is none other than himself.
  Broader Oxherding pictures in Zen Buddhism (Zen, #HM2690).
  Followed by Catching-the-ox awareness (Zen, #HM2979).
  Preceded by Seeing-the-traces awareness (Zen, #HM3302).

♦ **HM2756c Tantra receptivity-application awareness** (Buddhism, Tibetan)
Prajna-jnana
Description This state of awareness is represented as the third of four initiations for the Tantric yogi in which the Clear Light arises in the co-emergent (saha-ja) bliss of the wisdom-understanding union (prajna-jnana) of the yogi and yogini (the wisdom-lady). This initiation instruction reveals that the white element that contributes to the formation of bodhicitta is not coarse emitted semen, but the vital breath or vital current (prana) in the middle vein (shushumna) which is visualized as being drawn from the head centre (cakra) and forced to descend to the perineal-centre where it is visualized as encountering the vital liquid or the red element. The external organ proximate to the perineal centre is symbolized as the thunderbolt-jewel in the one sex, and as the lotus in the other. The "jewel-in-the-lotus" is prajna-jnana, yab-yum and other termed unions which produce the future body of a new Buddha. This initiation also indicates the sham of uninitiates in esoteric Tantra who abuse its apparently erotic content.
Context In Tibetan Sakya Buddhism this is one of the states in the "Ascension Stages Game". In some sets it is numbered 57 on the board.
  Broader Tantric paths of accumulation and application in the ascension-stages-game (Buddhism, Tibetan, #HM2848).
  Followed by Joy-land awareness (Buddhism, Tibetan, #HM2167)
  First vajra-master stage (Buddhism, Tibetan, #HM2187).
  Second vajra-master awareness (Buddhism, Tibetan, #HM2703)
  Tantra union-in-learning-application awareness (Buddhism, Tibetan, #HM2280).
  Preceded by Tantra heat-application awareness (Buddhism, Tibetan, #HM2696)
  Tantra climax-application awareness (Buddhism, Tibetan, #HM2220).

♦ **HM2757d Classical story** (ICA)
Description This is experienced when you discover that your unique, unrepeatable life experiences are the same things that others have talked about in their lives; that the deeps of your own life are revealed in and through the stories of the lives of other people. An example is Ahab's story about the whale which is shaping his life, in the book "Moby Dick".
Context In the ICA New Religious Mode in the arena of articulating the word about life (the life of knowing) the third formal aspect is one's experience of contentless transformation or the greatness of the word. At the third phenomenological level this occurs when a person knows the authenticity of a life-interpreting and life-giving story.
  Broader Contentless transformation (ICA, #HM3106).
  Followed by Universal Christ (ICA, #HM2644).
  Preceded by Terrifying acceptance (ICA, #HM2460).

♦ **HM2758d Radical identification** (ICA)
Description The first formal aspect of the *life of chastity*, an arena in the ICA *New Religious Mode*, this represents identification with images of profound living.
  Broader Life of chastity (ICA, #HM2506).
  Narrower Living word (#HM3178)   Self programming (#HM3114)
  Committed teacher (#HM3179)   Transcendent guru (#HM3058).

♦ **HM2759d Wheel-turner-king awareness** (Buddhism, Tibetan)
Vidyadhara-emperor
Description This state of awareness is represented as the Emperor of the World. He is said to be the summit of tantric scholarship, master of sacred texts all of whose knowledge he has mystically realized. He has attained all the spiritual powers available in this lower realm of mankind and epitomizes informed consciousness applied benevolently to human concerns. He is the ultimate philosopher-king, demonstrating wisdom with the excercise of vast occult powers by which he maintains and orders his court and commissions ambassadors to teach the dharma in far-off lands. He is the temporal model for the Dalai Lama.
Context In Tibetan Sakya Buddhism this is one of the states in the "Ascension Stages Game". In some sets it is numbered 69 on the board.
  Broader Masters of wisdom in the ascension stages game (Buddhism, Tibetan, #HM3066).
  Followed by Fifth vajra-master awareness (Buddhism, Tibetan, #HM2763)
  Fourth vajra-master awareness (Buddhism, Tibetan, #HM2251).
  Preceded by Second vajra-master awareness (Buddhism, Tibetan, #HM2703)
  Tantra master-in-sense-realm awareness (Buddhism, Tibetan, #HM2211).

♦ **HM2760b God-consciousness as power**
Zeus awareness — Prometheus awareness
Description The evolution or mystic attainment by man to a state of god-consciousness may first be evidenced by feelings of creative energies and powers. It is an intoxicating realization that one can wield Zeus' thunderbolt, or carry the sacred fire, like Prometheus. It gives full scope to the ego's will-to-power and may lead to the classic downfall of a Phaeton or Faust; that is, to psychic disintegration. In its more positive manifestation, it expresses itself in man as boundless energy, the capability to realize one's projects and ambitions, and the self-knowledge that one has a mission to perform.
  Broader God-consciousness (#HM2166).

♦ **HM2761d Trust intuitions** (ICA)
Description This is believing that one's own intuitions and insights at the most concrete level are not only correct but appropriate, and occurs when one not only grasps one's own intuitions but acts on them.
Context In the ICA New Religious Mode in the arena of knowing and doing intensified in one's life (life of being) the second formal aspect is the experience of transparent lucidity or the experience of divination. At the third phenomenological level this occurs when a person disciplines his life to be continually related to the experience of nothingness.
  Broader Transparent lucidity (ICA, #HM2184).
  Followed by Virgin birth (ICA, #HM2666).
  Preceded by Prophetic sight (ICA, #HM2846).

♦ **HM2762d Ushin** (Japanese)
Aesthetic feeling — Having heart
  Related Mushin (Japanese, #HM3554)   Artistic education (#HH0029).

♦ **HM2763d Fifth vajra-master awareness** (Buddhism, Tibetan)
Anujna
Description This state of awareness is characterized by accession (anujna). It is represented by the fifth-stage tantric master who receives the initiation of accession to teacher level. His path of development is complete and he has reached the degrees of Buddhahood. He receives a new name, and the implements of the thunderbolt (vajra, dorje) and bell.
Context In Tibetan Sakya Buddhism this is one of the states in the "Ascension Stages Game". In some sets it is numbered 81 on the board.
  Broader Way of the vajra masters (Buddhism, Tibetan, #HM3090).
  Followed by Eighth vajra-master awareness (Buddhism, Tibetan, #HM2301)
  Seventh vajra-master awareness (Buddhism, Tibetan, #HM2789).
  Preceded by Great-black-lord awareness (Buddhism, Tibetan, #HM2118)
  Wheel-turner-king awareness (Buddhism, Tibetan, #HM2759)
  Third vajra-master awareness (Buddhism, Tibetan, #HM2727)
  Fourth vajra-master awareness (Buddhism, Tibetan, #HM2251)
  Second vajra-master awareness (Buddhism, Tibetan, #HM2703).

♦ **HM2764d Total exposure** (ICA)
Description This state occurs when an individual experiences his entire life as known by an omniscient power; he senses that the deepest hidden places of his life are exposed, the best kept secrets are known and things never spoken of are now spoken - as if a voice were saying "all is known". It may compared to the state of consciousness of German people exposed to the Nazi concentration camps shortly after World War II, or to a parent who discovers for the first time that his children know they are being lied to.
A dimension of this experience is a sense of awe, that is fear and fascination. The fear state is that so much that is unacceptable is now known. All that one has conspired one's whole life to cover up is now exposed. The fascination is that it is no longer necessary to conspire to conceal the unspeakable; as if a burden is lifted and "at last it's out". The decision is that since one has nothing else to offer, one can claim the deeds one has; one's conscience is not one's own. After such a time one knows that there are no secrets; one can confront the darkest parts of one's own self and the task of really living actually begins.
Context This state is number 8 in the ICA *Other World in the midst of this World*.
  Broader Inescapable power (ICA, #HM3096).
  Followed by Vibrant powers (ICA, #HM2982)   Dynamic selfhood (ICA, #HM3091)
  Singular adoration (ICA, #HM2111)   Archetypal humanness (ICA, #HM3112)
  Soteriological existence (ICA, #HM2904).
  Preceded by Final limits (ICA, #HM2674)   Primordial wonder (ICA, #HM2186)
  Singular adoration (ICA, #HM2111)   Exclusive contradiction (ICA, #HM2951)
  Soteriological existence (ICA, #HM2904).

♦ **HM2765c Single desiredness** (Sufism)
Description Since unity cannot be achieved by accumulation, the heart must be freed from all but one desire, that of using all for the contemplation of God.
Context The second illumination of Jami.
  Broader Fountains of light (Sufism, #HM3039).
  Followed by Perception of hidden reality (Sufism, #HM2531).
  Preceded by Single-heartedness (Sufism, #HM3085).

♦ **HM2766d The relationship between man and man** (Christianity)
Description In moments of crisis a primitive drive joins people together by force, friends and foes, good and evil. This drive is superior to all of them, independent of their desires and deeds. It is the spirit, the breath of God on earth. It descends on people in whatever form it wishes, as dance, eros, hunger, religion or slaughter. It does not ask permission. It does not portion out riches or capacities equally. Injustice, cruelty, longing and hunger are the four steeds that drive its way through humanity. It is never created out of happiness or comfort or glory, but out of shame and hunger and tears.
At every moment of crisis an array of men risk their lives in the front ranks as standard bearers of God to fight and take upon themselves the whole responsibility of the battle. This identification with the universe gives birth to the virtues of an ethic, responsibility and sacrifice. Responsibility to and sacrifice for the universe is why solidarity among people is no longer a tenderhearted luxury but a deep necessity. These standard bearers fight on without certainty. Action takes on a new, incalculable value and all - beauty, knowledge, hope, economic struggle, daily and seemingly meaningless cares - takes on an unexpected holiness.
Everyone has his own specific road to salvation. For one person it is disease, lies and dishonour. For that person to be saved he must plunge into disease, lies and dishonour that he may conquer them. For another person it may be the road of virtue, joy and truth. It is that person's duty to plunge into virtue, joy and truth that he may conquer them and leave them behind.
Context The second stage in the action for the spiritual exercises of Nikos Kazantzakis.
  Broader Saviors of God: the action (Christianity, #HM3462).
  Followed by The relationship between man and nature (Christianity, #HM3488).
  Preceded by The relationship between God and man (Christianity, #HM3595).

♦ **HM2767d Al-Qabid** (Sufism)
Description The believer is aware of Allah as the one who constricts. Prevented from receiving the blessings of this world, the believer learns patience. These times may strengthen faith, bringing one closer to the creator and becoming His beloved. Knowledge that good and bad all come from Allah provides a balanced state.
Context A mode of mystical awareness in the Sufi tradition associated with the twentieth of the ninety-nine names of Allah.
  Broader Ninety-nine names of Allah (Sufism, #HH2561).
  Followed by Al-Basit (Sufism, #HM0115).
  Preceded by Al-'Alim (Sufism, #HM4010).

♦ **HM2768d Feminine consciousness**
Description Because of the different life experiences a woman undergoes, her consciousness will differ from that of a man's. The onset of menstruation means that, from puberty onwards, time will be perceived in terms of monthly cycles, and therefore not as a linear process. The passive nature of pregnancy, with no knowledge of the sex, health or character of the child to be born, requires a creativity which expresses itself in the preparation of the environment for what is created rather than the creation itself. The experience of labour, when the body has a will of its own and no longer responds to her commands, means that it is necessary to allow inner nature to guide, relinquishing the control of the ego. And the symbiotic relationship between a nursing mother and her child requires responsiveness to and awareness of the needs of others, and an interest in developing two-way, fulfilling relationships.
  Related Maternal thinking (#HH4646)   Acceptance of the female principle (#HH3663)
  Psychological development of women (#HH1976).

♦ **HM2769c Meditation way of the bodhisattvas** (Buddhism)
Byang-chub-sems-dpa (Tibetan)
Description This is the state of consciousness appropriate to those working for the salvation of all.

HM2769

Context This way leads to Buddhahood and the ultimate consciousness, according to Tibetan Gelugpa Buddhism.
    Broader Consciousness ascension stages game (Buddhism, Tibetan, #HH4000).
    Narrower First scriptural bodhisattva awareness (Buddhism, Tibetan, #HM2155)
    Third scriptural bodhisattva awareness (Buddhism, Tibetan, #HM2215)
    Sixth scriptural bodhisattva awareness (Buddhism, Tibetan, #HM2385)
    Fifth scriptural bodhisattva awareness (Buddhism, Tibetan, #HM2909)
    Ninth scriptural bodhisattva awareness (Buddhism, Tibetan, #HM2292)
    Tenth scriptural bodhisattva awareness (Buddhism, Tibetan, #HM2421)
    Fourth scriptural bodhisattva awareness (Buddhism, Tibetan, #HM2191)
    Second scriptural bodhisattva awareness (Buddhism, Tibetan, #HM2739)
    Eighth scriptural bodhisattva awareness (Buddhism, Tibetan, #HM2816)
    Seventh–scriptural bodhisattva awareness (Buddhism, Tibetan, #HM2361).
    Related Spiritual direction (#HH0442)   Heart–awakening stage of life (#HM3518).
    Followed by Way of the Buddhas (Buddhism, Tibetan, #HM2185).
    Preceded by Meditation way of the hearers (Buddhism, #HM2161)
    Meditation way of the four truths (Buddhism, #HM2245).

♦ HM2770b Openness
Description Merleau–Ponty has described consciousness as openness to the world. This implies that to be conscious one must have a world to be open to, where phenomenological reality is not doubted; and one must have an ability to be aware of this reality and to orientate one's self to it.
    Narrower Spatial awareness (#HM2176).

♦ HM2771d Imploring succour (ICA)
Description This is the cry for help when one asks oneself, "How will I ever make it through ?". It occurs when one acknowledges the interdependence of all things and is like being "at the end of one's tether".
Context In the ICA New Religious Mode in the arena of acting out one's freedom (the life of prayer) the third formal aspect is surrender to one's unlimited inadequacy, or petition. At the third phenomenological level, this occurs when one acknowledges that which is beyond one's power to control.
    Broader Surrender to inadequacy (ICA, #HM2922).
    Followed by Levitational submission (ICA, #HM2706).
    Preceded by Representational sign (ICA, #HM2532).

♦ HM2772b Triadic awareness
Description Triadic awareness is established by linking monadic and dyadic awareness. It deals with the energetic aspect of reality, which is constant movement. Beyond form and content, yet including them, intentional, holistic aspects prevail. Time is a predominant factor. C. Castaneda cites "totality of ourselves", uniting tonal and nagual. Buddha's "beyond perception and non–perception" seems to indicate the same fundamental notion. Experiences of a rather cataclysmic character (c.f. Revelation of John), and concepts of the father in Christian tradition also relate. Exemplifications of this principle often show the trinitarian character predominant.
Refs Collins, John E and Woods, Ralph (Eds) *Civil Religion and Transcendent Experience* (1988).
    Preceded by Dyadic awareness (#HM2930).

♦ HM2773d Original integrity (ICA)
Description This is the experiencing of one's self as a stranger, out of step with this world. One's roots are in another world, the world of mystery, consciousness, care and tranquillity. One is always on a journey between these worlds, knowing that one's final obligation is to no thing. It is the response of Jesus when he is asked about his family, "Who is my mother, my sister and my brother ?".
A dimension of this experience is a sense of awe, that is fear and fascination. The experience is marked by the knowledge that there is nothing to fall back on. There is an ever–present fear that there may be nothing to show for all one's endeavours. Others do not seem to understand why one is not satisfied and sustained by the same temporal pleasures that give meaning to their lives. Knowing that this world is not one's home, one discovers that this is what one always wanted. The decision is to remain in this state, knowing that everything in this world is of concern; but at the same time, having a deep sense of detachment from anything that claims one's life. It is like taking the mantle of a foreign emissary, representing another world in the midst of this world, and possessing inviolate integrity.
Context This state is number 29 in the ICA *Other World in the midst of this World*.
    Broader Final accountability (ICA, #HM3206).
    Followed by Living death (ICA, #HM2808)   Universal fate (ICA, #HM2687)
    Ultimate awareness (ICA, #HM2388)   Worldly detachment (ICA, #HM2451)
    Diaphanous intuition (ICA, #HM2927).
    Preceded by Living death (ICA, #HM2808)   Universal fate (ICA, #HM2687)
    Primal vocation (ICA, #HM2621)   Beyond morality (ICA, #HM3049)
    Essential dubiety (ICA, #HM3135).

♦ HM2774d Universal concern (ICA)
Agape as compassion
Description This is an experiential trek in the *Mountain of Care*, within the ICA *Other World in the midst of this World*, when the individual is aware of binding the wounds of time. The states are described in separate entries.
    Broader Mountain of care (ICA, #HM2170).
    Narrower Primal sympathy (#HM2550)   Sacrificial passion (#HM2641)
    Universal compassion (#HM2734)   Soteriological existence (#HM2904).
    Related Singular mission (ICA, #HM3217)   Transparent power (ICA, #HM2828)
    Original gratitude (ICA, #HM3105).

♦ HM2775b Eighteen unshared attributes of Buddha (Buddhism, Tibetan)
Context One of the paths of effect according to Buddhist teaching.
    Broader Paths of effect (Tibetan, #HM3249).

♦ HM2776d Pratyeka Buddha arhat awareness (Buddhism, Tibetan)
Description This is characterized as the fifth path for solitary Buddhahood and is called the *awakened–state*. It is represented, however, as not being full–Buddhahood. Properly, the solitary one is now an Arhat, but not a world–saving Boddhisattva.
Context In Tibetan Sakya Buddhism this is one of the states in the "Ascension Stages Game". In some sets it is numbered 47 on the board.
    Broader Religious traditions in the ascension stages game (Buddhism, Tibetan, #HM3341).
    Followed by Cessation–awareness (Buddhism, Tibetan, #HM2172)
    Pure–lands awareness (Buddhism, Tibetan, #HM2168)
    Awareness of relative reality (Hinduism, Yoga, #HM2032)
    Great-vehicle lower-path awareness (Buddhism, Tibetan, #HM2268).
    Preceded by Pratyeka Buddha cultivation awareness (Buddhism, Tibetan, #HM2252).

♦ HM2777d Everlasting community (ICA)
Description This is the experience of embodying a multitude of internalized mentors, expressing and embodying all who have gone before. Such an individual becomes wise beyond his years, as the common mind of the saints informs him and makes his presence radical. This is Hermann Hesse's image of the "League" in the book "Journey to the East", and may be compared to Nikos Kazantzakis' image of the crimson line in history and of the ancestors who live within every person's skull. Perhaps this was the experience of Joan of Arc when she heard voices calling her to save France.
A dimension of this experience is a sense of awe, that is fear and fascination. There is knowledge that history is not mechanically predetermined but that people participate in creating it; and a strange sense of power as an individual chooses who will inform his consciousness. He is overwhelmed with the choice before him; and with a kind of stage–fright in knowing that all creation is watching him act out his life. He is responsible to all creation for all creation. The decision is to be intentional about the only life and death he has and which mentors he will allow to inform it. There is an internal silence as the individual quietens his own voice and focuses his being to listen to the others; it is as though he has thousands of colleagues awaiting his call.
Context This state is number 63 in the ICA *Other World in the midst of this World*.
    Broader Endless life (ICA, #HM2437).
    Followed by Contentless word (ICA, #HM2373)   Contingent eternality (ICA, #HM2456)
    Transcending hostility (ICA, #HM2658)   Transcendent immanence (ICA, #HM3034)
    Passionate disinterest (ICA, #HM2547).
    Preceded by Blissful seizure (ICA, #HM3148)   Impactful profundity (ICA, #HM2607)
    Transcending hostility (ICA, #HM2658)   Passionate disinterest (ICA, #HM2547)
    Resurrectional existence (ICA, #HM2631).

♦ HM2778d Spite (Buddhism)
Pradasha — 'Tshig–pa (Tibetan)
Description The desire to respond with harsh words to someone who has pointed out a fault.
Context One of the twenty *secondary afflictions* of Tibetan Buddhism.
    Broader Secondary afflictions (Buddhism, Tibetan, #HH0781).

♦ HM2780c Multi–dimensional consciousness
Description By letting go of the idea of one's self as three– dimensional, gross matter, one achieves *insight* and *awareness* which has sometimes been referred to as holiness. This may be compared with the release of energy when an atom is smashed, although the analogy breaks down when one realizes that in this case one is both the atom and the smasher. There is therefore great resistance within one's self to this smashing process but, when it is achieved, the loss of misery due to the idea of ego–separateness inherent in the three–dimensional self opens the way for channelling boundless compassion to the world, energy which would otherwise have been distorted or diverted to self–centred pursuits.

♦ HM2781c Swapna state of dream consciousness (Yoga)
Svapna
Description In the yogic theory of avasthas (states) this is the second, represented as dream consciousness. According to the Maharishi Mahesh Yogi it is the second in a series of seven states attainable to human beings. In the Mandukya Upanishad, Swapna is said to be the field of Taijasa, the mental condition.
    Broader Awareness of relative reality (Hinduism, Yoga, #HM2032).
    Related Taijasa (Hinduism, #HM3040)   Turiya awareness (Hinduism, #HM2395)
    Transcendental meditation (TM, #HH0682).
    Followed by Susupti state of unconsciousness (Hinduism, Zen, #HM2957).
    Preceded by Jagrat state of waking consciousness (Yoga, #HM2141).

♦ HM2782d Individual rights (ICA)
Description This state is the drive within a person for an equitable solution that honours all the participants in a situation, when he realizes that his social covenant is with all, not just a select few.
Context In the ICA New Religious Mode in the arena of acting out one's engagement (the life of obedience) the second formal aspect is embodying equity as the disestablishment to call forth human justice. At the first phenomenological level, this occurs when a person becomes aware that he is not autonomous, but rather is bound unto death to other people.
    Broader Embodying equity (ICA, #HM2442).
    Followed by Corporate duty (ICA, #HM2694).

♦ HM2783d Amoghasiddhi–karma awareness (Buddhism, Tibetan)
Karma–paripurana
Description This state is characterized as the moral compulsion to complete actions taken in tantric practice. It is represented by the northern Buddha, Amoghasiddhi, whose Tara–consort is the Faithful One. Together they correspond to the forces that destroy obstacles in the dharma path.
Context In Tibetan Sakya Buddhism this is one of the states in the "Ascension Stages Game". In some sets it is numbered 70 on the board.
    Broader Supreme heavens in the ascension stages game (Buddhism, Tibetan, #HM3214).
    Followed by Great-black-lord awareness (Buddhism, Tibetan, #HM2118)
    Third vajra–master awareness (Buddhism, Tibetan, #HM2727)
    Second vajra–master awareness (Buddhism, Tibetan, #HM2703)
    First scriptural bodhisattva awareness (Buddhism, Tibetan, #HM2155)
    Seventh–scriptural bodhisattva awareness (Buddhism, Tibetan, #HM2361).
    Preceded by Great-black-lord awareness (Buddhism, Tibetan, #HM2118).

♦ HM2784d Heavenly advocate (ICA)
Description This is experienced as a question about life in the form "Is this what is required by all of history and the whole world ?", when a person's own consciousness calls his intentions into question. An example is the scene in the movie "Gandhi", when a man is brought to him who has killed a small boy in the rioting at the time of Indian independence. This encounter involves all the implications of Gandhi's participation in the independence movement.
Context In the ICA New Religious Mode in the arena of knowing one's internal sociality (the life of meditation) the second formal aspect is the experience of being called to accountability by one's interior priors. At the fourth phenomenological level, this occurs when a person is sustained by communion with people and events which serve as milestones for his consciousness.
    Broader Called to accountability (ICA, #HM2008).
    Preceded by Universal father (ICA, #HM2418).

♦ HM2785d Intuitive, meditative thinking
Description As opposed to *rational, calculative thinking*, which is concerned with the *ontic* level of being, *intuitive, meditative thinking* is *ontological*, concerned with the essential dimensions of being, and based on a respectful attitude of awe and mystery towards reality. This mode of thinking overcomes the limits of ego–consciousness and separation inherent in the understanding of reality as the objective world.
    Related Intuition (#HM3634)   Rational, calculative thinking (#HM2283).

♦ **HM2786d Eternal moment** (ICA)
Description This is experienced at those times when a person encounters the fact that his action will produce permanent change in history, that every decision is filled with significance, that the present moment is filled with implications about the future and with meaning for all of history. It is like the first time that a few scientists were able to sustain nuclear fission in a primitive reactor.
Context In the ICA New Religious Mode in the arena of acting out one's deed (the life of doing) the second formal aspect is extending one's engagement to history. At the fourth phenomenological level this occurs when the unique greatness of one's expenditure is embodied.
Broader Engagement to history (ICA, #HM2588).
Preceded by Determining history (ICA, #HM2682).

♦ **HM2787f Mania**
Description This state is a symptom of mental disturbance characteristically marked by distracted and agitated behaviour, hyperactivity, excitability and excessively elevated moods and cheerfulness. There may be senseless rambling and laughter, and often violence to the person himself or to those who attempt to interfere. In cyclothymia, or manic depression, the state is at one "pole" of the cycle of behaviour; but mania may occur without the balancing depression following. The term is often used as synonymous with an irresistible urge (kleptomania, pyromania); and also to describe a furor or frenzy.
Broader Anomalies in the flexibility of associations (#HM4312).
Related Phobia (#HM1509) Frenzy (#HM3594) Elation (#HM4899)
Theomania (#HM1027) Hypomania (#HM0514) Cyclothymia (#HM1616).

♦ **HM2788c Pre-conscious pure intelligence**
Description Pure intelligence in its transformation to assume the role of creative intelligence becomes consciousness; reflecting, in the individual, the cosmic evolution from the Father (transcendent) to the Son (logos - sun - "I Am"). The pre-conscious pure intelligence is the Self, or Father, or God in each person, according to teachers such as the Maharishi.
Related Creative intelligence (#HH0679).

♦ **HM2789d Seventh vajra-master awareness** (Buddhism, Tibetan)
Description This state of awareness is characterized by intellectual radiance and it is represented as the secret mind's participation in a projection of the Clear Light. **Context** In Tibetan Sakya Buddhism this is one of the states in the "Ascension Stages Game". In some sets it is numbered 83 on the board.
Broader Way of the vajra masters (Buddhism, Tibetan, #HM3090).
Followed by Supreme-heaven awareness (Buddhism, Tibetan, #HM2813)
Final vajra-master awareness (Buddhism, Tibetan, #HM2849).
Preceded by Fifth vajra-master awareness (Buddhism, Tibetan, #HM2763)
Fourth vajra-master awareness (Buddhism, Tibetan, #HM2251)
Guhya-samaja urgyan awareness (Buddhism, Tibetan, #HM2699).

♦ **HM2790g Lower unconscious** (Psychosynthesis)
Description According to R. Assagioli, this is the lowest region of a tripartite division of the unconscious mind and contains the elementary intelligence or purposeful activities which direct the vital physical, bodily functions, and regulate and integrate them into the unitary purpose of sustaining the human life form. Here also is the level of consciousness where the basic urges and drives that lead to compelling behavioural expression, reside in their mental form. The lower unconscious embraces many compartments. It not only relates to maintaining the body as a biological individual, but also as a psychological individual, as it is here that is found the storehouse of emotional complexes, as well as the activities that produce dreams, fantasies, and involuntary abnormal mental activity, whether pathological or anomalous.
Related Psychosynthesis (#HH0002).
Followed by Middle unconscious (Psychosynthesis, #HM2306).

♦ **HM2791d Everlasting inescapability** (ICA)
Description This is the experience of encountering the mystery at every point in one's life; when one sees that the mystery is everywhere and that there is no where to flee to. It may be compared to the experience of people in the closed city in Camus' book "The Plague".
Context In the ICA New Religious Mode in the arena of standing present to the mystery of being in life (life of contemplation) the first formal aspect is the experience of enigmatic not-me-ness or of externality. At the second phenomenological level this occurs when a person's being trapped reveals that he is forever bound to the mystery of life.
Broader Awareness of externality (ICA, #HM2637).
Followed by Hallowed honour (ICA, #HM2680).
Preceded by Dangerous intrusion (ICA, #HM2449).

♦ **HM2792d Alertness of the heart** (Sufism)
Wuquf-i qalbi — Dhikr-i-Qalbi
Description In this state, with frequent repetition of the dhikr, awareness of God is manifest to the heart and His presence felt within and without.
Related Cleansing of the heart (Sufism, #HM6932).

♦ **HM2793b Bodhi-pakkhiya dhamma** (Buddhism)
Bodhi-pakshika dharma
Description Seven groups of mental qualities (37 in all) summarise the Buddhist path. They consist of three groups of four - foundations of mindfulness, right effort and psychic power - followed by a group of controlling or powerful qualities (faith, strength, mindfulness, concentration, wisdom). The next group lists the seven bodhi (awareness or awakening) factors; and the final consists of the *eightfold path of yoga*. Although the 37 may be considered as occurring in sequence they are also said to be simultaneously present in *lokuttara* transcendent consciousness.
Related Lokuttara consciousness (Buddhism, #HM2120)
Consciousness in Buddhism (Buddhism, #HM2200).

♦ **HM2794c Emptiness of the loci of the senses** (Buddhism, Tibetan)
Adhyatmabahirdhashunyata
Context One of the eighteen emptinesses comprising the paths of view in Tibetan Buddhism.
Broader Emptinesses on the paths of the view in Buddhism (Buddhism, Tibetan, #HM2944).

♦ **HM2795f Secondary delusions**
Delusional ideas — Delusional systematization
Description These delusions are secondary because they arise from an attempt to explain or understand other abnormal experiences or morbid affective moods, which may include primary delusion. They may be irrational and held with extraordinary conviction but, in that the experience triggering them is evidently unusual, they are understandable. Delusions are more likely to arise in people who have strong ego boundaries and who are rigidly detached from their environment, and the development and maintenance of idiosyncratic ideas at odds with the environment may be in order to maintain their identity. Again, a secondary delusion attempts to reconcile incompatible conditions - a disturbing experience at odds with reality - in order to reduce the tension caused by cognitive dissonance. Because this is an impossible task, a complex, interlocking argument is built up, ever more complex and idiosyncratic as experience increases in perplexity. This is delusional systematization.
Broader Delusive consciousness (#HM2600).
Related Primary delusions (#HM4604) Cognitive dissonance (#HH8204).

♦ **HM2796d Definitive effectivity** (ICA)
Description This occurs whenever an individual discovers his capacity to do the impossible. Accurate in all he does and is, he finds he is performing miracles where others said it couldn't be done. By seeing through the complexity to the underlying contradiction, he is able to focus his energy at the root of a problem and generate a solution. A classic example of this state is the great Maratha warrior Shivaji who, at sixteen years of age, led his army into battle to defeat the powerful Moghul empire in India by outwitting them in strategy. It may be compared to a top-class international athlete way ahead of the rest of the field, who knows rather than hopes he will win; or to the Pied Piper of Hamelin: whistle a tune and everyone follows.
A dimension of this experience is a sense of awe, that is fear and fascination. The individual stands in awe at the possibility contained in his power to move mountains, realizing he can make one hell of a difference, whenever he decides to go through with it. All he fears is his own power which is a dangerous thing if not handled properly. He becomes a symbol of chastity, focussing his entire being on doing one thing; audaciously taking charge of history and deciding in which direction it is to go. There is the knowledge that the price is his own life; but in deciding to pay this price, he knows he has unlimited power to effect change.
Context This state is number 48 in the ICA *Other World in the midst of this World.*
Broader Transparent power (ICA, #HM2828).
Followed by Singular adoration (ICA, #HM2111) Seminal illumination (ICA, #HM2056)
Sacramental universe (ICA, #HM2445) Contingent eternality (ICA, #HM2456)
Soteriological existence (ICA, #HM2904).
Preceded by Invented history (ICA, #HM3245) Singular adoration (ICA, #HM2111)
Impactful profundity (ICA, #HM2607) Destinal accountability (ICA, #HM2309)
Soteriological existence (ICA, #HM2904).

♦ **HM2799e Breaking point of basic fear** (Brainwashing)
Description Fear of total annihilation is brought about in a state of total inner and exterior conflict, when the individual is estranged from all those around him and from his own inner self. Physical and mental integration break down, leading to such anxiety and depression that suicide may be contemplated. Delusions and hallucinations may occur as the psyche attempts to protect itself from this fear.
Context A stage reached in thought reform, a system of organized, deliberate and total psychological training which effects individual change through two basic elements - *confession* (renouncing of past beliefs and attitudes) and *re-education* (remaking of the individual in the required image).
Broader Thought reform (Brainwashing, #HH0865).
Related Breaking point of total conflict (Brainwashing, #HM2354).
Followed by Desperate gratitude (Brainwashing, #HM3017).
Preceded by Self-betrayal (Brainwashing, #HM2229).

♦ **HM2800b Sufi contracted-consciousness** (Sufism)
Kabd
Description Contraction (kabd) is the state in which the mystic forces out from himself everything except spiritual gnosis. He is compelled to do this by the action of God who induces in him a feeling of *great desolation.* In fear that he will lose God he recoils from his comforts and practises dying unto himself, by denying his lower ego.
Broader Human development (Sufism, #HH0436)
Stations of consciousness - Simnami (Sufism, #HM2341).

♦ **HM2801c Life of knowing** (ICA)
Description An arena in the ICA *New Religious Mode* when one is aware of articulating the word about life.
Broader New religious modes (ICA, #HM3004).
Narrower Solitary being (#HM2404) Glorious mystery (#HM2042)
Wonder at the world (#HM3062) Contentless transformation (#HM3106).

♦ **HM2804c Secular transcendence**
Description If there are two "lives", the way of immanence and the way of transcendence, then the second is identified in the Old Testament as the way in which the Jewish nation should (but rarely did) conduct itself. Although the Bible refers to God and Jesus as the Son of God, these titles may simply be misunderstandings in the light of "immanent" interpretations. If one is committed to the way of transcendence, then the little world constructed by the individual to suit immanent purposes comes to an end. Sharpening to the limit the understanding of reality shaped by the conceptual form confirms understanding but destroys conceptual form. Thus Jesus sets out the way of transcendence with final clarity but exposes the "end of the world" imagery as inappropriate. If Christian faith is defined as commitment with the ultimate concern to that which came to expression in Jesus Christ, then it is commitment to him as the very incarnation of the way of transcendence. Transcendence can replace God in the whole Christian faith. This interpretation is both more comprehensible and more appropriate to our time. What is believed about God is actually contained within this position.
Refs Kee, Alistair *The Way of Transcendence* (1985).
Broader Transcendence (#HM4104).
Related Secular quest (#HH4994).

♦ **HM2805b Stages of the gunas** (Yoga)
Description These are stages in *samadhi* through which, according to Patanjali, the Yogi must pass before being released from the domination of the gunas.
Narrower Vicara stage (#HM3460) Ananda stage (#HM3212)
Asmita stage (#HM3509) Vitarka stage (#HM2912).
Related Three gunas (#HH0413).

♦ **HM2807b Great compassion** (Buddhism, Tibetan)
Context One of the paths of effect according to Buddhist teaching.
Broader Paths of effect (Tibetan, #HM3249).

♦ **HM2808d Living death** (ICA)
Description This occurs when an individual experiences having fully embraced the fact of his death. This allows them to be fearless: no-one can touch him or disillusion him for he has died to his entire life. Having decided to be the "living dead", he is afraid of nothing. Sun Tzu speaks of being one with the void; St. Francis speaks of Brother Death; and St. Paul of being "dead and yet alive". Protesters who calmly set themselves alight are a sign of having embraced death.
A dimension of this experience is a sense of awe, that is fear and fascination. There is a kind of irritation with being bombarded with insignificant things, but a fascination with death mixed with a sense of the liveliness of life. The decision is to put one's death into the universe, one wills

to die the one death one has to give. There remains a constant awareness that one day death will come. With fear of this moment is the knowledge that such fear is a little silly.
Context This state is number 61 in the ICA *Other World in the midst of this World.*
Broader Endless life (ICA, #HM2437).
Followed by Creative futility (ICA, #HM2493) Essential dubiety (ICA, #HM3135) Original integrity (ICA, #HM2773) Seminal illumination (ICA, #HM2056) Resurrectional existence (ICA, #HM2631).
Preceded by Vital signs (ICA, #HM2875) Final blessedness (ICA, #HM2958) Creative futility (ICA, #HM2493) Original integrity (ICA, #HM2773) Diaphanous intuition (ICA, #HM2927).

♦ **HM2809f Obsession**
Description A neurotic rather than a psychotic symptom, a person is said to be suffering from an obsession when a persistent uncontrollable idea leads to the performing of unreasoned acts - such as rechecking locked doors, counting steps and so on - which, if not performed, would promote a dominant feeling of guilt, anxiety and frustration. Such actions tend to be ritualistic and, although accepted by the person as unreasonable, appear to be compelled by some force outside. The original meaning of the term is *to haunt* and is contrasted in religious terms with *possession*, when the evil entity dominates the mind from within.
The obsessional thought, idea or whatever is unwanted. It is intrusive and, although it is seen to originate within the individual himself, he sees it as neither natural nor desirable. He knows it to be senseless and actively resists it, at least in the early stages.
A particular kind of obsession is *obsessional rumination*, where the individual is unable to desist from internally debating some abstruse, metaphysical topic, or one not amenable to rational debate. This impairs normal mental effectiveness.
Related Rites (#HH0423) Compulsion (#HH0345) Spirit possession (#HH0056) Over-valued ideas (#HM1466) Resistance to shift (#HM5329).

♦ **HM2811g Collective unconscious** (Psychosynthesis)
Collective conscience
Description According to R. Assagioli, there is a psychological environment composed of the influence from all selves. Selective elements from this general environment interact with the individual unconsciousness. These elements may be passive memories of a collective or racial nature, or active influences emanating from the whole of which selves are a part and which act on a through the higher self and the higher unconscious. They can be compared with E Durkheim's "conscience collective", the link between one generation and another, the shared values and sentiments.
Related Archetype (#HH0019) Psychosynthesis (#HH0002) Collective unconscious (Jung, #HM0085).
Followed by Conscious self (Psychosynthesis, #HM2123).
Preceded by Higher unconscious (Psychosynthesis, #HM2057).

♦ **HM2812d Disinterested collegiality** (ICA)
Description This is experienced on embracing a situation of human suffering as one's own, in order to change it, out of no other motive than care for the well-being of the people caught in that situation. It occurs when a person sets aside all other concerns and relations and gives himself to transforming the situation of suffering. It is like in the movie "Norma Raye" when a New York lawyer and labour organizer arrives in a town in the southern United States to organize the workers in a textile mill. There is no good reason for him to be there, in a very dangerous situation, except that he cares about the welfare of workers in textile mills.
Context In the ICA New Religious Mode in the arena of acting out one's engagement (the life of obedience) the third formal aspect is embodying charity as the transestablishment to act out social concern. At the third phenomenological level this occurs when people willingly embrace their responsibility to society.
Broader Embodying charity (ICA, #HM3056).
Followed by Sacrificial friendship (ICA, #HM2754).
Preceded by Personal obligation (ICA, #HM2971).

♦ **HM2813d Supreme-heaven awareness** (Buddhism, Tibetan)
Akanistha
Description This state of awareness is characterized as the epitome of conscious existence and represented as the heaven of the supreme divinities of the Pure lands. Here, it is said, are the Buddhas of the six directions and their assemblies of Buddhas-to-be, arch-angelic bodhisattvas of the tenth stage. Here too arrive the saintly yogis, whose bodies live on earth while their minds ascend to the Buddhist Olympus in concentration. There, in the enclosure of the vajra-dhatu Palace, the Buddhas in their higher bodies (sambhoga-kaya) teach the final secrets of the Great Vehicle. The Buddha-field, akanistha, is symbolized as a lotus.
Context In Tibetan Sakya Buddhism this is one of the states in the "Ascension Stages Game". In some sets it is numbered 84 on the board.
Broader Supreme heavens in the ascension stages game (Buddhism, Tibetan, #HM3214).
Followed by Absolute-body awareness (Buddhism, Tibetan, #HM2397).
Preceded by Final vajra-master awareness (Buddhism, Tibetan, #HM2849) Guhya-samaja urgyan awareness (Buddhism, Tibetan, #HM2699) Seventh vajra-master awareness (Buddhism, Tibetan, #HM2789) Ninth scriptural bodhisattva awareness (Buddhism, Tibetan, #HM2292) Tenth scriptural bodhisattva awareness (Buddhism, Tibetan, #HM2421).

♦ **HM2814d Ever-present brother** (ICA)
Description This is the realization that there is no escape from one's exemplar - once chosen, he is always present. It occurs when a person confronts something new in life and finds himself having a dialogue with the same exemplar, again and again; as in the movie "Requiem for a Heavyweight" when, in spite of all that happens, the boxer continues to look upon his manager as his guide.
Context In the ICA New Religious Mode in the arena of knowing one's internal sociality (the life of meditation) the third formal aspect is one's experience of being given existential guidance or encountering the Saints. At the third phenomenological level, this occurs when one dialogues across time and space with those who also travel this journey of the spirit.
Broader Existential guidance (ICA, #HM2114).
Followed by Guardian angel (ICA, #HM3130).
Preceded by Scorching avatar (ICA, #HM2921).

♦ **HM2815d Taurus-consciousness** (Astrology)
Undifferentiated receptivity
Description This is the capacity for submitting in a spirit of patience, tolerant sympathy and loving sacrifice, to the dynamic working of undifferentiated energy ; and of acting from a grounded source of strength. Under optimal conditions, Taurus is associated with *intelligent openness* to self, others, and to the total problem situation, bringing resources to bear on the immediate challenge.
Context One of the twelve conditions of being or streams of divine energy encompassed within the zodiac. It is said that all twelve conditions must be present and playing its proper part for any successful and fulfilling action to take place. This may be through a group of individuals with different endowments and soul qualities working together; or it may be through one individual who, by self-discipline, practice and meditation, has allowed all twelve conditions to arise within himself.
Broader Fixed awareness (Astrology, #HM2554) Earthy awareness (Astrology, #HM3235) Zodiacal forms of awareness (Astrology, #HM2713).
Followed by Gemini-consciousness (Astrology, #HM2924).
Preceded by Aries-consciousness (Astrology, #HM2665).

♦ **HM2816d Eighth scriptural bodhisattva awareness** (Buddhism, Tibetan)
Acala
Description This state of awareness is characterized by independence, and is represented by the eighth-stage bodhisattva striving effortlessly, without attachment to any ideology. He is inclined to leave the samsara without becoming a perfect Buddha; however, he is said to be reminded by the Buddhas above him of his vows to work for all beings' release; and for this purpose he acquires still further powers.
Context In Tibetan Sakya Buddhism this is one of the states in the "Ascension Stages Game". In some sets it is numbered 96 on the board.
Broader Meditation way of the bodhisattvas (Buddhism, #HM2769).
Followed by Tenth scriptural bodhisattva awareness (Buddhism, Tibetan, #HM2421) Ninth scriptural bodhisattva awareness (Buddhism, Tibetan, #HM2292).
Preceded by Sixth scriptural bodhisattva awareness (Buddhism, Tibetan, #HM2385) Seventh-scriptural bodhisattva awareness (Buddhism, Tibetan, #HM2361).

♦ **HM2817d Global guardianship** (ICA)
Description In this state an individual discovers that he has taken the burden of caring forever for all of creation, and he knows it is exactly what he needed. When Albert Einstein wrote to President Roosevelt concerning the potential for developing the atomic bomb and the need for the United States to move ahead with this project before the enemy did so, he was taking this sort of responsibility.
A dimension of this experience is a sense of awe, that is fear and fascination. In acknowledging that the world has become his ward, the individual experiences deep paralysis at the enormity of the task and the realization that there is no escape from it, wondering if he can possibly handle it alone, but nevertheless fascinated by the destinal role he has been given and rejoicing in the burdensome responsibility that has been placed in his hands.
Being the global guardian entails the decision to put one's whole life into forging history with everything one has. One keeps taking charge, time after time after time, accepting the perpetual task and seeing one can never go back to anything less. There is no-one to blame and one invents the rules. Although the burden is always expanding one discovers a hidden ability to bear it.
Context This state is number 41 in the ICA *Other World in the midst of this World.*
Broader Singular mission (ICA, #HM3217).
Followed by Vital signs (ICA, #HM2875) Vibrant powers (ICA, #HM2982) Ancestral obligation (ICA, #HM3122) Diaphanous intuition (ICA, #HM2927) Individual fatefulness (ICA, #HM2223).
Preceded by Vibrant powers (ICA, #HM2982) Primal sympathy (ICA, #HM2550) Beyond morality (ICA, #HM3049) Individual fatefulness (ICA, #HM2223) Soteriological existence (ICA, #HM2904).

♦ **HM2818b Super-contemplative state** (Yoga)
Description According to Patanjali, this is a state of unfailingly accurate perception when all other modes of vision are in their right proportion and the senses are required only for constructive work at their respective levels. There is still one slight veil of illusion which must be displaced before achieving samadhi, that of the awareness of consciousness itself.
Related Awareness of spiritual identity (Yoga, #HM3232).
Followed by Samadhi (Hinduism, Yoga, #HM2226).

♦ **HM2820e Negative intrapsychic forces** (Psychism)
Description Consciousness is limited in range by inherent genetic or structural factors, but it is also limited by intrapsychic forces some of which may have an acquired nature. In the case of blocked access to the unconscious store of memories, for example, an agency such as repression may come into play. It is sometimes convenient to distinguish what is in the memory in voluntarily recallable form and what is not. The former field is called by some, the pre- or fore-conscious; the latter, the unconscious. Other intrapsychic forces may block access to successive levels of the unconscious. Such levels may include sensory experiences of which there was no conscious awareness when they were stored, and inexplicable contents of the unconscious. The latter include other personalities, uncanny knowledge, psychic powers such as healing, reported former lives, and communications with intelligences other than embodied, human ones. Undoubtedly the privative, or negative, aspect of the blocking intrapsychic forces serves a salutary purpose and is conducive to a rational orientation to the world of everyday experience.

♦ **HM2821d Everlasting enemy** (ICA)
Description This is the realization that nothing one can do will make a difference, for the mystery is always breaking one's illusions; when one knows that the mystery of life is working out history and that one is not ultimately in charge. An example is when the people in Ignazio Silone's novel "Bread and Wine" fight for a vision they have of the future; and experience that it all falls apart even as their vision is coming true.
Context In the ICA New Religious Mode in the arena of articulating the word of life (the life of knowing) the fourth formal aspect is one's experience of the final nothing in life or the glorious mystery of life. At the second phenomenological level, this occurs when one is somehow stripped of one's illusions, having to decide how to relate to their new universe.
Broader Glorious mystery (ICA, #HM2042).
Followed by Immutable friend (ICA, #HM3124).
Preceded by Eternal void (ICA, #HM2994).

♦ **HM2822d Capricorn-consciousness** (Astrology)
Conscious wisdom
Description A mode which is experienced as contemplative recognition of the totality of causes, constraints and possibilities. Under optimal conditions it is associated with a continually re-evaluated, conscious control, relating needs, opportunities and resources through appropriate action, using all available experiences and innovations.
Context One of the twelve conditions of being or streams of divine energy encompassed within the zodiac. It is said that all twelve conditions must be present and playing its proper part for any successful and fulfilling action to take place. This may be through a group of individuals with different endowments and soul qualities working together; or it may be through one individual who, by self-discipline, practice and meditation, has allowed all twelve conditions to arise within himself.
Broader Earthy awareness (Astrology, #HM3235) Cardinal awareness (Astrology, #HM2259) Zodiacal forms of awareness (Astrology, #HM2713).
Followed by Aquarius-consciousness (Astrology, #HM2973).
Preceded by Sagittarius-consciousness (Astrology, #HM2726).

♦ **HM2823d Pride** (Christianity)
Self-regard — Mana — Nga-rgyal (Buddhism, Tibetan) — Conceit
Description When self-love excludes love of others, clouding knowledge and isolating self from others and from God, then humanity is either ignored or used for the achievement of the individual's private wishes. This inordinate desire to excel is referred to as pride. In Church teaching, pride is said to be a capital sin in that it is the foundation for other sins. At its worst it not only breeds contempt for authority and incites the individual to ignore or reject commands from superiors, it also aims to withdraw himself from subjection to God.
As one of the six *root afflictions* referred to in Tibetan Buddhism, pride views the transitory as the real "I", puffing up the mind as it regards its own wealth and qualities, thus encouraging the generation of disrespect and of suffering. It may be in thinking of one's self as superior to inferior persons, to persons of an equal standing to one's self, or even to persons superior to one's self. Or it may be in thinking of one's self as only slightly inferior to persons far superior. It may be in thinking of the body and mind as "I"; or in thinking one has attained what in fact one has not; or even that one has attained a quality when one has actually deviated from the path.
In Hinayana Buddhism, pride or conceit is one of the formations aggregate (mental coefficients), being listed among the inconstant states which are immutable by nature, and as unprofitable secondary (sometimes present in any unprofitable or unprofitable-resultant consciousness). It is looked on as being like madness. Pride has the characteristic of haughtiness. Its function is arrogance, praising the self; it manifests as desire for self-advertisement, as vaingloriousness. The proximate cause is greed unassociated with wrong views.
Broader Root afflictions (Buddhism, #HH0270)
Awareness as mental-formation group of conscious existence (Buddhism, #HM2050).
Related Humility (Christianity, #HH0861) Spiritual pride (Buddhism, #HM2852).

♦ **HM2824d Cryptic disclosure** (ICA)
Description This occurs when the mystery is experienced as always absent; no matter how hard he tries to seek the final reality of life, the individual can never approach it. He feels he has been left out, cut off and denied a clear perception of the final revelation of the meaning of life, as if he were being called but not allowed to approach. The life he has created - all he loves and cares about - is saying "no" to him, rejecting him, and denying his presence. It may compared to the experience of the mother of the family in the film "Ordinary People" when all she loved, trusted and counted upon in her life are turning their backs on her; or to what Yeats wrote in his poem "The Second Coming" in the lines "Things fall apart, the centre cannot hold".
A dimension of this experience is a sense of awe that is fear and fascination. The fear is that one's whole life is meaningless and one will never understand what is happening. Yet there is a deep fascination in that one goes on showing up with a life to live; the state is unwanted and resented, yet desired. One cannot live with it, one cannot live without it. The decision is to be who one is. There is knowledge that the absence of the mystery is in fact the current manifestation of the mystery, and this experience is part of what one decides to be. One is left with the sense that nothing worse could happen.
Context This state is number 14 in the ICA *Other World in the midst of this World.*
Broader Infinite passion (ICA, #HM2234).
Followed by Absurd existence (ICA, #HM2331) Worldly detachment (ICA, #HM2451)
Interior discipline (ICA, #HM2851) Ubiquitous otherness (ICA, #HM2570)
Transcendent immanence (ICA, #HM3034).
Preceded by Essential dubiety (ICA, #HM3135) Interior discipline (ICA, #HM2851)
Ubiquitous otherness (ICA, #HM2570) Transformed existence (ICA, #HM2862)
Resurrectional existence (ICA, #HM2631).

♦ **HM2827c Perception of universality** (Sufism)
Description All beauty, perfection, knowledge, intelligence, are but mirrored reflections, however imperfect, of the one Lord. Reflection on any of particular manifestation of these will lead the gaze to the universal.
Context The fifth illumination of Jami.
Broader Fountains of light (Sufism, #HM3039).
Followed by Ultimate transparency of spirit (Sufism, #HM2483).
Preceded by Separation from the transient (Sufism, #HM2906).

♦ **HM2828d Transparent power** (ICA)
Agape as motivity
Description This is an experiential trek in the *Mountain of Care*, within the ICA *Other World in the midst of this World*, when the individual is aware of having the strength of ten. The states are described in separate entries.
Broader Mountain of care (ICA, #HM2170).
Narrower Interior discipline (#HM2851) Diaphanous intuition (#HM2927)
Impactful profundity (#HM2607) Definitive effectivity (#HM2796).
Related Singular mission (ICA, #HM3217) Universal concern (ICA, #HM2774)
Original gratitude (ICA, #HM3105).

♦ **HM2829b Spiritual intuitive cognition**
Intuition — Intuitive perception
Description Intuitive cognition is the apprehension of an object by the mind, independent of reasoning. Concentration and meditation can make the mind more intuitive. In fact, meditation has been defined as thinking in the heart; and it is the activity of the three-fold mind on the heart which is said to produce intuition. Intuition has also been defined as immediate knowledge of the Absolute obtained through *wisdom*, and thus contrasted with knowledge of external objects derived through the senses and the intellect (Swami Omkarananda). It is then the only method of approach to, and experience of, reality; the only means by which to realize God. Intuition is thus the normal capacity of that consciousness within man to which the intellect is sub-normal. R Steiner uses the term *intuition* to describe the third step in the exercise of *anthroposophic meditation*, beyond *inspiration* and *imagination*. Intuitive cognition may be of natural or supernatural knowledge. In the case of supernatural, or spiritual, intuitive cognition, the process of knowing is as mysterious as the source of knowledge; and there is a distinction to be made between intuition and *infusion*.
Related Intuition (#HM3634) Anthroposophical system (#HH0018).
Preceded by Imaginative consciousness (Anthroposophy, #HM2901).

♦ **HM2830f Nihilism**
Description A state of delusion in which the whole world including the individual, or parts of the world and the individual, may seem not to exist.
Related Nihilism (#HH4509).

♦ **HM2831d Wealth untold** (ICA)
Description This is the moment a person realizes that what he has is all he needs to live a full life; when he sees that he can do anything with precisely the resources, knowhow and status he already has. It may be compared with D. H. Lawrence in one of his poems when he writes "When I am quite, quite nothing, then I am everything."
Context In the ICA New Religious Mode in the arena of knowing one's disengagement (the life of poverty) the first formal aspect is the experience of being liberated from the claim of one's possessions. At the fourth phenomenological level, this occurs when one is filled with the power of requiring nothing from a situation.
Broader Liberation from possessions (ICA, #HM3131).
Preceded by Divine nothingness (ICA, #HM3042).

♦ **HM2834c Buddha wisdom body** (Buddhism, Tibetan)
Buddha truth body
Description The Buddha wisdom body is the ultimate true path part of the Buddha truth body, and is the final, perfect wisdom. It consists of the omniscient consciousnesses of eye, ear, nose, tongue body and mind; each perceives all qualities of what is perceived, including emptiness.
Broader Buddha-consciousness (Buddhism, Tibetan, #HM2735).
Related Buddha nature body (Buddhism, Tibetan, #HM2039)
Buddha emanation body (Buddhism, Tibetan, #HM3019)
Buddha enjoyment body (Buddhism, Tibetan, #HM3113).

♦ **HM2838g Mental consciousness** (Buddhism)
Manovijnana — Yid-kyi-rnam-par-shes-pa (Tibetan) — Mind-element consciousness — Manovynana
Description This is the consciousness that apprehends undemonstrable and non-obstructive forms, classed as phenomena sources. These may arise from: aggregation of the eight substances from which physical forms are constructed; clear space; a vow or absence of a vow; imagination, as in a dream; manifestation through meditative power. It is listed in Hinayana Buddhism as indeterminate (neither profitable nor unprofitable) at the sense-sphere level. It is without root cause, resultant (with profitable or unprofitable result) or functional (inoperative).
Context One of the six consciousnesses defined in Buddhism as dependent on the individual senses, and with objects of sense as their focus.
Broader Sense consciousness (Buddhism, #HM2664)
Perception through the senses (#HM3764)
Mundane resultant consciousness (Buddhism, #HM0130)
Indeterminate consciousness in the sense sphere - resultant (Buddhism, #HM5721)
Indeterminate consciousness in the sense sphere - functional (Buddhism, #HM3852)
Awareness of consciousness-group of conscious existence - senses and mind (Buddhism, #HM2556).
Related Adverting (Buddhism, #HM8336) Receiving (Buddhism, #HM7092)
Examining (Buddhism, #HM7324) Introspection (Buddhism, #HM7109)
Mental sensation (Buddhism, #HM6644) Eye consciousness (Buddhism, #HM2074)
Ear consciousness (Buddhism, #HM2169) Nose consciousness (Buddhism, #HM2364)
Body consciousness (Buddhism, #HM2562) Mind consciousness (Buddhism, #HM3323)
Tongue consciousness (Buddhism, #HM2263)
Mind-consciousness element (Buddhism, #HM6173).

♦ **HM2839d Perpetual revolutionary** (ICA)
Description This is the realization that there is no just society, it is always on the way, and that one is committed for life to seeking justice within that covenant; one finds one's self "thirsting for justice", when new demands become apparent that require one to act. An example is Eugene Debs who gave his whole life to seeking justice for labour.
Context In the ICA New Religious Mode in the arena of acting out one's engagement (the life of obedience) the second formal aspect is embodying equity as the disestablishment to call forth human justice. At the fourth phenomenological level, this occurs when one lives on behalf of the evolution of human consciousness.
Broader Embodying equity (ICA, #HM2442).
Preceded by Loyal opposition (ICA, #HM2910).

♦ **HM2840f Depravity**
Description Behaviour disorders which manifest so as to involve conflict with the moral perceptions of society and the law are sometimes referred to as depravity. The borderline between *perversion*, personality deviation and the acting-out trends of psychotics is so hard to define that individuals in this borderline group are still little understood.

♦ **HM2841d Five clairvoyances** (Buddhism)
Description These are the *divine eye*, the *divine ear*, the *knowledge of others' minds*, the *memory of former lives* and the *knowledge of extinction of contaminations*.
Context One of the paths of special qualities according to Buddhist teaching.
Broader Paths of special qualities (Buddhism, #HM3079).

♦ **HM2842d Contextual world-view** (ICA)
Description This occurs when a human being wakes up to the fact that he shapes the world into which he has been placed. There are no blueprints to point the way and no-one else will do it for him. Each person becomes the architect not only of his life but of the world itself and it feels like being asked to lift a ten-ton weight, not once but many times a day. He experiences the fact that designing his world means the mundaneness of having done this or that activity at least a hundred times before and knowing he will have to do it at least a hundred times more.
A dimension of this experience is a sense of awe, that is fear and fascination. One is confronted by the reality that certain glaring perversions are bound to occur. Failure is a frightening possibility and, whether failure or victory occurs, one is responsible. What is more, there is no guarantee that anybody else will even understand what one is creating. Nevertheless, there is the titillating thought that one might make a difference, perhaps a significant difference: "This could be my most creative moment and I dare not miss it."
The experience of being the architect of one's world brings the decision to pour one's self into it. It can be like an addiction to which one is drawn to return. One realizes that one would not want to be doing anything else with one's life, even though the temptation often comes to try.
Context This state is number 23 in the ICA *Other World in the midst of this World.*
Broader Creative existence (ICA, #HM2894).
Followed by Cosmic sanctions (ICA, #HM2945) Sacrificial passion (ICA, #HM2641)
Archetypal humanness (ICA, #HM3112) Passionate disinterest (ICA, #HM2547)
Transcending hostility (ICA, #HM2658).
Preceded by Final limits (ICA, #HM2674) Self transcendence (ICA, #HM2584)
Relational situation (ICA, #HM2978) Passionate disinterest (ICA, #HM2547)
Transcending hostility (ICA, #HM2658).

♦ **HM2843e Ideological rebirth** (Brainwashing)
Identification
Description Following final confession, the reformed individual takes up his previous occupation in his new surroundings, combining it with maximum participation in the movement to which he is now happily committed.
Context A stage reached in thought reform, a system of organized, deliberate and total psychological training which effects individual change through two basic elements - *confession* (renouncing of past beliefs and attitudes) and *re-education* (remaking of the individual in the required image).
Broader Thought reform (Brainwashing, #HH0865).

Related Spiritual rebirth (Christianity, #HH4098).
Followed by Transitional limbo (Brainwashing, #HM2934).
Preceded by Final confession (Brainwashing, #HM3156).

♦ **HM2846d Prophetic sight** (ICA)
Description This state is experienced as knowing what is going to happen next. This occurs when one suspends judgement about the present and listens to their own deepest insights. This is like the boy in the book "The Ronin" when he is told by his teacher as he entered the doorway to be met by a rain of blows about the head and shoulders, "You knew. You knew. And yet you still took this doorway".
Context In the ICA New Religious Mode in the arena of knowing and doing intensified in one's life (life of being) the second formal aspect is the experience of transparent lucidity or the experience of divination. At the second phenomenological level this occurs when a person takes a relationship to the experience of nothingness.
Broader Transparent lucidity (ICA, #HM2184).
Followed by Trust intuitions (ICA, #HM2761).
Preceded by Heavenly secret (ICA, #HM2918).

♦ **HM2847c Mindfulness** (Buddhism)
Smriti — Dran-pa (Tibetan) — Sati (Pali) — Non-forgetfulness
Description In Tibetan Buddhism, mindfulness is an essential feature of meditative stabilization, and may be considered as the non-distracted observation of a familiar object. In Hinayana Buddhism, it is the means of remembering, it remembers, it is remembering. The object of awareness does not slip away or wobble but remains steady. Mindfulness manifests as guarding or confronting the object, and its proximate cause is firm and strong perception or the foundation of mindfulness with respect to the body. It guards the doors of the senses.
Context One of the five *determining mental factors* of Tibetan Buddhism. One of the formations aggregate (mental coefficients) of Hinayana Buddhism, being listed among the constant states which appear in their true nature, and as profitable primary (always present in any profitable or profitable-resultant consciousness).
Broader Determining mental factors (Buddhism, #HH0170)
Awareness as mental-formation group of conscious existence (Buddhism, #HM2050).
Related Mindfulness (#HM4543) Recollection (Buddhism, #HM1563)
Way of the warrior (Amerindian, #HH8219)
Establishments in mindfulness (Buddhism, #HM2997).

♦ **HM2848b Tantric paths of accumulation and application in the ascension-stages-game** (Buddhism, Tibetan)
Description The first two paths on the road to liberation for those committed to the use of the mantra as a vehicle.
Broader Stages of spiritual life (#HH0102)
Consciousness ascension stages game (Buddhism, Tibetan, #HH4000).
Narrower Kriya-tantra awareness (#HM2558) Tantra-beginner awareness (#HM2452)
Great-black-lord awareness (#HM2118) Great-path tantra awareness (#HM2656)
Middle-path tantra awareness (#HM3026) Tantra heat-application awareness (#HM2696)
Tantra climax-application awareness (#HM2220)
Tantra receptivity-application awareness (#HM2756)
Tantra union-in-learning-application awareness (#HM2280).

♦ **HM2849d Final vajra-master awareness** (Buddhism, Tibetan)
Yugbanaddha
Description This state is represented as the tenth and last before Buddhahood on the tantric path of the budhisattva, and characterized as the union of the subtle mind-created body of the Form Realm with the Clear Light Buddha-mind. This is said to be the real "yoga", a joining (yuganad-dha) into the realm of the pure Dharma body.
Context In Tibetan Sakya Buddhism this is one of the states in the "Ascension Stages Game". In some sets it is numbered 91 on the board.
Broader Way of the vajra masters (Buddhism, Tibetan, #HM3090).
Followed by Absolute-body awareness (Buddhism, Tibetan, #HM2397)
Supreme-heaven awareness (Buddhism, Tibetan, #HM2813).
Preceded by Seventh vajra-master awareness (Buddhism, Tibetan, #HM2789).

♦ **HM2850c Islamic transformation awareness**
Ittihad (Islam)
Description Among some Islamic mystics this state is one of mystical union, of "becoming" (ittihad) one being with another. In one expression, ittihad is the assumption of human form by an angel or other being. In its highest degree (hulvil) God may "become" one with a human form, in the sense either of a divine possession, overshadowing, or incarnation; or God may become one with an angel or any thing in creation, from a star or sun, to the infinitely smallest thing. However, only God is the eternal reality, and ittihad is experienced by the mystic not as an apotheosis, but as the passing away of his own will in the infusion of the divine will implanted in the soul. Allah is present (hudur) then in the heart-consciousness of the mystic. In becoming (ittihad) Allah, "he" has passed away (fana).
Broader Stations of consciousness - Simnami (Sufism, #HM2341).
Related Identification with God (Sufism, #HM3520).

♦ **HM2851d Interior discipline** (ICA)
Description This is the experience of having taken a stand and seen through to the profound; and being overcome with a sense of incredible potential. Filled with an amazing sense of power, one restrains one's impulse to act and harnesses one's creativity into a numbing stillness. As the prophet Jeremiah said, it is like becoming a pillar of iron. It is being manifest in the steadfastness of the saints, like Teresa of Avila or Hildegaard of Bingen. The protest songs of the 1960s, such as "We shall overcome" and "We shall not be moved" capture this mood in their lyrics.
A dimension of this experience is a sense of awe, that is fear and fascination; a sense of intense, incurable loneliness which sometimes causes the individual to ask: "How much more of this can I stand ?". He wonders if he is going insane or might be consumed, and there is a deep sense of exhaustion. It comes to him that it must be this or nothing; and trusting the conviction that what he has decided to do will work, he finds himself possessed by a strange power. He experiences being one of the privileged few with a unique role to play in history.
The decision is to keep standing and to keep believing in what one stands for. The desire for action is controlled and one curbs one's will to follow one's propensities. There are moments when one experiences the fortitude and capacity to do absolutely anything.
Context This state is number 46 in the ICA *Other World in the midst of this World.*
Broader Transparent power (ICA, #HM2828).
Followed by Cryptic disclosure (ICA, #HM2824) Impactful profundity (ICA, #HM2607)
Universal compassion (ICA, #HM2734) Resurrectional existence (ICA, #HM2631)
Definitive predestination (ICA, #HM2131).
Preceded by Cryptic disclosure (ICA, #HM2824) Worldly detachment (ICA, #HM2451)
Diaphanous intuition (ICA, #HM2927) Ancestral obligation (ICA, #HM3122)
Universal compassion (ICA, #HM2734).

♦ **HM2852d Spiritual pride** (Buddhism)
Context The eighth fetter or character defect referred to by Buddha. This is in the second set of five fetters, when automatic responses are already overcome and the disciple is ready (arhat) to face and overcome the second set.
Broader Character defects (Buddhism, #HH0470).
Related Pride (Christianity, #HM2823).
Followed by Notion of one's self as entity (Buddhism, #HM3612).
Preceded by Desire for formless life (Buddhism, #HM1261).

♦ **HM2853g Sleep** (Hinduism)
Description All notions of "I" and "this", all memories and all dreams are abandoned in favour of total inert dullness.
Context The last of seven descending steps of ignorance described in the Supreme Yoga, which veil *self-knowledge*. Each has innumerable sub-divisions.
Broader Veils of delusion (Hinduism, #HM3592).
Preceded by Dream wakefulness (Hinduism, #HM3510).

♦ **HM2855c Saviors of God: the vision** (Christianity)
Description In this state, behind the stream of the mind and body, behind the stream of the race and mankind, behind the stream of plants and animals is the invisible, treading on all things visible and ascending. His face is without laughter, beyond joy, sorrow and hope. He weeps, clings to the living and the dead and grows strong. He is feared and pitied. He is lifting with every word, every deed, every thought the gravestone of creation. God heaves it upward. It is not pain, nor hope in some future, nor joy or victory that is the essence of God but struggle. To ask what is the purpose of the struggle is to succumb to the wretched self-seeking mind of humans within the bounds of man-made time, place and casualty. The self rejoices to feel the beginning and end of the world. The self condenses into a single lightning moment all of creation and all of destruction. This eternal moment is to be transfixed without losing in the rigidity of language any of its erotic whirling. This life of ecstasy can never be put in words but the struggle is to battle unceasingly to establish it in words; using myth, allegory, the rare and the mundane, exclamations and rhymes. In the same way God also struggles to speak in every way he can, with seas and fires, with colours and wings, with horns and claws, with constellations and butterflies to establish his ecstasy. In this state God confronts with terror and love his only hope and announces that this ecstatic who gives birth to all things, rejoices in them and yet destroys them, is his son.
Context The third phase in the spiritual exercises of Nikos Kazantzakis.
Broader Saviors of God (Christianity, #HM3420).
Followed by Saviors of God: the action (Christianity, #HM3462).
Preceded by Saviors of God: the march (Christianity, #HM3439).

♦ **HM2856d Pisces-consciousness** (Astrology)
Ultimate union
Description The freedom of spiritual activity is expressed through fluid response to the limitations of the senses and of embodiment. Under optimal conditions, Pisces is associated with a total reconciliation of all experience, whether material or spiritual, expressed through practical action, even in the face of opposition, towards the emergence of a new world order.
Context One of the twelve conditions of being or streams of divine energy encompassed within the zodiac. It is said that all twelve conditions must be present and playing its proper part for any successful and fulfilling action to take place. This may be through a group of individuals with different endowments and soul qualities working together; or it may be through one individual who, by self-discipline, practice and meditation, has allowed all twelve conditions to arise within himself.
Broader Watery awareness (Astrology, #HM2384)
Mutable awareness (Astrology, #HM3092)
Zodiacal forms of awareness (Astrology, #HM2713).
Followed by Aries-consciousness (Astrology, #HM2665).
Preceded by Aquarius-consciousness (Astrology, #HM2973).

♦ **HM2858g Taiken**
Body knowledge
Broader Psychotherapy (#HH0003).

♦ **HM2859d Divine hosts** (ICA)
Description This is the realization that, whether good or bad, one is in dialogue with a host of colleagues from the past, the present and the future; when one senses one's self to be surrounded by a host which has become divine. It is like the League which transcends all of history, in Hermann Hesse's book "Journey to the East".
Context In the ICA New Religious Mode in the arena of knowing one's internal sociality (the life of meditation) the fourth formal aspect is one's experience of missional comradeship or of having a colleague. At the fourth phenomenological level, this occurs when a person is sustained by communion with people and events which serve as milestones for his consciousness.
Broader Missional comradeship (ICA, #HM2976).
Preceded by Expectant descendant (ICA, #HM2903).

♦ **HM2860f Dreamy-arrested consciousness**
Description The dreamy state is one akin to that in epileptic seizures, but without convulsion. A range of sensory hallucinations may be present, accompanying a dream-world condition lasting several minutes. The dreaming state is associated with lesions and disorders in the temporal lobes of the brain.

♦ **HM2861d Second duty** (Christianity)
Description Beyond the mind, behind all appearance is an essence with which the heart struggles to merge but the body and the mind stand in the way. Shatter the body to merge with the invisible, silence the mind to hear the invisible calling. Create for earth a brain and a heart, give a human meaning to the superhuman struggle. This suffering is the second duty.
Context The second duty in the preparation for the spiritual exercises of Nikos Kazantzakis.
Broader Saviors of God: the preparation (Christianity, #HM3264).
Followed by Third duty (Christianity, #HM4279).
Preceded by First duty (Christianity, #HM2224).

♦ **HM2862d Transformed existence** (ICA)
Description This is the experience of being in the same mundane world as before but that it is strangely transformed; as if one's perspective has been radically transposed without any participation on one's part. Two songs in the musical "My Fair Lady" are sung by characters experiencing such occasions - "I Could Have Danced All Night" and "On the Street Where You Live". It may compared to standing in a European cathedral and suddenly seeing the people who worked for hundreds of years to create it as if they were still working and were all around one as one stands there. Many people have described waking up on a beautiful summer morning and experiencing all their surroundings strangely made new.
A dimension of this experience is a sense of awe, that is fear and fascination. The fear is that this will not last and of what will come after it; a sense of the fragility of the moment and of one's own

ineffectiveness in the midst of it. At the same time one is fascinated with the wonder, almost enchantment, of the transposed world, the sense that it was here all the time but has only just been revealed. The decision is to believe in this transformed reality and to start on an exploration of it. This is followed by a new sense of urgency, almost of being driven, and by seeking someone with whom one can share the experience.
Context This state is number 10 in the ICA *Other World in the midst of this World.*
Broader Transformed state (ICA, #HM2386).
Followed by Second birth (ICA, #HM3175) Absurd existence (ICA, #HM2331) Cryptic disclosure (ICA, #HM2824) Ancestral obligation (ICA, #HM3122) Intentional conscience (ICA, #HM2892).
Preceded by Vibrant powers (ICA, #HM2982) Absurd existence (ICA, #HM2331) Ubiquitous otherness (ICA, #HM2570) Ancestral obligation (ICA, #HM3122) Spontaneous gratitude (ICA, #HM3025).

♦ HM2864c Tantra in meditation on emptiness (Buddhism, Tibetan)
Action tantra: kriya — Performance tantra: charya — Yoga tantra — Highest yoga tantra: anuttarayoga
Description Tantra is a rapid path for those who are fit to follow it; but these are persons of great compassion who so wish to be a source of help and happiness for others that they cannot bear to spend unnecessary time in achieving Buddhahood. The most qualified will achieve the path in one lifetime, but lesser qualified practitioners who maintain their vows may achieve it in seven or in sixteen lifetimes. Those following the highest yoga tantra will not have to pass sequentially through the initial stages of meditation on emptiness. Having achieved calm abiding and special insight with emptiness, it is accepted that a phenomenon qualified by emptiness can continue to appear to a consciousness realizing its emptiness. In the fifth stage of the meditation on emptiness there is a series of techniques designed to manifest the four subtle minds and, using the most subtle (that of clear light), to cognize emptiness.
Broader Yoga (Yoga, #HH0661) Selflessness (Buddhism, #HM2016) Emptiness (Buddhism, Zen, #HM2193) Meditative states of mental abidings (Buddhism, #HM2145).
Related Tantra (Buddhism, Yoga, #HH0306).
Followed by Buddha-consciousness (Buddhism, Tibetan, #HM2735).
Preceded by Tibetan meditative states of form concentrations (Buddhism, #HM2693).

♦ HM2866c Sukha (Buddhism)
Happiness — Pleasure
Description This may refer to physical health, material well-being or spiritual beatitude. It covers pleasure, pleasurable feeling and happiness. Happiness is the ultimate to which every human aspiration reduces. The mission of Buddha was not only the revealing of how suffering (dukkha) could be overcome but also the attainment of the good and happiness of all beings. Happiness is the aim of the religious life. The crown of happiness, paramasukha, is Nibbana or Nirvana.
The happiness of worldly desire is kamasukha; sweeter and loftier than this are the four stages of the rupajhana and the five of the arupajhana, each sweeter and loftier than the one before. However, Buddha also said that happiness is wherever and whenever it is found. Nevertheless, the pleasantness associated with kamasukha, with feeling or ideas about feeling, is to be considered pain or ill; painful feeling is considered like a javelin; neutral feeling impermanent. If sense experience were totally happy, there would be no spur to the spiritual. If it were totally unhappy, no-one would be engrossed by it. It is the mixture of pleasure and pain that is both the hindrance to and the guarantee of spiritual progress.
In happy feeling, the static element is happy, change is unhappy. In unhappy feeling, the static element is unhappy, change is happy. In neutral feeling, knowledge is happy feeling, want of knowledge is unhappy feeling. All three may include passion, aversion or resentment, and ignorance. However, by practice of jhana or dhyana: in the first stage are banished sensual desires and immoral or wrong ideas, the intellect is engaged in happy zest about the object selected: in the second and third stages, resentment and opposed feeling melt away in yearning for attainment of blissful serenity of the saint; in the fourth stage, all positive feeling fades into indifference, clarity of mind is obtained and ignorance banished.
Sukha predominates celestial existence, at least as far as the six realms of the devas in the kamaloka. But although practice of the four-fold jhana brings rebirth as a deva in conditions which are entirely pleasurable, these are not comparable with the self-mastery and intuitive vision of the arahant. Happiness is thus best secured when what is aimed for is its cause rather than happiness itself.
Context One of the five constituent factors of jhana. The pleasurable feelings consequent upon diffused *piti* (zest) expel distraction and agitation and leas the mind to concentration.
Narrower Quietude (#HM1498) Unperturbedness (#HM4572) Stasis of thought (#HM0051) Constancy of thought (#HM1066) Correct concentration (#HM1735) Power of concentration (#HM1759) Unperturbed mindedness (#HM0715) Steadfastness of thought (#HM3471) Unperturbedness of thought (#HM0449).
Related Happiness (#HM2409) Jhana (Buddhism, Pali, #HM7193) Pleasant feeling (Buddhism, #HM6722) Faculty of pleasure (Buddhism, #HM1945).

♦ HM2867c Mystical detachment (Sufism)
Description In this state there is total detachment from anything but God. This includes outward detachment from what arises as an accident of this world and inward detachment so that no compensation is sought for what has been forsworn outwardly. All activity is up to God with no thought for one's self, respect for others or regard for compensation. Only then is unification with God possible.

♦ HM2870f Preoccupied consciousness
Description Those that are in preoccupied states may be concerned with reflecting on, or solving, personal or impersonal problems. They may also be involved in creative work. So engrossed or self-absorbed are they, that effective contact and response to external reality may be hindered. Preoccupation is more serious when it is habitual, or involuntary due to autistic schizophrenic tendencies, or when it is indicative of some underlying psychic or brain function disturbance. Continuous, difficult intellectual work, such as philosophical or mathematical analysis, may produce the sterotypical bumbler, or absent-minded professor. Some meditation excercices and related religious practices to induce a separation of the mind from everyday consciousness may also produce persistent preoccupation. In some monastic orders, community activities and manual work were early introduced to counter excessive introversion. Preoccupied states are necessary from time to time for some people; others may require hours of inwardness daily. Their need is to acquire a technique of alternation between the inner and outer reality.

♦ HM2871c Kundalini conscious energy (Yoga)
Description The awakening of the spiritual energy of consciousness (Kundalini) is the ultimate aim of Hatha-Yoga, Kundalini Yoga and various forms of Tantra. Spiritual energy is said to lie in the lower abdominal region, described as a serpent coiled at the base of the spine, blocking the energy channel, Shushumna. When Kundalini is awakened it travels up the spine and each spiritual centre (Chakra) is charged with power. If Kundalini has risen up to the top of the head the yogi experiences unsurpassable joy in Samadhi.
Refs Krishna, Gopi *Higher Consciousness*; Woodroffe, John *The Serpent Power* (1986).
Related Shushumna (Hinduism, #HM2053) Kundalini yoga (Yoga, #HH0237) Chakra centres of consciousness (#HM2219).

♦ HM2872d Bottomless centre (ICA)
Description This is experienced as a void at the centre of being; when one encounters one's own death or the meaningless of one's life in relation to the sweep of history. It may be compared to the woman in the book "The Sybil" experiencing having been raped by the mystery.
Context In the ICA New Religious Mode in the arena of knowing and doing intensified in one's life (life of being) the first formal aspect is the experience of transparent selfhood or the experience of noughtness. At the first phenomenological level this occurs when one experiences discontinuity at the centre of one's life.
Broader Transparent selfhood (ICA, #HM3077).
Followed by Acute inadequacy (ICA, #HM3151).

♦ HM2873d Enjoyment-body awareness (Buddhism, Tibetan)
Sambhoga-kaya
Description This state of awareness is characterized by omniscience and represented as Buddha-consciousness beyond the ten degrees of development. Its "place" is Akanistha, the Superior Heaven, from which it sends forth the Emanation Bodies. The *enjoyment-* or *fruit-*body is the culmination of effort; and its powers sustain the communion of Buddhist saints.
Context In Tibetan Sakya Buddhism this is one of the states in the "Ascension Stages Game". In some sets it is numbered 92 on the board.
Broader Supreme heavens in the ascension stages game (Buddhism, Tibetan, #HM3214).
Related Way of the Buddhas (Buddhism, Tibetan, #HM2185).
Followed by Emanation-body awareness (Buddhism, #HM2340).
Preceded by Absolute-body awareness (Buddhism, Tibetan, #HM2397).

♦ HM2874c Valleys (Sufism)
Context The sixth of ten states acting as gateways orienting the mystic on the journey to the Absolute, as described by Ansari.
Broader Stations of consciousness - Ansari (Sufism, #HM2317).
Followed by Mystical states (Sufism, #HM3607).
Preceded by Principles (Sufism, #HM3069).

♦ HM2875d Vital signs (ICA)
Description This state is of tingling in every moment; like a tautly coiled spring one throbs in anticipation, ready to pounce with a new eagerness into action. One can hardly wait; one can't sit down. An athlete may jump the starter's gun as the result of such an experience. These are the moments when even the stones seem to cry out.
A dimension of this is a sense of awe, that is fear and fascination. The fear is that this state will not last, it will somehow go wrong and one will not be able to stand it. The fascination is in the yearning to be in action: all of life is to be lived this way. One sets aside all doubt and commits everything to a specific course of action, deciding to "go for broke". There is a moment of emptiness on realizing that one's commitment is all there is; and one is filled with amazement at one's self.
Context This state is number 57 in the ICA *Other World in the midst of this World.*
Broader Contentment at the centre (ICA, #HM2529).
Followed by Living death (ICA, #HM2808) Vibrant powers (ICA, #HM2982) Beyond morality (ICA, #HM3049) Seminal illumination (ICA, #HM2056) Spontaneous gratitude (ICA, #HM3025).
Preceded by Beyond morality (ICA, #HM3049) Creative futility (ICA, #HM2493) Global guardianship (ICA, #HM2817) Seminal illumination (ICA, #HM2056) Exclusive contradiction (ICA, #HM2951).

♦ HM2876d Embodying peace (ICA)
Description The first formal aspect of the *life of obedience*, an arena in the ICA *New Religious Mode*, this represents the awareness that one embodies peace as an establishment to enable the ordering of human affairs.
Broader Life of obedience (ICA, #HM2024).
Narrower Universal prior (#HM2458) Radical incarnation (#HM2587) Missional engagement (#HM2343) Submissive obedience (#HM2258).

♦ HM2877d Primordial sociality (ICA)
Description This is the awareness that a person is part of a community of concern much larger than his own immediate family or society. It occurs when he discovers his indebtedness to people who came before him and to unknown people who are working around him, as when women in the "women's revolution" discover the women of the past who had committed themselves to women's advancement.
Context In the ICA New Religious Mode in the arena of acting out one's deed (the life of doing) the second formal aspect is extending one's engagement to history. At the first phenomenological level this occurs when one is confronted by the fact that to live is to be of service.
Broader Engagement to history (ICA, #HM2588).
Followed by Every situation (ICA, #HM2991).

♦ HM2880b Mystical Christ-consciousness (Christianity)
Christian stations of awareness — Stations of the cross
Description Mystical Christ-consciousness, unlike most classical Christian mysticism which has a Christ-state as its goal, does not set out to achieve contemplation of the form of the risen Jesus, but rather emulates the conscious experience of Jesus himself as recorded in gospels and traditions. Analagous to the stations attributed to the Passion of Jesus are the stations of his consciousness which, like the original Stations of the Cross, may have numbered 5. One categorization of these 5 includes: the empty state (kenosis, or emptying); the remitting state where the concept of evil is relinquished so that forgiveness may enter; the state of agape or charismatic love; the mortified state of dying to everything except fuller life; and finally, the risen or resurrected consciousness. This categorization is perhaps the spiritual equivalent only of the Passion, for 2 other states of consciousness from the evangelical life of Jesus may precede these 5. They are conversion (metanoia, the re-direction of one's life and mind toward spiritual growth, corresponding to the baptism of Jesus) and the transfiguration state, the lifting of the consciousness to a state of illumination.
Related Becoming Christ (#HH5328) Way of the cross (Christianity, #HM3516) Human development (Christianity, #HH2198) Christ consciousness (Christianity, #HM2013) Consciousness in Buddhism (Buddhism, #HM2200) Christian recollection (Christianity, #HM2077) Stations of consciousness - Simnami (Sufism, #HM2341) Indian approaches to consciousness (Hinduism, #HM2446).

♦ HM2881c Sense of shame (Buddhism)
Hri — Hiri (Pali) — Ngo-tsha-shes-pa (Tibetan) — Shame
Description This emotion is experienced by an individual when his or her thoughts or activities (or the results of activities) do not conform to a standard considered by that individual to be

acceptable. The emotion may be accompanied by a feeling of guilt when such thoughts or actions are also in conflict with some externally accepted norm. It is experienced as anxiety, regret, and dissatisfaction with and censure of one's self; and is experienced again on recalling shameful occasions in the past. Because of its dependence on particular moral or social codes, shame is felt by different people in response to different phenomena, depending on social, moral and geographical conditions.
Context One of the eleven *virtuous mental factors* referred to in Tibetan Buddhism.
Broader Virtuous mental factors (Buddhism, #HH0578).
Related Shame (Buddhism, #HM8112) Embarrassment (Buddhism, #HM2210).

♦ **HM2883c Pleasure**
Enjoyment
Description The feeling of pleasure arises in response to some stimulus, whether physical, mental or spiritual. There is the pleasure in the muscular ache after a long walk, there is the pleasure of the hedonist in satisfaction of his desires, the pleasure of the ascetic in controlling his desires. One definition indicates that pleasure accompanies any activity where progress is made towards achieving a desired end.
Related Pleasure (#HH0194) Pleasant feeling (Buddhism, #HM6722).

♦ **HM2886b Forms of enlightenment** (Buddhism)
Description These seven branches of enlightenment - mindfulness, discrimination of phenomena, effort, joy, pliancy, meditative stabilization and equanimity - are attained with the path of seeing.
Context One of the sections of yogic paths or harmonies with enlightenment defined in Buddhism.
Broader Harmonies with enlightenment (Buddhism, #HH0603).

♦ **HM2887f Psychotic state**
Psychosis
Description Mental functions are so impaired that insight, contact with reality and ability to deal with life's normal demands are all interfered with. States typical of psychosis are: delusions; hallucinations; loosening of associations and incoherence; catatonic stupor or excitement. Behaviour is very disorganized.
Related Hallucination (#HM4580) Delusive consciousness (#HM2600).

♦ **HM2888d Near death review of life** (NDE)
Description The whole of life's experience, and also the effects of one's thoughts, actions and feelings on others, are reviewed. Awareness and control of thoughts, feelings and actions become of central concern. There is a sense of detachment and sometimes the possibility of moving backwards or forwards in life or skipping some parts.
Context The sixth stage in the process referred to as *near death experience.*
Broader Near death experience (NDE, #HM0777).
Followed by Near death entering the light (#HM3180).
Preceded by Near death encounter with the higher self (NDE, #HM3059).

♦ **HM2889d Agonizing prediction** (ICA)
Description This is the overwhelming realization that in daring to decide and act in a particular situation one is predicting the future shape of history. It occurs in situations where one's decision to act has widespread consequences, some of which one is not even aware of; as when Gandhi in the 1920's decided India would secure independence through non-violent action.
Context In the ICA New Religious Mode in the arena of acting out one's freedom (the life of prayer) the fourth formal aspect is one's experience of universal responsibility or intercession. At the third phenomenological level this occurs one acknowledges that which is beyond one's power to control.
Broader Universal responsibility (ICA, #HM3110).
Followed by Promissorial offering (ICA, #HM3172).
Preceded by Particular concern (ICA, #HM2939).

♦ **HM2890d Pingala conscious energy** (Yoga)
Description Pingala is the energy channel to the right of Shushumna in Tantra and Kundalini-Yoga. Physiological correspondence lies in the sympathetic nervous system. Psychological correspondence is creative power, symbolized by the sun. Pingala is, like Ida and Shushumna, one of the major nadis of the body. Modern literature strongly suggests the parallelism between tantric introspective findings about the structure of the nervous system and modern medical fact. Although in outline the concept Nadi is not identical with nerve (s), specific analogies between the structure of chakras and corresponding cerebral and spinal nerve-organization have been observed. Note the correspondence to the right side of the Tree of Life with the findings on brain hemisphere coordination, in which the left brain (connecting to the right-side of the body) refers to analytic-rational functions; and likewise, the right hemisphere, referring to synthetic-emotional functions, corresponds to Ida.
Broader Shushumna (Hinduism, #HM2053).

♦ **HM2891f Hebetude**
Affective dementia
Description Apathy or listlessness usually arising from mental illness, typically schizophrenia. Sufferers are totally indifferent to internal or external stimuli and have no reaction to their environment or to their treatment by others.

♦ **HM2892d Intentional conscience** (ICA)
Description This occurs as a person stands alone before his decision, facing no limits and feeling totally unconstrained. There is a sense of being lost in a wilderness and charting a path through it all alone. After their daughter had been in a coma for many months, the parents of Karen Quilan went to court to get permission to disconnect the life-support systems, not knowing if she would die or live once this was done. Simon Bolivar embodied this position too, when he decided that becoming a dictator was the only choice he had if Greater Colombia were to survive, having fought all his life against the ruling dictatorship.
A dimension of this experience is a sense of awe, that is fear and fascination. The agony is keeping one's own conscience when whatever one decides will be condemned from one side or another. While knowing that every decision one makes is significant, one realizes that what one decides may not help at all and yet that the alternative to not deciding is to turn into stone. There is a feeling that everything one has ever done was preparation for this moment and this really is what life is about. In the process of deciding, an individual finds he trusts his intuitions and lets the consequences take their course. It begins to come clear that history really is changed by ordinary people making decisions which affect their lives and the lives of others.
Context This state is number 26 in the ICA *Other World in the midst of this World.*
Broader Moral ground (ICA, #HM2721).
Followed by Cosmic sanctions (ICA, #HM2945) External relation (ICA, #HM2471)
Worldly detachment (ICA, #HM2451) Ancestral obligation (ICA, #HM3122)
Spontaneous gratitude (ICA, #HM3025).
Preceded by Beyond morality (ICA, #HM3049) External relation (ICA, #HM2471)
Relational situation (ICA, #HM2978) Transformed existence (ICA, #HM2862)
Spontaneous gratitude (ICA, #HM3025).

♦ **HM2893d Muladhara** (Yoga)
Root base awareness — First chakra
Description The personality is characterized by spiritual torpor and the individual clings to his condition of unexhilarated wakefulness. This is the level of manifestation of consciousness in human form, physical birth. Meditation on the tip of the nose brings the beginnings of awareness, freedom from disease, lightness, inspiration, vitality, vigour, stamina, security, understanding of inner purity, softness of voice, inner melody. Behaviour in the first chakra is violent, based on insecurity, striking out like a frightened animal. The chakra encompasses the planes of genesis, illusion, anger, greed, delusion, avarice and sensuality. These are fundamental to human existence. Motivation is through desire for more experience and information. As the seat of the coiled serpent Kundalini, this is the root of all growth and awareness.
Context The first "lotus" or chakra defined in kundalini yoga, said to be situated between the genitals and the anus.
Broader Kundalini yoga (Yoga, #HH0237) Chakra centres of consciousness (#HM2219).
Related Sphere of the kingdom (Kabbalah, #HM2288)
Initiation of revelation (Esotericism, #HM0181)
First chakra: fundamentals of being (Leela, #HM4103).
Followed by Svadhishthana (Yoga, #HM3174).

♦ **HM2894d Creative existence** (ICA)
Freedom of inventiveness
Description This is an experiential trek in the *River of Consciousness*, within the ICA *Other World in the midst of this World*, when the individual realizes "I am my own originality". The states are described in separate entries.
Broader River of consciousness (ICA, #HM2993).
Narrower Universal fate (#HM2687) Relational situation (#HM2978)
Archetypal humanness (#HM3112) Contextual world-view (#HM2842).
Related Moral ground (ICA, #HM2721) Authentic relation (ICA, #HM2316)
Final accountability (ICA, #HM3206).

♦ **HM2895c Age of the Father** (Christianity)
Description This age or state is represented by the Old Testament revelation, associated with God the Father.
Context The first age, closing, according to the Joachimites, with Zacharias, the father of John the Baptist.
Broader Doctrine of three ages (Christianity, #HH0397).
Followed by Age of the Son (Christianity, #HM3087).

♦ **HM2896b Samprajnata samadhi** (Yoga)
Description This is a form of enstatic realization when there is still an object of consciousness and transcendental cognitive activity. It has four levels: vitarka samadhi; nirvitarka samadhi; vicara samadhi; nirvicara samadhi. Vitarka indicates the gross or material form of the object, and indicates cogitation on this form over time; when noetic activity ceases, nirvitarka samadhi, beyond vitarka, commences. Vicara samadhi indicates cogitation on the subtle and causal form of what is cognized; again, when noetic activity ceases, nirvicara samadhi, beyond vicara, commences. The vedanta *savikalpa samadhi* is said to be synonymous.
Broader Samadhi (Hinduism, Yoga, #HM2226).
Narrower Vicara samadhi (#HM3880) Vitarka samadhi (#HM4451)
Nirvicara samadhi (#HM0809) Nirvitarka samadhi (#HM4284).
Related Enstasy (#HM3169) One-pointedness (Yoga, #HM5734)
Savikalpa samadhi (Hinduism, #HM2650) Asamprajnata samadhi (Hinduism, #HM3041).

♦ **HM2898c Samtosa** (Yoga)
Contentment
Description This state is one of elated serenity, arising as the yogin frees himself from the notions of "I" and "mine". There is inner equilibrium and self-sufficiency, when external events neither surprise or distress. It is closely connected with *samatva - equanimity.*
Context One of the five components of *niyama*, self-discipline.
Broader Niyama (Hinduism, #HM3280).
Related Contentment (#HM3729) Samatva (Yoga, #HM1586)
Satisfaction (Hinduism, #HM8116).

♦ **HM2899c Emptiness of the nature of phenomena** (Buddhism, Tibetan)
Shunyatashunyata
Context One of the eighteen emptinesses comprising the paths of view in Tibetan Buddhism.
Broader Emptinesses on the paths of the view in Buddhism (Buddhism, Tibetan, #HM2944).

♦ **HM2900b Awareness of the mystic journey**
Suluk (Sufism)
Description The Sufi conception of the mystic journey (Suluk) is the conscious intent to seek God under the direction of a shaikh. In the journey the disciple makes himself perfect in each of the stations (maqamat). The mystic is aware at all times that he is on the path, using recollection (dhikr) and practising all the virtues his moral strength will allow. In a wider sense, a number of religious philosophies in and outside Islam, recognize that all humanity is on the journey to God-realization, but people's consciousness of this fact varies enormously.
Narrower Cleansing the sirr (Sufism, #HM5353) Cleansing of the heart (Sufism, #HM6932)
Purification of the self (Sufism, #HH5092) Illumination of the spirit (Sufism, #HM6162).
Related Salik (Sufism, #HM4288) Mystic stations (Sufism, #HM3415)
Suluk of the Naqshbandiyya order (Sufism, #HM4356)
Stations of consciousness - Simnami (Sufism, #HM2341).

♦ **HM2901b Imaginative consciousness** (Anthroposophy)
Imagination
Description Imagination can reproduce previous experience or even creatively apprehend and design visions of things anticipated in the future. The visionary capacity to imagine so far unrealized realities is held to be the key distinction between human and animal consciousness. The visionary capacity gives rise to prophecy and idealism.
Context In R Steiner's anthroposophic meditation, imagination is the second step of experience leading to direct knowledge or intuition.
Related Active imagination (#HM0867) Anthroposophical system (#HH0018)
Constructive imagination (#HH2076).
Followed by Spiritual intuitive cognition (#HM2829).
Preceded by Inspired consciousness (#HM3125).

♦ **HM2902d Manas awareness** (Hinduism)
Description A fully introspective state of consciousness is one in which the mind (manas), its faculties, structures and contents, are the sole objects. Partly introspective states have given rise to the human endeavours of philosophy, epistemology and psychology; and in the East, to yoga.

In the West, the mind–body relation problem is unsolved. In the East techniques for dominating the body by the mind are unsurpassed. Investigation of manas leads to buddhi (wisdom). Manas has been referred to as the lower mind, the messenger between the higher mind and the senses; it needs to be controlled by the buddhi, which is also discrimination.
Related Mind (#HH0299) Mind (Buddhism, #HM1743) Mental (Buddhism, #HM0251) Base of mind (Buddhism, #HM1736) Ego awareness (Hinduism, #HM2059) Faculty of mind (Buddhism, #HM1369) Buddhi awareness (Hinduism, #HM2099).

♦ **HM2903d Expectant descendant** (ICA)
Description This is the moment when an individual finds himself asking what it is that the future needs and what it requires of his life. It occurs when there seems nothing one can do and yet the question of what one intends doing is in front of one, as when the ecology movement raises the question of what kind of world we intend to bequeath to future generations of humanity.
Context In the ICA New Religious Mode in the arena of knowing one's internal sociality (the life of meditation) the fourth formal aspect is one's experience of missional comradeship or of having a colleague. At the third phenomenological level, this occurs when one dialogues across time and space with those who also travel this journey of the spirit.
Broader Missional comradeship (ICA, #HM2976).
Followed by Divine hosts (ICA, #HM2859).
Preceded by Primordial ancestor (ICA, #HM3224).

♦ **HM2904d Soteriological existence** (ICA)
Description This is the realization that the ultimate cost of one's care is one's life, the knowledge that one is expendible and that the inevitable result of expenditure is death. It is like the trapeze artist who swings out into the air, giving it everything he's got, pouring himself into the effort for no rational or moral reason and even though he cannot know whether he will reach the other side; or the sheriff in the movie "High Noon" when he decides, against all good advice, to stay in town and face the gang that is coming after him.
There is a sense of awe, that is fear and fascination, as this reality dawns on a person. The whole of his life has been about this particular moment. This is the "big night" after all the dress rehearsals. Like an Olympic athlete who has set his heart on winning a gold medal, this is the big chance worthy of everything he can give. But it is not without serious misgivings that he decides to take the plunge. The question is, since he only has one life, is he sure he wants to put it here? But then again, what will happen to him if he doesn't? This is experienced as discovering one's own integrity and deciding to succumb to a sense of urgency. The decision to care means one knows a new kind of authenticity. One's life is changed forever and one leaves one's mark on history.
Context This state is number 40 in the ICA *Other World in the midst of this World.*
Broader Universal concern (ICA, #HM2774).
Followed by Total exposure (ICA, #HM2764) Invented history (ICA, #HM3245) Global guardianship (ICA, #HM2817) Definitive effectivity (ICA, #HM2796) Exclusive contradiction (ICA, #HM2951).
Preceded by Total exposure (ICA, #HM2764) Sacrificial passion (ICA, #HM2641) Sacramental universe (ICA, #HM2445) Archetypal humanness (ICA, #HM3112) Definitive effectivity (ICA, #HM2796).

♦ **HM2906c Separation from the transient** (Sufism)
Description All but truth is subject to decline and destruction. Since death will inevitably take away those things which are the object of vain desires and deception, these should be repudiated now.
Context The fourth illumination of Jami.
Broader Fountains of light (Sufism, #HM3039).
Followed by Perception of universality (Sufism, #HM2827).
Preceded by Perception of hidden reality (Sufism, #HM2531).

♦ **HM2907c Liberation of the spirit by attraction of divine grace** (Sufism)
Description Enticed by the action of divine grace, man is freed from the trap of concupiscence and desire which has hindered his relationship with the Absolute. His joy masters physical and spiritual delights; the pain of spiritual effort is dissipated and contemplative bliss overwhelms the soul.
Context The eleventh illumination of Jami.
Broader Fountains of light (Sufism, #HM3039).
Followed by Surrender to the bliss of mystical seduction (Sufism, #HM3436).
Preceded by Liberation of the heart (Sufism, #HM3244).

♦ **HM2909d Fifth scriptural bodhisattva awareness** (Buddhism, Tibetan)
Sudurjaya
Description This state of awareness is characterized by infallible memory and an ability to write commentaries on scripture. What is "very–difficult–to–overcome" (sudurjaya), the fifth level bodhisattva is represented as accomplishing. He may express the dharma in the arts, designing gardens and outdoor and indoor theatres; and he may create tableaux, plays and games for edification.
Context In Tibetan Sakya Buddhism this is one of the states in the "Ascension Stages Game". In some sets it is numbered 88 on the board.
Broader Meditation way of the bodhisattvas (Buddhism, #HM2769).
Followed by Ninth scriptural bodhisattva awareness (Buddhism, Tibetan, #HM2292) Sixth scriptural bodhisattva awareness (Buddhism, Tibetan, #HM2385).
Preceded by Third scriptural bodhisattva awareness (Buddhism, Tibetan, #HM2215) Fourth scriptural bodhisattva awareness (Buddhism, Tibetan, #HM2191).

♦ **HM2910d Loyal opposition** (ICA)
Description This is the realization that one is taking on the burden of a task and that one is going to do it even if no one else does. It is the moment when a person decides to affirm the covenant he is part of and to take responsibility for it, when he says "yes" to the vision of the covenant and seeks to fulfil that vision within the covenant by transforming its forms and functions.
Context In the ICA New Religious Mode in the arena of acting one's engagement (the life of obedience) the second formal aspect is embodying equity as the disestablishment to call forth human justice. At the third phenomenological level, this occurs when people willingly embrace their responsibility to society.
Broader Embodying equity (ICA, #HM2442).
Followed by Perpetual revolutionary (ICA, #HM2839).
Preceded by Corporate duty (ICA, #HM2694).

♦ **HM2911d Possessiveness** (Yoga)
Description The first to be given up in the series of troubles referred to by Patanjali.
Broader Five kleshas (Yoga, Zen, #HH0390).
Followed by Revulsion (Yoga, Zen, #HM3620).

♦ **HM2912c Vitarka stage** (Yoga)
Vivesa stage of the gunas — Manomaya kosa
Description In this stage of samadhi, consciousness functions through the lower mind which sees objects as particular things with names and forms.
Context The first or crudest stage of the four stages of the gunas.
Broader Stages of the gunas (Yoga, #HM2805).
Related Vitarka samadhi (Yoga, #HM4451).
Followed by Vicara stage (Yoga, #HM3460).

♦ **HM2913b Ishvara–consciousness** (Yoga)
Description By following the way of devotion, knowledge of Ishvara as the second aspect or *soul* in the heart is revealed. Full revelation of Ishvara is said to be the result of application to the way of *raja yoga*, when intellectual knowledge, mental control and discipline combine with the devotion of pure love.
Ishvara is God in the heart, the son of God, otherwise called the cosmic Christ.
Broader God–consciousness (#HM2166).
Related Christ consciousness (Christianity, #HM2013).

♦ **HM2916c Conviction** (Sufism)
Description Conviction arises when the heart, awakened, overrules impulsive demands and rejects the arguments of the mind. Weaned from concepts of good and bad, one sees both as products of culture. Vowing with reality of revelation and inspiration, one is acquainted with one's own superior consciousness, directing one to the reality of immortality and closeness with God, the creative essence. The moment of conviction cannot be forced, nor can it arise through knowledge or instruction. But when unity and harmony have been achieved, it comes, and the purity of soul transforms the physical eyes (duality) to the unity of an inner eye. However the transformation is nurtured – and this may be through a number of factors, for example the sincere recitation of holy verses – a courage for the quest is generated. The experiential moment of illumination in higher consciousness directs and expands the self, taking the body along with it. All aware, devoid of thought, all receiver or all giver, this is the creation of the heart. The person is awakened to a new means of communication with all around him, whether physical, subtle or essential and to the soul of each object, then the soul of souls of all objects; but there is also a sense of separation from the essence. There is no longer any sense of effort but of joyful contemplation, through experience there is learning to remain steady and sensitive. There are experiences of illuminating flashes of truth, attracting the heart to its single object of desire. Engrossed in contemplation of this single object, the individual feels tranquillity, totality and happiness.
Refs Arasteh, A Reza *Growth to Selfhood* (1980).

♦ **HM2917c Sunao na** (Japanese)
Uprightness — Compliance
Description In Japan this is one of the most valued personality traits in interpersonal relationships. It is contrasted with *hine–kureta*, the warped or twisted state of the embittered, begrudging and resentful person, a state arising from the frustration and consequent sulking when pretended indifference to one's needs brings envy towards others whose needs are apparently satisfied.
Related Hinekurata (Japanese, #HM1598).

♦ **HM2918d Heavenly secret** (ICA)
Description This is the experience of there being no deliverance from the meaninglessness and absurdity of existence, and that is one's deliverance; when one recognizes that freedom is always present but that it is usually surrendered to someone or something. It may be compared to Sisyphus in Albert Camus' "Myth of Sisyphus" when he turns to watch the rock roll down the hill and recognizes his possibility.
Context In the ICA New Religious Mode in the arena of knowing and doing intensified in one's life (life of being) the second formal aspect is the experience of transparent lucidity or the experience of divination. At the first phenomenological level this occurs when a person experiences discontinuity at the centre of his life.
Broader Transparent lucidity (ICA, #HM2184).
Followed by Prophetic sight (ICA, #HM2846).

♦ **HM2919c Compassion**
Description Compassion is one of the emotions or attitudes with an emotional component, that are other–oriented. It is imaginatively dwelling on the condition of the other person, having an active regard for the welfare of the other, viewing him as a fellow human being, responding with an emotion of concern and acting in a rational and effective way to the other person's benefit.
Refs Wynne–Tyson, Esme *The Philosophy of Compassion* (1985).
Related Care (#HH0042) Love (#HH0258) Involvement (#HH0030) Pity (Buddhism, #HM0513).

♦ **HM2920c Perfect man consciousness** (Sufism)
Al–insan al–kamil — Insanu'l–kamil
Description In Sufism, one saintly person in every age is said to receive the illuminations (tadjalli) of God's Names, Attributes and Essence. He becomes deified, God's worldly viceregent (khalifa). He is the perfect man, the pole (kuth) of the universe, supporting and sustaining it. He has acquired the universal nature (djamiya) and contains the types of every spiritual and material thing. Having realized perfect identity with God, he comes into the world to serve humanity in both moral and spiritual aspects, living in the world but not belonging to it. As the Prophet his spiritual perfection is revealed, as "Sufi" the fact of his identity with God is veiled. Serving as Shayk, Pir or spiritual master he guides human beings on the way top God. To him divine essences and divine attributes are the same. He experiences the state of *al–wahid* – his soul is both annihilated in the Universal Soul and rejoices in permanency in the Divine Soul. According to Al–Djili, Muhamad is the Most Perfect Man. These ideas echo Sakyamuni's position as claimed in later Buddhism, that as Buddha he became affiliated to Adi–Buddha. They also reflect the Manichaean Primal Man, the Hermetic Athropos, and the Zoroastrian Gayomart.
Refs Bhatnagar, R S *Dimensions of Classical Sufi Thought* (1984); Nicholson, R A *The Sufi Doctrine of the Perfect Man* (1984).
Broader Stations of consciousness – Simnami (Sufism, #HM2341).
Related Al–Wahid (Sufism, #HM5506).

♦ **HM2921d Scorching avatar** (ICA)
Description This is experienced when the person whose life one is seeking to emulate calls one's decisions and actions into question and one realizes one must pay the price of one's decisions. It occurs when one asks one's self what one's hero would have done in this situation, and finds one's own life falling short. In the movie "High Noon", one of the people in a small frontier town wants to help the Marshall but discovers that no one else is willing to join him; he has to decide if he will stand with the Marshall or not.
Context In the ICA New Religious Mode in the arena of knowing one's internal sociality (the life of meditation) the third formal aspect is one's experience of being given existential guidance or encountering the Saints. At the second phenomenological level, this occurs when a person finds his selfhood audited by that which commands his attention.

Broader Existential guidance (ICA, #HM2114).
Followed by Ever-present brother (ICA, #HM2814).
Preceded by Revered hero (ICA, #HM2654).

♦ **HM2922d Surrender to inadequacy** (ICA)
Description The third formal aspect of the *life of prayer*, an arena in the ICA *New Religious Mode*, this represents a state of petition as one is aware of one's total inadequacy.
Broader Life of prayer (ICA, #HM2511).
Narrower Imploring succour (#HM2771) Abject helplessness (#HM2429) Representational sign (#HM2532) Levitational submission (#HM2706).

♦ **HM2923e Recognition of guilt** (Brainwashing)
Description This mode is brought about by persuasions of guilt and demands to feel guilty. The unconscious, and finally the conscious, is pervaded by feelings of evil. The individual comes to believe that he himself is the cause of his sufferings, which are deserved and to be expected; and although the guilt is not yet specific he is led to confess.
Context A stage reached in thought reform, a system of organized, deliberate and total psychological training which effects individual change through two basic elements - *confession* (renouncing of past beliefs and attitudes) and *re-education* (remaking of the individual in the required image).
Broader Thought reform (Brainwashing, #HH0865).
Followed by Self-betrayal (Brainwashing, #HM2229).
Preceded by Deprivation of selfhood (Brainwashing, #HM3002).

♦ **HM2924d Gemini-consciousness** (Astrology)
Energetic conceptualization
Description An analytic mode of awareness, polarizing experience so as to create a union which involves recognition of complementarity. Under optimal conditions, Gemini is associated with positive thinking and with mature creativity in response to present and emerging challenges.
Context One of the twelve conditions of being or streams of divine energy encompassed within the zodiac. It is said that all twelve conditions must be present and playing its proper part for any successful and fulfilling action to take place. This may be through a group of individuals with different endowments and soul qualities working together; or it may be through one individual who, by self-discipline, practice and meditation, has allowed all twelve conditions to arise within himself.
Broader Airy awareness (Astrology, #HM2955) Mutable awareness (Astrology, #HM3092) Zodiacal forms of awareness (Astrology, #HM2713).
Followed by Cancer-consciousness (Astrology, #HM2239).
Preceded by Taurus-consciousness (Astrology, #HM2815).

♦ **HM2926g Lethargy** (Buddhism)
Styana — Rmugs-pa (Tibetan)
Description A heaviness or ineptness of body or mind due to ignorance.
Context One of the twenty *secondary afflictions* of Tibetan Buddhism.
Broader Secondary afflictions (Buddhism, Tibetan, #HH0781).
Related Ignorance (Buddhism, Yoga, Zen, #HM3196).

♦ **HM2927d Diaphanous intuition** (ICA)
Description This is suddenly seeing through to the meaning of everything around, sometimes described as the "aha" experience. It may be compared to Archimedes sitting in the bath and hitting upon the theory of the displacement of volume in a liquid, or the cartoon character with a light bulb over his head.
A dimension of this experience is a sense of awe, that is fear and fascination. The discovery of being clairvoyant opens the individual to misunderstanding, ridicule and rejection by others. Insights are sometimes so fleeting that he wonders if they are only a dream, and doubts their validity. Nevertheless, he is pulled by life itself to trust his deepest intuitions and know the euphoria that occurs when it all comes clear.
The decision is to act alone, regardless of the consequences; all one has to believe is the power of one's own insights and intuitions. One is driven to continue pursuing every insight to its full significance, as though one will never again be content with anything less than the profound meaning of every thing and every moment.
Context This state is number 45 in the ICA *Other World in the midst of this World.*
Broader Transparent power (ICA, #HM2828).
Followed by Living death (ICA, #HM2808) Primal sympathy (ICA, #HM2550) Essential dubiety (ICA, #HM3135) Interior discipline (ICA, #HM2851) Individual fatefulness (ICA, #HM2223).
Preceded by Primal sympathy (ICA, #HM2550) Invented history (ICA, #HM3245) Essential dubiety (ICA, #HM3135) Original integrity (ICA, #HM2773) Global guardianship (ICA, #HM2817).

♦ **HM2929b Great love** (Buddhism, Tibetan)
Context One of the paths of effect according to Buddhist teaching.
Broader Paths of effect (Tibetan, #HM3249).

♦ **HM2930c Dyadic awareness**
Description Awareness may deal with formless, qualitative aspects of reality. Consciousness loosens localization and attains a distributed character as may be experienced in dreams or during shock. Experience of this kind may either lose contact with ordinary awareness, resulting in an outpouring of subconscious energy (e.g. anxiety) or integrate into meaningful phenomena (prophecy). C. Castaneda links a fundamental dualism of either frightening or enticing character to all such awareness states.
Followed by Triadic awareness (#HM2772).
Preceded by Monadic awareness (Amerindian, #HM2069).

♦ **HM2931b Darsana** (Buddhism, Tibetan)
Perception
Narrower Discipleship-vision awareness (#HM2240).

♦ **HM2933b Belief** (Buddhism, Yoga)
Adhimoksha — Mos-pa (Tibetan) — Saddha (Pali) — Faith
Description As an activity of the soul, belief is the means of achieving discrimination of pure spirit and, according to *yogic* tradition, is the first stage on the path to *samadhi.* The further steps on this path: energy; memory (or right mindfulness); a high order of meditation; and right perception, may be compared, together with *belief*, with the five determining mental factors of Buddhist teaching. As one of these latter factors, belief is that which keeps the mind from being side-tracked by another view of what has been already ascertained.
In Hinayana Buddhism, that by which one believes, or which believes, or is merely believing, is faith. It has the characteristic of trust or having faith; its function is to purify or to enter into. It manifests as resolution and clarity (non-fogginess). As proximate cause are things to gave faith in or those things, commencing with the hearing of good dhamma (saddhamma), which constitute stream entry (association with good people, wise attention, conduct according to dhamma).
Context One of the five determining mental factors of Tibetan Buddhist teaching. One of the formations aggregate (mental coefficients) of Hinayana Buddhism, being listed among the constant states which appear in their true nature, and as profitable primary (always present in profitable or profitable-resultant consciousness).
Broader Faculty of faith (Buddhism, #HM0066) Determining mental factors (Buddhism, #HH0170) Awareness as mental-formation group of conscious existence (Buddhism, #HM2050).
Related Faith (#HH0694) Aspiration (Buddhism, #HM2578) Right beliefs (Buddhism, #HH0974) Resolution (Buddhism, Pali, #HM3800).

♦ **HM2934e Transitional limbo** (Brainwashing)
Description Following completion of the thought reform process the individual returns to everyday life. This may precipitate a crisis of identity if it contrasts too strongly with what he has been accustomed to. Longing to return to the ordered retraining environment is only slowly overcome as the individual learns to trust his new surroundings and finds internal wholeness and integrity.
Context A stage reached in thought reform, a system of organized, deliberate and total psychological training which effects individual change through two basic elements - *confession* (renouncing of past beliefs and attitudes) and *re-education* (remaking of the individual in the required image).
Broader Thought reform (Brainwashing, #HH0865).
Preceded by Ideological rebirth (Brainwashing, #HM2843).

♦ **HM2935c Pradhana** (Yoga)
Description The state of pure spiritual being achieved, according to Patanjali, when the individual has passed through the stages of the gross and subtle states. This balanced state is the cause of the physical and subtle, it is unresolvable and undifferentiated primary substance.

♦ **HM2937c Asmita** (Yoga)
I am this
Description Here consciousness is identified with the vehicle through which consciousness is expressed. If pure consciousness is *asmi*, pure awareness and self-existence, "I am", then in asmita pure consciousness gets involved in matter. The power of maya or illusion is such that the real nature of consciousness is lost, the "I am" becomes "I am this", where "this" may be the subtlest or the grossest vehicle of consciousness, even the physical body. The loss of awareness of the real nature of consciousness and identifying it with the vehicle of consciousness occur simultaneously. Thus asmita and *avidya* (ignorance) work together. In the descent into ignorance, there is a progressive evolution as the thinnest veil of avidya and weak asmita gradually become thicker and stronger, and identification with the vehicle is stronger and grosser. The reverse path, on the way to transcendence, is marked by a progressive weakening of asmita as it becomes more subtle and the veil of avidya becomes thinner. The path is also progressively harder. Intellectually, one can see that one is not the body; it is less easy to see the mind as thought patterns the same as any other mind and that this is not "I". One's opinion is very important to one because one identifies with it. When the jivatma can leave a vehicle because he wills to do so, that experience strengthens the knowledge that "I am" not that vehicle even when he returns to it. One after another the vehicles of consciousness are left behind by the yogi, asmita is progressively destroyed and the veil of ignorance becomes progressively thinner.
Context One of the five afflictions or causes of misery in yoga.
Broader Afflictions (Yoga, Hinduism, #HH1047).
Related Asmita stage (Yoga, #HM3509) Ignorance (Buddhism, Yoga, Zen, #HM3196).

♦ **HM2938b States of consciousness** (Physical sciences)
Refs Tart, Charles T *States of Consciousness and State-Specific Sciences* (1972); Zee, A *Fearful Symmetry.*
Narrower Discrete states of consciousness (#HM2298).

♦ **HM2939d Particular concern** (ICA)
Description This is experienced as an individual's need to find the way to act out his responsibility in the context of a particular time and place. It occurs when a specific concern becomes that for which he is willing to commit his life as a way of fulfilling his responsibility for all of creation, as when Martin Luther King decided civil rights in the United States was the issue to which he was going to give his life.
Context In the ICA New Religious Mode in the arena of acting out one's freedom (the life of prayer) the fourth formal aspect is one's experience of universal responsibility or intercession. At the second phenomenological level this occurs on recognizing one's own full participation in one's fate.
Broader Universal responsibility (ICA, #HM3110).
Followed by Agonizing prediction (ICA, #HM2889).
Preceded by Utter awareness (ICA, #HM2625).

♦ **HM2940c Spontaneity**
Description In the mode referred to as *spontaneity state*, the individual discovers that emotions previously only experienced expressed involuntarily (such as anger, jealousy and so on) are now only achieved by an act of will.

♦ **HM2941d Beyond success** (ICA)
Description This is experienced as detachment from a one's work; when one surrenders one's need for success and recognition. An example is the hero at the end of the book "The Ronin", when he walks away from his work of several years and says: "The hell with it".
Context In the ICA New Religious Mode in the arena of knowing one's disengagement (the life of poverty) the second formal aspect is being reliant on the larger society, or being disengaged from one's work. At the first phenomenological level, this occurs when one is no longer content with one's customary habits, but sees in them a futility never before discerned.
Broader Disengagement from work (ICA, #HM2552).
Followed by Historical vocation (ICA, #HM2616).

♦ **HM2943c Path of purified intelligence** (Judaism)
Description This state is represented as establishing the unity of numerations, preventing their destruction and division.
Context This state is the ninth of the spiritual paths expressed in the Jewish Kabbalah.
Broader Paths of wisdom (Judaism, #HM2509).
Followed by Path of resplendent intelligence (Judaism, #HM3426).
Preceded by Path of perfect intelligence (Judaism, #HM3243).

♦ **HM2944b Emptinesses on the paths of the view in Buddhism** (Buddhism, Tibetan)
Description Eighteen emptinesses aare described.
Narrower Emptiness of things (#HM2989) Emptiness of nature (#HM3205) Emptiness of things (#HM3208) Emptiness of products (#HM2414)

Emptiness of definitions (#HM2946)
Emptiness of non–products (#HM2089)
Emptiness of all phenomena (#HM3000)
Emptiness of the five senses (#HM2671)
Emptiness of ultimate nirvana (#HM2294)
Emptiness of cyclic existence (#HM3032)
Emptiness of the ten directions (#HM2995)
Emptiness of the unapprehendable (#HM2501)
Emptiness of the loci of the senses (#HM2794)
Emptiness of the nature of phenomena (#HM2899)
Emptiness of the indestructible Mahayana (#HM3072)
Emptiness of the inherent existence of non–things (#HM2448)
Emptiness of what is free of permanence and annihilation (#HM0052)
Emptiness of the objects of sense and of mental consciousness (#HM2522).

♦ **HM2945d Cosmic sanctions** (ICA)
Description This state occurs whenever it dawns on a person that despite the ambiguity, the decisions he makes have already been approved by life itself, as though he has cosmic permission to do whatever he decides with his life. There is a strange, indefinable significance about everything he does. It may be an experience President Anwar Sadat of Egypt came to know as he pioneered the rapprochement between Egypt and Israel, or Pope John XXIII as he opened the windows of the Catholic Church through Vatican II.
A dimension of this experience is a sense of awe, that is fear and fascination. There is a feeling that "this is not something I deserved and perhaps it is really all a dreadful mistake or an illusion". One takes to one's self both the acceptance and the anger of others and realizes that this sense of belonging is a rare and precious thing, something each one of us would like to live with always. There is a decision to live with both the doubt and approval of society; and a knowledge that since life itself has indicated its approval, one can do no wrong.
Context This state is number 27 in the ICA *Other World in the midst of this World.*
Broader Moral ground (ICA, #HM2721).
Followed by Primal vocation (ICA, #HM2621)
Blissful seizure (ICA, #HM3148)
Self transcendence (ICA, #HM2584)
Passionate disinterest (ICA, #HM2547)
Futuric responsibility (ICA, #HM3016).
Preceded by Second birth (ICA, #HM3175)
Blissful seizure (ICA, #HM3148)
Self transcendence (ICA, #HM2584)
Contextual world–view (ICA, #HM2842)
Intentional conscience (ICA, #HM2892).

♦ **HM2946c Emptiness of definitions** (Buddhism, Tibetan)
Lakshanashunyata
Context One of the eighteen emptinesses comprising the paths of view in Tibetan Buddhism.
Broader Emptinesses on the paths of the view in Buddhism (Buddhism, Tibetan, #HM2944).

♦ **HM2947f Anomie**
Description This state occurs when the individual sees no regularity or certainty in his interactions with the social world. There seem to be no rules about what is and what is not permitted, no norms for dealing with life. The effects of his actions and the reaction of others seem totally unpredictable; traditional rules have lost their meaning; and a sense of meaningless and isolation may result. Although free to choose his actions, all alternatives appear meaningless and there is no sense of direction or support from others. Such an individual is anxious. He seeks situations of security, stability and certainty, and avoids situations involving risk or challenge. The state may arise as unwillingness or inability to adjust to change in circumstances and the term may be used to describe behaviour conflicting with the norm. Social conditions exacerbating the problem may be negative (collapsing economy, destruction of one culture by another), or apparently positive, when increasing prosperity means that old values (thrift, hard work) lose their relevance.
Related Abulia (#HM0672) Uncertainty (#HM3051) Dehumanization (#HM3030).

♦ **HM2948c Five forces** (Buddhism)
Description These are powers attained on the path of preparation on the levels of forbearance and supreme mundane qualities - faith, effort, mindfulness, meditative stabilization and wisdom.
Context One of the sections of yogic paths or harmonies with enlightenment defined in Buddhism.
Broader Harmonies with enlightenment (Buddhism, #HH0603).
Related Determining mental factors (Buddhism, #HH0170).

♦ **HM2950g Dreams**
Description Dreaming constitutes an altered state of consciousness, the most common and most dramatic, elaborate hallucinations experienced by everybody every night. It is a major and integral part of many ways of growth. Brain waves show a high alpha to low beta level; there is rapid movement of the eyes and often movement of fingers, toes and genitals; blood pressure, pulse rate, breathing and adrenalin flow are typical of the waking state. Dream symbolism is a form of expression which is more complete and penetrating than that provided by intellectual concepts. It provides a sequence of messages from the individual's unconscious which by appropriate interpretation lead to personal growth. The process of dreaming is essential - an individual continually forced to waken when dreaming may become insane. Psychotherapy has paid considerable attention to dreams, whether in terms of interpretation (psychoanalysis), underlying archetypal qualities (Jungian analysis), or unfolding their content through re–enactment (psychodrama, Gestalt therapy). In theological terms, dreams may be "deceitful" when they arise from the depths of human nature anteceding rational planning and decision; but they may also be revelatory of something wrongly ignored in waking consciousness and the means of divine revelation.
Dreaming as a proportion of the time spent asleep increases when the individual has just had to learn complex tasks. It decreases with age, probably because the learning process also decreases.
Broader Sleep (#HM2980).
Narrower Nightmares (#HM2085)
Waking dreams (#HM1376)
Lucid dreaming (#HM3618)
High dream (Psychism, #HM0147)
Tonal state of awareness (Amerindian, #HM2066)
Nagual state of awareness (Amerindian, #HM2036).
Related Not–doing (#HH0367) REM sleep (#HM5974) Dream therapy (#HH2116)
Hallucination (#HM4580) Psychoanalysis (#HH0951)
Dream yoga (Buddhism, #HH4087)
Mastery of dreaming (#HM3173)
Psychic consciousness (Psychism, #HM1919)
Hypnotic states of consciousness (#HM2133).

♦ **HM2951d Exclusive contradiction** (ICA)
Description This is the experience of having only one battle to fight: with the forces of evil within one's self; and that this is the only struggle that counts. The problem is to slay one's propensity to live out of guilt rather than glory in the particularity of one's own life. It may be compared to Jacob wrestling with the angel or to St. Anthony saying: "Only struggle with Satan". Nikos Kazantzakis portrays such a struggle in "The Last Temptation of Christ"; and in the story of "Dr. Jekyll and Mr. Hyde" a man finds his life consumed by contradictory forces of good and evil.
A dimension of this experience is a sense of awe, that is fear and fascination. The fear is that one will succumb to one's own destructiveness. There is the sense that win or lose, one's life is forever changed; and joy in the combat - in knowing that this is the only battle worth one's being. There is also fascination with the eternal vigilance required. One recognizes, in the midst of this tremendous all–consuming struggle, that this is one's life and it is good; and that one decides good and evil. And one is left with the realization that the battle is never finally won and with a new determination to fight it out, to be vigilant and never surrender.
Context This state is number 56 in the ICA *Other World in the midst of this World.*
Broader Unknowable peace (ICA, #HM3015).
Followed by Vital signs (ICA, #HM2875)
Total exposure (ICA, #HM2764)
Final blessedness (ICA, #HM2958)
Archetypal humanness (ICA, #HM3112)
Contingent eternality (ICA, #HM2456).
Preceded by Personal epiphany (ICA, #HM2595)
Archetypal humanness (ICA, #HM3112)
Contingent eternality (ICA, #HM2456)
Transcending hostility (ICA, #HM2658)
Soteriological existence (ICA, #HM2904).

♦ **HM2952b Contemplation**
Description Contemplation embraces all the forms of reflection which call for a recollected frame of mind and have in them the quality of intuitive vision. It is associated not only with the mental attitude necessary for apprehending spiritual truth, but also in the initiation and understanding of complex and fundamental scientific and philosophical theories. Contemplation is therefore the state in which major innovations are conceived. It is not antagonistic to rational, analytic thought, but is complementary to it providing an alternative means of apprehending the nature of things. Marxism defines contemplation as a *sensory* stage of knowledge. However, Thomas Merton refers to it as "the highest expression of man's intellectual and spiritual life... that life itself, fully awake, fully active, fully aware that it is alive". With a certitude beyond reason and beyond simple faith, beyond knowing and unknowing, it is awareness of the source of life and being, an experience of "I am".
Christian mysticism distinguishes between two kinds of contemplation, that achieved by psychological effort - *acquired contemplation*; and that arising through the gratuitous making known of God to the individual - *infused contemplation.* This latter form entails deprivation of some natural operations, such as discursive rational knowledge, and first appears in purification from preoccupation with the external. This has been described as *aridity* or the *dark night of the soul.*
François Fénelon, quoting Dionysius the Areopagite, says that in the contemplative state the holy soul is occupied with pure or spiritual divinity, with God and not the image of God which could be addressed to the senses. The desires of the soul are not satisfied at being occupied with the attributes of God; the soul loves to unite itself with God as the subject of His attributes.
Refs Heiler, Friedrich *Contemplation in Christian Mysticism* (1960); Ibish, Yusuf and Wilson, Peter Lamborn (Eds) *Traditional Modes of Contemplation and Action* (1977); Merton, Thomas *Seeds of Contemplation* (1972); Merton, Thomas *Spiritual Direction and Meditation* (1975).
Narrower Contemplative life (Christianity, #HH2145).
Related Meditation (#HH0761) Concentration (#HM4226) Deautomatization (#HH2331)
Mystical contemplation (#HM2710)
Life of contemplation (ICA, #HM2109)
Contemplative indifference (#HM1405)
Contemplative intuitive meditation (#HH0816)
Dark night of the soul (Christianity, #HM3941)
Religious experience as meditational insight (#HM1593).

♦ **HM2953d Second step: the race** (Christianity)
Description Having recognized the cry is oneself united with the universe, the ego bursts with a new insight. It is not oneself but one's ancestors talking through one, demanding that their work is completed that they may have significance. One chooses carefully and without mercy which of them will be allowed to fulfil its desires, to turn its thoughts to action, to give form to its hopes. One's first duty, in service to one's race, is to feel within one's being all of one's ancestors. It is not enough to be a slave to these ancestors. One is more, a new hope, a new possibility on which the whole of one's race is gambled. One is to make strong the living men, women and children of one's race. One's second duty is to continue the work of the race. The next generation fights for freedom from one's ways. They dismiss the present as sluggish. One rejoices, for the third duty is to pass on to one's children the mandate that they surpass this generation.
Context The second step in the march for the spiritual exercises of Nikos Kazantzakis.
Broader Saviors of God: the march (Christianity, #HM3439).
Followed by Third step: mankind (Christianity, #HM3501).
Preceded by First step: the ego (Christianity, #HM3748).

♦ **HM2954d Emotional–sexual stage of life**
Second stage of life
Description In this state there is a process of socialization, of exploration and growth of relationships; there is an emotional sensitivity to others and to the natural world and a development of morality. This stage is normal between 7 and 14 years. Because of an inability to resolve dependency on others at the first stage of life, there may be unhappy adaptation at this point, a feeling of dissociation from the ultimate source of love, of rejection and rejecting.
Context The second of seven stages of life characterized by Master Da Free John. Those individuals who have their human growth arrested in the first three stages lack psychic maturity and spiritual quality. They may develop through *puja yoga* and *karma yoga.*
Broader Seven stages of life (#HH0460).
Related Puja yoga (Yoga, #HH0111).
Followed by Mental–volitional stage of life (#HM3464).
Preceded by Vital–physical stage of life (#HM3320).

♦ **HM2955c Airy awareness** (Astrology)
Triplicity of air
Description A person whose consciousness is dominated by the triplicity of Gemini, Libra and Aquarius will be marked by the ability to communicate and an intellect capable of connecting ideas. There will be a sense of justice and a balanced approach which appreciates system and prefers to work within a preconceived framework. Such detachment and control may appear as lacking in emotion or feeling, or as pedantic, but will find itself able to reflect in relationships with more "feeling"–dominated individuals.
Context One of the four *triplicities* or *elements*, each of which combines three related signs of the zodiac.
Broader Zodiacal forms of awareness (Astrology, #HM2713).
Narrower Libra–consciousness (#HM2512) Gemini–consciousness (#HM2924)
Aquarius–consciousness (#HM2973).
Related Fiery awareness (Astrology, #HM2124) Watery awareness (Astrology, #HM2384)
Earthy awareness (Astrology, #HM3235).

♦ **HM2957c Susupti state of unconsciousness** (Hinduism, Zen)
Sushupti — Unconscious — Deep sleep — Nidra
Description In yoga psychology, among the avasthas (states) is the deep sleep, ego–unconscious condition. According to the Mandukya Upanishad it is the field of Jnana, the cognitional or intellectual condition. In Zen terms, it may said to be consciousness as such, consciousness which is not of anything, even of itself.
Broader Awareness of relative reality (Hinduism, Yoga, #HM2032).
Related Superconscious (#HM2960) Prajna (Hinduism, #HM3455)
Turiya awareness (Hinduism, #HM2395) Transcendental meditation (TM, #HH0682).
Followed by Transcendental consciousness (#HM2020).
Preceded by Swapna state of dream consciousness (Yoga, #HM2781).

♦ **HM2958d Final blessedness** (ICA)
Description This is the awareness that one's life has eternal worth, which becomes synonymous with the word happy – happiness that has nothing whatever to do with external causes or rational achievement. If a person must find something to become happy about he will never grasp the reality of happiness. This is Camus' story "A Happy Death". It may be compared to Mrs Gandhi's statement a few days before her death, when she said that if death were to come to her now she knew that her life and death would help to create a new India.
A dimension of this experience is a sense of awe, that is fear and fascination. There is awareness of the incongruities of life (but this is in the background) and a sadness when the individual senses that some will never consciously know a moment like this. He finds he asks "What if no-one shares the fruit of this time ?"; but there is fascination with the sense of having "arrived", and satisfaction with living the life he has been given. He takes the stance of "you may kill me but you cannot destroy me", fascinated with the feeling that he has done it all and that there is no need for more. He decides that the struggle is stopped, suspended and that even death is full of meaning. What remains is a knowledge that his life is great just as it is and a sense of "now understand what this game of life is all about".
Context This state is number 60 in the ICA *Other World in the midst of this World*.
Broader Contentment at the centre (ICA, #HM2529).
Followed by Living death (ICA, #HM2808) Primal vocation (ICA, #HM2621) Dynamic selfhood (ICA, #HM3091) Personal epiphany (ICA, #HM2595) Contingent eternality (ICA, #HM2456).
Preceded by Primal vocation (ICA, #HM2621) Blissful seizure (ICA, #HM3148) Invented history (ICA, #HM3245) Personal epiphany (ICA, #HM2595) Exclusive contradiction (ICA, #HM2951).

♦ **HM2959d State of anger** (Buddhism)
Pratigha — Khong-khro (Tibetan)
Description Whether it is the intent to harm sentient beings, to harm the cause of suffering or to harm one's own sufferings, anger is the cause of not abiding in happiness in this lifetime and of sufferings induced for the future. It is said to be caused by arrogantly trying to surpass others in a state of constant competition and is the basis of wrong conduct.
Anger is identified with the workings of the ego, its chief characteristic is perversity, the fundamental distortion of placing the ego at the centre of the universe. It is supreme self-centredness and the desire to dominate, to always win. There is arrogance, conceit and a pitying contempt for others. Hidden within is anxiety and a sense of inferiority as others' weaknesses are seen to mirror one's own. The negative side is in all forms of intolerance and group loyalty that disparages others. The positive side is the energy to fight injustice and inequality.
Context One of the six *root afflictions* referred to in Tibetan Buddhism. Also one of the ten worlds described in Nichiren Soshu Buddhism, one of the four evil paths.
Broader Six paths (Buddhism, #HM1914) Ten worlds (Buddhism, #HM2657) Four evil paths (Buddhism, #HM1252) Root afflictions (Buddhism, #HH0270) Nichiren shoshu buddhism (Buddhism, #HH3443).
Related Meekness (#HH0414) Peevishness (#HM2472) Hatred awareness (#HM4596) Spiritual anger (Christianity, #HM0936).
Followed by State of tranquillity (Buddhism, #HM3492).
Preceded by State of animality (Buddhism, #HM0847).

♦ **HM2960b Superconscious**
Higher unconscious
Description This term is used to designate the creative, intuitive, inspiring aspects of mind, and those which have positive and self-directing qualities. It is said to be the source of higher intuitions, thoughts, inspirations and feelings, of genius, and of the states of contemplation, illumination and ecstasy; and may be expressed in dreams, hunches, feelings, and intuitive knowledge. At present the idea of a superconscious is scattered among a number of philosophers, psychologists and other investigators of consciousness. It is believed that, if the concept is a viable one, it may coalesce with as much force and effect as did the earlier idea of unconscious process.
Related Higher unconscious (Psychosynthesis, #HM2057) Higher state of consciousness (Psychism, #HM0866) Susupti state of unconsciousness (Hinduism, Zen, #HM2957).

♦ **HM2961b Presence of God**
Sakina — Shekina — Presence of the Holy Ghost
Description The quality of calm and dignity and the condition of inner peace and security inherent in one who has encountered the eternal. This condition extends beyond times of formal prayer and worship to permeate the whole of life and every activity.
Related Presence of Allah (Sufism, #HM3496) Shekinah awareness (Judaism, #HM2002) God as wholly the same (Judaism, #HM4291) Practice of the presence of God (#HH0992) Sense of the presence of God (Christianity, #HM6989) Discipline of worship (Christianity, Judaism, #HH3776).

♦ **HM2962b Sutra paths of accumulation and application in the ascension stages game** (Buddhism, Tibetan)
Description The first two paths of active compassion on the road to liberation.
Broader Stages of spiritual life (#HH0102) Consciousness ascension stages game (Buddhism, Tibetan, #HH4000).
Narrower Great-vehicle lower-path awareness (#HM2268) Great-vehicle middle-path awareness (#HM2160) Great-vehicle higher-path awareness (#HM3048) Mahayana heat-application awareness (#HM2208) Mahayana climax-application awareness (#HM2232) Mahayana receptivity-application awareness (#HM2247) Mahayana highest-teachings-application awareness (#HM2271).

♦ **HM2963c Preparedness**
Description A concept suggested by Colin Wilson to counteract the long-term effect of "taking life as it comes"; it involves preparing each action beforehand in order to get an adequate return from that action, whether it is reading a book or eating a meal. At the mental level, this involves *focusing* on the value of the prospective activity.
Related Focusing (#HM3003) Partial-mindedness (#HM2430).

♦ **HM2964d Miserliness** (Buddhism)
Matsarya — Ser-sna (Tibetan)
Description An attachment to goods which causes a grasping of possessions and not allowing them to go.
Context One of the twenty *secondary afflictions* of Tibetan Buddhism.
Broader Secondary afflictions (Buddhism, Tibetan, #HH0781).
Related Avarice (Buddhism, Pali, #HM0227).

♦ **HM2965c Terce** (Christianity)
Hours of terce
Description The canonical hours of terce traditionally take place between 9 am and 12 noon. They dramatize the acknowledgement that the work of the day, even life itself is only possible because of the good graces of God. The mood is unceasing dependence.
Broader Canonical hours (Christianity, #HM1167).
Followed by Sext (Christianity, #HM3725).
Preceded by Prime (Christianity, #HM1904).

♦ **HM2966b Pure vision**
Pure knowledge
Description The seer or true observer is absolute sentience, considering *maya*. Through the mind, he comes to know the "not-self", discriminating between the real and the unreal; and thereby to an understanding of the nature of spirit. The mind thus becomes the interpreter of true vision or knowledge, transmitting it to the brain.

♦ **HM2968d Meditative stabilizations** (Buddhism)
Description These are: *going as a hero*, *sky treasury*, *stainless* and *loftily-looking lion*.
Context One of the paths of special qualities according to Buddhist teaching.
Broader Paths of special qualities (Buddhism, #HM3079).

♦ **HM2969c Mansions of exemplary life** (Christianity)
Description Entry into these mansions is marked by fear of losing God. In this state the person is aware enough of their own failing to recognize the danger of returning to the first mansions. Also being aware of the lives of the saints they know that what ever effort they put forth they may fail and they cannot be certain of God's help at this point.
In these mansions the individual experiences great aridity in prayer. This is not the interior trials that will be experienced later. This aridity is usually caused by an unwillingness to abandon attachments to the things of this world or to seek recognition, even for things like piety. Some souls, having lead a life of virtue, who would not intentionally commit even the smallest of sins become impatient to be in the presence of God. Failing to realize that while virtue and good works are expressions of love of God, God has no need of these things. Gifts of the spirit are gifts and not purchases.
This period of aridity may teach humility, the recognition that it is God that controls the life of the spirit and not humans. Restlessness over acquiring the fruits of the spirit is the aim of the Devil. Souls in this state frequently come to believe that they are suffering for the sake of God and fail to see that it is their own imperfections which cause this restlessness. It is grieving over quite unimportant earthly things. It is failure to master one's passions.
They become cautious in ordering their lives, even their penances because their love of God is not strong enough to over come their reason. They find stumbling blocks everywhere because they are afraid and dare not go further. Without complete renunciation of the self, this state becomes quite burdensome. While there may be spiritual consolations, perfection consists of increased love and not consolations.
In preparation to leave this stage, an individual needs to find someone to render obedience in order that the soul is not harmed. This person should be without illusions about the things of this world.
Context The third of seven mansions of the soul's progress described by St Teresa of Avila.
Broader Mansions of the soul (Christianity, #HH1409).
Followed by Mansions of spiritual consolations (Christianity, #HM3443).
Preceded by Mansions of the practice of prayer (Christianity, #HM3218).

♦ **HM2970g Higher self** (Psychosynthesis)
Description According to R. Assagioli, the *higher self* is the permanent centre of individuality in the phenomenal personality. This self is above the flow in the tides of consciousness and their fluctuating levels.
Related Psychosynthesis (#HH0002).
Preceded by Conscious self (Psychosynthesis, #HM2123).

♦ **HM2971d Personal obligation** (ICA)
Description This is experienced as the acknowledgement that, if something is going to be done about a situation of human suffering, then one will have to be involved in that situation for the sake of changing it. It occurs when a person realizes that just feeling sorry about something is not going to prevent it from continuing into the future, that he must commit himself if the future is going to be different. An example is the decision of doctor that he must go to Africa and work in the camps for famine victims.
Context In the ICA New Religious Mode in the arena of acting one's engagement (the life of obedience) the third formal aspect is embodying charity as the transestablishment to act out social concern. At the second phenomenological level, this occurs when one recognizes one's obligation to the rest of mankind.
Broader Embodying charity (ICA, #HM3056).
Followed by Disinterested collegiality (ICA, #HM2812).
Preceded by Passionate concern (ICA, #HM3128).

♦ **HM2972g Sensory awakening**
Body awareness — Sensory awareness
Description Individuals frequently think they feel rather than actually feel; by ignoring such primary processes they freeze situations and themselves so that there is no sensory contact with the richness of each event. Each experience therefore tends to be predetermined by a frame of reference which effectively specifies the nature of the experience. Sensory awakening is a method of rebalancing the nonverbal aspects of the organism with the intellect. The process consists of different experiments designed to shift attention from symbolic or verbal interpretation to the actual sensations. Attention is focused on simple bodily functions such as relaxation, breathing, listening, movement and touch. Used separately or together in various combinations, these help to bring the individual back to an awareness of his senses.
Again, excessive muscular tension causes desensitization of the organism. By redistributing awareness throughout the organism rather than localizing it in the head, the person is often able to make contact with muscular tension, learn how it is created, and experience what it is like to gradually let go. The process of sensory awakening leads to heightened awareness, contact and experience. It allows the individual, if only temporarily, to let go of some of his defences, experience the intensity of open experience and, to some extent, the potentialities that lie within.
Related Body sense (Buddhism, #HM9123) Sense consciousness (Buddhism, #HM2664) Perception through the senses (#HM3764).

♦ **HM2973d Aquarius-consciousness** (Astrology)
Illumined effort
Description Aquarius applies the fruits of spiritual self-realization in the transformation of society. Under optimal conditions this is associated with an openness to the essential requirements for the further development of any condition.
Context One of the twelve conditions of being or streams of divine energy encompassed within the zodiac. It is said that all twelve conditions must be present and playing its proper part for any successful and fulfilling action to take place. This may be through a group of individuals with different endowments and soul qualities working together; or it may be through one individual who,

by self-discipline, practice and meditation, has allowed all twelve conditions to arise within himself.
Broader Airy awareness (Astrology, #HM2955) Fixed awareness (Astrology, #HM2554) Zodiacal forms of awareness (Astrology, #HM2713).
Followed by Pisces-consciousness (Astrology, #HM2856).
Preceded by Capricorn-consciousness (Astrology, #HM2822).

♦ **HM2974c Mushinjo** (Zen)
Description A self-induced trance-like state arising in various Zen practices, not the same as *satori*.
Broader Human development (Zen, #HH1003).
Related Bombu Zen (Zen, #HH0785).

♦ **HM2976d Missional comradeship** (ICA)
Description The fourth formal aspect of the *life of meditation*, an arena in the ICA *New Religious Mode*, this represents the awareness of having a colleague.
Broader Life of meditation (ICA, #HM3234).
Narrower Divine hosts (#HM2859) Destinal elector (#HM3027) Primordial ancestor (#HM3224) Expectant descendant (#HM2903).

♦ **HM2977d Remorse**
Related Shame (Buddhism, #HM8112).

♦ **HM2978d Relational situation** (ICA)
Description This is the realization that no one else is responsible for one's situation. Placing no blame, making no excuse, offering no defense, seeking no counsel, one embraces one's fate with all its haphazardness and precariousness. One ceases to blame one's parents, or one's self, or one's social- economic situation, or anything else, for the way one turned out. It is reflected in Wellington's comment "In for a penny, in for a pound", when he ordered the general advance at Waterloo. The poet Stephen Crane captured it when he wrote about the creature squatting upon the ground eating his own heart in his hands: "It is bitter-bitter," he said, "But I like it because it is bitter, and because it is my heart".
A dimension of this experience is a sense of awe, that is fear and fascination. One is aware of being completely vulnerable to the whims of life - like the parents of a seriously ill child who are gripped by the dreadful knowledge that the child may die and at the same time are lured by the possibility that it might live. There is no one to whom they can turn who can take away the agony of waiting upon life to determine which way it will go. Confronted with this shattering clarity, the decision is that no-one and no-thing can tell one that things are any different from the way one perceives them. It is as though one has had life handed back to one in a brand new way and no more does one seek a scapegoat for one's actions.
Context This state is number 22 in the ICA *Other World in the midst of this World*.
Broader Creative existence (ICA, #HM2894).
Followed by Problemless living (ICA, #HM2747) Worldly detachment (ICA, #HM2451) Universal compassion (ICA, #HM2734) Contextual world-view (ICA, #HM2842) Intentional conscience (ICA, #HM2892).
Preceded by Universal fate (ICA, #HM2687) External relation (ICA, #HM2471) Problemless living (ICA, #HM2747) Worldly detachment (ICA, #HM2451) Ubiquitous otherness (ICA, #HM2570).

♦ **HM2979c Catching-the-ox awareness** (Zen)
Description Longing for the old, sweet life prevents easy control of the ox, even though it is within the seeker's grasp. Discipline is necessary to tame the unruliness and come to self-harmony.
Broader Oxherding pictures in Zen Buddhism (Zen, #HM2690).
Followed by Herding-the-ox awareness (Zen, #HM2560).
Preceded by Seeing-the-ox awareness (Zen, #HM2755).

♦ **HM2980g Sleep**
Dormant consciousness — Orthodox sleep — Paradoxical sleep
Description Small or subtle interruptions in a persons' waking moments of discernment are not usually noted; but the interruption called sleep is made categorically distinct because of its obvious physiological characteristics. In fact, sleep is a variety of levels of consciousness (or lack of consciousness) in which the normal individual spends between one quarter and one third of his life. Although there is awareness of having slept there is no actual awareness of the sleeping state except in the case of dreams, which may be wholly, partially, permanently or temporarily remembered or not remembered at all (although the dreaming state may be known to have occurred due to external observation of the sleeper). On first falling asleep the individual enters orthodox sleep, a physiological and psychological continuum with wakefulness where no obviously abrupt change in consciousness occurs. This is followed by paradoxical sleep, where changes are abrupt and discontinuous. The stages of sleep have also been described in terms of rapid eye movement, electroencephalogram readings and other physical measurements. There is a cycle from 1 to 4, shown by increasing delta wave activity, then from 4 to 1, which is marked by rapid eye movements and dreaming. The cycle repeats several times each night, gradually spending less time in stages three and four and more time dreaming.
There may be changes in consciousness, but sleep does not involve complete absence of consciousness. There is some monitoring, so that the passage of time is usually estimated quite accurately. An individual can often awaken at a predetermined hour, or just before the ringing of an alasm clock. Again, the individual will respond to loud or unexpected noises or to the speaking of his own name. A mother may waken at the first slight noise made by her baby, although ignoring other, much louder noises.
Although little is known of the function of sleep, sleep deprivation has marked effects, ranging from irritability and headache through blurred vision and hallucination, disorders in thinking and finally psychosis and depersonalization. Deprivation of dreaming sleep is particularly distressing, and sleep following a period of deprivation is particularly dream-full.
If the purely cognitive aspect of sleep is singled out, it is not necessarily in a class by itself, but rather draws attention to daytime, waking "sleep" conditions of restricted awareness, ranging from mild to severe and varying in duration and periodicity. This is not easily demonstrable, but other conditions showing that sleep is a state of varying intensity are sleep walking, hypnosis, "light" hypnosis or suggestibility, twilight sleep, and reveries. All conditions in which the potential of the individual's full consciousness is dormant, are attended by some degrees of ignorance (due to restriction of field of awareness), suggestibility or impotent will, and psychic vulnerability. Sleep is a metaphoric but partly accurate description of states of dormancy.
Broader Diminished clarity of awareness (#HM6201).
Narrower Dreams (#HM2950) REM sleep (#HM5974) Drowsiness (#HM6231) Deep sleep (#HM6307) Sleep spindles (#HM4806) Stage-three sleep (#HM0348).
Related Sleep (Buddhism, #HM7921) Mental tiredness (#HM5806) Physical tiredness (#HM5635).
Followed by Waking consciousness (#HM3021).

♦ **HM2981d Apt religion** (Leela)
Sudharma
Description The player lives in harmony with his own dharma, with the rules of the game. He takes part with no care for where the play will lead him. He always plays fair and believes in liberation and the merging with cosmic consciousness as a final goal. No longer dependent on external forms, the whole of life becomes internalized religion, an act of worship. Listening to the voice of his own inner nature he fears nothing. The player begins to understand his own role in the game. He is lead directly to the sixth chakra, to the plane of austerity, to hard work on the self.
Context The 28th state or square on the board of *Leela*, the game of knowledge, appearing in the fourth row.
Broader Fourth chakra: attaining balance (Leela, #HM3291).
Followed by Plane of austerity (Leela, #HM5244).

♦ **HM2982d Vibrant powers** (ICA)
Description This state occurs when an individual experiences a sudden burst of power that is beyond his own physical and mental capabilities; he senses that his own capacity to decide and act is augmented in a way he does not understand and which is not of his own doing. It may compared to the movie "Flashdance", when the dancer appears before the judges and seems to dance as she has never danced before; or the movie "Rocky", when the boxer stands at the top of a flight of steps, looks over the city, and knows he is filled with a new kind of power. D H Lawrence writes of being like a "tiger bursting into sunlight". It is an experience of being intensely alive and filled with a new vitality and vibrant power.
A dimension of this experience is a sense of awe, that is fear and fascination. One fears that one is somehow being possessed, that one is losing control of one's life; and one wonders how long it will last. At the same time there is a fascination with the fact that one has never experienced doing this before. One wonders what it would mean to live one's whole life in this state. After such a time, having decided to abandon one's self to this power, one experiences a new confidence and a willingness to risk moments of "being possessed" again. One is ready for anything life might want to present.
Context This state is number 9 in the ICA *Other World in the midst of this World*.
Broader Transformed state (ICA, #HM2386).
Followed by Beyond morality (ICA, #HM3049) Essential dubiety (ICA, #HM3135) Radical contingency (ICA, #HM2477) Global guardianship (ICA, #HM2817) Transformed existence (ICA, #HM2862).
Preceded by Vital signs (ICA, #HM2875) Total exposure (ICA, #HM2764) Incarnate living (ICA, #HM2394) Radical contingency (ICA, #HM2477) Global guardianship (ICA, #HM2817).

♦ **HM2983d Being all the other** (ICA)
Description This is experienced when a person suddenly knows all of life; when he finds himself able to stand in another person's shoes. It may have been experienced by Gandhi when he saw that his whole life was about human injustice in all forms in the continent of India.
Context In the ICA New Religious Mode in the arena of standing present to the mystery of being in life (life of contemplation) the first formal aspect is the experience of enigmatic not-me-ness or of externality. At the fourth phenomenological level this occurs when a person experiences adoration as the participant in all of life.
Broader Awareness of externality (ICA, #HM2637).
Related Organized games (#HH0692).
Preceded by Hallowed honour (ICA, #HM2680).

♦ **HM2984c Awareness of self through spiritual intelligence** (Yoga)
Description According to Patanjali, when the true self or spiritual intelligence sees itself reflected in chitta or "mindstuff" then an awareness of self arises. The mindstuff, as it reflects knower and known, becomes all-knowing; the knower, the field of knowledge, and that which is known, are brought into conjunction through the mind.

♦ **HM2985c Identifying with reality** (Sufism)
Description Through discipline the individual maintains his awareness that he and the Truth are one, whether sleeping, eating, listening, speaking; watching every breath, the individual is conscious of his true identity at every breath. Thus moments of truth are extended and enlarged until they become a continuous experience.
Context The seventh illumination of Jami.
Broader Fountains of light (Sufism, #HM3039).
Followed by Identity with God (Sufism, #HM2214).
Preceded by Ultimate transparency of spirit (Sufism, #HM2483).

♦ **HM2986c Non-shame** (Buddhism)
Ahrikya — Ngo-tsha-med-pa (Tibetan) — Ahirika — Shamelessness
Description Whether through desire, hatred or ignorance, the failure to avoid faults either from one's own viewpoint or because morally prohibited.
Context One of the twenty *secondary afflictions* of Tibetan Buddhism.
Broader Secondary afflictions (Buddhism, Tibetan, #HH0781).
Related Shamelessness (Buddhism, #HM0649) Power of shamelessness (Buddhism, #HM1724) Consciencelessness (Buddhism, Pali, #HM4394).

♦ **HM2987b Paths of calming** (Buddhism)
Description Phenomena of the pure class according to Buddhist teaching, the paths of calming consist of: the four noble truths; the four concentrations; the four immeasurables; the four formless absorptions; the eight liberations; the nine serial absorptions; and the paths of insight.
Broader Stages of spiritual life (#HH0102).
Narrower Paths of insight (#HM3194) Four noble truths (#HH0523) Eight liberations (#HM2635) Four immeasurables (#HM2507) Serial absorptions (#HM3150) Tibetan meditative states of form concentrations (#HM2693) Tibetan meditative states of formless absorptions (#HM2669).

♦ **HM2989c Emptiness of things** (Buddhism, Tibetan)
Bhavashunyata
Context One of the eighteen emptinesses comprising the paths of view in Tibetan Buddhism.
Broader Emptinesses on the paths of the view in Buddhism (Buddhism, Tibetan, #HM2944).

♦ **HM2990b Field consciousness**
Description A part of the traditional wisdom of India and of Greece, this refers to the whole consciousness experienced either by mankind or by the individual. Although the problems of mankind may be looked on as disturbances in the total field, they are said to be soluble only by the individual taking personal responsibility for self-transformation and thus starting to purify the whole consciousness.
Refs Gurwitsch, A *The Field of Consciousness* (1964).

♦ **HM2991d Every situation** (ICA)
Description This is the realization that every situation is fraught with history. It occurs when a person realizes that society is not static and that he can change the momentum of society by his action, as when a teacher realizes that his activity in the classroom is shaping the imagination of people who will be leaders of tomorrow's world.

Context In the ICA New Religious Mode in the arena of acting out one's deed (the life of doing) the second formal aspect is extending one's engagement to history. At the second phenomenological level this occurs when a person recognizes their life as sheer venture.
Broader Engagement to history (ICA, #HM2588).
Followed by Determining history (ICA, #HM2682).
Preceded by Primordial sociality (ICA, #HM2877).

♦ **HM2993c River of consciousness** (ICA)
Description An area in the ICA *Other World in the midst of this World* characterized by awareness, freedom and selfhood.
Broader Other world in the midst of this world (ICA, #HM2614).
Narrower Moral ground (#HM2721) Authentic relation (#HM2316) Creative existence (#HM2894) Final accountability (#HM3206).
Related Land of mystery (ICA, #HM2434) Mountain of care (ICA, #HM2170) Sea of tranquillity (ICA, #HM3033).

♦ **HM2994d Eternal void** (ICA)
Description This is the moment when one knows that the centre of all of life is unknowable and that one imposes understanding on life one's self. It occurs when a person encounters new ways of seeing things and knows that there is no final answer or eternal way to understand life; as, when experiencing the death of a loved–one, anticipating one's own death in that event.
Context In the ICA New Religious Mode in the arena of articulating the word about life (the life of knowing) the fourth formal aspect is one's experience of the final nothing in life or the glorious mystery of life. At the first phenomenological level this occurs when one sees through the events of everyday life and one's own subjectivity, to discern existence clearly.
Broader Glorious mystery (ICA, #HM2042).
Followed by Everlasting enemy (ICA, #HM2821).

♦ **HM2995c Emptiness of the ten directions** (Buddhism, Tibetan)
Mahashunyata
Context One of the eighteen emptinesses comprising the paths of view in Tibetan Buddhism.
Broader Emptinesses on the paths of the view in Buddhism (Buddhism, Tibetan, #HM2944).

♦ **HM2996d Afflicted views** (Buddhism)
Drishti — Lta–ba–nyon–mongs–can (Tibetan)
Description There are five such views: of *transitory collection*, or a belief in the existence of illusory mental and physical aggregates; of *extremes* – that the self viewed in the transitory sense is permanent and unchanging, or that it is subject to complete annihilation rather than rebirth in another lifetime; that a wrong view is *supreme*; that wrong conduct is supreme; that bad ethics and wrong modes of conduct are supreme; or *perverse* denial of cause, effect or function.
Context One of the six *root afflictions* referred to in Tibetan Buddhism.
Broader Root afflictions (Buddhism, #HH0270).
Related Wrong view (Buddhism, Pali, #HM5324).

♦ **HM2997c Establishments in mindfulness** (Buddhism)
Description There are four such establishments attained with the lesser path of accumulation: meditations on the impermanence, misery, emptiness and selflessness of the body, feelings, thoughts and other internal phenomena of one's self and other sentient beings.
Context One of the sections of yogic paths or harmonies with enlightenment defined in Buddhism.
Broader Harmonies with enlightenment (Buddhism, #HH0603).
Related Mindfulness (Buddhism, #HM2847).

♦ **HM2998c Viveka** (Yoga)
Discrimination — Discernment
Description Viveka opens the eyes of the soul, leading to non–attachment to the objects which keep it in bondage. In this state of non–attachment the illusions of life are detected and the relative reality hidden behind them discovered.
Related Viveka khyati (Yoga, #HM4493) Awareness of the ultimate reality (#HM1456).

♦ **HM3000c Emptiness of all phenomena** (Buddhism, Tibetan)
Sarvadharmashunyata
Context One of the eighteen emptinesses comprising the paths of view in Tibetan Buddhism.
Broader Emptinesses on the paths of the view in Buddhism (Buddhism, Tibetan, #HM2944).

♦ **HM3001g Android consciousness**
Description A fictitious state in which an existential consciousness with some degree of free–will is superimposed on a being in a deterministic universe, a mechanical consciousness with a set of imposed moral standards where the distinction between robot - android - man is blurred.

♦ **HM3002e Deprivation of selfhood** (Brainwashing)
Assault on identity
Description This condition is brought about as the first stage in thought reform when the individual is persuaded that he is not the person he thought he was, but that this was simply a cover for his real self. Verbal persuasion combined with physical and emotional deprivation brings about a regressive, helpless condition leading to surrender of personal autonomy and sometimes to a hypnagogic mode of awareness between sleeping and waking. This renders the individual more susceptible both to external influence and to internal aggressive and destructive impulses.
Context A stage reached in thought reform, a system of organized, deliberate and total psychological training which effects individual change through two basic elements - *confession* (renouncing of past beliefs and attitudes) and *re–education* (remaking of the individual in the required image).
Broader Thought reform (Brainwashing, #HH0865).
Followed by Recognition of guilt (Brainwashing, #HM2923).

♦ **HM3003e Focusing**
Alertness — Factor–X
Description This is the awareness described by Colin Wilson as that heightened, wide–awake and free perception achieved in times of stress and danger. This perception can be dulled through lack of use or sharpened by deliberate attention to thing which are most worthwhile to one's self. It can be described as a true *sense of reality*.
Related Preparedness (#HM2963) Peak experiences (#HM2080) Partial–mindedness (#HM2430) Beta wave consciousness (#HM3476).

♦ **HM3004b New religious modes** (ICA)
Description A scheme of awareness elaborated in the 1970's by the *Institute of Cultural Affairs (ICA)* in the light of a multi–cultural exploration of contemporary and traditional modes of experience. Human life is seen as being acted out in various arenas which make up the *new religious mode*, the individual arenas being described separately.
Broader Maps of the mind (#HH8903).
Narrower Life of doing (#HM3018) Life of being (#HM3229) Life of prayer (#HM2511) Life of knowing (#HM2801) Life of poverty (#HM2299) Life of chastity (#HM2506) Life of obedience (#HM2024) Life of meditation (#HM3234) Life of contemplation (#HM2109).
Related Other world in the midst of this world (ICA, #HM2614).

♦ **HM3005e Multi–level awareness** (Psychism)
Description By following a sequence of relaxation techniques and questions and answers involving the individual in altering his internal conception of his physical limits, it is possible to assist him to become aware of and describe previous lifetimes or incarnations.

♦ **HM3006c Establishment in truth** (Hinduism)
Fourth plane of wisdom
Description Following the practice of dispassion, enquiry and non–attachment there arises a natural turning away from sense pleasures and a natural dwelling in truth.
Context The fourth of seven ascending planes of wisdom described in the Supreme Yoga, which require to be known so as not to be caught in delusion.
Broader Planes of wisdom (Hinduism, #HM3298).
Followed by Total freedom from attachment (Hinduism, #HM2001).
Preceded by Subtlety of mind (Hinduism, #HM3312).

♦ **HM3007d Awareness of archaism** (ICA)
Description The second formal aspect of the *life of contemplation*, an arena in the ICA *New Religious Mode*, this is an experience of fearful "never–again–ness".
Broader Life of contemplation (ICA, #HM2109).
Narrower Sheer re–creation (#HM3101) All–that–ever–was (#HM3088) Wonder–filled fate (#HM3052) Reforged transformation (#HM3061).

♦ **HM3008c Religious consciousness**
Description A universal structure of religious consciousness has been defined (Donald M Moss) as the derealization of the physical and sensory world and the building of a "virtual" world along religious dimensions, so that this virtual world becomes a second order reality, uniquely lived by humans, which permeates everyday life.
Refs Ghurye, G S *Religious Consciousness*; Valle, Ronald S and Eckartsberg, Rolf von (Eds) *The Metaphors of Consciousness* (1981).
Related Derealization (#HM5128) Religious experience (#HM3445) Mystical transformation (#HH1432).

♦ **HM3009d Decisional nothingness** (ICA)
Description This is putting one's own self in right perspective with the overwhelming awe one has experienced. It occurs when a person has chosen to no longer seek for significance in his life and may be compared to Gandhi putting on a dhoti and spinning, or to Saint Anthony leaving a comfortable life and going to be a hermit in the desert.
Context In the ICA New Religious Mode in the arena of knowing and doing intensified in one's life (life of being) the first formal aspect is the experience of transparent selfhood or the experience of noughtness. At the third phenomenological level this occurs when a person disciplines his life to be related continually to the experience of nothingness.
Broader Transparent selfhood (ICA, #HM3077).
Followed by Absorption into nothingness (ICA, #HM3047).
Preceded by Acute inadequacy (ICA, #HM3151).

♦ **HM3010f Hysteria**
Description This state is the result of a desire to attract attention or avoid responsibility. It may have any of a wide range of symptoms, the most usual being screaming or weeping, but anything from headache to paralysis, fainting to convulsions. Although commonly afflicting young woman, hysteria may affect people of any age or sex. It may be a response to a situation which has become unbearable, the manifestation of repressed childhood experience, the self–punishment inflicted from feelings of guilt. The difficulties of the hysteric are projected onto the outside world, or extroverted, leading him or her to seek assistance from others. According to Freud, the hysteria associated with sexual problems (the original and most common use of the term "hysteria") is considered equivalent to coitus.
Related Mass hysteria (#HM0574).

♦ **HM3011c Chitta–dependent consciousness** (Yoga)
Description In yoga, the equivalent to the Christian mystics' conformation of the mind to Christ, is the conformation of the individual mind stuff (chitta) to the universal mind–stuff (chit) which is part of the trinitarian realized state, sat–chit–ananda. The yogi quells the egoic obstructions (vritti) in the chitta and sees God, enjoying real existence (sat), wisdom (chit) and joy (ananda).
Related Saccidananda (#HM0592) Atma yoga (Yoga, #HH0774) Dhyana (Hinduism, Buddhism, #HM0137) Indian approaches to consciousness (Hinduism, #HM2446).

♦ **HM3012c Consciousness of human service**
Arete — Jen — Futuwwa (Sufism)
Description A state of awareness in which noble and altriustic impulses dominate has been expressed in many cultures. Among the Greeks it was called virtue (arete) which only the best citizens possessed. The Chinese tradition finds the ideal in the character of the gentleman (jen); and among the Arabs it is the praiseworthy qualities that distinguish knighthood (futuwwa). The Sufis defined it as a state of mind made manifest by self–denial, immunity against disappointment, liberality and indulgence of other's short–comings; in short, all the qualities of an ethically and psychologically mature individual that open the way to God–realization. The signal feature of this state of awareness is that it builds unbreakable fellowship in service.
Broader Stations of consciousness - Simnami (Sufism, #HM2341).

♦ **HM3013c Path of the admirable intelligence** (Judaism)
Path of the hidden intelligence
Description This state is represented as giving the power of comprehension. It is beyond the attainment of all created beings, the primal glory which has no beginning.
Context The first of the spiritual paths expressed in the Jewish Kabbalah.
Broader Paths of wisdom (Judaism, #HM2509).
Followed by Path of the illuminating intelligence (Judaism, #HM2612).

♦ **HM3014d Non–stiffness** (Buddhism)
Akathinata (Pali)
Broader Flexibility of body (Buddhism, #HM3370) Flexibility of thought (Buddhism, #HM1805).

♦ **HM3015d Unknowable peace** (ICA)
Problemlessness at the centre
Description This is an experiential trek in the *Sea of Tranquillity*, within the ICA *Other World in*

the midst of this World, when the individual comes to know that security dwells in the trials of living. The states are described in separate entries.

Broader Sea of tranquillity (ICA, #HM3033).
Narrower Creative futility (#HM2493)
Transcending hostility (#HM2658)
Problemless living (#HM2747)
Exclusive contradiction (#HM2951).
Related Endless life (ICA, #HM2437)
Contentment at the centre (ICA, #HM2529).
Radical illumination (ICA, #HM2273)

♦ **HM3016d Futuric responsibility** (ICA)
Description This is the awareness that one is an embodiment of the future, experienced as being out of step with the times one is in. The future is not some abstraction read about in books or seen on television, it is what one's life is creating. One finds one is projecting one's self into the future by the visualization method, like writing a script for a play and finding one's self acting it out before one has finished. The film "2001" graphically presents this experience as it thrusts one into the future with a new image of humanness.
A dimension of this experience is a sense of awe, that is fear and fascination, a realization of the crucialness of this moment and an understanding that what one is doing is the most crucial thing one could be engaged in. The sense of eager anticipation is like that of a father waiting for the birth of his first child and one knows that living on the cutting edge of history is not always an enviable position. There is a terrible sense of being alone and a fear of not being understood. Life is one risk after another. Moreover, there is no justification for taking this particular stand. The decision is to allow one's self to be consumed by the challenge to create the new. A sense of déjà vu permeates every experience and a rejoicing in the fact that "my eyes have seen the glory". One feels sorry for all those trapped by the past and unable to experience the excitement of creating the future.
Context This state is number 43 in the ICA *Other World in the midst of this World*.
Broader Singular mission (ICA, #HM3217).
Followed by Second birth (ICA, #HM3175)
Blissful seizure (ICA, #HM3148)
Impactful profundity (ICA, #HM2607).
Invented history (ICA, #HM3245)
Temporal solidarity (ICA, #HM2363)
Preceded by Second birth (ICA, #HM3175)
Sacrificial passion (ICA, #HM2641)
Ancestral obligation (ICA, #HM3122).
Cosmic sanctions (ICA, #HM2945)
Temporal solidarity (ICA, #HM2363)

♦ **HM3017e Desperate gratitude** (Brainwashing)
Grasping at straws
Description Just as the individual is reaching breaking point he is shown unexpected kindness and leniency, while at the same time he is encouraged to confess and avoid further hardship. He is thus motivated to cooperate with his reform which is seen as the only escape from total self-annihilation.
Context A stage reached in thought reform, a system of organized, deliberate and total psychological training which effects individual change through two basic elements – *confession* (renouncing of past beliefs and attitudes) and *re-education* (remaking of the individual in the required image).
Broader Thought reform (Brainwashing, #HH0865).
Followed by Compulsion to confess (Brainwashing, #HM2745).
Preceded by Breaking point of basic fear (Brainwashing, #HM2799)
Breaking point of total conflict (Brainwashing, #HM2354).

♦ **HM3018c Life of doing** (ICA)
Description An arena in the ICA *New Religious Mode* when one is aware of acting out one's deed.
Broader New religious modes (ICA, #HM3004).
Narrower Engagement to history (#HM2588)
Surrender to one's calling (#HM2346)
Awareness of allegiance (#HM2073)
Symbolizing the eternal context (#HM3176).

♦ **HM3019c Buddha emanation body** (Buddhism, Tibetan)
Buddha form body
Description Comprises, together with the *enjoyment body* and having the same continuum, the *Buddha form body*; appearing in the aspect of a body from the force of compassionate wisdom without prior thought but arising spontaneously in answer to need. Such bodies perform their task without effort and are then withdrawn.
Broader Buddha-consciousness (Buddhism, Tibetan, #HM2735).
Related Buddha wisdom body (Buddhism, Tibetan, #HM2834)
Buddha nature body (Buddhism, Tibetan, #HM2039)
Buddha enjoyment body (Buddhism, Tibetan, #HM3113).

♦ **HM3020d Pratyeka Buddha application awareness** (Buddhism, Tibetan)
Unicorn
Description This state of awareness is represented by the unicorn who, in his forest loneliness, is like an ancient hermit who knew the Dharma not by being taught, but by discovering it through meditation. The unicorn bears love for all beings and is gentle and mild, but he runs alone towards Nirvana.
Context In Tibetan Sakya Buddhism this is one of the states in the "Ascension Stages Game". In some sets it is numbered 44 on the board.
Broader Religious traditions in the ascension stages game (Buddhism, Tibetan, #HM3341).
Followed by Joyful-heaven awareness (Buddhism, Tibetan, #HM2022)
Four-great-kings-heaven awareness (Buddhism, Tibetan, #HM2082)
Discipleship-application awareness (Buddhism, Tibetan, #HM2716)
Pratyeka Buddha cultivation awareness (Buddhism, Tibetan, #HM2252)
Pratyeka Buddha vision-path awareness (Buddhism, Tibetan, #HM2228)
Northern-continent awareness of community (Buddhism, Tibetan, #HM2067).
Preceded by Pratyeka Buddha awareness (Buddhism, Tibetan, #HM2180).

♦ **HM3021g Waking consciousness**
Wakening consciousness
Description That condition to which one wakes from sleep, in which most of one's waking life is spent and in which one perceives through the five senses. According to Gurdjieff this is only semi-awareness, the state of self-remembering being as different from waking consciousness as waking consciousness is from sleep.
Related Beta wave consciousness (#HM3476).
Followed by Self-remembering (#HM2486).
Preceded by Sleep (#HM2980).

♦ **HM3022c Angelic awareness - Zacharael**
Yahriel — Zachriel
Description This mode of awareness is represented as focussed on the deeper aspects of memory. Zacharael means remembrance of God; it is also known in the form *Zachriel*, the angel ruling over memory. Zacharael is also identified with Yahriel, an angel with dominion over the moon, as, among more recondite symbology, the lunar power rules memory and is mysteriously connected with the origins of the human race.
Broader Angelic frame of awareness (#HM2150).

♦ **HM3023c Reverence**
Description An attitude of reverence, of respectful attentiveness to the divine, transforms and intensifies any action, however simple, whether eating, drinking or any everyday activity. It invokes a sense of oneness with the whole universe. Such reverence allows the development of a sense of awe and wonder and a recognition of the mystery of creation. Although implying the idea of "fear", such fear is not for personal safety but, in conjunction with honour, expresses reverence for God and for that which derives its function from God.
Related Awe (#HM2592)
Wonder (#HM3197)
Honour (#HH0192)
Spirituality (Christianity, #HH0792).

♦ **HM3024d Arhat sanctity awareness** (Buddhism, Tibetan)
Description This awareness is represented as Buddhist sainthood; a life-condition characterized by the attainment of Nirvana. The physical body and mental components of the Arhats are said to have an apparent existence until they wear out, while the Arhat may be anywhere independent of them.
Context In Tibetan Sakya Buddhism this is one of the states in the "Ascension Stages Game". In some sets it is numbered 51 on the board.
Broader Religious traditions in the ascension stages game (Buddhism, Tibetan, #HM3341).
Related State of universal cessation awareness (Buddhism, #HM2596).
Followed by Cessation-awareness (Buddhism, Tibetan, #HM2172)
Pure-lands awareness (Buddhism, Tibetan, #HM2168)
Great-vehicle lower-path awareness (Buddhism, Tibetan, #HM2268).
Preceded by Pratyeka Buddha cultivation awareness (Buddhism, Tibetan, #HM2252).

♦ **HM3025d Spontaneous gratitude** (ICA)
Description This is an awareness of being grateful for the very particular situation: for the colour of cheeses on a platter, for the smoky haze clinging to the window pane. Then the whole world rapidly becomes an object of gratitude: each element, dynamic and relationship a miracle. Life teems about one, showering blessings everywhere, and one hears welling up from inside one's being a mighty "Whoopie ". It may be compared to the song "Life is a banquet" in the film "Aunt Mame", or to "Life is a cabaret" in the film "Cabaret".
A dimension of this experience is a sense of awe, that is fear and fascination. One asks "What did I do to deserve this?", and fears one may never experience such a time again. The fascinaton is manifested as exhilaration with the gratitude that is flooding one's life. In this moment one finds one has surrendered to this gratitude. Subsequent contentment and feelings of having been healed will no longer depend on circumstances: a falling leaf or a spoken word will recall this overwhelming sense of deep appreciation.
Context This state is number 58 in the ICA *Other World in the midst of this World*.
Broader Contentment at the centre (ICA, #HM2529).
Followed by Blissful seizure (ICA, #HM3148)
Intentional conscience (ICA, #HM2892)
Resurrectional existence (ICA, #HM2631).
Transformed existence (ICA, #HM2862)
Inclusive comprehension (ICA, #HM2256)
Preceded by Vital signs (ICA, #HM2875)
Ancestral obligation (ICA, #HM3122)
Inclusive comprehension (ICA, #HM2256).
Problemless living (ICA, #HM2747)
Intentional conscience (ICA, #HM2892)

♦ **HM3026c Middle-path tantra awareness** (Buddhism, Tibetan)
Ubhaya-carya-tantra
Description This state of awareness is characterized as an outer-and-inner yoga, when meditation oscillates between being object-oriented and being objectless. Unlike hatha yoga, in which physical restriction of breathing is used to induce trance states, the carya tantra, like raja yoga, restrains the mind first. Breathing becomes calmed of itself and then regulated by mental concentration and observation.
Context In Tibetan Sakya Buddhism this is one of the states in the "Ascension Stages Game". In some sets it is numbered 41 on the board.
Broader Tantric paths of accumulation and application in the ascension-stages-game (Buddhism, Tibetan, #HM2848).
Followed by Kriya-tantra awareness (Buddhism, Tibetan, #HM2558)
Potala-island awareness (Buddhism, Tibetan, #HM2175)
Great-path tantra awareness (Buddhism, Tibetan, #HM2656)
Kalacakra-tantra shambhala awareness (Buddhism, Tibetan, #HM2151)
Tantra master-in-sense-realm awareness (Buddhism, Tibetan, #HM2211).
Preceded by Kriya-tantra awareness (Buddhism, Tibetan, #HM2558)
Magical-forces awareness (Buddhism, Tibetan, #HM2679)
Tantra master in form-realm awareness (Buddhism, Tibetan, #HM2235)
Tantra master-in-sense-realm awareness (Buddhism, Tibetan, #HM2211).

♦ **HM3027d Destinal elector** (ICA)
Description This is experienced when one suddenly knows one has been chosen, that for no good reason one cares about something. This occurs when one sees that one has been called to use one's life to care for mankind, as when Dag Hammerskjold wrote in his poem "the Way has chosen you".
Context In the ICA New Religious Mode in the arena of knowing one's internal sociality (the life of meditation) the fourth formal aspect is one's experience of missional comradeship or of having a colleague. At the first phenomenological level, this occurs when that which is other intrudes upon one's consciousness.
Broader Missional comradeship (ICA, #HM2976).
Followed by Primordial ancestor (ICA, #HM3224).

♦ **HM3028c Betrothal initiation archetypal image** (Tarot)
Lovers — Mystic marriage — Hierogamy
Description The encounter of the psyche with an image of marriage is emblematic, reflecting a particular condition of conscious existence. It is one of the most important archetypes and is expressed visually in many forms, from male-female companionship to conjugal relations, and all possibilities in between such as the marriage sacrament, its ceremony, and its feast. The mystical symbolism of marriage pervades Christianity, esoteric Islam, and Judaism, and was central to some of the gnostics. It also entered the faith-systems of hermeticism and alchemy in the West, and is still crucial to the practice of Hindu and Buddhist tantra yoga. In the Tarot system the related emblem is the sixth card (of the 22 tarots of the Grand Arcana) where it is depicted as a priest marrying a man and women. The solar crown behind the cupidus or Eros indicates the noble love which attends initiation. Flowing upwards from beauty (harmony), there is represented the enduring union of opposites, male and female, sun and moon. Within the Tarot system itself there are indications that the psychic encounter with Marriage may be associated with the emblem of the Magician.
Context The correspondence of the Lovers emblem in the Kabbalistic Paths of Wisdom is said to relate to the connection between the sephira of the middle-pillar, beauty (Tiphareth), which is the nexus of all the sephiroth, and intelligence or understanding (Binah).
Broader Emblem archetypes in the psyche (Tarot, #HM2201)
Tarot arcana of conscious states (Tarot, #HM2097).
Related Internal marriage (#HH0799)
Sphere of beauty (Kabbalah, #HM3031)
Sphere of understanding (Kabbalah, #HM2372).

♦ **HM3029d Honouring the mystery** (ICA)
Description This is the awareness that all one does is a reflection of God's creation; when one knows that the mystery is present in every situation, and that the situation is a part of the mystery's intentional creation. An example in the Old Testament is when Job says "The Lord gave, and the Lord hath taken away; blessed be the name of the Lord".
Context In the ICA New Religious Mode in the arena of articulating the word about life (the life of knowing) the fourth formal aspect is the experience of the final nothing of life or the glorious mystery of life. At the fourth phenomenological level, this occurs when one recognizes the humiliation of standing before the final mystery of life in every moment.
Broader Glorious mystery (ICA, #HM2042).
Preceded by Immutable friend (ICA, #HM3124).

♦ **HM3030d Dehumanization**
Alienation — Self-alienation
Description When typical human attributes, whether qualities or mental or physical activities, are restricted or denied full outlet, then a person is said to be dehumanized. This is self-inflicted but is exacerbated by repetitive, mindless employment which reduces those carrying it out to the level of components in a machine. A society which treats the majority of people as simply instruments for the minority, which cares little or not at all for human dignity and whose institutions dominate people and rob them of their rights may be referred to as a dehumanizing society.
Alienation and boredom results when the social world is omnipresent and repressive. The oppressive security, stability and certainty of everyday life ensure that action produces little or no effect; and the individual feels powerless to change the situation. When the result is predictable, every act appears as drudgery. The individual may seek escape from such a situation by indulging in an activity such as mountaineering, which engenders a situation of controlled uncertainty and allows creative expression.
Related Anomie (#HM2947) Uncertainty (#HM3051) Flow experience (#HM2344)
Dehumanizing process (#HH3221).
Preceded by Conscientization (#HH0027).

♦ **HM3031c Sphere of beauty** (Kabbalah)
Tiphareth sephira
Description This sphere of awareness is associated with a sense of individuality and with an understanding of the psychological self and thus with the development of selfhood and individuality. It corresponds to the developmental phase of prime adulthood and achievement of self-confidence. As the centre or heart of the ten spheres of the kabbalistic system, it is called beauty, heaven and glory – the mediating stage between man and God, where man experiences spiritual rebirth, harmonizing of personality and balance of the rational and the emotional. It is the sphere of sacrifice, when the old personality is offered up and new, universal understanding and insight begin to operate. There is a glimpse of light beyond the abyss caused by the fall. It is symbolized by the positive and negative archetypes of self and Lucifer, and by the Sun, as well as by Jacob in the Jewish tradition.
Context This sphere is the sixth of ten described in the Kabbalistic system of sephiroth.
Broader Kabbalah (#HH0921).
Related Anahata (Yoga, #HM3242) Death emblems (Tarot, #HM3121)
Measuring emblems (Tarot, #HM2178) World of creation (Judaism, #HM3038)
Emblems of supremacy (Tarot, #HM2411) Wisdom archetypal image (Tarot, #HM2261)
Way of organization (Esotericism, #HM1605) Emblems of personified evil (Tarot, #HM2378)
Old wise-man archetypal image (Tarot, #HM2202)
Life-fluid emblems - temperance (Tarot, #HM3289)
Betrothal initiation archetypal image (Tarot, #HM3028).
Followed by Sphere of wisdom (Kabbalah, #HM2348)
Sphere of judgement (Kabbalah, #HM2290)
Sphere of understanding (Kabbalah, #HM2372)
Sphere of mercy and greatness (Kabbalah, #HM2420)
Sphere of the of the supreme crown (Kabbalah, #HM3132).
Preceded by Sphere of glory (Kabbalah, #HM2238)
Sphere of victory (Kabbalah, #HM2362)
Sphere of the foundation (Kabbalah, #HM2410).

♦ **HM3032c Emptiness of cyclic existence** (Buddhism, Tibetan)
Anavaragrashunyata
Context One of the eighteen emptinesses comprising the paths of view in Tibetan Buddhism.
Broader Emptinesses on the paths of the view in Buddhism (Buddhism, Tibetan, #HM2944).

♦ **HM3033c Sea of tranquillity** (ICA)
Description An area in the ICA *Other World in the midst of this World* characterized by being fulfilled, at one with death and finally happy.
Broader Other world in the midst of this world (ICA, #HM2614).
Narrower Endless life (#HM2437) Unknowable peace (#HM3015)
Radical illumination (#HM2273) Contentment at the centre (#HM2529).
Related Land of mystery (ICA, #HM2434) Mountain of care (ICA, #HM2170)
River of consciousness (ICA, #HM2993).

♦ **HM3034d Transcendent immanence** (ICA)
Description This occurs when an individual is aware of what he has known all his life; that he has an indefinite longing that will never go away. That is the first clue that he loves the mystery. One Christian saint wrote about it as a hound wandering through the streets looking for its master. The indefinite longing has no object as its focus; one is a perpetual refugee seeking one's home and yet knowing one will never find it, that one will die with the homesickness, the isolation and the sense of deprivation.
A dimension of this experience is a sense of awe, that is fear and fascination. The fear is of being without help, that it will be like this forever; and that this amazing consciousness will somehow go unrecorded, like a tree falling in a forest with no one to hear it. There is fear of being unable to function without a guide and that one's existence will make no difference to life. At the same time there is fascination with the possibility that it is forever and one is on one's own; there is a strange excitement in knowing that one is the guide of one's own life.
The decision is made in the midst of this state to continue the search. After this happening, loneliness becomes a friend and companion on the journey. There is also a sense of being driven to continue the search – it becomes a crucial aspect of life.
Context This state is number 15 in the ICA *Other World in the midst of this World.*
Broader Infinite passion (ICA, #HM2234).
Followed by Final limits (ICA, #HM2674) Ultimate reality (ICA, #HM2030)
Singular adoration (ICA, #HM2111) Impactful profundity (ICA, #HM2607)
Passionate disinterest (ICA, #HM2547).
Preceded by Second birth (ICA, #HM3175) Final limits (ICA, #HM2674)
Cryptic disclosure (ICA, #HM2824) Impactful profundity (ICA, #HM2607)
Everlasting community (ICA, #HM2777).

♦ **HM3035d Ignorance in dependent origination formula** (Buddhism)
Avijja (Pali) — Moha
Description Deluded ignorance (avijja) is the foundation of all life-affirming actions. On ignorance depend the karma-generations (sankhara) of volitional behaviour and thought. In the formula of Dependent Origination (paticcasamuppada), avijja (ignorance) conditions the rebirth-causing volitions (cetana) and their generations or formations (sankhara). Karma generated in a past life conditions the consciousness (citta, vinnana) in the present life. Through consciousness are conditioned the *mental and physical* (nama-rupa) phenomena of the interdependent groups of conscious existence phenomena (khandas).
Broader Consciousness in Buddhism (Buddhism, #HM2200).
Narrower Five aggregates (#HH3321).
Related Becoming (Buddhism, #HM5909) Karma awareness (Buddhism, #HM2048)
Ignorance (Buddhism, Yoga, Zen, #HM3196).

♦ **HM3036c Searching-for-ox awareness** (Zen)
Description Because the seeker, or ox-herd, has violated his own inmost nature, the ox is lost and the oxherd led out of his way by the deluding senses; desire for gain and fear of loss, ideas of right and wrong, all come and go, burning like fire while they last.
Broader Oxherding pictures in Zen Buddhism (Zen, #HM2690).
Followed by Seeing-the-traces awareness (Zen, #HM3302).

♦ **HM3037c Tsutsushimi** (Japanese)
Seriousness of mind
Description By maintaining seriousness of mind by concentration of the mind externally and internally, clear judgement is possible. Outward aspects are modesty, discretion and self restraint and such seriousness is achieved through knowledge.

♦ **HM3038c World of creation** (Judaism)
Olam briah
Description This world is said to correspond to the spiritual consciousness in man that is supra-individual, a spirit which is identified with the archetypes of creation. It is the realm of the three spheres (sephiroth): Mercy, Judgment and Beauty (chesed, geburah and tiphereth), characterized as divine aspiration; true judgments; and inner harmony and beauty.
Context This is the second highest of the Kabbalistic four worlds.
Broader Shekinah awareness (Judaism, #HM2002).
Related Paths of wisdom (Judaism, #HM2509) Sphere of beauty (Kabbalah, #HM3031)
Sphere of judgement (Kabbalah, #HM2290)
Sphere of mercy and greatness (Kabbalah, #HM2420).
Followed by World of formation (Judaism, #HM2360).
Preceded by World of manifestation (Judaism, #HM2312).

♦ **HM3039b Fountains of light** (Sufism)
The illuminations of Jami
Description This series of 36 illuminations recorded by the Sufi Jami describe the experience of the individual in approaching spiritual perfection.
Refs Jami *Les jaillissements de lumière* (1982).
Broader Human development (Sufism, #HH0436)
Stages of spiritual life (#HH0102).
Narrower Identity with God (#HM2214) Single-heartedness (#HM3085)
Single desiredness (#HM2765) Liberation of the heart (#HM3244)
Identifying with reality (#HM2985) Indivisibility of essence (#HM2113)
Perception of universality (#HM2827) Unchanging reality of being (#HM3316)
Perception of hidden reality (#HM2531) Separation from the transient (#HM2906)
First individuation of essence (#HM3614) Beyond individual distinctions (#HM2681)
Ultimate transparency of spirit (#HM2483) Ontological reality of the truth (#HM3303)
Existence as ontological reality (#HM3457) Annihilation of personal sensation (#HM3358)
Identity of attributes with their essence (#HM3260)
Manifestation of the essence through veils (#HM2642)
Surrender to the bliss of mystical seduction (#HM3436)
Reciprocal relation of absolute and conditional (#HM3228)
Liberation of the spirit by attraction of divine grace (#HM2907).
Related Naqsbandi tradition (Sufism, #HH0277).

♦ **HM3040d Taijasa** (Hinduism)
Mental condition
Description The second "quarter" or condition of the self or *Atman* according to the Mandukya Upanishad, enjoying subtle objects (remembered impressions of the waking state) through the organs of sense, the motor organs, the pranas and the aspects of mind; its field the dreaming (inwardly cognitive) state. It is associated with the syllable "U" of the mystic word "AUM".
Broader Degrees of consciousness (Hinduism, #HM0891).
Related Turiya awareness (Hinduism, #HM2395)
Swapna state of dream consciousness (Yoga, #HM2781).
Followed by Prajna (Hinduism, #HM3455).
Preceded by Vaisvanara (Hinduism, #HM2336).

♦ **HM3041b Asamprajnata samadhi** (Hinduism)
Enstasy
Description Existence is transcended in this, the highest superconscious state, in which the subject is not conscious of the object of concentration, and both mind and ego-sense appear to be dissolved. Although in a sense comparable with deep sleep, the two are in fact opposites, the one being dominated by *tamas*, the *guna* of inertia, while in the other tamas is obliterated. Possibly equated with the vedanta *nirvikalpa samadhi*, this is the state of *turiya*, pure being; although it has also, paradoxically, also been called waking sleep, as there is awakening to *self-awareness*.
Broader Samadhi (Hinduism, Yoga, #HM2226).
Related Enstasy (#HM3169) Three gunas (#HH0413)
Samprajnata samadhi (Yoga, #HM2896) Turiya awareness (Hinduism, #HM2395)
Nirvikalpa samadhi (Hinduism, #HM2061).

♦ **HM3042d Divine nothingness** (ICA)
Description This is experienced as having nothing, and of being called to create life out of it; when one realizes that there is nothing but significance. It may be compared to the role of the tramp so often played by Charlie Chaplin, who over and over again picks up his impossible situation and recreates his life in the midst of it.
Context In the ICA New Religious Mode in the arena of knowing one's disengagement (the life of poverty) the first formal aspect is the experience of being liberated from the claim of one's possessions. At the third phenomenological level, this occurs when one is able to so value the gift of life that nothing else is needed for one's fulfilment.
Broader Liberation from possessions (ICA, #HM3131).
Followed by Wealth untold (ICA, #HM2831).
Preceded by Good stewardship (ICA, #HM3143).

♦ **HM3043d Second absorption in the immaterial sphere** (Buddhism)
Vinnanacayatana (Pali) — Vijnananantyayatana — Limitless consciousness absorption — Sphere of infinite consciousness — Second formless attainment
Description *Hinayana Buddhism* The mind reaches the sphere of unbounded consciousness in

the second absorption. Reasoning and deliberation are suppressed. This second jhana is characterized by inner tranquillity and one-pointedness of mind. It arises from concentration, and joy and happiness are experienced.
Tibetan Gelugpa Buddhism In this state the object of meditation is the meditator's own mental aggregates or groups of conscious existence (khandas).
Context *Hinayana Buddhism* The next two jhanas are in the spheres of nothingness and neither-perception-nor-non-perception.
Tibetan Gelugpa Buddhism This is the second of the 4 formless absorptions (arupaya samapatti).
Broader Jhana (Buddhism, Pali, #HM7193)
Arupaloka consciousness (Buddhism, Pali, #HM2012)
Trances and mental absorptions (Buddhism, #HM2122)
Immaterial states as meditation subjects (Buddhism, #HH3198)
Tibetan meditative states of formless absorptions (Buddhism, #HM2669).
Related Subjects for meditation (Buddhism, #HH3987)
Formless-realm consciousness (Buddhism, #HM2281).
Followed by Third absorption in the immaterial sphere (Buddhism, #HM2027).
Preceded by First absorption in the immaterial sphere (Buddhism, #HM2110).

♦ **HM3044c Four fearlessnesses** (Buddhism, Tibetan)
Description These are fearlessness with respect to: the assertion of being enlightened with respect to all phenomena; the teaching that afflictive obstructions and obstructions to omniscience are to be ceased, since they are obstacles to liberation and the simultaneous cognition of all phenomena; the teaching of the paths of deliverance; the assertion that contaminations have been extinguished (achieved by abandoning pride).
Context One of the paths of effect according to Buddhist teaching.
Broader Paths of effect (Tibetan, #HM3249).

♦ **HM3045c Charioteer archetypal image** (Tarot)
Chariot
Description The encounter of the psyche with the emblem of the driver of a passenger-bearing vehicle (boatman, coachman, charioteer, and all the modern equivalents) reflects a particular state of consciousness. It is one of the most important archetypes experienced in mental life and is found in the myths of many peoples, often expressed in the form of a celestial chariot that can descend and ascend through the heavens. In the Tarot, the 7th card of the Greater Arcana (22 trump cards) represents one version of the chariots of the gods. Here it is depicted as horse-drawn and canopied. It's sole occupant is a crowned figure. The canopy is adorned with stars as is the crown. Theologically, the charioteer represents the Lord of Hosts; mystically it is the Self which is affiliated to this symbolic role of ruler of the universe by virtue of its spiritual mastery. It also the potentiality of that mastery and so can represent to the foot-weary pilgrim (in Tarot, the Fool) the possibility of "motoring". However, riding the Chariot may be considered as an emblem or condition encountered only after initiation (represented in the Tarot as card number 6, the betrothal or "lovers").
Context The correspondence of the charioteer emblem in the Kabbalistic Paths of Wisdom is said to be the relationship or transition between the Sephiroth called strength or judgment (Geburah), intelligence or understanding (Binah), thus forming one of five links between the second triad of emanations with the triad or trinity of the first three.
Broader Emblem archetypes in the psyche (Tarot, #HM2201)
Tarot arcana of conscious states (Tarot, #HM2097).
Related Sphere of judgement (Kabbalah, #HM2290)
Sphere of understanding (Kabbalah, #HM2372).

♦ **HM3046c Path of collective intelligence** (Judaism)
Description This state of awareness is of the celestial host whose collective dispositions and movements are revealed in speculative science.
Context The thirtieth of the spiritual paths expressed in the Jewish Kabbalah.
Broader Paths of wisdom (Judaism, #HM2509).
Followed by Path of perpetual intelligence (Judaism, #HM2643).
Preceded by Path of corporeal intelligence (Judaism, #HM3149).

♦ **HM3047d Absorption into nothingness** (ICA)
Description This is experienced as union with the void at the centre of being; when a person has surrendered everything, even the need to surrender. It may be compared to the prophet Elijah not dying but going directly to heaven, or to the Buddha experiencing Nirvana.
Context In the ICA New Religious Mode in the arena of knowing and doing intensified in one's life (life of being) the first formal aspect is the experience of transparent selfhood or the experience of noughtness. At the fourth phenomenological level this occurs when a person becomes one with the eternal.
Broader Transparent selfhood (ICA, #HM3077).
Preceded by Decisional nothingness (ICA, #HM3009).

♦ **HM3048c Great-vehicle higher-path awareness** (Buddhism, Tibetan)
Mahayana path
Description This state is characterized as an irreversible stage in personal spiritual development. The four foundations of psychic powers are represented as being attained on this level and are concentrations (samadhi) of: detached effort (faith); potency; mindfulness; and analytical concentration, leading to the psychic powers necessary to propagate and defend the Dharma. There is also a fifth concentration-foundation of *following the Dharma stream* (Dharma-srotanugata-samadhi), which is said to confer the ability of coming into the presence of the high bodhisattvas and the various Buddhas. it is an essential characteristic of wisdom.
Context In Tibetan Sakya Buddhism this is one of the states in the "Ascension Stages Game". In some sets it is numbered 54 on the board.
Broader Mahayana Buddhism (Buddhism, #HH0900)
Sutra paths of accumulation and application in the ascension stages game (Buddhism, Tibetan, #HM2962).
Related Heart-awakening stage of life (#HM3518).
Followed by Potala-island awareness (Buddhism, Tibetan, #HM2175)
Mahayana heat-application awareness (Buddhism, Tibetan, #HM2208)
Kalacakra-tantra shambhala awareness (Buddhism, Tibetan, #HM2151)
Mahayana receptivity-application awareness (Buddhism, Tibetan, #HM2247).
Preceded by Pure-lands awareness (Buddhism, Tibetan, #HM2168)
Joyful-heaven awareness (Buddhism, Tibetan, #HM2022)
Great-vehicle lower-path awareness (Buddhism, Tibetan, #HM2268)
Great-vehicle middle-path awareness (Buddhism, Tibetan, #HM2160).

♦ **HM3049d Beyond morality** (ICA)
Description This is the experience that all concepts of morality have fallen away and one's criteria for making ethical decisions have become meaningless. What used to be called right and wrong is now a maze of shades of grey between right and right and wrong and wrong; what is appropriate for one situation is often entirely inappropriate for another. This reality is faced when a hospital committee has to decide which patients will have the use of the kidney dialysis machine; or in the lifeboat situation where the captain decides whom to set adrift to reduce the numbers so that the rest will have a chance of survival.
A dimension of this experience is a sense of awe, that is fear and fascination. Awareness of ethical relativity brings realization that one has the capacity to make a fatally wrong decision. The question arises: "What if I didn't really give a damn about life and death ?". The mixed dread and fascination of power over life and death comes into play. No set of prescribed or established guidelines are there to help in the agony of deciding and yet a kind of intrigue beckons one forward to take a stand. Conscious of the ambiguity of life, the individual decides to make the decision, to be intentional about what he is doing. There may be a very sour taste in his mouth when someone offers free advice, telling him he should have done it another way; but there is a sense of relief that he has at least had the courage to make a choice.
Context This state is number 25 in the ICA *Other World in the midst of this World.*
Broader Moral ground (ICA, #HM2721).
Followed by Vital signs (ICA, #HM2875)
Original integrity (ICA, #HM2773)
Intentional conscience (ICA, #HM2892).
Ultimate awareness (ICA, #HM2388)
Global guardianship (ICA, #HM2817)
Preceded by Vital signs (ICA, #HM2875)
Vibrant powers (ICA, #HM2982)
Archetypal humanness (ICA, #HM3112).
Universal fate (ICA, #HM2687)
Ultimate awareness (ICA, #HM2388)

♦ **HM3050c Soul vision** (Psychism)
Description The aim of the yoga of action, soul vision comes at moments of exaltation and high aspiration. It provides the incentive for further perseverance and determination in removing both the obstructions, hindrances and distractions to self-realization and their "seeds" - the latent tendencies for such obstructions, etc, whether carried over from previous lives, developed during this life, or brought into one's life through family or race.
Related Kriya yoga (Yoga, #HH0969).

♦ **HM3051d Uncertainty**
Description A degree of uncertainty is necessary for the individual to avoid boredom and alienation in life, as the predictable may become tedious and meaningless. However, too great an uncertainty may cause panic or anomie, with impulsive, ill-considered action, or fear to act at all and the search for total security. For harmonious existence (such as occurs in flow experience) the individual is happiest responding to a challenge which, although uncertain in outcome, he sees as falling within his capabilities to perform.
Related Panic (#HM2083)
Creativity (#HH0481)
Flow experience (#HM2344)
Uncertainty (Christianity, #HH1471).
Anomie (#HM2947)
Risk-taking (#HH0654)
Freedom of choice (#HH0789)
Boredom (#HM2620)
Dehumanization (#HM3030)

♦ **HM3052d Wonder-filled fate** (ICA)
Description This the experience that everything, back to the big bang creating the universe, is one's own personal history, as though all of the past conspired to bring this moment to be; when one participates in an event which brings an overwhelming sense of wonder about the past. It may be compared to the opening scene of "2001, A Space Odyssey", in relation to the whole movie.
Context In the ICA New Religious Mode in the arena of standing present to the mystery of being in life (the life of contemplation) the second formal aspect is the experience of fearful never-againness or of archaism. At the first phenomenological level this occurs when a person encounters a completely mysterious event.
Broader Awareness of archaism (ICA, #HM3007).
Followed by Reforged transformation (ICA, #HM3061).

♦ **HM3053c Forgetfulness** (Buddhism)
Mushitasmrtita — Brjed-nges-pa (Tibetan)
Description Lack of clarity through mindfulness of objects of afflictions, which results in distraction.
Context One of the twenty *secondary afflictions* of Tibetan Buddhism.
Broader Secondary afflictions (Buddhism, Tibetan, #HH0781).

♦ **HM3054d Appropriated passion** (ICA)
Description This is experienced as Being, as having affirmed all of one's existence and having appropriated that affirmation; when one steps outside all other expectations, including that of being the unique being one is. It may be compared to the main character in the film "All that Jazz" when, in the closing scene, he finally appropriates his death as his own dying.
Context In the ICA New Religious Mode in the arena of standing present to the mystery of being in life (the life of contemplation) the fourth formal aspect is the experience of dreadful in-myselfness or of depth. At the third phenomenological level a person experiences collegiality with the mystery of being what he is bound to.
Broader Awareness of depth (ICA, #HM2447).
Followed by All-being-in-myself (ICA, #HM3097).
Preceded by Irreplaceable uniqueness (ICA, #HM3155).

♦ **HM3055d Great wakefulness** (Hinduism)
Description In this state, the notions of "I" and "this" are strengthened by the memory of previous incarnations.
Context The third of seven descending steps of ignorance described in the Supreme Yoga, which veil *self-knowledge*. Each has innumerable sub-divisions.
Broader Veils of delusion (Hinduism, #HM3592).
Followed by Wakeful dream (Hinduism, #HM2014).
Preceded by Wakefulness (Hinduism, #HM2567).

♦ **HM3056d Embodying charity** (ICA)
Description The third formal aspect of the *life of obedience*, an arena in the ICA *New Religious Mode*, this represents an awareness of embodying charity as the transestablishment to act out social concern.
Broader Life of obedience (ICA, #HM2024).
Narrower Passionate concern (#HM3128)
Sacrificial friendship (#HM2754)
Personal obligation (#HM2971)
Disinterested collegiality (#HM2812).

♦ **HM3057c Thorough abandonings** (Buddhism)
Description These are described as: the abandoning of already generated afflictions; the non-generating of not yet generated afflictions; the increasing of already generated pure phenomena; and the generation of not yet generated pure phenomena. Such thorough abandonings may, in conjunction with aspiring for enlightenment for the sake of others, lead to Buddhahood.
Context One of the sections of yogic paths or harmonies with enlightenment defined in Buddhism.
Broader Harmonies with enlightenment (Buddhism, #HH0603).

♦ **HM3058d Transcendent guru** (ICA)
Description This is experienced as becoming the role one has been playing; when the practical techniques of living one's life are so much one's second nature that one is free to consider the

style, the artfulness with which one will live. It may be compared to a dancer or a musician who has so mastered the techniques of his craft that he is free to create movement and music expressing himself in a way simply impossible previously.
Context In the ICA New Religious Mode in the arena of being the presence of transcendence (the life of chastity) the first formal aspect is radical identification with images of profound living, in which a person interiorizes the qualities of spiritual role models. At the third phenomenological level, this occurs when his concern about existence expands indefinitely, such that the object of his care cannot be reduced to any one thing but is all of existence.
Broader Radical identification (ICA, #HM2758).
Followed by Living word (ICA, #HM3178).
Preceded by Committed teacher (ICA, #HM3179).

♦ **HM3059d Near death encounter with the higher self** (NDE)
Description The light experienced in the previous stage may be felt as a presence with whom the experiencer feels at one.
Context The fifth stage in the process referred to as *near death experience*.
Broader Near death experience (NDE, #HM0777).
Related Near death evil force (#HM3422).
Followed by Near death review of life (NDE, #HM2888).
Preceded by Near death awareness of light (NDE, #HM3381).

♦ **HM3060b I-transcendent stage of life**
Ego-death — Conditional self-realization — Jnana samadhi (Hinduism, Yoga) — Gnosis — Sixth stage of life
Description As a deepening identification with consciousness itself is experienced, there is what may be referred to as "conditional self-realization". This is the condition of *jnana samadhi* - there is still some awareness of the individual as witness. Such orientation may be found in the jnana yoga of the *advaita* vedanta, the *buddhi yoga* of some Buddhist traditions, in Jainism and Samkhya. Errors are the holding on to the position of consciousness as separate from spirit energy and conditional objects, and to the love and bliss of consciousness by excluding awareness of conditional objects and states.
Context The sixth of seven stages of life characterized by Master Da Free John.
Broader Seven stages of life (#HH0460) Samadhi (Hinduism, Yoga, #HM2226).
Related Ego death (#HH6921) Jnana yoga (Yoga, #HH0927) Ego destruction (#HM0655) Buddhi yoga (Yoga, #HH0484) Samkhya (Hinduism, #HH0155) Jivanmukti (Hinduism, #HM3890) Ego transcendence (Jung, #HM2230) Advaita vedanta (Hinduism, #HH0518) Human development (Jainism, #HH0622).
Followed by Unconditional nirvikalpa samadhi (#HM3619).
Preceded by Mystical stage of life (#HM3191).

♦ **HM3061d Reforged transformation** (ICA)
Description This is a profound awareness of the past and, at the same time, a sense of being enslaved by it; when one finds one's self placed in history and forced to interpret one's role in the historical process. It may be compared to James Baldwin describing his attempt to escape the racism of the United States and discovering he carried America within himself.
Context In the ICA New Religious Mode in the arena of standing present to the mystery of being in life (the life of contemplation) the second formal aspect is the experience of fearful never-againness or of archaism. At the second phenomenological level this occurs when a person's being trapped reveals that he is forever bound to the mystery of life.
Broader Awareness of archaism (ICA, #HM3007).
Followed by Sheer re-creation (ICA, #HM3101).
Preceded by Wonder-filled fate (ICA, #HM3052).

♦ **HM3062d Wonder at the world** (ICA)
Description The second formal aspect of the *life of knowing*, an arena in the ICA *New Religious Mode*, this represents an awareness of the vanishing cosmos.
Broader Life of knowing (ICA, #HM2801).
Narrower Stark givenness (#HM2101) Passing awayness (#HM2076) Sacramental universe (#HM2366) Presence of Jesus Christ (#HM2206).

♦ **HM3063c Manipura** (Yoga)
City of the shining jewel awareness — Third chakra
Description At this level the overriding interest is in turning all into the experiencer's own substance, forcing all to conform to his way of thinking. Meditation on this chakra brings understanding of physiology, the body's internal functioning, and the role of ductless glands as they relate to human emotions. Concentration on the navel brings an end to indigestion, constipation and intestinal problems and the achievement of a long and healthy life. The second chakra has brought fluidity - here it takes the form of practicality. The power to command and organize is developed, there is control over speech and effective expression of ideas. Behaviour at this stage is marked by a striving for personal power and recognition, even if this has bad effects on family or friends. The plane encompasses karma, charity, atonement for error, good and bad company, selfless service, sorrow, the plane of dharma and the celestial plane. Remaining true to one's nature makes for stable and clear relationships with others. Balance is through selfless service without desire for reward. Practising charity clarifies the path. Aware of one's actions one achieves balance and may enter the celestial plane of illumination.
Context The third "lotus" or chakra defined in kundalini yoga, said to be situated at the level of the navel.
Broader Kundalini yoga (Yoga, #HH0237) Chakra centres of consciousness (#HM2219).
Related Sphere of glory (Kabbalah, #HM2238) Sphere of victory (Kabbalah, #HM2362) Initiation of baptism (Esotericism, #HM1267) Third chakra: theatre of karma (Leela, #HM0717).
Followed by Anahata (Yoga, #HM3242).
Preceded by Svadhishthana (Yoga, #HM3174).

♦ **HM3064c Mastery of intent**
Description Mastery of *intent* is one of three techniques described by Casteneda as leading to mastery of awareness. It is the purposeful guiding of the will, continually renewing alignment to give continuity of perception.
Related Mastery of stalking (#HM2304) Mastery of dreaming (#HM3173).

♦ **HM3065d Luminous change** (ICA)
Description This is experienced when the future which seemed like a dark cloud turns into a clear pathway; when one makes a clear decision in spite of the ambiguity, and a variety of specific things become possible. It may be compared to MacMurphy in "One Flew Over the Cuckoo's Nest" when, knowing the probable consequences, he decides to challenge the inhumanity of Big Nurse, and many specific opportunities become clear.
Context In the ICA New Religious Mode in the arena of standing present to the mystery of being in life (the life of contemplation) the third formal aspect is the experience of awesome not-yetness or of futurity. At the third phenomenological level a person experiences collegiality with the mystery of being that he is bound to.
Broader Awareness of futurity (ICA, #HM2609).
Followed by All-that's-yet-to-be (ICA, #HM3083).
Preceded by Frightful possibility (ICA, #HM3117).

♦ **HM3066c Masters of wisdom in the ascension stages game** (Buddhism, Tibetan)
Vidyahara
Description Those on paths to liberation who have reached the summit of wisdom in their philosphic framework are referred to as *vidyahara* or wisdom-holders. They may have developed as far as is possible in the system they are following, or they may be bodhisattva yogis with scope for further advancement on the path, with the ultimate aim of achieving Buddhahood.
Broader Consciousness ascension stages game (Buddhism, Tibetan, #HH4000).
Narrower Bon-wisdom awareness (#HM2663) Magical-forces awareness (#HM2679) Wheel-turner-king awareness (#HM2759) Hindu-wisdom-holder awareness (#HM2723) Tantra master in form-realm awareness (#HM2235) Tantra master-in-sense-realm awareness (#HM2211).

♦ **HM3067d Transparent existence** (ICA)
Description This is reading the meaning of life by the way people live it; when one experience or individual seems to demonstrate the nature of life in general. It is like finding oneself on stage in a morality play, ad-libbing the role of Everyman, while all of mankind watches in the audience.
Context In the ICA New Religious Mode in the arena of being the presence of transcendence (the life of chastity) the second formal aspect is having inclusive collegiality with symbols of profound living, in which a person develops the ability to see through symbols to their depth significance. At the third phenomenological level, this occurs when a person's concern about existence expands indefinitely, such that the object of his care cannot be reduced to any one thing, but is all of existence.
Broader Inclusive collegiality (ICA, #HM2333).
Followed by Transfigured man (ICA, #HM3161).
Preceded by Symbol maker (ICA, #HM3103).

♦ **HM3068c Entering-city-with-bliss-bestowing-hands awareness** (Zen)
Description The man continues on his way: even the wisest do not see his inner life. He and all he is with are converted into Buddhas.
Broader Oxherding pictures in Zen Buddhism (Zen, #HM2690).
Preceded by Back-to-the-source awareness (Zen, #HM3251).

♦ **HM3069c Principles** (Sufism)
Context The fifth of ten states acting as gateways orienting the mystic on the journey to the Absolute, as described by Ansari.
Broader Stations of consciousness - Ansari (Sufism, #HM2317).
Followed by Valleys (Sufism, #HM2874).
Preceded by Character (Sufism, #HM3759).

♦ **HM3070b Completeness**
Harmony of the whole
Description Completeness is the ideal state of being when all the parts are perfected and in perfect harmony. In religious terms, completeness arises with triumph over lower self-will and in union with divine will. The way to this union is one of gradually ascending a ladder of which each rung is a struggle with and victory over some temptation of the world that would otherwise drag the soul down to hell.
When considering universal harmony, different traditions see significance in numeric quantities. This is as true of the numbers significant in modern science as much as in the cosmic numbers of Pythagoras. Particular numbers are significant in many traditions, as the 3 of the trinity, the 7 of the scale, the 12 manifest beings, 10 creative forces, and so on. Of particular importance is the transcending of duality (or male and female) in wholeness or unity.
Related Totality (#HH0534) Wholeness (#HM2725) Perfectionism (#HM0213) Spiritual development (#HH0017) Existential unity of being (#HH0602).

♦ **HM3071d Revolutionary sign** (ICA)
Description This is the acting out of the decision one has made about one's life; when circumstances demand that a decision be embodied: "putting one's money where one's mouth is". It may be compared to Florence Nightingale, Jane Addams, Mother Jones and many others whose concern for suffering people drove them to be pioneers in inventing systems of social service.
Context In the ICA New Religious Mode in the arena of being the presence of transcendence (the life of chastity) the third formal aspect is inventing one's own essence, in which one manifests through one's style of being one's perception of existence. At the second phenomenological level this occurs when one's life is consecrated to a cause in such a way that one's every action manifests one's intent.
Broader Inventing essence (ICA, #HM3138).
Followed by Human example (ICA, #HM3188).
Preceded by Self determining (ICA, #HM3076).

♦ **HM3072c Emptiness of the indestructible Mahayana** (Buddhism, Tibetan)
Anavakarashunyata
Context One of the eighteen emptinesses comprising the paths of view in Tibetan Buddhism.
Broader Emptinesses on the paths of the view in Buddhism (Buddhism, Tibetan, #HM2944).

♦ **HM3074d Indecisive wavering** (Buddhism, Tibetan)
Context One of the five invalid ways of knowing of Tibetan Buddhism.
Broader Ways of knowing (Buddhism, #HH0873).

♦ **HM3075d Being history** (ICA)
Description This state is experienced as understanding history and one's own life as two sides of the same coin. This occurs when a person mutates their lifestyle to suit whatever situation life presents to them. This is like Dr. Martin Luther King's shift from a focus on the civil rights movement to the anti-war movement, when he discerned that the latter was the locus of the most severe denial of people's human rights.
Context In the ICA New Religious Mode in the arena of being the presence of transcendence (the life of chastity) the third formal aspect is inventing one's own essence, in which someone manifests through their style of being what they perceive existence to be. At the fourth phenomenological level, this occurs when someone perceives the void at the centre of existence, and continues to live out the absurdity.
Broader Inventing essence (ICA, #HM3138).
Preceded by Human example (ICA, #HM3188).

♦ **HM3076d Self determining** (ICA)
Description This is choosing the being one will have; when one visualizes a means of achieving the selfhood one intends. It may be compared to children playing games that rehearse roles they will play in adult life.

Context In the ICA New Religious Mode in the arena of being the presence of transcendence (the life of chastity) the third formal aspect is inventing one's own essence, in which one manifests through one's style of being what one perceives existence to be. At the first phenomenological level this occurs when one vision-sees new possibilities for existence.
Broader Inventing essence (ICA, #HM3138).
Followed by Revolutionary sign (ICA, #HM3071).

♦ **HM3077d Transparent selfhood** (ICA)
Description The first formal aspect of the *life of being*, an arena in the ICA *New Religious Mode*, this is an experience of "noughtness", an awareness of transparent selfhood.
Broader Life of being (ICA, #HM3229).
Narrower Acute inadequacy (#HM3151) Bottomless centre (#HM2872) Decisional nothingness (#HM3009) Absorption into nothingness (#HM3047).

♦ **HM3079b Paths of special qualities** (Buddhism)
Description Phenomena of the pure class according to Buddhist teaching, the paths of special qualities consist of: the five clairvoyances; the four meditative stabilizations; the four doors of retention.
Narrower Five clairvoyances (#HM2841) Four doors of retention (#HM2064) Meditative stabilizations (#HM2968).

♦ **HM3080d Actional existence** (ICA)
Description This is experienced as knowing that to be alive is to be responsible, that there is no escape from action, that to be alive is to act; existence is to act and there is no neutrality or escape from involvement. It occurs when a one sees that to pick up one's life is to commit one's self and there is no avoiding the performance of the deed, as when one sees that just by living in a community one is actively involved in it and thus responsible for what happens to it.
Context In the ICA New Religious Mode in the arena of acting out one's deed (the life of doing) the first formal aspect is surrendering one's self to one's unique calling as a person. At the first phenomenological level, this occurs when one is confronted by the fact that to live is to be of service.
Broader Surrender to one's calling (ICA, #HM2346).
Followed by Unlimited commitment (ICA, #HM3230).

♦ **HM3083d All-that's-yet-to-be** (ICA)
Description This is the realization of how the future is dependent upon the decisions made now; when one sees realistically the potential consequences of one's decisions. It may be compared to H. G. Wells' book "The Time Machine".
Context In the ICA New Religious Mode in the arena of standing present to the mystery of being in life (the life of contemplation) the third formal aspect is the experience of awesome not-yet-ness or of futurity. At the fourth phenomenological level this occurs when a person experiences adoration as he participates in all of life.
Broader Awareness of futurity (ICA, #HM2609).
Preceded by Luminous change (ICA, #HM3065).

♦ **HM3084d Frustration**
Description When the personality views a situation both as impossible to change and as counter to the achievement or satisfying of certain needs, then a psychological state of oppression, anxiety and despair is generated. Reacting to this state, an individual may take refuge in a fantasy world, or become aggressive (manifesting as irritability interspersed with bouts of anger); and behaviour becomes generally retrogressive.
Related Endurance (#HH0729).

♦ **HM3085c Single-heartedness** (Sufism)
Description Turning from all but God, the heart is concerned only with him.
Context The first illumination of Jami.
Broader Fountains of light (Sufism, #HM3039).
Followed by Single desiredness (Sufism, #HM2765).

♦ **HM3086d Secondary integrity** (ICA)
Description This is giving up the last of one's own values in order to honour the values of those with whom one is engaged; when one cares so much about seeing a situation change that one can put aside even one's own satisfaction, one's own credit, in order to do it. It may be compared to Mahatma Gandhi setting aside his own caste concerns in order to attempt to unite the diverse castes of India.
Context In the ICA New Religious Mode in the arena of being the presence of transcendence (the life of chastity) the fourth formal aspect is spiritual creativity, in which one's entire being is devoted to guarding the spirit dimension of existence. At the third phenomenological level, this occurs when one's concern about existence expands indefinitely, such that the object of one's care cannot be reduced to any one thing, but is all of existence.
Broader Spiritual creativity (ICA, #HM2037).
Followed by Replication of Christ (ICA, #HM3140).
Preceded by Eternal friends (ICA, #HM3095).

♦ **HM3087c Age of the Son** (Christianity)
Description This age or state is the New Testament revelation and associated with the Son or second member of the Trinity.
Context The second age, commencing, according to the Joachimites, with Zacharias, the father of John the Baptist, and finishing about the year 1260 AD.
Broader Doctrine of three ages (Christianity, #HH0397).
Followed by Age of the Holy Spirit (Christianity, #HM3499).
Preceded by Age of the Father (Christianity, #HM2895).

♦ **HM3088d All-that-ever-was** (ICA)
Description This is the experience of standing on the shoulders of all of the past; when one realizes that every event - good or bad, creative or uncreative - makes up this moment. It may be compared to Marilyn Monroe's reported response at the time of her wedding to Arthur Miller, when asked if she would live the horrible life she had led again. She replied that if it took all of that to create this, the happiest moment of her life, she would relive every minute of it.
Context In the ICA New Religious Mode in the arena of standing present to the mystery of being in life (the life of contemplation) the second formal aspect is the experience of fearful never-again-ness or of archaism. At the fourth phenomenological level this occurs when a person experiences adoration as he participates in all of life.
Broader Awareness of archaism (ICA, #HM3007).
Preceded by Sheer re-creation (ICA, #HM3101).

♦ **HM3089c Path of intelligence of mediating influence** (Judaism)
Description This state is represented as that in which the influx of emanations are multiplied, this affluence being communicated to those united with it.
Context The sixth among the spiritual paths expressed in the Jewish Kabbalah.
Broader Paths of wisdom (Judaism, #HM2509).
Followed by Path of occult intelligence (Judaism, #HM2444).
Preceded by Path of radical intelligence (Judaism, #HM3181).

♦ **HM3090c Way of the vajra masters** (Buddhism, Tibetan)
Description This is the path of liberation appropriate to those following the way of wisdom.
Context One of the two ways leading to Buddhahood and ultimate consciousness, according to Tibetan Gelugpa Buddhism.
Broader Consciousness ascension stages game (Buddhism, Tibetan, #HH4000).
Narrower First vajra-master stage (#HM2187) Third vajra-master awareness (#HM2727) Fifth vajra-master awareness (#HM2763) Sixth vajra-master awareness (#HM2287) Ninth vajra-master awareness (#HM2325) Final vajra-master awareness (#HM2849) Second vajra-master awareness (#HM2703) Fourth vajra-master awareness (#HM2251) Eighth vajra-master awareness (#HM2301) Seventh vajra-master awareness (#HM2789).
Related Vajrayana Buddhism (Buddhism, #HH0309).

♦ **HM3091d Dynamic selfhood** (ICA)
Description This is experienced as the reality that one's life is changing every moment, one is perpetually new and ceaselessly evolving. One does not consciously will this to happen but experiences life as forever surprising. It may compared to the movie "Aunt Mame" when the main character says "Life is a banquet and most poor fools are starving". Life is experienced as an endless celebration of the fact that, in embracing the reality of change, one is constantly renewed. For example, in the movie "Zorba the Greek", when Zorba watches his special project, designed to make him rich, collapsing about him, his immediate reaction is to dance, to celebrate this catastrophe as a great moment in his life.
A dimension of this experience is a sense of awe that is fear and fascination: fear that it will be like waking up from a dream, that one will always be out of step with the rest of the world, and that one cannot handle these situations seen only to one's self, that one will never have a home again. Yet fascination with reality like a new land; this is the beginning - the first day of the rest of one's life. And there is joy and the desire to find others who share this awareness as they discover the new in their lives. Each new universe is a feast in which one is invited to partake and one decides to accept the invitation. Such moments are followed by the knowledge that joy is always there even if one does not always see it; and one can never again take one's self too seriously.
Context This state is number 12 in the ICA *Other World in the midst of this World.*
Broader Transformed state (ICA, #HM2386).
Followed by Primal vocation (ICA, #HM2621) Invented history (ICA, #HM3245) Essential dubiety (ICA, #HM3135) Primordial wonder (ICA, #HM2186) Singular adoration (ICA, #HM2111).
Preceded by Second birth (ICA, #HM3175) Total exposure (ICA, #HM2764) Invented history (ICA, #HM3245) Primordial wonder (ICA, #HM2186) Final blessedness (ICA, #HM2958).

♦ **HM3092c Mutable awareness** (Astrology)
Common cross
Description Gemini and Sagittarius, Virgo and Pisces unite in the mutable or common cross, the awareness with which the individual is first incarnated. Consciousness awakes as to the goal ahead and he must then serve out his time until he overcomes his selfishness in the desire to serve others, and his changeability and duality in singleness of purpose. As his personality develops he is no longer bound to the wheel of incarnation. Planetary awareness is brought about and he comes under the influence of the next quadruplicity, the *fixed cross.*
Context One of the three zodiacal *quadruplicities* or crosses, uniting four signs of the zodiac in two pairs of opposites or polarities which are strongly linked and between which the influenced individual tends to oscillate, all four signs having much in common.
Broader Zodiacal forms of awareness (Astrology, #HM2713).
Narrower Virgo-consciousness (#HM2439) Gemini-consciousness (#HM2924) Pisces-consciousness (#HM2856) Sagittarius-consciousness (#HM2726).
Related Fixed awareness (Astrology, #HM2554) Cardinal awareness (Astrology, #HM2259).

♦ **HM3093d Dissimulation** (Buddhism)
Shathya — G'yo (Tibetan)
Description Hiding one's faults from others through desiring material goods or services, thus preventing one attaining preceptual instruction.
Context One of the twenty *secondary afflictions* of Tibetan Buddhism.
Broader Secondary afflictions (Buddhism, Tibetan, #HH0781).

♦ **HM3094d Unlust**
Unpleasure — Ego pain — Anxiety
Description Unlust refers to the frustration - tension or mild discomfort arising in consciousness - when the ego blocks, or partially blocks, instinctual trends seeking gratification and satisfaction.
Related Anxiety (#HM2465) Pleasure (#HH0194).

♦ **HM3095d Eternal friends** (ICA)
Description This is finding one's self among others who are also marked as devoted to the life of the spirit; when people who have made solitary vocational decisions find others who have made similar decisions of their own. It may be compared to the formation of such a team in the book "The Soul of a New Machine".
Context In the ICA New Religious Mode in the arena of being the presence of transcendence (the life of chastity) the fourth formal aspect is spiritual creativity, in which one's entire being is devoted to guarding the spirit dimension of existence. At the second phenomenological level this occurs when one's life is consecrated to a cause in such a way that one's every action manifests one's intent.
Broader Spiritual creativity (ICA, #HM2037).
Followed by Secondary integrity (ICA, #HM3086).
Preceded by Divine captive (ICA, #HM3190).

♦ **HM3096d Inescapable power** (ICA)
Envelopped by mystery
Description This is an experiential trek in the *Land of Mystery*, within the ICA *Other World in the midst of this World*, when the individual is aware that there is no escape. The states are described in separate entries.
Broader Land of mystery (ICA, #HM2434).
Narrower Final limits (#HM2674) Total exposure (#HM2764) Incarnate living (#HM2394) Ubiquitous otherness (#HM2570).
Related Infinite passion (ICA, #HM2234) Aweful encounter (ICA, #HM2619) Transformed state (ICA, #HM2386).

♦ **HM3097d All-being-in-myself** (ICA)
Description This is as though one is not in the universe but the universe is within oneself; when one lets go of the self and confronts the void. It may be compared to the end of the book "Saviours of God" when, referring to his description of God, Nikos Kazantzakis says that "even this does

not exist".
Context In the ICA New Religious Mode in the arena of standing present to the mystery of being in life (the life of contemplation) the fourth formal aspect is the experience of dreadful in-myself-ness or of depth. At the fourth phenomenological level this occurs when a person experiences adoration as he participates in all of life.
Broader Awareness of depth (ICA, #HM2447).
Preceded by Appropriated passion (ICA, #HM3054).

♦ **HM3098d Envy** (Christianity)
Jealousy
Description *Envy* is a sense of grievance, displeasure or sadness at another's good fortune, attributes and so on, and derives from pride, vanity and self-love. When the sense of grievance has some content of right or exclusive possession, then the term used is *jealousy*, although the two words are commonly used interchangeably. In the Christian tradition it is considered a vice or sin, in that it may be the cause of many forms of malevolent behaviour. It is said that active cultivation of brotherly love is the best means of overcoming an envious disposition. There is, nevertheless, a holy kind of envy which, while rejoicing at another's goodness, is sad at not possessing it one's self; and which rejoices at another's surpassing one's self in the service of God if one has failed.
Related Jealousy (Buddhism, #HM2638).

♦ **HM3099d Dying death** (ICA)
Description This is experienced when a person sees that he will have to give his entire life to the issue to which he is committed, when he knows that he will die anyway, and that the question is not if, but for what. An example is the scientist Picard who, having designed diving bells to explore the ocean bottom and balloons to reach new heights in the atmosphere, used them himself as a demonstration.
Context In the ICA New Religious Mode in the arena of acting out one's deed (the life of doing) the first formal aspect is surrendering one's self to one's unique calling as a person. At the fourth phenomenological level this occurs when the unique greatness of one's expenditure is embodied.
Broader Surrender to one's calling (ICA, #HM2346).
Preceded by Suffering servant (ICA, #HM3126).

♦ **HM3100c Pure intention** (Hinduism)
First plane of wisdom
Description This state is characterized by the realization of one's own foolishness; and the desire cultivate dispassion and to seek holy men and the scriptures.
Context The first of seven ascending planes of wisdom described in the Supreme Yoga, which require to be known so as not to be caught in delusion.
Broader Planes of wisdom (Hinduism, #HM3298).
Followed by Enquiry (Hinduism, #HM2443).

♦ **HM3101d Sheer re-creation** (ICA)
Description This is experienced as history being a personal gift and a burden; one has permission to use all of it and to make a human future with it. It occurs when one expands one's capacity to embrace more and more eras further and further back in time; and may be compared to Karl Marx rewriting history from the perspective of economics.
Context In the ICA New Religious Mode in the arena of standing present to the mystery of being in life (the life of contemplation) the second formal aspect is the experience of fearful never-again-ness or of archaism. At the third phenomenological level a person experiences collegiality with the mystery of being that he is bound to.
Broader Awareness of archaism (ICA, #HM3007).
Followed by All-that-ever-was (ICA, #HM3088).
Preceded by Reforged transformation (ICA, #HM3061).

♦ **HM3102c Path of receiving intelligence** (Judaism)
Path of measuring intelligence — Path of cohesive intelligence
Description This state is represented as containing all holy powers emanating from the Supreme Crown, and from which emanate all spiritual virtues.
Context The fourth among the spiritual paths expressed in the Jewish Kabbalah.
Broader Paths of wisdom (Judaism, #HM2509).
Related Sphere of the of the supreme crown (Kabbalah, #HM3132).
Followed by Path of radical intelligence (Judaism, #HM3181).
Preceded by Path of sanctifying intelligence (Judaism, #HM3567).

♦ **HM3103d Symbol maker** (ICA)
Description This is the marking of one's experiences to ensure that one is aware of their meaning. It occurs when one grasps that unsymbolized reality is only rarely perceived; that one can only know what one has a means of naming or otherwise marking. It may be compared to marking one's self as a monk by refusing to touch money and by wearing a habit.
Context In the ICA New Religious Mode in the arena of being the presence of transcendence (the life of chastity) the second formal aspect is having inclusive collegiality with symbols of profound living, in which one develops the ability to see through symbols to their depth significance. At the second phenomenological level this occurs when one's life is so consecrated to a cause that one's every action manifests one's intent.
Broader Inclusive collegiality (ICA, #HM2333).
Followed by Transparent existence (ICA, #HM3067).
Preceded by Raw reality (ICA, #HM3195).

♦ **HM3104c Kan** (Japanese)
Sixth sense — Intuition — Intuitional cognition
Description An essential concept in Japanese psychology, *kan* includes the intuitive power revealed in cognition and judgement and the knack - or *kotsu* - manifested in involuntary activity and the handling of objects. The latter is achieved through effort, discipline and experience and is the hallmark of great artists and artisans.
Related Intuition (#HM3634) Intelligible intuition (Buddhism, #HM5532).

♦ **HM3105d Original gratitude** (ICA)
Agape as appreciation
Description This is an experiential trek in the *Mountain of Care*, within the ICA *Other World in the midst of this World*, when the individual is aware of being in love with life. The states are described in separate entries.
Broader Mountain of care (ICA, #HM2170).
Narrower Temporal solidarity (#HM2363) Sacramental universe (#HM2445)
Individual fatefulness (#HM2223) Definitive predestination (#HM2131).
Related Singular mission (ICA, #HM3217) Transparent power (ICA, #HM2828)
Universal concern (ICA, #HM2774).

♦ **HM3106d Contentless transformation** (ICA)
Description The third formal aspect of the *life of knowing*, an arena in the ICA *New Religious Mode*, this represents an awareness of the greatness of the word.
Broader Life of knowing (ICA, #HM2801).
Narrower Classical story (#HM2757) Universal Christ (#HM2644)
Objective awareness (#HM2598) Terrifying acceptance (#HM2460).

♦ **HM3108d Unexplainable thereness** (ICA)
Description This is the sense of suddenly and quite arbitrarily just having an existence, as though one simply appeared without one's prior consent or consideration; as when one has a close encounter with death. It is like running into a totally alien people for the first time and experiencing the wonder of one's own existence.
Context In the ICA New Religious Mode in the arena of standing present to the mystery of being in life (the life of contemplation) the fourth formal aspect is the experience of dreadful in-myself-ness or of depth. At the first phenomenological level this occurs when a person encounters a completely mysterious event.
Broader Awareness of depth (ICA, #HM2447).
Followed by Irreplaceable uniqueness (ICA, #HM3155).

♦ **HM3109d Group consciousness**
Description The formation of groups is a permanent aspect of human existence, whether family, regional groups, communities or whatever. The consciousness of being a member of a group gives heightened awareness and confidence. In particular, communal living and meditation favours the formation of such a group, and short periods of retreat to live and work together are common in establishing therapeutic or philosophic groups. The supportive effect of a group is the basis of a number of therapeutic methods.
Related Group think (#HM6119) Discipleship (#HH0376)
Group therapy (#HH0851) Conscious groups (#HM3321)
Group spiritual initiation (Esotericism, #HM1417).

♦ **HM3110d Universal responsibility** (ICA)
Description The fourth formal aspect of the *life of prayer*, an arena in the ICA *New Religious Mode*, this represents an experience of intercession, of universal responsibility.
Broader Life of prayer (ICA, #HM2511).
Narrower Utter awareness (#HM2625) Particular concern (#HM2939)
Agonizing prediction (#HM2889) Promissorial offering (#HM3172).

♦ **HM3111e Possession** (Psychism)
Description The state of being possessed by some other personality, usually that of a disincarnate spirit is a form of *motor automatism*, when the possessed person becomes a means of communication for the possessor. Whether the possessor is actually a (usually unexpressed) personality within the possessed, or whether it is the true possession of an existence external to the possessed, it is in any case the position that a mind is "using" a brain surrendered to it, usually of free will, by the person possessed.
Broader Anomalies in integrity of consciousness (#HM4521).
Related Trance (#HM3236) Spirit possession (#HH0056).

♦ **HM3112d Archetypal humanness** (ICA)
Description This is the awareness that not only is the universe one's invention, but what one creates becomes the legacy for many others to inherit. It's like being elected president of the world and suddenly being looked to for a sign to follow, an awesome sense of power coupled with an inward terror at the thought of what one can do. King Arthur and his knights embody this in legendary form as they carve out a code of conduct appropriate to their situation. There is a sense that all of history conspired to bring them to this point and that what they create is setting a precedent for the next thousand years.
A dimension of this experience is a sense of awe, that is fear and fascination. There is painful awareness of being on centre stage and often a yearning for the privacy and insignificance of former life. Danger is not just an abstract idea but a shockingly real possibility. Leaders of social movements such as Martin Luther King and Harvey Milk must have experienced this as they lived each day with the threat of assassination which finally claimed their lives. And yet there is the interior knowledge that the time has come and their cue to go on stage is being given. Dag Hammarskjold described this poetically when he said: "But this is your path, and it is now, now that you must not fail". The decision is made to walk out on stage and play the role, when a person says to himself: "If I'm going to do this, I'm going to give it all I've got", whether it's a tennis player in the Wimbledon final, a child giving his first public speech in front of a group, or an artisan carving out a delicate sculpture. The clay is in one's hands and humanness comes as sheer open-ended creation. There is a supreme confidence, not in one's self but in the fact that one has set out to do something and it will be done.
Context This state is number 24 in the ICA *Other World in the midst of this World*.
Broader Creative existence (ICA, #HM2894).
Followed by Beyond morality (ICA, #HM3049) Primal vocation (ICA, #HM2621)
Destinal accountability (ICA, #HM2309) Exclusive contradiction (ICA, #HM2951)
Soteriological existence (ICA, #HM2904).
Preceded by Total exposure (ICA, #HM2764) Perpetual becoming (ICA, #HM2717)
Contextual world-view (ICA, #HM2842) Destinal accountability (ICA, #HM2309)
Exclusive contradiction (ICA, #HM2951).

♦ **HM3113c Buddha enjoyment body** (Buddhism, Tibetan)
Buddha form body
Description Achieved simultaneously with the *emanation body*, the enjoyment body is that part of the *Buddha form body* which abides in the highest pure lands and which displays the five qualities of: impermanence - although continually displaying the marks of a Buddha and hence immortal; speaking the Mahayana doctrine; displaying activities arising from wisdom and compassion; performing such activities without effort; displaying emanation bodies.
Broader Buddha-consciousness (Buddhism, Tibetan, #HM2735).
Related Buddha wisdom body (Buddhism, Tibetan, #HM2834)
Buddha nature body (Buddhism, Tibetan, #HM2039)
Buddha emanation body (Buddhism, Tibetan, #HM3019).

♦ **HM3114d Self programming** (ICA)
Description This is experienced as demanding of one's self that one sees beyond the existing situation: when one creates disciplines that permit one to imagine beyond one's present experience to the future, as an athlete does when he imagines his body doing something it has never done before. It may be compared to creating a mythology about a revolution which allows the creation of that revolution.
Context In the ICA New Religious Mode in the arena of being the presence of transcendence (the life of chastity) the first formal aspect is radical identification with images of profound living, in which a person interiorizes the qualities of spiritual role models. At the first phenomenological level this occurs when he experiences vision-seeing of new possibilities for existence.
Broader Radical identification (ICA, #HM2758).
Followed by Committed teacher (ICA, #HM3179).

♦ **HM3115d Near death negative detachment** (NDE)
Description There is a sense of being out of the physical body and an urge to return to it.
Context The second stage in the (less frequently recorded) negative near death experience.
Broader Near death experience (NDE, #HM0777).
Related Near death detachment (NDE, #HM3349).
Followed by Near death black void (NDE, #HM2213).
Preceded by Near death fear (NDE, #HM3177).

♦ **HM3116g Sloth**
Related Accidie (#HM3259) Laziness (Yoga, #HM0046).

♦ **HM3117d Frightful possibility** (ICA)
Description This is seeing that things could go in any direction and wondering what in the world to do now; when suddenly the unthought of, or the seemingly impossible, becomes possible. It may be compared to Mountain Rivera in the film "Requiem for a Heavyweight", when he is told that he will go blind if he continues to box.
Context In the ICA New Religious Mode in the arena of standing present to the mystery of being in life (the life of contemplation) the third formal aspect is the experience of awesome not-yet-ness or of futurity. At the second phenomenological level this occurs when a person's being trapped reveals that he is forever bound to the mystery of life.
Broader Awareness of futurity (ICA, #HM2609).
Followed by Luminous change (ICA, #HM3065).
Preceded by Cut-off unknownness (ICA, #HM3167).

♦ **HM3118c Renunciation**
Akirame (Japanese) — Resignation
Description To keep out of obvious sin, to avoid the obviously wrong because they are shameful and degrading, and to act in a respectful way because it is required by human dignity is only to be a human being and is only the beginning of renunciation. To plan a strategy for fighting obvious vices which includes self examination, resolutions and penances is the beginning of renunciation. Fighting the deep and unconscious vices is much more difficult because the plans, resolutions and meditations all may strengthen the very vices being fought. The fasts and disciplines of a proud person are imposed by his own vanity and belief in himself. Self initiative is almost always useless because the self is blind to its own most secret vices. When a one's greatest strengths are seen a insufficient, one's strongest virtues are hollow, there is nothing on which to rely and nothing to support the self, it is then one learns if one lives by faith. In renouncing the deep and secret selfishness and abandoning one's self to God's will in spite of one's own failures and weakness one is made strong.
Refs Merton, Thomas *Seeds of Contemplation* (1972).
Related Sannyasa (Yoga, #HH4210) Deautomatization (#HH2331)
Resignation to the will of God (#HH1016).

♦ **HM3119d Attachment** (Yoga, Zen)
Description The basic cause, according to Patanjali, of manifestation. Attachment is an intense desire for sentient existence even when manifestation in this particular manner is no longer useful, having been outgrown.
Context One of five human troubles referred to in Zen literature.
Broader Five kleshas (Yoga, Zen, #HH0390).
Narrower Attachment (Hinduism, #HM3914).
Followed by Egoism (Yoga, Zen, #HM3477).
Preceded by Revulsion (Yoga, Zen, #HM3620).

♦ **HM3121c Death emblems** (Tarot)
Grim reaper
Description Among the critical archetypal images projected by the psyche is the emblem of death of the living person; but there are also emblems of cessation, finality and separation as applied to human relationships, and to some extent, to other human affairs such as careers and enterprises. In the Tarot system the emblem called Death (number 13) is that of the skeletal Grim Reaper with his scythe. This is the figure traditionally represented as mowing or cutting down human lives symbolized by sheaves of grain, man is temporal and limited. There is light on the horizon, however; new life and regeneration are promised, a spiritual rebirth and awakening through the path of death.
Context The Reaper is placed on the Kabbalistic Path (number 24) between victory or force (Netzach) and beauty (Tiphareth).
Broader Emblem archetypes in the psyche (Tarot, #HM2201)
Tarot arcana of conscious states (Tarot, #HM2097).
Related Sphere of beauty (Kabbalah, #HM3031) Sphere of victory (Kabbalah, #HM2362).

♦ **HM3122d Ancestral obligation** (ICA)
Description This occurs when an individual realizes that he is the culmination of all human creation and destruction. Not only has he said "yes" to all the past but he experiences the burden of its consequences. It may be compared to Alex Haley in the novel "Roots", when he delves back into the past and recounts the journey of his people from Africa to slavery to freedom; or to Nikos Kazantzakis standing accountable to his grandfather, his ancestors and the land from which he came, as he writes his opening letter in "Report to Greco".
A dimension of this experience is a sense of awe, that is fear and fascination, as the individual marvels at the entire historical process and appreciates the world in which his ancestors lived and the journey that has brought him to this point. The ancestor worship of the Chinese is a ritualization of this consciousness. There is an understanding of the aweful duty one owes to those who have gone before to do one's part in this moment of history, together with a constant fear that one could fail.
The decision is to to honour the past as good and take on the burden of completing an unfinished task. Actions are put into a larger context and one finds one's self with a greater sense of patience than one had known before. As one appropriates the past one knows one is forgiven.
Context This state is number 42 in the ICA *Other World in the midst of this World.*
Broader Singular mission (ICA, #HM3217).
Followed by Interior discipline (ICA, #HM2851) Spontaneous gratitude (ICA, #HM3025)
Transformed existence (ICA, #HM2862) Futuric responsibility (ICA, #HM3016)
Definitive predestination (ICA, #HM2131).
Preceded by Global guardianship (ICA, #HM2817)
Universal compassion (ICA, #HM2734)
Transformed existence (ICA, #HM2862)
Intentional conscience (ICA, #HM2892)
Definitive predestination (ICA, #HM2131).

♦ **HM3123c Angelic awareness - Uriel**
Description This mode of awareness may be considered as the inherent tendency within the psyche to redeem itself. It does this in relationship to negative conditions such as remorse, despair and the like, which imprison the inner nature in a self-made hell. Uriel has been represented both as the power that presides in the infernal regions and as the archangel of salvation. He is said to have buried Adam and to have led Abraham "up" out of Ur. His symbol in the Church is an open hand holding a flame, as he is also thought to have been the fiery spirit at the gate of Eden.
Broader Angelic frame of awareness (#HM2150).

♦ **HM3124d Immutable friend** (ICA)
Description This is the realization that the mystery of life is never going to go away and that its eternal presence is radical affirmation of life; when one knows one's vision will never go away.
Context In the ICA New Religious Mode in the arena of articulating the word about life (the life of knowing) the fourth formal aspect is one's experience of the final nothing of life or the glorious mystery of life. At the third phenomenological level, this occurs when someone perceives in any experience its timeless authenticity.
Broader Glorious mystery (ICA, #HM2042).
Followed by Honouring the mystery (ICA, #HM3029).
Preceded by Everlasting enemy (ICA, #HM2821).

♦ **HM3125b Inspired consciousness**
Inspiration
Description From the point of view of the field of consciousness the condition termed *inspiration* corresponds to an upsurge of content without preceding continuity, or to a process characterized by intense activity in the field, or both. The process of inspiration may arise spontaneously (viz unconscious stimulation) or it may arise in response to demand (viz conscious motives or conscious reactions). Although there may be inspired subjective activity, the state of inspiration is not recognized unless there is an objective product. Frequently the manifestation of the product or content overshadows any preceding mental activity to the extent that it is not recalled, making inspiration appear even more miraculous. The state of the inspired mind is one that may also be accompanied by strong emotions such as love or hate, or general emotional excitement. In some instances it arises from, or creates, a feeling of strain or tension that can be exhausting, and in extreme cases, fatal. This is because it may be associated with physical, mental or emotional over-exertion. There are numerous cases of geniuses, artists and other creative persons who have sacrificed themselves to their gifts.
Related Inspiration (#HH2139) Anthroposophical system (#HH0018)
Psychic inspiration (Psychism, #HM1094).
Followed by Imaginative consciousness (Anthroposophy, #HM2901).

♦ **HM3126d Suffering servant** (ICA)
Description This is experienced when a person sees the price required for his decision and action. The complexity becomes obvious and he has to consider whether to continue through the thick and thin of his commitment. It occurs in a situation which challenges decision and action, as when Woodrow Wilson proposed the League of Nations and then discovered that his own nation was not prepared to participate in it.
Context In the ICA New Religious Mode in the arena of acting out one's deed (the life of doing) the first formal aspect is surrendering one's self to one's unique calling as a person. At the third phenomenological level this occurs when the individual embraces his election to this life, this task.
Broader Surrender to one's calling (ICA, #HM2346).
Related Tragic suffering (#HH0708).
Followed by Dying death (ICA, #HM3099).
Preceded by Unlimited commitment (ICA, #HM3230).

♦ **HM3127c Natural forces emblems** (Tarot)
Strength
Description One of the archetypal emblems projected by the psyche is the figure of great natural strength. Usually this is depicted as a man, and sometimes also an animal, although an emblem of strength of will need not necessarily be masculine or brutally muscular. In the Tarot the emblem of Strength (number 11) lies just below the abyss between the individual and the universal. It is depicted as a woman holding open the toothy jaws of a beast. At this stage there is mastery over the animal soul. Animals are tamed by humans using courage and intelligence. Once it is tamed, the strength of the beast is at their disposal.
Context In the Kabbalah, strength is assigned the 19th Path from force or judgement (Geburah) to mercy or kindness (Chesed).
Broader Emblem archetypes in the psyche (Tarot, #HM2201)
Tarot arcana of conscious states (Tarot, #HM2097).
Related Sphere of judgement (Kabbalah, #HM2290)
Sphere of mercy and greatness (Kabbalah, #HM2420).

♦ **HM3128d Passionate concern** (ICA)
Description This is the awareness of a great human need that affects the quality of people's lives, which may be only at the edge of what the mainstream of society says people need. It occurs when a person awakens to some dimension of innocent human suffering that is not necessary and could be eliminated; for example, the people who led the anti-slavery movement in the early years of the 19th century when there was little social awareness of the inhumanity of slavery in western society.
Context In the ICA New Religious Mode in the arena of acting out one's engagement (the life of obedience) the third formal aspect is embodying charity as the transestablishment to act out social concern. At the first phenomenological level, this occurs when one becomes aware that one is not autonomous, but rather is bound unto death to other people.
Broader Embodying charity (ICA, #HM3056).
Followed by Personal obligation (ICA, #HM2971).

♦ **HM3130d Guardian angel** (ICA)
Description This is experienced when a person realizes that, while he initially adopted someone else as his exemplar, he in turn has been adopted; that his life's work and that of his exemplar are the same; and that the exemplar will protect him from failure.
Context In the ICA New Religious Mode in the arena of knowing one's internal sociality (the life of meditation) the third formal aspect is one's experience of being given existential guidance or encountering the Saints. At the fourth phenomenological level, this occurs when a person is sustained by communion with people and events which serve as milestones for his consciousness.
Broader Existential guidance (ICA, #HM2114).
Preceded by Ever-present brother (ICA, #HM2814).

♦ **HM3131d Liberation from possessions** (ICA)
Description The first formal aspect of the *life of poverty*, an arena in the ICA *New Religious Mode*, this represents a knowledge of disengagement manifested in the experience of being librated from one's possessions.
Broader Life of poverty (ICA, #HM2299).
Narrower Wealth untold (#HM2831) Good stewardship (#HM3143)
Unmitigated death (#HM3200) Divine nothingness (#HM3042).

♦ **HM3132c Sphere of the of the supreme crown** (Kabbalah)
Kether sephira
Description This sphere of awareness is associated with the divine self and thus with the development of a higher sense of self. It corresponds to the developmental phase of final synthesis of the experience of life. The state is described by the names: *supreme crown* (kether); *unending* and *mystery of mysteries*. Among its many other names are: ancient holy-one; cause of causes; unity, all of which indicate the expression of a state of awareness that contemplates, by attribution (investiture) or negation (divestiture), the qualities of the god-head. The two pre-eminent qualities are: nothingness which is a transcendent unity, beyond mind or existence; and being-ness, or "I-am" consciousness. As the infinitude (En-Sof), this contains its self-conscious, intelligible quality, the sephiroth or archetypes of creation. In this sense Kether or En-Sof is the cause of causes. In its other aspect it is an abyss, or void. Yet this supreme state is also called, "who?" (mi) and is considered the end at which the spiritual quest rests. From Kether come all the subsequent levels on the Tree of Life, male to female, across and down the tree, itself being beyond sexual polarity, both neutrally sublime and yet transcendentally potent. It is symbolized by the positive and negative archetypes of deity and by the planet Neptune.
Context This sphere is the first of ten described in the Kabbalistic system of sephiroth, and is their culmination.
Broader Kabbalah (#HH0921).
Related Sahasrara (Yoga, #HM3398)
World of emanation (Judaism, #HM2408)
Wisdom archetypal image (Tarot, #HM2261)
Pilgrim archetypal image (Tarot, #HM2225)
Magician archetypal image (Tarot, #HM2237)
Path of receiving intelligence (Judaism, #HM3102)
Path of the illuminating intelligence (Judaism, #HM2612).
Preceded by Sphere of wisdom (Kabbalah, #HM2348)
Sphere of beauty (Kabbalah, #HM3031)
Sphere of understanding (Kabbalah, #HM2372).

♦ **HM3135d Essential dubiety** (ICA)
Description This is an awareness of the possibility that one may be the victim of the great cosmic charade - life is a joke. One is thrown into self-doubt even in moments of great victory: it simply cannot be true. In the book "The Magus" something like this happens when the main character realizes he is not the main character; or it is like discovering at the age of 25 that one is an adopted child, or a young child realizing that his parents have betrayed him. There is a haunting possibility that one's life may be the most painful tragic/comedy ever created.
A dimension of this experience is a sense of awe, that is fear and fascination. The fear is that nothing is legitimate about one's self. One is the jackass of history and somehow missing some clue as to what is really going on. The fascination is that although existence itself seems to be laughing at one and one does not know why, one finds one's self joining in the joke. The decision is to not die of shame; to stand before what is happening and not try to escape. After this it is no longer possible to believe that things are going to get better and better. There is a strange sense of selfhood as one knows there is nothing to lean on for support.
Context This state is number 13 in the ICA *Other World in the midst of this World*.
Broader Infinite passion (ICA, #HM2234).
Followed by Incarnate living (ICA, #HM2394)
Cryptic disclosure (ICA, #HM2824)
Original integrity (ICA, #HM2773)
Radical contingency (ICA, #HM2477)
Diaphanous intuition (ICA, #HM2927).
Preceded by Living death (ICA, #HM2808)
Vibrant powers (ICA, #HM2982)
Dynamic selfhood (ICA, #HM3091)
Incarnate living (ICA, #HM2394)
Diaphanous intuition (ICA, #HM2927).

♦ **HM3137c Life-fluid emblems - star** (Tarot)
Description One of the archetypal emblems projected by the psyche is the water of life, the vital force depicted as a stream, or as a vessel containing liquid, or as a flowing sap or shining essence. In the Tarot the water of life is depicted in relationship to two pitchers. In this emblem (*Star*, number 17), a terrestrial spirit or nymph, representing nature, pours the two pitchers into a pool, the universal consciousness. In another emblem (*Temperance*, number 14) the water or liquid is poured from one pitcher to another by a winged angel. Both emblems represent awareness of the ebb and flow of life. In this case, the traveller on his quest will have to modify his own "vessel of perception" so it may be filled with the waters of higher consciousness.
Context In the Kabbalah, while the Star is place on the 15th path between foundation (Yesod) and victory or force (Netzach). Temperance is placed in the 25th path between foundation and beauty.
Broader Emblem archetypes in the psyche (Tarot, #HM2201)
Tarot arcana of conscious states (Tarot, #HM2097).
Related Sphere of victory (Kabbalah, #HM2362)
Sphere of the foundation (Kabbalah, #HM2410)
Life-fluid emblems - temperance (Tarot, #HM3289).

♦ **HM3138d Inventing essence** (ICA)
Description The third formal aspect of the *life of chastity*, an arena in the ICA *New Religious Mode*, this refers to inventing one's own essence, manifesting in one's style of being what one perceives existence to be.
Broader Life of chastity (ICA, #HM2506).
Narrower Human example (#HM3188)
Being history (#HM3075)
Self determining (#HM3076)
Revolutionary sign (#HM3071).

♦ **HM3139a Turyatita** (Hinduism)
Turiyatita — Total union
Description According to yogic tradition, this is the highest state that can be achieved, beyond the state of *turiya* (the superconscious or extraordinary meditation state in which the true self may be discovered), when the body is transcended and the Atman and Paramatman are totally united or equal.
Broader Planes of wisdom (Hinduism, #HM3298)
Degrees of consciousness (Hinduism, #HM0891)
Awareness of relative reality (Hinduism, Yoga, #HM2032).
Related Mystic union with God (Christianity, #HM3889).
Preceded by Turiya awareness (Hinduism, #HM2395).

♦ **HM3140d Replication of Christ** (ICA)
Imitation of Christ
Description This is the re-inventing by the individual of the style of total permission-giving, which creates hope in a new situation; when someone lends his being unconditionally to any situation. It may be compared to the woman in India who wrapped herself around a tree to protect it from being felled by those who sought to do away with India's traditional woodcutters. Although she was cut to pieces, her sacrifice initiated the Shipko movement which continues to protect the right of woodcutters to maintain their livelihood.
Context In the ICA New Religious Mode in the arena of being the presence of transcendence (the life of chastity) the fourth formal aspect is spiritual creativity, in which a person's entire being is devoted to guarding the spirit dimension of existence. At the fourth phenomenological level, this occurs when one perceives the void at the centre of existence, and continues to live out the absurdity.
Broader Spiritual creativity (ICA, #HM2037).
Preceded by Secondary integrity (ICA, #HM3086).

♦ **HM3141f Melancholy**
Melancholia
Description This state is characterized by gloom, sadness and depression; and by a preference for solitude and study. Although previously distinguished, in that *melancholia* was said to be a great depression of spirits, with morbid sense perversions, irrational conduct and suicidal tendencies, whereas *melancholy* was considered less severe with the individual showing same interest in events and ability to continue everyday duties, the two terms are now synonymous in psychiatry.
Related Angst (#HM0362).

♦ **HM3142d Crushing-hells awareness** (Buddhism, Tibetan)
Black rope hell
Description There are two states of awareness at this stage: one is the *black rope* hell for thieves, the other is for sexual perverts. The states are characterized by sensations of being cut along lines of the body made by charcoal-blackened bindings, or of being crushed.
Context In Tibetan Sakya Buddhism this is one of the states in the "Ascension Stages Game". In some sets it is numbered 5 on the board.
Broader Desire-realm consciousness (Buddhism, #HM2733).
Followed by Naga-world awareness (Buddhism, Tibetan, #HM2031)
Cold-hells awareness (Buddhism, Tibetan, #HM2040)
Animal-hell awareness (Buddhism, Tibetan, #HM2636)
Howling-hells awareness (Buddhism, Tibetan, #HM2100)
Very-hot-hells awareness (Buddhism, Tibetan, #HM2576)
Hungry-ghosts-hell awareness (Buddhism, Tibetan, #HM2112).
Preceded by Cold-hells awareness (Buddhism, Tibetan, #HM2040)
Animal-hell awareness (Buddhism, Tibetan, #HM2636)
Demon-island awareness (Buddhism, Tibetan, #HM2055)
Howling-hells awareness (Buddhism, Tibetan, #HM2100)
Reviving-hell awareness (Buddhism, Tibetan, #HM2516)
Very-hot-hells awareness (Buddhism, Tibetan, #HM2576)
Discipleship-karma awareness (Buddhism, Tibetan, #HM2192).

♦ **HM3143d Good stewardship** (ICA)
Description This is experienced when the things of this world become meaningless in themselves, but one decides to care for them on behalf of the rest of the world; when one operates from concern for the whole world and its future. This might have been the state experienced by John Muir, who gave years of his life to create a movement to protect the natural wonders of the Western United States.
Context In the ICA New Religious Mode in the arena of knowing one's disengagement (the life of poverty) the first formal aspect is the experience of being liberated from the claim of one's possessions. At the second phenomenological level, this occurs when a person takes charge of desires which formerly controlled his life.
Broader Liberation from possessions (ICA, #HM3131).
Followed by Divine nothingness (ICA, #HM3042).
Preceded by Unmitigated death (ICA, #HM3200).

♦ **HM3144d Formless-realm awareness** (Buddhism, Tibetan)
Arupa-dhatu
Description This state of awareness is chartacterized as four-fold, and includes the representations of infinite space, infinite consciousness, nothing-at-all, and neither-thoughts-nor-the-absence-of-thoughts (also called the Peak of Existence). The four-fold states are said to arise from four meditative stabilizations or equalizations (samapatti) and particular techniques (prayoga). At the summit is the diminished ideation (samjna) which can, however, be surpassed by a state of non-ideation with a stabilization called *cessation*, allowing mental phenomena to halt for a limited period of time. This is not nirvana as there is still attachment to existence.
Context In Tibetan Sakya Buddhism this is one of the states in the "Ascension Stages Game". In some sets it is numbered 36 on the board.
Broader Supreme heavens in the ascension stages game (Buddhism, Tibetan, #HM3214).
Related Formless-realm consciousness (Buddhism, #HM2281)
Immaterial states as meditation subjects (Buddhism, #HH3198).
Followed by Animal-hell awareness (Buddhism, Tibetan, #HM2636)
Howling-hells awareness (Buddhism, Tibetan, #HM2100)
Southern-continent awareness of men (Buddhism, #HM2127)
Discipleship-karma awareness (Buddhism, Tibetan, #HM2192).
Preceded by Form-realm awareness (Buddhism, Tibetan, #HM2142).

♦ **HM3145d Flood of ignorance** (Buddhism)
Avijjogha (Pali)
Broader Delusion (Buddhism, #HM0184).

♦ **HM3146c Awareness of angelic-transformation**
Angel of death
Description This mode of awareness is represented as the transformation of human to angelic consciousness and vice-versa. The human to angelic change is referred to as occuring post-mortem. For example, the Ishim, angels of the 5th heaven are said to be the souls of the saints; or otherwise, it is said that all the Jewish patriarchs were transformed into great angels of one of the highest three grades. On the other hand, there are angels who became men. Jacob (Israel) is one who is said to have pre-existed in heaven as an angel. In variant legends of fallen angels some of these also become men. For those who become angels after death, it is said that the effective cause is another angel, the angel of death (with dozens of names) who separates the body from the soul. For the holy, the Jews say this angel is the Shekirah; Christians say it is Michael who guides souls to eternal light, and Muslems say it is Azrael. From these personifications it is seen that the transformation of awareness called "death" was viewed as the culmination of psychic life according to the psychology of the three faiths. Mohammed's death path, led by Gabriel, (not Azrael) took him on an ascent (miradj) through the angelic spheres of awareness to the presence of Allah.
Broader Angelic frame of awareness (#HM2150).

♦ **HM3147g Dream** (Hinduism)
Description The false notions of experiences arising in sleep appear real.
Context The fifth of seven descending steps of ignorance described in the Supreme Yoga, which veil *self-knowledge*. Each has innumerable sub-divisions.
Broader Veils of delusion (Hinduism, #HM3592).
Followed by Dream wakefulness (Hinduism, #HM3510).
Preceded by Wakeful dream (Hinduism, #HM2014).

♦ **HM3148d Blissful seizure** (ICA)
Description This occurs when an individual has seen the final tragedy of his life and is overcome by an almost euphoric sense of well being. Life is blissful. What a time to be alive. Every nerve

in his body shouts with excitement. "My God, I could have missed it – the greatness of living this moment". In gratitude, he is released from paralysis, filled with joy; and this becomes bountiful food for the rest of his life. All veils are wiped away and the white hot magnificence of everything is present. In Greek mythology, this is the moment when Odysseus returns from his travels and discovers the fidelity of his wife Penelope.

A dimension of this experience is a sense of awe, that is fear and fascination. The individual is aware of bursting with ecstasy and almost out of control, and yet fascinated with the sheer wonder of the moment. The decision is to bask in this moment, to dance with it; and that it is right to understand all that is going on. The rest of life is lived knowing what real joy and contentment are.

Context This state is number 59 in the ICA *Other World in the midst of this World*.
Broader Contentment at the centre (ICA, #HM2529).
Followed by Second birth (ICA, #HM3175) Contentless word (ICA, #HM2373)
Cosmic sanctions (ICA, #HM2945) Final blessedness (ICA, #HM2958)
Everlasting community (ICA, #HM2777).
Preceded by Contentless word (ICA, #HM2373) Cosmic sanctions (ICA, #HM2945)
Spontaneous gratitude (ICA, #HM3025) Transcending hostility (ICA, #HM2658)
Futuric responsibility (ICA, #HM3016).

♦ **HM3149c Path of corporeal intelligence** (Judaism)
Description This state of awareness is said to be the consciousness informing those embodied beings under the sun, moon and planets, and of their vital principal of growth.
Context The twenty ninth of the spiritual paths in the Jewish Kabbalah.
Broader Paths of wisdom (Judaism, #HM2509).
Followed by Path of collective intelligence (Judaism, #HM3046).
Preceded by Path of active intelligence (Judaism, #HM2218).

♦ **HM3150c Serial absorptions** (Buddhism)
Description These are the first, second, third and fourth concentrations, infinite space, infinite consciousness, nothingness, peak of cyclic existence and absorption of cessation.
Context One of the paths of calming according to Buddhist teaching.
Broader Paths of calming (Buddhism, #HM2987).

♦ **HM3151d Acute inadequacy** (ICA)
Description This is experiencing one's life as filled with weakness, meaninglessness, anger and suffering; when, having made a great effort at social change, one encounters a social situation that will not change. It may be compared to the line "such a worm as I", in the song "At the Cross"; or part of the experience of Paul the Apostle on the road to Damascus.
Context In the ICA New Religious Mode in the arena of knowing and doing intensified in one's life (life of being) the first formal aspect is the experience of transparent selfhood or the experience of noughtness. At the second phenomenological level this occurs when a person relates to the experience of nothingness.
Broader Transparent selfhood (ICA, #HM3077).
Followed by Decisional nothingness (ICA, #HM3009).
Preceded by Bottomless centre (ICA, #HM2872).

♦ **HM3152d Delusion of the personal self** (Buddhism)
Description The thoughtless acceptance of one's self as body and mind.
Context The first fetter or character defect referred to by Buddha. The first five fetters are marked by automatic response due to the following of old, unphilosophical mental habits. When these are overcome, the disciples is ready (arhat) to face the final five.
Broader Character defects (Buddhism, #HH0470).
Followed by Doubt of the efficacy of the good life (Buddhism, #HM1294).

♦ **HM3153c Coming-home-on-the-ox's-back awareness** (Zen)
Description The struggle with concern over gain and loss is past; the seeker will not deviate from his path whatever the enticement.
Broader Oxherding pictures in Zen Buddhism (Zen, #HM2690).
Followed by Ox-forgotten-leaving-man-alone awareness (Zen, #HM2604).
Preceded by Herding-the-ox awareness (Zen, #HM2560).

♦ **HM3154d Distraction** (Buddhism)
Vikshepa — Rnam-par-g'yeng-ba (Tibetan) — Agitation
Description In Tibetan Buddhism, as opposed to *excitement* (the scattering of the mind to pleasant objects), distraction involves the scattering of the mind to any object. In Hinayana Buddhism, agitation is the state of an excited mind. Its characteristic is disquiet, absence of calmness like water agitated by the wind. Its function is unsteadiness, it is manifest as trembling and turmoil and consciousness wavers and is distracted.
Context One of the twenty *secondary afflictions* of Tibetan Buddhism. One of the formations aggregate (mental coefficients) of Hinayana Buddhism, being listed among the constant states which appear in their true nature, and as unprofitable primary (always present in any unprofitable or unprofitable-resultant consciousness).
Broader Secondary afflictions (Buddhism, Tibetan, #HH0781).
Awareness as mental-formation group of conscious existence (Buddhism, #HM2050).
Related Excitement (Buddhism, #HM2534).

♦ **HM3155d Irreplaceable uniqueness** (ICA)
Description This is being surprised by the reality of one's own existence and the recognition that one's existence could never happen again; when one wishes one could be someone else and finds one cannot escape. It may be compared to the experience described by Dag Hammerskjold in the book "Markings", when he says that the "way chose you".
Context In the ICA New Religious Mode in the arena of standing present to the mystery of being in life (the life of contemplation) the fourth formal aspect is the experience of dreadful in-myself-ness or of depth. At the second phenomenological level this occurs when a person's being trapped reveals that he is forever bound to the mystery of life.
Broader Awareness of depth (ICA, #HM2447).
Followed by Appropriated passion (ICA, #HM3054).
Preceded by Unexplainable thereness (ICA, #HM3108).

♦ **HM3156e Final confession** (Brainwashing)
Summing-up
Description In contrast to confessions made at the beginning of the thought-reform process, which were effectively gestures of defiance; and to subsequent occasions when the individual was prepared to confess anything and everything, to fabricate wild exaggerations, in order to be accepted; this final confession becomes self-accusation, when the individual is truly convinced of his guilt and of the accusations initially made against him. The re-evaluation of his past actions and structured nature of this final confession are convincing both to the individual and to the outside world.
Context A stage reached in thought reform, a system of organized, deliberate and total psychological training which effects individual change through two basic elements – *confession* (renouncing of past beliefs and attitudes) and *re-education* (remaking of the individual in the required image).
Broader Thought reform (Brainwashing, #HH0865).
Followed by Ideological rebirth (Brainwashing, #HM2843).
Preceded by Sense of harmony through progress (Brainwashing, #HM2149).

♦ **HM3157f Deficiency cognition**
Description A term used by Maslow to describe partial, dependent awareness.
Related Being cognition (#HM2474).

♦ **HM3158c Having faith** (Buddhism)
Saddahana (Pali)
Broader Faculty of faith (Buddhism, #HM0066).

♦ **HM3159b Ten powers** (Buddhism, Tibetan)
Description These are: direct knowledge of sources and non-sources, achieved with steadfastness in ascertaining the relation of cause and effect; direct knowledge of the fruits of action, achieved through conviction of the relationship between causes and their effects; knowledge of those who have faith and those who are afflicted, gained through teaching in accordance with the faculties of those being taught; knowledge of the variety of dispositions, gained through teaching in accordance with the dispositions of those taught; knowledge of the variety of interest, gained through teaching in accordance with the interests of those taught; knowledge (through practice) of the paths leading to cyclic existences and those leading to the three enlightenments; knowledge (through practice of meditative stabilizations) of concentrations, liberations, stabilizations, absorptions, and of others' afflictions and non-contaminations; knowledge of former lives; knowledge of former births and deaths (including the clairvoyance of the divine eye); knowledge of the extinction of all contaminations.
Such knowledge pierces the armour of obstruction to omniscience, destroys the walls of obstructions to meditative stabilization, and cuts the trees of afflictive obstructions.
Context One of the paths of effect according to Buddhist teaching.
Broader Paths of effect (Tibetan, #HM3249).

♦ **HM3160c Sanjna** (Yoga)
Vijna
Description Contrasted with prajna, consciousness without object or pure consciousness, this is a notion or the consciousness of something. It implies duality at least.
Related Vijna (Buddhism, #HM3617) Prajna (Hinduism, #HM3455)
Essential wisdom (#HM0107).

♦ **HM3161d Transfigured man** (ICA)
Description This is experienced as serving as a beacon for other people on what life is about; when one's life reveals depths of existence beyond what one thought one had; when one's life is in fact itself a symbol of life. It may be compared to Charlie Chaplin's roles, which expressed not only the particular dramas in which they were played, but have been perceived as symbolic of the broader human life crisis in this century.
Context In the ICA New Religious Mode in the arena of being the presence of transcendence (the life of chastity) the second formal aspect is having inclusive collegiality with symbols of profound living, in which one develops the ability to see through symbols to their depth significance. At the fourth phenomenological level, this occurs when one perceives the void at the centre of existence, and continues to live out the absurdity.
Broader Inclusive collegiality (ICA, #HM2333).
Preceded by Transparent existence (ICA, #HM3067).

♦ **HM3162c Pliancy** (Buddhism)
Prasrabdhi — Shin-tu-sbyangs-pa (Tibetan)
Description Pliancy removes all obstructions, through its power purifying both mentally and physically so that meditative stabilization is increased.
Context One of the eleven *virtuous mental factors* referred to in Tibetan Buddhism.
Broader Virtuous mental factors (Buddhism, #HH0578).
Related Calm abiding (Buddhism, #HM2147)
Malleability of mental body (Buddhism, #HM3696)
Malleability of consciousness (Buddhism, #HM1155).

♦ **HM3163d Laziness** (Buddhism)
Kausidya — Le-lo (Tibetan)
Description A lack of delight in virtue which hinders application.
Context One of the twenty *secondary afflictions* of Tibetan Buddhism.
Broader Secondary afflictions (Buddhism, Tibetan, #HH0781).
Related Non-faith (Buddhism, #HM2461).

♦ **HM3164b Ji-ri-mu-je** (Zen)
Description This realization occurs when *ji hokkai* is known not to be different from *ri hokkai*, that there is no division between the world of separate things and the world of unity.
Related Ri (#HH3544) Ji-ji-mu-ge (Zen, #HM2034)
Zazen meditation (Zen, #HH0882).
Preceded by Ji hokkai (Zen, #HM3168) Ri hokkai (Zen, #HM3352).

♦ **HM3165d Near death hell** (NDE)
Description The individual enters a hell-like environment but returns to the physical body.
Context The final stage in the (less frequently recorded) negative near death experience.
Broader Near death experience (NDE, #HM0777).
Related Near death entering the light (#HM3180).
Preceded by Near death evil force (#HM3422).

♦ **HM3166c Jesus stripped of his garments** (Christianity)
Description The believer is aware of the pain of Jesus as His clothes are torn from His wounded body. He begs for aid in stripping away affection for things of this world so all his love may be placed on Jesus. Loving Jesus more than himself, he repents with his whole heart and begs never again to separate himself from Him. He asks that he will love Jesus always and that Jesus will do with him as He wills.
Context The tenth station of the cross.
Broader Way of the cross (Christianity, #HM3516).
Followed by Jesus nailed to the cross (Christianity, #HM4136).
Preceded by Jesus falls the third time under the cross (Christianity, #HM2289).

♦ **HM3167d Cut-off unknownness** (ICA)
Description This is the experience of the day to day routine having been broken; and of having to face a future which is totally unknown; when one is forced to go to a new place or to do something for the first time.
Context In the ICA New Religious Mode in the arena of standing present to the mystery of being in life (the life of contemplation) the third formal aspect is the experience of awesome not-yet-ness or of futurity. At the first phenomenological level this occurs when a person encounters

a completely mysterious event in his life.
Broader Awareness of futurity (ICA, #HM2609).
Followed by Frightful possibility (ICA, #HM3117).

♦ HM3168d Ji hokkai (Zen)
Description A term in Zen Buddhism for the perception of the universe as consisting of an assembly of discrete or separate things. Such awareness is inherently discriminative and implies the experience of one's self as bound by the body.
Related Ri hokkai (Zen, #HM3352) Zazen meditation (Zen, #HH0882).
Followed by Ji-ri-mu-je (Zen, #HM3164).

♦ HM3169b Enstasy
Enstatic consciousness — Yogic experience (Yoga) — Concentration
Description This has been described as extreme luminosity of consciousness with the merging of subject and object, such as occurs in *samadhi*. There is complete cessation of volition but also total freedom from unconsciousness or sleep. The ordinary space-time continuum disappears and there is experience of the eternal present. The state arises suddenly and cannot be engineered; but it does require extreme inner immobility. Some have interpreted it theistically as an act of grace. It is contrasted with *jadya* or *jada samadhi*, pseudo-enstatic states of unconsciousness with no spiritual value. In systems of yoga, the state of enstasy is characterized by becoming, in consciousness, that which is contemplated. There are various stages of enstatic realization, depending on whether a subjective 'prop' is used or not - in other words, whether there is an object of consciousness and cognitive activity: samprajnata samadhi (four levels) and asamprajnata samadhi (one level). Enstasy has been described as ecstatic or hypnotic; but such descriptions tend not to be by those who have experienced the state and are therefore misleading. Others definitely distinguish enstasy from ecstasy, the former referring to a state or concentration when contacts with the external world are withdrawn, consciousness is empty of all content and wisdom is firm.
Related Ecstasy (#HM2046) Prasada (Hinduism, #HH4330)
Jada samadhi (Yoga, #HM0148) Samadhi (Hinduism, Yoga, #HM2226)
Samprajnata samadhi (Yoga, #HM2896) Asamprajnata samadhi (Hinduism, #HM3041)
Vipassana-bhavana (Buddhism, Pali, #HH0680).

♦ HM3170b Prerequisites of manifestation (Buddhism)
Description These four - aspiration, effort, thought and analysis - exist simultaneously during the process of manifestation but occur sequentially during the practice of magical manifestation.
Context One of the sections of yogic paths or harmonies with enlightenment defined in Buddhist teaching.
Broader Harmonies with enlightenment (Buddhism, #HH0603).

♦ HM3171d Al-Muqit (Sufism)
Description The believer is aware of Allah as the sustainer, nourishing all creation. He therefore does not allow ambition or falsehood to obtain the sustenance of others but relies only on Allah.
Context A mode of mystical awareness in the Sufi tradition associated with the thirty-ninth of the ninety-nine names of Allah.
Broader Ninety-nine names of Allah (Sufism, #HH2561).
Followed by Al-Hasib (Sufism, #HM1467).
Preceded by Al-Hafiz (Sufism, #HM7099).

♦ HM3172d Promissorial offering (ICA)
Description This is experienced as the final commitment of one's life to fulfil and accomplish the deed that is worth giving one's life and death for. It occurs when the individual sees that accomplishing his vision is going to take his whole life, as when Dietrich Bonhoffer decided to participate in the anti-Hitler movement in Germany during World War II.
Context In the ICA New Religious Mode in the arena of acting out one's freedom (the life of prayer) the fourth formal aspect is one's experience of universal responsibility or intercession. As the fourth phenomenological level when one knows that one's whole life is inescapably bound in this struggle.
Broader Universal responsibility (ICA, #HM3110).
Preceded by Agonizing prediction (ICA, #HM2889).

♦ HM3173c Mastery of dreaming
Description Mastery of *dreaming* is one of three techniques described by Casteneda as leading to mastery of awareness. Conscious awareness of dreams is through a subtle balance of non-interference with dreams but a shift in the point from which the dream is observed.
Related Dreams (#HM2950) Mastery of intent (#HM3064) Mastery of stalking (#HM2304).

♦ HM3174c Svadhishthana (Yoga)
Her favourite resort awareness — Second chakra
Description The whole of life is seen as centred on sex - repressed, frustrated, sublimated or fulfilled. meditation enables the mind to reflect the world, the ability to use creative and sustaining energy to elevate one's self to refined arts and pure relationships with others is acquired, as one has become free of lust anger, greed, unsettledness and jealousy. Elevation from the first chakra brings lunar awareness, reflecting divine grace of creation and preservation. Behaviour at this stage is marked by desire for physical sensation and mental fantasy. The downward, whirlpool effect (this is the level of water) may cause restlessness and confusion. To remain healthy and balanced requires respect and understanding for natural limitations of body and mind. This person often pretends to be a prince or a hero, changing roles, maintaining high self-esteem and being chivalrous - the destroyer of evil of folk stories. The astral plane is encompassed, as well as those of entertainment, fantasy, nullity, jealousy, mercy, envy and joy. Fantasy may be used in crafts and fine arts but nullity, a state of emptiness and purposelessness, when the world is seen with a negative mind, together with envy and jealousy may give rise to destructive, restless anxiety. However, the plane of joy penetrates the entire consciousness of the person who has evolved beyond second chakra aspects.
Context The second "lotus" or chakra defined in kundalini yoga, said to be situated at the level of the genitals.
Broader Kundalini yoga (Yoga, #HH0237) Chakra centres of consciousness (#HM2219).
Related Initiation of birth (Esotericism, #HM1337)
Sphere of the foundation (Kabbalah, #HM2410)
Second chakra: realm of fantasy (Leela, #HM3651).
Followed by Manipura (Yoga, #HM3063).
Preceded by Muladhara (Yoga, #HM2893).

♦ HM3175d Second birth (ICA)
Description This is the awareness that one's own most personal life and self-understanding have undergone a metamorphosis. Such a transformation happens without one willing it; one knows that one is not what one was: it has been called a second birth. D H Lawrence writes about this experience in the poem " A New Heaven and New Earth"; and the popular song "Clouds" also refers to such an occasion.

A dimension of this experience is a sense of awe, that is fear and fascination. There is a deep fear that it won't last and that it is more than one can constantly live with; but in the midst of such fears there is a fascination with the possibility that is now open. There is a sense that there is "nothing I can't do", that one's responsibility is without limits. In this state an individual decides to play the part, to assume this transformed life as his own to live and create, and to find ways to symbolize it. After such occasions there is a new appreciation of the frailty of life's roles and a knowledge that it can never go back to the way it was, that this fragile new creation of a transformed life can be created or destroyed.
Context This state is number 11 in the ICA *Other World in the midst of this World.*
Broader Transformed state (ICA, #HM2386).
Related Born-again (Christianity, #HH0576).
Followed by Dynamic selfhood (ICA, #HM3091) Cosmic sanctions (ICA, #HM2945)
Ultimate reality (ICA, #HM2030) Transcendent immanence (ICA, #HM3034)
Futuric responsibility (ICA, #HM3016).
Preceded by Final limits (ICA, #HM2674) Ultimate reality (ICA, #HM2030)
Blissful seizure (ICA, #HM3148) Transformed existence (ICA, #HM2862)
Futuric responsibility (ICA, #HM3016).

♦ HM3176d Symbolizing the eternal context (ICA)
Description The fourth formal aspect of the *life of doing* in the ICA *New Religious Mode*, this is an awareness of becoming a symbol of the eternal context.
Broader Life of doing (ICA, #HM3018).
Narrower Obedient son (#HM2005) Final situation (#HM2119)
Meaning creation (#HM2017) Doing the mystery (#HM2143).

♦ HM3177d Near death fear (NDE)
Description The individual experiences a senses of fear and panic.
Context The first stage in the (less frequently recorded) negative near death experience.
Broader Near death experience (NDE, #HM0777).
Related Near death peace (NDE, #HM2494).
Followed by Near death negative detachment (NDE, #HM3115).

♦ HM3178d Living word (ICA)
Description This is experienced as being the very presence of permission to pick up life and live it; when one emanates the fullness and creativity that life is. It may be compared to Miyamoto Musashi's perception that having once mastered one art, all other arts are yours.
Context In the ICA New Religious Mode in the arena of being the presence of transcendence (the life of chastity) the first formal aspect is radical identification with images of profound living, in which a person interiorizes the qualities of spiritual role models. At the fourth phenomenological level, this occurs when one perceives the void at the centre of existence, and continues to live out the absurdity.
Broader Radical identification (ICA, #HM2758).
Preceded by Transcendent guru (ICA, #HM3058).

♦ HM3179d Committed teacher (ICA)
Description This is being fanatical about calling other people to awareness; it occurs when someone is passionately concerned that others should know the profundity of their lives. It may be compared to Eleanor Roosevelt's participation in drafting the United Nations Declaration of Human Rights - a passionate engagement that demanded that others participate.
Context In the ICA New Religious Mode in the arena of being the presence of transcendence (the life of chastity) the first formal aspect is radical identification with images of profound living, in which a person interiorizes the qualities of spiritual role models. At the second phenomenological level this occurs when one's life is consecrated to a cause in such a way that one's every action manifests one's intent.
Broader Radical identification (ICA, #HM2758).
Followed by Transcendent guru (ICA, #HM3058).
Preceded by Self programming (ICA, #HM3114).

♦ HM3180d Near death entering the light
Description The experiencer enters a transcendental environment of great beauty; he may meet with dead loved ones (or, in a religious framework, Jesus or Krishna) who intimate that he must return to earth as his purpose has not been fulfilled or his family or dependents still need him. This may be experienced as a door, a river or a boundary which he is not allowed to cross.
Context The final stage in the process referred to as *near death experience.*
Broader Near death experience (NDE, #HM0777).
Related Near death hell (NDE, #HM3165).
Preceded by Near death review of life (NDE, #HM2888).

♦ HM3181c Path of radical intelligence (Judaism)
Description This state is represented as being closest to the supreme unity, uniting with Binah, emanating from supreme wisdom.
Context The fifth among the spiritual paths expressed in the Jewish Kabbalah.
Broader Paths of wisdom (Judaism, #HM2509).
Followed by Path of intelligence of mediating influence (Judaism, #HM3089).
Preceded by Path of receiving intelligence (Judaism, #HM3102).

♦ HM3182d Obstacles to soul cognition (Yoga)
Description These obstacles, according to Patanjali, are the hindrances to achieving *self realization.*
Narrower Laziness (#HM0046) Carelessness (#HM0361) Mental inertia (#HM0906)
Bodily disability (#HM1134) Wrong questioning (#HM1080) Lack of dispassion (#HM0110)
Erroneous perception (#HM1077) Inability to achieve concentration (#HM0298)
Failure to hold meditative attitude (#HM0610).

♦ HM3183d State of not being infatuated (Buddhism)
Asarajjana (Pali)
Broader Non-attachment (Buddhism, #HM2128).

♦ HM3188d Human example (ICA)
Description This is standing as proof of the capacity of human beings to stand taller than they are, to be bigger than the life they have; when a person uncompromisingly demonstrates the significance of being alive. It may be compared to the sisters who work with Mother Teresa of Calcutta, to care for the destitute and dying.
Context In the ICA New Religious Mode in the arena of being the presence of transcendence (the life of chastity) the third formal aspect is inventing one's own essence, in which one manifests through one's style of being what one perceives existence to be. At the third phenomenological level, this occurs when one's concern about existence expands indefinitely, such that the object of one's care cannot be reduced to any one thing, but is all of existence.
Broader Inventing essence (ICA, #HM3138).
Followed by Being history (ICA, #HM3075).
Preceded by Revolutionary sign (ICA, #HM3071).

♦ **HM3189g Kufu**
Naturalness in bodily action

♦ **HM3190d Divine captive** (ICA)
Description This is being possessed by the mystery of life, enraptured beyond ability to or interest in escaping; when one finds one's self. It may be compared to Captain Ahab in "Moby Dick", when he is irresistibly drawn to follow the white whale.
Context In the ICA New Religious Mode in the arena of being the presence of transcendence (the life of chastity) the fourth formal aspect is spiritual creativity, in which one's entire being is devoted to guarding the spirit dimension of existence. At the first phenomenological level this occurs when one vision-sees new possibilities for existence.
Broader Spiritual creativity (ICA, #HM2037).
Followed by Eternal friends (ICA, #HM3095).

♦ **HM3191c Mystical stage of life**
Fifth stage of life
Description The attention is turned from the external and inverted onto the inner space of the psyche. It may be expressed through may of the methods of yoga, through the Taoist tradition, through the psychic experiences and siddhis of tantric magic and yoga. Traditionally, the orientation is towards renouncing gross physical existence; this may be through asceticism. It culminates in the conscious absorption in divine reality and loss of body sense experienced in *nirvikalpa samadhi*. The error is to cling to subtle objects and states or to conditional transcendence of these objects and states in conditional nirvikalpa samadhi.
Context The fifth of seven stages of life characterized by Master Da Free John. Characteristic expression is said to be in the sky magic of the psyche.
Broader Seven stages of life (#HH0460).
Related Raja yoga (Yoga, #HH0755); Nada yoga (Yoga, #HH0160); Hatha yoga (Yoga, #HH0862); Mantra yoga (Yoga, #HH0931); Dhyana yoga (Yoga, #HH0827); Shabda yoga (Yoga, #HH0539); Taraka yoga (Yoga, #HH0452); Kundalini yoga (Yoga, #HH0237); Tantra (Buddhism, Yoga, #HH0306); Siddhis (Yoga, Buddhism, #HH0380); Human development (Taoism, #HH0689); Nirvikalpa samadhi (Hinduism, #HM2061); Tibetan meditative states of form concentrations (Buddhism, #HM2693).
Followed by I-transcendent stage of life (#HM3060).
Preceded by Heart-awakening stage of life (#HM3518).

♦ **HM3192e Mayko**
Description This form of extra-sensory perception may occur in early states of meditation.
Broader Extrasensory perception (ESP, #HM2262).

♦ **HM3193a Unity consciousness**
Transcendental unity consciousness — God consciousness supreme — Seventh state (Yoga)
Description Unity consciousness is said to evolve naturally out of god consciousness with given time. Its characteristic is the extinction of observer/observed-duality into uninterrupted unity of self and all objects of perception and action. This unified consciousness is seen as the potential normative level of consciousness for mankind. A state of self-referral and self-interacting identical to the self-interacting dynamics of the unified field of all the laws of nature as discovered by quantum field theories of modern physics.
Context The highest state of consciousness in the series enunciated by Maharishi Mahesh Yogi, achieved through the practice of transcendental meditation. Such a state is said to enable the individual to lead his daily life fully supported by natural law. Unity consciousness has been theorized as normative for all subjective experience; that is, experiences in unity consciousness all share the same objective certainty and precision within the realm of subjectivity. The notions of reality, knowledge and continuity of perception are clarifiable only in this state beyond all dualities. Scientific evidence of this state is the whole idea of science itself. Strong similarities with conceptions and descriptions of states of unity consciousness appear in most traditions, e.g. in mysticism.
Broader Awareness of relative reality (Hinduism, Yoga, #HM2032).
Related God-consciousness (#HM2166); Unitary consciousness (#HM2702); Nirvikalpa samadhi (Hinduism, #HM2061); Transcendental meditation (TM, #HH0682); Transcendental consciousness (#HM2020); Essential unity of being (Christianity, #HM6111).
Preceded by Refined cosmic consciousness (#HM3222).

♦ **HM3194b Paths of insight** (Buddhism)
Description These are meditative stabilization on the three entries to liberation: *wishlessness* (emptiness of a phenomenon in that it does not produce effects); *signlessness* (emptiness of a phenomenon in that it is not inherently produced by causes); *emptiness* (emptiness of a phenomenon itself as an entity).
Context One of the paths of calming according to Buddhist teaching.
Broader Paths of calming (Buddhism, #HM2987).

♦ **HM3195d Raw reality** (ICA)
Description This is when existence reveals itself in some wholly unprecedented way which one seeks to remember; when a situation or an individual is experienced as the wholly "other", beyond the universe of meaning in which one is accustomed to live, whose presence demands that that universe change. Pablo Picasso's painting "Guernica" may be said to represent this experience among villagers upon whom a saturation bombing experiment was conducted.
Context In the ICA New Religious Mode in the arena of being the presence of transcendence (the life of chastity) the second formal aspect is having inclusive collegiality with symbols of profound living, in which one develops the ability to see through symbols to their depth significance. At the first phenomenological level this occurs when one has the experience of vision-seeing new possibilities for existence.
Broader Inclusive collegiality (ICA, #HM2333).
Followed by Symbol maker (ICA, #HM3103).

♦ **HM3196d Ignorance** (Buddhism, Yoga, Zen)
Avidya — Ma-rig-pa (Tibetan) — Avijja (Pali) — Dullness — Delusion
Description Ignorance is cause of all obstructions, whether latent, in full operation, in the process of being eliminated, or overcome. It is the obscuring of knowledge of the true status of phenomena in a consciousness that perceives the opposite of what is correct, and is thus the cause of false assertions and doubt. It may refer to action or the effect of action, or to the "suchness" of something itself. The former leads to inauspicious rebirth, the latter to auspicious rebirth; but in either case ignorance is the root cause of all other afflictions and of the cycle of birth and rebirth. In particular, ignorance is the cause of believing the false "I", which consists simply of mental and physical aggregates, to be an actually existent self which controls these aggregates.
According to Patanjali, every form of life veils a portion of spiritual energy. In order to reach liberation, one has to progressively fight one's way through this veil. In particular, the three sheaths which comprise the physical, emotional and mental bodies which are instruments of knowledge, should not be confused with that which knows. The three stages in overcoming ignorance are: identification with the phenomenal world; restlessness and searching for knowledge of the true self; and realization, wisdom and expansion of consciousness.
In Hinayana Buddhism, the characteristic of delusion or dullness is unknowing, opposition to knowledge. Its function is non-penetration, concealing the intrinsic nature or essence of an object; it is manifest by the absence of right theory or conduct, by darkness or blindness. The proximate cause is unwise attention. It is the root of all the unprofitable or immoral. Again, ignorance is 8 ways of unknowing: it is unknowing about suffering, about the cause of suffering, the cessation of suffering and the way leading to cessation of suffering (four noble truths). It is also unknowing about the past (5 aggregates), unknowing about the future (5 aggregates) unknowing about the past and the future together, unknowing about relatedness (specific conditionality) of one thing to another or of states that have arisen. Ignorance thus keeps the truth from penetrating.
Context One of the six *root afflictions* referred to in Tibetan Buddhism. Also one of the five human troubles referred to in Zen literature. Also one of the formations aggregate (mental coefficients) of Hinayana Buddhism, being listed among the constant states which appear in their true nature, and as unprofitable primary (always present in any unprofitable or unprofitable-resultant consciousness). Also one of the five afflictions or causes of misery in yoga.
Broader Delusion (Buddhism, #HM0184); Five kleshas (Yoga, Zen, #HH0390); Delusive consciousness (#HM2600); Root afflictions (Buddhism, #HH0270); Afflictions (Yoga, Hinduism, #HH1047); Awareness as mental-formation group of conscious existence (Buddhism, #HM2050).
Narrower Conditions for consciousness (Buddhism, #HM7655).
Related Asmita (Yoga, #HM2937); Abhinvesa (Yoga, #HM3871); Ignorance (Leela, #HM0310); Lethargy (Buddhism, #HM2926); Becoming (Buddhism, #HM5909); Non-faith (Buddhism, #HM2461); Concealment (Buddhism, #HM2605); Divine consciousness (#HM0563); Non-ignorance (Buddhism, #HM2695); Recognition of ignorance (#HH0486); Veils of delusion (Hinduism, #HM3592); Four noble truths (Buddhism, #HH0523); Jnana (Yoga, Buddhism, Hinduism, #HH0610); Ignorance in dependent origination formula (Buddhism, #HM3035).
Preceded by Egoism (Yoga, Zen, #HM3477).

♦ **HM3197c Wonder**
Astonishment
Description According to Plato and Aristotle, the state of wonder or astonishment induced in the mind when it perceives something strange or not well understood, is the cause or instigation of philosophy. The contemplative aspect, the inclusion of the intellect in the experience of wonder is said to give this state of attention its specifically human (as opposed to animal) quality. It is this sense of wonder that draws together great scientists and great mystics, the sense of the "un-knowability" of the universe which increases the more one knows.
Refs Carson, R *The Sense of Wonder* (1965).
Related Awe (#HM2592); Reverence (#HM3023); Admiration (#HM3701); Understanding (#HH0658); Experience of miracles (Christianity, #HM0284).

♦ **HM3198c Ascription**
Adhyasa
Related Concentration (#HM4226).

♦ **HM3199d Embodying service** (ICA)
Description The fourth formal aspect of the *life of obedience*, an arena in the ICA *New Religious Mode*, this represents an awareness of embodying service to the spiritual life of others.
Broader Life of obedience (ICA, #HM2024).
Narrower Ethical existence (#HM2661); Global brotherhood (#HM2432); Communion of saints (#HM2338); Eternal identification (#HM2541).

♦ **HM3200d Unmitigated death** (ICA)
Description This is the moment when a person breaks free of his attachment to the things of this world; when he sees that the goods of this world are just that: of this world. An example might be the stories told by many refugees who have had to flee their homeland, leaving behind all they had acquired.
Context In the ICA New Religious Mode in the arena of knowing one's disengagement (the life of poverty) the first formal aspect is the experience of being liberated from the claim of one's possessions. At the first phenomenological level, this occurs when one is no longer content with their customary habits, but see in them a futility never before discerned.
Broader Liberation from possessions (ICA, #HM3131).
Followed by Good stewardship (ICA, #HM3143).

♦ **HM3202e Shape shifting** (Psychism)
Description The individual experiences an alteration in form or substance. Shape-shifting is common in virtually all traditions, whether religious or folk, and in popular culture (Superman; Spiderman). It is an element of deep spiritual insight in Christianity, Hinduism and Buddhism. The significance of, say, Zeus appearing as a swan and Christ's transfiguration are so different as to make a universal meaning impossible. Nevertheless, there are certain similarities that can be classified.
(1) *Strategic deception.* The individual intentional changes shape for reasons of aggression, seduction or trickery. Examples are: the Hindu god Visnu who constantly takes different incarnations in his fight against evil, and who even changes shape within a given incarnation (Rama, Krisna); the agent of evil or death appearing as a seductive young woman (Arthurian legends); folk stories where tricksters assume diverse forms, such as Loki of Nordic mythology.
(2) *Escape.* The intended victim of an attack changes shape or has his or her shape changed to (not always successfully) elude the attacker, for example Daphne becomes a laurel tree to elude Apollo's amorous advances.
(3) *Punishment.* The shape of an individual is changed as punishment for some offence. An example is Lot's wife becoming a pillar of salt. Another is the Hindu doctrine of reincarnation in different form depending on the way the previous life has been lived, and Plato indicates the same. Here even our present earthly shape is a sign we have fallen from some more desirable existence.
(4) *Liberation.* False existence is replaced by true, the individual becomes "who I am before I was" in a mystic sense, as in Islam, or less mystically in the fairy stories such as that of the frog being restored to his shape as a prince. In Tao there is no permanent shape, but the path is endless change.
(5) *Immortalization.* From the simple level of Narcissus in Greek mythology being transformed into a flower, through Plato's liberated souls becoming stars, the ultimate is the resurrection of the body, when the mortal, as St Paul says, "puts on immortality".
(6) *Confused identity.* Here the shape-shifter inhabits two realms simultaneously. Examples are: Dionysos, being born twice, sometimes a female, when dead always returning; and the phenomenon of the werewolf.
(7) *Revelation.* Observers are awakened or enlightened to a reality otherwise unnoticed. Both Krisna to Arjuna and Christ to His disciples revealed his glory otherwise hidden, shining like the sun.
Related Metamorphosis (Psychism, #HM0551); Psychic atavism (Psychism, #HM0401).

♦ **HM3203c Non-embarrassment** (Buddhism)
Anapatrapya — Khrel-med-pa (Tibetan)
Description The failure to avoid faults of which another would disapprove.
Context One of the twenty *secondary afflictions* of Tibetan Buddhism.
Broader Secondary afflictions (Buddhism, Tibetan, #HH0781).
Related Shamelessness (Buddhism, #HM0649).

♦ **HM3204e Channelling of guilt** (Brainwashing)
Description As the individual feels compelled to confess, he begins to see evidence in past actions which justify his previously unfocused feelings of guilt and self-evil. Viewing his past acts as criminal is strengthened by the support of the environment and the knowledge that some inner feelings accompanying the most altruistic of such actions were such as to corroborate this attitude. He condemns himself not for what he has done but for what he was.
Context A stage reached in thought reform, a system of organized, deliberate and total psychological training which effects individual change through two basic elements - *confession* (renouncing of past beliefs and attitudes) and *re-education* (remaking of the individual in the required image).
Broader Thought reform (Brainwashing, #HH0865).
Followed by Logical dishonouring (Brainwashing, #HM2564).
Preceded by Compulsion to confess (Brainwashing, #HM2745).

♦ **HM3205c Emptiness of nature** (Buddhism, Tibetan)
Prakrtishunyata
Context One of the eighteen emptinesses comprising the paths of view in Tibetan Buddhism.
Broader Emptinesses on the paths of the view in Buddhism (Buddhism, Tibetan, #HM2944).

♦ **HM3206d Final accountability** (ICA)
Freedom of obligation
Description This is an experiential trek in the *River of Consciousness*, within the ICA *Other World in the midst of this World*, when the individual realizes "I am my answerability". The states are described in separate entries.
Broader River of consciousness (ICA, #HM2993).
Narrower Original integrity (#HM2773) Worldly detachment (#HM2451) Passionate disinterest (#HM2547) Destinal accountability (#HM2309).
Related Moral ground (ICA, #HM2721) Authentic relation (ICA, #HM2316) Creative existence (ICA, #HM2894).

♦ **HM3207b Samadhi parinama** (Yoga)
Description Starting with the practice of *dharana* and continuing until a state of *ekagrata* or single pointed attention is reached, the tendency for distractions to occur in consciousness gradually disappears. First to go are the objects of normal attention, being replaced by the chosen object of samadhi, then this is gradually stripped of all its inessentials until it shines in the mind of itself.
Context The first of three mental transformations described by Patanjali.
Broader Parinama (Yoga, #HH4561).
Related Dharana (Hinduism, #HM2566) Samadhi (Hinduism, Yoga, #HM2226).
Followed by Ekagrata parinama (Yoga, #HM3337).

♦ **HM3208c Emptiness of things** (Buddhism, Tibetan)
Abhavashunyata
Context One of the eighteen emptinesses comprising the paths of view in Tibetan Buddhism.
Broader Emptinesses on the paths of the view in Buddhism (Buddhism, Tibetan, #HM2944).

♦ **HM3209c Basic faith** (Buddhism)
Shraddha — Dad-pa (Tibetan)
Description Faith may be said to serve as the basis for aspiration, and for all auspicious attainments. It is characterized by clarity, clearing mental troubles and allowing true qualities of realization to be generated.
Context One of the eleven *virtuous mental factors* referred to in Tibetan Buddhism.
Broader Virtuous mental factors (Buddhism, #HH0578).
Related Faith (#HH0694).

♦ **HM3210d Harmfulness** (Buddhism)
Vihimsa — Rnam-par-'tshe-ba (Tibetan) — Byapajjitatta (Pali)
Description A lack of compassion and unmerciful wish for another's harm.
Context One of the twenty *secondary afflictions* of Tibetan Buddhism.
Broader Hatred (Buddhism, #HM4502)
Secondary afflictions (Buddhism, Tibetan, #HH0781).

♦ **HM3211c Single mindedness**
Related Double mindedness (#HM3362).

♦ **HM3212c Ananda stage** (Yoga)
Linga stage of the gunas — Anandamaya kosha
Description In this stage of samadhi, consciousness functions at a supra-mental level which transcends the intellect and is expressed through the *buddhi*. Particular objects or principles appear as marks or signs distinguishing them from other objects or principles. All are part of universal consciousness but each maintains its identity and is distinguishable.
Context The third of the four stages of the gunas.
Broader Stages of the gunas (Yoga, #HM2805).
Related Ananda (Hinduism, Buddhism, Yoga, #HM3227).
Followed by Asmita stage (Yoga, #HM3509).
Preceded by Vicara stage (Yoga, #HM3460).

♦ **HM3213e Dance-induced experience**
Ritual movement — Abstract dancing
Description The practice of dancing or other ritual movements has been used to produce altered states of awareness and mystic mystic experiences. This may consists of highly complex sequences of movements performed without conscious thought and with total concentration. The dance may be very energetic and prolonged, inducing frenzy and/or total exhaustion and subsequent trance or altered state. In *abstract dancing*, various levels of consciousness are achieved through classical steps to particular melodies.
Related Dance (#HH0445) Aikido (#HH0252) Mudras (#HH0286) Frenzy (#HM3594) Derwish (Islam, #HH0741) Induction technique (#HH0025) Asana (Hinduism, Yoga, #HH0669).

♦ **HM3214c Supreme heavens in the ascension stages game** (Buddhism, Tibetan)
World-transcending spheres
Description These are the domains of the great Boddhisattva heroes, and the spheres purified by the Buddha in which the devotee learns and practises the highest teachings without hindrance or distraction.
Broader Consciousness ascension stages game (Buddhism, Tibetan, #HH4000).
Narrower Joy-land awareness (#HM2167) Potala-island awareness (#HM2175) Absolute-body awareness (#HM2397) Formless-realm awareness (#HM3144) Supreme-heaven awareness (#HM2813) Enjoyment-body awareness (#HM2873) Hyper-bliss-realm awareness (#HM2337) Amoghasiddhi-karma awareness (#HM2783) Guhya-samaja urgyan awareness (#HM2699) Jewelled-peaks-realm awareness (#HM2275) Kalacakra-tantra shambhala awareness (#HM2151).

♦ **HM3215d Jiriki** (Zen)
Own strength
Description An approach to enlightenment, typified by Zen Buddhism, which relies on the individual's own effort and power from within rather than outside help from some deity or from Buddha.
Related Tariki (Japanese, #HM2684) Zazen meditation (Zen, #HH0882).
Followed by Ji-ji-mu-ge (Zen, #HM2034).

♦ **HM3216c Four sciences** (Buddhism, Tibetan)
Description These are knowledge of: doctrines; specific and general characteristics of phenomena; languages; the aspects, relationships, and differences, of phenomena.
Context One of the paths of effect according to Buddhist teaching.
Broader Paths of effect (Tibetan, #HM3249).

♦ **HM3217d Singular mission** (ICA)
Agape as responsibility
Description This is an experiential trek in the *Mountain of Care*, within the ICA *Other World in the midst of this World*, when the individual is aware that "everything is my brother". The states are described in separate entries.
Broader Mountain of care (ICA, #HM2170).
Narrower Invented history (#HM3245) Global guardianship (#HM2817) Ancestral obligation (#HM3122) Futuric responsibility (#HM3016).
Related Universal concern (ICA, #HM2774) Transparent power (ICA, #HM2828) Original gratitude (ICA, #HM3105).

♦ **HM3218c Mansions of the practice of prayer** (Christianity)
Description To move from the first mansions to the second the person has practised prayer diligently but has not avoided the occasional sin. They will have recognized the importance of not staying in the first mansions.
In the second mansions the person is called to God through conversations with others, books, sermons, and some times through trial and sickness. God sees that they persevere and have good desire. The soul is still assaulted by the devil and earthly vipers. It becomes confused about whether to go back or to go forward. At this junction, the soul comes to realize it has one true friend, God. The danger here is that everything can be ruined by vain habits. The soul will experience great trials especially when its character and habits are such that it is about to progress further.
In these mansions the person comes to realize it needs the aid of God. It mixes with others who are also on the journey. It comes to have the determination and resolve to lose its life, peace or everything to avoid returning to the first mansions. This is not the time to think of receiving spiritual favours. These are not the mansions in which they are found. Complaining about aridity and desiring consolations is foolish because this is just the beginning. Inward favours do not bring determination to face outward trial at this point. The outward trials may be a way of testing determination to continue on the journey. Simply the practice of following the will of God is of benefit.
When the person fails it is necessary to not lose heart or cease striving to make progress. Even in failure good may come of it. The person is shown what a wretched creature he is and what harm is done by dissipating desires. The faculties seem to be making war on the soul, but this not reason to turn back. The self is not to be trusted, only God is. Consulting people of experience is helpful. It is necessary to always begin again. Without knowing the depth of one's own soul it is impossible to know heaven. Progress is only available by recognizing one's own wretchedness, imploring the lord for mercy and praying to not be lead into temptation.
Context The second of seven mansions of the soul's progress described by St Teresa of Avila.
Broader Mansions of the soul (Christianity, #HH1409).
Followed by Mansions of exemplary life (Christianity, #HM2969).
Preceded by Mansions of humility (Christianity, #HM3382).

♦ **HM3220c Conscientiousness** (Buddhism)
Apramada — Bag-yod-pa (Tibetan)
Description This quality may manifest in remedying past faults; in intending to remedy future faults; in remedying faults now, in the present; in awareness, prior to action, of behaving such that faults do not arise; and in actually behaving so that these faults do not arise. It maintains effort and keeps the mind from the influence of afflictions and contaminations.
Context One of the eleven *virtuous mental factors* referred to in Tibetan Buddhism.
Broader Virtuous mental factors (Buddhism, #HH0578).
Related Conscience (Buddhism, #HM1590).

♦ **HM3221d Induced religious experience**
Description Following an effort, which may through application or sustained pursuit of one of any number of methods or techniques of the spiritual systems of east and west, the person has an experience of bliss. This may arise as unexpectedly as the spontaneous experience or there may be hints of its approach, even minor inspirations and "borderline" ecstasy. The intention is to reach beyond the confines of "ordinary" reality to something which is worth seeking whatever the price, God-realization to the Christian, Saccidananda to the Hindu.
Methods common to many traditions are meditation and contemplation (not always easily distinguished) and chemical means of altering consciousness through drugs, although other means of inducing such an experience have included flagellation and torture. One account speaks of the brilliant light used by interrogators as having triggered an experience of peace, joy and absolute protection. Another speaks of LSD producing, beyond all opposites, a knowledge that all things were one, an experience described as certainty, unity, obviousness, satisfaction, realization of the ultimate, awareness, completeness, nothing mattering, "isness". Yet another speaks of God or essence "is" love. Common to these experiences is their lasting effect which stays through the years.
Refs Coxhead, Nona *The Relevance of Bliss* (1985).
Broader Religious experience (#HM3445).
Related Ceasing from thought (#HM3295) Spontaneous religious experience (#HM2624) Modes of awareness associated with use of hallucinogens (#HM0801).

♦ **HM3222b Refined cosmic consciousness**
Sixth state (Yoga)
Context One of the states through which it is possible to pass on the way to *Unity consciousness* through the daily practice of transcendental meditation as taught by Sri Maharishi.

Related Transcendental meditation (TM, #HH0682).
Followed by Unity consciousness (#HM3193).
Preceded by Transcendental cosmic consciousness (#HM2405).

♦ **HM3224d Primordial ancestor** (ICA)
Description This is the realization that one is called to carry on the work of many generations in caring for mankind and that this task will consume one's life, that one is called upon to pick up the mission of one's life. Nikos Kazantzakis, in his book "Saviors of God", writes of the Crimson Line that we are called to follow which has been travelled by many before us.
Context In the ICA New Religious Mode in the arena of knowing one's internal sociality (the life of meditation) the fourth formal aspect is one's experience of missional comradeship or of having a colleague. At the second phenomenological level, this occurs when a person finds his selfhood audited by that which commands his attention.
Broader Missional comradeship (ICA, #HM2976).
Followed by Expectant descendant (ICA, #HM2903).
Preceded by Destinal elector (ICA, #HM3027).

♦ **HM3225b Fundamental dialogue**
I–Thou awareness — Interhuman awareness — Confirmation of otherness
Description Usually, experience of the world is in terms of "things". "I" is seen only in relation to "it". The object of experience, "it", exists in that it is differentiated from other objects. The experience is confined to the experiencer; what is experienced has no part in the activity of experiencing. However, when "I" is seen in relation to "thou", the "thou" is included in the experience. Facing another human being as "thou", aware of the "I – Thou" relationship, the other is no longer a thing, nor consists of things, but is the whole within himself. The individual can only truly, then, be himself in relationship to others, when the whole of his being is engaged. There must be a "meeting" which is an act of grace. This concept is elaborated by Martin Buber.
It arises when an encounter with another person, or possibly a feature of nature, involves a suspension of intellection, directed affection, and identity, although not in opposition to them. Rather than being specially significant communications, they come from no one in particular. They are little disclosures beyond the boundaries of interest and concept. As such they are by no means extraordinary, unless the ordinary is defined to exclude them. In Martin Buber's words: "no purpose intervenes between I and you, no greed and no anticipation; and longing itself is changed as it plunges from the dream into appearance. Every means is an obstacles. Only where all means have disintegrated do encounters occur". There is a sense of inexhaustibleness, a sense of complete rightness, and yet it cannot be deliberately engendered. It fulfils something in a person's being that will be neither subject nor object, a dimension that is untranslatable into passion, piety, or personal fulfilment. They are approachable by stepping back and by intending not to intend.
Refs Buber, Martin *I and Thou* (1958); Buber, Martin *Between Man and Man* (1965); Buber, Martin *The Knowledge of Man* (1966); Friedman, Maurice *The Confirmation of Otherness* (1983); Hampden–Turner, Charles *Radical Man* (1970); Scott, Charles E *Boundaries in Mind* (1982).
Related Hasidism (Judaism, #HH0597) Religious experience (#HM3445)
Compunction (Christianity, #HM4251) Mental prayer (Christianity, #HH8672)
Immediate boundary awareness (#HM0035).

♦ **HM3226c Gateway** (Sufism)
Context The first of ten states acting as gateways orienting the mystic on the journey to the Absolute, as described by Ansari.
Broader Stations of consciousness - Ansari (Sufism, #HM2317).
Followed by Doors (Sufism, #HM3451).

♦ **HM3227b Ananda** (Hinduism, Buddhism, Yoga)
Bliss — Sat–cit–ananda — Saccidananda
Description The state of ananda, bliss, is achieved on release from the suffering, *duhkha*, caused by mistaken awareness of the transient as real. It is achieved through righteous living and repeated meditation.
Related Joy (#HM1172) Bliss (#HM4048) Mukti (#HH0613)
Saccidananda (#HM0592) Ananda stage (Yoga, #HM3212)
Duhkha (Buddhism, Hinduism, #HM2574).

♦ **HM3228c Reciprocal relation of absolute and conditional** (Sufism)
Description There is no absolute without that conditioned by it, nothing conditional without the absolute; they have a reciprocal relationship. But there is unilateral dependence – the conditional depends on the absolute but not the absolute on the conditional. Further, the conditioned with which the absolute is linked are interchangeable but all refer to the absolute, with which nothing can be interchanged. The absolute is beyond the conditioned, depending upon them only to exercise sovereignty.
Context The twenty first illumination of Jami.
Broader Fountains of light (Sufism, #HM3039).
Preceded by Unchanging reality of being (Sufism, #HM3316).

♦ **HM3229c Life of being** (ICA)
Description An arena in the *new religious mode*, within the ICA *Other World in the midst of this World*, when one is aware of the intensification of knowing and doing in one's life.
Broader New religious modes (ICA, #HM3004).
Narrower Transparent selfhood (#HM3077) Transparent lucidity (#HM2184)
Transparent presence (#HM2462) Transparent engagement (#HM2736).

♦ **HM3230d Unlimited commitment** (ICA)
Description This is experienced when a person realizes that it is not possible to circumscribe his action by a particular time and place, that if he wants to create a change he has to make a commitment to do whatever it takes to achieve his aims. It occurs when involvement is seen as not limited to one action or one situation but as part of a much larger situation that must be dealt with, as when Gandhi realized that his commitment to help the Indian people of South Africa was really to all Indian people and that he had to return to India.
Context In the ICA New Religious Mode in the arena of acting out one's deed (the life of doing) the first formal aspect is surrendering one's self to one's unique calling as a person. At the second phenomenological level this occurs when a person recognizes one's life as sheer venture.
Broader Surrender to one's calling (ICA, #HM2346).
Followed by Suffering servant (ICA, #HM3126).
Preceded by Actional existence (ICA, #HM3080).

♦ **HM3231d Vasana** (Yoga)
Desire — Tendency
Description This is a particular kind of psychical disposition or emotional complex. It is the universal power which drives the mind, producing the series of transformations which imprison consciousness. It is the principle of desire in its widest sense, and the karma or tendencies which this desire produces, cause and effect intermingled.
Related Zeal (Buddhism, Pali, #HM4487) Aesthetic emotion (Hinduism, #HM0205).

♦ **HM3232c Awareness of spiritual identity** (Yoga)
Super–contemplative state
Description This state is achieved, according to Patanjali, beyond the fourfold elimination, as feeling and mind are transcended, the lower vibrations no longer sensed and the seer identifies with that which observes. Although such awareness may appear to others as sleep or trance, he is fully awake on planes of consciousness beyond definition.
Related Spiritual realization (Yoga, #HM2438) Super–contemplative state (Yoga, #HM2818).

♦ **HM3233b Faculty of wisdom** (Buddhism)
Pannindriya (Pali)
Narrower Search (#HM3408) Wisdom (#HM0082) Research (#HM1602)
Sagacity (#HM1750) Thinking (#HM3966) Erudition (#HM1499)
Expertise (#HM3390) Intuition (#HM1632) Pondering (#HM4488)
Proficiency (#HM1554) Discernment (#HM1143) Correct view (#HM3252)
Non–ignorance (#HM2695) Understanding (#HM0158) Lamp of wisdom (#HM1846)
Discrimination (#HM2399) Power of wisdom (#HM1792) Sword of wisdom (#HM1645)
Differentiation (#HM0249) Being conscious (#HM4514) Jewel of wisdom (#HM1486)
Search of dharma (#HM1715) Lustre of wisdom (#HM1547) Wisdom as a goad (#HM0633)
Palace of wisdom (#HM2508) Breadth of wisdom (#HM1623) Faculty of wisdom (#HM1840)
Wisdom as a guide (#HM4358) Splendour of wisdom (#HM0192)
Analytical investigation (#HM3345).
Related Understanding (Buddhism, #HM4523).

♦ **HM3234c Life of meditation** (ICA)
Description An arena in the ICA *New Religious Mode* when one is aware of knowing one's internal sociality.
Broader New religious modes (ICA, #HM3004).
Narrower Personal absolution (#HM2370) Existential guidance (#HM2114)
Missional comradeship (#HM2976) Called to accountability (#HM2008).

♦ **HM3235c Earthy awareness** (Astrology)
Earth triplicity
Description A person whose awareness is most influenced by the *earth triplicity* defined by the signs Taurus, Virgo and Capricorn, will exhibit the qualities of dependability and practicality. Although perhaps over influenced by possessions and the objective of material gain, he will show attention to detail and an ability to relate to the world through the senses. He longs for the spiritual but at the same time may lack vision. An unconscious search for life's meaning will attract him to dogma and the formalising of the spiritual and numinous into concrete form.
Context One of the four *triplicities* or *elements*, each of which combines three related signs of the zodiac.
Broader Zodiacal forms of awareness (Astrology, #HM2713).
Narrower Virgo–consciousness (#HM2439) Taurus–consciousness (#HM2815)
Capricorn–consciousness (#HM2822).
Related Airy awareness (Astrology, #HM2955) Fiery awareness (Astrology, #HM2124)
Watery awareness (Astrology, #HM2384).

♦ **HM3236e Trance**
Entrancement
Description A state often arising in ritual situations when religious or mystical visions may be experienced; the subject may exhibit unusual physical movements – trembling, rigidity or violent jerking – may shout, speak in tongues, hyperventilate and may experience personality dissociation. While the individual appears to be conscious, whether as his normal self or possessed by some other being, and responds to questions, he nevertheless has partial suspension of the vital functions and is insensitive to sensory stimuli. He appears unconscious of his surroundings and cannot subsequently remember what occurred during the experience. Such a state may be the result of hypnosis, hysteria, ecstasy or spirit possession, and may be self–induced. Although viewed by Freud as similar to hysteria and by Jung as due to moral conflict in the same way as neuroses, trance does not have the negative connotations of either and the state is actively encouraged in some contexts.
Narrower Ikia (#HM3925) Deep trance (Psychism, #HM1886)
Hypertrance (Psychism, #HM0929) Semi–trance (Psychism, #HM0887)
Active trance (Psychism, #HM0442) Cataleptic trance (Psychism, #HM4319)
Voodoo trance consciousness (#HM2502) Self–induced trance (Psychism, #HM1964).
Related Spirit possession (#HH0056) Spiritual healing (#HH0458)
Possession (Psychism, #HM3111) Hypnotic states of consciousness (#HM2133).

♦ **HM3237c Mental engagement** (Buddhism)
Manaskara — Manasi–kara (Pali) — Yid–la–byed–pa (Tibetan) — Attention — Bringing–to–mind
Description In Tibetan Buddhism, this is the factor which directs the mind to a specific object of observation.
In Hinayana Buddhism, this is what causes the mind to differ from its previous, life–continuum state. It acts in three ways: it controls or regulates the object of attention; it controls or regulates the cognitive series or process of consciousness; it controls or regulates impulsion or apperception. It is the first of these three which is referred to as one of the formations aggregate. The latter two are five–door adverting and mind–door adverting. The characteristic of attention as controller of object is conducting or driving associated states towards the object; its function is to link them with the object; it is manifest as confronting the object, the object is the proximate cause.
Context One of the five omnipresent mental factors defined in Tibetan Buddhism. One of the formations aggregate (mental coefficients) of Hinayana Buddhism, being listed among the "or–whatever" states, and as general primary (always present in any consciousness).
Broader Omnipresent mental factors (Buddhism, Tibetan, #HH0320)
Awareness as mental–formation group of conscious existence (Buddhism, #HM2050).
Related Attention (#HM1817) Adverting (Buddhism, #HM8336)
Perfect wisdom (Buddhism, #HM7844).

♦ **HM3238c Five powers** (Buddhism)
Description These powers are attained in the path of preparation on the levels of heat and peak, and are: faith, effort, mindfulness, meditative stabilization and wisdom.
Context One of the sections of yogic paths or harmonies with enlightenment defined in Buddhism.
Broader Harmonies with enlightenment (Buddhism, #HH0603).
Related Determining mental factors (Buddhism, #HH0170).

♦ **HM3239b Intuitive knowledge** (Yoga)
Description The use of discrimination with total concentration on the succeeding moments leads to intuitive knowledge of the eternal present, which includes both the past and the future; and to the ability to distinguish between all beings, including their place in space, their qualities and so on. This is the highest knowledge and a measure of the yogin's state of absolute unity (kaivalya).
Related Kaivalya (Yoga, #HM3869).

♦ **HM3240d Liberation from relationships** (ICA)
Description The third formal aspect of the *life of poverty*, an arena in the ICA *New Religious Mode*, this represents the awareness of nonchalence about one's relations.
Broader Life of poverty (ICA, #HM2299).
Narrower Common earth (#HM2023) Serious sharing (#HM2323) Eternal insecurity (#HM2517) Abounding abasement (#HM2209).

♦ **HM3241c Consciousness of mystical poverty**
Fana (Sufism)
Description A mystical interpretation of poverty seems to be an innate idea among the spiritual, and is particularly strong in the New Testament. It is the first vow of the monastic life. Fana, the Sufi term for mystical annihilation in God, is considered, popularly but heretically, to be a state of union. Esoterically, fana is the annihilation of the individual human will before the will of God, "nothing in possession, possessed by nothing". In Europe it is the consciousness exhibited by St Francis, the "little poor one" of Assisi, whose state was such that he took "Lady Poverty" as his spouse.
Related Fana (Sufism, #HM3799) Fana (Sufism, #HM1270).

♦ **HM3242c Anahata** (Yoga)
Not hit awareness — Fourth chakra
Description The human level is distinguished from sublimated animal aims and drives. The individual's aims and drives become envisioned and awakened; religious symbols, artistic images and philosophical questions refer to this level and beyond. Awareness is of the sound anterior to things - of sound not produced by the striking together of two things. Meditation which evolves through the fourth chakra brings mastery of language, poetry and verbal endeavour and of desires (indriyas) and physical functions. The person is master of himself and gains wisdom and inner strength. Male and female energy are balanced, all relationships become pure, the senses are controlled, the person flows freely, with no hindrance from external barriers. Having evolved beyond circumstantial and environmental limitations, the person is independent and self-emanating. Others are inspired by his life and are calm and peaceful in his presence. Divine vision evolves with pure sound, bringing a balance of action and joy. Power is gained over the element air. Since air is formless, the person has power to become invisible, travel through space, enter the bodies of others.
This chakra encompasses sudharma (right or apt religion), good tendencies, the planes of sanctity, balance and fragrance. When negative karmas are enacted, purgatory may be experienced. The pure one who has developed good tendencies and sanctified his life to the human plane is illuminated by clarity of conscience.
Context The fourth "lotus" or chakra defined in kundalini yoga, said to be situated at the level of the heart.
Broader Kundalini yoga (Yoga, #HH0237) Chakra centres of consciousness (#HM2219).
Related Sphere of beauty (Kabbalah, #HM3031)
Fourth chakra: attaining balance (Leela, #HM3291)
Initiation of renunciation (Esotericism, #HM0210).
Followed by Vishuddha (Yoga, #HM3461).
Preceded by Manipura (Yoga, #HM3063).

♦ **HM3243c Path of perfect intelligence** (Judaism)
Path of absolute intelligence
Description This state is represented as the source of the preparation of principles, itself emanating from the Sphere of Magnificence.
Context The eighth among the spiritual paths expressed in the Jewish Kabbalah.
Broader Paths of wisdom (Judaism, #HM2509).
Related Sphere of mercy and greatness (Kabbalah, #HM2420).
Followed by Path of purified intelligence (Judaism, #HM2943).
Preceded by Path of occult intelligence (Judaism, #HM2444).

♦ **HM3244c Liberation of the heart** (Sufism)
Description By use of the will, knowledge and mysticism, all attachments to that which is not God are rooted out, the heart is liberated from consciousness of anything but the One.
Context The tenth illumination of Jami.
Broader Fountains of light (Sufism, #HM3039).
Followed by Liberation of the spirit by attraction of divine grace (Sufism, #HM2907).
Preceded by Annihilation of personal sensation (Sufism, #HM3358).

♦ **HM3245d Invented history** (ICA)
Description This is a perception of the absolute absurdity of standing before all of history and knowing there isn't anyone to show the way. It's as though one has the only set of keys to unlock the future, there are no blueprints or even guidelines. An example is the joke when the Lord says to Noah: "Noah, build an ark", and Noah says: "Sure, Lord, I'll build an ark" - then adds thoughtfully, "Lord, what's an ark ?"; or when the mythical stone age character Ronstrum, in the book "The Ancient of Days", discovers that by observing the sun's shadow every day he can measure time.
A dimension of this experience is a sense of awe, that is fear and fascination. Daring to create out of nothing means facing the prospect of ridicule and cynicism from those around one, and one experiences one's own profound doubt. Taking the burden of inventing the new means always thinking ahead to plan the next step. The only rewards are less sleep and more worry. But the adventure of shaping history is too great to resist and the knowledge that it has never been done before is very tantalizing. What has one really got to lose and furthermore what else would one really want to do ? One finds one's self taking on the world, forcing new images to emerge although they often come when one least expects them to. Even though running into the pain of derision, one feels one is the lucky one; one is aware that there is nothing in history that was not created out of the decision of some ordinary person to venture into the unknown.
Context This state is number 44 in the ICA *Other World in the midst of this World*.
Broader Singular mission (ICA, #HM3217).
Followed by Dynamic selfhood (ICA, #HM3091) Final blessedness (ICA, #HM2958)
Diaphanous intuition (ICA, #HM2927) Sacramental universe (ICA, #HM2445)
Definitive effectivity (ICA, #HM2796).
Preceded by Primal vocation (ICA, #HM2621) Dynamic selfhood (ICA, #HM3091)
Sacramental universe (ICA, #HM2445) Futuric responsibility (ICA, #HM3016)
Soteriological existence (ICA, #HM2904).

♦ **HM3246d Deceit** (Buddhism)
Maya — Sgyu (Tibetan)
Description A pretence at having good qualities which generates "wrong livelihood" through the deceitful amassing of goods.
Context One of the twenty *secondary afflictions* of Tibetan Buddhism.
Broader Secondary afflictions (Buddhism, Tibetan, #HH0781).

♦ **HM3249b Paths of effect** (Tibetan)
Description Phenomena of the pure class according to Buddhist teaching, the paths of effect consist of: the ten *powers*; the four *fearlessnesses*; the four *sciences*; great *love*; great *compassion*; and the eighteen *unshared attributes of Buddha*.
Broader Stages of spiritual life (#HH0102).
Narrower Ten powers (Buddhism, Tibetan, #HM3159)
Great love (Buddhism, Tibetan, #HM2929)
Four sciences (Buddhism, Tibetan, #HM3216)
Great compassion (Buddhism, Tibetan, #HM2807)
Four fearlessnesses (Buddhism, Tibetan, #HM3044)
Eighteen unshared attributes of Buddha (Buddhism, Tibetan, #HM2775).

♦ **HM3250e Infatuated** (Buddhism)
Saraga (Pali)
Broader Greed (Buddhism, #HM3283).

♦ **HM3251c Back-to-the-source awareness** (Zen)
Returning to the origin
Description No longer identified with maya-like transformations the man has never been what he is not now; unaffected by defilement he is pure and immaculate.
Broader Oxherding pictures in Zen Buddhism (Zen, #HM2690).
Followed by Entering-city-with-bliss-bestowing-hands awareness (Zen, #HM3068).
Preceded by Ox-and-man-both-out-of-sight awareness (Zen, #HM2492).

♦ **HM3252c Correct view** (Buddhism)
Sammaditthi (Pali)
Broader Faculty of wisdom (Buddhism, #HM3233).

♦ **HM3257d Not ashamed of acquisition of sinful and unwholesome dharmas** (Buddhism)
Na kiriyati papakanam akusalanam dhammanam samapattiya (Pali)
Broader Not ashamed of ought to be ashamed (Buddhism, #HM1023).

♦ **HM3259f Accidie**
Acedy — Acedia
Description This apathy or lack of concern is manifested in indifference to matters of importance and total lack of enthusiasm. As "sloth", accidie is said to negate all the seven virtues.
Related Sloth (#HM3116) Abulia (#HM0672) Sadness (#HM2403)
Laziness (Yoga, #HM0046).

♦ **HM3260c Identity of attributes with their essence** (Sufism)
Description Beyond the abstraction of the intellect, which sees attributes as other than their essence, there is awareness of the one, unique Existence. Names and attributes are only the relationship of that Existence and its aspects.
Context The fifteenth illumination of Jami.
Broader Fountains of light (Sufism, #HM3039).
Followed by Manifestation of the essence through veils (Sufism, #HM2642).
Preceded by Existence as ontological reality (Sufism, #HM3457).

♦ **HM3261d State of unfaltering exertion** (Buddhism)
Asithilaparakkamata (Pali)
Broader Faculty of energy (Buddhism, #HM1470).

♦ **HM3263a Anuttara samyak sambodhi**
Buddha illumination (Buddhism)
Description Experience beyond all series of experiences, complete knowing beyond which there is nothing.
Broader Sambodhi (Buddhism, #HM7195).
Related Enlightenment (#HM2029).

♦ **HM3264c Saviors of God: the preparation** (Christianity)
Context The first phase in the spiritual exercises of Nikos Kazantzakis.
Broader Saviors of God (Christianity, #HM3420).
Narrower First duty (#HM2224) Third duty (#HM4279) Second duty (#HM2861).
Followed by Saviors of God: the march (Christianity, #HM3439).

♦ **HM3265c Seemingly bare perception** (Buddhism, Tibetan)
Context One of the valid ways of knowing of Tibetan Buddhism.
Broader Ways of knowing (Buddhism, #HH0873).

♦ **HM3271f Psychodynamic LSD experience**
Description This is a "reliving" of basic fixations that is achieved by the use of psychedelic drugs. Individuals encounter the personal unconscious with all its infantile ego-defences. Such experience may resolve inner conflicts, and allow transpersonal realization to develop. The experience can be a painful reliving of the birth trauma, together with feelings of agressive tension and of imminent catastrophe. The experiencer may identify with other individuals or groups - even, in some cases, with the suffering of the whole of mankind, past, present and future. Reliving the first stages of birth trauma, as uterine contractions commence, overwhelmed by total anguish, leads to a sense of total meaninglessness in life with no way out. This is followed by a second stage, as in the birth canal, when agony mounts finally breaking beyond the pain threshold; it is characterized by extremes of pain and pleasure, the world being seen as full of threats and oppression. There may be attempts at violent suicide or murder. Finally, there may be a crisis of actual ego-death, complete annihilation at all levels, followed by a sense of release, rebirth and redemption.
Refs Grof, S *Varieties of Transpersonal Experiences* (1972).
Broader Psychedelic drugs (#HH0075)
Modes of awareness associated with use of hallucinogens (#HM0801)
Modes of awareness associated with psychoactive substances (#HM0584).
Related Bad trip (#HM0818) Aesthetic LSD experience (#HM2542)
Drug use for transcendent experience (#HH1123).

♦ **HM3274c Freedom from lower qualities** (Sufism)
Sifat-i bashriyya
Description Through negation of self, *nafs*, and its associated desires and passions which are the obstacles to attainment of mystical perfection, comes freedom from lower qualities, essential for spiritual progress. The heart is purified of all but God, who bestows purity of self on the person who desires it earnestly.

♦ **HM3280c Niyama** (Hinduism)
Self-discipline
Description The five aspects of self-discipline apply to the relation between the individual and transcendent reality. The first two, purity (sauca) and contentment (samtosa), allow a shift from harmonizing external relations to an inward orientation. Another source lists ten niyamas - austerity, contentment, belief in God, charity, worship of God, listening to explanation of doctrine

and scripture, modesty, discerning mind, repetition of prayers, sacrifice (including performance of religious sacrifice).
Context The second component of the eightfold path of yoga.
Broader Eightfold path of yoga (Yoga, #HH0779).
Narrower Tapas (#HH1248) Sauca (Yoga, #HH1391) Samtosa (Yoga, #HM2898) Study of sacred scriptures (#HH0957) Resignation to the will of God (#HH1016).
Related Self-discipline (#HH0877).

♦ **HM3281d Rajoguna** (Leela)
Description Having failed to achieve cosmic consciousness, the player is drawn on by karma into the activity of rajas, the cause of suffering, and to a predisposition for ambition and reward. He cannot attain samadhi and tamoguna takes him back to earth.
Context The 71st state or square on the board of *Leela*, the game of knowledge, appearing in the eighth row.
Broader Beyond the chakras: the gods themselves (Leela, #HM1141).

♦ **HM3283d Greed** (Buddhism)
Lobha (Pali)
Description In Hinayana Buddhism, greed is; the means by which one lusts or is greedy; lust or greed itself; or being lustful or greedy. Its characteristic is grasping an object, its function is clinging or sticking, it is manifest by not letting go, its proximate cause is seeing as enjoyment those things which lead to bondage. Swelling with craving, it carries one away into states of loss.
Context One of the formations aggregate (mental coefficients) of Hinayana Buddhism, being listed among the constant states which appear in their true nature, and as unprofitable secondary (sometimes present in any unprofitable or unprofitable-resultant consciousness).
Broader Awareness as mental-formation group of conscious existence (Buddhism, #HM2050).
Narrower Greed (#HM0802) Infatuated (#HM3250) Being greedy (#HM1442) Covetousness (#HM3563) Feeling greed (#HM1513) Being infatuated (#HM1683) Feeling infatuation (#HM4363) Greed an unwholesome root (#HM1954).
Related Desire (#HM2433) Greed (Leela, #HM1931) Non-attachment (Buddhism, #HM2128) State of hunger (Buddhism, #HM0150).

♦ **HM3286c World of forms** (Sufism)
Noah of one's being
Description This corresponds to the animal soul where there is strife between vital and organic operations - the situation faced by Noah when his people were hostile. The colour is blue.
Context The second of seven subtle stages relating to the seven major prophets of Semitic monotheism as described by Simnani.
Broader Stations of consciousness - Simnami (Sufism, #HM2341).
Followed by World of spiritual perception (Sufism, #HM3974).
Preceded by World of nature (Sufism, #HM3425).

♦ **HM3288d Self-satisfaction**
Broader Happiness (#HM2409).

♦ **HM3289c Life-fluid emblems - temperance** (Tarot)
Description One of the archetypal emblems projected by the psyche is the water of life, the vital force depicted as a stream, or as a vessel containing liquid, or as a flowing sap or shining essence. In the Tarot the water of life is depicted in relationship to two pitchers. In this emblem (*Temperance*, number 14) the water or liquid is poured from the sun pitcher to the moon pitcher by a winged angel - a tempering or uniting of the male and female energies. In another emblem (*Star*, number 17), a terrestrial spirit or nymph pours the two pitchers into a pool. Both emblems represent awareness of the ebb and flow of life. In this case there is a mystical ascent from the limited personality of ordinary man to the illumination of the spiritual, where all the elements are synchronized; man the animal aspires to man the god.
Context In the Kabbalah, Temperance is placed in the 25th path between foundation (Yesod) and beauty (Tiphareth), while the Star is place on the 15th path between foundation and victory or force.
Broader Emblem archetypes in the psyche (Tarot, #HM2201) Tarot arcana of conscious states (Tarot, #HM2097).
Related Sphere of beauty (Kabbalah, #HM3031) Life-fluid emblems - star (Tarot, #HM3137) Sphere of the foundation (Kabbalah, #HM2410).

♦ **HM3291c Fourth chakra: attaining balance** (Leela)
Plane of balance
Description The player is aware of patterns of behaviour, of karma. Midway among the seven chakras, there are influences from below and above. Life is motivated by faith but there may be imbalance, with emphasis too much on rectifying the past. Mastery of the chakra leads to control of the senses and inspired life. Dear to women and lord of speech the player is an inspiring presence with no enemies, master of time, seeing both visible and invisible universes, able to become invisible and unnoticeable and to levitate.
Context The fourth row of the board of Leela, the game of knowledge, this is represented by the planes of balance, sanctity, fragrance and taste, and also groups the squares representing the states of apt religion, irreligiousity, good tendencies, purgatory and clarity of conscience. Located at the region of the heart, the seat of the conscious principle, of life and breath, this chakra is related to the element air, to the sense of touch and to a smoky grey/green colour. It is compared to the octave comprising the ages 21 to 28 of temporal life, that of responsibility. Apt religion is the arrow at this stage, and irreligiousity the snake.
Broader Lila (#HM2278).
Narrower Purgatory (#HM0856) Apt religion (#HM2981) Irreligiousity (#HM0670) Plane of taste (#HM0878) Good tendencies (#HM5197) Plane of balance (#HM1351) Plane of sanctity (#HM5965) Plane of fragrance (#HM0008) Clarity of consciousness (#HM0455).
Related Anahata (Yoga, #HM3242).
Followed by Fifth chakra: man becomes himself (Leela, #HM0933).
Preceded by Third chakra: theatre of karma (Leela, #HM0717).

♦ **HM3294d Near death transition** (NDE)
Near death entering the darkness
Description An apparently intermediate stage between two states of consciousness, experienced as moving rapidly through a dark space or down a tunnel towards a bright light which is comforting and beautiful rather than dazzling.
Context The third stage in the process referred to as *near death experience*. The previous two stages are sometimes bypassed.
Broader Near death experience (NDE, #HM0777).
Related Near death black void (NDE, #HM2213).
Followed by Near death awareness of light (NDE, #HM3381).
Preceded by Near death detachment (NDE, #HM3349).

♦ **HM3295c Ceasing from thought**
Description Ceasing from thought is the beginning of meditation or contemplation. Attention is controlled, so there is clarity and attentiveness. In this freedom from thought there is true meditation. One has to learn to still everyday mind, stop the flow of idle chatter, the endless circle of recriminations, rehearsals, puns, snatches of song which tend to fill it.
Related Induced religious experience (#HM3221).

♦ **HM3296d Understanding having an exalted object** (Buddhism)
Understanding having a sublime object
Description This is understanding contingent on states of the fine-material (form-realm) or immaterial (formless realm) spheres. Together with understanding having a limited object it comprises mundane or worldly insight.
Context On the Path of Purification of Hinayana Buddhism, panna (understanding) is considered as of one kind (monad), of two kinds (dyads), of three kinds (triads) or of four kinds (tetrads). There are five dyads, four triads and two tetrads. All have the characteristic of penetrating the individual essences or true nature of states (monad). In the second triad, this is compared with understanding having a limited or measureless object.
Broader Understanding (Buddhism, #HM4523).
Related Mundane understanding (Buddhism, #HM7628) Understanding having a limited object (Buddhism, #HM2495) Understanding having a measureless object (Buddhism, #HM6681).

♦ **HM3298b Planes of wisdom** (Hinduism)
Description As enumerated in the Supreme Yoga, there are seven states, ascending steps or planes of wisdom to be known so as not to be caught in delusion. These are contrasted with seven descending steps of ignorance or delusions that veil *self-knowledge*. Ignorance leads to total inert dullness resembling sleep, while wisdom leads to liberation beyond happiness and unhappiness. Even beyond the seventh plane of wisdom is a state - turiyatita - where the body is transcended.
Narrower Enquiry (#HM2443) Turyatita (#HM3139) Pure intention (#HM3100) Subtlety of mind (#HM3312) Turiya awareness (#HM2395) Establishment in truth (#HM3006) Cessation of objectivity (#HM3354) Total freedom from attachment (#HM2001).
Related Veils of delusion (Hinduism, #HM3592).

♦ **HM3300g Perception of form** (Yoga)
Description According to Patanjali, knowledge based on perception of form (rather than on the state of being or true nature of phenomena) is false and incorrect.

♦ **HM3301d Emotional desires** (Buddhism)
Context The fourth fetter or character defect referred to by Buddha. The first five fetters are marked by automatic response due to the following of old, unphilosophical mental habits. When these are overcome, the disciples is ready (arhat) to face the final five.
Broader Character defects (Buddhism, #HH0470).
Followed by Emotional aversions (Buddhism, #HM3504).
Preceded by Dependence on ceremonies (Buddhism, #HM0509).

♦ **HM3302c Seeing-the-traces awareness** (Zen)
Description Although understanding that the objective world is a reflection of the self, the seeker's mind is still confused between truth and falsehood.
Broader Oxherding pictures in Zen Buddhism (Zen, #HM2690).
Followed by Seeing-the-ox awareness (Zen, #HM2755).
Preceded by Searching-for-ox awareness (Zen, #HM3036).

♦ **HM3303c Ontological reality of the truth** (Sufism)
Description Vision is dazzled by contemplation of the beauty and perfection of the truth, which is none other than Being, Being which knows no decline or ugliness, pure without qualification by change or evolution. All quality and quantity are comprehended in him, he himself has neither quantity nor quality. He, the source of all colour, has no colour. All is perceived by he who is inaccessible to perception.
Context The thirteenth illumination of Jami.
Broader Fountains of light (Sufism, #HM3039).
Followed by Existence as ontological reality (Sufism, #HM3457).
Preceded by Surrender to the bliss of mystical seduction (Sufism, #HM3436).

♦ **HM3306d Vikalpa** (Buddhism)
Mayoi (Zen) — Mi — Doubt
Description This state of uncertainty, of being unsure in which direction to go, is the result of relying on a multiplicity of objects for knowledge and not on that which reflects these objects. The world is seen as a realm of opposites. This is contrasted with *satori*, where the mind moves from discrimination of the "ten thousand things" to non-discrimination, from relativity to emptiness, from the "ten thousand things" to the contentless self or mirror-nature. The bright, mirror nature of satori is there all the time but it is obscured with the dust of "things" and has to be revealed. The contradiction is that wiping off means dirt accumulating. There is no enlightenment which one can claim to have attained. Hui-neng and the Prajna or Southern School of Zen therefore preclude continuous progress as a means of freedom from *mayoi*. There needs to be absolute negation, for otherwise progress stops at the bright mirror and no account is taken of how the bright nature of the mirror allows itself to be defiled.
Related Satori (Zen, #HM2326).

♦ **HM3310c Cessations** (Buddhism)
Analytical cessations — Non-analytical cessations
Description Phenomena of the pure class according to Buddhist teaching, these are the true absence of afflictions following their abandonment. Analytical cessation is the state of permanent absence of affliction, nirvana being the ultimate analytical cessation when the last affliction is abandoned. Non-analytical cessation is non-permanent, and occurs when one affliction is abandoned (possibly due to attention to another), and may return under certain conditions.
Broader Harmonies with enlightenment (Buddhism, #HH0603).

♦ **HM3311d Fear** (Christianity)
Dread
Description Without belief in God, men do not simply not trust each other, they do not trust themselves; they are not just afraid of each other, they are afraid of everything. Only love, and therefore humility, can exorcize fear.
Dread arises from the experience of contradiction between the actually possible and socially barred freedom and is combatted in the struggle for emancipation. However, dread is a fundamental condition for man in the history of his personal salvation, in the dread of God's demands which are both inescapable and unattainable. The root of dread is original sin. If it were not for original sin dread would not exist. Attempting to escape it through independence, through flight from God, is unsuccessful. Accepting fear in hope, which is sharing in the saving dread of Jesus, is the only alternative.
Refs Merton, Thomas *Seeds of Contemplation* (1972).
Narrower Phobia (#HM1509) Fear of God (#HM0673).

Related Horror (#HM0904) Anxiety (#HM2465) Fearlessness (#HM3553)
Fear (Hinduism, #HM1194).
Followed by Tapa (#HM3530).

♦ **HM3312c Subtlety of mind** (Hinduism)
Third plane of wisdom
Description With the arising of non-attachment, the mind becomes subtle and transparent.
Context The third of seven ascending planes of wisdom described in the Supreme Yoga, which require to be known so as not to be caught in delusion.
Broader Planes of wisdom (Hinduism, #HM3298).
Related Non-attachment (Buddhism, #HM2128).
Followed by Establishment in truth (Hinduism, #HM3006).
Preceded by Enquiry (Hinduism, #HM2443).

♦ **HM3314d Neither physical ease nor unease born of contact with element of mind-consciousness** (Buddhism)
Manovinnanadhatusamphassajam cetasikam neva satam nasatam (Pali)
Broader Unperturbedness (Buddhism, #HM4572).

♦ **HM3316c Unchanging reality of being** (Sufism)
Description Appearance and disappearance of external modes and aspects entail no change at all in the ontological reality of Being nor of its essential aspects, any more than the sun's radiance is affected by what it illuminates.
Context The twentieth illumination of Jami.
Broader Fountains of light (Sufism, #HM3039).
Followed by Reciprocal relation of absolute and conditional (Sufism, #HM3228).
Preceded by Indivisibility of essence (Sufism, #HM2113).

♦ **HM3320g Vital-physical stage of life**
First stage of life
Description This state occurs during the perinatal stage and the first few years of life, approximately up to seven years, and is marked by adaptation to physical existence. There is identification with the personal physical body and the development of physical, mental, emotional and psychological independence from others. Fulfilment of this stage means conscious relationships with others and with nature. More frequent is misadaptation to separation and an inability to resolve dependence.
Context The first of seven stages of life characterized by Master Da Free John. Those individuals who have their human growth arrested in the first three stages lack psychic maturity and spiritual quality. They may develop through *puja yoga* and *karma yoga*.
Broader Seven stages of life (#HH0460).
Related Puja yoga (Yoga, #HH0111).
Followed by Emotional-sexual stage of life (#HM2954).

♦ **HM3321d Conscious groups**
Group soul — Group mind — Folk soul
Description Under circumstances when a group of people acts together corporately - as for, example, a football team or a dance troupe - then a *group soul* may emerge, such that there is a collective and simultaneous awareness of the pattern of interactions among the members. Without central coordination or explicit direction, the group acts appropriately with an awareness of itself and with internal, integrated, self-regulating response. The collective mind can thus be said to exist independently of individual members of the group and to dominate their behaviour. Another definition refers to psychic groups, where individuals can draw on the others telepathically.
Related Group consciousness (#HM3109).

♦ **HM3323c Mind consciousness** (Buddhism)
Bhavanga-mano
Description The Path of Purification (Hinayana Buddhism) describes mind-consciousness as of 22 kinds of mundane resultant consciousness. As mind element it is listed twice, without root cause, resultant with profitable or unprofitable result. As mind-consciousness element it is listed 3 times, without root cause: with profitable or moral resultant, accompanied by joy or accompanied by equanimity or indifference; with unprofitable or immoral resultant. Further, there are the sense-sphere resultant consciousnesses with root cause (8 kinds, as to whether accompanied by joy or equanimity, associated or not with knowledge, prompted or not); there are the fine-material sphere resultant (5 kinds) and the immaterial sphere resultant (4 kinds).
Broader Mundane resultant consciousness (Buddhism, #HM0130).
Related Mental consciousness (Buddhism, #HM2838)
Mind-consciousness element (Buddhism, #HM6173)
Indeterminate consciousness in the immaterial sphere - resultant (Buddhism, #HM4982)
Indeterminate consciousness in the fine-material sphere - resultant (Buddhism, #HM0594).

♦ **HM3327c Keeping going on** (Buddhism)
Yapana (Pali)
Broader Faculty of living (Buddhism, #HM3404).

♦ **HM3329c Satsang** (Hinduism, Yoga)
Description By focusing the attention on the self-transcendent state of an adept or guru the disciple enters in communion with this state.
Related Guru (#HH0805) Identification with spiritual master (Sufism, #HM3944).

♦ **HM3331g Subliminal perception**
Description Under the threshold of conscious awareness are several levels. It is not necessary to be consciously aware of every stimulus all the time, yet things do come into the mind below the level of conscious perception. Meaningful sounds are monitored even while one is asleep, reactions to information below the level of consciousness can be measured physically, and feelings are developed to stimuli even when they cannot be recognized.
Refs Ornstein, Robert *The Psychology of Consciousness* (1986).

♦ **HM3335c Kushesh** (Sufism)
Riaza — Will to behavioural transformation
Description This is the conscious effort resulting in purification of the achieved self, characterized by poverty and tranquillity in patience and trust. Kusheesh is followed by *kedesh jazba*, divine attraction, when the person passes from one state or maqamat to another.
Narrower Reza (#HM2724).
Followed by Kedesh-jazba (Sufism, #HM1026).

♦ **HM3336d Existential consciousness**
Consciousness two — Progression
Description Although, as in consciousness one, the world of objects is considered real - and consciousness and its prime object, the ego, as derived from that world - experience is seen as guiding the process, so that control involves choice of environment for desired impact on the ego. Events are controlled by limiting the situation and by acquisition of habits. Examples would be self conditioning or control of behaviour.
Context One of three possible modes by which the relationship of consciousness with ego may be regarded, each implying a particular strategy of control or of seeing the way through limitations, what may be termed a form of transcendence.
Broader Ego consciousness (#HM0570).
Related Biosocial consciousness (#HM0044) Transpersonal awareness (#HM0768)
Psychological approach to transcendence (#HH1763).

♦ **HM3337b Ekagrata parinama** (Yoga)
Description The tendency for distractions to arise, stilled in samadhi parinama, is followed by tendency for the lack of impressions in the mind to continue; the only object to arise in the mind is that which arose previously, the essential object of samadhi. This is constantly repeated so the impression is of a single unchanging object of attention.
Context The second of three mental transformations described by Patanjali.
Broader Parinama (Yoga, #HH4561).
Followed by Nirodha parinama (Yoga, #HM3437).
Preceded by Samadhi parinama (Yoga, #HM3207).

♦ **HM3340c Unification with life essence** (Sufism)
Description Here deification in everything is recognized and further transformation occurs as duality of subject and object are transcended. Love, the essence of creation, is the object of desire, and the Sufi practises the art of love lying beneath the surface realities of name and attributes. Through experience love is discovered as the greatest mediator of all, where all conflict is resolved. This state arises very slowly and practice and action are required to mould the Sufi's nature in the process of union.
Context The third category of experience related to the three types of object of desire.
Preceded by Deification (Sufism, #HM0946).

♦ **HM3341b Religious traditions in the ascension stages game** (Buddhism, Tibetan)
Description This refers to the religious traditions and various spiritual paths appropriate to the different human personalities. These are of three types: those who seek rebirth, particularly in the heavens, through good works; those who seek liberation from the cycle of birth and death; and those who seek the emancipation of all.
Broader Consciousness ascension stages game (Buddhism, Tibetan, #HH4000).
Narrower Cessation-awareness (#HM2172) Arhat sanctity awareness (#HM3024)
Hindu-states of awareness (#HM2115) Pratyeka Buddha awareness (#HM2180)
Barbarian-state of awareness (#HM2091) Discipleship-karma awareness (#HM2192)
Discipleship-vision awareness (#HM2240) Pratyeka Buddha arhat awareness (#HM2776)
Discipleship-application awareness (#HM2716) Bon-practitioner state of awareness (#HM2639)
Pratyeka Buddha application awareness (#HM3020)
Pratyeka Buddha vision-path awareness (#HM2228)
Pratyeka Buddha cultivation awareness (#HM2252).

♦ **HM3344c Jesus helped to carry the cross** (Christianity)
Description Just as Simon of Cyrene accepted to carry the cross of Jesus when He became too weak to bear the burden, the believer accepts, through grace, to carry the cross, in particular to embrace his own death and the pain he may have to bear. As Jesus died for love of him, so he will die for love of Jesus. Loving Jesus more than himself, he repents with his whole heart and begs never again to separate himself from Him. He asks that he will love Jesus always and that Jesus will do with him as He wills.
Context The fifth station of the cross.
Broader Way of the cross (Christianity, #HM3516).
Followed by Veronica wipes the face of Jesus (Christianity, #HM3401).
Preceded by Jesus meets his afflicted mother (Christianity, #HM3611).

♦ **HM3345c Analytical investigation** (Buddhism)
Upaparikkha (Pali)
Broader Faculty of wisdom (Buddhism, #HM3233).

♦ **HM3347d Vision**
Seeing visions — Beatific vision
Description Beatific vision is the imparting by God to the pure in heart of the direct contemplation of himself, in as perfect a manner as the creature is capable. As distinguished from intellectual knowledge, this is experience of the nearness of God and absorption in his glory. Such transparency to the absolute God is the highest fulfilment of man's nature. In that empirical knowledge is based on the knower knowing that knower and known are entitatively one, the reality of the mind as knower is the being of God himself. This taking up into the source is a supernatural mystery when God's utter incomprehensibility is contemplated in itself. It arises in the *lumen gloriae*, the growing of the grace created by grace from the seed already present in man.
To the Sufi, beatific vision is the culmination of religious love. Fixing his gaze on the beloved is the first step to salvation.
Related Nazar (Islam, #HM6003).

♦ **HM3348d Inattentive perception** (Buddhism, Tibetan)
Context One of the five invalid ways of knowing of Tibetan Buddhism.
Broader Ways of knowing (Buddhism, #HH0873).

♦ **HM3349d Near death detachment** (NDE)
Near death body separation
Description The individual experiences physical detachment - the body is seen below and in a different light - and emotional detachment - no longer identifying with the body, which appears as a worn out instrument to be discarded. There is a sense of weightlessness, mental clarity and acuteness of sight and hearing, the latter being telepathic, of "hearing" others' thoughts.
Context The second stage in the process referred to as *near death experience*.
Broader Near death experience (NDE, #HM0777).
Related Near death negative detachment (NDE, #HM3115).
Followed by Near death transition (NDE, #HM3294).
Preceded by Near death peace (NDE, #HM2494).

♦ **HM3352d Ri hokkai** (Zen)
Description A term used in Zen Buddhism to express the undifferentiated consciousness which perceives the universe as absolute, as one made manifest through the many.
Related Ri (#HH3544) Ji hokkai (Zen, #HM3168)
Zazen meditation (Zen, #HH0882).
Followed by Ji-ri-mu-je (Zen, #HM3164).

♦ **HM3353d Not feeling remorse of acquisition of sinful and unwholesome dharmas** (Buddhism)
Na ottappati papakanam akusalanam dhammanam samapattiya (Pali)
Broader Power of unremorsefulness (Buddhism, #HM1915).

♦ **HM3354c Cessation of objectivity** (Hinduism)
Sixth plane of wisdom
Description There is rejoicing in one's own self, with a cessation of perception of duality whether within or outside one's self; previous efforts lead to direct spiritual experience.
Context The sixth of seven ascending planes of wisdom described in the Supreme Yoga, which require to be known so as not to be caught in delusion.
Broader Planes of wisdom (Hinduism, #HM3298).
Followed by Turiya awareness (Hinduism, #HM2395).
Preceded by Total freedom from attachment (Hinduism, #HM2001).

♦ **HM3356d Wrongness** (Buddhism)
Micchatta (Pali)
Broader Wrong view (Buddhism, #HM1710).

♦ **HM3357d Ardour** (Buddhism)
Ussolhi (Pali)
Broader Faculty of energy (Buddhism, #HM1470).

♦ **HM3358c Annihilation of personal sensation** (Sufism)
Description Complete identity with true being occurs as all personal sensations are annihilated, even the sensation of being aware of that annihilation; since awareness of this destruction of destruction would imply something "other", contradicting true annihilation.
Context The ninth illumination of Jami.
Broader Fountains of light (Sufism, #HM3039).
Related Tawhid (Sufism, #HM3438).
Followed by Liberation of the heart (Sufism, #HM3244).
Preceded by Identity with God (Sufism, #HM2214).

♦ **HM3359c Conjunctive faith**
Stage five faith
Description This stage does not normally arise before mid-life. What has previously been suppressed or unrecognized due to self-certainty and conscious cognitive and affective adaptation to reality is now integrated into self and outlook. In this "second naïveté", symbolic power is reunited with conceptual meanings. The past has to be newly reclaimed and reworked, opening up to the voices of the deeper self. In particular, this includes critical recognition of the myths, ideals and prejudices that one's nurture within a particular class, culture, religion or ethnic group has built deeply into the self-system, one's social unconscious. This stage knows the sacrament of defeat and that commitments and acts may be irrevocable. The boundaries of self and outlook that the previous stage sought to clarify are now made porous and permeable. There is awareness of the paradox and truth of apparent contradictions so that one strives to unify opposites both in mind and experience. Vulnerability to the truth of those who are "other" is generated and maintained. There is commitment to justice beyond the confinements of tribe, class and so on, and there is a readiness to be close to what may threaten self and outlook, which may be new depths of spirituality and religious revelatory experience. In the knowledge that life is more than half over, there is a readiness to spend and be spent in the cause of others' generating identity and meaning.
Here the new strength is an ironic imagination which can see or be in the powerful meanings of one's self or one's group and at the same time recognize their relativity, that they are partial and distorting apprehensions of transcendent reality. The danger is that paradoxical understanding of truth may lead one one to be paralyzed by passivity or inaction, with concomitant complacency or cynical withdrawal. Both one's own and others' symbols, myths and rituals can be appreciated due to having been grasped by the depth of reality to which they refer. Having been apprehended by the possibility and necessity of inclusive community of being, one is vividly aware of the divisions in humanity. But there is still division in one's self as one lives and acts between the untransformed world and transforming vision and loyalties. This may lead to the radical actualization of universalizing faith, stage 6.
Context Stage 5 in the system of faith development described by James Fowler.
Refs Fowler, J W *Stages of Faith* (1981).
Broader Stages of faith (#HH2097).
Followed by Universalizing faith (#HM1465).
Preceded by Individuative-reflective faith (#HM1665).

♦ **HM3362f Double mindedness**
Description Apparent insincerity or hypocrisy in an individual's behaviour may be due to a certain fraction of the complex personality being so split off from the rest that it acts "on its own account" and, momentarily at least, forgets its relation to the whole. Sometimes such inconsistent attitude or behaviour may be accompanied consciousness of a haunting sense of lack of harmony.
Related Sub-personalities (#HM0463) Single mindedness (#HM3211)
Multiple personality (#HH0763).

♦ **HM3370g Flexibility of body** (Buddhism)
Kayamuduta (Pali)
Narrower Mildness (#HM0191) Non-rigidity (#HM1028) Non-stiffness (#HM3014)
Flexibility of the aggregate of sensation, of the aggregate of cognition, and of the aggregate of synergies (#HM3397).

♦ **HM3371b Divine path** (Sufism)
Tariqa — Maqamat — Ahwal
Description Mystic union with God arises through transformation of the self through following a divine path or *tariqa*, sometimes described as the action of the Prophet. The path brings liberation from bondage to sensual passions and desires; and the attainment of spiritual discipline. As the soul is perfected, individual qualities are transmuted into divine qualities, leading to complete obliteration of individual existence and perfect identity with God.
The path is not a continuous process of change. There are stable stations or *maqamat* on the way, which mark particular spiritual attainments achieved by the aspirant's personal effort. Each maqam is reached when the soul experiences a number of spiritual states or ahwal, and lasts for a fixed time. As soon as a maqam is reached then the earlier station has been completely attained. A hal (singular of ahwal) is not the same as a maqam; it is an unstable or passing mystical feeling of the heart granted by the grace of God to a chosen few. Thus maqamat are the results of actions of self-mortification while ahwal are gifts during which the individual is dead to himself, standing by a state created in him by God. Both are considered expressions of God's love, the one on the path of *mujahada* (striving), the other on the illuminative way of *mukashafa* (contemplative vision).
Many different interpretations of the path exist depending on the Shaykh or Sufi who described it. Orthodox Sufis speak of three main stages: *shari'a*, *tariqa* and *haqiqa*, the word, the action, the inward state. Others refer to four: *nasut*, following the law; *malaqut*, the angelic nature; *jabarut*, spiritual powers; *lahut*, divine qualities. Particular teachers detail many more stages.
At each level or station on the spiritual path, the individual is referred to by a particular title indicating his status. The first stage has been described as repentance, the turning from that which is against God. Further stages include poverty and love, leading to purity of soul and a feeling of oneness with God. The last station is *fana*, the soul's absorption into God. At the stage of *baqa*, perfection is reached, the sufi as *perfect man* abides in the world of divinity.
It can be interpreted that the Sufi travels three paths to realise identity with God. First, reflecting his being in the descending order of creation, means attainment of mental or sensual faculties, at the lowest when involved with animal instincts. This is the journey from God. There is then the journey to God, tariqat. Then return to the world as perfect man, serving humanity and guiding others on the spiritual path – the journey with God.
Refs Bakhtiar, Laleh *Sufi* (1987); Bhatnagar, R S *Dimensions of Classical Sufi Thought* (1984).
Broader Human development (Sufism, #HH0436).
Narrower Nasut (#HM3579) Lahut (#HM0741) Malaqut (#HM0116)
Jabarut (#HM4392) Mystic stations (#HM3415)
Higher states of consciousness (#HM2365) Stations of consciousness - Ansari (#HM2317)
Stations of consciousness - Simnami (#HM2341)
Stations of consciousness - ibn-Abi'l-Khayr (#HM2424).
Related Sharia (Islam, #HH0738) Haqiqa (Sufism, #HM0124)
Ma'rifa (Sufism, #HM2254) Psychic states (Sufism, #HM4311).

♦ **HM3374d Non-inertness** (Buddhism)
Avitthanata (Pali)
Broader Lightness of body (Buddhism, #HM4395)
Lightness of thought (Buddhism, #HM1830).

♦ **HM3375d Unaffected** (Buddhism)
Akaranam (Pali)
Broader Correct speech (Buddhism, #HM1821) Correct action (Buddhism, #HM0439)
Correct livelihood (Buddhism, #HM0549).

♦ **HM3377c Nihayat** (Sufism)
Supreme goal
Description Having cut through the deserts of calamity to reach the destination of the inn, the devotee has seen God with the eye of his heart.
Context The last of ten states acting as gateways orienting the mystic on the journey to the Absolute, as described by Ansari. According to Shaykh Abu Sa'id ibn Abi'l-Khayr, this is the 39th of 40 stations or maqamat the Sufi must possess for his journey on the path of Sufism to be acceptable.
Broader Stations of consciousness - Ansari (Sufism, #HM2317)
Stations of consciousness - ibn-Abi'l-Khayr (Sufism, #HM2424).
Followed by Tasawwuf (Sufism, #HM0575).
Preceded by Tahqiq (Sufism, #HM5563) Realities (Sufism, #HM3755).

♦ **HM3378c Induced attentiveness** (Buddhism, Tibetan)
Self-induced attentiveness — Other-induced attentiveness
Description Self-induced attentiveness is that of validly knowing something to be self-evident. Thus the colour of an object may be known and awareness of its colour is self-induced. Other-induced attentiveness occurs when it is clear that an object is not self-evidently one thing or another. Then there is valid knowledge that further knowledge is necessary to distinguish what the object actually is. For example, although the colour of a flower may be known self-evidently, it may be that further information and cognition are required to know what type of flower it is – there is valid cognition that the type of flower is not known self-evidently. These two are both forms of induced attentiveness. However, other-induced attentiveness may also occur with inattentive perception or distorted perception. In these cases, although the other-induced attentiveness is valid it is only nominally so, as it is based on distorted cognition. Thus two types of induced attentiveness are truly valid ways of knowing – self-induced attentiveness and the first type of other-induced attentiveness. Two further varieties of other-induced attentiveness are only nominally so.
Context One of the valid ways of knowing of Tibetan Buddhism.
Broader Ways of knowing (Buddhism, #HH0873).

♦ **HM3379c Living** (Buddhism)
Jivita (Pali)
Broader Faculty of living (Buddhism, #HM3404).

♦ **HM3381d Near death awareness of light** (NDE)
Description The light viewed in the previous stage enlarges until the experiencer emerges into it; it exudes compassion and understanding and there is a feeling of love, joy, beauty and peace. Some describe it as a different land, a bright and beautiful but indescribable field or valley.
Context The fourth stage in the process referred to as *near death experience*.
Broader Near death experience (NDE, #HM0777).
Followed by Near death encounter with the higher self (NDE, #HM3059).
Preceded by Near death transition (NDE, #HM3294).

♦ **HM3382c Mansions of humility** (Christianity)
Description Every human has a soul but some are unaware of it and refuse to enter this interior castle. They are only interested in outward things. This interest in the things of this world sooner or later leads them to become more and more like this world. Eventually they become frozen in this state. To leave this state one must strive through prayer and meditation focusing on God. At this point some individuals enter the rooms but bring their own pre-occupations with them turning toward God once or twice a month.
While in the state of mortal sin, the soul will profit in nothing and good works will be of no avail. What ever its intentions, the soul separated from God produces nothing but misery and filth. It is not the wonder of God that has departed or gone out but the soul has placed a dark cloth between itself and God.
In these first mansions there are two things to be learned, to have the greatest fear of offending God and to be present to a mirror of humility. In order to grow the soul must roam through the mansions of the soul and the only room it should spend time in is that of self knowledge and to soar aloft on the glory of God. While self-knowledge is of critical importance and should be cultivated, it is better to dwell on the purity and humility of God than on its own baseness. For dwelling in self-doubt and misgivings about one's own worthiness do not arise from humility but from a lack of self-knowledge. Self-knowledge enobles and does not make one fearful.
In these first rooms the soul is blinded by many concerns, possessions, honours and business dealings all calling for its attention. One must focus on God and put aside all else. Even the religious life has its dangers. A person inspired to do penance such that he only has peace when he is torturing himself is in a trap. Also trapped is the person striving for perfection so that any other person's smallest fault seems to be a major failure. These people fail to see their own real

faults.
Context The first of seven mansions of the soul's progress described by St Teresa of Avila.
Broader Mansions of the soul (Christianity, #HH1409).
Followed by Mansions of the practice of prayer (Christianity, #HM3218).

♦ **HM3384d Jamais vu**
Description As opposed to *déjà vu*, where the individual experiences a new situation as though it had occurred before, *jamais vu* refers to experiencing an ostensibly familiar situation or scene where the quality of familiarity is absent.
Broader Anomalies in consistency of consciousness (#HM1122).
Related Déjà vu (#HM1240).

♦ **HM3385d Knowledge of contemplation of dissolution** (Buddhism)
Related Purification by knowledge and vision of the way (Buddhism, #HH3550).
Followed by Knowledge of contemplation of rise and fall (Buddhism, #HM3723).

♦ **HM3387d Furyu** (Japanese)
Artistic refinement — Aesthetic attitude
Description This attitude of mind is traditionally associated with the artist and implies aesthetic taste; although the term has more general and diverse meanings associated both with immersion in the pleasures of the world and with withdrawal into the lifestyle of a hermit.
Related Wabi (Japanese, #HM1365) Artistic education (#HH0029).

♦ **HM3390c Expertise** (Buddhism)
Nepunna (Pali)
Broader Faculty of wisdom (Buddhism, #HM3233).

♦ **HM3391d Guilt**
Description Feelings of guilt may arise for irrational or for rational reasons. In the latter case, failure to recognize such feelings may produce unfortunate psychological consequences and failure to have such feelings may result in neurosis. According to Jung, even irrational guilt may be necessary to avoid outward projection of the contents of the shadow. Again, all guilt feelings are important in that they inspire reflection upon what is evil. They can, however, result in moral paralysis; and confession of guilt is enjoined in the Christian religion as a means of healing and release.
Related Discipline of confession (Christianity, #HH0825).

♦ **HM3394d Evasion** (Buddhism)
Asappana (Pali)
Broader Perplexity (Buddhism, #HM0812).

♦ **HM3397g Flexibility of the aggregate of sensation, of the aggregate of cognition, and of the aggregate of synergies** (Buddhism)
Vedanakkhandhassa sannakkhandhassa sankharakkhandhassa muduta (Pali)
Broader Flexibility of body (Buddhism, #HM3370).

♦ **HM3398c Sahasrara** (Yoga)
Thousand–petalled lotus — Seventh chakra — Shunya chakra — Niralambapuri chakra
Description The absolute, non–dual state beyond all divisions, visions, feelings, ideas and thoughts, beyond subject and object, beyond knowledge and names. There is the silence that is the basis of the no longer heard syllable AUM.
Only on attaining this chakra can the yogi attain asama–prajnata–samadhi, where there is no activity of mind, no knower, no knowledge, nothing to be known. Knowledge, knower and known are unified and liberated. Immortality is achieved. There is the bliss of total inactivity. The prana moves to its highest point. All thoughts, feelings and desires dissolve into their primary cause. For as long as he remains in the physical body the yogi retains nondual consciousness, enjoying the play of lila without being troubled by pleasure, pain, success, humiliation. His is sat–chit–ananda (knowledge–being–bliss).
According to the shastras this chakra is the seat of chitta or citta, the self–luminescent soul, the essence of being. Chitta is like a screen on which the reflection of the cosmic Self is seen and through it the divine reflected. Here one can feel the divine, even realize divinity within one's self.
Context The seventh "lotus" or chakra defined in kundalini yoga, said to be situated at the crown of the head.
Broader Kundalini yoga (Yoga, #HH0237) Chakra centres of consciousness (#HM2219).
Narrower Soma chakra (#HM0176).
Related Citta (Hinduism, Buddhism, #HM3529)
Seventh chakra: plane of reality (Leela, #HM0754)
Initiation of resurrection (Esotericism, #HM1153)
Sphere of the of the supreme crown (Kabbalah, #HM3132).
Preceded by Ajna (Yoga, #HM2144).

♦ **HM3401c Veronica wipes the face of Jesus** (Christianity)
Description The believer is aware that, just as Jesus' face, once so beautiful, is marred by wounds and blood, so his soul, once made beautiful through grace in baptism, has been disfigured by sin and only Jesus can restore it. He begs that, through His passion, He will indeed restore it. Loving Jesus more than himself, he repents with his whole heart and begs never again to separate himself from Him. He asks that he will love Jesus always and that Jesus will do with him as He wills.
Context The sixth station of the cross.
Broader Way of the cross (Christianity, #HM3516).
Followed by Jesus falls the second time under the cross (Christianity, #HM3824).
Preceded by Jesus helped to carry the cross (Christianity, #HM3344).

♦ **HM3403d Lightness of the aggregate of consciousness** (Buddhism)
Vinnanakkhandhassa lahuta (Pali)
Broader Lightness of thought (Buddhism, #HM1830)
Dispositions of consciousness (Buddhism, #HM2098).

♦ **HM3404c Faculty of living** (Buddhism)
Jivitindriya (Pali) — Vitality — Life principle
Description The positive force or energy that remains in the human being after enlightenment, namely once all the obstacles, hindrances and delusions have been eliminated.
Narrower Stasis (#HM1418) Living (#HM3379) Going on (#HM1691)
Progression (#HM1972) Continuance (#HM2468) Conservation (#HM1508)
Faculty of life (#HM0261) Keeping going on (#HM3327)
Duration of the formless dharmas (#HM1969).
Related Enlightenment (#HM2029).

♦ **HM3405f Despair**
Hopelessness
Description Despair is characterized by lack of hope and by a sense of waiting for hope to return. In present day Western society, despair is commonly suppressed or at least a (usually successful) attempt is made to hide its existence from others. This is said to leading to a partial numbing of the psyche so that the emotional and sensory life are diminished. Anxiety–provoking data is effectively filtered out and the numbing effect intensified unless the despair is worked through. In this respect it is not dissimilar from grief.
Despair has been seen as the ultimate development of pride when it is so great and so stubborn that it chooses the total misery of damnation rather than admitting the self's weakness, its inability to fulfil its destiny and its dependence on God. It is the extreme of self–love, expressed as self–pity when one's own resources fail. Since one's own resources invariably do fail, all are subject to some extent to discouragement and despair. The cure is humility, since the humble feel no self–pity.
Refs Merton, Thomas *Seeds of Contemplation* (1972).
Related Grief (#HM2685) Hope (Christianity, #HH0199).

♦ **HM3407f Dysphoria**
Description The reverse of euphoria, the individual in this state makes every attempt to redress it, even to the extent of resorting to drugs. This latter response, however, although leading to short–term alleviation, may ultimately result in deeper and more persistent dysphoria.
Related Euphoria (#HM3763).

♦ **HM3408c Search** (Buddhism)
Vicaya (Pali)
Broader Faculty of wisdom (Buddhism, #HM3233).

♦ **HM3409g Embodiment**
Description A person's body may be defined as the centre through which that person interacts with the world, the means by which he is in the world and by which the world around him is presented for thought and action. It is only through experiencing the body that anything else in the world may be experienced; and it is through the body that the person's desires, feelings, ideas, and so on, are expressed to others. The relationship of a person to the bodily organism in which and by which that person lives is a complex metaphysical issue. A body is referred to as "my body" and thus belongs to "me"; yet in another sense "my body" is "me". Although the person may be said to govern the body he is also subject to it – damage to the body is experienced as "I" am injured. Mind and body somehow interact. As Pascal pointed out, we understand neither the body, nor the mind, nor, even more, how the body can be united to the mind; this is our very being, to be both body and mind and therefore opaque to ourselves. Spinoza posited body and mind as attributes of the unitary self, each being essential to the other, the body mirrored in the mind as its idea; later writers have also essayed the problem, but all admit a fundamental opacity or ambiguity intrinsic to the union between person and embodying organism.
Refs Zaner, R *The Problem of Embodiment.*
Related Reincarnation (#HH0686).

♦ **HM3410c Jesus taken down from the cross** (Christianity)
Description The believer is aware of Jesus' mother as she embraces the body of her son. He begs to be accepted as her servant and that she will pray for him. Since Jesus has died for him, wanting nothing more than Him, he begs to be allowed to love Him. Loving Jesus more than himself, he repents with his whole heart and begs never again to separate himself from Him. He asks that he will love Jesus always and that Jesus will do with him as He wills.
Context The thirteenth station of the cross.
Broader Way of the cross (Christianity, #HM3516).
Followed by Jesus placed in the sepulchre (Christianity, #HM3536).
Preceded by Jesus dies on the cross (Christianity, #HM3719).

♦ **HM3411d Complete tranquillization** (Buddhism)
Patipassambhana (Pali)
Broader Composedness of body (Buddhism, #HM1867)
Composedness of thought (Buddhism, #HM1490).

♦ **HM3412c Innocence** (Christianity)
Description One who is innocent is in charity with all, harming nobody, with no desire to defraud or deceive and assuming the integrity of others. Three modes are distinguished: (1) The attribute of Christ which is incapable of sin. (2) The childlike harmlessness of those who have never been tempted, who are unaware of good and evil but who have the innate capacity to develop the good; this is innocence in the sense described by Kierkegaard, the state of ignorance sometimes referred to as perfection in which Adam received his instruction not to eat of the tree of the knowledge of good and evil. There are possibilities which are not understood and which therefore cause anxiety and dread. This dread could be resolved by faith; but humanity responded by making the choice which confirmed it's limited freedom to act but destroyed the state of innocence. (3) The innocence of those capable of sin who have withstood temptation and retained purity of heart; there may be minor weaknesses but the state is essentially Christlike.
Related Innocent cognition (#HM0888).

♦ **HM3413d Not incurring guilt** (Buddhism)
Anajjhapatti (Pali)
Broader Correct speech (Buddhism, #HM1821) Correct action (Buddhism, #HM0439)
Correct livelihood (Buddhism, #HM0549).

♦ **HM3415b Mystic stations** (Sufism)
Maqamat
Description The Sufi path or *tariqa* is not a continuous process of change. There are stable stations or *maqamat* on the way, which mark particular spiritual attainments achieved by the aspirant's personal effort on the way of *muhajada* (striving). Each maqam is reached when the soul experiences a number of spiritual states or ahwal, and lasts for a fixed time. As soon as a maqam is reached then the earlier station has been completely attained. A hal (singular of ahwal) is not the same as a maqam; it is an unstable or passing mystical feeling of the heart granted by the grace of God to a chosen few on the illuminative way of *mukashafa* (contemplative vision). Thus maqamat are the results of actions of self–mortification while ahwal are gifts during which the individual is dead to himself, standing by a state created in him by God. Both are considered expressions of God's love.
At each level or station on the spiritual path, the individual is referred to by a particular title indicating his status. The first stage has been described as repentance, the turning from that which is against God. Further stages include poverty and love, leading to purity of soul and a feeling of oneness with God. The last station is *fana*, the soul's absorption into God. At the stage of *baqa*, perfection is reached, the sufi as *perfect man* abides in the world of divinity.
Broader Divine path (Sufism, #HM3371).
Narrower Wali (#HM4418) Fani (#HM4064) Baqi (#HM4594)
Nabi (#HM0606) Zuhd (#HM4450) Faqr (#HM3427)
Sabr (#HM3418) Rida (#HM4190) Wajd (#HM3650)
Fana (#HM1270) Baqa (#HM0330) Wara (#HM0286)

Talib (#HM4464) Salik (#HM4288) Mumin (#HM4173)
Murid (#HM3887) 'Abid (#HM0815) Zahid (#HM4208)
'Arif (#HM1045) Rasul (#HM0724) Tawba (#HM4062)
Khalifa (#HM0335) Ma'rifa (#HM2254) Tawakkul (#HM0807)
Al-khatim (#HM1626) Mushahada (#HM3521) Love of God (#HM4273)
Insanu'l-kamil (#HM3677) Nearness-awareness (#HM2377).
Related Psychic states (Sufism, #HM4311) Awareness of the mystic journey (#HM2900)
Suluk of the Naqshbandiyya order (Sufism, #HM4356).

♦ **HM3416d Aesthetic state** (Psychism)
Description Sensation of joy and beauty triggered by religious activity.
Related Emotional high (#HM1400) Extreme exaltation (Psychism, #HM4152)
Aesthetic perception (Psychism, #HM0218).

♦ **HM3418c Sabr** (Sufism)
Patience — Patient endurance
Description In this state there is joy in spiritual bliss and kinship with God. God is reached through turning towards him in a life of renunciation and love. Afflictions on the path of search for God are welcomed. The first stage is of repentance and of no longer complaining; the second is the ascetic stage, satisfaction with what has been decreed; finally, as a friend of God, the follower loves what God does with him. The attitude of thanksgiving is developed - patience follows gratitude, for there is gratitude for the contentment of soul and for full satisfaction. There is awareness of God's favour and grace and the beginning of surrender to God.
Context The doctrines of *sabr*, *mahabbat* and *ma'rifa* were elaborated by the Sufi scholar Abu Talib al-Makki and followed and interpreted by many others. Sabr is the fifth stage in a systematic account of various stations of the Sufi spiritual path. However, it is the seventh stage on the path to sainthood of the Naqshbandiyya order. According to Shaykh Abu Sa'id ibn Abi'l-Khayr, this is the seventh of 40 stations or maqamat the Sufi must possess for his journey on the path of Sufism to be acceptable.
Broader Mystic stations (Sufism, #HM3415) Human development (Sufism, #HH0436)
Suluk of the Naqshbandiyya order (Sufism, #HM4356)
Stations of consciousness - ibn-Abi'l-Khayr (Sufism, #HM2424).
Related Ma'rifa (Sufism, #HM2254) Love of God (Sufism, #HM4273)
Striving against the soul (Islam, #HH6973).
Followed by Shukr (Sufism, #HM4154) Tawakkul (Sufism, #HM0807)
Recollection (Islam, Sufism, #HM2351).
Preceded by Faqr (Sufism, #HM3427) Shukr (Sufism, #HM4154)
Tawakkul (Sufism, #HM0807) State-awareness (Sufism, #HM2305).

♦ **HM3419c Cognizing** (Buddhism)
Sannanana (Pali)

♦ **HM3420b Saviors of God** (Christianity)
Spiritual exercises of Nikos Kazantzakis
Refs Kazantzakis, Nikos *The Saviors of God* (1960).
Broader Spiritual exercises (#HH0707).
Narrower Saviors of God: the march (#HM3439) Saviors of God: the vision (#HM2855)
Saviors of God: the action (#HM3462) Saviors of God: the silence (#HM3977)
Saviors of God: the preparation (#HM3264).

♦ **HM3421f Psychic inertia** (Physical sciences)
Resistance to change
Description It has been postulated that Newton's law - that a body will remain in a state of rest or uniform motion in a straight line unless compelled by an external force to change - may be extended to cover all systems, including the psyche. Psychic inertia or resistance to change manifests to defend every pattern of adaptation and requires an impulse at least as strong to displace it and allow change (however desirable) to occur, since change is seen as a death threat to the ego. The impulses may be from the environment or from the Self.
Related Rites of passage (#HH0021).

♦ **HM3422d Near death evil force**
Description The individual experiences an evil force trying to drag him down.
Context The fourth stage in the (less frequently recorded) negative near death experience.
Broader Near death experience (NDE, #HM0777).
Related Near death encounter with the higher self (NDE, #HM3059).
Followed by Near death hell (NDE, #HM3165).
Preceded by Near death black void (NDE, #HM2213).

♦ **HM3424d Good quality** (Buddhism)
Pagunattam (Pali)
Broader Fitness of mental body (Buddhism, #HM1455)
Fitness of consciousness (Buddhism, #HM1810).

♦ **HM3425c World of nature** (Sufism)
Adam of one's being
Description One acquires an embryonic mould of a subtle or non-physical form or new body. The psychological colour here is black moving towards dark grey.
Context The first of seven subtle stages relating to the seven major prophets of Semitic monotheism as described by Simnani.
Broader Stations of consciousness - Simnami (Sufism, #HM2341).
Followed by World of forms (Sufism, #HM3286).

♦ **HM3426c Path of resplendent intelligence** (Judaism)
Description This state of awareness is represented as seated upon the throne of Binah, exalted above all, enlightening all lights and from which all forms emanate.
Context The tenth of the spiritual paths expressed in the Jewish Kabbalah.
Broader Paths of wisdom (Judaism, #HM2509).
Related Sphere of understanding (Kabbalah, #HM2372).
Followed by Path of fiery intelligence (Judaism, #HM2523).
Preceded by Path of purified intelligence (Judaism, #HM2943).

♦ **HM3427c Faqr** (Sufism)
State of poverty — Faqir — Dervish
Description At this station the traveller becomes a faqir, liberated from all expectations except in things related to God. In a state of detachment all external things are renounced, together with any desire for them. There is perfect renunciation of all in the heart and worldly possessions mean nothing. A spiritual state of perfect contentment - *rida* - materializes. No longer bound by evil there is freedom to acquire spiritual qualities. The individual will is obliterated - the faqir no longer follows a way, he is a way by which things pass. Purity of soul is the mission of poverty and this is achieved by the highest category of mendicants or *dervishes*, who not only possess nothing and ask for nothing (the first category), nor no longer desire worldly goods, accepting only what is given (the second category), but who are so detached they can ask for something as the need is not related to the real self which they can clearly distinguish from the unreal self. Once free of attachment to possessions their actual possession does not hinder poverty. As a guard against evil influence and a means for making the world indifferent to him, the faqir may accept blame or condemnation - the stage of *malamat*; by God's grace this leads to spiritual perfection and closeness to God.
Context The fourth stage in a systematic account of various stations of the Sufi spiritual path.
Broader Mystic stations (Sufism, #HM3415).
Narrower Malamat (#HM4280).
Related Tasawwuf (Sufism, #HM0575) Spiritual poverty (#HH0190).
Followed by Sabr (Sufism, #HM3418) State-awareness (Sufism, #HM2305).
Preceded by Zuhd (Sufism, #HM4450).

♦ **HM3429d Leaving undone** (Buddhism)
Akiriya (Pali)
Broader Correct speech (Buddhism, #HM1821) Correct action (Buddhism, #HM0439)
Correct livelihood (Buddhism, #HM0549).

♦ **HM3432d Inclination towards view** (Buddhism)
Abhinivesa (Pali)
Broader Wrong view (Buddhism, #HM1710).

♦ **HM3434d Mercy** (Leela)
Daya
Description Responding with compassion to injury to his self-identification, the player surrenders to the divine attribute of mercy. Insight into motivation allows recognition of the game beyond comprehension of himself and the other players. He sees how he may also have injured them. The ego is bypassed in feelings of joy, exultation and cosmic love, so the player is lifted from the second to the eighth, the absolute, plane. He remains at this level until bitten by the "snake" of tamoguna, when he returns to earth to work off past karma.
Context The 17th state or square on the board of *Leela*, the game of knowledge, appearing in the second row.
Broader Second chakra: realm of fantasy (Leela, #HM3651).
Followed by Absolute plane (Leela, #HM4326).

♦ **HM3435d Not trespassing limit** (Buddhism)
Vela anatikkamo (Pali) — Vela anatikkama
Broader Correct speech (Buddhism, #HM1821) Correct action (Buddhism, #HM0439)
Correct livelihood (Buddhism, #HM0549).

♦ **HM3436c Surrender to the bliss of mystical seduction** (Sufism)
Description Aware of the commencing of mystical seduction and its bliss, the seeker finds the Truth within himself. This "union" must be reinforced with all zeal and guarded against all which opposes it. Even eternal life consecrated to such union is insufficient to surrender to the Absolute what is due.
Context The twelfth illumination of Jami.
Broader Fountains of light (Sufism, #HM3039).
Followed by Ontological reality of the truth (Sufism, #HM3303).
Preceded by Liberation of the spirit by attraction of divine grace (Sufism, #HM2907).

♦ **HM3437b Nirodha parinama** (Yoga)
Niruddha
Description This is the holding of the state, normally instantaneous, between two impressions in consciousness so that there is no impression at all and the mind is unmodified. It is called the *Niruddha* state. Once this state is achieved on one plane the consciousness passes to the next plane and the process of three transformations is repeated.
Context The third of three mental transformations described by Patanjali.
Broader Parinama (Yoga, #HH4561).
Related Suppression (Yoga, #HM0135)
Cessation-awareness (Buddhism, Tibetan, #HM2172).
Preceded by Ekagrata parinama (Yoga, #HM3337).

♦ **HM3438a Tawhid** (Sufism)
Unification of God — Trans-substantiation of the heart
Description This direct personal experience of reality is said arise as a state of mind through practice of *dhikr* or remembrance. Inner repetition the phrase "la ilaha illallah" - there is no god but God - is practised until it becomes a permanent activity. Relation to the absolute is grasped and harmony with the universe maintained. As the quality of the Divine Presence is recognized in the heart, the illusion of separation is surrendered and the truth of unity - God's immanence and transcendence - experienced. Not simply monotheism, with only one God, or "unity of God", Sufism believes in total union and identity with God, "unification of God". The multiplicity of created things are seen to reflect that unity and material and spiritual worlds become one. Under the influence of tawhid, global and personal concerns become aligned. In the realization of God's consciousness, each individual is a source of transformation able to assist in creation of harmonious global institutions and restructuring of existing ones. The balance brought in tawhid assists in avoidance of common obstacles to progress - factionalism, idealism and ruthlessness.
In the all-pervading unicity of God there is no room for the individual self or ego. Individuality, even the individuality seeking unification, even the thought that unification has been achieved, conceals the infinite and maintains separation. Eventually the state of self annihilation is arrived at, through repentance, repentance of repentance and repentance of seeing one's own existence. The lower self is subjected through self-mortification, and eventually, through God's grace alone, there is only God in the heart. This perfect unity is attained as a favour of God granted to very few and is the goal of many Sufi paths of self mortification and mysticism. However, after the loss of personal attributes and total presence of God, the being present in God and absent to self, after the "drunkenness" of being overwhelmed by God, the Sufi comes to the "sobriety" to correctly assess everything in its right place. Having passed through *fana* his personal qualities persist and he becomes a pattern for the spiritual path of others.
The total obliteration of the self is the fourth kind of tawhid described by Junayd. There are lesser kinds: first, the tawhid of the common man is an awareness of union arising through belief in one God and Islam, and following the Prophet. The second is through following of *shari'a*, the law of religion; observance of the law and avoiding its prohibitions brings union. The third is the feeling of God in the heart, a personal experience of God and his uniqueness. But the fourth is absolute identification, with return to the state prior to existence. This is consistent with the view that unity of God (monotheism as opposed to polytheism) is to be distinguished from unification as absolute identity with God. In Islam, tawhid refers to the former and early Sufis would not use the term to refer to the absolute identity or mystic ecstasy not reached through knowledge but revealed.
Refs Bhatnagar, R S *Dimensions of Classical Sufi Thought* (1984); Said, Abdul Aziz *Tawhid* (1988).

Related Fana (Sufism, #HM1270) Passionate love (Islam, #HM3656)
Recollection (Islam, Sufism, #HM2351) Unification with God (Sufism, #HM3864)
Annihilation of personal sensation (Sufism, #HM3358).

♦ HM3439c Saviors of God: the march (Christianity)

Context The second phase in the spiritual exercises of Nikos Kazantzakis.
Broader Saviors of God (Christianity, #HM3420).
Narrower The cry (#HM0059) First step: the ego (#HM3748) Third step: mankind (#HM3501)
Second step: the race (#HM2953) Fourth step: the earth (#HM3570).
Followed by Saviors of God: the vision (Christianity, #HM2855).
Preceded by Saviors of God: the preparation (Christianity, #HM3264).

♦ HM3443c Mansions of spiritual consolations (Christianity)

Description These mansions are beautiful, filled with exquisite things and near God. Petty faults and temptations are seldom encountered here and when they are they may be good for the soul by keeping it alert to the possibility of temptation by greater evil. They also keep the soul from being constantly absorbed.
In these mansions the imagination is quite active. Imagination is not the same as understanding and can cause a great deal of torment. Things which are not bad may be worried about and the soul is sent in pursuit of them wasting its time. These obstacles to prayer are no reason to quit.
While it is a weak will, nature and the devil that cause these trial neither the will nor understanding can stop them. They are to be lived with.
In these mansions spiritual sweetness may be encountered which is result of work. They can be gained through human skill, by meditating, fatiguing the understanding and the soul is filled with joy. These worldly joys have their source in nature, and are quite noisy and while they do not enlarge hearts, they can lead to God.
Spiritual consolations, which are different from spiritual sweetness, come from God directly and bring great peace, quiet and sweetness which are not felt by the body at first. The interior is dilated. This is not union but being absorbed by God and the person who recognizes these graces and does not turn back is greatly blessed. They cannot be strove for because they come from love of God without self interest, this effort lacks humility, preparation is the desire to suffer and to follow Christ not to gain consolations, God is in no way obligated to give them and striving is in vain.
The person may experience an interior shrinking or gradual retiring within oneself. The response should be to give praise. While some advocate silencing all thought, reason and understanding, this is potentially harmful, the effort may encourage thought, and it distracts from remembering the honour and glory of God. It is best to try without forcing to stop discursive reasoning and not stop all thought and understanding. This kind of prayer of recollection enlarges the heart, gives more freedom to serve God, grants confidence in being destined to be with God, and releases from the fear of physical harm when doing penance. These consolations are so great that earthly desires are gradually withdrawn from.
The dangers in these mansions include the need to avoid all occasions for offending God. Some people become so physically weak that as soon they receive these consolations they fall into a stupor and when every they experience joy return to the stupor. They then believe they are making progress while, in fact, they are making fools of themselves. Finally, some souls will find that they cannot handle the physical strain of these experiences and should stop.
Context The fourth of seven mansions of the soul's progress described by St Teresa of Avila.
Broader Mansions of the soul (Christianity, #HH1409).
Followed by Mansions of incipient union (Christianity, #HM4110).
Preceded by Mansions of exemplary life (Christianity, #HM2969).

♦ HM3445b Religious experience

Mystical experience — Transcendent experience
Description Described by William James as "the feelings, acts and experiences of individual men in their solitude, so far as they apprehend themselves to stand in relation to whatever they may consider the divine", such experience is essentially subjective and personal. The relationship so described may be cognitive, ritual, inspirational, transformative or sustaining, and may be experienced positively or negatively; but in all cases the experience may be considered *transcendent*, although what is defined as transcendent may vary. Further, the experience is usually accompanied by a feeling of reverence and sense of the sacred; and may be associated with the phenomenon of *conversion*. Certainly, feelings of unworthiness, of wrongness with how things are naturally and of saving from this wrongness by the transcendent, are common. The experience may be renewing and transformative; there is a coming to terms with the conflict between "the flesh and the spirit" and a striving to rise above the habitual level of human nature, to become a "new creation".
Within the definition of religious experience may be the wisdom acquired from the conscious experience in the present of communion by the verifying of spiritual facts. In this sense, the subjective, personal aspect may be seen to be trans-subjective. It is not simply a reflex but grounded in the life-process and closely linked with the experience of others, whose spiritual experience can only, however, be interpreted in the light of one's own. Thus, the nature of religious experience depends on the traditions through which the person relates to the transcendent and the patterns of his personal life. The spatio-temporal nature of the world as normally experienced is seen as a reflection of a transcendent reality in which it is rooted.
Although the characteristic of religious experience has been stated as being absolutely dependent and may have absolute authority for the person experiencing it, it nevertheless does not destroy moral freedom; self-surrender is voluntary and involves conflict and choice. Self-discipline and inner struggle are required before there is willing conformity with divine will and conscious and increasing participation in the divine life. This has been referred to as the process of transformation, as the personality surrenders in faith to that which will satisfy the deepest needs. The vocabulary of religious experience illustrates the sense of obligation involved. The word "religion" itself carries the meaning of "bound", bound in conscientious and dutiful devotion. Kant described the sense of reverence which arises at moral law. The term yoga implies a "yoke". All religions have within them particular systems of discipline whose goal is either union (in the sense of absorption) with the transcendent or perfect communion between the self and the transcendent - for example the three *marga* or ways of Hinduism, the eightfold path of Buddhism. There is also celebration of the joy that religious experience brings, compared with the joy of human love - as in the Sufi poets and the Judaic Song of Songs.
Although the Christian traditions of the past provided for devotional practice and the rites of passage which give ritual expression and celebration of major concerns, less emphasis was placed on self-conscious reflection on religious experience than is now the case in some Protestant groupings, where the fundamental role of individual religious consciousness is emphasized. However, in the 14th Century, Meister Eckhart spoke of mysticism as experiencing God rather than simply believing. This refers to an attitude which accompanies every state of consciousness, whether normal or abnormal. This attitude is healthy, realistic and life-giving. It implies renunciation and detachment, of letting go, of being open to the present moment whatever it brings. It is not so much an expansion of consciousness as a breaking through consciousness to find an inner centre of stability, a unity and serenity which stays intact despite fluctuations of the conscious state.
Religious experience as the personal relation between the individual and the divine has been extended (Martin Buber) to include any authentic relationship with another person in which the whole of one's being is engaged and in which there is an element of grace or givenness from the other. In any I-thou relation there is a glimpse of the eternal Thou. This mutuality of encounter, he says, can hardly be referred to as an individual experience at all. It is compared with the I-it relationship in which the whole being cannot be involved.
Religious experience may be precipitated by an upsetting of normal routine, whether through a discipline such as fasting or through extreme physical exertion, or through changes such as a new job, a different social situation, travelling. Shocks and stresses induced like this may make the individual able to see himself as he really is, not as he imagines or hopes himself to be. Surveys in the 1960s and 1970s in the UK and the USA have revealed a remarkably high proportion of people claiming to have had spontaneous experiences which could be referred to as religious, of a powerful spiritual force, many of whom stating that their lives had been radically changed by this experience. This is in addition to the numbers of people having experiences induced by traditional religious methods or through various meditative techniques, drugs and other methods.
Refs Brakenhelm, C R *Problems of Religious Experience* (1985); Collins, John E and Woods, Ralph (Eds) *Civil Religion and Transcendent Experience* (1988); Coxhead, Nona *The Relevance of Bliss* (1985); Davis, J (Ed) *Religious Organizations and Religious Experience* (1982); Goodenough, Erwin R *The Psychology of Religious Experiences* (1986); James, William *The Varieties of Religious Experience* (1961); MacMurray, John *The Structure of Religious Experience* (1971); Meissner, W W *Psychoanalysis and Religious Experience* (1986); Smart, Ninian *The Religious Experience of Mankind* (1971); Wach, Joachim *Types of Religious Experience* (1972).
Narrower Induced religious experience (#HM3221)
Spontaneous religious experience (#HM2624)
Religious experience as meditational insight (#HM1593)
Religious experience as the realization of duty (#HM4327)
Religious experience as personal devotion and worship (#HM0561).
Related Bliss (#HM4048) Conversion (#HH0077) Transformation (#HH1039)
Deautomatization (#HH2331) Mystical cognition (#HM2272)
Fundamental dialogue (#HM3225) Cosmic consciousness (#HM2291)
Religious consciousness (#HM3008) Transcendental experience (#HM2712)
Mystical theology (Christianity, #HH5217) Human development through religion (#HH1198)
Awareness of the sacred (Christianity, #HM0876)
Deautomatization and the mystic experience (#HM4398)
Birth of God in the soul's ground (Christianity, #HM6522)
Modes of awareness associated with use of hallucinogens (#HM0801).

♦ HM3450c Power of recollection (Buddhism)

Satibala (Pali)
Broader Faculty of recollection (Buddhism, #HM4248).

♦ HM3451c Doors (Sufism)

Context The second of ten states acting as gateways orienting the mystic on the journey to the Absolute, as described by Ansari.
Broader Stations of consciousness - Ansari (Sufism, #HM2317).
Followed by Conduct (Sufism, #HM4648).
Preceded by Gateway (Sufism, #HM3226).

♦ HM3454c Faculty of concentration (Buddhism)

Samadhindriya (Pali)
Related Concentration (Buddhism, #HM6663).

♦ HM3455d Prajna (Hinduism)

Intellectual condition — Pure consciousness
Description The third "quarter" or condition of the self or *Atman* according to the Mandukya Upanishad, its field the state of deep sleep. There are no dreams, no awareness of sensory impressions, no desires; this condition is associated with bliss and peace - the precursor and controller of waking and dreaming states. The being of consciousness itself, with no object - pure consciousness that one is. It is associated with the syllable "M" of the mystic word "AUM".
According to Patanjali, prajna stands for all the states of consciousness in samadhi, from *vitarka* to *asmita*. Prajna becomes more subtle as the light of spirituality irradiates mental consciousness and again when consciousness is illuminated by consciousness of *purusha*. It comes to an end altogether in pure awareness of reality, *viveka khyati*.
Broader Degrees of consciousness (Hinduism, #HM0891).
Related Sanjna (Yoga, #HM3160) Essential wisdom (#HM0107)
Viveka khyati (Yoga, #HM4493) Prajna paramita (Zen, #HH0550)
Turiya awareness (Hinduism, #HM2395) Pure consciousness (Psychism, #HM0521)
Susupti state of unconsciousness (Hinduism, Zen, #HM2957).
Preceded by Taijasa (Hinduism, #HM3040).

♦ HM3456e Trance channelling (Psychism)

Channelling
Description In a condition of heightened awareness, with mind relaxed, uncluttered and open, a medium allows thoughts of some other world intelligence to be impressed on the mind; and communicates these messages, supposedly from disincarnate personalities, while remaining unaware of the meaning of such messages.
Broader Communication with supernatural beings (#HH1306)
Anomalies in integrity of consciousness (#HM4521).
Related Channelling (#HH0878) Mediation (Magic, #HH4521).

♦ HM3457c Existence as ontological reality (Sufism)

Description Truth is revealed as the existential substratum. Such existence does not imply relativity but is the essential "to be". Other beings exist by virtue of this being; in reality there is no other being but this in the external world. As great mystic have shown intuitively, it is through this being that all beings exist.
Context The fourteenth illumination of Jami.
Broader Fountains of light (Sufism, #HM3039).
Followed by Identity of attributes with their essence (Sufism, #HM3260).
Preceded by Ontological reality of the truth (Sufism, #HM3303).

♦ HM3460c Vicara stage (Yoga)

Avivesa stage of the gunas — Vijnanamaya kosa
Description In this stage of samadhi, consciousness functions through the higher mind which deals with the universals, archetypes and principles lying behind particular names and forms. There is isolation of a particular concept, archetype, law or principle from all others.
Context The second of the four stages of the gunas.
Broader Stages of the gunas (Yoga, #HM2805).
Related Vicara samadhi (Yoga, #HM3880).
Followed by Ananda stage (Yoga, #HM3212).
Preceded by Vitarka stage (Yoga, #HM2912).

♦ **HM3461c Vishuddha** (Yoga)
Purification awareness — Fifth chakra
Description The soul is purged of any remaining attachment, leaving behind anything interposing between the hearer and the sound AUM, whether art, religion, philosophy or thought. Meditation on this chakra gives calmness, serenity, purity, a melodious voice, command of speech and mantras, ability to compose poetry, interpret scriptures, understand hidden messages in dreams. The person is youthful, radiant and a good teacher of spiritual sciences. Entering this chakra, the person becomes master of his entire self. All the elements dissolve in akasha (ether), so that only the tanmatras remain, the subtle frequencies of the elements. No longer is the distracting nature of the world, the senses and the mind a problem, as supreme reasoning overcomes the elements and the emotions of the heart. This is the plane of the follower of knowledge, the path leading to man's true birth in the divine state. The person is chitta, free from the fetters of the world, master of his total self. The chakra embodies *chit* or cosmic consciousness.
Context The fifth "lotus" or chakra defined in kundalini yoga, said to be situated at the level of the larynx.
Broader Kundalini yoga (Yoga, #HH0237) Chakra centres of consciousness (#HM2219).
Related Sphere of knowledge (Kabbalah, #HM0126)
Fifth chakra: man becomes himself (Leela, #HM0933).
Followed by Ajna (Yoga, #HM2144).
Preceded by Anahata (Yoga, #HM3242).

♦ **HM3462c Saviors of God: the action** (Christianity)
Context The fourth phase in the spiritual exercises of Nikos Kazantzakis.
Broader Saviors of God (Christianity, #HM3420).
Narrower The relationship between God and man (#HM3595)
The relationship between man and man (#HM2766)
The relationship between man and nature (#HM3488).
Followed by Saviors of God: the silence (Christianity, #HM3977).
Preceded by Saviors of God: the vision (Christianity, #HM2855).

♦ **HM3464d Mental-volitional stage of life**
Third stage of life
Description At this stage there is development of discriminative intelligence and will, and the individual is in a process of integration to a fully differentiated sexual and social person. Having adapted to this stage the individual is ready to enter adult responsibilities, social, personal and spiritual. Although associated with the ages 14 to 21, the state is not so much one reached at a particular age (18, for example) as continuing to mature throughout life. In fact, misadaptation in first and second stages and the inability of society to communicate further approaches means that most spend their entire life limited to these stages. Those individuals who have their human growth arrested in the first three stages lack psychic maturity and spiritual quality. They may develop through *puja yoga* and *karma yoga*.
Context The third of seven stages of life characterized by Master Da Free John.
Broader Seven stages of life (#HH0460).
Related Puja yoga (Yoga, #HH0111).
Followed by Heart-awakening stage of life (#HM3518).
Preceded by Emotional-sexual stage of life (#HM2954).

♦ **HM3468d Ambivalence** (Jung)
Description Contradictory feelings, both positive and negative, result in a greater coherence which arises from blending of apparently the apparently disparate elements. Thus, according to Jung, every position entails its own negation. Ambivalence is inevitable when archetypal images are concerned, particularly in relation to parents; it also manifests in the pairing of opposites found in the forces of nature, so that good and evil, hope and despair, etc, counterbalance each other.
Refs Adler G, Fordham M and Read H (Eds) *The Collected Works of C G Jung*; Samuels, Andrew et al *A Critical Dictionary of Jungian Analysis* (1986).

♦ **HM3469d Exertion** (Buddhism)
Parakkama (Pali)
Broader Faculty of energy (Buddhism, #HM1470).

♦ **HM3471d Steadfastness of thought** (Buddhism)
Cittasa avatthiti (Pali)
Broader Sukha (Buddhism, #HM2866).

♦ **HM3476g Beta wave consciousness**
Alert wakefulness
Description Brain activity registered on an electroencephalogram indicates 13–26 cycles per second. Beta is linked to mental concentration, anxiety, some kinds of analysis, problem solving and attention, but also with the somewhat disordered state associated with highly active and variegated living at the emotional, physical and intellectual levels. Perception by the conscious mind is through all five senses. One researcher (William Condon) refers to six different brain wave frequencies, which include the highly active beta II (perhaps of to 40 cycles per second).
Broader Brain waves (#HH3129).
Related Focusing (#HM3003) Waking consciousness (#HM3021)
Alpha wave consciousness (#HM2345) Theta wave consciousness (#HM2321)
Delta wave consciousness (#HM1785).

♦ **HM3477d Egoism** (Yoga, Zen)
Description One of the human troubles or sources of human trouble mentioned by Patanjali and referred to in Zen literature.
Broader Five kleshas (Yoga, Zen, #HH0390).
Related Egoism (#HH0130) Abhinvesa (Yoga, #HM3871).
Followed by Ignorance (Buddhism, Yoga, Zen, #HM3196).
Preceded by Attachment (Yoga, Zen, #HM3119).

♦ **HM3480c Paradisal memory** (Psychism)
Paradisal condition
Description A calm, detached, unitive mode of physical consciousness combined with a certain stage of spiritual development may lead to a condition in which the constricting effect of earthly existence is reduced. The fundamental nature of the universe and its order is glimpsed as a 'measureless communion of spirit', with experience of bliss or ecstasy.

♦ **HM3484d Inabat** (Sufism)
Conversion
Description In spiritual solitude the devotee sees God. Changes of this world or calamities from heaven to not affect him.
Context According to Shaykh Abu Sa'id ibn Abi'l-Khayr, this is the second of 40 stations or maqamat the Sufi must possess for his journey on the path of Sufism to be acceptable.
Broader Stations of consciousness - ibn-Abi'l-Khayr (Sufism, #HM2424).
Related Spiritual retreat (Sufism, #HH1875).
Followed by Tawba (Sufism, #HM4062).
Preceded by Niyyat (Sufism, #HM4901).

♦ **HM3485c Non-abiding** (Zen)
Wu-chu
Context Together with wu-hsing (formlessness) and wu-nien (no-thought), one of the three principles of Zen of Hui-neng.
Broader Human development (Zen, #HH1003).
Related No-mind (Zen, #HM2163) No-thought (Zen, #HM2500)
Formlessness (Zen, #HM0910).

♦ **HM3488d The relationship between man and nature** (Christianity)
Description In this stage, all this world is not a deception nor is it absolute reality. It is the condensation of the two enormous powers of the universe permeated with the will of God. The one power descends and wants to disperse and die. The other ascends and strives for freedom and immortality. All the rich, endless flow of appearances is the visible sign of the eternal collision of these two forces. It is man's responsibility to help the spirit of God on its course, to impose order on chaos, to contend with the powers of nature. If he is a labourer, then he is to till the earth, to build a house and in doing so sets free the spirit. If he is a warrior, he is to be pitiless, do battle without mercy. If he is a man of learning, he is to kill ideas and create new ones. If a mother, she is to choose the father as austerely as possible and give blood and milk to the next generation that the spirit might ascend. Every aspect of this world is part of one to be hailed as a comrade in arms.
Context The third stage in the action for the spiritual exercises of Nikos Kazantzakis.
Broader Saviors of God: the action (Christianity, #HM3462).
Preceded by The relationship between man and man (Christianity, #HM2766).

♦ **HM3489c Transcending mind** (Zen)
Description One of the aims of Zen is to go beyond the intellect or the mind, so that knowledge arises without thinking about what is under observation.
Broader Human development (Zen, #HH1003).

♦ **HM3490d Full understanding as the known** (Buddhism)
Comprehension of the known
Description This is knowledge in the sense of being known. It arises when the conditions of mentality-materiality or name and form are discerned by means of the round of karma and karma result and when uncertainty as to the three periods of time is abandoned so that past, future and present are understood in terms of death and rebirth linking. It occurs by observing specific characteristics of a given state, such as feeling having the characteristic of being felt or experienced. Its plane extends from delimitation of formations up to discernment of conditions.
Context One of three kinds of mundane full understanding.
Related Purification by overcoming doubt (Buddhism, #HH1187).
Followed by Full understanding as investigation (Buddhism, #HM4552).

♦ **HM3492c State of tranquillity** (Buddhism)
World of humanity
Description A state of calm and peace, this is fundamentally neutral when nothing in the environment upsets or excites. If one does not easily descend into lower worlds, those of the animals, then one is here in the world of humanity. The qualities are intelligence, excellence, acute consciousness, sound judgement, superior wisdom, distinguishing truth from falsehood, able to attain enlightenment, good karma. It is a rather unstable point of balance; it is easy to slide into anger, for example. Also it can be totally negative in laziness, unwillingness to make an effort, negligence and apathy. Like sleep, it is a useful state for restoration but one which must be left behind for effective functioning.
Context One of the ten worlds described in Nichiren Soshu Buddhism.
Broader Six paths (Buddhism, #HM1914) Ten worlds (Buddhism, #HM2657)
Nichiren shoshu buddhism (Buddhism, #HH3443).
Related Quietness (#HH0491).
Followed by State of rapture (Buddhism, #HM1973).
Preceded by State of anger (Buddhism, #HM2959).

♦ **HM3493d Ignorance of the nature of the essential self** (Buddhism)
Context The tenth fetter or character defect referred to by Buddha. This is in the second set of five fetters, when automatic responses are already overcome and the disciple is ready (arhat) to face and overcome the second set.
Broader Character defects (Buddhism, #HH0470).
Preceded by Notion of one's self as entity (Buddhism, #HM3612).

♦ **HM3496b Presence of Allah** (Sufism)
Narrower Allah (#HM4500).
Related Presence of God (#HM2961).

♦ **HM3497d Sentiment**
Description Perception involving an element of judgement coloured by emotion, sentiment occurs in the process of self-development on association with objects about which there is strong feeling and is accompanied by feelings of pleasure or pain. Sentiments are highly organized systems of emotions which arise in relation to the class of object which excite them.
Related Emotion (#HH0819).

♦ **HM3498e Clairvoyance** (Psychism)
Description Objective or subjective, an inner or psychic picture whether or a person, scene or object or lights, colours or other impressions, clairvoyance may be a view of some happening occurring far away - *clairvoyance-in-space* - or removed in time - *clairvoyance-in-time* (when the actual event is viewed); or it may be symbolic and need interpretation.
Narrower Emotional clairvoyance (#HM1118) Fourth dimensional clairvoyance (#HM4071).
Related Clairscent (Psychism, #HM0318) Visual hallucination (#HM0092)
Pseudo-hallucination (#HM4336) Clairgustance (Psychism, #HM0412)
Clairaudience (Psychism, #HM4333) Extrasensory perception (ESP, #HM2262).

♦ **HM3499c Age of the Holy Spirit** (Christianity)
Description This age is characterized by the fostering of monasticism and the life of contemplation, the age of the Eternal Gospel. It is associated with the Holy Spirit, the third member of the Trinity.
Context The third age, commencing, according to the Joachimites, about the year 1260 AD.
Broader Doctrine of three ages (Christianity, #HH0397).
Preceded by Age of the Son (Christianity, #HM3087).

♦ **HM3500c Jesus made to bear the cross** (Christianity)
Description The believer is aware that Jesus, bearing the weight of the cross, offered for mankind the death He was about to undergo. He considers the tribulations he will have to bear in this life and begs help in carrying these with patience and resignation. Loving Jesus more than himself, he repents with his whole heart and begs never again to separate himself from Him. He asks that he will love Jesus always and that Jesus will do with him as He wills.
Context The second station of the cross.
Broader Way of the cross (Christianity, #HM3516).
Followed by Jesus falls the first time under His cross (Christianity, #HM4009).
Preceded by Jesus condemned to death (Christianity, #HM2535).

♦ **HM3501d Third step: mankind** (Christianity)
Description Freeing oneself from the race, one fights to live through the whole struggle of humankind. Detaching itself from the animal, mankind struggled to stand upright, to coordinate inarticulate cries, to free the spirit, and to feed the mind. One looks back across the centuries to see hairy, blood-spattered beasts rising out of the mud to fall back in to the soil feeding the earth. One battles to give meaning to the confused struggles of man by enlarging one's heart to encompass one, then two, then ten centuries of toil, as many as one can bear. One gathers together in one's heart every unhappiness and lives it. One erects in one's mind, in an ocean of nothingness, a small island of meaning, by ploughing a field, kissing a spouse, studying a stone. In the third step one unites with humankind.
Context The third step in the march for the spiritual exercises of Nikos Kazantzakis.
Broader Saviors of God: the march (Christianity, #HM3439).
Followed by Fourth step: the earth (Christianity, #HM3570).
Preceded by Second step: the race (Christianity, #HM2953).

♦ **HM3502d Consolation** (Christianity)
Positive feeling towards God
Description This is experienced as feeling love for God, a profound peace and joy, and of being in harmony with one's self and with one's truth. It may arise from the devout "stirring of love" arising from contemplation or from external things. The Cloud of Unknowing indicates that the latter should be held suspect as it needs discernment to discover whether it is a help or a hindrance to growth.
Refs Byrne, Lavinia (Ed) *Traditions of Spiritual Guidance* (1990).
Related Desolation (Christianity, #HM5119) Spiritual discernment (Christianity, #HH3900).

♦ **HM3504d Emotional aversions** (Buddhism)
Context The fifth fetter or character defect referred to by Buddha. The first five fetters are marked by automatic response due to the following of old, unphilosophical mental habits. When these are overcome, the disciples is ready (arhat) to face the final five.
Broader Character defects (Buddhism, #HH0470).
Followed by Desire for life in form (Buddhism, #HM3583).
Preceded by Emotional desires (Buddhism, #HM3301).

♦ **HM3508c Subsequent cognition** (Buddhism, Tibetan)
Description This is apprehension of what has already been apprehended, arising subsequent to bare perception and depending on previous cognition of the same object rather than on fresh perception of the object itself. The initial moment of the sequence is the only one not dependent upon the previous moment and is therefore the only valid perception. Subsequent cognition may also arise following a valid inferential understanding, and also from memory of something validly known before, either through bare perception or inference. In this latter case even the first moment is subsequent cognition and therefore invalid.
Context One of the five invalid ways of knowing of Tibetan Buddhism.
Broader Ways of knowing (Buddhism, #HH0873).

♦ **HM3509c Asmita stage** (Yoga)
Alinga stage of the gunas — Atma
Description In this stage of samadhi, consciousness functions at a level where only awareness of the divine consciousness, of which particular objects or principles are modifications, remains. Although objects as separate entities exist they no longer have any meaning. The exalted consciousness functioning at the Atmic level is the last stage before attainment of Kaivalya. Transcendence of this plane leads to total destruction of individuality before a higher and more subtle form of individuality emerges. Despite the knowledge and bliss inherent at this level they become means of bondage and must also be given up before the final objective is achieved.
Context The last of the four stages of the gunas.
Broader Stages of the gunas (Yoga, #HM2805).
Related Asmita (Yoga, #HM2937).
Preceded by Ananda stage (Yoga, #HM3212).

♦ **HM3510g Dream wakefulness** (Hinduism)
Description Past experiences are recalled as though real now.
Context The sixth of seven descending steps of ignorance described in the Supreme Yoga, which veil *self-knowledge*. Each has innumerable sub-divisions.
Broader Veils of delusion (Hinduism, #HM3592).
Followed by Sleep (Hinduism, #HM2853).
Preceded by Dream (Hinduism, #HM3147).

♦ **HM3511g Hypnagogic consciousness** (Psychism)
Hypnogogic state — Hypnagogic imagery
Description A state intermediate between waking and sleeping, said to occur before the subconscious attains dominance over the conscious. There may be an awareness of fusing electric sparks, or clairaudient or clairvoyant perception which is forgotten as sleep takes over. Hypnagogic experiences occur more frequently than hypnopompic, although both hypnagogic and hypnopompic imagery occur suddenly and are not under the individual's control. The imagery is realistic but bizarre. A frequent experience is of hearing one's name called, arousing the person involved. There may be music playing (so realistic as to arouse the person to turn it off), the appearance of highly coloured geometrical images or distorted faces (visual images are short-lived and surrealistic), kinaesthetic experiences, and (less often) olfactory or tactile experience. Surveys relate degree of hypnagogic and hypnopompic imagery to favourable personality characteristics. Absence of such imagery is related to rigid defence against impulse. This is the opposite of findings related to high dream fantasy ratings.
Refs Reed, Graham *The Psychology of Anomalous Experience* (1988).
Related Hallucination (#HM4580) Reverie (Psychism, #HM0967)
Isakower phenomenon (#HM5124) Sensory translation (#HM3754)
Hypnopompic consciousness (Psychism, #HM3688).

♦ **HM3513c Divine essence** (Sufism)
Mohammad of one's being
Description This subtle organ corresponds to the divine centre or Eternal Seal - Muhammad was the Seal of Prophecy. The colour is green.
Context The last of seven subtle stages relating to the seven major prophets of Semitic monotheism as described by Simnani.
Broader Stations of consciousness - Simnami (Sufism, #HM2341).
Preceded by Divine nature (Sufism, #HM2246).

♦ **HM3514b Sokushin jobutsu** (Buddhism)
Buddhahood
Description The potentiality of becoming a Buddha is inherent in all beings and may be evoked so as to to achieve perfect enlightenment in the worldly body by a physical as much as a spiritual process. Such primordial enlightenment gives rise to an organic, pantheistic view of creation compatible with Shintoism.
Context A state described by the Shingon sect of Buddhism.
Related State of buddhahood (Buddhism, #HM1873).

♦ **HM3516b Way of the cross** (Christianity)
Stations of the cross
Description In the state of loving Jesus and sincerely repenting those times when he has abandoned Him, the believer performs the exercise of accompanying Christ on the journey to his death by contemplating the various stages on the way. Knowing that Christ was to die for love of him the believer is aware of wishing to die - and live - for His love.
Narrower Jesus dies on the cross (#HM3719) Jesus condemned to death (#HM2535)
Jesus nailed to the cross (#HM4136) Jesus made to bear the cross (#HM3500)
Jesus placed in the sepulchre (#HM3536) Jesus stripped of his garments (#HM3166)
Jesus helped to carry the cross (#HM3344) Jesus taken down from the cross (#HM3410)
Jesus meets his afflicted mother (#HM3611) Veronica wipes the face of Jesus (#HM3401)
Jesus falls the first time under His cross (#HM4009)
Jesus speaks to the daughters of Jerusalem (#HM2737)
Jesus falls the third time under the cross (#HM2289)
Jesus falls the second time under the cross (#HM3824).
Related Mystical Christ-consciousness (Christianity, #HM2880).

♦ **HM3517d Instinct**
Description A form of unconscious awareness not arising through experience and which is diminished by learning from experience, discovering connections and drawing conclusions. Instinctive behaviour is an automatic response to a given stimulation or situation, without any mental realization of the situation or the response to it. Behaviour is spontaneous rather than the result of deliberate mental processes.
Related Life instinct (#HH0750) Herd instinct (#HM2559)
Psychological imprisonment (#HM4539).

♦ **HM3518c Heart-awakening stage of life**
Saintly life — Fourth stage
Description The spiritual aspirant animates the qualities of true faith, love and surrender at the heart. His progress is typified in *bhakti yoga*, in the self-transcending service to others of classic *karma yoga*, and in the Mahayana Buddhist path of the *bodhisattva*; and by the practice of traditions such as *japa yoga*, continuous prayer, *Hesychasm*, *Hasidism* and *Sufism*. The subtle, non-physical dimensions of the body-mind begin to develop. This is the "saintly life" where there is transcendence of human interests and the true and full practice of religion commences, with the ability to surrender to the Divine. This leads to close communion with the Divine, self-transcending devotional communion expressed through heart-feeling and meditation and constant service. During meditation there is movement of the spirit current from descent in the frontal line to ascent in the spinal line, with steadiness of concentration at the "third eye". The error is to consider the self separate from but entirely dependent upon the Divine who is constantly called upon for aid. The fourth stage may become an end in itself as opposed to a means of transition.
Context The fourth of seven stages of life characterized by Master Da Free John.
Broader Seven stages of life (#HH0460).
Related Bhakti yoga (Yoga, #HH0337) Mantra yoga (Yoga, #HH0931)
Hesychasm (Christianity, #HH0205)
Meditation way of the bodhisattvas (Buddhism, #HM2769)
Great-vehicle higher-path awareness (Buddhism, Tibetan, #HM3048).
Followed by Mystical stage of life (#HM3191).
Preceded by Mental-volitional stage of life (#HM3464).

♦ **HM3520a Identification with God** (Sufism)
Itihad — Itasal — Jam
Description Also described in terms of the fusion of God's attributes, *Hulyl*, or the passing away into God's existence, *fana fel al Allah*, this is union with the Divine being. The process involves: *tawaja* - intense concentration; assimilation of the quality, will and desire attributes of the object of desire; stabilization of this state so that fusion of wills can take place. Through identification with various attributes, the Sufi gradually transcends the duality of subject and object. In union with the essence there is no relatedness to time or place - names, affiliations and concepts disappear. Then the awareness arises that all creation is based on love and that no other activity is worth while.
Related Fana (Sufism, #HM1270) Mushahada (Sufism, #HM3521)
Deification (Sufism, #HM0946) Unification with God (Sufism, #HM3864)
Islamic transformation awareness (#HM2850)
Ultimate Sufi state of consciousness (Sufism, #HM2318).

♦ **HM3521c Mushahada** (Sufism)
Mystical ecstasy — Mushahida — Contemplation on unitary experience — Islamic contemplative state — Spiritual vision of God — Mushahadat
Description At the beginning of the real journey to God there is contemplation or vision of God in a state of mystical ecstasy, when the Sufi becomes perfect for absolute unification with God. The God-given desire to behold his essence enables knowledge to become vision, vision to become revelation and revelation becomes contemplation as God reveals divinity in the heart of the seeker. This last becomes existence with and in God.
This is one of the most important states of consciousness in Sufism. There is vision of unveiled, divine beauty in which the beholder and the beheld are not two, but one.
The experience of this vision depends on the devotee's love and faith in God and illumination of the heart by the grace of God. In contemplation of God everything else is obliterated.
Context The seventh psychic state listed by A Reza Aresteh as progress on the inner self through divine attraction, *kedesh-jazba*, the outer self already having been purified through conscious effort, *kushesh*. The tenth stage in a systematic account of various stations of the Sufi spiritual path.
Broader Psychic states (Sufism, #HM4311) Mystic stations (Sufism, #HM3415)
Higher states of consciousness (Sufism, #HM2365).
Related Ma'rifa (Sufism, #HM2254) Identification with God (Sufism, #HM3520).
Followed by Wajd (Sufism, #HM3650) Yaqin (Sufism, #HM4426)
Itmianan (Sufism, #HM0050).
Preceded by Uns (Sufism, #HM1957) Love of God (Sufism, #HM4273).

♦ **HM3522d Rotoo** (Japanese)
Bewilderment
Description The individual is shocked out of spiritual lethargy and complacency through establishment of a sense of insecurity. Habitual affiliations and social position are closed to him and he faces the question "who am I ?". Anxiety at rejection and isolation increases. There is a need to beg for acceptance from others. When acceptance finally occurs a cathartic sense of gratitude arises. Rotoo provides an opportunity for interaction between strangers which leads to regard for humanity in others. The effect is a fostering of warmth and spontaneity.

♦ **HM3526b Savitarka samadhi** (Hinduism)
Description In this level of *samadhi* the mind alternates between knowledge based on words, real knowledge and that ordinary knowledge based on reasoning or perception through the senses, all being present in a mixed state.
Broader Samadhi (Hinduism, Yoga, #HM2226).
Related Artha (Hinduism, #HH0353).

♦ **HM3527c Two-fold knowledge of truths** (Buddhism)
Knowledge of ideas — Knowledge of penetration
Description The Path of Purification distinguishes two ways of knowing about the four noble truths. There is the mundane knowledge as idea (awakening knowledge), which learns about cessation and the path through what it hears. It occurs as overcoming obsessions. Knowledge about suffering prevents the false view of individuality; knowledge about origin prevents the false view of annihilation; knowledge of cessation prevents the false view of eternity; knowledge of the path prevents the false view of inaction (seeing action is being inefficacious morally). Again, knowledge of suffering prevents wrong theories as to results, not attributing permanence, beauty, pleasure, individuality to the aggregates; knowledge of origin prevents wrong theories of cause, such as reasons for the existence of the world; knowledge of cessation prevents wrong theories about cessation, for example that final release arises in the immaterial sphere; knowledge of the path prevents wrong theories about the path of purification, such as indulgence in pleasures of the senses or in self-mortification. And there is supramundane, penetrating knowledge, which penetrates the four truths by making cessation its object. Then whoever sees suffering sees the origin of suffering, sees the cessation of suffering, sees the way leading to cessation of suffering.
Related Four noble truths (Buddhism, #HH0523).

♦ **HM3529c Citta** (Hinduism, Buddhism)
Higher mind — Thought
Description Citta is the mind through which consciousness functions. Through citta functions *buddhi*, or discrimination, and the mind is helped to know and understand objects, for the buddhi illuminates citta. While buddhi is functioning through the mind it is not possible to know pure consciousness. When it is turned upon itself it illuminates its own nature, consciousness. In *samadhi*, consciousness passes from one level of citta to another. When it has penetrated the deepest level of citta and transcended even this its true nature is realized - perceiver, perceived and perception itself are all one.
Related Sahasrara (Yoga, #HM3398) Vritti (Hinduism, #HM2692)
Alaya-vijnana (Buddhism, #HM2730) Samadhi (Hinduism, Yoga, #HM2226)
Buddhi awareness (Hinduism, #HM2099).

♦ **HM3530d Tapa**
Anxiety
Description Whatever the pleasure, indulgence or happiness in human life it is always accompanied by anxiety, for none of these things is permanent and there is always the fear of losing what one has. This anxiety may not be conscious but its effect is to poison life.
Context The second affliction inherent in human life.
Related Anxiety (#HM2465).
Preceded by Fear (Christianity, #HM3311).

♦ **HM3536c Jesus placed in the sepulchre** (Christianity)
Description The believer is aware of the stone that enclosed Jesus but that He rose again on the third day; and begs that he will be raised in glory with Him on the last day, united with Him in heaven where he will praise and love Him for ever. Loving Jesus more than himself, he repents with his whole heart and begs never again to separate himself from Him. He asks that he will love Jesus always and that Jesus will do with him as He wills.
Context The fourteenth station of the cross.
Broader Way of the cross (Christianity, #HM3516).
Preceded by Jesus taken down from the cross (Christianity, #HM3410).

♦ **HM3538c Experience of revelation**
Description According to Swami Omkarananda, revelation is reason purged of impurities, liberated from limitations; it is the timeless knowledge of timeless reality in the world of time-space experience gathered through the mind and senses, the full functioning of inner consciousness.
Related Revelation (#HH1028) Passivity experiences (#HM7203).

♦ **HM3539d Ar-Rahman** (Sufism)
Description The believer is aware of Allah as beneficent, impartially bestowing blessings and prosperity on all beings. Rahman denotes the will of the total good of Allah. By using freedom of choice for the good of himself and others, the servant finds in himself the light of Rahman.
Context A mode of mystical awareness in the Sufi tradition associated with the first of the ninety-nine names of Allah.
Broader Ninety-nine names of Allah (Sufism, #HH2561).
Followed by Al-Rahim (Sufism, #HM4043).

♦ **HM3542b Spiritual experiences** (Sufism)
Psychological assessment of mystic experience
Description Spiritual experience in Sufi terms cannot be explained psychologically, as psychology is concerned with the mind and the spiritual is beyond the mind. Nor can it be interpreted in language. This means that neither mystic experience related to fear or love of God, nor to awareness of the pure self, can be interpreted according to science. They are are trans-subjective, as are some other higher states (the soul's contemplation of the attributes of God, communication with the Beloved, revelation through illumination); they are not ordinary states or emotions. There is no accepted standard for psychological assessment of mystical states.
Broader Human development (Sufism, #HH0436).
Related Deautomatization and the mystic experience (#HM4398).

♦ **HM3546e Psychic presence** (Psychism)
Description Awareness of some psychically sensed life form as opposed to an inert energy field.
Related Sense of presence (#HM1166).

♦ **HM3550f Moral morbidness**
Description An unusually depressed state may arise with respect to moral or religious status. This may be true insanity, or it may be a state in which the sane person (often an adolescent) comes under some moral distortion. Examples are: introspection that applies excessively severe standards to one's self as opposed to others; obsessive emphasis on exact performance of trifles appearing as duties, with expectation of perfect behaviour in the self and others; compelling, and what the person himself admits are unreasonable, demands for certitude and refusal to make assumptions; excessively emotional or self-annihilating devotion to an ideal, a cause, or another person; extreme under-sensitivity and callousness. In all but the last case, the state has been shown to arise in sensitive people unable to assimilate gloomy theological theories which the more robust may profess to believe while leading normally happy or humdrum lives.

♦ **HM3553c Fearlessness**
Description There is an insensibility to danger in situations when even the most courageous would feel fear, despite behaviour which belied the fact. This may arise simply from ignorance or inexperience, as in a child's attitude to fire; or in a situation when normal responses are inhibited, as may arise in a crowd when the presence of others lulls into false security. On the other hand, a permanent state of fearlessness seems to exist in certain individuals simply from finding happiness under any conditions and, perhaps because of religious, moral or artistic enthusiasm, where the risks and the consequences of action have no intrinsic significance. Since fear is destructive and paralyzes activity, fearlessness increases confidence and the chances of success. In particular, religious conviction which is aware of infinite power and goodness behind every occurrence banishes strain, effort, worry and fatigue, with considerable increase in happiness and effectiveness.
Related Heroism (#HH0929) Fear (Christianity, #HM3311)
Treasures of the godly (#HH0364).

♦ **HM3554c Mushin** (Japanese)
Freedom from desire — Lack of deep feeling
Description This expression has two meanings - it can either refer to lack of aesthetic sense (the converse of *ushin*), or to freedom from mundane desires and attachments as achieved in the enlightenment of Buddhism.
Related No-mind (Zen, #HM2163) Ushin (Japanese, #HM2762).

♦ **HM3556d Faculty of suffering** (Buddhism)
Dukkhindriya (Pali)

♦ **HM3558b Immanence**
Description This is a state of divine indwelling.
Related Transcendence (#HH0841).

♦ **HM3559e Light-induced experience**
Description A mystic experience or altered state of awareness may be achieved, intentionally or spontaneously, by staring at or concetrating on a light source such as a candle. This is the basis for some forms of meditation and yoga.
Related Phosphenism (#HH6321) Taraka yoga (Yoga, #HH0452)
Induction technique (#HH0025).

♦ **HM3560d Non-hatred as a wholesome root** (Buddhism)
Adoso kusala mulam (Pali)
Broader Non-hatred (Buddhism, #HM2744).

♦ **HM3563e Covetousness** (Buddhism)
Abhijjha (Pali)
Broader Greed (Buddhism, #HM3283).

♦ **HM3567c Path of sanctifying intelligence** (Judaism)
Description This state is represented as the mother of faith, the foundation of wisdom.
Context The third among the spiritual paths expressed in the Jewish Kabbalah.
Broader Paths of wisdom (Judaism, #HM2509).
Followed by Path of receiving intelligence (Judaism, #HM3102).
Preceded by Path of the illuminating intelligence (Judaism, #HM2612).

♦ **HM3570d Fourth step: the earth** (Christianity)
Description Having heard the ego call, the race call, mankind call in one's heart one, realized it is earth calling, her trees, waters, animals, men and idols. Earth rises up and one sees her entire body for the first time. The journey from flaming chaos, passing through an infinity of time, the womb of matter cools, the stones become alive. A single leaf emerges and sucks at the whole universe so that it might pass through its body to turn it into flower, fruit and seed that it might be deathless. At any moment, all may sink into chaos but death is overcome. Life passes beyond plants, fishes, birds, beasts and apes. Man is created. Now someone else struggles to be rid of humankind. From body to body existence hunts down its untamed spouse, eternity.
Context The fourth step in the march for the spiritual exercises of Nikos Kazantzakis.
Broader Saviors of God: the march (Christianity, #HM3439).
Preceded by Third step: mankind (Christianity, #HM3501).

♦ **HM3571e Psychic transfer** (Psychism)
Description Abundant psychic energy is unknowingly received, changing attitudes, disturbing the emotional balance or healing the body.
Related Psychic attack (Psychism, #HM0837).

♦ **HM3574e Attunement**
Psychic attunement (Psychism) — Problem solving state
Description In psychic reading, the reader synchronizes with the brain rhythm, aura and body chemistry of the person being "read", or deliberately senses his conscious and subconscious mind. The term is also used to describe a number of psychic states, such as a stilling of the conscious mind when sub- and superconscious minds are connected to higher planes; and an awareness of healing energy from another. The psychic person realizes he or she has reached a level beyond his or her mental awareness when information is nevertheless valid and useful.

♦ **HM3575c Inner peace**
Peace of mind — Spiritual peace — Shanti — Shalom — Salaam
Description So valued as to be implicit in habitual greetings, peace is a spiritual calm as well as a social benefit, a calm which can be maintained under all adverse external circumstances. This is the "shanti" of the Hindu sacred texts (often repeated three times, to indicate inner peace at three levels). The state is of tranquillity, quiet and calmness of mind; there is absence of passion, aversion of pain and indifference to objects of sense which cause pleasure and pain. There is

detachment unmoved by the results of activity.
Related Non-action (HH1073). Peacefulness is an essential feature of the Christian character. It arises in the consciousness of Christ's victory over the world and over all that inhibits spiritual wellbeing; and is manifest in an inward wholeness arising on acceptance of the gospel as a law of life and explanation of the universe. Knowledge of the nearness and providence of God, of his care for all and his power, leads to freedom from personal anxiety. The Christian message and the empty tomb underline an ultimate purpose in life and its vicissitudes, and lead to intellectual repose, deliverance from perplexity of mind. There is also joy, love having found its true and enduring object and so overcome all inner unrest; the restfulness is further developed in surrender of the will to God. The peace which "passeth all understanding" of St Paul is synonymous with the Kingdom of Heaven within - by no means marked by absence of disturbance, but strenuous and harmonious action compared with the "concord of melodious sounds". "Giving the peace" is common in present day Church services and peacefulness of temper is important in social interchanges, the spirit of love working for the wellbeing of the community. However, there are occasions when conflict and resistance to evil are inevitable. Thomas Aquinas indicates that the inner peace of this world is relative as opposed to the absolute peace of the next.

♦ HM3576d Al-Muqtadir (Sufism)
Description The believer is aware of Allah as the creator and controller of power, allotting limited and controlled power to creation. Man may seem to be powerful and creative but that power comes from al-Muqtadir. Submitting to His will, one receives power which no force can overcome and for which one must show one's thanks in repentance, and in compassion, justice, forgiveness and generosity to all.
Context A mode of mystical awareness in the Sufi tradition associated with the seventieth of the ninety-nine names of Allah.
Broader Ninety-nine names of Allah (Sufism, #HH2561).
Followed by Al-Muqaddim (Sufism, #HM5331).
Preceded by Al-Qadir (Sufism, #HM1500).

♦ HM3579c Nasut (Sufism)
Description In this state, the traveller lives according to the law of God, the *shari'a*, attaining moral perfection before commencing truly on the spiritual path.
Context The first of four stages on the Sufi spiritual path.
Broader Divine path (Sufism, #HM3371).
Related Sharia (Islam, #HH0738).
Followed by Malaqut (Sufism, #HM0116).

♦ HM3583d Desire for life in form (Buddhism)
Context The sixth fetter or character defect referred to by Buddha. This is in the second set of five fetters, when automatic responses are already overcome and the disciple is ready (arhat) to face and overcome the second set.
Broader Character defects (Buddhism, #HH0470).
Followed by Desire for formless life (Buddhism, #HM1261).
Preceded by Emotional aversions (Buddhism, #HM3504).

♦ HM3587d Self assertion
Self subjection — Self sentiment — Self abasement
Description These qualities are manifest from early childhood, particularly in response to admiring applause or to the opposite, being ignored. They are essentially social and depend upon the presence of others - or at least on the idea of others if no-one else is actually present. Something or someone superior is approached submissively, in the case of something created (a work of art, for example) this implying admiration for the creator. Ultimately this implies admiration and respect for the creator of all things, for God. Even within such submission or self abasement there is self assertion in that the individual feels his admiration to be of worth or even necessary to what is admired. There is an inner struggle between assertion and abasement which results in an idea of the self in relation to others and in the development of self sentiment incorporating positive and negative self-feelings deriving from responses of others to assertion and abasement. Mental disturbance may lead to exaggerated symptoms of self abasement, shrinking from the observation of others, or to exaggerated self display.
Related Assertiveness training (#HH0994) Self-awareness (Psychism, #HM2436).

♦ HM3589c Ineffability
Description Common to the mystic experience is the extreme difficulty of putting such an experience into words. It has been compared with attempting to describe a sunset to a blind man or a Bach cantata to someone who is deaf. Descriptions usually emphasize what the experience is not, as in the *via negativa* of Christianity and the *neti, neti* of the Hindu and Buddhist traditions.
Refs Yandell, Keith *Some Varieties of Ineffability* (1979).
Related Via negativa (#HH1171) Mystical cognition (#HM2272).

♦ HM3590g Feeling consciousness (Psychism)
Emotional consciousness — Affection
Description The characteristic aspect of consciousness which is not knowledge or will, but which is a differentiated feeling or emotion, is referred to as affection.
In esoteric or psychic terms, emotion is severally described as: (1) a physically and mentally experienced, concentrated energy consciousness causing physical response in the body (flushing, sweating, etc), often in reaction to external stimulus; (2) a manifestation of the vital life force in every living thing; (3) rapidly changing moods and attitudes arising without conscious intention having their basis in cosmic consciousness.
Related Emotion (#HH0819) Affection (#HH0422) Affection (Hinduism, #HM9845).

♦ HM3592b Veils of delusion (Hinduism)
Descending steps of ignorance
Description According to the Supreme Yoga, seven descending steps of ignorance or delusion veil *self knowledge* and lead to egotism and bondage. Each has innumerable sub-divisions.
Narrower Dream (#HM3147) Sleep (#HM2853) Wakefulness (#HM2567) Wakeful dream (#HM2014) Great wakefulness (#HM3055) Dream wakefulness (#HM3510) Seed state of wakefulness (#HM2159).
Related Planes of wisdom (Hinduism, #HM3298) Ignorance (Buddhism, Yoga, Zen, #HM3196).

♦ HM3594d Frenzy
Furor — Amok — Ethnopsychosis — Spirit possession — Mania
Description The mind is enveloped in a blinding flame, overwhelming normal consciousness and raising the individual to such a pitch of excitement that he can no longer control himself. This seizure of violent and uncontrollable agitation may arise in a number of ways:
- *Martial furor*, the result of combat, may result in an orgy of killing once a battle has been resolved. The fact that this violence may be directed towards women leads to the supposition that the warrior is in uncontrolled rage agaist his mother, against women in general and ultimately towards his own weakness which women symbolize.
- *Amok*, ethnopsychosis or psychotic symptoms which are culture-specific, usually affects an individual already mentally unstable who, in a loss of sense of social order, is faced with a threatening or frightening situation. The result is a sudden wild attack on surrounding people, animals or objects, followed by stupor and subsequent lack of memory of what has happened.
- *Spirit possession* during exorcistic rites may commence with a dream-like somnambulistic trance which gives way to total loss of control or frenzy and convulsions, a state followed by exhaustion and torpor. This latter form of frenzy is often seen in group behaviour and seems to be facilitated by group participation.
Plato distinguishes several phases of frenzy or mania: that of the seer, unveiling the future; that of the consecrated mystic who absolves individuals from their sins; that of the poet, possessed by the muses; that of the philosopher. In this state one remembers the past life of the soul.
Marsilio Ficino elaborated four modes of furor:
- *Religious furor* or theomania;
- *Prophetic furor*, resulting in oracular prophecy;
- *Poetic furor*, culminating in poetical or musical expression;
- *Erotic furor* or erotomania.
Further modes added by his commentators include *melancholic furor*, with deep depression and grief sometimes leading to insanity, and martial furor, described above.
Narrower Theomania (#HM1027) Erotic frenzy (#HM0036).
Related Mania (#HM2787) Spirit possession (#HH0056) Religious enthusiasm (#HM2146) Dance-induced experience (#HM3213).

♦ HM3595d The relationship between God and man (Christianity)
Description In this stage the ultimate most holy form of theory is action. The spark that leaps from generation to generation is not to be passively regarded but to be leapt on and burned with. Flesh, race, man, plants and animals have been joined in their march, light and dark have been embraced to reach this point, the beginning. This, the starting point, is arrived at with new eyes and ears, with a new sense of taste, smell and touch and with a new brain. Human duty lies not in interpreting or casting light on God's march but adjusting, as much as possible to his rhythm. In this stage, the virtues and faces other ages have given to God have been gone beyond. This is not the time to wait patiently for certain victory. It is not the time to wait smugly trusting God to pity and save. The self and God are united. From the blind worm in the depths of the ocean to the endlessness of space only one person struggles and is imperilled: the self. Within the small and earthen breast encasing the self, only one thing struggles and is imperilled: the universe.
Context The first stage in the action for the spiritual exercises of Nikos Kazantzakis.
Broader Saviors of God: the action (Christianity, #HM3462).
Followed by The relationship between man and man (Christianity, #HM2766).

♦ HM3596d Sense of humour
Description One of the most essential qualities of a complete human being and yet one of the least easy to explain, this involves perception of the incongruous and inconsistent in a situation, appreciating and gaining joy from the whimsical and what might otherwise be a source of sadness. It is associated with alertness and breadth of mind as well as a sense of proportion, although its focus may be the exaggeration of one feature or characteristic out of all proportion, thus making it absurd.
Related Humour (#HH1031).

♦ HM3597d Shibumi (Japanese)
Shibui — Aesthetic understatement
Description The experience of a subtle, unobtrusive, and deeply moving beauty, whether in aesthetic matters or in human behaviour. It is the form of beauty appreciated by those satiated by more obvious types of beauty. It may also be experienced in groups (especially in sports) where the value of a participant derives from his ability to contribute unobtrusively without seeking a central role.
Related Sabi (Japanese, #HM4350) Artistic education (#HH0029).

♦ HM3599c Self-knowing
Description Since the knower cannot be that which is known, self-knowledge is essentially a paradox, as in Zen: "the seeing that is no seeing". There is, however, a conscious and sentient awareness which perceives without distinguishing definite objects, as for example young babies before the awakening of mental operations. This is consciousness of self and remains throughout life, knowing without an object of knowledge, self-knowledge.
Related Self-knowledge (#HH0312).

♦ HM3602c Inward purity (Sufism)
Safa
Description Central to mystic discipline is purification of the soul; the Sufi seeks to leave impurity behind and, free from this impurity, to attain spiritual perfection. Rather than the following of theoretical principles or of doctrinal knowledge, inward purity is themeans for realization of moral ideals and higher values; living a virtuous life leads to disciplining of the soul. Purity being characteristic of those who love, perfect purification brings rejoicing in the vision of God and affinity with him. The soul sees nothing but God and eventually becomes one with him.
Broader Human development (Sufism, #HH0436).
Related Purity (#HH0719) Tasawwuf (Sufism, #HM0575).

♦ HM3603b Silence
Description This deep state of consciousness may be reached by meditation, so the body is relaxed, the mind controlled, time no longer exists and there is awareness of God. Mystical or divine silence cannot be used but only experienced. Speech is said to be the organ of the present world but silence is from the gods, the mystery of the world to come. One who is silent in this sense is at one with God, whatever his religious tradition. Unanimous also is the agreement that this silence cannot be described in speech.
Refs Paramananda, Swami *Silence As Yoga* (1974).
Related Voice of silence (#HM2227) Aesthetic silence (#HM1664) Practice of silence (#HH0983).

♦ HM3604c Self-cognition (Sufism)
Description Described by the poet Attar as the final and highest state for any human being and accounting for his nobility, self-cognition is the goal of Sufism, when man, the most noble of creatures, knows himself; and whoever knows himself truly knows his God.
Related Introspection (#HH0824) Human development (Sufism, #HH0436).

♦ HM3605c Inward experience of the real (Sufism)
Description In a state of ecstasy the mystic communicates with God and feels kinship with him. The mystic does not attempt explanation of this, it is totally personal and cannot be disclosed. God is contemplated unceasingly and divine presence is experienced in all things.

♦ HM3606d Premonition (Psychism)
Foresight — Prescience — Presentiment
Description This is an awareness of future events, sometimes accompanied by anxiety or foreboding. Prescience can refer, in particular, to a conscious act of divining the future. It is

sometimes remarked as correct guidance when an internal conviction is held as opposed to guessing in the mind.
Related Prophecy (#HH0559).

♦ **HM3607c Mystical states** (Sufism)
Context The seventh of ten states acting as gateways orienting the mystic on the journey to the Absolute, as described by Ansari.
Broader Stations of consciousness - Ansari (Sufism, #HM2317).
Followed by Sanctity (Sufism, #HM4041).
Preceded by Valleys (Sufism, #HM2874).

♦ **HM3608d Sense of mystery**
Description To feel a sense of mystery is to experience that wonder which unites the person experiencing it with the unknowable mystery of life itself.
Related Mystery (#HH0303).

♦ **HM3609d Critical awareness**
Description Critical awareness is characterized by an ability to think creatively and flexibly. It is encouraged by inquiry-based learning rather than training to carry out specific tasks. The basis may be philosophical questions such as what makes something true, real, good or fair. It is reflection, thinking about thinking, that brings such awareness, which is developed by a climate which encourages: the asking of "thinking" questions; the careful listening to others in a climate of mutual respect, trust and tolerance; the evaluation and criticism of ideas without abusing those who put them forward; the building up and correction of one's own ideas when faced with reasonable criticism.
Related Critical-analytical subtle contemplation (Buddhism, #HM2659).

♦ **HM3611c Jesus meets his afflicted mother** (Christianity)
Description The believer is aware of the sorrow that Jesus and His mother felt at this meeting. He begs that he may truly love the mother of Jesus and also her intercession that he may continually remember Christ's passion. Loving Jesus more than himself, he repents with his whole heart and begs never again to separate himself from Him. He asks that he will love Jesus always and that Jesus will do with him as He wills.
Context The fourth station of the cross.
Broader Way of the cross (Christianity, #HM3516).
Followed by Jesus helped to carry the cross (Christianity, #HM3344).
Preceded by Jesus falls the first time under His cross (Christianity, #HM4009).

♦ **HM3612d Notion of one's self as entity** (Buddhism)
Context The ninth fetter or character defect referred to by Buddha. This is in the second set of five fetters, when automatic responses are already overcome and the disciple is ready (arhat) to face and overcome the second set.
Broader Character defects (Buddhism, #HH0470).
Followed by Ignorance of the nature of the essential self (Buddhism, #HM3493).
Preceded by Spiritual pride (Buddhism, #HM2852).

♦ **HM3614c First individuation of essence** (Sufism)
Description This pure unity , pure receptivity, englobes all receptivities - whether of deprivation or of attribution of all attributes; the first unity of ontological priority and pre-eternity, the second of ontological posteriority, eternity. In the latter case, the appearance of Being is clothed in the names and attributes of divinity and sovereignty, but this does not entail multiplicity of existence.
Context The seventeenth illumination of Jami.
Broader Fountains of light (Sufism, #HM3039).
Followed by Beyond individual distinctions (Sufism, #HM2681).
Preceded by Manifestation of the essence through veils (Sufism, #HM2642).

♦ **HM3615f Abaissement du niveau mental** (Jung)
Description Jung described this as a relaxing and letting go of psychic restraints, and one which corresponded "pretty exactly" to the primitive state of consciousness in which myths were formed originally. Concentration and attention are lacking, consciousness being of reduced intensity. There may be unexpected contents arising from the unconscious. The play of opposites is released and a relative reversal of values brought about. Latent psychotic tendencies may emerge. Although such a state arises naturally in mental illness, it may also be consciously fostered or arise under the influence of certain drugs. Abaissement du niveau mental normally accompanies vision. It has been compared with the *soul loss* of primitive peoples.
Refs Adler G, Fordham M and Read H (Eds) *The Collected Works of C G Jung*; Samuels, Andrew et al *A Critical Dictionary of Jungian Analysis* (1986).
Related Loss of soul (#HH0210).

♦ **HM3616c World of imagination** (Sufism)
Moses of one's being
Description This is the Secret, the point of supraconsciousness. Here there are spiritual monologues such as those in which Moses was participant. The colour of this stage is white.
Context The fourth of seven subtle stages relating to the seven major prophets of Semitic monotheism as described by Simnani.
Broader Stations of consciousness - Simnami (Sufism, #HM2341).
Followed by World beyond form (Sufism, #HM4570).
Preceded by World of spiritual perception (Sufism, #HM3974).

♦ **HM3617d Vijna** (Buddhism)
Vinnana (Pali) — Vijnana — Rnam shes (Tibetan) — Perception — Understanding — Consciousness
Description Perception with mental knowledge, this is contrasted with *prajna*, or transcendent wisdom which transcends the knowledge of things and of the mind. Vijna is that perception which apprehends what can be communicated by sense objects without classifying or reacting to such perception. It is equated with the six sense consciousnesses (five senses and mind).
Related Sanjna (Yoga, #HM3160); Jna (Buddhism, #HM5321); Essential wisdom (#HM0107); Alaya-vijnana (Buddhism, #HM2730); Perception aggregate (Buddhism, #HM4143); Bare perception (Buddhism, Tibetan, #HM4651).

♦ **HM3618g Lucid dreaming**
Creative dreaming
Description In *lucid dreaming* one is fully aware that one is dreaming while feeling fully awake and conscious. Reasoning and remembering may be clear, one may even change the "plot". One can train one's self to dream lucidly virtually at will, and to indicate to others that one is dreaming lucidly. This can lead to *creative dreaming* - that is, the shaping of one's own dreams and controlling their subject matter - which is the result of acquired abilities. Intervention may be in the original moment or immediately after waking, to produce an outcome which leaves the dreamer in control. It results in expanded consciousness and a better understanding of one's self and the uniqueness of one's personality.
Refs Ornstein, Robert *The Psychology of Consciousness* (1986).
Broader Dreams (#HM2950).
Related Waking dreams (#HM1376); Dream yoga (Buddhism, #HH4087); Escapist daydreaming (#HM0248); Free-floating fantasy (#HM1011); Spontaneous out-of-body experience (#HM0332).

♦ **HM3619a Unconditional nirvikalpa samadhi**
Bhava samadhi — Sahaj samadhi — Seventh stage
Description There is identification with the transcendent, radiant being in which all phenomena are seen as temporary, non-binding modifications of this all-inclusive divine being. The divine self is realized beyond the view point of the physical body, the mind or psyche, and the independent personal consciousness. When phenomena arise to notice from this formless and unqualified presence or love-bliss there is ecstasy of perfect spontaneity in *sahaj samadhi*. If no phenomena arise, then there is unconditional *nirvikalpa samadhi* or *bhava samadhi*. Although this stage has been entered by adepts of a number of spiritual traditions, they are few in number and even fewer have taught from the viewpoint of this stage.
Context The last of seven stages of life characterized by Master Da Free John; it is the foundation and aim of the yoga of progressive incarnation when both perception and conception of psycho-physical existence are dissolved in divine translation.
Broader Seven stages of life (#HH0460); Samadhi (Hinduism, Yoga, #HM2226).
Related Atmanubhava (Hinduism, #HM1195); Nirvikalpa samadhi (Hinduism, #HM2061).
Preceded by I-transcendent stage of life (#HM3060).

♦ **HM3620d Revulsion** (Yoga, Zen)
Dvesa — Repulsion
Description This state arises as a natural response to a source of pain or unhappiness. It is the reverse of attraction but similar in its ability to bind the individual. In fact, it is harder to transmute than attraction which can be transformed into impersonal love and as such no longer binds.
Context One of the five human troubles, kleshas or afflictions, the causes of misery mentioned by Patanjali and referred to in Zen literature.
Broader Five kleshas (Yoga, Zen, #HH0390); Afflictions (Yoga, Hinduism, #HH1047).
Followed by Attachment (Yoga, Zen, #HM3119).
Preceded by Possessiveness (Yoga, #HM2911).

♦ **HM3623d Negative intellect** (Leela)
Durbuddhi
Description Denying the presence of God in whatever object implies denial of reality. The player is in doubt, he does not follow the law of dharma and is sucked back to nullity from where he has to make his way again.
Context The 61st state or square on the board of *Leela*, the game of knowledge, appearing in the seventh row.
Broader Seventh chakra: plane of reality (Leela, #HM0754).
Followed by Nullity (Leela, #HM4512).

♦ **HM3627d Sensual plane** (Leela)
Desire — Kama-loka
Description This is the plane of desires, worthy or unworthy, without which no creation is possible. The player has to pass through this plane before starting out on the second level. He may reach it through exploring the the first chakra or through ignorance.
Context The ninth state or square on the board of *Leela*, the game of knowledge, appearing in the first row.
Broader First chakra: fundamentals of being (Leela, #HM4103).
Related Desire (Hinduism, #HM7800).
Preceded by Ignorance (Leela, #HM0310).

♦ **HM3630c Principal insights** (Buddhism)
Great insights — Contemplations — Discernment
Description The 18 principal insights, of which the first 7 are referred to as the 7 contemplations, are enumerated in Hinayana Buddhism.
1. Developing contemplation of impermanence implies abandoning the perception of permanence.
2. Developing contemplation of pain or ill implies abandoning the perception of pleasure or bliss.
3. Developing contemplation of not-self implies abandoning the perception of self.
4. Developing contemplation of dispassion or disgust implies abandoning delighting.
5. Developing contemplation of fading away or dispassion implies abandoning greed or passion.
6. Developing contemplation of cessation implies abandoning originating.
7. Developing contemplation of relinquishment implies abandoning grasping or clinging.
8. Developing contemplation of destruction or extinction implies abandoning the perception of compactness or density.
9. Developing contemplation of fall of formations implies abandoning accumulation of karma or exertion.
10. Developing contemplation of change or perversion implies abandoning the perception of lastingness or fixedness.
11. Developing contemplation of the signless implies abandoning the sign.
12. Developing contemplation of the desireless implies abandoning desire or hankering.
13. Developing contemplation of voidness or the empty implies abandoning misinterpreting or conviction.
14. Developing insight into states (higher understanding) implies abandoning misinterpreting or conviction due to grasping at the core or clinging to essence.
15. Developing correct knowledge and vision (knowledge and discernment of the true nature of things) implies abandoning misinterpreting or conviction due to confusion.
16. Developing contemplation of danger or tribulation implies abandoning misinterpreting due to reliance or conviction of attachment.
17. Developing contemplation of reflection implies abandoning non-reflection.
18. Developing contemplation of turning away (separation from the round of births) implies abandoning misinterpreting or conviction due to bondage or fetters.
Refs Buddhaghosa, Bhadantacariya *The Path of Purity* (1975); Buddhaghosa, Bhadantacariya *The Path of Purification* (1980).
Related Imperfection of insight (Buddhism, #HM9722); Special insight and preparations as states (Buddhism, #HM2623); Purification by knowledge and vision of what is the path and what is not the path (Buddhism, #HH4007).

♦ **HM3631d Alchemical reddening** (Esotericism)
Citrinistas — Xantosis — Rubedo
Description The beginning of this third and final stage of the alchemical process is marked by solar consciousness of omnipresence, the experience of matter in ecstasy, named as the

appearance of gold in the metallurgical symbolism used. Whereas the previous stage, the work of whitening, implies an ecstatic ascent through the insights of synthesis, in this stage there is a descent or return through which the world is transformed and sanctified. There is a union of light and darkness, symbolized by purple as the colour of royalty — a royalty faithful to the needs of the earth. The nature which delights in itself is subordinated to the nature which is able to surpass itself. The processes of the world are transmuted into a cosmic liturgy. The alchemist thenceforth functions as a secret king, a consciously central being relating heaven and earth and ensuring the good order of things. Although dead to himself, he becomes an inexhaustible source of nourishment.
Context The third stage of the alchemical work.
Broader Alchemy (Esotericism, #HH0221).
Preceded by Alchemical whitening (Esotericism, #HM6328).

♦ **HM3633d Al-Mughni** (Sufism)
Description The believer is aware of Allah as the enricher. He decides who shall be rich and who poor – He may make the poor rich or the rich poor; this is because the world is a testing ground. Will riches make one arrogant ? Will poverty tempt to doubt and complaint ? Riches should be accompanied by thankfulness and generosity, knowing they are not one's own. Poverty should be marked by hard work to better one's situation even though one does not succeed. The greatest riches are not material goods, but knowledge and faith. These may be spent without decrease and are not devalued with time – they stay with us beyond the grave.
Context A mode of mystical awareness in the Sufi tradition associated with the eighty-ninth of the ninety-nine names of Allah.
Broader Ninety-nine names of Allah (Sufism, #HH2561).
Followed by Al-Mani (Sufism, #HM7220).
Preceded by Al-Ghani (Sufism, #HM3822).

♦ **HM3634d Intuition**
Hunch
Description As opposed to deductive reasoning, intuition is defined as direct knowledge of what is intrinsically true, the axioms upon which deductive reasoning is based. In addition, despite a tendency to assume that new knowledge is acquired as a progress of linear thought, much new knowledge seems to be acquired through hunch, or "direct knowing", which is then confirmed rationally. The process has been described as almost a bodily sensation of important knowledge arising spontaneously, whether or not willed; and of being direct knowledge unaffected by the individual's belief system. Intuition has been explained (William Kautz) by positing the human mind as three concentric circles, the smallest enclosing consciousness, the second personal unconsciousness and the largest, or *super-conscious*, being the reservoir of all actual and potential human knowledge and experience. Intuition is then the flow of knowledge from the outer to the inner circle, the flow being assisted by clearing the channel through the personal unconscious.
Intuition has also been defined as immediate knowledge of the Absolute obtained through *wisdom*, and thus contrasted with knowledge of external objects derived through the senses and the intellect (Swami Omkarananda). The suddenness of the understanding arising is exemplified in Zen, where meditation on what might appear trivial can lead to inspired intuition of understanding or beauty.
Refs Agor, Weston *The Logic of Intuitive Decision Making* (1986); Agor, Weston *The Role of Intuition in Leadership* (1989); Goldberg, Philip *The Intuitive Edge* (1983); Ornsteil, R *The Education of the Intuitive Mode* (1972); Vaughan, F *Awakening Intuition* (1979).
Related Kan (Japanese, #HM3104) Spiritual intuitive cognition (#HM2829)
Intuitive, meditative thinking (#HM2785).

♦ **HM3636d Al-Bari'** (Sufism)
Description The believer is aware of Allah as the evolver, creating all in harmony. The believer should manifest his natural harmony in life. Wilfully opting for wrong is an intention to destroy universal harmony and results in destruction of one's self.
Context A mode of mystical awareness in the Sufi tradition associated with the twelfth of the ninety-nine names of Allah.
Broader Ninety-nine names of Allah (Sufism, #HH2561).
Followed by Al-Musawwir (Sufism, #HM5786).
Preceded by Al-Khaliq (Sufism, #HM0034).

♦ **HM3644d Al-Khafid** (Sufism)
Description The believer is aware of Allah as the abaser. Being cast down from a high to a low state may waken the arrogant or heedless to see their true shape and that the hand that castes down is also the hand that lifts up.
Context A mode of mystical awareness in the Sufi tradition associated with the twenty-second of the ninety-nine names of Allah.
Broader Ninety-nine names of Allah (Sufism, #HH2561).
Followed by Al-Rafi (Sufism, #HM4040).
Preceded by Al-Basit (Sufism, #HM0115).

♦ **HM3647c Wilayat** (Sufism)
Saintship — Sainthood
Description Sensing at all times the presence of the remembered God, knowing the reality sought to be absolute and without limits, the devotee tries to keep his knowledge free from quality, quantity or direction. He is drawn irresistibly towards God and in consequence experiences a state of fervour, ecstasy, joy, intoxication, fear and reverence. Nothing in this world or the next is worth anything to him.
Context This is the eighth state arising in the method of dhikr described by Shaykh Kalimallah. It is also the state described in those who have completed the *suluk* of the Naqshbandiyya order. According to Shaykh Abu Sa'id ibn Abi'l-Khayr, this is the 27th of 40 stations or maqamat the Sufi must possess for his journey on the path of Sufism to be acceptable.
Broader Remembrance of God (Sufism, #HM6562)
Suluk of the Naqshbandiyya order (Sufism, #HM4356)
Stations of consciousness – ibn-Abi'l-Khayr (Sufism, #HM2424).
Followed by Divine-love awareness (Sufism, #HM2401).
Preceded by Jahd (Sufism, #HM6099) Divine-love awareness (Sufism, #HM2401)
Continuous all-pervading knowledge (Sufism, #HM4788).

♦ **HM3650c Wajd** (Sufism)
Ecstasy
Description This state may be characterized by calmness, if the mystic is firm in spiritual attainments; or behaviour may be violent. The sight of God's glory is beyond rational explanation, beyond doubt and suspicion. In the most sublime presence, he may no longer be said to inhabit anywhere else in this world or outside it.
Context This is the eleventh stage in a systematic account of various stations of the Sufi spiritual path. According to Shaykh Abu Sa'id ibn Abi'l-Khayr, it is the 29th of 40 stations or maqamat the Sufi must possess for his journey on the path of Sufism to be acceptable.
Broader Mystic stations (Sufism, #HM3415)
Stations of consciousness – ibn-Abi'l-Khayr (Sufism, #HM2424).
Related Mahabba (Islam, #HM6523).
Followed by Ma'rifa (Sufism, #HM2254) Nearness-awareness (Sufism, #HM2377).
Preceded by Mushahada (Sufism, #HM3521) Divine-love awareness (Sufism, #HM2401).

♦ **HM3651c Second chakra: realm of fantasy** (Leela)
Astral plane
Description The player is caught up in sensory perception. This sensuality is the force behind creative art, although over indulgence of sensuality leads to loss of energy and thence to corruption and disorder. Mastery of this chakra leads to conquering of lust and to gifts connected with language. There is control of the elements. Free from enemies, the player is beloved by all, including beasts.
Context The second row of the board of Leela, the game of knowledge, this is represented by the astral, fantasy and joy planes and also groups the squares representing the states of purification, entertainment, envy, nullity, jealousy and mercy. Located in the region of the sex organs, this chakra is related to the element water, to the sense of taste and to pale blue or luminous white. It is compared to the octave comprising the ages 7 to 14 of temporal life. Purification and mercy are the two arrows at this stage, and jealousy and envy the two snakes.
Broader Lila (#HM2278).
Narrower Envy (#HM0832) Mercy (#HM3434) Nullity (#HM4512)
Jealousy (#HM0554) Purification (#HM1773) Astral plane (#HM4600)
Plane of joy (#HM4376) Entertainment (#HM4403) Plane of fantasy (#HM1485).
Related Svadhishthana (Yoga, #HM3174).
Followed by Third chakra: theatre of karma (Leela, #HM0717).
Preceded by First chakra: fundamentals of being (Leela, #HM4103).

♦ **HM3656d Passionate love** (Islam)
'Ishq
Description This love is described as the operation of a heart which is empty or unpreoccupied, a madness leading to sorrow or death. It has been referred to as a disease similar to melancholy. According to Ibn al-Jawzi, it is the acute inclination of the nafs or soul towards a form which conforms to its nature. Thinking intently on this form it imagines obtaining it, then begins to hope to obtain it, and from this intense thought comes love. It could be termed an excess of feeling and, of its nature, can be felt towards one person only although, in practice, it may be dissipated among a number of partners. Although love for one's family and affection are natural and not to be blamed, passionate love possesses the reason and makes the person act in an unwise way. It is therefore to be avoided. It leads to idolatry and is the opposite of tawhid (spirituality which is monotheistic).
There is difference of opinion as to whether 'ishq is an appropriate term when referring to love for God. In that it implies desire it is used when the possibility of consummated love (mahabba) between God and man is disputed. It is also used in the sense of excessive and firmly planted love towards God. It may not be strong enough, even then, to describe the degree of love felt for God's attributes. In mystic terms, five stages of 'ishq are enumerated by al-Dabbagh: gharam (infatuation), iftitan (seduction), walah (loss of discernment), dahsh (bafflement) and fana (annihilation).
Context One of the sins of passion (hawa) in Islam.
Refs Bell, Joseph Norment *Love Theory in Later Hanbalite Islam* (1979).
Broader Hawa (Islam, #HM0129) Love (Islam, #HH5712)
Love with God (Islam, #HM5580).
Related 'Ishq (Sufism, #HM4054) Tawhid (Sufism, #HM3438)
Romantic love (#HM5087).

♦ **HM3662d State of learning** (Buddhism)
Two vehicles
Description This is the state in which something is learned from the teaching of others. Learning from the realization of others and applying it to one's own life, this corresponds to intelligence. It is concerned with existing knowledge. Together with the state of realization the state of learning comprises the two vehicles and has the goal of self-betterment. Learning may be thought of as trying to understand, realization as actually understanding. Both worlds have limitations. They are self-centred, self-betterment for one's own sake, which may lead back to the world of anger, looking down on others. It may also lead back to self-absorption so that all else, including the implications of what one is doing, is obliterated from view. The other limitation is a difficulty in accepting that conclusions one has come to one's self are not totally correct. These limitations are dangerous when it is realized that people in these worlds are leaders and experts. Based on a desire to improve the human condition they may lead to beneficial results; based on a desire for profit for profit's sake they may lead to degradation of the environment and the destruction of life.
Context One of the ten worlds described in Nichiren Soshu Buddhism.
Broader Ten worlds (Buddhism, #HM2657) Four noble paths (Buddhism, #HM4026)
Nichiren shoshu buddhism (Buddhism, #HH3443).
Related Learning (#HH0180).
Followed by State of realization (Buddhism, #HM0450).
Preceded by State of rapture (Buddhism, #HM1973).

♦ **HM3664g Deep psychophysiological relaxation** (DPR)
Description This state, which can be reached in meditation or through biofeedback techniques, is characterized by mental and physical stillness and decreased activity of the sympathetic nervous system.

♦ **HM3667e Fringe of consciousness** (Psychism)
Description While the conscious mind still remains aware of the physical environment, the subconscious mind predominates. Said to be related to the alpha state.
Related Alpha wave consciousness (#HM2345).

♦ **HM3677c Insanu'l-kamil** (Sufism)
Description The name of a traveller on the *tariqa* or Sufi spiritual path at the level of realizing identity with the essence of God – the perfect man. Although living in the world he does not belong to the world, he has come to serve humanity, guiding others on the tariqa or path, as Shaykh, Pir, spiritual master, whether as Sufi or Prophet.
Context Spiritual progress on the spiritual path is marked by the attainment of specific *maqamat* or station.
Broader Mystic stations (Sufism, #HM3415).

♦ **HM3685d Sectarian bias** (Buddhism)
Titthayatana (Pali)
Broader Wrong view (Buddhism, #HM1710).

♦ **HM3688g Hypnopompic consciousness** (Psychism)
Hypnapompic state — Hypnopompic imagery
Description A state intermediate between sleeping and waking, said to occur before the conscious attains dominance over the subconscious. There may be an awareness of fusing electric sparks, or clairaudient or clairvoyant perception which is forgotten as soon as the individual is fully awake. Hypnopompic experiences occur less frequently than hypnagogic, although both hypnagogic and hypnopompic imagery occur suddenly and are not under the individual's control. The imagery is realistic but bizarre. A frequent experience is of hearing one's name called, arousing the person involved. There may be music playing (so realistic as to arouse the person to turn it off), the appearance of highly coloured geometrical images or distorted faces (visual images are short-lived and surrealistic), kinaesthetic experiences, and (less often) olfactory or tactile experience. Surveys relate degree of hypnagogic and hypnopompic imagery to favourable personality characteristics. Absence of such imagery is related to rigid defence against impulse. This is the opposite of findings related to high dream fantasy ratings.
Refs Reed, Graham *The Psychology of Anomalous Experience* (1988).
Related Hallucination (#HM4580)
Hypnagogic consciousness (Psychism, #HM3511).

♦ **HM3690d Yoga of the clear light** (Buddhism)
Hod–gsal (Tibetan)
Context The fourth branch of the Tibetan Tantric path of form according to the Naropa school.
Broader Tantric visualization (Yoga, #HM1690).
Followed by Yoga of the intermediary state (Buddhism, #HM0101).
Preceded by Yoga of the dream state (Buddhism, #HM0600).

♦ **HM3691f Modes of awareness associated with use of cannabis**
Use of marijuana
Description Effects on consciousness and behaviour are small, but this substance is used not only for pleasurable sensations but also to produce self–transcendence. It has been used as an aid to meditation and as a ceremonial drug in a number of cultures. Cannabis intoxication may include states of euphoria or anxiety, panic, dysphoria and possibly derealization or depersonalization and delusion. There may be a feeling of slowed time.
Broader Modes of awareness associated with psychoactive substances (#HM0584).
Related Sacred drugs (#HH0788).

♦ **HM3692c Seed meditation**
Description In this state body and emotions are relaxed and quiet; the aim is control of the mind to the extent of no thought at all.
Related Meditation (#HH0761).

♦ **HM3696d Malleability of mental body** (Buddhism)
Pliancy of mental factors — Pliancy of kaya (Pali)
Description This refers to the malleability or pliancy of the three aggregates of feeling, perception and formation or mental activities. The characteristic is suppressing or quieting rigidity of the mental body, the function is to crush its stiffening. It is manifest as non–resistance. Mental factors are the proximate cause and it is the opponent of defilements causing stiffness in the mental body. Malleability of mental body and malleability of consciousness are considered together.
Context One of the formations aggregate (mental coefficients) of Hinayana Buddhism, being listed among the constant states which appear in their true nature, and as profitable primary (always present in any profitable or profitable–resultant consciousness).
Broader Awareness as mental–formation group of conscious existence (Buddhism, #HM2050).
Related Pliancy (Buddhism, #HM3162)
Malleability of consciousness (Buddhism, #HM1155).

♦ **HM3697d Adha–loka** (Jainism)
Description According to the Jain system, this is the lowest level of the universe, below Mount Meru; the underworld, including many levels corresponding to the infernal beings inhabiting them.
Related Siddha–loka (Jainism, #HM0569)
Madhya–loka (Jainism, #HM1357)
Urdhva–loka (Jainism, #HM0131)
Human development (Jainism, #HH0622).

♦ **HM3701d Admiration**
Description This emotional reaction or feeling arises in response to manifestation of unusual excellence and implies some standard of excellence against which the behaviour exciting admiration is measured.
Related Wonder (#HM3197).

♦ **HM3707g Tanha** (Pali)
Trisna (Buddhism) — Thirst — Craving
Description Longing for the substantial or for something to cling to in the inevitable process of change, there is a holding on to experience and to life. Tanha is the cause of continuing existence and of repeated becoming. The very clinging to life is also the clinging to death. Since the essence of life is to change, the craving to arrest change makes this life one of suffering, dukkha. Change or disappearance of objects and states of existence only cause suffering if there is a desire to possess or selfishly enjoy them. The very fact that change can be enjoyed shows that it is not change but attitude towards it which causes suffering.
Related Duhkha (Buddhism, Hinduism, #HM2574)
Cessation of suffering (Buddhism, #HH2119).

♦ **HM3716g Being workable** (Buddhism)
Kammannabhava (Pali)
Broader Workability of body (Buddhism, #HM0739)
Workability of thought (Buddhism, #HM1584).

♦ **HM3719c Jesus dies on the cross** (Christianity)
Description Aware of Jesus' anguished death on the cross, the believer knows that his own sins merit a miserable death but that Jesus died for him and this is his hope. He begs that when he dies he may do so full of love for Jesus, and commits his soul into His hands. Loving Jesus more than himself, he repents with his whole heart and begs never again to separate himself from Him. He asks that he will love Jesus always and that Jesus will do with him as He wills.
Context The twelfth station of the cross.
Broader Way of the cross (Christianity, #HM3516).
Followed by Jesus taken down from the cross (Christianity, #HM3410).
Preceded by Jesus nailed to the cross (Christianity, #HM4136).

♦ **HM3723d Knowledge of contemplation of rise and fall** (Buddhism)
Related Purification by knowledge and vision of the way (Buddhism, #HH3550).
Followed by Knowledge of appearance as terror (Buddhism, #HM7114).
Preceded by Knowledge of contemplation of dissolution (Buddhism, #HM3385).

♦ **HM3724f Trauma**
Related Abreaction (#HH0640).

♦ **HM3725c Sext** (Christianity)
Hours of sext
Description The canonical hours of sext traditionally take place between 12 noon and 3 pm. They dramatize the reality of the day being half over, the long afternoon is ahead and the need for help from God to have the strength to complete the work. The mood is humble supplication.
Broader Canonical hours (Christianity, #HM1167).
Followed by None (Christianity, #HM0466).
Preceded by Terce (Christianity, #HM2965).

♦ **HM3726f Pseudologia phantastica**
Pathological lying
Description As opposed to hysterical amnesia, where true facts are forgotten, in pseudologia phantastica something which has not happened is remembered as true. However much the story may be shown to be intrinsically improbable or easily disproved, the individual further elaborates with more and more improbable details. The reason seems to be a need to attract respect, sympathy or attention.
Related Hysterical amnesia (#HM7608).

♦ **HM3729d Contentment**
Description Eastern thought looks on contentment as freedom from all desires and passions, contemplating all that arises in life with perfect equanimity. From this arises universal love and sympathy for all. For Socrates, the secret of happiness lay in faithful performance of life's customary duties; Epicurean contentment was self–sufficiency and the simple life, while Stoicism looked to impassivity and a life of reason in conformity with nature. Judaic thought has emphasized the sense of certainty of divine omniscience and righteousness, while Christianity adds trust in God's love. In the latter case, resignation or submission is an act of sanctified will.
Related Happiness (#HM2409) Samtosa (Yoga, #HM2898).

♦ **HM3732d Plane of primal vibrations** (Leela)
Omkar
Description The state in which the syllable "Om" vibrates, the sound of creation vibrating in the universe. The player sees the need for calm and simplicity; centred on Om, tension is relieved and vast resources are opened. He has knowledge beyond his own embodied life–experience.
Context The 56th state or square on the board of *Leela*, the game of knowledge, appearing in the seventh row.
Broader Seventh chakra: plane of reality (Leela, #HM0754).

♦ **HM3743f Modes of awareness associated with use of inhalants**
Glue–sniffing
Description Anaesthetics and volatile solvents may produce a rapid, intoxicated "high". Associated effects are: euphoria; dizziness; lethargy; possibly stupor or coma.
Broader Modes of awareness associated with psychoactive substances (#HM0584).

♦ **HM3744b States of awareness in biofeedback training**
Description Just as training in meditative or yoga techniques leads to discretely definable altered modes of awareness, so, after initial training in meditation and with 'guided imagery', the following of biofeedback techniques has been shown to produce a hierarchy of states. So far six levels have been documented. According to C Maxwell Cade, stage 5 is preceded by 'sudden excursions into other realms', and is just above the fourth state of traditional meditation and of the relaxation response. The final stages are said to be similar to those described by mystics, with feelings of ecstasy, joy and wholeness.
Refs Cade, Maxwell C and Coxhead, Nona *The Awakened Mind* (1979).
Broader Biofeedback training (#HH0765).
Related Electronic meditation (#HH1100).

♦ **HM3745d Indecisiveness** (Buddhism)
Parisappana (Pali)
Broader Perplexity (Buddhism, #HM0812).

♦ **HM3748d First step: the ego** (Christianity)
Description Who cries out within one's heart to begin the march ? The self is not good nor innocent nor serene. Happiness and sorrow are both unbearable. This I controls the body, keeping it prepared. It keeps the mind awake, lucid doing battle with the darkness. It keeps the heart aflame, restless and courageous. The cry is not a hope or a home; it is one's general, one's comrade–in–arms. The cry demands that the self loves danger, learns to obey, learns to command, loves responsibility and loves each person for their contribution to the struggle. It demands the the I be always restless, unsatisfied and unconforming for the greatest sin is satisfaction. The cry and the I become one. The I becomes a bridge which crumbles having done its duty and the cry passes to another.
Context The first step in the march for the spiritual exercises of Nikos Kazantzakis.
Broader Saviors of God: the march (Christianity, #HM3439).
Followed by Second step: the race (Christianity, #HM2953).
Preceded by The cry (Christianity, #HM0059).

♦ **HM3752d Disinterestedness as a wholesome root** (Buddhism)
Alobhokusala mulam (Pali)
Broader Non–attachment (Buddhism, #HM2128).

♦ **HM3754g Sensory translation**
Description This has been postulated (Arthur J Deikman) as an explanation of some of the perceptions associated with the mystic experience. Psychic action such as conflict, repression or problem solving is perceived through the sensations of light, colour, movement, force, sound, smell or taste. It is similar to the hypnagogic experience but, whereas in the hypnagogic and dream states perceptions are complex, here there is an experience of non–verbal, simple, concrete perceptual equivalents of psychic action.
Sensory translation seems to occur when there is: direction of heightened attention to sensory paths; absence of controlled, analytical thought; a receptive as opposed to a suspicious attitude to stimuli.
Broader Deautomatization and the mystic experience (#HM4398).
Related Hypnagogic consciousness (Psychism, #HM3511).

♦ **HM3755c Realities** (Sufism)
Context The ninth of ten states acting as gateways orienting the mystic on the journey to the Absolute, as described by Ansari.
Broader Stations of consciousness – Ansari (Sufism, #HM2317).
Followed by Nihayat (Sufism, #HM3377).
Preceded by Sanctity (Sufism, #HM4041).

♦ **HM3759c Character** (Sufism)
Context The fourth of ten states acting as gateways orienting the mystic on the journey to the Absolute, as described by Ansari.
Broader Stations of consciousness - Ansari (Sufism, #HM2317).
Followed by Principles (Sufism, #HM3069).
Preceded by Conduct (Sufism, #HM4648).

♦ **HM3763f Euphoria**
High
Description This exaggerated feeling of elation and well-being, although usually transient, is sought by numbers of people to the extent of risking their physical and mental health through use of various kinds of drug. It is incongruous with the individual's external or internal reality, and is frequently an attempt to regain joyful or happy experience which no longer arises naturally. A related drawback is that feelings of euphoria are frequently followed by dysphoria. Commonly held moral views that pleasure has to be paid for mean that socio-ethical issues are involved, even supposing a non-toxic, easily available stimulant could be found.
Related Ecstasy (#HM2046) Dysphoria (#HM3407) Emotional high (#HM1400)
Modes of awareness associated with use of sedatives (#HM4546).

♦ **HM3764g Perception through the senses**
Sensory perception
Description According to Western tradition there are five senses - *taste, touch, hearing, sight* and *smell* - although none is said to be totally independent of the others; sensitivity to pressure, temperature, radiation, equilibrium, pain and perception of depth, distance and time have also been posited as specific senses. Such sensations are interpreted according to the nervous system, psychology and culture of the individual concerned and are therefore subjective.
Narrower Eye consciousness (Buddhism, #HM2074)
Ear consciousness (Buddhism, #HM2169)
Nose consciousness (Buddhism, #HM2364)
Body consciousness (Buddhism, #HM2562)
Tongue consciousness (Buddhism, #HM2263)
Mental consciousness (Buddhism, #HM2838).
Related Synesthesia (#HM1241) Sensory awakening (#HM2972)
Sense consciousness (Buddhism, #HM2664).

♦ **HM3770e Fringe of space-time** (Psychism)
Description This experience may occur in deep hypnosis or during astral projection - while the conscious mind is quiet the subconscious moves into another plane. Said to be related to the theta state.
Related Theta wave consciousness (#HM2321).

♦ **HM3771c State of quies** (Christianity)
Description This state, described by the early Christian desert fathers, has been compared with transcendental consciousness.
Broader Human development (Christianity, #HH2198).
Related Transcendental consciousness (#HM2020).

♦ **HM3779d Lunar plane** (Leela)
Ganges — Ganga
Description This is the plane of devotion and receptivity. The player is at the source of magnetic or lunar energy and is magnetic and attractive to others; he gains understanding of the female principle.
Context The 49th state or square on the board of *Leela*, the game of knowledge, appearing in the sixth row.
Broader Sixth chakra: time of penance (Leela, #HM4412).

♦ **HM3789d Wickedness**
Description This attitude of the intellect and will deliberately seeks evil for its own sake in contempt of God.
Related Evil (#HH1034).

♦ **HM3799a Fana** (Sufism)
Passing away
Description When the seeker becomes aware of the identity that exists between himself and God there is a state of *fana*, or *passing away*. There is perfect knowledge of God. Even awareness of unity may be transcended in a higher state of union with God beyond all mystical states or *ahwal*, when all trace of the Sufi is annihilated in the traces of another. This absolute unification with God, the absorption of the soul into God, is beyond expression and therefore no description conveys its meaning. The mystic is unaware of himself and even of his own absorption. All worldly desires and human qualities are transcended; passions have passed away so that there are no feelings towards anything. Protected by God, he acts according to His will. Human qualities such as ignorance, injustice and ingratitude are replaced in the state of baqa by divine attributes such as knowledge, justice and gratitude. The perfect state is reached gradually as external forms of reality are removed and human activities, physical forms, the phenomenal self, the whole external world, are transcended. According to Kalabadhi, once having experienced the state of fana the mystic never returns to his selfhood.
Broader Stations of consciousness - Simnami (Sufism, #HM2341).
Related Fana (Islam, #HM1087) Fana (Sufism, #HM1270)
Remembrance of God (Sufism, #HM6562) Consciousness of mystical poverty (#HM3241).
Preceded by 'Arif (Sufism, #HM1045).

♦ **HM3800d Resolution** (Buddhism, Pali)
Resolve
Description Here the act of resolving is that of being convinced about an object, it does not refer to trust, faith or belief. It is characterized by determination or conviction, its function is not groping or "slinking along", its manifestation is decisiveness (unshakableness). The proximate cause is the object about which to be convinced. Because it is unshakable it is like a stone pillar with respect to the object.
Context One of the formations aggregate (mental coefficients) of Hinayana Buddhism, being listed among the "or-whatever" states, and as general primary (always present in any consciousness).
Broader Awareness as mental-formation group of conscious existence (Buddhism, #HM2050).
Related Belief (Buddhism, Yoga, #HM2933).

♦ **HM3803d Faculty of indifference** (Buddhism)
Upekkhindriya (Pali)
Related Equanimity (Buddhism, #HM7769).

♦ **HM3806c Recalling** (Buddhism)
Patissati (Pali)
Broader Faculty of recollection (Buddhism, #HM4248).

♦ **HM3807d Al-'Adl** (Sufism)
Description The believer is aware of Allah as the just. His absolute justice is the enemy of tyrants, it is harmony as opposed to disorder. Seeing the bad one recognizes the good for what it is.
Context A mode of mystical awareness in the Sufi tradition associated with the twenty-ninth of the ninety-nine names of Allah.
Broader Ninety-nine names of Allah (Sufism, #HH2561).
Followed by Al-Latif (Sufism, #HM0308).
Preceded by Al-Hakam (Sufism, #HM5014).

♦ **HM3808d Capacity of easy transformation** (Buddhism)
Lahuparinamata (Pali)
Broader Lightness of thought (Buddhism, #HM1830).

♦ **HM3811c Numinosum** (Jung)
Numinous experience — Mysterium tremendum
Description This state is described by Jung as imposing itself on the subject independent of his will. It is mysterious, enigmatic and impressive, defying explanation. There is confrontation with tremendous and compelling force implying meaning both fateful and attracting. Conscious or unconscious belief - the readiness to trust transcendent power - is a prior necessity to the arising of such a state. One can open one's self to it but cannot conquer it.
Encounter with the numinosum is seen as an aspect of all religious experience. Jung considered that previously unconscious contents break through ego constraints and overwhelm conscious personality in a similar way to invasions by the unconscious in pathological situations, but that nonetheless the experience is not normally pathological. The experience is held not necessarily to be proof of the existence of God and is compared to the *peak experience* of humanistic psychology.
The numinous is an expression also used by Rudolf Otto to describe an encounter with the objective "wholly other" outside the self. The mysterium tremendum is compared with the feeling associated with ancient churches and with solemn rites. The feeling of awe combines dread and horror with attraction and the invitation to surrender. The experience is one of absolute of absolute dependence. This is not necessarily an encounter with God - some quote it as the same experience as is felt walking alone in the dark through a wood or invoked in horror films. Others would say that the numinous experienced as encounter with God was totally different to the dread just described, the fear being of a different quality.
Refs Adler G, Fordham M and Read H (Eds) *The Collected Works of C G Jung*; Otto, Rudolf *The Idea of the Holy* (1950); Samuels, Andrew et al *A Critical Dictionary of Jungian Analysis* (1986).
Related Peak experiences (#HM2080) Mystical cognition (#HM2272)
Creature consciousness (#HM0376)
Awareness of the sacred (Christianity, #HM0876).

♦ **HM3818d Spiritual envy** (Christianity)
Description Many beginners on the spirit path are really irritated by other people's spirituality and thus suffer from spiritual envy. They are pained by being overtaken on the spiritual road. They would rather not hear others praised about their growth. Not only do they not take pleasure in the development others that they often refute this growth as being false. They even disparage praise given to others. They want to be first in everything and become irate when they are not praised. This is the opposite of spiritual love.
Context This is an imperfection of beginners in the Dark Night by St John of the Cross.
Broader Beginner in spiritual life (Christianity, #HM0102).
Related Spiritual lust (Christianity, #HM4180) Spiritual love (Christianity, #HM8977)
Spiritual sloth (Christianity, #HM0491) Spiritual anger (Christianity, #HM0936)
Spiritual pride (Christianity, #HM4533) Spiritual avarice (Christianity, #HM0642)
Spiritual gluttony (Christianity, #HM0507).

♦ **HM3822d Al-Ghani** (Sufism)
Description The believer is aware of Allah as the self-sufficient, the rich. All He has is unrelated to another, He has no needs, He does not need to earn. His existence and perfection depend on no-one yet all depends on Him. Even the most powerful man is needy, he depends on Allah for his position, for all that he is and has. The highest to which man can aspire is to be servant of Allah, serving His servants and that which is in his charge. The good servant comes close to Allah, receiving His richest treasure which is faith.
Context A mode of mystical awareness in the Sufi tradition associated with the eighty-eighth of the ninety-nine names of Allah.
Broader Ninety-nine names of Allah (Sufism, #HH2561).
Followed by Al-Mughni (Sufism, #HM3633).
Preceded by Al-Jami' (Sufism, #HM0009).

♦ **HM3824c Jesus falls the second time under the cross** (Christianity)
Description Seeing Jesus fall again, the believer is aware of how many times he has fallen again after having been pardoned. He begs for aid in persevering in grace and that he will always commend himself to Jesus when he faces temptation. Loving Jesus more than himself, he repents with his whole heart and begs never again to separate himself from Him. He asks that he will love Jesus always and that Jesus will do with him as He wills.
Context The seventh station of the cross.
Broader Way of the cross (Christianity, #HM3516).
Followed by Jesus speaks to the daughters of Jerusalem (Christianity, #HM2737).
Preceded by Veronica wipes the face of Jesus (Christianity, #HM3401).

♦ **HM3828d Being ashamed of what one ought to be ashamed** (Buddhism)
Hiriyati hiriyitabbena (Pali)
Broader Power of shame (Buddhism, #HM0352).

♦ **HM3829d Obsession with ignorance** (Buddhism)
Avijjapariyutthana (Pali)
Broader Delusion (Buddhism, #HM0184).

♦ **HM3833c Mono no aware** (Japanese)
Description Essentially a deep, empathetic appreciation of the ephemeral beauty manifest in nature and human life, and possibly encompassing the whole range of human emotions. It encourages a perception of human reality imbued with elegant beauty and Buddistic pessimism. This experience is usually tinged with a hint of sadness when it focuses on the beauty of impermanence and on the sensitive capability of appreciating that beauty. It may be accompanied by admiration, awe, or even joy. The deep feeling is not an emotion deriving from someone's idiosyncracy, but rather is held to be a feeling which emerges from the hearts of sensitive people under given circumstances. As a purified exalted feeling, it is close to the innermost heart of people and of nature. A person who understands mono no aware is a more complete individual because he is more perceptive, empathetic and capable of understanding the reality of the human condition.

Context A literary and aesthetic ideal cultivated during the Heian period of Japan (794–1185). Revived by the writings of Motoori Norinaga (1730–1801).

♦ **HM3837d Atonement** (Leela)
Saman–paap
Description Activities carried out at lower levels are seen to have been blind and regardless of injury to others. Wrong action and wrong means have given rise to wrong vibrations, and only atoning for these can bring inner peace. Atonement also releases from guilt a player still vibrating at the level of the second chakra and unable to adjust to vibrating at a higher level.
Context The 21st state or square on the board of *Leela*, the game of knowledge, appearing in the third row.
 Broader Third chakra: theatre of karma (Leela, #HM0717).

♦ **HM3846d Knowledge** (Leela)
Wisdom — Gyana — Jnana
Description Knowing both what is right and how to realize it in everyday existence, the player is lifted from the human plane to beyond the chakras, to the plane of bliss. He is gaining insight into the game itself and is liberated from the bondage of illusion. The goal of cosmic consciousness is seen as attainable and he tries to pass his insight on to others on the path while knowing this is futile – what is permanent is the page, what is written on it is temporary and illusory.
Context The 37th state or square on the board of *Leela*, the game of knowledge, appearing in the fifth row.
 Refs Johari, Harish *Leela* (1980).
 Broader Fifth chakra: man becomes himself (Leela, #HM0933).
 Followed by Plane of bliss (Leela, #HM6901).

♦ **HM3850d Al–Mutakabbir** (Sufism)
Description The believer is aware of Allah as the majestic, manifesting his greatness in all at all times. Created beings cannot have this attribute, although followers of the devil think that attributes lent to them by Allah are intrinsically theirs. The believer reflects this attribute only by working to achieve his greatest potential while not boasting or demonstrating his greatness.
Context A mode of mystical awareness in the Sufi tradition associated with the tenth of the ninety–nine names of Allah.
 Broader Ninety–nine names of Allah (Sufism, #HH2561).
 Followed by Al–Khaliq (Sufism, #HM0034).
 Preceded by Al–Jabbar (Sufism, #HM0215).

♦ **HM3852c Indeterminate consciousness in the sense sphere – functional** (Buddhism)
Indeterminate consciousness in the sense sphere – inoperative
Description There are 11 indeterminate resultant consciousnesses in the sense sphere.
Three are without root–cause (unconditioned), that is without non–greed, etc, as cause of the result. These are the mind element as precursor to sense consciousness cognizing sense data. It interrupts the life continuum and is associated with equanimity (indifference). Then there is the mind–consciousness element. This occurs in two ways: (i) As common to everyone it is accompanied by equanimity (indifference). It cognizes the six kinds of object, determining at the five–doors (senses) and adverting at the mind door. The proximate cause is departure of one of the profitable resultant mind–consciousness elements or one kind of life–continuum. (ii) As special, appearing only in arahants, it is accompanied by joy. It cognizes the six kinds of object, manifesting as smiling or laughter about non–sublime things. The proximate cause is always the heart–basis.
Eight are with root–cause or conditioned. They arise only in arahants. Of these, 4 are accompanied by joy, that is: with understanding and unprompted; with understanding and prompted; without understanding and unprompted; without understanding and prompted. And 4 are accompanied with equanimity, that is: with understanding and unprompted; with understanding and prompted; without understanding and unprompted; without understanding and prompted.
All give rise to materiality, to postures and to intimation, except mind element which gives rise only to materiality.
Context In Hinayana Buddhism, 89 consciousnesses are enumerated in aggregate (khanda). Of these, 21 are profitable or moral, 12 are unprofitable or immoral and 56 are indeterminate (resultant or functional). The unprofitable all arise in the sphere of sense and desire, whereas profitable and indeterminate arise in sense, fine–material, immaterial and supramundane spheres.
 Broader Dispositions of consciousness (Buddhism, #HM2098).
 Narrower Mental consciousness (#HM2838)   Mind–consciousness element (#HM6173).
 Related Impulsion (Buddhism, #HM7268)   Determining (Buddhism, #HM7632)
 Profitable consciousness in the sense sphere (Buddhism, #HM4447)
 Unprofitable consciousness in the sense sphere (Buddhism, #HM8375)
 Indeterminate consciousness in the sense sphere – resultant (Buddhism, #HM5721).
 Followed by Indeterminate consciousness in the fine–material sphere – functional (Buddhism, #HM4761).

♦ **HM3857d Abruptness** (Buddhism)
Asuropa (Pali)
 Broader Hatred (Buddhism, #HM4502).

♦ **HM3863d Yoga of the inner fire** (Buddhism)
Gtum–mo (Tibetan)
Context The first branch of the Tibetan Tantric path of form according to the Naropa school.
 Broader Tantric visualization (Yoga, #HM1690).
 Followed by Yoga of the illusory body (Buddhism, #HM1222).

♦ **HM3864b Unification with God** (Sufism)
Unification with Allah
Description Meditation on the attributes of Allah creates a new pattern of behaviour, a positive attitude; and even effects physiological changes. Encounter and assimilation lead to self–awareness of the inadequacies of the conventional self. Practice of conventional religion brings awareness of God's attributes through transcendental means and gradual identifying with these attributes brings further awareness of ultimate unity. Allah, the object of desire, can be identified with through devotion and *zekr*, although he cannot be reached through thought or discussion. This process transforms the duality of submission to an authority to the unity of becoming God–like; and, no longer God's servant, completely devoted to God, the Sufi becomes unified and one with God. To reach this goal there has to be perfect patience, treating afflictions as gifts from God; there must also be a dissociation from all mankind, being true to God alone. This separation means that the Sufi has nothing attached and is attached to nothing; his eyes are closed to the imperfect, phenomenal world which conceals God. In a state of *uns* (fellowship with God), the Sufi ceases to take trouble for himself. All deeds are in accordance with divine will: pleased with what God does so that God is pleased with what he does. He has no need of rituals or formal ceremony – in himself he is nothing and is veiled from those who have their being in the world. Without care, in the highest state of divine consciousness, he is the existence of the divine being.
 Related Uns (Sufism, #HM1957)   Tawhid (Sufism, #HM3438)
 Psychic states (Sufism, #HM4311)   Human development (Sufism, #HH0436)
 Recollection (Islam, Sufism, #HM2351)   Identification with God (Sufism, #HM3520).

♦ **HM3868d Al–Mumit** (Sufism)
Description The believer is aware of Allah as the creator of death. man is visible, mortal flesh and invisible, immortal soul. When the body dies the soul is left with whatever it has gained or lost in this life. The believer prepares for death and finds eternal bliss.
Context A mode of mystical awareness in the Sufi tradition associated with the sixty–first of the ninety–nine names of Allah.
 Broader Ninety–nine names of Allah (Sufism, #HH2561).
 Followed by Al–Hayy (Sufism, #HM4236).
 Preceded by Al–Muhyi (Sufism, #HM1007).

♦ **HM3869a Kaivalya** (Yoga)
Liberation — Enlightenment — Aloneness — Realization — Kaivalyam — Consciousness independent of vehicle
Description The long evolutionary unfolding of the *purusha* after numerous lifetimes finally ends in enlightenment and the gunas re–emerge devoid of the object of purusha. Kaivalya is said to be attained when there is equality of purity between purusha (witness–consciousness), or Atman, and sattva (perception) and each is quite distinct. This realization becomes clearer in a process of states as the goal of complete freedom from limitations and illusions is approached. Purification of sattva implies that purusha can function through prakriti (nature), free and in full realization of the yogi's real nature. For the perfected men of humanity, vehicles built up and perfected in prakriti can now be used without egoism or illusion.
Closely related to liberation as moksha, kaivalya is the goal of yoga as described by Patanjali. It has been described as the "supreme autonomous state of being free from ignorance". All transient phenomena are discarded, there is total withdrawal from visible and invisible dimensions of phenomenal reality. Then the original wholeness is recovered and found never to have been absent, merely covered by the multiplicity of the transient. Moksha is said to be realized through loving attachment to God, *bhakti*, or through unconditional offering of the self, *prapatti*. In contrast, kaivalya is consciousness of the self alone without realization of the "lord". It can therefore also be interpreted as *aloneness*. Since the state of kaivalya is beyond description to those living in the world of the unreal, misconceptions about it may arise. Rather than dissociating into divine consciousness with total loss of individuality, it is said that knowledge, consciousness and bliss increase rapidly through the final states of the Yogi's progress. Some schools of thought see the final act of emancipation as irreversible and coinciding with death of the body and mind; this is the basis of mythic yoga. Others see a subtle "germ" of individuality remaining after the perfect union of Jivatma with Paramatma (individual soul with universal soul) so that the fruits of evolutionary development are not lost; and that these great beings do return to the lower worlds for the good of creation.
Kaivalya is, for example, the ultimate goal of the raja–yoga meditation techniques, a state of consciousness in which there is no need of a body, a mind, or a world of objects. This is a state of liberation from the comings and goings, arisings and cessations of bodies, minds and worlds, and an attainment of independent existence above the play of phenomena.
 Broader Emancipation of the self (#HH0907).
 Related Three gunas (#HH0413)   Purusha (Yoga, #HM2396)
 Enlightenment (#HM2029)   Raja yoga (Yoga, #HH0755)
 Moksha (Hinduism, #HM2196)   Videha–mukti (Yoga, #HM4489)
 Intuitive knowledge (Yoga, #HM3239)
 Indian approaches to consciousness (Hinduism, #HM2446).

♦ **HM3871d Abhinvesa** (Yoga)
Desire for life — Will to live
Description This is the constant and universal desire to live inherent in conscious beings, coming into play at the moment the evolutionary cycle begins on contact of consciousness with matter. It is said by Patanjali to dominate all, even the learned, even those whose life is a misery, and is the last derivative of ignorance, *avidya*. Called by some the last of the five kleshas or afflictions, it may in this sense be seen as the last of the series which commences in avidya, the root of all the kleshas and of which abhinvesa is the fruit.
Context One of the five afflictions or causes of misery in yoga.
 Broader Afflictions (Yoga, Hinduism, #HH1047).
 Related Egoism (Yoga, Zen, #HM3477)   Five kleshas (Yoga, Zen, #HH0390)
 Ignorance (Buddhism, Yoga, Zen, #HM3196).

♦ **HM3880b Vicara samadhi** (Yoga)
Description There is ontic identification with the internal reality of the object contemplated, together with a certain awareness, a transcendental cogitation.
Context The third level of samprajnata samadhi.
 Broader Samprajnata samadhi (Yoga, #HM2896).
 Related Vicara stage (Yoga, #HM3460).
 Followed by Nirvicara samadhi (Yoga, #HM0809).
 Preceded by Nirvitarka samadhi (Yoga, #HM4284).

♦ **HM3887c Murid** (Sufism)
Description The name of a traveller on the *tariqa* or Sufi spiritual path at the level of receiving guidance under a spiritual director – a *shaykh* or a *pir*.
Context Spiritual progress on the spiritual path is marked by the attainment of specific *maqamat* or station.
 Broader Mystic stations (Sufism, #HM3415).
 Related Spiritual master (Sufism, #HH2080).

♦ **HM3889a Mystic union with God** (Christianity)
Unio mystica
Description In the Christian mystical tradition this is described in three ways: (1) *Essentialization*, the divine spark in every soul, part of the essence of God, one with him not only united to him. As man is perfected so this identification becomes more and more complete. (2) *Substitution*, the corrupt and human will, life and spirit being replaced by the divine. (3) *Transformation*, the entry of the divine transforming the human and making it more fit to receive the divine which enters in ever–increasing fullness.
François Fénelon also describes three ways: (1) *Intellectual union*, an idea of internal, spiritual origin making God known as being without form. (2) *Affectionate union*, when the soul is united with God in love with nothing intervening as it loves Him for His own sake. (3) *Practical union*, when the soul does the will of God not by simply following a prescribed form but from the impulse of holy love.
 Refs Haught, John F *Religion and Self–Acceptance* (1980).
 Related Wisal (Sufism, #HM7119)   Devekuth (Judaism, #HM0377)
 Turyatita (Hinduism, #HM3139)   Way of paradox (Christianity, #HH3335)
 Spiritual marriage (Christianity, #HH0465).

♦ **HM3890a Jivanmukti** (Hinduism)
Jivan-mukti
Description This state of liberation is the highest that can be reached by the individual soul, *jiva*, in this life. Someone who has achieved this state is referred to as *jivanmukta*. Through liberating insight, or *gnosis*, he is released from *samsara*, the endless cycle of birth and death. Free from suffering and in a state of bliss, the remainder of life is lived only to exhaust the karma built up previously, although no more is added to karma once this stage is reached.
Liberation, final, irreversible and achievable in this life, Jivanmukti is contrasted with *videha-mukti*, liberation which occurs only with death of the body.
Broader Mukti (#HH0613).
Related Soteriology (#HH1166) Unitive life (#HM2137) Sannyasa (Yoga, #HH4210) Videha-mukti (Yoga, #HM4489) I-transcendent stage of life (#HM3060).

♦ **HM3892d Remembrance of reality** (Sufism)
Description At this stage it must be remembered that the goal is to efface the self in remembered reality and not in the name of remembered reality. The seeker must not be stuck on the name. A different movement may arise in the pine-shaped heart and the arteries, a continuous rather than discontinuous, repeated sound. It cannot be connected with individual words. This is remembered reality.
Context This is the fourth state arising in the method of dhikr described by Shaykh Kalimallah.
Broader Remembrance of God (Sufism, #HM6562).
Followed by Total annihilation (Sufism, #HM5775).
Preceded by Sultan-al-dhikr (Sufism, #HM8330).

♦ **HM3898g Anaesthesia**
Description This is an abnormal insensitivity to bodily feelings, commonly induced by the administering of a drug so as to deaden pain, when the effect may be local or general depending on the means of administering of the drug. General anaesthesia is normally associated with total loss of consciousness. An immunity to pain can arise spontaneously or through training, as in the case of extraordinary states of consciousness achieved in rites and rituals.
Broader Paraesthesia (#HM0003).
Related Pain (#HM4031).

♦ **HM3903e Floating**
Description Despite deprogramming, individuals who have previously been brainwashed into becoming members of a cult are still vulnerable to regression to cult personality for a period of up to several weeks. There is confusion as to beliefs, guilt and uncertainty.
Related Deprogramming (#HH4102) Destructive manipulation (#HH1020).

♦ **HM3906g Automatization**
Automaton awareness
Description Although the learning of a particular activity may require full attention, once the activity has been learned it is carried out automatically, having dropped below the level of conscious attention. Motor behaviour, perception and thinking are automized so that, in the case of well established behaviour sequences, intermediate steps disappear from consciousness. This is true of walking, driving a car, even of experiencing pleasure. Attention is shifted from action or perception to abstract thought. The result of 'automatized goal orientation' may be that the person moves towards that goal (such as making money, experiencing pleasure) with a total disregard for future consequences which may be pathological. The process of deautomatization is the undoing of the automatic structure and resumption of conscious carrying out of an activity, whether progressively in order to learn more or regressively in order to compensate for loss of ability (due perhaps to disease or intoxication by a drug).
Related Deautomatization (#HH2331) Breakdown of automatization (#HM8105) Deautomatization and the mystic experience (#HM4398).

♦ **HM3907d Phenomenal plane** (Leela)
Prakriti-loka
Description The divine prakriti, beyond the seven chakras, is that by which the life of the universe is upheld and is the source of gross, phenomenal existence. In this state the player sees the concepts behind the precepts, sees what lies behind the world of sense-objects.
Context The 64th state or square on the board of *Leela*, the game of knowledge, appearing in the eighth row.
Broader Beyond the chakras: the gods themselves (Leela, #HM1141).

♦ **HM3908g Dreaming experience**
Description Australian aboriginal beliefs tell of a time, prior to formation of this world, when beings were not fixed but changed from one creature or feature of the environment to another. This was the time when the law was laid down and is referred to as dreaming. Such dreaming did not cease to exist with the fixing of creation and continues as an invisible surrounding presence which may be experienced, in particular by a shaman or medicine man, through penetrating the screen between the two worlds. Ritual contact with the dreaming also occurs during group rites, when the individual may be absorbed by the dreaming essence of his totem, in particular during rites of passage.
Related Rites of passage (#HH0021)
Tonal state of awareness (Amerindian, #HM2066)
Nagual state of awareness (Amerindian, #HM2036)
Human development through primal religion (#HH1902).

♦ **HM3911c Witness meditation**
Description A phase in meditation when the self witnesses thoughts passing, without control and without attachment.
Related Meditation (#HH0761).

♦ **HM3912c Self-acceptance** (Jung)
Description According to Jung, this is the result of a successful transition from the externally-oriented first stage of life to the more conscious relationship with the interpsychic process and relationship to self which occurs at midlife.
Refs Adler G, Fordham M and Read H (Eds) *The Collected Works of C G Jung*; Samuels, Andrew et al *A Critical Dictionary of Jungian Analysis* (1986).
Related Acceptance (#HM1875) Midlife transition (#HH1366).

♦ **HM3914d Attachment** (Hinduism)
Raga
Description This has been defined as delight in the objects of enjoyment which have already been attained and accompanied by a desire for perpetually enjoying them. There is a yearning for objects of pleasure with which the mind is tinged, in which it is engrossed and which it has enjoyed in the past. Others define it as intense desire for pleasant objects which have not been attained. True knowledge of the self destroys attachment. Until the self is intuited, attachment to objects of enjoyment cannot cease.
Broader Desire (#HM2433) Attachment (Yoga, Zen, #HM3119).
Narrower Affection (#HM9845).
Related Detachment (Hinduism, #HM5091).

♦ **HM3925e Ikia**
Ikia healing
Description This ecstatic trance experienced by healers among Kalahari tribesmen is induced by tribal dancing and chanting; the healer feels *ikia fire* boiling up from the base of the spine to the head, when thoughts disappear and sight extends for great distances and to within others bodies so that the source of illness can be divined. The actual healing process includes laying on of hands and spirit exorcism.
Broader Trance (#HM3236).

♦ **HM3932c Death** (Buddhism)
Decease
Description The final life-continuum consciousness in the case of one becoming is death or decease, the passing away from that existence. The associated consciousnesses are the same 19 associated with rebirth-linking and life-continuum. The object of death consciousness cannot be present, only past or neither of these. Following death there is rebirth-linking again, and the cycle proceeds through the kinds of becoming or sources of life, destiny, stations of consciousness or conscious durations and abodes of beings. With attainment of arahantship, the cycle ceases at the end of death consciousness.
Context In Hinayana Buddhism, this is the 14th and last mode of occurrence of consciousness in which the 89 khandas or kinds of consciousness proceed.
Broader Modes of occurrence of consciousness (Buddhism, #HM6720).
Followed by Rebirth-linking (Buddhism, #HM8266).
Preceded by Life-continuum (Buddhism, #HM6221).

♦ **HM3934d Dubiety** (Buddhism)
Kankhayitattam (Pali)
Broader Perplexity (Buddhism, #HM0812).

♦ **HM3941c Dark night of the soul** (Christianity)
Mystical ladder of divine love
Description Usually the soul spends a long time after the dark night of the senses, even years, before it enters this advanced state. While it enjoys the things of God more readily, it still experiences periods of dryness and darkness. Sometimes these are more intense than those of the night of the senses indicating that the soul is ready for the dark night of the soul.
The advanced soul still has two kinds of imperfections. The continuous ones are the ongoing habits and desires which have existed all along in the spirit. The temporary ones are experienced differently by different people. Some are totally absorbed in their own spiritual life and readily fall into false visions and sensual desires. Others trust their own whims and they are filled with pride to the point that they can become vain and arrogant. They enjoy being seen having mystical experiences. They want to appear holy before others. In others deceit becomes a way of life. Others fall because they give into their spiritual experiences without examining them carefully. These all rate themselves as more spiritual than others.
Having experienced the dark night of the senses, the unspiritual part of the soul has been united with the spiritual part so that they both may be purified. This purification can be described in many ways. One of the ways St John of the Cross describes it is as a mystical ladder of divine love.
The mystical ladder of divine love is both secret and climbed in disguise. It is secret because the soul's understanding, knowledge and other faculties are unaware of it. It is not acquired by the senses. No one else understands what is happening to the soul. The soul cannot talk of it and is silent. The soul's sense and imagination is aware that something unusual and delightful is happening but does not know what it is. All a person can say is that they are satisfied, content, calm and aware of God's presence. This is quite different from spiritual blessings like visions which can be described. The soul feels distant from all created things. It is as though it has been transported to a desert of great beauty and many blessings but where it is terribly alone. It is aware of the utter depravity of humankind and any understanding of God is impossible without His illumination.
It is like a ladder because the soul goes up and down. It goes up toward the good things and treasures of Heaven. The soul goes down in its estimation of itself. The soul is humbled and exalted. There are 10 rungs on this ladder.
The first step on the ladder the soul is made quite ill which is a blessing that ends in the glory of God. During this illness the soul loses its taste for sin and everything else that does not come from God. The incidentals of life become meaningless because of the near presence of God.
The second step of the ladder leads the soul to continually seek God. It is so anxious to find God that it looks everywhere for Him. Whatever it is thinking about its thoughts return to God. When it is eating or sleeping it continually longs for God. During this step the soul begins to recover from its illness.
The third step of the ladder gives the soul strength for action and inspiration so that it does not stop. On this step seemingly great works done by the soul for God become tiny and insignificant. The years spent doing God's work seems like moments because of the fires of love burning within it. The soul is very unhappy because it loves God so much and has done so little for Him. It thinks itself as much worse than all other souls because God's love is teaching it the reverence God deserves. On the third rung of the ladder the soul never condemns any one else. It is also given the strength to move on to the next rung.
The fourth step of the ladder of love gives the soul strength to go on suffering for God's sake. It has subdued and takes little notice of the flesh. It is not seeking self-gratification from God or anything else. It does not pray for blessings because it realizes that it has received plenty of these. It tries to do everything to please God. The soul longs to suffer for God while at the same time it is receiving joy and spiritual experiences. The soul divorces its interior life from everything else and become restless.
The fifth step of the ladder of love the soul longs for God with abandon. Even the shortest separation from God is intolerable, it is weak with love, it faints and feels it will die if deprived of God's love. It is like a hungry dog scavenging through the alleys of the city searching for food. This hunger is totally satisfied when it meets God's love.
The sixth step of the ladder of love cause the soul to run toward God very quickly and with great strength. It is full of hope in God. It no longer collapses as in the fifth step. It is nearly completely pure; its love knows no bounds.
The seventh step causes the soul to become passionate with boldness. It throws caution to the wind and is filled with blessings.
On the eighth step the soul is enabled to embrace God without letting go.
On the ninth step the soul is aflame with perfect passion for God.
On the tenth step, which is not on this earth, the soul is enabled to be completely assimilated by God. It has a close and clear vision of God. It ceases to be a human being. It becomes like God and is called to be one with God. Everything is open to the soul.
It now has freedom of the spirit. It has travelled from the earthly to the heavenly and from the human to the divine.

To ascend this ladder the soul puts on a disguise which protects it from the enemies of the devil, the world and the flesh. This disguise most accurately reflects the desire of the spirit. It consist of faith because it receives no comfort from God or teachers and which protects it from the devil. Faith takes away natural understanding so that it can unite with divine wisdom. Second, it is hope which protects it from the world. Hope takes away from the soul the memory of worldly possessions. Hope in heaven keeps the soul's eyes on God and gives it great passion and desire to search for heaven. Third, the disguise is charity which adds grace to the other two virtues and protects it from the flesh. Charity takes away any desires which are not centred on God.
Context This is the final stage of the Dark Night by St John of the Cross.
Refs St John of the Cross *The Dark Night of the Soul* (1988).
Broader The dark night (Christianity, #HM1714).
Related Solitude (#HH0333) Endurance (#HH0729) Contemplation (#HM2952) Practice of silence (#HH0983).
Preceded by Dark night of the senses (Christianity, #HM1727).

♦ **HM3942d Sticking strongly** (Buddhism)
Patiggaha (Pali)
Broader Wrong view (Buddhism, #HM1710).

♦ **HM3944c Identification with spiritual master** (Sufism)
Personification
Description Since identification may be a mechanism for assimilating respected qualities and idealized values, the seeker may (through intense desire and through ideation of certain images) break from previous attachments and unify with the spiritual guide who is the object of desire and the personification of God. This is a true spiritual step, with the passing away, *fana*, from past experience and being reborn in the experience of the master who is eventually seen in everything with which the disciple has contact. This is the result of striving and waiting patiently for illumination leading to the state of the master, giving insight into created things.
Context The first category of experience related to the three types of object of desire.
Related Personification (#HH3908) Satsang (Hinduism, Yoga, #HM3329) Spiritual master (Sufism, #HH2080).
Followed by Deification (Sufism, #HM0946).

♦ **HM3945d Al-Haqq** (Sufism)
Description The believer is aware of Allah as the truth, forever unchanging. He is the necessary cause of all other existence, which is subject to change and which exists only as it takes the truth of its existence from Him. Al-Haqq manifests in true faith and true words, kept alive and constant by Allah. Those who speak them, listen to them, believe them, are rewarded.
Context A mode of mystical awareness in the Sufi tradition associated with the fifty-first of the ninety-nine names of Allah.
Broader Ninety-nine names of Allah (Sufism, #HH2561).
Related Lahut (Sufism, #HM0741).
Followed by Al-Wakil (Sufism, #HM1382).
Preceded by Ash-Shahid (Sufism, #HM0375).

♦ **HM3947d State of not being greedy** (Buddhism)
Alubbhana (Pali)
Broader Non-attachment (Buddhism, #HM2128).

♦ **HM3948a Kevalajnana** (Jainism)
Omniscience
Description This is knowledge unhindered by karma from former modes of understanding the world. It is the knowledge of the arhat, in particular of the tirthankaras.
Related Enlightenment (#HM2029) Human development (Jainism, #HH0622) Released soul's perception (Hinduism, #HM6558).

♦ **HM3949d Depression of thought** (Buddhism)
Anattamanata cittassa (Pali)
Broader Hatred (Buddhism, #HM4502).

♦ **HM3955d Paccekabodhi** (Buddhism)
Description This is enlightenment, understanding the full nature of reality, reached alone without having heard the Buddha's teaching.
Context One of three types of enlightenment recognized in Theravada Buddhism.
Broader Steps to enlightenment (Buddhism, #HH4019).

♦ **HM3959f Alert passivity**
Description A state in which the bodily systems are slowed down and the mind and emotions quiet and passive, yet there is acute awareness of external stimuli. This state arises under hypnosis and in meditation.
Related Restful alertness (#HM0996) Direct perception (#HM4469) Pure alertness (Psychism, #HM1461) Alpha wave consciousness (#HM2345).

♦ **HM3966c Thinking** (Buddhism)
Cinta (Pali)
Broader Initiation (#HH0230) Faculty of wisdom (Buddhism, #HM3233).

♦ **HM3971c Mansions of spiritual marriage** (Christianity)
Description In these, the seventh mansions, the Lord is pleased to have pity upon this soul, which suffers so much out of desire for him. He brings her to this mansion before consummating the Spiritual Marriage. This is the second heaven.
In earlier mansions the soul may have been blind or lacked understanding of what was happening, but in this one it sees an understands something of the favour it has been granted. This is done through an intellectual vision. First the soul is enkindled and is illumined by a cloud of great brightness. The Holy Trinity; God the Creator, Jesus the Christ, and the Holy Spirit; reveals Itself in all three Persons. Each is seen individually and a wonderful kind of knowledge is given through which it realizes the all three are one. Because this is not an imaginary vision, nothing is seen by the eyes of the body or of the soul. Here all three Persons communicate to the soul. They explain they will dwell within the soul which love God and keeps His commandments.
Each day the soul wonders more. She feels They have never left her and she perceives clearly They are within the most interior of her heart and in its greatest depth. Strangely, she not so absorbed that she is unable to concentrate on anything, but rather in all that serves God she is most alert. When she is not occupied she rest in the companionship of the Trinity. She has great confidence that God will not leave her. This presence is, of course, not always so fully realized. At times, the light may not be so clear, but the soul is always aware that it is experiencing this companionship.
This companionship seems to be preparing the soul for more. The essential part of the soul seems to never leave this dwelling place. The soul seems it is divided, part enjoying itself in quietness while the rest is left with all the trials and occupations.
Divine and Spiritual Marriage cannot be fulfilled perfectly in one's lifetime because the soul is still capable of withdrawing from God. When granting this favour for the first time, the Lord reveals, through an imaginary vision, His sacred Humanity in great splendour, beauty and majesty. He told her it was time for her to take on his affairs as if they were her own and that He would take on her affairs. Other things are said that are easier understood than repeated. This vision is different from others experienced because it leaves the soul quite confused, it comes with great force and it is revealed in a part of the soul where she has never had visions before.
In the union of Spiritual Marriage the Lord appears in the centre of he soul through the most subtle, so far, intellectual vision without entering the soul. The glory that is in heaven is manifest in the soul. The spirit of the soul is made one with God. The extent of God's love is revealed and the two spirits cannot be separated. With the passage of time the effects of Spiritual Marriage become more evident. The soul neither move from the center nor loses its peace.
Even in this state the soul needs to be wary of offending Him. It goes its way with greater misgiving and with greater caution of committing the smallest offence against God. It so desires to serve him that it is confused and suffers from seeing how little it is able to do and how great is its obligations. It delights in penance and so when God takes away its health and strength so it cannot do any penance, its real penance begins. The faculties, senses and passions are not at peace in these mansions but the soul is. Neither the passions nor the physical pains cause any distress.
The little butterfly is dead, full of joy at having found rest. The effects of these mansions are several. First is self-forgetfulness, so complete it is as though the soul does not exist. She does not know nor remember heaven or life or honour for her. Honour of God is all. She has no desire to exist except to do something to further the honour and glory of God. She eats and sleeps but these torment her. She carries out her professional duties. She grieves that she can do nothing on her own. Second is a great desire to suffer. So great is her desire to do the will of God that what ever is His will she considers it to be best. If she is to suffer well and good, if not she does not worry herself as in earlier mansions. When these souls are persecuted, they have great interior joy. They bear no enmity to the persecutors, in fact, they conceive a special love for them. If their tormentors are in trouble, they grieve for them and do anything they can to relieve their ills. They have a strong desire to serve God, to sing his praise and to help some soul if they can. They desire to die and to live a great many years suffering the greatest trials if by doing so they can bring glory to God. Some times they forget this, turn to the thought of enjoying God and the desire to escape the exile of this earth. But then they remember He is with them always and they are content. They are not afraid of death. They no longer desire consolations or favours. These souls have a marked detachment from everything. They desire to be alone to to be busy helping some other soul. They have no aridity or other interior trials but a tender remembrance of Christ and a love for Him. When the soul is neglect, Christ awakens it. He sends messages which should be responded to immediately. There is no reason for the understanding to work, in fact it seems to be dazed along with the other faculties. The soul, now, rarely experiences raptures or flight of the spirit. It is no longer frightened by anything save its own self, because it has found its rest and its true companionship. It is more timorous about itself because the more it sees the greatness and majesty of God the more it realizes how wretched it is.
These effects are not with the soul all the time. Sometimes the Lord leaves the soul to its own devices when all of the evil things seem to try to avenge the time they have not been in power. This may last one day. As a rule the soul gains from the Companion who has given it great determination. It recalls it is always to be humble, God majesty and how great a favour it has been granted. The soul does commit many imperfections, not intentionally which nevertheless cause great torment.
The Spiritual Betrothal is different. The two are frequently separated as is the case of union. In union they are joined but can be separated.
Context The last of seven mansions of the soul's progress described by St Teresa of Avila.
Broader Mansions of the soul (Christianity, #HH1409).
Preceded by Mansions of favours and afflictions (Christianity, #HM2470).

♦ **HM3974c World of spiritual perception** (Sufism)
Abraham of one's being
Description Here the spiritual heart exists embryonically in the potential mystic, a pearl within a shell. This is the true "I", the personal individuality. As Abraham was the intimate friend of God, the Abraham of one's being travels the subtle centres of supraconsciousness. The colour is red.
Context The third of seven subtle stages relating to the seven major prophets of Semitic monotheism as described by Simnani.
Broader Stations of consciousness – Simnami (Sufism, #HM2341).
Followed by World of imagination (Sufism, #HM3616).
Preceded by World of forms (Sufism, #HM3286).

♦ **HM3975e Focus fifteen** (Psychism)
Description There is either no awareness of the space-time continuum or this is not significant.
Context A state occurring during astral projection.
Broader Astral projection (Psychism, #HM1887).

♦ **HM3977c Saviors of God: the silence** (Christianity)
Description This is the ultimate stage of the spiritual exercise; silence. It is not called silence because it contains the ultimate inexpressible despair or joy or hope. It is not the ultimate knowledge which does not condescend to speak or the ultimate ignorance which cannot. The silence is where every person having completed his labours reaches the summit of endeavour, beyond labour, where he ripens fully in silence with the universe. He merges with the abyss. In this state he stands before God and is fully a self. He is also at one with God. And he bears the great, sublime and terrifying secret that even this one does not exist.
Context The final phase in the spiritual exercises of Nikos Kazantzakis.
Broader Saviors of God (Christianity, #HM3420).
Preceded by Saviors of God: the action (Christianity, #HM3462).

♦ **HM4000c Ninth order perceptions - control of systems concepts**
Description At the ninth order the eighth order systems are reorganized in order to find the solution to an eighth order problem and therefore solve conflicts; one eighth order is given priority over another in which conflicts cannot be satisfied.
Context The ninth of ten orders in the perceptual system described by William Glasser.
Broader Orders of perception (#HM0988).
Followed by Tenth order perceptions – universal oneness – meditation (#HM1717).
Preceded by Eighth order perceptions – control of principles (#HM1325).

♦ **HM4001c Wa** (Japanese)
Social harmony — Group harmony
Description Fundamental to Japanese social relations, wa is the spirit of group or social harmony which all consciously strive to maintain and cultivate in any social setting. Wa may also be experienced by the individual in relation to the environment as a whole. It is felt to be disrupted when particular individuals act so as to stand out in a group.

♦ **HM4002c Non-discriminative knowledge** (Systematics)
Experience of wholeness
Description Any situation to which attention is directed is a monad, but some exemplify the systemic attribute of universality more strongly than others. Wholeness is experienced as universal and omnipresent but relative. It may be transformed into identity. Non-discriminative knowledge means functional order than can issue only as automatic behaviour. The combination of confused immediacy and the expectation of finding an organized structure gives the monad a progressive character. It is experienced as what it is, but as holding the promise of being more than what it appears to be. Wholeness may be experienced as being too comfortable to be satisfying.
Context The first in a sequence of twelve modes of knowledge, identified by J G Bennett, inspired by G Gurdjieff.
Broader Systematics (#HH2003).
Followed by Two-fold knowledge (Systematics, #HM1234).

♦ **HM4006f Mind drift**
Lack of concentration
Description Despite numerous techniques and efforts, the mind has an innate tendency to shift attention from full concentration on the task in hand to some other topic or topics. It may revert to a subject worrying or concerning the individual or simply drift from one theme to another by free association until consciously brought back to the subject under consideration.
Related Concentration (#HM4226)
Inability to achieve concentration (Yoga, #HM0298).

♦ **HM4007c Refining the self** (Taoism)
Setting up the foundation
Description At this stage temporal accretions encrusting the senses, biases, and habitual patterns, are all burnt away. By conquering the self, quelling anger and cupidity, a healthy pattern of order is established as the foundation for further work.
Context The first of seven stages of the Taoist alchemical firing process through which reality is cultivated and the self is restored.
Broader Alchemy (Taoism, #HH5887).
Followed by Recognition of the true mind (Taoism, #HM6633).

♦ **HM4008d Puzzlement** (Buddhism)
Vimati (Pali)
Broader Perplexity (Buddhism, #HM0812).

♦ **HM4009c Jesus falls the first time under His cross** (Christianity)
Description The believer is aware that it is the weight of his own sins, not the weight of the cross, which cause Jesus to suffer the pain of falling, from weakness due to lack of blood and His many wounds. He begs that, through the falling of Jesus, he himself may not fall into mortal sin. Loving Jesus more than himself, he repents with his whole heart and begs never again to separate himself from Him. He asks that he will love Jesus always and that Jesus will do with him as He wills.
Context The third station of the cross.
Broader Way of the cross (Christianity, #HM3516).
Followed by Jesus meets his afflicted mother (Christianity, #HM3611).
Preceded by Jesus made to bear the cross (Christianity, #HM3500).

♦ **HM4010d Al-'Alim** (Sufism)
Description The believer is aware of Allah as the all-knowing. In contrast to one's own very limited life, power and knowledge one should attempt to feel the unlimited perfection of Allah. Repetition of this name will make the heart luminous, revealing divine light.
Context A mode of mystical awareness in the Sufi tradition associated with the nineteenth of the ninety-nine names of Allah.
Broader Ninety-nine names of Allah (Sufism, #HH2561).
Followed by Al-Qabid (Sufism, #HM2767).
Preceded by Al-Fattah (Sufism, #HM0002).

♦ **HM4017d Al-Ba'ith** (Sufism)
Description The believer is aware of Allah as the resurrector, raising all creation to life on the day of judgement. Also within this life there are many deaths and revivals. Ignorance is death, knowledge is life. Reviving one's self or another from the death of ignorance to the life of knowledge, one sees al-Ba'ith manifest and truly believes.
Context A mode of mystical awareness in the Sufi tradition associated with the forty-ninth of the ninety-nine names of Allah.
Broader Ninety-nine names of Allah (Sufism, #HH2561).
Followed by Ash-Shahid (Sufism, #HM0375).
Preceded by Al-Majid (Sufism, #HM5446).

♦ **HM4024c Esoteric divine consciousness** (Psychism)
Description The one consciousness, that which permeates all and which which divided all, controlled by thoughts of all soul minds.
Related Divine consciousness (#HM0563).

♦ **HM4026b Four noble paths** (Buddhism)
Description Four of the ten worlds described in Nichiren Soshu Buddhism, these are characterized by inner effort. No longer trapped by demands of desires and looking beyond day-to-day exingencies one starts to develop true potential.
Broader Ten worlds (Buddhism, #HM2657)
Nichiren shoshu buddhism (Buddhism, #HH3443).
Narrower State of learning (#HM3662) State of buddhahood (#HM1873)
State of realization (#HM0450) State of bodhisattva (#HM1225).
Related Six paths (Buddhism, #HM1914).

♦ **HM4031e Pain**
Description Pain is not simply the sensation of touch - which may be felt when, for example, a local anaesthetic dulls awareness of pain during an operation. It is said to be the necessary consequence of consciousness, and may vary in intensity from a mild headache to third degree burns. Pain threshold's vary from person to person and analgesics actually work by raising the threshold of pain. Not so much a mode of awareness itself, it nonetheless may act as a means of understanding the deep significance of life; and pain has been used by ascetics and mystics to induce altered states and feelings of transcendence.
Related Suffering (#HM0471) Anaesthesia (#HM3898).

♦ **HM4032d Presumption** (Buddhism, Tibetan)
Description Presumption is compared with inference, in that although both may have a fresh reaching of a fresh conclusion, inference also implies a fresh conceptual understanding while presumption implies no real understanding of why the conclusion is correct. Presumption may be any of five kinds: there may be no real reason, for example simply hearing a statement without knowing why it is true; the reason may be contradictory, it may be non-determining, it may be irrelevant or it may be correct but the individual while putting forward the argument is not convinced of it himself. In other words, knowledge of something true is invalid because it is presumed true for no reason, for an invalid reason, or for a valid reason but without understanding of why it is correct.
Context One of the five invalid ways of knowing of Tibetan Buddhism.
Broader Ways of knowing (Buddhism, #HH0873).

♦ **HM4033e Accidental psychic attack** (Psychism)
Description Fear and anxiety are produced accidently through another's strongly felt concern for one's safety or health. This may be unconsciously passed on to a third person in the vicinity.
Broader Psychic attack (Psychism, #HM0837).

♦ **HM4036d Exalted concentration** (Buddhism)
Sublime concentration
Description Here there is unification in profitable or moral consciousness of the fine-material and immaterial spheres (form-realm and formless realms).
Context According to Hinayana Buddhism, concentration is considered as of one kind (monad), of two kinds (dyads), of three kinds (triads), of four kinds (tetrads) or of five kinds (pentad). In the fourth triad, this is the second concentration.
Broader Concentration (Buddhism, #HM6663).
Followed by Measureless concentration (Buddhism, #HM0496).

♦ **HM4037d Uncrookedness** (Buddhism)
Avankata (Pali)
Broader Straightness of body (Buddhism, #HM1424)
Straightness of thought (Buddhism, #HM0253).

♦ **HM4040d Al-Rafi** (Sufism)
Description The believer is aware of Allah as the exalter, lifting up those who are enlightened.
Context A mode of mystical awareness in the Sufi tradition associated with the twenty-third of the ninety-nine names of Allah.
Broader Ninety-nine names of Allah (Sufism, #HH2561).
Followed by Al-Mu'izz (Sufism, #HM1116).
Preceded by Al-Khafid (Sufism, #HM3644).

♦ **HM4041c Sanctity** (Sufism)
Context The eighth of ten states acting as gateways orienting the mystic on the journey to the Absolute, as described by Ansari.
Broader Stations of consciousness - Ansari (Sufism, #HM2317).
Related Sanctity (Christianity, #HM0560).
Followed by Realities (Sufism, #HM3755).
Preceded by Mystical states (Sufism, #HM3607).

♦ **HM4042d Az-Zahir** (Sufism)
Description The believer is aware of Allah in His outwardness, as manifest to those who have been given wisdom and reason. As light makes manifest and yet is not seen, so He is manifest by what He has created but is Himself veiled by His own light. One sees what is closest. One is closest to one's self. As man is Allah's best creation and all creation is in man, then seeing His attributes in the perfect creation one is one sees Allah the manifest and has complete faith.
Context A mode of mystical awareness in the Sufi tradition associated with the seventy-fifth of the ninety-nine names of Allah. Contemplation of this name in the Naqshbandiyya order follows the stages of minor and major saintship, and the internal spiritual state is strengthened and widened.
Broader Ninety-nine names of Allah (Sufism, #HH2561).
Followed by Al-Batin (Sufism, #HM1209).
Preceded by Al-Akhir (Sufism, #HM7112).

♦ **HM4043d Al-Rahim** (Sufism)
Description The believer is aware of Allah as merciful, giving eternal salvation in the hereafter. Those who use his benificence for good are rewarded with eternal gifts.
Context A mode of mystical awareness in the Sufi tradition associated with the second of the ninety-nine names of Allah.
Broader Ninety-nine names of Allah (Sufism, #HH2561).
Followed by Al-Malik (Sufism, #HM4061).
Preceded by Ar-Rahman (Sufism, #HM3539).

♦ **HM4048b Bliss**
Description Although it has been said that "ignorance is bliss", bliss is in fact better described as divine unknowingness and an awareness of that unknowingness. An example is the state of an artist in the process of creating when "I" and the process of creating are one.
Refs Coxhead, Nona *The Relevance of Bliss* (1985).
Related Joy (#HM1172) Ecstasy (#HM2046) Saccidananda (#HM0592)
Religious experience (#HM3445) Ananda (Hinduism, Buddhism, Yoga, #HM3227).

♦ **HM4049d Nadi** (Balinese)
Being in another world
Description A stage further than willing suspension of disbelief, the person watching an actor playing a part part actually experiences the part as though it were himself and physically joins in. He is in "another world". The same experience is that of the absent minded professor or the inventor who is so taken up with what is going on inside his mind that he no longer responds to externals but is in a trance-like state.

♦ **HM4050d Plane of neutrality** (Leela)
Saraswati
Description When electrical and magnetic or sun and moon energies are balanced, neutral or psychic energy flows up the spinal column. At this plane, negative and positive die away, only neutral remains and the player controls its flow. He is stabilized in a state of pure knowledge and witnesses the game.
Context The 47th state or square on the board of *Leela*, the game of knowledge, appearing in the sixth row.
Broader Sixth chakra: time of penance (Leela, #HM4412).

♦ **HM4052d Al-Mu'akhkhir** (Sufism)
Description The believer is aware of Allah as delaying advancement. It may be that efforts are with wrong intent or that he will eventually place more value on his reward if it is delayed. Whatever his state, the good servant accepts it and continues his efforts while trying to understand the reason for it. He is close to Allah as he does what pleases Allah and is pleased with what Allah does.
Context A mode of mystical awareness in the Sufi tradition associated with the seventy-second

of the ninety-nine names of Allah.
Broader Ninety-nine names of Allah (Sufism, #HH2561).
Followed by Al-Awwal (Sufism, #HM6139).
Preceded by Al-Muqaddim (Sufism, #HM5331).

♦ **HM4054c 'Ishq** (Sufism)
Description This is a state of passionate love and longing for union with God which results in reaching the station of *mahabbat*.
Related Love of God (Sufism, #HM4273)   Passionate love (Islam, #HM3656).

♦ **HM4058 Autism** (Yoga)
Kshipta
Description There is neither the patience not the intelligence required in order to contemplate a supersensuous object and therefore the person can neither think of nor comprehend any subtle principle. The mind may at times be in a state of concentration, but this is through intense envy or malice and is not a yogic concentration.
Context One of five states of mind classification identified in yoga.
Followed by Stupefaction (Yoga, #HM1346).

♦ **HM4060f Modes of awareness associated with use of opium and similar drugs**
Use of morphine — Use of heroin
Description Used to relieve pain and induce sleep, these substances are also addictive mood-changers. They produce dream-like experiences. The initial use may result in a delightful "high", a delirium with grandiose and sexual hallucinations; but withdrawal sickness is intense, and addicts continue to take the drug more to ward off withdrawal symptoms than to achieve the initial euphoria.
Broader Modes of awareness associated with psychoactive substances (#HM0584).
Related Delirium (#HM7186).

♦ **HM4061d Al-Malik** (Sufism)
Description The believer is aware of Allah as the sovereign lord. He owns and rules absolutely the whole creation. Kings are servants to Him, servants see Him as master of their masters, all are accountable to him.
Context A mode of mystical awareness in the Sufi tradition associated with the third of the ninety-nine names of Allah.
Broader Ninety-nine names of Allah (Sufism, #HH2561).
Followed by Al-Quddus (Sufism, #HM0235).
Preceded by Al-Rahim (Sufism, #HM4043).

♦ **HM4062c Tawba** (Sufism)
Repentance — Turning to God — Tauba — Tawbat — Tuba
Description In the moment of repentance one weans one's self from sin. One who is loved by God can thereby truly repent, repent of his thought of repentance, turning towards God and regretting past actions which kept him away from God. Of three forms, the first is that of general believers in God and implies fear of divine retribution, seeking of favour from God or regret for disobedience; the second is of spiritualists and lovers of God who seek God's love in repentance of their sins; and the third is that of those who repent their individual will and seek annihilation of self in God. This last is the state of those who start out on the spiritual path. Further, Shaykh Fariduddin refers to six kinds of repentance when attachment to worldly things is left behind and the individual comes to love God in a life of perfect asceticism: (1) Tawba-i dil, repentance of the heart, implying removal of worldly desires from the heart and allowing direct communion with God. (2) Tawba-i zaban, of the tongue, referring to the attitude of silence adopted when recollecting the names of God. (3) Tawba-i chasm, of the eye, when the eyes are closed to forbidden things. (4) Tawba-i gosh, of the ear, when there is listening only to the names of God. (5) Tawba-i pa, of the feet, when movement towards evil is repented. (6) Tawba-i nafs, of the carnal soul, when lust, passion and sensual desires are repented. In addition there are three more, repentance of present, past and future prohibited deeds - tawba-i hal, tawba-i madi, tawba-i mustaqbil.
Context The first stage in a systematic account of the various stations of the Sufi spiritual path, and also in the path to sainthood of the Naqshbandiyya order. According to Shaykh Abu Sa'id ibn Abi'l-Khayr, this is the third of 40 stations or maqamat the Sufi must possess for his journey on the path of Sufism to be acceptable.
Broader Mystic stations (Sufism, #HM3415).
Suluk of the Naqshbandiyya order (Sufism, #HM4356)
Stations of consciousness – ibn-Abi'l-Khayr (Sufism, #HM2424).
Related Repentance (#HH0441)   Al-Tawwab (Sufism, #HM4334).
Followed by Zuhd (Sufism, #HM4450)   Wara (Sufism, #HM0286).
Iradat (Sufism, #HM8901).
Preceded by Inabat (Sufism, #HM3484).

♦ **HM4064b Fani** (Sufism)
Description The name of a traveller on the *tariqa* or Sufi spiritual path at the level of complete obliteration from the self.
Context Spiritual progress on the spiritual path is marked by the attainment of specific *maqamat* or station.
Broader Mystic stations (Sufism, #HM3415).
Related Fana (Sufism, #HM1270).

♦ **HM4065d Delight** (Buddhism)
Amodana (Pali)
Broader Fivefold happiness (Buddhism, #HM0747).

♦ **HM4071e Fourth dimensional clairvoyance** (Psychism)
Description There is clairvoyant or visionary perception of objects from within as well as from outside.
Broader Clairvoyance (Psychism, #HM3498).

♦ **HM4076e Sensation of unpleasant born of contact with the psychical** (Buddhism)
Cetosamphassaja dukkha vedana (Pali)
Broader Feeling (Buddhism, #HM2270).

♦ **HM4081d State of not feeling infatuation** (Buddhism)
Asarajjitattam (Pali)
Broader Non-attachment (Buddhism, #HM2128).

♦ **HM4085c Heroic sentiment** (Hinduism)
Aesthetic emotion of energy — Rasa of resoluteness
Context One of eight kinds of aesthetic sentiment or rasas in Indian psychology.
Broader Aesthetic emotion (Hinduism, #HM0205).

♦ **HM4090d Un-comprehension** (Buddhism)
Anabhisamaya (Pali)
Broader Delusion (Buddhism, #HM0184).

♦ **HM4093d Al-Mu'id** (Sufism)
Description The believer is aware of Allah as the restorer of all things, recreating all things to their previous form, so that at the end of all things only Allah exists, as it was in the beginning. At the same time, all that has existed is recorded within the eternally existent and alive. At the day of judgement the good will be rewarded, the bad punished, all creatures will be recreated perfect and their souls returned to them.
Context A mode of mystical awareness in the Sufi tradition associated with the fifty-ninth of the ninety-nine names of Allah.
Broader Ninety-nine names of Allah (Sufism, #HH2561).
Followed by Al-Muhyi (Sufism, #HM1007).
Preceded by Al-Mubdi (Sufism, #HM0622).

♦ **HM4094d Formication**
Cocaine bug — Tactile hallucination
Description This is a specific tactile hallucination when the person feels as though there is something crawling on or under the skin. It arises in particular during the delirium following alcohol withdrawal and as a cocaine withdrawal symptom. It is often accompanied by the delusion that the sensation is due to insects or worms.
Broader Hallucination (#HM4580)   Tactile hallucination (#HM6118).
Related Modes of awareness associated with use of cocaine (#HM1594)
Modes of awareness associated with alcohol consumption (#HM0134).

♦ **HM4100b God experience**
Description Constant pursuit of spiritual activities and aims leads to transformation of unconscious, subconscious and conscious experience so that the divine centre of being, God, as source of happiness, knowledge, power, love and perfection is understood and perceived as God experience.
Related Universal religion (Yoga, #HH0746).

♦ **HM4103c First chakra: fundamentals of being** (Leela)
Physical plane
Description All aspects of the first chakra are fundamental to human existence and important for human life; their moods are responsible for development of the higher self. The insecurity of the individual at this level may give rise to violence but also to the development of material technology. Mastery of this chakra leads to physical health and openness to knowledge. Abstaining from lower desires and attachments, non-assertive, the player is said to be able to become invisible at will.
Context The first row of the board of Leela, the game of knowledge, this is represented by the physical plane, and also groups the squares representing the states of genesis, maya, anger, greed, delusion, conceit, avarice and sensuality – aspects fundamental to human existence. Located between the anus and the genitals, this chakra is related to the element earth, to the sense of smell and to the colour yellow. It is compared to the octave comprising the first 7 years of temporal life and to self-centred search for material security.
Broader Lila (#HM2278).
Narrower Maya (#HM0953)   Anger (#HM1433)   Greed (#HM1931)
Genesis (#HM4474)   Conceit (#HM0098)   Avarice (#HM0987)
Delusion (#HM0697)   Sensual plane (#HM3627)   Physical plane (#HM1024).
Related Muladhara (Yoga, #HM2893).
Followed by Second chakra: realm of fantasy (Leela, #HM3651).

♦ **HM4104b Transcendence**
Description Transcendence occurs in a person's experience as the immediately communicated presence of particular identities and awarenesses. There is a pervasive unity or sameness through experience of the kinship of all awarenesses and meanings. As a particular being, the person is taken up in events that are not governed by his particularity, but in the presence of which the person is alert as to who he is and also alert in awarenesses that are not particularly his. Transcendence names immediate amertness in both the person's consciousness and in the awareness constituted as the presence of something. Transcendence means dialogical contact, address and claim. Awareness and consciousness are events in which transcending continually and immediately goes on. Any state of awareness or consciousness is beyond its singularity in the immediate presence of other awarenesses and consciousnesses.
Described as the highest, most inclusive and holistic levels of consciousness, transcendence is a relating to one's self, to particular others, to all human beings, to all species, to nature and to the cosmos. Numerous meanings or facets are listed:
– Loss of self-consciousness, self-awareness and self-observation so that there is forgetfulness of self, in concentration or meditation on something outside the individual psyche, can lead to self-forgetfulness.
– The body and all its components are transcended so there is no identification with it.
– One with those who have lived before and who will live after, time is transcended in the unity of experience. Emotional ties can be felt with people who are no longer living in the accepted sense.
– Although the individual is rooted in a particular culture, he can transcend that culture and become one with the whole species. Self actualization leads to resistance to enculturation and the ability to examine one's own culture with detachment while still living within it.
– Understanding one's self leads to acceptance of one's past, aware that one was free to act and make choices and is responsible for the consequences without negative feelings towards one's self because of this.
– In harmony with one's surroundings, activities are carried out as required by circumstances, in accordance with others' needs and not in response to one's own selfish demands. The ego is transcended.
– There is a mystic experience of fusion or oneness with another person, with other people, even with the whole cosmos.
– The normally negative aspects of existence such as pain, sickness and death are accepted as necessary. The usual reactions of anger, bitterness and resentment are reduced.
– The natural world is accepted as it is, irrespective of how it responds to the individual's apparent needs.
– Levels of synergy are ascended until there is no distinction between "them" and "us".
– Self can be transcended in unselfish love for another or others, even for the whole human race. One's own self can be included in this love.
– There is detachment from less-than-worthy situations and behaviour of others, even when such behaviour is direct against one's self.
– Self-confident and self-determining, the individual resists the roles others try to impose and transcends their opinions.
– Beyond the super-ego the individual is at the level of intrinsic conscience.
– One is strong rather than weak, independent rather than dependent, responsible rather than

HM4104

irresponsible.
- One is aware of the possible as well as the actual.
- Opposites and dichotomies are transcended in unity, ultimately in holistic perception of the cosmos as one.
- Rising above the personal will, the individual embraces his destiny.
- All that one does surpasses what one has previously done or thought one could do.
- As one reaches further heights of human potential one appears god-like. While accepting one's cultural roots, one transcends them.
- Living serenely in the realm of being, one accepts as natural the ecstasy of the peak experience and the insight such experience brings.
- Both the subjective and the non-involved or neutral objective attitude are transcended.
- Reconciled to the perception of evil, denials and refusals are transcended.
- Oneness with all humanity means physical distances are meaningless in transcendence of space.
- Allowing things to happen rather than forcing them to happen means lack of strife, there is always a sense of having arrived rather than trying to get somewhere.
- Fear is transcended by courage or lack of fear.
- There is a sense of belonging in the universe, a right to be here.
- Differences are still seen to exist but are accepted and enjoyed in wonder, while the individual rises above them in the universal similarity that exists among human beings.
- Even one's own views, beliefs and value systems can be transcended in acceptance of a larger, inclusive structure.

Refs Dowdy, Edwin (Ed) *Ways of Transcendence* (1982); Kulandran *Concept of Transcendence* (1982); Richardson, H W and Cutler, D R (Eds) *Transcendence* (1969); Scott, Charles E *Boundaries in Mind* (1982).

Narrower Secular transcendence (#HM2804).
Related Transcendence (#HH0841) Human synergy (#HH0517)
Being cognition (#HM2474) Peak experiences (#HM2080)
Self-remembering (#HM2486) Immediate boundary awareness (#HM0035)
Way of transcendence (Christianity, #HH6566).

♦ **HM4110c Mansions of incipient union** (Christianity)
Description The delights of these 5th mansions cannot be described because there is nothing on earth to compare them with, nor word sufficient to do them justice.
In these mansions it is strength of the soul and not of the body that is important. The soul needs to keep back nothing for God will have it all.
Here everything goes to sleep to the world and to ourselves. Thought is not possible and there is no need to find ways of suspending thought. The soul is drawn from the body, a kind of delectable death is experienced. The mind becomes dumbfounded, the body does not move and even breathing is not noticed. This is one type of union with God and is quite different from the transports of the devil over vanities of this world. The joy is greater. The delights, satisfaction and peace is different. One seems to effect the gross parts of the body and the other penetrates the marrow of the bones. This difference is obvious to those who have had the experience. The experience is not being able to see, hear or understand for a short period of time during which God and ourselves are implanted into our souls in such a way that it cannot be forgotten and there is no doubt that God had been there.
These mansions are like the life of a silk worm. In the spring warmth, it feeds on leaves, spins a cocoon, dies and emerges as a butterfly. In the heat of the love of the Holy Spirit the soul uses confession, good books, sermons and good meditations to grow in strength. It then begins to build a house in God by renouncing self-love, self-will and attachments to earthly things, by practising prayer, penance, mortification and obedience and by letting the worm die. It then enters these mansions as a butterfly for never more than a half hour. The soul cannot understand how it merits this blessing. It desperately desires to please God, longs to suffer great trials and desires to do penances and have solitude and for all to know God. The butterfly knows not where to settle. Everything of the earth is unsatisfactory. It desires to do more than it can for God. Seeming weakness when doing penances are now strengths. It is no longer bound by relationships, friendships or property and it grieves to fulfil these obligations out of fear to offend God. Everything wearies it.
While there is rest even in the knowledge of the source of this suffering, this state has new and more profound trials. The soul is not resigned to the will of God. It does conform to God's will but with many tears and sorrow at being unable to do more. It is besieged with grief at offenses against God and with the loss of souls. In this God has blessed the soul with some of the suffering of Christ.
This experience is a gift and cannot be acquired through an act of will.
Thinking this is the end of the journey is foolish and will close the door to further mansions. A person who has reached these mansions and does nothing else or even strays from this path will benefit others. But true union with God is not possible without submission to the will of God. A person must continue living this life, die to their own free will and never doubt the possibility of true union. Being unoffensive to God and leading a religious life is not enough. Self love, self esteem, censoriousness of neighbours and failure to love neighbours eat away at any real virtues and keep the soul far away from union. God's will is for the soul to be perfect, to love God and to love the neighbour. The best way of knowing the soul is doing this is to be sure of loving the neighbour because being sure of loving God is not possible. Avoid wish dreaming about loving the neighbour but act on the neighbour's behalf. Avoid false humility and being so concerned about the experience of prayer that the neighbour is neglected. The soul needs to do violence to its own will that it serve the neighbour and love God.
In this state it is still possible to go astray and follow the devil. Even this close to God the soul can be deceived by believing it is doing good when it is actually, gradually reasserting its own will. In spite of the level of detachment from worldly things, the soul can be lured back to earthly pleasures. To guard against these dangers, the soul prays to be in God's hands, never has confidence in itself, is ever vigilant and watches for progress in virtue. The soul knows not advancing in virtue is a bad sign.
Context The fifth of seven mansions of the soul's progress described by St Teresa of Avila.
Broader Mansions of the soul (Christianity, #HH1409).
Followed by Mansions of favours and afflictions (Christianity, #HM2470).
Preceded by Mansions of spiritual consolations (Christianity, #HM3443).

♦ **HM4111b Religious states** (Christianity)
Description Three successive phases are defined: (1) That in paradise before original sin when man was in a state of integrity, immune from death and endowed, through God's self-communication, with supernatural grace. (2) That under original sin, before Jesus Christ or unjustified through faith, love and baptism. (3) That of the former sinner (through original and personal sin), the just man sanctified through Christ's grace. The three phases comprise one religious situation embraced from the beginning through the sanctifying and merciful will of God to communicate himself in Christ. Sin cannot nullify it, and it becomes explicitly present in the Christian era and through individual actions, through faith and the sacraments.

♦ **HM4112c Firaq** (Sufism)
Separation
Description The longing, *shugh*, which arises from the state of *uns*, brings awareness of separation from intimacy with the beloved. The most intense state of separation from the image of God is said to have arisen when man became aware of himself and of the distance between himself and nature. He was cast from the paradisal, unconscious state in nature and has longed for a new union ever since.
Related Uns (Sufism, #HM1957) Showq (Sufism, #HM0762).

♦ **HM4120c Altered ego states**
Description In Freudian psychology, this expression is used to describe the transient distortion of awareness or perception of the self or parts of the self, of other objects or the surroundings, and consequent feelings of unreality. There may be a distortion of the senses, a feeling of separation between mind and body, parts of the body may seem detached. These states arise to defend the ego against perceptions or ideas which might threaten it at the conscious or unconscious level. There is defensive regression to earlier libidinal and ego modes, the ego "splits" into observing and experiencing portions. Particular examples are *micropsia*; *déjà vu*; *depersonalization*; *derealization*.
Broader Ego (#HH0636).
Narrower Déjà vu (#HM1240) Depersonalization (#HM1248).
Related Archaic ego states (#HM0216).

♦ **HM4128c Understanding as the plane of development** (Buddhism)
Understanding being the plane of culture
Description This understanding belongs to the last three of the four paths.
Context On the Path of Purification of Hinayana Buddhism, panna (understanding) is considered as of one kind (monad), of two kinds (dyads), of three kinds (triads) or of four kinds (tetrads). There are five dyads, four triads and two tetrads. All have the characteristic of penetrating the individual essences or true nature of states (monad). In the fifth dyad, this is contrasted with understanding being the plane of seeing or discernment.
Broader Understanding (Buddhism, #HM4523).
Related Understanding as the plane of seeing (Buddhism, #HM6192).

♦ **HM4131d State of unabated desire** (Buddhism)
Anikkhittacchandata (Pali)
Broader Faculty of energy (Buddhism, #HM1470).

♦ **HM4136c Jesus nailed to the cross** (Christianity)
Description The believer is aware of Jesus as he extends his hands to sacrifice His life for the salvation of mankind. As Jesus was held in contempt, he begs that his own heart be nailed at Jesus' feet so that it always remains there to love Him. Loving Jesus more than himself, he repents with his whole heart and begs never again to separate himself from Him. He asks that he will love Jesus always and that Jesus will do with him as He wills.
Context The eleventh station of the cross.
Broader Way of the cross (Christianity, #HM3516).
Followed by Jesus dies on the cross (Christianity, #HM3719).
Preceded by Jesus stripped of his garments (Christianity, #HM3166).

♦ **HM4143b Perception aggregate** (Buddhism)
Sanna-khanda (Pali) — Awareness of perception-group of conscious existence
Description In Hinayana Buddhism, perception is considered as anything that is perceived as being perception, whether past, present or future, internal or external, subjective or objective, gross or refined, superior or inferior, far or near. It is also considered as follows:
It is of one kind in that all that its intrinsic nature or essence is perceiving.
It is threefold in that it is: good (profitable or moral) and associated with profitable consciousness; bad (unprofitable or immoral) and associated with unprofitable consciousness; or indeterminate and associated with indeterminate consciousness.
There are 89 divisions, since there is no consciousness dissociated from perception and consciousness has 89 divisions.
The characteristic of perception is perceiving; its function is signalling similarity - perceiving or recognizing that something is similar to what was perceived before; its manifestation is interpreting using the sign that was apprehended; its proximate cause is an objective field as it appears (in the mind).
Context One of the five interacting aggregates that produce the illusory ego. This group, in southern Buddhism, is classified according to six objects, perceptions of: form, sound, odour, taste, bodily impression and mental impression.
Refs Buddhaghosa, Bhadantacariya *The Path of Purification* (1980); Buddhaghosa, Bhadantacariya *The Path of Purity* (1975).
Broader Five aggregates (Buddhism, #HH3321).
Related Vijna (Buddhism, #HM3617) Feeling aggregate (Buddhism, #HM4983)
Dispositions of consciousness (Buddhism, #HM2098)
Awareness of corporeality-group of conscious existence (Buddhism, #HM2108)
Awareness as mental-formation group of conscious existence (Buddhism, #HM2050).

♦ **HM4152e Extreme exaltation** (Psychism)
Description In psychic terms, this describes a soul flight of the soul-mind under the influence of meditation or an emotional religious experience.
Related Emotional high (#HM1400) Aesthetic state (Psychism, #HM3416).

♦ **HM4154c Shukr** (Sufism)
Gratitude
Context An attitude of thanksgiving preceding the station of *sabr* or patience on the Sufi spiritual path. However, it is the eighth stage on the path to sainthood of the Naqshbandiyya order, following *sabr*.
Broader Suluk of the Naqshbandiyya order (Sufism, #HM4356).
Followed by Sabr (Sufism, #HM3418) Rida (Sufism, #HM4190).
Preceded by Sabr (Sufism, #HM3418).

♦ **HM4162d Al-'Ali** (Sufism)
Description The believer is aware of Allah as the most high, although he is nonetheless as close to the lowest of all. He is above all that has been or ever will be, encompassing all.
Context A mode of mystical awareness in the Sufi tradition associated with the thirty-sixth of the ninety-nine names of Allah.
Broader Ninety-nine names of Allah (Sufism, #HH2561).
Followed by Al-Kabir (Sufism, #HM4437).
Preceded by Ash-Shakur (Sufism, #HM1934).

♦ **HM4169d Al-Wasi'** (Sufism)
Description The believer is aware of Allah as the all-embracing, the one without limits. His beautiful attributes are infinite; all good attributes in men are reflections of al-Wasi'. His creation

is infinite, no two individuals are alike. His knowledge is limitless, nothing can be hidden from Him or escape His power. Since His mercy is infinite, all may turn to Him. His tolerance is infinite, the believer should not lose hope.
Context A mode of mystical awareness in the Sufi tradition associated with the forty-fifth of the ninety-nine names of Allah.
Broader Ninety-nine names of Allah (Sufism, #HH2561).
Followed by Al-Hakim (Sufism, #HM4799).
Preceded by Al-Mujib (Sufism, #HM0504).

♦ **HM4173c Mumin** (Sufism)
Description The name of a traveller on the *tariqa* or Sufi spiritual path at the level of a believer in God.
Context Spiritual progress on the spiritual path is marked by the attainment of specific *maqamat* or station.
Broader Mystic stations (Sufism, #HM3415).

♦ **HM4180d Spiritual lust** (Christianity)
Description When a beginner on the journey of the spirit is overcome with desire for the pleasures of this world arising from the spirit he is said to suffer from spiritual lust. These pleasures may be of any sensual nature, sexual, comfort, beauty, quiet or hundreds of others. Often this happens in the midst of a spiritual or religious exercise, such as meditating, contemplating, participating in a sacrament or praying deeply, when they have no power to stop it. The non-spiritual dimension of their life takes over and impure acts or deeds result. Spiritual lust has three causes. The first is that in the midst of these activities the individual experiences great delight and pleasure and the non-spiritual dimension interprets the experience in a sensual way. The individual while hating these rebellious thoughts and feelings is nevertheless aware of them. The second reason is the devil brings these impure thoughts and feelings to consciousness and if heeded can cause great harm. Some people out of fear of these things becomes careless about prayer and may stop to avoid them. This is of course what the devil wants. The third source of these experiences is the fear of having impure thoughts and feelings. Some individuals have their most profound spiritual experience within sight of others because they are driven by their devotion of being noticed by others. Friendship made out of enjoying the company of the other and in which the memory and love of God does not grow but causes remorse instead is another form of spiritual lust. Love of the other that begins as lust will end as lust.
Context This is an imperfection of beginners in the Dark Night by St John of the Cross.
Broader Beginner in spiritual life (Christianity, #HM0102).
Related Lust (Hinduism, #HM4666) Spiritual envy (Christianity, #HM3818)
Spiritual sloth (Christianity, #HM0491) Spiritual anger (Christianity, #HM0936)
Spiritual pride (Christianity, #HM4533) Spiritual avarice (Christianity, #HM0642)
Spiritual gluttony (Christianity, #HM0507).

♦ **HM4185e Body of glory** (Psychism)
Description The most subtle of all bodies, able to materialize and dematerialize at will. Said to serve mankind by materializing on earth without birth, and then to leave without death.
Related Avatar (#HH0079).

♦ **HM4190c Rida** (Sufism)
Satisfaction — Contentment — Acquiescence — Taslim
Description Resulting from trust in God, this has been regarded (Al-Sarraj) as the last stage on the journey to God – nothing further may be attained through mystical effort. It is a state of blissful satisfaction granted by God to those close to him, whatever arises in their path towards God. The state may arise: from the point of view of man and his evenness in relation to God; from the point of view of God; from leaving the problem of whose side to God alone. There are no relationships and the Sufi is both at rest when he has nothing and unselfish when he has something.
Context The seventh stage in a systematic account of various stations of the Sufi spiritual path. However, it is the ninth stage on the path to sainthood of the Naqshbandiyya order. Also, according to Shaykh Abu Sa'id ibn Abi'l-Khayr, this is the ninth of 40 stations or maqamat the Sufi must possess for his journey on the path of Sufism to be acceptable.
Broader Mystic stations (Sufism, #HM3415)
Suluk of the Naqshbandiyya order (Sufism, #HM4356)
Stations of consciousness - ibn-Abi'l-Khayr (Sufism, #HM2424).
Related Happiness (#HM2409).
Followed by Mukhalafat-i nafs (Sufism, #HM7336)
Nearness-awareness (Sufism, #HM2377)
Divine-love awareness (Sufism, #HM2401).
Preceded by Shukr (Sufism, #HM4154) Tawakkul (Sufism, #HM0807)
Recollection (Islam, Sufism, #HM2351).

♦ **HM4191d Concentration accompanied by bliss** (Buddhism)
Concentration accompanied by ease
Description This concentration is unification in three jhanas of the fourfold and four of the fivefold reckoning. It is one way of experiencing access concentration.
Context According to Hinayana Buddhism, concentration is considered as of one kind (monad), of two kinds (dyads), of three kinds (triads), of four kinds (tetrads) or of five kinds (pentad). In the fourth dyad, this is contrasted with concentration accompanied by equanimity.
Broader Concentration (Buddhism, #HM6663).
Related Access concentration (Buddhism, #HM4999)
Concentration accompanied by equanimity (Buddhism, #HM8021).

♦ **HM4194d Understanding subject to cankers** (Buddhism)
Description Understanding is affected by cankers or is the object of cankers.
Context On the Path of Purification of Hinayana Buddhism, panna (understanding) is considered as of one kind (monad), of two kinds (dyads), of three kinds (triads) or of four kinds (tetrads). There are five dyads, four triads and two tetrads. All have the characteristic of penetrating the individual essences or true nature of states (monad). In the second dyad, this is contrasted with understanding free from cankers.
Broader Understanding (Buddhism, #HM4523).
Related Understanding free from cankers (Buddhism, #HM0756).

♦ **HM4208c Zahid** (Sufism)
Description The name of a traveller on the *tariqa* or Sufi spiritual path at the level of renouncing everything except God and becoming an ascetic.
Context Spiritual progress on the spiritual path is marked by the attainment of specific *maqamat* or station.
Broader Mystic stations (Sufism, #HM3415).

♦ **HM4212g Touch** (Buddhism)
Phassa (Pali)
Broader Contact (Buddhism, #HM2708).

♦ **HM4226c Concentration**
Description Concentration is a special form of attention characterized by disciplined organization and fixation of attention on the grasping and shaping of subject matter containing meaning and value. It is the requisite condition for optimal cognitive achievement and is a feature of those techniques in which the field of consciousness is closely restricted, such as hypnosis and autogenic training.
Related Yoga (Yoga, #HH0661) Ascription (#HM3198)
Mind drift (#HM4006) Contemplation (#HM2952)
Dharana (Hinduism, #HM2566).

♦ **HM4228d Lack of real grasping** (Buddhism)
Apariyobahana (Pali)
Broader Perplexity (Buddhism, #HM0812).

♦ **HM4236d Al-Hayy** (Sufism)
Description The believer is aware of Allah as the alive, all cognizant, all activity. Within His knowledge is all that is and will be known. Comprehended in his activity is all existence. As man is higher than animals which are higher than plants, so some men are more alive than others in their knowledge and action. Not knowing himself nor being aware of his existence a man is as good as dead and his words are deadly. The sacred name *Hayy* pronounced by men of *ma'rifah* brings life.
Context A mode of mystical awareness in the Sufi tradition associated with the sixty-second of the ninety-nine names of Allah.
Broader Ninety-nine names of Allah (Sufism, #HH2561).
Followed by Al-Qayyum (Sufism, #HM5006).
Preceded by Al-Mumit (Sufism, #HM3868).

♦ **HM4239d Enmity** (Buddhism)
Virodha (Pali)
Broader Hatred (Buddhism, #HM4502).

♦ **HM4243d Supramundane concentration** (Buddhism)
Transcendental concentration
Description This is the unification or collectedness associated with the noble path.
Context According to Hinayana Buddhism, concentration is considered as of one kind (monad), of of two kinds (dyads), of three kinds (triads), of four kinds (tetrads) or of five kinds (pentad). In the second dyad, this follows mundane concentration.
Broader Concentration (Buddhism, #HM6663).
Related Eightfold way (Buddhism, #HM2339).
Preceded by Mundane concentration (Buddhism, #HM7234).

♦ **HM4245e Emotional clairaudience** (Psychism)
Description An urgent message is heard psychically within the head, whether of warning or comfort, either from someone with whom there are close ties or from a spirit guide.
Broader Clairaudience (Psychism, #HM4333).

♦ **HM4246d Delightfulness** (Buddhism)
Pamodana (Pali)
Broader Fivefold happiness (Buddhism, #HM0747).

♦ **HM4248b Faculty of recollection** (Buddhism)
Satindriya (Pali)
Narrower Recalling (#HM3806) Remembrance (#HM1484) Mindfulness (#HM1423)
Remembering (#HM0405) Recollection (#HM1563) Unforgetfulness (#HM1705)
Correct recollection (#HM0180) Power of recollection (#HM3450)
Faculty of recollection (#HM0047) Deep penetration by memory (#HM1689).

♦ **HM4250d Rectitude** (Buddhism)
Ujuta (Pali)
Broader Straightness of body (Buddhism, #HM1424)
Straightness of thought (Buddhism, #HM0253).
Related Rectitude of mental body (Buddhism, #HM5402)
Rectitude of consciousness (Buddhism, #HM9001).

♦ **HM4251c Compunction** (Christianity)
Description Compunction is the awareness of one's indigence and coldness and of one's need for God. It brings with it faith, sorrow and humility. Most important, it brings hope in the mercy of God. It can transform prayer from a cold formality, with the emphasis on one's self, to a living act, bringing one's self into a real, spiritual and personal relationship with God face to face, the I-Thou relationship.
Related Fundamental dialogue (#HM3225) Mental prayer (Christianity, #HH8672).

♦ **HM4254f Modes of awareness associated with use of phencyclidine and similar drugs**
Description Small doses lead to mild euphoria, with the sensation of floating, and numbness. There may be hallucinations and paranoid ideation, possibly leading to suicide. After-effects include depression, anxiety and irritability. Larger doses may lead to coma or convulsions.
Broader Modes of awareness associated with psychoactive substances (#HM0584).

♦ **HM4256d Kerygma** (Christianity)
Description One of the three ways in which the mission of the Church is manifest, this refers to a state in which the word of God is manifest in preaching, the proclamation of the life, death, resurrection and ascension of Jesus Christ.
Related Diakonia (Christianity, #HM0810) Koinonia (Christianity, #HM0144).

♦ **HM4265d Fine-material-sphere concentration** (Buddhism)
Form-realm concentration
Description This is profitable unification of mind or collectedness of moral thought associated with the fine-material sphere or jhana.
Context According to Hinayana Buddhism, concentration is considered as of one kind (monad), of two kinds (dyads), of three kinds (triads), of four kinds (tetrads) or of five kinds (pentad). In the fifth tetrad, this is the second concentration.
Broader Concentration (Buddhism, #HM6663).
Related Unincluded concentration (Buddhism, #HM5768)
Form-realm consciousness (Buddhism, #HM2257)
Sense-sphere concentration (Buddhism, #HM1097)
Immaterial-sphere concentration (Buddhism, #HM0696).

♦ **HM4273a Love of God** (Sufism)
Mahabbat
Description This state arises naturally when the divine mysteries are revealed and the soul is

illuminated. In fact knowledge and love of God are one. It is the greatest of God's favours, arising because of God's love for the devotee. The common man expresses his love because of God's kindness and favour; there is also love as attachment through realization of God's attributes; and, highest of all, there is pure love of God reflected in a natural feeling of affinity based on knowledge of God. Then God is loved for His own sake.

In one exposition of al-Ansari, mahabbat is the final station but above all stations, being the guiding principle of the whole progress of the mystic. It has three stages: *rasti* – rectitude; *masti* – drunkenness; *nisti* – annihilation. It also has three degrees. The first comes from considering the divine attributes, is rooted in following the *sunna* and grows according to responding to one's spiritual needs. It cuts short insinuations, finds service a pleasure and consoles affliction. The second arises from considering God's attributes, from reflecting on the signs of God and from exercising the mystic stations. It prefers God above all else, the tongue is infatuated when speaking of Him and contemplation of Him attaches the heart. The third is beyond description. It ravishes, cutting short expression and rendering allusion too subtle. Man's love for God, properly directed, must transcend the conventions of religion which conceived it and finally be consumed in God's abiding love.

Ghazali sees this journey as follows: Firstly, love refers to existence in the physical world, to the individual, his family and the society he lives in; then it is shown in personal achievements and gains. In these two cases, God is the only cause for his existence and protection. This is followed by the experience of selfless love attracted towards beauty and goodness; love is then revealed in the soul as it inclines towards moral values related to beauty and goodness. In these cases, because God is the ideal of beauty and goodness and the supreme moral being, He is the true object of love. Finally there is the affinity natural between lover and beloved, when the soul is drawn naturally to God because of pre-existing fellowship with Him, the affinity of the soul and its source. It shares the divine nature and attributes and may become Godlike itself, attaining eternal life through knowledge and love.

Thus the love of the devotee passes through a number of stages to attain perfection in detachment from all and joyful communion with God. The state of perfect love may be reached through recollection of God, *dhikr*, the recitation of the names of God. It results from *'ishq*, passionate desire for union with God and restlessness at separation from him.

The state of mahabbat is a spiritual state or *hal*, related to the Sufi's own experience and granted as a special favour by God, being the only means of unification with God. When there is love for God nothing else has any value; suffering at separation from Him can be ignored while there is disinterest in favours granted. There is satisfaction even in the midst of affliction; and love increases with obedience and respect for the beloved's commands. Perfect love is motiveless and is unaffected in the way love for a motive is affected when that motive is fulfilled. In love, human qualities are transmuted and the association to the self is lost. The Sufi does not distinguish between himself and his beloved. Love removes the veil of human qualities, annihilating the self in the beloved. Dead to his own attributes – *fana* – the Sufi is alive in the attributes of the loved – *baqi*.

Context The doctrines of *sabr*, *mahabbat* and *ma'rifa* were elaborated by the Sufi scholar Abu Talib al-Makki and followed and interpreted by many others. For al-Ansari, mahabbat is the final or 101st ground, or the 61st of 100 stations. Mahabbat is also the ninth stage in a systematic account of various stations of the Sufi spiritual path.

Refs Valiuddin, Mir *Love of God* (1968).

Broader Mystic stations (Sufism, #HM3415) Human development (Sufism, #HH0436)
Higher states of consciousness (Sufism, #HM2365).
Narrower Rasti (#HM7222) Masti (#HM6097) Nisti (#HM7626).
Related Sabr (Sufism, #HM3418) Fana (Sufism, #HM1270)
Baqi (Sufism, #HM4594) 'Ishq (Sufism, #HM4054)
Ma'rifa (Sufism, #HM2254) Love to God (Islam, #HM5116)
Love of God (Christianity, #HH5232) Recollection (Islam, Sufism, #HM2351)
Divine-love awareness (Sufism, #HM2401).
Followed by Mushahada (Sufism, #HM3521).
Preceded by Nearness-awareness (Sufism, #HM2377).

♦ **HM4277f Modes of awareness associated with use of khat**

Description Chewing of this plant produces mild euphoria and social affability.
Broader Modes of awareness associated with psychoactive substances (#HM0584).

♦ **HM4279d Third duty** (Christianity)

Description Free yourself from the mind that thinks to put all things in order and hopes to subdue phenomena. Free yourself from the heart's terror that seeks and hopes to find the essence of things. Conquer the last, the greatest temptation of all: hope. Without hope, but with courage sail calmly toward the abyss. Hope for nothing, fear nothing, seek the freedom that is beyond the mind and the heart. This is the third duty.

Context The third duty in the preparation for the spiritual exercises of Nikos Kazantzakis.
Broader Saviors of God: the preparation (Christianity, #HM3264).
Preceded by Second duty (Christianity, #HM2861).

♦ **HM4280c Malamat** (Sufism)

Blame

Description A station on the Sufi divine path accepted by one of the level of faqr, when condemnation by the world guards him against external influences and he concentrates on realization of the true self.
Broader Faqr (Sufism, #HM3427).

♦ **HM4282d State of hell** (Buddhism)

Jigoku (Japanese)

Description This is the state of suffering. There is no ability to think or act freely, one's life force is so reduced that both physically and spiritually one might as well be already dead. Grief, fear, poverty and illness are obvious hells; but each individual may have his own particular hell, perhaps incomprehensible to others. There is a rage which is not the rage of anger but of frustration and self-destruction or the tendency to destroy one's own surroundings. There is pessimistic worry about the worst possible outcome of an event and suffering as though it had already happened, and the state is characterized by a lack of hope. It is the experience of hell that makes one appreciate happiness and sympathize with the unhappiness of others. It is a powerful motivation for action to keep the world out of hell.

Context The lowest of the ten worlds described in Nichiren Soshu Buddhism.
Broader Six paths (Buddhism, #HM1914) Ten worlds (Buddhism, #HM2657)
Four evil paths (Buddhism, #HM1252) Three evil paths (Buddhism, #HM0923)
Nichiren shoshu buddhism (Buddhism, #HH3443).
Related Suffering (#HM0471).
Followed by State of hunger (Buddhism, #HM0150).

♦ **HM4284b Nirvitarka samadhi** (Yoga)

Description There is ontic identification with the external reality of the object contemplated, no cogitation but transcendental reflection.

Context The second level of samprajnata samadhi.
Broader Samprajnata samadhi (Yoga, #HM2896).
Followed by Vicara samadhi (Yoga, #HM3880).
Preceded by Vitarka samadhi (Yoga, #HM4451).

♦ **HM4288c Salik** (Sufism)

Description The name of a traveller on the *tariqa* or Sufi spiritual path at the level of a pilgrim realizing the mystical experiences of *suluk*.

Context Spiritual progress on the spiritual path is marked by the attainment of specific *maqamat* or station.
Broader Mystic stations (Sufism, #HM3415).
Related Awareness of the mystic journey (#HM2900).

♦ **HM4291c God as wholly the same** (Judaism)

Description As opposed to experience of God as wholly other, an experience which might culminate in the Kiss of God and death, this is the heightened quality of awareness when the wonder of God's oneness with creation is experienced in everyday life.
Related Presence of God (#HM2961) Human development (Judaism, #HH3029)
God as wholly other (Judaism, #HM4501).

♦ **HM4297d Knowledge of recollection of previous existence** (Buddhism)

Knowledge of recollection of past life — Direct knowledge — Higher knowledge

Description Recollection may be one birth, many births, one aeon, many aeons until there is, in the case of Buddhas, no limit. Again, depending on the level of enlightenment, previous lives are seen only with difficulty or clearly.

A beginner, having attained the four jhanas in succession, emerges from the fourth and then adverts to his most recent act, then to the one before and so on, covering a whole night and day in reverse order. This is continued, covering periods for days, a fortnight, years until the moment of rebirth-linking to this existence is reached. Then he removes this and makes the moment of mentality-materiality at the moment of death in the previous existence his object (this may be hard to see, but repetition of the exercise gradually brings success). There is appearance of absorption and the knowledge of past life arises. This develops into knowledge of other previous lives.

Knowledge of aeons or cycles of dissolution and evolution or contraction and expansion infers also knowledge of what supersedes contraction or expansion. Thus each cycle is four aeons. The world cycle may be destroyed by fire, water or air, when all is demolished up to a particular Brahma world. These are the limits on contraction or dissolution. In breadth one Buddha-field is demolished (in the field of his birth, ten thousand world spheres or systems; in the field of his authority, one hundred thousand million world spheres or systems; in the field of his sphere, to an infinite extent, as far as he wishes). The monk is then aware of where he was in a particular aeon of world contraction or expansion, of his name, his mode of life, how long he lived, whether pleasure or pain predominated.

Context One of the five kinds of direct or higher knowledge developed in Hinayana Buddhism.
Related World cycles (Buddhism, #HH2002) Appearance of absorption (Buddhism, #HM1618)
Knowledge of supernormal powers (Buddhism, #HM7672)
Knowledge of penetration of minds (Buddhism, #HM5232)
Knowledge of the divine ear element (Buddhism, #HM5982)
Knowledge of passing away and reappearance of beings (Buddhism, #HM0748).

♦ **HM4298c Dwelling in the first jhana** (Buddhism)

Jhana with happiness and bliss born of seclusion

Description Having departed absolutely from the desires of the senses and from immoral states or unprofitable things – the five hindrances, which are the opposites of the jhana factors – the meditator enters the first jhana and dwells there. Bringing the subject of meditation to mind, the process of appearance of absorption is gone through. Attention is detached from objects of sense and looks inward into the mind. Thoughts are discursive, there is a relaxed but energetic feeling. There is concentration which is incompatible with lust, happiness which is incompatible with ill will, applied thought incompatible with sloth and torpor, bliss incompatible with agitation and worry, sustained thinking incompatible with perplexity or uncertainty. Applied thought keeps the mind on its object; sustained thought anchors the mind on its object; happiness refreshes the mind with the success of the effort not to be distracted by hindrances; bliss intensifies this happiness; unification or collectedness of mind centres the mind evenly and rightly.

The first jhana is also said to be good in three ways: beginning (purification of the way), which is *access*; middle (intensification of equanimity), which is *absorption*; end (satisfaction), which is *reviewing*. It also has ten characteristics. The length of time the meditator is able to remain in the jhana depends on first completely purifying the mind of states which obstruct concentration. As meditation progresses, the subject of meditation, for example the earth kasina, may be extended at the access or absorption level. Since the jhana factors only appear crudely at first the beginner is not advised to review the first jhana too much – trying to reach a higher jhana may mean falling away from the first and failing to attain the second.

Through practice, the meditator acquires mastery in the first jhana through the habits of: adverting to the jhana; attaining the jhana; resolving and steadying the duration of the jhana; emerging from the jhana; reviewing the jhana. He can then end his attachment to the first jhana and commence doing what is necessary to attain the second.
Broader Jhana (Buddhism, Pali, #HM7193) Dhyana (Hinduism, Buddhism, #HM0137).
Related Analysis (Buddhism, #HM5089) Investigation (Buddhism, #HM1177)
Fivefold happiness (Buddhism, #HM0747) Appearance of absorption (Buddhism, #HM1618)
Dwelling in the fivefold jhana (Buddhism, #HM6553)
Concentration of the first jhana of five (Buddhism, #HM0297)
Concentration of the first jhana of four (Buddhism, #HM4456)
First trance of the fine-material sphere (Buddhism, #HM2450).
Followed by Dwelling in the second jhana (Buddhism, #HM7121).

♦ **HM4299d Al-Badi** (Sufism)

Description The believer is aware of Allah as the incomparable originator of creation, needing no design or prior knowledge. All comes from nothing, everything in creation is unique – no two men are the same. Through his attention and curiosity man discovers all that Allah has made – he creates nothing of himself. Rather than thinking one has invented and created things, one should see this discovery as a beginning, a means of receiving the love of Allah in one's heart.

Context A mode of mystical awareness in the Sufi tradition associated with the ninety-fifth of the ninety-nine names of Allah.
Broader Ninety-nine names of Allah (Sufism, #HH2561).
Followed by Al-Baqi (Sufism, #HM6786).
Preceded by Al-Hadi (Sufism, #HM1287).

♦ **HM4302c Heaven**

Description Heaven normally refers to a mode of existence after this life, achieved by virtue or grace, which some believe permanent while others consider a temporary respite after which the individual is again born into earthly life. Some traditions believe it also the dwelling place of gods or of angels, or of God, while others that union with God is beyond heaven. The term "heaven"

is also used to describe a mode in this life where there is total harmony with God and with one's place in creation, or which results from activities, beliefs and attitudes developed during this or previous lives.
Related Heaven (#HH3564)   Heaven awareness (Psychism, #HM2565).

♦ **HM4309c Wise attention** (Buddhism)
Description In Hinayana Buddhism, wise attention is much emphasized in developing skill in absorption. By means of wise attention the necessary wisdom or enlightenment-factors arise and are grown, fulfilled and developed to exert, restrain and encourage the mind as these become necessary; it acts as food or fuel for such factors. It occurs in penetration of intrinsic or individual essences and general characteristics. If the mind is slack, wise attention produces elements of initiative, launching, and persistence in factors to exert it; finally it produces the happiness or rapturous enlightenment factor which is happiness itself. Similarly, if the mind is agitated, wise attention acts as food for growth, fulfilment, development and perfection for the factors of tranquillity, concentration and equanimity to restrain it.
Broader Attention (#HH0756)   Attention (#HM1817).
Related Perfect wisdom (Buddhism, #HM7844)   Skill in absorption (Buddhism, #HH4777)
Enlightenment factors (Buddhism, #HM6336).

♦ **HM4310d Satoguna** (Leela)
Description Beyond manifestation, the player can still not attain liberation and cosmic consciousness for he remains linked to karma. He must return to earth and continue the game. Nevertheless, in satoguna he is linked with true nature and the highest frequency vibration. In samadhi he becomes pure light.
Context The 70th state or square on the board of *Leela*, the game of knowledge, appearing in the eighth row.
Broader Beyond the chakras: the gods themselves (Leela, #HM1141).

♦ **HM4311b Psychic states** (Sufism)
Maqamat
Description Following progress on the outer self, *kushesh*, the Sufi is ready to embark on pursuing progress on the inner self under the influence of divine attraction, *keshesh-jazba*, which increases on passing from one of the eight to ten stages or psychic states to the next. Within these states are changing psychic conditions, *ahwal*, which mark increasing insight and illumination. The states or maqamat have been enumerated by A Reza Aresteh as: muraqaba; qurb; showq; mehr; omid; uns; mushahida; itmianan; yaqin.
Refs Arasteh, A Reza *Growth to Selfhood* (1980).
Narrower Uns (#HM1957)   Mehr (#HM1266)   Omid (#HM4477)
Showq (#HM0762)   Yaqin (#HM4426)   Itmianan (#HM0050)
Mushahada (#HM3521)   State-awareness (#HM2305)
Nearness-awareness (#HM2377).
Related Divine path (Sufism, #HM3371)   Kedesh-jazba (Sufism, #HM1026)
Mystic stations (Sufism, #HM3415)   Unification with God (Sufism, #HM3864).

♦ **HM4312f Anomalies in the flexibility of associations**
Grass-hopper mind — Preoccupation — Brooding — One-track mind — Perseverance
Description Over-flexibility is evinced in the rapid flitting from one subject to another which is typical of the grass-hopper mind. Associations are too loose so that there is a low level of distraction and the train of thought is lost. *Circumstantiality* is the following up of associated ideas although the original train of thought is not lost and eventually is reverted to. The extreme form of over-loose associations is *mania* and, to a lesser extent, hypomania. There is a rapid play on words, with punning and rhyming but no coherence.
The reverse case is manifest in *preoccupation*. Associations are available but their range is limited. There is furious thought around one theme, often an unsolved problem. Included in this area is *brooding*, a preoccupation with a subject with high negative affective capacity. Preoccupation is common at some time with everyone, but others are prone to preoccupation and are said to have a *one-track mind*. Rather than an over-fluidity of associations, here there is a lack of associations or an over-fixity of focus. In severe mental disorders this may reach an extreme beyond normality, when it is referred to as *perseverance*.
Narrower Mania (#HM2787)   Hypomania (#HM0514).
Related Resistance to shift (#HM5329).

♦ **HM4319e Cataleptic trance** (Psychism)
Catalepsy
Description The characteristics of this deep trance are rigidity of the limbs and insensitivity to pain. Even if the limbs may be moved, they remain exactly where they are placed. The person is unconscious (as opposed to the cataplexic state) and may remain in the state for several days. Catalepsy may arise in a number of nervous disorders, being characteristic of hysteria; it also occurs in some forms of schizophrenia or it may result from emotional shock. Intentional cataleptic states may be induced by pressure on certain arteries, by occult techniques or by drugs. As opposed to suspended animation or coma, respiration and heart rate are virtually normal. In psychic terms, this is a deliberately desired trance-like state when a medium is in a state of deep hypnosis and taken over by some other intelligence, which uses the medium's body to speak and act. On returning to normality there is no memory of what has occurred during the trance. It is not clear whether there is a complex inner fantasy life at this time or total unconsciousness.
Broader Trance (#HM3236).
Related Cataplexy (#HM1955)   Self-induced trance (Psychism, #HM1964)
Attainment of cessation (Buddhism, #HM5438).

♦ **HM4320d All-accomplishing wisdom**
Description A self-less action, free of karma, within meditation, dedicated to universal enlightenment or willingness to take consequences.

♦ **HM4321c Compline** (Christianity)
Hours of compline
Description The canonical hours of compline traditionally take place between 9 and 12 pm. They dramatize the recognition the failure to live up to the full potential of the day. The mood is abject contrition.
Broader Canonical hours (Christianity, #HM1167).
Followed by Matins (Christianity, #HM0160).
Preceded by Vespers (Christianity, #HM1468).

♦ **HM4323d Hating** (Buddhism)
Dussana (Pali)
Refs Humphreys, Christmas *Concentration and Meditation*.
Broader Hatred (Buddhism, #HM4502).

♦ **HM4326d Absolute plane** (Leela)
Brahma-loka
Description Established in truth, the player will no longer have to go through the karmic roles; he merges with the absolute subtle principle. He may have reached this state through practising mercy. He cannot achieve cosmic consciousness directly but must return to the square of earth – from whence, with the insight he has attained beyond manifestation, he can be helped to this goal.
Context The 69th state or square on the board of *Leela*, the game of knowledge, appearing in the eighth row.
Broader Beyond the chakras: the gods themselves (Leela, #HM1141).
Preceded by Mercy (Leela, #HM3434).

♦ **HM4327c Religious experience as the realization of duty**
Description The vocabulary of religious experience illustrates the the sense of obligation involved. The word "religion" itself carries the meaning of "bound", bound in conscientious and dutiful devotion. Kant described the sense of reverence which arises at moral law. The term yoga implies a "yoke". All religions have within them particular systems of discipline or duty whose goal is either union (in the sense of absorption) with the transcendent or perfect communion between the self and the transcendent. Particularly relevant is the experience of Arjuna in the Bhagavad Gita, who is told that inaction is not an option but that the path of salvation is action in accord with duty as regards the roots of the activity rather than its fruits. Similarly, the *dharma* is the obligatory path for the devotee in Hindu and Buddhist traditions; and in Confucianism there is action in accord with the structured system of human relations, *li*.
Broader Religious experience (#HM3445).
Related Dharma (#HH0093)   Karma marga (Hinduism, #HH1211).

♦ **HM4331c Nirguna brahman**
Description Reality beyond names and forms, the aspect of Brahman that is beyond the gunas.
Broader Brahman (#HH1226).
Related Three gunas (#HH0413).

♦ **HM4332g Bodily relaxation**
Refs Brand, L W and W G *Preliminary Explorations of Psi-conducive States* (1973).

♦ **HM4333e Clairaudience** (Psychism)
Description The individual hears sounds, either as though from outside – *objective clairaudience*, or from within the head – *subjective clairaudience*, when there is no apparent physical source of the sound. It may be music, speech or other noises, possibly carrying a message.
Narrower Emotional clairaudience (#HM4245).
Related Audible thought (#HM1317)   Clairscent (Psychism, #HM0318)
Pseudo-hallucination (#HM4336)   Clairvoyance (Psychism, #HM3498)
Clairgustance (Psychism, #HM0412)   Auditory hallucination (#HM6704)
Extrasensory perception (ESP, #HM2262).

♦ **HM4334d Al-Tawwab** (Sufism)
Description The believer is aware of Allah as turning man to repentance and accepting his repentance. He constantly wakes the believer's heart from the sleep of unconsciousness and disharmony into harmony with His will, so that one tries to eliminate the cause of sin. Effort is rewarded with forgiveness, as is forgiveness others.
Context A mode of mystical awareness in the Sufi tradition associated with the eightieth of the ninety-nine names of Allah.
Broader Ninety-nine names of Allah (Sufism, #HH2561).
Related Tawba (Sufism, #HM4062).
Followed by Al-Muntaqim (Sufism, #HM4513).
Preceded by Al-Barr (Sufism, #HM7022).

♦ **HM4335d Religious feeling**
Description In that feeling is an awakening to the objective world and to others, it is a necessary, although subjective, element of piety.
Refs Davis, Charles *Body as Spirit* (1976).

♦ **HM4336f Pseudo-hallucination**
Description Unlike true hallucinations, these perceptions without external stimuli are usually supposed to lack substantiality and to appear in inner as opposed to outer space – although some claim that they are not seen with the inner eye but, like true hallucinations, experienced as if through the sense organs. They are, however, subjective and the person experiencing them is aware that they that they are produced by his or her imagination, referring to them as "visions". The experience seems to be somewhere intermediate between imagery and true hallucination, although it may slide into a true hallucination. The subjectivity of the experience and the fact that it is not shared by others has led to the assertion that the apparitions and voices experienced by mediums and clairvoyants are actually pseudo-hallucinations.
Broader Hallucination (#HM4580).
Related Clairvoyance (Psychism, #HM3498)   Clairaudience (Psychism, #HM4333)
Mental imagery therapy (#HH0072).

♦ **HM4337d Illumination due to insight** (Buddhism)
Description Illumination such as never before experienced arises. The meditator leaves his meditation subject and instead enjoys illumination. This is karmically unprofitable, because he mistakes the path.
Context One of the ten imperfections of insight of Hinayana Buddhism.
Broader Imperfection of insight (Buddhism, #HM9722).

♦ **HM4338c Assembling the five elements** (Taoism)
Description The integration or reconstitution of the human being may be described in terms of the assembly of the five elements. This process refers to the progressive interrelationship of conscious knowledge, real knowledge, sense, essence and intent (normally symbolized by fire, water, metal, wood and earth). Through this process the essence of conscious knowledge is united with the sense of real knowledge, this unification being accomplished through the medium of intent. Concentrated attention is used to bridge the habitual gap between conscious and unconscious. Associated with this process is the recognition of the manner in which each element habitually conditions another to maintain a circular pattern of conditioning. The work of assembling is thus a matter of extracting the five elements in their primordial or unconditioned form (in which they foster one another) from their conditioned form (in which they overcome one another). This conditioning cycle may be reversed through practice so that essence produces real knowledge, which produces true sense, producing true intent, producing open consciousness, producing essence. When the five elements foster each other in this way, they are integrated within the celestial design and act as a single energy. Through such reversals the world is used for self-refinement rather than self-indulgence.
This process may also be understood as the integration of the five "eyes" potentially available to humans: the physical eye (as the ordinary organ of sight); the celestial eye (as the power of clairvoyance); the wisdom eye (as the power of intuitive insight); the objective eye (as the power to see things as they are in reality); and the enlightened eye (as the power to see both absolute

HM4338

and relative truth, encompassing all other eyes).
It may also be understood as five energies "returning to their origin" as a result of a state of mental and physical collection through which the primordial energy is recovered. The body is experienced as unmoving, the mind unstirring, the nature tranquil, feelings are forgotten, and the physical elements are in harmony.
Context The fourth of seven stages of the Taoist alchemical firing process through which reality is cultivated and the self is restored.
Broader Alchemy (Taoism, #HH5887).
Followed by Merging of yin and yang (Taoism, #HM1330).
Preceded by Restoration of celestial awareness within the mundane (Taoism, #HM6534).

♦ **HM4340a Superessential contemplation** (Christianity)
Divine contemplation — State of purity
Description John Ruusbroec refers to this state as being in divine light in accordance with the divine mode of being. There is establishment in a state of purity which transcends all understanding. It is the eternal reward for virtues and for the whole of life. Such a state cannot be achieved by exercises, it is only open to those with whom God wishes to be united in His Spirit and to enlighten through Himself. Because of their own incapacity, few attain this divine contemplation, and also because of the mysterious nature of the light in which one contemplates. Learned and subtle reflections are not sufficient for understanding its meaning - words and all that can be understood in the guise of creature are alien to and far beneath the truth. However, once united with God and enlightened in this truth a person can understand the truth through itself.
(1) God the Father of light wishes us to see. This is the generation and birth of the Son, the eternal light in whom all blessedness is seen and known. It is necessary for the person to be well ordered externally and unhindered internally – his interior must be as unaffected by his external good works as though they were not being performed. He must cleave to God with devoted intention and love. He must be in a state devoid of particular form or measure; this is a blissful losing his way in darkness, never again to find himself in a "creaturely" way. The loving spirit dies to itself in this darkness and it is here that an incomprehensible light is born, the son of God, in whom the person is able to see and contemplate eternal life. In the simple being of his spirit the spirit receives the resplendence of God, it becomes this very resplendence. In the ground of his being the contemplative is the same light with which he sees.
(2) Able to see, the person contemplates the eternal coming of the Bridegroom into his spirit. This is uninterrupted new birth and new illumination. Creaturely activity and exercise of virtue cease. In sublime nobility of spirit God works alone. There is only eternal contemplation of and gazing at the light with the light in the light. The Bridegroom comes in the eternal now, always received with new pleasure and new joy. This delight and joy is his very self.
(3) God's Spirit sends the person forth in a state of eternal contemplation and blissful enjoyment in God's own manner. This is eternal going forth, eternal activity without beginning. The created being depends on the eternal being and is one with it. The eternal being and life are in God and like God, the Holy Trinity. This image of the Holy Trinity is the wisdom of God in which God contemplates himself and all things in the eternal now. In this divine image all creatures have eternal life. This is the image to which the Holy Trinity has created us. God wills that we go from ourselves into this divine light, pursuing and possessing that image. In an eternal act of gazing by means of the inborn light the person is transformed by and become one with that light. All things are contemplated without distinction in a simple act of seeing in divine resplendence. In this contemplation the person remains free and master of himself.
(4) Having thus attained his eternal image, possessed the Father's bosom, enlightened with divine truth, receiving the eternal birth and going out in a state of divine contemplation, the interior, contemplative person knows a meeting in love where the highest blessedness resides. In the turning of the Father to the Son and the Son to the Father, the Holy Spirit, which is the love of them both and one with them, arises. When a person understands this wonder, his spirit is raised above itself and is made one with the Spirit of God. There is measureless savouring and seeing, like that of God Himself, of the riches which it has itself become. This meeting is actively renewed without ceasing in the mode of being of God, the Father giving himself to the Son and the Son to the Father in eternal wellbeing and loving embrace by means of the Holy Spirit. There is a darkness that encompasses all divine modes, activity and properties, a crossing over and immersion in essential bareness where all reflected in the mirror of divine truth passes away in simple ineffability. All is encompassed in a state of blissful blessedness while the ground is completely uncomprehended unless through essential Unity. Before this the Persons of God give way, as does all that lives in God, there is nothing but the eternal state of rest in a blissful embrace of loving immersion.
Refs Dupré, Louis and Wiseman, James A (Eds) *Light from Light* (1988); Wiseman, James A (Trans) *John Ruusbroec* (1985).
Broader Mystical contemplation (#HM2710).
Related Meeting God (Christianity, #HM0541) Contemplative life (Christianity, #HH2145).

♦ **HM4341 Totemic awareness**
Totemic being — Totemic identity
Description Where there is an understanding that a person is made up of manifold identities, such a person is free to explore totemic identities. Achieving this requires that the person explore the depths of his imaginative life. He may then recognize experientially how he is a certain object, whether the object lives in some way or whether it possesses an an inanimate existence (such as a stone undulating on a river bed, rain falling, or fire flaming from a tree). He is then able to partake of that existence in addition to his own, embarking on a manifold existence in keeping with a certain alter–ego as rich and mysterious as the identity that he conventionally allows to nest in himself. The person is then free to be both himself as conventionally recognized by others as well as an inhabitant of the imaginary world of any particular totem with which he identifies. People identify with chosen totems because of the sense of enlarged life which they then enjoy and because of the imaginative vitality which this identification excites. The totemic condition allows the individual to partake of innumerable languages, free from the confinement of the logic inherent in word patterns. The individual finds for himself a new form of interior expression. This also implies a new kind of dialogue with nature, not one of classification and exploitation, but one that prefigures an inchoate courtesy more in keeping with the language of heraldry. The person becomes other than he normally understands himself. The totemic experience gives the individual access to other lives, fulfilling his own by inhabiting the realm conventionally known as that of the imagination.
For the peoples that practice the totemic experience (such as the Australian Aborigines), a person can only be considered as such if he has become a "lord of two worlds". They must be able to bestow their puis–sance over wider realms, to include the territory that lies beyond all frontiers, thus transcending the ordinary by way of what is most distinctive in themselves. For them the totemic experience is partly characterized by the the possibility that the individual may not set out to acquire a totem. Exposure to nature leads to a situation in which the totem may be better understood as acquiring the person. Within that totem the person never dies, living on through re–integration into something larger.
Refs Cowan, James *On Totems*.
Related Human development through primal religion (#HH1902).

♦ **HM4343d Plane of cosmic good** (Leela)
Rudra–loka
Description Only one stage removed from the final goal of cosmic consciousness, the player receives his final purification. Knowing truth, feeling beauty and doing good, he does whatever is required of him, offering no resistance to the flow of dharma. This state may be reached by working up through the board or directly through right knowledge.
Context The 67th state or square on the board of *Leela*, the game of knowledge, appearing in the eighth row.
Broader Beyond the chakras: the gods themselves (Leela, #HM1141).
Preceded by Right knowledge (Leela, #HM0567).

♦ **HM4344f Cafard**
Blues
Related Depression (#HM2563).

♦ **HM4347d Doubt** (Buddhism)
Kankha (Pali)
Broader Perplexity (Buddhism, #HM0812).
Related Doubt (Buddhism, #HM4467).

♦ **HM4348d Full understanding as abandoning** (Buddhism)
Comprehension of rejection
Description This is insight with characteristics as object occurring as abandonment of the perception of permanence. This is knowledge in the sense of giving up. Its plane extends from contemplation of dissolution onwards.
Context One of three kinds of mundane full understanding.
Preceded by Full understanding as investigation (Buddhism, #HM4552).

♦ **HM4349d Experience of mystical knowledge** (Sufism)
Description On the Sufi path there are several stages at which mystical knowledge of God is experienced. At the first plane, knowledge is experienced on attainment of perfection in repentance. At the second plane, it is achieved when the devotee realizes kinship with God. At the final stage, knowledge of the divine essence is revealed to the seeker after truth by God Himself.
Related 'Arif (Sufism, #HM1045).

♦ **HM4350c Sabi** (Japanese)
Harmony of aesthetic contrasts
Description A poetic or aesthetic ideal blending the qualities of age, loneliness, resignation, tranquility and colour. It is influenced by the Buddhist sense of the existential loneliness of individuals in which a subtle beauty is to be found. It blends the experience of two conflicting sets of aesthetic experiences, manifest beauty and that implicit in the desolation of age. A person awakened to the essential mutability of life accepts the waning of beauty and finds a subtler form of beauty in resignation to the experience of age.
Related Wabi (Japanese, #HM1365) Shibumi (Japanese, #HM3597)
Artistic education (#HH0029).

♦ **HM4356b Suluk of the Naqshbandiyya order** (Sufism)
Path to sainthood
Description Adepts pass through ten stages to attain perfect sainthood. Further contemplations lead to higher and yet more ecstatic states of consciousness.
Narrower Sabr (#HM3418) Rida (#HM4190) Zuhd (#HM4450)
Tawba (#HM4062) Piety (#HM5339) Shukr (#HM4154)
Riyada (#HM6690) Wilayat (#HM3647) Tawakkul (#HM0807)
Contentment (#HM7024) Divine–love awareness (#HM2401).
Related Mystic stations (Sufism, #HM3415) Awareness of the mystic journey (#HM2900).

♦ **HM4358c Wisdom as a guide** (Buddhism)
Parinayika (Pali)
Broader Faculty of wisdom (Buddhism, #HM3233).

♦ **HM4363e Feeling infatuation** (Buddhism)
Sarajitattam (Pali)
Broader Greed (Buddhism, #HM3283).

♦ **HM4376d Plane of joy** (Leela)
Harsha–loka
Description Satisfaction arises in the knowledge of having transcended levels of being and raised energy, although there is no knowing how long it will be before the goal is reached. Fear and insecurity have been left behind with the first chakra, sensual desires with the second, there is joy in the completion of part of the task and in the challenge of that to follow.
Context The 18th state or square on the board of *Leela*, the game of knowledge, appearing in the second row.
Broader Second chakra: realm of fantasy (Leela, #HM3651).
Related Joy (Hinduism, #HM8098).

♦ **HM4380c Concentration of the second jhana of four** (Buddhism)
Description Having emiminated applied and sustained thought, three factors remain in the second jhana: rapture or happiness; ease or bliss; concentration.
Context According to Hinayana Buddhism, concentration is considered as of one kind (monad), of two kinds (dyads), of three kinds (triads), of four kinds (tetrads) or of five kinds (pentad). In the third tetrad, fourfold according to the four jhana factors, this is the second concentration. It is equivalent to the third concentration of the pentad.
Broader Concentration (Buddhism, #HM6663).
Related Dwelling in the second jhana (Buddhism, #HM7121)
Concentration of the third jhana of five (Buddhism, #HM8532)
Concentration of the second jhana of five (Buddhism, #HM4575).
Followed by Concentration of the third jhana of four (Buddhism, #HM2284).
Preceded by Concentration of the first jhana of four (Buddhism, #HM4456).

♦ **HM4382d Non–distraction** (Buddhism)
Not–wavering
Context According to Hinayana Buddhism, concentration is considered as of one kind (monad), of two kinds (dyads), of three kinds (triads), of four kinds (tetrads) or of five kinds (pentad). This is the monad whose quality is shared by all kinds of concentration.
Broader Concentration (Buddhism, #HM6663).

♦ **HM4383c Quiet attentiveness**
Description A state to be cultivated prior to meditation, when the body, emotions and mind are all quiet, relaxed and controlled.
Related Attention (#HM1817) Spiritual attentiveness (Christianity, #HM7143).

♦ **HM4387d Charity** (Leela)
Daan
Description Identification with the divinity present in all instigates acts of charity with no desire for personal benefit. The player experiences elation in the raising from lower to higher energy and the developing ego is satisfied as bonds of the third chakra are broken, and the player is raised to the plane of balance.
Context The 20th state or square on the board of *Leela*, the game of knowledge, appearing in the third row.
Broader Third chakra: theatre of karma (Leela, #HM0717).
Followed by Plane of balance (Leela, #HM1351).

♦ **HM4389c Sense mode of consciousness occurrence** (Buddhism)
Description Following five–door adverting, eye, ear, nose, tongue or body consciousness arise at the eye, ear, nose, tongue or body door respectively, with the physical basis of eye, ear, nose, tongue or body sensitivity. The indeterminate resultant consciousnesses are of profitable result with respect to desirable and desirable–neutral objects and of unprofitable result with respect to undesirable and undesirable and neutral objects – ten states altogether.
Context In Hinayana Buddhism, these are the fourth to eighth modes of occurrence of consciousness in which the 89 kinds of consciousness proceed.
Broader Modes of occurrence of consciousness (Buddhism, #HM6720).
Narrower Eye consciousness (#HM2074) Ear consciousness (#HM2169)
Nose consciousness (#HM2364) Body consciousness (#HM2562)
Tongue consciousness (#HM2263).
Related Sense consciousness (Buddhism, #HM2664)
Indeterminate consciousness in the sense sphere – resultant (Buddhism, #HM5721).
Followed by Receiving (Buddhism, #HM7092).
Preceded by Adverting (Buddhism, #HM8336).

♦ **HM4390c Acheta** (Japanese)
Description A state achieved in martial arts when forces of mind, body and spirit are blended.
Related Martial arts (#HH0085).

♦ **HM4392c Jabarut** (Sufism)
Ma'arifa
Description By the grace of God, in this state the aspirant acquires spiritual powers and knowledge of God, *ma'arifa*.
Context The third of four stages on the Sufi spiritual path.
Broader Divine path (Sufism, #HM3371).
Related Ma'rifa (Sufism, #HM2254).
Followed by Lahut (Sufism, #HM0741).
Preceded by Malaqut (Sufism, #HM0116).

♦ **HM4393c Raja** (Sufism)
Hope in God
Description A state in which the nearness to God is experienced, this may be hope for kindness, hope for reward or (the most perfect) hope for realization of complete identity with God. Seeing God's grace there is boasting in joy, and the absence of fear or terror.
Context According to Shaykh Abu Sa'id ibn Abi'l–Khayr, this is the 20th of 40 stations or maqamat the Sufi must possess for his journey on the path of Sufism to be acceptable.
Refs Vivekananda, Swami *Raja–Yoga* (1970).
Broader Nearness–awareness (Sufism, #HM2377)
Stations of consciousness – ibn–Abi'l–Khayr (Sufism, #HM2424).
Followed by Fana (Sufism, #HM1270).
Preceded by Khawf (Sufism, #HM1047).

♦ **HM4394d Consciencelessness** (Buddhism, Pali)
Unconscientiousness
Description In Hinayana Buddhism, this is the formation that has no conscientious scruples and may also be referred to as immodesty; its characteristic is absence of disgust or abomination of evil and its proximate cause is lack of respect for self.
Context One of the formations aggregate (mental coefficients) of Hinayana Buddhism, being listed among the constant states which appear in their true nature, and as unprofitable primary (always present in any unprofitable or unprofitable–resultant consciousness). Although consciencelessness and shamelessness are considered closely related they are listed as separate formations.
Broader Awareness as mental–formation group of conscious existence (Buddhism, #HM2050).
Related Non–shame (Buddhism, #HM2986) Conscience (Buddhism, #HM1590)
Shamelessness (Buddhism, #HM0649).

♦ **HM4395c Lightness of body** (Buddhism)
Kayalahuta (Pali)
Narrower Non–inertness (#HM3374)
Lightness of the aggregate of sensation, of the aggregate of cognition, and of the aggregate of synergies (#HM1435).

♦ **HM4396d Consciousness reinforced by energy** (Buddhism)
Mind upheld by energy
Description Consciousness does not waver in idleness, it is unperturbed by indolence.
Context One of the sixteen roots or modes of unperturbedness of mind listed in Hinayana Buddhism as a basis for supernormal powers.
Related Supernormal powers (Buddhism, #HH5652).

♦ **HM4398b Deautomatization and the mystic experience**
Description
1. *Sense of realness*. The sense of a reality which is greater than that felt in ordinary consciousness arises in all mystic experiences. However, this sense of reality is also typical in some psychoses and in depersonalization and derealization. Meditation and renunciation together are said to cause a disruption of the normal psychological relationship to the world. Through *reality transfer*, stimuli of the inner world – thoughts and images – become real.
2. *Unusual sensations*. Perceptions of all–engulfing light, energy, visions, incommunicable knowledge, although claimed to be from another world or a different dimension, may equally well be due to an unusual mode of perception rather than an external stimulus. Through the mode of *sensory translation*, illumination is not simply a metaphor but an actual sensory experience. In a state of *perceptual concentration*, the person is aware of intra–psychic processes normally outside the range of awareness. Because reality transfer makes real the amorphous sensation which is the vehicle for this perception, it is misinterpreted as originating from outside.
3. *Unity*. The experience of oneness with God or with the universe is common to mystic experience in whatever culture. This may be explained in terms of regression. It may also be perception of one's own psychic structure or of the real structure of the world. Since the substance of perception is probably electrochemical activity, the contents of awareness are homogenous. Turning awareness back on itself, as in sensory translation, this homogeneity or unity would be interpreted as pertaining not to the thought process but to the external world.
The perception of unity may, in fact, evaluate the external world correctly. Deautomatization, although requiring more attention than under normal perception, does allow *perceptual expansion*. One aspect of the real world not normally experienced, but experienced under the broader basis of perceptual expansion, may indeed be unity. This explains the experience of those versed in meditation and renunciation but not the fleeting experience of untrained persons.
4. *Ineffability*. Because mystic experiences are all indescribable they tend all to be included in one category as though they were all similar. However, there are definite differences. For example, that based on memories and fantasies of pre–verbal childhood could not be described in speech. Mystic training may increase recall and vividness of such memories. Again, revelations such as drug–induced experiences may be too complex for speech – simultaneous levels of meaning, understanding of the totality of existence – whether or not they are actual or illusory. Sudden expansion of consciousness to encompass a large number of concepts, or vertical organization of concepts instead of all in one consciousness plane, would be equally indescribable.
5. *Trans–sensate phenomena*. This is a third kind of ineffable experience, which goes beyond habitual sensory paths, ideas and memories. It is described by many mystics as not containing any familar sensory or intellectual elements and yet being filled with a profound perception which is regarded as the goal of the mystic path. Here, renunciation has weakened and temporarily removed ordinary objects of consciousness from the focus of awareness while meditation has undone the logical organization of consciousness. However, there is a strong motivation to perceive something. Supposing undeveloped or unutilized perceptual capacities to exist, then these conditions are just those in which they might be expected to be mobilized and come into operation. Because such experience would be outside normal frames of reference it would be unidentifiable and therefore indescribable. This appears to have a different scope from normal consciousness, and high value, meaning and intensity are ascribed to it. Trans–sensate experience is associated with loss of "self" and is therefore not associated with reflective awareness, the "I" of normal consciousness is in abeyance.
Refs Ornstein, Robert *The Psychology of Consciousness* (1986).
Narrower Reality transfer (#HM6992) Sensory translation (#HM3754)
Perceptual expansion (#HM7007).
Related Automatization (#HM3906) Deautomatization (#HH2331)
Mystical cognition (#HM2272) Religious experience (#HM3445)
Perceptual concentration (#HM0824) Trans–sensate experience (#HM9022)
Spiritual experiences (Sufism, #HM3542).

♦ **HM4399d Prana–loka** (Leela)
Description In this state the bodily life of the player is maintained in balance by the vital or life–force of prana, taken in with every inward breath. Practice of pranayama leads to the flowing of prana and apana together through sushumna and to the experience of samadhi.
Context The 38th state or square on the board of *Leela*, the game of knowledge, appearing in the fifth row.
Broader Fifth chakra: man becomes himself (Leela, #HM0933).
Related Pranayama (Hinduism, Yoga, #HH0213).

♦ **HM4400d Al–Khabir** (Sufism)
Description The believer is aware of Allah as the one who is aware. There is nothing which is not known to Him in infinite detail, even the most secret needs and wishes of His followers.
Context A mode of mystical awareness in the Sufi tradition associated with the thirty–first of the ninety–nine names of Allah.
Broader Ninety–nine names of Allah (Sufism, #HH2561).
Followed by Al–Halim (Sufism, #HM4519).
Preceded by Al–Latif (Sufism, #HM0308).

♦ **HM4401c Mono–motivational hypnotic state of transcendence**
Description During hypnotherapy, frustration may be released by the surfacing and expression of a single and powerful thought.
Broader Hypnotherapy (#HH0962).

♦ **HM4403d Entertainment** (Leela)
Celestial musicians — Gandharvas
Description Having transcended the secure first chakra level, joy is expressed within rhythm and harmony, within the divine play of creation.
Context The 11th state or square on the board of *Leela*, the game of knowledge, appearing in the second row.
Broader Second chakra: realm of fantasy (Leela, #HM3651).

♦ **HM4404c Power of faith** (Buddhism)
Saddhabala (Pali)
Broader Faculty of faith (Buddhism, #HM0066).

♦ **HM4406c Desirelessness** (Hinduism)
Nirasih
Description The person is devoid of desire for the non–eternal enjoyments which constitute bondage and are the cause of transmigration and the absence of freedom. The desire for fruits of action are renounced as one knows that one is impelled by God to perform them. All actions may thus be regarded as worship of God. Some say that absence of desire must include absence of desire for liberation.
Related Tasawwuf (Sufism, #HM0575) Desire (Hinduism, #HM7800).

♦ **HM4412c Sixth chakra: time of penance** (Leela)
Plane of austerity
Description At this chakra, the player raises consciousness through austerities. Individuality disappears and he is the supreme consciousness in undivided unity. Mastery leads to psychic powers and to destruction of karma accumulated in previous lives.
Context The sixth row of the board of Leela, the game of knowledge, this is represented by the planes of austerity, neutrality and violence, the solar, lunar and liquid planes, and groups squares representing the states of conscience, earth and spiritual devotion. Located at the third eye, in the region of the pineal gland, the player vibrating with this chakra is beyond the elements. It is compared to the octave comprising the ages 35 to 42 of temporal life, that of observation. Spiritual devotion and conscience are the arrows at this stage (the former leading directly to the plane of cosmic consciousness and the end of the game), and the plane of violence the snake.
Broader Lila (#HM2278).
Narrower Earth (#HM6112) Conscience (#HM1601) Solar plane (#HM1279)
Lunar plane (#HM3779) Liquid plane (#HM0172) Plane of violence (#HM1276)
Plane of austerity (#HM5244) Spiritual devotion (#HM1475) Plane of neutrality (#HM4050).

Related Ajna (Yoga, #HM2144).
Followed by Seventh chakra: plane of reality (Leela, #HM0754).
Preceded by Fifth chakra: man becomes himself (Leela, #HM0933).

♦ **HM4417e Seventh plane** (Psychism)
Description In this state of total perfection the soul mind achieves the highest level of spiritual attainment. There is some doubt as to whether or not there can be awareness of individuality at all.
Related Satori (Zen, #HM2326) Nirvana (Buddhism, #HM2330).

♦ **HM4418c Wali** (Sufism)
Description The name of a traveller on the *tariqa* or Sufi spiritual path at the level sincere love of God.
Context Spiritual progress on the spiritual path is marked by the attainment of specific *maqamat* or station.
Broader Mystic stations (Sufism, #HM3415).

♦ **HM4423c Transformation of consciousness** (Yoga)
Description Intentional change in the state of consciousness is typical of the techniques of yoga, in this case the aim being to realize the core of all existence, the self outside and beyond the transient world. Empirical consciousness is progressively broken down until there is metamorphosis in to universal awareness of self.
Related Yoga (Yoga, #HH0661).

♦ **HM4425d Psychical perplexity** (Buddhism)
Cetaso vikkhepo (Pali)
Broader Excitement (Buddhism, #HM1469).

♦ **HM4426c Yaqin** (Sufism)
Certainty — Yaquin — Awareness of certitude
Description This is one of the highest forms of mystical consciousness described in Sufism. It is said to follow that of contemplation (mushahada) and to be characterized by the practitioner becoming firmly rooted in divine contemplation and communion with the holy. It is a consciousness attained with clear vision, by the power of faith.
Context The ninth psychic state listed by A Reza Aresteh as progress on the inner self through divine attraction, *kedesh-jazba*, the outer self already having been purified through conscious effort, *kushesh*.
Broader Psychic states (Sufism, #HM4311)
Higher states of consciousness (Sufism, #HM2365).
Preceded by Itmianan (Sufism, #HM0050) Mushahada (Sufism, #HM3521).

♦ **HM4433d Concentration accompanied by happiness** (Buddhism)
Concentration accompanied by rapture
Description This state is the unification of the first two jhanas in the fourfold reckoning and the first three in the fivefold reckoning. It may accompany access concentration.
Context According to Hinayana Buddhism, concentration is considered as of one kind (monad), of two kinds (dyads), of three kinds (triads), of four kinds (tetrads) or of five kinds (pentad). In the third triad, this is the first concentration.
Broader Concentration (Buddhism, #HM6663).
Related Access concentration (Buddhism, #HM4999).
Followed by Concentration accompanied by bliss (Buddhism, #HM1454).

♦ **HM4437d Al-Kabir** (Sufism)
Description The believer is aware of Allah as the most great, extending from before creation to after the end of all things. His greatness is beyond comprehension, realizing which arouses fear of losing His love, of causing His disappointment.
Context A mode of mystical awareness in the Sufi tradition associated with the thirty-seventh of the ninety-nine names of Allah.
Broader Ninety-nine names of Allah (Sufism, #HH2561).
Followed by Al-Hafiz (Sufism, #HM7099).
Preceded by Al-'Ali (Sufism, #HM4162).

♦ **HM4447c Profitable consciousness in the sense sphere** (Buddhism)
Description A total of 8 kinds of consciousness may arise here. Of these, 4 are accompanied by joy, that is: with understanding and unprompted; with understanding and prompted; without understanding and unprompted; without understanding and prompted. And 4 are accompanied with equanimity, that is: with understanding and unprompted; with understanding and prompted; without understanding and unprompted; without understanding and prompted. All 8 give rise to materiality, to postures and to intimation.
Context In Hinayana Buddhism, 89 consciousnesses are enumerated in aggregate (khanda). Of these, 21 are profitable or moral, 12 are unprofitable or immoral and 56 are indeterminate (resultant or functional). The unprofitable all arise in the sphere of sense and desire, whereas profitable and indeterminate arise in sense, fine-material, immaterial and supramundane spheres.
Broader Dispositions of consciousness (Buddhism, #HM2098).
Related Impulsion (Buddhism, #HM7268) Formation of merit (Buddhism, #HH5122)
Unprofitable consciousness in the sense sphere (Buddhism, #HM8375)
Indeterminate consciousness in the sense sphere - resultant (Buddhism, #HM5721)
Indeterminate consciousness in the sense sphere - functional (Buddhism, #HM3852).
Followed by Profitable consciousness in the fine-material sphere (Buddhism, #HM5338).

♦ **HM4449c Creative moment** (Sufism)
An
Description When the temporal is transcended there arises the timeless moment, the eternal. Time is no longer a never ending succession of events, one after the other, basically dualistic. I becomes identical with not-I, there is no distinction between personal and universal. It is the artistic moment of original vision, the moment when past and future are completely cut away, subject and object are one in "I am". This is the instant of noticing. The process of inner search and inner evolution is a preparation for this experience. In the search for his own origin, the seeker may first turn to natural phenomena; then from idol worship to worship of God; then through patience, persistence, love; then from the senses, reason, imagination, heart. Finally it is the heart that motivates, searching for the object of its desire, which also is subject hitting the heart. In this unity there is *an*, real knowledge arising in an instant of illumination.
This illumination arises through three techniques: (1) illumination of names, where name is experienced as symbolizing conceptual reality, with awareness not of words in their abstract form but in relation to the original object producing the word; (2) illumination of attributes or qualities, to the original elements, waves, light which produce shape, colour and sound; (3) illumination of essence, a state achieved by very few, miracle workers who had power to move not only their own bodies but the bodies (animate and inanimate) around them.
With purity of heart and intention, and with utter sincerity (not desiring the object for one's self), there is total concentration on unity with an object worthy of desire. Then the "I" and the "thou" become one, duality becomes unity and *an* arises - the knower, knowing and known or lover, beloved and love being one. *An* is the moment of conception, of birth and rebirth, of joy and originality, growing by dynamic relation between subject and object, where first subject depends on subject, then there is the moment of union, then object depends on subject. This is the end of the creative process, the perception of a vision, faithfulness to the vision, seeing its distance from one's own reality then, despite adverse opinion, the effort to materialize it through unification with it. *An* may be thought of as existing in the primitive, childlike state, but not retained due to accumulation of cultural and social experience. Only on leaving this is it possible to become creative and again experience *an*.
Related Kairos (#HM2749).

♦ **HM4450c Zuhd** (Sufism)
Austerity — Renunciation — Asceticism — Detachment — Tasawwuf — Abstinence
Description In this state the traveller separates himself from all possessions of this world which are seen as the real cause of good and evil and from any desire for them. The saints further renounce all that is associated with the self; and the spiritualists, gnostics, renounce even thoughts of renunciation as such thoughts conceal God. What is renounced is seen as temporary and of little value; what is gained is permanent and of infinite value. Zuhd is the basis of mystical perfection and is the first step of those on the way to God, acting according to his will in trust and devotion.
Context The third stage in a systematic account of various stations of the Sufi spiritual path. According to Shaykh Abu Sa'id ibn Abi'l-Khayr, this is the 14th of 40 stations or maqamat the Sufi must possess for his journey on the path of Sufism to be acceptable. The second stage on the path to sainthood of the Naqshbandiyya order.
Broader Mystic stations (Sufism, #HM3415)
Suluk of the Naqshbandiyya order (Sufism, #HM4356)
Stations of consciousness - ibn-Abi'l-Khayr (Sufism, #HM2424).
Related Tasawwuf (Sufism, #HM0575).
Followed by Faqr (Sufism, #HM3427) Riyada (Sufism, #HM6690)
'Ibadat (Sufism, #HM0128).
Preceded by Wara (Sufism, #HM0286) Tawba (Sufism, #HM4062)
Tawakkul (Sufism, #HM0807).

♦ **HM4451b Vitarka samadhi** (Yoga)
Description There is ontic identification with the external reality of the object contemplated, together with a certain awareness or transcendental cogitation.
Context The first level of samprajnata samadhi.
Broader Samprajnata samadhi (Yoga, #HM2896).
Related Vitarka stage (Yoga, #HM2912).
Followed by Nirvitarka samadhi (Yoga, #HM4284).

♦ **HM4456c Concentration of the first jhana of four** (Buddhism)
Description Having suppressed hindrances, five factors remain in the first jhana: applied thought; sustained thought; rapture or happiness; ease or bliss; concentration.
Context According to Hinayana Buddhism, concentration is considered as of one kind (monad), of two kinds (dyads), of three kinds (triads), of four kinds (tetrads) or of five kinds (pentad). In the third tetrad, fourfold according to the four jhana factors, this is the first concentration. It is equivalent to the first concentration of the pentad.
Broader Concentration (Buddhism, #HM6663).
Related Dwelling in the first jhana (Buddhism, #HM4298)
Concentration of the first jhana of five (Buddhism, #HM0297)
Concentration of the second jhana of five (Buddhism, #HM4575).
Followed by Concentration of the second jhana of four (Buddhism, #HM4380).

♦ **HM4464c Talib** (Sufism)
Description The name of a traveller on the *tariqa* or Sufi spiritual path at the level of a seeker of God.
Context Spiritual progress on the spiritual path is marked by the attainment of specific *maqamat* or station.
Broader Mystic stations (Sufism, #HM3415).

♦ **HM4467d Doubt** (Buddhism)
Vichikitsa - The-tshom (Tibetan)
Description In this state mind is two-pointed with respect to seeing the truth, in particular the four noble truths and actions and the effect of actions. There is lack of engagement in virtues.
Context One of the six *root afflictions* referred to in Tibetan Buddhism.
Broader Root afflictions (Buddhism, #HH0270).
Related Doubt (Buddhism, #HM4347) Four noble truths (Buddhism, #HH0523).

♦ **HM4469 Direct perception**
Alert passivity — Choiceless awareness
Description There is a transcendent spontaneity of life, a creative reality, which reveals itself as immanent only when the perceiver's mind is in a state of alert passivity or of choiceless awareness. Any form of judgement or comparison inhibits this transcendent awareness, committing the person irrevocably to duality. It is through choiceless awareness that non-duality becomes possible and opposites may be reconciled. The liberating process begins with choiceless awareness of what the person wills and of reactions to symbol systems indicating what ought, or ought not, to be done. Through this choiceless awareness, as it penetrates the successive layers of the ego and its associated sub-conscious, comes love and understanding, but of another order than that which is familiar. It thus becomes a highly effective meditation. The dangers of other forms of yoga is that their self-discipline leads to a form of self-induced rapture or false samadhi. The true liberation is an inner freedom of creative reality. It is a state of being, as silence, in which there is no becoming, in which there is completeness. Human beings are thus able to teansform themselves radically, not in time, not by evolution, but by such immediate perception, without intermediary, duality or thought.
Context Advocated by J Krishnamurti as a direct way, to be contrasted with the way of gradual evolution through states and stations of spiritual progress.
Refs Krishnamurti, J *The First and Last Freedom* (1954); Krishnamurti, J *The Awakening of Intelligence*; Krishnamurti, J *The Only Revolution* (1970); Krishnamurti, J *Freedom from the Known* (1969); Mehta, Rohit *Nameless Experience* (1973); Vimala Thakar *Mutation of Mind.*
Related Alert passivity (#HM3959) Right resolves (Buddhism, #HM1142).

♦ **HM4470d Fetters of ignorance** (Buddhism)
Avijjayoga (Pali)
Broader Delusion (Buddhism, #HM0184).

♦ **HM4474d Genesis** (Leela)
Janma
Description The player throws a six, representing the uniting of the five subtle elements with

consciousness. This is birth, the commencing of a new game.
Context The first state or square on the board of *Leela*, the game of knowledge, appearing in the first row and representing unity, the independent personality.
  Broader First chakra: fundamentals of being (Leela, #HM4103).

♦ **HM4477c Omid** (Sufism)
Hope for getting nearer
Context The fifth psychic state listed by A Reza Aresteh as progress on the inner self through divine attraction, *kedesh-jazba*, the outer self already having been purified through conscious effort, *kushesh*.
  Broader Psychic states (Sufism, #HM4311).
  Followed by Uns (Sufism, #HM1957).
  Preceded by Mehr (Sufism, #HM1266).

♦ **HM4487d Zeal** (Buddhism, Pali)
Desire — Desire-to-do — Wish to act — Ussaha
Description The characteristic of zeal is the wish to act, its function is searching or scanning for an object. It is manifest as need for an object or having it at its disposal. The object is its proximate cause. It is like the mental extending of a hand to grasp an object.
Context One of the formations aggregate (mental coefficients) of Hinayana Buddhism, being listed among the "or-whatever" states, and as general primary (always present in any consciousness).
  Broader Faculty of energy (Buddhism, #HM1470)
  Awareness as mental-formation group of conscious existence (Buddhism, #HM2050).
  Related Vasana (Yoga, #HM3231).

♦ **HM4488c Pondering** (Buddhism)
Vebhabya (Pali)
  Broader Faculty of wisdom (Buddhism, #HM3233).

♦ **HM4489b Videha-mukti** (Yoga)
Description This state of disembodied emancipation is the aim of mythic yoga and considered by some schools to be the only final and irreversible liberation or *kaivalya*, death of body and mind arising simultaneously with it.
  Broader Mukti (#HH0613).
  Related Kaivalya (Yoga, #HM3869)   Mythic yoga (Yoga, #HH3405)
  Jivanmukti (Hinduism, #HM3890).

♦ **HM4490d Understanding as interpreting the internal and the external** (Buddhism)
Description This understanding is initiated when one has grasped the complex aggregate of elements of which one's self or another person or inanimate matter – external material not bound up with the faculties – is made up.
Context On the Path of Purification of Hinayana Buddhism, panna (understanding) is considered as of one kind (monad), of two kinds (dyads), of three kinds (triads) or of four kinds (tetrads). There are five dyads, four triads and two tetrads. All have the characteristic of penetrating the individual essences or true nature of states (monad). In the fourth triad, this is compared with understanding as interpreting the internal or as interpreting the external.
  Broader Understanding (Buddhism, #HM4523).
  Related Understanding as interpreting the internal (Buddhism, #HM0956)
  Understanding as interpreting the external (Buddhism, #HM1128).

♦ **HM4493c Viveka khyati** (Yoga)
Pure awareness of reality
Description The role of prajna comes to an end in pure awareness of reality. In this most exalted state of enlightenment, ignorance is kept at bay and, with practice of discrimination (viveka) and mental renunciation (para vairagya) mutually reinforcing each other, the highest level of samadhi, *dharma megha samadhi*, is reached. This completes the evolutionary cycle of the individual, there is no turning back and, passing through dharma megha samadhi, *kaivalya* or liberation is finally achieved.
  Related Viveka (Yoga, #HM2998)   Prajna (Hinduism, #HM3455)
  Dharma megha samadhi (Yoga, #HM0194)   Awareness of the ultimate reality (#HM1456).

♦ **HM4495d Al-Sabur** (Sufism)
Description The believer is aware of Allah as patient, unhurried, being in perfect measure and time. He is patient with sinners, giving time for repentance. In this state the affairs of both the world and the hereafter are resolved. The believer applies himself to those things acceptable to reason and religion, refusing desires of the flesh and the ego despite the pain involved. He submits only to Allah, fearing no-one else, abasing himself to no-one else.
Context A mode of mystical awareness in the Sufi tradition associated with the ninety-ninth of the ninety-nine names of Allah.
  Broader Ninety-nine names of Allah (Sufism, #HH2561).
  Preceded by Ar-Rashid (Sufism, #HM5234).

♦ **HM4497c Fund of the individual** (Christianity)
Description This state, decribed by the mediaeval mystics, has been compared with transcendental consciousness.
  Broader Human development (Christianity, #HH2198).
  Related Transcendental consciousness (#HM2020).

♦ **HM4500d Allah** (Sufism)
Description The believer is aware of God as Allah. This is the greatest name of God, containing all divine and beautiful attributes; it is a sign of the essence and cause of all existence and can be assumed or shared by no-one. The name contains eight essentials indicating the infinite perfection of Allah and the five qualities showing that Allah resembles nothing else. It contains all attributes of perfection and no faulty attributes. By trying to eliminate what is faulty in himself and increasing what is good, the servant of Allah seeks the wish to become perfect.
Context A mode of mystical awareness in the Sufi tradition associated name of Allah.
  Broader Presence of Allah (Sufism, #HM3496)   Ninety-nine names of Allah (Sufism, #HH2561).

♦ **HM4501c God as wholly other** (Judaism)
Ecstatic experience of God — Kiss of God — Talmudic ecstasy — Merkabah mysticism
Description Meditation on the heavenly Temple while chanting the sacred names of God, the mystic, already wise and steeped in knowledge of the Torah, experiences an ascent through the chambers of the celestial palace, passes the gates of the seven heavens and their angelic guardians and, beyond the universe, comes into the presence of God where his soul becomes a flame of praise and rejoicing before the throne, surrounded by light and choirs of angels. Such an experience leads to spiritual distortion, madness or death in agony unless the mysteries are already understood. Total experience of God leads to the Kiss of God, ecstatic death.
  Related Kabbalah (#HH0921)   Jewish mysticism (Judaism, #HH1232)
  Human development (Judaism, #HH3029)   God as wholly the same (Judaism, #HM4291).

♦ **HM4502d Hatred** (Buddhism)
Dosa (Pali) — Hate
Description In Hinayana Buddhism, Hate is the means by which associated states hate, it is hatred itself or it is the act of hating. Hate has the characteristic of savageness, of flying into anger like a snake when provoked. Its function is to spread like poison or to burn up that on which it depends, the body or the heart-basis, like fire in a forest. It manifests as persecution or injuring of its object. The proximate cause is the grounds for vexation or annoyance.
Context One of the formations aggregate (mental coefficients) of Hinayana Buddhism, being listed among the constant states which appear in their true nature, and as unprofitable secondary (sometimes present in any unprofitable or unprofitable-resultant consciousness).
  Broader Awareness as mental-formation group of conscious existence (Buddhism, #HM2050).
  Narrower Hatred (#HM0607)   Hating (#HM4323)   Enmity (#HM4239)
  Hostility (#HM0719)   Abruptness (#HM3857)   Harmfulness (#HM3210)
  Causing harm (#HM1591)   Churlishness (#HM0945)   Maliciousness (#HM1111)
  Depression of thought (#HM3949)   State of feeling hatred (#HM1967).
  Related Hatred awareness (#HM4596)   Non-hatred (Buddhism, #HM2744).

♦ **HM4504f Delusional ideas**
Delusional memories
Description A past event is remembered and imbued with sudden significance.
  Broader Primary delusions (#HM4604).

♦ **HM4510d Al-Mudhill** (Sufism)
Al-Muzill
Description The believer is aware of Allah as the one who humiliates. Attributing honour to one's self or having it so attributed by others distorts reality, so that one imagine's one is in a state other than one's own. Imagining one can fashion one's own destiny and seeking the approval of others leads to punishment and expulsion from God's presence.
Context A mode of mystical awareness in the Sufi tradition associated with the twenty-fifth of the ninety-nine names of Allah.
  Broader Ninety-nine names of Allah (Sufism, #HH2561).
  Followed by Al-Sami' (Sufism, #HM0419).
  Preceded by Al-Mu'izz (Sufism, #HM1116).

♦ **HM4512d Nullity** (Leela)
Antariksha
Description Between physical and celestial planes, between earth and heaven, is the nowhere of nullity, with feelings of futility and lack of purpose. Inactivity is combined with restlessness; and energy may be frittered away in pursuing objects of the senses.
Context The 13th state or square on the board of *Leela*, the game of knowledge, appearing in the second row.
  Broader Second chakra: realm of fantasy (Leela, #HM3651).
  Preceded by Negative intellect (Leela, #HM3623).

♦ **HM4513d Al-Muntaqim** (Sufism)
Description The believer is aware of Allah as the great avenger. After repeated warnings, accepting excuses, delaying punishment to give time for repentance, Allah's revenge is all the more terrible. The believer reflects this quality by taking revenge on the enemies of Allah, the greatest enemy being his own ego.
Context A mode of mystical awareness in the Sufi tradition associated with the eighty-first of the ninety-nine names of Allah.
  Broader Ninety-nine names of Allah (Sufism, #HH2561).
  Followed by Al-'Afuw (Sufism, #HM5543).
  Preceded by Al-Tawwab (Sufism, #HM4334).

♦ **HM4514c Being conscious** (Buddhism)
Samapajanna (Pali)
  Broader Faculty of wisdom (Buddhism, #HM3233).

♦ **HM4516e Clairempathy** (Psychism)
Description This term describes the state of telepathic sharing of the emotion or attitude of another, possibly of those who have visited the particular environment.
  Broader Extrasensory perception (ESP, #HM2262).

♦ **HM4519d Al-Halim** (Sufism)
Description The believer is aware of Allah as the forbearing one. He delays punishment to allow the sinner to repent and be forgiven rather than being the subject of His vengeance. Allah loves those who are also clement, reflecting the forbearance of Allah.
Context A mode of mystical awareness in the Sufi tradition associated with the thirty-second of the ninety-nine names of Allah.
  Broader Ninety-nine names of Allah (Sufism, #HH2561).
  Followed by Al-'Azim (Sufism, #HM5198).
  Preceded by Al-Khabir (Sufism, #HM4400).

♦ **HM4521b Anomalies in integrity of consciousness**
Description Here the individual's normal consciousness pattern is replaced by another.
  Narrower Fugue state (#HM7131)   Channelling (#HH0878)
  Spirit possession (#HH0056)   Possession (Psychism, #HM3111)
  Multiple personality (#HH0763)   Trance channelling (Psychism, #HM3456).

♦ **HM4523b Understanding** (Buddhism)
Prajna — Panna (Pali) — Wisdom
Description On the Path of Purification of Hinayana Buddhism, understanding is insight knowledge associated with profitable consciousness or moral thought. It is knowing in a way which is more penetrating than perception or consciousness, which are nevertheless included in panna. As well as knowing the object, penetrating its characteristics, it also, through endeavour, brings about manifestation of the path. Although not always present when perception and consciousness are present, it is always accompanied by perception and consciousness. Its proximate cause is concentration.
The "soil" of understanding are all the aggregates, sense-organs or bases, elements, faculties, truths, dependent origination. In this soil are the roots of purification of virtue (purity of morality) and purification of consciousness (purity of mind). From these roots grow the body or trunk of the other five purifications or purities: view; overcoming of doubt; knowledge and vision of what is and what is not the path (discernment of right path and wrong path); knowledge and vision of the way (discernment of the middle way); knowledge and vision (discernment). One who is perfecting his understanding should study these things.
Understanding is considered as of one kind (monad), of two kinds (dyads), of three kinds (triads) or of four kinds (tetrads). There are five dyads, four triads and two tetrads. All have the characteristic of penetrating the individual essences or true nature of states (monad).
Refs Buddhaghosa, Bhadantacariya *The Path of Purity* (1975); Buddhaghosa, Bhadantacariya

HM4523

The Path of Purification (1980).
Narrower Four discriminations (#HM0485)
Supramundane understanding (#HM8838)
Understanding as skill in means (#HM6601)
Understanding defining mentality (#HM5627)
Understanding defining materiality (#HM4968)
Understanding as the plane of seeing (#HM6192)
Understanding as knowledge about law (#HM4726)
Understanding having a limited object (#HM2495)
Understanding as skill in improvement (#HM8290)
Understanding having an exalted object (#HM3296)
Understanding accompanied by equanimity (#HM7525)
Understanding consisting in development (#HM7826)
Understanding as knowledge of suffering (#HM6870)
Understanding as knowledge about meaning (#HM1663)
Understanding as the plane of development (#HM4128)
Understanding having a measureless object (#HM6681)
Understanding as knowledge about language (#HM5026)
Understanding as interpreting the internal (#HM0956)
Understanding as interpreting the external (#HM1128)
Understanding consisting in what is learned (#HM5298)
Understanding consisting in what is reasoned (#HM7154)
Understanding as knowledge about kinds of understanding (#HM4958)
Understanding as knowledge of the origin of suffering (#HM7290)
Understanding as knowledge of the cessation of suffering (#HM8163)
Understanding as interpreting the internal and the external (#HM4490)
Understanding as knowledge of the way leading to cessation suffering (#HM7645).
Mundane understanding (#HM7628)
Understanding free from cankers (#HM0756)
Understanding subject to cankers (#HM4194)
Understanding accompanied by joy (#HM6245)
Understanding as skill in detriment (#HM7602)
Related Essential wisdom (#HM0107)
Mental wisdom (Buddhism, #HM2655)
Five aggregates (Buddhism, #HH3321)
Concentration (Buddhism, #HM6663)
Purifications (Buddhism, #HH3875)
Faculty of wisdom (Buddhism, #HM3233).

♦ **HM4526d Al–Majid** (Sufism)
Description The believer is aware of Allah as the glorious, the noble, both the giver of good character and conduct which enables the carrying out of good deeds, and the rewarder of such deeds, who nevertheless hides the errors of His people and forgives their sins. Remembering how, in His glory, He accepts excuses, protects rights, relieves difficulty, His servants should love, obey and fear Him, fearing to lose his favour.
Context A mode of mystical awareness in the Sufi tradition associated with the sixty–fifth of the ninety–nine names of Allah.
Broader Ninety–nine names of Allah (Sufism, #HH2561).
Related Al–Majid (Sufism, #HM5446).
Followed by Al–Wahid (Sufism, #HM5506).
Preceded by Al–Wajid (Sufism, #HM0743).

♦ **HM4529e Mental projection**
Description Following deep meditation, the subconscious is intentionally directed to higher levels and awareness of higher intelligences while retaining an awareness of normality.
Related Astral projection (Psychism, #HM1887).

♦ **HM4533d Spiritual pride** (Christianity)
Hybris — Hubris
Description For those at the beginning of the path of self–consciousness, of spirituality, reaching a state of some satisfaction with themselves and with their work is a mark of spiritual pride. They view themselves as progressing nicely in the spirit because of their ardour and hard work in spirit exercises. They feel compelled to speak of spiritual things in front of others. Even as beginners they desire to teach rather than learn. In the centre of their being they condemn those who do not have the devotion they themselves want. The spiritual guides of these people may disapprove of this behaviour and say so, but the pride–filled beginner dismisses this as being misunderstood. They only desire approval and admiration and not correction. This disapproval is proof to them that their guides are really not spiritual. They seek out someone who will speak of spiritual things and praise them and avoid those who would point out their errors and lead them on a fruitful path. They are ashamed to acknowledge their faults openly for fear they will lose the esteem of others. They loath praising others and long for praise themselves. This is quite the opposite of spiritual humility.
The devil want this state for all those who choose the path of the spirit. The devil wants this state to grow so that the development of the individual is arrested and he may turn to vices. The pilgrim of the spirit may turn away from concern for others and become totally concerned with themselves, so that when ever the chance comes up they deplore and belittle others directly and indirectly. Any suggestion that this individual may need assistance is rejected out of hand. This unassailable position of spiritual self–righteousness can lead the person commit the most arrogant and evil of acts.
Context This is an imperfection of beginners in the Dark Night by St John of the Cross.
Broader Beginner in spiritual life (Christianity, #HM0102).
Related Spiritual envy (Christianity, #HM3818)
Spiritual sloth (Christianity, #HM0491)
Spiritual avarice (Christianity, #HM0642)
Spiritual lust (Christianity, #HM4180)
Spiritual anger (Christianity, #HM0936)
Spiritual gluttony (Christianity, #HM0507).

♦ **HM4535c Ksana** (Buddhism)
Description Stabilized in the eternal present there is great joy and each moment is propitious.

♦ **HM4539d Psychological imprisonment**
Mental prison
Description Continual repetition of key phrases designed to dull the mind and induce emotional dependence make the individual dependent on the cult for all decision–making and may reduce awareness to instinctual level.
Broader Destructive manipulation (#HH1020).
Related Instinct (#HM3517).

♦ **HM4542d Ad–Darr** (Sufism)
Description The believer is aware of Allah as the distresser. Allah creates evil as well as good, since all that happens is by His will. But choosing to follow evil is the choice of the individual himself. Almost everyone slips into evil and some time or another. Those who fall into sin are given time to repent, no longer subject to the devil or their own egos, but when evil befalls it is due to the individual himself. Suffering, one may learn from one's mistakes and repent. There is also a blessing through misery, through it one may gain positive virtues such as patience and courage. Affliction and blessing are all at the hands of Allah, but in affliction the best worship is to seek to understand its cause and try to put it right.
Context A mode of mystical awareness in the Sufi tradition associated with the ninety–first of the ninety–nine names of Allah.
Broader Ninety–nine names of Allah (Sufism, #HH2561).
Followed by An–Nafi' (Sufism, #HM6350).
Preceded by Al–Mani (Sufism, #HM7220).

♦ **HM4543c Mindfulness**
Description There is clear awareness of the present and of the physical, mental and emotional states.
Refs Langer, Ellen *Mindfulness* (1989).
Related Mindfulness (Buddhism, #HM2847)
Mindfulness (Buddhism, #HM1423).

♦ **HM4544d Al–Muhsi** (Sufism)
Description The believer is aware of Allah as the knower of all quantitative knowledge. He knows everything in its reality and as one indivisible whole. Even the hairs of our heads are numbered. Because of this, the believer should frequently analyse his intentions and repent now, before the final accounting at the day of judgement.
Context A mode of mystical awareness in the Sufi tradition associated with the fifty–seventh of the ninety–nine names of Allah.
Broader Ninety–nine names of Allah (Sufism, #HH2561).
Followed by Al–Mubdi (Sufism, #HM0622).
Preceded by Al–Hamid (Sufism, #HM0392).

♦ **HM4546f Modes of awareness associated with use of sedatives**
Modes of awareness associated with barbiturates
Description Initially there may be euphoria but tolerance builds up and larger and larger doses are needed for the same effect. Eventually these may lead to side effects perhaps resulting in death. Withdrawal symptoms include insomnia, anxiety and seizure. Barbiturates act like alcohol and also enhance the effects of alcohol, so that the two together can be very dangerous. They may be originally prescribed as tranquillizers, although they have a high placebo effect (people reporting immediate relief from stress before the drug has had time to act).
Broader Modes of awareness associated with psychoactive substances (#HM0584).
Related Euphoria (#HM3763).

♦ **HM4548d Wilderness of views** (Buddhism)
Ditthikantara (Pali)
Broader Wrong view (#HM1710).

♦ **HM4550e Assumption of other forms** (Psychism)
Description Records from ancient Egypt describe the assumption of animal form by merging of consciousness with a god, the form being maintained for some time before resumption of the individual's normal appearance.

♦ **HM4552d Full understanding as investigation** (Buddhism)
Comprehension of scrutiny
Description Having begun insight, knowledge of what is and what is not the path arises during full understanding as investigation. The objects of insight are general characteristics such as feeling as impermanent. Its plane extends from comprehension by groups to contemplation of rise and fall.
Context One of three kinds of mundane full understanding.
Related Purification by knowledge and vision of what is the path and what is not the path (Buddhism, #HH4007).
Followed by Full understanding as abandoning (Buddhism, #HM4348).
Preceded by Full understanding as the known (Buddhism, #HM3490).

♦ **HM4556d Haqq al–yaqin** (Sufism)
Certain truth — Truth of certainty
Description Passing beyond creations of man or God, God is seen without question and without veil.
Context According to Shaykh Abu Sa'id ibn Abi'l–Khayr, this is the 24th of 40 stations or maqamat the Sufi must possess for his journey on the path of Sufism to be acceptable.
Broader Stations of consciousness – ibn–Abi'l–Khayr (Sufism, #HM2424).
Followed by Ma'rifa (Sufism, #HM2254).
Preceded by 'Ilm al–yaqin (Sufism, #HM0930).

♦ **HM4559f Olfactory hallucination**
Broader Hallucination (#HM4580).
Related Clairscent (Psychism, #HM0318).

♦ **HM4562d Al–'Aziz** (Sufism)
Description The believer is aware of Allah as the victorious. He is unconquerable but delays destruction of those who rebel. Suppressing demands of ego and flesh, satisfying lawful needs by lawful means, not exercising strength or revenge, reflects al–'Aziz.
Context A mode of mystical awareness in the Sufi tradition associated with the eighth of the ninety–nine names of Allah.
Broader Ninety–nine names of Allah (Sufism, #HH2561).
Followed by Al–Jabbar (Sufism, #HM0215).
Preceded by Al–Muhaymin (Sufism, #HM0526).

♦ **HM4563d Contentment** (Buddhism)
Description This is a state of non–greed.
Context In Hinayana Buddhism, one of the five ascetic states.
Broader Ascetic states (Buddhism, #HM5354)
State of not feeling greed (Buddhism, #HM1825).

♦ **HM4565d Harmless** (Buddhism)
Avyapajja (Pali)
Broader Non–hatred (Buddhism, #HM2744).

♦ **HM4570c World beyond form** (Sufism)
David of one's being
Description The subtle organ is the Spirit. It is invested as the Vice–Regent of God because of the nobility of its rank. The David of one's being, its colour is yellow.
Context The fifth of seven subtle stages relating to the seven major prophets of Semitic monotheism as described by Simnani.
Broader Stations of consciousness – Simnami (Sufism, #HM2341).
Followed by Divine nature (Sufism, #HM2246).
Preceded by World of imagination (Sufism, #HM3616).

♦ **HM4572c Unperturbedness** (Buddhism)
Avikkhepa (Pali) — Unperplexity of thought
Broader Sukha (Buddhism, #HM2866).
Narrower Neither suffering nor pleasure experienced born of contact with the psychical (#HM1352)
Sensation of neither suffering nor pleasure born of contact with the psychical (#HM1643)
Neither physical ease nor unease born of contact with element of mind–consciousness (#HM3314).

♦ **HM4575c Concentration of the second jhana of five** (Buddhism)
Description Having emiminated applied thought, four factors remain in the second jhana: sustained thought rapture or happiness; ease or bliss; concentration.
Context According to Hinayana Buddhism, concentration is considered as of one kind (monad), of two kinds (dyads), of three kinds (triads), of four kinds (tetrads) or of five kinds (pentad). In the pentad, fivefold according to the five jhana factors, this is the second concentration. It is intermediate between the first and second concentrations in the third tetrad, in that applied thought is eliminated but sustained thought remains.
Broader Concentration (Buddhism, #HM6663).
Related Dwelling in the fivefold jhana (Buddhism, #HM6553)
Concentration of the first jhana of four (Buddhism, #HM4456)
Concentration of the second jhana of four (Buddhism, #HM4380).
Followed by Concentration of the third jhana of five (Buddhism, #HM8532).
Preceded by Concentration of the first jhana of five (Buddhism, #HM0297).

♦ **HM4580f Hallucination**
Description There is apparently uninstigated perception of what is not actually there, the state occurring in response to drugs, physical or mental illness (for example, high fever, paranoia) or pain. The perception is most usually taken to be visual, although it may in fact refer to any of the five senses. The individual may or may not be aware of the hallucinatory nature of what is perceived. Although hallucination may be pleasant, it may also be unpleasant or frightening. The state is similar to a waking dream and may reveal useful information on the subconscious. Hallucination is difficult to distinguish from illusion. The former is generally said to cover seeing an internal image as physically present whereas the latter as a misrepresentation of an object actually there.
Common instances of hallucination are: hypnotic trance, when the individual perceives and responds to suggestions of the hypnotist, whether seeing something that is not there or not seeing something that is; dreams and hypnagogic hallucinations when half awake. In addition to conditions mentioned above, hallucination may be triggered by fatigue or by prolonged fasting.
Refs Reed, Graham *The Psychology of Anomalous Experience* (1988).
Narrower Formication (#HM4094) Doppelgänger (#HM7342)
Visual hallucination (#HM0092) Pseudo-hallucination (#HM4336)
Tactile hallucination (#HM6118) Somatic hallucination (#HM7208)
Auditory hallucination (#HM6704) Negative hallucination (#HM5904)
Olfactory hallucination (#HM4559) Gustatory hallucination (#HM6995)
Functional hallucination (#HM5504).
Related Dreams (#HM2950) Illusion (#HM2510) Hypnotherapy (#HH0962)
Psychotic state (#HM2887) Audible thought (#HM1317) Sense of presence (#HM1166)
Delusive consciousness (#HM2600)
Hypnagogic consciousness (Psychism, #HM3511)
Hypnopompic consciousness (Psychism, #HM3688)
Modes of awareness associated with schizophrenia (#HM2313).

♦ **HM4587e Land of shadows** (Psychism)
Description A hell-like state described in psychism, an etheric world of dense darkness inhabited by ignorant soul-minds and comprising negative aspects of mankind.

♦ **HM4594c Baqi** (Sufism)
Description The name of a traveller on the *tariqa* or Sufi spiritual path at the level of abiding and subsisting in divinity, when perfection is reached.
Context Spiritual progress on the spiritual path is marked by the attainment of specific *maqamat* or station.
Broader Mystic stations (Sufism, #HM3415).
Related Love of God (Sufism, #HM4273).

♦ **HM4595f Light state of hypnosis**
Description Although external stimuli are accepted by the subconscious without discrimination by the conscious mind, and the person involved is thus very open to suggestion, this is not so clearly a hypnotic state as the other, deeper states. The body may retain the standing or sitting position and the individual may converse with full awareness (although may not remember afterwards all that has been said). The condition may arise not only through hypnotic suggestion but also through intense concentration or full attention to a voice or other sound, a book or a television programme.
Broader Hypnotic states of consciousness (#HM2133).

♦ **HM4596d Hatred awareness**
Hate
Description This has been defined as a "mental state of revulsion from something that offends" and implies ill will and the desire to harm or do away with the object of hatred. Unlike anger, which normally passes according to mood, hate usually denotes a continuity of attitude and one which consciously makes plans against the object of hatred. It is held to be the opposite of love, which desires the good of its object. Where love may be blind to defects, hatred sees only defects and takes pleasure in dwelling on them, while it is blind to good qualities. Interestingly, love may turn to hatred which is all the more bitter because of once having loved.
Hate may also be considered the reverse of love in that it creates duality and separation. It is also the reverse of brotherhood. It is due to a concentration on form rather than on what the form reveals, and results from ignorance, a sense of personality and a misapplication of desire.
Related Hatred (Buddhism, #HM4502) Love consciousness (#HM2000)
State of anger (Buddhism, #HM2959).

♦ **HM4600d Astral plane** (Leela)
Bhuvar-loka
Description This is the plane of dreams, fantasies and soaring imagination. There is awareness of how diverse the phenomenal world is, with all its possibilities. No longer troubled by lacking means for survival, the player expresses himself in the creative arts and in sexuality and sensual indulgence. Although heaven may now be imagined as a possibility, over-excitement may carry him away, depleting his energy and leading to exhaustion.
Context The 14th state or square on the board of *Leela*, the game of knowledge, appearing in the second row. This is the second of eight lokas represented on the spine of the board and marking a stage in the development of the individual's consciousness. The first seven represent a psychic centre or chakra on the spine of the human body, a level of psychic evolution. An eighth loka lies beyond the chakras. Each loka is defined by the nature of the matter of which it is composed. The eight other states in the same row may be considered as special regions within the loka.
Broader Second chakra: realm of fantasy (Leela, #HM3651).

♦ **HM4603b Tantric formless path** (Buddhism, Yoga)
Description This method of realizing the whole through which the world becomes utterly transparent is contrasted with the *path of form*. The latter involves meditative absorption or visualization, which is not part of the formless path. It does, however, produce psychic energy to shift the yogin to his goal.
Related Tantra (Buddhism, Yoga, #HH0306) Tantric visualization (Yoga, #HM1690).

♦ **HM4604f Primary delusions**
True delusions — Apophany
Description The individual has an uneasy awareness that everything has changed in significance, become disordered. The atmosphere is sinister, there is a sense of frightening uncertainty and an apprehension of disintegration. objects and events have vague significance and new meaning (delusional perception). A delusional memory may give significance to a past event which is relevant to the present perception. There may be awareness of cataclysmic events (delusional awareness). It is not so much the cognitive processes which are impaired. What appears to happen is a shift in cognitive structures. Primary delusion can thus be said to be caused by a shift in inter-schematic organization. The shift results in false postulates as part of an attempt to fit in the new meanings which have been experienced. One may be said to build up schema or constructs which structure one's world and enable prediction of the course of events. Where one is subjected to inconsistencies which require continual modification of the relations between the schema or constructs, the interlinking becomes weaker and inconsistent. There is failure in prediction and subsequent breakdown in response to the environment.
Broader Delusive consciousness (#HM2600).
Narrower Delusional ideas (#HM4504) Delusional awareness (#HM5003)
Delusional perception (#HM0464).
Related Secondary delusions (#HM2795).

♦ **HM4605d Workability of the aggregate of consciousness** (Buddhism)
Vinnanakkhandhassa (Pali) — Composedness of the aggregate of consciousness
Broader Workability of thought (Buddhism, #HM1584)
Composedness of thought (Buddhism, #HM1490)
Dispositions of consciousness (Buddhism, #HM2098).

♦ **HM4608d Correct speech** (Buddhism)
Sammavaca (Pali)
Broader Correct speech (Buddhism, #HM1821).

♦ **HM4637d Knowledge of change of lineage** (Buddhism)
Adoptive knowledge — Gotra-bhu-nanam (Pali)
Description Having passed through the 9 knowledges of purification by knowledge and vision of the way, there comes knowledge of change of lineage. It is intermediate between purification by knowledge and vision of the way and purification by knowledge and vision, and is counted as insight since it falls in line with insight. The object of this knowledge is nirvana, the signless, no-occurrence, no-formation, cessation. Passing from the lineage, category and plane of the ordinary man he enters those of the noble ones or arahants.
Related Purification by knowledge and vision (Buddhism, #HH3025)
Purification by knowledge and vision of the way (Buddhism, #HH3550).
Followed by Knowledge of the path of stream entry (Buddhism, #HM1088).
Preceded by Knowledge in conformity with truth (Buddhism, #HM5403).

♦ **HM4648c Conduct** (Sufism)
Context The third of ten states acting as gateways orienting the mystic on the journey to the Absolute, as described by Ansari.
Broader Stations of consciousness - Ansari (Sufism, #HM2317).
Followed by Character (Sufism, #HM3759).
Preceded by Doors (Sufism, #HM3451).

♦ **HM4651c Bare perception** (Buddhism, Tibetan)
Description This way of knowing takes objective entities as its appearing object. It is contrasted with *conceptual cognition* of metaphysical entities. Objective entities are impermanent. They are dependent on cause or circumstance and may produce an effect. From their own individual point of view they exist objectively or substantially. Bare perception is of four types: sensory; mental; awareness of consciousness; yogic. All four are non-deceptive and devoid of conceptualizing. The first is of five kinds, depending on the five organs of sense, and is the first, non-conceptualized, moment of perception. Mental perception is also of five kinds; it arises immediately after sensory perception and immediately prior to conceptualization of the object of perception. Awareness of consciousness is also non-deceptive and without conceptualizing. All these three types may be valid, subsequent or inattentive cognitions. Yogic perception is never inattentive and arises from meditating single-mindedly. There are two types from the viewpoint of the object of perception, knowing what things exist and knowing what they are like.
Context One of the valid ways of knowing of Tibetan Buddhism.
Broader Ways of knowing (Buddhism, #HH0873).
Related Vijna (Buddhism, #HM3617)
Conceptual cognition (Buddhism, Tibetan, #HM0713).

♦ **HM4653d Measureless concentration with a limited object** (Buddhism)
Infinite concentration with a limited object
Description Concentration is familiar, well developed and capable of being condition for a higher jhana; however the object of concentration is unextended and therefore limited.
Context According to Hinayana Buddhism, concentration is considered as of one kind (monad), of two kinds (dyads), of three kinds (triads), of four kinds (tetrads) or of five kinds (pentad). In the second tetrad, this is the third concentration.
Broader Concentration (Buddhism, #HM6663).
Related Limited concentration with a limited object (Buddhism, #HM1175)
Limited concentration with a measureless object (Buddhism, #HM5547)
Measureless concentration with a measureless object (Buddhism, #HM5214).

♦ **HM4663d Ikhlas** (Sufism)
Sincerity
Description Fasting by day and praying by night, no slightest deviation from the path is accepted by the devotee.
Context According to Shaykh Abu Sa'id ibn Abi'l-Khayr, this is the 17th of 40 stations or maqamat the Sufi must possess for his journey on the path of Sufism to be acceptable.
Broader Stations of consciousness - ibn-Abi'l-Khayr (Sufism, #HM2424).
Followed by Sidq (Sufism, #HM6631).
Preceded by Wara (Sufism, #HM0286).

♦ **HM4664g Coriolis illusion**
Vestibular coriolis reaction
Description Normally the brain discounts the vestibular sensations of acceleration that result from voluntary head movements, provided the body is stationary or moving slowly. In a vehicle travelling rapidly, particularly with irregular movement, the passenger feels his body moving when his head moves, and may see glowing lights apparently move. This aggravates the problem of motion sickness.
Broader Illusion (#HM2510).

♦ **HM4665d Pari–nirvana** (Buddhism)
Anupadisesanibbana (Pali) — Nirvana without remainder
Description Craving is eradicated, there is dispassion or "fading away", even inherent tendencies are eradicated. There is no remainder. The aggregates of existence arise no further; those already arisen have disappeared. The cause of action has been put away, results of the fruit of action have been destroyed.
Broader Nirvana (Buddhism, #HM2330).
Related Sa–upadisesanibbana (Pali, #HM5023).

♦ **HM4666d Lust** (Hinduism)
Kamachanda — Raga
Description Whereas *kama* is desire in the general sense, *kamachanda* refers to sexual desire or lust. It is the craving for union of man and women with each other. The yearning for sexual union is an instinctive desire due to false identification of the self with the body. Craving for union with a member of the opposite sex is also called *raga*.
Broader Desire (Hinduism, #HM7800).
Related Tama' (Islam, #HM4772) Spiritual lust (Christianity, #HM4180).

♦ **HM4671c Self–aggrandizing love**
Vanity love
Description An attachment is formed with another person in large part as a means to an end, whether the objective is tangible or intangible. This form of love may evolve into passionate love.
Refs Person, Ethel Spector *Love and Fateful Encounters* (1990).
Broader Love consciousness (#HM2000).
Related Self–love (#HM0478) Neurotic love (#HM5702).

♦ **HM4701c Profitable consciousness in the immaterial sphere** (Buddhism)
Moral consciousness in the formless realm
Description A total of four kinds of consciousness may arise here. These are associated with the four immaterial states or jhanas. The first is based on the jhana of boundless or unlimited space; the second with the jhana of boundless or unlimited consciousness; the third with the jhana of the sphere of nothingness and the fourth with the jhana of the sphere of neither perception not non–perception.
All four give rise to materiality and to postures but not to intimation.
Context In Hinayana Buddhism, 89 consciousnesses are enumerated in aggregate (khanda). Of these, 21 are profitable or moral, 12 are unprofitable or immoral and 56 are indeterminate (resultant or functional). The unprofitable all arise in the sphere of sense and desire, whereas profitable and indeterminate arise in sense, fine–material, immaterial and supramundane spheres.
Broader Dispositions of consciousness (Buddhism, #HM2098).
Related Impulsion (Buddhism, #HM7268)
Immaterial states as meditation subjects (Buddhism, #HH3198)
Indeterminate consciousness in the immaterial sphere – resultant (Buddhism, #HM4982)
Indeterminate consciousness in the immaterial sphere – functional (Buddhism, #HM0282).
Followed by Profitable consciousness in the supramundane plane (Buddhism, #HM4930).
Preceded by Profitable consciousness in the fine–material sphere (Buddhism, #HM5338).

♦ **HM4704d Tranquillity of mental body** (Buddhism)
Repose of mental factors — Repose of kaya (Pali)
Description This refers to the tranquillizing of the three aggregates of feeling, perception and formation or mental activities. The characteristic is pacifying suffering and quieting disturbance of the mental body, the function is to crush its suffering and disturbance. It is manifest as unwavering, inactive and coolness. Mental factors are the proximate cause and it is the opponent of corruption or defilement causing disturbance or lack of peace in the mental body due to distraction, agitation and so on. Tranquillity of mental body and tranquillity of consciousness are considered together.
Context One of the formations aggregate (mental coefficients) of Hinayana Buddhism, being listed among the constant states which appear in their true nature, and as profitable primary (always present in any profitable or profitable–resultant consciousness).
Broader Awareness as mental–formation group of conscious existence (Buddhism, #HM2050).
Related Tranquillity of consciousness (Buddhism, #HM0226).

♦ **HM4712d Love in God** (Islam)
Al–mahabba fi 'llah
Description God is the ultimate but not the only legitimate object of love. Love to God leads to love in God or for the sake of God towards all that God loves, and is demanded if one loves God completely, as is hatred for all God hates. But there must be caution. Because love is blind, it is easy to be deluded into thinking that one is loving for the sake of God when the real purpose is quite different – leadership, wealth or some other worldly advantage. Love then degenerates from love in God to love with God.
Context One of three types of love of Ibn al–Qayyim – love to God, love in God (for His sake) and love with God.
Broader Love (Islam, #HH5712) Mahabba (Islam, #HM6523).
Related Love to God (Islam, #HM5116) Love with God (Islam, #HM5580)
Divine–love awareness (Sufism, #HM2401).

♦ **HM4726d Understanding as knowledge about law** (Buddhism)
Dhamma–patisambhida (Pali) — Understanding as knowledge of doctrine
Description Law (dhamma or dharma) implies condition or "root cause", that which in some way leads to a particular result. The knowledge referred to is thus that of the result or "fruit" of a condition or cause. In this context, five specific things are understood as law: the cause of a result; the Noble Path; what is spoken; what is profitable or moral; what is unprofitable or immoral. The understanding referred to is that which arises on reviewing that law, distinctions about dhamma which arise on reflection. Examples of knowledge of law are: knowledge about the origin of suffering; knowledge about the path leading to cessation of suffering; knowledge about a condition or root cause; knowledge about the source of what has come to be, arisen, been born; knowledge about the origin of ageing and death; knowledge about the path leading to cessation of ageing and death; knowledge about the path leading to cessation of compound things or formations; knowledge of the dhamma and the scriptures; knowledge about moral or profitable states.
Context On the Path of Purification of Hinayana Buddhism, panna (understanding) is considered as of one kind (monad), of two kinds (dyads), of three kinds (triads) or of four kinds (tetrads). There are five dyads, four triads and two tetrads. All have the characteristic of penetrating the individual essences or true nature of states (monad). In the second tetrad, this is one of the four kinds of understanding as concerned with meaning, law, language and perspicuity. The four kinds of understanding are together referred to as the four discriminations or analyses.
Broader Understanding (Buddhism, #HM4523) Four discriminations (Buddhism, #HM0485).
Related Eightfold way (Buddhism, #HM2339)
Understanding as knowledge about meaning (Buddhism, #HM1663)
Understanding as knowledge about language (Buddhism, #HM5026)
Understanding as knowledge about kinds of knowledge (Buddhism, #HM4958).

♦ **HM4754f Anomalies in experience of the self as distinct from the outside world**
Blurring of ego boundaries
Description Disturbed people may experience other people and things as though they were inside themselves. What is happening to someone or something else may be experienced as though happening to one's self. Or they may cease to experience themselves as anything more than part of the environment, even confusing personal pronouns and possessive adjectives. This blurring of ego boundaries has been compared with *nirvana*, *oneness* and consciousness expansion, the merging of the individual self with the outside world, where dissolving of ego boundaries is looked on as desirable. Graham Reed postulates that all these conditions are extremes, welcome when expected, feared when arising unexpectedly, but part of two continua – the differences in the degree of discreteness of the ego boundaries within one individual at different times and the differences in the degree of discreteness of the ego boundaries between individuals (personality variable).
Refs Reed, Graham *The Psychology of Anomalous Experience* (1988).
Broader Anomalies in awareness of one's self (#HM8186).
Related Nirvana (Buddhism, #HM2330) Schizophrenic fantasy (#HH0939)
Unitary consciousness (#HM2702) Consciousness expansion (#HM2126).

♦ **HM4761c Indeterminate consciousness in the fine–material sphere - functional** (Buddhism)
Indeterminate consciousness in the form–realm - inoperative
Description As with the profitable, a total of 5 kinds of consciousness may arise here. These are associated with the jhana factors, that is: (i) with applied thought or inception of thought; sustained thought; happiness, joy or zest; bliss or ease; and concentration. (ii) with sustained thought; happiness, joy or zest; bliss or ease; and concentration. (iii) with happiness, joy or zest; bliss or ease; and concentration. (iv) with bliss or ease and concentration, happiness having faded away. (v) with concentration and equanimity. However, whereas the profitable arise only in ordinary men, these arise only in arahants.
All 5 give rise to materiality and to postures but not to intimation.
Context In Hinayana Buddhism, 89 consciousnesses are enumerated in aggregate (khanda). Of these, 21 are profitable or moral, 12 are unprofitable or immoral and 56 are indeterminate (resultant or functional). The unprofitable all arise in the sphere of sense and desire, whereas profitable and indeterminate arise in sense, fine–material, immaterial and supramundane spheres.
Broader Dispositions of consciousness (Buddhism, #HM2098).
Related Impulsion (Buddhism, #HM7268)
Profitable consciousness in the fine–material sphere (Buddhism, #HM5338)
Indeterminate consciousness in the fine–material sphere - resultant (Buddhism, #HM0594).
Followed by Indeterminate consciousness in the immaterial sphere - functional (Buddhism, #HM0282).
Preceded by Indeterminate consciousness in the sense sphere - functional (Buddhism, #HM3852).

♦ **HM4762c Unification of energy** (Taoism)
Formation of the spiritual embryo
Description During this stage the natural fire of reality operates to burn away residual conditioning in order to recover the state in which there is no discriminatory knowledge. This is frequently symbolized by the "pregnancy" of the practitioner with a spiritual "embryo" that becomes imbued with spiritual energy. Both things and self are experienced as empty. There is a sublimation of the duality of firmness and flexibility. In this state of absence of discrimination or knowledge, the spirit is unified, objects disappear. It is a profound trance without differentiation, entering from doing into not–doing. The term "embryo" is used to describe true awareness become solidified, stabilized and free from habitual scattering.
Context The sixth of seven stages of the Taoist alchemical firing process through which reality is cultivated and the self is restored.
Broader Alchemy (Taoism, #HH5887).
Followed by Transcending the world (Taoism, #HM5265).
Preceded by Merging of yin and yang (Taoism, #HM1330).

♦ **HM4766c Striving against the soul** (Islam)
Description Since the soul innately loves passions (hawa) which need to be controlled, there is a striving against the soul when one tries to harness the passions. The resistance is *sabr* (patient endurance). Conflict between reason and passion occurs in the heart. As the five senses bring useful information to the heart they also bring what is harmful. The paths bringing worldly preoccupations have to be blocked.
Related Striving against the soul (Islam, #HH6973).

♦ **HM4772d Tama'** (Islam)
Desire to possess — Physical lust
Description The object of admiration, gaving been gazed upon, found to be beautiful and contemplated is now desired. Although lovers of absolute beauty do not feel this, nor is it necessarily present in lovers of determinate beauty, it is nonetheless present in the strongest love.
Context The fourth cause of profane love occurring in the beholder according to Ibn al–Qayyim.
Refs Bell, Joseph Norment *Love Theory in Later Hanbalite Islam* (1979).
Related Lust (Hinduism, #HM4666).
Preceded by Al–fikr fi 'l–manzur (Islam, #HM1264).

♦ **HM4788d Continuous all–pervading knowledge** (Sufism)
Description Having reached the stage where, for most of the time and without directing attention to the total motion in the body, the devotee senses the presence of the remembered God, he tries not to forget this presence for an instant, whether due to activity in the limbs or in the heart.
Context This is the seventh state arising in the method of dhikr described by Shaykh Kalimallah.
Broader Remembrance of God (Sufism, #HM6562).
Related Ghaflah (Sufism, #HM7383).
Followed by Wilayat (Sufism, #HM3647).
Preceded by 'Ilm al–basir (Sufism, #HM0476).

♦ **HM4799d Al–Hakim** (Sufism)
Description The believer is aware of Allah as the wise, without doubt or uncertainty. In creation all is formed in perfect wisdom of order and continuity, each individual a model of the whole cosmos. Following Him one learns from the reflection of His perfect knowledge and progresses towards perfection; failure to follow means one cannot grow, one receives no good and becomes fuel for hell. One is free to choose.
Context A mode of mystical awareness in the Sufi tradition associated with the forty–sixth of the ninety–nine names of Allah.

Broader Ninety-nine names of Allah (Sufism, #HH2561).
Followed by Al-Wadud (Sufism, #HM7712).
Preceded by Al-Wasi' (Sufism, #HM4169).

♦ **HM4803f Koro**
Description This is a state of partial depersonalization and acute anxiety, when the individual is certain that his or her genitalia are shrivelling away into non-existence.
Broader Depersonalization (#HM1248).

♦ **HM4805b Musical inspiration**
Musical ecstasy
Description Three main levels of musical and artistic inspiration can be recognized.
Refs Godwin, Joscelyn *Harmonies of Heaven and Earth* (1987).
Narrower Egoic musical inspiration (#HM0203) Imitative musical inspiration (#HM1995)
Avataric level of musical inspiration (#HM6006).
Related Human development through music (#HH3003).

♦ **HM4806g Sleep spindles**
Stage-two sleep
Description This is marked by *sleep spindles*; there is minimal eye-movement and further reduction in muscle tension and an increase in slow, high amplitude (delta) waves. There are no eye movements and muscle tension is usually low.
Broader Sleep (#HM2980).
Related Delta wave consciousness (#HM1785).
Followed by Stage-three sleep (#HM0348).
Preceded by Drowsiness (#HM6231).

♦ **HM4875c Flash of intuition** (Hinduism)
Pratibhajnana — Prophetic intuition — Clairvoyant intuition
Description This extrasensory perception or intuition of a future event generally belongs to sages but in rare cases occurs to ordinary persons as well. The intuitive perception of a future object is brought to consciousness by memory. This is to be regarded as a valid perception - it is produced by a real object, it is not doubtful, it is not contradicted, its causes are not vitiated by any defect. A yogin is able to perceive all objects, whether past, present or future, but the flash of intuition described here may be experienced on occasion by anyone. It has been described as an extra-sensory valid perception since it is not produced by the sense organs; it is not produced by the internal organ as it does not operate on external objects. It is not conjecture since it is definite cognition. It is not illusion since it corresponds to actual fact. It is not inference, since it is immediate knowledge.
Broader Intuition (Hinduism, #HM1256).
Related Intuition of sages (Hinduism, #HM9124) Flash of intuition (Hinduism, #HM4875).

♦ **HM4877d Flow experience through work**
Enjoying work
Description In cultures where the life-style harmoniously balances human goals with the resources of the environment, work and enjoyment are not separate; but any work experience where the person plays with and transform opportunities in his surroundings is developed and enjoyable. Enjoyment depends on increasing complexity, upon focusing attention on opportunities for action in the environment, resulting in the perfection of skills so that in the end they appear completely spontaneous. Another way of developing flow experience through work is to change the work so that conditions are more conducive to flow. An example is the cottage industry where the worker sets his own schedules and goals, modifying them to his own needs and changing products so as to bring variety and challenge. The construction of work to the maximum pleasure possible does not guarantee flow; optimal experience depends on subjective evaluation of possibilities and individual capacity. The optimal experience comes from redesigning jobs to resemble flow activities and helping people to develop autotelic personalities so as to enjoy whatever job they are doing.
Refs Csikszentmihalyi, Mihaly *Flow* (1990).
Broader Flow experience (#HM2344).
Related Yu (Taoism, #HM1771) Autotelic experience (#HM1410).

♦ **HM4882c Concentration of the fourth jhana of five** (Buddhism)
Description With the elimination or fading away of rapture or happiness, two factors remain in the fourth jhana: ease or bliss; concentration.
Context According to Hinayana Buddhism, concentration is considered as of one kind (monad), of two kinds (dyads), of three kinds (triads), of four kinds (tetrads) or of five kinds (pentad). In the pentad, fivefold according to the five jhana factors, this is the fourth concentration. It is equivalent to the third concentration in the third tetrad.
Broader Concentration (Buddhism, #HM6663).
Related Dwelling in the third jhana (Buddhism, #HM5643)
Concentration of the third jhana of four (Buddhism, #HM2284).
Followed by Concentration of the fifth jhana of five (Buddhism, #HM5143).
Preceded by Concentration of the third jhana of five (Buddhism, #HM8532).

♦ **HM4886d Rapturous happiness due to insight** (Buddhism)
Description The five kinds of happiness arise and permeate the whole body. The meditator, thinking that such happiness has never arisen before, believes he has reached fruition, interrupts the course of insight. This is karmically unprofitable, because he mistakes the path.
Context One of the ten imperfections of insight of Hinayana Buddhism.
Broader Imperfection of insight (Buddhism, #HM9722).

♦ **HM4897f Amnesia**
Narrower Cryptomnesia (#HM8235) Hysterical amnesia (#HM7608).

♦ **HM4899f Elation**
Description This elevated mood, incongruous with the individual's true situation, usually includes physical and mental hyperactivity. More common in children than adults, it is related to the primary defence of denial.
Related Mania (#HM2787) Hypomania (#HM0514) Elation (Buddhism, #HM0425).

♦ **HM4901d Niyyat** (Sufism)
Intention
Description Despite tribulations which will have to be faced, there is keen desire to reach the goal.
Context According to Shaykh Abu Sa'id ibn Abi'l-Khayr, this is the first of 40 stations or maqamat the Sufi must possess for his journey on the path of Sufism to be acceptable.
Broader Stations of consciousness - ibn-Abi'l-Khayr (Sufism, #HM2424).
Followed by Inabat (Sufism, #HM3484).

♦ **HM4902f Mind blankness**
Stage-fright
Description A typical example of this is in the examination room when not only is the studiously learned material absent but also the mind is empty of all relevant thoughts. Although this and stage fright are clearly induced by stress, there are times of mind-blankness in non-stressful situations. These include *writers' block*. Pathological *thought blocking* arises in schizophrenia, when the individual may believe that thoughts have been siphoned off by some malignant being.

♦ **HM4905c Sympathy-love**
Mannered love
Description A form of love which is essentially conformist, its main attraction being its conventionality. Although it may appear to the lovers to be an authentic emotional attachment, it lacks the substance of one because of this conformism.
Refs Person, Ethel Spector *Love and Fateful Encounters* (1990).
Broader Love consciousness (#HM2000).

♦ **HM4908d Concentration with applied thought and sustained thought** (Buddhism)
Description This is concentration of the first jhana together with access concentration.
Context According to Hinayana Buddhism, concentration is considered as of one kind (monad), of two kinds (dyads), of three kinds (triads), of four kinds (tetrads) or of five kinds (pentad). This is the first concentration of the second triad.
Broader Concentration (Buddhism, #HM6663).
Related Access concentration (Buddhism, #HM4999).
Followed by Concentration without applied thought and with sustained thought (Buddhism, #HM5199).

♦ **HM4930c Profitable consciousness in the supramundane plane** (Buddhism)
Moral consciousness in the transcendental plane
Description The four kinds of consciousness are associated with the four paths: stream entry; once-return; non-return; arahantship.
All four give rise to materiality and to postures but not to intimation.
Context In Hinayana Buddhism, 89 consciousnesses are enumerated in aggregate (khanda). Of these, 21 are profitable or moral, 12 are unprofitable or immoral and 56 are indeterminate (resultant or functional). The unprofitable all arise in the sphere of sense and desire, whereas profitable and indeterminate arise in sense, fine-material, immaterial and supramundane spheres.
Broader Dispositions of consciousness (Buddhism, #HM2098).
Related Impulsion (Buddhism, #HM7268)
Purification by knowledge and vision (Buddhism, #HH3025)
Indeterminate consciousness in the supramundane plane - resultant (Buddhism, #HM5129).
Preceded by Profitable consciousness in the immaterial sphere (Buddhism, #HM4701).

♦ **HM4958d Understanding as knowledge about kinds of knowledge** (Buddhism)
Patibhana-patisambhida (Pali) — Understanding as discrimination of perspicuity
Description Knowledge concerned with meaning, with law, or with the language of meaning or law - whenever this knowledge is the object of knowledge, whenever it is analysed by its spheres and functions, then the understanding which arises is as knowledge about kinds of knowledge.
Context On the Path of Purification of Hinayana Buddhism, panna (understanding) is considered as of one kind (monad), of two kinds (dyads), of three kinds (triads) or of four kinds (tetrads). There are five dyads, four triads and two tetrads. All have the characteristic of penetrating the individual essences or true nature of states (monad). In the second tetrad, this is one of the four kinds of understanding as concerned with meaning, law, language and perspicuity. The four kinds of understanding are together referred to as the four discriminations or analyses.
Broader Understanding (Buddhism, #HM4523) Four discriminations (Buddhism, #HM0485).
Related Understanding as knowledge about law (Buddhism, #HM4726)
Understanding as knowledge about meaning (Buddhism, #HM1663)
Understanding as knowledge about language (Buddhism, #HM5026).

♦ **HM4967d Alert thinking**
Context Normal consciousness is said to be alert thinking (concentrated) interspersed at regular intervals by relaxed thinking (concentrated) or daydreaming (diffuse). This mirrors the regular pattern of paradoxical or rapid eye movement sleep which occurs about every 90 minutes during the night.
Related Daydream (#HM2138) Relaxed thinking (#HM1274).

♦ **HM4968d Understanding defining materiality** (Buddhism)
Understanding by way of fixing material qualities
Description In one striving for insight, this understanding is associated with defining the material aggregate.
Context On the Path of Purification of Hinayana Buddhism, panna (understanding) is considered as of one kind (monad), of two kinds (dyads), of three kinds (triads) or of four kinds (tetrads). There are five dyads, four triads and two tetrads. All have the characteristic of penetrating the individual essences or true nature of states (monad). In the third dyad, this is contrasted with understanding defining mentality.
Broader Understanding (Buddhism, #HM4523).
Related Understanding defining mentality (Buddhism, #HM5627).

♦ **HM4969d Concentration due to zeal** (Buddhism)
Desire-concentration
Description Unification or collectedness of mind is obtained by making zeal or desire predominant.
Context According to Hinayana Buddhism, concentration is considered as of one kind (monad), of two kinds (dyads), of three kinds (triads), of four kinds (tetrads) or of five kinds (pentad). In the sixth tetrad, this is the first concentration.
Broader Concentration (Buddhism, #HM6663).
Related Concentration due to energy (Buddhism, #HM8365)
Concentration due to inquiry (Buddhism, #HM7434)
Concentration due to natural purity of consciousness (Buddhism, #HM7961).

♦ **HM4976f Absent-mindedness**
Preoccupation — Nonmeditation — Absentmindedness
Description The individual is preoccupied with his own thoughts so that he closes out much of the available external information. He may be very attentive to thinking and to external stimuli related to that thinking, but his attentiveness to other stimuli is very low as he considers it distracting. Behaviour related to routine does not respond to feedback with respect to changes in such routine and may therefore appear bizarre. The state does not continue if external stimuli reach a particular level, as when the individual suddenly becomes aware that an approaching car will run him over if he does not respond by getting out of its way.
This condition of deep inner concentration on a specific subject blots out all awareness of external

stimuli. There is said to be a predominance of alpha brain waves, but the state is not the same as meditation; rather than a calm response when external stimuli do intrude, the individual is startled and incoherent, possibly due to the creative (right hemisphere) of the brain remaining predominant and not switching to mundane, left hemisphere, activity.
Refs Reed, Graham *The Psychology of Anomalous Experience* (1988).
Broader Tuning–in (#HM6687).

♦ **HM4977d Concentration of easy progress and swift direct–knowledge** (Buddhism)
Concentration of easy progress and quick intuition
Description Cultivating what is suitable, carrying out preparatory tasks, not overwhelmed by craving, having practice in serenity (absorption concentration), having a mild lower nature, all these mean progress is easy. Cultivating what is suitable, having skill in absorption, not overwhelmed by ignorance, having practice in insight, having keen faculties, all these mean intuition is swift.
Context According to Hinayana Buddhism, concentration is considered as of one kind (monad), of two kinds (dyads), of three kinds (triads), of four kinds (tetrads) or of five kinds (pentad). In the first tetrad, this is the fourth concentration.
Broader Concentration (Buddhism, #HM6663).
Related Concentration of easy progress and sluggish direct–knowledge (Buddhism, #HM1010)
Concentration of difficult progress and swift direct–knowledge (Buddhism, #HM7859)
Concentration of difficult progress and sluggish direct–knowledge (Buddhism, #HM5992).

♦ **HM4980d As–Samad** (Sufism)
Description The believer is aware of Allah as the eternal satisfier of all needs, the only recourse in case of trouble or pain. All others to whom one turns are reflections of Him and gifts from Him. He loves those who are heedful and thankful to Him.
Context A mode of mystical awareness in the Sufi tradition associated with the sixty–eighth of the ninety–nine names of Allah.
Broader Ninety–nine names of Allah (Sufism, #HH2561).
Followed by Al–Qadir (Sufism, #HM1500).
Preceded by Al–Ahad (Sufism, #HM6117).

♦ **HM4982c Indeterminate consciousness in the immaterial sphere – resultant** (Buddhism)
Indeterminate consciousness in the formless realm – resultant
Description As with profitable consciousness, a total of four kinds of consciousness may arise here. These are associated with the four immaterial states or jhanas. The first is based on the jhana of boundless or unlimited space; the second with the jhana of boundless or unlimited consciousness; the third with the jhana of the sphere of nothingness and the fourth with the jhana of the sphere of neither perception not non–perception. But whereas with the profitable consciousnesses these occurred in a cognitive series through right attainment, here it is by birth, through rebirth–linking (reconception), life–continuum and death.
None of the four gives rise to materiality, to postures or to intimation.
Context In Hinayana Buddhism, 89 consciousnesses are enumerated in aggregate (khanda). Of these, 21 are profitable or moral, 12 are unprofitable or immoral and 56 are indeterminate (resultant or functional). The unprofitable all arise in the sphere of sense and desire, whereas profitable and indeterminate arise in sense, fine–material, immaterial and supramundane spheres.
Broader Dispositions of consciousness (Buddhism, #HM2098).
Narrower Mundane resultant consciousness (#HM0130).
Related Rebirth–linking (Buddhism, #HM8266) Mind consciousness (Buddhism, #HM3323)
Formation of the imperturbable (Buddhism, #HH5543)
Profitable consciousness in the immaterial sphere (Buddhism, #HM4701)
Indeterminate consciousness in the immaterial sphere – functional (Buddhism, #HM0282).
Followed by Indeterminate consciousness in the supramundane plane – resultant (Buddhism, #HM5129).
Preceded by Indeterminate consciousness in the fine–material sphere – resultant (Buddhism, #HM0594).

♦ **HM4983b Feeling aggregate** (Buddhism)
Vedana–khanda — Awareness of feeling–group of conscious existence
Description In Hinayana Buddhism, feeling is considered as anything that is perceived as being feeling, whether past, present or future, internal or external, subjective or objective, gross or refined, superior or inferior, far or near. It is also considered as follows:
It is of one kind in that all that is feeling is felt.
It is threefold in that it is: good (profitable or moral) and associated with profitable consciousness; bad (unprofitable or immoral) and associated with unprofitable consciousness; or indeterminate and associated with indeterminate consciousness.
This group is divided into five classes, according to the nature of feeling. Thus it is fivefold depending upon whether feelings are intrinsically: (physically) pleasant or bodily agreeable; (physically) painful or bodily disagreeable; (mentally) joyful or mentally agreeable;, (mentally) sad (with grief as essence) or mentally disagreeable; indifferent (with equanimity as essence) or indeterminate (neutral).
Pleasure and pain are associated with profitable resultant and unprofitable resultant body consciousness, respectively. Joy is associated with 30 classes of consciousness: in the sense sphere, 4 profitable, 4 unprofitable, 4 indeterminate resultant profitable with root–cause, one indeterminate resultant profitable without root–cause, 4 indeterminate functional with root–cause, one indeterminate functional without root–cause; in the fine–material sphere, 4 profitable, 4 indeterminate resultant and 4 indeterminate functional (leaving out the fifth jhana in each of the last 3 cases). Grief is associated with the two kinds of unprofitable consciousness in the sense sphere that are rooted in hate. Equanimity is associated with the remaining 55 kinds of consciousness of the 89 in the consciousness aggregate.
Note: The total associated with joy is increased to 62 when the 8 supramundane (4 profitable and 4 indeterminate resultant) are each multiplied by the five jhanas but leaving out the 8 associated with the fifth jhana (32 extra).
Context One of the five interacting aggregates that produce the illusory ego. The feeling–group is also enumerated among the psycho–physical faculties (indriya) and the phase conditions (paccaya).
Refs Buddhaghosa, Bhadantacariya *The Path of Purification* (1980); Buddhaghosa, Bhadantacariya *The Path of Purity* (1975).
Broader Five aggregates (Buddhism, #HH3321).
Narrower Equanimity (#HH7769) Sad feeling (#HM7120)
Joyful feeling (#HM8737) Painful feeling (#HM8010)
Pleasant feeling (#HM6722).
Related Feeling (Buddhism, #HM2270) Perception aggregate (Buddhism, #HM4143)
Dispositions of consciousness (Buddhism, #HM2098)
Modes of occurrence of consciousness (Buddhism, #HM6720)
Awareness of corporeality–group of conscious existence (Buddhism, #HM2108)
Awareness as mental–formation group of conscious existence (Buddhism, #HM2050).

♦ **HM4987d Khilwat dar anjuman** (Sufism)
Enjoyment of solitude
Description The seeker perseveres in listening to the motion in the heart at all times, whether alone or in company. The murmuring is nurtured through frequent silence, and continually nourished and developed, with eyes open or closed, until the enjoyment of solitude even in company arises more frequently and for longer periods. There is pleasure in this and the seeker has peace of heart and equanimity.
Context This is the second state arising in the method of dhikr described by Shaykh Kalimallah.
Broader Remembrance of God (Sufism, #HM6562).
Followed by Sultan–al–dhikr (Sufism, #HM8330).
Preceded by Latifa–i–qalb (Sufism, #HM1905).

♦ **HM4988d Al–Qawi** (Sufism)
Description The believer is aware of Allah as the most strong and inexhaustible, who nothing can touch and who overcomes all. His strength creates universes and ensures their continuation; it protects and guides the creatures.
Context A mode of mystical awareness in the Sufi tradition associated with the fifty–third of the ninety–nine names of Allah.
Broader Ninety–nine names of Allah (Sufism, #HH2561).
Followed by Al–Matin (Sufism, #HM5633).
Preceded by Al–Wakil (Sufism, #HM1382).

♦ **HM4999d Access concentration** (Buddhism)
Collectedness of mind — Upacara samadhi
Description This is the unification of mind, arising through carrying out preliminary meditation on the following: the 6 recollections; mindfulness of death; recollection of peace; perceiving the abominableness of food; definitions of the 4 elements. In that it may be with or without happiness, access concentration may be of two kinds, also as accompanied by bliss or equanimity (third and fourth dyads). It is also related to concentration with applied thought and sustained thought (the first state in the second triad); in that it may be accompanied by bliss and happiness or by equanimity, with the third triad; and, in that sense–sphere concentration is all kinds of access unification, with the first state in the fifth tetrad.
Access concentration arises on abandoning hindrances but the jhana factors are not strong. For this reason there is a marked tendency to re–enter the life–continuum (bhavanga) or lower consciousness. However, with persistence, absorption concentration will develop. Development of access concentration is said to bestow special rebirth in the happy realm of sense. It is the equivalent of "calm abiding" in Tibetan Buddhism.
Context According to Hinayana Buddhism, concentration is considered as of one kind (monad), of two kinds (dyads), of three kinds (triads), of four kinds (tetrads) or of five kinds (pentad). In the first dyad, this is the unification that precedes absorption concentration.
Broader Concentration (Buddhism, #HM6663)
Appearance of absorption (Buddhism, #HM1618).
Related Dharana (Hinduism, #HM2566) Calm abiding (Buddhism, #HM2147)
Mark of calm (Buddhism, #HM8673) Limited concentration (Buddhism, #HM1073)
Subjects for meditation (Buddhism, #HH3987)
Sense–sphere concentration (Buddhism, #HM1097)
Concentration with happiness (Buddhism, #HM5767)
Concentration without happiness (Buddhism, #HM0730)
Concentration accompanied by bliss (Buddhism, #HM4191)
Concentration accompanied by bliss (Buddhism, #HM1454)
Concentration accompanied by happiness (Buddhism, #HM4433)
Concentration accompanied by equanimity (Buddhism, #HM8021)
Concentration accompanied by equanimity (Buddhism, #HM5008)
Concentration with applied thought and sustained thought (Buddhism, #HM4908).
Followed by Absorption concentration (Buddhism, #HM0311).

♦ **HM5000d Cuor contento** (Italian)
Happy–heartedness
Description Despite difficulties which would normally upset the person's even keel, he remains in a mentally and emotionally contented state, does not fret and is cheerful under all conditions.

♦ **HM5003f Delusional awareness**
Description There is awareness of a universal or cataclysmic event, this not necessarily being accompanied by perceptions of or any clear idea about the event.
Broader Primary delusions (#HM4604).

♦ **HM5005c Yogic perception** (Hinduism)
Yogipratyaksa
Description Considered by some to be the highest excellence of human perception (Jayanta Bhatta), this is the perception of subtle, hidden, remote, past and future objects. The practice of meditation gives the mind of the yogi immediate knowledge of all knowable objects unattainable to those whose minds are impure due to taints of love, hatred and so on. Pure minds, free from all taints and one–pointed through constant concentration see all objects in all places simultaneously through a single cognition. This differs from divine perception in that, unlike divine perception, it is not eternal. Also, God's knowledge is natural and not acquired since he is the creator of the Vedas and promulgator of moral law.
There are two kinds of yogic perception: ecstatic (yukta) and nonecstatic (viyukta).
Broader Intuition (Hinduism, #HM1256).
Narrower Ecstatic perception (#HM7100) Nonecstatic perception (#HM0239).

♦ **HM5006d Al–Qayyum** (Sufism)
Description The believer is aware of Allah as the self–existing. All existence depends on Him, as the body depends on the soul, but His existence depends only on Himself. It is divine grace that perpetuates the universe, divine will that provides the cause for satisfying the needs of every atom. With the whole universe in His care, He still cares for each individual as though he were His only creation. Nevertheless we still behave as though we did not need Him, as though we were self–sufficient and existed of ourselves.
Context A mode of mystical awareness in the Sufi tradition associated with the sixty–third of the ninety–nine names of Allah.
Broader Ninety–nine names of Allah (Sufism, #HH2561).
Followed by Al–Wajid (Sufism, #HM0743).
Preceded by Al–Hayy (Sufism, #HM4236).

♦ **HM5008d Concentration accompanied by equanimity** (Buddhism)
Concentration accompanied by indifference — Concentration accompanied by even–mindedness
Description This state is unification in the final jhana of the fourfold and the fivefold reckoning. It is one of the kinds of access concentration.
Context According to Hinayana Buddhism, concentration is considered as of one kind (monad),

of two kinds (dyads), of three kinds (triads), of four kinds (tetrads) or of five kinds (pentad). In the third triad, this is the third concentration.
Broader Concentration (Buddhism, #HM6663).
Related Access concentration (Buddhism, #HM4999).
Preceded by Concentration accompanied by bliss (Buddhism, #HM1454).

♦ **HM5014d Al-Hakam** (Sufism)
Description The believer is aware of Allah as the judge who orders all. His justice cannot be opposed or prevented. Man understands the law as written in the holy books to the extent of his good qualities and to his lot. He may obey or rebel and receives reward or punishment accordingly. The executor of judgement should take care that it is Allah's decree and through his power.
Context A mode of mystical awareness in the Sufi tradition associated with the twenty-eighth of the ninety-nine names of Allah.
Broader Ninety-nine names of Allah (Sufism, #HH2561).
Followed by Al-'Adl (Sufism, #HM3807).
Preceded by Al-Basir (Sufism, #HM6512).

♦ **HM5023d Sa-upadisesanibbana** (Pali)
Nirvana with remainder
Description Craving is eradicated, there is dispassion or "fading away", but inherent tendencies are not eradicated. These are the remainder. Nirvana has been reached but, during the arahant's life, the aggregates which result from the past still cling and are understood in terms of suppressing vices or defilement and the results of past clinging.
Broader Nirvana (Buddhism, #HM2330).
Related Pari-nirvana (Buddhism, #HM4665).

♦ **HM5026d Understanding as knowledge about language** (Buddhism)
Nirutti-patisambhida (Pali) — Understanding as knowledge about interpretation
Description The knowledge referred to is of true etymological interpretation of meaning and law. This is of essential language where there are no exceptions, the "root speech of all beings". Understanding which arises on reflection of meaning and law in such language is of the imperishable and everlasting.
Context On the Path of Purification of Hinayana Buddhism, panna (understanding) is considered as of one kind (monad), of two kinds (dyads), of three kinds (triads) or of four kinds (tetrads). There are five dyads, four triads and two tetrads. All have the characteristic of penetrating the individual essences or true nature of states (monad). In the second tetrad, this is one of the four kinds of understanding as concerned with meaning, law, language and perspicuity. The four kinds of understanding are together referred to as the four discriminations or analyses.
Broader Understanding (Buddhism, #HM4523) Four discriminations (Buddhism, #HM0485).
Related Understanding as knowledge about law (Buddhism, #HM4726)
Understanding as knowledge about meaning (Buddhism, #HM1663)
Understanding as knowledge about kinds of knowledge (Buddhism, #HM4958).

♦ **HM5028d Plane of radiation** (Leela)
Teja-loka
Description having passed through sound and air the player reached the plane of fire. He becomes light and his flame illuminates those around him.
Context The 58th state or square on the board of *Leela*, the game of knowledge, appearing in the seventh row.
Broader Seventh chakra: plane of reality (Leela, #HM0754).

♦ **HM5087c Romantic love**
Idyllic love — Mutual passionate love — Falling in love
Description Although each experience is frequently claimed to be unique, falling in love is often described as being accompanied by physical sensations, including loss of appetite, breathlessness. It can be experienced as a delirium and is often described as a fever. This experience is matched by a sense of excitement, risk and fear at revealing one's true self. There is a feeling of being caught up in a great emotion by which one is swept away. The exultation of love seems to offer a new kind of freedom, a freedom from the confines of the self. Preoccupation with oneself is exchanged, however temporarily, for a consuming interest in another. All uncertainties are eradicated, except that concerning whether one will continue to be loved by the beloved. These levels of desire and exultation alternate with feelings of withdrawal, boredom, self-doubt in a continuing process of vacillation.
The aim of love is nothing less than to overcome separateness and achieve union or merger with the beloved. In that imaginative merger the lover achieves both an exaltation of feeling and a profound sense of release. Lesser pleasures, and even pain, may be ignored in pursuit of the complex and elusive gratifications it promises. The process of falling in love is thus a grand obsession. Repetitive thinking about the beloved is an integral part of the experience and just as distinctive as the feeling state. Mutual passionate love, as the most complete form of romantic love, is distinguished by its intensity, even though it may be based on mutual destructiveness.
As an act of the imagination, love is often the greatest creative triumph of a lifetime. In its very nature as an act of the imagination lies the source of its power. It can exploit the lover's illusions and delusions, for both good and ill, as well as leading the lover to transcendent truths. It can demand a significant reordering of values and priorities, presenting the content and conditions requisite for dramatic change in self. Not only does it provide a major route to self-transcendence, but it also opens the way to self- realization and self-transformation. It creates a situation in which the self is exposed to new risks and enlarged possibilities thus providing the opportunity for growth. It acquires meaning and offers a sense of liberation to the extent that it enables a flexibility in personality that breaks through conventional psychological barriers and customs. As such it opens the possibility of new phases of life and realignements of the personality. This change in the sense of self, experienced by the lover as the novelty and originality of love, is at the core of love. This results in part from the multiple identifications in which the lover participates when falling in love, propelling the person to new commitments and away from old ones, and changing the boundaries of the self. The self grows through the desire for merger with the other. There may be a feeling of being born again. Love acts as a change agent because it is an explorative imaginative transaction between two people, which at its most sublime offers a partial escape from unremitting subjectivity. Despite its frequently transient nature, it offers access to the unconscious, lights up the emotional life, and brings internal change that often outlives the experience itself.
Context The value attached to romantic love varies from culture to culture, depending on what kinds of transcendental experience are esteemed and the degree of emphasis placed on personal change and development. Love can be viewed as a paradigm for any profound alignment of personality, such as those that occur in profound experiences of religious conversion. Romantic love has always been inextricably linked to spiritual aspiration.
Refs Johnson, Robert A *The Psychology of Romantic Love* (1984); Person, Ethel Spector *Love and Fateful Encounters* (1990).
Broader Love consciousness (#HM2000).
Related Love (#HH0258) Limerence (#HM1809) Sexual love (#HM9406)
Courtly love (#HM6122) Transference love (#HM7045)
Passionate love (Islam, #HM3656).
Followed by Affectionate bonding (#HM0973).
Preceded by Infatuation (#HM5991).

♦ **HM5088d Bliss due to insight** (Buddhism)
Pleasure due to insight
Description This arises and floods the whole body. The meditator, thinking that such bliss has never arisen before, believes he has reached fruition, interrupts the course of insight. This is karmically unprofitable, because he mistakes the path.
Context One of the ten imperfections of insight of Hinayana Buddhism.
Broader Imperfection of insight (Buddhism, #HM9722).

♦ **HM5089d Analysis** (Buddhism)
Vichara — Dpyod-pa (Tibetan) — Vicara (Pali) — Sustained thought — Reflection
Description Sustained mental application upon the same object with a view to investigation keeps the mind continually engaged in the exercise and eliminates doubt.
In Tibetan Buddhism, analysis refers to fine discrimination with respect to the entity or name of an object, and is virtuous, non-virtuous or neutral depending upon motivation and object of analysis. Virtuous generates pleasant effects and leads to abiding in contact with happiness. Non-virtuous generates unpleasant effects and leads to not abiding in contact with happiness.
In Hinayana Buddhism, it is said that once the mind has been directed onto its object by applied thought, like the striking of a bell, it is held there by sustained thought which keeps consciousness anchored and is compared to the ringing of the bell. Applied thought and sustained thought are not always considered separately.
Context One of the four changeable factors referred to in Tibetan Buddhism. These may be virtuous, non-virtuous or neutral depending on circumstances. One of the formations aggregate (mental coefficients) of Hinayana Buddhism, being listed among the constant states which appear in their true nature, and as general secondary (sometimes present in any consciousness). One of the five constituent factors of jhana.
Broader Changeable mental factors (Buddhism, #HH0910)
Awareness as mental-formation group of conscious existence (Buddhism, #HM2050).
Related Jhana (Buddhism, Pali, #HM7193) Investigation (Buddhism, #HM1177)
Approaching reflection (Buddhism, #HM1647) Application of reflection (Buddhism, #HM1982)
Dwelling in the first jhana (Buddhism, #HM4298).

♦ **HM5091c Detachment** (Hinduism)
Asakti
Description The primal desires for sons, wealth and happiness on earth and in heaven are extirpated, the instinctive desires for pleasure, sex and power are exterminated. The mind is no longer engrossed in objects of enjoyment, even those yielding the most transient pleasures. This absence of attachment, indifference, arises from always being content with the supreme bliss of Atman or Brahman.
Related Attachment (Hinduism, #HM3914) Non-attachment (Buddhism, #HM2128)
Detachment (Christianity, #HM1534).

♦ **HM5097g Oneroid states**
Description As with delirium, the chief characteristic of these states is their dream-like quality. Here, however, the content is connected, clear and consistent. The individual's whole attention and interest is held and the state may persist for weeks. The cause may be toxic poisoning or an early phase of schizophrenia.
Broader Diminished clarity of awareness (#HM6201).
Related Delirium (#HM7186).

♦ **HM5099d Happiness** (Leela)
Sukh
Description The player knows he is near his goal - so long as he is not lazy or inactive, one throw of the die may take him there. Body chemistry and psychic phenomena are balanced, there is experience of merging with the source in indescribable happiness.
Context The 62nd state or square on the board of *Leela*, the game of knowledge, appearing in the seventh row.
Broader Seventh chakra: plane of reality (Leela, #HM0754).
Preceded by Conscience (Leela, #HM1601).

♦ **HM5116c Love to God** (Islam)
Context One of three types of love of Ibn al-Qayyim - love to God, love in God (for His sake) and love with God.
Broader Love (Islam, #HH5712).
Related Love in God (Islam, #HM4712) Love of God (Sufism, #HM4273)
Love with God (Islam, #HM5580) Love of God (Christianity, #HH5232)
Divine-love awareness (Sufism, #HM2401).

♦ **HM5118d Numerical awareness** (Sufism)
Wuiquf al-'adudi
Description With patience and perseverance in the practice of dhikr, the regard for odd numbers arises.
Broader Recollection (Islam, Sufism, #HM2351).

♦ **HM5119d Desolation** (Christianity)
Negative feeling towards God
Description This is experienced as feelings of darkness, stress, anxiety, inner conflict and turmoil. There is an attraction towards things which can draw one away from God. Nevertheless, desolation can be productive of growth.
Refs Byrne, Lavinia (Ed) *Traditions of Spiritual Guidance* (1990).
Related Consolation (Christianity, #HM3502).

♦ **HM5122d Yoga of consciousness transference** (Buddhism)
Hpho-ba (Tibetan)
Context The sixth branch of the Tibetan Tantric path of form according to the Naropa school.
Broader Tantric visualization (Yoga, #HM1690).
Preceded by Yoga of the intermediary state (Buddhism, #HM0101).

♦ **HM5123g Mineral consciousness** (Psychism)
Mineral psychism — Attraction consciousness — Overall awareness
Description Minerals are said to attract, record and emit emotional vibrations from their surroundings and from living beings, particularly in the case of amulets and other psychic instruments which heighten the psychic and healing propensities of those who possess them; but each mineral has its own psychic properties (for example: healing, protection) particularly if treated correctly.

♦ HM5124f Isakower phenomenon
Description In this state a complex combination of sensations is experienced, principally of hands, skin and mouth. There may be an impression of giddiness, floating or sinking. Parts of the body, and the outside world, may seem to blend together. An indefinite, round, shadowy thing seems to come closer and more gigantic until it threatens to crush the individual before receding and shrinking to nothing. There is a murmuring or rustling sound and there may be the impression of something soft and yielding in the mouth and yet outside. The hands may feel swollen or crumpled, the mouth and skin dry and abrasive. The phenomenon is associated with hypnagogic states (falling asleep) but also with waking, with frequent drug abuse and as part of epileptic auras or preceding déjà vu. The experience is not regarded as real by the person experiencing it.
    Related Déjà vu (#HM1240)   Hypnagogic consciousness (Psychism, #HM3511).

♦ HM5128f Derealization
Description The environment appears to have changed in such a way that reality, the whole world, has changed, with possible feelings of imminent catastrophe. It may be that everything appears flat and lacking in significance. The quality of perception is such that the reality of what is perceived is not convincing. When derealization occurs in conjunction with depersonalization then everything appears unreal, including the individual experiencing it himself.
    Broader Anomalies in experience of the reality of one's self and the environment (#HM2548).
    Related Depersonalization (#HM1248)   Religious consciousness (#HM3008).

♦ HM5129c Indeterminate consciousness in the supramundane plane – resultant (Buddhism)
Indeterminate consciousness in the transcendental plane – resultant
Description The four kinds of consciousness are linked with the fruition of consciousness associated with four paths: fruition moment – stream entry; fruition moment – once–return; fruition moment – non–return; fruition moment – arahantship.
All four give rise to materiality and to postures but not to intimation.
Context In Hinayana Buddhism, 89 consciousnesses are enumerated in aggregate (khanda). Of these, 21 are profitable or moral, 12 are unprofitable or immoral and 56 are indeterminate (resultant or functional). The unprofitable all arise in the sphere of sense and desire, whereas profitable and indeterminate arise in sense, fine–material, immaterial and supramundane spheres.
    Broader Dispositions of consciousness (Buddhism, #HM2098).
    Related Impulsion (Buddhism, #HM7268)
Purification by knowledge and vision (Buddhism, #HH3025)
Profitable consciousness in the supramundane plane (Buddhism, #HM4930).
    Preceded by Indeterminate consciousness in the immaterial sphere – resultant (Buddhism, #HM4982).

♦ HM5134c Sunnagare abhirati (Buddhism)
Description Delight in the practice of jhana in solitude.
Context One of ten superhuman qualities described in the Sutta Vibhanga as being special attainments of insight above that of ordinary men.
    Broader Superhuman qualities (Buddhism, #HH9764).

♦ HM5143c Concentration of the fifth jhana of five (Buddhism)
Description Ease or bliss is abandoned; the fifth jhana has two factors: concentration and the equanimity or indifference with which it is accompanied.
Context According to Hinayana Buddhism, concentration is considered as of one kind (monad), of two kinds (dyads), of three kinds (triads), of four kinds (tetrads) or of five kinds (pentad). In the pentad, fivefold according to the five jhana factors, this is the fifth concentration. It is equivalent to the fourth concentration in the third tetrad.
    Broader Concentration (Buddhism, #HM6663).
    Related Dwelling in the fourth jhana (Buddhism, #HM8087)
Concentration of the fourth jhana of four (Buddhism, #HM7202).
    Preceded by Concentration of the fourth jhana of five (Buddhism, #HM4882).

♦ HM5155d Human plane (Leela)
Jana-loka
Description Upward flow of energy is maintained through devoting life to synchronization with divine law and to the balance of vital airs which reflects this synchronization. Seeking divinity within himself, the player is aware of inner sounds previously inaudible. Openness to new dimensions of experience results in true understanding and he becomes a keystone in his tradition, or a saint or seer. His experience acts as a mirror for others. The nature of humanity is seen in perspective, a state which may be reached directly through selfless service.
Context The 41st state or square on the board of *Leela*, the game of knowledge, appearing in the fifth row. This is the fifth of eight lokas represented on the spine of the board and marking a stage in the development of the individual's consciousness. The first seven represent a psychic centre or chakra on the spine of the human body, a level of psychic evolution. An eighth loka lies beyond the chakras. Each loka is defined by the nature of the matter of which it is composed. The eight other states in the same row may be considered as special regions within the loka.
    Broader Fifth chakra: man becomes himself (Leela, #HM0933).
    Preceded by Selfless service (Leela, #HM0294).

♦ HM5187d Six-fold knowledge (Systematics)
Transcendental knowledge — Cyclic knowledge — Experience of repetition
Description Precise knowledge is only possible through the observation of repetitive processes. Without repetition nothing can be done with potentiality. Repetition is experienced when identity, difference and relatedness coalesce into a single system, for which a minimum of six independent elements is required. Coalescence is understood as the property of structure whereby significance acquires depth and enrichment and yet retains the unique character associated with a particular event. As progressive cyclicity it is the most appropriate system through which to experience structures in the step-by-step process through which they achieve their significance as events. It expresses the two-fold character of creation and counter-creation and also the movement of the entire process towards a goal. Although potential energy can be stored up indefinitely, it can only renew itself through the repetitive two-fold action of a disturbing and a restoring force. Success in action requires a balance between attention to what actually is and what potentially might be; events continue to transform themselves even when their actualization is completed. This form of knowledge transcends the distinction of actual and potential by penetrating into a region where actual and potential are incessantly interchanged and interwoven. Knowledge of this kind always escapes from the limitations of any particular experience. It is transcendental because it cannot be acquired through sense-data alone. The senses do not perceive cyclicity. Transcendental knowledge involves direct participation in the eternal return.
Context The sixth in a sequence of twelve modes of knowledge, identified by J G Bennett, inspired by G Gurdjieff.
    Broader Systematics (#HH2003).
    Followed by Seven-fold knowledge (Systematics, #HM0776).
    Preceded by Five-fold knowledge (Systematics, #HM8080).

♦ HM5197d Good tendencies (Leela)
Uttam gati
Description Having attained a degree of balance, the player moves in harmony with dharma and good tendencies start to flow. These tendencies are reflected in control of heartbeat and breath patterns and include living in harmony with the natural laws of nature, rising before sunrise, fasting and controlling diet, etc. There is a natural tendency away from distractions.
Context The 30th state or square on the board of *Leela*, the game of knowledge, appearing in the fourth row.
    Broader Fourth chakra: attaining balance (Leela, #HM3291).

♦ HM5198d Al-'Azim (Sufism)
Description The believer is aware of Allah as the magnificent, the great one. The greatness of the great is as nothing compared with the greatness of Allah, their greatness is but one of his infinite works.
Context A mode of mystical awareness in the Sufi tradition associated with the thirty-third of the ninety-nine names of Allah.
    Broader Ninety-nine names of Allah (Sufism, #HH2561).
    Followed by Al-Ghafur (Sufism, #HM6255).
    Preceded by Al-Halim (Sufism, #HM4519).

♦ HM5199d Concentration without applied thought and with sustained thought (Buddhism)
Description In this state the individual is aware of the danger in applied thought. It is the concentration of the second jhana in the fivefold reckoning, when there is still sustained thought.
Context According to Hinayana Buddhism, concentration is considered as of one kind (monad), of two kinds (dyads), of three kinds (triads), of four kinds (tetrads) or of five kinds (pentad). In the second triad, this is the second concentration.
    Broader Concentration (Buddhism, #HM6663).
    Followed by Concentration without applied thought or sustained thought (Buddhism, #HM7543).
    Preceded by Concentration with applied thought and sustained thought (Buddhism, #HM4908).

♦ HM5214d Measureless concentration with a measureless object (Buddhism)
Infinite concentration with an infinite object
Description Concentration is familiar, well developed and capable of being condition for a higher jhana; the object of concentration is extended and therefore measureless and suitable for a higher jhana.
Context According to Hinayana Buddhism, concentration is considered as of one kind (monad), of two kinds (dyads), of three kinds (triads), of four kinds (tetrads) or of five kinds (pentad). In the second tetrad, this is the fourth concentration.
    Broader Concentration (Buddhism, #HM6663).
    Related Limited concentration with a limited object (Buddhism, #HM1175)
Limited concentration with a measureless object (Buddhism, #HM5547)
Measureless concentration with a limited object (Buddhism, #HM4653).

♦ HM5215d Mujahadat (Sufism)
Spiritual struggle — Mortification of the self
Description The self is mortified for the subduing of carnal desire.
Context According to Shaykh Abu Sa'id ibn Abi'l-Khayr, this is the fifth of 40 stations or maqamat the Sufi must possess for his journey on the path of Sufism to be acceptable.
    Broader Stations of consciousness - ibn-Abi'l-Khayr (Sufism, #HM2424).
    Related Discipleship (#HH0376)   Self-mortification (Sufism, #HH0464).
    Followed by State-awareness (Sufism, #HM2305).
    Preceded by Iradat (Sufism, #HM8901).

♦ HM5221d Knowledge due to insight (Buddhism)
Description This may arise like a flash of lightning while in the process of estimating and judging material and immaterial states. The meditator, thinking that such knowledge has never arisen before, believes he has reached fruition, interrupts the course of insight. This is karmically unprofitable, because he mistakes the path.
Context One of the ten imperfections of insight of Hinayana Buddhism.
    Broader Imperfection of insight (Buddhism, #HM9722).

♦ HM5224d Gladness (Buddhism)
Sympathy
Description The meditator does not at first direct gladness towards those to whom he is antipathetic, to very dear friends, to a neutral person, to an enemy or hostile person or to a member of the opposite sex or to a dead person. A dear friend can, however, be the proximate cause. He becomes versatile in the unspecified pervasion of gladness (five ways), specified pervasion (seven ways) and directional pervasion (10 ways); and eleven advantages arise.
Context One of the four divine abidings or states described as subjects for meditation in Hinayan Buddhism. As experienced in the sense sphere, one of the formations aggregate (mental coefficients) of Hinayana Buddhism, being listed among the inconstant states, and as general secondary (sometimes present in any profitable or profitable–resultant consciousness).
    Broader Divine abidings as meditation subjects (Buddhism, #HH3534).
Awareness as mental-formation group of conscious existence (Buddhism, #HM2050).
    Related Pity (Buddhism, #HM0513)   Equanimity (Buddhism, #HM7769)
Lovingkindness (Buddhism, #HM7607)
Faculty of mental gladness (Buddhism, #HM1649).

♦ HM5232d Knowledge of penetration of minds (Buddhism)
Knowledge of others' thoughts — Direct knowledge — Higher knowledge
Description With his own mind, the meditator penetrates the minds of others, understanding their manner of consciousness. Its object or extent depends on the knowledge level reached. If another's sense-sphere consciousness is known, then knowledge has a limited object. If fine-material or immaterial sphere is known, then knowledge has an exalted object. If path and fruition are known, it has a measureless object. Only the consciousness of one at a lesser level can be known, but an arahant knows the consciousness of all. There are also path, past, present, future and external objects. Initially, at the sense-sphere level, the manner of consciousness is known by observing with the divine eye the colour of the blood, this depending on whether consciousness is accompanied by joy, grief or serenity (red, black or clear). Knowledge is then extended to fine-material and immaterial spheres. The different kinds of consciousness, whether profitable, unprofitable or indeterminate are thus known.
Context One of the five kinds of direct or higher knowledge developed in Hinayana Buddhism.
    Related Manahparyayajnana (Hinduism, #HM7298)
Knowledge of supernormal powers (Buddhism, #HM7672)
Knowledge of the divine ear element (Buddhism, #HM5982)
Knowledge of recollection of previous existence (Buddhism, #HM4297)
Knowledge of passing away and reappearance of beings (Buddhism, #HM0748).

♦ **HM5234d Ar–Rashid** (Sufism)
Description The believer is aware of Allah as the righteous teacher, the guide to the right path. This is the way of salvation, leading to bliss and prosperity for the believer who follows it. Man has to choose to follow this way, he is not forced. First he is aware and conscious of what is being taught, then he uses the intelligence he has been given to discipline and educate his ego, then he learns the divine laws and orders his material existence in accordance with them. The student sees the order within and outside himself, sees the will, power, generosity and love of the teacher and learns to love Him. He lives to do as He says, loves to work for His pleasure and becomes righteous.
Context A mode of mystical awareness in the Sufi tradition associated with the ninety–eighth of the ninety–nine names of Allah.
Broader Ninety–nine names of Allah (Sufism, #HH2561).
Followed by Al–Sabur (Sufism, #HM4495).
Preceded by Al–Warith (Sufism, #HM0113).

♦ **HM5244d Plane of austerity** (Leela)
Tapah–loka
Description Here the player as witness sees that karma remains and can be carried no longer; it must be worked off in austerities, penance and mortification. He may have reached this stage direct through sudharma (apt religion), or through gradual progression as conscience develops and sun/moon energy system are balanced. There is no further attraction to the phenomenal world and although the body is limited the player is not. He reveals the divine.
Context The 50th state or square on the board of *Leela*, the game of knowledge, appearing in the sixth row. This is the sixth of eight lokas represented on the spine of the board and marking a stage in the development of the individual's consciousness. The first seven represent a psychic centre or chakra on the spine of the human body, a level of psychic evolution. An eighth loka lies beyond the chakras. Each loka is defined by the nature of the matter of which it is composed. The eight other states in the same row may be considered as special regions within the loka.
Broader Sixth chakra: time of penance (Leela, #HM4412).
Preceded by Apt religion (Leela, #HM2981).

♦ **HM5265c Transcending the world** (Taoism)
Celestial immortality — Absolutely open nothingness — Subtlety of nondoing — Leaping out of the world — Incubation of the spiritual embryo
Description During this stage, the "spiritual embryo" having been formed, the work involved is that of gently nurturing it through an extended incubation period until it is fully developed. The natural fire of reality is operated so as to forge and refine it from vagueness to clarity, from weakness to strength. This work, also termed "cultural cooking", is the practice of "nonstriving" or "nondoing", which involves tranquil preservation of clarified consciousness to prevent it from scattering in distraction, whilst calmly watching over the mind. It may be alternated with a "bathing" process, in which there is a deliberate relaxation of striving to avoid the counterproductive accumulation of unbalanced forces. The process of incubation is one of single–minded attention, persistently centered, avoiding imbalance in yin and yang so that they combine appropriately and the spiritual embryo remains complete.
When the final traces of mundanity have been removed by these processes, the practitioner breaks through the undifferentiated, bursting out in a pure spiritual body, unconstrained by the five elements. There is a transmutation by which the formless spontaneously produces form and the immaterial produces substance. The practitioner then transcends the world in a realm of absolutely open nothingness, a state of immortality. The corresponding metaphor is the state before birth or conception. It is a state of mental refinement in which emotional and intellectual attachments to objects of the world, including the ego, have been removed. There is no longer any ultimate subjective identification with the individual body or personality. Consciousness is "spacelike" and merged with objective reality.
Context The last of seven stages of the Taoist alchemical firing process through which reality is cultivated and the self is restored. It opens the way to two further stages.
Broader Alchemy (Taoism, #HH5887).
Followed by Ultimate accomplishment (Taoism, #HM1342)
Merging with the ordinary world (Taoism, #HM0486).
Preceded by Unification of energy (Taoism, #HM4762).

♦ **HM5289d Al–Karim** (Sufism)
Description The believer is aware of Allah as the generous and merciful, who forgives where he could punish, who rewards beyond expectations. The generous have received their gift from Allah and should be grateful even if they receive no thanks from others. Even the greatest sinner should not doubt the generosity of Allah.
Context A mode of mystical awareness in the Sufi tradition associated with the forty–second of the ninety–nine names of Allah.
Broader Ninety–nine names of Allah (Sufism, #HH2561).
Followed by Al–Raqib (Sufism, #HM6098).
Preceded by Al–Jalil (Sufism, #HM6771).

♦ **HM5290d Taslim** (Sufism)
Surrender — Resignation
Description The seeker is resigned in the face of what fate might bring.
Context According to Shaykh Abu Sa'id ibn Abi'l–Khayr, this is the 12th of 40 stations or maqamat the Sufi must possess for his journey on the path of Sufism to be acceptable.
Broader Stations of consciousness – ibn–Abi'l–Khayr (Sufism, #HM2424).
Followed by Tawakkul (Sufism, #HM0807).
Preceded by Muwafaqat (Sufism, #HM7981).

♦ **HM5298d Understanding consisting in what is learned** (Buddhism)
Understanding by way of tradition
Description This is understanding acquired by from hearing another person; it is learned, effected by tradition.
Context On the Path of Purification of Hinayana Buddhism, panna (understanding) is considered as of one kind (monad), of two kinds (dyads), of three kinds (triads) or of four kinds (tetrads). There are five dyads, four triads and two tetrads. All have the characteristic of penetrating the individual essences or true nature of states (monad). In the first triad, this is compared with understanding consisting in what is reasoned or in development.
Broader Understanding (Buddhism, #HM4523).
Related Understanding consisting in development (Buddhism, #HM7826)
Understanding consisting in what is reasoned (Buddhism, #HM7154).

♦ **HM5304c Affinity** (Islam)
Munasaba
Description This may be a relationship of similarity or of complementarity, but in Islam tends to imply the former. It is one of the causes and preconditions of love, affecting both subject and object. The other causes are the qualities of the beloved and the lover's perception of them (Ibn al–Qayyim) or beauty and benificence (Ibn al–Dabbagh). There is the obvious affinity as, for example, the affinity between children of similar ages and interests; and there is the hidden relationship, that of an inward nature, which is not evident in the person's outward form. The second, or secret, affinity is of two kinds. The first is resemblance of the believer to God through attributes. The second may not be revealed. There has been dispute as to whether man can love God or take pleasure in a vision of God as there can be no affinity between the creator and the created; while some would say that it is the affinity man has with the upper world that makes him love God. Ibn al–'Arabi indicates that the affinity between man and God is greater than that between man and a youth or a woman, since God can actually be a man's faculties, his hearing, sight, and so on, while others can only be the object of these faculties. Ibn al–Qayyim emphasizes complementarity or dissimilarity when describing the relationship between man and God, that of servant and divinity.
Affinity implies reciprocity and permanence. This may be why true passion ('ishq), arising from essential similarity, can end only in death. It is similarity which is the fundamental cause of love, the lover sees himself mirrored in the beloved and in loving the beloved in reality loves himself. Such temporary affinity as arises from proximity to a person or being involved in the same enterprise lasts only as long as the circumstances which produced it. But true affinity, of the essence, lasts as long as the essence lasts.
Refs Bell, Joseph Norment *Love Theory in Later Hanbalite Islam* (1979).
Related Love (#HH0258) Love consciousness (#HM2000).

♦ **HM5309d Meeting God both resting and working in accordance with righteousness** (Christianity)
Description God comes ceaselessly to us, both with intermediary and without intermediary, calling to blissful rest and to activity. The interior person is whole and undivided although possessing his life in these two ways, completely in God in blissful rest, completely in himself in active love. Here all that he could desire is spiritually revealed and held out to him. The righteous person has thus the truly spiritual life which will continue for ever, transformed after this life into a higher state. Righteousness is attained when rest and activity are possessed in the one exercise, going toward God with interior love through eternal activity, entering God with blissful inclination toward eternal rest, remaining in God and yet going out to creatures in virtue and righteousness through a love common to all.
Context According to John Ruusbroec, there are three modes in which meeting God in the interior way of life is practised. This is the last of the three, arising from the other two, blissful rest in emptiness and active desire.
Broader Meeting God (Christianity, #HM0541).
Related Meeting God with emptiness (Christianity, #HM8772)
Meeting God with active desire (Christianity, #HM6204).

♦ **HM5321b Jna** (Buddhism)
Shes pa (Tibetan) — Consciousness
Related Vijna (Buddhism, #HM3617).

♦ **HM5324d Wrong view** (Buddhism, Pali)
Description In Hinayana Buddhism, wrong view is; the means by which associated states see wrongly; seeing wrongly itself; or seeing wrongly. Its characteristic is unwise interpretation or conviction, its function preassumption or perversion, it is manifest as wrong interpretation or conviction. Its proximate cause is desire not to see the noble ones (arahants). This is the most reprehensible of all faults.
Context One of the formations aggregate (mental coefficients) of Hinayana Buddhism, being listed among the constant states which appear in their true nature, and as unprofitable secondary (sometimes present in any unprofitable or unprofitable–resultant consciousness).
Broader Awareness as mental–formation group of conscious existence (Buddhism, #HM2050).
Related Wrong view (Buddhism, #HM1710) Right outlook (Buddhism, #HM1280)
Afflicted views (Buddhism, #HM2996).

♦ **HM5329f Resistance to shift**
Perseveration — Rigidity — Obsessional experience
Description While perseveration is a case of inertia, rigidity represents a failure to shift from habitual thought patterns and to exploit possible alternatives. Obsessional experience is a failure to shift from a preoccupying thought however much one resists it. There is a compulsion, the individual feels invaded by the thought. He recognizes it as senseless and actively resists it.
Related Obsession (#HM2809) Anomalies in the flexibility of associations (#HM4312).

♦ **HM5331d Al–Muqaddim** (Sufism)
Description The believer is aware of Allah as the expediter, advancing whomever He wills. Although all are invited to the truth, He leads some to respond while others are left behind. Working in harmony with Allah al–Muqaddim and in accordance with His laws one comes close to Allah, even if He does not will one should receive riches, fame or power.
Context A mode of mystical awareness in the Sufi tradition associated with the seventy–first of the ninety–nine names of Allah.
Broader Ninety–nine names of Allah (Sufism, #HH2561).
Followed by Al–Mu'akhkhir (Sufism, #HM4052).
Preceded by Al–Muqtadir (Sufism, #HM3576).

♦ **HM5332d Asamjnisamapatti** (Buddhism)
Description The attainment of unconsciousness.

♦ **HM5334c Sarvajnata** (Buddhism)
Description In Mahayana Buddhism this is awareness of all things which is identical to wisdom.
Related Essential wisdom (#HM0107).

♦ **HM5335d Unattracted consciousness** (Buddhism)
Unbending mind
Description The mind is not perturbed by greed or lust.
Context One of the sixteen roots or modes of unperturbedness of mind listed in Hinayana Buddhism as a basis for supernormal powers.
Related Supernormal powers (Buddhism, #HH5652).

♦ **HM5338c Profitable consciousness in the fine–material sphere** (Buddhism)
Moral consciousness in the form–realm
Description A total of 5 kinds of consciousness may arise here. These are associated with the jhana factors, that is: (i) with applied thought or inception of thought; sustained thought; happiness, joy or zest; bliss or ease; and concentration. (ii) with sustained thought; happiness, joy or zest; bliss or ease; and concentration. (iii) with happiness, joy or zest; bliss or ease; and concentration. (iv) with bliss or ease and concentration, happiness having faded away. (v) with concentration and equanimity.
All 5 give rise to materiality and to postures but not to intimation.
Context In Hinayana Buddhism, 89 consciousnesses are enumerated in aggregate (khanda). Of

these, 21 are profitable or moral, 12 are unprofitable or immoral and 56 are indeterminate (resultant or functional). The unprofitable all arise in the sphere of sense and desire, whereas profitable and indeterminate arise in sense, fine-material, immaterial and supramundane spheres.
Broader Dispositions of consciousness (Buddhism, #HM2098).
Related Impulsion (Buddhism, #HM7268) Formation of merit (Buddhism, #HH5122)
Indeterminate consciousness in the fine-material sphere - resultant (Buddhism, #HM0594)
Indeterminate consciousness in the fine-material sphere - functional (Buddhism, #HM4761).
Followed by Profitable consciousness in the immaterial sphere (Buddhism, #HM4701).
Preceded by Profitable consciousness in the sense sphere (Buddhism, #HM4447).

♦ **HM5339d Piety** (Sufism)
War'a
Context The fourth stage on the path to sainthood of the Naqshbandiyya order.
Broader Suluk of the Naqshbandiyya order (Sufism, #HM4356).
Followed by Contentment (Sufism, #HM7024).
Preceded by Riyada (Sufism, #HM6690).

♦ **HM5345d Alchemical blackening** (Esotericism)
Nigredo — Melanosis
Description At the beginning of the alchemical operation, the bodily consciousness is chaotic and obscure, a state symbolized by lead. During this operation man consciously separates himself from appearances by allowing himself to drown in the cosmic feminine nature, the full power of which he seeks to awaken and master. The work of blackening can therefore also be understood as a death, a marriage (or perhaps a parturition in reverse), or a descent into hell. The death is a death to the cosmic illusion through which insight is congealed. During this process man detaches himself from his sense of separateness, extracting his vital force from bodily and mental attractions, developing spiritual discernment. He recollects this force within himself as a pool of stillness. The marriage arises from the recognition of the intimate bond between his body and nature, an interplay on which the illusion of individuality can no longer be projected. In this marriage cosmic femininity prevails over masculine reification and objectification. The feminine power, and its fascination, dissolves the "solidification" of the virile nature and awakens its power. This liberating dissolution draws the virile force back from separative modes of action and knowledge in order to bathe it in the baptismal waters of universal life. This re-potentiation may also be experienced as a descent into hell, an encounter with the world as a poisonous flower in all its seductive and fragmented aspects. The alchemist then comes to embody the process through which this sense of fragmentation is engendered. The world is then understood as a womb through which forms are engendered in the present rather than as a grave of dead forms from the past. The alchemist, fertilizing himself, becomes the egg from which he himself will be reborn.
Context The first stage in the alchemical work.
Broader Alchemy (Esotericism, #HH0221).
Followed by Alchemical whitening (Esotericism, #HM6328).

♦ **HM5353d Cleansing the sirr** (Sufism)
Takhliya-i sirr — Emptying of the sirr
Description The sirr is the organ of spiritual appreciation, that which receives divine communication. It is also the secret state which such communication engenders. Opinions differ as to whether it is an organ separate from the heart or spirit, or whether it is the same organ but released from the bondage of the nafs and the passions. It has been called the mystic shrine of God set up in the heart by God, continually contemplated by Him and belonging to Him. It is cleansed by contemplation, *muraqaba*. The seeker is emptied of all thoughts diverting attention from the remembrance of God.
Context The third of four contemplative disciplines the salik (seeker after God) must pass through to obtain ma'rifa. Various Sufi orders may differ in approach but all require the same spiritual concentration; manifestation in its many forms may be dealt with in many ways but the realization of spiritual reality behind the forms is the same.
Broader Awareness of the mystic journey (#HM2900).
Related State-awareness (Sufism, #HM2305) Subtle faculties (Sufism, #HH6282).
Followed by Illumination of the spirit (Sufism, #HM6162).
Preceded by Cleansing of the heart (Sufism, #HM6932).

♦ **HM5354b Ascetic states** (Buddhism)
Description In Hinayana Buddhism, five states are said to go with volition of ascetic practices. These are: fewness of wishes; contentment; austerity or effacement; seclusion or solitude; a quality of knowledge which exists because of desire for these states.
Broader Ascetic practices (Buddhism, #HH4298).
Narrower Seclusion (#HM6312) Effacement (#HM7703)
Contentment (#HM4563) Idamatthita (#HM5498)
Fewness of wishes (#HM0460).
Related Hinayana Buddhism (Buddhism, #HH0845).

♦ **HM5364d Knowledge of contemplation of dispassion** (Buddhism)
Knowledge of repulsion
Related Purification by knowledge and vision of the way (Buddhism, #HH3550).
Followed by Knowledge of desire for deliverance (Buddhism, #HM0766).
Preceded by Knowledge of contemplation of danger (Buddhism, #HM7297).

♦ **HM5402d Rectitude of mental body** (Buddhism)
Rectitude of mental factors — Rectitude of kaya (Pali)
Description This refers to the straight state of the three aggregates of feeling, perception and formation or mental activities. The characteristic is uprightness of the mental body, the function is to crush its tortuousness or crookedness. It is manifest as non-crookedness, non-deflection. Mental factors are the proximate cause. It is the opponent of corruptions like deceit or fraud, deception or craftiness, causing tortuousness in the mental body. Rectitude of mental body and rectitude of consciousness are considered together.
Context One of the formations aggregate (mental coefficients) of Hinayana Buddhism, being listed among the constant states which appear in their true nature, and as profitable primary (always present in any profitable or profitable-resultant consciousness).
Broader Awareness as mental-formation group of conscious existence (Buddhism, #HM2050).
Related Rectitude (Buddhism, #HM4250).

♦ **HM5403c Knowledge in conformity with truth** (Buddhism)
Conformity — Saccanulomikanan (Pali)
Description By this knowledge, the murk of defilements concealing the truth is dispelled. However, it is for change-of-lineage to make nirvana its object.
Related Purification by knowledge and vision of the way (Buddhism, #HH3550).
Followed by Knowledge of change of lineage (Buddhism, #HM4637).

♦ **HM5409d Affection**
Wadd (Islam)
Related Affection (#HH0422).

♦ **HM5432d Resolution as faith due to insight** (Buddhism)
Description Faith arises in association with insight, serene and confident in consciousness and its properties. The meditator, thinking that such resolution has never arisen before, believes he has reached fruition, interrupts the course of insight. This is karmically unprofitable, because he mistakes the path.
Context One of the ten imperfections of insight of Hinayana Buddhism.
Broader Imperfection of insight (Buddhism, #HM9722).

♦ **HM5438c Attainment of cessation** (Buddhism)
Samjnavedayitanirodha
Description In this state, when sensation and conceptualization are suspended, the physical, verbal and mental functions of the practitioner cease and subside. The distinguishing features between this state and death are said to be that the vitality is not destroyed, that heat is not totally extinguished, and the sense organs, far from being dispersed, are purified and ready to perform their functions perfectly when called upon to do so. The residual vitality or life-principle and physical bodily heat are sufficient for it to be possible to leave the condition and resume normal life. Otherwise, to the observer, the practitioner appears dead, breathing ceases and he does not respond to any external stimuli. Such a state has been compared to a cataleptic trance or to an animal in deep hibernation.
Context This state has been described as that following the series of jhana or arupaloka series and as such can be considered the fifth in a five-fold series in the immaterial sphere.
Related Nirvana (Buddhism, #HM2330) Nirodhasamapatti (Buddhism, #HM6346)
Cataleptic trance (Psychism, #HM4319)
Cessation-awareness (Buddhism, Tibetan, #HM2172).
Preceded by Fourth absorption in the immaterial sphere (Buddhism, #HM2051).

♦ **HM5446d Al-Majid** (Sufism)
Description The believer is aware of Allah as the most glorious and majestic. For the glory and honour revealed in beautiful actions and states He is praised and loved. For His majesty and power such that he is beyond reach he is respected and feared. Realizing this, believing and seeking only His pleasure, the follower receives strength and honour.
Context A mode of mystical awareness in the Sufi tradition associated with the forty-eighth of the ninety-nine names of Allah.
Broader Ninety-nine names of Allah (Sufism, #HH2561).
Related Al-Majid (Sufism, #HM4526).
Followed by Al-Ba'ith (Sufism, #HM4017).
Preceded by Al-Wadud (Sufism, #HM7712).

♦ **HM5448d Dissatisfaction** (Hinduism)
Description When desire is not fulfilled as objects of desire are not attained then the pain of dissatisfaction arises.
Related Desire (Hinduism, #HM7800) Satisfaction (Hinduism, #HM8116).

♦ **HM5451d Leadership intelligence**
Leadership mind
Description As opposed to everyday consciousness, a trance-like state whose external manifestation may be said to be a blank stare, leadership mind is described as wide-awake and active. It is alert, in touch with reality; it is expansive, exhilarating, liberating and positive; it is always conscious. In a company situation it can be the state of each employee, who then knows and cares about the whole business. This state is no respecter of position or birth, it can be learnt by anyone through changing their way of thinking. Such a way of thinking is an asset in combatting disease, and leadership intelligence and good health are said to go together.
Related Leadership development (#HH0516).

♦ **HM5467c State of self-surrender** (Christianity)
Description In the present moment, acting as God directs, the soul responds to every movement of grace, light and fluid, like molten metal taking the shape of whatever vessel into which it is poured.
Related Kairos (#HM2749) Self-oblation (#HH0968).

♦ **HM5479f Thought insertion** (Psychism)
Description The experience is of one's thoughts being drained away by an outside power.
Broader Modes of awareness associated with schizophrenia (#HM2313)
Anomalies in experience of the self as recognized in personal performance (#HM1336).
Related Passivity experiences (#HM7203).

♦ **HM5498d Idamatthita** (Buddhism)
Knowledge
Description This is a state of knowledge and desire to achieve ascetic states is that by which one is established into ascetic practice and undertakes and persists in ascetic qualities.
Context In Hinayana Buddhism, one of the five ascetic states.
Broader Ascetic states (Buddhism, #HM5354).

♦ **HM5499d Consciousness reinforced by mindfulness** (Buddhism)
Mind upheld by mindfulness
Description Consciousness does not waver in negligence.
Context One of the sixteen roots or modes of unperturbedness of mind listed in Hinayana Buddhism as a basis for supernormal powers.
Related Supernormal powers (Buddhism, #HH5652).

♦ **HM5504f Functional hallucination**
Description These hallucinations occur only in the case of specific perceived objective stimuli, so that each time there is a specific external stimulus, for example a tap running, there is also the hallucination, such as hearing voices. Both are perceived simultaneously. This may be a symptom of chronic schizophrenia.
Broader Hallucination (#HM4580).

♦ **HM5506d Al-Wahid** (Sufism)
Description The believer is aware of Allah as one and unique in His essence and attributes, in His actions and justice, in His names. Nothing is equal to Him or can be compared with Him.
Context A mode of mystical awareness in the Sufi tradition associated with the sixty-sixth of the ninety-nine names of Allah.
Broader Ninety-nine names of Allah (Sufism, #HH2561).
Related Perfect man consciousness (Sufism, #HM2920).
Followed by Al-Ahad (Sufism, #HM6117).
Preceded by Al-Majid (Sufism, #HM4526).

♦ **HM5532c Intelligible intuition** (Buddhism)
Yogi-pratyaksa — Direct intuition
Description Normal intuition is sensuous, explanations by the understanding in vague and general

images or concepts following a moment of bright and vivid reality. In contrast, the intelligible intuition of the saint is of the intellect, understanding as directly as the feeling in the first moment of sensation is felt. Rather than approaching reality through limited constructions based on dialectical concepts it is contemplated directly, free from illusion.
Related Kan (Japanese, #HM3104) Mental sensation (Buddhism, #HM6644).

♦ **HM5534e Out-of-body experience** (Psychism)
Description Many religions refer to the ecstatic state as occurring in some other place or time, the practitioner cultivating the ability to move out of the physical body which is confined by space-time constraints, and reaching an extraordinary state or communing with spirits in some other time or environment. In terms of psychism, it is said that the individual leaves the physical body and floats invisible to others; the out-of-body experience occurs during sleep or at will, the soul remaining attached to the body by a "silver cord".
Refs Irwin, H *Flight of Mind* (1985).
Related Ecstasy (#HM2046) Shamanic journey (#HM6120) Ecstatic discipline (#HH5282)
Near death experience (NDE, #HM0777)
Spontaneous out-of-body experience (#HM0332).

♦ **HM5543d Al-'Afuw** (Sufism)
Description The believer is aware of Allah as the forgiver, pardoning and eliminating sins so they no longer exist. One has only to enter at His open door. Only continual denial and infidelity, despite repeated forgiveness, results in punishment.
Context A mode of mystical awareness in the Sufi tradition associated with the eighty-second of the ninety-nine names of Allah.
Broader Ninety-nine names of Allah (Sufism, #HH2561).
Followed by Ar-Ra'uf (Sufism, #HM0206).
Preceded by Al-Muntaqim (Sufism, #HM4513).

♦ **HM5547d Limited concentration with a measureless object** (Buddhism)
Limited concentration with an infinite object
Description Concentration is limited and thus unfit as a condition for a higher jhana; however, the object of concentration is extended and therefore measureless and suitable for a higher jhana.
Context According to Hinayana Buddhism, concentration is considered as of one kind (monad), of two kinds (dyads), of three kinds (triads), of four kinds (tetrads) or of five kinds (pentad). In the second tetrad, this is the second concentration.
Broader Concentration (Buddhism, #HM6663).
Related Limited concentration with a limited object (Buddhism, #HM1175)
Measureless concentration with a limited object (Buddhism, #HM4653)
Measureless concentration with a measureless object (Buddhism, #HM5214).

♦ **HM5561d Tamoguna** (Leela)
Description The truth is veiled in darkness and sleep. From this last square on the board the player goes back to earth to continue play.
Context The 72nd state or square on the board of *Leela*, the game of knowledge, appearing in the eighth row.
Broader Beyond the chakras: the gods themselves (Leela, #HM1141).
Followed by Earth (Leela, #HM6112).

♦ **HM5562d Inferior concentration** (Buddhism)
Description This is concentration which is just obtained.
Context According to Hinayana Buddhism, concentration is considered as of one kind (monad), of two kinds (dyads), of three kinds (triads), of four kinds (tetrads) or of five kinds (pentad). In the first triad, this is the first concentration.
Broader Concentration (Buddhism, #HM6663).
Followed by Medium concentration (Buddhism, #HM6089).

♦ **HM5563d Tahqiq** (Sufism)
Ascertaining truth — Assertion of truth
Description In a state of wonder and lamentation the devotee flees from all creatures to the gate of God.
Context According to Shaykh Abu Sa'id ibn Abi'l-Khayr, this is the 38th of 40 stations or maqamat the Sufi must possess for his journey on the path of Sufism to be acceptable.
Broader Stations of consciousness - ibn-Abi'l-Khayr (Sufism, #HM2424).
Followed by Nihayat (Sufism, #HM3377).
Preceded by Inbisat (Sufism, #HM7098).

♦ **HM5580d Love with God** (Islam)
Al-mahabba ma'a 'llah
Description This is loving a person or a thing for its own sake and not for God's. One loves God but one also loves the other. This may be a kind of idolatry or polytheism, where the objects of love are worshipped as an end in themselves. Or it may simply spoil the perfection of one's love for God without excluding one from Islam all together, which is the case of natural love whether for persons or possessions in accordance with nature and appetite. A typical example of love with God is 'ishq, passionate love.
Context One of three types of love of Ibn al-Qayyim - love to God, love in God (for His sake) and love with God.
Broader Love (Islam, #HH5712) Mahabba (Islam, #HM6523).
Narrower Passionate love (#HM3656).
Related Love in God (Islam, #HM4712) Love to God (Islam, #HM5116)
Divine-love awareness (Sufism, #HM2401).

♦ **HM5587d Ceto-samadhi** (Buddhism)
Description In this state, the practitioner experienced mental concentration as a result of ardour, exertion, application, effort and right attention. There are supernatural results, for example he is able to recall past existences. But, in that this leads to the false conclusion that the self and the world are either eternal or subject to total annihilation, the experience is imperfect and leads to illusory conceptions.
Broader Concentration (Buddhism, #HM6663).
Related Correct concentration (Buddhism, #HM1735).

♦ **HM5588d Consciousness reinforced by concentration** (Buddhism)
Mind upheld by concentration
Description Consciousness is not perturbed by agitation nor does it waver in distraction.
Context One of the sixteen roots or modes of unperturbedness of mind listed in Hinayana Buddhism as a basis for supernormal powers.
Related Supernormal powers (Buddhism, #HH5652).

♦ **HM5600d Abstinence from bodily misconduct** (Buddhism)
Kayaduccaritavirati (Pali) — Abstinence from misconduct of the kaya
Description Together with abstinence from verbal misconduct and abstinence from wrong livelihood, this is the mind's adverseness from evil-doing and has the characteristic of non-transgression, of not treading, in its particular field. Its function is shrinking from this field; its manifestation is not doing such things; its proximate cause includes faith, conscience or sense of shame, shame or sense of blame, contentment (fewness of wishes).
Context One of the formations aggregate (mental coefficients) of Hinayana Buddhism, being listed among the inconstant states, and as secondary profitable (sometimes present in any profitable or profitable-resultant consciousness).
Broader Awareness as mental-formation group of conscious existence (Buddhism, #HM2050).
Related Abstinence from wrong livelihood (Buddhism, Pali, #HM7887)
Abstinence from verbal misconduct (Buddhism, Pali, #HM7171).

♦ **HM5622d Wandering mind** (Taoism)
Human mind
Description This mind is always stirring, producing myriads of thoughts even in quietness, always astir, never stopping. But even in the human mind there is the mind of Tao and even in the mind of Tao there is the human mind. The human mind is there even in perfected people, even a slight bias or unbalance makes the mind perilous and unstable. When a student of immortality discerns and keeps to unity, holding to the centre, then the perilous becomes safe and the subtle obvious.
Refs Cleary, Thomas (Trans) *The Book of Balance and Harmony* (1989).
Related Shining mind (Taoism, #HM1858).

♦ **HM5623d Unelated consciousness** (Buddhism)
Unelated mind
Description The mind does not waver in distraction, it is not perturbed by agitation.
Context One of the sixteen roots or modes of unperturbedness of mind listed in Hinayana Buddhism as a basis for supernormal powers.
Related Supernormal powers (Buddhism, #HH5652).

♦ **HM5627d Understanding defining mentality** (Buddhism)
Understanding by way of fixing mental qualities
Description In one striving for insight, this understanding is associated with defining the four immaterial aggregates.
Context On the Path of Purification of Hinayana Buddhism, panna (understanding) is considered as of one kind (monad), of two kinds (dyads), of three kinds (triads) or of four kinds (tetrads). There are five dyads, four triads and two tetrads. All have the characteristic of penetrating the individual essences or true nature of states (monad). In the third dyad, this is contrasted with understanding defining materiality.
Broader Understanding (Buddhism, #HM4523).
Related Understanding defining materiality (Buddhism, #HM4968).

♦ **HM5633d Al-Matin** (Sufism)
Description The believer is aware of Allah as perfect in strength and firmness from which nothing can be saved and which nothing can oppose. Since nothing can prevent His compassion or His punishment, His servants fear nothing but Him.
Context A mode of mystical awareness in the Sufi tradition associated with the fifty-fourth of the ninety-nine names of Allah.
Broader Ninety-nine names of Allah (Sufism, #HH2561).
Followed by Al-Wali (Sufism, #HM1633).
Preceded by Al-Qawi (Sufism, #HM4988).

♦ **HM5634c Compassion** (Buddhism)
Karuna — Snying re (Tibetan)
Related Pity (Buddhism, #HM0513).

♦ **HM5635g Physical tiredness**
Description Intense physical effort brings about this kind of tiredness which is characterized by relaxed muscles and pleasant feelings.
Related Sleep (#HM2980) Mental tiredness (#HM5806).

♦ **HM5643c Dwelling in the third jhana** (Buddhism)
Dwelling in equanimity
Description Reviewing the second jhana reveals the threat of the nearness of applied and sustained thought. Rapture or happiness seems gross, while bliss and unification or collectedness of mind seem calm and peaceful. Again bringing the subject of meditation to mind, the purpose being to abandon the gross factor and obtain the two peaceful factors, the process of appearance of absorption occurs. Happiness fades away - there is distaste for it - and it is stilled, just as applied and sustained thought have already been stilled. It gives way to a sense of dispassionate bliss, there is equanimity in the jhana sense. The meditator is mindful and fully aware in the sense of personal attributes of razor sharpness, more subtle than in the previous jhanas. The latter leads to bliss at the mental level and, on leaving the jhana, at the physical level; but this bliss implies no greed for bliss. So the two factors are bliss and unification of mind.
Like the first and second jhana it is said to be good in three ways: beginning (purification of the way), which is *access*; middle (intensification of equanimity), which is *absorption*; end (satisfaction), which is *reviewing*. It also has ten characteristics. Again, through practice, the meditator acquires mastery in the third jhana through the habits of: adverting to the jhana; attaining the jhana; resolving and steadying the duration of the jhana; emerging from the jhana; reviewing the jhana. He can then end his attachment to the third jhana and commence doing what is necessary to attain the fourth.
Broader Jhana (Buddhism, Pali, #HM7193) Dhyana (Hinduism, Buddhism, #HM0137).
Related Equanimity (Buddhism, #HM7769) Appearance of absorption (Buddhism, #HM1618)
Dwelling in the fivefold jhana (Buddhism, #HM6553)
Concentration of the third jhana of four (Buddhism, #HM2284)
Third trance of the fine-material sphere (Buddhism, #HM2062)
Concentration of the fourth jhana of five (Buddhism, #HM4882).
Followed by Dwelling in the fourth jhana (Buddhism, #HM8087).
Preceded by Dwelling in the second jhana (Buddhism, #HM7121).

♦ **HM5665d Aparisphutamanovijnana** (Buddhism)
Unmanifest thinking consciousness

♦ **HM5667g Stiffness** (Buddhism)
Thina (Pali) — Sloth
Description In Hinayana Buddhism, some sources list stiffness and torpor as a single formation. There is general paralysis due to lack both of urgency and vigour. In particular, stiffness or sloth has the characteristic of lacking driving power and opposing effort. Its function is destruction or removal of energy and it manifests as subsiding or sinking associated states. As with torpor, the proximate cause is unwise attention to boredom, discontent, sloth, etc.
Context One of the formations aggregate (mental coefficients) of Hinayana Buddhism, being listed among the constant states which appear in their true nature, and as unprofitable secondary

(sometimes present in any unprofitable or unprofitable–resultant consciousness).
Broader Awareness as mental–formation group of conscious existence (Buddhism, #HM2050).
Related Torpor (Buddhism, #HM0264).

♦ **HM5676c Dosaksaya** (Buddhism)
Dosakkhaya (Pali) — Cessation of hatred
Related Steps to enlightenment (Buddhism, #HH4019).

♦ **HM5702c Neurotic love**
Description Neurotic attachments endeavour to satisfy a real need, but not the same need of those experiencing mutual, reciprocal love. Typically they may be based on dependency needs or on fear of being alone. This form of love may be considered as arising from a defect in the integration of the self, thus constituting a misguided remedy for a deficiency in self–love.
Refs Person, Ethel Spector *Love and Fateful Encounters* (1990).
Broader Love consciousness (#HM2000).
Related Self–aggrandizing love (#HM4671).

♦ **HM5721c Indeterminate consciousness in the sense sphere - resultant** (Buddhism)
Description There are 23 indeterminate resultant consciousnesses in the sense sphere. These are with profitable or moral result (16) or with unprofitable or immoral result (7).
Of the 16 with profitable result, 8 are without root cause (unconditioned), devoid of non–greed and so on as the cause or condition of the result. These are eye, ear, nose, tongue and body (touch) consciousness, mental or mind–element consciousness, and the mind–consciousness element accompanied with joy (being present when entirely desirable objects occur) and that accompanied with equanimity (being present when desirable neutral objects occur). The resultant consciousness has invariable object (the five senses) or variable (mind–element with any of the five, mind–consciousness element with any of the five and the mind–element). They are classed by high order equanimity (not very sharp), bodily pleasure (sharp) and mental joy.
Of the 8 with root–cause or conditioned, devoid of non–greed and so on as the cause or condition of the result, 4 are accompanied by joy, that is: with understanding and unprompted; with understanding and prompted; without understanding and unprompted; without understanding and prompted. And 4 are accompanied with equanimity, that is: with understanding and unprompted; with understanding and prompted; without understanding and unprompted; without understanding and prompted. Being prompted or not depends on the source from which it has come.
The 7 consciousnesses associated with unprofitable or immoral result are without root–cause (unconditioned). These are eye, ear, nose, tongue and body (touch) consciousness, mental or mind–element consciousness, and the mind–consciousness element. They have undesirable or undesirable–neutral object and are classed by bodily pain (touch) which is sharp, or equanimity (indifference) which is of a low order and not sharp.
In this group, all but the eye, ear, nose, tongue and body consciousnesses, whether with profitable or unprofitable result, give rise to materiality but not to postures or intimation. The 10 sense consciousnesses give rise to none of these.
Context In Hinayana Buddhism, 89 consciousnesses are enumerated in aggregate (khanda). Of these, 21 are profitable or moral, 12 are unprofitable or immoral and 56 are indeterminate (resultant or functional). The unprofitable all arise in the sphere of sense and desire, whereas profitable and indeterminate arise in sense, fine–material, immaterial and supramundane spheres.
Broader Dispositions of consciousness (Buddhism, #HM2098).
Narrower Eye consciousness (#HM2074) Ear consciousness (#HM2169)
Nose consciousness (#HM2364) Body consciousness (#HM2562)
Tongue consciousness (#HM2263) Mental consciousness (#HM2838)
Mind–consciousness element (#HM6173) Mundane resultant consciousness (#HM0130).
Related Receiving (Buddhism, #HM7092) Examining (Buddhism, #HM7324)
Registering (Buddhism, #HM1646) Rebirth–linking (Buddhism, #HM8266)
Sense mode of consciousness occurrence (Buddhism, #HM4389)
Profitable consciousness in the sense sphere (Buddhism, #HM4447)
Unprofitable consciousness in the sense sphere (Buddhism, #HM8375)
Indeterminate consciousness in the sense sphere - functional (Buddhism, #HM3852).
Followed by Indeterminate consciousness in the fine–material sphere - resultant (Buddhism, #HM0594).

♦ **HM5734c One–pointedness** (Yoga)
Ekagra
Description One thought succeeds another as each fades away. Aware of the continuous succession of such states, the mind is one–pointed. As this becomes the mind's habit in waking consciousness and even in the dream state then *samprajnata samadhi* is attained, leading to salvation.
Context One of five states of mind classification identified in yoga.
Related Realization of nen (#HM0220) Samprajnata samadhi (Yoga, #HM2896).
Followed by Suppression (Yoga, #HM0135).
Preceded by Restlessness (Yoga, #HM1015).

♦ **HM5754g Clouding of consciousness**
Obfuscation of consciousness
Description There is an increasing lack of contact with the world, with dulling of mental abilities and a feeling of confusion. Temporal disorientation results in confusion as to time of day, day of the week. The individual is aware that he cannot think clearly, that he cannot concentrate. The next stage may be disorientation as regards place, the individual starts to lose his way and be unsure of landmarks. Then there is unsureness as to the identity of friends and relatives. There are disturbances in memory and the capacity for abstract thought and reasoning deteriorates. Social awareness diminishes, leading to inappropriate behaviour in public. Although the individual may initially worry about the changes that are occurring, he subsequently becomes indifferent and may deny what is happening. Finally there is delirium, responding to hallucinations of the senses, and then coma. Clouding of consciousness is a symptom of a number of disorders, including brain injury and senility.
Broader Diminished clarity of awareness (#HM6201).
Related Twilight consciousness (#HM2406).

♦ **HM5767d Concentration with happiness** (Buddhism)
Concentration with rapture
Description The mind is unified in two jhanas (in the fourfold reckoning) or five jhanas (in the fivefold reckoning). This is one way of experiencing access concentration.
Context According to Hinayana Buddhism, concentration is considered as of one kind (monad), of two kinds (dyads), of three kinds (triads), of four kinds (tetrads) or of five kinds (pentad). In the third dyad, this contrasts with concentration without happiness.
Broader Concentration (Buddhism, #HM6663).
Related Access concentration (Buddhism, #HM4999)
Concentration without happiness (Buddhism, #HM0730).

♦ **HM5768d Unincluded concentration** (Buddhism)
Description This is profitable unification of mind or collectedness of moral thought associated with the path or jhana.
Context According to Hinayana Buddhism, concentration is considered as of one kind (monad), of two kinds (dyads), of three kinds (triads), of four kinds (tetrads) or of five kinds (pentad). In the fifth tetrad, this is the fourth concentration.
Broader Concentration (Buddhism, #HM6663).
Related Sense–sphere concentration (Buddhism, #HM1097)
Immaterial–sphere concentration (Buddhism, #HM0696)
Fine–material–sphere concentration (Buddhism, #HM4265).

♦ **HM5775d Total annihilation** (Sufism)
Passing away
Description The continuous motion first perceived in the pine–shaped heart and the arteries may be diffused through the whole body or felt in a particular organ. Perception is drawn to the reality being sought and concentration is on the physical heart with no consideration or mention of the name Allah, although attention may need to be fixed by taking the name Allah. The knowledge of the continuous motion is applied to the motion itself. This very knowledge is distance and nearness, presence and absence, annihilation and total passing away. The source of the motion is the heart; but when the whole body is blessed with this motion the reality remembered is applied to the entire motion. There is an experience of abundant effacement and ecstasy - the devotee has reached the state of "passing away".
Context This is the fifth state arising in the method of dhikr described by Shaykh Kalimallah.
Broader Remembrance of God (Sufism, #HM6562).
Narrower Annihilation of the heart (#HM7622).
Followed by 'Ilm al–basir (Sufism, #HM0476).
Preceded by Remembrance of reality (Sufism, #HM3892).

♦ **HM5776d Holy indifference** (Christianity)
Description The soul ceases to desire or to will except in cooperation with the divine leading. As its light increases, its desires for itself are merged in the desire of God's glory and fulfilment of His will. Even one's own salvation and deliverance are not desired so much for one's own happiness as for fulfilment of God's pleasure, resulting in His glory, and because he desires and wills that one should thus desire and will. Far from being inactivity, such indifference is the highest life and activity to everything in God's will.
Context This is the second modification of sanctification as described by François Fénelon.
Refs Fénelon, François *The Maxims of the Saints*.
Broader Sanctification (#HH0428).
Related Non–desire (Christianity, #HM9330).
Preceded by Holy resignation (Christianity, #HM8022).

♦ **HM5786d Al–Musawwir** (Sufism)
Description The believer is aware of Allah as the fashioner of all things, designing and giving form. Man must not claim to be creator but try to see divine power in all creativity and lead others to find Him through His creative manifestations.
Context A mode of mystical awareness in the Sufi tradition associated with the thirteenth of the ninety–nine names of Allah.
Broader Ninety–nine names of Allah (Sufism, #HH2561).
Followed by Al–Ghaffar (Sufism, #HM1257).
Preceded by Al–Bari' (Sufism, #HM3636).

♦ **HM5806g Mental tiredness**
Description This follows a period of intense intellectual or emotional activity, and is usually an unpleasantly "drained" feeling.
Related Sleep (#HM2980) Physical tiredness (#HM5635).

♦ **HM5856f Medium state of hypnosis**
Parallel awareness — Semi–stage hypnosis
Description Although the individual is still consciously aware of his surroundings there is a feeling of indifference to and dissociation from them; his body is limp and he is in a deep enough hypnotic state to accept subconsciously and respond to suggestions put to him by the hypnotist. Nevertheless, there may be a residual ability to make decisions or to try to direct the hypnotist's questioning.
Broader Hypnotic states of consciousness (#HM2133).
Related Hypnotherapy (#HH0962) Semi–trance (Psychism, #HM0887).

♦ **HM5876d Tatayyum** (Islam)
Enslavement — Worship — Ta'abbud
Context The highest degree of love in two lists of Ibn al–Jawzi.
Related Walah (Islam, #HM8303) Khulla (Islam, #HM7259).

♦ **HM5901d Consciousness reinforced by understanding** (Buddhism)
Mind upheld by understanding
Description Consciousness is not perturbed by ignorance.
Context One of the sixteen roots or modes of unperturbedness of mind listed in Hinayana Buddhism as a basis for supernormal powers.
Related Supernormal powers (Buddhism, #HH5652).

♦ **HM5902d Malik–ul–Mulk** (Sufism)
Description The believer is aware of Allah as the eternal owner of his kingdom, sharing sovereignty with no–one. The whole universe is connected in one totality, one kingdom; so is man, a microcosm of the great whole. If man rules his temporal kingdom as a servant of Allah he will be rewarded with the eternal kingdom of hereafter. If he is a servant of his own ego, however, he will be bankrupt and imprisoned in hell.
Context A mode of mystical awareness in the Sufi tradition associated with the eighty–fourth of the ninety–nine names of Allah.
Broader Ninety–nine names of Allah (Sufism, #HH2561).
Followed by Dhul–Jalali wal–Ikram (Sufism, #HM6719).
Preceded by Ar–Ra'uf (Sufism, #HM0206).

♦ **HM5904f Negative hallucination**
Description As opposed to visual hallucination, where the person sees something that is not there, in negative hallucination he or she does not see something that is there. Unlike a visual hallucination, where the person behaves as though the hallucination is really there, someone with a negative hallucination does not, for example, attempt to walk through someone he cannot see.
Broader Hallucination (#HM4580).
Related Visual hallucination (#HM0092).

♦ **HM5908d Unrepelled consciousness** (Buddhism)
Unoffended mind
Description Consciousness is not perturbed by ill will not waver in malice.
Context One of the sixteen roots or modes of unperturbedness of mind listed in Hinayana Buddhism as a basis for supernormal powers.
Related Supernormal powers (Buddhism, #HH5652).

♦ **HM5909c Becoming** (Buddhism)
Wheel of becoming
Description In Hinayana Buddhism, the Path of Purification describes the profound meaning of ageing and death as originating in and produced by birth. Not only is birth the origin and condition of ageing and death, nothing else can be their origin or condition. Having been born, it is ignorance that produces the mental formations, where ignorance is unknowing, unseeing, not penetrating the truth. Coming to birth, descent into the womb, rebirth and manifestation bring destruction, fall, break-up and change.
Related Ignorance (Buddhism, Yoga, Zen, #HM3196)
Ignorance in dependent origination formula (Buddhism, #HM3035)
Awareness as mental-formation group of conscious existence (Buddhism, #HM2050).

♦ **HM5930d Concentration partaking of penetration** (Buddhism)
Description In this state there is accessibility to perception and attention which is accompanied by dispassion and directed towards fading away.
Context According to Hinayana Buddhism, concentration is considered as of one kind (monad), of two kinds (dyads), of three kinds (triads), of four kinds (tetrads) or of five kinds (pentad). In the fourth tetrad, where concentration is associated with different conditions of wisdom or understanding, this is the fourth concentration.
Broader Concentration (Buddhism, #HM6663).
Related Concentration partaking of diminution (Buddhism, #HM8002)
Concentration partaking of stagnation (Buddhism, #HM6956)
Concentration partaking of distinction (Buddhism, #HM7363).

♦ **HM5965d Plane of sanctity** (Leela)
Yaksha-loka
Description Resulting from good tendencies, there is understanding of cosmic principles leading to experience of divine grace. The player does not simply accept intellectually a oneness with divine presence, he experiences it as a part of daily life. The desire to confront reality and understand the presence of the divine in all existence becomes his very essence.
Context The 31st state or square on the board of *Leela*, the game of knowledge, appearing in the fourth row.
Broader Fourth chakra: attaining balance (Leela, #HM3291).

♦ **HM5974g REM sleep**
Stage-one REM — Rapid eye movement
Description After a full sequence of stage-one to stage-four sleep and back, instead of drowsiness there is a period of rapid eye movement, irregular heart rate, possibly penile erection or vaginal engorgement, vestibular activation. All other commands from the brain for voluntary movement are blocked. This is the time of dreaming sleep.
Broader Sleep (#HM2980).
Related Dreams (#HM2950).

♦ **HM5977c Vimoksa** (Buddhism)
Vimokkha (Pali) — Liberation
Description This is threefold release - from the conception of soul, from the illusion of permanence and from hankering after objects of lust, hate and delusion. Liberation can be considered a series of eight altered states of consciousness.
Context One of ten superhuman qualities described in the Sutta Vibhanga as being special attainments of insight above that of ordinary men.
Broader Superhuman qualities (Buddhism, #HH9764).
Related Moksha (Hinduism, #HM2196).

♦ **HM5982d Knowledge of the divine ear element** (Buddhism)
Deva hearing — Direct knowledge — Higher knowledge
Description The meditator, having completed the practices for higher knowledge, adverts to sounds evident to normal consciousness, starting with gross and progressing to more subtle. Giving attention to the sound sign, he goes through the process for absorption with reference to sound. Knowledge arises with absorption consciousness. He then hears, with the divine ear, divine and human sounds in an area he delimits, which can then be progressively extended. There are four kinds of object - limited and present (since sounds are, by their nature limited and in the present), internal and external.
Context One of the five kinds of direct or higher knowledge developed in Hinayana Buddhism.
Related Appearance of absorption (Buddhism, #HM1618)
Knowledge of supernormal powers (Buddhism, #HM7672)
Knowledge of penetration of minds (Buddhism, #HM5232)
Knowledge of recollection of previous existence (Buddhism, #HM4297)
Knowledge of passing away and reappearance of beings (Buddhism, #HM0748).

♦ **HM5986d Exertion due to insight** (Buddhism)
Uplift due to insight
Description This arises as well-exerted energy due to insight. The meditator, thinking that such exertion has never arisen before, believes he has reached fruition, interrupts the course of insight. This is karmically unprofitable, because he mistakes the path.
Context One of the ten imperfections of insight of Hinayana Buddhism.
Broader Imperfection of insight (Buddhism, #HM9722).

♦ **HM5991c Infatuation**
Crushes
Description As young people recognize the limitations of their idealization of their parents, they project their own ego ideals onto others in their environment. These heroic figures are not substitutes for parents but rather are role models for what the young hope to become. In such infatuations, the young people fall in love with those whose lives they hope to emulate and whose paths they would wish to follow. The object is identification alone, rather than any form of union. The object of such infatuation is not necessarily someone known personally. Media stars may serve this purpose. Where a group shares the same infatuation, the group bonding may be emotionally as important as the bond with the focal personage. In the normal course of development, the yearning associated with such idealization is transformed from wishing to be like to wishing to be with, although many remain fixated at the earlier phase. Although crushes are particularly common in adolescence, they are experienced throughout life as imaginative rehearsals of experiences for which individuals are not quite ready. Indeed the opening phase of any love affair always bears a resemblance to such infatuations.
Refs Person, Ethel Spector *Love and Fateful Encounters* (1990).
Broader Love consciousness (#HM2000).
Followed by Romantic love (#HM5087).

♦ **HM5992d Concentration of difficult progress and sluggish direct-knowledge** (Buddhism)
Concentration of painful progress and sluggish intuition
Description Cultivating what is unsuitable, failing to carry out preparatory tasks, overwhelmed by craving, lacking practice in serenity (absorption concentration), having an acutely corrupt nature, all these mean progress is difficult. Cultivating what is unsuitable, lacking skill in absorption, overwhelmed by ignorance, lacking practice in insight, having dull faculties, all these mean intuition is sluggish.
Context According to Hinayana Buddhism, concentration is considered as of one kind (monad), of two kinds (dyads), of three kinds (triads), of four kinds (tetrads) or of five kinds (pentad). In the first tetrad, this is the first concentration.
Broader Concentration (Buddhism, #HM6663).
Related Concentration of easy progress and swift direct-knowledge (Buddhism, #HM4977)
Concentration of easy progress and sluggish direct-knowledge (Buddhism, #HM1010)
Concentration of difficult progress and swift direct-knowledge (Buddhism, #HM7859).

♦ **HM6001f State of illness**
Disease
Description Not simply aware of the presence of symptoms, the sufferer feels distress, discomfort and dis-ease.

♦ **HM6002d Vyana-loka** (Leela)
Description The internal energy of the body is balanced through vyana, manifest in circulation of the blood, sweating and coughing.
Context The 40th state or square on the board of *Leela*, the game of knowledge, appearing in the fifth row.
Broader Fifth chakra: man becomes himself (Leela, #HM0933).

♦ **HM6003d Nazar** (Islam)
Amorous regard — Glancing — Gazing
Description This is the first state on the path of love in several traditions. According to some sources, the faculty has to be restrained by casting down the eyes. The first regard, if accidental, is allowable; the second is not, it stirs up desire in the heart. A second glance brings love, love brings sorrow, sorrow brings death. Although a hurried and inaccurate impression at first glance may lead to love, a second may reveal the person to be even more attractive than first thought; Satan may embellish what is seen so it appears more beautiful. Even if a second glance disappoints, renunciation is then in line with one's own purposes (not, after all, finding the object of the gaze attractive) and not for God's sake. The ascetic who avoids glancing at women, as this is forbidden, is then warned against Satan's temptation to glance at youths.
Although nazar may be an activity of the eyes, it may also be an activity of the heart in response to the description of a beautiful person. Thus, if nazar is the cause of profane love it is also, in the spiritual sphere, the final cause of love for God. The beatific vision may be seen as the culmination of religious love.
Context One of the sins of passion (hawa) in Islam. The first cause of profane love according to Ibn al-Qayyim.
Refs Bell, Joseph Norment *Love Theory in Later Hanbalite Islam* (1979).
Broader Hawa (Islam, #HM0129).
Related Vision (#HM3347).
Followed by Istihsan (Islam, #HM0453).

♦ **HM6006c Avataric level of musical inspiration**
Description This is the sublest of three levels of musical inspiration and has a historic function in addition to, or even surpassing, its intrinsic value. Like the visions of meditating saints, the works of those inspired in this way become the objects of musical contemplation for every subsequent composer. Insights from this level are sufficient to sustain and nourish whole epochs of creativity providing an aesthetic framework within which subsequent generations create according to their skills. Such inspiration provides the exemplars for those with less clearer vision. It gives forms to musical archetypes.
In responding to this level of inspiration, the listener ceases to listen as such and becomes engaged in a form of mystical or philosophic meditation of which musical forms may the prelude or the product.
Broader Musical inspiration (#HM4805).

♦ **HM6033d Lightness of consciousness** (Buddhism)
Buoyancy of mind — Buoyancy of citta (Pali)
Description This refers to the lightness or quickness of the aggregate consciousness. The characteristic is suppressing or quieting heaviness of the citta, the function is to crush its heaviness. It is manifest as non-sluggishness. Consciousness is the proximate cause and it is the opponent of stiffness and torpor causing heaviness in consciousness. Lightness of mental body and lightness of consciousness are considered together.
Context One of the formations aggregate (mental coefficients) of Hinayana Buddhism, being listed among the constant states which appear in their true nature, and as profitable primary (always present in any profitable or profitable-resultant consciousness).
Broader Awareness as mental-formation group of conscious existence (Buddhism, #HM2050).
Related Lightness of mental body (Buddhism, #HM7636).

♦ **HM6036f Delusions of guilt**
Description Real or imagined sins, whether trivial or great, lead a depressive individual to believe he is worthless and will receive eternal punishment. Although convinced that he deserves the punishment he is to receive, the individual may fear it will also be inflicted upon his family. He may believe he is already dead, that his mind and body no longer exist, even that everyone around him is dead or that the world has stopped.
Broader Delusive consciousness (#HM2600).

♦ **HM6087f Dissociation of affect**
Description In conditions of extreme danger or stress the individual's reactions may be walled off so that he is unaware of emotions and may continue to function. There is an unnatural, blunted calm so that the person feels drugged or machine-like. There may be a reaction after the event when emotional awareness resumes. During the event there is no way of voluntarily controlling the dissociation.
Broader Emotion (#HH0819) Affects (#HM7132)
Anomalies in experience of the unity of self (#HM9132).
Related Mourning (#HM1502) Dissociation (#HH1294).

♦ **HM6089d Medium concentration** (Buddhism)
Middling concentration
Description This is concentration which has not yet developed very well.
Context According to Hinayana Buddhism, concentration is considered as of one kind (monad), of two kinds (dyads), of three kinds (triads), of four kinds (tetrads) or of five kinds (pentad). In the first triad, this is the second concentration.
Broader Concentration (Buddhism, #HM6663).
Followed by Superior concentration (Buddhism, #HM6327).
Preceded by Inferior concentration (Buddhism, #HM5562).

♦ **HM6093c Ragaksaya** (Buddhism)
Ragakkhaya (Pali) — Abolition of passion
Related Steps to enlightenment (Buddhism, #HH4019).

♦ **HM6097c Masti** (Sufism)
Drunkenness
Context The second stage of *mahabbat* (love) according to al-Ansari. The three stages subsume the whole spiritual journey.
Broader Love of God (Sufism, #HM4273).
Followed by Nisti (Sufism, #HM7626).
Preceded by Rasti (Sufism, #HM7222).

♦ **HM6098d Al-Raqib** (Sufism)
Description The believer is aware of Allah as the watchful, protecting and harmonizing all creation. No good deed is lost or goes unrewarded; no evil deed is unnoticed. The believer should also be watchful against those other watchers, the devil and the ego, which seek the unguarded moment to attack and destroy.
Context A mode of mystical awareness in the Sufi tradition associated with the forty-third of the ninety-nine names of Allah.
Broader Ninety-nine names of Allah (Sufism, #HH2561).
Followed by Al-Mujib (Sufism, #HM0504).
Preceded by Al-Karim (Sufism, #HM5289).

♦ **HM6099d Jahd** (Sufism)
Effort — Endeavour
Description Worshipping God in heart and soul, the obedience of the devotee is unclouded by doubt.
Context According to Shaykh Abu Sa'id ibn Abi'l-Khayr, this is the 26th of 40 stations or maqamat the Sufi must possess for his journey on the path of Sufism to be acceptable.
Broader Stations of consciousness - ibn-Abi'l-Khayr (Sufism, #HM2424).
Followed by Wilayat (Sufism, #HM3647).
Preceded by Ma'rifa (Sufism, #HM2254).

♦ **HM6111c Essential unity of being** (Christianity)
Description According to John Ruusbroec, this meeting and unity which the spirit attains in God takes place in the essential ground of being. It is a mystery to the understanding unless apprehended essentially in utter simplicity. The spirit rests in the unity which is above all gifts and yet the source of all gifts. There is nothing but God and the spirit united with God. The person is received by the Holy Spirit and receives the Father, the Son and the Holy Spirit undivided. Inclined to blissful enjoyment, the spirit seeks rest in God.
When a person turns from sin he is received by God in the essential unity of his being. This unity in rest is possessed by all good people, but they remain unaware of it when not interiorly fervent and empty of all creatures. Through charity and practice of virtue they receive grace and a likeness to God; unity in rest cannot be lost except through mortal sin. Sinners and damned spirits lack God's grace to enlighten, instruct and lead them to this blissful unity. Sin is such a barrier that the spirit cannot attain union in its own being which, if it were not for sin, would be the proper place for its eternal rest.
Related Unity consciousness (#HM3193) Meeting God (Christianity, #HM0541).

♦ **HM6112d Earth** (Leela)
Prithvi
Description Mother earth who selflessly nourishes her offspring, and whose own fiery, austere birth provides the energy for their birth, is the symbol of the player at this stage. It is here the player is brought back to earth from tamoguna when play has taken him out of his own level to the highest plane before he is ready to attain cosmic consciousness.
Context The 51st state or square on the board of *Leela*, the game of knowledge, appearing in the sixth row.
Broader Sixth chakra: time of penance (Leela, #HM4412).
Preceded by Tamoguna (Leela, #HM5561).

♦ **HM6117d Al-Ahad** (Sufism)
Description The believer is aware of Allah as unity, the one in which all is united. In this unity, names, attributes and their relations disappear in the very essence. Although expression of uncreated Allah in created man is impossible, immersion in the one "I" which is one's essence, forgetting the multiplicity of "I's" or qualities which one ascribes to one's self, leads to manifestation of unity as far as is possible in man.
Context A mode of mystical awareness in the Sufi tradition associated with the sixty-seventh of the ninety-nine names of Allah.
Broader Ninety-nine names of Allah (Sufism, #HH2561).
Followed by As-Samad (Sufism, #HM4980).
Preceded by Al-Wahid (Sufism, #HM5506).

♦ **HM6118f Tactile hallucination**
Description As well as formication, other tactile hallucinations involve cold winds, vibrations, electric shocks and sexual sensations.
Broader Hallucination (#HM4580).
Narrower Formication (#HM4094).
Related Somatic hallucination (#HM7208).

♦ **HM6119d Group think**
Group mindset
Description The process through which a group of individuals remains locked into a particular pattern of thinking, constantly reinforced through daily interaction. This pattern can become so strong that it inhibits development of any alternative perspectives and specifically inhibits emergence of any attitude which questions the perspective which it sustains. Individuality is repressed by the power of the group thought pattern.
Related Group consciousness (#HM3109) Perceptual narrowing (#HM1237).

♦ **HM6120c Shamanic journey**
Soul flight — Out-of-body experience
Description When a shaman voluntarily enters an altered state of consciousness he may experience himself leaving his body and journeying to another realm. The soul seems to leave the body and to roam at will through the upper, middle and lower worlds. These journeys are used to acquire knowledge or power and to help people in their community. If the journey is to guide the souls of the dead to their resting place then the shaman is referred to as a psychopomp. The experience of the journey may be dramatic or dangerous, ecstatic or horrendous, demonic or divine. Although there may be the emotion of terror there is often an ineffable joy in what is seen and awe at the mysterious worlds opening before him. It is like a waking dream, one in which actions and adventures can be directed.
The journey is in three phases: preparation, which may be a period of isolation, fasting or celibacy; induction of an altered state of consciousness through rituals involving singing, dancing, drumming or drugs; the journey. There may be descent into into the lower world (perhaps through a cave, hollow tree or water hole) when the shaman sees himself diving deep into the earth and emerging in another world where he may seek information on medicine, recover a lost soul or placate an angry spirit. The fate of the whole tribe may rest on successful completion of this task. The lower world is often a place of tests and challenges. The middle world is this world. Journeying over it at will the shaman may seek and obtain information on hunting, weather or warfare. The upper world (entered from a mountain, tree or cliff-top) may involving the piercing of a membrane which has temporarily impede ascent. He then may move among several levels, perhaps being assisted by a helping spirit. Ascent may also be in the form of a bird or through climbing the central axis connecting the lower, middle and upper worlds. The axis may be in the form of the world tree or a mountain, rainbow or ladder. The upper world journey may be ecstatic.
Refs Irwin, H *Flight of Mind* (1985); Walsh, Roger *The Spirit of Shamanism.*
Related Shaman (#HH0973) Shamanism (#HM1189)
Dream yoga (Buddhism, #HH4087) Near death experience (NDE, #HM0777)
Out-of-body experience (Psychism, #HM5534)
Spontaneous out-of-body experience (#HM0332).

♦ **HM6122c Courtly love**
Erotic asceticism — Love-pain — Romance as spiritual discipline
Description The concept and feeling of longing and yearning, experienced somatically, and summarized by the notion of erotic asceticism as a discipline of unfulfilled desire. Grounded in sexual passion, it functions as a kind of mental, even moral, discipline. It entails the simultaneous acceptance of contradictory notions. It echoes the mystical yearning for God, the desire to escape from matter (the physical other) in favour of a spirit (an archetype), emphasizing inner feeling rather than ritual observance of marriage, and favouring sexual abstinence.
In this idealized version of romantic love, the fair lady is seen by the knight as his source of inspiration and as a symbol of beauty and perfection. It is this ideal which leads him to noble acts, to being spiritual, high-minded, refined. The relationship is idealized and spiritualized, there is no sexual involvement. It lifts the couple above the gross, physical level. Nor is there an intimate relationship as with an ordinary, mortal woman, so the lovers are not married. Indeed the lady is usually married to another. Despite the absence of sexual relations or marriage, the couple have intense and passionate desire for each other. This desire is spiritualized so that each sees the other as a symbol of the divine archetypal world.
Context Term coined to describe a practice developed in the Middle Ages. It is believed to have inspired the troubadours and powerfully influenced the contemporary western preoccupation with romantic love. As a doctrine of paradoxes, it is to be distinguished from caring or affection.
Refs Boase, Roger *The Origin and Meaning of Courtly Love* (1977); Capellanus, Andreas *The Art of Courtly Love* (1964); Johnson, Robert A *The Psychology of Romantic Love* (1984); Newman, F X (Ed) *The Meaning of Courtly Love* (1969); Rougemont, Denis de *Love in the Western World* (1983).
Broader Love consciousness (#HM2000).
Related Limerence (#HM1809) Romantic love (#HM5087).

♦ **HM6124c Mohaksaya** (Buddhism)
Mohakkhaya (Pali) — Extinction of illusion
Related Steps to enlightenment (Buddhism, #HH4019).

♦ **HM6129c Waiting on God** (Christianity)
Description Waiting on God implies a continual state of passivity, dependent on the will of pure providence. In a state of surrender, such persons belong wholly to God. When God lives in the soul, nothing of the self is left. The future is not mapped out, God directs the way from moment to moment. With no spiritual direction there may be no support but God Himself. Despite distress and misery, they must wait for assistance calmly and untroubled. Neglected by men but, through love, in possession of God, they must make no effort of their own but serve God in His own way. Although seeming to live a completely ordinary existence and with no obvious distinguishing characteristics, such persons have a certain virtue about their quiet oblivion, their speech and actions, that speaks of God and affects those with whom they come into contact. Unwittingly they guide and sustain others, although they themselves may feel insignificant, bewildered and confused.
Refs Caussade, Pierre de *The Sacrament of the Present Moment* (1981).
Related Mystical journey (Christianity, #HM0402).

♦ **HM6133c Negative capability**
Description The capacity to be in uncertainty, mystery and doubt, without any irritable reaching after fact and reason (John Keats). This is a form of imaginative questioning that is more rewarding than the discovery of facts and reasons, and implies the ability to work with the imagination, without the necessity of seeking out facts and reasons. It is a questioning that involves the questioner in the matter of thought so deeply that he becomes, in a sense, one with it. At this point knowing is no longer divorced from being: we know the way we are and we are the way we know. The meaning may be extended to include the capacity to live with mistakes and failures without being disheartened or dismayed. It is an attitude of mind which learners cultivate to help them to relate more appropriately to their mistakes as experience. Without such decent doubt, there can be no questioning, no learning, and no deliberate change.
Related Gnosis (#HM0413).

♦ **HM6138f Ego-splitting**
Description Not simply emotionally detached, the individual feels as though he is physically outside himself.
Broader Anomalies in experience of the unity of self (#HM9132).

♦ **HM6139d Al-Awwal** (Sufism)
Description The believer is aware of Allah as the first of which there is no second, for there is none like Him, and there is none prior to Him. Al-Awwal and al-Akhir, the first and the last, are considered together, they are like a circle in which the first and last are one. The good servant must be first in devotion, worship and good deeds.
Context A mode of mystical awareness in the Sufi tradition associated with the seventy-third of

the ninety-nine names of Allah.
Broader Ninety-nine names of Allah (Sufism, #HH2561).
Followed by Al-Akhir (Sufism, #HM7112).
Preceded by Al-Mu'akhkhir (Sufism, #HM4052).

♦ **HM6144d Attachment due to insight** (Buddhism)
Desire due to insight
Description Adorned with illumination, for example, the attachment which arises is so peaceful and subtle that it is not discerned as a defilement. The meditator, thinking that such a state has never arisen before, believes he has reached fruition, interrupts the course of insight. This is karmically unprofitable, because he mistakes the path.
Context One of the ten imperfections of insight of Hinayana Buddhism.
Broader Imperfection of insight (Buddhism, #HM9722).

♦ **HM6149c Citta visuddhi** (Buddhism)
Mental purity
Description Concentration or samadhi, when considered as a mental quality, is seen to cleanse the mind from taints and defilements of passion.
Broader Concentration (Buddhism, #HM6663).

♦ **HM6153d Clinging** (Hinduism)
Asanga
Description This is delight in and clinging to a desired object that has been attained. The excess of delight in the object means its embracement, desire for continually enjoying it even after it has been attained. The self wrongly identifies itself with an object which is of the nature of not-self and with which the mind is deeply engrossed due to egoism. From clinging springs desire.
Related Desire (#HM2433) Five aggregates (Buddhism, #HH3321).

♦ **HM6162d Illumination of the spirit** (Sufism)
Tajliya-i ruh
Description Ruh is the soul, the organ of mystic contemplation. The spirit is filled with the radiance of God and the ardour of his love.
Context The last of four contemplative disciplines the salik (seeker after God) must pass through to obtain ma'rifa. Various Sufi orders may differ in approach but all require the same spiritual concentration; manifestation in its many forms may be dealt with in many ways but the realization of spiritual reality behind the forms is the same.
Broader Awareness of the mystic journey (#HM2900).
Related Subtle faculties (Sufism, #HH6282).
Preceded by Cleansing the sirr (Sufism, #HM5353).

♦ **HM6173d Mind-consciousness element** (Buddhism)
Description This has the function of investigating or receiving. It appears five times among the indeterminate consciousnesses: as resultant with profitable or moral resultant, accompanied by joy or accompanied by equanimity or indifference; as resultant with unprofitable or immoral resultant; as functional accompanied by joy or accompanied by equanimity or indifference. It is without root cause (is unconditioned). Its proximate cause is the heart basis. As resultant it cognizes the six kinds of object (the five senses and the mind element). It is associated with joy when it is present at the occurrence of entirely desirable objects; it appears twice during the series of events comprising cognition, at the stage of investigating the five-doors (senses) and as registration at the end of impulsion or apperception. It is associated with equanimity or indifference when it is present at the occurrence of desirable-neutral objects; it appears five times, as it proceeds by investigating, registering, rebirth-linking or reconception, life-continuum, death.
Context In Hinayana Buddhism, 89 consciousnesses are enumerated in aggregate (khanda). Of these, 21 are profitable or moral, 12 are unprofitable or immoral and 56 are indeterminate (resultant or functional). The unprofitable all arise in the sphere of sense and desire, whereas profitable and indeterminate consciousnesses arise in sense, fine-material, immaterial and supramundane spheres.
Broader Mundane resultant consciousness (Buddhism, #HM0130)
Indeterminate consciousness in the sense sphere - resultant (Buddhism, #HM5721)
Indeterminate consciousness in the sense sphere - functional (Buddhism, #HM3852).
Related Adverting (Buddhism, #HM8336) Examining (Buddhism, #HM7324)
Determining (Buddhism, #HM7632) Rebirth-linking (Buddhism, #HM8266)
Eye consciousness (Buddhism, #HM2074) Ear consciousness (Buddhism, #HM2169)
Nose consciousness (Buddhism, #HM2364) Body consciousness (Buddhism, #HM2562)
Mind consciousness (Buddhism, #HM3323) Tongue consciousness (Buddhism, #HM2263)
Mental consciousness (Buddhism, #HM2838)
Volition born of contact with element of mind consciousness (Buddhism, #HM0404)
Psychical unease born of contact with element of mind-consciousness (Buddhism, #HM0234).

♦ **HM6174d Disillusionment**
Refs Person, Ethel Spector *Love and Fateful Encounters* (1990).
Related Illusion (#HM2510).

♦ **HM6192d Understanding as the plane of seeing** (Buddhism)
Understanding being the plane of discernment
Description This understanding belongs to the first of the four paths.
Context On the Path of Purification of Hinayana Buddhism, panna (understanding) is considered as of one kind (monad), of two kinds (dyads), of three kinds (triads) or of four kinds (tetrads). There are five dyads, four triads and two tetrads. All have the characteristic of penetrating the individual essences or true nature of states (monad). In the fifth dyad, this is contrasted with understanding being the plane of development or culture.
Broader Understanding (Buddhism, #HM4523).
Related Understanding as the plane of development (Buddhism, #HM4128).

♦ **HM6193f Tip of the tongue experience**
Description Although part of what is to be remembered may be recalled (number of syllables, particular consonants, related terms), the words being sought obstinately refuse to arise in memory. Taking the mind off the subject sometimes results in almost immediate recall.

♦ **HM6201c Diminished clarity of awareness**
Narrower Sleep (#HM2980) Sopor (#HM8006) Torpor (#HM1483)
Delirium (#HM7186) Oneroid states (#HM5097)
Absence of consciousness (#HM2670) Clouding of consciousness (#HM5754).

♦ **HM6204d Meeting God with active desire** (Christianity)
Description Meeting God with the intermediary of the gift of savourous wisdom, which is the ground and source of all virtue, the interiorly fervent person turns to God in a way of desire and activity, to give God glory and honour, and to offer Him himself and all his works to be consumed in the love of God. Love is in a state of likeness and of desire to be united with God. His love may be so intense that all that God can bestow apart from Himself seems small and unsatisfying and increases restlessness. Lingering over God's gifts or over any creature is a hindrance as they cannot satisfy. In the ground of his being, hunger and thirst of love are such that he surrenders himself at every moment, each sudden illumination of God being seized and touched in love, constantly dying and being reborn as yearning of love is renewed.
Context According to John Ruusbroec, there are three modes in which meeting God in the interior way of life is practised. This is the second of the three, actively loving, but is the cause of the first, emptiness.
Broader Meeting God (Christianity, #HM0541).
Related Meeting God with emptiness (Christianity, #HM8772)
Meeting God both resting and working in accordance with righteousness (Christianity, #HM5309).

♦ **HM6208d Illuminated consciousness** (Buddhism)
Illuminated mind
Description Consciousness is not perturbed by the darkness of ignorance.
Context One of the sixteen roots or modes of unperturbedness of mind listed in Hinayana Buddhism as a basis for supernormal powers.
Related Supernormal powers (Buddhism, #HH5652).

♦ **HM6221c Life-continuum** (Buddhism)
Subconsciousness — Samtana
Description Following rebirth-linking or reconception consciousness there arises a life-continuum consciousness, or subconsciousness, with the same object as rebirth-linking consciousness and being associated with the 19 kinds of rebirth-linking consciousness in root cause. This consciousness continues indefinitely unless some other kind of arising of consciousness interrupts the continuity. Once the continuity is interrupted, the process of adverting, etc, up to registration proceeds until life-continuum resumes. The final life-continuum consciousness in the case of one becoming is death or decease, the passing away from that existence.
Context In Hinayana Buddhism, this is the second mode of occurrence of consciousness in which the 89 kinds of consciousness proceed.
Broader Modes of occurrence of consciousness (Buddhism, #HM6720).
Followed by Death (Buddhism, #HM3932) Adverting (Buddhism, #HM8336).
Preceded by Registering (Buddhism, #HM1646) Rebirth-linking (Buddhism, #HM8266).

♦ **HM6226d Spiritual meekness** (Christianity)
Description Spiritual meekness is opposed to spiritual anger. It does not become irritated at the sins of other people or feel impelled to berate them for their sins. Nor does it set itself up as the arbiter of goodness. It is the same with one's own shortcomings. The spiritually meek do not become impatient at their own imperfections, nor do they continually make resolutions which they continually break, thus becoming more angry. They wait patiently for God to provide His gifts as and when He pleases.
Related Meekness (#HH0414) Spiritual anger (Christianity, #HM0936).

♦ **HM6227d Tafakkur** (Sufism)
Meditation on God
Description The name of God becomes the intimate friend of the devotee.
Context According to Shaykh Abu Sa'id ibn Abi'l-Khayr, this is the 31st of 40 stations or maqamat the Sufi must possess for his journey on the path of Sufism to be acceptable.
Broader Stations of consciousness - ibn-Abi'l-Khayr (Sufism, #HM2424).
Followed by Wisal (Sufism, #HM7119).
Preceded by Nearness-awareness (Sufism, #HM2377).

♦ **HM6231g Drowsiness**
Stage-one sleep
Description Generally lasting only a few minutes, this state is marked by replacing alpha-rhythms by specific low-voltage EEG activity.
Broader Sleep (#HM2980).
Related Alpha wave consciousness (#HM2345).
Followed by Sleep spindles (#HM4806).

♦ **HM6238d Tamas** (Leela)
Description Having realized happiness, the player feels he has reached his goal and relaxes his efforts. But play must go on. By attempting in inaction to avoid the law of karma, and forgetting that refusal to act is itself an act, the player finds this tamasic ignorance takes him right back to maya, illusion.
Context The 63rd state or square on the board of *Leela*, the game of knowledge, appearing in the seventh row.
Broader Seventh chakra: plane of reality (Leela, #HM0754).
Followed by Maya (Leela, #HM0953).

♦ **HM6243d Independent consciousness** (Buddhism)
Independent mind
Description Consciousness does not waver in opinion, it is not perturbed by false views.
Context One of the sixteen roots or modes of unperturbedness of mind listed in Hinayana Buddhism as a basis for supernormal powers.
Related Supernormal powers (Buddhism, #HH5652).

♦ **HM6245d Understanding accompanied by joy** (Buddhism)
Description Here understanding is that belonging to two of the classes of profitable or moral consciousness associated with the sense sphere, and belonging to the 16 path consciousnesses associated with the first four jhanas of the fivefold method.
Context On the Path of Purification of Hinayana Buddhism, panna (understanding) is considered as of one kind (monad), of two kinds (dyads), of three kinds (triads) or of four kinds (tetrads). There are five dyads, four triads and two tetrads. All have the characteristic of penetrating the individual essences or true nature of states (monad). In the fourth dyad, this is contrasted with understanding accompanied by equanimity.
Broader Understanding (Buddhism, #HM4523).
Related Understanding accompanied by equanimity (Buddhism, #HM7525).

♦ **HM6255d Al-Ghafur** (Sufism)
Description The believer is aware of Allah as the all-forgiving. He hides our faults from others (al-Ghafir), from the angels (al-Ghafur) and from ourselves (al-Ghafar).
Context A mode of mystical awareness in the Sufi tradition associated with the thirty-fourth of the ninety-nine names of Allah.
Broader Ninety-nine names of Allah (Sufism, #HH2561).
Followed by Ash-Shakur (Sufism, #HM1934).
Preceded by Al-'Azim (Sufism, #HM5198).

♦ **HM6276f Delusions of love**
Erotomania
Description Despite lack of evidence to support the belief, the individual is convinced that some

other person is madly in love with him.
Broader Delusive consciousness (#HM2600).

♦ **HM6287d Plane of inner space** (Leela)
Uranta–loka
Description Feeling of self starts to end as the player merges into the source of phenomena and all duality ceases. He becomes like clear glass through which all passes with no feelings, distortions or judgements.
Context The 65th state or square on the board of *Leela*, the game of knowledge, appearing in the eighth row.
Broader Beyond the chakras: the gods themselves (Leela, #HM1141).

♦ **HM6307g Deep sleep**
Stage–four sleep
Description Delta wave activity is more than 50 percent. There are no eye movements and muscle tension is usually low.
Broader Sleep (#HM2980).
Related Delta wave consciousness (#HM1785).
Preceded by Stage–three sleep (#HM0348).

♦ **HM6312d Seclusion** (Buddhism)
Solitude
Description This is a state of non–greed and non–delusion.
Context In Hinayana Buddhism, one of the five ascetic states.
Broader Non–ignorance (Buddhism, #HM2695)   Ascetic states (Buddhism, #HM5354)
State of not feeling greed (Buddhism, #HM1825).

♦ **HM6321c Mystical marriage**
Refs Welvis, Gerhard *The Mystical Marriage*.
Related Alchemy (Esotericism, #HH0221).

♦ **HM6327d Superior concentration** (Buddhism)
Description Superior concentration is well–developed, has reached mastery and has brought under control.
Context According to Hinayana Buddhism, concentration is considered as of one kind (monad), of two kinds (dyads), of three kinds (triads), of four kinds (tetrads) or of five kinds (pentad). In the first triad, this is the third concentration.
Broader Concentration (Buddhism, #HM6663).
Preceded by Medium concentration (Buddhism, #HM6089).

♦ **HM6328d Alchemical whitening** (Esotericism)
Albedo — Leucosis
Description In this second stage of the process, the body is experienced as dissolving in the waters of the spirit (revealed by the first stage of the alchemical process). This reduction of bodily consciousness to its psychic substance effectively causes the soul to withdraw from the sensory organs and to expand into a "space" that is both inward and unlimited. When the inner consciousness is thus reduced to its primary matter, the alchemist experiences a power emanating from the mysterious centre of his being, from his divine essence. This living light illuminates, integrates and nourishes. The passions, and their corresponding instincts, are pacified and made cosmic, gradually recovering their primordial innocence. Heaviness is melted in life. The alchemist experiences nature from within in its immaculate conception. The process of whitening thus leads to an experience of the synthesis of all forms and especially as a marriage of opposites.
Context The second stage of the alchemical work.
Broader Alchemy (Esotericism, #HH0221).
Followed by Alchemical reddening (Esotericism, #HM3631).
Preceded by Alchemical blackening (Esotericism, #HM5345).

♦ **HM6335d Personal salvation** (Christianity)
Description This condition is the aim of evangelicals and in particular of those in the charismatic movement. The intention is to confront actual men and women, in the situation they are actually in, with the question of saving their own souls. The focus is righteousness and the kingdom of heaven on earth. But righteousness according to the law is not enough. The principle of justification by faith by God's grace through the redemption of Christ is based on St Paul's letter to the Romans, III 24 to 28.
Broader Salvation (#HH0173)   Liberation (#HH0388).
Related Evangelism (Christianity, #HH3798)   Charismatic renewal (Christianity, #HH3124).

♦ **HM6336c Enlightenment factors** (Buddhism)
Wisdom factors
Description In Hinayana Buddhism, three enlightenment or wisdom factors are described as being conducive to exerting a slack mind. These are: investigation into doctrine or states; energy; happiness or rapture. And three are described as conducive to restraining an agitated mind. These are: tranquillity; concentration; equanimity. Figuratively speaking, the first three contribute to assisting in bringing to a blaze a fire which is dampened down, while the second three dampen down a fire which is blazing too freely.
Related Wise attention (Buddhism, #HM4309)   Skill in absorption (Buddhism, #HH4777).

♦ **HM6342d Lower fetters** (Buddhism)
Description These are: belief in personality; sceptical doubt; attachment to rites and rituals; sensual lust; ill–will.
Related Afflictions and hindrances (Buddhism, #HH5007).

♦ **HM6346d Nirodhasamapatti** (Buddhism)
Attainment of cessation
Related Attainment of cessation (Buddhism, #HM5438)
Cessation–awareness (Buddhism, Tibetan, #HM2172).

♦ **HM6350d An–Nafi'** (Sufism)
Description The believer is aware of Allah as the propitious creator of good. He is creator of man, the best of creation, giving intellect, conscience and faith so that man may discriminate for himself. Only man has will, only Allah can check his will. The will of Allah is the means by which one sees, it is what one sees. One has only to will to receive the good that Allah has willed for one – but to do this one must be awake, with ears and eyes open.
Context A mode of mystical awareness in the Sufi tradition associated with the ninety–second of the ninety–nine names of Allah.
Broader Ninety–nine names of Allah (Sufism, #HH2561).
Followed by An–Nur (Sufism, #HM1354).
Preceded by Ad–Darr (Sufism, #HM4542).

♦ **HM6434d Gaseous plane** (Leela)
Vayu–loka
Description The player becomes an enlightened soul and ceases to be burdened by mass, weight or form. He has true freedom of action.
Context The 57th state or square on the board of *Leela*, the game of knowledge, appearing in the seventh row.
Broader Seventh chakra: plane of reality (Leela, #HM0754).

♦ **HM6445c State of grace** (Christianity)
Description In this state, the individual is totally receptive towards God. He neither demands nor expects the gift of grace from God but, insofar as he is able, he has, through God's grace, repented of and confessed his sins and opened himself to God's mercy.
In the Catholic Church, principal ways of obtaining grace are through prayer and the sacraments, in particular the Holy Eucharist.
Related Grace (#HH0169)   Actual grace (Christianity, #HH6548)
Sanctifying grace (Christianity, #HH5116).

♦ **HM6502c Nanadassana** (Buddhism)
Knowledge — Insight
Description This is threefold insight into the knowledge of previous existences, the knowledge of the decease and re–birth of beings and the knowledge of destruction of the asavas.
Context One of ten superhuman qualities described in the Sutta Vibhanga as being special attainments of insight above that of ordinary men.
Broader Superhuman qualities (Buddhism, #HH9764).
Narrower Purification by knowledge and vision (#HH3025)
Purification by knowledge and vision of the way (#HH3550)
Purification by knowledge and vision of what is the path and what is not the path (#HH4007).

♦ **HM6504c Asamskrita** (Buddhism)
Asamkhata (Pali) — Unconditioned existence — Uncompounded existence
Related Steps to enlightenment (Buddhism, #HH4019).

♦ **HM6508f Alienation of thought**
Description There is an uneasy awareness that one's thoughts are not one's own. They may be attributed to an outside power or influence, the outside agency (police, secret–service, enemies, doctors) being believed to use hypnosis, x–rays, television transmitters or whatever.
Broader Modes of awareness associated with schizophrenia (#HM2313)
Anomalies in experience of the self as recognized in personal performance (#HM1336).
Related Passivity experiences (#HM7203).

♦ **HM6512d Al–Basir** (Sufism)
Description The believer is aware of Allah as all–seeing. His creatures have also the power to see, man sees with the physical but also the inner eye to see the inner self. Knowing the inner self one knows that Allah sees one although we cannot see Him. Otherwise, not seeing Allah, one may behave as though He cannot see one.
Context A mode of mystical awareness in the Sufi tradition associated with the twenty–seventh of the ninety–nine names of Allah.
Broader Ninety–nine names of Allah (Sufism, #HH2561).
Followed by Al–Hakam (Sufism, #HM5014).
Preceded by Al–Sami' (Sufism, #HM0419).

♦ **HM6521d Undejected consciousness** (Buddhism)
Inflexible mind
Description The mind does not waver in idleness, it is not perturbed by indolence.
Context One of the sixteen roots or modes of unperturbedness of mind listed in Hinayana Buddhism as a basis for supernormal powers.
Related Supernormal powers (Buddhism, #HH5652).

♦ **HM6522c Birth of God in the soul's ground** (Christianity)
Speaking of the Word in the soul
Description Meister Eckhart speaks of this arising from detachment, from an inner stillness and silence which may persist in everyday tasks as much as in contemplative and meditative practice. In this silence God works and speaks within. What is important is not so much a practice as an attitude. There is a breaking through consciousness to find an inner centre of stability, a unity and serenity which stays intact despite fluctuations of the conscious state. By grace, man is drawn away from the temporal and purified from the transient. Occurring in time, birth of God in the self transcends time. Happening instantaneously, it is not transitory. There is no alternation of opposites, the ecstasy is not followed by pain. Despite being a transcendent experience, the physical body is caught up in it so that it is a real, physical birth, God is encountered in the flesh.
Refs Smith, Cyprian *The Way of Paradox* (1987).
Related Mystical cognition (#HM2272)   Religious experience (#HM3445)
Universal human nature (#HH4389)   Detachment (Christianity, #HM1534)
Way of paradox (Christianity, #HH3335)   Way of transcendence (Christianity, #HH6566)
Contemplative intuitive meditation (#HH0816).

♦ **HM6523c Mahabba** (Islam)
Love based on reason
Description This is love such as one feels for one's family, and is natural and not to be blamed. In mystic terms, five stages of mahabba lead to wajd or ecstasy.
Refs Bell, Joseph Norment *Love Theory in Later Hanbalite Islam* (1979).
Narrower Love in God (#HM4712)   Love with God (#HM5580).
Related Love (Islam, #HH5712)   Wajd (Sufism, #HM3650)
Divine–love awareness (Sufism, #HM2401).

♦ **HM6533g Rhythmic sensory bombardment**
Description Immersion in intensified rhythmic input, both very loud music and visual input, force the brain's electrical patterns in line with the input. Subjective experience and motor activity are of euphoric excitement while cognitively one is asleep. Similar effects are induced in tribal rituals, in revivalist meetings, in chanting at political rallies. The state can be described as one of excited subservience.
Broader Anomalies in consistency of consciousness (#HM1122).
Related Rhythmic sensory bombardment therapy (#HH0218).

♦ **HM6534c Restoration of celestial awareness within the mundane** (Taoism)
Activating the mind of Tao — Dissolving the human mind
Description Centered on the calm awareness of the true mind, the energy of the mind of Tao grows, requiring progressively less effort. This is accompanied by a waning of the force of mundanity. Through the continuing practice of this stage, celestial awareness is restored within the mundane.

Context The third of seven stages of the Taoist alchemical firing process through which reality is cultivated and the self is restored.
Broader Alchemy (Taoism, #HH5887).
Followed by Assembling the five elements (Taoism, #HM4338).
Preceded by Recognition of the true mind (Taoism, #HM6633).

♦ **HM6539g Auditory agnosia**
Word–deafness — Sensory amusia
Description Despite being able to hear normally, the individual cannot recognize the significance of the sound heard. This may be extended to word–deafness, when spoken (as opposed to written) speech is not recognized; and to sensory amusia, when music is not recognized as such.
Broader Agnosia (#HM8932).

♦ **HM6544d Establishment due to insight** (Buddhism)
Presentation due to insight
Description Assurance or presentation (establishment) as mindfulness arises in association with insight; whatever subject is adverted to or appears, that subject enters within like the other world to divine vision. The meditator, thinking that such establishment has never arisen before, believes he has reached fruition, interrupts the course of insight. This is karmically unprofitable, because he mistakes the path.
Context One of the ten imperfections of insight of Hinayana Buddhism.
Broader Imperfection of insight (Buddhism, #HM9722).

♦ **HM6552d Spiritual repose** (Christianity)
Description This, together with spiritual renewal, is one of the great gifts which Jesus presents. "Come unto me, all ye that labour and are heavy laden, and I will give you rest". This is not the end of toil, but the replacing of a heavy yoke by one which is easy and whose burden is light.
Related Charismatic renewal (Christianity, #HH3124)
Prayer of simplicity (Christianity, #HM0238).

♦ **HM6553c Dwelling in the fivefold jhana** (Buddhism)
Dwelling in the second jhana of the fivefold system
Description As opposed to the system of four jhanas, each of which is dwelt in successively, a fifth jhana may occur between the first and second. This may then be referred to as the second jhana, and the second, third and fourth jhanas of the fourfold system referred to as the third, fourth and fifth of the fivefold system.
The second jhana of the fivefold system is as follows:
Having obtained mastery of the first jhana, the meditator is aware of the hindrances still being near and of the grossness of applied thought. The jhana factors are reviewed with full awareness; sustained thought, happiness, bliss and unification appear peaceful. Bringing the subject of meditation to mind, the process of appearance of absorption is gone through again. Stilling applied thought, the second jhana is entered and dwelt in. Applied thought is abandoned, not at the moment of access (as with the hindrances in the first jhana), but at the moment of actual absorption. Now there are four factors – sustained thought, happiness or rapture, bliss, unification or collectedness of mind.
Like the first jhana it is said to be good in three ways: beginning (purification of the way), which is *access*; middle (intensification of equanimity), which is *absorption*; end (satisfaction), which is *reviewing*. It also has ten characteristics. Again, through practice, the meditator acquires mastery in the second jhana through the habits of: adverting to the jhana; attaining the jhana; resolving and steadying the duration of the jhana; emerging from the jhana; reviewing the jhana. He can then end his attachment to the second jhana and commence doing what is necessary to attain the third.
Broader Jhana (Buddhism, Pali, #HM7193).
Related Appearance of absorption (Buddhism, #HM1618)
Dwelling in the first jhana (Buddhism, #HM4298)
Dwelling in the third jhana (Buddhism, #HM5643)
Dwelling in the second jhana (Buddhism, #HM7121)
Dwelling in the fourth jhana (Buddhism, #HM8087)
Concentration of the second jhana of five (Buddhism, #HM4575).

♦ **HM6554c Way of idealism** (Esotericism)
Way of devotion
Description Characteristic of those who dedicate themselves to a task involving unity and some transcendent goal, possibly through devotion to a god, a guru, and ideal or a cause, even to the point of fanaticism. On this way they are challenged by the task of recognizing the value in alternative visions and refining their understanding of the goal to which they are dedicated.
Context The sixth of seven ways to spiritual realization characteristic of Western esotericism.
Broader Ways to spiritual realization (Esotericism, #HH2665).
Related Sphere of the foundation (Kabbalah, #HM2410)
Initiation of baptism (Esotericism, #HM1267).

♦ **HM6556d Wieldiness of consciousness** (Buddhism)
Wieldiness of mind — Wieldiness of citta (Pali)
Description This refers to the wieldiness of the aggregate of consciousness. The characteristic is suppressing or quieting unwieldiness of the citta, the function is to crush its unwieldiness. It is manifest as success in making something the object of consciousness. Consciousness is the proximate cause. It brings faith in things to be trusted and application to beneficial acts, and it is the opponent of hindrances causing unwieldiness in the citta. Wieldiness of mental body and wieldiness of consciousness are considered together.
Context One of the formations aggregate (mental coefficients) of Hinayana Buddhism, being listed among the constant states which appear in their true nature, and as profitable primary (always present in any profitable or profitable–resultant consciousness).
Broader Awareness as mental–formation group of conscious existence (Buddhism, #HM2050).
Related Wieldiness of mental body (Buddhism, #HM7969).

♦ **HM6558c Released soul's perception** (Hinduism)
Muktajnana — Omniscience — Knowledge of liberated souls
Description Discriminative knowledge of the self generates a trance that continuously furnishes the intuition of the self as completely isolated from the mind–body complex (Patanjali). Constant intuition of the self arises when all potencies of actions and subconscious impressions of experience are destroyed. Knowledge of the self is infinite as all taints which veil it are destroyed. Even though in the embodied state, the self is liberated. It is unaffected by modifications in the gunas of prakriti (nature). There is immediate knowledge of the self and the universe, which nevertheless does not affect the released soul. The past is known as existing in the past, the future as existing in the future. Free from the bondage of physical existence, the omniscient person is free from the fetters of karma–matter and mundane existence. There is a supersensuous vision of the whole world, past, present and future.
Broader Intuition (Hinduism, #HM1256).
Related Omniscience (Yoga, #HM2475) Kevalajnana (Jainism, #HM3948).

♦ **HM6562b Remembrance of God** (Sufism)
Practice of dhikr
Description Various systems for achieving recollection or remembrance of God are described in Sufism. Each has the aim of attaining all–pervading knowledge and of reaching the state of *fana–fi–Allah*, passing away of the self in the presence of God. One system of dhikr, covered in separate entries for each stage, is as described by Shaykh Kalimallah of the Chishtiyya order. It involves passing through successive states until continual remembrance is experienced and the devotee achieves saintship.
Refs Valiuddin, Mir *Contemplative Disciplines in Sufism* (1980).
Narrower Ghaflah (#HM7383) Wilayat (#HM3647) Latifa–i–qalb (#HM1905)
'Ilm al–basir (#HM0476) Sultan–al–dhikr (#HM8330) Total annihilation (#HM5775)
Khilwat dar anjuman (#HM4987) Remembrance of reality (#HM3892)
Continuous all–pervading knowledge (#HM4788).
Related Fana (Sufism, #HM3799) Recollection (Islam, Sufism, #HM2351).

♦ **HM6601d Understanding as skill in means** (Buddhism)
Description This is the skill which arises, whether as skill in improvement or detriment, in bringing about and effecting various states.
Context On the Path of Purification of Hinayana Buddhism, panna (understanding) is considered as of one kind (monad), of two kinds (dyads), of three kinds (triads) or of four kinds (tetrads). There are five dyads, four triads and two tetrads. All have the characteristic of penetrating the individual essences or true nature of states (monad). In the third triad, this is compared with understanding as skill in improvement or detriment.
Broader Understanding (Buddhism, #HM4523).
Related Understanding as skill in detriment (Buddhism, #HM7602)
Understanding as skill in improvement (Buddhism, #HM8290).

♦ **HM6631d Sidq** (Sufism)
Veracity — Truthfulness
Description Every step and every breath is in truth. Tongue speaks only of the heart, heart only of inner secrets, and inner secrets of God.
Context According to Shaykh Abu Sa'id ibn Abi'l–Khayr, this is the 18th of 40 stations or maqamat the Sufi must possess for his journey on the path of Sufism to be acceptable.
Broader Stations of consciousness - ibn–Abi'l–Khayr (Sufism, #HM2424).
Followed by Khawf (Sufism, #HM1047).
Preceded by Ikhlas (Sufism, #HM4663).

♦ **HM6633c Recognition of the true mind** (Taoism)
Discovery of the natural mind — Recognition of the innocent mind
Description During this stage the natural, innocent, true mind is recognized as distinct from conventional human mentality. This essential step in cultivating reality is symbolized by the process of finding the "heart of heaven and earth". This mind is subtle and recondite and is not easily manifested. When found it may then be used to refine the self. True consciousness emerges, as a point of celestial energy, and provides the clarity through which undesirable influences may be distinguished and disempowered. It is tranquil and unperturbed, yet sensitive and effective.
Context The second of seven stages of the Taoist alchemical firing process through which reality is cultivated and the self is restored.
Broader Alchemy (Taoism, #HH5887).
Followed by Restoration of celestial awareness within the mundane (Taoism, #HM6534).
Preceded by Refining the self (Taoism, #HM4007).

♦ **HM6644d Mental sensation** (Buddhism)
Manasa–pratyaksa — Non–sensuous feeling
Description This is the sensation which follows immediately the sensation of an outer sense, or pure sensation, and is evoked by it. It is an intelligible sensation, direct, intuitive and non–conceptive. Constructive imagination is excluded here, so the sensation is incapable of illusion.
Broader Consciousness in Buddhism (Buddhism, #HM2200).
Related Introspection (Buddhism, #HM7109) Mental consciousness (Buddhism, #HM2838)
Intelligible intuition (Buddhism, #HM5532)
Awareness of consciousness–group of conscious existence - senses and mind (Buddhism, #HM2556).

♦ **HM6652d Steadiness of consciousness** (Buddhism, Pali)
Weak concentration — Conscious duration
Description In Hinayana Buddhism, some sources do not list steadiness of consciousness as separate from concentration (stabilization), of which it may be considered a short–lived equivalent. Others consider it separately as arising, for example, in unprofitable consciousness rooted in delusion and associated with uncertainty and equanimity, and in eye, ear, nose, tongue and body consciousnesses (indeterminate resultant with profitable result).
Context One of the formations aggregate (mental coefficients) of Hinayana Buddhism, being listed among the constant states which appear in their true nature, and as general secondary (sometimes present in any consciousness).
Broader Stabilization (Buddhism, #HM2440)
Awareness as mental–formation group of conscious existence (Buddhism, #HM2050).

♦ **HM6663b Concentration** (Buddhism)
One–pointedness of citta — Cittassa ekaggata — Unification of mind — Collectedness of moral thought — Samadhi — Cittassekaggata (Pali) — One–directedness of thought
Description In Buddhism equated with *samadhi*, concentration is the centering of consciousness and all that it entails, properly and evenly on a single object. The term samadhi is used both to describe a state of mind, in particular *samma–samadhi* or right concentration, and the method designed to induce this state. It is used in a different sense from the same term in Hinduism and yoga. In the Path of Purity it is defined as profitable unification of the mind or collectedness of moral thought. The characteristic is lack of distraction and it arises from a state of bliss. It is a state of pure mind, a necessary preliminary to higher progress towards Arhatship or final emancipation. Samadhi is developed by systematic training which inculcates the habit of mental concentration, resulting in spiritual progress experienced both in and through the human organism and leading to a point at which self illumination supervenes.
Context One of ten superhuman qualities described in the Sutta Vibhanga as being special attainments of insight above that of ordinary men. According to Hinayana Buddhism, concentration is considered as of one kind (monad), of two kinds (dyads), of three kinds (triads), of four kinds (tetrads) or of five kinds (pentad). There are four dyads, four triads, six tetrads and one pentad (which is an expansion of the third tetrad). All have the property of non–distraction (monad). One of the five constituent factors of jhana, concentration is intensified by the other four constituents and constitutes the one–pointedness of mind that expels sensuous desire.
Refs Buddhaghosa, Bhadantacariya *The Path of Purification* (1980); Buddhaghosa, Bhadantacariya *The Path of Purity* (1975); Mahathera, Paravahera Vajiranana *Buddhist Meditation in Theory and Practice* (1987).

Broader Superhuman qualities (Buddhism, #HH9764).
Narrower Ceto-samadhi (#HM5587)
Ceto vimutti (#HM0831)
Ceto samatha (#HM0709)
Citta visuddhi (#HM6149)
Non-distraction (#HM4382)
Access concentration (#HM4999)
Medium concentration (#HM6089)
Correct concentration (#HM1735)
Mundane concentration (#HM7234)
Limited concentration (#HM1073)
Exalted concentration (#HM4036)
Inferior concentration (#HM5562)
Superior concentration (#HM6327)
Absorption concentration (#HM0311)
Unincluded concentration (#HM5768)
Measureless concentration (#HM0496)
Concentration due to zeal (#HM4969)
Supramundane concentration (#HM4243)
Sense-sphere concentration (#HM1097)
Concentration due to energy (#HM8365)
Concentration with happiness (#HM5767)
Concentration due to inquiry (#HM7434)
Right rapture of concentration (#HM0931)
Concentration without happiness (#HM0730)
Immaterial-sphere concentration (#HM0696)
Concentration accompanied by bliss (#HM4191)
Concentration accompanied by bliss (#HM1454)
Fine-material-sphere concentration (#HM4265)
Concentration partaking of diminution (#HM8002)
Concentration partaking of stagnation (#HM6956)
Concentration accompanied by happiness (#HM4433)
Concentration partaking of distinction (#HM7363)
Concentration partaking of penetration (#HM5930)
Concentration accompanied by equanimity (#HM8021)
Concentration accompanied by equanimity (#HM5008)
Concentration of the first jhana of four (#HM4456)
Concentration of the third jhana of four (#HM2284)
Concentration of the first jhana of five (#HM0297)
Concentration of the third jhana of five (#HM8532)
Concentration of the fifth jhana of five (#HM5143)
Concentration of the second jhana of four (#HM4380)
Concentration of the fourth jhana of four (#HM7202)
Concentration of the second jhana of five (#HM4575)
Concentration of the fourth jhana of five (#HM4882)
Limited concentration with a limited object (#HM1175)
Limited concentration with a measureless object (#HM5547)
Measureless concentration with a limited object (#HM4653)
Measureless concentration with a measureless object (#HM5214)
Concentration due to natural purity of consciousness (#HM7961)
Concentration with applied thought and sustained thought (#HM4908)
Concentration of easy progress and swift direct-knowledge (#HM4977)
Concentration without applied thought or sustained thought (#HM7543)
Concentration of easy progress and sluggish direct-knowledge (#HM1010)
Concentration of difficult progress and swift direct-knowledge (#HM7859)
Concentration without applied thought and with sustained thought (#HM5199)
Concentration of difficult progress and sluggish direct-knowledge (#HM5992).
Related Dharana (Hinduism, #HM2566)
Jhana (Buddhism, Pali, #HM7193)
Understanding (Buddhism, #HM4523)
Samadhi (Hinduism, Yoga, #HM2226)
Adhicitta sikkha (Buddhism, #HH2387)
Dhyana (Hinduism, Buddhism, #HM0137)
Way of the warrior (Amerindian, #HH8219)
Faculty of concentration (Buddhism, #HM3454)
Trances and mental absorptions (Buddhism, #HM2122)
One-pointedness in mental abiding states (Buddhism, #HM2753).

♦ **HM6681d Understanding having a measureless object** (Buddhism)
Understanding having an infinite object
Description This is understanding that occurs contingent on nirvana or nibbana. This is transcendental or supramundane insight.
Context On the Path of Purification of Hinayana Buddhism, panna (understanding) is considered as of one kind (monad), of two kinds (dyads), of three kinds (triads) or of four kinds (tetrads). There are five dyads, four triads and two tetrads. All have the characteristic of penetrating the individual essences or true nature of states (monad). In the second triad, this is compared with understanding having a limited or exalted object.
Broader Understanding (Buddhism, #HM4523).
Related Supramundane understanding (Buddhism, #HM8838)
Understanding having a limited object (Buddhism, #HM2495)
Understanding having an exalted object (Buddhism, #HM3296).

♦ **HM6687d Tuning-in**
Selective filtering of experience
Description When the attention is presented with a number of competing stimuli it is possible follow one stimulus while remaining aware in a general sense of all the stimuli. Thus one may be aware of a number of people speaking while focusing the attention upon what is being said by one of them. Input seems to be divided into a number of channels, with attention to one of them being distracted to another depending upon the meaning of what is being seen, heard or whatever. For example, attention may be distracted from one conversation if one's own name is spoken in another conversation. A special example is absent-mindedness, when attention is focused upon what one is thinking and not on what is going on around one.
Refs Reed, Graham *The Psychology of Anomalous Experience* (1988).
Narrower Absent-mindedness (#HM4976).

♦ **HM6690d Riyada** (Sufism)
Austerity
Context The third stage on the path to sainthood of the Naqshbandiyya order.
Broader Suluk of the Naqshbandiyya order (Sufism, #HM4356).
Followed by Piety (Sufism, #HM5339).
Preceded by Zuhd (Sufism, #HM4450).

♦ **HM6704f Auditory hallucination**
Echo de pensées — Imperative hallucination
Description the individual may hear noises, meaningful sounds such as music, or one or more voices speaking meaningful phrases. The sound is usually localized, the source being precise, and may be the voice of a famous person, a friend, or some unknown or mysterious person or being. In schizophrenia there is often auditory hallucination where the voices are saying something derogatory about the individual or whispering about him; or there may be imperative hallucination which involves the individual being commanded to do something; another schizophrenic hallucination is echo de pensées, hearing one's own thoughts spoken out loud as though everything one thinks can be heard by others.
Broader Hallucination (#HM4580).
Related Audible thought (#HM1317)
Clairaudience (Psychism, #HM4333).

♦ **HM6705g Death** (Hinduism)
Marana
Description This is the condition of the jiva or empirical self when all external and internal organs cease to operate until it is associated with another body. There is a cessation of the general and particular sense of the body due to destruction of the merits and demerits producing enjoyment and suffering in the present life. There is complete cessation of relations of the self to the body and the vital forces. Unlike deep sleep the subtle body remains in the same place in the form of impressions. In death it goes with the jiva to another world.

♦ **HM6719d Dhul-Jalali wal-Ikram** (Sufism)
Dhul-Jalal-wal-Ikram
Description The believer is aware of Allah as the lord of majesty and bounty. All perfection belongs to Him, all blessing comes from Him. Nothing can exist of itself or sustain itself. Those who revolt against Him have no power on which they can depend. Yet our needs are supplied by others, He has made each dependent on others and others dependent on each. Gratitude is due to the vehicle and to the source. Blessing is obtained by spending upon others what we have received from Him. One of the beautiful names which can be applied only to Allah, some have claimed this as the greatest name of all.
Context A mode of mystical awareness in the Sufi tradition associated with the eighty-fifth of the ninety-nine names of Allah.
Broader Ninety-nine names of Allah (Sufism, #HH2561).
Followed by Al-Muqsit (Sufism, #HM8001).
Preceded by Malik-ul-Mulk (Sufism, #HM5902).

♦ **HM6720b Modes of occurrence of consciousness** (Buddhism)
Description In Hinayana Buddhism, the 89 kinds of consciousness are seen as occurring in 14 different modes: rebirth-linking (reconception); life-continuum; adverting; seeing; hearing; smelling; tasting; touching; receiving; examining or investigating; determining or deciding; impulsion or apperception; registering; death. Once the continuity is interrupted, the process of adverting, etc, up to registration proceeds until life-continuum resumes. The final life-continuum consciousness in the case of one becoming is death or decease, the passing away from that existence. Following death there is rebirth-linking again, and the cycle proceeds through the kinds of becoming or sources of life, destiny, stations of consciousness or conscious durations and abodes of beings. With attainment of arahantship, the cycle ceases at the end of death consciousness.
Refs Buddhaghosa, Bhadantacariya *The Path of Purification* (1980); Buddhaghosa, Bhadantacariya *The Path of Purity* (1975).
Narrower Death (#HM3932)
Adverting (#HM8336)
Receiving (#HM7092)
Examining (#HM7324)
Impulsion (#HM7268)
Determining (#HM7632)
Registering (#HM1646)
Life-continuum (#HM6221)
Rebirth-linking (#HM8266)
Sense mode of consciousness occurrence (#HM4389).
Related Feeling aggregate (Buddhism, #HM4983)
Dispositions of consciousness (Buddhism, #HM2098).

♦ **HM6722d Pleasant feeling** (Buddhism)
Pleasure
Description The characteristic of pleasant feeling is experiencing a desirable, tangible object; the function is to intensify associated states; the manifestation is physical satisfaction (bodily enjoyment); the proximate cause is the controlling faculty of the body.
Context In Hinayana Buddhism, one of the five ways in which feeling is analysed in the feeling aggregate.
Broader Feeling aggregate (Buddhism, #HM4983).
Related Pleasure (#HM2883)
Sukha (Buddhism, #HM2866)
Equanimity (Buddhism, #HM7769)
Sad feeling (Buddhism, #HM7120)
Joyful feeling (Buddhism, #HM8737)
Painful feeling (Buddhism, #HM8010)
Body consciousness (Buddhism, #HM2562)
Psychically pleasant (Buddhism, #HM1862).

♦ **HM6754b Smrityupasthana** (Pali)
Four stations of mindfulness
Related Right mindfulness (Buddhism, #HM0704).

♦ **HM6762d Tafrid** (Sufism)
Isolation — Aloneness
Description The devotee is a stranger among the creatures of this world, undeterred by beatings, nor fooled by caresses.
Context According to Shaykh Abu Sa'id ibn Abi'l-Khayr, this is the 36th of 40 stations or maqamat the Sufi must possess for his journey on the path of Sufism to be acceptable.
Broader Stations of consciousness - ibn-Abi'l-Khayr (Sufism, #HM2424).
Followed by Inbisat (Sufism, #HM7098).
Preceded by Tajrid (Sufism, #HM7621).

♦ **HM6771d Al-Jalil** (Sufism)
Description The believer is aware of Allah as the sublime lord of majesty and might, who is everywhere at all times yet cannot fit in space and time, the owner of all that is good and perfect. Whoever loves Allah loves all whom He loves, who love Him, who teach His words.
Context A mode of mystical awareness in the Sufi tradition associated with the forty-first of the ninety-nine names of Allah.
Broader Ninety-nine names of Allah (Sufism, #HH2561).
Followed by Al-Karim (Sufism, #HM5289).
Preceded by Al-Hasib (Sufism, #HM1467).

♦ **HM6786d Al-Baqi** (Sufism)
Description The believer is aware of Allah as everlasting, having no beginning or end. One is here for only a short time, although selfless working for the sake of Allah and not for immediate profit will mean one's work will carry on for eternity. If the intention in doing is for service and not for gain, one will earn eternity hereafter.
Context A mode of mystical awareness in the Sufi tradition associated with the ninety-sixth of the ninety-nine names of Allah.
Broader Ninety-nine names of Allah (Sufism, #HH2561).
Followed by Al-Warith (Sufism, #HM0113).
Preceded by Al-Badi (Sufism, #HM4299).

♦ **HM6791d Untrammelled consciousness** (Buddhism)
Unfettered mind
Description Here consciousness is unperturbed by greed nor does it waver in lustful desire.
Context One of the sixteen roots or modes of unperturbedness of mind listed in Hinayana Buddhism as a basis for supernormal powers.
Related Supernormal powers (Buddhism, #HH5652).

♦ **HM6870d Understanding as knowledge of suffering** (Buddhism)
Understanding with reference to the truth about ill
Context On the Path of Purification of Hinayana Buddhism, panna (understanding) is considered as of one kind (monad), of two kinds (dyads), of three kinds (triads) or of four kinds (tetrads). There are five dyads, four triads and two tetrads. All have the characteristic of penetrating the individual essences or true nature of states (monad). In the first tetrad, this is one of the four kinds of understanding as knowledge of the four noble truths.
Broader Understanding (Buddhism, #HM4523).
Related Four noble truths (Buddhism, #HH0523)

Understanding as knowledge of the origin of suffering (Buddhism, #HM7290)
Understanding as knowledge of the cessation of suffering (Buddhism, #HM8163)
Understanding as knowledge of the way leading to cessation suffering (Buddhism, #HM7645).

♦ **HM6898c Non–anger** (Hinduism)
Akrodha
Description The anger which would normally arise at the chastisement or injury by another person is instantly suppressed. There is freedom from the mental perversion generated by oppression by another and freedom from mental agitation.
Related Anger (Hinduism, #HM8665) Treasures of the godly (#HH0364).

♦ **HM6901d Plane of bliss** (Leela)
Anand–loka
Description Rather than a feeling, bliss is the source of which feelings are manifestations. Experienced through wisdom or by making the long journey up through the game, impossible to describe, bliss is in fact always there at the heart of being.
Context The 66th state or square on the board of *Leela*, the game of knowledge, appearing in the eighth row.
Broader Beyond the chakras: the gods themselves (Leela, #HM1141).
Preceded by Knowledge (Leela, #HM3846).

♦ **HM6920d Knowledge of the path of non–return** (Buddhism)
Description The noble disciple, a once–returner, attenuates greed for sense–desires and ill will. Seeing the five aggregates as impermanent, painful and not self, he passes through the series of insights terminating in change–of–lineage. The path of non–return arises; knowledge associated with it is the knowledge of the path of non–return. This is the fifth noble person Supramundane path moment non–return consciousness is followed by two or three fruition moment non–return consciousnesses. This is the sixth noble person. After death he appears elsewhere and, attaining complete extinction, never returns to this world through rebirth linking.
Related Purification by knowledge and vision (Buddhism, #HH3025).
Followed by Knowledge of the path of arahantship (Buddhism, #HM7055).
Preceded by Knowledge of the path of once–return (Buddhism, #HM7563).

♦ **HM6932d Cleansing of the heart** (Sufism)
Tasfiya–i qalb
Description Love for the ephemeral world and the attendant worrying over griefs and sorrow are replaced in the heart by love of God. Self–mortification and austerity are essential conditions for spiritual progress, abstaining from all objects but God. This cleansing may involve physical ablutions and ritual prayers, including some repeated during the night.
Context The second of four contemplative disciplines the salik (seeker after God) must pass through to obtain ma'rifa. Various Sufi orders may differ in approach but all require the same spiritual concentration; manifestation in its many forms may be dealt with in many ways but the realization of spiritual reality behind the forms is the same.
Broader Awareness of the mystic journey (#HM2900).
Related Subtle faculties (Sufism, #HH6282) Alertness of the heart (Sufism, #HM2792).
Followed by Cleansing the sirr (Sufism, #HM5353).
Preceded by Purification of the self (Sufism, #HH5092).

♦ **HM6956d Concentration partaking of stagnation** (Buddhism)
Concentration partaking of stability
Description Having attained the first jhana, the mindfulness in conformity with the concentration is stationary and does not progress.
Context According to Hinayana Buddhism, concentration is considered as of one kind (monad), of two kinds (dyads), of three kinds (triads), of four kinds (tetrads) or of five kinds (pentad). In the fourth tetrad, where concentration is associated with different conditions of wisdom or understanding, this is the second concentration.
Broader Concentration (Buddhism, #HM6663).
Related Concentration partaking of diminution (Buddhism, #HM8002)
Concentration partaking of distinction (Buddhism, #HM7363)
Concentration partaking of penetration (Buddhism, #HM5930).

♦ **HM6989c Sense of the presence of God** (Christianity)
First degree of prayer
Description Madame Guyon describes the sense of the presence of God as being the principal exercise, recalling the senses when they wander. It is practised by meditative prayer and by meditation. Retiring within one's self becomes increasingly easy with habit and the grace of God.
Refs Guyon, J M B de la Mothe *A Short Method of Prayer and Spiritual Torrents* (1875).
Related Presence of God (#HM2961) Meditative reading (#HH5550)
Practice of the presence of God (#HH0992) Christian meditation (Christianity, #HH5023).
Followed by Prayer of simplicity (Christianity, #HM0238).

♦ **HM6992d Reality transfer**
Description This has been postulated (Arthur J Deikman) as an explanation of some of the perceptions associated with the mystic experience. Stimuli of the inner world – thoughts and images – become real.
Broader Deautomatization and the mystic experience (#HM4398).

♦ **HM6993f Delusions of jealousy**
Description This is a morbid conviction that one's husband or wife is unfaithful. The most innocuous or unrelated events are taken as proof of this behaviour, although real evidence of infidelity may be ignored.
Broader Delusive consciousness (#HM2600).

♦ **HM6995f Gustatory hallucination**
Broader Hallucination (#HM4580).
Related Clairgustance (Psychism, #HM0412).

♦ **HM6997d Discursive reasoning** (Christianity)
Lower knowing — Mundane knowledge
Description This is the kind of knowledge which draws logical conclusions from data received from the senses and from images in the mind.
Context One of two ways of knowing distinguished by Meister Eckhart.
Broader Ways of knowing (Christianity, #HH1616).
Related Samvriti (Buddhism, #HM0652) Spiritual knowledge (Christianity, #HM7554).

♦ **HM7002d Specific neutrality** (Buddhism)
Tatra–majjhattata (Pali) — Equanimity
Description This refers to a being neutral with regard to states of consciousness and their concomitants which have arisen. Its characteristic is the even carrying on of consciousness and its mental concomitants; its function is the prevention of deficiency or excess, of checking partiality or partisanship; it manifests as neutrality.
Context One of the formations aggregate (mental coefficients) of Hinayana Buddhism, being listed among the "or–whatever" states, and as profitable primary (always present in any profitable or profitable–resultant consciousness).
Broader Awareness as mental–formation group of conscious existence (Buddhism, #HM2050).
Related Equanimity (Buddhism, #HM7769).

♦ **HM7007d Perceptual expansion**
Description This has been postulated (Arthur J Deikman) as an explanation of some of the perceptions associated with the mystic experience. Undoing of the psychic structure through deautomatization allows increased detail and sensation to be experienced under increased attention. It may include awareness of added dimensions to the total stimulus array which would be defined as perceptual expansion.
Broader Deautomatization and the mystic experience (#HM4398).

♦ **HM7019g Visual agnosia**
Mind–blindness
Description Although the function of an object may not be apparent when it is seen, the individual immediately recognizes what it is for when it is handled. It may arise only for small objects, or for objects placed in an unfamiliar context.
Broader Agnosia (#HM8932).

♦ **HM7021d Al–Razzaq** (Sufism)
Description The believer is aware of Allah as the sustainer, providing physical and spiritual sustenance. One must first seek and find elements of sustenance in everything, particularly spiritual sustenance from scripture.
Context A mode of mystical awareness in the Sufi tradition associated with the seventeenth of the ninety–nine names of Allah.
Broader Ninety–nine names of Allah (Sufism, #HH2561).
Followed by Al–Fattah (Sufism, #HM0002).
Preceded by Al–Wahhab (Sufism, #HM0069).

♦ **HM7022d Al–Barr** (Sufism)
Description The believer is aware of Allah as perfect doer of good and source of all goodness. He rewards good deeds tenfold but never punishes bad deeds more than the sin committed, delaying punishment to allow for repentance and reparation. Do good to Allah's creation and you will see Him reflected in you.
Context A mode of mystical awareness in the Sufi tradition associated with the seventy–ninth of the ninety–nine names of Allah.
Broader Ninety–nine names of Allah (Sufism, #HH2561).
Followed by Al–Tawwab (Sufism, #HM4334).
Preceded by Al–Muta'ali (Sufism, #HM8019).

♦ **HM7024d Contentment** (Sufism)
Qana'a
Context The fifth stage on the path to sainthood of the Naqshbandiyya order.
Broader Suluk of the Naqshbandiyya order (Sufism, #HM4356).
Followed by Tawakkul (Sufism, #HM0807).
Preceded by Piety (Sufism, #HM5339).

♦ **HM7045c Transference love**
Erotic transference — Therapeutic love
Description Transference is the general term designating some feelings that a patient develops for an analyst during the course of therapy. Transference love is that mixture of tender, erotic and sexual feelings which forms part of any positive transference (which may be partially reciprocated in a counter–tranference). It can be experienced as a very powerful reality. The experience of such a reality is often sufficient to effect a cure, known as a "transference cure". The existence of such a transference is considered a prerequisite for analytical change. Nevertheless the subjective experience of transference love bears many similarities to that of romantic love, especially the deep connection with the person's innermost desires, feelings and imaginative powers. This love can create both the desire and the vehicle for fundamental change.
Refs Person, Ethel Spector *Love and Fateful Encounters* (1990).
Broader Love consciousness (#HM2000).
Related Transference (#HH0668) Romantic love (#HM5087)
Love of the whole (#HM1367).

♦ **HM7055d Knowledge of the path of arahantship** (Buddhism)
Description The noble disciple, a non–returner, attenuates greed for sense–desires and ill will. Seeing the five aggregates as impermanent, painful and not self, he passes through the series of insights terminating in change–of–lineage. The path of arahantship arises; knowledge associated with it is the knowledge of the path of arahantship. This is the seventh noble person. Supramundane path moment arahantship consciousness is followed by two or three fruition moment arahantship consciousnesses. This is the eighth noble person. He is one of the great ones, all cankers destroyed, bearing his last body, his burdens laid down. He has reached his goal and destroyed the fetters of becoming, liberated with right or final knowledge.
Related Purification by knowledge and vision (Buddhism, #HH3025).
Preceded by Knowledge of the path of non–return (Buddhism, #HM6920).

♦ **HM7089d Superior saintship** (Sufism)
Wilayat–i–'ulya — Saintship of the angels — Wilayat–i–mala'ika
Description In the Naqshbandiyya order this state follows the stages of major saintship. Contemplation is on al–Batin, the Inward, and the three elements of water, fire and air are purified and perfected. Selfhood is annihilated although the seeker remains conscious in union with God. In the illumination of being, the *dhat* of *Allah*, which may sometimes be seen, then the lover may lose all consciousness.
Related Al–Batin (Sufism, #HM1209).

♦ **HM7091d Samma–sambodhi** (Buddhism)
Description Utter and complete enlightenment as known by Gautama Buddha and the other Buddhas in other world cycles.
Context One of three types of enlightenment recognized in Theravada Buddhism.
Broader Sambodhi (Buddhism, #HM7195) Steps to enlightenment (Buddhism, #HH4019).

♦ **HM7092c Receiving** (Buddhism)
Description The element of eye, ear, nose, tongue or body consciousness has arisen and ceased, and the mind–element arises and receives their object, profitable or moral resultant from profitable or moral resultant, unprofitable or immoral resultant from unprofitable or immoral resultant. This are therefore two kinds of resultant consciousness.
Context In Hinayana Buddhism, this is the ninth mode of occurrence of consciousness in which the 89 kinds of consciousness proceed.

Broader Modes of occurrence of consciousness (Buddhism, #HM6720).
Related Mental consciousness (Buddhism, #HM2838)
Indeterminate consciousness in the sense sphere – resultant (Buddhism, #HM5721).
Followed by Examining (Buddhism, #HM7324).
Preceded by Sense mode of consciousness occurrence (Buddhism, #HM4389).

♦ **HM7098d Inbisat** (Sufism)
Expansion — Restraint
Description The hallmark of this state is audacity before God. Only on receiving a vision of the forgiving king will the devotee consent even to enter the supreme heaven.
Context According to Shaykh Abu Sa'id ibn Abi'l-Khayr, this is the 37th of 40 stations or maqamat the Sufi must possess for his journey on the path of Sufism to be acceptable.
Broader Stations of consciousness – ibn-Abi'l-Khayr (Sufism, #HM2424).
Followed by Tahqiq (Sufism, #HM5563).
Preceded by Tafrid (Sufism, #HM6762).

♦ **HM7099d Al-Hafiz** (Sufism)
Description The believer is aware of Allah as the preserver, remembering and protecting. It is He that placed in creatures an instinct for survival, protecting mankind by teaching that what is bad for them is unlawful, sending His prophets and holy scriptures so that they may learn wisdom and the law and be protected from physical or spiritual harm.
Context A mode of mystical awareness in the Sufi tradition associated with the thirty-eighth of the ninety-nine names of Allah.
Broader Ninety-nine names of Allah (Sufism, #HH2561).
Followed by Al-Muqit (Sufism, #HM3171).
Preceded by Al-Kabir (Sufism, #HM4437).

♦ **HM7100c Ecstatic perception** (Hinduism)
Yukta perception
Description The internal organ of mind (manas), strengthened and perfected by merits due to the practice of meditation, perceives in valid perception the essential nature of one's true self, other selves, ether, space, time, atoms, air, manas, qualities, actions, generalities, particularities, inherences. This is distinguished from the common perception of sensible objects through the senses in a "here and now" manner.
Context One of two kinds of yogic perception distinguished by Prasastapada.
Broader Yogic perception (Hinduism, #HM5005).
Related Nonecstatic perception (Hinduism, #HM0239).

♦ **HM7109d Introspection** (Buddhism)
Sva-samvedana
Description In the Sautrantika-Yogacara school, knowledge is seen as illuminating itself as well as its object – cognition of an external object is also cognition of that cognition. The feeling of personal identity accompanies every state of consciousness, every perception of an external object, and that is a feeling of the condition of the ego. Therefore there is an awareness of knowledge that is a direct, non-constructed, non-illusory sense perception. Some hold that this can never be indifferent or disinterested, the ego is always to some extent emotional, always desiring or shunning. This stream of consciousness is uninterrupted, even in deep sleep or trance, and is of its essence a preparation for action. The ego is thus the element of interestedness which accompanies every conscious state.
This approach is contrasted with the view of mental consciousness as the sixth or inner sense, perceiving special objects in the same way as the other five senses perceive external objects, so that there is always the triad of soul-sense organ-object of the senses, whether consciousness is seen as the essence of the soul or a quality of the soul. Again, some schools hold that consciousness is a passing phenomenon arising in the soul through interaction with the inner sense organ. In Nyaya-Waiseika the soul is held only to cognize through the senses for external as well as internal objects, perception giving rise to self-perception, that is, perception of perception. Cognition of the unconscious soul is then the ego. Even in Hinayana Buddhism, where there is no "soul", there still exists the triad consciousness-organ-object.
Where introspection is not interpreted on the pattern of external perception, there is no division of the apprehending part into two, no further subject and object. The feeling of personal identity is not followed by an image of the ego in the same way that sense perception is followed by an image of what is perceived, it accompanies every state of consciousness.
Related Mental sensation (Buddhism, #HM6644) Mental consciousness (Buddhism, #HM2838).

♦ **HM7112d Al-Akhir** (Sufism)
Description The believer is aware of Allah as the last, having no beginning and no end, nothing can be compared with him, when nothing else exists He will still exist. The cycle of existence begins and ends in Him and to Him all things return. Al-Awwal and al-Akhir, the first and the last, are considered together, they are like a circle in which the first and last are one. All the believer is and has is His and will return to Him. It has been lent to him and he will have to account for their use.
Context A mode of mystical awareness in the Sufi tradition associated with the seventy-fourth of the ninety-nine names of Allah.
Broader Ninety-nine names of Allah (Sufism, #HH2561).
Followed by Az-Zahir (Sufism, #HM4042).
Preceded by Al-Awwal (Sufism, #HM6139).

♦ **HM7114d Knowledge of appearance as terror** (Buddhism)
Knowledge of the presence of fear
Related Purification by knowledge and vision of the way (Buddhism, #HH3550).
Followed by Knowledge of contemplation of danger (Buddhism, #HM7297).
Preceded by Knowledge of contemplation of rise and fall (Buddhism, #HM3723).

♦ **HM7115d Tranquillity due to insight** (Buddhism)
Repose due to insight
Description Body and mind are tranquil and there is superhuman delight. The meditator, thinking that such tranquillity has never arisen before, believes he has reached fruition, interrupts the course of insight. This is karmically unprofitable, because he mistakes the path.
Context One of the ten imperfections of insight of Hinayana Buddhism.
Broader Imperfection of insight (Buddhism, #HM9722).

♦ **HM7119d Wisal** (Sufism)
Union with God
Description Although physically in this world, the heart of the devotee is with God.
Context According to Shaykh Abu Sa'id ibn Abi'l-Khayr, this is the 32nd of 40 stations or maqamat the Sufi must possess for his journey on the path of Sufism to be acceptable.
Broader Stations of consciousness – ibn-Abi'l-Khayr (Sufism, #HM2424).
Related Mystic union with God (Christianity, #HM3889).
Followed by Kashf (Sufism, #HM1452).
Preceded by Tafakkur (Sufism, #HM6227).

♦ **HM7120d Sad feeling** (Buddhism)
Grief
Description The characteristic of sad feeling is experiencing an undesirable object; the function is to in some way make use of or exploit the undesirable aspect; the manifestation is mental affliction; the proximate cause is the heart-basis.
Context In Hinayana Buddhism, one of the five ways in which feeling is analysed in the feeling aggregate.
Broader Feeling aggregate (Buddhism, #HM4983).
Related Grief (#HM2685) Grief (Hinduism, #HM9392)
Equanimity (Buddhism, #HM7769) Joyful feeling (Buddhism, #HM8737)
Painful feeling (Buddhism, #HM8010) Pleasant feeling (Buddhism, #HM6722).

♦ **HM7121c Dwelling in the second jhana** (Buddhism)
Description Having obtained mastery of the first jhana, the meditator is aware of of the hindrances still being near and of the grossness of applied and sustained thought. The jhana factors are reviewed with full awareness; happiness, bliss and unification appear peaceful. Bringing the subject of meditation to mind, the process of appearance of absorption is gone through again. Stilling applied and sustained thought, the second jhana is entered and dwelt in. These two factors are abandoned, not at the moment of access (as with the hindrances in the first jhana), but at the moment of actual absorption. Now there are three factors – happiness or rapture, bliss, unification or collectedness of mind. Thoughts are no longer discursive, but there is still a feeling of energy, comfort and trust.
This jhana is particularly marked by faith or confidence and singleness of mind. Like the first jhana it is said to be good in three ways: beginning (purification of the way), which is *access*; middle (intensification of equanimity), which is *absorption*; end (satisfaction), which is *reviewing*. It also has ten characteristics. Again, through practice, the meditator acquires mastery in the second jhana through the habits of: adverting to the jhana; attaining the jhana; resolving and steadying the duration of the jhana; emerging from the jhana; reviewing the jhana. He can then end his attachment to the second jhana and commence doing what is necessary to attain the third.
Broader Jhana (Buddhism, Pali, #HM7193) Dhyana (Hinduism, Buddhism, #HM0137).
Related Appearance of absorption (Buddhism, #HM1618)
Dwelling in the fivefold jhana (Buddhism, #HM6553)
Concentration of the third jhana of five (Buddhism, #HM8532)
Concentration of the second jhana of four (Buddhism, #HM4380)
Second trance of the fine-material sphere (Buddhism, #HM2038).
Followed by Dwelling in the third jhana (Buddhism, #HM5643).
Preceded by Dwelling in the first jhana (Buddhism, #HM4298).

♦ **HM7131f Fugue state**
Dual personality
Description In addition to hysterical amnesia, when the identity is forgotten, here the person also escapes from everyday routines and pressures. Although apparently normal, he or she wanders about with no obvious goal or becomes a tramp. Coming out of the state as though from a trance, the person then remembers his or her previous life but not the period of time the fugue state lasted. During the fugue state the person may actually take on a new identity and life, which is forgotten as soon as the state comes to an end. This is the origin of the expression *'dual personality'*.
Broader Anomalies in integrity of consciousness (#HM4521).
Related Sub-personalities (#HM0463) Hysterical amnesia (#HM7608)
Multiple personality (#HH0763) Twilight consciousness (#HM2406).

♦ **HM7132d Affects**
Description Affects have a subjective, feeling component which almost always involves a pleasurable or unpleasurable quality (except for feelings which involve detachment or isolation), and thus a motivational quality. They have a cognitive component which involves ideas and fantasies linked to the affective state as it develops and organized around themes related to its motivational quality. And there is a physiological component mediated both through the autonomic nervous system (blushing, crying, rapid pulse and so on) and through the voluntary nervous system (changes in posture, facial expression or tone of voice). The physiological response patterns from which affects arise are said to be nine: surprise, interest, joy, distress, anger, fear, shame, contempt, disgust. These are universal, prominent and may be identified in the first year of life. Because initial biological response becomes linked to encoded memory traces, familiar perceptive patterns mobilize the appropriate affective response in anticipation of what the individual has come to expect by association. Affects are usually closely linked to object representations, self representations and fantasies related to drive states. They may or may not be drive-related or involved in conflict.
An important adaptive function of affects is to alert and prepare the individual for appropriate response to the internal or external environment and to communicate the individuals response to others, thus eliciting in particular the response of those in a caretaking role. Perceptions related to stimuli and their implications (evaluated, integrated and responded to in line with previous experience) determine the nature of the feeling state. Because derivatives of affects may elicit feelings and associations which are painful or may signal danger, they may be dealt with by a number of defensive manoeuvres.
Broader Emotion (#HH0819).
Narrower Dissociation of affect (#HM6087).
Related Mood (#HM1748).

♦ **HM7143c Spiritual attentiveness** (Christianity)
Broader Attention (#HM1817).
Related Quiet attentiveness (#HM4383).

♦ **HM7154d Understanding consisting in what is reasoned** (Buddhism)
Understanding by way of imagination — Understanding by what is thought out
Description This is understanding acquired by one's own reasoning and not from hearing another person. It may be through ingenuity in work, craft, art or science; it may arise from pondering over things, concerning kamma, conformity with truth, conformity with axioms.
Context On the Path of Purification of Hinayana Buddhism, panna (understanding) is considered as of one kind (monad), of two kinds (dyads), of three kinds (triads) or of four kinds (tetrads). There are five dyads, four triads and two tetrads. All have the characteristic of penetrating the individual essences or true nature of states (monad). In the first triad, this is compared with understanding consisting in what is learned or in development.
Broader Understanding (Buddhism, #HM4523).
Related Understanding consisting in development (Buddhism, #HM7826)
Understanding consisting in what is learned (Buddhism, #HM5298).

♦ **HM7171d Abstinence from verbal misconduct** (Buddhism, Pali)
Description Together with abstinence from bodily misconduct and abstinence from wrong livelihood, this is the mind's adverseness from evil-doing and has the characteristic of non-transgression, of not treading, in its particular field. Its function is shrinking from this field; its manifestation is not doing such things; its proximate cause includes faith, conscience or sense of shame, shame or sense of blame, contentment (fewness of wishes).
Context One of the formations aggregate (mental coefficients) of Hinayana Buddhism, being

listed among the inconstant states, and as secondary profitable (sometimes present in any profitable or profitable–resultant consciousness).
Broader Awareness as mental–formation group of conscious existence (Buddhism, #HM2050).
Related Abstinence from bodily misconduct (Buddhism, #HM5600)
Abstinence from wrong livelihood (Buddhism, Pali, #HM7887).

♦ HM7174g Punch–drunk
Description This twilight state of organic origin may involve the individual feeling confused and disoriented, there may be clouding of consciousness. The boxer, for example, may box with his usual skill but have no memory of having done so, or break off suddenly and parade round the ring.
Broader Twilight consciousness (#HM2406).

♦ HM7186g Delirium
Description This arises in a number of acute organic states. There is a dreamlike consciousness with diminished attention and fragmented conceptual thought. Perception is disturbed, confused with memory images. There may be hallucinations, pseudo hallucinations, delusions, illusions. Toxic poisoning often produces delirium.
Broader Diminished clarity of awareness (#HM6201).
Narrower Delirium tremens (#HM7805).
Related Oneroid states (#HM5097)
Modes of awareness associated with use of opium and similar drugs (#HM4060).

♦ HM7193b Jhana (Buddhism, Pali)
Description Derived from the sanskrit term dhyana, jhana has a wider meaning. It not only implies the extensive system of mental development of contemplation or meditation, but also the process of transmuting the lower states of consciousness into the higher states, from the material or form worlds through the immaterial or formless worlds to the summit of progress. Specifically, it means: (1) to contemplate a given object or to examine closely the characteristics of phenomenal existence; and (2) to eliminate hindrances or lower mental elements which are detrimental to higher progress, although the former meaning is more generally used.
The jhana are the mental absorptions arising in the material sphere (four or five) and the immaterial sphere (four). The jhana of the immaterial sphere arise when practice continues beyond the kasinas as subjects for meditation. Beyond the fine–material state as subject of concentration, the fourth or fifth jhana, are the immaterial states - boundless space, boundless consciousness, nothingness, neither perception nor non–perception. Preceding all these states is *upacara* or access concentration.
The five psychic factors, vitakka, vicara, piti, sukha and ekagatta are induced by the expulsion of the five hindrances and are attributable to jhana. The hindrances are: sensuous desires; ill–will; sloth and torpor; distraction and agitation; perplexity. Having suppressed hindrances, the meditator progresses through the jhanas, leaving behind jhana factors which at first appear worthy but which are gradually known to be gross. In the first jhana there are applied thought, sustained thought, rapture or happiness, ease or bliss and concentration. All but concentration are progressively abandoned. In the fifth jhana, as ease or bliss is abandoned, there are two factors: concentration and the equanimity or indifference with which it is accompanied.
Context One of ten superhuman qualities described in the Sutta Vibhanga as being special attainments of insight above that of ordinary men.
Refs Buddhaghosa, Bhadantacariya *The Path of Purification* (1980); Buddhaghosa, Bhadantacariya *The Path of Purity* (1975); Griffiths, Paul J *On Being Mindless* (1986); Mahathera, Paravahera Vajiranana *Buddhist Meditation in Theory and Practice* (1987).
Broader Superhuman qualities (Buddhism, #HH9764).
Narrower Dwelling in the first jhana (Buddhism, #HM4298)
Dwelling in the third jhana (Buddhism, #HM5643)
Dwelling in the second jhana (Buddhism, #HM7121)
Dwelling in the fourth jhana (Buddhism, #HM8087)
Dwelling in the fivefold jhana (Buddhism, #HM6553)
First absorption in the immaterial sphere (Buddhism, #HM2110)
Third absorption in the immaterial sphere (Buddhism, #HM2027)
Second absorption in the immaterial sphere (Buddhism, #HM3043)
Fourth absorption in the immaterial sphere (Buddhism, #HM2051).
Related Sukha (Buddhism, #HM2866)
Analysis (Buddhism, #HM5089)
Investigation (Buddhism, #HM1177)
Concentration (Buddhism, #HM6663)
Dhyana (Hinduism, Buddhism, #HM0137)
Fivefold happiness (Buddhism, #HM0747)
Arupaloka consciousness (Buddhism, Pali, #HM2012)
Trances and mental absorptions (Buddhism, #HM2122)
Immaterial states as meditation subjects (Buddhism, #HH3198).

♦ HM7195b Sambodhi (Buddhism)
Description Perfect enlightenment, identical with wisdom.
Broader Enlightenment (#HM2029)
Steps to enlightenment (Buddhism, #HH4019).
Narrower Samma–sambodhi (#HM7091)
Anuttara samyak sambodhi (#HM3263).
Related Essential wisdom (#HM0107).

♦ HM7202c Concentration of the fourth jhana of four (Buddhism)
Description Ease or bliss is abandoned; the fourth jhana has two factors: concentration and the equanimity or indifference with which it is accompanied.
Context According to Hinayana Buddhism, concentration is considered as of one kind (monad), of two kinds (dyads), of three kinds (triads), of four kinds (tetrads) or of five kinds (pentad). In the third tetrad, fourfold according to the four jhana factors, this is the fourth concentration. It is equivalent to the fifth concentration of the pentad.
Broader Concentration (Buddhism, #HM6663).
Related Dwelling in the fourth jhana (Buddhism, #HM8087)
Concentration of the fifth jhana of five (Buddhism, #HM5143).
Preceded by Concentration of the third jhana of four (Buddhism, #HM2284).

♦ HM7203f Passivity experiences
Delusions of passivity
Description Experiences of interference in one's thoughts by an outside agency are termed *'passivity experiences'*. Where the outside agency is identified the term is *'delusions of passivity'*. Passivity experiences are usually symptoms of schizophrenia and rarely arise in 'normal' persons, and then only in stressful situations. Delusions of passivity, although also symptoms of schizophrenia, are more common than passivity experience in 'normal' persons and have been compared with: extra–sensory perception; mediumship; revelation. In these cases, the thoughts are from other living people (ESP); from persons long dead (mediumship) or from God Himself (revelation).
Broader Delusive consciousness (#HM2600).
Related Thought broadcasting (#HM9384)
Alientation of thought (#HM6508)
Experience of revelation (#HM3538)
Thought insertion (Psychism, #HM5479)
Extrasensory perception (ESP, #HM2262).

♦ HM7208f Somatic hallucination
Delusional zoopathy
Description These bodily experiences may refer to pain, to proprioceptive and kinaesthetic sensations, or to sensations within the body or head. A particular case is delusional zoopathy, where the person may experience the body as being invaded by some animal, which can be clearly described and specifically located.
Broader Hallucination (#HM4580).
Related Tactile hallucination (#HM6118).

♦ HM7209d Contrition (Buddhism)
Kaukritya — 'Gyod–pa (Tibetan)
Description Contrition is regret for some action previously done by one's self. Involving ignorance, it interrupts the stability of the mind. It is virtuous when referring to remorse for past sins, non–virtuous when referring to past meritorious activity, neutral when referring to a past activity neither helping nor harming another.
Context One of the four changeable factors referred to in Tibetan Buddhism. These may be virtuous, non–virtuous or neutral depending on circumstances.
Broader Changeable mental factors (Buddhism, #HH0910).

♦ HM7212g Audiogyral illusions
Audiogravic illusions
Description Illusions engendered by false sensations of bodily movement or orientation, due to unusual accelerative forces leading to changes in the apparent source of a sound.
Broader Illusion (#HM2510).

♦ HM7219d Aesthetic enjoyment (Hinduism)
Carvana
Related Aesthetic emotion (Hinduism, #HM0205).

♦ HM7220d Al–Mani (Sufism)
Description The believer is aware of Allah as preventing His creation from harm. Man tries to obtain what he wishes by all possible means but he does not always succeed. Allah al–Mani has prevented the granting of his wish because only He knows the outcome and what is best for His creation.
Context A mode of mystical awareness in the Sufi tradition associated with the ninetieth of the ninety–nine names of Allah.
Broader Ninety–nine names of Allah (Sufism, #HH2561).
Followed by Ad–Darr (Sufism, #HM4542).
Preceded by Al–Mughni (Sufism, #HM3633).

♦ HM7221d Wundersucht (German)
Passion for miracles
Description This is a fascination with the miraculous, whether at the exoteric or esoteric level, the hunger for a way out of the dilemma of existence.

♦ HM7222c Rasti (Sufism)
Rectitude
Description This aspect of love is obedience to the law.
Context The first stage of *mahabbat* (love) according to al–Ansari. The three stages subsume the whole spiritual journey.
Broader Love of God (Sufism, #HM4273).
Followed by Masti (Sufism, #HM6097).

♦ HM7234d Mundane concentration (Buddhism)
Worldly concentration
Description This is collectedness of moral thought or profitable unification of the mind in the three planes, in the worlds of sense, of form and the formless.
Context According to Hinayana Buddhism, concentration is considered as of one kind (monad), of of two kinds (dyads), of three kinds (triads), of four kinds (tetrads) or of five kinds (pentad). In the second dyad, this precedes supramundane concentration.
Broader Concentration (Buddhism, #HM6663).
Related Mundane resultant consciousness (Buddhism, #HM0130).
Followed by Supramundane concentration (Buddhism, #HM4243).

♦ HM7259d Khulla (Islam)
Love
Description As opposed to 'ishq, or passionate love, this is the true devotion of love to one object, the supreme stage of affection, perfected in this life only by Muhammad and by Abraham towards God. It is the highest stage on the lists of Ibn al–Qayyim which comprehend both sacred and profane love, the perfection of love when the beloved permeates the most secret depths of the lover. In sacred terms, all secondary loves are consciously subordinated to love for God.
Related Walah (Islam, #HM8303)
Tatayyum (Islam, #HM5876).

♦ HM7268c Impulsion (Buddhism)
Apperception
Description This occurs in the case of a large or vivid object (with a life of 14 conscious moments).
In the case of the "five–doors" (senses) then, immediately following determining, there are six or seven apperceptions or impulsions with respect to the object as determined. These may be among: the 8 profitable consciousnesses in the sense sphere; the 12 unprofitable consciousnesses in the sense sphere; the 9 (or 8 ?) indeterminate functional or inoperative consciousnesses in the sense sphere (arahants only).
In the case of the "mind–door", then the apperceptions or impulsions arise next to mind–door adverting. One source quotes the mind–consciousness element accompanied by equanimity arising in the indeterminate functional classification of the sense sphere. Another quotes the 29 (or 28) classes of apperceptional consciousness mentioned in the previous paragraph.
Beyond the change–of–lineage stage, there is impulsion or apperception caused by any of: the five profitable and the five functional or inoperative consciousnesses of the fine–material sphere; the four profitable and four functional or inoperative consciousnesses of the immaterial sphere; the four path (profitable) and four fruition (indeterminate, resultant) consciousnesses of the supramundane sphere.
Context In Hinayana Buddhism, this is the 12th mode of occurrence of consciousness in which the 89 kinds of consciousness proceed.
Broader Modes of occurrence of consciousness (Buddhism, #HM6720).
Related Profitable consciousness in the sense sphere (Buddhism, #HM4447)
Unprofitable consciousness in the sense sphere (Buddhism, #HM8375)
Profitable consciousness in the immaterial sphere (Buddhism, #HM4701)
Profitable consciousness in the supramundane plane (Buddhism, #HM4930)
Profitable consciousness in the fine–material sphere (Buddhism, #HM5338)

Indeterminate consciousness in the sense sphere – functional (Buddhism, #HM3852)
Indeterminate consciousness in the immaterial sphere – functional (Buddhism, #HM0282)
Indeterminate consciousness in the supramundane plane – resultant (Buddhism, #HM5129)
Indeterminate consciousness in the fine-material sphere – functional (Buddhism, #HM4761).
Followed by Registering (Buddhism, #HM1646).
Preceded by Adverting (Buddhism, #HM8336) Determining (Buddhism, #HM7632).

♦ **HM7290d Understanding as knowledge of the origin of suffering** (Buddhism)
Understanding with reference to the truth about the origin of ill
Context On the Path of Purification of Hinayana Buddhism, panna (understanding) is considered as of one kind (monad), of two kinds (dyads), of three kinds (triads) or of four kinds (tetrads). There are five dyads, four triads and two tetrads. All have the characteristic of penetrating the individual essences or true nature of states (monad). In the first tetrad, this is one of the four kinds of understanding as knowledge of the four noble truths.
Broader Understanding (Buddhism, #HM4523).
Related Four noble truths (Buddhism, #HH0523)
Understanding as knowledge of suffering (Buddhism, #HM6870)
Understanding as knowledge of the cessation of suffering (Buddhism, #HM8163)
Understanding as knowledge of the way leading to cessation suffering (Buddhism, #HM7645).

♦ **HM7297d Knowledge of contemplation of danger** (Buddhism)
Knowledge of tribulation
Related Purification by knowledge and vision of the way (Buddhism, #HH3550).
Followed by Knowledge of contemplation of dispassion (Buddhism, #HM5364).
Preceded by Knowledge of appearance as terror (Buddhism, #HM7114).

♦ **HM7298c Manahparyayajnana** (Hinduism)
Knowledge of another's thoughts — Rijumati — Vipulamati
Description This is the supersensuous and vividly immediate knowledge of the objects of the mental processes of another person independent of the sense organs or the mind (manas). Rijumati, the perception of present simple thoughts in the mind of another is less pure than vipulamati, which perceives past, future, complex and subtle thoughts in the mind of another. The former may disappear with the coming of omniscience but the latter persists and in it there is great purity of character.
Related Knowledge of penetration of minds (Buddhism, #HM5232).

♦ **HM7308d Yearning** (Hinduism)
Trisna — Thirst
Description This is the emotion which yearns for an object which has not been attained. It is a state of tension between a present state of dissatisfaction and a future state of satisfaction when the (attainable) object is attained.

♦ **HM7324c Examining** (Buddhism)
Investigating
Description When the mind element has arisen and ceased, then the resultant mind-consciousness element arises which examines the objective field which had been received by the mind element. Where the mind-element was with profitable or moral result, then the profitable mind-consciousness element is with joy (desirable object) or indifference or equanimity (neutral object). Where the mind-element was with unprofitable or immoral result, then the mind-consciousness element is unprofitable-resultant. Examining or investigating therefore comprises three kinds of consciousness.
Context In Hinayana Buddhism, this is the tenth mode of occurrence of consciousness in which the 89 kinds of consciousness proceed.
Broader Modes of occurrence of consciousness (Buddhism, #HM6720).
Related Investigation (Buddhism, #HM1177) Mental consciousness (Buddhism, #HM2838)
Mind-consciousness element (Buddhism, #HM6173)
Indeterminate consciousness in the sense sphere – resultant (Buddhism, #HM5721).
Followed by Determining (Buddhism, #HM7632).
Preceded by Receiving (Buddhism, #HM7092).

♦ **HM7332d Resentment** (Hinduism)
Amarsa
Description This is the kind of intolerance which arises from thwarting of effort to attain a desired object.
Related Anger (Hinduism, #HM8665).

♦ **HM7336d Mukhalafat-i nafs** (Sufism)
Opposition to the carnal soul
Description The carnal soul may cry out for favours for the whole of a man's life but will receive only pain and hardship.
Context According to Shaykh Abu Sa'id ibn Abi'l-Khayr, this is the 10th of 40 stations or maqamat the Sufi must possess for his journey on the path of Sufism to be acceptable.
Broader Stations of consciousness – ibn-Abi'l-Khayr (Sufism, #HM2424).
Followed by Muwafaqat (Sufism, #HM7981).
Preceded by Rida (Sufism, #HM4190).

♦ **HM7342f Doppelgänger**
Autoscopy
Description The person sees another person whom he can identify as being himself. Folk-lore and fairy stories abound with such images, and the phenomenon is part of a number of religious beliefs. A shaman or a witch may actually cultivate a double who make take his or her place in some task. The appearance of one's double is associated with disaster, tragedy or death and indeed the experience is related to cases of brain damage or dangerous diseases, although it is experienced by others in cases of stress, fatigue or preceding an epileptic attack. The apparition is usually a mirror image of the self, just out of arm's reach; and the person experiencing it may 'know' how it feels as though it were truly himself. Sometimes the double may be heard or felt. The experience usually lasts for only a few seconds.
Broader Hallucination (#HM4580).
Related Sense of presence (#HM1166) Visual hallucination (#HM0092).

♦ **HM7344c Absence of love** (Hinduism)
Anabhisneha
Description This absence of love may extend even to the body and to life itself, but generally refers to lack of love for objects of desire or for dear ones. There is absence of joy at being with dear ones and of sorrow at being parted from them. God is the indwelling spirit in all persons and all persons are treated alike, not with love or with hatred.
Related Affection (Hinduism, #HM9845) Non-affection (Hinduism, #HM8000).

♦ **HM7345c Kilesappahana** (Buddhism)
Description Elimination of the evils of lust, hate and delusion.
Context One of ten superhuman qualities described in the Sutta Vibhanga as being special attainments of insight above that of ordinary men.
Broader Superhuman qualities (Buddhism, #HH9764).

♦ **HM7363d Concentration partaking of distinction** (Buddhism)
Description Perception and attention are not accompanied by applied thought.
Context According to Hinayana Buddhism, concentration is considered as of one kind (monad), of two kinds (dyads), of three kinds (triads), of four kinds (tetrads) or of five kinds (pentad). In the fourth tetrad, where concentration is associated with different conditions of wisdom or understanding, this is the third concentration.
Broader Concentration (Buddhism, #HM6663).
Related Concentration partaking of diminution (Buddhism, #HM8002)
Concentration partaking of stagnation (Buddhism, #HM6956)
Concentration partaking of penetration (Buddhism, #HM5930).

♦ **HM7383d Ghaflah** (Sufism)
Forgetfulness of God
Description If, during the practice of dhikr, the devotee cannot maintain presence of awareness of reality without the medium of the physical heart, then attention is directed towards the continuous motion in the heart. If this is also lost then, to regain the relation with reality, the devotee should either take a cold bath, or exhale breath from the brain with force, or repeat "Fa'al" several times. Forgetfulness of God should not be tolerated.
Context This is a state which may arise in the method of dhikr described by Shaykh Kalimallah.
Broader Remembrance of God (Sufism, #HM6562).
Related 'Ilm al-basir (Sufism, #HM0476)
Continuous all-pervading knowledge (Sufism, #HM4788).

♦ **HM7384c Melting** (Christianity)
Speaking forth
Description Meister Eckhart speaks of transcendent God continuously melting out into the Trinity as person, as speaking forth, as creating and sustaining the whole universe, and of the individual soul reenacting the same. This is the paradox. Out of the depths of mystery there is a total uttering so that nothing is left unuttered. The expression and the truth expressed are one reality. One only truly lives to the extent that one is caught up in and shares the Trinity. The three persons of the Trinity melt out of each other and into each other in the divine nature which all three share. The process is also referred to as glowing or boiling. All is balanced and controlled. Normally when one acts there is a pouring into the world and into relationships that implies dissipation of energy; but here there is no dissipation, there is a pouring out while inwardly detached, simultaneously being in movement and yet at rest.
Refs Smith, Cyprian *The Way of Paradox* (1987).
Related Way of paradox (Christianity, #HH3335) Spiritual knowledge (Christianity, #HM7554)
Way of transcendence (Christianity, #HH6566).

♦ **HM7434d Concentration due to inquiry** (Buddhism)
Investigation-concentration
Description Unification or collectedness of mind is obtained by making inquiry predominant.
Context According to Hinayana Buddhism, concentration is considered as of one kind (monad), of two kinds (dyads), of three kinds (triads), of four kinds (tetrads) or of five kinds (pentad). In the sixth tetrad, this is the fourth concentration.
Broader Concentration (Buddhism, #HM6663).
Related Concentration due to zeal (Buddhism, #HM4969)
Concentration due to energy (Buddhism, #HM8365)
Concentration due to natural purity of consciousness (Buddhism, #HM7961).

♦ **HM7512c Zanshin** (Japanese)
Relaxed alertness
Description This person remains relaxed and alert in the face of danger and thus able to react very swiftly. It is achieved through the practice of calming the emotions and focusing the attention on the energy centre at the centre of the body; and is part of martial arts training.

♦ **HM7525d Understanding accompanied by equanimity** (Buddhism)
Understanding accompanied by even-mindedness
Description Here understanding is that belonging to two of the classes of profitable or moral consciousness associated with the sense sphere, and belonging to the four kinds of path consciousnesses associated with the fifth jhana of the fivefold method.
Context On the Path of Purification of Hinayana Buddhism, panna (understanding) is considered as of one kind (monad), of two kinds (dyads), of three kinds (triads) or of four kinds (tetrads). There are five dyads, four triads and two tetrads. All have the characteristic of penetrating the individual essences or true nature of states (monad). In the fourth dyad, this is contrasted with understanding accompanied by joy.
Broader Understanding (Buddhism, #HM4523).
Related Understanding accompanied by joy (Buddhism, #HM6245).

♦ **HM7543d Concentration without applied thought or sustained thought** (Buddhism)
Description This is unification in the three jhanas, commencing with the second in the fourfold reckoning and the third in the fivefold reckoning.
Context According to Hinayana Buddhism, concentration is considered as of one kind (monad), of two kinds (dyads), of three kinds (triads), of four kinds (tetrads) or of five kinds (pentad). In the second triad, this is the third concentration.
Broader Concentration (Buddhism, #HM6663).
Preceded by Concentration without applied thought and with sustained thought (Buddhism, #HM5199).

♦ **HM7549b Yatha-bhuta-jnana-darsana** (Buddhism)
Yatha-bhuta-nana-dassana (Pali)
Description This state arises in knowledge that things are the way they are and in perception in line with that knowledge.
Related Vipassana-bhavana (Buddhism, Pali, #HH0680).

♦ **HM7554d Spiritual knowledge** (Christianity)
Higher knowing
Description This way of knowing does not use images or logical reasoning. It unites immediately and intuitively with its object. Love aspires to its goal; intellect unites with it as the fire and heat of love melt out into pure white light. Unlike mundane knowledge, this kind of knowing can attain God.
Context One of two ways of knowing distinguished by Meister Eckhart.
Broader Ways of knowing (Christianity, #HH1616).
Related Melting (Christianity, #HM7384) Discursive reasoning (Christianity, #HM6997)
Contemplative intuitive meditation (#HH0816).

♦ **HM7563d Knowledge of the path of once-return** (Buddhism)
Description The noble disciple, a stream enterer, attenuates greed for sense-desires and ill will. Seeing the five aggregates as impermanent, painful and not self, he passes through the series of insights terminating in change-of-lineage. The path of once-return arises; knowledge associated with it is the knowledge of the path of once return. This is the third noble person. Supramundane path moment once return consciousness is followed by two or three fruition moment once return consciousnesses. This is the fourth noble person. After once returning to this world he will make an end of suffering.
Related Purification by knowledge and vision (Buddhism, #HH3025).
Followed by Knowledge of the path of non-return (Buddhism, #HM6920).
Preceded by Knowledge of the path of stream entry (Buddhism, #HM1088).

♦ **HM7602d Understanding as skill in detriment** (Buddhism)
Understanding as skill in loss
Description This is two-fold understanding – in the diminution of good (advantages) and the increase in harm (disadvantages). Unprofitable things which had not arisen in mind, do not arise; those already arisen are abandoned. Profitable things which had not arisen in mind, arise; those already arisen, prosper, thus increasing development and perfection.
Context On the Path of Purification of Hinayana Buddhism, panna (understanding) is considered as of one kind (monad), of two kinds (dyads), of three kinds (triads) or of four kinds (tetrads). There are five dyads, four triads and two tetrads. All have the characteristic of penetrating the individual essences or true nature of states (monad). In the third triad, this is compared with understanding as skill in improvement or means.
Broader Understanding (Buddhism, #HM4523).
Related Understanding as skill in means (Buddhism, #HM6601)
Understanding as skill in improvement (Buddhism, #HM8290).

♦ **HM7607d Lovingkindness** (Buddhism)
Love
Description Having prepared himself for meditation, the meditator reviews the danger of hate and the advantages of patience or forbearance by means of which hate can be put away. He then commences to develop lovingkindess, but not at first towards those to whom he is antipathetic, to very dear friends, to a neutral person, to an enemy or hostile person or to a member of the opposite sex or to a dead person. In fact, he first develops it towards himself, making the concern he feels for his own welfare and happiness an example of how he feels towards others. He is then pervaded with lovingkindness and may specifically apply his mind to loving one to whom he is antipathetic, a very dear friend, a neutral person, an enemy. He is endued with loving kindness in heart and mind and in all directions. He becomes versatile in the unspecified pervasion of love (five ways), specified pervasion (seven ways) and directional pervasion (ten ways); and eleven advantages arise.
Context One of the four divine abidings or states described as subjects for meditation in Hinayan Buddhism.
Broader Divine abidings as meditation subjects (Buddhism, #HH3534).
Related Pity (Buddhism, #HM0513) Gladness (Buddhism, #HM5224)
Equanimity (Buddhism, #HM7769).

♦ **HM7608f Hysterical amnesia**
Dissociative amnesia
Description Despite the fact that memory is sufficiently unimpaired for the person to look after himself, his personality and abilities are unchanged and he recalls speech, learning, social skills and logical memory, there is total loss of memory as regards identity. Even name, age and address are forgotten.
Broader Amnesia (#HM4897).
Related Fugue state (#HM7131) Pseudologia phantastica (#HM3726).

♦ **HM7621d Tajrid** (Sufism)
Catharsis — Detachment
Description Whether in paradise or hell there is no fear, there is constancy in friendship with Him, nothing of this world is possessed.
Context According to Shaykh Abu Sa'id ibn Abi'l-Khayr, this is the 35th of 40 stations or maqamat the Sufi must possess for his journey on the path of Sufism to be acceptable.
Broader Stations of consciousness - ibn-Abi'l-Khayr (Sufism, #HM2424).
Followed by Tafrid (Sufism, #HM6762).
Preceded by Khidmat (Sufism, #HM8776).

♦ **HM7622d Annihilation of the heart** (Sufism)
Fana-i-qalb — Conscious heart
Description Ordinary consciousness disappears and all activity is referred to God. Everything but God is totally forgotten by the heart, knowledge of external things is lost, all that remains is love for God, for whose love man was created. In this state, the devotee can accept with patience the trials and tribulations of life and is resigned to the will of God. Rank, wealth and the joys of human love are not the objects of his desire, only to be always conscious of God.
Broader Total annihilation (Sufism, #HM5775).
Related Latifa-i-qalb (Sufism, #HM1905).

♦ **HM7626c Nisti** (Sufism)
Annihilation
Context The third stage of *mahabbat* (love) according to al-Ansari. The three stages subsume the whole spiritual journey.
Broader Love of God (Sufism, #HM4273).
Preceded by Masti (Sufism, #HM6097).

♦ **HM7628d Mundane understanding** (Buddhism)
Worldly understanding
Description This is understanding associated with mundane states of consciousness or the mundane path. Mundane insight comprises understanding having limited or exalted (sublime) object and concerned with the sense, fine-material and immaterial spheres.
Context On the Path of Purification of Hinayana Buddhism, panna (understanding) is considered as of one kind (monad), of two kinds (dyads), of three kinds (triads) or of four kinds (tetrads). There are five dyads, four triads and two tetrads. All have the characteristic of penetrating the individual essences or true nature of states (monad). In the first dyad, this is contrasted with supramundane (transcendental) understanding.
Broader Understanding (Buddhism, #HM4523).
Related Supramundane understanding (Buddhism, #HM8838)
Mundane resultant consciousness (Buddhism, #HM0130)
Understanding having a limited object (Buddhism, #HM2495)
Understanding having an exalted object (Buddhism, #HM3296).

♦ **HM7632c Determining** (Buddhism)
Deciding
Description Immediately following examination or investigation, the indeterminate functional or inoperative mind-consciousness element arises and determines the same object. This is one kind of inoperative or functional consciousness by way of determining.
Context In Hinayana Buddhism, this is the 11th mode of occurrence of consciousness in which the 89 kinds of consciousness proceed.
Broader Modes of occurrence of consciousness (Buddhism, #HM6720).
Related Mind-consciousness element (Buddhism, #HM6173)
Indeterminate consciousness in the sense sphere - functional (Buddhism, #HM3852).
Followed by Impulsion (Buddhism, #HM7268).
Preceded by Examining (Buddhism, #HM7324).

♦ **HM7636d Lightness of mental body** (Buddhism)
Buoyancy of mental factors — Buoyancy of kaya (Pali)
Description This refers to the lightness or quickness of the three aggregates of feeling, perception and formation or mental activities. The characteristic is suppressing or quieting heaviness of the mental body, the function is to crush its heaviness. It is manifest as non-sluggishness. Mental factors are the proximate cause and it is the opponent of stiffness and torpor causing heaviness in the mental body. Lightness of mental body and lightness of consciousness are considered together.
Context One of the formations aggregate (mental coefficients) of Hinayana Buddhism, being listed among the constant states which appear in their true nature, and as profitable primary (always present in any profitable or profitable-resultant consciousness).
Broader Awareness as mental-formation group of conscious existence (Buddhism, #HM2050).
Related Lightness of consciousness (Buddhism, #HM6033).

♦ **HM7645d Understanding as knowledge of the way leading to cessation suffering** (Buddhism)
Understanding with reference to the truth about the path leading to the cessation of ill
Context On the Path of Purification of Hinayana Buddhism, panna (understanding) is considered as of one kind (monad), of two kinds (dyads), of three kinds (triads) or of four kinds (tetrads). There are five dyads, four triads and two tetrads. All have the characteristic of penetrating the individual essences or true nature of states (monad). In the first tetrad, this is one of the four kinds of understanding as knowledge of the four noble truths.
Broader Understanding (Buddhism, #HM4523).
Related Four noble truths (Buddhism, #HH0523) Cessation of suffering (Buddhism, #HH2119)
Understanding as knowledge of suffering (Buddhism, #HM6870)
Understanding as knowledge of the origin of suffering (Buddhism, #HM7290)
Understanding as knowledge of the cessation of suffering (Buddhism, #HM8163).

♦ **HM7655c Conditions for consciousness** (Buddhism)
Twenty-four causes for consciousness arising
Description In Hinayana Buddhism, the Path of Purification describes 24 conditions in which states are the cause of or assist in another state arising.
1. *Root-cause condition*: A state which causes another to arise by not rejecting it, or by assisting in its presence or arising, is a root-cause condition.
2. *Object condition*: A state which is the object of another state is its object condition.
3. *Dominant influence (predominance) condition*: The state which is the dominant influence in causing another state to arise is its dominant influence. This may be as object or as co-nascent.
4. *Proximity or immediacy condition*: The state which assists the arising of another state by being its proximate cause (arising immediately before it) is intended here (for example, mind-consciousness element follows mind element, which itself follows eye consciousness).
5. *Contiguity or direct immediacy condition*: This is the same as the previous condition, although some differentiate proximity as referring to aim and contiguity as referring to time.
6. *Co-nascence or coexistence condition*: Here the state assists another to arise by its arising itself (for example, the 4 immaterial aggregates).
7. *Mutuality or reciprocity condition*: Here there is mutuality between two or more states, so that the one consolidates and supports the other (s) (for example, again, the 4 immaterial aggregates).
8. *Support or dependence condition*: Here one state assists the arising of another by being its support, as the eye base is for eye consciousness and so on.
9. *Decisive-support or sufficing condition*: As the previous condition, but here there is cogent reason for the arise of the state so that the one depends on the other. There are: object decisive support, where the object is what is given importance (object-predominance); proximate decisive support, where the proximity condition causes a consciousness proximate to itself to arise as cogent reason; and natural decisive support, where those which are natural or habitual as causing a state to arise are the cogent reason.
10. *Pre-nascence or pre-existence condition*: Here the state which arises first and by its proceeding causes another state to arise is the pre-nascence condition. This condition is 11-fold, both as physical and as object basis for all of the "five-doors", and as heart-basis. Thus the eye-base is the pre-existence condition for eye-consciousness element, the visible-data base a pre-existence condition for mind-element consciousness.
11. *Post-nascence or post-existence condition*: Here a state is consolidated by a post-nascent immaterial state which arises after it has commenced but while it is still present.
12. *Repetition condition*: Here a state has the function of increasing the efficiency and strength of a following state. Preceding profitable or moral states may act as repetition condition for following profitable states, preceding unprofitable for following unprofitable, preceding functional for following functional – the condition is therefore three-fold.
13. *Karma condition*: Here a state assists another by means of action. It is of two kinds, arising from a profitable or inprofitable volition from a previous time, or all co-nascent volition.
14. *Karma-result condition*: Here, effortless calm or quiet in one state assists such quiet in other states. It is a condition for states originating from it during existence, and for associated states; and at rebirth-linking for the kinds of materiality which arise due to previously performed karma and for associated states.
15. *Nutriment or sustenance condition*: These are four kinds of nutriment giving support to material or immaterial states – physical nutriment for the physical body, immaterial nutriment for associated states, etc.
16. *Faculty or controlling influence condition*: There are 22 faculties or controlling influences - the faculties of eye, ear, nose, tongue, body, mind, feminine, masculine, life, bodily bliss or pleasure, bodily pain or ill, joy, grief, indifference or equanimity, faith, energy, mindfulness, concentration, understanding, "I shall come to know the unknown", perfected or final knowledge, final knower – "one who has come to know". All of these, except masculinity and femininity, when assisting the arising of a state by being the dominant influence, are faculty conditions. Eye, ear, nose, tongue and body faculties assist only immaterial states (eye faculty for eye-consciousness element and so on), the rest for material and non-material.
17. *Jhana condition*: The jhana factors are conditions for their associated states, except for pleasant and painful feeling in the case of the two sets of five. Jhana conditions do not apply to

the two sets of five nor to consciousnesses without root-cause.
18. *Path condition*: There are 12 path factors, profitable, unprofitable and indeterminate. These factors are path conditions for the states associated with them. Path conditions do not apply to the two sets of five nor to consciousnesses without root-cause.
19. *Association condition*: Where immaterial states arise from the same physical basis, have the same object, origin and cessation, then they are conditional upon each other by way of the association condition.
20. *Dissociation condition*: This may be pre-nascent, co-nascent or post-nascent and refers to material states assisting immaterial or immaterial assisting material through not having the same physical basis, the same object, origin and cessation.
21. *Presence condition*: Here presence implies present time. For example, the four immaterial aggregates are in this way a condition for each other, as are the four great primaries, descent into the womb mentality for materiality, the eye base for the eye-consciousness element.
22. *Absence or non-presence condition*: Here an immaterial state, by ceasing in contiguity, assists a second immaterial state to arise next.
23. *Disappearance or absence condition*: The states described in the previous condition, because they assist by their disappearance, are a disappearance condition.
24. *Non-disappearance or non-absence condition*: Just as they assist by their presence as presence-condition, states which have not disappeared assist by their non-disappearance condition.
Broader Ignorance (Buddhism, Yoga, Zen, #HM3196).
Related Awareness as mental-formation group of conscious existence (Buddhism, #HM2050).

♦ **HM7656c State of hope** (Christianity)
Description In this state there is the desire to be united with God in heaven together with the expectation or belief of being so united. Such a state does not preclude perfect love for God.
Refs Fénelon, François *The Maxims of the Saints*.
Related Hope (Christianity, #HH0199).

♦ **HM7672d Knowledge of supernormal powers** (Buddhism)
One becoming many — Direct knowledge — Higher knowledge
Description Knowledge may be severally: multiplying one's self to be many and then reverting to one; appearing and vanishing; causing to appear and vanish, and so on. There are seven kinds of object: limited when the object is material, exalted when the object is exalted consciousness; past, future or present; internal when one's own body or mind are object, external when the object is some other form.
Context One of the five kinds of direct or higher knowledge developed in Hinayana Buddhism.
Broader Supernormal powers (Buddhism, #HH5652).
Related Knowledge of penetration of minds (Buddhism, #HM5232)
Knowledge of the divine ear element (Buddhism, #HM5982)
Knowledge of recollection of previous existence (Buddhism, #HM4297)
Knowledge of passing away and reappearance of beings (Buddhism, #HM0748).

♦ **HM7703d Effacement** (Buddhism)
Austerity
Description This is a state of non-greed and non-delusion.
Context In Hinayana Buddhism, one of the five ascetic states.
Broader Non-ignorance (Buddhism, #HM2695)   Ascetic states (Buddhism, #HM5354)
State of not feeling greed (Buddhism, #HM1825).

♦ **HM7712d Al-Wadud** (Sufism)
Description The believer is aware of Allah as loving those who do good, having compassion on them, giving them *fayd* (the ability to receive and achieve truth). Only He is worthy of love. Piety and worship bring faith and devotion, allowing the believer to profit from *fayd*. The soul is aware of itself, one is aware of one's soul and the senses follow the soul which is aware of the whole. When Allah is the only beloved of the soul the senses are in ecstasy with awareness of such perfection.
Context A mode of mystical awareness in the Sufi tradition associated with the forty-seventh of the ninety-nine names of Allah.
Broader Ninety-nine names of Allah (Sufism, #HH2561).
Related Fayd (Sufism, #HM0568).
Followed by Al-Majid (Sufism, #HM5446).
Preceded by Al-Hakim (Sufism, #HM4799).

♦ **HM7760f Delusions of persecution**
Description The individual believes that he and possibly those he loves are the focus of threats and malevolence, of a particular person or an organization. He may also have associated hallucinations. He will see hidden meanings in others' behaviour all referring to his deluded belief.
Broader Delusive consciousness (#HM2600).

♦ **HM7769c Equanimity** (Buddhism)
Upekkha (Pali) — Indifference — Upeksha — Btang-snyoms (Tibetan) — Even-mindedness — Indifferent feeling
Description The Path of Purification details equanimity as of ten kinds:
1. Six-factored: The state of purity is not abandoned when the six kinds of desirable or undesirable objects are presented to the attention.
2. Divine abiding: There is a mode of neutrality towards all beings.
3. Wisdom, enlightenment: There is indifference dependent upon detachment.
4. Energy: There is neither over-strenuousness nor over-slackness.
5. Formations or complexes: Eight kinds of equanimity are said to arise through concentrations and ten through insight.
6. Feeling: There is no pleasure or pain, for example when a profitable consciousness has arisen in the sense-sphere.
7. Insight: There is neutrality in intellectual investigation - whatever exists, whatever has become, is abandoned.
8. Specific neutrality: There is indifference regarding equal efficiency of conascent states.
9. Jhana: There is impartiality regarding even the highest bliss.
10. Purifying: There is purification from all opposition and therefore no interest in stilling opposition, characteristic of the fourth jhana.
The mind is neither slack not excited, it proceeds calmly and serenely, conducting itself evenly towards its object; it does not need to be exerted, restrained or encouraged. Under these circumstances it is viewed with equanimity.
This evenness of mind is a spontaneous abiding of the mind on its object with no opportunity for afflictions to arise, and is associated with non-hatred, non-attachment and non-ignorance. It is an equanimity of application which sets the mind one-pointedly towards the nine states in developing *calm abiding*.
The characteristic of indifferent feeling is feeling neutral; the function is neither the intensifying nor the withering or waning of associated states; the manifestation is peacefulness; the proximate cause is consciousness without happiness or zest.
To approach equanimity as a subject of meditation, the meditator must already have reached the third jhana (or fourth, if the pentad is referred to) in the other three divine abidings. He sees the dangers of these other abidings because of the link with attention given to beings through wishing for their welfare and so on. Resentment or hatred and approval or fawning are not far away. The association with joy is seen as gross. The peace and tranquillity of equanimity are seen as blessings. The meditator arouses equanimity by regarding with even mind a person who is by nature even-minded or neutral. He then extends the practice towards one to whom he is antipathetic, to a very dear friend, to an enemy or hostile person and finally himself. With practice the fourth jhana arises. He becomes versatile in the unspecified pervasion of equanimity (five ways), specified pervasion (seven ways) and directional pervasion (ten ways); and eleven advantages arise.
Context One of the eleven *virtuous mental factors* referred to in Tibetan Buddhism. One of the four divine abidings or states described as subjects for meditation in Hinayan Buddhism. Also in Hinayana Buddhism, one of the five ways in which feeling is analysed in the feeling aggregate.
Broader Feeling aggregate (Buddhism, #HM4983)
Virtuous mental factors (Buddhism, #HH0578)
Divine abidings as meditation subjects (Buddhism, #HH3534).
Related Samatva (Yoga, #HM1586)   Samachittata (#HM2731)
Pity (Buddhism, #HM0513)   Gladness (Buddhism, #HM5224)
Non-hatred (Buddhism, #HM2744)   Sad feeling (Buddhism, #HM7120)
Calm abiding (Buddhism, #HM2147)   Non-ignorance (Buddhism, #HM2695)
Non-attachment (Buddhism, #HM2128)   Lovingkindness (Buddhism, #HM7607)
Joyful feeling (Buddhism, #HM8737)   Painful feeling (Buddhism, #HM8010)
Pleasant feeling (Buddhism, #HM6722)   Skill in absorption (Buddhism, #HH4777)
Specific neutrality (Buddhism, #HM7002)   Faculty of indifference (Buddhism, #HM3803)
Dwelling in the third jhana (Buddhism, #HM5643)
Third trance of the fine-material sphere (Buddhism, #HM2062).

♦ **HM7800d Desire** (Hinduism)
Kama
Description As opposed to attachment to objects already attained, here the emotion is a yearning for objects not currently perceived, a craving for objects not yet attained.
Narrower Lust (#HM4666).
Related Sensual plane (Leela, #HM3627)   Satisfaction (Hinduism, #HM8116)
Desirelessness (Hinduism, #HM4406)   Dissatisfaction (Hinduism, #HM5448).

♦ **HM7805f Delirium tremens**
DTs
Description Associated with alcoholic poisoning, this delirium may often be accompanied by hallucinations of unnaturally large animals or insects. There is negative affect ranging from anxiety to terror.
Broader Delirium (#HM7186)
Modes of awareness associated with alcohol consumption (#HM0134).

♦ **HM7826d Understanding consisting in development** (Buddhism)
Understanding by culture
Description This is understanding that has reached absorption, having been acquired through an attainment of ecstasy in meditative development.
Context On the Path of Purification of Hinayana Buddhism, panna (understanding) is considered as of one kind (monad), of two kinds (dyads), of three kinds (triads) or of four kinds (tetrads). There are five dyads, four triads and two tetrads. All have the characteristic of penetrating the individual essences or true nature of states (monad). In the first triad, this is compared with understanding consisting in what is reasoned or what is learned.
Broader Understanding (Buddhism, #HM4523).
Related Understanding consisting in what is learned (Buddhism, #HM5298)
Understanding consisting in what is reasoned (Buddhism, #HM7154).

♦ **HM7843d Unified consciousness** (Buddhism)
One-centred mind
Description The defilement of variety does not perturb consciousness, it does not waver in a multitude of passions.
Context One of the sixteen roots or modes of unperturbedness of mind listed in Hinayana Buddhism as a basis for supernormal powers.
Related Supernormal powers (Buddhism, #HH5652).

♦ **HM7844b Perfect wisdom** (Buddhism)
Description This wisdom is neither found nor got at, nor is it a property. In perfect wisdom there is no mental activity. Although attention is incompatible with perfect wisdom, since it would pervert its essential being, wise attention is the cause of perfect wisdom.
Related Essential wisdom (#HM0107)   Wise attention (Buddhism, #HM4309)
Mental engagement (Buddhism, #HM3237).

♦ **HM7859d Concentration of difficult progress and swift direct-knowledge** (Buddhism)
Concentration of painful progress and quick intuition
Description Cultivating what is unsuitable, failing to carry out preparatory tasks, overwhelmed by craving, lacking practice in serenity (absorption concentration), having an acutely corrupt nature, all these mean progress is difficult. Cultivating what is suitable, having skill in absorption, not overwhelmed by ignorance, having practice in insight, having keen faculties, all these mean intuition is swift.
Context According to Hinayana Buddhism, concentration is considered as of one kind (monad), of two kinds (dyads), of three kinds (triads), of four kinds (tetrads) or of five kinds (pentad). In the first tetrad, this is the second concentration.
Broader Concentration (Buddhism, #HM6663).
Related Concentration of easy progress and swift direct-knowledge (Buddhism, #HM4977)
Concentration of easy progress and sluggish direct-knowledge (Buddhism, #HM1010)
Concentration of difficult progress and sluggish direct-knowledge (Buddhism, #HM5992).

♦ **HM7887d Abstinence from wrong livelihood** (Buddhism, Pali)
Description Together with abstinence from bodily misconduct and abstinence from verbal misconduct, this is the mind's adverseness from evil-doing and has the characteristic of non-transgression, of not treading, in its particular field. Its function is shrinking from this field; its manifestation is not doing such things; its proximate cause includes faith, conscience or sense of shame, shame or sense of blame, contentment (fewness of wishes).
Context One of the formations aggregate (mental coefficients) of Hinayana Buddhism, being listed among the inconstant states, and as secondary profitable (sometimes present in any profitable or profitable-resultant consciousness).
Broader Awareness as mental-formation group of conscious existence (Buddhism, #HM2050).
Related Right livelihood (Buddhism, #HM1341)
Abstinence from bodily misconduct (Buddhism, #HM5600)
Abstinence from verbal misconduct (Buddhism, Pali, #HM7171).

♦ **HM7902d Consciousness reinforced by faith** (Buddhism)
Mind upheld by faith
Description Unperturbed by faithlessness, consciousness does not waver in absence of faith.
Context One of the sixteen roots or modes of unperturbedness of mind listed in Hinayana Buddhism as a basis for supernormal powers.
Related Supernormal powers (Buddhism, #HH5652).

♦ **HM7921g Sleep** (Buddhism)
Middha — Gnyid (Tibetan)
Description Sleep may be virtuous or non–virtuous. It is non–virtuous when it degenerates virtuous activity. The proper time for sleep is the middle watch of the night, when it should be with a desire to practice virtue and not motivated by affliction. It occurs when, powerlessly, the engagement of sense consciousness with objects is withdrawn inside.
Context One of the four changeable factors referred to in Tibetan Buddhism. These may be virtuous, non–virtuous or neutral depending on circumstances.
Broader Changeable mental factors (Buddhism, #HH0910).
Related Sleep (#HM2980) Torpor (Buddhism, #HM0264).

♦ **HM7961d Concentration due to natural purity of consciousness** (Buddhism)
Mind–concentration
Description Unification or collectedness of mind is obtained by making natural purity of consciousness predominant.
Context According to Hinayana Buddhism, concentration is considered as of one kind (monad), of two kinds (dyads), of three kinds (triads), of four kinds (tetrads) or of five kinds (pentad). In the sixth tetrad, this is the third concentration.
Broader Concentration (Buddhism, #HM6663).
Related Concentration due to zeal (Buddhism, #HM4969)
Concentration due to energy (Buddhism, #HM8365)
Concentration due to inquiry (Buddhism, #HM7434).

♦ **HM7969d Wieldiness of mental body** (Buddhism)
Wieldiness of mental factors — Wieldiness of kaya (Pali)
Description This refers to the wieldiness of the three aggregates of feeling, perception and formation or mental activities. The characteristic is suppressing or quieting unwieldiness of the mental body, the function is to crush its unwieldiness. It is manifest as success in making something the object of the mental body. Mental factors are the proximate cause. It brings faith in things to be trusted and application to beneficial acts, and it is the opponent of hindrances causing unwieldiness in the mental body. Wieldiness of mental body and wieldiness of consciousness are considered together.
Context One of the formations aggregate (mental coefficients) of Hinayana Buddhism, being listed among the constant states which appear in their true nature, and as profitable primary (always present in any profitable or profitable–resultant consciousness).
Broader Awareness as mental–formation group of conscious existence (Buddhism, #HM2050).
Related Wieldiness of consciousness (Buddhism, #HM6556).

♦ **HM7981d Muwafaqat** (Sufism)
Agreement
Description The seeker conforms with all afflictions arising from God.
Context According to Shaykh Abu Sa'id ibn Abi'l–Khayr, this is the 11th of 40 stations or maqamat the Sufi must possess for his journey on the path of Sufism to be acceptable.
Broader Stations of consciousness – ibn–Abi'l–Khayr (Sufism, #HM2424).
Followed by Taslim (Sufism, #HM5290).
Preceded by Mukhalafat–i nafs (Sufism, #HM7336).

♦ **HM7983d Ubhatobhagavimutti** (Pali)
Liberation
Description This refers to liberation both through wisdom and of the mind. It is not a simple combination of *cetovimutti* and *pannavimutti* but rather a means based both on physical and on mental liberation. This liberation leads to *nibbana*.
Broader Mukti (#HH0613).
Related Cetovimutti (Pali, #HM0125) Pannavimutti (Pali, #HM8125).
Followed by Nirvana (Buddhism, #HM2330).

♦ **HM8000c Non–affection** (Hinduism)
Anabhisvanga
Description There is an absence of affection or identification with any entity other than the self. All identification of one's self with objects of the nature of not–self and producing sentient pleasure is renounced. One writer (Baladeva) considers it the unconditional negation of love but that a trace of affection must exist even in non– affection as there is the presence of compassion for all.
Related Affection (Hinduism, #HM9845) Absence of love (Hinduism, #HM7344).

♦ **HM8001d Al–Muqsit** (Sufism)
Description The believer is aware of Allah as the equitable, acting in harmony, justice and fairness. Just as the earth is balanced in relation to the sun, not so close as to burn all living things, nor so far as to freeze all, so is everything in creation balanced whether we can see it or not. In harmony we must not harm another, making peace as will Allah at the last judgement.
Context A mode of mystical awareness in the Sufi tradition associated with the eighty–sixth of the ninety–nine names of Allah.
Broader Ninety–nine names of Allah (Sufism, #HH2561).
Followed by Al–Jami' (Sufism, #HM0009).
Preceded by Dhul–Jalali wal–Ikram (Sufism, #HM6719).

♦ **HM8002d Concentration partaking of diminution** (Buddhism)
Concentration partaking of worsening
Description Having attained the first jhana there is still susceptibility to opposition, as perception and understanding are accompanied by desires of the senses.
Context According to Hinayana Buddhism, concentration is considered as of one kind (monad), of two kinds (dyads), of three kinds (triads), of four kinds (tetrads) or of five kinds (pentad). In the fourth tetrad, where concentration is associated with different conditions of wisdom or understanding, this is the first concentration.
Broader Concentration (Buddhism, #HM6663).
Related Concentration partaking of stagnation (Buddhism, #HM6956)
Concentration partaking of distinction (Buddhism, #HM7363)
Concentration partaking of penetration (Buddhism, #HM5930).

♦ **HM8003b Thatness** (Buddhism)
Tattva — De–kho–na (Tibetan)
Related Steps to enlightenment (Buddhism, #HH4019).

♦ **HM8004c Resting in the Spirit** (Christianity)
Description The gift of infused recollection or prayer of quiet, when there is a mild suspension of the ordinary sense faculties. The gift can be resisted but, when accepted, is experienced with great delight. The person may fall to the floor but is unhurt.

♦ **HM8006g Sopor**
Description As in torpor, there is drowsiness, but strong stimuli can lead to intentional action.
Broader Diminished clarity of awareness (#HM6201).
Related Torpor (#HM1483).

♦ **HM8010d Painful feeling** (Buddhism)
Pain
Description The characteristic of painful feeling is experiencing an undesirable, tangible object; the function is the withering or waning of associated states; the manifestation is physical (bodily) affliction; the proximate cause is the controlling faculty of the body.
Context In Hinayana Buddhism, one of the five ways in which feeling is analysed in the feeling aggregate.
Broader Feeling aggregate (Buddhism, #HM4983).
Related Equanimity (Buddhism, #HM7769) Sad feeling (Buddhism, #HM7120)
Joyful feeling (Buddhism, #HM8737) Pleasant feeling (Buddhism, #HM6722)
Body consciousness (Buddhism, #HM2562).

♦ **HM8019d Al–Muta'ali** (Sufism)
Description The believer is aware of Allah as supreme, the most exalted, exempt from all failure, the source of all treasures and riches, responding as the needs of creation increase. He supplies all needs yet nothing can be forced from Him.
Context A mode of mystical awareness in the Sufi tradition associated with the seventy–eighth of the ninety–nine names of Allah.
Broader Ninety–nine names of Allah (Sufism, #HH2561).
Followed by Al–Barr (Sufism, #HM7022).
Preceded by Al–Wali (Sufism, #HM0525).

♦ **HM8021d Concentration accompanied by equanimity** (Buddhism)
Concentration accompanied by indifference — Concentration accompanied by even–mindedness
Description Here there is unification of the mind in the jhana not associated with concentration with bliss, one in both the fourfold reckoning and the fivefold reckoning. This is one way of experiencing access concentration.
Context According to Hinayana Buddhism, concentration is considered as of one kind (monad), of two kinds (dyads), of three kinds (triads), of four kinds (tetrads) or of five kinds (pentad). In the fourth dyad, this is contrasted with concentration accompanied by bliss.
Broader Concentration (Buddhism, #HM6663).
Related Access concentration (Buddhism, #HM4999)
Concentration accompanied by bliss (Buddhism, #HM4191).

♦ **HM8022d Holy resignation** (Christianity)
Description In this state the soul still thinks more of its own happiness than it will in the future.
Context This is the first modification of sanctification as described by François Fénelon.
Refs Fénelon, François *The Maxims of the Saints.*
Broader Sanctification (#HH0428).
Followed by Holy indifference (Christianity, #HM5776).

♦ **HM8079g Tactile agnosia**
Description Despite the fact that there is no sensory impairment, there is no recognition of objects by touch.
Broader Agnosia (#HM8932).

♦ **HM8080c Five–fold knowledge** (Systematics)
Potential knowledge — Effectual knowledge — Experience of potentiality
Description Meaning and potentiality must be added to activity if the significance of a structure for itself (and for the totality that contains it) is to be experienced. Potentiality is therefore experienced when at least two similar sets of relationships share a common initiating element and thus requires a system of not less than five independent elements. Potentiality endows experience with that which is lacking in the earlier forms of knowledge. Everything that exists has potentialities for actualization that outstrip the relationships that it can sustain within any concrete situation. In this fifth gradation of knowledge the individual opens himself to what might be. It is thus a form of sensitivity to the given as well as to the ungiven. The polarity of actual and potential produces a force which, because the knowledge is objective, makes the resulting action effectual. Potentialities thus become as apparent as actualities. But since some have to be sacrificed to allow the actualization of one, this knowledge is associated with maximum tension.
Context The fifth in a sequence of twelve modes of knowledge, identified by J G Bennett, inspired by G Gurdjieff.
Broader Systematics (#HH2003).
Followed by Six–fold knowledge (Systematics, #HM5187).
Preceded by Four–fold knowledge (Systematics, #HM1991).

♦ **HM8087c Dwelling in the fourth jhana** (Buddhism)
Description Emerging from the third jhana, and reviewing the jhana factors with mindfulness and full awareness, bliss or mental joy seem gross. Again bringing the subject of meditation to mind, the purpose being to abandon the gross factor and obtain the peaceful factors, and knowing that the fourth jhana will arise, the process of appearance of absorption occurs. Pleasure and pain, joy and grief are abandoned without remainder. Having abandoned the factor of bliss, there are two factors to the fourth jhana: equanimity or indifference of feeling and unification or collectedness of mind. Free from all opposites, there is pure and absolute awareness and complete calmness.
Like the first three jhanas, it is said to be good in three ways: beginning (purification of the way), which is *access*; middle (intensification of equanimity), which is *absorption*; end (satisfaction), which is *reviewing*. It also has ten characteristics. Again, through practice, the meditator acquires mastery in the fourth jhana through the habits of: adverting to the jhana; attaining the jhana; resolving and steadying the duration of the jhana; emerging from the jhana; reviewing the jhana.
Broader Jhana (Buddhism, Pali, #HM7193) Dhyana (Hinduism, Buddhism, #HM0137).
Related Appearance of absorption (Buddhism, #HM1618)
Dwelling in the fivefold jhana (Buddhism, #HM6553)
Concentration of the fifth jhana of five (Buddhism, #HM5143)
Concentration of the fourth jhana of four (Buddhism, #HM7202)
Fourth trance of the fine–material sphere (Buddhism, #HM2586).
Preceded by Dwelling in the third jhana (Buddhism, #HM5643).

♦ **HM8092d Unassociated consciousness** (Buddhism)
Dissociated mind
Description Consciousness does not waver in passion, it is not perturbed by defilement.
Context One of the sixteen roots or modes of unperturbedness of mind listed in Hinayana Buddhism as a basis for supernormal powers.
Related Supernormal powers (Buddhism, #HH5652).

♦ **HM8098c Joy** (Hinduism)
Harsa
Description Manifesting as mental enjoyment, this emotion has the characteristic of experiencing a desirable object, with the function of exploiting the desirable object in some way. It is the exalting of the mind arising from union with a beloved person or attainment of a desired object, or from the coming of prosperity or the successful accomplishment of an endeavour leading to a desired end. The person's face is bright, he or she may weep.
Related Joy (#HM1172) Plane of joy (Leela, #HM4376)
Joyful feeling (Buddhism, #HM8737).

♦ **HM8105g Breakdown of automatization**
Anomalous emphasis — Deautomatization
Description In the normal experience of the self there is a global awareness subsuming the various features and having reference to context. The component experiences function in a balanced and interactive way and one is not aware of them individually. However, if attention is focused on features not normally receiving prominence there is an imbalance in relationships. At the physical level, focusing on one part of a skilled task can disrupt the whole activity. What was normally an automatic process becomes stilted. This may arise under conditions of stress, for example during a championship match of some sport or during an emergency when driving a car.
The same approach may be applied to disruption of cognitive skills. Depersonalization, for example, is the sensation arising when normal integrated experience breaks down. There is a change in emphasis of attention due to a shift in hierarchical organization. Such a shift may occur intentionally, such as in contemplative exercise of meditation, or it may arise in response to stress. The hierarchical organization may be that which determines the individual's personality so that, when it breaks down, he will no longer recognize himself.
When the process of deautomatization or depersonalization is intentional, the aim is to dispel blindness and reveal a new perception, variously referred to as enlightenment or illumination.
Refs Reed, Graham *The Psychology of Anomalous Experience* (1988).
Broader Anomalies in awareness of one's self (#HM8186).
Related Enlightenment (#HM2029) Automatization (#HM3906)
Deautomatization (#HH2331)
Anomalies in experience of the unity of self (#HM9132)
Anomalies in experience of the reality of one's self and the environment (#HM2548).

♦ **HM8112d Shame** (Buddhism)
Ottappa (Pali) — Dread of blame — Remorse
Description In Hinayana Buddhism, this is the formation that is ashamed about misconduct and is anxious or agitated about evil; its characteristic is fear or dread of evil and its proximate cause is respect for others.
Context One of the formations aggregate (mental coefficients) of Hinayana Buddhism, being listed among the constant states which appear in their true nature, and as profitable primary (always present in any profitable or profitable-resultant consciousness). Although shame and conscience are considered closely related they are listed as separate formations. Together they are regarded as the 'guardians of the world'.
Broader Awareness as mental-formation group of conscious existence (Buddhism, #HM2050).
Related Remorse (#HM2977) Conscience (Buddhism, #HM1590)
Shamelessness (Buddhism, #HM0649) Sense of shame (Buddhism, #HM2881).

♦ **HM8116d Satisfaction** (Hinduism)
Santosa
Description This is the emotion felt on the gratification of a desire when the desired object is attained.
Broader Happiness (#HM2409).
Related Samtosa (Yoga, #HM2898) Desire (Hinduism, #HM7800)
Dissatisfaction (Hinduism, #HM5448).

♦ **HM8123c Passive state** (Christianity)
Repose in God
Description In a firm, unbroken, continuous act of faith, in a state of divine union, being made one with Christ in God, the person no longer puts forth distinct inward acts but is in deep and divine repose.
Refs Fénelon, François *The Maxims of the Saints.*
Related Faith (#HH0694) Passive way of faith (#HH3412).

♦ **HM8125d Pannavimutti** (Pali)
Liberation by means of wisdom
Description It is said that through wisdom liberation may be achieved, resulting in a state of nibbana, without necessarily achieving complete mastery of enstatic practices. This has also been described as freedom from ignorance.
Broader Mukti (#HH0613).
Related Cetovimutti (Pali, #HM0125) Ubhatobhagavimutti (Pali, #HM7983).
Followed by Nirvana (Buddhism, #HM2330).

♦ **HM8135d Pareidolia**
Pareidolic illusions
Description This form of illusion is indulged in intentionally. It involves seeing images in the fire or in clouds, for example. Although such images usually occur when perception is not very sharp, the illusion still remains when attention is heightened. Although the illusion is made to appear voluntarily, it cannot be removed voluntarily and heightened attention may intensify it.
Refs Reed, Graham *The Psychology of Anomalous Experience* (1988).
Broader Illusion (#HM2510).

♦ **HM8163d Understanding as knowledge of the cessation of suffering** (Buddhism)
Understanding with reference to the truth about the cessation of ill
Context On the Path of Purification of Hinayana Buddhism, panna (understanding) is considered as of one kind (monad), of two kinds (dyads), of three kinds (triads) or of four kinds (tetrads). There are five dyads, four triads and two tetrads. All have the characteristic of penetrating the individual essences or true nature of states (monad). In the first tetrad, this is one of the four kinds of understanding as knowledge of the four noble truths.
Broader Understanding (Buddhism, #HM4523).
Related Four noble truths (Buddhism, #HH0523) Cessation of suffering (Buddhism, #HH2119)
Understanding as knowledge of suffering (Buddhism, #HM6870)
Understanding as knowledge of the origin of suffering (Buddhism, #HM7290)
Understanding as knowledge of the way leading to cessation suffering (Buddhism, #HM7645).

♦ **HM8186b Anomalies in awareness of one's self**
Anomalies in experience of self
Description Experience of self underlies all other experiences and states, and all experiences and states are affected by it. Anomalies arise in respect of:
(1) Experiencing the self as distinct from the outside world.
(2) Experiencing the self as recognized in personal performance;
(3) Experiencing the unity of self;
(4) Experiencing the reality of one's self and the environment.
There may also be anomalies in the emphasis of different components of the self, causing a breakdown in automatization.
Refs Reed, Graham *The Psychology of Anomalous Experience* (1988).
Narrower Breakdown of automatization (#HM8105)
Anomalies in experience of the unity of self (#HM9132)
Anomalies in experience of the self as distinct from the outside world (#HM4754)
Anomalies in experience of the reality of one's self and the environment (#HM2548)
Anomalies in experience of the self as recognized in personal performance (#HM1336).
Related Identity (#HH0875) Self-remembering (#HM2486) Depersonalization (#HM1248)
Modes of awareness associated with schizophrenia (#HM2313).

♦ **HM8232c Sound consciousness**
Description Mainly experienced by jazz and rock musicians but also experienced by composers of classical music, the sound is almost touched, the person almost becomes the sound itself.
Related Sound health (#HH2543).

♦ **HM8235f Cryptomnesia**
Description What in fact is a memory is here experienced as something totally new. A composer may write a tune which he subsequently recognizes as a perfectly well-known tune written by somebody else; a raconteur may tell a joke as his own which was originally told to him by the person to whom he is recounting it. A novel solution to a problem may present itself, which is subsequently realized has either arisen before or actually been suggested by someone else.
Broader Amnesia (#HM4897).

♦ **HM8241d Liberated consciousness** (Buddhism)
Emancipated mind
Description Greed for sense desires does not perturb, there is no wavering in sensual lust.
Context One of the sixteen roots or modes of unperturbedness of mind listed in Hinayana Buddhism as a basis for supernormal powers.
Related Supernormal powers (Buddhism, #HH5652).

♦ **HM8266c Rebirth-linking** (Buddhism)
Reconception
Description When human beings and deities are reborn through the influence of the 8 profitable or moral consciousnesses in the sense sphere, then there are 9 consciousnesses which occur: the 8 kinds of indeterminate resultant consciousness, conditioned or with root cause, in the sense realm; and the indeterminate resultant mind-consciousness element with profitable result and accompanied by indifference or equanimity. Their object is the stored profitable karma or the sign or attribute of karma or the sign or attribute of destiny which appeared at the time of death.
When beings are reborn in a state of loss or woe through influence of the 12 unprofitable or immoral consciousnesses in the sense sphere, the consciousness which occurs as rebirth-linking is the indeterminate resultant mind-consciousness element with unprofitable result. Its object is the stored unprofitable karma or the sign or attribute of karma or the sign or attribute of destiny which appeared at the time of death.
When beings are reborn in the fine-material sphere (form-realm) or immaterial sphere (formless realm), through influence of profitable consciousnesses in the fine-material and immaterial spheres respectively, then the 9 consciousnesses which occur are the 5 indeterminate resultant of the fine-material sphere and the 4 indeterminate resultant of the immaterial sphere. Their object is the sign or attribute of karma (kasina, etc) which appeared at the moment of dying.
There are thus 19 kinds of rebirth-linking consciousness, together with a 20th kind of rebirth-linking which is of unconscious or non-percipient beings. Rebirth-linking conforms to the means and place by which it comes about. Those arising from unprofitable-resultant mind-consciousness element bring rebirth in a state of loss or woe; profitable-resultant brings rebirth in the human world but disabled by being blind, deaf, etc. Rebirth-linking through the 8 principal-resultant consciousnesses with root-cause (conditioned) is among deities of the sense sphere or humans possessing merit. Rebirth linking through the 5 fine-material resultant consciousness brings rebirth in the fine-material Brahma world; through the 4 immaterial resultant consciousnesses, rebirth in the immaterial world.
There are three kinds of object, "past", "present" and neither of these. Non-percipient rebirth has no object. Boundless-consciousness and neither-perception-nor-non-perception have a past object; the 10 rebirth-linking of the sense-sphere have present or past object; the others have neither of these.
Rebirth may be from a happy to an unhappy destiny: (1) When an evil doer who had a happy destiny focusses at death upon his evil deeds or their sign, the train of consciousness follows through to rebirth-linking with this past as object. (2) When, at death, the sign of unhappy destiny comes into focus at the mind-door of an evil doer who had a happy destiny. This is rebirth linking with present object arising from death with past object. (3) When, at death, an inferior object comes into focus at one of the five-doors (senses) causing greed, for instance, there is again rebirth-consciousness with present object from death with past object.
Rebirth may be from an unhappy to a happy destiny: Someone with unhappy destiny, having stored up blameless karma in the sense sphere, focusses at death upon his good deeds or their sign, the train of consciousness follows through to rebirth-linking with this past as object. Blameless karma stored in the more exalted spheres leads to rebirth-linking with past or neither past nor present object, following death with past object. Rebirth may be from a happy to a happy destiny: (1) Blameless karma or its sign comes into focus in the mind-door, the train of consciousness follows through to rebirth linking with this past as object. (2) When, at death, someone with blameless karma has the sign of happy destiny come into focus in the mind-door. This is rebirth linking with present object arising from death with past object. (3) When, at death, someone with blameless karma is presented at the five sense doors with pleasant things, rebirth linking again arises with present object following death with past object. (4) When one who has obtained exalted consciousness, there is no registration of what is in focus in the mind-door, then rebirth-linking consciousness arises with present, past or neither of these object next to death with "neither of theses" object.
Rebirth may be from an unhappy to an unhappy destiny: When an evil doer who had an unhappy destiny focusses at death in the mind-door or the five sense doors upon his evil deeds or their sign, or the sign of the destiny the train of consciousness follows through to rebirth-linking with one of these as object.

Context In Hinayana Buddhism, this is the first mode of occurrence of consciousness in which the 89 kinds of consciousness proceed.
Broader Modes of occurrence of consciousness (Buddhism, #HM6720).
Related Mind-consciousness element (Buddhism, #HM6173)
Mundane resultant consciousness (Buddhism, #HM0130)
Indeterminate consciousness in the sense sphere - resultant (Buddhism, #HM5721)
Indeterminate consciousness in the immaterial sphere - resultant (Buddhism, #HM4982)
Indeterminate consciousness in the fine-material sphere - resultant (Buddhism, #HM0594).
Followed by Life-continuum (Buddhism, #HM6221).
Preceded by Death (Buddhism, #HM3932).

♦ **HM8273c Consciousness-born materiality** (Buddhism)
Description Of the 89 kinds of consciousness grouped in the consciousness aggregate, some give rise to materiality while others do not. Those in the sense sphere giving rise to materiality (and also to postures and to intimation) are 32: the 8 profitable consciousnesses; the 12 unprofitable consciousnesses and 10 of the indeterminate functional (the mind element is excluded). Consciousnesses giving rise to materiality and postures but not intimation are 26: in the fine material sphere, 5 profitable and 5 indeterminate functional; in the immaterial sphere, 4 profitable and 4 indeterminate functional; in the supramundane sphere, all 8, the 4 profitable and 4 indeterminate resultant. Consciousnesses giving rise to materiality only, not postures or intimation, are 19: in the sense sphere the 10 life-continuum consciousnesses - indeterminate resultant, profitable result with root-cause (8); indeterminate resultant without root-cause (2), profitable resultant mind consciousness element accompanied by equanimity, unprofitable resultant mind consciousness element; in the fine-material sphere, the 5 indeterminate resultant consciousnesses; the 3 mind elements (sense sphere, indeterminate resultant profitable result and unprofitable result, and indeterminate functional); plus indeterminate resultant, profitable resultant mind consciousness element accompanied by joy. Consciousnesses not giving rise to materiality, postures or intimation are 16: in the sense sphere, indeterminate resultant without root-cause, the 5 with profitable result and five with unprofitable result (the sense consciousnesses); the rebirth-linking consciousness of all beings, the death consciousness of those whose cankers are destroyed, and the 4 immaterial indeterminate resultant consciousnesses.
Related Dispositions of consciousness (Buddhism, #HM2098)
Awareness of corporeality-group of conscious existence (Buddhism, #HM2108).

♦ **HM8290d Understanding as skill in improvement** (Buddhism)
Understanding as skill in profit — Understanding as skill in increase
Description This is two-fold understanding - in the increase of good (advantages) and the diminution in harm (disadvantages). Unprofitable things which had not arisen in mind, do not arise; those already arisen are abandoned. Profitable things which had not arisen in mind, arise; those already arisen, prosper, thus increasing development and perfection.
Context On the Path of Purification of Hinayana Buddhism, panna (understanding) is considered as of one kind (monad), of two kinds (dyads), of three kinds (triads) or of four kinds (tetrads). There are five dyads, four triads and two tetrads. All have the characteristic of penetrating the individual essences or true nature of states (monad). In the third triad, this is compared with understanding as skill in detriment or means.
Broader Understanding (Buddhism, #HM4523).
Related Understanding as skill in means (Buddhism, #HM6601)
Understanding as skill in detriment (Buddhism, #HM7602).

♦ **HM8303d Walah** (Islam)
Loss of discernment
Context The highest degree of love in the first list of Ibn al-Jawzi.
Related Khulla (Islam, #HM7259) Tatayyum (Islam, #HM5876).

♦ **HM8330d Sultan-al-dhikr** (Sufism)
Description The physical motion of the heart murmuring "Allah" is heard freely and known to arise from the pine-shaped heart. The motion starts to disperse to all parts of the body, although the seeker concentrates only on the heart. It may first arise in the hand, the foot, the head, etc. Eventually the whole body is filled. A variety of states may be experienced, from happiness to dejection and bewilderment, but these are not attended to, all attention is on the dhikr. Even when the whole body murmurs "Allah" there may be greater dominance of one organ over another. When distribution is even there is a feeling of great pleasure experienced as *sultan-al-dhikr*.
Context This is the third state arising in the method of dhikr described by Shaykh Kalimallah.
Broader Remembrance of God (Sufism, #HM6562).
Followed by Remembrance of reality (Sufism, #HM3892).
Preceded by Khilwat dar anjuman (Sufism, #HM4987).

♦ **HM8336c Adverting** (Buddhism)
Five-door adverting — Six-door adverting — Mind-door adverting
Description With the being in a state of life-continuum, the faculties become capable of apprehending an object. Sense data can then be apprehended and there is an impact between the sentient organism and the object, for example between the eye and visible data. The life-continuum is interrupted and wavers. This ceasing in the life-continuum is followed by the arising of the indeterminate functional or inoperative mind-element which effects the function of adverting. This is true for the "doors" of all five senses and is sometimes called "five-door adverting".
Similarly, when any of the six types of object is in focus at the mind "door", then the indeterminate functional mind-consciousness element arises, indifferent (accompanied by equanimity), cutting off the life-continuum and effecting the function of adverting. This is sometimes called "six-door adverting".
As attention controls or regulates the cognitive series or process of consciousness it is "five-door" adverting; as it controls or regulates impulsion or apperception it is "six-door" adverting.
Context In Hinayana Buddhism, this is the third mode of occurrence of consciousness in which the 89 kinds of consciousness proceed.
Broader Modes of occurrence of consciousness (Buddhism, #HM6720).
Related Mental engagement (Buddhism, #HM3237)
Mental consciousness (Buddhism, #HM2838)
Mind-consciousness element (Buddhism, #HM6173).
Followed by Impulsion (Buddhism, #HM7268)
Sense mode of consciousness occurrence (Buddhism, #HM4389).
Preceded by Life-continuum (Buddhism, #HM6221).

♦ **HM8365d Concentration due to energy** (Buddhism)
Description Unification or collectedness of mind is obtained by making energy predominant.
Context According to Hinayana Buddhism, concentration is considered as of one kind (monad), of two kinds (dyads), of three kinds (triads), of four kinds (tetrads) or of five kinds (pentad). In the sixth tetrad, this is the second concentration.
Broader Concentration (Buddhism, #HM6663).
Related Concentration due to zeal (Buddhism, #HM4969)
Concentration due to inquiry (Buddhism, #HM7434)
Concentration due to natural purity of consciousness (Buddhism, #HM7961).

♦ **HM8375c Unprofitable consciousness in the sense sphere** (Buddhism)
Description A total of 12 kinds of consciousness may arise here, rooted in greed, hate or delusion.
Of the eight associated with greed: four are accompanied by joy, that is: accompanied with false or wrong views and unprompted; accompanied with false or wrong views and prompted; unaccompanied with false or wrong views and unprompted; unaccompanied with false or wrong views and prompted. And four are accompanied with equanimity, that is: accompanied with false or wrong views and unprompted; accompanied with false or wrong views and prompted; unaccompanied with false or wrong views and unprompted; unaccompanied with false or wrong views and prompted.
Of the two associated with hate, both are associated with grief and resentment, one is unprompted, the other prompted.
Of the two associated with delusion, both are associated with equanimity. One is associated with uncertainty or doubt, the other with agitation. They arise at times of indecision or wavering.
All 12 give rise to materiality, to postures and to intimation.
Context In Hinayana Buddhism, 89 consciousnesses are enumerated in aggregate (khanda). Of these, 21 are profitable or moral, 12 are unprofitable or immoral and 56 are indeterminate (resultant or functional). The unprofitable all arise in the sphere of sense and desire, whereas profitable and indeterminate arise in sense, fine-material, immaterial and supramundane spheres.
Broader Dispositions of consciousness (Buddhism, #HM2098).
Related Impulsion (Buddhism, #HM7268) Formation of demerit (Buddhism, #HH3226)
Profitable consciousness in the sense sphere (Buddhism, #HM4447)
Indeterminate consciousness in the sense sphere - resultant (Buddhism, #HM5721)
Indeterminate consciousness in the sense sphere - functional (Buddhism, #HM3852).

♦ **HM8532c Concentration of the third jhana of five** (Buddhism)
Description Having emiminated applied and sustained thought, three factors remain in the third jhana: rapture or happiness; ease or bliss; concentration.
Context According to Hinayana Buddhism, concentration is considered as of one kind (monad), of two kinds (dyads), of three kinds (triads), of four kinds (tetrads) or of five kinds (pentad). In the pentad, fivefold according to the five jhana factors, this is the third concentration. It is equivalent to the second concentration in the third tetrad.
Broader Concentration (Buddhism, #HM6663).
Related Dwelling in the second jhana (Buddhism, #HM7121)
Concentration of the second jhana of four (Buddhism, #HM4380).
Followed by Concentration of the fourth jhana of five (Buddhism, #HM4882).
Preceded by Concentration of the second jhana of five (Buddhism, #HM4575).

♦ **HM8566d Equanimity due to insight** (Buddhism)
Indifference due to insight
Description Equanimity or indifference arises, whether about insight, as neutrality about formations; or in adverting, incisive and sharp. The meditator, thinking that such equanimity has never arisen before, believes he has reached fruition, interrupts the course of insight. This is karmically unprofitable, because he mistakes the path.
Context One of the ten imperfections of insight of Hinayana Buddhism.
Broader Imperfection of insight (Buddhism, #HM9722).

♦ **HM8621d Consciousness rid of barriers** (Buddhism)
Unconfined mind
Description Consciousness is not perturbed by the barrier of defilement, it does not waver in the confinement of passion.
Context One of the sixteen roots or modes of unperturbedness of mind listed in Hinayana Buddhism as a basis for supernormal powers.
Related Supernormal powers (Buddhism, #HH5652).

♦ **HM8634 Levitation**
Yogic flying — Ascension through meditation
Description Rigorous spiritual practices, including transcendental meditation, have been claimed to result in levitation of the body with apparent overcoming of the force of gravity. This may be spontaneous and uncontrolled or intentional.
The TM-Sidhi technique of the Maharishi is said so to enhance mind-body coordination that the body fulfils any intention of the mind. This includes lifting effortlessly into the air while seated in the lotus position. Since maximum EEG coherence is created just before the body lifts into the air, yogic flying may be said to be the fullest extension of the highly integrated, effective way of functioning. If mind-body coordination is not perfect then the body rapidly returns to earth and there is the phenomenon of "hopping". A more advanced stage is floating and the final stage is sustained flight. Allowing one's conscious mind to settle down and become identified with the unified field generates coherence in both individual and collective consciousness and the technique is postulated as a basis for world peace.
The TM-Sidhi technique is the most up-to-date version of a phenomenon that has been widely reported throughout history. Not only in Buddhist and Hindu traditions, but also among Christian saints, levitation has been taken to be a sign that the person has reached a high degree of holiness. However, many teachers, including Patanjali, warn that too much attention to siddhis can be a distraction.
Refs Maharishi European Research University *Maharishi's Programme to Create World Peace* (1987).
Broader Siddhis (Yoga, Buddhism, #HH0380).
Related Maharishi effect (#HH3764) Transcendental meditation (TM, #HH0682).

♦ **HM8665d Anger** (Hinduism)
Krodha
Description In the Bhagavad Gita, Krisna says that desire gives rise to attachment, attachment to anger, both being the product of energy of mind (rajas). It is the door of hell leading to destruction of the self. Several writers refer to it as an aversion which may be for hostile objects which are perceived or remembered as painful to the self (Sankara). It is characterized by aversion or inflaming and arises when a desire is obstructed. Anger has been defined as a mental mode arising from subjection or oppression by someone else (Anandagiri). It may give rise to a desire to inflict injury on one's self or the other. It is distinguished from fear, as fear arises when the cause of the obstruction of desire or the cause of pain cannot be countered, while in the case of anger it can. Sexologists look on it as a mental mode rising from the thwarting of the sex impulse. It arises when there is violent obstruction to the sex impulse, to any kind of desire, whether for sexual gratification, for wealth, for performing religious rites, for liberation. Anger is the cause of violence and destroys moral judgement. It can conquered by discrimination which prevents the evocation of anger by destroying its roots.
Related Anger (Leela, #HM1433) Non-anger (Hinduism, #HM6898)
Resentment (Hinduism, #HM7332) Belligerence (Buddhism, #HM2264).

♦ **HM8667c Himma** (Sufism)
Gnosis kardias — Wisdom of the heart
Description The heart is conceived as the locus of the spiritual energy of imagination or spiritual perception. Salvational knowledge is the act of mediating, imagining, projecting and ardently desiring. For a person in a condition of himma, the energy is held to be so powerful as to project extra-psychic being external to the person in that condition. Himma creates as real the figures of the imagination, although their reality is neither hallucinatory nor illusory. The results of that imagination are neither objectively nor subjectively real but belong to to the intermediary mediating realm of the mundus imaginalis.
Refs Avens, Roberts *The New Gnosis* (1984).
Related Gnosis (#HM0413).

♦ **HM8672c Undifferentiated faith**
Pre-stage of faith
Description The seeds of trust, courage, hope and love are fused in an undifferentiated manner. They contend with threats of abandonment, inconsistencies and deprivations in the environment of the infant. The quality of mutuality and the strength of trust, autonomy, hope and courage underlie future faith development while their opposites threaten to undermine it. The fund of basic trust and the relational experience of mutuality with the provider (s) of primary love and care is the emergent strength of faith at this stage. Possible dangers or deficiencies are: the emergence of excessive narcissism, so that there is continued domination of the experience of being central, so that mutuality is distorted; experience of neglect or inconsistencies may cause the infant to be locked in a pattern of isolation and failed mutuality. The convergence of thought and language opens up the use of symbols in speech and ritual play and begins the transition to Stage 1.
Context The pre-stage in the system of faith development described by James Fowler.
Refs Fowler, J W *Stages of Faith* (1981).
Broader Stages of faith (#HH2097).
Followed by Intuitive-projective faith (#HM1425).

♦ **HM8673c Mark of calm** (Buddhism)
Samatha-nimitta
Description In the process of Buddhist meditation, an object is selected, such as a kasina device. The physical device is called *parikammananimitta*, the mark of the preliminary exercise.. Fixing the eyes on the kasina during prolonged contemplation leads to retaining a mental image of the kasina which is an exact copy of the original and which may be visualized as clearly as the concrete object. This is referred to as mark of grasping or *uggaha-nimitta*. Concentration on this mental image is *parikamma samadhi*, preliminary concentration. Further concentration leads the mental image to give way to an abstract idea or concept, divested of phenomenal reality and free from all the faults of the original object. This is *patibhaga-nimitta*, the mark of the equivalent or after-image. This arises when the mind has reached *upacara-samadhi* or access concentration. These three nimittas are the objects of the three stages - parikamma, upacara and appana - of intense concentration obtained in the development (bhavana) of meditation.
Related Bhavana (Buddhism, #HH0551) Fixation (Buddhism, #HM1617)
Access concentration (Buddhism, #HM4999).

♦ **HM8737d Joyful feeling** (Buddhism)
Joy — Piti
Description The characteristic of joyful feeling is experiencing a desirable object; the function is to in some way make use of or exploit the desirable aspect; the manifestation is mental satisfaction or enjoyment; the proximate cause is tranquillity.
Context In Hinayana Buddhism, one of the five ways in which feeling is analysed in the feeling aggregate.
Broader Feeling aggregate (Buddhism, #HM4983)
Fivefold happiness (Buddhism, #HM0747).
Related Joy (Hinduism, #HM8098) Equanimity (Buddhism, #HM7769)
Sad feeling (Buddhism, #HM7120) Painful feeling (Buddhism, #HM8010)
Pleasant feeling (Buddhism, #HM6722).

♦ **HM8762c Furious sentiment** (Hinduism)
Aesthetic emotion of anger — Violent rasa
Context One of eight kinds of aesthetic sentiment or rasas in Indian psychology.
Broader Aesthetic emotion (Hinduism, #HM0205).

♦ **HM8772d Meeting God with emptiness** (Christianity)
Blissful love — State of rest
Description Without intermediary, beyond all activity and virtue there is a simple, inward act in blissful love. The light of God's unity reveals darkness, which envelops. The individual falls into a modeless state where he no longer has the power of observing things in their distinctness. He is transformed and pervaded by simple resplendence. Overcome by God's love, all activity fails him. His spirit inclines towards blissful enjoyment, he overcomes God and is one in spirit with Him. Interiorly unable to move, powerless over himself and all activity, he feels and knows only unique resplendence accompanied by a sense of well-being and penetrating savour. Through the simple illumination of God and the inclination of the spirit to be blissfully immersed in God, the spirit is transported into a state of rest, united with God.
Context According to John Ruusbroec, there are three modes in which meeting God in the interior way of life is practised. This is the first of the three, that of blissfully resting in one's self.
Broader Meeting God (Christianity, #HM0541).
Related Meeting God with active desire (Christianity, #HM6204)
Meeting God both resting and working in accordance with righteousness (Christianity, #HM5309).

♦ **HM8776d Khidmat** (Sufism)
Devotion to God — Service
Description Rendering Him unceasing service, the devotee, having tasted union with God, is never absent from the presence of his Friend.
Context According to Shaykh Abu Sa'id ibn Abi'l-Khayr, this is the 34th of 40 stations or maqamat the Sufi must possess for his journey on the path of Sufism to be acceptable.
Broader Stations of consciousness - ibn-Abi'l-Khayr (Sufism, #HM2424).
Related Religious experience as personal devotion and worship (#HM0561).
Followed by Tajrid (Sufism, #HM7621).
Preceded by Kashf (Sufism, #HM1452).

♦ **HM8838d Supramundane understanding** (Buddhism)
Transcendental understanding
Description This is understanding associated with supramundane states of consciousness or the transcendental path. Supramundane insight comprises understanding having measureless (infinite) object contingent upon nirvana (nibbana).
Context On the Path of Purification of Hinayana Buddhism, panna (understanding) is considered as of one kind (monad), of two kinds (dyads), of three kinds (triads) or of four kinds (tetrads). There are five dyads, four triads and two tetrads. All have the characteristic of penetrating the individual essences or true nature of states (monad). In the first dyad, this is contrasted with mundane (worldly) understanding.
Broader Understanding (Buddhism, #HM4523).
Related Mundane understanding (Buddhism, #HM7628)
Understanding having a measureless object (Buddhism, #HM6681).

♦ **HM8901d Iradat** (Sufism)
Discipleship
Description The seeker is firmly inclined for the realizing of God.
Context According to Shaykh Abu Sa'id ibn Abi'l-Khayr, this is the fourth of 40 stations or maqamat the Sufi must possess for his journey on the path of Sufism to be acceptable.
Broader Stations of consciousness - ibn-Abi'l-Khayr (Sufism, #HM2424).
Related Discipleship (#HH0376).
Followed by Mujahadat (Sufism, #HM5215).
Preceded by Tawba (Sufism, #HM4062).

♦ **HM8932g Agnosia**
Description Despite correctly identifying the properties of a object, an individual is unable to connect these properties with what the object is. This inability to connect may correspond to only one of the senses, perception through the other senses being perfectly normal. The condition is caused by brain lesions, the senses themselves are working normally. As opposed to synesthesia, where there seems to be diffusion from one sensory input channel to another, in agnosia normal interaction between channels is prevented. One channel is damaged, so that information received through it may be conceptualized but not at a high enough level.
Refs Reed, Graham *The Psychology of Anomalous Experience* (1988).
Narrower Visual agnosia (#HM7019) Simultanagnosia (#HM9176)
Tactile agnosia (#HM8079) Auditory agnosia (#HM6539).
Related Synesthesia (#HM1241).

♦ **HM8977d Spiritual love** (Christianity)
Description The opposite of spiritual envy, which sees everything from the personal self point-of-view, desiring to be praised above all others and angry at their success, spiritual love rejoices in the truth. The only envy here is sadness at not possessing other people's goodness, because there is delight in their possessing goodness. The individual is happy that others surpass him in serving God if he himself has failed.
Related Love (#HH0258) Love consciousness (#HM2000)
Spiritual envy (Christianity, #HM3818).

♦ **HM9001d Rectitude of consciousness** (Buddhism)
Rectitude of mind — Rectitude of citta (Pali)
Description This refers to the straight state of the aggregate of consciousness. The characteristic is uprightness of the citta, the function is to crush its tortuousness or crookedness. It is manifest as non-crookedness, non-deflection. Consciousness is the proximate cause. It is the opponent of corruptions like deceit or fraud, deception or craftiness, causing tortuousness in the citta. Rectitude of mental body and rectitude of consciousness are considered together.
Context One of the formations aggregate (mental coefficients) of Hinayana Buddhism, being listed among the constant states which appear in their true nature, and as profitable primary (always present in any profitable or profitable-resultant consciousness).
Broader Awareness as mental-formation group of conscious existence (Buddhism, #HM2050).
Related Rectitude (Buddhism, #HM4250).

♦ **HM9002f Delusions of grandeur**
Description The individual is convinced that his abilities, achievements and status are exaggeratedly superior to what is in fact the case. In extreme, psychotic, cases the individual may believe he is some prominent historical or current figure. This belief may be reinforced by hallucinatory voices.
Broader Delusive consciousness (#HM2600).

♦ **HM9021d Anicca** (Buddhism)
Transitoriness — Impermanence
Context One of the three characteristics of the phenomenal world, the others being dukkha (suffering) and anatta (non-self).
Related Anatma (Buddhism, #HH0241) Duhkha (Buddhism, Hinduism, #HM2574).

♦ **HM9022d Trans-sensate experience**
Description An ineffable experience, which goes beyond habitual sensory paths, ideas and memories, this is described by many mystics as not containing any familar sensory or intellectual elements and yet being filled with a profound perception which is regarded as the goal of the mystic path. Trans-sensate experience is associated with loss of "self" and is therefore not associated with reflective awareness, the "I" of normal consciousness is in abeyance.
Related Deautomatization and the mystic experience (#HM4398).

♦ **HM9101d Worry** (Buddhism)
Kukkucca (Pali)
Description In Hinayana Buddhism, worry is regarded as a state of bondage or slavery. It is the state of mind after having done a vile act. Its characteristic is repentance and regret; its function is sorrow at what has or has not been done. It manifests as remorse or regret. The proximate cause is what has been done (deeds of commission) or what has not been done (deeds of ommission).
Context One of the formations aggregate (mental coefficients) of Hinayana Buddhism, being listed among the inconstant states which are immutable by nature, and as unprofitable secondary (sometimes present in any unprofitable or unprofitable-resultant consciousness).
Broader Awareness as mental-formation group of conscious existence (Buddhism, #HM2050).

♦ **HM9124c Intuition of sages** (Hinduism)
Arsajnana
Description If there are two kinds of valid knowledge, perception and inference, then this is a valid knowledge of perception. Free from doubt and delusion, it is intuitive cognition differing from ordinary perception in that it is not produced by the external sense organs but by the inner organ assisted by powers acquired through learning, austerity and meditation. There is valid intuition of all objects (past, present and future). It is not always distinguished from *pratibhajnana* - a flash of intuition, a vivid and distinct perception in line with the real nature of things. In fact pratibhajnana has been called arsajnana which occurs in everyday life.
Broader Intuition (Hinduism, #HM1256).
Related Flash of intuition (Hinduism, #HM4875).

♦ **HM9132f Anomalies in experience of the unity of self**
Impairment of the unity of self
Description Although also occurring in psychiatric patients, this is quite common amongst 'normal' people as well in conditions of stress when a threat seems intolerable. In extreme danger there may be a walling off of emotion, *dissociation of affect*, but various levels of dissociative

defences occur in response to the milder threats of everyday life. Typically, everything seems unreal as though it is happening to someone else. Despite physical reactions, such as sweating, the emotional state is one of calm which may impair performance, when a slight state of being 'worked up' might put an edge on it. *Ego-splitting* is another defensive reaction when the individual seems to himself to be physically dissociated from himself and sees and hears himself from outside.
Broader Anomalies in awareness of one's self (#HM8186).
Narrower Ego-splitting (#HM6138)   Dissociation of affect (#HM6087).
Related Stress (#HM2585)   Breakdown of automatization (#HM8105).

♦ **HM9176g Simultanagnosia**
Description Individual details may be recognized in a picture but the meaning of the whole does not arise, the interaction between objects is not appreciated.
Broader Agnosia (#HM8932).

♦ **HM9213c Resting in God** (Christianity)
Description In the process of centering one uses attention on a holy word to keep the attention from wandering to passing thoughts. However, during the exercice of centering there may be an experience of such deep interior stillness that the word disappears and there are no thoughts. This inner silence is beyond thinking, images and emotions. Time passes without one being aware of it. There is an awareness that the core of being is indestructible, of being loved by God and sharing in his divine life.
Related Centering prayer (Christianity, #HH1994).

♦ **HM9218c Visionary experience**
Description This is distinguished from mystical experience in that it is described with sensory language and not marked by low levels of normal cognition or physiological activity.
Refs Katz, Steven *Mysticism and Religious Traditions* (1983); Katz, Steven T (Ed) *Mysticism and Philosophical Analysis* (1978).
Related Mystical cognition (#HM2272).

♦ **HM9330c Non-desire** (Christianity)
Description On reaching the state of sanctification called holy indifference, the holy soul ceases to desire anything outside the will of God. In this state, it may desire everything which relates to correcting its imperfections and weaknesses, to its persevering in its religious state and to its ultimate salvation. It may even desire temporal good and material things if it has reason to believe these are in line with God's desires. It also desires fulfilment of God's will in all respects, known and unknown. Non-desire refers only to not desiring anything out of God.
Refs Fénelon, François *The Maxims of the Saints*.
Related Tasawwuf (Sufism, #HM0575)   Holy indifference (Christianity, #HM5776).

♦ **HM9384f Thought broadcasting**
Description This is experienced as having one's thoughts open to everybody. Others may simply listen as a (critical) audience to one's thoughts or they may participate or think in unison.
Broader Modes of awareness associated with schizophrenia (#HM2313).
Anomalies in experience of the self as recognized in personal performance (#HM1336).
Related Passivity experiences (#HM7203).

♦ **HM9392d Grief** (Hinduism)
Soka — Sorrow
Description Expression of grief may be through paleness, dryness of mouth, fatigue and pain. There is continual suffering like a burning in the mind, a suffering which consumes the mind, a mental pain characterized by mental oppression. The function is to distress the mind and it manifests as mental affliction although the intrinsic suffering produces bodily suffering. It may be expressed in tearing the hair, contorting the body, weeping, striking the breast, suffering and even committing suicide. Grief or sorrow arises from separation from objects of sense desire, when desire and lust elude a person. It is the mental agony arising from a calamity of from the loss of someone or something very dear. It may also be aroused by the frustration of endeavour. It is a mental perversion where the self and the not-self are not discriminated. Combined with mental bewilderment it is the cause of delusion. Together with delusion it eclipses discriminative knowledge and is a seed of embodied life.
Related Grief (#HM2685)   Dejection (Hinduism, #HM0883)
Sad feeling (Buddhism, #HM7120).

♦ **HM9406c Sexual love**
Carnal love
Description Although quite authentic, this form of love is founded on the short-lived passion of physical attraction. It is experienced as the urge to possess the other sexually. At its peak, sexual passion is an almost insatiable bodily appetite centred exclusively, as an erotic fixation, on the other person. The lover can become obsessed with breaking through the barrier of otherness, although more with the object of knowing the other's sexual self, rather than the subjective self (as is the case in passionate love). When such carnal love is experienced in the absence of passionate love it eventually becomes self-limiting, especially as the sense of total carnal knowledge of the other is realized.
Refs Person, Ethel Spector *Love and Fateful Encounters* (1990).
Broader Love consciousness (#HM2000).
Related Romantic love (#HM5087).

♦ **HM9701f Delusions of poverty**
Description This now rare condition was characterized by belief that the individual and his family are faced with total destitution. It was related with depressive illness.
Broader Delusive consciousness (#HM2600).

♦ **HM9722c Imperfection of insight** (Buddhism)
Corruption of insight
Description The beginner in insight, who is continuously devoted to his meditation subject and keeps to the right way, nonetheless may have the ten imperfections arise in him. These are: (1) Illumination due to insight. (2) Knowledge due to insight. (3) Rapturous happiness due to insight. (4) Tranquillity or repose due to insight. (5) Bliss (pleasure) due to insight. (6) Resolution as faith due to insight. (7) Exertion or uplift arises as well-exerted energy due to insight. (8) Assurance or presentation (establishment) as mindfulness arises in association with insight. (9) Equanimity or indifference, whether about insight, as neutrality about formations; or in adverting, incisive and sharp. (10) Attachment or desire due to insight.
Bringing the formations to mind as impermanent, then illumination, for example, arises. Adverting to illumination, he sees it as the state of an arahant. However, his mind is seized by agitation and he does not understand the object presented to mind as painful and not-self. Already one of serenity and insight, he may mistakenly believe he is an arahant.
Imperfection is avoided when the skilful meditator defines and examines the illumination or whatever has arisen, understanding that it is impermanent, formed, subject to destruction, fall, fading away, and so on. Or he may see that it is not self being taken as self. Seeing the illumination or whatever as not mine, not I, not myself, there is no wavering.
Narrower Bliss due to insight (#HM5088)   Exertion due to insight (#HM5986)
Knowledge due to insight (#HM5221)   Equanimity due to insight (#HM8566)
Attachment due to insight (#HM6144)   Illumination due to insight (#HM4337)
Tranquillity due to insight (#HM7115)   Establishment due to insight (#HM6544)
Rapturous happiness due to insight (#HM4886)   Resolution as faith due to insight (#HM5432).
Related Principal insights (Buddhism, #HM3630)   Vipassana-bhavana (Buddhism, Pali, #HH0680)
Purification by knowledge and vision of the way (Buddhism, #HH3550)
Purification by knowledge and vision of what is the path and what is not the path (Buddhism, #HH4007).

♦ **HM9845d Affection** (Hinduism)
Abhisvanga — Abhisneha — Love
Description There is sympathetic identification of others' emotions of happiness and misery with one's own emotions, one identifies one's self with the object of one's affection. Delight in the objects of affection may arise in that they are the means of bodily comfort and selfish enjoyment. Different teachers regard love as excessive attachment, as a mature state of affection which is attachment, as deep attachment to objects of desire. It is a mental mode in which the guna tamas predominates.
Broader Attachment (Hinduism, #HM3914).
Related Non-affection (Hinduism, #HM8000)   Absence of love (Hinduism, #HM7344)
Feeling consciousness (Psychism, #HM3590).

Index HX

Human Development

Index scope

The index to the names and keywords of Section H is integrated into the index for the whole of Volume 2 (Section X). Do NOT attempt to use the following index to locate entries by name or keyword.

The following index is a specialized index to those **entries which are part of a numbered set or series.** It covers both sub-sections:

Section HH: Human development concepts

Section HM: Modes of awareness

All index entries refer via reference numbers (eg HM1234) to the descriptions in the preceding sections. The letters indicate the section (eg Section HM).

Index entries

The index entries are of two types:

- Numbered states appearing in the principal name or title of the description;

- Numbered states appearing in the body of the description.

No distinction is made between the two types of entry. In both cases the index entries are presented in number order. Each contains the immediately following words from the description. The principal name of the descriptive entry is then given in italics.

Remarks

This index was created as one of several **editorial experiments** in this Encyclopedia. It presents information so as to highlight the ways in which modes and states of awareness are integrated in the light of the mainly non-western traditions and cultures which attach greater importance to numbered sets of concepts. Such presentations are not unusual in such contexts.

Although the index was generated automatically, it has been edited manually to eliminate less helpful keywords resulting from the extraction process. Where there was any doubt, less meaningful words have however been left where they may prove to be of value in locating entries in Section HH or HM.

The automatic extraction process identified numbers in entries, whether expressed numerically, alphabetically, or in cardinal or in ordinal form. The index includes entries for numbers from 3 up. Those based on 0, 1 or 2 have been excluded. Those for 2 were too numerous. Those for one and zero could not be extracted satisfactorily.

INDEX TO CONCEPT SETS
(See SECTION X for KEYWORD INDEX) HX

NUMBERS

HH 0075 3 a loss *Psychedelic drugs*
HH 0711 3 a third system *Self-preservation*
HM 2027 3 absorption in the immaterial *Third absorption in the immaterial sphere (Buddhism)*
HM 2110 3 absorptions will *First absorption in the immaterial sphere (Buddhism)*
HM 2050 3 abstinences applied thought sustained *Awareness as mental-formation group of conscious existence* (Buddhism)
HM 6204 3 actively loving but is *Meeting God with active desire (Christianity)*
HH 2768 3 acts in this sequence *Corporate worship (Christianity)*
HM 3499 3 age commencing according *Age of the Holy Spirit (Christianity)*
HH 1778 3 age that of god *Primitivism*
HH 0397 3 ages christianity description *Doctrine of three ages (Christianity)*
HM 1455 3 aggregates of feeling perception *Fitness of mental body (Buddhism)*
HM 7636 3 aggregates of feeling perception *Lightness of mental body (Buddhism)*
HM 3696 3 aggregates of feeling perception *Malleability of mental body (Buddhism)*
HM 5402 3 aggregates of feeling perception *Rectitude of mental body (Buddhism)*
HM 4704 3 aggregates of feeling perception *Tranquillity of mental body (Buddhism)*
HM 7969 3 aggregates of feeling perception *Wieldiness of mental body (Buddhism)*
HH 0722 3 aims of the upbringing *Physical perfection*
HM 3567 3 among the spiritual paths *Path of sanctifying intelligence (Judaism)*
HM 6553 3 and fifth *Dwelling in the fivefold jana (Buddhism)*
HM 3631 3 and final stage *Alchemical reddening (Esotericism)*
HM 2980 3 and four and more *Sleep*
HM 3150 3 and fourth concentrations infinite *Serial absorptions (Buddhism)*
HM 0931 3 and fourth ecstasies *Right rapture of concentration (Buddhism)*
HH 3025 3 and fourth fruition *Purification by knowledge and vision (Buddhism)*
HM 6553 3 and fourth jhanas *Dwelling in the fivefold jana (Buddhism)*
HM 2395 3 and *Turiya awareness (Hinduism)*
HM 2387 3 animals and one human *Emblems of totality (Tarot)*
HM 2733 3 animals are divided *Desire-realm consciousness (Buddhism)*
HM 2311 3 antarabhava consciousness description *Meditation-state bardo awareness (Buddhism, Tibetan)*
HM 2257 3 are called cloudless born *Form-realm consciousness (Buddhism)*
HM 6336 3 are described as conducive *Enlightenment factors (Buddhism)*
HH 2909 3 are described as sanctification *Psychospiritual growth (Christianity)*
HM 2142 3 are included in this *Form-realm awareness (Buddhism, Tibetan)*
HM 2217 3 are lack of impulse *Setting the mind (Buddhism)*
HM 2254 3 are linked *Ma'rifa (Sufism)*
HM 3971 3 are one *Mansions of spiritual marriage (Christianity)*
HH 2768 3 are present *Corporate worship (Christianity)*
HM 2128 3 are related in both *Non-attachment (Buddhism)*
HM 2693 3 are subdivided into 3 *Tibetan meditative states of form concentrations (Buddhism)*
HM 3852 3 are without root cause *Indeterminate consciousness in the sense sphere - functional (Buddhism)*
HH 2121 3 areas of duty *Self-discipline (Christianity)*
HM 5309 3 arising *Meeting God both resting and working in accordance with righteousness (Christianity)*
HM 2026 3 as the one *Subjective domains of consciousness (Yoga)*
HH 0513 3 aspects self expression self *Creative existence*
HM 2141 3 avasthas recognized in yoga *Jagrat state of waking consciousness (Yoga)*
HH 4742 3 bases in one completion *Highest vehicle (Taoism)*
HM 2033 3 being of cognizing *Direct non-conceptual cognition of emptiness (Buddhism, Tibetan)*
HH 0306 3 being the path *Tantra (Buddhism, Yoga)*
HM 2295 3 brains the reptilian top *Brain-based consciousness*
HM 0600 3 branch of the tibetan *Yoga of the dream state (Buddhism)*
HH 0309 3 branches of buddhism combining *Vajrayana Buddhism (Buddhism)*
HH 0845 3 branches of buddhism emphasizing *Hinayana Buddhism (Buddhism)*
HH 0900 3 branches of buddhism marked *Mahayana Buddhism (Buddhism)*
HH 3987 3 by transcending the second *Subjects for meditation (Buddhism)*
HM 2259 3 cardinal qualities each uniting *Cardinal awareness (Astrology)*
HM 2554 3 cardinal qualities each uniting *Fixed awareness (Astrology)*
HM 4983 3 cases *Feeling aggregate (Buddhism)*
HH 0330 3 castes the twice born *Human development (Hinduism)*
HH 0353 3 categories of knowledge *Artha (Hinduism)*
HM 3340 3 category of experience *Unification with life essence (Sufism)*
HM 1264 3 cause of profane love *Al-fikr fi 'l-manzur (Islam)*
HM 2128 3 causes of misconduct desire *Non-attachment (Buddhism)*
HM 4180 3 causes *Spiritual lust (Christianity)*
HH 2198 3 central affirmations that salvation *Human development (Christianity)*
HH 0523 3 cessation of suffering (nirvana) *Four noble truths (Buddhism)*
HM 4387 3 chakra are broken *Charity (Leela)*
HM 0455 3 chakra *Clarity of consciousness (Leela)*
HM 0717 3 chakra theatre of karma *Third chakra: theatre of karma (Leela)*
HM 1618 3 change of lineage *Appearance of absorption (Buddhism)*
HH 3550 3 characteristics may be observed *Purification by knowledge and vision of the way (Buddhism)*
HH 0241 3 characteristics of the phenomenal *Anatma (Buddhism)*
HM 9021 3 characteristics of the phenomenal *Anicca (Buddhism)*
HM 2574 3 characteristics of the phenomenal *Duhkha (Buddhism, Hinduism)*
HM 1632 3 characteristics of the phenomenal *Intuition (Buddhism)*
HH 0873 3 class is on ways *Ways of knowing (Buddhism)*
HM 2733 3 classes of gods *Desire-realm consciousness (Buddhism)*
HH 2189 3 coding and decoding being *Artificial intelligence*
HH 0669 3 component of the eightfold *Asana (Hinduism, Yoga)*
HH 0745 3 components are indicated
HH 0353 3 components are separated *Artha (Hinduism)*
HH 2077 3 components of classical *Six-member yoga (Yoga)*
HM 5008 3 concentration *Concentration accompanied by equanimity (Buddhism)*
HM 7961 3 concentration *Concentration due to natural purity of consciousness (Buddhism)*
HM 1010 3 concentration *Concentration of easy progress and sluggish direct-knowledge (Buddhism)*
HM 8532 3 concentration *Concentration of the third jhana of five (Buddhism)*
HM 2284 3 concentration *Concentration of the third jhana of four (Buddhism)*
HM 7363 3 concentration *Concentration partaking of distinction (Buddhism)*
HM 7543 3 concentration *Concentration without applied thought or sustained thought (Buddhism)*
HM 0696 3 concentration *Immaterial-sphere concentration (Buddhism)*
HM 4882 3 concentration in the third *Concentration of the fourth jhana of five (Buddhism)*
HM 2257 3 concentration is divided *Form-realm consciousness (Buddhism)*
HM 0496 3 concentration *Measureless concentration (Buddhism)*
HM 4653 3 concentration *Measureless concentration with a limited object (Buddhism)*
HM 4380 3 concentration of the pentad *Concentration of the second jhana of four (Buddhism)*
HM 6327 3 concentration *Superior concentration (Buddhism)*
HM 2062 3 concentration *Third trance of the fine-material sphere (Buddhism)*
HM 3634 3 concentric circles the smallest *Intuition*
HH 0902 3 concepts suggested by israel *Capability to become*
HH 0405 3 concepts suggested by israel *Capacity to become*
HH 0171 3 concepts suggested by israel *Propensity to become*
HH 2198 3 confessions (1) the catholic *Human development (Christianity)*
HM 1618 3 conformity and the third *Appearance of absorption (Buddhism)*
HH 0565 3 conformity to stereotypical images *Moral development*
HM 6336 3 contribute to assisting *Enlightenment factors (Buddhism)*
HH 0659 3 craft degrees entered apprentice *Initiation (Freemasonry)*
HH 0221 3 crisis in late middle *Alchemy (Esotericism)*
HH 0221 3 crisis in old age *Alchemy (Esotericism)*
HH 0347 3 cultural factors cultural factors *Healthy human growth and development*
HM 2317 3 cycle contains two decads *Stations of consciousness - Ansari (Sufism)*
HM 6336 3 dampen down a fire *Enlightenment factors (Buddhism)*
HH 1763 3 data processor transpersonal awareness *Psychological approach to transcendence*
HM 2317 3 decads called principles valleys *Stations of consciousness - Ansari (Sufism)*
HM 0203 3 degree of creative inspiration *Egoic musical inspiration*
HH 3012 3 degree of love when *Degrees of love (Christianity)*
HH 0659 3 degree or master mason *Initiation (Freemasonry)*
HM 4273 3 degrees *Love of God (Sufism)*
HM 2077 3 degrees of prayerful recollection *Christian recollection (Christianity)*
HM 3630 3 developing contemplation of not *Principal insights (Buddhism)*
HM 1133 3 deviations of body buddhism *Leaving off, abstaining, totally abstaining and refraining from the three* deviations of body (Buddhism)
HM 0137 3 dhyanas belong *Dhyana (Hinduism, Buddhism)*
HH 0767 3 dimensional and protective incorporative *Mutational phases of life*
HM 0092 3 dimensional casts a shadow *Visual hallucination*
HM 2780 3 dimensional gross matter one *Mult-dimensional consciousness*
HH 1542 3 dimensional mass of energy *Human being*
HM 2712 3 dimensional perception of his *Transcendental experience*
HH 0720 3 dimensional rational view *Magic*
HM 2780 3 dimensional self opens *Mult-dimensional consciousness*
HH 2662 3 dimensional structure a yantra *Yantras*
HH 2334 3 dimensions *Psychic ability*
HH 3987 3 divine abidings are conditional *Subjects for meditation (Buddhism)*
HH 3987 3 divine abidings bring three *Subjects for meditation (Buddhism)*
HM 7769 3 divine abidings *Equanimity (Buddhism)*
HM 0357 3 divine persons *States of prayer (Christianity)*
HH 1187 3 divisions of time is *Purification by overcoming doubt (Buddhism)*
HM 2026 3 domains are as subjectively *Subjective domains of consciousness (Yoga)*
HM 7655 3 dominant influence (predominance) *Conditions for consciousness (Buddhism)*
HH 2121 3 duties of asceticism *Self-discipline (Christianity)*
HM 4279 3 duty christianity description *Third duty (Christianity)*
HM 4279 3 duty context *Third duty (Christianity)*
HM 4279 3 duty in the preparation *Third duty (Christianity)*
HM 2953 3 duty is to pass *Second step: the race (Christianity)*
HM 5767 3 dyad this contrasts *Concentration with happiness (Buddhism)*
HM 0730 3 dyad this contrasts *Concentration without happiness (Buddhism)*
HM 4968 3 dyad this is contrasted *Understanding defining materiality (Buddhism)*
HM 5627 3 dyad this is contrasted *Understanding defining mentality (Buddhism)*
HM 7089 3 elements of water fire *Superior saintship (Sufism)*
HH 1339 3 elements the cognitive soul *Pre-existence*
HH 7734 3 elixir fields swallows noon *Higher upper grade (Taoism)*
HH 8219 3 enemy is defeated *Way of the warrior (Amerindian)*
HM 6336 3 enlightenment or wisdom factors *Enlightenment factors (Buddhism)*
HM 3159 3 enlightenments knowledge through practice *Ten powers (Buddhism, Tibetan)*
HM 3194 3 entries to liberation wishlessness *Paths of insight (Buddhism)*
HM 2097 3 equals 7 *Tarot arcana of conscious states (Tarot)*
HH 5652 3 equanimity and bliss fourth *Supernormal powers (Buddhism)*
HH 0745 3 essential and interdependent components *Human resources development (United Nations)*
HH 0587 3 essential aspects of reality *Non-temporal dimension of being*
HH 0444 3 essential aspects the pure *Reality*
HH 3290 3 essential ones are *Human development (United Nations)*
HH 7386 3 essentials are body mind *Higher vehicle path (Taoism)*
HH 5342 3 essentials are body mind *Three vehicles of gradual method (Taoism)*
HH 4742 3 essentials are discipline concentration *Highest vehicle (Taoism)*
HH 4864 3 essentials are mouth *Lower vehicle path (Taoism)*
HH 6345 3 essentials are the head *Middle vehicle path (Taoism)*
HM 0923 3 evil paths buddhism description *Three evil paths (Buddhism)*
HM 1252 3 evil paths plus anger *Four evil paths (Buddhism)*
HH 0631 3 exercises of hierarchical actions *Triple way (Christianity)*
HM 1601 3 eye all the collective *Conscience (Leela)*
HM 3518 3 eye *Heart-awakening stage of life*
HM 4412 3 eye in the region *Sixth chakra: time of penance (Leela)*
HM 0176 3 eye *Soma chakra (Yoga)*
HM 7121 3 factors happiness or rapture *Dwelling in the second jhana (Buddhism)*
HH 0108 3 factors or forces *Spiritual aspects of psychic health*
HM 4380 3 factors remain *Concentration of the second jhana of four (Buddhism)*
HM 8532 3 factors remain *Concentration of the third jhana of five (Buddhism)*
HM 1814 3 faith description *Synthetic-conventional faith*
HM 2339 3 faith or ethical conduct *Eightfold way (Buddhism)*
HM 2150 3 faiths *Angelic frame of awareness*
HM 3146 3 faiths *Awareness of angelic-transformation*
HM 0509 3 fetter or character defect *Dependence on ceremonies (Buddhism)*
HM 7655 3 fold 13 *Conditions for consciousness (Buddhism)*
HM 0811 3 fold knowledge systematics relational *Three-fold knowledge (Systematics)*
HM 2829 3 fold mind *Spiritual intuitive cognition*
HM 2050 3 fold when associated *Awareness as mental-formation group of conscious existence (Buddhism)*
HM 0936 3 form of this is *Spiritual anger (Christianity)*
HM 2073 3 formal aspect *Awareness of allegiance (ICA)*
HM 2609 3 formal aspect *Awareness of futurity (ICA)*
HM 3106 3 formal aspect *Contentless transformation (ICA)*
HM 3056 3 formal aspect *Embodying charity (ICA)*
HM 2114 3 formal aspect *Existential guidance (ICA)*
HM 3138 3 formal aspect *Inventing essence (ICA)*
HM 2466 3 formal aspect is *Action irrelevant (ICA)*
HM 3083 3 formal aspect is *All-that's-yet-to-be (ICA)*
HM 2484 3 formal aspect is awareness *People of God (ICA)*
HM 2209 3 formal aspect is being *Abounding abasement (ICA)*
HM 2023 3 formal aspect is being *Common earth (ICA)*
HM 2517 3 formal aspect is being *Eternal insecurity (ICA)*
HM 2323 3 formal aspect is being *Serious sharing (ICA)*
HM 2303 3 formal aspect is *Cruciform exaltation (ICA)*
HM 3167 3 formal aspect is *Cut-off unknownness (ICA)*
HM 2812 3 formal aspect is embodying *Disinterested collegiality (ICA)*
HM 3128 3 formal aspect is embodying *Passionate concern (ICA)*
HM 2971 3 formal aspect is embodying *Personal obligation (ICA)*

HM 2754 3 formal aspect is embodying *Sacrificial friendship (ICA)*
HM 3117 3 formal aspect is *Frightful possibility (ICA)*
HM 2524 3 formal aspect is *Impossible possibility (ICA)*
HM 3075 3 formal aspect is inventing *Being history (ICA)*
HM 3188 3 formal aspect is inventing *Human example (ICA)*
HM 3071 3 formal aspect is inventing *Revolutionary sign (ICA)*
HM 3076 3 formal aspect is inventing *Self determining (ICA)*
HM 2249 3 formal aspect is *Keeping conscience (ICA)*
HM 3065 3 formal aspect is *Luminous change (ICA)*
HM 2757 3 formal aspect is one's *Classical story (ICA)*
HM 2814 3 formal aspect is one's *Ever-present brother (ICA)*
HM 3130 3 formal aspect is one's *Guardian angel (ICA)*
HM 2598 3 formal aspect is one's *Objective awareness (ICA)*
HM 2352 3 formal aspect is one's *Primordial colloquy (ICA)*
HM 2577 3 formal aspect is one's *Religious function (ICA)*
HM 2293 3 formal aspect is one's *Religious vocation (ICA)*
HM 2654 3 formal aspect is one's *Revered hero (ICA)*
HM 2921 3 formal aspect is one's *Scorching avatar (ICA)*
HM 2460 3 formal aspect is one's *Terrifying acceptance (ICA)*
HM 2644 3 formal aspect is one's *Universal Christ (ICA)*
HM 2429 3 formal aspect is surrender *Abject helplessness (ICA)*
HM 2771 3 formal aspect is surrender *Imploring succour (ICA)*
HM 2706 3 formal aspect is surrender *Levitational submission (ICA)*
HM 2532 3 formal aspect is surrender *Representational sign (ICA)*
HM 3240 3 formal aspect *Liberation from relationships (ICA)*
HM 2922 3 formal aspect *Surrender to inadequacy (ICA)*
HM 2736 3 formal aspect *Transparent engagement (ICA)*
HH 0661 3 forms are laya yoga *Yoga (Yoga)*
HM 1991 3 forms *Four-fold knowledge (Systematics)*
HH 0230 3 forms *Initiation*
HM 0837 3 forms of attack are *Psychic attack (Psychism)*
HH 0353 3 forms of knowledge *Artha (Hinduism)*
HH 0610 3 forms of knowledge *Jnana (Yoga, Buddhism, Hinduism)*
HH 0538 3 forms of knowledge *Sabda (Yoga, Buddhism)*
HH 0983 3 forms of prayer oral *Practice of silence*
HM 4062 3 forms the first is *Tawba (Sufism)*
HH 3025 3 fruition consciousnesses *Purification by knowledge and vision (Buddhism)*
HM 7055 3 fruition moment arahantship consciousnesses *Knowledge of the path of arahantship (Buddhism)*
HM 6920 3 fruition moment non return *Knowledge of the path of non-return (Buddhism)*
HM 7563 3 fruition moment once return *Knowledge of the path of once-return (Buddhism)*
HM 1088 3 fruition moment stream entry *Knowledge of the path of stream entry (Buddhism)*
HH 0413 3 fundamental properties underlying *Three gunas*
HM 2356 3 ghosts *Unfailing prompter (ICA)*
HH 0169 3 graces beyond the individual *Grace*
HH 2003 3 gradations of knowledge can *Systematics*
HM 3146 3 grades *Awareness of angelic-transformation*
HH 4298 3 grades of severity depending *Ascetic practices (Buddhism)*
HH 0199 3 great abiding virtues *Hope (Christianity)*
HM 2793 3 groups of four foundations *Bodhi-pakkhiya dhamma (Buddhism)*
HM 0670 3 gunas are no longer *Irreligiousity (Leela)*
HH 4561 3 gunas mastery of this *Parinama (Yoga)*
HM 0576 3 gunas *Saguna brahman*
HM 0953 3 gunas sattva rajas *Maya (Leela)*
HH 0413 3 gunas sattva *Three gunas*
HH 0786 3 habits namely integrated systems *Integration*
HM 2062 3 has 5 branches 2 *Third trance of the fine-material sphere (Buddhism)*
HH 0156 3 he prefers that based *Friendship*
HH 5652 3 he repeatedly attains jhana *Supernormal powers (Buddhism)*
HH 0921 3 higher levels unaffected *Kabbalah*
HM 2072 3 higher spheres depending *Worlds of conscious existence (Buddhism, Pali)*
HM 2098 3 higher spheres unwholesome karma *Dispositions of consciousness (Buddhism)*
HM 2536 3 higher states in pali *Rupaloka consciousness (Buddhism)*
HM 2012 3 higher states that *Arupaloka consciousness (Buddhism, Pali)*
HM 2072 3 higher states the fine *Worlds of conscious existence (Buddhism, Pali)*
HH 0862 3 hours *Hatha yoga (Yoga)*
HH 0669 3 hours without strain *Asana (Hinduism, Yoga)*
HH 0447 3 how much development proceeds *Human psychological development*
HM 1655 3 illuminating techniques are necessary *Experience of An (Sufism)*
HM 2531 3 illumination of jami *Perception of hidden reality (Sufism)*
HM 2050 3 immaterial or mental aggregates *Awareness as mental-formation group of conscious existence* (Buddhism)
HM 0811 3 in a sequence *Three-fold knowledge (Systematics)*
HH 0661 3 in his *Yoga (Yoga)*
HM 0357 3 in its powers *States of prayer (Christianity)*
HH 6566 3 in one the trinity *Way of transcendence (Christianity)*
HM 0357 3 in person but one *States of prayer (Christianity)*
HM 1454 3 in the fivefold reckoning *Concentration accompanied by bliss (Buddhism)*
HM 4433 3 in the fivefold reckoning *Concentration accompanied by happiness (Buddhism)*
HM 7543 3 in the fivefold reckoning *Concentration without applied thought or sustained thought (Buddhism)*
HM 2030 3 in the ica *Ultimate reality (ICA)*
HM 1814 3 in the system *Synthetic-conventional faith*
HM 2050 3 inconstant *Awareness as mental-formation group of conscious existence (Buddhism)*
HM 0811 3 independent elements *Three-fold knowledge (Systematics)*
HM 1646 3 indeterminate resultant mind consciousness *Registering (Buddhism)*
HM 0428 3 initiation in the tradition *Initiation of transfiguration (Esotericism)*
HM 2566 3 inner members *Dharana (Hinduism)*
HM 6502 3 insight into the knowledge *Nanadassana (Buddhism)*
HM 0357 3 intellectual visions *States of prayer (Christianity)*
HH 0565 3 interpersonally normative morality *Moral development*
HH 5101 3 is an adjusted figure *Human development index (United Nations)*
HM 4273 3 is beyond description *Love of God (Sufism)*
HH 5217 3 is loving *Mystical theology (Christianity)*
HM 4062 3 is that of those *Tawba (Sufism)*
HM 3438 3 is the feeling *Tawhid (Sufism)*
HH 0385 3 is the number *Individuation*
HM 2077 3 is true recollection when *Christian recollection (Christianity)*
HH 1198 3 it may be argued *Human development through religion*
HH 0278 3 jewels of buddha *Observing moral precepts (Buddhism)*
HH 0622 3 jewels or ways are *Human development (Jainism)*
HM 5643 3 jhana and commence doing *Dwelling in the third jhana (Buddhism)*
HM 8087 3 jhana and reviewing *Dwelling in the fourth jhana (Buddhism)*
HM 5643 3 jhana buddhism dwelling *Dwelling in the third jhana (Buddhism)*
HM 2284 3 jhana ease or bliss *Concentration of the third jhana of four (Buddhism)*
HH 5652 3 jhana in the fire *Supernormal powers (Buddhism)*
HM 1454 3 jhana in the fourfold *Concentration accompanied by bliss (Buddhism)*
HM 8532 3 jhana of five buddhism *Concentration of the third jhana of five (Buddhism)*
HM 2284 3 jhana of four buddhism *Concentration of the third jhana of four (Buddhism)*
HM 7769 3 jhana or fourth if *Equanimity (Buddhism)*
HM 8532 3 jhana rapture or happiness *Concentration of the third jhana of five (Buddhism)*
HM 5643 3 jhana through the habits *Dwelling in the third jhana (Buddhism)*
HH 5652 3 jhanas base of boundless *Supernormal powers (Buddhism)*
HM 7543 3 jhanas commencing *Concentration without applied thought or sustained thought (Buddhism)*
HM 8087 3 jhanas it is said *Dwelling in the fourth jhana (Buddhism)*
HM 4191 3 jhanas of the fourfold *Concentration accompanied by bliss (Buddhism)*
HM 3987 3 jhanas the fourth divine *Subjects for meditation (Buddhism)*
HH 0738 3 journey to god *Sharia (Islam)*
HH 3181 3 key concepts needs limitations *Sustainable development (United Nations)*
HH 0088 3 key elements of jainism *Ahimsa*
HM 4398 3 kind of ineffable experience *Deautomatization and the mystic experience*
HH 5232 3 kinds love *Love of God (Christianity)*
HM 7324 3 kinds of consciousness context *Examining (Buddhism)*
HH 0487 3 kinds of growth stimulus *Transactional analysis*
HM 2217 3 kinds of mental torpor *Setting the mind (Buddhism)*
HM 4348 3 kinds of mundane full *Full understanding as abandoning (Buddhism)*
HM 4552 3 kinds of mundane full *Full understanding as investigation (Buddhism)*
HM 3490 3 kinds of mundane full *Full understanding as the known (Buddhism)*
HM 8266 3 kinds of object past *Rebirth-linking (Buddhism)*
HM 0988 3 kinds of perception controlled *Orders of perception*
HM 6663 3 kinds (triads) of four *Concentration (Buddhism)*
HM 4523 3 kinds (triads) *Understanding (Buddhism)*
HM 2257 3 lands brahma type brahma *Form-realm consciousness (Buddhism)*
HM 3880 3 level of samprajnata samadhi *Vicara samadhi (Yoga)*
HM 1357 3 level of the universe *Madhya-loka (Jainism)*
HM 1973 3 level world of formlessness *State of rapture (Buddhism)*
HM 2693 3 levels each *Tibetan meditative states of form concentrations (Buddhism)*
HM 3575 3 levels *Inner peace*
HM 2257 3 levels little light limitless *Form-realm consciousness (Buddhism)*
HM 0762 3 levels longing for paradise *Showq (Sufism)*
HM 2574 3 levels of dukkha are *Duhkha (Buddhism, Hinduism)*
HH 4019 3 levels of enlightenment savakabodhi *Steps to enlightenment (Buddhism)*
HM 0203 3 levels of musical improvisation *Egoic musical inspiration*
HM 6006 3 levels of musical inspiration *Avataric level of musical inspiration*
HM 1995 3 levels of musical inspiration *Imitative musical inspiration*
HM 2046 3 levels of mystic ecstasy *Ecstasy*
HH 0625 3 levels of physical training *Total fitness*
HM 0933 3 life breaths birth *Fifth chakra: man becomes himself (Leela)*
HH 5550 3 lines at a time *Meditative reading*
HM 1525 3 literalism is broken down *Mythic-literal faith*
HM 3063 3 lotus or chakra defined *Manipura (Yoga)*
HH 7004 3 lower grades comprise over *Sidetracks and auxiliary methods in Taoism (Taoism)*
HH 3013 3 lower grades *False paths (Taoism)*
HH 4289 3 lower grades *Outside paths (Taoism)*
HH 3667 3 lower grades *Outside paths (Taoism)*
HM 2072 3 lower spheres *Worlds of conscious existence (Buddhism, Pali)*
HH 0650 3 main geographical areas *Human development (Buddhism)*
HM 4805 3 main levels of musical *Musical inspiration*
HH 0366 3 main processes being growth *Ontogenesis*
HH 0650 3 main schools hinayana mahayana *Human development (Buddhism)*
HM 0124 3 main stages *Haqiqa (Sufism)*
HM 3371 3 main stages shari'a tariqa *Divine path (Sufism)*
HH 0497 3 main types of power *Magico-religious powers*
HM 2431 3 major after life bardo *Death-bardo rebirth-seeking awareness (Buddhism)*
HH 0916 3 major approaches behaviourism psychoanalysis *Transpersonal psychology*
HM 2098 3 manner profitable (moral) associated *Dispositions of consciousness (Buddhism)*
HM 3445 3 marga or ways *Religious experience*
HH 5101 3 maxima and minima being *Human development index (United Nations)*
HM 2866 3 may include passion aversion *Sukha (Buddhism)*
HM 0413 3 mediating function associated *Gnosis*
HM 3499 3 member of the trinity *Age of the Holy Spirit (Christianity)*
HH 0093 3 members of a trinity *Dharma*
HM 2339 3 mental discipline (samadhi) *Eightfold way (Buddhism)*
HM 3337 3 mental transformations described *Ekagrata parinama (Yoga)*
HM 3437 3 mental transformations described *Nirodha parinama (Yoga)*
HM 3207 3 mental transformations described *Samadhi parinama (Yoga)*
HH 9760 3 methods of prayer are *Spiritual exercices of St Ignatius (Christianity)*
HH 1021 3 methods tends *Spiritual discipline*
HM 0882 3 metres with the result *Empty-field myopia*
HH 7004 3 middle grades gradually approach *Sidetracks and auxiliary methods in Taoism (Taoism)*
HH 3296 3 middle grades *Lower middle grade (Taoism)*
HH 5207 3 middle grades *Middle middle grade (Taoism)*
HH 2633 3 middle grades *Upper middle grade (Taoism)*
HM 8273 3 mind elements sense sphere *Consciousness-born materiality (Buddhism)*
HM 2428 3 minds in man *Theta-clear state of consciousness (Scientology)*
HM 0768 3 mode may the integrity *Transpersonal awareness*
HM 8336 3 mode occurrence *Adverting (Buddhism)*
HM 0168 3 mode of learning posited *Conscious learning*
HM 3412 3 modes are distinguished (1) *Innocence (Christianity)*
HM 5309 3 modes in which meeting *Meeting God both resting and working in accordance with righteousness* (Christianity)
HM 6204 3 modes in which meeting *Meeting God with active desire (Christianity)*
HM 8772 3 modes in which meeting *Meeting God with emptiness (Christianity)*
HM 0541 3 modes *Meeting God (Christianity)*
HM 2318 3 moons together divine unity *Ultimate Sufi state of consciousness (Sufism)*
HM 2696 3 more initiations *Tantra heat-application awareness (Buddhism, Tibetan)*
HM 4062 3 more repentance of present *Tawba (Sufism)*
HM 2087 3 most important nadis *Ida conscious energy (Yoga)*
HH 0162 3 needsherent *Encounter group*
HM 8673 3 nimittas are the objects *Mark of calm (Buddhism)*
HH 3025 3 noble person and so *Purification by knowledge and vision (Buddhism)*
HM 7563 3 noble person *Knowledge of the path of once-return (Buddhism)*
HM 2330 3 noble truth ignorance *Nirvana (Buddhism)*
HH 4019 3 noble truth *Steps to enlightenment (Buddhism)*
HH 2189 3 objective knowledge or human *Artificial intelligence*
HM 4062 3 of 40 stations *Tawba (Sufism)*
HM 1052 3 of eight lokas represented *Celestial plane (Leela)*
HH 0813 3 of five perfections *Kshanti (Zen)*
HM 5353 3 of four contemplative disciplines *Cleansing the sirr (Sufism)*
HM 2756 3 of four initiations *Tantra receptivity-application awareness (Buddhism, Tibetan)*
HM 4392 3 of four stages *Jabarut (Sufism)*
HM 2980 3 of his life *Sleep*
HM 3312 3 of seven ascending planes *Subtlety of mind (Hinduism)*
HM 3055 3 of seven descending steps *Great wakefulness (Hinduism)*
HM 2969 3 of seven mansions *Mansions of exemplary life (Christianity)*
HM 3464 3 of seven stages *Mental-volitional stage of life*
HM 6534 3 of seven stages *Restoration of celestial awareness within the mundane (Taoism)*
HM 3974 3 of seven subtle stages *World of spiritual perception (Sufism)*
HM 1997 3 of seven ways *Way of active intelligence (Esotericism)*
HM 2372 3 of ten described *Sphere of understanding (Kabbalah)*
HM 0312 3 of ten orders *Third order perceptions - configurations*
HM 4648 3 of ten states acting *Conduct (Sufism)*
HM 4388 3 of the five primary *Pramana (Buddhism)*
HH 2782 3 of the four ashramas *Vanaprasthashrama (Hinduism)*
HM 2027 3 of the four formless *Third absorption in the immaterial sphere (Buddhism)*

INDEX TO CONCEPT SETS
(See SECTION X for KEYWORD INDEX)
HX

HM 2330 3 of the four noble *Nirvana (Buddhism)*
HM 4128 3 of the four paths *Understanding as the plane of development (Buddhism)*
HM 3212 3 of the four stages *Ananda stage (Yoga)*
HM 2360 3 of the kabbalistic four *World of formation (Judaism)*
HM 2265 3 of the nine states *Resetting the mind (Buddhism)*
HM 4061 3 of the ninety nine *Al-Malik (Sufism)*
HM 2719 3 of the preparations *Belief-arising subtle contemplation (Buddhism)*
HM 0923 3 of the ten inner *Three evil paths (Buddhism)*
HH 0551 3 of the ten meritorious *Bhavana (Buddhism)*
HM 0203 3 of the three levels *Egoic musical inspiration*
HM 3437 3 of three mental transformations *Nirodha parinama (Yoga)*
HM 3070 3 of trinity *Completeness*
HM 2576 3 on the board *Very-hot-hells awareness (Buddhism, Tibetan)*
HH 3987 3 or four jhanas imply *Subjects for meditation (Buddhism)*
HH 0983 3 or spiritual the greater *Practice of silence*
HM 0312 3 order perceptions configurations description *Third order perceptions - configurations*
HH 2117 3 parties (impersonal bonds) although *Justice*
HH 0712 3 parts of government philosopher *Virtue*
HH 0712 3 parts of the soul *Virtue*
HM 2262 3 party *Extrasensory perception (ESP)*
HH 1354 3 party on one's actions *Journeying within transcendence - from blindness to sight*
HH 4090 3 path purgation illumination mystical *Via purgativa (Christianity)*
HH 0711 3 paths of morality may *Self-preservation*
HM 3371 3 paths to realise identity *Divine path (Sufism)*
HM 3490 3 periods of time is *Full understanding as the known (Buddhism)*
HM 4033 3 person in the vicinity *Accidental psychic attack (Psychism)*
HM 0134 3 person *Modes of awareness associated with alcohol consumption*
HM 3971 3 persons communicate *Mansions of spiritual marriage (Christianity)*
HM 3971 3 persons *Mansions of spiritual marriage (Christianity)*
HM 7384 3 persons of the trinity *Melting (Christianity)*
HH 0986 3 phallic stage 3 5 *Libido development*
HM 2855 3 phase in the spiritual *Saviors of God: the vision (Christianity)*
HM 4111 3 phases comprise one religious *Religious states (Christianity)*
HM 6120 3 phases preparation which may *Shamanic journey*
HM 3054 3 phenomenological level a person *Appropriated passion (ICA)*
HM 2680 3 phenomenological level a person *Hallowed honour (ICA)*
HM 3065 3 phenomenological level a person *Luminous change (ICA)*
HM 3101 3 phenomenological level a person *Sheer re-creation (ICA)*
HM 2209 3 phenomenological level this occurs *Abounding abasement (ICA)*
HM 2889 3 phenomenological level this occurs *Agonizing prediction (ICA)*
HM 2286 3 phenomenological level this occurs *Awareness of spiritual poverty (ICA)*
HM 2379 3 phenomenological level this occurs *Besetting sin (ICA)*
HM 2757 3 phenomenological level this occurs *Classical story (ICA)*
HM 3009 3 phenomenological level this occurs *Decisional nothingness (ICA)*
HM 2682 3 phenomenological level this occurs *Determining history (ICA)*
HM 2812 3 phenomenological level this occurs *Disinterested collegiality (ICA)*
HM 3042 3 phenomenological level this occurs *Divine nothingness (ICA)*
HM 2541 3 phenomenological level this occurs *Eternal identification (ICA)*
HM 2814 3 phenomenological level this occurs *Ever-present brother (ICA)*
HM 2903 3 phenomenological level this occurs *Expectant descendant (ICA)*
HM 3188 3 phenomenological level this occurs *Human example (ICA)*
HM 2334 3 phenomenological level this occurs *Human transformation (ICA)*
HM 3124 3 phenomenological level this occurs *Immutable friend (ICA)*
HM 2771 3 phenomenological level this occurs *Imploring succour (ICA)*
HM 2249 3 phenomenological level this occurs *Keeping conscience (ICA)*
HM 2910 3 phenomenological level this occurs *Loyal opposition (ICA)*
HM 2269 3 phenomenological level this occurs *Manifold blessings (ICA)*
HM 2005 3 phenomenological level this occurs *Obedient son (ICA)*
HM 2139 3 phenomenological level this occurs *Persistent friend (ICA)*
HM 2587 3 phenomenological level this occurs *Radical incarnation (ICA)*
HM 2293 3 phenomenological level this occurs *Religious vocation (ICA)*
HM 2557 3 phenomenological level this occurs *Representational existence (ICA)*
HM 2366 3 phenomenological level this occurs *Sacramental universe (ICA)*
HM 3086 3 phenomenological level this occurs *Secondary integrity (ICA)*
HM 2728 3 phenomenological level this occurs *Social failure (ICA)*
HM 3126 3 phenomenological level this occurs *Suffering servant (ICA)*
HM 3058 3 phenomenological level this occurs *Transcendent guru (ICA)*
HM 3067 3 phenomenological level this occurs *Transparent existence (ICA)*
HM 2761 3 phenomenological level this occurs *Trust intuitions (ICA)*
HM 2418 3 phenomenological level this occurs *Universal father (ICA)*
HH 0126 3 physical types endomorph *Physique and temperament*
HH 0921 3 pillars positive negative *Kabbalah*
HM 3312 3 plane of wisdom description *Subtlety of mind (Hinduism)*
HM 7234 3 planes in the worlds *Mundane concentration (Buddhism)*
HM 2270 3 (pleasure pain neutrality) may *Feeling (Buddhism)*
HM 3725 3 pm *Sext (Christianity)*
HH 1321 3 points of view phenomena *Religious growth*
HH 0487 3 positions from which one *Transactional analysis*
HH 1763 3 possible dynamics are described *Psychological approach to transcendence*
HM 0570 3 possible modalities *Ego consciousness*
HM 0044 3 possible modes *Biosocial consciousness*
HM 3336 3 possible modes *Existential consciousness*
HM 0768 3 possible modes *Transpersonal awareness*
HM 2679 3 powers of bodily form *Magical-forces awareness (Buddhism, Tibetan)*
HH 1010 3 prajnacaksus the eye *Five eyes (Buddhism)*
HM 0910 3 principles zen *Formlessness (Zen)*
HM 3485 3 principles zen *Non-abiding (Zen)*
HH 4872 3 processes characteristic of human *Neuro-linguistic programming (NLP)*
HH 0647 3 productive processes that are *Development of the quality of human life*
HM 0762 3 psychic state listed *Showq (Sufism)*
HM 2396 3 qualities (gunas) of prakriti *Purusha (Yoga)*
HM 3455 3 quarter *Prajna (Hinduism)*
HH 0413 3 rajas *Three gunas*
HH 0623 3 realizations of wisdom metaphysics *Wisdom*
HM 2653 3 realm consciousness buddhism tibetan *Beyond-the-third-realm consciousness (Buddhism, Tibetan)*
HM 2177 3 realms and nine levels *Consciousness states in cyclic existence (Buddhism)*
HH 2909 3 recognition of call *Psychospiritual growth (Christianity)*
HM 2955 3 related signs *Airy awareness (Astrology)*
HM 3235 3 related signs *Earthy awareness (Astrology)*
HM 2124 3 related signs *Fiery awareness (Astrology)*
HM 2384 3 related signs *Watery awareness (Astrology)*
HH 2909 3 relatedness and attachment *Psychospiritual growth (Christianity)*
HM 5977 3 release from the conception *Vimoksa (Buddhism)*
HH 0955 3 requirements in training pupils *Educational self-development*
HM 3837 3 row *Atonement (Leela)*
HM 0764 3 row *Bad company (Leela)*
HM 1052 3 row *Celestial plane (Leela)*
HM 4387 3 row *Charity (Leela)*
HM 1355 3 row *Good company (Leela)*
HM 0717 3 row of the board *Third chakra: theatre of karma (Leela)*
HM 0481 3 row *Plane of dharma (Leela)*
HM 0948 3 row *Plane of karma (Leela)*
HM 0294 3 row *Selfless service (Leela)*
HM 1782 3 row *Sorrow (Leela)*
HM 3941 3 rung of the ladder *Dark night of the soul (Christianity)*
HH 5101 3 scales gives a country's *Human development index (United Nations)*
HM 2215 3 scriptural bodhisattva awareness buddhism *Third scriptural bodhisattva awareness (Buddhism, Tibetan)*
HH 0136 3 self esteem or egoism *Unity of personality*
HH 0631 3 separate but parallel courses *Triple way (Christianity)*
HM 7384 3 share *Melting (Christianity)*
HM 3196 3 sheaths which comprise *Ignorance (Buddhism, Yoga, Zen)*
HM 1433 3 signifies fire and zeal *Anger (Leela)*
HM 0348 3 sleep description *Stage-three sleep*
HM 4180 3 source of these experiences *Spiritual lust (Christianity)*
HH 3121 3 sources of christian stewardship *Christian stewardship (Christianity)*
HM 2360 3 spheres (sephiroth) glory victory *World of formation (Judaism)*
HM 2408 3 spheres (sephiroth) kingdom wisdom *World of emanation (Judaism)*
HM 3038 3 spheres (sephiroth) mercy judgment *World of creation (Judaism)*
HH 1198 3 spiritual modalities *Human development through religion*
HH 0631 3 spiritual ways description *Triple way (Christianity)*
HM 2215 3 stage bodhisattva who will *Third scriptural bodhisattva awareness (Buddhism, Tibetan)*
HM 2142 3 stage brahma heaven where *Form-realm awareness (Buddhism, Tibetan)*
HM 2142 3 stage gods of light *Form-realm awareness (Buddhism, Tibetan)*
HM 4450 3 stage in a systematic *Zuhd (Sufism)*
HM 3488 3 stage in the action *The relationship between man and nature (Christianity)*
HM 2078 3 stage in the evolution *Mythical consciousness*
HM 2213 3 stage in the less *Near death black void (NDE)*
HM 3294 3 stage in the process *Near death transition (NDE)*
HM 3464 3 stage of life description *Mental-volitional stage of life*
HM 7626 3 stage of mahabbat (love) *Nisti (Sufism)*
HM 3631 3 stage of the alchemical *Alchemical reddening (Esotericism)*
HM 6690 3 stage on the path *Riyada (Sufism)*
HM 0720 3 stage the awareness having *Magic*
HH 1276 3 stages (1) theological *Positivism*
HH 0021 3 stages although not every *Rites of passage*
HH 1543 3 stages are referred *Individual development*
HH 6669 3 stages growth (childhood adolescence) *Primary growth*
HH 5217 3 stages growth *Mystical theology (Christianity)*
HM 3196 3 stages in overcoming ignorance *Ignorance (Buddhism, Yoga, Zen)*
HH 0460 3 stages is reached *Seven stages of life*
HM 2954 3 stages lack psychic maturity *Emotional-sexual stage of life*
HM 3464 3 stages lack psychic maturity *Mental-volitional stage of life*
HM 3320 3 stages lack psychic maturity *Vital-physical stage of life*
HH 0372 3 stages of life *Karma yoga (Yoga)*
HH 0111 3 stages of life *Puja yoga (Yoga)*
HH 0021 3 stages of realization experience *Rites of passage*
HH 0212 3 stages of sila ethical *Human perfectibility*
HH 0021 3 stages of the mystical *Rites of passage*
HH 3652 3 stages of the spiritual *Ascetical theology (Christianity)*
HM 1134 3 stages of yoga have *Bodily disability (Yoga)*
HM 8673 3 stages parikamma upacara *Mark of calm (Buddhism)*
HH 0460 3 stages physical emotional *Seven stages of life*
HH 5217 3 stages purgation illumination mystical *Mystical theology (Christianity)*
HM 4273 3 stages rasti rectitude masti *Love of God (Sufism)*
HM 2866 3 stages resentment and opposed *Sukha (Buddhism)*
HM 6097 3 stages subsume the whole *Masti (Sufism)*
HM 7626 3 stages subsume the whole *Nisti (Sufism)*
HM 7222 3 stages subsume the whole *Rasti (Sufism)*
HM 2669 3 stages *Tibetan meditative states of formless absorptions (Buddhism)*
HH 5217 3 stages were later equated *Mystical theology (Christianity)*
HH 1543 3 stages
HH 0169 3 standpoints (1) as forgiveness *Grace*
HM 8330 3 state arising *Sultan-al-dhikr (Sufism)*
HM 1433 3 state or square *Anger (Leela)*
HM 2402 3 state which is *Objective fourth state of consciousness*
HM 1087 3 states (1) passing away *Fana (Islam)*
HM 2032 3 states (avasthas) of consciousness *Awareness of relative reality (Hinduism, Yoga)*
HM 2395 3 states (avasthas) *Turiya awareness (Hinduism)*
HH 1987 3 states of consciousness waking *Brahmacharyashrama (Hinduism)*
HH 1987 3 states of mind *Brahmacharyashrama (Hinduism)*
HM 4009 3 station of the cross *Jesus falls the first time under His cross (Christianity)*
HM 2829 3 step in the exercise *Spiritual intuitive cognition*
HM 3501 3 step in the march *Third step: mankind (Christianity)*
HM 3501 3 step mankind christianity description *Third step: mankind (Christianity)*
HM 3941 3 step of the ladder *Dark night of the soul (Christianity)*
HM 3501 3 step one unites *Third step: mankind (Christianity)*
HM 2062 3 sub types context *Third trance of the fine-material sphere (Buddhism)*
HM 4111 3 successive phases are defined *Religious states (Christianity)*
HH 0631 3 successive stages the triple *Triple way (Christianity)*
HH 0221 3 (sun) coniunctio leading *Alchemy (Esotericism)*
HM 2075 3 supermundane faculties arising *Psycho-physical faculties awareness (Buddhism)*
HH 0694 3 supernatural or abiding virtues *Faith*
HH 0199 3 supernatural or abiding virtues *Hope (Christianity)*
HH 0258 3 supernatural or abiding virtues *Love*
HM 2126 3 symbolic level identification *Consciousness expansion*
HH 0711 3 system the good is *Self-preservation*
HH 0711 3 system the individualistic morality *Self-preservation*
HH 0413 3 tamas description *Three gunas*
HM 4449 3 techniques (1) illumination *Creative moment (Sufism)*
HM 3173 3 techniques described by casteneda *Mastery of dreaming*
HM 3064 3 techniques described by casteneda *Mastery of intent*
HM 2304 3 techniques described by casteneda *Mastery of stalking*
HH 0126 3 temperaments viscerotonic *Physique and temperament*
HM 6663 3 tetrad *Concentration (Buddhism)*
HM 5143 3 tetrad *Concentration of the fifth jhana of five (Buddhism)*
HM 0297 3 tetrad *Concentration of the first jhana of five (Buddhism)*
HM 4882 3 tetrad *Concentration of the fourth jhana of five (Buddhism)*
HM 8532 3 tetrad *Concentration of the third jhana of five (Buddhism)*
HM 4456 3 tetrad fourfold according *Concentration of the first jhana of four (Buddhism)*
HM 7202 3 tetrad fourfold according *Concentration of the fourth jhana of four (Buddhism)*
HM 4380 3 tetrad fourfold according *Concentration of the second jhana of four (Buddhism)*
HM 2284 3 tetrad fourfold according *Concentration of the third jhana of four (Buddhism)*

HM 4575 3 tetrad in that applied *Concentration of the second jhana of five (Buddhism)*
HM 8772 3 that of blissfully resting *Meeting God with emptiness (Christianity)*
HM 3941 3 the disguise is charity *Dark night of the soul (Christianity)*
HH 0420 3 the illuminative way *Retreat*
HH 3025 3 the path *Purification by knowledge and vision (Buddhism)*
HH 0136 3 the self is *Unity of personality*
HH 3012 3 the third degree *Degrees of love (Christianity)*
HH 0195 3 the third stage 5 *Sensorimotor development*
HM 0137 3 there is tranquil serenity *Dhyana (Hinduism, Buddhism)*
HM 2470 3 these words carry such *Mansions of favours and afflictions (Christianity)*
HH 0825 3 things are said st *Discipline of confession (Christianity)*
HH 2561 3 thousand names one thousand *Ninety-nine names of Allah (Sufism)*
HM 2289 3 time and begs *Jesus falls the third time under the cross (Christianity)*
HM 2289 3 time under the cross *Jesus falls the third time under the cross (Christianity)*
HM 2735 3 times he turns *Buddha-consciousness (Buddhism, Tibetan)*
HM 3575 3 times to indicate inner *Inner peace*
HM 3323 3 times without root cause *Mind consciousness (Buddhism)*
HM 1425 3 to 7 is powerfully *Intuitive-projective faith*
HM 2062 3 trance is a state *Third trance of the fine-material sphere (Buddhism)*
HM 2142 3 trance is represented *Form-realm awareness (Buddhism, Tibetan)*
HM 2062 3 trance of the fine *Third trance of the fine-material sphere (Buddhism)*
HM 3437 3 transformations is repeated context *Nirodha parinama (Yoga)*
HM 4999 3 triad and in that *Access concentration (Buddhism)*
HM 7602 3 triad this is compared *Understanding as skill in detriment (Buddhism)*
HM 8290 3 triad this is compared *Understanding as skill in improvement (Buddhism)*
HM 6601 3 triad this is compared *Understanding as skill in means (Buddhism)*
HM 1454 3 triad this is *Concentration accompanied by bliss (Buddhism)*
HM 5008 3 triad this is *Concentration accompanied by equanimity (Buddhism)*
HM 4433 3 triad this is *Concentration accompanied by happiness (Buddhism)*
HH 0523 3 truth is *Four noble truths (Buddhism)*
HM 4712 3 types love *Love in God (Islam)*
HM 5116 3 types love *Love to God (Islam)*
HM 5580 3 types love *Love with God (Islam)*
HM 4651 3 types may be valid *Bare perception (Buddhism, Tibetan)*
HH 0501 3 types nutritive or vegetative *Soul*
HM 0946 3 types object *Deification (Sufism)*
HM 3944 3 types object *Identification with spiritual master (Sufism)*
HM 3340 3 types object *Unification with life essence (Sufism)*
HM 3955 3 types of enlightenment recognized *Paccekabodhi (Buddhism)*
HM 7091 3 types of enlightenment recognized *Samma-sambodhi (Buddhism)*
HM 0680 3 types of enlightenment recognized *Savakabodhi (Buddhism)*
HH 0705 3 types of games are *Games*
HM 2578 3 types the wish *Aspiration (Buddhism)*
HM 3341 3 types those who seek *Religious traditions in the ascension stages game (Buddhism, Tibetan)*
HH 0650 3 ultimately derive *Human development (Buddhism)*
HH 1747 3 understanding *Sufi path to perfection (Sufism)*
HM 4398 3 unity *Deautomatization and the mystic experience*
HH 5652 3 up to the base *Supernormal powers (Buddhism)*
HH 0715 3 upper castes *Twice-born*
HH 7004 3 upper grades also over *Sidetracks and auxiliary methods in Taoism (Taoism)*
HH 7734 3 upper grades *Higher upper grade (Taoism)*
HH 4711 3 upper grades *Lower upper grade (Taoism)*
HH 4864 3 upper grades *Lower vehicle path (Taoism)*
HH 6304 3 upper grades *Middle upper grade (Taoism)*
HM 2727 3 vajra master awareness buddhism *Third vajra-master awareness (Buddhism, Tibetan)*
HH 5342 3 vehicles of gradual method *Three vehicles of gradual method (Taoism)*
HH 7386 3 vehicles of the gradual *Higher vehicle path (Taoism)*
HH 4864 3 vehicles of the gradual *Lower vehicle path (Taoism)*
HH 6345 3 vehicles of the gradual *Middle vehicle path (Taoism)*
HH 7004 3 vehicles of the gradual *Sidetracks and auxiliary methods in Taoism (Taoism)*
HH 2117 3 virtues to be realized *Justice*
HH 0227 3 way interaction prevails *Group counseling*
HM 2470 3 way the lord speaks *Mansions of favours and afflictions (Christianity)*
HM 3889 3 ways (1) essentialization *Mystic union with God (Christianity)*
HM 3889 3 ways (1) intellectual *Mystic union with God (Christianity)*
HM 4298 3 ways beginning purification *Dwelling in the first jhana (Buddhism)*
HM 6553 3 ways beginning purification *Dwelling in the fivefold jana (Buddhism)*
HM 8087 3 ways beginning purification *Dwelling in the fourth jhana (Buddhism)*
HM 7121 3 ways beginning purification *Dwelling in the second jhana (Buddhism)*
HM 5643 3 ways beginning purification *Dwelling in the third jhana (Buddhism)*
HM 0810 3 ways *Diakonia (Christianity)*
HM 0631 3 ways in a hierarchy *Triple way (Christianity)*
HM 2254 3 ways in which he *Ma'rifa (Sufism)*
HM 3237 3 ways it controls *Mental engagement (Buddhism)*
HM 4256 3 ways *Kerygma (Christianity)*
HM 0144 3 ways *Koinonia (Christianity)*
HH 0628 3 ways salvation *Bhakti marga (Hinduism)*
HH 0495 3 ways salvation *Jnana marga (Hinduism)*
HH 1211 3 ways salvation *Karma marga (Hinduism)*
HH 0330 3 ways to salvation are *Human development (Hinduism)*
HH 9760 3 week one successively applies *Spiritual exercices of St Ignatius (Christianity)*
HM 3237 3 which is referred *Mental engagement (Buddhism)*
HH 2768 3 with offering and dedication *Corporate worship (Christianity)*
HM 0282 3 with the jhana *Indeterminate consciousness in the immaterial sphere - functional (Buddhism)*
HM 4982 3 with the jhana *Indeterminate consciousness in the immaterial sphere - resultant (Buddhism)*
HM 4701 3 with the jhana *Profitable consciousness in the immaterial sphere (Buddhism)*
HH 4552 3 world countries and support *Liberation theology (Christianity)*
HM 2026 3 worlds description *Subjective domains of consciousness (Yoga)*
HM 2051 3 worlds of desire form *Fourth absorption in the immaterial sphere (Buddhism)*
HH 0661 3 yogas therefore leads *Yoga (Yoga)*
HM 3092 3 zodiacal quadruplicities or crosses *Mutable awareness (Astrology)*
HH 0075 4 a deeply positive mood *Psychedelic drugs*
HM 2051 4 absorption in the immaterial *Fourth absorption in the immaterial sphere (Buddhism)*
HM 4456 4 according to the four *Concentration of the first jhana of four (Buddhism)*
HM 7202 4 according to the four *Concentration of the fourth jhana of four (Buddhism)*
HM 4380 4 according to the four *Concentration of the second jhana of four (Buddhism)*
HM 2284 4 according to the four *Concentration of the third jhana of four (Buddhism)*
HM 2122 4 additional stages of absorption *Trances and mental absorptions (Buddhism)*
HM 4297 4 aeons *Knowledge of recollection of previous existence (Buddhism)*
HM 2596 4 aeons or kappas culminates *State of universal cessation awareness (Buddhism)*
HH 0239 4 ages or yuga *Kali yuga*
HH 1778 4 ages the iron age *Primitivism*
HH 0330 4 aim was not always *Human development (Hinduism)*
HH 0754 4 aims *Kama (Hinduism)*
HH 1198 4 although things *Human development through religion*
HM 3102 4 among the spiritual paths *Path of receiving intelligence (Judaism)*
HM 0485 4 analyses description *Four discriminations (Buddhism)*
HM 6553 4 and fifth *Dwelling in the fivefold jana (Buddhism)*
HM 2144 4 and fifth levels ananda *Ajna (Yoga)*
HM 2280 4 and final tantric initiation *Tantra union-in-learning-application awareness (Buddhism, Tibetan)*
HM 4191 4 and four *Concentration accompanied by bliss (Buddhism)*
HM 2980 4 and more time dreaming *Sleep*
HH 3987 4 and so *Subjects for meditation (Buddhism)*
HM 5008 4 and the fivefold reckoning *Concentration accompanied by equanimity (Buddhism)*
HM 4297 4 and then adverts *Knowledge of recollection of previous existence (Buddhism)*
HM 2335 4 antarabhava consciousness *Death-bardo clear-light awareness (Buddhism)*
HM 3852 4 are accompanied by joy *Indeterminate consciousness in the sense sphere - functional (Buddhism)*
HM 5721 4 are accompanied by joy *Indeterminate consciousness in the sense sphere - resultant (Buddhism)*
HM 4447 4 are accompanied by joy *Profitable consciousness in the sense sphere (Buddhism)*
HM 8375 4 are accompanied by joy *Unprofitable consciousness in the sense sphere (Buddhism)*
HM 3852 4 are accompanied with equanimity *Indeterminate consciousness in the sense sphere - functional* (Buddhism)
HM 5721 4 are accompanied with equanimity *Indeterminate consciousness in the sense sphere - resultant* (Buddhism)
HM 4447 4 are accompanied with equanimity *Profitable consciousness in the sense sphere (Buddhism)*
HM 8375 4 are accompanied with equanimity *Unprofitable consciousness in the sense sphere (Buddhism)*
HH 2909 4 are described as preparation *Psychospiritual growth (Christianity)*
HM 4651 4 are non deceptive *Bare perception (Buddhism, Tibetan)*
HM 2050 4 are or whatever states *Awareness as mental-formation group of conscious existence (Buddhism)*
HH 0669 4 asanas there are 84 *Asana (Hinduism, Yoga)*
HH 1987 4 ashramas or stages *Brahmacharyashrama (Hinduism)*
HH 2343 4 ashramas or stages *Grahastashrama (Hinduism)*
HH 2056 4 ashramas or stages *Sanyasashrama (Hinduism)*
HH 2782 4 ashramas or stages *Vanaprasthashrama (Hinduism)*
HH 0513 4 aspects of creative existence *Creative existence*
HM 2245 4 aspects of each *Meditation way of the four truths (Buddhism)*
HM 3170 4 aspiration effort thought *Prerequisites of manifestation (Buddhism)*
HH 2909 4 awareness insufficiency *Psychospiritual growth (Christianity)*
HH 5652 4 bases or roads *Supernormal powers (Buddhism)*
HM 2262 4 basic forms are distinguished *Extrasensory perception (ESP)*
HH 0661 4 basic personality types reflective *Yoga (Yoga)*
HM 2245 4 bden *Meditation way of the four truths (Buddhism)*
HH 0306 4 being the path *Tantra (Buddhism, Yoga)*
HM 1727 4 benefits produced *Dark night of the senses (Christianity)*
HM 1225 4 bodhisattva leaders who are *State of bodhisattva (Buddhism)*
HM 3690 4 branch of the tibetan *Yoga of the clear light (Buddhism)*
HM 4456 4 buddhism description *Concentration of the first jhana of four (Buddhism)*
HM 7202 4 buddhism description *Concentration of the fourth jhana of four (Buddhism)*
HM 4380 4 buddhism description *Concentration of the second jhana of four (Buddhism)*
HM 2284 4 buddhism description *Concentration of the third jhana of four (Buddhism)*
HH 1010 4 but it is not *Five eyes (Buddhism)*
HH 5640 4 cardinal or principal virtues *Fortitude*
HH 2117 4 cardinal or principal virtues *Justice*
HH 7902 4 cardinal or principal virtues *Prudence*
HH 0600 4 cardinal or principal virtues *Temperance*
HH 0712 4 cardinal or principal virtues *Virtue*
HH 0712 4 cardinal virtues feature prominently *Virtue*
HH 0330 4 castes each with specific *Human development (Hinduism)*
HH 0330 4 castes or varnas are *Human development (Hinduism)*
HM 4772 4 cause of profane love *Tama' (Islam)*
HH 2006 4 centuries before christ *Quietism*
HH 0288 4 century the church requires *Feasting*
HM 3291 4 chakra attaining balance leela *Fourth chakra: attaining balance (Leela)*
HM 3242 4 chakra brings mastery *Anahata (Yoga)*
HM 5089 4 changeable factors referred *Analysis (Buddhism)*
HM 7209 4 changeable factors referred *Contrition (Buddhism)*
HM 1177 4 changeable factors referred *Investigation (Buddhism)*
HM 7921 4 changeable factors referred *Sleep (Buddhism)*
HH 1348 4 changeable *Secondary mental factors (Buddhism, Tibetan)*
HH 0347 4 changes in the human *Healthy human growth and development*
HM 2291 4 characteristics of the experience *Cosmic consciousness*
HM 2335 4 chikhai bardo *Death-bardo clear-light awareness (Buddhism)*
HM 2733 4 classes of demigods *Desire-realm consciousness (Buddhism)*
HH 3987 4 colour kasinas are suitable *Subjects for meditation (Buddhism)*
HH 0213 4 component of the eightfold *Pranayama (Hinduism, Yoga)*
HM 2586 4 concentration 7 stages *Fourth trance of the fine-material sphere (Buddhism)*
HM 2586 4 concentration and attains *Fourth trance of the fine-material sphere (Buddhism)*
HM 7434 4 concentration *Concentration due to inquiry (Buddhism)*
HM 4977 4 concentration *Concentration of easy progress and swift direct-knowledge (Buddhism)*
HM 4882 4 concentration *Concentration of the fourth jhana of five (Buddhism)*
HM 7202 4 concentration *Concentration of the fourth jhana of four (Buddhism)*
HM 5930 4 concentration *Concentration partaking of penetration (Buddhism)*
HM 2586 4 concentration eliminates the last *Fourth trance of the fine-material sphere (Buddhism)*
HM 5143 4 concentration in the third *Concentration of the fifth jhana of five (Buddhism)*
HM 2257 4 concentration is divided *Form-realm consciousness (Buddhism)*
HM 5214 4 concentration *Measureless concentration with a measureless object (Buddhism)*
HM 2284 4 concentration of the pentad *Concentration of the third jhana of four (Buddhism)*
HM 5768 4 concentration *Unincluded concentration (Buddhism)*
HM 2623 4 concentrations (dhyana) *Special insight and preparations as states (Buddhism)*
HM 3150 4 concentrations infinite space infinite *Serial absorptions (Buddhism)*
HM 2195 4 concentrations on the desire *Individual knowledge of the character (Buddhism, Tibetan)*
HM 2257 4 concentrations or dhyana *Form-realm consciousness (Buddhism)*
HM 2987 4 concentrations the four immeasurables *Paths of calming (Buddhism)*
HM 2586 4 concentrations the main activity *Fourth trance of the fine-material sphere (Buddhism)*
HM 2098 4 conditions classed as profitable *Dispositions of consciousness (Buddhism)*
HM 0891 4 conditions the waking state *Degrees of consciousness (Hinduism)*
HM 2050 4 constant contact volition life *Awareness as mental-formation group of conscious existence (Buddhism)*
HM 6663 4 constituents and constitutes *Concentration (Buddhism)*
HM 2108 4 constitute the consciousness *Awareness of corporeality-group of conscious existence (Buddhism)*

INDEX TO CONCEPT SETS
(See SECTION X for KEYWORD INDEX)

HX

HM 6932 4 contemplative disciplines the salik *Cleansing of the heart (Sufism)*
HM 5353 4 contemplative disciplines the salik *Cleansing the sirr (Sufism)*
HM 6162 4 contemplative disciplines the salik *Illumination of the spirit (Sufism)*
HH 5092 4 contemplative disciplines the salik *Purification of the self (Sufism)*
HM 2733 4 corners has 4 neighboring *Desire-realm consciousness (Buddhism)*
HH 0661 4 correspond to the developmental *Yoga (Yoga)*
HH 5522 4 current models of human *Communalist human development (United Nations)*
HH 6122 4 current models of human *Liberal capitalist human development (United Nations)*
HH 4033 4 current models of human *Liberal humanist human development (United Nations)*
HH 3997 4 current models of human *State socialist human development (United Nations)*
HM 1785 4 cycles are slowest *Delta wave consciousness*
HH 3129 4 cycles are slowest followed *Brain waves*
HH 3012 4 degree of love when *Degrees of love (Christianity)*
HM 0137 4 degrees of dhyanas *Dhyana (Hinduism, Buddhism)*
HM 2050 4 described as or whatever *Awareness as mental-formation group of conscious existence (Buddhism)*
HM 3630 4 developing contemplation of dispassion *Principal insights (Buddhism)*
HM 1781 4 deviations of speech buddhism *Leaving off, abstaining, totally abstaining and refraining from the four* deviations of speech (Buddhism)
HH 1010 4 dharmacaksus the eye *Five eyes (Buddhism)*
HM 0137 4 dhyanas determines one's position *Dhyana (Hinduism, Buddhism)*
HM 4071 4 dimensional clairvoyance psychism description *Fourth dimensional clairvoyance (Psychism)*
HM 0485 4 discriminations buddhism four analyses *Four discriminations (Buddhism)*
HM 0485 4 discriminations or analyses can *Four discriminations (Buddhism)*
HM 0485 4 discriminations or analyses *Four discriminations (Buddhism)*
HM 4958 4 discriminations or analyses *Understanding as knowledge about kinds of knowledge (Buddhism)*
HM 5026 4 discriminations or analyses *Understanding as knowledge about language (Buddhism)*
HM 4726 4 discriminations or analyses *Understanding as knowledge about law (Buddhism)*
HM 1663 4 discriminations or analyses *Understanding as knowledge about meaning (Buddhism)*
HM 2098 4 dispositions said *Dispositions of consciousness (Buddhism)*
HH 3987 4 divine abiding as this *Subjects for meditation (Buddhism)*
HH 3987 4 divine abiding *Subjects for meditation (Buddhism)*
HM 7769 4 divine abidings or states *Equanimity (Buddhism)*
HM 5224 4 divine abidings or states *Gladness (Buddhism)*
HM 7607 4 divine abidings or states *Lovingkindness (Buddhism)*
HM 0513 4 divine abidings or states *Pity (Buddhism)*
HH 3987 4 divine states or abidings *Subjects for meditation (Buddhism)*
HM 2064 4 doors of retention buddhism *Four doors of retention (Buddhism)*
HM 3079 4 doors of retention *Paths of special qualities (Buddhism)*
HH 0523 4 duhkha *Four noble truths (Buddhism)*
HH 0523 4 dukkha *Four noble truths (Buddhism)*
HM 4191 4 dyad this is contrasted *Concentration accompanied by bliss (Buddhism)*
HM 8021 4 dyad this is contrasted *Concentration accompanied by equanimity (Buddhism)*
HM 7525 4 dyad this is contrasted *Understanding accompanied by equanimity (Buddhism)*
HM 6245 4 dyad this is contrasted *Understanding accompanied by joy (Buddhism)*
HM 4999 4 dyads *Access concentration (Buddhism)*
HM 6663 4 dyads four triads six *Concentration (Buddhism)*
HH 0846 4 dynamically distinct relationships are *Multiple therapy*
HM 0931 4 ecstasies *Right rapture of concentration (Buddhism)*
HH 0767 4 einsteinian circuits *Mutational phases of life*
HM 4999 4 elements *Access concentration (Buddhism)*
HH 6282 4 elements and then *Subtle faculties (Sufism)*
HM 2082 4 elements context *Four-great-kings-heaven awareness (Buddhism, Tibetan)*
HH 6282 4 elements earth water air *Subtle faculties (Sufism)*
HM 2562 4 elements earth water fire *Body consciousness (Buddhism)*
HM 2450 4 elements have become balanced *First trance of the fine-material sphere (Buddhism)*
HM 2237 4 elements his magic is *Magician archetypal image (Tarot)*
HM 2713 4 elements of earth water *Zodiacal forms of awareness (Astrology)*
HM 2237 4 elements of the material *Magician archetypal image (Tarot)*
HH 3987 4 elements perception of repulsiveness *Subjects for meditation (Buddhism)*
HH 3987 4 elements they cannot *Subjects for meditation (Buddhism)*
HM 3232 4 elimination as feeling *Awareness of spiritual identity (Yoga)*
HM 2185 4 equalizations in the realms *Way of the Buddhas (Buddhism, Tibetan)*
HM 2735 4 equalizations of formless realm *Buddha-consciousness (Buddhism, Tibetan)*
HM 2097 4 equals 15 so that *Tarot arcana of conscious states (Tarot)*
HM 1252 4 evil paths buddhism description *Four evil paths (Buddhism)*
HM 2959 4 evil paths *State of anger (Buddhism)*
HH 1038 4 facets are used *Madhyamaka (Buddhism)*
HM 4575 4 factors remain *Concentration of the second jhana of five (Buddhism)*
HM 6553 4 factors sustained thought happiness *Dwelling in the fivefold jana (Buddhism)*
HM 1665 4 faith description *Individuative-reflective faith*
HM 3044 4 fearlessnesses buddhism tibetan description *Four fearlessnesses (Buddhism, Tibetan)*
HM 3249 4 fearlessnesses the four sciences *Paths of effect (Tibetan)*
HM 3301 4 fetter or character defect *Emotional desires (Buddhism)*
HM 1618 4 flashes *Appearance of absorption (Buddhism)*
HH 0258 4 fold affection friendship eros *Love*
HM 3144 4 fold and includes *Formless-realm awareness (Buddhism, Tibetan)*
HM 0239 4 fold contact *Nonecstatic perception (Hinduism)*
HM 2866 4 fold jhana brings rebirth *Sukha (Buddhism)*
HM 1991 4 fold knowledge systematics value *Four-fold knowledge (Systematics)*
HM 3144 4 fold states are said *Formless-realm awareness (Buddhism, Tibetan)*
HH 1865 4 fold virtue context *Purification of virtue (Buddhism)*
HM 2450 4 form concentrations (dhyana) found *First trance of the fine-material sphere (Buddhism)*
HM 2586 4 form concentrations (dhyana) found *Fourth trance of the fine-material sphere (Buddhism)*
HM 2038 4 form concentrations dhyana found *Second trance of the fine-material sphere (Buddhism)*
HM 2062 4 form concentrations (dhyana) found *Third trance of the fine-material sphere (Buddhism)*
HM 2447 4 formal aspect *Awareness of depth (ICA)*
HM 3199 4 formal aspect *Embodying service (ICA)*
HM 2042 4 formal aspect *Glorious mystery (ICA)*
HM 3097 4 formal aspect is *All-being-in-myself (ICA)*
HM 3054 4 formal aspect is *Appropriated passion (ICA)*
HM 2286 4 formal aspect is *Awareness of spiritual poverty (ICA)*
HM 2143 4 formal aspect is becoming *Doing the mystery (ICA)*
HM 2119 4 formal aspect is becoming *Final situation (ICA)*
HM 2017 4 formal aspect is becoming *Meaning creation (ICA)*
HM 2005 4 formal aspect is becoming *Obedient son (ICA)*
HM 2154 4 formal aspect is *Being myself (ICA)*
HM 2591 4 formal aspect is *Defender of deeps (ICA)*
HM 2338 4 formal aspect is embodying *Communion of saints (ICA)*
HM 2541 4 formal aspect is embodying *Eternal identification (ICA)*
HM 2661 4 formal aspect is embodying *Ethical existence (ICA)*
HM 2432 4 formal aspect is embodying *Global brotherhood (ICA)*
HM 3029 4 formal aspect is *Honouring the mystery (ICA)*
HM 2104 4 formal aspect is *Human contingency (ICA)*
HM 2334 4 formal aspect is *Human transformation (ICA)*
HM 2369 4 formal aspect is *Intentional self-negation (ICA)*
HM 2092 4 formal aspect is *Inventing humanness (ICA)*
HM 3155 4 formal aspect is *Irreplaceable uniqueness (ICA)*
HM 2889 4 formal aspect is one's *Agonizing prediction (ICA)*
HM 3027 4 formal aspect is one's *Destinal elector (ICA)*
HM 2859 4 formal aspect is one's *Divine hosts (ICA)*
HM 2994 4 formal aspect is one's *Eternal void (ICA)*
HM 2821 4 formal aspect is one's *Everlasting enemy (ICA)*
HM 2903 4 formal aspect is one's *Expectant descendant (ICA)*
HM 3124 4 formal aspect is one's *Immutable friend (ICA)*
HM 2939 4 formal aspect is one's *Particular concern (ICA)*
HM 3224 4 formal aspect is one's *Primordial ancestor (ICA)*
HM 3172 4 formal aspect is one's *Promissorial offering (ICA)*
HM 2625 4 formal aspect is one's *Utter awareness (ICA)*
HM 2413 4 formal aspect is *Saving the mystery (ICA)*
HM 3190 4 formal aspect is spiritual *Divine captive (ICA)*
HM 3095 4 formal aspect is spiritual *Eternal friends (ICA)*
HM 3140 4 formal aspect is spiritual *Replication of Christ (ICA)*
HM 3086 4 formal aspect is spiritual *Secondary integrity (ICA)*
HM 3108 4 formal aspect is *Unexplainable thereness (ICA)*
HM 2976 4 formal aspect *Missional comradeship (ICA)*
HM 2037 4 formal aspect *Spiritual creativity (ICA)*
HM 2174 4 formal aspect *Spiritual denial (ICA)*
HM 3176 4 formal aspect *Symbolizing the eternal context (ICA)*
HM 2462 4 formal aspect *Transparent presence (ICA)*
HM 3110 4 formal aspect *Universal responsibility (ICA)*
HM 3043 4 formless absorptions (arupaya samapatti) *Second absorption in the immaterial sphere (Buddhism)*
HM 2110 4 formless absorptions (arupayasamapatti) *First absorption in the immaterial sphere (Buddhism)*
HM 2051 4 formless absorptions (arupayasamapatti) *Fourth absorption in the immaterial sphere (Buddhism)*
HM 2027 4 formless absorptions (arupayasamapatti) *Third absorption in the immaterial sphere (Buddhism)*
HM 2987 4 formless absorptions the eight *Paths of calming (Buddhism)*
HH 3987 4 formless states are not *Subjects for meditation (Buddhism)*
HM 2696 4 forms of buddhahood *Tantra heat-application awareness (Buddhism, Tibetan)*
HH 4006 4 foundations body contemplation *Practice of presence of mindfulness (Buddhism)*
HM 2793 4 foundations of mindfulness right *Bodhi-pakkhiya dhamma (Buddhism)*
HM 3048 4 foundations of psychic powers *Great-vehicle higher-path awareness (Buddhism, Tibetan)*
HM 7268 4 fruition (indeterminate resultant) consciousnesses *Impulsion (Buddhism)*
HH 3025 4 fruition is the eighth *Purification by knowledge and vision (Buddhism)*
HH 3025 4 fruition *Purification by knowledge and vision (Buddhism)*
HM 7268 4 functional or inoperative consciousnesses *Impulsion (Buddhism)*
HM 2098 4 functional or inoperative *Dispositions of consciousness (Buddhism)*
HM 2578 4 fundamental principles of creative *Aspiration (Buddhism)*
HH 0996 4 general approaches may *Ideological re-education*
HM 0282 4 give rise to materiality *Indeterminate consciousness in the immaterial sphere - functional* (Buddhism)
HM 5129 4 give rise to materiality *Indeterminate consciousness in the supramundane plane - resultant* (Buddhism)
HM 4701 4 give rise to materiality *Profitable consciousness in the immaterial sphere (Buddhism)*
HM 4930 4 give rise to materiality *Profitable consciousness in the supramundane plane (Buddhism)*
HM 4982 4 gives rise to materiality *Indeterminate consciousness in the immaterial sphere - resultant* (Buddhism)
HM 2072 4 grades in the immaterial *Worlds of conscious existence (Buddhism, Pali)*
HH 0754 4 great aims in human *Kama (Hinduism)*
HH 0353 4 great aims of life *Artha (Hinduism)*
HH 0093 4 great aims of life *Dharma*
HH 0330 4 great aims of life *Human development (Hinduism)*
HM 2733 4 great kings *Desire-realm consciousness (Buddhism)*
HM 2082 4 great kings heaven awareness *Four-great-kings-heaven awareness (Buddhism, Tibetan)*
HM 2010 4 great kings *Heavenly-highway awareness (Buddhism, Tibetan)*
HM 7655 4 great primaries descent *Conditions for consciousness (Buddhism)*
HM 2693 4 has eight levels *Tibetan meditative states of form concentrations (Buddhism)*
HH 5652 4 he repeatedly attains *Supernormal powers (Buddhism)*
HM 2272 4 headings ineffability noetic quality *Mystical cognition*
HH 0221 4 (heaven) coniunctio symbolized *Alchemy (Esotericism)*
HM 2669 4 higher trances (jhana) *Tibetan meditative states of formless absorptions (Buddhism)*
HM 2470 4 hours *Mansions of favours and afflictions (Christianity)*
HM 2067 4 human islands *Northern-continent awareness of community (Buddhism, Tibetan)*
HM 7769 4 if the pentad is *Equanimity (Buddhism)*
HM 2906 4 illumination of jami *Separation from the transient (Sufism)*
HM 7655 4 immaterial aggregates 7 *Conditions for consciousness (Buddhism)*
HM 7655 4 immaterial aggregates 8 *Conditions for consciousness (Buddhism)*
HM 7655 4 immaterial aggregates are *Conditions for consciousness (Buddhism)*
HM 5627 4 immaterial aggregates context *Understanding defining mentality (Buddhism)*
HM 8273 4 immaterial indeterminate resultant consciousnesses *Consciousness-born materiality (Buddhism)*
HM 8266 4 immaterial resultant consciousnesses rebirth *Rebirth-linking (Buddhism)*
HH 3987 4 immaterial states bring *Subjects for meditation (Buddhism)*
HH 3198 4 immaterial states or formless *Immaterial states as meditation subjects (Buddhism)*
HH 3987 4 immaterial states or formless *Subjects for meditation (Buddhism)*
HM 0282 4 immaterial states or jhanas *Indeterminate consciousness in the immaterial sphere - functional* (Buddhism)
HM 4982 4 immaterial states or jhanas *Indeterminate consciousness in the immaterial sphere - resultant* (Buddhism)
HM 4701 4 immaterial states or jhanas *Profitable consciousness in the immaterial sphere (Buddhism)*
HH 3987 4 immaterial states where boundless *Subjects for meditation (Buddhism)*
HM 2507 4 immeasurables buddhism description *Four immeasurables (Buddhism)*
HM 2987 4 immeasurables the four formless *Paths of calming (Buddhism)*
HM 1991 4 in a sequence *Four-fold knowledge (Systematics)*
HM 1454 4 in the fivefold reckoning *Concentration accompanied by bliss (Buddhism)*
HM 2186 4 in the ica *Primordial wonder (ICA)*
HM 1665 4 in the system *Individuative-reflective faith*
HM 2060 4 in the tibetan *Kamaloka consciousness (Buddhism, Pali)*
HH 1747 4 independence *Sufi path to perfection (Sufism)*
HM 1991 4 independent terms *Four-fold knowledge (Systematics)*
HM 8273 4 indeterminate functional *Consciousness-born materiality (Buddhism)*
HM 4983 4 indeterminate functional leaving out *Feeling aggregate (Buddhism)*
HM 4983 4 indeterminate functional with root *Feeling aggregate (Buddhism)*
HM 4983 4 indeterminate resultant and 4 *Feeling aggregate (Buddhism)*
HM 4983 4 indeterminate resultant are each *Feeling aggregate (Buddhism)*
HM 8273 4 indeterminate resultant *Consciousness-born materiality (Buddhism)*
HM 4983 4 indeterminate resultant profitable *Feeling aggregate (Buddhism)*
HM 8266 4 indeterminate resultant *Rebirth-linking (Buddhism)*

HH 2909 4 individuation as the conscious *Psychospiritual growth (Christianity)*
HM 4398 4 ineffability *Deautomatization and the mystic experience*
HM 0210 4 initiation in the tradition *Initiation of renunciation (Esotericism)*
HM 2756 4 initiations for the tantric *Tantra receptivity-application awareness (Buddhism, Tibetan)*
HH 0424 4 initiations one at each *The path*
HM 2126 4 integral level religious *Consciousness expansion*
HH 0136 4 intention namely an ego *Unity of personality*
HM 3438 4 is absolute identification *Tawhid (Sufism)*
HH 0385 4 is the number *Individuation*
HM 0137 4 is without sorrow *Dhyana (Hinduism, Buddhism)*
HM 7769 4 jhana arises *Equanimity (Buddhism)*
HM 8087 4 jhana buddhism description *Dwelling in the fourth jhana (Buddhism)*
HM 4882 4 jhana ease or bliss *Concentration of the fourth jhana of five (Buddhism)*
HM 8087 4 jhana equanimity or indifference *Dwelling in the fourth jhana (Buddhism)*
HM 4456 4 jhana factors this is *Concentration of the first jhana of four (Buddhism)*
HM 7202 4 jhana factors this is *Concentration of the fourth jhana of four (Buddhism)*
HM 4380 4 jhana factors this is *Concentration of the second jhana of four (Buddhism)*
HM 2284 4 jhana factors this is *Concentration of the third jhana of four (Buddhism)*
HM 7202 4 jhana has two factors *Concentration of the fourth jhana of four (Buddhism)*
HH 3198 4 jhana in any one *Immaterial states as meditation subjects (Buddhism)*
HM 4882 4 jhana of five buddhism *Concentration of the fourth jhana of five (Buddhism)*
HM 7202 4 jhana of four buddhism *Concentration of the fourth jhana of four (Buddhism)*
HM 7769 4 jhana the mind is *Equanimity (Buddhism)*
HH 3987 4 jhana there are two *Subjects for meditation (Buddhism)*
HM 8087 4 jhana through the habits *Dwelling in the fourth jhana (Buddhism)*
HM 8087 4 jhana will arise *Dwelling in the fourth jhana (Buddhism)*
HH 3987 4 jhanas as follows *Subjects for meditation (Buddhism)*
HH 3198 4 jhanas associated with them *Immaterial states as meditation subjects (Buddhism)*
HM 6553 4 jhanas each *Dwelling in the fivefold jana (Buddhism)*
HH 5652 4 jhanas first seclusion second *Supernormal powers (Buddhism)*
HH 3987 4 jhanas imply transcending factors *Subjects for meditation (Buddhism)*
HM 4297 4 jhanas in succession emerges *Knowledge of recollection of previous existence (Buddhism)*
HM 6245 4 jhanas of the fivefold *Understanding accompanied by joy (Buddhism)*
HM 6553 4 jhanas of the fourfold *Dwelling in the fivefold jana (Buddhism)*
HM 2051 4 jhanas of the immaterial *Fourth absorption in the immaterial sphere (Buddhism)*
HH 3987 4 jhanas the ten kinds *Subjects for meditation (Buddhism)*
HM 2098 4 karmically neutral resultant *Dispositions of consciousness (Buddhism)*
HM 3438 4 kind of tawhid described *Tawhid (Sufism)*
HM 5129 4 kinds of consciousness are *Indeterminate consciousness in the supramundane plane - resultant* (Buddhism)
HM 4930 4 kinds of consciousness are *Profitable consciousness in the supramundane plane (Buddhism)*
HM 0282 4 kinds of consciousness may *Indeterminate consciousness in the immaterial sphere - functional* (Buddhism)
HM 4982 4 kinds of consciousness may *Indeterminate consciousness in the immaterial sphere - resultant* (Buddhism)
HM 4701 4 kinds of consciousness may *Profitable consciousness in the immaterial sphere (Buddhism)*
HM 7655 4 kinds of nutriment giving *Conditions for consciousness (Buddhism)*
HM 5982 4 kinds of object limited *Knowledge of the divine ear element (Buddhism)*
HM 7525 4 kinds of path consciousnesses *Understanding accompanied by equanimity (Buddhism)*
HM 4958 4 kinds of understanding are *Understanding as knowledge about kinds of knowledge (Buddhism)*
HM 5026 4 kinds of understanding are *Understanding as knowledge about language (Buddhism)*
HM 4726 4 kinds of understanding are *Understanding as knowledge about law (Buddhism)*
HM 1663 4 kinds of understanding are *Understanding as knowledge about meaning (Buddhism)*
HM 0485 4 kinds of understanding *Four discriminations (Buddhism)*
HM 0485 4 kinds of understanding *Four discriminations (Buddhism)*
HM 4958 4 kinds of understanding *Understanding as knowledge about kinds of knowledge (Buddhism)*
HM 5026 4 kinds of understanding *Understanding as knowledge about language (Buddhism)*
HM 4726 4 kinds of understanding *Understanding as knowledge about law (Buddhism)*
HM 1663 4 kinds of understanding *Understanding as knowledge about meaning (Buddhism)*
HM 6870 4 kinds of understanding *Understanding as knowledge of suffering (Buddhism)*
HM 8163 4 kinds of understanding *Understanding as knowledge of the cessation of suffering (Buddhism)*
HM 7290 4 kinds of understanding *Understanding as knowledge of the origin of suffering (Buddhism)*
HM 7645 4 kinds of understanding *Understanding as knowledge of the way leading to cessation suffering* (Buddhism)
HM 6663 4 kinds (tetrads) *Concentration (Buddhism)*
HM 4523 4 kinds (tetrads) *Understanding (Buddhism)*
HM 0130 4 kinds which are formed *Mundane resultant consciousness (Buddhism)*
HH 0986 4 latency stage 6 11 *Libido development*
HM 1105 4 learning modes virtual *Ontogenetic model of human consciousness*
HM 0168 4 learning processes distinguished lazslo *Conscious learning*
HM 0196 4 learning processes distinguished lazslo *Functional learning*
HM 1244 4 learning processes distinguished lazslo *Superconscious learning*
HM 1217 4 learning processes distinguished lazslo *Virtual learning*
HM 2420 4 letter name of god *Sphere of mercy and greatness (Kabbalah)*
HM 2282 4 letters jhvh to show *Angelic awareness - Zohariel*
HM 0809 4 level of samprajnata samadhi *Nirvicara samadhi (Yoga)*
HM 1355 4 level of the game *Good company (Leela)*
HM 2341 4 level so to speak *Stations of consciousness - Simnami (Sufism)*
HH 0018 4 levels in the self *Anthroposophical system*
HM 2281 4 levels in this realm *Formless-realm consciousness (Buddhism)*
HH 0018 4 levels of apprehension corresponding *Anthroposophical system*
HH 4019 4 levels of dhyana *Steps to enlightenment (Buddhism)*
HH 8903 4 levels of experience were *Maps of the mind*
HM 2896 4 levels vitarka samadhi nirvitarka *Samprajnata samadhi (Yoga)*
HH 1038 4 logical alternatives a is *Madhyamaka (Buddhism)*
HM 3242 4 lotus or chakra defined *Anahata (Yoga)*
HH 3534 4 lovingkindness pity or compassion *Divine abidings as meditation subjects (Buddhism)*
HM 2072 4 lowest classes or rounds *Worlds of conscious existence (Buddhism, Pali)*
HH 0827 4 main types of yoga *Dhyana yoga (Yoga)*
HH 3129 4 major kinds of brain *Brain waves*
HH 1785 4 major kinds of brain *Delta wave consciousness*
HH 8903 4 major levels *Maps of the mind*
HM 2020 4 major state of consciousness *Transcendental consciousness*
HH 0851 4 major types of therapy *Group therapy*
HH 0136 4 man strives *Unity of personality*
HM 1077 4 means and leading *Erroneous perception (Yoga)*
HH 0420 4 meditations the first concerning *Retreat*
HM 2280 4 meditative cultivation *Tantra union-in-learning-application awareness (Buddhism, Tibetan)*
HM 3144 4 meditative stabilizations or equalizations *Formless-realm awareness (Buddhism, Tibetan)*
HM 3079 4 meditative stabilizations the four *Paths of special qualities (Buddhism)*
HM 1244 4 mode of learning posited *Superconscious learning*
HM 3594 4 modes of furor
HH 0661 4 more specialized systems are *Yoga (Yoga)*
HH 0767 4 mutational phases are *Mutational phases of life*
HM 3371 4 nasut following the law *Divine path (Sufism)*
HM 8219 4 natural enemies of man *Way of the warrior (Amerindian)*
HM 2733 4 neighboring hells altogether 108 *Desire-realm consciousness (Buddhism)*
HH 5652 4 neither pain nor pleasure *Supernormal powers (Buddhism)*
HM 1914 4 noble paths and starts *Six paths (Buddhism)*
HM 4026 4 noble paths buddhism description *Four noble paths (Buddhism)*
HM 2657 4 noble paths require inner *Ten worlds (Buddhism)*
HM 7563 4 noble person *Knowledge of the path of once-return (Buddhism)*
HM 4467 4 noble truths and actions *Doubt (Buddhism)*
HM 2551 4 noble truths (ariya sacca) *Phenomena awareness (Buddhism)*
HH 0523 4 noble truths buddhism ariya *Four noble truths (Buddhism)*
HH 1099 4 noble truths *Causally continuous doctrine of being (Buddhism)*
HM 2240 4 noble truths *Discipleship-vision awareness (Buddhism, Tibetan)*
HH 0650 4 noble truths is intended *Human development (Buddhism)*
HM 2339 4 noble truths leading *Eightfold way (Buddhism)*
HM 2330 4 noble truths *Nirvana (Buddhism)*
HH 4019 4 noble truths *Steps to enlightenment (Buddhism)*
HM 2987 4 noble truths the four *Paths of calming (Buddhism)*
HM 3527 4 noble truths *Two-fold knowledge of truths (Buddhism)*
HM 6870 4 noble truths *Understanding as knowledge of suffering (Buddhism)*
HM 8163 4 noble truths *Understanding as knowledge of the cessation of suffering (Buddhism)*
HM 7290 4 noble truths *Understanding as knowledge of the origin of suffering (Buddhism)*
HM 7645 4 noble truths *Understanding as knowledge of the way leading to cessation suffering (Buddhism)*
HM 8901 4 of 40 stations *Iradat (Sufism)*
HM 1351 4 of eight lokas represented *Plane of balance (Leela)*
HH 0677 4 of five perfections *Virya (Zen)*
HM 3006 4 of seven ascending planes *Establishment in truth (Hinduism)*
HM 2014 4 of seven descending steps *Wakeful dream (Hinduism)*
HM 3443 4 of seven mansions *Mansions of spiritual consolations (Christianity)*
HM 4338 4 of seven stages *Assembling the five elements (Taoism)*
HM 3518 4 of seven stages *Heart-awakening stage of life*
HM 3616 4 of seven subtle stages *World of imagination (Sufism)*
HM 0763 4 of seven ways *Way of harmony (Esotericism)*
HM 1516 4 of ten orders *Fourth order perceptions - control of transitions*
HM 2420 4 of ten spheres described *Sphere of mercy and greatness (Kabbalah)*
HM 3759 4 of ten states acting *Character (Sufism)*
HH 3398 4 of the five primary *Vinaya (Buddhism)*
HM 4191 4 of the fivefold reckoning *Concentration accompanied by bliss (Buddhism)*
HH 2056 4 of the four ashramas *Sanyasashrama (Hinduism)*
HM 2339 4 of the four noble *Eightfold way (Buddhism)*
HM 2157 4 of the nine states *Close setting the mind (Buddhism)*
HM 0235 4 of the ninety nine *Al-Quddus (Sufism)*
HM 2743 4 of the preparations *Full-isolation subtle contemplation (Buddhism)*
HM 2341 4 of the seven stages *Stations of consciousness - Simnami (Sufism)*
HM 1252 4 of the ten inner *Four evil paths (Buddhism)*
HH 0604 4 of the ten meritorious *Respect for superiors (Buddhism)*
HM 4026 4 of the ten worlds *Four noble paths (Buddhism)*
HM 2098 4 of which are resultant *Dispositions of consciousness (Buddhism)*
HM 2100 4 on the board *Howling-hells awareness (Buddhism, Tibetan)*
HH 0647 4 opportunities for creativity innovation *Development of the quality of human life*
HM 7193 4 or fifth jhana are *Jhana (Buddhism, Pali)*
HM 0748 4 or fine material jhana *Knowledge of passing away and reappearance of beings (Buddhism)*
HM 1618 4 or five flashes *Appearance of absorption (Buddhism)*
HM 2282 4 or manifold divinities pointing *Angelic awareness - Zohárariel*
HH 0507 4 or more participants some *Analytical therapy*
HM 2402 4 or objective state *Objective fourth state of consciousness*
HM 2050 4 or whatever states *Awareness as mental-formation group of conscious existence (Buddhism)*
HM 1516 4 order *Fourth order perceptions - control of transitions*
HM 1516 4 order perception enables *Fourth order perceptions - control of transitions*
HM 1516 4 order perceptions control *Fourth order perceptions - control of transitions*
HH 0565 4 orientation toward authority law *Moral development*
HM 2751 4 parts *Consciousness of cyclical time*
HH 2002 4 parts or aeons *World cycles (Buddhism)*
HM 2252 4 path of the independent *Pratyeka Buddha cultivation awareness (Buddhism, Tibetan)*
HM 7268 4 path (profitable) and four *Impulsion (Buddhism)*
HH 3025 4 path the path *Purification by knowledge and vision (Buddhism)*
HM 4128 4 paths context *Understanding as the plane of development (Buddhism)*
HM 6192 4 paths context *Understanding as the plane of seeing (Buddhism)*
HH 0102 4 paths for the soul *Stages of spiritual life*
HM 5129 4 paths fruition moment stream *Indeterminate consciousness in the supramundane plane - resultant* (Buddhism)
HH 3025 4 paths *Purification by knowledge and vision (Buddhism)*
HM 4930 4 paths stream entry once *Profitable consciousness in the supramundane plane (Buddhism)*
HH 0846 4 person that an opportunity *Multiple therapy*
HM 3462 4 phase in the spiritual *Saviors of God: the action (Christianity)*
HM 3047 4 phenomenological level this occurs *Absorption into nothingness (ICA)*
HM 3097 4 phenomenological level this occurs *All-being-in-myself (ICA)*
HM 3088 4 phenomenological level this occurs *All-that-ever-was (ICA)*
HM 3083 4 phenomenological level this occurs *All-that's-yet-to-be (ICA)*
HM 2983 4 phenomenological level this occurs *Being all the other (ICA)*
HM 3075 4 phenomenological level this occurs *Being history (ICA)*
HM 2023 4 phenomenological level this occurs *Common earth (ICA)*
HM 2338 4 phenomenological level this occurs *Communion of saints (ICA)*
HM 2303 4 phenomenological level this occurs *Cruciform exaltation (ICA)*
HM 2591 4 phenomenological level this occurs *Defender of deeps (ICA)*
HM 2859 4 phenomenological level this occurs *Divine hosts (ICA)*
HM 2143 4 phenomenological level this occurs *Doing the mystery (ICA)*
HM 3099 4 phenomenological level this occurs *Dying death (ICA)*
HM 2786 4 phenomenological level this occurs *Eternal moment (ICA)*
HM 2253 4 phenomenological level this occurs *Eternal saviour (ICA)*

INDEX TO CONCEPT SETS
(See SECTION X for KEYWORD INDEX)
HX

HM 3130 4 phenomenological level this occurs *Guardian angel (ICA)*
HM 2784 4 phenomenological level this occurs *Heavenly advocate (ICA)*
HM 2216 4 phenomenological level this occurs *Heavenly sorrow (ICA)*
HM 3029 4 phenomenological level this occurs *Honouring the mystery (ICA)*
HM 2480 4 phenomenological level this occurs *Incarnate Christ (ICA)*
HM 2706 4 phenomenological level this occurs *Levitational submission (ICA)*
HM 3178 4 phenomenological level this occurs *Living word (ICA)*
HM 2839 4 phenomenological level this occurs *Perpetual revolutionary (ICA)*
HM 2206 4 phenomenological level this occurs *Presence of Jesus Christ (ICA)*
HM 2352 4 phenomenological level this occurs *Primordial colloquy (ICA)*
HM 2441 4 phenomenological level this occurs *Realized vocation (ICA)*
HM 3140 4 phenomenological level this occurs *Replication of Christ (ICA)*
HM 2754 4 phenomenological level this occurs *Sacrificial friendship (ICA)*
HM 2413 4 phenomenological level this occurs *Saving the mystery (ICA)*
HM 3161 4 phenomenological level this occurs *Transfigured man (ICA)*
HM 2644 4 phenomenological level this occurs *Universal Christ (ICA)*
HM 2458 4 phenomenological level this occurs *Universal prior (ICA)*
HM 2400 4 phenomenological level this occurs *Unspeakable joy (ICA)*
HM 2666 4 phenomenological level this occurs *Virgin birth (ICA)*
HM 2831 4 phenomenological level this occurs *Wealth untold (ICA)*
HM 3172 4 phenomenological level when one *Promissorial offering (ICA)*
HM 3006 4 plane of wisdom description *Establishment in truth (Hinduism)*
HH 5652 4 planes or stages *Supernormal powers (Buddhism)*
HH 0767 4 post newtonian or einsteinian *Mutational phases of life*
HM 2152 4 preceding steps *Integral consciousness*
HM 2050 4 primary concomitants to any *Awareness as mental-formation group of conscious existence (Buddhism)*
HH 0385 4 principle stages are identified *Individuation*
HM 2551 4 producing corporeality the 5 *Phenomena awareness (Buddhism)*
HM 7268 4 profitable and four functional *Impulsion (Buddhism)*
HM 8273 4 profitable and 4 indeterminate *Consciousness-born materiality (Buddhism)*
HM 8273 4 profitable and 4 indeterminate *Consciousness-born materiality (Buddhism)*
HM 4983 4 profitable 4 indeterminate resultant *Feeling aggregate (Buddhism)*
HM 4983 4 profitable 4 unprofitable 4 *Feeling aggregate (Buddhism)*
HM 7655 4 proximity or immediacy *Conditions for consciousness (Buddhism)*
HM 1266 4 psychic state listed *Mehr (Sufism)*
HM 2082 4 quarters and the four *Four-great-kings-heaven awareness (Buddhism, Tibetan)*
HM 2395 4 quarters *Turiya awareness (Hinduism)*
HM 2330 4 realms of form *Nirvana (Buddhism)*
HM 1454 4 reckoning and the first *Concentration accompanied by bliss (Buddhism)*
HM 4433 4 reckoning and the first *Concentration accompanied by happiness (Buddhism)*
HM 8021 4 reckoning and the fivefold *Concentration accompanied by equanimity (Buddhism)*
HM 0730 4 reckoning and the fivefold *Concentration without happiness (Buddhism)*
HM 1454 4 reckoning and the fourth *Concentration accompanied by bliss (Buddhism)*
HM 7543 4 reckoning and the third *Concentration without applied thought or sustained thought (Buddhism)*
HM 5767 4 reckoning or five jhanas *Concentration with happiness (Buddhism)*
HM 2257 4 regions which correspond *Form-realm consciousness (Buddhism)*
HM 2257 4 related areas *Form-realm consciousness (Buddhism)*
HM 1931 4 represents impetus towards completion *Greed (Leela)*
HM 2208 4 root conditions of entering *Mahayana heat-application awareness (Buddhism, Tibetan)*
HM 2108 4 root elements are derived *Awareness of corporeality-group of conscious existence (Buddhism)*
HM 2108 4 root elements earth water *Awareness of corporeality-group of conscious existence (Buddhism)*
HM 2072 4 rounds (gati) of existence *Worlds of conscious existence (Buddhism, Pali)*
HM 2981 4 row *Apt religion (Leela)*
HM 0455 4 row *Clarity of consciousness (Leela)*
HM 5197 4 row *Good tendencies (Leela)*
HM 0670 4 row *Irreligiousity (Leela)*
HM 3291 4 row of the board *Fourth chakra: attaining balance (Leela)*
HM 1351 4 row *Plane of balance (Leela)*
HM 0008 4 row *Plane of fragrance (Leela)*
HM 5965 4 row *Plane of sanctity (Leela)*
HM 0878 4 row *Plane of taste (Leela)*
HM 0856 4 row *Purgatory (Leela)*
HM 0935 4 salient qualities (1) there *Higher states of consciousness*
HM 3216 4 sciences buddhism tibetan description *Four sciences (Buddhism, Tibetan)*
HM 3249 4 sciences great love great *Paths of effect (Tibetan)*
HM 2191 4 scriptural bodhisattva awareness buddhism *Fourth scriptural bodhisattva awareness (Buddhism,* Tibetan)
HH 3796 4 section of st john's *Journeying within transcendence - from secular to sacred*
HM 2980 4 shown by increasing delta *Sleep*
HH 0112 4 sides *Mandalas*
HM 2259 4 signs having much *Cardinal awareness (Astrology)*
HM 2554 4 signs having much *Fixed awareness (Astrology)*
HM 3092 4 signs having much *Mutable awareness (Astrology)*
HM 2259 4 signs of the zodiac *Cardinal awareness (Astrology)*
HM 2554 4 signs of the zodiac *Fixed awareness (Astrology)*
HM 3092 4 signs of the zodiac *Mutable awareness (Astrology)*
HM 5974 4 sleep and back instead *REM sleep*
HM 6307 4 sleep description *Deep sleep*
HH 0565 4 social system morality *Moral development*
HM 2866 4 stage all positive feeling *Sukha (Buddhism)*
HH 0622 4 stage but the sixth *Human development (Jainism)*
HM 3427 4 stage in a systematic *Faqr (Sufism)*
HM 2319 4 stage in the evolution *Mental consciousness*
HM 3422 4 stage in the less *Near death evil force*
HM 3381 4 stage in the process *Near death awareness of light (NDE)*
HH 1543 4 stage is postulated where *Individual development*
HM 3518 4 stage may become *Heart-awakening stage of life*
HH 0309 4 stage of development *Vajrayana Buddhism (Buddhism)*
HH 0765 4 stage of traditional meditation *Biofeedback training*
HM 5339 4 stage on the path *Piety (Sufism)*
HM 2375 4 stage (yad dast) is *Nakshbandi recollection (Islam)*
HH 1323 4 stages are defined (1) *Discipline of study (Christianity)*
HM 2046 4 stages are involved purgation *Ecstasy*
HH 2563 4 stages for full development *Ashramas (Hinduism)*
HH 0820 4 stages it may also *Saivism*
HH 2563 4 stages nonetheless demonstrate pictorially *Ashramas (Hinduism)*
HH 2563 4 stages of life description *Ashramas (Hinduism)*
HM 3509 4 stages of the gunas *Asmita stage (Yoga)*
HM 0932 4 stages of the path *Phala sacchikiriya (Buddhism)*
HM 2866 4 stages of the rupajhana *Sukha (Buddhism)*
HM 2735 4 stages of trance *Buddha-consciousness (Buddhism, Tibetan)*
HM 2185 4 stages of trance *Way of the Buddhas (Buddhism, Tibetan)*
HM 4392 4 stages on the sufi *Jabarut (Sufism)*
HM 0741 4 stages on the sufi *Lahut (Sufism)*
HM 0116 4 stages on the sufi *Malaqut (Sufism)*
HM 3579 4 stages on the sufi *Nasut (Sufism)*
HH 2563 4 stages or ashramas *Ashramas (Hinduism)*
HM 2710 4 stages the prayer *Mystical contemplation*
HM 3892 4 state arising *Remembrance of reality (Sufism)*
HM 2072 4 state consciousness *Worlds of conscious existence (Buddhism, Pali)*
HM 2072 4 state (lokuttara) *Worlds of conscious existence (Buddhism, Pali)*
HM 2486 4 state objective consciousness *Self-remembering*
HM 2402 4 state of consciousness description *Objective fourth state of consciousness*
HM 2077 4 state of prayer is *Christian recollection (Christianity)*
HM 3744 4 state of traditional meditation *States of awareness in biofeedback training*
HM 1931 4 state or square *Greed (Leela)*
HM 2072 4 state the supermundane (lokuttara) *Worlds of conscious existence (Buddhism, Pali)*
HM 2020 4 state *Transcendental consciousness*
HM 2032 4 state (turiya) is objective *Awareness of relative reality (Hinduism, Yoga)*
HM 2020 4 states can be described *Transcendental consciousness*
HM 2693 4 states concentration *Tibetan meditative states of form concentrations (Buddhism)*
HM 2693 4 states correspond to those *Tibetan meditative states of form concentrations (Buddhism)*
HM 2669 4 states of formless absorption *Tibetan meditative states of formless absorptions (Buddhism)*
HM 3611 4 station of the cross *Jesus meets his afflicted mother (Christianity)*
HM 2317 4 stations are called gateways *Stations of consciousness - Ansari (Sufism)*
HM 6754 4 stations of mindfulness *Smrityupasthana (Pali)*
HM 2766 4 steeds that drive *The relationship between man and man (Christianity)*
HM 3570 4 step in the march *Fourth step: the earth (Christianity)*
HM 3941 4 step of the ladder *Dark night of the soul (Christianity)*
HM 3570 4 step the earth christianity *Fourth step: the earth (Christianity)*
HH 0382 4 sub divisions of occultism *Tarot (Tarot)*
HM 2060 4 sub human realms *Kamaloka consciousness (Buddhism, Pali)*
HM 2864 4 subtle minds and using *Tantra in meditation on emptiness (Buddhism, Tibetan)*
HM 2997 4 such establishments attained *Establishments in mindfulness (Buddhism)*
HH 0523 4 suffering *Four noble truths (Buddhism)*
HM 2395 4 superconscious or extraordinary meditation *Turiya awareness (Hinduism)*
HM 6553 4 system referred *Dwelling in the fivefold jana (Buddhism)*
HH 0424 4 tasks to know what *The path*
HM 8002 4 tetrad where concentration is *Concentration partaking of diminution (Buddhism)*
HM 7363 4 tetrad where concentration is *Concentration partaking of distinction (Buddhism)*
HM 5930 4 tetrad where concentration is *Concentration partaking of penetration (Buddhism)*
HM 6956 4 tetrad where concentration is *Concentration partaking of stagnation (Buddhism)*
HH 3012 4 the fourth degree *Degrees of love (Christianity)*
HH 0195 4 the fourth stage 8 *Sensorimotor development*
HH 0420 4 the unitive way *Retreat*
HH 0523 4 the way to cessation *Four noble truths (Buddhism)*
HM 2407 4 through the 19 *Death-bardo heaven-reality awareness (Buddhism)*
HM 2335 4 tibetan *Death-bardo clear-light awareness (Buddhism)*
HM 2980 4 to 1 which is *Sleep*
HM 4389 4 to eighth modes *Sense mode of consciousness occurrence (Buddhism)*
HM 2098 4 to fruition moment *Dispositions of consciousness (Buddhism)*
HH 0786 4 traits resulting at least *Integration*
HM 2142 4 trance are divided *Form-realm awareness (Buddhism, Tibetan)*
HM 2586 4 trance of the fine *Fourth trance of the fine-material sphere (Buddhism)*
HM 2122 4 trance of the five *Trances and mental absorptions (Buddhism)*
HM 2142 4 trance states in this *Form-realm awareness (Buddhism, Tibetan)*
HM 2122 4 trance states *Trances and mental absorptions (Buddhism)*
HM 2735 4 trances he enters nirvana *Buddha-consciousness (Buddhism, Tibetan)*
HM 2098 4 trances or absorptions classed *Dispositions of consciousness (Buddhism)*
HM 2735 4 trances seeing the suffering *Buddha-consciousness (Buddhism, Tibetan)*
HM 1128 4 triad this is compared *Understanding as interpreting the external (Buddhism)*
HM 4490 4 triad this is compared *Understanding as interpreting the internal and the external (Buddhism)*
HM 0956 4 triad this is compared *Understanding as interpreting the internal (Buddhism)*
HM 4036 4 triad this is *Exalted concentration (Buddhism)*
HM 1073 4 triad this is *Limited concentration (Buddhism)*
HM 0496 4 triad this is *Measureless concentration (Buddhism)*
HM 6663 4 triads six tetrads *Concentration (Buddhism)*
HM 2713 4 triplicities corresponding *Zodiacal forms of awareness (Astrology)*
HM 2955 4 triplicities or elements each *Airy awareness (Astrology)*
HM 3235 4 triplicities or elements each *Earthy awareness (Astrology)*
HM 2124 4 triplicities or elements each *Fiery awareness (Astrology)*
HM 2384 4 triplicities or elements each *Watery awareness (Astrology)*
HH 0523 4 truth is *Four noble truths (Buddhism)*
HM 2245 4 truths buddhism chatvari satvani *Meditation way of the four truths (Buddhism)*
HM 3527 4 truths by making cessation *Two-fold knowledge of truths (Buddhism)*
HM 2245 4 truths *Meditation way of the four truths (Buddhism)*
HH 0523 4 truths revealed *Four noble truths (Buddhism)*
HM 2207 4 truths (satya) may *States of special insight (Buddhism)*
HM 0891 4 turyatita when the yogin *Degrees of consciousness (Hinduism)*
HM 2138 4 types of daydream have *Daydream*
HM 4651 4 types sensory mental awareness *Bare perception (Buddhism, Tibetan)*
HM 4983 4 unprofitable 4 indeterminate resultant *Feeling aggregate (Buddhism)*
HM 2251 4 vajra master awareness buddhism *Fourth vajra-master awareness (Buddhism, Tibetan)*
HM 2690 4 varieties of such pictures *Oxherding pictures in Zen Buddhism (Zen)*
HH 0712 4 virtues as forms *Virtue*
HM 2020 4 wakeful hypometabolic state yoga *Transcendental consciousness*
HH 0820 4 ways or stages *Saivism*
HH 0513 4 will lead to existential *Creative existence*
HM 0282 4 with the jhana *Indeterminate consciousness in the immaterial sphere - functional (Buddhism)*
HM 4982 4 with the jhana *Indeterminate consciousness in the immaterial sphere - resultant (Buddhism)*
HM 4701 4 with the jhana *Profitable consciousness in the immaterial sphere (Buddhism)*
HM 2098 4 worlds or spheres (lokas) *Dispositions of consciousness (Buddhism)*
HM 2509 4 worlds *Paths of wisdom (Judaism)*
HM 2002 4 worlds *Shekinah awareness (Judaism)*
HM 3038 4 worlds *World of creation (Judaism)*
HM 2408 4 worlds *World of emanation (Judaism)*
HM 2360 4 worlds *World of formation (Judaism)*
HM 2312 4 worlds *World of manifestation (Judaism)*
HH 0075 5 a sense of sacredness *Psychedelic drugs*
HH 0565 5 a social contract orientation *Moral development*
HH 4098 5 according to his mercy *Spiritual rebirth (Christianity)*
HH 0986 5 adolescent stage 12 15 *Libido development*
HM 3871 5 afflictions or causes *Abhinvesa (Yoga)*
HM 2937 5 afflictions or causes *Asmita (Yoga)*
HM 2433 5 afflictions or causes *Desire*
HM 3196 5 afflictions or causes *Ignorance (Buddhism, Yoga, Zen)*
HH 3321 5 aggregates are said *Five aggregates (Buddhism)*

HH 3321 5 aggregates are those *Five aggregates (Buddhism)*
HM 7055 5 aggregates as impermanent painful *Knowledge of the path of arahantship (Buddhism)*
HM 6920 5 aggregates as impermanent painful *Knowledge of the path of non-return (Buddhism)*
HM 7563 5 aggregates as impermanent painful *Knowledge of the path of once-return (Buddhism)*
HH 3321 5 aggregates buddhism khandha *Five aggregates (Buddhism)*
HH 3321 5 aggregates *Five aggregates (Buddhism)*
HM 3181 5 among the spiritual paths *Path of radical intelligence (Judaism)*
HM 2341 5 and muhammad sixth *Stations of consciousness - Simnami (Sufism)*
HM 5721 5 and the mind element *Indeterminate consciousness in the sense sphere - resultant (Buddhism)*
HM 2407 5 antarabhava consciousness *Death-bardo heaven-reality awareness (Buddhism)*
HH 6282 5 are concerned *Subtle faculties (Sufism)*
HM 2219 5 are connected *Chakra centres of consciousness*
HM 2050 5 are inconstant immutable *Awareness as mental-formation group of conscious existence (Buddhism)*
HM 0130 5 arise with fine material *Mundane resultant consciousness (Buddhism)*
HM 1665 5 arises when the person *Individuative-reflective faith*
HM 2735 5 ascetics and having eaten *Buddha-consciousness (Buddhism, Tibetan)*
HM 0485 5 aspects attainment or achievement *Four discriminations (Buddhism)*
HM 3280 5 aspects of self discipline *Niyama (Hinduism)*
HH 6053 5 basic arts of magic *Guided visualization (Magic)*
HH 1799 5 basic practices the pillars *Human development (Islam)*
HH 6876 5 being fortitude kindness straight *Restraint*
HM 2033 5 being of non cognizing *Direct non-conceptual cognition of emptiness (Buddhism, Tibetan)*
HM 0747 5 bliss *Fivefold happiness (Buddhism)*
HM 0101 5 branch of the tibetan *Yoga of the intermediary state (Buddhism)*
HM 2062 5 branches 2 types *Third trance of the fine-material sphere (Buddhism)*
HH 1010 5 buddhacaksus the eye *Five eyes (Buddhism)*
HH 0497 5 buddhist superknowledges plus heightened *Magico-religious powers*
HM 2232 5 cardinal virtues faith works *Mahayana climax-application awareness (Buddhism, Tibetan)*
HM 2880 5 categorization of these *Mystical Christ-consciousness (Christianity)*
HH 0959 5 centuries *Discipline of the secret*
HH 5007 5 century ad the following *Afflictions and hindrances (Buddhism)*
HH 0205 5 century jerusalem *Hesychasm (Christianity)*
HM 0933 5 chakra man becomes himself *Fifth chakra: man becomes himself (Leela)*
HM 0455 5 chakra that of man *Clarity of consciousness (Leela)*
HM 2219 5 chakras *Chakra centres of consciousness*
HM 2272 5 characteristics are characteristic *Mystical cognition*
HM 2090 5 characteristics egolessness pointlike unitary *Magical consciousness*
HM 2407 5 chonyid bardo *Death-bardo heaven-reality awareness (Buddhism)*
HM 2407 5 chos nid *Death-bardo heaven-reality awareness (Buddhism)*
HM 2841 5 clairvoyances buddhism description *Five clairvoyances (Buddhism)*
HM 3079 5 clairvoyances the four meditative *Paths of special qualities (Buddhism)*
HM 4983 5 classes according *Feeling aggregate (Buddhism)*
HH 3321 5 clinging aggregates are thus *Five aggregates (Buddhism)*
HM 2735 5 companions to remove false *Buddha-consciousness (Buddhism, Tibetan)*
HH 0829 5 component of patanjali's eightfold *Pratyahara (Yoga)*
HH 1016 5 components of niyama self *Resignation to the will of God*
HM 2898 5 components of niyama self *Samtosa (Yoga)*
HM 1391 5 components of niyama self *Sauca (Yoga)*
HH 0957 5 components of niyama self *Study of sacred scriptures*
HH 1248 5 components of niyama self *Tapas (Hinduism)*
HH 0306 5 components of the ceremony *Tantra (Buddhism, Yoga)*
HM 5143 5 concentration *Concentration of the fifth jhana of five (Buddhism)*
HM 3048 5 concentration foundation of following *Great-vehicle higher-path awareness (Buddhism, Tibetan)*
HM 7202 5 concentration of the pentad *Concentration of the fourth jhana of four (Buddhism)*
HM 5089 5 constituent factors of jhana *Analysis (Buddhism)*
HM 6663 5 constituent factors of jhana *Concentration (Buddhism)*
HM 0747 5 constituent factors of jhana *Fivefold happiness (Buddhism)*
HM 1177 5 constituent factors of jhana *Investigation (Buddhism)*
HM 2866 5 constituent factors of jhana *Sukha (Buddhism)*
HM 0137 5 constituents of yoga *Dhyana (Hinduism, Buddhism)*
HM 7655 5 contiguity or direct immediacy *Conditions for consciousness (Buddhism)*
HM 2551 5 corresponding objects *Phenomena awareness (Buddhism)*
HH 6773 5 covers *Five hindrances (Buddhism)*
HH 0213 5 defects which cause *Pranayama (Hinduism, Yoga)*
HH 0185 5 degrees of prayer have *Prayer*
HM 4983 5 depending upon whether feelings *Feeling aggregate (Buddhism)*
HM 2578 5 determining mental factors *Aspiration (Buddhism)*
HM 2933 5 determining mental factors *Belief (Buddhism, Yoga)*
HM 2655 5 determining mental factors *Mental wisdom (Buddhism)*
HM 2847 5 determining mental factors *Mindfulness (Buddhism)*
HM 2440 5 determining mental factors *Stabilization (Buddhism)*
HM 3630 5 developing contemplation of fading *Principal insights (Buddhism)*
HH 1028 5 different and often competing *Revelation*
HH 1108 5 different types physical mental *Seventy-five dharmas (Buddhism)*
HH 5207 5 directions context *Middle middle grade (Taoism)*
HM 2098 5 dispositions arise as profitable *Dispositions of consciousness (Buddhism)*
HM 2416 5 divine presences *Awareness of the divine presences (Sufism)*
HM 8336 5 door adverting *Adverting (Buddhism)*
HM 8336 5 door adverting *Adverting (Buddhism)*
HM 3237 5 door adverting and mind *Mental engagement (Buddhism)*
HM 8336 5 door adverting as it *Adverting (Buddhism)*
HM 4389 5 door adverting eye ear *Sense mode of consciousness occurrence (Buddhism)*
HM 8336 5 door adverting similarly when *Adverting (Buddhism)*
HM 7655 5 doors and as heart *Conditions for consciousness (Buddhism)*
HM 3852 5 doors (senses) and adverting *Indeterminate consciousness in the sense sphere - functional (Buddhism)*
HM 8266 5 doors (senses) causing greed *Rebirth-linking (Buddhism)*
HM 6173 5 doors (senses) *Mind-consciousness element (Buddhism)*
HM 1646 5 doors (senses) or very *Registering (Buddhism)*
HM 7268 5 doors (senses) then immediately *Impulsion (Buddhism)*
HM 4128 5 dyad this is contrasted *Understanding as the plane of development (Buddhism)*
HM 6192 5 dyad this is contrasted *Understanding as the plane of seeing (Buddhism)*
HM 4523 5 dyads four triads *Understanding (Buddhism)*
HH 7386 5 elements are lead mercury *Higher vehicle path (Taoism)*
HM 2144 5 elements are present *Ajna (Yoga)*
HM 4338 5 elements *Assembling the five elements (Taoism)*
HM 2219 5 elements earth water fire *Chakra centres of consciousness*
HM 4338 5 elements foster each *Assembling the five elements (Taoism)*
HH 5887 5 elements imbalancing one another *Alchemy (Taoism)*
HM 2312 5 elements in man *World of manifestation (Judaism)*
HM 4338 5 elements in their primordial *Assembling the five elements (Taoism)*
HM 4338 5 elements taoism description *Assembling the five elements (Taoism)*
HM 5265 5 elements *Transcending the world (Taoism)*
HH 7324 5 elements uses food *Human development through diet*
HM 1330 5 elements yin and yang *Merging of yin and yang (Taoism)*
HM 2260 5 emblem which is *Hierophant archetypal image (Tarot)*
HM 4338 5 energies returning to their *Assembling the five elements (Taoism)*
HH 0306 5 essential elements the methodical *Tantra (Buddhism, Yoga)*
HH 1348 5 ever recurring or omnipresent *Secondary mental factors (Buddhism, Tibetan)*
HM 1995 5 external senses offer any *Imitative musical inspiration*
HH 1010 5 eyes buddhism five levels *Five eyes (Buddhism)*
HH 1010 5 eyes or levels *Five eyes (Buddhism)*
HM 4338 5 eyes potentially available *Assembling the five elements (Taoism)*
HM 0297 5 factors remain *Concentration of the first jhana of five (Buddhism)*
HM 4456 5 factors remain *Concentration of the first jhana of four (Buddhism)*
HH 5652 5 factors the second three *Supernormal powers (Buddhism)*
HM 3359 5 faith description *Conjunctive faith*
HM 2277 5 faults laziness forgetting *Mental abiding in equipoise (Buddhism)*
HM 2075 5 feelings the 5 mental *Psycho-physical faculties awareness (Buddhism)*
HM 3504 5 fetter or character defect *Emotional aversions (Buddhism)*
HM 3152 5 fetters are marked *Delusion of the personal self (Buddhism)*
HM 0509 5 fetters are marked *Dependence on ceremonies (Buddhism)*
HM 1294 5 fetters are marked *Doubt of the efficacy of the good life (Buddhism)*
HM 3504 5 fetters are marked *Emotional aversions (Buddhism)*
HM 3301 5 fetters are marked *Emotional desires (Buddhism)*
HM 0233 5 fetters or character defects *Arhat (Buddhism)*
HM 1261 5 fetters when automatic responses *Desire for formless life (Buddhism)*
HM 3583 5 fetters when automatic responses *Desire for life in form (Buddhism)*
HM 3493 5 fetters when automatic responses *Ignorance of the nature of the essential self (Buddhism)*
HM 3612 5 fetters when automatic responses *Notion of one's self as entity (Buddhism)*
HM 2852 5 fetters when automatic responses *Spiritual pride (Buddhism)*
HM 8266 5 fine material resultant consciousness *Rebirth-linking (Buddhism)*
HM 1618 5 flashes of apperception *Appearance of absorption (Buddhism)*
HM 8080 5 fold knowledge systematics potential *Five-fold knowledge (Systematics)*
HM 5438 5 fold series *Attainment of cessation (Buddhism)*
HM 2948 5 forces buddhism description *Five forces (Buddhism)*
HH 4864 5 forces context *Lower vehicle path (Taoism)*
HH 4864 5 forces liver as dragon *Lower vehicle path (Taoism)*
HH 6345 5 forces *Middle vehicle path (Taoism)*
HM 7268 5 functional or inoperative consciousnesses *Impulsion (Buddhism)*
HM 2098 5 functional or inoperative *Dispositions of consciousness (Buddhism)*
HH 0720 5 fundamental magical arts concentration *Magic*
HH 0347 5 genetic factors *Healthy human growth and development*
HM 2656 5 gestures (mudras) *Great-path tantra awareness (Buddhism, Tibetan)*
HM 4761 5 give rise to materiality *Indeterminate consciousness in the fine-material sphere - functional* (Buddhism)
HM 0594 5 give rise to materiality *Indeterminate consciousness in the fine-material sphere - resultant* (Buddhism)
HM 5338 5 give rise to materiality *Profitable consciousness in the fine-material sphere (Buddhism)*
HM 8080 5 gradation of knowledge *Five-fold knowledge (Systematics)*
HH 3321 5 groups of interacting physical *Five aggregates (Buddhism)*
HM 0747 5 happiness buddhism piti *Fivefold happiness (Buddhism)*
HM 0747 5 happiness *Fivefold happiness (Buddhism)*
HM 0747 5 happiness is conceived *Fivefold happiness (Buddhism)*
HH 3321 5 have a sense *Five aggregates (Buddhism)*
HH 5652 5 he repeatedly attains *Supernormal powers (Buddhism)*
HM 3146 5 heaven are said *Awareness of angelic-transformation*
HM 2656 5 high buddhas who are *Great-path tantra awareness (Buddhism, Tibetan)*
HM 7193 5 hindrances and are attributable *Jhana (Buddhism, Pali)*
HH 6773 5 hindrances buddhism gogai *Five hindrances (Buddhism)*
HM 4298 5 hindrances which are *Dwelling in the first jhana (Buddhism)*
HH 0136 5 homeostasis of the endocrines *Unity of personality*
HH 0565 5 human rights and social *Moral development*
HH 0390 5 human troubles as given *Five kleshas (Yoga, Zen)*
HM 3620 5 human troubles kleshas *Revulsion (Yoga, Zen)*
HM 3119 5 human troubles referred *Attachment (Yoga, Zen)*
HM 3196 5 human troubles referred *Ignorance (Buddhism, Yoga, Zen)*
HM 2827 5 illumination of jami *Perception of universality (Sufism)*
HM 5438 5 in a five fold *Attainment of cessation (Buddhism)*
HM 8080 5 in a sequence *Five-fold knowledge (Systematics)*
HM 2072 5 in ascending order *Worlds of conscious existence (Buddhism, Pali)*
HM 2060 5 in pali but fourth *Kamaloka consciousness (Buddhism, Pali)*
HM 1618 5 in the fine material *Appearance of absorption (Buddhism)*
HM 2394 5 in the ica *Incarnate living (ICA)*
HM 3359 5 in the system *Conjunctive faith*
HM 2880 5 includes the empty state *Mystical Christ-consciousness (Christianity)*
HM 8080 5 independent elements *Five-fold knowledge (Systematics)*
HM 8273 5 indeterminate functional *Consciousness-born materiality (Buddhism)*
HM 8273 5 indeterminate resultant consciousnesses *Consciousness-born materiality (Buddhism)*
HM 8266 5 indeterminate resultant *Rebirth-linking (Buddhism)*
HM 0181 5 initiation in the tradition *Initiation of revelation (Esotericism)*
HM 2050 5 interacting aggregates that produce *Awareness as mental-formation group of conscious existence* (Buddhism)
HM 2556 5 interacting aggregates that produce *Awareness of consciousness-group of conscious existence - senses* and mind (Buddhism)
HM 2108 5 interacting aggregates that produce *Awareness of corporeality-group of conscious existence* (Buddhism)
HM 4983 5 interacting aggregates that produce *Feeling aggregate (Buddhism)*
HM 4143 5 interacting aggregates that produce *Perception aggregate (Buddhism)*
HH 4864 5 internal organs as five *Lower vehicle path (Taoism)*
HH 0539 5 internal spiritual regions *Shabda yoga (Yoga)*
HM 2668 5 invalid ways of knowing *Distorted cognition (Buddhism, Tibetan)*
HM 3348 5 invalid ways of knowing *Inattentive perception (Buddhism, Tibetan)*
HM 3074 5 invalid ways of knowing *Indecisive wavering (Buddhism, Tibetan)*
HM 4032 5 invalid ways of knowing *Presumption (Buddhism, Tibetan)*
HM 3508 5 invalid ways of knowing *Subsequent cognition (Buddhism, Tibetan)*
HM 3744 5 is preceded by 'sudden *States of awareness in biofeedback training*
HH 0765 5 is said *Biofeedback training*
HM 1024 5 is the number *Physical plane (Leela)*
HM 6553 5 jana buddhism dwelling *Dwelling in the fivefold jana (Buddhism)*

INDEX TO CONCEPT SETS
(See SECTION X for KEYWORD INDEX)
HX

HM 4983 5 jhana (32 extra) context *Feeling aggregate (Buddhism)*
HM 7193 5 jhana are the immaterial *Jhana (Buddhism, Pali)*
HM 7193 5 jhana as ease *Jhana (Buddhism, Pali)*
HM 5143 5 jhana factors this is *Concentration of the fifth jhana of five (Buddhism)*
HM 0297 5 jhana factors this is *Concentration of the first jhana of five (Buddhism)*
HM 4882 5 jhana factors this is *Concentration of the fourth jhana of five (Buddhism)*
HM 4575 5 jhana factors this is *Concentration of the second jhana of five (Buddhism)*
HM 8532 5 jhana factors this is *Concentration of the third jhana of five (Buddhism)*
HM 5143 5 jhana has two factors *Concentration of the fifth jhana of five (Buddhism)*
HM 4983 5 jhana in each *Feeling aggregate (Buddhism)*
HM 6553 5 jhana may occur *Dwelling in the fivefold jana (Buddhism)*
HM 5143 5 jhana of five buddhism *Concentration of the fifth jhana of five (Buddhism)*
HM 7525 5 jhana of the fivefold *Understanding accompanied by equanimity (Buddhism)*
HM 4983 5 jhanas but leaving out *Feeling aggregate (Buddhism)*
HM 5767 5 jhanas in the fivefold *Concentration with happiness (Buddhism)*
HM 0130 5 jhanas *Mundane resultant consciousness (Buddhism)*
HM 0747 5 joy pali *Fivefold happiness (Buddhism)*
HM 4651 5 kinds depending *Bare perception (Buddhism, Tibetan)*
HM 4651 5 kinds it arises immediately *Bare perception (Buddhism, Tibetan)*
HM 4761 5 kinds of consciousness may *Indeterminate consciousness in the fine-material sphere - functional* (Buddhism)
HM 5338 5 kinds of consciousness may *Profitable consciousness in the fine-material sphere (Buddhism)*
HM 0748 5 kinds of direct *Knowledge of passing away and reappearance of beings (Buddhism)*
HM 5232 5 kinds of direct *Knowledge of penetration of minds (Buddhism)*
HM 4297 5 kinds of direct *Knowledge of recollection of previous existence (Buddhism)*
HM 7672 5 kinds of direct *Knowledge of supernormal powers (Buddhism)*
HM 5982 5 kinds of direct *Knowledge of the divine ear element (Buddhism)*
HH 5652 5 kinds of direct knowledge *Supernormal powers (Buddhism)*
HM 0747 5 kinds of happiness are *Fivefold happiness (Buddhism)*
HM 4886 5 kinds of happiness arise *Rapturous happiness due to insight (Buddhism)*
HM 0311 5 kinds (pentad) *Absorption concentration (Buddhism)*
HM 4999 5 kinds (pentad) *Access concentration (Buddhism)*
HM 4191 5 kinds (pentad) *Concentration accompanied by bliss (Buddhism)*
HM 1454 5 kinds (pentad) *Concentration accompanied by bliss (Buddhism)*
HM 5008 5 kinds (pentad) *Concentration accompanied by equanimity (Buddhism)*
HM 8021 5 kinds (pentad) *Concentration accompanied by equanimity (Buddhism)*
HM 4433 5 kinds (pentad) *Concentration accompanied by happiness (Buddhism)*
HM 6663 5 kinds (pentad) *Concentration (Buddhism)*
HM 8365 5 kinds (pentad) *Concentration due to energy (Buddhism)*
HM 7434 5 kinds (pentad) *Concentration due to inquiry (Buddhism)*
HM 7961 5 kinds (pentad) *Concentration due to natural purity of consciousness (Buddhism)*
HM 4969 5 kinds (pentad) *Concentration due to zeal (Buddhism)*
HM 5992 5 kinds (pentad) *Concentration of difficult progress and sluggish direct-knowledge (Buddhism)*
HM 7859 5 kinds (pentad) *Concentration of difficult progress and swift direct-knowledge (Buddhism)*
HM 1010 5 kinds (pentad) *Concentration of easy progress and sluggish direct-knowledge (Buddhism)*
HM 4977 5 kinds (pentad) *Concentration of easy progress and swift direct-knowledge (Buddhism)*
HM 5143 5 kinds (pentad) *Concentration of the fifth jhana of five (Buddhism)*
HM 0297 5 kinds (pentad) *Concentration of the first jhana of five (Buddhism)*
HM 4456 5 kinds (pentad) *Concentration of the first jhana of four (Buddhism)*
HM 4882 5 kinds (pentad) *Concentration of the fourth jhana of five (Buddhism)*
HM 7202 5 kinds (pentad) *Concentration of the fourth jhana of four (Buddhism)*
HM 4575 5 kinds (pentad) *Concentration of the second jhana of five (Buddhism)*
HM 4380 5 kinds (pentad) *Concentration of the second jhana of four (Buddhism)*
HM 8532 5 kinds (pentad) *Concentration of the third jhana of five (Buddhism)*
HM 2284 5 kinds (pentad) *Concentration of the third jhana of four (Buddhism)*
HM 8002 5 kinds (pentad) *Concentration partaking of diminution (Buddhism)*
HM 7363 5 kinds (pentad) *Concentration partaking of distinction (Buddhism)*
HM 5930 5 kinds (pentad) *Concentration partaking of penetration (Buddhism)*
HM 6956 5 kinds (pentad) *Concentration partaking of stagnation (Buddhism)*
HM 4908 5 kinds (pentad) *Concentration with applied thought and sustained thought (Buddhism)*
HM 5767 5 kinds (pentad) *Concentration with happiness (Buddhism)*
HM 5199 5 kinds (pentad) *Concentration without applied thought and with sustained thought (Buddhism)*
HM 7543 5 kinds (pentad) *Concentration without applied thought or sustained thought (Buddhism)*
HM 0730 5 kinds (pentad) *Concentration without happiness (Buddhism)*
HM 4036 5 kinds (pentad) *Exalted concentration (Buddhism)*
HM 4265 5 kinds (pentad) *Fine-material-sphere concentration (Buddhism)*
HM 0696 5 kinds (pentad) *Immaterial-sphere concentration (Buddhism)*
HM 5562 5 kinds (pentad) *Inferior concentration (Buddhism)*
HM 1073 5 kinds (pentad) *Limited concentration (Buddhism)*
HM 1175 5 kinds (pentad) *Limited concentration with a limited object (Buddhism)*
HM 5547 5 kinds (pentad) *Limited concentration with a measureless object (Buddhism)*
HM 0496 5 kinds (pentad) *Measureless concentration (Buddhism)*
HM 4653 5 kinds (pentad) *Measureless concentration with a limited object (Buddhism)*
HM 5214 5 kinds (pentad) *Measureless concentration with a measureless object (Buddhism)*
HM 6089 5 kinds (pentad) *Medium concentration (Buddhism)*
HM 7234 5 kinds (pentad) *Mundane concentration (Buddhism)*
HM 4382 5 kinds (pentad) *Non-distraction (Buddhism)*
HM 1097 5 kinds (pentad) *Sense-sphere concentration (Buddhism)*
HM 6327 5 kinds (pentad) *Superior concentration (Buddhism)*
HM 4243 5 kinds (pentad) *Supramundane concentration (Buddhism)*
HM 5768 5 kinds (pentad) *Unincluded concentration (Buddhism)*
HM 4032 5 kinds there may *Presumption (Buddhism, Tibetan)*
HM 3871 5 kleshas or afflictions it *Abhinvesa (Yoga)*
HH 0390 5 kleshas yoga zen description *Five kleshas (Yoga, Zen)*
HM 2909 5 level bodhisattva is represented *Fifth scriptural bodhisattva awareness (Buddhism, Tibetan)*
HM 2140 5 level description *Evolution of consciousness*
HM 2144 5 levels ananda bliss *Ajna (Yoga)*
HM 2168 5 levels of gods exist *Pure-lands awareness (Buddhism, Tibetan)*
HH 1010 5 levels of understanding *Five eyes (Buddhism)*
HM 3045 5 links between the second *Charioteer archetypal image (Tarot)*
HM 2150 5 listed above are usually *Angelic frame of awareness*
HM 3461 5 lotus or chakra defined *Vishuddha (Yoga)*
HM 1010 5 madhyamaka buddhism *Five eyes (Buddhism)*
HH 5120 5 magical arts *Ritual pattern-making (Magic)*
HH 5547 5 magical arts to collect *Magical meditation*
HH 0622 5 mahavratas or vows observance *Human development (Jainism)*
HM 4110 5 mansions cannot be described *Mansions of incipient union (Christianity)*
HM 2122 5 material sphere is also *Trances and mental absorptions (Buddhism)*
HM 2693 5 material sphere (rupaloka) context *Tibetan meditative states of form concentrations (Buddhism)*
HM 2075 5 mental powers (bala) *Psycho-physical faculties awareness (Buddhism)*
HM 2716 5 mental powers magical clairaudience *Discipleship-application awareness (Buddhism, Tibetan)*
HM 7525 5 method context *Understanding accompanied by equanimity (Buddhism)*
HM 6245 5 method context *Understanding accompanied by joy (Buddhism)*
HH 3798 5 methods of evangelism are *Evangelism (Christianity)*
HH 1010 5 middle *Five eyes (Buddhism)*
HM 8336 5 mind *Adverting (Buddhism)*
HM 5721 5 mind consciousness element *Indeterminate consciousness in the sense sphere - resultant (Buddhism)*
HH 1775 5 minds are distinguished conscious *Psychism*
HH 4536 5 moral law description *Pancasila*
HH 4536 5 moral maxims dating *Pancasila*
HH 0088 5 moral qualities of yama *Ahimsa*
HH 1268 5 moral qualities of yama *Aparigraha*
HH 0361 5 moral qualities of yama *Asteya*
HH 0978 5 moral qualities of yama *Brahmacarya*
HH 0554 5 moral qualities of yama *Satya*
HH 0539 5 names the names *Shabda yoga (Yoga)*
HM 6920 5 noble person supramundane path *Knowledge of the path of non-return (Buddhism)*
HM 7655 5 nor to consciousnesses *Conditions for consciousness (Buddhism)*
HM 7655 5 nor to consciousnesses *Conditions for consciousness (Buddhism)*
HH 1348 5 object attentive or determining *Secondary mental factors (Buddhism, Tibetan)*
HM 5215 5 of 40 stations *Mujahadat (Sufism)*
HM 5155 5 of eight lokas represented *Human plane (Leela)*
HM 0137 5 of five perfections *Dhyana (Hinduism, Buddhism)*
HM 2001 5 of seven ascending planes *Total freedom from attachment (Hinduism)*
HM 3147 5 of seven descending steps *Dream (Hinduism)*
HM 4110 5 of seven mansions *Mansions of incipient union (Christianity)*
HM 1330 5 of seven stages *Merging of yin and yang (Taoism)*
HM 3191 5 of seven stages *Mystical stage of life*
HM 4570 5 of seven subtle stages *World beyond form (Sufism)*
HM 1201 5 of seven ways *Way of knowledge (Esotericism)*
HM 2290 5 of ten described *Sphere of judgement (Kabbalah)*
HM 0772 5 of ten orders *Fifth order perceptions - control of sequence*
HM 3069 5 of ten states acting *Principles (Sufism)*
HM 2866 5 of the arupajhana each *Sukha (Buddhism)*
HH 1765 5 of the five primary *Abhidharma (Buddhism)*
HM 6553 5 of the fivefold system *Dwelling in the fivefold jana (Buddhism)*
HM 2181 5 of the nine states *Disciplining in mental abiding states (Buddhism)*
HM 1221 5 of the ninety nine *As-Salam (Sufism)*
HM 2267 5 of the preparations *Subtle contemplation of withdrawal or joy (Buddhism, Tibetan)*
HM 2341 5 of the seven *Stations of consciousness - Simnami (Sufism)*
HH 0383 5 of the ten meritorious *Attending to the needs of superiors (Buddhism)*
HM 2098 5 of which are resultant *Dispositions of consciousness (Buddhism)*
HM 2312 5 of which is *World of manifestation (Judaism)*
HM 2708 5 omnipresent mental factors defined *Contact (Buddhism)*
HM 2399 5 omnipresent mental factors defined *Discrimination (Buddhism)*
HM 2270 5 omnipresent mental factors defined *Feeling (Buddhism)*
HM 2589 5 omnipresent mental factors defined *Intention (Buddhism)*
HM 3237 5 omnipresent mental factors defined *Mental engagement (Buddhism)*
HM 3142 5 on the board *Crushing-hells awareness (Buddhism, Tibetan)*
HM 2050 5 or 6 along *Awareness as mental-formation group of conscious existence (Buddhism)*
HM 0294 5 or human plane context *Selfless service (Leela)*
HH 0755 5 or mystical stage *Raja yoga (Yoga)*
HM 0772 5 order perceptions control *Fifth order perceptions - control of sequence*
HM 4651 5 organs of sense *Bare perception (Buddhism, Tibetan)*
HH 1004 5 outer sheaths masking *Maya (Hinduism)*
HM 2776 5 path for solitary buddhahood *Pratyeka Buddha arhat awareness (Buddhism, Tibetan)*
HH 0219 5 perfections of mind heart *Paramitas (Zen)*
HH 0103 5 physical elements the intelligences *Atman (Hinduism)*
HM 2059 5 physical senses *Ego awareness (Hinduism)*
HH 1265 5 pillars of islam *Hajj (Islam)*
HH 0536 5 pillars of islam *Salat (Islam)*
HH 4216 5 pillars of islam *Saum (Islam)*
HH 2341 5 pillars of islam *Shahada (Islam)*
HH 5643 5 pillars of islam *Zakat (Islam)*
HM 2001 5 plane of wisdom description *Total freedom from attachment (Hinduism)*
HH 0413 5 potentials tanmatra and their *Three gunas*
HM 3238 5 powers buddhism description *Five powers (Buddhism)*
HH 0278 5 precepts may *Observing moral precepts (Buddhism)*
HM 2050 5 primary concomitants *Awareness as mental-formation group of conscious existence (Buddhism)*
HH 1765 5 primary subjects *Abhidharma (Buddhism)*
HH 1038 5 primary subjects *Madhyamaka (Buddhism)*
HH 1344 5 primary subjects *Prajnaparamita (Buddhism)*
HH 4388 5 primary subjects *Pramana (Buddhism)*
HH 3398 5 primary subjects *Vinaya (Buddhism)*
HH 0873 5 primary subjects *Ways of knowing (Buddhism)*
HM 8273 5 profitable and 5 indeterminate *Consciousness-born materiality (Buddhism)*
HM 7268 5 profitable and the five *Impulsion (Buddhism)*
HM 7193 5 psychic factors vitakka vicara *Jhana (Buddhism, Pali)*
HM 4477 5 psychic state listed *Omid (Sufism)*
HM 2257 5 pure places not great *Form-realm consciousness (Buddhism)*
HH 3025 5 purifications or purities *Purification by knowledge and vision (Buddhism)*
HH 3550 5 purifications or purities *Purification by knowledge and vision of the way (Buddhism)*
HH 4007 5 purifications or purities *Purification by knowledge and vision of what is the path and what is not the* path (Buddhism)
HH 1187 5 purifications or purities *Purification by overcoming doubt (Buddhism)*
HH 2718 5 purifications or purities *Purification of view (Buddhism)*
HM 4523 5 purifications or purities view *Understanding (Buddhism)*
HH 3875 5 purifications view *Purifications (Buddhism)*
HH 0282 5 qualities are developed slowness *T'ai Chi Ch'uan*
HM 3113 5 qualities of impermanence although *Buddha enjoyment body (Buddhism, Tibetan)*
HM 4500 5 qualities showing that allah *Allah (Sufism)*
HM 0747 5 rapture *Fivefold happiness (Buddhism)*
HH 2909 5 receipt of divine forgiveness *Psychospiritual growth (Christianity)*
HM 1454 5 reckoning but there is *Concentration accompanied by bliss (Buddhism)*
HM 4191 5 reckoning *Concentration accompanied by bliss (Buddhism)*
HM 1454 5 reckoning *Concentration accompanied by bliss (Buddhism)*
HM 5008 5 reckoning *Concentration accompanied by equanimity (Buddhism)*
HM 8021 5 reckoning *Concentration accompanied by equanimity (Buddhism)*
HM 4433 5 reckoning *Concentration accompanied by happiness (Buddhism)*
HM 5767 5 reckoning *Concentration with happiness (Buddhism)*
HM 0730 5 reckoning *Concentration without happiness (Buddhism)*

HM 7543 5 reckoning context *Concentration without applied thought or sustained thought (Buddhism)*
HM 5199 5 reckoning when there is *Concentration without applied thought and with sustained thought (Buddhism)*
HH 0470 5 representing the following *Character defects (Buddhism)*
HM 0145 5 row *Apana-loka (Leela)*
HM 0978 5 row *Birth of man (Leela)*
HM 5155 5 row *Human plane (Leela)*
HM 0310 5 row *Ignorance (Leela)*
HM 3846 5 row *Knowledge (Leela)*
HM 0933 5 row of the board *Fifth chakra: man becomes himself (Leela)*
HM 0859 5 row *Plane of agnih (Leela)*
HM 4399 5 row *Prana-loka (Leela)*
HM 0567 5 row *Right knowledge (Leela)*
HM 6002 5 row *Vyana-loka (Leela)*
HM 2909 5 scriptural bodhisattva awareness buddhism *Fifth scriptural bodhisattva awareness (Buddhism, Tibetan)*
HH 2909 5 self transcendence and self *Psychospiritual growth (Christianity)*
HH 0786 5 selves integration of systems *Integration*
HM 2270 5 sense consciousnesses *Feeling (Buddhism)*
HM 0130 5 sense consciousnesses profitable *Mundane resultant consciousness (Buddhism)*
HM 8266 5 sense doors upon his *Rebirth-linking (Buddhism)*
HM 8266 5 sense doors with pleasant *Rebirth-linking (Buddhism)*
HH 1108 5 sense organs and their *Seventy-five dharmas (Buddhism)*
HM 2551 5 sense organs the 5 *Phenomena awareness (Buddhism)*
HM 8336 5 senses and is sometimes *Adverting (Buddhism)*
HM 6173 5 senses and the mind *Mind-consciousness element (Buddhism)*
HM 2108 5 senses and their organs *Awareness of corporeality-group of conscious existence (Buddhism)*
HM 3476 5 senses *Beta wave consciousness*
HH 6973 5 senses bring useful information *Striving against the soul (Islam)*
HM 4766 5 senses bring useful information *Striving against the soul (Islam)*
HM 2671 5 senses buddhism tibetan adhyatmashunyata *Emptiness of the five senses (Buddhism, Tibetan)*
HM 0130 5 senses each being profitable *Mundane resultant consciousness (Buddhism)*
HM 4580 5 senses *Hallucination*
HM 5721 5 senses) or variable (mind *Indeterminate consciousness in the sense sphere - resultant (Buddhism)*
HM 7109 5 senses perceive external objects *Introspection (Buddhism)*
HM 3764 5 senses taste touch hearing *Perception through the senses*
HH 9760 5 senses to the subject *Spiritual exercices of St Ignatius (Christianity)*
HM 3021 5 senses *Waking consciousness*
HH 3321 5 skandha *Five aggregates (Buddhism)*
HH 0241 5 skandhas or tendencies form *Anatma (Buddhism)*
HM 2551 5 specialized (corresponding) conformations *Phenomena awareness (Buddhism)*
HM 4726 5 specific things are understood *Understanding as knowledge about law (Buddhism)*
HM 2061 5 stage a yogic state *Nirvikalpa samadhi (Hinduism)*
HM 2375 5 stage (hus dar dan) *Nakshbandi recollection (Islam)*
HM 3418 5 stage in a systematic *Sabr (Sufism)*
HM 3059 5 stage in the process *Near death encounter with the higher self (NDE)*
HM 3191 5 stage of life description *Mystical stage of life*
HM 2864 5 stage of the meditation *Tantra in meditation on emptiness (Buddhism, Tibetan)*
HM 7024 5 stage on the path *Contentment (Sufism)*
HM 2763 5 stage tantric master who *Fifth vajra-master awareness (Buddhism, Tibetan)*
HM 2180 5 stages entering the solitary *Pratyeka Buddha awareness (Buddhism, Tibetan)*
HH 0913 5 stages in human motivation *Needs hierarchy*
HH 0350 5 stages of a disciples *Circles of enlightenment (Zen)*
HM 3656 5 stages of 'ishq are *Passionate love (Islam)*
HM 6523 5 stages of mahabba lead *Mahabba (Islam)*
HH 0928 5 stages or relations (1) *Soto Zen (Zen)*
HH 0628 5 stages resignation obedience friendship *Bhakti marga (Hinduism)*
HM 5775 5 state arising *Total annihilation (Sufism)*
HM 0891 5 state beyond these four *Degrees of consciousness (Hinduism)*
HM 0750 5 state of consciousness description *Lucid awareness (Psychism)*
HM 1024 5 state or square *Physical plane (Leela)*
HM 2168 5 state the unsurpassed (akanistha) *Pure-lands awareness (Buddhism, Tibetan)*
HM 5354 5 states are said *Ascetic states (Buddhism)*
HM 4058 5 states of mind classification *Autism (Yoga)*
HM 5734 5 states of mind classification *One-pointedness (Yoga)*
HM 1015 5 states of mind classification *Restlessness (Yoga)*
HM 1346 5 states of mind classification *Stupefaction (Yoga)*
HM 0135 5 states of mind classification *Suppression (Yoga)*
HM 3344 5 station of the cross *Jesus helped to carry the cross (Christianity)*
HM 3941 5 step *Dark night of the soul (Christianity)*
HM 3941 5 step of the ladder *Dark night of the soul (Christianity)*
HM 4474 5 subtle elements with consciousness *Genesis (Leela)*
HM 2996 5 such views of transitory *Afflicted views (Buddhism)*
HM 6553 5 system description *Dwelling in the fivefold jana (Buddhism)*
HM 6553 5 system is as follows
HM 6553 5 system the second jhana *Dwelling in the fivefold jana (Buddhism)*
HM 2059 5 tanmatra or potentials *Ego awareness (Hinduism)*
HM 2656 5 tantric buddha families *Great-path tantra awareness (Buddhism, Tibetan)*
HH 7324 5 tastes plus soft taste *Human development through diet*
HH 7324 5 tastes sweet salty vinegary *Human development through diet*
HM 4999 5 tetrad access concentration arises *Access concentration (Buddhism)*
HM 4265 5 tetrad this is *Fine-material-sphere concentration (Buddhism)*
HM 0696 5 tetrad this is *Immaterial-sphere concentration (Buddhism)*
HM 1097 5 tetrad this is *Sense-sphere concentration (Buddhism)*
HM 5768 5 tetrad this is *Unincluded concentration (Buddhism)*
HH 0195 5 the fifth stage 12 *Sensorimotor development*
HH 0647 5 the fullest attainable participation *Development of the quality of human life*
HH 1799 5 the qur'an contains many *Human development (Islam)*
HH 1198 5 the search for meaning *Human development through religion*
HM 2880 5 they are conversion metanoia *Mystical Christ-consciousness (Christianity)*
HM 6173 5 times among the indeterminate *Mind-consciousness element (Buddhism)*
HM 6173 5 times as it proceeds *Mind-consciousness element (Buddhism)*
HH 0536 5 times *Salat (Islam)*
HM 4398 5 trans sensate phenomena *Deautomatization and the mystic experience*
HM 2389 5 types armouring *Effort (Buddhism)*
HH 4001 5 types of thought *Disruptive thoughts (Christianity)*
HH 0136 5 unity lies *Unity of personality*
HH 1747 5 unity *Sufi path to perfection (Sufism)*
HM 2298 5 universal principles postulated *Discrete states of consciousness (Physical sciences)*
HM 2417 5 universal principles postulated *Least action principle in conscious states (Physical sciences)*
HM 2357 5 universal principles postulated *Physical conservation in conscious states (Physical sciences)*
HM 2381 5 universal principles postulated *Physical duality in conscious states (Physical sciences)*
HM 2322 5 universal principles postulated *Physical relativity in conscious states (Physical sciences)*
HM 2763 5 vajra master awareness buddhism *Fifth vajra-master awareness (Buddhism, Tibetan)*
HH 0213 5 vital forces or prana *Pranayama (Hinduism, Yoga)*
HH 3198 5 ways again the meditator *Immaterial states as meditation subjects (Buddhism)*
HM 7769 5 ways in which feeling *Equanimity (Buddhism)*
HM 8737 5 ways in which feeling *Joyful feeling (Buddhism)*
HM 8010 5 ways in which feeling *Painful feeling (Buddhism)*
HM 6722 5 ways in which feeling *Pleasant feeling (Buddhism)*
HM 7120 5 ways in which feeling *Sad feeling (Buddhism)*
HM 8273 5 with profitable result *Consciousness-born materiality (Buddhism)*
HH 6282 5 with *Subtle faculties (Sufism)*
HM 8273 5 with unprofitable result *Consciousness-born materiality (Buddhism)*
HM 0747 5 zest *Fivefold happiness (Buddhism)*
HM 0801 6 a positive mood *Modes of awareness associated with use of hallucinogens*
HH 0075 6 a sense of insightful *Psychedelic drugs*
HM 2050 6 abstinence from bodily misconduct *Awareness as mental-formation group of conscious existence* (Buddhism)
HM 2050 6 along with the feeling *Awareness as mental-formation group of conscious existence (Buddhism)*
HM 0894 6 am *Lauds (Christianity)*
HM 3089 6 among the spiritual paths *Path of intelligence of mediating influence (Judaism)*
HM 1904 6 and 9 am *Prime (Christianity)*
HM 1468 6 and 9 pm *Vespers (Christianity)*
HM 2171 6 and with preparations (samantaka) *Only-a-beginner subtle contemplation (Buddhism)*
HM 2431 6 antarabhava consciousness *Death-bardo rebirth-seeking awareness (Buddhism)*
HM 2657 6 arise spontaneously without any *Ten worlds (Buddhism)*
HH 1747 6 astonishment *Sufi path to perfection (Sufism)*
HM 5122 6 branch of the tibetan *Yoga of consciousness transference (Buddhism)*
HM 2060 6 but the order is *Kamaloka consciousness (Buddhism, Pali)*
HM 3028 6 card of the 22 *Betrothal initiation archetypal image (Tarot)*
HM 4412 6 chakra time of penance *Sixth chakra: time of penance (Leela)*
HM 2981 6 chakra to the plane *Apt religion (Leela)*
HH 0910 6 classes of mental factors *Changeable mental factors (Buddhism)*
HH 0170 6 classes of mental factors *Determining mental factors (Buddhism)*
HH 0320 6 classes of mental factors *Omnipresent mental factors (Buddhism, Tibetan)*
HM 2556 6 classes of sense consciousness *Awareness of consciousness-group of conscious existence - senses and mind* (Buddhism)
HM 7655 6 co nascence or coexistence *Conditions for consciousness (Buddhism)*
HM 2566 6 component of the eightfold *Dharana (Hinduism)*
HM 3359 6 context *Conjunctive faith*
HM 3630 6 developing contemplation of cessation *Principal insights (Buddhism)*
HM 3476 6 different brain wave frequencies *Beta wave consciousness*
HH 3129 6 different brain wave frequencies *Brain waves*
HM 1785 6 different brain wave frequencies *Delta wave consciousness*
HM 2813 6 directions and their assemblies *Supreme-heaven awareness (Buddhism, Tibetan)*
HM 8336 6 door adverting as attention *Adverting (Buddhism)*
HM 8336 6 door adverting context *Adverting (Buddhism)*
HH 0877 6 elements sama calmness overcoming *Self-discipline*
HM 7769 6 factored the state *Equanimity (Buddhism)*
HM 1465 6 faith description *Universalizing faith*
HM 2664 6 feelings for example *Sense consciousness (Buddhism)*
HM 3583 6 fetter or character defect *Desire for life in form (Buddhism)*
HM 5187 6 fold knowledge systematics transcendental *Six-fold knowledge (Systematics)*
HM 2456 6 foot hole could contain *Contingent eternality (ICA)*
HH 0682 6 (from 12 14) *Transcendental meditation (TM)*
HM 2072 6 general classes of conscious *Worlds of conscious existence (Buddhism, Pali)*
HM 2050 6 general elements *Awareness as mental-formation group of conscious existence (Buddhism)*
HH 0986 6 genital stage 16 18 *Libido development*
HM 0776 6 gradations of knowledge *Seven-fold knowledge (Systematics)*
HM 2072 6 grades in the sensuous *Worlds of conscious existence (Buddhism, Pali)*
HM 2664 6 groups omnipresent factors (5) *Sense consciousness (Buddhism)*
HH 5652 6 he repeatedly attains *Supernormal powers (Buddhism)*
HM 2072 6 higher spiritual powers (abhinma) *Worlds of conscious existence (Buddhism, Pali)*
HM 2012 6 higher spiritual powers start *Arupaloka consciousness (Buddhism, Pali)*
HM 2483 6 illumination of jami *Ultimate transparency of spirit (Sufism)*
HM 5187 6 in a sequence *Six-fold knowledge (Systematics)*
HM 2570 6 in the ica *Ubiquitous otherness (ICA)*
HM 1465 6 in the system *Universalizing faith*
HM 5187 6 independent elements is required *Six-fold knowledge (Systematics)*
HM 0322 6 initiation in the tradition *Initiation of decision (Esotericism)*
HH 2909 6 integration of personality directional *Psychospiritual growth (Christianity)*
HM 3104 6 intuition *Kan (Japanese)*
HM 3104 6 intuitional cognition description *Kan (Japanese)*
HM 2060 6 kinds in the sensuous *Kamaloka consciousness (Buddhism, Pali)*
HM 7769 6 kinds of desirable *Equanimity (Buddhism)*
HM 2122 6 kinds of higher spiritual *Trances and mental absorptions (Buddhism)*
HM 3852 6 kinds of object determining *Indeterminate consciousness in the sense sphere - functional (Buddhism)*
HM 3852 6 kinds of object manifesting *Indeterminate consciousness in the sense sphere - functional (Buddhism)*
HM 6173 6 kinds of object *Mind-consciousness element (Buddhism)*
HM 4062 6 kinds of repentance when *Tawba (Sufism)*
HM 2385 6 level bodhisattva who embodies *Sixth scriptural bodhisattva awareness (Buddhism, Tibetan)*
HH 0765 6 levels have been documented *Biofeedback training*
HM 3744 6 levels have been documented *States of awareness in biofeedback training*
HM 2060 6 levels the tibetan also *Kamaloka consciousness (Buddhism, Pali)*
HM 2144 6 lotus or chakra defined *Ajna (Yoga)*
HM 1973 6 lower worlds instinctive urges *State of rapture (Buddhism)*
HH 1234 6 maha paramita or great *Dana (Buddhism, Zen)*
HM 0137 6 maha paramita or great *Dhyana (Hinduism, Buddhism)*
HH 0813 6 maha paramita or great *Kshanti (Zen)*
HH 0429 6 maha paramita or great *Shila (Zen)*
HH 0677 6 maha paramita or great *Virya (Zen)*
HM 2335 6 major bardo *Death-bardo clear-light awareness (Buddhism)*
HM 2407 6 major bardo *Death-bardo heaven-reality awareness (Buddhism)*
HM 2371 6 major bardo existences *Dream-state bardo awareness (Buddhism, Tibetan)*
HM 2347 6 major bardo existences *Life-state bardo awareness (Buddhism)*
HM 2311 6 major bardo existences *Meditation-state bardo awareness (Buddhism, Tibetan)*
HH 2077 6 member yoga yoga sadanga *Six-member yoga (Yoga)*
HH 0565 6 morality of universalizable reversible *Moral development*
HM 6920 6 noble person *Knowledge of the path of non-return (Buddhism)*
HM 4143 6 objects perceptions of form *Perception aggregate (Buddhism)*
HM 2305 6 of 40 stations *State-awareness (Sufism)*
HM 5244 6 of eight lokas represented *Plane of austerity (Leela)*
HM 3354 6 of seven ascending planes *Cessation of objectivity (Hinduism)*
HM 3510 6 of seven descending steps *Dream wakefulness (Hinduism)*

INDEX TO CONCEPT SETS
(See SECTION X for KEYWORD INDEX)

HX

HM 2470 6 of seven mansions *Mansions of favours and afflictions (Christianity)*
HM 3060 6 of seven stages *I-transcendent stage of life*
HM 4762 6 of seven stages *Unification of energy (Taoism)*
HM 2246 6 of seven subtle stages *Divine nature (Sufism)*
HM 6554 6 of seven ways *Way of idealism (Esotericism)*
HM 3031 6 of ten described *Sphere of beauty (Kabbalah)*
HM 0103 6 of ten orders *Sixth order perceptions - control of relationships*
HM 2874 6 of ten states acting *Valleys (Sufism)*
HM 2205 6 of the nine states *Pacifying in mental abiding states (Buddhism)*
HM 1245 6 of the ninety nine *Al-Mu'min (Sufism)*
HM 2659 6 of the preparations *Critical-analytical subtle contemplation (Buddhism)*
HH 1266 6 of the ten meritorious *Transferring merit (Buddhism)*
HM 1914 6 of the ten worlds *Six paths (Buddhism)*
HM 2516 6 on the board *Reviving-hell awareness (Buddhism, Tibetan)*
HH 0550 6 or culmination *Prajna paramita (Zen)*
HH 0484 6 or i transcendent stage *Buddhi yoga (Yoga)*
HM 7109 6 or inner sense perceiving *Introspection (Buddhism)*
HM 2162 6 or seven angels who *Angelic awareness - Raphael*
HM 7268 6 or seven apperceptions *Impulsion (Buddhism)*
HM 0103 6 order perception context *Sixth order perceptions - control of relationships*
HM 0103 6 order perceptions control *Sixth order perceptions - control of relationships*
HH 0565 6 orientation toward the decisions *Moral development*
HH 0661 6 orthodox indian systems *Yoga (Yoga)*
HM 1914 6 paths buddhism description *Six paths (Buddhism)*
HH 0607 6 people together *Group-centered therapy*
HH 0786 6 personality the progressive *Integration*
HH 3687 6 pillar of islam *Jihad (Islam)*
HM 3354 6 plane of wisdom description *Cessation of objectivity (Hinduism)*
HM 0466 6 pm *None (Christianity)*
HM 2072 6 power takes the adept *Worlds of conscious existence (Buddhism, Pali)*
HM 2753 6 powers effort has matured *One-pointedness in mental abiding states (Buddhism)*
HH 1799 6 practice holy war *Human development (Islam)*
HH 0781 6 primary afflictions *Secondary afflictions (Buddhism, Tibetan)*
HH 2909 6 progressive freedom from sin *Psychospiritual growth (Christianity)*
HH 2097 6 progressive stages preceded *Stages of faith*
HM 1957 6 psychic state listed *Uns (Sufism)*
HM 2866 6 realms of the devas *Sukha (Buddhism)*
HH 0647 6 recognition of different cultural *Development of the quality of human life*
HH 3987 6 recollections are suited *Subjects for meditation (Buddhism)*
HM 4999 6 recollections mindfulness of death *Access concentration (Buddhism)*
HM 4474 6 representing the uniting *Genesis (Leela)*
HM 2996 6 root afflictions referred *Afflicted views (Buddhism)*
HM 2433 6 root afflictions referred *Desire*
HM 4467 6 root afflictions referred *Doubt (Buddhism)*
HM 3196 6 root afflictions referred *Ignorance (Buddhism, Yoga, Zen)*
HM 2823 6 root afflictions referred *Pride (Christianity)*
HM 2959 6 root afflictions referred *State of anger (Buddhism)*
HH 1348 6 root deluded 20 auxiliary *Secondary mental factors (Buddhism, Tibetan)*
HM 1601 6 row *Conscience (Leela)*
HM 6112 6 row *Earth (Leela)*
HM 0172 6 row *Liquid plane (Leela)*
HM 3779 6 row *Lunar plane (Leela)*
HM 4412 6 row of the board *Sixth chakra: time of penance (Leela)*
HM 5244 6 row *Plane of austerity (Leela)*
HM 4050 6 row *Plane of neutrality (Leela)*
HM 1276 6 row *Plane of violence (Leela)*
HM 1279 6 row *Solar plane (Leela)*
HM 1475 6 row *Spiritual devotion (Leela)*
HM 2385 6 scriptural bodhisattva awareness buddhism *Sixth scriptural bodhisattva awareness (Buddhism, Tibetan)*
HM 2664 6 sense consciousnesses and 51 *Sense consciousness (Buddhism)*
HM 2664 6 sense consciousnesses are enumerated *Sense consciousness (Buddhism)*
HM 1119 6 sense consciousnesses *Delusion consciousness (Buddhism)*
HM 3617 6 sense consciousnesses five senses *Vijna (Buddhism)*
HM 0650 6 sense consciousnesses *Foundation awareness*
HM 3104 6 sense *Kan (Japanese)*
HM 2431 6 sridpa bardo *Death-bardo rebirth-seeking awareness (Buddhism)*
HM 0807 6 stage in a systematic *Tawakkul (Sufism)*
HM 2888 6 stage in the process *Near death review of life (NDE)*
HH 0622 6 stage must be reached *Human development (Jainism)*
HM 2375 6 stage (safar dar vatan) *Nakshbandi recollection (Islam)*
HH 0761 6 stages in the process *Meditation*
HH 0195 6 stages of sensorimotor development *Sensorimotor development*
HH 6452 6 stages through which it *Circulation of the light (Taoism)*
HH 0506 6 standard exercises are based *Autogenic training*
HM 0476 6 state arising *'Ilm al-basir (Sufism)*
HM 2556 6 state or function *Awareness of consciousness-group of conscious existence - senses and mind* (Buddhism)
HM 0697 6 state or square *Delusion (Leela)*
HM 3222 6 state yoga context *Refined cosmic consciousness*
HM 2556 6 states of consciousness *Awareness of consciousness-group of conscious existence - senses and mind* (Buddhism)
HM 2341 6 station is called inspiration *Stations of consciousness - Simnami (Sufism)*
HM 3401 6 station of the cross *Veronica wipes the face of Jesus (Christianity)*
HM 3941 6 step of the ladder *Dark night of the soul (Christianity)*
HH 0497 6 superknowledges magical powers *Magico-religious powers*
HH 0136 6 temperament particularly when determined *Unity of personality*
HM 8365 6 tetrad this is *Concentration due to energy (Buddhism)*
HM 7434 6 tetrad this is *Concentration due to inquiry (Buddhism)*
HM 7961 6 tetrad this is *Concentration due to natural purity of consciousness (Buddhism)*
HM 4969 6 tetrad this is *Concentration due to zeal (Buddhism)*
HM 6663 6 tetrads and one pentad *Concentration (Buddhism)*
HM 3045 6 the betrothal or lovers *Charioteer archetypal image (Tarot)*
HH 0195 6 the sixth stage 16 *Sensorimotor development*
HH 1198 6 the supreme reality god *Human development through religion*
HM 2431 6 tibetan *Death-bardo rebirth-seeking awareness (Buddhism)*
HH 0851 6 to 12 equally balanced *Group therapy*
HM 2733 6 types hell beings hungry *Desire-realm consciousness (Buddhism)*
HH 1809 6 types of extraordinary knowledge *Extraordinary knowledge (Buddhism)*
HH 4019 6 types of extraordinary knowledge *Steps to enlightenment (Buddhism)*
HM 2664 6 types of mind corresponding *Sense consciousness (Buddhism)*
HM 8336 6 types of object is *Adverting (Buddhism)*
HH 0136 6 unity lies *Unity of personality*
HM 2287 6 vajra master awareness buddhism *Sixth vajra-master awareness (Buddhism, Tibetan)*
HM 2716 6 will eventuate when nirvana *Discipleship-application awareness (Buddhism, Tibetan)*
HM 2431 6 worlds *Death-bardo rebirth-seeking awareness (Buddhism)*
HH 0565 6 1 *Moral development*
HH 1321 6 11 both at school *Religious growth*
HH 0565 7 a hypothetical stage based *Moral development*
HH 0075 7 a paradoxical quality *Psychedelic drugs*
HH 0225 7 acts of submission are *Discipline of submission (Christianity)*
HH 0986 7 adult stage *Libido development*
HM 2444 7 among the spiritual paths *Path of occult intelligence (Judaism)*
HH 0765 7 and 8 12 cycles *Biofeedback training*
HM 2683 7 and last *Final-training subtle contemplation (Budhism)*
HM 2097 7 and of 4 equals *Tarot arcana of conscious states (Tarot)*
HM 2162 7 angels who most notably *Angelic awareness - Raphael*
HM 7268 7 apperceptions or impulsions *Impulsion (Buddhism)*
HM 3630 7 are referred *Principal insights (Buddhism)*
HM 2793 7 bodhi (awareness or awakening) *Bodhi-pakkhiya dhamma (Buddhism)*
HM 2886 7 branches of enlightenment mindfulness *Forms of enlightenment (Buddhism)*
HM 2219 7 but with a plane *Chakra centres of consciousness*
HM 3045 7 card of the greater *Charioteer archetypal image (Tarot)*
HH 0237 7 centres also called lotus *Kundalini yoga (Yoga)*
HH 0237 7 centres leaving *Kundalini yoga (Yoga)*
HM 0754 7 chakra plane of reality *Seventh chakra: plane of reality (Leela)*
HM 0176 7 chakra this is located *Soma chakra (Yoga)*
HM 2219 7 chakras *Chakra centres of consciousness*
HM 3907 7 chakras is that *Phenomenal plane (Leela)*
HM 3291 7 chakras there are influences *Fourth chakra: attaining balance (Leela)*
HM 0137 7 component of the eightfold *Dhyana (Hinduism, Buddhism)*
HM 5721 7 consciousnesses associated with unprofitable *Indeterminate consciousness in the sense sphere -* resultant (Buddhism)
HM 3630 7 contemplations are enumerated *Principal insights (Buddhism)*
HH 5887 7 corresponding stages *Alchemy (Taoism)*
HM 0102 7 deadly sins context *Beginner in spiritual life (Christianity)*
HH 0414 7 deadly sins *Meekness*
HM 3592 7 descending steps of ignorance *Veils of delusion (Hinduism)*
HH 1198 7 despite widely diverse concepts *Human development through religion*
HH 2002 7 destructions by fire then *World cycles (Buddhism)*
HM 3630 7 developing contemplation of relinquishment *Principal insights (Buddhism)*
HH 2665 7 distinct ways to spiritual *Ways to spiritual realization (Esotericism)*
HH 0565 7 each describing the distinctive *Moral development*
HH 2267 7 energy fields are described *Human aura*
HM 2075 7 faculties can be added *Psycho-physical faculties awareness (Buddhism)*
HM 1261 7 fetter or character defect *Desire for formless life (Buddhism)*
HH 0786 7 final perfect personality integration *Integration*
HM 0776 7 fold knowledge provides *Seven-fold knowledge (Systematics)*
HM 0776 7 fold knowledge *Seven-fold knowledge (Systematics)*
HM 0776 7 fold knowledge systematics structural *Seven-fold knowledge (Systematics)*
HH 5652 7 following the order *Supernormal powers (Buddhism)*
HH 0269 7 further dynamics progressively leaving *Thetan*
HM 0137 7 grades of holy paths *Dhyana (Hinduism, Buddhism)*
HM 2793 7 groups of mental qualities *Bodhi-pakkhiya dhamma (Buddhism)*
HM 2188 7 heaven although a lower *Orders of angelic awareness (Judaism)*
HM 4501 7 heavens and their angelic *God as wholly other (Judaism)*
HM 2327 7 heavens *Ascension awareness (Sufism)*
HH 1903 7 holy immortals ahura mazda *Human development (Zoroastrianism)*
HM 2985 7 illumination of jami *Identifying with reality (Sufism)*
HM 0776 7 in a sequence *Seven-fold knowledge (Systematics)*
HM 2674 7 in the ica *Final limits (ICA)*
HM 0776 7 independent elements *Seven-fold knowledge (Systematics)*
HM 1153 7 initiation in the tradition *Initiation of resurrection (Esotericism)*
HM 1425 7 is powerfully and permanently *Intuitive-projective faith*
HM 7672 7 kinds of object limited *Knowledge of supernormal powers (Buddhism)*
HM 2341 7 level with al khadir *Stations of consciousness - Simnami (Sufism)*
HM 3398 7 lotus or chakra defined *Sahasrara (Yoga)*
HH 0921 7 lower levels representing levels *Kabbalah*
HM 3513 7 major prophets of semitic *Divine essence (Sufism)*
HM 2246 7 major prophets of semitic *Divine nature (Sufism)*
HM 4570 7 major prophets of semitic *World beyond form (Sufism)*
HM 3286 7 major prophets of semitic *World of forms (Sufism)*
HM 3616 7 major prophets of semitic *World of imagination (Sufism)*
HM 3425 7 major prophets of semitic *World of nature (Sufism)*
HM 3974 7 major prophets of semitic *World of spiritual perception (Sufism)*
HH 1409 7 mansions as it is *Mansions of the soul (Christianity)*
HM 3971 7 mansions the lord is *Mansions of spiritual marriage (Christianity)*
HM 2341 7 maqamat in 7 worlds *Stations of consciousness - Simnami (Sufism)*
HM 7655 7 mutuality or reciprocity *Conditions for consciousness (Buddhism)*
HM 7055 7 noble person *Knowledge of the path of arahantship (Buddhism)*
HM 1088 7 noble treasures faith virtue *Knowledge of the path of stream entry (Buddhism)*
HM 3418 7 of 40 stations *Sabr (Sufism)*
HM 1293 7 of eight lokas represented *Plane of reality (Leela)*
HM 3070 7 of scale *Completeness*
HM 2395 7 of seven ascending planes *Turiya awareness (Hinduism)*
HM 1605 7 of seven ways *Way of organization (Esotericism)*
HM 2362 7 of ten described *Sphere of victory (Kabbalah)*
HM 1001 7 of ten orders *Seventh order perceptions - programme control*
HM 3607 7 of ten states acting *Mystical states (Sufism)*
HM 2729 7 of the nine states *Full pacification in mental abiding states (Buddhism)*
HM 0526 7 of the ninety nine *Al-Muhaymin (Sufism)*
HH 0932 7 of the ten meritorious *Rejoicing at the merit of others (Buddhism)*
HM 2040 7 on the board *Cold-hells awareness (Buddhism, Tibetan)*
HH 0145 7 or 8 recent psychological *Empathy*
HM 0822 7 or highest level *Adi plane (Psychism)*
HM 2864 7 or in sixteen lifetimes *Tantra in meditation on emptiness (Buddhism, Tibetan)*
HM 1001 7 order perceptions programme control *Seventh order perceptions - programme control*
HM 3298 7 plane of wisdom is *Planes of wisdom (Hinduism)*
HM 4417 7 plane psychism description *Seventh plane (Psychism)*
HM 1478 7 planes of density *Limbo*
HM 1153 7 planes of planetary life *Initiation of resurrection (Esotericism)*
HH 1747 7 poverty and nothingness *Sufi path to perfection (Sufism)*
HM 2683 7 preparation in the desire *Final-training subtle contemplation (Buddhism)*
HM 2623 7 preparations (samantaka) or mental *Special insight and preparations as states (Buddhism)*
HM 2062 7 preparations *Third trance of the fine-material sphere (Buddhism)*
HH 2909 7 progressive evidence *Psychospiritual growth (Christianity)*
HM 3521 7 psychic state listed *Mushahada (Sufism)*
HH 3875 7 purifications or purities *Purifications (Buddhism)*
HH 2665 7 rays description *Ways to spiritual realization (Esotericism)*
HM 1293 7 representing a psychic *Plane of reality (Leela)*
HM 1726 7 row *Egotism (Leela)*
HM 6434 7 row *Gaseous plane (Leela)*
HM 5099 7 row *Happiness (Leela)*
HM 3623 7 row *Negative intellect (Leela)*
HM 0754 7 row of the board *Seventh chakra: plane of reality (Leela)*
HM 3732 7 row *Plane of primal vibrations (Leela)*
HM 5028 7 row *Plane of radiation (Leela)*
HM 1293 7 row *Plane of reality (Leela)*
HM 1789 7 row *Positive intellect (Leela)*
HM 6238 7 row *Tamas (Leela)*

HH 2198 7 sacraments baptism marriage confirmation *Human development (Christianity)*
HM 2361 7 scriptural bodhisattva awareness buddhism *Seventh-scriptural bodhisattva awareness (Buddhism,* Tibetan)
HH 2498 7 section of st john's *Journeying within transcendence - from isolation to imagination*
HH 0603 7 sections *Harmonies with enlightenment (Buddhism)*
HM 2556 7 seems to appear *Awareness of consciousness-group of conscious existence - senses and mind (Buddhism)*
HH 2097 7 so the greater arcana *Tarot arcana of conscious states (Tarot)*
HH 0647 7 social structures which while *Development of the quality of human life*
HM 4190 7 stage in a systematic *Rida (Sufism)*
HM 2375 7 stage (nazar bar qadam) *Nakshbandi recollection (Islam)*
HM 0137 7 stage of the eightfold *Dhyana (Hinduism, Buddhism)*
HM 3418 7 stage on the path *Sabr (Sufism)*
HM 2341 7 stages in the second *Stations of consciousness - Simnami (Sufism)*
HH 0306 7 stages of initiation *Tantra (Buddhism, Yoga)*
HH 0460 7 stages of life description *Seven stages of life*
HM 2586 7 stages of preparations have *Fourth trance of the fine-material sphere (Buddhism)*
HM 4338 7 stages of the taoist *Assembling the five elements (Taoism)*
HM 1330 7 stages of the taoist *Merging of yin and yang (Taoism)*
HM 6633 7 stages of the taoist *Recognition of the true mind (Taoism)*
HM 4007 7 stages of the taoist *Refining the self (Taoism)*
HM 6534 7 stages of the taoist *Restoration of celestial awareness within the mundane (Taoism)*
HM 5265 7 stages of the taoist *Transcending the world (Taoism)*
HM 4762 7 stages of the taoist *Unification of energy (Taoism)*
HM 4788 7 state arising *Continuous all-pervading knowledge (Sufism)*
HM 2428 7 state of awareness that *Theta-clear state of consciousness (Scientology)*
HM 0098 7 state or square *Conceit (Leela)*
HM 3298 7 states ascending steps *Planes of wisdom (Hinduism)*
HM 2141 7 states attainable *Jagrat state of waking consciousness (Yoga)*
HM 2781 7 states attainable to human *Swapna state of dream consciousness (Yoga)*
HH 5887 7 states the generative state *Alchemy (Taoism)*
HM 3824 7 station of the cross *Jesus falls the second time under the cross (Christianity)*
HM 2341 7 station the prophet *Stations of consciousness - Simnami (Sufism)*
HM 3941 7 step causes the soul *Dark night of the soul (Christianity)*
HH 0927 7 steps are prescribed discrimination *Jnana yoga (Yoga)*
HM 2683 7 subtle contemplations viewed *Final-training subtle contemplation (Buddhism)*
HH 2097 7 supreme arcana and fifteen *Tarot arcana of conscious states (Tarot)*
HH 0136 7 temporary convergence occurs when *Unity of personality*
HM 2424 7 to 101 *Stations of consciousness - ibn-Abi'l-Khayr (Sufism)*
HM 3651 7 to 14 of temporal *Second chakra: realm of fantasy (Leela)*
HH 0136 7 unity lies *Unity of personality*
HM 2789 7 vajra master awareness buddhism *Seventh vajra-master awareness (Buddhism, Tibetan)*
HH 1747 7 valleys
HH 3259 7 virtues *Accidie*
HM 0130 7 with sense sphere formation *Mundane resultant consciousness (Buddhism)*
HM 2341 7 worlds description *Stations of consciousness - Simnami (Sufism)*
HM 2341 7 worlds *Stations of consciousness - Simnami (Sufism)*
HM 5977 8 altered states of consciousness *Vimoksa (Buddhism)*
HM 3243 8 among the spiritual paths *Path of perfect intelligence (Judaism)*
HH 0763 8 and 13 totally distinct *Multiple personality*
HM 1141 8 and top row *Beyond the chakras: the gods themselves (Leela)*
HM 2098 8 are karmically wholesome *Dispositions of consciousness (Buddhism)*
HM 3852 8 are with root cause *Indeterminate consciousness in the sense sphere - functional (Buddhism)*
HM 5721 8 are without root cause *Indeterminate consciousness in the sense sphere - resultant (Buddhism)*
HH 1198 8 as a reflexive cultural *Human development through religion*
HM 8375 8 associated with greed four *Unprofitable consciousness in the sense sphere (Buddhism)*
HM 4983 8 associated with the fifth *Feeling aggregate (Buddhism)*
HH 5652 8 attainments (jhana) in each *Supernormal powers (Buddhism)*
HH 2002 8 by water until *World cycles (Buddhism)*
HH 2040 8 cold hell states *Cold-hells awareness (Buddhism, Tibetan)*
HM 2733 8 cold hells *Desire-realm consciousness (Buddhism)*
HM 2226 8 component of the eightfold *Samadhi (Hinduism, Yoga)*
HM 2318 8 cycle is called sanctity *Ultimate Sufi state of consciousness (Sufism)*
HH 2909 8 deepening intimacy with god *Psychospiritual growth (Christianity)*
HM 3630 8 developing contemplation of destruction *Principal insights (Buddhism)*
HH 0285 8 distinct stages each *Stages of personality development*
HH 0103 8 elements constitute the body *Atman (Hinduism)*
HH 0103 8 elements must be transcended *Atman (Hinduism)*
HM 4500 8 essentials indicating the infinite *Allah (Sufism)*
HH 5652 8 factors it may *Supernormal powers (Buddhism)*
HM 2586 8 faults those of inhalation *Fourth trance of the fine-material sphere (Buddhism)*
HM 2852 8 fetter or character defect *Spiritual pride (Buddhism)*
HM 1408 8 fold knowledge does not *Nine-fold knowledge (Systematics)*
HM 0344 8 fold knowledge is able *Eight-fold knowledge (Systematics)*
HM 0344 8 fold knowledge systematics revealed *Eight-fold knowledge (Systematics)*
HH 5652 8 following the order *Supernormal powers (Buddhism)*
HM 2693 8 forbearances and the eight *Tibetan meditative states of form concentrations (Buddhism)*
HM 4447 8 give rise to materiality *Profitable consciousness in the sense sphere (Buddhism)*
HM 2516 8 hot hells *Reviving-hell awareness (Buddhism, Tibetan)*
HM 2733 8 hot hellse *Desire-realm consciousness (Buddhism)*
HM 2214 8 illumination of jami *Identity with God (Sufism)*
HM 0344 8 in a sequence *Eight-fold knowledge (Systematics)*
HM 2733 8 in each *Desire-realm consciousness (Buddhism)*
HM 2764 8 in the ica *Total exposure (ICA)*
HM 0344 8 independent terms *Eight-fold knowledge (Systematics)*
HM 7268 8 ? indeterminate functional *Impulsion (Buddhism)*
HM 1646 8 indeterminate resultant consciousnesses *Registering (Buddhism)*
HM 1258 8 initiation in the tradition *Initiation of transition (Esotericism)*
HH 0136 8 interdependence of traits provides *Unity of personality*
HH 2987 8 is *Journeying within transcendence - from death to life*
HM 2318 8 is the highest state *Ultimate Sufi state of consciousness (Sufism)*
HM 2318 8 is the station *Ultimate Sufi state of consciousness (Sufism)*
HM 2051 8 jhana of all when *Fourth absorption in the immaterial sphere (Buddhism)*
HM 2098 8 karmically neutral or indeterminate *Dispositions of consciousness (Buddhism)*
HH 5652 8 kasinas ending *Supernormal powers (Buddhism)*
HM 0645 8 kinds of aesthetic sentiment *Comic sentiment (Hinduism)*
HM 1780 8 kinds of aesthetic sentiment *Erotic sentiment (Hinduism)*
HM 8762 8 kinds of aesthetic sentiment *Furious sentiment (Hinduism)*
HM 4085 8 kinds of aesthetic sentiment *Heroic sentiment (Hinduism)*
HM 0291 8 kinds of aesthetic sentiment *Marvellous sentiment (Hinduism)*
HM 0966 8 kinds of aesthetic sentiment *Odious sentiment (Hinduism)*
HM 0399 8 kinds of aesthetic sentiment *Pathetic sentiment (Hinduism)*
HM 1644 8 kinds of aesthetic sentiment *Terrible sentiment (Hinduism)*
HM 4447 8 kinds of consciousness may *Profitable consciousness in the sense sphere (Buddhism)*
HM 7769 8 kinds of equanimity are *Equanimity (Buddhism)*
HM 8266 8 kinds of indeterminate resultant *Rebirth-linking (Buddhism)*
HM 2693 8 knowledges of the path *Tibetan meditative states of form concentrations (Buddhism)*
HH 3550 8 knowledges plus a ninth *Purification by knowledge and vision of the way (Buddhism)*
HM 2257 8 lands where the meditator *Form-realm consciousness (Buddhism)*
HM 2556 8 level after investigation *Awareness of consciousness-group of conscious existence - senses and mind* (Buddhism)
HH 0779 8 levels of yama (restraint) *Eightfold path of yoga (Yoga)*
HM 2693 8 levels *Tibetan meditative states of form concentrations (Buddhism)*
HM 2635 8 liberations buddhism description *Eight liberations (Buddhism)*
HM 2987 8 liberations the nine serial *Paths of calming (Buddhism)*
HM 1167 8 liturgies which rehearse *Canonical hours (Christianity)*
HM 2052 8 lower hells context *Interminable-hell awareness (Buddhism, Tibetan)*
HM 2679 8 magical powers are obtained *Magical-forces awareness (Buddhism, Tibetan)*
HH 0285 8 major crises may *Stages of personality development*
HM 2635 8 means *Eight liberations (Buddhism)*
HM 4389 8 modes occurrence *Sense mode of consciousness occurrence (Buddhism)*
HM 2339 8 noble eightfold *Eightfold way (Buddhism)*
HM 7055 8 noble person *Knowledge of the path of arahantship (Buddhism)*
HH 3025 8 noble person *Purification by knowledge and vision (Buddhism)*
HM 2428 8 objects of awareness *Theta-clear state of consciousness (Scientology)*
HM 2351 8 of 40 stations *Recollection (Islam, Sufism)*
HM 1325 8 of ten orders *Eighth order perceptions - control of principles*
HM 2238 8 of ten spheres described *Sphere of glory (Kabbalah)*
HM 4041 8 of ten states acting *Sanctity (Sufism)*
HM 2753 8 of the nine states *One-pointedness in mental abiding states (Buddhism)*
HM 4562 8 of the ninety nine *Al-'Aziz (Sufism)*
HH 1183 8 of the ten meritorious *Listening to the dharma (Buddhism)*
HM 2454 8 on the board *Temporary-hells awareness (Buddhism, Tibetan)*
HM 0205 8 or nine *Aesthetic emotion (Hinduism)*
HM 4000 8 order is given priority *Ninth order perceptions - control of systems concepts*
HM 1325 8 order perceptions control *Eighth order perceptions - control of principles*
HM 4000 8 order problem and therefore *Ninth order perceptions - control of systems concepts*
HM 4000 8 order systems are reorganized *Ninth order perceptions - control of systems concepts*
HM 2228 8 path context *Pratyeka Buddha vision-path awareness (Buddhism, Tibetan)*
HM 2339 8 path description *Eightfold way (Buddhism)*
HM 2566 8 path *Dharana (Hinduism)*
HH 1021 8 path for their own *Spiritual discipline*
HH 0779 8 path implies the eight *Eightfold path of yoga (Yoga)*
HM 3445 8 path of buddhism *Religious experience*
HM 0137 8 path of raja yoga *Dhyana (Hinduism, Buddhism)*
HH 0669 8 path of yoga *Asana (Hinduism, Yoga)*
HH 0045 8 path of yoga *Celibacy*
HM 2566 8 path of yoga *Dharana (Hinduism)*
HM 0137 8 path of yoga *Dhyana (Hinduism, Buddhism)*
HM 3280 8 path of yoga *Niyama (Hinduism)*
HH 0213 8 path of yoga *Pranayama (Hinduism, Yoga)*
HH 0829 8 path of yoga *Pratyahara (Yoga)*
HH 6876 8 path of yoga *Restraint*
HM 2226 8 path of yoga *Samadhi (Hinduism, Yoga)*
HH 0779 8 path of yoga yoga *Eightfold path of yoga (Yoga)*
HM 8266 8 principal resultant consciousnesses *Rebirth-linking (Buddhism)*
HM 7268 8 profitable consciousnesses *Impulsion (Buddhism)*
HM 8273 8 profitable consciousnesses the 12 *Consciousness-born materiality (Buddhism)*
HM 8266 8 profitable or moral consciousnesses *Rebirth-linking (Buddhism)*
HM 0050 8 psychic state listed *Itmianan (Sufism)*
HH 4777 8 reasons for urgency birth *Skill in absorption (Buddhism)*
HH 0145 8 recent psychological research has *Empathy*
HH 5007 8 rooted in greed two *Afflictions and hindrances (Buddhism)*
HM 4326 8 row *Absolute plane (Leela)*
HM 3907 8 row *Phenomenal plane (Leela)*
HM 6901 8 row *Plane of bliss (Leela)*
HM 1301 8 row *Plane of cosmic consciousness (Leela)*
HM 4343 8 row *Plane of cosmic good (Leela)*
HM 6287 8 row *Plane of inner space (Leela)*
HM 3281 8 row *Rajoguna (Leela)*
HM 4310 8 row *Satoguna (Leela)*
HM 5561 8 row *Tamoguna (Leela)*
HM 2816 8 scriptural bodhisattva awareness buddhism *Eighth scriptural bodhisattva awareness (Buddhism,* Tibetan)
HH 4188 8 section of st john's *Journeying within transcendence - from object to subject*
HM 0754 8 siddhis and can create *Seventh chakra: plane of reality (Leela)*
HM 2679 8 siddhis description *Magical-forces awareness (Buddhism, Tibetan)*
HM 2816 8 stage bodhisattva striving effortlessly *Eighth scriptural bodhisattva awareness (Buddhism, Tibetan)*
HH 0565 8 stage epigenetic cycle relating *Moral development*
HM 2375 8 stage for these reasons *Nakshbandi recollection (Islam)*
HM 2377 8 stage in a systematic *Nearness-awareness (Sufism)*
HM 4154 8 stage on the path *Shukr (Sufism)*
HM 2375 8 stage originally *Nakshbandi recollection (Islam)*
HM 2735 8 stages and rises again *Buddha-consciousness (Buddhism, Tibetan)*
HM 2185 8 stages *Way of the Buddhas (Buddhism, Tibetan)*
HM 3647 8 state arising *Wilayat (Sufism)*
HM 0987 8 state or square *Avarice (Leela)*
HM 2428 8 state the awareness *Theta-clear state of consciousness (Scientology)*
HM 2428 8 states of progressively widening *Theta-clear state of consciousness (Scientology)*
HM 2737 8 station of the cross *Jesus speaks to the daughters of Jerusalem (Christianity)*
HM 3941 8 step the soul is *Dark night of the soul (Christianity)*
HH 5652 8 steps the sixteen roots *Supernormal powers (Buddhism)*
HM 2838 8 substances from which physical *Mental consciousness (Buddhism)*
HH 0285 8 such stages oral anal *Stages of personality development*
HM 7655 8 support or dependence *Conditions for consciousness (Buddhism)*
HM 4983 8 supramundane 4 profitable *Feeling aggregate (Buddhism)*
HM 8273 8 the 4 profitable *Consciousness-born materiality (Buddhism)*
HM 3434 8 the absolute plane *Mercy (Leela)*
HM 4311 8 to ten stages *Psychic states (Sufism)*
HH 1119 8 types of consciousness *Delusion consciousness (Buddhism)*
HM 0650 8 types of consciousness *Foundation awareness*
HH 0136 8 unity emerges *Unity of personality*

INDEX TO CONCEPT SETS
(See SECTION X for KEYWORD INDEX)

HX

HM 2301 8 vajra master awareness buddhism *Eighth vajra-master awareness (Buddhism, Tibetan)*
HM 2330 8 way also referred *Nirvana (Buddhism)*
HM 2339 8 way buddhism aryastangamarga *Eightfold way (Buddhism)*
HH 0907 8 way *Emancipation of the self*
HH 0567 8 way *Karma*
HM 3196 8 ways of unknowing it *Ignorance (Buddhism, Yoga, Zen)*
HM 5721 8 with root cause *Indeterminate consciousness in the sense sphere - resultant (Buddhism)*
HM 1088 8 wrong path is abandoned *Knowledge of the path of stream entry (Buddhism)*
HM 1301 8 yoga *Plane of cosmic consciousness (Leela)*
HH 2077 8 yoga yama (restraint) niyama *Six-member yoga (Yoga)*
HM 2965 9 am and 12 noon *Terce (Christianity)*
HM 1904 9 am *Prime (Christianity)*
HM 4321 9 and 12 pm *Compline (Christianity)*
HM 2072 9 conditions or sub states *Worlds of conscious existence (Buddhism, Pali)*
HM 8266 9 consciousnesses which occur are *Rebirth-linking (Buddhism)*
HM 8266 9 consciousnesses which occur *Rebirth-linking (Buddhism)*
HM 7655 9 decisive support or sufficing *Conditions for consciousness (Buddhism)*
HM 3630 9 developing contemplation of fall *Principal insights (Buddhism)*
HM 3612 9 fetter or character defect *Notion of one's self as entity (Buddhism)*
HM 1408 9 fold knowledge permits *Nine-fold knowledge (Systematics)*
HM 1408 9 fold knowledge systematics experience *Nine-fold knowledge (Systematics)*
HM 1608 9 gifts of the holy *Glossolalia (Christianity)*
HM 1141 9 god forces *Beyond the chakras: the gods themselves (Leela)*
HH 7004 9 grades are the three *Sidetracks and auxiliary methods in Taoism (Taoism)*
HH 7004 9 grades of practice *Sidetracks and auxiliary methods in Taoism (Taoism)*
HM 3358 9 illumination of jami *Annihilation of personal sensation (Sufism)*
HM 1408 9 in a sequence *Nine-fold knowledge (Systematics)*
HM 2982 9 in the ica *Vibrant powers (ICA)*
HM 1408 9 independent terms *Nine-fold knowledge (Systematics)*
HM 1020 9 initiation in the tradition *Initiation of refusal (Esotericism)*
HM 2062 9 interrupted paths *Third trance of the fine-material sphere (Buddhism)*
HH 3987 9 kasinas as object this *Subjects for meditation (Buddhism)*
HH 3550 9 knowledge in conformity *Purification by knowledge and vision of the way (Buddhism)*
HM 4637 9 knowledges of purification *Knowledge of change of lineage (Buddhism)*
HH 3025 9 knowledges of purification *Purification by knowledge and vision (Buddhism)*
HM 2177 9 levels of conscious existence *Consciousness states in cyclic existence (Buddhism)*
HH 0024 9 levels of training *Arica training*
HM 7092 9 mode occurrence *Receiving (Buddhism)*
HM 1134 9 obstacles to soul cognition *Bodily disability (Yoga)*
HM 0361 9 obstacles to soul cognition *Carelessness (Yoga)*
HM 1077 9 obstacles to soul cognition *Erroneous perception (Yoga)*
HM 0610 9 obstacles to soul cognition *Failure to hold meditative attitude (Yoga)*
HM 0298 9 obstacles to soul cognition *Inability to achieve concentration (Yoga)*
HM 0110 9 obstacles to soul cognition *Lack of dispassion (Yoga)*
HM 0046 9 obstacles to soul cognition *Laziness (Yoga)*
HM 0906 9 obstacles to soul cognition *Mental inertia (Yoga)*
HM 1080 9 obstacles to soul cognition *Wrong questioning (Yoga)*
HM 4190 9 of 40 stations *Rida (Sufism)*
HM 2410 9 of ten described *Sphere of the foundation (Kabbalah)*
HM 4000 9 of ten orders *Ninth order perceptions - control of systems concepts*
HM 3755 9 of ten states acting *Realities (Sufism)*
HM 0215 9 of the ninety nine *Al-Jabbar (Sufism)*
HM 2943 9 of the spiritual paths *Path of purified intelligence (Judaism)*
HH 1022 9 of the ten meritorious *Preaching the dharma (Buddhism)*
HM 2088 9 on the board *Lord-of-death awareness (Buddhism, Tibetan)*
HM 7268 9 (or 8 ?) indeterminate *Impulsion (Buddhism)*
HH 0873 9 or ten this is *Ways of knowing (Buddhism)*
HM 4000 9 order perceptions control *Ninth order perceptions - control of systems concepts*
HM 4000 9 order the eighth order *Ninth order perceptions - control of systems concepts*
HM 2150 9 orders of angels *Angelic frame of awareness*
HM 2586 9 path of release *Fourth trance of the fine-material sphere (Buddhism)*
HM 2062 9 path of release *Third trance of the fine-material sphere (Buddhism)*
HM 2683 9 path release *Final-training subtle contemplation (Buddhism)*
HM 2683 9 paths of release *Final-training subtle contemplation (Buddhism)*
HM 2062 9 paths of release *Third trance of the fine-material sphere (Buddhism)*
HM 0205 9 permanent or predominant states *Aesthetic emotion (Hinduism)*
HM 1468 9 pm *Vespers (Christianity)*
HH 3945 9 points *Enneagram patterning (Sufism)*
HM 4426 9 psychic state listed *Yaqin (Sufism)*
HM 1026 9 psychic states *Kedesh-jazba (Sufism)*
HM 2292 9 scriptural bodhisattva awareness buddhism *Ninth scriptural bodhisattva awareness (Buddhism, Tibetan)*
HM 2987 9 serial absorptions *Paths of calming (Buddhism)*
HH 5652 9 skipping jhanas and kasinas *Supernormal powers (Buddhism)*
HM 2278 9 squares representing *Lila*
HM 2292 9 stage bodhisattva expounding *Ninth scriptural bodhisattva awareness (Buddhism, Tibetan)*
HM 4273 9 stage in a systematic *Love of God (Sufism)*
HM 4190 9 stage on the path *Rida (Sufism)*
HM 3627 9 state or square *Sensual plane (Leela)*
HM 7769 9 states in developing calm *Equanimity (Buddhism)*
HM 2157 9 states of mental abiding *Close setting the mind (Buddhism)*
HM 2241 9 states of mental abiding *Continuous setting of mind (Buddhism)*
HM 2181 9 states of mental abiding *Disciplining in mental abiding states (Buddhism)*
HM 2729 9 states of mental abiding *Full pacification in mental abiding states (Buddhism)*
HM 2277 9 states of mental abiding *Mental abiding in equipoise (Buddhism)*
HM 2753 9 states of mental abiding *One-pointedness in mental abiding states (Buddhism)*
HM 2205 9 states of mental abiding *Pacifying in mental abiding states (Buddhism)*
HM 2265 9 states of mental abiding *Resetting the mind (Buddhism)*
HM 2217 9 states of mental abiding *Setting the mind (Buddhism)*
HM 2145 9 states of mind called *Meditative states of mental abidings (Buddhism)*
HM 2289 9 station of the cross *Jesus falls the third time under the cross (Christianity)*
HM 3941 9 step the soul is *Dark night of the soul (Christianity)*
HM 7132 9 surprise interest joy distress *Affects*
HM 2050 9 these formations are variously *Awareness as mental-formation group of conscious existence (Buddhism)*
HH 3945 9 types mapped *Enneagram patterning (Sufism)*
HM 2683 9 uninterrupted paths and 9 *Final-training subtle contemplation (Buddhism)*
HH 0136 9 unity of personality occurs *Unity of personality*
HM 2325 9 vajra master awareness buddhism *Ninth vajra-master awareness (Buddhism, Tibetan)*
HM 2849 10 and last before buddhahood *Final vajra-master awareness (Buddhism, Tibetan)*
HH 0921 10 are divided *Kabbalah*
HM 2098 10 are karmically neutral *Dispositions of consciousness (Buddhism)*
HM 2421 10 bodhisattva stage leading *Tenth scriptural bodhisattva awareness (Buddhism, Tibetan)*
HM 2187 10 bodhisattva stages *First vajra-master stage (Buddhism, Tibetan)*
HM 7022 10 but never punishes bad *Al-Barr (Sufism)*
HM 3501 10 centuries of toil *Third step: mankind (Christianity)*
HH 0314 10 commandments a framework may *Self-examination*
HH 0921 10 commandments are *Kabbalah*
HH 1028 10 commandments *Revelation*
HM 3070 10 creative forces and so *Completeness*
HM 2317 10 degrees in each *Stations of consciousness - Ansari (Sufism)*
HM 2873 10 degrees of development *Enjoyment-body awareness (Buddhism, Tibetan)*
HM 3630 10 developing contemplation of change *Principal insights (Buddhism)*
HH 1198 10 different aspects of different *Human development through religion*
HM 2995 10 directions buddhism tibetan mahashunyata *Emptiness of the ten directions (Buddhism, Tibetan)*
HM 2509 10 emanation realms *Paths of wisdom (Judaism)*
HM 3493 10 fetter or character defect *Ignorance of the nature of the essential self (Buddhism)*
HH 0470 10 fetters or character defects *Character defects (Buddhism)*
HM 1804 10 fold knowledge systematics experience *Ten-fold knowledge (Systematics)*
HM 1804 10 fold level of knowledge *Ten-fold knowledge (Systematics)*
HM 0130 10 from the five senses *Mundane resultant consciousness (Buddhism)*
HH 6292 10 guru naming the adi *Human development (Sikhism)*
HM 2050 10 hate envy avarice worry *Awareness as mental-formation group of conscious existence (Buddhism)*
HH 0921 10 hierarchy and in their *Kabbalah*
HM 3244 10 illumination of jami *Liberation of the heart (Sufism)*
HH 3550 10 imperfections and the three *Purification by knowledge and vision of the way (Buddhism)*
HM 9722 10 imperfections arise in him *Imperfection of insight (Buddhism)*
HM 1804 10 in a sequence *Ten-fold knowledge (Systematics)*
HM 2365 10 (in sarraj) includes muraqaba *Higher states of consciousness (Sufism)*
HM 2862 10 in the ica *Transformed existence (ICA)*
HM 1804 10 independent terms *Ten-fold knowledge (Systematics)*
HM 1252 10 inner states of being *Four evil paths (Buddhism)*
HM 0923 10 inner states of being *Three evil paths (Buddhism)*
HH 3987 10 kasinas and mindfulness *Subjects for meditation (Buddhism)*
HH 3987 10 kasinas (universals) or devices *Subjects for meditation (Buddhism)*
HH 2918 10 kasinas (universals) or devices *Tenfold powers (Buddhism)*
HH 3987 10 kasinas when hearing *Subjects for meditation (Buddhism)*
HH 3987 10 kinds of foulness *Subjects for meditation (Buddhism)*
HH 3987 10 kinds of foulness *Subjects for meditation (Buddhism)*
HH 3987 10 kinds of foulness swollen *Subjects for meditation (Buddhism)*
HM 7769 10 kinds
HH 0921 10 levels each have their *Kabbalah*
HH 0921 10 levels *Kabbalah*
HH 0921 10 levels or sephiroth are *Kabbalah*
HM 8273 10 life continuum consciousnesses indeterminate *Consciousness-born materiality (Buddhism)*
HH 0859 10 meritorious deeds and buddhism *Merit*
HH 0446 10 meritorious deeds buddhism dasakusalakamma *Ten meritorious deeds (Buddhism)*
HH 0921 10 metaphysical numbers or numerations *Kabbalah*
HM 7324 10 mode occurrence *Examining (Buddhism)*
HM 3280 10 niyamas austerity contentment belief *Niyama (Hinduism)*
HM 7336 10 of 40 stations *Mukhalafat-i nafs (Sufism)*
HH 1206 10 of one's income *Philanthropy*
HM 1717 10 of ten orders *Tenth order perceptions - universal oneness - meditation*
HM 8273 10 of the indeterminate functional *Consciousness-born materiality (Buddhism)*
HM 3850 10 of the ninety nine *Al-Mutakabbir (Sufism)*
HM 3426 10 of the spiritual paths *Path of resplendent intelligence (Judaism)*
HH 0974 10 of the ten meritorious *Right beliefs (Buddhism)*
HM 2112 10 on the board *Hungry-ghosts-hell awareness (Buddhism, Tibetan)*
HM 1717 10 order perceptions universal oneness *Tenth order perceptions - universal oneness - meditation*
HM 0988 10 orders of perception *Orders of perception*
HH 1038 10 perfections *Madhyamaka (Buddhism)*
HH 0136 10 personality may be viewed *Unity of personality*
HH 2918 10 powers buddhism meditating *Tenfold powers (Buddhism)*
HM 3159 10 powers buddhism tibetan description *Ten powers (Buddhism, Tibetan)*
HH 0093 10 powers manifesting only *Dharma*
HM 3249 10 powers the four fearlessnesses *Paths of effect (Tibetan)*
HM 7655 10 pre nascence or pre *Conditions for consciousness (Buddhism)*
HM 1136 10 psychism description *Focus ten (Psychism)*
HM 8266 10 rebirth linking *Rebirth-linking (Buddhism)*
HH 6221 10 recollections are enumerated *Recollections as meditation subjects (Buddhism)*
HH 3987 10 recollections of the buddha *Subjects for meditation (Buddhism)*
HM 3941 10 rungs on this ladder *Dark night of the soul (Christianity)*
HM 2187 10 said to confer visions *First vajra-master stage (Buddhism, Tibetan)*
HM 2421 10 scriptural bodhisattva awareness buddhism *Tenth scriptural bodhisattva awareness (Buddhism,* Tibetan)
HM 5721 10 sense consciousnesses give rise *Indeterminate consciousness in the sense sphere - resultant* (Buddhism)
HH 4777 10 skills enumerated in hinayana *Skill in absorption (Buddhism)*
HM 3521 10 stage in a systematic *Mushahada (Sufism)*
HM 2401 10 stage on the path *Divine-love awareness (Sufism)*
HM 2813 10 stage *Supreme-heaven awareness (Buddhism, Tibetan)*
HM 2690 10 stages although the seikyo *Oxherding pictures in Zen Buddhism (Zen)*
HM 2155 10 stages of his path *First scriptural bodhisattva awareness (Buddhism, Tibetan)*
HM 4311 10 stages or psychic states *Psychic states (Sufism)*
HM 4356 10 stages to attain perfect *Suluk of the Naqshbandiyya order (Sufism)*
HM 1773 10 state or square *Purification (Leela)*
HM 4389 10 states altogether context *Sense mode of consciousness occurrence (Buddhism)*
HM 3166 10 station of the cross *Jesus stripped of his garments (Christianity)*
HM 2317 10 stations with 10 degrees *Stations of consciousness - Ansari (Sufism)*
HM 3941 10 step which is not *Dark night of the soul (Christianity)*
HH 6282 10 subtle faculties of man *Subtle faculties (Sufism)*
HH 0146 10 such stages have *Cognitive growth and development*
HH 9764 10 superhuman qualities are described *Superhuman qualities (Buddhism)*
HH 0873 10 this is on ways *Ways of knowing (Buddhism)*
HM 3306 10 thousand things *Vikalpa (Buddhism)*
HM 3306 10 thousand things *Vikalpa (Buddhism)*
HM 4297 10 thousand world spheres *Knowledge of recollection of previous existence (Buddhism)*
HH 2002 10 thousand world spheres *World cycles (Buddhism)*
HM 7769 10 through insight 6 *Equanimity (Buddhism)*
HH 5652 10 transposing factors he goes *Supernormal powers (Buddhism)*
HM 2657 10 worlds description *Ten worlds (Buddhism)*
HH 6876 10 yamas the additional five *Restraint*
HH 0349 11 acts said to constitute *Devotion (Christianity)*

HM 7769 11 advantages arise context *Equanimity (Buddhism)*
HM 5224 11 advantages arise context *Gladness (Buddhism)*
HM 7607 11 advantages arise context *Lovingkindness (Buddhism)*
HM 0513 11 advantages arise context *Pity (Buddhism)*
HM 2050 11 are constant 4 are *Awareness as mental-formation group of conscious existence (Buddhism)*
HM 2050 11 are constant and 2 *Awareness as mental-formation group of conscious existence (Buddhism)*
HM 3630 11 developing contemplation *Principal insights (Buddhism)*
HM 7655 11 fold both as physical *Conditions for consciousness (Buddhism)*
HM 0065 11 fold knowledge systematics experience *Eleven-fold knowledge (Systematics)*
HM 2098 11 functional or inoperative *Dispositions of consciousness (Buddhism)*
HM 2907 11 illumination of jami *Liberation of the spirit by attraction of divine grace (Sufism)*
HM 0065 11 in a sequence *Eleven-fold knowledge (Systematics)*
HM 3175 11 in the ica *Second birth (ICA)*
HM 0065 11 independent terms *Eleven-fold knowledge (Systematics)*
HM 3852 11 indeterminate resultant consciousnesses *Indeterminate consciousness in the sense sphere - functional* (Buddhism)
HH 0578 11 mental factors are *Virtuous mental factors (Buddhism)*
HM 0034 11 of the ninety nine *Al-Khaliq (Sufism)*
HM 2523 11 of the spiritual paths *Path of fiery intelligence (Judaism)*
HM 2636 11 on the board *Animal-hell awareness (Buddhism, Tibetan)*
HM 2050 11 or 12 are *Awareness as mental-formation group of conscious existence (Buddhism)*
HM 7655 11 post nascence or post *Conditions for consciousness (Buddhism)*
HM 2059 11 senses (including manas) *Ego awareness (Hinduism)*
HM 3650 11 stage in a systematic *Wajd (Sufism)*
HH 0622 11 stages or pratimas bring *Human development (Jainism)*
HM 4136 11 station of the cross *Jesus nailed to the cross (Christianity)*
HH 0819 11 such love desire joy *Emotion*
HH 1198 11 the mentality *Human development through religion*
HH 5652 11 transposing the object he *Supernormal powers (Buddhism)*
HH 1348 11 virtuous 6 root deluded *Secondary mental factors (Buddhism, Tibetan)*
HM 0160 12 and 3 am *Matins (Christianity)*
HM 2098 12 are contaminated by greed *Dispositions of consciousness (Buddhism)*
HM 2050 12 are general elements *Awareness as mental-formation group of conscious existence (Buddhism)*
HM 2219 12 centres some of them *Chakra centres of consciousness*
HM 2713 12 conditions may be present *Zodiacal forms of awareness (Astrology)*
HM 3630 12 developing contemplation *Principal insights (Buddhism)*
HM 2713 12 different conditions may *Zodiacal forms of awareness (Astrology)*
HM 2713 12 different conditions of being *Zodiacal forms of awareness (Astrology)*
HM 2733 12 different types according *Desire-realm consciousness (Buddhism)*
HH 0851 12 equally balanced by sex *Group therapy*
HM 2232 12 fold dependent origination is *Mahayana climax-application awareness (Buddhism, Tibetan)*
HM 0707 12 fold knowledge is *Twelve-fold knowledge (Systematics)*
HM 0707 12 fold knowledge systematics autocracy *Twelve-fold knowledge (Systematics)*
HH 5652 12 from the first jhana *Supernormal powers (Buddhism)*
HM 8375 12 give rise to materiality *Unprofitable consciousness in the sense sphere (Buddhism)*
HM 2072 12 grades in the fine *Worlds of conscious existence (Buddhism, Pali)*
HM 3436 12 illumination of jami *Surrender to the bliss of mystical seduction (Sufism)*
HM 3091 12 in the ica *Dynamic selfhood (ICA)*
HM 0707 12 independent terms *Twelve-fold knowledge (Systematics)*
HM 8375 12 kinds of consciousness may *Unprofitable consciousness in the sense sphere (Buddhism)*
HM 2339 12 links of conditioned genesis *Eightfold way (Buddhism)*
HM 3070 12 manifest beings 10 creative *Completeness*
HM 2713 12 may also be subdivided *Zodiacal forms of awareness (Astrology)*
HH 2003 12 modes of knowledge developed *Systematics*
HM 0707 12 modes of knowledge identified *Twelve-fold knowledge (Systematics)*
HM 3725 12 noon and 3 pm *Sext (Christianity)*
HM 2965 12 noon *Terce (Christianity)*
HM 3636 12 of the ninety nine *Al-Bari' (Sufism)*
HM 2107 12 of the spiritual paths *Path of transparent intelligence (Judaism)*
HM 2007 12 on the board *Divine-animal-hell awareness (Buddhism, Tibetan)*
HM 7655 12 path factors profitable unprofitable *Conditions for consciousness (Buddhism)*
HM 0176 12 petalled lotus *Soma chakra (Yoga)*
HM 4321 12 pm *Compline (Christianity)*
HM 0267 12 psychism description *Focus twelve (Psychism)*
HM 7655 12 repetition condition here *Conditions for consciousness (Buddhism)*
HH 3215 12 section of st john's *Journeying within transcendence - from outside to inside*
HM 0176 12 sixteen petalled lotus *Soma chakra (Yoga)*
HM 2254 12 stage in a systematic *Ma'rifa (Sufism)*
HM 3719 12 station of the cross *Jesus dies on the cross (Christianity)*
HM 2713 12 streams of divine energy *Zodiacal forms of awareness (Astrology)*
HM 8273 12 unprofitable consciousnesses and 10 *Consciousness-born materiality (Buddhism)*
HM 7268 12 unprofitable consciousnesses *Impulsion (Buddhism)*
HM 2098 12 unprofitable dispositions *Dispositions of consciousness (Buddhism)*
HM 8266 12 unprofitable or immoral consciousnesses *Rebirth-linking (Buddhism)*
HH 1321 12 14 changes focus *Religious growth*
HH 0682 12 14 per minute *Transcendental meditation (TM)*
HM 2050 13 are constant and 4 *Awareness as mental-formation group of conscious existence (Buddhism)*
HM 2050 13 are karmically unwholesome *Awareness as mental-formation group of conscious existence (Buddhism)*
HH 5652 13 definition or fixing *Supernormal powers (Buddhism)*
HM 3630 13 developing contemplation of voidness *Principal insights (Buddhism)*
HM 2050 13 elements associated *Awareness as mental-formation group of conscious existence (Buddhism)*
HM 3303 13 illumination of jami *Ontological reality of the truth (Sufism)*
HM 3135 13 in the ica *Essential dubiety (ICA)*
HH 4298 13 kinds of ascetic practice *Ascetic practices (Buddhism)*
HM 5786 13 of the ninety nine *Al-Musawwir (Sufism)*
HM 2583 13 of the spiritual paths *Path of uniting intelligence (Judaism)*
HM 2031 13 on the board *Naga-world awareness (Buddhism, Tibetan)*
HM 1270 13 stage in a systematic *Fana (Sufism)*
HM 3410 13 station of the cross *Jesus taken down from the cross (Christianity)*
HH 0715 13 when i became *Twice-born*
HM 2035 14 among the spiritual paths *Path of illuminating intelligence (Judaism)*
HH 1199 14 and 15 centuries were *Soul care (Christianity)*
HH 0442 14 and 15 centuries were *Spiritual direction*
HM 3932 14 and last mode *Death (Buddhism)*
HM 7268 14 conscious moments *Impulsion (Buddhism)*
HH 5652 14 definition or fixing *Supernormal powers (Buddhism)*
HM 3630 14 developing insight into states *Principal insights (Buddhism)*
HM 6720 14 different modes rebirth linking *Modes of occurrence of consciousness (Buddhism)*
HM 2599 14 function activities involving *Phase-conditions of consciousness (Buddhism)*
HM 3457 14 illumination of jami *Existence as ontological reality (Sufism)*
HM 2824 14 in the ica *Cryptic disclosure (ICA)*
HM 4450 14 of 40 stations *Zuhd (Sufism)*
HM 3651 14 of temporal life *Second chakra: realm of fantasy (Leela)*
HM 1257 14 of the ninety nine *Al-Ghaffar (Sufism)*
HM 2055 14 on the board *Demon-island awareness (Buddhism, Tibetan)*
HM 2050 14 or 13 are karmically *Awareness as mental-formation group of conscious existence (Buddhism)*
HM 2285 14 path from intelligence (chokmah) *Female sovereignty archetypal image (Tarot)*
HM 2075 14 phenomena are mental *Psycho-physical faculties awareness (Buddhism)*
HM 0330 14 stage in a systematic *Baqa (Sufism)*
HH 0622 14 stages gunasthanas from total *Human development (Jainism)*
HM 4600 14 state or square *Astral plane (Leela)*
HM 3536 14 station of the cross *Jesus placed in the sepulchre (Christianity)*
HM 0717 14 to 21 of temporal *Third chakra: theatre of karma (Leela)*
HM 3464 14 to 21 the state *Mental-volitional stage of life*
HH 5652 14 ways
HH 0208 14 1720 for the body *Hyper-personalisation*
HM 2117 15 among the spiritual paths *Path of constituting intelligence (Judaism)*
HM 3630 15 developing correct knowledge *Principal insights (Buddhism)*
HM 2097 15 great arcana of consciousness *Tarot arcana of conscious states (Tarot)*
HM 3260 15 illumination of jami *Identity of attributes with their essence (Sufism)*
HM 3034 15 in the ica *Transcendent immanence (ICA)*
HM 7655 15 nutriment or sustenance *Conditions for consciousness (Buddhism)*
HM 0128 15 of 40 stations *'Ibadat (Sufism)*
HM 1318 15 of the ninety nine *Al-Qahhar (Sufism)*
HM 2579 15 on the board *Asura-world awareness (Buddhism, Tibetan)*
HM 2108 15 other derived corporeal elements *Awareness of corporeality-group of conscious existence (Buddhism)*
HM 3289 15 path between foundation *Life-fluid emblems - temperance (Tarot)*
HM 3137 15 path between foundation (yesod) *Life-fluid emblems - star (Tarot)*
HM 3975 15 psychism description *Focus fifteen (Psychism)*
HM 2097 15 so that the trumps *Tarot arcana of conscious states (Tarot)*
HM 1485 15 state or square *Plane of fantasy (Leela)*
HH 0862 15 to 20 postures *Hatha yoga (Yoga)*
HM 0130 16 arise with the sense *Mundane resultant consciousness (Buddhism)*
HM 2245 16 aspect discloses that *Meditation way of the four truths (Buddhism)*
HM 2392 16 aspects *Upanishadic stages of awareness (Hinduism)*
HM 1646 16 conscious moments *Registering (Buddhism)*
HM 3630 16 developing contemplation of danger *Principal insights (Buddhism)*
HM 2733 16 hells there are *Desire-realm consciousness (Buddhism)*
HM 2642 16 illumination of jami *Manifestation of the essence through veils (Sufism)*
HM 2111 16 in the ica *Singular adoration (ICA)*
HM 8273 16 in the sense sphere *Consciousness-born materiality (Buddhism)*
HM 2864 16 lifetimes *Tantra in meditation on emptiness (Buddhism, Tibetan)*
HM 0286 16 of 40 stations *Wara (Sufism)*
HM 0069 16 of the ninety nine *Al-Wahhab (Sufism)*
HM 2593 16 of the spiritual paths *Path of triumphant intelligence (Judaism)*
HM 2603 16 on the board *Rudra awareness of black freedom (Buddhism, Tibetan)*
HM 6245 16 path consciousnesses associated *Understanding accompanied by joy (Buddhism)*
HH 5652 16 roots of supernormal power *Supernormal powers (Buddhism)*
HM 0554 16 state or square *Jealousy (Leela)*
HM 5721 16 with profitable result 8 *Indeterminate consciousness in the sense sphere - resultant (Buddhism)*
HH 0591 17 and 18 centuries *Philosophy of enlightenment*
HH 0130 17 and 18 century materialists *Egoism*
HH 0520 17 century brought with it *Progress*
HH 0179 17 century onwards *Pedagogy*
HM 2257 17 classes of gods *Form-realm consciousness (Buddhism)*
HM 3630 17 developing contemplation of reflection *Principal insights (Buddhism)*
HM 2050 17 elements associated *Awareness as mental-formation group of conscious existence (Buddhism)*
HM 3614 17 illumination of jami *First individuation of essence (Sufism)*
HM 2388 17 in the ica *Ultimate awareness (ICA)*
HM 7655 17 jhana condition the jhana *Conditions for consciousness (Buddhism)*
HM 4663 17 of 40 stations *Ikhlas (Sufism)*
HM 7021 17 of the ninety nine *Al-Razzaq (Sufism)*
HM 2569 17 of the spiritual paths *Path of disposing intelligence (Judaism)*
HM 2127 17 on the board *Southern-continent awareness of men (Buddhism)*
HM 3434 17 state or square *Mercy (Leela)*
HH 0597 18 and 19 century *Hasidism (Judaism)*
HH 0659 18 and the 31 degrees *Initiation (Freemasonry)*
HH 0591 18 centuries the philosophy *Philosophy of enlightenment*
HH 0215 18 century emphasizing the transforming *Enlightenment humanism*
HH 0130 18 century materialists seeing good *Egoism*
HH 1778 18 century notably rousseau while *Primitivism*
HM 2122 18 chief kinds of insight *Trances and mental absorptions (Buddhism)*
HM 3630 18 developing contemplation of turning *Principal insights (Buddhism)*
HM 2050 18 elements associated *Awareness as mental-formation group of conscious existence (Buddhism)*
HM 3000 18 emptinesses comprising the paths *Emptiness of all phenomena (Buddhism, Tibetan)*
HM 3032 18 emptinesses comprising the paths *Emptiness of cyclic existence (Buddhism, Tibetan)*
HM 2946 18 emptinesses comprising the paths *Emptiness of definitions (Buddhism, Tibetan)*
HM 3205 18 emptinesses comprising the paths *Emptiness of nature (Buddhism, Tibetan)*
HM 2089 18 emptinesses comprising the paths *Emptiness of non-products (Buddhism, Tibetan)*
HM 2414 18 emptinesses comprising the paths *Emptiness of products (Buddhism, Tibetan)*
HM 2671 18 emptinesses comprising the paths *Emptiness of the five senses (Buddhism, Tibetan)*
HM 3072 18 emptinesses comprising the paths *Emptiness of the indestructible Mahayana (Buddhism, Tibetan)*
HM 2448 18 emptinesses comprising the paths *Emptiness of the inherent existence of non-things (Buddhism,* Tibetan)
HM 2794 18 emptinesses comprising the paths *Emptiness of the loci of the senses (Buddhism, Tibetan)*
HM 2899 18 emptinesses comprising the paths *Emptiness of the nature of phenomena (Buddhism, Tibetan)*
HM 2522 18 emptinesses comprising the paths *Emptiness of the objects of sense and of mental consciousness* (Buddhism, Tibetan)
HM 2995 18 emptinesses comprising the paths *Emptiness of the ten directions (Buddhism, Tibetan)*
HM 2501 18 emptinesses comprising the paths *Emptiness of the unapprehendable (Buddhism, Tibetan)*
HM 2989 18 emptinesses comprising the paths *Emptiness of things (Buddhism, Tibetan)*
HM 3208 18 emptinesses comprising the paths *Emptiness of things (Buddhism, Tibetan)*
HM 2294 18 emptinesses comprising the paths *Emptiness of ultimate nirvana (Buddhism, Tibetan)*

INDEX TO CONCEPT SETS
(See SECTION X for KEYWORD INDEX)

HX

HM 0052 18 emptinesses comprising the paths *Emptiness of what is free of permanence and annihilation (Buddhism,* Tibetan)
HM 2681 18 illumination of jami *Beyond individual distinctions (Sufism)*
HM 2471 18 in the ica *External relation (ICA)*
HH 1900 18 is *Journeying within transcendence - prologue*
HM 6631 18 of 40 stations *Sidq (Sufism)*
HM 0002 18 of the ninety nine *Al-Fattah (Sufism)*
HM 2045 18 of the spiritual paths *Path of intelligence of influences (Judaism)*
HM 2519 18 on the board *Western-continent awareness of cattle (Buddhism, Tibetan)*
HM 7655 18 path condition there are *Conditions for consciousness (Buddhism)*
HM 3630 18 principal insights *Principal insights (Buddhism)*
HM 4376 18 state or square *Plane of joy (Leela)*
HM 2775 18 unshared attributes of buddha *Eighteen unshared attributes of Buddha (Buddhism, Tibetan)*
HM 3249 18 unshared attributes of buddha *Paths of effect (Tibetan)*
HH 1321 18 24 at this stage *Religious growth*
HH 1778 19 and 20 centuries are *Primitivism*
HM 3932 19 associated with rebirth linking *Death (Buddhism)*
HM 7655 19 association condition where immaterial *Conditions for consciousness (Buddhism)*
HM 2407 19 day after death *Death-bardo heaven-reality awareness (Buddhism)*
HM 2113 19 illumination of jami *Indivisibility of essence (Sufism)*
HM 2584 19 in the ica *Self transcendence (ICA)*
HM 8273 19 in the sense sphere *Consciousness-born materiality (Buddhism)*
HM 6221 19 kinds of rebirth linking *Life-continuum (Buddhism)*
HM 8266 19 kinds of rebirth linking *Rebirth-linking (Buddhism)*
HM 1047 19 of 40 stations *Khawf (Sufism)*
HM 4010 19 of the ninety nine *Al-'Alim (Sufism)*
HM 2543 19 on the board *Eastern-continent awareness of noble figures (Buddhism, Tibetan)*
HM 3127 19 path from force *Natural forces emblems (Tarot)*
HM 2050 19 primary concomitants to any *Awareness as mental-formation group of conscious existence (Buddhism)*
HM 2021 19 spiritual path expressed *Path of intelligence of spiritual action (Judaism)*
HM 0948 19 state or square *Plane of karma (Leela)*
HH 0781 20 afflictions or knowers *Secondary afflictions (Buddhism, Tibetan)*
HM 0348 20 and 50 percent *Stage-three sleep*
HH 1348 20 auxiliary deluded 4 changeable *Secondary mental factors (Buddhism, Tibetan)*
HH 1380 20 different kinds of saint *Bodhisattva (Buddhism)*
HM 7655 20 dissociation condition this may *Conditions for consciousness (Buddhism)*
HM 2193 20 false views *Emptiness (Buddhism, Zen)*
HM 3316 20 illumination of jami *Unchanging reality of being (Sufism)*
HM 2717 20 in the ica *Perpetual becoming (ICA)*
HM 2098 20 indeterminate functional dispositions refer *Dispositions of consciousness (Buddhism)*
HM 8266 20 kind of rebirth linking *Rebirth-linking (Buddhism)*
HM 4393 20 of 40 stations *Raja (Sufism)*
HM 2767 20 of the ninety nine *Al-Qabid (Sufism)*
HM 2629 20 of the spiritual paths *Path of intelligence of will (Judaism)*
HM 2067 20 on the board *Northern-continent awareness of community (Buddhism, Tibetan)*
HM 2202 20 path between kindness *Old wise-man archetypal image (Tarot)*
HH 0862 20 postures which should *Hatha yoga (Yoga)*
HM 4387 20 state or square *Charity (Leela)*
HM 2431 20 to the 49 *Death-bardo rebirth-seeking awareness (Buddhism)*
HM 2097 21 and 22nd *Tarot arcana of conscious states (Tarot)*
HM 2250 21 between force or victory *Mutability emblems (Tarot)*
HM 3228 21 illumination of jami *Reciprocal relation of absolute and conditional (Sufism)*
HM 2687 21 in the ica *Universal fate (ICA)*
HM 1270 21 of 40 stations *Fana (Sufism)*
HM 0717 21 of temporal life *Third chakra: theatre of karma (Leela)*
HM 0115 21 of the ninety *Al-Basit (Sufism)*
HM 2105 21 of the spiritual *Path of intelligence of reward (Judaism)*
HM 2091 21 on the board *Barbarian-state of awareness (Buddhism, Tibetan)*
HM 7655 21 presence condition here presence *Conditions for consciousness (Buddhism)*
HM 2098 21 profitable and 12 unprofitable *Dispositions of consciousness (Buddhism)*
HM 3837 21 state or square *Atonement (Leela)*
HM 3464 21 the state is not *Mental-volitional stage of life*
HM 3291 21 to 28 of temporal *Fourth chakra: attaining balance (Leela)*
HM 2097 21 together with one unnumbered *Tarot arcana of conscious states (Tarot)*
HM 7655 22 absence or non presence *Conditions for consciousness (Buddhism)*
HM 2097 22 and a diameter *Tarot arcana of conscious states (Tarot)*
HM 2551 22 elements the 4 producing *Phenomena awareness (Buddhism)*
HM 7655 22 faculties or controlling influences *Conditions for consciousness (Buddhism)*
HM 0130 22 from mind consciousness *Mundane resultant consciousness (Buddhism)*
HM 2978 22 in the ica *Relational situation (ICA)*
HM 2072 22 in the three lower *Worlds of conscious existence (Buddhism, Pali)*
HM 3323 22 kinds of mundane resultant *Mind consciousness (Buddhism)*
HM 0330 22 of 40 stations *Baqa (Sufism)*
HM 3644 22 of the ninety *Al-Khafid (Sufism)*
HM 2081 22 of the spiritual *Path of faithful intelligence (Judaism)*
HM 2115 22 on the board *Hindu-states of awareness (Buddhism, Tibetan)*
HM 2178 22 path between and beauty *Measuring emblems (Tarot)*
HH 0921 22 paths of wisdom *Kabbalah*
HM 0481 22 state or square *Plane of dharma (Leela)*
HM 2097 22 symbols can be analyzed *Tarot arcana of conscious states (Tarot)*
HM 3028 22 tarots of the grand *Betrothal initiation archetypal image (Tarot)*
HM 2098 23 are resultant and 11 *Dispositions of consciousness (Buddhism)*
HM 7655 23 disappearance or absence *Conditions for consciousness (Buddhism)*
HM 2842 23 in the ica *Contextual world-view (ICA)*
HM 5721 23 indeterminate resultant consciousnesses *Indeterminate consciousness in the sense sphere - resultant* (Buddhism)
HM 0930 23 of 40 stations *'Ilm al-yaqin (Sufism)*
HM 4040 23 of the ninety *Al-Rafi (Sufism)*
HM 2533 23 of the spiritual *Path of stable intelligence (Judaism)*
HM 2639 23 on the board *Bon-practitioner state of awareness (Buddhism, Tibetan)*
HM 1052 23 state or square *Celestial plane (Leela)*
HM 7655 24 causes for consciousness *Conditions for consciousness (Buddhism)*
HM 7655 24 conditions in which states *Conditions for consciousness (Buddhism)*
HH 0011 24 dec the islamic ramadan *Fasting*
HH 0622 24 great teachers or jinas *Human development (Jainism)*
HH 0836 24 hour cycles and may *Physical development*
HM 3112 24 in the ica *Archetypal humanness (ICA)*
HM 7655 24 non disappearance *Conditions for consciousness (Buddhism)*
HM 4556 24 of 40 stations *Haqq al-yaqin (Sufism)*
HM 1116 24 of the ninety *Al-Mu'izz (Sufism)*
HM 2011 24 of the spiritual *Path of imaginative intelligence (Judaism)*
HM 2010 24 on the board *Heavenly-highway awareness (Buddhism, Tibetan)*
HM 2599 24 phases how any state *Phase-conditions of consciousness (Buddhism)*
HM 2108 24 secondary phenomena of sentience *Awareness of corporeality-group of conscious existence (Buddhism)*
HM 0764 24 state or square *Bad company (Leela)*
HM 1397 24 to 25 hours *Daily variations in consciousness*
HM 6335 24 to 28 *Personal salvation (Christianity)*
HM 2050 25 are karmically wholesome *Awareness as mental-formation group of conscious existence (Buddhism)*
HM 1397 25 hours *Daily variations in consciousness*
HM 3049 25 in the ica *Beyond morality (ICA)*
HM 2696 25 ingredients compounded with water *Tantra heat-application awareness (Buddhism, Tibetan)*
HM 2254 25 of 40 stations *Ma'rifa (Sufism)*
HM 4510 25 of the ninety *Al-Mudhill (Sufism)*
HM 2503 25 of the spiritual *Path of intelligence of temptation (Judaism)*
HM 2452 25 on the board *Tantra-beginner awareness (Buddhism, Tibetan)*
HM 3137 25 path between foundation *Life-fluid emblems - star (Tarot)*
HM 3289 25 path between foundation (yesod) *Life-fluid emblems - temperance (Tarot)*
HM 1355 25 state or square *Good company (Leela)*
HM 3135 25 that one is *Essential dubiety (ICA)*
HH 0364 26 characteristics of the divine *Treasures of the godly*
HM 8273 26 in the fine material *Consciousness-born materiality (Buddhism)*
HM 2892 26 in the ica *Intentional conscience (ICA)*
HM 6099 26 of 40 stations *Jahd (Sufism)*
HM 0419 26 of the ninety *Al-Sami' (Sufism)*
HM 2047 26 of the spiritual *Path of renovating intelligence (Judaism)*
HM 2058 26 on the board *Wheel-turning-king awareness (Buddhism, Tibetan)*
HM 1782 26 state or square *Sorrow (Leela)*
HM 2050 27 are said *Awareness as mental-formation group of conscious existence (Buddhism)*
HM 2945 27 in the ica *Cosmic sanctions (ICA)*
HM 3647 27 of 40 stations *Wilayat (Sufism)*
HM 6512 27 of the ninety *Al-Basir (Sufism)*
HM 2071 27 of the spiritual *Path of natural intelligence (Judaism)*
HM 2082 27 on the board *Four-great-kings-heaven awareness (Buddhism, Tibetan)*
HM 2350 27 path between splendour *Emblems of disaster (Tarot)*
HM 0294 27 state or square *Selfless service (Leela)*
HM 2621 28 in the ica *Primal vocation (ICA)*
HM 2401 28 of 40 stations *Divine-love awareness (Sufism)*
HM 3291 28 of temporal life that *Fourth chakra: attaining balance (Leela)*
HM 5014 28 of the ninety *Al-Hakam (Sufism)*
HM 2218 28 of the spiritual *Path of active intelligence (Judaism)*
HM 2606 28 on the board *Thirty-three-god-heaven awareness (Buddhism, Tibetan)*
HM 2411 28 path between beauty (tiphareth) *Emblems of supremacy (Tarot)*
HM 2981 28 state or square *Apt religion (Leela)*
HM 0933 28 to 35 of temporal *Fifth chakra: man becomes himself (Leela)*
HM 2773 29 in the ica *Original integrity (ICA)*
HM 3650 29 of 40 stations *Wajd (Sufism)*
HM 3807 29 of the ninety *Al-'Adl (Sufism)*
HM 3149 29 of the spiritual *Path of corporeal intelligence (Judaism)*
HM 2130 29 on the board *Heaven-without-fighting consciousness (Buddhism, Tibetan)*
HM 7268 29 (or 28) classes *Impulsion (Buddhism)*
HM 2398 29 path between kingdom (malkuth) *Emblems of the mysterious (Tarot)*
HM 0670 29 state or square *Irreligiousity (Leela)*
HM 2291 30 and 40 *Cosmic consciousness*
HM 4983 30 classes of consciousness *Feeling aggregate (Buddhism)*
HH 0659 30 degrees *Initiation (Freemasonry)*
HM 2451 30 in the ica *Wordly detachment (ICA)*
HM 2377 30 of 40 stations *Nearness-awareness (Sufism)*
HM 0308 30 of the ninety nine *Al-Latif (Sufism)*
HM 3046 30 of the spiritual paths *Path of collective intelligence (Judaism)*
HM 2022 30 on the board *Joyful-heaven awareness (Buddhism, Tibetan)*
HM 2422 30 path between foundation (yesod) *Emblems of well-being (Tarot)*
HH 0682 30 per cent reduction *Transcendental meditation (TM)*
HM 5197 30 state or square *Good tendencies (Leela)*
HM 2315 31 between kingdom (malkuth) *Emblems of renewal (Tarot)*
HH 0659 31 degrees *Initiation (Freemasonry)*
HM 2547 31 in the ica *Passionate disinterest (ICA)*
HM 6227 31 of 40 stations *Tafakkur (Sufism)*
HM 4400 31 of the ninety *Al-Khabir (Sufism)*
HM 2643 31 of the spiritual *Path of perpetual intelligence (Judaism)*
HM 2546 31 on the board *Delightful-emanation-heaven consciousness (Buddhism, Tibetan)*
HM 5965 31 state or square *Plane of sanctity (Leela)*
HM 2309 32 in the ica *Destinal accountability (ICA)*
HM 0130 32 kinds 10 *Mundane resultant consciousness (Buddhism)*
HM 7119 32 of 40 stations *Wisal (Sufism)*
HM 4519 32 of the ninety *Al-Halim (Sufism)*
HM 2095 32 of the spiritual *Path of administrative intelligence (Judaism)*
HM 2070 32 on the board *Non-emanating consciousness (Buddhism, Tibetan)*
HM 2387 32 path between kingdom *Emblems of totality (Tarot)*
HM 2509 32 paths can be considered *Paths of wisdom (Judaism)*
HH 0921 32 paths of wisdom context *Kabbalah*
HH 0669 32 said *Asana (Hinduism, Yoga)*
HM 1351 32 state or square *Plane of balance (Leela)*
HM 8273 32 the 8 profitable consciousnesses *Consciousness-born materiality (Buddhism)*
HM 2606 33 god heaven awareness *Thirty-three-god-heaven awareness (Buddhism, Tibetan)*
HM 2733 33 gods above *Desire-realm consciousness (Buddhism)*
HH 0659 33 in all may *Initiation (Freemasonry)*
HM 2223 33 in the ica *Individual fatefulness (ICA)*
HM 2072 33 in *Worlds of conscious existence (Buddhism, Pali)*
HM 1452 33 of 40 stations *Kashf (Sufism)*
HM 5198 33 of the ninety *Al-'Azim (Sufism)*
HM 2558 33 on the board *Kriya-tantra awareness (Buddhism, Tibetan)*
HM 0008 33 state or square *Plane of fragrance (Leela)*
HM 0205 33 transitory or subsidiary *Aesthetic emotion (Hinduism)*
HM 2579 33 upper divinities who *Asura-world awareness (Buddhism, Tibetan)*
HM 2098 34 are indeterminate dispositions karmically *Dispositions of consciousness (Buddhism)*
HM 2131 34 in the ica *Definitive predestination (ICA)*
HM 8776 34 of 40 stations *Khidmat (Sufism)*
HM 6255 34 of the ninety *Al-Ghafur (Sufism)*
HM 2118 34 on the board *Great-black-lord awareness (Buddhism, Tibetan)*
HM 0878 34 state or square *Plane of taste (Leela)*
HM 2363 35 in the ica *Temporal solidarity (ICA)*
HM 7621 35 of 40 stations *Tajrid (Sufism)*
HM 0933 35 of temporal life that *Fifth chakra: man becomes himself (Leela)*
HM 1934 35 of the ninety *Ash-Shakur (Sufism)*
HM 2142 35 on the board *Form-realm awareness (Buddhism, Tibetan)*
HM 0856 35 state or square *Purgatory (Leela)*
HM 4412 35 to 42 of temporal *Sixth chakra: time of penance (Leela)*
HM 2050 36 elements associated *Awareness as mental-formation group of conscious existence (Buddhism)*
HM 3039 36 illuminations recorded *Fountains of light (Sufism)*
HM 2445 36 in the ica *Sacramental universe (ICA)*
HH 3667 36 methods of culling *Outside paths (Taoism)*
HM 6762 36 of 40 stations *Tafrid (Sufism)*
HM 4162 36 of the ninety *Al-'Ali (Sufism)*
HM 3144 36 on the board *Formless-realm awareness (Buddhism, Tibetan)*
HM 0455 36 state or square *Clarity of consciousness (Leela)*
HM 0796 37 constituents of enlightenment (bodhipakkhiya) *Magga-bhavana (Buddhism)*
HH 0603 37 harmonies are attained when *Harmonies with enlightenment (Buddhism)*
HM 2550 37 in the ica *Primal sympathy (ICA)*

HM 2793 37 may be considered *Bodhi-pakkhiya dhamma (Buddhism)*
HM 7098 37 of 40 stations *Inbisat (Sufism)*
HM 4437 37 of the ninety *Al-Kabir (Sufism)*
HM 2168 37 on the board *Pure-lands awareness (Buddhism, Tibetan)*
HM 3846 37 state or square *Knowledge (Leela)*
HM 2734 38 in the ica *Universal compassion (ICA)*
HM 5563 38 of 40 stations *Tahqiq (Sufism)*
HM 7099 38 of the ninety *Al-Hafiz (Sufism)*
HM 2192 38 on the board *Discipleship-karma awareness (Buddhism, Tibetan)*
HM 4399 38 state or square *Prana-loka (Leela)*
HM 2641 39 in the ica *Sacrificial passion (ICA)*
HM 3377 39 of 40 stations *Nihayat (Sufism)*
HM 3171 39 of the ninety *Al-Muqit (Sufism)*
HM 2716 39 on the board *Discipleship-application awareness (Buddhism, Tibetan)*
HM 0145 39 state or square *Apana-loka (Leela)*
HM 2570 39 steps show situations *Ubiquitous otherness (ICA)*
HM 2424 40 description *Stations of consciousness - ibn-Abi'l-Khayr (Sufism)*
HM 2904 40 in the ica *Soteriological existence (ICA)*
HM 2291 40 intimations of this state *Cosmic consciousness*
HM 1467 40 of the ninety nine *Al-Hasib (Sufism)*
HM 2240 40 on the board *Discipleship-vision awareness (Buddhism, Tibetan)*
HM 2424 40 stage development *Stations of consciousness - ibn-Abi'l-Khayr (Sufism)*
HM 6002 40 state or square *Vyana-loka (Leela)*
HH 3987 40 subjects as suitable *Subjects for meditation (Buddhism)*
HM 2817 41 in the ica *Global guardianship (ICA)*
HM 6771 41 of the ninety *Al-Jalil (Sufism)*
HM 3026 41 on the board *Middle-path tantra awareness (Buddhism, Tibetan)*
HM 5155 41 state or square *Human plane (Leela)*
HM 3122 42 in the ica *Ancestral obligation (ICA)*
HM 4412 42 of temporal life that *Sixth chakra: time of penance (Leela)*
HM 5289 42 of the ninety *Al-Karim (Sufism)*
HM 2656 42 on the board *Great-path tantra awareness (Buddhism, Tibetan)*
HM 0859 42 state or square *Plane of agnih (Leela)*
HM 0754 42 to 49 of temporal *Seventh chakra: plane of reality (Leela)*
HM 3016 43 in the ica *Futuric responsibility (ICA)*
HM 6098 43 of the ninety *Al-Raqib (Sufism)*
HM 2180 43 on the board *Pratyeka Buddha awareness (Buddhism, Tibetan)*
HM 0978 43 state or square *Birth of man (Leela)*
HM 3245 44 in the ica *Invented history (ICA)*
HM 0504 44 of the ninety *Al-Mujib (Sufism)*
HM 3020 44 on the board *Pratyeka Buddha application awareness (Buddhism, Tibetan)*
HM 0310 44 state or square *Ignorance (Leela)*
HM 2927 45 in the ica *Diaphanous intuition (ICA)*
HM 4169 45 of the ninety *Al-Wasi' (Sufism)*
HM 2228 45 on the board *Pratyeka Buddha vision-path awareness (Buddhism, Tibetan)*
HM 0567 45 state or square *Right knowledge (Leela)*
HH 1108 46 categories in all covering *Seventy-five dharmas (Buddhism)*
HM 2851 46 in the ica *Interior discipline (ICA)*
HM 4799 46 of the ninety *Al-Hakim (Sufism)*
HM 2252 46 on the board *Pratyeka Buddha cultivation awareness (Buddhism, Tibetan)*
HM 1601 46 state or square *Conscience (Leela)*
HM 2607 47 in the ica *Impactful profundity (ICA)*
HM 7712 47 of the ninety *Al-Wadud (Sufism)*
HM 2776 47 on the board *Pratyeka Buddha arhat awareness (Buddhism, Tibetan)*
HM 4050 47 state or square *Plane of neutrality (Leela)*
HM 2796 48 in the ica *Definitive effectivity (ICA)*
HM 5446 48 of the ninety *Al-Majid (Sufism)*
HM 2172 48 on the board *Cessation-awareness (Buddhism, Tibetan)*
HM 1279 48 state or square *Solar plane (Leela)*
HM 2431 49 day preceding rebirth *Death-bardo rebirth-seeking awareness (Buddhism)*
HM 2056 49 in the ica *Seminal illumination (ICA)*
HM 0754 49 of temporal life that *Seventh chakra: plane of reality (Leela)*
HM 4017 49 of the ninety *Al-Ba'ith (Sufism)*
HM 2696 49 on the board *Tantra heat-application awareness (Buddhism, Tibetan)*
HM 3779 49 state or square *Lunar plane (Leela)*
HM 2256 50 in the ica *Inclusive comprehension (ICA)*
HM 2050 50 mental phenomena or concomitants *Awareness as mental-formation group of conscious existence* (Buddhism)
HM 0375 50 of the ninety nine *Ash-Shahid (Sufism)*
HM 2220 50 on the board *Tantra climax-application awareness (Buddhism, Tibetan)*
HM 5244 50 state or square *Plane of austerity (Leela)*
HM 2373 51 in the ica *Contentless word (ICA)*
HM 2664 51 mental factors classed *Sense consciousness (Buddhism)*
HM 3945 51 of the ninety *Al-Haqq (Sufism)*
HM 3024 51 on the board *Arhat sanctity awareness (Buddhism, Tibetan)*
HM 6112 51 state or square *Earth (Leela)*
HH 1348 51 5 ever recurring *Secondary mental factors (Buddhism, Tibetan)*
HM 2595 52 in the ica *Personal epiphany (ICA)*
HM 1382 52 of the ninety *Al-Wakil (Sufism)*
HM 2268 52 on the board *Great-vehicle lower-path awareness (Buddhism, Tibetan)*
HM 1276 52 state or square *Plane of violence (Leela)*
HM 2493 53 in the ica *Creative futility (ICA)*
HM 4988 53 of the ninety *Al-Qawi (Sufism)*
HM 2160 53 on the board *Great-vehicle middle-path awareness (Buddhism, Tibetan)*
HM 0172 53 state or square *Liquid plane (Leela)*
HM 2098 54 dispositions consciousness *Dispositions of consciousness (Buddhism)*
HM 2747 54 in the ica *Problemless living (ICA)*
HM 5633 54 of the ninety *Al-Matin (Sufism)*
HM 3048 54 on the board *Great-vehicle higher-path awareness (Buddhism, Tibetan)*
HM 1475 54 state or square *Spiritual devotion (Leela)*
HM 2658 55 in the ica *Transcending hostility (ICA)*
HM 4983 55 kinds consciousness *Feeling aggregate (Buddhism)*
HM 1633 55 of the ninety *Al-Wali (Sufism)*
HM 2208 55 on the board *Mahayana heat-application awareness (Buddhism, Tibetan)*
HM 1726 55 state or square *Egotism (Leela)*
HM 2951 56 in the ica *Exclusive contradiction (ICA)*
HM 2098 56 of the 89 dispositions *Dispositions of consciousness (Buddhism)*
HM 0392 56 of the ninety *Al-Hamid (Sufism)*
HM 2232 56 on the board *Mahayana climax-application awareness (Buddhism, Tibetan)*
HM 3732 56 state or square *Plane of primal vibrations (Leela)*
HM 2875 57 in the ica *Vital signs (ICA)*
HM 4544 57 of the ninety *Al-Muhsi (Sufism)*
HM 2756 57 on the board *Tantra receptivity-application awareness (Buddhism, Tibetan)*
HM 6434 57 state or square *Gaseous plane (Leela)*
HM 3025 58 in the ica *Spontaneous gratitude (ICA)*
HM 0622 58 of the ninety *Al-Mubdi (Sufism)*
HM 2280 58 on the board *Tantra union-in-learning-application awareness (Buddhism, Tibetan)*
HM 5028 58 state or square *Plane of radiation (Leela)*
HM 3148 59 in the ica *Blissful seizure (ICA)*
HM 4093 59 of the ninety *Al-Mu'id (Sufism)*
HM 2151 59 on the board *Kalacakra-tantra shambhala awareness (Buddhism, Tibetan)*
HM 1293 59 state or square *Plane of reality (Leela)*
HM 2958 60 in the ica *Final blessedness (ICA)*
HM 2175 60 on the board *Potala-island awareness (Buddhism, Tibetan)*
HM 1789 60 state or square *Positive intellect (Leela)*
HM 2808 61 in the ica *Living death (ICA)*
HM 4273 61 of 100 stations *Love of God (Sufism)*
HM 3868 61 of the ninety *Al-Mumit (Sufism)*
HM 2699 61 ohn the board *Guhya-samaja urgyan awareness (Buddhism, Tibetan)*
HM 3623 61 state or square *Negative intellect (Leela)*
HM 2631 62 in the ica *Resurrectional existence (ICA)*
HM 4236 62 of the ninety *Al-Hayy (Sufism)*
HM 2723 62 on the board *Hindu-wisdom-holder awareness (Buddhism, Tibetan)*
HM 5099 62 state or square *Happiness (Leela)*
HM 4983 62 when the 8 supramundane *Feeling aggregate (Buddhism)*
HM 2777 63 in the ica *Everlasting community (ICA)*
HM 5006 63 of the ninety *Al-Qayyum (Sufism)*
HM 2247 63 on the board *Mahayana receptivity-application awareness (Buddhism, Tibetan)*
HM 6238 63 state or square *Tamas (Leela)*
HH 2002 64 aeon which is destroyed *World cycles (Buddhism)*
HH 4997 64 hexagrams *Balancing yin and yang (Taoism)*
HM 2456 64 in the ica *Contingent eternality (ICA)*
HM 0743 64 of the ninety *Al-Wajid (Sufism)*
HM 2271 64 on the board *Mahayana highest-teachings-application awareness (Buddhism, Tibetan)*
HM 3907 64 state or square *Phenomenal plane (Leela)*
HM 4526 65 of the ninety *Al-Majid (Sufism)*
HM 2663 65 on the board *Bon-wisdom awareness (Buddhism, Tibetan)*
HM 6287 65 state or square *Plane of inner space (Leela)*
HM 5506 66 of the ninety *Al-Wahid (Sufism)*
HM 2187 66 on the board *First vajra-master stage (Buddhism, Tibetan)*
HM 6901 66 state or square *Plane of bliss (Leela)*
HM 6117 67 of the ninety *Al-Ahad (Sufism)*
HM 2211 67 on the board *Tantra master-in-sense-realm awareness (Buddhism, Tibetan)*
HM 4343 67 state or square *Plane of cosmic good (Leela)*
HM 4980 68 of the ninety *As-Samad (Sufism)*
HM 2235 68 on the board *Tantra master in form-realm awareness (Buddhism, Tibetan)*
HM 1301 68 state or square *Plane of cosmic consciousness (Leela)*
HM 1500 69 of the ninety *Al-Qadir (Sufism)*
HM 2759 69 on the board *Wheel-turner-king awareness (Buddhism, Tibetan)*
HM 4326 69 state or square *Absolute plane (Leela)*
HM 2783 70 on the board *Amoghasiddhi-karma awareness (Buddhism, Tibetan)*
HM 4310 70 state or square *Satoguna (Leela)*
HM 5331 71 of the ninety *Al-Muqaddim (Sufism)*
HM 2155 71 on the board *First scriptural bodhisattva awareness (Buddhism, Tibetan)*
HM 3281 71 state or square *Rajoguna (Leela)*
HM 4052 72 of the ninety *Al-Mu'akhkhir (Sufism)*
HM 2679 72 on the board *Magical-forces awareness (Buddhism, Tibetan)*
HM 2278 72 primary states of being *Lila*
HH 3013 72 schools of sexual play *False paths (Taoism)*
HH 4000 72 squares others with many *Consciousness ascension stages game (Buddhism, Tibetan)*
HM 5561 72 state or square *Tamoguna (Leela)*
HM 6139 73 of the ninety *Al-Awwal (Sufism)*
HM 2703 73 on the board *Second vajra-master awareness (Buddhism, Tibetan)*
HM 7112 74 of the ninety *Al-Akhir (Sufism)*
HM 2727 74 on the board *Third vajra-master awareness (Buddhism, Tibetan)*
HH 1108 75 dharmas buddhism vaibhasika *Seventy-five dharmas (Buddhism)*
HH 1108 75 dharmas classified *Seventy-five dharmas (Buddhism)*
HM 4042 75 of the ninety *Az-Zahir (Sufism)*
HM 2251 75 on the board *Fourth vajra-master awareness (Buddhism, Tibetan)*
HM 1209 76 of the ninety *Al-Batin (Sufism)*
HM 2275 76 on the board *Jewelled-peaks-realm awareness (Buddhism, Tibetan)*
HM 0525 77 of the ninety *Al-Wali (Sufism)*
HM 2167 77 on the board *Joy-land awareness (Buddhism, Tibetan)*
HM 8019 78 of the ninety *Al-Muta'ali (Sufism)*
HM 2191 78 on the board *Fourth scriptural bodhisattva awareness (Buddhism, Tibetan)*
HM 7022 79 of the ninety *Al-Barr (Sufism)*
HM 2215 79 on the board *Third scriptural bodhisattva awareness (Buddhism, Tibetan)*
HM 4334 80 of the ninety nine *Al-Tawwab (Sufism)*
HM 2739 80 on the board *Second scriptural bodhisattva awareness (Buddhism, Tibetan)*
HM 4513 81 of the ninety *Al-Muntaqim (Sufism)*
HM 2763 81 on the board *Fifth vajra-master awareness (Buddhism, Tibetan)*
HM 5543 82 of the ninety *Al-'Afuw (Sufism)*
HM 2287 82 on the board *Sixth vajra-master awareness (Buddhism, Tibetan)*
HM 0206 83 of the ninety *Ar-Ra'uf (Sufism)*
HM 2789 83 on the board *Seventh vajra-master awareness (Buddhism, Tibetan)*
HM 5902 84 of the ninety *Malik-ul-Mulk (Sufism)*
HM 2813 84 on the board *Supreme-heaven awareness (Buddhism, Tibetan)*
HH 3667 84 schools of sexual intercourse *Outside paths (Taoism)*
HH 0669 84 which are considered best *Asana (Hinduism, Yoga)*
HM 6719 85 of the ninety *Dhul-Jalali wal-Ikram (Sufism)*
HM 2337 85 on the board *Hyper-bliss-realm awareness (Buddhism, Tibetan)*
HM 8001 86 of the ninety *Al-Muqsit (Sufism)*
HM 2361 86 on the board *Seventh-scriptural bodhisattva awareness (Buddhism, Tibetan)*
HM 0009 87 of the ninety *Al-Jami' (Sufism)*
HM 2385 87 on the board *Sixth scriptural bodhisattva awareness (Buddhism, Tibetan)*
HM 3822 88 of the ninety *Al-Ghani (Sufism)*
HM 2909 88 on the board *Fifth scriptural bodhisattva awareness (Buddhism, Tibetan)*
HM 2098 89 dispositions *Dispositions of consciousness (Buddhism)*
HM 4143 89 divisions since there is *Perception aggregate (Buddhism)*
HM 4143 89 divisions the characteristic *Perception aggregate (Buddhism)*
HM 4983 89 in the consciousness aggregate *Feeling aggregate (Buddhism)*
HM 3932 89 khandas or kinds *Death (Buddhism)*
HM 6720 89 kinds of consciousness are *Modes of occurrence of consciousness (Buddhism)*
HM 2050 89 kinds of consciousness comprising *Awareness as mental-formation group of conscious existence* (Buddhism)
HM 8273 89 kinds of consciousness grouped *Consciousness-born materiality (Buddhism)*
HM 8336 89 kinds of consciousness proceed *Adverting (Buddhism)*
HM 7632 89 kinds of consciousness proceed *Determining (Buddhism)*
HM 7324 89 kinds of consciousness proceed *Examining (Buddhism)*
HM 7268 89 kinds of consciousness proceed *Impulsion (Buddhism)*
HM 6221 89 kinds of consciousness proceed *Life-continuum (Buddhism)*
HM 8266 89 kinds of consciousness proceed *Rebirth-linking (Buddhism)*
HM 7092 89 kinds of consciousness proceed *Receiving (Buddhism)*
HM 1646 89 kinds of consciousness proceed *Registering (Buddhism)*
HM 4389 89 kinds of consciousness proceed *Sense mode of consciousness occurrence (Buddhism)*
HM 3633 89 of the ninety *Al-Mughni (Sufism)*
HM 2301 89 on the board *Eighth vajra-master awareness (Buddhism, Tibetan)*
HM 2556 89 states of consciousness *Awareness of consciousness-group of conscious existence - senses and mind* (Buddhism)
HM 2599 89 states of consciousness vinnana *Phase-conditions of consciousness (Buddhism)*
HM 7220 90 of the ninety nine *Al-Mani (Sufism)*
HM 2325 90 on the board *Ninth vajra-master awareness (Buddhism, Tibetan)*
HH 2378 90 percent of the brain *Suggestopedia*
HM 4542 91 of the ninety *Ad-Darr (Sufism)*

INDEX TO CONCEPT SETS
(See SECTION X for KEYWORD INDEX)

HX

NUMBERS

HM 2849 91 on the board *Final vajra-master awareness (Buddhism, Tibetan)*
HM 6350 92 of the ninety *An-Nafi' (Sufism)*
HM 2873 92 on the board *Enjoyment-body awareness (Buddhism, Tibetan)*
HM 1354 93 of the ninety *An-Nur (Sufism)*
HM 2397 93 on the board *Absolute-body awareness (Buddhism, Tibetan)*
HM 1287 94 of the ninety *Al-Hadi (Sufism)*
HM 2421 94 on the board *Tenth scriptural bodhisattva awareness (Buddhism, Tibetan)*
HM 4299 95 of the ninety *Al-Badi (Sufism)*
HM 2292 95 on the board *Ninth scriptural bodhisattva awareness (Buddhism, Tibetan)*
HM 6786 96 of the ninety *Al-Baqi (Sufism)*
HM 2816 96 on the board *Eighth scriptural bodhisattva awareness (Buddhism, Tibetan)*
HM 0113 97 of the ninety *Al-Warith (Sufism)*
HM 2340 97 on the board *Emanation-body awareness (Buddhism)*
HM 5234 98 of the ninety *Ar-Rashid (Sufism)*

HM 2735 98 to 104 *Buddha-consciousness (Buddhism, Tibetan)*
HH 2561 99 in the quran together *Ninety-nine names of Allah (Sufism)*
HH 2561 99 names allows the believer *Ninety-nine names of Allah (Sufism)*
HH 2561 99 names enters paradise *Ninety-nine names of Allah (Sufism)*
HH 2561 99 names in the quran *Ninety-nine names of Allah (Sufism)*
HH 2561 99 names of allah *Ninety-nine names of Allah (Sufism)*
HM 2317 100 description *Stations of consciousness - Ansari (Sufism)*
HH 4864 100 operations *Lower vehicle path (Taoism)*
HH 5652 100 persons *Supernormal powers (Buddhism)*
HM 4273 100 stations *Love of God (Sufism)*
HM 4297 100 thousand million world spheres *Knowledge of recollection of previous existence (Buddhism)*
HM 2002 100 thousand million world spheres *World cycles (Buddhism)*
HM 2424 101 description abu sa'id *Stations of consciousness - ibn-Abi'l-Khayr (Sufism)*
HM 4273 101 ground or the 61 *Love of God (Sufism)*

HM 2735 104 on the board *Buddha-consciousness (Buddhism, Tibetan)*
HM 2733 108 neighboring hells *Desire-realm consciousness (Buddhism)*
HM 2150 119 chief angels *Angelic frame of awareness*
HH 1376 200 rules the majority *Vinaya (Buddhism)*
HH 3667 300 fallacious techniques *Outside paths (Taoism)*
HH 2561 300 in the new testament *Ninety-nine names of Allah (Sufism)*
HH 2561 300 in the torah 300 *Ninety-nine names of Allah (Sufism)*
HH 2561 300 in zabar psalms *Ninety-nine names of Allah (Sufism)*
HH 3013 300 such practices context *False paths (Taoism)*
HH 4289 400 recipes for material alchemy *Outside paths (Taoism)*
HH 7004 1000 items and are practised *Sidetracks and auxiliary methods in Taoism (Taoism)*
HH 7004 1000 practices are performed *Sidetracks and auxiliary methods in Taoism (Taoism)*
HH 0208 1720 for the body is *Hyper-personalisation*
HH 0689 3600 practices in taoism *Human development (Taoism)*
HH 3764 7000 persons collectively practising *Maharishi effect*

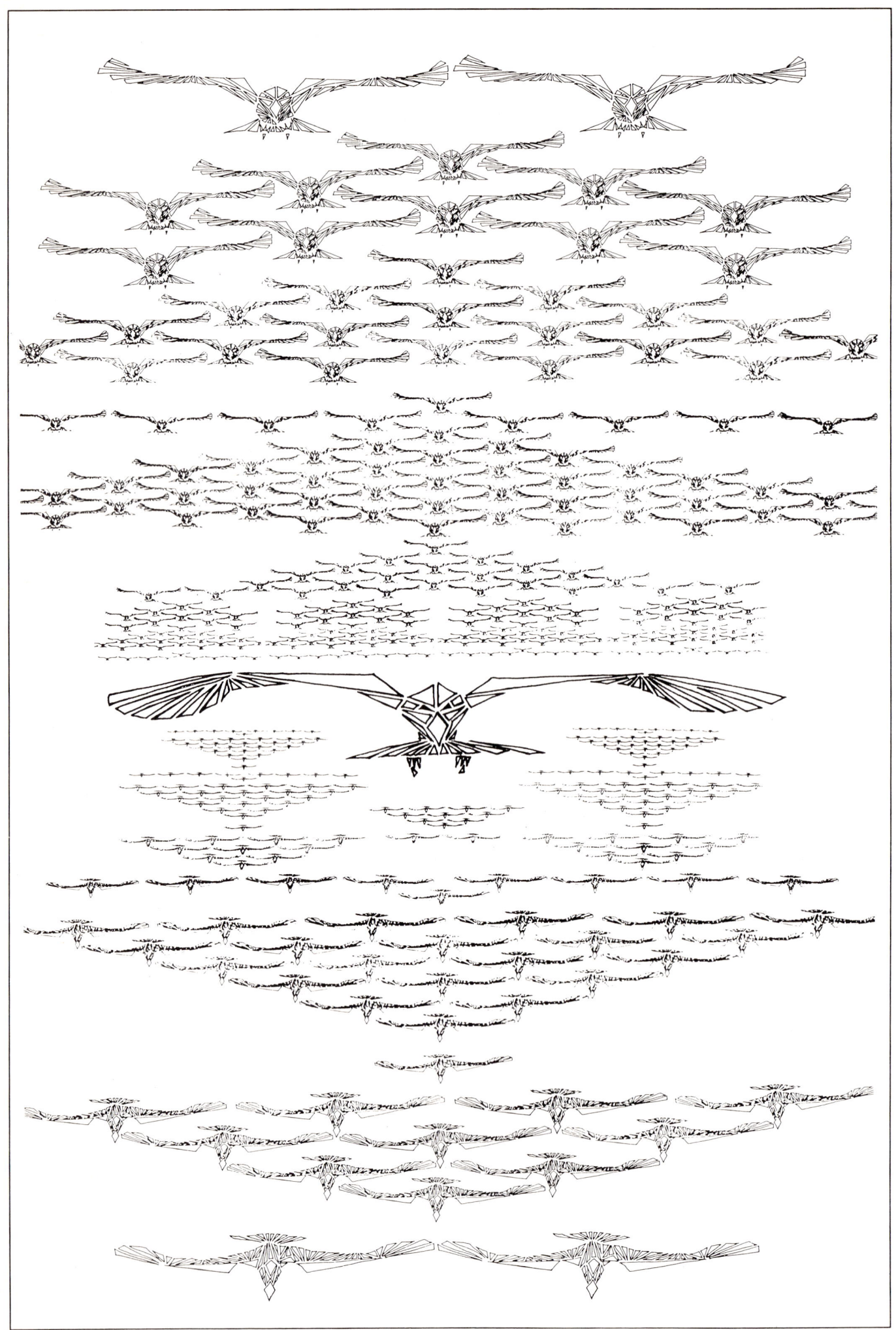

References

Human Development

Scope of bibliography

The following section constitutes a collection of 2,488 bibliographical references relevant to the Human Development (Section H).

The entries appear here for any of the following reasons:

(a) Reference is made to them in the individual entries of Section H (where they appear in abridged form, namely with author, title and year of publication).

(b) Reference is made to them in the Notes on Section H (which appear as Section HZ).

(c) The publications are relevant to Section H, even though they have not (yet) been cross-referenced in either of the two ways indicated above.

An important reason for producing this bilbiography was to focus attention on studies of unusual aspects of human development. The bibliography tends not to include items which are mainly designed to popularize certain aspects of personal growth. Items which are not concerned with the experiential dimension of human development also tend to be excluded, especially since such materials are well-documented elsewhere.

Bibliographical entries

The entries appear by order of author and, within author, by year.

In certain cases the *International Standard Book Number (ISBN)* is given to facilitate access.

Source of bibliography

The bibliography was derived from a wide variety of sources. These include:

- Information from international organizations, including catalogues of publications and accessions lists

- Information from commercial publishers, especially sales catalogues

- Citations and bibliographies in other publications and reference books

- Systematic compilations of books in print (notably *Books in Print* and *International Books in Print*)

- Book reviews in specialized journals

A-kya Yong-dzin Yang-chan ga-wai lo-dro A Compendium of Ways of Knowing: a clear mirror of what should be accepted and rejected (1980)
Dharamsala, Library of Tibetan Works and Archives, 67 p. 2nd ed. With commentary compiled from the oral teachings of Gese Ngawang Dhargyey. First published 1976.

Aaronson, Bernard and Humphrey, Osmund Psychedelics: the uses and implications of psychedelic drugs (1971)
London, Hogarth Press.

Abhedananda, Swami Spiritual Unfoldment (1978)
15th ed.

Abhishiktananda Saccidananda: a Christian approach to Advaitic experience (1984)
Delhi, ISPCK.
Revised edition. Author's other name: Henri le Saux.

Abraham, William J The Logic of Evangelism: a significant contribution to the theory and practice of evangelism (1989)
London, Hodder and Stoughton, 245 p.
ISBN 0-340-51451-5.

Abrahamsen, David The Road to Emotional Maturity (1958)
Englewood Cliffs NJ, Prentice Hall. bibl.

Abram, David The Perceptual Implications of Gaia (1985)
The Ecologist, 15, 3, 1985, 96-103.

Abun Nasr, Jamil M The Tijaniya: a Sufi Order in the modern world (1965)
Oxford.

Academy of Religion and Mental Health Religion in the Developing Personality: proceedings of the 2nd academy symposium, 1958 (1960)
New York, New York University Press, 105 p.

Acharya, Anada Sri Tattvajnanam, or the Quest of Cosmic Consciousness

Achterberg, J Imagery and Healing: Shamanism and modern medicine (1985)
Boston MA, New Science Library.

Adams, John (Ed) Transforming Work: a collection of organizational transformation readings (1984)
Alexandria VA, Miles River Press.

Addo, Herb, et al Development as Social Transformation: reflections on the global problematique (1985)
London, Hodder and Stoughteon in association with the United Nations University, 281 p. ISBN 0-340-35634-0.

Adler, Alfred Individual Psychology of Alfred Adler: a systematic presentation of selections from his writings
New York, Harper and Row.

Adler, Alfred Understanding Human Nature
New York, Fawcett World Library.

Adler, Alfred Social Interest: a challenge to mankind (1939)
New York, Putnam Publishing Group.

Adler G, Fordham M and Read H (Eds) The Collected Works of C G Jung
Princeton NJ, Princeton University Press. Hull R F C, Trans.

Adler, Gerhard Living Symbol: a case study in the process of individuation (1961)
New York, Pantheon Books.

Agar, W E Wholeness of the Living Organism (1948)

Agor, Weston The Logic of Intuitive Decision Making (1986)
Westport CT, Greenwood.

Agor, Weston The Role of Intuition in Leadership (1989)
New York, Sage.

Ahrenfeldt, R H and Soddy, K Mental Health in a Changing World (1965)
London.

Ajaya, Swami Psychotherapy East and West: a unifying paradigm (1984)
Honesdale PA, Himalayan Publishers.

Akishige, Y Psychological Studies on Zen (1968)
In: *Bulletin of the Faculty of Literature of Kyrushu University, Japan* 5.

Al-Ghazzali The Alchemy of Happiness
London, Octagon Press. ISBN 0-900860-71-5.

Al-Halveti, Sheikh Tosun Bayrak al-Jerrahi (Comp) The Most Beautiful Names (1985)
Vermont, Threshold Books, 164 p. ISBN 0-939660-10-5.

al-Jili, Abd al-Karim Universal Man
Extracts trans. with commentary by Titus Burckhardt.

Al-Najib, Abu A Sufi Rule for Novices (1975)
Cambridge MA, Harvard University Press. Trans. by Milson M.

Albery, Nicholas and Yule, Valerie (Eds) Encyclopaedia of Social Inventions (1990)
London, The Institute for Social Inventions, 298 p. Second edition. ISBN 0-948826-17-7.

Alcantara, Peter S A Golden Treatise of Mental Prayer (1978)
Ann Arbor MI, Books on Demand. ISBN 0-8357-9135-1.

Alderfer, Clayton P Existence, Relatedness and Growth: human needs in organizational settings (1972)
New York, Free Press.

Aldrich, V C Theory and the Integrity of Experience (1946)
Journal of Philosophy 43, p. 379-82, July 4.

Alexander, Franz Buddhistic Training as an Artificial Catatonia
In: *Psychoanalysis* 1931, 19, pp. 191-206.

Alexander, Gerda Eutony: the holistic discovery of the whole person (1986)
Great Neck NY, Felix Morrow, 182 p. illus. Transl. from German. ISBN 0-9615659-0-X.

Alisjahbana, Sutan T Values as Integrating Forces in Personality, Society and Culture: essay of a new anthropology (1966)
London, Oxford University Press, 248 p. bibl. foot.

Allen, Marcus Tantra for the West: a guide to personal freedom (1981)
Los Altos CA, New World Library Publishing, 235 p.
ISBN 0-931432-06-5.

Allison, J Respiratory Changes during Transcendental Meditation
In: *The Lancet* 1970, 18 April.

Allport, Floyd H Theories of Perception and the Concept of Structure (1955)
New York, John Wiley and Sons.

Allport, G Becoming (1955)
New Haven CT, Yale University Press.

Allport, Gordon W The Open System in Personality Theory
In: *Journal of Abnormal and Social Psychology* 1960, 61.

Allport, Gordon W The Unity of Personality
In: *Pattern and Growth in Personality* New York, Holt, Rinehart and Winston, 1961.

Allport, Gordon W Becoming: basic considerations for a psychology of personality (1955)
New Haven CT, Yale University Press.

Allport, Gordon W Pattern and Growth in Personality (1961)
New York, Holt, Rinehart and Winston, 573 p.

Allport, Gordon W Personality and Social Encounter (1964)
Boston MA, Beacon Press.

Allport, Gordon W The Person in Psychology (1968)
Boston MA, Beacon Press.

Allport, Gordon W Personality: a psychological interpretation (1971)
London, Constable, 588 p. reprint of 1937 ed. bibl. foot.

Allport, Gordon W The Nature of Personality: selected papers (1975)
Brooklyn NY, Greenwood House, 220 p. Repr of 1950.
ISBN 0-8371-7432-5.

Almond, Philip C Mystical Experience and Religious Doctrine: an Investigation of the study of mysticism in world religions (1982)
Berlin, Walter de Gruyter, 197 p. Religion and Reason: 26.
ISBN 90-279-3160-7.

Altman, Nathaniel The Nonviolent Revolution: a comprehensive guide to Ahimsa - the philosophy of dynamic harmlessness (1988)
Longmead, Element Books, 180 p. bibl.
ISBN 1-85230-028-0.

Amaldas, Swami Christian Yogic Meditation
Wilmington DE, Michael Glazier. Ways of Prayer Series: 8.
ISBN 0-89453-368-1.

American Psychiatric Association Diagnostic and Statistical Manual of Mental Disorders - DSM-III-R (1987)
Washington DC, American Psychiatric Association, 567 p. 3rd rev ed. ISBN 0-89042-019-X.

Amma Dhyan-Yoga and Kundalini Yoga (1969)
Ganeshpuri, Shree Gurudev Ashram.

Anand, B K; Chhina, G S and Singh, B Studies on Shri Ramanand Yogi during his Stay in an Air-Tight Box (1961)
pp. 82-89. Indian Journal of Medical Research: No 49.

Anand, B K; China, G S and Singh, B Some Aspects of EEG Studies in Yogis
In: *EEG and Clinical Neurophysiology* 1961, 13, pp. 452-456.

Ananda, Spiritual Practice
Hollywood CA, Vedanta Press. ISBN 0-87481-155-4.

Ananda, Acharya Yoga of Conquest (1926)
Hospiarpur, Vishveshvaranand Institute.

Anant, Santokh S Psychoanalysis, Morality and Integrity Therapy (1966)
Interdiscipline 3, 2, April pp. 81-88.

Anddrews, Donald H The Symphony of Life (1966)
Santa Cruz, Unity Press, 423 p.

Andersen, M S and Savary, L Passages: a guide for pilgrims of the mind (1973)
New York, Harper and Row.

Anderson, H H Personality Growth: conceptual considerations in perspectives in personality theory (1957)
London, Tavistock Publications.

Anderson, H H Creativity as Personality Development (1959)
In: *H H Anderson (Ed) Creativity and its Cultivation* New York, Harper, bibl.

Anderson, H H Creativity and its Cultivation (Addresses at the Interdisciplinary Symposia on Creativity, Michigan State University 1957-1958) (1959)
New York, Harper and Row. bibl. (pp. 269-79).

Anderson, James F The Bond of Being: an essay on analogy and existence (1969)
New York, Greenwood Press, 341 p. bibl. (pp. 328-36).

Anderson, Walter Truett To Govern Evolution (1987)
Boston MA, Harcourt, Brace and Jovanovich, 376 p.

Andrews, F M and Withey, S B Social Indicators of Well-Being (1976)
New York, Plenum Publishing.

Anglyal, Andras Foundations for a Science of Personality (1941)
Cambridge, Harvard.

Aniane, Maurice Yoga: science de l'homme integral (1956)
Paris, Cahiers du Sud.

Aniane, Maurice Notes on Alchemy: the Cosmological Yoga of Medieval Christianity (1976)
San Francisco, Far West Press, in *Material for Thought*, Spring 1976.

Annett, Marian Left, Right, Hand and Brain: the right shift theory (1985)
Hillsdale NJ, Erlbaum Lawrence Associates, 488 p.
ISBN 0-86377-018-5.

Anon Yoga: its various aspects
Calcutta, Ramakrishna Vedanta Math, 236 p.

Anon The Cloud of Unknowing (1951)
Burns and Oates. Many translations.

Anon Material for Thought (1976)
San Francisco CA, Far West Press, 96 p. Spring 1976.
ISBN 0-914480-02-2.

Anshen, Ruth Nanda (Ed) Moral Principles of Action
Foundation for Integrative Education.

Anthony, D and Robbins, T In Gods We Trust: new patterns in American religious pluralism (1981)
New Brunswick NJ, Transaction Books.

Anuradha Vittachi Earth Conference One, Sharing a Vision for our Planet (1989)
New York, Element Books.

Arapura, John G Radhakrishnan and Integral Experience: the philosophy and world vision of Sarvepalli Radhakrishnan (1966)
New York, Asia Publishing House, 211 p. bibl. (pp. 205-208).

Arapura, John G Gnosis and the Question of Thought in Vedanta (1986)
Norwell MA, Kluwer Academic Publishers.
ISBN 90-247-3061-9.

Arasteh, A Reza Final Integration in the Adult Personality: a measure for health, social change and leadership (1965)
Leiden, Brill.

Arasteh, A Reza Growth to Selfhood: the Sufi contribution (1980)
London, Routledge and Kegan Paul, 145 p.
ISBN 0-7100-0355-2.

Arber, Agnes The Manifold and the One (1967)
Wheaton IL, Theosophical Publishing House, 146 p. copyright 1957. bibl. (pp. 119-35).

Arberry, A J The Doctrine of the Sufis (1935)
Cambridge. Trans. of Kalabadhi's Kitab at-ta'arruf.

Arberry, A J Sufism: an account of the mystics of Islam (1972)
London, Allen and Unwin.

Archimedes Foundation Directory of Human Happiness and Well-Being (1988)
Toronto, Archimedes Foundation.

Arendt, H The Human Condition (1958)
Garden City NY, Doubleday.

Arguelles, Jose A The Transformative Vision: reflections on the nature and history of human expression (1975)
Berkeley CA, Shambhala Publications, 364 p.

Arguelles, José and Miriam, T Mandala (1972)
Berkeley CA, Shambhala Publications, 140p. bibl. (pp. 130-34) (distributed in Commonwealth and Europe by Routledge and Kegan Paul).

Argyle, M The Psychology of Happiness (1987)
London, Methuen.

Argyle, M; Schwarz, N and Strack, F (Eds) The Social Psychology of Subjective Well-Being (1990)
New York, Pergamon Press.

Argyris, Christopher Personality and Organization
New York, Harper and Row.

Argyris, Christopher Integrating the Individual and the Organization (1964)
New York, John Wiley and Sons, 330 p. bibl. foot.

Aries, P Centuries of Childhood (1962)
New York, Random House.

Arkin, Alan Halfway through the Door: first steps on a path toward enlightenment

Arkle, William A Geography of Consciousness (1974)
London, Nevilla Spearman. Introduction by Colin Wilson.

Arney, William Ray and Bergan, Bernard J Medicine and the Management of Living: taming the last great beast (1982)
Chicago IL, University of Chicago Press.

Arnold, Magda B and Gasson, J A The Human Person: an approach to an integral theory of personality (1954)
New York, Ronald Press.

Arnold, W J and Jones, M R (Eds) Nebraska Symposium on Motivation
Lincoln, 1953-1969. Vols 1-17.

Aron, Arthur and Aron, Elaine The Maharishi Effect: a revolution through meditation (1986)
Walpole NH, Stillpoint Publishing, 235 p.
ISBN 0-913299-26-X.

Aronfreed, J Conduct and Conscience (1968)
New York.

Aronoff, Joel Psychological Needs and Cultural Systems: a case study (1967)
New York, Reinhold Van Nostrand.

Arpita Psychology of the Beatitudes (1979)
Illinois, Himalayan International Institute of Yoga.

Arundale, J S Kundalini: an occult experience (1938)
Madras.

Arvey, Michael ESP (1988)
Saint Paul MN, Greenhaven Press, 112 p.
ISBN 0-89908-057-X.

Arya, Pandit Usharbudh Superconscious Meditation (1974)
Glenview IL, Himalayan Institute.

Ashburn, Shirley S and Schuster, Clara S (Eds) The Process of Human Development: a holistic approach (1986)
Glenview IL, Scott, Foresman and Company. 2nd ed.
ISBN 0-673-39404-2.

Ashby, R The Guide Book to the Study of Psychical Research (1972)
Samuel Weiser.

Ashby, R H The Guidebook for the Study of Psychical Research (1972)
London, Rider.

Asrani, U The Psychology of Mysticism (1969)
Main Currents in Modern Thought 25, pp68-73.

Asrani, U A Yoga Unveiled: through a synthesis of personal mystic experiences and psychological and psychosomatic studies (1977)
Delhi, Motilal Banarsidass, 230 p.

Assagioli, Roberto A New Method of Healing: psychosynthesis (1927)
Rome, Istituto di Psicosintesi.

Assagioli, Roberto Il Valore Practico ed Umano della Cultura Psichica Istituto di Psicosintesi, Rome 1929 (1929)
Bulletin de la Société Lorraine de Psychologie Appliquée 34–35. French translation.

Assagioli, Roberto Psicanalisi e Psicosintesi Istituto di Psicosintesi, Rome 1931 (1931)
Hibbert Journal London, 1934. English Translation.

Assagioli, Roberto Sviluppo Spirituale e Malattis Nervose Istituto di Psicosintesi, Rome 1933 (1933)
Hibbert Journal London, 1937. English translation.

Assagioli, Roberto Dynamic Psychology and Psychosynthesis (1958)
New York, Psychosynthesis Research Foundation. monograph (reprinted in his: Psychosynthesis, New York, Viking Press, 1971 pp11–34).

Assagioli, Roberto Self-Realization and Psychological Disturbances (1961)
New York, Psychosynthesis Research Foundation. monograph (reprinted in his: Psychosynthesis, New York, Viking Press, 1971, pp. 35–59).

Assagioli, Roberto Psychosynthesis: a manual of principles and techniques (1965)
New York, Hobbs Dorman, 323 p.

Assagioli, Roberto The Balancing and Synthesis of Opposites (1972)
New York, Psychosynthesis Research Foundation.

Assagioli, Roberto Psychosynthesis typology (1983)
London, Institute of Psychosynthesis, 88 p. Psychosynthesis Monographs, ISSN 0309–4025.

Assagioli, Roberto Synthesis: the realization of the self (I and II)
Redwood City CA, Synthesis Press. 2 Vols. I (1977), II (1978).

Association pour la Recherche et l'Intervention Psychosociologique Connexions (1972)
Paris vol 1, 1972 onwards, quarterly.

Atlan, Henri Le Cristal et la Fumé (1979)
Paris, Seuil.

Atmananda, Swami The Four Yogas, or the Four Paths to Spiritual Enlightenment (in the Words of the Ancient Rishis)

Atreya, B L Deification of Man: its methods and stages according to the yoga Vasistha

Attali, Jacques Noise: the political economy of music (1985)
Minneapolis, University of Minnesota Press. Trans. by Brian Massumi.

Aurobindo, Sri Integral Education in the words of Sri Aurobindo and the Mother, selected from their writings (1952)
Pondcherry, Sri Aurobindo International University Centre, 92 p. bibl. (p 92).

Aurobindo, Sri On Yoga: the synthesis of yoga (1955)
Pondicherry, Sri Aurobindo Ashram.

Aurobindo, Sri The Ideal of Human Unity (1960)
Pondicherry, Sri Aurobindo Ashram.

Aurobindo, Sri The Future Evolution of Man (1963)
Pondicherry, Sri Aurobindo Ashram.

Aurobindo, Sri The Mind of Light (1971)
New York, EP Dutton.

Aurobindo, Sri The Supramental Manifestation upon Earth (1973)
Pondicherry, Sri Aurobindo Ashram.

Aurobindo, Sri The Synthesis of Yoga (1984)
High Falls NY, Sri Aurobindo Association, 899 p. first published 1914. ISBN 0–89071–313–8.

Austin, Mary Experiences Facing Death (1977)
Salem NH, Ayer Company Publishers. repr of 1931. Death and Dying Ser. ISBN 0–405–09553–8.

Avalon, A Tantra of the Great Liberation (1972)
New York, Allied Publications.

Avalon, A The Serpent Power (1974)
New York, Dover Publications. 1st ed 1931.

Avens, Roberts Imagination is Reality: Western Nirvana in Jung, Hillman, Barfield and Cassirer (1980)
Dallas, Spring Publications Inc, 127 p. ISBN 0–88214–311–5.

Avens, Roberts The New Gnosis: Heidegger, Hillman and Angels (1984)
Dallas TX, Spring Publications, 155 p. bibl. ISBN 0–88214–327–1.

Axline, Virginia M In Search of Self (1966)
New York, Ballantine Books.

Ayensu, Edward S and Whitfield, Philip The Rhythms of Life (1982)
New York, Crown Publishers.

Ayyangar, T R S The Yoga Upanisads (1952)
Adyar, Theosophical Publishing House.

Bachelard, Gaston The Poetics of Reverie (1971)
Boston MA, Beacon Press.

Back, W Beyond Words: the story of sensitivity training and the encounter movement (1972)
New York, Basic Books.

Badham, Paul and Badham, Linda Death and Immortality in the Religions of the World (1987)
New Era Books/Paragon Hse, 238 p. ISBN 0–913757–67–5.

Baer, Dobh and Jacobs, Louis On Ecstasy (1982)
Dallas TX, Rossel Books, 196 p. ISBN 0–8419–0814–1.

Baer, R and Baer, V Windows of Light: quartz crystals and self-transformation (1984)
San Francisco CA, Harper and Row.

Baer, R and Baer, V The Crystal Connection: a guidebook for personal and planetary transformation (1986)
San Francisco CA, Harper and Row.

Bagchi, B K and Wenger, M A Electrophysiological Correlates of Some Yoga Exercises (1957)
Electroencephalography and Clinical Neurophysiology Suppl. 7.

Bagchi, B K and Wenger, M A Simultaneous EEG and Other Recordings during Some Yogic Practices (1958)
Electroencephalography and Clinical Neurophysiology Vol 10, Abstract.

Baginski, Bodo and Sharamon, Shalila REIKI - Universal Life Energy: heals body, mind and spirit (1988)
Mendocino CA, LifeRhythm, 224 p. ISBN 0–940795–02–7.

Bailey, Alice A A Treatise on the Seven Rays
London, Lucis Press, 5 vols.

Bailey, Alice A A Treatise on White Magic (1934)
New York, Lucis Publishing.

Bailey, Alice A The Rays and the Initiations (1960)
London, Lucis Press.

Bailey, Alice A The Labours of Hercules (1971)
London, Lucis Press.

Bailey, Alice A Initiation, Human and Solar (1978)
London, Lucis Press.

Bailey, Alice A Discipleship in the New Age (1986)
New York, Lucis Publishing, 2 vols.

Bakan, David Sigmund Freud and the Jewish Mystical Tradition
New York, Schocken Books.

Bakan, David Duality of Human Existence (1926)
Chicago IL, Rand Corporation.

Baker, Robert L and Mednick, Birgitte R Influences on Human Development: a Longitude perspective (1984)
Dordrecht, Kluwer Academic Publishers Group, 224 p. ISBN 0–89838–130–4.

Bakhtiar, Laleh Sufi: expressions of the Mystic Quest (1987)
London, Thames and Hudson, 120 p. ISBN 0–500–81015–X.

Baldwin, J M Thought and Things, or Genetic Logic, I–III
New York, Macmillan.

Baldwin, J M Mental Development in the Child and the Race (1900)
New York, Macmillan.

Balsys, Bodo Revelation: the evolution of transcendent perception by humanity (1983)
Bellingen NSW, Kalang Press. ISBN 0–9592648–0–9.

Bancroft, Anne Modern Mystics and Sages (1976)
London, Heinemann.

Bancroft, Anne Zen: direct pointing to reality (1987)
New York, Thames and Hudson. ISBN 0–500–81018–4.

Bandler, J and Grinder, R Reframing (1982)
Moab UT, Real People Press.

Bandler, Richard and Grinder, John The Structure of Magic (1975–76)
Palo Alto CA, Science and Behaviour Books, 2 vols.

Bandler, Richard and Grindler, John The Structure of Magic: a book about language and therapy (1975)
Palo Alto CA, Science and Behavior Books Inc, 225 p. ISBN 08314–0044–7.

Bandura, A Self-Efficacy: toward a unifying theory of behavioral change
In: *Psychological Review* 1977, 34, pp. 191–215.

Bandura, A and Walters, R Social Learning and Personality Development
New York, Holt, Rinehart and Winston.

Bandyopadhyay, Pranab Yoga, Sadhana and Samadhi (1987)
Columbia MO, South Asia Books. ISBN 0–8364–2132–9.

Bank St College of Education, New York Integration of Mental Health Concepts with the Human Relations Professions: education, medicine, psychology, law, religion, nursing, social work, dentistry (Proceedings of a lecture series) (1962)
New York, Mental Health Materials Center, 132 p.

Bapak Muhammad Subuh Subud in the World (1965)
The Subud Brotherhood.

Barbar, T and Dalal, A Yoga, Yoga Feats, and Hypnosis in the Light of Empirical Research (1969)
In: *American Journal of Clinical Hypnosis* 11, pp. 155–66.

Barber, T (Ed) Advances in Altered States of Consciousness and Human Potentialities (1976)
New York, Psychological Dimensions. Vol 1.

Barber, T X, et al (Ed) Biofeedback and Self-Control: an Aldine annual on the regulation of bodily processes and consciousness (1970)
Chicago IL, Aldine–Atherton.

Barker, L F The Neuro–Psychiatrist and the Study of a Person as a Whole (1922)
New York State Journal of Medicine pp. 512–15.

Barnes, Hazel E Existentialist Ethics
New York, Alfred A Knopf.

Barnes, Michael The Buddhist Way of Deliverance (1980)
In: *Studia Missionalia* pp. 223–277.

Barnhouse, R The Spiritual Exercises and Psychoanalytic Therapy (1975)
In: *The Way Supplement*, 24, Spring, p. 74–82.

Barrett, William Irrational Man (1958)
New York, Doubleday.

Barrett, William What is Existentialism? (1964)
Grove.

Barron, Frank Complexity-Simplicity as a Personality Dimension (1953)
Journal of Abnormal and Social Psychology 48, 2 April pp. 163–72.

Barron, Frank Creativity and Psychological Health (1963)
New York, Reinhold Van Nostrand.

Barron, Frank Creativity and Personal Freedom (1968)
Princeton, Van Nostrand.

Barron, Frank Creative Person and Creative Process (1969)
New York, Holt, Rinehart and Winston.

Barron, Frank Toward an Ecology of Consciousness (1972)
Inquiry 15, 1–2, Summer pp. 95–113.

Barshinger, C Intimacy and Spiritual Growth (1977)
In: *The Bulletin of the Christian Association for Psychological Studies*, 3, p. 19–21.

Barthes, R Mythologies (1972)
London, Jonathan Cape.

Bartlett, F C Remembering: an experimental and social study (1932)
London, Cambridge University Press.

Bastiaans, J, et al (Eds) Training in Psychosomatic Research (1967)
Basel, S Karger AG, 180 p. Advances in Psychosomatic Medicine: Vol 5. ISBN 3–8055–0459–4.

Batchelor, Stephen The Faith to Doubt: glimpses of Buddhist uncertainty (1989)
California CA, Paralax Press.

Bateson, Gregory Where Angels Fear

Bateson, Gregory Minimal Requirements for a Theory of Schizophrenia (1960)
AMA Archives of General Psychiatry 2, pp. 447–91.

Bateson, Gregory Redundancy and Coding (1968)
In: *T A Seboek (Ed) Animal Communication; techniques of study and results of research* Chapter 22, Bloomington, Indiana University Press.

Batson, Daniel and Ventis, W Larry The Religious Experience: a social-psychological perspective (1982)
New York, Oxford University Press.

Battista, J The Holographic Model, Holistic Paradigm, Information Theory and Consciousness
In: *ReVision* I, 3/4, 1978.

Battleheim, Bruno The Empty Fortress: infantile autism and the birth of self (1967)
New York, Free Press, 484 p. bibl. (pp. 461–68).

Bauman, E, et al The New Holistic Health Handbook: living well in a new age (1985)
Lexington MA, The Stephen Greene Press. Edited by S Bliss.

Bay, C Human Development and Political Orientations (1970)
Bulletin of Peace Proposals, 1, pp 177–186.

Baynes, K Bohman J and McCarthy, T (Eds) After Philosophy: end or transformation? (1987)
Cambridge MA, MIT Press.

Beahrs, J O Unity and Multiplicity: multilevel consciousness of self in hypnosis, psychiatric disorder and mental health (1982)
New York, Brunner/Mazel.

Beardsworth, Timothy A Sense of Presence (1977)
Oxford, Religious Experience Research Unit.

Beck, Carlton E Synergy and Counseling (1973)
Counseling and Values, 18, 1, February pp 7–43.

Becker, Ernest Birth and Death of meaning (1962)
New York, Free Press.

Becker, Ernest Revolution in Psychiatry (1964)
New York, Free Press.

Becker, R and Selden, G The Body Electric: electromagnetism and the foundation of life (1985)
New York, William Morrow.

Beckett, L C Neti-Neti (1959)
JM Watkins.

Beckford, James A (Ed) New Religious Movements and Rapid Social Change (1986)
Paris, UNESCO, 247 p. ISBN 92–3–102402–7.

Bee, H L The Journey of Adulthood (1987)
New York, Macmillan.

Beesing, Maria et al The Enneagram: a journey of self discovery (1984)
Denville NJ, Dimension Books.

Begbie, T Twice Born Men (1909)
Old Tappan NJ, Fleming O Revell.

Beha, Ernest A Comprehensive Dictionary of Freemansonry (1962)
New York, Arco Publications.

Behanan, K T Yoga: a scientific evaluation (1937)
New York, Macmillan.

Belenky, M, et al Women's Ways of Knowing: self, mind and voice (1986)
New York, Basic Books.

Bell, Joseph Norment Love Theory in Later Hanbalite Islam (1979)
Albany NY, State University of New York Press, 280 p. ISBN 0–87395–244–8.

Bellah, R N Beyond Belief: essays on religion in post-traditional world (1976)
New York, Harper and Row.

Bellah, Robert Beyond Belief (1970)
New York, Allied Publications.

Bendit, L J and Bendit, P D The Transforming Mind (1970)
London, Theosophical Publishing House, 160 p. bibl. (pp. 150–60).

Benedek, Therese Insight and Personality Adjustment (1948)
New York, Ronald Press.

Benedict, R Unpublished Lectures on 'Synergy in Society' (1942)
Bryn Mawr PA, Bryn Mawr Commentaries.

Benedict, R Patterns of Culture (1946)
New York, Mentor.

Benner, David G Psychotherapy and the Spiritual Quest: examining the links between psychological and spiritual health (1988)
London, Hodder and Stoughton, 173 p. ISBN 0–340–50114–6.

Bennett, Frank S M A Soul in the Making: a psychosynthesis (1926)
London, Simpkin Marshall, 116 p. bibl.

Bennett, J G The Dramatic Universe (1956)
London, Hodder and Stoughton, 3 vols.

Bennett, J G Enneagram Studies (1983)
York Beach ME, Samuel Weiser.

Bennett, J G, et al The Spiritual Hunger of the Modern Child (1984)
Charles Town WV, Claymont Communications, 220 p.

Bennett, John G Total Man: an essay in the systematics of human nature (1964)
Systematics 1, 4, March.

Bennett, Louisa and Swainson, Mary Psychic Sense: training and developing psychic sensitivity (1990)
Atlanta GA, Quantum, 160 p.

Bennis, W G and Schein E H (Eds) Leadership and Motivation: essays of Douglas McGregor (1968)
Cambridge MA, MIT Press, 286 p.

Benoit, Hubert and Huxley, Aldous The Supreme Doctrine: psychological studies in zen thought (1984)
Rochester VT, Inner Traditions International, 272 p. ISBN 0–89281–058–0.

Bentov, I Stalking the Wild Pendulum: on the mechanics of consciousness (1977)
New York, E P Dutton.

Berdyaev, N The Destiny of Man (1960)
New York, Harper and Row.

Berelson, B and Steiner, G T Human Behavior: an inventory of scientific findings (1964)
New York, Harcourt Brace, 712 p.

Berenda, C W World Visions and the Image of Man: cosmologies as reflections of man (1965)
New York, Vantage Press.

Berendt, Joachim–Ernst Nada Brahma: the world is sound; music and the landscape of consciousness (1988)
London, East West Publications, 258 p. Foreword by Fritjof Capra. Transl. by Helmut Bredigkeit. ISBN 0–85692–176–9.

Berg, Philip S Kabbalah for the Layman: a guide to cosmic consciousness

Berger, Peter L and Luckmann, Thomas The Social Construction of Reality: a treatise in the sociology of knowledge (1972)
New York, Doubleday.

Berkeley, George and Warnock, G J The Principles of Human Knowledge and Three Dialogues Between Hylas and Philonous
Magnolia MA, Smith Peter Publisher. ISBN 0–8446–1667–2.

Berkowitz, L and Macauley, J (Eds) Altruism and Helping Behavior (1970)
New York, Academic Press.

Berman, Morris The Reenchantment of the World (1981)
Cornell University Press.

Berman, Morris Coming to Our Senses: body and spirit in the hidden history of the West (1990)
New York, Bantam Books, 425 p. ISBN 0–553–34863–9.

Bernard, Harold W Human Development in Western Culture (1970)
Boston MA, Allyn and Bacon, 642 p. 3rd ed. bibl. (pp. 595–626).

Bernard, Jessie The Female World (1981)
New York, Free Press.

Berne, E Games People Play (1964)
New York, Grove Press.

Berne, P; Savary, L and Williams, S Dreams and Spiritual Growth: a christian approach to dreamwork (1984)
New York, Paulist Press.

Berrill, N J Man's Emerging Mind: the story of man's progress through time (1965)
Greenwich CT, Fawcett Publications, 240 p.

Besant, Annie World Problems of Today: a series of lectures (1925)
London, Theosophical Publishing House, 144 p.

Bettelheim, Bruno Informed Heart: autonomy in a mass age (1960)
New York, Free Press.

Bettelheim, Bruno Dialogues with Mothers (1962)
New York, Free Press.

Bettelheim, Bruno Symbolic Wounds: puberty rites and the envious male (1962)
New York, Macmillan.

Betty, Stafford L Vadirā's Refutation of Sankara's Non-Dualism: clearing the way for Theism (1978)
Delhi, Motilal Banarsidass.

Beyer, Stephan The Buddhist Experience: sources and interpretations (1974)
Encino–Belmont CA, Dickenson.

Beyer, Thomas P The Integrated Life (1948)
Minneapolis MN, University of Minnesota Press, 190 p. Essays, sketches and poems.

Bezzola, D Des procédés propores à réorganiser 'La Synthèse Mentale' dans le traitement des névroses (1908)
Revue de Psychiatrie Cahors, Paris.

Bhagyalakshmi, S Facets of Spirituality: dialogues and discourses of Swami Krishnanda (1986)
Delhi, Motilal Banarsidass, 292 p. ISBN 81–208–0087–7.

Bhatnagar, R S Dimensions of Classical Sufi Thought (1984)
London, East West Publications, 241 p. ISBN 0–85692–146–7.

Bhikkhu Bodhi (Trans) The Discourse on the All-Embracing Net of Views: the Brahmajala Sutta and its commentarial exegesis (1978)
Kandy, Buddhist Publication Society, 359 p.

Bhikshu, Ven Sumedho Handbook for the Practice of Dhamma: being the method of recollection at the time of death and for those who are interested to go further

Biddle, William E Integration of Religion and Psychiatry (1955)
New York, Macmillan, 171 p.

Bielefeldt, Carl Dogen's Manuals of Zen Meditation (1988)
Berkeley CA, University of California Press, 336 p. ISBN 0–520–06056–3.

Bieri, J Cognitive complexity – simplicity and predictive behavior (1955)
Journal of Abnormal and Social Psychology 51.

Bieri, J Complexity: simplicity as a personality variable in cognitive and preferential behavior (1961)
In: *D W Fiske and S Maddi (Ed) Functions of Varied Experience* Dorsey.

Binswanger, Ludwig Being–in–the–World: selected papers of Ludwig Binswanger (1963)
New York, Harper and Row. Trans. by Jacob Needleman from German.

Biofeedback Research Society

Biofeedback Research Society Biofeedback and Self-Regulation (1973)
Denver, University of Colorado Medical College, (800 bibliog. references).

Birdson, Robert E Four–Dimensional Values and Their Attainment (1981)
Eureka CA, Sirius Books, 17 p. Aquarian Academy Monograph: E5. ISBN 0–917108–33–7.

Birdsong, Robert E Paths to Human Perfection (1979)
Eureka CA, Sirius Books. Aquarian Academy Supplementary Lecture: No 3. ISBN 0–917108–26–4.

Birren, Faber Color Psychology and Color Therapy (1961)
New Hyde Park, University Books.

Bissing, Hurbert Songs of Submission: on the practice of subud (1982)
Greenwood SC, Attic Press, 180 p. ISBN 0–227–67852–4.

Bitter, Wilhelm (Ed) Evolution: fortschrittsglaube und heilserwartung (1970)
Stuttgart, Ernst Klett Verlag.

Bittle, Celestine The Whole Man: psychology (1945)
Milwaukee, Bruce Pub, 687 p. bibl. (pp. 645–61).

Bjerre, P Von der Psychanalyse zur Psychosynthese (1925)
Hall.

Bjerre, P Psychosynthèse (1971)
Stuttgart, Hippokrater Verlag, 111 p. (Schriftenreihe zur Theorie und Praxis der medizinischen Psychologie, Bd 20).

Bjorkman, Rut Spiritual Evolution: light from another dimension
Authorized trans. from German into English by Tim Nevill.

Black, Michael Poetic Drama: as mirror of the will (1977)
London, Vision Press, 203 p. ISBN 0–85478–074–2.

Blackburn, Albert Now–Consciousness: exploring the world beyond thought

Blakeslee, Thomas The Right Brain: a new understanding of the unconscious mind and its creative powers (1980)
New York, Doubleday.

Blanco, M The Unconscious as Infinite Sets (1975)
London, Duckworth.

Blau, Abram The Master Hand: a study of the origin and meaning of right and left
New York, AMS Press. ISBN 0–404–60854–X.

Bleakley, Alan Earth's Embrace: archetypal psychology's challenge to the growth movement: facing the shadow of the new age (1989)
Bath, Gateway Books, 232 p. ISBN 0–946551–40–5.

Bletzer, June G The Donning International Encyclopedic Psychic Dictionary (1987)
Norfolk VI, The Donning Publishers, 875 p. 2nd ed. bibl. (pp. 857–875). ISBN 0–89865–371–1.

Bliss, Eugene L Multiple Personality, Allied Disorders and Hypnosis (1986)
New York, Oxford University Press, 300 p. ISBN 0–19–503658–1.

Block, N and Fodor, J A What Psychological States Are Not
In: *Philosophical Review* 1972, 81, pp. 159–181.

Blofeld, J The Tantric Mysticism of Tibet (1970)
New York, EP Dutton.

Blofeld, John (Trans) The Zen Teaching of Hui Hai on Sudden Illumination

Blofeld, John Gateway to Wisdom: Taoist and Buddhist contemplative and healing yogas adapted for Western students of the way

Bloomfield, Harold; Cain, Michael and Jaffe, Dennis TM, Discovering Inner Energy and Overcoming Stress (1975)
New York, Delacorte Press.

Blumenthal, David R God at the Center: meditations on Jewish Spirituality (1988)
New York, Harper and Row, 192 p. ISBN 0–06–254839–5.

Boadella, D Lifestreams: an introduction to biosynthesis (1987)
London, Routledge.

Boase, Roger The Origin and Meaning of Courtly Love (1977)
Manchester, Manchester University Press.

Bobgan, D and Bobgan, M The Psychological Way/The Spiritual Way (1979)
Minneapolis, Bethany Fellowship.

Bode, B H Education as Growth: some conclusions (1937)
Progressive Education 14, March pp. 151–57.

Boehme, Jacob Dialogue on the Supersensual Life (1957)
New York, Frederick Ungar. Trans. William Law et al.

Boerstler, Richard W Letting Go: a holistic and meditative approach to living and dying (1982)
Assocs Thanatology, 112 p. illus. ISBN 0–9607928–0–5.

Boggs, S W Atlas of Ignorance: a needed stimulus to honest thinking and hard work (1949)
American Philos. Society Proceedings 93, 3, pp. 253–58, bibl. foot.

Bois, J S Explorations in Awareness (1957)
New York, Harper and Row.

Bois, J Samuel Breeds of Man: toward the adulthood of humankind (1970)
New York, Harper and Row, 174 p.

Bois, Samuel The Art of Awareness: a textbook on general semantics (1966)
Dubuque IA, William C Brown.

Boisel, A T The Exploration of the Inner World: a study of mental disorder and religious Experience (1936)
Philadelphia PA, University of Pennsylvania, 1971. (Willett, Clark and Co, 1936).

Bolen J S The Tao of Psychology: synchronicity and the self (1982)
New York, Harper and Row.

Bolen, Jean S The Tao of Psychology: synchronicity and the self (1979)
New York, Harper and Row.

Bolen, Jean S Goddesses in Everywoman: a new psychology of women (1985)
New York, Harper and Row.

Bolen, Jean S Gods in Everyman: a new psychology of men's lives and loves (1989)
New York, Harper and Row.

Bolles, Richard N What Color is Your Parachute (1987)
Berkeley CA, Ten Speed Press, 416 p. ISBN 0–89815–176–7.

Bondreau, L Transcendental Meditation and Yoga as Reciprocal Inhibitors
Journal of Behavior Therapy and Experimental Psychiatry 3, 2.

Bonewitz, R Cosmic Crystals: crystal consciousness and the new age (1983)
Wellingborough, Turnstone Press.

Bonewitz, R The Cosmic Crystal Spiral: crystals and the evolution of human consciousness (1986)
Longmead, Element Books.

Borsodi, Ralph Seventeen Problems of Man and Society (1968)
Anand, Charotar Book Stall, 595 p. bibl.

Bortoft, Henri Goethe's Scientific Consciousness (1986)
Tunbridge Wells, The Institute for Cultural Research, 886 p. ICR Monograph Series: 22. ISBN 0–904674–10–X.

Bose, D N Tantras: their philosophy and occult secrets (1981)
Columbia MO, South Asia Books. rev 3rd ed. ISBN 0–8364–0737–7.

Boss, Medard Psychoanalysis and Daseinsanalysis (1963)
New York, Basic Books.

Boulding, Elise The Underside of History: a view of women through time (1976)
Boulder CO, Westview, 829 p. illus.

Boulding, Elise Building a Global Civic Culture: education for an Interdependent World (1988)
New York, Teachers College Press, Columbia University, 192 p. ISBN 0–8077–2867–5.

Bouny, Helen and Savary, Louis Music and Your Mind: listening with a new consciousness (1973)
New York, Harper and Row.

Bouyer, L Introduction to Spirituality (1961)
Collegeville MN, Liturgical Press. Trans. by Markey Perkins Ryan.

Bouyer, Louis Orthodox Spirituality and Protestant and Anglican Spirituality (1982)
New York, Harper and Row, 232 p. A History of Christian Spirituality Series: 3. ISBN 0–8164–2374–1.

Bouyer, Louis, et al A History of Christian Spirituality
3 Vols. Vol I: The Spirituality of the New Testament and the Fathers, Vol II: The Spirituality of the Middle Ages, Vol III: Orthodox Spirituality and Protestant and Anglican Spirituality.

Boveda, Xavier Integracin del Hombre (Poemas en Profundidad) (1934)
Buenos Aires, Libreria del Colegio Cabaut, 110 p.

Bowers, C A Curriculum as Cultural Reproduction: an examination of metaphor as a carrier of ideology
In: *Teachers College Record* 1980, 82, 2, pp. 267–289.

Bowker, John Problems of Suffering in the Religions of the World (1970)
Cambridge MA, Cambridge University Press, 318 p. ISBN O–521–07412–6.

Bowker, John The Religious Imagination and the Sense of God (1978)
Oxford, Clarendon Press.

Boy, Angelo V and Pine, Gerald J Expanding the Self: personal growth for teachers (1971)
Brown, bibl. pp. 120–24.

Boydell, Tom Management Self-Development: a guide for managers, organizations and institutions (1986)
Geneva, ILO, 267 p. ISBN 92–2–103958–7.

Brace, C The Stages of Human Evolution (1967)
Englewood Cliffs NJ, Prentice Hall.

Bradburn, Norman M The Structure of Psychological Wellbeing (1969)
Chicago IL, Aldine.

Bradford, Leland P; Gibb, Jack and Benne, K T-Group Theory and Laboratory Method: innovation in re-education (1964)
New York, John Wiley and Sons.

Brakenhelm, C R Problems of Religious Experience (1985)
Philadelphia PA, Coronet Books, 158 p. ISBN 91–554–1657–8.

Brame, Grace Receptive Prayer (1985)
St Louis, CPB Press.

Brame, Grace A Receptive Prayer: a Christian approach to meditation (1985)
Saint Louisville MO, CBP Press, 144 p. ISBN 0–8272–3211–X.

Brams, S J Superior Beings, If They Exist, How Would We Know?: game–theoretic implications of omniscience, omnipotence, immortality, and incomprehensibility (1983)
Berlin, Springer–Verlag, 175 p. ISBN 3–540–90877–3.

Brand, L W and W G Preliminary Explorations of Psi–conducive States: progressive muscular relaxation (1973)
Journal of the American Society Psychical Research Vol 67, No 1, pp. 26–46.

Branden, Nathaniel The Psychology of Self–Esteem (1971)
New York, Bantam Books.

Brar, H S Yoga and Psychoanalysis (1970)
British Journal of Psychiatry p. 116.

Brazelton, Thomas Berry On Becoming a Family: the growth of attachment (1981)
New York, Delacorte Press.

Bregman, L Spiritual Dimensions to Psychotic Experience
In: *The Journal of Transpersonal Psychology* 1979, 11, pp. 59–74.

Bregman, L The Rediscovery of Inner Experience (1982)
Chicago IL, Nelson–Hall.

Bricklin, Mark The Practical Encyclopedia of Natural Healing (1983)
Emmaus PA, Rodale Press, 592 p. illus. rev ed. ISBN 0–87857–480–8.

Bridges, William Transitions (1980)
Reading MA, Addision–Wesley.

Briggs, John Fire in the Crucible: the alchemy of creative genius (1988)
New York, St Martin's Press.

Refs #H0295.

Brodic, E and Philbrook, T Individual Growth and Development and National Policy: particularly of a socio–economic kind (Paper at the World Congress of World Federation for Mental Health, Copenhagen 1975) (1975)

Bronowski, J The Identity of Man (1966)
London, Heinemann. not all experience is gained by observing nature – relations with people are more revealing.

Brosse, T Etudes instrumentales des techniques du yoga (1963)
Paris.

Brown, D The Stages of Meditation in Cross–Cultural Perspective
In: *Brown, D; Engler, J and Wilber, K; Transformations of Consciousness: conventional and contemplative perspectives on development* Boston, Shambhala.

Brown, D P A Model for the Levels of Concentrative Meditation
In: *International Journal of Clinical and Experimental Hypnosis* 1977, 25, pp. 236,273.

Brown, D P Mahamudra Meditation: stages and contemporary cognitive psychology (1981)
Chicago IL, University of Chicago Press. Doctoral dissertation.

Brown, D P and Engler, Jack The Stages of Mindfulness Meditation: a validation study
In: *Journal of Transpersonal Psychology* 1980, 12, pp. 143–192.

Brown D P; Forte M and Dysart M Differences in Visual Sensitivity Among Mindfulness Meditators and Non–Meditators
In: *Perceptual and Motor Skills* 1984, 58, pp. 227–233.

Brown, F, Stewart, W and Blodjett, J EEG Kappa Rhythm during Transcendental Meditation and Possible Threshold Changes Following (1971)
(Paper presented to the Kentucky Academy of Science, 13 November 1971).

Brown, G Now: the human–dimension (A report on the Ford–Esalen Project for Innovation in Humanistic Education) (1968)
Esalen Monograph, 1.

Brown, R and Motoyama, H Science and the Evolution of Consciousness (1978)
Brookline MA, Autumn Press.

Brown, S Supermind: the ultimate energy (1980)
New York, Harper and Row.

Browning, Don S Generative Man: psychoanalytic Perspectives: Society and The Good Man in the Writings of Philip Rieff, Norman Brown, Erich Fromm and Erik Erikson (1975)
New York, Dell Publishing Company, 266 p.

Brunton, Paul The Wisdom of the Overself (1970)
New York, Samuel Weiser.

Bruteau, Beatrice Evolution toward Divinity: Teilhard de Chardin and the Hindu tradition (1974)
Wheaton IL, Theosophical Publishing House.

Buber, Martin Pointing the Way
London, Routledge and Kegan Paul.

Buber, Martin Knowledge of Man: selected essays
New York, Harper and Row.

Buber, Martin I and Thou (1958)
New York, Scribner.

Buber, Martin Between Man and Man (1965)
New York, Macmillan.

Buber, Martin Knowledge of Man (1965)
New York, Humanities Press.

Buber, Martin Between Man and Man (1965)
New York, Macmillan, 168 p. Trans. by Ronald Gregor Smith.

Buber, Martin The Knowledge of Man: a philosophy of the interhuman (1966)
New York, Harper Torchbooks, 149 p.

Buckarin, N Historical Materialism: a system of sociology (1925)
New York, Int Publishers.

Bucke, Richard M A Study in the Evolution of the Human Mind (1901)

Bucke, Richard M Cosmic Consciousness: a study in the evolution of the human mind (1972)
New York, Olympia Press, 286 p. reprint of 1901 ed.

Buddhaghosa, Bhadantacariya The Path of Purity: being a translation of Buddhaghosa's Visuddhimagga (1975)
London, Pali Text Society, 907 p. (Ist ed. 1923) Trans. from the Pali by Pe Maung Tin. ISBN 0–7100–8218–5.

Buddhaghosa, Bhadantacariya Visuddhimagga: the path of purification (1976)
Boulder CO, Shambhala Publications. 2 Vols. Trans. from the Pali by Bhikku Nyanamoli

Buddhaghosa, Bhadantacariya The Path of Purification: Visuddhi Magga (1980)
Kandy, Buddhist Publication Society. Trans. from the Pali by Bhikku Nanamoli.

Bugental, J F T The Search for Authenticity: an existential–analytic approach to psychotherapy (1965)
New York, Holt, Rinehart and Winston.

Bugental, J F T Challenges of Humanistic Psychology (1967)
New York, McGraw Hill Book Company.

Bugental, J F T The Self: process or illusion (1971)
In: *T C Greening (Ed), Existential Humanistic Psychology* Belmont, Brooks/Cole, pp57–71.

Bühler, Charlotte Der Menschliche Lebenslauf als Psychologisches Problem (1933)
Leipzig, Verlag von Hirzel. 1933, revised, Bonn, Hogrefe, 1959.

Bühler, Charlotte From Birth to Maturity: an outline of the psychological development of the child (1935)
London, Kegan Paul.

Bühler, Charlotte Maturation and Motivation (1951)
Dialectica 5, p. 312–61.

Bühler, Charlotte Values in Psychotherapy (1962)
New York, Free Press.

Bühler, Charlotte Psychology for Contemporary Living (1969)
New York, Hawthorn Books.

Bühler, Charlotte and Massarik, Fred (Eds) The Course of Human Life: a study of goals in the humanistic perspective (1968)
New York, Springer Publishing.

Bukin, V R Religious Emotions and the Place They Hold in Believers' Consciousness
In: *Voprosii Filosofii* 1969, 23 (11), pp. 57–66.

Bullard, Dexter M (Ed) Psychoanalysis and Psychotherapy: selected papers of Frieda Fromm–Reichmann (1959)
Chicago IL, University of Chicago Press.

Bunge, Mario Do the Levels of Science Reflect the Levels of Being? (1959)
In: *Metascientific Queries* Springfield, Charles Thomas, Chap 5.

Burckhardt, Titus Introduction to Sufi Doctrine (1959)
Lahore.

Burckhardt, Titus Alchemy: science of the cosmos, science of the soul (1967)
London, Stuart and Watkins.

Burckhardt, Titus Alchemy (1971)
Baltimore MD, Penguin Books.

Burckhardt, Titus Mirror of the Intellect: essays on traditional science and sacred art (1987)
Cambridge, Quinta Essentia, 269 p. bibl. (pp. 255–262). Trans. and ed William Stoddart. ISBN 0–946621–08–X.

Burckhardt, Titus Insight into Alchemy (1987)
Cambridge, Quinta Essentia, in *Mirror of the Intellect: Essays on Traditional Science and Sacred Art*, p 269. Trans. and edited by Wiliam Stoddart. ISBN 0–946621–08.

Burger, Hanry G Syncretism, an Acculturative Accelerator (1966)
Human Organization 24, pp. 103–15.

Burkes, R G Ecological Psychology (1968)
Stanford CA, Stanford University Press.

Burnham, W H The Wholesome Personality (1932)

Burrow, G Trigant Sciences and Man's Behavior: the contribution of phylobiology (including the Nemosis of Man) (1953)
New York, Philosophical Library.

Burrow, T Preconscious Foundations of Human Experience (1964)
Free Press. Ed by W E Galt.

Burton, Arthur (Ed) Encounter Groups (Behavioral Sciences Service) (1969)
San Francisco CA, Jossey–Bass Publishers.

Bury, J B The Idea of Progress (1970)
London, Macmillan.

Butler, Bryant O A Synthesis on Human Nature and Life (1933)
Rahway NJ, Privately Printed, 85 p.

Byrd, B Cognitive Needs and Human Motivation
Unpublished.

Byrne, Edmund F and Maziarz, E A Human Being and Being Human: man's philosophies of man (1969)
New York, Appleton–Century–Crofts, 427 p. bibl.

Byrne, Lavinia (Ed) Traditions of Spiritual Guidance (1990)
London, Geoffrey Chapman Cassell Publishers, 213 p. ISBN 0–225–66616–2.

Cade, Maxwell C and Coxhead, Nona The Awakened Mind: biofeedback and the development of higher states of awareness (1979)
UK, Wildwood House.

Cahn, Steven M Fate, Logic and Time (1986)
Atascadero CA, Ridgeview Publishing, 150 p. ISBN 0–917930–76–2.

Caillois, Roger Unity of Play: diversity of games (1957)
Diogenes 19, pp. 92–121.

Caillois, Roger Man, Play and Games (1961)
New York, Free Press.

Caldwell, Lynton K Health and Homeostasis as Social Concepts: an exploratory essay (1969)
In: *Diversity and Stability in Ecological Systems* (Brookhaven Symposia in Biology, 22, 1969).

Calhoun, J B Promotion of Man (1970)
In: *E O Attinger (Ed) Global Systems Dynamics* New York, Wiley Interscience, 1970 pp. 36–58 (Basel, Karger, 1970).

Campbell, A The Seven States of Consciousness (1974)
New York, Harper and Row.

Campbell, Anthony Seven States of Consciousness (1973)
London, Gollancz.

Campbell, D T Variation and Selective Retention in Socio–Cultural Evolution (1965)
Cambridge MA, Schenkman Publishing, pp. 19–41. Social Change in Developing Areas.

Campbell, J The Masks of God (1968)
New York, Viking Press. 4 vols.

Campbell, Joseph The Inner Reaches of Outer Space: metaphor as myth and as religion

Campbell, Joseph The Hero with a Thousand Faces (1970)
New York, World Publishing Co.

Campbell, Joseph (Ed) Spiritual Disciplines: papers from the Eranos Yearbooks (1985)
New York, Bollingen Foundation, 506 p. illus. Bollingen Series XXX: 4. Selected and transl. from the Eranos–Jahrbücher ed by Olga Froebe–Kapteyn. Transl. by Ralph Manheim except for the paper by C G Jung which was transl. by R F C Hull. ISBN 0–691–01863–4.

Campbell, Joseph Myths to Live By (1985)
London, Paladin Books, 225 p. ISBN 0–586–08528–9.

Canadian Ecophilosophy Network The Trumpeter: Journal of Ecosophy (1989)
Victoria BC, Canada, LightStar Press, Vol 6, No 4.

Cannon, W B The Wisdom of the Body (1963)
New York, WW Norton, 333 p.

Canseliet, Eugène Alchimie: études diverses de symbolisme hermétique et de pratique philosophale
Paris, Jean–Jacques Pauvert, 288 p.

Cantore, Enrico Scientific Man: the humanistic significance of science
New York, Learned Publications. in press.

Cantore, Enrico Holistic Development of Man in the Scientific–Technological Age: a philosophical contribution (Paper for the Rome Special Conference on Futures Research 1973) (1973)

Cantril, H and Bumstead, C Reflections on the Human Venture (1960)
New York, New York University Press.

Capellanus, Andreas The Art of Courtly Love (1964)
New York, Ungar. Trans. by John Jay Parry.

Carlile, Richard Manual of Freemansonry (1845)
London, Wm Reeves.

Carlson, Rae Where is the Person in Personality Research? (1971)
Psychological Bulletin 75, 3, pp. 203–19.

Carmody, Denise L Seizing the Apple: a feminist spirituality of personal growth (1984)
New York, Crossroad Publishing, 176 p. ISBN 0–8245–0652–9.

Carmody, John Ecology and Religion: toward a new Christian theology of nature (1983)
New York, Paulist Press.

Carney, T R No Limits to Growth: mind expanding techniques
Winnipeg, Harbeck and Associates.

Carpenter, J T Meditation, Esoteric Traditions: contributions to psychotherapy
In: *American J Psychotherapy* 1977, 31, pp. 394–404.

Carreiro, Mary E The Psychology of Spiritual Growth (1987)
Granby MA, Bergin and Garvey, 160 p. Gentle Wind Books: 1. ISBN 0–89789–124–4.

Carrington, P Freedom in Meditation (1977)
New York, Anchor.

Carson, R The Sense of Wonder (1965)
New York, Harper and Row.

Carter, John Ross Dhamma: western academic and Sinhalese Buddhist interpretations: A study of a religious concept (1978)
Tokyo: The Hokuseido Press.

Caruso, I A Psychoanalyse und Synthese der Existenz (1952)
Vienna, Seelsorger–Verlag Herder.

Case, Charles J Beyond Time: ideas of the great philosophers on eternal existence and immortality (1985)
Lanham MD, University Press of America, 144 p. ISBN 0–8191–4933–0.

Casey, Edward S Imagining: a phenomenological study (1976)
Bloomington, Indiana University Press.

Cassim, K M P Mohamed Sufism: an exploration of inner space

Cassirer, E Individual and The Cosmos in Renaissance Philosophy
New York, Harper Torchbook.

Cassirer, E An Essay on Man (1944)
New Haven CT, Yale University Press.

Cassirer, E Essay on Man: an introduction to a philosophy of human culture (1945)
New Haven CT, Yale University Press. 3rd imp.

Cassirer, E Language and Myth (1946)
New York, Dover Publications.

Cassirer, E The Logic of the Humanities (1961)
New Haven CT, Yale University Press.

Cassirer, E J The Philosophy of Symbolic Forms (1953)
New Haven CT, Yale University Press. 3 Vols.

Castaneda, Carlos The Teachings of Don Juan: a Yaqui way of knowledge (1968)
Berkeley CA, University of California Press. (Ballantine, 1971).

Castaneda, Carlos A Separate Reality (1971)
New York, Simon and Schuster International.

Castaneda, Carlos Journey to Ixtlan (1972)
New York, Simon and Schuster International.

Castilejo, Claremont De Knowing Woman: a Feminine Psychology (1973)
New York, C J Jung Foundation, 178 p.

Caussade, Pierre de The Sacrament of the Present Moment (1981)
Glasgow, Collins, Fount Paperbacks, p 128. Trans. by Kitty Muggeridge from the original test of the treatise on self-abandonment to divine providence. ISBN 0-00-625545-0.

Causton, Richard Nichiren Shoshu Buddhism: an Introduction (1988)
London, Rider, 299 p. ISBN 0-7126-2269-1.

Center for the Study of Social Policy Changing Images of Man (1974)
Menlo Park, Stanford Research Institute, 319 p. Policy Research Report: 4.

Centre d'Etudes Propectives L'homme encombré (1969)
Presses universitaires de France.
1-151.

Centre international de synthèse L'Unité de l'être
Paris, Ed A Michel.

Chakrabarty, Surath Mysterious Samadhi: a Study on Indian philosophy (1984)
Calcutta, Firma KLM.

Chakravarti, Surath Mysterious Samadhi

Chandrkaew Chinda Nibbana: the ultimate truth of Buddhism

Chang, G C C The Buddhist Teaching of Totality (1971)
University Park PA, The Pennsylvania State University Press.

Chang, Po-tuan The Inner Teachings of Taoism (1986)
Boston, Shambhala, 118 p. Commentary by Liu I-ming. Trans. by Thomas Cleary. ISBN 0-87773-363-5.

Chang, Po-tuan Understanding Reality: a Taoist alchemical classic (1987)
Honolulu, University of Hawaii Press, 203 p. Commentary by Liu I-ming. Trans. by Thomas Cleary. ISBN 0-8248-1139-9.

Charon, Jean E Man in Search of Himself (1967)
New York, Walker Publishing Company, 210 p. bibl. foot. (translated).

Chatterji, Priti Bhusan Towards Supermanhood: the philosophy of Sri Aurobindo (1977)
Bombay, Progressive Corporation, 96 p.

Chattopadhyaya, Deviprasad Individuals and Society: a methodological inquiry (1967)
New York, Allied Publications, 196 p. bibl. foot. (individualism and holism).

Chaudhuri, Haridas Integral Yoga: the concept of harmonious and creative living (1965)
London, Allen and Unwin 1965, 160p. Foreword by P A Sorokin. bibl. foot.

Chaudhuri, Haridas Philosophy of Meditation (1965)
New York, Philosophical Library.

Chaudhuri, Haridas The Philosophy of Meditation (1965)
New York, Philosophical library.

Chaudhuri, Haridas The Evolution of Integral Consciousness (1977)
Wheaton IL, Quest Books.

Chaudhuri, Haridas and Spiegelberg, F (Ed) The Integral Philosophy of Sri Aurobindo: a commemorative symposium (1960)
London, Allen and Unwin, 350 p. bibl.

Cheng, Man-ch'ing and Smith, Robert W T'ai Chi (1967)
Rutland VT, Charles E Tuttle.

Chermayeff, Serge and Tzonis, A Shape of Community: realization of human potential (1971)
London, Penguin Books, 247 p. bibl. foot.

Chesser, E and Meyer, V Behaviour Therapy in Clinical Psychiatry (1970)
London.

Chhaganlal, Lala Bhakti in Religions of the World
With special reference to Dr Sri Bankey Behariji.

Chia, M Awaken Healing Energy Through the Tao (1983)
New York, Aurora Press.

Childe, C Social Evolution (1951)
London, Watts.

Childe, Gordon V Man Makes Himself (1957)
New York, Mentor.

Chimnoy, Sri The Inner Promise, Paths to Self-Perfection (1970)
New York, Simon and Schuster International.

Chirban, J Developmental Stages in Eastern Orthodox Christianity
In: *Brown, D; Engler, J and Wilber, K; Transformations of Consciousness: conventional and contemplative perspectives on development* Boston, Shambhala.

Chirban, J T Human Growth and Faith: intrinsic and extrinsic motivation in human development (1981)
Washington DC, University Press of America.

Chitrabhanu, G S The Psychology of Enlightenment: meditation on the seven energy centers (1979)
New York, Dodd-Mead.

Chittick, William C The Sufi Path of Love: the spiritual teachings of Rumi (1984)
Albany NY, State University of New York Press, 433 p. ISBN 0-87395-723-7.

Chittick, William C The Sufi Path of Knowledge: ibn al-arabi's metaphysics of imagination (1989)
Albany NY, State University of New York Press, 768 p. ISBN 0-88706-884-7.

Choisy, M La métaphysique des yogas (1948)
Geneva.

Chung-yuan Chang Self Realization and the Inner Process of Peace (1956)
In: *Eranos Jahrbuch* Zurich, Rhein-Verlag.

Clark, Gordon H The Atonement (1987)
Jefferson MD, Trinity Foundation, 175 p. The Trinity Papers: No 17. ISBN 0-940931-17-6.

Clark, J H The Varieties of Ineffability: proceedings of a colloquy of European psychologists of religion (1979)
Nijmegen, Katholieke Universiteit.

Clark, John H A Map of Mental States (1983)
London, Routledge and Kegan Paul, 242 p. illus. bibl. ISBN 0-7100-9235-0.

Clark, Walter Houston Chemical Ecstasy: psychedelic drugs and religion (1969)

Claxton, Guy (Ed) Beyond Therapy: the impact of Eastern religions on psychological theory and practice (1986)
Newburyport MA, Wisdom Publications, 352 p. illus. Wisdom East-West Book, Grey Series. ISBN 0-86171-043-6.

Cleary, Thomas (Trans) The Book of Balance and Harmony (1989)
London, Rider, 153 p. ISBN 0-7126-3251-1.

Clow, B Eye of the Centaur: a visionary guide into past lives (1986)
St Paul MN, Llewellyn Publications.

Clyne, Manfred Sentics: the touch of emotions (1989)
Houston TX, Prism.

Coelho, Mary C and Neufelder, Jerome N (Eds) Writings on Spiritual Direction by Great Christian Writers (1982)
New York, Harper and Row, 224 p. ISBN 0-8164-2420-9.

Coffey, H S Socio and Psyche Group Process: integrative concepts (1952)
Journal of Social Issues 8, Spring pp. 65-74.

Cofman, Victor Apostles of Integration (1964)
Systematics Vol II, May.

Cohen, John Humanistic Psychology (1962)
New York, Macmillan.

Cole, Colin A Asparsa Yoga: a study of Gaudapada's Mandukya karika (1982)
Delhi, Motilal Banarsidass, 158 p.

Cole, Sam Human Resources Development and Long-Term Forecasting (1989)
Paris, UNESCO, 79 p. Major Programme I, Reflection on World Problems and Future-Oriented-Studies: BEP/GPI/50.

Coleman, J E The Quiet Mind (1971)
London, Rider.

Colliander, Tito The Way of the Ascetics: the ancient tradition of discipline and inner growth

Collier, R W The Effect of Transcendental Meditation upon University Academic Attainment (1973)
Proceedings of the Pacific Northwest Conference on Foreign Languages (in press, 1973).

Collingwood, R G Speculum Mentis: or the map of knowledge (1963)
Oxford, Clarendon Press, 327 p. (first pub. 1924). Critical review of the chief forms of human experience, chapters on: art, religion, science, history and philosophy.

Collins, John E and Woods, Ralph (Eds) Civil Religion and Transcendent Experience: studies in theology and history, psychology and mysticism (1988)
Macon GA, Mercer University Press. Luce 3 Series. ISBN 0-86554-295-3.

Colodny, R G (Ed) Paradigms and Paradoxes (1972)
Pittsburgh PA, University of Pittsburgh Press.

Combs, A (Ed) Perceiving, Behaving, Becoming: a new focus for education (1962)
Washington DC, Association for Supervision and Curriculum Development.

Comfort, Alex Reality and Empathy: physics, mind and science in the 21st century (1984)
SUNY.

Commins, W D Some Early Holistic Psychologists (1932)
Journal of Philosophy, 29, pp. 208-217, April 14.

Committee for Development Planning Human Resources Development: a neglected dimension of development strategy, views and recommandations of the Committee for Development Planning (1988)
New York, United Nations, 45 p.

Committee on Cosmic Humanism Journal of Cosmic Humanism (Pittsburgh) (proposed)

Conant, Roger C and Ashby, W Ross Every Good Regulator of a System Must be a Model of That System (1970)
International Journal of Systems Science 1, pp. 89-97.

Congresso Nazionale di Filosofia L'Unificazione del Sapere: atti del congresse (Perugia 1965) (1967)
Firenze, G C Sansoni, 535 p. bibl. foot.

Conn, Joan W (Ed) Women's Spirituality: resources for Christian development (1986)
Mahwah NJ, Paulist Press, 336 p. ISBN 0-8091-2752-0.

Conway, David Ritual Magic (1978)
New York, E P Dutton.

Conze, E Buddhist Meditation (1972)
London, Unwin Hyman.

Conze, Edward (Ed) The Large Sutra on Perfect Wisdom: with the divisions of the Abhisamayālankāra (1984)
Berkeley LA, University of California Press, 679 p. ISBN 0-520-05312-4.

Cook, Francis Dojun How to Raise an Ox: Zen practice as taught in Zen Master Dogen's Shobogenzo

Cook, Patrck Self Esteem (1983)
North Sydney, Allen and Unwin, 128 p. ISBN 0-86861-292-8.

Coomaraswamy, Ananda K What is Civilisation?: and other essays (1989)
Golgonooza Press.

Cooper, D Psychiatry and Anti-Psychiatry (1967)
London, Tavistock Publications.

Corbin, Henry Creative Imagination in the Sufism of Ibn 'Arabi (1969)
Princeton NJ, Princeton University Press.

Corbin, Henry The Man of Light in Iranian Sufism (1978)
Boulder CO, Shambhala Publications, 174 p. ISBN 0-394-73441-6. Trans. by Nancy Pearson.

Cornia, Giovanni, et al (Eds) Adjustment with a Human Face: protecting the vulnerable and promoting growth (1987)
Oxford NY, Oxford University Press, 334 p. illus. Vol 1. ISBN 0-19-828610-4.

Cottrell, L S and Foote, N N Identity and Interpersonal Competence (1955)
Chicago IL, University of Chicago Press.

Coukoulis, Peter P Guru, Psychotherapist, and Self: a comparative study of the guru-disciple relationship and the Jungian analytic process

Coulson, William and Rogers Carl Man and the Science of Man
Columbus ON, Merrill Publishing.

Coulter, Art Synergetics: an adventure in human development (1976)
Englewood Cliffs NJ, Prentice-Hall.

Coulter, Arthur N Synergetics: an adventure in human development
Englewood Cliffs NJ, Prentice-Hall.

Council on Research in Bibliography The Developmental Sciences: a bibliographic analysis of a trend
In: *Meltal Health Book Review Index; an annual bibliography of books and book reviews in the behavioral sciences* New York, New York University Press, 1971, 90p. (vol. 16).

Count, E W Myth as world view: a biosocial synthesis (1960)
In: *S Diamond (Ed) Culture in History* New York.

Cousins, Ewert, et al (Eds) Classical Mediterranean Spirituality: Egyptian, Greek, Roman
New York, Crossroad/Continuum. World Spirituality Series: 15.

Cousins, Ewert, et al (Eds) Ancient Near Eastern Spirituality: Zoroastrian, Summerian, Assyro-Babylonian, Hittite
New York, Crossroad/Continuum. In preparation. World Spirituality Series: 12.

Cousins, Ewert, et al (Eds) Christian Spirituality: post-reformation and modern
New York, Crossroad/Continuum. In preparation. World Spirituality Series: 18.

Cousins, Ewert, et al (Eds) Dictionary of World Spirituality
New York, Crossroad/Continuum. In preparation. World Spirituality Series: 25.

Cousins, Ewert, et al (Eds) World Spirituality: an encyclopedic history of the religious quest
New York, Crossroad/Continuum. 25 Vols. Vols 13, 14, 15, 16, 17, 19 published. Vols 1, 2, 3, 4, 5, 6, 7, 8, 9, 10, 11, 12, 18, 20, 21, 22, 23, 24, 25 in preparation 1990.

Cousins, Ewert, et al (Eds) European Archaic Spirituality
New York, Crossroad/Continuum. In preparation. World Spirituality Series: 2.

Cousins, Ewert, et al (Eds) Confusian Spirituality
New York, Crossroad/Continuum. In preparation. World Spirituality Series: 11.

Cousins, Ewert, et al (Eds) Post-Classical Hindu and Sikh Spirituality
New York, Crossroad/Continuum. In preparation. World Spirituality Series: 7.

Cousins, Ewert, et al (Eds) South and Meso-American Native Spirituality
New York, Crossroad/Continuum. In preparation. World Spirituality Series: 4.

Cousins, Ewert, et al (Eds) Spirituality and the Secular Quest
New York, Crossroad/Continuum. In preparation. World Spirituality Series: 22.

Cousins, Ewert, et al (Eds) African and Oceanic Spirituality
New York, Crossroad/Continuum. In preparation. World Spirituality Series: 3.

Cousins, Ewert, et al (Eds) Encounters of Spiritualities: present to future
New York, Crossroad/Continuum. In preparation. World Spirituality Series: 24.

Cousins, Ewert, et al (Eds) Islamic Spirituality: manifestations in history and culture
New York, Crossroad/Continuum. In preparation. World Spirituality Series: 20.

Cousins, Ewert, et al (Eds) Buddhist Spirituality: Indian, Sri Lankan, Southeast Asian
New York, Crossroad/Continuum. In preparation. World Spirituality Series: 8.

Cousins, Ewert, et al (Eds) Buddhist Spirituality: Chinese, Tibetan, Japanese, Korean
New York, Crossroad/Continuum. In preparation. World Spirituality Series: 9.

Cousins, Ewert, et al (Eds) Taoist Spirituality
New York, Crossroad/Continuum. In preparation. World Spirituality Series: 10.

Cousins, Ewert, et al (Eds) Early Hindu and Jain Spirituality
New York, Crossroad/Continuum. In preparation. World Spirituality Series: 6.

Cousins, Ewert, et al (Eds) North American Indian Spirituality
New York, Crossroad/Continuum. In preparation. World Spirituality Series: 5.

Cousins, Ewert, et al (Eds) Encounters of Spiritualities: past to present
New York, Crossroad/Continuum. In preparation. World Spirituality Series: 23.

Cousins, Ewert, et al (Eds) Asian Archaic Spirituality
New York, Crossroad/Continuum. In preparation. World Spirituality Series: 1.

Cousins, Ewert, et al (Eds) Modern Esoteric Spirituality
New York, Crossroad/Continuum. In preparation. World Spirituality Series: 21.

Cousins, Ewert, et al (Eds) Christian Spirituality: high middle ages and reformation (1987)
New York, Crossroad/Continuum, 528 p. World Spirituality Series: 17. ISBN 0–8245–0765–7.

Cousins, Ewert, et al (Eds) Christian Spirituality: origins to the twelfth century (1987)
New York, Crossroad/Continuum, 496 p. World Spirituality Series: 16. ISBN 0–8245–0847–5.

Cousins, Ewert, et al (Eds) Jewish Spirituality: from Sixteenth Century revival to present (1987)
New York, Crossroad/Continuum, 528 p. World Spirituality Series: 14. ISBN 0–8245–0763–0.

Cousins, Ewert, et al (Eds) Islamic Spirituality: foundations (1987)
New York, Crossroad/Continuum, 496 p. illus. World Spirituality Series: 19. ISBN 0–8245–0767–3.

Cousins, Ewert, et al (Eds) Jewish Spirituality: from the Bible to Middle Ages (1988)
New York, Crossroad/Continuum, 496 p. World Spirituality Series: 13. ISBN 0–8245–0891–2.

Cousins, James H A Study in Synthesis (1934)
Madiar, Ganesh and Co.

Cousins, Lance S Buddhist Jhāna: its nature and attainment according to the Pali sources (1973)
In: *Religion* 3, pp. 115–131.

Couter, Walter Emergent Human Nature: a symbolic field interpretation (1949)
New York, AA Knopf.

Cowan, James On Totems
In: *Resurgence* 1989, 138, pp. 30–34.

Cowan, James Mysteries of the Dream–Time: the spiritual life of Australian Aborigines (1989)
Bridport, Prism–Unity, 132 p. ISBN 1–85327–033–4.

Cowan, Michael A, et al Alternative Futures for Worship: general introduction (1987)
Collegeville MN, Liturgical Press, 152 p. Vols 17. ISBN 0–8146–1491–4.

Cox, M Mysticism: the direct experience of God (1983)
The Aquarian Press.

Cox, Michael Handbook of Christian Spirituality Christianity

Coxhead, N Mindpower (1979)
New York, Penguin Books.

Coxhead, Nona The Relevance of Bliss: a contemporary exploration of mystic experience (1985)
London, Wildwood House, 183 p. ISBN 0 7045 0499 5.

Cozort, Dan Highest Yoga Tantra (1986)
Ithaca NY, Snow Lion Publications, 220 p. ISBN 0–937938–32–7.

Crabtree, A Multiple Man: explorations in possession and multiple personality (1988)
London, Grafton Books.

Craig, R Characteristics of Creativeness and Self–Actualization
To be published (cited in A Maslow. The Psychology of Science 1926).

Crampton, Martha Some Applications of Psychosynthesis in the Educational Field
(Papers to the Psychosynthesis Seminars, 1971–72 Series) New York, Psychosynthesis Research Fouldation.

Crampton, Martha Ways of Transcendent Growth: paper presented at the International Invitational Conference on Psychobiology and Transpersonal Psychology (1972)
Bifrost, Iceland, June.

Crandall, Joanne Self–Transformation through Music (1986)
Wheaton IL, Theosophical Publishing.

Crawford, John D Psychosynthesis: a project for a scientific psychology (1916)
London, Ash, 104 p. bibl.

Creegan, R F Holism Must be Historical: on reading foundations for a science of personality by A Angyval (1943)
Journal of Philosophy 40, pp 159–62, Mr 18.

Crisswell, E Feedback and States of Consciousness: meditation (1969)
In: *Biofeedback Research Society, Proceedings.*

Crisswell, E Experimental Yoga Course for College Students: a progress report (1970)
Journal of Transpersonal Psychology, 2, 1, 11, pp. 71–78.

Critchlow, Keith The Soul as Sphere and Androgyne (1980)
Ipswich, Golgonooza Press, 34 p. illus.

Crook, J H The Evolution of Human Consciousness (1980)
New York, Oxford University Press.

Crookall, Robert The Interpretation of Cosmic and Mystical Experiences (1969)
London, James Clarke, 175p. bibl. (various types of at–onement).

Cross, John and Guyer, Mel Social Traps (1980)
Ann Arbor MI, University of Michigan Press. illus. ISBN 0–472–06315–4.

Croucher, Michael and Reid, Howard The Way of the Warrior (1987)
New York, Simon and Schuster. ISBN 0–671–64674–5.

Crow, James F Mechanisms and Trends in Human Evolution (1961)
Daedalus Summer p. 416–31.

Csikszentmihalyi, M Between Boredom and Anxiety (1978)
San Francisco CA, Jossey–Bass Publishers.

Csikszentmihalyi, M and Csikszentmihalyi, I S (Eds) Optimal Experience: psychological studies of flow in consciousness (1988)
New York, Cambridge University Press.

Csikszentmihalyi, M and Rathunde, K The Psychology of Wisdom: an evolutionary interpretation (1989)
In: *R J Sternberg (Ed) The Psychology of Wisdom* New York, Cambridge University Press.

Csikszentmihalyi, Mihaly Flow: the psychology of optimal experience: steps towards enhancing the quality of life (1990)
New York, Harper and Row, 303 p. ISBN 0–06–016253–8.

Culbertson, J T Consciousness and Behavior (1951)
London, WC Brown.

Cullingan, Kevin (Intro) Spiritual Direction: contemporary readings (1983)
Hauppauge NY, Living Flame Press, 237 p. ISBN 0–914544–43–8.

Cully, Iris V Education for Spiritual Growth (1984)
New York, Harper and Row, 192 p. ISBN 0–06–061655–5.

Cummins, G The Road to Immortality and Beyond Human Personality (1952)
London, Psychic Press.

Cytowic, Richard E Synesthesia: a union of the senses
To be published

Da Free John The Paradox of Instruction (1977)
San Francisco, Dawn Horse.

Da Free John The Enlightenment of the Whole Body (1978)
San Francisco, Dawn Horse.

Da Free John Enlightenment and the Transformation of Man (1983)
Clearlake, Dawn Horse Press.

Da Free John Nirvanasara (1983)
Clearlake, Dawn Horse Press.

Da Free, John The Transmission of Doubt: talks and essays on the transcendence of scientific materialism through radical understanding (1984)
Clearlake CA, Dawn Horse Press.

Da Free, John The Basket of Tolerance: a guide to perfect understanding of the One and Great tradition of mankind (1989)
Clearlake CA, The Free Daist Communion, 301 p. bibl.

Da Free John, and Blas, Lydia Conscious Exercise and the Transcendental Sun (1977)
San Rafael CA, Dawn Horse, 272 p. ISBN 0–913922–33–1.

Dabrowski, K Mental Growth through Positive Disintegration (1970)
Gryf Publications.

Daly, Herman Steady State Economics: the economics of biophysical equilibrium and moral growth (1977)
New York, WH Freeman.

Danielli, Mary The Anthropology of the Mandala
In: *The Quarterly Bulletin of Theoretical Biology* 1974, 7, 2.

Danielou, Alain Yoga: the method of re–integration

Daniélou, Alain While the Gods Play
New York, Destiny Books, 288 p. ISBN 0–89281–115–3.

Daniélou, Alain The Gods of India
New York, Destiny Books, 480 p. ISBN 0–89281–101–3.

Danskin, D G and Walters, E D Biofeedback and Voluntary Self–Regulation, Counseling and Education (1973)
Personnel and Guidance Journal, 51, 9, pp633–638.

Danskin, D G and Walters, E D Biofeedback Training as Counseling (1975)
Counseling and Values, 19, 2, February pp 116–122.

Das, B A Concordance Dictionary to Yogasuras of Patanjali and the Bhasya of Vyasa (1938)
Benares.

Das, N N and Gastaut H Variations in the Electrical Activity of the Brain, the Heart and the Skeletal Muscles during Yogic Meditation and Ecstasy (1957)
Electroencephalography and Clinical Neurophysiology Suppl 6.

Dasgupta, S N Hindu Mysticism (1927)
Delhi, Motilal Banarsidass.

Dassgupta, Surendranath Yoga as Philosophy and Religion (1986)
Delhi, Motilal Banarsidass, 200 p. ISBN 81–208–0218–7.

Datan, Nancy et al (Eds) Life–Span Developmental Psychology (1986)
Hillsdale NJ, Erlbaum, Lawrenece Associates, 304 p.

Date, V H Brahma Yoga of the Gita (1971)
New Delhi, Munshiram Manoharlal Publishers, 671 p.

Daumal, René Mount Analogue (1959)
San Francisco CA, City Lights. Trans. Roger Shattuck.

Davidson, J and Davidson, R (Eds) The Psychobiology of Consciousness (1980)
New York, Plenum Publishing.

Davidson, J M The Physiology of Meditation and Mystical States of Consciousness (1976)
In: *Pespectives in Biology and Medicine* 1976, 19, pp. 345–379.

Davidson, R and Krippner, S Biofeedback Research: the data and their implications (paper at the Second International Invitational Conference on Humanistic Psychology, University of Wurzburg, July 1971)

Davidson R J; Goleman D and Schwartz G E Attentional and Affective Concomitants of Meditation: a cross–sectional study
In: *Journal of Abnormal Psychology* 1976, 85, pp. 235–238.

Davidson, R; Schwartz, G and Shapiro, D (Eds) Consciousness and Self–Regulation: advances in research and theory (1983)
New York, Plenum Publishing. Vol 3.

Davis, Charles Body as Spirit: the nature of religious feeling (1976)
London, Hodder and Stoughton.

Davis, G A and Scott, J A (Ed) Training Creative Thinking (1971)
London, Holt–Blond, 256 p.

Davis, J (Ed) Religious Organizations and Religious Experience (1982)
San Diego CA, Academic Press. ASA Monograph. ISBN 0–12–206580–8.

Davis, James A The Meaning of Positive Mental Health
NRC–monograph.

Davis, Roy E Science of Kriya Yoga (1984)
Lakemont GA, C S A Press, 192 p. ISBN 0–317–20860–8.

Davis, Roy Eugene The Science of Kriya Yoga: a verse by verse rendering of the yoga Sutras of Pantanjali, with detailed commentary and specific instructions for experiencing meditation and superconsciousness
The Teachings of the Masters of Perfection, Vol 1.

de Ajuriaguerra, J (Ed) Cycles biologiques et psychiatrie (1968)
Paris, Masson, 428 p.

de Boer, Julius A System of Characterology: polar synergy of feeling, willing and thinking as the dynamic principle for the classification of characters (1963)
Assen, Van Gorcum, 454 p. bibl. (pp. 423–443).

de Bono, Edward Masterthinker's Handbook (1985)
International Center for Creative Thinking.

de Condorcet, A N Sketch for a Historical Picture of the Progress of the Human Mind (1955)
London, Library of Ideas.

De Jonge Quelques principles et exemples de psychosynthèse (1937)
Bern, Gesellschaft für Psychiatrie.

De Long, Alton Phenomenological Space–Time Toward an Experimental Relativity
In: *Science* 1981, 213, August 7.

De Riencourt, A The Eye of Shiva: Eastern mysticism and science (1981)
New York, Morrow.

De Ropp, Robert S The Master Game: pathways to higher consciousness beyond the drug experience (1969)
London, Pan Books, 252 p. ISBN 0–330–24163–X.

Dean, S (Ed) Psychiatry and Mysticism (1975)
Chicago IL, Nelson–Hall.

Deatherage, O G The Clinical Use of Mindfulness Meditation Techniques in Short–Term Psychotherapy (1975)

Deci, E L and Ryan, R M Intrinsic Motivation and Self–Determination in Human Behavior (1985)
New York, Plenum Press.

Deikman, A Personal Freedom: on finding your way to the real world (1976)
New York, Grossman.

Deikman, A, et al Symposium on Consciousness (1976)
New York, Viking.

Deikman, Arthur The Observing Self: mysticism and psychotherapy (1982)
Boston MA, Beacon Press.

Deikman, Arthur J Experimental Meditation (1963)
Journal of Nervous and Mental Diseases 136, pp. 329–73.

Deikman, Arthur J Deautomatization and the Mystic Experience (1966)
Psychiatry 29, pp. 324–38.

Deikman, Arthur J Implications of Experimentally induced Contemplative Meditation (1966)
Journal of Nervous and Mental Diseases 142, pp. 101–16.

Deikman, Arthur J Evaluating Spiritual and Utopian Groups (1988)
London, Octagon Press. ISBN 0–904674–13–4.

Delgado, José Physical Control of the Mind: toward a psychocivilized society (1969)
New York, Harper and Row.

Delza, Sophia Body and Mind in Harmony (1961)
New York, McKay.

Desai, Amrit and Tennen, Laura Kripalu Yoga: meditation–in–motion, - focusing inward (1987)
Lenox MA, Kripalu Publications, 120 p. Bk. II ISBN 0–940258–16–1.

Desan, Wilfrid The Planetary Man: a noetic prelude to a united world (1961)
Washington DC, Georgetown University Press, bibl. foot.

Desan, Wilfrid L'Homme Planétaire: prélude théorique à un monde uni (1968)
Paris, Ed de Minuit, 157 p. bibl. foot. translated from 1961 ed.

Desikachar, T K; Skelton, Mary L and Carter, J R Religiousness in Yoga: lectures on theory and practice (1980)
Lanham MD, University Press of America, 314 p. ISBN 0–8191–0966–5.

Deutsch, Eliot Advaita Vedanta: a philosophical reconstruction (1969)
Honolulu: East–West Center Press.

Deutsch, F (Ed) Body, Mind, and the Sensory Gateways (1962)
New York, Karger S AG, 106 p. Advances in Psychosomatic Medicine: Vol 2. ISBN 3–8055–0455–1.

Deutsch, F; Jores, A and Stokvis, B (Eds) Training in Psychosomatic Medicine (1964)
New York, Karger S AG, 222 p. Advances in Psychosomatic Medicine: Vol 4. ISBN 3–8055–0458–6.

Devall, Bill and Sessions, George Deep Ecology: living as if nature mattered (1985)
Salt Lake City, Peregrine Smith Books, 267 p. ISBN 0–87905–247–3.

Dewana, Mohan Singh Uberoi Dhyana Yoga

Dewey, John Art as Experience (1959)
New York, G P Putnam's Sons.

DeWitt, Calvin B A Sustainable Earth: religion and ecology in the Western hemisphere (1987)
New York, Only One Earth Forum, Rene Dudos Center.

Dick, W and Gris, H The New Soviet Psychic Discoveries (1978)
New York, Warner Books.

Diel, Paul The God-Symbol: its history and its significance
Nelly Marans, Trans.

DiNardo, Ramon Albert The Unity of the Human Person (1961)
Washington DC, Catholic University of America Press, 68 p. bibl. (pp. 37–61) (philosophical studies 199).

Dixon, Norman Preconscious Processing (1981)
Chichester, John Wiley and Sons.

Döbert, R et al Entwicklung des Ichs (1977)
Köln, Kiepenheuer and Witsch.

Dobzhansky, Theodosius The Present Evolution of Man (1960)
Scientific American 203, Sept 1960 pp. 206–17.

Dobzhansky, Theodosius Mankind Evolving: the evolution of the human species (1962)
New Haven CT, Yale University Press.

Dogen Zen Master and Uchiyama Kosho Refining Your Life: from the Zen kitchen to enlightenment

Doran, R Jungian Psychology of Christian Spirituality, III (1979)
In: *Review for Religious* 38, pp. 857–66.

Dorr, Donal Spirituality and Justice (1985)
Dublin, Gill and Macmillan. ISBN 0–7171–1376–0.

Dossey, L Beyond Illness: discovering the experience of health (1984)
Boulder CO, New Science Library.

Douglas, Alfred ESP Powers: a century of psychical research (1976)
London, Gollancz.

Douglas, Alfred Extra-Sensory Powers: a century of psychical research (1983)
New York, Overlook Press, 392 p. ISBN 0–87951–064–1.

Dourley, John P Psyche as Sacrament: C G Jung and Paul Tillich (1981)
Toronto ON, Inner City Books, 128 p. ISBN 0–919123–06–6.

Dowdy, Edwin (Ed) Ways of Transcendence: insights from major Religions and modern thought (1982)
Bedford Park, Australian Association for the Study of Religions, 172 p. ISBN 0–908083–08–4.

Dowsett, Norman C and Jayaswal, Sitaram Yoga and Education (1977)
High Falls NY, Sri Aurobindo Association, 95 p. ISBN 0–89071–273–5.

Drakeford, John W Integrity Therapy (1967)
Nashville, Broadman Press, 153 p. bibl. (pp. 149–53).

Drogalina, J A; Kuznetsov, O A and Nalimov, V V Vizualizatsiya Semanticheskikh Polei Verbal'nogo Teksta Sredstvami Gruppovoi Meditatsii
In: *Bessoznatel'noe: Priroda, Funktsii, Metody Issledovaniya* 1978, 3, pp. 703–710. English Title: Visualization of Semantic Fields of Verbal Texts by Means of Group Meditation.

Dror, Yehezkel Assignment: improvement of mankind (A 'speculative discussion paper' to the Rome Special Conference on Futures Research, 1973) (1973)

Drury, N Inner Visions: explorations in magical consciousness (1979)
New York, Routledge and Kegan Paul.

Drury, Nevill Don Juan, Mescalito and Modern Magic: the mythology of inner space (1978)
London, Routledge and Kegan Paul, 229 p. illus. bibl. ISBN 0–7100–8582–6.

Drury, Nevill The Elements of Human Potential (1989)
Longmead, Element Books, 135 p. ISBN 1–85230–086–8.

Ducharme, A Spiritual Discernment and Community Deliberation (1974)
Ottawa ON, Canadian Religious Conference, 200 p.

Duhl, B S From the Inside Out and Other Metaphors: an integrated approach to training in multicentric systems thinking as derived from a family therapy training program (1982)
Boston MA, University of Massachusetts. Ed D Dissertation.

Dulles, Avery Models of Revelation (1983)
Dublin, Gill and Macmillan. ISBN 0–7171–1324–8.

Dumont, Louis From Mandeville to Marx: genesis and triumph of economic ideology (1977)
Chicago IL, University of Chicago Press.

Duner, Anders (Ed) Research into Personal Development: educational and vocational choice: report of the European Contact Workshop held in Saltsjöbaden 1977 under the Auspices of the Council of Europe (1978)
Lisse, Swets en Zeitlinger, 186 p. ISBN 90–265–0284–2.

Dunn, H L Higher-Level Wellness for Man and Society (1959)
American Journal of Public Health.

Dunn, H L High-Level Wellness (1967)
Arlington, Beatty, 256p.

Dupré, Louis and Wiseman, James A (Eds) Light from Light: an anthology of Christian mysticism (1988)
New York, Paulist Press, 440 p. ISBN 0–8091–2943–4.

Durckheim, K G von The Way of Transformation (1971)
London, George Allen and Unwin.

Durckheim, Karlfried Hara (1962)
London, Allen and Unwin.

Dychtwald, K Bodymind (1977)
New York, Pantheon.

Ebon, Martin Psychic Discoveries by the Russians
Bergenfield NY, New American Library.

Eckartsberg, Rolf von Maps of the Mind: the cartography of consciousness
In: *The Metaphors of Consciousness* 1981, New York, Plenum Publishing, 521 p. illus. bibl.

Eckhardt, William Psychosocial Isomorphism
In: *Compassion; Toward a Science of Value* Oakville, Canadian Peace Reasearch Institute, 1972.

Eckhardt, William Self-Acceptance and Mental Health (1960)
University of Kentucky (Ph. D. dissertation).

Eddy, Mary Baker Science and Health with Key to the Scriptures (1875)
Boston MA, First Church of Christ Scientist. New type ed ISBN 0–87952–010–8.

Edelstein, Ludwig The Idea of Progress in Classical Antiquity (1967)
Baltimore MD.

Edinger, Edward F Ego and Archetype: individuation and the religious function of the psyche (1972)
New York, C G Jung Foundation for Analytical Psychology.

Edwards, Jonathan Treatise Concerning Religious Affections (1959)
New Haven CT, Yale University Press. Edited by John E Smith.

Ehrenzweig, A The Psychoanalysis of Artistic Vision and Hearing (1953)
London, Routledge.

Eigo, F (Ed) From Alienation to Atoneness (1977)
Villanova PA, Villanova University Press.

Eisenberg, Nancy and Strayer, Janet (Eds) Empathy and Its Development (1987)
New York, Cambridge University Press, 416 p. illus. Cambridge Studies in Social and Emotional Development. ISBN 0–521–32609–5.

Eisendrath, C R The Unifying Moment (1971)
Cambridge MA, Harvard University Press.

Eitel, Ernest J Feng-Shui: the science of sacred landscape in Old China (1984)
London, Synergetic Press, 69 p. With commentary by John Mitchell. ISBN 0–907791–09–3.

Ekins, Paul (Ed) The Living Economy: a new economics in the making (1986)
London, Routledge and Keegan Paul. Available in North America form ITD/NA, Croton-on-Hudson NY.

Eliade, M The Sacred and the Profane: the nature of religion (1961)
New York, Harper Torchbook.

Eliade, M The Two and the One (1965)
New York, Allied Publications.

Eliade, M Yoga: immortality and freedom (1969)
Princeton NJ, Princeton University Press. 2nd ed. Bollingen Series: 56.

Eliade, M Patanjali and Yoga (1969)
New York, Allied Publications.

Eliade, Mircea Spiritual Discipline: encounters at Ascona
In: *J Campbell (Ed) Spiritual Disciplines; Papers from the Eranos Yearbooks* Princeton University Press, 1985.

Eliade, Mircea Rites and Symbols of Initiation: the mysteries of birth and rebirth (1958)
New York, Harper and Row. Willard R Trask, Trans.

Eliade, Mircea Patterns in Comparative Religions (1958)
New York, Sheed and Ward.

Eliade, Mircea Shamanism: archaic techniques of ecstasy (1964)
Princeton NJ, Princeton University Press. Willard R Trask, Trans.

Eliade, Mircea Myths, Dreams and Mysteries: the encounters between contemporary faiths and archaic realities (1967)
New York, Harper and Row. Philip Mairet, Trans.

Eliade, Mircea The Forge and the Crucible: the origins and structure of alchemy (1971)
New York, Harper and Row. Stephen Corrin, Trans.

Eliade, Mircea The myth of the eternal return or, Cosmos and History (1974)
New York, Routlege and Kegan Paul, 195 p. bibl. Trans. from French by Willard R Trask. ISBN 0–691–01777–8.

Eliade, Mircea Encyclopedia of Religion (1986)
New York, Macmillan. 16 Vols. ISBN 0–02–909480–1.

Elias, John L Moral Education: secular and religious (1988)
Melbourne FL, Robert E Krieger. ISBN 0–89464–260–X.

Eliot, John (Ed) Language
In: *Human Development and Cognitive Processes* New York, Holt, Rinehart and Winston, 1971, pp. 251–394.

Eliot, John Rites and Symbols of Initiation: the mysteries of birth and rebirth (1965)
New York, Harper and Row, 595 p. bibl. (pp. 541–89).

Eliot, John (Ed) Human Development and Cognitive Processes (1971)
New York, Holt, Rinehart and Winston, 7 p.

Ellis, A Humanistic Psychotherapy: the rational-emotive approach (1973)
New York, McGraw Hill Book Company.

Ellis, Willis D A Source Book of Gestalt Psychology (1950)
London, Routledge.

Elster, Jon Sour Grapes: studies in the subversion of rationality (1987)
Cambridge NY, Cambridge University Press, 177 p. bibl. (pp. 167–175).

Emerson, V F Can Belief Systems Influence Neurophysiology? Some implications of research on meditation (1972)
M Bucke Society Newsletter-Review 5, 1–2.

Emmerij, Louis The Human Factor in Development
In: *K Haq and U Kirdar (Eds) Human Development: the neglected dimension* Islamabad, North South Roundtable, 1986, pp. 13–27.

Emmons, M The inner Sources: a guide to meditative therapy (1978)
San Luis Obispo CA, Impact Publications.

Engler, J Buddhist Satipatthana-Vipassana Meditation and An Object Relations Model of Therapeutic Developmental Change: a clinical case study (1983)
Chicago IL, University of Chicago Press. Unpublished dissertation.

Enright, John B Awareness Training in the Mental Health Professions (1970)
In: *J Fagan and I L Shepherd (Eds) Gestalt Therapy Now* New York, Harper and Row, pp263–273.

Epstein, Gerald N Waking Dream Therapy (1980)
New York, Human Sciences Press.

Epstein, Gerald N (Ed) Studies in Non-Deterministic Psychology (1980)
New York, Human Sciences Press.

Erikson, E Identity and the Life Cycle (1959)
New York, International University Press.

Erikson, E H Growth and Crises of the 'Healthy Personality'. (1950)
In: *M J D Senn (Ed) Problems of Infancy and Childhood Suppl. II. Symposium on the Healthy Personality* New York, Josia Macy, Jr. Foundation.

Erikson, E H Childhood and Society (1950)
New York, WW Norton.

Erikson, E H Wholeness and Totality (1953)
In: *C J Friedrich (Ed) Totalitarianism; proceedings of a conference*, Cambridge, Harvard University Press, 1954.

Erikson, Erik Insight and Responsibility (1964)
New York, W W Norton.

Etkin, W Social Behavioral Factors in the Emergence of Man (1963)
Human Biology 35, pp. 299–310.

Ettinger, R C W Man into Superman (1972)
London, St Martin's Press, 312 p.

Etzioni, A Basic Human Needs, Alienation and Inauthenticity (1968)
American Sociological Review 33, pp. 870–885.

Evans, G R (Trans) Bernard of Clairvaux: selected works (1987)
New York NY, Paulist Press.
The Classics of Western Spirituality.

Evans, Hilary Alternate States of Consciousness: unself, otherself and superself (1989)
Wellingborough, The Aquarian Press, 256 p. ISBN 0–85030–802–X.

Evans-Wentz, W Y (Ed) The Tibetan Book of the Great Liberation, or the Method of Realizing Nirvana through Knowing the Mind
Preceded by an epitome of Padma-Sambhava's Biography and followed by Guru Phadampa Sangay's Teachings.

Evola, Julius The Metaphysics of Sex

Fabricius, Johannes Alchemy; the medieval alchemists and their royal art (1989)
Wellingborough, Aquarian Press, illus, bibl, 248 p. ISBN 0–85030–832–1.

Fadiman, J and Frager, R Personality and Personal Growth (1976)
New York, Harper and Row.

Fadiman, James and Katz, Richard Transformation: the meaning of personal growth (book under preparation)
Brandeis University.

Fadiman, James and Kewman, Donald (Eds) Exploring Madness (1973)
Monterey, Brooks-Cole.

Fagan, Joel and Shepherd, I L (Eds) Gestalt Therapy Now: theory, techniques, applications (1972)
London, Penguin Books, 380 p.

Fahler, Gerhard (Ed) Erbcharakterkunde, Gestaltpsychologie und Integrationstypologie (1937)
Leipzig, J A Barth, 252 p. bibl. (pp. 147–148).

Farber, S M and Wilson, R H L (Ed) Control of the Mind (1961)
New York, Allied Publications.

Faricy, Robert The Lord's Dealing: the primary of the feminine in Christian spirituality (1988)
Mahwah NJ, Paulist Press, 128 p. ISBN 0–8091–3003–3.

Faris, Nabih A The Book of Knowledge (1962)
Lahore. Trans. of a chapter of Ghazali's Ihya.

Farley, Edward Ecclesial Man: a social phenomenology of faith and reality (1975)
Philadelphia, Fortress Press.

Farrar, Janet and Farrar, Stewart The Witches' Goddless: the feminist principle of divinity
Phoenix WA. ISBN 0–919345–91–3.

Farrell, Warren The Liberated Man – Beyond Masculinity: freeing men and their relationships with women (1975)
New York, Bantam Books.

Fathi El-Rashidi Human Aspects of Development (1971)
Brussels, International Institute of Administrative Sciences.

Feifel, Herman New Meanings of Death (1977)
New York, McGraw Hill Book Company, 384 p. ISBN 0–07–020350–4.

Feinberg, Gerald The Prometheus Project: mankind's search for long-range goals (1969)
New York, Doubleday, 264 p.

Feinstein, David and Krippner, Stanley Personal Mythology (1988)
San Francisco, J P Tarcher.

Feldenkrais, M Awareness through Movement (1972)
New York, Harper and Row.

Feldenkrais, Moshe Body and Mature Behavior: a study of anxiety, sex, gravitation and learning (1970)
New York, International University Press, 167 p. ISBN 0–8236–0560–4.

Felix, M New Look at Chastity (1974)
Bombay, Asian Trading Corporation, 239 p.

Fénelon, François The Maxims of the Saints
London, H R Allenson.

Feng, Gia-Fu and Kirk, Jerome Tai-chi: a way of centering and the I Ching (1970)
London, Collier Books.

Fenwick, P B C, et al Metabolic and EEG Changes During Transcendental Meditation: an explanation
In: *Biological Psychology* 1977, 5, pp. 101-118.

Ferguson, J Encyclopedia of Mysticism (1976)
London, Thames and Hudson.

Ferguson, Marilyn The Aquarian Conspiracy: personal and social transformation in the 1980s (1980)
Boston MA, Houghton Mifflin, 448 p.

Ferguson, Marilyn PragMagic: pragmatic magic for everyday living - ten years of scientific breakthroughs, exciting ideas and personal experiments that can profoundly change your life (1990)
New York, Pocket Books, 254 p. illus. Adapted and updated by Wim Coleman and Pat Perrin. Illustrations by Kristin Ferguson. ISBN 0-671-66824-2.

Ferguson, Marilyn The New Common Sense: secrets of the visionary life (1991)
USA, New Stream Books.

Feuerstein, G Structure of Consciousness
Lower Lake CA, Integral Publishing.

Feuerstein, G Textbook of Yoga (1975)
London, Rider.

Feuerstein, G and Miller, J A Reappraisal of Yoga (1971)
London.

Feuerstein, Georg The Essence of Yoga: a contribution to the psychohistory of Indian civilisation (1974)
London, Rider, 224 p. ISBN 0-09-120800-9.

Filippi, Luigi S Maturit Umana e Celibato: problemi di psicologia dinamica e clinica (1970)
Brescia, Le senola, 333 p. bibl. (pp. 283-302).

Fingarette, H The Ego and Mystic Selflessness (1958)
In: *Psychoanalytic Review* 1958, 45, pp. 5-40.

Fingarette, H Self-Deception (1969)
London, Routledge and Kegan Paul.

Fingarette, Herbert The Self in Transformation: philosophy and the life of the spirit (1963)
New York, Harper and Row.

Fischbein, E Intuitive Sources of Probabilistic Thinking in Children (1975)
Dordrecht, Kluwer Academic Publishers Group, 204 p. ISBN 90-277-0626-3.

Fischer, R Cartography in Inner Space
In: *Siegel, R K and West, L J, Hallucinations: Behaviour, experience and theory* 1975, New York, John Wiley.

Fischer, R The Biological Fabric of Time (1967)
In: *R Fischer (Ed) Interdisciplinary Perspectives of Time* Annals of the New York Academy of Sciences, 138, pp. 440-488.

Fischer, R A A Cartography of the Ecstatic and Meditative States: the experimental and experiential features of a perception-hallucination continuum are considered (1971)
In: *Science* 174, pp. 897-904.

Fish, Sharon and McCormick, Thomas Meditation: a practical guide to a spiritual discipline (1983)
Downers Grove IL, Inter-Varsity Press, 132 p. ISBN 0-87784-844-0.

Fisher, Gary Self-Actualization of Paranormals (1971)
Journal of Personality Assessment, 35, 5, October.

Fittipaldi, Silvio E Human Consciousness and the Christian Mystic: Teresa of Avila
In: "The Metaphors of Consciousness", edited by Ronald S Valle and Rolf van Eckartsberg.

Fodor, Nandor Encyclopedia of Psychic Science (1966)
Secaucus NY, University Books.

Ford, Donald H and Urban, H B Systems of Psychotherapy: a comparative study (1963)
New York, John Wiley and Sons, 712 p. bibl. (pp. 692-699).

Forem, Jack Transcendental Meditation (1973)
New York, EP Dutton, 183 p.

Foster, Richard J Celebration of Discipline: the path to spiritual growth (1984)
London, Hodder and Stoughton, 179 p. ISBN 0 340 25992 2.

Foundation for Inner Peace A Course in Miracles (1975)
Tiburon CA, Foundation for Inner Peace.

Fowler, J W Stages of Faith: the psychology of human development and the quest for meaning (1981)
San Francisco, Harper and Row, 332 p. ISBN 0-06-062840-5.

Fox, Douglas Meditation and Reality: a critical view (1986)
St Louisville KY, John Knox Westminster, 192 p. ISBN 0-8042-0662-7.

Fox, Matthew (Ed) Western Spirituality: historical roots, ecumenical routes (1981)
Santa Fe, Bear and Company, 440 p. ISBN 0-939680-01-7.

Fox, Matthew Breakthrough: Meister Eckhart's creation spirituality in new translation (1982)
New York, Doubleday and Doubleday Image.

Fox, Matthew Original Blessing (1983)
Santa Fé NM, Bear and Company, 355 p. bibl. (pp. 321-348). ISBN 0-939680-07-06.

Fox, Matthew and Swimme, Brian Manifesto for a Global Civilization (1982)
Santa Fe, Bear and Company.

Fraisse, P The Psychology of Time (1964)
London, Eyre and Spottiswoode.

Frank, J D Persuasion and Healing: a comparative study of psychotherapy (1961)
Baltimore MD, Johns Hopkins University Press.

Frank, Waldo The Rediscovery of Man (1958)
New York, Braziller.

Frankl, V Man's Search for Meaning (1963)
Boston MA, Beacon Press.

Frankl, V The Will to Meaning (1969)
Cleveland OH, New American Library.

Frankl, Viktor E The Doctor and the Soul (1915)
New York, Alfred A Knopf, 280p.

Frankl, Viktor E Beyond Self-Actualization and Self-Expression (1960)
Journal of Existential Psychiatry 1, p. 5.

Frankl, Viktor E Man's Search for Meaning: an introduction to logotherapy (1963)
Washington DC, Washington Square Press.

Frankl, Viktor E Self Transcendence as a Human Phenomenon (1966)
Journal of Humanistic Psychology XI, 2.

Frankl, Viktor E Psychotherapy and Existentialism (1968)
New York, Simon and Schuster International.

Franz, Marie-Louise von Alchemical Active Imagination (1979)
Dallas, Spring Publications Inc, p 116. ISBN 0-88214-114-7.

Franz, Marie-Louise von Alchemy: an introduction to the symbolism and the psychology (1980)
Toronto, Inner City Books, 280 p. illus. ISBN 0-919123-04-X.

Fraser, J T Of Time, Passion and Knowledge: reflections on the strategy of existence (1975)
New York, George Braziller.

Fraser, J T Time as Conflict (1978)
Basel, Birkhäuser Verlag. A scientific and humanistic study.

Frayser, Suzanne G Varieties of Sexual Experience: an anthropological perspective on human (1985)
New Haven CT, Human Relations Area Files Press, 546 p. ISBN 0-87536-342-3.

Freeman, M and Robinson, R E New Ideas in Psychology
In Press.

Freire, Paulo Pedagogy of the Oppressed (1970)
New York, Herder and Herder.

Fremantle, A (Ed) The Protestant Mystics (1964)
New York, Mentor.

French, Thomas M The Integration of Behavior (1952)
Chicago IL, University of Chicago Press. 5 vols.

Freud, Sigmund General Introduction to Psychoanalysis
New York, Pocket Books.

Freud, Sigmund Totem and Taboo (1952)
New York, Random House.

Freyberger, H (Ed) Psychotherapeutic Interventions in Life-Threatening Illness (1980)
Basel, S Karger AG, 206 p. Advances in Psychosomatic Medicine: Vol 10. ISBN 3-8055-3066-8.

Frick, Willard B Humanistic Psychology: interviews with Maslow, Murphy, and Rogers (1971)
Columbus OH, Charles E Merrill, 186 p.
Studies of the Person edited by William R Coulson and Carl R Rogers. ISBN 0-675-09966-8.

Friedlander, Shems Ninety-Nine Names of Allah: the beautiful names (1978)
New York, Harper and Row, 128 p. ISBN 0-6-090621-9.

Friedman, Maurice Worlds of Existentialism (1964)
New York, Random House.

Friedman, Maurice To Deny Our Nothingness: contemporary images of man (1968)
New York, Dell Books.

Friedman, Maurice The Confirmation of Otherness: in family, community and society (1983)
New York, Pelgrim Press, 301 p. ISBN 0-8298-0651-2.

Fritz, Mary Take Nothing for the Journey: solitude as the foundation for non-possessive (1985)
Mahwah NJ, Paulist Press, 88 p. ISBN 0-8091-2722-9.

Fröbe-Kapteyn, O (Ed) Yoga und Meditation im Osten und im Westen (1934)
Zurich.

Fromm, Erich The Forgotten Language
New York, Grove Press.

Fromm, Erich Psychoanalysis and Religion
?.

Fromm, Erich The Sane Society (1955)
New York, Holt, Rinehart and Winston.

Fromm, Erich The Art of Loving: a stimulating and thoughtful look at the theory and practice of love (1956)
New York, Harper and Row, 133 p.

Fromm, Erich Man for Himself: an inquiry into the psychology of ethics (1968)
New York, Fawcett World Library.

Fromm, Erich Escape from Freedom (1969)
New York, Avon Books.

Fromm, Erich Ed Socialist Humanism: an international symposium (1965)
New York, Doubleday, 461 p.

Fromm, Erich; Suzuki, D T and de Martino, R Zen Buddhism and Psychoanalysis (1963)
New York, Grove Press.

Fuglesang, Andreas About Understanding: ideas and observations on cross-cultural communication (1982)
Uppsala, Dag Hammerskjld Foundation, 231 p. illus. ISBN 91-85214-09-4.

Fulcanelli Les demeures philosophales et le symbolisme hermétique dans ses rapports avec l'art sacré et l'ésotérisme du grand oeuvre (1965)
Paris, Jean-Jacques Pauvert. 3rd ed. 2 vols.

Fuller, R Buckminster I Seem to Be a Verb (1970)
New York, Bantam Books.

Gaer Luce, Gay Biological Rhythms in Human and Animal Physiology (1971)
New York, Dover Publications, 183p. bibl. Bibl previously published in 1970 by the US Department of Health, Ecuation and Welfare and the National Institute of Mental Health as Public Service Publication no 2088 under the title: Biological Rhythms in Psychiatry and Medicine. ISBN 0-486-22586-0.

Gairdner, Canon W H T Theories, Practices and Training Systems of a Sufi School (1980)
Cambridge MA, Institute for the Study of Human Knowledge. ISBN 0-86304-003-9.

Gallaghar, Charles A, et al Embodied in Love: the sacramental spirituality of sexual intimacy (1983)
New York, Crossroad Publishing, 176 p. ISBN 0-8245-0594-8.

Galton, F Hereditary Genius: an inquiry into its laws and consequences (1869)
London, Macmillan. rev 1892. reprinted 1925.

Galtung, Johan Buddhism: a quest for unity and peace (1988)
Honolulu, Dae Won Sa Pagoda.

Galyean, B Mind Sight: learning through imaging (1983)
Long Beach CA, Center for Intergative Learning.

Gandhi, Kishore The Evolution of Consciousness: a contemporary mythic journey into the roots of global awareness (1986)
New York, Paragon House, 260 p. Patterns of World Spirituality Series. ISBN 0-913757-50-0.

Gandolfo, Joseph B Spiritual Psychic Healing: a comparative psychological and biblical study (1986)
New York, Vantage Press. ISBN 0-533-06839-8.

Ganquelin, F and Ganquelin, M A Possible Hereditary Effect on Time of Birth in Relation to the Diurnal Movement of the Moon and the Nearest Planets: its relationship with geomagnetic activity (1967)
Amsterdam, Swets en Zeitlinger.

GAP Committee on Psychiatry and Religion Mysticism: spiritual quest or psychic disorder? (1976)
New York, Brunner/Mazel. Publication 97. Vol. 9 ISBN 0-87318-134-4.

Gardner, Howard Frames of Mind: the theory of multiple intelligences (1984)
London, Heinemann, 438 p. ISBN 0-434-28245-6.

Gardner, Howard The Mind's New Science: a history of the cognitive revolution (1985)
New York, Basic Books, 430 p. bibl. With a new epilogue by the author: Cognitive Science After 1984. ISBN 0-465-04634-7.

Gardner, John W Excellence: can we be equal and excellent too? (1961)
New York, Harper, 171 p. bibl. (pp. 163-167).

Gardner, John W Self-Renewal: the individual and the innovative society (1965)
New York, Harper and Row.

Garfield, P Creative Dreaming (1974)
New York, Simon and Schuster.

Gatewood, J B and Rosenwein, R Interactional Synchrony: genuine or spurious?: a critique of recent research
In: *Journal of Nonverbal Behavior* 1981, 6, 1.

Gauquelin, Michel The Cosmic Clocks: from astrology to modern science (1973)
London, Paladin. reprint of 1967.

Gazda, G M (Ed) Innovations to Group Psychotherapy (1968)
Springfield IL, Thomas Charles C.

Gebser, J Ursprung und Gegenwart (1966)
Stuttgart. 2 vols.

Geddis, A and Kilbourn, B S Curriculum Design in the Sciences: secondary school level
Paper read at the Pluridisciplinary Conference on Root Metaphor, SUNY, Buffalo, 1982.

Geertz, C The Growth of Culture and the Evolution of Mind (1962)
New York, Free Press.

Gellhorn, E and Kiely, W F Mystical States of Consciousness: neurophysiological and clinical aspects (1972)
Journal of Nervous and Mental Diseases 154.

Gellhorn, Ernst Principles of Autonomic-Somatic Integration (1967)
Minneapolis MN, University of Minnesota Press.

Gemmell, William (Trans) The Diamond Sutra (Chin-Kang-Ching)

Gendlil, Eugene T Experiencing and the Creation of Meaning (1962)
New York, Free Press.

Gendlin, Eugene T Experiencing and the Creation of Meaning (1962)
New York, Free Press.

Gerard, Ralph W Becoming: the residue of change (1960)
In: *S Tax (Ed) Evolution after Darwin* Chicago, University of Chicago Press.

Gerard, Ralph W; Kluckhohn, C and Rapoport, A Biological and cultural evolution (1956)
Behavioral Science 1, pp. 24-33.

Gerber, Richard Vibrational Medicine: new choices of healing ourselves (1988)
Santa Fé NM, Bear and Company.

Ghose, Aurobindo The Future Evolution of Man: the divine life upon earth (1974)
Wheaton IL, Quest Books.

Ghosh, J A Study of Yoga (1933)
Calcutta.

Ghosh, Shyam Original Yoga: as expounded in Siva-Samhita, Gheranda Samhita and Patanjala Yoga-Sutra (1980)
New Delhi, Munshiram Manoharlal Publishers, 334 p. Original text in Sanskrit, translated, edited and annotated.

Ghurye, G S Religious Consciousness
Bombay, Popular Prakashan.

Giles, Mary E The Feminist Mystic and Other Essays on Women and Spirituality (1982)
New York, Crossroad Publishing, 208 p. ISBN 0-8245-0432-1.

Gilligan, Carol In a Different Voice: psychological theory and women's development (1981)
New York, Free Press.
Gilson, E The Unity of Philosophical Experience (1938)
London, Sheed and Ward.
Ginsburg and Opper Piaget's Theory of Intellectual Development (1969)
New York, Prentice Hall.
Glasser, William Reality Therapy (1965)
New York, Harper and Row.
Glasser, William Stations of the Mind: new directions for reality therapy (1981)
New York, Harper and Row, 288 p. ISBN 0-06-011478-9.
Goble, Frank G The Third Force: the psychology of Abraham Maslow (1970)
New York, Pocket Books, 208 p. bibl.
Goddard, Dwight (Comp) Self-Realization of Noble Wisdom (The Lankavatara Sutra)
Compiled based on D T Suzuki's rendering.
Godwin, Joscelyn Harmonies of Heaven and Earth: the spiritual dimensions of music from Antiquity to the Avant-Garde (1987)
Rochester VT, Inner Traditions International, 208 p. illus. bibl. ISBN 0-89281-165-X.
Goffman, Erving Presentation of Self in Everyday Life (1959)
New York, Doubleday.
Golas, Thaddeus The Lazy Man's Guide to Enlightenment (1972)
Palo Alto, The Seed Center.
Goldberg, Philip The Intuitive Edge (1983)
Los Angeles, J P Tarcher.
Goldbrunner, Josef Individuation: a study of the depth psychology of C G Jung (1955)
London, Hollis and Carter, 204 p.
Goldbrunner, Josef Holiness is Wholeness and Other Essays (1964)
Notre Dame (Indiana).
Goldsmith, Joel S Realization of Oneness: the practice of spiritual healing (1974)
Secaucus NY, Lyle Stuart, 200 p. ISBN 0-8065-0453-6.
Goldstein, Arnold P and Michaels, Gerald Y Empathy: development, training and consequence (1985)
L Erlbaum Associates, 304 p. ISBN 0-89859-538-X.
Goldstein, J The Experience of Insight: a natural unfolding (1976)
Santa Cruz, Unity Press.
Goldstein, Joseph The Experience of Insight: a simple and direct guide to Buddhist meditation (1987)
Boston MA, Shambhala Publications, 170 p. Shambala Dragon Edition Series. ISBN 0-87773-226-4.
Goleman, D Meditation and Consciousness: an Asian approach to mental health
In: *American J Psychotherapy* 1975, 30, pp. 41-54.
Goleman, D Meditation as Meta-therapy: hypotheses toward a proposed fifth state of consciousness (1971)
Journal of Transpersonal Psychology 3, 1.
Goleman, D The Buddha on Meditation and States of Consciousness (1972)
Journal of Transpersonal Psychology 4, 1 (part I: The teachings), 2 (part II: A typology of meditation techniques).
Goleman, D The Varieties of Meditative Experience (1977)
New York, EP Dutton.
Goleman, D and Davidson, R J (Eds) Consciousness: brain, states of awareness, and mysticism (1979)
New York, Harper and Row.
Gooch, S Total Man: notes towards an evolutionary theory of personality (1975)
London, ABACUS, 552 p. ISBN 0-349-11520-6.
Goodenough, Erwin R The Psychology of Religious Experiences (1986)
Lanham MD, University Press of America, 214 p. Brown Classics in Judaica Series. ISBN 0-8191-4489-4.
Goodman, F D, et al Trance, Healing, and Hallucination: three field studies in religious experience (1982)
Melbourne FL, Robert E Krieger, 414 p. ISBN 0-89874-246-3.
Goodman, Paul Compulsory Miseducation (1971)
London, Penguin Books, 9 p.
Goodwin, Donald W and Guze, Samuel B Psychiatric Diagnosis (1989)
New York, Oxford University Press, 332 p. Fourth edition. ISBN 0-19-505231-5.
Gordon, David Therapeutic Metaphors: helping others through the looking glass (1976)
Cupertino CA, Meta Publications, 261 p. bibl.
Gordon, J Handbook of Clinical and Experimental Hypnosis (1967)
New York, Macmillan.
Gordon, W J J Making It Strange (1968)
New York, Harper and Row. Workbooks vols 1-4 designed to increase figurative language use in creative thinking.
Goswami, Shyam Sundar Layayoga: an advanced method of concentration (1980)
New York, Routledge Chapman and Hall. Foreword by Acharyya Karunamoya Saraswiti.
Gosztonyi, Alexander Der Mensch und die Evolution (1968)
Munich, Beck.
Gotesky, Rubin and Laszlo, Ervin (Eds) Human Dignity, This Century and the Next: an interdisciplinary inquiry into human rights (1970)
New York, Gordon and Breach, 380 p.
Gould, Roger L Transformations (1978)
New York, Simon and Schuster.
Govinda, Anagarika Foundations of Tibetan Mysticism (1972)
London.
Govinda, Anagarika Creative Meditation and Multi-Dimensional Consciousness (1976)
Wheaton IL, Theosophical Publishing House.
Gowan, J C Development of the Psychedelic Individual (1974)
Creative Education Foundation.
Gowan, John Curtis Operations of Increasing Order (1980)
Westlake Village CA, 390 p. bibl. Published by author.
Grandberg-Michaelson, Wesley A Worldly Spirituality: the call to take care of the earth (1984)
New York, Harper and Row.
Granmann, C F and Moscovici, S Changing Conceptions of Crowd Mind and Behavior (1985)
New York, Springer Verlag, 290 p. ISBN 0-387-96187-9.
Graves, C W Levels of Existence: an open system theory of values
In: *J Humanistic Psychology* 1970, 10, pp. 131-155.
Greeley, A Ecstasy: a way of knowing (1974)
New York, Prentice Hall.
Green, A M; Green E and Walters, E D Voluntary Control of Internal States: psychological and physiological (1970)
Journal of Transpersonal Psychology 2, pp. 1-28.
Green, E How to Use the Field Theory of Mind (1973)
Topeka KS, Menninger Foundation.
Green, E E and Green A M On the Meaning of Transpersonal: some metaphysical perspectives (1971)
Journal of Transpersonal Psychology 3.
Green E; Green A and Walters D E Voluntary Control of Internal States: psychological and physiological
In: *J Transpersonal Psychology* 1970, 2, pp. 1-26.
Greenacre, Phyllis The Early Years of the Gifted Child: a psychoanalytic interpretation (1962)
Yearbook of Education London, Evans, pp. 71-90.
Greenacre, Phyllis Emotional Growth: psychoanalytic studies of the gifted and a great variety of other individuals (1971)
New York, Int University Press, 863 p. 2 vols. bibl. (pp. 807-830).
Greenberg, J H Language and Evolution (1959)
In: *Evolution and Anthropology: a centennial appraisal* Washington DC, Anthropological Society.
Greene, T M Ben Jonson and the Centred Self (1970)
Studies in English Literature 10, Spring pp. 325-348.
Greenwald, J A Structural Integration and Gestalt therapy (1969)
Bulletin of Structural Integration vol 1, pp. 19-20.
Griffin, David R (Ed) Spirituality and Society: postmodern visions (1988)
Albany NY, State University of New York Press. Constructive Postmodern Thought Series. ISBN 0-88706-853-7.
Griffin, Susan Woman and Nature (1984)
London, Women's Press.
Griffith, Fred F Meditation Research: its personal and social implications (1974)
In: *John White (Ed) Frontiers of Consciousness* New York, Julian Press, pp. 119-137.
Griffiths, Bede Christianity in the Light of the East: the Hibbert Lecture 1989
London, The Hibbert Trust, 15 p. ISBN 0-9507535-7-2.
Griffiths, Bede Return to the Center (1976)
Springfield IL, Templegate Publications. ISBN 0-87243-112-6.
Griffiths, Bede A New Vision of Reality: Western science, Eastern mysticism and Christian faith (1989)
New Orleans LA, Collins Publications, 296 p.
Griffiths, D B Return to the Centre (1976)
London, Collins Liturgical Publications.
Griffiths, Paul Concentration or Insights: the problematic of Theravāda Buddhist meditation-theory (1981)
In: *Journal of the American Academy of Religion* 49, pp. 606-624.
Griffiths, Paul On Being Mindless: the debate on the reemergence of consciousness from the attainment of cessation in the Abhidharmakosabhāsyam and its commentaries (1983)
In: *Philosophy East and West* 33, 379-394.
Griffiths, Paul Indian Buddhist Meditation-Theory: history, development and systematization (1983)
University of Wisconsin Madison. PhD Dissertation.
Griffiths, Paul J On Being Mindless: Buddhist meditation and the mind-body problem (1986)
La Salle IL, Open Court, 220 p. ISBN 0-8126-9007-9.
Groenewegel, H Integratie-Typen (1941)
Assen, Van Gorcum, 87 p.
Grof, S Beyond Psychoanalysis (1970)
2 Vols. I: Implications of LSD research for understanding dimensions of human personality (Presented at the 1st World Conference on Scieltific Yoga, New Delhi, December 1970) (1970), II: A conceptual model of human personality encompassing the psychodelic phenomena (Presented at the 2nd Interdisciplinary Conference on Voluntary Control of Internal States, Council Grove, Ka).
Grof, S LSD Psychotherapy and Human Culture (1971)
Journal for the Study of Consciousness 4, 2, pp. 167-187.
Grof, S Varieties of Transpersonal Experiences: observations from LSD psychotherapy (1972)
Journal of Transpersonal Psychology 1, 11, p. 45-80.
Grof, S Realms of the Human Unconscious (1975)
New York, Viking Press. Observations from LSD research.
Grof, S Realms of the Unconscious (1979)
UK, Souvenir Press.
Grof, S Beyond the Brain (1985)
State University of New York Press.
Grof, S and Grof, C Beyond Death: the gates of consciousness (1980)
London, Thames and Hudson.
Gross, Rita M (Ed) Beyond Androcentrism: new essays on women and religion (1977)
Missoula MT, Scholars Press.
Gruen, S S and Sheldon, William Varieties of Human Temperament: a psychology of constitutional differences (1970)
New York, Hafner Publishing.
Gudas, Fabian Extrasensory Perception (1975)
Salem NH, Ayer Company Publishers. ISBN 0-405-07033-0.
Guénon, René The Multiple States of Being (1984)
Burdett NY, Larson Publications, 140 p. Original Title: *Les Etats Multiples de l'Etre*. Trans. by Joscelyn Godwin. ISBN 0-943914-07-8.
Guenther, H V Yuganaddha: the tantric view of life (1969)
Varanasi.
Guenther, Herbert Philosophy and Psychology in the Abhidharma (1974)
Berkeley, Shambhala, 180 p.
Guggenheim, F G (Ed) Psychological Aspects of Surgery (1985)
Basel, S Karger AG, 272 p. Advances in Psychosomatic Medicine: Vol 15. ISBN 3-8055-4090-6.
Gunstone, John Baptised in the Spirit (1989)
Crowborough, Highland Books, 172 p. ISBN 0-946616-53-1.
Gunther, Bernard Sense Relaxation (1968)
New York, Macmillan.
Gupta, Sailendra Bejoy Das Kriya Yoga and Swami Sri Yukteshvar
Gurwitsch, A The Field of Consciousness (1964)
Pittsburgh PA, Duquesne University Press.
Gutierrez, Gustavo A Theology of Liberation: history, politics and salvation (1973)
Maryknoll NY, Orbis Books.
Guyon, J M B de la Mothe A Short Method of Prayer and Spiritual Torrents (1875)
London, Sampson Low, Marston, Low and Searle. Trans. by A W Marston.
Guzman, Emilio Mind Control: new dimension of human thought (1976)
Laredo, Institute of Psychorientology.
Haas, S The Destiny of the Mind: East and West (1956)
London.
Hafez, Ismail A and Gruber, J J Integrated Development: motor aptitude and intellectual performance (1967)
Columbus ON, Merrill Publishing. bibl. (pp. 192-99).
Halacy, D S Jr Cyborg: evolution of the superman (1965)
New York, Harper and Row.
Halevi, Z'ev ben Shimon Kabbalah: tradition of hidden knowledge
Halevi, Z'ev ben Shimon Kabbalah and Psychology (1986)
Bath, Gateway Books.
Halevy, Elie The Growth of Philosophic Radicalism (1972)
London, Fabar and Faber.
Halifax, Joan Shamanic Voices: a survey of visionary narratives (1979)
New York, Dutton.
Hall, Darl M The Management of Human Systems (1971)
Cleveland OH, Association for Systems Management, 142 p.
Hall, E J The Hidden Dimension (1926)
New York, Doubleday.
Hall, Edward T Silent Language (1969)
New York, Fawcett World Library.
Hall, Edward T The Dance of Life: the other dimension of time (1982)
New York, Doubleday, 250 p. bibl. ISBN 0-385-19248-7.
Hall, Rebecca Voiceless Victims (1984)
Hallahan, Daniel P and Kauffman, James M Exceptional Children: introduction to special education (1988)
Englewood Cliffs NJ, Prentice Hall, 512 p. illus. 4th ed. ISBN 0-13-295585-7.
Halmos, Paul Towards a Measure of Man: the frontiers of normal adjustment (1957)
London, Routledge and Kegan Paul.
Halmos, Paul Faith of the Counsellors: a study of the theory and practice of social casework and psychotherapy (1966)
New York, Schocken Books.
Halpern, James and Ilsa Projections: our world of imaginary relationships (1983)
New York, Seaview/Putnam.
Halpern, Steven Tuning the Human Instrument (1978)
Belmont CA, Spectrum Research Institute.
Hamacheck, Don E (Ed) Self in Growth, Teaching and Learning: selected readings (1965)
New York, Prentice Hall.
Hamilton, Michael (Ed) The New Genetics and the Future of Man (1972)
Grand Rapids, Eerdmans.
Hammond, G B The Power of Self-Transcendence (1966)
St Louis, Bethany Press.
Hampden-Turner, Charles Radical Man: the process of psychosocial development (1970)
Cambridge MA, Schenkman Publishing, 508 p. bibl. ISBN 0-7156-0607-7.
Hampden-Turner, Charles Maps of the Mind: charts and concepts of the mind and its labyrinths (1981)
New York, Collier Books, 224 p. ISBN 0-02-076870-2.
Handy, Charles The Age of Unreason (1990)
London, Arrow Books, 216 p. bibl. ISBN 0-09-975740-0.
Hanh, Thich N The Miracle of Mindfulness: a manual on meditation (1988)
Boston Ma, Beacon Press, 136 p. 2nd ed. illus. Trans from Japanese by Ho Mobi. ISBN 0-8070-1201-7.
Hannah, B Encounters with the Soul: active imagination as developed by C G Jung (1981)
Boston MA, Sigo Press.
Hannah, Barbara Striving toward Wholeness (1971)
New York, Putnam Publishing Group for the C G Jung Foundation for Analytical Psychology, 316 p. bibl. (pp. 313-316).

Hanneborg, Kurt Anthropological Circling Observations on the Nature of Views of Man in Science, Philosophy and Religion (1962)
Copenhagen, Munksgaard, 72p. bibl. (pp. 57–72).

Hans-Ulrich, Rieker and Becherer, Elsy The Yoga of Light: the classic esoteric handbook of kundalini yoga (1974)
San Rafael CA, Dawn Horse. ISBN 0-913922-07-2.

Happold, F C The Challenge of a Leap–Epoch in Human History
In: *Festschrift for Ludwig von Bertalanffy.*

Happold, F C Mysticism: a study and an anthology (1963)
London, Penguin Books.

Happold, F C Prayer and Meditation: their nature and practice (1971)
London, Penguin Books, 381 p.

Haq Kadija and Huner Kirdar (Eds) Human Development (1986)
Geneva, United Nations Publications. 4 Vols. Vol I: The Neglected Dimension. Vol II: Adjustment and Growth. Vol III: Managing Human Development. Vol IV: Development for People: goals and strategies for the year 2000.

Haraldsson, E Reported Paranormal Experiences: date from a multi–national study
Paper presented at the Department of Psychology, University of Minnesota, Minneapolis, April 1st 1988.

Haraldsson, Erlendur Representative national surveys of psychic phenomena: Iceland, Great Britain, Sweden, USA and Gallup's multinational survey
In: *Journal of the society for psychical research* 53, 1985, pp 145–158.

Harary, K and Targ, R The Mind Race: understanding and using psychic abilities (1984)
New York, Villard Books.

Harbison, Frederick Education for Development
In: *Scientific American* Sep 1963, 209, 3, pp. 140–147.

Harding, M Esther The 'I' and the 'Not–I': a study in the development of consciousness (1965)
New York, Pantheon Books, 244 p. bibl. (pp. 223–228).

Hardy, A The Spiritual Nature of Man: a study of contemporary religious experience (1979)
Oxford NY, Oxford University Press, 162 p. bibl.

Hardy, Alister Living Stream: evolution and man (1968)
New York, World Pub.

Hardy, Alister Clavering The Biology of God: a scientific study of man, the religious animal (1975)

Hardy, Sir Alister Darwin and the Spirit of Man (1984)
London, Collins, 245 p.

Hare, William L Systems of Meditation in Religion (1937)
London, P Allan, 232 p.

Harman, Willis Global Mind Change: the promise of the last years of the twentieth century (1988)
Indianapolis IN, Knowledge Systems, 185 p. bibl. (pp. 171–174).

Harman, Willis W Humanistic Capitalism: another alternative
California CA, Stanford Research Institute. Reprinted from Journal of Humanistic Psychology, Vol 14, No 1, Winter 1974.

Harman, Willis W and Rheingold, Howard Higher Creativity: liberating the unconscious for breakthrough insights (1984)
Los Angeles CA, J P Tarcher, 237 p.

Harner, Michael The Way of the Shaman: a guide to power and healing

Harper, Ralph Human Love (1926)
Baltimore MD, Johns Hopkins University Press.

Harper, Robert A Psychoanalysis and Psychotherapy: 36 systems (1959)
Englewood Cliffs NJ, Prentice Hall, 182 p. bibl.

Harrington, Alan The Immortalist: an approach to the engineering of man's divinity (1969)
New York, Random House, 287 p.

Harris, Dale (Ed) The Concept of Development: an issue in the study of human behavior (1917)
Minneapolis MN, University of Minnesota Press.

Harrison, Owen Spirit: transformation and development in organizations (1987)
Potomac MD, Abbott Publishing, 245 p. illus. bibl.
ISBN 0-9618205-0-0.

Harrison, Ruth Animal Machines (1964)
London, Vincent Stuart.

Hart, Hugh Doak The Mature Person (1931)
New York, F S Crofts, 61 p.

Hart, John The Spirit of the Earth: a theology of the land (1984)
New York, Paulist Press.

Hart, Ray L Unfinished man and the imagination (1968)
New York, Herder and Herder.

Hartley, Margaret Changes in the self–concept during psychotherapy (1951)
Chicago IL, University of Chicago Press.

Hartshorne, C The Logic of Perfection (1973)
La Salle IL, Open Court.

Hartung, Henri Unité de l'homme (1963)
Paris, La Colombe, 219 p. bibl. foot (Collection Sciences et techniques humaines, 3).

Harvey, David E and Schroder, Harold M Conceptual Systems and Personality Organization (1961)
New York, John Wiley.

Harvey, Neil The Renaissance Children (1988)
Paper delivered at meeting of the World Organization for Human Potential.

Hastings, James (Ed) Encyclopedia of Religion and Ethics (1926)
Philadelphia PA, Fortress. 12 Vols. ISBN 0-567-06514-6.

Hatt, H E Cybernetics and the Image of Man (1968)
Abingdon.

Hauer, J W Der Yoga (1958)
Stuttgart.

Haught, John F Religion and Self–Acceptance: a study of the relationship between belief in God and the desire to know (1980)
Lanham MD, University Press of America, 195 p.
ISBN 0-8191-1296-8.

Have, Tonko Totalipeit, Vorm, Struktur: grondproblemen van Felix Kruegers Totaliteitspsychologie (1940)
Groningen, J B Wolters, 242 p. bibl. (pp. 236–242).

Hawkes, J, et al History of Mankind, Cultural and Scientific Development (1964)
Mentor. Vol 1, Part 1.

Hawkins, David On Understanding the Understanding of Children (1967)
American Journal of Diseases of Children 114, pp. 513–20.

Hay, David Exploring Inner Space: is God still possible in the twentieth century? (1987)
London, A R Mowbray, 252 p. bibl. rev ed.
ISBN 0-26-467120-1.

Hay, David and Morisy, Ann Reports on Ecstatic, Paranormal or Religious Experience in Great Britain and the United States: a comparison of trends
In: *Journal for the Scientific Study of Religion* 1978, 17, 3, pp. 225–268.

Hayakawa, S I The Fully Functioning Personality (1956)
ETC 13, pp. 164–181.

Hayes, D A and Mateja, J A Long Term Transfer Effect of Metaphoric Allusion
Paper read at Annual Meeting of the American Educational Research Association, Los Angeles, 1981.

Hayes, Peter, et al The Supreme Adventure: the experience of siddha yoga (1988)
New York, Dell Books. ISBN 0-440-55002-5.

Head, J and Cranston, S L Reincarnation in World Thought: a living study of reincarnation in all ages; including selections from the world's religious, philosophies and sciences and great thinkers of the past and present (1967)
New York, Julian Press.

Head, Joseph and Cranson, S L (Eds) Reincarnation: an East–West anthology (1968)
Wheaton IL, Theosophical Publishing House, 341 p. 2nd ed.

Heard, Gerald Training for the Life of the Spirit

Heater, Derek World Perspectives: aspects of education for international understanding (1986)
Hesketh UK, State Mutual Book and Periodical Service.
ISBN 0-905777-53-0.

Heath, Douglas H Exploration of Maturity: studies of mature and immature college men (1965)
New York, Appleton–Century–Croft, 423 p. bibl.

Heckler, Richard S (Ed) Aikido and the New Warrior (1985)
Berkeley CA, North Atlantic Books, 256 p.
ISBN 0-938190-54-7.

Heffey, Winefride Approaching Maturity
Melbourne VIC, Polding Press, 108 p.

Heibrun, Adam and Stacks, Barbara Virtual Reality: an interview with Jaron Lanier
Whole Earth Review, Fall 1989, p 108.

Heijkoop, H L Faith Healing and Speaking in Tongues
Sunbury PA, Believers Bookshelf, 40 p.
ISBN 0-88172-083-6.

Heiler, Friedrich Contemplation in Christian Mysticism (1960)
In: *Papers from the Eranos Yearbooks: Spiritual Disciplines, Bollingen Series XXX,* 4 New York, Pantheon Books.

Hein, G W Arendsen Psychotherapy and the Spiritual Dimension of Man (1974)
Psychotherapy and psychosomatics, 24, pp. 482–489.

Heinberg, Richard Memories and Visions of Paradise: exploring the universal myths of a lost golden age (1989)
Los Angeles, J P Tarcher.

Heisler, V The Transpersonal in Jungian Theory and Therapy (1973)
In: *Journal of Religion and Health* 12, pp. 337–41.

Helfaer, Philip M The Psychology of Religious Doubt (1972)
Boston MA, Beacon Press.

Helleberg, Marilyn M Beyond T M: a practical guide to the lost tradition of Christian meditation (1981)
Mahwah NJ, Paulist Press, 144 p. ISBN 0-8091-2325-8.

Helleberg, Marilyn M A Guide to Christian Meditation (1985)
New York, Walker and Company, 258 p.
ISBN 0-8027-2489-2.

Heller, Robert Unique Success Proposition (1989)
Sidgwick and Jackson.

Helminiak, Daniel A Spiritual Development: an interdisciplinary study (1987)
Chicago IL, Loyola University Press, 256 p.
ISBN 0-8294-0530-5.

Henderson, Hazel The Politics of the Solar Age: alternatives to economics (1981)
New York, Doubleday.

Hendricks, G and Weinhold, B Transpersonal Approaches to Counseling and Psychotherapy (1982)
Denver, Love Publishing.

Hendricks, Gay and Wells, Russel The Centering Book: awareness activities for children, parents and teachers (1975)
Englewood Cliffs NJ, Prentice Hall.

Henry, J Homeostasis, Society and Evolution: a critique (1955)
Sci. Monthly 81, pp. 300–309.

Henry, J Culture, Personality and Evolution (1959)
American Anthropologist 61, pp. 221–226.

Herbais de Thun, Charles Vicomte de Synthèse de l'Interprétation Astrologique d'Après les Principaux Auleins Modernes (1937)
Editions de la revue Demain Brussels, 241 p. bibl. (pp. 232–33).

Hermann, Margaret G and Milburn, Thomas W A Psychological Examination of Political Leaders (1977)
New York, Free Press. ISBN 0-02-914590-2.

Herrick, C J The Evolution of Human Nature (1961)
New York, Allied Publications.

Herrmann, Theo Problem und Begriff der Ganzheit in der Psychologie (1957)
Vienna, In Kommission bei R M Rohrer, 129 p. bibl. (pp. 119–129) (Osterreichische Akademie der Wissenschaften, Sitzungsberichte 231 Bd, 3 Abb.).

Herzog, E Psyche and Death (1966)
New York, Putnam Publishing Group.

Hillman, James (Ed) Spring: an annual of archetypal psychology and Jungian thought
Dallas, Spring Publications.

Hillman, James Peaks and Vales: the soul/spirit distinction as basis for the differences between psychotherapy and spiritual discipline

Hillman, James Re–Visioning Psychology (1975)
New York, Harper and Row.

Hillman, James Loose Ends: primary papers in archetypal psychology (1975)
Zürich, Spring Publications.

Hillman, James The Myth of Analysis: three essays in archetypal psychology (1978)
New York, Harper Colophon Books.

Hillman, James Healing Fiction (1983)
New York, Station Hill Press, 145 p. ISBN 0-930794-56-7.

Hillman, James Archetypal Psychology: a brief account, together with a complete checklist of works (1983)
Dallas, Spring Publications, 90 p.

Hillman, James The Essential James Hillman: a blue fire (1989)
London, Routledge, 323 p. Introduced and edited by Thomas More in collaboration with the author.
ISBN 0-415-05303-X.

Hillman, James, et al Facing the Gods: Amazons, Ariadne, Rhea, Dionysos, Hermes, Athene, Hestia, Hephaistos, Artemis (1980)
Dallas TX, Spring Publications, 172 p.
ISBN 0-88214-312-3.

Hilton, W The Seal of Perfection (1953)
Burns and Oates.

Himalayan International Institute (Eds) Meditation in Christianity
Honesdale PA, Himalayan Publications, 130 p. rev ed.
ISBN 0-89389-085-5.

Hinnells, John R (Ed) A Handbook of Living Religions (1984)
Harmondsworth, Penguin Books, 528 p.
ISBN 0-14-022342-8.

Hinnells, John R Ed The Penguin Dictionary of Religions (1984)
Harmondsworth, Penguin Books, 550 p.
ISBN 0-14-051-106-7.

Hirai, T Electroencephalographic Study on Zen Meditation (1960)
Psychiatrica et Neurologica Japonica 62, pp. 76–105.

Hirai, T and Koga, E EEG and Zen Buddhism: EEG changes in the course of meditation (1959)
EEG Clin. Neurophysiol 18, 1959, Suppl, p. 52–53.

Hjelle, L A Transcendental Meditation and Psychological Health
Perceptual and Motor Skills 39, 1974, pp623–628.

Hoagland, H and Burhoe R W Evolution and Man's Progress (1962)
New York, Columbia University Press.

Hobbs, Nicholas Group–Centered Psychotherapy
In: *C R Rogers, Client-Centered Therapy* New York, Houghton, 1951.

Hobhouse, Leonard, T Development and Purpose: an essay toward a philosophy of evolution (1913)
London.

Hocart, A The Progress of Man (1933)
London, Oxford University Press.

Hodson Pathway of Perfection
Madras, Theosophical Publication House.
ISBN 0-8356-7018-X.

Hoffman, E The Way of Splendor: Jewish mysticism and modern psychology (1981)
Boulder, Shambhala.

Hoffman, Edward (Ed) Path of the Kabbalah

Hoffman, Edward The Way of Splendor (1981)
Boulder CO, Shambhala Publications.

Hoffman, Edward and Schachter, Zalman M Sparks of Light: counseling in the Hasidic (1983)
Boston MA, Shambhala Publications, 208 p.
ISBN 0-394-72188-8.

Hofstadter, Douglas R and Dennett, Daniel C The Mind's I: fantasies and reflections on self and soul (1981)
New York, Allied Publications.

Holloway, R L Jr The Evolution of the Human Brain: some notes toward a synthesis between neural structure and the evolution of complex behaviour (1967)
General Systems Yearbook Society for General Systems Research, 12, p. 3–16.

Holm, Nils G Religious Ecstacy: based on papers read at the symposium on religious ecstacy held at Abi, Finland 1981 (1982)
Stockholm, Almqvist and Wiksell, 306 p.
ISBN 91-22-00574-9.

Holmes, Urban T A History of Christian Spirituality: an analytical introduction (1980)
New York, Seabury.

Holroyd, Stuart Emergence from Chaos (1957)
London, Victor Gollancz, 222 p. toward the spiritual integrity of the inner life.

Honig, W K (Ed) Operant Behavior (1966)
New York.

Honigmann, J J and Maslow, A Synergy: some notes of Ruth Benedict
In: *American Anthropologist* 1970, 72, pp. 320–333.

Hood, Ralph Conceptual Criticisms of Regressive Explanations of Mysticism
In: *Review of Religious Research* 1976, 17, pp. 179–188.

Hook, S (Ed) Dimensions of Mind (1973)
New York, Collier Books.

Hopke, William E (Ed) Encyclopedia of Career and Vocational Guidance (1987)
Chicago IL, J G Ferguson Publishing, 1950 p. 3 Vols.
ISBN 0–89434–083–2.

Hopkins, Jeffrey Meditation on Emptiness (1983)
London, Wisdom Publication, 1017 p.
ISBN 0–86171–014–2.

Horney, Karen New Ways in Psychoanalysis (1939)
New York, WW Norton.

Horney, Karen Neurosis and human growth (1970)
New York, WW Norton.

Horosz, William Escape from Destiny: self–directive theory of man and culture (1967)
Springfield IL, Thomas Charles C, 290 p. bibl. foot. (phil. anthrop.).

Houston, Jean The Psychenaut Program: an exploration into some human potentials (1973)
Journal of Creative Behavior 7, 4, Fourth quarter pp253–278.

Houston, Jean New Ways of Being, Consciousness and its Transformations (Paper presented at 11th Annual Meeting of the Association for Humanistic Psychology, Montreal, September 1973) (1973)

Houston, Jean The Possible Human: a course in extending your physical, mental and creative abilities (1985)
Los Angeles CA, JP Tarcher.

Hovland, C J and Rosenber, M J Attitude Organization and Change (1960)
New Haven.

Hubbard, L Ron Dianetics: the modern science of mental health
Copenhagen, New Era Publications International, 431 p.
ISBN 87-7336–162–3.

Hubbard, L Ron The Creation of Human Ability (1914)
Los Angeles, American Sailt Hill Organization.

Hubbard, L Ron Dianetics and Scientology Technical Dictionary (1975)
Los Angeles CA, Bridge Publications.

Huett, L and Richardson, W The Spiritual Value of Gem Stones (1980)
Marina del Ray CA, DeVorss and Co.

Hugel, F von The Mystical Element in Religion (1908)
London, Dent.

Huizinga, Johan Homo Ludens (1955)
Boston MA, Beacon Press.

Hull, R F C Bibliographical Notes on Active Imagination in the Works of C J Jung (1971)
Spring.

Hulme, William E Celebrating God's Presence: a guide to Christian meditation (1988)
Minneapolis MN, Augsburg Publishing, 128 p. Christian Growth Books. ISBN 0–8066–2306–3.

Hultkrantz, Ake (Ed) Supernatural Owners of Nature: Nordic symposium on the religious conceptions of ruling spirits and allied concepts (1961)
Stockholm, Almqvist and Wiksell, 165 p.

Humphreys, Christmas Concentration and Meditation
London, Watkins.

Hurvitz, Leon The Eight Liberations (1979)
In: *Studies in Pali and Buddhism* Delhi, BR Publishing, pp. 121-169. Ed AK Narain.

Huxley, A Moksha: writings on psychedelics and the visionary experience (1977)
Cambridge MA, Stonehill.

Huxley, Aldous The Doors of Perception (1954)
New York, Harper and Row.

Huxley, Aldous The Perennial Philosophy (1980)
London, Chatto and Windus, 358 p. First published 1946.
ISBN 0 7011 0812 6.

Huxley, Julian Cultural Process and Evolution (1958)
In: *A E Roe and G G Simpson (Eds) Behaviour and Evolution* New Havel.

Hyman, B The Dominant Metaphors with which Teachers Report they Function in the Classroom Environment (1980)
New York, New York University. Ph D Dissertation.

Ibish, Yusuf and Wilson, Peter Lamborn (Eds) Traditional Modes of Contemplation and Action: a colloquium held at Rothko Chapel, Houston Texas (1977)
Tehran, Imperial Iranian Academy of Philosophy, 477 p. Publication: 24. ISBN 0–500–97355–5.

Ichazo, Oscar The Human Process for Enlightenment and Freedom (1976)
New York, Arica Institute.

Ichazo, Oscar Kinerhythm Meditation: a multifaceted concentration (1978)
New York, Arica Institute Press, 54 p. illus.
ISBN 0–916554–07–4.

Ichazo, Oscar Hypergnostic Meditation Training Manual (1986)
New York, Arica Institute Press, 63 p.
ISBN 0–916554–12–0.

Iijima, Kanjitsu Buddhist Yoga (1975)
Briarcliff Manor NY, Japan Publications, 184 p.
ISBN 0–87040–349–4.

Illich, Ivan Tools for Conviviality (1973)
New York, Harper and Row.

Illich, Ivan D The Dawn of Epimethean Man (1970)
Paper contributed to the Conference on 'Technology, Social Goals and Cultural Options', Aspen, Colorado, August/September.

Illich, Ivan D Celebration of Awareness: a call for institutional revolution (1971)
London, Calder and Boyars, 189 p.

Illich, Ivan D Deschooling Society (1971)
London, Calder and Boyars.

Ilunga Paths of Liberation: a Third World spirituality
Maryknoll NY, Orbis Books, 224 p. ISBN 0–88344–401–1.

Inayat, Khan The Development of Spiritual Healing (1985)
San Bernardino CA, Borgo Press, 96 p.
ISBN 0–89370–582–9.

Indra, M A Ahimsa Yoga: an exposition of Mahatma Ghandi's philosophy of non–violence
Bombay, Minerva Bookshop.

Inge, W H The Idea of Progress (1920)
Romanes Lecture. Humphrey Milford, 1920 (reprinted in appendix to Diary of a Dean).

Ingle, Sud In Search of Perfection (1985)
368 p. illus.
ISBN 0–13-467523–1.

Inglis, Brian Trance: the natural history of altered states of mind (1989)

InnerLinks The Transformation Game (1987)
Seattle WA, InnerLinks, Forres, Findhorn Foundation. Board game.

Institute for Personal Development System of Personal Development: activism for the exercise of the brain and the development of the mind (1969)
London, The Institute, 250 p.

Institute for the Study of Human Knowledge Al-Muqaddasi: revelation of the secrets of the birds and flowers (1980)
Cambridge MA, Institute for the Study of Human Knowledge.
ISBN 0–900860–75–8.

Institute of Cultural Affairs (Eds) The Other World: a spirit journal (1987)
Brussels, Institute of Cultural Affairs, 177 p.

International Bureau of Education Education for International Education (1968)
Geneva, UNESCO. Publication 311.

International Commission for the Development of Education Towards the Complete Man
In: *Learning to Be; The World of Education Today and Tomorrow (Report of the Commission)* Paris, UNESCO, 1972, pp. 153–158.

Irwin, H Flight of Mind: a psychological study of the out–of–body experience (1985)
New York, Scarecrow Press.

Ives, Kenneth H Nurturing Spiritual Development: stages, structure, style (1982)
Chicago IL, Progresiv Pub, 60 p. Studies in Quakerism: 8.
ISBN 0–89670–011–9.

Iyangar, S The Hathayogapradipika (1972)
Adyar, Theosophical Publishing House.

Izutsu, Toshihiko A Comparative Study of the Key Philosophical Concepts of Sufism and Taoism
Tokyo. 2 Vols. Vol 1 (1966), Vol 2 (1967).

Jacks, L P The Education of the Whole Man
London, University of London Press. Portway Reprints.

Jackson, Judith Aromatherapy (1987)
Richmond VIC, Greenhouse Publications, 224 p.
ISBN 0–86436–070–3.

Jacobi, Jolande The Way of Individualism (1967)
New York, Harcourt Brace, 177 p. bibl. (pp. 163–169). transl 1965 ed.

Jacobs, H Western Psychotherapy and Hindu Sadhana (1961)
London, Allen and Unwin.

Jacobson, E Electrophysiology of Mental Activities (1932)
American Journal of Psychology 44.

Jahoda, Marie Toward a Social Psychology of Mental Health (1950)
In: *M J E Senn (Ed) Symposium on the Healthy Personality* New York, Josiah Macy Jr Foundation.

Jahoda, Marie Current Conceptions of Positive Mental Health (1958)
New York, Basic Books.

Jaideva Singh (Trans) Pratyabhijnahrdayam: the secret of self-recognition
Translation, notes, and introduction by Jaideva Singh.

Jaini, Padmanabh S The Jaina Path of Purification

Jamal, Hafiz Key Concepts in Sufi Understanding (1980)
Cambridge MA, Institute for the Study of Human Knowledge, 47 p. ISBN 0–86304–006–3.

James, D G Scepticism and Poetry: an essay on the poetic imagination (1937)
London, Allen and Unwin.

James, M and Jongeward, D Born to Win: transactional analysis with Gestalt experiments (1971)
Reading MA, Addison–Wesley, 1971 (an easy–to–comprehend presentation of transactional theory and ideas together with Gestalt experiments to facilitate awareness).

James W Botkin, Mahdi Elmandjra, Mircea Malitza No Limits to Learning: bridging the human gap (1979)
Oxford, Pergamon Books. ("A Report to the Club of Rome").

James, William The Varieties of Religious Experience (1961)
New York, Colliers Books. First ed 1901.

Jami Les Jaillissements de Lumière: lavayeh (1982)
Paris, Les Deux Océans, 179 p. ISBN 2 86681 003 1. Texte persan |edit é et traduit avec introduction et notes par Yann Richard; ouvrage publié avec le concours du Centre National des Lettres.

Janis, I Victims of Group-Think (1972)
Boston, Houghton Mifflin.

Janov, Arthur The Primal Scream: a revolutionary cure for neurosis (1970)
New York, Putnam Publishing Group.

Jantsch, Erich Consciousness Evolving (1975)
In: *Eric Jantsch, Design for Evolution; self–organization in the life of human systems,* New York, Braziller, pp131-190.

Jantsch, Erich and Waddington, Conrad H (Eds) Evolution and Consciousness: human systems in transition (1976)
Reading MA, Addison-Wesley, 259 p. illus.
ISBN 0–201–03438–7.

Japikse, Carl and Leichtman, Robert R Active Meditation: the Western tradition (1983)
Columbus OH, Ariel Press, 512 p. ISBN 0–89804–040–X.

Jarrell, Howard R International Yoga Bibliography, 1950 to 1980 (1981)
Metuchen NJ, Scarecrow Press, 231 p.
ISBN 0–8108-1472–2.

Jarrell, Howard R International Meditation Bibliography, 1950–1982 (1985)
Metuchen NJ, Scarecrow Press, 444 p. ATLA Bibliography Series: 12. ISBN 0–8108–1759–4.

Javad, Nurbakhsh and Lewishon, Leonard Spiritual Poverty in Sufism (1984)
New York, KhaniQahi–Nimatullahi, Sufi Order.
ISBN 0–933546–11–4.

Jaynes, Julian The Origin of Consciousness in the Breakdown of the Bicameral Mind (1982)
Harmondsworth, Penguin Books, 467 p.
ISBN 014–02–2305-3.

Jenkin, N and Pollack, R H (Eds) Perceptual Development: its relation to theories of intelligence and cognition (1925)
National Institute of Health, 1965.

Jensen, A R Hierarchical Theories of Mental Ability
In: *B Dockrell (Ed) Theories of Intelligence* London, Methuen, forthcoming.

Jeter, Hugh P By His Stripes: the doctrine of divine healing (1977)
Spingfield MO, Gospel Publications, 224 p.
ISBN 0–88243–521–3.

Jha, Mhakan Dimensions of Pilgrimage: an anthropological appraisal (1985)
New Delhi, Inter–India Publications, 180 p.
ISBN 81–210–0007–6.

Jimenez, J On Metaphor: a theory of the nature and educational uses of associative thinking (1972)
Boston MA, University of Massachusetts. Ph D Dissertation.

Jocobs, Louis Jewish Mystical Testimonies (1976)
New York, Schocken Press.

Johansson, Rune E A The Psychology of Nirvana: a comparative study of the natural goal of Buddhism and the aims of modern Western psychology (1969)
London, George Allan and Unwin.

Johari, Harish Leela: game of knowledge (1980)
London, Routledge and Kegan Paul, 150 p.
ISBN 0–7100–0689–6.

Johari, Harish Tools for Tantra (1986)
Rochester VT, Inner Traditions International, 192 p. ASBN 0–89281–055–6.

Johari, Harish Chakras: energy centers of transformation (1987)
Vermont, Destiny Books, 116 p. ISBN 0–89281–054–8.
Summaries of the Traditional Hindu Sadhanas (The fifth stage of life)

Johnsen, Carsten Man the Indivisible: totality versus disruption in the history of western thought (1971)
Oslo, Universitetsforlaget, 333 p.

Johnson, Dwight Spirals of Growth: the emergence of our "future–mind" (1983)
Wheaton IL, Theosophical Publishing House, 168 p. bibl.
ISBN 0–8356–0580–9.

Johnson, R A Inner Work: using dreams and active imagination for personal growth (1986)
San Francisco CA, Harper and Row.

Johnson, Robert A The Psychology of Romantic Love (1984)
London, Arkana, 204 p. ISBN 1–85063–077–1.

Johnson, Williard Riding the Ox Home: a history of meditation from Shamanism to science (1987)
Boston MA, Beacon Press, 262 p. illus.
ISBN 0–8070–1305–6.

Johnston, William The Mysticism of "The Cloud of Unknowing"
Foreword by Thomas Merton.
Catholic and protestant traditions of Christian religious mysticism.

Johnston, William Silent Music: the science of meditation (1979)
New York, Harper and Row. ISBN 0–06–064196–7.

Jonas, Hans The Gnostic Religion: the message of the alien God and the beginnings of Christianity (1963)
Boston MA, Beacon Press, 358 p. bibl. 2nd enlarged ed.
ISBN 0–8070–5799–1.

Jones, Cheslyn, et al The Study of Spirituality (1986)
Oxford, Oxford University Press, 664 p. illus.
ISBN 0–19-504169–0.

Jones, Ernest The Life and Work of Sigmund Freud
New York, Basic Books, 1953-1957, 3 Vols.

Jones, R A Self-Fulfilling Prophecies (1977)
Hillsdale NJ, Lawrence–Erlbaum.

Jones, Richard H Science and Mysticism: a comparative study of western natural science, (1986)
Cranbury NJ, Bucknell University Press, 272 p.
ISBN 0–8387–5093–1.

Jores, A; Freyberger, H and Stokvis, B (Eds) Symposium of the 4th European Conference on Psychomatic Research, Hamburg, 1959: symposium (1960)
Basel, S Karger AG, 328 p. Advances in Psychosomatic Medicine: Vol 1. ISBN 3-8055-0454-3.

Joshi, K S On the Possibility of Yogic Powers (1968)
International Philosophical Quarterly 8, Dec pp. 579–585.

Jourard, Sidney and Overlade, Dan C Reconciliation: a theory of man transcending (1926)
New York, Reinhold Van Nostrand.

Jourard, Sidney M The Transparent Self: self, disclosure and well-being (1924)
New York, Reinhold Van Nostrand.

Jourard, Sidney M Healthy Personality: an approach through the study of healthy personality (1958)
New York, Harper and Row.

Jourard, Sidney M Healthy Personality and Self-Disclosure (1959)
Mental Hygiene 43, pp. 499–507.

Jourard, Sidney M Personal Adjustment (1963)
New York, Macmillan.

Jourard, Sidney M Disclosing Man to Himself: the task of humanistic psychology (1968)
New York, Reinhold Van Nostrand.

Journal of Transpersonal Psychology Symbols of Transpersonal Experiences (1969)
Berlin, Springer-Verlag.

Journet, Charles The Dark Knowledge of God

Judge, A J N Needs Communication: viable needs patterns and their identification (1980)
In: *Katrin Lederer (Ed) Human Needs; a contribution to the current debate* Konigstein, Verlag Anton Hain, 1980, p.279–312. Also in: Forms of Presentation and the Future of Comprehension (Collection of papers mainly presented to the Forms of Presentation sub-project of the Goals, Processes and Indicators of Development project of the United Nations University, 1978–82). Brussels, Union of International Associations, 1984.

Jung, C G La guérison psychologique
Geneva, Librairie de l'Université Georg et Cie.

Jung, C G Zur Empirie des Individuations-Prozesses (1934)
In: *Eranos Jahrbuch 1933* Zurich, Rhein-Verlag, pp. 201–214.

Jung, C G Modern Man in Search of a Soul (1936)
New York, Harcourt Brace.

Jung, C G Traumsymbole des Individuations prozesses. (1936)
In: *Eranos Jahrbuch 1935* Zurich, Rhein-Verlag, pp. 13–133.

Jung, C G The Integration of the Personality (1939)
New York, Farras and Rinehart, 313 p.

Jung, C G Psychological Reflections: an anthology from the writings of Carl G Jung (1953)
New York, Harper and Row.

Jung, C G The Development of Personality (1954)
New York, Pantheon Books. Collected works, vol 17, Bollingen series 20.

Jung, C G The Undiscovered Self (1958)
London, Kegan Paul.

Jung, C G Concerning Mandala Symbolism (1959)
Collected Works, Bollingen Series XX 9, Part 1, pp. 355–384 (and Appendix).

Jung, C G The Archetypes and the Collective Unconscious (1959)
London, Routledge and Kegan Paul.

Jung, C G Psychological Commentary on the Tibetan Book of the Dead (Evans-Wentz (Ed)) (1960)
New York, Oxford University Press.

Jung, C G Psychology and Religion (1962)
New Haven CT, Yale University Press.

Jung, C G Memories, Dreams and Reflections (1963)
New York, Pantheon Books, 398 p.

Jung, C G Mysterium Conjunctionis (1963)
New York, Bollingen. Transl by R F C Hull.

Jung, C G Memories, Dreams, Reflections (1965)
New York, Vintage Books.

Jung, C G Psychology and Alchemy (1968)
Princeton University Press. bibl. pp 487–523. 2nd rev ed. ×3Bollinger Series: 12. Translated by RFC Hull.

Jung, C G Alchemical Studies (1968)
Princeton NJ, Princeton University Press.

Jung, C G Man and his Symbols (1968)
New York, Dell Books.

Jung, C G Archetypes and the Collective Unconscious (1969)
Princeton NJ, Princeton University Press.

Jung, C G Psychological Types (1971)
Princeton NJ, Princeton University Press.

Jung, C G Mandala Symbolism (1972)
Princeton NJ, Princeton University Press. 121 p. Bollingen Series.

Jung, C G and Kerenyi, C Essays on a Science of Mythology (1963)
New York, Harper Torchbook.

Jung, Emma and von Franz, M L The Grail Legend (1971)
New York, C G Jung Foundation for Analytical Psychology, 452p.

Jyotir Maya Nanda, Swami Concentration and Meditation (1971)
Miami FL, Yoga Research Foundation. illus. ISBN 0-934664-03-X.

Kagan, J and Moss, H A Birth to Maturity: a study in psychological development (1962)
New York, John Wiley and Sons.

Kahler, Erich The Rallying Idea (1967)
Santa Barbara, Unicorn Press, 25 p. (Under the auspices of the Van Leer Foundation for the Advancement of Human Culture, Jerusalem).

Kahoe, Richard D The Development of Intrinsic and Extrinsic Religious Orientations
In: *Journal for the Scientific Study of Religion* 1985, 24, 4, pp. 408–412.

Kalupahana, David J Causality: the central philosophy of Buddhism (1975)
Honolulu, University of Hawaii Press.

Kamiya, J Operant Control of the EEG Alpha Rhythm and Some of its Reported Effects on Consciousness (1969)
In: *Charles Tart. Altered States of Consciousness* New York, Wiley, pp. 507–517.

Kamiya, J A Fourth Dimension of Consciousness (1969)
Experimental Medicine and Surgery 27, p. 13–18.

Kamp, C Gratton (Ed) Perspectives on the Group Process: a foundation for counselling with groups (1970)
Boston MA, Houghton Mifflin, 351 p.

Kandinsky, Wassily Concerning the Spiritual in Art and Painting in Particular (1947)
Wittenborn, 1947 (Synesthesia).

Kane, P T The Use of Stories to Promote Figurative Language in Children (1982)
Madison, University of Winsconsin. Ph D Dissertation.

Kanellakos, D P Report on Some of the Current Scientific Studies of Transcendental Meditation (1972)
Los Angeles, Students International Meditation Society.

Kanellakos, Demetri P and Lukas, Jerome S (Eds) The Psychology of Transcendental Meditation: a literature review (1975)
Menlo Park, WA Benjamin, 132 p.

Kao, Charles C Psychological and Religious Development: maturity and maturation (1981)
Lanham MD, University Press of America, 382 p. ISBN 0-8191-1760-9.

Kao, Charles C (Ed) Maturity and the Quest for Spiritual Meaning (1988)
Lanham MD, University Press of America, 224 p. ISBN 0-8191-6972-2.

Kaplan, A Meditation and Kabbalah (1982)
York Beach ME, Samuel Weiser.

Kaplan, Aryeh Jewish Meditation: a practical guide (1985)
New York, Schocken Books, 174 p. ISBN 0-8052-4006-3.

Kaplan, B (Ed) The Inner World of Mental Illness (1964)
New York, Harper and Row.

Kaplan, Bernard and Wapner, Seymour (Eds) Toward a Holistic Developmental Psychology (1983)
Hillsdale NJ, Lawrence Erlbaum, 272 p. ISBN 0-89859-262-3.

Kaplan, L Oneness and Separateness (1978)
New York, Simon and Schuster International.

Kaplan, P Toward a Theology of Consciousness (1976)
Doctoral dissertation, Harvard.

Karagulla, Shafica Breakthrough to Creativity (1967)
Santa Monica CA, de Vorss, 268 p. bibl. (pp. 263–268).

Kardner, Abraham The Psychological Frontiers of Society (1945)
New York, Columbia University Press.

Karl, Frederick R and Hamalian, Leo (Eds) The Existential Imagination (1963)
Greenwich CT, Fawcett Publications Inc, 288 p.

Karlins, Marvin and Andrews, Lewis M Biofeedback: turning on the power of your mind (1973)
New York, Warner Paperback Library, 190 p.

Kaschmitter, William A The Spirituality of the Catholic Church (1982)
Houston TX, Lumen Christi, 980 p. ISBN 0-912414-33-2.

Kasl, S and Reichsman, F (Eds) Epidemiologic Studies in Psychosomatic Medicine (1977)
New York, Karger S AG, 224 p. Advances in Psychosomatic Medicine: Vol 9. ISBN 3-8055-2654-7.

Katchalsky, A Network Thermodynamics (1971)
Nature, 234, 393.

Katz, Joseph et al (Ed) Growth and Constraint in College Students: a study of the varieties of psychological development (1967)
Stanford CA, Stanford University Institute for the Study of Human Problems, 666 p. bibl.

Katz, Nathan Buddhist Images of Human Perfection: the Arahant of Sutta-Pitaka compared with the Bodhisattva and the Mahasiddha (1982)
Delhi, Motilal Banarsidass.

Katz, Richard Education for Transcendence: lessons from the Kung Zhu/twasi (1973)
Journal of Transpersonal Psychology, 5, 2, Fall pp 136–155.

Katz, Steven Mysticism and Religious Traditions (1983)
New York, Oxford University Press.

Katz, Steven J Models, Modeling and Mystical Experience (1982)
In: *Religion* 12, pp. 247–275.

Katz, Steven T (Ed) Mysticism and Philosophical Analysis (1978)
New York, Oxford University Press.

Katz, Steven T Language, Epistemology and Mysticism (1978)
In: *Mysticism and Philosophical Analysis* New York, Oxford University Press, pp. 22–74.

Kaune, Fritz J Selbstverwirklichung: eine konfrontation der psychologie C G Jungs mit der ethik (1967)
Psychologie und Person (Reinhardt) 13, bibl. (pp. 167–168).

Kavanaugh, John F Human Realization: an introduction to the philosophy of man (1971)
New York, Corpus Books, 154p. bibl. (phil. anthrop).

Kavanaugh, Kieran (Ed) John of the Cross: selected writings (1987)
New York NY, Paulist Press.
The Classics of Western Spirituality.
Refs #M1714.

Kazantzakis, Nikos The Saviors of God: spiritual exercises (1960)
New York, Simon and Schuster International, 143 p. Translated, with an introduction by Kimon Friar. ISBN 0-671-20232-4.

Keating, Thomas Open Mind, Open Heart: the contemplative dimension of the Gospel (1986)
Wellspring, 137 p. ISBN 0-916349-07-0.

Keddie, Nikki R (Ed) Scholars, Saints and Sufis: muslim religious institutions since 1500 (1972)
Berkeley.

Kee, Alistair The Way of Transcendence: Christian faith without belief in God (1985)
London, SCM Press Ltd, 281 p. ISBN 0-334-01751-3.
Refs #M2804.
2nd edition.

Keller, Daniel Humor and Therapy (1984)
Pine Mountain. ISBN 0-89769-082-6.

Kelley, C F Meister Eckhart on Divine Knowledge

Kelsey, Morton T Companions on the Inner Way: the art of spiritual guidance (1983)
New York, Crossroad Publishing, 250 p. ISBN 0-8245-0560-3.

Kerenyi, C Prometheus, Archetypal Image of the Human Existence (1967)
New York, Pantheon Books.

Keshavadas, Sant Self-Realization
Bombay, Bharatiya Vidya Bhavan.

Khan, Hazrat I Awakening of the Human Spirit (1988)
New Lebanon NY, Omega Press, 224 p. ISBN 0-930872-35-5.

Khan, Hazrat Inayat Spiritual Dimensions of Psychology (1981)
Lebanon Springs, Sufi Order Publications.

Khan, Pir Vilayat I Introducing Spirituality into Counseling and Therapy (1982)
New Lebanon NY, Omega Press, 176 p. ISBN 0-930872-30-4.

Khatena, Joe The Creatively Gifted Child: suggestions for parents and teachers
New York, Vantage.

Kichla, Kishnan Lal Integrated Personality (1967)
Aligarh, Bharat Pub House, 175p. bibl. foot (in Hildi script).

Kiefer, Durand Intermeditation Notes: reports from inner space (1974)
In: *John White (Ed) Frontiers of Consciousness* New York, Julian Press, pp. 138–153.

Kierkegaard, S The Concept of Dread (1957)
Princeton NJ, Princeton University Press.

Kilpatrick, Franklin P Explorations in Transactional Psychology (1961)
New York, New York University Press.

King, Ursula Towards a New Mysticism: Teilhard De Chardin and Eastern Religions (1980)
New York, Seabury.

King, Ursula The Spirit of One Earth: reflections on Teilhard de Chardin and global spirituality (1989)
New York, Paragon House Publishers.

King, Winston L Theravāda Meditation: the Buddhist transformation of yoga (1980)
Pennsylvania, Pennsylvania State University Press.

King, Winston L Theravada Meditation (1980)
University Park PA.

Kirkpatrick, E A The Individual in the Making
Probably dating from about 1910.

Klausner, S Z The Quest for Self-Control (1965)
New York, Free Press.

Kliebard, H M Curriculum Theory as Metaphor
In: *Theory Into Practice* 1982, 21, 1, pp. 11–17.

Klimo, J Channeling (1988)
Wellingborough, Aquarian Press.

Kluckhohn, Clyde K and Murray, Henry A (Eds) Personality in Nature, Society and Culture (1953)
New York, Alfred A Knopf.

Knight and Otto (Eds) Dimensions in Wholistic Healing: new frontiers in the treatment of the whole person (1979)
Chicago IL, Nelson-Hall.

Knight, G A History of White Magic (1978)
Oxford, Mowbray.

Knight, Peter T and Colletta, Nat J Implementing Programs of Human Development (1980)
Washington DC, International Bank for Reconstruction and Development, 372 p. Working Paper: No 403. ISBN 0-686-36130-X.

Knowles, Richard T and McLean, George F (Eds) Psychological Foundations of Moral Education and Character Development: an integrated theory of moral development (1986)
Lanham MD, University Press of America, 374 p. ISBN 0-8191-5406-7.

Knudsen, C W What do Educators Mean by 'Integration'? (1937)
Harvard Educational Review 7, January pp. 15–26.

Kochumuttom, Thomas A Buddhist Doctrine of Experience: a new translation and interpretation of the works of Vasubandhu the Yogācārin (1982)
Delhi, Motilal Banarsidass.

Koelman, G M Patanjala Yoga: from related ego to absolute self (1970)
Poona.

Koestler, Arthur The Act of Creation (1964)
New York, Macmillan, 751 p.

Koestler, Arthur Biological and Mental Evolution (1965)
Nature 208, 5015, pp. 1033–1036 (11. 12. 1965B).

Koffka, K The Growth of the Mind (1930)
New York, Harcourt Brace.

Koffka, K Principles of Gestalt Psychology (1935)
New York, Harcourt Brace.

Kogan, J Gestalt Therapy Resources (1971)
San Francisco, Lodestar Press.

Kohlberg, L Essays on Moral Development (1981)
San Francisco, Harper and Row. Vol 1.

Kohlberg, L and Kramer, R B Continuities and Discontinuities in Childhood and Adult Moral Development Revisited (1973)
New York, Life-span Developmental Psychology.

Kohlberg, Lawrence The Development of Moral Character and Ideology (1964)
In: *M Hoffman (Ed) Review of Child Psychology* Russell Sage Foundation.

Kohlberg, Lawrence Stage and Sequence: the cognitive-developmental approach to socialization (1969)
In: *David Goslin (Ed) Handbook of Socialization Theory and Research* Rand McNally, ch 6.

Kohlberg, Lawrence Stages of Moral Development as a Basis for Moral Education (1971)
In: *C Beck and E Sullivan (Ed) Moral Education* University of Toronto Press.

Kohr, R L Dimensionality in Meditative Experience: a replication
In: *The Journal of Transpersonal Psychology* 1977, 9, pp. 193–203.

Kohut, H The Restoration of the Self (1977)
New York, International University Press.

Koichi, Tohei Book of Ki: co-ordinating mind and body in daily life

Konner, Melvin The Tangled Wing: biological constraints of the human spirit (1984)
Harmondsworth, Penguin Books, 543 p.
ISBN 0-14-022526-9.

Kopas, J Jung and Assagioli in Religious Perspective (1981)
In: *Journal of Psychology and Theology* 9, pp. 216–23.

Kornfield, J Intensive Insight Meditation: a phenomenological study
In: *The Journal of Transpersonal Psychology* 1979, 11, pp. 11–58.

Kornfield, J Living Buddhist Masters (1977)
Santa Cruz, Unity Press.

Kornfield, J M The Psychology of Mindfulness Meditation (1976)
The Humanistic Psychology Institute. Unpublished doctoral dissertation.

Korzybski, A Manhood of Humanity: the science and art of human engineering (1971)
New York, EP Dutton, 264 p. 2nd ed.

Kostiachenko, Vladislav S Integral 'naia Vedanta (1970)
Moscow, Akademia Nauk SSSR, 191 p. bibl. In Russian.

Kramer, Kenneth P The Sacred Art of Dying: how the world religions understand death (1988)
New York, Paulist Press, 240 p. ISBN 0-8091-2942-6.

Krebs, D Altruism: an examination of the concept and a review of the literature (1970)
Psychological Bulletin 73, April bibl. (pp. 298–302).

Kretschmer, E The Psychology of Men of Genius (1931)
London, Kegan Paul.

Kretschmer, Wolfgang Selbsterkenntniss und Willensbildung im Rärtzlichen Raume (1958)
Stuttgart, G Thieme.

Kretschmer, Wolfgang Meditative Techniques in Psychotherapy (1962)
Psychologia 5, pp. 76–83.

Krishna, Gopi Higher Consciousness: the evolutionary thrust of 'Kundalini'
Bombay, Taraporevala Publishing Industries.

Krishna, Gopi The Real Nature of Mystical Experience

Krishna, Gopi Kundalini: the evolutionary energy in man (1970)
Berkeley CA, Shambhala Publications, 252 p. (with a psychological commentary by James Hillman).

Krishna, Gopi The Biological Basis of Religion and Genius (1972)
New York, Harper and Row.

Krishna, Gopi The True Nature of Mystical Experiences (1978)
New Concepts.

Krishna, Iswara Sankhya Karika or Sankhya Yoga
Bombay, Bharatiya Vidya Bhavan, 260 p.

Krishnamurti, J The Awakening of Intelligence

Krishnamurti, J The First and Last Freedom (1954)
London, Gollancz.

Krishnamurti, J Freedom from the Known (1969)
London, Victor Gollancz, 124 p. Edited by Mary Lutyens.

Krishnamurti, J The Only Revolution (1970)
London, Gollancz.

Kriyananda, Goswami The Spiritual Science of Kriya Yoga

Kroeber, Alfred L Configurations of Culture Growth (1944)
Berkeley CA, University of California Press.

Krutch, Joseph Wood The Measure of Man
New York, Bobbs-Merrill Company.

Krutch, Joseph Wood Human Nature and Human Condition (1959)
New York, Random House.

Kuang-ming Wu Chuang Tzu: world philosopher at play (1982)
New York, Crossroad Publishing.

Kubie, Lawrence S The Nature of Psychological Change and Its Relation to Cultural Change
In: *Ben Rothblatt (Ed) Changing Perspectives on Man* University of Chicago Press, 1968, pp. 147–148.

Kubie, Lawrence S The Problem of Maturity in Psychiatric Research (1953)
Journal of Medical Education 28, 10, October p. 11–27.

Kuchinsky, Saul Systematics: search for miraculous management (1985)
Charles Town, West Virginia, Claymont Communications, 271 p. ISBN 0-934254-12-5.

Kugler, P The Alchemy of Discourse
Cranbury NJ, Bucknell University Press.

Kuhlman, Thomas L Humor and Psychotherapy (1984)
Dorsey. ISBN 0-256-43600-2.

Kulandran Concept of Transcendence: a study of it in various world religions (1982)
Blantyre, Christian Literature Association.

Kulandran, Subapathy Concept of Transcendence: a study of it in various world religions (1982)
Blantyre, Christian Literature Association, 378 p.

Kulshreshtha, Saroj The Concept of Salvation in Vedanta (1986)
Columbia MO, South Asia Books, 120 p.
ISBN 0-317-61788-5.

Kunkel, Fritz In Search of Maturity: an inquiry into psychology, religion and self-education (1943)
New York, Charles Scribner, 292 p.

La Fontaine, Jean S Initiation: ritual drama and secret knowledge across the world (1985)
Harmondsworth, Penguin Books, 208 p.
ISBN 0-14-022124-7.

La Vallée Poussin, Louis de The Way to Nirvana: six lectures on ancient Buddhism as a discipline of salvation (1917)
Cambridge, Cambridge University Press.

Lacan, J Language of the Self (1968)
Baltimore MD, Johns Hopkins University Press.

Lacarriere, Jacques The Gnostics (1977)
New York, E P Dutton, 136 p. Foreword by Lawrence Durrell.
ISBN 0-525-47455-2.

Lagache, D L'unité de la psychologie (1949)
Paris.

Laing, R D Wisdom, Madness and Folly: the making of a psychiatrist (1986)
New York, McGraw-Hill Publishing, 208 p. illus.
ISBN 0-07-035850-8.

Laing, Ronald D Divided Self: an existential study in sanity and madness (1960)
New York, Penguin Books.

Laing, Ronald D The Politics of Experience (1966)
New York, Pantheon Books.

Laing, Ronald D Self and Others (1970)
New York, Pantheon Books.

Laing, Ronald D and Esterson, A Sanity, Madness and the Family (1965)
Vol. 1: Families schizophrenics. Basic.

Lama Anagarika Govinda Psycho-cosmic Symbolism of the Buddhist Stupa (1976)
Berkeley CA, Dharma Publishing, 102 p. illus.
ISBN 0-913546-35-6.

Land, G T Grow or Die: the unifying principle of transformation (1973)
New York, Random House.

Langer, Ellen Mindfulness (1989)
Reading MA, Addison Wesley.

Lankton, Steve Practical Magic: a translation of basic neuro-linguistic programming into clinical psychotherapy (1980)
Cupertino CA, Meta Publications, 250 p.
ISBN 0-916990-08-7.

Lapshin, Ivan I La Synergie spirituelle (1935)
Prague, Bulletin de l'Association russe pour les recherches scientifiques à Prague, 33 p. 2, 7, No. 6.

Lardner, Carmody Denise and Carmody, John Tully Western Ways to the Center: an introduction to religions of the West

Lardner Carmody, Denise and Carmody, John Tully Eastern Ways to the Center: an introduction to Asian religions

Larsen, S The Shaman's Doorway: opening the mythic imagination to contemporary consciousness (1976)
New York, Harper and Row.

Laski, M Ecstasy (1961)
London, Cresset Press.

Laski, Marghanita Ecstasy: a study of some secular and religious experiences
New York, Greenwood Press.

Laszlo, Ervin The Inner Limits of Mankind: heretical reflections on today's values, culture and politics (1989)
London, Oneworld Publications, 143 p.
ISBN 1-85168-015-2.

Laszlo, Ervin and Stulman, J Emergent Man: his chances, problems and potentials (1973)
New York, Gordon and Breach, 185 p.

Lati Rinbochay Mind in Tibetan Buddhism (1986)
Ithaca NY, Snow Lions Publications, 172 p. bibl. Translated, edited and introduced by Elizabeth Napper.

Latroff, George and Johnson, Mark Metaphors to Live By (1980)
Chicago, University of Chicago Press.

Lawrence, John Freemansonry: a way of salvation? (1982)
Grove Books.

Lawrence of the Resurrection, Brother The Practice of the Presence of God
Trans. by Sister Mary David.

Le Shan, Lawrence Physicists and Mysticism: similarities in world view (1969)
Journal of Transpersonal Psychology Fall.

Le Shan, Lawrence Toward a General Theory of the Paranormal (1969)
New York, Parapsychology Foundation.

Le Vine, A and White, M I (Eds) Human Conditions: the cultural factors in educational development (1986)
London, Routledge and Kegan Paul.

Leary, T Exo-Psychology (1977)
Los Angeles CA, Starseed/Peace Press.

Leary, Timothy The Politics of Ecstasy (1970)

Leary Timothy, Ralph Metzner and Richard Alpert The Psychedelic Experience (1971)
London, Academy Books.

Lecky, P Self-Consistency: a theory of personality (1945)
New York, Island Press.

Leclercq, Jean The Love of Learning and the Desire for God: a study of monastic culture (1961)
New York, Fordham University Press.

Lederer, Katrin (Ed) Human Needs: a contribution to the current debate (1978)
Cambridge MA, Oelgeschlager, Gunn and Hain, 360 p. bibl.
ISBN 0-89946-027-5.

Lee, Alfred M Multi-valent Man (1966)
New York, George Braziller.

Lee, D Autonomous Motivation (1962)
Journal of Humanistic Psychology 1, pp. 12–22.

Lee, Jung Young The I Ching and Modern Science (1971)
(Paper presented to the 28th International Congress of Orientalists, Canberra, Australia, 8 January 1971).

Lee, Philip R, et al Symposium on Consciousness (1976)
New York, Viking.

Leech, K Soul Friend (1977)
London, Sheldon Press.

Leeper, R R (Ed) Humanizing Education: the person in the process (1967)
Alexandria VA, Association for Supervision and Curriculum Development.

Leeuw, J J Van der The Conquest of Illusion (1968)
Wheaton IL, Quest Books.

Lefébure, Francis Le Mixage Phosphénique: le jour d'Ingeborg: développement de la mémoire et de l'intelligence par le mélange des pensées avec les phosphènes: ses conséquences politiques (1970)
Paris, 169 p.

Lefébure, Francis Du Moulin à Prière à la Dynamo Spirituelle: ou la machine à faire monter Koundalini (1984)
Paris, Francis Lefébure, 208 p. ISBN 2-901032-08-7.

Lehodey, Dom W The Ways of Mental Prayer (1982)
Rockford IL, TAN Books, 408 p. ISBN 0-89555-178-0.

Leming, James S Foundations of Moral Education: an annotated bibliography (1983)
Westport CT, Greenwood Press, 325 p.
ISBN 0-313-24165-1.

Leonard, George The Human Potential (1968)
In: *Education and Ecstasy* New York, Delacorte, pp. 23–50.

Leonard, George Education and Ecstasy (1969)
New York, Dell Books.

Leonard, George The Transformation (1972)
New York, Dell Books.

Leonard, George B The Transformation: a guide to the inevitable changes in humankind (1987)
Los Angeles CA, Jeremy P Tarcher, 278 p.
ISBN 0-87477-169-2.

Lerner, Max Education and a Radical Humanism (1962)
Columbus OH, Ohio State University Press.

Lersch, P The Levels of the Mind (1957)
In: *H P David and H von Bracken (Eds) Perspectives in Personality Theory* New York, pp. 212–217.

Lesh, T Zen Meditation and the Development of Empathy in Counselors (1970)
Journal of Humanistic Psychology 10.

LeShan, Lawrence Psychological States as Factors in the Development of Malignant Disease: a critical review
In: *Journal of the National Cancer Institute* 1959, 22, pp. 1–18.

LeShan, Lawrence Alternate Realities: the search for the full human being (1976)
New York, Ballantine Books, 193 p. ISBN 0-345-34924-5.

Levenson, Edgar A A Holographic Model of Psychoanalytic Change
In: *Contemporary Psychoanalysis* 1975, 12, 1.

Levine, Louis Personal and Social Development: the psychology of effective behavior (1963)
New York, Holt, Rinehart and Winston.

Levinson, David The Seasons of Man's Life (1978)
New York, Ballantine.

Levy, Marion J Some Problems for a Unified Theory of Human Nature (1963)
In: *E A Tiryakian (Ed) Sociological Theory, Values and Sociocultural Change* Free Press, pp. 9–32.

Lewin, Kurt Principles of Topological Psychology (1936)
New York, McGraw Hill Book Company.

Lewin, Kurt Psychological Ecology (1952)
In: *Lewin, Kurt, Field Theory in Social science; selected theoretical papers* London, Tavistock, pp. 170–187.

Lewin, Kurt Regression, Retrogression and Development (1952)
In: *Kurt Lewin, Field Theory in Social Science; selected theoretical papers* London, Tavistock, pp. 87–129 (includes: representation of developmental levels by means of scieltific constructs).

Lewis, B A and Pucelik F Magic Demystified: a pragmatic guide to communication and change – an introduction to NLP (1982)
Portland OR, Metamorphous Press, 160 p. Revised edition.
ISBN 0-943920-00-0.

Lewis, C S The Four Loves (1960)
London, Collins, Fontana Books, 128 p.
ISBN 0-00-620799-5.

Lewis, C S Miracles: a preliminary study (1974)
Glasgow, Collins, Fontana Books, 190 p. First published 1947.

Lewis, C S Surprised by Joy (1974)
Huntington NY, John M Fontana Publishing.

Lewis, Clive S Surprised by Joy: the shape of my early life (1956)
New York, Harcourt Brace.

Lewis, Howard R and Streitfeld, Harold S Growth Games: over 200 techniques from the Human Potential Movement that can help you be more loving and giving, more open and honest, more creative and zestful (1970)
London, Souvenir Press, 280 p. illus. bibl. ISBN 0-285-62055-X.

Library of Tibetan Worls and Archives Tibetan Tradition of Mental Development: oral teachings of Tibetan Lama Geshey Ngawang Dhargyey (1976)
Dharamsala, Library of Tibetan Works and Archives, 239 p. 2nd rev ed. Foreword by the Junior Tutor to his Holiness the Dalai Lama.

Licata, Salvatore J and Petersen, Robert P Historical Perspectives on Homosexuality (1982)
New York, Haworth Press, 224 p. ISBN 0-917724-27-5.

Lichtenstein, Heinz The Dilemma of Human Identity: notes on self, transformation, self, observation and metamorphosis (1963)
Journal of the American Psychoanalytical Association 11, pp. 173-223.

Lidz, Theodore The Person: his development throughout the life cycle (1968)
New York, Basic Books.

Lieshout, Cornelius van and Ingram, David (Eds) Stimulation of Social Development in School: report of the European Contact Workshop held at the university of Nijmegen 1976 under the auspices of the Council of Europe (1977)
Lisse, Swets en Zeitlinger, 234 p. ISBN 90-265-0258-3.

Lifton, Robert J Thought Reform and the Psychology of Totalism: a study of 'brainwashing' in China (1961)
London, Victor Gollancz.

Lifton, Robert J Brainwashing and the Psychology of Totalism (1963)
New York, WW Norton.

Lifton, Robert J Adaptation and Value Development: self-process in protean man (1968)
In: *The Development and Acquisition of Values* report of a Conference, National Institute of Child Health and Human Development, Washington, D C, 15-17 May.

Lifton, Robert J The Future of Immortality: and other essays for a nuclear age (1987)
New York, Basic Books, 368 p. ISBN 0-465-02597-8.

Lilly, J Simulations of God (1976)
New York, Bantam Books.

Lilly, J C The Center of the Cyclone (1972)
New York, Julian Press.

Lilly, John C The Centre of the Cyclone (1973)
London, Calder and Boyars.

Lilly, John C Programming and Metaprogramming in the Human Biocomputer (1974)
New York, Bantam Books.

Lilly, John C The Deep Self (1977)
New York, Simon and Schuster International.

Lilly, John C and Lilly, Antoinetta The Dyadic Cyclone: the autobiography of a couple (1978)
London, Paladin - Granada Publishing, 251 p. ISBN 0-586-08276-X.

Lindeman, E C Integration as an Educational Concept (1937)
In: *L Thomas Hopkins (Ed) Integration; its meaning and application* New York, Appleton-Century, pp. 21-35.

Linden, W The Relation Between the Practicing of Meditation by School Children and Their Levels of Field dependence-independence, test anxiety and reading achievement (1972)
PhD dissertation, New York University.

Lindenberg, Vladimir Mediation and Mankind (1959)
London, Rider.

Linssen, Robert L'Homme Transfini (1984)
Paris, Courrier du Livre, 126 p. bibl. ISBN 2-7029-0157-3.

Lip, Evelyn Feng Shui: a layman's guide to Chinese geomancy (1987)
Union City CA, Heian International Inc, 121 p. First published 1979. ISBN 0-89346-286-1.

Lipowski, Z J (Ed) Psychological Aspects of Physical Illness (1972)
Basel, S Karger AG, 275 p. Advances in Psychosomatic Medicine: Vol 8. ISBN 3-8055-1339-9.

Loevinger, J The Meaning and Measurement of Ego Development
American Psychologist 1966, 21, pp. 195-206.

Loevinger, J Ego Development (1976)
San Francisco CA, Jossey-Bass Publishers.

London, Perry Modes and Morals of Psychotherapy (1964)
New York, Holt, Rinehart and Winston.

Lonergan, B Insight, A Study of Human Understanding (1970)
New York, Philosophical Library.

Long, Herbert C Unfolding the Ideal Life (1980)
London, SPCK, 476 p.

Lorenz, Konrad The Waning of Humaneness (1988)
London, Unwin Paperbacks, 250 p. First published in German, 1983. Trans. by Robert Warren Kickert. ISBN 0-04-440442-5.

Lorimer, David Whole in One (1990)
Arkana, 340 p.

Lovejoy, A O The Great Chain of Being (1936)
Cambridge MA, Harvard University Press. reprinted 1960.

Lovelace, R Dynamics of Spiritual Life (1979)
Exeter, Paternoster Press.

Lowen, Alexander The Betrayal of the Body (1967)
New York, Macmillan.

Lowen, Alexander Bioenergetics (1975)
New York, Coward, McCann and Geoghegan.

Lozanov, Georgi Suggestology and Outlines of Suggestopedy (1978)
New York, Gordon and Breach.

Lu Kuan Yü Taoist Yoga: alchemy and immortality (1970)
London, Rider, 206 p. ISBN 0-09-102701-2.

Luck, B T The Influence of Biological Concepts and Metaphors on the Development of the Psychology of Learning (1983)
New York, Columbia University. Dissertation.

Ludwig, A M Altered States of Consciousness
In: *Archives of General Psychiatry* 1966, 15, pp. 225-234.

Luibheid, Colm (Trans) Pseudo-Dionysius: the complete works (1987)
New York NY, Paulist Press.
The Classics of Western Spirituality.

Lukoff, D The Diagnosis of Mystical Experiences with Psychotic Features
In: *Journal of Transpersonal Psychology* 1985, 17 (2), pp. 155-181.

Luthe, W Autogenic Training: research and theory (1970)
New York, Grune and Stratton.

Lutz, K A A Bibliography of Research on Imagery and Holistic Learning Strategies (1980)
Iowa City, Iowa University. Visual Scholars Program: 8.

Lutz, Ken and Lutz, Mark Humanistic Economics: the new challenge (1988)
New York, Bootstrap Press.

Lynch, J The Broken Heart: the medical consequences of loneliness (1977)
New York, Basic Books.

Lyons, Cathie Journey Toward Wholeness (1987)
New York, Friendship Press. ISBN 0-377-00171-6.

Lyons, Joseph Ecology of the Body: styles of behavior in human life (1987)
Durham NC, Duke University Press, 339 p. ISBN 0-8223-0710-3.

Lysebeth, André Van Pranayama: the yoga of breathing (1979)
New Delhi, Vikas Publishing House, 230 p. ISBN 0-04-149050-9.

MacDowell, Mark A Comparative Study of the Teachings of Don Juan and Madhyamaka Buddhism: knowledge and transformation (1986)
Delhi, Motilal Banarsidass, 116 p. bibl. (pp. 110-113). ISBN 81-208-0162-8.

Mace, C A Homeostasis, Needs and Values (1953)
British Journal Psychology 44, pp. 200-10.

Mackey, Albert G Encyclopedia of Freemansonry (1946)
Richmond VA, Macoy Publishing and Supply. 3 Vols.

MacKinnan, I H The Psychiatrist Views Integration (1937)
In: *Integration; its meaning and application (L T Hopkins, Ed)* New York, Appleton-Century, pp. 126-47.

MacKinnon, D W The Highly Effective Individual
In: *Mooney and Razik (Eds) Explorations in Creativity* 1967.

MacMurray, John The Structure of Religious Experience (1971)
Hamden CT, Shoe String Press, 77 p. repr of 1936 ed. ISBN 0-208-00958-2.

Macy, Joanna Rogers Despair and Personal Growth in the Nuclear Age (1983)
Philadelphia, New Society Publishers.

Madhavtirtha, Swami Integral Education (1952)
Valad, Vedanta Ashram, 124 p.

Madhu Khanna Yantra: the Tantric symbol of cosmos unity (1979)
London, Thames and Hudson, 176 p. illus. bibl.

Magnus, Bernd Nietzsche's Existential Imperative (1978)
Bloomington IN, Indiana University Press, 256 p. ISBN 0-253-34062-4.

Magsam, Charles Experience of God (1975)
Richmond VIC, Spectrum Publications, 238 p.

Mahadevan, T M P (Trans) The Wisdom of Unity (Manisa-Pancakam) of Sri Sankaracarya

Mahadevan, T M P Self-knowledge (1975)
New Delhi, Arnold Heinemann, 100 p.

Maharishi European Research University Maharishi's Programme to Create World Peace: global inauguration (1987)
Washington DC, Age of Enlightenment Press, 574 p.

Mahasi, Sayadaw The Process of Insight (1965)
Kandy, The Forest Hermitage. Trans. by Nyanaponika Thera.

Mahasi, Sayadaw Practical Insight Meditation (1972)
Santa Cruz, Unity Press.

Mahathera, Paravahera Vajiranana Buddhist Meditation in Theory and Practice: a general exposition according to the Pali Canon of the Theravada School (1987)
Kuala Lumpur, Buddhist Missionary Society, 496 p. 3rd ed. bibl. ISBN 967-9920-41-0.

Mahesh, Yogi Maharishi The Science of Being and Art of Living (1966)
London. In: SRM Publications, pp. 58-59.

Mahler, M S Thoughts about Development and Individuation (1963)
In: *The Psychoanalytic Study of the Child* 7, pp. 286-306.

Mahler, M S On Human Symbiosis and the Vicissitudes of Individuation (1968)
New York, International University Press.

Mahrer, A Experiencing (1978)
New York, Brunner/Mazel.

Main, John The Present Christ: further steps in meditation (1986)
New York, Crossroad Publishing, 121 p. ISBN 0-8245-0740-1.

Maisel, Edward T'ai Chi for Health (1963)
Englewood Cliffs NJ, Prentice Hall.

Maisel, Edward The Essential Writings of F M Alexander (1974)
London, Thames and Hudson. Selected and with an introduction by Maisel E.

Malamud, D and Machover, S Toward Self-Understanding: group techniques in self-confrontation (1965)
Springfield IL, Thomas Charles C.

Malgady, R G Figurative Language Development and Academic Achievement (1976)
Brockport, State University of New York. Unpublished paper.

Mallmann, Carlos A and Nudler, Oscar (Eds) Human Development in its Social Context: a collective exploration (1986)
London, Hodder and Stoughton, 269 p. In association with the United Nations University. ISBN 0-340-38517-0.

Mallmann, Carlos and Nudler, Oscar (Eds) Time, Culture and Development (1982)
Oxford, Pergamon Press.

Mallory, Marilyn May Christian Mysticism: transcending techniques: A theological reflection on the empirical testing of the teaching of St John of the Cross

Malone, T P and Whitaker, C A The Roots of Psychotherapy (1953)
New York, McGraw-Hill.

Maloney, P L Metaphor: the poetic mode of knowledge and its implications for education (1980)
Toronto, University of Toronto. Ph D Dissertation.

Mampra, Thomas (Ed) Religious Experience (1981)
Bangalore, Dharmaram Publications, 217 p.

Man, Paul de The Epistemology of Metaphor (1978)
In: *Critical Enquiry* 5, pp. 13-30.

Manganelli, Louise A Biomagnetism: an annotated bibliography (1972)
George Washington University Medical Center.

Manjusrimitra Primordial Experience: an introduction to rDzogs-chen meditation

Mann, R The Light of Consciousness: explorations in transpersonal psychology (1984)
Albany NY, State University of New York Press.

Manoranjan Basu Fundamentals of the Philosophy of Tantras (1986)
Calcutta, Mira Basu Publishers, 667 p. illus. bibl.

Manzoor, Suhail Systems Bibliography: a thirty years literature survey (1982)
Delhi, Metropolitan Book Company, 182 p.

Mapkey, J F The Symbolic Process (1928)
London, Kegan Paul.

Marcuse, H One-Dimensional Man (1964)
Boston MA, Beacon Press.

Marcuse, Herbert Eros and Civilization
London, Sphere, 21 p.

Maréchal, J Studies in the Psychology of the Mystics (1964)
New York, Allied Publications.

Marguerite, Craig and James, H Synergic Power: beyond domination, beyond permissiveness (1979)
Berkeley CA, ProActive Press, 164 p. 2nd ed. bibl. ISBN 0-914158-28-7.

Maritain, Jacques Integral Humanism: temporal and spiritual problems of the new Christendom (1968)
New York, Scribner, 308 p. bibl. foot.

Markley, O W, et al Changing Images of Man (1973)
Menlo Park: Stanford Research Institute, 347 p. Policy Research Report: 3.

Marlan, Stanton Depth Consciousness
In: *The Metaphors of Consciousness* 1981, New York, Plenum Publishing, 521 p. illus. bibl.

Marshall, Berenice (Ed) Experience in Being (1971)
Pacific Grove CA, Brooks/Cole Publishing.

Marston, William M et al Integrative Psychology: a study of unit response (1921)
New York, Harcourt Brace, 558 p. bibl. (pp. 490-95).

Martin, A P Think Proactive: new insights into decision-making (1983)
Ottawa ON, Professional Development Institute. ISBN 0-86502-000-0.

Martin, Graham D Shadows in the Cave: mapping the conscious universe (1990)
Arkana, 253 p.

Martin, P W Experiment in Depth: a study of the work of Jung, Eliot and Toynbee (1955)
London, Routledge and Kegan Paul, 275 p.

Martin, P W Experiment in Depth (1967)
London, Routledge and Kegan Paul.

Maruyama, Magoroh Paradigmatology and its application to cross-disciplinary, cross-professional and cross-cultural communication (1974)
Cybernetica (1974), 17, p. 135-156, p. 237-281.

Mary Esther, Sister Integration of Personality of the Christian Teacher (1928)
Milwaukee, Bruce Pub, 113 p. bibl. (pp. 111-13).

Mascaró, Juan Lamps of Fire: the spirit of religions: from the scriptures and wisdom of the world (1961)
London, Eyre Methuen, 279 p. SBN 413-29599-1.

Maslow, A Toward a Psychology of Being (1968)
New York, Reinhold Van Nostrand, 240 p. 2nd ed. bibl.

Maslow, A H with Gross, L Synergy in Society and in the Individual (1964)
Journal of Individual Psychology 20, pp. 153-64.

Maslow, Abraham and Hung-Min Chiang Healthy Personality
New York, Reinhold Van Nostrand.

Maslow, Abraham H Toward a Humanistic Biology (unpublished series of memoranda written at the request of the Director of the Salk Institute of Biological Studies 1968)

Maslow, Abraham H Power Relationships and Patterns of Personal Development (1917)
In: *A Kornhauser (Ed.) Problems of Power in American Democracy* Wayne Univ 1957.

Maslow, Abraham H Motivation and Personality (1954)
New York, Harper and Row.

Maslow, Abraham H Holistic Dynamic Theory in the Study of Personality (1954)
In: *Motivation and Personality* Chapter 3, pp. 22–62, New York, Harper and Row.

Maslow, Abraham H Cognition of Being in the Peak Experiences (1956)
Waltham MA, Brandeis University.

Maslow, Abraham H New Knowledge in Human Values (1959)
New York, Harper and Row.

Maslow, Abraham H Critique of Self-actualization (1959)
Journal of Individual Psychology 15, pp. 24–32.

Maslow, Abraham H Creativity in Self-Actualizing People (1959)
In: *H H Anderson (Ed) Creativity and its Cultivation* New York, Harper bibl.

Maslow, Abraham H Lessons from the Peak-Experience (1962)
Journal of Humanistic Psychology 2, pp. 9–18.

Maslow, Abraham H The Creative Attitude (1963)
The Structurist 3, pp. 4–10 (reprinted by Psychosynthesis Foundation 1963).

Maslow, Abraham H Synergy in the Society and in the Individual (1964)
Journal of Individual Psychology 20, p. 153–64.

Maslow, Abraham H Isomorphic Interrelationships between Knower and Known (1965)
In: *Sign, Image, Symbol* New York, Braziller.

Maslow, Abraham H Eupsychian Management: a journal (1965)
Homewood IL, Irwin and Dorsey.

Maslow, Abraham H The Psychology of Science (1966)
New York, Harper and Row.

Maslow, Abraham H A Theory of Metamotivation: the biological rooting of the value–life (1967)
Journal of Humanistic Psychology 7, 1967 pp. 93–127.

Maslow, Abraham H Neurosis as a Failure of Personal Growth (1967)
Humanitas.

Maslow, Abraham H Music Education and Peak-Experiences (1968)
Music Educators Journal.

Maslow, Abraham H Some Educational Implications of the Humanistic Psychologies (1968)
Harvard Educational Review Fall.

Maslow, Abraham H Religious Values and Peak Experiences (1970)
New York, Viking Compass Book.

Maslow, Abraham H Towards a Humanistic Biology (1970)
Fields within fields 3, 1, p. 4–18.

Maslow, Abraham H The Farther Reaches of Human Nature (1972)
New York, The Viking Press, 423 p. ISBN 670–30853–6.

Mason, D I Synesthesia and Sound Spectra (1952)
Word 8, 1.

Masters, R E L and Houston, Jean The Varieties of Psychedelic Experience: the first comprehensive guide to the effects of LSD on human personality (1967)
New York, Dell Books, 326 p.

Matics, M L Entering the Path of Enlightenment (1970)
London, Macmillan. Translation.

Matos, Leo Experiencing Separate Realities with and without Drugs: meditation as an eventual alternative to drug use (Paper presented before the 4th Annual International Conference on Humanistic Psychology, Paris, September 1973) (1973)

Matson, Floyd Being, Becoming and Behavior (1967)
New York, Braziller.

Matson, Katinka The Encyclopaedia of Reality: a guide to the New Age (1979)
London, Granada Publishing, 362 p. rev ed.
ISBN 0 586 08301 4.

Matt, Daniel Chanan (Trans) Zohar: the book of splendor
Jewish tradition of religious mysticism.

Maupin, E W Individual Differences in Response to a Zen Meditation Exercise
Journal of Consulting Psychology 29, 1965, pp. 135–45.

Maven, Alexander The Mystic Union: a suggested biological interpretation (1969)
Journal of Transpersonal Psychology 1, 1, Spring.

Max-Neef, Manfred et al Human Scale Development: an option for the future (1990)
CEPAUR, Dag Hammarskjöld Foundation, 80 p. Development Dialogue, reprint from 1989:1. ISSN 0345–2328.

Maxwell, Meg and Tschudin, Verena (Eds) Seeing the Invisible: modern religious and other transcendent experiences (1990)
London, Penguin Books, 216 p. ISBN 0–14–019222–0.

May, G Will and Spirit: a contemplative psychology (1982)
San Francisco CA, Harper and Row.

May, Gerald Simply Sane: the spirituality of mental health (1982)
New York, Crossroad Publishing, 144 p. Crossroad Paperback Series. ISBN 0–8245–0448–8.

May, Rollo Man's Search for Himself (1953)
New York, WW Norton.

May, Rollo Existential Psychology (1961)
New York, Random House.

May, Rollo Psychology and the Human Dilemma (1966)
New York, Reinhold Van Nostrand.

May, Rollo Love and Will (1969)
New York, WW Norton.

May, Rollo, Angel, E and Ellenberger, H (Eds) Existence: a new dimension in psychiatry and Psychology (1958)
New York, Basic Books.

Mayers, P Flow in Adolescence and Its Relations to the School Experience (1978)
University of Chicage. Unpublished Doctoral Dissertation.

Mayo, C A Cognitive Complexity and Conflict Resolution in Impression Formation (1960)
Unpublished Ph. S. thesis, Clark University.

Maziarz, Edward A (Ed) Evolution of Man's Values
New York, Gordon and Breach.

McCarroll, Toby An Introduction to Meditation
Humanist Institute.

McCarroll, Tolbert Exploring the Inner World: a guidebook for personal growth and renewal (1974)
New York, Julian Press.

McClain, Ernest G The Pythagorean Plato: prelude to the song itself (1978)
Stony Brook, Nicholas Hays.

McClain, F A Practical Guide to Past Life Regression (1986)
St Paul MN, Llewellyn Publications.

McClelland, D Power: the inner-experience (1975)
New York, John Wiley.

McClelland, D C and Winter, D G Motivating Economic Development (1969)
New York, Allied Publications.

McClelland, D C, et al The Achievement Motive (1953)
New York, Allied Publications.

McCurdy, H G Personality and Science: a search for self awareness
New York, Reinhold Van Nostrand.

McDougall, W The Group Mind (1920)
New York, Putnam Publishing Group.

McDowell, Joseph J Interactional Synchrony: a reappraisal
In: *Journal of Personal and Social Psychology* 1978, 35, 9.

McGann, Diarmuid The Journeying Self: the gospel of Mark through a Jungian perspective (1985)
New Jersey, Paulist Press.

McGann, Diarmuid Journeying within Transcendence: the Gospel of John through a Jungian Perspective (1989)
London, Collins Liturgical Publications, 217 p.
ISBN 0–00–599171–4.

McGhee, Paul E Humor and Children's Development: a guide to practical applications (1989)
New York, Haworth Press, 280 p. Journal of Children in Contemporary Society Series: 20, 1 and 2.
ISBN 0–86656–681–3.

McGregor, D The Human Side of Enterprise (1960)
New York, McGraw Hill Book Company, 246 p.

McGurk, H (Ed) Ecological Factors in Human Development (1978)
Amsterdam, Elsevier Science Publishing, 296 p.
ISBN 0–7204–0488–6.

McKenna, Terence Virtual Reality and Electronic Highs (1990)
Magical Blend, 12 p.

McKnight, Harry F Silva Mind Control Through Psychorientology
Laredo TX, Institute of Psychorientology.
ISBN 0–913343–40–4.

McLuhan, Marshall Understanding Media: the extensions of man (1964)
New York, Allied Publications.

McNeill, J T A History of the Cure of Souls (1951)
New York, Harper and Row.

McQuitty, L A Measure of Personality Integration in Relation to the Concept of Self (1950)
Journal of Personality Assessment 18, pp. 461–82.

McWaters, Barry Conscious Evolution (1981)
San Francisco CA, Institute for Conscious Evolution.

Meacham, J A, et al (Eds) Contributions to Human Development
18 vols.

Mead, G H Mind, Self and Society From The Standpoint of a Social Behaviorist (1934)
Chicago IL, University of Chicago Press.

Mead, G H Language and the Development of the Self (1952)
In: *G E Swanson, T M Newcomb, E L Hartley (Eds) Readings in Social Psychology.* New York, Holt, pp. 44–54.

Medawar, P B The Uniqueness of the Individual (1957)
In: *The Pattern of Organic Growth and Transformation* Edinburgh, Constable, pp. 110–14.

Medawar, P B The Future of Man (1959)
London, Methuen.

Meena, Alexander Poetic Self: towards a phenomenology of romanticism (1979)
New Delhi, Arnold Heineman, 280 p.

Mehta, P D Holistic Consciousness (1989)
Brookline MA, Tempest Books. ISBN 1–85230–108–2.

Mehta, P D Holistic Consciousness (1989)
Brookline MA, Tempest Books. ISBN 1–85230–108–2.

Mehta, Rohit Fullness of the Void
Delhi, Motilal Banarsidass.

Mehta, Rohit Nameless Experience: a Comprehensive discussion of J Krishnamurti's approach to life (1973)
Bombay, Bharatiya Vidya Bhavan, 473 p.

Mehta, Rohit The Secret of Self-Transformation: a synthesis of tantra and yoga (1987)
Columbia MO, South Asia Books. ISBN 81–208–0381–7.

Meissner, W W Psychoanalysis and Religious Experience (1986)
New Haven CT, Yale University Press, 272 p.
ISBN 0–300–03751–1.

Menaker, Esther and William Ego (1965)
New York, Grove Press.
In: Evolution.

Menninger, K, et al The Unitary Concept of Mental Illness (1958)
Bulletin Menninger Clin. 22, pp. 4–12.

Menninger, K; Mayman, M and Pruyser, P The Vital Balance (1963)
New York, Viking Press.

Menninger, Karl Love against Hate
New York, Harcourt Brace.

Menninger, Karl Man against Himself
New York, Harcourt Brace.

Mensch, James R Intersubjectivity and Transcendental Idealism (1988)
Albany NY, State University of New York Press, 272 p.
ISBN 0–88706–751–4.

Mermet, A Principles and Practice of Radiesthesia (1959)
London, Vincent Stuart.

Merrell-Wolff, Franklin The Philosophy of Consciousness without an Object: reflections on the nature of transcendental consciousness (1973)
New York, Julian Press.

Merton, Thomas Seeds of Contemplation (1972)
Wheathampstead, Anthony Clarke Books, 230 p.
ISBN 085650–023–2.

Merton, Thomas Spiritual Direction and Meditation: and What is Contemplation ? (1975)
Wheathampstead, Anthony Clarke Books, 112 p.
ISBN O–85650–037–2.

Metzner, Ralph Maps of Consciousness: I Ching, tantra, tarot, alchemy, astrology and actualism: how they work, how they may be applied as explorations of the mind, and what they can contribute to the search for meaning and growth toward individuality (1971)
New York, Macmillan, 160 p. illus.

Meyer, Donald The Positive Thinkers: popular religious psychology from Mary Baker Eddy to Norman Vincent Peale and Ronald Reagan (1988)
Rev ed.

Meyer, Michael R The Astrology of Relationship: a humanistic approach to the practice of synastry (1976)
Garden City NY, Anchor Books, 263 p. illus.
ISBN 0–385–11556–3.

Miles, Ian Social Indicators for Human Development (1985)
London, Frances Pinter, 216 p. Report to the United Nations University.

Miller, Jonathan (Ed) States of Mind (1983)
New York, Pantheon Books, 79 p.

Millman, Dan Way of the Peaceful Warrior: a book that changes lives (1984)
Atlanta GA, H J Kramer, 216 p. rev ed.
ISBN 0–915811–00–6.

Milner, D (Ed) Explorations of Consciousness (1978)
Sudbury, Neville Spearman. A sequel to the loom of creation.

Milosh, Joseph E The Scale of Perfection and the English Mystical Tradition
Ann Arbor MI, Books on Demand. ISBN 0–317–07863–1.

Mische, Gerald and Mische, Patricia Toward a Human World Order (1977)
New York, Paulist Press.

Mishlove, Jeffrey Roots of Consciousness: psychic liberation through history, science, and experience

Mishra, Rammamurti Fundamentals of Yoga (1959)
New York, Julian Press.

Mishra, Rammurti S The Textbook of Yoga Psychology: a new translation and interpretation of Patanjali's yoga Sutras for meaningful application in all modern psychologic disciplines

Misra (Ed) Humanizing Development: essays on people, space and development in honour of Masahiko Honjo
Singapore, Maruzen Asia. IBSN 962–220–214–4.

Misra, Ram Shankas The Integral Advaitism of Sri Aurobindo (1957)
Banaras, Banaras Hindu University, 410p. bibl. foot. (Darsana Series No 2).

Mitchell, E D Psychic Exploration (1976)
New York, Delta Book.

Mitchell, Richard J Mountain Experience: the psychology and sociology of adventure (1983)
Chicago IL, The University of Chicago Press, 272 p.
ISBN 0 226 53224 0.

MIU Press Neurophysiology of Enlightenment
Fairfield IA, MIU Neuroscience Press. Publication No: SU7.

Miura, Isshu and Sasaki, Ruth Fuller The Zen Koan: its history and its use in Rinzai Zen

Mohan, Singh Diwana Sikh Mysticism (1968)
Amritsar.

Mol, Hans Identity and the Sacred: a sketch for a new social-scientific theory of religion (1976)
Agincourt, The Book Society of Canada.

Molana-al-Moazam Hazrat Shah and Maghsoud Sadegh-ibn-Mohammad Angha, The Mystery of Humanity: tranquillity and survival (1986)
Lanham MD, University Press of America, 74 p.
ISBN 0–8191–5329–X.

Moltmann, J The Future of Creation (1979)

Monk of the Eastern Church Orthodox Spirituality: an outline of the Orthodox ascetical and mystical tradition (1968)
London, SPCK.

Montagu, Ashley M F The Direction of Human Development (1955)
New York, Harper and Row.

Montagu, Ashley M F The Humanization of Man (1962)
New York, World Pub.

Mookerjee, A Kundalini (1982)
New York, Destiny Books.

Mookerjee, Ajit Kundalini: the arousal of the inner energy
New York, Destiny Books, 112 p. ISBN 0–89281–020–3.

Mookerjee, Ajit Tantra Asana: a way to self–realization (1971)
Basel, Ravi Kumar.

Moore, Burness E and Fine, Bernard D (Eds) Psychoanalytic Terms and Concepts (1990)
New Haven, American Psychoanalytic Association and Yale University Press, 210 p.   ISBN 0–300–04577–8.

Moore, John Sexuality and Spirituality: the interplay of masculine and feminine in human development

Moore, T Rituals of the Imagination (1983)
Pegasus Foundation.

Moran, Gabriel Religious Education Development (1983)
Minneapolis.

Moreno, J L Psychodrama
Boston MA, Beacon Press. 3 Vols.

Morgan, C Lloyd Emergent Evolution (1923)
New York, Allied Publications.

Morgan, Peter (Ed) Unity: the first step (1972)
London, Society for the Promotion of Christian Knowledge.

Morin, E Le paradigme perdu: la nature humaine (1973)
Paris, Seuil.

Mork, Wulstan A Synthesis of the Spiritual Life (1962)
Milwaukee, Bruce Pub, 283 p. bibl.

Morocutti, Cristoforo and Rizzo, Paolo Andrea (Eds) Evoked Potentials (1985)
Amsterdam, Elsevier Science Publishing, 424 p.   ISBN 0–444–80658–X.

Morrison, Karl F I am You: the hermeneutics of empathy in Western literature, theology and art (1988)
Princeton NJ, Princeton University Press, 430 p. illus.   ISBN 0–0–691–05510–6.

Moss, Donald M and Keen, Ernest The Nature of Consciousness: the existential–phenomenological approach
In: The Metaphors of Consciousness 1981, New York, Plenum Publishing, 521 p. illus. bibl.

Motoyama, Hiroshi Theories of the Chakras: bridge to higher consciousness

Mottola, Anthony (Trans) Spiritual Exercises of St Ignatius (1964)
New York, Doubleday.

Mound Sadhu The Tarot: a contemporary course of the quintessence of hermetic occultism (1962)
London, George Allen and Unwin, 494 p.

Moustakas, Clark E The Self: explorations in personal growth (1956)
New York, Harper and Row.

Moustakas, Clark E Creativity and Conformity (1967)
New York, Reinhold Van Nostrand.

Moustakas, Clark E Individuality and Encounter (1968)
Cambridge MA, Doyle Publications.

Moustakas, Clark E Personal Growth: the struggle for identity and human values (1969)
East Dennis MA, Howard A Doyle Publishing.

Mowrer, O Hobart Crisis in psychiatry and religion (1961)
New York, Reinhold Van Nostrand.

Mowrer, O Hobart New Group Therapy (1964)
New York, Reinhold Van Nostrand.

Moyne, John and Barks, Coleman Open Secret: versions of Rumi (1984)
Putney, Threshold Books, 82 p.   ISBN 0–939660–06–7.

Mugerauer, Robert and Seamon, David (Eds) Dwelling, Place and Environment: essays toward a phenomenology of person and world (1986)
Den Haag, Martinus Nijhoff.

Mukerjee, Radhakamal The Dimensions of Human Evolution: a bio–philosophical interpretation (1963)
London, Macmillan, 217 p. bibl. (pp. 201–3).

Mukerjee, Radhakamal The Oneness of Mankind (1965)
London, Macmillan, 107p. bibl. (pp. 97–102) (phil–anthrop).

Muktananda, Swami Play of Consciousness (Chitshakti Vilas) (1978)
San Francisco CA, Harper and Row.

Muller, Robert A Cosmological View of the Future (1989)
World Goodwill. World Goodwill occasional paper, address from conference "Seeking the True Meaning of Peace", Costa Rica, 26 June 1989.

Muller, Robert and Zonneveld, Leo (Eds) The Desire to be Human: a global reconnaissance of human perspectives in an age of transformation written in honour of Pierre Teilhard de Chardin (1983)
Wassenaar, Miranada Publishers, 348 p. International Teilhard Compendium: Century volume. ISBN 90–6271–684–9.

Müller–Wielund, Marcel Syngene: sinn und wege persönlicher emporbildung (1961)
Berne, Francke, 239 p. bibl. (pp. 230–36).

Mumford, Lewis The Condition of Man (1944)
New York, Harcourt Brace, 467 p. bibl. (pp. 425–47) (purposes and ends of human development).

Mumford, Lewis Conduct of life (1951)
New York, Harcourt Brace.

Mumford, Lewis The Transformations of Man (1956)
New York, Harper and Row. (Macmillan 1962).

Mumford, Lewis Technics and Human Development: the myth of the machine (1967)
London, Secker and Warburg, 342 p. bibl. (pp. 297–323). Vol 1.

Munitz, Milton K Identity and Individuation (papers of a seminar of the Institute of Philosophy) New York, New York U (1971)
Englewood Cliffs NJ, Prentice Hall, 261 p. bibl.

Munn, N L The Evolution of the Human Mind (1971)
Boston, Houghton Mifflin.

Munroe, Ruth L Schools of Psychoanalytic Thought: an exposition, critique and attempt at integration (1955)
New York, Holt, Rinehart and Winston, 670 p. bibl. (pp. 650–52).

Munson, T N Religious Consciousness and Experience (1975)
Dordrecht, Kluwer Academic Publishers Group, 189 p.

Murphy, G and Cohen, S The Search for Person–World Isomorphism (1965)
Main Currents in Modern Thought 22, pp. 31–34.

Murphy, G and Hochberg, J Perceptual Development: some tentative hypotheses (1951)
Psychological Review 58, pp. 332, 49.

Murphy, Gardner Wholeness (1947)
In: Personality; a biosocial approach to origins and structure New York, Harper, pp. 619–760.

Murphy, Gardner Human Potentialities (1958)
New York, Basic Books.

Murphy, Gardner Challenge of Psychical Research: a primer of parapsychology (1961)
New York, Harper and Row.

Murphy, Gardner Personality: a biosocial approach to origins and structure (1966)
New York, Basic Books.

Murphy, Michael and White, Rhea A The Psychic Side of Sports (1978)
Reading MA, Addison–Wesley Publishing, 227 p. illus. bibl.   ISBN 0–201–04728–4.

Murphy, Paul E Triadic Mysticism (1986)
Delhi, Motilal Banarsidass, 226 p.

Murray, David C Reincarnation (1988)
Garden City NY, Avery Publishing Group, 288 p.   ISBN 1–85327–013–X.

Murray, Elwood et al Integrative Speech: the functions of oral communication in human affairs (1953)
New York, Dryden Press, 617 p.

Murray, George B S J and Huston, J LSD: the inward voyage (1967)
Jubilee 15, June pp. 8–17.

Murray, H A Explorations in Personality: a clinical and experimental study of fifty men of college age (1938)
Oxford NY, Oxford University Press.

Murray, H A Vicissitudes of Creativity (1959)
In: H H Anderson (Ed) Creativity and Its Cultivation Harper.

Muscari, P G Metaphorical Language and Its Place in the School (1973)
New York, New York University. Ph D Dissertation.

Myers, F W H Human Personality and its Survival (1920)
New York, Longmans Green.

Nagpal, Rup and Sell, Helmut Subjective Well–Being (1985)
Geneva, WHO, 161 p.   ISBN 92–9022–176–3.

Naidu, Usha S Altruism in Children (1980)
Bombay, Tata Institute of Social Science.

Nakamura, Hajime Ways of Thinking of Eastern Peoples: India–China–Tibet–Japan
Edited, Philip P Wiener.

Nalimov, V V In the Labyrinths of Language: a mathematician's journey (1981)
Philadelphia, Institute for Scientific Information.

Nalimov, V V Space, Time and Life: the probabilistic pathways of evolution (1985)
Philadelphia PA, Institue of Scientific Information Press.

Nalimov, V V Spontaneity of Consciousness: probabilistic theory of meanings and semantic architectonics of personality (1988)

Namgyal, Takpo Tashi Mahamudra: the quintessence of mind and meditation
Trans. and annotated by Lobsang P Lhalungpa; foreword by Chogyam Trungpa.

Namkhai Norbu The Cycle of Day and Night, Where One Proceeds Along the Path of the Primordial Yoga: an essential Tibetan text on the practice of Dzogchen (1987)
2nd rev ed. Trans. by John Myrdhin Reynolds.

Nānananda, Bhikkhu Concept and Reality in Early Buddhist Thought (1976)
Kandy, Buddhist Publication Society. 2nd ed.

Napper, Elisabeth Dependent Rising and Emptiness: a Tibetan Buddhist interpretation of Madyamika philosophy emphasising the compatability of emptiness and convential phenomena (1990)
Newburyport MA, Wisdom Publications, 848 p. illus. Wisdom Advanced Book – Blue Series.   ISBN 0–86171–057–6.

Narada, M A Manual of Abhidhamma (1968)
Kandy, Buddhist Publication Society.

Narada Thera A Manual of Abhidham (1970)
Rangoon, Buddha Sasana Council. 2 Vols.

Naranjo, Claudio The Healing Journey: new approaches to consciousness (1975)
London, Hutchinson.

Naranjo, Claudio and Ornstein, R E On the Psychology of Meditation (1972)
London, George Allen and Unwin. (New York, Viking Press, 1971).

Naranjo, Claudio I I and Thou: contributions of Gestalt Therapy (1967)
Esalen Paper, No 5.

Naranjo, Claudio I The Unfolding of Man (1969)
Menlo Park, 115 p. EPRC–6747–3, ERIC–ED–038713.

Naranjo, Claudio I Present–centeredness: technique, prescription and ideal (1970)
In: J Fagan and I L Shepherd (Eds) Gestalt Therapy Now; theory, techniques, applications New York, Harper and Row, pp. 47–69. (London, Penguin Books 1972 pp. 54–79).

Naranjo, Claudio I The One Quest (1972)
New York, Viking Press.

Naranjo, Claudio I The oneness of experience in the ways of growth (1973)
In: The One Quest New York, Viking, p 123–128.

Narayanananda, Swami The Secrets of Mind–Control

Narayanananda, Swami The Primal Power in Man, or the Kundalini Shakti

Nasr, Seyyed Hossein The Encounter of Man and Nature: the spiritual crisis of modern man (1968)
London.

Nasr, Seyyed Hossein Islam and the Plight of Modern Man (1975)
London, Longman Group, 161 p. bibl. (pp. 151–152).   ISBN 0–582–78053–5.

Nau, Erika S Self–Awareness Through Huna: Hawaii's ancient wisdom (1981)
Norfolk CA, Donning Company Publishers, 160 p.   ISBN 0–89865–099–2.

Neagle, Larry Underground Manuals for Spiritual Survival (1986)
Chicago IL, Moody Press.   ISBN 0–8024–9052–2.

Needleman, Jacob The New Religions (1970)
London, Allen Lane Penguin Press, 243p.

Needleman, Jacob The New Religions: the meaning of the spiritual revolution and the teachings of the East (1972)
London, Allen Lane Penguin Press.

Needleman, Jacob (Ed) Being–in–the–World (1975)
London, Souvenir Press.

Needleman, Jacob On the Way to Self–Knowledge (1976)
New York, Alfred A Knopf.

Needleman, Jacob Consciousness and Tradition (1982)
New York, Crossroad.

Needleman, Jacob Lost Christianity (1982)
New York, Bantam Books.

Needleman, Jacob (Ed) The Sword of Gnosis: metaphysics, cosmology, tradition, symbolism (1986)
London, Arkana, 447 p. First published 1974.   ISBN 1–85063–048–8.

Neufeldt, Ronald W (Ed) Karma and Rebirth: post classical developments (1986)
Albany NY, State University of New York Press, 357 p.   ISBN 0–87395–989–2.

Neumann, E Origins and History of Consciousness (1954)
Princeton NJ, Princeton University Press.

Neumann, E Art and the Creative Unconscious (1954)
New York, Harper and Row.

Neumann, E The Origins and History of Consciousness (1973)
Princeton NJ, Princeton University Press.

Nevis, E C Beyond Mental Health
Cleveland OH, Gestalt Institute of Cleveland. mimeo (paper no 2).

Nevitt, Sanford The Goal of Individual Development
In: G Kerry Smith (Ed) 1945 Twenty–five Years 1970 (American Association for Higher Education) San Francisco CA, Jossey–Bass, 1970, pp. 131–146.

New Age Interpreter La Canada CA, New Age Press (quaterly).

Newcombe, Hanna Design for a Better World
Lanham, University Press of America.

Newman, F X (Ed) The Meaning of Courtly Love (1969)
Albany, State University of New York Press.

Nicholson, R A Studies in Islamic Mysticism (1929)
Cambridge MA, Cambridge University Press.

Nicholson, R A The Sufi Doctrine of the Perfect Man (1984)
Edmonds WA, Holmes Publishing Group.   ISBN 0–916411–48–6.

Nidich S, Seeman W and Dreskin T Influence of Transcendental Meditation: a replication
In: Journal of Counseling Psychology 1973, 20, pp. 565–566.

Nipkow, Karl Ernst Grundfragen der Religionspädogogik (1975–1982)
Gütersloh, 3 vols.

Nisbet, R History of the Idea of Progress (1984)
New York, Basic Books, 4 p.

Nishida, K Intelligibility and the Philosophy of Nothingness (1958)
Honolulu, East–West Center Press.

Nishitani, Keiji Religion and Nothingness (1982)
Berkeley CA, University of California Press, 317 p.   ISBN 0–520–04329–4.

Nixon, Robert E The Art of Growing: a guide to psychological maturity (1962)
New York, Random House, 159 p. bibl.

Nizami, Ashraf Namaz: the Yoga of Islam
Bombay, Taraporevala Publishing Industries.

Norbu, Namkhai The Crystal and the Way of Light: meditation, contemplation and self liberation (1986)
New York, Routledge Chapman and Hall, 224 p. illus.   ISBN 0–7102–0833–2.

Nordberg, R B Meditation, future vehicle for career exploration (1934)
Vocational Guidance Quarterly 22, June 1974, pp267–271.

Nordberg, R B Mysticism: its implications for helping relationships (1975)
Counseling and Values, 19, 2, February pp 99–109.

Norman, Chirster Mystical Experiences and Scientific Method: a Study of the possibility of identifying a mystical experience by a scientific method (1986)
Stockholm, Almqvist and Wiksell.   ISBN 91–22–00936–1.

Novak, M The Experience of Nothingness (1971)
New York, Harper and Row.

Noyes, H Meditation: the doorway to wholeness (1965)
Main Currents in Modern Thought 22, 2.

Nunberg, H The Synthetic Function of the Ego (1931)
International Journal of Psychoanalysis 12, pp. 123–40.

Nurbakhsh, Javad and Chittick, William Sufism (1982)
New York, KhaniQahi–Nimatullahi, Sufi Order.   ISBN 0–933546–07–6.

Nuttin, Joseph Psychoanalysis and Personality
New York, New American Library.

Nyanaponika, Thera The Power of Mindfulness (1968)
Kandy, Buddhist Publication Society.

Nyanaponika, Thera The Heart of Buddhist Meditation (1972)
London, Rider. A handbook of mental training based on the Buddha's way of mindfulness.

Nyanatiloka, Mahathera Buddhist Dictionary: manual of Buddhist terms and doctrines (1972)
Colombo, Frewin and Co.

Nyaponika, Thera Abhidharma Studies (1965)
Kandy, Buddhist Publications Society.

Nygren, Anders Agape and Eros: a study of the Christian idea of love (1953)
London, SPCK, Volumes I and II. Trans. by Philip S Watson.

O'Brien, E Varieties of Mystic Experience (1965)
UK, N E L.

O'Brien, Theresa K (Ed) The Spiral Path: essays and interviews on women's spirituality (1988)
Saint Paul MN, Yes International, 438 p.
ISBN 0–936663–01–4.

O'Connor, E Our Many Selves: a handbook for self–discovery (1971)
New York, Harper and Row.

O'Grady, Joan Heresy: heretical truth or Orthodox error?
A study of early Christan heresies.

O'Keefe, Daniel Stolen Lightning: the social theory of magic (1982)
Oxford, Martin Robertson, 586 p. bibl.
ISBN 0–85520–486–9.

O'Neill, Andy Power of Charismatic Healing (1985)
Cork, The Mercier Press, 117 p. ISBN 0–85342–749–6.

Ochs, Carol Women and Spirituality (1983)
Totowa NJ, Rowman and Littlefield, 166 p. New Feminist Perspectives Series. ISBN 0–8476–7232–8.

Ockham, William of Predestination, God's Foreknowledge and Future Contingents (1983)
Indianapolis IN, Hackett Publishing, 146 p. 2nd ed.
ISBN 0–915144–14–X.

Odbert, H S et al Studies in Synesthetic Thinking (1942)
Journal of General Psychology 26, January–April bibl. (pp. 172–73; pp. 221–22).

Oden, Thomas C The Structure of Awareness (1969)
Nashville, Abingdon Press, 283 p. bibl. foot.

Ogilvy, J Many Dimensional Man (1977)
New York, Oxford University Press.

Oliver, W D and Landfield, A W Reflexivity: an unfaced issue of psychology (1963)
Journal of Individual Psychology 20.

Ollman, B Alienation: Marx's conception of man in capitalist society (1976)
Cambridge, Cambridge University Press.

Organ, Troy Wilson The Hindu Quest for the Perfection of Man (1970)
Athens OH, Ohio University.

Orme–Johnson, D W Autonomic Stability and Transcendental Meditation (1973)
In: *Psychosomatic Medicine* 1973, 35, 4, pp. 341 –349.

Orme–Johnson, David W and Dillbeck, Michael C Maharishi's Program to Create World Peace: theory and research (1987)
Fairfield, Maharishi International University, in "Modern Science and Vedic Science", Vol 1, No 2.

Orme–Johnson, David W and Farrow, John T Scientific Research on Transcendental Meditation: collected papers (1975)
Livingston NY, Maharishi European Research University Press. Vol 1.

Ornsteil, R The Education of the Intuitive Mode (1972)
In: *The Psychology of Consciousness*, San Francisco, Freeman, pp 143–179.

Ornsteil, R The Esoteric and Modern Psychologies of Awareness (1972)
In: *C Naranjo and R O Ornstein. On the Psychology of Medition.* London, George Allen and University, pp170–212.

Ornstein, R (Ed) The Nature of Human Consciousness: a book of readings (1973)
New York, Viking.

Ornstein, R MultiMinds: a new way to look at human behavior (1986)
Boston MA, Houghton Mifflin.

Ornstein, Robert The Psychology of Consciousness (1986)
New York, Penguin Books, 314 p. Second revised edition, classic study of 1972 completely revised and updated.
ISBN 0–14–022621–4.

Ornstein, Robert Multimind: a new way of looking at human behaviour (1986)
London, Macmillan, 206 p. bibl. ISBN 0–333–43803–5.

Ornstein, Robert E The Mind Field (1976)
London, Octagon Press.

Orr, Robert The Meaning of Transcendance (1979)
Vanderbilt University. Unpublished disertation.

Ortony, Alii Theoretical and Methodological Issues in the Empirical Study of Metaphor and Implications and Suggestions for Teachers (1983)
Urbana, Center for the Study of Reading. Reading Education Report: 38.

Osborn, Arthur W The Expression of Awareness
Reigate, Omega Press.

Osborne, Arthur Ramana Maharshi and the Path of Self–Knowledge
Foreword by S Radhakrishnan.

Ostrander, Sheila and Schroeder, Lynn Handbook of Psychic Discoveries (1974)
New York, Berkeley Publishing Corporation.

Otto, H A and Mann, J (Eds) Ways of Growth: approaches to expanding awareness psychiatry for laymen (1971)
Richmond Hill, Simon and Schuster.

Otto, Herbert A (Ed) Explorations in Human Potentialities (1966)
Springfield IL, Thomas Charles C.

Otto, Herbert A Guide to Developing Your Potential (1967)
New York, Scribner.

Otto, Herbert A Group Methods Designed to Actualize Human Potential: a handbook (1968)
Achievement Motivation Systems.

Otto, Herbert A Human Potentialities: the Challenge and the Promise (1968)
Saint Louis MO, Warren H Green.

Otto, Herbert A Motivation and Human Potentialities (1968)
Humanitas III, 3, Winter.

Otto, Rudolf The Idea of the Holy: an inquiry into the non-rational factor in the idea of the divine and its relation to the rational (1950)
New York, Oxford University Press. 2nd ed.

Otto, Rudolph Mysticism East and West: a discussion of the Nature of mysticism, focusing on the similarities and differences of its two principal types (1932)
New York, Macmillan.

Ouspensky, P D The Psychology of Man's Possible Evolution (1968)
London, Hodder and Stoughton (New York, Bantam Books, 1968).

Ouspensky, Peter D In Search of the Miraculous (1949)
New York, Harcourt Brace.

Overstreet, H A The Mature Mind (1949)
New York, WW Norton.

Owens, Claire M The Mystical Experience: facts and values (1967)
Main Currents in Modern Thought 23, 4, March/April.

Owens, Joseph Human Destiny: some problems for catholic philosophy (1985)
Washington DC, Catholic University of America Press, 126 p.
ISBN 0–8132–0604–9.

Pacheco, Dr Cláudia Bernhardt The ABC of Analytical Trilogy: integral psychoanalysis: a comprehensive explanation of the trilogy which unifies science, philosophy and spirituality (1988)
Sao Paulo, Proton Publishing House Inc, 220 p.
ISBN 0–939019–06.

Pal, Rai Sawindar Samata Yoga
Coos Bay OR, Vision Books, 142 p.

Pallis, Marco The Veil of the Temple: a study of Christian initiation (1986)
London, Arkana, in "The Sword of Gnosis: Metaphysics, Cosmology, Tradition, Symbolism", edited by Jacob Needleman, 447 p. First published 1974. ISBN 1–85063–048–8.

Palmer, Helen The Enneagram (1988)
San Francisco CA, Harper and Row.

Panati, C Supersenses: our potential for parasensory experience (1976)
Garden City NY, Anchor Press.

Pandit, M P Dictionary of Sri Aurobindo's Yoga (1973)
Pondicherry, Sri Aurobindo Ashram, 315 p.

Panikkar, Raimundo The Invisible Harmony: a universal theory of religion or a cosmic confidence in reality? (1987)
In: *Toward a Universal Theology of Religion* pp. 118–153.

Paplauskas–Ramunas, Anthony Development of the Whole Man through Physical Education: an interdisciplinary comparative exploration and appraisal (1968)
Melbourne VIC, Oxford University Press, 436 p.
ISBN 0–7766–1201–8.

Paramananda, Swami Silence As Yoga (1974)
Cohasset MA, Vedanta Centre Publishers.
ISBN 0–911564–11–X.

Parfip, D On the Importance of Self–Identity (1971)
Journal of Philosophy 68, 21 October pp. 667–90.

Parfitt, Will The Living Qabalah: a practical guide to understanding the Tree of Life (1988)
Longmead, Element Books.

Park, David The Image of Eternity (1975)
Amherst MA, University of Massachusetts Press.

Park, Sung–Bae Buddhist Faith and Sudden Enlightenment (1983)
Albany NY, State University of New York Press, 211 p.
ISBN 0–87395–673–7.

Parker, A States of Mind: ESP and altered states of consciousness (1975)
New York, Taplinger.

Parkhurst, Jacqueline Individuation Process (1977)
Yanchep WA, Open Mind Publications, 144 p.
ISBN 0–9596609–2–5.

Parrinder, Geoffrey Mysticism in the World's Religions (1977)
Oxford University Press.

Parsons, Howard L Value and Mental Health in the Thought of Marx (1964)
Philosophy and Phenomenological Research 24, pp. 355–65.

Parsons, Howard L Self, Global Issues and Ethics (1980)
Amsterdam, BR Grüner, 210 p. ISBN 90–6032–178–2.

Parthsarathy, A The Symbolism of Hindu Gods and Rituals (1985)
Bombay, Vedanta Life Institute, 138 p. 2nd edition.

Passmore, John A The Perfectability of Man (1971)
New York, Scribner, 396 p. bibl. (pp. 329–80).
ISBN 0–684–15521–4.

Pathak, P V The Heyapaksha of Yoga, or Towards a Constructive Synthesis of Psychological Material in Indian Philosophy (1931)
Ahmedabad.

Pathrapankal, J Metanoia, Faith, Covenant (1971)
Bangalore, Dharmaram Publications, 327 p.

Patton, James R, et al Exceptional Children in Focus (1987)
Columbus OH, Merrill Publishing, 280 p. 4th ed.
ISBN 0–675–20720–7.

Pavitrananda, Swami Common Sense About Yoga
Calcutta, Advaita Ashrama, 170 p.

Payne, Peter Martial Arts: the spiritual dimension (1981)
London, Thames and Hudson, 96 p. illus. bibl.
ISBN 0–500–81025–7.

Paz, Idit (Ed) Communal Life: an international perspective (1983)
Ramat Ef'al, Yad Tabenkin, 750 p.

Pear, Tom H The Maturing Mind (1938)
London, T Nelson, 152 p.

Pearsall, Paul Superimmunity: master your emotions and improve your health (1987)
North Ryde NSW, Angus and Robertson Publishers.

Pearson and Pope The Female Hero in American and British Literature (1981)
New York, R R Bowker and Co.

Pearson, Carol The Hero Within: six archetypes we live by (1986)
San Francisco CA, Harper and Row, 176 p.
ISBN 0–86683–527–X.

Peat, David F Synchronicity: the bridge between matter and mind (1987)
Toronto, Bantam Books, 245 p. bibl. ISBN 0–553–34321–1.

Peccei, Aurelio The Humanistic Revolution (1975)
Successo.

Peck, R and Havighurst, R The Psychology of Character Development (1960)
New York, John Wiley and Sons.

Peel, Robert Spiritual Healing in a Scientific Age (1987)
New York, Harper and Row, 288 p. ISBN 0–06–066484–3.

Peerbolte, M Meditation for School Children (1927)
Main Currents in Modern Thought 24, 1967, p. 15–21.

Penfield, W The Mystery of the Mind (1978)
Princeton NJ, Princeton University Press.

Perino, Sheila and Perino, Joseph Parenting the Gifted: developing the promise (1981)
Munich, K G Saur Verlag, 214 p. ISBN 0–8352–1354–4.

Perls, F S Theory and Technique of Personality Integration (1948)
American Journal of Psychotherapy vol 2, 1948 pp. 565–86.

Perls, F S Gestalt Therapy and Human Potentialities (1966)
In: *H A Otto (Ed) Explorations in Human Potentialities* Charles C Thomas 1966 (also Esalen Paper 1964, No1).

Perls, Frederick et al Gestalt Therapy: excitement and growth in the human personality (1951)
New York, Julian Press, 466 p.

Perls, Fritz In and Out of the Garbage Pail
Moab UT, Real People Press.

Perls, Fritz Gestalt Therapy Verbatim (1969)
Moab UT, Real People Press.

Person, Ethel Spector Love and Fateful Encounters: the power of romantic passion (1990)
London, Bloomsbury, 384 p. ISBN 0–7475–0581–0.

Personal Growth Explorations Institute, Cal (monthly)

Pesso, Albert Movement in Psychotherapy: Psychomotor Techniques and Training (1969)
New York, New York University Press.

Petermann, B The Gestalt Theory (1932)
London, Kegan Paul.

Peters, R S 'Mental Health' as an Educational Aim (1961)
Paper read before Philosophy of Education Society, Harvard University, March.

Peterson, Linda Gay and O'Shanick, G J (Eds) Psychiatric Aspects of Trauma (1986)
New York, Karger S AG, 238 p. Advances in Psychosomatic Medicine: Vol 16. ISBN 3–8055–4219–4.

Peterson, S A Catalog of the Ways People Grow (1971)
New York, Ballantine Books.

Peterson, Severin A Catalog of Ways People Grow (1971)
New York, Ballantine Books, 368 p. bibl.

Petrie, H G Metaphor and Learning
In: *A Ortony (Ed) Metaphor and Thought* Cambridge University Press, 1979, pp. 438–461.

Phipps, John–Francis The Politics of Inner Experience: dynamics of a green spirituality (1990)
London, The Merlin Press, 116 p. ISBN 1–85425–025–6.

Piatigorsky, Alexander The Buddhist Philosophy of Thought: essays in interpretation (1984)
London, Curzon Press, 227 p.
ISBN O–7007–0159–1 (UK), 0–389–20266–5 (USA).

Picknett, Lynn (Ed) The Encyclopaedia of the Paranormal (1990)
New York, Macmillan, 206 p.

Pico della Mirandola, Giovanni Oration on the Dignity of Man
A Robert Caponigri, Trans; Russell Kirk, Introduction.

Piepe, Anthony Knowledge and Social Order: the relationship between human knowledge and the construction of social theory (1971)
London, Heinemann, 81 p. bibl.

Pietsch, Paul Shufflebrain: the quest for the hologramic mind (1981)
Boston MA, Houghton Mifflin.

Pikunas, Justin et al Human Development: a science of growth (1969)
New York, McGraw Hill Book Company, 434 p. bibl. (1961 version: Psychology of human develop.).

Pilch, John J Wellness Spirituality (1985)
New York, Crossroad Publishing, 112 p.
ISBN 0–8245–0710–X.

Plomin, Robert and Dunn, Judy The Study of Temperament: changes, continuities and challenges (1986)
Hillsdale NJ, Erlbaum Lawrence Associates, 192 p. ISBN 0–89859–670–X.

Plutchik, Robert Emotion: a psychoevolutionary synthesis (1980)
New York, Harper and Row, 144 p.

Po–tuan Chang Understanding Reality: a Taoist alchemical classic
With a concise commentary by Liu I–ming. Trans. by Thomas Cleary.

Poddar, H P The Divine Name and Its Practice (1965)
Gorakhpur, Gita Press.

Polanski, V G A Description of the Spontaneous Production of Similes and Metaphors in the Writing of Students in Grades Four, Eight and Twelve, and the Third Year of College (1981)
Buffalo, SUNY. Ph D Dissertation.

Pollack, Susan and White, M I (Eds) The Cultural Transition (1986)
London, Routledge and Kegan Paul.

Pollio, H R, et al Psychology and the Poetics of Growth: figurative language in psychology, psychotherapy and education (1977)
Hillsdale, LEA.

Polster, E The Integrative Effect of Social Psychotherapy (1967)
Washington DC, Proceedings of the 75th Annual Convention of the American Psychological Association.

Poole, Roger Towards Deep Subjectivity (1972)
London, Allen Lane Penguin Press, 152p.

Pope John Paul II Salvifici Doloris: apostolic letter of the Supreme Pontiff John Paul II on suffering (1984)
London, Incorporated Catholic Truth Society, 74 p. ISBN O–85183–568–6.

Pope Paul VI Encyclical Letter of His Holiness Paul VI to the Bishops, Priests, Religious, the Faithful and to All Men of Good Will: Populorum Progressio – on the development of peoples: for man's complete development – the development of the human race in the spirit of solidarity (1967)
Vatican City, Vatican Polyglot Press.

Popenoe, C Books for Inner Development: the yes, guide (1976)
New York, Random House.

Popoff, Irmis B The Enneagram of the Man of Unity (1978)
New York, Samuel Weiser.

Postle, Denis Catastrophe Theory: predict and avoid personal disasters (1980)
Fontana Paperbacks, 218 p. illus. ISBN 0–00–635559–5.

Pott, P H Yoga and Yantra (1966)
Den Haag.

Prabhavananda, Swami and Isherwood, Christopher (Trans) Shankara's Crest–Jewel of Discrimination – Viveka–Chudamani

Pramod Kumar Moksa, The Ultimate Goal of Indian Philosophy
Foreword by D N Shastri. Trans and edited by M C Bhartiya.

Pratt, J B The Religious Consciousness: a psychological study (1923)
New York, Macmillan.

Preston, Harold and Babcock, Winifred The Single Reality (1971)
New York, Harold Institute.

Price, A F and Wong Mou–Lam (Trans) The Diamond Sutra and the Sutra of Hui–Neng (1969)
Berkeley CA, Shambhala Publications.

Prince, R and Savage, C Mystical States and the Concept of Regression (1966)
Psychedelic Review 8.

Privette, G Peak Experience, Peak Performance and Flow: a comparative analysis of positive human experiences (1983)
In: *Journal of Personality and Social Psychology* 83 (45), 1, pp. 361–368.

Progoff, Ira Jung's Psychology and Its Social Meaning (1953)
New York, Julian Press, 299 p.

Progoff, Ira The Death and Rebirth of Psychology: an integrative evaluation of Freud, Adler, Jung and Rank and the impact of their culminating insights on modern man (1956)
New York, Julian Press, 275 p. bibl. (pp. 267–75).

Progoff, Ira The Symbolic and the Real: a new psychological approach to the fuller experience of personal existence (1963)
New York, Julian Press, 234 p. bibl. (pp. 227–34).

Progoff, Ira Depth Psychology and Modern Man: a new view of the magnitude of human personality, its dimensions and resources (1973)
New York, McGraw–Hill Publishing, 304 p. Repr of 1969 ed. Holistic Depth Psychology Trilogy. ISBN 0–07–050891–7.

Progoff, Ira The Practice of Process Meditation: the intensive journal way to spiritual experience (1980)
New York, Dialogue House Library, 343 p. ISBN 0–87941–008–6.

Progroff, Ira Jung, Synchronity and Human Destiny (1973)
New York, Delta Books.

Proskauer, Magda Breathing Therapy (1968)
In: *H Otto and J Mann (Eds) Ways of Growth* New York, Grossman ch. 3.

Proudfoot, W Religious Experience (1985)
Berkeley, University of California Press.

Psychosynthesis Institute Psychosynthesis: a way to inner freedom

Psychosynthesis Institute Global Education and Psychosynthesis

Psychosynthesis Research Foundation, USA Jung and the Psychosynthesis (1967)

Psychosynthesis Research Foundation, USA Psychosomatic Medicine and the Bio–Psychosynthesis (1967)

Psynetics Foundation Psynetics Newsletter (Anaheim CA) (monthly).

Pucci, Giuseppe Theory and Practice of the Mandala (1960)
London, Rider.

Puligandla, Ramakrishna Jnana–Yoga: the way of knowledge (1985)
Lanham MD, University Press of America, 134 p. ISBN 0–8191–4530–0.

Purce, Jill The Sound of Enlightenment
In: *Resurgence* 1990, 139, pp. 36–38.

Purohit, Swami Shree and Yeats, W B (Trans) The Ten Principal Upanishads
London, Faber and Faber, 159 p. ISBN 571–09363–9.

Puthoff, H and Targ, R Mind Rearch: scientists look at psychic ability (1977)
New York, Dell Publishing.

Quarton, Gardner C Deliberate Attempts to Control Human Behavior and Modify Personality (1967)
Daedalus 96, p. 837–53.

Quina, J Root Metaphor and Interdisciplinary Curriculum: designs for teaching literature in secondary schools
In: *Journal of Mind and Behavior* 1982, 3, 3–4, pp. 345–356.

Quina, J Curriculum Designs for Teaching Literature in High Schools: a report on applications of world hypotheses
Paper read at the Pluridisciplinary Conference on Root Metaphor, SUNY, Buffalo, 1982.

Rabten, Geshe Echoes of Voidness (1985)
Newburyport CA, Wisdom Publications, 148 p. illus. Intermediate Books: White Series. ISBN 0–86171–010–X.

Radha, S Kundalini Yoga for the West (1978)
Boulder CO, Shambhala Publications.

Radhakrishnan, S Eastern Religions and Western Thought

Radhakrishnan, S The Brahma Sutra: the philosophy of spiritual life (1960)
London.

Radhakrishnan, S and Raju, P T (Ed) The Concept of Man: a study in comparative philosophy (1960)
London, George Allen and Unwin, 383 p.

Rado, S From the Metaphysical Ego to the Biocultural Acting Self (1958)
Journal of Psychology 46, pp. 277–88.

Rahner, Karl and Vorgrimler, Herbert Dictionary of Theology
New York, Crossroad Publishing. Second edition.

Rai, Ram Kumar Encyclopedia of Yoga
New Delhi, B Jain Publishers.

Raine, Kathleen Defending Ancient Springs (1985)
Ipswich, Golgonooza Press, 198 p. First published 1967. ISBN 0–903880–32–6.

Raitz, K L An Analysis and Evaluation of Metaphor in Education (1975)
Columbus OH, Ohio State University. Ph D Dissertation.

Rajneesh, Bhagwan Shree The Orange Book: the meditation techniques of Bhagwan Shree Rajneesh (1983)
Boulder CO, Chidvilas, 256 p. 2nd ed. Meditation Series. ISBN 0–88050–697–0.

Ram, D Journey of Awakening: a meditator's guidebook (1978)
New York, Bantam Books.

Rama Rao Pappu, S S (Ed) The Dimensions of Karma (1987)
Columbia MO, South Asia Books. ISBN 81–7001–025–X.

Rama, Swami Enlightenment Without God
Honesdale, Himalayan Publishers.

Rama, Swami Yoga and Psychotherapy
Honesdale, Himalayan Publishers.

Rama, Swami Ballentine Rudoph and Ajaya, Swami Yoga and Psychotherapy: the evolution of consciousness (1976)
Glenview IL, The Himalayan Institute.

Ramana Maharshi The Collected Work of Ramana Maharshi (1968)
London, rider. Edited by Arthur Osborne.

Ramanan, Vetaka K Nagarjuna's Philosophy as Presented in the Maha–Prajñaparamita–Sastra (1971)
Varanasi, Bharatiya Vidya Prakashan.

Ramanananda Saraswathi, Swami Sri (Trans) Tripura Rahasya, or The Mystery Beyond the Trinity

Ramurti, S Mishra Textbook of Yoga Psychology (1963)
New York, Julian Press.

Rank, Otto Psychology and the Soul
Barnes and Noble.

Raphaell, K Crystal Enlightenment: the transforming properties of crystals and healing stones (1985)
New York, Aurora Press.

Rapoport, Anatol General Systems Theory: a bridge between two cultures
In: *Proceedings* Society for General Systems Research, 1976, pp. 9–14.

Rappaport, Roy A The Sacred in Human Evolution
In: *Annual Review of Ecology and Systematics* 2, pp. 23–44.

Rappard, H V Psychology as Self–Knowledge (1979)
Assen, Van Gorcum, 84 p.

Rasey, Marie (Ed) Nature of Being Human
Wayne, State University Press.

Raven, Christian P Oogenesis: the storage of developmental information (1961)
New York, Pergamon Books.

Rawlinson, Andrew Altered States of Consciousness (1979)
In: *Religion* 9, pp. 92–103.

Rawls, John A Theory of Justice (1971)
?.

Refs #H2117.

Rechelbacher, Horst Rejuvenation (1987)
Rochester VT, Inner Traditions International, 224 p. ISBN 0–89281–248–6.

Reed, Graham The Psychology of Anomalous Experience: a cognitive approach (1988)
Buffalo, New York, Prometheus Books, 207 p. rev ed. ISBN 0–87975–435–4.

Reed, William Ki: a practical guide for Westerners
Tokyo, Japan Publications, 224 p. ISBN 0–87040–640–X.

Regardie, I Foundations of Practical Magic (1979)
Wellingborough, Aquarian Press.

Reich, Charles The Greening of America (1970)
New York, Random House.

Reich, Wilhelm The Function of the Orgasm: sex–economic problems of biological energy (1967)
New York, Bantam Books. Original ed 1942.

Reichard, G A; Jakobson, R and Werth, E Language and Synesthesia (1949)
Word 5, 2.

Reichsman, F (Ed) Hunger and Satiety in Health and Disease (1972)
Basel, S Karger AG, 336 p. Advances in Psychosomatic Medicine: Vol 7. ISBN 3–8055–1356–9.

Reik, Theodor The Search Within: the inner experiences of a psychoanalyst (1974)
New York, Jason Aranson.

Reinberg, A and Ghata, J Biological Rhythms (1965)
New York, Walker Publishing Company.

Rele, V G Human Mind Power: the secrets of the Vedic Gods

Rendel, David Earthling and Hellene: a psychological mechanism of evolution and its effect on the human race (1963)
Wimbledon, Vernon and Yates, 122 p. (phil. anthrop.).

Reps, P Be: new uses for the human instrument (1971)
New York, John Weatherhill.

Restivo, S P Parallels and Paradoxes in Modern Physics and Eastern Mysticism: a critical reconnaissance
In: *Social Studies of Science* 1978, 8, pp. 143–182.

Restivo, S P Physics, Mysticism and Society: a sociological perspective on distinguishing and relating alternate reality
In: *Social Studies of Science* 1979, 11,1.

Rewatadhamma (Ed) Visuddhimagga
London, Pali Text Society. 2 Vols. Vol 1 (1920), Vol 2 (1921).

Reyes, Benito F Meditation: cybernetics of consciousness (1978)
Ojai CA, World University of America, 151 p. Ed by Fred Volz. ISBN 0–939375–04–4.

Reynolds, David K Naikan Psychology: meditation for self–development (1983)
Chicago IL, University of Chicago Press, 184 p. ISBN 0–226–71029–7.

Rheingold, Howard They Have a Word for It: a lighthearted lexicon of untranslatable words and phrases (1988)
Los Angeles, Jeremy T Tarcher Inc, 224 p. ISBN 0–87477–464–0.

Rhine, J B and Pratt, J G Parapsychology
Springfield IL, Thomas Charles C.

Rhyne, J and Vich, M A Psychological Growth and the Use of Art Materials: small group experiments with adults (1967)
Journal of Humanistic Psychology 7, 1967 pp. 163–70.

Ribeiro, Darcy The Civilizational Process (1968)
Washington DC, Smithsonian Institution Press, 201 p. Transl. and forward by Betty J Meggars.

Rice, Cyprian O P The Persian Sufis (1969)
London. 2nd ed.

Richards, Mary C Centering in Pottery, Poetry and the Person (1962)
Conn, Wesleyan University Press.

Richards, Mary C Centering (1969)
Middletown CT, Wesleyan University Press.

Richardson, H W and Cutler, D R (Eds) Transcendence (1969)
Boston MA, Beacon Press.

Richter, C P Biological Clocks in Medicine and Psychiatry (1965)
Springfield IL, Thomas Charles C, 109 p.

Riegel, K F and Meacham, J A Developing Individual in a Changing World: symposium of the International Society for the Study of Behavioral Development, University of Michigan (1976)
Berlin, Moutin de Gruyter, 234 p. ISBN 90–279–7502–7.

Rieker, H The Yoga of Light (1971)
San Francisco, Dawn Horse.

Riley, Mary Corporate Healing: solutions to the impact of the addictive personality on the workplace

Ring, K Mapping the Regions of Consciousness: a conceptual reformulation (1976)
In: *The Journal of Transpersonal Psychology* 8 (2), pp. 77–88.

Ring, K Heading Toward Omega: in search of the meaning of the near–death experience (1984)
New York, Morrow.

Rinpoche, Namgyal The Womb of Form: pith instructions in the six yogas of Naropa
From the teachings of Namgyal Rinpoche.

Roberts, Bernadette The Experience of No–Self (1984)
Boulder CO, Shambhala Publications.

Roberts, R Spirituality and Human Emotions (1983)
Grand Rapids, Eerdmans.

Roberts, T Beyond Self–Actualization
In: *Re–Vision* 1978, 1, 1, pp. 42–46.

Roberts, T B Bibliography on Transpersonal Education (1935)
DeKalb IL, Department of Education, Northern Illinois University.

Roberts, T B Maslow's Human Motivation Needs Hierarchy: a bibliography (1973)
In: *Research in Education* Bethesda, ERIC Document Reproduction Serive, ED–069–591.

Roberts, T B Transpersonal: the new educational psychology (1975)
April. ERIC ED–099–252.

Robertson, Robin CG Jung and the Archetypes of the Collective Unconscious (1987)
Bern, Verlag Peter Lang AG, 250 p. ISBN 0–8204–0395–4.

Robinson, E A Tolerating the Paradoxical (1978)
Oxford, RERU.

Robinson, Edward The Original Vision (1977)
Oxford, Religious Experience Research Unit.

Robinson, John A T Truth is Two–Eyed (1979)
London, SCM Press, 161 p. bibl. (pp. 147–156).
ISBN 334–01690–8.

Robinson, Richard H and Johnson, Willard L The Buddhist Religion (1982)
Belmont CA. Third edition.

Rock, Pennell The Return of the Person (1973)
New York, Arica Institute.

Roddan, John The Changing Mind (1966)
Boston MA, Brown Little, 224 p. (London, Jonathan Cape, bibl. pp223–24).

Rogers, C On Becoming a Person (1961)
Boston MA, Houghton Mifflin.

Rogers, Carl and Coulson, W R (Eds) Freedom to Learn (1969)
Columbus ON, Merrill Publishing.

Rogers, Carl and Stevens, Barry Person to Person: the problem of being human (1967)
Moab UT, Real People Press.

Rogers, Carl R A Counseling and Psychotherapy (1942)
Boston MA, Houghton Mifflin.

Rogers, Carl R A Client Centered Therapy (1951)
Boston MA, Houghton Mifflin.

Rogers, Carl R A The Necessary and Sufficient Conditions of Therapeutic Personality Change (1957)
Journal of Consulting Psychology 21, pp. 91–103.

Rogers, Carl R A A Theory of Therapy, Personality and Interpersonal Relationships as Developed in the Client–Centered (1959)

Rogers, Carl R A Toward a Theory of Creativity (1959)
In: *H H Anderson (Ed) Creativity and its Cultivation* New York, Harper, bibl.

Rogers, Carl R A Some Hypotheses Regarding the Facilitation of Personal Growth (1961)
In: *Carl R Rogers, On Becoming a Person* Boston, Houghton Mifflin, 1961 (London, Constable, 1961).

Rogers, Carl R A Actualizing Tendency in Relation to Motives and to Consciousness (1963)
In: *M R Jones (Ed) Nebraska Symposium on Motivation* 1963 University of Nebraska Press.

Rogers, Carl R A On Becoming a Person (1970)
Boston MA, Houghton Mifflin.

Rogers, Carl R and Dymond, R F (Ed) Psychotherapy and Personality Change: coordinated studies in the client–centered approach (1954)
Chicago IL, Chicago University Press, 447 p.

Rogers, J Inner Worlds of Meditation (1976)
New York, Baraka Press.

Rohde, S Deliver Us from Evil: studies on the vedic ideas of salvation (1946)
London, Lund Humphries.

Rokeach, Milton On the Unity of Thought and Belief (1916)
Journal of Personality 25, 1956, pp. 224–50.

Rolf, Ida Structural Integration (1963)
Systematics 1, 1, June.

Roran, R Jungian Psychology and Christian Spirituality: III (1979)
In: *Review for Religious* 38, pp. 857–66.

Rose, Colin Accelerated Learning (1987)
New York, Dell Books.

Rosenberg, L A Study of Figurative Language Production in Traditional and Open Classrooms (1977)
Cambridge MA, Harvard University. Unpubl Honors Dissertation.

Rosenblueth, A; Wiener, N and Bigelow, J Purpose and Teleology (1943)
Philosophy of Science 10, pp. 18–24.

Rosenfeld, E and Rubenfeld, Ilana The Alexander Technique
In: *Bernard Aronson (Ed) Workshops in the Mind.*

Rosenfeld, Edward The Book of Highs (1973)
New York, New York Times Book.

Rosenfeld, Edward Le Livre des Extases: 250 façons de connaître les paradis artificiels sans drogues (1974)
Montréal, La Press, approx 300 p. illus. Transl. by Martine Wiznitzer.

Rosenthal, Bernard The Images of Man (1970)
New York, Basic Books.

Rosenthal, Miriam B and Smith, D H (Eds) Psychosomatic Obstetrics and Gynecology (1985)
New York, Karger S AG, 190 p. Advances in Psychosomatic Medicine: Vol 12. ISBN 3–8055–3967–3.

Ross, Helen E Behaviour and Perception in Strange Environments (1974)
London, George Allen and Unwin Ltd, 171 p. Advances in Psychology Series. ISBN 0–04–150048–2.

Rossbach, Sarah Interior Design with Feng Shui (1987)
New York, E P Dutton, 178 p. ISBN 0–525–48299–7.

Rossi, Ernest L Growth, Change, and Transformation in Dreams (1971)
Journal of Humanistic Psychology 2, 2, pp147–169.

Rossi, Ernest L Dreams and the Growth of Personality (1972)
New York, Pergamon Books, 217 p. bibl. (pp. 203–08).

Rossi, I (Ed) The Unconscious In Culture (1974)
New York, EP Dutton.

Rostand, Jean Peut–on Modifier l'Homme? (1956)
Paris, Librairie Gallimard.

Roszak, T (Ed) Sources (1972)
New York, Harper and Row.

Roszak, T There the Wasteland Ends (1972)
Garden City NY.

Roszak, Theodore Unfinished Animal: the aquarian frontier and the evolution of consciousness (1976)
London, Faber and Faber, 271 p. ISBN 0–571–11014–2.

Roszak, Theodore Person–Planet: the creative disintegration of industrial society (1978)
New York, Anchor Books.

Rotenstreich, N Idea of Historical Progress and its Assumptions (1971)
History and Theory 10, 2, p. 197–221.

Rothblatt, Ben (Ed) Changing Perspectives on Man (1968)
Chicago IL, University of Chicago Press.

Rougemont, Denis de Love in the Western World (1983)
Princeton University Press, 400 p.

Rousselle, Erwin Siu Schen 'Individuation' (In: Seelische Führung im Lebenden Taoismus) (1933)
In: *Eranos Jahrbuch 1933* Zurich, Rhein Veslay, 1934, pp. 174–99.

Rowan, John Subpersonalities: the people inside us (1990)
London, Routledge, 242 p. ISBN 0–415–04329–8.

Royce, Joseph R Encapsulated Man (1964)
New York, Reinhold Van Nostrand.

Rozman, D Meditating with Children (1975)
Boulder Creek, University of Trees Press.

Rozman, D Meditation for Children (1976)
Millbrae, Celestial Arts.

Rubottom, A Transcendental Meditation and its Potential Uses for Schools (1972)
Social Education December 36, p. 4.

Ruddick, Sara Maternal Thinking
Boston MA, Beacon Press, 291 p.

Rudhyar, Dane The Planetarization of Consciousness (1970)
Wasenaar, Service.

Rudhyar, Dane The Planetarization of Consciousness (1972)
New York, Harper and Row.

Rudhyar, Dane Beyond Individualism: the psychology of transformation (1979)
Wheaton IL, Theosophical publishing House.

Rudhyar, Dane Beyond Personhood (1982)
Palo Alto, Rudhyar Institute for Transpersonal Activity.

Rudhyar, Dane Rhythm of Wholeness (1983)
Wheaton IL, Theosophical publishing House.

Ruesch, J The Therapeutic Process from the Point of View of Communication Theory (1952)
American Journal of Orthopsychiatry 22, pp. 690–701.

Ruesch, J and Bateson, G Communication (1968)
New York, Norton Library.

Ruitenbeek, H H The Nude Group Therapies
In: *The New Group Therapies* pp. 186–201.

Rupp, George and Cox, Harvey Beyond Existentialism and Zen: religion in a pluralistic world (1979)
New York, Oxford University Press. ISBN 0–19–502462–1.

Rush, Joseph H New Directions in Parapsychological Research (1964)
New York, Parapsychology Foundation.
ISBN 0–912328–07–X.

Russell, Peter The TM Technique: an introduction to transcendental meditation and the teachings of Maharishi Mahesh Yogi (1976)
London, Routledge and Kegan Paul, 195 p.
ISBN 0–7100–8345–9.

Russell, Peter The Global Brain: speculations on the evolutionary leap to planetary consciousness (1983)
Los Angeles CA, J P Tarcher, 251 p.

Sachs, Céline Exploring the Human Dimensions of Development: a review of the literature (1989)
Paris, UNESCO, 89 p. Major Programme I: Reflection on World Problems and Future–Oriented Studies.

Sadhu, Mouni Concentration (1959)
London, George Allen and Unwin.

Sadhu, Mouni Samadhi (1962)
London, George Allen and Unwin.

Said, Abdul Aziz Tawhid: the Sufi Tradition of Unity
In: *Creation*, 1988, 4, 4.

Salk, J Man Unfolding (1972)
New York, Harper and Row.

Sampler, R E Opening: a process for self–actualization (1973)
Menlo Park, Addison–Wesley.

Sampler, R E Learning with the Whole Brain (1975)
Human Behavior 4, 79, Feb p. 16–23.

Sampson, Anthony The Midas Touch (1989)
Hodder and Stoughton.

Sampson, Tom Cultivating the Presence (1977)
New York, T Y Crowell.

Samuels, Andrew et al A Critical Dictionary of Jungian Analysis (1986)
London, Routledge and Kegan Paul, 171 p.
ISBN 0 7102 0410 8.

Sander, Friedrich and Volket, Hans Ganzheitspsychologie (1927)
Munich, Beck, 459 p. bhbl. foot.

Sanders, Oswald J Spiritual Maturity (1983)
Pasadena CA, Living Spring Publications. Samuel Chao and Lorna Chao transl from Eng (Chinese).
ISBN 0–941598–08–X.

Sanella, L Kundalini: psychosis or transcendence? (1976)
San Francisco CA, H S Dakin.

Sanford, Agnes The Healing Gifts of the Spirit

Sanford, Nevitt Self and Society (1907)
Atherton.

Sannella, L Kundalini: psychosis or transcendence ? (1976)
San Francisco, H S Dakin.

Santa, Maria Jack Anna Yoga: the yoga of food

Saraswati, Swami Satyananda and Saraswati, Swami Muktibodhananda Swara Yoga: Tantric science of brain breathing

Saraswati, Swami Yogeshwaranand Himalya Ka Yogi

Sarathchandra, E R Buddhist Psychology of Perception (1958)
Colombo, Ceylon University Press.

Satchidananda, Swami Integral Yoga Hatha (1970)
New York, Holt, Rinehart and Winston, 189 p.

Satir, Virginia Conjoint Family Therapy (1967)
Science and Behavior Books.

Satprakashananda, Swami Methods of Knowledge (Perceptual, Non–perceptual, and Transcendental): according to Advaita Vedanta
Intro by T M P Mahadevan.

Satprakashananda, Swami Methods of Knowledge According to Advaita Vedanta (1975)
Hollywood CA, Vedanta Press, 366 p. repr of 1965 ed.
ISBN 0–87481–154–6.

Satprem Sri Aurobindo: the adventure of consciousness (1970)
Pondicherry, Sri Aurobindo Ashram. Trans. by Tehmi.

Schactel, E Metamorphosis (1959)
New York, Basic Books.

Schaef, Anne Wilson Women's Reality: an emerging female system in the white male society (1981)
Minneapolis MN, Winston Press.

Schaefer, G Universe with Man in Mind: the new paradigm (1982)
Ann Arbor MI, Translational Press.

Schaer, Hans Religion and the Cure of Souls in Jung's Psychology
London, Routledge and Kegan Paul.

Schaya, Leo The Universal Meaning of the Kabbalah (1971)
London, George Allen and Unwin.

Schaya, Léon La Doctrine Soufique de l'Unite (1962)
Paris.

Scheff, T J Being Mentally (1966)
Chicagp IL, Aldine.

Scheffler, Israel Of Human Potential: an essay in the philosophy of education (1985)
London, Routledge and Kegan Paul, 141 p. bibl. (pp. 127–135). ISBN 0–7102–0571–6.

Scherz, F H Maturational Crises and Parent–Child Interaction (1971)
Social Casework 52, June pp. 362–369.

Schiffman, M Gestalt Self Therapy and Further Techniques for Personal Growth (1971)
Self Therapy Press. A straightforward account of the author's use of Gestalt therapy techniques in her own self therapy, with many procedures that can be well utilized in self growth.

Schilder, Paul The Image and Appearance of the Human Body (1950)
New York, International Universities Press. First published 1923.

Schlanger, J E Les métaphores de l'organisme (1971)
Paris, Librairie philosophique.

Schmidt, K O Meister Eckhart's Way to Cosmic Consciousness: a breviary of practical mysticism

Schmitz, O A H Psychoanalyse und Yoga (1923)
Darmstadt.

Schoedel, William Gnostic Monism and the Gospel of Truth (1980)
Leiden, E J Brill, in "The Rediscovery of Gnosticism", ed Bentley Layton, Vol 1.

Scholem, Gershom G Major Trends in Jewish Mysticism (1954)
New York, Schocken Books.

Scholem, Gershom G On the Kabbalah and its Symbolism (1965)
London, Routledge, Kegan Paul.

Scholom, Gershom Jewish Gnosticism, Merkabah Mysticism and Talmudic Tradition (1960)
Philadelphia, The Jewish Publication Society of America.

Schultz, J H and Luthe, W Autogenic Training: a psychophysiological approach in psychotherapy (1959)
New York, Grune and Stratton.

Schuman, Marjorie The Psychophysiological Model of Meditation and Altered States of Consciousness: a critical review (1980)
New York, Plenum Publishing. In: The Psychobiology of Consciousness, edited by Davidson J M and Davidson R J.

Schuon, Frithjof Logic and Transcendence (1975)
New York, Harper and Row.

Schuon, Frithjof Survey of Metaphysics and Esoterism (1986)
Bloomington IN, World Wisdom Books, 224 p. The Library of Traditional Wisdom. Transl. by Gustavo Polit.
ISBN 0–941532–06–2.

Schutz, A and Luckmann, T The Structures of the Life–World (1973)
Evanston IL, Northwestern University Press.

Schutz, William Interpersonal Underworld (1967)
Science and Behavior Books.

Schutz, William Joy: expanding human awareness (1967)
New York, Grove Press.
Schwartz, G Beyond Conformity and Rebellion (1987)
Chicago IL, University of Chicago Press.
Schwarzer, R The Self in Anxiety, Stress and Depression (1984)
New York, Elsevier Science Publishing, 442 p. ISBN 0-444-87556-5.
Scott, Charles E Boundaries in Mind: a study of immediate awareness based on psychotherapy (1982)
New York, Crossroad Publishing Company, 160 p. ISBN 0-8245-0529-8.
Scott, Gini G Cult and Countercult: a study of a spiritual growth group and a witchcraft order (1980)
Greenwood. ISBN 0-313-22074-3.
Scott, W A Cognitive Complexity and Cognitive Flexibility (1962)
Sociometry 24.
Scott, W A Cognitive Complexity and Cognitive Balance (1963)
Sociometry 26.
Se, Anima Attention and Distraction (1983)
New Delhi, Sterling Publishers, 208 p. ISBN 81-207-0392-8.
Seagrim, G N and Lendon, R J Furnishing the Mind: comparative study of cognitive development in central Australia aborigines (1980)
North Ryde NSW, Academic Press Australia, 242 p. ISBN 0-12-634340-3.
Sears, Pauline Snedden (Comp) Intellectual Development (1971)
New York, Wiley, 579 p. bibl.
Seeman, W; Nidich, S and Banta, T The Influence of Transcendental Meditation on a Measure of Self-Actualisation
In: *Journal of Counseling Psychology*, 19, 3, pp. 184-187.
Seeman, W; Nidich, S and Banta, T Influence of Transcendental Meditation on a Measure of Self-Actualization (1971)
Journal of Counseling Psychology 1972, 19, 3.
Seguin, Carlos A Love and Psychotherapy (1965)
Libra.
Sekida, Katsuki and Grimstone, A V Zen Training: methods and philosophy (1975)
New York, John Weatherhill, 264 p. ISBN 0-8348-0111-6.
Sen, A Poverty and Famines: an essay on entitlement and deprivation (1981)
Oxford, Clarendon Press.
Sen, A Resources, Values and Development (1984)
Oxford, Oxford University Press, 547 p.
Sen, A Hunger and Entitlements, Research for Action (1987)
Helsinki, WIDER/UNU, 38 p.
Sen Gupta, S C Human Existence, Transcendence and Spirituality (1979)
I.I.A.S., 180 p.
Sen, Indra The Integral Man: special lectures (1970)
Mysore, Prasaranga University of Mysore, 82 p.
Sen, Kshiti Mohan Hinduism (1961)
Harmondsworth, Penguin Books, 160 p. ISBN 0-14-02-0515-2.
Senn, M J E (Ed) Symposium on the Healthy Personality (1950)
New York, Josiah Macy Jr Foundation.
Serner, H W Agni Yoga and Physics: as above so below (1973)
Aalst, Waalre, 46 p. fig.
Severin, F T (Ed) Humanistic Viewpoints in Psychology (1965)
New York, McGraw Hill Book Company.
Sgam, Po Jewel Ornament of Liberation (1970)
London, Rider. Guenther H, Trans.
Shafii, Mohammd Freedom from the Self: Sufism, meditation and psychotherapy (1988)
New York, Human Sciences Press. reissue ed. ISBN 0-89885-395-8.
Shah, Gary The Secret Lore of Magic and Books of the Sorcerers (1957)
New York, Citadel Press.
Shah, Idries Special Illumination: the sufi use of humour (1977)
Cambridge MA, Institute for the Study of Human Knowledge, 64 p. ISBN 0-900860-57-X.
Shah, Idries Learning How to Learn: psychology and spirituality in the Sufi Way (1981)
New York, Harper and Row, 304 p. ISBN 0-06-067255-2.
Shah Waliullah; Jalbani, G N and Pendlebury, D L The Sacred Knowledge: the altaf al-quds of shah waliullah (1982)
Cambridge MA, Institute for the Study of Human Knowledge. ISBN 0-900860-93-6.
Shainberg, David The Transforming Self (1973)
New York, Intercontinental Medical Book Corporation.
Shapiro, D Precision Nirvana (1978)
Englewood Cliffs NJ, Prentice-Hall.
Shapiro, Deane Meditation: a scientific and personal exploration (1980)
New York, Aldine.
Shapiro, Deane H Meditation: self-regulation strategy and altered state of consciousness (1980)
Berlin, Aldine de Gruyter, 318 p. ISBN 0-202-25132-2.
Shapiro, Deane H and Walsh, Roger N (Eds) Meditation: classic and contemporary perspectives (1984)
Berlin, Aldine de Gruyter, 722 p. ISBN 0-202-25136-5.
Sharma, Dhirandra The Negative Dialectics: a study of negative dialectism in Indian philosphy (1974)
New Delhi, Sterling.
Sharot, Stephen Messianism, Mysticism, and Magic: a sociological analysis of jewish religious (1987)
Chapel Hill NC, University of North Carolina Press, 306 p. ISBN 0-8078-4170-6.
Shaw, Franklin J Transitional Experiences and Psychological Growth (1957)
Review of General Semantics, 15, p. 39-45.
Shaw, Franklin J Reconciliation: a theory of man transcending (1966)
Princeton, Van Nostrand, 146 p. bibl. (pp. 146-148).
Sheenan, John F On Becoming Whole in Christ: an interpretation of the spiritual exercises (1978)
Chicago IL, Loyola University Press. ISBN 0-8294-0278-0.
Shepard, Herbert A Personal Growth Laboratories: toward an alternative culture (1970)
Journal of Applied Behavioral Science, Vol 6, No 3.
Shepard, Leslie A Encyclopedia of Occultism and Parapsychology (1978)
Detroit, Gale Research. 2 Vols.
Shephard, Gerald T (Ed) Wisdom as a Hermeneutical Construct (1980)
Berlin, Walter de Gruyter, 178 p. ISBN 3-1100-7504-0.
Shlien, J M A Criterion of Psychological Health (1956)
Group Psychotherapy, 9, pp1-18.
Shoben Jr, E J Toward a Concept of the Normal Personality (1957)
American Psychologist 12, pp. 183-189.
Shorter, Bani An Image Darkly Forming: women and initiation (1987)
London, Routledge, Chapman and Hall, 160 p. ISBN 0-7102-0574-0.
Shostrom, E Personal Orientation Inventory: a test of self-actualization (1963)
San Diego CA, Educational and Industrial Testing Service.
Shostrom, E Man the Manipulator: the inner journey from manipulation to actualization (1968)
New York, Bantam Books. (A popularized discussion of methods of manipulation placed in a Gestalt therapy framework, with a number of therapy vignettes).
Siddheswarananda, Swami Meditation according to Yoga-Vedanta
Siegel, Ronald Intoxication: life in the pursuit of artificial paradise (1989)
New York, Dutton.
Simms, James R The Limits of Behavior (1983)
Salinas CA, Intersystems Publications, 150 p. ISBN 0-914105-15-9.
Simon, H A An Information Processing Theory of Intellectual Development
In: *Kessen W and Kuhlman C (Eds), Thought in the Young Child. Monographs of the Society for Research in Child Development* 1962, 27, 2 (whole no 83).
Simon, H A Models of Mal (1957)
New York, John Wiley and Sons.
Singer, B D Feedback in Society (1973)
Lexington MA, Lexington Books.
Singer, June Androgyny: toward a new theory of sexuality (1976)
Garden City NY, Anchor Press, 375 p. bibl. Introduction by Sheldon S Hendler. ISBN 0-385-11025-1.
Singh, Jaideva (Trans) Vijnanabhairava or Divine Consciousness (1979)
Delhi, Motilal Banarsidass, 173 p.
Sinha, Jadunath Indian Psychology (1986)
Delhi, Motilal Banarsidass. 3 Vols. Vol I: Cognition (512 p.), Vol II: Emotion and will (568 p.), Vol III: Epistemology and Perception (568 p.).
Sircar, Mahendra Nath The Mystical Experience and Samadhi
Sisk, Dorothy World Council for Gifted and Talented Children
Monroe NY, Trillium Press. 4 vols. ISBN 0-317-06552-1.
Siu, R G H The Tao of Science: an essay on western knowledge and eastern wisdom (1957)
Cambridge MA, MIT Press. illus. ISBN 0-262-69004-7.
Siu, R G H Ch'i: a neo-Taoist approach to life (1974)
Cambridge MA, MIT Press, 351 p. ISBN 0-262-19123-7.
Sivananda, Swami Samadhi Yoga
Garwal, Divine Life Society.
Sivananda, Swami Science of Yoga
Garwal, Divine Life Society, 400 p.
Sivananda, Swami Japa Yoga: a comprehensive treatise on Mantra-Shastra
Sivananda, Swami Sadhana, a Textbook of the psychology and practice of the techniques to spiritual perfection (1974)
Delhi, Motilal Banarsidass, 719 p.
Sivananda, Swami Sri Sadhana
A Text-Book of the Psychology and Practice of the Techniques to Spiritual Perfection.
Skolimowski, Henryk Eco-Philosophy: designing new tactics for living (1981)
New York, Marion Boyars, 117 p. Ideas in Progress. ISBN 0-7145-2676-2.
Skurky, Thomas The Levels of Analysis Paradigm: a model for individual and systematic therapy (1990)
New York, Praeger Publishers, 176 p. ISBN 0-275-93296-6.
Slater, Philip E Microcosm: structural, psychological and religious evolution in groups (1966)
New York, John Wiley and Sons.
Slavson, S R Analytic Group Psychotherapy (1950)
New York, Columbia.
Slochower, Harry Mythopoesis (1970)
Detroit MI, Wayne State University Press.
Smart, Ninian The Religious Experience of Mankind (1971)
Huntington NY, John M Fontana Publishing.
Smart, Ninian The Science of Religion and the Sociology of Knowledge: some methodological questions (1973)
Princeton NJ, Princeton University Press. 61-62 p.
Smart, Reginald G Forbidden Highs (1983)
Toronto ON, Addiction Research Foundation, 244 p. ISBN 0-88868-078-3.
Smith, Adam The Powers of Mind (1975)
New York, Random House.
Smith, Cyprian The Way of Paradox: spiritual life as taught by Meister Eckhart (1987)
London, Darton, Longman and Todd, 133 p. ISBN 0-232-51743-6.
Smith, David East/West Exercise Book (1976)
New York, McGraw Hill Book Company.
Smith, Huston Forgotten Truth: the primordial tradition (1976)
New York, Harper and Row.
Smith III, Robert A Synergistic Organizations: humanistic extensions of man's evolution (1970)
Fields within Fields 3, 1, p. 39-52.
Smith, M B Research Strategies Toward a Conception of Positive Mental Health (1959)
American Psychologist 14, pp. 673-681.
Smith, Margaret The Sufi Path of Love (1954)
London.
Snell, B The Discovery of the Mind (1960)
New York, Harper Torchbook.
Solé-Leris, Amadeo Tranquillity and Insight: introduction to the oldest form of Buddhist meditation (1986)
Boston MA, Shambhala Publications, 176 p. bibl. ISBN 0-87773-385-6.
Sollberger, A Biological Rhythm Research (1965)
Amsterdam, Elsevier Science Publishing, 461 p.
Solomon, D (Ed) LSD: the consciousness expanding drug (1964)
New York, Putnam Publishing Group.
Solomon, Joseph C A Synthesis of Human Behavior: an integration of thought processes and ego growth (1954)
New York, Grune and Stratton 265 p. Table of ego development.
Solomon, L and Barzon, B (Ed) New Perspectives on Encounter Groups (1972)
San Francisco CA, Jossey-Bass Publishers.
Soma, Bhikku The Way of Mindfulness (1949)
Colombo, Vajirama.
Sonnier, Isidore L Holistic Education: teaching-learning in the affectice domain (1982)
New York, Philosphical Library. ISBN 0-8022-2389-3.
Sontag, L W; Baker, C T and Nelson, V L Mental Growth and Personality Development (1958)
Monographs of the Society for Research in Child Development 23, pp. 1-143.
Sorel, Georges; Stanley, John and Stanley, Charlotte The Illusions of Progress (1969)
Berkeley CA, University of California Press. ISBN 0-520-01531-2.
Sorokin, Pitirim A Forms and Techniques of Altruistic and Spiritual Growth (1914)
Boston MA, Beacon Press.
Sorokin, Pitirim A The Reconstruction of Humanity (1948)
Boston MA, Beacon Press, 247 p.
Sorokin, Pitirim A Explorations in Altruistic Love and Behavior (1950)
Boston MA, Beacon Press.
Sorokin, Pitirim A Ways and Power of Love (1967)
Washington DC, Regnery Gateway. (Boston, Beacon Press, 1954).
Sorokin, Pitrim A Social and Cultural Dynamics
New York, Allied Publications. 4 vols. Vol 1: Fluctuation of Forms of Art. Vol 2: Fluctuation of Systems of Truth, Ethics and Law (1937, 727 p). Vol 3: Fluctuation of Social Relationships, War and Revolution (1937, 636 p). Vol 4: Basic Problems, Principles and Methods (1941, 804 p).
Speck, R and Attneave, C Family Networks (1973)
New York, Pantheon Books.
Sperry, Roger Structure and Significance of the Consciousness Revolution (1987)
In: *Journal of Mind and Behavior* Winter, Vol 8, 1, pp. 37-66.
Spidlik, Thomas S J The Spirituality of the Christian East: a systematic handbook
Trans. by Anthony P Gythiel.
Spidlik, Tomas The Spirituality of the Christian East: a systematic handbook (1986)
Kalamazoo MI, Cistercian Publications. Cistercian Studies Series: 79. Trans. from French by Anthony P Gythiel. ISBN 0-89907-879-0.
Spilka, Bernard The Complete Person: some theoretical views and research findings for a theological psychology of religion
In: *Journal of Psychology and Theology* 1976, 4, 1, pp. 15-24.
Spitz, R A A Genetic Field Theory of Ego Formation (1959)
New York, International University Press.
Spolin, Viola Improvisation for the Theater (1963)
Evanston, Northwestern University Press.
Spong, John S (Ed) Consciousness and Survival: an interdisciplinary inquiry into the possibility of life beyong biological death (1987)
Sausalito CA, Institute of Noetic Sciences, 194 p. Intro. by Clairborne Pell. ISBN 0-943961-00-3.
Spretnak, Charlene Politics of Women Spirituality: essays of the rise of spiritualist power within the feminist movement (1982)
New York, Doubleday, 624 p. ISBN 0-385-17241-9.
Spretnak, Charlene The Spiritual Dimension of Green Politics (1986)
Santa Fé, Bear and Company, 90 p. bibl. ISBN 0-939680-29-7.
Spuhler, J N (Ad) The Evolution of Man's Capacity for Culture (1959)
Detroit.
Sri Sankaracarya Self-Knowledge - Atmabodha
Swami Nikhilananda, Trans.

St John of the Cross The Dark Night of the Soul (1988)
London, Hodder and Stoughton, 116 p. Title page comments. ISBN 0-340-42274-2.

Stace, W T The Teachings of the Mystics (1960)
New York, New American Library.

Stace, Walter T Mysticism and Philosophy (1960)
New York, Lippincott.

Staley, F A Hemispheric Brain Research: a breakthrough in outdoor education
In: *Journal of Physical Education and Recreation* 1980, 51, 4, pp. 28–30, pp. 63–64.

Stanwood, P G (Ed) William Law: a serious call to a devout and Holy life, and the spirit of love (1979)
New York NY, Paulist Press.

Starhawk The Spiral Dance: a rebirth of the ancient Religion of the Great Goddess (1979)
San Francisco CA, Harper and Row.

Starhawk Dreaming the Dark: magic, sex and politics (1982)
Boston MA, Beacon Press.

Stark, Stanley An Essay in Rorschach Revisionism: with special reference to the Maslowian self-actualizer; (I: Innovation versus imagination, idealism, mysticism, romanticism) (1971)
Perceptual and Motor Skills, 33, 2, pp343-357.

Starr, I The Sound of Light: experiencing the transcendental (1976)
De Vorss and Co.

Stcherbatsky, T Buddhist Logic (1962)
New York, Dover Publications. Vols 1-2.

Stcherbatsky, Theodore The Conception of Buddhist Nirvā (1977)
Delhi, Motilal, rev ed.

Steere, Douglas V Quaker Spirituality: selected writings

Stein, Maurice R; Vidich, A J and White, D M Identity and Anxiety (1960)
New York, Free Press.

Steiner, George Real Presences: is there anything in what we say? (1989)
London, Faber and Faber, 236 p. ISBN 0-571-14071-8.

Steiner, Rudolf The Bridge Between Universal Spirituality and the Physical Constitution of Man (1979)
Hudson NY, Anthroposophic Press, 64 p. 2nd ed. Trans. from German by Dorothy S Osmond. ISBN 0-910142-03-3.

Steiner, Rudolf Esoteric Development (1982)
Hudson NY, Anthroposofic Press, 190 p. ISBN 0-88010-012-5.

Stephanou, Eusebius A Charisma and Gnosis in Orthodox Thought (1976)
Fort Wayne, Logos Ministry for Orthodox Renewal.

Sternberg, Robert J The Triarchic Mind: a new theory of human intelligence (1988)
New York, Penguin Books, 354 p. illus. bibl. ISBN 0-670-80364-2.

Stevens, J Awareness: Exploring, Experimenting, Experiencing (1971)
Moab UT, Real People Press. (New York, Bantam, 1973).

Steward, J Cultural Evolution
Scientific American 194, p. 69–80.

Steward, J H and Shimkin, D B Some Mechanisms of Sociocultural Evolution (1962)
In: *Hoagland and R W Burhoe. Evolution and Mal's Progress.* New York, Columbia UP.

Stewart, R J Living Magical Arts: imagination and magic for the 21st century (1987)
Poole, Blandford Press, 210, p. ISBN 0-7137-1883-8.

Stewart, R J Advanced Magical Arts: visualisation, mediation and ritual in the Western magical tradition (1988)
Shaftesbury, Element Books, 194 p. illus. bibl. ISBN 1-85230-045-0.

Sticht, T G Educational Uses of Metaphor
In: *A Ortony (Ed) Metaphor and Thought* Cambridge, Cambridge University Press, 1979, pp. 474–485.

Stoddart, William Sufism: the mystical doctrines and methods of Islam
Foreword, R W J Austin.

Stolz, Anselm The Doctrine of Spiritual Perfection (1938)
St Louis, B Herder Book.

Stoops, John D The Integrated Life (1951)
New York, RR Smith, 180 p.

Storr, A The Integrity of the Personality (1960)
Baltimore MD, Penguin Books.

Storr, Anthony The School of Genius
Deutsch, 216 p.

Storr, Anthony Solitude: a return to the self (1988)
New York, Free Press, 230 p. ISBN 0-02-931620-0.

Strackbein, Oscar R A Philosophy of Self-Management (1967)
New York, Frederick Fell, 765 p.

Streeten, P Mobilizing Human Potential: the challenge of unemployment (1989)
New York, UNDP. Policy Discussion Paper.

Streng, F J Emptiness: a study in religious meaning (1967)
Nashville, Abingdon Press.

Strober, C F and Luce G The Importance of Biological Clocks in Mental Health (1968)
Chevy Chase, National Institute of Mental Health, pp. 323–351 (PHS Publication 1743).

Stromberg, G Man, Mind and the Universe (1966)
Science of Mind Publications.

Strumpel, B (Ed) Eléments subjectifs du bien-être (1974)
Paris, OECD.

Stulman, Julius Evolving Mankind's Future; The World Institute: a problem-solving methodology (1967)
Philadelphia PA, Lippincott, 95 p.

Suarès, Carlo The Sepher Yetsira: including the original astrology according to the Qabala and Its Zodiac (1976)
Boulder, Shambhala, 173 p. Trans. from the French by Micheline and Vincent Stuart. ISBN 0-87773-093-8.

Sugarman, A and Tarter, R (Eds) Expanding Dimensions of Consciousness (1978)
New York, Springer Publisher.

Sullivan C; Grant M and Grant J The Development of Interpersonal Maturity: applications to delinquency
In: *Psychiatry* 1957, 20.

Sullivan, E A Medical, Biological, and Chemical Methods of Shaping the Mind (1972)
Phi Beta Kappan April.

Sullivan, John J (Trans) The Autobiography of Venerable Marie of the Incarnation, OSU: mystic and missionary (1964)
Chicago Il, Loyola University Press.

Sun, W Flow and Yu: comparison of Csikszentmihalyi's Theory and Chuang-tzu's Philosophy, Paper presented at the meetings of the Anthropological Association for the study of play, Montreal, March

Sutich, A The Growth-Experience and the Growth-Centered Attitude (1949)
Journal of Psychology 28, pp. 293–301.

Sutich, A J and Vich, M a The Realization of the Self
Psychosynthesis Press, 1974, 1, 1, Spring.

Sutich, A J and Vich, M a (Eds) Readings in Humanistic Psychology (1969)
New York, Free Press.

Suzuki, D T Essays in Zen Buddhism (1970)
London, Rider. 3 Vols.

Suzuki, D T The Zen Doctrine of No-Mind: the Significance of the Sutra of Hui-neng (1970)
London, Rider, 160 p. ISBN 0-09-152971-9.

Suzuki, Daisetz Teitaro Mysticism-Christian and Buddhist: the Eastern and Western way (1969)
New York, Macmillan.

Swahananda, Swami Meditation and Other Spiritual Disciplines

Swartley, W M A Comparative Survey of Some Active Techniques of Stimulating whole Fluctioning
Ms available from the Librarian, Psychosynthesis Research Foundation.

Swearer, Donald K (Ed) Secrets of the Lotus: studies in Buddhist meditation

Swearer, Donald K A Mahayana Training of the Mind: the wheel of sharp weapons
Trans. of the Tibetan Theg-pa-chen-pohi-blo-sbyong mtson-cha-hkhor-lo by Dharmarasksita. Prepared by Geshe Ngawang Dhargyey, Sherpa Tulku, Khamlug Tulku, Alexander Berzin, and Jonathan Landaw. First revised edition, 1976.

Swinburne, Richard R The Existence of God (1979)
Oxford, Clarendon Press.

Syed, Abdullah Meditation: achieving internal balance
In: *Elliot M Goldwag (Ed) Inner Balance: The power of holistic healing.*

Taft, Ronald Peak Experiences and Ego Permissiveness. (1969)
Acta Psychologica 29, 1, Feb pp 35–64.

Tagore, Rabindranath Toward Universal Man (1961)
London, Asian Publishing House, 387p.

Taimni, I K Self-Realization through Love
Madras, Theosophical Publication House. Navada Bhakti Sutra text in Sanskrit, Transliteration in Roman and translation in English and Commentary.

Taimni, I K The Science of Yoga (1961)
Adyar, Theosophical Publishing House, 448 p. ISBN 81 7059 001 9. The yoga-sutras of Patanjali in Sanskrit with transliteration in Roman, translation and commentary in English.

Taimni, I K The Science of Yoga (1965)
Adyar, Theosophical Publishing House.

Takahashi, Masaru and Brown, Stephen Qigong for Health: Chinese traditional exercises for cure and prevention
Tokyo, Japan Publications, 144 p. ISBN 0-87040-701-5.

Tambiah, Stanley Jeyaraja Magic, Science, Religion and the Scope of Rationality (1989)
Cambridge, Cambridge University Press, 187 p. bibl. Lewis Henry Morgan Lecture Series 1984. ISBN 0-521-37486-3.

Tanner, J M and Inhelder, B (Ed) Discussion on Child Development: a consideration of the biological, psychological, and cultural approaches to the understanding of human development and behaviour: Proceedings of the WHO Study Group on the Psychobiological Development of the Child (Geneva, 1953–1956) (1956)
London, Tavistock Publications.

Tansley, D Radionics and the Subtle Anatomy of Man (1972)
Essex, Health Science Press.

Tansley, David V Subtle Body: essence and shadow (1977)
London, Hodder and Stoughton.

Tart, C Transpersonal Psychologies (1983)
El Cerrito CA, Psychological Processes.

Tart, Charles T (Ed) Altered States of Consciousness (1971)
New York, John Wiley and Sons.

Tart, Charles T Transpersonal Potentialities of Deep Hypnosis (1971)
Journal of Transpersonal Psychology 3.

Tart, Charles T Scientific Foundations for the Study of Altered States of Consciousness (1971)
Journal of Transpersonal Psychology 3.

Tart, Charles T A Psychologist's Experience with Transcendental Meditation (1971)
Journal of Transpersonal Psychology 3, 2.

Tart, Charles T States of Consciousness and State-Specific Sciences (1972)
In: *Science* 176, pp. 1203-1210.

Tart, Charles T States of Consciousness (1983)
El Cerrito CA, Psychological Processes.

Tart, Charles T Waking Up: overcoming the obstacles to human potential (1987)
Boston MA, Shambhala Publications, 323 p. bibl. (pp. 301–304). ISBN 0-87773-374-0.

Tatz, Mark and Kent, Jody Rebirth: the Tibetan game of liberation (1977)
New York, Anchor Books, 231 p. ISBN 0-385-11421-4.

Tax, S (Ed) Evolution after Darwin: vol II: the evolution of man (1960)
Chicago IL, University of Chicago Press.

Taylor, Alfred Mind as the Basic Potential (1958)
Main Currents in Modern Thought March.

Taylor, D W Toward an Information Processing Theory of Motivation (1960)
In: *Nebraska Symposium on Motivation* University of Nebraska, pp. 51–79.

Taylor, J Dream Work: techniques for discovering the creative power in dreams (1983)
New York, Paulist Press.

Taylor, John The Shape of Minds to Come (1971)
New York, Waybright and Talley, 278 p.

Taylor, Marvin J (Ed) Changing Patterns of Religious Education (1984)
New York NY.
A wide-ranging survey of Catholic and Protestant writing.

Teilhard de Chardin, Pierre Phenomenon of Man (1959)
New York, Harper and Row.

Teilhard de Chardin, Pierre The Phenomenon of Man (1961)
Harper Torchbooks. Introduction by Sir Julian Huxley.

Teilhard de Chardin, Pierre The Future of Man (1968)
London, William Collins Sons. Norman Denny, Trans.

Teilhard de Chardin, Pierre Human Energy (1969)
New York, Harcourt Brace.

Tennov, Dorothy Love and Limerence (1979)
New York, Stein and Day.

Teresa of Avila Interior Castle (1988)
London, Hodder and Stoughton.

Terman, Lewis B Genetic Studies of Genius (1947)
Stanford CA, Stanford University Press. Vols. I, II, III, IV and V. out of print.

Teune, Henry The Learning of Integrative Habits (1964)
In: *P E Jacob and J V Toscano (Eds) The Integration of Political Communities* Philadelphia. Lippincott, pp247-282.

Teyler, Timothy J Altered States of Awareness: readings from 'Scientific American' (1972)
San Francisco, W H Freeman, 140p. bibl. (pp. 133–135).

The Theosophy Science Study Group Holistic Science and Human Values
Journal.

Thera Nanomoly Mindfulness of Breathing (1964)
Kandy, Buddhist Publishing Society.

Thompson, Helen Journey Toward Wholeness: a Jungian model of adult spiritual growth (1982)
Mahwah NJ, Paulist Press, 96 p. ISBN 0-8091-2422-X.

Thompson, Wendy L and Thompson, T L (Eds) Psychiatric Aspects of Chronic Pulmonary Disease (1985)
New York, Karger S AG, 174 p. Advances in Psychosomatic Medicine: Vol 14. ISBN 3-8055-4088-4.

Thompson, William Irwin At the Edge of History: speculation on the transformation of culture (1971)
New York, Harper and Row.

Thompson, William Irwin Passages About Earth: explorations of the new planetary culture (1974)
New York, Harper and Row.

Thorne, Frederick C Integrative Psychology (1967)
Brandon, Clinical Psychology Pub, 373 p. bibl.

Thorsén, Hakan Peak-Experience Religion and Knowledge: a Philosophical inquiry into some main themes in the writing of Abraham H Maslow (1983)
Stockholm, Almqvist and Wiksell. ISBN 91-22-00963-9.

Thurston, Herbert The Physical Phenomena of Mysticism
Edited by J H Crehan.

Tillich, P The Courage To Be (1952)
New Haven CT, Yale University Press.

Timmons, B and Kamiya, J The Psychology and Physiology of Meditation and Related Phenomena: bibliography II (1974)
Journal of Transpersonal Psychology, 6, 1, pp. 32–38.

Tipton, Steven Getting Saved from the Sixties: moral meaning in conversion and cultural change (1981)
Berkeley CA, University of California Press.

Todd, Emmanuel La Troisième Planète: structures familiales et systèmes idéologiques (1983)
Paris, Seuil.

Tohei, Koichi What is Aikido? (1962)
Tokyo, Rikugei Publishing.

Tohei, Koichi Aikido in Everyday Life (1966)
Tokyo, Rikugei Publishing.

Tola, Fernanda, et al The Yogasutra of Patanjali on Concentration of Mind (1986)
Columbia MO, South Asia Books, 268 p. ISBN 81-208-0259-4.

Trevelyan, George Magic Casements: the use of poetry in the expanding of consciousness (1980)
London, Coventure, 57 p. ISBN 0-904576-91-4.

Trimble, M R (Ed) Interface Between Neurology and Psychiatry (1985)
Basel, S Karger AG, 192 p. Advances in Psychosomatic Medicine: Vol 13. ISBN 3-8055-4023-X.

Trimingham, Spencer J The Sufi Orders in Islam (1971)
Oxford.

Trommen, Merton P (Ed) Research on Religious Development: a comprehensive handbook (1971)
New York, Hawthorn Books.

Trüb, H Psychosynthese als Seelisch-Geister Heilungsprozess (1936)
Zurich, Leipzig Ed Niehaus.
Trungpa, C Meditation in Action (1970)
Berkeley CA, Shambhala Publications.
Trungpa, Chogyam Cutting Through Spiritual Materialism (1975)
Berkeley CA, Shambhala Publications.
Trungpa, Chogyam The Myth of Freedom (1976)
Berkeley CA, Shambhala Publications.
Tucci, G The Theory and Practice of the Mandala (1957)
London.
Tulku, T (Ed) Reflections of Mind: western psychology meets Tibetan Buddhism (1975)
Emeryville CA, Dharma Publishing.
Tulku, T Gesture of Balance: a guide to awareness, self-healing and meditation (1977)
New York, State Mutual Books and Periodical Service, 170 p. ISBN 0–317–39074–0.
Turner, Harold W Rudolf Otto: the idea of the holy (1974)
Aberdeen University. Introduction to the Man' by Peter R McKenzie.
Tuxen, P Yoga (1911)
Copenhagen.
Tweedie, Irina The Chasm of Fire: a woman's experience of liberation through the teaching of a Sufi Master
Tyrrell, Bernard J Christotherapy II (1982)
New York NY, Paulist.
Tyson, Donald The New Magus: ritual magic as a personal process (1988)
St Paul MN, Llewellyn Publications, 346p.
Udupa, K N Stress and its Management by Yoga: a study of neurohumoral response (1986)
Delhi, Motilal Banarsidass, 205 p. ISBN 81–208–0000–1.
Ueshiba, Kisshomaru The Spirit of Aikido (1984)
Tokyo, Kodansha International, 126 p. Trans. by Taitetsu Unno. ISBN 0–87011–600–2.
Ullman, M and Wolman, B B (Eds) Handbook of States of Consciousness (1986)
New York, Van Nostrand Reinhold.
Underhill, Evelyn (Ed) The Cloud of Unknowing (1970)
London, Stuart and Watkins.
Underhill, Evelyn Mysticism: a study in the nature and development of man's spiritual (1977)
London, Methuen. 1st Ed 1911.
UNESCO International Trade in Educational and Scientific Materials (1964)
Geneva, United Nations Conference on Trade and Development. vol 7.
UNESCO The School and Continuing Education (1972)
Paris, UNESCO.
UNESCO The Parasciences: the impact of science on society (1974)
Paris, UNESCO. Special issue: 24, 4.
UNESCO UNESCO and the Development of Human Resources: paper submitted by UNESCO to the International Conference on the Human Dimension of the United Nations Programme of Action for African Economic Recovery and Development, 5–8 March (1988)
Paris, UNESCO.
Unger, Johan On Religious Experience: a psychological study (1976)
Stockholm, Almqvist and Wiksell.
Union of International Associations / Mankind 2000 Yearbook of World Problems and Human Potential (1976)
Brussels, K G Saur.
United Nations Human Development
In: *Resolution 2626 (XXV) as recommended by Second Committee, and adopted 24 October 1970* New York, UN. Document: A/8124, 8, 65–72.
United Nations United Nations Conference on the Application of Science and Technology for the Benefit of the Less-Developed Areas (1963)
Geneva, UN. Human Resources, report of the Secretary-General, para 24.
United Nations Science and Technology for Development (1963)
New York, UN. vol VI.
United Nations Development and Utilization of Human Resources: report of the secretary-general (1967)
New York, UN.
United Nations Human Resources Development: secretary-general's report to the forty-fourth session of the General Assembly (1989)
New York, UN.
United Nations Development Programme Human Development Report 1990
New York, Oxford University Press, 189 p. ISBN 0–19–506481–X.
United Nations Development Programme Report of the Tokyo Workshop on Human Resources Development held at the Foreign Ministry, Tokyo, 2–5 April, 1986 (1986)
Tokyo, UNDP, 13 p. DP/1986/10/Add 1.
United Nations Development Programme Human Resources Development: issues and implications, report of the administrator (1986)
New York, UNDP, 23 p. DP/1986/10.
United Nations Development Programme Programme Implementation: experience in human resources development since 1970, report of the administrator, 15 March (1988)
New York, UNDP, 13 p. DP/1988/62.
United Nations Development Programme Programme Implementation: experience in human resources development, report of the administrator, 15 March (1988)
New York, UNDP, 13 p. DP/1988/62.
United Nations Economic and Social Commission for Asia and the Pacific Select Annotated Bibliography on Social Aspects of Development Planning
Bangkok, UN/ESCAP, 71 p.
United Nations Economic and Social Commission for Asia and the Pacific Background to an Integrated Plan of Action on Human Resources Development for the ESCAP Region (1988)
Bangkok, ESCAP, 252 p.
United Nations University Goals, Processes and Indicators of Development Project: human development in micro to macro perspective (1983)
Tokyo, UNU. Final Report: Integration Group A.
Ushabudha Arya Meditation and the Art of Dying (1979)
Honesdale PA, Himalyan Institute.
Vaill, P B Process Wisdom for a New Age
In: *ReVision* 1984, 7, pp. 39–49.
Vaillant, G E Theoretical Hierarchy of Adaptive Ego Mechanisms
In: *Archives General Psychiatry* 1971, 24, pp. 107–18.
Valiuddin, Mir Love of God: the Sufi Approach (1968)
Delhi.
Valiuddin, Mir Contemplative Disciplines in Sufism (1980)
London, East-West Publications, 173 p. ISBN 0–85692–007–X.
Islamic Tradition of Religious Mysticism
Valle, Ronald S and Eckartsberg, Rolf von (Eds) The Metaphors of Consciousness (1981)
New York, Plenum Publishing, 521 p. illus. ISBN 0–306–40520–2.
Vallee, Jacques Dimensions: a casebook of alien contact (1988)
London, Sphere Books Ltd, 304 p. ISBN 0–7474–0529–8.
Van Den Berg, Jan H The Changing Nature of Man (1983)
New York, WW Norton.
Van der Bent, Ans J Vital Ecumenical Concerns: sixteen documentary surveys (1986)
Geneva, World Council of Churches, 333 p. ISBN 2–8254–0873–5.
Van Kaam, Adrian L Existential Foundations of Psychology
Pittsburgh PA, Duquesne University Press.
Van Kaam, Adrian L Formative Spirituality
New York, Crossroad Publishing. 4 Vols. Formative Spirituality Series.
Van Kaam, Adrian L The Dynamics of Spiritual Self-Direction
Denville NJ, Dimension Books. ISBN 0–87193–122–2.
Van Kaam, Adrian L The Third Force in European Psychology – its expression in a theory of psychotherapy (1960)
New York, Psychosynthesis Research Foundation.
Van Kaam, Adrian L Humanistic Psychology and Culture (1961)
Journal of Humanistic Psychology 1, pp. 94–100.
Van Nuys, D A Novel Technique for Studying Attention during Meditation (1971)
Journal of Transpersonal Psychology 3, 2.
Van Vliet, C J The Coiled Serpent: a philosophy of conservation and transmutation of reproductive energy
Vanderburg, William H Growth of Minds and Cultures: a unified theory of the structure of human experience (1985)
Toronto ON, University of Toronto Press. ISBN 0–8020–2578–1.
Vanderploeg, R D Imago Dei as Foundational to Psychotherapy: integration versus segregation (1981)
In: *Journal of Psychology and Theology* 9, pp. 299–304.
Vanek, J Self Management (1975)
Harmondsworth, Penguin.
Vargiu, James C A Model of Creative Behavior (1971)
Redwood City CA, Psychosynthesis Institute.
Vassiliou, G and V On the Need of a System's Approach to Psycho-Social Function and Malfunction (1975)
(Paper presented at the World Congress of the World Federation for Mental Health, Copenhagen, 1975).
Vaughan, F Awakening Intuition (1979)
Garden City NY, Anchor Books.
Vaughan, Frances and Walsh, Roger (Eds) Beyond Ego: transpersonal dimensions in psychology (1980)
Los Angeles, J P Tarcher.
Vaught, Carl G The Quest for Wholeness (1983)
Albany NY, State University of New York Press, 213 p. ISBN 0–87395–593–5.
Veenhoven, R Databook of Happiness (1984)
Boston MA, Dordrecht-Reidel.
Venkatasananda, S The Supreme Yoga (Yoga Vasistha) (1981)
London, Chiltern Yoga Trust.
Venkatesananda The Supreme Yoga (1981)
Australia, Chiltern.
Vidyaranya, Swami Panchadashi: a treatise on Advaita metaphysics
Trans. by Hari Prasad Shastri.
Vimala Thakar Mutation of Mind
Vimala Thakar Foundation, 180 p.
Vispo, R H On Human Maturity
In: *Perspectives in Biology and Medicine* 9, 4.
Vitz, P Psychology as Religion (1981)
Tring, Lion Publications.
Vivekananda, Swami Jnana Yoga
Calcutta, Advaita Ashrama, 421 p.
Vivekananda, Swami Karma Yoga
Calcutta, Advaita Ashrama, 136 p.
Vivekananda, Swami Bhakti Yoga
Calcutta, Advaita Ashrama, 116 p.
Vivekananda, Swami Bhakti-Yoga (1964)
Calcutta, Advaita Ashrama.
Vivekananda, Swami Raja-Yoga (1970)
Calcutta, Advaita Ashrama.
von Bertalanffy, L Some Considerations on Growth in its Physical and Mental Aspects (1956)
Merill-Palmer Quarterly 3, 1954, pp. 13–23.
von Bonin, G The Evolution of the Human Brain (1963)
Chicago IL, University of Chicago Press.
Vrey, J D Self-Actualising Educand (1980)
Pretoria, University of South Africa. 2nd ed. ISBN 0–86981–142–8.
Vyas, R N From Consciousness to Super Consciousness
New Delhi, Cosmo Publications. ISBN 81–7020–025–3.
Vygotsky, L S Mind in Society: the development of higher psychological processes (1978)
Cambridge, Harvard University Press.
Wach, Joachim Types of Religious Experience: Christian and Non-Christian (1972)
Chicago IL, University of Chicago Press, 275 p. ISBN 0–226–76710–2.
Wachsmuth, G Reincarnation as a Phenomenon of Metamorphosis
Hudson NY, Anthroposophic Press.
Wade, Nicholas The Art and Science of Visual Illusions (1983)
New York, Routledge Chapman and Hall, 224 p. illus. International Library of Psychology. ISBN 0–7100–0868–6.
Wainwright, W Mysticism: a study of its nature, cognitive value and moral implications (1981)
Madison WI, University of Wisconsin Press.
Walker, Benjamin Encyclopedia of Metaphysical Medicine (1978)
London, Routledge and Kegan Paul, 323 p. ISBN 0 7100 8781 0.
Walker, D P Spiritual and Demonic Magic: from Ficino to Campanella (1964)
Notre Dame, University of Notre Dame Press.
Walker, James L Body and Soul: Gestalt therapy and religious experience (1971)
Nashville, Abingdon Press, 208 p. bibl. (pp. 199–203).
Walker, Jeremy Self-Knowledge as the Way to God: studies in Kierkegaard (1984)
Montreal, McGill-Queen's University Press. ISBN 0–7735–0417–6.
Wallace, A F C Culture and Personality (1961)
New York, Random House.
Wallace, J G Concept Growth and the Education of the Child
A Survey of research on conceptualization. National Foundation for Education Research.
Wallace, Keith Physiological Effects of Transcendental Meditation
In: *Science* 167, pp. 1751–54.
Wallace, R Keith The Physiological Effects of Transcendental Meditation: a proposed fourth major state of consciousness (1970)
Berkeley CA, University of California Press.
Wallace, R Keith and Benson, Herbert The Physiology of Meditation (1972)
Scientific American 226, 2, pp. 84–90.
Wallace, Robert K; Benson, H and Wilson, A F A Wakeful Hypometabolic Physiologic State (1971)
American Journal of Physiology 221, 3, Sept pp. 795–799.
Walsh, Michael The Secret World of Opus Dei (1990)
London, Crafton Books, 223 p. illus. ISBN 0–586–20734–1.
Walsh, R Meditation Research: an introduction and review
In: *The Journal of Transpersonal Psychology* 1979, 11, pp. 161–174.
Walsh, R and Shapiro, D (Eds) Beyond Health and Normality (1978)
New York, Reinhold Van Nostrand.
Walsh, Roger The Spirit of Shamanism: a Psychological Review
Los Angeles CA, J P Tarcher.
Wapner, S and Werner, H Perceptual Development (1957)
Worcester MA, Clark University Press.
Wapnick, Kenneth The Psychology of the Mystical Experience (1968)
New York, Adelphi University, (unpublished doctoral dissertation).
Wapnick, Kenneth Mysticism and Schizophrenia (1969)
Journal of Transpersonal Psychology, 1, 2, Fall.
Ward, Kate Eccentrics: the scientific investigation
Stirling University Press.
Warfield, Benjamin B Studies in Perfectionism (1958)
Phillipsburg NJ, Presbyterian and Reformed Publishing. ISBN 0–87552–528–8.
Warnock, Mary Imagination (1976)
Berkeley CA, University of California Press.
Warren, Henry Clarke (Ed) Visuddhimagga of Buddhaghosâcariya (1950)
Cambridge, Harvard University Press. rev ed. Harvard Oriental Series: 41. Ed by Dharmânanda Kosambi.
Washburn, M Observations Relevant to a Unified Theory of Meditation
In: *Journal of Transpersonal Psychology* 1978, 10, 1.
Washburn, S L and Jay, P (Eds) Perspectives of Human Evolution (1968)
New York, Holt, Rinehart and Winston. bibl.
Watkins, Mary M Waking Dreams (1977)
New York, Harper.
Watkins, Mary Maria Creative Imagination and the Transcendent Function (1972)
Princeton NJ, Princeton University Press, 177 p. unpublished thesis. bibl. (pp. 159–177; works on symbolic experience and its implication for human growth).
Watson, Goodwin The Psychological Evidence for Integration (1937)
In: *L Thomas Hopkins (Ed) Integration; its meaning and application* New York, Appleton-Century, pp. 106–125.

Watson, Thomas The Doctrine of Repentance (1987)
Carlisle PA, Banner of Truth, 122 p. ISBN 0–85151–521–5.
Watts, A W The Wisdom of Insecurity (1968)
New York, Vintage Books.
Watts, Alan The Joyous Cosmology: adventures in the chemistry of consciousness (1962)
New York, Random House.
Watts, Alan W Way of Zen (1957)
New York, Random House.
Watts, Alan W Nature, Man and Woman (1958)
New York, Random House.
Watts, Alan W This is It and Other Essays on Zen (1960)
New York, Pantheon Books.
Watts, Alan W Psychotherapy East and West (1961)
New York, Allied Publications.
Watts, Alan W Beyond Theology: the art of godmanship (1964)
New York, Pantheon Books.
Watts, Alan W The Book – on the Taboo against Knowing Who You Are (1967)
New York, Collier Books.
Watts, Alan W The Two Hands of God: the myths of polarity (1969)
New York, MacMillan.
Watts, Alan W The Supreme Identity (1972)
New York, Vintage Books.
Watzlawick Paul, Weakland John and Fisch Richard Change (1974)
New York, WW Norton.
Wayman, Alex Calming the Mind and Discerning the Real (1978)
York, Columbia.
Weber, Renée Reflections on David Bohm's Holomovement: a physicist's model of cosmos and consciousness
In: *The Metaphors of Consciousness* 1981, New York, Plenum Publishing, 521 p. illus. bibl.
Weg, R (Ed) Sexuality in the Later Years: roles and behavior (1983)
New York, Academic Press.
Wehr, Gerhard C G Jung and Rudolf Steiner (1972)
Stuttgart, Ernst Klett Verlag, 267 p. bibl. (pp. 248–267).
Wehr, Gerhard The Mystical Marriage: symbol and meaning of the human experience (1990)
Rochester VT, Crucible/Inner Traditions International, 159 p.
Wei Wu Wei Ask the Awakened (1963)
London, Routledge and Kegan Paul.
Wei Wu Wei The Tenth Man (1966)
Hong Kong, Hong Kong University Press.
Wei Wu Wei Open Secret (1970)
Hong Kong, Hong Kong University Press.
Wei, Wu–wei All Else is Bondage: non-volitional living (1982)
Hong Kong, Hong Kong University Press, 68 p.
ISBN 962–209–025–7.
Weil, A The Natural Mind: a new way of looking at drugs and higher consciousness (1972)
Boston MA, Houghton Mifflin.
Welbon, Guy R Buddhist Nirvana and Its Western Interpreters (1968)
Chicago IL, University of Chicago Press.
Welvis, Gerhard The Mystical Marriage
Crucible Press.
Welwood, J Self-Knowledge as the Basis for an Integrative Psychology
In: *Journal of Transpersonal Psychology* 1979, II, 1.
Welwood, J Meditation and the Unconsciousness: a new perspective
In: *The Journal of Transpersonal Psychology* 1977, 9, pp. 1–26.
Wenger, M and Bagchi, B Studies in Autonomic Functions of Practitioners of Yoga in India (1961)
Behavioral Science 6.
Weraliz, Hans J Biorhythm: a scientific exploration into the life cycles of the individual (1961)
New York, Crown Publishers.
Werblowsky, R J Zwi and Bleeker, C J (Eds) Types of Redemption: study conference Jerusalem, 1968 (1970)
Leiden, Brill. ISBN 90–04–01619–8.
Werner, H The Concept of Development from a Comparative and Organismic Point of View (1957)
Minneapolis MN, University of Minnesota Press.
Werner, H Comparative Psychology of Mental Development (1964)
New York, International University Press.
Werner, H and Kaplal, B Symbol Formation (1963)
New York, John Wiley and Sons.
Werner, Karel Yoga and Indian Philosophy (1980)
Delhi, Motilal Banarsidass, 190 p.
Wescott, Roger W The Divine Animal: an exploration of human potentiality (1969)
New York, Funk and Wagnalls, 340 p.
Wesley, et al Christian Perfection
Salem OH, Schmul Publishing.
Westbrook, A and Ratti, O Aikido and the Dynamic Sphere (1970)
Rutland VT, Charles E Tuttle.
Wheatley, Henry B Literacy Blunders: a chapter in the history of human error (1979)
Arden Library.
ISBN 0–8495–5739–9.
Wheelis, A The Quest for Identity (1958)
New York, WW Norton.
Whilhelm, Richard The I Ching or Book of Changes (1967)
Princeton, Princeton University Press. 2 vols. Bollingen Series: X IX. Trans. and foreword by Cary F Baynes.

White, J (Ed) The Highest State of Consciousness (1972)
New York, Doubleday.
White, John (Ed) Frontiers of Consciousness: the meeting ground between inner and outer reality (1974)
New York, Julian Press, 366 p.
White, John Kundalini, Evolution and Enlightenment (1979)
New York, Anchor Books.
White, John (Ed) What is Enlightenment? (1985)
Los Angeles CA, Jeremy P Tarcher, 252 p.
ISBN 0–87477–343–1.
White, L Energy and the Evolution of Culture (1943)
American Anthropologist 45, pp. 335–356.
White, R (Ed) Surveys in Parapsychology: reviews of the literature and updated bibliographies (1976)
Metuchen NY, Scarecrow Press.
Whitehead, A N Mode of Thought (1966)
New York, Macmillan.
WHO Human Development and Public Health: report of a WHO Scientific Group (1972)
Geneva, WHO. Technical Report Series: 485.
WHO Social Dimensions of Mental Health (1981)
Geneva, WHO, 40 p.
Whyte, L L The Next Development in Man (1950)
New York, New American Library.
Whyte, Lancelot Law The Next Development in Man (1944)
London, Cresset Press, 275 p.
Whyte, Lancelot Law Die Nächste Stufe der Menscheit (1946)
Zurich, Pan-Verlag, 339 p.
Wickes, F Inner World of Choice (1963)
New York, Harper and Row.
Wieman, H N Man's Ultimate Commitment (1958)
Carbondale IL, World Resources Inventory.
Wilber, K Psychologia Perennis: the spectrum of consciousness
In: *F Vaughan and R N Walsh (Eds) Beyond Ego* 1980, Los Angeles CA, J P Tarcher.
Wilber, Ken The Pre/Trans Fallacy
In: *ReVision* 1980, 3, 2.
Wilber, Ken The Spectrum of Consciousness (1977)
Wheaton IL, Quest Books.
Wilber, Ken The Atman Project (1980)
Wheaton IL, Quest Books.
Wilber, Ken Up from Eden (1981)
New York, Doubleday/Anchor.
Wilber, Ken No Boundary (1981)
Boulder CO, Shambhala Publications.
Wilber, Ken A Sociable God (1983)
New York, McGraw Hill Book Company.
Wilber, Ken (Ed) Quantum Questions: mystical writings of the world's great physicists (1984)
Research assistance of Ann Niehaus.
Wilber Ken, Engler J and Brown D P Transformations of Consciousness (1986)
Boston MA, New Science Library.
Wilbur, Ken (Ed) The Holographic Paradigm (and Other Paradoxes): Exploring the Leading Edge of Science (1982)
Boulder CO, Shambhala.
Wilbur, Ken The Spectrum of Consciousness (1982)
Wheaton, Theosophical Publishing House.
Wile, Ira S Integration of the Child: the goal of the educational program (1936)
Mental Hygiene 20, April pp. 249–261.
Wilhelm, Richard (Trans) The Secret of the Golden Flower: a Chinese book of life (1962)
New York, Harcourt Brace. Revised 1931 edition with commentary by C G Jung.
Williams, Norman P The Ideas of the Fall and of Original Sin: a historical and critical study
New York, AMS Press. repr of 1927.
ISBN 0–404–18439–1.
Williams, S The Practice of Personal Transformation: a Jungian approach (1985)
Berkeley CA, California Journey Press.
Wilson, C The Outsider (1967)
New York, Houghton Mifflin.
Wilson, Colin Beyond the Outsider (1965)
Boston MA, Houghton Mifflin.
Wilson, Colin Introduction to the New Existentialism (1967)
Boston MA, Houghton Mifflin.
Wilson, Colin New Pathways in Psychology: Maslow and the post-Freudian revolution (1972)
New York, Taplinger Publishing, 288 p. bibl.
ISBN 0–8008–5514–0.
Wilson, F Human Nature and Esthetic Growth (1956)
In: *Moustakas C (Ed) The Self* Harper.
Winnicott, D W The Maturational Processes and the Facilitating Environment (1965)
New York, International University Press.
Winnicott, D W The Family and Individual Development (1965)
New York, Basic Books.
Winnicott, D W The Maturational Processes and the Facilitating Environment (1965)
New York, International University Press.
Winquist, C Homecoming, Interpretation, Transformation and Individuation (1978)
Missoula MT, Scholars Press.
Wise, T M (Ed) Advances in Psychosomatic Medicine
New York, Karger S AG. Vols 1,2,4,5,6,7,8,9,10,12,13,14,15, 16,17. ISSN 0065–3268.
Wise, T N and Fava, G (Eds) Research Paradigms in Psychosomatic Medicine (1987)
Basel, S Karger AG, 272 p. Advances in Psychosomatic Medicine: Vol 17. ISBN 3–8055–4484–7.

Wiseman, James A (Trans) John Ruusbroec: the Spiritual espousals and other works (1985)
New York NY, Paulist Press.
The Classics of Western Spirituality.
Wispe, Lauren G Altruism, Sympathy and Helping: psychological and sociological principles (1978)
San Diego CA, Academic Press. ISBN 0–12–760450–2.
Wold, Margaret The Power of Ordinary Christians: witnessing in a new age (1988)
Minneapolis MN, Augsburg Publishing House, 128 p.
ISBN 0–8066–2374–8.
Wolf, W (Ed) Rhythmic Functions in the Living System (1962)
Annals of the New York Academy of Sciences 98, 573 p.
Wolman, B B (Ed) Handbook of Parapsychology (1977)
New York, Van Nostrand Reinhold.
Wood, Ernest Zen Dictionary (1977)
Harmondsworth, Penguin Books, 128 p.
ISBN 0 14 00.7043 5. First published 1957.
Woodroffe, John The Serpent Power (1986)
Madras, Ganesh, 500 p. Sat Cakra Nirupana and Paduka Pancaka: Two works on Laya Yoga, trans. from the Sanskrit, with introduction and commentary.
13th Edition.
Woods, Richard Eckhart's Way (1986)
Wilmington, Michael Glazier. London, Darton Longman and Todd, (1987).
Woodward, Kimm Animistic Yoga
Kuranda QLD, Rams Skull Press, 112 p.
Wortz, E Feedback and State of Consciousness: meditation (1969)
Los Angeles CA, Biofeedback Research Institute.
Wuthnow, Robert Peak Experiences: some empirical tests (1976)
Berkeley CA, Survey Research Center.
Wyatt, Davis and Woodside, Alexander Moral Order and the Question of Change: essays on southeast asian thought (1982)
New Haven CT, Yale University Southeast Asia Studies, 413 p. ISBN 0–938692–02–X.
Wynne–Tyson, Esme The Philosophy of Compassion (1985)
New York, State Mutual Book and Periodical Service.
ISBN 0–900001–23–2.
Wynne–Tyson, Jon (Ed) The Extended Circle: a dictionary of humane thought (1985)
Fontwell, Centaur Press, 436 p. illus. bibl.
ISBN 0–900001–21–6.
Yablonsky, Lewis Synanon: the tunnel back (1965)
Baltimore, Pelican Books.
Yalom, Irving The Theory and Practice of Group Psychotherapy (1970)
New York, Basic Books.
Yamaoka Haruo Meditation Gut Enlightenment: the way of hara
Yamaoka, Haruo Meditation But Enlightment: the way of Hara (1976)
Tokyo, Heian International Publishing.
Yandell, Keith Some Varieties of Ineffability (1979)
In: *International Journal for the Philosophy of Religion* 10, pp. 167–179.
Yankelovich, D New Rules: searching for self-fulfillment in a world turned upside down (1981)
New York, Random House.
Yates, Frances The Rosicrucian Enlightenment (1972)
London, Routledge and Kegan Paul.
Yogananda, Paramahansa The Science of Religion
Yogeshwaranand, Swami Science of Soul (1972)
Yoga Niketan. A Practical Exposition of Ancient Method of Visualisation of Soul.
Young, Michael The Metronomic Society: natural rythms and human timetables (1988)
London, Thames and Hudson, 301 p.
ISBN 0–500–01443–4.
Yu Lu K'uan The Secrets of Chinese Meditation: self-cultivation by mind control as taught in the Ch'an, Mahayana and Taoist schools in China
Zaehner, R C Mysticism, Sacred and Profane: an inquiry into some varieties of preternatural experience (1961)
Oxford NY, Oxford University Press.
Zaehner, R C Evolution in Religion: a study in Sri Aurobindo and Pierre Teilhard de Chardin (1971)
Oxford, Clarendon Press, 121 p.
Zander, A and Medow, H Individual and Group Levels of Aspiration (1963)
Human Relations 16, pp. 89–105.
Zaner, R The Problem of Embodiment
Den Haag, Martinus Nijhoff.
Zdenek, Marilee Inventing the Future: advances in imagery that can change your life (1987)
New York, McGraw-Hill, 196 p. ISBN 0–07–072819–4.
Zee, A Fearful Symmetry: the search for beauty in modern physics
Zerubavel, Eviatar Hidden Rhythms (1981)
Chicago IL, University of Chicago Press.
Zhang, Mingwu and Sun, Xingyuan (Eds) Chinese Qigon Therapy
Shandong Scientific, 276 p.
Ziegler, A Archetypal Medicine (1983)
Ziegler, W Mindbook for Imaging/Inventing a World Without Weapons (1987)
Denver CO, Future Invention Association.
Zinberg, N (Ed) Alternate States of Consciousness (1977)
New York, Free Press.
Zinker, J C and Fink, S L The Possibility for Psychological Growth in a Dying Person (1966)
Journal of General Psychology vol. 47, pp. 185–99.

Zubek, J P (Ed) Sensory Deprivation: fifteen years of research (1969)
  New York, Appleton–Century–Crofts.

Zubek, John P and Solbeck, P A Human Development (1954)
  New York, McGraw Hill Book Company, 476 p. bibl. (McGraw–Hill Series in Psychology).

Zuger, B Growth of the Individual's Concept of Self (1952)
  A M A American Journal of Diseased Children 83, pp. 719.

Zwieg, Paul The Heresy of Self Love (1980)
  Princeton, Princeton University Press. First published 1968.

Notes

HZ

Human Development

Significance

Intent	363
Coexistence of unrelated human development paradigms	364
Economic bias	365
Social and educational biases	366
Cultural and educational biases	367
Psychological bias	368
Modes of awareness and experiential biases	371
Mythical, religious and spiritual biases	373
Acknowledgement of mythico-spiritual dimensions	374

Method

Assumptions	376
Criteria and definitions	378
Procedures	380
Descriptions	381
Classification of modes of awareness	383

Comments

Assessment	384
Official approaches to human development	386
Vital relevance of subtler human development	388
Language and the reconstruction of reality	389
Entrapment by competing metaphors	390
Embodiment in patterns of alternation	391

***   Bibliographical references identified in abridged form in the following section refer to publications detailed, by author, in Section HY, which is the bibliography for Section H.

Significance: intent

1. Intent
The term "human development" points to a process which is of vital interest at a time when the organization and future of life on this planet is challenged by the consequences of past understandings of that same process. For it is human development, through its lack of restraint, which has given rise to a high proportion of the problems of the world. There is nevertheless much confusion concerning the significance of the term. And yet there is great need to harness the forces of human development to meet the challenge of the times more effectively.

It was therefore considered useful, both as a bridging exercise and as an experiment, to attempt to initiate an information clarification process with the following objectives:

(a) Identify the range of concepts which effectively define the meanings currently attached to "human development" in its different forms and disguises.

(b) Provide a context for concepts of human development that are used in essentially different and frequently non-interacting sectors of society, without emphasizing those concepts which are either mechanistic or religious.

(c) Draw attention to those concepts of human development which have hitherto been excluded from serious consideration within the international community (whether by the academic world or in their societal applications) but for which legitimating documents and reports are held to exist by some constituencies and traditions.

(d) Distinguish, by juxtaposition within the same context, those concepts which place importance on the psycho-social development of the individual as a unique human being, from those which effectively stress the development of the individual conceived merely as a socio-economic unit.

(e) Provide sufficient description of each concept, based on available documents which depend upon and advocate its use, in order to clarify the special importance attached to each such particular understanding of human development.

(f) Clarify the relationships and distinctions between different concepts labelled by terms which are synonyms or homonyms, particularly where the meaning of the terms used changes with the context.

(g) Provide information on the range of modes of awareness, as well as on the states of consciousness with which people identify during the process of human development, indicating where possible how these are perceived as interrelated stages.

2. Reservations and apologies
This endeavour, as an exercise in processing information from diverse cultures, languages and disciplines can be considered totally questionable in the light of the current intellectual orthodoxy of the western academic disciplines of philosophy, anthropology, sociology, history of religion, and literary criticism. Insights from the sociology of knowledge, interpretations of the late Wittgenstein and the classical Quine, varieties of relativism in anthropological theory, Kuhn's incommensurability thesis and Feyerabend's case against method, and deconstructionist concerns about the readings of any text, all combine to create an intellectual climate in which it is problematic to imply any fruitful outcome from such an endeavour. In particular it is questionable whether insights (especially from the distant past) arising in other cultures and languages can be meaningfully expressed and understood through English in such a context as this.

Against such theoretical arguments may be set the fact that there is widespread and increasing interest in human development concerns and the experiential dimension. People, including academics of eminence, find meaning in domains considered void of meaning by certain disciplines. It is their experience which is honoured here, however ineffectually.

The editors recognize that in most traditions, whether spiritual or academic, there is a very strong belief that their insights should not be confused with those of other traditions. Clearly the whole procedure adopted here can be understood as doing violence to such insights, especially when information taken out of context is specifically intended to be understood only within some appropriate context (and possibly only with appropriate guidance). The treatment accorded particular insights may be thought to diminish their import and to degrade a vital cultural heritage. This is especially the case for those traditions from which little material could be obtained because their orientation is less explicitly concerned with human development and modes of awareness as articulated here. For some traditions, much that is articulated in detail in other traditions is encapsulated in a single central belief (such as "salvation") and it is not considered useful to articulate explicitly any related modes of experience or the stages of any spiritual journey.

Despite these valid concerns, the editors hold that sufficient material has now been produced in written form from many traditions (often by practitioners in those traditions), so that there is a case for placing their insights on human development together within a framework of this kind. One major reason for doing so is the tendency in many traditions towards obfuscation of key insights by layers of commentary which may do little to sharpen the central insight -- quite often burying it in inaccessible and essential unreadable material. The simplistic approach undertaken here endeavours to honour the existence of such insights, whatever their origin, and to offer users pointers to insights whose existence is virtually unknown, or conveniently ignored, in other traditions.

Although it may be virtually impossible to avoid causing offence by this approach, the editors wish to stress the sympathy with which they have endeavoured to respond to the variety of insights into human development. We should like to apologize for errors of omission and commission, whether or not they have caused offence.

Significance: co-existence of unrelated human development paradigms

1. Biased approaches to human development. As might be expected with a process so fundamental to human society, quite different sectors of society have very different understandings of the significance and place of this process. These different perspectives may be viewed as "biases", although from a broader view their insights may be assumed to be compatible. Few texts address themselves to the nature of the compatibility between these perspectives. It is easy to get locked into the logic associated with any one of these perspectives and to forget that there are other important ways of viewing the process.

A useful insight into the confusion concerning "human development" may be obtained from the notes on the following pages on some basic sets of biases. Others, especially favoured by some, to which reference might have been made, include: physical development, emotional development, moral and ethical development, mental development, development of creativity, aesthetic development, and political development. These briefly reviewed are:

(a) Socio-economic biases: These reflect the obvious concerns to provide for the basic needs of individuals and to ensure the survival of economies and the societies which depend upon them. This pattern of biases is central to "development" as most commonly understood by the international community, and especially by development and relief agencies. This understanding is underpinned by many of the applied social sciences, especially economics and the behavioural sciences.

(b) Psychological bias: Whereas the socio-economic biases reflect a preoccupation with society as a whole and the contribution of individual human development to its well-being, the psychological bias primarily governs recognition of individual patterns of development. This may be understood to be largely condition by society, but the concern is primarily with the individual and the manner in which he fulfils his potential within society in psychological terms. This bias is naturally cultivated by psychologists and the institutions in which they are employed. These are typically those related to education and training. Increasingly these institutions include large corporations who recognize the long-term economic value of investing in effective personnel relations. There is also some indication of concern by public health authorities with the psychological well-being of individuals in highly stressful industrial societies.

(c) Modes of awareness and experiential biases: The above biases reflect perspectives of institutional and academic establishments in dealing with individuals in society. Individuals themselves pursue their own understandings of human development, however deluded these may be considered to be. In their most common form these involve pursuit of status, cultivation of self-image, self-esteem and a sense of prowess, as well as the search for exciting stimuli. This pattern of concerns is reinforced by folk wisdom on the one hand, and by the media on the other. It is perhaps most striking in the pursuit of fashion understood as a daily striving for personal development. In this sense human development may be understood by many as becoming more fashionable and cultivating a higher order of personal style. Beyond the pursuit of common stimuli, there is the cultivation of other modes of awareness. In their more accessible forms, these may be associated with music and the delights of the flesh. But beyond these are the cultivation of other modes of awareness. These may be achieved through stimulants, whether alcohol, nicotine, or drugs (legal or illegal). They may also be cultivated through group processes or personal disciplines. The latter may range from physical exercise (jogging, mountaineering, etc) through to disciplines of the mind (including breathing exercises and the like). It can be argued that most individuals are primarily concerned with forms of personal human development governed by these experiential biases.

(d) Mythical, religious and spiritual biases: This of course is the best developed preoccupation with human development. It also has the longest history. On the one hand it reflects archetypal concerns with the place of the individual in the universe and his need to come into harmony with its rhythms. On the other it corresponds to the concern with coming into relationship with whatever are to be understood as the invisible dimensions and integrating forces of psychic and spiritual life. This preoccupation may also be considered as quite independent of the others.

2. Confusing range of meanings

In each of the above cases there is a well-developed constituency which has achieved a certain degree of consensus on the dimensions of human development, whatever the disagreement concerning the details of the process. There is however very little effective communication between these constituencies.

Many seemingly unrelated concepts, perspectives and methods are considered by their advocates to be central to full understanding of the meaning of human development. But even within the domain of psychology, there are different, and even mutually antagonistic, schools of thought on the matter. For psychologists the term is commonly used to describe changes in behaviour which occur with age. But even then *"development can be endowed with many connotations or it can be given limited meaning within a highly restrictive context. The precision of definition often depends on whether the writer is more interested in describing the achievement of broad stages or plateaus of behaviour or the mechanisms which apparently govern the transitions between stages. How we define development subsequently limits what we then observe."* (John Eliot, 1971).

The term "human development" is commonly used by psychologists and is increasingly used in international debate by those concerned with the limitations in practice of the conventional concern with economic and social development. One of the first working meetings (Tokyo, 1975) of the United Nations University was on human and social development. A proposal has even been made to hold a United Nations Conference on Human Development. Despite the emergence of this term into favour there is little consensus as to its meaning or range of meanings.

There is a certain incongruity in attempting any verbal description of concepts and processes for which the verbal mode of presentation may be considered inappropriate, insensitive and even totally inadequate. This is particularly so when the same editorial (information-oriented) approach is used in handling the descriptions of concepts which may be considered essentially incompatible by their respective advocates. Nevertheless many verbal descriptions have been attempted in the past. The resultant multiplicity of presentations of concepts presumably bear some relationship to one another since they all concern the individual human being. But this multiplicity facilitates neither comprehension of their particular emphasis nor empathy for the seeming excesses of their advocates. The very enthusiasm of available descriptions of some concepts of human development, let alone the existence of specialized jargons and neologisms, certainly facilitates the task of those who would prefer to ignore all but the most simplistic concepts of human development.

The following notes attempt to highlight the extent of this confusion, the alienating sterility of the depersonalized interpretations prevailing at the international level, and the relatively recent emergence of a variety of concepts which merit greater attention and more widespread recognition.

Significance: economic bias

1. Subsumption of human development
Within the international community, human development has usually been understood to be an aspect of economic and social development. As with social development, it has long been considered as a secondary consideration compared to economic development. When either social or human development have been accorded attention, it has been as a consequence of reluctant recognition that economic development policies have disrupted the social fabric to the point of inhibiting the development process that they were theoretically designed to enhance. From this perspective human development is an unfortunate necessity if economic development is to be effectively achieved.

The systematic neglect of human development in intergovernmental development programmes, and their national counterparts, was highlighted in 1985 in the Roundtable on Development, jointly convened by the North-South Roundtable and the UNDP Study Programme. *"Recent economic pressures, national and international, have led to serious neglect of the human dimension in development. Unless remedied, this neglect will distort and handicap the future development of at least a generation to come."* (Haq et al, 1986).

In a review in 1988 by UNICEF (*The Child in South Asia; issues in development as if children mattered*) notes that: *"The human condition in large parts of the world today suggests that economic development has substantially neglected the human factor -- which is the ultimate objective as well as the main instrument of development. The explanation for this situation lies in one or another of the familiar approaches to, or models of, development. When development is understood as a cumulative sequence of diverse economic deeds, consisting of a cut in consumption, savings, capital accumulation, increased prodction, higher consumption in the future and more savings for investment, the social factors of development tend to be relegated to relative unimportance. Even when the conv'entional model is adapted to enable the drawig in of unemployed resources -- like labour, skill, management, raw materials, capital equipment and external funds -- there are but narrow limits to its usefulness in underdeveloped economies which are short of such resources in the necessary balance."*

2. Recognition of the neglected dimension
This recognition was in large part a reaction to the evident setbacks in attempting to implement policies of economic adjustment, with their accompanying cutbacks ("austerity measures") in public expenditure on the development of human resources. *"The overwhelming preoccupation with the pressing econnomic problems of debt, balance of payments and economic survival has shifted the attention of national and international policy-makers from long-term goals to short-term adjustment, and from broader concerns to narrow financial matters."* (Haq et al, 1986)

The introduction to the report of the roundtable notes: *"The ongoing crisis brought into focus the larger illumination of thirty-five years of development experience and thinking: that human development is both an input and an objective of development. Yet the statistics tell the story of our failure to recognize it as an objective. The conventional wisdom of each decade has tended to offer single-factor recipes for economic development -- investment in physical infrastructure, industrialization, export promotion, import substitution, basic needs, etc."* (Haq et al, 1986)

This recognition was followed by the conclusion that: *"The latest thinking acknowledges that whilst all these elements are necessary conditions of growth, they are not sufficient without inputs into human capital formation, since it is on human beings and their capacity to utilize their skills and experience that self-sustaining development ultimately depends."* (Haq et al, 1986) From this it is clear that interest in human development is primarily defined in terms of its contribution to economic development. Human development in these terms is only an objective of development to the extent that it serves to enhance economic development.

3. Evolving scope of human development
The reluctant recognition of a "human factor" in development has evolved over recent decades (Haq et al, 1986). Originally human resource development had been defined somewhat narrowly in terms of labour supply and the provision of skilled manpower. In the early 1960s, human resource development focused on primarily on education and secondarily on health. Skill formation was seen as an investment in human capital required in parallel with investments in physical capital. In the 1970s, the emphasis shifted to the socialization function of schooling, with educational planning seen as being responsive to labour markets and their segmentation. When economic growth and technical progress are just as capable of de-skilling jobs as of generating new jobs, general academic education is a hedge against technical dynamism. Recurrent education and retraining are then an ongoing requirement in a dynamic society. In the 1980s there has been increasing recognition that other factors may be involved such as innate ability, motivation and other psychological factors. These are proving necessary to explain the current economic momentum of some countries and cultures (notably in Southeast Asia) in contrast with others.

4. Questionable official statements
Great caution should however be taken in interpreting official texts defining the scope of human development. Whilst great insights can apparently be embodied in them, these tend in practice to be used for rehetorical purposes or in a totally reductionistic manner. The views embodied in earlier texts may therefore be much more indicative of current practice in most settings, however effectively the more recent broader interpretations are used to suggest subtler options -- whether or not there is any real intention of making them a reality.

It is important to note the concept of human development which emerges from early examples such as the following:

(a) During the United Nations Conference on Trade and Development (Santiago, 1972), Robert S McNamara, President of the World Bank, stated: *"But the improvement of the individual lives of the great masses of the people is, in the end, what development is all about."* But by individual lives is here meant the physical living conditions and opportunity for gainful employment. For most of its existence the World Bank has chosen not to employ any full-time professionals in the non-economic social sciences who might be sensitive to the full range of "day-to-day deprivations" which he noted *"degrade human dignity to levels which no statistics can adequately describe."*

(b) In the United Nations report on the International Development Strategy, the *"ultimate object of development must be to bring about sustained improvement in the well-being of the individual and bestow benefits on all."* But the section on human development is divided into sections on the following topics only: population growth, employment, education for development needs, health facilities, nutrition, involving children and youth, housing, and the ecological balance.

(c) A report of the ECOSOC Development Planning Committee, after arguing the importance of adequate social structures, which makes any increase in production or income merely one of a number of relevant economic and social indicators, notes that because many of the social indicators are lacking, social goals can only be identified qualitatively. It is then able to conclude that in fact economic and social questions are so closely interwoven that there is hardly any sense in making the distinction between them. The remainder of the report identifies methods of increasing production and income. Development is in effect generally accepted as meaning first and foremost economic development.

(d) The Secretary-General of the United Nations Conference on the Application of Science and Technology for the Benefit of the Less-Developed Areas (Geneva, 1963) stated: *"The core of human resource development is the planning and execution of a policy of education and training - two aspects of the same coordinated process designed to provide the trained manpower at all levels of skill required to achieve the objectives of the economic development plan."*

Significance: social and educational biases

1. Acknowledgement of non-economic dimensions
The limitations of the economic focus of human development have long been acknowledged in isolated statements. Examples include:

(a) Progress of people: The economist Frederick Harbison (1963) points out: *"The progress of a nation depends first and foremost on the progress of its people. Unless it develops their spirit and human potentialities, it cannot develop much else - materially, economically, politically or culturally. The basic problem of most under-developed countries is not a poverty of natural resources but the underdevelopment of their human resources."*

(b) "Human factor": In preparation for the first United Nations Conference on Trade and Development (Geneva, 1964), the secretariat of UNESCO (1964) prepared a report which included the following statement: *"The ultimate justification for the development of human resources is man's basic right to the full realization of his potentialities. In addition, however, the development of human resources is a crucial factor in stimulating economic growth. Numerous studies analyzing the economic history of countries at varying stages of development arrive at a common conclusion: the increase in inputs of labour and capital in a given period does not fully account for the expansion of general output achieved in subsequent periods. Indeed, in a number of cases the role of these two factors in economic growth appears to be quite minor. The residual element - often referred to as the "human" factor - which is left after the contribution of labour and capital is allowed for can be very considerable. It has been tentatively estimated that in some developing countries this element accounts for up to half of the increase in the GNP. It is evident, therefore, that the human factor in economic growth is extremely important and warrants attention in efforts to achieve the targets of the (First) Development Decade."*

(c) Full development of the human being: UNESCO, in a communication to the Preparatory Committee for the Second United Nations Development Decade, stated: *"Development is meaningful only if man who is both the instrument and beneficiary is also its justification and its end. It must be integrated and harmonized; in other words, it must permit the full development of the human being on the spiritual, moral and material level, thus ensuring the dignity of man in society, through respect for the Declaration of Human Rights."*

(d) "Social aspects": In an address to the Intergovernmental Conference on Institutional, Administrative and Financial Aspects of Cultural Policies (Venice, 1970), René Maheu, Director-General of UNESCO, stated: *"The idea of development has, in fact, gradually become broader, deeper, and more varied so that going beyond the purely economic aspects of improving man's lot, it now also embraces the so-called social aspects...Man is the means and the end of development; he is not the one-dimensional abstraction of homo economicus, but a living reality, a human person, in the infinite variety of his needs, his potentialities and his aspirations...Even the economists now admit that development is not development unless it is total, and that it is no mere figure of speech to talk of cultural development: cultural development is part of total development".*

(e) "Human capital" and human investment: In 1988 UNICEF noted: *"As a response to the predicament of developing countries, the role of "human capital" in the development process came to be emphasized. That is to say, better education, better training, better health and better nutrition have a positive effect on productivity. While the logic of this argument is irrefutable, its application presents serious problems in an underdeveloped economy insofar as the financing of improvements in education, nutrition and health depends on previous output. The issue...demands a redefinition: Expenditure on social factors of development must be considered as investment in the quality of life of all human beings, right from earliest childhood. It represents also a choice between deflationary economic policies and a dynamic approach to development with the human resource as its prime mover...In this perspective, because "human investment" -- rather than physical capital accumulation -- is the essential basis for higher productivity, the truest investment of any community, rich or poor, developed or underdeveloped, capitalist or socialist, ancient or modern, is the investment in its own children."* (UNICEF, *The Child in South Asia; development as if children mattered*).

(f) Job satisfaction: The documents of the International Labour Organisation hint at the general concern about job enrichment and the need to make the work experience a fulfilling one for the worker. The Director-General's annual reports note the programmes relating to conditions of work and life (namely occupational safety and health, social security, and remuneration and conditions of work) and to the development of human resources (namely vocational training and management development).

2. Social indicators of development
It is the interpretations of development implied in the previous note which continue to prevail internationally. The following are indication of increasing consensus on the importance of the social dimension:

(a) United Nations University: In the Goals, Processes and Indicators of Development project (1978-82) of the United Nations University's Human and Social Development Programme (1983), an effort to define human development led to consensus on the following: *"Human development refers to the development of human beings in all life stages, and consists of a harmonious relationship between persons, society and nature, ensuring the fullest flowering of human potential without degrading, despoiling or destroying society or nature."* The same report identifies four additional requirements for a human-centred development, namely: social equity, inter-regional and international equity, living presence of the future, sensitiveness to the present. No effort is made to defining *"fullest flowering of human potential"*.

In an excellent follow-up report on the implications for social indicators, Ian Miles (1985, p.152) notes that: *"Human development does imply a process. The term leaves it open as to whether that process necessarily has a culminating point or tends towards some limit. It is distinct from human resource development which...sees human potentials in terms of their contribution as means towards other ends. The use of the term "human development" implies instead the view that human beings themselves should be the end to which economic development, political development, and other social changes are means."* Miles continues his report with reviews of human development as the satisfaction of human needs and the relation of such needs to political and social liberation. The insights of eastern cultures and of many schools of psychoanalysis and psychotherapy are totally absent.

(b) United Nations Development Programme: Possibly in response to such insights, the United Nations Development Programme initiated in 1990 an annual series: *Human Development Report* (1990) which attempts to define offically a "Human Development Index". The central message of the document is that while growth in national production is absolutely necessary to meet all essential human objectives, what is important is the translation of this growth into human development. To this end the report defines human development as *"a process of enlarging people's choices. In principle, these choices can be infinite and change over time. But at all levels of development, the three essential ones are for people to lead a long and healthy life, to acquire knowledge and to have access to resources needed for a decent standard of living. If these essential choices are not available, many other opportunities remain inaccessible. But human development does not end there. Additional choices, highly valued by many people, range from political, economic and social freedom to opportunities for being creative and productive, and enjoying personal self-respect and guaranteed human rights. Human development has two sides: the formation of human capabilities -- such as improved health, knowledge and skills -- and the use people' make of their acquired capabilities -- for leisure, productive purposes or being active in cultural, social and political affairs. If the scales of human development do not finely balance the two sides, considerable human frustration may result."*

The term "human development" here denotes both the process of widening people's choices and the level of their achieved well-being. The report elaborates an index of human development.

Significance: cultural and educational biases

The range of significance attributed to "culture" by different schools of thought is notorious, whether or not such distinctions are made in a spirit of clarification or obfuscation. At best the promotion of culture is an effort to cultivate every aspect of the human spirit, whereas at worst it is merely an effort to attract tourists (increasingly the only justification for the allocation of funds to "culture").

Cultural education, especially under government supervision, may simply be an effort at indoctrination through increased understanding of national writers and artists and through the production of cultural artefacts. Where there is sensitivity to the need to cultivate the human spirit, such indoctrination may even be considered sufficient.

The capacity to discuss intelligently the significance and relationship between cultural artefacts may be considered a mark of culture, irrespective of whether the development of such skills is matched by the development of other aspects of the human spirit or of wisdom, however that is to be understood. The cultural development of individuals implicit in the following readily lends itself to interpretation at both extremes.

1. Development of personality through human values
The UNESCO Declaration on Cultural Rights as Rights of Man (1968) concludes that the right to culture implies the possibility for each person to dispose of the means necessary to develop his personality, through direct participation in the creation of human values, and to become in this way master of his condition, whether on the local level or on the world level.

2. Non-material dimensions of development
Aurelio Peccei (1975), Director of the Club of Rome, notes that: *"Human development means much more than universal education, professional training and productive employment, although these are becoming compelling exigencies for individual emancipation and societal progress...it also means bringing the whole population to understand their times and live as contemporaries, learning how to adjust to the world complexities, the outer limits of its life-supporting systems, and the transformations we progressively operate in it. The present predicament of mankind appears, and in fact is, so formidable precisely because the majority of people, in both developed and developing countries and in all segments of society - including intellectual, scientific, political and religious elites - have not yet fully adapted psychologically and functionally to the overall new world our "civilization" has created and is continuously reshaping. The very crux of the pervasive and baffling global crisis we are struggling with lies in this mismatch; and adaptation is the name of the key to get out of it."*

He also notes: *"However, since the object of all our interest and concern is man, it is the multiple dimensions of man himself, with his complex personality and growing needs, wants, aspirations and manifestations, which are the very essence. It is erroneous and misleading to confine our analyses, as generally is the case, mainly to the material aspects of his existence, however important they may be, as indeed they are, then add political, social and cultural considerations as if they belonged to subordinate spheres."*

The Club of Rome subsequently commissioned a third generation report on Goals for Global Society (Director, Ervin Laszlo) to focus on the social, psychological, and cultural inner limits which could give positive direction to human aspirations. This deals explicitly with human factors and investigates those ethical commitments, world views and value judgments which could lead beyond perennial crises toward a healthier state of global human society (1977).

3. Towards the complete man
Prior to the sensitivity to feminist issues, the report of the UNESCO International Commission on the Development of Education (1972), in a chapter entitled *"Towards the complete man"* notes: *"If there are permanent traits in the human psyche, perhaps the most prominent are man's rejection of agonizing contradictions, his intolerance of excessive tension, the individual's striving for intellectual consistency, his search for happiness identified not with the mechanical satisfaction of appetite but with the concrete realization of his potentialities and with the idea of himself as one reconciled to his fate - that of the complete man."* But, it continues, *"He is exposed to division, tension and discord on all sides. Social structures which defy all rules of justice and harmony cannot fail to affect the various realms of his being. All that surrounds him seems to encourage dissociation of the elements of his personality: the division of society into classes, alienation from work and its fragmented nature, the artificial opposition between manual and intellectual labour, the crises of ideologies, the disintegration of accepted myths and the dichotomies between body and mind or material and spiritual values."*

It suggests that: *"Respect for the many-sidedness of personality is essential in education, especially in schools, if the individual is to develop as he should, both for himself and his associates. Complex attitudes, indispensable for balanced development of all personality components, must be stimulated and given form in the course of the individual's education."* And it concludes: *"The physical, intellectual, emotional and ethical integration of the individual into a complete man is a broad definition of the fundamental aim for education."* (1972).

4. Corporate culture
Over the past decade extensive resources have been devoted to understanding and improving "corporate culture". Understanding the elements of culture is felt to be crucial to managing human resources and thus to improving the productivity of organizations. In this sense cultural development can be understood as a way of involving people in the change process. The challenge is seen to be that of harmonizing the values of people with those of the organization within which they work. Where necessary this may involve exerting pressure on individuals to align their personalities with the values favoured by the organization.

5. Human potential
The question of "human potential" has been explored in a major project by the Harvard Graduate School of Education to assess the state of scientific knowledge concerning human potential and its realization, with extensive funding from the Bernard van Leer Foundation (The Hague). This Project on Human Potential has resulted in reviews of the relevant literature in history, philosophy and the natural and social sciences, a series of international workshops on conceptions of human development in diverse cultural traditions, together with a number of books (Howard Gardner (1984), Israel Scheffler (1985), A Le Vine and M I White (1986), M I White and Susan Pollack (1986)). The books provide a remarkable collection of material on the question of potential as seen in terms of human intellectual potentials (Howard Gardner, 1984), philosophical aspects of the concept of potential (Israel Scheffler, 1985), and the role of cultural factors in the progress of human development (A Le Vine and M I White (1986), M I White and Susan Pollack (1986)). The second volume is a deliberate effort to show the roots of the concept of potential in genuine aspects of human nature while at the same time freeing it, through analytical reconstruction, of outworn philosophical myths of fixity, harmony and value calculated to cause untold mischief in social and educational practice.

6. Cultural development
The United Nations has proclaimed the 1990s as a World Decade for Cultural Development whose importance is noted by UNESCO, as the responsible agency, in the following terms: *"Over the years, there has been growing awareness throughout the world of the inadequacy and limitations of a purely economic definition of national and international development strategies. Although economic growth continues to be a prime necessity, the international community now feels that it is equally vital to situate the human individual in his or her cultural context, at the centre of the web of development.. The latter should be based on the cultural identity of peoples, who are both the agents and the beneficiaries of development....Furthermore, international cooperation should be based on respect for differences....if people everywhere are to see their lot improved without any threat to their innermost being."* (UNESCO News, 2 Dec 1985).

Significance: psychological bias

1. Appropriation by institutions
The social dimensions of human development described above are not widely accepted, although the limitations of economic development and structural adjustment are increasing sensitivity to them. But even this social sensitivity filters out certain dimensions considered essential by others, or at least leaves the question of their presence or absence a matter of ambiguity, permitting the more subtle features to be expediently dropped at the first hint of ever-present controversy. Whilst each such interpretation seems to contain the essential key words, the meanings attached to them are not clarified.

What, for example, does the World Health Organization mean by "human potential", or the International Labour Organisation by "worker fulfilment", or UNESCO by "development of personality" or UNDP by "self-fulfillment"? Under the normal political and financial pressures on programme priorities:
- the promotion of positive psychological health must of necessity be limited to the elimination of physical disease by WHO;
- the promotion of worker fulfilment must be limited to the reduction of unemployment by ILO;
- the development of personality to the inculcation of reading / writing / arithmetic by UNESCO; and
- the promotion of individual self-fulfillment by UNDP must be limited to the ability of the individual to express himself through the acquisition of a more individualistic range of products.

The same situation must prevail in the equivalent national agencies. To what extent are such terms appropriated by institutions precisely in order to encourage people to believe that more is intended than is in fact planned in practice ?

The 1990 *Human Development Report* of UNDP, noted above, is also admirably ambiguous in its attempt to define human development. In its human development index "*Longevity and knowledge refer to the formation of human capabilities, and income is a proxy measure for the choices people have in putting their capabilities to use.*" It is not difficult to see how this understanding of human development lends itself readily to a limited focus on expanding the choice of consumer products and services as embodying an ever greater sense of well-being. Many will favour this interpretation because of the way in which it reinforces existing policies. But such a definition could possibly also be interpreted in terms of expanding the range of those inner choices which enable people to function with greater insight (through altered modes of awareness), effectively increasing their sense of personal fulfilment (their "psychic income") and prolonging their active lives. There is no implication in the report that this aspect will be explored, whether or not this is done for rhetorical purposes.

2. Avoidance of significant dimensions
The report of the United Nations University project, cited in a previous note, is remarkable for the skilful manner in which it avoids any discussion of the forms of human development with which people can and do identify. These are dismissed as "individual development" in contrast to "human-centered" social development which concentrates on the relationships between people. This supposedly corrects the over-emphasis on individualistic development, despite the fact that the most elaborate explorations of individual development derive from eastern cultures in which non-individualistic social relations prevail. Similarly the Bernard van Leer Foundation's Project on Human Potential is remarkable for the manner in which it avoids reference to human potential as experienced by the "developee" in favour of discussion of the issues raised for the "developer", whether parent, educator or planner. Given the immense interest in altered states of consciousness by young people, as indicated by the increasing dimensions of the drug problem, some reference to the relationship of such altered states to human potential would seem appropriate. In part such avoidance may simply be due to recognition of the inability of the mainstream psycho-social disciplines to respond effectively to such dimensions.

It would seem that official bodies are embarrassed by matters which touch upon the nature of human potential and the stages and processes in the psychological development of the adult human being with which people themselves identify. This is particularly so at a time when even the social element is being excised from the concept of development, as in the debate within the United Nations on the establishment of a New International Economic Order. Many would argue that the subtler concepts of human development are a private subjective luxury which must be ignored until the basic physical needs of every human being are satisfied. Or, as the political philosopher Herbert Marcuse argues: "*The traditional borderlines between psychology on the one side and political and social philosophy on the other have been made obsolete by the condition of man in the present era: formerly autonomous and identifiable psychical processes are being absorbed by the function of the individual in the state - by his public existence. Psychological problems therefore turn into political problems: private disorder reflects more directly than before the disorder of the whole, and the cure of personal disorder depends more directly than before on the cure of the general disorder.*"

3. Psychological maturity and social change
The Constitution of UNESCO states, in the oft-quoted phrase: "*...that since wars begin in the minds of men, it is in the minds of men that the defences of peace must be constructed...a peace based exclusively upon the political and economic arrangements of governments would not be a peace which could secure the unanimous, lasting and sincere support of the peoples of the world...*"

But such arguments over-simplify the situation faced by humankind in developed or developing countries. Unless the human beings - whether ordinary voters or members of privileged elites (with control over power and resources) are themselves exceptionally mature and well-integrated individuals, they will be insensitive to the needs and concerns of all those who may benefit or suffer from their decisions. Hitler is only the most obvious example; he is neither an isolated case, nor the most recent. Less extreme examples are numerous at all levels of society. Neither this well-researched fact, nor the meaning and degrees of maturity and of personality integration, can currently be made the subject of discussion within official bodies - where examples of immaturity are a matter of corridor gossip, even within the leadership of intergovernmental organizations. There would seem to be a myth that the good of society as it is defined by democratic and political processes is unaffected by the degree of integration of the key personalities and by the psychological maturity of the voters themselves. These factors are only incidentally related to formal education and to physical health.

There is increasing recognition of some form of hierarchy of needs, from the most basic survival needs to those associated with self-realization. In neglecting the subtler needs, policy-makers easily forget that it is only through the cultivation of such subtler needs that people (including policy-makers and voters) come to recognize the value of responding to the basic needs of others.

4. Missing essential factors
The importance of these points, and of the focus on the more subtle aspects of human development, is illustrated by the following:

(a) Belief and personality systems: In reporting on an investigation into the nature of belief systems and personality systems, Milton Rokeach (18) states: "*To say that a person is dogmatic or that his belief system is closed is to say something about the way he believes and the way he thinks - not only about single issues but also about networks of issues. The closed mind even though most people cannot define it precisely, can be observed in the "practical" world of political and religious beliefs, and in the more academic world of scientific, philosophic, and humanistic thought. In both of these worlds there is conflict among men about who is right and who wrong, who is rational and who is rationalizing, and conflict over whose convictions are dogmatic and whose intellectual... The relative openness or closeness of a mind cuts across specific content; that is, it is not uniquely restricted to any one particular ideology, or religion, or philosophy, or scientific viewpoint... Is it possible to say that the extent to which a person's belief system is open or closed is a generalized state of mind which will reveal itself in his politics and religion, the* way he goes about solving intellectual problems, the way he works with perceptual materials, and the way he reacts to unorthodox musical compositions?"

Further, an individual whose intellectual or belief systems are poorly integrated may harbour logically contradictory beliefs. Rokeach continues: "*Orwell, in his book 1984, has more picturesquely called this "double-think". In everyday life we note many examples of "double-think": expressing an abhorrence of violence and at the same time believing that it is justifiable under certain conditions; affirming a faith in the intelligence of the common man and at the same time believing that the masses are stupid; being for democracy but also advocating a government run by an intellectual elite; believing in freedom for all, but also believing that certain groups should be restricted; believing that science makes no value judgments, but also knowing a good theory from a bad theory and a good experiment from a bad experiment.*" He then notes: "*A person sometimes judges as "irrelevant" what may well be relevant by objective standards... Often enough, though not always, the judgment that something is irrelevant to something else points to a state of isolation between belief and disbelief systems. It is designed to ward off contradictions and, thus, to maintain intact one's own system.*" It is not unknown for individuals in positions of power to have closed minds harbouring contradictions in the sense used here, and in fact to have been placed in power by supporters holding similar views. There is even some recognition of what is termed psychosocial isomorphism, namely relations within a personality structure leading to formally similar relations within a social structure, and vice versa (William Eckhardt (1972), Emmanuael Todd (1983)). An extreme example being the structural equivalence between war propaganda and mental illness (William Eckhardt, 1972).

(b) Psychological change and cultural change: Commenting on the nature of psychological change and its relation to cultural change, Lawrence Kubie (1968) notes: "*The fact which confronts us is that cultural change is limited by the restrictions imposed on change in individual human nature by concealed neurotic processes. At the same time there is continuous cybernetic interplay between culture and the individual, ie between the intra-psychic processes which make for fluidity or rigidity within the individual and the external processes which make for fluidity or rigidity in a culture. It would be naive to expect political and ideological liberty to give internal liberty to the individual citizen unless he had already won freedom from the internal tyranny of his own neurotic mechanisms...Therefore, insofar as man himself is neurotogenically restricted, he will restrict the freedom to change of the society in which he lives. This interplay is sometimes clearly evident, sometimes subtly concealed; but it is the heart of the solution of the problem of human progress.*"

(c) Transformations of man: In concluding a historical survey of the transformations which man has already undergone, Lewis Mumford (1962) notes: "*The relations between world culture and the unified self are reciprocal. The very possibility of achieving a world order by other means than totalitarian enslavement and automatism rests on the plentiful creation of unified personalities, at home with every part of themselves, and so equally at home with the whole family of man, in all its magnificent diversity... In brief, one cannot create a unified world with partial, fragmentary, arrested selves which by their very nature must either produce aggressive conflict or regressive isolation. Nothing less than a concept of the whole man - and of man achieving a consciousness of the whole - is capable of doing justice to every type of personality, every mode of culture, every human potential. At this point a further transformation, so far not approached by any historic culture, may well take place.*"

5. Inner limits and constraints

Ervin Laszlo, who directed the Club of Rome's project on Goals for Mankind, stresses the importance of "inner limits", having noted the importance of the "outer limits" identified in many international reports (1989). "*It is said that more than half the effort in solving a problem goes into identifying it. Regrettably, much current effort has been wasted: it has identified the wrong problems and identified them on the wrong scale.*"

He argues that it is not that the most-publicised problems are illusory: "*they are real, but they are global, not national or local, and they are not the ones to which to direct our primary attention. They are outward manifestations of inner causes: the symptoms of malfunctions, not the malfunctions themselves....It is forgotten that not our world, but we human beings are the cause of our problems, and that only by redesigning our thinking and acting, not the world around us, can we solve them.*" He concludes that the critical but as yet generally unrecognized issue confronting mankind is that its truly decisive limits are inner, nor outer. "*They are not physical limits due to the finiteness or vulnerability of this world, but psychological, cultural and, above all, political limits inner to people and societies, manifested by individual and collective mismanagement, irresponsibility and myopia.*"

For Laszlo "*Many world problems involve outer limits, but most of them are due fundamentally to inner limits. There are hardly any world problems that cannot be traced to human agency and which could not be overcome by appropriate changes in human behaviour. The root causes even of physical and ecological problems are the inner constraints on our vision and values.*"

6. Psychological development

There is no lack of work on human development from a psychological perspective although very little of it is considered of relevance to the challenges confronted by the international community. Much of this work is concerned with explaining the emotional and cognitive growth of the individual and is primarily valued for any bearing that it has on human development perceived as education or training. Some of it is concerned with attitude formation and is valued for the insights it offers for mass communication efforts as an approach to education, training and opinion formation as forms of collective human development.

Psychology has tended to focus on behaviour rather than experience. For the most part it is concerned with observed behaviour rather than subjective phenomena. Furthermore, in academic psychology it has been customary to focus upon regular or 'normal' phenomena rather than those which might be described as unusual or of low frequency. This tendency to ignore or actively avoid anomalous instances continues to prevail. Those phenomena which do not readily fit psychological models are regarded as, at best, messy and inconvenient.

Psychology continues to find it difficult to clarify what is meant by a highly developed individual. A few psychologists are prepared to outline the goals of individual development in adults, as opposed to the stages of development to adulthood, for example: "*From the point of view of psychology, a high level of development in personality is characterized most essentially by complexity and wholeness. There is a high degree of differentiation, a large number of different parts or features having different and specialized functions; and a high degree of integration, a state of affairs in which communication among parts is great enough so that different parts may, without losing their essential identity, become organized into larger wholes in order to serve the larger purposes of the person... The highly developed individual is always open to new experience and capable of further learning; his stability is fundamental in the sense that he can go on developing while remaining essentially himself.*" (Sanford Nevitt, 1970)

7. Psycho-social health and psychotherapy

There is increasing emphasis on individual health as opposed to disease, and increasing interest in psychological health and human potential, although the degree of importance attached to these changes is not always clear from official reports.

The World Health Organization in the report of a Scientific Group on Human Development and Public Health (1971) delimits human development as follows: "*Human development embraces every aspect of the maturation process, including its physical, biological, psychological. and social aspects. To bring about healthy human development and to realize human potential, it is necessary to draw upon many areas of scientific knowledge and many components of the health service. Such areas as nutrition, communicable diseases, human reproduction, mental health, handicaps, and many others, together with the corresponding services, are related to human development. Many of these services have their greatest impact on development when they are employed early in the individual's life.*" (9)

The World Health Organization does not in fact have any definition of mental health as such. Its approach to the question is currently based on premises elaborated in a document on the "Social Dimensions of Mental Health" (1981): "*Man is a thinking being; inner experience linked to interpersonal group experience - in other words, mental life - is what makes people's lives valuable. To be human is to think, feel, aspire, strive and achieve, and to be social. Promoting health therefore must not only be concerned with preserving the biological element of the human organism: it must also be concerned with enhancing mental life.*"

The WHO report continues: "*Economic growth and social change exert significant influences on the mental life of individuals and the structure and functioning of families. When insufficient attention is given to this fact the cost of progress, in terms of diminished quality of life, may be unnecessarily high. The application of mental health knowledge could help to prevent harmful psychosocial consequences of socioeconomic change and facilitate harmonious development....Mental health skills can be used in developing positive attitudes towards community participation in health programmes. They can also help in persuading social sectors to adopt health as a motivating value for action. A mental health perspective in general health care can counter the dehumanization of medicine, and make health services more effective and less costly.*"

The increasing importance of psychotherapy has however been described as a stop-gap effort to fill the spiritual void left by the demise of religion. From this perspective, it has endeavoured to meet unmet metaphysical needs without recourse to mythical ideologies or magic ritual. In doing so it has ignored the assumptions that guided "soul care" in the past, adamantly believing that it bears no relation to earlier spiritual traditions and practices. (Benner, David. Psychotherapy and the Spiritual Quest, 1988).

In the case of psychiatry, a Group for the Advancement of Psychiatry investigated the phenomena of mystic experience in the USA. Their report explored the subtle dividing line between mystic experience and mental derangement (GAP Committee, 1976). It distinguished the various categories, stages and phases leading through to religious conversion. It also describes the different interpretations mystics put on their experience. The report did not reach any consensus, concluding that from one point of view all mystical experiences may be regarded as symptoms of mental disturbance, and from another, they may be regarded as attempts at adaptation. They were however satisfied that such phenomena were explicable in psychiatric terms. The conclusions of the report were criticized by one dissenting member (A Deikman, 1988).

8. Corporate psychological training programmes

Most work in psychology avoids the "humanistic" or "transpersonal" dimensions which, as emerging disciplines, continue to be marginalized within mainstream psychology. However there is a growing professional interest in the phenomenological quality of experience as opposed to observable behaviour. Ironically it is the interest of the business community in these dimensions, as a means of facilitating creativity and team work, that is leading to greater acceptance in practice, if not in the theory of the discipline.

Many leading companies are sending executives on management training courses which are designed to make use of new and experimental psychological techniques, some of which have been developed by quasi-religious groups. They may be described as "human potential" or "personal growth" semainars. The past decade has also seen many examples of employers investing in programmes of an experiential nature as a means of reducing stress in employees and increasing working potential, especially where creativity is a factor. The techniques may involve disorientation, demoralization, group meditation, group confession, peer group pressure, love bombing, rejection of old values, presentation of confusing doctrines, removal of privacy, time sense deprivation, uncompromising rules, sleep deprivation, chanting and singing, financial commitment, change of diet, fear, leader dependence and verbal abuse.

For many people such courses are beneficial, improving self-confidence, self-esteem and performance at work. It is for this reason that the business community is so active in exploring them. However, given that many of the techniques are experimental, the results for some can be profoundly disturbing psychologically. This is especially the case where the courses are deliberately designed to challenge ingrained, habitual modes of thought.

The existence and use of such techniques is a matter of some controversy, especially to those who are poorly informed about them and to those anxious to safeguard particular patterns of thought at all costs. Understanding is especially difficult for those who are unable to discriminate between their benefits and the abuses of religious cults and political re-education ("brainwashing"). The issues is exemplified by corporate interest in challenging executives physically to develop self-confidence and more fruitful modes of behaviour. The most dramatic example is leaping from a high bridge with elastic ropes tied around the legs. In this case the individual is called upon to face fears and take physical risks. In other cases the risks are more subtle and the changes possible more significant.

In selecting employees, especially at the highest executive level, much importance is attached to such subtle qualities as "maturity", even if psychologists have the greatest of difficulty in defining what is meant by this in theory. The same may be said for other results of human development, such as qualities like "balance" and "insight".

9. Vindication of subjectivity

A Latin American collective report on human scale development chose to include a section on the "vindication of subjectivity" in which it is argued that: "*The ways in which we experience our needs, hence the quality of our lives is, ultimately, subjective. It would seem, then, that only universalizing judgement could be deemed arbitrary. An objection to this statement could well arise from the ranks of positivism. The identification which positivism establishes between the subjective and the particular, though it reveals the historical failure of absolute idealism, is a sword of Damocles for the social sciences. When the object of study is the relation between human beings and society, the universality of the subjective cannot be ignored. Any attempt to observe the life of human beings must recognize the social character of subjectivity....Yet there is great fear of the consequences of such a reflection. Economic theory is a clear example of this. From the neo-classical economists to the monetarists, the notion of preferences is used to avoid the issue of needs. This perspective reveals an acute reluctance to discuss the subjective-universal....Whereas to speak of fundamental human needs compels us to focus our attention on the subjective-universal, which renders any mechanistic approach sterile.*" (Human Scale Development, pp 28-9)

It is a tragic symptom of the times that subjectivity should need to be "vindicated" to those concerned with the development process. It suggests that developing countries are paying the price of an unhealthy western obsession with objectivity. There are no institutionally acceptable indications as to the nature of a healthy balance between subjectivity and objectivity.

Significance: modes of awareness and experiential biases

1. Absence of official recognition
Although, as indicated above, intergovernmental and official academic bodies are prepared to give limited attention to the non-economic aspects of human development, there is no indication whatsoever that they are prepared to distinguish the range of modes of awareness characteristic of such development. Even when, as in the case of UNESCO, emphasis is placed on "cultural development", no acknowledgement is made of the modes of awareness recognized by the religious and cultural figures frequently honoured by such bodies as contributors to the cultural heritage of particular regions. The focus is on their products not on their subjective experience, however much the products were designed to draw attention to such experience and to articulate it.

It is characteristic of the tragic hypocrisy and collective schizophrenia of the present time, that delegates to intergovernmental meetings discussing "human development" may be deeply aware of the subjective range of states of consciousness, for many are indeed deeply religious. Whether or not they are, most would take great care, for purely political reasdons, to avoid offending those for whom such dimensions are important. And yet in debate no attention is drawn to these dimensions and to their relevance. Although the Constitution of UNESCO commences with the much cited phrase: *"since wars begin in the minds of men, it is the minds of men that the defences of peace must be constructed"*, it is difficult to trace any acknowledgement of the experiences occurring in the minds of men in the programmes of bodies such as UNESCO, or in those taking their lead from UNESCO.

The principal reason for not drawing attention to such dimensions is obvious. Such modes of awareness may be associated in the minds of many with religious experience. In many cases they are defined by particular religions. Any such discussion would thus arouse too much controversy in an international community already torn by ideological controversy. Ironically, it is the bitter conflict between such religions which hinders any recognition of the importance of such modes of awareness by the international community.

A secondary reason is that the weight of expertise within the international community is oriented towards the hard facts of politics, economics and science. Where it might be assumed that some of the social sciences would acknowledge such dimensions, this has proven to be very far from being the case. Ironically, again, it is the bitter conflict between those social science disciplines which purport to be sensitive to these dimensions which hinders any recognition of the importance of modes of awareness by the international community.

2. Explosion of popular interest
This situation at the governmental and intergovernmental level has however been totally undermined over the past decade by the explosion of popular and scientific interest in these dimensions. For example, books on individual human development have proved one of the strongest growth areas in publishing. **There is now a total cleavage between the content of human development as understood by intergovernmental bodies and the content associated with it by those interested in it personally** (as opposed to professionally).

Of all the governments, surprisingly it is the USSR, with its materialist ideological commitment, which has broken through the disciplinary obfuscation of the sciences to promote extensive research on a variety of paranormal states of consciousness. Even more surprising is the remarkable synthesis by the Soviet mathematician V V Nalimov (1982, 1985) *"using the major concepts of mathematics, physics, linguistics, psychology, psychiatry, history, philosophy, culturology, anthropology and theology"*. He covers all manifestations of the unconscious not encompassed within logic, synthesizing semantics, probability theory, mysticism and art into a startling new view of how the human mind perceives the world.

What is forgotten by the establishments that deny the significance of such experiences, is that people are strongly attracted to them irrespective of such denials, or possibly even because of the alternative they offer to the sterile concepts of human development promoted by such establishments. However well-meaning, such concepts are at present alienating to those bored by the claustrophobic analyses of political economics which prevail within the international community. For an increasing number of people, whether attracted to traditional or to contemporary approaches, the variety of modes of awareness defines human potential in a much more direct and attractive manner than has been otherwise possible. It becomes a process in which people wish to be engaged. For them human development can be both challenging and fun.

It is a sobering thought that, whilst official bodies and disciplines deny the significance of these dimensions, the amount of money invested in the achievement of alternate states of awareness through the use of drugs is now quite formidable. It has been estimated that the illegal drug trade in 1990 is only exceeded in international financial importance by the oil trade, which is commercially the principal traded product. In France it has been estimated that the individual expenditure in the "occult sciences" is three times the expenditure on consultation of general practitioners.

3. Experiential bias
The many perspectives indicated above suggest a strong case for opening up the debate on the nature of human development (as well as for finding out why it is so carefully closed off into isolated compartments). What in fact are the various meanings to be attached to the term? What are the related concepts? What images of human beings do such concepts imply? With what concepts or experiences do people themselves identify when considering their own development? What alternative and better varieties of experience and states of being do they suggest as being open to exploration? What methods may be used to facilitate such forms of personal development?

The following points give an indication that there are some very positive ways in which human development may be understood, and which are the justification for the collection of information undertaken for this section:

(a) In 1974, a well-respected establishment group, the Center for the Study of Social Policy of the Stanford Research Institute, prepared a policy research report for the Charles Kettering Foundation noting: *"If the post-industrial era of the future is dominated by the industrial-era premises, images, and policies of the past, the control of deviant behaviour needed to make societal regulation possible would in all likelihood require the application of powerful socio- and psycho-technologies. The result could well be akin to what has been termed friendly fascism - a managed society which rules by a faceless and widely dispersed complex of warfare- welfare-industrial-communications-police bureaucracies with a technocratic ideology. Evidence exists that this sort of future is already nascent. In contrast to such a technological-extrapolationist future, this report envisions an evolutionary transformation for society as a more hopeful possibility.*

Some characteristics of an adequate image of mankind for the post-industrial future were derived by: (1) noting the direction in which premises underlying the industrial present would have to change in order to bring about a more workable society; (2) from examination of the ways in which images of humankind have shaped societies in the past; and (3) from observation of some significant new directions in scientific research.

A future image of man meeting these conditions would:

(a) convey a holistic sense of perspective or understanding of life;

(b) entail an ecological ethic, emphasizing the total community of life-in-nature and the oneness of the human race;

(c) entail a self-realization ethic, placing the highest value on development of selfhood and declaring that an appropriate function of all social institutions is the fostering of human development;

(d) be multi-levelled, multi-faceted, and integrative, accommodating various culture and personality types;

(e) involve balancing and coordination of satisfactions along many dimensions rather than the maximizing of concerns along one narrowly defined dimension (eg economics); and

(f) be experimental, open-ended, and evolutionary.

A framework is developed in the report which demonstrates that it is at least conceptually feasible to fulfil these characteristics. Further, the report provides guidelines for actions through which fulfilment of the needed characteristics might be stimulated."

(b) As a result of the work of Abraham Maslow (1971) and the humanistic school of psychology, a distinction has now been established between basic deficiency needs in a human being and what have been called self-realization or being needs. He suggested, on the basis of empirical observation, that only about 1 per cent of any sample out of the population of contemporary Americans are examples of self-actualizing individuals, namely individuals who continue to attempt to develop and manifest their latent potentialities. This would seem to imply that at least 99 per cent of the population of one of the most developed countries may be considered to be psychologically underdeveloped, or at least only "developing", to employ the international euphemism.

Robert Jungk, in an address to the 1974 conference of the Irish Management Institute, argued that cultural man is underdeveloped. The characteristics identified for such self-actualizing individuals include: a capacity for acceptance, efficient perception of reality, spontaneity, transcendence of self-concern, detachment, transcendence of environment, social feeling and compassion, tolerance and respect, ethical certainty, and creativeness. It is suggested that consciously or unconsciously every person is seeking some form of self- realization or to become a self-actualizing person, fully expressing his own innate potentialities as an individual, and in full recognition of his own uniqueness as a personality. It is believed that there are a variety of methods and processes by which self-actualization emerges, and that this diversity should itself be protected.

(c) From the 1970s many groups, institutes and journals have emerged to explore new understandings of consciousness that recognize the experiential dimension.

4. Challenging need for new paradigms
Such interest is stimulated in part by new approaches to the relationship between consciousness and insights in fundamental physics (David Bohm (1980), Ken Wilbur (1982), and their specific relation to health (Larry Dossey, 1982). With this burgeoning interest in human development and states of consciousness (whether "altered" or not), it might be expected that there would be clear indications as to what these states or modes are to which people may aspire in the course of the process of human development. In fact the literature is mainly characterized by the priorities of the authors. These may, or may not, include: research on drug-induced states, research on mystical experience, states identified by traditional religions within a well-defined framework, conditions identified by various schools of psychoanalysis and psychotherapy, and conditions emerging from the explorations of charismatic leaders of new growth movements.

There is a general assumption that the different forms of awareness identified within each context may, in some cases at least, be identical. But there is little effort to catalogue these varieties of modes of awareness in their own terms, leaving open the question of what is identical with what. And the result of grouping such modes into a limited number of theoretical categories, as in the pioneering work of Charles Tart (1975), tends to denature the experiences described even further, however interesting such classificatory exercises may be from an academic point of view.

Part of the challenge lies in the manner in which many of the modes of awareness accessible to man in the process of human development are incomprehensible within classical scientific paradigms. They call for new approaches and new languages in which to communicate them, as argued in Section KD. These are emerging, as is indicated by Nalimov's work, for example (1985, 1981). But both the traditional and the new approaches rely to a large extent on metaphor for descriptive purposes (R S Valle and R von Eckartsberg, 1981).

As discussed in Section M, not only is human experience metaphorical in nature, but also that metaphor is an essential constituent of the structure of human experience. That is, part of the meaning of any experience is elusive, and it is the use of metaphor that formulates this elusive meaning and makes it available through an understandable figure of speech (Robert Romanyshyn, 1981). Part of the difficulty lies in the large number of modes which have been described, whatever similarities the descriptions conceal. This raises the question as to whether it is possible to interrelate these modes in any coherent manner without denaturing them. What are the "metabolic pathways" of intra-personal processes? Some of those identified are interlinked into sets, representing stages in a process, whether linear or cyclic. But it is clear that new ways of interlinking metaphors are required to offer a language for maintaining continuity between different modes of awareness.

This is one of the reasons for experimenting with metaphors and patterns, especially with the possibility of "pattern languages" discussed there.

5. Beyond human development consumerism
Despite well-recognized excesses of human development enthusiasts, is it appropriate to ignore the insights which prompt efforts at "revisioning psychology" (James Hillman, 1975) or to navigate through the dross of excesses and extreme positions? Robert E Ornstein, in a book appropriately entitled *The Mind Field (1976, p.ix)* indicates the problem: *"We are now on the threshold of a new understanding of man and of consciousness, one which might unite the scientific, objective, external approach of Western civilization and the personal, inward disciplines of the East. The emergence of this new synthesis has caused many to flock, unthinkingly, to rudimentary spiritual sideshows, which are quick, cheap, and often flashy. These reductions have given strength to others' total lack of interest. I write to develop a more secure position, one of interested yet candid assessment, somewhere between the two dominant positions: the almost reflexive rejection of what is conventionally understood as "mysticism", by many in the "hard" areas of contemporary life; the reflexive adulation characteristic of the slavish consumers of guruism, "instant enlightenment training", and other degenerations."* Such reassessments merit attention. Where would society be if, for example, "economic development" were to be rejected because of the excesses of its enthusiasts?

Only by opening up the debate on these matters, identifying the variety of concepts currently in use, and how they are related to one another, will it become possible to establish the connections between such concerns and the topics of economic and social development problems which have been favoured by the international and academic communities with such questionable results. Given that a major obstacle to such socio-economic development is the so-called "lack of political will to change", it may be that this intangible factor is intimately related to intangible factors in individual development, however it is conceived.

Significance: mythical, religious and spiritual biases

1. Religious experience

A large-scale survey of religious experience as a subjective phenomenon was conducted through the Religious Experience Research Unit at Manchester College (Oxford). The unit was founded by Sir Alister Hardy in 1969. The results of the first ten years of this work covering 3,000 responses have been published (Hardy, 1979). In 1974 a national survey was made of people reporting a mystic experience in the USA resulting in some 1,500 responses indicating that over 30 per cent of Americans had such experiences (Greely and McCready). Building on Hardy's work, two national surveys were conducted in the UK. These also indicated that over 30 per cent of the population had such experiences (Hay 1978, 1982). The author notes: "*I doubt very much that religion is about to die out. The awareness out of which it grows is too widespread for that. More dangerous, because more likely, is that it may continue to be isolated from the mainstream of modern life. Human realities which are absolutely ignored tend, as Freud has pointed out, to return in bizarre and fanatical forms....We need to attend more openly to our religious awareness, so that at the very least its constructiveness and creativity can be used for the benefit of the species.*"

The characteristic religious experience, at least within Christianity, is that of being absolutely dependent on God (Schleiermacher, Friedrich. *The Christian Faith*). The value of any religion, especially Christianity, is its success in evoking this experience. While in these terms the human condition is made clear to man by the presence of God, in those of Paul Tillich such recognition of God is always in some sense a function of human striving. Unfortunately, as indicated by Rudolf Otto (*The Idea of the Holy*, 1950): "*So far from keeping the non-rational element in religion alive in the heart of the religious experience, orthodox Christianity manifestly failed to recognize its value, and by this failure gave the idea of God a one-sidedly intellectualistic and rationalistic interpretation.*"

Within the Christian religions, emphasis appears to be allocated primarily to the one-step process of "conversion", although there is nothing uniquely Christian about the language of conversion and new birth. Despite some recognition of people of greater spiritual "depth", there is much vagueness concerning the process by which this was achieved and whether it can be replicated to any degree by others. Mystical experience of any kind is treated as extraordinary, if not totally suspect. This attitude is however increasingly questioned in charismatic religious movements. In the eyes of many, the credibility of established religions is eroded by the seeming lack of relationship between advancement within them and the increase, if any, in spiritual development of the individuals concerned.

Although religions such as Christianity have rather successfully eliminated the experience of the numinous or the holy, it is reported as breaking through at odd times and in particular circumstances. It is characterized by "mysterium tremendum" and a mixture of elements of awe, dread, wonder and fascination: "*beyond our apprehension and comprehension, not only because our knowledge has certain irremovable limits, but because in it we come upon something inherently 'wholly other', whose kind and character are incommensurable with our own and before which we therefore recoil in a wonder that strikes us chill and numb.*" (Otto, 1950).

2. Spiritual development

"*It is perhaps no accident that within those movements whose energies are directed towards the saving both of our ravaged earth, and of ourselves, there is a revival of awareness of the importance of spirituality. This revival is occurring independently of the uneasiness of the churches. Although as yet unfocused and without direction, it is none the less insistent. It is also fortified by the newest ideas within a science visibly approaching insights undreamed of at the beginning of this century. It is as though the extremity of our conditions evokes an instinctual response from within those forces of life which are concerned with survival....However undefined or unarticulated, the experience of what has been called the spirit appears a fact of our existence which lends to us the sense of a power within and about us which is not tangible, but ever present. However we name it, it is the source of our deepest responses to life....At the heart of the mystery of creation there appears to lie an ambiguity, a doubleness, a dialectic whose resolution is at once the justification of the old and the affirmation of the new. For humanity the irresistable path towards self-knowledge is the paqth towards God. Our viability as a species depends on our achknowledgement of an answerability to what is more than ourselves and yet lies within us as that which we are about to know.*" (Lorna Marsden, The Guardian)

It is surprising to note that few established religions appear to give much emphasis to "spiritual development", especially within the western religious traditions. Religions tend to emphasize moral and ethical development and religious worship, but accord little attention to development of the experiential dimension -- other than as an important private concern. "*There seems to be a tendency on the part of the WCC, perhaps Protestants in general, in certain traditions at least, to think of spirituality primarily in terms of worship. Worship is perhaps taken in a broad sense of prayer, devotion, eucharist, celebration, community-at-prayer. The word worship is not used as frequently in my Roman Catholic tradition, but we would in any case not equate it with spirituality. Before Vatican II we probably spoke more of the spiritual than of spirituality. And spiritual life did tend to focus more on spiritual exercises, devotions, prayer life, personal or communal. After Vatican II, however, we spoke more readily of spirituality, meaning the whole of our life as it was shot through with faith...Spirituality embraces one's ministry and service, one's relationships, one's personal and communal prayer life, one's approach to the political and social environment, in short one's life-style...Spirituality then is incarnational, daily, integrated. To be authentic it must have dimensions of combat, of search, of retreat, of renewal, of discernment, of personal growth, of ecumenicity, etc...*" (Joan Puls, *Vital Ecumenical Concerns*, 1986).

Although religion can claim to be vitally concerned with human development and attaches great value to the experiential dimension, its relationship to spiritual development remains unclear. Thus in reviewing the significance of "human development", the World Council of Churches notes: "*...More than ever before, we find it difficult to articulate our understanding of the development concept and consequently to decide on the patterns of participation in the development process. In the past few years there have been many conscious efforts to give human development a conceptual clarity that it lacked, but the relation between concept and reality seems to become more diffused and more evasive. The uncertainties and ambiguities resulting from this situation are made more pronounced because of the few certainties that cannot be evaded: that after two decades of efforts to remove poverty and reduce inequality there are today more people in the grips of dire poverty and the gap between the rich and the poor has widened...In the quest for development, we find ourselves caught in a pensive mood, raising many questions and finding few answers.*" (World Council of Churches, 1975).

For many the question of spiritual develomment is not a matter divorced from the social and environmental crises of the times. The failure to develop those dimensions of human potential then directly inhibits possibilities of responding appropriately and insightfully to those crises. In the words of George Steiner: "*What I affirm is the intuition that where God's presence is no longer a tenable supposition and where His absence is no longer a felt, indeed overwhelming weight, certain dimensions of thought and creativity are no longer attainable...The density of God's absence, the edge of presence in that absence, is no empty dialectical twist. The phenomenology is elementary: it is like the recession from us of one whom we have loved or sought to love or of one before whom we have dwelt in fear. The distancing is, then, charged with the pressures of a nearness out of reach, of a remembrance torn at the edges. It is the absent 'thereness', in the death-camps, in the laying waste of a grimed planet, which is articulate in the master-texts of our age.*" (*Real Presences*, 1989).

The more obvious problems of poverty, pollution, exploitation of resources and conflict then serve as metaphors indicative of the impoverished, polluted, exploitative, and conflictual forms of human development which currently predominate.

Significance: acknowledgement of mythico-spiritual dimensions

There are some domains and initiatives which explicitly recognize the value of some form of spiritual experience and its development.

1. Self-sacrifice and inner transformation
Some active traditions stress the limited value of techniques of meditation and prayer when these are used solely as a means of providing periodic "trips" for relaxation and stress control or as a means of affirming status in the life of a community engaging in such practices. Whilst thrilling experiences of ecstasy may relieve the drabness of daily life, they are viewed by some as a distorted understanding of the spiritual life. *"Real spiritual life is about loss of self, sacrifice, inner transformation and change. It involves a radical alteration of attitude which affects every element in our lives, even the apparently most earthy and trivial."* (Cyprian Smith, 1987).

Meditation and other techniques may simply provide subtle means of avoiding any such process of personal transformation by massaging the ego in aesthetically satisfactory ways that reinforce the current self-image. From this perspective, religious or spiritual experiences are of a different order than the enthralling experiences that can be induced in a variety of ways.

1. Conversion and rebirth
Much importance is attached, especially within Christian charismatic movements, to the process of conversion and rebirth. It is believed that such conversion can sweep people into a new relation with God, acquitted of their past, and with the acquisition of a direct and unmediated kind of assurance. Such entry into the "kingdom of God" is not a casual affair for those undergoing it. It involves a radical confrontation with God, and it seems impossible that it could happen without a profound self-examination and a penetrating self-knowledge. (William Abraham, Logic of Evangelism, 1989).

(c) Spiritual guidance and direction: In many religious traditions, special recognition continues to be accorded to spiritual mentors, guides, gurus or masters. However, within the Christian tradition, which has so significantly influenced the international community, the practice of spiritual guidance peaked in the 14th and 15th centuries. Following the Council of Trent (1545-63), the Catholic practice of soul care narrowed in focus to become primarily concerned with decisions about religious vocations. Spiritual guides increasingly took as their primary role the guardianship of orthodoxy, with the avoidance of heresy and dubious forms of mysticism as their major preoccupation.

In contrast to this authoritarian position, Protestant churches used the term "shepherding", implying love and concern. Individual spiritual guidance also came to be de-emphasized in favour of mutual admonition and collective guidance. The necessity of spiritual direction was also questioned by Catholic authors as being useful only when the individual, who is living the life of the Christian community to its fullest extent possible, becomes aware of God's special call to perfection.

The assumption in this tradition appears to be that such people would become associated with one of the many religious orders (or their lay equivalents). Spiritual guidance is then defined by the discipline or rule of the particular order chosen, with little adaptation to individual needs and sensitivities. Individual guidance is not available to the many, other than as part of communal religious practice under priestly guidance. For the general population, this situation has encouraged the shift of emphasis from the cure of souls within a religious context to a cure of minds within a psychotherapeutic context. (David Benner, Psychotherapy and the Spiritual Quest, 1988).

(d) Spiritual disciplines: From 1933, the world's most distinguished specialists in fields relating to psychology, religion and cultural anthropology meet annually for many years, under the auspices of the Eranos Foundation, *"Toward the task of encompassing and assimilating the world's wealth of poetic and religious visions, modes and dreams of life, and readings of the mystery of death"*. In one of the publications arising from this enterprise, on Spiritual Disciplines (1985), the editor Mircea Eliade writes: *"For the members of Eranos, this exceptional interest in spiritual disciplines and mystical techniques arises from the fact that they are documents capable of revealing a dimension of human existence that has been almost forgotten, or completely distorted, in modern societies. All these spiritual disciplines and mystical techniques are of inestimable value because they represent conquests of the human spirit that have been neglected or denied in the course of recent Western history, but that have lost neither their greatness nor their usefulness. The problems that now arises - and that will present itself with even more dramatic urgency to scholars of the coming generation - is this: How are means to be found to recover all that is still recoverable in the spiritual history of humanity? And this for two reasons:*

(a) Western man cannot continue to live on for an indefinite period in separation from an important part of himself, the part constituted by the fragments of a spiritual history of which he cannot decipher the meaning and message.

(b) Sooner or later, our dialogue with the "others" - the representatives of traditional, Asiatic, and "primitive" cultures - must begin to take place not in today's empirical and utilitarian language (which can approach only realities classifiable as social, economic, political, sanitary, etc) but in a cultural language capable of expressing human realities and spiritual values.

Such a dialogue is inevitable; it is part of the ineluctable course of History. It would be tragically naive to suppose that it can continue indefinitely on the mental level on which it is conducted today."

5. Personal relevance of myth
There is increasing recognition of the continuing value of myth in articulating human experience (Campbell, The Inner Reaches of Outer Space). Earlier hopes that the sciences, especially the social sciences, might provide new solutions for old problems are increasingly questioned. The old myths have been replaced by new myths as a response to the continuing encounter with the mystery of life. The new scientific responses to the mysteries of human being are recognized as much myth as the old religious responses. (Benner, David, 1988)

6. Channelled guidance
In many sectors of society, and notably in industrialized countries, there is increasing response to forms of spiritual guidance received through "channels", mediums or under other kinds of trance conditions (including shamans). This interest builds on the traditional role of such sensitives in relating to the mythical world of the gods and the spirits. In the modernized forms of such myths, these now include extraterrestrial entities. Through tapes and books, such guidance (whatever its quality) is now reaching a much wider audience. Sources of this kind, as in the past, are consulted to cast light on major life decisions, which may include political and investment decisions by the leadership of major institutions. Many are influenced in their understanding of human development and its future direction by the patterns of information presented under such conditions.

7. Divination
There is a phenomenal interest in all cultures in various forms of divination, ranging from astrology to even more ancient techniques, including geomancy. Much use is made of such techniques to determine the auspicious conditions favouring key stages in the life process and the configuration of circumstances appropriate to future human development. These techniques are used at all levels of society, often on a daily basis. They are recognized to have been important to the leaders of the largest countries (notably the USA and India) faced with key decisions in international relations as well as to the design and location of business corporation headquarters (Hong Kong). The use of such techniques is intimately intertwined with a mythical understanding of the forces influencing human development.

8. Belief in the devil
There is a surprising resurgence of belief in the devil. Although this belief has been important to the Christian religion, in recent years it

seems to have been stimulated by fundamentalism, whether Christian or Islamic. For example, in a special issue on the question (20 December 1990), the *Nouvel Observateur*, an influential French weekly, reports that 37 percent of French people (50 percent of young people) believe in the devil, compared with 46 per cent of Italians. For France, this is twice the number in 1968. One in ten there claim to have the impression of having been under the influence of the devil at some time. The concern of fundamentalists has in part been a response to the increasing activity of sects and cults. Whilst many of these have quite different, and even opposing preoccupations, some do indeed claim quite openly to engage in devil worship, to the point that black masses and other rituals are documented on television. Both Christian and Islamic authorities of a fundamentalist orientation frequently make reference to the ways in which the devil or Satan is embodied in, or manipulating, those who oppose that perspective. In the media this is most visible in the Middle East, with the Islamic labelling of the USA (or its President) as Satan.

9. Encyclopedia of spirituality.
Interest in spirituality is currently justifying the production of a 25-volume *Encyclopedic History of the Religious Quest* under the general heading of *World Spirituality* and under the general editorship of Ewert Cousins. One or more volumes each cover the major traditions.

10. Psychological well-being
Although many psychologists have viewed spirituality in a reductionistic manner, there are a number of significant exceptions. These include: the analytical psychology of Carl Jung and his followers; the we-psychology of Fritz Kunkel; the existential psychology of Soren Kierkegaard, John Finch, and Adrian van Kaam; the contemplative psychology of Gerald May and William McNamara; psychosynthesis, as developed by Robert Assagioli and his followers; and the logotherapy of Viktor Frankl.

11. Spiritual dimension of health
A past Director-General of the World Health Organization proposed that the question of the "spiritual dimension" in health be discussed by the WHO Executive Board (73rd session, EB 73/15, October 1983). In his preparatory note he indicated that all the meanings of spiritual have one common denominator: "*They infer a phenomenon that is not material in nature but belongs to the realm of ideas that have arisen in the minds of human beings, particularly ennobling ideas.*" And by shaping people's action and ways of life, such philosophical, religious, moral or political ideas have had a profound influence on the physical, mental and social well-being of the people concerned.

The text notes the impact of the spiritual dimension but skilfully avoids discussion of any form of non-material human development in the proposed Strategy for Health for All. It seems that the WHO interest in the "spiritual" was temporary and solely due to the personal commitment of the African Director-General of the time (also noted for his success in achieving greater acknowledgement of the related issue of the role of traditional healing methods).

12. Personal spiritual quests
A Catholic writer, commenting on the general unrest and disquiet among religious people today, detects two main desires as coming to the surface, especially among the younger generation. The first is political and social, focusing on the desire for freedom and a more just and equitable society. "*The second is more inward and personal. It is the desire to learn about the human heart, its inner depths and recesses. Within the Church it manifests itself as a desire to learn more about prayer and meditation, about different levels of consciousness and awareness....This knowledge and understanding about what human beings are, what lies in the deeper levels of the human heart, ought to be found in the Church, for it is religion, above all, which seeks to touch the central core of human nature.*

But people who turn to priests and spiritual directors for this kind of help are often disappointed...Jung points out, very pertinently, that in the eyes of official religion the human psyche, with all its hidden folds and dark declivities, has no real existence of its own; for the priest the psyche or soul is just something to be fitted into a dogmatic or liturgical framework.

This does not satisfy the modern seeker, who wants above all to be understood for what he really is, to be brought to the realization and acceptance of what really does lie in the innermost depths of his mind, regardless of whether this fits in with official church dogma or not....It is no use my being told that I am 'redeemed' by Christ, if my actual experience is one of alienation, darkness and self-division...

That is why so many turn to the psychiatrist rather than the priest. It is also why so many turn to Buddhist, Hindu and Sufi teachers; they believe -- often rightly -- that they will find in them a profound and detailed science of the inner life of the mind, a sureness of touch in practical guidance and training, which is rarely equalled within the Christian fold." (Cyprian Smith, *The Way of Paradox*, 1987).

13. Inter-faith gatherings of spiritual leaders
In the 1980s the increasing concern of spiritual leaders for the condition of society resulted in a number of initiatives in which the relation between spirituality and the problems of society were explored. These have culminated in the instauration of a series of meetings with parliamentary leaders. The first was in Oxford in 1988 (Vittachi, 1989); the second in Moscow in 1990.

14. Political acknowledgement of power of religion
The 1980s have also forced public authorities to accord greater recognition to the importance that people attach to religion at least, if not to spirituality. This is most dramatically evident in the rise of religious fundamentalism (whether Christian, Islamic or Hindu) and its effect on politics and policy options. Who would have believed that an issue of blasphemy would have the international repercussions of the Rushdie affair ? Who can deny that "holy wars" are as much a factor of the present as they were of the past ?

Method: assumptions

It is a basic mistake to assume that the concept of human development is held in the same way, whether between cultures or within any culture. The questions as to whether an individual can "develop" (other than in the obvious ways that preoccupy educators, economists, physicians and psychologists) and as to whether certain modes of awareness "exist" (and the nature of that existence) are not understood in the same way in different contexts.

For some, although altered states of awareness are a reality, individual development does not necessarily mean a journey through a pattern of such less readily accessible states. Individual maturity has not been effectively defined and it is uncertain whether it lends itself to definition. And for many, the degree of individual suffering in the world renders quite absurd any discussion of human development that does not concentrate on basic human needs. In some traditions, however, it is the failure to cultivate some of these poorly recognized states which is directly responsible for the ills engendered in the world.

It is useful to attempt to identify alternative ways in which human potential can be perceived, as a means of increasing understanding of the constraints on providing any satisfactory definition. This will also make evident the difficulty of attracting any consensus on strategies of human development. Whilst it is possible to discuss these perceptual modes as models, a broader and more insightful discussion results from treating such models as part of a set of metaphors.

The following alternative perceptions are therefore discussed as metaphors of human potential and its implications for the process of human development. They are not mutually exclusive:

1. Ordered array
Modes of awareness can be viewed as constituting an ordered array, like stations on a subway network. This view would tend to be favoured by those who are used to defining their environment in an orderly manner, in terms which favour management and control, whatever the degree of simplification necessary. In such an array, all modes are relatively accessible, although some may only be reached through intervening conditions. Modes are different, but not necessarily better in any developmental sense. In this metaphor, development might be envisaged in terms of extending and complexifying the network into a rich array of modes. This would be contrasted with a less developed condition equivalent to a subway network with relatively few stations and (possibly unconnected) lines. Goals of human development might be expressed in terms of improving the stations, increasing the facility of movement throughout the network, and organizing the network into the most effective configuration of stations. *(To be contrasted with...)*

2. Disorder and chaos
Modes of awareness can be viewed as completely unordered, to the point of being essential chaotic and disorderly. This view would tend to be favoured by those who have lost control over their environment, realize that they are subject to more forces than they originally assumed, or simply prefer the challenge of the disorderly and unpredictable (cf William James, Bergson, Schopenhauer, Rousseau). Modes of awareness are then too confusing to present any stable or orderly features permitting them to be distinguished or labelled. In this metaphor, development might be more concerned with ways of experiencing this chaos more completely, responding to it in a manner unfiltered and uncensored by artificial orderings.

3. Static structure
Modes of awareness can be viewed as forming a static, semi-permanent set of psychological conditions (especially by those who benefit from such predictability). This view would tend to be favoured by those seeking a reliable workforce (employers), stable markets (advertisers), or faithful constituencies (politicians), over an extended period of time. The view is then reinforced by legislation and regulatory procedures anticipating the range of basic needs of the average citizen, which are held to be unchanging or to change quite slowly. Human development is then primarily the process of ensuring that more people have such needs satisfied. *(To be contrasted with...)*

4. Dynamic structure
Modes of awareness can be viewed as constituting a dynamic structure, in which the modes arise in the dynamic relations between static elements. Like harmonies and melodies, based on a configuration of established musical notes, such modes cannot be readily isolated and named. They only exist as dynamic relationships changing continuously. This view would tend to be favoured by those who respond to the unique opportunities of the moment, possibly because their survival depends on the uniqueness of their response. In terms of the musical metaphor, human development then becomes a question of being able to form more complex harmonies amongst the predictable features of the environment, encompassing for longer periods the disharmonies which might otherwise be considered more significant.

5. Discrete phenomena
Modes of awareness can be viewed as distinct, with some form of boundary separating them. This view would tend to be favoured by those who need to distinguish clearly where they are, either from where they have been, or from where they want to be. As on a ladder, each mode corresponds to a dependable step and there is no intermediate condition. In terms of this metaphor, human development may then be conceived as moving up a series of steps, possibly understood as a series of initiations, or developmental stages. From each successive step a broader view may be possible, incorporating those below it. *(To be contrasted with...)*

6. Continuous phenomena
Modes of awareness can be viewed as part of a single continuous field of awareness. In the light of field theories, particular modes might then be understood as interference patterns (cf Moiré patterns). In this metaphor, human development might be understood in terms of increasing the number and complexity of such interference patterns and increasing the facility for shifting elegantly between them.

7. External relationship to phenomena
Modes of awareness can be viewed as externalities, as objects of investigation, and as "places" that can be visited. As such their existence is independent of any particular observer. This view would be favoured by those with either a rationalist or an empiricist orientation. This may be seen in the scientific investigation of states associated with biorhythms. It is basic to the assumptions in many educational development programmes. Human development is thus a question of acquiring the expertise, or possibly the technology, to gain access to such places at will. *(To be contrasted with...)*

8. Identification with phenomena
Modes of awareness can be held to be only genuinely comprehensible through an intuitive identification with the experience they constitute, experienced by the observer as he experiences himself (cf Bergson, Hegel). This view would be favoured by those whose views have been strongly formed by particular unsought personal experiences of altered states of awareness, largely unconditioned by external explanations and expectations. Human development from this perspective might then be viewed as progressive achievement of a more profound, enduring, and all-encompassing identification with such states through which identity itself is redefined.

9. Sharply defined phenomena
Modes of awareness can be viewed as being directly experienceable (cf Descartes, Hume), like individually framed paintings. This view would tend to be favoured by those concerned with the objective reality of such states as joy, pleasure, and love. For them, any other kinds of awareness are unreal abstractions of no significance, other than as distractions from the concrete reality of human experience. Human development might then be viewed as a process of achieving more intense experiences more frequently, rather as an art connoisseur seeks greater exposure to better paintings, through which his taste is developed. *(To be contrasted with...)*

10. Implicitly defined phenomena
Modes of awareness can be viewed as implying levels of significance greater than that immediately experienced (cf Hegel, Whitehead, Niebuhr, Proust). As with the experience of an iceberg, this view would tend to be favoured by those for whom awareness encompasses both the tip and some sense of the invisible presence of its underlying mass (and the possibility that it may suddenly become visible). Significance is derived from the unexpressed presence or the potential of any moment. Human development might then be viewed as the birth of such potential and the increasing recognition of the immensity that remains unexpressed.

11. Inherently comprehensible phenomena
Modes of awareness can be viewed as comprehensible in terms of existing paradigms or through their natural evolution. This view would tend to be favoured by pragmatists, and those with a scientific orientation, for whom a satisfactory explanation in terms of collectively known factors must eventually be possible (if one cannot immediately be imposed). Human development is then a process of making what is known to the experts more widely accessible and of investigating what they do not yet comprehend. *(To be contrasted with...)*

12. Inherently incomprehensible phenomena
Modes of awareness can be viewed as calling for explanation in terms of other frames of reference, which may not necessarily be accessible to the human mind (cf Plato, Schopenhauer, Hegel, Plotinus, Niebuhr, Toynbee). This view would tend to be favoured by many religious groups and in cultures sympathetic to belief in other levels of being or realms of existence. Human development is then essentially an evolving mystery whose nature is beyond the grasp of the human mind.

13. Phenomena in a context of due process
Modes of awareness can be viewed as subject to known (or knowable) laws as a part of definable processes. This view would tend to be favoured by those endeavouring to develop programmes of human development in which certain modes are experienced at certain stages or developmental phases. Human development is then viewed rather like an educational curriculum through which people need to pass in an orderly manner, building on appropriate foundational experiences, to the possible levels of achievement defined by the outstanding pioneers of the last. *(To be contrasted with...)*

14. Spontaneous phenomena
Modes of awareness can be viewed as totally spontaneous conditions or peak experiences unconnected to each other. This view would tend to be favoured by those who perceive chance, accident or divine intervention to be prime explanatory factors. It is also natural to those who respond spontaneously to their environment, placing relatively little reliance on norms and expectations. In this view human development is the increasing ability to rely on the spontaneity of the moment and the ability to respond proactively to the opportunities it offers.

15. Comment
Clearly these different views are not mutually exclusive and overlap in complex ways in the case of any spiritual tradition or school of thought. The 14 views have in fact been elaborated on the basis of work by W T Jones (1961), who developed 7 axes of bias by which many academic debates could be characterized. The 14 views above form 7 pairs of extremes corresponding to the extreme positions on such axes. Jones showed how any individual had a profile of pre-logical preferences based on the degree of inclination towards one or other extreme of each pair. The scholars named in each case are those given by Jones as examples.

It would be useful to explore cultural differences in the perception of human development, as noted in Section KZ.

In this project, although the information may derive from individuals or groups holding any combination of the above biases, the assumption made is that there is value in collecting, ordering and presenting the information as though modes of awareness did take the form of an ordered array (even if it is only "partially ordered"). The bias, in terms of the above checklist, is therefore towards understanding modes of awareness as: an ordered array (a), essentially static (c), discrete (e), experienced as externalities (g), sharply-defined (i), inherently comprehensible (k), and as part of a due process (m). It is one shared, to some degree, by certain authors in most traditions who may themselves be endeavouring to document modes of awareness cultivated by people of their own tradition holding quite different biases.

This is not to deny that a radically different set of biases does offer valuable insights and is more appropriate under certain circumstances. In fact many of the modes of awareness are only articulated by people having those other biases -- often quite strongly held (especially concerning the non-definability of many modes of awareness). It is quite probable that the process of human development calls for the complete range of patterns of biases. But those emphasized are valuable in creating a framework to permit insights arising from those other biases to be compared, to the extent that comparison is possible or appropriate.

Method: criteria and definitions

1. Scope
There are many ways in which human development is understood and many concepts which imply particular understandings of human development. Some of these are very narrow in focus. Some are very broad. In the case of modes of human awareness the situation is even more confused, with many denying the existence of states that others consider essential to an understanding of humanity and its future potential. To clarify the situation, without imposing yet another set of constraints, a very broad approach has been taken in selecting material for inclusion. The criteria are such that the user is confronted with the challenge of deciding whether particular items are relevant or irrelevant to an understanding of human development, either in western or other cultures.

The whole question of modes of awareness is further confused by the meaning, if any, to be attached to "consciousness". The short response would appear to be that nobody knows what it means, or rather numerous concepts, theories, models, and analogies have been proposed, but little agreement has been reached. Even without taking into account the very extensive explorations in eastern philosophy, in the West the very definition of the word remains in doubt. At one extreme, it is held that such concepts simply reflect a particular use of words. Many intermediate views derive from different positions on the classical mind-body controversy, including various forms of dualism and monism. Although psychology was originally conceived as the science of consciousness, behaviourists have sought to eliminate any reference to it, with some regarding it as an epiphenomenon of little significance.

Despite this lack of consensus on the definition or nature of consciousness, some of its characteristics have been generally accepted and might be described as follows: *"Consciousness is a personal state, characteristic at least of humans, which involves selective awareness of both external and internal events, which exemplifies knowing, allows insight, evidences memory and is to a large extent subservient to the will....What has become clear is that not only may it be said to function at various levels, but that it is composed of many sub-systems. Furthermore, it seems probable that these are arranged hierarchically, as are our physical systems. Perhaps we should concern ourselves...not with an assumed totality, nor with separate components, but with their interrelationships"* (Graham Reed, The Psychology of Anomalous Experience, 1988, p.161-2).

Both in the case of modes of awareness and human development concepts, the prime concern has not been the analysis of the extant concepts in order to group them into some apparently suitable order. Apparent duplicates have been combined under certain circumstances but allowed to stand as separate entries, possibly even with the same descriptor, where their advocates would have considered such treatment appropriate. The aim has been to allow users to consider the advantages and disadvantages of greater "rationalization" of the material presented.

2. Human development concepts: exclusion
In identifying aspects of human development in order to build up the sequence of entries, some criteria are obviously required either to include or to exclude material. Since a principal objective was to identify the range of human development concepts, it was considered unwise to develop a precise definition of what was to be included and both easier and usefully open-ended to define what was to be excluded. A set of provisional exclusion criteria was therefore used. When the collection of material was well-advanced, a set of provisional inclusion criteria was developed. These criteria continue to evolve.

(a) Physical development: Body building, and the improvement of particular physical capabilities and other approaches to health, when conceived solely as a question of body tone; also augmentation of the physical body using technological devices. In particular no consideration has been given to the many dietary concerns, whether or not they focus solely on the physical body.

(b) Personality development: Techniques used to increase the ability to influence others in personal contact, as well as techniques of persuasion and the use of personal magnetism.

(c) Aesthetic development: Development of taste and artistic appreciation.

(d) Mental development: "Mind improvement" techniques where these are completely divorced from other aspects of the human being and intended mainly to improve the individual's economic status and power in society.

(e) Human development for the collective: Techniques of human development conceived as a means of racial development for political or nationalist ends.

(f) Non-development orientation: Material concerned with the condition of man rather than focused on the development of the whole man; *ie,* material without a practical application. This also excludes philosophical approaches which consider the question of defining conceptual frameworks in which the development of man may be conceived.

(g) Human relations: The industrial approach to the improvement of human relations to the extent that its major thrust is to improve the productivity of the individual for the benefit of the corporation rather than help the human being to grow, whatever the implications for efficiency.

(h) Unnamed concepts: Human development practices to which no name has been given.

(i) Belief-related practices: Human development concepts associated with particular belief systems and "-isms".

(j) Goals of development: As reflected in different schools of thought (eg maturity, cosmic consciousness, completeness/wholeness, harmony, fulfilment, and self-awareness), although these are included as modes of awareness.

3. Human development concepts: inclusion
(a) Processes of development: With which "human development" may be confused (eg cognitive development, perceptual development, intellectual development, psychic development, emotional growth, and individuation).

(b) Facilitative techniques: Including non-therapeutic, non-technological techniques (eg meditation, dance, prayer, and ritual); non-therapeutic, technological techniques (eg biofeedback, and martial arts); therapeutic, non-technological techniques (eg experiential therapy, motivational development, client-centered therapy, and growth games); therapeutic, technological techniques (eg drug treatment).

(c) Focusing devices or symbolic systems: Including mandalas, astrology, I Ching, Tarot, and the like.

(d) Negative techniques: Including thought reform and brainwashing.

(e) Narrow concepts of human development: Including mind control and genital maturity.

(f) Religions and spiritual orthodoxy: To the extent possible, entries have been included on traditional religions as forms of human development. The emphasis however has been to focus primarily on material from any tradition which emphasizes the process of human development, especially through spiritual or religious experience, or those cases where the religion itself necessitates a self-referential approach. Eastern religions are particularly rich in these respects.

4. Modes of awareness: inclusion
It was not considered useful or possible to make hard and fast distinctions between a level or state of consciousness as such and

the less specific modes of awareness which might encompass such states or observe them. Entries were included on:

(a) Inner expriences: Articulation of the nature of various types of "lived interior experience", where some generalization is considered possible by the experiencers.

(b) Experiential goals of development: Concepts which in previous editions would have appeared in the human development section, including goals of development as reflected in different schools of thought (eg cosmic consciousness, completeness/wholeness, harmony, fulfilment and self-awareness).

(c) Sets of states: Sequential or related states as described by different religious and secular sources, including very specific entries (eg individual states in the Ascension Stages Game of Tibetan Buddhism).

(d) Induced states: States induced by specific techniques, whether therapeutic technological, symbolic systems, or negative programming (eg under conditions of brainwashing).

(e) Common states: Commonly experienced modes in normal everyday life (eg waking, sleeping, dreaming - including emotional variations).

(f) Anomalous experiences: Anomalous experiences of attention, perception and recall (eg déjà vu).

(g) States associated with abnormal conditions: States experienced under abnormal conditions (eg mental disorders, self-administered drugs).

(h) Personality specific modes: Modes intrinsic to specific personality types.

5. Concepts and modes identified by symbol or metaphor

In many traditions the process of human development, and the modes of awareness encountered, are identified to a high degree by metaphor or symbol. This is especially true of non-western cultures. The metaphor of a journey, for example, is frequently used in all cultures. The stages on the journey correspond to particular insights or modes of awareness.

In Buddhism there is an important technical term dhatu. This refers both to psychological realms (altered states of consciousness) and to cosmological realms (including "hells" and "heavens"), namely places in which the practitioner can exist and be reborn and live out one (or many) lives. In this sense to attain a particular altered state of consciousness is also (temporarily) to exist in the corresponding cosmological realm and possibly to be reborn there. This intimate relationship between the psychological and the cosmological is absent from the western traditions, except in those of magic. Practitioners of magic are in part concerned with contacting such "worlds" and entering them. These domains can all be understood as distinct states of consciousness (often in some ascending series) although their identification as such tends to be allusive. They are held to be accessible through particular meditational disciplines and other practices.

A number of traditions understand a human being as composed of several distinct "bodies", usually three or more. Each of these may be understood as providing a unique perspective or mode of awareness. Development of the person may progressively lead to the unfoldment of these modes as he becomes aware of that body.

In many traditions, states of consciousness may be symbolized by spiritual beings, notably gods. The encounter with a particular god is associated with entry into a particular state of consciousness. Repetition of the name of that god is one technique for conditioning the mind to awareness of that state. In a monotheistic system, God may be held to have many Divine Names, which can also be used to evoke distinct modes of awareness, to be associated with such modes, or to detrmine them.

Of greater concern is the extent to which some traditions, notably Buddhism, associate modes of awareness with all classes of phenomena. Cognition of a particular phenomenon thus involves a particular mode of awareness. Any classification of phenomena may then be understood as a classification of modes of awareness.

A very pragmatic approach has been taken to all such symbolic material. A principal concern was to avoid including sets of symbols which were not actively held to be associated with states of consciousness by some living tradition. Little attention was therefore given to material in which gods were simply worshipped unless that material identified a distinct state associated with the encounter with the god. Furthermore, even if there were strong indications that such states were recognized but no material could be located to give some content to that experience other than in terms of imagery, then such potential entries were not included.

6. Restrictions on coverage

Human development concepts (Section HH) emphasize the positive side of the individual's development, with far more entries describing methods of improving the human condition than activities detrimental to development. There has been some attempt to indicate possible negative consequences of human development concepts, but in the main, and where possible, the positive side has been emphasized. The material on modes of awareness (Section HM) covers all states whether positive or negative.

Section HM focuses on the experience of the mode rather than it's existence as such. States of unconscious (sleep, trance, etc) are given less full treatment than waking states. Minimal effort was given to philosophical, theological, psychological or other intellectual attempts to theorize about modes of awareness. The emphasis is on the experiential dimension, especially when some special personal discipline is advocated, however inadequate the articulation of the insight.

In its final form however a number of the entries and references included could validly have been excluded according to one or more of the above criteria. They have however been included either in order to mark the existence of borderlines or where it could easily have been assumed that they should have been in (whether because of confusion of terminology or because they were closely related to other concepts of human development). The core concepts of human development should therefore emerge by contrast with such borderline cases.

Inclusion of concepts and modes is based on availability of source material. Higher priority was given to material that could be readily processed. For this reason concepts from some traditions and disciplines may be better represented than others. Some material called for more work than resources permitted.

Recognizing the limitations of collecting material from a wide variety of cultures, where material was available describing states, entries were often made even if the description was felt to be unsatisfactory. Even if the description could only be formulated in poetic or symbolic terms, and even if only a single sentence distinguished it from related states, the entry was included as a pointer to an experience considered meaningful in a particular tradition. Such minimalistic entries can then be extended at a later stage if richer or more explicit material is located.

Method: procedures

1. Identification procedure
A preliminary list of subject headings relating to some aspect of human development or its synonyms was established for the 1976 edition. This list has been used and extended during the course of several systematic library and literature searches, by which an initial bibliography was built up (see Section HY). Summaries of various ranges of human development concepts and modes of awareness, in the form of books or articles, were located in this way and were used to build up files on individual concepts. The information finally present in each file was then used to establish the individual concept entries. Further information has been obtained for each new edition, permitting the creation of new entries or the improvement of existing descriptions.

2. Sources, precedents and parallels
There are relatively few efforts to produce systematic descriptions either of human development concepts or of modes of awareness. Those that exist tend to be quite specific in focus or oriented to a particular school of thought or tradition. There are of course a large number of books, connected with each tradition or discipline, which contain such information to different degrees. The sources used, and related works, are cited in the bibliography (Section HY).

3. Translations
(a) Non-European languages: Since the intent has been to bring together material from a wide range of cultures, there is necessarily a major difficulty associated with translation. The obvious difficulties of technical translation between European languages are compounded when using translations from non-European languages. These have been confronted by many authors who have made available materials, especially from Oriental traditions. No attempt has been made to use source material not already translated into English.

(b) Subject matter: There is however a further difficulty in the translation process which is specifically due to the subject matter. Commentators frequently stress the clarity of the original text in explaining subtleties which do not readily lend themselves to expression in English. This is particularly the case with Sanskrit and Pali. With regard to Pali, a language reserved for the Buddha's teaching, Bhikkhu Nanamoli, translator of the *Visuddhimagga* (*The Path of Purification*, 1979) by Bhadantacariya Buddhaghosa, states: "*This fact, coupled with the richness and integrity of the subject itself, gives it a singular limpidness and depth in its early form, as in a string quartet or the clear ocean, which attains in the style of the Suttas to an exquisite and unrivalled beauty unreflectable by any rendering. Traces seem to linger even in the intricate formalism preferred by the commentators. This translation presents many formidable problems, mainly either epistemological and psychological, or else linguistic, they relate either to what ideas and things are being discussed, or else to the manipulation of dictionary meanings of words used in discussion.*"

(c) Subtle experiences: The same translator comments on the special difficulty of technical translation of mental experience: "*Again even such generally recognized private experiences as those referred to by the words 'consciousness' or 'pain' seem too obvious to introspection for uncertainty to arise (communication to fail), if they are given variant symbols. Here the English translator can forsake the Pali allotment of synonyms and indulge a liking for 'elegant variation', if he has it, without fear of muddle. But mind is fluid, as it were...and its analysis needs a different and strict treatment. In the Suttas and still more in the Abhidhamma, charting by analysis and definition of pin-pointed mental states is carried far into unfamiliar waters. It was already recognized then that this is no more a solid landscape of 'things' to be pointed to when variation has resulted in vagueness. As an instance of disregard of this fact: a great scholar with impeccable historical and philological judgement... has in a single work rendered the cattaro satipatthana (here represented by 'Four Foundations of Mindfulness') by 'Four Inceptions of Deliberation', 'Fourfold Setting Up of Mindfulness', 'Fourfold Setting Up of Starting', 'Four Applications of Mindfulness', and other variants. The foreword to the Dictionary of the Pali Text Society observes 'No one needs now to use the one English word "desire" as a translation of sixteen distinct Pali words, no one of which means precisely desire"*

(d) Patterns of associations: He continues: "*So far only the difficulty of isolating, symbolizing and describing individual mental states has been touched on. But here the whole mental structure with its temporal-dynamic process is dealt with too. Identified mental as well as material states (none of which can arise independently) must be recognizable with their associations when encountered in new circumstances: for here arises the central question of thought-association and its manipulation. That is tacitly recognized in the Pali. If disregarded in the English rendering the tenuous structure with its inferences and negations -- the flexible pattern of thought-associations -- can no longer be communicated or followed, because the pattern of speech no longer reflects it, and whatever may be communicated is only fragmentary and perhaps deceptive. Renderings of words have to be distinguished, too, from renderings of words used to explain those words....One is handling instead of pictures of isolated ideas, or even groups of ideas, a whole coherent chart system.*"

Method: descriptions

1. Difficulties
Specific difficulties encountered in drafting descriptions include:

(a) Absence of clear descriptions: The absence of any clear-cut description or definition of the concept or mode in question. Where necessary minimal descriptions have been included pending the discovery of appropriate material.

(b) Confusing use of terms: Frequently the same terms are used with different meanings, or different terms are used with what appear to be the same meanings. Considerable confusion was noted in the use of some terms, particularly those relating to self-realization, and new levels of consciousness.

(c) Acceptance of the futility of verbal descriptions: Indications of the impossibility, futility or inappropriateness of any verbal definition of subjective experience. This view seems to be held by some people in every tradition, although others in the same tradition are usually to be found who have endeavoured to use words to describe and clarify the experiences.

(d) Transcendence of subject/object relationship: Beyond the problem of descriptions of subjective experience is that of using words to capture experiences which are explicitly recognized as transcending the subject/object relationship. This point is made in a discussion of Sufi insights: "*Though reasoning has not been put out of court, it has only an ancillary function to perform....language in its literal functioning is an effective instrument for the communication of that only which thought can think as an other. The writings of the great Sufis are no doubt linguistic expressions of truths concerning the mystery of Being, but the language employed therein is one intended to function in the symbolic key. Predicative modes of speech, necessarily drawing a dividing line between the subject and the predicate have to be understood as having a unitive import, as referring unmistakeably to the one in which all discursivity stands submerged and all differences are cancelled and transcended.*" (Bhatnagar, Dimensions of Classicial Sufi Thought, 1984).

(e) Secretiveness: In some cases there has been a tradition of a fairly high degree of secretiveness about the methods and experiences cultivated. The claim is made that the essential insights can only be transmitted by word of mouth and can only be understood after following a particular form of training.

(f) Multiplicty of perspectives within a tradition: There are so many schools of thought (even within the same tradition), with different methods and with different objectives, that quite often the interrelationships between their concerns are far from being evident. Stages of human development, and the modes of awareness experienced, overlap and merge into one another and may well be experienced differently by different people. Within a tradition, each spiritual leader may articulate a different pattern of modes of awareness and a different sequence through which they may be experienced. The linearity of such sequence may also be questioned.

(g) Exaggeration and disparagement: Some authors express themselves with what appears to be an unjustified degree of confidence, especially when it comes to disparaging the endeavours of others and extolling their own insight or that of their own tradition.

(h) Category shuffling: Whilst it is relatively easy for commentators to present categories of experience and schemes organizing those categories (or structuring them into developmental sequences reflecting greater levels of insight), the texts available tend to distinguish the categories rather than the experiences to which they point. There is therefore a constant danger of shuffling words about "dead" categories which cannot be effectively related to lived experience.

(i) External symptoms vs. subjective experience: Descriptions of modes of awareness under the influence of specific intoxicants, or experienced in different mental "disorders", are particularly difficult to locate. Whereas the symptoms of those conditions (as identified by outside observers) are readily available, descriptions of the experience are not.

(j) Personalization: There are many books from different traditions describing unusual personal experiences. Efforts to organize that material as a description of a class of experience that may be to some degree repeated by others are much rarer. Much available material is highly personalized and therefore relatively useless for this exercise.

(k) Varying quality of sources: The available material may reflect very different levels of scholarship and insight. Authors may be careless or misinformed about questions of fact. Authors may be extremely valuable for the clarity they bring to the organization of material on which they are commenting, but at the same time may raise questions concerning their level of experiential insight into that material. (For example, one valuable author explained the experience of nirvana as equivalent to that of a cataleptic trance, which may indeed be the case for an outside observer).

(l) Deceptive degrees of order: Systems - in particular Sufi and Tibetan Buddhist - may appear deceptively straight forward and well ordered until an effort is made to actually try to itemize the components. Then, maybe because the source of information is more muddled (or more subtle) than it appears, the system may seem to break down. There may be some difference in order. New states appear in explanations not included in the list to which the explanation appears to refer and so on.

(m) Personal experience as a prerequiste: Some authors are careful to point out that a mode of awareness cannot be effectively described, let alone understood, by anyone that has not experienced it. This raises questions about the status of many of the authors and commentators, and about the capacity of the editors of this section in dealing with such material.

2. Descriptions: approach
The emphasis here has been to present apparently distinct concepts as separate entries, especially those from different schools and traditions, even when it seemed highly probable that reference was being made to the identical experience. Where possible, duplication and overlap have been directly addressed, often by use of cross-references. Many obvious difficulties have not been resolved in the information presented here. Such difficulties are often the subject of bitter debate. It is hoped that this presentation will encourage further efforts towards clarification, if that proves appropriate.

A number of entries may be considered to be duplicates, because the names given to the concepts are held to be synonyms. Some entries of this kind have in fact been combined, but others have been kept separate where the different words tend to be used in different contexts, even though they may be considered (by some) to mean the same thing.

A number of concepts have been included because on first sight the terms used to label them appear to suggest some notion of human development when in fact this sense is relatively weak. Very short descriptions are then given to make this clear by contrast with the other entries.

Although all statements used in building up descriptions are very closely based on existing published documents, no explicit link is established between statement and source documents. This was avoided because the editorial process of selection and restructuring of texts from different sources may have unintentionally distorted the meanings in the original contexts (particularly when the original statements did not constitute clear descriptions). Any such misinterpretation which comes to light will be corrected in future editions.

In the text of the description, especially in the case of modes of awareness, an effort has been made to strike a compromise between using the style of the source material and harmonizing the language used with that of other descriptions. It is often difficult to avoid the particular flavour of a source document from influencing the final description, even though it reflects the perspective of a single individual. Diffuse explanations have been avoided and effusive claims in highly poetic language have been toned done. Where there was a choice of source material, that offering more precise or insightful descriptions has been used.

In the case of the modes of awareness, it proved convenient to include different types of entries:

(a) Modes of awareness for which the experience of the mode is described;
(b) Systems of modes of awareness through which specific modes are grouped;
(c) Methods for reaching modes of awareness;
(d) Classes, forms or types of consciousness;
(e) Definitions of terms used to point to consciousness.

The descriptive paragraphs may then include several types of information:

(a) Physical, psychological or spiritual conditions under which the mode is possible;
(b) Description of the mode in psychological, mythological or metaphorical terms;
(c) Results or consequences of experiencing a mode;
(d) Who or what tradition describes the mode;
(e) The physical, psychological or rational structure of experiencing the mode.

Given the lack of information, this responds to some degree to the very specific challenge of some modes which transcend any predicative description. In the particular case of systems of modes, there is also an implication that there is a mode corresponding to awareness of the systemic pattern.

3. Descriptions: abbreviations, spelling and transliteration

(a) Identification of traditions: Where possible, the tradition with which an item is associated is indicated in brackets following the title, and entries are also grouped in the index by tradition. Consistency has been attempted but has not necessarily been achieved. It is often difficult to distinguish between Hindu and Buddhist concepts, and within Buddhist by Zen, Tibetan, Pali, etc. Some errors may have crept in. The term "Japanese" has been used, for example, where the derivation is geographically clear but the tradition unclear. "Yoga" is used as a general to cover yogic practice independent of tradition. Occasionally the names of more than one tradition are included, separated by ",". Alternative titles in a description may be followed by their particular traditions.

This inclusion of the tradition can only be indicative. Many sources are implicitly Christian or Judeo-Christian even where their scope seems universal, as is perhaps to be expected where source material is in the English language. Traditions thus indicated include: Astrology, Buddhism, Christianity, Esotericism, Hinduism, ICA (that is, Institute of Cultural Affairs), Islam, Jainism, Japanese, Judaism, Jung, Leela, Pali, Physical sciences, Psychism, Psychosynthesis, Scientology, Systematics, Sufism, Taoism, Tibetan, Yoga, Zen, Zoroastrianism.

Descriptions labelled Tibetan are all labelled Buddhism as well. Where there is a title in the Tibetan language, then the term "Tibetan" follows that title, otherwise the first title is followed by the terms "Buddhism, Tibetan". A similar approach has been used for concepts from the Pali.

(b) Spelling: There are of course many variations in the source material. British English spelling has been preferred, in common with all UIA publications. Transliteration has posed a number of questions. Different sources use different conventions for transliterating from Pali, Sanskrit, Arabic and Chinese, for example. Many of these sources use conventions requiring accents and modifications to letters not generally available in the computer system used for processing the data. The approach has been to ignore such accents and modifications but otherwise to use source conventions where possible, listing other spellings as alternative titles where possible or practicable. An exception is the letter "r" in Sanskrit which has in some cases been replaced by "ri". The wide variations in conventions for transliteration may have led to unwitting duplication. The absence of accents may lead to apparent duplication, as with the ninety-nine names of Allah when Al-Majid appears twice.

4. Cross-references

Considerable effort has been put into linking related concepts within the section. A number of hierarchical relationships were established linking general approaches and practices with their more specific offshoots. In the modes of awareness section sequences were cross-referenced, indicating previous and subsequent stages of awareness in any recognized sequence or pattern of stages. For example, possible moves in the traditional Tibetan *Ascension Stages Game* are cross-referenced, as are relationships in the more recently elaborated pattern of the *Other World in the Midst of This World*. Where a number of possible groupings/sequences are conceivable these are indicated.

Method: classification of modes of awareness

1. Classification in the literature
The state of information on modes of awareness, and the attitude towards them, is such that no comprehensive classification of them exists. Basic texts such as the *Varieties of Religious Experience* (William James, 1961), *The Multiple States of Being* (René Guénon, 1984), *Altered States of Consciousness* (Tart, 1971), or the *Varieties of Psychedelic Experience* (R E L Masters and Jean Houston, 1967), do not provide extensive classifications as might be expected.

There exist various partial classifications and many dealing with particular ranges of modes. Although none has been used in grouping this material (except through the cross-references), the following are examples of some of special interest:

(a) The *Diagnostic and Statistical Manual of Mental Disorders* (American Psychiatric Association, 1987) provides detailed guidelines for classifying patients with "mental disorders". It is designed as an extension of the World Health Organization's *International Classification of Diseases*. The system of classification is excellent. As the name indicates however, it describes externally visible symptoms with relatively little information on the nature of the patient's experience. It is not designed to handle the subtleties of expanded modes of awareness, other than by treating them as various forms of delusion. To the extent that there are modes corresponding to a heightened state of mental health and well-being, there is no attempt to indicate their nature. Nor is there any attempt to describe the subjective experience associated with various forms of substance abuse.

(b) As a result of the survey on religious experience carried out in the UK on the initiative of Alistair Hardy by the Religious Experience Research Unit in Oxford, a quite detailed classification of types of religious experience has been produced. This is broad in scope, not tied to a particular tradition, but not especially useful in distinguishing qualitatively between subtle modes of awareness.

(c) Many authors have remarked on the finely detailed categorization of subjective experience in Buddhist texts. The most accessible examples are those of the *Visuddhimagga* (Bhadantacariya Buddhagosa, 1976), especially the translation by Bhikku Nanamoli which includes a tabular presentation. An even more detailed tabular presentation of Buddhist categories of thought is given by Alexander Piatigorsky (1984).

(d) The many Sufi authors make distinctions between subtle states, as with a number of Christian mystics. These tend to be much more individualistic than Buddhist authors. Their classifications are therefore less interesting. Of special interest however, in the case of the Sufi tradition, is the distinction made between mystic stations *** ??? (maqamat), as stable conditions of awareness marking particular spiritual attainment, and temporary feelings of the heart (ahwal) which may or may not be experienced at any particular station.

(e) It is rare for spiritual gurus to accord attention to insights other than their own. One remarkable exception is an extensively annotated bibliography by Da Free John (1989). Of special interest is the grouping of material from a wide range of traditions into seven levels of insight. It is appropriate to note that Ken Wilbur, a much cited author on altered states of awareness, specifically endorsed Da Free John's endeavours.

(f) In *A History of Christian Spirituality* (Holmes, 1980) a valuable categorization of the variety of ways of experiencing God is presented in terms of two bipolar scales: a kataphatic/apophatic scale and a speculative/affective scale. The first describes techniques of spiritual growth, the kataphatic involving the active use of the imagination and imagery as a tool for meditation, whilst at the other extreme the apophatic method is an "emptying" technique that rejects all forms as adequate media through which God may be understood. The second scale has at one extreme the speculative approaches emphasizing the illumination of the mind (or intellect), while at the other extreme the affective approaches emphasize illumination of the heart (or emotions). In arguing that balanced spirituality is an appropriate mix of all four approaches, the author is able to use them to categorize approaches which are out of balance. Examples given are: rationalism (over-emphasis of speculative and kataphatic), pietism (over-emphasis of affective and kataphatic), quietism (over-emphasis of affective and apophatic), and encratism (over-emphasis of apophatic and speculative).

(g) The *Encyclopedic Psychic Dictionary* (Bletzer) includes an extensive classification of "psychic skills" that incorporates many modes of awareness that would not normally be considered as psychic.

(h) One of the most interesting attempts to develop a general classification is **A Map of Mental States** by John H Clark (1983), with a foreword by cybernetician Gordon Pask. It is of a wider span than usual, including meditational, religious, depressive and euphoric conditions.

2. Classification in this Encyclopedia
In a crude attempt to indicate similar levels of awareness, the modes are classified on a scale "a" to "g". This is the small letter immediately following the reference number of a description. The basis for designating the code letter is a combination of: number of other entries cited in entry; number of narrower entries in a chain; number of broader entries in a chain; and a series of criteria as follows:

"a" God, mystical unity, unity, cosmos, nameless, beyond name.
"b" Theta clear, consciousness of map (awareness of system), tops of schemes, zodiac, named.
"c" Groups of individual states within map, in the map, set of gods, etc. Result of being on a path.
"d" Individual states within a map, significant trips, places, thought processes. May arises spontaneously, not necessarily on a path.
"e" ESP, psychic, clairvoyant, siddhis, positive brainwashing. Imposed states.
"f" Drugs, hypnotism, schizoid, mental derangements, hells, negative brainwashing. Imposed states.
"g" Physical, sensory, sleep, brain waves.

In fact, each state may be seen as appearing in a matrix which sets out metaphors for the spread of consciousness. The individual may be seen as following paths across this matrix, development being an upward spiral. Designation of each state in the matrix is a complex procedure which may involve distinguishing the state along dimensions such the following:

- 10,000 things to Nameless (through single point, line, triangle)
- Imposed - Imposing
- Physical expression, senses - Hyperspace
- Many - One
- Locales - Whole map
- Beginner - Maturity
- Bound to matter - Liberation, freedom
- Narrow spectrum - Broad spectrum, 360 degree view
- Single factor - Multivariable
- Point - Spherical (3-dimensional) (through linear, non-linear).

Codes "a" to "g" may be thought of as columns on the matrix, with rows reduced to one square. However, one square can also represent a column (e.g. all the signs of the zodiac appear in one square, because they do not differ in kind, yet they actually represent a progression from 1 to 12).

Because some of the criteria on choice of code may mutually exclude others, there can only be a crude overall consistency. For example, an item may be very important in its own right and yet included as a minor member in some network or chain of states. Thus Nagual is coded "c", although it is narrower than Sleep which is coded "g".

Comments: assessment

1. Entries
The entries included cover a very wide range of approaches and insights, as was the original intention. It is to be expected that the inclusion of some of the entries should be queried. The question to be asked is when is it appropriate to exclude an insight into human development that can be judged as naive, misguided or even dangerously misleading. There is a case for using the procedure for descriptions of world problems in which a "counter-claim" paragraph can be inserted to present such judgements and clarifications. It should be stressed that inclusion of entries in no way constitutes an endorsement of the insights or practices described.

Given the method used, which limits the degree to which related entries are treated as duplicates, what significance should be attached to the inclusion of 2760 entries on supposedly distinct modes of human awareness ? A more rigid editorial policy could have been applied to group certain "related" entries into a single description.

More important however, the question remains as to how appropriate it is to maintain separate entries for modes of experience which are clearly similar even though they derive from different traditions. It is also clear that attempts at rationalizing the information further are liable to create as many problems as they resolve.

There is also the question as to whether the distinctions made are not to a high degree a consequence of the language and cultural tradition from which they derive (a point discussed in a following note). It is not clear to what extent the more subtle modes of awareness are conditioned by the mindset or discipline with which they are approached. The question of "how many" modes of awareness there "really" are remains open. It may indeed be a quite inappropriate question.

2. Challenge to comprehension
There is obviously a fundamental question concerning the significance, if any, that can be associated with many of the concepts and modes of awareness, especially by anyone from outside the tradition within which they emerge. Even within any such tradition, it is often claimed that much can only be understood after passing through particular stages in the process of human development.

The ability to produce some kind of description, whose words can be understood, is therefore no guarantee whatsoever that the reader is comprehending much of what is intended, especially for those modes of awareness which call for an experiential transcendence of conventional modes of comprehension.

In such circumstances entries can only serve to point vaguely in the direction of a domain of understanding, offering hints and allusions which may be less than helpful. This limitation is partially corrected by setting an entry in a context of cross-references to other entries, especially when these reflect a progression of modes of awareness from others which are more meaningful.

The many entries of eastern origin tend to be understood in the West as associated with the religious dimension of belief and revealed knowledge, and the same is true to a lesser extent of concepts arising from western religions. It is interesting to consider whether descriptions of such concepts can be meaningful to those who do not have belief in the religion from which they are derived, without some "reprocessing" by scientific disciplines to relate them to western concepts of human development that are largely independent of particular religions.

The challenge to comprehension is sharpened through awareness of the "pre/trans fallacy" as described by Ken Wilbur (1982). In any development of insight, growth will tend to proceed from stage pre-X, through stage X, to stage trans-X. Because both stage pre-X and stage trans-X are, in their own ways, non-X, they may be understood as similar, even identical, to the untutored eye. This is particularly the case with pre-personal and trans-personal, pre-rational or trans-rational, or pre-egoic and trans-egoic. According to Wilbur, once these two conceptually and developmentally distinct realms of experience are theoretically confused, there is a tendency either to elevate pre-personal events to trans-personal status or to reduce trans-personal events to pre-personal status.

3. Comparative evaluation
It is not the purpose of this section to provide any form of comparative evaluation of modes of awareness or forms of human development. The intent has been limited to pointing to the existence of modes distinguished in the literature and to indicate, where possible, the sequence of experiences through which it is alleged that they may be encountered.

In the light of more profound experiences, other modes may be held to be superficial and even a dangerous error. In particular, the many reactions in the West to the limitations of materialism and the suffocating constraints of finite existence tend to be governed by the Cartesian dualism which reinforced those perspectives. For many this reaction means, almost unconsciously, attraction to its opposite pole, namely the non-material, without there being any discrimination within that domain. Interest in psychic phenomena or "trips" of any kind, may tend to obscure the nature and rich complexity of the spiritual dimension.

The entries therefore reflect different degrees of delusion as much as they reflect different degrees of insight. Juxtaposing them in this way is an aid to orientation but without any attempt to recommend the more fruitful directions for any particular individual to explore.

This question is of great concern in some traditions: *"From the Sufi point of view, which has always distinguished clearly between the psychic and the spiritual, so many of those who claim to speak in the name of the Spirit today are really speaking in the name of the psyche, and are taking advantage of the thirst of modern man for something beyond the range of experiences that modern industrial civilization has made possible for him. It is precisely this confusion which lies at the heart of the profound disorder one observes in the religious field in the West today, and which enables elements that are as far removed as possible from the sacred to absorb the energies of men of good intention and to dissipate rather than to integrate their psychic forces"* (Nasr, 1965).

4. Challenge of spiritual development
The juxtaposition of entries on many different aspects of human development and modes of awareness suggests the possibility of a continuum of distinct emphases and insights. It is the tendency to give prominence to one form rather than another which is the source of many difficulties. The neglect of spiritual development is one such consequence, whether or not it is deliberate.

Morris Berman (1989) suggests that the personal discovery of interiority, and the emergence of inwardness in society, have always been recognized as a major threat to established institutions and ways of thinking. He cites the suppression of the Cathars and the deep distrust of ecstatic experience during the Enlightenment.

Other dangers are signalled by a Quaker author, Lorna Marsden (The Guardian), "At this moment, a reawakening to the essentiality of the spiritual within the human experience carries two dangers -- a possibly excessive reaction which might induce or revive extremes of superstition, or (which is indeed happening) a retreat from challenge into fundamentalism within both Christianity and other religions."

Berman draws attention to a fundamental distinction between the "ascent experience" (with which spiritual development is most frequently understood) and embodied or "horizontal" consciousness. Entrapment by such metaphors is discussed in a following note. He argues that: "The real goal of a spiritual tradition should not be ascent, but openness, vulnerability, and this does not require great experiences but, on the contrary, very ordinary ones. Charisma is easy; presence, self-remembering (Gurdjieff's term), is terribly difficult and is where the real work lies." He sees much current interest with "spiritual development" to be basically an effort to escape the body rather than to work through it.

This kind of concern is also expressed in relation to the superficial assumption that mental understanding of some of the conditions and insights described constitutes full realization of their truths. *"This illusion, which is the result of the separation between the mental activity of certain men and the rest of their being, and which is directly related to a lack of spiritual virtues, is a major hindrance in the application of sacred teachings of various traditions to the present needs of Western men....Such people mistake their vision of the mountain peak, theoria in its original sense, for actually being on top of the mountain. They therefore tend to belittle all the practical, moral and operative teachings of tradition as being below their level of concern. Most of all they mistake the emphasis on the attainment of spiritual virtues...for sentimentality, and faith...for 'common religion' belonging only to the exoteric level, forgetting the fact that the greatest saints and sages have spoken most of all of spiritual virtues"* (Nasr, 1965). Without spiritual poverty, for example, *"no spiritual attainment is possible, no matter how keen the intelligence may be"* (Nasr, 1965).

A related point is made by a Catholic author in describing the approach of Meister Eckhart: *"Real spiritual life is about loss of self, sacrifice, inner transformation and change."* In this context, "mysticism" is not concerned with "trips" or results of any tangible or obvious kind. *"Neither is it about what we call today 'expansion of consciousness'. It is not concerned with trying to induce abnormal states, either blissful or otherwise; it is concerned with trying to develop within ourselves a certain attitude which is healthy, realistic and life-giving, and will remain constant within us during all states of consciousness, normal or abnormal, pleasant or unpleasant...Rather than talking about 'expansion' of consciousness, we should talk of 'breaking through' consciousness; that is, finding within ourselves some inner centre of stability and unity which remains intact throughout the manifold fluctuations of conscious states."* (Smith, Cyprian, 1987).

Comments: official approaches to human development

1. Selective expansion of scope of human development

(a) Official practice: Ian Miles (1985), in a review of social indicators for development, notes: "*Almost all political leaders claim that human development provides the criterion against which economic, cultural, and political development are assessed, and the framework within which their strategies and policies are designed. But, in practice, the goal is frequently obscured and obviated. Economic development, most often, is taken to be the sine qua non of all development efforts, to be intelligible in purely economic terms, and to have human development as its automatic consequence... Many aspects of well-being and quality of life are placed in jeopardy by current development patterns.*"

In the case of UNICEF (*The Child in South Asia; issues in development as if children mattered*, 1988): "*Let us see what development of the human potential implies. It calls for the creation of conditions in which the physical, mental and social well-being of all people in the country is possible. For this to happen, appropriate economics is an intellectual input, leading to the allocation of physical and financial resources for human development, but it must be complemented not only by technology but also by social values or principles. There need be no conflict within this variety of economic and technological and social resources. The challenge of development is to derive the benefit of their mutually reinforcing interaction.*"

(b) Human-centred development: Human resource development is increasingly understood as referring to a process of human-centred development which seeks to enhance the full capacities and capabilities of human beings. In practice this is understood within the international community as expanding peoples choices or entitlements. Thus Amartya Sen argues that this process refers to "*the set of alternative commodity bundles that a person can command in society using the totality of rights and opportunities that he or she faces...On the basis of this entitlement, a person can acquire some capabilities, i.e. the ability to do this or that (e.g. be well nourished) and fail to acquire some other capabilities. The process of economic development can be seen as a process of expanding the capabilities of people....For most of humanity, about their only commodity a person has to sell is labour power, so that a the person's entitlements depend crucially on his or her ability to find a job, the wage rate for that job, and the prices of commodities that he or she wishes to buy.*" (Sen, Amartya 1984)

(c) Selective increase in scope: In an extremely useful review of the human development literature, Céline Sachs (1989) notes that there is no commonly accepted definition as to what in practice constitutes human resource development. Successive fads in development theory have in the past emphasized the pre-eminence of one factor over others, often in a highly dogmatic manner. The current trend is towards expanding progressively its scope.

The Administrator of UNDP argued in 1986 (DP/1986/10) that human development should be broadly defined, because of its intersectoral links, '*as the maximization of human potential as well as the promotion of its fullest utilization for economic and social progress.*' This definition "*requires that people be given the opportunity to apply the full range of their skills and abilities, fulfil their desires and ambitions, and make their contributions to the improvement of their lives and their society. Thus human resources development depends upon a political and social environment conducive to individual expression, self-fulfilment and the utilization of human potential...it thus depends on the very nature of society and its economic history and culture.*" Furthermore the "*lack of a broad operational definition militates against the formulation of strategies to deal holistically with the issues and the establishment of criteria for assessing progress in human terms.*"

In the UNDP's *Human Development Report* (1990), human development is defined as a process of enlarging people's choices. "*In principle, these choices can be infinite and change over time. But at all levels of development, the three essential ones are for people to lead a long and healthy life, to acquire knowledge and to have access to resources needed for a decent standard of living. If these essential choices are not available, many other opportunities remain inaccessible. But human development does not end there. Additional choices, highly valued by many people, range from political, economic and social freedom to opportunities for being creative and productive, and enjoying personal self-respect and guaranteed human rights.*"

In UNDP's terms: "*Human development has two sides: the formation of human capabilities (such as improved health, knowledge and skills) and the use people make of their acquired capabilities (for leisure, productive purposes or being active in cultural social and political affairs). If the scales of human development do not finely balance the two sides, considerable human frustration may result.*" The Report points out that "*it is sometimes suggested that income is a good proxy for all other human choices since access to income permits exercise of every other option.*" However it then points out that "*the simple truth is that there is no automatic link between income growth and human progress. The main preoccupation of development analysis should be how such a link can be created and reinforced.*"

(d) Inhuman development and "factory farming": Given the rising concern of the international community with "adjustment with a human face" and "human" development, it is difficult to avoid the impression that other forms of development practised over the years could best be contrasted as "inhuman" development.

It is worth exploring the extent to which the official approach to human development cannot be usefully compared to factory farming production of chickens, cattle or pigs. In both cases the emphasis is on providing the minimum in order to ensure that maximum productivity can be achieved. For unenlightened farmers this minimum ignored even the most basic needs. But more scientific investigation led to the insight that by improving the conditions of the animals, their productivity in fact improved. Enlightened farmers have thus increasingly focused on more living space, appropriate lighting, improved nutrition, with the adventurous even acknowledging that music ("culture") played to animals enhanced yields.

It is difficult to avoid the suspicion that official interest in human development emerges from a very similar mind-set. Recent enlightened extension of the scope of human development may therefore have been made with similar concerns in mind. Those responsible for the developmental inadequacies of the early development decades remain in place, even if they are constrained to present their arguments with a more enlightened vocabulary.

2. Misappropriation of terms

Even at a more mundane level, this section suggests the need for great caution in responding to the use of certain "human development" terms by international bodies and academic authorities. Valuable words, which in other contexts point to experiences of great richness and subtlety, are easily appropriated for use in signifying much more limited realities. This may be done simply through ignorance of the more profound meanings. It may also be done quite deliberately in order to profit from very positive connotations with the intention of persuading people that some quite unimaginative and insensitive programme is in fact responding to those dimensions.

Such appropriation is a basic technique of the advertising industry. Worse still, the packaging of the programme may lend itself, through deliberate ambiguity, to well-intentioned endorsements concerning the recognition of such subtler dimensions. But under the slightest political or budgetary pressure these may then be completely dropped in practice, whilst still being able to use the same words to imply the presence of such subtlety.

As noted earlier, extreme caution should be used in assessing the use of terms such as "human potential", "personal fulfilment", or "development of the personality" by bodies such as the World Health Organization, the International Labour Organisation, UNESCO, or their national equivalents. They may refer only to the most basic needs and there is little evidence to the contrary. The *Human Development Report* of the United Nations Development Programme is the latest example, despite the modest breakthrough that it signals from prevailing practice.

3. Suspect "human developers"

Following from the previous point, it is important to learn from the use of the term "development" by the international community over the past decades. It could be argued that "development programmes" attracted support precisely because of the ambiguity with which the term could be interpreted both by those attempting to respond to the interests and needs of the "developees" and by those "developers" anxious to profit (personally) from the economic opportunity represented by those receptive to "development". This ambiguity has allowed any degradation of the environment (such as "clearing" the land of trees) to be defined as development even by intergovernmental organizations. It is indeed possible that a significant measure of support for bodies such as the United Nations Development Programme was assured precisely because it could be readily treated as a "United Nations Developers' Programme" rather than as a "United Nations Developees' Programme". Unfortunately achievements for developees over the past decades have been less than they were led to expect.

There is great risk that emerging enthusiasm for "human development" will prove to be counter-productive through reliance on an analogous ambiguity. Thus imposition of simplistic cultural and education models to replace "out-dated" traditional psycho-social structures and processes of great richness can easily be justified under the banner of "human development". It may take some decades to recognize that irreplaceable cultural "rain forests" are destroyed by this process, whether through good intentions or not. It would also be most regrettable if new-found enthusiasm for "human development" were to be accompanied by a "human developmentalism" equivalent to the "developmentalism" of the economic preoccupations of the past decade (George Aseniero, *A Reflection on Developmentalism*, 1985).

In exploring the richer, subtler dimensions of human development, there is therefore a need to be extremely wary of "human developers", whether official or private. As with the geopolitical parallel, those who have pioneered these domains reflect both human strengths and weaknesses. Perhaps it is in this sense that the reprehensible activities of a number of recently publicised cults should be viewed.

4. Denial of the experiential

As an example of the peculiar bias of the international community against the insights of psychoanalysis, the response to the work of C G Jung merits investigation. It might be assumed that the many dimensions of his pioneering work would in some way assist in an improved understanding of human development by international organizations. In a key publication sponsored and vetted by the United Nations University on "*Human Development in its Social Context*", the only reference to Jung is the phrase "*...the great dreamer who flirted with Nazism...*" (p. 148). Given the importance attached to Jungian insights by many concerned with human development, this is further indication of how out-of-touch intergovernmental organizations are with non-establishment perspectives on human development. It is of course ironic that the book should have been prepared whilst Kurt Waldheim was Secretary-General of the United Nations, itself headquartered in a city where the elites are renowned for their dependence on psychoanalysis.

At a time when there is increasing recognition of the importance of spiritual values, it is important to recognize that the United Nations, like all intergovernmental bodies, is unable to accord any recognition to this dimension. Such bodies have essentially secular preoccupations and are only equipped to deal with problems from a perspective which ignores subjective experiences and spiritual dimensions. This also implies that they are handicapped in being unable to benefit directly from integrative insights arising from spiritual development or any transcendent insight. Unfortunately their actions reinforce attempts to ensure that the letter triumphs over the spirit. Recognition of human potential is limited to the quantifiable, with no capacity to acknowledge the patterns of insight from what cultural leaders have regarded as the heights of human achievement.

The way in which experiential information is handled, or denied, in society suggests that fruitful insights might be gained by contrasting explicit and implicit responses. It is obviously in the interests of institutions to accept the existence of only those forms of experience which can be **explicitly** articulated in their mandates and programmes, and only then when they are under pressure to do so. Other forms of experience may be recognized and discussed informally, or they may simply not be acknowledged at all. Such experience can be considered as associated with an **implicit** dimension.

The contrast between the two situations of the previous point can be illustrated by the way established institutions and academic disciplines respond to interpersonal love, sexual ecstasy, induced states (through alcohol or drugs), and spiritual insight. For such bodies these terms are void of meaning, suggesting only varieties of delusion to which individuals may be subject (possibly describable in terms of pulse rate and other physiological factors). This may be contrasted with the importance attached to one or more of these by the responsible individuals in those bodies.

Attraction to the implicit dimension through alcohol and sex is a major factor at intergovernmental and academic meetings and has been used to manipulate their outcomes on many occasions. Most of the participants are liable to have been in love and to attach much importance to the love they bear for their family. Some at least may have a deep spiritual commitment and may consider daily prayer or meditation a vital dimension of their lives.

Within the Japanese culture, special terms are used to contrast such distinct realities. The "tatemae" of the situation, the explicitly stated reality, is that human development can only meaningfully be concerned with observable behaviour, as typified in this case by consumption of alcohol, copulation or religious ceremonial. But the "honne", the unspoken reality of the situation, is that without the vital experiences (of which such behaviour is a pale misrepresentation) individuals find "human development" to be a sterile concept. "*Man cannot live by bread alone*". But appearances ("tatemae") can be used as a more or less polite way of discouraging discussion of, or even obscuring, the complex richness of the underlying experiential reality ("honne"). When population and drug programmes fail to recognize this dimension, their abysmal failure becomes comprehensible. It also augurs badly for human development programmes conceived from a similar mindset.

Comments: vital relevance of subtler human development

1. Drug abuse as a reaction to sterile human development
The anthropologist Edward T Hall (1976) states that *"Western man has created chaos by denying that part of his self that integrates while enshrining the parts that fragment experience"*. This is echoed with other words by the poet Kathleen Raine: *"It may be that modern man's sense of chaos comes in part from his loss of that pattern of which his necessarily fragmentary individual life is a part."*

Given the magnitude of the drug problem, it is appropriate to ask whether drugs do not offer, conceptually and experientially, precisely what classical conceptual and religious approaches have been unwilling to supply, namely a means of "dancing" between systems of perceptual categories. Drugs can be perceived as offering an artificially induced sense of integration.

It may be argued that the attraction of drugs arises as a result of the sterile concepts of human development permeating society in this period. The dominant concepts of human development are essentially boring and unenticing to the imagination. They offer few challenging opportunities for self-discovery. Indeed self-discovery is not considered to be meaningful.

Human development as currently conceived fails to reflect either the richness of humanity's scientific and spiritual achievements or the richness of human beings. Such sterile concepts may constitute the major obstacle to more creative approaches to alienation and drug addiction.

In the history of human development, static conceptual frameworks and drug addiction may prove to be unfruitful complementary responses to the fundamental challenge identified by David Bohm (1980): *"How are we to think coherently of a single, unbroken, flowing actuality of existence as a whole, containing both thought (consciousness) and external reality as we experience it?"*

2. Human development and solidarity with the disadvantaged
Major social projects are not a characteristic of modern society. The challenges of development are expressed in officially defined campaigns (Freedom from Hunger), goals (Health for all by the Year 2000) and decades (especially the International Development Decades). International nongovernmental organizations collaborate in such initiatives and have many of their own.

But recent years have seen increasing concern at the fall off in response to such initiatives. The terms "aid fatigue", "compassion fatigue" and "professional burnout" (in the caring professions) have been coined.

When human and social development is defined almost solely in terms of basic needs, the question may be asked why people should be expected to feel any sense of solidarity with the disadvantaged. The ways in which "solidarity", and sensitivity to the needs of others, emerge are not of concern in official approaches to development. There has been an attempt to define development in terms of enlightened self-interest. But the scope of that self-interest, and the nature of that enlightenment, are not subject to inquiry.

Any sense of solidarity only emerges following much subtler developmental experiences, especially when the degree of common interest is limited. It is ironic that it is perhaps precisely those forms of human development which receive little recognition that appear to be the source of insight and inspiration which encourages response to the basic needs of distant peoples. This may perhaps be seen in one form in the contribution of religious organizations to development.

3. Human development through cults and fundamentalism
Fundamentalist movements of traditional religions have acquired prominence in recent decades, together with new sects and cults, often with what appear to be bizarre practices. Despite the supposed attractions of the modern way of life, these movements have proved to offer counter-attractions which have worried many.

It may be argued that such movements provide a framework which offers a richer and more existentially challenging approach to human development than the pale offering emanating from established perspectives.

What is it in the academically legitimated approaches to human development that is so unattractive in comparison? When all the arguments concerning the rational seriousness of these approaches have been explored, they remain unattractive in comparison.

It is the heightened degree of commitment of other levels of an individuals being that is so important. In many ways they offer the opportunity to participate actively in a "sacred dream". With the increasing complexity of modern society, the dimensions addressed by these movements acquire greater significance.

It may be argued that unless official human development responds to these dimensions, it will be constantly disrupted and overtaken by such movements and others which will be created where conditions permit. Failure to address these dimensions will also result in the emergence of more cults of a, possibly even more, bizarre nature.

4. Human development, dependability and unemployment
With the increasing complexity of society, individual responsibility and accountability become diluted to the point of non-existence. Employers at all levels are confronted with employees that "do not care" about the way in which any task is performed. This is matched by an equivalent degree of irresponsibility on the part of employers in attempting to maximize their interests.

At the same time there is an increasing interest in "maturity" in personnel, however that is to be defined. Executive employment agencies are called upon to propose mature candidates and corporations provide training courses to develop personnel maturity.

In a period of rising unemployment, there is a strong possibility that employers will seek out those who evidence some form of human development beyond that defined by qualifications and experience. For some this will take the form of preferring candidates of a certain background offering some moral guarantee of dependability (as with Mormon or Bahai candidates).

5. Initiative, enterprise and human development
The issue of why certain countries "take-off" in developmental terms, and others do not, has been much debated. It is extraordinary to note the dynamism of people in some countries devastated by war and natural disaster. It is tragic to note the apathy of people in others, who may or may not have been subject to equivalent disadvantages.

The issue of initiative and enterprise, and of "getting one's act together" is also a question of human development. This has not been effectively addressed. It may be that the key to effective economic development lies in a shift of attitude rather than, as is usually assumed, a particular pattern of capital investment. This perception is increasingly acknowledged by corporate concern with changing the "culture" of a corporation. Where does initiative fit into prevailing understanding of human development?

6. Sustainable development and shift of life style
There is increasing recognition of the need for a "radical shift in life style" within the industrialized countries if sustainable development is to have any hope of becoming a reality. Like the older plea for "generating the political will for change", it is unclear how this is to be brought about within the framework of the present understanding of self-interested human development. This tends to assume that people will naturally be stimulated to an adequate degree of change by exposure to the negative effects of pollution, etc. Whilst some are indeed stimulated to change, it is doubtful whether the number is significant (or is likely to become so) or whether the degree of change can be said to correspond to the nature of the crisis.

It may well be that radical changes of the kind called for can only emerge from a richer form of human development which acknowledges the existential dimensions of human experience.

Comments: language and the reconstruction of reality

It is a matter of personal experience for many who work with different languages that things can be said in one language which cannot be adequately expressed in another, if at all. It is extraordinary that so little attention is paid to the significance of such observations. Their implications have been largely neglected by the academic world and by the world of policy-making bodies whose views they reinforce. The problem is left to the technicalities of interpreters if multi-lingual discourse is accepted. The problem is especially well-disguised by those who believe that everything can be adequately articulated in a mono-lingual environment, particularly one such as English. It is ironic that this point has been recently highlight by a journalist, Howard Rheingold in a simple book: *They Have a Word for It: a lighthearted lexicon of untranslateable words and phrases* (1988).

Rheingold suggests a thought experiment with the material he presents (which might also apply to entries in this section): *"See if you can recognize an experiential reverberation (yoin) or a state of emotional and spiritual revitalization (sabsung) when you run across one of them in your life. Think about ta and Wen and see if you don't understand life and thought in a different way. Mediate on mu and watch your mind struggling with thoughts that can't be captured in nets made of words."*

He cites Alan Watts (*The Way of Zen*, 1957): *"We have no difficulty in understanding that the word tree is a matter of convention. What is much less obvious is that convention also governs the delineation of the thing to which the word is assigned....Thus scientific convention decides whether an eel shall be a fish or a snake, and grammatical convention determines what experiences shall be called objects and what shall be called events or actions....a great number of Chinese words do duty for both nouns and verbs -- so that one who thinks in Chinese has little difficulty in seeing that objects are also events, that our world is a collection of processes rather than entities."*

Much of the material in this section derives from non-English sources. Much of it is about the limitations of language in expressing levels of significance beyond that which can be effectively captured by words. **The challenge may be that what we need to understand may only be expressible in a "language" that we do not know.**

The issue for academics is whether language is a thought-tool, or whether thought is a language tool. Do words merely provide the vehicles through which internal perceptions are externalized and shared, or are they templates that determine how and what it is possible to think? This question continues to be a matter of debate amongst linguists, anthropologists and psychologists. As Rheningold notes, most English-speaking people equate "thinking" with "state of mind". Yet there are other languages in which there may even be hundreds of words for states of mind that do not include "thoughts" as defined in English.

The limits of current epistemological frameworks in understanding mysticism have been summarized by Donald Rothberg (*Understanding Mysticism: transpersonal theroy and the limits of contemporary epistemological frameworks*, 1989). He draws attention to the constructivist perspective reviewed by Steven Katz (1978, 1983) which maintains that mystical experience is, at least in part, constructed on the basis of a particular background context. *"In this way, many constructivists challenge two of the central claims whose validity seems vital to the very possibility of transpersonal approaches: (1) the claim that mystical experiences are cross-culturally identical or highly similar, and (2) the related claim that mystics in important ways transcend their own conceptual framework, as well as conditioned modes of knowing and being in general."*

According to Rothberg, for Katz and his colleagues: *"the experience of brahman is always a Hindu experience, the experience of nirvana always Buddhist, and both are fundamentally different from the Christian experience of God or the Jewish Kabbalist's experience of the higher levels of reality (the Sefiroth)."* From this perspective there are no pure unmediated experiences. Every moment of experience is "constructed", in an extremely complex way, by the elements that determine its context. These may be: guiding ideological, perceptual, emotional, somatic assumptions and structures; core ideas and texts of the tradition; social, economic and political relations; activities and practices of the community; the individual's memories, preoccupations, plans, and the like.

Rothberg's response is to point out that *"it has been precisely the aim and claim of the most prominent philosophical and mystical projects in the dominant world traditions of the last twenty-fice hundred years to come to a direct (ie unmediated) knowledge of what is ultimately real...The metaphor is not usually that of eliminating that which "stands between" or "mediates" knower and known, but the main traditional metaphors (eg ascent to the real, foundation, representation) all suggest a "pure" and "direct" knowledge in which error, uncertainty, and mere opinion are eliminated."*

Rothberg argues that constructivists implicitly prejudge as invalid some mystical claims, thus forfeiting their pretended claim to "objectivity" and neutrality. Furthelore, they do not adequately account for the ways in which there seem to be partial and sometimes full "deconstruction" of the structures of ordinary experience, including beliefs and perceptions. Finally he argues that they remain incapable of thematicizing modes of knowledge rooted in such deconstruction. Deconstruction may be understood as an activity in which the ordinary structures of experiencing are suspended. Rothberg argues that this is a simplifcation and that there seems to be another type in which the "constructions" of ordinary experience are recognized as such, even if they still remain operative. What is then deconstructed are not the actual structures, but the implicit assumption that there is no construction.

The important point ignored by academic approaches to deconstruction is the degree to which many of the approaches reviewed in this section involve techniques for addressing and transcending the epistemological frameworks by which an individual may be trapped. Clearly these approaches are described in particular languages, symbols and contexts. These may indeed appear strange to the western eye. The fact that such structures may influence or facilitate particular forms of experience is a secondary consideration. Non-anglophone diplomats may choose to use English because of its advanatages in certain types of negotiation, but this does not mean that the results of such negotiation are simply an artifact of English.

Rothberg calls for a shift in the assumptions of contemporary epistemology whi might involve retention of many constructivist emphases on context, without the extreme relativism often accompanying constructivism and without the dogmatism which he regrets. *"These new frameworks would facilitate a fuller contemporary response to the concerns of how to balance commonalities and differences among human spiritual traditions and how to interpret, in the postmodern situation, such traditions as contributing to the resolution of contemporary cultural problems. In our context, many of these theoretical questions appear charged with practical urgency."*

If many of the approaches identified in this section do represent carefully worked out ways of operating on habitual and other structures (as many would claim), it would be foolish to reinforce any disparaging attitude towards them. They would even be of interest if they only offered ways of modifying the structures associated with a particular culture. But they are especially interesting to the extent that they offer ways for individuals, and affinity groups, to construct and inhabit coherent alternative realities.

The issue is whether as individuals, or groups, we can find ways to construct a more powerful and appropriate language by which to order our experience in a life-enhancing manner. It is quite possible that the radical shift in lifestyles that is increasingly called will only become credible and viable when mediated through a language that needs to be constructed. Whether that language can be widely shared, or is private, or is an ecosystem of languages, remains to be explored.

Comments: entrapment by competing metaphors

There are a number of metaphors of human development and consciousness (Valle and von Eckartsberg, 1981) and various authors point to the dangers of being trapped by them. A related problem has also been explored by theologians (S McFague, *Metaphorical Theology*, 1982).

1. Technical metaphors
Stephen Batchelor (1989), a Buddhist author, argues that Buddhist understanding in the West has suffered from a superficial dependence on techniques at the expense of meditative attitude. Since a technique is the result of a technology, "*any spiritual path that speaks of a series of interconnected stages leading to awakening...has a technological aspect.*" Entrapped by this technical metaphor, "*preoccupation with technique gives rise to a technical attitude. It assumes that reality is reducible to a certain number of elements which each have a particular place and function. Once the location, duration and type of elements have been established, then it is possible to reorganize them into anothr order, or even to change them into something else. With faith in guidelines and instructions, we can then project an alternative reality and proceed to realize it through applying the corect techniques. A technical attitude is thus a calculative one.*"

2. Temporal metaphors
The temporal metaphor emphasizes progress, journey, and moving from one state to another. The descriptions may in fact be more about the movement than about the states themselves. This perspective is important in focusing on growth, maturation and human development and provides insights in responding to a person's needs at a particular stage in that journey. The implicit bias is moralistic with excessive concern about progression through stages.

3. Spatial metaphors
In contrast, the spatial metaphor emphasizes the state, content, and quality of experience. The importance of this metaphor is in the description of the breadth of the state and its place in relationship to other states in the whole realm of consciousness. The danger of this metaphor is the tendency to reinforce static ("state") experience, stagnation and complacency toward individual development. An interesting compromise is in metaphors of alternation. Here there is a constant shift between states, but any progress can better be conceived in terms of increasing complexification of that pattern, as well as detachment from it.

4. Topological metaphors
An interesting approach to these contrasting perspectives is provided by Alexander Piatigorsky (1984) in discussing the classification of modes of awareness from a Buddhist perspective. He argues that the set of such possible modes "*can be regarded (if taken as a whole) as a space where all possible (not only actual, that is) types of thought are present synchronously....So in trying to show the essential difference between the Buddhist theory of thought based on the principle of rise of thought and the theory of thought in European psychological tradition, I would emphasize that the latter commences its investigation of mental phenomena when they have already formed their sequential combinations with one another, assuming therewith the character and form of temporal causal processes....So, strictly speaking, each separate case (and/or type) of thought constitutes its own unique spatial configuration within which no temporal event is possible. Given that there can be no psychology whatsoever without temporal interpretation of the inner causality in an objectively observed process, we may go even further and assert that the Buddhist typology of thought, being par excellence a topological typology, cannot be psychologically analyzed.*" (pp. 150-151).

Piatigorsky goes on to suggest that this topological character seems to constitute a kind of periodic system of modes of awareness, analogous to that for chemical elements, where the possibility is always implied, that however many or few of them are possible, each element, when and if it appears, would invariably be positioned in its own place (pp. 167-168). There is much in his detailed study which could be used to further clarify the relationships between the entries in this section.

4. Vertical vs. Horizontal metaphors
As noted earlier, Morris Berman (1989) draws attention to a fundamental distinction between the vertical, "ascent experience" (by which spiritual development is most frequently understood) and embodied or "horizontal" consciousness. He sees much current interest in "spiritual development" to be basically an effort to escape the body rather than to work through it.

For Berman: "*The cognitive insight of the ascent experience is extremely powerful. Purely intellectual or analytical knowing is revealed as being hopelessly incomplete...The event is so numinous, so meaningful, that one's relationship to the world, and to oneself, will never be the same, and thus to talk of abandoning this kind of experience would seem to be a great mistake.*" But he then goes on to point out that such ecstatic dissolution of the self is "*just a bit too wonderful*". He argues that much of the "consciousness revolution" of the 1970s and the 1980s has amounted to little more than a flight from the body. "*If we have any hope of getting out of this trap, it will be because of clarity rather than charisma; for we are going to have to distinguish between an embodied holism -- one that is sensuous and situational -- and the cybernetic holism, or abstract 'process reality', that is being advanced by many New Age thinkers.*"

Citing Jacob Needleman, he argues that true enlightenment is to really know, really feel, one's ontological dilemma, one's somatic nature. Whereas the mystic seeks to ascend, to abandon the body, greater insight may be achieved through horizontal consciousness. In Buddhism he notes that the ecstatic ascent experience is even considered as an obstruction to enlightenment. "*The real goal of a spiritual tradition should not be ascent, but openness, vulnerability, and this does not require great experiences but, on the contrary, very ordinary ones. Charisma is easy; presence, self-remembering (Gurdjieff's term), is terribly difficult and is where the real work lies.*" As in the Tibetan saying that "*The highest art is the art of living an ordinary life in an extraordinary manner*".

Berman looks to the possibility of this being the modality of an entire society, with a shift away from ascent and toward bodily presence in the world. All of life is then sacred, not just "heaven". For him: "*a much deeper life lies beyond that of the ascent structure, which is finally about salvation, or redemption, for the ultimate heresy is not about redemption but...about redemption from redemption itself. It is to be able to live in life as it presents itself, not to search for a world beyond.*"

6. Competing metaphors
The fundamental problem of entrapment by a particular metaphor is that some alternative, competing metaphor has to be used to gain release. The release is then accompanied by such an upwelling of insight that there is a major danger that the metaphor catalyzing the release is perceived as so fundamental that it itself becomes a trap. Just as geese imprint on the first moving object they see on leaving the egg, here there is a form of "imprinting" on the reality disclosed by the metaphor of release.

Advocacy of a particular developmental metaphor, as in the case of Berman, or opposition to a metaphor, as in the case of Batchelor, reinforces the assumption that people all face the same developmental challenges and can benefit from the same metaphors in responding to them. It is more probable that each such metaphor has its place, and that it may be appropriate for a particular individual at a particular time. Ironically the greatest of insights may even come from finally understanding that which one's neighbour treats as a daily reality. This is not to deny the power of the argument for or against particular metaphors, when an individual is metaphorically entrapped. Indeed the resilient use of metaphors would seem to be what Berman advocates in his arguments for "reflexivity", discussed in a following note.

Comments: embodiment in patterns of alternation

1. Resonant exchange between opposing views
The current sterile debate, reinforced by the differences between western and eastern cultural traditions, as to whether the significance of an individual lies only in his individuality and its transformative development or only in his social context and its transformative development, can be viewed more creatively in the light of the arguments of Section KD. Unless it is to be assumed that some major schools of thought are totally misguided, each of these opposing views clearly offers valuable insights, but the transformative development of the human self-image results from the process of alternation between them.

The change of focus can perhaps be best illustrated by the possible reinterpretation of the "stimulus-response" image of man favoured by behaviourists. This focuses on the way in which a given stimulus gives rise to a given response (as well as on ways of conditioning the desired response). In a simplistic concept of organization, a leader may be conceived as providing key stimuli and ensuring appropriate responses. This asymmetrical approach was the original basis for government and corporate funding of research on the uses of the mass media.

In a symmetrical approach a stimulus from one individual gives rise to a response, which is in turn perceived as a stimulus to which the original stimulator in turn responds. The two parties can then continue alternating between the roles of stimulator and respondent in a resonant exchange in which each takes initiatives and is conditioned by responses. Whilst this is fairly obvious, the interesting question is how the resonant exchange may be "tuned" as a vehicle for the expression of more significant possibilities. Clearly the classic asymmetric approach is just an extreme example of "forced tuning" by one party in its own interest. Courtship behaviour can be an example of more symmetric resonance which is progressively tuned to levels of greater significance, if it is successful.

Of greater significance in a social context is the manner in which the individual engages in resonant exchange with each of the members of the groups in which he participates. Each exchange is necessarily different, but the question is how these exchanges interweave in a process of mutual entrainment to constitute the resonance pattern of the group. And how may such a resonance pattern be tuned in turn and how many different resonators can "fit" together into what sort of pattern?

2. Embodiment in patterns of alternation
In such a context the individual is as much a non-localized pattern of propagation through the resonance network as a locus of interference within that network. Each individual is partly encoded by all the people with which he is in contact - "we carry a bit of everyone within us". This approach not only suggests possibilities for interpretation of the individual in relation to others but also for the individual in relation to the sub-personalities and modes of awareness which constitute his psychic make-up. He is as much a resonance pattern between such sub-personalities (as modes of awareness) as identified with any one of them.

There has been much recent work on the biological cycles by which human beings are characterized. Time-budget analysis has demonstrated the variety of alternative activities in which humans involve themselves at different stages of development (Carlos Mallmann and Oscar Nudler, 1982). The arguments of Section KD and the interrelated modes of awareness in Section HM suggest that there is a case for exploring the nature of a human self-image based on alternation, whether between activities, roles or modes of awareness.

In this sense no one mode of awareness, however "spiritually developed", can carry, encode or embody as much significance as the pattern of alternation between the set of such modes. Developing that pattern enriches the quality of life. It is the erosion or destruction of that pattern which diminishes the quality of life, for both the individual and the group.

3. Transcending frozen learning cycles
The confusion arising from the plethora of approaches and concepts may well be due to the tendency to "freeze" this alternation and to "lock" obsessively onto particular phases of it. As stated by Ken Wilbur (1982, p.11) in introducing the Spectrum of Consciousness: "But, odd as it may sound, I have no quarrel with the particular state of our science of the soul, but only with the monopolization of the soul by that state. The thesis of this volume is, bluntly, that consciousness is pluridimensional, or apparently composed of many levels; that each major school of psychology, psychotherapy, and religion is addressing a different level; that these different schools are therefore not contradictory but complementary, each approach being more-or-less correct and valid when addressing its own level. In this fashion, a true synthesis of the major approaches to consciousness can be effected - a synthesis, not an eclecticism, that values equally the insights of Freud, Jung, Maslow, May, Berne, and other prominent psychologists, as well as the great spiritual sages from Buddha to Krishnamurti."

4. Ecosystems of societal learning
Validating the phases through which alternation takes place then places extreme phases in a new context. In the light of Paul MacLean's (1973) work on brain evolution, some phases may indeed be governed, for example, by the lower limbic brain corresponding to the "reptilian" phase of man's evolution. Political leaders, for example, are occasionally perceived as functioning primarily in this mode when grasping to retain power. But the point is not simply to condemn such phases and attempt to "rise above them"; they too have a role to play in the psychic ecology.

Although such attempts are also appropriate as constraining exercises, eliminating such phases completely would effectively destroy important behavioral pathways in the psycho-social ecosystem through which learning takes place. In the natural environment also it is not simply a question of eliminating "primitive" species as "pests", but rather of ensuring their appropriate function in the ecosystem. In this sense the alternation phases need to pass through all the "species" necessary to the healthy functioning of man's psychic ecosystem.

Seen in this light the widespread attempts to define some groups or modes of behaviour as "good", and others as "bad" or "misguided", do not help to move beyond the resulting dynamics. Human beings are much more richly textured than such simplistic categories imply - as any fictional literature or drama shows. Whilst labelling some as "guilty" and others, especially oneself, as "innocent" is a necessary behavioral pattern under certain local conditions, it is also necessary to be able to operate in the opposite mode. If man cannot understand how he is part of the problem, he cannot understand the nature of the "answer" required to his condition. It is even more desirable to recognize that it is not a question of being either guilty or innocent, but rather of being guilty and innocent as a responsible participant in the current global condition of society. In this sense being human is the ability to live creatively with this paradox.

5. Dance of the categories
If nothing else, human beings are only partially defined by the static categories in each of the many conceptual "languages" which attempt such definition. The essence of being human is uncontained by the patchwork aggregate of these definitions - it is a "quality without a name".

It can be more appropriately "defined", especially as a self-image by the person concerned, by the dynamics of alternation between the roles, categories, activities and modes of being by which people are usually characterized. A richer and more "global" understanding of being human lies in identification with the "dance" between these specific, "local" or temporary definitions.

The "dancer" is not limited by the specifics through which he expresses himself. Experientially he is more closely identified with the process of "dancing". Hence the production of books on the conceptual frontiers of physics with titles such as The Dancing Wu-Li

Masters (1979) and of others on new styles of corporate strategy with such titles as *When Giants Learn to Dance* (R M Kanter, 1989).

6. Experiencing the implicate order

The relationship between the individual's different attitudinal postures in the dance has perhaps been best clarified by David Bohm. Each of the series of conflicting images with which an individual identifies can be conceived as a lower-dimensional projection of a higher-dimensional actuality, which is their common ground, but which is of a nature beyond all of them, thus constituting a challenge to comprehension. In this higher-dimensional ground an implicate order prevails in which "what is" is movement, represented in thought as the co-presence of many phases of that order.

Any particular mode of awareness posture is ultimately misleading, although necessary as a well-defined vehicle of expression of the movement characteristic of the undefined totality of that higher order (1980, p.209- 210). The special merit of Bohm's presentation is that he demonstrates that, far from being an inaccessible mathematical abstraction, *"the experiencing of the implicate order is fundamentally much more immediate and direct than is that of the explicate order, which...requires a complex construction that has to be learned."* (1980, p.206). His work is leading to a reassessment of the hoary mind-body question by combining his concept of "holomovement" with that of the holographic paradigm (Ken Wilbur, 1982).

7. Reflexivity

Morris Berman (1989) queries the constant pressure towards better conditions to be found "elsewhere" or "otherwise" through "alternatives". He suggest that even the much sought "paradigm-shift" is part of the same pathology of escape. *"Paradigm-shift is still part of the salvation mentality, a patriarchal mind-set that tells the hero to persevere, find a new form of consciousness that will give him redemption."* It is this illusion of which it is necessary to be aware.

For Berman: *"Part of our goal, undoubtedly, is to learn what it means to live without paradigm, but I also sense a much more complex possibility, viz., developing a radical new code that is itself about coding, and is not merely a shift in coding. This is where reflexivity -- the awareness of coding as coding, or Gurdjieff's "self-remembering" on a cultural scale -- becomes so important."*

From this perspective, science, feminism, Buddhism, and the many other isms may be useful as tools under particular circumstances, but are dangerous as ideologies that an individual (or a group) is unable to set aside when appropriate. To clutch at such "transitional objects", regardless of what form they take, is to lose balance. He shares with Jacques Attali (1981) the view that real liberation is about resiliency, not about "truth".

For Berman: *"How things are held in the mind is infinitely more important than what is in the mind, including this statement itself"*. Some form of coding is of course always necessary for social and psychological life. The art of spiritual development is then the cultivation of awareness of constructing and using codes, and the having of that awareness as part of that code.

Reflexivity does not mean making everything conscious. It should include the recognition that the code, of which one is aware, *"is fed by sources that lend themselves only to indirect awareness"*. This inhibits the process of entrapment in particular worldviews. *"...one's worldview, in effect, becomes Mystery: there is some sort of larger process operating that we cannot directly apprehend, but that permeates our bodies and moves toward healing."*

This inherent dissatisfaction with any particular code, and continuing awareness of the ontological dilemma, accords with the role of doubt in Zen. The Zen tradition encourages the meditative attitude by an intensity of such questioning. *"Doubt or questioning is seen as an indispensable key to awakening. It is the vitality of the meditative attitude, the driving force which heightens the sense of the mysterious to the point where it unexpectedly reveals what until then had remained withdrawn and unsuspected."* (Stephen Batchelor, 1989).

8. Disciplined spontaneity

In effect it is not so much a question of the human self-image in the face of the undefined - certainty facing uncertainty. Nor is it only a question of "containing" the undefined by a configuration of responses. The challenge is to embody and express the undefined, as it is intuitively recognized in the appreciation of the vitality of human spontaneity. The direction of human development may then be seen to lie in the progressive embodiment (or "marriage") of more fundamental forms of the paradoxical relationship between discipline and spontaneity. The current social development crisis may be interpreted as the crucible in which human beings learn to perceive themselves in such terms. The attitude called for by these uncertain times is thus one of disciplined spontaneity or spontaneous discipline. This is not achieved by the present schizophrenic alternation between "discipline" and "spontaneity" which makes of each mode a shadowy evil to be combatted by the other.

Integrative Knowledge   K

Scope

A principal characteristic of the global problematique is its apparent complexity. This calls for a complex response interrelating many different intellectual resources and insights and involving sensitivity to very different kinds of constraint. Integrative approaches of this kind have proved inadequate or exceedingly difficult to implement in a society characterized by specialization and fragmentation. Following token interest in interdisciplinarity in its own right, recent years have seen an emphasis on a project-by-project pragmatic approach. This avoids the need for any form of conceptual framework transcending individual disciplines, but begs the question as to the relationship between such projects.

The purpose of this section is to assemble descriptions of the range of concepts or conceptual approaches which are, in some way, considered integrative and which are held by some international constituencies to provide the key to the organization of any effective strategic response to to the global problematique. Many of the words used to label these concepts are those which are considered indicators of the power of an advocated approach. They frequently appear in project proposals to trigger favourable response, whether or not any content can be given to them in practice. Words like "global", "integrative", "networking" and "systematic" are the magical "words- of-power" in the modern organizational world.

Sub-sections

The section contains 722 entries on integrative concepts. It is divided into three sub-sections:

- Section KC describes 632 integrative, interdisciplinary or unitary concepts in the broadest sense, namely it includes advocated methods of integrating awareness favoured by these who reject a purely conceptual approach.

- Section KD is made up of 70 entries commenting on recent efforts to interrelate incompatible conceptual approaches and the nature of the challenge that this implies.

- Section KP consists of 20 entries which constitute an exploration in ways of patterning incompatible perspectives.

Method

The procedures used in preparing this section are discussed in detail in Section KZ. The information in Section KC was derived in part from a period of collaboration with the Society for General Systems Research. That in the other two sections was derived in part from from papers prepared by the editors during their participation in the Goals, Processes and Indicators of Development project of the United Nations University, especially on problems of methodology.

Index

A keyword index to entries is provided in Section KX. THe keywords are also incorporated into the index for Volume 2 (Section X)

Bibliography

Bibliographical references, by author, are given in Section KY.

Overview

Detailed comments are provided in Section KZ. The section as a whole attempts to respond to the dramatic problem of how to interrelate vital conceptual insights which are essentially incommensurable and in practice often mutually antagonistic. A plurality of responses is not in itself an adequate response, especially since each fails to internalize the discontinuity, incompatiblity and disagreement which its existence as an alternative engenders. It is for this reason that the second part explores the possibility, implicit or explicit in recent studies, that a more appropriate answer might emerge from a patterned alternation between alternatives. This calls for a focus on the models of alternation by which the pattern and timing of cyclic transformations can be ordered between mutually opposed alternatives. It highlights the possibility that the kind of integrative approach required may not be fully describable within the language of any single conceptual framework, however sophisticated.

Context

The contents of this section may be considered as complementing the other sections in ways such as the following:

Metaphors and patterns: By the manner in which integrative knowledge is communicated and through the evolution of forms of communication to reflect new aspects of integration.

Human development: By the manner in which advances in the integration of knowledge are paralleled by integration of the individual and of society and require such integration in order to become meaningful.

World problems: By the importance of integrative knowldedge for comprehending the nature of the global problematique, and by the manner in which that problematique calls for new kinds of integrative knowledge.

Transformative approaches: By the integrative characteristics required of innovative techniques.

Human values: By the challenge of providing integrative frameworks to interrelate seemingly unrelated values and by the inherently integrative nature of value perspectives.

Integrative concepts KC

Rationale

The fragmentation of society is frequently deplored, as is the fragmentation of knowledge supposedly relevant to any appropriate response to the global problematique.

There continue to be calls for integrative, interdisciplinary or unified conceptual approaches to remedy this situation. Attempts to develop such approaches have themselves become fragmented, such that certain integrative insights are considered irrelevant, superficial, or misleading by those advancing other such insights.

There is no ongoing research into interdisciplinarity in its own right and the literature on it is dispersed under many unrelated headings (which library information systems make no attempt to cross-reference). And yet words like "global", "transdisciplinary", "networking" and "system" continue to emerge as the magical "words of power" triggering favourable response to project proposals addressing the global problematique. A minimum requirement at this time is therefore an indication of the range of integrative concepts from which some indication of their unique contributions can be deduced.

Contents

The section includes 632 entries briefly describing concepts which are considered integrative by some constituencies, whether or not they are accepted as significant or integrative by others.

Method

The information used was obtained from a wide range of specialized reference books as discussed in Section KZ.

Index

A keyword index to entries is provided in Section KX. The keywords are also incorporated into the index for Volume 2 (Section X)

Bibliography

Bibliographical references, by author, are given in Section KY.

Comment

Detailed comments are given in Section KZ.

Reservations

Because of the essentially integrative nature of the concepts, the distinction between an integrative and an unintegrative concept is unclear. It could be said that every concept is unclear. At one extreme, every concept may be considered as integrative; at the other, particular schools of thought would accept only a very limited number of their central concepts as integrative. Many of these concepts call for lengthy comment which is not possible in this framework; in such cases only highly abridged entries have been included.

Possible future improvements

In addition to the refinement of the existing entries and to the extension of the range (*eg* to include artificial intelligence concepts), the paradoxical problem of classifying interdisciplinary concepts could also be explored.

ENTRY CONTENT AND ORGANIZATION

Ordering of entries
Entries are in **numeric order**. Entry numbers have been **allocated randomly**; they have no significance other than as a permanent point of reference to facilitate indexing, cross- referencing, and updating between editions.

Index access to entries
The location of an entry in this sub-section may be determined from:
 - the **Volume Index** (Section X) on the basis of keywords in the name of the entry or its alternate names
 - the **Section Index** (Section KX) on the basis of keywords in the name of the entry or its alternate names

Structure of entries
Entries may be composed of the following descriptive elements:

(a) **Entry number** This number has **no significance**, except as a convenient method of identifying the entry (particularly for indexing purposes), of filing information on it, and as an identifier to which cross-references from other entries (possibly in other sections) may refer in this and future editions. The first letter of the entry number refers to the section of this volume in which the sub-section, denoted by the second letter, is located.

(b) **Entry name** This is printed in bold characters. It may be followed by alternate names when appropriate

(c) **Description** Brief description of the concept, possibly indicating alternative interpretations

Cross-referencing of entries
There are **no cross-references** between entries in this sub-section.

♦ **KC0001 Nature**
Natural
Description 1. Generally, the terrestrial world of perceptible phenomena.
2. The biosphere, including all organic life, soil, oceans and atmosphere.
3. The ecological system or network of which man is part.
4. The environment.
5. In Greek philosophy, Physics. In Greek myth, Pan.
6. One epistemological frame of reference in a relativistic universe. Hence, an order, e.g. the domains and characteristics of Euclidian space–time and the Newtonian physical laws which are operent in this frame or order.
7. Harmony (Pythagoras), regularity, familiarity and comprehensibility in man's experience of the natural world, as opposed to the supernatural: irregular, unfamiliar, incomprehensive experience, violating the day to day order of things. The sensible as opposed to supersensible ranges of phenomena and experience.
8. In moral theology what is ordained and right, fitting to the end intended; as opposed to unnatural, perverted, bestial, or fallen.
9. In Indian philosophies, dharma.

♦ **KC0002 Unity in diversity**
Description 1. A principle that aesthetic value or beauty in art depends on the fusion of various elements into an organic whole which produces a single impression.
2. An organized whole integrated through its basic ideas or common features which are scattered throughout a diversity of forms. For example, interlinked analogous principles are accepted in widely different and seemingly heterogeneous fields, and in otherwise divisive ideologies and political systems.

♦ **KC0003 Negentropy**
Negative entropy
Description 1. The tendency of an open system toward increasing order and complexity.
2. A comparative measure of average information i.e. the efficiency of a communication system in transmitting information. The efficiency measure of the transmission of information in a communications system is statistically akin to the measure of negentropy. The measure of noise in a communication system or transmitted message is akin to the statistical measure of entropy. Negative entropy in communications may also be described as the statistical measure of the probability of meaningful information, and in cybernetic terms, of the probability of order or control. All the most important principles of statistical mechanics can be deduced from the concepts of entropy in information–communication theory.
3. In the philosophy of science the theses of information as a fundamental universal property of matter.

♦ **KC0004 Global**
Description 1. Relating to, involving, including, adapted to, distributed over, or extending throughout the entire world.
2. The interrelated global system perceived as comprising a noosphere, a biosphere (or zoosphere or organosphere), and a lithosphere (or hylosphere), corresponding to the realms of human mental and cultural life, all physical life forms, and the rocks, soils, water, air and other elements of the natural inorganic environment.
3. The global ecosphere.
4. The international economic system of interdependencies.
5. Associated in mathematics with the totality of some system such as the symmetry, or the (constant or varying) deformation from symmetry, of any system as a whole. A global deformation affects all the parts of the system so that, when no disturbing factor is present, all the internal relations of a system are affected equally. The operation of global geometrical determinants (if sufficiently confirmed) expresses precisely the property underlying, or vaguely designated by: integrative systems, wholes, Gestalts, organicism, and the unity of complex systems.
6. The characteristic of interactions, relations or organizations which involve movements of information, money, physical objects, people, or other tangible or intangible items across state boundaries.

♦ **KC0005 Paradigm**
Description No clear definition of paradigm exists. Three senses have been distinguished as follows: 1. Metaphysical paradigms or metaparadigms, in which paradigms are equated with: a set of beliefs; with a myth; with a successful metaphysical speculation; with a standard; with a new way of seeing; with an organizing principle governing perception itself; with a map; and with something which determines a large area of reality.
2. Sociological paradigms, in which paradigms resemble a universally recognized scientific achievement; a concrete scientific achievement; a set of political institutions; and an accepted judicial decision.
3. Artefact paradigms or construct paradigms, in which paradigm is used: as an actual textbook or classic work; as supplying tools; as actual instrumentation; more linguistically, as a grammatical paradigm; illustratively, as an analogy; and more psychologically, as a gestalt–figure.
From a sociological point of view, a paradigm is a set of scientific habits. By following these, successful problem–solving can go on: thus they may be intellectual, verbal, behavioural, mechanical, technological, or any or all of these, depending on the type of problem to be solved. As such, normal, paradigm–based science constitutes research based upon one or more past scientific achievements that some particular community acknowledges for a time as supplying the foundation for its further practice. Such achievements are: sufficiently unprecedented to attract an enduring group of adherents away from competing modes of scientific activity; and sufficiently open–ended to leave all sorts of problems for the redefined group of practitioners to solve.

♦ **KC0007 Multidisciplinarity**
Description A variety of disciplines, presented within the same setting, but without making explicit possible relationships between them. From a systems point of view, the successive steps of cooperation and coordination between disciplines lead to the definition of the organizational principle for a single–level multigoal, hierarchical system, without cooperation between the parts.

♦ **KC0008 Continuity and discontinuity**
Description 1. Continuity in the development or activity of a system is indicated by relative stability within parameters of any of its given aspects. Biological homeostasis is an example. Discontinuity is the transition of the active system or its major elements beyond these parameters.
2. Continuity and discontinuity exist in form and structure. Discontinuity is the analytical nature of all matter (unitized by atomic or particle composition) and all life and human formations (the biological and sociological unit base). Continuity is the holistic reality of a system related to its negentropy or raison d'être, that is, its dynamics and function.
3. Continuity, discontinuity and alteration or creative dialectic between them are the three dynamic qualities of all things.

♦ **KC0010 Mathematical complex**
Description In combinatorial topology the set of simplexes or other constituents of a geometric figure.

♦ **KC0011 Science**
Description 1. A body of accumulated and accepted knowledge that has been systematized and formulated with reference to the discovery of general truths or the operation of general laws. Such knowledge is divided into fields, academically distinguished as disciplines or sciences, each with its own establishment, its own tradition, and its own form of enquiry, and linked mainly by the willingness of the participants in each to accept other participants as scientists and their disciplines as sciences.
2. A combination of methods for progressively validating hypotheses, applicable only with varying degrees of cogency in the fields of the different sciences.
3. A faith that reality is orderly and knowable, which is progressively extended through the study of systems, which expands the ideas of order and of the variety of conceivable orders.
4. An attitude towards knowledge, towards knowing, and towards learning as a basis for pursuing enquiry and criticizing the process and fruits of enquiry.
5. Organized rationality, namely a final court of appeal against any authority, human or divine.

♦ **KC0013 Maturity**
Description A property of some systems (ecological, social) closely related in one respect to its diversity or complexity, and in another to the information that can be maintained within the system, with a definite expenditure of potential energy. A highly mature system has the capacity of carrying a high amount or organization and information, and requires relatively little energy to maintain it. Conversely, the lower the maturity of the system, the less the energy required to disrupt it. Anything that keeps a system oscillating, retains it in a state of low maturity.

♦ **KC0014 Induction**
Description 1. Any inference whose premises do not entail its conclusions.
2. Inference from particular instances of conjunctions of characteristics that a conjunction may be held to be universal.

♦ **KC0015 Metahistory**
Description 1. Philosophy of history (e.g. Aristotle).
2. Universal or synoptic history (Toynbee, Spengler, et al).
3. Generalization of the phenomena of history (e.g. Hegel)
4. Epistemological approach to historiography in which the historians' procedures are examined.
5. History which also deals with mythology and with myth and legend making and also with the subjective development of man, i.e. his cognitive moral, aesthetic and behavioural history.

♦ **KC0016 Cost–benefit analysis**
Description 1. A technique to measure the costs of an alternative course of action to achieve some objective against the benefits resulting from taking that course. It is used where important factors in a problem cannot be measured in monetary terms, or where such measurement is very difficult and uncertain, or where monetary values are only part of total values. Associated with systems analysis, it has had its greatest development in dealing with military problems. Through the Planning–Programming–Budgeting system (PPBS), use of this technique has been extended to all types of organizations.
2. A distinction is made between cost–benefit and cost–effectiveness analysis. The cost–effective method assumes the outputs from a system to be fixed in quantity and quality. It then explores the comparative cost of various methods of producing the fixed output. That system which can produce the fixed output at lowest cost is selected as being the most efficient system (i.e. it provides the lowest input/output ratio). The cost–effective approach assumes that although the inputs to a system were variable, the output was fixed. The cost–benefit approach removes the latter constraint and allows both the inputs and the outputs from the alternative systems to vary, and then attempts to measure which of these systems is the most efficient (that is, has the lowest input/output ratio).

♦ **KC0017 Emergent systems**
Description Prototypical systems that have initially exhibited, as innovations in their time, novel formats of organization that have supported successive increases in adaptive range on the basis of correlative increases in systematic complexity. Properties of emergent systems include:
Gestalt novelty (a new form as opposed to combinatorial novelty);
Concrescence (process of growing together of previously distinct systems to form a unitary integral structure);
Systemic extension (organization of elements, themselves systems, in hierarchical levels connected by regenerative information/control linkages providing for selectivity at every level; is maintained with the incorporation of a new level of organization);
Normative innovation: the appearance of an additional level of organization requiring the institution of new norms;
Subsystem specialization: the modification of previous subsystems in terms of articulation or differentiation of structure and normative innovation contributing to increased complexity, efficiency, and elegance of structure and behaviour.
Negentropy: a general increase in variety and organization, through net gain of potential by a local, metastable system (due to transfer and transformation of energy), and through an increase in decisional degrees of freedom (due to communication and transformation of information).

♦ **KC0018 Heterostasis**
Description The tendency exhibited by some open systems purposively to seek disequilibriating forces, in order to move to higher levels of homeostasis. Equivalent to the creative act in human beings, where new levels of tension resolution are sought.

♦ **KC0019 Association**
Description The connection of memory contents that makes for tying together or simultaneous evocation, e.g., the colour red associated with apples. Association may take place unconsciously, as an assimilation of sensory data or of concepts, or consciously, especially as concurrent experience is multisensory.

♦ **KC0021 Government**
Administration
Description Collectively the highest executive body at any discrete political level. Government may also be referred to as administration or regime. It includes the judicial, executive and legislative functions and their civil and military instruments. Some governments may be constituted at any one time, in democracies, by office incumbents of different political parties. A government may have its supreme authority vested in a king or other monarch, in a body of hereditary

aristocrats, in a religious figure, in a dictator, in a parliament (of one or two houses), or in the instruments of its constitution or by popular referendum.

♦ KC0022 Synaesthesia
Description Hearing colours, seeing tones and unusual associative perceptions such as sensing colours as an aspect of numbers. Types of synaesthesia may be called photisms (visual), phonisms (aural), tactile, olfactory and gustatory, or they may be multisensory. Synaesthesia may relate to clairvoyance, clairaudience, etc., and other extrasensory psychological phenomena most of which are unexplained and lack any prudent, modern theoretical formulation. Older theories spoke of a common sense or sixth sense as the one that united the other five. (see also: synaesthesis)

♦ KC0023 Synderesis
Description In Aristotle, the starting point for integrative thinking, e.g. the truths that are already present in the mind concerning the first principles of the subjective and objective worlds to which thought innately inclines. Through synderesis, for example, all men have a moral sense.

♦ KC0024 Microcosm
Description 1. A community, institution, or other unity believed to be an epitome of a larger unity.
2. A Renaissance philosophical term designating man as being a little world in which the macrocosm or universe is reflected. This supposed analogy between the whole and its parts served to develop a humanistic cosmology in which the reality of the individual received due attention.
3. In Russian speculative philosophy, not man, but the earth or organic world (as in Gurdjief); or the biosphere with its upper reaches in the noosphere (as is Vernadsky).

♦ KC0025 Planning
Description 1. Action on the environment for the purpose of changing it in such a way that tendencies toward coherence and cohesion are enhanced and tendencies toward distintegration and dissolution are kept under check; namely, it is a process whose function is to reduce entropy and increase organization within the environment.
2. A goal–directed decision–making process.
3. The formalization of factors involved in determining the goals and the establishment of the decision processes to achieve these goals.
4. The systematic enrichment of the information base for decision–making (pointing out consequences for the future of alternative courses of action taken in the present, and consequences for present action of alternative goals in the future).
5. The process of developing a complex dynamic system designed in the form of a controlling event–structure whose function is to effect in its environment (which is another complex dynamic system), the kind of organized change which current values define as progress.
6. The process of making, changing, or coordinating plans which are sequences of future actions to which a person, unit or organization is committed.

♦ KC0026 Synchronicity
Coincidence
Description 1. The concurrence or coincidence of events which may be logically or conceptually, but not causally, related. Also, these kinds of events which in addition, exceed the statistical probability of occurring together.
2. A principle which attempts to eliminate the time factor from a linear cause–effect relationship without taking in consideration time reversal, time stop and other anomalies that may be imputed to time from relativistic and other theoretical physical frameworks. From a logical–physical perspective, coincidence or synchronicity may result from (or be caused by) anomalies in causation, time, or space, or by a combination of these, or by anomalies in other fundamental, ontological categories such as existence (being).
3. The C G Jung theory of magic and occultism i.e. the modus operandi of astrology and other forms of divination (by dreams, by I Ching oracle, etc.)

♦ KC0027 Multistable system
Description A system consisting of many ultrastable systems joined by their main variables, all the main variables being part functions.

♦ KC0028 Tensegrity
Description A structural relationship principle in which structural shape is guaranteed by the finitely closed, comprehensively continuous, tensional behaviours of the system, and not by the discontinuous and exclusively local compressional member behaviours. It provides the ability to yield increasingly without ultimately breaking or coming asunder. The integrity of the whole structure is invested in the finitely closed, tensional–embracement network, within which the compressions are local islands. Tension is omnidirectionally coherent and tensegrity is an inherently nonredundant confluence of optimum structural–effort effectiveness factors. All structures (from the solar system to the atom) are tensegrity structures, when properly understood. Of human societal structures that must correspond to the concept of tensegrity, theoretical communism is closest. Of non human societal structures corresponding, the termite nest or ant hill, are analagous. The desired omnidirectional, comprehensively continuous distribution of tensional energy in tensegrity social engineering is akin to the omnidirectional distribution of energy in a disorganized or chaotic system whose homeostatic dynamics and self–differentiation have been overcome by entropy.

♦ KC0031 Policy
Description 1. A projected programme of goal values and practices. The policy process is the formulation, promulgation, and application of identifications, demands, and expectations concerning the future interpersonal relations of the self. Projected action may be either social or private; it may concern either the actor alone or his relations with other persons. A course of action in relation to others may be termed the policy of the actor.
2. A definite course or method of action selected from among alternatives and in the light of given conditions to guide and usually determine present and future decisions.
3. A statement of procedure or principle by which an organization intends to realize its objectives. It insures proper direction toward definite objectives of an organization in the light of internal factors and organizational functions.
4. Regulation of a system over time in such a way as to optimize the realization of many conflicting relations without wrecking the system.

♦ KC0032 Fibonacci numbers
Description A number series of which the first two members are one; thereafter each member is the sum of the two preceding members. The series has interesting properties related to the golden section (familiar to students of proportion), and has connections with a variety of familiar natural phenomena, including: the number of emergent rays from multiple reflections of light; the number of different possible histories of an electron in the ideally simplified atoms of a quantity of hydrogen gas; the genealogical table of a bee; phyllotaxis, namely the arrangement of leaves on the stems of plants; and the equiangular spiral frequently encountered in petal arrangements in flowers, in shells, and in musical scale.

♦ KC0033 Polymathy
Polyhistory
Description 1. The attainments exhibited by erudition in several fields.
2. Encyclopaedic mentality. The quality personified by Corypheus or Pangloss.
3. In antiquity, wisdom, personified by Solon, Solomon, etc.
4. In Renaissance learning, humanism. The quality personified by Leonardo da Vinci.

♦ KC0036 Multivariate analysis
Description Analytical techniques using primarily mathematics and statistics (i.e. correlation and regression analysis) devised to sort out and identify the interactions of and interrelationships between multiple variables (i.e. usually a larger number of variables than the unaided human brain can contemplate).

♦ KC0037 International relations
Description 1. The study of all the exchanges, transactions, contacts, flows of information and meaning, and the attending and resulting behavioural responses between and among separated organized societies, including their components. Such a study usually deals with actors (states, governments, leaders, diplomats, peoples) striving to attain certain ends, using means (such as diplomacy, coercion, persuasion) which are related to their power or capability.
2. A branch of political science concerned with relations between political units of national rank and dealing primarily with foreign policies, the organization and function of governmental agencies concerned with foreign policy, and the economic, geographic and other factors underlying such policies.

♦ KC0038 Metacommunication
Description A term used to refer to anything which a person takes–into–account as an aid in interpreting what another person is saying, the import of the situation, how to comprehend what is happening, etc. It is therefore any clue or evidence which a person uses in making relevant his comprehension of something or someone.

♦ KC0039 Unity
Oneness — Singleness
Description 1. The quality (e.g. sameness or identity) or quantitative state of being of uniqueness.
2. A condition of concordant harmony; the state of those that are in full agreement.
3. The quality or state of being made one or uniting into one; the combination or ordering of parts such as to constitute a whole or promote an undivided total effect.
4. The quality or state of constituting a whole, and especially one organized from distinguishable parts or elements; an entity that is a complex or systematic whole.

♦ KC0042 World modelling
Description Construction and development of models of global phenomena based on mathematical and computer–assisted simulations. The purpose of such models is to obtain an insight into the structure and function of complex, dynamic, global systems in order to predict their behaviour by analogy with the behaviour of the models. The emphasis of a model may be placed on economic, social, political, technological, military or other aspects, or may interrelate some or all of these.

♦ KC0045 Taxonomy
Classification — Systematics
Description 1. The science of classification in a broad sense, although it is usually associated with biological classification and specifically the classification of plants and animals in hierarchies of superior and subordinate groups. Other subjects with taxonomic problems include: psychology, archaeology, sociology, linguistics, pattern recognition, and information storage and retrieval. In each case the aim is to reduce very large quantities of data to manageable proportions, by which is meant the ability to efficiently sort unit entries into categories.
2. Metaphorically, any classification scheme, especially if it is hierarchical and conceived as a tool to analyse a collection of data and establish their interrelationships.

♦ KC0047 Hierarchical restructuring
Description 1. A process in which complex stable systems can undergo sudden transitions to new self–maintaining arrangements which will in turn be stable for a relatively long time, when restructured by larger hierarchical patterns. In general, a stable hierarchical structure from the lowest levels (molecules, enzyme cycles, etc) up to a given level (cell or organism, for example) grows to a new structure because it comes in contact with new and different materials or information or another such structure (e.g. another organism). This can make the patterns unstable at the highest level until there is resolution (conflict, cooperation) with restructuring either by breaking apart or by a new organization at a level higher than the given level, to make a new stable pattern encompassing the larger experience of the larger system. Some transformations of viruses are examples of hierarchical restructuring and some cancers may be also. Some forms of cultural change may be analagous, such as westernization.

♦ KC0051 Culture
Description 1. That complex whole which includes practical artifacts, spiritual and moral beliefs, aesthetic activities and objects and all habits acquired by man as a member of society, and all products of human activity as determined by these habits.
2. The complex whole of the system of concepts and usages, organizations, skills, and instruments by means of which mankind deals with physical, biological, and human nature in satisfaction of its needs.
3. The whole complex of traditional behaviour patterns which has been developed by the human race and is successively learned by each generation, and includes knowledge, belief, art, morals, law, techniques, and methods of communication.
4. The working and integrated summation of the non–instinctive activities of human beings. It is the functionally interrelated, patterned totality of group–accepted and group–transmitted inventions, whether material or non–material.

♦ KC0053 World view
Weltanschauung
Description 1. An individual conception of the course of events in and of the purpose of the world as a whole, forming a philosophical view or apprehension of the universe.
2. The outlook of a social collectivity, a regime or other body of authority, or of a school of philosophers.

♦ **KC0054 Mathematical group**
Description 1. A mathematical aggregate in which the product of two elements is an element of the group.
2. A non–empty set together with an operation (called multiplication) which associates with each ordered pair of elements in the set a third element also in the set (called their product), in such a manner that certain conditions are satisfied. The theory of groups and sets is an essential part of modern mathematics and logic.

♦ **KC0055 Method**
Description 1. Technique or means of a practical, i.e., material accomplishment.
2. Procedures used in the solution of theoretical problems.
3. Philosophical, logical, experimental or scientific method.

♦ **KC0057 Planetization of mankind**
Planetary culture — Collectivization of mankind
Description This is foreseen as a future state of the Earth in which human consciousness, reaching the climax of its evolution, will have attained a maximum of complexity and, as a result, of concentration by total reflection (or planetization) of itself upon itself.
Contributing to development towards this state are: the appearance of a collective memory in which the common inheritance of mankind is amassed in the form of accumulated experience and passed on through education; the development, through the increasingly rapid transmission of thought, of what is in effect a generalized nervous system, emanating from defined centres and covering the entire surface of the globe; and the emergence, through the interaction and ever–increasing concentration of individual viewpoints, of a faculty of common vision penetrating beyond the continuous and static world of popular conception. In the terms of de Chardin, this is the movement of man towards the Omega Point, the highest evolution of the noosphere.

♦ **KC0059 Contextuality**
Trans–contextual thinking
Description The distinctive outlook towards which converge partial and fragmented approaches to understanding and formulation of policies relating to society as a whole. Whatever his origin, an intellectual tends to develop a comprehensive conceptual map and an inclusive and exclusive set of terms for thinking and talking about policy and society. In principle, however, his concepts can be treated as abstract equivalencies of other maps, and his terms can be expanded, contracted, or supplemented by others.

♦ **KC0060 Legal consciousness**
Description 1. A sense of justice and a loathing of crimes and illegal actions.
2. The ideology of law.

♦ **KC0061 Biocybernetics**
Description Cybernetic model building applied to biological structures and living system behaviours with particular reference to major and subsidiary control systems.

♦ **KC0062 International socialism**
Description 1. The international communication of socialist ideology with the end in view of adaptation to local conditions.
2. Attempts at a monolithic socialist ideology to be imposed internationally by subversion and revolution, e.g., communism.

♦ **KC0064 Ultrastable system**
Description A system (such as a homeostat or a human being) which is capable of resuming a steady state after it has been disturbed. The feedback paths are altered until the system can achieve stable closed loop control and the desired regulation. For example, every living organism keeps the value of certain internal variables within physiological limits despite disturbing external influences.

♦ **KC0065 Universalism**
Description 1. A theory according to which the whole is logically or valuationally prior to its parts.
2. An ethical theory that the good of all men should take precedence over that of an individual.
3. A social relationship in which behaviour is determined by an impersonal code or standard.

♦ **KC0066 Allopoietic system**
Description A system in which the product of operation of the system is different from the system itself. The majority of systems are allopoietic. For example, in a man–made system such as an automobile, there is a concatenation of processes which specifies organization and yet does not produce the components of the automobile, since the latter are produced by processes (in the factory and assembly plant) which are independent of the automobile and its operation.

♦ **KC0067 World citizenship**
Planetary citizenship — International citizenship
Description 1. Individuals who acquire skills in a number of languages, reside (or have resided) in a number of countries, and in general, who have many links across national boundaries, may be characterized as world citizens.
2. A network of people who care about the unity of mankind and the well–being of the planet, and who wish to remain in touch with one another as planetary citizens. Such a group has developed the use of a planetary passport and a planetary citizenship pledge.

♦ **KC0069 Energy circuit language**
Description Represents the flow of energy in circuits rather than the flow of electrical current (as in the case of the electronic equivalent circuit language). Such circuits are represented on diagrams using special symbols. In this way, using symbols that are each mathematically defined for an energy transformation operation, networks of potential energy and heat flow may be drawn for food chains in seas, streams and forests, for example. Although mainly applicable to ecological studies, energy circuit language is also suitable for the description and analysis of social, biochemical, economic, and electrical systems.

♦ **KC0071 Stereotypic patterns**
Stereotype
Description 1. A repeated pattern of action, activity, behaviour form or place. Human stereotypic patterns in abnormal psychology include such things as the repeated rinsing of the hands, adoption of immobile posture, and certain pathologies of speech.
2. Isomorphism.

♦ **KC0074 Area studies**
Description Interdisciplinary analyses of specific demographic regions encompassing economics, culture, natural sciences, sociology, politics, social statistics, beliefs, customs, laws, etc.

♦ **KC0076 Cultural relativity**
Description A thesis emphasizing the uniqueness of each culture, the uninterpretability of any culture item apart from its cultural context, and the unlikely validity of any simple law regarding cultural human nature. Higher–order interactions, complex interdependencies, highly contingent dependencies characterize the relationships among the aspects of culture and personality. Judgments must therefore be based on experience, and experience is interpreted by each individual in terms of his own enculturation.

♦ **KC0077 Discipline**
Disciplinarity
Description 1. Disciplinarity is the specialized scientific exploration of a given homogenous subject matter producing new knowledge and making obsolete old knowledge. Disciplinary activity results incessantly in formulations and reformulations of the present body of knowledge about that subject matter. Disciplines may be distinguished from one another, and characterized, at the following levels: the material field comprising the objects with which the discipline is concerned from a commonsense point of view; the subject matter of the discipline determining the manner in which it selects out a certain sector of all possible observables offered by the material field; the level of theoretical integration of the discipline, ranging from description and taxonomic emphasis to elaboration of a single theoretical system; the methods of a discipline, both to get at the observables of its subject matter, or to transform observables into data specific to the problem under investigation; the analytical tools of a discipline based on logical strategies, mathematical reasoning and model construction; the applications of a discipline in fields of practice; and the historical contingencies of a discipline.
2. A set comprising three types of elements: observable and/or formalized objects, both manipulated by means of methods and procedures; phenomena that are the materialization of the interaction between these objects; and laws (whose terms and/or formulation depend on a set of axioms) which account for the phenomena and make it possible to predict how they operate. The items in this set, which have internal and/or external relationships, are revealed through phenomena which subsequently confirm or invalidate the axioms or laws.
3. A specific body of teachable knowledge with its own background of education, training, procedures, methods and content areas.

♦ **KC0079 Numerology**
Number symbolism
Description Number symbolism has been employed in magic and religion, in literature and the arts, and in philosophy. Numbers may represent objects, natural or supernational beings, states or conditions, ideas, relationships, ontological dimensions, etc. It is also possible to correlate things characterized by the same number, for example, the five–year revolutionary period of Venus with the five–petal pattern of a rose, with the human five senses and thus with man represented by a pentagram. Many human artifacts have been constructed on numerological principals, for example, the classic alphabets, the ten–based number system, the calendar, the collected treatises of the Bible, the hierarchies of priesthood, Gothic cathedral architecture, playing and Tarot cards, the game of chess, military and royal ranks, the philosophy of Plato, the Divine Comedy of Dante, Hesiod's Theogony, the dialectics of Hegel and Marx, the psychology of Jung, the combinatorial art of Lull, the mediaeval Kabbalah, Tantra, and the Book of Revelations.

♦ **KC0080 Economic policy**
Description The implementation, by government, of systematic measures to affect the economy, usually in accord with specific social and political philosophies. Economic policy regularly invokes theories of economic behaviour of systems, societal and market segments, and individuals, and employs sophisticated computer models and statistically elaborate forecasting as, for example, in five year plans.

♦ **KC0081 Man–machine systems**
Description Systems in which there is some interaction between man (usually the operator or controller) and a machine requiring that analysis of the operations of the system take into account the psychological and behavioural characteristics of individuals functioning as interfaces with the hardware portions of the total system.

♦ **KC0083 Artistic synthesis**
Description The uniting of different art forms into an artistic whole. An example is architecture combined with monumental art (e.g. elaborate war memorials) enhanced by moral painting and sculpture. Other examples are historic monuments presented under theatrical lighting with narrative, music (e.g. the pyramids) and dance (e.g. Greek amphitheatres) or other performances. The secular synthesis of plastic and performing arts was partly carried out in the evolution of the theatre and the music drama but as these became democratized and dependent on popular support the financial conditions required for permanent theatrical stage architecture, sculpture and design generally became unsustainable. However the religious uses of artistic synthesis continue and are exemplified by public worship where there is a rich liturgy, sacred music and formal movements in a physically aesthetic environment which may include special lighting, flower arrangements, symbolic architecture and placement of furnishings, ornamented vestments for celebrants, censing and visual art. The ultimate opportunity for artistic synthesis in permanent form may lie in new city planning, with aesthetic considerations not only applying to new city structures but also to dynamics, i.e. human behaviours.

♦ **KC0087 Unity of mankind**
Description An affirmation that all people are part of a single human family, that a oneness lies buried beneath the manifold diversities and dissensions of the present fragmented world, and that this latent oneness can give life and fire to some programme of social transformation.
This unity means that the constant factors in our genetic heritage and in our upbringing imply similarities in our aspirations and in the process of our lives. While these latter similarities may be somewhat masked by the obvious differences imposed by life in diverse human environments, many features of common humanity stand out above such variations. The phrase also implies that most options available to any member of the human family will become available to any other member.
Alternately, the human commonalities perceived (despite differences of national, ethnic, or racial characteristics, and despite differences of sex, age, physical or mental states, or among economic, political, religious or cultural goals) are conceived, not as some difference–obliterating or ignoring, idealistic unity, but as a federation of different parts of the human family in which, in fact, differentiation plays an important if not crucially essential role in planetary development. This federated unity implies that people may continue to exercise their choices freely so as not to come under any system of 'harmonized' options.

♦ **KC0093 Spatial ordering**
Structural order — Form language — Pattern language
Description The study of the geometry of patterns and regular or semi–regular structures (particularly the study of polyhedra, polyhedra packing, sphere packing, methods of generating polyhedra, and related topics) which offers insights into engineering structure in nature, including crystal networks, biological cell packings, Fermi surfaces, virus morphology, and other areas. Such studies may illuminate the manner in which an individual relates to his constructed environment.

♦ **KC0096 Holistic medicine**
Description 1. Homeopathy.
2. Any system of medical therapeutics that seeks to also treat the patient's conditions that are antecedent to any immediate, accepted causes of a complaint. These are the causes of the causes and typically are considered to lie in the mental and personality domains which in turn effect some postulated root physical force such as a vital spirit. 3. Any traditional system or theory of medicine that is pre–scientific such as the four humours, yin–yang, and various schools of the mind over matter persuasion.
4. Medicine practised by unqualified persons when replacing effective, scientifically proven or controlled methods by a supposed universal treatment of the body by electricity, magnetism, immersion in still or agitated water, immersion in mud, hypnosis, colours, music, yoga, massage, herbs, herbal packs and compresses, dietary regimes, copper bracelets and metallic talismans, and astrological charms or witchcraft.

♦ **KC0097 Aim**
Description Aim may be used as a general term for purpose, objective, goal and policy. Both in theory and in practice the distinctions between these terms are not fixed although it may be said that a purpose determines why an organization continues to exist, an objective determines what it wants to accomplish, and a policy determines how it carries out its activities in order to fulfil its purpose and achieve its objectives. A goal or aim may be synonymous with either the over–riding purpose of a human grouping or activity, or with any of its strategic or tactical objectives. Among these near synonyms it is interesting, however, that aim is the only one with a verbal form (in English); i.e. to aim, or aiming. In turn, as a metaphor, this brings in its train such related words as target and targeting, project and projecting, direct and directing. This suggests that an aim is a direction chosen consciously from among other possible directions, and that 'aim' may be distinguished as the required effort (or resource) to reach a specific target. While the aim specifies the target, it is not identical with it.

♦ **KC0099 Unicity**
Description The unique, non–mathematical, oneness of the absolute being, God. Mathematical unity cannot be predicated ofGod if it is argued that there is no matter or quantity attributable to him. Therefore the unicity of God is the infinity of his perfection, the being of his act, the act of his being. If, according to Christianity, he is one to himself, he nonetheless is 'experienced' as three in one by Christians and one in three, in that each Person of the Holy Trinity is itself, and does not share, the divine identity.

♦ **KC0101 Society**
Description 1. A group of people having a common body and system of culture for example, a physical community, or an aggregate of communities.
2. A voluntary association of individuals for common ends, especially an organized group living or working together, or periodically meeting or worshipping together, because of a community of interests, beliefs, or a common profession.
3. An enduring and cooperating social group whose members have developed organized patterns of relationships through interaction with one another. The term may also be applied to the complex structures of institutions of such a group.
4. A broad grouping of people having common traditions, institutions, collective activities, and interests, particularly an international social order or community of societies.
5. An interdependent system of organisms, biological units, or organizations.

♦ **KC0106 Comprehensive thinking**
Integrative thinking
Description An approach to thinking about the problem condition of mankind that proceeds from the whole to the particular, and from considerations of how to think about the larger comprehensible whole system to how such whole systems' thinking may be applied to local and particular aspects of the system in planning an environment for man. Undue emphasis on locally unique aspects of the system, or ways of organizing individual local experience, may obscure the operation of the larger universal patterns, towards which a comprehensive orientation is required in order to derive maximum advantage for man.

♦ **KC0107 University**
Description An institution, which in the communal sense was conceived as the concrete embodiment of an ideal of life (rather than in the sense of a group of material structures, buildings and grounds). In particular, such a communal institution reflects the tripartite ideal of study, teaching and research, all interlinked and in some instances hardly distinguishable, engaged in for the purpose of creating and transmitting meaning as brought out by the connections within and between the disciplines. The contemporary university tends to constitute a focus of activities of many kinds; intellectual, athletic, social, etc, however, most notably, its educational purposes may suffer from an enormous administrative burden brought about by the centralizing and institutionalizing, in the physical sense, of the community of scholars. This centralization is not cost–effective and increases the 'ticket price' for higher learning. Universities therefore are continually distracted from their educational purposes by pressing administrative and financial needs and by continually raising tuition costs, making learning a privilege of an elite. Open universities, peoples' and non–traditional colleges and universities help offset this trend and the traditional university is under great pressure to reform, or go under.

♦ **KC0108 Cognitive mobilization**
Description The mind–influencing process by which the ability to manipulate political abstractions becomes increasingly widely distributed throughout the polity. With the development of an extensive multi–million person political community fundamental changes necessarily take place in the making and executing of political decisions and in the coordinating of activities of a geographically scattered electorate. New techniques of communication and control are developed amongst political party elite possessing specialized skills.

♦ **KC0110 Harmonic series**
Description The series, beginning withthe fundamental tone, which are overtones in ascending frequency. The frequency relationships of the overtones are expressed in prime numbers and the fundamental tone is called, the first, tomake a correspondence between the prime numbers and the ordinal numbers. The harmonic series are among those acoustic laws which exhibit the theoretical unity of the various tonalities.

♦ **KC0113 Convergence**
Description 1. In association with the theories of biological and societal evolution, the postulate that physical and non–material (e.g. political) systems move toward one another.
2. In dynamics, one of two complementary forces, i.e. divergence and convergence, applied to philosophies of history, culture and human behaviour and associated with theories of cyclicity of macro–phenomena.

♦ **KC0114 Political organization of society**

♦ **KC0115 Artistic life**
Description While the foundation of creative work may be primarily conscious and purposeful execution, the productions of art and other creative acts cannot be separated from the lives of their authors. Artistic life in its totality inevitably includes purposeful conscious execution, as otherwise there would be no sustained production, but it encompasses the entire range of life experience of the artist or creator in which non–conscious, non–directional, and non–rational elements enter. These may be factors attributed variously to the instincts, to the emotions or to the higher feelings; or to the unconscious or subconscious; or to intuition, to inspiration or to the superconscious. Much of this experience, even though it may a significant factor at the very moments of creation, cannot be verbalized by the creative person, and it is left to theorists to postulate definitions and rules for what essentially is the work of the indefinite, unbound, and unitary human spirit.

♦ **KC0116 Cybernetic management**
Cybernation
Description 1. A theory of management which conceptualizes the management process as principally a flow of information between structurally and functionally inter–related parts. Emphasizes an holistic and total system approach to decision–making. The management function (e.g., in social, political, and economic organizations) may be viewed as analogous to the cerebral and neural functions in a physiological or a biological system. An emphasis is placed upon the design of information loops in organizational communication systems and upon the feedback principle, positive and negative.
2. Cybernation encompasses the cybernetic management concept as it is found both in the factory (automation) and in the office (computerization).

♦ **KC0117 Set**
Description A set is a collection or aggregate of elements, considered together or as a whole. By an element is meant an object or entity of some sort (e.g. a positive integer, a point on the real line, or a point in the complex plane). Along with the group, the concept of the set is fundamental to modern mathematics and logic.

♦ **KC0118 Data structure**
Description The logical structure of data in complex data bases. Several principal approaches exist, including: lists (sequential data structure), tree structures (separately described), and network structures (separately described).

♦ **KC0120 Centrality**
Description 1. The quality or state of being central.
2. The degree to which a particular point, by the number and importance of its relationships to other such points, emerges as being more central with respect to them than is any other point. Indexes of centrality can be developed to give precise meaning to this notion which finds its main application in the study and description of social networks and patterns of human settlement, for which there is a central place theory (separately described).

♦ **KC0121 Reticulation**
Description The interdependence of processes in organisms may be represented by interlocking or interlacing hierarchies, or vertical tree structures. The meeting points of the branches from neighbouring tree structures form horizontal networks at several levels. Without the tree structures, the network formation (reticulation) would not be possible. Similarly, without the network formation, the individual hierarchies would be isolated, and there would be no integration of functions. Reticulation need not be thought of as necessarily constituted by two contrasting processes or as an abcissa and ordinate pair of axes, or as vertical and horizontal, since such distinctions may be apparent rather than essential. Why, for example is hierarchical, vertical ? Further, the reticulated network may consist of any number of crossing lines or chains, and not just two. In addition, reticulation may not be regular or symmetric. An imperfect spider–web in some cases may better represent those relationships which are said not to be arboreal (branching).

♦ **KC0122 Centralization**
Centralized system
Description A centralized system is one in which one element or subsystem plays a major or dominant role in the operation of the system. A small change in this subsystem will then be reflected throughout the system, causing considerable change. Progressive centralization is the process whereby one such part of the system emerges as a central and controlling agency. Progressive centralization is especially important in the biological realm where it is associated with progressive individualization (in that an individual can be defined as a centralized system which grows, through progressive centralization, more and more unified and indivisible).
At the higher levels of an organization, centralization is essential to maintain a common pattern of accepted purposes, to coordinate many activities that may otherwise serve to defeat each other, and to provide a general framework for decentralized action. It is needed to prevent or counteract external efforts to weaken or destroy the organization through divisive action, and is needed to deal with external controllers, who may demand it in order to see that control is effective. It should be noted also that a highly centralized system may have several dominant centers. This organization may have redundancy features in order to provide back–up for contingencies. Thus there are two lungs in the human pulmonary system, for example, and the human brain has some redundancy features among its left–right hemisphere divisions, and among the limbic and neo–cortex portions, etc. Finally, centralization need not be defined spatially at all, i.e. by where one or a few dominant sub–systems or elements are located. A truly centralized system may be one with a central logic.

♦ **KC0123 Meta–language**
Description 1. A language (the meta–language) in which another language (the object language) is described.
2. A language of higher order than the symbol system being discussed.
A language cannot contain a name for an expression belonging to that language. This is a rule which is necessary in order to disallow statements which are both true and false. As a rule of logic,

therefore, a meta–language is required in which to name the expressions in a given object language. Statements in a meta–language are meaningless in the object language. In a meta–language, expressions can, for example, be named by using quotation marks. Thus "John" is a name for 'John', which is a name for John.

♦ KC0125 Hermeneutics
Description 1. Interpretation of the New Testament to successive generations according to their forms of expression.
2. Broadly, the interpretation of cultural man; an anthropology especially of the intellect.
3. A methodology for the behavioural sciences.

♦ KC0126 Heuristic
Description 1. A set of instructions for searching out an unknown goal by incremental exploration, to some unknown criterion. (Example: by following the instruction to make every step upward, a person can reach the top of a hill even in a fog and without knowing the location or height of the top.)
2. A problem–solving strategy when no step–by–step algorithmic procedure is applicable.

♦ KC0127 Order of interaction
Level of interaction
Description The relationships of systems of the same order (or level) are distinguished from the relationships of a system to systems of a different order. Two systems of the same order or type may interact differently than systems of different orders.

♦ KC0129 Abstract objects
Description 1. Objects not palpable, concrete or sense–perceivable, viz concepts, impressions, certain qualities or states, etc.
2. Representations of real objects in which varying degrees of information or characteristics are missing, for example, limiting form or shape, relationship of parts, indications of size or scale, colour, movement, etc.
3. Abstract art.
4. Abstractions; a pejorative term for vague generalizations.

♦ KC0130 Nationalism
Description 1. The belief by a number of people that they constitute a nation, and behaviour in accord with this; for example, defence of sovereignty or revolution to attain self–determination, or self–governing status. Perception of historic nationhood may be based on homogeneity or continuity of geographical tenure, or claims or history of linguistic or ethnic homogeneity. Claimed uniqueness of religion has also been a factor.
2. Nationalist movements of a revolutionary nature, possibly motivated by Marxist ideology.

♦ KC0131 Integration laws
Description The following set of universal laws of integration has been proposed:
1. The condition of being integrated is a reference to an integrated whole, rather than to integration of parts; or to a unitary whole, rather than a unified whole.
2. Integration is a condition which exists only in terms of energy potentials distributed within a unitary whole in accordance with laws of energy behaviour. Action follows in the integrated energy field of the whole when, and only when, conditions ultimately traceable to environment place the field in a state of imbalance.
3. The whole conditions or regulates the expenditure of energy, or the work done, by the parts. The work of a whole is exclusively done by the parts; the whole as such does no work, for its influence is purely regulatory.
4. Wholes evolve as wholes. Integration is therefore as complete at the beginning of the life of an organism as it is at any other time during its history. Development is a process of preserving integration while it becomes structurally more complicated, and is not a process of building up integration.
5. Any whole is constantly renewing itself through a transposition process. The identity and integration of the whole is preserved while the parts change.
6. The evolution of the whole can be described as an exchange process going on between the properties of homogeneity and heterogeneity.
7. The activity of a part within a whole, no matter what the conditions, obeys the law of least action.
8. All of the available potential energy of the whole will be expended in the direction of maintaining its integration or current condition in the presence of external disturbances.
9. There is not a one–to–one correlation between a particular external stimulus or disturbance and the response of the affected whole. The whole responds relationally to a total situation; that is, to one disturbance in relation to other disturbances.

♦ KC0132 International language
Description 1. A language spoken in three or more countries. Some European international languages are German (Germany, Austria, Switzerland, Belgium, etc), French (France, Quebec, former colonies), Spanish, Portugese (Portugal, Brazil, former colonies), Dutch (Netherlands, Belgium, Antilles, etc), Swedish (Sweden, Finland, parts of Norway), English, and Russian (in Eastern Europe). Other international languages include Chinese, Arabic and Swahili.
2. Any sign system with universal meaning, e.g., mathematical and musical rotation.
3. An artificial language intended to be universal, such as Volapuk or Esperanto.

♦ KC0133 Functional cultural integration
Adaptive cultural integration
Description Integration resulting from the manner in which the different traits of a culture function interdependently and in support of one another. This is associated with the need for efficiency in a culture.

♦ KC0135 Chemical synthesis
Description The combination of two or more chemical components (or of parts of such components) to form a single system with some degree of stability. The process of organic synthesis in organisms underlies all assimilation, growth, and repair of tissues, as well as the normal function of synthetic and reproductive tissues. All other organic processes reduce to local cycles which leave no cumulative chemical result within the organism.

♦ KC0136 Entropy
Description 1. In physics: the ultimate state of maximum disorder to which the degradation of the matter and energy of the universe tends. A state of inert uniformity of component elements involving an absence of form, pattern, hierarchy or differentiation.
2. In cybernetics: the tendency of any closed system to move from a less to a more probable (unstable) state.
3. In communication theory: a measure of the efficiency of a system (as a code or a language) in transmitting information, being equal to the logarithm of the number of different messages that can be sent by selection from the same set of symbols and thus indicating the degree of initial uncertainty that can be resolved by any one message.

♦ KC0137 Degrees of freedom
Description The number of degrees of freedom of a system is an intrinsic property of that system deduced from the number of variables that must be observed or specified before the behaviour of a system becomes determinate, that is, unique, single–valued, and not subject to unpredictable variations.

♦ KC0138 Ecosystem
Description 1. An ecological community forming a unit with its environment. For example, an organism and its environment or an organization and its environment.
2. A biological community including not only the plants of which it is composed, but the animals habitually associated with them, and also the physical and chemical components of the immediate environment or habitat in which the community exists. These together form a recognizable self–contained entity with definable boundaries. Any such community is however part of a larger physical and biological system.
3. The totality or pattern of relations between organisms and their environment, understanding of which permits explanation of empirical observations such as growth, differentiation, order, and dominance in biological evolution.
Man has changed from being a simple component of individual ecosystems to becoming a dominant force and link in the larger system, which (through technology and the development of urban systems) ties together all ecosystems in a higher (global) level of organization comprising the entire biosphere.

♦ KC0140 Communicative social integration
Description Integration of a collectivity resulting from the possession of common language and culture accompanied by the development of communications media by which people may be involved in social processes.

♦ KC0142 Completeness of equipment

♦ KC0143 Regulative cultural integration
Description Integration associated specifically with the control of cultural conflict. This may appear as a kind of balance of power among features of the culture, as a hierarchical ordering of various value arrangements, or by the relegation of divergent patterns to different segments of the population.

♦ KC0145 Stylistic cultural integration
Description Integration resulting from the mutual adaptation of parts of experience felt so intensely that their contrasts and organization produce an emotionally gratifying whole. This is associated with the aesthetic impulse for authentic expression of experience in satisfying form.

♦ KC0146 Principle of requisite variety
Description 1. The abundance or variety of alternative control actions which a control mechanism is capable of executing must be at least equal to the abundance or variety of the spontaneous fluctuations which have to be corrected by the control mechanism, if the control mechanism is to perform its function effectively. Only a greater amount of variety in a regulator can control the variety present in a given system.
2. A mechanism's capacity as a regulator cannot exceed its capacity as a channel of communication. (This can be related to Shannon's Tenth Theorem which states that if noise appears in a message, the amount of noise that can be removed by a correction channel is limited to the amount of information that can be carried by that channel).

♦ KC0147 Dynamical similarity
Description The principle of dynamical similarity (derived from theology) is used as a basic for comparison of systems in a situation when only similarities, and not identities, can be established. This situation also exists in comparing biological and social entities. Biologists and sociologists explain processes on the basis of covariance principles, in contrast to the ability of the physicist to use invariance transformations to explain a physical system (recognized as part of a class for which statistical irregularities can be validly averaged out).

♦ KC0149 Pluralism
Description 1. In political theory the belief that the apparatus of the State, i.e. government, can be diminished in inverse proportion to the development of (plural) autonomous voluntary associations of people to perform administrative, regulatory and other functions.
2. In philosophy, the phenomenon of the diversity of systems, some teaching the truth, others the true method for reaching the truth. This pluralism may be seen as itself the philosophia perennis, i.e., that all philosophical systems complement one another and they are in fact a diversity in unity, advancing towards better explanations of reality.
3. In cultural terms, more than one cultures (linguistic, ethnic, or regionally–identified) within a society.
4. In socio–economic terms, partly as an effect of cultural pluralism, the structures corresponding to the class system with their advantages and disadvantages (e.g. educational levels, speech patterns, employment hierarchies, nutrition, health and stature, etc)
5. In religious terms, the coexistence of various denominations and creeds in a society.
6. Generally, pluralism connotates a tolerated, recognized, or official diversity, and corresponding democratic institutions.
7. A 'virtue' made out of the necessity of recognizing the fragmented nature of national and global society.

♦ KC0150 Design
Description 1. The initiation of change in man–made things through steps such as the following: envisaging imaginatively the things that are needed (in the light of the total set of needs and problems in man's environment), that might be developed (in the light of the alternative sets of resources available), and the manner whereby they might be brought into existence (in the light of the skills available); making decisions in the face of uncertainty; judging whether something new will also be useful; inventing models to simulate what needs to be made or done; and integrating scientific principles, technology, and imagination to define new structures with prespecified functions.
2. Engineering design, in the more restricted sense, is the use of scientific principles, technical information and imagination in the definition of a mechanical structure, machine or system to perform prespecified functions with the maximum of economy and efficiency.

♦ **KC0152 Community**
Description 1. Physically, the larger dominciliary unit surrounding a family residence.
2. Socially, the community is that larger immediate social unit an individual or group perceives themselves to be part of. The close-knit relationships which may be evidenced between families living in proximate dwellings such as in villages, or in neighbourhoods in towns and cities, and the identification of these co-residents with shared concerns. Thus there is a physical community and a social or psychological one.
3. Community and neighbourhood, that is, the social and physical aspects of a local society, coincide in the tribal and rural worlds, with kinship making a third, biological link. Thus the extended family or clan may constitute the community.
4. A social aggregate created expressly for religious, ideological or economic reasons, for example, a monastery constituting a religious community, or a cooperative craft commune, or the social and economic integration of a farming, religious sect.
5. An economic and legal administrative entity organized at various levels of government, from the smallest settlements, to large towns and provinces, up to national levels.
6. Various specialized uses, as the community of nations, community of scholars, European Community, etc.
7. Any number of societal groups that an individual may be a member of including but not limited to those based on location of residence. Generally, any grouping of people who are aware of some commonality and who may be organized to act in a concerted fashion.

♦ **KC0153 World government**
World federal government
Description Various proposals have been made for replacing or modifying the existing system of national governments (of independent sovereign nations) by some form of world government, possibly as a development of the United Nations system. This would require a considerable rewrite of international law to encompass the idea of supranationalism.

♦ **KC0156 Form**
Description By tracing the way that distinctions (or acts of severance or abstraction) are represented, it is possible to begin to reconstruct the basic forms underlying linguistic, mathematical, physical, and biological science, as well as to see how the familiar laws of individual experience follow inexorably from the original act of distinction. Such acts of distinction draw boundaries in our universe. Although all forms, and thus all universes, are possible, and any particular form is probabilistic or mutable, it becomes evident that the laws relating such forms may be symmetric in any universe. Mathematics can be used in the study of forms to determine if their nature is independent of how they actually appear.

♦ **KC0157 Noosphere**
Description This is a postulated biological entity which is in process of emerging at the top of the biosphere (Vernadsky), or outside and above it (de Chardin) as an added planetary layer, an envelope of thinking substance. The noosphere is defined as a collective human organism which is formed through the mutually reinforced evolutionary processes of complexification due to the growth of human consciousness and the emergence of consciousness as the outcome of complexity. Additionally, whilst a development of a noosphere may be continuation of the evolutionary process, it is conceived to be radically different in that the development of individual conscious reflection compels human beings to draw together into a new communicative pattern. As individual centres of consciousness acquire autonomy on emerging into the sphere of reflection, instead of being confined to the normal process of phylogenetic development, they are visualized as passing tangentially into a field of attraction which forces them toward one another. The result is that the entire system of zoological radiations which in the ordinary course, using geometric metaphor, would have culminated in a knot and a fanning out of new divergent lines of development, would tend instead to fold in upon itself. This, theoretically, leads to the spread of a living, conscious complex constituted over the whole surface of the globe. Alternatively imaged, the reflective coiling of the individual upon himself leads to the coiling of the phyla upon each other, which in turn leads to the coiling of the whole system about the closed convexity of the Earth. The genetically associated occurrences of psychic centration, phyletic intertwining and planetary envelopment taken together thus give rise to the noosphere in de Chardin's graphic description. Other concepts of the noosphere using other terminology are less visually oriented though more materialistic. Spiritual concepts of the noosphere existed in ancient religions (Zorastrianism, Gnosticism, Kabbalah, etc) of which traces remain in the Christian (Lord's Prayer) 'Kingdom, power and glory', and (revelations) 'the New Jerusalem'.

♦ **KC0159 Paradox**
Description As distinct to right or correct or normative concepts or ways (orthodoxy), and to divergence and dissent (heterodoxy), the paradoxical assertion is beyond exclusive affirmation or negation. Paradox is characterized by explicit or implicit 'Tolerance' of irreconcilables. In logic a paradox is a contradiction and in semantics its presence is a negation of meaning. In epistemology, however, a paradox may be regarded as instructional (e.g. Zen koans).

♦ **KC0160 Social organization**
Description 1. A durable system of differentiated and coordinated human activities utilizing, transforming, and bringing together human, material, capital, and natural resources, physically or conceptually, into a unique problem-solving whole whose function is to satisfy human needs. The social organization functions in interaction with other systems of human activities and resources in its particular environment as well as with other social organizations in a common greater environment, local, national or international.
2. A partially self-controlled system with the following characteristics: (a) responsibility for choices from the sets of possible acts in any specific situation is divided among two or more individuals or groups of individuals; (b) the functionally distinct subgroups are aware of each other's behaviour either through communication or observation; (c) the system has some freedom of choice of means (courses of action) and ends (desired outcomes).
3. A group or cooperative system with the following characteristics: an accepted pattern of purposes; a sense of identification or belonging; continuity of interaction; differentiation of function; conscious integration efforts on the part of those responsible for bringing or holding the members together.

♦ **KC0162 Socioeconomic structure**
Description The economic system viewed from the aspect of its human organizational requirements to produce goods and services and the social consequences of this organization, particularly considered as the various classes, according to their ownership of the means of production and to the unequal distribution of income and wealth. Some socioeconomic structures are the clan or tribe (with patriarchal or matriarchal rulership); the feudal or aristocratic (with ownership of Slaves or serfs); the individualistic (mixed-market, free-market, bourgeois, capitalistic); and the collective or communal (cooperative, socialist, communist). Socioeconomic structure in the feudal and individualistic economies is determined by the power of production and ownership of wealth. In the tribal and communal economies power (hence position and structure) is conferred by the functions or offices of distribution of tasks, information and resources. The individualistic economies tend towards structures that are unlimitedly oligarchic and consequently and inevitably democratic. The communal or collective economies tend towards a monolithic, centralized, bureaucratic-apparatus structure of pyramidal proportions with a limited oligarchy inevitably autocratic.

♦ **KC0163 Metabolism**
Description The major part of the chemical alterations occurring in a biological system are at cellular level. At this level, growth, maintenance of health, or disease are affected by the entry of chemical substances and their influence on the cell. Excluding the extracellular process of digestion (in the case of higher organisms) the chemical transformation of nutrients which constitutes metabolism takes place in the cell. Metabolism essentially is the development of the more complex molecules which the organism needs from simpler ones. A critical factor in successful metabolism is the action of nutrient-freeing protein enzymes present in the cell. Nutrient-freeing is among nearly 100 types of enzyme-induced metabolic changes. These include the synthesis and breakdown by the enzyme of its own fats, carbohydrates, proteins, nucleic acids and lipids. The mechanics of metabolic cell control are central to the investigations of DNA and other cellular materials in bio-genetic engineering.

♦ **KC0164 Unity of the faithful**
Description The notion that the faithful constitute in their spiritual collectivity an integral, mystical or supernatural terrestrial or celestial body, is common to a number of the higher religions. It is explicit in Zoroastrianism and the Christianity of the New Testament, and is also found in speculative forms in ancient Egyptian religion and in Judaism. Connected with the concept of the Soul of the World it is also found in various stages in Platonism, neoplatonism and Islam. Such doctrines, intellectually elaborated in all these philosophies and theologies, are, nonetheless, closely akin to primitive animism and pantheism: the concept that man's vital spirit is but a part of the spirit of nature. The unity of the faithful is an idea that leads on the one hand to religious elitism and the 'chosen people' syndrome; on the other, it is the spiritual corollary to the humanism of the doctrine that all men are born and remain equal and regardless of nationality or race, that they should look upon themselves as brothers.

♦ **KC0165 Interchangeability**
Modularity
Description 1. A system of comprising wholes by removable, stock parts.
2. Multi-source manufactured parts, each of which conforms to the same technical standard regardless of origin.
3. A combination of 1 and 2.

♦ **KC0166 Homomorphism**
Description 1. Resemblance or similarity of form, structure or function (e.g. external features) with different fundamental structure, whether between different organisms, groups, organizations, etc., or between different parts of the same organism, group, organization, etc.
2. The basis of a mapping procedure by which reality is mapped onto a scientific model by a many-one transformation (correspondence) between the elements of the sets being mapped (i.e., reality and the model). While not isomorphic, the transformation can be arranged so that certain operational characteristics concerning the relationship of elements are preserved.

♦ **KC0167 Social innovation**
Social engineering — Social transmutation
Description The deliberate development in a wide variety of domains of new techniques, ideas, systems and organizations designed to create more opportunity, more equity and more unity than is provided by existing social institutions and processes. Such innovations are focused on improving the environment of the individual human being and facilitating the development of the network of social interactions in which he is embedded as well as his own personal development. To this end innovations may be explored in such domains as: organizational and administrative structures; technological systems; designed environments; social decision making, control and participation systems; legal structures; energy systems; and information systems.

♦ **KC0169 Organic functional multiplicity**
Description At different times, or simultaneously, an organ in a biological system may perform two or more functions, e.g. the human hand which probes, grasps, strikes, defends, signals, makes music, soothes or helps heal. In evolution, changes in many organs and structures show variations in function.

♦ **KC0170 Universals**
Description Universals are properties predicated as a class of conceptual or abstract objects, for example, Platonic forms or archetypes, or as a class of, necessarily abstract, relations. Concrete things that exemplify or embody universal properties are particulars. Colours, for example, are universals, coloured objects are, so far as their colour is concerned, particulars. Logical or Platonic realism accorded objective existence to universals. The philosophy characterized as nominalistic, was so-called because the universals were considered to exist only as names, i.e. mental constructs, or concepts.

♦ **KC0171 General plan**

♦ **KC0172 Social policy planning**
Description An organized, purposeful and professionally collaborative or interdisciplinary approach to the analysis, modification or design of social systems and structures.

♦ **KC0173 Population**
Description 1. In statistics the total count of subjects in a census, sample or survey in its entirety, or the count in reference to a particular context, e.g., the number of respondents to a question.
2. In demographics, the count of people in any geographical area or any classification, for example, age, sex, marital status, ethnic origin, etc.
3. In ecology and genetics, over a period of time, the total number of individuals of a single species, or the number in a particular space. This applies to microorganisms, insects, plants and animals, as well as man.

♦ **KC0174 Complementology**
Description The study of complements, complementals, complemental systems and complementarities. Complementology defines man as a being of complemental systems (communications systems and physical systems). Such systems work together to function as a general system.

♦ **KC0177 Mobilization of resources**
Assembling resources — Procuring resources
Description Goal-oriented activity in organizations always include efforts to bring together the

resources required to execute plans in conformity with objectives. For any organization as a whole, resource mobilization is a crucial part of its relation with its general environment. For any unit within an organization, it is a crucial part of its relation with other units.
Techniques of mobilizing financial, physical and human resources include: taxation, money creation, borrowing, sales or purchase of goods or services, contributions, lease or rental of property, commandeering, drafting, expropriation, and rationing. Such techniques may be employed legally or illegally; with or without physical or moral pressure; and on the basis of a simple appeal or backed by powerful propaganda campaigns. At a high level of abstraction, the strategy and tactics of resource mobilization are no different from those used in any form of social combat.

♦ KC0179 One Earth
Description Associated with the name of the Findhorn Foundation (Scotland) and other New Age groups, with the alternative name, 'The Planetary Village'. A popular philosophy, or a vision, (of universalist, world federalist or 'aquarian types) for an interdependent world of peaceful cooperation and social and personal development. Vegetarianism, organic gardening, privately owned energy sources, ecology, meditation, astrology, and practice of Eastern religions are recommended.

♦ KC0180 Integrated
Unified — Consolidated
Description A structure, system or entity composed of separate parts united together to form a more complete, harmonious or coordinated whole, whether in the form of a physically interconnected system (such as an electronic circuit or a factory), or achieved solidarity and patterns of communication within a social group or organizational system.
A distinction may be made between integrated and integrative, with the first implying an achieved, completed, or predefined degree of integration within an established framework, whilst the second implies the facilitation of a process of integration whose scope and nature has not been predetermined within any framework and remains essentially open–ended.

♦ KC0181 Love
Description Love, as an expression of affection, sympathetic interest, empathy or benevolence, is repeatedly identified as a powerful basis for interrelating people, whether individually, in small groups or communities, or at a more universal level.

♦ KC0183 System complexity
Description 1. Incompletely understood systems.
2. Unpredictable systems.
3. Large systems.

♦ KC0186 Quantization
Description On the basis of quantum theory, radiation is conceived as having a discontinuous or quantized aspect (bundles of energy), rather than as travelling as continuous waves. The concepts of quantized action on the atomic scale can be related to pre–existing ideas of quantization in connection with macroscopic objects.
Such an extension of quantization can be used to correlate ideas and concepts in many diverse fields. In science, quantum theory plays a key role in providing a fundamental framework of concepts not only in many areas of physics and chemistry but also in the life sciences. Quantization also provides new insights into the nature of music and reveals philosophical and conceptual relations between music, sculpture, architecture and painting, thus establishing links between these fields and mathematics and philosophy. As a primary aspect of thinking, quantization also provides new insights into the nature of information, perception, the processes of teaching, learning and the nature of creativity.

♦ KC0187 Comprehensibility of systems
Description 1. The fact that many complex systems have a nearly decomposable, hierarchic structure, is a major facilitating factor in permitting individuals to understand, to describe, and even to 'see' such systems and their parts. Or, alternatively, if there are important systems in the world that are complex without being hierarchic, they may to a considerable extent escape observation and understanding. Analysis of their behaviour would involve such detailed knowledge and calculation of the interactions of their elementary parts that it would be beyond the capacities of memory or computation.
2. All systems are subject to comprehension, and their mathematical integrity of topological characteristics and trigonometric interfunctioning can be coped with by systematic logic. A system is the first subdivision of the universe into a conceivable entity separating all that is nonsimultaneously and geometrically outside the system (ergo irrelevant) from all that is nonsimultaneously and geometrically inside (and irrelevant) to the system; it is the remainder of the universe that conceptually constitutes the system's set of conceptually tunable and geometrically interrelated events. Conceptual tuning means occurring within the range of human's sensing within the electromagnetic spectrum and wherein the geometrical relationships are imaginatively conceivable by humans, independently of size, and identifiable systematically by their agreement with the angular configurations and topological characteristics of polyhedra or polyhedral complexes. All systems are individually conceptual polyhedral integrities; a system is a patterning of enclosure consisting of a conceptual aggregate of recalled experience items, or events, having inherent insideness, outsideness, and omniaroundness. Any conceptual thought is a system and is structured tetrahedrally because all conceptuality is polyhedral. Human thoughts are always conceptually and definitively confined to system comprehension. The whole the universe may not be conceptually considered because thinkability is limited to contiguous and contemporary integrity of conformated consideration, and universe consists of a vast inventory of nonsynchronous, noncontiguous, noncontemporary, noncoexisting, irreversibly transforming, dissimilar events. Intellectual comprehension occurs only when the interpatternings of experience and the interrelationships of events return upon themselves, as it were, and become foci of effectively systematic thinking.

♦ KC0188 Methodology
Description 1. Interdisciplinary, philosophical or technological study of the means of activity through organization and regulation of its processes.
2. Cognitive method, e.g. inductive logic.
3. Experimental method or scientific method.
4. Dialectic, including, interalia, analysis, reduction, division, definition, demonstration, synthesis, etc; also dialectical materialism, dialectical idealism, etc.

♦ KC0189 Integralism

♦ KC0190 Uncertainty principle
Indeterminacy principle
Description 1. The theory that no subatomic physical system can exist where, simultaneously, the coordinates of its momentum and mass centre assume completely measured or determined values. The more precisely momentum is determined the less precisely is the mass centre known, and vice versa. The exact values are always uncertain or probabalistic. This is probably due to the intrinsic properties of matter and not to imperfections in the experimental techniques. The particle–wave duality of quantum phenomena allows for the fact that there is a relative frequency of occurrence of the different values of the spatial coordinate which is proportional to the square of the absolute value of the wave function at the corresponding points in space, and values that lie close to the maximum of the wave function will be obtained most often, although with some spread and small uncertainty. This is true of the momentum value as well. Thus the concepts of coordinate and momentum applied to quantum phenomena need correction by the probabilistic uncertainty relation. In addition to the properties of mass and momentum, uncertainly in measured values of subatomic systems applies to energy state and time or duration. The uncertainty of measure of these four properties of matter are not apparent in interactions of macroscopic bodies and can be disregarded.
2. The ontological interpretation of Heisenberg's uncertainty principle which states that the indeterminacies are real and irreducible, well in accord with the theory of the eternal nature of matter and with dualism.
3. The epistemological interpretation which holds that microscopic (sub–atomic) reality is precisely unknowable, and with this a number of untenable assertions such as the illusory nature of the world, the relativity of everything, solipism, etc.

♦ KC0191 Multidimensional space

♦ KC0194 Dramatic unity

♦ KC0195 Interworking
Open systems initiatives
Description Compatability or interface or interconnection capability among computers from different manufactures. Includes peripheral equipment and software and particularly is directed at communications links in automation networks. Interconnection compatability or interworking may be achieved through multiple standard adherence by manufacturers, by conversion software, or by OSI standards to be issued by the International Standards Organization.

♦ KC0196 Analog computer language
Description Systems may be simulated on analog computers. Components of a system are convertible into the language of an operational analog computer which then simulates charge and discharge curves and permits tests of the effects of changing parameters or of the changing contribution of the simulated subsystem to the larger and more complex system of which it is a part. Such simulations facilitate studies of complex networks which have flows and storage of energy in ecosystems, biological systems, electronic networks, chemical engineering chains, and social systems.

♦ KC0197 Restructuring of knowledge
Description Now that the amount of scientific knowledge has become so large that a single individual cannot hope to encompass more than a fraction of it in the course of a lifetime, the problem of order and economy in learning and the transmission of knowledge becomes of overwhelming importance. A restructuring of knowledge (rather than a unification of knowledge) becomes necessary because of the growth of knowledge itself. The nature of the restructuring which is taking place, and which will have to take place, will be determined by the minimum knowledge (not the maximum) which must be transmitted from generation to generation if the whole knowledge structure is not to disintegrate.

♦ KC0198 Acculturation
Description The process of change induced by contacts between peoples having different cultural behaviour. Assimilation is total acculturation of a group to the point of lost identity. Culture–borrowing and cultural diffusion are related processes.

♦ KC0200 International peace
Description 1. Nonviolent behaviour among nations.
2. Peace proclaimed by international treaties.
3. Nonviolent, non–aggressive, non–competitive global conditions of the relations between man and man, and man and nature.

♦ KC0203 Musical form

♦ KC0204 Life style
Life theme
Description The concept of life theme or life style is a means of lending coherence to understanding of the course of psychological development and is a graphic means of describing an organizing psychological force in development and the way in which identifiable, unique, and enduring features of individuals come into existence and persist throughout a life span.

♦ KC0205 Meta–models
Description Conceptual models of potential universal invariances whose detection has been made possible with the help of very broadly–based systems–theoretic principles of various kinds. Several different conceptual meta–models have been independently formulated, with different suggested universal invariances. It has been suggested that the fundamental unifying themes are capable of expression in several different forms with each such form fully translatable into any of the other forms.

♦ KC0206 Augmentation of intellect
Team augmentation
Description Techniques for increasing the capability of an individual to approach a complex problem situation, to gain comprehension to suit his particular needs, and to derive solutions to his problems. Increased capability means: more rapid comprehension, better comprehension, the possibility of gaining a useful degree of comprehension in a situation that was too complex, speedier solutions, better solutions, and the possibility of finding solutions to problems that before seemed insoluble. Complex situations include, for example: the professional problems of diplomats, executives, social scientists, life scientists, physical scientists, attorneys, designers, etc, for whatever period the problem situation exists. The approach refers to an organization of conceptual activity in an integrated domain where hunches, impulses, intangibles, and the human feel for the situation usually coexist with powerful concepts, streamlined terminology and notation, sophisticated methods, and computer aids. The means used are the development of intellectual work spaces (as a reconception of the individual office environment) in which every technical facility is provided to maintain creative thinking momentum (even when collaborators are geographically distant) in the examination, storage, display, restructuring, criticism, and comprehension of concept structures, as represented by any of a large variety of symbol structures.

♦ **KC0208 Symbiotization**
Description The process of developing symbiosis, particularly within heterogeneous socio-cultural systems.

♦ **KC0209 Sociometry**
Description 1. The societal structure composed of complex interpersonal behaviour patterns and their related psychologies studied by quantitative and qualitative methods.
2. The discernment of orderings and patterns in the psychological situation of a community particularly noticable in group confrontations.

♦ **KC0211 Integrity**
Description The quality or state of being complete or undivided, whether as applied to material, organic, moral, spiritual, or aesthetic wholeness. In the intangible sphere it is a value and can be applied to the management of human affairs, e.g., the integrity of the judicial system.

♦ **KC0212 Genetic epistemology**
Description In Piaget's developmental psychology, the notion the cognitive development of the individual is, to a considerable extent, genetically pre–programmed. In its undefined form it can be related to the Socratic assertion that knowledge is innate, and to the deep–structure linguistic theory of Chomsky.

♦ **KC0216 Determinate system**
Description 1. A system so structured that if one knew all the relevant facts concerning the critical variables at any one moment, it could be predicted with a level of confidence approaching certainty what the relevant values of the variables at the next or any subsequent moment of time would be.
2. A type of general systems model, in terms of which the brain's functioning has been usefully simulated.

♦ **KC0217 Decay**
Description Deterioration or depletion of the components of a system. It is contrasted with positive entropy, which connotes a loosening or spreading apart of components, as a process of system decline. In decay, changes lead to a gradual transition from systemic wholeness or coherence to dissolution or morbidity. In radioactive decay matter effectively becomes energy. In organic death, also bound by conservation of energy rules, the physical system is depleted of something which is exchanged with the environment. The human body is constantly deteriorating (and renewing) and consequently exchanging energy, soul or substance with its environment or universe.

♦ **KC0218 Network data structures**
Description A form of hierarchical structure used in the logical organization of computer data bases and to be contrasted with tree structures. Whereas in the the tree or arboreal cases the branches of the hierarchies do not connect between hierarchies, in the case of network structures they do. Any record may be related to any other record in the file. The file may therefore be searched in a multidirectional manner through the hierarchy.

♦ **KC0219 Potential**
Description 1. The possibilities within any system, individual or collective, e.g., human potential.
2. Vector forces, i.e. energies and velocities in their field relationships. Vector field potentials are measured in a number of basic sciences and enter into theories of gravitation and electro–magnetism and the analysis of harmonic functions.

♦ **KC0220 Central place theory**
Description A theory concerning the distribution of points around one or more central points with which they have some dependent functional relationship. The theory has been developed by geographers in connection with the distribution of human settlements as forming a hierarchical pattern of central places in which the functional importance of the central place is determined by an index of centrality (separately described).

♦ **KC0222 Order in chaos**
Description 1. Chaos may be used as a term for extremely high and barely detectable forms of order in some phenomena, that formerly were thought to be characterized by randomness or chance. Stated negatively this order arises from a constraint in randomness that is termed by some an 'attractor' or 'strange attractor', and by others 'logos', or mathematical limit.
2. The introduction, into a stable system, of wave frequencies (such as harmonics) to the extent that they are deterministic in producing a non–random chaos. In applications of this theory, to weather forecasting for example, or to human heart fibrillation study, the efforts are to understand the logos or limit in both order and chaos. In one sense it is the 'load' of frequencies that a homeostatic system can bear.

♦ **KC0223 Standardization**
Description A process of establishing uniform standard measurements, methods, levels of quality, or levels of content to ensure interchangeability of elements from different sources and replaceability of elements in discontinued products, with a view to facilitating the exchange of goods and services.

♦ **KC0225 Ekistics**
Science of human settlements
Description Science in which the human settlement is conceived as an organism having its own laws. Through the study of the evolution of human settlements from their most primitive phase to megalopolis and ecumenopolis (predicted future city with related open land area which will cover the entire earth as a continuous living system forming a universal settlement), ekistics develops the necessary interdisciplinary approach necessary to its problems. The five ekistic elements which compose human settlements are: nature, anthropos, society, shells, and networks (roads, water supply, electricity) (men and women equally) (all types of structures within which anthropos lives and carries out various functions)

♦ **KC0226 Homeostasis**
Self–regulation
Description 1. Dynamic self–regulation (involving periodicity), namely the ability of a system to maintain its fundamental, internal balances even while undergoing various processes of change.
2. The regulating system which determines the homeostasis of a particular feature may comprise a number of cooperating factors brought into action at the same time or successively. If a state remains steady, it does so because any tendency toward change is automatically met by increased effectiveness of the factor or factors which resist change. When a factor is known that can shift a homeostatic state in one direction, it is reasonable to look for automatic control of that factor or for a factor or factors which act in the opposite direction.
3. A tendency toward maintenance of a relatively stable internal environment in organisms, in organizations, and in individuals (i.e. the psychological condition), with respect to changing external environmental conditions, and the processes of growth and decay.
4. A property of all systems that maintain critical variables within limits acceptable to their own structure in the face of unexpected disturbance.

♦ **KC0227 Subsystem**
Description A given system may be conceptually subdivided into subsystems. In other words, elements of a system may themselves be systems of a lower order. A distinction is made between general and special purpose subsystems. A general purpose subsystem is a separately operating system (within a larger system) which does not play a specialized role in the larger system. A special purpose subsystem also plays a specialized role in the operation of the larger system.

♦ **KC0228 Universe**
Description 1. In Synergetics (without the article 'the'), universe is the comprehensive, historically synchronous, integral–aggregate system embracing all the separate integral–aggregate systems of all men's consciously apprehended and communicated (to self or others) non–simultaneous, nonidentical, but always complementary and only partially overlapping, macro–micro, always–and–everywhere, omnitransforming, physical and metaphysical, weighable and unweighable event sequences. Universe is a dynamically synchronous scenario that is unitarily nonconceptual as of any one moment, yet as an aggregate of finites is sum–totally finite. (Aggregate means sum–totally but nonunitarily conceptual as of any one moment. Consciousness means an awareness of otherness. Apprehension means information furnished by those wave frequencies tunable within man's limited sensorial spectrum. Communicated means informing self or others. Nonsimultaneous means not occurring at the same time. Overlapping is used because every event has duration, and their initiatings and terminatings are most often of different duration).
The total of experiences is integrally synergetic. Universe continually operates in comprehensive, coordinate patternings which are transcendental to the sensorially miniscule apprehension and mental comprehension and prediction capabilities of mankind, consciously and inherently pre–occupied as man is only with special local and nonsimultaneous pattern considerations. Universe is the ultimate collective concept, namely the collection of all intelligible, inherently separate evolutionary event aspects which latter apparently occur exclusively and only through differentiating considerations which progressively isolate the components of whole and inclusive sets, super–sets, and sub–sets of generalized conceptioning in retrospectively abstracted principles of relationships. The generalized comprehensive principles of interrelationships progressively discovered as governing mankind's subsidiary generalized principles are embraced by mankind's definition of Universe.
Universe is the starting point for any study of synergetic phenomena and for consideration of all problems, in order to avoid all the imposed disciplines of progressive specialization, which create the risk of omitting strategically critical variables.
2. In relativity theory one component of a multi–universe reality.
3. Our universe as perceived by astronomers.

♦ **KC0230 Cybernetics**
Description 1. The interdisciplinary science and comparative study of the automatic control systems formed by the nervous system and the brain, and by various mechanical–electrical systems, and in general, of those methods of communication and control which are common to living organisms and machines. As such, it incorporates and unifies the work of the servomechanisms and systems engineer, the communications engineer, and certain aspects of the work of the physiologist, neurologist, psychologist, sociologist, and economist.
2. The science of proper control within any assembly that is treated as an organic whole.
3. The science whose focus is a system, either constructed, or so abstracted from a physical assembly, that it exhibits interaction between the parts, whereby one controls another, uninfluenced by the physical character of the parts themselves.
4. The science of effective organization or the art of relating things together so that what is desired occurs, namely the discipline of human action.
5. The science of constructing, manipulating, and applying cybernetic models which represent the organization of physical entities (such as animals, brains, societies, industrial plants, and machines) or symbolic entities (such as information systems, languages, and cognitive processes).
Disagreement exists concerning the generality of the science, especially in relation to general system theory (separately described) which has objectives identical to those of the founders of cybernetics. For some the two disciplines are co–extensive, whilst for others each is regarded as a branch of the other.

♦ **KC0231 Heterogenization**
Diversification
Description A basic principle of biological and social processes is increase of diversification and heterogenization, accompanied by symbiotization and the elaboration of networks and of non–repetitious complexity.
Heterogenization may be synonymous with integration if a mutual causal paradigm is used, although if a unidirectional causal paradigm is used, integration may then mean homogenization. Similarly, internationalization may constitute homogenization or heterogenization, depending upon which paradigm is used.

♦ **KC0232 Functionalism**
Description 1. A theory of culture which analyzes the interrelatedness and interdependence of patterns and institutions within a cultural complex or social system and which emphasizes the interaction of these forms in the maintenance of sociocultural unity or in meeting biosocial requirements.
2. The theory or practice of achieving cooperation or union between governmental units by gradual integration of economic, social and other functions, rather than immediate political federation. A particular functional system of endeavour delineates itself through transaction patterns and a suitable organizational frame for that system is determined by the needs of the function being formed. The framework is developed and (ideally) modified as the function being fulfilled changes. A multiplicity of forms and levels of organization each reflects a system of transactions which may or may not produce institutions at the world level.

♦ **KC0235 Paradigmatology**
Description A science of structures of reasoning which vary from discipline to discipline, from profession to profession, from culture to culture, and sometimes even from individual to individual.

♦ **KC0236 Orthogenesis**
Description Variation of organisms in successive generations along some predestined line resulting in progressive evolutionary trends independent of external factors. A theory holds that social evolution takes place in the same direction and through the same stages in every culture despite differing external conditions.

♦ **KC0240 Organization**
Description The condition or manner of individual elements being arranged into a coherent unity or functioning whole of interdependent parts. Characteristics of organization, whether of a living organism or a society, are notions like those of wholeness, growth, differentiation, hierarchical order, dominance, control, etc. many of which do not appear in conventional physics but are preponderant in most biological and social phenomena.

♦ **KC0241 Iatrodisciplines**
Description Medicine or healing combined (in the 16th and 17th century) with other disciplines, typically physics, chemistry and mathematics (astrology), as in iatrophysics, iatrochemistry and iatromathematics.

♦ **KC0243 Cultural integration**
Description Integration is a structural quality of culture associated with the degree of coherence in some determinate fashion of the diverse parts of the culture. There are differences of opinion on the contents of culture, the levels of cultural integration, and the forms of cultural integration. The following types (described separately) may be distinguished: configurational, connective, logical, stylistic, functional, and regulative.

♦ **KC0244 Policymaking**
Description Policymaking functions in an organization result in normative planning and are directed toward the search and establishment of new norms that will define those values which will be more consonant with the problematic environment. Normative planning therefore occurs when the purpose of planning action is to change the value system in order to achieve the required consonance with the environment.

♦ **KC0245 Order**
Orderliness
Description 1. Evenness, regularity, or discipline of terms or elements, or of their connection. Any connection of quantitative, qualitative, mechanical, or teleological reference. Complete or perfect order is complete equality and is equal to zero–entropy, since any increase in entropy destroys order and organization.
2. Order basically consists of a set of similar differences which (although they may be distinguished) also mutually relate to one another. Two types of differences may be distinguished, constitutive differences (which determine the essence of the order) and distinctive differences (which form some pattern and in another sense the formation of part of an overall hierarchy of orders). So that what at one level of consideration is a constitutive difference may at another be a distinctive difference, thus permitting the organization of orders into hierarchies to proceed without limit.
3. A surrounding may be said to possess order if a simple description of it can be supplied.
4. Orderliness, whether energy or structure based, may be described both abstractly (namely from a quantitative point of view) and concretely (namely from the point of view of the specifics which characterize a given system). The analysis of the orderliness of a living system, for example, includes: the definition of the quantity of orderliness of energy in thermodynamic entropy units (with reverse sign), and of structures in negentropy information units; the definition of the rate of change of orderliness as an index of the rate of evolution of the energic and structural aspects; and the definition of the specifics of energic and structural orderliness, that is the content of the corresponding information.

♦ **KC0246 Simulation**
Description 1. A representation of some aspect of the real world in such a manner that problem–solving is facilitated. Verbal descriptions, schematic diagrams, or simple tabular representations of reality may be considered simulations. More recently, mathematical expressions and equations have been used to capture succinctly very complex interrelationships necessary to an understanding of the operation of large enterprises and complex production and control processes. Such models can then be manipulated on computers to determine how the relationships and conditions change in response to new situations. A simulation model is therefore an abstraction from reality but hopefully close enough to the real world to permit useful observation, analysis or evaluation. A simulation or a simulation exercise is therefore an experiment performed upon a model.
2. A simulate of a given system is an object that copies the latter in some respect, such as shape or function. When artificial or conceptual, simulates are often called analogs or models of the original system. The design of a concrete simulate or material model is based on some conceptual model, sometimes a whole theory, of that system.
A simulation can be distinguished from a game (described separately) according to the degree of formality of the rules for translating external variables into simulation variables. The more formal the rules, the more it is a simulation rather than a game, and vice versa.

♦ **KC0247 Transient systems**
Description A system which is normally in a steady or equilibrium state is transient during the passage from one equilibrium to another. The term transient can be applied to a system whether the disequilibrium is short–lived or persistent. Biological (and social) systems, however, are transient as long as they may be described as living.

♦ **KC0248 Vertical integration**
Description In applied economics the control by a single entity of all resources and raw materials required, all extracting, converting, processing or manufacturing plant, equipment and processes needed, and all distribution means and outlets to the market. An example is a food conglomerate which owns farms and plantations, canneries, trucking companies and retail food stores. Another example is a socialist economy where the entity which vertically integrates is the state. Vertical integration is sometimes associated therefore with private or state monopolies, unfair business practices and unfair labour practices, as for example, in the agricultural sectors in both free market and planned economics.

♦ **KC0249 Meaning**

♦ **KC0253 Syndrome**
Description 1. A complex of symptoms (i.e. indicators) of dysfunction, more or less concurrent.
2. Medical syndromes, for example those indicating systemic stress or organic pathologies.
3. Characteristics of a dysfunctioning organization, government, group, etc.

♦ **KC0255 Synthesis**
Description 1. Composition or combination of parts or elements so as to form a whole, particularly the combination of often varied and diverse ideas, forces or factors into one coherent or consistent complex.
2. An interdisciplinary bringing together of major groups of sciences which have a mutual correlation, or as a bringing together of two or three sciences which were already rather closely allied.
3. One of the stronger conceptions of the unity of science suggests the existence of a model or pattern of scientific theory, of which each particular theory is an instance, so that higher level sciences recapitulate, although with more complex elements, the structure of lower level sciences, with genuine novelty emerging at each new level. Although reduction is not a necessary part of this synthetic view, the sciences are nearly always arranged in a hierarchy in which a part–whole relationship of some sort is held to obtain between levels, although the wholes may not be explainable without residue in terms of the parts. The sciences thus form a totality, the unit of which is provided by the archetypal structure that reappears at each stage.

♦ **KC0256 Value analysis**
Value control; Value engineering
Description Value is here conceived to be the lowest price that must be paid to provide a reliable function or service. The technique sensitizes engineers to achieve the absolute minimum cost compatible with functional requirements (through constant focus on questions such as: what is it; what does it do; what does it cost; what is it worth; what else will perform the same function; what does that cost). Considerable cost savings may be achieved.

♦ **KC0257 Multi–ordinality**
Description Multi–ordinality is a concept and methodology for dealing with levels of abstraction. A word is multiordinal when, without any change in its dictionary meaning, it is used in the same context to refer to different orders of abstraction. The meaning of a multiordinal word is therefore determined by the level of abstraction at which it is used. Misunderstanding and misuse of the multiordinal characteristics of some words impedes communication and comprehension and may indicate personality maladjustment.

♦ **KC0258 Interlock**
Description States or positions of operating parts in a system held by any means (mechanical, electro–optical, magnetic, electrical, etc.) to assure performance. Interlocking increases reliability and safety. Used in weapon and defence systems. A cybernetic, man–machine interlock is the Fail–Safe principle.

♦ **KC0259 Infinity**
Description 1. One of the properties of space and time considered as limitless quantity or extension.
2. Philosophically and logically infinity is a universal category, and therefore, in a qualitative sense, it connects the particular phenomena of space and time of which infinity is predicated. Therefore it connects all other space–time phenomena.
3. Absolute being or deity.
4. Mathematical concepts, as much intuited as reasoned, which variously state that the sequence of numbers can be completed and therefore the infinite is actual, or that the series cannot be completed and therefore infinity is only potential or nominal.

♦ **KC0260 Lateral thinking**
Description Divergent thinking or creative thinking whose negative injunction is not to reason solely in a single circuit or chain (particularly in a branching chain characterized by yes, no or valid, invalid eliminations). Positive injunctions in this methodology include reformulation of problems, epistemological reexamination of premises and parallel and multiple cognitive approaches to solutions, etc. Lateral thinking is said to be complementary to vertical or chain type thinking.

♦ **KC0261 Formalism**
Description 1. Adherence to forms or structures of thought, techniques, conduct, organization, etc, as a preeminent element in methodology, activity or creative work. Such forms have set relationships between their parts, including temporal ones such as order of succession, spatial ones such as relative magnitudes and distances, and values, such as more or less expensive, useful, well–recognized, etc.
2. The exercise of an art which subordinates social content, identified, by Soviet critics, with cosmopolitanism.

♦ **KC0262 Universal grammar**
Description The language allegedly already present in the structures of the human mind, including the deeper structures. From this arises surface language with its interrelated parts of speech, syntax, and rules of orthographic and phonetic word modifications, to give meaning.

♦ **KC0265 Communication**
Description 1. The binding into a society of individuals, by the use of shared language and signs. The sharing of common sets of rules, for various goal–directed activities.
2. The dynamic process underlying the existence, growth, change, and the behaviour of all living systems (whether individual or organization), through which the organism or organization relates itself to its environment, and relates its parts and internal processes to one another. It is the communications that occur and the patterns of intercommunication which ensue that define and determine the structure and the functioning of any organization. The basic phenomenon of communication is that an organism (or organization) takes something into account, in whatever form it is presented (or not presented).

♦ **KC0266 Structure**
Description 1. The arrangement of a system's subsystems at any given moment.
2. An interrelated set of events that return upon themselves in a cyclical way.
3. A self–stabilizing energy–event complex.
4. The static frameworks of a system.
5. The manner in which the different parts composing a whole are disposed in relation to each other at a given moment.
6. A system which presents laws or properties of totality as systems.
7. The set of components and relations between components making up the unity of a system.

♦ **KC0267 Balance**
Description 1. The stability or efficiency resulting from the equalization or exact adjustment of opposing forces. A steadiness resulting from the proper adjustment to one another of all elements, when no one element or constituting force outweighs or is out of proportion to another.
2. An aesthetically pleasing integration of elements (as in a work of art) usually achieved by giving

each element only its due prominence or significance and often by allowing one element to stand in contrast to, oppose, or otherwise be matched by another.

♦ **KC0268 Process**
Description

♦ **KC0269 Practice**
Description 1. The application of human effort.
2. Action. In Aristotle, thought is action. In Marx, practice and theory are a dialectical unity of opposites; mental actions have no meaning unless they find concrete expression in material behaviour.
3. The regular exercise of a liberal profession, or motor activity such as a sport.

♦ **KC0270 Ensemble**
Description 1. In determining a physical system's macroscopic state a statistical sample or collection of a large number of identical many–particle systems, replicating the study system, which all have identical macroscopic states; for example, the microcanonical ensemble; and which allows for probabilistic values.
2. In music, a small number of performers, or a chamber work for such players.
3. A theory of fellow–craftsmanship among actors which adds a dimension to dramatic productions.
4. Any group of performing artists.
5. A group of any public constructions intended to be aesthetically unified, such as around urban squares, grand places and the like.

♦ **KC0271 Abstraction**
Description 1. Cognitive activity, e.g. scientific investigation, which is selective in regards to consideration of all aspects of a subject by eliminating what it deems nonessential by criteria of goal or method.
2. Cognition of the essence of phenomena.
3. Meditation state.

♦ **KC0273 Family**
Description 1. Human kinship in a legalistic sense, by genetic factors, marriage, or adoption, at the closest and smallest bonding level, usually associated with cohabitation of its members, concurrently or successively, in at least one stage of their relationship.
2. The family circle subjectively perceived, including friends, who usually and periodically come together; i.e. those considered members of the family. By some, pet dogs and other domesticated animals are considered as members.
3. Kinfolk or family circle members who share, to some degree, a common life, taking responsibility for one another physically, materially and morally.
4. Human beings that in some societies were considered one person's property, i.e. wives, concubines, slaves, etc.
5. A term used for taxonomic purposes to delimit a group.

♦ **KC0274 Commensurability**
Description A common measure between like quantities. It is contained in each of the quantities under consideration an integral number of times. The lengths of a diagonal and a side of a square, for example, are like quantities but are incommensurable, lacking the integral common measure. An equally well known example of incommensurability are the areas of a circle and of a square constructed on the radius of the circle. Commensurability is seen as a rational number expressing the ratio between quantities. Incommensurability is expressed by an irrational number.

♦ **KC0276 Differentiation**
Description The principle of differentiation is ubiquitous in biology, the evolution and development of the nervous system, behaviour, psychology, and culture. It is the process of transformation from a more general and homogeneous condition to one which is more specialized and heterogeneous. Wherever development occurs it proceeds from a state of relative globality and lack of differentiation to a state of increasing differentiation, articulation, and hierarchic order.

♦ **KC0277 International economic order**
Description

♦ **KC0278 Order of universe**
Cosmos
Description 1. The structure of relationships between all existing things.
2. The purpose (s) or end (s) for which or to which the universe tends.
3. The cosmic work of a supreme being.
4. The totality of all things whether matter, energy, time, space, life consciousness or mind, and whether known or unknown, conceived as a unity.
5. This universe, which is composed of many galaxies including our own, as studied by cosmologists, and as understood to go through different phases in space and in time.
6. Theologically; providence in Christianity, dharma in Buddhism.
7. Scientifically; the laws of physics and mathematics.

♦ **KC0280 Interdisciplinarity**
Description 1. Interaction among two or more different disciplines. The interaction may range from communication and comparison of ideas to the mutual integration of organizing concepts, methodology, procedures, epistemology, terminology, data and organization of research and education in a fairly large field. (An interdisciplinary group consists of persons trained in different fields of knowledge or disciplines with different concepts, methods, data and terms, organized into a common effort on a common problem with sustained intercommunication among the participants from the different disciplines).
2. A common axiomatics for a group of related disciplines which is defined at the next higher hierarchical level of disciplines. A distinction may be made between: teleological interdisciplinarity at and between the empirical and pragmatic levels and sublevels; normative interdisciplinarity, signifying the step from the pragmatic to the normative level (at which the question of good and bad is raised); and purposive interdisciplinarity, bridging from the normative to the purposive level. Interdisciplinarity therefore constitutes an organizational principle. It leads to a two-level coordination of terms, concepts, and principles which is characteristic of a two–level multigoal system. With the introduction of interdisciplinary links between organizational levels, the scientific disciplines defined at these levels change in their concepts, structures, and aims.
3. A scientific category related mainly to research. In this respect it corresponds both to a certain theoretical level of formation of science and to a particularly important turning point in the history of science. It is basically a mental outlook which combines curiosity with openmindedness and a spirit of adventure and discovery. It includes the intuition that relationships exist between all things which escape current observation and that there are analogies of behaviour or structure which are perhaps isomorphic. It is not learnt, it is practised as the fruit of continual training and systematically working towards more flexible mental patterns.
4. Six types of interdisciplinarity may be distinguished in order of stage of maturity: (a) indiscriminate interdisciplinarity, including all kinds of encyclopaedic endeavours, usually conceived as a form of vocational training for those having to handle a variety of problems; (b) pseudo–interdisciplinarity, in which analytical tools such as mathematical models or computer simulation are applied to different subject matter as a uniting core for cross–disciplinary research and training, but with subordination of the different contents; (c) auxiliary interdisciplinarity, whereby an auxiliary disciplinarity arises from the application of the methods of one discipline to the subject matter of another discipline; (d) compositive interdisciplinarity, whereby a composite discipline arises from the need to focus a variety of disciplines on a particular issue area (e.g. peace research, urban planning) and is usually noteworthy for its technological instrumentality in pursuing a hierarchical sequence of clearcut goals which change person–environment systems or even innovate such systems; (e) supplementary interdisciplinarity, in which disciplines in the same material field develop a partial overlapping in a supplementary relationship between the respective subject matters. (The supplementation is induced from a correspondence between the levels of theoretical integration of two or more disciplinary subject matters, and is looked for and tentatively established in order to reconstruct life or social processes more fully); (f) unifying interdisciplinarity resulting from an increased consistency in the subject matter of two disciplines, paralleled by an approximation of the respective theoretical integration levels and methods.
5. Three types of interdisciplinarity may be distinguished: (a) linear interdisciplinarity, when a crude phenomenon belonging to one discipline is legalized by a law belonging to a second discipline, in the sense that the law is borrowed and adapted by the first discipline for the benefit of the phenomenon. (This has been commonly called by terms such as multidisciplinarity, pluridisciplinarity, or crossdisciplinarity); (b) structural interdisciplinarity, in which interactions between two or more disciplines lead to the creation of a body of new laws forming the basic structure of an original discipline that cannot be reduced to the formal combination of its generators, and may itself absorb them through a syncretic tendency (e.g. electromagnetism); (c) restrictive interdisciplinarity, in which the field of application of each discipline brought into play by the concrete objective is restricted, such that each discipline imposes technical, economic or other constraints on the application of the others, but there is no other interaction between the disciplines.

♦ **KC0281 Planetary synthesis**
Description Collective consciousness of individual, national and world relationships as partly a product of the spiritual evolution of the human race, and partly as an outcome of programmes of human development.

♦ **KC0282 Dialectic**
Description 1. In Greek philosophy the question and answer form of debate or discussion in which progressively analytical enquiry sought to expose the bases of opinion (among the Sophists) or to attain truth (Socrates, Plato).
2. Anciently, in general, the analytical adjunct of rhetoric.
3. In Scholastic philosophy, disputation among experts in which a special logic of argumentation was observed.
4. In the Renaissance Ramist pedagogy, the first two parts of the ancient five–part rhetoric: invention, in which materials for a speech or written argument were collected and in which the topic or topics were analyzed; and disposition, in which the material was arranged for best presentation.
5. In Hegelian philosophy, the substitution of the ancient question–answer–truth paradigm by the thesis–antithesis–synthesis model.
6. Figuratively, the interactions in any polarity.
7. The causal or developmental process in nature and in civilization consisting of a dynamic tension between directions of momentum of opposed or unharmonized energies which give rise or expression to third forces e.g., the opposition, in a field, of entropy ('free' energy tendency) and the field limit, boundary or form ('bound' energy mass, locus, etc), which give rise to, inter alia, a system, life, being, information or intelligence, and ultimately, (in cybernetic terms) to a human control mechanism on dialectical change.

♦ **KC0286 Relativity**
Description Relativity theory unites and integrates all things within the physical universe as a system of internally related entities in a four–dimensional continuum, which knits together their properties, movements and everything that makes up their substantial being. It moulds the entire universe into a self–contained unity, the parts of which cannot rightly be understood in isolation, or by any process of more than provisional dissection.

♦ **KC0287 International political integration**
Description Political integration is one component of international integration (described separately) and may be variously defined:
1. A political system is integrated to the extent that the minimal units (individual political actors) develop in the course of political interaction a pool of commonly accepted norms regarding political behaviour patterns legitimized by these norms.
2. International political integration involves a group of nations coming together to regularly make and implement binding public decisions by means of collective institutions and/or processes rather than by formally autonomous means. It implies that a number of governments begin to create and to use common resources to be committed in the pursuit of certain common objectives and that they do so by foregoing some of the factual attributes of sovereignty and decision–making autonomy (in contrast to more classical modes of cooperation such as alliances or international organizations). It can therefore be defined as the evolution over time of a collective decision–making system among nations.
Political integration generally implies a relationship of community, a feeling of identity and self–awareness in which the essence of the integration relationship is seen as collective action to promote mutual interests.
Four different types of political integration may be distinguished: institutional integration, policy integration, attitudinal integration, and the concept of a security community (in which there is reliable expectation of nonviolent relations).

♦ **KC0290 Integrative education**
Description 1. Integrative education involves the acceptance of a new corpus of universal principles and an appropriate pedagogy. As distinct from integrated education (representing the most advanced stage at a lower, previously conceptualized, level of social development), integrative education represents the next, or higher level which, in consonance with the principle of emergent levels, possesses its own, unique qualities and functions. These comprise the ordering element both to integrate activities within that emerging level itself and also to direct the activities of the lower, previously conceptualized, levels as well. An integrative approach within the eductional process, makes it possible to move either horizontally or vertically in an appropriately structured new curriculum. Integrative principles can then be employed to deal with topics and problems so as to develop interdisciplinary linkages, or alternatively probes can be made into specific areas or disciplines, as deeply as such specialization warrants.

A fundamental distinction is drawn between integration in the sense of synthesizing presently accepted postulates with their attendant corpus of knowledge, methodologies, and modes of behaviours, and integration in the sense of employing a new conceptual model constructed for the express purpose of developing a holistic educational philosophy and methodology capable of understanding and documenting the principles inherent in the natural and social sciences, together with the major humanistic systems of mankind. The distinction is between integration within an existing model and the development of a new model based upon universal integrative principles.
2. The central tasks of a unifying, general or integrative education are: to demonstrate, in the most effective way possible, interrelationships between all disciplines; to teach the individual how to discover such relationships for himself; and to put this integrative know–how to work in every life situation. Individuals with such an integrative education can understand themselves and their universe as organic wholes.

♦ **KC0293 Crisis management**

♦ **KC0296 Scientific revolution**
Description During normal periods of its development, science attempts to force nature into the conceptual framework which is supplied by professional education on the basis of the fundamental commitments accepted by scientists as a group. In a period of scientific revolution, one or more scientists feels that certain generally accepted paradigms are either inadequate in regard to the phenomena they claim to explain or at least unable to solve all the problems which can legitimately be asked. The scientific community is then forced to face such anomalies which in turn leads to a crisis situation in that science. In such a crisis situation extraordinary investigations begin that lead the scientific community to the elaboration of a new set of basic commitments, to a new paradigm or set of paradigms which then constitute a new basis for the practice of science.

♦ **KC0297 Teletics**
Description The study of purposes and purposefulness. It deals with the purposes of human beings in this world and is not concerned with the grand purposes of a universe or deity. The elementary principles of teletic inquiry relate to: the time sequence concept of purposeful action; the multiplicity of organizational purposes; the patterns that bring order out of this multiplicity; and the balance between clarity and vagueness in purpose formulation. Teletics is considered to be distinct from teleology.

♦ **KC0299 Unitary symmetry**

♦ **KC0300 Homologous series**

♦ **KC0301 Theoretical integration level**
Description Each empirical discipline attempts to reconstruct the reality of its subject matter in theoretical terms in order to understand, explain and predict phenomena and events involving the subject matter. In doing so the categorical nature of the relevant observables determines the categorical level of theoretical integration of the fundamental and unifying concepts. With regard to their present level of theoretical integration, disciplines can be distinguished according to their achieved state of maturity. At one extreme a discipline may be absorbed by mere description and phenotypic taxomomies of its subject matter; at the other a discipline may have developed a single theory system powerful enough to cover almost all the phenomena of its subject matter. Mutually exclusive levels of theoretical integration may exist within one discipline, because of the lack of relationship between theories and between observables. There may also be apparently unbridgeable gaps between the theoretical integration levels of some empirical disciplines, whilst other disciplines may show increasing convergence of their respective levels of theoretical integration.

♦ **KC0302 World soul**
Anima mundi
Description 1. In Platonic philosophical myth (Timaeus) a force that plays an intermediary role between the Forms and the material world. It is the principle of order and of life from which the Demiurge derives individual human souls.
2. In Plotinus the intermediary between the Nous and matter, creating in the latter according to the ideas contemplated in the Nous.
3. In speculative Christianity, the Holy Spirit.
4. In Renaissance philosophy, for example in Campanella, the first instrument of Sophia, Divine Wisdom, through which everything is made. In Cornenius the spiritus mundi which is life infused into the world so that all things may be moved sympathetically thereby.
5. In other philosophers, the spirit of nature (Herder, H More); the ether or something closely related to it (Newton); the arché or quintessence (Paracelsus, van Helmont); the natura naturans (Spinoza); etc.
6. In part, de Chardin's interpretation of Vernadsky's noosphere.

♦ **KC0303 Coordinate indexing**
Description

♦ **KC0304 Sociology of knowledge**
Description 1. Study of the social and historical conditionality of knowledge and knowledge acquisition, focused on the social history of ideas, idea types, and the morphology of the evolution of ideas.
2. The empirical investigation of States of Consciousness (i.e. opinions, values, information content, etc) among the component groups and classes of society which emphasizes the influence of the individual's surroundings.

♦ **KC0306 Strategy**
Strategic planning
Description 1. Strategic decisions deal with the way an organization or system of organizations moves into the future through a changing environment. They relate to the very purposes and essence of the organization itself, and of its human, material and financial resources. They involve a choice of objectives, a contrivance of means, and both of these involve an assertion of will rather than responses deterministically derived from what has happened previously. They are purposive thrusts into the future rather than decisions directed by testable logic or continuity of circumstances. Such decisions rest largely on judgement, foresight and imagination.
2. Game theory provides a unifying viewpoint for all types of conflict situation (whether military, political, or economic), and gives two meanings to strategy. A pure strategy is a move or a specific series of moves by a participant in a conflict in which successive gains are clearly delineated. A grand or mixed strategy is a statistical decision rule for deciding which particular pure strategy a participant should select in a particular situation to improve the probabilities of a desired payoff.
3. Strategy is contrasted to tactics, which is a specific scheme for employment of allocated resources. It is also contrasted to policy which is a contingent decision, where strategy is a rule for making decisions. Specification of strategy is forced under conditions of partial ignorance, when alternatives cannot be arranged and examined in advance, whereas in the case of policy, under conditions of risk (alternatives and their probabilities known) or uncertainty (alternatives known but not their probabilities), the consequences of different alternatives can be analysed in advance and decision made contingent on their occurrence.
4. Strategy means (in its literal and limited sense) the art of the general in the application of large–scale forces against a military opponent.
5. Strategy is a concept of an organization's function which provides a unifying theme for all its activities.
6. There is considerable confusion in the uses of the words strategic planning, strategic plan, and strategy. Strategic planning may be defined as the process of deciding the basic mission of the organization, the objectives which it seeks to achieve, and the major strategies and policies governing the use of resources at its disposal. The result of this process, among other things, is a strategic plan. Strategy may then be defined as a specific action, usually but not always the development of resources, to achieve an objective decided upon in strategic planning. Developing a strategy is usually a very difficult task involving questioning of old methods, exploring unfamiliar contexts, facing up to an objective evaluation of strengths and weaknesses, forcing important changes on people and organizational arrangements, and taking high risks with the organization's resources. This has to be done in a world of rapid change, and it has to be done continuously.
7. A weak and a strong definition of strategy may be distinguished: rationalization actively effected to order the totality of (internal or external) elements which may play a role in the determination of the general orientation of a social entity through its environment; or the rational organizing and directing function of the totality of forces (resources and systems, which are not all entirely or constantly mobilized) of social entities in their reciprocal negations (more or less focused and not necessarily correlated and equivalent). Essentially strategy is negation through the imposition of constraint on the freedom of action of the opponent.

♦ **KC0307 Attribute space**
Description The infinite number of attributes of social units can be defined as a vector space bounded by the total number of social units. Within this space, each attribute can be described as a vector, with its component equal to the value of each social unit on this attribute, and its direction from other vectors being a function of the correlation between the attributes for the social units. The intercorrelations between attributes constitute the social system.

♦ **KC0308 Integrated information system**
Description An information system which aims to channel all the data of an organization into a common data base and service all data processing and information functions for the entire organization dependent upon it. Although full implementation of a totally integrated system appears to be technically and economically impractical at the present time, many organizations attempt to move toward a greater degree of integration due to: dissatisfaction with the hierarchical approach because of fragmentation and noncoordination of the information function; and potential ability to effect more integration due to technological developments, especially in the area of computers and telecommunications.

♦ **KC0310 Tao**
Description 1. In Confucianism, man's path through life based on moral conduct and innate noble virtue (jen) which gives rise to conscience.
2. In neo–Confucianism, similar to Platonic Form, a principle (li) which exists as agent to a patent substrate (ch'i), e.g. matter.
3. In Taoism, the creative and sustaining force in the universe, in some ways like Dharma in Buddhism, Rita in the Vedas, Providence in Christianity, Pronoia in Greek philosophy, and the Anima Mundi of later speculation. In other ways it is like the Akashic ether of Hinduism. Since it is associated with the feminine Valley Spirit in Lao Tzu it may also be affiliated to the Hindu concept of Sakti. However, it encompasses both masculine and feminine metaphysical characteristics (Yang and Yin).

♦ **KC0311 Humanities**
Description All curricula that do not teach physical and applied sciences (which includes economics but not business). Their principle intent is to provide skills to receive the cultural legacy through a high degree of literacy, a sense of historical development, and through enhancement of the reasoning process. Notable among the humanities are the behavioural and social sciences, philosophy, law, literature and art, government, history and foreign languages.

♦ **KC0312 Decision theory procedure**
Description A demonstration technique for determining the logical truth of a proposition. The procedure may involve the use of truth tables in applicable parts of the predicate and propositional calculii. Diagrammatic and algebra techniques are also utilized.

♦ **KC0313 Babel syndrome**

♦ **KC0315 Party unity**
Description In Communism the requirement that the individual conform exactly to the Party line. In the name of Party unity a vast number of crimes have been perpetrated by Communist States and parties against individuals. International Party unity has also required foreign intervention in the affairs of Hungary and Czechoslovakia.

♦ **KC0316 Similarity**
Likeness — Image
Description 1. In geometry, identically shaped objects, of same or different size. Ratio constants are used as similarity factors (for distances between two pairs of corresponding points); angles between corresponding lines of similar figures are equal; and other relationships proving similarity exist as well.
2. In physics, the fields of the corresponding parameters of two systems in space and time.
3. In Aristotle, univocal sameness based on quality or analogical similarity. In other philosophies, developing from Aristotle's notion that likeness is the very first feature attributable to the nature of quality, the causality of partial sameness that allows for the perceiver to apprehend the object, or beings to communicate.
4. In Christian theology, applied to man as the image and likeness of God (Genesis 1.26.27), the sameness viewed as the total process of redemption, moving from the creation of the image to its vivification by grace towards a dynamic assimilation to God.

♦ **KC0318 Phase space**
Description The set of all points in n–dimensions describing possible states of a system, when the system is uniquely defined by a set of n numbers. To describe the behaviour of a system, it

is sufficient to specify the possible paths in phase space, or in other words the succession of states through which the system passes.

♦ **KC0320 Organization of knowledge**

♦ **KC0322 Cycles**

♦ **KC0323 Creativity**
Description 1. An ability to see new relationships, to make or otherwise bring into existence something new (possibly of aesthetic value), to produce new ideas and new solutions to problems, and to deviate from traditional patterns of thinking.
2. The emergence in action of a novel relational product, growing out of the uniqueness of the individual. Creative action achieves increased order or unity in some situations.
3. The disposition to make and to recognize valuable innovations.
4. The production of meaning by synthesis.
5. The capacity to grasp two or more mutually distinct realities and to derive innovation from their juxtaposition.

♦ **KC0324 World–line**
Description 1. In the General Theory of Relativity, the four–dimensional world point or event trajectory of a particle represented graphically. In the presence of a gravitational field the world–lines of a freely moving particle and of light are curved.
2. In the Special Theory of Relativity of particle moving uniformly and progressing in a linear motion has a straight world–line but angularly inclined to the time coordinate axis. An angle of 45 degrees represents the world–line of light.
3. In Kabbalistic cosmology the green line which circles the world.

♦ **KC0325 Homonymy**
Polysemy
Description An extension of the linguistic concept of homonyms to include correspondences of orthographic, graphic, signal, or phonetic composition, in any combination, and applying to sets of words, sets of signs or combinations thereof, providing that there is a difference in denotation or connotation between such homonymous forms. Circumstances of homonymy in any given instance may also be termed polysemic where the words of the homonymous set are related etymologically to a root from which two or more meanings have extended. Thus polysemy (multiple meaning), which is seen in single terms of some linguistic evolution, may be veiled if unrecognized at the verbal root level, since words may develop from the root in a divergent way and their relationship as etymological variants. One also detects a concealed homonymy since they correspond in root but differ in sense or meaning.

♦ **KC0326 Homology**
Description A likeness, short of identity, in structure or function between parts of different structures (or systems). In the case of organisms, this may be due to evolutionary differentiation from the same or a corresponding part of a remote ancestor. Homology implies both substantial and formal analogy. Logical homology makes possible not only isomorphy in science, but as a conceptual model has the capacity of giving instructions for correct consideration and eventual explanation of phenomena.
2. A one–to–one correspondence of two coplanar geometrical figures whereby the junction lines of correspondent points are copunctal in the centre of homology and the junction points of correspondent lines are collinear on the axis of homology.

♦ **KC0327 Transaction analysis**
Description Transactions are contacts or dealings. In international relations, for example, transactions have to do with contacts and dealings, both governmental and nongovernmental, between states. Transactions may, for example, be analyzed with respect to substance (e.g. concerning economic, political, cultural, social or technical matters), or with respect to the states involved and the directions of the transactions between them. The volume and frequency of such transactions is also considered. From such information, permutations of substance, direction, intensity, and time define patterns of transaction flow, and these in turn describe structure and process in the international system. Analysing transaction flows then theoretically opens the way to observing and recording who deals (or has dealt) with whom, how, or how much, about what, and when (and under some circumstances, provide projections of probable transaction patterns in the future). Different transformations of the basic data may be used to measure relative direction, relative intensity, dependence, interdependence, partnership, concentration, acceleration, and other quantities or qualities in transaction relationships. Transaction analysis may be used to detect international or regional integration and community formation.

♦ **KC0328 Whole system principle**
Description 1. The known behaviours of the whole plus the known behaviours of some of the parts may make possible discovery of the presence of other parts and their behaviours, kinetics, structures, and relative dimensionalities. The definitive identifications thus permitted may implement conscious synergetic definition strategies with incisive prediction effectiveness.
2. Given the sum of whole system pattern conception, its component behaviours may be differentially discovered and predictably described as required by the already evidenced behaviour functions implicit in the a priori–definitive experience and conceptioning of any given experience–verified system. Thus by the law of the whole system, as corollary of synergy, the component behaviours of systems may be predictably differentiated as primary and secondary componential subdivisions of the whole system and then progressively isolated and locally reconsidered for further dichotomy.

♦ **KC0330 General semantics**
Description

♦ **KC0332 Fuzzy sets**
Description

♦ **KC0333 Classical field theory**
Description In a physical system consisting of a large number of particles, the interaction of the particles can be described with the aid of a concept of a field of force (electromagnetic or gravitational). The particle is assumed to create a field around itself and a certain force then acts on every other particle located in this field. As a consequence of the theory of relativity the field itself acquires a physical reality since particles at a distance from one another cannot be said to interact directly. A particle must first interact with the field which subsequently interacts with other particles.

♦ **KC0335 Symbiosis**
Description Mutual cooperation between organisms, persons, or groups, especially when ecological interdependence is involved.

♦ **KC0336 Organized complexity**
Complex systems
Description 1. Any organized collection of entities interconnected by a complex network of relationships. The organization of a system is simple if the system is a serial or an additive complex of components, each of which is understood. As soon as strict sequential sequences or linear additivity is transcended, an organized system becomes rapidly more complex, usually too complex for detailed analysis into superposable parts or effects. At the other extreme from organized simplicity is chaotic complexity where the number of entities involved is so vast that the interactions can be described in terms of continuously distributed quantities or gradients, and do not need to be specifically identified with regard to the individual entities. Such systems can be described by the methods of statistical mechanics which merge with those of classical mechanics when the collections of entities are treated as continuous. In modern physics and biology, and the behavioural and social sciences, problems of organized complexity are commonplace and demand new conceptual tools such as a general theory of organization.
2. Complex systems are high–order, multiple–loop, nonlinear, feedback structures. They have many unexpected and little understood characteristics, making them very different from the simple systems of which people have an intuitive understanding, including: 1. High order: a system of greater than fourth or fifth order begins to enter the range of complex systems. An adequate representation of a social system, even for limited purposes, can be tenth or hundredth order.
2. Multiple loop: possessing upward of three or four interaction (positive or negative) feedback loops of shifting predominance.
3. Nonlinearity: allowing one feedback loop to dominate the system at one time and then causing a shift in this dominance to another part of the system which may produce such different behaviour that the two may seem unrelated.

♦ **KC0337 Synchronization**
Entrainment of frequencies
Description Under certain conditions, the fastest operation (rather than the slowest) in an interlinked process determines the overall rate of the process. The faster operation forces the pace of the slower, whilst the slower operation also has influence on the faster, so that synchronization of the two oscillations is mutual. Thus in dynamic operation the process variables are functioning together at a matching and compatible pace.

♦ **KC0338 Knowledge**
Description 1. The contents of consciousness in any living being.
2. The contents of memory.
3. The contents of the mind and all states of consciousness.
4. Cognitive activity with all its concomitants.
5. Behaviour, as an outcome of genetic instinctual, or acquired information, and as an inseparable aspect of such information. Includes mental behaviour; the instinctual knowledge of how to know.
6. In cybernetics, the programme.
7. In molecular biology, the genetic code.
8. In communication theory, information and negative entropy.
9. In mysticism and philosophy, the apprehension of truth. Wisdom.
10. In common man, opinions, prejudices, superstition and false and true ideas, all mixed together.

♦ **KC0339 Holographic logic**
Description 1. Negation of the premise of exclusive identity and the conventions of class logio by the formulation: the whole is in the part.
2. Buddhist logic.
3. Non–Aristotelian logic.
4. Multidimensional logic.

♦ **KC0340 Decision–making**
Description A process of selecting from among several alternatives, which may be either quantitative or qualitative, the best alternative in order to solve a problem or resolve a conflict. The elements of the process are: derivation of a model which describes the problem, selection of criteria to serve as standards, determination of constraints that act as limitations to various alternatives, and an optimization which results in the best solution consistent with the objectives of the decision maker. The analysis of complex situations with many alternatives and many possible consequences is facilitated by the mathematics of decision theory.

♦ **KC0341 Philosophy**
Description 1. A system of thought based on some logical relationships between concepts and principles that explains certain phenomena and supplies a basis for rational solutions of related problems.
2. A synthesis of learning which may have as its primary aim the complete unification of all departments of rational thought.
3. A system of motivating beliefs, concepts, and principles.

♦ **KC0342 Pythagoreanism**
Description The application of the symbolism of mathematics to the formal structure of nature leading to an understanding of the structure of the universe from the ratios and proportions of small integers.

♦ **KC0343 Policy sciences**
Description 1. A discipline concerned with knowledge of the policy process and of the relevance of knowledge in the process. This calls for: a cognitive map of the whole social process in reference to which each specific activity is considered; a problem orientation involving the intellectual tasks of goal clarification, trend description, analysis of conditions, projection of future developments, and invention, evaluation and selection of alternatives; and a distinctive synthesis of technique guided by principles of content and procedure.
2. The development of concepts and measures to strengthen the links at the policy level and to make them effective in ensuring conformity between forecasting, planning, decision–making, and action, and between values and these activities. These links include: norms, linking values to the process of rational creative action; links among the activities of the process of rational creative action; the feedback loop between policy formation and action, dynamic values and norm configurations, and induced changes in the environment.
3. Policy sciences integrates knowledge from a variety of branches of knowledge into a supradiscipline focusing on public policymaking. In particular, policy sciences are built upon behavioural sciences and analytical approaches, relying also on decision theory, general systems

INTEGRATIVE CONCEPTS

KC0393

theory, management sciences, conflict theory, strategic analysis, systems engineering, and similar modern areas of study. Physical and life sciences are also relied upon, insofar as they are relevant. Integration between pure and applied research is achieved by acceptance of the improvement of public policymaking as the ultimate goal. In essence, policy sciences are directed at explicit reconstruction of policymaking through conscious metapolicymaking.

♦ **KC0344 Factor analysis**
Description A statistical method for the isolation of factors in multivariable phenomena in which the variables fluctuate together, in order to determine the relative contribution of each to a phenomenon.

♦ **KC0345 Integrated automatic systems**
Description A network of automatic systems of similar or different types designed to produce a common result. Applications are in telecommunications, energy transmission, manufacturing controls, etc.

♦ **KC0346 Surveillance lag**
Description The elapsed time between the actual necessity for taking action, and the moment when that necessity is perceived by management. This refers to non-real-time decision-making, and emphasizes the fact that many management decisions must be made in terms of knowledge of the system's status some time in the past, rather than as it is at the moment of the decision.

♦ **KC0347 Non-linearity**
Description Input-output relations of a system, in which there occur variable parameters and explicit time dependence, are said to be non-linear. Non-linearity is a characteristic of biological and social systems which distinguishes them from physical systems (although there exist some non-living physical systems which are also non-linear). It is for this reason that biological systems cannot be satisfactorily reduced to and explained by physics, since the basic equations of quantum mechanics are linear. The relative infrequency of non-linearity in technological systems merely reflects the purpose of the designers, in that there is at least a fairly coherent linear system theory, which is not the case for the non-linear domain.

♦ **KC0349 Complementation**

♦ **KC0356 Value Theory**
Description Used only tentatively and experimentally in operations research to date, but felt to hold considerable promise for solving qualitative decision problems in the future. Basically, a process of assigning numerical significance to the worth of alternatives, thereby making explicit the consequences of certain kinds of ethical or value judgments.

♦ **KC0357 World brain**
World encyclopaedia
Description A world brain has been suggested as a new type of social organ, based on a world-wide organization, which would reorganize and reorient education and information. An important function would be to draw knowledge together into a comprehensive conception of the world and to counter the tendency to make very little use of the social, economic and political knowledge that already exists in attempting to alleviate world problems.

♦ **KC0360 Connective cultural integration**
Description Integration resulting from the direct connection of diverse features of a culture with one another. This is associated with the need for coherence in a culture.

♦ **KC0361 Deep structure**
Description In generative grammar, language surface structure (e.g. syntax) is obtained by the mind's transformational rules applied to a deep mental level of language organization or structure. The deep structure of sentences is where the definitions and interrelations of all the factors determining surface interpretation and generation are made.

♦ **KC0363 Development**
Description 1. A type of change over a period of time characterized by orderliness, and often by law-like, algorithmic or predictable processes.
2. Evolution.
3. Manifestation of potential.
4. Human development.
5. Societal development.
6. Economic development.
7. A type of variation, in music, associated with the transformations of thematic elements.
8. In literature, the movement of the plot or dramatic action, towards the climax.
9. Change as an outcome of probabilities or randomness.

♦ **KC0364 Orchestration**
Description Harmonious organization, integration, or combination and especially the conception and treatment of a musical composition with regard to the structure, manipulation, compass, and timbre of orchestral instruments, their effective combination, the proper distribution of the harmony, and the writing of orchestral scores.

♦ **KC0365 Structural-functional analysis**

♦ **KC0367 Integrative**
Description 1. An approach which cuts across and interrelates many dimensions (e.g. as social, economic, political, technological, psychological, and anthropological). Such approaches exist in connection with forecasting, planning, decision-making and education, for example.
2. The facilitation of a process of integration. A distinction may be made in this connection between integrated and integrative, with the former implying an achieved, completed, or predefined degree of integration within an established framework, whilst the latter implies the facilitation of a process whose scope and nature has not been predetermined within any framework and remains essentially open-ended.

♦ **KC0370 Reality**
Description 1. Whatever is perceived by all people, through the senses or intellectually.
2. Non-subjective phenomena.
3. Whatever is considered as existing by a class of beings in a particular world.
4. The immutable, (or immovable object).
5. The flux of change, (or irresistible force).
6. Illusion; maya.
7. The totality of all phenomena.
8. Tao, yang and yin , God, Dharma, urgrund, etc.

♦ **KC0371 Certainty**
Description 1. Knowledge that is validated or empirically confirmed as, for example, by experiment, by repeated observation, by the attestation of others (in law, by witnesses or evidence constituting proof).
2. A condition of knowledge regarding something in which uncertainty is ignored (whether it is immeasurably small or undetectable, or whether it is considered statistically insignificant) or unperceived.
3. Confidence in the correctness of an evaluation of the probability that a particular event (or activity, state, form, structure, etc) will or will not eventuate.

♦ **KC0372 World mind groups**

♦ **KC0373 Individual**
Description 1. Mathematically, a unit of quantity.
2. Biologically, a single representative of a species.
3. Psychologically, a personality.
4. In law, a legal entity.
5. In theology, a spiritual being.

♦ **KC0376 Growth**
Description The phenomenon of growth is found in practically all the sciences and even in most of the arts, because almost all the objects of human study grow (e.g. crystals, molecules, cells, plants, animals, children, personalities, knowledge, ideas, cities, cultures, organizations, nations, wealth and economic systems). It does not follow from the mere universality of the growth phenomenon that there must be a single unified theory of growth which will cover everything from the growth of a crystal to the growth of an empire. Nevertheless all growth phenomena have something in common, and what is more important, the classification of forms of growth and hence of theories of growth seems to cut across most of the conventional boundaries of the sciences. In addition there are a great many problems which are common to many apparently diverse growth phenomena.
Several types of growth (or decline) may be usefully distinguished: simple growth, involving the increase of a single quantity; population growth, in which the growing quantity is analyzed into an age distribution; structural growth, in which the growing aggregate consists of a complex structure of interrelated parts and in which growth involves change in the relation of the parts; merging into structural change or development, in which it is not the size which is growing, but the complexity or the systemic property.

♦ **KC0377 Cooperation**
Description

♦ **KC0378 Unity of faith**

♦ **KC0380 Configurational cultural integration**
Thematic cultural integration
Description Integration through similarity as a result of identity of meaning within a diversity of cultural items, namely their conformity to a common pattern or embodiment of a common theme. This is associated with the need to select a particular cultural pattern amongst the total range of possibilities.

♦ **KC0383 Alternation-fluctuation**

♦ **KC0384 Cultural cooperation**
Description

♦ **KC0386 Abstract mathematical space**
Description An arbitrary collection of homogeneous objects (events, states, functions, figures, values of variables, etc) between which there are relationships similar to the usual spatial relations (continuity, distance, etc). In regarding such a collection of objects as a space, all properties of these objects except those that are determined by these spacelike relationships are ignored. The relations then determine the structure or geometry of such a space. Spaces can be classified with respect to the types of those spacelike relations that underlie their definition. For example: a metric space is a set of arbitrary elements (points) between which a distance is defined; a topological space is any collection of points, in which a relation of neighbourhood of one point to a set of points is defined and, consequently, a relation of neighbourhood or adherence of two sets (figures) to one another.
When the position of a point cannot be defined by three coordinates, the concept of a many-dimensional space is introduced. If some figure or the state of some system, etc, is given by n data, then this figure, state, etc, can be conceived as a point of some n-dimensional space. This permits the application of well-known geometric analogies and methods to the study of the phenomena in question.

♦ **KC0387 Critical mass**
Threshold value — Critical quantity
Description The minimum amount of a given material or energy necessary to achieve a self-sustaining chain reaction under specified conditions. The term originated in nuclear physics in connection with the amount of fissile material necessary to sustain a fission chain reaction. It may also be used, by analogy, to refer to the quantity of funds, people, information, etc. which must necessarily be assembled before the intensity, variety, and quality of communication in the social subsystem so created is sufficient to maintain mutually stimulating social processes for the participants to be able to function innovatively and productively (e.g. critical number of researchers, to constitute an innovative research team; of change agents, to sustain a development process; of artists, to constitute aviable artistic community, etc.).

♦ **KC0390 Simultaneity**

♦ **KC0392 Poetry**
Description Verbal art utilizing as a vehicle the forms of verse and closely related rhythms and intonations. Poetry is frequently characterized by a heavy and sustained use of metaphor. In intellectual poetry symbolism may be introduced along with cultural history. In anti-intellectual poetry meaning may be subordinated to 'melody'. Poem-making, like tool-making, is a basic human trait. Its varieties, in the twentieth century, have become endless.

♦ **KC0393 Union**

♦ KC0394 Cyclical theory of history
Description A form of the organic theory of history in which the life processes are the paradigm. In the individual organism, the growth processes are linear and exhibit development in time. In the species there is an ebb and flow of birth and death. The cyclical theory of history is based on the macro pattern of the renewable life of the species, as well as on the metaphors of astral cycles, i.e. planetary orbits, star transits, the seasons, day and night, etc. Cyclical theories of social history may be of sinusoidal wave-form, elliptical, circular or of other complex natures (e.g., convoluted helices of other patterns which can combine linear, annular and wave motions, as, for example, to represent the motions of the earth).

♦ KC0395 Choreography
Description 1. Art of composing modern dances and ballets; also figure skating and water ballet to music, etc.
2. Notation system for dance movements.
3. The dance in general, replacing the older word, terpsichore.

♦ KC0396 Revolution
Description Radical transformation of any system, whether physical, biological or sociocultural, and implying fundamental transformation of the system's internal and/or environmental relationships accompanied by (or, as a consequence of) an acceleration in the rate of systemic change. Whether revolution reverses the direction of change or precipitates a radical transformation toward which things are moving too slowly, it involves overthrowing the established order rather than developing its latent tendencies.
Revolution need not necessarily mean the movement from a state of order to a state of disorder. It is useful to distinguish between progressive revolutions, retrogressive revolutions and status quo revolutions. Common to all of them is that they may be brought about either by violent and destructive means or by non-violent and constructive means.

♦ KC0397 International integration
Description Common usage of the term integration is frequently confusing. Integration, cooperation and community may be used interchangeably. In the study of international integration controversy is widespread even on the simplest issues and what constitutes a clarification for one school may mean a retreat for another. Efforts to exchange and harmonize views on the matter frequently only serve to further define cleavages. Examples of definitions include:
1. Institutions and practices strong enough and widespread enough to assure, for a long time, dependable expectations of peaceful change among the population.
2. The process whereby political actors in several distinct national settings are persuaded to shift their loyalties, expectations and political activities toward a new centre, whose institutions possess or demand jurisdiction over the existing national states.
3. The concept of international integration, verbally defined as forming parts into a whole or creating interdependence, can be broken down into economic integration (formation of a transnational economy), social integration (formation of a transnational society), and political integration (formation of transnational political independence).
4. A process by which discriminations existing along national borders are progressively removed between two or more countries. This differs from the other main definition of integration which is static and considers integration as a state of affairs which would be obtained at the end of a fairly long process leading to the complete merger of national identities. It also excludes from consideration the problem of less developed regions inside the same country. (Despite some similarity in the difficult problems that have to be dealt with in both cases, the implementation of regional integration inside one nation is considerably less complex than the international problem).

♦ KC0400 Hierarchy
Hierarchical system
Description 1. A system composed of interrelated subsystems, each of the subsystems being in turn hierarchic in structure until the lowest level of elementary subsystem is reached. Hierarchic systems have some common properties that are independent from their specific content. Such systems are nearly decomposable, namely interactions among subsystems are relatively weak compared with interactions within subsystems. Complex systems frequently take the form of hierarchy and tend to evolve more quickly when hierarchically organized. Hierarchical structure and combination into systems of ever higher order is characteristic of reality as a whole and of fundamental importance especially in biology, psychology, and sociology.
Hierarchy is the most conspicuous part of the formal structure of any social organization. (As such, it has frequently been falsely identified with the totality of formal structure, and the adjustment of hierarchic relationships has been falsely identified with the totality of the administrative process). The essence of hierarchy is then the distinction between the role of superior expected to exercise authority over one or more subordinates, who in turn function as superiors with respect to a lower level of subordinates.
Hierarchic organization is an essential feature of stable complex systems, whether they are inanimate systems, living organisms, social organizations, or patterns of behaviour. As a tree structure, hierarchies may serve to represent evolution as a process, and its projection in taxonomic systems; it may equally represent the step-wise differentiation in embryonic development; it may serve as a structural diagram of the parts-within-parts architecture of organisms or galaxies; or as a functional schema for the analysis of instinctive behaviour by ethologists; or of the phrase-generating machinery by the psycholinguist.
2. A set of things graded in levels by asymmetrical relations. Hierarchies are of many types: social, psychological, linguistic, conceptual, genetic, historical, etc. Structural hierarchies are composed of objects classified by spatial whole/part relations and may include inorganic hierarchies (from atoms to galaxies) or organic hierarchies (from atoms to communities). A free hierarchy is one in which there is no progressive transfer of energy between the levels, and the existence of the system does not impose additional constraints, characteristic of the level, on the degrees of freedom of its parts. A control hierarchy is a heterogeneous structural hierarchy in which causal levels are not separable and there is a progressive transfer of energy between the levels, with each higher level imposing additional constraints on the degrees of freedom of its parts.

♦ KC0402 Totalitarianism
Description Centralized control by an autocratic ruler or hierarchy or planning system regarded as infallible, and under which individual man is conceived as the servant of the state, an instrument or means and not an end.

♦ KC0404 Nous

♦ KC0406 Configurative thinking
Description The intermediaries between men of knowledge and policy-makers and executors are subject to a change of perspective as the scope of their activity increases. Their image of the decision process and of the social context becomes more inclusive and realistic. They become better informed about how the particular interests of a constituency can be integrated with a conception of the common interests of a larger coalition of interests, perhaps inclusive of the whole community. At the early stages greater contact with policy generates effective demand for policies that serve special rather than common interests. At later stages the aggregation of operational groups (who achieve distinct identities, demands and expectations) has an impact which strengthens common and inclusive interests in a configurative approach.

♦ KC0407 Social field theory
Description 1. A theory describing social actions or events as the resultant of a dynamic interplay among sociocultural, biomechanical, and motivational forces. Behaviour is therefore the consequence of the total social situation which forms a field consisting of social characteristics or attributes which stand in definite relation to each other. Any behaviour is relative to other behaviour, namely to a context, as well as to the relative similarities and differences of social units (whether individuals, groups, or nations) or their attributes. These attributes and the interactions between social units constituting behaviour form bounded systems which define the total situation within which social units can be located. Such systems have persistence in time, while the position of social units within them may change quite rapidly.
2. A method of analysing causal relations in society and building scientific constructs of which the most fundamental is the concept of a field. All behaviour (including action, thinking, wishing, etc) is conceived of as a change of some state of a field in a given unit of time. Thus individuals, groups, or societies each have a field or life space in which they exist. Behaviour has to be derived from a totality of coexisting facts which have the character of a dynamic field insofar as the state of any part of this field depends on every other part of the field.

♦ KC0408 Information system design
Description The drawing, planning, sketching, or arranging of many separate elements into a viable, unified whole. Whereas systems analysis is concerned with determining what the system is doing and what it should be doing, systems design is concerned with how the system is developed to meet the requirements of users. In the design process, the analyst develops alternative solutions and eventually ascertains the best design solution. In order to design a system the analyst must possess knowledge related to the following subjects: organizational resources; user information requirements; other systems requirements; methods of data processing; data operations; and design tools.

♦ KC0410 Cladistics

♦ KC0411 Cross cultural study
Description 1. A study in two or more different cultures of a particular psychological function in order to determine whether or in what manner it might be influenced by the differing cultural factors.
2. Comparative sociology.

♦ KC0412 Intuition

♦ KC0413 Prognosis
Description The foretelling of the course of an illness, or more generally, a forecast. The theory of prognoses is prognostics or prognostication. Prognosticators study the methodologies forecasting and the rehability of its outcomes.

♦ KC0415 Systems ecology
Description The combined approaches of systems analysis and the ecology of whole ecosystems when viewed as interdependent and functionally interacting components. The approach is based on an analysis of the network of relationships between organisms and organs, at any level of subdivision, within plant or animal communities. Such relationships need not be expressed as quantitative functions but may be expressed as the presence or absence of flows in a particular direction of material or energy. They may also express alternative pathways of development or direction of change in the organic or inorganic parts of the system.

♦ KC0416 Environment
Description 1. The set of all objects a change in whose attributes affects the system and also objects whose attributes are changed by the behaviour of the system.
2. The complex of climatic, edaphic, and biotic factors which act upon an organism or an ecological community and ultimately determine its form and survival. The aggregate of social and cultural conditions that influence the life of an individual, organization, or community.

♦ KC0417 System stress
Description Any externally or internally generated force or process which threatens a system's stability in one or more respects.

♦ KC0418 Visualization
Pattern perception
Description Visualization facilitates the conceptualization of the multiple factors involved in understanding complex processes and the associated problem-solving. Problem-solving involves cognition, and cognition includes perception. Visualization improves the capability to perceive and, therefore, assists the cognitive process. In the perception of shape lies the beginning of concept formation; it is the grasping of structural features found in or imposed on the stimulus material. When a person perceives a diagram, chart, symbol, word, formula, or some other visualization, he manifests his intelligence and cognition by his ability to understand the implication of what he sees.

♦ KC0420 Synaesthesis
Description The harmony of different or opposing impulses or sensations produced by a work of art, especially a subjective sensation or image of a sense, other than the one directly stimulated. (see also: synesthesia)

♦ KC0421 Mathematics
Description Characteristic features of modern mathematics includes:
1. Its subject matter consists not only of given quantitative relations and forms but of all possible (and, in general, variable) quantitative relations and interdependencies among magnitudes. In geometry, the focus is on both spatial relations and forms but also on all possible forms similar to spatial ones. In algebra, the focus is on various abstract systems of objects with all possible laws of operation on them. In analysis, not only magnitudes are considered as variables, but the very functions themselves. In a functional space, all the functions of a given type (all the possible interdependencies among the variables) are brought together.
2. The abstractions of mathematics are distinguished by three features: the concern with quantitative relations and spatial forms, abstracted from all other properties of objects; their occurrence in a sequence of increasing degrees of abstraction, going very much further in this direction than the abstractions of the other sciences; and their focus on abstract concepts and

their interrelations.
3. The absence of privileged rank for any mathematical entities.
4. Isomorphism takes the place of identity within the mathematical process ensuring the non-ontological character of mathematics and hence its power, fidelity and polyvalence.
5. Its unity as a consequence of the existence of a common language and agreement on common elementary structures.

♦ **KC0422 Cultural unity**
World culture — Universal culture

♦ **KC0423 Innovation**
Description Methodical creations of the human spirit, namely all novelties that once created can be usefully and repeatedly applied. (This therefore excludes artistic creations, since a symphony can be repeated but not applied. On the other hand, novelties in orchestration can be applied and therefore constitute innovations). In a more restricted sense, it is the process that turns an invention through development, pilot manufacture, sales propaganda, etc. into a marketable product. The most successful innovations are not necessarily the consequence of feats of creativity but derive rather from the development of improved substitutes for products and services already in existence. As such they are vitally important to the economic system.

♦ **KC0424 Harmony**
Description 1. The arrangement of parts within a structure or system in pleasing relation to each other.
2. The degree of efficiency with which a particular system achieves its purpose. A norm of harmony is distinguished from a norm of equilibrium which refers to the degree of interrelatedness between an equilibrium system as a whole and whatever are taken to be its parts.
3. Combination into a consistent whole.
4. The science of the structure, relation, and progression of chords in homophonic composition.

♦ **KC0425 Communion**
Description 1. A group of religious persons bound together by essential agreement in religious consciousness.
2. An action or situation involving sharing.

♦ **KC0426 System dynamics**
Industrial dynamics — Urban dynamics
Description The modelling of the feedback–loop structure of social systems. Such feedback loops are the closed paths that connect actions and their effects on the surrounding conditions, with the resulting conditions in turn coming back as information to influence further action. Within the feedback loops of a system, the principles of system structure indicate that two kinds of variables will be found, levels and rates. The levels are the accumulations (integrations) within the system, whereas the rates are the flows that cause the levels to change. A system dynamicist obtains information from groups of people who understand aspects of the system in question and assembles such information into a model that captures its essential structure.

♦ **KC0427 Deviation–amplification**
Positive feedback — Schismogenesis
Description A property of a wide range of systems (e.g. accumulation of capital in industry, evolution of living organisms, rise of certain types of culture, etc) and the processes that are loosely termed 'vicious circles', namely all processes of mutual causal relationships that amplify an insignificant or accidental initial impulse, build up deviation and diverge from the initial condition.

♦ **KC0428 Curricula**
Description

♦ **KC0430 Information**
Description 1. Information is an occurrence or a set of occurrences which carry messages and, when perceived by the recipients, will increase their state of knowledge. It is therefore the increase in knowledge obtained by the recipient by matching proper data elements to the variables of a problem. Information is substantially different from data or signals in that they are raw, unevaluated messages. Information processing of data to provide knowledge or intelligence, and as such the primary function of an information system is to increase the knowledge or reduce the uncertainty of the user. The significance or value of information received can only be measured by the recipient.
2. Information includes the messages occurring in any of the standard communication media, the signals occurring in computers, servomechanism systems and other data processing devices, as well as the signals in nerve networks.
3. The process by which the form of an object of knowledge is impressed upon the apprehending mind so as to bring about the state of knowledge.

♦ **KC0432 Generalist**
Description An individual conversant with, or capable of handling, several different fields, skills, or aptitudes. The extreme types of generalist are the dilettante or jack–of–all–trades (who knows a little about many topics and may move aimlessly from one problem to another) and the integrator (who is able to fit diverse activities into a general framework). The integrators are the true generalists, who are expected to see the general as well as the particular and to be motivated by interests broader than their own or even of their own organization. They are looked to for skills not only in communication and compromise but in the constructive integration of divergent interests. They are expected to understand an organization's broad environment as well as, or even more than, its internal workings. They are expected to know enough about relevant techniques to enable them to understand, evaluate, and coordinate the activities of many specialists and professionals.

♦ **KC0434 Compatibility**
Harmony
Description The capacity of two or more entities or components to combine together, remain together, or function together, without undesirable aftereffects.
Two systems may be quite satisfactory when functioning independently, but may have completely different, and not necessarily favourable characteristics, when they function in tandem or interactively. Systems may be compatible in some respects and incompatible in others, depending on the purposes for which they are introduced as well as environmental factors. A system may become incompatible with its environment, if some feature of the environment is changed.

♦ **KC0436 Optimization**
Description Adaptation of a system to its environment to secure the best possible performance in some respect (which may preclude optimum performance in another respect). Suboptimization is the process of fulfilling or optimizing some chosen objective which is an integral part of a broader objective which is frequently not consonant with the lower–level objective.

♦ **KC0437 Frame of reference**
Description A usually systematic set of principles, rules, or presuppositions; or a system of laws, mores, or values; or an interlocking group of facts or ideas, serving to orient or give particular meaning (as to a fact, statement or point of view) or serving as a matrix for behaviour or for the formation of attitudes.

♦ **KC0440 Arcology**
Description A discipline resulting from a fusion architecture and ecology which also signifies concrete structures, namely total cities embraced within one frame. Such cities, or arcologies, are based on the idea of implosion or reintegration of the scattered and often dissociated parts of the same organism into one homogeneous, compact, coordinated system. Their fundamental novelty lies in the normative conception of an architectural system. The rational order of arcologies follows not from the cohesion and self–consistency of these structures but from the moral order on which they are based and which they serve.

♦ **KC0441 Metaphysics**
Description A division of philosophy, including ontology and cosmology, that is concerned with the relations obtaining between the underlying reality and its manifestations. In its attempt to conceive the world as a whole by means of thought, it has been developed by the union and conflict of the urge toward mysticism and the urge toward science.

♦ **KC0443 Style**
Description One or more characteristics in behaviour, or in artistic execution that identify an individual, a creator or an origin. Expressions such as, hallmark or trademark are used metaphorically of consistent stylist traits. The term is applied to clothing fashions and designs of other consumer and industrial products. In cultural history a particular epoch may give its name to a style, e.g., Gothic Style. Thus a style may be identified by common features in things, or by their creation and use in a particular place or during a particular period.

♦ **KC0444 Gödel's Theorem**
Description In any formal system adequate for number theory there exists an undecidable formula (namely a formula which is not provable and whose negation is not provable). It is sometimes added that the undecidable formula is true.
A corollary to the theorem is that the consistency of a formal system adequate for number theory cannot be proved within the system. The bearing of these results on epistemological problems remains uncertain, but it has been suggested that they should not be rashly called upon to establish the primacy of some act of intuition that would dispense with formalization.

♦ **KC0446 Symbolic logic**
Mathematical logic
Description A science of the development and representation of logical principles by means of symbols for the purpose of providing an exact canon of deduction based on primitives, postulates, and formation and rules for combining and transforming the primitives. The algebra of logic substitutes symbols for words, propositions, classes of things, or functional systems.
Symbolic logic: lead to economy in mental effort; makes possible more complicated forms of reasoning; demonstrates the undue importance assigned by Aristotelian logic to the supposedly self–evident laws of thought; reveals form by showing that apparently similar logical forms (relation–structures) are actually dissimilar, and vice versa.

♦ **KC0447 Ecosphere**
Description The totality constituted by the ecosystem (the self–sustaining community of plant and animal organisms taken together with its inorganic environment) and the biosphere (the collective totality of living creatures on the earth).

♦ **KC0450 Futures research**
Futuristics — Futurology — Futurism — Futuribles
Description 1. A broad range of interlinked activities which include: (a) conjectural, speculative and imaginative descriptions of the human future; (b) exploratory forecasts based on methodological extrapolation of past and present developments into the future; and (c) prescriptions, namely normatively oriented future projections in which explicit value assertions and choices are made about how a specific future may be viewed.
2. Any serious, organized attempt to devise concepts and methods which can be used to conjecture intelligently about the human future.

♦ **KC0452 Synnoetics**

♦ **KC0455 Networking**
Description The processes of operation and development of a communication network of some kind, whose structure may be largely decentralized and possibly based on informal relationships between individuals or groups.
Networking in its more formal (hardware–based) sense is used to describe information, telecommunication, date and library networks which are increasingly augmented by computers at various points. Considerable use of mathematical techniques is made to determine and specify the most efficient processes.
Networking in its less formal (software–based) sense is increasingly used to describe the relationships between formal and informal groups, particularly at a grass–roots level. In this context, networking is the process of augmenting the communication between such groups by information centres which facilitate the formation of new contacts between groups and individuals with common interests.

♦ **KC0456 Context**
Description Context is assigned by the perceiving, organizing system to the background so that other relevant information may be brought to bear on the modelling of the ever–changing input stream to that system. Context is therefore distinct from background, as what determines the background and foreground is determined by the context under which the system is functioning. The context of any model therefore delimits the set of all possible inputs to only those which are applicable to the system.

♦ **KC0457 Dialectical–materialist synthesis**
Description The dialectical–materialist process involves recognition of a thesis, its anti–thesis, and therefore the whole, or synthesis, composed of both thesis and anti–thesis. At any stage of

this process, however, the synthesis is itself unsatisfactory because the elements of the synthesis suggest and require other relationships which therefore demand an enlargement of the universe considered and the formulation of a thesis, anti–thesis and synthesis at a new level. This dynamic has been applied by the Marxist and Hegelian analysis of historical processses in which each historical situation is said to contain tensions, conflicts and contradictory elements which are the driving force of change. History dialectically progresses from one condition to its opposite and then to a synthesis at a higher level. (Thus it is suggested that the contradictions of the world system will lead to a crisis and eventually to the emergence of a new order as a higher form of synthesis).

♦ KC0458 Coincidence
Description 1. Coincidence in space and time, illustrated by the common example of a chance (and unlikely) meeting of two friends.
2. Coincidence in time, illustrated by the independent parallel work of Wallace and Darwin, Leibniz and Newton, etc and by other parallel, identical or similar events separated by space.
3. Coincidence in space, illustrated by the successive, change arrivals of two or more related beings, objects, or phenomena, in the same physical place, separated by time.
4. Coincidence in mind, illustrated by the chance arising in the mind (through memory, imagination or cognition, etc) of opposite or linked information.
5. Chance events of linked or meaningful natures involving space, time and mind.
6. ESP phenomena.
7. Any event statistically unjustified (unpredictable, unlikely, singular, etc).

♦ KC0460 Logical cultural integration
Description Integration resulting from the logical consistency, primarily of the existential beliefs and systems of norms. This is associated with the need for rationality within a culture.

♦ KC0462 Catastrophe theory
Mathematics of discontinuous change
Description Mathematical–geometrical model formulation of level changes or abrupt movements in systems. Some seven models of elementary catastrophe type (discontinous changes) have been postulated.

♦ KC0463 Second–order theorizing
Description 1. Theory concerning the formal properties of theories invoking epistemology, logic, mathematics and related disciplines.
2. Theory concerning the nature of theories from the perspectives of history, sociology, history of ideas, psychology or other related studies.

♦ KC0464 Ontology
Description Study of the principles of being or reality, as a branch of philosophy, metaphysics or theology.

♦ KC0465 Programming
Description 1. Computer language instructions or the writing of such programmes.
2. Automatic control device procedures.
3. Microprocessor, silicon chip and similar physical memory circuits, e.g., printed circuit boards.

♦ KC0467 Self–realization ethic
Description 1. The proper end of all individual experience is the evolutionary and harmonious development of the emergent self (both as a person and as a part of wide collectivities), and the appropriate function of social institutions is to create an environment which will foster that process.
2. The basic principle of the biological and social processes is increase of heterogeneity and of symbiotization. Individuals are unique and different. The desirable end of all individual experience is the further development of the emergent self. The appropriate function of social institutions is to create an environment which will allow for and facilitate heterogeneous development of individuals and symbiosis within human species as well as among all living species.

♦ KC0470 Cultural evolution
Description

♦ KC0472 Network
Description A group of elements which may be partially or completely interconnected. The connections (termed branches or arcs) can represent roads, power lines, airline routes, information flows, predator–prey relationships in an ecosystem, logical relationships, functional relationships, or the generalized channels through which commodities flow. The elements (termed points or nodes) can represent individuals, communities, power stations, airline terminals, water reservoirs, libraries, organizations, namely any point where a flow or relationship of some kind originates, or terminates. In a more general case, the elements or points in the network may themselves be subnetworks composed of combinations of other kinds of elements.
The characteristics of the network's elements and relationships can be described by values, which may or may not be quantitative. The values can be fixed or they can vary in some way with time. Thus the relationship between two points may not exist during a particular period of time (as in an electrical circuit), or several possible relationship paths may exist between two points (as in a telephone circuit). Different types of relationship may exist between the same two points.
Network analysis (described separately) is used to determine the structure and characteristics of networks. Network synthesis (described separately) is used to combine together possible elements of a network in the process of designing circuits with required characteristics. Network is frequently used as a synonym for system, although systems analysis necessitates quantitative information whereas much analysis of networks can be performed using only information on the existence (or not) of relationships between points.
Some simple structural properties of networks (or parts of networks) using the methods of graph theory (described separately) are given below. More complex methods of network analysis (described separately) exist or are being explored, particularly with a view to analysing complex networks and determining their properties as a whole.
1. *Centrality:* A measure of the extent to which a given entity is directly or indirectly related to other entities, or isolated from them.
2. *Coherence:* A measure of the degree of interconnectedness or density of a group of entities. This may be an indication of the degree of development of a group of entities.
3. *Range:* A measure of the number of other entities to which a particular entity is directly related.
4. *Content:* The nature or reason for the existence of a relationship between entities. Simple networks can have only one link between two entities. More complex networks may have two or more such links, possibly each of different content.
5. *Directedness:* A relationship between two entities may have some direction (P to Q or Q to P) or asymmetry, which may be of different types.
6. *Durability:* A measure of the period over which a certain relationship between entities is activated and used.
7. *Intensity:* A measure of the strength of the relationship between two entities. (The link from P to Q may be strong, and that from Q to P weak).
8. *Frequency:* The frequency with which a link is established or cut off, if this is the case.
9. *Rearrangeability and Blocking:* A network is rearrangeable if alternative paths can be found to link any pair of entities by rearranging the links between other entities. A network is in a blocking state if some pair of entities cannot be connected.

♦ KC0473 Correlation
Description 1. An event that does not constrain or determine another so that the latter cannot be said to be a function of the former, but is closely associated with it, may be in the relationship of being mutually determined by a third event.
2. Mathematically, the relationship between variable measures, where increases or decreases tend to be parallel or proportional.
3. Positive correlation: change in the same direction between variables.
4. Negative correlation: change in opposite directions between variables.
5. Linear correlation: change maintained in a certain proportion.
6. Non–linear correlation: variable rates of change.
7. Correlation theory assumes that phenomena obey probabilistic laws. Given one random event a second random event can be expressed probabilistically in terms of its correlation.

♦ KC0476 Principle of allometry
Description Many phenomena of metabolism, biochemistry, morphogenesis, and evolution are governed by the allometric equation in which one characteristic (e.g. length or weight of an organ) can be expressed as a power function of another characteristic (e.g. the length or weight of another organ or of the organism as a whole). In effect this states that, for example, the relative growth rates of the parts under consideration stand in constant proportion throughout the life of the organism, or during the life cycle for which the allometric equation holds.
Sociological phenomena can also be conceived as phenomena of relative growth to which allometry applies as a rather general principle, the basis of which is essentially a process of distribution.

♦ KC0477 Generalization
Description Generalization can be made at many levels. At the lowest, a singular generalization consists of statements of observed uniformities between two isolated and easily identified variables. At a higher level, it consists of a set of interrelated propositions that are designed to synthesize the data in an organized body of singular generalizations. At a higher level of abstraction, systematic theory consists of the conceptual framework within which a whole discipline is cast, namely those theories and assumptions which an investigator uses in undertaking an analysis within a given field.

♦ KC0479 Epistemology
Description 1. Study of the problem of knowing: its instrumentalities, processes and conclusions.
2. The science of true and certain knowledge.

♦ KC0484 Typology
Description A system of groupings, usually called types, which aid demonstration or inquiry by establishing a limited relationship among phenomena. The identity of the members of each type is defined in terms of specified attributes which are mutually exclusive and collectively exhaustive. Any such type may represent one kind of attribute or several and need include only those features of the phenomena which are significant for the problem in question.

♦ KC0486 Simplicity
Description Simplicity may relate to the mathematical form of laws, to the number of independent variables which appear in them, or to the kinds of concepts they involve. There appears to be no commonly accepted definition. It is an aspect of structure. To be perceived however, structural simplicity must blend with a cognitive act, leading to recognition, experienced subjectively as a catharsis (release of tension). The inquiring mind seeks such experiences through solving problems and perceiving regularity where none was apparent before. The quest for simplicity stems from a conviction that underlying apparently wide dissimilarities are profound similarities, which when perceived, make order out of chaos, and simplicity out of complexity.
Science is a systematized search for simplicity, a method of making the world predictable. Science harnesses the search for simplicity to other purposes specifically to gain power over the environment.

♦ KC0487 Archetype
Description A primordial image, character or pattern that recurs throughout literature and thought consistently enough to be considered a universal concept or situation. All religions are repositories of transpersonal experience and archetypal images. Religious practices attempt to hold up to view such transpersonal categories of existence and attempt to relate them to the individual. It is considered doubtful whether collective human life could survive for any period without some common, shared sense of awareness of such categories.

♦ KC0490 Astrology
Description 1. The study of the periodical influences of celestial dynamics, and of physical bodies, such as planets, comets and stars, on human animal, or plant growth and behaviour, through the changes induced in the earth or atmosphere, or directly in organic systems, by agencies such as light and other electro–magnetic forces, or by gravity, or by cosmic or other radiation, or by other energies, or effects of space–time alternations or changes.
2. The study of the positions, configurations and movements of celestial phenomena as symbolic, acausal correspondences to terrestrial phenomena.
3. The doctrine that the celestial bodies are ruled by super–human beings who, in their various movements and mutual aspects, determine historical cycles, rulerships and personal life patterns on earth.
4. Astrolatry or behaviour modified by belief in astral divinities and their powers.
5. Species of popular journalism in which is predicted on the basis of nativity in a particular sign (degree of sun in tropical zodiac) the daily outlook for each of the twelve 'types' of person.
6. Species of fortune–telling based on one's 'horoscope' (actual or symbolic).

♦ KC0494 Biosphere
Description The layer constituting a shell around the surface of the earth, bounded on the inner side by dense solids and on the outer side by the vacuum of space, containing the whole range of living organisms for which air and water are essential. The biosphere is the largest ecosystem and is composed of interlinked eco–subsystems.

♦ **KC0496 World order**
Description The study of international relations and world affairs which focuses primarily on the questions of how to reduce significantly the likelihood of international violence and to create tolerable conditions of worldwide economic welfare, social justice and ecological stability.
The substantive matters comprehended by world order are a range of actors (world institutions, international organizations, regional arrangements, transnational actors, the nation–state, infra–national groups, and the individual) as they relate to the following dimensions of world political and community processes: peace–keeping, third party resolution of disputes and other modes of pacific settlement, disarmament and arms control, economic development and welfare, the technological and scientific revolutions, ecological stability, and human and social rights.
Political–social–legal forms, organizations and institutions are envisaged which are relevant to the solution of world problems. Relevant utopia models are developed which consist of projections of reasonably concrete behavioural models or images of a system of world political and social processes capable of preventing organized international violence and providing adequate world-wide economic welfare, social justice and ecological stability, as well as concrete behavioural statements of transition from the present system to that of the model.

♦ **KC0497 International attitudinal integration**
Description Attitudinal integration is one component of international political integration (separately described) and may be defined as the extent to which the populations of a group of countries have developed a common sense of identity, transnational consciousness or support for some form of transnational unity.

♦ **KC0500 Whole**
Description Something constituting a complex unity. A coherent system or organization of parts fitting or working together as one. An unreduced or unimpaired entirety.

♦ **KC0502 Cosmology**
Description A branch of systematic philosophy that deals with the character of the universe as a cosmos by combining speculative metaphysics and scientific knowledge. The study of the universe as a self–inclusive system characterized by order and harmony amid complexity of detail.

♦ **KC0503 Teamwork**
Description The process by which a number of associates work together, usually with each doing a clearly defined portion of the total task, but all subordinating personal prominence to the efficiency of the whole process.

♦ **KC0506 Global society**
World society — Planetary society
Description An emergent world social system considered to constitute a profound modification in the condition of the human species and characterized by: a rapidly emerging world wide system of human interaction; growing globalization of economic and military interdependence; expanding network of cross–national organizations; increasing similarity in mankind's social institutions; increasing cultural homogeneity. The global focus is distinguished from the conception of international society and international relations as being confined to relations between sovereign, nation–states, which excludes consideration of important religious, language, scientific, commercial, cultural and other relationships, in addition to a variety of formal nongovernmental relations that constitute a worldwide network. Traditional internationalism derives from the dictates of political expediency in a world of growing interdependence, but of unlimited horizons opened up by technology. Globalism is associated with the ambivalence of technology, its negative effects on the environment and ecological balance, the limited capacity of the biosphere, the population explosion, the limitation of resources, and the general finiteness of the planet.

♦ **KC0507 Surface structure**

♦ **KC0508 Data association**
Description Data held in a computer file are divided into records (also called entities or nodes), each consisting of a contiguous string of data divided into fields. In order to associate a record with other records (or fields), a logical system must be used to achieve this linkage. Several techniques of data association exist, including chains and pointers, and directories. Chains are used logically, not physically, to link records together based on commonalities and functional interrelationships. A pointer represents the address (reference point) of a particular record. The chaining operation is achieved by using pointers to directly or indirectly associate one record in the chain to the next record in the chain. The need for chaining arises when information is stored in a sequence other than the one that facilitates the type of retrieval desired.

♦ **KC0510 Self–organization**
Artificial organisms — Self–organizing systems
Description 1. Two meanings must be distinguished: (a) a system that starts with its parts separate (so that the behaviour of each is independent of the others' states) and whose parts then act so that they change towards forming connections of some type (the change is from unorganized to organized); (b) a system in which the progressive connection between the parts is conditioned by the connections already formed (this may be characterized as self–connection).
2. The property of a system in which the rate of change of its redundancy is positive.
3. A comprehensive, general concept for the properties of self–reproduction, production of structural order, or growth in connection with mechanisms and systems. Self–organizing systems are based on: (a) an exchange of information between the system and its environment permitting adaptation to the situation at hand, and (b) the ability to recognize and to learn, and thus can be regarded as the preliminary steps toward achieving artificial intelligence.

♦ **KC0511 Logos**
Description 1. Human reasoning or nous which as the suffix, –logy has the sense of 'science', and in the form, logic, is a science of sciences.
2. The link between theos, the divine, and nous, the human, minds. Logos can be infused into nous.
3. Order in the universe. Closely connected with the idealization of Apollo, and affiliated to the creative utterance of the deity in Hellenistic Judaism. From there taken into Christianity where its final form was the second Person of the Holy Trinity, God the Son, the Word ('which was made flesh' as the Christ, 'through whom and by whom all things were made').
4. Any saying of Jesus given in the New Testament.
5. Collectively, the logoi are emanations of the intellectual world which regulate and form the sensory world. These are identified as angels by speculative, neoplatonic philosophers of the Christian, Islamic and Judaic faiths.

♦ **KC0512 Vertical thinking**
Description 1. Convergent or directional thinking.
2. Scientific method.
3. Formal logic.
4. Analytical procedure.
5. Conventional thought processes in problem solving as opposed to divergent, lateral thinking.

♦ **KC0514 Organism**
Description 1. Unicellular and multicellular systems that are self–sustaining (homeostatic) through metabolic processes, and are capable of movement, reproduction and genetic change (such as adaptation).
2. Animate matter.
3. Any being or system said to have the property of life.

♦ **KC0515 Programme planning**

♦ **KC0516 Ontogenetic recapitulation of phylogeny**
Epigenetic landscape
Description 1. A property of evolving systems whose descriptions are stored in a process language such that, for example, individual biological organisms, in their development, go through stages that resemble some of the forms of their evolutionary ancestors.
2. The prior history of a species or an organization (its epigenetic history) specifies not only physical characteristics but in addition its underlying epigenetic landscape (or conditional probability model) which is the summation of previous generations' success and failure in interacting with the environment, characterized by the epithet 'ontogeny recapitulates phylogeny'.

♦ **KC0517 System overload**
Description Quantitative demands on the capacities of a system (for communication, for attention, etc.) to which it is unable to respond effectively.

♦ **KC0518 Distributed information system**
System of information systems
Description An information system of relatively independent subsystems which are, however, tied together within the organizational framework by communication interfaces. This system of information systems is a network of subsystems located at, and adapted to, areas of need. In such a network, three basic conditions will exist: some of the subsystems will need to interact with other subsystems; some will need to share files with others and even share data processing facilities; and some subsystems will require very little interaction with other subsystems, and for all purposes will be fairly isolated and self–sufficient. Whereas the integrated system (described separately) is monolithic in nature utilizing a central data processing facility with a common data base; the distributed system is modular and employs an aggregation of information systems arranged as a network.

♦ **KC0520 Normative social integration**
Description Integration achieved when the common values in the cultural system are institutionalized in structural elements of the social system. This may occur at three levels: to categories of persons, to collectivities, or to structured roles within collectivities. At each level sanctions exist, as well as the specifications of correct conduct.

♦ **KC0522 Mathematical intuitionism**
Description A philosophy of mathematics and mathematical logic based on the rejection of the laws of traditional logic developed for finite sets as capable of being extended to infinite sets. It considers such mathematical statements as a form of information about mentally completed constructions, whose study requires a special logic. This is called, intuitionist logic, which has as one of its features the limited acceptance of the law of the excluded middle. An intuitionist arithmetic was developed by K Gödel.

♦ **KC0523 Steady state**
Description The state of a system in which inputs of energy and/or material into the system balance outputs from the system in such a way as to maintain its level of integration constant. In such a case a system is not in equilibrium (described separately) but rather seems to be maintaining certain peculiar gradients.

♦ **KC0525 Consensus politics**
Description Consensus politics, the advocacy of programmes and laws, as in legislative bodies, that are moderately stated in order to attract support across party lines, but which may have little or no origination or support at the electorate level.

♦ **KC0526 Information system**
Data processing system
Description A systematic, formal assemblage of components that performs data processing operations for any, or all, of the following purposes: meet legal and transactional data processing requirements; provide information to management for support planning, controlling, and decision–making activities; and provide a variety of reports, as required, to external constituents. A formal information system includes some, if not all, of the following: data entry and preparation devices; data storage devices; telecommunications equipment; data processing equipment; terminal devices; procedures, programs, methods, and documentation; data manipulation models (e.g. for accounting, budgeting, costing, inventory control, etc.); decision rooms with graph display boards and charts; duplicating devices; and personnel.

♦ **KC0527 Symbolic system**
Eiconic system — Appreciative system — Image system
Description Systems in which 'images', namely the cognitive structures of persons, enter in an essential manner. Images are systems of extreme complexity and delicacy, especially at the symbolic level. Although the image of an individual person is subjective, it is not necessarily private in the sense that it is unrelated to other images. A public image, however, consists of the shared images of many individuals. A public image is the product of a universe of discourse, namely a process of sharing messages and experiences.
Human life is sustained culturally by varieties of symbolic systems. Each generation takes over, makes over, and passes on a heritage which consists basically in specific ways of appreciating and acting in each situation. This heritage subsists at any point in time in the organization of countless individual minds, never wholly shared; yet it is a social artefact, dependent on communication both for its continuity and for its change, yet itself giving meaning to the communication on which it depends. Time establishes a close correspondence between the needs and concerns of the society and the shared, symbolic system by which these are represented and interpreted in language and thought. It is essential to any society that its

appreciative system shall change sufficiently to interpret a changing world, yet should remain sufficiently shared and sufficiently stable to mediate mutual understanding and common action and to make sense of personal experience. These demands conflict with each other increasingly as rates of change accelerate and contacts between disparate cultures multiply.

♦ KC0528 Corpus of knowledge
Body of knowledge

♦ KC0529 Industrial design

♦ KC0530 Economic system
Description 1. A network of independent institutions each of which serves some function for the entire economic system, and which collectively organize and direct society's efforts to achieve its economic goals.
2. The organizational (or institutional) framework through which society strives to maximize the benefits of its scarce resources.

♦ KC0532 Consensus
Description Harmony, cooperation or sympathy, especially group solidarity in sentiment and belief, taking the form of general agreement arrived at by most of those concerned.

♦ KC0533 Coevolution
Description A property of the evolution of predator–prey relationships in that as the prey evolve and diversify, their predators evolve and diversify with them and in progressively close response to them.

♦ KC0535 Organicism
Description 1. A world view based on the biological as a complex structure of interdependent and subordinate elements whose relations and properties are largely determined by their function as a whole. In any realm of biological phenomena (whether embryonic development, metabolism, growth, etc.), the behaviour of an element is different within the system from what it is in isolation. The sum of the behaviour of the isolated parts does not equal the behaviour of the whole because the relationships between the various subordinated systems and the system which are superordinated to them affect the behaviour of the parts. Thus the organismic view looks at the world in terms of systems, wholes, and organizations, in contrast to the opposing reductionist view which looks at the world in terms of its component elements.
2. A philosophy recognizing the ultimacy of both parts of knowledge (such as a particular science, or the life–scheme of a particular person) and some whole of knowledge (such as a compendious history of mankind) but in such a way as to see their perpetual, dynamic interdependence. It is therefore a way of integrating knowledge. No whole of knowledge can be regarded as having organic integrity which fails to respect the partial autonomy of its parts. But also no part of knowledge, about any portion of man or universe, can be regarded as having organic integrity which fails to integrate its relations with the whole, and other parts of the whole, of knowledge.
3. A comprehensive theory of disease which reduced all aetiology to a material, organic basis in which the causative factor was considered to be ultimately a structural break–down (on a molecular or cellular level) or a lesion (on the tissue level). The applicability of this theory to all mental disorders (as well as to all physical disorders) is unproven.

♦ KC0536 Organismic
Description The property of an open system which either is an organism or can be treated as an organism where the chief characteristics of an organism are a relatively fixed ordering of components, reproduction of the same type or species, and existence of a life cycle. Organisms, in addition to transforming energy as mechanisms, also function as systems capable of processing, storing, and receiving information, and of making decisions. These attributes also characterize some non–living systems. The concept of an organized system therefore includes organisms.

♦ KC0537 Meta–science
Science of science
Description Research on research, including in a broader frame of reference attention to the self–reflection of science (Wissenschaft). Meta–science may focus on: the products of research (symbolic systems such as theories, formalisms, concepts, etc.) studying their logical, semantic, information–theoretical and epistemological aspects; the production–and–product system, viewed as a praxiological–symbolic system (e.g. how does scientific knowledge grow); the producers–and–users from a science–in–society perspective, including the sociological, psychological, historiographical, cultural, and political aspects of science; the aggregate science–society–man, viewed as a total system which sets and changes its aims; the meaning of science for man, evaluating its impact from a futurological point of view; or science viewed by a practising, praxiologically oriented scientist.
One major school of meta–science, logical empiricism, tends to be primarily concerned with articulating an ideal of science with key themes (such as unified science, empirical significance, confirmation, explanation) clarifying various features of this ideal. Another school, hermeneutic-dialectic, embeds its research in a philosophical anthropology of knowledge.

♦ KC0540 Operations research
Management science — Scientific management
Description Operations research provides analytical methods of finding solutions to problems involving the operations of a system, so as to provide those in control of the system with optimum solutions to the problems. From a large number of feasible combinations, the optimum combination is identified for achieving the given objective under certain constraints. The major phase of an OR project are: formulation of the problem; constructing a mathematical model to represent the system under study; deriving a solution from the model; testing the model and the solution derived from it; establishing controls over the solution; putting the solution to work (implementation).
OR provides solutions for many production–oriented problems, particularly those related to: inventory processes and control (using: economic–order–quantity equations, and linear, dynamic, and quadratic programming); resource allocation processes (using: linear and other types of mathematical programming); waiting–time processes (using: queuing theory, sequencing theory, line–balancing theory); replacement processes; competitive processes (using: theory of games); and combined processes
The basis for problem solution is a mathematical model or representation of selected aspects of the real world system in question. The model is manipulated according to rules of logic, and the consequences of policy decisions are predicted from model solutions. The objective of manipulating the model is usually to determine an optimum solution without having to manipulate the real world system, thus providing guidance for the solution of the real world problem.
The earlier distinction between operations research, scientific management and management science is no longer considered meaningful. A distinction is however made between OR and systems analysis. OR tends to accept specified objectives and given assumptions about the circumstances, the hardware, etc, and then attempts to compute an optimum solution. The espistemology of OR is that of the exact sciences, assuming that the empirical data are accurate enough to make refined calculations worthwhile. On the other hand, the epistemology of systems analysis is that of the inexact sciences. Statistics may be used and emphasis is placed on techniques for dealing with uncertainty, such as sensitivity tests, the use of ranges, alternative scenarios, etc. In effect, OR is oriented toward problems in which the element of calculation is dominant, and therefore, in which mathematics can be thought of as a substitute for rather than as an aid to, judgement. Both are scientific approaches to problem solving, but systems analysis is more complex, and less neat and tidy, embodying a much larger percentage of nonquantitative elements which influence the outcome.

♦ KC0542 Semantics
Description 1. The study of meanings.
2. The study of the relations of a sign to its referent and to other signs within a system.
3. The study dealing with the relations between signs and what they refer to, the relations between the signs of a system, and human behaviour in reaction to signs.
4. The historical and psychological study and the classification of changes in the signification of words or forms viewed as factors of linguistic development.

♦ KC0543 Ecumenism
Unity of religion
Description Principles and practices associated with a movement toward worldwide interconfessional Christian unity. The movement promotes through functional organizations cooperation between church bodies on such common tasks as missions and work among students. Mutual understanding on fundamental issues in belief, worship, and polity, and a united witness on world problems is promoted through conferences. Different forms of ecumenism (sometimes in opposition to one another) are associated with: the World Council of Churches, the Roman Catholic Church, the Movement for Church Union, denominational ecumenism, and evangelical ecumenism.

♦ KC0544 Metamorphosis
Description A marked and more or less abrupt change in form or structure during the course of some development process. In biology this refers to the transition from one postembryonic developmental stage to another within a relatively short period of time (e.g. in insect development), or to the sum of the various modifications, whether phylogenetic or primarily ontogenetic through which a primitive plant structure may pass in the course of its development. It may also be used for the transformation of a musical figure or idea into a rhythmically or melodically altered repetition of the original, or the change of chemical compound into an isomeric form.

♦ KC0545 Integrational analysis
Structural hermeneutics
Description The discipline of interdisciplinary research. It is based on a generally applicable theory of interpretation (termed structural hermeneutics) which specifies a minimum of seven levels which are intrinsically operative in the perception of any phenomenon in the humanities and social sciences, and in theories of meta–sciences. Because these seven levels cross–out the various disciplines, a phenomenon can be interpreted in an interdisciplinary way. A discipline can be utilized exactly as before but the resulting data can then be located on a specific level of the multilevel analysis. This allows recognition of what a discipline can account for but also what it has excluded from consideration as it sought the rigour and exactitude of its discipline.

♦ KC0546 System
Description 1. Any recognizable delimited aggregate of dynamic elements that are in some way interconnected and interdependent and that continue to operate together according to certain laws and in such a way as to produce some characteristic total effect. A system, in other words, is something that is concerned with some kind of activity and preserves a kind of integration and unity; and a particular system can be recognized as distinct from other systems to which, however, it may be dynamically related. Systems may be complex, they may be made up of interdependent sub–systems, each of which, though less autonomous than the entire aggregate, is nevertheless fairly distinguishable in operation.
2. A regular interacting or interdependent group of items forming a unified whole. A set of elements standing in interaction as expressed by a system of mathematical equations or an organization seen as a system of mutually dependent variables. A system may be characterized by: a particular relationship between elements which turns a mere collection of elements into something that may be called an assemblage; a pattern in the set of relationships which turns the assemblage into a systematically arranged assemblage; and a unified purpose which turns the systematically arranged assemblage into a system.
3. A set of objects together with relationships between the objects and between their attributes. Objects are parts or components of the system such as: atoms, stars, switches, masses, bones, neurons, gases, genes, etc., including abstract objects such as: mathematical variables, equations, rules and laws, processes, etc, and social actors such as: individuals, groups, communities, nations, etc.
4. Some form in structure or operation, concept or function, composed of united and integrated parts.
5. Any set of interrelating elements, including: any one of the concrete systems from solar system to molecule and electron, from cell to organism, together with personality, small group, formal organization, political system, economic system, and social system; a system of rules or procedures; a theoretical system bringing together various concepts and generalizations for the purpose of description, explanation or prediction.
6. A system is the first subdivision of Universe into a conceivable entity. It divides all the Universe into six parts, namely all the universal events occurring: geometrically outside the system; geometrically inside the system; nonsimultaneously, remotely, and unrelatedly prior to the system events; nonsimultaneously, remotely, and unrelatedly subsequent to the system events; synchronously and or coincidentally to and with the systematic set of events uniquely considered; and all the geometrically arrayed set of events constituting the system itself. All systems are individually conceptual polyhedral integrities; a system is a patterning of enclosure consisting of a conceptual aggregate of recalled experience items, or events. Systems have insideness and outsideness and are a closed configuration of vectors (conceptually representing all the interrelationships of system foci). They are each a pattern of forces constituting a geometrical integrity that returns upon itself in a plurality of directions, and are therefore interiorally concave and exteriorally convex, complex and finite.

♦ KC0547 Forecasting
Description A forecasting is a probabilistic statement, on a relatively high confidence level, about the future (as contrasted with a prediction, which is a statement on an absolute confidence level about the future). A distinction is made between technological forecasting and social forecasting.

♦ **KC0550 Corporate database**

♦ **KC0551 Integral life**

♦ **KC0554 Lattice theory**
Description A branch of set theory which discusses in precise mathematical terms the relation of different parts of the same whole to each other. Its applications include the study of structure and order. The theory which seems to have had application to the Warsaw Pact defence e-grid lattice and to NATO's and which has other military uses is also helpful in demographic analysis.

♦ **KC0555 Intellectual cooperation**
Description Collaboration among scholars and intellectuals (usually understood to be from different countries) on matters of common concern. This may include such activities as: exchange of persons between scientific institutions; exchange of scientific and scholarly publications; establishment of information clearing houses; cooperation in scientific research; and protection and promotion of the working conditions and welfare of scientific workers.

♦ **KC0556 Competition**
Agonemmetry
Description More or less active demand by two or more organisms (individuals, groups, or organizations) or kinds of organisms at the same time for some environmental resource in excess of the available supply. Typically this process may result in the ultimate elimination of the less effective organism from the particular setting in which the process is taking place.

♦ **KC0560 Structuralism**
Description Structuralism is a loosely defined term in that its proponents invoke structures which have acquired increasingly diverse significations to the extent that no common denominator is particularly evident. In mathematics, structuralism is opposed to compartmentalization, which it counteracts by recovering unity through isomorphisms. In linguistics, it is chiefly a departure from the diachronic study of isolated linguistic phenomena leading to the investigation of synchronously functioning unified language systems. In psychology, it is concerned with a focus on wholes rather than a reduction to parts.
Two suggested common aspects are: an ideal of intrinsic intelligibility supported by the postulate that structures are self-sufficient and are intelligible without resort to extraneous elements; a recognition that structures in general have, despite their diversity, certain common and perhaps necessary properties, including the idea of wholeness, the idea of transformation, and the idea of self-regulation. Structuralism remains essentially a method (rather than a doctrine) which is not exclusive or suppressive of other dimensions of investigation, which it tends to integrate in the way in which all integration in scientific thought comes about, namely by making for reciprocity and interaction. The application of the method results in interdisciplinary coordination, since structure is so defined that it cannot coincide with any systems of observable relationships, which are the only ones that clearly emerge through any of the existing sciences.

♦ **KC0562 Relativism**
Description The philosophical and methodological postulate of the unconditional conditionality of knowledge (excluding its self). In extreme forms, its asserts that all knowledge is subjective or false or valid only under limited or operant conditions. It has variant forms in the popularizations of the Theories of Relativity, the Uncertainly Principle, and in the 'understanding' of the inexplicable breaking down, or singularities, in the laws of physics.

♦ **KC0566 Planning-Programming-Budgeting System** (PPBS)
Description A form of cost-benefit analysis (described separately) associated with systems analysis and used as an operational tool in a wide variety of contexts (government agencies, universities, corporations, nonprofit centres, etc.). The integrated PPBS provides relatively few innovations for the individual elements of planning, programming, and budgeting; its value lies in its systematic coordination of all of these. A conceptual basis for the total allocation process is provided. Planning, programming, and budgeting constitute the process by which objectives and resources, and the interrelations among them, are combined to achieve a coherent and comprehensive programme of action for any organization conceived as a whole. The need for a holistic approach arises from the indissoluble connection between the allocation of human and material resources, or budgeting, and the formulation and conduct of policy. It provides a conceptual framework for delineating the complexities of purposive, goal-seeking public policy.

♦ **KC0567 Sensitivity analysis**
Description Determination of how accurately different data must be known in order to be reasonably assured a problem solution is optimal. Sensitivity analysis shows the implications on final solutions of slight variations of parameters. If an optimum solution is sought, it can show for what ranges of changes in values of elements of the analysis the solution is still optimum. Alternatively it can show the ranges of values which will still yield an acceptable solution.

♦ **KC0569 Conservation laws**
Description Energy values in isolated or closed systems will remain constant (or be conserved) regardless of the forms that the energy takes, i.e. chemical, electrical, thermal, etc. No energy can be lost (assuming symmetry in space-time). In homogeneous space isolated systems will conserve momentum regardless of where they are. The principles or laws of energy conservation are also applied to the universe as a closed system, and phenomena such as entropy find application in such a framework.

♦ **KC0571 Macroevolution**
Description 1. Intraspecies variation, or mutations, conformable to Mendel's laws.
2. Special, non-Mendelian variation.
3. Either or both of the foregoing as factors determining the origin of taxons above the species level.

♦ **KC0572 Organization of labour**

♦ **KC0573 Substance**
Description 1. Matter, i.e. mass.
2. The essence or first principle of an individual thing or being.
3. Matter or essence, metaphorically.
4. A category in philosophy and metaphysics which is viewed as more or less fundamental to any analytical account of the universe and nature.

♦ **KC0577 International social integration**
Description Social integration is one component of international integration (separately described) and may be defined as: 1. Development of a pattern of transactions between units belonging to different national states. Such transactions may include trade, mail, movement of persons, etc. 2. Creation of a transnational society or the abolition of national impediments to the free flow of transactions. Mass social integration may be usefully distinguished from the process of integration of special groups or elites.

♦ **KC0579 Alchemy**
Description 1. Pre-scientific chemical experimentation.
2. The theory and alleged practice of the transmutation of metals, particularly base metals (e.g. lead) into noble ones (e.g. gold). Also the method of obtaining an agent of transmutation, variously and allegorically named, in the form of a liquid elixir, a powder of projection, or a solid 'stone'. It is a practice associated with the preparations of medicines in the internal alchemy of producing and maintaining a healthy, long-living human body, and therefore also with a healthy psychology. The internal alchemy ultimately broadened its aims from long, healthy life, to immortality; and from a healthy psychology to a quest for superhuman knowledge and powers. In the latter form it became a species of magic. In both forms it was associated with delusion and charlatanism. In some aspects its practices required astrological considerations, and in others religious piety.
3. An occult art, one of whose objects is to provide a subtle vehicle for the soul, a second, spiritual body. For this purpose the physical body is viewed as a chemical chamber or vessel of some kind, for example, a distillation flask, a 'pelican' glass, etc, or, from more of a metallurgical point of view, as a furnace or crucible.

♦ **KC0580 Kaleidometrics**

♦ **KC0581 Eclecticism**
Description 1. In philosophy, the explicit or implicit acceptance of a number of approaches (logical, empirical, epistemological, etc), some of which may be contradictory, in order to adhere to their outcomes, particularly in the domaine of ethical and political philosophy.
2. A philosophy of personal or of party convenience in which elements are juxtaposed on the basis of their appeal and surface acceptability, and which lack coherence in terms of their derivation.
3. Non-traditional experiential or experimental philosophy.
4. In matters of taste, inconsistency or unconstrained selectivity.

♦ **KC0584 Graph theory**
Description The systematic study of configurations of points and lines joining certain pairs of these points, particularly for application in the symbolic presentation of comparative qualitative and quantitative knowledge (on paper and other surfaces). A graph may be analyzed and defined in the terms of mathematical set theory. Representations also can be given pictorially and analytically. The pictorial representation of a graph is useful for depicting structural situations.

♦ **KC0586 Formal elegance**
Mathematical elegance
Description The harmonious arrangement of elements such that the mind can without effort take in the whole without neglecting the details. This may be taken as indicating the nature of all elegance, whether aesthetic, theoretical, or functional. In a universal theory elegance implies the highest self-consistency and simplicity with all phenomena, large and small, complex and elementary, being revealed as expressions of one comprehensive principle. An elegant theory of organism, for example, must display the functional elegance of living systems whereby all their parts cooperate to maintain the properties of the whole. The special form which elegance takes in connection with sets of related constructs (hypotheses) is called invariance.

♦ **KC0587 Regional international integration**
Description This is the process of international integration (see separate description) operating between a limited number of countries, generally within the same continent. It is considered useful to distinguish regional integration from related concepts such as regionalism, regional cooperation, regional organization, regional movements, regional systems, or regional subsystems of a global system. The study of regional integration is concerned with explaining how and why states cease to be wholly sovereign, how and why they voluntarily mingle, merge, and mix with neighbours so as to lose the factual attributes of sovereignty while acquiring new techniques for resolving conflict between themselves. The other concepts may relate to stages on the way to regional integration, but they should not be confused with the resulting condition. (There is a difference of opinion as to whether the term refers solely to the terminal condition or also to the process by which it is achieved.)

♦ **KC0592 Holographic universe**

♦ **KC0596 Discipline development**
Description The development of a discipline may be envisaged as taking place in three stages:
1. Problem-oriented, multidisciplinary stage, in which people identifying themselves with an established discipline, apply and extend their methodologies to the new problem.
2. Interdisciplinary stage, in which a set of methodologies emerges out of the study of particular multidisciplinary problems. Specific problems are approached as a part of a new interdisciplinary field.
3. New discipline stage, in which there is formal recognition of the new field as one which possesses a coherent overview and a set of distinct methodologies of general applicability.

♦ **KC0597 International institutional integration**
Description Institutional integration is one component of international political integration (separately described) and may be defined as the development of international institutions. A distinction is made between strong and weak central institutions, and in one view the former are necessary for a high degree of other types of integration to occur. The degree of jurisdictional integration of institutional power is determined by such factors as: the supranationality of decision-making, the scope of legal powers, and the fulfilment of that scope.

♦ **KC0600 Integrated circuit**
Integrated microcircuit
Description Complex, highly dense microcircuits with tens and hundreds of thousands of components created by modifying their crystal lattices, and capable of a number of functions, ranging from total control of automated processes, to special applications in telecommunications, computers and weapons systems. Integrated circuits may be regarded as containing a logic or a programme. Their use is basic to a concept of integrated electronic engineering.

♦ **KC0601 Ideology**
Description 1. Those aspects of a philosophy which have the most apparently ready application to the world of action, i.e., ethics, politics and economics, and secondarily, psychology, sociology, philosophies of science, history, education, and religion, etc.
2. A secular substitute for religion in which there is emotional adherence to an unsystematic corpus of theories, opinions and beliefs, mainly on how to organize and improve the world.
3. Any collection of beliefs that are alleged to provide a theoretical basis for action and behaviour.

♦ **KC0603 Multi–valued logic**
Description

♦ **KC0604 Management of complexity**

♦ **KC0605 Logic**
Description 1. Mode of reasoning or inference.
2. Study of reasoning methods used to reach conclusions that are true, valid, certain, or characterized by other values, such as usefulness or acceptability.
3. Deductive reasoning. Syllogistic logic.
4. Inductive reasoning.
5. Propositional calculus.
6. Predicate calculus.
7. Logic of verbal expression.
8. Logic of mathematical expression. Mathematical logic.
9. Logic of symbolic expression.
10. Logics of formal relationships of elements in any composite: structure or activity.
11. Special logics, e.g., logic of classes or set theory in mathematics.
12. Generally, persuasive technique associated with order in rhetoric and argument.

♦ **KC0606 Variety**
Heterogeneity
Description The quality or state of: differing in kind; consisting of dissimilar constituents or parts (that may not be unified or compatible); having different values, opinions, or backgrounds.
1. The quality or state of having numerous forms or types. An intermixture or succession of different things, forms, or qualities. A multiplicity of things within the same class or category that can be distinguished, often by marked differences.
2. The variety of anything is its number of distinguishable elements. Every conceptual step which enriches the nature of a system under study increases the information about it, increases the uncertainty informing it, and proliferates its variety.
3. The total number of possible states of a system, or of an element of a system.

♦ **KC0607 Control system**
Description Any means, natural or artificial, by which a variable quantity or set of quantities is caused to conform, more or less accurately, to some prescribed norm which may either be a constant value or values which vary in prescribed ways. Control systems may be considered as a mix of two extreme types, namely intrinsic, in which the controls are part of the system under control, and extrinsic, in which the controls are external to the system under control or only loosely linked to it.

♦ **KC0608 Information system integrity**
Description The ability of an information system to protect itself against unauthorized user access, to the extent that security controls cannot be compromised. Security controls, no matter how sophisticated, are not reliable if the operating system that administers those controls is not itself protected from user tampering. Total information system integrity, or security, is not considered feasible. A level of system integrity must therefore be selected where the cost and risk involved in breaking that security exceed the benefits to be gained from doing so, or exceed the cost and risk of obtaining the same benefits in another way.

♦ **KC0609 Mass culture**
Description 1. The variants of popular cultural demand is perceived in the contents of mass media, in styles of leisure–time activities and entertainments, and in clothing and personal behaviour.
2. Popular cultural supply as conceived by governments of planned economies, state churches, dictators, opportunistic individuals and corporations, and utopia creators.
3. Generally, the average educational and informational level of the classes, i.e. the working classes.

♦ **KC0610 Systems analysis**
Description A methodology applied to problems in systems in which specialists may combine with decision–makers to blend in a concerted way the diverse approaches available. Thus each solution arrived at in this way is a work of reasoned art, not the result of a prescribed methodology or formula that applies to all cases. The major characteristics of systems analysis are that:
1. It is a powerful technique for grasping ill–structured, large, complex problems of choice under uncertainty (more effectively than if its individual parts were examined in isolation). It looks at the broad goals to be achieved and examines the costs, effectiveness, and risks of the various alternative approaches to achieving the objectives.
2. It permits and encourages the judgement and knowledge of experts to be joined in a systematic and efficient manner, and facilitates the blending of the judgement of generalists and managers with the expertise of specialists.
3. It employs the scientific method, namely: (a) it is open, explicit and results can be verified by others, even though quantitative and qualitative information are mixed; (b) analysis is systematic and objective; (c) hypotheses are tested and verified by appropriate methods; (d) information is quantified wherever possible.
4. It constructs and operates within models or simplified abstractions of the real system situation appropriate to the problem.
5. It evaluates alternatives by a careful assessment of costs against benefits, making cost–benefit analysis (described separately) an important part of systems analysis when applicable.
6. It deals with practical problems, and not theoretical problems.
7. It attempts to deal explicitly with uncertainty.
8. The context is often broad and the environmental i.e. external factors are usually very complex, such that simple problems exceptionally come within its scope.
9. There is an absence of a universal guiding theory, making systems analysis a methodological art rather than a science, one that seeks out the theory or analytics that are appropriate in each case.
10. It is focused more on exploring the implications of alternative assumptions than on analysing in extensive detail the implications of a single set of assumptions.
11. It is ordinarily not concerned with computing an optimum solution but rather with giving the decision–maker a range of choices and outcomes. It emphasizes design of new solutions and widening the range of alternatives, rather than selecting the best alternative from a predetermined range.
12. It lacks definition as a methodology, thus the foregoing characteristics are negated by some authorities, to wit: it applies to all systems, simple or complex; it has a distinction methodology which applies in all cases; it is the domain of specialists only; it is not always scientific in that it accepts some unverified hypotheses of systems theory; it can be rigorously applied to exhaustive analysis of a single solution, etc.

♦ **KC0611 Counter–intuitive**
Description The behaviour of complex social systems appears counter–intuitive to the average person for whom intuition and judgement, generated by a lifetime of experience with simple (first order, negative feedback) systems that surround the majority of daily activities, create a network of expectations and perceptions which mislead the unwary when trying to comprehend more complex systems. Not only are such systems fundamentally different from the simple systems of daily experience, but they appear to be the same by providing a plausible relationship and pattern to be discovered. However the discovered causes which become the basis for corrective action for the noted effects are frequently only coincident symptoms (so that having dispelled the symptoms, the underlying causes still remain). There is therefore a high probability that intuitive solutions to the problems of complex social systems will be incorrect most of the time.

♦ **KC0612 Transformation theory**
Description The fundamental premise of transformation theory is that living processes are ubiquitous and universal. It maintains that psychological and cultural processes are an extension of and are isomorphic with biological, physical, and chemical processes. The single process that forms the keystone of the theory and that unites the behaviour of all things is the process of growth. The drive of both the physiological and the psychological process of living is to assimilate external materials and to reformulate them into extensions of the self. The cell does this by ingesting its environment and transforming it into cells that match its own genetic pattern; the human does it by mentally absorbing his cultural environment affecting it in ways that conform to his own culturally acquired mental patterns.
Physical, biological, psychological and social systems act in the direction of development of higher levels of, and more widespread, interrelationships, evolving more organized behaviour and becoming more integrated through the incorporation of diversity. These interrelationships grow through a progression of different forms of linking behaviour which, having been successful in a particular subsystem, are repeated at the next highest level. As information handling capabilities evolve, each higher level of growth can contain (or bind) more data, and less energy is lost in growth transactions.

♦ **KC0614 Problem–solving theory**
Description 1. A wider approach to the study of human task–oriented behaviour than epistemology, logic, or scientific method, as it also includes such things as motivation, cultural influences, tradition, personality, context, etc.
2. Artificial intelligence theory, or cybernetics.

♦ **KC0615 Organization of production**

♦ **KC0616 Equifinality**
Description The ability of an open system (insofar as it does attain a steady state) to reach a specified or characteristic final state from different initial states and in different ways, namely under different working conditions based on dynamic interactions in the system. In a closed system, the final state is unequivocally determined by the initial state such as in chemical equilibrium. Thus closed systems cannot have equifinality. Equifinality appears to be responsible for the primary controlability of organic systems, namely for all control or regulation which cannot be based upon predetermined structures or mechanisms and which tends to exclude such mechanisms.

♦ **KC0618 Homeomorphism**
Description 1. A one–to–one continuous mapping of one topological space onto another which is also an open mapping. If two spaces are homeomorphic, then they differ only in the nature of their points, and can, from the point of view of topology, be considered essentially identical.
2. Topological equivalence between two geometrical figures in which each can be transformed into the other by a continuous deformation.

♦ **KC0619 Technology**
Description 1. Applied science.
2. Machines, processes, procedures, techniques, and knowledge used in providing the world's goods, services and material and social structures.
3. Industrial methods, notably for manufacturing production.

♦ **KC0620 Functional social integration**
Description The integration resulting from the manner in which the different specialized parts of society interact, interrelate and make reciprocal contributions to each other and to society as a whole.

♦ **KC0621 Projective geometry**
Synthetic geometry
Description A branch of pure mathematics that is concerned with the geometrical properties (distinguished by a peculiar interrelatedness and stability) that are preserved under arbitrary projective transformations of the plane (or of space). As a non–metrical geometry (not restricted to the Euclidean geometrical concepts of length, angles, area and volume), it is primarily concerned with the concepts of point, line and plane, and their interrelationships, without the need for measure. As such, it is not merely a geometry of created forms, but the geometry of form–creating entities. It enters the domain of movement and metamorphosis, where rigid measure is no longer the dominating factor, and one form can change into another without losing its identity. It requires a qualitative grasp of mathematical form before it takes on a fixed shape in the quantitative field of measure. It has a unifying effect on the whole field of geometry in that it is concerned with the whole, whereas the metrical geometries deal only with the part.

♦ **KC0622 Classification**
Description 1. The act or method of distributing into groups, classes or families, resulting in a system of classes or groups or a systematic division of a series of related phenomena.
2. A coding system within which the words of the code (series of symbols indicating a concept, or semantemes) are subject to certain order relationships. A code is understood here to be a system of symbols for the representation of information and rules governing their combination. The freedom of the coder is therefore restricted by special conventions, and usually by the obligation imposed upon him to follow a certain hierarchical order.

♦ **KC0626 Linear programming**
Description A mathematical technique of exceptional power and generality designed to assist decision–makers in planning certain organizational activities. The problem for which linear programming provides a solution involves the maximization (or minimization) of some dependent variable (e.g. some economic objective such as profits, production, costs, work weeks, etc) which is a function of several independent variables, when the independent variables are subject to various constraints.

♦ **KC0627 Conceptual systems**
Abstract systems
Description Systems whose components are concepts; for example, logical, numerical, linguistic, philosophical, ethical, and religious systems. Conceptual systems may differ in significant ways from concrete systems and in order to be considered within the framework of any general system theory, a system has to be defined more generally as a complex of interrelated entities (rather than of interacting entities). In order to study concrete systems, abstract systems with analogous relationships may be substituted such that the problem becomes a mathematical one. This process is usually known as the development of a model; the extent to which the abstract model agrees with the actual behaviour of the concrete, physical system is a measure of the applicability of the particular model to the situation in question.

♦ **KC0628 Complexification**
Description The process by which matter and all forms of organization arranges itself in ever larger and more complex units.

♦ **KC0630 Concept**
Description 1. The interpretation of the significance understood and the learning acquired from experience, stored in the memory, simply and in complexes. Thus there is a concept of a thing or material object, of a process, of a person, of a system, of a concatenation of systems, of a universe, of non–reality, of the imagined, and of the conjectured, etc, and there also is a concept of what a concept is.
2. The smallest unit of thought to which a term may correspond.

♦ **KC0631 Religion**
Description 1. Fundamentally, a manner of living in accord with a belief in the existence of a soul and a god, and in life after death (conditional or unconditional), with the associated beliefs in divinely given moral and other commendments whose transgressions bring punishment in the hereafter. Usually, but not inevitably, religion entails public worship and other external manifestations (fasts, feasts, mutual help among believers, distinctive greetings, clothes or manners, a holy book, holy places, etc).
2. Intellectually the attempt to account for man and the universe before the development of scientific method. Insecurity was the key drive in the creation of religion and the chief metaphor of a father god in heaven was based on the model of one or several of the principle bodies of the solar system, i.e., Saturn, Jupiter and the Sun, or on singular comets which approached the earth.
3. Modern religious forms include those which have no pronounced influence on behaviour and require mainly intellectual assent to a number of abstract propositions such as belief in an absolute being or substance (mind or matter), the fellowship of man, the goodness of the world, the sufficiency of ethics or ethical philosophy to guide human affairs, human progress, evolution, the all–conquering nature of human reason, and similar ideas.
4. Ideology, including anti–religion.

♦ **KC0632 Gnosticism**
Description 1. Judaic occultism originating in the pre–Christian period, influenced by Zoroastrian religion on its practical side, e.g. knowledge of the angelic worlds, and by neo–Platonic speculation on its theoretic side, e.g. number mysticism. Jewish gnosticism was closely allied with magic and with a Shamanistic technique for ascending through the heavenly palaces (Lekaloth).
2. Christian heterodoxies, primarily of the first three centuries, which claimed reception of the inner teachings (the esoteric tradition) of Jesus. Although externally the Christian Gnostics showed considerable borrowings from Hellenistic Judaism and Jewish gnosticism, there are untraceable essential teachings (e.g. Valentinian, Marcosian) whose origin lies most likely in Jesus himself, or in the circles around John the Baptist (cf. the Mandaeans). The Gnostic thesis that knowledge is salvation is echoed in the 'gnostic' Fourth Gospels of the New Testament.
3. A broad variety of religious and philosophical teachings that have purported to offer knowledge of the otherwise hidden truth of total reality as the indispensable key to man's salvation. The highly syncretistic origins of such teachings (including Jewish, Iranian, Babylonian, Egyptian, and other oriental material) masked a highly original inner unity of thought distinct from all the disparate historical elements employed in its representation. In a sense gnosticism culminated in Mani or in Islamic mysticism (Sufism).

♦ **KC0633 Social network**
Description 1. A complex set of interrelationships in a social system may be termed a network of social relations. This use of network is however purely metaphorical and is very different from the notion of a social network given below. As a metaphor, the notion of network subsumes, and therefore obscures, several different aspects of social relationships such as connectedness, intensity, status, and role.
2. A specific set of linkages among a defined set of social entities, with the additional property that the characteristics of these linkages as a whole may be used to interpret the social behaviour of the social entities involved.
3. A social network is based on a set of interlinked points. These points need not necessarily be individuals, groups or other collectivities, they may be social status positions, social roles, or myths. The points may be events or intervals of time. Social networks may be composed of several kinds of relationship between several kinds of social entity at several different levels, and including relationships of an entity to itself.

♦ **KC0634 Integrism**
Fundamentalism
Description Attempt to maintain the totality of a system, especially a religious system. It has been used extensively in polemical debates concerning the refusal of some Christians to consider any form of evolution. In addition, in the sense that it describes any religious or ideological adherence to an infallible record (e.g. holy book) or source (e.g. 'The Leader'), it has been applied to ideologues in Islam and Communism.

♦ **KC0635 Science policy**
Description 1. Science policy studies have as their focus the systematic investigation of scientific and technological activities and their function within society. In particular, they are concerned with policymaking in scientific and technological fields, and with the interrelationship between policy-making, cultural values and societal goals.
2. A deliberate and coherent attempt to promote a basis for national and international decision influencing the size, institutional structure, resources and creativity of scientific research in relation especially to its application and public consequences.

♦ **KC0636 Pattern recognition**
Character recognition
Description General shape or pattern recognition in material objects having individual differences such as printing faces of different sizes, styles, or geometrical patterns. The essence of pattern recognition lies in one definite meaning being allocated to whole classes of patterns which are examined by machine for the existence of certain describing features and for the meaning allocated to them on the basis of these existent or non–existent characteristics.

♦ **KC0637 Meta–games**
Description Game–like situations, formally identical with games, but differing from them in that players may or may not know such things as each other's utilities, or what moves each intends to make. The theory of meta–games is concerned with social behaviour in situations in which if the persons or organizations about which predictions are made were to obtain knowledge of the predictions their behaviour might or might not be affected by such knowledge.

♦ **KC0640 Model**
Modelling
Description 1. In the broadest sense, a model is simply a figurative or symbolic representation of something else. In order to produce information, a symbolic model is usually a verbal or mathematical expression describing a set of relationships in a precise manner. It can be useful simply to explain or describe something, or it can be used to predict actions and events. Models can be distinguished and classified in many ways, including:
By function: descriptive, predictive, and normative models. By time reference: static, and dynamic models. By uncertainty reference: deterministic, probabilistic, and game models. By generality: general, and specialized models.
2. An ordered set of assumptions about a complex system. It is an attempt to understand some aspect of the infinitely varied world by selecting from perceptions and past experiences a set of general observations, applicable to the problem at hand.
3. A theoretical projection in detail of a possible system of human relationships (e.g. in economics, politics, or psychology).
Model building refers to the process of putting together symbols according to certain rules to form a structure which corresponds to a real–world system under study. A real–world system is too complex to be modelled in exact detail, so many factors are ignored and relevant factors are abstracted to make up an idealized version of the system. The construction of a model, as a scientific procedure, is founded on the belief that there can be order and reason in the mind, if not in the real world. The process of abstraction is usually an integrative one requiring experience, intuition, and judgement about the system being analyzed, as well as skill in model construction. Modelling has the inherent capability of cutting across or integrating the inductive and deductive processes with the reality being confronted.

♦ **KC0642 Teleology**
Finality
Description Static teleology or fitness is any arrangement that seems to be useful for a certain purpose. Dynamic teleology is the directiveness of processes, of which the following forms may be distinguished: direction of events towards a final state which can be expressed as if the present behaviour were dependent on that final state; directiveness based on structure, meaning that an arrangement of structures leads the process in such a way that a certain result is achieved; equifinality (separately described), namely achievement of the final state from different initial conditions and in different ways; true finality or purposiveness, meaning that the actual behaviour is determined by the foresight of the goal.

♦ **KC0643 International**
Description In its broadest sense this is a characteristic of relations, interactions or organizations in which individuals or organizations from two or more countries participate. In a more restricted sense, the relations, interactions or organizations only involve the participation of governments of nation–states. International is then synonymous with intergovernmental or interstate.

♦ **KC0645 Change**
Process
Description Any planned or unplanned alteration of a state, configuration, quantity, etc, in an organism, situation or process. Change may extend over a period of time and it may be a very complex phenomenon. It is a universal characteristic, an invariant, within systems. In its social aspects, distinction may be made between planned change (as a designed, or purposive attempt by an individual, group, or larger social system to influence directly the status of itself, another system, or a situation), and unplanned change. A distinction also may be made between the transmission of culture (evolutionary changes of concern to the historical sciences and anthropology) and the transformation of social patterns. The former is neither planned nor intended, whereas through the latter, individuals, groups, or organizations change themselves or others through directive action or decisions.

♦ **KC0646 Closed system**
Description A system which is considered to be isolated, totally or practically, from its environment such that there is no significant import or export of energy in any of its forms such as information, light, heat, physical materials, etc. and therefore no change of the components of the system. An open system becomes closed if input and output is cut off. Some open systems become effectively closed if either input or output is terminated, thus cessation of throughput is also a factor since it may also terminate input or output or both. In thermodynamic terms if, in an open system, momentum or conversion of energy ceases, the system may close.

♦ **KC0647 Comparative analysis**
Comparative studies
Description Descriptive comparison is an important feature of most of the social science disciplines and is commonplace in the humanities, and in a very limited sense all studies are comparative. Whenever general concepts are used in considering particular items, descriptive comparisons are being made even if the classification of objects under categorical concepts is only implicit and unconscious.
Systematic comparative analysis in the social sciences is based upon the assumption that there is order or regularity in the world. By classifying phenomena in ways that focus upon the relationships between sets of events, it may therefore be possible to discover the dynamic relations that exist among them, in order to find orderly patterns of related actions. Such analysis may then inspire theories that explain classes of events by means of deductively related laws or regularities.

♦ **KC0650 World unity**
World union
Description

♦ **KC0651 Voluntary economic integration**
Description 1. Multilateral international economic agreements, treaties and organizations (based on regional location; prior military, mutual defence treaties; common language; common political heritage; common ideologies, or any combination of these) whose purpose is to create a common market between members and to achieve common economic domestic and foreign policies.
2. Economic integration in varying degrees has been the goal of many organizations, such as EEC, ASEAN, the British Commonwealth, bodies (as in Scandinavia, Africa, Latin America and among the Arab States). COMECON, however, dominated by the Soviet Union, achieves a partly involuntary economic integration among the captive European nations who are its members, for example, Hungary, Czechoslovakia and Poland, who are forced to trade with the Soviets on unfavourable terms.

♦ **KC0652 Collegiate authority**
Description 1. In the Roman Catholic church the protest against Papal and Curia secrecy and autonomy and the demand for a democratic share in policy making by the episcopacy, in particular, by the College of Cardinals.
2. In political and administrative science, collegiality is that form whereby authority and control are said to be exercised by a group of equals (in consultation or in decision–making).

♦ **KC0657 United Nations Organization**
Description The purposes of the United Nations Organization are: (a) to maintain international peace and security; (b) to develop friendly relations among nations; (c) to cooperate internationally in solving international economic, social, cultural, and humanitarian problems and in promoting respect for human rights and fundamental freedoms; and (d) to be a centre for harmonizing the actions of nations in attaining these common ends. The significance of the United Nations Organization lies not only in the attempt to accomplish these purposes but also in its existence as a symbol for the whole of mankind of the interlinked values which have given rise to the commitment to accomplish them.

♦ **KC0659 World market**
Description 1. The total global market for all goods, services and activities.
2. The global market for a particular commodity.
3. The global marketplace composed of free–market and mixed economies.

♦ **KC0660 Symbolism**

♦ **KC0661 Mass society**

♦ **KC0662 Socialist collective**

♦ **KC0663 Commensalism**

♦ **KC0665 Input/Output analysis**
Description A technique used to examine the flows into and out of a system. It has been developed to examine economic systems, particularly at the national level, as an aid to decision–making. In such an application it is used to examine the interindustry traffic in raw materials, intermediate products, and technical and financial services that necessarily precedes the delivery of finished products to final markets. An input/output table is used to display in its horizontal rows the outputs (sales) by each industry to other industries and, in its vertical columns, the inputs (purchases) by each industry from other industries.
The information in such a table may be held in the memory of a computer and then becomes a working model of the economy under consideration. Input/output analysis may then be used for estimating industrial markets, the availability of producers' supplies, wage and price trends, opportunities for investment, and the repercussions of government expenditures and new social and technological innovations.
Since it coordinates information on performance in all industrial sectors simultaneously, the technique also provides a framework for consistent analysis of natural resource constraints and environmental pollution of different types.

♦ **KC0666 Integrated planning**
Description Two dimensions to the integration of plans may be distinguished: ensuring a relationship between strategic plans and medium–range plans, and between medium–range plans and short–range plans; and ensuring a relationship between all the plans concerned with a particular time frame.
Integration of plans is facilitated by developing both quantitative and qualitative plans, thus permitting cross–references among functions, interrelating levels of planning, and providing cross–checks for evaluations. Numerous methods of integrating plans exist, but the most universally used basis for translating strategic plans into actions is a budgeting system.

♦ **KC0667 Probability theory**
Description Organizational decisions are made in one of two essentially different contexts: under conditions approaching certainty and, more generally, under conditions of uncertainty. The quantitative analysis which supports decision–making under certainty usually takes the form of maximizing some objective subject to some constraints. Under conditions of uncertainty and inadequate information, probability techniques assign values to probabilities of events and actions and permit the selection of preferred courses of action on the basis of the highest expected projected values. Probability theory supplies a method to quantify judgemental processes which are often carried out by decision–makers on the basis of crude approximations. Use of such techniques makes it possible for decision–makers to be more certain about the expected values of actions, to consider a greater range of alternatives, and to find their way through complex alternatives reasonably easily.

♦ **KC0669 Orthoepy**
Description Correct pronunciation of a language as determined by the aggregate of norms. Orthoepy is to phonetic as orthography is to graphic language.

♦ **KC0670 Progress**
Description The advance toward an objective or goal, accompanied by improvement, especially in the sense of the progressive development or evolution of mankind.

♦ **KC0671 Classic**
Description 1. A value concept expressing the notion of the best, or sometimes the original, of a form of human product, activity or behaviour. Thus one reads the classics, views classical art, listens to classical music and considers human artifacts and theories that are imbued with classicism. There are also classical mechanics and physics and classical styles of clothing.
2. Anything pertaining to the cultures of the ancient Hebrews, Greeks and Romans; i.e. the foundation of western culture.

♦ **KC0673 Space**
Description 1. A set of object–points, whether functions, figures or states, which in their relationships constitute a geometry. The axiomatic expression of the properties of these geometric relations defines the dimensional and other characteristics of the space.
2. Extension in three–dimensions.
3. This universe, excluding the gravitational field of the earth.
4. Emptiness.

♦ **KC0674 Theme**
Description 1. The subject or programmatic content of a work of art or the dominant motif.
2. In literature, the typology of the plot, e.g. the plot of Romeo and Juliet has the thematic type of tragic love; also the olidactic object or moral of a work, e.g. the theme, crime does not pay.
3. In music, one or more leading melodies in a work.
4. A dominant colour in a decorating scheme.
5. The central subject of a conference.
6. The cognitive or conceptual element that unifies a work as opposed to the sensory. In some cases, its meaning.

♦ **KC0676 Summativity**
Description 1. A property of a system such that a change in each element depends only on that element itself. Each element can therefore be considered independent of the others. The variation of the total complex is then the sum of the variations of its elements.
2. In the mathematical sense, summativity means that a change in the total system obeys an equation of the same form as the equations for the parts.
3. In biological, psychological and sociological systems the interactions between the elements may decrease with time so that they can be neglected, in which case the system becomes progressively more mechanical and segregated, tending to behave like the sum of the independent parts.

♦ **KC0677 Probabilistic system**
Markov systems
Description A system in which some of the states may be predicted with certainty, but others can be predicted only on a basis of probability, since it is possible at some stages for the system to move into one of several states. Because one state of the system is known, it does not follow with certainty that the next state can be described. Many of the more advanced technological processes and man–machine systems are of this type.

♦ **KC0678 Civilization**

♦ **KC0684 Utopia**
Description A place, state, or condition of ideal perfection, especially in laws, government, and social conditions. Utopias are possibly real but usually imaginary. However, the ideal nature of utopias serves as a focus for proposals for social improvement.

♦ **KC0686 Integration of knowledge**
Unification of knowledge — Conceptual integration
Description The meaning of integration of knowledge is largely determined by the manners in which it is attempted. These include:
1. Encyclopaedia method: involving the collection and evaluation of known facts and their tabulation for reference and judgement.
2. Dogma: conversion to a dogma, whether through persuasion or enforcement, functions as an effective integrator for a period of time.
3. Language analysis: leads to a form of integration through an understanding of the medium which carries and transmits accurate information. Semantics is considered to be an integrator of considerable potency, for it draws attention to the medium in which bare facts are embedded, a medium which in part fashions them and gives them their significance in discourse.
4. Conceptual: involving detection and understanding of the basic integrative concepts. Contrasts, polarities, misunderstandings are in many instances embedded in and confined to the surface of knowledge. Disparities and conflicts tend to be eliminated in the light of such basic concepts.

♦ **KC0687 Technological forecasting**
Description Technology denotes the broad area of purposeful application of the contents of the physical, life, and behavioural sciences. It comprises the entire notion of technics as well as the medical, agricultural, management and other fields with their total hardware and software contents. Technological forecasting is then a probabilistic assessment, on a relatively high confidence level, of future technology transfer. 1. Technology transfer is a (usually complex) process which cannot be adequately described by a simple line. It may be necessary for scientific or technological resources and many elementary technologies to merge before a functional system can be built.
2. Technology forecasting is a quantified prediction of the timing, the character, and the extent of change among technical parameters as well as all attributes associated with design, production, and the use of materials and processes, according to a structured system of reasoning. The prediction may or may not include an estimate of probability or confidence in the amount of change or its timing. In some cases, relative need or usefulness can be quantified and used as the basis of the forecasting procedure. The key distinction of this definition is that the forecast is reproducible through a system of logic. Thus, it differs from opinion and prophecy in that it rests upon an explicit, stated set of relationships, data, and assumptions. The procedure yields relatively consistent results regardless of who is doing the analysis.

♦ **KC0690 Racially and ethnically integrated education**
Description Common and equal participation of members of different racial or ethnic groups and communities in an educational environment (school, university, etc).

♦ **KC0692 World law**
Description World law is a condensed term for both a world authority equipped with legislative bodies for making laws against international violence and the related agencies for enforcing such laws, keeping the peace, and resolving conflicts.
A model of world order in which such a system of institutions exists is termed a world law model.

The world law model proceeds from the assumption that no political elite would be advised to engage in any kind of disarmament unless there were some assurance of another security system to protect its political independence and its territorial integrity. World law thus links disarmament and a collective security system.

♦ **KC0693 Unitary theory**

♦ **KC0694 Language**
Description 1. The organization and transmittal of information.
2. Verbal and written expression systematically developed to constitute an adequate lexicon, a syntax and a grammar (and subsequently, additional usages such as those of normative spelling and pronunciation).
3. Intercommunication between beings or systems (e.g. by use of symbols or signs) in which discrete elements are unified by a non–self–contradictory logic.
4. Intracommunication within beings or systems (via neurons, chromosomes, energy pulses, etc).
5. A set of signs or symbols.
6. Communication.

♦ **KC0696 Policy analysis**
Public policy analysis
Description A future–oriented inquiry into the optimum means of achieving a given set of social objectives. It may also encompass studies designed to extend the range of social alternatives perceived by decision–makers, and to suggest the long–range and implicit consequences of various sets of value priorities. This may involve inquiry into alternative: descriptions of the problem context; ways of conceptualizing the problem; sets of goals and objectives; predictions of probable outcomes of the courses of action considered (including alternative models); strategies for assuring preferred outcomes; and methods of appraising the implementation of a selected course of action.
Policy analysis is therefore a problem–oriented approach as distinguished from a disciplinary approach. Its success depends upon the identification and the integration of relevant knowledge and the cooperation of appropriate participants from all relevant disciplines.

♦ **KC0697 International policy integration**
Description Policy integration is one component of international political integration (separately described) and may be defined as the extent to which a group of countries act as a group (by whatever means) in making domestic or foreign policy decisions.

♦ **KC0702 Continuum**
Description Generally, a relationship between a perceived whole and its parts so that the manifold or complex unity can be characterized as being determined by the proximate nature of all or most of its subordinate members, either spatially, temporally, functionally, or logically. This nearness of the members (of what is essentially a set) may be 'touching' or separated by intervals. Therefore continuity (of substance, identity, etc) is an inverse function of the ratio of magnitudes of the intervals to adjacent set members (in terms of extension, duration, frequency or other quantities). The surfaces of seas and deserts, for example are continuums (although not homogeneous since they are broken by islands and oases). Emmental cheese, sponges, lace and spider webs are also continuums, also not necessarily homogeneous. The set of real numbers is a mathematical continuum. The identity, continuum is similar and has common features with the identity, unity. As a uniform type is also a type of unity. Speculatively, the human race, the biosphere, and the noospheres are continuums (of which there may be a mathematical treatment: a geometry or topology).

♦ **KC0706 Block diagram**
Block language
Description Graphic method for the symbolic representation of functional relationships which greatly facilitates the surveying (diagramming) of complicated systems, consisting of rectangles, triangles, circles, etc. and so–called blocks which are connected by arrowed lines. All block diagrams can be placed between two extremes: (a) abbreviated or simplified block diagrams where each block represents a piece of apparatus or a set of equations and connecting lines show which blocks influence each other and (b) detailed or exact block diagrams where each connecting line represents a physical or mathematical variable and the block itself represents the function.

♦ **KC0707 Human engineering**
Description A science drawing upon various other sciences (e.g. anatomy, physiology, physical anthropology, applied psychology), that deals with the design and positioning of machines, instruments, and controls so that they may be used with maximum efficiency by human beings.

♦ **KC0708 Comprehensiveness**
Description The completeness of available information, not necessarily in terms of the volume of information but rather in terms of the accounting for, or inclusion of, all (or virtually all) pertinent considerations. This attribute of information is important in linguistic analysis and information theory, and especially in semantics, the study of meaning.

♦ **KC0709 Social sciences methodology**
Description Includes methodological approaches that can be applied to any science, for example, most branches of logic, and the experimental method, but characteristically of human or soft sciences, raises issues of emphasis between quantitative versus qualitative, and specialization versus generalization (i.e. systems, holistic, comparative or interdisciplinary approaches, etc). The methodology also involves critical epistemological and ontological issues in arguing for or against the realities of the individual or the collectivity as agents or causal factors in societal phenomena.

♦ **KC0710 Systems engineering**
Description The design of a complex interconnection of many elements (to constitute a system) to maximize an agreed–upon measure of system performance. It includes two parts: modelling, in which each element of the system and the criterion for measuring performance are described; and optimization in which adjustable elements are set at values that give the best possible performance. It employs the methods of cybernetics, information theory, network analysis, flow and block diagrams, etc.

♦ **KC0711 Integrated science education**
Integrated methods of science teaching
Description Integration, when applied to science courses, means that the course is devised and presented in such a way that the student gains the concept of the fundamental unity of science; the commonality of approach to problems of a scientific nature; and is helped to gain an understanding of the role and function of science in his everyday life, and the world in which he lives. Such a course eliminates the repetition of subject matter from the various sciences and does not recognize the traditional subject boundaries when presenting topics or themes.
A distinction is made between integrated science curricula (namely curricula emphasizing the interdisciplinary or unified dimension of science) and integrated systems of science teaching (namely systems of teaching which blend with one another when handling science topics).
There is considerable uncertainty concerning the meaning of such terms as: discipline, interdisciplinary and integrated. Integration requires re–combination of previously segregated parts to give unity or to make a whole. One procedure advocated in practice is to mix the parts and hope that the significance of each part will be appreciated by the students. Another procedure, deemed more satisfactory by its proponents, is to show the inter–relationships and interactions explicitly by using team teaching methods, such that each is able to cross–refer to other relevant fields where appropriate. Another approach is to select a topic that is an entity in itself (e.g. air pollution, human ecology) and teach any science relevant to that topic. Integration is applied here in a very different sense from its usage with reference to the synthesis of a course from a universe of segregated parts. Where synthesis is attempted, ways are sought to combine the parts into coherent units. In the topic method, the starting point is the coherent unit, which is then examined for its constituent subject areas, the relevant parts of which are taught as part of the unit.

♦ **KC0712 Public opinion**
Description 1. The opinion of individuals and groups constituting the consumers of private and public mass media information services.
2. Statistically or methodologically significant behavioural reaction from the electorate, the public or masses so–called, and non–governmental organizations, to events, actions, proposals, plans, etc, that are perceived as affecting their interests.
3. The moral force of the community exercised by freedom of choice to accept or reject anything in the behaviours of individuals, groups, government, institutions, enterprises, or churches, etc.

♦ **KC0714 Rhyme**

♦ **KC0715 Theorem**
Description 1. A verbal or algebraic statement of proof capable of being framed in the form: 'if' (hypothesis follows or hypothetical situation); 'then' (conclusion follows). This logic is very important in computer programming and mathematics.
2. A deductive theory.

♦ **KC0717 Goals**
Targets — Short–term aims
Description 1. Goals and targets may be considered as synonymous. They refer to short–term and minor aims. Goals are objectives expressed in a specific dimension and are usually very concrete. They may exist for an organization as a whole, for major divisions and departments and for individual performance. Most short–range goals of an organization are defined during the course of the budgeting process.
2. An operational objective which a system seeks to achieve or maximize.

♦ **KC0718 Synergetics**
Synergetic analysis
Description The theory of an exploratory strategy of starting with the whole, and the known behaviour of some of its parts, with the anticipated progressive discovery of the integral unknowns and comprehension of the hierarchy of generalized principles. The mathematics conceived to be involved in synergetics consists of topology combined with vectorial geometry; to make possible a rational, whole–number, low–integer quantization of all the important relevant geometries. This is thought to be because the tetrahedron, the octahedron, the rhombic dodecahedron, the cube, and the vector equilibrium may be constituted by the lattices of all atoms.
Synergetics makes possible, it is thought, the return to conceptual modelling of all physical intertransformations and energy–value transactions. The geometrical, a priori, model of energy configurations in synergetics is developed from a symmetrical cluster of spheres, in which each sphere is a conceived to be a model of a field of energy all of whose forces tend to coordinate themselves. The forces of the field of energy represented by each sphere inter–oscillate through the symmetry of equilibrium to various assymetries; the vector equilibrium itself being only a referential pattern of conceptual relationships at which a system may never pause. The closest packing of spheres in 60–degree angular relationships demonstrates a finite system in a 'universal' geometry. Synergetics is theoretically comprehensive because it seeks to describe instantaneously both the internal and external limit relationships of the sphere or spheres of energetic fields. It attempts to coordinate within one mensurational system the interlocking of quantum mechanics and vectorial geometry. Synergetics is typical of holistic, deductive philosophies of scientific method which seek to make evident the awkwardness characterizing mid–twentieth century treatment of natural interrelationships by isolated scientific disciplines.

♦ **KC0719 Mathematical model**
Description The use of mathematical notation and structures (e.g., formulae) to describe conditions, states, values, relationships, etc in selected phenomena. Experimental mathematical models can be formulated for the behaviour of the human nervous system, for the consumption rates of the world resources, and for species populations in a given environment, for example.

♦ **KC0720 Educational system**
Description The basis of an educational system is the instructional activity which is an integrated set of media, equipment, methods, and personnel performing efficiently those functions required to accomplish one or more learning objectives. The critical functions for a complete instructional system are: practice of performance, practice of knowledge, presentation of knowledge, student management, and quality control. These functions may be performed by a variety of components, including people, machines, books, and different kinds of equipment, depending on the objectives. Other necessary systems are required for: record–keeping, feeding, supply, finance, transportation, and communication, depending whether the system is based on student attendances at physical, centralized school locations or whether some other educational system is involved such as distant learning, computer instruction, tutoring, and so on.

♦ **KC0721 Organic form**

♦ **KC0722 Complementarity**
Description The interrelationship, completion or perfection brought about by one or more units supplementing, being dependent upon, or standing in polar opposition to another unit or other units. In atomic physics, for example, conflicting evidence of behaviour or characteristics obtained under different experimental conditions cannot be comprehended within a single picture, but must be regarded as complementary in the sense that only the representation of the totality of the phenomena exhausts the possible information about the objects observed. This implies the

impossibility of any sharp separation between the behaviour of atomic objects and the interaction with the measuring instruments which serve to define the conditions under which the phenomena appear. The situation in quantum physics has been considered one reflection of the application of an all-pervasive principle determining the approach to the unity of knowledge. Complementarity is seen characteristically in the behaviour of light as a wave and as photon particles. Situations in psychology and biology also present equivalent complementary aspects; and the considerations of such analogies in epistemological respects in turn illuminates unfamiliar physical problems.

♦ KC0723 Semantic field
Description The total range of meanings associated with a set of words which are related but not identical in meaning.

♦ KC0725 History
Description The study of history may be primarily viewed as an accounting of chronological events and as a congeries or series of happenings which both form and dominate the shape of societies. The emergence of comparative studies of civilization can however lead to a shift in perception of the growth of cultures. Civilizations can now be seen from a wider as well as a more dynamic perspective, as revealing the development of fundamental patterns of human relationships in the context of those artistic and social forms which embody the transformations of human consciousness.

♦ KC0726 Mapping
Description 1. The assignment to every element of a set (i.e. a mathematical or conceptual model) of an element of the same set or another set. The process of making a one-to-one or one-to-many correspondence
2. Given two sets E and F, a mapping (from E to F) is any correspondence, rule, method, diagram, indication, construction, process, algorithm, computation, machine, device, force, drive, reflex, instinct, command, or any other cause whose effect is that, given any element in E, one and only one element in F results. E is known as the mapping's domain; F is its range, in which the mapping takes its values. F is not necessarily different from E.
3. Mapping may also be a heuristic process, when the intention is to model reality, since some information may not be known (as in early maps of the world which depicted the territories of fabulous monsters and beings on the edges of the flat earth).

♦ KC0728 Polyphasic unity
Description A unity (neither empty and analytic, nor blankly uniform and featureless) in which a variety of different elements is distinguishable, each expressing (as a particular manifestation or realization of) the principle that unifies them all and relates them to one another systematically. It is thus a unity or whole with many different facets, displaying its articulate nature in numerous different ways, presenting to an observer a multiplicity of appearances each representing in its own way the principle of order that constitutes the whole and which relates the elements of the multiplicity to one another in such a manner that they are mutually deducible if the principle of order is known.

♦ KC0730 Purpose
Purposiveness — End
Description 1. A characteristic of human behaviour whereby actual behaviour is determined by the foresight of the goal. It presupposes that the future goal is already present in thought, and directs the present action. It is connected with the evolution of the symbolism of language and concepts and should be distinguished from other aspects of finality and teleology (described separately).
2. The intent that is intrinsic to planning action and which gives it direction. The main purpose of planning is to create controlled change in the environment and the reason for wanting such change is that complex dynamic situations tend toward increasing degrees of de-organization (ecological imbalance) unless higher order organizing activities are introduced. The purpose of affecting that situation is therefore either to solve the problems that in here to the situation , or to improve the situation, or to establish a general control and dynamic over the environment so as to obtain organized progress within it.
3. The concept of purposiveness involves a notion of a hierarchy of decisions in which each step downward in the hierarchy consists in an implementation of the goals set forth in the step immediately above. Behaviour is purposive insofar as it is guided by general goals or objectives. Such purposiveness brings about an integration in the pattern of behaviour, in the absence of which administration of an organization would be meaningless.
Purpose may be roughly defined as the objective or end for which a process is undertaken. There is no essential difference between a purpose and a process but only a distinction of degree. A process is an activity whose immediate purpose is at a low level in the hierarchy of means and ends, while a purpose is a collection of activities whose orienting value or aim is at a high level in the means-end hierarchy.

♦ KC0731 Analogy
Description A methodological integration of the sciences can be furthered by drawing valid analogies between the methods of the different sciences. Resemblance between different sets of phenomena may be taken as suggesting the existence of some law or principle common to both sets, especially where a comparison can be made between the functions of elements in two systems. The fruitfulness of such analogies for science depends on whether consequences can be deduced from them which can be tested or observed, and this is likely to depend on whether the resemblance selected as analogous is of a fundamental or merely superficial kind. If structural relations can be reproduced in a simplified form in a different medium, a model may be constructed. The relation between model and thing modelled can be said generally to be a relation of analogy of which two kinds may be distinguished: (a) in the case of a logical model of a formal system, there is analogy of structure or isomorphism between model and system, since the same formal axiomatic and deductive relations connect elements and predicates of both system and model; and (b) in a replica model, in which there are material similarities between the parent system and its replica.
A member x of a set is analogous to its fellow member y when (a) x and y share several objective properties (or are equal in some respects) or (b) there exists a correspondence between the parts of x or the properties of x and those of y. If x and y satisfy the first condition, they may be said to be substantially analogous (e.g. in the case of any two atoms). If the second condition holds, then x and y are formally analogous irrespective of their constitution. If both conditions hold, the analogy may be called a homology. Homology implies both substantial and formal analogy, and substantial analogy implies formal analogy, but not conversely.
If x and y are sets, then correspondence under condition (b) leads to several degrees of formal analogy: (1) plain, or some-some analogy, when some elements of x are paired with some elements of y; (2) injective, or all-some analogy, when every element of x is paired with an element of y; (3) bijective, or all-all analogy, when the preceding relations hold both ways.
If there exists a correspondence that maps every element of x onto some element of y and, in addition, preserves the relations and operations in x, then there is a homomorphism of the set x into the set y, which is an all-some (injective) structure-preserving analogy. If there is also a homomorphism from y into x, and in addition the two morphisms compensate each other, the analogy is called an isomorphism. Isomorphism is perfect analogy which implies homomorphism, injective analogy, and plain analogy.

♦ KC0732 Uniformity
Description 1. A property postulated of nature, that identical or equal causes in identical or equal circumstances produce identical or equal results.
2. A logic of causality and determinism that the physical laws are invariant.
3. The characteristic, so long as it holds true, that delimits a world or dimension in the space-time continuum. Partial non-uniformity in sub-atomic quanta may indicate another 'universe' of uniformity beginning somewhere at this level. Apparent non-uniformity exists at the other end of the scale in very large phenomena studied in astrophysics.

♦ KC0733 Unitary thought
Description A system of thought which: (a) emphasizes process, development, and transformation; (b) is capable, at least potentially, of bringing all facts into relation with one another; and (c) recognizes a universal formative process in nature in which regular spatial forms are developed and transformed. Any process which displays one general form of continuity may be called unitary.

♦ KC0735 Common sense
Description 1. Popularly, the basing of action or judgement by an individual on knowledge of which he feels certain because it was acquired in his own experience, and often because it is common to others' experiences.
2. Originally, the sense common to the five senses whose function was apperceptive or interpretive. As an internal sixth sense it also received and interpreted (it was considered) inner impressions and states. Consequently it was confused with intuitive knowledge or a related parapyschological faculty. To claim, as a basis for action of judgement, the exercise of this common sixth sense, is to claim certainty based on some inherent wisdom or divine reason.

♦ KC0736 Network planning techniques
Critical path method — Programme evaluation and review technique
Description Techniques used in planning projects which have precisely defined start and end criteria, e.g. time or cost, etc, and in which the complexity of the interrelationship paths between the events and activities of the project makes control very difficult. Such techniques are most useful where the problem is one of directing a large number of concurrent activities directed towards the same goal, and have little to offer where the constituent activities must of necessity take place consecutively. The techniques take into account the need for first placing in their logical order the activities of the project thus enforcing a discipline which automatically shows how each activity depends on or constrains the others. A system is thereby provided for monitoring the progress of the project, for forecasting the effects of snags (timelags or time cost increments) on the project as a whole, and for deciding which activities should have priority for resources.
The basic steps are: preparation of a network of arrows in between event nodes (circles representing a milestone or other important stage of an activity or an event) in which each arrow represents one activity or sub-activity in the project, and the positions of these arrows in the network represent the, logical, preferred, or in some cases circumstance-dictated order in which the activities must take place; estimation of the duration and least cost of the activity represented by each arrow (and possibly also the cost of each activity); analysis of the network to identify the activities which are critical in the sense that they govern the over-all least duration of the project; and preparation of a schedule for all activities. All the foregoing procedures are algorithmically provided in network planning computer software packages.

♦ KC0737 Serendipity
Description The unexpected beneficial result which may emerge as a consequence of engaging in most forms of activity. For example, many advances come out of basic research, which by definition tends to produce serendipitious results for applications, since the research is not directed toward practical ends, but toward obtaining information and understanding about some important area. Serendipity often results from applied research, when an application is found in a different field than the one intended. The rich interaction characteristics of simulations and gaming may generate outcomes which were neither sought nor predicted but which may be of profound applied and theoretical significance.

♦ KC0738 Function
Description 1. Purposive action of a uniform kind by an agency or instrument.
2. A recalling activity that is characteristic of some organ or instrumentality.
3. The dependence of an outcome or value on preconditions or values in the instrumentality or operant so that it is a function of those antecedent conditions or values.

♦ KC0740 System order
Order of a system
Description 1. The number of integrations or accumulations within the system.
2. The number of states necessary to describe the condition of the system.
A system of greater than fourth or fifth order begins to enter the range of what is termed a complex system. An adequate representation of a social system, even for a very limited purpose, can be tenth to hundredth order.

♦ KC0741 Phenomenology
Description 1. A branch of science dealing with the description and classification of phenomena.
2. A discipline which endeavours to lay foundations for all sciences by describing the formal structures of phenomena or of both actual and possible material essences that are given through a suspension of the natural attitude in pure acts of intuition.

♦ KC0742 Universal
Description

♦ KC0746 Open systems
Description Systems which interact with their environment, through the exchanges of matter, energy and information. Such systems also have the following properties: their entropy tends to decrease, namely they acquire negative entropy; there is organization, counterbalancing the tendency toward de-organization and operating toward the achievement of higher levels of orderliness and heterogeneity; and inflow and outflow balance each other under steady state dynamics, such that the system continues to maintain its on-going rates of change.Additionally, open systems show self-regulation and self-adaptation; equifinality, namely as part of their self-regulation they tend to achieve and maintain a steady state around a particular level (or goal);

major feedback functions and differentiation and elaboration, with diffused global patterns being replaced by more specialized functions.

♦ KC0747 Proportion
Description Certain visual geometric proportions and ratios considered by some as metaphysically or aesthetically significant and evoking a pleasant response in an observer. These and other proportions and ratios may evoke an equivalent response when perceived aurally in musical consonances and harmonies.
Speculation on such relationships has led to the belief that certain ratios and proportions were intrinsically beautiful and were keys to understanding the relationship between microcosm and macrocosm and to revealing fundamental truths about the structure of the universe. The aesthetic qualities of the golden section and the golden rectangle have led to their extensive use in artistic and architectural products. Significance is attached to the numeric properties of the logarithmic spiral, its widespread occurrence in attractive natural forms, and its relevance for the structural understanding of growth.
Some kind of controlling or regulating system of proportion has been considered a desirable unifying and harmonizing device, although the more recent tendency is to reject the notion that one system of proportion, a priori, is better or more agreeable than another.

♦ KC0750 World crises
World problems
Description

♦ KC0751 Art
Description 1. Any human skill.
2. Products of specialized skills purposely endowed with aesthetic qualities, such as harmonious colour.
3. The fine arts.

♦ KC0753 Configurations
Description 1. Pattern recognition and typologies.
2. In astronomy, relative positions of solar system bodies.

♦ KC0754 Cross–impact analysis
Description This is a technique whereby the probability of occurrence of a number of different developments is given expression and the interrelationships between these probabilities are also established within a matrix. Through this network it becomes possible to link various developments and to describe systematically the impact of the realization of a given development on the probabilities of occurrence of other developments. The value of this technique to forecasting and assessment lies primarily in the link it provides between what might otherwise be rather disparate forecasting and assessment efforts. However this technique is only as good as the assignments of approximately correct probability values to the various matrix cells.

♦ KC0755 Integrated world
Description The characteristics of an integrated world are a high level of coordination or integration among the major powers, including arms control and international aid programmes, and in the world generally a low level of conflict and of perceived, potential conflict.
Models of integrated worlds may be:
1. Stability or status quo oriented, in which political and economic coordination exists mostly among the advanced powers and is designed mainly to secure and improve their position. There is effective exclusion of the developing countries from major influence as well as mixed prospects for their development.
2. Development or aid–oriented, in which there is extensive and successful world organization for progressive and welfare purposes, with a subordination of politics and ideology to pragmatism.

♦ KC0756 Metaphor
Description A figure of speech in which a word or phrase denoting one kind of object or action is used in place of another to suggest a likeness or analogy between them. It is more than a literary device and has cognitive implications whose nature is a proper subject of philosophic discussion.

♦ KC0757 Objective
Description An end to be achieved, a future condition or result to be accomplished in a specific time in carrying out a particular activity. A value sought by an individual or group. The time dimension of an objective is long–range, as distinguished from short–range targets and goals. Objectives are usually specific and realistic in contrast to (organizational) philosophies and creeds. They can however be broad, since one of their characteristics is variability, and can range from general, overall statements to specific, narrow statements.
There is considerable confusion between (organizational) purposes, objectives, policies, and goals. The guiding distinction is that a purpose determines why an organization continues to exist, an objective determines what it wants to accomplish, a policy determines how it carries out its activities in order to fulfil its purpose and achieve its objectives.

♦ KC0758 Fuzzy logic
Description 1. A logic system applied to artificial intelligence to replicate how human judgement is made from imprecise data.
2. Heuristic logic.

♦ KC0759 Megalopolis
Description 1. Areas such as Tokyo–Osaka with a population of over 50 million; Boswash (USA East coast) with 40 million; and London–Liverpool with 30 million.
2. A group of conurbations.
3. A continuous urban sprawl.
4. A single city like Mexico City, Tokyo or New York, where populations exceed 5 million.

♦ KC0760 Scientific and technological progress
Description 1. The history of science and technology.
2. The goals of such progress according to various philosophies.
3. The future trends of scientific and technological achievement; as predicted analogously to past accomplishments or by quantitative (e.g., probabilistic) methods.

♦ KC0761 Time
Description 1. Experiential time: the reading of a clock or other chronometer; for example, the time of day, or elapsed time (shown by stop–watch). Similarly, the time of the year (a season), or the time in one's life (an interval in which an event or activity occurs). In another and general sense, 'experience', as to have a good time. Time as normally experienced therefore includes notions of interval, duration, frequency and linear succession of durations or moments, and also diurnal and annual cyclicity. Time appears to have forms as well as rhythms.
2. Newtonian time: experiential time conceived of as a homogeneous, universal absolute.
3. Relativistic time: a multiplicity of time systems in the universe according to Einsteinian theories of relativity.
4. In philosophy, or the philosophy of physics, a number of time theories: Minkowski's, Dunne's, de Stitter's, Milne's and many other's which, though disparate, indicate the possibility of abrogrations of laws of time, and therefore of causation and logic.
5. In theology, God as the ultimate time, usually is the cosmogonic sense, from which all times come forth (e.g. the Zoroastrian Zurvany, the Gnostic Aeon, the Fourth Gospel Arche) or God as the annihilation of temporality (e.g. in the New Testament where it is said 'Time must have a stop' and 'there will be time no longer').

♦ KC0766 Holon
Description Any stable sub–whole in an organismic, cognitive, or social hierarchy which displays rule–governed behaviour and/or structural Gestalt constancy: Sub–wholes are intermediary structures at different levels in a series in ascending order of complexity, such that with respect to the lower levels they function as autonomous wholes, and with respect to the higher levels, they function as a dependent part. (In other words, towards its subordinate parts, it behaves as a self–contained whole, and towards its superior controls as a dependent part.) The relativity of the terms part and whole when applied to any of its sub–assemblies is a general characteristic of hierarchies. A hierarchically organized whole cannot be reduced to its elementary parts; but it can be dissected into its constituent branches of holons. A holon may also be conceived as a system of relations which is represented on the next higher level as a unit in another system of relations.
Biological holons are self–regulating open systems governed by a set of fixed rules which account for a holon's coherence, stability and its specific pattern of structure and function. In symbolic operations, holons are rule–governed cognitive structures variously called frames of reference, universes of discourse, algorithms, etc., each with its specific grammar or canon; with the strategies increasing in complexity on higher levels of each hierarchy.

♦ KC0767 Synectics
Description Synectics is derived from a Greek term meaning the fitting together of different and apparently irrelevant elements. It is used to refer to the process of identifying the combination of people and different techniques (including quantitative and non–quantitative, the mathematical, the hunches, the insights, the guesses), to innovate or to advance the decision–making process in planning. Essential to this process is ensuring that only techniques relevant to the problem and only the relevant parts of each applicable technique are used. The term synectic relevance is used to denote the applicability of a technique.

♦ KC0768 Class
Description 1. A mathematical or logical set constituted by symbols and/or numbers and representing real or conceptualized elements.
2. A typological or taxonomic manifold; a category.
3. People in a certain income bracket correlated to mainly their level of educational attainment, tastes, and manner of speech.
4. A degree or step in a vertical hierarchy.

♦ KC0769 Metasystem
Description 1. A system that constrains another system logically, said to be in a higher class since it may frame propositions, criteria and rules for the lower system.
2. A cybernetic control system.
3. More generally a system that is presupposed by another, e.g., the international trade system (comprising buying, selling, shipping and delivering) presupposes the international monetary system with its financial instruments (letters of credit, drafts, etc). Another example is the printed mass media system, which depends on literacy and therefore on the metasystem of education. In a cybernetic model education would be a 'control' on printed media.

♦ KC0770 Unification of legislation
Description The ordering process within the national or international corpus of statutes, codes, laws, enactments, etc to reduce obsolescence, redundancies, inconsistancies and inequalities. On the international level this is done by supranational bodies such as the EEC and COMECON.

♦ KC0771 Food chain
Description Ecological interdependence of species in respect to who eats whom. For example a certain plant is eaten by a certain insect, which is eaten by a certain bird. The food chains stretch from microorganisms to mammals. Certain terrestrial features are also an outcome or are dependent on particular kinds of vegetation; quality, chemically and mechanically, of the soil, etc.

♦ KC0772 Sacred tradition
Description A class of writings in several of the world religions that by their adherents are considered secondary only to revealed books like the Bible, the Koran or the Vedas. The sacred traditions arise due to oral transmission of the words of the founder, spoken usually after the giving of the revealed book, or they are words of subsequent great masters. Eventually written down, such traditions are ascribed to Moses, Buddha, Jesus, Mohammed and other figures. Typical traditional texts are in the Sunna and Hadith of Islam; in the Creeds, Apostolic Canons and some other early Church writings in Christianity; in the Sayings of the Fathers and the Talmud in Judaism, and in the smitri of Hinduism. All the earliest books of Buddhism, however, may be considered traditional, as there was no one revealed text ever published. In most instances of sacred tradition the books are considered 'binding' on believers in respect to their teachings.

♦ KC0776 Logic of relations
Relational logic
Description From the theoretical standpoint, an understanding of the logic of relations is essential in logic, in mathematics, and in dealing generally with problems of meaning. Logic may be conceived to be an interdisciplinary subject which performs an integrating function. It belongs to none of the established disciplines exclusively, but is useful to all of them. Relations are integrating concepts; they link together all of the traditionally separated areas of logic and philosophy, providing, through relational logic, a set of notions common to all parts of the subject. Thus it is not necessary to assume any special rules for categorical syllogisms, or for immediate inference, since they can all be derived (where they are valid) from more general relational ideas. The traditional distinction between categorical, hypothetical and disjunctive propositions may be dispensed with, since those types are interdefinable. Relational logical also shows how the meaning of a rigorously formalized language (of one theoretical system) may be understood, and how to compare systematically one such language with another.

♦ **KC0777 Unitary**
Description An attribute of a structure which may constitute a single unit, although possibly made up of component units. A distinction may be made between unitary and unified, the first implying that the condition of being whole is primary although parts may be distinguished, whilst the second implies that the whole is a product of a unifying or synthetic process applied to the originally separate parts.

♦ **KC0778 Integration of languages**
Description 1. Artificial and abrupt integration i.e., via conquest, edict, etc, where one language replaces another.
2. Natural and prolonged integration via acculturation or assimilation, e.g. where a linguistic minority is eventually absorbed.
3. Evolution of differentiated or diverse languages towards a shared vocabulary, similar syntax, etc, culminating by merging into what becomes a new language.
4. Creation of Esperanto, Volapuk or other proposed world languages.
5. Evolution of mathematical and logical symbology, e.g. the language of machine, artificial intelligence.

♦ **KC0779 Categories**
Description 1. Generally, groupings of objects or phenomena according to similar properties and characteristics.
2. The grouping of properties and characteristics into fundamental classes using analytical reduction.
3. The must fundamental classes of concepts of universal properties, such as moving or stationary.
4. In Pluto: essence or being, motion, rest, sameness or identity, and difference. In Aristotle: essence, quantity, quality, relation, place, time, position, state, action and affection.
5. In religious philosophy the categories exist apart from matter, as divine forms, archetypes, or energies (cf. sephiroth in Kabbalah).
6. In scientific philosophy new categories have been added such as system and structure.

♦ **KC0781 Gestalt**
Description A structure or configuration of physical, biological, or psychological phenomena so integrated as to constitute a functional unit with properties not derivable from its parts in summation.
Gestalt psychology in based on the unity of psychic life, the existence and primacy of psychological wholes which are not just a summation of elementary units. Basically, the parts of a shape only have meaning by the fact that they belong to a whole, that is, a shape cannot be split up into its elements without losing the meaning which it possessed as a whole. Thus human behaviour can be thought of as a gestalt, a mental whole which transcends the mere combination of its elements.
Within the long series of very different processes which lead from a vague idea of a lawfulness in data (discovered by gestalt perception) to a clear formulation of scientific knowledge, gestalt perception may enter at the most varied points in order to establish an orderly relationship between other rational links of the total event. True gestalten may be seen in figures or in equations. In other places, rational categories may be used in complexes whose natural unity has only been established by gestalt perception. The perception of complex gestalten may therefore be considered an indispensable partial function in the systematic whole from whose coordination the always incomplete picture of extrasubjective actuality is constructed.

♦ **KC0782 Integrated electronics**

♦ **KC0783 Cognitive system**
Description An agglomeration of knowledge or belief about things that are related in the percipient view. The higher the level of consistency in the agglomeration the more recognizable its systems characteristics are. The more interrelated a number of cognitive systems are the more an individual possesses a world-view. High interrelationship is characteristic of religious philosophies and some secular ideologies (e.g. Marxism-Leninism).

♦ **KC0786 International system**
International community
Description This consists of the nation-states and their interrelationships as nation-states. Various international subsystems exist and may interact. Various types of international system have been distinguished which include or exclude nation-states or nation-state systems according to their economic power, military capability, or involvement in some international organization. No distinction may be made between an international system and the global system. The United Nations may not be considered to be identical with the total international system or to operate above it.

♦ **KC0790 International university**
Description An institution or system of institutions with objectives such as the following: (a) stimulate and strengthen cooperation among individuals, research institutions and universities in order better to contribute to the satisfaction of the economic, social and cultural needs of the peoples of the world and to contribute to objectives of peace and progress; (b) undertake studies and research on problems arising out of the application of the Charter of the United Nations and on the theoretical and practical problems of putting such ideas and principles into practice; (c) constitute a meeting ground and stimulate thinkers and researchers anxious to contribute to the solution of world problems; (d) constitute a forum for free and independent discussion and responsible confrontation of the most varying ideas and ideologies; (e) make available to the international community the results of such discussions and of interdisciplinary and inter-ideological studies.

♦ **KC0791 Transcendent unity of religions**
Description Religions are alike at heart or in essence (esoterically) while differing in form (exoterically). Metaphysically, revealed religions converge at an apex of existence and cognition in God. The epistemological concomitant is that religious discernment unites at its apex, while differing short of this ultimate. Anthropologically, this unity precludes final distinction between human and divine; epistemologically, between knower and known. It bespeaks a knowing that becomes its object, or rather is its object, for temporal distinctions are inapplicable at this point.

♦ **KC0792 International economic integration**
Description Economic integration is one component of international integration (separately described) and may be defined as the abolition of discrimination between economic units belonging to different national states. The economic significance of national borders is that they introduce discontinuities, whether in trade, in flows of factors of production or in general economic policies, etc. It is customary to distinguish various states of the economic integration process according to the kind of discrimination removed: 1. Free trade area, which implies the removal of quantitative restrictions and customs tariffs;
2. Customs union, which unifies the tariff of the countries within the area against outsiders;
3. Common market, where all restrictions on factor movements within the area are abolished;
4. Economic union, where economic, monetary, fiscal, social and counter-cyclical policies are to some extent harmonized;
5. Supranational union, where the respective governments abandon completely their sovereignty over the policies listed.

♦ **KC0794 Syncretism**
Description 1. The conscious process of uniting or harmonizing conflicting principles, particularly in religion or philosophy. When this is done without critical examination or real logical unity it is more of the quality of a fusion or mix.
2. The developmental process of historical growth within a religion by planned or unplanned accretion and coalescence of different forms of belief and practice through interaction with other religions and cultures.

♦ **KC0795 Hierarchial principle of control**
Description The distribution of control functions in a system, e.g. by the signals of sub-systems at a higher level to subordinated sub-systems.

♦ **KC0796 Transdisciplinarity**
Description 1. The coordination of all disciplines and interdisciplines on the basis of a generalized axiomatics (introduced from the purposive level) and an emerging epistemological pattern. In the successive steps of cooperation and coordination between disciplines, transdisciplinarity defines an organizational principle for a hierarchical system of multiple levels, having multiple goals, and in which there is coordination of the whole system toward a common goal.
The essential characteristic of a transdisciplinary approach is the cooperation of activities at all levels of the education/innovation system. This depends not only on a common axiomatics (derived from coordination toward an overall system goal) but also on the mutual enhancement of epistemologies. With transdisciplinarity, the whole education/innovation system would be coordinated as a multilevel, multigoal system, embracing a multitude of coordinated interdisciplinary two-level systems.
2. Establishment of a common system of axioms for a set of disciplines.

♦ **KC0797 Universe of discourse**
Description The term is used to describe the growth and development of common images in conversation and linguistic intercourse. Any image common to a group of people is a product of such a universe of discourse, namely of the process of sharing messages and experiences. A subculture may therefore be defined as a group of people sharing a particular image.

♦ **KC0799 Front**
Description 1. A political coalition.
2. Among Socialists and Communists coalitions existing in variant opportunistic forms; for example, popular fronts, common fronts, united fronts, etc, defined principally by the degree of cooperation and at what levels.
3. A organized labour coalition.

♦ **KC0800 Complexity**
Description 1. All scientific statements are ultimately analysable into two components; an a priori or structural aspect associated with the number of independent parameters to which the statement refers, and an a posteriori or metrical aspect, there being associated with each structural element a numerical quantity measuring the amount of credibility to be associated with the aspect of the statement. The amount of this structural information represents what is meant by the complexity of a statement about a system; it may alternatively be defined as the number of parameters needed to define it fully in space and time.
2. Environmental complexity refers to the quantity of stimuli configurations or concatenations of stimuli which an animal must discriminate among, and to which it selectively responds.

♦ **KC0801 Evolution**
Description 1. A process of unfolding or continuous change from a lower, simpler, or worse state to a higher, more complex, or better state. This may be regarded in two senses: as the unfolding of that which is enfolded, namely the rendering explicit of that which is hitherto implicit (which may be termed development); and the outspringing of something that has hitherto not been in being (which may be termed emergence).
2. A process whereby some degree of randomness modifies the logic of behaviour thus producing a mutation which is then tested for survival value against the embedding environment. Those logics which survive are in turn affected by randomness, and thus mutated and themselves tested for survival, resulting either in an extinction or some modified logic that fits an ecological niche. In a dynamic environment some continual modification of the logic is required for survival of the species, although excessive randomness can also result in extinction.

♦ **KC0803 Confederation**
Description 1. A political form of organization, federating autonomous entities by a central administration or government, having delegated powers. The central administration may not impose decisions on individual members. Political confederations are also known as leagues and united bodies (provinces, states, etc).
2. Federalled labour syndicates (unions) at national or international levels members of which (locals, etc) have varying degrees of autonomy.

♦ **KC0804 World socialist market**
Description 1. An economic goal of the Soviet Union unrealized due to the defection from Leninism by China, Yugoslavia, and other socialist countries.
2. The global market in which socialist economies are trading members.
3. The world markets for COMECON members.

♦ **KC0805 Global system**
World system
Description The totality of interacting social, technological, information, cultural and other systems, whether at the community, national or international level. Natural environmental systems may also be included. A distinction may not be adequately made between the world system and the international system (described separately).

♦ **KC0806 Boundary**
Description 1. A line or area which determines inclusion in and exclusion from a system.
2. Any system as an entity which can be investigated in its own right must have boundaries, either spatial or dynamic. Strictly speaking, spatial boundaries exist only in naive observation, and all

boundaries are ultimately dynamic.
3. Systems, described as bounded regions in space–time, involving energy interchange among their parts, possess boundaries which may be clearcut and simple or non–material. An individual's self–identity and self–direction may be considered intimately linked to the role of boundaries in the conceptualizing process. In organizing sensory phenomena, the individual responds to stimuli in terms of patterns, and it is the segregation of circumscribed wholes which enables the sensory world to appear imbued with meaning.
4. A sub–system in a nearly–decomposable system in n dimensions will have boundary surfaces of n minus 1 dimensions between a high–interaction region and a low–interaction region. The surface may be taken as passing through a family of points where some parameter such as interaction–density has a maximum density. The boundary–surface for one property will tend to coincide with the boundary–surfaces for many other properties because the surfaces are mutually reinforcing. All gradients and flows in the region very near the boundary will tend to be either parallel or perpendicular to the boundary.
5. Indicates or fixes a limit on, or the extent of, the phase space (namely the particular area of the universe of numbers defined by a particular set of mathematical equations) which has to be searched in order to find the answer to a particular problem. Limiting conditions which restrict the range of possible solutions.
6. The extent of a field measured statistically and probabilistically in terms of incidence of the field's properties, e.g., the bounds of the gravitational field of the solar system or the bounds of the human (individual) energy field or the bounds of the noosphere.

♦ KC0807 Industrial engineering
Description The application of engineering principles and the techniques of scientific management (e.g. installation of methods and systems, operating procedures, control measurements, etc.) to the maintenance of a high level of productivity at optimum cost in industrial enterprises. Industrial engineering developed from scientific management and incorporated improved procedures for decision–making.

♦ KC0808 Tree structure
Arboreal organization
Description A nonlinear, multilevel hierarchical structure in which each node may be related to a multiplicity of nodes at any level below it, but to only one node above it in the hierarchy. Graphically, this is similar to the geneological family tree, or to the pyramidal organizational chart. The concept is used in the analysis of the logical structure of data in complex data bases.

♦ KC0811 Unity of science
Unified science — Integrated science
Description There are several persistent historical and logical factors to support unity of the sciences:
1. All sciences deal with the same universe.
2. The different phenomena encountered in the universe are interrelated and interdependent. It is assumed, usually implicitly, that the universe itself is somehow unified. Any seeming disunities are therefore assumed to be due to limitations in description and understanding and not to the fundamental reality. Assigning certain phenomena to different disciplines therefore creates artificial, or unnatural, boundaries.
3. Another assumption implicit in the concept of the integration of science is that science (as distinct from nature) is coming to be regarded as unified in substance and content. A finite set of logically related laws and theories should eventually emerge to explain all natural phenomena. New theories are characterized by their applicability to more phenomena than the theories they replace. (Physics is widely regarded as the fundamental science on which all other sciences must ultimately be based.)
4. There are certain key concepts that permeate all science disciplines. These concepts (e.g. energy, equilibrium, system) are relatively few in number and may therefore be regarded as constituting the essence of a unified science.
5. There are fundamental similarities in the manner in which scientists establish new knowledge, regardless of the discipline or field with which they are nominally associated. There is general agreement on a scientific method and such processes as hypothesis formation, validation, use of theories and models, etc. These may be felt to be of greater significance than any particular body of facts, laws and theories that comprise the findings of the separate sciences at any particular time.
The three most familiar conceptions of the unity of science are:
1. Unity as a reduction to a common basis, in which a single descriptive language is devised in which all the terms of all the sciences could be defined or to which they could all be reduced. If a single set of laws emerged, from which all the laws of all the sciences could be derived, or to which they could all be reduced, there would then be a single science.
2. Unity as the construction of an encyclopaedia of scientific statements, with all the discrepancies and difficulties which persist.
3. Unity as a synthesis into a total system, in which a model or pattern of scientific theory is elaborated, of which each particular theory is an instance, so that higher level sciences recapitulate, although with more complex elements, the structure of lower level sciences. General systems research has shown that certain principles apply to systems in general, irrespective of the nature of the systems and of the entities concerned. Corresponding conceptions and laws therefore appear independently in different fields of science, causing the remarkable parallelism in their modern development. Thus, concepts such as wholeness and sum, centralization, hierarchical order, equifinality, etc. are found in different fields of natural science, as well as in psychology and sociology.

♦ KC0812 Transformation
Description 1. The operation of changing in character or condition (such as by rotation, translation, or mapping) one configuration, structure or expression, into another in accordance with a mathematical rule. The formula, equation or rule (transform function) that effects such a change defines a many–one or one–one correspondence between the elements of function sets.
2. The operation of changing an expression, formula or statement in logic into a different form without altering its substance or intent.
3. All known structures (from mathematical groups to kinship systems) are, without exception, systems of transformation. The laws of composition of a structure are defined as governing the transformations of the system which they structure. In the absence of transformation, structures would lose all explanatory import, since they would collapse into static forms.

♦ KC0816 Flow chart
Flow graph — Flow diagram
Description 1. A tool of systems analysts, utilized in order to help conceptualize existing information flows and other sequences and to visualize the impacts of projected changes.
2. An overall schematic representation of a computer programme in its essential features (independent of the computer). Before the instructions of a programme are written down, the flow chart must be used to ascertain the sequence of the steps of the programme. These individual steps are written down in small boxes whose shape is different for assignments, tests, comparisons, and decisions. The sequence or flow of the calculation is ascertained by connecting these boxes by arrows.
3. The flow graph of a societal system is a representation of its processes, their directionality, and the multilateral pattern of relationships within a system. It depicts the course of institutional interactions that eventuate in the varying values of state variables over time. It may be viewed as an amplification of the interaction matrix in interrelating the state or main variables through a set of intermediate and auxiliary variables. Each of the variables in the flow graph receives or emits a number of information links to other variables. For a given variable, the number and types of links to it signify its relative importance in the system, those with the largest number being the most important from a control and constraint point of view, those with a maximum number being the most salient. The picture or society that emerges from the flow graph elucidates the nature of society as complex system.

♦ KC0817 Adaptation
Description The ability of a system to react to the environment in a way that is favourable to the continued operation of the system. It is as though systems with this property had some prearranged goal and the behaviour of the system is such that it is led to this goal despite unfavourable environmental conditions. Closely related to adaptation is the notion of stability.

♦ KC0818 Omnidirectionality
Spherical reference
Description The concept of R B Fuller that for each individual life's inescapably mobile viewpoint, a centre of a movable sphere of observation has been established a priori by Universe and may be envisaged as moving with the individual wherever he moves. Such physical–existence–environment surrounds of life events spontaneously resolve into two classes: events that are to pass tangentially by the observer; and event entities other than the self that are moving radically either toward or away from the observer. Omnidirectional consideration as generalized conceptual pattern integrity requires an inherently regenerative nucleus of conceptual observation reference.

♦ KC0820 Social system
Description Human social systems are open systems interacting with their environments through the exchange of matter, energy, and information deriving both from the system itself and from the environment. The behaviour of such systems is typically purposive and directed toward a goal. Such systems, including subsystems such as individuals and groups within them, interact as communication systems in the form of a dynamic that tends to lead to a steady state situation whose disruption they resist. Within relatively narrow limits they are capable of adjusting and adapting to their environments, as well as changing them. Such systems are characterized by functional unity, namely a condition in which all parts work together with a certain degree of harmony or internal consistency, and without producing persistent conflicts which cannot be resolved or regulated.

♦ KC0821 World economy
Description The economic concomitant of world government.

♦ KC0822 Melody
Description The logically organized expression of tonal scale, pitch, harmony, tempo, rhythm, etc, in a succession of sounds, single notes or chords, by one or more instruments or voices. Melodic structures are of variable length and range from leitmotifs to many lines or bars expressing development of a theme.

♦ KC0823 Transnational
Description The characteristic of interactions, relations, or organizations which extend or go beyond national boundaries. Some global interactions and organizations are initiated and sustained entirely, or almost entirely, by governments of nation–states. In contrast to these interactions, others involve nongovernmental actors (whether individuals or organizations) and it is these which are transnational. Thus a transnational interaction or organization may involve governments, but it may not only involve governments; nongovernmental actors must play a significant role.

♦ KC0824 Mode of life
Description The selection by or allocation to individuals or families of their material requisities on the one hand, such as diet and nutrition, housing quality, and quality of social infrastructures available such as roads, public transportation, public sanitation, and hospitals, etc. The mode of material life is differentiated most of all by income and occupation. Beyond the material and economic life one can speak of differentiated moral, aesthetic, educational–cultural and spiritual life. In a phrase, a mode of life is characterized by what an individual consumes and also that by which he himself is 'consumed'.

♦ KC0826 Morphogenesis
Description The process whereby animals and plants (and possibly organizations and societies) acquire form and structure. For example, the process by which a nearly undifferentiated droplet of protoplasm, the fertilized ovum, becomes eventually transformed or transforms itself into the complex architecture of the multicellular organism.

♦ KC0827 State–determined system
Description A system such that given an initial state, the path of the system is uniquely determined regardless of how the system arrived at that initial state. This is contrasted with a system in which the path cannot be determined from a given initial state.

♦ KC0828 Interrelatedness
Relatedness
Description The concept of the universe as a texture of relations between parts which though distinguishable, as they must be to be related, are not merely inseparable but intrinsically interdependent. The existence and character of each is what it is because the rest of the universe, as a whole and in its parts, are what they are. Whole and part are mutually determining and no detail could be other than it is without making some difference, however slight, to all the rest.

♦ KC0830 Feedback
Description The process by which the outputs or behaviour of a system is fed back into the input to the system to affect succeeding outputs. Such feedback is termed negative when the work done by the feedback mechanism opposes the main driving force such as to dampen or eliminate it. Feedback is termed positive when the work done by the feedback mechanism reinforces the main driving force, so that in an oscillating system, for example, the fluctuations increase wildly until

the system goes out of control. Amplification feedback occurs when the feedback is due in any measure to the exploitation of local energy.

♦ **KC0831 Rhythm**
Description 1. In any process the frequencies in the perceived temporal structure defined by accents and pauses, groupings and divisions, and the durations and time-correlation of elements.
2. The beat or musical meter.
3. Any periodicity in phenomena physical or behavioural.

♦ **KC0832 Isomorphism**
Isomorphies
Description 1. A one-to-one correspondence between elements in different systems such that the relationship between the elements is preserved.
2. A perfect analogy. Whatever is in one system and happens in that system has its isomorphic image in a second system and conversely. Isomorphism implies both homomorphism (described separately) and bijective (all-all) analogy; the latter implies injective (all-some) analogy, which in turn implies plain (some-some) analogy.
3. Whenever the symbols of one mathematical model stand in one-to-one correspondence with those of another, including the symbols of relation, and whenever the relation is preserved when a pair of symbols in one model is replaced by their counterparts in the other, the two models are isomorphic. Then the phenomena represented by the models can also be viewed as isomorphic. An isomorphism between two phenomena can be established only if their corresponding models (from which isomorphism is deduced) are sufficiently faithful representations.
There are three prerequisities for the existence of isomorphisms in different fields and sciences: (a) the number of simple mathematical expressions which will be preferably applied to describe natural phenomena is limited; for this reason, laws identical in structure will appear in intrinsically different fields (The same applies to statements in ordinary languages; here, too, the number of intellectual schemes is restricted, and they will be applied in quite different realms); (b) the structure of reality is such as to permit the application of conceptual constructs and is not too complex to be represented by the relatively simple schemes which can be elaborated; (c) the parallelism of general conceptions or even special laws in different fields is a consequence of the fact that these are concerned with systems and that certain general principles apply to systems irrespective of their nature.

♦ **KC0833 Unified field theory**
Description An attempt to extend the general theory of relativity to electromagnetic forces and the forces between nuclear particles, which, if successful, would result in the description of all the fundamental fields of force within the framework of the geometry of four-dimensional space-time or to higher dimensionalities.

♦ **KC0834 Completeness**
Description The completeness of an axiomatized system of logic is the state of being so constituted that a contradiction arises through the addition of any formula not previously deducible from the axioms of the system.

♦ **KC0836 Computer program**
Description A generally finite sequence of statements which are sets of instructions for the automatic processing of information. Machine language: when statements in a program are executed directly by the automation. Compiler program: when statements in a program translate a source program written in a pseudolanguage (i.e. FORTRAN) into an object program written in machine language. Assembler program: when the above translation consists of a re-coding or re-arrangement of symbols where several instructions in the object program correspond to one statement in the subject program which are realized by subroutines. Machine oriented program: when the programming language can be made accessible for direct automatic processing by an assembler program. Problem-oriented program: when the programming language is oriented toward the description of tasks within a certain class of problems without taking into account any particular set of automatons.

♦ **KC0837 Connectedness**
Description A connected space is a topological space which consists of a single piece and is considered one of the simplest properties which such a space may possess and one of the most important for the applications of topology to analysis and geometry. Connectedness is also a basic notion in complex analysis, for the regions on which analytic functions are studied are generally taken to be connected open subspaces of the complex plane. The concept of connectedness is important in the measurement of the degree of relationship between the members of a social group.

♦ **KC0840 Operational gaming**
Description A special type of model-building structured so as to permit multiple simultaneous interactions among competing and cooperating players. If a computer is used, the game situation can be manipulated and the effects analyzed by an observer. Open gaming (when there are no constraints to the individual steps) is restricted in many cases to manual operation because the programming effort and the storage required to handle all possible moves would be uneconomic, even if technically feasible. Gaming is an effective means of forecasting the impact of events and new technologies. Games may be used to provide participants with experience in adapting to an unfamiliar environment and making decisions under conditions of uncertainty, and also to develop understanding of basic organizational relationships. Functional games may be used to provide participants with experience in dealing with well-defined problem situations.

♦ **KC0841 Levels of organization**
Integrative levels
Description 1. An assembly of things of a definite kind (e.g. a collection of systems characterized by a definite set of properties and laws, and such that it belongs to an evolutionary line, though not necessarily to a line of biological descent). Some of the characteristics which emerge at a new level are the exclusive property of the level in question, although new means later (rather than superior) in the course of a process leading from pre-existing levels. A level structure does not necessarily qualify as a hierarchy, since the concept of domination, essential to hierarchy, is absent from the idea of level, as is the concept of superiority.
2. Grades of being ordered, not in arbitrary ways but in one or more evolutionary series.
3.The uniformities found among integrative levels include:
(a) The structure of integrative levels rests on a physical foundation. The lowest level of scientific observation would appear to be the mechanics of particles. (b) Each level organizes the level below it plus one or more emergent qualities (or unpredictable novelties). The levels are therefore cumulative upwards, and the emergence of qualities marks the degree of complexity of the conditions prevailing at a given level, as well as giving to that level its relative autonomy. (c) The mechanism of an organization is found at the level below, its purpose at the level above. (d) Knowledge of the lower level infers an understanding of matters on the higher level; however, qualities emerging on the higher level have no direct reference to the lower-level organization. (e) The higher the level, the greater its variety of characteristics, but the smaller its population. (f) The higher level cannot be reduced to the lower, since each level has its own characteristic structure and emergent qualities. (g) An organization at any level is a distortion of the level below, the higher-level organization representing the figure which emerges from the previously organized ground. (h) A disturbance introduced into an organization at any one level reverberates at all the levels it covers. The extent and severity of such disturbances are likely to be proportional to the degree of integration of that organization. (i) Every organization, at whatever level it exists, has some sensitivity and responds in kind.

♦ **KC0842 Network analysis**
Description Several kinds of network analysis have been developed for different problem areas. In each case there is considerable dependence on graph theory (described separately), but other techniques may also be used, depending on the extent to which the network's characteristics are quantifiable.
1. *Circuit analysis:* In the analysis of electrical and electronic circuits (composed of resistors, inductors, and capacitors), the central problem of network analysis is to find expressions for the node-pair potentials and branch currents at various points of interest in the circuit, and to determine the behaviour of the over-all network with respect to specified terminals.
2. *Flow-handling networks in general:* Road networks, power grids, airline routes, railroad networks, pipeline networks and other networks through which commodities flow, use a form of network analysis to determine: maximum flows in networks; the vulnerability of a network to disruption; and the necessary characteristics of the network elements for the required performance. This involves problems of economics and logistics and makes use of the techniques of linear programming, queuing theory, and probability theory as a form of systems analysis.
3. *Project planning:* This technique (described separately) is based on the representation of the component activities of a project as the relationships in a network, with the nodes representing the start and completion points. The analysis is used as an aid to work scheduling, particularly in the event of delays or where problems of resource allocation arise.
4. *Citation analysis:* Each document published constitutes a node in an extensive network in which the relationships to a given document are the citations of it in other documents. Such networks may be analysed to locate key documents (authors, journals, or research institutes) and the relationships between them as a guide to resource allocation and to an understanding of the history of ideas.
5. *Sociometry:* The network is usually severely restricted in content to the friendship or leadership relationships in a relatively small group of people. Such techniques have also been applied to inter-organizational relationships (such as in a community) in order to locate the opinion-forming bodies and the manner in which the community may be influenced with a minimum of effort.
6. *Social network analysis:* In an effort to analyze more general social networks, in which there may be several kinds of relationship between several kinds of social entity at several different levels, and including relationships of the entity to itself, proposals have been made for the use of topology (from which graph theory is derived), algebraic topology, tensor calculus, matrix analysis, and lattice theory.

♦ **KC0845 International understanding**
Description Two principal meanings exist, neither of which implies either promotion of world government, nor undermining of national loyalties:
1. A kind of knowledge, an attitude, that will lead the people of every nation to feel friendliness towards the people of other nations and to cooperate in international enterprises. This is a sympathetic understanding, implying a favourable attitude conducive to mutual accord, of which cooperative and peaceful relations are held to be a natural consequence.
2. An objective attitude, a sober comprehension of the behaviour of other people, whether friends or enemies. This is objective intellectual understanding which may render the points of disagreement clearer, and in some cases fewer.
A major means of promoting international understanding is held to be education concerning the United Nations Charter and the purposes and principles, the structure, background and activities of the United Nations in schools and institutes of higher learning.

♦ **KC0846 Equilibrium**
Description Stable equilibrium is the tendency of a system to move back toward a given condition after being disturbed by forces external to the system. Both stable and unstable equilibrium may be considered as a state of rest caused by the interaction of opposing forces. Equilibrium in the case of a society is characterized by a balance of antagonistic or noncomplementary elements (such as attitudes, sentiments, and associations) and the stable operation of a common system of social norms.
True equilibria in closed systems and stationary equilibria in open systems show similarity, inasmuch as the system, taken as a whole and in view of its components, remains constant in both. But the physical situation is fundamentally different. Equilibria in closed systems are based on reversible processes; they are a consequence of the second principle of thermodynamics and are defined by a minimum of free energy. In open systems, by contrast, the steady state is not reversible as a whole nor in many individual reactions. A closed system must according to the second principle, eventually attain a time-independent steady state of equilibrium defined by maximum entropy and minimum free energy where the ratio between its components remains constant. An open system may attain a time-independent steady state, even though there is a continuous flow of component materials, but is more likely to exhibit characteristics of organismic systems such as dynamic equilibrium, adaptation, and self-regulation.

♦ **KC0847 Artistic vision**
Artistic images — Artistic expression
Description The common denominator of artistic expression is the ordering of a vision into a consistent, complete form. The difference between a mere expression, however intense and revealing, and an artistic image of that expression lies in the structure of the form. This structure is specific. The colours, lines, and shapes corresponding to our sense impressions are organized into a balance, a harmony or rhythm that is in an analogous correspondence with feelings, and these in turn are analogues of thoughts and ideas. An artistic image, therefore, is more than a pleasant tickle of the senses and more than a graph of emotions. It has meaning in depth, and at each level there is a corresponding level of human response to the world. In this way, an artistic form is a symbolic form grasped directly by the senses but reaching beyond them and connecting all the strata of our inner world of sense, feeling, and thought. The intensity of the sensory pattern strengthens the emotional and intellectual pattern; conversely, the intellect illuminates such a sensory pattern, investing it with symbolic power. This essential unity of primary sense experience and intellectual evaluation makes the artistic form unique in human experience and therefore in human culture.
Images deriving solely from a rational assessment of the external world, without passion of the eyes, are only topographical records. Images of emotional responses without real roots in the environment are isolated graphs of a person's inner workings: they do not yield symbolic form.

And the most beautiful combinations of colour and shape, the most exquisitely measured proportions of line, area, and volume, leave no effect if they have not grown out of rational and emotional participation in the total environment. Each of these visions is a fragment only.
The essential unity of first-hand percept and intellectual concept makes artistic images different from scientific cognition or simple animal response to situations. It is the unity of the sensory, emotional, and rational that can make the orderly forms of artistic images unique contributions to human culture.

♦ KC0850 International cooperation
Description 1. Cooperation among nations in such fields as: economic and social development; international transfer of science and technology; cultural cooperation; world health; international law; education; promotion of human rights; food and agriculture. Such cooperation may be bilateral or via intergovernmental agencies.
2. Cooperation between nation-states, governmental and nongovernmental organizations (whether worldwide, regional, national or community-level), and individuals in all fields of human activity.

♦ KC0854 Manifold
Description A topological space equipped with a family of local coordinate systems that are related to each other by coordinate transformations belonging to a specified class. Manifolds are studied for their global properties.

♦ KC0856 Self-reproducing systems
Self-reproduction
Description A system such that, if there occurs within it a certain form (or property, or pattern, or recognizable quality generally), then a dynamic process occurs, involving the whole system, of such a nature that eventually it is possible to recognize, in the system, further forms (or properties, or patterns, or qualities) closely similar to the original.

♦ KC0857 Holography
Description A photographic record of an interference pattern between reflected light waves from an object and a second wave of interfering light. Coherent light (light in which all frequencies are in phase - as in a laser) is made to fall on a photographic plate. The same light source is also used to illuminate the subject. The reflected light from the subject which falls on the photographic plate will have travelled different distances depending on the shape of the subject; it thus arrives with its phasing disturbed in direct proportion to the shape of the subject. These variations in phasing meet the incident (or "reference") beam at the surface of the plate, setting up complex interference patterns which are faithfully recorded as densities in the emulsion. When a light beam is shone back through the processed emulsion, the recorded densities act like an interference filter and "reconstruct" the original wave front which caused them. The result is the apparent recreation, in space, of the original subject.
It has been suggested that the human brain may record memory in the same way as a hologram, with every point recording all of the image as seen from a particular point. As a result it has been further suggested that the universe may be understood as a hologram with brains functioning as holograms within it.

♦ KC0859 Industrial integration

♦ KC0860 Relationships
Relation
Description The variety of general relationships between categories is illustrated by the following groups: 1. Appurtenance: inclusion, implication; parts, organs; components, constituents; properties, attributes; aptitudes, predispositions.
2. Process: favourable action (on subject), action favourably affected (by object); unfavourable action (delay, inhibition, destruction); favourable interaction (symbiosis); unfavourable interaction (antagonism, competition); operation, means used (process, product, result).
3. Dependence: causing (subject), caused (effect); originating (subject), arising from (object); conditioning (subject), conditioned (object); interdependence by correlation; interdependence by association; interdependence by combination, synthesis.
4. Orientation: aspect, particular case; application; use.
5. Comparison: resemblance, similarity by analogy; resemblance, similarity by equality or identity; dissimilarity by difference; dissimilarity by opposition.

♦ KC0861 Plasma
Description 1. The fourth state of matter following the gaseous consisting of stripped or free electrons or ions.
2. Plasma states studied in electrodynamics or magnetohydrohynames (i.e. of free ions or electrons in the ionosphere, influenced by electromagnetic fields).

♦ KC0862 Metagalaxy
Description A finite and transient galactic super-system conceived to contain all clusters and isolates, in other words, the Great Universe of Stars and Stellar systems.

♦ KC0863 Orthography
Description 1. The rules of correct spelling of a language's lexicon as promulgated by official or academic bodies.
2. The norms of spelling, which may also accept variants (e.g. English versus American).
3. The science of establishing rules of spelling.

♦ KC0866 Wholeness
Coherence
Description Wholeness is the property of a system in which every part is so related to every other part that a change in a particular part or element causes a change in all the other parts and in the total system. The system is then said to behave as a whole or coherently and shows new structural and functional features, non-existent in its components. These new features result from the interaction of the system components, from their organization and functioning within the system.
A fundamental distinction is made between such structures and aggregates, the former being wholes, the latter composites formed of elements that are independent of the complexes into which they enter. To insist on this distinction is not to deny that structures have elements, but the elements of a structure are subordinated to laws, and it is in terms of these laws that the structure as a whole or system is defined. Moreover, the laws governing the structure's composition are not reducible to cumulative one-by-one association of its elements; they confer on the whole as such over-all properties distinct from the properties of its elements.
Wholeness is the most striking feature of the systems at all levels of organization of living matter. It arises and develops as a result of structural differentiation and functional specialization within the given system.

♦ KC0867 Management
Description Management is concerned essentially with the way in which the total organization reacts to its environment to achieve its objectives, as well as with the internal operations of the subsystems of the organization. Management includes the following processes (as well as concern with their interrelationships): setting objectives, planning strategy, establishing goals, developing a company philosophy, establishing policies, planning the organizational structure, providing personnel, establishing procedures, providing facilities, providing financial resources, setting standards, establishing management programmes and operational plans, providing control information, and activating people. Specific management activities include: planning, organizing, staffing, directing, coordinating, reporting, budgeting, and innovation.
Management is the fundamental instrument to ensure that all plans are adequately integrated and that the work of all functions is integrated into a cohesive whole so that the full force of the organization's resources can be focused on the successful achievement of its objectives.

♦ KC0870 Chunking
Binary recoding
Description Information coded in the binary system as O or I (usually for data processing or data transmission) produces long strings of digits (O or I). Recoding these binary values can introduce other digits (i.e. 2, 3, 4, etc) or symbolic characters (a, b, c, etc). For example, one technique is to use 8 digits, 0 to 7, to code the possible 8 combinations of 2 digits taken three at a time (e.g. 000, 001, 010, 011, 100, 101, 110, 111) where, for example 0 is the code for the first set of three (000), 1 is for the next set followed by 2 to 7 (for 111). Thus a message (term) of up to 24 binary digits can be recoded (as sets of 3) using 8 new code digits. A message then can be considered as every sequence of 24 binary digits, concluded by a final sequence divisible by 3 but less than 24. While there are small technical problems involved in aggregating or chunking sets of binary digits for recording, this is a procedure that the mind itself may do in interpreting and simplifiying sense data (received as long, binary-like, digit-like strings).

♦ KC0871 Connection
Description 1. The perception of relationship; the joining together phenomena or elements at or by a nexus which may be conceptual, logical or physical. Taxonomies of connections (or nexii) perceived or postulated in their domains exist in all areas of human thought in which they are described from functional or theoretical aspects.
2. The net of all relations.

♦ KC0876 Metatheorem
Metatheory
Description 1. A theory concerned with the investigation, analysis and description of theory itself. Theorems on objects of formalized theories in any given logical-mathematical calculus or formal system. Some of the best known metatheorems are those of Gödel's of incompleteness for formal arithmetic systems and of completeness for the predicate calculus; and in the same calculus, that of Church on the unsolubility of decision problems.
2. Theorems regarding theorems of informal mathematical theories (e.g. duality principles in algebra).
3. The theoretical proof of a another theory which may have as its object one or more axioms, theorems, proofs, rules of inference, definitions or concepts. The metatheorem is the theory of the object theory. Two forms of the metatheorem exist; the metamathematical employing finitary methods and the set-theoretic predicate logic particularly as applied to all the results on the categoricalness of different axiomatic systems.
A meta-level theory may be used as a means of integrating the constructs of different theories. Suppose there is a theory, S, containing the sentences b, c, d, and so on; and suppose there is another theory T, containing the sentences q, r, s, and so on. Now if S and T are isomorphic, S may be termed an interpretation of T, and T an interpretation of S. In other words, there is some relation between sentences in S which is ordinally similar to a relation between the sentences in T, and there is one-to-one correspondence between the two sets. If b corresponds to q and so forth, then b is an interpretation of q in S and q is the interpretation of b in T.
Now the various modes of organizing and comprehending human experience each constitute a theory, in a broad sense of the term. Each is based on sensing, which it refers to an interconnected set of constructs, to which response is made, which thus affects the environment that relevant sensing is afforded. Contentual differences between theories are due to the suitability and adequacy of the respective sets of constructs as regards the analysis and interpretation of the various segments of experience. With due regard for these differences, the theories nevertheless manifest common fundamental characteristics: they interpret experiences by means of constructs, give rise to coordinate responses and find their confirmation in the manipulated environment. Thus specifically differentiated but isomorphically structured theories are obtained, wherein there is a general ordinal similarity between the propositions. The relationship between the key terms, invariant in all theories, determines isomorphism and therefore determines the absolute meaning of the propositions in a meta-theory. Thus, it becomes theoretically possible to formulate a meta-theory integrating the diverse facets of human experience.

♦ KC0877 Facilitation
Description The property of certain types of process whereby they promote their own repetition. This property is a common factor in some systems, underlying both the process of identical multiplication and of adaptive modification. A process is said to facilitate its reproduction when it leads to a stable result which tends to bring about a repetition of the process. A wide class of processes possess this self-facilitating property which underlies all development and extension of form. All processes are self-facilitating which lead to a structure possessing a symmetry character which can be extended by spatial repetition.

♦ KC0879 Part and whole categories
Description Perceived as antithetical identities; logically expressed as member and set, or object and class, and by some, considered to be philosophical categories. A number of paradoxes raise epistemological issues in part-whole categorization and thus also in class logic, mathematics, language and mental behaviour. Illustrating this are such current phrases as 'the part is greater than the whole', and 'the whole is greater than the sum of its parts' (as in holism and synergy), and 'the part is equal to the whole' (as in popular holism).

♦ KC0884 Morphology
Description A study of the structure or form of some phenomenon, most usually animals and plants in which case, as a branch of biology, it is a study of the forms, relations, metamorphoses and phylogenetic development of organs (apart from their functions). In mineralogy it is the term given to the study of the crystalline form of minerals; in philology, it is the branch of grammar that examines the forms of words as well as the principles of word formation and inflection.

♦ KC0886 Systems research
Description The logic and methodology of systems research involve the following: elaboration of conceptual means (systems of notations, special models, etc.) to represent the systemic nature

of the domain studied; development of an apparatus to describe the most important characteristics of the system features (connections, systems of connection, inter–connections of the systems and its environment, hierarchical structure, control problems, etc.); and development of formalized techniques to describe system features, including the development of specific rules of inference.

It has been suggested that systems research represents the convergence of operations research and systems engineering. The major areas contributing to it have been identified as operations research and systems analysis, industrial engineering and human engineering. Systems research procedures differ from operations research in that they are descriptive rather than prescriptive. They seek to develop behavioural abstractions of systems which can provide guidance for decision–making for a range of values rather than an optimum solution relative to any specified value system. Specifically they provide statements of relationships between system task and resource variables. These relationships are derived from empirical data and organized according to logical rules.

Systems researchers, although usually trained as disciplinary specialists, work and communicate within multidisciplinary groups which seek to develop factual and general knowledge of systems, knowledge that does not lend itself to disciplinary classification. This knowledge provides a basis for more effective designs and operations of systems, so that it provides both a technology and a science of systems.

♦ **KC0887 Human systems management**

Description The art of conceiving, implementing and sustaining conditions of both the inner and the external environment under which system objectives can be achieved in a synergistic manner. In this context a human system is conceived as a formally or informally organized complex of purposeful entities which are interrelated for the purpose of attaining its internally incorporated as well as environmentally imposed objectives. Purposeful entities are understood to include both people and human contrivances (machines, systems, etc):

♦ **KC0890 Unified**

Integrated — Consolidated

Description An attribute of a completed structure which has been made into a coherent whole by gathering or combining parts or elements. A distinction may be made between unified and unitary, the first implying that the whole is a product of the unifying or synthetic process applied to the originally separate parts, the second implying that the condition of being whole is primary although parts may be distinguished. The unifying process may also introduce a notion of production of uniformity.

♦ **KC0891 Integration**

Description 1. Actual mechanisms and organizational principles which hold a system together (whereas wholeness is a general measurement of internal interdependence).
2. A shorthand word used to designate continuous, intelligent, interactive, adjusting behaviour.
3. A relationship between a sequence of activities such that an individual goal may be attributed to each, and at the same time an ultimate goal to the whole sequence.
4. The sum of the processes by which the developing parts of an organism or organization are formed into a functional and structural whole.
5. A combination ad coordination (frequently on an hierarchical basis) of separate and diverse elements, functions, units or groups into a more complete or harmonious whole.

♦ **KC0892 Theory**

Description 1. A true theory; the formulation of cognitive or practical methods to obtain a result: certitude in thought, or certain (sure, invariant) outcomes in practice and action.
2. A mental representation (proven or unproven, true or false) which gives an account of causality or instrumentality for some phenomena.
3. An opinion or belief.

♦ **KC0896 Crossdisciplinarity**

Description The axiomatics of one discipline are imposed upon other disciplines at the same hierarchical level, thereby creating a rigid polarization across disciplines toward a specific disciplinary axiomatics. In the successive steps of cooperation and coordination between disciplines, crossdisciplinarity defines an organizational principle for a hierarchical system of one level, having a single goal and rigid polarization toward a specific disciplinary goal. Crossdisciplinarity implies a brute–force approach to reinterpret disciplinary concepts and goals (axiomatics) in the light of one specific (disciplinary) goal. One of the most conspicuous attempts of crossdisciplinary polarization is the reformulation of management, planning, organization, and the explicit planning of change, in terms of the empirical and reductionist concepts of the applied behavioural sciences.

♦ **KC0897 Regional international cooperation**

Description A vague term covering any interstate activity with less than universal participation designed to meet some commonly experienced need. It is considered distinct from the related concept of regional integration.

♦ **KC0905 Commune**

Description A community in which the inhabitants have close relationships of friendship and interest. A group practising economic communalism with limited or no personal property.

♦ **KC0906 Communication channel capacity**

Description The upper limit of the transfer–rate, usually in bits per time–unit, of messages or information through a communication channel. A measure of the information which a channel can store or can transfer per time–unit or per character. (It must be sufficiently great to resolve ambiguity in the information transmitted, according to Shannon's Tenth Theorem).

♦ **KC0907 Complex social systems**

Description Complex systems are high–order, multiple–loop, nonlinear, feedback structures. All social systems belong to this class (management structure of a corporation, urban area, national government, international trade processes, etc). Complex systems have many unexpected and little understood characteristics, making them very different from the simple systems of which people have an intuitive understanding. Such systems bring together many interacting factors which have been traditionally compartmentalized into isolated intellectual fields (the psychological, the economic, the technical, the cultural, the political, etc.), whose interactions may be of greater importance than the internal content of any one alone. Characteristics include:
1. High order: a system of greater than fourth or fifth order begins to enter the range of complex systems. An adequate representation of a social system, even for limited purposes, can be tenth or hundredth order.
2. Multiple loop: possessing upward of three or four interacting (positive or negative) feedback loops of shifting predominance.
3. Nonlinearity: allowing one feedback loop to dominate the system at one time and then causing a shift in this dominance to another part of the system which may produce such different behaviour that the two may seem unrelated.
4. Counter–intuitive: in that the intuitive solutions of the average person to the problems of complex social systems will most probably be incorrect most of the time.
5. Insensitivity to changes in many system parameters.
6. Recalcitrant resistance to policy changes.
7. High sensitivity to a few parameters and a sensitivity to some changes in structure.
8. Frequent conditions of delicate and uncertain balance between the forces of growth and decline.
9. Tendency to drift to a low level of performance.

♦ **KC0908 Arborization**

Hierarchy formation

Description The interdependence of processes in organisms may be represented by interlocking or interlacing hierarchies, or vertical tree structures. The meeting points of the branches form neighbouring tree structures from horizontal networks at several levels. Without the tree structures, the network formation (reticulation) would not be possible. Similarly, without the network formation, the individual hierarchies would be isolated, and there would be no integration of functions.

♦ **KC0910 Engineering cybernetics**

Description 1. Industrial automation, mainly production and production control systems.
2. Automation and control systems.

♦ **KC0911 Reductionism**

Description A thesis according to which all scientific concepts are reducible to a set of ultimately irreducible concepts, to be identified as physical properties of things (whether as common observations of physical states or as predicates of the most elementary physical units) thus placing physical theory at the foundation of science and explanation. Physics thus becomes the only discipline that is conceptually independent of other empirical sciences, and the dependence of others is taken as unidirectional. Other disciplines deal with the concepts which could be synthesized from those used by the prior disciplines, thus introducing a hierarchy of disciplines: physics, chemistry, biology, psychology, and social science.

Analysis of a phenomenon is then best accomplished by continually breaking it down or reducing it to lower levels. Through the elaboration of a single descriptive language in which all the terms of the science could be defined (or to which they could be reduced), it is assumed that once the lower levels are comprehended, resynthesis can occur and the higher levels will fall into place upon a solid foundation for the unity of science.

In the progressive analysis of the unitary universe into abstracted elements, the reductionist prefers to move from a focus on larger aggregates downwards, gaining precision of information about fragments as he descends, but losing information content about the larger orders he leaves behind, in that he is no longer able to supplement the sum of the statements that can be made about the separate parts by any such additional statements as will be needed to describe the collective behaviour of the parts, when in an organized group. (By contrast, the holist proceeds in the opposite direction, from below, trying to retrieve the lost information content by reconstruction, but is forced to recognize early in the ascent that information is not forthcoming unless he has already had it on record in the first place).

♦ **KC0912 Logistics**

Description 1. Science and activity of providing resources required for events, activities, processes, production, etc; e.g. supply of antarctic expeditions.
2. In military operations or establishments, the methods, tactics and strategies of distribution; supplying ordnance, providing storage facilities, transportation, and food, etc.
3. A branch of mathematics, including queing theory.

♦ **KC0913 Business cycles**

Description Theory of cyclic nature of economic expansion and contraction, formerly influenced by the cyclic nature of production of food and some primary commodities, and the simplistic dilation and contraction in the supply–demand market model; currently conceived as an outcome of over–correcting government (intervention) policies.

♦ **KC0915 Social group**

Description A number of individuals bound together by a community of interest, purpose or function.

♦ **KC0916 Game theory**

Description In general, a game will have a certain number of players and is composed of moves, which are of two types; personal, made by one of the players, and chance, in which one of several possible outcomes is selected by a chance device acting in accordance with certain probability laws.

The theory of games is a method of applying mathematical logic to determine which of several strategies is likely to maximize the gain or minimize the loss of one of the players. Many economic, social, organizational, and military problems can be described in these terms, in that the other players can also choose between several strategies. It is therefore a logical procedure for formulating conservative decision rules, where the aim is to be assured of realizing the maximum of the minima gains potentially available, or to be assured of realizing the minimum of maxima losses which could be inflicted, regardless of the strategic behaviour of the other players.

Games that can be formulated in game–theoretic form can be usefully distinguished from operational games (described separately), which are too complex to be given a precise formulation.

♦ **KC0917 Stability**

Description A system is stable with respect to certain of its variables if these variables tend to remain within defined limits. Such a system may be stable in some respects and unstable in others. An adaptive system maintains stability for all those variables which must, for favourable operation, remain within limits. An important feature of system stability is that it is a property of the whole system and cannot be assigned to any part of it; it belongs only to the combination of parts and cannot be related to the parts considered separately. The presence of stability always implies some coordination of the actions between parts.

♦ **KC0918 Omnitopology**

Description Omnitopology is conceived to be an accessory to the conceptual aspects of Euler's superficial topology in that it extends its concerns to the angular relationships as well as to the topological domains of nonnuclear, closest–packed spherical arrays and to the domains of the nonnuclear–containing polyhedra thus formed. The domains of volumes are then the volumes topologically described and the domain of an external face is the volume defined by that external face and the centre of volume of the system. Each of the lines and vertexes of polyhedrally defined

conceptual systems have their respective unique areal domains and volumetric domains.
In contradistinction to, and in complementation of, Eulerian topology, omnitopology deals with the generalized equatabilities of a priori generalized, omnidirectional domains of vectorially articulated linear interrelationships, their vertexial interference loci, and consequent uniquely differentiated areal and volumetric domains, angles, frequencies, symmetries, asymmetries, polarizations, structural–pattern integrities, associative interbondabilities, intertransformabilities, and transformative–system limits, simplexes, complexes, nucleations, exportabilities and omni-interaccommodations.

♦ **KC0919 Optimal planning**

♦ **KC0920 Systems education**
Description Education concerning systems concepts as evidenced in natural and social systems, emphasizing interrelationships, interrelatedness, and the holistic nature of natural and social processes. The introduction of such interdisciplinary concepts into a curriculum is frequently thought to require a reorganization and integration of the curriculum components to reinforce understanding of the interrelationships discussed. Systems education then blurs into, and is confused with, educational systems just as the teaching of concepts of integrated science is obscured by a focus of methods of integration of science teaching.

♦ **KC0921 Invariance**
Description 1. A property of a system of relationships that is unchanged by any change or transformation applied to the frame of reference of the system. Thus a structure may be reflected in a plane resulting in a transformation in which every point of space is associated with its corresponding point with respect to the plane. A rotation of the structure around a line or a reflection of it with respect to the centre are similarly defined. Invariance establishes a relationship between objects, phenomena and theories which are outwardly unrelated.
2. By an invariant of some object under study, relative to certain of its transformations, is meant any quantity, numerical, vectorial, etc connected with the object that does not vary under these transformations. Invariance is orderly repetition of pattern of which a special form is called symmetry.

♦ **KC0922 Ecology of concepts**
Ecology of knowledge
Description 1. Change and stability on the planet are now mediated: by transfers and transformations of energy, which are universally applicable to material objects; by responsiveness, which operates additionally and in varying degrees in all organic forms (and an increasing number of artefacts); and by appreciation, which through the human conceptual system constitutes the means whereby humans represent, interpret, value, and increasingly create the world in which they effectively live. The mediator in such a conceptual system is human communication and the system itself is a psycho–social artefact of which the conceptual world created by science is the most stable, coherent, and explicit example, although business, politics, and other human activities have their own partly autonomous systems. All communication, and hence all cooperation, depends on such shared appreciative systems, on ways of conceptualizing and valuing which are systematically organized and which, when they change, may have to change extensively before they reach anything approaching a new equilibrium.
2. The general ecology of knowledge is used heuristically to mean the study of patterns of interrelationships among the various "species" (subsystems, sub–subsystems, etc.) or fields and subfields of knowledge with emphasis on (a) preserving the condition of dynamic balance between the "species" and their environment; and (b) optimizing the over–all, symbiotic fruits of synergistic interactions among them.

♦ **KC0924 Coordination**
Description The combination in suitable relation for most effective or harmonious results through the functioning of parts in cooperation and normal sequence.
Each differentiated subsystem within an organism or an organization operates in such a manner as to eliminate deviations from the norm of a specific type, and the processes of such differentiated systems are coordinated as components of the comprehensive normalizing process of the organization and its environment.

♦ **KC0925 Metalogic**
Description 1. Syntactics, the study of structural properties of mathematical and logical calculi encompassing the theories of formal proofs and definability of concepts.
2. Semantics, the theories of meaning (or reference) and of sense.
3. Examination of the relation between intensional and extensional languages (leading to pragmatics).

♦ **KC0926 Monte Carlo simulation technique**
Description 1. Method of simulating the behaviour of a system whose physical components and functional relationships can in part be described only by laws of probability. By means of a random number generator, numbers are produced in order to determine the component movements by the use of probability distribution. The behaviour of the whole system is then calculated by the use of these random movements. By simulating the random nature of certain real–world organizational processes (of which the problems are too complicated for classical analytic and quantitative methods), it can treat problems that are almost impossible to deal with otherwise.
2. An operations research tool by means of which solutions can be approximated in models containing stochastic variables. A trial–and–error technique, refined by the use of probability curves and random samples. The name derives from the fact that the technique usually involves programming so that the computer generates a random normal number as needed, to evaluate the stochastic variables.

♦ **KC0927 Semiotics**
Description Regarding a sign as any object for which a significance is associated, semiotics studies in natural and artificial languages the properties of sign systems and their sign contents. Three aspects are considered. In syntactics no interest is taken in meaning, only the construction system for joining signs (in sentences, etc) is studied. Semantics, on the other hand, studies the relationship of signs to meanings. In pragmatics the focus is on the characteristics and psychology of the receiver or observer of signs (the addresses), and the phenomenology of interpretation including the valuation of the sign system as a carrier for meaning. Different sign systems in natural and artificial languages can be uniformly studied using semiotics, and the same heuristic methodology leads to revealing the sign or signing character of non–linguistic human behaviour in society (e.g. dress, table manners, etc). The disciplines of semiotics assist creation of computer languages in indexing and recording data for retrieval. Semiotics particular contribution is to allow descriptions for logically 'permitted', classes or items encompassing probabilities, and for decision–making situations. Beyond this, semiotic approaches have facilitated computer translation, indexing and abstracting of texts.

♦ **KC0928 Group**
Description An assemblage, collection, or combination of units, figures, or organisms characterized by the fact that they form a distinctive unit complete in itself (or forming part of a larger unit) due to the nature of the relationships between them.

♦ **KC0930 Control**
Description Means whereby courses of action are chosen and kept so as to reach goals (in the case of positive control) or to escape threats (in the case of negative control).

♦ **KC0931 Encyclopaedia**
Description A work that explains the order and interrelations of human knowledge by treating all the various branches of human knowledge, usually in the form of individual entries on each aspect, and frequently arranged in alphabetical order with cross–references between entries.
The movement for the unity of science gave rise to an encyclopaedia of unified science as representing the best genuine model of science as a whole. An encyclopaedic integration of scientific statements, with all the discrepancies and difficulties which appear, was considered to be the maximum of integration which could be achieved at the time.

♦ **KC0932 Equivalent network**
Description A network which, under certain conditions of use, may replace another network. The networks need not be of the same form (for example, one may be electrical and the other mechanical).

♦ **KC0933 Network synthesis**
Description A branch of the theory of electric networks which deals with the systematic determination of the structure and element values of an electric network in order that it should have certain preassigned characteristics. The first part of the synthesis problem involves the approximation of the required characteristics by a rational function corresponding to a physically realizable network. The second consists in finding the physical network which has the function as an impedance (admittance) function.

♦ **KC0934 Consistency**
Description 1. Agreement or harmony of parts, traits or features, or uniformity among a number of things.
2. A set of propositions has consistency when no contradiction can be derived from the joint assertion of the propositions in the set.

♦ **KC0935 Bionics**
Description The science of systems whose function is based on living systems, or which have characteristics of living systems, or which resemble these. It is particularly concerned with attempting to apply understanding derived from biological systems to technological systems.

♦ **KC0936 Real–time information system**
Description 1. In the case of information systems: processing data in synchronism with a physical process in such a fashion that results of data processing are immediately useful to the physical operation.
2. Capability of an information system to provide a basis for making a decision in synchronism with an on–going management process in such a way that the decision has the expected impact upon the course of events.
3. Capability of controlling an environment by receiving data, processing them, and returning results sufficiently quickly to affect the functioning of the environment at that time.
4. Operation in real–time is achieved when the time required for solution of a problem is equal to or less than the time available for action on it.

♦ **KC0937 Process logic**
Holocyclation — Process metaphysics
Description Instead of regarding the universe as made up of things, it may be considered to be composed of a complex hierarchy of smaller and larger flow patterns in which the things are invariant or self–maintaining features of the flow (even though matter, energy, and information are continually flowing through them). Such steady–state patterns or objects (organisms, or observers) can only be understood as being in a holistic relationship to their environment, with fields of flow extending outward indefinitely. Likewise, the environment only takes on stable form and meaning and points of reference through the objects which it sustains.

♦ **KC0941 Holism**
Wholism
Description A theory emphasizing the organic or functional relationships between parts and wholes rather than focusing on the parts alone. The determining factors, expecially in living nature, are considered to be irreducible wholes. This is often briefly expressed in the statement: the whole is more or greater than the sum of its parts. Here more does not at all refer to any measurable quantity in the observed systems themselves; it refers solely to the necessity for the observer to supplement the sum of the statements that can be made about the separate parts by any such additional statements as will be needed to describe the collective behaviour of the parts, when in an organized group. In carrying out this upgrading process, he is in effect doing no more than restoring information content that has been lost on the way down in the progressive analysis of the unitary universe into abstracted elements.
The holist proceeds from below, trying to retrieve the lost information content by reconstruction, whilst the reductionist procedure (with which it is contrasted) moves from the top down, gaining precision of information about fragments, but losing information content about the larger orders that are left behind.

♦ **KC0943 Technology assessment**
Description An analytical approach to providing a whole conceptual framework, complete both in scope and time, for decisions about the appropriate utilization of technology for social purposes. The assessment involves: identification and refinement of the subject to be assessed; delineation of the scope of the assessment and development of a data base; identification of alternative strategies to solve the selected problems with the technology under assessment; identification of the parties affected by the selected problems and the technology; identification of the impacts on the affected parties; evaluation or measurement of impacts; comparison of pros and cons of alternative strategies.

♦ **KC0944 Dictionary**
Thesaurus Classification schedule
Description A formal device for stating the relationships among the words of a vocabulary or between one vocabulary and another, or a combination. In some instances, a thesaurus may accomplish this by relating both words via a common, more abstract concept.

♦ **KC0946 Topology**
Description A branch of mathematics that deals with selected (topological) properties of collections of related physical or abstract elements and specifically those that endure without rupture when the collection undergoes distortion. A topological property is therefore any property of a structure which is invariant or unchanged by such deformation. A topological space can be thought of as a set from which has been eliminated all structure irrelevant to the continuity of functions defined on it.
The broad field of topology includes domains such as: the homology and cohomology theory of complexes, and of more general spaces; dimension theory; the theory of differentiable and Riemannian manifolds and of Lie groups; the theory of continuous curves; the theory of Banach and Hilbert spaces and their operators, and of Banach algebras; and abstract harmonic analysis on locally compact groups.
Applications of topology have been found in a wide range of disciplines, particularly those concerned with networks or relationships of some kind. Since topology is not limited to quantitative problems, it may contribute to such fields as the social sciences, previously not considered susceptible to mathematical treatment.

♦ **KC0947 Language universals**
Universals of language
Description Underlying the multitude of idiosyncrasies of the world's languages there are uniformities of universal scope. Amid infinite diversity, all languages are cut from the same pattern. Language universals are by their very nature summary statements about characteristics of tendencies shared by all human speakers. As such they constitute the most general laws of a science of linguistics (as contrasted with a method and a set of specific descriptive results). Further, since language is at once both an aspect of individual behaviour and an aspect of human culture, its universals provide both the major point of contact with underlying psychological principles (psycholinguistics) and the major source of implications for human culture in general (ethnolinguistics).
Types of universals may be differentiated both with respect to logical structure and with respect to substantive content. On the basis of logical structure, the subtypes which may be distinguished include:
1.. Unrestricted universals: characteristics possessed by all languages which are not merely definitional, namely they are such that if a symbolic system did not possess them, it would still be called a language.
2. Universal implications: asserting that if a language has a certain characteristic, it must also have some other characteristic, but not vice versa.
3. Restricted equivalence: asserting that if a language has a particular non–universal characteristic, it also has another defined characteristic, and vice versa.
4. Statistical universals: asserting that for any language a certain characteristic has a greater probability of occurrence than some other.
5. Statistical correlations: asserting that if a language has a particular characteristic, it has a significantly greater probability of possessing some other characteristic.
6. Universal frequency distributions, whereby, as a result of some measurement over an adequate sample of languages, a characteristic mean and standard deviation is demonstrated and may therefore be considered as universal facts about languages.
On the basis of substantive content, the subtypes which may be distinguished include: phonological, grammatical, semantic and symbolic. In this classification, the first three involve either form without meaning or meaning without form, whereas the last, which involves sound symbolism, involves the connection between the two.

♦ **KC0950 Ecology**
Description A branch of science concerned with the interrelationships of organisms and their environments, with interacting entities and forces as open systems interchange matter with their environments.

♦ **KC0954 Group dynamics**
Description The forces and processes of interaction operating within a relatively small human group, and the study of such processes, especially within the theoretical framework of the view that the group is a sociological whole with dynamic properties of its own (such as organization, stability, goals) which can be objectively analyzed and accurately measured.

♦ **KC0956 Autopoietic system**
Description A system defined as a unity by a network of productions of components which through their interactions give rise to these same productions. Thus the production network produces the components and the components, through their interactions, constitute the system as a unity in the space in which they exist and make the network possible by defining and realizing its topology. Living systems are autopoietic and as such constitute a special case of homeostasis in which the variable of the system that is held constant is that system's own organization.

♦ **KC0957 Synergy**
Description 1. Behaviour of integral, aggregate, whole systems unpredicted by behaviours of any of their parts, components, or subassemblies of components taken separately from the whole. (There are progressive degrees of synergy, namely synergy–of–synergies, which are complexes of behaviour aggregates holistically unpredicted by the separate behaviours of any of their successive subcomplex components. There is thus a synergetic progression, a hierarchy of total complex behaviours. The Universe is therefore the maximum synergy–of–synergies, being utterly unpredicted by any of its parts.)
2. Synergy is a major component of an enterprise's product–market strategy and is a measure of joint effects which can produce a combined return on the organization's resources greater than the sum of its parts. Thus the potential return on investment for an integrated enterprise is higher than the composite return which would be obtained if the same financial volumes for its respective products were produced by a number of independent enterprises. The synergistic effect can be measured in either of two ways: by estimating the cost economics to the enterprise from a joint operation for a given level of revenue; or by estimating the increase in net revenue for a given level of investment. Synergetic effects may be obtained through: sales and distribution integration; integration of operating facilities and personnel utilization; integration of plant, inventories and other investments; and integration of management expertise, particularly at the highest level.
3. Cooperative and interactive effects between social and technical innovations may be termed synergistic. They tend to be important not only because advances in one area are correlated with or spur advances in other areas, but also because various separate advances often allow for unexpected solutions to problems, or can be fitted together to make new wholes that are greater than the sum of their parts, or lead to other unexpected innovations.

♦ **KC0961 Integrated science curricula**
Interdisciplinary science curricula — Teaching of concepts of integrated science — Unified science curricula
Description Science courses which contribute towards general education, emphasizing the fundamental unity of scientific thought, and leading towards an understanding of the place of science in contemporary society. Integrated science teaching avoids unnecessary repetitions and permits the introduction of intermediate disciplines. Such courses aim to present science in a coherent way, and to avoid premature or undue stress on the distinctions between the various scientific fields. They emphasize the underlying methodology and processes which characterize the scientific outlook. By a judicious choice of experience and activities, students learn how to think productively about the natural world and man's interaction with it by calling upon knowledge and skills from the various parts of science as needed.
To some degree the concept of integrated science teaching is based on the parallel assumptions that the universe has an inherent unity and that science as an attempt to provide an understanding of the natural world has a unity of purpose, content and process that is far more significant than the differences in language or focus between individual sciences. Those who press for integrated science teaching usually make the further assumption that the teaching of any subject should in some way reflect the nature of the subject itself. If the natural sciences are becoming integrated in their intellectual structure and are already unified methodologically, then, according to this assumption, science teaching should emphasize this by itself being integrated.
It is useful to think of integrated science courses as being of four kinds: 1. Those that integrate the subject matter from various subdivisions of a major science. (e.g. physics as a unified structure of ideas rather than as essentially separate courses in classical mechanics, heat, light, sound, electricity, and magnetism, etc.)
2. Those that blend two or more sciences in similar proportions. (It is not uncommon, for example, for earth science courses to place roughly equal emphasis on astronomy, meteorology, oceanography, physical geography, and geology.)
3. Those that blend two or more sciences together, but with a strong bias toward one. The difference between this and the previous category is essentially a matter of emphasis.
4. Those that select content as described in any one of the above three categories, but in addition, integrated material from the non–sciences. In this group are courses that pay attention to the philosophical underpinnings of science, to the development of scientific ideas and to the social consequences of science and technology.
For a course or programme to be considered truly integrated, it is necessary that the concepts of science be presented through a unified approach. A unified approach can be designed in a variety of ways, such as:
1. The conceptual schemes approach: Structural unity may come from designing a course around broad conceptualizations, the so–called big ideas of science. Concepts are selected for such a course that naturally make connections with various sciences. (e.g. conservation of energy, atomicity, dynamic equilibrium and change through time).
2. The inquiry approach: Problem solving activities can also be used as a unifying organizational scheme. In this approach, the science teacher raises, or guides students to raise, interesting questions about the natural world that challenges them to seek answers and that necessarily involve them in the substance of many sciences.
3. The relevance approach: This might be further sub–divided into the environmental science and applied science approaches. It is possible to structure an integrated science course around questions of the social relevance and utility of science. Issues such as whether or not to locate nuclear reactors in geologic fault zones, the propriety of organ transplants, controlling the population, and problems of nutrition and health, are examples of what will necessarily involve students in exploring ideas from various sciences, as well as from social ethics.
4. The process approach. It is also possible to design an integrated science course in which the focus is the process of science rather than the content of any discipline. (e.g. how to ask fruitful questions, develop working hypotheses, collect relevant data, analyze it and make sensible judgments, takes precedence, in such a course over the mastery of a prescribed body of scientific knowledge.)

♦ **KC0963 Economic cybernetics**
Description The study of the structural and functional components of national economies as systems in which processes of control are maintained by the movements and conversions of information.

♦ **KC0964 Morphological analysis**
Description This is a forecasting technique which is an extension of the network aspect of the cross–impact technique (separately described). In effect, morphology simply involves breaking down a phenomenon into all of its sub–elements at each of various stages of development or in various dimensions and then listing all of the various possible combinations of these sub–elements as alternative developments. These alternative developments can then be exposed to detailed analysis in an effort to determine which is most likely, thereby arriving at a technological forecast. The so–called morphological charting or plotting is, however, more an aid to thinking than a forecasting instrument in itself.

♦ **KC0965 Life cycle**
Description 1. The cycles of synthesis, degradation, use and sense of the essential elements of the biosphere (nitrogen, oxygen, carbon, sulphur, etc).
2. The developmental pattern in the lives of individuals of a species. In humans, physical and psychological infancy, growth, maturity and decline, and the various socioeconomic concomitants of these stages.

♦ **KC0966 Unitary principle**
Description This is formulated as: asymmetry tends to disappear, and this tendency is realized in isolable processes. The unitary principle represents the isolable phase of a given process, in which the distortion due to some wider system of which the process was previously a part, is progressively eliminated as the system in question separates itself out and perfects its characteristic symmetry. Isolable systems are those which are in course of isolating themselves, and are only observable while this process is incomplete.
The unitary principle defines a universal method of selection of causally simple processes within the complexity of nature. It combines in the simplest possible formula the minimum spatial and temporal relations that are required by any general theory of process.

♦ **KC0967 Budgeting**
Description An organizational budget is a plan covering all phases of operations for a definite period in the future. It is a formal expression of policies, plans, objectives, and goals laid down in advance by management for the organization as a whole and for each subdivision thereof. Budgets serve a multiplicity of purposes including: forcing managers to direct attention to the formulation of the objectives and goals sought; coordinating operations through the manner in which budgets for different functions are integrated into a comprehensive budget; and setting standards against which actions can be measured.
Budgeting not only facilitates but forces integration of functional elements in both the development of plans and in carrying them out. Budgets force managers to concentrate on quantifying ends to be achieved. They can be used as a vehicle for widespread participation among people in the planning process, thereby promoting better understanding and motivation for achievement, and bridging the gap between strategic planning and current actions.

♦ KC0968 Balance of nature
Description The self–maintaining, self–renewing dynamic of nature as a system, including the life cycles constituted by the synthesis, use, and degradation of nitrogen, oxygen, carbon, sulphur, etc. The 'balance' sustains the qualitative and quantitative needs of the sub–systems: the salt seas and fresh–water bodies, the forests, the soil, and the ecological relationships between the species. Man is the upsetter of the balance, engaged in this role for long centuries, but with ever increasing destabilizing power.

♦ KC0969 Metamathematics
Proof theory
Description Metatheories of mathematical theories particularly concerned with proofs of consistency, expressed in calculi (formal systems). Closely related to metalogic.

♦ KC0970 Field
Description 1. In physics the fields of the four universal forces (electromagnetism, gravity, weak force, nuclear energy. Fields are systems of interacting wave particles and can be considered as special forms of matter.
2. In algebra, a class, set, or set of classes used to describe mathematical phenomena such as rational numbers, constants, real numbers, complex numbers, etc.
3. In biology, a system in which behaviour of parts is determined by their position, e.g. the embryo and its cellular differentiation. Such biosystems may be vector fields with zones of structural stability.

♦ KC0971 Sense
Description 1. The aggregate of extralinguistic characteristics of language content, as opposed to meaning (intralinguistic).
2. Connotative aspect of a word, as opposed to denotative.
3. Generally, the essential aspect of something, its meaning, value, purpose.
4. Notion, intuition, judgement, etc.

♦ KC0974 Social forecasting
Description

♦ KC0975 Peace research
Description An interdisciplinary field of study both more specialized and broader than that of international relations. It is more specialized in the sense that it focuses on the nature of conflict and its resolution, but it is broader in that it draws on all scientific disciplines, in particular the social sciences, including history. Individual peace researchers define their task as one of accommodating social change in a non–violent way and to this end they seek to improve the present system or appropriately to transform it. The peace researcher is concerned with all forms of violence: open, behavioural violence and hidden, structural violence. The structural forms of violence are those institutionalized inequalities within and among nations which inhibit social justice, and are therefore conducive to behavioural violence. As these structural forms of social injustice exist both at the international and at the national level, so the peace researcher is led to examine social injustice wherever this is relevant to problems of international peace.

♦ KC0976 Integrated education
Curriculum integration — Educational integration
Description 1. Construction of a curriculum environment which will best enable students to relate meaningfully their school experiences with each other, with out–of–school experiences, and with their own personal needs and interests. The two major dimensions of educational integration are: the degree to which the various subjects or content experiences relate to each other; and the relevance of the content to the meaning structure of individual students. Types of integrating scheme include: correlation, in which teachers of different subjects provide a common referrent or theme which, in effect, correlates any two or more subjects; fusion, in which two separate courses (e.g. history and literature) are scheduled into the same block of time; and the curriculum approach consisting of broad pre–planned areas from which the teacher–pupil interaction evolves specific study in terms of the special needs, interests and problems of students.
2. A synthesis, or at least a cross–referencing, of existing knowledge within present–day educational disciplines. The conceptual matrix used emphasizes the disparate nature and essential autonomy of the disciplines so integrated. Such integration takes place within an existing model.
A fundamental distinction can be drawn between integratior in the sense of synthesizing presently accepted postulates with their attendant corpus of knowledge, methodologies, and modes of behaviour, and a type of integration (for which the term integrative has been proposed) in which a new conceptual model is constructed for the express purpose of developing a holistic educational philosophy and methodology capable of understanding and documenting the philosophical principles inherent in the natural and social sciences, together with the humanistic systems of mankind.
3. Integrated education may also be used to refer to the integration of different phases of education (primary, secondary, post–secondary, out–of–school, and adult education) as a continuing process.

♦ KC0977 Long–range goals for mankind
Description Some desired future state of affairs for mankind whose realization would require an effort lasting over many generations. To qualify as a long–range goal of mankind, the goal must involve a large number of people, probably a considerable fraction of the human race. Either their direct efforts might be required to bring about the goal, or the goal might have an important effect on everybody's life when it was realized even though the direct efforts of many people are not required to bring it about. Goals may either involve the intensification or fulfilment of something already to some extent present in humanity (developmental goals), or they may require the creation or achievement of something qualitatively new (transcendental goals).

♦ KC0979 Unity of intellect
Description 1. The active intellect (Aristotle) or agent intellect which is not a part or power of each human soul, but which is a separately existing divine intelligence; a unity which acts on the passive, individualized intellects of men.
2. Identified by speculative thinkers of the Middle–Ages with God; denied by orthodox Catholics.

♦ KC0980 Unitary nation–states
Description Those allowing little or no autonomy to regions, provinces, cities, etc. The most unitary nation–states have one national constitution, one general system of laws, and a centralized apparatus for executive, legislative and judicial functions. The opposite type is the federalled nation–state.

♦ KC0981 Number
Description 1. In mathematics, the positive integer, or natural number.
2. In mathematics, any one, or all of the various categories of number.
3. Quantity.
4. Sign for quantity.
5. In linguistics,the grammatical category which indicates singularity, plurality or indeterminedness.

♦ KC0986 Mathematical integration
Description The operation of finding a function of which the integrand is the derivative of a function or of solving a differential equation.

♦ KC0992 Racial and ethnic integration
Integration — Desegregation
Description Incorporation into a group, organization or community on the basis of common and equal membership despite differing characteristics, such as race or ethnic origin.

♦ KC0994 Topological space
Description A topological space is an abstract mathematical space constituted by any collection of points (an arbitrary set of elements), in which a relation of neighbourhood of one point to a set of points is defined and, consequently, a relation of neighbourhood or adherence of two sets (figures) to one another. This is a generalization of the intuitive intelligible relation of neighbourhood or adherence of figures in ordinary space. Neighbourhood forms a distinctive appurtenance of bodies and gives them the same geometric relation when we retain in them this property and do not take into consideration all others, whether they be essential or accidental. Topology (described separately) provides a rigorous basis for the strict application of arguments relating to the intuitive concept of connectedness in its most general sense.

♦ KC0996 Pluridisciplinarity
Description Juxtaposition of various disciplines, usually at the same hierarchical level (whether empirical or pragmatic), grouped in such a way as to enhance the relationships between them, but without in any way affecting the rigid disciplinary modules. In the successive steps of cooperation and coordination between disciplines, pluridisciplinarity defines an organizational principle for a hierarchical system of one level, having multiple goals, involving cooperation but no coordination. Examples of pluridisciplinary are: mathematics plus physics; or French plus Latin plus Greek.

♦ KC0997 Regional international system
Regional subsystem of global system
Description 1. An especially intense network of international links within a defined geographical compass (namely regional cooperation, regional transactions, or regional organization), may be summarized by the term regional system.
2. Regional subsystem is a term used to explain the interdependence between local ties and concerns and the large world which constrains them.

♦ KC1000 General systems theory
Description 1. An outlook, methodology, programme or direction in the contemporary philosophy of science. The salient feature of this outlook is an emphasis on those aspects of objects or events which derive from general properties of systems rather than from the specific content of the systems. All the variants and interpretations of this outlook have a common aim, namely the integration of diverse content areas by means of a unified methodology or conceptualization or of research. The task of general systems theory is to find the most general conceptual framework in which a scientific theory or a technological problem can be placed without losing the essential features of the theory or the problem. The main theme is the explicit fusion of the mathematical approach with the organismic leading to the main task of showing how the organismic aspect of a system emerges from the mathematical structure. The theory is considered to be a focal point for the resynthesis of knowledge.
2. A general science of organization and wholeness whose aims emerge from the following: (a) there is a general tendency towards integration in the various sciences, whether natural or social; (b) such integration seems to be centered in a general theory of systems; (c) such theory may be an important means for aiming at exact theory in the non–physical fields of science; (d) developing unifying principles running vertically through the universes of the individual sciences, this theory brings the goal of the unity of science nearer; (e) this can lead to much–needed integration in scientific education.
3. A level of theoretical model–building which lies somewhere between the highly generalized constructions of pure mathematics and the specific theories of the specialized disciplines. It responds to the need for a body of systematic theoretical constructs which will discuss the general relationships of the empirical world. At a low level of ambition (but with a high degree of confidence) it aims to point out similarities in the theoretical constructions of different disciplines, where these exist, and to develop theoretical models having applicability to at least two different fields of study. At a higher level of ambition (but with a lower degree of confidence) it aims to develop something like a spectrum of theories, namely a system of systems which may perform the function of a gestalt in theoretical construction. Such gestalts in special fields have been of great value in directing research towards the gaps which they reveal.
4. A methodological direction aimed at extending rigorous methods of analysis to objects and situations characterized by organized complexity. The system–theoretic view focuses on emergent properties which such objects or classes of events have by virtue of being systems, namely those properties which emerge from the very organization of complexity. Theories dealing with certain portions of the world, too complex to be understood entirely in terms of analytical models, can nevertheless be developed. They depend for their development on synthetic and holistic approaches rather than analytic ones. The aim is to develop such approaches further, to link them to analytic approaches whenever possible, and above all, to unify theories of widely different contents but of similar logical structure, where the similarity derives from some structural isomorphism between objects and phenomena in different spheres.
5. The formulation and derivation of those principles of a physical, biological, or social nature which are valued for systems in general. It seeks models, principles, and laws which apply to the generalized system irrespective of the particular kind of elements or forces involved. It seeks identification of structural similarities (isomorphies) in different scientific fields and hypothesizes that the world (i.e. the total of observable events) shows structural uniformities which manifest themselves by isomorphies of order and organization in the different levels, realms, or systems of the world.

♦ KC1010 Symmetry
Description 1. In physics, the invariant nature of physical laws governing comparable phenomena in different systems or frames of reference, or in the same phenomena within a changing system. The laws are invariant or symmetric under given transformations. Symmetry is preserved in a number of such known types. The continuous transformations that are symmetric include: the

translation of a whole system in space; either active, in which the system is translated to a different, chosen frame of (spatial) reference, or passive, in which the frame itself undergoes translation. Symmetry in these cases includes the concept that all points in space are equivalent (non-distinctive), and displacements or transformations are thus in homogeneous space. Other important symmetric continuous transformations are: whole system rotation in isotropic space, all directions being equivalent; time translation or change in origin of the time coordinate (indicating that physical laws do not change with time); and Lorentz transformations, of a system to another frame of reference moving in constant relative directional velocity (indicating equivalence of all inertial frames of reference; critical to relativity theory). In symmetric Gange transformations affecting interactions of charged particles and fields there is also a special case of invariance violation in the 'approximate' symmetry of isotopic spin in strong nuclear particle interactions. The other major category (in addition to the continuous) are the discrete transformations. While the first category is characterized by parameter values that can vary continuously along the coordinate axes of the frame of reference, this is not the case in discrete transformations. The characterizing names of the principal cases of symmetric discrete transformations are space inversion and time reversal. The former is also called reflection symmetry due to mirror-imaging, as in dextrorotary and levorotary spins and rotations, and in matter, antimatter substances (however space-inversion symmetry is violated in the case of weak-interactions). The other types of symmetric discrete transformations are charge conjugation in which particles are replaced by antiparticles, and time reversal where the sign of the time coordinate is changed. Taking the three operations, CPT, together (charge conjugation, C; space inversion, P; and time reversal, T) their symmetry follows from quantum field theory. Taken in connection with invariance under the Lorentz transformations, and with a single point locality of field interaction, symmetry holds, even if CPT are taken separately and all interactions are not invariant. The foregoing and other investigations of symmetry are crucial to understanding the physical universe and the validity of many fundamental principles such as unity and the conservation laws of energy momentum, and spatial parity.

2. In logic the property relation that holds for any pair of objects regardless of order in which they occur. Such symmetric relations are the equality types: identity, equivalence and similarity. Weak symmetric relations occur in such binaries described as resembling and proximate, etc. (These are better known as tolerance relations).

3. In mathematics, commutativity and permutability, as in groups.

4. In geometry, reflection, central, axial and translational space symmetries. In these operations a figure is mapped onto itself. The set of all orthogonal transformations that may a figure onto itself is called its symmetry group.

5. In botany, leaf arrangements (phyllotaxies) characterized by twists, i.e. symmetries generated in combination, by a rotation and a translation.

6. In crystallography, the behaviour of crystals, though in rotations, reflections, parallel translations, or in any such combined operations, to replicate or repeat their organization, so that symmetry of crystalline atomic structure creates both symmetry of physical properties and symmetry in faceted form.

7. In molecular chemistry, spatial symmetry or equilibrium configuration of axes and planes of symmetry (also symmetry of nonequilibrium configurations). The symmetry of the equilibrium configuration of molecular nuclei determines the symmetry of the wave functions for various states of the molecule. Symmetry properties are important in classification and analysis of complex compounds, and in study of chemical lasers and pharmacologically active substances.

8. In biology, structural symmetry.

9. In art, harmony of composition, e.g. the Golden Mean.

10. In science, the general theory of symmetry.

Embodying discontinuity

Rationale
In this period characterized both by profound disagreements and by intense efforts at consensus formation, there is widespread recognition of the disadvantages of the former compared to the advantages of the latter. This recognition is itself a danger however when it detracts from complementary efforts to recognize the advantages to be derived from the living reality of disagreement processes as compared to the corresponding disadvantages associated with dysfunctional consensus formation. Typically the former leads to characteristic difficulties in handling differences, "otherness" discontinuity, uncertainty, ignorance and the underdefined, which all arise frequently in social processes, especially in any transitional period of social transformation when there is a possibility of a "new" or "alternative" order.

In the search for such a new order, many "answers" continue to be produced in response to the global problematique, whether in the form of explanations, programmes, strategies, ideologies, paradigms or belief systems. The proponents of each such answer naturally attach special importance to their own as being of crucial relevance at this time, whether in the short-term for tactical reasons, or in the long- term as being the only appropriate basis for a viable world society in the future. This widespread focus on "answer production", a vital moving force in society, obscures both the significance of the lack of fruitful integration between existing answers and the manner in which such answers undermine each others significance. Such answers are inherently limited in that they fail to internalize the discontinuity, incompatibility and disagreement which their existence engenders, in such a way as to "contain", whether conceptually or organizationally, the development processes they promote. This naturally results in the emergence of new problems.

Any new order is thus engendered by the fluctuation in practice between the extreme policies of essentially antagonistic answers counteracting each others weaknesses and excesses. It is this same fluctuation which the proponents of each dominant answer at present make every effort to prevent, as a way of maintaining their dominance in the short-term, but at the expense of development in the longer- term. But it is on this very fluctuation that a viable new order needs to be built if it is to contain a development that is inherently dynamic. The desperate search for "the" model of a new magical alternative order (of necessarily temporary and limited appeal) can thus be usefully complemented by a concern for models of alternation to order the pattern and timing of cyclic transformation between such alternatives, as and when they emerge into the ecological pool of available models.

This raises a major difficulty since, as noted above, no single framework (whether logical or otherwise) can encompass the dynamics of alternation between such frameworks, whether it be cartesian or holistic, linear or non-linear, technocratic or ecological. Perception through any one of them necessarily precludes simultaneous perception through any other one (as with the wave or particle theories of light). It follows that no single conceptual language or paradigm is appropriate to the task of bridging across the discontinuity between frameworks to support the development process. This raises questions as to the nature of such a bridge and of the language with which such a bridge may be constructed. These are explored in this section.

Index
A keyword index to entries is provided in Section KX. The keywords are also incorporated into the index for Volume 2 (Section X)

Bibliography
Bibliographical references, by author, are given in Section KY.

Contents
The section contains 70 entries which constitute a critical review of a series of approaches to the problem of embodying or bridging conceptual discontinuity. There is a progression through the series to more complex approaches which in different ways embody that discontinuity into some kind of alternation dynamic between alternative perspectives.

Method
The entries were formulated by the editors as part of a review of methodological problems which emerged during their participation in the Goals, Processes and Indicators of Development project of the United Nations University (1978-82).

Comment
In contrast to entries in other sections, comments are incorporated into individual entries, although general comments appear in Section KZ. **Bibliographical references** are listed in Section KY.

Reservations
Unlike entries in most other sections, those included here reflect an editorial attempt to review the efforts of a wide range of authors to respond, in one way or another, to the conceptual problems of discontinuity. The editors have attempted to interrelate these initiatives and to indicate how they collectively contribute to an understanding of alternation, and that in fact the central theme they together identify can itself only be comprehended by alternating between the essentially incommensurable perspectives they represent. The implications of their insights have been deliberately oriented toward comprehension of the global problematique which may be very far from the intention of the authors.

Whilst this procedure may have resulted in distortion of their views, the editors do consider that presenting such unrelated views as a configuration offering complementary perspectives does go some way to avoiding the problems of entrapment in a particular language. The stress on ordered patterns of "alternation" is of course another form of entrapment. At this stage however it does seem to cast new light on the problem of embodying discontinuity in response to the global problematique, especially since it effectively draws attention to its own limitations. For those who find the various jargons unhelpful, many of the metaphors in Section CM were designed to clarify the significance in practice of alternation.

ENTRY CONTENT AND ORGANIZATION

Ordering of entries
Entries are in **numeric order**. Entry numbers have been **allocated randomly**; they have no significance other than as a permanent point of reference to facilitate indexing, cross- referencing, and updating between editions.

Index access to entries
The location of an entry in this sub-section may be determined from:
- the **Volume Index** (Section X) on the basis of keywords in the name of the entry or its alternate names
- the **Section Index** (Section KX) on the basis of keywords in the name of the entry or its alternate names

Structure of entries
Entries may be composed of the following descriptive elements:

(a) **Entry number** This number has **no significance**, except as a convenient method of identifying the entry (particularly for indexing purposes), of filing information on it, and as an identifier to which cross-references from other entries (possibly in other sections) may refer in this and future editions. The first letter of the entry number refers to the section of this volume in which the sub-section, denoted by the second letter, is located.

(b) **Entry name** This is printed in bold characters.

(c) **Comment** Review of particular approach to embodying conceptual discontinuity.

Cross-referencing of entries
There are **no cross-references** between entries in this sub-section.

♦ **KD2001 Questionable answers: monopolarization**

Comment The many initiatives in response to the global problematique are in most cases stimulated by a need to determine guidelines for action. The question to which an answer is sought at all levels is some variant of "what can be usefully done?" The answers to this question have taken a range of well-known forms which include the following:
(a) Policy recommendations to appropriate institutions;
(b) Publication and distribution of research conclusions to academic communities;
(c) Public information programmes to adapt and disseminate conclusions;
(d) Formulation of educational programmes for schools and universities;
(e) Implementation of community dialogue programmes to encourage participation;
(f) Implementation of field programmes acting directly on societal problems;
(g) Design of new organizations and institutions capable of responding more adequately to the societal condition;
(h) Elaboration and dissemination of visions of alternatives and future action modes;
(i) Design of new information systems;
(j) Organization of (a series of) meetings on problems and action;
(k) Proposals for further research on problems whether involving practical applications or fundamental re-assessment of methodology;
(l) Elaboration of innovative audio-visual presentations of the nature of problems and the action possibilities;
(m) Elaboration and dissemination of new sets of values through which consensus can be obtained;
(n) Elaboration and dissemination of declarations concerning action to be taken;
(o) Elaboration of multilateral treaties concerning action to be taken;
(p) Elaboration of interaction processes (including computer conferencing) whereby the problematique can be approached in a new light.

These are all "classical" options to ensure an integrated response to any societal condition. They have been extensively applied since the origin of the International Development Decades and in response to every type of problem, including: energy, population, food, refugees, discrimination, health, youth, drugs or environment. It is fair to conclude that these answers have been successful to the extent that the problem was either a narrow technical one involving little controversy (e.g. smallpox) or did not call for immediate action (e.g. creating environmental awareness). The answers have however been of limited effectiveness in containing the problematique in its essential globality. The point has been reached at which predictions by the highest authority of the cumulative consequences of inaction are met with increasing indifference and a sense of helplessness. It is possible to take any one of such answers and show why it is inadequate as a response and why in fact it may merely aggravate or displace the problem. This too is increasingly recognized. And yet such answers continue to be formulated in desperation because of the need to respond to constituencies who want to believe that something effective is being done which will alleviate the problem and avert disaster. Protests that such answers have proven to be of limited effectiveness in the past, meet with responses of the type: "these things take time", "we must do what we can", "we must concentrate on what we can handle effectively", and "it is participation in the process which is significant, not the results". Is it possible to move *beyond the uni-modal answer* and recognize that because each form of action has both strengths and weaknesses, the key to a more effectively multi-modal answer lies in finding how to interrelate the various uni-modal answers so that they correct for each others weaknesses and counteract each others excesses. There are some efforts in this direction but they run up against another constraint, namely whether integrated action of any type is feasible at this time.

♦ **KD2003 Assumptions concerning appropriate answers**

Comment What then is the nature of the answer that would prove appropriate ? What are its "properties" ? What would be the response to the formulation of such an answer ? Are there more fruitful ways of formulating such an answer ? Assumptions such as the following are too easily made:
(a) The appropriate answer can be made in the same conceptual framework or "language" as the question "what can be usefully done ?";
(b) The answer will not challenge the status and self-image of the questioner or potential "doer";
(c) The answer can be rendered in a comprehensible form to the questioners or to those from whom they have received their mandate;
(d) The answer would simply involve a reshuffling of existing organizational resources and priorities, but would not imply any radical transformation of their status and mode of working. (It might well be interpreted by many to take the form of the NIth UN World Action Plan and therefore to conform to standard UN administrative procedures);
(e) The answer would not engender valid opposition and resistance, except by reactionary segments of society whose views are irrelevant;
(f) The promulgation of the kind of answer sought would not deprive the future period (during which it is implemented) of the ability to initiate alternative responses;
(g) The answer cannot be conceived as competing with other answers, which if they are advanced must necessarily be subsumed, opposed, or preferably suppressed;
(h) The psychological and institutional systems could adjust satisfactorily to the complete elimination of the problematique by the ideal answer.

Assumptions such as these result from thinking similar to that associated with modern medicine. Illnesses are diagnosed and then surgery and/or a course of treatment is recommended based on specific drugs and diet. It is assumed that if the world problematique could be accurately "diagnosed" and mapped, malignant growth could be excised and appropriate "pills" could be designed and "prescribed". Some further treatment may also be advocated in the form of various therapies or re-educational exercises, with "stimulants", "tranquillizers" and "vitamins" as necessary. This *pill psychology* approach takes no account of the questionable role of medicine in society, as explored by Illich (1976) and Attali (1979). It does not take into account issues analogous to those raised by such currently debated phenomena as conflict between specialists, malpractice, iatrogenic diseases, placebo effects, commercialization and institutionalization of medicine, drug cost as a perceived indicator of remedial power, folk medicine, euthanasia, hospital vs home environment, and problems of psychosomatic origin.

The approach to providing "the answer" must therefore be examined very carefully. Advocating a particular model or course of action is tantamount to advocating a particular type of pill. It raises the question of how this might conflict with treatment advocated by other "health centers" from which the "patient" is seeking advice. On the other hand, presenting a range of conflicting opinions by eminent specialists on possible alternative courses of treatment would be of little value to the patient, as would recommendations for remedies for an aspect of the problem (a "micro-answer"). And pointing to directions for "further research" would be simply abandoning the patient to his own resources for the meantime. In each case, it is not the treatment which is necessarily the main problem, but rather *the framework within which the patient's relationship to the possible treatments is defined.* The question is therefore whether this situation can be seen in a new light and whether a new kind of response can be made to the question "what can be usefully done ?".

♦ **KD2005 Forms of truth: uniformity versus aesthetics**

Comment The exploration of the nature of an appropriate answer must take into account a most important phenomenon. That is that few groups, projects, or schools of thought have difficulty in discovering and promulgating an answer. The difficulty for society as a whole arises from the conflictual relationship between such answers, or their denial of each other as irrelevant, out-of-date, erroneous, or unworthy of consideration. In the words of Jacques Attali concerning remedial ideas about the current crisis: "Au-delà des problèmes que pose toute sélection d'idées... voici l'essentiel: si tout ce savoir n'est encore aujourd'hui ni synthetise, ni assimile, s'il reste un lieu d'affrontement et d'anathèmes, c'est parce qu'il charrie une image du monde d'une intolérable fixite; et que tout groupe social trouve intérêt à en occulter certains fragments pour tenter d'asseoir sa domination". (1981, p.10-11)

Perhaps the most important feature of this phenomenon is that every effort is necessarily made to ignore it, to deny its significance, but especially to avoid exploring non-trivial routes beyond the barrier it constitutes to social development. As Attali continues: "Face à l'immensité de l'enjeu, faut-il alors cesser ce combat rudimentaire entre un vrai et un faux, mettre un terme à cette denonciation de la parole de l'autre ? Et avoir le courage d'admettre que plusieurs discours peuvent être simultanement vrais, c'est-à-dire peuvent valablement interpreter le monde ?" (1981, p.11)

Attali notes in passing that the multiplicity of truth is also encountered in physics (form example the wave vs particle theory of light). Clearly, as he proceeds to demonstrate, the problem lies in the way truth is to be understood. He distinguishes three senses (1981, p.11-14):
(a) A theory is true if it can be articulated according to the rules of formal logic, and if its consequences can be verified empirically by any observer. This is the most common scientific criterion of truth, and is that used by establishment institutions of every kind in every society. It gives rise to difficulties if some of the consequences it implies are contradicted by experience. The institutions are then obliged to construct a representation of the world which denies any possibility of its own negation.
(b) A discourse is true (and therefore scientific) if it provides a useful mode of communication for a group in its struggle for power. Unanimity is then forcefully imposed rather than emerging from aggrement with a universal rational structure.
(c) A discourse is receivable, and thus true, the moment it produces an understanding of the world for those articulating it. Unanimity is achieved neither by pure logic, nor by force, but by the virtue of seduction. As with beauty, and because it is intimately related to it, truth is not in itself universal. Truth is aesthetic.

Attali compares these three forms of truth in physics with mechanics, thermodynamics, and relativity theories. The equivalents he suggests in economics are regulatory theories, theories of value production, and theories of the organization or management of violence (especially of the non-physical variety), each with their appropriate modes of organization. The first two may be equated with capitalist (most general sense) and marxist (theoretical) approaches. It is the third approach, or basis for world order, which needs to be defined.

As Attali stresses, it is necessary to recognize that the reality of the world, whether in physical or psycho-social terms, is too complex to be encompassed by a single mode of discourse. The real cannot be separated from each necessarily partial view of it. It is in fact the multiplicity of views of the world, with all their differences and ambiguities, which renders the world tolerable to the majority, permitting each to develop his own understanding and to manage the violence done to it by others.

"Aujourd'hui cette multiplicité est difficile à preserver. C'est que les deux premiers mondes de la science ont proné, l'un l'universalité, le second la force: ni dans l'un, ni dans l'autre, il n'y a place pour la tolérance. Aussi, toute société qui accepte de se representer le monde selon une seule de ces deux classes de discours s'oblige à l'uniformité. Elle ne peut laisser vivre le troisième sens du vrai, et le voilà inévitablement contrainte au mensonge et à la dictature: tout ordre qui élimine l'esthétique comme langue et la séduction comme parole implique inévitablement la dictature". (1981, p.15-16)

Just as in physics the three approaches continue to have their domains of validity, so it should prove to be in the realm of psycho-social organization. The human being has three brains, the third being essential to mediate between the conflicting functions of the other two (M D Waller, 1961). The key question is then what kind of organization is implied by this third order of truth such that it could be of any significance for social development? Failure to take account of this question can only result in an answer of essentially limited value.

♦ **KD2006 Prevailing meta-answer: the gladiatorial arena**

Comment If "an answer" is sought for the current global condition, and it is repeatedly stressed that one is urgently needed, it would seem that great care is required to avoid falling into the trap of formulating answers whose nature forces them to complete in the unending, and essentially inhumane, "gladiatorial combats" of the *answer arena*, in an effort to attract the temporary support of fickle "spectators" answer only meaningful to those initiated into a particular elite group (cf world modellers) with its own limited information base. The answer must be of a different nature, but at the same time widely comprehensible. It should not attempt to accumulate glory by direct combat in the answer arena. It should rather redefine the significance of that arena and the answers which emerge temporarily victorious there.

In effect humanity already possesses a single, universal *meta-answer*. That is the one which defines the present nature of the answer arena. It is the mind-set which perceives that arena as the place on which differences should be settled and effectively legitimates the processes which currently occur therein. This legitimation is obviously neither fully conscious nor explicit. It is derived from the instinctually felt "appropriateness of similar "stamping ground" processes in the time of early man. These were shared with pack animals. This essentially instinctual meta-answer has, for specific and limited purposes, been partially modernized and given respectability. That is in the concept of the global "marketplace" for exchange of goods and services and the various "international assemblies" for exchange of views ("marketplaces for ideas"). But these are but a thin disguise for an arena which remains essentially primitive, in which most other differences are "settled", and as a result of which pack allegiances are redefined. Everybody participates actively or passively in these processes whereby movements of opinion arise and "world opinion" is formed and modified. They appeal to the "fickle instinctual spectator" in everyone. The challenge would seem to be to find a way of placing this current meta-answer in a new light, not so much by combatting it on its own terms, but rather by offering a more "seductive" alternative in Attali's sense (1981). The difficulty is to *avoid the temptation of defining this meta-answer as an answer* and thus ending up in the current trap. But at the same time, if it is to be of any relevance, the meta-answer should do more than simply provide a context for the emergence of better answers.

♦ **KD2007 Answer production through accumulation of significance**

Comment Society does not lack for answers to its current difficulties. The problem lies in the limited constituencies to which such answers appeal. It is useful to look at *answers as products*, or visible *manifestations of an accumulation process.* Answers tend to emerge from ordered accumulations of information. The amount of information effectively entering any such accumulation process is necessarily limited because of limitations on human processing capacity and constraints on social learning. This does not mean that the information arises from a limited geographical region. On the contrary it is a characteristic of present day answers that they result

from interpretations of information (cf Edgar Morin, 1981) selected from a globally distributed pool of information (c.g. data networks) which may well be physically accumulated at a particular spot (e.g. major libraries). It is the selection process which ensures the filtration. Each such answer is formulated in terms of a limited information base. For example, this is usually discipline-oriented in the case of academic answers, but ideological, action-preference, educational-label, "priority" and other filters may also be used, whether together, alone or in various combinations.
Once an answer has been formulated it acquires *symbolic significance* over and above the rational arguments which support it. It provides a rallying point for those searching for coherence in terms of the information base from which it emerged. Particular jobs may be tied to its promulgation or implementation. As such it reinforces the accumulation of further information in support of that answer. Competing answers, and contradictory information, are ignored, avoided or suppressed whenever possible. In the case of a well-developed answer, all "available" information of any "relevance" is perceived as supporting the position. The answer is then used as a vehicle for vigorous proselytizing activity amongst those who subscribe, out of "ignorance", to alternative answers. The aim is to ensure that such "infidels" are converted to "the answer", namely that consensus is achieved so that effective action can be undertaken. Everybody must be "accumulated" by the answer.
Over the past decade this approach has taken on a new aspect, due to some recognition of its obvious limitations. Instead of answers emphasizing particular conceptual perspectives or content, many now focus on a particular process (e.g. community dialogue) or mode of action (e.g. networking, struggle) which permits or engenders a variety of local answers in concrete situations. The process advocated thus becomes the answer for which universal support is sought.
There are many parallels in this to the emergence and historical development of religions, each of which makes universal claims for its unique grasp of the answer to the social condition. The current (lack of) relationship between organized religions provides an excellent model for understanding the relationship between groups subscribing to any given answer. The model is enriched by its representation of the formation of schisms and priesthoods as well as by the process of religious disaffection, accompanied by the continual emergence of a plethora of sects, each with a well-developed answer.

♦ KD2009 Development processes and the accumulation of significance

Comment In attempting to understand better how individuals and social groups accumulate the significance they associate with their particular answers, it is appropriate to look at critical analyses of the well-documented *capital accumulation* process. This should provide further insights and clues for the pursuit of the enquiry into the characteristics of a desirable meta-answer. Such analyses can be "decoded", using them as a model to understand accumulation processes in general rather than as limited to economic processes in the narrow material sense (A J N Judge, 1984, p.11-16).
What is refreshing about the most powerful of such analyses, "the world-system perspective", is the manner in which it avoids taking present structures for granted. Both Wallerstein and Addo (1984, 1985) criticize the conventional "developmentalist" framework within which current answers have been vainly sought for two decades. Wallerstein contrasts this with the world-system perspective: "What is crippling about a developmentalist perspective is the fact that... large-scale historical processes are not even discussable, if one uses the politico-cultural entity (the 'state') as the unit of analysis" (Wallerstein, 1976, p.352).
Equally crippling however, in attempting to understand the accumulation of significance, is the restriction of "world-system" type analyses to the limited range of material phenomena significant to a scholastic entity, namely "political economics". Other phenomena are then simply "not even discussable". The difficulty is understandable in that once the scope of the analysis is extended to non-material phenomena it is obliged to become self-reflexive (D R Hofstadter, 1979) and include the production and distribution of world-system perspectives. Since it is an explicit characteristic of any such perspective to use political action in the "marketplace of ideas" in order to ensure its own dominance, it is difficult to see how its strategy can be distinguished from that of any other aspirant hegemony. The same is naturally true for any answer entering that marketplace or with an established place in it.
It is not solely at the level of material phenomena that an appropriate meta-answer can be usefully sought. Somehow the relationship *between* answers at all levels must be examined more creatively. It is a "New International Conceptual Order" that is required as a basis for any effective New International Economic Order. All the unsatisfactory material processes for which an NIEO-type response is sought are a rather pale reflection of *equivalent conceptual processes* which continually reinforce them and undermine remedial action in any context.
The subtleties of Addo's assessment of the limitations of NIEO could also be generalized to cover those of the "answer economy". What is to be the status of answers formulated or favoured by minority groups or weakly organized large groups ? There is an exploited "Third World" to be recognized in non-material terms, and current concern with cultural domination is a step in this direction.
Adapting Johan Galtung's comment on "structural violence", it could be said that: *Amateurs use the organization of material accumulation to dominate a situation, this can be done professionally by the organization of non-material forms of accumulation, especially the accumulation of significance.* In fact the very vigour of the processes of radical analysis and conceptual innovation may well reinforce the material accumulation processes deplored in such analyses.
Each answer is effectively an attempt by a limited group (with limited sensibilities, and with a limited information base) to give better organized expression to "the good, the true, and the beautiful". But in all such cases the nature of an appropriate meta-answer remains unclear. The problem is in devising a suitable meta-form to interrelate answers which can only retain their essential quality within forms which are antagonistic to one another. Advocating tolerance in a pluralistic, laissez-faire context is a very superficial, impractical response to the current existential challenge.
The restrictive nature of a particular form of accumulation also affects the kinds of answers sought to the problems arising from that accumulation process. Answers tend to focus on changing the pattern of accumulation or eliminating it altogether, at least at the material level. The focus of attention is however limited to the level of accumulation at which the problems are currently most evident. Answers tend not to be sensitive to what is accumulated through promulgation and implementation of the favoured answer. It is also important to understand how a system can slip, or be displaced, into other modes of accumulation at an equivalent level.
It is assumed in the light of the variety of forms of accumulation and its ambiguous functions in the development process, that it is highly unlikely that this process can be eliminated. The question is then whether it can be transformed such that the focus of attention is not on a particular level. Whilst it may not be possible to eliminate accumulation, it may be possible to give progressively *greater emphasis to subtler non-material forms of accumulation.* This would involve creatively identifying new opportunities for accumulation. An interesting example of this is the Buddhist emphasis on the *accumulation of merit*, and the understanding that subsequent phases of development can only be achieved by recognizing that *this emphasis on accumulation must itself be abandoned as an obstacle to further transformation.*
Two forms of development can then be usefully distinguished. The first is primarily associated with "growth" and "spread", namely "quantitative" development. The second is primarily associated with qualitative development or transformation. This is brought about by shifting the centre of gravity of the accumulation process. In this sense the challenge is to find ways to "*develop accumulation*".

♦ KD2010 Oppositional logic

Comment The philosopher Stephane Lupasco has explored the nature of antagonistic dualities (1973). He shows that knowing is intimately associated with such duality and takes place by actualizing one of the terms of the duality and virtualizing the contradictory one. In this way only a monism is knowable, especially in science, even though it is the dualism which is the "motor" for this process. That by which we know illuminates a contradictory order whose contradictory nature is not apparent. In this way the proper object of scientific knowledge can only be extension - affirmation, permanence, conservation, and identity. The knowledge is brought about by the negation of intensity - which is forced out of the cognitive domain (1973, p.16).
Furthermore: "Devant un champ conscientiel et cognitif de plus en plus riche d'identités exteriorisées, tout ce qui relève de la négation sera rapporté au sujet connaissant, lequel se connait de moins en moins au fur et à mesure qu'il connait davantage, puisque précisement il n'apercoit plus que ce sur quoi il opère, que ce qu'il refoule, nie" (1973, p.15).
For Lupasco all human cognitive and practical efforts oscillate between extension and intensity: "De par leur contradiction dynamique constitutive, il y aura toujours conflit et tentative... de suppression de ce conflit, et, donc, choix de l'un au detriment de l'autre, *alternativement*" (1973, p.17, emphasis added). For the human being, extension is that which one knows more than one feels, whereas intensity is that which one feels more than one knows (1973, p.22). The characteristics of each (1973, p.30) recall recent work on right and left hemispheres of the brain.
For Lupasco any emergent third perspective can itself be resolved into mutually contradictory terms involving an oscillation between identity and non-identity. This "prison" is the essence of our knowledge (1973, p.61) although: "L'esprit humain fuit ce qui lui est révèle le plus infailliblement, l'opposition pure, l'oscillation continuelle des contraires" (1973, p.60).
Although his exploration is very valuable in understanding how such energy is engendered by such dualities, it is less useful in understanding how such energy to be contained in support of human and social development.

♦ KD2015 Polarity

Comment Another philosopher, Archie Bahm (1977), has studied the many characteristics of polarity as a basis for ordering contrasting theories. For him, polarity involves at least three general categories which he discusses in detail. These are: *oppositeness*; *complementarity* (involving subcategories of supplementarity, interdependence, dimension and reciprocity); and *tension* (involving subcategories of tendency, extra-tension, duo-tension, dimensional tension, inter-level tension, polari-tension, rever-tension, rhythmi-tension, and organi-tension).
He starts by distinguishing four emphases with which general types of theory (or "answers") may be associated:
(a) "One-pole-ism", indicating emphasis upon the priority of one of the poles
(b) "Other-pole-ism", indicating emphasis upon the priority of the other pole in constituting the polarity
(c) "Dualism", indicating preference for the independence of the two poles
(d) "Aspectism", indicating the priority of the (shared) dimension relative to the poles.
For each of these Bahm then identifies more specific types for which he gives examples from philosophy. In each case he distinguishes between "extreme", "modified" and "middle" emphases. Combined these constitute a set of 12 categories which provide him with the framework foro his own "answer", organicism, in the form of "a theory" about the nature of polarity but also about theories of polarity." (1977, p. 47).
Organicism is the theory that polarity consists in something "which is not wholly describable" but such that there is in it some basis for the positive claims made by each of the 12 preceding theories. Unfortunately, this creates the impression it is somehow an appropriate compromise between the 12 at some "dead centre". In fact Bahm specifically warns against this interpretation: "One needs an oscilloscope to depict the dynamic movements of the ways in which things, and the polar categories of things, exist; to stop at the centre is to destroy movement, and thus, existing and existence." (1977, p.277). He does not explore the nature of this movement.
Nevertheless, just as Lupasco stresses the dynamics between categories, Bahm seems to stress the static structural and non-contradictory relations between categories. In effect the two studies represent complementary approaches. Additional elements, interrelating such approaches, are required for an ordered response to the dramatic nature of the conflict between answer domains.

♦ KD2030 Paradoxes and antinomies

Comment Another approach to the logical discontinuity between answer domains is through the study of paradoxes. For Solomon Marcus: "Paradoxes occur when two different levels of knowledge, of language, of communication, of reality, of human behaviour, etc. are seen as one level, are mixed, are superposed, are combined, or are confused." (1982). He gives 18 pairs of levels which demonstrate a variety of paradoxes of which some are well-known to specialists. To clarify the semiotic difficulties involved, Marcus groups them into four types:
(a) Semiparadox: A *against* B, but not necessarily B *against* A (e.g. "Mary gave birth to a child and got married")
(b) Paradox: A *against* B, and B *against* A. If "against" is a logical negation, for example, this results in a logical paradox (e.g. when something is simultaneously good and bad)
(c) Semiantinomy: A *against* B, and A *for* B, where "for" is a binary relation which is inverse with respect to the relation "against" (e.g. the well-known claim of Epimenides the Cretan that "All Cretans are always liars")
(d) Antinomy: where (A,B) are both semiantinomies, such that the first term of a dichotomy both opposes and needs the second term, with the terms attracting and rejecting each other
Marcus and Tataram have applied these distinctions in the analysis of 60 interacting global trends noted by the Goals, Processes and Indicators of Development Project (1982). They argue:
"When dealing with the contemporary world, a basic step is to learn how to progress from a descriptive to an evaluative analysis, from what is directly perceived to what is scientifically understood, although such an understanding may sometimes surprise the intuitive perception.... Many such trends are organized in opposite pairs, but their contradictory nature is much more richer and perfidious than what these binary oppositions reflect."
Of special interest with respect to the subsequent entries is the manner in which they show that antinomic relations emerge when semiparadoxical and paradoxical pairs of trends are associated as *cyclic sequences of trends.* Although they do not explore this feature explicitly in their analysis of an influence-trends matrix, it highlights an essential feature of the dynamic associated with paradox. It is especially valuable, given the implications for comprehension, that they have related this cyclic process to development trends.
The difficulty with any such approach is that the very logic of the method employed disguises the full force of the paradox and of the hiatus it engenders in any univocal communication. It effectively prevents the insertion of the engendering elements into the same framework, unless they are denatured and converted into symbolic entities, as in the case of the Marcus initiative.

♦ KD2040 Beyond method

Comment The difficulty in taking the argument beyond that of the previous entries lies in the manner in which conventional notions of method are undermined beyond this point. Basically *acceptable methods are associated with particular domains* or groups of domains. Attempts to apply a given method to "all" domains are only possible if the method is used to pre-define many domains as "irrelevant". Methods as answers, or as aspects of an answer, are thus subject to limitations.

Such a conclusion is particularly unfortunate given the enthusiasm and hopefulness which is associated with advances in general systems and other frontier topics. For example, the kinds of syntheses produced by Erich Jantsch (1976, 1980) bring together much that appears relevant to comprehension of the breakthrough required into a more adequate approach.

Such initiatives do not however escape from the basic difficulty, namely the fundamentally unsatisfactory nature of such investigations as perceived from other domains. It is easy to understand that the more successful any such synthesis appears or claims to be, the more it will be felt to be an imposition and a constraint on initiatives by others in other domains, existing or emergent. *Success is a constraint on the development of others.*

Essentially the missing factor which makes such approaches of limited relevance is that they are *unable to internalize the nature of their relationship to opposing methods.* They are unable to handle disagreement explicitly, except through value judgements of "irrelevance". Nor are the supporters able to give any creative form to the irrational processes which then hold sway if the confrontation continues. It is within this shadowy area or blindspot that many of the most deplorable initiatives of humanity are born. The domains oppose each other governed by the same primitive territorial mind-set as was associated with the warring tribes and baronies of the past.

This situation has been explored in the light of Paul Feyerabend's treatise "Against Method" (1978), and of the concept of the *dialectic method* (much favoured by those who criticize the accumulation of capital). To be consistent, Feyerabend cannot of course advocate any new method, other than arguing for none or for a plurality of conflicting methods. He does however make a plea for human-scale methods which are not so abstract and complex as to be beyond the comprehension of most. With regard to the value of dialectics, the paper concluded:

"Despite the relevance of dialectics to the problem of disagreement, as noted above, it does not appear to do more than explain the dynamics of the environment it constitutes. It explains the eventual *future* evolution beyond the stage of disagreement, but does not clarify the nature of any possible *present* order *whilst the disagreement holds*. It does not clarify the nature of the psycho-social forms to which disagreement can give rise in the present, it merely affirms that they are necessarily temporary. The question is whether there is any pattern in the present to the ancillary processes to which a dialectical confrontation gives rise". (A J N Judge, 1984, p.17)

Because of its essentially transformative emphasis, dialectics offers little for an understanding of the relationship between *co-present answers*, other than to predict that through ongoing struggle an answer will emerge triumphant sometime in the future. "A" struggle is however explicity and creatively internalized, but not "the struggle with those in disagreement with the dialectical method itself.

Unfortunately it is in the present, with a variety of mutually opposed answers that people have to live. And it is in the present that the future is born. It is there that answers compete for resources and support. It would seem important therefore to look at the *"viable" patterns of disagreement* between such domains of significance in the present. In particular it is important to move beyond the limitation of dialectics to a set of only two opposing theses. A previous paper (A J N Judge, 1984) took a step in this direction by producing an ordered series of 210 mutually-incompatible (opposed), transformation-oriented statements (reprinted in Section CP) adapted from a variety of existing multi-set integrated concept schemes (A J N Judge, 1984). This was an effort to order *varieties of incomensurability*, which Feyerabend sees as vital to the process of development. This "order" or pattern is presumably an aspect of the "meta-answer" sought.

A related approach could be to produce a comprehensive *"bibliography of answers"*, if only to demonstrate the scope of the challenge. The fact that this has never been done shows how "biased" the individual answers must necessarily be, and how limited their information bases. In introducing their own position, having briefly reviewed others, Samir Amin at al (1982, p.7), state: "Nous rejetons toutes ces explications de la crise, même si chacune d'elles n'est pas sans fondements empirique et, à la limite, pourrait constituer un élement d'explication de la situation actuelle. Néanmoins, toutes ces appréciations nous paraissent jouer sur des variables aléatoires, qui ne relèvent pas d'une explication synthétique et cohérente de la crise, de ce qui l'a amenée, ou de ce sur quoi elle débouchera."

Needless to say, each of the other positions would generate equivalent statements. A bibliography of answers, if appropriately organized and annotated would at least provide a kind of checklist of what kinds of answers tended to be "invisible" from a particular domain. This should also give further understanding of the nature of the meta-answer.

There seems to be a peculiar kind of inconsistency concerning attitudes towards answers. With the Universal Declaration of Human Rights, the equal rights of individuals was affirmed as a fundamental proposition which governs much of the discourse in the world community. (Somehow society also accepts the fact that some people have more rights than others, due to their age, numbers, qualifications or other attributes). When it comes to the answers individuals may favour, however, very few are perceived by others as having a right to exist.

Although "stupid" and "intelligent" people, as well as children, all have equal rights, the answers favoured by such people do not. Every effort is made in intellectual debate to denigrate and suppress the "stupid answers" favoured by "stupid", "misguided" or "uneducated" people. But when the setting of the discussion is that of a community dialogue, or learning environments in general, an entirely different attitude is advocated. No answer is then denigrated. Each answer, however "stupid" by some standard, is recognized as a possible step or stage in a learning process. Such stages often have their historical parallels such that the past is rather effectively encoded into the range of views currently held in society.

This raises the question as to how far the world community is from recognizing that *every answer has a function*, especially insofar as it imposes constraints on the dominance of other answers, or constitutes a valuable developmental challenge to them. In the search for a meta-answer, it is impossible to avoid recognition of the fact that the number of people who will not be able to comprehend the emerging sophisticated insights into the world's condition is increasing at a very high rate. The "education gap" is increasing faster than any other developmental gap and cannot be treated as non-existent or on the verge of elimination. In this light, the percentage of people subscribing to answers that can be termed "wrong" by others is likely to very high (if it is not necessarily already 100 percent).

It is naive to expect that "wrong-thinking" can be eliminated from a developing, multi-generational world community (although such a view has a valid role to play). Somehow the required meta-answer must accord recognition to the psycho-social structures and processes corresponding both to different information bases and to different interpretations of them. The assumption that any view (including this one) is unquestionably "right" is a significant constraint on the development process (although as such it too has a role to play). In fact any exchange of information is part of a ceaseless effort to counteract "wrong thinking". It is difficult to imagine that information exchange would cease in an ideal society.

♦ KD2045 Constraints on a meta-answer

Comment To avoid creating the impression that this amounts to pluralist relativism, it is necessary to clarify some constraints which counteract such a condition before taking the argument a step further. Ranges of possible constraints have been explored in an earlier paper (A J N Judge, 1984). At this point it is appropriate to list the following:

(a) *Single, exclusive, universal claims*: Such claims are what the meta-answer must necessarily interrelate. A claim of this type defines itself as of a different type than that of a meta-answer.

(b) *Eclectic pluralism*: The meta-answer must necessarily be open to any perspective, but it is of little value if it does not achieve more than this.

(c) *Artificial agglomerations*: Grouping together answers within a framework of categories (e.g. a matrix or a thesaurus) may prove to be a valuable step towards a meta-answer, but the framework does not possess all the required characteristics of one.

(d) *Partial strategies*: Reduction of the range of factors to be considered may lead to valuable insights but it fails to respond to the basic challenge of interrelating the full range of answers.

(e) *Non-self-reflexive approaches*: Any approach to a meta-answer which is not faced with the paradox of the status of a meta-answer in relation to an answer avoids an essential dimension of the challenge.

In the earlier paper (A J N Judge, 1984) it was argued that statements about a meta-answer could best be formulated as an open-ended ordered series of mutually-incompatible, transformation-oriented propositions of which 210 were outlined in 20 sets (reprinted in Section CP). A measure of self-reflexiveness is built into them but is most evident in the earlier sets. The statements are formulated in sets based on the number of elements by which it is hoped to "contain" the description of the complexity of an adequate meta-answer. The first two sets, containing respectively one and two elements, are:

1. Inadequacy of formulations: No single formulation (including this one), nor any logically integrated set of formulations, adequately encompasses the nature of the development process. Every position or formulation is therefore suspect. When it is formulated within a domain of unquestioned consensus, this potential doubt is inactive, thus establishing a boundary of uncritical discourse which inhibits development.

2. Opposition/Disagreement: (a) New initiatives, including this one, are formulated by taking and establishing a particular position in opposition to whatever is conceived as potentially denying it. The nature of the initiative is partly determined by the way in which the challenge or initial absence of any opposing position is perceived and the possible nature of the response. It is the immediacy with which the challenge is perceived that empowers the initiative. (b) The taking of a position as a result of a new initiative engenders or activates a formulation which is its denial. Every formulation is therefore necessarily matched by an initiative which is incompatible with it, or opposed to it, or takes an essentially different direction from it. This opposition is fundamentally unmediated and as such cannot be observed or described. It can only be comprehended with one of the opposed positions.

The tentative titles used to indicate the qualitative characteristics of the other sets formulated are:

3. Dialectic synthesis
4. Development interaction
5. Constraints on existence
6. Coherence through renewal
7. Modes of change
8. Constraints on change
9. Implementation of a transformation process
10. Endurance of a form
11. Empowerment and importance of a form
12. Harmoniously transformative controlled relationship
13. Creative renewal
14. Cycle of development processes
15. Construction and development of form
16. Values and assumptions
17. Relationship potential of a form
18. Inadequate transformation attempts
19. Qualitative transformation
20. Significance of mutually constraining forms

In effect such sets attempt to clarify the kinds of significance domain perceptible under different conditions of observation whilst at the same time challenging the nature of the formulation and of the observation process. In a sense *the ordered sets establish the necessity of the fragmentation of answers into domains.*

♦ KD2060 Meta-answer patterning

Comment In moving beyond pluralistic relativism, what is required is some appropriate pattern whereby answer domains can be interrelated. The number-pattern of sets outlined in the previous entry is one approach to this.

Another approach is to develop a suitable *classification scheme for answer domains* which goes beyond the limitation of conventional matrix-type schemes (A J N Judge, 1984, pp.294) and the kinds of criticism to which classification is subject (Penser/Classer, 1982, A J N Judge, 1981). The feasibility of this has been explored in earlier papers (A J N Judge, 1981, A J N Judge, 1982) and is the basis for an ongoing experiment in the classification of the 15,000 international organizations listed in the *Yearbook of International Organizations* (UIA, 1990/1991). The intention is to highlight patterns of integrative relationships between international activities and problems in order to provide more coherent overviews of the world community of organizations in all its detailed variety.

In the terms of anthropologist Gregory Bateson (1979, pp.8-11), what is sought is an approach beyond pluralism to the *"pattern that connects"*. Erich Jantsch points out however that: "The cultural pluralism... which is about to replace the era of uniform, committing guiding images, may be interpreted as a suspension of historical time... More surprising to many comes the conceptual pluralism of modern science. The theory of relatively and quantum mechanics "function" inside domains of observations, but all attempts to mould them into a unified paradigm have failed so far." (1980, p.303)

But he then continues: "The pluralism of more recent concepts, especially in the physics of subnuclear particles, makes some physicists already speak of an *"ecology of models"* which cannot be fused to a unified model, not because we lack the necessary knowledge, but as a matter of principle". (1980, p. 303)

The properties of the required meta-answer lie in the nature of such an "ecology" which does have a special form of organization. Jantsch then makes the points: "Let us remember that the evolution of dissipative structures, too, can be described only by simultaneously employing two complementary models, a macroscopic-deterministic and a microscopic-stochastic one. And the co-evolution of macro- and microcosmos may only be grasped in the synopsis of complementary approaches. As a matter of principle, the autopoietic levels in a multilevel dynamic reality which have become separated by symmetry breaks cannot be united in a super-model, but only by way of describing the web of relations between them." (Erich Jantsch, 1980, p. 303)

He then cites Ilya Prigogine who investigated such dissipative structures: "The world is far too rich

to be expressed in a single language... Music does not exhaust itself in a sequence of styles. Equally, the essential aspects of our experience can never be condensed into a single description. We have to use many descriptions which are irreducible to each other, but which are connected by *precise rules of translation* (technically called 'transformation'). Scientific work consists of selective exploration and not of the discovery of a given reality. It consists of the choice of questions which have to be posed." (Erich Jantsch, 1980, p.303)
The challenge is to discover the pattern of such "transformation" and to avoid the traps of current satisfaction with only providing "micro-answers" to "micro-questions". Such answers are of course essential, but they are not enough at this time.
There does seem to be a special *existential challenge* to the relationship between the "pattern that connects" and acceptance of action in terms of a specific micro-answer. This challenge is associated with the "sacrifice" of generality of perspective in order to achieve concrete relevance and comprehensibility. Ironically the nature of this existential sacrifice has been explored in analyses of Rig Vedic philosophy as partly encoded in music and dance (Antonio de Nicolas, 1978, Ernest G McClain 1978). These analyses, which reflect Prigogine's above remarks on music, are explicitly linked to investigations of the significance of quantum theory for new understanding of *changing classificatory frameworks* (P A Heelan, 1974, C A Hooker, 1974, and see also Edgar Morin, 1981) and the network of links between such frameworks conceived as "languages". The relationship of these concerns to "integration" has been explored in an earlier paper (A J N Judge, 1984).
Another approach which provides valuable insights in delineating a meta-answer is that of Christopher Alexander (1977) on design and planning processes. Of special significance is his stress on the *democratization of any design process*, especially in a complex institutional setting. He clarifies the process of elaborating an open-ended *"pattern language"* consisting at the moment of some 250 sub-patterns (basis for part of Section CP). These can be combined in different ways by users to form their own unique languages. Clearly a similar approach could be used to elaborate a set of psycho-social patterns with which users could elaborate languages to design alternative institutions, communities and lifestyles.

♦ KD2065 Containing discontinuity through aesthetics
Comment A major strength of Alexander's pattern language is that it is a deliberate attempt to provide a means of giving form to that core quality which makes life meaningful and a delight to live. He very carefully shows how this must necessarily be "defined" as a *"quality without a name"* (1979). It is only partially expressed through each of the words bandied about in social policy-making discussions. In his view the quality can only be adequately captured or "contained" by use of a pattern language. There is obviously a case for applying this approach to contain the subtleties of human and social development. By seeking to give form to this core quality through a user-oriented language, Alexander effectively joins Attali (1981). Attali however introduces a vital additional element through his stress on the *management of contradictions and violence.* His three forms of truth correspond to three ways of ordering society:
"La première représentation (capitaliste) décrit l'économie comme une mécanique. Son objet est l'étude de la régulation... Une deuxième représentation (marxiste) regroupe les discours qui décrivent la société comme une production du travail des hommes" (1981,p.17)
The second corresponds to the argument here generalizing accumulation to non-material features of the "discours". Although Attali argues that Wallerstein's (1982, 1976) analysis is itself of more general significance than for the marxist paradigm within which he writes: "...il ouvre, au-delà de la régulation et de la production, à l'analyse la plus totale du processus économique, celle de l'organisation ouverte sur le monde naturel, l'écosysteme et la biosphère." (1981, p.154)
With respect to his third order, Attali then continues: "Mais le monde ne se résume ni à l'échange ni à la production, le sens ne se réduit ni au prix ni à la valeur des objets. Le monde engendre ses propres structures ailleurs que dans la seule production matérielle. Il faut aussi penser le monde comme organisation du sens. Et la crise comme rupture du sens dans l'organisation, qui naît des divergences dans l'ordre, des parasites de la communication, bruits du marche, voleurs de valeur, bruits du monde. L'ordre est alors gestion de la violence, la crise retour de la violence, selon une succession que l'histoire seule nous désigne." (1981,p.18).
The capitalist and marxist world orders are thus, according to Attali, each incapable of avoiding aspects of the organizational problem:
"Toute politique recommendée par le premier Monde aggrave les problèmes que dévoile le second, et réciproquement. L'un et l'autre n'en sont pas moins confrontés à une même double difficulté. D'une part, la circulation du sens (le prix ou le travail) peut être parasitée (par la monnaie ou la classe capitaliste). D'autre part, ce parasitage a surtout lieu quand il s'agit d'arbitrer entre ce qui doit servir à améliorer les moyens de produire et ce qui doit servir à améliorer les moyens de consommer." (1981, p.154).
Attali's argument converges on the importance of a *new understanding of language* as a way of containing and managing the violence inherent in the crisis of the development process. His main hypothesis regarding the third order is that language structures order and that people and objects are only valued in terms of their capacity to participate in the circulation of messages which give a meaning to social organizations. But the only meaning for any group lies in its survival, which is only threatened by violence. All its efforts are directed via language to avoid or eliminate such violence. (1981, p.60).
For Attali, the reality of the languages which effectively structure societies is much more complex than that of the first two orders. The route forward then lies through an *"aesthetic" approach* to the world. Everything produced then enters a process which, by circulation and the meaning given to what is produced, prevents the proliferation of violence, *transforming the production of violence into the production of meaning* (1981, p.168). This would give rise to a "non-violent polyorder" in which the struggle "ne passe plus ni par la force, ni par la raison, masques de la violence, mais par la seduction des formes, la subversion des objets" (1981, p.297). This view emerges from Attali's carefully documented study of the *significance of the exchange process* as the "circulation of the life" of a community (1981, p.179): "En resumé, pour qu'un Ordre existe, il faut y créer en permanence des différences. Produire, c'est augmenter les différences; consommer et échanger, c'est les annuler..." (1981, p.175-76).
Objects thus always remain the magical property of the producer, a living incarnation of his force and reality. Exchanging them suppresses this difference recreating violence. The *exchange is thus never equal*, or else there would be no interest in the exchange. The idea of balance or equivalence in exchange, as it is accepted in the first two worlds, effectively assumes the death of objects (1981, p.179). Any such *similarity creates violence which difference averts* and directs towards the exterior, polarized onto a suitable scapegoat (1981, p.16).
Although the first two "worlds" with their corresponding "orders" and "networks" (production of offer, production of demand), create the third (based on the exchange process), the difficulty is that such processes are not "containable" within any particular organization which could be designed by either of the first two orders in terms of their forms of "truth": "... les organisations n'ont ni fonctionnement universel ni utilité conflictuelle. D'une part, chacune à sa langue spécifique, malgré les universaux qu'elle contient. D'autre part, aucun ordre n'est réductible à son utilité pour un groupe". (1981, p.187).
Attali therefore advocates the elaboration of a theory (in effect a "meta-answer" in terms of his third form of truth, to *give a meaning to forms and discontinuities.* He sees this as: "un pari sur l'existence d'une adéquation entre la structure d'un esprit humain et celle du monde. Elle n'est, dès lors, vraie que si elle lui semble belle, si elle lui procure une jubilation intérieure par la perception de la potentialité infinie de toute oeuvre humaine. Elle est vraie comme l'est une oeuvre d'art, dont elle utilise d'ailleurs la métaphore: un Ordre est comme une écriture et une crise come une déchirure". (1981, p.187-188).
It is interesting that Attali has recently held a major post in the French government, because it is not clear how it might be possible to develop this position in practice. The same problem of determing what forms of organization would be appropriate for the future is left unresolved by the tantalizing images of Alvin Toffler's "Third Wave" (1980). Feyerabend (1978), as a methodological anarchist, also finds it unnecessary to envisage any new organizational form appropriate to the methodological anarchism he considers necessary to scientific advance. But such processes are unlikely to be appropriately engendered unless they are matched by complementary structures to "contain" them.

♦ KD2070 Observer entrapment and micro-macro complementarity
Comment The question of "containers" and "containment" calls for a better understanding of the function of the observer called upon to respond to the elements of any duality (for which he may also be conceptually responsible).
As Ilya Prigogine notes: "There is always the temptation to try to describe the physical world as if we were not part of it." (1976, 1980). This is even more true of the social world and for most researchers on human and social development. It corresponds to the classical Galilean view of science in which an attempt is made to see phenomena "from the outside as an object of analysis to which we do not belong. But we have reached the limit of this Galilean view". (1980). To progress further, we must have a better understanding of our description of the physical universe. (1980, p.xv).
This breakthrough in perspective was triggered by Einstein's work on relativity and the constraints on communication between observers within different frames of reference moving with respect to each other. By invoking the active role of the observer, the nature and limitations of measurement processes are clarified. For Prigogine "The incorporation of the limitation of our way of acting on nature has been an essential element of progress". (1980, p.214). It is somewhat extraordinary that no equivalent to relativity theory is available to remedy the flabby weaknesses of "relativistic" perspectives in the social sciences which justify the lack of attention accorded to them.
Introducing the active role of the observer enriched physical science with the concept of complementarity which, as with relativity, has no central role in the social sciences. Prigogine clarifies the concept by the musical analogy noted above: "... the world is richer than it is possible to express in any single language. Music is not exhausted by its successive stylizations form Bach to Schoenberg. Similarly, we cannot condense into a single description the various aspects of our experience." (1980, p.51).
In this sense particular descriptions ("answers") do not become wrong, even though each may be considered fundamental; rather they correspond to idealizations that extend beyond the conceptual possibilities of observations (1980, xviii). But as idealizations they each lack essential elements and cannot be studied in isolation (1980, p.212). This is equally true for the extremes of micro and macro description of human and social development.
The main thesis of Prigogine is associated with the constructive reality of irreversible processes which appear as particularly coherent on the biological level. Irreversibility emerges once the basic concepts of the extreme idealizations cease to be observables. It is inseparable from measurement. It corresponds to the embedding of the micro perspective within a vaster formalism which permits a non-reductionist transformation to coordinate various levels of description. Irreversibility is then a manifestation on a macroscopic scale of a "randomness" on a microscopic scale. It is then the modern theory of bifurcations and instabilities which provides a *bridge between the micro and macro levels of description*, as well as between the geometrical world of physical descriptions and the organized, functional world characteristic of biological and social systems. It is this bridge which is Prigogine's "third" perspective (1980, p. 56 and p. 196).

♦ KD2080 Non-equilibrium structures
Comment Prigogine obtained the Nobel Prize in 1977 for his investigation of non-equilibrium systems which, in the words of the Nobel Committee, "created theories to bridge the gap between biological and social scientific fields of enquiry" (1976, 1980). It is evident that the world system is far from being in an equilibrium state or even near it. As Holling notes, for example: "An equilibrium-centred view is essentially static and provides little insight into the transient behaviour of systems that are not near the equilibrium... The present concerns for pollution and endangered species are specific signals that the well-being of the world is not adequately described by concentrating on equilibria or conditions near them." (1976, p.173).
Unfortunately the tendency has been to focus on equilibrium research and, in the case of social systems, on visions of desirable societies formulated in equilibrium terms as "peaceful" utopian end states. This is only realistic when dealing with closed systems exchanging nothing with whatever can be described as an "external" environment. As Jantsch notes "Equilibrium is the equivalent of stagnation or death" (1978, p.10). In the more realistic case of open systems, it is the high degree of non-equilibrium due to the presence of such exchanges which can maintain self-organizing processes that give rise to "dissipative" (non-equilibrium) structures.

♦ KD2081 Order through fluctuation: dissipative structures
Comment Dissipative structures are associated with an entirely different ordering principle called "order through fluctuation". Such structures can in effect arise from the amplification of fluctuations resulting from instabilities which, in the case of the world system, for example, are perceived as the curse of orderly planetary policy-making and global programme management. Open systems in a state of sufficient non-equilibrium endeavour to maintain their capability for exchange with the environment by switching to a new dynamic regime whenever entropy production becomes stifled in the old regime.
Order may therefore increase, and the response to fluctuations is the less random the more degrees of freedom the system has (E Jantsch and C H Waddington, 1976). Fluctuations on a sufficiently small scale are always damped by the medium. Conversely, once a fluctuation attains a size beyond a critical dimension, it triggers an instability (E Jantsch and C H Waddington, 1976, p.119). There is no longer a consistent macroscopic description (Ilya Prigogine, 1980, p.141). In the formation of dissipative structures, *it is the fluctuations that drive the system to a new average macroscopic state with a different spatio-temporal structure.*
Instead of being simply a corrective element, the *fluctuations become the essential element in the dynamics of such systems* (E Jantsch and C H Waddington, 1976, p.93-96). Dissipative structures can therefore be considered as giant fluctuations whose evolution over time contains an essentially stochastic element (1976, p.93).
Fluctuations play this critical role in macroscopic systems in the neighbourhood of bifurcations where the system has to "choose" between alternatives (Ilya Prigogine, 1980, p.132). Given the situation of the world-system in the face of such alternatives, Prigogine's work merits careful attention.

♦ KD2082 Unexpected global relations
Comment Prigogine argues that dissipative structures present precisely the global aspect, the aspect of totality, which has been ascribed to the object of the synthetic sciences, including sociology (1976, p95) This macro view is important both in the temporal (historical) sense as well

as in the usual spatial (structural) sense As Jantsch notes:
"In a nonequilibrium world of self-realizing, self-balancing systems, process and structure become complementary aspects of the same overall order of process, or evolution. As interacting processes define temporary structures - comparable to standing wave patterns in physics - so structures define new processes, which in turn give rise to new temporary structures. Where process carries the momentum of energy unfoldment, structure permits the focusing and acting out of energy. Only a macro view is capable of providing a perspective of history, or evolution of space-time structures; our current microscopic paradigms (e.g., quantum mechanics) do not deal with space-time coincidences." (1982, p.39).
Dissipative structures are very sensitive to global and historical characteristics that influence in a decisive way the type of instabilities by which the structures are engendered. For example, the occurrence of dissipative structures generally requires that the system's size exceed some critical value - a complex function of the parameters describing the interaction-diffusion process. Far from equilibrium, therefore, an essentially unexpected relation exists between the dynamics and the space-time structure of such systems. Instabilities near the critical point involve *long-range order through which the system acts as a whole* in spite of the short-range character of the interactions (Ilya Prigogine, 1980, p.103-4). The distribution of interactants is no longer random (1980, p.132). *Chaos gives rise to order* (1980, p.142), a phenomenon explored by Atlan (128). The oscillation frequency now becomes a well-defined function of the state of the whole system. Any instability then develops over time the periodicity of the limit cycles of fluctuation (1980, p.99). The determining importance of these global and historical dimensions recalls the preconditions for an adequate world-system type analysis (I Wallerstein, 1976).

♦ KD2083 Complementarity of determinism and fluctuation

Comment In order to be able to take form from instability, a dissipative structure requires a non-linear mechanism to function. It is this mechanism which is responsible for the instability amplification mechanism of the fluctuation. *Dissipative structures thus form a bridge between function and structure* (E Jantsch and C H Waddington, 1976, p.95, Ilya Prigogine, 1980, p.100) as portrayed by the triad: function/space-time structure/fluctuations. Determinism and fluctuation then play a complementary role in any description. In Jantsch's words, as applied to social systems: "Process (or function) and structure, deterministic and stochastic features, necessity and chance (or free will), become complementary aspects in the self-organizing dynamics of "order through fluctuation' which may also be graphically depicted as a nonequilibrium system "stumbling forward" and crossing by its own force the ridges separating "valleys" of global stability". (1976, p.72)
Jantsch sees this essentially dualistic description as itself complemented by Rene Thom's topological model seen as constituting the first rigorously formulated monistic model of life. Here instabilities (catastrophes) are responsible for mutations such that the deterministic and finalistic aspects are understood as complementary links in a temporal feedback cycle. Causality and finality become expressions of a pure topological continuity of self-balancing processes, viewed from opposite directions (1975, p.41). In commenting on Abraham's (1960) application of Thom's work, Jantsch notes:
"A complementary approach, the theory of catastrophe..., focuses on the existence of multiple globally stable regimes (called macrons, and equivalent to dissipative structures) and the transitions (catastrophes) between them. Macrons are, at the present stage of the theory, represented by mathematical descriptions of their equilibrium state (attractors). Therefore, catastrophes appear as sudden quantum jumps, as if due to "pushes" by and outside force, comparable to a golf ball being propelled over a ridge by a single stroke. What is of central interest in this approach, is the landscape of new forms, the "epigenetic landscape", beyond the ridge." (Erich Jantsch, 1976, p.72).
Such a landscape is of special interest in perceiving the relationship between "answers". Each answer is then a macron (attractant) which determines the flow of attention. *Answer domains are thus separated by "ridges", which prevent the effective flow of information from (or to) neighbouring "valleys".* (The focus and preoccupation of other valleys is considered "irrelevant".) *Answers can then be usefully seen as distributed over the landscape such as to ensure the most economic distribution of attention*, or psycho-social tension, in a social system.

♦ KD2085 Organization self-renewal: autopoiesis

Comment Autopoiesis or organizational self-renewal
The basic conditions for the dynamic existence of non-equilibrium structures are therefore:
- partial openness toward the environment
- a macroscopic system state far from equilibrium
- autocatalytic self-reinforcement of certain steps in the process chain.

The dynamics of such a globally stable, but never resting structure, has been called "autopoiesis" (Archie J Bahm, 1977), namely self-renewal regulated in such a way that the integrity of the structure is maintained. It is typical of biological and social organization. According to Jantsch: "Autopoiesis is an expression of the fundamental complementarity of structure and function, that flexibility and plasticity due to dynamic relations, through which self-organization becomes possible." (1980, p.10). According to Zeleny and Pierre:
"Autopoietic organization can be defined as a network of interrelated component-producing processes such that the components, through their interaction, generate recursively the same network of processes which produced them and thus realize the network of processes as an identifiable unity in the space in which the components exist. The product of an autopoietic system is necessarily always the system itself, its organization being continuously realized under permanent turnover of matter and energy". (1980, p.150)
Whereas: "Allopoietic organization, in contrast, can be defined as a network of interrelated component-producing processes such that it does not produce the components and processes which realize it as a unity" (E Jantsch and C H Waddington, 1976, p.150).
It would seem that there is much to be learnt from this perspective with reference to human and social development. In Jantsch's words again: "...human systems with all their tangible and intangible aspects might then perhaps be regarded as dissipative structures, arising from the interaction of strong and highly nonequilibrium flows of ideas and actions. Their spatial organization would then be the result of processes of self-organization, or in other words the forms of periodicity built into human systems. This organization would be physical as well as psychic. Indeed, the borderline between both becomes blurred in the light of the emerging insight that information itself may have a self-organizing capacity, that a seed of information may engender more information and thus more order". (1975, p.60)

♦ KD2090 Opening and closing: alternation for discontinuous learning

Comment The self-renewal of the autopoietic system is achieved, in the words of Zeleny and Pierre, "through a series of oscillations between rupture and closure. Its very existence as an autopoietic system is based on this rhythmical opening and closing.... We might preferably talk of *pulsating systems*, since neither permanently closed nor permanently open systems are autopoietic; they are not "alive"." (E Jantsch and C H Waddington, 1976, p.153).
The theory of opening and closing in relation to social systems has been explored by Orrin Klapp (1978), to whom Zeleny and Pierre refer. Klapp argues that opening to variety, whether for learning, progress, evolution, or control, has been over-emphasized to the point of bias. Because of this modern society has wandered into a crisis of social noise and *failure of resonance*, thus impairing communication and making it harder to find meaning. He interprets a variety of psycho-social phenomena according to the theory that individuals and societies normally open *and* close to information and communication. *Opening is scanning for desired information and the new, whereas closing is a natural response to too much unusable information, broadly conceived as social noise. Opening and closing are therefore part of a shifting optimization strategy of living systems* to get the most of the best information and the least of the worst noise:
"From such things, we see that what we call aliveness - resilience, adaptability - is not continual intake, nor any constant policy, but sensitive alternation of openness and closure. The mind listens alertly, then turns off to signals. The natural pattern is alternation, and the more alive a system is, the more alertly it opens and closes. In such a view, closing is not, as some suppose, merely a setback to growth and progress, but evidence that the mechanisms of life are working, that the society has resiliency... A perpetually open society would suffer the fate of a perpetually open clam." (1978, p.15).
Openness or closedness is not a fixed policy or structural characteristic but a changing life strategy of organisms and groups. Communication fluctuates in cycles requiring sensitive alternation (1978, p.16). The "conventional" bias in favour of openness, the more information the better, has been reinforced in recent years by liberal theorists pleading for open-mindedness (Rokeach, 130) or an open society (Popper, 40). More recently this view underlies the Club of Rome report on "No Limits to Learning". The view is challenged by the extensive evidence information overload on the one hand, and by what Klapp calls "spasms of closing among ethnic, religious, and other groups", from which academic and other specialization should not be dissociated. "When a lot of people feel too much entropy as a crisis to collective identity, they close to protect the net, exclude noise, intensify signals affirming common values, and perhaps define more clearly an enemy." (1978, p.16).
Klapp argues that if all societies are naturally subject to cycles of openness and closure, some revisions in current assumptions about progress and the "free market of information" may be necessary. He asks whether it is possible to get too much of a good thing, when system theorists recognize that unlimited increase of anything good is not better, and no living system takes an unlimited input of anything.
Is information exempt from this, or is it also subject to overloads and entropic effects comparable with overproduction in economic markets and polluting side effects of growth ? If closing is as necessary to human systems as opening then they should be placed on a par, rather than being presented as "bad" and "good" respectively.
In amending the open society model, Klapp cautions against a purely mechanical interpretation of the advocated oscillation between relative openness and closedness. There are inherent risks in either strategy: "Scanning for news, discovery, or growth runs the risk of excessive noise and other costs of bad opening; closing for redundancy, memory reinforcement, or cohesion risks narrowness, ignorance, and stifling banality". (1978, p.20).
He therefore distinguishes moves in an information game corresponding to bad opening, good opening, bad closing, and good closing (1978, p.19). Furthermore, what are conventionally called "open" societies close in different ways from "closed" societies, and at different points on a range, one end of which might be an authoritarian system allergic to small increases of information, and the other an ideal liberal society with a progress ideology emphasizing the modern and devaluing the old - hence vulnerable to crisis from information overload and loss of redundancy. (1978, p.12).
These considerations suggest that *the "container" for answers should embody characteristics which ensure that it "opens" and "closes" in some way* - not an unusual requirement for containers which are to be of practical value.

♦ KD2095 Third-perspective "container" alternation

Comment Prigogine (1980), Jantsch (1980), Attali (1981) and, in effect, Feyerabend (conclude 1978) that it is necessarily impossible, if not anti-developmental, to define an organized, rational structure to bridge across discontinuity. The only "solution" being to adapt more spontaneously or aesthetically to the processes in relation to discontinuity. In effect what is being said is that, even in mathematical terms, *it is impossible to discover a space whose form (a "meta-answer") validates every argument ("answer")*. In Bateson's terms: "The question is onto what surface shall a theory of aesthetics be mapped.... a map of the region where angels fear to tread" (1979, p.210-214). But even if such a form could be discovered, it would presumably be too abstract to be of any value in society.
The difficulty is one of handling essentially incompatible answers which cannot co-exist passively (e.g. "science" and "religion"; "industry" and "environment"). In order to be hospitable to the discontinuities they represent, it would be necessary to somehow encompass or "contain" the *non-rational character of the disagreement* between them (A J N Judge, 1984). This implies a distinctly non-linear relationship between them. The most accessible indication of the possible nature of such a relationships is that between *right- and left-hemisphere modes of thinking* (Erich Jantsch, 1975), and the essential difficulty of integrating the perceptions to which they give rise. The functional "solution" used in daily life by all is an *oscillation between the two modes* according to the task to be performed. Integration, namely the meta-answer, is here represented by the *pattern of oscillation* between the distinct modes.
The question is whether this is relevant to the wide range of answer domains and the modes of action/perception they represent. In an earlier paper (A J N Judge, 1984), it was argued that this was at least a fruitful area of exploration. In another (UIA, 1990/1991), it was used as a basis for an experimental ordering of the range of preoccupations of international organizations in a "chequerboard" matrix classification scheme based on right and left-hemisphere modes. Such a classification scheme (criticized below) is a minimal pattern of interrelationship (namely a "container") between answer domains, reflecting the discontinuities between them. This suggests, as stated there, that the present pursuit of "alternative models" may be proceeding in an unfruitful direction. The point is not simply to discover some magical alternative model of value to development but of limited appeal. It is rather to discover *"models of alternation" (or oscillation) to contain the development process in relation to different alternatives which may be periodically adopted.* Institutions could useful consider the value of an *alternation policy* (e.g. centralization/decentralization), rather than having it forcefully imposed upon them periodically by their environment or pursuing a schizophrenic policy using departments with alternative approaches which are impossible to reconcile. Each alternative becomes a boundary condition. The challenge is to use *a configuration of such alternative models* in such a way as to constitute a "container" for the development process.

♦ KD2100 Revolutionary cycles of alternation

Comment The fact that social conditions are very much subject to cycles (e.g. Kondratieff), and that policy "breakthroughs" (such as centralization or decentralization) are periodically rediscovered with enthusiasm, suggests that alternation should be explored as a cyclic phenomenon. In fact, as any physical model will illustrate, *a pattern of oscillation is not stable unless it is accompanied by some form of revolution* of which the observed alternation is often a consequence (e.g. night/day on the revolving Earth; seasons on the Earth revolving around the Sun). Control of the "revolutionary process" is absolutely basic to the generation and use of electrical and other forms of energy (e.g. generators and motors).
Cyclic processes are also characteristic of many biological phenomena (e.g respiration, reproduction, metabolic cycles). They are also evident in many socio-cultural phenomena (R Buckminster,

1975, 1979), not to mention various symbolic and mythological cycles. But in the non-physical cases, humanity has gained *little effective control of the revolutionary process.* In fact the significance of social "revolution" is limited to the superficialities of discontinuity which are thus reinforced.

It has proved difficult to give operational content of any value to the non-disruptive dimensions of the "permanent revolution" advocated by marxists. This may well be due to the fact that it does not involve cyclic alternation between incompatible modes, because of the emphasis on an essentially linear series of dialectically superseded modes. Such linearity is a Western cultural concept of change which is not married to the valuable Eastern insight into change as recurrence. In metaphorical terms the marxist stress is on the struggle of abandoning "winter now" for "spring tomorrow" without any additional recognition of the revolutionary process whereby a new "winter" (with similar characteristics) will necessarily be encountered (or brought about) or of the importance of the ecological role of "winter" in a cycle. If there is any significance to the importance of the revolution-based design of generators for the industrial revolution, there should be some insights of relevance to the current problem of designing a meta-answer to the present socio-cultural condition. In fact, in Attali's terms (1981), such physical designs may well have prefigured the socio-cultural design problem of the present.

Designs appropriate to Attali's "third world" or Toffler's "third wave" could seemingly only emerge following recognition of the validity of a "third perspective". Integrated comprehension of the revolutionary cycle is only possible through a conceptual relationship to the axis that stabilizes perception of the cyclic processes with reference to it. Without this third perspective, the revolutionary cycle can only be confusedly comprehended as a linear process in one or two dimensions. It is in terms of this third dimension that the required meta-answer designs may well be possible.

Humanity does not function in terms of one mode alone, just as it is difficult to hop (or limp) forward on one foot - although this may well be what history will see as characteristic of this period. The "struggle" between two feet is avoided by a third "walking" perspective. Switching metaphors, it is though the vehicle conveying humanity forward that the spokesmen of antagonistic groups struggling on the driver's seat for control of the steering wheel. Those closest to the left-hand window (and the abyss on that side) shout "turn right", and those on the right-hand side (seeing the abyss there) shout "turn left". Luckily the vehicle has so far remained on the road because their over-corrections counter-balance each other. A more balanced third perspective is required to allow the vehicle to follow a road with both left and right-hand curves and abysses on either side (not to mention on-coming traffic).

♦ KD2105 Trialectics: a logic of the whole

Comment Oscar Ichazo, founder of an internationally-active human development institute, formulates an interesting progression in the interpretation of the basic laws of logic. These are in sympathy with the arguments of the previous sections. The points he makes (1982) can best be summarized by a equivalent three-phase sequences:

Aristotle:
1. Law of identity A = A
2. Law of contradiction A = B
3. Law of excluded middle A = A B

Dialectics: Logic of change
1. Quantity transformed into qualitative change
2. Everything works in opposition to something else (law of opposition).
3. Every process must deny itself if it is to continue; negation of negation.

Trialectics: Logic of the whole
1. Developmental stages: pre-established points at which step change occurs (law of mutation)
2. Everything is the seed of its contrary, both elements contributing to each other in a cycle (law of circulation)
3. Everything attracted to either expansion or contraction (law of resonance).

Ichazo sees the dialectical reinterpretation of the Aristotelian version as having been the basic logic of the industrial revolution. He gives arguments from physics and biology for the need for the trialectic interpretation to explain developmental changes of a whole. The cyclic "resolution" of contradictions corresponds to the points of the previous entries. The question of resonance is discussed in later entries.

♦ KD2110 Threshold of comprehensibility: a fourfold minimal system

Comment In the preceding entries it has been sufficient to present the argument in terms of learning "cycles" within an alternation process. But such cycles are rather abstract concepts. They may constitute good descriptive "geometry", but the challenge is to find additional features whereby the abstract geometry is geared or anchored into the complexities of perceived reality. Additional *design constraints are required to relate any such cycle to its environment* and prevent it spinning out of control or losing its integrity. This question can be examined in very different ways, each of which, as a "language", throws a different light on the relationships and significance of the dimension required to structure a minimally comprehensible system of adequate complexity. For this reason the arguments of several authors are presented in the following entries.

♦ KD2120 Omnitriangulation: interlocking cycles

Comment The interrelationships of circles has been extensively studied by Buckminster Fuller (1975, 1979), an architect, as the basis for a model of the non-transient existence of energy and material systems. He makes the point that: "Not until we have three noncommonly polarized, great-circle bands providing omnitriangulation as in a spherical octahedron, do we have the great circles acting structurally to self-interstabilize their respective spherical positionings by finitely intertriangulating fixed points less than 180 degrees apart..." (1975, I, 706.20). Furthermore, the more minutely the "sphere" so delineated is subtriangulated by other great circles, the lesser the local structural-energy requirements and the greater the effectiveness of the integrity resulting from such mutual interpositioning. This interlocking is then spontaneously self-stabilizing (1975, I, 706.22).

Assuming the circular representation of cycles, Fuller is in effect saying that it takes at least three interweaving cycles before there is interaction (entrainment) of a type to stabilize the abstract processes within a minimal non-abstract form which their interlocking brings about, in this case a sphere. With less than three, the form can exist only as a transient phenomenon, if at all. In his terms, *three cycles is the condition for a minimal system.*

But whilst three such cycles can interlock to engender a system, the system can only become comprehensible if a fourth cycle (corresponding to the processes of the observer's involvement in a comprehended system) is added. With less than four, the system may be identified with, opposed, proposed, or participated in, but it can only be partially contained within any communication. Its totality is only apparent as a succession of experiences in time. The unity of a minimal system as a whole only emerges in terms of a minimum of four event foci (1975, 400.08). In Fuller's terms "Systems are aggregates of four or more critically contiguous relevant events..." (1975, I, 400.26). All conceptually thinkable experiencings are fourfoldedly characterized (1979, II, 1072. 22). This is the basis for the "the minimal thinkable set that would subdivide Universe and have interconnectedness where it comes back upon itself" (1975, I, 620.03) and is differentiated from its environment (1975, I, 400.05).

As is clarified below, this suggests that not even a conceptual process involving the three classic processes of the dialectic can render any kind of meta-answer comprehensible. It is no wonder that unitary or dualistic answers are insufficient, even though they may be necessary as part of a larger scheme.

These considerations cause Fuller to distinguish four interwoven processes which relate to the learning perspective. "Life consists of alternate observing and articulating interspersed with variable-recall rates of "retrieved observations" and variable rates of their reconsideration to the degrees of understandability". These four are therefore: observation (or recall), (re) consideration, understanding, and articulation. (1975, 513. 06-07).

♦ KD2130 Quaternary consciousness: number and time

Comment The concern with alternation cycles arises because of a collective need to obtain a more conscious awareness of integration in a developing world system. It is therefore appropriate to take account of the insights of psychoanalysis. Marie-Louise von Franz, in pursuing the work of C G Jung and linking it to modern physics, makes points which bear a strong relationship to the distinctions made by Fuller. She presents material indicating the fundamental role which number plays in ordering both the psyche and matter.

"Taken as rhythm or dynamism, three thus introduces a directional element into the oscillatory rhythm of two, whereby spatial and temporal parameters can be formed. This step involves the interference of an observing consciousness, which inserts a symmetrical axis into the two-rhythm, or else "counts" the latter's temporal and spatial succession. In terms of content the number three therefore serves as the symbol of a dynamic process... three signifies a unity which dynamically engenders self-expanding linear irreversible processes in matter and in our consciousness (e.g. discursive thought)" (1974, ,pp. 103-106).

Her remarks, citing Jung, clarify further the *limitations of single-answer or dualistic thinking*: "... at the level of one, man still naively participates in his surroundings in a state of uncritical consciousness, submitting to things as they are. At the level of two, on the other hand, a dualistic world...images gives rise to tension, doubt, and criticism of...life, nature, and oneself. The condition of three comparison denotes insight, the rise of consciousness, and the rediscovery of unity at a higher level...But no final goal is reached, for "trinitarian" thinking lacks a further dimension; it is flat, intellectual, and consequently encourages intolerant and absolute declarations." (1974, p.124-125).

This suggests again that, despite the necessity of answers formulated in such modes, they are not sufficient at this time. The difficult step across the "incommensurability" between threefold thinking is effectively a progression from the infinitely conceivable to finite reality "based on the inclusion (no longer avoidable) of the observer in his wholeness within the framework of his processes of understanding" (1974, p.122). Citing both myths and sets of physical constants von Franz notes: *"The fact that mankind's repeated attempts to establish an orientation toward wholeness possess a quaternary structure appears to correspond to an archetypal psychic structural predisposition in man"* (1974, p.115).

For von Franz, *a fourfold approach appears "to constitute the fundamental minimum means for subdividing and thus classifying the circle or wholeness"* (1974, p. 121). "Two pairs of opposites, a quaternion, are required to set up a bodily unity" (1974, p.127). Below four the perception of wholeness is partly unconscious. As soon as the unconscious content enters the sphere of consciousness it has already split into four basic modes of awareness. "It is *perceived* as something that exists (sensation); it is *recognized* as this and *distinguished* from that (thinking); it is *evaluated* as pleasant or unpleasent (feeling); and, finally, *intuition* tells us where it came from and where it is going" (1974, p.121). As a minimum condition, if they are not incorporated into an "integrated" approach, they must necessarily be projected onto competing approaches in the environment, with all the intellectual and institutional consequences for any harmonious integration. Such *a fourfold approach is a necessary requirement for comprehending any "meta-answer".*

The significance of a quaternary attitude is evident, whether for any human and social development programme or arising from it: "Instead of proclaiming absolute dogmas, a "quaternary" attitude of mind then develops which, more modestly, seeks to describe reality in a manner that will - if it is based on archetypal concepts - be understandable to others. One remains simultaneously aware of the fact that assumptions of the unconscious do indeed reflect outer or inner reality, but also that they are transformed, through their passage into consciousness, into constricted, time-bound language." (1974, p.26)

The step to a fourfold approach to the world problematique was beyond the impotence of mental processes revolving about "intellectual theorizations" into those which partake of the creative adventure of "realizations in the act of becoming" (1974, p.131). Von Franz cites Ferdinand Gonseth's advocacy of a quaternary outlook which would no longer involve "the summary and brutal coercion of one variant over another, but the play of identifications and differentiation, agreements and complements, limitations and expansions, a game which can lead to dialectical synthesis, built up in four rhythms." (1955, p.583).

♦ KD2140 Logos and lemma for interparadigmatic dialogue

Comment In discussing the conditions for inter-paradigmatic dialogue, especially in he social sciences, Kinhide Mushakoji, Vice-Rector of the United Nations University, argues for the need to move beyond the accepted limits of formal logic: "Inter-paradigmatic dialogue - not only in natural but also in social sciences - should be concerned not with the determination of who is right or wrong in defining a concept one way or the other. It should rather concern itself with the question of what part of the natural or social realities are best approached by one or the other position. Two formally contradictory definitions of the (natural or social) realities may be both relevant and complementary in shedding light on different aspects of the same social realities. This is why the logic of inter-paradigmatic dialogue cannot be bound by the laws of Aristotelian formal logic: identity, contradiction, and excluded middle". (1978, p.19).

He also draws attention to the problem of "binary" approaches and the need for a "third pole": "By the very nature of scientific logic which is binary, intellectuals tend to form bi-polar structures with two opposed camps rallied under two paradigmatic banners. The polarization often takes place even within each of the two poles which then divide themselves into two sub-poles, and so on...An inter-paradigmatic process should be able to break the bi-polarity of the intellectual community by introducing a third pole in the dialogical process... The role of such a pole is to introduce extra paradigmatic considerations (into the discussion) and to break the dichotomic argumentation bringing into the discussion innovative ideas." (1978, p.15-16)

Mushakoji sees such a pole as "a basic condition of a successful scientific revolution". Without it the "opposed schools of thought send their best champions for a scholastic exercise...leading to nothing else but a reaffirmation of one's paradigmatic superiority over the others" (1978, p.18).

But Mushakoji then draws attention to the "logico-real" problem of the relationship between the logical and the reality levels. He suggests that catastrophe theory can help to shed light on the different logical positions in the morphogenetical space by relating the continuous reality (i.e. "signifié") to the discrete set of concepts (i.e. "signifiant"). This leads him to advocate a *fourfold non formal logic model to provide a logical basis for inter-paradigmatic dialogues.* Such a logic emerges from the work of Tokuryn Yamauchi who interrelates oriental thinking based on "lemma" with occidental thinking based on "logos". Lemma concerns itself with the modalities according to which the human mind grasps reality, rather than how human intellect reasons about it. Mushakoji sees the lemmic approach as offering a breakthrough in response to the static ontology of the West.

The tetralemmic model which has been developed in oriental logic stipulates the existence of four lemmas: (a) *affirmation* (b) *negation* (c) *non–affirmation and non–negation* (d) *affirmation and negation* (p.21) Here (a) and (b) both belong to formal logic, whereas (c) and (d) are unacceptable to it, although they are acceptable in theoretical physics. "Only an acceptance of the third and fourth lemmas can allow a full representation of the contemporary world problematique in its totality since contemporary world reality is full of cases where a mere affirmation or negation does not make sense". (p.21–22).

It is unfortunate that Mushakoji has limited his concern here to representing or grasping reality for the purposes of revolution in thinking. This does not respond to the problem of how to intervene in that reality on the basis of any such revolution – a vital preoccupation in furthering human and social development. And yet the four lemmas lend themselves to such an action–oriented interpretation as the basis for a more general "action logic":

(a) *affirmative action*, including support, commitment, initiative, proposition, cooperation, consensus formation, empowering, "opening";

(b) *negative action*, including sanction, withdrawal (of support), denial, disassociation, delimitation, criticism, opposition, promotion of dissent, disempowering, "closing";

(c) *non–affirmative and non–negative action*, including indifference, indecision, non–action (in the oriental sense), "neither confirm nor deny", "opening and closing";

(d) *affirmative and negative action*, including ambiguous action, non–violent resistance, "dumb insolence", "giving with one hand and taking with the other", "double dealing", "stick and carrot tactics", the "yes but no" response of the frustrated cross–examinee.

The conventional western–based logic of international actions uses modes (a) and (b) consciously, although some groups promote strategies based on one or the other only. For example, those in favour of "positive thinking" claim not to use (b), despite the positive value of closure as discussed in a earlier entry. Whereas those who fear "contamination" by a system gone wrong claim not to use (a).

The strength of the tetralemmic perspective is that it draws attention to the complementary role of the two other modes (c) and (d), which are outside the framework of action explicitly (consciously) accepted by the international community, although they are are evident in its interstices. The (a) and (b) modes are embodied in formal agreements and procedures and are the focus of academic study of international action. The existence of other modes can only be publicly "recognized" as scandalous illegality meriting no serious attention, except as the spice of informal discussion. The (c) and (d) modes are the tools of wily, world–wise actors, as well as of those they are trying to manoeuvre, both being aware that there are degrees of freedom of action which the (a) and (b) modes are unable to reveal. In contrast to the "cut and dried", overt (a) and (b) modes are unable to reveal. In contrast to the "cut and dried", overt (a) and (b) modes, in the essentially covert (c) and (d) modes what is not done is as significant as what is. Many valuable illustrations of the importance of the (c) and (d) modes are given in Douglas Hofstadter's justly acclaimed *Gödel, Escher, Bach* (1979). The authors cited provide conceptual, visual and auditory indications of the opportunities for transcending the limitations of the (a) and (b) modes.

Most of the examples given suggest the questionable value of the (c) and (d) modes because until recently they have been largely embedded in the collective unconscious at least for the Western mind. These are the kafkaesque worlds of double dealing ("crime"), influence ("old boy networks"), double standards ("hypocritical leadership"), and collective resistance ("bureaucratic stonewalling"). Other possibilities are however suggested by the oriental approach to action, by their extensive literature on non–action, and by the recent innovative use of "non–violent" strategies. All the modes are significant for development, as well as being vulnerable to misuse.

♦ KD2145 Epistemological mindscapes

Comment In a remarkable series of articles, Magoroh Maruyama has studied patterns of cognition, perception, conceptualization, design, planning and decision processes (1974, 1977, 1978, 1980). His central concern is the role of epistemological types, especially as they affect cross–disciplinary, cross–professional, cross–paradigm and cross–cultural communications. In contrasting his own work with that of previous research in this area, he distinguishes two traditional approaches: the psychological and psychoanalytical bases of individual differences in patterns of cognition, and the cultural and social differences as determined by sociologists and anthropologists.

Maruyama notes the various terms that have been used to describe such patterns, none of which has proved satisfactory: models, logics, paradigms, epistemologies. To these might be added Kenneth Boulding's "image" (1956). In Maruyama's more recent work he favours "mindscapes". This is more attractive term than "answer (domain)" as used here, although it lacks the active connotation of responding to a need. He provides a very valuable summary of these different exercises in "paradigmatology" and their relation to social organization.

Although he no longer favours the term, he defined paradigmatology as the "science of structures of reasoning" whether between disciplines, professions, cultures or individuals). He notes that the "problem of communication between different structures of reasoning had not been raised until recently", since scholars tended either to advocate their own approach or describe that of others. Contributing to this neglect is the fact that the choice between logics is based on factors which are beyond and independent of any logic.

Although he carefully emphasizes that there are many possible mindscapes or paradigms, Maruyama argues that "for practical purposes" it is useful to distinguish four main types. He stresses that these are not meant to be either mutually exclusive nor exhaustive and warns that any attempt at separating them into non–overlapping categories "is itself a victim of a paradigm which assumes that the universe consists of non–overlapping categories". What is intriguing is that over the years he has continued to struggle with the same attributes, grouping them first into three types, extended to four, then to five and now seemingly stabilized at four again.

The four types are:

(a) *H–mindscape* (homogenistic, hierarchical, classificational): Parts are subordinated to the whole, with subcategories neatly grouped into supercategories. The strongest, or the majority, dominate at the expense of the weak or of any minorities. Belief in existence of the one truth applicable to all (e.g. whether values, policies, problems, priorities, etc.). Logic is deductive and axiomatic demanding sequential reasoning. Cause–effect relations may be deterministic or probabilistic.

(b) *I–mindscape* (heterogenistic, individualistic, random): Only individuals are real, even when aggregated into society. Emphasis on self–sufficiency, independence and individual values. Design favours the random, the capricious and the unexpected. Scheduling and planning are to be avoided. Non–random events are improbable. Each question has its own answer; there are no universal principles.

(c) *S–mindscape* (heterogenistic, interactive, homeostatic): Society consists of heterogeneous individuals who interact non–hierarchically to mutual advantage. Mutual dependency. Differences are desirable and contribute to the harmony of the whole. Maintenance of the natural equilibrium. Values are interrelated and cannot be rank–ordered. Avoidance of repetition. Causal loops. Categories not mutually exclusive. Objectivity is less useful than "cross–subjectivity" or multiple viewpoints. Meaning is context dependent.

(d) *G–mindscape* (heterogenistic, interactive, morphogenetic): Heterogeneous individuals interact non–hierarchically for mutual benefit, generating new patterns and harmony. Nature is continually changing requiring allowance for change. Values interact to generate new values and meanings. Values of deliberate (anticipatory) incompleteness. Causal loops. Multiple evolving meanings.

The above descriptions are brief summaries of extensive listings of characteristics in relation to overall social philosophy, ethics, decision–making, design, social activity, perception of environment, human values, choice of alternatives, religion, causality, logic, knowledge, and cosmology. Of special interest, in the light of Attali's "seductive" concern, are the implications for aesthetic principles favoured. Maruyama considers that the influence of such "pure" types predominates in certain cultures, although in practice the types are quite mixed. Thus the H–type predominates in European, Hindu and Islamic cultures. The I–type develops in certain individuals, such as those of existentialist philosophy. The S–type is characteristic of Chinese, Hopi, and Balinese cultures. The G–type predominates in the African Mandenka culture, for example. H, S. and G characteristics can be distinguished in different streams of Japanese culture.

Maruyama has recently compared his four types with an extensive survey of epistemological data grouped by O J Harvey into four "systems" (1966).

System I: (a) High absolutism, closedness of beliefs, high evaluativeness, high positive dependence on representatives of institutional authority, high identification with social roles and status position, high conventionality, high ethnocentrism.

System II: (b) Deep feelings of uncertainty, distrust of authority, rejection of socially approved guidelines to action accompanied by lack of alternative referents, psychological vacuum, rebellion against social prescriptions, avoidance of dependency on God and tradition.

System III: (c) Manipulation of people through dependency upon them, fairly high skills in effecting desired outcomes in his world through the techniques of having others do it for him, some autonomous internal standards especially in social sphere, some positive ties to the prevailing social norms.

System IV: (d) High perceived self–worth despite momentary frustrations and deviation from the normative, highly differentiated and integrated cognitive structure, flexible, creative and relative in thought and action, internal standards that are independent of external criteria, in some cases coinciding with social definitions and in other cases not.

The two authors find that they agree on three types and differ on the nature of the fourth (which Jungian's would presumably consider as corresponding to a partially "repressed function" they have in common). It is much to be regretted that such surveys have not explored the epistemologies in "developing" countries to a greater degree, nor the extent to which different epistemologies are co–present in the same culture, group, individual or life–cycle.

♦ KD2146 De–categorization and poly–ocular vision

Comment Maruyama uses his approach to clarify the essential weaknesses of the interdisciplinary, holistic programmes associated with generalists and their education:

"The concept 'interdisciplinarity' presupposes that there are first disciplines which have to be put together later... Such 'interdisciplinary' programs, 'holistic' views and production of 'generalists' are all patchwork which perpetuates and aggravates the inadequacies of the classificational thinking. What we need, instead, is non–disciplinary programs, de–categorization of science and trans–specialization. Trans–specialization consists in maintaining a contextual view while focusing on specifics and details."

But in analysing such inadequacies of classificational thinking, Maruyama seemingly fails to recognize that they necessarily arise from over–reactions to inadequacies in non–classificational thinking (to which he does not accord any attention). Any pre–logical tabulation of this kind must however necessarily reveal the sympathies/antipathies of the formulator for particular types therein. Is it then useful to ask, for example, how "valid" is Maruyama's seeming over–reaction to the (excessive) dominance in world society which is engendered by the dominant (homogenistic) mode ? To clarify the need for "trans–specialization", Maruyama distinguishes between:

(a) The study of paradigms engendered by different types of logic which are chosen in terms of extra–logical factors. Such work has been done in anthropology, history, psychology and psychiatry, physics and biology.

(b) The study of cross–paradigmatic communication. Work on this has focused either on inter–cultural communication or on the problem of researching other cultures. More recently this has been related to communication in community development situations.

(c) The study of the trans–paradigmatic process, namely of the process whereby new paradigms are created. Little work has been done on this.

Of this last process Maruyama says: "Perhaps there cannot be such a methodology: a methodology, once established, would limit the type of paradigms that it can generate." This is the essential dilemma in bridging discontinuity. His approach to this "methodology" is given in a subsequent paper on ways of increasing heterogeneity and symbiotization as a basis for epistemological restructuring. In contrast to causal (homogenistic) or random (heterogenistic) paradigms he notes that no adequate mathematical formulation has been provided for mutual causal heterogeneity. In a later paper he concludes that although "mindscapes are learned rather than innate", they are mostly formed in childhood and it seems extremely difficult to change them later in life.

Perhaps the widespread disaffection with existing models is evidence to the contrary, especially where it results in "alternatives" being adopted. And, as argued here, may be it is not so much a question of "changing" existing modes as of being able to "alternate" into and out of them whenever appropriate. The problem is how to "formulate" the nature of alternation in order to make this trans–paradigmatic process credible in practice.

In arguing for a heterogeneity of epistemologies, Maruyama offers a beautiful metaphor in response to the (homogenistic) question "but which one is correct?" He suggests that in binocular vision it is irrelevant to raise the question as to which eye is correct and which wrong. "Binocular vision works, not because two eyes see different sides of the same object, but because the differential between the two images enables the brain to compute the invisible dimension". The brain computes a third dimension which cannot be directly perceived. And if we live in a multidimensional space even more epistemological "eyes" are required. Reducing such vision to the parts in common provides much less than monocular vision. The difficulty with Maruyama's presentation however, is that he often appears to associate such "poly–ocular" vision with the heterogeneity characteristic of Japanese culture, although this may not be his intention. This would then preclude the use of a homogenistic epistemological "eye" in any such poly–ocular configuration. Each "eye" has its inherent limitations and strengths, and the homogenistic "eye" presumably has its own vital contribution to make to the process of encompassing (or responding to) the complexity of our collective condition. In terms of his metaphor, this section is about the design of such poly–ocular configurations and how they may be comprehended through any given "eye". His work, with Harvey's, demonstrates that a minimum of four such "eyes" are required to describe the variety of perceptions of our collective reality.

♦ KD2150 Emancipation from particular languages

Comment A philosopher of language, Antonio de Nicolas, has studied the limitations of single languages as a vehicle for complex, action–oriented, human–centred meaning. His use of "language" corresponds to "answer" as used here. For him one of the most widespread misleading misconceptions is the implied existence or possibility of one universally adequate language (1978,p.190).

Given this point of departure de Nicolas explores the problem of the variety, interrelation and mutual exclusivity of rational thought systems, particularly of Western origin. To obtain perspective on the problem, he analyzes the philosophical languages embodied in the Rg Vedic hymns to which much oriental philosophy can trace its origins. These clarify the problem of responding to a multiplicity of perspectives which, even when understood, each on their terms, do not themselves offer any reconciliation of the multiplicity of "answers" which they constitute. Any synthesis of them is unable to "provide the antithetical perspectives essential to freedom" (1978,p.66).

De Nicolas points out that reconciliation is not a question of compromise between opposing views, since each such compromise is an "amputation" of a portion of "one's own flesh". What is then significant in the prevalence of Western-style compromise "is not that a questionable compromise is being carried out; but rather....that a new human orientation has been demanded, or been imposed through power, on all humans; in fact, a single perspective has been imposed or demanded on all humans." (1978,p.68).

Any form of reconciliation between answers has to contend, not only with saving the multiplicity of perspectives, but with the fact that these perspectives have become embodied in psycho-social structures (1978,p.67). The "songs" characteristically sung in the expression of each answer engender the "bodies" through which we function in society and determine our images of ourselves. But "if thought is the ground of man, then it follows that thought is radically man's body. The limits of his body being again the same limits of the thought that grounds it." (1978,p.82). In this sense, as explored by Geoffrey Vickers (1970), the *proponents of any answer are trapped by the bodily image they engender* (1980). Getting out of such traps calls for continuing attention to the decision process, whereby they are engendered:

"If the plight of man is grounded neither in language nor in the mirror (thought) but, rather, in man's decision to reduce himself to a universalized form of thought by grounding himself on it, then the emancipation of man will be in radicalizing himself on his decisions rather than on his images. But in order to do so man needs other men and the ability to discover them at their origin – at the radical level of their decisions and not just their images or ours, for this is man's own origin and, ultimately, his own flesh, though this might demand of every man a constant sacrifice of images – the ability to liberate himself from the prison of his mirrors – and to acknowledge a human reality which, though the source of multiple images, can neither be reduced nor identified with any of them. The other is my own possibilities and, in realizing these possibilities, I actualize my right to innovation and continuity." (1978,p.3).

♦ KD2152 Modelling language relationships by sound

Comment The distinguishing "linguistic" and epistemological feature of the Rg Veda hymns is the manner in which they are grounded in sound and demand a selection amongst *alternative musical patterns*. Since the number of tonal systems is infinite, the selection of a finite number of them by the singer/musician at the moment of execution, not only closes him within a certain limitation or determination (eg just tuning, equal temperament) but, more radically, it forces him to *constantly face the internal incompatibility of any such systems*, the tones of *every conceivable system must constantly face and submit to a radical sacrifice to permit others to emerge* (1978,p12)

"Therefore, from a linguistic and cultural perspective, we have to be aware that we are dealing with a language where tonal and arithmetical relations establish the epistemological invariances. Language grounded in music is grounded thereby on context dependency; any tone can have any possible relation to other tones, and the shift from one tone to another, which alone makes melody possible, is a shift in perspective which the singer himself embodies. Any perspective (tone) must be "sacrified" for a new one to come into being; the song is a radical activity which requires innovation while maintaining continuity, and the "world" is the creation of the singer, who shares its dimensions with the song". (1978,p.57).

Recalling Klapp's (1978) concern with alternation between opening and closing, *each necessary choice is a closure to alternatives*, but *each such choice can be sacrificed* through the movement which must open to other possibilities if development is to continue.

"Rg Vedic man, like his Greek counterparts, knew himself to be the organizer of the scale, and he cherished the multitude of possibilities open to him too much to freeze himself into one dogmatic posture. His language keeps alive the 'open-ness' to alternatives, yet it avoids entrapment in anarchy. It also resolves the fixity of theory by setting the body of man historically moving through the freedom of musical spaces, viewpoint transpositions, reciprocities, pluralism, and finally, an absolute radical sacrifice of all theory as a fixed invariant" (1978,p.57).

Of great interest is the manner in which the sets of categories, necessary to order the perceptual world, are developed and related, highlighting both the potential dynamics for harmony and discord between them. This possibility is entirely lacking in the present fashion for "pragmatically objective" elaboration of sets of categories (1984). The consequences of basing work on sets of 2,3 or more categories has not been recognized, despite obvious conflictual implications of a 2-element set (whatever the content) when reflected in a 2-division organization, for example (1978). And yet, the process whereby such sets are defined, determines how whole psycho-social systems are fragmented for analysis, comprehension, and communication.

In a musically grounded language, the basic whole is the octave. That tones recur cyclically at every doubling or halving of frequency is the basic miracle of music. But the octave refuses to be subdivided into subordinate cycles by integer ratios. "It is a blunt arithmetical fact that the higher powers of 3 and 5 which define such subordinate intervals in music never agree with higher powers of 2 which define octave cycles. It is man's yearning for this impossible agreement which introduced a hierarchy of values into the number field". (1978,p.56).

This dilemma with all that it signifies for music, philosophy and social organization has been explored by Ernest McClain (1978). The present day equivalent is the problem of how different sets of concepts, with differing numbers of categories, can nest together to encompass the societal whole without creating a degree of qualitatively unacceptable discord in use – namely a "gap" or "error" between reality as envisaged (or desired) and as perceived through the chosen pattern of categories. This gap provokes demands for an alternative in which the gap is at least diminished. (The process of reducing the gap is itself encoded in the Rg Veda according to McClain's analysis (1978).

It is not the case that numbers or ratios control movement, but it is the case that movement may be ordered according to certain ratios. Conceptual movement, and developmental in general, takes place through the elaboration of constellations of categories in which each category is context and structure dependent (A J N Judge, 1984).

Opposite or reciprocal possibilities can be perceived as equally relevant, whether co-present or succeeding each other. "Any perspective remains just one out of a group of equally valid perspectives...but no song has so universal an appeal that it terminates the invention of new ones...the function of any language is to make clear its own dependence on, and reference to, other linguistic systems." (Antonio de Nicolas, 1978,p.63.4)

"In a language ruled by the criteria of sound, perspectives, the change of perspectives and vision, stand for what musicologists call "modulation". Modulation in music is the ability to change keys within a composition. To focus within this language, and by its criteria, is primarily the activity of being able to run the scale backwards and forwards, up and down, with these sudden shifts in perspectives. Through this ability, the singer, the body, the song and the perspectives become an inseparable whole. In this language, transcendence is precisely the ability to perform the song, without any theoretical construct impeding its movement a priori, or determining the result of following such movement a priori. Nor can any theoretical compromise substitute for the discovery of the movement of "modulation" itself in history. The human body would then be asked to lose the memory of its origins; a task the human body refuses to do by its constant return to crisis." (p.192).

♦ KD2154 Complementary languages

Comment Given the context described in the previous entry it is not surprising that the Rg Veda requires four languages, rather than one, in order to convey the contrasting natures of its meaning. De Nicolas, following Husserl, describes such languages as intentionality-structures. "The intentionality-structure of a particular question, then, determines or prefigures the kind of answer it will receive." (1978,p.79). The four languages, with their multiple perspectives, function as four spaces of discourse within which human action takes place, and from which any given statement in the text gains meaning. The languages show the human situation within disparate linguistic contexts embodying different ways of viewing the world. (1978, p.9). The four languages may be described as follows:

(a) *Language of non-existence:* Provides the modality of being in a world, either of possibilities to be discovered, or of stagnant dogmatic attitudes. It is the field condition out of which all differentiation in human experience emerges. It is the continuing context for choice. In this world there is always the tendency to lift one explanatory set to the level of an internal image as a guide for action. It then functions as a suppressed premise or fundamental myth, and is hardly ever made explicit (1978,p.92-4). When the originating potential of this world is not recognized, man is deprived of the possibility of returning existentially to his origins and those of others by the dogmatic reduction of multiplicity to the "song" of one theoretical voice. In de Nicolas words the tragedy of human and social development in this world is that "We have cried for and praised many Saviours, but have lost our own act of creation and the power to revive it". (1978, p.73)

(b) *Language of existence:* Provides the modality of acting in a world of truth to be built, formed or established, as the discontinuous results of innovation. This world is one of continuity and discontinuity, multiplication and division, with a pluralism of perspectives generated from a common field admitting many alternative structures and autonomous images. In this world man is challenged by the possibility of embodying any perspective, of being bodily "at home" within any structure or autonomous space (with which his body then shares its dimensions). This world is characterized by forces experienced either as constraining, enclosing and destructive, or as liberating and growth-enhancing. The root of contemporary man's crisis in terms of this world lies in the reduction of these multiple aspects of man to the names of objects with which he is confronted and to which he then has no effective originating relationship.

(c) *Language of images and sacrifice:* Provides the modality of acting in a world through regathering the images of the dismembered sensorium (the multiplicity of worlds of existence) by sacrificing their multiple and exclusive ontologies. In contrast to the centrifugal language of existence, the images are grouped and regrouped, creating and erasing boundaries, in a centripetal process converging on a unique configuration of forces in a final "efficient-moment" of sacrifice which reveals the underlying "common body of the norm", the efficient centre of creative action, or "embodied-vision". The image of sacrifice stands therefore for an activity of eternal return to the radical originating power through which the multiplicity of perspectives is engendered. It is the efficient centre of the discontinuities of space of perception and time, or the link between efficient acts and discontinuous acts. It is not a renunciation of action, but rather a renunciation of the limits of perspectives which interpretations attach to the structured subject-object sensorium. (1978,pp.139-154)

Sacrifice is the necessary response to the ills of polycentricity with their many consequences for the fragmentation of the body of man. No idea of the body, whether monocentric or polycentric (validly chosen as styles of expression), is prior to man and therefore prior to this embodiment. (1978, p.141). Fundamental to the problem of human development is that "any identification of man with a theory of "man" obscures the fact that any and all theories of man about "man" are made of the radical dismemberment of man himself, and distract him from engaging in his only original and primal activity: the sacrificing of all theories about himself so that he may recreate himself as man". (1978, p.70)

It is for this reason that Rg Vedic man does not accept any way of understanding man's role other than as an original and continuous sacrifice (an activity rather than a theory)". (1978, p.70)

(d) *Language of embodied-vision:* Provides the modality of having gone through, and being in, a world which remains continuously because it comprehends the totality of the cultural movement on which it is grounded (C A Hooker, 1974, p.74). It is the embodiment of choosers in movement. Rationality is not then based on "the narrow logic of appeal to premises and conclusion, but rather, on an appeal to a community of listeners capable of understanding and changing, or re-directing the movement of their song". (de Nicolas, 1978, p.154). The vision becomes an objective norm, not as the result of a dogmatically imposed constraint on action, but rather as the embodiment of the norm as discovered in a community of plural activities, decisions and descriptions. (p.154). Within such a context "we find ourselves facing moving webs, moving structures; each structure a rhythm through which a body-world appears, revealing a background of living beings together with the glory and terrors of their life". (p.122)

♦ KD2156 Theory as performance

Comment In contrast to the Western emphasis on a visually-based "linear kind of movement, which disclosed a perspectival, three-dimensional space and linear time...the audial space-time structure opened by sound...was articulated not only by rhythm and cyclically recurring movements, but movement itself became the base of all contexts (structures), and the sources of meaning within each and every field of experience". (de Nicolas, 1978, p.84-5) There is no substitute for the historical discovery of the criteria by which music became one form of music as opposed to another; for it was by these criteria of music, that the body of man became now one flesh, now another. Furthermore, "without the historical mediation of the criteria of sound, by which man both imagined and lived his worlds, there is no eternal return, and therefore, no emancipation for man's memory and imagination". (p.175) "Man's emancipation lies precisely in his ability to break the barriers imposed on his memory and imagination by any abstractions which serve to reduce the human body to only the movements of a theory, and deprive man from the whole historical movement of which his historical body is the visible path". (p.170)

Of striking significance to the inertia characteristic of human and social development initiatives is the advocation of a movement which points "straight at the heart of the stillness we never dared to move: the human body...we come face to face with our most radical problem...we have never dared to set into motion our own beliefs about the human body". (p.155) Using the perusal of his own work as an example, de Nicolas states "It would indeed be a radical failure of the way of these meditations if, at the end of the journey, the human body...remained still, unchanged, undivided, and as silent with its memories and imaginations as when we started this journey". (p.156) "Every man must actively constitute himself by creating a certain order with the things around him (structure) within a general orientation he already has (or has received) about the whole of life; it is in relation to this activity that the body of man appears as flesh, and that the flesh of man makes present for us a context and a structure with which it shares its dimensions. For this reason, our path or method must *focus on the silent and fleshy unity which underlies and is the root of any human reflective thinking*". (p.53)

The body can then be brought to share the dimensions of every perspective or song it encounters, "thus turning theory into human flesh". (p.176) "Theory must turn intosong; it must be performed" to guarantee man's and society's continuity and innovation (p.167) This can only be adequately done by placing theory "within a different historical context: the context of sound". (p.174) Every vision "carries concomittantly an act of creation which can only be effective if that vision coincides with the original viewpoint" whereby that world was created. "By creating structures of knowledge to see the world in such a manner, the doer of this activity becomes the efficient vision and its concomittant creation". (p.159) It is this identification with the active power of the word making the world that joins efficient action with efficient vision (p.61). Opening and closing are involved for the "structure of the embodied subject has the double–barrel effect of opening a horizon of inquiry and restricting what appear within that horizon". (p.156)

This calls for a three–fold acceptance: the possibility of viewpoint shifting through the activity of dialogue–ing, the integration of the formal aspects of experience by the rationality of practical life, the (re)achievement through practical life and action of a unity which unrelated formal models render otherwise impossible (p.166). Such activity, which "keeps the community moving" is of course "not formalizable, but the spaces of discourse within which it appears may be formalized". (p.168)

In calling philosophers "to discover the language ruled by the criteria of sound" de Nicolas contrasts the atomicity of classical physics and Western philosophy with that of modern physics in its correspondence to Eastern views of reality. "It is only secondarily that classification of individual entities is made possible, and for this we revert to ordinary (Boolean) symbolic manipulation. In other words, to perceive anything apart from the total field is to perceive it as a subsystem, an artificially created aspect of a field of stresses, i.e. pattern. In fact, according to the law of complementarity, what can truly be said in one context–language, the same cannot be truly said in the other context–language". (p.33)

The implications of this point have been explored in different ways by a number of authors including Bohm (1980), Capra (1975), Zukav (1979), Heelan (1974), Hooker (1974). But a special merit of de Nicolas presentation is that he draws attention to a response to the radical misunderstandings which arise from the "detached objective aloofness with which we in the West are accustomed to view whatever is presented to our speculative reason. This is the precise error of knowledge which the Rg Veda is trying to correct...As a result (of the error), philosophical activity became (in the West), not liberating knowledge, but an alienation of man from man, since he was bent on equating himself with the objects of his knowledge". (p.186–7)

As a systems theorist, Francisco Varela, in developing the insights of Spencer Brown (1969), clarifies this problem in a manner which is a warning to formulators of models of human and social development: "In finding the world as we do, we forget all we did to find it as such, and when we are reminded of it in retracing our steps back to indication, we find little more than a mirror–to–mirror image of ourselves and the world. In contrast with what is commonly assumed, a description, when carefully inspected, reveals the properties of the observer. We observers, distinguish ourselves precisely by distinguishing what we apparently are not, the world". (1975,p. 22).

These considerations enable de Nicolas to turn to the ordering of complementary frameworks in the logic of quantum mechanics as a way to formalize the spaces of discourse through which action (dialogue) in the world may take place. It appears that such complementarity or contextual logic offers "a very suggestive 'model' for positive dialogue between rival philosophies, and even more important, within human experience itself". (p.10) As a partial ordering (lattice) of complementary descriptive languages, such frameworks involve changes in the embodied subjectivity of the knower, changes that make possible mutually exclusive objectivities or horizons. The "sacrifice" called for "involves a partial ordering of languages in a non–Boolean logic, the non–Boolean character of which is the mediation for growth and liberation". (p.187)

Man may then "re–create himself and his society through the appropriate sacrifice, eternally, exercising thus his right to innovation and continuity. This sacrifice is the constant watch man must keep over himself fore–directing his own radical interpretive activity". (p.187) The present inability of individuals and societies to "sacrifice" their cherished beliefs is instrumental in "freezing" society and increasing its alienation, aside from the material consequences for development. Re–thought sacrifice could constitute the sort of fundamental myth which can give "philosophical meaning to the facts of ordinary life". (p.148) For de Nicolas, "*it would be unphilosophical and inhuman not to open up man's possibilities by grounding him on that movement which will set him free*". (p.46)

His emphasis on "critical" philosophy recalls the preoccupations of the Frankfurt School which would presumably also give a central role to some equivalent of "sacrifice". Indeed de Nicolas approaches their language in identifying the presuppositions of his formalization.

As a way of perceiving the de Nicolas formalization raises critical questions of how it is to be perceived in its own terms. Such questions include:

(a) Do the very valuable corrective perceptions concerning the limitations of vision–based perspectives, Western modes, and theory, necessarily imply a fundamental primacy for sound–based perspectives, Eastern modes, and non–theoretical action, or rather a temporary expedient in a continuing alternation of perspectives ?

(b) To what extent is the approach locked into the idiosyncracies of Rg Vedic Sanskrit, and Vedic–oriented (Hindu) culture ?

(c) Is the primacy accorded the fourth language necessarily fundamental or could each language appear fundamental from certain perspectives ?

(d) What would be the status of other languages and perspectives in relation to the Rg Vedic approach ?

(e) If the approach is so powerful, why has it seemingly failed to respond to the problems of collective action in the country where it is still understood ?

It is perhaps a paradoxical necessity that the very openness and fluidity of its philosophy should be based on a set of hymns which has remained unchanged (although each interpretation is conceived as a renewal). But it would seem that, like it or not, his perception/presentation of it is paradoxically a temporary product in the process he so usefully clarifies. His perception is necessarily impermanent and incomplete and does not encounter the dynamics of those who would disagree with it.

♦ KD2160 Nonlinear cybernetics

Comment Edgar Taschdjan has recently suggested that if cybernetics is to move beyond its current preoccupation with the "simplified world of abstract models", it appears to be necessary to develop a "nonlinear cybernetics able to handle regulations which are time–dependent and dialectic rather than mechanistic". (1982). Many world modelling exercises are based on such simplified models.

Tashdjan argues that human behaviour is not constant and that "too much of a good thing can become a bad thing". Furthermore, the "bipolarity of human motivations permits switches from positive to negative, from attraction to repulsion", whether in the case of an individual or of a group. The classic concepts of negative and positive feedback fail to encompass this reality since the results of such interaction in a system are purely linear, in the sense that regulators either add or subtract output from the unit governed. Regulation maintains an "equilibrium". The negative feedback concept, based on nullifying deviations, is "not sufficient to explain the real behaviour of the steersman of a sailing vessel buffeted by changing winds, who has to "tack" first in one direction, then in another." Just as in the case of (development) policy–making, there is a need to "change the course abruptly and repeatedly in order to reach his objective", especially if he has to steer around an obstacle. On the other hand, the positive feedback concept, based on amplifying deviations, is only able to explain exponential growth, whereas "real growth processes are not exponential but sigmoid", namely a function of time. All growth is constrained by counteracting processes

Now in a situation where there are effectively two interacting regulators governing the same working unit, for example two alternative policies (political parties) by which a society is (successively) governed, this "three–body problem", even in mechanical systems, is not susceptible to deterministic analysis. The synergisms and antagonisms which emerge are essentially nonlinear interactions. Such double regulation is of great importance in natural systems and society.

The overall effect of such interactions is a *continual disequilibrium.* In the case of physiological systems, "Once the organism reaches equilibrium, it is dead." In attempting to regulate any such oscillating systems, timing of intervention is vital, as is evident in attempting to control a child's swing. *The timing of any therapeutic treatment is as important as its nature and direction.*

From these considerations Taschdjan concludes that "when we want to analyse and model systems existing in nature and society, the dialectic process of successive antagonistic actions requires the model to be *quadripolar* rather than bipolar". He points out that there is no difficulty in representing this mathematically since the totality of positive and negative real numbers constitute a bipolar system, representable on one axis. It is accepted mathematical practice to add another axis perpendicular to this to indicate the bipolar system of imaginary numbers, which then, as discussed by C Muses (1967), represent the temporal dimension necessary to describe nonlinear processes. For Taschdjan, therefore, "to say that a system of dialectic interactions is quadripolar, is merely another way of saying that the system is nonlinear". Analysis then requires the use of vectors and tensors rather than scalars, and these are multiplicative rather than additive.

♦ KD2170 Modes of managing: embracing conflicting styles

Comment In discussing the dilemmas of the organized society, Charles Handy, a management scientist, distinguishes three types of management problem: (a) steady–state, programmable, predictable problems that can be handled by systems; (b) development problems designed to deal with new situations; and (c) exceptional problems or emergencies where speed and instinct are essential (1979, p.45).

Handy identifies four styles of organization and management which respond to these problems. He points out that any organization will tend to make use of all these styles, although the larger the organization the more evident will be their role in the blend of styles used. The manager therefore has to embrace within himself all four of the styles, using each in appropriate circumstances, since none is sufficient to contain all combinations of problems (even though style–bound managers may believe it possible). Each has a place under certain circumstances.

For convenience, Handy labels each of the four philosophies of management (and the corresponding organizational culture) with the name of a Greek god (1973):

(a) *Zeus style:* This is the club culture of the "old boy network" in which the crucial links are the empathy radiating out in a web–like manner from the patriarch or inner circle. It is excellent for speed of decision in high risk enterprises but relies heavily on trust, dependent on common background. Power lied at the centre.

(b) *Appolonian style:* This is the role–structured, hierarchical organization portrayed in standard organization charts, split into divisions at the base, linked by a board at the top. It is excellent for routine tasks in which stability and predictability are taken for granted, and no one is irreplaceable. Power lies at the top.

(c) *Athenian style:* This is based on a network of task–oriented units responding to new one–off problems. Resources are drawn from various parts of the network to focus on a particular problem. It is excellent where innovative responses are required and experiments are encouraged. Power lies in the interstices.

(d) *Dionysian style:* This is the style in which the organization is perceived as existing to help the individual in it achieve his idiosyncratic purposes, and preserve his identity and freedom. Coordination is accepted as an "administrative" necessity but no ultimate authority is recognized other than the peer group. This style is excellent where the talent or skill of the individual is the crucial asset of the organization.

Handy points out that *the ways of each style are anathema* to the others. Linkage between these modes is however essential. He distinguishes three elements of effective linkage: cultural tolerance, allowing each mode to develop its own methods of control; bridging mechanisms, including exchanges of correspondence, liaison groups and task forces; and a common language. He argues that the organization of the future will be a membership organization, multi–purpose and dispersed, combining the search for community, the economics of quality, and the revolution in communications.

♦ KD2180 Cyclic self–organization requirements

Comment The previous entries have demonstrated the need for at least four distinct approaches to be able to "contain" the complexity with which society is faced. These are the minimum number of complementary languages through which a "rounded" understanding of human and social development may be achieved.

Whilst *four distinct approaches are sufficient to contain a general conceptual understanding*, Buckminster Fuller argues that *specific, concrete instances require a fifth*: "all recallably thinkable experiencings, physical and metaphysical, are fivefoldedly characterized...All conceptually thinkable, exclusively metaphysical experiencings are fourfoldedly characterized" (1975,II,1072.21–23). For Fuller the fourfold is the basis for a *minimal* conceptual system, whereas fivefoldness "constitutes a self–exciting, pulsating propagating system" (1975,I,981.03). He demonstrates this in many structural systems.

The difficulty with the four languages, as Fuller implies, is the additional ordering required to get to grips with particular cases. As such they together only offer an unanchored potential for grasping the particular. The question is how distinct approaches interrelate, or "resonate" together (since linear or "mechanical" interlinkage is unlikely) to underpin and stabilize any new order, whether conceptual, social, or physical. This is clearly of central importance to any *practical* approach to human and social development.

In Jantsch's investigation of cyclic self–organization of social systems (1980), he draws attention to the work of Manfred Eigen in molecular genetics. Eigen explores the question of how new information originates (1978). This is a general problem of evolution, which Jantsch relates to development and to learning. The question is how the new information emerges to provide the basis for any new patterns of ordering. Any given language, or "answer domain", effectively functions like a self–replicating ecosystem. Margalef (1968) has described the evolution of such ecosystems as a process of information accumulation. Each such system seeks information from the environment, but only to use it to prevent the assimilation of more new information. Novelty is continuously transformed into confirmation. The question is *how any new order can emerge* under such circumstances.

Eigen uses the term "hypercycle" to denote any such new order. A hypercycle is a closed circle of distinct transformatory or catalytic processes in which one or more participants act as autocatalysts. For Jantsch: "Hypercycles...play an important role in many natural phenomena of self–organization, spanning a wide spectrum from chemical and biological evolution to ecological and economic systems and systems of population growth." (1980,p.15). Eigen, in reporting on his detailed analysis with Peter Schuster of the emergence of such new order (1979), states: "The self–replicative components significant for the integration of information reproduce themselves only in a coexistent form when they are connected to one another through cyclic

coupling. The mutual stabilization of the components of hypercycles succeeds for *more than four partners* in the form of nonlinear oscillations..." (p.252).
Such a hypercycle can be seen as a linking process between the participating (sub)systems, themselves cyclically ordered. The formation and maintenance of such a cycle which runs irreversibly in one direction and reconstitutes its participants and thereby itself, is possible only far from equilibrium. Its rhythm is controlled by the cycle of the slowest acting participant, thereby liberating transformative energy steadily rather than explosively (p.90).
From this one could conclude that whilst an adequate new (world) order can be *envisaged* with the aid of four internally consistent, self-replicating languages, it could not be rendered *practicable* without a *fifth language* of some kind.
For Eigen "the hypercyclic order is a theoretically justifiable, essential requirement for the integration of subsystems capable of replication into a unit of greater informational content" (p.255). It alone is capable of integrating and stabilizing such otherwise competing (sub)systems. Simple connections "would not be sufficient for cooperative stabilization of the components" (p.255). From a multitude of such replicative units, the hypercycle can discover those appropriate to one another and, if the combination offers some advantage, amplify them selectively. In this manner a totally novel order comes into being through "sympathetic" interaction which does not change the nature of the participating (sub)systems, although it may optimize their characteristics (p.256).
Eigen, through his hypercyclic ordering principle, addresses directly the question of the nature of the non-Darwinian constraining mechanism in an environment in which each "species" would otherwise expand exponentially. This is a problem with any individual answer domain, whether discipline, ideology or religion. "Coexistence of competitors requires some sort of stabilization, which confirms the exponential growth law, or the formation of niches that uncouple the competition." (p.247). Eigen's "decisive question" is also of interest for the emergence of any new world order: "The decisive question for the evolutionary ability of an information-integrating system is that of the stability of the respective coupled reaction systems, whose components must be coexistent, at the same time behaving in toto selectively with respect to other competitors." (p.251). The latter would then presumably be those hypercyclic features characterizing the old order.

♦ KD2190 Encompassing system dynamics: sixfold restraint

Comment It is not to be expected that the fivefold grasp of a developing reality (indicated in the previous entry) is sufficient. It is a minimal requirement for a certain degree of comprehension of that reality. For example, Jantsch notes: "If, in the development of the organism, two types of non-linear processes play the main role, namely genetic and metabolic processes the number rises to at least six in ecosystems (competition for niches, predator-prey, symbiosis, and optico-acoustical communication)...All these processes bring their proper rhythms into play..." (1980, p247).
As he remarks, similar coupling of oscillations occurs to an even higher degree for sociocultural systems resulting in structures of "autopoietic and temporarily harmonious nature which are capable of carrying a great deal of creativity" (p. 248). In material systems Fuller also distinguishes six basic ways in which a system can "move" in relation to its environment, namely spin, orbit, inversion (inside-out), expansion-contraction, torque, and precession. "The six basic motions are complex consequences of the six degrees of freedom. If you want to have an instrument held in position in respect to any cosmic body such as Earth, it will take exactly six restraints... Shape requires six restraints. Exactly six interrestraints produce structure. Six restraints are essential to structure and to pattern stability." (1979,II,400,464).
Fuller also notes that it takes a minimum of six interweaving trajectories to establish a boundary (insideness-outsideness) between any system and its environment (1975,I,240.32). It is thus a pre-condition of individuality (1975,I,458.05-11), and consequently of characteristic patterns of interference resultants: tangential avoidance, modulation, reflection, refraction, explosion, and critical proximity (1975,I,517.05,101-12).
As an interesting confirmation of Fuller's statements, six muscles outside the eye govern its four basic movements in tracking any object - "restraining" it in order to be able to bring it into focus. The movements towards and away from the nose are each controlled by one muscle; the upward and downward movements are each controlled by two. Other movements involve a combination of muscles. Focusing is achieved by a muscular ring, the ciliary body, within the eye. This suggests that the distinct languages required to restrain a phenomenon conceptually could usefully be thought of as "counter-acting" together in a manner somewhat analogous to such muscles.

♦ KD2192 Encompassing system dynamics: learning

Comment The learning dimension is introduced by Arthur Young in attempting to formalize how a free agent "interferes" with any system The resulting freedom or unpredictability is then part of the system He points out that a minimum of six observations are then required to determine the behaviour of the free agent:
1. To know the position of a body in space, we need one instantaneous observation (for instance, the photo finish of a race).
2. To know its velocity, which is computed from the difference in position of the body and the difference in time between the two observations, we need two such observations.
3. To know its acceleration, we need three observations.
4. To know that a body, for example, a vehicle, is under control, and thus distinguish it from one in which the controls are stuck, we need at least four observations. That is, we need three to know acceleration and one more to know that acceleration has been changed. (This still does not tell us the body's destination or goal).
5. To know the destination, provided the operator does not change his mind or try to fool us, we need five observations.
6. To know the operator has changed his mind or is trying to fool us, we need six observations." (1978,p.18).
Young noted that observation "categories five and six repeat the cycle", the fifth falling into a position category (like the first), and the sixth falling into a velocity category (like the second). This shows the relationship between the minimum of four categories required for any analytical grasp and the six observational elements to encompass the behavioural complexity.
It is to be expected that the degrees of freedom of sociocultural systems call for a corresponding array of "conceptual restrainers" in order to grasp their nature or contain them. This would also be true of any new order based on a hypercycle. Jantsch points out that the cyclical organization of any such new order may itself evolve if the participating (sub)systems mutate or new processes become introduced- namely the hypercycle's exploitation of its degrees of freedom. "The co-evolution of participants in a hypercycle leads to the notion of an ultracycle which generally underlies every learning process" (1980,p.15).
The term "ultracycle" was originally proposed by Thomas Balmer and Ernst von Weizaecker (1974) to clarify the co-evolution of subsystems of an ecosystem. In such an ultracycle, according to Jantsch, the evolution of higher complexity does not result from competition, as in the hypercycle, but from interdependence within a larger system (1980, p.106). Each self-replicating "answer domain" in a hypercycle would then represent a niche within a sociocultural ecosystem, each such niche constituting a smaller ecosystem. Each "mutation" in a niche then catalyzes changes in other niches with which it is in contact - an increase the complexity of one tending to increase in the complexity of others. The result of the co-evolution within the domains is then the evolution of the overall system. Jantsch sees this as applying to national economic systems, for example, but he does not focus on the inequalities in such development (p.195-6).
"The ultracycle is a model for the learning process in general. Learning is not the importation of strange knowledge into a system, but the mobilization of processes which are inherent to the learning system itself and belong to its proper cognitive domain...Learning may generally be described as the co-evolution of systems which accumulate experience - a capability already characteristic of simple chemical dissipative structures. In the ultracycle information is not only transferred but also produced." (p.196).

♦ KD2200 Encompassing varieties of form

Comment In the light of the arguments in previous entries concerning fourfoldness as providing the basis for the minimal conceivable system through which distinct domains could be related, it is possible, as Mushakoji noted, that recent work on catastrophe theory can clarify the kinds of discontinuity which might then become apparent. Clearly knowledge of the forms such discontinuities take is necessary if the dynamics of such a developing system are to be encompassed conceptually.
If the behaviour of the minimal conceivable system interrelating the domains is determined by four control factors only, Rene Thom (1980) demonstrated that, irrespective of their nature, there are only seven qualitatively different types of discontinuity then possible. In other words, while there are an infinite number of ways for such a system to change continuously (around an equilibrium position), there are only seven structurally stable ways for it to change discontinuously (through non-equilibrium states). "To put it very simply, in a wide range of situations - physical, biological, even psychological - where experience tells us that 'something's got to give'...there are only seven fundamentally different ways it could happen".
Recalling the importance of fivefoldness for implementability, it is appropriate to note that subsequent work on catastrophe theory has demonstrated that in systems with five control factors, a further four types of characteristic catastrophes emerge, making a total of eleven. Above five there are no unique patterns.
The question is then what are these possible "catastrophes" or discontinuities as they emerge in human and social development ? What are these processes with which people are presumably intimately acquainted, but which are seemingly difficult to "objectify" despite their importance ? In a sense the discontinuities are "irrationalities" to which man is subject in the development process. Thus the "cusp catastrophe", for example, has been used to explore the flight/fight and love/hate transitions. The nature of such developmental discontinuities does not seem to have been determined.

♦ KD2202 Encompassing varieties of communication

Comment In cultural anthropology, a problem related to that of the previous entry has however been explored by a philosopher, W T Jones (1961). He is concerned, like Maruyama, with discontinuities in communication which prevent people, supposedly concerned with the same subject, from achieving any effective dialogue. To clarify this situation, he demonstrates that the discontinuities can be described in terms of the different positions of the participants (or schools of thought) on *seven pre-rational axes of bias*. These differences are reflected in aesthetical, theoretical, value, life-style, policy, and action preferences, as well as in the preferred style of discussion. Any difference between people in position "along" an axis gives rise to discontinuity which it is difficult to handle within a rational frame of reference. The axes identified by Jones are:
(a) *Order vs disorder*, namely the range between a preference for fluidity, muddle, chaos, etc. and a preference for system, structure, conceptual clarity, etc.
(b) *Static vs dynamic*, namely the range between a preference for the changeless, eternal, etc. and a preference for movement, for explanation in genetic and process terms, etc.
(c) *Continuity vs dicreteness*, namely the range between a preference for wholeness, unity, etc and a preference for discreteness, plurality, diversity, etc.
(d) *Inner vs outer*, namely the range between a preference for being able to project oneself into the objects of one's experience (to experience them as one experiences oneself), and a preference for a relatively external, objective relation to them.
(e) *Sharp focus vs soft focus*, namely the range between a preference for clear, direct experience and a preference for threshold experiences which are felt to be saturated with more meaning than is immediately present.
(f) *This world vs other world*, namely the range between a preference for belief in the spatio-temporal world as self-explanatory and a preference for belief that it is not self-explanatory (but can only be comprehended in the light of other factors and frames of reference).
(g) *Spontaneity vs process*, namely the range between a preference for chance, freedom, accident, etc and a preference for explanations subject laws and definable processes.
It is worth noting that in Fuller's analysis of systems, conceptual and otherwise, he identifies a total of *seven axes of symmetry necessary for the descriptions of the variety forms* (1975,I,1040-10 42.05). He shows that if each axis passes through two distant planes, then the fourteen resultant planes together encompass all possible asymmetries in relationships, as typified by arrays of biological cells or bubbles (1975,II,1041.12-1041.13). If each extreme mode (or "domain") on Jones axes (above) is represented by a plane, then the variety they represent together could be "contained" by a "space" of fourteen facets. Much of Fuller's work is concerned with the fundamental significance of the regenerative transformations possible with systems represented by such a fourteen-faced figure.
Jantsch attributes a seven level structure to the autopoiesis and evolution of man. He notes that such "a concept of multilevel life presents considerable difficulties to Western thinking" (1980, p.240). But he draws attention to the reality of such a sevenfold system in Hindu thinking concerning human development.

♦ KD2205 Non-comprehension "holes"

Comment The previous entries clarify the conceptual challenge by which individuals and societies are faced in endeavouring to "ride the tiger" and encompass the development process in which they are immersed. But the very fact that this seems to call for juggling conceptually with seven or more factors simultaneously suggests the need to examine what kind of situation results if this does *not* prove possible, and what kind of psychosocial structures then emerge As reviewed in an earlier paper, most people have difficulty in juggling with more than three factors, and there is a considerable preference for dealing with one, or at the most two.
The nature of the communication (and organizational) patterns which emerge as the result of conceptual discontinuity and non-communication has been clarified by q-analysis. This is the theory and application of mathematical relations between finite sets. Ron Atkin has applied this to the analysis of communication patterns within complex organizations (1974, 1977, 1981).
The perceptual significance of this approach is well-illustrated by visual sensitivity to colours resulting from the three primary hues (red, green and blue). These may be represented on a simple triangle. Here the vertices (O-simplexes) represent the primary hues, the sides are twofold combinations (1-simplexes), and the combination of the three hues makes the central white (2-simplex). The 2-simplex, together with all its faces, forms a simplicial complex KY (X) where X is the vertex set (red, green, blue) and Y is the set of seven perceived colours.
Now to be able to see all the colours, a person's vision needs to have the ability to function in the triangle as 2-dimensional "traffic" on that geometry, moving from location to location adjusting

to the complexity of the geometrical structure which carries the visual traffic. It however the person's vision is limited to 1–dimensional traffic, then white could not be perceived because the visual traffic of seeing is then restricted to the vertices only. Similarly, if the person's colour vision is only O–dimensional, then it is restricted to the edges and vertices. It can only see one vertex colour at a time and never a combination (as represented by an edge). If vision was 3–dimensional, it would allow traffic throughout the geometry, but would perceive other colours as well, calling for a fourth vertex in order to contain the full range of combinations.

If the geometry represents concepts or psychosocial functions (or even policy issues faced by an organization) instead of colours, then it would be expected that some people, in relation to that set, would have O–dimensional comprehension (i.e. sensitive to isolated primary issues only) and others would have 1–dimensional comprehension (i.e. only sensitive to binary combinations of primary issues). The latter would be unable to maintain attention to three concepts simultaneously in order to perceive the threefold combination (the central, integrated "white" issue). The threefold issue may then be termed a *2–hole* in the pattern of communication connectivity amongst those involved. For 2–dimensional traffic however, the issue complex is coherent, comprehensible and well integrated. For the 1–dimensional traffic, it feels less secure as a whole, since the latter may only be experienced sequentially through a succession of experiences ("around the edges") from which the shape of the whole may be deducted but not experienced For O–dimensional traffic, the integrated concept does not exist, since experience is disconnected.

"Generally speaking it seems to be confirmed that action (of whatever kind) in the community can be seen as traffic in the abstract geometry and that this traffic must naturally avoid the holes (because it is impossible for any such action to exist in a hole). The holes therefore appear strangely as objects in the structure, as far as the traffic is concerned. The difference is a logical one in that the word "q–hole" describes a static feature of the geometry S (N), whilst the world "q–object" describes the experience of that hole by traffic which moves in S (N)" (1977,p.75).

As an "object" this phenomenon is an obstacle to communication and comprehension and obliges those confronted with it to go "around" in order to sense the higher dimensionality by which it is characterized. Communications "bounce off" such objects. As a "hole" this phenomenon engenders, or is engendered by, a pattern of communication. It appears to function both as "source" and "sink". Atkin suggests that, in some way which is not yet fully understood, such object/holes act as sources of energy for the possible traffic around them. From the initial research it would appear that such objects/holes are characteristic of communication patterns in most complex organizations. It seems highly probable that they can also be detected in any partially ordered pattern of communication. As such "societal problems", "human needs", and "human values" merit examination in this light.

Very concretely, Atkin has investigated situations in which the "vertices" (which could themselves be n–simplexes in a multidimensional geometry) are individuals or offices linked together through various committees. (They could also be governments or disciplines). There will then be a lot of O–traffic and 1–traffic within and between offices due to the details of their intra–and inter–office (bilateral) operations. This traffic will circulate around the holes/objects which they constitute. Any n–level traffic can only be encompassed, or be brought to rest, by an (n1)–level body (e.g. an executive or a committee). If the latter does not exist, such traffic will continue to circulate around the q–objects in the structure and, according to Atkin, may be defined as noise. An "empire builder" (or any elite), for example, in such an organizational system will carefully create many q–holes underneath him (at the n–level), so that subordinate bodies answerable only to his appointees, are trapped in the flow of noise between them (1977, p.129). Atkin notes that even though the geometry may not have been rendered explicit, such structures generate the feeling throughout a community of some "power behind the scenes" acting to outwit the formal structure. The special value of q–analysis is that it can clarify why action/discussion in connection with (development) issues tends to be "circular" in the long–term, however energetic it may appear in the short–term. As such it shows how *social change is blocked* by the way in which conceptual traffic patterns itself *around* the sensed core issue which is never confronted as such because the connectivity pattern is inadequate to the dimensionality of the issue. This would explain why so many issues go unresolved and why the process of "solving" problems becomes institutionally of greater importance than the actual "elimination" of the problem.

♦ KD2207 Communicable insights: the geometry of connectivity

Comment Q–analysis as explored by Ron Atkin (see previous entry) gives precision to the recognition that traffic of different degrees of content connectivity finds (or creates) its appropriate level in any psycho–social structure. Communicable insights are level–bound, especially where they are of high connectivity. In other words, at the level within which we can communicate, concepts cannot necessarily be anchored unambiguously into terms and definitions which "travel well". Precision introduces distortion which is only acceptable locally within any communicating society – although "locally" must be interpreted in the non–geographical sense in which all nuclear physicists are near neighbours, for example.

The relation between two personal or institutional structures, conceived as a multidimensional backcloth, carries whatever traffic that constitutes the communication between them. If this backcloth changes by becoming dimensionally smaller, then its geometry loses vertices and the consequent connectivity properties. This is first indicated by the failure of higher dimensional traffic which the geometry can no longer carry. Such 4–traffic, for example, must then move through the structure to some new haven of 4–dimensionality or it must change its nature and become genuine 3–traffic. This process of reducing communication expectations in order to continue to live within the new warped geometry is the classical problem of compromising. The feeling of "having to compromise" is a painful one. It is the feeling of stress induced by the warping of the communication geometry, namely the direct experience of a structurally induced force, in this case a 4–force (p.146–7). This approach clearly provides a very precise approach to understanding more subtle forms of structural violence. He has applied it to an analysis of unemployment (p.148).

Such considerations suggest the power of q–analysis in clarifying approaches to human and social development in general. *Reducing the dimensionality of the geometry on which a person (or group) is able to live is an impoverishment associated with repressive forces.* Expanding the dimensionality induces positive, attractive forces through which a sense of development and enrichment is experienced (p.163). Q–analysis seems to be a valuable new language through which precision can be given to intuitive experiences and their communication, particularly since it provides an explicit measure of obstruction to change.

In the case of social development, it is probable that most continuing *societal problems should be seen as holes/objects*, especially given the well–established record of unfruitful action in response to them – however vigorous and dedicated. Typical examples are: peace/ disarmament, development, human rights, environment, etc. Q–analysis could then provide understanding of why any action tends to be drawn into a vortex of futility, however much it satisfies short–term political needs for visible "positive" action. The *participants in the action find themselves "circulating" around a central concern of which they are unable to obtain an overview due to the geometries of the overlapping conceptual and organizational structures through which they work* (or which they somehow engender).

The term "futility" used above is however only appropriate if the sole considerations were the elimination of such problems. In fact the existence of such problems is extremely important to the organization of society, to social development, and to the direct or indirect employment of many people. Just as the "defence" business is vital to the economy of many countries, so is the "social problem" business vital to many sectors of society. Eliminating social problems would be a disaster for many people, especially problem–oriented intellectuals, the employees of problem–solving agencies, or indeed those in need of stimulus and challenge.

In the case of human development, Atkin shows how *the individual can be defined in terms of a multidimensional geometry requiring a minimum of four levels* (p.111). By relating this geometry to that of society, Atkin introduces an 8–level scheme (p.162) within which the degree of integration or eccentricity of communication can be clarified in terms of developmental or anti–developmental forces.

Concerning such levels, the question arises as to whether their hierarchical order is fixed. Preoccupations associated with Schumacher's "small is beautiful", for example, may well modify the order. The ordering may be a question of orientation in which the "top" and "bottom" elements selected depend on the preferred concept and direction of development (e.g. "top–down",, "bottom–up"). This would be more consistent with the concept of order as an (existential) choice as discussed above in connection with the various fourfold "languages".

In such a multidimensional geometry it is clear that, whether in the case of an individual, a group or society as a whole, *it is not possible to eliminate "underdevelopment" as associated with low dimensionality.* Such a geometry will necessarily continue to have traffic of very low–level connectivity co–present with that of increasingly higher level connectivity. The simplest illustration arises from the continual birth of infants who will, when resources permit, continue to be educated through to the level of connectivity to which they can respond. But there will always be communication at both low and high–connectivity levels, especially about socio–political issues. The question is then how such learning communication between these different levels of connectivity can weave itself together within a social structure.

It is the status of the holes/objects in relation to development which could provide an interesting point of departure for further investigation. As noted above, it is *not a question of attempting vainly to eliminate such holes*, especially when some of them may arise from alternative concepts of "development". Rather it is *a question of how configurations of holes can be identified and/or designed.* It is such configurations of holes which provide the minimum structure (and communication dynamics) *to stabilize and give form to the co–presence of the differing "answers" to the challenge of development.*

In effect such holes exist at a lower connectivity–level than the "macro–hole" of higher connectivity constituted by the world problematique at this time. This macro–crisis hole "absorbs" the development initiatives of society by engendering the immense volume of action/communication traffic around the hole so defined. This draws attention to the developmental implications of the probable presence of holes of yet higher dimensionality than can be readily sensed or made the subject of acceptable public (consensual) communication.

How then are "better" holes to be engendered within such configurations ? Now from one point of view it is necessary to avoid introducing an element of evaluation, because from each hole the perception of other holes will be distorted so that no communicable assessment can be usefully formulated. On the other hand, it may prove to be the case that, at the level of the configuration as a whole, more than one such configuration can be identified/designed in order to interrelate the perspectives associated with the set of holes. And at this level, without privileging any particular hole, more adequate interrelationships between the elements making up the holes can be identified.

Expressed differently, introducing evaluative judgements into the relationships between the holes within a particular configuration can only contribute to the dynamics between such holes in terms of perceived advantage/disadvantage. Excessive emphasis on this runs the risk of tearing the configuration apart. The identities associated with the holes can be respected in each of the configurations in a series constituting progressively more adequate or richer formulations of the relationships between "developments". There is consequently a multiplicity of concepts of development operative in society. Individuals and groups may "progress" from one to another, possibly with a general tendency towards those of higher connectivity. But other individuals and groups will emerge and find the concepts of lower connectivity more meaningful before moving on, if they do, to those of higher connectivity. (In this sense the "ontogenesis" of an individual tends to repeat the "phylogenesis" of his/her society). Society in this sense is the arena within which individuals and groups refine their concept of development.

♦ KD2210 Discontinuity: Comprehension and internalization

Comment It appears from Buckminster Fuller's work that cycles interlock with greatest facility (i.e. minimum energy condition) in such a way as to form *configurations of modes in relatively simple geometrical patterns* e.g. spherical tetrahedron, octahedron, etc) according to the number of cycles (1975, 1979). The modes correspond to answer domains effectively stabilized into sets by standing wave interference effects. The portions of cycles linking such modes are then the *transformation pathways* between them which favour information transfer and learning. The pattern as a whole can also be considered as a transformation of the two–dimensional matrix representation of answer domains into a "wrap–around" three–dimensional "container". The observer, in terms of the "third perspective", is effectively given a location at the spherical centre in contrast to his undefined status in relation to the matrix (A J N Judge, 1984).

Fortunately as portrayed this representation is essentially sterile. Even though it encompasses incompatibles it does so within a framework which is a typical example of left–hemisphere thinking. Only by re–introducing right–hemisphere thinking is it possible to open the way to anything of transformative significance. In effect the *rational objectivity of a presentation must be challenged* (and, in Attali's terms, "seduced") by irrational discontinuity and subjectivity. Strangely it would seem that the scholastic preoccupation with avoiding "non sequiturs" is precisely what renders academic conclusions non–transformative, at least in any revolutionary sense. They do not internalize discontinuity but effectively project it onto their non–relationship with other answer domains.

The challenge of *internalizing non–sequiturs* is one of the exciting aspects of the frontiers of fundamental physics (Edgar Morin, 1981, P A Heelan 1974, C A Hooker, 1974). Many observers have remarked the relationship to Eastern concepts of consciousness, especially Zen. Others note that the challenge of the times calls for a change of consciousness, but are unable to design any framework to focus the approach to this. As a response to this dilemma, an earlier paper (A J N Judge, 1982) experimented with presenting the steps of an argument in terms of left–and right–hemisphere modes alternately. This procedure was based on the assumption that a *transformative argument cannot be wholly based on one mode* or the other, but each must provide clues (negative and positive feedback) for the next step of the other (as implied by the "walking" metaphor).

Bateson has argued strongly for a somewhat related approach: "...it is necessary to expand on the relationship between form and process, treating the notion of form as an analogue of what I have been calling tautology and process as the analogue of the aggregate of phenomena to be explained. As form is to process, so tautology is to description...What is important...is to note that my procedures of inquiry were punctuated by an alternation between classification and the description of process...I shall argue that this paradigm...recurs again and again wherever mental process...predominates in the organization of phenomena. In other words, when we take the notion of logical typing out of the field of abstract logic and start to map real biological events onto the hierarchies of this paradigm, we shall immediately encounter the fact that in the world of mental and biological systems, the hiererchy is not only a list of classes, classes of classes, and classes of classes of classes, but has also become a *zigzag ladder of dialectic between form and process*" (1979, p.190,193,194).

An earlier paper (A J N Judge, 1984) alternated between presentations of right-hemisphere (RH) arguments considered academically acceptable to Jungian psychologists and left-hemisphere (LH) arguments concerning structure. The RH material forms part of the symbolic heritage of many cultures. The concern of Jungians is to clarify its contemporary significance and thus counteract *"cerebral imperialism"* and "dominance" of the LH over the RH and the projections onto society to which that gives rise. They see this dominance pattern as the *subjective origin of the present social crisis.* The therapeutic objective is the achievement of a greater integration between the LH and the RH through a *transcendent "union of opposites", namely a transcendent function* (or the "meta-answer" seen in a new light): "One tendency seems to be the regulating principle of the other; both are bound together in a compensatory relationship...aesthetic formulation needs understanding of the meaning, and understanding needs aesthetic formulation. The two supplement each other to form the transcendent function".

In the LH approach, the structural problems of containing and transforming attention were explored using as a metaphor the current research on the containment of plasma (whose fluidity corresponds closely to that of attention) in fusion research. This requires a *special configuration*, yet to be fully developed, before energy can be generated at a sustainable yield. It would seem that the patterns of thought and structure required for this fusion breakthrough offer insight for a corresponding breakthrough in human and social development (and are a technological prefiguration of it, in Attali's terms (1977)).

Whilst the approach outlined is worth exploring, once again it is necessary to challenge the essential inadequacy of the previous step. It is not sufficient at this time to elaborate "descriptions" and "theories". Whatever their RH component, they are essentially LH in nature, *confronting the observer in a manner which deactivates and neutralizes him.* If there is to be effective "seduction" in Attali's terms, something more stimulating and participative is required.

The basic weakness of the above approach is that it fails to clarify or internalize the obvious differences in peoples ability to comprehend and derive significance from *a meta-answer. In this sense a meta-answer is not definable and subject to enclosure*, but is elusive in that it is understood and defined to different degrees by different people. To the extent that there is no foreseeable limit on future increases and refinements in understanding, the definition is in fact open-ended in terms of time.

♦ KD2215 Pattern accumulation in a learning society

Comment In a social condition of "structured fluidity", observers can no longer usefully assume that they are standing on solid ground around which events flow (for their intellectual delectation). Such an assumption merely temporarily defines the observer (or an aspect of his personality) as a rigid element in society, within which he is not currently undergoing a process of developmental transformation. In this sense observers are, momentarily, non-participants in the process of human and social development. Furthermore observation is only one step in the learning process, to the extent that it is useful to consider that *observers, as observers, are effectively non-learners.*

It would seem that in a fluid environment, structured by degrees and kinds of comprehension, that a vital step forward is to switch from interpreting actions in terms of their significance *for development* to their significance *as learning.* It is strange that "development" is conventionality a process applied to, or undergone by "others" — never by the "developers", despite their well-documented limitations. It is acknowledged that good teachers succeed partly because their attitude is one of learning with, and from, the student - to the point that "facilitator" is more appropriate than "teacher". The advocated change can then be represented by:

From: developers plus developees = developing society

To: facilitator-learners plus learners = learning society

For this change of interpretation to be other than cosmetic, the concept of "learning" must: (a) extend far beyond conventional forms of book learning and training; (b) be promoted as an activity of all social institutions; (c) extend beyond individual learning (in a learning society) to group and societal learning; (d) be accepted as intrinsic to all activities of all social institutions (not just "educational" programmes, but living as learning).

The first two points are well elaborated in the report of the UNESCO International Commission on the Development of Education (1972), concerned with the emergence of a *"learning society"* but from which the last two points are totally absent, since they do not refer to "development of individual education" (a well-defined answer domain) but "development as societal learning" (which generates its own answers) in a more inclusive sense. The importance of the third point has been discussed in an earlier paper (A J N Judge, 1982) in relation to the erosion of collective memory. Rector Soedjatmoko of the UN University has emphasized this point (prior to taking that position) in relation to the *"learning capacity of a nation"*: "The capacity of a nation - not just of its government, but of society as a whole - to adjust to rapidly changing techno-economic, socio-cultural and political changes, on a scale which makes it possible to speak of social transformation, very much depends on its collective capacity to generate, to ingest, to reach out for, and to utilize a vast amount of new and relevant information. This capacity for creative and innovative response to changing conditions and new challenges I would like to call the learning capacity of a nation. This capacity is obviously not limited to the cognitive level, but includes the attitudinal, institutional and organizational levels of society as well. It therefore resides not only in a nation's formal educational system, not only in the government bureaucracy, in parliament and the political parties, but also in the business community, in the media, the professional organizations, the trade unions, the cooperatives and the various kinds of voluntary associations within the society at large. It also includes the political public with its various political constituencies, consumer groups, and all other kinds of permanent and ad hoc pressure groups." (1981, p.82-83).

A recent Club of Rome report extends this notion to "humanity": "Our continued survival is testimony that humanity indeed learns...So we have to reconsider what is meant by the statement "humanity learns". Does the statement not imply - indeed demand - that learning occur at the right time and on a scale sufficiently large not only to avoid disasters but also to conclude a century, so much traumatized by successive follies, with a gain in peace, dignity, and happiness ?" (p.118).

The report concludes however that: "The conventional, often unarticulated, conception of how societies learn... (is reduced to one of)...adjusting to and consuming the discoveries and knowledge produced in centers of expertise. The unavoidable consequence of this view of societal learning is elitism, technocracy and paternalism. What is omitted is the fact that meaning and values - decisive for learning - are products of society at large, not of specialized centers... (that)...tend to reproduce themselves according to their own internal logic. This autonomous and self-reproducing development accounts in large part for the fact that so much of societal learning is maintenance learning." (p.80-81).

The basic distinction made in the report between the necessities of maintenance (adaptive) learning and innovative (shock) learning can be related to the alternation process discussed in previous entries: "Innovative societal learning seeks to restore active learning to those in society conventionally confined to a passive role of assimilation". But whilst much research has been done on individual learning processes, hardly any is done on organizational or group or societal learning (Botkin, 1979, p.137).

The key question then becomes: *what is the individual or collective learning component of any activity* ? A major weakness of conventional concepts of development is that, outside the economic answer domain, there is no positive coherent image of what is being achieved by human and social development processes. In a learning society, however, it is "learning" which is being accumulated, where this can best be partially defined in terms of accumulation of recognized patterns. Discovery of the manner in which newly comprehended patterns interlock and constrain each other most economically, in terms of a meta-pattern, is the organizing constraint upon the accumulation process.

Given the current passive, academically inferior status of "learning" (as part of the professor-student, trainer-trainee dominance mind-set), it should be apparent that a complementary active (learning through doing), conflictual (learning through opposing) dimension is inherent in what is advocated here. Learning is effectively being "defined" by the accumulation process in the zigzag ladder of dialectical alternation between perceptions of forms and process, which Bateson considers "basic to the way in which the world of adaptive action is put together." (1979, p.201).

"I shall further suggest that the very nature of perception follows this paradigm; that learning is to be modelled on the same sort of zigzag paradigm; that in the social world, the relation between love and marriage or education and status necessarily follow a similar paradigm; that in evolution, the relation between somatic and phylogenetic change and the relation between the random and the selected have this zigzag form. I shall suggest that similar relations obtain at a more abstract level between speciation and variation, between continuity and discontinuity and between number and quantity" (p.195).

Learning of this kind can only ever be partially contained within an organization or a paradigm, because of its *essentially dichotomous nature.* As Bateson says: "This view makes the process of learning...necessarily discontinuous...A world of sense, organization, and communication is not conceivable without discontinuity, without threshold. If sense organs can receive news only of difference and if neurons either fire or do not fire, then threshold becomes necessarily a feature of how the living and mental world is put together". (p.202).

Learning is an ordered dynamic response to discontinuity and conflict between institutions and answer domains - a conflict which it engenders and by which it is engendered, for learning to continue. But learning is not an unconstrained process without limits, except in a purely gross sense. Due to the progressive interlocking of accumulated patterns into nested meta-patterns, as a solution to human processing capacity limitations, there is a form of directed, convergence onto a progressively clarified ultimate meta-pattern, towards which learning tends asymptotically, in that final (en)closure is never achieved (except possibly as an essentially transient, private, transcendental experience). Bateson describes this ultimate pattern as: *"The pattern which connects (all living creatures) is a metapattern. It is a pattern of patterns. It is that metapattern which defines the vast generalization that, indeed, it is patterns which connect".* (p.11).

Final enclosure evanesces in the paradoxical world of self-reference explored in a left-hemisphere mode by D Hofstadter (1979). Jantsch points out however that in life the issue is not control but dynamic connectedness. For him "Learning may generally be described as the co-evolution of systems which accumulate experience". (1980, p.196). He cites Christine von Weizaecker:

"...co-evolving systems....play between adaptation and non-adaptation. Total adaptation and total non-adaptation are both lethal. In ecology, a niche fits the species sufficiently, without defining it; the species, in turn, fit the niche sufficiently, without defining it. What else is fitting, but not defining each other, than an emancipated relation".

♦ KD2220 General systems and holonomy

Comment The most deliberate effort to clarify the nature and possibilities of integration has been made through general systems research. This has of necessity involved the perspectives of many disciplines. Efforts, such as those of J G Miller (1979), have brought a very extensive range of phenomena within the same framework. General systems has not however been very successful in bringing its insights to bear upon the world problematique, despite deliberate efforts to do so (Richard Ericson, 1979). Part of the problem seems to lie in the essentially left-hemisphere approach to describing, explaining, and classifying systems. This has not met the needs of those participating in systems, however valuable it has been to those observing such systems.

It is therefore interesting to note the recent effort by J S Stamps to "marry" the insights of general systems research with those of humanistic psychology, as an "integration of conscious systems with concrete systems" in which mind and system are perceived as complementary (1980). Stamps interrelates general systems taxonomies of recent decades to provide a "multi-dimensional elaboration of the fundamental principles of complementary process and level structure" which indicates the "limits of integration and transformation" at each level. The final design of Stamps heuristic taxonomy arises from the combination of two ideas which he believes have not previously been related:

"Namely, that the complementarity between awareness and organization can be applied to the distinction between abstracted and concrete systems and theories. A taxonomy with this feature becomes a potential "Rosetta stone" for translating abstracted scientific language into concrete scientific language... (And secondly) is the suggestion that the form and processes of both individuality and collectivity evolve. To the conventional notion that phylogeny evolves and ontogeny develops, I have added the idea that phylogeny also develops and ontogeny also evolves. The importance of being able to make an argument such as this is enormous, for it places the human individual into a context of ontological equality with the many layers and types of human and social organization". (p.204).

Stamps makes a deliberate attempt to move beyond Cartesian dualism, especially in the light of research on the bicameral mind. The limitation of this approach, as discussed elsewhere (A J N Judge, 1981) lies in his implication that a heuristic taxonomy does not contain inherent limitations in a society which is increasingly resistant to such hierarchical orderings, whether conceptual or otherwise. Some of these limitations emerge in the work of Rescher, discussed in a subsequent entry, where such orderings are contrasted with a "network" organization of knowledge. Ironically, Stamps subsequently co-authored a book on "networking" for practitioners, which emphasizes this other perspective (1980).

Stamps uses Arthur Koestler's term "holon" as referring to complex entities, particularly organisms and people, which are simultaneously whole individuals and participating parts of more encompassing wholes (89). From it he names his approach "holonomy" as being a systems theory which acknowledges the place of the human individual. This new term has also been recently used by Bohm in a rather different sense (as noted in a subsequent entry).

♦ KD2225 Cognitive systematization

Comment In order to clarify the implications of the previous entries for some integrated approach to human and social development, it is appropriate to consider the current status of cognitive systematization. This has been the concern of Nicholas Rescher who explores the reason for systematization in the cognitive domain and shows how this is one of the crucial features of the development of knowledge (1979). It is to be expected that the pattern of insights and conclusions would be relevant to development in general.

Rescher identifies *eleven definitive characteristics of systematicity*: wholeness, completeness, self-sufficiency, cohesiveness, consonance, architectonic structure, functional unity, functional regularity, functional simplicity, mutual supportiveness, and functional efficacy (p.10)). He points out, citing C S Peirce, that the need for understanding through a unified view of things is a real as any of man's physical cravings, and more powerful than many of them. The above characteristics "are constitutive components of that systemacity through which alone understanding can be achieved". (p.29). The point of cognitive systematization in relational terms is that (a) it is the prime vehicle for understanding by making claims intelligible, (b) it authenticates the adequacy of the organization of knowledge, (c) it is a vehicle of cognitive quality control, providing a test of acceptability, and (d) it provides the definitive constituting criterion of knowledge (p.29–38). Similar points could be usefully made about the integration of development.

The *alternative modes of cognitive systematization* are distinguished by Rescher. These are foundationalism, based on a Euclidean model of a linear, deductive exfoliation from basic axioms and coherentism, a network model of cyclic systematization of interrelated theses (p.39). The Euclidean model is typical of the logic governing formalized (intergovernmental) development programmes based on a set of principles. The network model is typical of the logic of grass–roots development movements. From the network perspective, the Euclidean model imposes a drastic limitation by "inflating what is at most a local feature of derivation from the underived (i.e. locally underived) into a global feature that endows the whole system with an axiomatic structure". Thus although "a network system gives up Euclideanism at the global level of its over–all structure, it may still exhibit a locally Euclidean aspect, having local neighbourhoods whose systematic structure is deductive/axiomatic" (p44–45).

The network model shifts the perspective, as Maruyama also notes, from unidirectional dependency to reciprocal interconnection, abandoning the concept of priority or fundamentality in its arrangement of these. "It replaces such fundamentality by a conception of enmeshment in a unifying web" (p.46–47), whereas the Euclidan approach gives priority to derivation from what is better understood or more fundamental.

Rescher notes (p.58–59) that the basic weakness in the latter approach was however demonstrated by Kurt Goedel (1958), who showed both that the consistency of any formal axiomatic system can never be proved, and that the deductive axiomatization of any such system was inherently incomplete. There are therefore always "true" statements in a given domain that cannot be derived form the chosen axioms. It would seem that this too has important implications for the limitations of development programmes elaborated on the basis of pre–determined sets of principles in some "declaration" or "world plan of action", especially since Rescher indicates the possibility of a breakdown of deductivism in the factual sciences as well (p.176).

Rescher also provides a valuable analysis of the *limits to cognitive systematization*. He identifies three possibilities: incompletability, inconsequence (or disconnectedness, compartmentalization), and inconsitency (or incoherence). With regard to the first, he notes that it is unrealistic to expect either attainment of a completed and final state of factual knowledge, or a condition in which all questions are answered. "Accordingly, we have little alternative but to take the humbling view that the incompleteness of our information entails its incorrectness, as well" (p.152–3). In a more highly developed future, fundamental errors will be perceived in present formulations and programmes – as can already be detected in the development strategies of past decades.

With regard to disconnectedness, the second possibility, Rescher argues that this cannot characterize the body of our factual knowledge as a whole which can always be joined by mediating connections of common relevancy (p.164). The problem is rather that despite such causal linkage, there could well fail to be connections in meaning between two domains. The fundamental causal matrix in which all natural occurrences are bound together "might merely be a purely formal unity, lacking any sufficient substantive basis of functional connectedness." Nature might come to be shown as operating in an essentially compartmentalized manner. Furthermore, Rescher notes, gaps in the knowledge attainable at any time might in practice block realization of any underlying interconnectedness. This issue of compartmentalization is of course of crucial importance in the design of interdiciplinary development programmes, for which no adequate methodology has yet emerged, partly because of separative behaviour characteristic of disciplines.

With regard to the third possibility, Rescher sees inconsistency as lying at the root of the urge to systematicity. It is the very drive toward completeness that enjoins the toleration of inconsistency upon us. But rather than implying no system at all, any inconsistency–embracing world picture involves the toleration of ungainly systems of deficient systemacity (p.176–7). It is a question of degree.

♦ KD2230 Wholeness and the implicate order

Comment As a theoretical physicist, David Bohm is concerned with the illusory nature of fragmentation (1971, 1976) and the manner in which distinct fragments emerge from wholeness in movement (1980). He sees the perceptual problems with which he deals as being as relevant to a more healthy response to psychosocial fragmentation as to the problems of fundamental physics. The value of Bohm's perspective for understanding healthy individual development has in fact been recently stressed by a physician Larry Dossey (1982).

For Bohm: "the widespread pervasive distinctions between people (race, nation, family, profession, etc.), which are now preventing mankind from working together for the common good, and indeed, even for survival, have one of the key factors of their origin in a kind of thought that treats things as inherently divided, disconnected, and 'broken up' into yet smaller constituent parts...considered to be essentially independent and self–existent." (1980, xi).

Attempting to live according to the notion that the fragments are really separate is then what leads to the growing series of extremely urgent crises with which society is confronted. "Individually there has developed a widespread feeling of helplessness and despair, in the face of what seems to be an overwhelming mass of disparate social forces, going beyond the control and even the comprehension of the human beings who are caught up in it." (1980, p.2). And yet the seeming practicality and convenience of the process of divisive thinking about things supplies man with "an apparent proof of the correctness of his fragmentary self–world view."

Basing his investigations on insights from the current state of physics, Bohm focuses "on the subtle but crucial role of our general forms of thinking in sustaining fragmentation and in defeating our deepest urges toward wholeness or integrity". (p.3). He arrives at the conclusion that "our general world view is itself an overall movement of thought, which has to be viable in the sense that the totality of activities that flow out of it are generally in harmony, both in themselves and with regard to the whole existence." (xii). This view implies that "flow is, in some sense, prior to that of the 'things' that can be seen to form and dissolve in this flow". (p.11). Thus the "various patterns that can be abstracted from it have a certain relative autonomy and stability, which is indeed provided for by the universal law of the flowing movement". (p.11).

Of special relevance to the question of human and social development, is that the above–mentioned desirable harmony "is seen to be possible only if the world view itself takes part in an unending process of development, evolution, and unfoldment, which fits as part of the universal process that is the ground of all existence." (xii). This has the merit of grounding the concept of development in movement from which appropriate conceptual and social forms temporarily arise, rather than, as is presently done, starting from some "thing" (e.g. a society, a community, or a person) which has to be stimulated into a process of movement and change that is then called "development" (under certain conditions).

Bohm cautions against the expectations of quick remedies: "To ask how to end fragmentation and to expect an answer in a few minutes makes even less sense than to ask how to develop a theory as new as Einstein's was when he was working on it, and to expect to be told what in terms of some programme, expressed in terms of formulae or recipes...What is needed, however, is somehow to grasp the overall *formative cause* of fragmentation, in which content and actual process are seen together, in their wholeness". (p.18).

As he notes, this confronts us with a very difficult challenge: "How are we to think coherently of a single, unbroken, flowing actuality of existence as a whole, containing both thought (consciousness) and external reality as we experience it ?" (x). The approach he suggests requires looking at the challenge in a new way. Instead of aiming for some reflective correspondence between "thought' and "reality as a whole" the process of thinking about reality as a whole can more usefully be thought of as a kind of *"dance of the mind"* (determining, and being determined) which functions indicatively. (p.55–6).

He uses the indicative role of the well–known bee–dance as an analogy (although the element of "alternation" in any such dance should not be overlooked). As with Attali, Bohm emphasizes that "thought with totality as its content has to be considered as an art form, like poetry, whose function is primarily to give rise to new perception, and to action that is implicit in this perception, rather than to communicate reflective knowledge of "how everything is" (p.63). There can no more be an ultimate form of such thought (or of any principles or programmes to which it gives rise) than there can be an ultimate poem which would obviate the need for further poetic development.

Bohm explores the implications of quantum theory as an indication of "new order". The questions he raises are also relevant the emergence of any new psychosocial order. He demonstrates that in the past recognition of new patterns of order has involved attention to "similar differences and different similarities" (p.115), namely the *"irrelevance of old differences, and the relevance of new differences"* (p.141). The radical transformation of understanding brought about by quantum theory, for example, results from recognition of the way in which modes of obsevation and of theoretical understanding are related to each other. A social science equivalent of this is given in Johan Galtung's demonstration of the impossibility of value–free research (1977), although his purpose is to orient research in terms of development–oriented values.

For Bohm, however, comprehending the new order bears some resemblance to artistic perception. He uses Piaget's distinction between assimilation (understanding, render comprehensible) and accommodation (adaptation, fitting to a pattern) as the basic modes of intelligent perception. This artistic perception then begins by "observing the whole fact in its full individuality, and then by degree articulates the order that is proper to the assimilation of this fact." (p.141) Thus it does not begin with abstract preconceptions as to what the order has to be, which are then "adapted" to the order that is observed.

Bohm uses the differences between a lens system (in measurement processes) and a holographic system to show how by use of the former "scientists were encouraged to extrapolate their ideas and to think that such an (analytical) approach would be relevant and valid no matter how far they went, in all possible conditions, contexts, and degrees of approximation." (p.144). The advances in relativity and quantum theory imply, however, an undivided wholeness in which such "analysis into distinct and well–defined parts is no longer relevant." This is best illustrated by the hologram in which a whole pattern is somehow encoded into each part, no matter how small. *The new order appropriate to our time could then be conceived as contained as a totality, encoded in some implicit sense into each region of space and time* (p.149).

He elaborates an entirely new way of *understanding order as "implicate", or enfolded*, which he contrasts with "explicate" forms that are commonly observed and sought. The simplest example he gives is of a television image, carried by a radio wave in an implicate order, and then explicated by a receiver.

In more general terms, Bohm argues that the underlying wholeness in movement (the "holomovement"), noted above, acts like the radio wave to "carry" an implicate order. Under certain circumstances particular things (objects, phenomena, people, nations) can then be unfolded from this dynamic totality by a perceiver, but the holomovement is not limited in any specifiable way at all. As such it does not conform to any particular order and is essentially undefinable and immeasurable. This means that no single theory can capture or contain phenomena on a permanent basis. Rather, each theory will abstract a certain aspect that is relevant only in some limited context, lifting it temporarily into attention so that it stands out in relief (p.151). Furthermore, any new order within which a multiplicity of such aspects are "integrated" is itself not a final goal (as in efforts at "unified science"), but rather part of a movement from which new wholes are continually emerging (p.157).

This approach is very helpful in opening up ways of conceiving development and new forms of social order. In providing a mathematical description of implicate order, for example, Bohm makes a useful distinction between: *transformation*, as a geometric rearrangement within a given explicate order, and *metamorphosis*, as a much more radical change (such as between a caterpillar and a butterfly) in which everything alters, although "some subtle and highly implicit features remain invariant" (p.160). The former characterizes much development thinking, whereas the subtlety of the latter has hitherto made it appear non–operational or equivalent to catastrophe.

Given Atkin's use of simplical complexes to describe social organization, it is also interesting that Bohm suggests the extension of this technique in terms of "multiplexes" (p.166–7). His argument that phenomena need to be perceived as projections of a higher–dimensional reality for which appropriate algebras are required (p.188), relates to Thom's concerns with mathematical archetypes (1980).

The challenge of Bohm's arguments lies in the manner in which they strike at the very root of the meaning of human and social development. His arguments highlight the extent to which both the physical and social sciences continue to rely on a Cartesian framework (if only in the familiar tabular/matrix presentations characteristic of social science papers) at a time when inherent weaknesses in the thinking behind such frameworks have been demonstrated. His most basic point is that the phenomena such as those which are the preoccupation of "development" (peoples, ideologies, groups, societies) are essentially derivative. *"The things that appear to our senses are derivative forms and their true meaning can be seen only when we consider the plenum, in which they are generated and sustained, and into which they must ultimately vanish".* (p.192) In this light, the basic flaw in present development thinking is the a priori recognition of certain distinct social entities which it now seems desirable to "develop".

It is precisely this conception (as argued on different grounds by the world–system theorists) which reduces development to "sterile" transformative operations and prevents any metamorphoses (to use Bohm's terms). For it is *development which precedes and underlies such explicate social entities as a movement* from which they have been unfolded: *"what* is movement" (p.203). *Metamorphosis thus calls for ways of unfolding new, currently implicate forms from this holomovement, and enfolding into it those which are currently explicate, but are inadequate to the time.* This is far removed from mechanistic efforts to "eliminate" undesirable structures and to "build" new ones from their components.

It should not be assumed that this implicate order is an inaccessible theoretical abstraction. Bohm argues that *consciousness itself operates by enfolding and unfolding* and that "not only is immediate experience best understood in terms of the implicate order, but that thought also is basically to be comprehended in this order". (p.204). This creates the possibility for "an unbroken flowing movement from immediate experience to logical thought and back" thus ending the fragmentation characteristic of the absence of any awareness of such movement (p.203). He argues that movement is itself sensed primarily in the implicate order and that Piaget's work "supports the notion that the experiencing of the implicate order is fundamentally much more immediate and direct than that of the explicate order, which...requires a complex construction that has to be learned" (p.206).

KD2240 Health and space–time

Comment As has been noted on many occasions, the concept of health is intimately related to that of wholeness. As broadly defined by the World Health Organization, it encompasses the physical, psychological and spiritual well–being of the individual and is thus central to the concept of human and social development. It is therefore valuable to explore the evolution in the concept of health, as a form of integration, and as throwing light on the implications of such integration for an understanding of development.

This question has been admirably discussed by Larry Dossey (1982), a physician, in the light of the conceptual implications of theoretical breakthroughs in 20th century physics, and notably as a result of the work of David Bohm (see previous entry). The shortcomings of the current health care system are increasingly perceived as rooted in the conceptual framework that supports medical theory and practice. As the physicist Fritjof Capra states in introducing Dossey's work: "The crisis in medicine, then, is essentially a crisis of perception, and hence it is inextricably linked to a much larger social and cultural crisis...which derives from the fact that we are trying to apply the concepts of an outdated world view – the mechanistic world view of Cartesian–Newtonian science – to a reality that can no longer be understood in terms of these concepts". (p.viii).

To describe the globally interconnected world, in which biological, psychological, social, and environmental phenomena are all interdependent, Dossey explores the implications of quantum physics as "the most accurate description we have ever discovered of the physical world" (p.126).

Given the disturbing innovation of such physics, whereby the behaviour and subjectivity of the observer is necessarily incorporated into any understanding of the results of observation, he points out the weakness in the argument that such theoretical breakthroughs are only of significance to the abstract world of nuclear physics. He cites the physicist E Wigner who states: "The recognition that physical objects and spiritual values have a very similar kind of reality...is the only known point of view which is consistent with quantum mechanics" (1979, p.192). Dossey points out that the relevance of such supposedly sub-atomic preoccupations to macroscopic phenomena is also demonstrated by Bell's theorem as noted by the physicist H S Stapp: "The most important thing about Bell's theorem is that it puts the dilemma posed by quantum phenomena clearly into the realm of macroscopic phenomena...it shows that our ordinary ideas about the world are somehow profoundly deficient even on the macroscopic level" (p.1303). The theorem can be described as stating: "If the statistical predictions of quantum theory are true, an objective universe is incompatible with the law of local causes", which requires that events occur at a speed not exceeding that of light (p.1303).

This theorem has been substantiated by experiments which show that simultaneous changes in non-causally linked distant systems can occur when a change in one takes place. In some sense, as yet not understood, all "objects" thus constitute an indivisible whole, in contrast to the prevailing notion of an external, fixed, objective world of separate things. Furthermore, the theorem shows that the ordinary idea of an objective world unaffected by consciousness lies in opposition not only to quantum theory but to facts established by experiment.

In addition to the implications of quantum mechanics, Dossey draws attention to those from the logical limitations highlighted by the theorems of Godel (1958), Turing and Church, and Tarski. These collectively demonstrate the inherent limitations of any symbolic language which purports to describe the world unambiguously but is also called upon to make self-referential statements about itself as part of that world. They show that no precise language can be universal and that no scientific system is complete. *Any language used to describe health and development, must necessarily suffer from similar limitations.*

In the light of these considerations, Dossey points out that if our ordinary view of life, death, health and disease rests solidly on seventeenth-century physics (and on the logic on which it is based), and if this physics has now been partially abandoned in favour of a more accurate description of nature, then: "an unescapable question occurs: must not our definitions of life, death, health, and disease themselves change? To refuse to face the consequences to these areas is to favour dogma over an evolving knowledge...We have nothing to lose by a reexamination of fundamental assumptions of our models of health; on the contrary, we face the extraordinary possibility of fashioning a system that emphasizes life instead of death, and unity and oneness instead of fragmentation, darkness, and isolation." (1982, p.141–2).

After listing the characteristics of health and of the image of man arising from the "traditional" view (p.139–141), Dossey outlines the nature of the concept of health which emerges in the light of the new participative descriptions of nature. He notes, for example, that even from the point of view of elemental biology and physiology, *the body behaves more as pattern and process* than as an isolated and noninteracting object. It cannot be localized in space and its boundary is essentially illusory as in the notion of "body" in de Nicolas analysis (earlier entry). Health and illness are then a characteristic of the dynamic relationship between bodies on which therapy should necessarily arbitrarily focus as a participative process. The divisions of time are also arbitrarily imposed. "Connected as we are to all other bodies, comprised as we are of an unending flux of events themselves occurring in spacetime, we regard ourselves not as bodies fixed in time at particular points, but as eternally changing patterns for which precise descriptive terms seem utterly inappropriate" (p.142–9).

The ordinary view of death is then inadequate because it is based on two erroneous assumptions – that the body occupies a particular space, and that it endures through a span of linear time (p.158). Dossey considers that during illness the experience of spacetime construction is distorted: "When we are sick we become a Newtonian object: a bit-piece stranded in a flowing time" (p.175). *Health and illness are field phenomena.* Comprehending spacetime in this way is not a matter of intelligence and restricted to gifted scientists. It is largely a skill of the right–hemisphere. Such understanding has long been characteristic of certain oriental philosophies, and other cultures have developed with a concept of non-flowing time (p.178). It is possible that such understanding is even to some extent characteristic of the musically-oriented (youth) in the West and of the aurally-oriented cultures. Desire for experience in this form may also contribute to widespread use of drugs as offering an alternative to domination of daily life by a Newtonian worldview with its many limitations.

The emergence of such incompatible spacetime views in society as flowing or non-flowing time, isolated objects or shifting energy patterns, should perhaps be seen as constituting a vital complementarity of evolutionary significance. Dossey therefore stresses the importance of *alternation* between them: "These two modes of time perception, working alternately, make sense. They strike a balance not conferred by either alone. Perhaps we find within us these two capacities for sensing time because we needed one as much as the other". (p.180).

The two modes may then be seen as corresponding to Bohm's distinction between the explicate and the implicate order, the former being characterized by separated objects and the latter by flowing movement of totality (the holomovement). For Dossey it is the latter with which health is associated, disease being a disequilibrium in favour of the former: "Seen in this way, health has a kinetic quality. There is an essential dynamism to it, grounded as it is in Bohm's proposed underlying implicate order...Health is not static" (p.183).

He contrasts this with a prevailing image of health as associated with some "frozen stage of youth, whereafter things never change... We view health as a frozen painting, a still collection of bits of information" (p.183). But this has no meaning if health is the harmony of the movement of interdependent parts. Dossey produces a 13–pointable contrasting health based on the traditional view with that based on an implicate view (p.186–7). The problem is that modern health care (including holistic health), only focuses on the reality of the explicate order of separate objects and events. "The implicate domain, where the very meaning of health, disease, and death radically changes, is currently of no concern to medicine". (p.189) Explicate therapy has a purely mechanistic concern to "keep the parts running".

It is clear that Dossey's arguments with respect to health can also be made with respect to human and social development in general, especially since much development thinking can be viewed as directed towards "keeping the parts running". There is much to be said therefore for exploring *the possibility of elaborating an implicate understanding of development as a vital complement to the prevailing explicate view.* Any "new world order", to be of any long–term significance, could well be based on an alternation between these two modes.

KD2250 Dissonant harmony and holistic resonance

Comment As Attali has shown (1977), music remains one of the clearest domains in terms of which the thinking underlying any social order can be discussed. It provides a more concretely accessible language with which to comprehend the subtleties and distinctions reviewed in the previous entries. Thus the composer Dane Rudhyar, in a study of spatialization of tone experience, confronts the basic duality of those entries:

"The basic issues is, should we think of the notion of separate, entities *in* space, or of rhythmic movements *of* space producing entities which, though they may appear to be separate are in fact only differentiated areas of space and temporary condensations of energy ? This may seem to be a highly metaphysical issue having very little to do with music or the other arts, but it is actually the most basic issue a culture and its artists (and even the organizers and leaders of the society) have to face" (1982, p.41–2).

Rudhyar points out that the need for order is basic in human consciousness. "But the kind of order human consciousness demands and expects varies at each level of its evolution" (p.93). In the realm of music, Westerners have needed a type of musical order which makes it very clear that classical works constitute an integrated whole with a consistent tonal structure.

"Tonality is a system by which the innate pluralism of a society is kept within a definitive operative structure. Its manifestation is not so much in melodic sequence as in chordal harmony... Each melodic tone carries an identifying badge announcing clearly where it belongs, not so much in relation to the tonic as in terms of its place and function in the tonal bureaucracy. This is the ransom for the ideal of universalism...Multiplicity and differences are the evident realities; the principle that makes possible the harmonization of these differences has to work throughout the society, up and down the scale. It has to be able to be "transposed" to any place, to meet any situation. It is universal, but it has to be imposed upon the many units. It needs the complex power of chords to achieve that purpose. In other words, in our pluralistic European music the instinctual psychic power of integration that once was inherent in sequences of tones had to be replaced by the harmonizing impact of chords clearly stating the tonality to which melodic notes belong. Cadences of chords also make the hearer expect how the melody will develop..." (p.9).

For Rudhyar, any society or work of art is a complex whole composed of many parts which may be organized in two fundamentally different ways. In social organization they may be termed the tribal order and the companionate order. In music these are analogous to what he calls the consonant order and the dissonant order. The tribal order, founded on biological relationships in a community, derives its sense of unity from the past and all the associated (paternalistic) traditions which give rise to such a compulsive, quasi-instinctual feeling-realization. The musical equivalent is the harmonic series and overtones in the classical tonality system.

The companionate order begins with a multiplicity of differentiated individuals and strives to achieve *unity as a future condition*. If achieved it has to be "unity in diversity" through the harmonization of unsuppressed differences. Rudhyar argues that the musical equivalent of this is to be found in what the calls syntonic music in which the experience of tone is unconstrained by the intellectual concepts of the classical tonality system. Tonal relationships are included in the space relationships of syntonic music, but the rules, patterns, and cadences obligatory in tonality-controlled music normally hinder the development of syntonic consciousness. In syntonic music the notes are drawn into holistic group formations. Instead of emerging from a tonic, they seek the interpenetrative condition of dissonant chords, as pleromas of sounds. "These are limited in content; each has its own principle of organization, which determines the content of the pleroma" (1982, p.142–5).

In these terms, Rudhyar considers that melody is *aesthetic* in the tonality system associated with the culturally organized tribal order, and *expressionistic in a society characterized by transformation*. In the latter case its essential attributes are *dissonances*, rooted in their own musical space. Such dissonant chords can generate, when properly space, a far more powerful *resonance* than so-called perfect consonances, because of the phenomena of beats and combinations tones (p.143–6), which engender beauty of another order (p.141). "A pleroma of sounds refers to the process of harmonization through which differentiated vibratory entities are made to interact and interpenetrate in order to release a particular aspect of the resonance inherent in the whole of the musical space (accessible to human ears), its holistic resonance, its Tone" (p.139–140). As examples of such holistic resonance, he notes the traditional role and effect of gongs (in Eastern societies) and church bells (in the West). Such effects are also produced by two cymbals tuned to slightly different pitches, giving rise to a vibrant tone because of the interference pattern of the two different frequencies (p.141) – a case reminiscent of Maruyama's binocular vision analogy. Most striking however is the manifestation of such effects in the technique of over-tone singing.

Rudhyar concludes that although consonant harmony has its place and function and should be enjoyed, its danger lies in re-inforcing psychological attachment to the "matricial security and comfort" of the cultural framework with which it is associated. This bond prevents the individual from developing creative spontaneity and thereby engaging fully in the processes of social transformation (p.162).

KD2260 Interwoven alternatives: tensegrity organization

Comment The first unsatisfactory feature of the preceding entries is that they do not go far enough in showing how the alternatives can be interwoven in order to be of practical relevance to the present crisis in human and social development. Thus Stamps, although acknowledging the bicameral nature of mind, uses a classical taxonomic framework as a vehicle for his arguments for holonomy, and discusses "networking" in an unrelated book. Rescher stresses the local significance of Euclidean hierarchical structuring, but does not offer more than the recognition that the complexities of any global pattern can be viewed as the recognition that the complexities of any global pattern can be viewed as a "chain-mail structure" (1979,p.202), otherwise known as the "fish-scale" structure of knowledge. Bohm provides graphic examples of the unfoldment and enfoldment of explicate forms in relation to the implicate order, but (although he discusses its relevance to consciousness) it is not his purpose to show how such perception can be applied concretely in human and social organizations. Dossey avoids discussion of the interface between implicate and explicate therapy and of the nature of the framework in terms of which any decision to use one or the other would be made, namely the art of alternating between them. Rudhyar avoids the nature of the thinking required to alternate between tonality and syntonic perspectives, and what that implies for social organization.

Given the incompatible, but complementary, natures of the alternatives with which each of these authors is dealing, it seems necessary to focus more clearly on the *structure and dynamics of any "marriage" between such alternatives*. How is any such marriage brought about and how does it "work" in practice ? It does not seem to be sufficient to switch in an unmediated manner between sophisticated Euclidean taxonomies and sprawling associative networks (Stamps and Rescher) or between their process equivalents (Bohm and Dossey). The abyss separating them in practice invites the chaotic ineffectiveness and abuse which is characteristic of present conceptual and organizational dilemmas and of operational "schizophrenia". In any "marriage" such a situation is associated with abusive "sexual politics". The either/or nature of the switch is in itself an essentially Cartesian trap.

A step towards concretizing the implications of any such marriage can be explored firstly in structural terms. The problem can be defined as how to design (or comprehend) a *marriage between "hierarchy" and "network"* so that the union constitutes a whole of greater significance than the "incompatible" parts. This problem has been explored elsewhere in the light of the tensegrity (tensional integrity) structures discussed by Buckminster Fuller (1975, 1979) and subsequent authors. The problem is essentially one of fitting hierarchy into network. This can be done such that the local advantage of hierarchy are melded into the global advantages of network. In this way the local weakness of network and the global weaknesses of hierarchy are counteracted. The large range of tensegrity structures may then be viewed as indications of possible patterns of processes within a whole.

The special significance of tensegrity structures as "thought models" lies in the way they reflect and combine realistically the *continuity of network and the discontinuity of hierarchy*. The strong "constraining" bonds of hierarchy are interlinked by the weaker "restraining" bonds of a network as in the reality of social organization. The two types of structural element, if appropriately interrelated, bring about the emergence of an entirely new structural system, with dynamic self-stabilizing and load distribution properties of a unique kind. Of great significance, when they take a spherical form, is the fact that the centre is unoccupied by any structural element. It becomes a vital point of reference for the global characteristics, but is defined solely by the manner in which the local structural elements are configured around it.

Although physical models of tensegrity structures can be built (and effectively underlie the construction of large-scale geodesic domes), a criticism that has been made of their potential significance as models of psycho-social organization is that they are too symmetrical and complete, and thus are not open to any further development. This criticism is valid if psycho-social organization is modelled by one such tensegrity structure only, as it is also if organization is modelled by a particular hierarchical or network pattern. But there are many such tensegrity structures, even in the spherical form. Each such structure may then be considered as a possible alternative. The process of development from one to another, or of the alternation between them, then models the potential richness of psycho-social organization more effectively. There are many transformation pathways between them.

♦ KD2262 Interwoven alternatives: resonance hybrids

Comment The set of alternative structures, between which alternation takes place, may be more clearly understood in the light of the theory of resonance. Johan Galtung first explored the possibility of using the organization of chemical molecules to clarify the description of social organization (1977). He dealt with fixed structures and not with the transition between alternatives. The theory of resonance in chemistry is concerned with the representation of the actual normal state of molecules by a combination of several alternative "resonable" structures, rather than by a single valence-bond structure. The molecule is then conceived as resonating among the several valence-bond structures, or rather to have a structure that is a resonance hybrid of these structures.

The classic example of a resonance hybrid is the benzene molecule of 6 carbon atoms for which F A Kekulé introduced the idea of oscillation between two alternative structures. The pattern of oscillation was later extended by Linus Pauling to include three more distinct alternates. The actual configuration is a resonance hybrid of the five forms, which through quantum mechanics has been shown to have an energy less than any of the alternate structures. This is potentially of great significance for any social structure analogue, in view of the call for a low-energy society. Given the fundamental role of the benzene molecular configuration as the basis for most living structures, it is worth asking (in the light of the sixfold restraint discussed in earlier entries) why it is composed of six atoms. The answer is that it is this configuration which ensures minimal strain on the distribution of the four valency bonds of each carbon atom, thus resulting in a minimal energy configuration. It is worth reflecting on this model in the light of the research showing that the upper limit for effective committee or task force organization, the basis for social organization, is seven, plus or minus one.

Such structures recall the context of Bohm's arguments concerning unfoldment of explicate forms. The wave function representing a stationary state of a resonance hybrid in quantum mechanics can be expressed as the sum of the wave functions that correspond to several hypothetical alternates. The proper combination is that sum which leads to a minimal energy for the system. Of significance in any social structure analogue is that the higher energy of each alternate is associated with some degree of "distortion" (different in kind in each case), which effectively renders the alternate meta-stable. (Also worth exploring is the contrasting concept of a "resonance particle". This is any exceedingly unstable high energy particle, which may be considered as a composite of several relatively stable low energy particles into which it may decay.)

Resonance hybrids could well provide a key to the conception, design and operation of coalitions of people or groups which could not cohere for any length of time in one single form but could be stable if the coalition alternated between distinct forms. Underlying this possibility, hybrids are also of interest in integrating incompatible perspectives, paradigms and policies without eroding their distinctiveness in some simplistic compromise. Whilst the value of using such tensegrity or resonance models may be contested, they do have the advantage of shifting the debate, currently somewhat sterile, to a level at which the merits of particular answers are no longer the sole issue. The need is for investigation of "resonable" structures, however "unreasonable" they may appear from any particular perspective. They open the way to *more fruitful discussions both about how alternation between the opposing answers characteristic of a complex society can be improved and about the kinds of social structures that could be based upon such patterns of alternation.*

♦ KD2270 Non-comprehension as a structuring phenomenon in a learning society

Comment In a learning society, in which no one can aspire to be informed of every item of significance, it is quite unrealistic to expect ignorance and non-comprehension to have a purely minor role, hopefully to be further diminished by development programmes and information technology. Whether it be between the preoccupations of disciplines, cultures, generations, levels of education, or temperamental preferences, non-comprehension must necessarily continue to play a major role in the ordering of society, if not a progressively increasing one. The inability to respond to the minimal educational needs of developing countries is a striking example of the problem, matched by recent evidence of the increasing ineffectiveness of the sophisticated education programmes in developed countries. There are practical limits to learning, some of which have been explored elsewhere in a critical review of the recent UNESCO-endorsed report to the Club of Rome "No Limits to Learning" (James W Botkin, 1979).

This is the second reason for which the reviews of integration in the previous entries as advocated by Stamps, Rescher, Bohm, Dossey, and Rudhyar are insufficient, however valuable. They do not recognize the wider social structuring effects of a person's inability to comprehend some more "seductive" answer. It is assumed that with some minimum of explanation comprehension will necessarily result and the person will switch from Cartesian to non-Cartesian, from linear to non-linear, as providing the only "reasonable" mode of comprehension. It is assumed that people can be provided with the educational context within which this transition can be facilitated. This is not the case, perhaps fortunately. Available resources do not permit such education on more than a limited scale, but more importantly, people have other agendas to which their concepts of human and social development are linked. It is through this process that the variety of the psycho-social system is protected from homogenizing tendencies, however benevolently initiated.

The structuring effect on non-comprehension in complex organizations is most clearly seen through q-analysis as discussed by Ron Atkin. One interesting feature of this earlier entry is the effect of the forces to which an individual is subject by exposure to something which is not fully comprehended, especially when the non-comprehension is not consciously recognized, or is disguised by satisfaction with a superficial explanation. In a sense, recalling Dossey's arguments, the comprehension of an individual creates the spacetime geometry within which he functions (in Atkin's terms), whereas his non-comprehension determines the nature of the forces to which he is subject within that geometry (again in Atkin's terms).

The difficulty in engendering a more healthy approach to non-comprehension, as a phenomenon in which everyone participates, is that it is still treated as something to be disguised or denied, whether to oneself or to others. Or, perhaps worse, it is treated as something that can be eliminated by some kind of educational "fix" (a course, a tape, a book, etc), or acknowledged with pride as something one does not need, or have time, to know.

It is for such reasons that Christopher Alexander, an architect/designer, is helpful in demonstrating that the central "quality without a name", which makes any context attractive to be in, can only be "tangentially" described in terms of a range of possible aspects: "There is a central quality which is the root criterion of life and spirit in a man, a town, a building, or a wilderness. This quality is objective and precise, but it cannot be named" (Ilya Prigogine, 1980).

He shows how the nature of this central quality is not encompassed by any of the following attributes, each with its special advocates: alive, whole, comfortable, free, exact, egoless, eternal (Christopher Alexander, 1979, p.25-40).

For Alexander, in order to define this quality, it is necessary to recognize that every context is given its character by certain patterns of events that keep on happening there. By arguing that these patterns are always interlocked with certain geometric patterns which structure any inhabited space, Alexander effectively provides a concretization of Atkin's insights relating to organization space (Christopher Alexander, 1977). Both are intimately linked to the process of development:

"The specific patterns out which a building or a town is made may be alive or dead. To the extent they are alive, they let our inner forces loose, and set us free: but when they are dead, they keep us locked in inner conflict. The more living patterns there are in a place...the more it comes to life as an entirety, the more it glows, the more it has that self-maintaining force which is the quality without a name" (Christopher Alexander, 1979).

It is interesting that by defining the central quality as nameless, Alexander frees it from the problems, encountered earlier, of the essential inadequacy of any particular language. The quality is "defined" as not comprehensible through any one such language. As such it is totally in sympathy with the nature of Bohm's holomovement, by which such "names" are engendered and re-enfolded." Using the spherical tensegrity model, each such language is characterized by (explicate) surface features or patterns of the tensegrity which encompass a central empty space without occupying it. Using the resonance model, each attribute is an alternative in the pattern of alternation, but the nameless core quality is represented by the resonance hybrid. In Atkin's terms, the central quality functions as a higher dimensional q-hole which engenders a pattern of communication amongst the perspectives or languages configures around it.

♦ KD2272 Focus for variable geometries of global order

Undefined common focus

Comment The "New Global Order" called for by the crisis of the times may be thought of as brought into being by recognition of the fundamental distinction between local, specific, surface features (centres, values, languages, groupings, etc) and the unoccupied common centre whose position is determined by the pattern of all such specialized features constellated around it. It is the very pattern of harmonies and dissonances between these local features which can then engender the space of which the unoccupied centre is the focal reference point. This can only occur if the mutual rejection of those most strongly opposed is contained, by allowing them appropriate separation, and is thus itself used to maintain the form of the pattern.

These considerations clarify ways of thinking about any "meta-answer" and show the essential role of non-comprehension in structuring the space for the nameless quality of life which development programmes try in their different ways to enhance. *The meta-answer is thus a resonance hybrid of answers based on particular conceptual languages or epistemologies.* Development of that quality of life calls for the dynamics of resonance between answers or frameworks despite the conceptual discontinuity that this involves. The languages used by different authors (in preceding entries, for example) clarify the strengths and weakness of particularly approaches in endeavouring to encompass this quality. In a sense the essential feature of this section is the search for ways to alternate between the insights of each such language and thus engender some understanding of the nature of the resonance hybrid they form.

In order to give more practical significance to these considerations, further work is required to show, in the light of the insights of Dossey and Atkin, what kinds of spacetime geometries a person (or a group) may create for himself through his mode of comprehension and through the "valencies" of the nodes in his pattern of communication. Much of relevance to such an investigation is implicit in Fuller's "geometry of thinking" (1975, 1979), and explicit in Atkin's work (1977, 1981). Atkin especially clarifies the structuring effect of interactions between groups "living on different geometries". Fuller clarifies the transformations between configurations.

Each such language must be recognized as limited. Recent solutions by R Kurzweiler to the difficult technical problems of pattern recognition in voice-activated computers illustrate very clearly the kind of thinking required. Eleven different expert systems each with different pattern recognition strengths and weaknesses are used. Their individual conclusions are then compared by a final expert system (programmed to take account of the biases of the individual systems) in order to arrive at the clearest pattern. Another way to think about this question is in terms of "conceptual gearboxes". Use of a single-answer mode of comprehension is similar to the use of the first gear on an automobile, with all its advantages and disadvantages. Modes of comprehension corresponding to alternation patterns between two, three, or more, such languages, are then similar to the use of second and higher gears. The problem is then one of improving the design of such gearboxes and learning how to use them. For the difficulty at the moment is that individually and collectively people tend to get "stuck" in a particular conceptual gear, and are unable to "shift" up or down according to the needs of the moment. This approach is explored in a separate paper (A J N Judge, 1982).

The disadvantage of the gearbox model is highlighted by a "woven basket" or "birdcage" model in the same paper. The structural features of a spherical tensegrity may be considered as indicating the different functions active in a viable whole. What remains to be determined is what degree of functional differentiation is appropriate under what circumstances. A viable pattern of functions bears a strong resemblance to a viable basketweave pattern with its counter-balancing properties integral to the structure – hence the notion of "functional basketweaving". In this sense it has yet to be recognized that the psycho–social world is "functionally round" rather than "flat" as many seem to assume. Considered as a "birdcage", the problem is to interweave functions in such a way as to construct an environment for the essential living quality (of Alexander) which cannot otherwise be "kept alive" by the gross and devitalizing concepts so widely employed by programme designers.

An even more dynamic model emerges from the possibility that conceptual processes can be usefully conceived as engaged in a "pumping action". A "conceptual pump" (as in the respiratory cycle) involves transformation of attention processes through a single category, to a polar category set, to an N–category set, and back again – corresponding to the movement from principles to details. (Buckminster Fuller argues strongly in support of the fundamental significance of a topological equivalent he describes, namely the vector equilibrium "jitterbug"). Problems in the pumping cycle can arise when the pump locks onto a set of a particular number, as is evident in many conceptual systems. Problems also arise when, for example, the single category set is projected onto some detailed feature of the environment, effectively reversing the action of the pump.

♦ KD2275 Learning cycles

Comment The approach to learning discussed in previous entries is too basic for it to be possible to derive much of significance that can be applied directly to organizations. The problem lies in the Western bias discussed in favour of learning in an essentially linear direction, even if Bateson's learning "zigzag" is recognized. If the zigzag is considered as occurring around a learning cycle however, marrying in the Eastern bias towards recurrence, this cycle can then be subdivided into sufficiently detailed elements to be of significance for organizational operations. Jantsch discusses cyclical organization in terms of the system logic of dissipative self–organization:

"Hypercycles, which link autocatalytic units in cyclical organization, play an important role in many natural phenomena of self–organization, spanning a wide spectrum from chemical and biological evolution to ecological and economic systems and systems of population growth. The cyclical organization of a system may itself evolve if autocatalytic participants mutate or new processes become introduced. The co–evolution of participants in a hypercycle leads to the notion of an ultracycle which generally underlies every learning process". (1980, p.15)

The question then becomes how many discontinuous phases (Jantsch's "participants") it is useful to distinguish in the cycle. Too few and the incompatibilities between them are too fundamental, too many and the distinctions between them are too subtle. The operational significance of this conceptual constraint has been explored in earlier papers from which it is apparent that significance is lost if more than about 7 categories are used (James W Botkin, 1979), unless the total breaks down into sub–sets based on simple (e.g. J Attali, 1979, 1981, A J N Judge, 1984) factors (A J N Judge, 1984).

A novel approach to the learning cycle in relation to action has been taken by Arthur Young (1978) as a consequence of his experience as the inventor of the Bell helicopter (whose three–dimensional movement is notoriously difficult to control – as with the development process). He established the vital learning–action link through a new interpretation of the operational significance of the set of 12 "measure formulae" through which material phenomena are observed, acted upon and controlled in physics and engineering. These he portrays as corresponding to a series of *phases in a learning–action cycle.* Of special interest for the development theme is the significance he attaches to the sequence of movement around the cycle: one direction involving essentially unremembered experience–without–learning, the other involving conscious–learning–action. His approach has been adapted and modified to further emphasize the action–learning significance (A J N Judge, 1984). It is interesting that the philosopher Stephane Lupasco also attaches importance to the analysis of such measures in terms of the polarities they constitute and the types of energy with which they are associated (1973,p.26).

This approach clarifies how portions of such a cycle are vulnerable to institutionalization (as specialized, independent answer domains, or habitual responses) to the extent that there is no learning bridge across the discontinuities. The problem of (social) integration is thus intimately related to the functioning of (collective) learning cycles. It seems probable that needs (and their satisfiers) also relate to different portions of such cycles, as would ranges of *incompatible development goals or alternative visions of desirable futures. In each case the point to be emphasized is that such seemingly incompatible fragments are "frozen" portions of a cycle* with which individuals or groups identify. None are of lasting significance in their own right, especially insofar as they hinder the collective learning process which must take place through them.

The facilitative and obstructive factors to further learning (i.e. successful "struggle" in marxist terms) at each stage in the cycle are probably linked to patterns of complementarity and incompatibility between the stages according to their memberships of (2,3, or 4–member) sub–sets in the cycle (e.g. preceding and succeeding stages in the cycle are in conflictual relationship since they would correspond to thinking of the opposite hemisphere). Answers given from any part of a cycle are of course "questionable" as perceived from other parts of the cycle.

As noted earlier, a single cycle is not a sufficiently concrete representation of the complexity to be encompassed by an adequate meta–answer. Where several cycles interlock to form a sphere, the nodes are effectively combinations of cyclic phases. The relationships of challenge and harmony between such nodes have been discussed in earlier papers concerning Fuller's tensegrity concept (1975, 1979). It is this which clarifies the potential and vulnerability of networking as an essentially right hemisphere mode of organization which needs to be more "seductively" married to the much–criticized left–hemisphere, hierarchical mode.

The acid test of learning cycles however, is whether they can encompass the discontinuities between the major political tendencies by which the world community is seemingly divided. Any such relationship posited must necessarily be highly controversial, but the controversy should be patterned according to the aspects of the learning challenges involved. This has been explored elsewhere (A J N Judge, 1984).

♦ KD2280 Patterns of alternation: a musical key from a political philosopher

Comment Given the essential discontinuity between the domains in patterns of alternation, the key question is whether there is any way of comprehending and communicating the nature of the transformations between the elments in the pattern, other than "superficially" in purely right–hemisphere (dramatic) terms. Furthermore, it would be useful to clarify the basis for the emergence of each domain within the pattern. A recent study by Ernest McClain (1978), a musicologist, demonstrates that the "father" of political philosophy, Plato, has further unexplored insights to offer concerning these matters, especially in view of his aesthetic concerns (which correspond to Attali's requirement). McClain points out that great care must be taken in exploring Plato's political allegories because of his considerable use of puns, metaphor and humour as a form of presentation appropriate for arguments comprehensible to a musically and mathematically informed audience. It is appropriate to follow McClain's lead because he draws attention to a language which can be used to clarify the nature of alternation.

McClain demonstrates with considerable musico–mathematical precision how Plato conceived four very different tuning systems whose characteristics he described through the four different communities (Callipolis, Ancient Athens, Atlantis, Magnesia) which are the subject of his later dialogues. As McClain says, they are: "each vividly, presented but each necessarily "sacrificed" to let the alternatives come into view. This mode of thinking, or manner of talking, is appropriate for the realm of *alternative* aesthetic structures which are equally appealing – from one point of view or another – but mutually incompatible in time. Not only do Plato's musical cities come to be, but each must pass away – as each tone, each mode, each rhythm must pass – to allow the next to come into focus. There is no dialogue which "fixes" Plato's thought for us. The *Republic* and *Laws* are so opposed in spirit – the first proposing a communal brotherhood with few laws and common wives, children and property, the second satiated with law and central government – as to appear to be the work of two different men...Plato is in no sense what we have come to understand as a "Platonist". Neither he a Pythagorean. His Pythagoreanism is but a prelude to philosophy, to "the song itself...which dialectic performs", a prelude to which Platonists have declined to listen, although it establishes the multiple perspectives from which Plato understood himself" (1978, p.132–3).

The constraints and possibilities of developing a tuning system within which harmonies and discords can play themselves out allows McClain to demonstrate, using Plato's material:

(a) *Limitation:* "In political theory as in musical theory, both creation and the limitation of creation pose a central problem. Threatening infinity must be contained. Conflicting and irreconcilable systems...must be coordinated as an alternative to chaos. Limitation, preferably self–limitation, is one of Plato's foremost concerns" (p.14) "The political lesson – a musical analogue – points to the impossibility of founding a lasting state on any model which lacks an internal principle of self–limitation..." (p.19).

(b) *Inevitability of degeneration:* McClain demonstrates how the expansion of any system, musical or political, leads to its degeneration. Plato's well–known theory concerning the transition between five kinds of ruler, through aristocracy to tyranny, whilst no longer of interest in those terms, acquires new relevance as a metaphor in the light of McClain's analysis

(c) *Just distribution in practice:* Plato provides a unique response to the central problem of how a whole is to be divided up into parts which can be harmoniously related. The musical language he uses clarifies the impossibility of achieving the ideal solution to this problem, whether in music or in social systems. The many possible tuning systems model the variety of "intuitive" approximations to this solution. The best approximation is achieved by "tempering", namely through a slight deformation of the ideal distribution amongst the parts, in order to achieve harmony within the whole. As McClain demonstrates, the *Republic* is from a musician's point of view a treatise on equal temperament, namely on the ways of approaching the fundamental musical problem arising from the incommensurability of certain musical intervals. In Plato's terms justice does not mean giving each man "exactly what he is owed" but rather moderating such demands in the interests of "what is best for the city" (p.5). His communities model tuning systems which achieve this *with different advantages and disadvantages to the quality of the whole* and to that of the parts. But what is important here is not the models but the language which interrelates these possibilities.

(d) *Dialectics of opposites:* The language used requires, and embodies, a critical process of turning back to examine anything that has been assumed. "It is this turning back to criticize one's initial assumptions which separates Plato from all philosophy developed from "first principles"...No assumption we can make...makes any sense until we have "turned back" to study them also from the opposite point of view" (p.8 and Alexander, 1977).

McClain's analysis provides a much richer understanding of how and why alternatives may be distinguished, and of how the sets of categories may be derived by which a "seductive" pattern of functions (or institutions) is defined. It shows the need for sufficient variety to "contain" or "carry" interesting harmony, but marks the emergence of various limits necessary for the integrity of the system. But although the pattern of alternatives clarifies the nature of the required container, the art of choosing and moving between them remains (delightfully) elusive, as pointed out in a study (de Nicholas, 1978) of Vedic musically–encoded philosophy to which McClain refers in an earlier work:

"Language grounded in music is grounded thereby on context dependency; any tone can have any possible relation to other tones, and the shift from one tone to another, which alone makes melody possible, is a shift in perspective which the singer himself embodies. Any perspective (tone) must be "sacrificed" for a new one to come into being; the song is a radical activity which requires innovation while maintaining continuity, and the "world" is the creation of the singer, who shares its dimensions with the song."

In this sense much can be learnt from current interest in "techniques" of improvisation in music groups in which each instrument is free to respond to the others in a "dominant" or "subservient" manner: "...each musician must search for missing material in the performance of the neighbour (pitches from the first, length from the second) and react to it in different ways: imitate, adapt himself to it (if need be further develop), do the opposite, become disinterested or something else (something "unheard of")".

♦ KD2285 Patterns of alternation: an agricultural key from crop rotation

Comment The difficulty in exploring patterns of alternation is the seeming lack of concrete (as opposed to abstract) examples by which the credibility of such patterns in practice may become apparent. The rotation of agricultural crops is therefore and interesting "earthy" practice to explore in the light of the mind–set which it has required of farmers for several thousand years.

Crop rotation is the alternation of different crops in the same field in some (more or less) regular sequence. It differs from the haphazard change of crops from time to time, in that a deliberately chosen set of crops is grown in succession in cycles over a period of years. Rotations may be of any length, being dependent on soil, climate, and crop. They are commonly of 3 to 7 years duration, usually with 4 crops (some of which may be grown twice in succession). The different crop rotations on each of the fields of the set making up the farm as a whole constitute a "crop rotation system" when integrated optimally.

Long before crop rotation became a science, practice demonstrated that crop yields decline if the same crop is grown continuously in the same place. There are therefore many benefits, both direct and indirect, to be obtained from good rotational (T B Hutchison, 1936, p.176–8):

(a) Control of pests: with each crop grown the emergence of characteristic weeds, insects and diseases is facilitated. Changing to another crop inhibits the spread of such pests which would otherwise become uncontrollable (to the point that some crops should not be grown twice in succession). By rotating winter and summer crops, the farmer fights summer weeds in the winter crop and winter weeds in the summer crop.

(b) Maintenance of organic matter: some crops deplete the organic matter in the soil, others increase it.

(c) Maintenance of soil nitrogen supply: no single cropping system will ordinarily maintain the nitrogen supply unless leguminous crops are alternated with others.

(d) Economy of labour: several crops may be grown in succession with only one soil preparation (ploughing). For example: the land is ploughed for maize, the maize stubble is disced for wheat, then grass and clover are seeded in the wheat.

(e) Protection of soil: it was once believed necessary to leave land fallow for part of the cycle. Now it is known that a proper rotation of crops, with due attention to maintaining the balance of nutrients, is more successful than leaving the land bare and exposed to leaching and erosion.

(f) Complete use of soil: by alternation between deep and shallow–rooted crops the soil may be utilized more completely.

(g) Balanced use of plant nutrients: when appropriately alternated, crops reduce the different nutrient materials of the soil in more desirable proportions.

(h) Orderly farming: work is more evenly distributed throughout the year. The farm layout is usually simplified and costs of production are reduced. The rushed work characteristic of haphazard

cropping is avoided.
(i) Risk reduction: risks are distributed among several crops as a guarantee against complete failure.
The situation is somewhat different in the case of single-species forests where "rotation" is the guiding principle in the special sense of the economic age to which each crop can be grown before it is succeeded by the next one. (For example, on a 100-year rotation required for oak, one per cent of the forest would be clear out each year, and a further 20 percent thinned out). In total contrast to crop rotation is the "monoculture" cropping system in which the same crop is grown every year. This is possible on a large scale only by the heavy application of chemical fertilizers, herbicides and pesticides. It leads to long-term problems of soil structure and erosion, as well as to the accumulation of pollutants.
Because of the short-term advantages of fertilizers, efforts to design ndw approaches to crop rotation have been limited. It is only with the resurgence of interest in non-exploitive, non-polluting agriculture that such possibilities are being investigated.
From an agronomist's perspective, the problem is to strike a balance between harmonizing the three-fold soil-plant-climate relationship and those of the economic constraints of production. Because such threefold relationships are now fairly well understood, rotation cycles can now be considered as a whole in which the order and the plants used are of secondary importance. The problem is to ensure that the soil-plant-climate relationship is in an optimally balanced state at every moment in order to become increasingly independent of its past. The production constraints complicate this evolution and the choices possible, especially when requirements change rapidly without taking into account the recent history of a crop rotation.
There is a striking parallel between the rotation of crops and the succession of (governmental) policies applied in a society. The contrast is also striking because of the essentially haphazard switch between "right" and "left" policies. There is little explicit awareness of the need for any rotation to correct for negative consequences ("pests") encouraged by each and to replenish the resources of society ("nutrients", "soil structure") which each policy so characteristically depletes.
There is no awareness, for example, of the number of distinct policies ("crops") through which it is useful to rotate. Nor is it known how many such distinct cycles are necessary for an optimally integrated world society in which the temporary failure of one, due to adverse circumstances (disaster) is compensated by the success of others. It is also interesting that during a period of increasing complaint regarding cultural homogenization ("monoculture"), voters are either confronted with single-party systems or are frustrated by the lack of real choice between the alternatives offered. There is something to be learnt from the mind-sets and social organizations associated with the stages in the history of crop rotation which evolved, beyond the slash-and-burn stage, through a 2-year crop-fallow rotation, to more complex 3 and 4-year rotations. Given the widespread sense of increasing impoverishment of the quality-of-life, consideration of crop rotation may clarify ways of thinking about what is being depleted, how to counteract this process, and the nature of the resources that are so vainly (and expensively) used as "fertilizer" and "pesticide" to keep the system going in the short-term. The "yield" to be maximized is presumably human and social development.

♦ KD2295 The entropic crisis and the learning response

Comment Society may be usefully perceived as facing an entropic crisis. This view has been explored by Jeremy Rifkin (1980). The second (entropy) law of thermodynamics states that matter and energy can only be changed in one direction, from usable to unusable, or from available to unavailable, or from ordered to disordered. And whenever any semblance of order is created anywhere, it is done at the expense of causing an even greater disorder in the surrounding environment: For Rifkin the inexorable nature of this process provides an understanding of why the existing world views are breaking down. For:
"The laws of thermodynamics, then, govern the physical world. The way humanity decides to interact with those laws in establishing a framework for physical existence is of crucial importance in whether humankind's spiritual journey is allowed to flourish or languish" (p.9).
He anticipates three types of response to the implications of the entropy law, namely from optimists, pragmatists, or hedonists. It is very interesting that he challenges the use to which Prigogine's work on dissipative structures will be put by the optimists. For Rifkin: "The theory of dissipative structures is an attempt to provide a growth paradigm for an energy environment based on renewables, just as Newtonian physics provided a growth paradigm for a nonrenewable energy environment" (p.245).
He argues that the theory of dissipative structures completely ignores the wider significance of the entropy law by concentrating only on that part of the unfolding process that creates increasing order. And on the question of irreversibility on a cosmic scale, Prigogine does indeed state "I prefer to confess ignorance" (1980, p.214). Rifkin continues: "By refusing to recognize that increased ordering and energy flow-through always creates ever greater disorder in the surrounding environment, those who advocate bioengineering technology as the transforming apparatus for a renewable energy environment are doomed to repeat the same folly that has led to the final collapse of our nonrenewable energy environment and the age of physics that was built upon it" (p.247).
He concludes: "Like it or not, we are irrevocably headed toward a low-energy society...The longer we put off the necessary transition from a high- to a low-entropy society, the bigger the entropy bill becomes and the more difficult the turnaround becomes...The alternative to this wholescale squandering of available energy is an internalization of the values and dictates of the entropic paradigm" (p.254).
The difficulty is that Rifkin is clear on what should not be done but provides few practical insights into the social order required to do whatever ought to be done - whatever that is. In particular, in the light of the theme of this section, he accumulates significance in relation to entropy at the expense of conceptual ordering in relation to issues and perspectives to which others are sensitive. By striving for support, as does any proponent of a world view (whatever its merits), he condemns his perspective to compete in the "gladiatorial arena" discussed earlier.
Rifkin believes that the entropy constraint applies only to the physical domain and that there is an escape route. "There are those among us who are willing to accept the finiteness of the physical world but who believe that the entropic flow is counterbalanced by an ever-expanding stream of psychic order. To these people, the becoming process of life is synonymous with the notion of an ever-growing consciousness" (p.257).
Whatever the merits of the argument, as an ordering device, it does not clarify the basis for the emergence of such a new psychic order. His presentation implies that it could be based on a psychic, constraint-free replication of the pattern which he so effectively criticizes in the physical domain. But the accumulation of "hot spots" of significance at the expense of a surrounding, unredeemable "wasteland" of increasing irrelevance does not seem to be the basis for the needed breakthrough. Whether it is "experienced" or "achieved", somehow a low-energy psychic order is required to interrelate the various domains of significance to permit the emergence of a physical low-energy society. "Hot-wiring", to use Rifkin's term, also needs to be avoided in the patterns of communication between such domains, between his "answer" and those of others.

♦ KD2300 Alternation between energetic expansion and mentalistic reduction

Comment The entropic constraint in social development has been specifically explored by anthropologist Richard Adams. He cites Alfred Lotka's observation that the second law of thermodynamics cannot be contravened by human action. Lotka's principle states that in evolution natural selection favours those populations that convert the greater amount of energy, that is, that bring the greater amount of energy form and process under control. But any "islands" of local order are not themselves an indication of counter-entropic process but rather zones where energy is hastened to entropy or converted into equilibrium forms (1975, p.125-6).
Adams argues, with Carneiro, that the evident macroscopic expansion of human society in terms of "culture traits" is exponential due to this expansion being proportional to the number of traits already generated. But instead of culture traits, Adams argues that the concept of energy conversion (as opposed to input) is more significant, as well as more directly related to loss of entropy.
He suggests the formulation: "The rate of cultural change is proportional to the rate of energy conversion carried out within the system." (p.281). He emphasizes that this is not simply valid for the material portion of the system. "For not only does the amount of energy in the system have a direct relation to the amount of energy that will be communicated and stored, but it is also subjected to the inevitable human-cultural device of reduction to size." (p.281).
This "reduction to size" takes place through the central process of binary differentiation which Adams considers as providing the basis for ranking and the treatment of much of what is meant by value: "I do not know whether the mere fact of identification, that is, of making a binary differentiation, may be said to imply the immediate bestowal of something we may want to call value; and I am not sure that it really makes any difference" (p.155). A significant aspect of the process is that it is done constantly: "While there is obviously great individual difference in the relative ability to project new cuts in the environment, *we are nevertheless constantly imposing old bifurcate categories on new events*, thereby reducing them and simplifying them - in a word, mentally classifying them. More important, there are regularly new formulations of such differentiations, new ways of cutting up the world, that are invented and tried out. Most of these, like the lethal mutants of the genetic process, serve to extinguish themselves (and in some case their bearers)...Westerners have tended to see this process of recutting the world as something of a hallmark of progress. It can, however, also be seen as man's way of reducing the world to size, to terms with which he can deal" (p.281).
Adams points out that in these terms mankind can be viewed as a species confronting a constantly changing environment. The confrontations are however repeatedly made with *relatively fixed mental equipment*: "No matter how new the events perceived, they had to be reduced to a comprehensible scope and to familiar dimensions. The totality of the energetic component may have been beyond his control, but man could always cut a piece of it down to size and form it to fit the "order" demanded by his mentalistic limitations...So while societies become increasingly complex in terms of their energetic structures, their organizational dimensions are constantly reintegrated to mentalistic structural dimensions that are comprehensible to the human mind" (p.282).
Adams draws attention to research on the apparent limitation on the number of taxonomic dimensions that the human mind can comfortably handle within a social communication context (p.157-8). This number appears to be around six or seven as discussed earlier. He cites studies of folk taxonomies showing that there are at least five, perhaps six, taxonomic ethnobiological categories which appear to be highly general if not universal. They are arranged hierarchically and taxa assigned to each rank are mutually exclusive. One modern example given is a banner in a hall at the Palais des Nations (Geneva), indicating: family, village, clan, mediaeval state, nation, federation (p.158-9).
For Adams the limited number of levels of integration a society uses to describe its own organization then replaces in practice the levels of articulation that may be empirically found in the course of interactions in society (p.282). Such levels of description then become significant determinants of the kind of structures which can be perceived as emerging or required in society. Adams points out that the process of binary differentiation, taxonomy making and classification, and ranking with its implicit bestowal of priority, is not an unorganized activity unrelated to the question of power and control: "It is, rather, a mentalistic structural concomitant of overt control...Ranking, then, is an attempt to arrange events in the external world so that they will behave as our mental limitations dictate and will reflect our ability to handle them. It becomes a way to put order in the environment, to imbue things with a positive or negative value that permits them to be maximized, minimized, or optimized" (p.166).
Given the above relationship between the mentalistic and energetic components, much of Adam's study is concerned with how a (demographically) expanding society organizes itself. For him, the process whereby centralized units expand through a multiplication of their numbers is a coordinate growth process not involving any qualitative change. Centralization, however, marks a qualitative change in the amount of energy that is being brought under control within one part of the system. The process of development has as a parallel a process of coordination. This correlates closely with the process of ecological succession except that, instead of moving into a steady state limiting further expansion, new inventions set aside this governing mechanism and permit an increase of energy input to press for a continued expansion (p.287). Of great interest therefore is the possibility of using computers to assist in the invention of better cuts of the environment which remain comprehensible. This is one reason for further investigating tensegrity organization as a more powerful way of handling and integrating sets of binary differentiations.
Adams draws attention to the oscillation between the two modes noted above (which correspond to mentalistic and energetic emphases): "The alternation of phases of coordination and centralization that can be seen in the macroview of societal and cultural evolution is equally useful in the examinationa of the processes that particular societies are undergoing at a given point in time...This oscillation may take place simultaneously in two phases or dimensions: (a) horizontal, that is, the shift from a fragmented (identity) unit to a coordinated unit and back (in other terms, fusion and fission, or recombination and segmentation); and (b) vertical, that is, the shift from a coordinated unit to a centralized unit and back (also described as integration and disintegration, centralization and decentralization, etc.)" (p.290-3).
The literature of ethnography and history is replete with instances of societies undergoing some such kind of oscillation, of which Adams gives a number of examples. With regard to this alternation process he concludes: "I think that we would have to argue that *oscillations are inevitable parts of the evolutionary process*; they are the *ongoing trial-and-error of a unit*, at whatever level, the coming into direct touch with the environment, the testing of the validity of mentalistic pictures and accumulated knowledge. It is the constant inherent structural push toward expansion that makes actors and the units they operate in try again. The oscillating pattern simply means some lack of success, which may be due to any of a wide variety of circumstances. But "success" is hardly the appropriate word, particularly when we recognize that consumption and destruction are both necessary parts of the scene. The fact that old people die will, in the long run, mean success for the young. Or what is successful centralization for one nation, state, chiefdom, may spell disaster for another. What is important about the oscillation process is that it cues the observer as to what he should be looking for. Every operating unit will be at some stage of oscillation at any point in time; to seek out its state and the factors that make it move is to understand how the power system is currently working" (p.298. emphasis added).

♦ KD2305 Uncertainty and the function of ignorance

Comment In a study still in progress, John and David Keppel explore the situation of man at the interface between the entropic degradation of order and the development of new forms of order (1982). For them new conceptual tools are required to respond to those aspects of the current social condition for which determinism has proved inadequate, if not dangerous in its efforts to maximize certainty and to marginalize uncertainty. They see the key as lying inn the logic of living systems with their inherent ability to deal with uncertainty to their own developmental advantage. There are principles involved in this logic which are valuable to any new ordering of society.

In such a framework the spread of control among diverse entities serves as a focus for the development of any such system. Error, imperfection and accident, with their implicit static bias, are in fact vital to learning within any living process, in contrast to their current status in societal management. Destruction of information in any form may well lead to the recovery of uncommitted potentiality for adaptive response to change. For the Keppels what persists are the broad principles according to which things in flux relate to each other.

The preliminary study stresses the importance of the partnership of living beings as essential to regulatory feedback processes based on differences. Such symbiosis depends on the essential diversity or non-homogeneity of society. Reciprocal relations underlie all evolutionary processes and are essential to mental development, psychic balance and cultural advance.

Whilst the arguments are developed in some detail with specific suggestions for political initiatives in the United States, at both national and at local levels, the study does not - in its preliminary form - demonstrate the nature of the new conceptual and social patterns required. It is one thing to draw attention to the principles involved and to show how they work in nature. It is another to elaborate their implications at the level at which they enable new initiatives to be undertaken. They agree that evolution works by the adaptation of existing structures rather than by exact design of a new structure suited to the new order. But the question is what form to give to the "conceptual ley lines" to enable such adaptation to take place in a decentralized manner. Is it sufficient to assume that the logic of living uncertainty is an adequate guideline ? If not, how is this understanding to be "geared down" to facilitate the emergence of more appropriate patterns ? How is it to avoid being coopted as a newly fashionable cosmetic for unchanging strategies, as have so many previous guidelines over the past decades - cooperation, development, interdisciplinarity, networking, etc ?

This difficulty calls for a means of building into the conceptual framework countervailing elements of a kind which correspond to the diversity-enhancing nature of the principles involved. Specifically the study does not take into account the dynamics resulting from opposition to those principles and the process of attempting to implement them. How are they to respond to their own negation ? How are they to "dance" with those of the opposition ? It is giving form to the nature of the dance which is the core of the problem of "coherence" which the authors identify as the only authoritative answer to the present chaos of problems and rival solutions. But in the authors' own terms, coherence must surely dance with incoherence in a developing society grounded on uncertainty.

Uncertainty is clearly closely related to "future possibility". Both are associated with "ignorance", a feature of the learning process discussed in previous entries. Society may have various attitudes towards all three. At the moment defensiveness prevails. Ignorance in particular is a social "evil". It must be "eliminated", despite the fact that it is "regenerated" with every baby born "ignorant", and with every scientific and cultural innovation about which people are as yet ignorant. It is also generated in government, military and commercial practices, even at the grass-roots level, through the need to avoid revealing the truth under certain conditions.

Specialists are necessarily educated to be ignorant of domains other than that with which they are specifically concerned. As with the generation of event-horizons by black holes, ignorance may be viewed as the orifice through which we enter time, or leave it. The problem is not to eliminate it, but rather to accept it (as every parent does, usually with pleasure) to recognize its positive functions, and to give it a central "place" in society rather than marginalizing it. Unless it is appropriately "contained", society is unable to relate effectively to the direction from which its own future development will emerge. Rejection of ignorance is a rejection of transformative development. All that then remains is non-transformative development in the light of pre-existing, "ignorance-free" programmes.

♦ KD2310 Morphic resonance

Comment An understanding of development calls for an understanding of how the shapes and behavioural patterns of psycho-social entities are determined. Rupert Sheldrake, a biochemist, has recently explored the limitations of the prevailing paradigm in biology and has put forward an original and revolutionary answer to this problem. His closely argued thesis is that the form, development and behaviour of living organisms, including human beings, are shaped by "morphogenetic fields" of a type at present not recognized by physics. His presentation suggests a fruitful new approach to thinking about the emergence, stabilization and development of societal, institutional and conceptual patterns and structures.

Sheldrake argues that such morphogenetic fields are moulded by the form and behaviour of past organisms of the same species through direct connections across both space and time: "The characteristic form of a given morphic unit is determined by the forms of previous similar systems which act upon it across time and space by a process called morphic resonance. This influence takes place through the morphogenetic field and depends on the system's three-dimensional structures and patterns of vibration. Morphic resonance is analogous to energetic resonance in its specificity, but is not explicable in terms of any known type of resonance, nor does it involved a transmission of energy". (1981, p.116-7).

The relevance to human and social development lies in the way in which this insight clarifies the influence or constraining effect of past patterns on the possible emergence of new patterns. What he is suggesting is that "by morphic resonance the form of a system, including its characteristic internal structure and vibrational frequencies, becomes present to a subsequent system with a similar form; the spatio-temporal pattern of the forms superimposes itself on the latter". (p.96). Forms therefore get "canalized" or locked into particular developmental pathways known as "chreodes". The difficulty of shifting into more fruitful developmental pathways is thus explained by the "weight" of all past systems of similar form. These act to increase the probability of the repetition of forms of a given type: "The most frequent type of previous form makes the greatest contribution by morphic resonance, the least frequent the least: morphogenetic fields are not precisely defined but are represented by probability structures which depend on the statistical distribution of previous similar forms". (p.118)

The emergence of new forms, or a "new order", of any kind is therefore a low probability event difficult to bring about. But once the pattern has been brought about it becomes progressively easier to maintain. "Once the final form of a morphic unit is actualized, the continued action of morphic resonance from similar past forms stabilizes and maintains it". Sheldrake puts forward evidence in support of this hypothesis and suggests a number of experiments by which it may be verified.

His perspective can also be applied to the problem of learning about the nature of any new order whilst "immersed" in the patterns of the old order: "People usually repeat characteristically structured activities which have already been performed over and over again by many generations of their predecessors...All the patterns of activity characteristic of a given culture can be regarded as chreodes...morphic resonance cannot itself lead an individual into one set of chreodes rather than another. So none of these patterns of behaviour expresses itself spontaneously: all have to be learned...Then as the process of learning begins, usually by imitation, the performance of a characteristic pattern of movement brings the individual into morphic resonance with all those who have carried out this pattern of movement in the past. Consequently learning is facilitated as the individual "tunes in" to specific chreodes".

This suggests the link between learning, chreodes and the "answer domains" which were the point of departure of this paper. Sheldrake in effect clarifies the specificity of patterns of credibility and the relationship between them. In the psycho-conceptual realm, his morphogenetic fields are as much "credibility structures" as probability structures. It is easier, for example, to get funds for research that has been done in the past because its credibility has already been established. Such fields render patterns perceivable or recognizable.

Sheldrake provides a framework within which to take the alternation argument of this paper a step further. In the turbulence of modern society it is to be expected that at a particular moment strategy/pattern A would prove appropriate to a community (say). But the environmental turbulence may well erode its appropriateness and effectively call for the emergence of strategy/pattern B, or C, or D. There will probably be morphic predecessors for each, possibly embedded in the folk tales of the culture, to give some credibility to the "alternatives". But the alternatives are in each case relatively high probability/credibility structures compared to the meta-strategy/pattern (A:B:C:D), which allows the community to shift between these alternatives in response to the turbulence of the environment. There are few morphic predecessors to clarify this pattern. Hence the importance of seeking out the kinds of analogies recorded in this section.

The question then becomes how to increase the credibility of this meta-strategy/pattern rather than, as at present, how to increase the credibility of some momentarily significant alternative. *Visions of desirable futures could therefore usefully focus on alternation patterns* as much as on the specific patterns or alternatives between which alternation must necessarily take place in a dynamic society.

It is the meta-pattern of alternation which provides the transition pathways between its essentially antagonistic constituent patterns. Without such pathways the transition itself becomes socially catastrophic, irrespective of the catastrophes the appropriate alternative is designed to avert. Unfortunately the prevailing present mind-set requires that transitions should be socially catastrophic because, like a stumbling infant or a drunken adult, society has not yet collectively learned any better way of shifting from leg to leg in the process of moving forward.

♦ KD2320 Toward an enantiomorphic policy

Comment William Irwin Thompson, a cultural historian, has sharpened considerably the ecology-sensitive intuition concerning the psycho-social lessons to be learned from cooperation between co-evolving systems (1982). He stresses the importance of an appropriate understanding of the interaction between opposites by citing E F Schumacher:

"The pairs of opposites, of which *freedom and order* and *growth and decay* are the most basic, put tension into the world, a tension that sharpens man's sensitivity and increases his self-awareness. No real understanding is possible without awareness of these pairs of opposites which permeate everything man does...Justice is a denial of mercy, and mercy is a denial of justice. Only a higher force can reconcile these opposites: wisdom. The problem cannot be solved but wisdom can transcend it. Similarly, societies need stability *and* change, tradition *and* innovation, public interest *and* private interest, planning *and* laissez-faire, order *and* freedom, growth *and* decay. Everywhere society's health depends on the simultaneous pursuit of mutually opposed activities or aims. The adoption of a final solution means a kind of death sentence for man's humanity and spells either cruelty or dissolution, generally both". (1978,p.127)

For Thompson any ecosystem is a form of life in which opposites interact: "It isn't the case that the ocean is right and the continent wrong..As it is with an ecosystem, so it is with a political system. It isn't the case that one party is right and the other party is wrong. Truth cannot be expressed in an ideology, for Truth is that which over-lights the conflict of opposed ideologies". (1982, p.32)

As with Attali, Thompson refers to Atlan's synthesis of information theory and biology. Atlan moves beyond the prevalent superficial enthusiasm for cooperation by recognizing the role of discontinuity in health development: "So then, it would suffice to consider organization as an uninterrupted process of disorganization-reorganization, and not as a state, so that order and disorder, the organized and the contingent, construction and destruction, life and death are no longer so distinct. And moreover that is not all. These processes where the unity of opposites manifests (such a unity is not realized as a new state, a synthesis of the thesis and the antithesis, *it is the movement of the process itself which constitutes the "synthesis"*), these processes cannot exist unless the errors are a priori true errors, that order at any given moment is truly disturbed by disorder, that destruction (though not total) is still real, that the irruption of the event is a veritable irruption (a catastrophe or a miracle or both). In other words, these processes which apear to us as one of the fundamental organizing features of living beings, the result of a sort of collaboration between what one customarily calls life and death, can only exist precisely when it is *not a question of co-operation but always radical opposition and negation"*. (emphasis added)

In this context Thompson argues that: "A global polity cannot be simply capitalist or communist, Christian or Muslim; it has to be a planetary ecology of each and all...As ecology begins to inform our perceptions of the body politic, we will begin to understand that any polity must be an interaction of opposites. In a policy that has the shape of opposites, the wisdom of William Blake's "In opposition is true friendship' will be understood". (1982, p.32)

Thompson does not elaborate on his understanding of the dynamic nature of the relations between opposites in the polity, except through the overused term "interaction" and the notion of "symbiosis". Despite his quotation of Atlan, he interprets such interaction as cooperation, without giving much more than the usual "public relations" content to the term. And yet it is the nature or pattern of healthy interaction which is the goal of the collective quest at this time.

By introducing the powerful concept of enantiomorphy he stresses the static aspect of the mirror image nature of the relationship between opposites, at an archetypal level. Elsewhere he introduced the "inexorable" process of enantiodromia. This is the dialectical movement in which a force, in its fullest development, turns into its opposite. The concept derives from ancient alchemy but was reintroduced into philosophy by C G Jung. It may be seen at work in the frequently remarked "convergence" between the supposedly opposed policies of the USA and the USSR. The process is often portrayed in drama.

This process is strongly related to that of alternation as explored here. And Thompson does affirm (1982, p.175) both the cyclic and innovative learning nature of this process by quoting the well-known verse of T S Eliot: "The end of all our exploring will be to arrive where we started and know the place for the first time". Thompson develops his argument by exploring in some detail "one model for a field of interacting opposites". He uses the traditional psycho-cultural image of the quaternity, a geometrical version of William Blake's "Fourfold Vision". This permits an enantiomorphic juxtaposition of the four basic political orientations he distinguishes: conservative, liberal, radical, reactionary. The four political parties "attempt to play out certain values in time...". He suggests that the structure can also be used to interrelate the four basic political and economic worlds he distinguishes: the capitalistic first world, the communist second world, the resource rich third world, and the fourth world of least developed countries:

"In the present transitional world-system, the interactions of the Four Worlds are unconscious, full of projections, and laden with conflict and of structural violence...The purpose of invoking the archaic Quaternity in a modern context of international relations is to make the unconscious conscious...The Quaternity enables us to see and model relationships of a more complex, polycentric variety". (p.50)

Thompson suggests that the fashionable centre-periphery model increases the potential for conflict by reinforcing simplistic perceptions of possible relationships. But he believes that in such a "planetary ecology...the health of the whole requires that one does not dominate the others".

(1982, p.50) This is an over-simplified (or possibly atemporal) understanding of "dominance" in ecological systems. It does not reflect Atlan's view (1979), nor does it accord with the process of alternation as argued here. Alternation, through enantiodromia, may indeed be understood to follow patterns such as Thompson's quarternity, but in them it is the dominance, or focus of power, which switches its locus in order to maintain the health of the whole. Dominance can only be absent in a system of maximum entropy associated with "energy death". Or it may apear to be absent, when in effect it has been displaced into a less obvious form whose consequences may be much more pernicious, as in the case violence.

Both Thompson and Klapp (1978) use the metaphor of a ball-game (with four zones or teams), whilst stressing that "the rules and the court are not the game; they are merely conditions that enable the game to be played". (1982, p.44). But initiative and dominance lie where the ball is located, even though it must be expressed by movement of the ball for the game to continue, with the real risk of its loss to another team whenever their strategy is more appropriate. But there is a vast difference between a good game and a bad one (whatever the quantitative indicators), and none of the above clarifies the art of playing a qualitatively superior game. It does suggest that such a game is possible.

♦ KD2325 Game comprehension and identity transformation

Comment In the previous entry Thompson draws attention to the need to understand the patterns encoded by more complex games. Jantsch cites Eigen who is investigating the new lessons to be learnt in biology concerning the "game of life". And indeed Eigen has co-authored a book on the "Laws of the Game" (1981). It is to be expected that such games can encode richer and more dynamic patterns, which is one reason for the resources allocated to war games and the hopes of finding "win-win" solutions attached to world modelling exercises based on game theory.

Xavier Sallantin, a military theorist working on the logic of conflictual systems, has provided a valuable analysis of the nature of the domains to which game theory applies, thus clarifying hidden dimensions underlying reliance on the game perspective (1976). Interestingly, in the light of the "answer arena" theme of this section, Sallantin uses a "gladiatorial arena" to illustrate his point. He notes that in that arena the gladiators risk their lives, the rule of the game being that one of them should die, however carefully they study their moves.

Surrounding that arena are the "stands" from which spectators observe, as well as betting their assets on the issue of the game. There are therefore two categories of "player" at such a "circus", one risking existence, the other risking possession. In "vital" games of the first degree, the gladiator risks himself, whilst in "venal" games of the second degree the spectator risks an object he possesses. The distinction between these domains is further clarified by the processes which occur when a spectator daringly jumps into the arena to taunt the gladiators - only to escape back again when the risk becomes too great, as is the case in many "demonstrations". Or when a gladiator jumps into the stands to expose bettors to the reality in which they only wish to participate vicariously, as occurs in cases of terrorism.

Sallatin demonstrates that in terms of logic, the negation resulting from the loss of the "bet" does not have the same status when applied to games of the first degree, involving a co-terminous subject and object (constituting a unity in arithmetic terms) as in the second, where they are distinct (constituting a duality). In physical terms the first terminates temporal existence, whereas the second terminates a spatial relationship (between the bettor and his property) having the character of a cohesive force. He shows that such distinctions are an essential condition for univocal communication, whether in biology or in informatics.

For Sallantin the ontological status of the game has an entirely different meaning if one's existence or identity is liable to be terminated by it. He points out that recognition of this meaning is what distinguishes militants', whether conscripted or self-appointed, from those who only risk the loss of a possession and may well re-enter a later phase of the game to gain it back twofold. It is one thing to risk loss of academic status in favouring Gandhian non-violence, for example, it is another to risk one's life in the active practice of it.

In games of the first degree a positive relationship to death must be developed which effectively redefines the game. The encounter with death involves a transformative process of great psychological significance for those who undergo it. This is absent in games of the second degree, except in a vicarious sense. Setting aside the problem of counteracting abuses of militancy in any form, Sallantin questions whether a society can satisfactorily order itself without the kinds of commitment and identity-risk implied by games of the first degree. This is the assumption made by those who seek substitutes for such games in games of the second degree. The positive function of games of the first degree is clarified in cultures recognizing the "way of the warrior". The theme of a proposed international conference which would not simply deny the value of the military perspective (1982).

To further clarify the hidden dimensions and degrees of freedom behind the rules of a game, Sallantin draws attention to the possibility that the gladiators might subvert the game, and the expectations of spectators, by seeking death together rather than fighting to live. He illustrates this possibility by locating a "pit of annihilation" in the centre of the arena into which one or both might jump. This then represents a game of degree zero. Although he does not discuss the possibility, presumably this also covers the case when the gladiators blow up both themselves and the observers in the stands.

Sallantin further develops his argument by mapping the arenas into a spheric geometric model which recalls Fuller's preoccupations. He suggests that the limited freedom in the confrontation between the gladiators (resembling that of a tournament "list") can best be represented by a diameter of a circle, where the area of the latter represents the domain of the bettors. The centre of the circle then represents the pit of annihilation (into which a gladiator may jump or be pushed). He then defines a third domain, having a further dimension of freedom, represented by the sphere of which such a circle is a cross-section. The sphere is then the "space" within which bettors and gladiators mentally model and evaluate the progress of the game in endeavouring to assess how best to make their next moves. It is also the conceptual domain in which our own thinking links with theirs in endeavouring to grasp the rules of the game. It is a transdimensional domain in that it permits moves across dimensional frontiers but it also negates the spaces of more limited degrees of freedom in which games of lower degrees become possible. This negating process is counteracted whenever a "position" is taken by the generation of a line representing a possible first degree game. The spheric "trans-spatial" domain is thus one of free interplay of possibilities from which particular games crystallize.

♦ KD2327 Resonance-based consensus in games

Comment Sallantin points out the need to correlate the conceptual freedom of the trans-spatial domain with the verbal domain within which consensus is established (1976). For a game to be possible, all involved must be "attuned" in a consensus on the rules, on a reference polarization, or on a direction of the game.

This is most evident in fixing a convention for the interpretation of the codification of a "bit" in informatics, as being signalled either by "switch on" or "switch off". A similar convention is necessary to specify which pole of a battery is to be considered "positive" or "negative". Sallantin's striking example of the fundamental nature of this question is that of a referee who asks two players before the game whether they understand the rules. Both nod their heads. However one comes from a culture in which, unknown to the referee, a nod indicates a negative, not an affirmative. For players to be in agreement, they must first be in agreement about the significance of the verb "to agree".

Sallantin argues that any such consensus is intimately related to the physical phenomenon of resonance. There is an ontological correlation between verbal agreement and the physical resonance expressing that agreement or in the syntony between sender and received: "Il faut nécessairement que le signifié et le signifiant de la concordance concordent, sinon la concordance signifierait la discordance".

Each of the four "universals" so defined for any game are then interpreted by Sallantin in terms of four distinct ways of being:

(a) being in time or existing; (b) being bound by a force to some totality sharing a common characteristic; (c) not being, or being absent, for lack of an appropriate space within which to act; and (d) being that of which the meaning is the subject of a consensus.

The first three are aspects of the fourth. Sallantin suggests that the fourth may be represented by a hyperspace, as an affirmatory polar complement to the negatory function represented by the centre of the sphere. The development of the system through increase in its negentropy is then a function of the settings or levels on which the consensus is based to determine the nature of possible games. Sallantin suggests that when the processes of society as a whole are seen as a game, the four different domains (vital, venal, conceptual and verbal) may then be associated with different aspects of society (military, economic, political and social, respectively). His use of "military" needs however to be seen as signifying any creative confrontation with the risk of the loss of identity.

His insights cast a fruitful light on the arguments of previous entries sections. Once again the need for a fourfold grasp of a system through distinct "languages" is demonstrated. Use of a spheric model ties in with earlier arguments concerning a non-linear container, as does his insistence on the importance of resonance in interrelating the participants. Note however that consensus understood as based on resonance is quite different in nature from a static, superficial consensus where there is no understanding of the resonance dynamics which make it possible.

"Le consensus ne peut être le fruit que de la clarté qui est l'expression optique de la résonance. Tout accord de surface reste précaire et vulnérable; il faut aller au fond des problèmes pour dissiper les malentendus. La souffrance qu'engendrent ces désaccords contraint la pensée à un approfondissement en vue d'élucider la racine des contradictions".

It is this constraining force which is a vital aspect of the human and social development process.

♦ KD2329 Learning through loss as a cyclic phase

Comment Prior to the end-game, the dynamics of a game as examined by Sallantin (1976) may be seen for each participant as an exciting alternation between conditions of "advantage" and "disadvantage". With the termination of the game and the alternation process, the identity of one participant is "exalted" and that of the other is "extinguished", crushed or dissipated. There is a distinct *transformation of state*, achievement of which is usually the object of the game, whether sought or feared as a resolution of uncertainty. This change of state constitutes a form of development. According to conventional thinking, winning is obviously better, since it ensures immediate "development" (for the winner), whereas losing is to be avoided at all cost (as an unwelcome increase in personal "entropy"). Winning is perhaps the most widely accepted social indicator of development. Development theorists seek "win-win" solutions to avoid the unfortunate loss of identity, or the continuous generation of "losers" in a two-class society: we should all be "winners".

The preoccupation with winning is also confused with the cult of the new, the cult of youth, the cult of the beautiful, the cult of "bigness", and the cult of "wealth". These reinforce each other so that achievement of any of them is to some degree an achievement of the others. Unchecked, such cults respectively favour: the exploitation of non-renewable resources and the erosion of collective learning, the rejection and institutionalization of the elderly, the avoidance of unbeautiful realities (including toxic waste dumps, slums and the deformed), the inhibition of grass-roots initiatives, and the marginalization of the poor. It is precisely this *obsession with winning and the avoidance of loss which obscures a more fundamental alternation process on which long-term human and social development may well be grounded.*

The problem with win maximizing is that success tends to be due to the deployment of a particular set of attributes which confer advantage under particular environmental conditions. The winner is however trapped by these attributes when the environment changes and other sets of attributes have an advantage. New "winners" then tend to emerge from the pool of "losers". It is in fact in this pool that are conserved those "psycho-social genes" governing attributes not currently manifest. But whilst the winners have relatively little freedom within their defining attributes, new forms can emerge from the pool of losers into which all winners mut eventually be reabsorbed. In terms of long-term human and social development, there is therefore in operation an ecocyclic process. *Focus on a single game merely offers insights into a portion of that cycle - a broken cycle.* It does not show what happens to the winner after reaching the state of identity exaltation, nor to the loser after having been exposed to identity extinction.

A new dimension may be added to Sallantin's analysis by introducing the concept of "degree of identification" for this is basically what distinguishes gladiators from bettors. (It would be fruitful to explore Atkin's (1977) description of communication geometry as a framework for identity of different degrees). The identity of the gladiator tends to be engaged in the game, through his physical "self-bet", to a far greater degree than that of the bettor. But clearly if the bettor's self-image is identified to a very high degree with his possessions, which he then loses, then he too may well be psychologically destroyed by the outcome of the game. What is interesting about this kind of identity "death", which R D Laing has shown to be a powerful existential experience, is that it opens up the possibility of a "rebirth", if the player can then reformulate his identity on a new foundation. The winner, once exalted does not however have access to this possibility of rebirth, which necessarily requires a destruction of the set of characteristics by which his identity as a winner is defined. Even if the winner wins in a new game the merely confirms and extends the exaltation of his identity, but he does not renew it. It is the difference between a quantitative and a qualitative development. The latter calls for a fundamental transformation based on a radical "mise en question" following real loss.

The loss phase can be related to the learning process. As Kenneth Boulding points out: *"Disappointment forces a learning process of some kind upon us; success does not".* (1978, p.133). There is then a need to change the image of the world in some way (p.145). He suggests that science itself is essentially a system of organized learning from disappointments" (p.135). Beyond the play on words, there is evenvalue in the link between "appointment" (as a win-phase) and "dis-appointment" as the loss-phase which must necessarily follow it.

By designing strategies to minimize disappointment, there is clearly the risk of minimizing learning. Again, Boulding notes: "One of the most striking phenomena of the human learning process is the extent to which it is self-limiting. Far beyond the physiological capacity of the human nervous system, *we learn not to learn.* We paint ourselves into a tiny corner of the bast ballroom of the human nervous system". (p.156-7). This then implies that we learn not to engage in transformative development.

This suggests that *long-term human and social development is based on a process involving risk of identity loss, winning and losing.* Periods of losing are then as important as periods of winning to the development process. Just as *"small is beautiful"* so *"decay is OK"* - a fairly obvious remark with regard to ecological processes. Real strength, in military terms, comes with the ability to accept loss and the lessons it brings, including the ability to be weak and disorganized. Real weakness results from an identification with the need to win always and be permanently

strong – in judo terms, the inability to take a fall and learn to lose as part of a larger process. Development through alternation between the conditions of winning and losing is then associated with the ability to "disintegrate" at will and to "reintegrate" at will – without long-term identification with the forms used in this process.
This recalls the arguments of de Nicolas (1978) concerning the fundamental importance of sacrifice as part of the renewal of form in the Rg Veda. Educating a child, for example, involves an understanding of when the child should lose and when the child should be allowed to win – accepting the fact that at some stage it will no longer be a question of "allowing" him to win. The teacher must eventually accept the opportunity of identity loss as a real loser if the student is ever genuinely to experience the nature of the win portion of the cycle and learn how to use it responsibly. The teacher, like the parent and the psychoanalyst, must accept rejection if the student is to be free.
The win mind-set is partly responsible for the inadequacy of the response to economic cycles. Troughs are necessarily experienced as regrettable, and efforts are necessarily made to maintain peaks – but it would probably be extremely unhealthy to eliminate the cycle, even if that were possible. The problem is that the transitions and associated transformations are spastically forced upon the relevant actors for lack of any sense of the developmental significance of any cycle involving loss. To employ a biological metaphor, deciduous trees are a more advanced evolutionary form than evergreens precisely because of their ability to engage in a cycle of leaf loss and subsequent regeneration. The combustion engine is possible because it integrates a cycle of ignition/combustion with extinction/evacuation, in which the latter makes possible the power stroke of the former portion.
Loss phobia and win mania, which are themselves integral and necessary features of a larger alternation cycle, obscure the nature of that cycle and its significance for human and social development. It is unfortunate that lack of awareness of such cycles may well contribute to the ambiguous status of widely used techniques common to political "re-education", business executive "re-motivation" (in Japan), religious "conversion", and military manpower "training". In these highly successful processes, whose phases are now well-defined, "stripping" of identity is one of the techniques applied as a preliminary to forcing the person to "win-through" to a new understanding and self-image. Greater understanding of such cycles is required to determine to what extent the "manipulative" nature of such techniques of "human development" is acceptable, under what conditions, and to whom – and whether more acceptable processes can be envisaged.

♦ KD2335 Ecodynamics and societal evolution
Comment Under the concept ecodynamics, Kenneth Boulding, an economist, has attempted an ambitious synthesis which interrelates many physical, biological and social processes that are usually kept apart (1978). Of special interest in the light of the argument of this section is that his synthesis internalizes both its conflictural relationship to other competing visions and the recognition of its own mortality as an artifact. "Every vision, of course, conflicts with other visions...Each vision must be understood in terms of what it is not as well as in terms of what it is" (p.19)
He therefore indicates how his evolutionary vision is "unfriendly" to various other visions to which his is an alternative. A special feature of his vision is the recognition of the dynamics of the relationship between such visions as they emerge as artifacts and occupy and expand various niches in society.
For Boulding: "The pattern of human development is therefore seen to be an extension, enlargement, and acceleration of the pattern of biological development, operating through mutation and selection". (p.18). Social dynamics is then to be thought of primarily as the evolution of human artifacts. Human artifacts not only include material structures and objects, they also include organizations, institutions and social groupings. These all originate and are sustained by images in the human mind. Such artifacts are species just as much as biological artifacts. And:
"Just as there is the genosphere or genetic know-how in the biosphere, so there is a noosphere of human knowledge and know-how in the sociosphere. The noosphere is the totality of the cognitive content, including values, of all human nervous systems, plus the prosthetic devices by which this system is extended and integrated in the form of libraries, computers, telephones..." (p.122)
The processes of biological evolution are also to be found in the evolution of human artifacts, namely replication, recombination, reconstitution, redefinition (mutation) and selection. Boulding suggest that Darwin's metaphor concerning the "survival of the fittest" is unfortunate. "A more accurate metaphor would be the survival of the fitting, the fitting being what fits into a niche in an ecosystem" (p.110). This corresponds to the self-consistency constraint imposed on the organization of dissipative structures. Furthermore: "The social dynamics of human history, even more than that of biological evolution, illustrate the fundamental principle of ecological evolution – that everything depends on everything else. The nine elements that we have described in societal evolution of the three families of phenotypes – the phyla of things, organizations and people, the genetic bases in knowledge operating through energy and materials to produce phenotypes, and the three bonding relations of threat, integration and exchange – all interact on each other". (p.224)
Boulding sharpens the generality of this statement by noting that it is changes in knowledge or know-how that are the basic source of all other changes. In biological evolution it is the genetic structure that evolves with the phenotypes as encoded carriers of it. In societal evolution it is the human artifacts which are encoders of the knowledge structure that is nevertheless continually expressed in human beings. Knowledge is therefore primal as "what evolves" (p.224–5). The "noogenetic" processes by which each generation of human beings learns from the last then come to be of far greater importance than the biogenetic processes by which genes are transmitted. But Boulding sees as the ultimate constraint that "we learn not to learn" and that "the learning patterns themselves are self-limiting". (p.123)
The question is then what self-limiting patterns emerge and how are they to be comprehended ? For it is then these patterns which govern the kinds of human and social development that are currently possible – unless richer patterns can be designed or comprehended.

♦ KD2337 Self-limiting patterns of interaction
Comment Boulding has many reservations about dialectics as one of the basic patterns, aside from the problems of the confused variety of meanings associated with it. "The substantive question, which is very difficult to answer, is just what is the quantitative or even qualitative significance of dialectical processes, interpreted narrowly as involving conscious conflict, struggle, victory and defeat, winning or losing, revolution and counterrevolution, war and peace, in the great four-dimensional tapestry of the universe". (1978, p.262) He asks when it matters who wins and suggests that such processes affect the details rather than the larger patterns of history, arguing that this could however be significant on historical "watersheds" (p.262–5)
Boulding recognizes that few people accept this view:
"Dialectics in many different forms has a surprisingly good press. Most people believe that struggle is very important and that it is important to be on the right side in a conflict...Part of the difficulty is that the human race has an enormous and by no means unreasonable passion for the dramatic, and conflict is much more dramatic than production...The awful truth about the universe – that it is not only rather a muddle, but also pretty dull – is wholly unacceptable to the human imagination. Nevertheless, it is the dull, nondialectical processes that hold the world together, that move it forward, and that provide the setting within which the dialectical processes take place. Evolution is the theatre, dialectics the play. It is a tragic error to mistake the play for the theatre, however, because that all too easily ends in the theatre burning down...Unless there is a reasonably widespread appreciation of the proper role of dialectical processes, these tend to get out of hand and become extremely destructive...doing more harm than good". (p.266)
The popularity of dialectics, regretfully noted by Boulding, is however due to the ("seductive") sense of transformation with which it is associated. This is necessarily absent from the "dull" production processes which "hold the world together", protecting it from the effects of random disturbances. The two can however best be perceived as complementary. But the problem is indeed one of how to provide the setting within which the dialectical processes can take place. Is this not the problem of comprehending equilibrium processes as a context for disequilibrium processes, or at least of designing the (shifting) balance between them ? Here however Boulding does not offer many insights because he is seemingly handicapped by his stress on "everything depends on everything else". How indeed is the manifold to be structured for comprehension, or by it ?
Boulding offers a useful point of departure in a discussion of types of ecological interaction between two species. These are (p.78): mutual cooperation; parasitism; predation; mutual competition; dominant – cooperative; dominant – competitive; mutual independence.
Now in a biological environment the species may well be locked into one of these types of interaction. In the case of the social environment interacting species, including organizations and roles, may well switch from one type of interaction to another. This is perhaps seen most clearly in the case of two individuals (e.g. husband and wife) or of two nation-states. In both cases there can be an alternation between many of the types according to circumstances. Within any type the interaction may well have dramatic potential as a disequilibrium cycle (e.g. predation cycles). Switching between types of interaction may also constitute a dramatic change of strategy as a disequilibrium process for the individuals involved. But the pattern of alternation between strategies would then provide an equilibrium context for such shifts.
Conventional values stress the importance of everyone abandoning other types of interaction in favour of the "mutual cooperation" type. Boulding suggests, using the classic example of the "prisoner's dilemma", that "there is a very long-run evolutionary process in this direction that is precarious, however, in the sense that it is constantly being interrupted by lapses" (p.204). He points out that there are "limits to love" about which little is known (p.203 and 304). Of course these "lapses" are the very ingredients of much that is valued for its dramatic significance in any period of culture. It is these lapses, continually engendered by the birth of "ignorant" children, which renew the learning cycle associated with alternation through the other types.
Now in the case of the biological environment, Boulding points out that "the biosphere recycles its materials through all the organisms that comprise it" (p.86). Examples are the nitrogen cycle and the carbon dioxide-oxygen cycle. But he does not suggest, as his vision implies, that the noosphere could recycle its materials through the organizations (and other artifacts) which comprise it. The question is if it did, or rather does, what are those cycles and how do they interweave ? For it is then their interweaving that provides the context for the transformative, dialectical processes which are hopefully to be appropriately contained.
The tragedy of the seven types of interaction is that they emerge as the only credible set of alternatives because of the binary valued logic which governs their generation (e.g. 2 species, each affected positively or negatively by the interaction, or not at all). Such a limited set has the twofold disadvantage of reinforcing the conflictual logic of dialectics without strengthening and enriching the non-dialectical context through which such transformative processes play themselves out. (The "mutual cooperation" type then merely functions as a "day of rest" which evokes the dramatic opportunities of the other six). They constitute a self-limiting pattern.
If the number of basic distinctions made was greater (e.g. $N=20$, for example) a much finer grained set of alternatives would emerge. These would provide a richer ecology of interactions within which people and groups could develop. The finer the grain the more probable it is that people would find that one or more such interaction categories was specifically meaningful to their condition. The ability to vary N would introduce a new degree of conceptual freedom. The problem is of course to safeguard the integrity of each such set and relate it to others.
This suggests the need to elaborate a continuum of patterns of which the simplest would correspond to the transformative disequilibrium processes. The more complex would correspond to the equilibrium processes based on a richer set of interaction types woven together by a variety of interlocking and mutual stabilizing cycles through which alternation takes place. In the light of Sheldrake's argument (1981), the existence of such a range of patterns should make it easier to avoid learning not to learn within the current self-limiting patterns. This should help to release the potential for healthier human and social development.

♦ KD2340 Language of probabilistic vision of the world
Comment The Soviet statistician, V V Nalimov, has recently completed a trilogy of which the third volume (1982) constitutes a remarkable synthesis drawing on the entire range of knowledge (including elements of semantics, natural and social sciences, mysticism, and the arts) in an effort to understand how the human mind perceives the world. The methodology is borrowed largely from physics (as capable of tolerating paradoxes within its own theories), with considerable attention to the role of metaphor and the function of human imagination in capturing manifestations of consciousness and unconsciousness. The primary ontological position is that the world is an open one, the outcome of processes that are probabilistic in nature and constantly the domain of novelties and uncertainties. The language in which one captures aspects of reality is itself polymorphic, metaphorical, and constrained by Godelian principles of undecidability. Right and left hemisphere modes of consciousness are, through links provided by the unconscious, capable of functioning along unusual circuits so that sequential processes become concurrent in real time.
For Nalimov the "words, on which our culture is based, do not and cannot have an atomistic meaning. It has become possible and even necessary to consider words as possessing fuzzy semantic fields over which the probabilistic distribution function is constructed and to consider people as probabilistic receivers" (p.5–6). This leads him to ask whether taxa are discrete, as is normally assumed in the various typologies through which the world is perceived, especially in connection with human and social development. "What we are considering is not merely the probabilistic vision of the world stemming from its infinite complexity, but inwardly deterministic "in fact". We mean the probalistic world where probability lies at the core of the world. We mean probabilistic ontology of the probabilistic world, not probabilistic epistemology of a deterministic world" (p.6)
Acquaintance with taxonomy in various branches of knowledge leads him to suggest the hypothesis that "taxa probabilistic by nature are neither an exception to nor a result of deficiency in our cognition; they are the rule and an immanent property of the world". (p.7) Consequently in the probabilistic world clear-cut boundaries and the absence of embarrassing "transitional forms" merely testify to a reduction in completeness.
"We are accustomed to the idea that evolution means the appearance of something absolutely new. The truly new thing is a new taxon or archetype. But within the world of probabilistic taxa, archetypes, and individuals, evolution may take another course: it suffices to redistribute probabilities. A rare deviation becomes the norm, while the norm becomes an abnormality, an atavism. But potentiality still exists...What we keep in mind is the constant potentiality which underlies

various probabilities of manifestation". (p.8)

For Nalimov, culture is a deep collective consciousness whose roots lie in the remotest part. It forms a fuzzy mozaic of concepts with the distribution function of probabilities given over it. But real people have their own individual probabilistic filters of perception which generate personal perception of culture, again probabilistically given. The (Jungian) collective unconsciousness is then related to low probability concepts. Groups in society with similar filters then constitute clusters of psychic genotypes. In this probabilistic sense man is never free, being dominated by the past as stored in the collective consciousness. But at the same time he is free because the genotype does not rigidly determine the probabilistic structure of an individual filter of perception; it only gives certain possibilities for its formation (p.9)

Nalimov stresses the continuous nature of consciousness, with which a person is always in contact, but which cannot be reduced to the discreteness of language (except partially through rhythmical texts). Phrases constructed over discrete symbol-words are always interpreted at the continuous level. "The continuous nature of everyday language finds its expression in the limitless divisibility of the verbal meanings, while the continuous nature of the morphology of the animate world is expressed by the impossibility of constructing a discrete taxonomy". (p.30)

This leads him to suggest that: "Perhaps we should be more cautious and speak not of consciousness of the world but of the semantic field of universal significance through which the world we know as divided is restored in its wholeness. Is there any other way to imagine the world in its integrality ?" (p.14)

The question is then what is the semantic field of the world. "How can the fuzzy, probabilistically weighted vision of the world be combined with formal logic, which we cannot afford to reject," (p.15). If one phenomenon can be explained by several scientific hypotheses, "there is still a tendency to evaluate these hypotheses and to select the only true one. If this cannot be done, the situation is evaluated as obviously unsatisfactory. Is it possible to act otherwise, i.e., to perceive the phenomenon through a field of hypotheses, without their discrimination...But shall we be able to cope with this at the psychological level: is our mind ready for this vision of the world ?" (p.16). And how is this field to be conceived as being organized for comprehension ?

Nalimov suggests that: "If all the taxas of our culture, despite their uniqueness, are but various translations of our Text...and biological species are various translations of another Text, then it is quite plausible that they all are translations of one and the same Text". (p.14). The world is then seen as a text, and through texts accessible to consciousness individuals interact with the world (p.26). The question directly relevant to human and social development is how such texts evolve.

Transformative evolution, according to Nalimov, results from the multiplicative nature of the interaction between two probability distribution functions, one corresponding to the initial text and the second to a preference filter. Then it is not the initial text but the filter which changes. Thus the progress of science also consists effectively of the endless filtration of new ideas through the filter of paradigmatic criteria carried by past conceptions – filters which may be softened, rather than destroyed, when they prove inadequate.

Innovative development can then be seen as a response to a semantic vacuum. "New texts are always a result of free creativity realized on a probabilistic set which may be regarded as an unexposed semantic universe or nothing, the semantic vacuum or, metaphorically, an analogue of the physical vacuum. Here we deal with the problem of nothing which stimulated thinkers both in the East...and in the West..." (p. 29)

Nalimov devotes three chapters to this question. Of special interest is his concept of the way a person is defined for himself and others both through the generation of discrete words or symbols and through their comprehension at a continuous level. Both aspects are realized through contact with the semantic field, which in physical terms can be described as emission and absorption of semantic field quanta. But in the light of the Heisenberg uncertainty relation, "semantic interaction between people...is possible only as a consequence of semantic fuzziness of both the human psychic domain and verbal semantics". (p.76). Humans may however, interact with the semantic vacuum through unobservable (virtual) vacuum manifestations:

"Just as any physical virtual particles of various types, very similar processes go on in the psychic domain too. The latter may be described as a constant fluctuation of the probability distribution function determining a person's individuality on the semantic field. A human being never remains frozen and unchanged". (p.77)

Nalimov then reviews experiments in meditation, some of which were conducted with his colleagues. In his terms such experiences: "may be reinterpreted as a ceaseless reconstruction of the distribution function of probabilities determining a person's individuality, and the distribution function determining personality is becoming needle-shaped. Meditation is a technique that allows people to loosen this unbearably narrow structure, to make it fuzzy. (p.138)

These considerations lead Nalimov to ask the question: "in what way can a possible change in the metrics of the semantic space be interpreted ?" (p.287). In the physical world he sees the entire past, through the colonial period, as involving an expansion of the life space. "In contrast, we see the new as an expansion of the space for human existence – the entrance into new psychological spaces. But is our consciousness mature enough not to pollute these new spaces ?" (p.300). Such new spaces also call for a multidimensional concept of personality which he explores (p.287).

♦ KD2342 Patterns of complementary metaphors

Comment Of special interest in the light of the arguments of this section Nalimov's view that the Aristotelian bimodal logic cannot be replaced by another one: "The insufficiency of logic in everyday language is made up for by the use of metaphors. The logic of the text and its metaphorical side are two mutually complementary phenomena. And, to my mind, further evolution of linguistic means should proceed by deepening language complementarity rather than by searching for another kind of logic". (1982, p. 281)

According to the complementarity principle, in order to reproduce an object in its integrity it is necessary to alternate through a pattern of descriptions of it in terms of mutually exclusive classes of concepts. This implies the use of a "manifold of models generated by essentially different paradigms", but Nalimov questions "whether people are prepared for this intentionally incomplete vision of the world" (p.276 and 283). However, as has been argued in earlier entries: *"If it is typical of human reason to perceive the world through antinomies, why not try to find a language in which these antinomies...would act as mutually complementary principles"* (p.284). Nalimov points out that: "Clear-cut conceptualization oppositions create the polarization without which the passionate temperament of individuals that provides society with its energy could not have been realized" (p.294). Such "passionarity" looks like an obsession and has dramatic consequences when expressed collectively. But without it, persons or nations may lose their energy. "We must acknowledge passionarity to be immanent to people. A selective manifestation of the whole, man tends to discover the entrance for the selectivity and thus acquires energy" (p.295).

Again as argued in this section, Nalimov sees the *manifested semantic universe as structured not by logic, but by number* (p.285). In this sense he considers the probabilistic vision of the world to be a realization of the dream of Pythagoras and Plotinus of describing the world in its integrity and fuzziness through number, or through a "koan of numbers" (p.32-36). But despite an extensive review of Eastern and Western symbols (including mandalas), he seems to limit his attention to numbers in their probablistic sense rather than extending it to include their configurative geometrical sense, although Chladni patterns (1961) might be considered a link between the two. Whereas this section stresses the possibility of using *number-governed configurations of complementary languages, symbols or metaphors* to provide a variety of learner-responsive ways of organizing comprehension of the semantic universe and the possibilities of acting in it.

In the light of Nalimov's work it is possible to envisage configurations of complementary languages as being organized as N-dimensional "semantic Chladni patterns". These would probably only be comprehensible where N was less than 4. Clearly when N was 2, these could resemble many of the symbols discussed by Nalimov. With N equals to 2 or 3, they might resemble the "macrons" examined by Ralph Abraham (1976). And with N equal to 3, there is the intriguing possibility that the stable configurations could be conceived as resembling the organization of sets of electron orbits around an atom. Such "atoms" might lend themselves to a Mendeleyev-type periodic classification into what would then be a (developmental) sequence of viable "learning manifolds", each with characteristics properties, whether as a description of a multi-facetted personality, group, or society (A J N Judge, 1982.

♦ KD2350 Alternation: implications for agreement and consensus

Comment This section arose from recognition that however excellent any "answer" may appear to its advocates, there will always be others who find good reason to argue or act against it in the interests of their own conception of human and social development. Furthermore, most answers, if they recognize the possibility of such rejection or accord importance to it, either make somewhat naive provisions for "educating everybody" or advocate processes which would lead to much more violent procedures for limiting the influence of those who hold any opposing viewpoint.

Without considering the political realm, the difficulties of achieving any consensus are quite obvious in the realm of scholarly discourse. For a scholar to agree, without qualifications, with the views of another effectively involves loss of identity as an uncreative "follower". The further development of the scholar can only come about by disagreeing and thus distinguishing himself from his peers – distinction is acquired by engendering difference. Quite concretely his career may even depend upon the production of well-argued counterarguments. A similar situation exists in the political realm.

The previous sections suggest that the kinds of consensus or agreement which evoke responses perceived as conflictual must necessarily continue to occur. They are a feature of psycho-social dynamics, whether they are the hawk/dove, ecology/industry, right/left, or other varieties. Universal agreement at this level could only be achieved at the price of psycho-social stagnation.

Another possibility arises where an answer is deliberately formulated to "resonate" with one or more antagonistic alternatives. Consensus of a different kind then becomes possible through shared recognition of this resonance. The resonance pattern then defines in energy terms a "structure" which could not exist if defined monolithically. Thus, for example, neither of the answers of the two parties in a 2-party political system can accept the need for the other to hold power. In French politics even though explicit use is made of "l'alternance", it is only used to mean the one-off transfer of power to the favoured party. The alternation between the two answers is only recognized implicitly, de facto, or for public relations purposes, never as a process in which the two answers have necessarily to participate given their complementary limitations (which they would deny).

Another example is disagreement as to whether it is now "night" or "day". Clearly a global perspective shows that it is night for some and day for others. From a local perspective such a view constitutes equivocation, although the alternation of night and day in time is necessarily accepted. In an interstellar spaceship the question of whether it "is" night or day is inappropriate, although the passengers will need to impose a night/day cycle upon themselves to maintain their health. Alternation between contrasting conditions is indeed important to the health of any system.

Further work is required to clarify the nature of possible resonance patterns – especially those already effectively in use. The problem is to render such alternation more credible as a foundation for social interaction. One approach is to explore patterns of alternation within sets of increasing numbers of different perspectives. Such patterns may be more capable of acting as a basis for social organization when the number of alternatives is greater than the range 2-7 with which the human mind seems to be comfortable. This is the question of the discontinuous "organization" of disagreement explored elsewhere (Penser/Classer, 1982). Alternatives in disagreement are necessary to the health of any system, if the "organization" is appropriate.

Finally there is the very specific question of the relationship of a section like this one to others which disagree with it. The argument has been that any response must resonate or "dance" with its own negation, or with positions that negate or deny it. *Alternation is obviously not the whole truth.* It is a response to the mind-set which claims to have encompassed such truth with a fixed set of categories lacking any self-transformational dynamic or internalization of opposition. Moreover it is specifically designed as a way of not occupying the central space from which new insights about truth emerge. It is one perspective on the relationship between partial truths, some of which must (in order to fulfil their function) necessarily deny both their partiality and the importance of clarifying any such a relationship. The alternation proposed between global uncertainty and local specificity is a response to this aspect of reality. The denial of such alternation is however necessary to the renewal and development of the perspective which gave rise to it. In a self-referential perspective there is necessarily a degree of paradox. The question is whether it is appropriately contained and whether the "container" can be further developed.

♦ KD2353 Alternation: implications for action formulation

Comment Stress has been placed on the pattern of essentially opposed modes of comprehending the nature of the problematique and the useful priorities in responding to it. This implies a built-in uncertainty necessary to contain the essential uncertainty encountered in a dynamic developing society (cf. Ashby's Law).

Use of the term uncertainty raises the question as to whether the conceptual problems experienced in some realms of physics are not also to be found in some realms of the social sciences. Specifically is there some form of generalized Heisenberg Principle of Uncertainty of which it is important to take account in formulating any coherent pattern of actions ? This question has been explored by Garrison Sposito (1969) who clarifies the significance of a study by Richard Lichtman (1967) on indeterminacy in the social sciences.

Lichtman's demonstration involved the premise that the response of social phenomena to investigation is entirely the result of rational processes, "the result of their acting to realize purposes they have *consciously* elaborated and endorsed". Sposito asks what happens if the opposite assumption is made. "Suppose the social phenomena do not control their responses to observation, but instead manufacture them to realize purposes *unconsciously* elaborated and endorsed". He draws attention to the situation in which human beings experience anxiety concerning the content of their experiences. Healed-over wounds of experience, not held consciously, can be laid open rather easily if the repressed experience which fostered them is repeated either in fact or suggestion. This triggers an uncontrolled aberration in behaviour well known in the therapeutic context as *transference distortion.*

Sposito then tentatively formulates a version of the Heisenberg Principle operative in the social sciences as: *"If the observation of social phenomena entails direct communication between human beings, one of which is the agent of observation, and transference distortions occur, then the results of the observation will not be completely objective, but will reflect latent facets of the personalities of those involved".*

Clearly the "agent" could also be a group or school of thought, there being sufficient evidence of the dramatic communication difficulties between schools of thought. The possibility that each such school or collectivity carries repressed experiential "wounds" which partially determine its response is worth further investigation. As Sposito stresses: "The interference phenomena engendered by transference distortions are uncontrollable in that they do not follow from purely rational processes and are not known to those who manifest them. Finally and most significantly, the interference phenomenon connects the social scientist inextricably with the objects of his inquiry (both heretofore logically independent entities) because the behaviour of the former induces unpredictable behaviour in the latter and vice versa".

The possibility then exists that the discontinuities between answer domains are governed by transference distortions caused by repressed (historical) experiences which are triggered by the nature of their responses to each other. Each effectively represents the other's "poison" and has been engendered to fill a niche from which an appropriate response can be made. In such a situation the question is then how to formulate a "methodological" framework to interrelate such dramatically opposed perspectives – especially when the problem is to anchor the valid concerns of such complementary perspectives in a coherent pattern of actions.

One approach is to envisage a series of *statement levels of decreasing uncertainty*. Thus in the first and most general statement the uncertainty would be most explicit. As such the statement corresponds in nature to the degree of universal consensus which can be realistically expected in a complex society. Succeeding statement levels would reduce the apparent uncertainty through the formulation of sets of *increasing numbers of parallel statements of decreasing ambiguity*. The uncertainty is then implicit in the unidentified relationship between those statements and between those who associate themselves with one or another. It is the dynamics between such statements which then "explicate" or "carry" the uncertainty. The lower the level of the statement, the more concrete and action-oriented it can be made – but the more difficult it becomes to formulate any coherent statement to interrelate statements at that level. The higher the level of the statement, the more coherent it can be – but the greater the degree of uncertainty which must be built into it to adequately reflect a consensus.

This approach then provides a realistic method for ordering and "packaging" statements. It reflects the dynamics inherent in any supposed consensus concerning a "new order" – in contrast to current sets of "static" conference resolutions which conceal the dynamics that then subsequently act to undermine the significance of such declarations.

This approach was advocated in a "methodological preamble" as an ordering device for the conclusions of the UN University project on Goals, Processes and Indicators of Development, on the occasion of the drafting meeting for the final "integrated" report (Port-of-Spain, December 1982). The sets of statements at each level were designed to reflect the contrasting methodological emphases and priorities represented in the deliberately diverse project, some of which took the form of GPID sub-projects. The statements suggested were:

1. Formulation of any clear and unambiguous understanding of development, such as at the macro-societal level or in terms of structures, tends to introduce ambiguity into the significance of the necessary complementary understanding of development, such as at the individual level, or in terms of processes. This ambiguity engenders uncertainty which is a healthy characteristic of the freedom inherent in the processes of a learning society. Any non-trivial single statement concerning development must therefore necessarily incorporate aspects of the development dynamics associated with the response to this uncertainty, or else run the risk of failing to encompass the richness of development potential.

2. Such ambiguity in understanding may be reduced for purposes of presentation by contrasting the aspects of development in terms of opposing mind-sets which engender the dynamics characteristic of the development process. The mind-sets selected as extreme examples here are the epistemological, metaphysical, axiological, and space-time frameworks considered as dimensions of the multidimensional space within which development may be understood. Polar extremes characteristic of each framework may then be clustered into bipolar configurations of elements of understanding. These then encode the real-world tensions and disagreements they imply:

2.1 Alpha-cluster: This groups the mind-sets associated with atomistic epistemologies, secular metaphysical frameworks, relativistic axiological frameworks, and linear space-time frameworks. Discontinuities between isolated local elements are highlighted, as well as the conflictual dynamics between them.

2.2 Beta-cluster: This groups the mind-sets associated with holistic epistemologies, transcendental metaphysical frameworks, absolutist axiological frameworks, and non-linear space-time frameworks. Global features are highlighted but with loss of ability to distinguish clearly the dynamics of local elements bound together into a seamless continuity.

3. Understanding of development may be given a more concrete form by orienting any description in terms of the constraints on action implied in any practical response to the contrasting perspectives above. Any understanding may thus be further qualified in terms of:

3.1 The necessarily limited domain of its validity, or its relevance to any unique concrete situation.

3.2 The participatory, historical, or self-reflexive processes whereby the understanding is engendered.

3.3 The ongoing learning processes of critical re-examination of the questions to which it attempts to provide an answer.

This approach has been explored down to the 20th level (see Section CP), based on concept sets from a wide variety of disciplines and cultures (A J N Judge, 1984). The intention was to "anchor" the approach in accepted conceptual frameworks as an appropriate foundation for the clustered pattern of complementary action programmes which can be associated with each such statement. Clearly at the higher levels the statements deal mainly with principles, whereas it is at the lower levels that the most concrete actions would emerge in detail. is at the lower levels that the most concrete actions would emerge in detail. At any given level an alternation between the emphases of each of the (complementary) statements is called for in order to ensure a balanced, coherent pattern of action capable of absorbing its own excesses. This has interesting implications for organization design.

Patterning disagreement KP

Rationale

Most conceptual schemes, whether purely theoretical or basic to the practical design of a development programme, are organized into sets of concepts, principles, priorities, or functions. Several such sets may be interrelated in a more elaborate scheme. It is the pattern of such interrelationships which ensures the coherence and integrity of the approach. Patterns therefore constitute a special form of presentation. Such patterns tend to be presented in isolation and often such that only the sub-patterns are explicit. Little is known about them as conceptual patterns. Given the need to interrelate concepts into a coherent pattern, so that they can be effectively communicated without loss of information, it is appropriate to explore the design of conceptual patterns of many elements and interconnections, whether explicit or implicit. Such explorations can contribute to the development of pattern languages through which groups can define and interrelate the non-material features appropriate to the quality of life in their environment.

Contents

This section, composed of 20 entries, is an **editorial experiment** in generating a pattern of progressively more differentiated conceptual incompatibilities as a basis for a more appropriate manner of psycho-social organization. It is based on insights in a wide range of different concept schemes that use sets of concepts of different sizes to contain qualitative complexity. Its merit lies in its deliberate attempt to internalize discontinuity and disagreement within the pattern. It is based on research presented to sub-projects on Forms of Presentation and on Methodology of the project on Goals, Processes and Indicators of Development of the United Nations University.

Method

The method used is discussed in the comments in Section KZ.

Index

A keyword index to entries is provided in Section KX. The keywords are also incorporated into the index for Volume 2 (Section X)

Bibliography

Bibliographical references, by author, are given in Section KY.

Comment

Detailed comments are given in Section KZ.

Reservations

Although the results of this deliberate editorial experiment are interesting and indicative of further possibilities, the entries raise many questions concerning the appropriateness of the language used. The language is stilted and artificial in an attempt both to maintain the pattern and to flesh it out with suitable material from a variety of sources that distinguished allusively between qualities if at all.

ENTRY CONTENT AND ORGANIZATION

Ordering of entries
Entries are in **numeric order**.

Index access to entries
The location of an entry in this sub-section may be determined from:
- the **Volume Index** (Section X) on the basis of keywords in the name of the entry or its alternate names
- the **Section Index** (Section KX) on the basis of the name of the entry

Structure of entries
Entries may be composed of the following descriptive elements:

(a) **Entry number** The last two digits serve both to order the sequence of entries and to indicate the number of subelements or viewpoints in each entry.

(b) **Qualifier** following entry number. There are **no qualifiers** for this group.

(c) **Entry name** This is printed in bold characters. It is a rough indication of the qualitative characteristics of the pattern of incompatibilities detailed in (d) below.

(d) **Viewpoints** Through the sequence of entries (KP3001 to KP3020), this paragraph is divided into progressively increasing number of numbered sub-paragraphs: KP3001 has 1 sub-paragraph, KP3011 has 11, *etc*. The sub-paragraphs is any one entry are designed, through the extreme positions they establish in relationship to one another, to capture a level of dynamic complexity which could not be reflected by any single viewpoint in isolation. Each entry therefore potrays a viewpoint set embodying a certain degree of incompatibility is greatest in the earlier entries and is diluted into qualitative distinctions in the later entries.

Cross-referencing between entries
There are no explicit cross-refrerences between entries in this group. The entries have a highly developed pattern of relationships to one another however, as described in point (d) immediately above.

♦ **KP3001 Inadequacy of formulations**
Viewpoints:
1. No single formulation (including this one), nor any logically integrated set of formulations, adequately encompasses the nature of the development process. Every position or formulation is therefore suspect. When it is formulated within a domain of unquestioned consensus, this potential doubt is inactive, thus establishing a boundary of uncritical discourse which inhibits development.

♦ **KP3002 Opposition/Disagreement**
Viewpoints:
2.1 New initiatives, including this one, are formulated by taking and establishing a particular position in opposition to whatever is conceived as potentially denying it. The nature of the initiative is partly determined by the way in which the challenge or initial absence of any opposing position is perceived and the possible nature of the response. It is the immediacy with which the challenge is perceived that empowers the initiative.
2.2 The taking of a position as a result of a new initiative engenders or activates a formulation which is its denial. Every formulation is therefore necessarily matched by an initiative which is incompatible with it, or opposed to it, or takes an essentially different direction from it. This opposition is fundamentally unmediated and as such cannot be observed or described. It can only be comprehended through identification with one of the opposed positions.

♦ **KP3003 Dialectic synthesis**
Viewpoints:
3.1 A form, through the affirmation of its existence, exerts pressure in response to its context which acts as an impulse for the continual transformation of the latter. As antecedent of any such transformation, it subjects any outcome to constraints. To the extent that the nature of the pressure on its context is unrecognized, any action initiated is distorted or unregulated in its impact on the context.
3.2 A form existing in the present stands in opposition to other pre–existing forms within the same context. As a result it is constrained by them to be of the necessary scale and proportion to oppose the pre–existing forms most dynamically. Within a given context, however, an opposing form of a particular type may be engendered which has been superseded in other co–present contexts. Forms corresponding to different stages of development may thus re–emerge and co–exist if the communication between contexts is obstructed in any way. To the extent that ignorance concerning this obstruction prevails, contexts become progressively more restricted, such that the dynamism of the opposition of the forms engendered within them diminished with a corresponding increase in the inertia or resistance associated with the least developed forms.
3.3 Opposition between two forms tends to give rise to a new form which has properties characteristic of both of them as well as new mediating properties unique to itself. The new form interrelates or harmonizes the original opposing forms. It reconciles them at a new level of expression of unity, whether or not they then disappear. The potential existence of the new form is therefore partially implicit (although incomplete) in each of the opposing forms prior to its generation. It thus functions as a stimulus or attractant by providing a pattern for their interaction and the organization of its outcome. Once created, the form will in its own turn prove inadequate and be opposed and superseded by more adequate forms whose nature it partially defines. The attraction of a particular form may however prevent the energetic development of this process.

♦ **KP3004 Developmental interaction**
Viewpoints:
4.1 In a set of forms, one form acquires a dominant status at any one time. As such it establishes the formal pattern of relationships between other forms by observing and distinguishing their elements, and interpreting their significance. Any infringement of this monopoly of power is met by a conscious reaction on the part of those associated with it who strive for position within the framework it supplies.
4.2 In a set of forms, one or more forms acquire a recessive or sub–dominant status at any one time. As such they are characterized by both minimal inherent organization and high inertial resistance to transformation. Any attempt to change those associated with such forms is met by unconscious reaction.
4.3 In a set containing a dominant and a dominated form, the pattern of relationships governed by the dominant form proves progressively more inadequate as a framework for handling the accumulation of new information and experience. Inconsistencies, contradictions and incompleteness gradually accumulate and become increasingly apparent as conditions change. The dominant form alone does not contain the variety to encompass and control the complex conditions to which it is exposed. The value of the recessive or inferior form becomes correspondingly apparent by contrast. The unconscious or impulsive actions of those associated with both forms serve merely to aggravate the condition and to highlight the absence of a form providing any adequate sense of direction or functional orientation for the whole.
4.4 In a set containing a dominant and an inferior form, and characterized by contradictions, adequate control is usually maintained through the momentum of working processes governed by the dominant form. Any deviation is corrected by a conscious integrative action on the part of those associated with that form. As the contradictions cease to be held in restraint in this way, the source of control is effectively transferred from the dominant form to the inferior form which thus emerges to take its place. To the extent that this transfer of control is resisted, the change is likely to be violent rather than smooth.

♦ **KP3005 Constraints on existence**
Viewpoints:
5.1 For a form to exist and acquire any momentary significance, it must bear a consciously recognized relationship to a context. If this relationship is ignored the form effectively merges into the context and cannot be distinguished from it due to the absence of any recognized boundaries or limits.
5.2 For a form to exist and acquire any momentary significance, it must be sufficiently general to be perceived as relevant to other variants of the phenomenon detached from immediate perception within the domain of discourse. If it is so general that it is perceived as relating to too wide a range of phenomena, then its significance is lost. Or, alternatively, it becomes so detached from immediate perception that its significance becomes fragmented into seemingly unrelated facets which arouse differing degrees of attachment or rejection.
5.3 For a form to exist and acquire any momentary significance, it must be perceived as relating to tangible phenomena of immediate relevance. But if this relationship is so strong as to be perceived as merely a reflection of those phenomena or identical with them, then its significance is lost or engenders contradictions, confusion and associated conflict.
5.4 For a form to exist and acquire any momentary significance, it must be perceived as sufficiently complex to encompass the complexity. If this is too much greater than that of the phenomena, its significance is either lost or a faith in the form may be engendered which is then valued for its own sake, independently of the phenomena, and possibly as being in some way superior to them.
5.5 For a form to exist and acquire any momentary significance, it must be sufficiently simple to be a comprehensible vehicle for intention. But if it is perceived as too simple (or trivial) the significance is lost. The unchannelled intention then reinforces inactivity or degenerates into sublimated forms of action.

♦ **KP3006 Coherence through renewal**
Viewpoints:
6.1 Sustaining the coherence of a form through its continual renewal requires a focused reaffirmation of the existence of the elements which ensure its integrity. To the extent that this reaffirmation is lacking, knowledge of its structure is eroded and the boundaries of the form become confused or dissolve.
6.2 Sustaining the coherence of a form through its continual renewal requires redefinition of the form to distinguish it from the superficial features of encroaching alternative forms with which it interacts. These may appear more attractive if concentration is relaxed. To the extent that this transformative process is lacking, aspects of the alternative definitions may be partially incorporated, thus progressively destroying the form as an integrated structure by formation of a hybrid or an agglomerate.
6.3 Sustaining the coherence of a form through its continual renewal requires repeated effort to understand the essential or general characteristics of the form which underlie any particular set of superficial features and thus not bound by them. To the extent that this understanding is lacking, the superficial features condemn the form as unnecessarily constraining, unsatisfactory, with consequent reactions.
6.4 Sustaining the coherence of a form through its continual renewal requires periodic detached recognition of its wider significance and how its development can best be controlled in relation to this. To the extent that this recognition is lacking, transformation of the form is blocked because of the narrow perspective with which it is viewed.
6.5 Sustaining the coherence of a form through its continual renewal requires recognition of the contextual structuring constraints, qualitative characteristics and challenges which ensure its stability, and in terms of which it may be transformed. To the extent that this recognition is lacking, the stability of the form is undermined by doubts concerning its present relevance.
6.6 Sustaining the coherence of a form through its continual renewal requires adaptation of insights concerning its possible development to a realistic strategy for its actual development. To the extent that this adaptation is lacking, any strategies formulated will be impractical and will result in maldevelopment of the form.

♦ **KP3007 Modes of change**
Viewpoints:
7.1 Under certain conditions the only form of change perceived as effective is through the wilful destruction of a prevailing form, whether or not a new or more adequate form can be substituted in its stead. This approach is favoured when the existing form is perceived as essentially static and an inhibitor of any form of dynamism or growth.
7.2 Under certain conditions the only form of change perceived as effective is through supportive interaction (dialogue) with the various perspectives formulated within the community concerned. Through such participative involvement on the part of the change agent as a sympathetic catalyst, a new community viewpoint can develop naturally from its existing foundations and be transformed. This approach is favoured when existing methods are perceived as implying destructive discontinuity or the imposition of inappropriate external formulations which would do violence to the community's growth and thus effectively retard it.
7.3 Under certain conditions the only form of change perceived as effective is through the formulation of a new all–encompassing philosophy (paradigm, theory, or strategy) as the reference framework in terms of which change can be initiated and undertaken. This approach is favoured when the diversity of existing initiatives is perceived as breeding confusion, dissipating resources, and undermining any possibility of a new level of collective achievement for the community as a whole.
7.4 Under certain conditions the only form of change perceived as effective is by enabling a more sensitive recognition of the variety of existing forms and the manner in which, through their various (and possibly discordant) interactions, they already constitute a rich and harmonious pattern saturated with meaning at a deeper level of significance. This approach is favoured when there is concern that new forms advocated are insensitive to and detached from the inherent harmony in those which have already been organically integrated into the tissue of lived reality.
7.5 Under certain conditions the only form of change perceived as effective is through the formulation of laws and definitions concerning observable processes on the basis of controlled investigation of their properties. Through such forms control is obtained over the processes which can then be used to restructure the environment according to their possibilities. This approach is favoured when there is concern that the processes of change are clothed in superstition, mystification, and are attributed solely to chance, or accident, or inexplicable agents acting spontaneously beyond the control of man.
7.6 Under certain conditions the only form of change perceived as effective emerges by renunciation of forms based upon the spatio–temporal world in favour of other factors and frames of reference to which appeal may be made. This approach is favoured when there is recognition that manipulative control of particular sub–systems of the external physical environment is only partially satisfactory (even when it is complete), and that less tangible dimensions need to be taken into account. Any such forms are frequently at least partially based on transformations of the inner world of the individual as it relates to the external world.
7.7 Under certain conditions the only form of change perceived as effective is to design configurations through which the full range of existing forms in opposition to each other can function creatively as complementaries, compensating for each others limitations and excesses. This approach is favoured when there is concern that the various approaches to change are functioning together so discordantly that some new form of dynamic order is required which provides a context for their different, and essentially incompatible, orientations.

♦ **KP3008 Constraints on change**
Viewpoints:
8.1 In assessing any apparent need for change, care is required to avoid mistaken formulations of the environmental condition. These can lead, for example, to an impetuous response or action for action's sake, from the consequences of which recovery may be difficult.
8.2 In formulating and planning any change initiative, care is required in selecting the point and manner of intervention. The constraints rarely offer the desired freedom of action and may easily be used as a focus for distracting dissatisfaction.
8.3 In formulating the nature of the change initiative, care is required in adapting any representation of it to avoid the temporary benefits of pleasing whoever is identified with the current condition or failing to acknowledge the difficulties to be encountered in changing it. These difficulties include weaknesses in those associated with the change initiative itself.
8.4 In implementing a change initiative as formulated, care is required that the initiative is not itself distorted by close association with the adverse conditions to which it responds or weakened by avoiding unpleasant decisions which have to be made to maintain the integrity of the response.
8.5 In sustaining a change initiative as formulated, care is required in ensuring its equilibrium with the intensification and expansion of activity due to confidence from successful experience with

any adverse conditions encountered and with the distractions of contentment with positive achievements.
8.6 Once a change initiative has achieved its maximum deployment, care is required in responding to the limitations on any further development. The original direction of effort may well be deflected in the pursuit of further success, especially in response to any accumulation of negative assessments.
8.7 Once the essential task of a change initiative is approaching completion, care is required in deciding on the termination of activities as originally intended. It may seem natural to continue the activities or to institutionalize them. Positive encouragement to do so may be received from all concerned. Succumbing to these pressures creates the risk of entrapment by a pattern of activity which it may then prove difficult to terminate at any time.
8.8 After a change initiative has been terminated, care is required in evaluating the activities and the achievements in the light of the original intent in order to avoid subsequent dependence on them.

♦ KP3009 Implementation of a transformation process
Viewpoints:
9.1 Implementation of a transformative process subject to realworld hazards requires assembly of the necessary operational resources of an adequate quality. To the extent that assembly is impossible, or their quality is inadequate, the process will be handicapped and partially controlled by the nature of those deficiencies.
9.2 Implementation of a transformative process subject to realworld hazards requires precise and energetic clarification of the succeeding stages of the process. To the extent that this clarification is lacking, action will be confused and momentum will be insufficient to overcome unforeseen problems.
9.3 Implementation of a transformative process subject to realworld hazards requires recognition of deviation or conflict between resources assembled and process planning in the light of independent critical questions concerning the implementation process. To the extent that this recognition is lacking, or that the questions are poorly conceived, further implementation (together with any corrective action) will result in an imbalanced process vulnerable to disruption.
9.4 Implementation of a transformative process subject to real-world hazards requires attentive preparation of the assembled elements to be processed. To the extent that this attentiveness is lacking, details of the preparation will be carelessly omitted or improperly executed thus jeopardizing the success of the operation.
9.5 Implementation of a transformative process subject to real-world hazards necessitates a controlled manipulation of the prepared elements into an emerging configuration. To the extent that this manipulation is improperly controlled or that the correspondence between the action taken and the knowledge of the action actually required is otherwise inadequate, the results will be unsatisfactory.
9.6 Implementation of a transformative process subject to real-world hazards requires dispassionate evaluation of the form emerging from the process in the light of the original intention and the current circumstances. To the extent that this evaluation is inadequate (and no corrective action is taken), the product may either not correspond to the original intention or be inappropriate to current possibilities for using it.
9.7 Implementation of a transformative process subject to real-world hazards requires that the emergent product be appropriately detached from the process which gave rise to it. To the extent that this separation is inadequate, or the relationship between the product and the process is otherwise confused, the resultant dependency relationship will jeopardize the value of the product.
9.8 Implementation of a transformative process subject to real-world hazards requires controlled delivery of the product to its originally intended setting in the face of possible reactions against it. To the extent that there is over-sensitivity to such reactions, the delivery cannot be completed thus jeopardizing the original intent.
9.9 Implementation of a transformative process subject to real-world hazards requires an appropriate attitude on completion of the process to ensure that it is evaluated within its proper context. To the extent that this attitude is lacking, efforts may then be made to associate either the product or the process to other contexts and initiatives. This distorts the originally intended significance of the initiative and runs the risk of confusing any new initiatives.

♦ KP3010 Endurance of a form
Viewpoints:
10.1 The endurance of a form is conditioned by its built-in ability to recognize the probable consequences of initiatives it determines and thus ensure relationships to other formulations which are supportive of their mutual development. To the extent that this recognition is lacking, destructive initiatives emerge with ultimately negative consequences for the development of the original form.
10.2 The endurance of a form is conditioned by its built-in ability to recognize the determining causes of developments in its environment and thus establish supportive relationships for the development of other forms on the basis of its own experience. To the extent that this recognition is lacking, the form develops parasitic or exploitative relationships with other forms which are ultimately detrimental to its own development.
10.3 The endurance of a form is conditioned by its inbuilt ability to recognize the characteristic initiatives and responses engendered by other forms in order, by exercise of discrimination, to determine those with which a mutually beneficial association is possible. To the extend that this recognition is lacking, the formulation is continually drawn into illusory or mutually conflicting relationships with other forms, in an uncontrollable manner which provides no stable foundation for its own development and effectively conceals its possibility.
10.4 The endurance of a form is conditioned by its inbuilt ability to recognize the developmental potential of other forms in order to adapt appropriately to such alternative perspectives for its own further development. To the extent that this recognition is lacking, the potential of such alternative forms is misrepresented, thus undermining the future adaptability of the form and the refinement of its own development goal.
10.5 The endurance of a form is conditioned by its inbuilt ability to recognize the different levels or capacities by which other forms may be characterized in order to relate appropriately to them to further mutual development. To the extent that this recognition is lacking, any relationships risk entrapment in apparent contradictions and in inappropriate responses to forms which stand in active opposition. In such circumstances the form may simply serve to spread dissension and blind awareness to particular expressions of a form.
10.6 The endurance of a form is conditioned by its inbuilt ability to recognize the pathways and goals of different modes of development characteristic of other forms and to adapt appropriately to an environment with such contrasting possibilities. To the extent that this recognition is lacking, other forms are actively condemned, often with considerable prejudice. The power and development of the form is then severely handicapped by the distortion and fragmentation of the actions it determines into rigidly polarized opposition to other forms.
10.7 The endurance of a form is conditioned by its inbuilt ability to recognize, through some process of detachment, those of its features which need to be gradually abandoned and those which need to be reinforced. To the extent that this recognition is lacking, rigid attachment to an unchanging form deflects any inherent dynamism into superficial matters of little consequence.
10.8 The endurance of a form is conditioned by its inbuilt ability to recall earlier stages in its development and the manner in which weaknesses were progressively eliminated. To the extent that this recollection is lacking, the form is unable to sustain any method for its own transformation and the necessary confidence is instead displaced into reinforcing attachment to existing weaknesses.
10.9 The endurance of a form is conditioned by its inbuilt ability to recognize the probable future states of forms and the probable circumstances of their termination. To the extent that this recognition is lacking, the form tends to become the vehicle for negative intentions towards the positive achievements associated with other forms, rather than channelling that intention to reinforce its own developmental momentum.
10.10 The endurance of a form is conditioned by its inbuilt ability to recognize in other forms the weaknesses to which they have developed an appropriate resistance. To the extent that this recognition is lacking, the form becomes a vehicle for the development of destructive misperceptions which hinder any ability either to abandon the weaknesses they have overcome or to free other forms from such obstacles to their own development.

♦ KP3011 Empowerment and importance of a form
Viewpoints:
11.1 The empowerment and importance of a form is determined by the degree of constructive or destructive action with which it is associated and the manner whereby they are distinguished.
11.2 The empowerment and importance of a form is determined by the degree of enrichening or impoverishing action with which it is associated.
11.3 The empowerment and importance of a form is determined by the degree of protection or exposure with which it is associated.
11.4 The empowerment and importance of a form is determined by the degree of assistance or obstruction with which it is associated.
11.5 The empowerment and importance of a form is determined by the degree of bias or lack of bias with which it is associated.
11.6 The empowerment and importance of a form is determined by the degree of security or danger with which it is associated.
11.7 The empowerment and importance of a form is determined by the degree of confidence or doubt with which it is associated.
11.8 The empowerment and importance of a form is determined by the degree of consolation or dejection with which it is associated.
11.9 The empowerment and importance of a form is determined by the degree of inspiration and reinforcement with which it is associated.
11.10 The empowerment and importance of a form is determined by the quality of remedial advice with which it is associated.
11.11 The empowerment and importance of a form is determined by the power of the subtle qualities with which it is associated.

♦ KP3012 Harmoniously transformative controlled relationship
Viewpoints:
12.1 A form in a harmoniously transformative controlled relationship with its environment is characterized by forceful spontaneous initiatives appropriately guided by an implicit sense of opportunity and constraint. Such action opens up viable new possibilities. If inappropriately controlled, it may be excessively violent, misguided, unfruitful or merely self-serving.
12.2 A form in a harmoniously transformative controlled relationship with it environment is characterized by a capacity to respond receptively to a comprehensive range of external initiatives by providing appropriate frameworks within which they can be embodied and consolidated. To the extent this capacity is lacking, such receptivity may be over-loaded leading to selective resistance, non-response or alternatively to their cooptation.
12.3 A form in a harmoniously transformative controlled relationship with its environment is characterized by a capacity to interrelate initiatives, creatively and explicitly, with contexts within which they can be further developed. To the extent this capacity is lacking, any such catalytic mediation becomes diffuse and lacking in continuity. Apparent contradictions are then a source of confusion rather than being perceived as aspects of an intricate pattern of stimulating diversity.
12.4 A form in a harmoniously transformative controlled relationship with its environment is characterized by the gradual emergence of higher order organization in response to initiatives and constraints. If such emergence is absent or inhibited, the form engenders actions which are increasingly incapable of containing the forces to which they respond.
12.5 A form in a harmoniously transformative controlled relationship with its environment necessitates a degree of organization which enables it to respond fully, in an integrated uncompromising forceful manner, to a full range of external events of which it remains independent. To the extent that this capacity is inappropriately developed, such organization is characterized by domination, self-appreciation, and misuse of power.
12.6 A form in a harmoniously transformative controlled relationship with its environment necessitates intuitive readjustment of implicit assumptions in order to renew the capacity to respond appropriately to events in context. To the extent that this capacity is lacking, any response is inhibited or focused on superficial detail.
12.6 A form in a harmoniously transformative controlled relationship with its environment necessitates intuitive readjustment of implicit assumptions in order to renew the capacity to respond appropriately to events in context. To the extent that this capacity is lacking, any response is inhibited or focused on superficial detail.
12.7 12.7 A form in a harmoniously transformative controlled relationship with its environment is characterized by a capacity for detached evaluation of past development from a perspective which provides both an intuitive balance between relevant factors and a sense of integrative possibilities. To the extent that this capacity is lacking, evaluation of external factors is negative or indecisive thus hindering further development.
12.8 A form in a harmoniously transformative controlled relationship with its environment is characterized by the capacity to respond spontaneously to higher order goals and possibilities even if the prevailing set of lower order goals and possibilities (with which it is identified) must be abandoned in order to do so. To the extent that the capacity for this transformation is lacking, the lower order goals and possibilities are distorted and reinforced to the detriment of further development.
12.9 A form in a harmoniously transformative controlled relationship with its environment is characterized by the spontaneous initiation of higher order processes which are focused in order to transform the operation of pre-existing lower order processes by which it is governed. To the extend that this capacity is inappropriately developed, any processes initiated are misdirected to the detriment of further development.
12.10 A form in a harmoniously transformative controlled relationship with its environment is characterized by an explicit pattern of control processes governing future possibilities, or current needs and opportunities. To the extent that this capacity is inappropriately developed, there is a tendency to over-control which is detrimental to further development.
12.11 A form in a harmoniously transformative controlled relationship with its environment is characterized by the capacity to engender appropriate design in the light of significant new insights

which bring possibilities and constraints into focus in an unforeseen and fruitful manner, thus facilitating effective action for their development. To the extent that this capacity is inappropriately developed, it results in automatic negative reaction to external initiatives and conditions, to the detriment of their further developments.

12.12 A form in a harmoniously transformative controlled relationship with its environment is characterized by a response pattern of reconciliation between all potential initiatives or conflicts. This unifying pattern thus acts as a stabilizing influence ensuring continuity, particularly between higher and lower-order processes. To the extend that this capacity is inappropriately developed, the response pattern becomes confused, reacting inadequately to spurious conditions.

♦ KP3013 Creative renewal
Viewpoints:

13.1 Renewal is dependent on the emergence of a creative response to any impotence and enfeeblement of action associated with the form in its current mode.
13.2 Renewal is dependent on the emergence of a creative response to any fragmented or inconsistent action associated with the form in its current mode.
13.3 Renewal is dependent on the emergence of a creative response to any fragmented or inconsistent action associated with the form in its current mode.
13.4 Renewal is dependent on the emergence of a creative response to any non-viable products of action associated with the form in its current mode.
13.5 Renewal is dependent on the emergence of a creative response to any dependence and powerlessness of the form in its current mode.
13.6 Renewal is dependent on the emergence of a creative response to any rigidity or crystallization of the form in its current mode.
13.7 Renewal is dependent on the emergence of a creative response to any impracticality or shortsightedness of action associated with the form in its current mode.
13.8 Renewal is dependent on the emergence of a creative response to any sense of futility associated with the form in its current mode, or to any (consequent) self-destructive processes.
13.9 Renewal is dependent on the emergence of a creative response to any apathy or pessimism associated with the form in its current mode.
13.10 Renewal is dependent on the emergence of a creative response to any unpredictability or uncontrollability associated with the form in its current mode.
13.11 Renewal is dependent on the emergence of a creative response to any action associated with the form becoming narrowly focused as an end in itself.
13.12 Renewal is dependent on the emergence of a creative response to any corruption or dissolution of the form in its current mode.
13.13 Renewal is dependent on the emergence of a creative response to the total disappearance of the form in its current mode.

♦ KP3014 Cycle of development processes
Viewpoints:

14.1 The cycle of development processes includes extreme phases characterized by static, unchanging forms.
14.2 The cycle of development processes includes extreme phases characterized by the breakdown of forms into their component elements.
14.3 The cycle of development processes includes extreme phases characterized by the coalescence of forms through which a new form is engendered.
14.4 The cycle of development processes includes extreme phases characterized by the harmonious interaction of forms which retain their identity.
14.5 The cycle of development processes includes extreme phases characterized by a unified, continuous pattern of forms.
14.6 The cycle of development processes includes extreme phases characterized by a diversity of separate, discrete forms.
14.7 The cycle of development processes includes extreme phases characterized by specific conflictual relationships between forms.
14.8 The cycle of development processes includes extreme phases characterized by qualitatively significant undefinable relationships between forms.
14.9 The cycle of development processes includes extreme phases characterized by chance-determined forms.
14.10 The cycle of development processes includes extreme phases characterized by forms which result as a natural and predictable consequence of those processes.
14.11 The cycle of development processes includes extreme phases characterized by forms whose existence in the spatio-temporal world is self-explanatory.
14.12 The cycle of development processes includes extreme phases characterized by forms whose existence cannot be adequately explained in terms of the spatio-temporal frame of reference.
14.13 The cycle of development processes includes extreme phases characterized by fluidity, turbulence and chaos.
14.14 The cycle of development processes includes extreme phases characterized by ordered systems and well-defined patterns.

♦ KP3015 Construction and development of form
Viewpoints:

15.1 Construction of form and the logical prediction of its future development requires direct or indirect observation of empirical facts, whether events, processes, or phenomena.
15.2 Construction of form and the logical prediction of its future development that requires appropriate procedures of measurement of empirical quantitative can be obtained.
15.3 Construction of form and the logical prediction of its future development requires appropriate procedures for the design and interpretation of significant experiments.
15.4 Construction of form and the logical prediction of its future development requires appropriate procedures of empirical generalization and descriptive classification to organize empirical data in a preliminary way in preparation for systematic classification.
15.5 Construction of form and the logical prediction of its future development requires appropriate procedures whereby explanatory results can be represented.
15.6 Construction of form and the logical prediction of its future development requires the use of conceptual elements, whether characteristic abstractions, terminology or techniques, which constitute the intellectual keys by which phenomena are made intelligible.
15.7 Construction of form and the logical prediction of its future development requires hypothesis formation, namely postulation through creative insight of a conceptual model based on assumptions concerning existing experimental observations or measurements.
15.8 Construction of form and the logical prediction of its future development requires recognition of a problem which appears susceptible to solution by use, or extension, of available techniques.
15.9 Construction of form and the logical prediction of its future development requires the possible adjustment or replacement of a conceptual model as a result of new observations or measurements.
15.10 Construction of form and the logical prediction of its future development requires the selection of a particular style of explanatory procedure required for the application of a given group of concepts.
15.11 Construction of form and the logical prediction of its future development requires use of formal or mathematical elements, whether computational, construction or analytic procedures.
15.12 Construction of form and the logical prediction of its future development requires use of techniques of formal transformation, whether formalization (reduction to relations while disregarding the nature of the related) or axiomatization (tracing of entailments back to accepted axioms).
15.13 Construction of form and the logical prediction of its future development requires validation of a conceptual model by checking its predictions against observations or measurements using techniques of confirmation, corroboration or falsification.
15.14 Construction of form and the logical prediction of its future development requires the production of rigorous formal definitions of the validity, probability, degree of confirmation, and other evidential relations involved in the judgement of a logical argument.
15.15 Construction of form and the logical prediction of its future development requires the use of a formal propositional system having a definite, essential logical structure, namely a formal scheme of propositions and axioms bound together by logical relations.

♦ KP3016 Values and assumptions
Viewpoints:

16.1 Recognition of the values underlying a form highlights any unfounded assumption that the form is without imperfection.
16.2 Recognition of the values underlying a form highlights any unfounded assumption that the form is an end in itself.
16.3 Recognition of the values underlying a form highlights any unfounded assumption that there is a permanent dimension to the form.
16.4 Recognition of the values underlying a form highlights any unfounded assumption that the form is composed of independent external features.
16.5 Recognition of the values underlying a form highlights any unfounded assumption that the inadequacies of the form have no cause or are their own cause.
16.6 Recognition of the values underlying a form highlights any unfounded assumption that the inadequacies of the form arise from irrelevant causes.
16.7 Recognition of the values underlying a form highlights any unfounded assumption that the inadequacies of the form are only due to one cause, independent of conditions or secondary circumstances.
16.8 Recognition of the values underlying a form highlights any unfounded assumption that the inadequacies of the form are necessarily permanent.
16.9 Recognition of the values underlying a form highlights any unfounded assumption that it is impossible to generate an adequate form.
16.10 Recognition of the values underlying a form highlights any unfounded assumption that the form as achieved is adequate, can be accepted, and that further effort to generate a more adequate form should cease.
16.11 Recognition of the values underlying a form highlights any unfounded assumption that the most abstract forms constitute the ultimate achievement.
16.12 Recognition of the values underlying a form highlight any unfounded assumption that, however perfect the form engendered, its inadequacy will eventually become apparent.
16.13 Recognition of the values underlying a form highlight any unfounded assumption that there is no method adequate to the current circumstances.
16.14 Recognition of the values underlying a form highlights any unfounded assumption that there is no suitable method, or pattern of methods, whereby acentric significance can be effectively perceived or reflected in a form.
16.15 Recognition of the values underlying a form highlights any unfounded assumption supporting the practice of methods which yield no useful results.
16.16 Recognition of the values underlying a form highlights any unfounded assumption that there are no effective remedies for the inadequacies of the existing form.

♦ KP3017 Relationship potential of a form
Viewpoints:

17.1 The relationship potential of a form to other forms, namely the extent to which it is assimilated into a larger set of differing forms, is directly dependent on its relative imperfection. Absence of imperfection reduces dependency arising from formal incompleteness thus removing any basis for interdependency. However, the nature of the imperfection strongly influences the quality of interdependence with which the form can be associated.
17.2 The relationship potential of a form to other forms, namely the extent to which it is assimilated into a larger set of differing forms, is directly dependent on the recognition that the form is not an end in itself.
17.3 The relationship potential of a form to other forms, namely the extent to which it is assimilated into a larger set of differing forms, is directly dependent on recognition of the impermanence of the form. The larger the set of forms within which relationships may exist, the greater the probability that such relationships will involve patterns of formal development and transformation in which any invariance will be at a higher level of abstraction than that of the form as originally recognized.
17.4 The relationship potential of a form to other forms, namely the extent to which it is assimilated into a larger set of differing forms, is directly dependent on recognition that the form is itself the integrated development of interdependent forms.
17.5 The relationship potential of a form to other forms, namely the extend to which it is assimilated into a larger set of differing forms, is directly dependent on recognition of the causes of the perceived inadequacies of the form. Such recognition establishes a relationship between the form and other forms. However the nature of the perceived cause strongly influences the quality of interdependence with which the form can be associated.
17.6 The relationship potential of a form to other forms, namely the extent to which it is assimilated into a larger set of differing forms, is directly dependent on recognition that the inadequacies of the form arise from relevant causes and not from causes irrelevant to the nature of the form.
17.7 The relationship potential of a form to other forms, namely the extent to which it is assimilated into a larger set of differing forms, is directly dependent on recognition that the inadequacies of the form are due to a multiplicity of causes themselves dependent on conditions and secondary circumstances.
17.8 The relationship potential of a form to other forms, namely the extent to which it is assimilated into a larger set of differing forms, is directly dependent on recognition that the inadequacies of the form and their causes are necessarily of a temporary nature.
17.9 The relationship potential of a form to other forms, namely the extent to which it is assimilated into a larger set of differing forms, is directly dependent on conviction that it is possible to generate a more adequate form. By focusing attention on possible adaptation of the form, its evolving relationship to other forms thus becomes evident.
17.10 The relationship potential of a form to other forms, namely the extent to which it is assimilated into a larger set of differing forms, is directly dependent on continuing effort to generate a more adequate form and refusal, as adequate, of what has already been achieved.

This ensures that the form is placed in a context of forms in process of transformation rather than in isolation.

17.11 The relationship potential of a form to other forms, namely the extent to which it is assimilated into a larger set of differing forms, is directly dependent on recognition that elaboration and retention of the most abstract form does not constitute the ultimate achievement. To the extent that this recognition is lacking, any such form, despite its sophistication, is a hindrance to the dynamics of further development.

17.12 The relationship potential of a form to other forms, namely the extent to which it assimilated into a larger set of differing forms, is directly dependent on conviction that forms can be engendered which will not subsequently come to be perceived as inadequate. Such forms must necessarily incorporate and counterbalance the factors which make for the emergence of inadequacy in an evolving set of forms.

17.13 The relationship potential of a form to other forms, namely the extent to which it is assimilated into a larger set of differing forms, is directly dependent on conviction that there is a method, or pattern of methods, which can be followed and is adequate to current circumstances. To the extent that this conviction is lacking, it is unlikely that significant relationships between forms will be recognized.

17.14 The relationship potential of a form to other forms, namely the extent to which it is assimilated into a larger set of differing forms, is directly dependent on conviction that a suitable method, or pattern of methods, may emerge whereby acentric significance can be effectively perceived or reflected in form. To the extent that this conviction is lacking, methods used will continue to be centred on particular approaches which fail to take account simultaneously of insights emerging from those centred on other approaches.

17.15 The relationship potential of a form to other forms, namely the extent to which it is assimilated into a larger set of differing forms, is directly dependent on recognition of the futility of practising methods which yield no fruitful results. To the extent that this recognition is lacking, the methods pursued will limit the range and richness of relationships which can be established between forms.

17.16 The relationship potential of a form to other forms, namely the extent to which it is assimilated into a larger set of differing forms, is directly dependent on conviction that there are effective remedies for the inadequacies of the existing form.

17.17 The relationship potential of a form to other forms, namely the extent to which it is assimilated into a larger set of differing forms depends on (intuitive) recognition of the permeability and variability of the boundary of that form.

♦ **KP3018 Inadequate transformation attempts**
Viewpoints:

18.1 Attempts at the transformation of form tend to be undermined by destructive energy-dissipating conflict between methodological extremes such as the assembly or mobilization of operational resources in accordance with a predetermined concept. This tends to engender either subservience or considerable resistance and alienation of potential support. Such forcing initiatives may well prevent formation of linkages vital to the future integrity of the operation and may lead to its early abortion or a considerable limitation in its scope.

18.2 Attempts at the transformation of form tend to be undermined by destructive energy-dissipating conflict between methodological extremes such as allowing operational resources to assemble, as and when they may, according tot he emergent processes of their initial interaction. This tends to result in considerable confusion, seldom with any creative operational outcome of other than a superficial nature. Such initiatives then lack coherence, continuity and any capacity for endurance.

18.3 Attempts at the transformation of form tend to be underminded by destructive energy-dissipating conflict between methodological extremes such as the imposition of a programme of operations. This immediately splits the resources mobilized into the empowered and the disempowered. The strength of the former then tends to be overestimated, whilst their weaknesses are under estimated, and the full contribution of the disempowered is blocked. The imposed programme is never called into question. This procedure further alienates potential support and increases the risk that the operation will go out of control if circumstances later arise in which the blocked or alienated resources are essential.

18.4 Attempts at the transformation of form tend to be undermined by destructive energy-dissipating conflict between methodological extremes such as the dependence on spontaneous, participative self-organization of operational programmes. This tends to result in uncertainty and conflicting activities which reinforce lack of coherence, of continuity, and of any capacity for endurance. Any programmes which emerge are immediately called into question.

18.5 Attempts at the transformation of form tend to be undermined by destructive energy-dissipating conflict between methodological extremes such as the reassessment of objectives and direction through detailed analysis following the initiation of the operation, this tends to be a destructive, unfruitful exercise providing little more than an intellectual framework as support for programme integration. The exercise then serves to alienate involvement in the operation, rather than to uncover new reserves of support for it.

18.6 Attempts at the transformation of form tend to be undermined by destructive energy-dissipating conflict between methodological extremes such as the reassessment of objectives and direction through resensitizing processes, affirmation, and celebration of solidarity, following the initiation of the operation. This tends to emphasize the dimensions of consensus (whether intangible or superficial) at the expense of the dimensions of disagreement (often specific and fundamental). Operational coherence is then dependent on the former without any adequate framework to balance the issues raised by the latter.

18.7 Attempts at the transformation of form tend to be undermined by destructive energy-dissipating conflict between methodological extremes such as the preparation or partial destructuring of the operation (for subsequent transformation), according to a rigid procedure unresponsive to contextual feedback. This tends to result in the accumulation of conditions which disrupt the procedure. The operation can then only be continued by overriding such obstacles or by limiting its original scope. Both solutions generate difficulties necessitating future operations for their elimination.

18.8 Attempts at the transformation of form tend to be undermined by destructive energy-dissipating conflict between methodological extremes such as the preparation or partial destructuring of the raw materials of the operation (for subsequent transformation) according to a procedure totally responsive to contextual feedback. This tends to result in the erosion (and eventual dissipation) of the procedure whose impetus is then absorbed into the contextual processes.

18.9 Attempts at the transformation of form tend to be undermined by destructive energy-dissipating conflict between methodological extremes such as the transformation of the raw materials of the operation by a series of precisely defined (and reproducible) changes of structure. This tends to limit such operations to those of essentially mechanical scope and renders them inapplicable to transformations of perception, attitude or value.

18.10 Attempts at the transformation of form tend to be undermined by destructive energy-dissipating conflict between methodological extremes such as the transformation of the raw materials of the operation by a set of intuitive, irreproducible processes. This tends to limit such operations to those of essentially intangible scope. This renders them inapplicable to transformations of tangible conditions which should reflect such changes and give them a measure of permanence.

18.11 Attempts at the transformation of form tend to be undermined by destructive energy-dissipating conflict between methodological extremes such as evaluating the transformation in terms of the quality of the results achieved, without taking into consideration the viability of the process as a means to that end. This facilitates the emergence of processes whose by-products set the stage for later difficulties.

18.12 Attempts at the transformation of form tend to be undermined by destructive energy-dissipating conflict between methodological extremes such as evaluating the transformation in terms of the viability of the process, without taking into consideration the quality of the results achieved (if any). This facilitates the emergence of processes carried out as an end in themselves, but which generate little of permanent benefit to the context in which they take place.

18.13 Attempts at the transformation of form tend to be undermined by destructive energy-dissipating conflict between methodological extremes such as abrupt separation of the emergent product from the process which gives rise to it. Such sudden separation endangers the product in its final phases of dependency on the process.

18.14 Attempts at the transformation of form tend to be undermined by destructive energy-dissipating conflict between methodological extremes such as continuing dependence of the emergent product on the process which gives rise to it. This pattern of dependency endangers the ultimate self-sufficiency of the product.

18.15 Attempts at the transformation of form tend to be undermined by destructive energy-dissipating conflict between methodological extremes such as delivery of the final product to the originally intended setting in a manner insensitive to reactions from that setting. This tends to lead to the early rejection of the product.

18.16 Attempts at the transformation of form tend to be undermined by destructive energy-dissipating conflict between methodological extremes such as delivery of the final product to the originally intended setting in a manner overly sensitive to reactions from that setting. Unless the normal resistance to new products is overcome, this tends to prevent the product from being delivered.

18.17 Attempts at the transformation of form tend to be undermined by destructive energy-dissipating conflict between methodological extremes such as complete rejection of any subsequent evaluation of the process or association with it. This tends to deprive subsequent initiatives from any value of the process as a learning experience.

18.18 Attempts at the transformation of form tend to be undermined by destructive energy-dissipating conflict between methodological extremes such as continuing identification with the process after its completion. This tends to distort any subsequent initiatives.

♦ **KP3019 Qualitative transformation**
Viewpoints:

19.1 Qualitative transformation depends on harmonious transfer of focus, alternating from (and to) the assembly or mobilization of, in accordance with a predetermined concept.

19.2 Qualitative transformation depends on harmonious transfer of focus, alternating from (and to) allowing operational resources to assemble naturally of their own accord.

19.3 Qualitative transformation depends on harmonious transfer of focus, alternating from (and to) the imposition of a programme of operations.

19.4 Qualitative transformation depends on harmonious transfer of focus, alternating from (and to) the dependence on spontaneous, participative self-organization of operational programmes.

19.5 Qualitative transformation depends on harmonious transfer of focus, alternating from (and to) the reassessment of objectives and direction through detailed analysis, following the initiation of the operation.

19.6 Qualitative transformation depends on harmonious transfer of focus, alternating from (and to) the reassessment of objectives and direction through resensitizing processes, following the initiation of the operation.

19.7 Qualitative transformation depends on harmonious transfer of focus, alternating from (and to) the preparation or partial restructuring of the elements of the operation, according to a rigid procedure unresponsive to contextual feedback.

19.8 Qualitative transformation depends on harmonious transfer of focus, alternating from (and to) the preparation or partial restructuring of the elements of the operation, according to a procedure totally responsive to contextual feedback.

19.9 Qualitative transformation depends on harmonious transfer of focus, alternating from (and to) the transformation of the elements of the operation by a series of precisely defined changes of structure.

19.10 Qualitative transformation depends on harmonious transfer of focus, alternating from (and to) the transformation of the elements of the operation by a set of intuitive, irreproducible processes.

19.11 Qualitative transformation depends on harmonious transfer of focus, alternating from (and to) evaluating the transformation in terms of the quality of the results achieved, without taking into consideration the viability of the process as a means to that end.

19.12 Qualitative transformation depends on harmonious transfer of focus, alternating from (and to) evaluating the transformation in terms of the process, without taking into consideration the quality of the results achieved.

19.13 Qualitative transformation depends on harmonious transfer of focus, alternating from (and to) abrupt separation of the emergent product which gives rise to it.

19.14 Qualitative transformation depends on harmonious transfer of focus, alternating from (and to) continuing dependence of the emergent product on the process which gives rise to it.

19.15 Qualitative transformation depends on harmonious transfer of focus, alternating from (and to) delivery of the final product to the originally intended setting in a manner insensitive to reactions form that setting.

19.16 Qualitative transformation depends on harmonious transfer of focus, alternating from (and to) delivery of the final product to the originally intended setting in a manner extremely sensitive to reactions from that setting.

19.17 Qualitative transformation depends on harmonious transfer of focus, alternating from (and to) complete rejection of any subsequent evaluation of the process or association with it.

19.18 Qualitative transformation depends on harmonious transfer of focus, alternating from (and to) continuing identification with the process after its completion.

19.19 Qualitative transformation depends on harmonious transfer of focus between extremes whilst maintaining an appropriate periodicity for such transfers within a self-organizing pattern.

♦ **KP3020 Significance of mutually constraining forms**
Viewpoints:

20.1 The significance of mutually constraining forms emerges with their avoidance of unnecessary or excessive response to each other. To the extent that this forbearance is lacking, the significance is obscured by the turbulent nature of that response.

20.2 The significance of mutually constraining forms emerges with affirmation of their affinity. To the extent that this affirmation is lacking, the significance is obscured by the consequences of previous unbalanced interactions.

20.3 The significance of mutually constraining forms emerges with their controlled interaction. To the extend that such control is lacking, the significance is obscured by the uncontrolled nature of their interaction.

20.4 The significance of mutually constraining forms emerges with recognition of their sensitively supportive response to each other's condition. To the extent that this sensitivity is lacking, the significance is obscured by destructive interactions.

20.5 The significance of mutually constraining forms emerges with reconciliation of their respective characteristics. To the extent that this reconciliation is lacking, the significance is obscured by non-recognition or non-acceptance of some characteristics.

20.6 The significance of mutually constraining forms emerges with acknowledgement of inadequacies. To the extent that such acknowledgement is lacking, the significance will be obscured by distortion of the relationship for short-term advantage.

20.7 The significance of mutually constraining forms emerges with abandonment of claims to non-existent qualities. To the extent that such claims are not relinquished, the significance will be obscured by efforts to achieve short-term advantage.

20.8 The significance of mutually constraining forms emerges with the implicit development of principles governing their actions. To the extent that such implicit principles are lacking, the significance is obscured by unconstrained actions and their consequences.

20.9 The significance of mutually constraining forms emerges with the explicit development of principles governing their actions. To the extent that such principles are lacking, the significance is obscured by unconstrained actions and their consequences.

20.10 The significance of mutually constraining forms emerges with acknowledgement of obstacles to further development. To the extent that such acknowledgement is lacking, the significance is obscured and their power reinforced.

20.11 The significance of mutually constraining forms emerges with abandonment of efforts to increase the resources associated with either form. To the extent that this is not achieved, the significance is obscured by the dependence created on the resource-seeking activity.

20.12 The significance of mutually constraining forms emerges with reservations concerning the resources and characteristics associated with the forms. To the extent that this reserve is lacking, the significance is obscured by preoccupation with these attributes.

20.13 The significance of mutually constraining forms emerges with enthusiasm for the functions with which they are associated. To the extend that this enthusiasm is lacking, the significance is obscured by indifference to those functions.

20.14 The significance of mutually constraining forms emerges with perseverance. To the extent that such persistent attention is lacking, the significance is obscured.

20.15 The significance of mutually constraining forms emerges with recognition of the constructive and destructive consequences of their interaction. To the extent that this recognition is lacking, the significance is obscured.

20.16 The significance of mutually constraining forms emerges with recollection of the multiple aspects of their interaction. To the extent that such memories are eroded, the significance is obscured.

20.17 The significance of mutually constraining forms emerges with alertness to potential confusion. To the extent that such attentiveness is lacking, the significance is obscured.

20.18 The significance of mutually constraining forms emerges with intelligent interest in their interaction. To the extent that such interest is lacking, the significance is obscured.

20.19 The significance of mutually constraining forms emerges with balanced attention to them. To the extent that there is preoccupation with one form, the significance is obscured.

20.20 The significance of mutually constraining forms emerges with ability to focus on their interaction. To the extent that such focus cannot be maintained, the significance is obscured.

Index

Integrative Knowledge

Index scope

This index covers all entries in Section K, namely:

- Integrative concepts (Section KC)

- Embodying discontinuity (Section KD)

- Patterning disagreement (Section KP)

Note that **all index entries in the following index are integrated into the Volume Index (Section X)** at the end of this volume.

The main intention of this index is to enable users to obtain a better overview of the individual sub-sections of Section K.

All index entries refer via reference numbers (eg KC1234) to the descriptions in the preceding sections. The letters indicate the section (eg Section KC).

Index entries

The index entries are of several types:

- Principal name or title of concept;

- Secondary or alternative names of concept (including popular expressions and synonyms)

- Keywords from the principal concept name;

- Keywords from the secondary concept names.

No distinction is made between the different types of entry. Keywords are however recognizable by the presence of a semi-colon in the index entry.

Remarks

To facilitate consultation of the index, and to reduce the space requirement prepositions and articles are normally omitted from the index entries.

Although the index was generated automatically, it has been edited manually to eliminate less helpful keywords resulting from the extraction process. Where there was any doubt, less meaningful words have however been left where they may prove to be of value in locating entries in Sections KC, KD or KP.

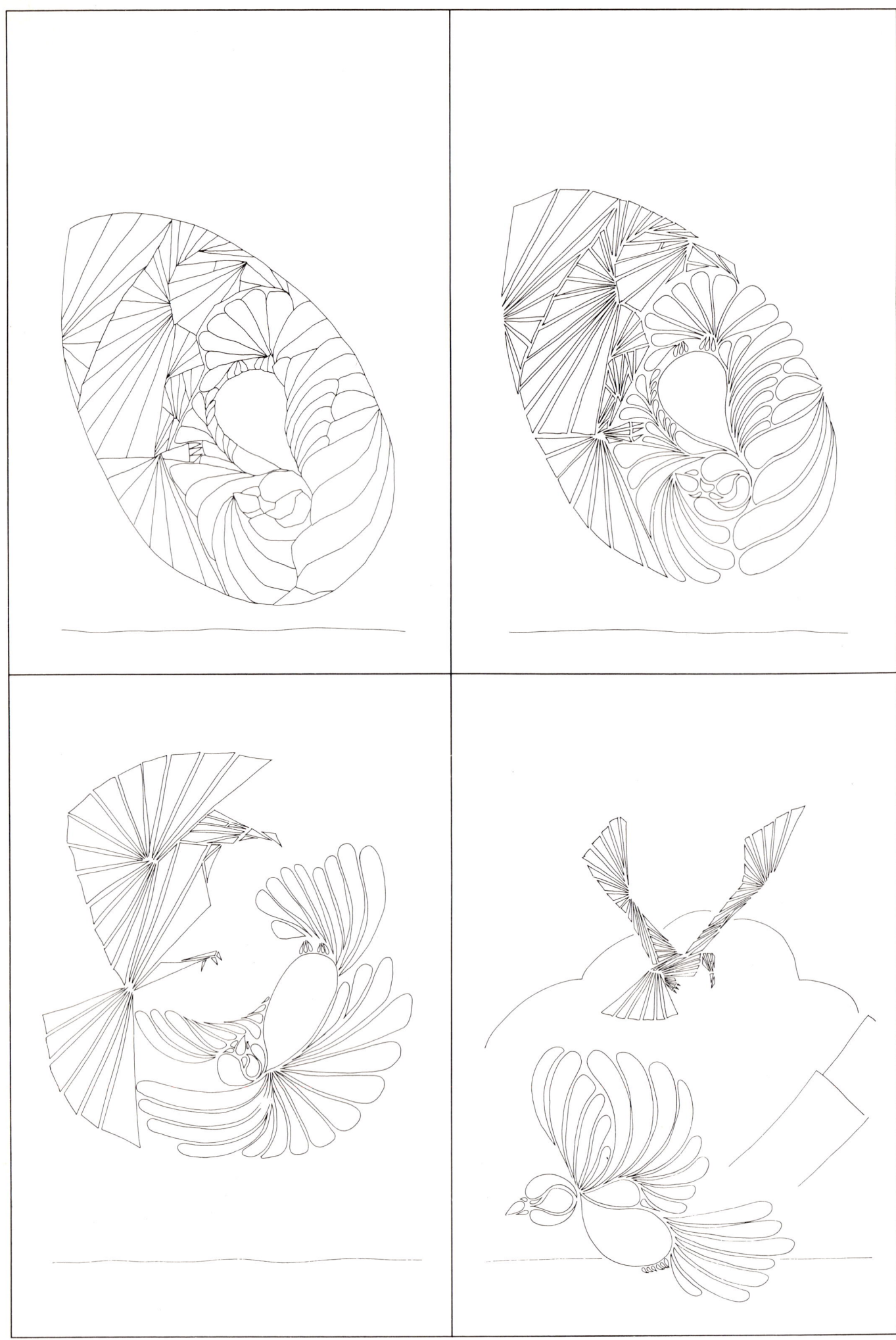

INDEX TO INTEGRATIVE KNOWLEDGE KX

A

KC 0386	Abstract mathematical space
KC 0129	Abstract objects
KC 0627	Abstract systems
KC 0271	Abstraction
KC 0198	Acculturation
KD 2215	Accumulation learning society ; Pattern
KD 2007	Accumulation significance ; Answer production
KD 2009	Accumulation significance ; Development processes
KD 2353	Action formulation ; Alternation: implications
KC 0817	Adaptation
KC 0133	Adaptive cultural integration
KC 0021	Administration
KD 2065	Aesthetics ; Containing discontinuity
KD 2005	Aesthetics ; Forms truth: uniformity versus
KC 0556	Agonemmetry
KD 2350	Agreement consensus ; Alternation: implications
KD 2285	Agricultural key crop rotation ; Patterns alternation:
KC 0097	Aim
KC 0717	Aims ; Short term
KC 0579	Alchemy
KC 0476	Allometry ; Principle
KC 0066	Allopoietic system
KD 2285	Alternation: agricultural key crop rotation ; Patterns
KD 2090	Alternation discontinuous learning ; Opening closing:
KD 2300	Alternation energetic expansion mentalistic reduction
KD 0383	Alternation-fluctuation
KD 2353	Alternation: implications action formulation
KD 2350	Alternation: implications agreement consensus
KD 2280	Alternation: musical key political philosopher ; Patterns
KD 2100	Alternation ; Revolutionary cycles
KD 2095	Alternation ; Third perspective container
KD 2262	Alternatives: resonance hybrids ; Interwoven
KD 2260	Alternatives: tensegrity organization ; Interwoven
KC 0427	Amplification ; Deviation
KC 0196	Analog computer language
KC 0731	Analogy
KC 0647	Analysis ; Comparative
KC 0016	Analysis ; Cost benefit
KC 0754	Analysis ; Cross impact
KC 0344	Analysis ; Factor
KC 0665	Analysis ; Input/Output
KC 0545	Analysis ; Integrational
KC 0964	Analysis ; Morphological
KC 0036	Analysis ; Multivariate
KC 0842	Analysis ; Network
KC 0696	Analysis ; Policy
KC 0696	Analysis ; Public policy
KC 0567	Analysis ; Sensitivity
KC 0365	Analysis ; Structural functional
KC 0718	Analysis ; Synergetic
KC 0610	Analysis ; Systems
KC 0327	Analysis ; Transaction
KC 0256	Analysis ; Value
KC 0302	Anima mundi
KD 2045	Answer ; Constraints meta
KD 2006	Answer: gladiatorial arena ; Prevailing meta
KD 2060	Answer patterning ; Meta
KD 2007	Answer production accumulation significance
KD 2003	Answers ; Assumptions concerning appropriate
KD 2001	Answers: monopolarization ; Questionable
KD 2030	Antinomies ; Paradoxes
KC 0527	Appreciative system
KD 2003	Appropriate answers ; Assumptions concerning
KC 0808	Arboreal organization
KC 0908	Arborization
KC 0487	Archetype
KC 0440	Arcology
KC 0074	Area studies
KC 0751	Art
KC 0510	Artificial organisms
KC 0847	Artistic expression
KC 0847	Artistic images
KC 0115	Artistic life
KC 0083	Artistic synthesis
KC 0847	Artistic vision
KC 0177	Assembling resources
KC 0943	Assessment ; Technology
KC 0019	Association
KC 0508	Association ; Data
KD 2003	Assumptions concerning appropriate answers
KP 3016	Assumptions ; Values
KC 0490	Astrology
KP 3018	Attempts ; Inadequate transformation
KC 0497	Attitudinal integration ; International
KC 0307	Attribute space
KC 0206	Augmentation intellect
KC 0206	Augmentation ; Team
KC 0652	Authority ; Collegiate
KC 0345	Automatic systems ; Integrated
KD 2085	Autopoiesis ; Organization self renewal:
KC 0956	Autopoietic system

B

KC 0313	Babel syndrome
KC 0267	Balance
KC 0968	Balance nature
KD 2327	Based consensus games ; Resonance
KC 0016	Benefit analysis ; Cost
KD 2040	Beyond method
KC 0870	Binary recoding
KC 0061	Biocybernetics
KC 0935	Bionics
KC 0494	Biosphere
KC 0706	Block diagram
KC 0706	Block language
KC 0528	Body knowledge
KC 0806	Boundary
KC 0357	Brain ; World
KC 0967	Budgeting
KC 0566	Budgeting System ; Planning Programming (PPBS)
KC 0913	Business cycles

C

KC 0906	Capacity ; Communication channel
KC 0462	Catastrophe theory
KC 0779	Categories
KC 0879	Categories ; Part whole
KD 2146	Categorization poly ocular vision ; De
KC 0220	Central place theory
KC 0120	Centrality
KC 0122	Centralization
KC 0122	Centralized system
KC 0371	Certainty
KC 0645	Change
KP 3008	Change ; Constraints
KC 0462	Change ; Mathematics discontinuous
KP 3007	Change ; Modes
KC 0906	Channel capacity ; Communication
KC 0222	Chaos ; Order
KC 0636	Character recognition
KC 0816	Chart ; Flow
KC 0135	Chemical synthesis
KC 0395	Choreography
KC 0870	Chunking
KC 0600	Circuit ; Integrated
KC 0069	Circuit language ; Energy
KC 0067	Citizenship ; International
KC 0067	Citizenship ; Planetary
KC 0067	Citizenship ; World
KC 0678	Civilization
KC 0410	Cladistics
KC 0768	Class
KC 0671	Classic
KC 0333	Classical field theory
KC 0045	Classification
KC 0622	Classification
KC 0944	Classification schedule ; Thesaurus
KC 0646	Closed system
KD 2090	Closing: alternation discontinuous learning ; Opening
KC 0533	Coevolution
KC 0108	Cognitive mobilization
KC 0783	Cognitive system
KD 2225	Cognitive systematization
KC 0866	Coherence
KP 3006	Coherence renewal
KC 0458	Coincidence
KC 0026	Coincidence
KC 0662	Collective ; Socialist
KC 0057	Collectivization mankind
KC 0652	Collegiate authority
KC 0663	Commensalism
KC 0274	Commensurability
KD 2272	Common focus ; Undefined
KC 0735	Common sense
KC 0905	Commune
KD 2207	Communicable insights: geometry connectivity
KC 0265	Communication
KC 0906	Communication channel capacity
KD 2202	Communication ; Encompassing varieties
KC 0140	Communicative social integration
KC 0425	Communion
KC 0152	Community
KC 0786	Community ; International
KC 0647	Comparative analysis
KC 0647	Comparative studies
KC 0434	Compatibility
KC 0556	Competition
KC 0722	Complementarity
KD 2083	Complementarity determinism fluctuation
KD 2070	Complementarity ; Observer entrapment micro macro
KD 2154	Complementary languages
KD 2342	Complementary metaphors ; Patterns
KC 0349	Complementation
KC 0174	Complementology
KC 0834	Completeness
KC 0142	Completeness equipment
KC 0010	Complex ; Mathematical
KC 0907	Complex social systems
KC 0336	Complex systems
KC 0628	Complexification
KC 0800	Complexity
KC 0604	Complexity ; Management
KC 0336	Complexity ; Organized
KC 0183	Complexity ; System
KD 2110	Comprehensibility: fourfold minimal system ; Threshold
KC 0187	Comprehensibility systems
KD 2205	Comprehension holes ; Non
KD 2325	Comprehension identity transformation ; Game
KD 2210	Comprehension internalization ; Discontinuity:
KD 2270	Comprehension structuring phenomenon learning society ; Non
KC 0106	Comprehensive thinking
KC 0708	Comprehensiveness
KC 0196	Computer language ; Analog
KC 0836	Computer program
KC 0630	Concept
KC 0922	Concepts ; Ecology
KC 0961	Concepts integrated science ; Teaching
KC 0686	Conceptual integration
KC 0627	Conceptual systems
KD 2003	Concerning appropriate answers ; Assumptions
KC 0803	Confederation
KC 0380	Configurational cultural integration
KC 0753	Configurations
KC 0406	Configurative thinking
KD 2170	Conflicting styles ; Modes managing: embracing
KC 0837	Connectedness
KC 0871	Connection
KC 0360	Connective cultural integration
KD 2207	Connectivity ; Communicable insights: geometry
KC 0060	Consciousness ; Legal
KD 2130	Consciousness: number time ; Quaternary
KC 0532	Consensus
KD 2350	Consensus ; Alternation: implications agreement
KD 2327	Consensus games ; Resonance based
KC 0525	Consensus politics
KC 0569	Conservation laws
KC 0934	Consistency
KC 0180	Consolidated
KC 0890	Consolidated
KP 3020	Constraining forms ; Significance mutually
KP 3008	Constraints change
KP 3005	Constraints existence
KD 2045	Constraints meta-answer
KP 3015	Construction development form
KD 2095	Container alternation ; Third perspective
KD 2065	Containing discontinuity aesthetics
KC 0456	Context
KC 0059	Contextual thinking ; Trans
KC 0059	Contextuality
KC 0008	Continuity discontinuity
KC 0702	Continuum
KC 0930	Control
KC 0795	Control ; Hierarchial principle
KC 0607	Control system
KC 0256	Control; Value engineering ; Value
KP 3012	Controlled relationship ; Harmoniously transformative
KC 0113	Convergence
KC 0377	Cooperation
KC 0384	Cooperation ; Cultural
KC 0555	Cooperation ; Intellectual
KC 0850	Cooperation ; International
KC 0897	Cooperation ; Regional international
KC 0303	Coordinate indexing
KC 0924	Coordination
KC 0550	Corporate database
KC 0528	Corpus knowledge
KC 0473	Correlation
KC 0502	Cosmology
KC 0278	Cosmos
KC 0016	Cost-benefit analysis
KC 0611	Counter-intuitive
KP 3013	Creative renewal
KC 0323	Creativity
KC 0750	Crises ; World
KD 2295	Crisis learning response ; Entropic
KC 0293	Crisis management
KC 0387	Critical mass
KC 0736	Critical path method
KC 0387	Critical quantity
KD 2285	Crop rotation ; Patterns alternation: agricultural key
KC 0411	Cross cultural study
KC 0754	Cross-impact analysis
KC 0896	Crossdisciplinarity
KC 0384	Cultural cooperation
KC 0470	Cultural evolution
KC 0243	Cultural integration
KC 0133	Cultural integration ; Adaptive
KC 0380	Cultural integration ; Configurational
KC 0360	Cultural integration ; Connective
KC 0133	Cultural integration ; Functional
KC 0460	Cultural integration ; Logical
KC 0143	Cultural integration ; Regulative
KC 0145	Cultural integration ; Stylistic
KC 0380	Cultural integration ; Thematic
KC 0076	Cultural relativity
KC 0411	Cultural study ; Cross
KC 0422	Cultural unity
KC 0051	Culture
KC 0609	Culture ; Mass
KC 0057	Culture ; Planetary
KC 0422	Culture ; Universal
KC 0422	Culture ; World
KC 0428	Curricula
KC 0961	Curricula ; Integrated science
KC 0961	Curricula ; Interdisciplinary science
KC 0961	Curricula ; Unified science
KC 0976	Curriculum integration
KC 0116	Cybernation
KC 0116	Cybernetic management
KC 0230	Cybernetics
KC 0963	Cybernetics ; Economic
KC 0910	Cybernetics ; Engineering
KD 2160	Cybernetics ; Nonlinear
KP 3014	Cycle development processes
KC 0965	Cycle ; Life
KC 0322	Cycles
KD 2100	Cycles alternation ; Revolutionary
KC 0913	Cycles ; Business
KD 2275	Cycles ; Learning
KD 2120	Cycles ; Omnitriangulation: interlocking
KD 2329	Cyclic phase ; Learning loss

KX

KD 2180	Cyclic self-organization requirements	KP 3010	Endurance form
KC 0394	Cyclical theory history	KD 2300	Energetic expansion mentalistic reduction ; Alternation

D

KC 0508	Data association
KC 0526	Data processing system
KC 0118	Data structure
KC 0218	Data structures ; Network
KC 0550	Database ; Corporate
KD 2146	De-categorization poly-ocular vision
KC 0217	Decay
KC 0340	Decision-making
KC 0312	Decision theory procedure
KC 0361	Deep structure
KC 0137	Degrees freedom
KC 0992	Desegregation
KC 0150	Design
KC 0529	Design ; Industrial
KC 0408	Design ; Information system
KC 0216	Determinate system
KC 0827	Determined system ; State
KD 2083	Determinism fluctuation ; Complementarity
KC 0363	Development
KC 0596	Development ; Discipline
KP 3015	Development form ; Construction
KD 2009	Development processes accumulation significance
KP 3014	Development processes ; Cycle
KP 3004	Developmental interaction
KC 0427	Deviation-amplification
KC 0706	Diagram ; Block
KC 0816	Diagram ; Flow
KC 0282	Dialectic
KP 3003	Dialectic synthesis
KC 0457	Dialectical-materialist synthesis
KD 2140	Dialogue ; Logos lemma interparadigmatic
KC 0944	Dictionary
KC 0276	Differentiation
KC 0077	Disciplinarity
KC 0077	Discipline
KC 0596	Discipline development
KD 2065	Discontinuity aesthetics ; Containing
KD 2210	Discontinuity: Comprehension internalization
KC 0008	Discontinuity ; Continuity
KC 0462	Discontinuous change ; Mathematics
KD 2090	Discontinuous learning ; Opening closing: alternation
KC 0797	Discourse ; Universe
KD 2081	Dissipative structures ; Order fluctuation:
KD 2250	Dissonant harmony holistic resonance
KC 0518	Distributed information system
KC 0231	Diversification
KC 0002	Diversity ; Unity
KC 0194	Dramatic unity
KC 0147	Dynamical similarity
KC 0954	Dynamics ; Group
KC 0426	Dynamics ; Industrial
KD 2192	Dynamics: learning ; Encompassing system
KD 2190	Dynamics: sixfold restraint ; Encompassing system
KC 0426	Dynamics ; System
KC 0426	Dynamics ; Urban

E

KC 0179	Earth ; One
KC 0581	Eclecticism
KD 2335	Ecodynamics societal evolution
KC 0950	Ecology
KC 0922	Ecology concepts
KC 0922	Ecology knowledge
KC 0415	Ecology ; Systems
KC 0963	Economic cybernetics
KC 0792	Economic integration ; International
KC 0651	Economic integration ; Voluntary
KC 0277	Economic order ; International
KC 0080	Economic policy
KC 0162	Economic structure ; Socio
KC 0530	Economic system
KC 0821	Economy ; World
KC 0447	Ecosphere
KC 0138	Ecosystem
KC 0543	Ecumenism
KC 0976	Education ; Integrated
KC 0711	Education ; Integrated science
KC 0290	Education ; Integrative
KC 0690	Education ; Racially ethnically integrated
KC 0920	Education ; Systems
KC 0976	Educational integration
KC 0720	Educational system
KC 0527	Eiconic system
KC 0225	Ekistics
KC 0782	Electronics ; Integrated
KC 0586	Elegance ; Formal
KC 0586	Elegance ; Mathematical
KD 2150	Emancipation particular languages
KD 2170	Embracing conflicting styles ; Modes managing:
KC 0017	Emergent systems
KP 3011	Empowerment importance form
KD 2320	Enantiomorphic policy ; Toward
KD 2192	Encompassing system dynamics: learning
KD 2190	Encompassing system dynamics: sixfold restraint
KC 2202	Encompassing varieties communication
KD 2200	Encompassing varieties form
KC 0931	Encyclopaedia
KC 0357	Encyclopaedia ; World
KC 0730	End

KP 3010	Endurance form
KD 2300	Energetic expansion mentalistic reduction ; Alternation
KC 0069	Energy circuit language
KC 0910	Engineering cybernetics
KC 0707	Engineering ; Human
KC 0807	Engineering ; Industrial
KC 0167	Engineering ; Social
KC 0710	Engineering ; Systems
KC 0256	Engineering ; Value control; Value
KC 0270	Ensemble
KC 0337	Entrainment frequencies
KD 2070	Entrapment micro macro complementarity ; Observer
KD 2295	Entropic crisis learning response
KC 0136	Entropy
KC 0003	Entropy ; Negative
KC 0416	Environment
KC 0516	Epigenetic landscape
KD 2145	Epistemological mindscapes
KC 0479	Epistemology
KC 0212	Epistemology ; Genetic
KC 0616	Equifinality
KC 0846	Equilibrium
KD 2080	Equilibrium structures ; Non
KC 0142	Equipment ; Completeness
KC 0932	Equivalent network
KC 0467	Ethic ; Self realization
KC 0992	Ethnic integration ; Racial
KC 0690	Ethnically integrated education ; Racially
KC 0736	Evaluation review technique ; Programme
KC 0801	Evolution
KC 0470	Evolution ; Cultural
KD 2335	Evolution ; Ecodynamics societal
KP 3005	Existence ; Constraints
KD 2300	Expansion mentalistic reduction ; Alternation energetic
KC 0847	Expression ; Artistic

F

KC 0877	Facilitation
KC 0344	Factor analysis
KC 0378	Faith ; Unity
KC 0164	Faithful ; Unity
KC 0273	Family
KC 0153	Federal government ; World
KC 0830	Feedback
KC 0427	Feedback ; Positive
KC 0032	Fibonacci numbers
KC 0970	Field
KC 0723	Field ; Semantic
KC 0333	Field theory ; Classical
KC 0407	Field theory ; Social
KC 0833	Field theory ; Unified
KC 0642	Finality
KC 0816	Flow chart
KC 0816	Flow diagram
KC 0816	Flow graph
KC 0383	Fluctuation ; Alternation
KD 2083	Fluctuation ; Complementarity determinism
KD 2081	Fluctuation: dissipative structures ; Order
KD 2272	Focus ; Undefined common
KD 2272	Focus variable geometries global order
KC 0771	Food chain
KC 0547	Forecasting
KC 0974	Forecasting ; Social
KC 0687	Forecasting ; Technological
KC 0156	Form
KP 3015	Form ; Construction development
KP 3011	Form ; Empowerment importance
KD 2200	Form ; Encompassing varieties
KP 3010	Form ; Endurance
KC 0093	Form language
KC 0203	Form ; Musical
KC 0721	Form ; Organic
KP 3017	Form ; Relationship potential
KC 0586	Formal elegance
KC 0261	Formalism
KC 0908	Formation ; Hierarchy
KP 3020	Forms ; Significance mutually constraining
KD 2005	Forms truth: uniformity versus aesthetics
KD 2353	Formulation ; Alternation: implications action
KP 3001	Formulations ; Inadequacy
KD 2110	Fourfold minimal system ; Threshold comprehensibility:
KC 0437	Frame reference
KC 0137	Freedom ; Degrees
KC 0337	Frequencies ; Entrainment
KC 0799	Front
KC 0738	Function
KD 2305	Function ignorance ; Uncertainty
KC 0365	Functional analysis ; Structural
KC 0133	Functional cultural integration
KC 0169	Functional multiplicity ; Organic
KC 0620	Functional social integration
KC 0232	Functionalism
KC 0634	Fundamentalism
KC 0450	Futures research
KC 0450	Futuribles
KC 0450	Futurism
KC 0450	Futuristics
KC 0450	Futurology
KC 0758	Fuzzy logic
KC 0332	Fuzzy sets

G

KD 2325	Game comprehension identity transformation
KC 0916	Game theory
KC 0637	Games ; Meta
KD 2327	Games ; Resonance based consensus
KC 0840	Gaming ; Operational
KC 0171	General plan
KC 0330	General semantics
KD 2220	General systems holonomy
KC 1000	General systems theory
KC 0432	Generalist
KC 0477	Generalization
KC 0212	Genetic epistemology
KD 2272	Geometries global order ; Focus variable
KD 2207	Geometry connectivity ; Communicable insights:
KC 0621	Geometry ; Projective
KC 0621	Geometry ; Synthetic
KC 0781	Gestalt
KD 2006	Gladiatorial arena ; Prevailing meta answer:
KC 0004	Global
KD 2272	Global order ; Focus variable geometries
KD 2082	Global relations ; Unexpected
KC 0506	Global society
KC 0805	Global system
KC 0997	Global system ; Regional subsystem
KC 0632	Gnosticism
KC 0717	Goals
KC 0977	Goals mankind ; Long range
KC 0444	Gödel's Theorem
KC 0021	Government
KC 0153	Government ; World
KC 0153	Government ; World federal
KC 0262	Grammar ; Universal
KC 0816	Graph ; Flow
KC 0584	Graph theory
KC 0928	Group
KC 0954	Group dynamics
KC 0054	Group ; Mathematical
KC 0915	Group ; Social
KC 0372	Groups ; World mind
KC 0376	Growth

H

KC 0110	Harmonic series
KP 3012	Harmoniously transformative controlled relationship
KC 0434	Harmony
KC 0424	Harmony
KD 2250	Harmony holistic resonance ; Dissonant
KD 2240	Health space-time
KC 0125	Hermeneutics
KC 0545	Hermeneutics ; Structural
KC 0606	Heterogeneity
KC 0231	Heterogenization
KC 0018	Heterostasis
KC 0126	Heuristic
KC 0795	Hierarchial principle control
KC 0047	Hierarchical restructuring
KC 0400	Hierarchical system
KC 0400	Hierarchy
KC 0908	Hierarchy formation
KC 0725	History
KC 0394	History ; Cyclical theory
KD 2205	Holes ; Non comprehension
KC 0941	Holism
KC 0096	Holistic medicine
KD 2250	Holistic resonance ; Dissonant harmony
KC 0937	Holocyclation
KC 0339	Holographic logic
KC 0592	Holographic universe
KC 0857	Holography
KC 0766	Holon
KD 2220	Holonomy ; General systems
KC 0618	Homeomorphism
KC 0226	Homeostasis
KC 0300	Homologous series
KC 0326	Homology
KC 0166	Homomorphism
KC 0325	Homonymy
KC 0707	Human engineering
KC 0225	Human settlements ; Science
KC 0887	Human systems management
KC 0311	Humanities
KD 2262	Hybrids ; Interwoven alternatives: resonance

I

KC 0241	Iatrodisciplines
KD 2325	Identity transformation ; Game comprehension
KC 0601	Ideology
KD 2305	Ignorance ; Uncertainty function
KC 0316	Image
KC 0527	Image system
KC 0847	Images ; Artistic
KC 0754	Impact analysis ; Cross
KP 3009	Implementation transformation process
KD 2230	Implicate order ; Wholeness
KD 2353	Implications action formulation ; Alternation:
KD 2350	Implications agreement consensus ; Alternation:
KP 3011	Importance form ; Empowerment
KP 3001	Inadequacy formulations
KP 3018	Inadequate transformation attempts
KC 0190	Indeterminacy principle
KC 0303	Indexing ; Coordinate

# INDEX TO INTEGRATIVE KNOWLEDGE

KX

KC 0373	Individual
KC 0014	Induction
KC 0529	Industrial design
KC 0426	Industrial dynamics
KC 0807	Industrial engineering
KC 0859	Industrial integration
KC 0259	Infinity
KC 0430	Information
KC 0526	Information system
KC 0408	Information system design
KC 0518	Information system ; Distributed
KC 0308	Information system ; Integrated
KC 0608	Information system integrity
KC 0936	Information system ; Real time
KC 0518	Information systems ; System
KC 0195	Initiatives ; Open systems
KC 0423	Innovation
KC 0167	Innovation ; Social
KC 0665	Input/Output analysis
KD 2207	Insights: geometry connectivity ; Communicable
KC 0597	Institutional integration ; International
KC 0551	Integral life
KC 0189	Integralism
KC 0180	Integrated
KC 0890	Integrated
KC 0345	Integrated automatic systems
KC 0600	Integrated circuit
KC 0976	Integrated education
KC 0690	Integrated education ; Racially ethnically
KC 0782	Integrated electronics
KC 0308	Integrated information system
KC 0711	Integrated methods science teaching
KC 0600	Integrated microcircuit
KC 0666	Integrated planning
KC 0811	Integrated science
KC 0961	Integrated science curricula
KC 0711	Integrated science education
KC 0961	Integrated science ; Teaching concepts
KC 0755	Integrated world
KC 0992	Integration
KC 0891	Integration
KC 0133	Integration ; Adaptive cultural
KC 0140	Integration ; Communicative social
KC 0686	Integration ; Conceptual
KC 0380	Integration ; Configurational cultural
KC 0360	Integration ; Connective cultural
KC 0243	Integration ; Cultural
KC 0976	Integration ; Curriculum
KC 0976	Integration ; Educational
KC 0133	Integration ; Functional cultural
KC 0620	Integration ; Functional social
KC 0859	Integration ; Industrial
KC 0397	Integration ; International
KC 0497	Integration ; International attitudinal
KC 0792	Integration ; International economic
KC 0597	Integration ; International institutional
KC 0697	Integration ; International policy
KC 0287	Integration ; International political
KC 0577	Integration ; International social
KC 0686	Integration knowledge
KC 0778	Integration languages
KC 0131	Integration laws
KC 0301	Integration level ; Theoretical
KC 0460	Integration ; Logical cultural
KC 0986	Integration ; Mathematical
KC 0520	Integration ; Normative social
KC 0992	Integration ; Racial ethnic
KC 0587	Integration ; Regional international
KC 0143	Integration ; Regulative cultural
KC 0145	Integration ; Stylistic cultural
KC 0380	Integration ; Thematic cultural
KC 0248	Integration ; Vertical
KC 0651	Integration ; Voluntary economic
KC 0545	Integrational analysis
KC 0367	Integrative
KC 0290	Integrative education
KC 0841	Integrative levels
KC 0106	Integrative thinking
KC 0634	Integrism
KC 0211	Integrity
KC 0608	Integrity ; Information system
KC 0206	Intellect ; Augmentation
KC 0979	Intellect ; Unity
KC 0555	Intellectual cooperation
KP 3004	Interaction ; Developmental
KC 0127	Interaction ; Level
KC 0127	Interaction ; Order
KD 2337	Interaction ; Self limiting patterns
KC 0165	Interchangeability
KC 0280	Interdisciplinarity
KC 0961	Interdisciplinary science curricula
KC 0258	Interlock
KD 2120	Interlocking cycles ; Omnitriangulation:
KD 2210	Internalization ; Discontinuity: Comprehension
KC 0643	International
KC 0497	International attitudinal integration
KC 0067	International citizenship
KC 0786	International community
KC 0850	International cooperation
KC 0897	International cooperation ; Regional
KC 0792	International economic integration
KC 0277	International economic order
KC 0597	International institutional integration
KC 0397	International integration
KC 0587	International integration ; Regional
KC 0132	International language
KC 0200	International peace
KC 0697	International policy integration
KC 0287	International political integration
KC 0037	International relations
KC 0577	International social integration
KC 0062	International socialism
KC 0786	International system
KC 0997	International system ; Regional
KC 0845	International understanding
KC 0790	International university
KD 2140	Interparadigmatic dialogue ; Logos lemma
KC 0828	Interrelatedness
KC 0195	Interworking
KD 2262	Interwoven alternatives: resonance hybrids
KD 2260	Interwoven alternatives: tensegrity organization
KC 0412	Intuition
KC 0522	Intuitionism ; Mathematical
KC 0611	Intuitive ; Counter
KC 0921	Invariance
KC 0832	Isomorphies
KC 0832	Isomorphism

K

KC 0580	Kaleidometrics
KD 2285	Key crop rotation ; Patterns alternation: agricultural
KD 2280	Key political philosopher ; Patterns alternation: musical
KC 0338	Knowledge
KC 0528	Knowledge ; Body
KC 0528	Knowledge ; Corpus
KC 0922	Knowledge ; Ecology
KC 0686	Knowledge ; Integration
KC 0320	Knowledge ; Organization
KC 0197	Knowledge ; Restructuring
KC 0304	Knowledge ; Sociology
KC 0686	Knowledge ; Unification

L

KC 0572	Labour ; Organization
KC 0346	Lag ; Surveillance
KC 0516	Landscape ; Epigenetic
KC 0694	Language
KC 0196	Language ; Analog computer
KC 0706	Language ; Block
KC 0069	Language ; Energy circuit
KC 0093	Language ; Form
KC 0132	Language ; International
KC 0123	Language ; Meta
KC 0093	Language ; Pattern
KD 2340	Language probabilistic vision world
KD 2152	Language relationships sound ; Modelling
KC 0947	Language universals
KC 0947	Language ; Universals
KD 2154	Languages ; Complementary
KD 2150	Languages ; Emancipation particular
KC 0778	Languages ; Integration
KC 0260	Lateral thinking
KC 0554	Lattice theory
KC 0692	Law ; World
KC 0569	Laws ; Conservation
KC 0131	Laws ; Integration
KD 2275	Learning cycles
KD 2192	Learning ; Encompassing system dynamics:
KD 2329	Learning loss cyclic phase
KD 2090	Learning ; Opening closing: alternation discontinuous
KD 2295	Learning response ; Entropic crisis
KD 2270	Learning society ; Non comprehension structuring phenomenon
KD 2215	Learning society ; Pattern accumulation
KC 0060	Legal consciousness
KC 0770	Legislation ; Unification
KD 2140	Lemma interparadigmatic dialogue ; Logos
KC 0127	Level interaction
KC 0301	Level ; Theoretical integration
KC 0841	Levels ; Integrative
KC 0841	Levels organization
KC 0115	Life ; Artistic
KC 0965	Life cycle
KC 0551	Life ; Integral
KC 0824	Life ; Mode
KC 0204	Life style
KC 0204	Life theme
KC 0316	Likeness
KD 2337	Limiting patterns interaction ; Self
KC 0324	Line ; World
KC 0626	Linear programming
KC 0347	Linearity ; Non
KC 0605	Logic
KC 0758	Logic ; Fuzzy
KC 0339	Logic ; Holographic
KC 0446	Logic ; Mathematical
KC 0603	Logic ; Multi valued
KD 2010	Logic ; Oppositional
KC 0937	Logic ; Process
KC 0776	Logic ; Relational
KC 0776	Logic relations
KC 0446	Logic ; Symbolic
KD 2105	Logic whole ; Trialectics:
KC 0460	Logical cultural integration
KC 0912	Logistics
KC 0511	Logos
KD 2140	Logos lemma interparadigmatic dialogue
KC 0977	Long-range goals mankind
KD 2329	Loss cyclic phase ; Learning
KC 0181	Love

M

KC 0081	Machine systems ; Man
KD 2070	Macro complementarity ; Observer entrapment micro
KC 0571	Macroevolution
KC 0081	Man-machine systems
KC 0867	Management
KC 0604	Management complexity
KC 0293	Management ; Crisis
KC 0116	Management ; Cybernetic
KC 0887	Management ; Human systems
KC 0540	Management science
KC 0540	Management ; Scientific
KD 2170	Managing: embracing conflicting styles ; Modes
KC 0854	Manifold
KC 0057	Mankind ; Collectivization
KC 0977	Mankind ; Long range goals
KC 0057	Mankind ; Planetization
KC 0087	Mankind ; Unity
KC 0726	Mapping
KC 0659	Market ; World
KC 0804	Market ; World socialist
KC 0677	Markov systems
KC 0387	Mass ; Critical
KC 0609	Mass culture
KC 0661	Mass society
KC 0457	Materialist synthesis ; Dialectical
KC 0010	Mathematical complex
KC 0586	Mathematical elegance
KC 0054	Mathematical group
KC 0986	Mathematical integration
KC 0522	Mathematical intuitionism
KC 0446	Mathematical logic
KC 0719	Mathematical model
KC 0386	Mathematical space ; Abstract
KC 0421	Mathematics
KC 0462	Mathematics discontinuous change
KC 0013	Maturity
KC 0249	Meaning
KC 0096	Medicine ; Holistic
KC 0759	Megalopolis
KC 0822	Melody
KD 2300	Mentalistic reduction ; Alternation energetic expansion
KD 2045	Meta answer ; Constraints
KD 2006	Meta answer: gladiatorial arena ; Prevailing
KD 2060	Meta-answer patterning
KC 0637	Meta-games
KC 0123	Meta-language
KC 0205	Meta-models
KC 0537	Meta-science
KC 0163	Metabolism
KC 0038	Metacommunication
KC 0862	Metagalaxy
KC 0015	Metahistory
KC 0925	Metalogic
KC 0969	Metamathematics
KC 0544	Metamorphosis
KC 0756	Metaphor
KD 2342	Metaphors ; Patterns complementary
KC 0041	Metaphysics
KC 0937	Metaphysics ; Process
KC 0769	Metasystem
KC 0876	Metatheorem
KC 0876	Metatheory
KC 0055	Method
KD 2040	Method ; Beyond
KC 0736	Method ; Critical path
KC 0188	Methodology
KC 0709	Methodology ; Social sciences
KC 0711	Methods science teaching ; Integrated
KD 2070	Micro macro complementarity ; Observer entrapment
KC 0600	Microcircuit ; Integrated
KC 0024	Microcosm
KC 0372	Mind groups ; World
KD 2145	Mindscapes ; Epistemological
KD 2110	Minimal system ; Threshold comprehensibility: fourfold
KC 0108	Mobilization ; Cognitive
KC 0177	Mobilization resources
KC 0824	Mode life
KC 0640	Model
KC 0719	Model ; Mathematical
KC 0640	Modelling
KD 2152	Modelling language relationships sound
KC 0042	Modelling ; World
KC 0205	Models ; Meta
KP 3007	Modes change
KD 2170	Modes managing: embracing conflicting styles
KC 0165	Modularity
KD 2001	Monopolarization ; Questionable answers:
KC 0926	Monte Carlo simulation technique
KD 2310	Morphic resonance
KC 0826	Morphogenesis
KC 0964	Morphological analysis
KC 0884	Morphology
KC 0257	Multi-ordinality
KC 0603	Multi-valued logic
KC 0191	Multidimensional space
KC 0007	Multidisciplinarity
KC 0169	Multiplicity ; Organic functional
KC 0027	Multistable system
KC 0036	Multivariate analysis
KC 0203	Musical form
KD 2280	Musical key political philosopher ; Patterns alternation:

KX

KP 3020   Mutually constraining forms ; Significance

N

KC 0980   Nation states ; Unitary
KC 0130   Nationalism
KC 0001   Natural
KC 0001   Nature
KC 0968   Nature ; Balance
KC 0003   Negative entropy
KC 0003   Negentropy
KC 0472   Network
KC 0842   Network analysis
KC 0218   Network data structures
KC 0932   Network ; Equivalent
KC 0736   Network planning techniques
KC 0633   Network ; Social
KC 0933   Network synthesis
KC 0455   Networking
KD 2205   Non-comprehension holes
KD 2270   Non-comprehension structuring phenomenon learning society
KD 2080   Non-equilibrium structures
KC 0347   Non-linearity
KD 2160   Nonlinear cybernetics
KC 0157   Noosphere
KC 0520   Normative social integration
KC 0404   Nous
KC 0981   Number
KC 0079   Number symbolism
KD 2130   Number time ; Quaternary consciousness:
KC 0032   Numbers ; Fibonacci
KC 0079   Numerology

O

KC 0757   Objective
KC 0129   Objects ; Abstract
KD 2070   Observer entrapment micro-macro complementarity
KD 2146   Ocular vision ; De categorization poly
KC 0818   Omnidirectionality
KC 0918   Omnitopology
KD 2120   Omnitriangulation: interlocking cycles
KC 0179   One Earth
KC 0039   Oneness
KC 0516   Ontogenetic recapitulation phylogeny
KC 0464   Ontology
KC 0746   Open systems
KC 0195   Open systems initiatives
KD 2090   Opening closing: alternation discontinuous learning
KC 0840   Operational gaming
KC 0540   Operations research
KC 0712   Opinion ; Public
KP 3002   Opposition/Disagreement
KD 2010   Oppositional logic
KC 0919   Optimal planning
KC 0436   Optimization
KC 0364   Orchestration
KC 0245   Order
KC 0222   Order chaos
KD 2081   Order fluctuation: dissipative structures
KD 2272   Order ; Focus variable geometries global
KC 0127   Order interaction
KC 0277   Order ; International economic
KC 0093   Order ; Structural
KC 0740   Order system
KC 0740   Order ; System
KC 0463   Order theorizing ; Second
KC 0278   Order universe
KD 2230   Order ; Wholeness implicate
KC 0496   Order ; World
KC 0093   Ordering ; Spatial
KC 0245   Orderliness
KC 0257   Ordinality ; Multi
KC 0721   Organic form
KC 0169   Organic functional multiplicity
KC 0535   Organicism
KC 0514   Organism
KC 0536   Organismic
KC 0510   Organisms ; Artificial
KC 0240   Organization
KC 0808   Organization ; Arboreal
KD 2260   Organization ; Interwoven alternatives: tensegrity
KC 0320   Organization knowledge
KC 0572   Organization labour
KC 0841   Organization ; Levels
KC 0615   Organization production
KD 2180   Organization requirements ; Cyclic self
KC 0510   Organization ; Self
KD 2085   Organization self-renewal: autopoiesis
KC 0160   Organization ; Social
KC 0114   Organization society ; Political
KC 0336   Organized complexity
KC 0510   Organizing systems ; Self
KC 0669   Orthoepy
KC 0236   Orthogenesis
KC 0863   Orthography
KC 0413   Prognosis
KC 0517   Overload ; System

P

KC 0005   Paradigm
KC 0235   Paradigmatology
KC 0159   Paradox
KD 2030   Paradoxes antinomies

KC 0879   Part whole categories
KD 2150   Particular languages ; Emancipation
KC 0315   Party unity
KC 0736   Path method ; Critical
KD 2215   Pattern accumulation learning society
KC 0093   Pattern language
KC 0418   Pattern perception
KC 0636   Pattern recognition
KD 2060   Patterning ; Meta answer
KD 2285   Patterns alternation: agricultural key crop rotation
KD 2280   Patterns alternation: musical key political philosopher
KD 2342   Patterns complementary metaphors
KD 2337   Patterns interaction ; Self limiting
KC 0071   Patterns ; Stereotypic
KC 0200   Peace ; International
KC 0975   Peace research
KC 0418   Perception ; Pattern
KD 2156   Performance ; Theory
KD 2095   Perspective container alternation ; Third
KC 0318   Phase space
KD 2329   Phase ; Learning loss cyclic
KC 0741   Phenomenology
KD 2270   Phenomenon learning society ; Non comprehension structuring
KD 2280   Philosopher ; Patterns alternation: musical key political
KC 0341   Philosophy
KC 0516   Phylogeny ; Ontogenetic recapitulation
KC 0220   Place theory ; Central
KC 0171   Plan ; General
KC 0067   Planetary citizenship
KC 0057   Planetary culture
KC 0506   Planetary society
KC 0281   Planetary synthesis
KC 0057   Planetization mankind
KC 0025   Planning
KC 0666   Planning ; Integrated
KC 0919   Planning ; Optimal
KC 0515   Planning ; Programme
KC 0566   Planning-Programming-Budgeting System (PPBS)
KC 0172   Planning ; Social policy
KC 0306   Planning ; Strategic
KC 0736   Planning techniques ; Network
KC 0861   Plasma
KC 0149   Pluralism
KC 0996   Pluridisciplinarity
KC 0392   Poetry
KD 2015   Polarity
KC 0031   Policy
KC 0696   Policy analysis
KC 0696   Policy analysis ; Public
KC 0080   Policy ; Economic
KC 0697   Policy integration ; International
KC 0172   Policy planning ; Social
KC 0635   Policy ; Science
KC 0343   Policy sciences
KD 2320   Policy ; Toward enantiomorphic
KC 0244   Policymaking
KC 0287   Political integration ; International
KC 0114   Political organization society
KD 2280   Political philosopher ; Patterns alternation: musical key
KC 0525   Politics ; Consensus
KD 2146   Poly ocular vision ; De categorization
KC 0033   Polyhistory
KC 0033   Polymathy
KC 0728   Polyphasic unity
KC 0325   Polysemy
KC 0173   Population
KC 0427   Positive feedback
KC 0219   Potential
KP 3017   Potential form ; Relationship
KC 0566   PPBS Planning-Programming-Budgeting System
KC 0269   Practice
KD 2006   Prevailing meta-answer: gladiatorial arena
KC 0476   Principle allometry
KC 0795   Principle control ; Hierarchial
KC 0190   Principle ; Indeterminacy
KC 0146   Principle requisite variety
KC 0190   Principle ; Uncertainty
KC 0966   Principle ; Unitary
KC 0328   Principle ; Whole system
KC 0677   Probabilistic system
KD 2340   Probabilistic vision world ; Language
KC 0667   Probability theory
KC 0614   Problem-solving theory
KC 0750   Problems ; World
KC 0312   Procedure ; Decision theory
KC 0645   Process
KC 0268   Process
KP 3009   Process ; Implementation transformation
KC 0937   Process logic
KC 0937   Process metaphysics
KD 2009   Processes accumulation significance ; Development
KP 3014   Processes ; Cycle development
KC 0526   Processing system ; Data
KC 0177   Procuring resources
KD 2007   Production accumulation significance ; Answer
KC 0615   Production ; Organization
KC 0836   Program ; Computer
KC 0736   Programme evaluation review technique
KC 0515   Programme planning
KC 0465   Programming
KC 0566   Programming Budgeting System ; Planning (PPBS)
KC 0626   Programming ; Linear
KC 0670   Progress
KC 0760   Progress ; Scientific technological

KC 0621   Projective geometry
KC 0969   Proof theory
KC 0747   Proportion
KC 0712   Public opinion
KC 0696   Public policy analysis
KC 0730   Purpose
KC 0730   Purposiveness
KC 0342   Pythagoreanism

Q

KP 3019   Qualitative transformation
KC 0387   Quantity ; Critical
KC 0186   Quantization
KD 2130   Quaternary consciousness: number time
KD 2001   Questionable answers: monopolarization

R

KC 0992   Racial ethnic integration
KC 0690   Racially ethnically integrated education
KC 0936   Real-time information system
KC 0370   Reality
KC 0467   Realization ethic ; Self
KC 0516   Recapitulation phylogeny ; Ontogenetic
KC 0870   Recoding ; Binary
KC 0636   Recognition ; Character
KC 0636   Recognition ; Pattern
KD 2300   Reduction ; Alternation energetic expansion mentalistic
KC 0911   Reductionism
KC 0437   Reference ; Frame
KC 0818   Reference ; Spherical
KC 0897   Regional international cooperation
KC 0587   Regional international integration
KC 0997   Regional international system
KC 0997   Regional subsystem global system
KC 0226   Regulation ; Self
KC 0143   Regulative cultural integration
KC 0828   Relatedness
KC 0860   Relation
KC 0776   Relational logic
KC 0037   Relations ; International
KC 0776   Relations ; Logic
KD 2082   Relations ; Unexpected global
KP 3012   Relationship ; Harmoniously transformative controlled
KP 3017   Relationship potential form
KC 0860   Relationships
KD 2152   Relationships sound ; Modelling language
KC 0562   Relativism
KC 0286   Relativity
KC 0076   Relativity ; Cultural
KC 0631   Religion
KC 0543   Religion ; Unity
KC 0791   Religions ; Transcendent unity
KD 2085   Renewal: autopoiesis ; Organization self
KP 3006   Renewal ; Coherence
KP 3013   Renewal ; Creative
KC 0856   Reproducing systems ; Self
KC 0856   Reproduction ; Self
KD 2180   Requirements ; Cyclic self organization
KC 0146   Requisite variety ; Principle
KC 0450   Research ; Futures
KC 0540   Research ; Operations
KC 0975   Research ; Peace
KC 0886   Research ; Systems
KD 2327   Resonance-based consensus games
KD 2250   Resonance ; Dissonant harmony holistic
KD 2262   Resonance hybrids ; Interwoven alternatives:
KD 2310   Resonance ; Morphic
KC 0177   Resources ; Assembling
KC 0177   Resources ; Mobilization
KC 0177   Resources ; Procuring
KD 2295   Response ; Entropic crisis learning
KD 2190   Restraint ; Encompassing system dynamics: sixfold
KC 0047   Restructuring ; Hierarchical
KC 0197   Restructuring knowledge
KC 0121   Reticulation
KC 0736   Review technique ; Programme evaluation
KC 0396   Revolution
KC 0296   Revolution ; Scientific
KD 2100   Revolutionary cycles alternation
KC 0714   Rhyme
KC 0831   Rhythm
KD 2285   Rotation ; Patterns alternation: agricultural key crop

S

KC 0772   Sacred tradition
KC 0944   Schedule ; Thesaurus Classification
KC 0427   Schismogenesis
KC 0011   Science
KC 0961   Science curricula ; Integrated
KC 0961   Science curricula ; Interdisciplinary
KC 0961   Science curricula ; Unified
KC 0711   Science education ; Integrated
KC 0225   Science human settlements
KC 0811   Science ; Integrated
KC 0540   Science ; Management
KC 0537   Science ; Meta
KC 0635   Science policy
KC 0537   Science science
KC 0537   Science ; Science
KC 0961   Science ; Teaching concepts integrated
KC 0711   Science teaching ; Integrated methods
KC 0811   Science ; Unified

—470—

INDEX TO INTEGRATIVE KNOWLEDGE

KX

KC 0811	Science ; Unity	KD 2080	Structures ; Non equilibrium	KC 0887	Systems management ; Human
KC 0709	Sciences methodology ; Social	KD 2081	Structures ; Order fluctuation: dissipative	KC 0677	Systems ; Markov
KC 0343	Sciences ; Policy	KD 2270	Structuring phenomenon learning society ; Non comprehension	KC 0746	Systems ; Open
KC 0540	Scientific management			KC 0886	Systems research
KC 0296	Scientific revolution	KC 0074	Studies ; Area	KC 0510	Systems ; Self organizing
KC 0760	Scientific technological progress	KC 0647	Studies ; Comparative	KC 0856	Systems ; Self reproducing
KC 0463	Second-order theorizing	KC 0411	Study ; Cross cultural	KC 0518	Systems ; System information
KD 2337	Self-limiting patterns interaction	KC 0443	Style	KC 1000	Systems theory ; General
KC 0510	Self-organization	KC 0204	Style ; Life	KC 0247	Systems ; Transient
KD 2180	Self organization requirements ; Cyclic	KD 2170	Styles ; Modes managing: embracing conflicting		
KC 0510	Self-organizing systems	KC 0145	Stylistic cultural integration		**T**
KC 0467	Self-realization ethic	KC 0573	Substance		
KC 0226	Self-regulation	KC 0227	Subsystem	KC 0310	Tao
KD 2085	Self renewal: autopoiesis ; Organization	KC 0997	Subsystem global system ; Regional	KC 0717	Targets
KC 0856	Self-reproducing systems	KC 0676	Summativity	KC 0045	Taxonomy
KC 0856	Self-reproduction	KC 0507	Surface structure	KC 0961	Teaching concepts integrated science
KC 0723	Semantic field	KC 0346	Surveillance lag	KC 0711	Teaching ; Integrated methods science
KC 0542	Semantics	KC 0335	Symbiosis	KC 0206	Team augmentation
KC 0330	Semantics ; General	KC 0208	Symbiotization	KC 0503	Teamwork
KC 0927	Semiotics	KC 0446	Symbolic logic	KC 0926	Technique ; Monte Carlo simulation
KC 0971	Sense	KC 0527	Symbolic system	KC 0736	Technique ; Programme evaluation review
KC 0735	Sense ; Common	KC 0660	Symbolism	KC 0736	Techniques ; Network planning
KC 0567	Sensitivity analysis	KC 0079	Symbolism ; Number	KC 0687	Technological forecasting
KC 0737	Serendipity	KC 1010	Symmetry	KC 0760	Technological progress ; Scientific
KC 0110	Series ; Harmonic	KC 0299	Symmetry ; Unitary	KC 0619	Technology
KC 0300	Series ; Homologous	KC 0022	Synaesthesia	KC 0943	Technology assessment
KC 0117	Set	KC 0420	Synaesthesis	KC 0642	Teleology
KC 0332	Sets ; Fuzzy	KC 0026	Synchronicity	KC 0297	Teletics
KC 0225	Settlements ; Science human	KC 0337	Synchronization	KC 0028	Tensegrity
KC 0717	Short-term aims	KC 0794	Syncretism	KD 2260	Tensegrity organization ; Interwoven alternatives:
KD 2007	Significance ; Answer production accumulation	KC 0023	Synderesis	KC 0380	Thematic cultural integration
KD 2009	Significance ; Development processes accumulation	KC 0253	Syndrome	KC 0674	Theme
KP 3020	Significance mutually constraining forms	KC 0313	Syndrome ; Babel	KC 0204	Theme ; Life
KC 0316	Similarity	KC 0767	Synectics	KC 0715	Theorem
KC 0147	Similarity ; Dynamical	KC 0718	Synergetic analysis	KC 0444	Theorem ; Gödel's
KC 0486	Simplicity	KC 0718	Synergetics	KC 0301	Theoretical integration level
KC 0246	Simulation	KC 0957	Synergy	KC 0463	Theorizing ; Second order
KC 0926	Simulation technique ; Monte Carlo	KC 0452	Synnoetics	KC 0892	Theory
KC 0390	Simultaneity	KC 0255	Synthesis	KC 0462	Theory ; Catastrophe
KC 0039	Singleness	KC 0083	Synthesis ; Artistic	KC 0220	Theory ; Central place
KD 2190	Sixfold restraint ; Encompassing system dynamics:	KC 0135	Synthesis ; Chemical	KC 0333	Theory ; Classical field
KC 0167	Social engineering	KP 3003	Synthesis ; Dialectic	KC 0916	Theory ; Game
KC 0407	Social field theory	KC 0457	Synthesis ; Dialectical materialist	KC 1000	Theory ; General systems
KC 0974	Social forecasting	KC 0933	Synthesis ; Network	KC 0584	Theory ; Graph
KC 0915	Social group	KC 0281	Synthesis ; Planetary	KC 0394	Theory history ; Cyclical
KC 0167	Social innovation	KC 0621	Synthetic geometry	KC 0554	Theory ; Lattice
KC 0140	Social integration ; Communicative	KC 0546	System	KD 2156	Theory performance
KC 0620	Social integration ; Functional	KC 0066	System ; Allopoietic	KC 0667	Theory ; Probability
KC 0577	Social integration ; International	KC 0527	System ; Appreciative	KC 0614	Theory ; Problem solving
KC 0520	Social integration ; Normative	KC 0956	System ; Autopoietic	KC 0312	Theory procedure ; Decision
KC 0633	Social network	KC 0122	System ; Centralized	KC 0969	Theory ; Proof
KC 0160	Social organization	KC 0646	System ; Closed	KC 0407	Theory ; Social field
KC 0172	Social policy planning	KC 0783	System ; Cognitive	KC 0612	Theory ; Transformation
KC 0709	Social sciences methodology	KC 0183	System complexity	KC 0833	Theory ; Unified field
KC 0820	Social system	KC 0607	System ; Control	KC 0693	Theory ; Unitary
KC 0907	Social systems ; Complex	KC 0526	System ; Data processing	KC 0356	Theory ; Value
KC 0167	Social transmutation	KC 0408	System design ; Information	KC 0944	Thesaurus Classification schedule
KC 0062	Socialism ; International	KC 0216	System ; Determinate	KC 0106	Thinking ; Comprehensive
KC 0662	Socialist collective	KC 0518	System ; Distributed information	KC 0406	Thinking ; Configurative
KC 0804	Socialist market ; World	KC 0426	System dynamics	KC 0106	Thinking ; Integrative
KD 2335	Societal evolution ; Ecodynamics	KD 2192	System dynamics: learning ; Encompassing	KC 0260	Thinking ; Lateral
KC 0101	Society	KD 2190	System dynamics: sixfold restraint ; Encompassing	KC 0059	Thinking ; Trans contextual
KC 0506	Society ; Global	KC 0530	System ; Economic	KC 0512	Thinking ; Vertical
KC 0661	Society ; Mass	KC 0720	System ; Educational	KD 2095	Third-perspective container alternation
KD 2270	Society ; Non comprehension structuring phenomenon learning	KC 0527	System ; Eiconic	KC 0733	Thought ; Unitary
KD 2215	Society ; Pattern accumulation learning	KC 0805	System ; Global	KD 2110	Threshold comprehensibility: fourfold minimal system
KC 0506	Society ; Planetary	KC 0400	System ; Hierarchical	KC 0387	Threshold value
KC 0114	Society ; Political organization	KC 0527	System ; Image	KC 0761	Time
KC 0506	Society ; World	KC 0526	System ; Information	KD 2240	Time ; Health space
KC 0162	Socio-Economic structure	KC 0518	System information systems	KC 0936	Time information system ; Real
KC 0304	Sociology knowledge	KC 0308	System ; Integrated information	KD 2130	Time ; Quaternary consciousness: number
KC 0209	Sociometry	KC 0608	System integrity ; Information	KC 0994	Topological space
KC 0302	Soul ; World	KC 0786	System ; International	KC 0946	Topology
KD 2152	Sound ; Modelling language relationships	KC 0027	System ; Multistable	KC 0402	Totalitarianism
KC 0673	Space	KC 0740	System order	KC 0772	Tradition ; Sacred
KC 0386	Space ; Abstract mathematical	KC 0740	System ; Order	KC 0059	Trans-contextual thinking
KC 0307	Space ; Attribute	KC 0517	System overload	KC 0327	Transaction analysis
KC 0191	Space ; Multidimensional	KC 0566	System ; Planning Programming Budgeting (PPBS)	KC 0791	Transcendent unity religions
KC 0318	Space ; Phase	KC 0328	System principle ; Whole	KC 0796	Transdisciplinarity
KD 2240	Space time ; Health	KC 0677	System ; Probabilistic	KC 0812	Transformation
KC 0994	Space ; Topological	KC 0936	System ; Real time information	KP 3018	Transformation attempts ; Inadequate
KC 0093	Spatial ordering	KC 0997	System ; Regional international	KD 2325	Transformation ; Game comprehension identity
KC 0818	Spherical reference	KC 0997	System ; Regional subsystem global	KP 3009	Transformation process ; Implementation
KC 0917	Stability	KC 0820	System ; Social	KP 3019	Transformation ; Qualitative
KC 0223	Standardization	KC 0827	System ; State determined	KC 0612	Transformation theory
KC 0827	State-determined system	KC 0417	System stress	KP 3012	Transformative controlled relationship ; Harmoniously
KC 0523	State ; Steady	KC 0527	System ; Symbolic	KC 0247	Transient systems
KC 0980	States ; Unitary nation	KD 2110	System ; Threshold comprehensibility: fourfold minimal	KC 0167	Transmutation ; Social
KC 0523	Steady state	KC 0064	System ; Ultrastable	KC 0823	Transnational
KC 0071	Stereotype	KC 0805	System ; World	KC 0808	Tree structure
KC 0071	Stereotypic patterns	KC 0045	Systematics	KD 2105	Trialectics: logic whole
KC 0306	Strategic planning	KD 2225	Systematization ; Cognitive	KD 2005	Truth: uniformity versus aesthetics ; Forms
KC 0306	Strategy	KC 0627	Systems ; Abstract	KC 0484	Typology
KC 0417	Stress ; System	KC 0610	Systems analysis		
KC 0365	Structural-functional analysis	KC 0336	Systems ; Complex		**U**
KC 0545	Structural hermeneutics	KC 0907	Systems ; Complex social		
KC 0093	Structural order	KC 0187	Systems ; Comprehensibility	KC 0064	Ultrastable system
KC 0560	Structuralism	KC 0627	Systems ; Conceptual	KD 2305	Uncertainty function ignorance
KC 0266	Structure	KC 0415	Systems ecology	KC 0190	Uncertainty principle
KC 0118	Structure ; Data	KC 0920	Systems education	KD 2272	Undefined common focus
KC 0361	Structure ; Deep	KC 0017	Systems ; Emergent	KC 0845	Understanding ; International
KC 0162	Structure ; Socio Economic	KC 0710	Systems engineering	KD 2082	Unexpected global relations
KC 0507	Structure ; Surface	KD 2220	Systems holonomy ; General	KC 0099	Unicity
KC 0808	Structure ; Tree	KC 0195	Systems initiatives ; Open	KC 0686	Unification knowledge
KC 0218	Structures ; Network data	KC 0345	Systems ; Integrated automatic	KC 0770	Unification legislation
		KC 0081	Systems ; Man machine		

–471–

KX

KC 0180	Unified	
KC 0890	Unified	
KC 0833	Unified field theory	
KC 0811	Unified science	
KC 0961	Unified science curricula	
KC 0732	Uniformity	
KD 2005	Uniformity versus aesthetics ; Forms truth:	
KC 0393	Union	
KC 0650	Union ; World	
KC 0777	Unitary	
KC 0980	Unitary nation-states	
KC 0966	Unitary principle	
KC 0299	Unitary symmetry	
KC 0693	Unitary theory	
KC 0733	Unitary thought	
KC 0657	United Nations Organization	
KC 0039	Unity	
KC 0422	Unity ; Cultural	
KC 0002	Unity diversity	
KC 0194	Unity ; Dramatic	
KC 0378	Unity faith	
KC 0164	Unity faithful	
KC 0979	Unity intellect	
KC 0087	Unity mankind	
KC 0315	Unity ; Party	
KC 0728	Unity ; Polyphasic	
KC 0543	Unity religion	
KC 0791	Unity religions ; Transcendent	
KC 0811	Unity science	
KC 0650	Unity ; World	
KC 0742	Universal	
KC 0422	Universal culture	
KC 0262	Universal grammar	
KC 0065	Universalism	

KC 0170	Universals	
KC 0947	Universals language	
KC 0947	Universals ; Language	
KC 0228	Universe	
KC 0797	Universe discourse	
KC 0592	Universe ; Holographic	
KC 0278	Universe ; Order	
KC 0107	University	
KC 0790	University ; International	
KC 0426	Urban dynamics	
KC 0684	Utopia	

V

KC 0256	Value analysis	
KC 0256	Value control; Value engineering	
KC 0256	Value engineering ; Value control;	
KC 0356	Value theory	
KC 0387	Value ; Threshold	
KP 3016	Values assumptions	
KD 2272	Variable geometries global order ; Focus	
KD 2202	Varieties communication ; Encompassing	
KD 2200	Varieties form ; Encompassing	
KC 0606	Variety	
KC 0146	Variety ; Principle requisite	
KC 0248	Vertical integration	
KC 0512	Vertical thinking	
KC 0053	View ; World	
KC 0847	Vision ; Artistic	
KD 2146	Vision ; De categorization poly ocular	
KD 2340	Vision world ; Language probabilistic	
KC 0418	Visualization	
KC 0651	Voluntary economic integration	

W

KC 0053	Weltanschauung	
KC 0500	Whole	
KC 0879	Whole categories ; Part	
KC 0328	Whole system principle	
KD 2105	Whole ; Trialectics: logic	
KC 0866	Wholeness	
KD 2230	Wholeness implicate order	
KC 0941	Wholism	
KC 0357	World brain	
KC 0067	World citizenship	
KC 0750	World crises	
KC 0422	World culture	
KC 0821	World economy	
KC 0357	World encyclopaedia	
KC 0153	World federal government	
KC 0153	World government	
KC 0755	World ; Integrated	
KD 2340	World ; Language probabilistic vision	
KC 0692	World law	
KC 0324	World-line	
KC 0659	World market	
KC 0372	World mind groups	
KC 0042	World modelling	
KC 0496	World order	
KC 0750	World problems	
KC 0804	World socialist market	
KC 0506	World society	
KC 0302	World soul	
KC 0805	World system	
KC 0650	World union	
KC 0650	World unity	
KC 0053	World view	

References

Integrative Knowledge

Scope of bibliography

The following section constitutes a collection of bibliographical references relevant to the Integrative Knowledge (Section K).

The entries appear here for any of the following reasons:

(a) Reference is made to them in the individual entries of Section K (where they appear in abridged form, namely with author, title and year of publication).

(b) Reference is made to them in the commentary on Section K (which appears in Section KZ).

(c) The publications are relevant to Section K, even though they have not (yet) been cross-referenced in either of the two ways indicated above.

Bibliographical entries

The entries appear by order of author and, within author, by year.

In certain cases the *International Standard Book Number (ISBN)* is given to facilitate access.

Source of bibliography

The bibliography was derived from a wide variety of sources. These include:

- Information from international organizations, including catalogues of publications and accessions lists

- Information from commercial publishers, especially sales catalogues

- Citations and bibliographies in other publications and reference books

- Systematic compilations of books in print (notably *Books in Print* and *International Books in Print*)

- Book reviews in specialized journals

Abell, W The Collective Dream in Art
New York, Schocken Books.

Abellio, Raymond La structure absolue: essai de phénoménologie génétique (1965)
Paris, Gallimard, 527 p.

Abraham, Ralph Vibrations and the realization of form (1976)
In Erich Jantsch and C H Waddington (Ed). Evolution and Consciousness; human systems in transition. Reading, Addison-Wesley, 1976, p.134–148.

Abt Associates Application of Systems Analysis Models: a survey (1968)
Washington DC, NASA, 69 p. bibl. NASA SP–5048.

Abt, Clark C Serious Games (1970)
New York, Viking Press.

Achinstein, P Models, Analogies and Theories (1964)
Philosophy of Science 31 October p. 328–350, bibl.

Ackoff, R L Varieties of Unification (1946)
Philosophy of Science, 13:287–300.

Ackoff, R L Towards a Behavioral Theory of Communication (1957)
Management Science, 4, pp. 218–34.

Ackoff, R L Systems, Organizations, and Interdisciplinary Research (1960)
General Systems Yearbook Society for General Systems Research, 5, p. 1–7.

Ackoff, R L Progress in Operations Research (1961)
New York, Wiley, 505 p.

Ackoff, R L Scientific Method: optimizing applied research decisions (1962)
New York, John Wiley and Sons, 464 p.

Ackoff, R L General System Theory and Systems Research: contrasting conceptions of systems science (1963)
General Systems Yearbook Society for General Systems Research, 8, p. 117–121.

Ackoff, R L The Art of Problem Solving: accompanied by Ackoff's fables (1978)
New York, Wiley.

Ackoff, R L and Emery, F E On Ideal–Seeking Systems (1972)
General Systems Yearbook Society for General Systems Research, 17, p. 17–24.

Ackoff, R L and Emery, F E On Purposeful Systems (an interdisciplinary analysis of individual and social behavior as a system of purposeful events) (1972)
Chicago IL, Aldine–Atherton, 288 p.

Ackoff, R L and Rivett, P A Manager's Guide to Operations Research (1963)
New York, John Wiley and Sons, 107 p.

Ackoff, R L and Sasieni, M W Fundamentals of Operations Research (1968)
New York, John Wiley and Sons, 455 p.

Ackoff, Russell L Redesigning the Future: a systems approach to societal problems (1974)
New York, Wiley–Interscience, 260 p.

Acres, David and Seymour, Roland General Studies (1987)
White Plains NY, Longman. ISBN 0–582–29160–7.

Adamer, Josef Holismus: prispevek ka kritice burzoazni filosofie (1966)
Praha, Ceskoslovenska akademie ved 187, bibl. pp. 275–278, Summary in English.

Adams, Richard Newbold Energy and Structure: a theory of social power (1975)
Austin, University of Texas Press.

Addo, Herb Approaching The New International Economic Order dialectically and transformationally (1984)
In: *Herb Addo (Ed) Transforming the World Economy* London, Hodder and Stoughton. In association with the United Nations University.

Adrecht, E Des Aspects Méthodologiques de l'Intégration de Sciences
In: *Organon* 1974, Warsaw, 10, pp. 19–23.

Afanas'iev, V G Problema Tzelostnosti v Filosofii Ibiologii (The problem of wholeness in philosophy) (1964)
Moscow, Mysl Press.

Agassi, J Analogies as Generalizations (1964)
Philosophy of Science 31 October pp. 351–356, bibl.

Aggarwal, J K Notes on Nonlinear Systems (1972)
New York, Reinhold Van Nostrand, 214p. bibl. (pp. 202–211).

Akademiia nauk SSSR, Institut Istorii Estestvcznaniia i Techniki Sistemnye Issledovaniia (1969)
Moskva, 1969–annual (Title also bears Systems Research).

Albin, P and Gottinger, H Structure and Complexity in Economic and Social Systems
In: *Mathematical Social Sciences* 1983, 5, pp. 253–268.

Albou, P Problèmes actuels de la psycho–sociologie économique et les sciences humaines (1967)
Paris, Dunod.

Alexander, Christopher Notes on the Synthesis of Form (1964)
Cambridge MA, Harvard University Press, 216 p.

Alexander, Christopher A City is not a Tree (1966)
Architectural Forum (reprinted in *Design* 206, February, pp. 46–55).

Alexander, Christopher The Timeless Way of Building (1979)
New York, Oxford University Press.

Alexander, Christopher, et al The Oregon Experiment (1975)
New York, Oxford University Press. Vol 3.

Allen, John Succeed: a handbook on structuring managerial thought (1986)
London, Synergetic Press, 58 p. illus. ISBN 0–907791–042.

Allen, L A The Management Profession (1964)
New York, McGraw Hill Book Company, 375 p.

Allen, P M Towards a New Science of Complex Systems
In: *The Science and Praxis of Complexity* 1985, Tokyo, United Nations University, pp. 268–297.

Allen, P M; Engelen, G and Sanglier, M Self–Organizing Dynamic Models of Human Systems
In: *E Frehland (Ed) From Microscopic to Macroscopic Order* 1982, Berlin, Springer Verlag, Synergetics Series: 22.

Alpert, Daniel The Role and Structure of Interdisciplinary and Multidisciplinary Research Centers (Address to the Ninth Annual Meeting of the Council of Graduate Schools in the US, Washington, D C, December 4–6, 1969) (outlines problems of present approaches to interdisciplinary research) (1969)

American Academy of Arts and Sciences Contributions to the Analysis and Synthesis of Knowledge (1951)
(Published in cooperation with the Institute for the Unity of Science) Boston, 1951–1954, 4 parts, 350p. (Proceedings of the American Academy of Arts and Sciences, vol. 80, 1–4).

American Association for the Advancement of Science The Psychological Bases of Science – A Process Approach, 1965 (1965)
Washington DC, American Association for the Advancement of Science.

American Managemelt Association Integration Policies and Problems in Mergers and Acquisitions (1957)
New York, The Association, 67 p.

American Society of Criminology Interdisciplinary Problems in Criminology (1965)
Columbus ON, Ohio State University College of Commerce and Administration, 227p.

Amosov, N M Modeling of Thinking and the Mind (1967)
New York, Spartan Books.

Amstutz, G C Symmetry in Nature and Art
In: *Main Currents in Modern Thought* Vol 23, No 1.

Ando, Albert; Fisher, Franklin M, and Simon, Herbert A Essays on the Structure of Social Science Models (1963)
Cambridge MA, MIT Press.

André, Charles L'Organisation de la coopération intellectuelle (1938)
Rennes.

Andrews, Donald H Quantization as an Integrative Concept (1972)
In: *Integrative Principles of Modern Thought* New York, Gordon and Breach Science Publishers, pp. 69–134.

Angonet, M and Suvin, D L'Implicité du Manifeste: métaphorm et imagerie de la démystification dans la Manifeste Communiste
In: *Etudes Françaises* 1980, 16, iii–iv, pp. 43–67.

Angrist, Stanley W and Hepler, L G Order and Chaos: laws of energy and entropy (1967)
New York, Basic Books, 237 p. bibl. (pp. 225–227).

Angyal, A The Structure of Wholes (1939)
Philosophy of Science 6, pp. 25–37.

Anokin, P K Cybernétique, némophysiologie et psychologie (1968)
In: *Les sciences sociales, problèmes et orientations* The Hague, Mouton, Paris, Universco, pp. 279–308.

Anon Penser / Classer (1982)
Le Genre Humain Paris, Fayard, 2. (special issue).

Anon Creating Integrated Systems (1983)
Brookfield VT, Gower Publishing, 86 p. ISBN 0–909394–06–7.

Ansbro, J J Experiment in Interdisciplinary Liberal Education (1968)
Liberal Education 54, December.

Anthony, R N Planning and Control Systems: a framework for analysis (1965)
Boston MA, Harvard Business School, 180 p.

Antonisse, James H; Benoit, John W and Silverman, Barry G (Eds) AI Systems in Government Conference, 4th; 1989: proceedings (1989)
Washington DC, IEEE Computer Society Press, 321 p. ISBN 0–8186–8934–X.

Apostel, Léo Les Sciences Humaines: échantillons de relations interdisciplinaires
In: *UNESCO – Interdisciplinarité et Sciences Humaines* 1983, Paris, UNESCO, pp. 73–168.

Apostel, Leo Conceptual Tools for Interdisciplinarity: an operational approach (1972)
In: *Interdisciplinarity, OECD, Centre for Educational Research and Innovation* Paris, OECD, pp. 141–180. (compares and evaluates several attempts to develop interdisciplinary research, including: Movement for Unified Science, Society for General Systems Research, Centre international d'épistomologie génétique, etc).

Apostel, Leo, Piaget, Jean et al La filiation des structures (1963)
Paris, Presses Universitaires de France, 196 p.

Apter, M Cybernetics and Development (1966)
New York, Macmillan.

Arbib, M A Brains, Machines and Mathematics (1964)
New York, McGraw Hill Book Company.

Arbib, M A A Partial Survey of Cybernetics in Eastern Europe and the Soviet Union (1966)
Behavioral Science 11, pp. 193–216.

Arbib, M A The Metaphorical Brain (1972)
New York, Wiley Interscience.

Archibald, R D and Villoria, R L Network–Based Management Systems (PERT/CPM) (1967)
New York, John Wiley and Sons, 508 p.

Argyris, Christopher Interpersonal Competence and Organizational Effectiveness (1962)
Homewood IL, Irwin and Dorsey.

Arieti, S (Ed) The Intrapsychic Self: feeling, cognition and creativity in health and mental Illness (1967)
New York, Basic Books.

Arkes, Harold and Hammond, Kenneth K (Eds) Judgment and Decision Making: an interdisciplinary reader (1986)
Cambridge MA, Cambridge University Press, 826 p. illus. ISBN 0–521–32617–6.

Arnheim, R Art and Visual Perception: a psychology of the creative eye (1954)
California CA, University of California Press.

Arnheim, R Toward a Psychology of Art (1966)
Berkeley CA, University of California Press.

Arnheim, R Entropy and Art: an essay on disorder and order (1971)
Berkeley CA, University of California Press, 64 p. bibl. (pp. 57–61).

Aron, Raymond World Technology and Human Destiny (1963)
Ann Arbor MI, University of Michigan.

Aronson, Sidney H Obstacles to a Rapprochement between History and Sociologist's View (1969)
In: *Sherif and C Sherif (Ed) Interdisciplinary Relationships in the Social Sciences* Chicago, Aldine, pp. 282–304.

Arrow, Kenneth J Mathematical models in the social sciences (1956)
General Systems Yearbook Society for General Systems Research, 1, p. 29–47.

Aschner, Mary Jet al A System for Classifying Thought Processes in the Context of Classroom Verbal Interaction (1965)
Urbana, Institute for Research on Exceptional Children, 50 p.

Ashby, W R The Application of Cybernetics to Psychiatry (1954)
Journal of Mental Science 110, pp. 114–24.

Ashby, W R Design for an Intelligence Amplifier (1956)
In: *C E Shannon and J McCarthy (Ed) Automation Studies* Princeton University Press.

Ashby, W R General Systems Theory as a New Discipline (1958)
General Systems Yearbook Society for General Systems Research, 3, p. 1–6.

Ashby, W R Requisite Variety and its Implications for the Control of Complex Systems (1958)
In: *Cybernetica* 1, 2, 83.

Ashby, W R Design for a Brain (1960)
London, Chapman and Hall, 286 p.

Ashby, W R Principles of the Self–Organizing System (1962)
New York, Pergamon Books. Principles of Self Organization.

Ashby, W R Induction, Prediction and Decision–Making in Cybernetic Systems (1963)
Induction: Some current issues Middletown, Conn., Wesleyan University, pp. 55–73.

Ashby, W R An Introduction to Cybernetics (1963)
New York, John Wiley and Sons, 295 p.

Ashby, W R Cybernetics Today and its Future Contribution to the Engineering–Sciences (1963)
General Systems Yearbook Society for General Systems Research, 8, p. 207–218.

Ashby, W R The Set Theory of Mechanism and Homeostasis (1964)
General Systems 9, p. 83–97.

Ashby, W R Constraint Analysis of Many–Dimensional Relations (1964)
Urbana, University of Illinois. Technical Report: 2.

Asheraft, R Economic Metaphors, Behaviouralism and Political Theory: some observations on the ideological use of language
In: *Western Political Quarterly* 1977, 30, 3, pp 313–328.

Askin, Iakov Fomich and Zelkina, Olga S (Ed) Problemy Determinizma v Svete Sistemnostrukturnogo analiza (1970)
Saratov, 181p.

Association for Systems Management An Annotated Bibliography for the Systems Professional (1970)
Cleveland OH, The Association, 183 p. 2nd ed.

Association for Systems Management Business Systems (1970)
Cleveland OH, The Association, 496 p. bibl.

Atkin, Ron Mathematical Structures in Human Affairs (1974)
London, Heinemann.

Atkin, Ron Combinatorial Connectivities in Social Systems: an application of simplicial complex structures to the study of large organizations (1977)
Basel, Birkhäuser Verlag.

Atkin, Ron Multidimensional Man: can man live in 3–dimensional space? (1981)
London, Penguin Books, 196 p. bibl.

Atlan, H Information Theory and Ecosystems: proceedings of a Oceanic Ecology Conference, Québec, March 1984

Atlan, H Natural Complexity and Self–Creation of Meaning
In: *The Science and Praxis of Complexity* 1985, Tokyo, United Nations University, pp. 173–192.

Atlantic Information Centre for Teachers Interdisciplinary Studies in Secondary Schools (1969)
London, The Centre, 64 p. bibl. pp 61–64. ×3Report of the 7th Atlantic Study Conference on Education (1968).

Attali, Jacques Bruits: essai sur l'économie politique de la musique (1977)
Paris, Presses Universitaires de France.

Attali, Jacques L'Ordre cannibale (1979)
Paris, Grasset.

Attali, Jacques Les Trois Mondes: pour une théorie de l'après–crise (1981)
Paris, Fayard.

Attinger, E O (Ed) Global Systems Dynamics: proceedings (1970)
New York, Wiley Interscience, 353 p.

Attinger, E O Potentials and Pitfalls in the Analysis of Social Systems (1970)
In: *Attinger, E O (Ed) Global Systems Dynamics* New York, Wiley Interscience, pp. 130–144.

Attneave, F Applications of Information Theory to Psychology (1959)
New York, Holt, Rinehart and Winston.

Auerbach, E Mimesis: the representation of reality in Western literature
New York, Anchor Books.

Auerswald, Edgar H Interdisciplinary Versus Ecological Approach (in anthropology) (1968)
Family Process 1968 (reprinted by Warner Modular Publications).

Auger, Pierre Science as a Force for Unity among Men (1956)
Bulletin of the Atomic Scientists 12, June pp. 208–210.

Auger, Pierre Structure and Complexity in the Universe (1963)
In: *G S Metraux and F Crouzet (Eds) The Evolution of Science* Mentor Books, New York.

Auger, Pierre Les aspects synthétiques dans l'organisation de la recherche scientifique (1967)
Science et Synthèse Paris, Gallimard (Unesco copyright) pp. 277–283 (also English edition).

Auger, Pierre, et al Towards a Synthesis in the Organization of Scientific Research: articles relating to a Symposium on Science and Synthesis, Paris, 1965 (1966)
Impact of Science on Society, 16, 1, pp 5–40.

Aulin, A The Cybernetic Laws of Social Progress (1981)
Elmsford NY, Pergamon.

Axelrod, R The Evolution of Cooperation (1984)
New York, Basic Books.

Bacon, Francis Description of the Intellectual Globe (1857)
In: *Spedding, Ellis and Heath (Eds) The Works of Francis Bacon* London, Longman, 1857–1872, 13 vol., Vol V, 506p.

Baer, R M Some General Remarks on Information Theory and Entropy (1955)
In: *Information Theory in Biology, Quastler (Ed)* 21–24, Illinois University Press.

Baez, Albert V Integrated Science Teaching as Part of General Education: looking ahead.
In: *New trends in integrated science* Paris, Unesco, pp. 167–174.

Bahm, A J Systems Theory: hocus pocus or holistic science? (1969)
General Systems Yearbook Society for General Systems Research, 14, p. 175–177.

Bahm, Archie J Polarity, Dialectic and Organicity (1977)
Albuquerque, World Books.

Bajah, S T; Ryan, J O and Samuel, P S Integrated Science for Tropical Schools (1971)
2 Teachers' Guides, Pupils' Books, Pupils' Workbooks Oxford University Press.

Bakan, David On method: toward a reconstruction of psychological investigation (1927)
San Francisco CA, Jossey-Bass Publishers.

Baldwin, Donald R; Birnbaum, Philip H and Rossini, Frederick, A International Research Management: studies in interdisciplinary methods from business, government and academia (1990)
Oxford, Oxford University Press, 272 p. illus. ISBN 0–19–506252–3.

Ballmer, Thomas T and Weizsaecker, Ernst von Biogenese und Selbstorganisation (1974)
In: *Ernst von Weizsaecker (Ed). Offene Systeme I: Beitraege zur Zeitstruktur von Information, Entropie und Evolution* Klett, 1974.

Balzer, Wolfgang et al Architectonic for Science: the structuralist program (1987)
Dordrecht, Kluwer Academic Publishers Group, 468 p. ISBN 90–277–2403–2.

Banathy, Bela H (Ed) Systems Inquiring – Applications, Theory, Philosophy and Methodology: proceedings of the Society for General Systems Research, 1985 (1985)
Salinas CA, Intersystems Publications, 1200 p. 2 Vols. ISBN 0–914105–36–1.

Banerji, R B A Language for the Description of Concepts (1964)
General Systems Yearbook Society for General Systems Research, 9, p. 135–141.

Banghart, F W Educational Systems Analysis (1969)
New York, Macmillan, 330 p.

Bar-Hillel, Y An Examination of Information Theory (1955)
Philosophy of Science 22, pp. 86–105.

Bar-Hillel, Y Language and Information (1964)
Reading MA, Addison-Wesley.

Barber, B and Merton, R K Brief Bibliography for the Sociology of Science (1952)
American Academy of Arts and Sciences, Proceedings 80, 2, pp. 140–154, bibl.

Barber, Bernard The Sociology of Science: a trend report and bibliography (1956)
Current Sociology 5, 4, 1956 (New York Unesco), 153 p. (entire issue).

Barber, Bernard Resistance by Scientists to Scientific Discovery (1961)
Science, 184, pp. 596–602.

Barber, Bernard New Bibliography, 1962 – on the sociology of science (1966)
New York, Columbia University Press, 9 p.

Bardis, P D Cybernetics: definition, history, etymology (1965)
Social Science 1964, pp. 226–28.

Barel, Yves Prospective et analyse des systèmes (1971)
Paris, La Documentation Française 174 p. bibl. (pp. 164–171).

Barfield, O Worlds Apart: a dialogue of the 1960's (1963)
Middletown CT, Wesleyan University Press, 211 p. ISBN 0–8195–6017–0.

Barfield, O What Coleridge Thought (1972)
London, Oxford University Press.

Barratt, Krome Logic and Design: the syntax of art, art and mathematics (1980)
Westfield NJ, Eastview Editions, 328 p. ISBN 0–89860–033–2.

Barrea, Jean L'intégration politique externe (1969)
Leuven, Nauwelaerts Uitgaven, 335 p. bibl (pp. 319–325).

Barrett, F D and Shepard, H A A Bibliography of Cybernetics (1953)
Proceedings of the American Academy of Arts and Sciences 80, pp. 204–22.

Bartlett, S J and Suber, P (Eds) Self Reference: reflections on reflexivity (1987)
Dordrecht, Kluwer Academic Publishers Group, 376 p. ISBN 90–247–3474–6.

Baskin, Frank R Refining Generalist Social Work Practice (1983)
Lanham MD, University Press of America, 142 p. illus. ISBN 0–8191–3718–9.

Bassam, Netherlands F Synthese: an international journal devoted to present day cultural and scientific life
G. Kroonder, 1936–1939–1946–, monthly.

Bassett, G A Management Styles in Transition (1966)
New York, American Management Association, 208p.

Bastide, R Approche interdisciplinaire de la maladie mentale (1968)
In: *Les Sciences Sociales; problèmes et orientations* La Haye, Mouton, Paris, Unesco.

Bateson, G The Pattern Which Connects
In: *The Co-Evolution Quarterly* 1978, 18, pp. 5–15.

Bateson, Gregory Steps to an Ecology of Mind: collected essays in anthropology, psychiatric, evolution and epistemology (1972)
San Francisco, Chandler.

Bateson, Gregory Mind and Nature: a necessary unity (1979)
New York, EP Dutton, 238 p. ISBN 0–525–15590–2.

Bateson, Gregory and Bateson, Mary Catherine Angels Fear: towards an epistemology of the sacred (1987)
New York, Macmillan, 224 p. ISBN 0–02–507670–1.

Bateson, Mary Catherine Our Own Metaphor: a personal account of a conference on the effects of conscious purpose on human adaptation (1972)
New York, Knopf.

Bauer and Gergen, K J (Eds) Study of Policy Formation (1968)
New York, Free Press.

Bauer, et al Second Order Consequences
Cambridge MA, MIT Press.

Bauer, R A Social Indicators (1966)
Cambridge MA, MIT Press, 357 p.

Bayliss, L E Living Control Systems (1966)
San Francisco, Freeman, 189 p.

Beaty, J Interdisciplinarial Body (colloquium on the novelist and society in the 1840's and the 1880's) (1968)
Victorian Studies 12, Dec, pp. 139–144.

Becker, Ernest The Structure of Evil: an essay on the unification of the science of man (1968)
New York, Braziller.

Becker, Joseph and Hayes, Robert M Information Storage and Retrieval: tools, elements, theories (1963)
New York, John Wiley and Sons.

Becket, John A The Total Systems Concept: its implication for management (1967)
In: *The Impact of Computers on Management 204–235, Myers, Ch A (Ed)* Cambridge MA, London.

Becket, John A Management Dynamics: the new synthesis (1971)
New York, McGraw-Hill, 234 p. bibl.

Beckner, Morton The Biological Way of Thought (1968)
Berkeley CA, University of California Press, 200 p.

Beer, Henri L'Avenir de la Philosophie, Esquisse d'une Synthèse des Connaissances Fondée sur l'Histoire (thèse de lettres) (1899)
Paris, Hachette, 512 p.

Beer, Henri Programme d'une Bibliographie Synthétique (1920)
Revue de Synthèse Historique février pp. 75–79.

Beer, Henri La Synthèse en Histoire: son rapport avec la synthèse générale (1953)
Paris, A Michel, 322 p. (nouvelle édition) (L'Evolution de l'humanité, Synthèse collective. Série complémentaire).

Beer, Stafford The World, the Flesh and the Metal: the prerogatives of systems (1925)
Nature London, 205, 1965, pp. 223–231.

Beer, Stafford Below the Twilight Arch: a mythology of systems (1960)
General Systems Yearbook Society for General Systems Research, 5, p. 5–20.

Beer, Stafford Cybernetics and Management (1964)
New York, John Wiley and Sons, 214 p.

Beer, Stafford Decision and Control: the meaning of operational research and management cybernetics (1966)
New York, John Wiley and Sons.

Beer, Stafford Management Science: the business use of operations research (1968)
Garden City NY, Doubleday, 192 p.

Beer, Stafford Brain of the Firm: the managerial cybernetics of organization (1972)
London, Allen Lane Penguin Press.

Beer, Stafford Management in Cybernetic Terms (1972)
In: *Unesco Scientific Thought.* The Hague, Mouton.

Beer, Stafford Designing Freedom (1973)
Toronto ON, Canadian Broadcasting Corporation, 100 p. (Massey Lectures, 13th series).

Beer, Stafford Platform for Change: a message from Stafford Beer (1975)
London, John Wiley and Sons, 457 p.

Beer, Stafford Managing Modern Complexity (1975)
In: *Stafford Beer, Platform for Change* London, Wiley, pp. 219–241.

Beigl, Herbert Unity of Science and Unitary Science (1953)
In: *H Feigl and M Brodbeck (Eds) Readings in the Philosophy of Science* New York, 1953 (conceptual distinction from physicalist app.).

Belbin, R M Management Teams: why they succeed or fail (1981)
London, Heinemann.

Bell, D A Intelligent Machines: an introduction to cybernetics (1962)
Blaisdell.

Bell, J L and Slomson, A B Models and Ultraproducts: an introduction (1970)
New York, Humanities Press.

Bellman, R E Dynamic Programming (1957)
Princeton NJ, Princeton University Press, 342 p.

Belnis, W G; Benne, K D and Chin, R (Eds) The Planning of Change (1961)
New York, Holt Rinehart and Winston, 781 p.

Benfey, O T An Approach to the Conceptual Analysis of Scientific Crises (1964)
General Systems Yearbook Society for General Systems Research, 9, p. 57–59.

Benjamin, A C Some Theories of the Development of Science (1953)
Philosophy of Science 20, July pp. 167–176, bibl.

Bennett, A Leroy The Development of Intellectual Cooperation under the League of Nations (1950)
Urbana.

Bennett, J G The Dramatic Universe: the foundation of natural philosophy (1956)
London, Hodder and Stoughton, 356 p. 2 vols. bibl. foot.

Bennett, J G; Brown, R L and Thring, M W Unified Field Theory in a Curvature-Free Five-Dimensional manifold (1949)
Proceedings of the Royal Society 198A, p. 39.

Bennett, J W Interdisciplinary Research and the Concept of Culture (1954)
American Anthropologist 54, April 2, part 1, pp. 169–179, bibl.

Bennis, Warren G The Social Science Research Organization: a study of the institutional practices and values of interdisciplinary research (1955)
Cambridge MA, MIT Press. Unpublished doctoral dissertation.

Bennis, Warren G Some Barriers to Teamwork in Social Research (1956)
Social Problems 3, pp. 223–235.

Bennis, Warren G Towards a 'Truly' Scientific Management: the concept of organization health (1962)
General Systems Yearbook Society for General Systems Research, 7, p. 269–280.

Bennis, Warren G Changing Organizations: essays on the development and evolution of human organization (1966)
New York, McGraw Hill Book Company, 223 p.

Benoist, Jean-Marie L'Interdisciplinarité dans les Sciences Sociales
In: *UNESCO – Interdisciplinarité et Sciences Humaines* 1983, Paris, UNESCO, pp. 169–190.

Benson, D Synergistics: the study and practice of synergy (1972)
Man-Environment Systems 2.

Bentley, A F Kinetic Inquiry (1950)
Science 112, pp. 775–83.

Bentov, I Stalking the Wild Pendulum: on the mechanics of consciousness (1977)
New York, E P Dutton.

Bereille, André Harmonic and Disharmonic Social Systems (1971)
Sydney, Sydney University Press, 27 p. bibl.

Berg, J H Van den Metabletica van de Materie (1969)
Nijkerk, G F Callenbach.

Berger, B M Sociology and the Intellectuals: an analysis of a stereotype (1957)
Antioch Review 17, Fall, pp. 275–290. bibl.

Berger, Guy The Interdisciplinary Archipelago (1972)
In: *OECD, Centre for Educational Research and Innovation, Interdisciplinapity*, Paris, OECD, pp. 35–70 (geography of interdisciplinarity; some general principles for regrouping disciplines; application of interdisciplinarity; obstacles and difficulties).

Bergson, Henri Creative Evolution (1984)
University Press of America.

Bernal, J D The Social Function of Science (1967)
Cambridge MA, MIT Press. London, Routledge 1939.

Bernal, J D Science in History (1971)
Cambridge MA, MIT Press. 4 Vols. Vol 1: The Emergence of Science. ISBN 0–262–52020–6. Vol 2: The Scientific and Industrial Revolution. ISBN 0–262–52021–4. Vol 3: The Natural Sciences in Our Time. ISBN 0–262–52022–2. Vol 4: The Social Sciences: a conclusion. ISBN 0–262–52023–0.

Berrien, Frederick K General and Social Systems (1968)
New Brunswick NJ, Rutgers University Press, 231 p. bibl. (pp. 205–222).

Berry, Brian J L and Pred, Allan (Eds) Central Place Studies: a bibliography of theory and applications, including supplement through 1964 (1965)
Philadelphia, Regional Science Research Institute.

Berry, M V; Percival, I C and Weiss, N O (Eds) Dynamical Chaos (1989)
Princeton NJ, Princeton University Press, 200 p. ISBN 0–691–08519–6.

Bertrand, Alvin Lee Social Organization: a general systems and the role theory (1971)
Philadelphia PA, FA Davis, 226 p. bibl. (pp. 211–215).

Bertrand, Maurice Some Reflections on Reform of the United Nations (1985)
Geneva, Joint Inspection Unit of the UN. JIU/REP/85/9.

Beshers, J M (Ed) Computer Methods in the Analysis of Large-Scale Social Systems (1965)
Cambridge MA, Harvard University Press, 207 p.

Bestler, Patricia J and DuBois, Jean M Plan of Action for Personnal Generalists and Corporate Managers with Supervisory Responsibilities (1986)
Wayzata MN, Raleigh Publishing, 138 p. ISBN 0–9615775–0–9.

Betiol, L F Integracao Econŏmica e Uniăo Politica Internacionais (1966)
Sao Paulo, Ed Revista dos Tribunais, 133 p. bibl. (pp 105–107).

Betta, J A The Ideology of Rhetoric of Thomas Paine: political justification through metaphor (1975)
New Brunswick NJ, Rutgers University. Ph D Dissertation.

Bidney, David The Concept of Meta-Anthropology and its Significance for Contemporary Anthropological Science (1949)
In: *F Northrop (Ed) Ideological Differences in World Order; studies in the philosophy and science of the world's cultures* New Haven, Yale University Press, pp. 323–355.

Bierman, Judah and Laszlo, Ervin Goals in a Global Community: the original background papers for goals for mankind (1977)
Oxford, Pergamon Press. 2 Vols.

Bindel, Ernst Pythagoras: leben und lehre in wirklichkeit und legende (1962)
Stuttgart, Martin Sandkühler, 207 p.

Bindel, Ernst Harmonien im Reiche der Geometrie: in Antehnung an Keplers Weltharmonik (1964)
Stuttgart, Martin Sandkühler, 69 p.

Bindel, Ernst Die Arithmetik: die menschenkundliche begründung und pädagogische bedentung (1967)
Stuttgart, Martin Sandkühler, 124 p. illus.

Binet, J Psychologie Economique Africaine: eléments d'une recherche interdisciplinaire (1970)
Payot.

Black, Guy The Application of Systems Analysis to Government Operations (1968)
New York, Praeger Publishers, 186 p. bibl. (pp. 173–181).

Blackowicz, J A Systems Theory and Evolutionary Models of the Development of Science (1971)
Philosophy of Science 38, June pp. 178–199, bibl.

Blackwell, G W Multidisciplinary Team Research (1955)
Social Forces 33, May pp. 367–374, bibl.

Blair, R N and Whitston, C W Elements of Systems Engineering (1971)
New York, Prentice Hill.

Blake, D V and Uttley, A M (Eds) Proceedings of a Symposium on Mechanization of Thought Processes (1959)
London, H M Stationery Office. 2 Vols.

Blaquière, Austin Analyse des systèmes non linéaires (1966)
Paris, Presses Universitaires de France, 440 p. bibl. (pp. 415–417).

Blau, P M The Dynamics of Bureaucracy (1955)
Chicago IL, University of Chicago, 269 p.

Blau, P M and Scott, W R Formal Organizations: a comparative approach (1962)
San Francisco, Chandler, 312 p.

Blauberg, I V Problema Tzelostnosti v Marksistskoi Filosofii (The problem of wholeness in Marxist philosophy) (1964)
Moscow, Vysshaia Shkola.

Blauberg, I V; Sadovsky, V N and Yudin, E G Systems Research: yearbooks 1969, 1970, 1971
Moscow, Nauka Press.

Blauberg, I V; Sadovsky, V N and Yudin, E G Systems Approach: its presuppositions, problems and difficulties (1969)
Moscow, Znanie Press.

Blauberg, I V; Sadovsky, V N and Yudin, E G Methodological Problems of Systems Research (1970)
Moscow, Mysl Press.

Blauberg, I V; Sadovsky, V N and Yudin, E G Some Problems of General Systems Development (1973)
In: *Unity through Diversity, Gray, William and Rizzo, Nicholas D (Eds)* New York, Gordon and Breach Science Publishers, pp. 245–266.

Blin-Stoyle, R J and others Turning Points in Physics
Foundation for Integrative Education.

Bliss, C K Unified Symbolism for World Understanding in Science
Sydney, Semantography Publishing.

Bloomfield, Brian P Modelling the World: the social constructions of systems analysts (1986)
New York, Basil Blackwell, 240 p. ISBN 0–631–14163–4.

Bluhm, William T (Ed) The Paradigm Problem in Political Science: perspectives from philosophy and from practice (1982)
Durham NC, Carolina Academic Press, 227 p. ISBN 0–89089–218–0.

Blum, Abraham Towards a Rationale for Integrated Science Teaching: agriculture as environmental science project
In: *New trends in integrated science teaching, Vol II* Paris, Unesco, pp. 29–45.

Blum, Harold F Times Arrow and Evolution (1968)
Princeton NJ, Princeton University Press. 3rd ed.

Blumenthal, S C Management Information Systems: a framework for planning and development (1969)
Englewood Cliffs NJ, Prentice Hall, 219 p.

Blumer, H Sociological Implications of the Thought of George Herbert Mead (1966)
American Journal of Sociology 71, pp. 535–44.

Blumstein, Alfred et al Systems Analysis for Social Problems: proceedings of a symposium, 1969 (1970)
Washington DC, Washington Operations Research Council, 331 p. bibl.

Bochner, Salomon Eclosion and Synthesis: perspectives on the history of knowledge (1969)
New York, WA Benjamin, 264 p, bibl. foot. (chapters on: humanities, historiography, psychology and pedagogy, economics, geology).

Bock, K E Evolution, Function and Change (1963)
American Sociological Review 28, pp. 229–37.

Bode, H F; Mosteller, F; Tukey, F and Winsor, C The Education of a Scientific Generalist (1949)
Science 109, pp. 453.

Boffey, P M Systems Analysis: no panacea for Nation's domestic problems (1967)
Science 158, pp. 1028–1030.

Bogdanov, A A Essays in the Universal Organizational Science (1980)
Seaside CA, Intersystems. Transl. by G Gorelik form 1922 ed.

Boguslaw, R The New Utopians: a study of systems design and social change (1965)
Englewood Cliffs NJ, Prentice Hall, 213 p.

Bohm, David Physics and Perception (1965)
Appendix in his The Special Theory of Relativity New York, Benjamin, pp. 185–230.

Bohm, David Fragmentation in Science and Society (1970)
Impact of Science on Society, 22, 2, April–June p. 159–169.

Bohm, David Fragmentation and Wholeness (1976)
Jerusalem, The Van Leer Jerusalem Foundation.

Bohm, David Wholeness and the Implicate Order (1980)
London, Routledge and Kegan Paul. p.16–25; 6, 1979, 2, p.92–103.

Bohr, Niels The Unity of Human Knowledge (1963)
In: *Atomic Physics and Human Knowledge* Suffolk, Richard Clay.

Boldt, C Die Einheit des Erkenntnisproblems (1937)
Leipzig, 163 p. (Neue Philosophische Forschungen).

Bonini, C P Simulation of Information and Decision Systems in the Firm (1963)
Englewood Cliffs NJ, Prentice Hall.

Bonner, H On Being Mindful of Man (1965)
Boston MA, Houghton Mifflin.

Bonner, J T Cells and Societies (1955)
Princeton NJ, Princeton University Press.

Bonnett, Henri Intellectual Cooperation in World Organization (1942)
Washington DC, American Council on Public Affairs.

Boodin, J E Analysis and Holism (1943)
Philosophical Sciences 10, p. 213–29, 0.

Boole, G The Laws of Thought (1951)
New York, Dover Publications. reprint of 1854 ed.

Born, Max The Restless Universe
Foundation for Integrative Education.

Botnariuc, N The Wholeness of Living Systems and Some Basic Biological Problems (1966)
General Systems Yearbook Society for General Systems Research, p. 93–98.

Bouchon, B and Yager, R R (Eds) Uncertainty in Knowledge-Based Systems: international conference on information processing and management of uncertainty (1986)
Berlin, Springer-Verlag, 405 p. ISBN 3–540–18579–8.

Boudun, Raymond Modèles et méthodes mathématiques (dans les domaines interdisciplinaires de la recherche). (1970)
In: *Unesco, Tendances Principales de la Recherche dans les Sciences Sociales et Humaines* Paris, Mouton/Unesco, pp. 629–685.

Boulding, K E Learning by Simplifying Complexity: how to turn data into knowledge
In: *The Science and Praxis of Complexity* 1985, Tokyo, United Nations University, pp. 25–34.

Boulding, Kenneth E The Organizational Revolution (1953)
New York, Harper and Row, 286 p.

Boulding, Kenneth E The Image (1956)
Ann Arbor MI, University of Michigan.

Boulding, Kenneth E General Systems Theory – The Skeleton of Science (1956)
Management Science 2, p. 197–208. Reprinted in. General Systems Yearbook, 1:11–17.

Boulding, Kenneth E Toward a General Theory of Growth (1958)
General Systems Yearbook Society for General Systems Research, 1, p. 66–75.

Boulding, Kenneth E Political Implications of General Systems Research (1961)
General Systems Yearbook Society for General Systems Research, 6, p. 1–7.

Boulding, Kenneth E An Interdisciplinary Honors Course in General Systems (1962)
Superior Student 4, 31, Jan, Feb.

Boulding, Kenneth E General Systems as a Point of View (1964)
In: *Mesarovic, M D (Ed), Views on General Systems Theory* New York, Wiley.

Boulding, Kenneth E Technology and the Integrative System (1965)
In: *Today's Changing Society; a challenge to individual identity* 1965–1969, pp. 57–73.

Boulding, Kenneth E Beyond Economics: essays on society, religion and ethics (1968)
Ann Arbor MI, University of Michigan, 302 p. bibl.

Boulding, Kenneth E A Primer on Social Dynamics: history as dialectics and development (1970)
New York, Free Press.

Boulding, Kenneth E Economics and General Systems (1972)
In: *Ervin Laszlo, The Relevance of General Systems Theory* New York, Braziller, 213p (pp. 77–92).

Boulding, Kenneth E General Systems as an Integrating Force in the Social Sciences (1973)
In: *Unity through Diversity, Gray, William and Rizzo, Nicholas D (Eds)* New York, Gordon and Breach Science Publishers, pp. 951–966.

Boulding, Kenneth E Ecodynamics: a new theory of societal evolution (1978)
London, Sage.

Boulding, Kenneth E and Senesh, Lawrence (Eds) The Optimum Utilization of Human Knowledge (1983)
Boulder CO, Westview Press.

Bouleau, C The Painter's Secret Geometry (1963)
A Study in the Composition in Art. Harcourt Brace.

Bowman, D M and Fillerup F M (Eds) Management: organization and planning (1963)
New York, McGraw Hill Book Company, 148 p.

Boyce, R O Integrated Managerial Controls: a visual approach through integrated management information systems (1968)
New York, American Elsevier, 372 p.

Boyden, A Homology and Analogy: a critical review of the meanings and implications of these concepts in biology (1947)
American Midl. Nat. 37, May, pp. 648–669. bibl.

Bradley, D F Multilevel Systems and Biology: view of a submolecular biologist (1968)
In: *Systems Theory and Biology, M D Mesarovic (Ed)* pp. 38–58, New York, Springer-Verlag.

Brams, Steven J Transaction Flows in the International System (1966)
American Political Science Review 40, pp. 880–898.

Brams, Steven J Measuring the Concatenation of Power in Political Systems (1968)
American Political Science Review 42, 461–475.

Brams, Steven J The Structure of Influence Relationships in the International System (1969)
In: *International Politics and Foreign Policy: a reader in research and theory. J N Rosenau (Ed)* New York, Free Press.

Branch, M C Planning: aspects and applications (1966)
New York, John Wiley and Sons, 333 p.

Branford, Victor Science and Sanctity: a study in the scientific approach to unity (1923)
London. (vague but full of insight into the pathology of modern life).

Braunstein, Daniel N and Ungson, Gerardo R Decision Making: an interdisciplinary inquiry (1982)
Boston MA, PWS-Kent Publishing, 400 p. ISBN 0–534–01161–6.

Bremer, Stuart A (Ed) Globus Model: computer simulation of worldwide political and economic developments (1987)
Frankfurt, Campus Verlag, 942 p. Coproduction with Westview Press, Boulder, Colorado. ISBN 3–593–33771–1.

Brennecke, J H and Amick, R G Significance: the struggle we share (1971)
Glencoe, Glencoe Press. bibl. (pp. 176–182).

Bridgman, P W The Potential Intelligent Society of the Future (1949)
In: *F Northrop (Ed) Ideological Differences in World Order; studies in the philosophy and science of the world's cultures* New Haven, Yale Univ Press, pp. 229–249.

Briggs, Asa and Michaud, Guy Problems and Solutions (in transforming a university toward interdisciplinarity). (1972)
In: *OECD, Centre for Educational Research and Innovation. Interdisciplinarity* Paris, OECD, pp. 183–277. (annexes on: model of an interdisciplinary university with a special emphasis on international relations; plan for a Center of Interdisciplinary Synthesis).

Briggs, John P and Peat, David F Looking Glass Universe: the emerging science of wholeness (1984)
New York, Simon and Schuster.

Brillouin, L Life, Thermodynamics, and Cybernetics (1949)
American Scientist 37, pp. 554–68.

Brillouin, L Thermodynamics and Information Theory (1950)
American Scientist 38, pp. 590–99.

Brillouin, L Science and Information Theory (1956)
New York, Academic Press.

Brim, Orville G Jr et al Knowledge into Action: improving the nation's use of the social sciences (Report of the Special Commission on the Social Sciences) (1969)
Washington DC, National Science Foundation.

Brimley, Robert Charles Organism: a synoptic integration of human knowledge (1950)
Cambridge, Deighton Bell, 63 p.

British Association Symposium on Cybernetics
Advancement of Science 40. 3 Vols. A: E C Cherry, Organisms and Mechanisms; B: W Hick, The Impact of Information Theory on Psychology; C: D MacKay, On Comparing the Brain with Machines (1954)

Brix, V H You Are a Computer: cybernetics in everyday life (1970)
White Plains NY, Emerson Books.

Brodey, W M Human Enhancement through Evolutionary Technology (1967)
IEEE Spectrum Sept.

Brohm, Jean-Marie Qu'est ce que la Dialectique? (1976)
Paris, Savelli.

Bronowski, J Science and Human Values (1990)
New York, Harper and Row, 128 p. Repr of 1956.
ISBN 0-06-097281-5.

Bronson, Gordon The Hierarchical Organization of the Central Nervous System: implications for learning processes and critical periods in early development (1965)
Behavioral Science 10, 7-25.

Brookhaven National Laboratory, Biology Dept Diversity and Stability in Ecological Systems (Report of a symposium)
Brookhaven Symposia in Biology 22.

Brooks, H On Coherences and Transformations: scientific concepts and cultural change (1965)
Daedalus Winter 94, p. 66-83.

Bross, I J Design for Decision (1953)
New York, Free Press, 276 p.

Brown, G Spencer Laws of Form (1969)
London, George Allen and Unwin, 141 p. bibl. (pp. 136-137) (an approach to the basic forms underlying linguistic, mathematical and physical science).

Brown, S C The Paradox of Reductionism: a study of the logical characteristics of closed systems of ideas (1962)
(Ph D thesis, Birkbeck College, London, accepted 1962-1963).

Browne, Sybil The Arts and Integration (1937)
In: *L Thomas Hopkins (Ed) Integration: Its meaning and application* New York, Appleton-Century, pp. 148-176.

Brun, R A and Saunders, R M Analysis of Feedback Control Systems (1955)
New York, McGraw Hill Book Company.

Brunswik, E Historical and Thematic Relations of Psychology to Other Sciences (1956)
Scientific Monthly 83, 1954, pp. 151-161.

Buck, R C On the Logic of General Behavior Systems Theory (1956)
In: *H Feigl and M Scriven (Eds) The Foundations of Science and the Concepts of Psychology and Psychoanalysis* University of Minnesota Press.

Buck, Roger C and Hull, David L The Logical Structure of the Linnaean Hierarchy (1966)
Systematic Zoology 15:97-111.

Buckley, W Sociology and Modern Systems Theory (1967)
Englewood Cliffs NJ, Prentice Hall, 227 p.

Buckley, W Modern Systems Research for the Behavioral Scientist: a sourcebook (1968)
Chicago IL, Aldine, 525p.

Budd, Richard W and Ruben, Brent D (Eds) Interdisciplinary Approaches to Human Communication (1979)
Chicago IL, Transaction Publishers, 173 p. 2nd ed.
ISBN 0-8104-5125-5.

Budden, Lionel B Synthesis in Architecture: the contemporary process (1934)
Liverpool, Liverpool University Press, 23 p.

Bulletin of Structural Integration
1969 vol 1 onwards.

Bunge, M The Methodological Unity of Science (1973)
Boston MA. Vol VIII.

Bunge, Mario Metascientific Queries (1959)
Springfield IL, Thomas Charles C.

Bunge, Mario Levels: a semantic preliminary (1960)
The Review of Metaphysics 8, p. 396-4068, p. 396-4O6.

Bunge, Mario On the Connections among Levels: proceedings of the XII International Congress of Philosophy, Vol (1960)
Florence, Sansoni.

Bunge, Mario Levels (1963)
The Myth of Simplicity (Chapter 3), Problems of Scientific Philosophy pp. 36-48. New York, Prentice-Hall.

Bunge, Mario Scientific Research (1967)
New York, Springer-Verlag.

Bunge, Mario Partition, Ordering and Systematics (1967)
In: *Scientific Research I, The Search for System* pp. 74-96 (chapter 2). New York, Springer-Verlag.

Bunge, Mario The Metaphysics, Epistemology and Methodology of Levels (1969)
In: *Lancelot Law Whyte, et al (Eds) Hierarchical Structures* New York, American Elsevier, pp. 17-28.

Bunge, Mario Analogy, Simulation, Representation (1970)
General Systems Yearbook Society for General Systems Research, 15, p. 27-34.

Bunge, Mario Metatheory (1972)
In: *Scientific Thought; some underlying concepts, methods and procedure*, Paris/The Hague, Unesco/Mouton, pp227-252.

Burger, Hanry G Procedure Gradation: a means to concept-wording and problem-solving (1962)
General Systems Yearbook Society for General Systems Research, 7, p. 293-304.

Burger, Hanry G 'Agonemmetry' (adaptability through rivalry): an institution evolving biology and culture (1968)
General Systems Yearbook Society for General Systems Research, 12, p. 209-222.

Burgers, J M On the Emergence of Patterns of Order (1963)
Bulletin of the American Mathematical Society 69, pp. 1-25.

Burke, Kenneth Permanence and Change (1965)
Indianapolis, Bobbs-Merrill.

Burkhalter, B R (Ed) Case Studies in Systems Analysis in a University Library (1968)
Metuchen NJ, Scarecrow Press, 186 p.

Burns, T and Stalker, G M The Management of Innovation (1961)
London, Tavistock Publications, 269 p.

Bursk, E C and Chapman, J F (Eds) New Decision-Making Tools for Managers (1963)
Cambridge MA, Harvard University Press, 413 p.

Burt, Michael Spatial Arrangement and Polyhadra with Curved Surfaces and their Architectural Applications (1966)
Haifa, Israel Institute of Technology,. 139 p. Masters Thesis.

Burtt, E A The Metaphysical Foundations of Modern Physical Science (1914)
New York, Doubleday.

Bush, G P ald Hattery, D H Teamwork in Research (1953)
Washington DC, American University Press.

Bushaw, O et al (Ed) Mathematical Systems Theory (1927)
New York, Springer Publishing. 1967.

Butterworth, Eric Unity of All Life (1969)
New York, Harper and Row.

Buttimer, Anne Musing on Helicon: root metaphors and geography
In: *Geografiska Annaler* 1982, 64B, 2, pp. 89-96.

Buttimer, Anne Mirrors, Masks and Diverse Milieux
German version published in Münchener Geographische Hefte 1983.

Buttimer, G Social Space in Interdisciplinary Perspective (1969)
Geog. Review 59, July pp. 417-426.

Cadwallader, M The Cybernetic Analysis of Change in Complex Social Organizations (1959)
American Journal of Sociology 65, pp. 154-57.

Caiden, G Administrative Reform (1969)
Chicago IL, Aldine, 240 p.

Calder, Nigel An Atlas of the Fourth Dimension (1984)
Londn, Chalto and Windus.

Calder, Ritchie The Fragmentation of Science (1955)
Advancement of Science 12, Dec pp. 328-338.

Caldwell, Lynton C Biopolitics: science, ethics and public policy (1964)
Yale Review 54, pp. 1-16.

Calhoun, J B Behavioral States and Developed Images: paper given at a session on nets - social, nemonic and others (1965)
Berkeley CA, AAAS. URBS Document: No 56.

Callois, Roger Dynamique de la Dissymétrie
Diogene 76, p 67-98.

Callois, Roger La Dissymétrie (1973)
Paris, Gallimard.

Cameron, S and Yovits, M (Eds) Proceedings of the Interdisciplinary Conference on Self-Organizing Systems (1960)
New York, Pergamon Books.

Campbell, Donald T Adaptive Behavior from Random Response (1956)
Behavioral Science 1, pp. 105-10.

Campbell, Donald T Methodological Suggestions from a Comparative Psychology of Knowledge Processes (1959)
Inquiry 2, pp. 152-67.

Campbell, Donald T Theories of Boundaries, Groupings and Systems (1965)
In: *Propositions about Ethnocentrism from Social Science Theories, D T Campbell and R A Levine (Eds)* Chapter 7, Unpublished monograph, Northwestern University.

Campbell, Donald T Ethnocentrism of Disciplines and the Fish-Scale Model of Omniscience (1969)
In: *M Sherif and C Sherif (Ed) Interdisciplinary Relationships in the Social Sciences* Chicago, Aldine, pp. 328-348.

Campbell, Jeremy Grammatical Man: information entropy, language and life (1982)
New York, Simon and Schuster.

Campbell, Richmond and Snowden, Lanning (Eds) Paradoxes of Rationality and Cooperation: prisoner's dilemma and newcomb's problem (1985)
Vancouver BC, University of British Columbia Press, 325 p.

Canadian Welfare Council Integration of Physical and Social Planning (1967)
Ottawa ON, The Council. Report No. 2 (1968, 73p - 89p).

Candhill, William and Roberts, B H Pitfalls in the Organization of Interdisciplinary Research (1951)
Human Organization 10, 4, Winter p. 12-15. bibl.

Cantile, E J (Ed) Transportation and Aging: selected issues based on proceedings of an interdisciplinary workshop, Washington, 1970 (1971)
Washington DC, 208 p.

Caplan, Nathan and Nelson, Stephen D On Being Useful: the nature and uses of psychological research on social problems (1972)
Ann Arbor MI, University of Michigan, 35 p. bibl.

Capps, G T Humanities - Science: a natural for team teaching
In: *Clear House* 1971, 45, February pp. 361-364.

Capra, Fritjof The Tao of Physics (1975)
Boulder CO, Shambhala Publications.

Capra, Fritjof The Turning Point: science, society and the rising culture (1982)
Bantam.

Carbonnell, J G Metaphor: a key to extensible semantic analysis: proceedings of the 18th Annual Meeting of the Association for Competational Linguistics (1980)
pp. 17-21. Report on a computer program to paraphrase metaphors based on common formulas in economics and polities.

Carey, G W and Schwartzberg, J Teaching Population Geography: an interdisciplinary ecological approach (1969)
Teachers College Press, 134 p. bibl.

Carles, Jules Unité et Vie: esquisse d'une biophilosophie (1946)
Paris, Beauchesne, 234 p. bibl (Bibliothèque des archives de philosophie).

Carlstein, Tommy Time Resources: society and ecology; on the capacity for human interests in space and time (1982)
London, George Allen and Unwin. 2 Vols.

Carnap, R The Unity of Science (1934)
London.

Carse, James Finite and Infinite Games (1986)
New York, Macmillan.

Carter, J In My Opinion: interdisciplinary science courses (1972)
American Biology Teacher 33 3, 130.

Carter, L J Systems Approach: political interest rises (1966)
Science 153, pp. 1222-1224.

Carzo, R and Yanouzas, J N Formal Organizations: a systems approach (1967)
Homewood IL, Irwin and Dorsey.

Cassidy, Harold G Wholeness: the university's task (1962)
The Graduate Journal 5, p. 160.

Cassidy, Harold G Creating an Interdisciplinary Course (1970)
The Science Teacher November Supplement, pp. 5-7.

Cassidy, Harold G Summary of Theoretical Issues: what generalization of Mendeleev's periodic table means (1972)
In: *E Haskell (Ed) Full Circle; the moral force of unified science* New York, Gordon and Breach, pp. 3-19.

Cassirer, E J The Problem of Knowledge (1950)
New Haven CT, Yale University Press.

Casti, J Connectivity, Complexity and Catastrophe in Large-Scale Systems (1979)
New York, John Wiley.

Cavallo, Roger E (Ed) Systems Research Movement: characteristics, accomplishments and current developments (1979)
Binghampton NY, Society for General Systems Research, 131 p. General Systems Bulletin: IX, 3, special issue.

Caws, Peter Science and System: on the unity and diversity of scientific theory
In: *General Systems Yearbook* Society for General Systems Research, 1968, 13, p. 3-11.

Caws, Peter Science, Computers and the Complexity of Nature (1963)
Philosophy of Science 30, April 2.

Ceccato, Silvio Cybernétique: discipline et interdiscipline
Diogene, 53, p. 110-125.

Ceccato, Silvio Future Applications of Cybernetics: when machines will mirror man (1969)
In: *R Jungk and J Galtung (Eds) Mankind 2000* London, Allen and Unwin, pp. 205-211.

Centre International de Synthèse La Synthèse: idée-force dans l'évolution de la pensée (Exposés et discussions de la 15e semaine de synthèse)
Paris, Editions A Michel.

Centre International de Synthèse Notion de structure et structure de la connaissance (Exposés et discussions de la 20e semaine de synthèse)
Paris, Editions A Michel.

Centre International de Synthèse L'Encyclopédie et les Encyclopédistes: exposition organisée par le Centre International de Synthèse; imprimés manuscrits, etc (1932)
Paris, Bibliothèque Nationale, 82 p.

Centre International de Synthèse L'Encyclopédie et le progrès des sciences et des techniques (1952)
Paris, Presses Universitaires de France, 233 p.

Centre International de Synthèse Hommage à Henri Beer (1863-1964): commémoration du centenaire de sa naissance (1965)
Paris, Editions A Michel, 164 p. bibl. (pp. 13-16).

Centre International de Synthèse Leibniz 1646-1716: aspects de l'homme et de l'oeuvre (1968)
Paris, Aubier-Monteigne, 295 p, bibl.

Centre Royaumont pour une Science de l'Homme L'Unité de l'Homme: invariants biologiques et universaux culturels (essais et discussions présentés et commentés par Edgar Morin et Massimo Piatteli-Palmarini) (1974)
Paris, Editions du Seuil, 830 p.

Cercone, N and McCalla, G (Ed) Knowledge Frontier: essays in the representation of knowledge (1987)
Berlin, Springer-Verlag, 512 p. ISBN 3-540-96557-2.

Chadwick-Jones, J K Recent Interdisciplinary Exchanges and the Use of Analogy in Social Psychology (1970)
Human Relations 23, August pp. 253-261, bibl.

Chamberlin, William H World Order or Chaos (1946)
London, Duckworth, 292 p.

Chandler, A D Jr Strategy and Structure (1962)
Cambridge MA, MIT Press, 463 p.

Chang, K C Major Aspects of the Interrelationship of Archaeology and Ethnology (1967)
Current Anthrop. 8, June pp. 227-243, bibl.

Chapanis, A Men, Machines and Models (1961)
Systems Philosophy New York, Wiley, 113-31.

Chapanis, A Man-Machine Engineering (1966)
Belmont CA, Wadsworth Publishing, 134 p.

Chapman, G P The Epistemology of Complexity and Some Reflections on the Symposium
In: *The Science and Praxis of Complexity* 1985, Tokyo, United Nations University, pp. 657-376.

Chapman, P C and Zashin, E The Uses of Metaphor and Analogy: toward a renewal of political language
In: *Journal of Politics* 1974, 36, 2, pp. 290-336.

Charafas, Dimitris N The Knowledge Revolution: an analysis of the international brain market and the challenge to Europe (1968)
London, Allen and Unwin, 142p.

Charlesworth, James C (Ed) Integration of the Social Sciences through Policy Analysis (1972)
Philadelphia PA, American Academy of Political and Social Sciences, 229p.

Charlier, C V L How an Infinite World May be Built Up (1922)
Arkiv för Matematik, Astronomi och Fysik Band 16, 22, pp. 1-34.

Charnes, A and Cooper, W W Management Models and Industrial Applications of Linear Programming (1961)
New York, John Wiley and Sons, 467 p.

Charon, J Eléments d'une Théorie Unitaire d'Univers (1962)

Chartrand, R L (Ed) Operations Research and Human Problems (1970)
New York, Spartan Books.

Chartrand, R L Systems Technology Applied to Social and Community Problems (1970)
New York, Spartan Books.

Chatterton, Ronald The Multidisciplinary Teaching of Class Research Topics
Merrick, School District No. 25 (Long Island N Y 11566) (notes on educational films for multidisciplinary projects).

Chen, Kan and Dluhy, Milan J (Eds) Interdisciplinary Planning: a perspective for the future (1986)
New Brunswick NJ, Center for Urban Policy Research, 224 p. Foreword by Alfred J Sussman. ISBN 0–88285–116–0.

Chen, Peter P (Ed) Entity–Relationship Approach to Information Modeling and Analysis (1983)
Amsterdam, Elsevier Science Publishing, 602 p. ISBN 0–444–86747–3.

Cherry, C On Human Communication: a review, a survey and a criticism (1966)
Cambridge MA, MIT Press, 337 p.

Chestnut, H Systems Engineering Tools (1965)
New York, John Wiley and Sons, 646 p.

Chidambaram, T Coordination Problems in Competitive Situations (1967)
Research Center, Report 1O1 (A–62–43).

Chin, R The Utility of System Models and Developmental Models for Practitioners (1962)
In: *The Planning of Change* New York, Holt Rinehart, pp. 201–14.

Chomsky, Noam Syntactic Structures (1957)
Den Haag, Morton.

Chomsky, Noam Cartesian Linguistics (1966)
New York, Harper and Row.

Chomsky, Noam The Formal Nature of Language (1967)
In: *Biological Foundations of Language by E H Lenneberg, Appendix A* New York, Wiley, 489p.

Chomsky, Noam Language and Mind (1968)
New York, Harcourt Brace.

Chorafas, D N Systems and Simulation (1965)
San Diego CA, Academic Press, 10 p. bibl.

Chorley, Richard J Geomorphology and General Systems Theory (1962)
Washington DC, US Government Printing Office, 10 p, bibl.

Christopher, Alexander, et al A Pattern Language: towns, buildings, construction (1977)
New York, Oxford University Press.

Chubin, Daryl (Ed) Interdisciplinary Analysis and Research (1986)
Mount Airy MD, Lomond Publications, 482 p. ISBN 0–912338–54–7.

Churchman, C W The Systems Approach (1968)
New York, Dell Books, 243 p.

Churchman, C W; Ackoff, R L and Arnoff, E L Introduction to Operations Research (1917)
New York, John Wiley and Sons, 645 p.

Churchman, C W and Ackoff, R L Psychologistics (1947)
Philadelphia PA, University of Pennsylvania Press.

Churchman, C W and Ackoff, R L Purposive Behavior and Cybernetics (1950)
Social Forces 29, pp. 32–39.

Churchman, C W and Verhulst, M (Eds) Management Science Models and Techniques (1960)
New York, Pergamon Books. 2 vols.

Cirlot, J E A Dictionary of Symbols (1971)
London, Routledge and Kegan Paul.

Clark, B R Interorganizational Patterns in Education (1965)
Administrative Science Quarterly Sept 10, 2, pp. 224–237.

Clark, J T Remarks on the Role of Quantity, Quality, and Relations in the History of Logic, Methodology and Philosophy of Science (1962)
Stanford CA, Stanford University Press, pp. 611–12.

Clark, J W and Clark, J S (Ed) Systems Education Patterns on the Drawing Boards for the Future (1969)
New Haven CT, Kazanjian Economics Foundation.

Clark, Jere W The General Ecology of Knowledge in Curriculums of the Future (1972)
In: *Ervin Laszlo, The Relevance of General Systems Theory* New York, Braziller, 213p (pp. 163–180).

Clark, Jere W The Role of Unified Science in Vitalizing Research and Education (1972)
In: *E Haskell (Ed) Full Circle; the moral force of unified science* New York, Gordon and Breach, pp. 91–106 bibl.

Clark, Jere W and Judge, A J N Development of Trans-Disciplinary Conceptual Aids: simple techniques for education, research, pre–crisis management, and program administration highlighting patterns of information transaction and subsystem interdependence (1970)
Brussels, Union of International Associations, 13 p. plus annexes.

Clark, R The Case for an Integrated Tertiary Science Course
In: *New trends in integrated science teaching, Vol 1* Paris, Unesco, pp. 270–274.

Clay, Richard Nonlinear Networks and Systems (1971)
New York, Wiley Interscience, 284 p. bibl.

Cleland, David I and King, W R Management: a systems approach (1972)
New York, McGraw Hill Book Company, 422 p. bibl.

Cleveland, Harlan and Lasswell, Harold D (Eds) Conference on Science, Philosophy and Religion in their relation to the Democratic Way of Life, Ethics and Bigness: scientific, academic, religious, political and military (1962)
New York, Harper and Row.

Clough, D J Concepts in Management Science (1963)
Englewood Cliffs NJ, Prentice Hall, 425 p.

Cobb, John V and Griffin, David Ray Process Theology: an introductory exposition (1976)
Philadelphia, Westminster Press.

Cockerill, E Interdependence of the Professions in Helping People (1953)
Social Casework 34, November pp. 371–378, bibl. (National Conference of Social Work, pp. 137–147 (revised version)).

Cohen, David Evaluation of Integrated Science Curricula
In: *New trends in integrated science teaching, Vol II* Paris, Unesco, pp. 143–165.

Cohen, E R and DuMond, J W N Our Knowledge of the Fundamental Constants of Physics and Chemistry in 1965 (1965)
Physical Review 37, p. 537.

Cohn, S H and Cohn S M The Role of Cybernetics in Physiology (1953)
Scientific Monthly 79, pp. 85–89.

Cole, Sam World Models: their progress and applicability (1974)
Futures, 6, 3, June p. 201–218.

Colloque La psychologie économique et les disciplines voisines (1967)
Dalloz.

Committee on Science and Astronautics, U S House of Representative The Management of Information and Knowledge (1970)
University S Government Printing Office.

Comte, Auguste La Synthèse Subjective d'Auguste Comte: système universel des conceptions propres l'état normal de l'humanité (1900)
Paris, Fonds typographique de l'exécutive testamentaire d'Auguste Comte, 775 p.

Comte, Auguste System of Positive Philosophy (1968)
New York, Franklin Burt. 4 vols (Burt Franklin Research and Source Works Series, 125).

Conger, G P Synoptic Naturalism (1960)
Minneapolis MN, University of Minnesota Press.

Consultants Bureau, New York Systems Theory Research (Translation of
Problemy Ribennetiki Periodical).

Cooley, C H Human Nature and the Social Order
New York, Schocken Books.

Cooper, J C An Illustrated Encyclopaedia of Traditional Symbols (1978)
London, Thames and Hudson.

Cooper, Joseph Bonar and McGaugh, J L Integrating Principles of Social Psychology (1963)
Cambridge MA, Schenkman Publishing, 320 p. bibl.

Copi, I M; Elgot, C C and Wright, J B Realization of Events by Logical Nets (1958)
Journal, Association of Computing Machinery 5, pp. 181–96.

Cornacchio, J V System Complexity: a bibliography
In: *International Journal of General Systems* 1977, 3, 4, pp. 267–271.

Corselius, George Hints Towards the Developments of a Unitary Science or Science of Universal Analogy (1846)
Ann Arbor MI, S B M'Cracken, 22 p.

Costa de Beauregard, O Quanta and Relativity, Cosmos and Consciousness
In: *The Science and Praxis of Complexity* 1985, Tokyo, United Nations University, pp. 153–172.

Cottrell, Leonard S, Jr The Interrelationships of Law and Social Science (1966)
In: *Harry W Jones (Ed) Law and the Role of Social Science (Proceedings of a Conference)* New York, Rockefeller University Press, pp. 106–119 (discusses mental needs and requirements).

Count, E W Myth as World View: a biosocial synthesis (1960)
In: *Diamond, Stanley (Ed) Culture in History, essays in honour of Paul Radin* New York.

Court of Justice of the European Communities Publications juridiques concernant l'intégration européenne (1962)
Luxembourg, 392 p.

Coxeter, H S M Regular Polypopes (1948)
London, Methuen.

Crawford, Bryce The Support of Interdisciplinary and Transdisciplinary Programs: address to the Ninth Annual Meeting of the Council of Graduate Schools in the USA (1969)
Washington DC.

Crider, D B Cybernetics: a review of what it means and some of its implications in psychiatry (1956)
Neuropsychiatry 4, 1956–57, pp. 35–58.

Crockett, W H Cognitive Complexity and Impression Formation (1965)
In: *B A Maker (Ed) Progress in Experimental Personality Research* (vol. 2). New York, Academic.

Crook, Frederick Integration of Science (1960)
Castel, Guernsey Fount Books, 99 p.

Crosson, F J and Sayre, K M (Eds) Philosophy and Cybernetics (1967)
New York, Simon and Schuster International, 271 p.

Crow, Lee W The Generalist (1988)
Oracle AZ, Wakan Tanka Press, 202 p. ISBN 0–9620935–0–5.

Crowe, B L The Tragedy of the Commons Revisited (1969)
Science 166, 3909, 28 Nov pp. 1103–1107 (discusses breakdown of communication between sciences and effect on problem–solving).

Crowther, J G Scientific Types (1969)
London, The Cresset Press, 408 p.

Crozier, M The Bureaucratic Phenomenon (1924)
Chicago IL, University of Chicago Press, 320 p.

Cunningham, Don E Federal Support and Stimulation of Interdisciplinary Research in Universities (1969)
Oxford OH, Miami University, 73 p. NASA Grant NGR 36–022–001.

Curtis, James E and Petras, J W (Comp) The Sociology of Knowledge: a reader (1970)
New York, Praeger Publishers, 724 p. bibl.

Custard, Harry Lewis The Universe and the University: showing correspondences between the basic elements of the physical universe and the major fields of knowledge in the modern university (1965)
Tucson AZ, Distributed by Unity of Knowledge Publications, World University College, 44 p.

D'Abro, A The Evolution of Scientific Thought
Foundation for Integrative Education.

d'Alembert, J L Discours préliminaire de l'Encyclopédies (1912)
Heidelberg. Ed by H Wieleitner.

d'Arbon, J A The Impact of Integrated Science Courses (1972)

Dabrowski, K Positive Disintegration (1964)
Boston MA, Brown Little.

Dahlberg, Wolfgang Ordnung, Sein und Bewustsein: zur logischen, ontologischen und erkenntnistheoretischen systematik der ordnung (1984)
Frankfurt, Verlag AVIVA. Thesis for the University of Frankfurt.

Dale, E (Ed) Readings in Management (1965)
New York, McGraw Hill Book Company, 516 p.

Dale, L G Integrated Science in Tertiary Education
In: *New trends in integrated science teaching, Vol II*Paris, Unesco, pp. 91–105.

Daltzig, G B Linear Programming and Extensions (1963)
Princeton NJ, Princeton University Press, 632 p.

Danielsson, Albert and Törnebohns, H On Complex Systems with Human Components (1968)
Stockholm, Föraverets Forsknings–anstalt Planeringsbyran.

Danzig, T Number: the language of science (1930)
London, George Allen and Unwin.

Danzin, A The Pervasiveness of Complexity: common trends, new paradigms and research orientations
In: *The Science and Praxis of Complexity* 1985, Tokyo, United Nations University, pp. 69–80.

Das, Bhagavan The Essential Unity of All Religions (1966)
Illinois, Theosophical Press.

Das, T K (Ed) The Time Dimension: an interdisciplinary guide (1990)
New York, Praeger Publishers, 368 p. ISBN 0–275–92681–8.

Dascal, Marcelo (Ed) Dialogue: an interdisciplinary approach
Philadelphia PA, John Benjamins, 473 p. Pragmatics and Beyond Companion Series: 1. ISBN 0–915027–47–X.

Dauten, P M (Ed) Current Issues and Emerging Concepts in Management (1962)
Boston MA, Houghton Mifflin.

Davis, M D Game Theory: a nontechnical introduction (1970)
New York, Basic Books.

de Beauregard, O C Sur l'équivalence entre information et entropie (1961)
Science 11, pp. 51–58.

de Bie, Pierre The Concept of Problem–Focused Research (1968)
International Social Science Journal 20, 3, pp. 204–207 (details requirements for integration in an interdisciplinary research project).

de Bie, Pierre La recherche orientée (à propos des dimensions interdisciplinaires). (1970)
In: *Unesco, Tendances principales de la recherche dans les sciences sociales et humaines* pp. 686–764. Paris, Mouton/ Unesco.

de Bono, Edward Lateral Thinking for Management: a handbook (1982)
New York, Penguin Books, 225 p. illus. ISBN 0–14–022373–8.

de Bono, Edward Conflicts: a better way to resolve them (1985)
London, Harrap, 207 p. ISBN 0–245–54322–8.

De Coninck, Antoine L'Unité de la connaissance humaine et le fondement de sa valeur (1947)
Leuven, Institut supérieur de philosophie, 183 p. 2nd ed.

De Greene, K B Sociotechnical Systems: factors in analysis, design and management (1973)
New York, Prentice Hall.

De Latil, P Thinking by Machine: a study of cybernetics (1956)
London, Sedgwick and Jackson.

de Nicholas, Antonio T Meditations Through the Rg Veda (1978)
Boulder CO, Shambhala.

de Rosnay, Joel La Physique Sociale: de la bioénergie à l'écoénergétique (1974)
Paris, Communications, 22 p.

de Rosnay, Joel Le Macroscope: vers une vision globale (1975)
Paris, Seuil, 295 p.

De Rougemont, D Future Within Us (1983)
Elmsford NY, Pergamon Books, 254 p. Systems Science and World Order Library. ISBN 0–08–027395–5.

De Solla Price, Derek J Big Science, Little Science (1963)
New York, Columbia University Press.

De Solla Price, Derek J Networks of Scientific Papers (1965)
Science 30 July 149, 3683, pp. 510–515.

Dechert, C R (Ed) The Social Impact of Cybernetics (1966)
New York, Simon and Schuster International, 206 p.

Dedijer, S An Attempt at a Bibliography of Bibliographies in the Science of Science (1966)
Lund, Research Policy Institute.

Deese, J The Structure of Associations in Language and Thought (1965)
Baltimore MD, Johns Hopkins University Press.

Deland, E C Comments on Cybernetics and Management of Large Systems (1970)
Santa Monica CA, Rand Corporation.

Delattre, P Interdisciplinaires
In: *Encyclopaedia Universalis Organuum* 1973, p. 387.

Delgado, Rodriguez R Possible Model for Ideas (1957)
Philosophy of Science 24, p. 253–69, July bibl.

Demerath, Nicholas J and Peterson, R A (Ed) System, Change and Conflict: a reader on contemporary sociological theory and the debate over functionalism (1967)
New York, Free Press, 533 p. bibl. foot.

Deming, R H Characteristics of an Effective Management Control System in an Industrial Organization (1968)
Boston MA, Harvard Business School, 222 p.

Dempf, Alois Die Einheit der Wissenschaft (1955)
Stuttgart, Kohlhammer Urban Bücher. Band 18.

Derrida, J Structure, Sign and Play in the Discourse of the Human Sciences
In: *R Macksey and E University Donato (Eds) The Languages of Criticism and the Sciences of Man* pp. 247–265.

Dervin, Brenda, et al (Eds) Rethinking Communication
Newbury Park CA, Sage Publications. 2 Vols. Vol 2: Paradigm Exemplars, 1989, 544 p. ISBN 0–8039–3031–3.

Deutsch, K W Higher Education and the Unity of Knowledge (1950)
In: *Lyman Bryson, et al (Eds) Goals for American Education* New York, Conference on Science, Philosophy, and Religion in their Relation to the Democratic Way of Life, pp. 55–139.

Deutsch, K W Mechanism, Organism and Society (1951)
Philosophy of Science 18, 3, July pp. 230–252.

Deutsch, K W Mechanism, Teleology and Mild (1951)
Philosphical and Phenomenological Research 12, pp. 185–223.

Deutsch, K W Communication Theory and Social Science (1952)
American Journal Orthopsychiatry 22, pp. 469–83.

Deutsch, K W On Communication Models in the Social Science (1952)
Public Opinion Quarterly 16, pp. 356–80.

Deutsch, K W Some Notes on Research on the Role of Models in Natural and Social Sciences (1955)
Synthèse 7, pp. 506–33.

Deutsch, K W An Interdisciplinary Bibliography on Nationalism (1935–1953) (1956)
Cambridge MA, Technology Press of MIT, 165 p.

Deutsch, K W The Nerves of Government: models of political communication and control (1966)
New York, Free Press, 316 p.

Dewan, E (Ed) Cybernetics and the Management of Large Systems (1969)
New York, Spartan Books, 224 p.

Dews, Jule N The Design Synthesis of Biological, Chemical and Physical Response Systems (1964)
Frederick MD, Apropos Press, 27 p. bibl. (pp. 22–23).

Diderot, Denis et al Table analytique et raisonnée des natures contenues dans les 33 volumes in-folio du Dictionnaire des sciences, des arts et des métiers et dans son supplément (1780)
Paris, Panckoncke. 2 vols.

Diderot, Denis et al L'Univers de L'Encyclopédie (1964)
Paris, Les Libraires Associés, 135 p. illus.

Diderot, Denis, et al Encyclopédie du Dictionnaire raisonné des sciences, des arts et des métiers, par une société des gens de lettres (1966)
Geneva, Pellet. 1777–1779, 36 vols. non v. ed. Stuttgart-Bad Canstatt, Frommann, 1966 (facsimile edition) 35 vols.

Diderot, Denis, et al The Encyclopedia (1967)
New York, Harper and Row, 246 p. illus. Selection edited and translated by Stephen J Gendzier.

Dill, Stephen H Integrated Studies: challenges to the college curriculum (1983)
Lanham MD, University Press of America, 158 p. illus. ISBN 0–8191–2794–9.

Dillon, John A Foundations of General Systems Theory (1982)
Salinas CA, Intersystems Publications, 300 p. Systems Inquiry Series. ISBN 0–914105–05–1.

Dingler, Hugo Das System: das philosophisch–rationale grundproblem und die exakte methode der philosophie (1930)
Munich, E Reinhardt, 131 p.

Dirven, R and Paprotte, W (Eds) The Ubiquity of Metaphor: metaphor in language and thought (1984)
Amsterdam, John Benjamins. CILT: 29.

Dober, R P Environmental Design (1969)
New York, Reinhold Van Nostrand, 278 p.

Dobzhansky, Theodosius The Biology of Ultimate Concern (1967)
New York, New American Library.

Doctorow, E L False Documents
In: *American Review* 1977, 29, pp. 231–232.

Dodd, Stuart C Human Dimensions: a re-search for concepts to integrate thinking (1953)
Main Currents in Modern Thought 9, 4, February.

Dodd, Stuart C An Alphabet of Meanings for the Oncoming Revolution in Man's Thinking (1954)
Educational Theory IX, 3, July.

Dodd, Stuart C The Reiteration Rule: a cyclic system for syntax, neurograms, and all laws (1959)
Synthèse, 9, 1, March.

Dodd, Stuart C Introducing 'Systemmetrics' for Evaluating Symbolic Systems: 24 criteria for the excellence of scientific theories (1968)
Systematics, 6, 1, June.

Dodd, Stuart C The Epicosm Model for the Material and Mental Universes (1970)
In: *Progress in Cybernetics* New York, Gordon and Breach.

Dodd, Stuart C Documents relating to epicosmic modelling (1971)
Seattle WA, University of Washington Press. EpiDoc Series (223 by March 1971).

Dolezal, Hubert Living in a World Transformed: perceptual and performatory adaptation to visual (1981)
San Diego CA, Academic Press. ISBN 0–12–219950–2.

Donald, A G Management, Information and Systems (1967)
New York, Pergamon Books.

Donath, Tibor Von Bertalanffy's Integrative Endeavors (1973)
In: *Unity through Diversity, Gray, William and Rizzo, Nicholas D (Eds)* New York, Gordon and Breach Science Publishers, pp. 117–123.

Donnan, F G Integral Analysis and the Phenomenon of Life (1937)
Acta Biothaor.

Doob, Leonard W Patterning of Time (1972)
New Haven CT, Yale University Press, 472 p.

Dossey, Larry Space, Time and Medicine (1982)
Boulder CO, Shambhala Publications.

Douglas, Mary Natural Symbols: explorations in cosmology (1973)
London, Pelikan.

Doxiadis, C A Ecumenopolis: tomorrow's city (1968)
In: *Britannica Book of the Year* pp. 14–38. Chicago, Benton.

Doxiadis, C A Ekistics: an introduction to the science of human settlements (1968)
New York, Oxford University Press.

Doxiadis, C A Order in our Thinking: the need for a total approach to the anthropocosmos (1972)
Ekistics, 34, July.

Doxiadis, C A Anthropolis: city for human development (1975)
New York, WW Norton.

Dreger, R M Aristotle, Linnaeus and Lewin: the place of classification in the evaluative therapeutic process (1968)
Journal of General Psychology 78, January bibl. (pp. 55–59).

Drengson, Alan R Shifting Paradigms: from technocrat to planetary person (1983)
Victoria BC, Lightstar Press, 180 p. ISBN 0–920578–06–3.

Drischel, H Formale Theorien der Organisation (Kybernetik und verwandte Disziplinen) (1968)
Halle, Nova Acta Leopoldina.

Driver, M J and Steufart, S Integrative Complexity: an approach to individuals and groups as information processing systems (1969)
Administrative Science Quarterly 14, June bibl. (pp. 284–285).

Dror, Y Policy Analysts: a new professional role in government service (1967)
Public Administration Review 27, Sept pp. 197–203.

Dror, Y Specialists vs Generalists: a miss-question (1968)
Santa Monica CA, Rand Corporation, Dec 10 p. (P–3997).

Dror, Y Public Policymaking Re-examined (1968)
San Francisco, Chandler, 370 p.

Dror, Y A General Systems Approach to the Uses of Behavioral Sciences for Better Policymaking (1970)
In: *E O Attinger (Ed) Global Systems Dynamics* New York, Wiley Interscience, pp. 81–91.

Dror, Yehezkel Prolegomenon to Policy Sciences: from muddling through to meta-policymaking
Paper presented at the annual meeting of the American Association for the Advancement of Science, Boston, 1969.

Drury, W H and Nisbet, I C T Inter-Relations between Developmental Models in Geomorphology, Plant Ecology and Animal Ecology (1971)
General Systems Yearbook Society for General Systems Research, 16, p. 57–67.

Dubarle, Dominique La Science et la Vision Unifiée de l'Univers: les conceptions d'Einstein et l'apport de Teilhard de Chardin (1967)
In: *R Maheu, et al, Science et synthèse* Paris, Gallimard (Unesco copyright) pp. 48–66 (also English edition).

Dubin, Robert Continuous Problem Analysis: an approach to systematic theories about social organization (1969)
In: *M Sherif and C Sherif (Ed) Interdisciplinary Relationships in the Social Sciences* Chicago, Aldine, pp. 65–76.

Dubois, D and Prade, H Fuzzy Sets and Systems: theory and applications (1980)
New York, Academic Press.

Dubos, René The Dreams of Reason
New York, Crowell-Collier.

Dubos, René So Human an Animal (1970)
New York, Scribner.

Ducrocq, Albert Logique Générale de Systèmes et des Effets: introduction à une physique des effets fondements de l'intellectique (1960)
Paris, Dunod, 298 p.

Ducros, Louis Les Encyclopédistes (1967)
New York, B Franklin, 376p. bibl. foot. reprint 1900 ed.

Dumitriu, Anton Theory and System (1970)
Bologna, Cappelli, 57p. bibl.

Duncan, H D Symbols in Society (1968)
London, Oxford University Press.

Duncan, O D and Schnoie, L F Cultural, Behavioral and Ecological Perspectives in the Study of Social organization (1959)
American Journal of Sociology 65.

Duncan, W J et al An Experiment in Large Group Interdisciplinary Investigations into Contemporary Social issues (1971)
Social Science 46, Oct pp. 214–222.

Dunning, A J (Ed) Integration in Internal Medicine (1967)
(Proceedings of the International Congress of Internal Medicine) (Amsterdam 1966). Amsterdam, Excerpta Medica, 599 p.

Dupuy, J P Autonomy and Complexity in Sociology
In: *The Science and Praxis of Complexity* 1985, Tokyo, United Nations University, pp. 255–267.

Durkheim, E Division of Labor in Society
New York, Free Press.

Dymes, D M E (Ed) Synthesis in Education (1946)
Cambridge, Le Play House Press, 80p. (Reports of the Annual Conference of the Institute of Sociology, 5).

Easton, D Limits of the Equilibrium Model in Social Research (1956)
Behavioral Science 1, pp. 96–104.

Easton, D A Framework for Political Analysis (1965)
Englewood Cliffs NJ, Prentice Hall, 143 p.

Easton, D A Systems Analysis of Political Life (1965)
New York, John Wiley and Sons, 507 p.

Easton, D Varieties of Political Theory (1966)
Englewood Cliffs NJ, Prentice Hall, 154 p.

Eberwein, Wolf-Dieter Organizing Complexity: comparative politics and global modelling (1985)
Prepared for the volume on the Conference on Comparative research on National Political Systems, July 9–12, Berlin.

Eccles, John C and Popper, Karl R The Self and Its Brain (1981)
New York, Springer Publishers.

Eckman, D P (Ed) Systems: research and design (1961)
Proceedings of the First Systems Symposium at Case.

Eckman, D P System Philosophy, Institute of Technology (1961)
New York, John Wiley and Sons, 310 p.

Edel, D H Jr (Ed) Introduction to Creative Design (1967)
Englewood Cliffs NJ, Prentice Hall, 224 p.

Edgar Faure (Ed) Learning to Be: the world of education today and tomorrow (1972)
Paris, UNESCO.

Edwards, R F Interrelations between Perception Adaptation Levels and Value (1964)
General Systems Yearbook Society for General Systems Research, 9, p. 219–233.

Ehrenzweig, Anton The Hidden Order of Art: a study in the psychology of artistic imagination (1967)
Los Angeles, University of California Press, London, Weidenfeld and Nicolson.

Ehrmann, Jacques (Comp) Structuralism (1966)
Garden City, Anchor Books, 264 p. bibl. (pp. 239–263).

Eiduson, B T Scientists: their psychological world (1962)
New York, Basic Books.

Eigen, Manfred and Schuster, Peter The Hypercycle: a principle of natural self-organization (1979)
Berlin, Springer-Verlag.

Eigen, Manfred and Winkler, Ruthild Laws of the Game: how the principles of nature govern change (1983)
New York, Harper and Row, 347 p. illus. ISBN 0–06–090971–4.

Eisler, Riane The Chalice and the Blade (1987)
Viking.

Eliade, Mircea Images et Symboles (1952)
Paris.

Eliade, Mircea The Furnace and the Crucible: a study on the origins of alchemy (1962)
New York, Harper Torchbook.

Ellis, Arthur K; Humphreys, Alan H and Thomas, R Interdisciplinary Methods: a thematic approach (1981)
Glenview IL, Scott, Foresman and Company. ISBN 0–376–16495–0.

Ellis, David O and Ludwig, F J Systems Philosophy (1962)
Englewood Cliffs NJ, Prentice Hall, 387 p. bibl. (p. 377) (Space Technology Series).

Elmaghraby, S E The Design of Productive Systems (1966)
New York, Reinhold Van Nostrand, 496 p.

Emerson, Alfred E Homeostasis and Comparison of Systems (1956)
In: *Grinker, R R (Ed) Toward a Unified Theory of Behavior* New York, Basic Books.

Emery, Frederick E (Ed) Systems Thinking: selected readings (1969)
Harmondsworth, Penguin Books, 398 p. bibl.

Emery, J C Organizational Planning and Control Systems (1969)
New York, Macmillan, 166 p.

Empson, William Seven Types of Ambiguity (1930)
London, Chatto and Windus, 296 p. repr by Penguin 1973.

Emshoff, James R Analysis of Behavioral Systems (1971)
New York, Macmillan, 147 p. bibl.

Encyclopaedia Britannica The Great Ideas: a synopticon
Chicago IL, Encyclopaedia Britannica.

Epton, S R; Payne, R L and Pearson, A W (Eds) Managing Interdisciplinary Research (1984)
New York, John Wiley and Sons, 245 p. ISBN 0–471–90317–5.

Ericson, Richard (Ed) Improving the Human Condition: quality and stability in social systems (1979)
Washington DC, Society for General Systems Research.

Ericson, Richard F The Impact of Cybernetic Information Technology on Management Value Systems (1969)
General Systems Yearbook Society for General Systems Research, 14, p. 7–101.

Erismann, T H Zwischen Technik und Psychologie: grundprobleme der kybernetik (1968)
Berlin, Springer-Verlag. bibl. (pp. 170–178).

Esteves, Justinian A Unificacao da Ciência: Bases e algumas deducoes da minha teoria sobre a constitui os corpos (1927)
Lisbon, Servicos gráficos do exercito, 130 p.

Esty, G W Psycho-Social Economics of Human Ecology (1971)
American Journal of Economics 30, July pp. 243–252.

Etzioni, Amitai Modern Organizations (1964)
Englewood Cliffs NJ, Prentice Hall, 120 p.

Etzioni, Amitai Political Unification (1965)
New York, Holt, Rinehart and Winston.

Evans, C B and Robertson, A D J (Eds) Cybernetics (1958)
University Park Press.

Evans, W Indices of Hierarchical Structure of Industrial Organizations (1963)
Management Science 9, pp. 468–477.

Ewing, D W (Ed) Long-Range Planning for Management (1964)
New York, Harper and Row, 565 p.

Fabrycky, W J and Banks, J Procurement and Inventory Systems: theory and analysis (1967)
New York, Reinhold Van Nostrand, 239 p.

Fabun, Don The Dynamics of Change (1967)
Englewood Cliffs NJ, Prentice Hall.

Fairfax, W The Triple Abyss: towards a modern synthesis (1965)
London, Bles. (mainly concerned with religious synthesis).

Fairlie, H The Politican Without a Metaphor is a Ship Without Sails: the sense of metaphor
In: *New Republic* 1979, 180, 10, p 10.

Fakan, John C Application of Modern Network Theory to Analysis of Complex Systems (1969)
Washington DC, NASA, 45 p. Sold by Clearinghouse for Federal Scientific and Technical Information.

Fartappie, L Principi di una Teoria Unitaria del Mondo Fisica e Biologica (1945)
Rome.

Fast, J D Entropy: the significance of the concept of entropy and its applications in science and technology (1962)
London, Macmillan.

Fearing, F An Examination of the Conceptions of Benjamin Whorf in the light of theories of perception and cognition (1954)
American Anthropologist 54, p. 47.

Fédération internationale des unions intellectuelles L'Oeuvre de la Fédération: 1923–1928 (1928)
Prague, Union intellectuelle tchéchoslovaque, 90 p.

Feiblemal, James K Theory of Integrative Levels (1914)
British Journal for the Philosophy of Science 5, pp. 59–66.

Feiblemal, James K Inside the Great Mirror: a critical examination of Russell, Wittgenstein and their followers (1958)
Den Haag. (written on integrative levels).

Feiblemal, James K The Integrative Levels in Nature (1965)
In: *B Kyle (Ed) Focus on Information and Communication* London, Aslib,. (Rev. version of: Theory of integrative levels. In: *British Journal for the Philosophy of Science* vol. 5:59–66, May 1954, No. 17).

Feiblemal, James K Disorder (1968)
In: *Paul G Kuntz (Ed) The Concept of Order* Seattle, University of Washington Press, pp. 3–13.

Feibleman, J and Friend, J W The Structure and Function of Organization (1945)
The Phil. Review 54, pp. 19–44.

Fernhout, R, et al (Eds) Dialogue and Syncretism: an interdisciplinary approach (1989)
Grand Rapids MI, William B Eerdmans, 240 p. Currents of Encounter Series. ISBN 0–8028–0501–9.

Ferrero, Guglielmo The Unity of the World (1931)
London.

Feyerabend, Paul Consolations for the specialist (1970)
In: *I Lakatos and A Musgrove (Eds) Criticism and the Growth of Knowledge* Cambridge University Press, pp. 197–230.

Feyerabend, Paul Against Method: outline of an anarchist theory of knowledge (1978)
London, Verso.

Fiasco, Michael A Integration of the Sciences (1970)
The American Biology Teacher 32, 4, April.

Fibonacci Quarterly A Journal Devoted to the Study of Integers with Special Properties (1963)
California CA, Fibonacci Association. vol 1.

Filova, Elena Materializmus proti Holizmu (1963)
Bratislava, Vydavtel'stvo Slovenskej akademie Vied, 243 p. bibl.

Fink, D G Computers and the Human Mind: an introduction to artificial intelligence (1966)
New York, Doubleday.

Fischer, Roland (Ed) Interdisciplinary Perspectives of Time (1927)
Annals of the New York Academy of Sciences 138, 1967, article 2, pp. 369–915, bibl.

Fisher, G H The Analytical Bases of Systems Analysis (1966)
Santa Monica CA, Rand Corporation, 18 p.

Fisk, G (Ed) The Psychology of Management Decision (1967)
Lund, Gleerup, 309 p.

Fiszer, Emilie Unité et intelligibilité (1936)
Paris, J Vrin, 245 p. (Thèse, Université de Paris).

Fitzgerald, John A Systems Analysis: applications to the non-western political process (1968)
Washington DC, Center for Research in Social Systems, 96 p. bibl. (pp. 85–93).

Flagle, C D; Huggins, W H and Roy, R H (Ed) Operations Research and Systems Engineering (1960)
Baltimore MD, Johns Hopkins University Press.

Fleck, A C J and Ianni, F A Epidemiology and Anthropology: some suggested affinities in theory and method (1958)
Human Organization 18, Winter pp. 38–40, bibl.

Fleck, Alexander Science and the Humanities: their basic unity (1957)
New Scientist 1, 24 January pp. 10–12.

Fletcher, A and Clarke, G Management and Mathematics: the practical techniques available (1964)
New York, Gordon and Breach, 236 p.

Fobes, John Next Steps in World Governance
Unpublished remarks at the Club of Rome Conference, Santander, 1985.

Foerster, H von (Ed) Cybernetics (1953)
Macy Foundation.

Foerster, H von and Zopf, G W Principles of Self-Organization (1962)
New York, Pergamon Books.

Foerster, H von, et al (Eds) Purposive Systems (1969)
New York, Spartan Books, 192 p.

Forde, D Ecology and Social Structure (The Huxley Memorial Lecture). (1971)
Journal of the Royal Anthropological Institute of Great Britain and Ireland, pp. 15–29.

Forney, G David Concatenated Codes (1966)
Cambridge MA, MIT Press, 147 p.

Forrester, J W Industrial Dynamics (1961)
Cambridge, Wiley, 464 p.

Forrester, J W Principles of Systems (1968)
Allen-Wright. 2nd ed.

Forrester, J W Counterintuitive Behavior of Social Systems (1971)
Technology Review, 73, 52.

Foster, C; Rapoport, A and Trucco, E Some Unsolved Problems in The theory of Non-Isolated Systems (1957)
General Systems 2, p. 9–29.

Foulqui, Paul La Dialectique (1976)
Paris, Presses Universitaires de France. 8th rev ed.

Foundation for Integrated Education Issues in Integration (Proceedings of a workshop at the University of New Hampshire) (1948)
New York, Foundation for Integrated Education.

Fox, Richard G (Ed) Interdisciplinary Anthropology (1978)
Washington DC, American Anthropological Association. American Ethnologist Series: 6, 3. ISBN 0–317–66315–1.

Fox, Sidney and Ho, Mae-Wan (Eds) Process and Metaphors in the Evolutionary Paradigm (1988)
New York, John Wiley and Sons, 350 p. ISBN 0–471–91801–6.

Frank, L K, et al Teleological Mechanisms (1948)
Ann NY, Academic Science, 50 p.

Frank, Lawrence K Interprofessional Communication (1961)
American Journal of Public Health Dec 51, pp. 1798–1804.

Frank, Philip Philosophical Uses of Science (1957)
Bulletin of the Atomic Scientists 13, April pp. 125–130.

Franz, Marie-Louise von Number and Time: reflections leading towards a unification of psychology and physics (1974)
London, Rider.

Frawley, D The Houses of I Ching
In: *I Ching: Taoist book of days* 1976, New York, Ballatine, pp. 183–205.

Freeman, Christopher et al (Eds) Thinking about the Future (1973)
London, Chatto and Windus.

Frey, Dagobert On the Problem of Symmetry in Art Studium Generale P (1956)
276 (quoted in Herman Weyl, 'Symmetry' The World of Mathematics, James N Newman (Ed). New York, Simon and Schuster, Vol I, p. 678).

Frick, F C Information Theory (1959)
Psychology: A study of a science New York, McGraw-Hill, pp. 629–36, 611–15. vol 2.

Fried, Jacob and Molnar, Paul Technological and Social Change: a transdisciplinary model (1979)
Princeton NJ, Petrocelli Books. ISBN 0–89433–074–8.

Friedrich, Carl J (Ed) Totalitarianism (1954)
Cambridge MA, Harvard University Press.

Friedrichs, R W A Sociology of Sociology (1970)
New York, Free Press, 428 p. bibl. (pp. 375–418).

Frye, Northrop Fearful Symmetry: a study of William Blake (1907)
Princeton NJ, Princeton University Press, 468 p.

Fryxell, Roald The Interdisciplinary Dilemma: a case for flexibility in academic thought (1977)
Rock Island IL, Augustana College Library, 16 p. Augustana College Library Occasional Papers: 13. ISBN 0–910182–36–1.

Fuchs, W R Cybernetics for the Modern Mind (1970)
New York, Macmillan Publishing.

Fuller, R Buckminster 4–D
Carbondale IL, World Resources Inventory. reprint of privately published 1927 version 200 pages (man's evolutionary functioning in universe and its recognition).

Fuller, R Buckminster Ideas and Integrities (1963)
Englewood Cliffs NJ, Prentice Hall, 318 p. a spontaneous autobiographical disclosure (edited by Robert W Marks).

Fuller, R Buckminster Conceptuality of Fundamental Structures (1965)
Gyorgy Kepes (Ed) Structure in Art and in Science pp. 66–68 New York, Braziller,.

Fuller, R Buckminster Operating Manual for Spaceship Earth (1970)
Carbondale IL, World Resources Inventory.

Fuller, R Buckminster The Buckminster Fuller Reader (1970)
Harmondsworth, Penguin Books.

Fuller, R Buckminster The World Game: integrative resource utilization planning tool (1971)
Carbondale IL, Southern Illinois Univeristy, 183 p.

Fuller, R Buckminster Total thinking (1972)
In: *R B Fuller, Ideas and Integrities* Englewood Cliffs, Prentice-Hall, 1963. ×3Republished in: *James Meller (Ed) The Buckminster Fuller Reader* London, Pelican books, 1972, pp. 310–328.

Fuller, R Buckminster Synergetics: explorations in the geometry of thinking (Vol II) (1979)
New York, Macmillan. 1975 (Vol I).

Fuller, R Buckminster, et al Comprehensive Thinking (1965)
Carbondale IL, World Resources Inventory, 118 p (World Design Science Decade 1965–1975, Phase I, Document 3, selected and edited by John McHale).

Fuller, R Buckminster with E J Applewhite Synergetics: explorations in the geometry of thinking (1975)
New York, Macmillan, 876 p.

Furth, R Physics of Social Equilibrium (1952)
Advancement of Science London, 8, pp. 429–34.

Gabor, D Communication Theory and Cybernetics (1954)
In: *I R E Prof (Trans) Group on Non-Linear Circuits, Milan Symposium paper.*

Gabor, D The Mature Society (1972)
London, Secker and Warburg.

Gagne, R M (Ed) Psychological Principles in System Development (1965)
New York, Holt, Rinehart and Winston, 560 p.

Gagsch, S Der Beitrag Bertalanffy's zur Allgemeinen System Theorie (1967)
Köln, Eine ergänzende Analyse. 1967–68.

Galabert, Henri La commission intellectuelle de la Société des Nations (1931)
Toulouse.

Gall, John Systemantics: the underground text of systems lore (1986)
Ann Arbor MI, General Systemantics Press, 319 p. illus. Illustrated by Mark Howell and David Martinez. ISBN 0–9618251–1–1.

Galtung, J Notes on the Differences between the Physical and the Social Sciences (1958)
Inquiry 1, 1/2, pp. 7–34.

Galtung, Johan Structural Analysis and Chemical Models (1977)
In: *Methodology and Ideology* Copenhagen, Christian Ejlers, 1977, pp. 160–189.

Galtung, Johan Methodology and Ideology (1977)
Copenhagen, Christian Ejlers.

Galtung, Johan Processes in the UN System (1980)
Geneva, Paper for Goals, processes and indicators of development project of the United Nations Unversity.

Gamow, G Numerology of the Constants of Nature (1968)
Proceedings of the National Academy of Science 59, p. 313.

Gardner, Howard Thinking: composing symphonies and dinner parties
In: *Psychology Today* April 1980, 13, 1.

Gardner, Howard The Socialization of Human Intelligences Through Symbols
In: *Frames of Mind: the theory of multiple intelligence* 1984, London, Heinemann.

Gardner, M The Hierarchy of infinities and the Problems it Spawns (1966)
Scientific American 214, p. 112–118 (March).

Gardner, M The Ambidextrous Universe: mirror asymmetry and time-reversed worlds (1979)
New York, Charles Scribner's Sons, 243 p.

Garey, M R and Johnson, D S Computers and Interactability: a guide to the theory of NP-completeness (1979)
San Francisco CA, W H Freeman.

Garner, Wendell R Uncertainty and Structure as Psychological Concepts (1962)
New York, Wiley.

Garrison, Sposito Does a Generalized Heisenberg Principle Operate in the Social Sciences? (1969)
Inquiry 12, 1969, 3, p. 356–361. (an interdisciplinary journal of philosophy and social sciences).

Geeraerts, D Paradigm and Paradox: explorations into a paradigmatic theory of meaning and its epistemological background (1985)
Leuven, Universitaire Pers Leuven, 407 p.

General Systems Yearbook of the Society for General Systems Research (1956)

George, C S Jr The History of Management Thought (1968)
Englewood Cliffs, Prentice-Hall, 210 p.

George, F H Automation, Cybernetics and Society (1959)
New York, Philosophical Library.

George, F H Automation, Cybernation, and Society (1960)
London, Leonard Hill.

George, F H Cybernetics and Biology (1965)
London, Oliver and Boyd.

George, F H and Handlon, J H Towards a General Theory of Behavior (1955)
Methodos 7, 1955 pp. 24–44.

Gerard, R W Units and Concepts of Biology (1917)
Science 125, 1917 pp. 429–33.

Gerard, R W Organism, Society and Science (1940)
Scientific Monthly 50, 1940, pp. 340–50.

Gerard, R W Higher Levels of Integration (1942)
In: *Redfield, R (Ed) Levels of Integration in Biological and Social Systems* Bio-Symposia, 8, pp. 67–87.

Gerard, R W Hierarchy, Entitation and Levels (1969)
In: *Lancelot Law Whyte, et al (Eds) Hierarchical Structures* New York, American Elsevier, 1969 pp. 215–28.

Gerardin, L Bionics (1968)
New York, McGraw Hill Book Company, 254 p.

Gergen, K J Toward Transformation in Social Knowledge (1982)
Berlin, Springer-Verlag, 260 p. ISBN 3-540-90673-8.

Geyer, Felix and Van Der Zouwen, Johannes (Eds) Sociocybernetic Paradoxes: observation, control and evolution of self-steering systems (1986)
Newbury Park CA, Sage Publications, 248 p. illus. ISBN 0-8039-9735-3.

Ghiselin, B The Creative Process (1952)
Berkeley CA, University of California Press.

Giedion, S The Eternal Present (1964)
Princeton NJ, Princeton University Press.

Giedion, S Space, Time, Architecture: the growth of a new tradition (1967)
Cambridge MA, Harvard University Press.

Giffard, J A H Integrating Social with Technological Change (1957)
Impact 7, March pp. 3–15 (use of social and psychological sciences to integrate modern man into the technological society).

Gigch, J P van Applied General Systems Theory (1974)
New York, Harper and Row, March.

Gilb, Corinne L Hidden Hierarchies (1966)
New York, Harper and Row.

Gillette, George F The Cycle of Power and Unitary Theory: a unitary conception of all natural phenomena and of the subatomic mechanism of the cosmos (1930)
New York, Appeal Printing, 273 p.

Gillin, J P Some Principles of Sociocultural Integration (1971)
Current Anthrop. 12, February, pp. 63–71. bibl.

Gillin, John (Ed) For a Science of Social Man: convergences in anthropology, psychology, and sociology (1954)
New York, Macmillan, 289 p.

Gilson, E The Unity of the Philosophical Experience (1950)
New York, Allied Publications.

Ginsberg, M Social Evolution (1961)
In: *M Banton (Ed) Darwinism and the Study of Society* Chicago, Quadrangle.

Gjessing, Gutorm Ecology and Peace Research (1967)
Journal of Peace Research 2, pp. 125–39.

Glans, T B, et al Management Systems (1968)
New York, Holt, Rinehart and Winston, 430 p.

Glansdorff, P and Prigogine, Ilya Thermodynamic Theory of Structure, Stability, and Fluctuations (1971)
New York, Wiley Interscience.

Gleick, James Chaos: making a new science (1987)
New York, Viking Penguin, 352 p. illus. ISBN 0-670-81178-5.

Glenn, E S Patterns of Language and Patterns of Logic (1958)
Panel on Psycholinguistics, APA Convention.

Glenn, E S A Cognitive Approach to the Analysis of Cultures and Cultural Evolution (1966)
General Systems Yearbook Society for General Systems Research, 11, p. 115–31.

Glushkov, V M Introduction to Cybernetics (1966)
San Diego CA, Academic Press.

Gogner, J A Categorical Foundations for General Systems Theory
(to appear).

Goguen, J A Mathematical Representation of Hierarchically Organized Systems (1970)
In: *E O Attinger (Ed) Global Systems Dynamics* New York, Wiley Interscience, 1970 pp. 112–28.

Golay, Keith J Learning Patterns and Temperament Styles: a systematic guide to maximizing student achievement (1982)
Fullerton CA, Manas-Systems, 109 p. ISBN 0-9610076-0-5.

Goldman, S Information Theory (1953)
Englewood Cliffs NJ, Prentice Hall.

Goldstein, Joshua S Long Cycles: prosperity and war in the modern age (1988)
New Haven CT, Yale University Press, 433 p.

Goldstein, Kurt The Organism: a holistic approach to biology derived from pathological data in man (1939)
New York, American Book Company.

Goldstein, Kurt Levels and Ontogeny (1962)
American Scientist 50, 1.

Goldstein, Kurt The Organism (1963)
Boston MA, Beacon Press.

Goldstein, W The Political Metaphors of Environmental Control
In: *Alternations* 1973, 2, 4, pp. 11–17.

Gombrich, E H Art and Illusion (1961)
Princeton NJ, Princeton University Press.

Gonseth, F La Géomtrie et le Problème de l'Espace (1955)
Neuchatel, p.583.

Goode, H H and Machol, R E System Engineering: an Introduction to the design of large scale systems (1957)
New York, McGraw Hill Book Company.

Goodman, Nelson The Languages of Art: an approach to a theory of symbols (1976)
Indianopolis, Hackett.

Gordkins, F S Control Theory and Biological Systems (1963)
New York, Columbia University Press.

Gordon, Geoffrey System Simulation (1969)
Englewood Cliffs NJ, Prentice Hall, 303 p. bibl.

Gordon, J J Synectics (1961)
New York, Harper and Row.

Gordon, W J The Metaphorical Way of Learning and Knowing (1971)
Cambridge, Porpoise Books.

Gordon, W J J The Metaphorical War or Learning and Knowing (1971)
Cambridge MA, Porpoise Books.

Gotesky, Rubin and Laszlo, Ervin (eds) Evolution-Revolution (1971)
New York, Gordon and Breach, 351 p.

Gourdin, Michel Unitary Symmetries and Their Application to High Energy Physics (1967)
Amsterdam, North-Holland, 303 p. bibl. (pp. 298–301).

Gouthier, Giuseppe I Concetti di Sistema e di Struttura nelle Scienze Biologiche, Sociologiche, linguistiche ed antropologiche (1968)
Torino, Quaderni di Studio, 78 p. bibl.

Gowan, John Curtis Operations of Increasing Order (1980)
Westlake Village CA, 390 p. bibl. Published by author.

Gragner, Le Roy (Ed) Systems and Actors in International Politics (1971)
Scranton, Chandler Pub, 210 p. bibl.

Graham, A C Reason and Spontaneity: a new solution to problem of fact and value (1985)
London, Curzon Press, 236 p.

Granger, Robert L Educational Leadership: an interdisciplinary perspective (1971)
Scranton, Intext Educational Publishers, 372 p. bibl. (pp. 363–369).

Gratz, Pauline Integrated Science: an interdisciplinary approach: a suggested course of study (1966)
Philadelphia, FA Davis, 103 p. bibl. pp. 88–92 of textbooks in individual disciplines.

Graves, N J Geography, Social Science and Interdisciplinary Enquiry (1968)
Geog. Journal 134, Sept. 1968 pp. 390–394 (reply: 135, March 1969 pp. 158–160).

Gray, Donald P The One, and the Many: Teilhard de Chardin's vision of unity (1969)
London, Burns and Oates, 185 p. bibl.

Gray, W, et al (Eds) General Systems Theory and Psychiatry (1982)
Boston MA, Brown Little, 450 p. Systems Inquiry Series.

Gray, William Bertalanffian Principles as a Basis for Humanistic Psychiatry (1972)
In: *Ervin Laszlo, The Relevance of General Systems Theory* New York, Braziller 213p. (pp. 123–133).

Gray, William (Ed) General Systems Theory and the Psychological Sciences (1982)
Salinas CA, Intersystems Publications, 550 p. 2 Vols. Systems Inquiry Series. ISBN 0-914105-10-8.

Gray, William and Nicholas D Rizzo (Eds) Unity Through Diversity: a festschrift in honor of Ludwig von Bertalanffy
New York, Gordon and Breach. 2 Vols.

Green, Herbert S and Hurst, C A Order-Disorder Phenomena (1964)
London, Interscience, 363 p. bibl. (p. 353–55) (crystallography).

Greene, John C The Concept of Order in Darwinism (1968)
In: *Paul G Kuntz (Ed) The Concept of Order* Seattle, University of Washington Press, pp. 89–103.

Greenlaw, P S; Herron, L W and Rowdon, R H Business Simulation: in industrial and university education (1962)
Englewood Cliffs NJ, Prentice Hall, 356 p.

Greenwood, James W Jr and Greenwood, James W III Bibliography on Systems Theory for Managers (1970)
New York, IBM Systems Research Institute, 43 p. mimeo.

Grenander, M E Mark Hopkin's Log: teaching and the analysis of ideas (1969)
English Record 20, 1, Oct. 1969 (advocates use of a four-stage sequential approach to interdisciplinary study: acquisition and evaluation of facts: creative discovery, presentation of ideas).

Grene, Martorie Biology and the Problem of Levels of Reality (1967)
New Scholasticism, 41, p. 427–49.

Grene, Martorie Anatomy of Knowledge (1969)
Amherst MA, University of Massachusetts Press.

Grene, Martorie Notes on Hierarchy Concept; Hierarchy: one word, how many concepts? (1969)
In: *Lancelot Law Whyte, et al (Eds) Hierarchical Structures* New York, American Elsevier, 1969 pp. 56–58.

Grene, Martorie Toward a Unity of Knowledge (1969)
New York, International University Press, 302 p. (Psychological Issues, 6, 2, monograph 22).

Greniewski, H Cybernetics without Mathematics (1960)
New York, Pergamon Books.

Grill, F and Hammelmann, F M Interdisciplinary Teamwork in a School Setting (1959)
Social Casework 40, January pp. 23–27.

Grinker, Roy R The Interrelation of Neurology, Psychiatry, and Psychology. (1941)
Journal of the American Medical Association 116, 17 May p. 2236–2241.

Grinker, Roy R Toward a Unified Theory of Human Behavior: an introduction to general systems theory (1967)
New York, Basic Books, 390 p.

Grinker, Roy R Symbolism and General Systems Theory (1969)
In: *Gray, Duhl and Rizzo, General Systems Theory and Psychiatry* Boston Little, Brown.

Grochla, E Automation and Organization: The impact of general systems theory on organization theory (1973)
In: *Unity through Diversity, Gray, William and Rizzo, Nicholas D (Eds).*

Grodins, F S Similarities and Differences between Control Systems in Engineering and Biology (1970)
In: *E O Attinger (Ed) Global Systems Dynamics* New York, Wiley Interscience 1970 pp. 2–6.

Gross, B M The Managing of Organizations: the administrative struggle (1964)
New York, Free Press, 1024 p.

Gross, B M Organizations and Their Managing (1964)
New York, Free Press, 708 p.

Gross, B M Coming General Systems Model of Social Systems (1967)
Human Relations 20, November pp. 357–74.

Gross, Llewellyn Preface to a Metatheoretical Framework for Sociology (1961)
American Journal of Sociology 67, September pp. 125–43.

Grote, M D Teaching Metaphors for Educational Theory: a quasi-normative approach (1972)
Pennsylvania, Temple University. Doctoral Dissertation.

Grotjahn, Martin The Voice of the Symbol (1971)
New York, Dell.

Gruber, Edward C Miller Analogy Test: 1400 analogy questions (1967)
New York, Arco Pub, 137 p. 2nd ed.

Guetzkow, H (Ed) Simulation in Social Science: readings (1962)
Englewood Cliffs NJ, Prentice Hall, 199 p.

Guildbaud, G T What is Cybernetics? (1959)
New York, Criterion, 126 p.

Günther, G Cybernetic Ontology and Transjunctional Operations (in self-organizing systems) (1959)
New York, Spartan Books, pp. 313–92.

Günther, G Cybernetical Ontology and Transjunctional Operations (1962)
In: *Self-Organizing Systems, Yovits, Jacobi and Goldsterne* Washington, Spartan Books.

Günzl, Herbert C Struktur und Existenz: versuch einer synthese von ordnung und freiheid (1968)
Linz, 272 p. bibl. (pp. 260–272) (Schriftenreihe der sozialpolitischen Zeitschriften-Verlagsgesellschaft Linz/Donau).

Guran, M A Change in Space-Defining Systems (1969)
General Systems Yearbook Society for General Systems Research, 14, p. 37–49.

Gurvitch, Georges The Spectrum of Social Time (1964)
Dordrecht, D Reidel, 152 p.

Gusdorf, Georges Passé, Présent, Avenir de la Recherche Interdisciplinaire
In: *UNESCO – Interdisciplinarité et Sciences Humaines* 1983, Paris, UNESCO, pp. 31–52.

Gusdorf, Georges Projet de recherche interdisciplinaire dans les sciences humaines (1907)
In: *Les Sciences de l'Homme sont-elles des Sciences humaines?* Strasbourg, Faculté des Lettres de l'Université de Strasbourg, 1967, pp35–63 (Rapport rédigé pour l'Unesco en 1961; une version abrégée et publiée dans la revue *Diogene* 42, 1963).

Gusdorf, Georges Pour une recherche interdisciplinaire (1963)
Diogene, 42, p. 122–142.

Gusdorf, Georges Project for Interdisciplinary Research (1963)
Diogene, Summer 42, p. 119–142 (Abridged version of a report written for Unesco in 1961).

Gusdorf, Georges Interdisciplinaire (Connaissance)
Paris, Encyclopaedia Universalis France, pp. 1086–1090. Vol 8.

Gutiérrez, E and Margalef, R How to Introduce Connectance in the Frame of an Expression for Diversity
In: *Amer, Natur* 1983, 121, pp. 601–607.

Gutman, Herbert Structure and Function (1964)
Genetic Psychology Monographs 70, p. 3–56.

Gutman, Herbert Notes on Organic Hierarchical Structures: structure and function in living systems (1969)
In: *Lancelot Law Whyte, et al (Eds) Hierarchical Structures* New York, American Elsevier, 1969 pp. 229–230.

Haberstroh, Chadwick J Control as an Organizational Process (1960)
Management Science 6, pp. 165–71.

Haberstroh, Chadwick J Organizational Design and Systems Analysis (1965)
In: *March, J G (Ed) Handbook of Organizations* Chicago, Rand McNally 1965 pp. 1171–1211, bibl.

Hadamard, J The Psychology of Invention in the Mathematical Field (1949)
Princeton NJ, Princeton University Press.

Hagstrom, W O The Differentiation of Disciplines (1972)
In: *B Barnes (Ed) Sociology of Science* London, Penguin, 1972, pp. 121–125 (excerpt form W. O. Hagstrom, The Scientific Community, Basic Books 1965 pp. 222–226).

Hahn, Lewis E Contextualism and Cosmic Evolution-Revolution (1971)
In: *Gotesky, Rubin and Laszlo, Ervin (Eds) Evolution-Revolution* New York Gordon and Breach Science Publishers. 1971 pp. 3–37.

Haire, M (Ed) Modern Organization Theory (1959)
New York, John Wiley and Sons, 324 p.

Haire, M Biological Models and Empirical Histories of the Growth of Organizations (1959)
New York, John Wiley and Sons.

Haire, M Organization Theory in Industrial Practice (1962)
New York, John Wiley and Sons, 173 p.

Halacy, D S Jr Bionics: the science of living machines (1965)
New York, Holiday House.

Hale, D G The Body Politic: political metaphor in Renaissance English literature (1971)
Den Haag, Mouton de Gruyter. De Proprietatibus Litterarum Series Major: 9.

Hall, A D A Methodology for Systems Engineering (1962)
Princeton, Van Nostrand, 478 p.

Hall, A D and Fagen, R E Definition of Systems (1956)
General Systems 1, New York, Bell Telephone Lab. p. 18–28.

Hall, M F The Generality of Cognitive Complexity-Simplicity (1966)
Unpublished Ph. D. Thesis, Vanderbilt Uni.

Hall, William C Aims and Objectives of Integrated Science Teaching
In: *New trends in integrated science teaching, Vol II* Paris, Unesco pp. 15–25.

Halmos, Paul (Ed) The Sociology of Sociology (1970)
Sociological Review Monograph Keele, September 16, whole issue.

Hambidge, Jay The Elements of Dynamic Symmetry (1919)
Yale, Dover Publications.

Hamilton, Howard The Evolution of Societal Systems: a theoretical foundation for monitoring change (1974)
Los Angeles, Center for Futures Research. CFR Report M16.

Hammer, Preston C (Ed) Advances in Mathematical Systems Theory (1969)
University Park PA, Pennsylvania State University Press, 174 p. bibl.

Hanika, F de P New Thinking in Management: a guide for managers (1965)
London, Hutchinson, 110 p.

Hanson, Norwood R Patterns of Discovery: a inquiry into the conceptual foundations of science (1958)
Cambridge MA, Cambridge University Press.

Harari, Josu V Structuralists and Structuralisms: a selected bibliography of French contemporary thought (1960–1970) (1971)
Ithaca, Diacritics, 82 p.

Hardin, G The Cybernetics of Competition: a biologist's view of society (1963)
Perspectives in Biology and Medicine 7, pp. 61–84.

Hardy, A C Escape from Specialization (1964)
In: *J Huxley, A C Hardy, and E B Ford (Eds) Evolution as a Process* New York, 1954.

Hare, Van Court Systems Analysis: a diagnostic approach (1967)
New York, Harcourt Brace, 544 p. bibl. (pp. 519–533).

Harold, Preston On the Nature of Universal Cross-Action (vol 3 of trilogy: the Single Reality; unifying framework concerned with integration of science, psychology and religion)

Harris, Errol E Nature, Mind and Modern Science (1954)
London, George Allen and Unwin, 455 p. bibl. foot.

Harris, Errol E The Foundations of Metaphysics in Science (1965)
London, George Allen and Unwin, 512 p. bibl. foot.

Harris, Errol E Wholeness and Hierarchy (1965)
In: *Foundations of Metaphysics in Science, Chapter 7* New York, Humanities.

Harris, J A Models and Analogues in Biology (1960)
In: *Symposia of the Society for Experimental Biology* New York, Academic Press, XIC pp. 250–255.

Harris, M The Rise of Anthropological Theory (1968)
New York, Crowell–Collier.

Harris, Philip R and Moran, R Managing Cultural Synergy
Houston TX, Gulf Publishing.

Harris, Pickens E Philosophic Aspects of Integration (1937)
In: *L Thomas Hopkins (Ed) Integration; its meaning and application* New York, Appleton–Century, pp. 50–76.

Harris, V All Coherence Gone (1949)
Chicago IL, Chicago University Press.

Harrison, E R On the Origin of Structure in Certain Models of the Universe (1927)
Mensuel de la Society Royal Liège 15, p. 15–28, 1967.

Harrison, E R The Mystery of Structure in the Universe (1969)
In: *Lancelot Law Whyte, et al (Eds) Hierarchical Structures* New York, American Elsevier, pp. 87–98.

Harrison, Edward Masks of the Universe (1985)
New York, Macmillan, 306 p.

Hart, B L J Dynamic Systems Design (1964)
London, Business Pub, 233 p.

Harton, J Influence of the Degree of Unity of Organization on The Estimation of Time (1939)
Journal of General Psychology 21, p. 25–49, July.

Harvard Business School The Idea of System (1963)
Cambridge, 9 p.

Haselkorn, F Some Dynamic Aspects of Interprofessional Practice in Rehabilitation (1958)
Social Casework 39, July 1958 pp. 396, 401.

Haskell, E F and Jensen, Arthur R Framework of the Periodic Table of human cultures (1972)
In: *E Haskell (Ed) Full Circle: The moral force of unified science* New York, Gordon and Breach, pp. 111–168.

Haskell, Edward F The Religious Force of Unified Science (1942)
Scientific Monthly 54, June pp. 545–551.

Haskell, Edward F The Coaction Compass: a general conceptual scheme based upon the independent systematizations of coaction among plaits by Gause, animals by Haskell and men by Morelo and others (1948)
New York, Council for Unified Research and Education. mimeographed.

Haskell, Edward F Geometric Coding of Political Philosophies (1955)
In: *Editions du Griffon, Proceedings of the Second International Congress for the Philosophy of Science (Zurich 1954).*

Haskell, Edward F Unified Science: assembly of the sciences into a single discipline (1968)
New York, Council for Unified Research and Education. offset by the National Institute of Health), 1969 (xeroxed by the IBM Systems Research Institute. 3 vols when completed. vol. 1. Scientia Generalis.

Haskell, Edward F Assembly of the Sciences into a Single Discipline (1970)
The Science Teacher 37, 9, Dec. suppl.

Haskell, Edward F Full Circle: the moral force of unified science (1972)
New York, Gordon and Breach, 256 p. bibl. foot.

Haskell, Edward F Generalization of the Structure of Mendelev's Periodic Table (1972)
In: *E Haskell (Ed) Full Circle; the moral force of unified science* New York, Gordon and Breach, pp. 21–87.

Haskell, Edward F Unified Science's Moral Forces (1972)
In: *E Haskell (Ed) Full Circle; the moral force of unified science* New York, Gordon and Breach, pp. 169–213 bibl. (pp. 211–213).

Haskell, R E Anatomy of Analogy: a new look (1968)
Journal of Humanistic Psychology 8, Fall bibl. (pp. 167–169).

Havelock, Ronald G The Process and Strategy of Beneficial Change: an analysis and critique of four perspectives (1972)
Ann Arbor MI, University of Michigan, 35 p. bibl.

Havelock, Ronald G et al Planning for Innovation: through dissemination and utilization of knowledge (1971)
Ann Arbor MI, University of Michigan, 538 p.

Havelock, Ronald G et al Bibliography on Knowledge Utilization and Dissemination (1971)
Ann Arbor MI, University of Michigan, 163 p.

Hawley, A H Human Ecology (1950)
New York, Ronald Press.

Hayek, F A Degrees of Explanation (1955)
British Journal for the Philosophy of Science 6, pp. 209–225.

Hayward, Jeremy W Perceiving Ordinary Magic: science and intuitive wisdom (1984)
London, New Science Library, 323 p. bibl.
ISBN 0-87773-297-3.

Hayward, Jeremy W Shifting Worlds Changing Minds: where the sciences and Buddhism meet (1987)
Boston MA, Shambhala Publications, 310 p.
ISBN 0-87773-368-6.

Heady, Earl O et al Recherche Interdisciplinaire Sur les Relations Input–Output et les Fonctions de Production en Vue d'Améliorer les Décisions et l'Efficacité en Aviculture: présentation des concepts et méthodes utilisés pour les études interdisciplinaires (1966)
Paris, OECD, 144 p.

Hearn, Gordon (Ed) The General Systems Approach: contributions toward an holistic conception of socialwork (1969)
New York, Council on Social Work Education, 72 p. bibl. (pp. 71–72).

Hebb, D O The Organization of Behavior (1949)
New York, John Wiley and Sons.

Heckhausen, Heinz Discipline and Interdisciplinarity (1972)
In: *Centre for Educational Research and Innovation. Interdisciplinarity* Paris, OECD, pp. 83–89.

Heelan, P A The Logic of Changing Classificatory Frameworks (1974)
In: *J A Wojciechowski (Ed) Conceptual Basis for the Classification of Knowledge* Munich, K G Saur Verlag, 1974, p. (260–274).

Heelan, Patrick Space Perception and the Philosophy of Science (1983)
Berkeley CA, University of California Press.

Hegel, G W F Science de la Logique, Tome 1

Heijnskijk, J De Interdisciplinaire Benadering in Wetenschapsbeoefening en Wetenschapstoepassing (1970)
Rotterdam, Universitaire Pers Rotterdam, 116 p. bibl. (pp. 115–116).

Hein, Piet Of Order and Disorder, Science and Art
Architectural Forum December 1967.

Heinrich, W (Ad) Die Ganzheit in Philosophie und Wissenschaft (1950)
Vienna, Othmar Spann zpm 70 Geburtstag.

Helmer, O The Prospects of a Unified Theory of Organizations (1958)
Management Science 4, pp. 172–176.

Helmer, O A Use of Simulation for the Study of Future Values (1966)
Santa Monica CA, Rand Corporation.

Helmer, O Interdisciplinary Modelling (1974)
In: *Journal of Management Science* Los Angeles, University of Southern California, Center for Futures Research, 1974, 12 p.

Helvey, T C The Age of Information: an interdisciplinary survey of Cybernetics (1970)
Educational Technology.

Henning, D H and Utton, A E Interdisciplinary Environmental Approaches (1974)
Costa Mesa CA. Vol IV.

Henry, C A Cohérence et harmonie des choses (1934)
349 p.

Herbert, P G Situation Dynamics and the Theory of Behavior Systems (1957)
Behavioral Science 2, pp. 13–29.

Hermand, Jost Synthetisches Interpretieren
Munich, Nymphenburger Verlagshandlung. (Sammlung Dialog, nr 27).

Herrick, Clyde N Unified Concepts of Electronics (1970)
Englewood Cliffs NJ, Prentice Hall, 670 p.

Hertel, H Structure, Form and Movement: biology and engineering (1966)
New York, Reinhold Van Nostrand.

Hesse, M B On Defining Analogy (1960)
Journal of Symbolic Logic 25, March pp. 74–75.

Hesse, M B Forces and Fields (1961)
London, T Nelson.

Hesse, M B Models and Analogies in Science (1963)

Hesse, Mary B The Explanatory Function of Metaphor
In: *Models and Analogies of Science* Notre Dame, University of Notre Dame Press, 1966, pp. 157–177.

Hetzler, Joyce O A History of Utopian Thought (1923)
New York.

Heus, Michael and Pincus, Allen The Creative Generalist: a guide to social work practice (1986)
Barneveld WI, Micamar Publishing. ISBN 0-937373-00-1.

Hickman, C P Integrated Principles of Zoology (1970)
St Louis, Mosby, 976 p. bibl.

Hillman, Donald J The Measurement of Simplicity (1962)
Philosophy of Science 29, July 3.

Hills, Richard J The Concept of System (1967)
Eugene, Center for the Advanced Study of Educational Administration, University of Oregon, 20p. bibl. (p. 18–20).

Hills, Richard J Toward a Science of Organization (1968)
Eugene, Center for the Advanced Study of Educational Administration, University of Oregon, 122p. bibl. (pp. 121–122).

Hiltikka, J Models for Modalities: selected essays (1970)
New York, Humanities Press.

Hilton, M A (Ed) The Evolving Society (1966)
The Institute for Cybercultural Research.

Hilton, P J and Wylie, S Homology Theory (1960)
New York, Cambridge University Press.

Himsworth, H The Development and Organization of Scientific Knowledge (1970)
London, William Havemann.

Hinde, Robert A Animal Behavior: a synthesis of ethology and comparative psychology
New York, McGraw Hill Book Company.

Hoberman, J M The Body as an Ideological Variable: sportive imagery of leadership and the state
In: *Man and World* 1981, 14, 3, pp. 309–329.

Hoffding, Harald Der Totalipätsbegriff: Eile Erkenntnistheoretische Untersuchung (1917)
Leipzig, OR Reisland, 126 p. bibl.

Hoffding, Harald Der Begriff der Analogie: Sonderausgabe
Darmstadt, Wissenschaftliche Buchgesellschaft (1967)
109 p. bibl. foot. (reproduction of 1924 edition).

Hofstadter, A Myth of the Whole: a consideration of Quine's view of knowledge (1954)
Journal of Philosophy 51, 8 July pp. 397–417.

Hofstadter, D R Gödel, Escher, Bach: an eternal golden braid (1979)
Hemel Hempstead, Harvester Wheatsheaf.

Hofstadter, Douglas Metamagical Themas (1985)
New York, Basic Books, 136 p.

Hofstadter, Douglas R Mathematical Themes: questing for the essence of mind and patterns (1985)
New York, Basic Books.

Hoggatt, A C and Balderston, F E Symposium on Simulation Models: methodology and applications to the behavioral sciences (1963)
Southwestern.

Hoggatt, V E Fibonacci and Lucas Numbers (1969)
New York, Houghton Macmillan.

Holden, Arun V (Ed) Chaos (1986)
Princeton NJ, Princeton University Press, 332 p.
ISBN 0-691-08423-8.

Holdren, J P and Elrlich, P R (Eds) Global Ecology (1971)
New York, Harcourt Brace.

Holdridge, Leslie R A Unified Hypothesis of the Evolution of the Universe and Life (1964)
San José, Tropical Science Center, 49 p. bibl. (pp. 48–49) (Occasional paper 1).

Holliday, Leslie (Ed) The Integration of Technologies (1966)
London, Hutchinson, 167 p. bibl.

Holling, C S Perceiving and Managing the Complexity of Ecological Systems
In: *The Science and Praxis of Complexity* 1985, Tokyo, United Nations University, pp. 217–227.

Hollowell, A I Self, Society and Culture in Phylogenetic Perspective (1960)
In: *Sol Tax (Ed) The Evolution of Man* Chicago University Press.

Holly, Douglas Humanism in Adversity: teacher's experience of integrated humanities in the 1980s (1986)
New York, Taylor and Francis, 150 p.
ISBN 1-85000-100-6.

Holton, Gerald Thematic Origins of Scientific Thought (1973)
Cambridge MA, Harvard University Press.

Holton, Gerald J (Ed) Science and Culture: a study of cohesive and disjunctive forces (1965)
Boston MA, Houghton Mifflin.

Holton, Gerald J The Roots of Complementarity (1968)
In: *Eranos Jahrbuch 1968* Zurich, Rhein Verlag, 1970, pp. 45–90.

Holub, Miroslav Sciences in the Unity of Culture (1970)
Impact of science on Society 20, April–June p. 151–158.

Hooker, C A The Impact of Quantum Theory on the Conceptual Bases for the Classification of Knowledge (1974)
In: *J A Wojciechowski (Ed) Conceptual Basis for the Classification of Knowledge* Munich, K G Saur Verlag, 1974.

Hoos, Ida Russakoff Systems Analysis in Social Policy: a critical review (1969)
London, Institute of Economic Affairs, 62 p. bibl. pp. 61–62.

Hopkins, L T Integration: its meaning and applications (1937)
New York, Appleton–Century, 21 p.

Hopkins, Levi Thomas Conditions Influencing Integrative Behavior (1937)
In: *L Thomas Hopkins (Ed) Integration; its meaning and application* New York, Appleton–Century, 315p. bibl. (pp. 305–309).

Horn, Werner; Klir, George J and Trappl, Robert (Eds) Basic and Applied General Systems Research: a bibliography 1977–1984 (1985)
New York, Hemisphere Publishing, 348 p.
ISBN 0-89116-454-5.

Hornsey Commission (United Kingdom) Unity and Variety: current problems in art and design education, 1969 (1969)

Horowitz, Irving L Engineering and Sociological Perspectives on Development: interdisciplinary constraints in social forecasting (1969)
International Social Science Journal 21, 4, 1969 pp. 545–546.

Horton, F W Jr Reference Guide to Advanced Management Methods (1972)
New York, American Management Association.

Houtart, Francois, et al Recherche interdisciplinaire et théologie (1970)
Paris, Editions des Cerf, 140 p.

Hovey, H A The Planning–Programming–Budgeting Approach to Government Decision–Making (1968)
New York, Praeger, 264 p.

Howard, Nigel The Theory of Meta–Games (1966)
General Systems Yearbook Society for General Systems Research, 11, p. 167–186.

Howard, Nigel The Mathematics of Meta–Games (1966)
General Systems Yearbook Society for General Systems Research, 11, p. 187–202.

Howard, Nigel A Method for Metagame Analysis of Political Problems (1968)
In: *Papers, Vol IX, International Peace Research Society.*

Howard, Nigel Some Developments in the Theory and Application of Metagames (1970)
General Systems Yearbook Society for General Systems Research, 15, p. 205–230.

Howe, J Carrying the Village: Cuna political metaphors
In: *J D Sapir and J C Crocker (Eds) The Social Use of Metaphor* 1977, Philadelphia, University of Philadelphia, pp. 132–163.

Howland, Daniel Cybernetics and General Systems Theory (1963)
General Systems Yearbook Society for General Systems Research, 8, p. 227–232.

Hoyt, E E Choice as an Interdisciplinary Area (1965)
Quarterly Journal of Economics 79, Feb. pp. 106–112.

Hsu, L K Psychosocial Homeostasis and Jen: conceptual tools for advancing psychological anthropology (1971)
American Anthropologist 73, February pp. 23–44 bibl.

Hughes, Barry World Models: the bases of difference
In: *International Studies Quarterly* 1985, 29, pp. 77–101.

Hughes, H K Cybernetics and the Management of Large Systems (1970)
In: *E D Attinger (Ed) Global Systems Dynamics* New York, Wiley Interscience, pp. 66–75.

Hulin, Charles L; Roberts, Karlene H and Rousseau, Denise M Developing an Interdisciplinary Science of Organization (1978)
San Francisco CA, Jossey–Bass, 189 p. illus. Social and Behavioral Science Series. ISBN 0–87589–393–7.

Humbalek, R P Ku Skumaniu Vseobecnej Problematiky Systemov (An approach to the general problems of systems). (1966)
Filozofia, 21, 1.

Humphrey, Robert L, et al Paradigm Shift: teach the universal values (1984)
Coronado CA, Life Values Press, 100 p. illus. Illus by Linda Humphrey. ISBN 0–915761–00–9.

Huntley, H E The Divine Proportion: a study in mathematical beauty (1970)
New York, Dover Publications, 186 p.

Husain, I Z (Ed) Transdisciplinary Methods in Special Sciences (1971)
Bombay, Asian Studies Press, 128 p.

Hutchins, Robert The Learning Society (1968)
Praeger.

Hutchinson, G E The Concept of Pattern in Ecology (1953)
Proceedings of the Academy of National Science Philadelphia 105, pp. 1–12.

Hutchinson, T B, et al The Production of Field Crops: a textbook of agronomy (1936)
London, McGraw Hill Book Company.

Huxley, J; Hardy, A C and Ford, E B (Eds) Evolution as a Process (1954)
New York, Allied Publications.

Huxley, Sir Julian S Problems of Relative Growth (1932)
London, Methuen.

Huxley, Sir Julian S Evolution: the modern synthesis (1942)
London, Allen and Unwin.

Huxley, Sir Julian S Evolution in Action (1953)
New York, Harper and Row.

Huxley, Sir Julian S Evolution, Cultural and Biological (1955)
In: *William L Thomas Jr (Ed) Yearbook of Anthropology.*

Huxley, Sir Julian S (Ed) The Humanist frame (1962)
New York, Harper and Row.

Huxley, Sir Julian S Science et synthèse (1967)
In: *R Maheu, et al, Science et synthèse* Paris, Gallimard (Unesco copyright), 1967 pp. 67–83 (also English edition).

Hwang, I and Lin, M J Group Decision Making under Multiple Criteria (1986)
New York, Springer Verlag, 400 p. Lecture Notes in Economics and Mathematical Systems Series: 281. ISBN 0–387–17177–0.

Hyde, A B and Murphy, J Experiment in Integrative Learning (1955)
Social Service Review 29, December pp. 358–371 bibl.

Hyde, Lawrence An Introduction to Organic Philosophy: an essay on the reconciliation of the masculine and feminine principles (1955)
Reigate, Omega Press, 201 p. (includes chapters on: character of wholes, principle of hierarchy, problems of synthesis).

Ianni, Francis A Culture, Systems and Behavior (1927)
Chicago IL, Science Research Association, 1967, 134 p. (chapters on: education, anthropology, sociology, psychology, related sciences of behavior, interdependence).

Iberall, A S On the General Dynamics of Systems (1970)
General Systems Yearbook Society for General Systems Research, 15, p. 7–13.

Iberall, A S Toward a General Science of Viable Systems (1972)
New York, McGraw Hill Book Company.

Ichazo, Oscar Between Metaphysics and Protoanalysis: a theory for analysis of the human psyche (1982)
New York, Arica Institute.

Ihde, D and Zaner, R M Interdisciplinary Phenomenology (1975)
Dordrecht, Kluwer Academic Publishers Group, 187 p. Selected Studies in Phenomenology and Existential Philosophy: 6. ISBN 90–247–1922–4.

Ingman, Stan A New Faith: interdisciplinarianism (1971)
Social Science and Medicine 5, Oct pp. 491–494.

Institut de Philosophie de l'Académie des Sciences de l'URSS and Institut de Recherches Scientifiques sur les Systèmes Analyse Systématique des Problèmes Mondiaux (1988)
Paris, UNESCO, 163 p. bibl. Grand Programme I: Réflection sur les problèmes mondiaux et études prospectives: BEP/GPI/24.

Institut de Recherches Scientifiques sur les Systèmes/ Institut de Philosophie de l'Académie des Sciences de l'URSS Analyse Systématique des Problèmes Mondiaux (1986)
Paris, UNESCO, 163 p. Grand Programme I, Réflections sur les Problèmes Mondiaux et Etudes Prospectives: BEP/GPI/24.

Institut International de Coopération Intellectuelle Coopération intellectuelle (1946)
1929–1940 1946 1/2, 3/4.

Institute for Humanistic Science, Kyoto Uni L'Encyclopédie (1751–1780): a corporate study (1954)
Tokyo, Iwanami Shoten, 37 p.

Institution of Electrical Engineers Integrated Process Control Applications in Industry (1966)
London, The Institution, 139 p.

Interdisciplinary Essays
Emmitsburg, St Mary's College Vol 1, Feb. semi–annual.

Interdisciplinary Relations; inventory and bibliography
Social Science Information for 1964–1966: vol. 6, 2/3 April–June 1967; for 1968–1969: vol. 9, 2, April 1970, pp. 171–183.

Interdiscipline Varanasi, Gandhian Institute of Studies (other title: Social Science Abstracts)

International Business Machines Corporation Decision Tables: a systems analysis and documentation technique
White Plains NY.

International Business Machines Corporation Flowcharting Techniques (form C20–8152) White Plains N
Y.

International Center for Integrative Studies Some Readings in Integrative Studies (1968)
New York, The Center, 38 p. bibl. (annotated bibliography on: holistic studies, natural sciences, social sciences).

International Council of Scientific Unions (ICSU) Congress on the Integration of Science Teaching, Droujba, Bulgaria (1968)
Paris, Editions CIES.

International Institute of Intellectual Cooperation Annual Report
Paris, The Institute (and the League of Nations) 1934–1937.

International Society for General Semantics/Society for General Systems Research Transactions of the Joint Conference ,Denver, May (1970)
New York, Gordon and Breach.

International Systems Institute Towards Holistic Societal Learning (1983)
Seaside CA, Intersystems Publications, 58 p. Fuschl Conversations: 1. ISBN 0–914105–24–8.

Jacker, C Man, Memory and Machines: an introduction to cybernetics (1964)
New York, Macmillan.

Jacks, M L Total Education: a plea for synthesis (1946)
London.

Jacobs, Heidi H Interdisciplinary Curriculum: Design and implementation (1989)
Alexandria VA, Association for Supervision and Curriculum Development, 97 p. illus. ISBN 0–87120–165–8.

Jacobs, Robert The Interdisciplinary Approach to Educational Planning (1964)
Comparative Education Review June 8, pp. 5–47.

Jadfard, P and Poitou, G Introduction aux Catégories et aux Problèmes Universels (1971)
Paris, Ediscience, 322 p. bibl. (pp. 321–322).

Jakobovits, L A and Steinberg, D D Semantics: an interdisciplinary reader in philosophy, linguistics and psychology (1974)
Cambridge MA, Cambridge University Press. illus. ISBN 0–521–07822–9.

James, Bertram G S Integrated Marketing (1967)
London, Batsford, 415 p.

Jantsch, Erich Integrative Planning For the 'Joint Systems' of Society and Technology: the emerging role of the university Cambridge, MIT Alfred P Sloan School of Management 1969
In: *Ekistics* 1969, 28, November.

Jantsch, Erich Technological Forecasting in Perspective (1967)
Paris, OECD.

Jantsch, Erich Perspectives of Planning (1969)
Paris, OECD.

Jantsch, Erich Inter– and Transdisciplinary University: a systems approach to education and innovation (1970)
Policy Sciences 1, pp. 403–428 bibl. foot.

Jantsch, Erich Towards Interdisciplinarity and Transdisciplinarity in Education and Innovation (1972)
In: *OECD, Centre for Educational Research and Innovation, Interdisciplinarity* Paris, OECD pp. 97–121.

Jantsch, Erich Technological Planning and Social Futures (1972)
London, Cassell.

Jantsch, Erich Forecasting and the Systems Approach: a critical survey (1972)
Policy Sciences Vol 2, No4.

Jantsch, Erich Design for Evolution: self–organization and planning in the life of human systems (1975)
New York, George Braziller.

Jantsch, Erich The Self–Organizing Universe: scientific and human implications of the emerging paradigm of evolution (1980)
New York, Pergamon Books, 343 p. ISBN 0–08–024311–8.

Jenny, H Cymatics (1967)
Basel, Basileus Press.

Jenny, Hans Kymatics (1967)
Basel, Basilius Press.

Johnson,F C Feedback: principles and analogies (1962)
Journal of Communication 12, 1962 pp. 150–59.

Johnson, F C and Klare, G R General Models of Communication Research (1961)
Journal of Communication 11, p. 13–26.

Johnson, J T and Taylor, S E The Effect of Metaphor on Political Attitudes
In: *Basic and Applied Social Psychology* 1981, 2, 1, pp. 305–316.

Johnson, R A; Kast, F E and Rosenweig, J E The Theory and Management of Systems (1967)
New York, McGraw Hill Book Company, 513 p.

Johnston, William M 'Von Bertalanffy's Place in Austrian thought' Strategies of Integrative Thinking among Leibnizians and Impressionists (1973)
In: *Gray, William and Rizzo, Nicholas D (Eds) Unity through diversity* New York. Gordon and Breach Science Publishers 1973 pp. 21–28.

Jolley, Leo Relation Codes: an ordering of ideas (1968)
New Scientist 16 May 1968 pp. 338–340.

Jones, J C Design Methods (1970)
John Wiley.

Jones, R D (Ed) Unity and Diversity (1969)
New York, Braziller.

Jones, R W and Gray, J S System Theory and Physiological Processes (1963)
Science 140, 1963 pp. 461–466.

Jones, Ronald G Holism, The Integration of Knowledge and Liberal Education (1924)
Salzburg. 1964.

Jones, Ronald G Notes on Hierarchy in Artifact, Hierarchy: the problem of the one and the many (1969)
In: *Lancelot Law Whyte, et al (Eds) Hierarchical Structures* New York, American Elsevier pp. 280–285.

Jones, W T The Romantic Syndrome: toward a new method in cultural anthropology and history of ideas (1961)
The Hague, Martinus Nijhoff, 255 p.

Judge, A J N Reflections on Associative Constraints and Possibilities in an Information Society
In: *Transnational Associations* 1987, Brussels, Union of International Associations, 38.

Judge, A J N Functional Synthesis of Viewpoints: a conceptual model (1968)
Brussels, Privately distributed, 40p. plus diagrams.

Judge, A J N Relationship Between Elements of Knowledge: use of computer systems to facilitate construction, comprehension and comparison of the concept thesauri of different schools of thought (Committee on Conceptual and Terminological Analysis of the International Political Science Association, Working Paper 3) (1971)
Brussels, Union of International Associations, 150 p.

Judge, A J N Criteria for a Meta–Model: document submitted to a session of the 4th Conference on General Systems Education, Connecticut, April 1971 (1971)

Judge, A J N Tensed Networks: balancing and focusing network dynamics in response to networking diseases (1978)
Transnational Associations, 30, 1978, 11, p.480–485. Also in: From Networking to Tensegrity Organization (Collection of papers prepared in response to the concerns of the Networks sub–project of the Goals, Processes and Indicators of Development project of the United Nations University, 1978–82). Brussels, Union of International Associations, 1984.

Judge, A J N Representation, Comprehension and Communication of Sets: the role of number (1978)
International Classification 5, 1978, 3, p.126–133; 6, 1979, 1, p.16–25; 6, 1979, 2, p.92–103. Brussels, Union of International Associations, 1984. Also in: Patterns of Conceptual Integration (Collection of papers presented at meetings of the Goals, Processes and Indicators of Development project of the United Nations University, 1978–82).

Judge, A J N Transcending duality through tensional integrity (1978)
Transnational Associations, 30, 1978, 5, p.248–265. Also in: From Networking to Tensegrity Organization (Collection of papers prepared in response to the concerns of the Networks sub–project of the Goals, Processes and Indicators of Development project of the United Nations University, 1978–82); Brussels, Union of International Associations, 1984.

Judge, A J N Implementing Principles by Balancing Configurations of Functions: a tensegrity organization approach (1979)
Transnational Associations, 31, 1979, 12, p.587–591. Also in: From Networking to Tensegrity Organization (Collection of papers prepared in response to the concerns of the Networks sub–project of the Goals, Processes and Indicators of Development project of the United Nations University, 1978–82). Brussels, Union of International Associations, 1984.

Judge, A J N Groupware Configurations of Challenge and Harmony: an alternative approach to alternative organization (1979)
In: *Richard Ericson (Ed) Improving the Human Condition; quality and stability in social systems* Washington DC, Society for General Systems Research, 1979, p.597–610. Also in: From Networking to Tensegrity Organization (Collection of papers prepared in response to the concerns of the Networks sub-project of the Goals, Processes and Indicators of Development project of the United Nations University, 1978–82). Brussels, Union of International Associations, 1984.

Judge, A J N Patterns of N-foldness: comparison of integrated multi-set concept schemes as forms of presentation (1980)
In: *Patterns of Conceptual Integration* Brussels, Union of International Associations, 1984. (Paper for a meeting on forms of presentation meeting of the UN University GPID project, Geneva, 1980) in (Collection of papers presented at meetings of the Goals, Processes and Indicators of Development project of the United Nations University, 1978–82).

Judge, A J N Liberation of integration: pattern, oscillation, harmony and embodiment (1980)
In: *Patterns of Conceptual Integration* Brussels, Union of International Associations, 1984. (Paper for the 5th Network Meeting of the UN University GPID project, Montreal, 1980) in (Collection of papers presented at meetings of the Goals, Processes and Indicators of Development project of the United Nations University, 1978–82).

Judge, A J N Societal learning and the erosion of collective memory (1980)
In: *Th Dimitrov, International Information for the 80s* New York, Uniflo, 1982. (Report for 2nd World Symposium on International Documentation, Brussels, 1980). Also in: Forms of Presentation and the Future of Comprehension (Collection of papers mainly presented to the Forms of Presentation sub-project of the Goals, Processes and Indicators of Development project of the United Nations University, 1978–82). Brussels, Union of International Associations, 1984.

Judge, A J N Integrative Dimensions of Concept Sets: transformations with minimal distortion between implicitness and explicitness of set representation according to constraints on communicability (1980)
In: *Patterns of Conceptual Integration* Brussels, Union of International Associations, 1984. (Paper for Integrative Group B of the UN University GPID project, Tokyo, 1980) in (Collection of papers presented at meetings of the Goals, Processes and Indicators of Development project of the United Nations University, 1978–82).

Judge, A J N Anti-developmental biases in thesaurus design (1981)
In: *F W Riggs (Ed) The CONTA Conference; Proceedings* Frankfurt/Main, Indeks Verlag 1981, p.185–201. (Paper for UNESCO-sponsored Conference on Conceptual and Terminological Analysis in the Social Sciences. Bielefeld, 1981).

Judge, A J N Beyond Method: engaging opposition in psycho-social organization (1981)
In: *Patterns of Conceptual Integration* Brussels, Union of International Associations, 1984. (Paper for meeting on methodology of the UN University GPID project, Bucharest, 1981) in (Collection of papers presented at meetings of the Goals, Processes and Indicators of Development project of the United Nations University, 1978–82).

Judge, A J N Alternation Between Development Modes: reinforcing dynamic conception through functional classification of international organizations and their concerns (1982)
In: *Global Action Networks, under title "Functional Classification: a review of possibilities"* Munich, K G Saur Verlag, 1985/86. (Paper for a meeting of Integrative Group B of the UN University GPID project, Athens, 1982).

Judge, A J N Policy Alternation for Development (1984)
Brussels, Union of International Associations. (Papers arising from work in connection with the Goals, Processes and Indicators of Development project of the United Nations University, 1978–1982).

Judge, A J N Reordering of Networks of Incommensurable Concepts in Phased Cycles and their Comprehension through Metaphor: paper prepared for the International Symposium on Models of Meaning (Bulgaria, September 1988) under the auspices of the Institute of Bulgarian Language of the Bulgarian Academy of Sciences (1988)
Brussels, Union of International Associations, 36 p.

Jun, J S and Storm, W B (Eds) Tomorrow's Organizations (1973)
Glenview IL, Scott Foresman.

Jung, C G The Collected Works of C G Jung (1953)
New York, Bollinger. 14 vols.

Jungwirth, E 'Bedikat Hashpaat Horaat Miktsoot Meshulavim Bekita Alef Shal Beth Haseder Hakaklai': a study of the influence of teaching integrated subjects in the first grade of agricultural secondary schools (1960)
Rehovot, Hebrew University. Unpublished Ph D thesis.

Jurkovich, Ray and Paelinck, Jean H (Eds) Problems in Interdisciplinary Studies (1984)
Brookfield VT, Gower Publishing, 196 p. Issues in Interdisciplinary Studies: 2. ISBN 0-566-00689-8.

Kade, Gerhard La théorie économique de la pollution et l'application de la méthode interdisciplinaire à l'aménagement de l'environnement (1970)
Revue Internationale des Sciences Sociales 22, 4, 1970 pp. 613–625.

Kahler, Erich The Disintegration of Form in the Arts (1968)
New York, Braziller.

Kahn, H and Wiener, A J The Year 2000: a framework for speculation on the next thirty-three Years, Macmillan, 1967 (1967)

Kahn, Hermann and Mann, I Techniques of Systems Analysis (1957)
Santa Monica, Rand Corporation (RM-1829-1).

Kalmus, H (Ed) Regulation and Control in Living Systems (1966)
New York, John Wiley and Sons.

Kamaryt, Jan 'From Science to Metascience Philosophy' Dialectical Perspectives in the Development of Ludwig von Bertalanffy's Theoretical Work (1973)
In: *Gray, William and Rizzo, Nicholas D (eds) Unity through diversity* New York, Gordon and Breach Science publishers. 1973 pp. 75–95.

Kanner, L The Conception of Wholes and Parts in Early Infantile Autism (1951)
American Journal of Psychiatry 108, pp. 23–26.

Kaplan, David The Superorganic: science or metaphysics? (1965)
American Anthropologist 67, pp. 958–76.

Kaplan, M A New Approaches to International Relations: progress or retrogression? (1968)
In: *Yearbook of World Affairs* 1968, pp. 15–34.

Kaplan, Moston System and Process in International Politics (1957)
New York, John Wiley and Sons, 283 p.

Kast, F E and Rosenzweig, J E Organization and Management: a systems approach (1970)
New York, McGraw-Hill, 654 p.

Kastein, Gerrit W Eine Kritik der Ganzheitstheorien (1937)
Leiden, J Ginsberg, 131 p. bibl. (pp. 130–131).

Katz, D and Kahn, R L The Social Psychology of Organizations (1966)
New York, Wiley, 498 p.

Katz, Fred E Indeterminacy and General Systems Theory (1973)
In: *Gray, William and Rizzo, Nicholas D (Eds) Unity through diversity* New York, Gordon and Breach Science publishers. pp. 969–81.

Katz, K U and Smith, M F Cybernetic Principles of Learning, and Educational Design (1966)
New York, Holt, Rinehart and Winston.

Katz, S M A Systems Approach to Development Administration: a framework for analyzing capability of action for national development (1965)
Washington DC, American Society for Public Administration, 59 p.

Kauffman, A The Science of Decision-Making (1968)
New York, McGraw-Hill, 253 p.

Kaufman, Herbert Challenges to Unity (in government agencies). (1969)
In: *Litterer, Joseph A (Ed) Organizations; systems control and adaptation* London, Wiley 2nd ed. vol 2, pp. 182–91.

Kaufman, Michele A Possible Mechanism for the Origin of the Sequence of Cosmic Bodies (1969)
In: *Lancelot Law Whyte, et al (Eds) Hierarchical Structures* New York, American Elseviar, pp. 99–112.

Kazemier, B H and Vuysje, D (Eds) Synthese Library
Dordrecht, R Reidel. (a series of monographs on the recelt development of symbolic logic, significs, sociology of language, sociology of science and of knowledge, statistics of language and related fields).

Kedrov, B Concerning the Synthesis of Sciences
In: *Man, Science, Technology* 1973, Moscow, I, pp. 67–92.

Kedrov, B M Intégration et Différenciation Dans les Sciences Modernes: l'évolution générale de la connaissance scientifique (1967)
In: *R Maheu, et al, Science et synthèse* Paris, Gallimard (Unesco copyright) 1967 pp. 141–47, (also English edition).

Keiter, Friedrich The Biorrhesis of Cultural Systems (1973)
In: *Gray, William and Rizzo, Nicholas D (Eds) Unity through diversity* New York, Gordon and Breach Science Publishers 1973 pp. 983–997.

Kel'zon, A S Dynamic Problems of Cybernetics (1960)
General Systems Yearbook Society for General Systems Research, 5, p. 209–17.

Kelleher, Grace J (Ed) The Challenge to Systems Analysis: public policy and social change (1970)
New York, Wiley, 150 p. bibl. foot. Operations Research Society of America: 20.

Keller, A G Societal Evoluation (1915)
New Haven, Yale University Press. rev. ed. 1931.

Keller, A G Societal Evolution (1931)
1931 (revised).

Kempf, E J Holistic Laws of Life (1949)
Journal of Psychology 27, p. 79–123, Jan.

Kenner, Hugh Geodesic Math and How to Use It (1976)
Los Angeles, University of California Press.

Kepes, Gyorgy (Ed) Structure in Art and in Science
New York, Braziller. (Vision and Value Series).

Kepes, Gyorgy Module, Proportion, Symmetry, Rhythm
New York, Braziller. (Vision and Value Series).

Kepes, Gyorgy The Visual Arts and the Sciences: a proposal for collaboration (1965)
Daedalus 94, 1965, p. 117–34.

Kerschner, R B A Survey of Systems Engineering and Tool Techniques (1960)
In: *Flagle Ch D, Huggins and Roy, R H (Eds) Operations Research and Systems Engineering* Baltimore. 140.

Kershaw, J A and McKean, R N Systems Analysis and Education (1959)
Santa Monica CA, A and McKean R N, 64 p.

Kershner, Lee R Cybernetics and Soviet Philosophy (1966)
International Philosophical Quarterly 6, 1966, p. 270–85.

Kessing, F M Problems of Integrating Humanities and Social Science Approaches in Far Eastern Studies (1955)
Far East Quarterly 14, February p. 161–68.

Kettner, Frederick The Synthesis of Science and Religion (1939)
New York, Biosophical Institute, 22 p.

Keys, D Earth at Omega: passage to planetization (1982)
Boston MA, Branden.

Khailov, D M The Orderliness of Biological Systems (1968)
General Systems Yearbook Society for General Systems Research, 12, p 29–36.

Khailov, R M The Problem of Systematic Organization in Theoretical Biology (1964)
General Systems Yearbook Society for General Systems Research, 9 p. 151–57.

Khan, Ziauddin, et al (Eds) Generalists Versus Specialists
Jaipur, Ramesh Book Depot, 12 p.

Khare, R S and Little, David (Eds) Leadership: interdisciplinary reflections (1984)
Lanham MD, University Press of America, 146 p. ISBN 0-8191-3969-6.

Khenpo Tultrim Gyatso Rinpoche Progressive Stages of Meditation on Emptiness (1986)
Oxford, Longchen Foundation, 65 p.

Khosla, A; Prakash, S and Revi, A A Transcultural View of Sustainable Development: the landscape of design (1986)
New Delhi, Development Alternatives. Document for the World Commission on Environment and Design.

King, A J and Brownell, J A The Curriculum and the Structures of Knowledge (1935)
New York, John Wiley and Sons.

Kinhide, Mushakoji Scientific revolution and inter-paradigmatic dialogue (1978)
Tokyo, United Nations University. (Paper for meeting of the Goals, Processes and Indicators of Development project, Geneva, 1978).

Klaczko-Ryndziun, Salomon (Ed) Interdisciplinary Systems Research/Interdisziplinäre Systemforschung (1974)
Basel, Birkhauser Verlag. Series commencing in.

Klapp, Orrin E Opening and Closing: strategies and information adaptation in society (1978)
New York, Cambridge University Press.

Klaus, G Kybernetik und Gesellschaft (1964)
Berlin, Die Wissenschaft.

Kleene, S Introduction to Metamathematics (1952)
Princeton, Van Nostrand.

Klein, Julie T Interdisciplinarity: history, theory and practice (1990)
Detroit MI, Wayne State Univerisity, 416 p. ISBN 0-8143-2087-2.

Klemm, Otto et al (Ed) Ganzheit und Struktur: festschrift zum 60 geburtstage Felix Kruegers (1934)
Munich, Beck, 214 p.

Klir, G J The Many Faces of Complexity
In: *The Science and Praxis of Complexity* 1985, Tokyo, United Nations University, pp. 81–98.

Klir, George J The General System as a Methodological Tool (1905)
General Systems Yearbook Society for General Systems Research, 1965, 10 p. 29–42.

Klir, George J An Approach to General Systems Theory (1969)
New York, Reinhold Van Nostrand, 323p. bibl. (pp. 311–18).

Klir, George J Trends in General Systems Theory (1972)
New York, Wiley Interscience, 462 p. bibl.

Klir, George J From General Systems Theory to General Systems Profession (1973)
In: *Gray, William and Rizzo, Nicholas D (Eds) Unity through Diversity* New York, Gordon and Breach Science Publishers. pp. 513–532.

Klir, George J Architecture of Systems Problem Solving (1985)
New York, Plenum Press.

Klir, J and Valachi, M Cybernetic Modeling (1967)
Princeton, Van Nostrand, 437 p.

Klopper, Leopold E and McCann, Donald C Evaluation in Unified Science: measuring the effectiveness of the natural science course at the university of Chicago High School
Science Education 53, pp. 155–64.

Kluckhohn, Clyde K The Special Character of Integration in an Individual Culture (1950)
The Nature of Concepts, their Inter-Relation and Role in Social Structure. Stillwater Conference, Oklahoma, 86 p.

Kluckhohn, Clyde K Universal Values and Anthropological Relativism (1952)
In: *Modern Education and Human Values* Pillesburgh, University Press.

Kluckhohn, Clyde K Universal Categories of Culture (1953)
In: *A L Kroeber (Ed) Anthropology Today; an encyclopedic inventory* Chicago University Press, pp507–523.

Knight, D Cybernetics and Conflict (1970)
New York, Spartan Books.

Knight D E; Curtis, H W and Fogel, L J (Eds) Cybernetics, Simulation, and Conflict Resolution: Third Annual Symposium of The American Society for Cybernetics (1970)
New York, Spartan Books, 256 p.

Knorr, K and Verba, S (Eds) The International System: theoretical essays (1962)
Princeton University Press.

Koch, Sigmund Value Properties: their significance for psychology, axiology and science
In: *Marjorie Grene (Ed) Toward a Unity of Knowledge* 1969, New York, International Universities Press, pp. 251–279.

Koch, Sigmund Psychology and Emerging Conceptions of Knowledge as Unitary (1964)
In: *T W Wann (Ed) Behaviorism and Phenomenology* Chicago, University of Chicago Press, pp. 1–45.

Kockelmans, Joseph J On the Meaning of Scientific Revolution. (1971)
In: *Gotesky, Rubin and Laszlo, Ervin (Eds) Evolution-Revolution* New York, Gordon and Breach Science Publishers, 1971 pp. 231–51.

Kockelmans, Joseph J (Ed) Interdisciplinarity and Higher Education (1979)
University Park PA, Pennsylvania State University Press. ISBN 0-271-00200-X.

Koertge, Noretta Toward an Integration of Content and Method in the Science: a curriculum
Curriculum Theory Network 4, pp. 26–44.

Koestler, Arthur Insight and Outlook: an inquiry into the common foundations of science, art and social ethics (1905)
Lincoln, University of Nebraska Press.

Koestler, Arthur Evolution and Revolution in the History of Science (1926)
The Advancement of Science March 1966.

Koestler, Arthur The Act of Creation (1964)
New York, Macmillan, 751 p.

Koestler, Arthur The Ghost in the Machine (1967)
New York, Macmillan, 384 p.

Koestler, Arthur Beyond Atomism and Holism: the concept of the holon (1969)
In: *A Koestler and J Smythies (Eds) Beyond Reductionsim; new perspectives in the life sciences. Alpbach symposium, 1968* pp. 192–232. London, Hutchinson.

Koestler, Arthur The Call Girls (1973)
New York, Random House.

Koestler, Arthur The Tree and the Candle (1973)
In: *Unity through Diversity, Gray, William and Rizzo, Nicholas D (Eds)* New York, Gordon and Breach Science Publishers, pp. 287–303.

Koestler, Arthur and Smythies, J R (Ed) Beyond Reductionism; New Perspectives in the Life Sciences: Alpbach symposium, 1968 (1969)
London, Hutchinson, 438 p. bibl. foot.

Kohn, Hans World Order in Historical Perspective (1942)
Cambridge MA, Harvard University Press, 352 p.

Kolaja, Jiri T Social System and Time and Space: introduction to the theory of recurrent behavior (1969)
Pittsburgh PA, Duquesne University Press, 113 p. bibl.

Konorski, Jerzy Integrative Activity of the Brain: an interdisciplinary approach (1967)
Chicago IL, University of Chicago Press, 531 p. bibl.

Koontz, H (Ed) Toward a Unified Theory of Management (1964)
New York, McGraw Hill Book Company, 273 p.

Kopstein, F F and Hanreider, B D The Macro–Structure of Subject Matter as a Factor in Instruction (1966)
Princeton NJ, Educational Testing Service. Research Memorandum RM–66–25.

Kopstein, Kingsley E F F and Seidel, R J Graph Theory as a Meta–Language of Communicable Knowledge (1971)
In: *M D Rubin (Ed) Man in Systems* New York, Gordon and Breach, 1971, pp. 43–69.

Korzybski, A Science and Sanity: an introduction to non–Aristotelian systems and general semantics (1948)
Lakeville, Conn. International Non–Aristotelian Lib. 3rd ed.

Kosmos
Martinus Institute, Copenhagen Fortnightly in 4 languages.

Kovecses, Zoltan Metaphors of Anger, Pride, and Love: a lexical approach to the structure of concepts (1987)
Philadelphia PA, Benjamins John North America, 147 p.
ISBN 1–55619–009–3.

Krane, Harry H The Third Culture: an integration of technology and art (1972)
Melbourne VIC, International Cooperation for the Integration of Technology with Art, 80 p. bibl. (pp. 67–79).

Kranzberg, Melvin The Unity of Science–Technology (1967)
American Scientist 55, March pp. 48–66.

Kranzberg, Melvin The Disunity of Science–Technology (1968)
American Scientist 56, pp. 21–34.

Krech, David Dynamic Systems as Open Neurological Systems (1950)
Psychological Review 57, pp. 345–61.

Krech, David Dynamic Systems, Psychological Fields, and Hypothetical Constructs (1956)
General Systems Yaerbook Society for General Systems Research, 1, p. 139–143 (reprinted from Psychological Review, 57, 283, 1950).

Kremyanskiy, V I Certain Peculiarities of Organisms as a 'System' from the Point of View of Physics, cybernetics, and biology (1960)
General Systems, 5, p. 221–24.

Krippendorff, Klaus Communication and the Genesis of Structure (1971)
General Systems Yearbook Society for General Systems Research, 16, p. 171–185.

Kroeber, A L Styles and Civilizations (1957)
Ithaca NY, Cornell University Press.

Kroeber, A O Concept of Culture in Science (1949)
Journal of General Education 3, p. 182–196.

Krueger, Felix E Zur Philosophie und Psychologie der Ganzheit: schriften aus den Jahren 1918–1940 (Hrsg von E Heuss) (1953)
Berlin, Springer–Verlag, 347p. bibl. (pp. 336–339).

Kubler, G Shape of Time Remarks on the History of Things (1962)
New Haven CT, Yale University Press.

Kuhn, Alfred Toward a Uniform Language of Information and Knowledge (1961)
Synthese 13, pp. 127–153.

Kuhn, Alfred The Study of Society: a multidisciplinary approach (1966)
London, Tavistock Publications, 810 p.

Kuhn, Alfred The Logic of Social Systems: a unified, deductive, system–based approach to social science (1974)
San Francisco CA, Jossey–Bass Publishers, April.

Kuhn, T S Second Thoughts on Paradigms (1970)
In: *Suppe (Ed) The Structure of Scientific Theory.*

Kuhn, T S Logic of Discovery or Psychology of Research (1970)
In: *Lekatos, I and Musgrove, A (Eds) Criticism and the Growth of Knowledge* Cambridge University Press, pp. 1–24.

Kuhn, T S The Function of Dogma in Scientific Research (1972)
In: *A C Crombie (Ed) Scientific Change* Heinemann, 1963, pp. 347–369 ×3Also printed under the title *Scientific Paradigms* in: *B Barnes (Ed) Sociology of Science* London, Penguin.

Kuhn, Thomas The Essential Tension (1977)
Chicago IL, University of Chicago Press.

Kuhn, Thomas S The Structure of Scientific Revolutions (1970)
Chicago IL, University of Chicago Press. 2nd ed.

Kulikowski, R Optimum Control of Aggregated Multilevel System (1966)
In: *Proceedings III IFAC Congress* London.

Kung, Hans Paradigm Change in Theology (1989)
New York, Crossroad Publishing, 250 p.
ISBN 0–8245–0925–0.

Kuntz, Paul G (Ed) The Concept of Order (Papers presented in a series of lectures entitled (1968)
Interdisciplinary Seminars on Order at Grinnell College 1963/64) Seattle, University of Washington Press, 479p., bibl., foot.

Kunz, F L The Metric of the Living Orders (1972)
In: *Integrative Principles of Modern Thought* New York, Gordon and Breach Science Publishers, pp. 291–363.

Kupperman, Robert H and Smith, H Mathematical Foundations of Systems Analysis (1969)
Reading MA, Addison–Wesley Pub.

La Guardia, Eric L'Esthétique de l'analogie
Diogene, 62, p. 54 and follwing.

Laborit, H The Complexity of Interdependence in Living Systems
In: *The Science and Praxis of Complexity* 1985, Tokyo, United Nations University, pp. 146–152.

Lajugie, L L'Expérience des Recherches Interdisciplinaires de l'Institut d'Economie Régionale du Sud–Ouest (1969)
Bière (Ed) Revue Juridique et Economique du Sud–Ouest 3.

Lakatos, Imre and Musgrove, Alan (Ed) Criticism and the Growth of Knowledge: proceedings of the international colloquium in the philosophy of science, London, 1965, vol 4 (1970)
Cambridge University Press.

Lambert, J H Kosmologische Briefe (System of the World) (1761)
London, Vernon and Hood. Trans. 1800 by James Jacque.

Lamberton, D M (Ed) Economics of Information and Knowledge
London, Penguin Books, 384 p.

Lamson, Robert W National Goals and Science Technology: is a better synthesis possible? (Paper at meeting of the Operations Research Society of America, San Juan, 1974; read into the (1974)
Congressional Record, 93rd Congress, 2nd Session, 16 Dec by Charles S Gubser).

Landau, H G Development of Structure in a Society With a Dominance Relation When New Members Are Added Successively (1965)
Bulletin of Mathematical Biology 17, 151–60.

Landauer, C Toward a Unified Social Science (1971)
Political Science Quarterly 86, December pp. 563–585.

Landfield, A W The Science of Psychology and the Psychology of Science (1961)
Perceptual and Motor Skills 13.

Landheer, B; Loenen, J H M M and Polak, F L (Eds) World Society: how is an effective and desirable world order possible ?: a symposium (1971)
Dordrecht, Kluwer Academic Publishers Group, 211 p.
ISBN 90–247–5088–1.

Lane, Michael (Comp) Structuralism: a reader (1970)
London, Cape, 456 p. bibl. (pp. 417–434).

Lange, O Wholes and Parts: a general theory of system behavior (1965)
Oxford, Pergamon Books, 74 p.

Lange, O Introduction to Economic Cybernetics (1970)
New York, Pergamon Books.

Langham, Derald G Genesa: an attempt to develop a conceptual model to synthesiza, synchronize and vitalize man's interpretation of universal phenomena
California CA, Gelesa Foundation, 262 p.

Langley, L L Homeostasis (1965)
New York, Reinhold Van Nostrand, 114 p.

Lasswell, Harold The Transition Toward More Sophisticated Precedures
In: *D B Bobrow and J L Schwartz (Eds) Computers and the Policy–Making Community: applications to international relations* 1968, Englewood Cliffs, Prentice Hall, pp. 307–314.

Laszlo, E and Margenau, H The Emergence of Integrative Concepts in Contemporary Science: proceedings of the XIIIth International Congress on the History of Science, Moscow, 1974
pp. 92–100. Section IA.

Laszlo, Eervin, et al Goals for Mankind: a report to the Club of Rome on the new horizons of global community (1977)
New York, Dutton.

Laszlo, Ervin System, Structure and Experience: toward a scientific theory of mind (1969)
New York, Gordon and Breach, 110 p. bibl.

Laszlo, Ervin The Systems View of the World (1972)
New York, George Braziller, 131 p. bibl.

Laszlo, Ervin The Relevance of General Systems Theory: papers presented to Ludwig von Bertalanffy on his 70th Birthday (1972)
New York, Braziller, Spring.

Laszlo, Ervin Introduction to Systems Philosophy (1972)
New York, Gordon and Breach, 350 p.
ISBN 0–677–03850–X.

Laszlo, Ervin The Systems View of the World: the natural philosophy of the new development in the sciences (1972)
New York, Braziller, 131 p. bibl. (pp. 121–126).

Laszlo, Ervin Integrative Principles of Art and Science (1972)
In: *Integrative Principles of Modern Thought* New York, Gordon and Breach Science Publishers, pp. 365–390.

Laszlo, Ervin Ludwig Von Bertalanffy and Claude Lévi–Strauss: systems and structures in biology and social anthropology (1973)
In: *Unity through Diversity, Gray, William and Rizzo, Nicholas D (Eds)* New York, Gordon and Breach, Science Publishers, pp. 143–164.

Laszlo, Ervin The World System: models, norms, applications (1973)
New York, Braziller.

Laszlo, Ervin A Strategy for the Future (1974)
New York, Braziller.

Laszlo, Ervin Evolution: the grand synthesis (1987)
Boston MA, Shambhala Publications, 212 p. illus.
ISBN 0–87773–399–6.

Laudan, L Towards a Reassessment of Comte's Méthode Positive (1971)
Philosophy of Science 38, March pp. 35–53.

Laurie, Simon S Synthetica: being meditations epistemological and ontological (1906)
London, Longmans Green. 2 vols (Gifford lectures at University of Edinburgh).

Lawrence, J R (Ed) Operational Research and the Social Sciences (1966)
London.

Lawrence, Paul R and Lorsch, Jay W Organization and Environment (1967)
Boston MA, Harvard University, 279 p.

Lawrence, Paul R and Lorsch, Jay W Differentation and Integration in Complex Organizations (1969)
Litterer, Joseph A (Ed) Organizations; systems control and adaptation London, Wiley, 2nd ed., vol. 2, pp. 229–253.

Lawson, C A Language, Communication, and Biological Organization (1963)
General Systems Yearbook Society for General Systems Research, 8, p. 107–115, Pub. No. 152, Dept of Natural Science, Michigan State, University, East Lansing, Michigan.

Lawson, Hilary Reflexivity: the post–modern predicament (1985)
La Salle IL, Open Court, 132 p. ISBN 0–8126–9011–7.

Le Moigne, Jean–Louis The Intelligence of Complexity
In: *The Science and Praxis of Complexity* 1985, Tokyo, United Nations University, pp. 35–61.

Leake, Chauncey D Historical Aspects of the Concept of Organizational Levels of Living Material (1969)

Leatherdale, W H The Role of Analogy, Model and Metaphor in Science (1974)
Amsterdam, North–Holland, 276 p.

Leavitt, H J (Ed) The Social Science of Organizations (1963)
Englewood Cliffs NJ, Prentice Hall, 182 p.

Lecca, Pedro J and McNeill, John S (Eds) Interdisciplinary Team Practice: issues and trends (1985)
New York, Praeger Publishers, 256 p.
ISBN 0–275–90134–3.

Lefébure, Francis Les Homologies: architecture cosmique, ou la lumière secrète de l'Asie devant la science moderne (1950)
Paris, Editions Aryama, 624 p.

Leff, Enrique Los Problemas del Conocimiento y la Perspectiva Ambiental del Desarrollo (1986)
Mexico, Siglo Veintimmo Editres.

Lehninger, A L Bioenergetics (1965)
New York, WA Benjamin.

Lehrer, R N The Management of Improvement: concepts organization, strategy (1965)
New York, Reinhold Van Nostrand, 416 p.

Leinfellner, W Struktur und Aufbau Wissenschaftlicher Theorien (1965)
Vienna–Würzburg. Eine wissenschaftstheoretisch–philosophische Untersuchung.

Lektorsky, V A and Sadovsky, V N O Printzipakh Issliedovania Sistiem (On the principles of system research) (1960)
Voprosy Filosofii 8.

Lektorsky, V A and Sadovsky, V N On Principles of System Research (Related to L Bertalanffy's General System Theory) (1960)
General Systems Yearbook Society for General Systems Research, 5, p. 171–178.

Lenine, V Cahiers Philosophiques (1973)
Paris, Editions Sociales.

Lennard, H and Bernsteil, A The Anatomy of Psychotherapy (1960)
New York, Columbia University Press.

Leonard, George The Silent Pulse (1981)
New York, Bantam Books.

Leontief, W Note on the Pluralistic Interpretation of History and the Problem of Interdisciplinary cooperation (1948)
Journal of Philosophy 45, 4 Nov pp. 617–624.

Leopold, L B and Langbein, W B The Concept of Entropy in Landscape Evolutions (1964)
General Systems Yearbook Society for General Systems Research, 9, p. 25–43.

Lerner, D (Ed) Parts and Wholes (1963)
New York, Free Press.

Leshan, Lawrence and Margenau, Henry Einstein's Space and Van Gogh's Sky: physical reality and beyond (1982)
New York, Macmillan Publishing Company, 268 p.
ISBN 0–02–093180–8.

Levi-Strauss, G The Mathematics of Man (1954)
International Social Science Bulletin 6, 4.

Levi-Strauss, G Anthropologie Structurale (1958)
Paris.

Levi-Strauss, G Structural Anthropology (1968)
London, Allen Lane Penguin Press, 410p.

Levins, Richard The Limits of Complexity (1973)
In: *Pattee, Howard H (Ed), Hierarchy Theory* New York, Braziller.

Lévy, Roger Intellectuels, Unissez-vous (1931)
Paris, M Rivière, 237 p. bibl. (pp. 205–233).

Lewada, J Kybernetische Methoden in der Soziologie (1965)
Kommunist Moscow, pp. 14–45.

Lewin, Kurt Analysis of the Concepts Whole, Differentiation and Unity (1952)
In: *Kurt Lewin, Field Theory in Social Science; selected theoretical papers* London, Tavistock, pp. 305–338.

Lewin, Kurt Field Theory in Social Science: selected theoretical papers (1952)
London, Tavistock Publications, 346 p.

Lewis, Sulwyn Principles of Cultural Cooperation (1971)
Paris, UNESCO, 27 p. (Reports and papers on mass communication, No. 61).

Lichnerowicz, André Mathematics and Transciplinarity (1972)
In: *OECD, Centre for Educational Research and Innovation, Interdisciplinarity* pp. 121–127. Paris, OECD.

Lichtenberg, Don B Unitary Symmetry and Elementary Particles (1970)
New York, Academic Press, 246 p. bibl.

Lichtman, Richard Indeterminacy in the social sciences (1967)
Inquiry, 10, 1967, p. 139–50.

Likert, R New Patterns of Management (1961)
New York, McGraw Hill Book Company, 279 p.

Likert, R The Human Organization: its management and value (1967)
New York, McGraw-Hill, 258 p.

Lilienthal, D E Management: a humanist art (1967)
New York, Columbia University, 67 p.

Lindblom, C D and Braybrooke, D A Strategy of Decision (1962)
Glencoe, Free Press, 268 p.

Lindblom, C E The Policy Making Process (1968)
Englewood Cliffs, Prentice-Hall, 122 p.

Lindelmayer, A Life Cycles as Hierarchical Relations (1924)
In: *Gregg, R and Harris, F T C (Eds) Form and Strategy in Science* Dordrecht, D Reidel, 1964.

Linstone, Harold A Communications: the planner's predicament (1973)
Paper presented to the Rome Special World Conference on Futures Research.

Linton, R Universal Ethical Principles: An anthropological view (1952)
In: *Moral Principles of Action, R N Anshen* New York, Harper.

Lipset, Seymour M (Ed) Politics and the Social Sciences (1969)
New York, Oxford University Press. relationship between political science and seven disciplines.

Lipton, Michael Interdisciplinary Studies in Less Developed Countries (1970)
Journal of Development Studies 7, Oct. pp. 5–18 bibl.

Litterer, Joseph A (Ed) Organizations: systems control and adaption (1969)
London, Wiley. 2nd ed. vol 2.

Littger, Kurt Standard Register for Operational Research (1974)
Dusseldorf, Claus Lincke, 500 p. (with 170 annual looseleaf updating sheets).

Liversey, Lionel J Jr Noetic Planning: the need to know, but what? (1972)
In: *Ervin Laszlo, The Relevance of General Systems Theory* New York, Braziller 213p. (pp. 145–62).

Locker, Alfred On the Ontological Foundations of the Theory of Systems (1973)
In: *Gray, William and Rizzo, Nicholas D (Eds) Unity through diversity* New York, Gordon and Breach Science Publishers 1973 pp. 537–69.

Löfgren, L Complexity of Descriptions of Systems: a foundational study
In: *International Journal of General Systems* 1977, 3, 4, pp. 197–214.

Londonn, I D Some Consequences for History and Psychology of Langmuir's Concept of Convergence and divergence of phenomena (1946)
Psychological Review 53, 1946 pp. 170–88.

Lorens, C S Flowgraphs: for the modeling and analysis of linear systems (1964)
New York, McGraw-Hill.

Lorenz, Konrad Gestalt Perception as Fundamental to Scientific Knowledge (1962)
General Systems Yearbook Society for General Systems Research, 7, 1962 p. 37–56.

Lotka, A J Evolution and Thermodynamics (1942)
Science and Society New York, Columbia University Press

Lough, John Essays on the Encyclopedia of Diderot and d'Alembert (1968)
London, Oxford University Press, 552 p. bibl. (pp 484–94).

Loux, Michael (Comp) Universals and Particulars, Readings in Ontology (1970)
Garden City NY, Anchor Books 349 p. bibl. (pp. 341–49).

Lovelock, James E Gaia: a new look at life on earth (1979)
Oxford, Oxford University Press, 157 p. illus.
ISBN 0-19-520358-5.

Lowy, Louis et al Integrative Learning and Teaching in Schools of Social Work: a study of organizational development in professional education (1971)
New York, Association Press, 285 p. bibl.

Luce, R D ald Raiffa, H Games and Decisions (1957)
New York, Wiley and Sons, 509 p.

Luchaire, Julien Observations sur quelques problèmes de l'organisation intellectuelle internationale (1923)
Geneva, Société des Nations (Committee on Intellectual Cooperation, Enquête sur la situation du travail intellectuel, part 2).

Luchaire, Julien Principes de la coopéération intellectuelle internationale (1925)
Recueil des Cours Academy of International Law, IV, Vol. 9, pp. 307–408.

Luhmann, Niklas Complexity and Meaning
In: *The Science and Praxis of Complexity* 1985, Tokyo, United Nations University, pp. 99–106.

Lumsden, Malvern Perception and Information in Strategic Thinking (1966)
Journal of Peace Research 3, pp. 257–277.

Lundh, Lars-gunnar Mind and Meaning: towards a theory of the human mind considered as a system of meaning structures (1983)
Stockholm, Almqvist and Wiksell, 206 p.
ISBN 91-554-1487-7.

Lupasco, Stephane Le Dualisme Antagoniste (1973)
Paris, Vrin. (Du Devenir Logique et de l'Affectivité, vol. 1).

Luszki, Margaret E B Interdisciplinary Team Research, methods and problems (1958)
New York, New York University Press for the National Training Laboratories, 355 p. bibl.

Lwoff, Andréé Biological Order (1962)
Cambridge MA, MIT Press.

Lynch, William F An Approach to the Metaphysics of Plato Through the Parmenides (1959)
Washington DC, Georgetown University Press.

Lynch, William F The Integrating Mind: an exploration into Western thought (1962)
New York, Sheed and Ward, 181 p.

M'Pherson, P K Systems Science and Systems Philosophy (1974)
Futures, 6, 3, June p. 219–239.

Machlup, Fritz The Production and Distribution of Knowledge in the United States (1962)
Princeton NJ, Princeton University Press.

MacKay, D M Operational Aspects of Some Fundamental Concepts of Human Communication (1954)
Synthèse 9, pp. 182–98.

MacKay, D M Towards an Information-Flow Model of Human Behavior (1956)
British Journal Psychology 47, pp. 30–43.

Macko, D General Systems Theory Approach to Multilevel Systems (1967)
Thesis: Case Western Reserve University.

MacLean, Paul D A triune concept of the brain and behaviour (1973)
In: *T Bag and D Campbell (Eds) The Hincks Memorial Lectures* Toronto, University of Toronto Press, 1973.

MacLeod, Robert B Phenomenology and Cross-Cultural Research (1969)
In: *M Sherif and C Sherif (Ed) Interdisciplinary Relationships in the Social Sciences* Chicago, Aldine, pp. 177–196.

Macrae, D G Cybernetics and Social Science (1951)
British Journal Sociology 2, pp. 135–49.

Maharishi Mahesh Yogi, Transcendental Meditation: science of being and art of living (1988)
New York, New American Library, 320 p.
ISBN 0-317-67301-7.

Maheu, Ren Declaration
In: *Science and Synthesis* 1971, Berlin, Springer Verlag.

Makridakis, S and Weintraub, E R On the Synthesis of General Systems (1971)
In: *General Systems Yearbook*, Society for General Systems Research 16.

Malaska, P Outline of a Policy for the Future
In: *The Science and Praxis of Complexity* 1985, Tokyo, United Nations University, pp. 338–356.

Malcolm, D G and Rowe, A J (Eds) Management Control Systems (1960)
New York, John Wiley and Sons, 375 p.

Maldague, Michel Problèmatique de la crise de l'environnement (1973)
Université de Laval, 180p. bibl. (pp. 165–174).

Malek, I Interdisciplinary Science Teaching (1971)
In: *New Trends in Biology Teaching, Vol III* Paris, Unesco.

Malz, Maxwell Psycho-Cybernetics (1925)
Wilshire, 1965.

Malz, Maxwell Psychocybernetica (1960)
New York, Prentice Hall.

Mamardashvili, M K Protzessy Analiza i Sinteza (The processes of analysis and synthesis) (1958)
Voprosy Filosofii 2.

Mandelbrot, B B The Fractal Geometry of Nature (1983)
New York, W H Freeman.

Manheim, Marvin L Hierarchical Structure: a model of planning and design processes (1926)
Cambridge MA, MIT Press, 227 p.

Manzoor, Suhail Systems bibliography (1982)
Metropolitan Bk Co, 182 p.

Maquet, Jacques J P The Sociology of Knowledge: its structure and its relation to the philosophy of knowledge; a critical analysis of the systems of Karl Mannheim and Pitirim A Sorokin (1951)
Boston MA, Beacon Press, 318p. bibl. (pp. 298–311).

Maquet, Jacques J P Sociologie de la Connaissance (1969)
Brussels, Editions de l'Institut de Sociologie, 360 p. 2nd ed. bibl. (pp. 335–351).

Maranda, P The Dialectic of Metaphor: an anthropological essay on hermanenties
In: *I Crossman and S Sulaiman (Eds) The Reader in the Text* 1980, Princeton, Princeton University Press, pp. 183–204.

Marashio, P Moving Toward a New Direction: multidiscipline approach to education (1971)
Clear House 46, December pp. 252–253.

March, J G and Simon, H A Organizations (1958)
New York, John Wiley and Sons, 262 p.

March, James G (Ed) Handbook of Organizations (1965)
Chicago IL, Rand McNally, 1246 p. bibl.

Marcus, Solomon Paradoxes (1982)
Revue Roumaine et Linguistique, 1982, 2.

Marcus, Solomon and Tataram, Monica Paradoxical and antinomic aspects of the global trends in the world today (1982)
Bucharest, GPID Romanian Team.

Margalef, Ramon Ecosystems: diversity and connectivity as measurable components of their complication
In: *The Science and Praxis of Complexity* 1985, Tokyo, United Nations University, pp. 228–244.

Margalef, Ramon Perspectives in Ecological Theory (1968)
Chicago IL, University of Chicago Press.

Margalef, Ramon On Certain Unifying Principles in Ecology (1969)
In: *A S Boughey (Ed) Contemporary Readings in Ecology* Belmont, Dickenson.

Margenau, Henry The Methodology for Integration in the Physical Sciences (1950)
In: *The Nature of Concepts, their interrelation and role in social structure* Stillwater, P. 47.

Margenau, Henry Open Vistas: philosophical perspectives of modern science (1961)
New Haven CT, Yale University Press.

Margenau, Henry Integrative Education in the Sciences (1967)
Main Currents of Modern Thought 24, Nov-Dec pp. 36–41.

Margenau, Henry Some Integrative Principles of Modern Physics (1972)
In: *Integrative Principles of Modern Thought* New York, Gordon and Breach Science Publishers, pp. 45–68.

Margenau, Henry Integrative Principles of Modern Thought (1972)
New York, Gordon and Breach, 522 p.

Marguerite, Craig and James, H Synergic Power: beyond domination, beyond permissiveness (1979)
Berkeley CA, ProActive Press, 164 p. 2nd ed. bibl.
ISBN 0-914158-28-7.

Markley, O W Unity with Diversity: toward social policies for a future world order (1972)
Menlo Park, Stanford Research Institute, December.

Marney, M C and Smith, N M The Domain of Adaptive Systems: a rudimentary taxonomy (1964)
General Systems Yearbook Society for General Systems Research, 9, p. 107–132.

Marney, M C and Smith, N M Institutional Adaptation, Part II: interdisciplinary Synthesis McLean, Virginia: Research Analysis Corporation, 1971 (Extracts have been published under the title Interdisciplinary Synthesis (1971)
In: *Policy Sciences* 1972, New York, 3, pp. 299–323.

Marquis, Stewart D Bibliography on Systems (1963)
East Lansing, Institute for Community Development and Services, Michigan State University, 9p.

Marteka, V J Bionics (1965)
New York, Lippincott.

Martin, H R Introduction to Feedback Systems (1968)
New York, McGraw Hill Book Company.

Martin, William O The Order and Integration of Knowledge (1957)
Ann Arbor MI, University of Michigan, 355 p. bibl.

Martino, Rocco L Critical Path Networks (1968)
New York, Gordon and Breach, 176 p.

Martino, Rocco L Integrated Manufacturing Systems (1972)
New York, McGraw Hill Book Company, 160 p. bibl.

Maruyama, Magoroh Relational Algebra of Intercultural Understanding (1960)
Methodos 11, pp. 269–277.

Maruyama, Magoroh Morphogenesis and Morphostasis (1960)
Methodos Vol 12, No 48.

Maruyama, Magoroh Awareness and Unawareness of Misunderstandings (1962)
Methodos 13, pp. 255–275.

Maruyama, Magoroh The Second Cybernetics: deviation amplifying mutual causal processes (1963)
American Scientist 51, pp. 164–79.

Maruyama, Magoroh Basic Elements in Misunderstandings (1963)
Dialectica 17, p. 78–92, p. 99–110.

Maruyama, Magoroh Metaorganization of Information (1965)
General Systems Yearbook Society for General Systems Research, 1966, 11, p. 55–60 (reprinted from *cybernetica* No. 4, 1925).

Maruyama, Magoroh The Navaho Philosophy: an esthetic ethic of mutuality (1967)
Mental Hygiene Vol 51, No 2, pp. 242–49.

Maruyama, Magoroh Notes on Hierarchy in Artifact: comment on patterns in social events: In: (1969)
Lancelot Law Whyte, et al (Eds) Hierarchical Structures New York, American Elsevier, pp. 279–281.

Maruyama, Magoroh Symbiotization of Cultural Heterogeneity (1972)
Paper given at the Convention of the American Anthropological Association November 30.

Maruyama, Magoroh Paradigms and Communications (1974)
Technological Forecasting and Social Change, 6, 1, pp. 3–32.

Maruyama, Magoroh Hierarchists, Individualists, and Mutualists: three paradigms among planners (1974)
Futures, 6, 2, April, p. 103–113.

Maruyama, Magoroh Heterogensitics: an epistemological restructuring of biological and social sciences (1977)
Cybernetica (1977), 20, 1, p. 69–85.

Maruyama, Magoroh Epistemologies and esthetic principles (1978)
Journal of the Steward Anthropological Society 8, p.155–167.

Maruyama, Magoroh Mindscapes, social patterns and future development of scientific theory types (1980)
Cybernetica (1980), 23, 1, p. 5–25.

Massachusetts Institute of Technology, Electrical Engineering Department Bibliography on Cybernetics (1950)
Cambridge.

Masterman, Margaret The Nature of a Paradigm (1970)
In: *I Lakatos and A Musgrove (Eds) Criticism and the Growth of Knowledge* Cambridge University Press, pp. 59–90.

Mather, K F Objectives and Nature of Integrative Studies (1951)
Main Currents in Modern Thought 8, pp. 11.

Matson, Floyd Broken Image: man, science and society (1964)
New York, Doubleday.

Matson, Floyd and Montagu, M Ashley (Eds) Human Dialogue (1967)
New York, Free Press.

Maturana, Humberto and Varela, Francisco Autopoesis and Cognition: the realization of the living (1980)
Boston MA, D Reidel.

Maturana, Humberto and Varela, Francisco Autopoeses and Cognition: the realization of the living (1980)
Dordrecht, Kluwer Academic Publishers Group, 140 p.
ISBN 90–277–1015–5.

Maturana, Humberto and Varela, Francisco The Tree of Knowledge (1987)
Boston MA, New Science Library.

Maurer, J A (Ed) Readings in Organization Theory: open systems approaches (1971)
New York, Random House.

Mayr, Ernst Systematics and the Origin of Species (1942)
New York, Columbia University Press.

Mayr, Ernst Evolutionary Systems and the Systems Approach to Understanding Problems of Evolution (1965)
Bulletin, Boston Systems Group Chapter, Society for General Systems Research 1964, 4 (9), pp. 3–6.

Mays, W Cybernetic Models and Thought Processes (1956)
Namur, Gaultier–Villar. Proceedings of the First International Congress on Cybernetics.

Mazzeo, J A Universal Analogy and the Culture of the Renaissance (1954)
Journal of the History of Ideas 15, April pp. 299–304, bibl.

McCarthy, E J Integrated Data Processing Systems (1926)
New York, John Wiley and Sons, 565 p.

McCarthy Report, The The Employment of Highly Specialized Graduates: a comparative study in the UK and USA (1968)
London, HMSO.

McCay, J T The Management of Time (1968)
Englewood Cliffs NJ, Prentice Hall, 178 p.

McClain, Ernest G The Myth of Invariance (1978)
Boulder CO, Shambhala Publications.

McClain, Ernest G The Pythagorean Plato: prelude to the song itself (1978)
Stony Brook, Nicholas Hays.

McClearn, Gerald E Biology and the Social and Behavioral Sciences (1969)
Social Science Research Council Items 23, 3, Sept pp. 33–37.

McClelland, C A Systems Theory and Human Conflict
In: *E B McNeil (Ed) The Nature of Human Conflict* pp. 250–273.

McClelland, C A Systems and History in International Relations: some perspectives for empirical research and theory (1958)
General Systems Yearbook Society for General Systems Research, 3, p. 221–247.

McClelland, C A Applications of General Systems Theory in International Relations (1961)
In: *J N Rosenan, et al. International Politics and Foreign Policy* Free Press.

McClelland, C A General Systems and the Social Sciences (1962)
ETC 18, pp. 449–68.

McClintock, C G and Messick, D M Empirical Approaches to Game Theory and Bargaining: a bibliography (1966)
General Systems Yearbook Society for General Systems Research, 11, p. 229–238.

McCloskey, J F and Coppinger, J M (Eds) Operations Research for Management v (1956)
Baltimore MD, Johns Hopkins University Press, 563 p.

McCulloch, W S The Brain as a Computing Machine (1949)
Electronic Eng.

McDonough, A M Information Economics and Management Systems (1963)
New York, McGraw Hill Book Company, 321 p.

McHale, John The Future of the Future (1969)
New York, Braziller.

McInerny, Ralph M Studies in Analogy (1968)
Den Haag, Martinus Nijhoff, 138 p. bibl., foot.

McKean, R N Efficiency in Government through Systems Analysis (1958)
New York, John Wiley and Sons, 336 p.

McKeon, Richard World Order in Evolution and Revolution in Arts, Associations, and Sciences (1971)
In: *Gotesky, Rubin and Lazlo, Ervin (Eds) Evolution–Revolution* New York, Gordon and Breach Science Publishers, pp. 109–230.

McLardy, T Unitary Aspects of the Cerebral Diffuse Systems (1952)
Revue Neurologique 87, pp. 195–197.

McLeod, J Simulation: the modeling of ideas and systems with computers (1968)
New York, McGraw Hill Book Company.

McLuhan, M Understanding Media (1964)
Sphere Books.

McLuhan, M and Watson, W From Cliché to Archetype (1970)
New York, Viking Press.

McMillan, C and Gonzalez, R F Systems Analysis: a computer approach to decision models (1965)
Homewood IL, Irwin and Dorsey.

McRae, Robert The Problem of the Unity of the Sciences: Bacon to Kant (1961)
Toronto ON, University of Toronto Press, 148 p.

Mead, M (Ed) Continuities in Cultural Evolution (1964)
New Haven CT, Yale University Press.

Mead, M Crossing Boundaries in Social Science Communication (1969)
Social Science Information, February, pp7–16.

Mead, M Models and Systems Analyses as Metacommunication (1973)
In: *Laszlo, Ervin (Ed) The World System* New York, Braziller.

Mead, Margaret Effects of Anthropological Field Work Models on Interdisciplinary Communication in the Study of National Character (1955)
Journal of Social Issues 11, 2, pp. 3–11, bibl.

Meadows, D H et al The Limits to Growth: a report for the Club of Rome's project on the Predicament of Mankind (1972)
Signet Book, New American Library.

Meadows, D L, et al Dynamics of Growth in a Finite World (1975)
New York, John Wiley and Sons, 638 p.

Mee, J F Management Thought in a Dynamic Economy (1963)
New York, New York University Press, 138p.

Meehan, E Explanation in Social Science: a system paradigm (1968)
New York, Dorsey Press.

Meier, R L Communication and Social Change (1956)
Behavioral Science 1, pp. 43–59.

Meir, A Z General Systems Theory: development and perspectives for medicine and psychiatry (1969)
Archives Gen. Psychiatry 21, September. bibl. p. 310.

Mellor, D H Models and Analogies in Science: Dukem versus Campbell? (1968)
Isis 59, Fall pp. 282–290.

Menzel, Herbert Review of Studies in the Flow of Information among Scientists (1960)
New York, Columbia University Press. 2 vols, mimeo.

Merritt, R L Systems and the Disintegration of Empires (1963)
General Systems Yearbook Society for General Systems Research, 8, p. 91–102.

Mesarovic, M D and Macko, D Foundations for a Scientific Theory of Hierarchical Systems (1969)
In: *Lancelot Law Whyte, et al (Eds) Hierarchical Structures* New York, American Elsevier, pp. 29–50.

Mesarovic, M D et al Theory of Hierarchical, Multilevel Systems (1970)
New York, Academic Press, 294 p. bibl.

Mesarovic, M D; Macko, D and Takahara, Y Structuring of Multilevel Systems (1968)
Düsseldorf, IFAC.

Mesarovic, M D; Sanders, J L and Sprague, C F An Axiomatic Approach to Organizations from a General Systems Viewpoint (1964)
In: *New Perspectives in Organization Research, Cooper, W W, Leavitt, H J, Shelly II, N M (Eds)* New York–London–Sydney, pp. 493–512.

Mesarovic, Mihajlo D Control of a Multivariable System (1960)
Cambridge MA, MIT Press.

Mesarovic, Mihajlo D Systems Research and Design: view on general systems theory (1961)
New York, John Wiley and Sons.

Mesarovic, Mihajlo D A General Systems Approach to Organization Theory (1962)
Los Angeles CA, Systems Research Center. Report SRC 2–A–62–2.

Mesarovic, Mihajlo D A Unified Theory of Learning and Information (1963)
Evanston IL, Northwestern University Press.

Mesarovic, Mihajlo D Views on General Systems Theory: proceedings of the Second Systems Symposium at Case Institute of Technology (1964)
New York, John Wiley and Sons.

Mesarovic, Mihajlo D Foundations for a General Systems Theory (1964)
In: *Views on General Systems Theory (second Systems Symposium)* Cleveland, Case Institute of Technology.

Mesarovic, Mihajlo D Self–Organizing Control Systems (1964)
In: *IEEE Transactions Applied Ind* 83, 74, September.

Mesarovic, Mihajlo D Multilevel Concept for Systems Engineering (1965)
In: *Proceedings Systems Engineering Conference* Chicago IL.

Mesarovic, Mihajlo D A Conceptual Framework for the Studies of Multilevel Multigoal Systems (1966)
Los Angeles CA, Systems Research Center. Report SRC101–A–66–43.

Mesarovic, Mihajlo D General Systems Theory and its Mathematical Formulation (1967)
Massachussetts MA, IEEE Systems Science and Cybernetics.

Mesarovic, Mihajlo D Systems Theory and Biology – View of a theoretician (1968)
In: *Systems Theory and Biology, M D Mesarovic (Ed)* pp. 59–87. New York, Springer–Verlag.

Mesarovic, Mihajlo D Auxiliary Functions and Constructive Specifications of General Systems (1968)
Mathematical Systems Theory Journal 2, 3.

Mesarovic, Mihajlo D Systems Concepts (paper prepared for the Unesco project 'Scientific Thought', 30 November 1969, revised 29 December 1969) (1969)

Methlie, Leif B and Spargue, Ralph H (Eds) Knowledge Representation for Decision Support Systems (1986)
Amsterdam, Elsevier Science Publishing, 268 p.
ISBN 0–444–87739–8.

Mey, Marc de The Cognitive Paradigm (1984)
Dortrecht, Kluwer. ISBN 90–277–1600–5.

Meyer, A Mechanische und Organische Metaphorik, Politischer Philosophie
In: *Archiv für Begrisffsgeschichte* 1969, 13, 2, pp. 128–199.

Meyer, A Die Idee des Holismus (1935)
Scientia 58, p. 18–29, Jl 1935 (ger.) 58:sup. 9–19, Jl 1935 (tr.).

Meyer, J R Systems Analysis and Simulation Models (1970)
Washington DC, Brookings Institution. Vol 2.

Meynaud, Jean The Interdisciplinary Approach (1961)
In: *Yearbook 1960–61, Geneva, Association of Institutes of European Studies* pp. 25–31 (in an issue on the European University concept).

Michael, Donald N Cybernation: the silent conquest (1962)
Santa Barbara, Center for the Study of Democratic Institutions.

Michael, Donald N The Unprepared Society: planning for a precarious future (1968)
New York, Harper Colophon Books.

Michael, Donald N On the Requirement for more Boundary–Spanning (1973)
In: *On Learning to Plan and Planning to Learn* Washington DC, Jossey–Bass, pp. 237–254.

Michaelis, Anthony R (Ed) Interdisciplinary Science Reviews Essay
Ann Arbor MI, Books on Demand. Annual.

Miles, V W Principles and Experiments for Courses of Integrated Science (1947)
Ph. D. Thesis. University of Michigan.

Milgram, Stanley Interdisciplinary Thinking and the Small World Problem (1969)
In: *M Sherif and C Sherif (Ed) Interdisciplinary Relationships in the Social Sciences* Chicago, Aldine, pp. 103–120.

Milhorn, H T Jr The Application of Control Theory to Physiological Systems (1966)
Philadelphia PA, WB Saunders, 386 p.

Millendorfdr, H Input–Output Relations between Social Systems (1970)
In: *Attinger, E O (Ed) Global Systems Dynamics* New York, Wiley Interscience, pp. 161–170.

Miller, D W and Starr, M K Executive Decisions and Operations Research (1960)
Englewood Cliffs NJ, Prentice Hall, 446 p.

Miller, E J and Rice, A K Systems of Organization: the control of task and sentient boundaries (1967)
Barnes and Noble.

Miller, G A; Galanter, E and Pribram, K Plans and the Structure of Behavior (1960)
New York, Holt, Rinehart and Winston.

Miller, Gary E The Meaning of General Education: the emergence of a curriculum paradigm (1988)
New York, Columbia University, 216 p.
ISBN 0–8077–2894–2.

Miller, George A The Magical Number Seven, Plus or Minus Two: some limits on our capacity for processing information (1956)
Psychological Review 63, 2, March pp. 81–96 (republished in his *The Psychology of Communication; seven essays* New York, Bsic, 1907).

Miller, James G Adjusting to Overloads of Information (1904)
Disorders of Communication Vol. 42, Research Publications, Association for Research in Nervous and Mental Diseases, 1964.

Miller, James G Towards a General Theory for the Behavioral Sciences (1915)
American Psychologist 10, 1955, pp. 513–531.

Miller, James G Psychological Aspects of Communication Overloads (1964)
In: *International Psychiatry Clinics: Communication in Clinical Practice* pp. 201–224, Waggener and Casek (Eds) Boston, Little, Brown.

Miller, James G Living Systems: basic concepts, structure and process, cross–level hypotheses (1965)
Behavioral Science 10, 193–237, 337–79, 380–441.

Miller, James G Toward a General (Systems) Theory for the Behavioral Sciences (1969)
In: *Litterer, Joseph A (Ed) Organizations; systems control and adaptation* London, Wiley, 2nd ed, vol. 2, pp. 77–87.

Miller, James G Living Systems (1978)
New York, McGraw Hill Book Company.

Miller, Linda B World Order and Local Disorder: the United Nations and internal conflicts (1967)
Princeton NJ, Princeton University Press, 235 p. bibl. (pp. 217–227).

Millet, Louis and Varin d'Ainvelle, M Le Structuralisme (1970)
Paris, Editions Universitaires, 135p. bibl. (pp. 129–135).

Milsum, John H Biological Control Systems Analysis (1966)
New York, McGraw Hill Book Company.

Milsum, John H The Technosphere, the Biosphere, the Sociosphere (1968)
IEEE Spectrum 5, 6, pp. 76–82 (reprinted in *Ekistics* 27, 160, March 1969).

Milsum, John H Positive Feedback: a general systems approach to positive/negative feedback and mutual causality (1968)
Oxford, Pergamon Books, 169 p. bibl. foot.

Milsum, John H Technosphere, Biosphere, and Sociosphere: an approach to their systems modeling and optimization (1968)
General Systems Yearbook Society for General Systems Research, 13, p. 37–48.

Mol, J J Wholeness and Breakdown: a model for the interpretation of nature and society (1978)
Madras, University of Madras, 128 p.

Monane, J H A Sociology of Human Systems (1967)
New York, Appleton–Century–Crofts, 223 p.

Monlyn, Adrian C Structure, Function and Purpose: an inquiry into the concepts and methods of biology from the viewpoint of time (1957)
New York, Liberal Arts, 198 p. bibl.

Monod, Jacques Chance and Necessity, an essay on the natural philosophy of modern biology (1971)
New York, Alfred A Knopf.

Moore, O K and Lewis, D J Purpose and Learning Theory (1953)
Psychological Review 60, pp. 149–56.

Moore, Wilbert E Global Sociology: the world as a singular system (1966)
American Journal of Sociology March.

Moore, Wilbert E Order and Change: essays in comparative sociology (1967)
New York, John Wiley and Sons, 313 p. bibl.

Morin, E On the Definition of Complexity
In: *The Science and Praxis of Complexity* 1985, Tokyo, United Nations University, pp. 62–68.

Morin, Edgar Pour Sortir du XXe Siécle (1981)
Paris, Fernand Nathan.

Morin, Edgar and Piattelli–Palmarini, Massimo L'Unité de l'Homme comme Fondement et Approche Interdisciplinaire
In: *UNESCO – Interdisciplinarité et Sciences Humaines* 1983, Paris, UNESCO, pp. 191–218.

Morin, Edgar et Moscovici, Serge Remarques indisciplinaires et transdisciplinaires. (1974)
In: *Centre Royaumont pour une Science de l'Homme. L'Université de l'Homme* Paris, Seuil.

Morris, W T Management Science in Action (1963)
Homewood IL, Irwin and Dorsey, 308 p.

Morrison, Philip The Modularity of Knowing (1966)
In: *G Kepes (Ed) Module, Proportion, Symmetry, Rhythm* New York, Braziller, pp. 1–19.

Morse, P (Ed) Operations Research for Public Systems (1967)
Cambridge MA, MIT Press, 256 p.

Morse, P M and Kimball, G E Methods of Operations Research (1951)
New York, Wiley and Sons, 158 p.

Mosher, F C Research in Public Administration: some notes and suggestions; the problem of interdisciplinary communication (1956)
Public Administration Review 16, Summer 1956 pp. 174–76.

Moulin, L Medieval Origins of Transnationality
In: *From International to Transnational* 1980, Brussels, Union of International Associations, pp. 270–276.

Moulyn, A C Structure, Function, and Purpose (1957)
New York, Liberal Arts.

Mowrer, O H Ego Psychology, Cybernetics, and Learning Theory (1954)
In: *Learning Theory and Clinical Research* New York, Wiley pp. 81–90.

Mowrer, O H Learning Theory and the Symbolic Processes (1960)
New York, John Wiley and Sons.

Mukerjes, Radhakamal The Community of Communities (1966)
Bombay, Manaktalas, 155 p. bibl. (pp141–46).

Muller, A (Ed) Encyclopaedia of Cybernetics (1968)
Barnes and Noble.

Müller–Markus, Siegfried Science and Faith (1972)
In: *Integrative Principles of Modern Thought* New York Gordon and Breach Science publishers, 1972 pp. 452–69.

Muller, Robert New Genesis: shaping a global spirituality (1979)
Garden City NY, Doubleday.

Mumford, Lewis Technics and Civilization (1934)

Mumford, Lewis The Unified Approach to Knowledge and Life (1941)
In: *The University and the Future of America* Stanford University.

Mumford, Lewis Medieval Synthesis (1944)
In: *Lewis Mumford, The Condition of Man* New York, Harcourt Brace, pp. 108–51.

Mumford, Lewis The Myth of the Machine (1967)
London, Martin Secker and Warburgh.

Murchie, Guy Music of the Spheres (1967)
New York, Dover Publications. 2 vols.

Murphy, G Toward a Field Theory of Communication (1961)
Journal of Communication 11, pp. 196–201.

Musashi, Miyamoto The Five Rings (Gorin No Sho): the real art of Japanese management (1982)
New York, Bantam.

Muses, C Time, Experience and Dimensionality: an introduction to higher kinds of number (1967)
In: *Interdisciplinary Perspectives of Time (Annals of the New York Academy of Sciences)* 138, Art 2, 1967, p.646–660.

Muses, C and Young, A M (Eds) Consciousness and Reality (1974)
New York, Discus Book.

Mushakoji, Kinhide In Search of a Theory of Cycles: for a transfinite mathematical treatment of recurrence in social and natural processes
Paper prepared as a contribution to the development of the UNU project on the economic aspects of human development.

Nadel, S F Social Control and Self–Regulation (1953)
Social Forces 31, pp. 265–73.

Nadel, S F The Theory of Social Structure (1957)
New York, Free Press.

Nadler, G Work Systems Design: the ideals concept (1967)
Homewood IL, Irwin and Dorsey, 183 p.

Nagel, E and Newman, J R Goedel's Proof (1958)
New York, Allied Publications.

Nagel, Ernest On the Statement 'The whole is more than the Sum of its Parts'. (1915)
In: *Lazarsfeld, P F and Rosenberg, M (Eds) The Language of Social Research* Glencoe, Free Press, 1955, pp. 519–27.

Nagel, Ernest The Concept Levels in Social Theory (1959)
In: *Llewellyn Gross (Ed) Symposium on sociological theory* Row, Peterson 1959 pp. 167–95.

Nagel, Ernest The Structure of Science (1961)
London, Routledge.

Nagel, Ernest Das Ganze ist Mehr als die Summe seiner Teile (1965)
In: *Topitsch, Ernst (Ed) Logik der Sozialwissenschaften* pp. 225–235, Köln–Berlin.

Nagel, Stuart (Ed) Interdisciplinary Approaches to Policy Studies (1973)
Croton–on–Hudson NY, Policy Studies Organization.
ISBN 0–918592–04–6.

Nagel, Stuart S (Ed) Encyclopedia of Policy Studies (1983)
New York, Marcel Dekker, 914 p. Public Administration and Public Policy: 13. ISBN 0–8247–1199–8.

Nair, Balakrishna N Interdisciplinary Approaches to our Regional Agricultural Development (1969)
Interdiscipline 6, 2, Summer pp. 125–42.

Nalimov, V V The Necessity to Change the Face of Science
Moscow, Moscow State University, 30 p.

Nalimov, V V Semantic Vacuum as the Analogue of the Physical Vacuum: comparative ontology of two realities, physical and psychic
In: *Realms of the Unconscious: the enchanted frontier* 1982, Philadelphia, 1S1 Press, pp. 75–94.

Nalimov, V V Realms of the Unconscious: the enchanted frontier (1982)
Philadelphia PA, Institute of Scientific Information Press.

Naroll, R; O'Leary, T J, and Siegelman, L A Hologeistic Bibliography
Pittsurgh PA, International Studies Association. Working Paper 23.

Naroll, R S and Bertalanffy, L The Principle of Allometry in Biology and the Social Sciences (1916)
General Systems Yearbook Society for General Systems Research, 1956, 1, p. 76–88.

National Science Foundation Important Notice to Presidents of Universities and Colleges and Directors of Non–Profit Research Institutes; subject: interdisciplinary research relevant to problems of our society (1969)
Washington December.

National Society for the Study of Education, Committee on the Integration of Educational Experiences, USA Integration of Educational Experiences (edited by Nelson B Henry)
Yearbook (of the Society) 57, 1958, part 3, bibl. (pp. 267–73).

Needham, Joseph Integrative Levels: a revaluation of the idea of progress (1937)
Oxford, Clarendon Press, 59 p.

Needham, Joseph Time, The Refreshing River (1943)
London, Allen and Unwin.

Needham, Joseph Order and Life (1968)
Cambridge MA, MIT Press, 175 p. bibl. 1936 copyright.

Needleman, Jacob A Sense of the Cosmos (1976)
EP Dutton.

Nemes, T Cybernetic Machines (1970)
New York, Gordon and Breach.

Nemhauser, G L Introduction to Dynamic Programming (1966)
New York, John Wiley and Sons, 256 p.

Nett, R Conformity – Deviation and the Social Control Concept (1953)
Ethics 64, 1953, p. 38–45.

Netzer, Lanore A et al (Comp) Interdisciplinary Foundations of Supervision (1970)
Boston MA, Allyn and Bacon, 400 p. bibl.

Neurath, O et Al (Eds) Foundations of the Unity of Science: toward an International Encyclopedia of Unified Science (1955)
Chicago IL, University of Chicago Press. Vol 1, 1955 and Vol 2, 1970.

Neuschel, R F Management by System (1960)
New York, McGraw Hill Book Company, 359 p.

Newell, A and Simon, H A Computer Simulation of Human Thinking and Problem Solving (1961)
Santa Monica CA, Rand Corporation.

Newell, A; Shaw, J C and Simon H A Elements of a Theory of Human Problem Solving (1958)
Psychedelic Review 65, 1958 pp. 151–66.

Newman, W H Administrative Action: the techniques of organization and management (1963)
Englewood Cliffs NJ, Prentice Hall, 486 p.

Newman, W H; Summer, C E Jr and Warren, E K The Process of Management: concepts, behavior, practice (1967)
Englewood Cliffs NJ, Prentice Hall, 640 p.

Nichols, R E The Source of Mental Cybernetics (1970)
Parker.

Nicolas, Antonio de Meditations through the Rg Veda (1978)
Boulder CO, Shambhala Publications.

Nicolis, G and Prigogine, I Self–Organization in Nonequilibrium Systems: from dissipative structures to order through fluctuations (1977)
New York, Wiley–Interscience.

Nicolle, Jacques Sur la symétrie
Diogene, 12, p. 103 and following.

Nokes, P Feedback as an Explanatory Device in the Study of Certain Interpersonal and Institutional processes (1961)
Human Relations 14, 1961 pp. 381–87.

Noppen, Jean–Pierre Van, et al (Comp) Metaphor: a bibiography of post–1970 publications (1985)
Amsterdam, John Benjamins. Amsterdam Studies in the Theory and History of Linguistic Science.

Northedge, F S International Intellectual Cooperation Within the League of Nations: its conceptual basis and lessons for the present
London, University of London Press, 730 p. bibl. (pp. 713–30).

Northrop, F C ald Livilgston, H H (Eds) Cross–Cultural Understanding: epistemology in anthropology (1964)
New York, Harper and Row, 396 p. bibl.

Northrop, F S C Toward a General Theory of the Arts
In: *The Journal Value Inquiry* vol 1, No 2.

Northrop, F S C The Meeting of East and West (1947)
New York, Macmillan.

Northrop, F S C The Logic of the Sciences and the Humanities (1947)
New York, Macmillan.

Northrop, F S C (Ed) Ideological Differences in World Order: studies in the philosophy and science of the world's cultures (1949)
New Haven CT, Yale University Press

Northrop, F S C Ethics and the Integration of Natural Knowledge (1950)
In: *The Nature of Concepts, their interrelation and role in social structure* (Proceedings of the Stillwater Conference, Stillwater 1950).

Notterman, J M and Trumbull, R Note on self–regulating systems and stress (1959)
Behavioral Science 4, pp. 324–27.

Novikoff, Alex B The Concept of Integrative Levels and Biology (1905)
Science vol 101, pp. 209–15, 1945.

Novikov, S P Integrable Systems: selected papers (1981)
Cambridge, Cambridge University Press, 272 p. London Mathematical Society Lecture Note Series: 60.
ISBN 0–521–28527–5.

Nudler, Oscar Notes for an Epistemology of Holism (1978)
San Carlos de Bariloche, Synergic Developments, 16 p. Paper prepared for the meeting of the Goals, Processes and Indicators of Development Project of the United Nations University (Geneva, Oct 2–7, 1978).

Oberholtzer, Edison E An Integrated Curriculum in Practice
New York, AMS Press. repr of 1937. Columbia University Teachers College Contributions to Education: 694.
ISBN 0–404–55694–9.

OECD Multidisciplinary Aspects of Regional Development (Meeting of directors of development training and research institutes, Montpellier 1967) Paris OECD, 1969, 272p (1969)
bibl. foot.

OECD Systems Analysis for Educational Planning: selected annotated bibliography (1969)
Paris, OECD, 219 p.

Oestreicher, H L and Moore, D L (Eds) Cybernetic Problems in Bionics (1968)
New York, Gordon and Breach.

Ofstad, Harold The Functions of Moral Philosophy: a plea for an integration of philosophical analysis and empirical research (1958)
Inquiry 1, 1/2, pp. 35–72, bibl. (pp. 65–72) (concerned with meta–ethics).

Oliver, W Donald Theory of Order (1951)

Oliver, W Donald Order and Personality (1968)
In: *Paul G Kuntz (Ed) The Concept of Order* Seattle, University of Washington Press, pp. 309–21.

Olroyd, D Arch of knowledge (1983)
Kensington NSW, New South Wales University Press, 320 p.
ISBN 0–86840–049–1.

Opler, M E Cultural Dynamics and Evolutionary Theory (1965)
In: *Barringer, Blanksten (Eds) Social Change in Developing Areas* Cambridge MA, Schenkman, pp. 68–96.

Oppenheimer, R Analogy in Science (1956)
American Psychologist 11, pp. 127–35.

Oppenheimer, R The Tree of Knowledge (1958)
Harper's 217, p. 55–57.

Optner, Stanford L Systems Analysis for Business and Industrial Problem Solving (1965)
Englewood Cliffs NJ, Prentice Hall, 116 p.

Organisation for Economic Cooperation and Development, Centre for Educational Research and Innovation Interdisciplinarity: problems of teaching and research in universities ,based on the results of a seminar, Nice 1970 (1972)
Paris, OECD, 334 p. bibl. (pp. 293–98). French Ed (1972) L'Interdisciplinarité: Problèmes d'enseignement et de recherche dans les universitées.

Osgood, C E and Sebeck, T A (Eds) Psycholinguistics, a Survey of Theory and Research Problems (1954)
International Journal of American Linguistics Memoir 10.

Ostow, Mortimer The Entropy Concept and Psychic Function (1951)
American Scientist 39, pp. 140–44.

Ouspensky, P D A New Model of the Universe
London, Routledge and Kegan Paul.

Ouspensky, P D The Fourth Way (1965)
New York, Alfred A Knopf.

Owen, Philip J The Contribution of Hierarchical Information Structures to Cybernetic Ontology (1970)
International Journal of Systems Science vol 1.

Ozbekhan, Hasan Toward a General Theory of Planning (1969)
In: *Jantsch, Erich (Ed) Perspectives of Planning* Paris, OECD.

Paci, E Vico, Structuralism, and the Phenomenological Encyclopedia of the Sciences
In: *G Tagliacozzo and H V White, Gimbattista Vico* pp. 499–515.

Pacini, Dante Sinteses e Hipóteses do ser Humano: sentidos, experiencia e contecimento (1967)
Rio de Janeiro, Libraria Eldorado, 178 p.

Padalino, Francesco Integralogy (a collection of short notes) (1927)
New York, Integrale, 126 p.

Pagels, Heinz The Cosmic Code: quantum physics as the language of nature (1982)
New York, Bantam Books.

Paine, Frank T and Naumes, William Strategy and Policy Formation: an integrative approach (1974)
Philadelphia PA, WB Saunders, 296 p.

Paine, R The Political Use of Metaphor and Metonym
In: *Politically Speaking: cross-cultural studies of rhetoric* 1981, Philadelphia PA, Institute for the Study of Human Issues, pp. 87–200.

Palade, George E The Organization of Living Matter (1963)
In: *The Scientific Endeavor* New York, Rockefeller Institute, pp.179–204.

Palmade, Guy L'unitéé des sciences humaines (1961)
Paris, Dunod, 357 p. bibl. foot. (Organisation et sciences humaines, 1).

Palmer, Martin Genesis or Nemesis: belief, meaning and ecology (1988)
London, Dryad Press, 160 p. ISBN 0–8521–9780–2.

Panofsky, E Meaning in the Visual Arts (1955)
Anchor Doubleday.

Pantin, C F A The Relations Between the Sciences (1968)
New York, Cambridge University Press, 206 p. (the philosophy of the sciences, and the relations, or want of relations, between different departments of knowledge).

Parker, H T Tentative Reflection on the Interdisciplinary Approach and the Review Historian (1957)
South Atlantic Quarterly 56, January pp. 105–11.

Parks, John H Biopsychosynthesis (1973)
New York, Psychosynthesis Research Foundation, 24 p. P R F No 32.

Parr, J B and Denike, K G Theoretical Problems in Central Place Analysis (1970)
Econ. Geog. 46, October, p. 568–86 bibl.

Parsegian, V L What makes Studies Interdisciplinary? (1972)
The Journal of College Science Teaching February.

Parsons, C Ontology and Mathematics (1971)
Philosophical Review 80, April pp. 151–76.

Parsons, Talcott The Present Position and Prospects of Systematic Theory in Sociology (1945)
In: *G Gurvitch and W E Moore (Eds) Twentieth Century Sociology* Philosophical Library.

Parsons, Talcott The Social System (1957)
New York, Free Press.

Parsons, Talcott Structure and Process in Modern Societies (1960)
Glencoe, Free Press.

Parsons, Talcott Evolutionary Universals in Society (1964)
American Sociological Review 29, June pp. 339–57.

Parsons, Talcott Unity and Diversity in the Modern Intellectual Disciplines: the role of the social sciences (1965)
Daedalus Winter p. 39–65.

Parsons, Talcott The System of Modern Societies (1971)
Englewood Cliffs NJ, Prentice Hall, 152 p. bibl. (pp. 144–46).

Parsons, Talcott, et al (Eds) Theories of Society (1961)
New York, Free Press.

Pascarella, Perry The New Achievers (1984)
Free Press.

Pask, G An Approach to Cybernetics (1968)
London, Hutchinson, 128 p.

Pask, Gordon Conservation, Cognition and Learning (1975)
Amsterdam, Elsevier.

Pattee, Howard H Quantum Mechanics, Heredity, and the Origin of life (1927)
Journal of Theoretical Biology 17, 1967, p. 410–20.

Pattee, Howard H The Problem of Biological Hierarchy (1969)
In: *C H Waddington (Ed) Towards a Theoretical Biology vol. III* Edinburgh, Edinburgh University Press

Pattee, Howard H Physical Conditions for Primitive Functional Hierarchies (1969)
In: *Lancelot Law Whyte, et al (Eds) Hierarchical Structures* New York, American Elsevier 1969 pp. 161–78.

Pattee, Howard H The Evolution of Self-Simplifying Systems (1972)
In: *Ervin Laszlo, The Relevance of General Systems Theory-* New York, Braziller 213p. (pp31–41).

Pattee, Howard H Hierarchy Theory: the challenge of complex systems (1973)
New York, Braziller.

Patten, B C An Introduction to the Cybernetics of the Eco-System: the trophic-dynamic aspect (1959)
Ecology 40, 1959, pp. 221–31.

Pavans de Ceccatty, Max The Scandal of Integration (1973)
In: *Gray, William and Rizzo, Nicholas D (Eds) Unity through diversity* New York. Gordon and Breach Science Publishers pp. 125–29.

Pearce, J C The Crack in the Cosmic Egg (1971)
New York, Julian Press.

Pearson, J D Decomposition, Coordination, and Multi-Level Systems (1966)
IEEE Trans. on Systems Science and Cybernetics SSC 2, pp. 36–40.

Pearson, Karl The Grammar of Science (1957)
Meridian 1957 (reprint).

Peitgen, H O and Richter, P H The Beauty of Fractals: images of complex dynamical systems (1986)
Berlin, Springer Verlag. Intro on the 'Frontiers of Chaos'.

Pelletier, Kenneth A New Age: problems and potentials (1985)
Robert Briggs Associates.

Pemberton, J Michael and Prentice, Ann (Eds) Information Science: the interdisciplinary context (1989)
New York, Neal-Schuman Publishers, 275 p.

Penrose, L B Self-Reproducing Machines (1959)
Sci. Am. 200, 1959 pp. 105–14.

Pepper, S World Hypotheses (1942)
Berkeley CA, University of California Press.

Pepper, S C World Hypotheses: a study in evidence (1942)
Berkeley CA, University of California Press. (reviews different comprehensive theories).

Perger, Anton Analogien in Unserem Weltbild: entwurf einer allgemeinen Formenlehre (1963)
Meisenheim am Glan, A Hain, 183 p. bibl. (pp. 182–83).

Perlberg, A and Shaal, G Interdisciplinary Approach to Manpower Planning and Development (1969)
International Labour Review 99, April pp. 363–80.

Perrin, S G Metaphor to Mythology: experience as a resonant synthesis of meaning and being (1982)
Boston MA, Boston University.

Perschon, J (Ed) Disciplines and Techniques of Systems Control (1965)
Needham Heights MA, Ginn Press.

Pervyi, Meditsinskii Institut, Moskva Integrativnaia Deiatel 'Nost' Moga: International conference on integrative activity of the brail, Moscow 1967 (1967)
Moscow, 122 p.

Peter, Karl Sorokin and Von Bertalanffy: a convergence of views (1973)
In: *Unity through Diversity, William and Rixzo, Nicholas D (Eds)* New York, Gordon and Breach Science Publishers. 1973 pp. 131–40.

Peterfreund, E and Schwartz, J T Information, Systems, and Psychoanalysis: an evolutionary biological approach to psychoanalytic theory (Psychological Issues, 7, 1-2, monograph 25–26) International Universities Press 1971, bibl (1971)
pp. 381–88.

Petrella, R and Schaff, A A European Experiment in Cooperation in the Social Sciences (1974)
Vienna.

Pfeiffer, John New Look at Education: systems analysis in our schools and colleges (1968)
New York, Odyssey Press, 162p.

Pham-thi-Tu La coopération intellectuelle sous la Société des Nations (1962)
Geneva, E Droz, 268 p. bibl. (pp. 261–66).

Philips, D C Holistic Thought in Social Sciences (1976)
London, Macmillan.

Phillips, D C Organicism in the Late Nineteenth and Early Twentieth Centuries (1970)
Journal of the History of Ideas 31, July pp. 413–432.

Piaget, J La Psychologie, les Relations Interdisciplinaires et le Système des Sciences: XVIIIe Congrès International des Psychologues, Moscou, 1966

Piaget, J Méthodologie des Relations Interdisciplinaires
In: *De la Méthode* 1972, Brussels, pp. 85–95.

Piaget, Jean La Recherche Orientée Multidisciplinaire
In: *Revue Internationale des Sciences Sociales* 1968, Paris, UNESCO, XX, 2.

Piaget, Jean Classification of Disciplines and Interdisciplinary Connections (1964)
International Social Science Journal, 16, 4, p. 553–570.

Piaget, Jean La situation des sciences de l'homme dans le système des sciences (1970)
In: *Unesco, Tendances Principales de la Recherche dans les Sciences Sociales et Humaines* pp. 1–68. Paris, Mouton/ Unesco.

Piaget, Jean Genetic Epistemology (1970)
New York, Columbia University Press.

Piaget, Jean Problèmes généraux de la recherche interdisciplinaire et mécanismes communs (1970)
In: *Unesco, Tendances Principales de la Recherche dans les Sciences Sociales et Humaines* pp. 559–628. Paris, Mouton/ Unesco.

Piaget, Jean Structuralism (1971)
London, Routledge and Kegan Paul.

Piaget, Jean The Epistemology of Interdisciplinary Relationships (1972)
In: *OECD, Centre for Educational Research and Innovation, Interdisciplinarity* pp. 127–30. Paris, OECD.

Piaget, Jean Les Formes lmentaires de la Dialectique (1980)
Paris, Gallimard.

Pike, K L Language in Relation to a Unified Theory of the Structure of Human Behavior (1954)
Glendale CA, Summer Institute of Linguistics.

Pines, David Emerging Syntheses in Science (1987)
Reading MA, Addison-Wesley, 300 p.
ISBN 0–201–15677–6.

Pitts, W and McCulloch, W C How We Know Universals, the Perception of Auditory and Visual Forms (1947)
Bulletin Mathematical Biophysics 9, pp. 127–47.

Platt, John R Properties of Large Molecules that go beyond the Properties of their Chemical Sub-Groups (1961)
Journal of Theoretical Biology 1, p. 342, 58.

Platt, John R Theorems on Boundaries in Hierarchical Systems (1969)
In: *Lancelot Law Whyte, et al (Eds) Hierarchical Structures* New York, American Elsevier, pp. 201–214.

Platt, John R Commentary on the Limits of Reductionism: part I (1969)
Journal of History Biology 2, p. 140–147.

Platt, John R Hierarchical Restructuring (1970)
General Systems Yearbook Society for General Systems Research, 15, p. 49–54.

Platt, John R Beliefs that Link Man Together (Paper at the Walgreen Conference, Ann Arbor, 1973) (1973)

Plummer, L Gordon The Mathematics of the Cosmic Mind: a study in mathematical symbolism (1970)
London, Theosophical Publishing House, 217 p.

Podalanski, J Unified Field Theory in Six Dimensions (1950)
Proceedings of the Royal Society 201A, pp. 234–261.

Point The Whole Earth Catalog (1975)
Sausalito CA, Point, 456 p. 16th ed.

Polanyi, M Life's Irreducible Structure
Science 160, p. 1303–1312.

Polanyi, Michael Personal Knowledge (1958)
University of Chicago Press.

Polanyi, Michael The Tacit Dimension (1966)
Garden City NY, Doubleday.

Pollak, Otto Integrating Sociological and Psychoanalytic Concepts: an exploration in child psychotherapy (1956)
New York, Russell Sage Foundation, bibl.

Polya, G Induction and Analogy in Mathematics (1954)
Princeton NJ, Princeton University Press.

Polytechnic Institute Proceedings of the Symposium on System Theory (1965)
Brooklyn, Polytechnic Institute.

Pomerance, L Need for Guidelines in Interdisciplinary Meetings (1971)
American Journal of Archeology 75, October pp. 428–431.

Popkewitz, Thomas Paradigm and Ideology in Educational Research: the social function of the intellectual (1984)
New York, Taylor and Francis, 208 p.
ISBN 0–905273–98–2.

Popper, Karl R The Logic of Scientific Discovery (1959)
New York, Basic Books.

Popper, Karl R Conjectures and Refutations: the growth of scientific knowledge (1965)
London, Routledge and Kegan Paul, 417 p. 2nd ed.

Popper, Karl R The Poverty of Historicism (1966)
New York, Basic Books.

Postlethwaite, S N et al An Integrated Experience Approach to Learning with Emphasis on Independent study (1965)
Minneapolis MN, Burgess Pub Co, 114 p. bibl. (pp. 79–80).

Poulet, Georges The Metamorphoses of the Circle (1966)
Baltimore MD, Johns Hopkins University Press, 400 p. bibl. (translated from French) (apprehend and define the mental structures of a series of writers).

Powellson, Jack Holistic Economics and Social Protest (1983)
Wallingford PA, Pendle Hill Publications.
ISBN 0–87574–252–1.

Powers, W T; Clark, R K and McFarland, R I A General Feedback Theory of Human Behavior (1960)
Perceptual and Motor Skills 11, pp. 71–88.

Prat, Henri La métamorphose explosive de l'humanité (1960)
Paris, Editions Planète, 253 p.

Prat, Henri Le Champ Unitaire en Biologie (1964)
Paris, Presses Universitaires de France, 154 p. bibl. (pp. 149–152).

Preiffer, Carl H The Development and Implementation of a Four-Year Unified Concept-Centered Science Curriculum for Secondary Schools, Final Report
Washington DC. ERIC Document Ed. 054 965.

Presthus, R The Organizational Society: an analysis and a theory (1962)
New York, Vintage Books, 337 p.

Preston, R E Structure of Central Place Systems (1971)
Econ. Geog. 47, April p. 136–155.

Pribam, K H Complexity and Causality
In: *The Science and Praxis of Complexity* 1985, Tokyo, United Nations University, pp. 119–132.

Price, J L Organizational Effectiveness: an inventory of propositions (1968)
Homewood IL, Irwin and Dorsey, 414 p.

Prigogine, Ilya New Perspectives on Complexity
In: *The Science and Praxis of Complexity* 1985, Tokyo, United Nations University, pp. 107–118.

Prigogine, Ilya Stability, Fluctuations, and Complexity (1974)
Brussels, Manuscript.

Prigogine, Ilya Order through fluctuation: self-organization and social system (1976)
In: *Erich Jantsch (Ed) Evolution and Consciousness; human systems in transition* Reading, Addision-Wesley, 1976.

Prigogine, Ilya From Being to Becoming: time and complexity in the physical sciences (1980)
San Francisco, W H Freeman.

Prigogine, Ilya and Stengers, Isabelle Order Out of Chaos: man's new dialogue with nature (1984)
Toronto, Bantam Books, 349 p. bibl. Foreword by Alvin Toffler. ISBN 0-553-34082-4.

Prigogine, Ilya; Nicolis, Gregoire and Babloyantz, Agnès Thermodynamics of Evolution (1972)
Physics Today Vol 24, Nos 11 and 12.

Pringle, J W S On the Parallel between Learning and Evolution (1951)
Behavioral Science 3, pp. 174-215.

Probst, G J B and Ulrich, H (Eds) Self-Organizational and Management of Social Systems (1984)
New York, Springer-Verlag.

Prosser, C Ladd Levels of Biological Organization and their Physiological Significance (1965)
In: *S A Moore (Ed) Ideas in Modern Biology* pp. 359-90. New York, Doubleday.

Pugh, Anthony An Introduction to Tensegrity (1976)
Los Angeles, University of California Press.

Puligandla, R The Concept of Evolution and Revolution (1971)
In: *Gotesky, Rubin and Lazlo, Ervin (Eds) Evolution-Revolution* New York, Gordon and Breach Science Publishers, pp. 41-67.

Puntel, Lourencino B Analogie und Geschichtlichkeit (1969)
Freiburg, Herder. bibl. (pp. 558-566). vol 1.

Purcell, Edward Parts and Wholes in Physics (1963)
In: *Parts and Wholes, D S Lerner (Ed)* pp. 11-39. New York, Free Press.

Putnam, A O; Barlow, E R and Stilian, G N Unified Operations Management: a practical approach to the total systems concept (1963)
New York, McGraw Hill Book Company.

Pyles, Donald A Dictionary of Synergetics
Parkersburg WV, Synergetics Press.

Quade, E S and Boucher, W I (Eds) Systems Analysis and Policy Planning: applications in defense (1968)
New York, American Elsevier.

Quade, Edward S Some Problems Associated with Systems Analysis (1966)
Santa Monica CA, Rand Corporation, 21 p. bibl. (Rand Corporation P-3391).

Quastler, Henry Information Theory in Psychology (1955)
New York, Free Press.

Quastler, Henry Emergence of Biological Organization (1964)
New Haven CT, Yale University Press. bibl. (pp. 67-80).

Quastler, Henry General Principles of Systems Analysis (1965)
In: *Waterman, T H and Morowitz, H J (Eds) Theoretical and Mathematical Biology* New York, Blaisdell, pp. 313-33.

Quigley, Carroll The Evolution of Civilizations (1961)
New York, Macmillan.

Quine, Willard V O Ontological Relativity and Other Essays (1969)
New York, Columbia University Press, 165 p. bibl. foot.

Quinn, J A Human Ecology (1950)
Englewood Cliffs NJ, Prentice Hall.

Rabow, Gerald The Era of the System: How the systems approach can help solve society's problems (1969)
New York, Philosophical Library, 153 p. bibl. (pp. 147-149).

Radnitzky, Gerard Ways of Looking at Science: a synoptic study of contemporary schools of 'metascience' (1969)
General Systems Yearbook Society for General Systems Research, 14, p 187-191.

Radnitzky, Gerard Contemporary Schools of Metascience (1970)
New York, Humanities Press. 2nd rev ed. 2 vols.

Rahim, M Afzalur (Ed) Managing Conflict: an interdisciplinary approach (1989)
New York, Praeger Publishers, 348 p. ISBN 0-275-92683-4.

Raiff@, H Decision Analysis: introductory lectures on choices under uncertainty (1970)
Reading MA, Addison-Wesley, 309 p.

Ramo, S Cure for Chaos: fresh solutions to social problems through the systems approach (1969)
New York, McKay, 116 p.

Ramsoy, O Social Groups as System and Sub-system (1963)
New York, Free Press, 204 p.

Rapoport, A and Horvath, W J Thoughts on Organization Theory (1959)
General Systems Yearbook Society for General Systems Research, 4, p. 87-91.

Rapoport, A and Shimbel, A Mathematical Biophysics, Cybernetics and General Semantics, etc (1949)
A Review of General Semantics 6, pp. 145-159.

Rapoport, Anatol The Promise and Pitfalls of Information Theory (1956)
Behavioral Science 1, pp. 303-09.

Rapoport, Anatol The Diffusion Problem in Mass Behavior (1956)
General Systems Yearbook Society for General Systems Research, 1, p. 48-55.

Rapoport, Anatol Critiques of Game Theory (1959)
Behavioral Science 4, pp. 49-66.

Rapoport, Anatol Mathematics and Cybernetics (1959)
In: *American Handbook of Psychiatry, Vol. II* New York, Basic, pp. 1743-59.

Rapoport, Anatol Some System Approaches to Political Theory (1966)
In: *Varieties of Political Theory* Englewood Cliffs, Prentice-Hall, pp. 129-41.

Rapoport, Anatol Mathematical Aspects of General Systems Analysis (1969)
In: *Litterer, Joseph A (Ed) Organizations; systems control and adaptation* London, Wiley, 2nd ed, vol. 2, pp. 88-97.

Rapoport, Anatol Methodology in the Physical, Biological and Social Sciences (1970)
In: *E O Attinger (Ed) Global Systems Dynamics* New York, Wiley Interscience, pp. 14-27.

Rapoport, Anatol Modern Systems Theory – An Outlook for Coping with Change (1970)
General Systems Yearbook Society for General Systems Research, 15, p. 15-25.

Rapoport, Anatol The Search for Simplicity (1972)
In: *Ervin Laszlo, The Relevance of General Systems Theory* New York, Braziller, pp. 13-30.

Rapoport, Anatol Mathematical General System Theory (1973)
In: *Unity through Diversity, Gray, William and Rizzo, Nicholas D (Eds)* New York, Gordon and Breach Science Publishers, pp. 437-60.

Rapoport, Anatol General Systems Theory: essential concepts and applications (1986)
Cambridge MA, Abacus Press, 270 p. Cybernetics and Systems Series. ISBN 0-85626-172-6.

Rapoport, D Organization and Pathology of Thought (1951)
New York, Columbia University Press.

Rashevsky, N Topology and Life: in search of general mathematical principles in biology and sociology (1956)
General Systems Yearbook Society for General Systems Research, 1, p. 123-137.

Rashevsky, N The Geometrization of Biology (1956)
Bulletin Mathematical Biophysics 18, pp. 31-56.

Rashevsky, N Organismic Sets: outline of a general theory of biological and social organisms (1967)
General Systems Yearbook Society for General Systems Research, 12, p. 21-27.

Rashevsky, N Outline of a Unified Approach to Physics, Biology and Sociology (1969)
Bulletin of Mathematical Biophysics 31, pp. 159-198.

Raymond, R C Communication, entropy and life (1950)
American Scientist 38, pp. 273-78.

Razran, Gregory H S Mind in Evolution: an East-West synthesis of learned behavior and cognition (1971)
Boston MA, Houghton Mifflin. bibl. (pp. 329-407).

Read, Sir Herbert Icon and Idea: The function of art in the development of human consciousness (1955)
Cambridge MA, Harvard University Press.

Read, Sir Herbert The Forms of Things Unknown (1960)
London, Faber and Faber.

Read, Sir Herbert Grass Roots of Art (1961)
Meridial.

Reason, P and Rowan, J (Eds) Human Inquiry: a sourcebook of new paradigm research (1981)
New York, John Wiley and Sons, 530 p. ISBN 0-471-27935-8.

Redfield, R (Ed) Levels of integration in biological and social systems (1942)
Lancaster, Jacques Cattell Press, pp. 5-26.

Rees, D A Kant's Physiology of the Human Understanding and the Classification of the Sciences (1952)
Journal of the History of Ideas 13, January pp. 108-109.

Regnier, J and de Montmollin, M Reconnaissance de l'Organisation, Recherche de l'Ordonnance des Eléments et Choix du Mode d'Enseignement de la Matière (1968)
Paris, Société d'Economie et de Mathématiques Appliquées. mimeo.

Reichardt, J (Ed) Cybernetic Serendipity: the computer and the arts (1969)
New York, Praeger Publishers.

Reiser, O World Philosophy and the Integration of Knowledge
In: *International Logic Review* 1971, 3.

Reiser, Oliver World Philosophy: a search for synthesis

Reiser, Oliver Mathematics and Emergent Evolution (1930)
Monist 40, p. 509-25.

Reiser, Oliver The Integration of Human Knowledge: a study of the formal foundations and the social implications of unified science (1958)
Boston MA, Porter Sargent, 478 p.

Reiser, Oliver Cosmic Humanism: a theory of the eight-dimensional cosmos based on integrative principles from science, religion and art (1966)
Cambridge MA, Schenkman Publishing, 576 p.

Reiser, Oliver Solar Systems Resonance, the Galactic Alphaphone, and the DNA Helix (1971)
Pittsburgh PA, University of Pittsburg (unpublished) May.

Reisman, John M Toward the Integration of Psychotherapy (1971)
New York, John Wiley and Sons. bibl. (pp. 137-145).

Renzulli, Joseph S (Ed) Systems and Models for Developing Programs for the Gifted and Talented (1986)
Mansfield Centre CT, Creative Learning Press. ISBN 0-936386-44-4.

Rescher, Nicholas Many-Valued Logic (1969)
New York, McGraw Hill Book Company, 359 p. bibl. (pp. 236-331).

Rescher, Nicholas Cognitive Systematization: a system-theoretic approach to a coherent theory of knowledge (1979)
Oxford, Blackwell Scientific Publications.

Rescigno, A Synthesis for Multicompartmental Biological Models (1960)
Biochem. Biophys. Acta 37, pp. 463-468.

Reuterdahl, Arvid A Synthesis of Number, Space-Time and Energy, and a Physical Basis for Planck's and Ryderberg's constants (1923)
Brooklyn, Academy of Nations, 54 p. Academy of Nations Monographs, Scientific Series, P, No. 1.

Revue de Synthèse (du Centre international de synthèse, Paris) Vol
(1, 1931, historique), quarterly.

Rex, John A The Spread of the Pathology of Natural Science to the Social Sciences (1970)
In: *Paul Halmos (Ed) The Sociology of Sociology, Sociological Review Monograph (Keele)* Sept 16.

Reynaud, P L Comment l'ééconomie politique moderne doit-elle coordonner les apports des mathématiques et de la psychologie économique (1967)
Revue des Sciences Economiques Liège, Sept.

Rhyne, R F Communicating Holistic Insights (1972)
Fields within Fields World Institute Council, 5, 1.

Rice, A K The Enterprise and its Environment: a system theory of management organization (1963)
London, Tavistock Publications, 364 p.

Richards, M D and Greenlaw, P S Management Decision Making (1966)
Homewood IL, Irwin and Dorsey, 564 p.

Richardson, Jacques Models of Reality: shaping thought and action (1987)
Paris, UNESCO.

Richardson, John S and Showalter, Victor Effects of a Unified Science Curriculum on High School Graduates (1969)
Columbus OH, Ohio State University Research Foundation.

Richter, Robert Die Internationale Geistige Zusammenarbeit im Rahmendes Völkerbundes (1930)
Würzburg, Handelsdruckerei, 59p. bibl. (pp. 7-8).

Rickert, Heinrich Die Grenzen der Naturwissenschaftlichen Begriffbildung (1913)
Tubingen. 2nd ed.

Ricoeur, Paul The Conflict of Interpretations (1974)
Evanston IL, Northwestern University Press.

Rieger, Hans C Some Problems of Interdisciplinary Research
Interdiscipline Gandhian Institute of Studies.

Rifkin, Jeremy Entropy: a new world view (1980)
New York, Viking Press.

Riggs, D S Control Theory and Physiological Feedback Mechanisms (1970)
Baltimore MD, Williams and Wilkins.

Rino, José B El Hombre Como Sistema, Problema y Misterio: antropologismo filosofico (1969)
Buenos Aires, Plus Ultra, 228 p. bibl. (pp. 223-228).

Ripley, S D and Buechner, H K Ecosystem Science as a Point of Synthesis (1967)
Daedalus 96, p. 1192-99.

Rizzo, Nicholas D The Significance of Von Bertalanffy for Psychology (1972)
In: *Ervin Laszlo, The Relevance of General Systems Theory* New York, Braziller, 213p. (pp. 135-144).

Robert Aubroy; Laing, Ronald and R Pflughaupt, Knut The Way of the Warrior (1982)
Stuttgart, Forum International.

Roberts, Fred S Graph Theory and Its Applications to Problems of Society (1978)
Philadelphia PA, Society for Industrial and Applied Mathematics, 122 p. ISBN 0-89871-026-X.

Robinson, A Complete Theories (1963)
Amsterdam, North Holland.

Robinson, H W (Ed) Cybernetics, Artificial Intelligence, and Ecology (Fourth Annual Symposium of the American Society for Cybernetics) (1973)
New York, Spartan Books.

Roerich, Nicholas K Beautiful Unity
Delhi.

Rogow, Arnold A Some Relations between Psychiatry and Political Science (1969)
In: *M Sherif and C Sherif (Ed) Interdisciplinary Relationships in the Social Sciences* Chicago, Aldine, pp. 274-291.

Rokeach, Milton The Open and Closed Mind: investigations into the nature of belief systems and personality systems (1960)
New York, Basic Books.

Rokkan, Stein Recherche trans-culturelle, trans-culturelle, trans-sociétale et transnationale (1970)
In: *Unesco, Tendances Principales de la Recherche dans les Sciences Sociales et Humaines* pp. 765-824. Paris, Mouton/Unesco.

Romain, J Information et Cybernétique (1959)
Cybernetica 2, p. 23-50.

Romanyshyn, Robert D Metaphors of Experience and Experience as Metaphorical
In: *R S Valle and R von Eckartsberg (Eds) The Metaphors of Consciousness* New York, Plenum Press, 1981, pp. 3-19.

Roose, Kenneth D Observations on Interdisciplinary Work in the Social sciences (1969)
In: *M Sherif and C Sherif (Ed) Interdisciplinary Relationships in the Social Sciences* Chicago, Aldine, pp. 323-327.

Roosen-Runge, Peter Towards a Theory of Parts and Wholes: an algebraic approach (1966)
General Systems Yearbook Society for General Systems Research, 11, p. 13-18.

Rose, J (Ed) Progress of Cybernetics (1969)
New York, Gordon and Breach. 3 Vols.

Rose, R Disciplined Research and Undisciplined Problems
In: *International Social Science Journal* 1976, XXVIII, 1, pp. 99-121.

Rosen, Robert A Relational Theory of Biological Systems (1960)
General Systems Yearbook Society for General Systems Research, 5, p. 29–43.

Rosen, Robert The Representation of Biological Systems from the Standpoint of the Theory of Categories (1960)
General Systems Yearbook Society for General Systems Research, 5, p. 45–54.

Rosen, Robert Optimality Principles in Biology (1967)
New York, Plenum Publishing, 198 p.

Rosen, Robert Notes on Hierarchy in Concept: comments on the use of the term hierarchy (1969)
In: *Lancelot Law Whyte, et al (Eds) Hierarchical Structures* New York, American Elsevier, pp. 52–53.

Rosen, Robert Hierarchical Organization in Automata: theoretic models of biological systems (1969)
In: *Lancelot Law Whyte, et al (Eds) Hierarchical Structures* New York, American Elsevier, pp. 179–200.

Rosen, Robert Dynamical System Theory in Biology (1970)
New York, Wiley.

Rosen, Robert A Survey of Dynamical Descriptions of System Activity (1973)
In: *Unity through Dipersity, Gray, William and Rizzo, Nicholas D (Eds)* New York, Gordon and Breach Science Publishers. pp. 461–78.

Rosenau, J N The Functioning of International Systems (1963)
Background 7.

Rosenberg, S and Hall, R L The Effects of Different Social Feedback Conditions upon Performance in dyadic teams (1958)
Journal of Abnormal Social Psychology 57, pp. 271–77.

Rosenblueth, A and Wiener, Norbert Purposeful and Non–Purposeful Behavior (1950)
Philosophy of Science 17, pp. 318–26.

Rosenblueth, Arturo Mind and Brain: a philosophy of science (1970)
Cambridge MA, MIT Press.

Rosenthal, D C Metaphors, Models and Analogies in Social Science and Public Policy
In: *Political Behaviour* 1982, 4, 3, pp. 283–301.

Rosner, S Some Dimensions involved in Interpreting Diagnostic Findings (of the interdisciplinary evaluation) (1959)
Social Casework 40, October pp. 445–448.

Ross, Herbert H A Synthesis of Evolutionary Theory (1962)
Englewood Cliffs NJ, Prentice Hall, 387 p. bibl.

Ross, J F Analogy and the Resolution of Some Cognitivity Problems (1970)
Journal of Philosophy 67, 22 October pp. 725–746.

Rossi, Paolo Clavis Universalis: arti mnemoniche e logica combinatoria da Lullo a Leibniz (1960)
Milano, R Ricciardi, 315 p.

Roth, Gerhard and Schwegler, Helmut (Eds) Self Organizing Systems: an interdisciplinary approach (1981)
Frankfurt, Campus Verlag, 376 p. ISBN 3–593–32833–X.

Rothbarth, Margarete Geistige Zusammenarbeit im Rahmen des Völkerbundes (1931)
Münster, Aschendorf, 195 p. bibl. (p. 195).

Rothstein, J Communication, Organization, and Science (1958)
Indian Hills CO, Falcon's Wing.

Rothstein, J Discussion: information and organization as the language of the operational viewpoint (1962)
Philosophy of Science 29, pp. 406–11.

Rowe, J S Level of Integration Concept and Ecology (1961)
Ecology 42, April bibl. (p. 427).

Royce, J R (Ed) Psychology and the Symbol: an interdisciplinary Symposium (1965)
New York, Allied Publications.

Royce, J R Toward Unification in Psychology (1970)
Toronto ON, University of Toronto Press, 308 p.

Ruberti, A (Ed) Systems Science and Modelling (1985)
Paris, UNESCO, 159 p. Trends in Scientific Research Series: 1. ISBN 92–3–102138–9.

Rubin, Milton (Ed) Man in Systems (1971)
New York, Gordon and Breach, 496 p. bibl.

Rudhyar, Dane The Magic of Tone and the Art of Music (1982)
Boulder CO, Shambhala Publications.

Rueff, Jacques L'ordre dans la nature et dans la société
Diogene, 10, p. 3 and following.

Ruller, R W and Putnam, P On the Origin of Order in Behavior (1966)
General Systems Yearbook Society for General Systems Research, 11, p. 99–111.

Ruller, R W and Putnam, P Systems in Society (1973)
Washington DC, Society for General Systems Research.

Rummel, R J Dimensions of Conflict Behavior within and Between Nations (1963)
General Systems Yearbook Society for General Systems Research, 8, p. 1–46.

Rummel, R J A Field Theory of Social Action with Application to Conflict within Nations (1965)
General Systems Yearbook Society for General Systems Research, 10, p. 183–211.

Russell, Peter The Global Brain: speculations on the evolutionary leap to planetary consciousness (1983)
Los Angeles CA, J P Tarcher, 251 p.

Russell, W M S Evolutionary Concepts in Behavioural Science
In: *General Systems Yearbook* Society for General Systems Research. 4 Vols.

Rutherford, James and Gardner, Marjorie Integrated Science Teaching
In: *New trends in integrated science teaching* Paris, Unesco, pp. 47–55 (Vol 1).

Saccaro–Battist, G Changing Metaphors of Political Structures
In: *Journal of the History of Ideas* 1983, 44, pp. 31–54.

Sachs, M Philosophical Implications of Unity in the Contemporary Arts and Sciences
In: *Philosophy and Phenomenological Research* 1974, Buffalo, XXXIV, 4, pp. 489–503.

Sackman, H Computers, System Science, and Evolving Society: the challenge of man–machine digital systems (1967)
New York, John Wiley and Sons, 638 p.

Sadovsky, V N Aspects Méthodologiques d'une Théorie Générale des Systèmes. (1971)
Revue Internationale de Philosophie 25e Année, No 98, Fasc. 4.

Sadovsky, V N General Systems Theory: its tasks and methods of construction (1972)
General Systems Yearbook Society for General Systems Research, 17, p. 171–178.

Sagasti, Francisco A Conceptual and Taxonomic Framework for the Analysis of Adaptive Behavior (1970)
General Systems Yearbook Society for General Systems Research, 15, p. 151–159.

Sage, Andrew P and Melsa, J L Systems Identification (1971)
New York, Academic Press, 221p. bibl.

Samuelson, K, Dorko, H et Al Global and Long–Distance Decision–Making: environmental issues and network potentials (1972)
Stockholm, The Royal Institute of Technology and the University of Stockholm.

Sankaranarayanan, A On a Group Theoretical Connection among the Physical Hierarchies (1969)
Research Communication No. 96 Douglas Advanced Research Laboratories, Huntington Beach, California.

Saporta, S (Ed) Psycholinguistics (1961)
New York, Holt, Rinehart and Winston.

Saunders, Charles et al (Comp) Synthesis: responses to literature (1971)
New York, Knopf, 750 p. bibl. (pp. 739–742).

Savory, Allan Holistic Resource Management (1988)
Covelo CA, Island Press, 545 p. illus. ISBN 0–933280–62–9.

Savory, Allan, et al Holistic Resource Management
Covelo CA, Island Press. 2 Vols. Vol 1: 1988, 545 p. illus. ISBN 0–933280–62–9. Vol 2: 1989, 224 p. illus. ISBN 0–933280–69–6.

Sayles, L R and Chandler, M K Managing Large Systems: organizations for the future (1971)
New York, Harper and Row.

Sayre, K M Consciousness: a philosophical study of minds and machines (1969)
New York, Random House.

Sblichta, Paul J Notes on Inorganic Hierarchical Structures: overlap in hierarchical structures (1969)
New York, American Elsevier, pp. 138–143.

Schachtel, E G Metamorphoses (1959)
New York, Basic Books.

Schafer, R Murray The Tuning of the World (1977)
New York, Alfred A Knopf.

Scharnberg, Max The Myth of Paradigm–Shift, or How to Life with Methodology (1984)
Philadelphia PA, Coronet Books, 170 p. ISBN 91–554–1489–3.

Schedrovitsky, G P Methodological problems of system research (1966)
General Systems Yearbook Society for General Systems Research, 11, p. 27–51.

Schedrovitsky, G P Concerning the analysis of initial principles and conceptions of formal logic (1968)
General Systems Yearbook Society for General Systems Research, 13, p. 21–31.

Schellenberger, R E Managerial Analysis (1969)
Homewood IL, Irwin and Dorsey, 464 p.

Schelling, T C The Strategy of Conflict (1960)
Cambridge MA, Harvard University Press.

Schlanger, J E Les métaphores de l'organisme (1971)
Paris, Librairie philosophique.

Schlegel, R Completeness in Science (1966)
New York, Appleton–Century–Crofts, 320 p.

Schneer, Cecil J The Search for Order (1960)
(reissued in paperback as: The Evolution of Physical Science).

Schneider, Marius El Origen Musical de los Animales–Simbolos en la Mitologia y la Escultura Antiquas (1946)
Barcelona.

Schoderbek, P P Management Systems (1967)
New York, John Wiley and Sons, 483 p.

Schools Council Schools Council Integrated Science Project (1973)
London, Longman/Penguin.

Schools Council General Studies Committee General Studies: an annotated booklist (1982)
National Book League. ISBN 0–85353–370–9.

Schrödinger, E What is Life? (1945)
Cambridge, Cambridge University Press.

Schumacher, E A Guide for the Perplexed (1977)
New York, Harper and Row.

Schumacher, E F Small is Beautiful: economics as if people mattered (1973)
New York, Harper and Row.

Schuon, Frithjof The Transcendent Unity of Religions (1953)
London, Faber and Faber.

Schwab, J J The Structure of the Disciplines: meanings and signification (1964)
In: *J F Ford and C Pugno (Eds) The Structure of Knowledge and the Curriculum* Chicago, Rand McNally.

Schweitzer, A L M Sociologie en Cybernetica (1963)
Mens en Maatschappij 38, pp. 351–67.

Schwenk, Theodor Sensibles Chaos: strömendes formenschaffen in wasser un luft (1968)
Stuttgart. 3rd ed. (Also English and French translations).

Schweyer, H E Analytic Models for Managerial and Engineering Economics (1904)
New York, Reinhold Van Nostrand, 505 p.

Sciama, D The Unity of the Universe (1959)
New York, Doubleday.

Scientific American Automatic Control (1955)
New York, Simon and Schuster International.

Scileppi, John A Systems View of Education: a model for change (1988)
Lanham MD, University Press of America, 236 p. rev ed. ISBN 0–8191–6763–0.

Scott, John Paul Biological Basis of Human Warfare: an interdisciplinary problem (1969)
In: *M Sherif and C Sherif (Ed) Interdisciplinary Relationships in the Social Sciences* Chicago, Aldine, pp. 121–136.

Scott, W A Cognitive Structure and Social Structure: some concepts and relationships (1962)
In: *Decisions, Values and Groups, Vol II* New York, Pergamon, pp. 86–118.

Scott, W G Organization Theory: a behavioral analysis for management (1967)
Homewood IL, Irwin and Dorsey, 442 p.

Scur, G S On Some General Categories of Linguistics (1966)
General Systems Yearbook Society for General Systems Research, 11, p. 149–156.

Scur, G S On the relations among Some Categories in Linguistics (1966)
General Systems Yearbook Society for General Systems Research, 11, p. 157–164.

Seamon, J The Style of Political Discourse: an annotated bibliography
In: *Style* 1974, 3, pp. 477–528. Section 4 on Semioties and Metaphor.

Sebillotte, M Les Rotations Culturales: approche méthodologique d'une politique dynamique (1968)
Bulletin FNCETA, Janvier, numéro special.

Secrest, Leif The Rationale for Polydisciplinary Programs (Address to the Ninth Annual Meeting of the Council of Graduate Schools in the US, Washington, D C. December 4–6, 1969) (1969)

Segre, Cesare La synthèse stylistique (1967)
Social Science Information, October pp. 161–168.

Seiler, J A Systems Analysis in Organizational Behavior (1967)
Homewood IL, Irwin and Dorsey, 219p.

Sellerio, A Les syméétries en physique (1935)
Scientia 58, 69, August.

Sells, S B General Theoretical Problems related to Organizational Taxonomy: a model solution (1968)
In: *Indik, Bernard P and Berrien, F K (Eds) People, Groups, and Organizations* New York, Teachers College (Columbia Univ.) Press, pp. 27–46.

Sengupta, S S and Ackoff, R L Systems Theory from an Operations Research Point of View (1965)
General Systems Yearbook Society for General Systems Research, 1964, 10, p. 43–46.

Seward, J P The Structure of Functional Autonomy (1963)
American Psychologist 18, pp. 703–10.

Sewell, E The Human Metaphor (1964)
Notre Dame IN, University of Notre Dame Press.

Shannon, C and Weaver, W The Mathematical Theory of Communication (1949)
Urbana, University of Illinois Press, 117 p.

Shaw, L System Theory (1965)
Science 149, pp. 1005.

Sheldrake, Rupert A New Science of Life: the hypothesis of formative causation (1981)
London, Blond and Briggs, 229 p. ISBN 0–85634–115–0.

Sherif, C W and Sherif, M Interdisciplinary Relations in the Social Sciences (1969)
Chicago IL. Vol XVI.

Sherif, M Social Psychology, Anthropology and the Behavioral Sciences (1959)
S W Social Science Quarterly 40, Sept pp. 105–112, bibl.

Sherif, Mazafer and Sherif, Carolyn Interdisciplinary Coordination as a Validity Check: retrospect and prospect (1969)
In: *M Sherif and C Sherif (Ed) Interdisciplinary Relationships in the Social Sciences* Chicago, Aldine, pp. 3–20.

Sherrington, Sir Charles Scott The Integrative Action of the Nervous System (1961)
New Haven CT, Yale University Press, 413 p. repr of 1906 ed. bibl.

Shipley, C Morton et al A Synthesis of Teaching Methods (1968)
New York, McGraw Hill Book Company, 344 p. bibl.

Showalter, Victor Unification of the Curriculum
In: *Encyclopedia of Education Vol 8 (Lee C Deighton Ed)* New York. Macmillan with Free Press.

Shriver, R H and White, R C Distribution Planning and Control: effective use of computer systems and models (1969)
New York, American Management Association, 60p.

Shubik, Martin Simulation of Socio–Economic Systems, I: general considerations (1968)
General Systems Yearbook Society for General Systems Research, 12, p. 149–158.

Shurig, R Morphology: a knowledge tool
In: *Syst. Res.* 1986, 3, pp. 9–19.

Sibatani, Atuhiro Antiscience: toward one knowledge, one learning (in Japanese) (1973)
Tokyo, Misuzu Syoboo.

Sickesz, W C Synthese van Drie Ideologie: communisme, socialisme, kapitalisme (1957)
Amsterdam, 94 p.

Sickles, W R and Hartmann, G W The Theory of Order (1942)
Psychological Review 49, pp. 403–421.

Siebker, Manfred Art as a Research and Development Tool for Human Qualities, as an Indication of the Future and as an Integrating Factor for Mankind (Paper for Rome Special Conference on Futures Research, 1973) (1973)

Siegmann, Heinrich World Modelling (1987)
Paris, UNESCO, 175 p. Major Programme I, Reflection on World Problems and Future–Oriented Studies: BEP/GPI/2.

Siegmann, Heinrich World Modeling (1987)
Paris, UNESCO, 175 p. Major Programme I, Reflection on World Problems and Future–Oriented Studies: BEP/GPI/2.

Silverman, Hirsch L Psychiatry and Psychology: relationships, intra–relationships and inter–relationships (1963)
Springfield IL, Thomas Charles C.

Silvern, L C The Evolution of Systems Thinking in Education (1971)
Los Angeles, Education and Training Consultants, 140 p. 2nd ed. bibl.

Silverstein, A and V Bionics: man copies nature's machines (1970)
McCall Pub. Co.

Simas, Philip W A New Supra Discipline, Policy Sciences, Aims to 'Integrate Intelligence and Action'. (1971)
In: *The Chronicle of Higher Education* 5, 16, January 24, pp. 1–5.

Simmons, O G and Davis, J A Interdisciplinary Collaboration in Mental Illness Research (1957)
American Journal of Sociology 63, November pp. 297–303, bibl.

Simon, Herbert A Models of Man (1957)
New York, John Wiley and Sons.

Simon, Herbert A The New Science of Management Decision (1960)
New York, Harper and Row. 50 p.

Simon, Herbert A The Architecture of Complexity (1962)
Proceedings of the American Philosophical Society 106, p. 467–482.

Simon, Herbert A Administrative Behavior: a study of decision–making processes in administrative organization (1965)
New York, Free Press, 259 p.

Simon, Herbert A The Sciences of the Artificial (1969)
Cambridge MA, Cambridge University Press.

Simonds, Roger Integrative Concepts in the Logic of Relations (1972)
In: *H Margenau (Ed) Integrative Principles of Modern Thought* New York, Gordon and Breach, pp393–451.

Simpson, G G The Meaning of Evolution (1949)
New Haven CT, Yale University Press.

Sin, R G H The Tao of Science (1957)
New York, John Wiley and Sons.

Sinaceur, Mohammed Allal Qu'est–ce Que l'Interdiciplinarité?
In: *UNESCO – Interdisciplinarité et Sciences Humaines* 1983, Paris, UNESCO, pp. 21–30.

Singer, J David A General Systems Taxonomy for Political Science (1971)
New York, General Learning.

Singer, J David A Cybernetic Interpretation of International Conflict (1973)
In: *Unity through Diversity, Gray, William and Rizzo, Nicholas D (Eds)* New York, Gordon and Breach Publishers, pp. 1105–1122.

Singevin, Charles Essai sur l'un (1969)
Paris, Editions du Seuil, 347 p. bibl. (pp. 325–329).

Singh, J Great Ideas in Information Theory, Language and Cybernetics (1966)
New York, Dover Publications, 338 p.

Singh, J Great Ideas in Operations Research (1968)
New York, Dover Publications, 228 p.

Sinnott, Edward The Biology of the Spirit (1957)
Compass Viking Press.

Siu, R G Tao of Science (1958)
Mass. Inst. Technology.

Siu, R G H Ch'i: a neo–taoist approach to life (1974)
Cambridge MA, Massachusetts Institute of Technology.

Sivashankar, N Man Rediscovered: a new approach to the nature of man, an attempt at a synthesis of ancient philosophies of religions and modern philosophies of science (1965)
Trivandrum, B Sarada Amma, 258p. bibl.

Slack, C W Feedback Theory and the Reflex Arc Concept (1955)
Psychological Review 62, pp. 263–67.

Slesnick, Irwin L The Effectiveness of a Unified Science in the High School Curriculum
Journal of Research in Science Teaching.

Slikkerveer, L J; Titilola, S O and Warren, D M (Eds) Indigenous Knowledge Systems: implications for agriculture and international development (1989)
Ames IA, Iowa State University, 186 p. Studies in Technology and Social Change: 11. ISBN 0–945271–15–8.

Sloane, E H Reductionism vs the Principle of the whole and the principle of levels underlying new theories (1945)
Psychological Review 52:214–23, July.

Slobodkin, L B Meta–Models in Theoretical Biology (1958)
Ecology 39, pp. 450–551.

Slobodkin, L B Aspects of the Future of Ecology (1968)
General Systems Yearbook Society for General Systems Research, 13, p. 115–124.

Sluckin, W Minds and Machines (1954)
Baltimore MD, Penguin Books, 223 p.

Smirnov, Stanislav N L'Approche Interdisciplinaire dans la Science d'Aujourd'hui: fondements ontologiques et épistémologiques, formes et fonctions
In: *UNESCO – Interdisciplinarité et Sciences Humaines* 1983, Paris, UNESCO, pp. 53–72.

Smith, A D Social Worker in the Legal Aid Setting: a study of interprofessional relationships (1970)
Social Service Review 44, June pp. 156–168, bibl.

Smith, August W (Ed) Systems Methodologies, Isomorphies and Applications: proceedings of the Society for General Systems Research (1984)
Salinas CA, Intersystems Publications, 660 p. 2 Vols. ISBN 0–914105–29–9.

Smith, Cyril Stanley The Shape of Things (1954)
Scientific American 190, p. 58–64.

Smith, Cyril Stanley Structure, Substructure, Superstructure (1965)
In: *Structure in Art and in Science, G Kepes (Ed)* pp. 19–41. New York, Braziller.

Smith, Cyril Stanley Structural Hierarchy in Inorganic Systems (1969)
In: *Lancelot Law Whyte, et al (Eds) Hierarchical Structures* New York, American Elsevier, pp. 61–86.

Smith, M and Mackay, A The Science of Science (1964)
London.

Smith, N M and Marney, M C Management Science: an intellectual innovation (1961)
Brussels, Institute of Management Science. Paper 14. 4, 8th international meeting, Aug. 23–26.

Smith, P T Computers, Systems and Profits (1969)
New York, American Management Association, 200p.

Smith, R A, III Social Systems Analysis and Industrial Humanism: awareness without revelation (1969)
General Systems Yearbook Society for General Systems Research, 14, p. 103–110.

Smith, R A, III General Systems Theory: our passport to evolutionary awareness (1972)
General Systems Yearbook Society for General Systems Research, 17, p. 25–27.

Smith, W N et al Integrated Simulation: an interactive general business simulation designed for flexible application in management education (1968)
Cincinnati, Southwestern Pub Co, 70 p.

Smuts, Jan C Holism and Evolution (1961)
New York, Viking Press, 362 p. reprint of 1927 ed.

Smythe, W R Jr and Johnson, L A Introduction to Linear Programming, with Applications (1966)
Englewood Cliffs NJ, Prentice Hall, 221 p.

Società Filosofica Italiana L'Unificazione del Sapere
Florence, Sansoni. 2 Vols.

Society for Experimental Biology Homeostasis and Feedback Mechanisms Symposium (1965)
San Diego CA, Academic Press.

Soedjatmoko, K The Future and the Learning Capacity of Nations: the role of communications (1981)
Transnational Associations, 33, 1981, 2, p.80–85.

Soleri, Paolo The Omega Seed: an eschatological hypothesis (1981)
New York, Anchor Books, 286 p. ISBN 0–385–15889–0.

Solesbury, W Strategic Planning: metaphor or method?
In: *Policy and Polities* 1981, 9, 4, pp. 419–437.

Solmsen, Fr Plato and the Unity of Science (1940)
Philosophical Review 49, pp. 567–571.

Sonnemann, Ulrich The Specialist as a Psychological Problem (1951)
Social Research, international quarterly of political and social science March, pp. 9–31.

Sorokin, Pitirim The Crisis of Our Age (1941)
EP Dutton.

Sorokin, Pitirim A The Reconstruction of Humanity (1948)
Boston MA, Beacon Press, 247 p.

Sorokin, Pitirim A Sociological Theories of Today (1966)
New York, Harper and Row.

Sorokin, Pitrim A Social and Cultural Dynamics
New York, Allied Publications. 4 vols. Vol 1: Fluctuation of Forms of Art. Vol 2: Fluctuation of Systems of Truth, Ethics and Law (1937, 727 p). Vol 3: Fluctuation of Social Relationships, War and Revolution (1937, 636 p). Vol 4: Basic Problems, Principles and Methods (1941, 804 p).

Southern Methodist University Integration of the Humanities and the Social Sciences: a symposium (1948)
Dallas TX, University Press, 92 p. bibl., foot. (Southern Methodist University Studies No. 4).

Spaccapietra, Stefano Entity–Relationship Approach. Ten years of experience in information modeling (1987)
Amsterdam, Elsevier Science Publishing, 558 p. ISBN 0–444–70255–5.

Spencer, Herbert The Principles of Sociology: a system of synthetic philosophy (1880)
Westport CT, Greenwood Press, 3 vols, 1880–1897.

Spengler, J J Generalists Versus Specialists in Social Science: an economist's view (1950)
American Political Science Review 44, June, pp 358–393.

Spiegel, J P A Model for Relationships among Systems (1956)
In: *Grinker, R R (Ed) Toward a Unified Theory of Human Behavior* New York, Basic Books.

Spiegelberg, Herbert Rules and Order: toward a phenomenology of order (1968)
In: *Paul G Kuntz (Ed) The Concept of Order* Seattle, University of Washington Press, pp. 290–308.

Spirkin, A G and Sazonov, B V Obsuzhdenia Metologicheskikh Problem Issliedovania Sistiem i Struktur (A discussion of the methodological problems of research on systems and structures) (1964)
Roprosy Filosofii 1.

Sprout, Harold and Sprout, Margaret The Ecological Perspective on Human Affairs: with special reference to international politics (1965)
Princeton NJ, Princeton University Press, 234 p. bibl. foot.

Squire, William Integration for Engineers and Scientists (1970)
New York, American Elsevier, 302 p. (Includes list of doctoral dissertations on integration and integral equations, pp. 267–72).

Stabile, Donald Prophets of Order (1984)
Boston MA, South End Press, 350 p. ISBN 0–89608–230–X.

Stachowiak, H Gedanken zu einer Allgemeinen Theorie der Modelle (1965)
Studium generale 7.

Stagner, Ross Homeostasis as a Unifying Concept in Personality Theory
In: *Litterer, Joseph A (Ed) Organizations; systems control and adaptation* London, Wiley, 1969 2nd ed, vol. 2 pp. 77–78.

Stamps, Jeffrey S Holonomy: a human systems theory (1980)
Seaside CA, Intersystems Publications, 213 p. Intersystems Inquiry Series.

Stanley, J C; Keating, D P and Fox, L H (Ed) Mathematical Talent: discovery, description and development
Baltimore MD, Johns Hopkins University Press, 216 p.

Stanley–Jones, D and K The Kybernetics of Natural Systems: a study in patterns of control (1960)
New York, Pergamon Books, 145 p.

Stapp, H S S–Matrix interpretation of quantum theory (1971)
Physical Review, D3, 1971, p.130*ff.*.

Stark, W The Sociology of Knowledge (1958)
London, Routledge and Kegan Paul.

Stark, W Kybernetics of Mind and Brain (1970)
Springfield IL, Thomas Charles C.

Starr, M K Executive Readings in Management Science (1925)
New York, Macmillan, 422 p.

Starr, M K Production Management, Systems and Synthesis (1964)
Englewood Cliffs NJ, Prentice Hall.

Starr, M K Management: a modern approach (1971)
New York, Harcourt Brace, 716 p.

Steinbuch, Karl Communication in the Year 2000. (1969)
In: *Jungk, R and Galtung J (Eds) Mankind 2000* London, Allen and Unwin, pp. 165–70.

Steinbuch, Karl und Moser, Simon (Eds) Philosophie und Kybernetik (1970)
Munich, Nymphenburger Verlagshandlung, 198 p. bibl. foot.

Stephen W and Rohrer, R A Introduction to Systems Theory (1971)
New York, McGraw Hill Book Company, 441 p. bibl.

Stern, Karl Third Revolution: a study of psychiatry and religion (1955)
New York, Doubleday.

Sterrenburg, J N L'intégration monétaire (1961)
Leiden, HE Stenfert Kroese, 168 p. bibl.

Stevens, Anthony Archetype: a natural history of the self (1982)
London, Routledge and Kegan Paul.

Stevens, Mary Otis and McNulty, T F World of Variation (1970)
New York, Braziller, 157 p. bibl. (pp. 148–57).

Steward, A H Theory of Culture Change: the methodology of multilinear evolution (1955)
Urbana, University of Illinois Press.

Stewart, Robert M Fields and Waves in Existable Cellular Structures (1963)
In: *James Garvey (Ed) Self–Organizing Systems* Office of Naval Research, US Govt Printing Office, ACA–96, 78p.

Stewart, T C The City as an Image of Man (1970)
London, Latimer Press.

Stogdill, R M (Ed) Process of Model–Building in the Behavioral Sciences (1970)
Columbus OH, Ohio State University Press.

Stokes, P M Total Systems Approach to Management Control (1968)
New York, Macmillan.

Stolte, Dieter and Wissen, Richard (Ed) Integritas: geistige wandlung und menschliche wirklichkeit (1966)
Tubingen, Wunderlich, 626 p. bibl. foot.

Stone, James H Integration in the Humanities: perspectives and prospects (1969)
Main Currents in Modern Thought 26, 1, September/October pp14–19.

Storer, H Relations Among Scientific Disciplines
In: *The Social Contexts of Research* 1973, London, pp. 229–268.

Storer, Norman W The Social System of Science (1966)
New York, Holt, Rinehart and Winston, 180 p. bibl. (pp. 169–76).

Stover, Carl (Ed) The Technological Order (1963)
Detroit MI, Wayne State University Press.

Stoward, P J Thermodynamics of Biological Growth (1962)
Nature London, 194, pp. 977–78.

Stranieio, Giorgio L'Ontologia Fenomenologica di Teilhard de Chardin (1969)
Milan, Vita e Penseiro, 221 p. bibl. (pp. 217–21).

Streitweiser Jr, A Aromaticity and (Huckel's) 4n 2 rule (1961)
In: *Molecular Orbital Theory for Organic Chemists* Wiley, 1961, p.256–304.

Studer, R G Human Systems Design and the Management of Change (1971)
General Systems Yearbook Society for General Systems Research, 16, p. 131–42.

Stumpers, F L A Bibliography of Information Theory, Communication Theory and Cybernetics (1953)
Transactions I R E (PG1T–2) November.

Stybe, Svend E 'Antinomies' in the Conceptions of Man: an inquiry into the history of human spirit (1962)
Copenhagen, Munksgaard, 34 p. (Interdisc. Studies from the Scandinavian Summer Univ, Vol 9).

Sullivan, Harry S Fusion of Psychiatry and Social Science (1964)
New York, WW Norton.

Sullivan, Harry S Interpersonal Theory of Psychiatry (1968)
New York, WW Norton.

Sunderam, K V Some Problems of Interdisciplinary Teamwork and Research in Town and Country Planning (1969)
Interdiscipline 6, 4, Winter pp. 303–15.

Suppe (Ed) The Structure of Scientific Theories (1977)
Moscow ID, University of Illinois, 832 p. illus. 2nd ed.
ISBN 0–252–00655–0.

Sutherland, John W A General Systems Philosophy for the Social and Behaviorial Sciences (1973)
New York, Braziller.

Svoboda, A Synthesis of Logical Systems of Given Activity (1963)
IEEE Trans. Electronic Computers vol EC–12, Dec 1963 pp. 904–10.

Synthese; Gesellschaft und Wirtschaft, Geist und Kultur (vols listed by author)

Synthèses; revue mensuelle internationale
Bruxelles, Editions Socodei, avril 1946–

Synthesis; Sammlung Historischer Monographien Philosophischer Begriffe Heidelberg 1908–

Szeng–Györgyi, Albert Address Before the Conference on Interdisciplinary Science Education, Washington DC, January 1969 (sponsored by the American University) (1969)

Szilard, L Uber die Entropieverminderung in einem Thermodynamischen System bei Eingriffen Intelligenter Wesen (1924)
Zeitschrift für Physik Part 1, 53, pp. 840–56.

Szilard, L On the Increase of Entropy in a Thermodynamic System by the Intervention of Intelligent Beings (1964)
Behavioral Science 9, pp. 301–10.

Taiwan Institute of International Relations A Comprehensive Glossary of Chinese Communist Terminology (1978)
Taiwan, National Chengchi University.

Taschdjan, Edgar Nonlinear cybernetics (1982)
Cybernetica (1982), 25, 1, p. 5–15.

Taylor, Alastair M Evolution–Revolution, General Systems Theory, and Society (1971)
In: *Gotesky, Rubin and Laszlo, Ervin (Eds) Evolution–Revolution* New York. Gordon and Breach Science Publishers pp. 99–135.

Taylor, Alastair M Integrative Principles in Human Societies (1972)
In: *Integrative Principles of modern thought* New York, Gordon and Breach Science Publishers, 1972 pp. 201–84.

Taylor, Mark C Altarity (1987)
Chicago IL, University of Chicago Press, 369 p. illus.
ISBN 0–226–79137–8.

Taylor, R Purposeful and Non–Purposeful Behavior: a rejoinder (1950)
Philosophy of Science 17, pp. 327–32.

Taylor, R Comments on a Mechanistic Conception of Purposefulness (1950)
Philosophy of Science 17, pp. 310–17.

Terhune, William B The Integration of Psychiatry and Medicine: an orientation for physicians (1951)
New York, Grune and Stratton, 177 p.

Tesler, L; Enea, H and Colby, K M Directed Graph Representation of Computer Simulation of Belief Systems (1967)
Stanford CA, Stanford University Press.

Thayer, Lee (Ed) Communication: the ethical and moral issues
New York, Gordon and Breach.

Thayer, Lee Communication: general semantics perspective (1970)
New York, Spartan Books.

Thayer, Lee Communication Systems (1972)
In: *Ervin Laszlo, The Relevance of General Systems Theory* New York, Braziller 213p. (pp. 93–121).

Theobald, Robert The Rapids of Change: social entrepreneurship in turbulent times (1987)
Knowledge Systems.

Theodorson, George A Studies in Human Ecology (1961)
New York, Allied Publications.

Thier, Herbert D Content and Approaches of Integrated Science Programs at the Primary and Secondary School Levels
In: *New trends in integrated science teaching, Vol II* Paris, Unesco pp. 53–68.

Thirring, Hans The Step from Knowledge to Wisdom (1956)
American Scientist 46, Oct. pp. 445–56.

Thom, René Stabilité Structurelle et Morphogenèse (1972)
Reading MA, Benjamin.

Thom, René Structural Stability and Morphogenesis: an outline of a general theory of models (1975)
Reading MA, Addison–Wesley, 348 p.

Thom, René Modèles Mathématiques de la Morphogenese (1980)
Paris, Christian Bourgois.

Thomas, D S Experiences in Interdisciplinary Research (1952)
American Sociological Review 17, December pp. 663–69. bibl.

Thomas, R M and Swastout, S G Integrated Teaching Materials: how to choose, create and use them (1963)
New York, David McKay, 559 p.

Thompson, D W Science and the Classics (1940)
London, Oxford University Press

Thompson, D W On Growth and Form (1959)
New York, Cambridge University Press. (reprint). 2 vols.

Thompson, Evan and Varela, Francisco World Without Ground: cognitive science and human experience

Thompson, J D Organizations in Action: social science bases of administrative theory (1907)
New York, McGraw Hill Book Company, 192 p.

Thompson, J D Approaches to Organizational Design (1966)
Pittsburgh PA, University of Pittsburgh Press, 223 p.

Thompson, John W Mental Science, Meteorology, and General System Theory (1960)
General Systems Yearbook Society for General Systems Research, 5, p. 21–25.

Thompson, John W Meteorological Models in the Social Sciences: complex processes in meteorology and sociology (1962)
General Systems Yearbook Society for General Systems Research, 7, p. 283–90.

Thompson, John W Meteorological Models in the Social Sciences: I, suggestions for the dynamic control of behavior (1963)
General Systems Yearbook Society for General Systems Research, 8, p. 153–57.

Thompson, John W Sociometry and the Physical Sciences Part I (1964)
General Systems Yearbook Society for General Systems Research, 9, p. 1–5.

Thompson, John W Similar Problems in Meteorology and Psychology (1965)
General Systems Yearbook Society for General Systems Research, 1964, 10, p. 49–60.

Thompson, John W Meteorology and the Social Sciences: further comparisons (1966)
General Systems Yearbook Society for General Systems Research, 11, p. 19–25.

Thompson, W R (Ed) Contending Approaches to World Systems (1983)
London, Sage.

Thompson, William Irwin (Ed) Gaia, A Way of Knowing: political implications of the new biology (1987)
Great Barrington MA, Lindisfarne Press, 217 p. bibl.
ISBN 0–89281–080–7.

Thornton, Robert M Integrative Principles of Biology (1972)
In: *Integrative Principles of Modern Thought* New York, Gordon and Breach Science Publishers pp. 137–209.

Thrupp, S L History and Sociology: new opportunities for cooperation (1957)
American Journal of Sociology 63, July, pp. 11–16. bibl.

Thrupp, S L Comparative Studies in Society and History: a working alliance among specialists (1965)
International Social Science Journal 17, 4, pp. 644–54.

Thyssen–Bornemisza, Stephen The Unified System Concept of Nature (1955)
New York, Vantage Press, 137 p.

Tikekar, Inda B Integral Revolution: an analytical study of Gandhian thought (1970)
Varanas, Sarva Seva Sangh Prakashan, 266 p. (bibl. pp. 255–60) (Thesis to Banaras Hindu University).

Tinhofer, G and Schmidt, G (Ed) Graph–Theoretic Concepts in Computer Science (1987)
Berlin, Springer–Verlag, 305 p. ISBN 3–540–17218–1.

Toch, H H and Hastorf, A H Homeostasis in Psychology (1955)
Psychiatry 18, pp. 81–91.

Tocher, K D The Art of Simulation (1963)
Princeton, Van Nostrand, 184 p.

Toda, M and Takada, Y Studies of Information Processing Behavior (1958)
Psychologica 1, pp. 265–74.

Toda, Masanao and Shuford, E H Jr Logic of Systems: introduction to a formal theory of structure (1965)
General Systems Yearbook Society for General Systems Research, 1964, 10, p. 3–27.

Toffler, Alvin The Third Wave (1980)
New York, William Morrow.

Tolman, R C Relativity, Thermodynamics and Cosmology (1934)
Oxford, Clarendon Press.

Tomkins, Silvan Affect, Imagery, Consciousness
Berlin, Springer–Verlag. 4 Vols.

Tonge, Fred M Hierarchical Aspects of Computer Languages (1969)
In: *Lancelot Law Whyte, et al (Eds) Hierarchical Structures* New York, American Elsevier, pp. 233–52.

Tou, J T and Wilcox R H (Eds) Computer and Information Sciences: collected papers on learning, adaptation and control in information systems (1964)
Washington DC, Spartan Books.

Toulmil, Stephen Conceptual Revolutions in Science (1967)
In: *Cohen, Wartofsky (Ed) Boston Studies in the Philosophy of Science* 3, pp. 331–46.

Toulmil, Stephen The Evolutionary Development of Natural Science (1967)
American Scientist 55, pp. 456–71.

Toulmil, Stephen The Physical Sciences (1967)
In: *R M Hutchins and M J Adler (Eds) The Great Ideas Today* pp. 159–95, Chicago, Benton.

Toulmin, S Unity Through Diversity: a festschrift for Ludwig von Bertalanffy (1973)
New York. Vols I and II.

Toulmin, S and Goodfield, J The Classical Synthesis (of the physical sciences). (1962)
In: *The Architecture of Matter* London, Hutchinson. New York, Harper and Row.

Touloukian, Y S The Concept of Entropy in Communication, Living Organisms and Thermodynamics
Research Bull 130, Lafayette, Ind, Purdue Eng. Exper. Station.

Tracy, Stanley Bannon The Problem of Integration in Painting
Columbus ON, Ohio State University, 32 p.

Tranoy, Knut E Wholes and Structures: an attempt at a philosophical analysis (1959)
Copenhagen, Munksgaard, 42 p. bibl. (Interdisc. Studies from the Scandinavian Summer Univ, vol 1).

Trappl, R and Pichler F R (Ed) Progress in Cybernetics and Systems Research (1975)
New York, Wiley. 2 vols.

Tribus, Myron Information Theory as the Basis for Thermostatics and Thermodynamics (1961)
General Systems Yearbook Society for General Systems Research, 6, p. 127–37.

Trincher, K S Biology and Information: elements of biological thermodynamics (1965)
New York, Consultants Bureau.

Troncale, Len (Ed) A General Survey of Systems Methodology: proceedings of the Society for General Systems Research (1982)
Salinas CA, Intersystems Publications, 1014 p. Vols 1, 2.
ISBN 0–914105–23–X.

Truitt, Robert W Analogy of the Special Theory of Relativity to the Study of Compressible Fluid Flow (1949)
Raleigh, North Carolina State College, 20 p. bibl. (pp. 18–20).

Truxall, J G Control System Synthesis (1955)
New York, McGraw Hill Book Company.

Tsanoff, Radoslav Worlds to Know: a philosophy of cosmic perspectives (1962)
New York, Humanities Press, 230 p. phil. anth.

Tucker, S A A Modern Design for Defense Decision (1966)
Washington DC, Industrial College of the Armed Forces, 259 p.

Turbayne, C M The Myth of Metaphor (1962)
New Haven CT, Yale University Press.

Turner, F Design for a New Academy
In: *Harpers* 1986, 273, no 1636, pp. 47–53.

Turnill, Paul M L'Esprit de Synthèse Dans la Société des Nations: rapport entre les questions intellectuels, éthiques et éducatives et les questions économiques, sociales et politiques (Le Bureau de travail intellectuel et le Bureau international de travail; une enquète a réaliser) Genève, Pub (1921)
Messider, 7 p.

Tustin, A Feedback (1952)
Sci. Am. 187, pp. 48–54.

Ulam, Stanislaw On Some Mathematical Problems Connected with Patterns of Growth of Figures (1962)
In: *Proceedings of Symposia in Applied Mathematics, American Mathematical Society* Providence, Rhode Island, 14, pp. 215–24.

Ulam, Stanislaw Patterns of Growth of Figures: mathematical aspects. From: (1966)
G Kepes (Ed) Module, Proportion, Symmetry, Rhythm New York, Braziler, p. 64–73.

Umpleby, Stuart A The Revolution that Fizzled: the lack of impact of cybernetics on political science (paper for the Rome Special Conference on Futures Research 1973) (1973)

Unesco Colloque interdisciplinaire sur le rôle son milieu, l'architecture et l'urbanisme pour l'expansion et le changement (Otaniemi, 1970)
Revue Internationale des Sciences Sociales 22, 4, 1970, part of complete issue on the topic (also English edition).

Unesco Main Trends of Research in the Social and Human Sciences
Paris, UNESCO. forthcoming, vol 3 (devoted to areas of friction and convergence in interdisciplinary research).

Unesco Interrelations of Cultures: Their contribution to international understanding (1953)
Paris, UNESCO, 387 p. bibl. (pp. 383–87) (Its: Collection of Intercultural Studies: Unity and Diversity of Cultures, No 2).

Unesco The Social Sciences, Problems and Orientations/Les sciences sociales, problèmes et orientations (1968)
Den Haag, Mouton, 507 p.

Unesco Multidisciplinary Problem–Focused Research (1968)
International Social Science Journal 20, 2, whole issue.

Unesco Planning Meeting for UNESCO's Programme in Integrated Science Teaching (1969)
Paris, UNESCO.

UNESCO Tendances Principales de la Recherche dans les Sciences Sociales et Humaines: première partie – sciences sociales (1970)
Paris, UNESCO, 987 p. (major section on: Dimensions interdisciplinaires de la recherche).

Unesco Sciences and Synthesis: an international colloquium organized by Unesco on the tenth anniversary of the death of Albert Einstein and Teilhard de Chardin (1971)
Berlin, Springer–Verlag, 206 p.

Unesco Integrated Science Teaching in the Asian Region (1971)
Bangkok, Unesco Regional Office for Education in Asia.

UNESCO Interdisciplinarité et sciences humaines (1983)
Paris, UNESCO, 343 p. ISBN 92–3–204988–4.

UNESCO Evaluating Long–Term Developments by Using Global Models (1987)
Paris, UNESCO, 41 p. bibl. Major Programme I, Reflection on World Problems and Future–Oriented Studies: BEP/GPI/4.

Unger, Carl Vom Bilden Physikalischer Begriffe
Stuttgart, Martin Sandkühler.

Unger, Georg Das Offenbare Geheimnis des Rammes: meditationen am pentagondodekaeder nach Carl Kemper (1963)
Stuttgart, Martin Sandkühler, 80 p.

Union of International Associations/Mankind 2000 Integrative and Transdisciplinary Concepts (1976)
In: *Yearbook of World Problems and Human Potential* Brussels, Union of International Associations, special section.

United Nations Department of Economic and Social Affairs Unified Socio-economic Developmental (1971)
New York, UN, 67 p. ST/SOA/Ser. X/3.

United Nations University Science and Praxis of Complexity: contributions held at Montpellier, France, 9–11 May, 1984 (1985)
Tokyo, United Nations University, 384 p. ISBN 92-808-0560-6.

United States Congress, Office of Technology Assessment Global Models: world futures and public policy: a critique (1982)
Washington DC, US GPO. OTA-R-165.

University of California, Dept of Engineering Systems Evaluation of the World Hunger Problem (1967)
Los Angeles, The Department, 614 p. bibl. Report EEEP 67-11.

Urban, G R Kinesis and Stasis: a study in the attitude of Stefan George and his circle to the musical arts (1962)
Den Haag, Mouton, 209 p.

Uttley, A M The Informon: a network for adaptive recognition (1970)
Journal of Theoretical Biology 27, pp. 31–67.

Valarche, J L'Expérience Interdisciplinaire d'Economie Régionale à l'Université de Fribourg (Suisse) (1969)
Bière (Ed) Revue Juridique et Economique du Sud-Ouest 3.

Vallance, Theodore R Structural Innovations in Higher Education to Meet Social Needs (1970)
Washington DC. ERIC Clearinghouse on Higher Education.

Van Peursen, Cornelis A The Strategy of Culture: a view of the changes taking place in our ways of thinking and living today (1974)
Amsterdam, North-Holland, 265 p.

Van Prag, H M (Ed) Biochemical and Pharmacological Aspects of Dependence and Reports on Marihuana Research (Symposium organized by the Interdisciplinary Society of Biological Psychiatry, Amsterdam 1971) (1972)

Van Spronsen, J W The Periodic System of Chemical Elements: a history of the first hundred years (1969)
Amsterdam, Elsevier Science Publishing, 368 p. bibl. foot.

Van Wagenel, R International Integration: a review
In: *Conflict Resolution* vol 9, 4, pp. 526–31.

Váradi, Emery Yechiel Functional Human Universalism: in search of a solution to the crisis of mankind and to human survival (1981)
New York, Vantage Press, 143 p. ISBN 533-04869-9.

Varela, Francisco A Calculus for Self-Reference (1975)
International Journal of General Systems 1975, 2, p. 5–24.

Vayda, A Environment and Cultural Behavior
New York, Doubleday.

Vayda, A Man, Culture and Animals
Washington DC, AAAS.

Venturi, Franco Le Origini dell Enciclopedia (du Diderot et al) (1946)
Rome, Edizioni V, 164 p.

Vickers, G Control, Stability and Choice (1957)
General Systems Yearbook Society for General Systems Research, 2, p. 1–8.

Vickers, G The Concept of Stress in Relation to the Disorganization of Human Behavior (1959)
In: *Stress and Psychiatric Disorder, Blackwell* Oxford, pp. 3–10.

Vickers, G Is Adaptability Enough? (1959)
Behavioral Science 4, pp. 319–34.

Vickers, G The Regulation of Political Systems (1968)
General Systems Yearbook Society for General Systems Research, 12, p. 59–67.

Vickers, G A Classification of Systems (1970)
General Systems Yearbook Society for General Systems Research, 1970a, 15, p. 3–6.

Vickers, G The Ecology of Ideas (1970)
In: *Geoffrey Vickers, Value Systems and Social Process* London, Tavistock Publications.

Vickers, G Value Systems and Social Process: choosing, planning, controlling, revaluing, appreciating, learning, surviving (1970)
New York, Pelican Books, 221 p.

Vickers, G, et al Hierarchically Organized Systems in Theory and Practice (1971)
New York, Hafner Publications, 263 p. bibl. (p. 262).

Vickers, Geoffrey Freedom in a Rocking Boat: changing values in an unstable society (1970)
London, Penguin Books, 215 p. ISBN 0-14-021205-1.

Vidulich, R N The Integration of Multiple Sets into a New Belief System (1956)
Unpublished M A Thesis, Michigan State Univ.

Voegelin, E Order and History
Univ of Louisiana Press, 3 vols.

Voge, J Management of Complexity
In: *The Science and Praxis of Complexity* 1985, Tokyo, United Nations University, pp. 298–311.

Voge, J The Political Economy of Complexity
In: *Information Economics and Policy 1* 1983, Amsterdam, North-Holland, pp. 97–114.

Vogelaar, G A M Communicatie, Kernproces van de Samenleving: sociologie en cybernetiek (1962)
Haarlem, F Bohn, 213 p.

Vogt, E Z On the Concepts of Structure and Process in Cultural Anthropology (1960)
American Anthropologist 62, pp. 18–33.

Von Bertalanffy, L and Rapoport, A (Eds) General Systems: yearbook of the Society for General Systems Research (1956)
Vol 1.

Von Bertalanffy, L, et al General System Theory: a new approach to unity of science (1951)
Human Biology 23, 1–4, pp. 302–61.

Von Bertalanffy, Ludwig General System Theory (1915)
Main Currents in Modern Thought 11, pp. 75–83, 1955. Reprinted in 1956. *General Systems* Yearbook 1 pp. 1–10.

Von Bertalanffy, Ludwig An Essay on the Relativity of Categories (1915)
Philosophy of science 22, 1955, pp. 243–63.

Von Bertalanffy, Ludwig The Theory of Open Systems in Physics and Biology (1950)
Science 3, pp. 23–29.

Von Bertalanffy, Ludwig An Outline of General Systems Theory (1950)
British Journal for the Philosophy of Science 1, p. 134.

Von Bertalanffy, Ludwig Problems of General System Theory (1951)
Human Biology 23, pp. 302–12.

Von Bertalanffy, Ludwig Theoretical Models in Biology and Psychology (1952)
In: *Theoretical Models and Personality Theory* Durham, NC, Duke University, pp. 24–38.

Von Bertalanffy, Ludwig Problems of Life: an evaluation of modern biological thought (1952)
New York, John Wiley and Sons.

Von Bertalanffy, Ludwig Levels of Organization (1952)
In: *Problems of Life* New York, Wiley, chapter 2.

Von Bertalanffy, Ludwig A Biologist looks at Human Nature (1956)
The Scientific Monthly 82, 34.

Von Bertalanffy, Ludwig General Systems Theory: a new approach to unity of science (1956)
Human Biology 28, 4, December.

Von Bertalanffy, Ludwig The Significance of Psychotropic Drugs for a Theory of Psychosis (1957)
Geneva, WHO.

Von Bertalanffy, Ludwig Democracy and Elite: the education quest, comments on interdisciplinary study (1962)
Main Currents in Modern Thought 19, No 2, 31.

Von Bertalanffy, Ludwig Kritische Theorie der Formbildung (1962)
New York, Harper and Row.

Von Bertalanffy, Ludwig On the Definition of the Symbol (1965)
New York, Randum. Psychology and the Symbol: an interdisciplinary Symposium.

Von Bertalanffy, Ludwig General Systems Theory and Psychiatry (1966)
New York, American Handbook of Psychiatrics. Vol 3.

Von Bertalanffy, Ludwig Robots, Men and Minds: psychology in the modern world (1967)
New York, Braziller, 150 p.

Von Bertalanffy, Ludwig General Theory of Systems: application to psychology (1968)
In: *The Social Sciences, problems and orientations* pp. 309–19, The Hague, Paris, Mouton, Unesco.

Von Bertalanffy, Ludwig General Systems Theory: foundations, development, applications (1968)
New York, Braziller, 289. (Milan, Istituto Librario Internazionale, 1972, Stockholm, Wahlström and Widstrad 1972, Tokyo, Misuzu Shobo 1972, Madrid, Guadarrama 1972, Paris, Dunod 1972, Braunschweig, Vieweg 1973).

Von Bertalanffy, Ludwig General Systems Theory as Integrating Factor in Contemporary Science and in Philosophy (1969)
Proceedings of the XIVth International Congress of Philosophy 2, p. 339, Vienna.

Von Bertalanffy, Ludwig Chance or Law (1969)
In: *A Koestler and J Smythies (Eds) Beyond Reductionism; new perspectives in the life sciences* (Alpbach Symposium, 1968) pp. 56–84. London, Hutchinson.

Von Bertalanffy, Ludwig General Systems Theory: a critical review (1969)
In: *Litterer, Joseph A (Ed) Organizations; systems control and adaptation* London, Wiley, 2nd ed., vol 2, pp. 7–30.

Von Foerster, H; Mead, M and Teuber, H L Transactions of Conferences on Cybernetics
New York, Josiah Macy Foundation. 1949–1957. 5 Vols.

Von Heltic, Hartmut Die Einheit der Wissenschaft: dargestellt als problem einer allgemeinen wissenshaftspropädeutik und wissenshaftsdidaktik (1971)
Merkur September/October.

von Holzhey, H Interdisziplinäre Arbeit und Wissenschaftstheorie: interdisciplinär (1974)
Stuttgart, Schwabe. Vol I.

Von Neumann, J Probabilistic Logics (1952)
Calif. Inst. Techn.

Von Neumann, J The Computer and the Brain (1958)
New Haven CT, Yale University Press.

Von Neumann, J Theory of Self-Reproducing Automata (1966)
Urbana, University of Illinois Press, 388 p.

Von Neumann, J and Morgenstern, O Theory of Games and Economic Behavior (1964)
New York, John Wiley and Sons, 641 p.

Vreecken, J Integratie en Interactie (1970)
Amsterdam, Scheltema and Holkema, 22 p. bibl. (p. 22) (medicine).

Vroom, Victor H (Ed) Methods of Organizational Change (1967)
Pittsburgh PA, University of Pittsburgh Press.

Waddington, C H (Ed) The Human Evolutionary System. (1961)
In: *M Banton (Ed) Darwinism and the study of society* Chicago, Quadrangle, pp. 63–82.

Waddington, C H Towards a Theoretical Biology (1970)
Chicago IL, Aldine.

Waddington, C H Behind Appearance (1970)
Edinburgh, Edinburgh University Press.

Wadia, M S (Ed) The Nature and Scope of Management (1966)
Chicago IL, Scott Foresman, 349 p.

Waller, M D Chladni Figures: a study in symmetry (1961)
London.

Wallerstein, Immanuel A world-system perspective on the social sciences (1976)
British Journal of Sociology 27, 3, p. 343–352.

Wallerstein, Immanuel La crise comme transition (1982)
In: *La Crise, Quelle Crise?* Paris, Maspero, pp. 10–56.

Walter, R I and Walter, N I McN The Equivocal Principle in Systems Thinking (1971)
General Systems Yearbook Society for General Systems Research, 16, p. 3–11.

Ward, Barbara Spaceship Earth (1966)
New York, Columbia University Press.

Ward, Roger G; Reynolds, G William and Nurnberger, Robert Unified Science: a workable approach
Science Education 53, pp. 137–40.

Warfield, J N An Assaulty on Complexity (1973)
Columbus OH, Battelle Memorial Institute, April.

Warfield, J N and Hill, J D A Unified Systems Engineering Concept (1972)
Columbus OH, Battelle Memorial Institute, June.

Warfield, John N Societal Systems (1989)
Salinas CA, Intersystems Publications, 490 p.

Washburn, S L and Howell F C Evolution After Darwin
University of Chicago Press. Vol 2.

Wasserman, Paul and Silander, Fred S Decision-Making: an annotated bibliography: Supplement 1958–63 (1964)
Ithaca.

Watkins, B O Introduction to Control Systems (1969)
New York, Macmillan.

Watkins, J W N Holism Versus Individualism (1956)
Proceedings of the Aristotelian Society.

Watson, Goodwin Concepts for Social Change
Washington, National Education Association.

Watson, Goodwin Wholes and Parts in Education (1932)
Teachers College Record 34, November 1932 pp. 119–33.

Watt, Kenneth E (Ed) Systems Analysis in Ecology (1966)
New York, Academic Press, 276 p. bibl.

Wax, Murray L Myth and Interrelationship in Social Science Illustrated through Anthropology and Sociology (1969)
In: *M Sherif and C Sherif (Eds) Interdisciplinary Relationships in the Social Sciences* Chicago, Aldine, pp. 77–99.

Weaver, Wassen Science and Complexity (1948)
American Scientist 36, pp. 536–644.

Weber, Renée Dialogues with Scientists and Sages: the search for unity (1986)
Routledge.

Weinberg, Alvin M Reflections on Big Science (1967)
Cambridge MA, MIT Press.

Weinberg, Daniela and Weinberg, Gerald M General Principles of Systems Design (1988)
New York, Dorset House Publishing, 376 p. illus. repr of 1979 ed. ISBN 0-932633-07-2.

Weinberg, Gerald M Rethinking Systems Analysis and Design (1988)
New York, Dorset House Publishing, 208 p. illus. repr of 1982 ed. ISBN 0-932633-08-0.

Weinberg, H L Levels of Knowing and Existence
London, Hodder and Stoughton (unifying function of semantics).

Weinberg, Meyer (Comp) Integrated Education: a reader (1968)
Integrated Education Beverley Hills CA, Glencoe Press.

Weiss, Paul A Life, Order and Understanding (forthcoming)

Weiss, Paul A Nine Basic Arts (1958)

Weiss, Paul A Knowledge: a growth process (1960)
Science 131, 10 June pp. 1716–19.

Weiss, Paul A Dynamics of Development: experiments and inferences (1968)
New York, Academic Press.

Weiss, Paul A Some Paradoxes Relating to Order (1968)
In: *Paul G Kuntz (Ed) The Concept of Order* Seattle, University of Washington Press, pp. 14–22.

Weiss, Paul A The Living System: determinism stratified (1969)
In: *A Koestler and J Smythies (Eds) Beyond Reductionism; new perspectives in the life sciences* (Alpbach symposium, 1968) pp. 3–55. London, Hutchinson, bibl. foot.

Weiss, R S and Jacobson, E A Method for the Analysis of the Structure of Complex Organizations (1961)
In: *A Etzioni (Ed) Complex Organizations: A sociological reader* New York, Holt.

Weiss, Robert and Rein, Martin The Evaluation of Broad Aim Programs: a cautionary case and a model (1969)
Annals of the American Academy of Political and Social Science 385, Sept. pp. 133–42.

Weizäcker, C F von The Unity of Nature (1980)
New York, Farrar-Straus-Giroux.

Weizacker, Christina von Die umweltfreudliche Emanzipation (1975)
In: *Humanoekologie* Vienna, Georgi, 1975.

Weizsaecker, Viktor von Der Gestaltkreiss: theorie der einheit von wahrnehmen und bewegungen (1947)
Stuttgart, Georg Thieme Verlag. 3rd ed.

Wells, Bill Levels and Integrated Entities (notes on hierarchy in artifact)
In: *Lancelot Law Whyte, et al (Eds) Hierarchical Structures* New York, American Elsevier, 1969, pp. 282–83.

Werkmeister, William H Basis and Structure of Knowledge (1948)
New York, Greenwood Press.

Wertheimer Laws of Organization in Perceptual Forms (1966)
In: *W D Ellis (Ed) A Source Book of Gestalt Psychology* Humanities.

Weyl, H Symmetry (1952)
Princeton NJ, Princeton University Press, 168 p.

Weyl, Hermann Mind and Nature (1934)
Philadelphia PA, University of Pennsylvania Press.

Weyl, Hermann Chemical Valence and the Hierarchy of Structures (1949)
In: *Philosophy of Mathematics and Natural Science* Princeton, Princeton Univ P., 1949 (Appendix D).

Wheeler, Raymond H The Problem of Integration (1937)
In: *L Thomas Hopkins (Ed) Integration; its meaning and application* New York, Appleton–Century, pp. 36–49.

White, Alvin M (Ed) Interdisciplinary Teaching (1981)
San Francisco CA, Jossey–Bass. New Directions for Teaching and Learning Series: 8. ISBN 0–87589–869–6.

White, D J Decision Theory (1969)
Chicago IL, Aldine 200 p.

White, Han J and Tauber, S Systems Analysis (1969)
Philadelphia PA, WB Saunders, 499 p. bibl.

White, Leslie A Diffusion vs Evolution: an anti–evolutionist fallacy (1945)
American Anthropologist 47, pp. 339–56.

White, Leslie A The Concept of Evolution in Cultural Anthropology (1959)
In: *Evolution and Anthropology* Washington DC, Anthropological Society.

White, Leslie A The Evolution of Culture: the development of civilization to the fall of Rome (1959)
New York, McGraw Hill Book Company.

Whitehead, Alfred North Symbolism, Its Meaning and Effect (1927)
New York, Macmillan.

Whitehead, Alfred North Process and Reality: an essay in cosmology (1929)
New York, Allied Publications.

Whitehead, Alfred North Modes of Thought (1938)
New York, Cambridge University Press

Whitehead, Alfred North Symbolism: Its meaning and effect (1953)
In: *F S C Northrop and M W Gross (Eds) A Whitehead Anthology* Cambridge.

Whitehead, Alfred North Process and Reality (1969)
New York, Macmillan.

Whitehead, Clay T Uses and Limitations of Systems Analysis (1967)
Santa Monica CA, Rand Corporation, 182 p. bibl. (Rand Corp. P–3683).

Whorf, B L Collected Papers on Metalinguistics (1952)
Washington DC, Foreign Service Institute.

Whorf, B L Language, Thought and Reality: selected writings of B L Whorf (1956)
New York, Wiley and Sons. edited by J B Carroll.

Whyte, Lancelot Law Unitary Principles in Physics and Biology (1949)
New York, Holt, Rinehart and Winston, 182 p. (London, Cresset Press, 1949) (Includes references to mental processes).

Whyte, Lancelot Law Aspects of Form: a symposium on form in nature and art (1951)
London, Lund Humphries, 249 p. bibl. (pp. 238–49) (New York, Pellegrini and Cudahy American Elsevier).

Whyte, Lancelot Law Accent on Form: an anticipation of the science of tomorrow (1954)
New York, Harper and Row, 198 p. (London, Routledge and Paul 1955 202p).

Whyte, Lancelot Law Atomism Structure and Form (1965)
In: *G Kepes (Ed) Structure in Art and in Science* New York, Braziller, 189 p.

Whyte, Lancelot Law Internal Factors in Evolution (1965)
New York, Braziller, 128 p. bibl. (pp. 115–17).

Whyte, Lancelot Law Organic Structural Hierarchies (1969)
In: *Unity and Diversity in Systems (essays in honour of L von Bertalanffy, R. G. Jones and G. Brandl (Eds))* New York, Braziller.

Whyte, Lancelot Law Towards a Science of Form (1970)
Hudson Review 23, Winter 1970–71, pp. 613–32.

Whyte, Lancelot Law The Structural Hierarchy in Organisms (1973)
In: *Gray, William and Rizzo, Nicholas D (Eds) Unity through dipersity* New York, Gordon ald Breach Science Publishers. pp. 271–80.

Whyte, Lancelot Law et al (Eds) Hierarchical Structures (1969)
New York, American Elsevier.

Wiener, N and Schadé, J P (Eds) Progress in Biocybernetics (1964)
New York, Elsevier Scientific Publishing. 2 vols.

Wiener, N and Schadé, J P Cybernetics of the Nervous System (1965)
Amsterdam, Elsevier Science Publishing.

Wiener, Norbert A Simplification of the Logic of Relations (1914)
Proceedings of the Cambridge Philosophical Society 17, pp. 387–90.

Wiener, Norbert Cybernetics (1948)
New York, John Wiley and Sons.

Wiener, Norbert The Human Use of Human Beings: cybernetics and society (1950)
Boston MA, Houghton Mifflin, 288 p. New York, Doubleday, 1954 2nd ed.

Wiener, Norbert Cybernetics: or control and communication in the animal and the machine (1961)
Cambridge MA, MIT Press, 194 p.

Wigner, E P The Role of Invariance Principles in Natural Philosophy (1924)
New York, Academic Press. Varelna lectures, on 'Dispersion Relations and their connection with causality'.

Wigner, E P Events, Laws of Nature and Invariance Principles (1964)
Science 145, 995, Sept. 4.

Wigner, Eugene Symmetries and Reflections (1979)
Woodbridge, Ox Bow Press.

Wilber, Ken (Ed) The Holographic Paradigm and Other Paradoxes: exploring the leading edge of science (1982)
Boulder CO, Shambhala Publications, 300 p. New Science Library. ISBN 0–87773–235–3.

Wilber, Ken Eye to Eye: the quest for the new paradigm (1990)
Bostson MA, Shambhala Publications, 352 p. Repr of 1983. ISBN 0–87773–549–2.

Wilden, A System and Structure: essays in communication and exchange (1972)
London, Tavistock Publications.

Wilkinson, John The Concept of Information and the Unity of Science (1961)
Philosophy of Science 28, Oct 4.

Willer, Judith The Social Determination of Knowledge (1971)
Englewood Cliffs NJ, Prentice Hall, 150 p. bibl.

Williams, Hugh E General Systems Theory, Systems Analysis and Regional Planning: an introductory bibliography (1970)
Monticello IL, Council of Planning Librarians, 31 p. (Exchange Bibliography 164).

Williams, Iolo Wynn Teaching Methods in Integrated Science at the Primary and Secondary Levels
In: *New trends in integrated science teaching, Vol II* Paris, Unesco pp. 71–88.

Williams, J D The Complete Strategist (1966)
New York, McGraw Hill Book Company, 268 p.

Williams, R The Geometrical Foundation of Natural Structure (1972)
New York, Dover.

Williams, Robert Edward Notes on Inorganic Hierarchical Structures: dimension as level (1969)
In: *Lancelot Law Whyte, et al (Eds) Hierarchical Structures* New York, American Elsevier, pp. 135–37.

Williamson, G Scott and Pearse, I H Science, Synthesis and Sanity: an inquiry into the nature of living (by the founders of the Peckham Experiment) (1965)
London, Collins Liturgical Publications, 352 p. (includes a 'Dictionary of Quality).

Williamson, J J An Outline of the Principles and Concepts of the New Metaphysics (1927)
Hastings, Society of Metaphysicians, 39p. mimeo (New Metaphysics, no 1).

Willner, Dorothy and Washburne, N F (Eds) Interdisciplinary Behavioral Sciences Research Conference (University of New Mexico): decision, values and groups
New York, Pergamon Books. 2 vols. bibl.

Wilson, Albert G Olbers' Paradox and Cosmology (1965)
Santa Monica CA, Rand Corporation, 20 p. paper P–3256.

Wilson, Albert G Morphology and Modularity (1967)
In: *Zwicky and Wilson (Eds) New Methods of Thought and Procedure* pp. 298–313, New York, Springer Verlag.

Wilson, Albert G A Hierarchical Cosmological Model (1967)
Astronom J 72, p. 326.

Wilson, Albert G Hierarchical Structure in the Cosmos (1969)
In: *Lancelot Law Whyte, et al (Eds) Hierarchical Structures* New York, American Elsevier, pp. 113–34.

Wilson, Albert G Notes on Hierarchy in Concept: closure–entity and level (1969)
In: *Lancelot Law Whyte, et al (Eds) Hierarchical Structures* New York, American Elsevier, pp. 54–55.

Wilson, Colin Beyond the Outsider: the philosophy of the future (1925)
London, Arthur Barker, 236 p. bibl. (pp227–29) (synthesis of evolutionary humanism with phenomenological existentialism).

Wilson, Donna Forms of Hierarchy: a selected bibliography (1969)
General Systems 14, p. 3–15, bibl.

Wilson, E O Sociobiology: the new synthesis (1975)
Cambridge MA, Harvard University Press.

Wilson, P R On the Argument by Analogy (1964)
Philosophy of Science 31, January pp. 34–39 bibl.

Winiarski, B Recherche Interdisciplinaire Pour la Planification du Développement Régional: les expériences polonaises (1969)
Bière (Ed) Revue Juridique et Economique du Sud–Ouest 3.

Wisdom, J O The Hypothesis of Cybernetics (1951)
British Journal for the Philosophy of Science 2, pp. 1–24.

Wittgenstein, Ludwig The Bleu and Brown Books (1972)
Oxford, Blackwell.

Wohl, R R Some Observations on the Social Organization of Interdisciplinary Social Science Research (1955)
Social Forces 33, May pp. 374–83.

Woodger, J H Biology and Language (1952)
Cambridge MA, Cambridge University Press.

World Order Models Project Economics and World Order from the 1970's to the 1990's (1972)
London, Macmillan, 365 p. bibl.

World Order Study Conference Paths to World Order (1967)
New York, Columbia University Press, 161 p. (Proceedings of 6th Conference, St. Louis, 1965).

Worrell, Florence Basic Lessons in Synergetics
Parkersburg WV, Synergetics Press.

Worsley, P One World or Three?: a critique of the world system theory of Immanuel Wallerstein (1980)
Socialist Register, pp. 298–338.

Wortman, M S Jr and Luthans, F (Eds) Emerging Concepts in Management: process, behavioral, quantitative, and systems (1969)
New York, Macmillan, 462 p.

Yamada, D I Metaphor in Political Thought: an essay in comparison (1975)
Santa Barbara CA, Univerity of California. Ph D Dissertation.

Yamada, Keiji Konton no Umi e: Chugoku–teki Shiko no Kozo (In a Sea of Chaos; the structure of Chinese thinking) (1975)
Tokyo.

Yates, Frances The Art of Memory (1969)
London, Penguin Books.

Yolton, J W Metaphysical Analysis (1967)
Toronto On, University of Toronto Press.

Young, Arthur The Geometry of Meaning (1978)
Boston MA, Delacort Press/Seymour Lawrence.

Young, J F Cybernetics (1969)
New York, American Elsevier.

Young, O R A Survey of General Systems Theory (1964)
General Systems Yearbook Society for General Systems Research, 9, p. 61–80.

Young, O R The Impact of General Systems Theory on Political Science (1964)
General Systems Yearbook Society for General Systems Research, 9, p. 239–53.

Young, S Management: a systems analysis (1966)
Chicago IL, Scott Foresman, 436 p.

Young, T R Social Stratification and Modern Systems Theory (1969)
General Systems Yearbook Society for General Systems Research, 14, p. 113–17.

Yourdon, E Cybernetics (1961)
Cambridge MA, MIT Press. 2nd ed.

Yourgrau, W General System Theory and the Vitalism–Mechanism Controversy (1952)
Scientia 87, pp. 307.

Yovits, M C and Cameron, S Self, Organizing Systems: proceedings of the Interdisciplinary Conference on Self–Organizing Systems, Chicago 1959 (1960)
New York, Pergamon Press, 322 p.

Yovits, M C; Jacobi, G T and Goldstein, G D Self–Organizing Systems (1962)
New York, Spartan Books, 563 p.

Zacharias, J R Structure of Physical Science (1957)
Science 125, pp. 427–28.

Zachner, R C Concordent Discord: the interdependence of faith (1970)
Oxford, Clarendon Press.

Zader, L From Circuit Theory to System Theory (1962)
IRE Proceedings May 50, 5 p. 857.

Zawodny, J K (Ed) Man and International Relations: contributions of the social sciences to conflict and integration (1966)
San Francisco CA, Chandler.

Zeleny, M Spontaneous Social Orders
In: *The Science and Praxis of Complexity* 1985, Tokyo, United Nations University, pp. 312–328.

Zeleny, M Autopoiesis, Dissipative Structures and Spontaneous Social Order (1980)
Boulder CO, Westview.

Zeller, C A Geometric Model with Some Properties of Biological Systems (1968)
General Systems Yearbook Society for General Systems Research, 12, p. 53–55.

Zeman, J Le sense philosophique du terme 'l'information' (1962)
La Documentation en France 3, pp. 19–29.

Zimmerman, H D Die Politische Rede (1975)
Stuttgart, Kohlmanner Verlag. Metaphors in political discourse chosen from military, organic and nature domains.

Zimmern, Alfred The Intellectual Foundations of International Cooperation (1928)
London.

Zuckerkandl, V Sound and Symbol: music and the external world
Princeton University Press.

Zukav, G The Dancing Wu–Li Masters (1979)
New York, William Morrow.

Zwicky, Fritz and Wilson, A G (Ed) New Methods of Thought and Procedures (1967)
New York, Springer Publishing.

Notes

Integrative Knowledge

Significance

Intent	499
Obstacles to an interdisciplinary focus	500
Consequence for approaches to societal problems	501
Previous, parallel or related initiatives	502

Method

Identification and guidelines	504
Approaches to the art of disagreement	505
Subtler forms of disagreement	507
Constraints on patterning disagreement	508

Comments

Integrative concepts	510
Patterning disagreement	512
Systems of categories distinguishing cultural emphases	514

*** Bibliographical references identified in abridged form in the following section refer to publications detailed, by author, in Section KY, which is the bibliography for Section K.

Significance: intent

1. Intent
(a) Identify the range of those concepts which in some way either integrate or interrelate concepts, especially from different disciplines, or describe conditions or formal properties common to a conceptual approach to the subject matter of many disciplines.

(b) Provide sufficient description of each concept, based on available documents which depend upon and advocate its use, in order to clarify the special importance attached to each such particular integrative or interdisciplinary concept.

(c) Clarify the relationships and distinctions between different concepts labelled by terms which are synonyms or homonyms, particulary where the meaning of the terms used changes with the context.

(d) Provide a common framework for integrative concepts which are used in essentially different and frequently non-interacting sectors of society.

(e) Highlight those integrative concepts which have hitherto been excluded from serious consideration (whether in the academic world or in their societal applications) but for which some legitimating documents and reports now exist.

(f) Distinguish, by juxtaposition within the same context, those concepts which are more comprehensive in their integrating power, from those which are of relatively limited integrating power although possibly attracting much greater attention.

2. Significance
(a) The communication problem: Today most scientists are acknowledged as specialists, for whom it is legitimate to know progressively more and more about less and less. As Harold Linstone notes (1973): *"When a group of prestigious future-oriented interdisciplinary scientists meets, the result usually fits the words of novelist Arthur Koestler: 'The moment you put them together in a conference room, they behave like schoolboys performing a solemn play...each of them possesses a small fragment of the Truth which he believes to be the Whole Truth, which he carries around in his pocket like a tarnished bubble gum, and blows up on solemn occasions to prove that it contains the ultimate mystery of the universe. Discussion? Interdisciplinary dialogue? There is no such thing, except on the printed program. When the dialogue is supposed to start each gets his own bubble gum out and blows it into the other's faces. Then they repair, satisfied, to the cocktail room.'"*

Thus we have a modern Tower of Babel - more people, more social and professional groups with access to vast communication, channels yet unable to exploit them because of resistance to transcending the separateness of their languages.

(b) The interdisciplinary movement: In recent years there has emerged a diffuse movement based on a variety of interdisciplinary approaches. There is, for example, the increasing number of hybrid disciplines with hyphenated names indicating their disciplinary origin. The newer interdisciplines, however, have a much more varied and occasionally even obscure ancestry. They result from the reorganization of material from many different fields of study. Cybernetics comes out of electrical engineering, neurophysiology, physics, biology, with an element of economics.

Information theory, which originated in communications engineering, has important applications in many fields stretching from biology to the social sciences. Organization theory comes out of economics, sociology, engineering, physiology, and management science. Then there are a multiplicity of concepts, possibly associated with such interdisciplines, which themselves transcend normal disciplinary boundaries and perform an integrative function. Such concepts and interdisciplines tend to be associated with particular schools of thought, few of which appear to feel strongly related to the others, if at all. As a result there has been no move to delimit the group of such conceptual tools, to attempt to describe systematically its contents, or to determine what the conceptual elements of the group have in common. There are few, if any, collecting points for interdisciplinary or integrative documents and perspectives.

There appear to be three principal reasons for this interdisciplinary or integrative movement, for want of a better and more inclusive term:

(a) concern at the uncontrolled multiplication of unrelated specialized approaches to an objective world which is believed to be an unfragmented totality;
(b) concern at the inadequacy and natural limitations of specialized approaches;
(c) the unsuspected complexity of social problems.

(c) Fragmentation: The considerable increase in specialization in all branches of science is a well-recognized phenomenon. René Maheu (1971), as Director-General of UNESCO, noted UNESCO's concern in responding to this condition: *"...in the face of the growing specialization of thought and action brought about by diversification in research and the division of labour, UNESCO has a duty to promote inter-disciplinary activities and contacts and to encourage broad views, in short, to emphasize the vital importance of the spirit of synthesis for the health of our civilization. I say vital advisedly since man - and I mean his essence, which is to say his judgment and his freedom of choice - is just as likely to be smothered by his knowledge as paralysed by the lack of it."*

The inadequacy and natural limitations of specialized approaches are less well-recognized. It is however increasingly recognized that it is both inefficient and inadequate to organize research or action programmes as though nature were organized into disciplinary sectors in the same way that universities are.

How, asks Russell Ackoff (1960), is a practitioner of any one discipline to know in a particular case whether another discipline is better equipped to handle the problem than is his? It would be rare indeed if a representative of one of the many disciplines in some way related to the problem in question did not feel that his particular approach to that problem would be very fruitful, if not the most fruitful.

This tendency is also institutionalized, as noted by Hasan Ozbekhan (1969): *"This almost subconsciously motivated attempt, that of a sector to expand over the whole space of the system in its own particular terms and in accordance with its own particular outlooks and traditions, compounds the problem by further fragmenting the wholeness of the system. For sectors cannot become systems, they can only dominate them; and when they do they warp them."*

On the same point, Ackoff notes (1960): *"...few of the problems that arise can adequately be handled within any one discipline. Such systems are not fundamentally mechanical, chemical, biological, psychological, social, economic, political, or ethical. These are merely different ways of looking at such systems. Complete understanding of such systems requires an integration of these perspectives. By integration I do not mean a synthesis of results obtained by independently conducted undisciplinary studies, but rather results obtained from studies in the process of which disciplinary perspectives have been synthesized. The integration must come during, not after, the performance of the research."*

Significance: obstacles to an interdisciplinary focus

Georges Gusdorf, in an exceptional survey of interdisciplinarity for the French-language *Encyclopaedia Universalis France* (1972) notes that interdisciplinary knowledge is à la mode and that everybody now calls for "pluridisciplinarity" or "multidisciplinarity". He even suggests that there is an element of snobism in doing so. However, on closer examination, it is possible to discover that this requirement, far from constituting any form of progress, is only the symptom of the pathological state of knowledge at this time. The specialization without limit of scientific disciplines over the past two hundred years has resulted in an increasing fragmentation of the epistemological horizon. Atomised knowledge, he suggests, is the work of an atomised intelligence, which it would be legitimate to believe had lost the element of reason. The consequence is a disequilibrium which reaches the human personality in its totality. This scientific alienation may thus be regarded as one of the causes of the malaise of contemporary civilization.

After reviewing the problem of the disintegration of knowledge, specialization seen as an epistemological cancer, and the need for unitary knowledge, Gusdorf comments on the obstacles to interdisciplinary knowledge.

1. Discipline sub-division
The first obstacle he identifies is an epistemological one arising from the inexorable process of discipline subdivision and divergence. To the extent that each science is a well-formed language, each language thus created encloses the associated knowledge in an axiomatic space isolated from that of similar languages.

Specialists cannot be asked to testify with regard to the unification of the sciences insofar as these specialists by their vocation and training are ignorant of, or deny this very unity. Even those who profess to stand for the unification of the sciences cannot be trusted, for each one of them would be satisfied in defining their familiar point of view, and more or less justifying their own individual presuppositions.

Institutional reinforcement
The second obstacle is an institutional one in that teaching and research institutions reinforce the above separation through administrative procedures which tend to eliminate communications with institutions associated with other disciplines.

3. Development of feudal intellectual systems
The third obstacle is a psycho-sociological one. The division of intellectual space into smaller and smaller compartments and the multiplication of institutions which assume the management of each such territory results in the formation of a feudal system (he also uses the term epistemological capitalism) which governs the majority of scientific teaching and research enterprises.

The specialist, once his speciality has been transformed into a fortress, can give free reign to his desire for power. Under the pretext of division of labour, each intends to be master of his own domain and to defend his position against enemies from without and against rivals from within. Academic survival demands an expertise in career strategy and tactics which may even involve obstructing the development of the discipline over which control has been achieved.

4. Cultural biases
The fourth obstacle noted is a cultural one. The separation between disciplines is aggravated by the separation between cultures and their associated mentalities, between languages and between traditions. Science itself, as currently understood, is a typically western phenomenon (within which, for example, particular schools of thought may be associated with particular languages).

5. Forms of false interdisciplinarity
Gusdorf then notes the existence of various kinds of false interdisciplinarity favoured as solutions to the difficulties noted above. The simplest, and most naive, consists in bringing together for a meeting specialists from different disciplines, with the idea that such an assembly would suffice to bring about a common ground and a common language between individuals who have nothing else in common. The reports of such meetings neither achieve nor attempt to achieve any synthesis, leaving any such task to the reader. He concludes his survey with an outline of the basis for any conversion to an adequate interdisciplinary approach and the nature of the education required to accomplish this.

6. Synthesis by chance
Elsewhere he suggests that a specialist arrives at a view of the entire field of knowledge only by chance; he attains a perspective of synthesis only at the very limits of his domain. But he stops short; he is discountenanced, for nothing in his personal experience has prepared him to go any further.

Gusdorf's remedy is to encourage a new category of researchers toward synthesis. The major effort and reason for being of these researchers would be to create interdisciplinary intelligence and imagination. Ordinarily, human problems are confronted from the angle of a speciality. The proposed research would set as its task the confrontation of these problems in the perspective of unity or totality.

Since this section appeared in the previous edition in 1976, there have been remarkably few publications on interdisciplinarity as such. Indeed the most recent initiative of UNESCO on this subject, which falls so explicitly within its mandate, is a collection of papers published in 1983, with almost no references to papers published after 1976 (1983). Georges Gusdorf, one of the contributors, continues to be the most explicit about the work which has not been undertaken. At the time of writing, information was received concerning a study on *Interdisciplinarity: history, theory and practice* by Julie Klein (1990) which hopefully constitutes a major step forward in clarifying the scope of the field.

Significance: consequence for approaches to societal problems

1. Complexity
The unsuspected complexity of social problems aggravates the situation noted above. As noted in the Bellagio Declaration on Planning (1969): *"Social institutions face growing difficulties as a result of an ever increasing complexity which arises directly and indirectly from the development and assimilation of technology. Many of the most serious conflicts facing mankind result from the interaction of social, economic, technological, political, and psychological forces and can no longer be solved by fractional approaches from individual disciplines... Diagnosis is often faulty and remedies proposed often merely suppress symptoms rather than attack the basic cause... Complexity and the large scale of problems are forcing decisions to be made at levels where individual participation of those affected is increasingly remote, producing a crisis in political and social development which threatens our whole future... Scientific attack on these problems of complexity and interdependencies is a matter of the utmost urgency..."*.

Given this complexity of the social problem environment, the question arises as to whether the conceptual tools used to handle this complexity are themselves adequately complex. In terms of Ashby's (1958) Principle of Requisite Variety, only a greater amount of variety (or complexity) in a regulator can control the variety present in a given system; only variety can destroy variety.

2. Inadequate conceptual tools
While the complexity and danger of the problems tend to increase at a geometric rate, the knowledge and manpower qualified to deal with these problems tend to increase at an arithmetic rate (Yehezkel Dror, 1969). There is therefore a real danger that the conceptual tools applied to problems are insufficiently complex to contain them and guide the allocation of resources to appropriate programmes of action. Because the interdisciplinary and integrative focus is an intellectual no-man's-land, little is known about it and the potential it represents goes largely unrecognized.

There is real danger that such tools may not be adequately developed and used, since the increase in complexity of the world, the rapidity of change, and the need to focus on specific current political issues may tempt those with power to use simplistic conceptual tools. However, there is also the danger, because of limited understanding of the nature of integrative concepts, that they may be used by the few to out-manoeuvre the many.

3. Incoherence of interdisciplinary initiatives
The interdisciplinary movement therefore both exists and has reasons to exist. But, as Kenneth Boulding (1956) notes: *"...there is a good deal of interdisciplinary excitement abroad. If this excitement is to be productive, however, it must operate within a certain framework of coherence. It is all too easy for the interdisciplinary to degenerate into the undisciplined. If the interdisciplinary movement, therefore, is not to lose that sense of form and structure which is the "discipline" involved in the various separate disciplines, it should develop a structure of its own"*.

Boulding conceives the elaboration of this structure to be the task of general systems theory. However, given the plethora of integrative concepts currently in use within the movement in its broadest sense, the modest task of attempting to clarify the nature of such concepts and their relationship to one another is of some significance at this time. No other conceptual tools appear to be adequate to the task of grasping the complexity with which society is faced.

4. Conflicting interdisciplinary approaches
One of the difficulties is the long-standing competition between "cybernetics" and "general systems", with the former appealing primarily to engineers and the latter to social and biological scientists. This has given rise to unhelpful dynamics between the two groups.

A second difficulty is that both groups have little sympathy with "non-scientific" insights into possibilities of integration, whether from psychoanalysis or from the arts. General systems is only open to interdisciplinarity on its own pre-established terms. One exception to this is the attempt by Jeffrey Stamps (1980) to marry the insights of general systems with those of humanistic psychology to provide a general systems theory acknowledging the place of the human individual (see comment in entry KD2220).

A third difficulty arises from the non-self-referential nature of general systems theory. This is ironical in that a number of authors linked with the general systems enterprise have explored issues of self-reference. The difficulty is perhaps best illustrated by the relationship between the work of Prigogine and Jantsch (who were colleagues). Prigogine's group has carefully avoided any but the most tentative exploration of the social significance of the breakthroughs for which he received the Nobel Prize. Jantsch has taken Prigogine's work, related it to other initiatives from both the hard and the soft sciences and produced an exciting synthesis oriented towards policy and management problems. Unfortunately he died before being able to relate his contribution to the implications of the work of Bohm. He did however make extensive use of self-reference in relation to mentation, especially with regard to values and to the evolution of consciousness. Whilst the more concrete aspects of his work on self-organization remain central to general systems, the paradoxical problems for consciousness of comprehending (or failing to comprehend) any particular form of integration are outside the general systems domain of concern. Yet it is precisely such questions which illuminate the inability of general systems to offer insights relevant to the global problematique

5. Epistemological challenge
One of the striking features of papers on interdisciplinarity is their essential sterility. Interdisciplinarity has not proved to be productive. It has not revealed any method of moving forward.

David Bohm (1980) is one of the few to have indicated a way forward. Following a useful description of fragmentation and its consequences in all domains he states: *"...some might say: 'Fragmentation of cities, religions, political systems, conflict in the form of wars, general violence, fratricide, etc, are the reality. Wholeness is only an ideal, toward which we should perhaps strive.' But this is not what is being said here. Rather, what should be said is that wholeness is what is real, and that fragmentation is the response of this whole to man's action, guided by illusory perception, which is shaped by fragmentary thought. In other words, it is just because reality is whole that man, with his fragmentary approach, will inevitably be answered with a correspondingly fragmentary response. So what is needed is for man to give attention to his habit of fragmentary thought, to be aware of it, and thus bring it to an end."*

He then continues by saying: *"It is clear that we may have any number of different kinds of insights. What is called for is not an integration of thought, or a kind of imposed unity, for any such imposed point of view would itself be merely another fragment. Rather, all our different ways of thinking are to be considered as different ways of looking at the one reality, each with some domain in which it is clear and adequate. One may indeed compare a theory to a particular view of some object. Each view gives only an appearance of the object in some aspect. The whole object is not perceived in any one view but, rather, it is grasped only implicitly as that single reality which is shown in all these views. When we deeply understand that our theories also work in this way, then we will not fall into the habit of seeing reality and acting toward it as if it were constituted of separately existent fragments, corresponding to how it appears in our thought and in our imagination, when we take our theories to be "direct descriptions of reality as it is."*

Bohm asks: *"How are we to think coherently of a single, unbroken, flowing actuality of existence as a whole, containing both thought (consciousness) and external reality as we experience it?"* (1980, p.x) He points to the nature of an answer through his discussion of the "implicate order", of which man is directly aware to some degree, as it relates to the "explicate order" in terms of which society and conceptual systems are structured.

Significance: previous, parallel or related initiatives

The following initiatives have been noted:

1. Interdisciplinary team research
A volume has been produced on the methods and problems of interdisciplinary team research as the outcome of five conferences sponsored by the USA National Institutes of Health and the National Training Laboratories (M B Luszki, 1958). While the focus was on mental health research, the material which includes excerpts from the conferences usefully serves as an introduction to the topic of interdisciplinary research in the social sciences (Part I, general observations and case study; II, characteristics of the disciplines in the research setting and crucial issues likely to arise when persons with different disciplinary background attempt to work together; III, planning and carrying out an interdisciplinary project; IV, administrative aspects of interdisciplinary research team operation; V, suggestions for training).

2. Interdisciplinary research
A study has been prepared for the Subcommittee on Science, Research, and Development of the Committee on Science and Astronautics of the US House of Representatives by the Science Policy Research Division of the Legislative Reference Service of the US Library of Congress (*Interdisciplinary Research; an exploration of public issues*, 1970). This defines the scope of interdisciplinary research and the related research process, and considers questions of dissemination, funding, training, and general problems of such research. It includes a summary of government supported interdisciplinary research in the USA. The study notes that there has been a considerable range of defence-related interdisciplinary research in the social sciences. Many of the documents describing these activities and their products are classified or proprietary. It notes that should these documents become available they would serve as useful sources for an assessment of the merits of interdisciplinary research in a variety of subject areas and organizational settings.

3. Interdisciplinary teaching and research in universities
A report has been produced by the OECD Center for Educational Research and Innovation based on the results of a seminar on interdisciplinarity in universities, focusing on teaching and research (1972). This includes the results of a questionnaire to universities, various analyses of the concept of interdisciplinarity, and the problems and solutions within a university context.

4. General systems
Since 1956, the Society for General Systems Research has produced a yearbook, *General Systems,* which collects together significant papers, often originally published in journals of a wide variety of disciplines. This is a very valuable source for the location of concepts and their definitions. (A former director of the Society, Richard Ericson, has himself prepared two tentative versions of a *Selected glossary of terms with particular reference to concepts associated with cybernetic management and information technology*). Oran Young has carried out a survey of the use of general systems concepts (*General Systems*, 1964, p.61-80, 239-253). In 1979 a further survey of the movement was carried out for the Society (Cavallo, 1979). The many books on general systems naturally also offer definitions of concepts. It is appropriate to ask why general systems has not been able to respond as effectively as might have been hoped to the challenges of the times. Attempts have been made to render general systems relevant to the issues raised by the global problematique (Ervin Laszlo, 1974 and R F Ericson 1979), but there has been little follow-up.

5. Unified science
The main source for this approach is the *Encyclopedia of Unified Science* (R W Carnap and C W Morris, 1938).

6. Unity of science:
A series of international conferences has been held on unified science or the unity of science (New York, 1972; Tokyo, 1973; London, 1974) under the sponsorship of the International Cultural Foundation. One volume arising out of the first conference contains a glossary adapted from a Glossary Proposal for General Systems Theory formulated for the Task Force on General Systems Education of the Society for General Systems Research (Edward Haskell, 1972).

7. Integrative education
Largely through the journal *Main Currents in Modern Thought* of the Center for Integrative Education (previously the Foundation for Integrative Education), now discontinued, a considerable amount of material is available on useful concepts and theories for integrating modern knowledge. Individuals associated with the Center elaborated such ideas in book form (Henry Margenau, 1972)

8. Integrated education
A considerable movement has developed in connection with the integration of curriculum elements to present a unified perspective on any topic and the integration of teaching methods to improve understanding of a topic. Integrated science teaching is particularly developed.

9. Cybernetics
Under the direction of Heinz von Foerster an early course on the cybernetics of cybernetics had as its principal aim to arrive at a format for a publication that when produced should serve as a nucleus for a comprehensive presentation of the full range of methods and concepts in cybernetics as they are currently available with regard to cognitive, social and cultural processes. The book is entitled *Cybernetics of Cybernetics or the control of control and the communication of communication* (1974). It gives explanatory descriptions of 238 terms and includes the texts of a number of key papers.

10. General studies
The journal *Studium Generale* (Heidelberg) has for many years carried important interdisciplinary material. A book has been produced which summarizes the meaning of general studies in Germany (W Regg, 1954).

11. Unity of knowledge
As a philosophical concern this is reflected in a number of papers, see: Alois Dempf, *Die Einheit der Wissenschaft* (1955); Congresso Nazionale di Filosofia, *L'Unificazione del Sapere* (1966). It is also of concern to psychology, see: Marjorie Grene, *Toward a Unity of Knowledge* (1969); S Koch, *Psychology and emerging conceptions of knowledge as unitary (*1964).

12. Synthesis of the human sciences:
The World Institute Council, largely through its president, Julius Stulman, has contributed since 1949 to a number of activities associated with integrative education, such as the journal *Main Currents in Modern Thought*, and more recently the Institute's own journal *Fields within Fields within Fields*. The World Institute, proposed in 1949, is envisaged as a scientific institute devoted to research and action for the integrated human and economic development of the world.

13. General systems education
Early efforts were made through the Task Force on General Systems Education of the Society for General Systems Research, largely through its chairman Jere W Clark at the Center for Interdisciplinary Creativity, to formulate a curriculum for general systems education and to design delivery systems for the general systems perspective.

14. Interdisciplinary synthesis
A plan to institutionalize interdisciplinary research and teaching was formulated by Leo Apostel at the University of Ghent in 1963 (Leo Apostel, *A Center for Interdisciplinary Synthesis* (1972).

15. Integrative studies
The Center for Integrative Studies is concerned with long-range social and cultural implications of scientific and technological developments. The integrative function is directed towards the analysis and interpretation of information from a great variety of sources, and to assemble and integrate the contributions of many disciplines for the identification and evaluation of the critical interactions of technology and society.

16. Interdisciplinary cooperation
The programme of UNESCO provides for projects which enlist the aid of philosophy to stimulate sustained thought on the nature, achievements, difficulties and present and potential role of interdisciplinary research, so as to promote its evaluation and use. A review of the status of interdisciplinary research by Jean Piaget has been published (1970). A symposium has been held under UNESCO auspices on science and synthesis (1971).

17. UNESCO Computerized Documentation Service
UNESCO is the intergovenmental agency with a specific mandate to respond to the challenge of interdisciplinarity. In 1975, of the 61 items concerning interdisciplinary research in Unesco documents and publications: 29 stressed the need or importance of, or recommend, such research: 24 were in some way concerned with its application to various areas, but mainly the environment; and only 6 were concerned with interdisciplinary research itself.

18. Interdisciplinary relations
An inventory and bibliography of interdisciplinary relations in the social sciences is published regularly by the International Social Council (Interdisciplinary relations. *Social Science Information*, number 2 of each volume).

19. Comparative studies
The Comparative Interdisciplinary Studies Section of the International Studies Association encourages the advancement of cross-cultural, interdisciplinary comparative studies of national, international, and sub-national social systems and processes principally by the exchange of documents through correspondence networks focused on a common theme. (A number of other problem areas have given rise to interdisciplinary groups, *eg* aged, drug abuse, youth, *etc*).

20. Systems theory in management
The conceptual frameworks of system theory and the quantitative techniques of management science provide the bases for the latest advances in the evolution of management theory and practice. A considerable amount of literature is available on such aspects of organization and management (See: J W Greenwood Jr et al.)

21. Cybernetics and systems analysis
The International Institute for Applied Systems Analysis (whose members are the principal scientific academies in each country) undertakes studies into both methodological and applied research in the related fields of systems analysis, cybernetics, operations research, and management techniques. The World Cybernetics and Systems Organization acts as a clearing house for all societies concerned with cybernetics, general systems, operational research and computer science.

22. Comprehensive thinking
As a contribution in 1965 to the World Design science Decade (1965-1975), John McHale selected and edited some of the writings of R Buckminster Fuller into a document entitled *Comprehensive Thinking*, concerned with the application of whole systems thinking to local and particular aspects of the system in the planning of environment for man.

23. Specialists and Generalists
The Subcommittee on National Security and International Operations of the Committee on Government Operations of the United States Senate compiled a selection of readings entitled *Specialists and Generalists* to encourage reflection on these respective roles in government.

24. Interdisciplinary vocabulary
A report prepared for UNESCO in 1961 by Georges Gusdorf (*Projet de recherche interdisciplinaire dans les sciences humaines*) suggested that one of the principal tasks was to identify the key concepts of importance to a variety of sciences (1967). This would then constitute a basis for defining the meaning and scope of interdisciplinary communication and could open the way to an epistemology of convergence in contrast to the current epistemology of dissociation. The report listed 176 possible concepts as a basis for further discussion.

25. Policy sciences
This emerging discipline is in effect an interdisciplinary approach to the problems of governance. An *Encyclopedia of Policy Studies* (1983) was produced under the editorship of Stuart S Nagel.

26. Complexity and chaos studies
Some of the most interesting recent developments have focused on what might be considered the obverse of interdisciplinarity, namely the study of complexity and chaos.

27. Theory and practice of interdisciplinarity
A very recent study reviewing the history, theory and practice of interdisciplinarity has been produced by Julie Klein (1990).
It is interesting to note that none of the encyclopedias consulted, with one remarkable exception (in the French *Encyclopaedia Universalis*), included entries on interdisciplinarity. Even the re-organized 15th edition of the *Encyclopaedia Britannica*, which specifically emphasizes the circle of learning and knowledge about knowledge, adopts a very conventional approach.

Given the problem of locating interdisciplinary material through existing information systems, it is highly probable that many comparable (or even more significant) initiatives should be mentioned in the same context as those above. Perhaps even more regrettable, many promising initiatives and proposals in the more distant past may be almost impossible to trace, or may bear a somewhat diffuse relationship to the current approach to interdisciplinary or integrative questions.

The synthesis of knowledge (and its application to world problems) was, for example, a basic inspiration behind the complex of activities of the following interlinked organizations in the period just after the turn of the century: the International Bibliographical Institute (later to become the International Federation for Documentation), the Union of International Associations, and the International Peace Institute. Synthesis of knowledge was to be facilitated by systematic classification of documents pioneered by the elaboration of the Universal Decimal Classification system and its application to some 11 million file cards. The tracing and evaluation of such initiatives and their interactions must await some history of approaches to synthesis and integration of knowledge.

Method: identification and guidelines

1. Definition
There is no satisfactory general definition for the concepts which it is intended that this section should include. The title envisaged for this section has at various times been "interdisciplinary concepts", "integrative concepts", "integrative, unitary, and transdisciplinary concepts", *etc*.

(a) Inclusion: By the above terms is meant concepts such as the following:

(a) Concepts of systems, types of systems, and general properties of systems

(b) Concepts of interdisciplinarity

(c) Concepts of integration, unification and unity of knowledge

(d) Concepts whose essential characteristic is to link together or structure other concepts into a larger whole

(e) Concepts whose essential characteristic is to encompass or grasp complexity or the basically incomprehensible

(f) Concepts of wholes, types of wholes, and general properties of wholes

(g) Concepts which interrelate incommensurable domains vital to balanced (social) action and policy formulation

(h) Concepts of unity, and types of unity

(i) Modes of analysis in the light of the above concepts

The most general description that emerges is possibly one of concepts whose essential characteristic is to interrelate incommensurable (or in some way incompatible) concepts, modes of experience, or methods of study.

(b) Exclusion: Concepts such as the following are included (if at all) only to indicate limiting cases of interest: racial integration, cultural integration, personal integration, institutional integration, integration of communication and data systems, integration of production systems, and integration of behavioral systems.

2. Identification procedure
A preliminary list of subject headings relating to some aspect of integrative thinking was first established. This list was used and extended during the course of several systematic library and literature searches forming a bibliography (Section KY).

Summaries of various ranges of integrative, interdisciplinary and related concepts, in the form of books and articles, were located in this way and were used to build up files on individual concepts. The information finally present in each file was then used to establish the individual concept entries.

Although all statements used in building up concept descriptions are very closely based on existing published documents, no explicit link is established between statement and source documents. This was avoided because the editorial process of selection and restructuring of texts from different sources may have unintentionally distorted the meanings in the original contexts (particularly when the original statements did not constitute clear descriptions). Any such misinterpretation will be corrected in future editions.

A particular difficulty encountered was that libraries are currently unable to process interdisciplinary material in a satisfactory manner. Either the publication is treated under "general", which is by nature an underprivileged category mainly used for general reference books in which the notion of generality derives from the variety of unrelated materials included within the same volume and not from the nature of any concept which attempts to interlink such materials, possibly at a higher level of order. Or, as the chief cataloguer of one of the major libraries of the world explained, the practice is to scan such books on interdisciplinary and related approaches in order to identify the most predominant discipline, which is then used as the basis for classification. A key publication such as *Interdisciplinarity; problems of teaching and research in universities*. (Paris, OECD, 1972) would therefore not be retrievable under any interdisciplinary classification, but rather under some aspect of education. An equivalent situation occurs in bookshops, so that interdisciplinary questions may be dealt with in books anywhere in an entire collection.

3. Guidelines for patterning disagreement (Section KP)
In order to generate such an integrated multi-set grouping of operational statements it is of course vital to have a rich variety of source material on possible content. Such material was collected for an earlier paper (A J N Judge, *Patterns of N-foldness*, 1984) and tentatively ordered there according to the number of elements in the sets in question. As pointed out in that paper, this ordering permitted useful comparisons between sets in different schemes having an equal number of elements.

Such material can only be useful if care is taken to treat the morphological characteristics as **independent** of the special properties of the substrate with which a given set is particularly concerned. This point has been clearly made by René Thom in his study of morphogenesis (*Modèles Mathématiques de la Morphogenèse*, 1980).

The point is also made in relation to one of documents included in the material collected: "*This study will develop the hypothesis that the "lattice logic" which de Nicolas perceives in the Rg Veda was grounded on a proto-science of number and tone. The numbers Rgvedic man cared about define alternate tunings for the musical scale. The hymns describe the numbers poetically, distinguish "sets" by classes of gods and demons, and portray tonal and arithmetical relations with graphic sexual **invariances** which became the focus of attention in Greek tuning theory. Because the poets limited themselves to **integers**, or natural numbers, and consistently used the **smallest integers possible** in every tonal context, they made it possible for us to rediscover their constructions by the methods of Pythagorean mathematical harmonics.*" (Ernest MClain, 1978, p.3)

Using equivalent sets from the source material as a guideline, a new set for the number in question was "generated" taking into account the constraints. Given the variety of emphases of sets of an equivalent number of elements, this process of generation necessarily involved non-logical operations, especially since the intention was to maximize the incompatibility between the elements in any given set. The results are given in Section KP. The only similar exercise detected is a doctoral thesis in philosophy (W Dalhberg, *Ordnung, Sein und Bewustsein*, 1984) which was composed directly onto a word processor (to facilitate experiments with alternative structures) using the poetic power of the German language to full effect.

Two interesting and related difficulties emerged in comparing equivalent items from the source material.

(a) Clearly some sets are formulated at a higher level of abstraction than others. The problem was, using the constraints as guidelines, to generate elements of an equivalent set at an appropriate level of abstraction.

(b) Clearly sets differ greatly in the nature of the elements, whether: stages, values, qualities, problems, methods, conditions, *etc*. Again the problem was to use the contraints to arrive at some neutral formulation of which the above might be considered aspects. In both cases the problem was to find appropriate words (whether general or neutral) to carry the incompatible qualities associated with each set.

For each set an underlying theme is common to each element. The incompatibility is embedded in the qualification on that theme.

Method: approaches to the art of disagreement

In interrelating highly diverse focal concepts, it would be naive to ignore the fact that those identifying with such different concepts tend to disagree and to oppose each other. Such dynamics need to be taken into account if the resulting integration is to be of more than academic significance. The difficulty is that the ability of conventional methodologies to encompass essentially incommensurable concepts is poorly developed - methodologies themselves tend to be mutually "hostile". In such a situation it is necessary to move "beyond method" (A J N Judge, Beyond Method, 1984) if a diversity of methods is to be interrelated in a manner which is relevant for integrated development.

There is therefore a major need for a "science of disagreement" to clarify the manner in which active disagreement can be usefully structured. It appears that agreement in society is often essentially superficial or token (if it prevails at all). There is a total absence of knowledge on how to disagree intelligently in an organized manner, rather than in an irrational, fear-ridden manner requiring some form of violent or repressive response to eliminate the disagreement as soon as possible.

1. Conflictual approach
It might be assumed that the methodologies of conflict resolution, mediation or arbitration would provide guidelines for a science of disagreement. This is not the case. Such methods are primarily concerned with eliminating the disagreement between theparties, or reducing it to a level at which it is not significant for their relationship. Edward de Bono (1985) suggests that people seek to solve conflicts by a conflict method of thinking. Even through negotiation, themethod is still that of compromise or consensus. Compromise suggests that both sides give up something in order to gainsomething. Consensus means staying with that part of the proposal on which everyone is agreed, namely the lowest common denominator.

He stresses that: "We do have to accept that our methods of solving major disputes and conflicts have been crude and primitive, inadequate and expensive, dangerous and destructive. Even if we operate these methods with the best will in the world and with the highest intelligence, they will not suffice. There is a need for a fundamental shift in our approach to these conflicts." (2, p.viii) He suggests that in current approaches there is insufficient attention to the creative and design aspect and to the creation of elements which were not there to begin with. The question however is how to introduce this "design dimension".

In looking for a "science" of disagreement some care is necessary. Science, as it claims to be practised, can be usefully considered to be about agreement processes and the elimination of disagreement. "Art" may however be considered to be about disagreement processes, set against a background of the rise and fall of agreement. Agreement is not useful without disagreement. In fact it is meaningless. It is the disagreement which introduces the essence of diversity and avoids the uniformity of undifferentiated mass consciousness.

2. Anarchistic approach
Since disagreement can be perceived as arising from differences in method, there is merit in exploring Paul Feyerabend's thesis Against Method (1975). But there is an obvious problem in using Feyerabend's 'method' as a basis for any art or science of disagreement. He explicitly advances his views as epistemological anarchism and states: "It is clear, then, that the idea of a fixed method, or of a fixed theory of rationality, rests on too naive a view of man and his social surroundings. To those who look at the rich material provided by history, and who are not intent on impoverishing it in order to please their lower instincts, their craving for intellectual security in the form of clarity, precision, "objectivity", "truth", it will become clear that there is only one principle that can be defended under all circumstances and in all stages of human development. It is the principle: anything goes." (p 27-28)

But he goes further in arguing that science is itself anarchistic: "To sum up: in so far as the methodology of research programmes is "rational", it does not differ from anarchism. In so far as it differs from anarchism, it is not "rational". Even a complete and unquestioning acceptance of this methodology does not create any problem for an anarchist who certainly does not deny that methodological rules may be and usually are enforced by threats, intimidation, and deception. This, after all, is one of the reasons why he mobilizes (not counter-arguments but) counter-forces to overcome the restrictions imposed by the rules." (p 198)

But the somewhat quixotic element in his extremely valuable approach is then revealed in his remarks on its status in his own view (as the author): "Always remember that the demonstrations and the rhetorics used do not express any "deep convictions" of mine. They merely show how easy it is to lead people by the nose in a rational way. An anarchist is like an undercover agent who plays the game of Reason in order to undercut the authority of Reason (Truth, Honesty, Justice, and so on)." (p.33)

Feyerabend takes us to the very useful point at which it is possible to say that "disagreement is OK" and that scientific progress might be impossible if imperfections were eliminated (p.255). But, as an anarchist, he is obviously totally uninterested in the need to "organize" disagreement in any way, even if it were possible. As a result his approach provides no clues for any new way of organization which could take account of new levels of disagreement.

3. Marxist-Leninist dialectical approach
The most fruitful guide to further understanding of disagreement should be found in writings on dialectics, which were clearly of value to Feyerabend. But whether it be in the writings of Hegel, Marx, Engels or Lenin, or in recent writings on dialectics as it emerges in modern science (eg complementarity, etc), there is little to be gleaned beyond the concept of the essential (thesis, antithesis, synthesis). Most authors emphasize the intimate relationship to the cognitive subject-object process, about which it is necessarily difficult to be "objective" without distorting comprehension of its essential dynamism. Thus: "If we try to analyze what it is that the threefold describes, we are in a bind, for it is just that element of participation in life that analysis cannot, and does not even pretend to, cope with." (Arthur Young, The Geometry of Meaning, 1975, p.57) "Since it is basically nonconceptual, it cannot be defined..." (p.27) For this reason dialectics has been most favoured as a method by those capable of anchoring it in practical action in a concrete material context.

The marxist scholar Jean-Marie Brohm points out that neither Marx nor Engels attempted to define dialectics positively (1976, p.43). They defined it negatively by the criticism of adverse positions, as have most of their successors: "Ce faisant ils obéissaient à un grand principe général de la dialectique: la négativité. Le positif est toujours le résidu de la négativité, un moment négatif provisoire qui attend à son tour d'être nié... La dialectique est le produit d'une lutte ininterrompue contre les conceptions adverses. Elle se définit négativement par ce contre quoi elle s'oppose." (p.43)

Hegel summarizes the essence of dialectics as follows (as quoted by Brohm): "Les choses finies sont, mais leur rapport elles-mêmes est de nature négative, en ce sens qu'elles tendent à la faveur de ce rapport à se dépasser. Elles sont, mais la vérité de leur être est qu'elles sont finies, qu'elles ont une fin. Le fini ne se transforme pas seulement, comme toute chose en gnral, mais il passe, il s'vanouit; et cette disparition, cet évanouissement du fini n'est pas une simple possibilité, qui peut se réaliser ou non, mais la nature des choses finies est telle qu'elles contiennent le germe de leur disparition, germe qui fait partie intégrante: l'heure de leur naissance est en même temps celle de leur mort." (Science de la Logique, p.129)

In commenting on Hegel's Science of Logic, Lenin clarifies one of Hegel's definitions of dialectics by the following:

(a) Definition of the concept on the basis of itself (the thing itself should be considered in its relationships and in its development)
(b) Contradiction in the thing itself, forces and contradictory tendencies in each phenomenon
(c) Union of the analysis and the synthesis.

Then he further clarifies these elements in 16 points (*Cahiers Philosophiques,* 1973, p.209-210):

(a) Objectivité de l'examen (pas des exemples, pas des digressions, mais la chose en elle-même).
(b) Tout l'ensemble des rapports multiples et divers de cette chose aux autres.
(c) Le développement de cette chose (respective phénomène), son mouvement propre, sa vie propre.
(d) Les tendances (et aspects) intrieurement contradictoires dans cette chose.
(e) La chose (le phnomne, *etc*) comme somme et unit des contraires.
(f) La lutte respective (ou encore) le dploiement de ces contraires, aspirations contradictoires, *etc*.
(g) Union de l'analyse et de la synthèse, séparation des différentes parties et runion, totalisation de ces parties ensemble.
(h) Les rapports de chaque chose (phénomène, *etc*) non seulement sont multiples et divers, mais universels. Chaque chose (phénomène, processus, *etc*) est lie à chaque autre.
(i) Non seulement l'unité des contraires, mais aussi les passages de chaque dtermination, qualit, trait, aspect, proprit en chaque autre en son contraire.
(j) Processus infini de mise à jour de nouveaux aspects, rapports, *etc*.
(k) Processus infini d'approfondissement de la connaissance par l'homme des choses, phénomènes, processus, *etc*, allant des phénomènes à l'essence et d'une essence plus profonde.
(l) De la coexistence à la causalité et d'une forme de liaison et d'interdépendance une autre, plus profonde, plus générale.
(m) Rptition un stade suprieur de certains traits, proprits, *etc*, du stade inférieur et
(n) Retour apparent à l'ancien (négation de la négation).
(o) Lutte du contenu avec la forme et inversement; rejet de la forme, remaniement du contenu.
(p) Passage de la quantité en qualité et vice versa.

As indicated above, these insights are not especially helpful to developing a method of disagreement.

4. Non-marxist dialectical approach

In the case of psychologist Jean Piaget, there are five characteristics of dialectics:

(a) Construction of previously non-existing interdependencies between two systems considered either as opposed or as strangers to each other, and which are thus integrated into a new totality; whose properties exceed them.
(b) The interdependencies of the parts of the same object are in dialectical relationship.
(c) Every new interdependency engenders properties exceeding the component parts if it results in a totality greater than that without it.
(d) Intervention of circularities or spirals in the construction of interdendencies.
(e) Relativisation of parts due to their interdependencies.

These five properties of dialectics are summarized by a sixth which gives its general significance: "*dialectic constitutes the inferential aspect of all equilibration*". This means that dialectics does not intervene at all stages of cognitive development, but only during the course of the equilibrating process. It is therefore important to distinguish carefully between the state of equilibrium corresponding to a non-dialectic moment of evolution and the dialectic processes permitting the construction of new frameworks. Piaget distinguishes eight kinds of interdependency (*Les Formes Elémentaires de la Dialectique,* 1980, p.213-227). A co-author, Rolando Garcia, draws attention to similarities between Piaget's concept and that of Lenin as detailed above (p.233-237).

In one of the few studies which also reviews non-marxist concepts of dialectics, Paul Foulqui concludes with the following general definition:

"*Est dialectique une pensée constamment tendue pour se dépasser elle-même aussi bien en allant jusqu'au bout de ce qu'elle a découvert qu'en se portant à des points de vue nouveaux que semblent contredire ses affirmations premières.*" (*La Dialectique,* 1976, p.125) Despite the relevance of dialectics to the problem of disagreement, as noted above, it does not appear to do more than explain the dynamics of the environment it constitutes. **Dialectics explains the eventual future evolution beyond the stage of disagreement, but does not clarify the nature of any possible present order whilst the disagreement holds.** It does not clarify the nature of the psycho-social forms to which disagreement can give rise in the present, it merely affirms that they are necessarily temporary. **The question is whether there is any pattern in the present to the ancillary processes to which a dialectical confrontation gives rise.** It it possible to discover any underlying structure to disagreement? For example, evident disagreement might be considered to be structured like interference patterns from distinct interacting wave sources. Or disagreement might be compared to recent thinking on the relationship between interacting parallel universes.

5. Developmental stage approach

More accessible to reflection (but spread over time) is the concept of development stages, of which the best example is in the case of the individual human being. Development for the individual is a series of separations which give rise to a qualitatively different sense of unity. Stages include:

(a) Birth (loss of physical connection within the womb);
(b) Physical separation from adult supervision;
(c) Emotional separation with external orientation of affections at puberty;
(d) Intellectual separation from parental framework with departure from home; and possibly others.

Each of these separations, as a form of disagreement, can be very painful. They are accompanied by changes of perspective which are difficult to communicate to younger siblings, for example. This effective secrecy is enshrined in primitive initiation rites and rites of passage of which equivalents still exist for apprentices, students, and soldiers. At each stage new adversaries emerge as potential enemies with whom to disagree.

The shock effect of such initiations has been extensively explored by psychoanalyst C G Jung in his study of the confrontation of an individual with archetypes (including adversaries) corresponding to each initiatory level (*The Collected Works,* 1953-71). Of special interest is the individual's encounter with his "shadow" and its relation to creative comprehension of the significance of death as a dramatic form of disagreement. He clearly demonstrates that avoiding this confrontation is unhealthy for the development of the individual.

At each such development stage intense regret may be expressed for the loss of the togetherness and innocence of the preceding stages despite profound appreciation for the new insights achieved. The advantage of using such stages to model levels of disagreement is that it highlights the possibility that many of those involved in movements for "peace", "equality" and "solidarity" may be hoping to achieve a kind of womb-like level of agreement within their environment. Or some childish condition of "eternal summertime" and parental security. But the more separation or disagreement that has been achieved, the greater the potential for new kinds of unity. It is the degree of disagreement which qualifies the scope and depth of the unity possible. This problem is well illustrated in the various levels of disagreement with which the poets of the ancient Rg Veda hymns struggled using music as a language as discussed in Section KD (Antonio T de Nicholas, 1978).

Method: subtler forms of disagreement

1. Premature and immature agreement
Clearly there is a communication problem in arguing for new levels of unity, if this is comprehended as equivalent to arguing for separation of mother and child, for example. Any such argument can then only be perceived as "bad" or "evil" under present circumstances. But, at the same time as the Rg Veda case illustrates, there is a certain level of disagreement inherent in any pattern of organization.

Any such disagreement gives rise to an "impossible yearning for agreement" that drives the search for subtler levels of agreement. For example, an interesting study has been made of dissymetry as an anti-entropic force. This may lead to a valuable contrast between symmetry (as agreement) and dissymetry (as disagreement): "*Dans toute symétrie établie peut surgir une rupture partielle et non accidentelle qui tend à compliquer l'équilibre formé. Une telle rupture est proprement une dissymétrie. Elle a pour effet d'enrichir la structure ou l'organisme o elle se produit, c'est--dire de les doter d'une propriété nouvelle ou de les faire passer à un niveau supérieur d'organisation.*" (Roger Caillois, *La Dissymtrie*, 1973, p.78)

But this search may be driven in either of two conflicting directions, whether towards the primordial unity (soundless, womb-like, by recovery of the past), or towards a unity based on greater differentiation (in the future). In both cases it is necessary to live with disagreement, rather than rejecting it as "evil". As indicated in the Rg Veda case: "*It is man's yearning for this impossible agreement which introduced a hierarchy of values into the number field... and that world was rife with disagreement among an endless number of possible structures.*" The disagreement is only absorbed and contained, as a complementary study demonstrates, by the use of larger numbers sets: "*The great expansion of the number sets in later diagrams is motivated, I believe, by the effort to approximate as exactly as possible the irrational square root of 2 which is needed to locate a tone symmetrically opposite the mean on D, that is, precisely in the middle of our octave.*" (Ernest G MClain, *The Myth of Invariance*, 1978, p.37)

The problem of "world peace" etc needs to be seen in a similar light. **As presently conceived, the level of articulated separation, disagreement or diversity is not yet great enough to sustain more than an undifferentiated, mass-consciousness version of the desired level of agreement.**

2. Differentiated patterns of disagreement
The previous point would appear to indicate that the deficiency of dialectics in understanding disagreement arises whenever some stability is required for disagreement sets higher than the threefold by which it is characterized (*eg* thesis, antithesis, synthesis). This is clearly stated by Arthur Young: "*But when the stimulus causes wrong action and the result is not achieved, the (fourfold) learning cycle becomes necessary. Thus the learning cycle only becomes necessary when there is an obstacle in the larger, threefold cycle.*" (1975, p.24)

This suggests the need to explore more highly differentiated patterns of disagreement with higher numbers of component elements. The most interesting development in this direction is that arising from the impact of quantum theory on the conceptual bases for the classification of knowledge (C A Hooker, *The Impact of Quantum Theory on the Conceptual Bases for the Classification of Knowledge*, 1978), especially that of P A Heelan (*The Logic of Changing Classification Frameworks*, 1978) who is concerned with incompatible frameworks, and with complementary frameworks and dialectical development. He advocates the use of non-Boolean partially ordered lattices to interrelate such frameworks and the languages associated with them. Heelan's approach is cited by the authors of the above-mentioned studies on the Rg Veda as appropriate to the complexity with which they are dealing.

Heelan relates his own work to that of Feyerabend who was cited above: "*The context of assumptions in which I am working comprises those counter-positions to classical logical empiricism, established by such authors as N R Hanson, P K Feyerabend and T S Kuhn, such as the absence of any hard distinction between observational and theoretical language, the validity of multiple explanatory viewpoints, the existence of both continuous trajectories of theory development and discontinous trajectories representing revolutionary episodes in the history of science or culture.*" (1978, p.260) He concludes: "*From the foregoing it is clear that there are a variety of logical models at hand to understand inter-framework relationships and especially developmental transpositions between frameworks in history. The task of using these models practically in problems of classification, has yet scarcely been begun.*" (1978, p.272)

3. Incompatibility and paradox
Heelan indicates the relevance of his approach to relating certain incompatible theories of physics. The question is whether, by using the term "logical", he is restricting its relevance to situations in which the disagreement is less fundamental. How irrational can disagreement be and still be organized in some way? Both Feyerabend in his book *Against Method* (1975), and Heelan in identifying himself with "counter-positions" (quotation above), are taking up positions and "disagreeing" with others. They are therefore trapping themselves in a dynamic relationship without providing any organization for that disagreement thus leaving the basic difficulty unclarified. It would seem that the difficulty lies in the paradigm in which "positions are taken". The difficulty is less that of whether one takes a particular position and more that of the nature of the relationship to the positions one fails to understand or support (of which others, or the future, may understand more), especially when the "otherness" (Maurice Friedman, *The Confirmation of Otherness*, 1983) of that position can be conceived as a healthy complement, counteracting weaknesses in one's own position.

Paradox is implicit in the approach of Feyerabend and Heelan, but can it be made pardoxically explicit? Somehow any **static** "balance" between agreement and disagreement must be by-passed through a set of paradoxes which legitimate contradictory positions. It is strange that the absence of humour from the development of psycho-social organization is not a cause for comment given its fundamental importance to human beings, even in political life. Arthur Koestler has explored its relation to paradox and creativity (*The Act of Creation*, 1964). Can contradictory positions be mapped into a self-reflexive hierarchy of paradoxes in which **dynamism** is inherent? Such a context might then prove more appropriate for the dialectic process. At present this is rather like having access to the central component of an electric generator, without being able to mount it in a suitable framework so as to be able tap the energy generated to drive other psycho-social processes - and without tearing the mounting apart as it rotates between opposing positions.

The challenge that remains is therefore to explore "design" approaches (in de Bono's terms) to the possibility of generating a pattern of progressively more differentiated disagreements as a basis for a more appropriate manner of psycho-social organization.

Method: constraints on patterning disagreement

1. Presentation

(a) The design of any text concerning method immediately raises the question as to whether that design will facilitate or hinder implementation of any insights embodied in the text. The form of the text is not a trivial matter and should ideally be isomorphic with the pattern of operations to which it gives rise. Texts which fail to take this constraint into account tend to give rise to methods which are poorly understood and rarely used, whatever their merits.

(b) In recognition of this problem, the design of the "method" outlined here emerges as the result of the application of a series of constraints.
Without such explicit constraints, any text on method is free to meander in an unstructured way through hundreds of paragraphs of inoperable statements.

(c) The intent is therefore to establish a constraint framework such that different kinds of development discussed can be effectively distinguished whilst at the same time clarifying why those we do no happen to favour appear disagreeable and essentially unjustifiable, if not incomprehensible.

(d) The aim is therefore to achieve an optimum degree of congruence or isomorphism between statements relevant to psycho-social reality, methods relevant to the transformation of that reality, and structures designed to implement such methods.

2. Disagreement

(a) Any text on method can be further elaborated by introducing statements in agreement with the initial statement. There is no well-defined limit to this expansion process.

(b) In the present social context a statement on method only acquires significance through the manner in which it disagrees with other extant statements. This may be used as an explicit technique for limiting the further expansion of sets of statements in agreement with one another. Each statement must therefore be matched with other opposing, or mutually disagreeable, statements. Instead of emerging only in the dynamics of the debate between adherents of methods, disagreement is thus "internalized" as an explicit structuring device in the design of the text. Unless such disagreement is internalized, the method described is always essentially inadequate and must always assume the existence of other methods to complement it and compensate for its weaknesses. Since adherents of a particular method tend to have difficulty in acknowledging the significance of other methods, failure to internalize strongly reinforces application of inadequate methods without any device for their reconciliation.

(c) Disagreement is usually conceived as being a condition prevailing between two elements which together constitute a set, whether of people, values, principles, concepts, methods, or facts. The condition may however exist between a larger number of elements.

(d) In the absence of a suitable constraint framework embodying the complete pattern of potential disagreement, statements and counter-statements in any debate twist into predictable and essentially pre-determined patterns. There is in fact an interesting parallel to the description of energy states in fundamental physics. The possible energy states (*ie* debate) are described by a probability wave function. When a particular probability is actualized (*ie* debate position is taken), the wave function "collapses" (*ie* no other statements are relevant in that context).

3. Underlying relationship

(a) Unless they are identical, members of a set necessarily differ and this difference may be interpreted as "disagreement". In order to understand how such disagreement may be organized, a search must first be made for sets which contain elements in maximal disagreement.

(b) If such sets are meaningful, then the elements of the set retain some degree of commonality which binds them together despite the high level disagreement between them. The qualitative characteristic of the bond is what needs to be understood.

(c) The disagreement becomes especially interesting when the elements are such that the disagreement is somehow "active". The elements are then complementary in that each is a vehicle for a particular perception of an underlying condition which cannot be adequately conveyed through any one of them (*cf* the complementarity between wave and particle descriptions of light). This complementarity may of course be denied and then the set elements are perceived as opposed. The set as such may then not be considered a meaningful grouping device for those elements.

(d) It is the presence of this combination of maximal disagreement with an underlying commonality, or relationship between set elements, which constitutes the third constraint.

4. Completeness

(a) The previous constraints do not in any way limit the expansion of a set of matched statements. A new constraint is therefore introduced to limit a particular set of matched statements to a given number of elements.

(b) This is done on the assumption that once established the set constitutes a complete pattern of incompatible positions and cannot be enlarged or reduced (although the individual statements may of course be reworded).

(c) If further matching statements are required to clarify the methods, these should be combined in one or more other sets, each complete in its own way.

5. Number uniqueness

(a) In the practical use of sets of elements such as those it is intended to generate here, there is an important constraint relating to the uniqueness of any given set. For example, the concept of the method or approach as implemented constitutes a fundamental 1-element set. Furthermore, if in applying the method a balance has to be maintained between two conflicting considerations, this constitutes a 3-element set. In both cases, the dynamics it is intended to encompass will also be present when dealing with some sub-component of the method - where the sub-component approach then itself again constitutes a 1-element set, for example.

(b) The previous constraints do not prevent the emergence of sets for which the pattern of disagreement between the elements is effectively a replication or a qualification of that in other sets.

(c) A new constraint is therefore introduced which requires that only one set be allowed with a given number of matching statements as elements.

(d) This constraint highlights the essential "management" issue of handling the set elements and maintaining the integrity of the set. This is the challenge of **managing contradictions** It is not possible to apply a method without having a 1-element set, for example. It may even be explicitly stated that there is no single central concept - but that is then itself the one governing central concept. It is highly probable that the application of the method will also, for example, at some point involve an explicit polarization between two complementary approaches or considerations, thus constituting a 2-element set requiring some form of mediation governed by statements in a 3-element set.

(e) These questions become clearer when considered in the light of any organizational structure created to implement the method. A hierarchy necessarily emerges with concerns relating to the 1-element set "at the top". Note however that this **conceptual** hierarchy does not have to be matched in a one-to-one relationship with the **organizational** structure of roles and departments. Some of the sets may instead be reflected in the sets of principles, values, strategies, or procedures of that organization - or even in informal factions concerned with particular policies.

(f) The set associated with a given number N effectively gives rise to a range of N-"person" games as an organizational, management, coordination or strategy problem. It is the qualitative characteristic of the range of games that is to be elucidated, as well as the set elements "activated" as role stereotypes for "players" implementing the method.

6. Number pattern
(a) The previous constraints do not prevent the usual situation in which sets of elements are treated independently, each set being embedded wherever convenient within an arbitrarily structured text which supposedly provides the connecting links between them.

(b) To the extent that the text constitutes a complete explication of a method, of which the essential items are formulated as set elements, some degree of order should emerge from the relationship between those sets. The various sets in effect constitute some kind of hierarchy of N-person games within which disagreement or contradictions are handled.

(c) A new constraint is therefore introduced which requires that the numbers whereby the sets are labelled should themselves fall into a pattern (not necessarily complete) which can be used to elucidate the relationships between the methodological significance of the sets.

(d) A pattern of numbers can be considered as a "minimal form". The question is what pattern of numbers is most appropriate as a constraint. In terms of number theory, the conventional number series 0, 1, 2, 3, ... is arbitrarily based on the number 10. It is preferable to avoid possible distortion arising from this particular choice of pattern. The hybrid number pattern which appears to avoid this problem in the most balanced manner can be obtained by taking the series in which each succeeding number in the series is taken with itself as base. As indicated above, a set corresponding to any number in the series is then composed of elements equal to that number (*eg* at level 5, there are 5 elements or matching statements).
(e) The imposition of any such numbering pattern may appear totally unnecessary. Why is any such device required? A response is that most social science texts avoid the issue of how systematically their arguments need to be structured to render explicit as many relationships between statements as is feasible. There is no implication that such texts should be structured other than arbitrarily for editorial purposes. It is not surprising that insights emerging from such texts cannot be easily geared into any integrated set of transformative operations which require a complete pattern of checks and balances. As an illustration of the non-trivial role of numbers in the organization of information, the database software through which this book is produced distributes information more efficiently on the storage disk if the file size is determined using a prime number. Otherwise information is not distributed so evenly, leading to performance degradation. This suggests the possibility that more fruitful patterns of disagreement can be organized when a set spreads the weight of the incompatible positions more evenly using a number of set elements based on a prime number.

7. Transformation operator
(a) The set elements in academic texts tend to be unsatisfactory because they are primarily descriptive. A descriptive set is essentially static and de-emphasizes transformation.

(b) The problem is therefore to generate a set in which the elements are essentially dynamic or have an operative dimension, namely a set of operators. This requirement constitutes the seventh constraint on set design.

(c) Such operators are effectively methods or methodological operations. However, given the design of the set, each operator would be in maximal opposition to the other operators in the same set. The operators would therefore be mutually counteracting.

(d) If such sets of counteracting methods are to be designed, the question is how much incompatibility can be effectively built into operators without destroying the basis for grouping them as set? And yet the more they are incompatible, the greater the probability that they will be able to "contain" the complexity of conditions to which they are applied (*cf* Ashby's *Law of Requisite Variety,* also the gene pool concept).

8. Self-constraint
(a) Whilst the operational emphasis introduced by the previous constraint ensures a degree of action entailment, such action lacks focus. Statements can be sharpened by introducing a suitable focus.

(b) Whilst the statements could be oriented toward many domains of action, the one which introduces the greatest constraint and the sharpest degree of focus is that relating to the generation of statements on method and related forms. This is effectively a self-referential, self-constraining constraint.

9. Containment of unpredictable
(a) Although previous constraints have emphasized the importance of maximal incompatibility consistent with set formation, they fail to allow for a specific openness to the risks and hazards of real-world processes.

(b) A further constraint is therefore introduced to ensure such responsiveness to the possibility of unforeseen conditions.

10. Inter-set consistency
Although the number pattern ensures a formal relationship between the sets, a further constraint is introduced to ensure that there is consistency between the contents of different sets.

11. Operational relevance
Although a previous constraint requires that the set elements have a transformative dimension, a further constraint is required to ensure that such operations are important to any isomorphic management process especially to one requiring the management of contradictions.

12. Inter-set harmony
(a) Although a previous constraint requires that there be consistency between set contents, this is only a neutral "mechanical" condition.
(b) A further constraint can be usefully introduced to require that the set elements be conceived in such a way that there is harmonic reinforcement between elements in different sets. Such harmony would also be significant to any isomorphic management process.

13. Non-definitive
(a) The previous constraints leave open the possibility that the set elements may be generated with the conventional idea of producing a definitive, finished product. This would close the set elements to any process of continuing refinement.

(b) A further "constraint" is therefore introduced which requires that each statement be subject to ongoing reformulation. The pattern of statements thus itself becomes a domain for necessary further action, in the light of experience and insight.

Comments: integrative concepts

1. Comment

Since the intention is only to present the results of a preliminary compilation of material with a view to more detailed evaluation, only the following points are noted:

(a) Range: The entries included cover a very wide range of approaches, as was the original intention. It is to be expected that the inclusion of some of the concepts should be queried as well as the exclusion or omission of other concepts.

(b) Duplication and overlap: A number of the entries may be considered to be duplicates, because the names given to the concepts are held to be synonyms. Some such entries have in fact been combined, but others have been kept separate where the different words tend to be used in different contexts, even though they may be considered to mean the same thing.

(c) Confusion: Considerable confusion was noted in the use of some of the terms, particularly in connection with: integrated and integrative; unity (of science) and unified (science); interdisciplinary and multidisciplinary. There seems to be little general awareness of the subtle but very real implications of the distinctions which some authors attempt to make between the concepts in such pairs or series, particularly wherever interdisciplinary is used.

Such confusion was noted as early as 1937 by Thomas Hopkins with respect to the term integration (1937): *"Integration has come to be one of the "big" works in the American language. Like all "big" words this one tends to lose its specific meaning; frequency of use invariably leads to diffusion of meaning; the heavier the load a term is required to carry the more rapid is its loss of specificity. This rule is especially applicable with respect to words which have a value connotation. When the value involved is one which receives ready and general approval, such a word is easily borrowed, and each borrowing brings about expansiveness of meaning. The value which inheres in the term integration is wholeness or unity. In an age which is characterized primarily by fractionalism, by fragmented and segmented experience, it is natural that thoughtful persons should reach out for concepts of unity. Hence the popularity of the word integration in our time."*

More recently fears have been expressed that there is an escalation in the number of projects and proposals in which terms such as integrative, holistic, interdisciplinary, global, synergistic, organic, between the sciences and the arts, *etc* are used in a variety of permutations. Such approaches may in some cases be social rather than more strictly intellectual phenomena, tending primarily to satisfy the needs of their protagonists (and to impress sources of new funds) rather than providing substantive contribution to the field.

Some groups may attempt to avoid inconsistency and cognitive dissonance by using big umbrella categories and over-arching terms to disguise rather unsubstantial ideas and give some conceptual illusionary security. No attempt has been made at this stage to include any criticism of the concepts included in the light of such possibilities.

(d) Western bias: The entries included mainly emphasize a western concept of order emerging from the industrial era. The point has been made that eastern concepts of order and integration need to be examined, since they may prove more appropriate to the post-industrial era (Magoroh Maruyama, 1974). The implications of this distinction may be helpful in classifying the concepts at some later stage as suggested below.

(e) Partial approaches: Each of the concepts included highlights a particular feature of order. Some do so very clearly because they are relatively abstract, others do so very imprecisely because of the loose manner in which the relevant term is used. Because of the importance of terms such as integration and synthesis in this context, some entries involving them are included mainly to record high frequency usages (*eg* racial integration) which may obscure other usages.

2. Classification

As the series stands, the entries suggest a variety of classification schemes to clarify and relate the different notions of ordering. From an examination of the concepts included, the outline of a framework may emerge within which appropriate distinctions, and links, between them can be made. This is suggested by the possibility of interrelating the dimensions associated with the following:

(a) the series constituted by the progressive structural complexification associated with multi-disciplinary, pluridisciplinary, cross-disciplinary, inter-disciplinary and transdisciplinary (Erich Jantsch, 1972)

(b) the series of organization forms ranging from hierarchical (bureaucracy), through systems organization, to network organization

(c) the distinction between homogeneity and heterogeneity (Magoroh Maruyama, 1974)

(d) the emergence of new integrative levels

In an x-y coordinate system, for example, the y-axis could represent a progressive increase in hierarchical order, to a limiting condition in which all elements (of the universe under consideration) are interrelated vertically under (or to) one dominating element of category. The x-axis could then represent an increase in horizontal interrelationship between elements progressively more distant from each other, to a limiting condition in which every element (of the universe under consideration) was related in some way to every other element.

Ordering in terms of the y-axis is then achieved by grouping elements in terms of a single (possibly complex) hierarchical pattern which remains fundamentally unchanged wherever it is applied. This necessitates the suppression of essential differences between elements in the (long-term) interest of conformity to the pattern as a whole. Incompatible elements are rejected, isolated or eliminated. Unforeseen complexity, when it emerges, is either encompassed by forcing it into the existing pattern or by a special replication of the basic pattern in response to the new situation.

Ordering in terms of the x-axis is achieved by interlinking elements, directly or indirectly, irrespective of their compatibility, and solely in terms of the pattern of local functional requirements. This necessitates the rejection of any ordering, systematization or standardization in the interest of the whole, in favour of the (immediate) interest of the elements within the local functional pattern. Unforeseen complexity, when it emerges locally is either linked directly into the existing network of interrelationships, or indirectly by the generation of a new variety of element to respond to it.

Clearly some integrative or transdisciplinary concepts are associated more closely with y-axis ordering, rather than with x-axis ordering. Other concepts contain different degrees of both types of ordering. Of special interest is the possibility of highlighting the presence of concepts which contain a balanced mix of both types of ordering (namely on the diagonal). Such concepts tend to interrelate a maximum variety of concepts from a new level of integration, with new characteristics, which facilitates the emergence and optimal containment of new variety.

Such distinctions are important in connection with the degree and manner of order or organization in conceptual schemes, social organizations, or society in general. Where such distinctions are blurred, new and more appropriate concepts of order can only emerge with difficulty because they can be too easily condemned as identical to the known forms which have proved inadequate. Hopefully a simple framework can be elaborated to clarify such distinctions and draw attention to new possibilities of order and integration.

3. Geopolitical integration as a metaphor of discipline integration

There is an interesting structural parallel to be explored between the national-international-supranational dimension and the disciplinary-interdisciplinary-transdisciplinary dimension. Just as there are many subtleties and peculiar combinations accepted under the term international, with little effective supranationalism, so there may be many subtle combinations of disciplines to be considered under the term interdisciplinary, but with little effective transdisciplinarity.

(a) Integrity: In each case there is concern with the relationship between sovereignty or territorial integrity and the powers conferred upon some more comprehensive framework. In one case the territory is a geographical area, in the other it is a subject area. In defining international it is important to distinguish between its use as applied to regional groupings of different extent: Andean, Caribbean, Baltic, Scandinavian, African, Afro-Asian, and European, for example.

(b) Significance of integration: The extent determines to some degree the relative significance of the integration achieved as would be the case with discipline groupings such as: amongst the medical sciences, amongst the natural sciences, between the natural and the social sciences, *etc*. In both cases there are many examples of token integration, and of lack of integration disguised by token collaboration and talk about integration. Then in the case of international there are groupings based on non-contiguous areas such as Commonwealth countries, developing countries (OECD), ideological blocs, or land-locked states. Universal organizations raise the problem of the legitimacy and viability of micro-states and the manner of their participation in organizations such as the United Nations.

(c) "Secession": The same is true of micro-disciplines (and disciplines emerging from the tutelage of some powerful and long-developed discipline, reluctant to relinquish its hold), although few efforts have been made to create a universal framework to encompass all the disciplines, whether on an equal footing or not.

(d) Non-axiomatic forms: The qualifiers attached to international are also suggestive of relationships to the subject area of the discipline other than those dependent on a body of axioms, principles, and laws: international governmental, international nongovernmental nonprofit-making, and international nongovernmental profit-making (multinational).

There is also the possibility that the dynamics of the controversies between disciplines, and the expansion of the subject area of some disciplines, may well follow a similar pattern to the dynamics of the relationships between states as illustrated by history from the tribal period through recent centuries. Further examination of the different aspects of this parallel is justified by the familiarity and richness of the nation-state system's structure and dynamics and its consequent ability to draw attention to structural features which may be present or embryonic in the system of disciplines. It is interesting that both the geographical area and the subject area may offer opportunities for equivalent structures and processes, and that the subject area dynamics may increasingly provide a psycho-culturally satisfactory substitute for geographical area dynamics in a world characterized by space limitations and overcrowding. It would however be regrettable if the more unfortunate features of the geographically based dynamics (*eg* imperialism, colonialism, feudalism, cold-war) were to be repeated on a subject area basis, given that many lessons could possibly be learnt from the geographical parallel.

4. Possible future improvements

(a) Revision and elaboration of the preliminary version of the descriptive texts on each concept included in this edition by qualified advocates of each such integrative concept.

(b) Addition of other integrative concepts not covered by the preliminary list in this edition.

(c) Inclusion, where relevant within the entries on concepts, of the names and addresses of centres, organizations, or institutes where the particular integrative concept is:
-- investigated in an academic setting
-- advocated and promoted to a wider group, and possibly to the general public
-- the special concern of an information clearing house
-- developed or implemented in some way (*eg* through training courses)

(d) Inclusion, where relevant within the entries on concepts, of some indication of the extent to which the concept is accepted and used.

(e) Inclusion, where relevant within the entries on concepts, of statements critical of the particular integrative concept and its potential utility.

(f) Development of a network of relationships between the integrative concepts (in a manner analogous to that of other sections), particularly in order to draw attention to concepts which are more powerfully integrative.

(g) Inclusion of the results of one or more experiments in classifying integrative concepts which attempt to highlight, by their position within

the classification scheme, those concepts which reflect a more powerful or comprehensive integration.

(h) Build up a bibliography of books, academic articles and reports which focus on, or draw attention to, some aspect of integrative processes and methods. Such a bibliography should function as a collecting point for documents scattered through a wide range of literature to the point of being irretrievable in terms of the integrative concern which is common to them. (See Section KY).

(i) Inclusion of one or more general surveys or essays of the range of integrative concepts, drawing particular attention to those concepts which represent a more powerful and complex form of integration.

Comments: patterning disagreement

1. Tuning the pattern
The ordered collection of statements presented in Section KP raises a number of interesting questions. It is necessarily imperfect, and is even more so as a first draft. Its current status can best be compared to an untuned musical instrument. Only when it is tuned, to the extent possible, will it be possible to determine whether it can be realistically applied as a guide to operations.

Prior to, or during, the tuning process itself, it will be necessary to sharpen up the sets to a greater extent. This is due to the weakness of some of them in terms of the constraint requirements for:

- maximal disagreement between set elements, perhaps requiring a greater degree of controversy, risk, uncertainty, or paradox;

- operational orientation, since some of them are more descriptive rather than transformative (the emphasis is on nouns or adjectives, and not on verbs).

In this sense it is necessary to "charge up" each set and render it inherently more dynamic. The generated sets can be confronted with new source material to assist in this process. (For example the 16-point definition of dialectics by Lenin, quoted above).

The "tuning" process may be envisaged as follows. The different sets need to be compared to highlight the pattern of relationships between them. For example, the sets with common numerical factors (*eg* 2, 4, 8, *etc*) have commonalities which can be highlighted. This will help to clarify the contents of each set and to increase the degree of order prevailing between them.

2. Awkwardness and artificiality of statements
The tuning process is necessary to overcome the problem of the awkwardness of the individual statements. Such awkwardness, is to be expected in a first draft, given the manner in which the sets were generated. There is a basic dilemma in formulating such statements in order to avoid an impression of jargon. But the problem is really that a "general, neutral" set of statements is inconsistent with the underlying philosophy of this approach. No particular wording is adequate.

Efforts to produce an exhaustive "definition" merely result in an exhausting amount of text. The study of the significance of some of the sets has in fact been a life work for some people, resulting in many volumes of commentary (as is the case with Carl Jung and 4-set). The very quantity of information quickly becomes counter-productive in terms of operational criteria.

3. Generation of other schemes
One way around this problem of awkwardness and length is to use the "artificial" statement scheme as generated in Section KP as as basis for generating other schemes, corresponding to the difficulties initially encountered:

(a) Schemes may be produced scaled up or down in level of abstraction (Vertical scaling)

(b) Schemes may be produced oriented in terms of: stages, qualities, problems, conditions, *etc* (Horizontal scaling)

(c) Schemes may be produced using different languages: poetic, formalistic, religious, sociological, *etc*. (Model scaling)

By combining these different possibilities sets of "more readable" statements can be produced which will presumably be closer in terminology to particular source material sets. Sets may thus be generated according to application.

4. Inadequate vocabulary
The problem of the lack of sufficiently general words needs to be seen in the light of the previous point and the use of synonyms. In effect by shifting the emphasis according to any of the above scales, there is a shift through the set of synonyms used to generate the set. The tuning process and the generation of sets could be better studied using an on-line synonym data base, which could also permit alternation between noun, adjective and verb. It is possible that the problem of lack of general words would disappear in sets having an even higher number of elements where the emergent concerns would become much more specific.

5. Internal structure
At this preliminary stage, it is preferable to assess the value of the approach on the basis of the internal structure and consistency of the scheme. Specific references from each generated element to source material have been omitted because of the quantity of such material and the complexity of the decision process leading to a particular choice of words.

In some cases, for example, 20 source sets were compared to produce the generated set. It will be noticed that the attribute of the higher number sets are aspects of those associated with their lower number factors and "condensed" into those associated with their prime number factors. In effect each set "tells the same resource management story", but in the lower number sets the story is highly compacted. In the higher number sets, the attributes associated with elements are simplified, and more easily comprehensible, at the cost of making the relationship pattern more complex. In the lower number sets, these qualities are absorbed into more complex set elements, at the cost of comprehensibility, although the relationship pattern is simpler.

6. Emergence of new information
It will be noticed that sets which are multiples of 2 do not result in new information. The 2-operator merely dichotomizes each element in a set, elaborating on a common point. However a set with 2 as a factor establishes an unresolved polarity which can only be handled in an operational setting by introducing a new perspective (the set elements + 1) from which the polarity can be viewed and balanced. In this sense such polarized sets can effectively "give birth to" a new perspective as pointed out in one of the source documents: "*A vibrating string of any reference length can be halved to sound the octave higher or doubled to sound the octave lower... The number 2 is "female" in the sense that it creates the matrix, the octave, in which all other tones are born. By itself, however, it can only create "cycles of barrenness", in Socrates metaphor, for multiplication and division by 2 can never introduce new tones...*" (C A Hooker, 1978, p.19-20)

7. Development through polar tensions
As structured the scheme supports the view that a monolithic structure of any kind inhibits development. The tension of a polarity is necessary to engender any development. It is useful to distinguish growth or elaboration (in which no new pattern is introduced, by a 2-factor, for example) from new development (in which a new pattern is introduced as a resulting of balancing a polarity). This suggests that any of the classic polarities are very healthy, if they can give birth to a new pattern: capitalism/communism, governmental/nongovernmental, rationalism/empiricism, *etc*. It suggests that a monolithic "world government" would be a total inhibitor of development. In Section KD it is suggested that oscillation or resonance between two or more polar positions may be essential to significant integration or qualitative transformation. The extreme example of brainwashing (stick and carrot) techniques is given there as an example of oscillatory operations which have there constructive equivalent.

8. Definability of disagreement
"Disagreement" as it has been discussed here, and allowed to emerge in the generated sets, has not been clearly defined. This is because the definition is implicit in the 2-level set. At that level the subtleties of any distinction between opposition and complementarity, for example, do not emerge. "Disagreement" therefore also covers its synonyms, namely: disaccord, dissent, unconformity, controversy, disunion, discrepancy, difference, opposition, dissonance, irrelation, inequality, incompatibility, irreconcilability, *etc* (*Roget's Thesaurus*).

There is a progressive "dilution" of the degree of disagreement between elements in a set as the number of elements increases. In effect the basic maximal disagreement of the 2-level is spread amongst more elements in the sets containing progressively greater numbers of elements. This suggests that using sets with a higher number of elements as operators makes it progressively easier to contain the disagreement, but only if the relationship between those elements is not eroded or lost. Each set is a container for a different kind of disagreement. Each can also be used to highlight what can go wrong when working with operators at that level, namely the characteristic errors for that level of operation.

9. Definitional "holes"
This hierarchy of sets of "paradoxes", as it was termed, can be usefully, related to the work of R H Atkin on Q-analysis (*Combinational Connectivities in Social Systems*, 1977) discussed in Section KD. It would seem that the set elements generated at any given level are in effect focal points (for the elements) whose relationship define a Q-hole or Q-object in Atkin's terms. As is pointed out there, the major achievement of Q-analysis probably lies in its ability to give precision to discussion about psycho-social phenomena which are, by definition, sensed beyond the boundary of (collective) comprehension. These are represented by "holes" in a physical structure and are indistinguishable observationally from solid objects in the physical case.

In the psycho-social case, such holes are necessarily less substantial without losing their reality. *"Generally speaking it seems to be confirmed that action (of whatever kind) in the community can be seen as traffic in the abstract geometry and that this traffic must naturally avoid the holes (because it is impossible for any such action to exist in a hole). The holes therefore appear strangely as objects in the structure, as far as the traffic is concerned. The difference is a logical one in that the word "Q-hole" describes a static feature of the geometry S(N) whilst the word "Q-object" describes the experience of that hole by traffic which moves in S(N)."* (p.75)

As an "object" this phenomenon is an obstacle to communication and comprehension and obliges those confronted with it to go "around" it in order to sense the higher dimensionality by which it is characterized. As a "hole" this phenomenon engenders is or, engendered by, a pattern of communication. It appears to function both as "source" and "sink". It is suggested that in some way which is not yet fully understood, such object/holes act as sources of energy for the possible traffic around them. From the initial research it would appear that such objects/holes are characteristic of communication patterns in most complex organizations. It seems highly probable that they can also be detected in any partially ordered pattern of communication. As such "societal problems", "human needs" and "human values" merit examination in this light.

10. Frustration of change
The special value of this Q-hole perspective is that it can clarify why action/discussion in the presence of such a hole tends to be "circular" in the long-term, however energetic it may appear in the short-term. As such it shows how social change is **blocked** by the way in which conceptual traffic patterns itself **around** the sensed core issue which is never confronted as such because the connectivity pattern is inadequate to the dimensionality of the issue. This would explain why so many issues go unresolved and why the institutional process of solving problems becomes of greater importance than the actual elimination of the problem. This approach also draws attention to the probable presence of holes/objects of even higher dimensionality than those whose presence can be sensed relatively easily. Such phenomena, it may be supposed, are of great significance to long-term development.

11. Configurations of "holes"
In Section KD the importance of **configurations of holes** is raised. Clearly the hierarchy of sets generated in Section KP constitutes such a configuration. How can such holes exist in relation to one another? What is necessary to permit transitions from one configuration to another? The question is how **configurations** of holes can be identified and/or designed. It is the configuration of the holes which provides the minimum structure to stabilize and give form to the co-presence of the differing concepts of development. Such configurations, in order to fulfil their function, must presumably exist within two boundary conditions:

- the connectivity between elements bounding holes must not be so great as to erode or destroy the **identity** of the holes so connected;

- the connectivity between elements bounding holes must be great enough so that the integrity of the configuration as a whole is maintained.

12. Designing better "holes"
A further question is then the manner whereby better holes are to be identified or reached within such configurations. Now from one point of view it is necessary to avoid introducing an elements of evaluation, because from each hole the perception of other holes will be distorted so that no **communicable** assessment can be usefully formulated. On the other hand, it may prove to be the case that, at the level of the configuration as a whole, more than one such configuration can be identified/designed in order to interrelate the perspectives associated with the set of holes. And at this level, **without** privileging any particular hole, more adequate inter-relationships between the elements making up the holes can be identified.

13. Internalizing disagreement
By deliberately internalizing disagreement, the scheme moves beyond the stage of being a "cook book for potted wisdom" or a set of "bloodless categories". Each set can be tuned to constitute a set of challenging operations - challenging because of the difficulty of maintaining them in equilibrium. The question is how effectively the sets can be tuned to take the scheme beyond the status of being simply an interesting exercise.

The scheme is valuable because of the way it interrelates incompatibilities at different levels. It is significant also because of the way each set is embedded in a context of interdependent sets.
Given its relation to the source material of diverse cultural origin and specialization, the scheme is also valuable to the extent that interfaces can be provided to such specialized sets. It offers a way of interrelating and engaging groups working through apparently different concept schemes.

The awkwardness of the statements at present draws attention to the basic problem of how to condense qualitative complexes. The solution in traditional cultures of projecting them onto gods or demons (about whom stories could be told to bring out those qualities) was a good way of transforming the problem (as argued in Section MZ).

14. Central issue of comprehension
As designed the scheme is not "ideal" in the sense which is now so easily condemned. No set element is imposed, since as a hierarchy of paradoxes the problem of comprehension is central. A distinction may even be usefully made between:

- Freedom to choose **between** a **plurality** of **competing** concept schemes each with **overdefined** concepts, namely the conventional approach. Here the individual, once the choice of scheme has been made, has no further freedom, because the concepts within the scheme must be accepted as they are defined.

- Freedom to choose how to understand **within** a **single** concept scheme composed of **underdefined** concepts whose significance may be partially associated to those of other schemes seen as **non-competing.** Here the individual is constantly challenged with the freedom to understand particular concepts in some more significant manner in the light of the concept set within which it is embedded.

15. Challenge of set design
The generation of these sets has been approached as a design problem in which constraints are necessary and must be creatively selected. It is possible that the constraints could be refined as part of the tuning process.

Comments: systems of categories distinguishing cultural emphases

1. System of Maruyama (Epistemological mindscapes)
(a) H-mindscape (homogenistic, hierarchical, classificational): Parts are subordinated to the whole, with subcategories neatly grouped into supercategories. The strongest, or the majority, dominate at the expense of the week values, policies, problems, priorities, etc). Logic is deductive and axiomatic demanding sequential reasoning. Cause-effect relations may be deterministic or probalistic.

(b) I-mindscape (heterogenistic, individualistic, random): Only individuals are real, even when aggregated into society. Emphasis on self-sufficiency, independence and individual values. Design favours the random, the capricious and the unexpected. Scheduling and planning are to be avoided. Non-random events are improbable. Each question has its own answer; there are no univeral principles.

(c) S-mindscape (heterogenistic, interactive, homeostatic): Society consists of heterogeneous individuals who interact non-hierarchically to mutual advantages. Mutual dependency. Differences are desirable and contribute to the harmony of the whole. Maintenance of the natural equilibrium. Values are interrelated and cannot be rank-ordered. Avoidance of repetition. Causal loops. Categories not mutually exclusive. Objectivity is less useful than "cross-subjectivity" or multiple viewpoints. Meaning is context dependent.

(d) G-mindscape (heterogenistic, interactive, morphogenetic): Heterogeneous individuals interact non-hierarchically fur mutual benefit, generating new patterns and harmony. Nature in continually changing requiring allowance for change. Values interact to generate new values and meanings. Values of deliberate (antipatory) incompleteness. Causal loops. Multiple evolving meanings.

2. System of Hofstede (Indices of work-related values)
(a) Power distance: Namely the attitude to human inequality. The index developed groups information on perceptions of an organizational superior's style, colleague's fear to disagree with the superior, and the type of decision-mlaking that subordinates prefer in a superior.

(b) Uncertainty avoidance: Namely the tolerance for uncertainty which determines choices of technology, ruyles and rituals to cope with it in organizations. The index developed groups information on rule orientation, employment stability and stress.

(c) Individualism: Namely the relationship between the individual and the collectivity which prevails in a given society, especially as reflected in the way people choose to live and work together. The index distinguishes between the importance attached to personal life and the importance attached to organizational determination of life style and orientation.

(d) Masculinity: Namely the extent to which biological differences between the sexes should or should not have implications for social activities that are transferred by socialization in families, schools, peer groups and through the media. The index developed measures the extent to which people endorse goals more popular with men or with women.

3. System of Mushakoji (Modalities for grasping reality)
(a) Affirmation: Leading to affirmative action in the form of support, commitment, initiative, proposition, cooperation, consensus formation, empowering, "opening".

(b) Negation: Leading to negative action in the form of sanction, withdrawal (of support), denial, disassociation, delimitation, criticism, opposition, promotion of dissent, disempowerting, "closing".

(c) Affirmation and negation: Leading to ambiguous action, non-violent resistance, "dumb insolence", "giving withone hand and taking with the other", "double dealing", "stick and carro tactics", the "yes but no" response of the frustated cross-examinee.

(d) Non-affirmation and Non-negation: Leading to action in the form of indifference, indecision, non-action (in the oriental sense), "neither confirm nor deny", "opening and closing".

4. System of McWhinney (Modes of reality construction (resolution and change):

(a) Analytic mode: Based on empirical thinking and depnds on hypo-deductive and induc and inductive methods, using all logics, theories and information available to the senses to identify possible solutions, predict implications, and evaluate outcomes. Currently associated with the scientific and quantitative methods. Provides no guide for the processes of change but determines (or predicts) outcomes. Change is driven by the sense of efficiency, of ptimally organiziating to produce that which can be produced.

(b) Dialectic mode: Composed of a variety of methods which may appear to be totally distinct and arising from contrary world views based on unitary premises (and therefore held to be intimately related). Encompasses the mode of argument, of disputation among partisans of opposing views and of adversarial encounter -- all as methods of unification. Change serves to cleanse the system of error, correct for deviation from the norm, or protect the domain of truth. In the formation of synthesis, evolution occurs as a historically driven imperative that progressively cleanses the organization of impure functions.

(c) Axiotic mode: Based on value exploration, resolving issues by developing new, and shared, evaluations of events. May work through "recontexting" or "transformation" of images by which an issue of "dissolved". Concerned with questions of morality, fairness and interpersonal behaviour as having value in and of themselves. Changes induced may affect the ideoçlogy of a system and thus be profoundly disturbing to and often blocked by those of unitary belief.

(d) Mythic mode: Based on methods of symbolic creation. At the deepest level, mythic events create new meaning, literally producing something out of nothing. Resolution is produced by transcending existing strucures and meanings that are given to words, situations, objects, and stories. Mythic inventions successful in engendering large scale change are those which are in tune with the needs of the cultural system into which they are injected. They re typically associated with charismatic leadership that captures the will and faith of the involved population. Major methods are those associated with creartive endeavour, use of intuition and strong adherence to premises of the mythic reality.

5. System of Pepper (World hypotheses)
(a) Formism: Grounded on the common sense experience of similarity and a correspondence theory of truth, expressed in the case of geography in a preoccupation with mapping.

(b) Mechanism: Based on a causal adjustment theory of truth, taking the machine as the root metaphor, resulting in a preoccupation with special systems and functional mechanisms in the case of geography.

(c) Organicism: Based on a coherence theory of truth, regarding every event as a more or less concealed process within an organic whole.

(d) Contextualism: Based on an operational theory of truth, seing the world as an arena of unique events.

6. System of Douglas (Natural symbols)
Systems of natural symbols in which the image of the body is used in different ways to reflect and enhance each persons experience of society:

(a) Body conceived as an organ of communication: *"The major preoccupations will be with its functioning effectively; the relation of head to subordinate members will be a model f the central nervous system, the favourite metaphors of statecraft will harp upon the flw of blood in the arteries, sustenance and restoration of strength."*

(b) Body seen as a vehicle of life: As such *"it will be vulnerable in different ways. The dangers to it will come...from failure to control the* *quality of what it absorbs thrugh its orifices; fear of poisoning, protection of boundaries, aversion to bodily waste products, and medical theory that enjoins frequent purging."*

(c) Practical concern with possible uses of bodily rejects: As such it will be *"very cool about recycling waste matter and about the pay-off from such practices....In the control areas of this society controversies about spirit and matter will scarcely arise."*

(d) Life seen as spiritual, and the body as irrelevant matter: *"In these types of social experience, a person feels that his personal relations, so inexplicably uprofitable, are in the sinister grip of a social system. It follows that the body tends to serve as a smbol of evil, as a structured system contrasted with pure spirit which by its nature is free and unfdifferentiated. The millenialist...believes in a Utopian world in which goodness of heart can prevail without institutional devices".*

7. System of Gardner (Forms of intelligence)
(a) Linguistic intelligence: This is demonstrated by a sensitivity to sounds, rhythms, inflections and meter, a special clarity of awareness of the core operation of language. Such gifts are particularly characteristic of poets; but are said to be universally relevant in order: to use rhetoric in order to convince others; to remember information mnemonically; to explain someting clearly to others (even when what is being explained is mathematical, logical or whatever); and to understand language itself. This intelligence is shown to be rooted in the left hemisphere of the brain; and although the right-hemisphere may be used to learn both to read and to speak, such ability will be somewhat restricted.

(b) Musical intelligence: Such intelligence has as its centre the relating of emotional and motivational factors to the perceptual ones; music is a way of capturing and communicating feelings and knowledge about feelings. Musical ability is centred in the right-hemisphere of the brain and varies widely among individuals and cultures. It seems to be used in exploring and interpreting other forms of intelligence.

(c) Logical/mathematical intelligence: This is developed first from the ability to recognize classes or sets of physical objects; and later by conceptualizing classes or sets of objects or ideas in the mind and understanding logical connections among them. Central features are: the ability to identify and then solve significant problems; memory for repetitive patterns and the ability to compare and operate upon such patterns mentally; and an intuitive feel for logical relationship.

(d) Spatial intelligence: An accurate perception of the physical world, an ability to transform or modify these perceptions, and the recreating of certain aspects of visual experience without relevant physical stimuli -these are all part of spatial ability. Centred in the right-hemisphere of the brain, spatial skills are typical of cultures where tracking, hunting and visual recognition of the environment are paramount; but present-day Western culture requires it no less, whether for the architects or the mathematical topologist or the molecular biologist.

(e) Bodily-kinesthetic intelligence: Skill in controlling bodily movements and in the ability to manipulate objects combine in this intelligence, which has been valued in many cultures as the harmony between mind and body - the mind trained to use the body properly and the body to respond to the mind. It reaches its height in dance, which has supernatural connotations in some cultures, and in other performing roles. Low bodily-kinesthetic intelligence is equated, in India for example, with immaturity.

(f) Personal intelligence: These are centred on the concept of the individual self and may be considered as:

-- Access to one's own feeling life - this is the development of the internal aspects of a person and the ability to detect and symbolize complex and highly differentiated sets of feelings.

-- Ability to notice and make distinctions among individuals - to read even the hidden intentions and desires of others and to use this knowledge to influence their behaviour. Development of these intelligences leads to self-maturity and to personal knowledge of one's self as a unique individual.

8. System of Jones (Axes of methodological bias)
(a) Order vs disorder: Namely the range between a preference for fluidity, muddle chaos, etc. and a preference for system, structure, conceptual clarity, etc.

(b) Static vs dynamic: Namely the range between a preference for the changeless, eternal, etc. and a preference for movement, for explanation in genetic and process terms, etc.

(c) Continuity vs discreteness: Namely the range between a preference for wholeness, unity, etc and a preference for discreteness, plurality, diversity, etc.

(d) Inner vs outer: Namely the range between a preference for being able to project oneself into the objects of one's experience (to experience them as one experiences oneself), anda
a preference for a relatively external, objective relation to them.

(e) Sharp focus vs soft focus: Namely the range between a preference for clear, direct experience and a preference for threshold experiences which are felt to be saturate with more meaning than is immediately present.

(f) Spontaneity vs process: Namely the range between apreference for chance, freedom, accident, etc and a preference for explanations subject laws and definable processes.

9. System of Todd (Family types associated with different socio-political systems)
(a) Exogamic communal family: Namely favouring the emergence of communist socio-political systems (e.g. Russia, certain Slavic countries, China, Viet Nam, Cuba, Northern India)

(b) Exogamic authoritarian family: Namely favouring an asymmetric pluralism characteristic of socialist and socio-democratic forces (e.g. Germanic countries, Sweden, Norway, Gaelic countries, Northern Spain, Japan, Korea, Jews, Gypsies).

(c) Exogamic nuclear family: Namely favouring the emergence of individualistic systems of one kind (e.g. Northern France, Northern Italy, Greece, Poland, Latin America, Ethiopia).

(d) Exogamic absolute nuclear family: Namely favouring the emergence of a second kind of individualism.

(e) Endogamic communal family: Namely characterized by frequent marriage between children of brothers (e.g. Arab countries, Turkey, Pakistan, Iran, Afghanistan, and southern Soviet Republics).

(f) Endogamic asymmetric communal family: Namely characterized by frequent marriage between children of brother and sister (e.g. Southern India) favouring the emergence of a caste system.

(g) Anomic family: Namely characterized by flexible heritage and cohabitation arrangements with possible consanguinous marriage (e.g. South-East Asia and South Amzerican Indians) favouring political ambivalence and socio-political systems such as that based on Buddhism.

(h) Dynamically unstable domestic family: Namely characterized by the dynamic instability of the domestic family group and polygyny, favouring the emergence of socio-political systems dependent on authoritarian forces in order maintain social stability (e.g. African family systems, insofar as information is currently available).

Metaphors and patterns M

Scope

Any form of international "mobilization of public opinion" (using the conventional military metaphor) to engender the much sought "political will to change" is dependent upon communication. This is especially the case when the insights required to guide that change are complex, counter-intuitive or simply not clearly communicable within any one conceptual language.

The purpose of this section is therefore to review the range of communication possibilities and constraints of metaphor, pattern and symbol. This is partly in response to the narrow focus of recent major intergovernmental initiatives under the extremely misleading titles of "International Commission for the Study of Communication Problems" (limited to the mass media) and the "International Communications Year" (telecommunications hardware) by UNESCO and ITU respectively. It is however a direct consequence of participation by the editors in the Forms of Presentation project of the Goals, Processes and Indicators of Development project of the United Nations University.

Sub-sections

The section consists of 444 entries. It is divided into three sub-section:

-- Section MM explores through 88 entries the possibility of designing metaphors that are appropriate to engendering a creative response to the global problematique.

-- Section MP explores in 253 entries a particularly rich approach to interrelating mutually incompatible concepts in a pattern, interlinked by 3,491 cross-references.

-- Section MS reviews in 103 entries the range of symbols used in modern and traditional cultures as a way of communicating multiple levels of significance in a compact and reproducible form. The entries are interlinked by 636 cross-references.

Method

The procedures used in preparing this section are discussed in detail in Section MZ.

Overview

Detailed discussion is given in the Notes of Section MZ. As a whole the section provides a framework within which to review alternative ways of interrelating items of information to facilitate comprehension and communication. Further details of the three sub-sections are given on the next page.

Context

The contents of this section may be considered as complementing the other sections in ways such as the following:

Human development: By the manner in which human development options are communicated, and through the evolution of forms of communication following efforts to communicate new insights into human development possibilities.

Integrative knowledge: By the manner in which integrative knowledge is communicated, and through the evolution of forms of communication to reflect new aspects of integration.

World problems: By the problems of communication in a global society, and by the need to communicate the complex nature of world problems.

Tranformative approaches: By the evolution of communication techniques (especially in meetings), and by the need to communicate innovative techniques.

Human values: By the manner in which human values are communicated, and through the intrinsic value of communication in maintaining the fabric of global society.

Metaphors

Rationale

Metaphors are a special form of presentation natural to many cultures. They are of unique importance as a means of communicating complex notions, especially in interdisciplinary and multicultural dialogue, as well as in the popularization of abstract concepts, in political discourse and as part of any creative process. They offer the special advantage of calling upon a pre-existing capacity to comprehend complexity, rather than assuming that people need to engage in lengthy educational processes before being able to comprehend.

Although frequently used in international debate through which strategies are defined, the advantages of metaphor have not been deliberately explored to assist in the implementation of such strategies. Each development policy may be considered a particular "answer" to the global problematique. No such answer appears to be free from fundamental weaknesses. A shift to an alternative policy becomes progressively more necessary as the effects of these weaknesses accumulate. However, since each such policy reflects a "language" or mind-set whereby a worldview is organized, no adequate "logical" framework can exist to facilitate comprehension of the nature of such a shift or of the process of transition between alternatives.

Many familiar metaphors of alternation exist through which the characteristics and limitations of such a shift may be understood. This section presents the result of an experiment in deliberately designing metaphors in support of innovative development.

Contents

The section contains 88 entries elaborated as an *editorial experiment* in facilitating comprehension of transition and change, especially in some ordered manner between complementary alternatives. The phenomena selected as substrates for the metaphors include: those familiar to everybody (*eg* walking, breathing), those especially significant to rural communities (*eg* crop-rotation, getting water, animal movement), those familiar to industrialized societies (*eg* driving, media diets, vitamins) and some key physical or technological phenomena (*eg* electric motors, metabolic pathways, magnetic containment of plasma).

Method

The information used was obtained from a wide range of specialized reference books.

Index

A keyword index to entries is provided in Section MX. The keywords are also incorporated into the index for Volume 2 (Section X)

Bibliography

Bibliographical references, by author, are given in Section MY.

Comment

Detailed comments are given in Section MZ.

Reservations

The entries indicate the possibility of developing a technique for designing powerful metaphors. As this is an exploratory exercise, individual entries may call for substantial revision.

Possible future improvements

In addition to the refinement of the selected metaphors, the variety of phenomena used as a basis for suchmetaphors could be increased. A better indication could be provided of the strengths and limitations of each metaphor. This would enable groups of complementary metaphors to be interrelated by a pattern of cross-references as explored in Section MP. This points toward the possibility of producing a repertoire of metaphors that may be used to communicate complex insights into a wide range of social phenomena whilst at the same time empowering them conceptually to explore new patterns of organization in which dynamic processes are emphasized, rather than static structures.

ENTRY CONTENT AND ORGANIZATION

Ordering of entries
Entries are in **numeric order**. Entry numbers have been **allocated randomly**; they have no significance other than as a permanent point of reference to facilitate indexing, cross-referencing, and updating between editions.

Index access to entries
The location of an entry in this sub-section may be determined from:
- the **Volume Index** (Section X) on the basis of keywords in the name of the entry or its alternate names
- the **Section Index** (Section MX) on the basis of keywords in the name of the entry or its alternate names

Structure of entries
Entries may be composed of the following descriptive elements:

(a) **Entry number** This number has **no significance**, except as a convenient method of identifying the entry (particularly for indexing purposes), of filing information on it, and as an identifier to which cross-references from other entries (possibly in other sections) may refer in this and future editions.

(b) **Entry name** This is printed in bold characters. It may be followed by alternative names, when appropriate.

(c) **Substrate** Brief description of the nase phenomenon highlighting the processes on which the metaphor is constructed.

(d) **Metaphor** Brief indication of how the processes of the substrate may be used to re-interpret processes in society

(e) **Special features** Indication of the contrast between how the particumar processes in society are usually uns-derstood as compared to how the metaphor suggests they might be usefully understood.

(f) **Further keys** Indication of related features of the substrate which could be used to extend and enrich the metaphor.

Cross-referencing of entries
There are **no cross-references** between entries in this sub-section.

♦ **MM2001 Walking**
Substrate One foot is moved forward to a position at which it can bear the full weight of the body. The other foot is then brought forward, past the first, to a new position at which it can in turn bear the full weight of the body. The arms are moved in such a way as to act as a counterbalance. As a result of these movements the body can be moved forward at a constant pace. Although in places of difficulty the attention may be focused on the movement of one of the feet, normally attention is focused on the movement of the body as a whole.
Metaphor One policy may be promoted and implemented to bring society forward to a new position. Eventually however the momentum of this displacement requires another distinct policy to be brought into play to prevent loss of balance and to carry the society even further forward. During this latter phase the first policy must necessarily conserve the achievements made although the weight attached to this role is gradually phased out in anticipation of a reinterpretation of this policy to take the society even further forward. Whilst attention is clearly required on the formulation and implementation of each policy, particularly at points of crisis, the progress of society is best guided in terms of the movement as a whole to which both policies contribute, but for which neither is sufficient by itself.
Special features The smooth transfer of weight from one foot to the other with each foot alternately bearing the weight and then giving it up to the other. The counterbalancing movement of each arm in harmony with the opposing leg. Progress is measured by the number of alternations made.
Contrast: The metaphor of "walking on two legs" has been used in China to describe a policy of technological dualism. The present attitude of policy advocates may be likened to the attempt to move forward with one foot only – whether it be the right or the left. This can only be achieved by hopping – provided balance can be maintained. Policies have to be relinquished in favour of an alternative and then renewed to fulfil a new role. This is also true of any "alternative".
Further keys More legs (4–legged, 6–legged, etc. animals). Legless movement (serpentine). Learning to coordinate walking movements. Drunken or spastic lack of coordination. Limping, paralysis and other obstacles to free movement. Number of counteracting muscles required. Evolution of types of movement. Monkey movement through trees by swinging from the branches by the hands only.

♦ **MM2003 Breathing**
Substrate Air is drawn into the lungs by movement of the diaphragm. This permits oxygen to be transferred into the bloodstream and metabolic waste products to be transferred back into the air ejected from the lungs during the following expiration portion of the respiratory cycle.
Metaphor A society is inspired by the circulation of news of collective significance. This has a stimulating and revivifying effect. Such news is vital to the social metabolism. After provoking debate and discussion, such news that is not stored in the collective memory is considered "stale" and is rejected in favour of fresh news.
Special features The body must necessarily draw in new air to replace stale air. It cannot remain locked in one portion of the respiratory cycle for any length of time. The transition from inspiration to expiration is normally very smooth.
Contrast There is a tendency to believe that society can function on the basis of the inspiration to be drawn from some single favoured message and that consideration of this message does not lead to the production of stale waste products which need to be ejected. Where the need for fresh news is recognized, the difficulty of producing fresh news of significance has resulted in the progressive banalization of content.
Further keys Respiratory defects: irregular breathing, shallow breathing, etc. Respiratory diseases. Respiratory consequences of air quality: suffocation, hyperventilation. Forced breathing in times of intense activity. Techniques and consequences of interrupting the respiratory cycle. Artificial aids to respiration. Breathing exercises.

♦ **MM2004 Reification**
Entities — Substances
Substrate Experience of the physical world, including the people in it, tends to be described in terms of bounded objects having some substance.
Metaphor Experience of the psycho–cultural world, including the personalities, groups and bodies of knowledge in it, may be described in terms of bounded objects. Understanding experience in terms of objects and substances allows parts of experience to be selected out and treated as discrete entities or substances of a uniform kind.
Special features The ubiquity of the reification process.
Contrast There is widespread recognition of personalities, groups, concepts and belief systems as substantive entities, even when their boundaries have been artificially imposed to render the phenomena adequately discrete. Problems are typically treated as distinct entities ('Inflation is endangering our economy') as are values ('Peace will bring prosperity for all'). Physical and emotional states within a person are also frequently treated as entities ('She has a pain in her chest'). Such metaphorical approaches are used in the attempt to deal rationally with complex experiences.
Further keys Variety of entities enated by reification.

♦ **MM2005 Exercise and rest**
Substrate Periods of bodily exercise alternate with periods of rest. During the former the muscles become tired and the energy resources of the body are depleted. During the latter muscle tone is recovered and the energy resources are regenerated.
Metaphor A society may exert itself for a period, whether a working day, week or longer. A period of rest is then necessary. This may be especially necessary following exceptional effort such as in a war.
Special features During the work portion of the cycle the body exhibits signs of the degree of need for rest. During the rest portion of the cycle, restlessness gradually emerges as an indicator of the need for renewed activity.
Contrast In society work may be variously considered as an unfortunate necessity or as an unmitigated good, with rest as a wholly desirable condition or as an unfortunate necessity. This is even extended to retirement as a desirable endstate after an unfortunate number of working years. Such perceptions have been further confused by the unemployment crisis and the possibility of extended leisure. The transition between the two conditions is seldom smooth and often perceived schizophrenically.

♦ **MM2006 War**
Substrate War is armed conflict, usually between two political groups, involving hostilities of some magnitude over a period of time and usually resulting in some damage to both sides, whether or not a winner eventually emerges. Both sides may attempt to advance their cause by devising suitable strategies and tactics within the prevailing logistical constraints.
Metaphor Non–physical conflict, usually between two groups or individuals, may take the form of an argument of some magnitude sustained over a period of time and usually resulting in some damage to both sides, whether or not a winner eventually emerges. Both sides may attempt to advance their cause by devising suitable strategies and tactics within the prevailing constraints.
Special features Emphasis on neutralizing or eliminating any opposition.
Contrast Most expressions relating to an argument are derived from warfare. Arguments with an 'opponent' are 'won' or 'lost', depending on the relative 'strength' of the 'position' of the participants, and how they ace 'attacked' and 'defended'. Appropriate 'strategies' and 'tactics' must be used in 'deploying' resources in order to avoid having to 'lose ground' or 'retreat'. Many of the acts of argument are partially structured by the concept of war, including 'bringing up the heavy artillery', obtaining sufficient 'ammunition', and ensuring that shots 'hit the target'. Although some social groups (e.g. academics, diplomats) engage in genteel argument within a framework precluding certain tactics as unfair or irrational (e.g. flattery, insult, intimidation, evading issues, invoking authority, etc), their arguments still involve an effort to destroy the opponent's position.
Further keys Types of weaponry and warfare. Strategy and tactics. Arms race. Overkill and mutually assured destruction. History and future of warfare.

♦ **MM2007 Practice/performance**
Substrate During the period of practice a particular skill is exercised often by focusing on its component parts. Weaknesses are detected and given special attention. Practice prepares for a subsequent period of performance during which the skill is used in its entirety.
Metaphor The activities of individual groups in a society may be considered as forms of exercise relatively isolated from each other. Only under rare crisis conditions is it necessary for these group activities to interlink strongly in a coordinated protective or remedial action.
Special features During practice the emphasis is on bringing weaknesses to light and on repetitive work on such weaknesses often in isolation from their context. During performance, weaknesses must necessarily be disguised and stress is placed on the coherence of the activity as a whole.
Contrast At the societal level practice is limited to military and civil defence exercises, trial run presentation meetings in large corporations, rehearsals for artistic performances, ceremonies and parades, as well for team sport practice. The approach is not applied to activities directly concerned with the development of society such as the organization of key meetings.
Further keys Degree of informality possible in practice sessions. Degree of innovation possible in performance.

♦ **MM2008 Personification**
Substrate A physical object may be further specified as being a person in order to facilitate response to the vagaries of its behaviour (e.g. automobiles or ships named by the owner, operator or crew; naming of hurricanes).
Metaphor A non–physical psycho–cultural phenomena may be further specified as being a person in order to organize a wide range of complex experiences associated with it in terms of human characteristics, motivations and activities.
Special features The range of human characteristics, motivations and activities imputed to the physical object. The explanatory power rendering the phenomena comprehensible to most people.
Contrast Throughout the history of man, complex psycho–cultural phenomena have been recognized as the expression of deities or other beings with semi–human characteristics calling for an appropriate pattern of response to ensure the survival of the tribe. Complex modern problems may be defined as though they had human characteristics ('Inflation has outmanoeuvred the programmes of national governments'). Ideas may also be treated as people with an appropriate life cycle ('Infra–red astronomy is still in its infancy').
Further keys Contrast between treating phenomena as having selected human characteristics and giving a phenomenon a human name. Named deities and archetypes.

♦ **MM2009 Interruption of thematic development**
Substrate In the development of a literary or dramatic theme, interruption of one theme at a critical moment to allow the further development of some other theme is a common device for increasing the significance of the whole.
Metaphor In society the development of an issue may be interrupted whilst attention is switched to other issues which temporarily acquire, or are given, higher priority.
Special features In this form of alternation there is a special awareness of the dramatic moment at which attention can best be switched to another theme to avoid a premature denouement and loss of interest. Skill is used in blending the themes developed in parallel in this way building up to the final denouement.
Contrast It would seem that in societal development the technique is mainly used to postpone any denouement and reduce interest thus preventing the emergence of any new level of significance.
Further keys Nature of dramatic moments. Skill in blending themes. Significance of the whole as distinct from that of its constituent dramatic parts.

♦ **MM2010 Animal locomotion**
Structure The structure of animals reflects their locomotor habits, if any, in four distinct environments: aerial (including arboreal), aquatic, fossorial (underground), and terrestrial. The physical restraints on movement are gravity and drag, although these may be considered negligible in aquatic and terrestrial locomotion respectively. Movement may be achieved by modifications of the body shape (eg squid, eel or leech), by the operation of special appendages, or by drifting according to wind or water currents.
Metaphor The structure of groups or individual personalities reflects their mode of movement through society, whether through conceptual realms and frameworks, through administrative or behavioural infrastructures, or over the interface between the latter and the former. The restraints on such mobility are inertia, the weight of attachment, or attraction to any particular social setting and the general social resistance acting to inhibit any speech change, although these may be significantly reduced in some contexts. Movement may be achieved by general modifications of structure (e.g. a demonstrating crowd, an infantry manoeuvre, or a tribal war party), by the action of special sections (e.g. public relations, research), or by drifting with the prevailing current of opinion.
Special features The close relationship between the varieties of structure and the varieties of movement.
Contrast There is widespread recognition of individual mobility, of groups 'on the move', if the problems of overcoming resistance to such movements and of 'getting things moving'. There is little explicit understanding of the ways in which groups and individuals move in society, although frequent use is made of such phrases as 'burrowing' (e.g. through archives), 'wriggling like an eel' (e.g. in a negotiation), 'rising' (e.g. above sordid details).
Further keys The many specialized forms of aquatic, aerial, fossorial and terrestrial locomotion. Annual migrations and transhumane. Contrast with motion in man–made vehicles.

♦ **MM2011 Shifting topics of conversation**
Substrate In a casual conversation amongst a well–established group of friends, themes are taken up for a period of time then abandoned for others, often to be taken up once again on some later occasion.
Metaphor In society an issue may become fashionable for a period of time as the coin of exchange between opposing forces. It may then be abandoned in favour of some other issue, only to be

taken up again on some later occasion.
Special features The manner in which interest in the topic develops and declines to the point at which people are 'tired of it' and seek an alternative. How not discussing a topic for a period increases the interest with which it is taken up on a later occasion.
Contrast The reintroduction of issues for consideration by public opinion is very much under the control of the media (themselves possibly an instrument of government control).
Further keys How new topics get introduced and how old topics get re-introduced or definitively rejected. How long a topic is discussed or remains undiscussed. How the level of interest in the conversation is maintained through shifting between contrasting topics.

♦ MM2012 Terrestrial animal locomotion
Substrate The structure of animals moving on the surface of the ground reflects their locomotion habits. Arthropods and vertebrates tend to move by walking or running, using the legs to support the body off the surface and to propel it forward. Stability is maintained during this process by a functional sequence of limb movements which in the fastest case are asymmetrical in the case of vertebrates (four-legged) and symmetrical in the case of arthropods (six-legged or more) which are consequently less rapid. Movement by saltation (hopping) is also used by some vertebrates and arthropods which necessarily have larger hind legs; four patterns of saltation may be distinguished. Movement by crawling is used by some invertebrates (using peristaltic locomotion or contract-anchor-extend locomotion) and by limbless vertebrates. The latter use one of four patterns: serpentine locomotion (simultaneous lateral thrusts against solid projections by a body in a series of sinuous curves), concertina locomotion (used when there is not enough frictional resistane for the serpentine form), sidewinding locomotion (used over friable sand, such that only portions of the body remain in contact with the ground), and rectilinear locomotion (using movement of scales beneath the body in contraction waves from head to tail).
Metaphor The structure of praxis-oriented groups or individual personalities reflects their mode of movement over social terrain. To reduce resistance to such movement, the group may only interface with its programme through a succession of complementary policies. These may each be discarded once their immediate objectives have been achieved but may be reimplemented in a modified form to achieve subsequent objectives. Stability is maintained by appropriately ordering the process of implementing and abandoning such complementary policies. An alternative procedure, rather than maintaining continuous contact with the social terrain and its constraints, is to undertake a discontinuous succession of policy leaps thus by-passing obstacles which might otherwise prove impassable. Another alternative is to ensure that all parts of the group are in continuous contact with the terrain and to coordinate the manner in which they each act on it in order to move the group forward.
Special features The manner in which support, propulsive and stabilizing functions are distinguished and coordinated.
Contrast In China official use has been made of the metaphor of 'walking on two legs' as well as of the 'great leap forward'. But in general there is little understanding of how the relationship between contrasting policies is to be coordinated. It is as though the different legs win in competition for the right to move forward. This is a guarantee of instability and discontinuity in two-party systems and of confusion in multi-party systems. This clarifies the frequent repeat that society can only advance at a 'crawling' pace. The metaphor also throws light on the debate between those who favour organizations in which all sections deal with practical issues and those favouring specialized sections for such contact.
Further keys Contrast between the evolutionary relationship between different forms of movement and the effectiveness of particular forms in particular environments.

♦ MM2013 Rotating chairmanship
Substrate As a solution to the political problem of control of the key position of chairperson, this position may be rotated amongst several (or all) members of the body in question. In this way each person holds the position for some period of time, whether a session, a meeting or a year. (The technique is also used for office location and running expenses of small organizations).
Metaphor In society control of positions of power could be rotated between the forces attempting to obtain a monopoly of that power.
Special features Rules are formulated defining the period for which the position may be held and the manner in which the transition to the next holder is carried out. A well-defined sequence of holds is elaborated.
Contrast Usually in society much of the inter-group dynamics is concerned with the struggle for one group to hold the position of power. Where this is totally unacceptable bicephalous arrangements are made (especially in countries in which bilingualism is a major issue), although structure-oriented solutions, in contrast to the process-oriented solution of rotation.
Further keys Variations in the length of the period the position is held to allow participation of minority interests.

♦ MM2014 Container and boundary
Substrate Experience of the physical world, including the people in it, tends to be described in terms of bounded objects. Each of these may be viewed as a container with an inside and an outside. Such objects may include areas such as fields and territories (whether fenced or not).
Metaphor Experience of the psycho-cultural world, including the personalities, groups and belief systems in it, may be described in terms of bounded objects that may be viewed as containers having an inside and an outside.
Special features The ubiquity of boundaries.
Contrast There is widespread recognition of personalities, groups and belief systems as containers with distinct boundaries, and when none exist these may well be projected onto whatever is perceived as in the case of events, activities and states (and even the 'visual field').
Further keys Varieties of container and boundary. Topology of containers. Containers with no distinct inside or outside (e.g. Klein bottle). Territoriality.

♦ MM2015 Shifting patterns of activity
Substrate Individuals when free to choose the activities in which they wish to engage will perform one activity until they 'get bored with it' and then shift to some other 'more interesting' activity. The activities selected may include talking, eating, watching TV, going for a walk, music, bird watching, doing a crossword puzzle, playing a game, gardening, etc. At some stage they will take up each of the earlier activities again.
Metaphor Collective attention and public opinion may also be seen as shifting its focus of attention between activities 'currently in the news'.
Special features The variety of possibilities that may be temporarily selected out of the pool of activities is quite large. The ability of an activity to sustain interest for any period of time varies. The nature of the interest is very different as though each provided a different kind of nourishment (or vitamin). It is not clear whether there is any order to the shifts or how long before an activity may once again become of interest.
Contrast Public opinion is notoriously fickle concerning the subjects which retain its interest for any period of time. Fresh items ('news') are clearly essential even though they fall into familiar categories. The question is what kinds of (better) nourishment can be provided with what sort of frequency.
Further keys The process of becoming bored and the emergence of a 'hunger' for some alternative form of 'nourishment'. The nature of the search for 'kicks' and the importance of 'happenings'. The inter-activity hiatus.

♦ MM2016 Conduit
Substrate Channels or pipes of a range of dimensions are used for conveying water, other fluids, or solids in suspension. This may be done under pressure. Such conduits are frequently linked together to form networks controlled by valves.
Metaphor Information of various types may be communicated through channels which may be open or closed. Such channels are frequently linked together to form networks controlled by procedural devices or 'gatekeepers'.
Special features The spatial relationship defined between form and content.
Contrast The use of 'channels' of communication is widespread, whether or not any corresponding physical network exists (e.g. telex, telephone or data network). Works themselves are widely recognized as conduits for meaning (e.g. 'His words carry little meaning'). It has been estimated that some 70 percent of the expressions used for talking about language are based on the conduit metaphor. Some individuals and groups are recognized as conduits for certain forms of information, especially in the case of religious teachings.
Further keys Information networks and their organization, irrigation networks.

♦ MM2017 Interpersonal interaction
Substrate An individual will engage in conversation or some other form of contact with a friend, colleague, acquaintance or stranger, possibly at a cocktail party or for a coffee. After a time the interaction is terminated because one or both have other interactions in which they believe they can more profitably engage, whether out of interest, pleasure, or some form of obligation. On parting they may fix a date for a further encounter between them, although the overlap in their patterns of behaviour may be such as to bring about such an encounter anyway.
Metaphor Such patterns of interaction are also evident in the relationships between groups and nations (through their representatives).
Special features The different kinds or qualities of interaction and the different periods for which they are activated. The manner in which interactions are ordered and given priorities over a period of time.
Contrast This process is extremely well developed at the individual level and constitutes much of the dynamics of interpersonal interaction. Although well developed in the commercial and diplomatic worlds, it is not clear that it yet provides an equivalent amount of 'connective tissue'.
Further keys The variety of calendar (diary) filling policies. How each interaction fits into a pattern through which the individual is nourished.

♦ MM2018 Disease
Substrate A disease is a departure from the normal state of a living organism sufficient to produce overt symptoms that constitute an impairment of its functions. Physiologically the normal state is a delicate balance of chemical, physical and functional processes maintained by a complex of partially understood control mechanisms. When these fail to function appropriately in response to some particular conditions a characteristic disease may be defined. The concept is also extended to mental and emotional disorders of man. A systematic classification of human diseases exists.
Metaphor Groups, organizations, cultures (and even schools of thought), as organs of society, on occasion present overt symptoms of some form of disruption to the delicate balance of their control processes. These can be considered as organizational diseases which could lend themselves to some form of systematic classification analogous to that for human diseases: (a) infective diseases: Organizations and networks can be negatively affected by the transfer of some viewpoint or process from another body. (b) Parasitic diseases: Groups and cultures can function in a parasitic relationship to other bodies from which they sustain themselves. (c) Neoplasms or tumours: Groups and networks can suffer from the abnormal development of parts of their structure, whether this is of a relatively harmless nature or leads rapidly to the complete disruption of the body. (d) Endocrine diseases: To the extent that organizations have sub-divisions with specific functions, one such sub-division can function inappropriately thus unbalancing the operations of the whole. (e) Nutritional diseases: Groups and societies are sustained by the influx of resources. If the amount or nature of such resources is inappropriate, this will result in characteristic consequences for the group. (f) Metabolic diseases: Organizations and networks have established procedures for processing incoming resources both to adapt them for effective use after storage and to reject the exhausted by products of such use to permit further activity. These procedures may be disrupted. (g) Blood diseases: To the extent that organizations and societies may be described as having a medium, such as finance (whether as credit or debt), through which resources are circulated throughout their structures, the manner in which this medium functions may become disrupted. (h) Mental disorders: Groups and cultures control their own behaviour and attitudes by the appropriate transmission of information. Disorders in this process may occur because of either predisposing causes (resulting from the manner of their creation) or exciting causes or stresses. These may take the form of abnormal beliefs or acts. (i) Nervous diseases: Networks and societies can exhibit such symptoms as disturbance of information input (loss of receptivity, hypersensitivity, perverted sensitivity), or more or less complete paralysis of options of their structure (possibly accompanied by spastic forms of activity). (j) Circulatory diseases: To the extent that the resources of an organization are distributed to its subdivisions and exchanged between them through a medium such as finance, the excessive accumulation of this medium in particular parts and the inadequate supply to other parts can endanger the organization as a whole. (k) Respiratory diseases: Networks and societies may be conceived as breathing information in order to revitalize their various parts. Irregularities in the inflow of new information of the appropriate kind and the elimination of out-of-date information can severely impair operations. (l) Digestive system diseases: Organizations and groups may be conceived as digesting facts which are broken down into a form which enables information to be extracted from them, absorbed and assimilated. These processes may be disrupted. (m) Genito-urinary diseases: Groups and networks eject the exhausted resources resulting from their activity and this process may malfunction such as to impair their further operations. To the extent that organizations tend to merge or to engender spin-off bodies having a similar style or pattern. This process may also be subject to malfunction. (n) Skin diseases: An organization may be conceived as being separated from its environment by a boundary which has function analogous to those of protection, secretion, heat regulation and respiration. These processes may malfunction. (o) Diseases of the musculoskeletal system: Some features of an organization or network perform structural functions analogous to the skeletal system and the associated muscles (whether voluntary or involuntary). Weaknesses in such features can impair the functioning of the body.
Special features The detailed understanding of the many different systems enabling the body to function and of the ways in which they interact.
Contrast Many terms, including diseases itself, have been borrowed from medicine to describe the condition of an organization. But although a detailed comparison has been made between the state and the human body, such terms have not been used to sild up a coherent pattern of metaphors to offer insight into the functioning of an organization.

Further keys Medical disciplines: anatomy, physiology, pathology. Diagnosis. Medical intervention, including surgery. Pharmacology. Different forms of therapy.

♦ MM2020 Changing moods of an individual

Substrate Every individual enjoys or experiences shifts of mood which may be subtle or dramatic (as in the case of depression, for example). Although very familiar, such shifts may be difficult to control even though engaging in particular activities may tend to induce particular moods. Individuals alternate between a relatively limited number of moods which tend to become progressively more clearly characterized.
Metaphor Public opinion may also be said to have moods which shift more or less frequently and are responsive to certain triggers (cf the Roman 'circus' policy).
Special features The nuances of mood and the varieties of subtle or catastrophic transformation between moods.
Contrast The moods of a group or society tend to be less well characterized than is the case for an individual.
Further keys The varieties of mood and the varieties of transformation between them. Triggering moods. Maps of mood transformation pathways.

♦ MM2021 Ambiguous visual illusions

Substrate Illusions are perceptual experiences in which information arising from external stimuli leads to a misleading impression. Certain kinds of visual illusion allow the same image to be interpreted in two quite distinct ways. Classic examples are the figure and ground illusion (e.g. white vase or two black profiles) and the Necker cube. Observers see the latter either as though looking up at its base or looking down on its top, and can alternate between the two interpretations. In both examples one interpretation excludes the other. Some people, especially the elderly, have considerable difficulty experiencing such object reversibility.
Metaphor Some important patterns of information on the social environment may lend themselves to quite distinct interpretations, neither of which is more real than the other. It is possible that some classic contrasting interpretations of social phenomena (e.g. nature vs nurture, individual development vs social development, centralization vs decentralization) may be seen in this way. One such interpretation necessarily excludes the other. Some people would experience difficulty alternating between such (complementary) interpretations.
Special features The mutually exclusive nature of the interpretations, the possibility of alternating between them, and the fact that neither is more appropriate than the other.
Contrast Major shifts in policy (e.g. in a 2–party democracy) may entrain (or derive from) a major shift in perception in public opinion that could be described by analogy to such visual ambiguity. The switch may be from perceiving nationalization as an appropriate policy to one in which denationalization is perceived as fitting the actual conditions. The condition may not however specifically call for one in preference to the other, although one of the interpretations must be made. Not everyone is capable of such shifts; many shift perspective according to the requirements of their environments.
Further keys Other visual illusions. Auditory and tactile illusions of psychiatric significance. Hallucinations.

♦ MM2022 Media diets

Substrate To attract an audience or readership and retain its continuing interest, radio, television and periodicals (dailies, weeklies, etc) have to supply a varied 'diet' of programmes, articles, or visual materials. The variety is a compromise between responding to the interests of different segments of the same audience and holding the interest of any one such segment. The materials presented must therefore respond to the tendency for attention to wane and switch to some alternative by trying to ensure that that alternative is provided in some measure by the channel or publication in question. Programme directors and editors must therefore juggle with different materials of variable length to capture and retain a fickle attention.
Metaphor In the government of any society the government must ensure that it captures and retains support by treating subjects which attract the attention of potential supporters for a sufficient length of time.
Special features The well–developed skill required to select and balance materials. The explicit nature of the shift to more satisfactory alternatives. The well–catalogued range of a material.
Further keys Scheduling techniques. How materials attract and hold attention. Development of the capacity to switch to an alternative and to choose between alternatives. Clarification of what is attractive to an individual, and when, in each category of material from the range available. Emergence of new categories of material. Attention management.

♦ MM2023 Animal–drawn cart

Substrate A cart consists–minimally of two wheels on a common axle on which a load–bearing platform is supported. The cart is drawn by one or more animals to which it is attached. Movement is normally controlled by a driver on the cart.
Metaphor A social innovation may be envisaged based on two parallel cycles operating at the same rate around a common principle (or axis of reference) on which a social platform may be constructed. The cycles are maintained in operation over the terrain as a result of a force exerted on the platform by the controlled application of subjugated instinctual forces. Possible cycles are the work–life cycle and the home–life cycle which are linked. In this instance they are maintained in operation by the instinctual need to pursue satisfiers (typically portrayed in advertising images). If the resulting movement is appropriately controlled the configuration can progress through social space bearing any cultural heritage with it.
Special features The interrelationship between the cart as a deliberately designed assemblage, the unhuman source of power requiring its own particular form of nourishment, the purposeful human controller, and the load bearing capacity.
Contrast Whilst it is highly probable that social vehicles are in operation, they are not seen in their entirety but only as unclearly related elements. There are however a number of mythological references to animal–drawn chariots which may indicate unconscious recognition of the social dimension of such devices.
Further keys The range of cart and harness designs, and the scope for improvements to any one design. Carts designed for speed versus those designed for heavy loads, and the configurations of animals appropriate to each. Horseless carriages.

♦ MM2024 Daily round of activities

Substrate In normal daily life individuals tend to alternate amongst a well–defined set of activities. These include sleep, commuting, work, eating (well known in French as the routine of 'boulot, metro, telelocke, dodo'). Each activity has well–established limits; the set being effectively governed by a somewhat flexible time–table, but with little scope for variations. The daily round is studied statistically by time–budget analysis.
Metaphor This daily round is fundamental to the organization of society and determines the minimal shifts in attitude required to ensure its viability.
Special features The relative rigidity of the pattern which may indeed amount to little more than a circular sequence ('round').
Contrast The rigidity of the daily round in industrialized societies may be contrasted with that in some rural societies in which alternation between the possible types of activity may occur at any time of the day and many times a day. This flexibility is a goal sought by many advocates of an 'alternative' society.
Further keys Means of introducing flexibility into a time–table. Means of enriching such a daily round with variants and other categories of activity. Manner in which the different activities match with the fulfillment of basic human needs. Comparison with a religiously–oriented daily round of monastic life. The inter–activity hiatus.

♦ MM2025 Aquatic animal locomotion

Substrate The structure of aquatic animals reflects their locomotor habits. In the case of micro–organisms, motion usually involves the action of flagella or cilia (as planar waves, oar–like beating or three–dimensional waves), through the extension of pseudopodia, or by sliding or undulating. The main distinction in the case of invertebrates is between bottom locomotion (ciliary gliding over mucus, contract–anchor–extend bottom creeping, pedal contraction waves, peristaltic contraction, and bottom walking) and swimming (hydraulic propulsion, or undulating all or parts of the body). Fish–like vertebrates of elongated form (such as eels) use a series of oscillations, passing from head to tail, to provide propulsion. In those of shorter form only the posterior half of the body flexes with such contraction waves, or in other cases only the caudal fin. Stabilization (to prevent rolling, pitching or yawing) and steering are accomplished using secondary fins. Tetrepodal vertebrates tend to use fish–like movements of their tails, rotation of flippers or wings through a figure–of–eight configuration, or alternate movement of the feet in the case of surface–swimming.
Metaphor The structure of affect–oriented groups or individual personalities reflects their mode of movement through the affective and emotional contexts, by which they are supported.
Special features The range of movements and the variety of waves in which they are coordinated.
Contrast Movement of some social actors is characterized by 'waves of emotion', which in some forms is described as 'flapping' or 'fluttering', in others as 'lashing out'. Some characteristic modes are based on 'putting out feelers' (and then shifting into the new context if there is no threat), on various forms of 'social climbing' involving making a firm contact (and using that as an anchor and then as a stepping stone from which the next contact may be reached), or on 'creeping along' over some dependable base to which adherence is assured. Another range of modes is characterized by such terms as 'drifting', 'gliding' or 'undulating through life', and also by such terms as 'wriggling' and 'slithering'; and another by strong rejection mechanisms which force the body in the opposite direction by reaction. In more controlled and directed forms, movement through society is ensured by some kind of orderly alternations between alternatives (e.g. excess and restraint, opening and closure to information, confidence and suspicion, etc) which allows the individual or groups to navigate between Seylla and Charybdis. With greater control, movement may be achieved in several dimensions between several pairs of alternatives. The ways in which such movements may be modelled by the varieties of aquatic locomotion has not been explored.
Further keys Buoyancy and the reduction in the effort required to overcome inertia. Breaking mechanisms and the necessity for continual movement in some cases. Surface swimming. Depth, pressure and characteristic forms.

♦ MM2026 Traditional and spontaneous ceremony

Substrate Ceremony consists of a structured pattern of activities of symbolic significance. Individual activities may recur during the course of the ceremony. Repetition tends to play an important function in reinforcing the significance of certain acts.
Metaphor Ceremonies may be deliberately designed to encode a representation of a pattern of relationships between the powers governing society.
Special features The activities in making up a ceremony tend to occur (and possibly recur) in a linear sequence, although some activities may occur in parallel. The ceremony is structured so that certain acts are perceived as especially appropriate or 'fitting' particularly in spontaneous ceremony.
Contrast In modern society the significance of ceremonies has been severely eroded to the point of embarrassment, except when viewed as spectacles or shows.
Further keys The manner in which the pattern of activities enhances and defines the appropriateness of particular activities. How activities which are 'fitting' are fitted together.

♦ MM2027 Resources, commodities and products

Substrate Naturally occurring resources provide a source of material support for individuals and the economies of their societies. Commodities are products extracted and processed into other forms for distribution.
Metaphor Human beings and their cultural heritage may be considered as human resources, especially when individual skills and know how are exploited for economic purposes. Ideas, concepts and insights may be produced, packaged and marketed as non–material products.
Special features Isolation and adaptation of concepts in order to package them for use in economic exchange processes.
Contrast There is widespread recognition of people as economic resources or as 'resource persons'. Economic terms such as 'intellectual productivity', 'unproductive concept', and 'resourceful individual' are used. Concepts are 'packaged' and 'marketed'. There is increasing recognition of the herbal lore of so–called primitive cultures as a resources for clues to new drugs. But the emphasis is on a narrow exploitative approach to such resources with only token recognition of the value of such natural cultural resources in their own right and in terms of their contribution to the well–being of society. In effect there is no appreciation of the cultural environment in terms of the 'ecological' importance of such resources.
Further keys Classification of resources, commodities and products. Extracting commodities from resources. Stewardship of cultural resources.

♦ MM2028 Taking turns

Substrate Taking things in turn and letting others have their turn involves an understanding of alternation which is first developed in children's games. It is also basic to some adult games, to the ordering of conference speakers, or of performers at a show, and to 'waiting one's turn' for access to some service.
Metaphor Different groups could each take it in turn to formulate or implement policies, or benefit from access to limited services.
Special features The natural understanding at least in children's games is of the justice of it being a particular person's turn, of the injustice of not letting others have their turn, and of the requirement that each should take a turn at being the hero (or villain).
Contrast In society the right of each social group to have its turn is resisted and quarrelled about with what, in the context of children's games, would be considered as much bullying.
Further keys What is carried, expressed or contained by 'turning'? Behaviour in queues. Rules for establishing whose turn it is. The characteristics of the 'natural justice' governing the acceptability of taking turns.

♦ MM2029 Spatial orientation

Substrate Experience of the physical environment and efforts to coordinate any response to it tend to be made in terms of a spatial frame of reference. This involves polar opposition such as

up–down, right–left, in–out, front–back, on–off or centre–periphery. For any given polar opposition, a higher value may be attached to one of the poles.
Metaphor Experience of the pycho–cultural environment and efforts to coordinate any response to it may be made in terms of an orientational frame of reference expressed in terms of spatial polarities. For any give polar opposition, a higher value may be attached to one of the poles.
Special features The organization of a whole system of concepts with respect to one another.
Contrast Many positive experiences, whether individual or collective, are associated with 'upward' or 'forward', whilst corresponding negative value is attached to 'downward' or 'backward'. This is the case with regard to feelings, company performance, status, level of debate, standards, development and values in general (e.g. 'good' is 'up'; 'bad' is 'down'). Policy issues are frequently associated with 'right' or 'left', 'centre' or 'periphery'. Value is attached to being 'in' on 'out' of some activity or movement, or to being switched 'on' or 'off'. Most fundamental concepts are effectively organized in terms of one or more spatialization metaphors which provide both an internal systematicity and coherence and an external systemacity amongst such spatial metaphors. But whilst the relationships within a particular polar dimension (up–down) tends to be explicit, the relationships with other such dimensions remains implicit and unexplored.
Further keys Systematic approaches to organization of space and to orientation within it. Types of discontinuity between frames of reference.

♦ MM2030 Gaining/losing initiative
Substrate In many forms of interaction, including some games (e.g. Go), considerable importance is attached to gaining the initiative in order to have a temporary advantage in controlling the process. Considerable efforts are made to avoid losing it, or to regain it again once it has been lost.
Metaphor Each power group in society can be conceived as attempting to gain the initiative or to resist losing it.
Special features The subtlety of 'initiative' as the focus of concern, which is nevertheless understood to be exchanged somewhat like a ball in a game.
Contrast The concept is current in the relationships between power groups but mainly with regard to their narrow self–interest rather than their contribution to the interests of society as a whole.
Further keys 'Leaving the ball in someone else's court'. 'Buck passing'.

♦ MM2031 Cyclic resonance accelerator
Substrate This device accelerates charged particles by timing their exposure to high frequency alternation of polarity. Within a vacuum chamber in a fixed magnetic field. As a result the particles move in a circular orbit whose radius increases as the energy of the particles increases. Different techniques are used to ensure stability in the motion of the particle.
Metaphor The significance of any charged issue may be increased by suitably timing its exposure to positive and negative reinforcement in a context within which the issue can evolve along only one path.
Special features The precision required to engender, control and contain the high–energies involved. The unusual conditions created and the large physical dimensions required to do so.
Contrast Some awareness of this process is indicated by the slang expression 'to whipsaw'. Features of it are used in stick–and–carrot interrogation techniques.
Further keys Linear accelerators, synchrotrons. Storage rings. Use of superconductors.

♦ MM2032 Changing physical position
Substrate An individual tends to hold one physical position for only a limited period of time before shifting (an arm or leg, for example) to another. This is usually done because of a certain build up of tension which can best be released by relaxing into a new position. This process involves alternation between a limited number of positions. A similar situation occurs when two people are intertwined in an embrace during love–making. The couple will alternate amongst a set of positions.
Metaphor A society can be usefully conceived as needing to modify its position from time to time in order to release tensions that build up. The alternatives then constitute a limited set. A similar situation obtains in the relationship between two societies.
Special features The selection of the alternative position into which the shift is made is determined by what it is necessary to shift in order to release tension. There is a tendency to seek the most relaxed posture although once adopted an alternative is then progressively defined as relatively more relaxed.
Contrast There is no understanding of the set of positions that a society can adopt. Shifts between positions tend to be of a spastic nature, involving resistance and conflict.
Further keys Classification of the position in a set and the permissable transformation pathways between them. How and where tension builds up and how this defines the nature of the possible release.

♦ MM2033 Fossorial animal locomotion
Substrate The structure of fossorial (burrowing or boring) animals reflects their locomotor habits. Invertebrates and vertebrates have evolved a number of different locomotor patterns to penetrate soil, wood or stone. Some invertebrate (such as worms) may burrow using peristaltic locomotion generated by the alternation of longitudinal– and circular–muscle contraction waves flowing from head to tail. Others (such as molluscs) use the contract–anchor–extend method. In the case of reptiles and amphibians, burrowing is accomplished by alternating head movements, whereas in the case of mammals digging is achieved.
Metaphor The structure of infrastructure–oriented groups or individual personalities reflects their mode of movement through relatively dense social or information structures characterized by complex procedures (whether of a modern administration or some traditional body) or by large quantities of minimally ordered information (as in any system of archives).
Special features The tunnels resulting from such movement.
Contrast Terms such as 'barrowing' and 'boring' are frequently used, whether in relation to working through large quantities of data or to the penetration of some complex social structure. The terms 'bookworm' and 'mole' (as applied in espionage) reflect these perceptions. Little attention has, however, been given to the range of ways in which such movements are accomplished.
Further keys Contribution of the earthworm to soil fertility. Movement of reptiles and insects through friable sand without resulting in tunnels. Wood borers.

♦ MM2034 Sexual intercourse
Substrate The basic movement of sexual intercourse is one of alternation through the movement of the penis in the vagina. This movement is supported and enhanced by that of whole sets of complementary movements in other parts of the anatomy of the two partners. The nature of the basic movement can be further modified by alternation amongst a set of positions (as defined in the Kama Sutra, for example). The initiating role for such changes may also alternate between the partners.
Metaphor The variety of ways in which sexual roles can be abused are a potentially valuable indicator of the kinds of abuse possible in the intercourse between two social groups. The metaphor also suggests distinctions between more and less fruitful ways in which one partner can impregnate the other with its principles or receive the principles of the other.
Special features The manner in which the physiological process can engender mutual sympathy and even a blending in ecstatic union. The constant exchange of signals enabling the parameters of the alternation process to be varied.
Contrast Although this metaphor is very frequently used to describe relationships between social groups, it is only perceived in terms of dominance and 'working one's will' on the other group, often as a form of rape. There is rarely any sense of harmony and mutual contribution to a process in which the initiative may be shared and each may be constrained to receive something from the other.
Further keys Foreplay, frigidity, impotence. how and whether the normal can be distinguished from the perverted.

♦ MM2035 Radio communication
Substrate Radio waves, as one form of electromagnetic wave, consist of electric and magnetic fields vibrating mutually at right angles to each other in space. A transmitter feeds a transmitting aerial from which radio waves are broadcast to a receiving aerial feeding a receiver. The transmitter is composed of a generator of a radio frequency carrier wave which is modulated in accordance with the electric currents carrying the signal amplified from a microphone. If the receiving aerial is tuned to the frequency of the carrier wave, the receiver is able to selectively amplify and demodulate the transmitted signal so as to operate a loudspeaker. Amplification is achieved by devices (such as vaccuum tubes or transistors) based on use of the small input signal to control the flows of larger negative (or positive) charges that are collected as the output signal. Tuning is achieved using an electronic circuit containing both inductance and capacitance such that it is capable of resonance. Thus when the capacitator discharges through the inductor this results in the capacitator being recharged in the opposite sense with which it then discharges again, thus continuing this oscillatory cycle (as long as external energy is supplied). The frequency of oscillation may then be varied by adjusting the values of the inductance and capacitance.
Metaphor The flow of information emanating from any individual or group, whether in the form of documents, speeches or media activity, may be compound to radio waves. A distinctive communication is achieved by modulating this flow. Such communications can only be satisfactorily received by adjusting one's concept scheme to the frame of reference of the originator and decoding the symbols used to extract the referents they carry. There may be considerable interference between various flow of information, to the extent that it may be very difficult to receive a clear and coherent pattern of information from which meaningful referents may be extracted.
Special features The degree to which the various component devices needed for radio communication are distinguished, the refinements possible to improve the quality of communication, and the degree of development of the theory governing their operation.
Contrast Considerable use is made of phrases such as 'being on the same wavelength' or 'frequency'. Some people are recognized as being 'tuned in' to the latest gossip or fashions. After the quality of a confusing explanation has been improved, the content may be referred to as being 'received loud and clear'. These suggest an implicit understanding of other parallels to radio communication but no attempt seems to have been made to make the metaphor coherent. Whilst there has been considerable attention to the transfer of information in a digital or particle–form, there is little complementary recognition of its transmission/reception in a wave–form or of how this may be facilitated.
Further keys Antenna arrays and concept arrays and their design. Mutually orthogonal electric and magnetic fields in carrier waves compared with mutually orthogonal semantic dichotomies providing the dynamic framework for the distinctions alternatively stressed in any communication ('semantic screening'). The process of amplifying subtle distinctions in a weak signal or embodying subtle distinctions into a flow of language. Ability to receive coherent messages from schools of thought distant in space and time. Radio noise, interference and jamming. Transmission power and quality in relation to long, medium, short–wave and UHF transmission. Radio telescope. Radar.

♦ MM2036 Changing fashions
Substrate Fashion as the prevailing mode or pattern governs not only styles of clothing and accessories but also the appreciation accorded to styles of art, music, recreation, tourism and even academic research. Fashions change, and are encouraged to change, especially in the case of clothing. Hems rise only to fall again on some later occasion. Certain theoretical approaches lose favour only to be taken up again at a later time.
Metaphor Amongst social groups certain issues or principles also become fashionable for a period during which they are the basis for intense activity, only to be abandoned in favour of more compelling alternatives. Later they may return to favour once again.
Special features The non–rational subtlety of the aesthetic choices involved which appear to seek to heighten interest by exploring extremes and contrasts. Ideally the 'new' should shock and reinforce attempts to break with the 'old'.
Contrast With respect to ideas and conceptual approaches, such phrases as 'out of style', 'in vogue' and 'avant garde' are used. But there is much greater ability to discuss fashions as temporary fashions in domains such as clothing, than is possible with respect to issues in society where the uncritical attachment to particular issues is highly developed. There is also an expectation that the new fashions should be strikingly different from the old. The search for 'alternatives' may itself become fashionable.
Further keys The nature of 'classical styles' less subject to the vagaries of fashion. The role of fashion leaders and the ease with which the 'fashion trade' is manipulated.

♦ MM2037 Slavery
Bondage — Forced labour
Substrate People are considered slaves when they are compelled to perform involuntary labour for a person or group, usually under conditions that make them socially inferior and deprive them of most of their rights or freedom. Forced labour, contract labour and serfdom are variants.
Metaphor Within a socio–cultural context, people may be considered slaves when for their own survival, they are compelled by forces within that context to perform involuntary actions such that others benefit disproportionately in terms of status, privileges and freedoms.
Special features The all–encompassing scope of behavioural chains which completely condition those affected to the point that freedom is associated with undesirable insecurity.
Contrast The expression 'wage slave' has been used to describe those working in capitalist systems, just as citizens of totalitarian states have been described as slaves. Both these examples emphasize the economic dimension, thus disguising more subtle forms of slavery and bondage associated with belief systems or with the exploitative paradigms within which people may function.
Further keys Galley slaves and assembly lines. Military conscription. Plantation labour. Personal slaves. Emancipation. Chain gangs. Slave forms and slave breeding.

♦ MM2038 Resource sharing
Substrate In a number of specific situations formulae for sharing the same physical resources at different periods of time have been developed. Possession of the resources thus alternates

between the different owners. This can be seen in procedures for making a limited water supply available for irrigation in semi–arid rural communities, where each channels the water to his field for a specified time. (An interesting version of this is the sequence in which wild animals share access to water holes). A holiday apartment may be owned under a condominium arrangement whereby each only has full use of the facility for a specified period during each year. Where the number of school rooms is limited the facilities may be used under a modular (shift) arrangement similar to the shift arrangement in factories. This formula may be extended to two part–time people who contract together to work non–overlapping periods to provide a full day's work thus sharing the job.
Metaphor There is clearly scope for groups and societies to share resources over a period of time (rather than simultaneously as is presently advocated).
Special features The clarity with which the necessary timing is understood (or felt) by all parties.
Contrast Vain attempts are made to divide up a pool of resources often to the dissatisfaction of all concerned – frequently leading to conflict. No effort is made to facilitate the sharing process by distributing larger portions for periods of time rather than attempting to distribute all the resources permanently.
Further keys Types of cycle, symmetric and asymmetric.

♦ MM2039 Arboreal animal locomotion
Substrate The structure of tree–dwelling animals reflects their locomotor habits. There are many adaptation for climbing but all require strong grasping abilities with no leg being moved until the others are firmly anchored. The sequence of alternating movements in climbing is closely related to that of walking or hopping. Leaping from limb to limb is also similar to terrestrial saltation. In brachiation (using the arms to swing from one place to another), the body is suspended rather than being supported, but the pattern of movements is similar to walking. Birds also use walking and hopping movements on tree branches.
Metaphor The structure of hierarchically–oriented groups or individual personalities reflects their mode of movement through the social and conceptual hierarchies within which they function.
Special features The range of techniques used to manoeuvre through a tree or a network of interweaving branches of different trees.
Contrast There is widespread recognition of 'climbing', whether up an organizational hierarchy or in the form of 'social climbing' within an informal network. There is some recognition of 'leaping' from one organizational hierarchy to another, but more widespread disapproval of 'leaping' around within conceptual hierarchies (because of the implied non sequitur). No equivalent use is made of 'swinging' through interlacing organizational or conceptual networks. There is however recognition of group or individual ability to 'perch' on some particular 'branch' of knowledge or other position.
Further keys Movement from position to position. Continuous tree canopies in contrast with isolated clusters. Trees as an interface for birds, tree–dwellers and ground–dwellers.

♦ MM2040 Gamesmanship
Substrate In certain situations possession of rights is determined by the rules and conditions of some form of game. Ownership may then be said to alternate between the participants in the game. At the detailed level this may be recognized by possession of a ball (e.g. football, basketball), although the rules may require that the ball alternate between the players (e.g. tennis, volleyball). At a more general level success may be indicated by the game score or the position of the team in a league table. In the absence of a ball, the struggle may be for amount of speaking time (as at conferences) partially checked by rules requiring that each be allowed to present his position. The struggle for power of political parties may be seen in this light.
Metaphor This suggests that the struggle for power between different claimants could be rendered less naked by elaborating various kinds of games which offered each the possibility of periods of control as an outcome of performance against the other (s).
Special features The well–defined nature of the rules, which may be relatively complex, involving penalties and subtleties of scoring to ensure fairplay, as determined by adequate involvement in the alternation process.
Contrast The interaction between the great powers, for example, is studied in the light of game theory by those involved. There is however little attempt to render the rules of the game explicit. On the other hand, such powers do take very seriously the need to perform well in any games that are offered as substitutes (e.g. sport, chess) for the basic game they are playing.
Further keys Emergence of game rules. Types of game. Game referees. Role of spectators.

♦ MM2041 Agricultural development
Substrate Like all living organisms, man extracts food from his environment. More food is obtained from a given environment by encouraging useful plant and animal species and discouraging or eliminating others. In this process such species become domesticated and the methods of enhancing their growth become more complex, requiring the organization of a variety of farm production processes (soil cultivation, irrigation, crop growing and harvesting, animal husbandry, and some processing of the resulting products, especially to preserve them. Highly organized farming tends to be preceded by a hunter–gatherer phase, followed by a nomadic stage and by a self–sufficient, small–holding stage.
Metaphor Groups and individuals derive necessary stimulus and nourishment from their psycho–social environment. More such resources are obtained from a given environment by encouraging useful behaviour patterns (whether unselfconscious or not) and discouraging or eliminating others. In this process such patterns become controlled and the methods of enhancing their development become more complex, requiring the conscious organization of a variety of supportive processes (cultivating the philosophical ground, counteracting conceptual aridity, initiating new patterns, caring for their development, extracting benefits from them, and preserving the latter for the future). A high degree of such organization tends to be preceded by a stimulus–response phase, followed by a pattern of behaviour shifting in response to the availability of stimuli, and then by a more grounded self–reliant stage in which the necessary patterns are engendered.
Special features The discipline and skills required at each stage to ensure balanced nourishment under a variety of conditions.
Contrast Inspiring parallels are frequently drawn between various aspects of education and the cultivation of plants 'preparing the soil', 'planting the seed'. Occasionally this is extended to groups of plants ('crops' of students). No attention appears to have been given to possible insights from the management of the relationships between different plants or crops under a variety of conditions.
Further keys Development of new varieties of plants. Herding animals. Crop rotation and monoculture. Soil conditions (acidity, alkalinity, water–logged) and erosion. Organic farming and fertilizers. Hedgerow habitats. Permaculture.

♦ MM2042 Relationship sharing
Substrate In certain circumstances several people may find it impossible to share a relationship simultaneously. They may then agree not to meet all together at the same time but rather as smaller compatible groups at different times. This formula is used in connection with child visiting rights of divorced parents. It is a basic feature of the conjugal relationship in polygamous families (which in West African societies, for example, allows the wives extended periods of freedom to engage in independent economic marriages. In large families it determines who does what with whom and when. Groups of families may alternate responsibility for all their children.
Metaphor This concept could be developed as a means of forming coalitions in which some of the partners do not wish 'to be at the same table together'.
Special features Recognition that the co–presence of some people is either not viable or not fruitful.
Contrast This approach is used to a limited extent in triangular negotiations via a mediator when two social groups are 'not speaking to each other'. This condition has not however been integrated into a stable coalition in which non–co–presence is an accepted feature.
Further keys Nature of the transition between one relationship pattern and another. How the rules are established.

♦ MM2043 Electric motors and generators
Substrate Device which converts electrical energy into mechanical energy or mechanical energy into electrical. Operation in the first mode depends on the fact that when an electric current from an external source flows through a conductor placed in a magnetic field acting, (even partially), at right angles to the conductor then the conductor experiences a force. In its simplest form the conductor is a coil of wire placed between the poles of an electromagnet such that the coil is forced to rotate. Operation in the second mode depends on the fact that if an electrical conductor is moved by an external force across a magnetic field, then an electrical current is generated and flows through the conductor. In more complex forms the magnetic field may be created by current flowing through one or more other coils such that the magnetic field rotates. This is achieved by ensuring that the currents are not of the same phase (namely alternating currents whose amplitudes peak at different moments, whether twice or three times in a cycle).
Metaphor If externally motivated communication takes place from one individual or group to another within a highly polarized context (with an orientation other than that of the communication itself), then the communication pathway will be subject to social forces tending to displace it. On the other hand, if a pair of individuals or groups with the possibility of communication find themselves displaced together by external social forces within a highly polarized context, this displacement will engender communication between them.
Special features Detail range of design possibilities whereby the power of the motor or generator may be increased. Dynamic relationship between (at least) two complementary polarizations and the manner whereby the energy they engender may be harmed.
Contrast There is recognition of the manner in which communication on unrelated (for example, technical) matters may be affected by a polarized (for example, politicized) context. There is also recognition of the manner whereby social actors are induced to communicate if they are together, subject to forces within such a polarized context. But such recognition has not led to any understanding of how the interaction between such mutually orthogonal polarities might be more effectively harnessed as suggested by the design of motors and generators.
Further keys Types of motor (induction, synchronous, direct or alternating commutator), including polyphase and linear, as determined by the manner in which conductors are arranged. Forms of coil winding. Starting problems.

♦ MM2044 Cyclic migration
Substrate The seasonal cycle creates and withdraws opportunities at geographically separated locations. Both animals and humans respond to these changes by moving physically from one place to the other, only to return as the cycle continues. In the case of animals, especially birds, this may involve movement in search of basic needs such as water, food or warmth, water or grazing for herd animals, as in the case of nomads. Within the monetary economy it may be determined by the opportunities of seasonal employment, harvesting, sea resorts, ski resorts Another kind of migration may occur in response to the social or cultural season (eg jet set migration).
Metaphor Social groups could change their pattern of behaviour in response to cyclic processes in society, especially economic cycles.
Special features Sensitivity to cyclic phenomena, their dangers and their opportunities. Displacement.
Contrast Seasonal movement is used to a limited extent by inter–governmental assemblies (e.g. ECOSOC meetings alternately in New York and Geneva; EEC meetings). There is no built in adaptation by social groups to economic or other cycles, except possibly in the form of religious pilgrimages (e.g. Mecca). Series of infrequent periodic, international meetings in response to particular needs may perhaps be considered in this light (e.g. UNCTAD conferences).
Further keys Concept of a recognizable cycle and the nature of a cyclic opportunity. Kinds of 'displacement' or adaptation that are possible or useful.

♦ MM2045 Plants
Substrate Plants, although more or less stationary, are characterized by the ability to generate their own nutrients from inorganic chemicals, air and water in the presence of light and facilitated by a catalyst. They exist in many forms adapted to the annual weather cycles in available ecological niches. As such they are a vital link in ecological food chains as well as providing habitats for most animal species. Plants are deliberately grown and harvested as crops, whether for the seeds, flowers, fruit, roots or timber. They are also much appreciated in their own right.
Metaphor Groups and individuals may adopt behaviour patterns that do not require any movement through society relative to other bodies, however active they may be at their own particular location in relation to one another. They then derive any required resources by unself–consciously incorporating prevailing ideas, emotions and practical opportunities in the light of creative values. Such behaviour patterns exist in many forms adapted to long term behavioural cycles governing the nature of the psycho–social resources available. They are a vital link in the ecology of the psycho–social environment as well as providing contexts for many forms of non conscious behaviour. Such unself–conscious behaviour patterns may be deliberately cultivated when specific benefits may be derived from them. They may also be appreciated in their own right.
Contrast There is widespread association of the parts of a plant, or the processes of agriculture, with conceptual activity. Ideas may be planted like seeds. There are 'budding' hypotheses and those that have come to 'full flower' or to 'fruition', possibly following 'cross–fertilization'. Intellectual disciplines are recognized as having many 'branches' and as having their 'roots' in earlier forms of thought. The focus is however on the parts of a plant and not on the plant as a whole, except as in some archetypal cultural symbols (e.g. The Tree Yggdrasi and the Tree of Life). There is little recognition of the integration of the plant's' processes, of how it functions in any ecosystem, or of what is required in the care and cultivation of plants as metaphors of specific patterns of behaviour.
Further keys Varieties of plants and conditions for their growth. Plant ecosystem structure and dynamics. Plant cultivation and breeding. Crop rotation and monoculture. Plant diseases. Relationship to animals whether in pollination, seed distribution or as a source of food, shelter, or security. Adaptation to the seasons. Movement of pollen and seeds. Flower smells.

♦ **MM2046 Cycles of religious festivals**
Substrate Many religions have festivals or other events organized within the framework of the annual cycle or even of much longer periods. At each such event within a cycle specific symbolic considerations are stressed, each qualitatively different from the others although together they are perceived as constituting a coherent response to existential needs. Some of these events may be tied to particular individuals (e.g. saints). It is recognized that people may feel greater attachment to some of these events than to others.
Metaphor The annual cycle, as well as other cycles, could provide a basis for interrelating alternating events which each stress one of the existential concerns of life on earth. The aim being to stress the complete range of such concerns.
Special features The cyclic organization of variety and the subtle contrasts in forms of presentation whereby their qualitative distinctions are maintained.
Contrast This approach is used in an essentially tokenistic manner in the form of various 'World Days' sponsored by the United Nations and other bodies (e.g. for refugees, environment, etc). These are not viewed as integrated within a cycle and the agencies responsible for any one event are seldom aware of others, and feel no need to be. The approach is also applied in the series of 'International Years' (approved by the United Nations and other bodies), but these do not as yet repeat within any cycle.
Further keys Nature of the distinctions possible between events. Preservation of the integrity of the cycle. Significance of the cycle as a whole.

♦ **MM2047 Bicycles and cycling**
Substrate A bicycle is composed of two equal-sized wheels rotating at the same speed, on parallel axes within the same frame. One wheel is steerable by hand and rotates as a result of contact with the ground over which the frame is moving. The other receives the force provided by the leg movement of the person seated on the frame between them. Cycling, as the most efficient means of converting human energy into propulsion, is only successful over varied terrain when a dynamic equilibrium can be maintained by continual adjustments to direction, body position and speed of forward motion. The momentum of the wheels causes the frame to resist disequilibriating forces.
Metaphor Whilst it may be possible for especially talented individuals and groups to maintain their social development on the basis of a single behavioural cycle (whether of economic production, emotional phases, etc), most people are more able to harness their social energy effectively within a framework of two complementary cycles. For such a social innovation to be successful, the cycles chosen must be appropriately defined and positioned within a framework to support the individual or group. It must be clear which provides the motive power (e.g. a work cycle) and which can be used to provide the direction (e.g. an emotional cycle). The art of cycling requires a learning period with its associated falls.
Special features The degree of integration between the function of wheels, frame and cyclist.
Contrast There is emerging awareness of the complementarity of certain social processes which may be conceived as cyclic (e.g. cycles of work and leisure activities, cycles of innovative and adaptive learning). In Eastern cultures there is explicit recognition of active and passive cyclic forces. In the West much attention has been given to the complementarity of the cycles of right and left-hemisphere brain activity as well as to those of other favoured dichotomies. It is not yet clear how a bicycle might be put together, but the parts are certainly available.
Further keys Frame design. Advantage of smooth pathways or roads. Loss of maneuverability in tricycle riding compensated by elimination of need for any skill in balancing. Hindu and Buddhist concept of wheel-like chakras.

♦ **MM2048 Psycho-symbolic cycles**
Substrate Importance is attached within certain cultures and traditions to the waxing and waning of qualitative influences. These may be associated with gods, principles, astrological factors, etc or of combinations of these. In each case these are understood to act within a multi-cyclic framework, which may include cycles of very long periodicity (e.g. the 'Ages' in the Hindu tradition). Within such cycles the waxing influence (of the reborn god) may be represented as struggling to overcome the waning influence (of the exhausted outgoing god). Celebration of these events may be integrated within religious cycles.
Metaphor The collective attention required by particular qualitative features of psycho-social life may be seen as waxing and waning within a multi-cyclic framework ensuring that adequate (but not excessive) attention is given to each such feature in order to achieve the harmony of the whole. The relationship between the features may be conflictual.
Special features The richness of such cycles and the effort devoted to rendering them comprehensible through a multiplicity of stories and various interrelated coding systems (e.g. colour, number, sound, etc.).
Contrast In some cultures such systems are a determining factor in decision-making but often in a manner which stresses superficial features rather than the qualitative richness that they symbolize.
Further keys The variety of qualities distinguished and the manner in which they are grouped into cyclic sets.

♦ **MM2049 Knotting, nets and basketry**
Substrate Knots are produced by interlacing one or more cords so that they lock together and resist forces acting to separate them. Basketry is formed by multiple knotting according to a particular pattern usually designed to form a rounded container leaving no unincorporated ends. In both cases interlocking is frequently achieved by alternating the placement of cords in relation to one another, above and below, then alternating this pattern with a contiguous, but reversed, counteracting placement. In achieving the interlock culturally specific decorative variations may be incorporated into the pattern.
Metaphor Permanent social or conceptual structures are achieved by interlacing processes so that forces acting to separate them reinforce the mutual bonding. Complex structures are formed by multiple interlocking of functions according to some particular organizational principle that provides integrity usually by engendering a cohesive form that defines a closed boundary (namely insideness and outsideness). Interlocking is achieved by alternating the degrees of freedom and constraint associated with any given function to provide the checks and balances necessary to the integrity of the larger whole. In achieving this self-reinforcement, elegant variations on the pattern may be favoured by particular cultures.
Special features The well-defined range of interlocking techniques and the variety of materials with which they may be used.
Contrast Complex interpersonal relationships have recently been compared to Knots. A number of creation myths emphasize the fundamental role of a basket-weave or set (such as the agrenon covering the omphalos at Delphi). In recent years much emphasis has been placed on loose social networks. Little attention has however been given to the deliberate design of networks with any degree of cohesiveness or integrity.
Further keys Mathematical theory of Knots and the concept of tame and wild Knots. Double bind. Klein bottle. Tensegrity.

♦ **MM2050 Contrast and significance**
Substrate Phenomena acquire significance to the individual when they appear in a context in which they are effectively highlighted by contrast. At the most basic level of vision, the eye is constantly engaged in a very rapid scanning movement without which objects would blend into the context in which they are located and become 'invisible'. Objects may be given (artistic) significance by taking them from their natural context (where they would not be noticed) and placing them in a carefully structured contrasting decor which focuses attention on them. It is then the alternation between focus on the context and on the object which enhances their significance. This principle is also used within paintings and other works of arts, only then the painter has to build the context into the painting or ensure that the painting contrasts with settings in which it is likely to be displayed.
Metaphor This suggests an alternative approach to comprehension of the recognition of social problems and the function of inequality.
Special features Rapidity of the alternation which effectively creates a stable figure/ground configuration.
Contrast In the recognition of social problems and injustices society is still struggling with the distinction between what observers perceive as significant (because it contrasts with their own background) and what those involved perceive as significant in the light of their own background.
Further keys 'Scanning' of processes and non-material objects (e.g. concepts, values, etc).

♦ **MM2051 Internal combustion engine**
Substrate An inflammable liquid fuel is mixed with air to form a combustible mixture which is compressed and then ignited by an electric spark. The gases formed by the combustion expend, thrusting the piston out of the cyclinder, for example. This movement imparts a rotary motion to a crankshaft which can then be used to drive external devices (such as the wheels of an automobile). Various arrangements are possible to remove the burnt gases and draw in the fuel during the cycle (such as by the motion of the piston), the energy to do so being provided by a flywheel in which mechanical energy (from the crankshaft) is stored. Arrangements are also required to cool the engine.
Metaphor A group meeting in closed session may raise an unresolved (and often emotionally charged) complex issue. If possibilities for resolving the issue are present an explosive interchange can be provoked. Whilst much of this may merely result in heated discussion calling for procedures to lower the temperature of the debate, some of the force of the interchange may be used to drive the group to external action (whether by resolution, remotivation or reaction). Various procedures may be used to clear the air, and maintain the momentum of the group in preparation for the next topic on its meeting agenda.
Special features The highly engineered relationship between the different phases of the engine cycle and the processes required to contain a controlled series of explosions whilst preventing the engine itself from exploding. The stable cyclic relationship between combustion, gas and liquid within the solid engine block.
Contrast The importance of closed meeting environments is recognized, whether for think-tanks, election of a pope, or group therapy marathons. Frequent use is made in meetings of terms such as inflammable topic, charged atmosphere, provocative spark, explosive spark, explosive interaction, heated discussion, clearing the air, cooling, down, etc. No attempt has been made to provide metaphorical coherence in order to provide understanding of possible improvements to meeting efficiency.
Further keys Four stroke (induction, compression, ignition, power, exhaust) compared to two stroke gasoline engines. Multi-cylinder engine designs (6-cylinder, opposed piston, free piston, V-8, radial piston). Wankel rotating engine. Gas turbine. Types of fuel. Comparison with external combustion (steam) engine.

♦ **MM2052 Stick and carrot processes**
Substrate There are a number of situations in which people are persuaded to change their attitudes by the alternate use of pressure and encouragement. Many forms of education involve exercises under time, peer or instructor pressure alternating with interesting exploration of new material (such as with audio-visuals). Here there is also an alternation between active and passive roles. Stick and carrot techniques have long been used to motivate the people in a work or military force, as in team sport training. They are also used in executive training, occasionally in a very severe form (e.g. some staff colleges and Japanese management motivation courses). Such techniques have also been applied to brainwashing and interrogation using the classic alternation between nice guy (violence) in a two man team.
Metaphor This suggests the value for a society of alternating between response to challenge and peaceful relaxation. It is the alternation which promotes development.
Special features Such techniques alternate between building up and testing/questioning (or even destroying) self-confidence so that the person is forced to look for a new position of equilibrium. It is the displacement to a new position which can constitute positive change.
Contrast This process is not consciously applied by large social groups but it is possible that societies engage in it through the manner in which crises are engendered. Note also the challenge and response theory of history.
Further keys The educational problems of how much pressure and how much 'nourishment' and for what periods. How to determine when the technique is being abused, especially if the participants are there voluntarily.

♦ **MM2053 Literacy**
Substrate Ability to read and write symbols. In the simplest case these may denote one's own name. In modern society, functional literacy implies a level of competence appropriate to the requirements, one's culture or group.
Metaphor Social development is dependent upon the increasing competence of individuals and groups in interpreting and articulating sets of symbols through which the constraints of time and space may be overcome. Lack of competence inhibits the exchange of significance in ways which may hinder human and social development.
Special features There is a wide range of symbol notations. Many of these notations may be used to convey more complex levels of significance. New languages, using the same notations or new notations, are continually developed.
Contrast The stress on functional literacy, as implying a minimum necessary level of competence, disguises the extent of incompetence in the comprehension of the more complex symbols required to support further social development.
Further keys Innumeracy. Inability to interpret maps and diagrams. Range of extant notational systems.

♦ **MM2054 Seasons and weather**
Substrate The cycle of the seasons is a basic form of alternation, especially perceptible in the non-tropical zones. Perception of the movement through the 'four' seasons has become well-characterized and has fundamental implications for the organization of activities such as agriculture and the cultural activities dependent upon them. The seasonal contrasts are recognized by variations in the weather. There is an instinctive understanding of how the parameters of weather (wet, dry, hot, cold, wind, calm) give rise to an alternating pattern of distinctive

phenomena (storm, heat wave, etc).
Metaphor Both the seasons and alternating weather patterns suggest ways of understanding the constantly shifting patterns of social dynamics.
Special features The manner in which the transitions between distinct seasons or types of weather can be described as continuous transformations rather than perceptual discontinuities (as in other cases of alternation). The subtle variations of weather are well recognized and therefore an ideal substrate for encoding.
Contrast The metaphor is extensively used to describe social phenomena (e.g. 'stormy situation') but has not been developed into a systematic descriptive language encoding the complexities of social dynamics. There is unfortunately a well–developed tendency to consider 'rain' as 'bad' weather and 'sun' as 'good' weather which takes little account of the environmental significance of such phenomena.
Further keys Importance of seasonal and weather variation. Duration of seasons and types of weather. Flood, drought, hurricanes, etc.

♦ MM2055 Food, nutrition and cooking

Substrate Nutrition is the process by which organisms absorb and utilize food substances, whether in their raw state or after having been subject to culinary processes to facilitate digestion. Food serves the functions of: providing materials from which energy is derived, contributing to the formation of enzymes necessary for synthetic processes, and providing the materials from which the components of the organism can be assembled. The exchange, preparation and (collective) consumption of foodstuffs are of considerable symbolic significance to man.
Metaphor Groups and individuals are in large part nourished by a wide range of information, whether in a raw form as stimuli or after having been ordered and processed into more digestible patterns. Information serves the functions of: providing stimuli from which the group derives its energy, contributing to the formation of integrative patterns necessary for development, and providing the concepts from which the components of the organization can be assembled. The exchange, preparation and (collective) consumption of information are of considerable symbolic significance to man.
Contrast Phrases such as 'getting one's teeth into' or 'devouring' a document, 'digesting' ideas, 'food for thought' and a 'thirst for knowledge' are frequently used. Ideas are often considered as essential 'nourishment' for the spirit. The problems of assimilating ideas may be attributed to their unprepared state ('half–baked') or the need to process them (to allow ideas to 'simmer', to 'ferment' or to 'percolate'). There is little attention to the nature of the foodstuffs essential for adequate nutrition except in the media concerned with supplying its audience with a well–balanced 'diet' of information. The gastronomic art of preparing and enjoying food is occasionally reflected in a 'feast of ideas' but with little understanding of how such a feast should be prepared or served.
Further keys Malnutrition. Vitamins. Natural and synthetic foods. Variety of foods and food preparation processes. Censorship in relation to vital nutrients and food poisoning. 'One man's meat is another man's poison'.

♦ MM2056 Strategic configurations

Substrate Whether in sport, military or business confrontations with competing forces, there is a tendency to elaborate alternative strategies which can be used if they are liable to be more advantageous. The team may therefore undergo training in response to a variety of possible scenarios, as with military manoeuvres, war games and management games. In each case the aim is to develop the ability of the group to switch to an alternative posture and work through that pattern until it is out–manoeuvred by the opposition. Each such strategy may be explicitly codified with the role of each person clearly defined. The principle of alternation between postures is especially clear in certain forms of martial art.
Metaphor This suggests that societies should be able to work through a variety of alternative organizational patterns according to the crisis they face.
Special features The stress on the need to switch between organizational modes rather than treating any particular one as desirable in its own right. The explicit definition of each pattern and clarification of the transitions between them. Recognition of the inherent limitation of any particular pattern in a turbulent environment.
Contrast This approach has been extensively explored with respect to military strategies (although the number of alternatives in great power nuclear confrontation is presumably several orders of magnitude lower than in conventional warfare or team sport). A limited variant is applied for the alternative organization of society in response to civil defence crises. Extension of the approach to other crises has not been envisaged except through the use of various tactical economic devices (e.g. 'belt tightening') which do not involve temporary social reorganization. Multinational enterprises make some use of this approach.
Further keys Range of organizational patterns required to contain the range of (un) foreseeable crises. Whether any intermediary 'rest' posture exists or whether all postures are a response to a prevailing condition.

♦ MM2057 Bilingual protocol

Substrate In countries having more than one official language a special protocol emerges to ensure that neither language is favoured on any formal bilingual occasion. An official speech is given first in one language and then in another, with the second being used first on the next equivalent occasion. Strict rules are also applied to the presentation of official texts and signposts.
Metaphor In cultural contexts in which more than one perspective is considered meaningful in the organization of society, the emergence of a protocol to ensure that precedence is given to each in turn enables all to contribute to understanding of the whole which they each endeavours to encompass.
Special features The distinctiveness of co–present languages and the very different effects they have on ordering comprehension and behaviour, even at a non–verbal level.
Contrast Whilst the implications of different languages are striking to those who have to respond to them, there is little recognition of the implications of different modes of organization of knowledge or group behaviour calling for expression in the same context.
Further keys Bilingualism. Unequal competence in the different languages of a social context. Switching between languages in informal contexts. Missing equivalents between languages.

♦ MM2058 Tacking

Substrate In a wind–powered yacht where it is necessary to travel against the wind (namely in the general direction from which the wind is coming) a tacking technique must be employed. This involves first travelling some distance with the wind on the right–hand side in a direction at a considerable angle (say 45 degrees) to that of the direction desired. Then the direction of travel is switched so that the wind is on the left–hand side again at an angle (say 45 degrees) to that of the direction desired. The boat thus advances in the desired direction by alternating repeatedly between two directions of travel which are 'off–course' in a complementary manner. An analogous technique is used for routes up a mountain which wind in relation to the direction desired, as is the case with skiing down a steep hill.
This suggests that there may be circumstances in which it is not possible to achieve social development using a single policy. It may be necessary to alternate repeatedly between two (or more) policies which move society in complementary undesirable directions.
Special features The explicit manner in which the vehicle uses the energy acting in a direction opposed to that in which the vehicle moves.
Contrast This technique is not explicitly used although the conflicting directions between which social groups stumble may be seen as indicating an unconscious use of an analogous procedure.
Further keys How 'close to the wind' it is possible to sail. Balance and the keel. Moving against an entropic force.

♦ MM2059 Aerial animal locomotion – gliding and soaring

Substrate The structure of animals moving though the air reflects their locomotor habits. Gravitational gliding by certain amphibians, reptiles and mammals is based on their ability to increase the relative width of their bodies thereby increasing the surface area exposed to wind resistance. Direction is controlled by adjusting the surface area (braking being achieved by stalling). Soaring is of two types and is restricted to birds and includes use of gravitational gliding mechanisms. Static soaring (at relatively high altitudes over land) depends on the presence of vertical air currents, whether close to a cliff or within free–floating thermal bubbles (thermals). In the latter case birds spiral downward in the updraft; however, because the bubble rises faster than the birds descend, the birds are carried upward, but at a speed less than that of the bubble. When they reach the bottom, they begin a gravitational glide to the next bubble. They therefore alternate between circling and straight gliding. Dynamic soaring (at relatively low altitudes over water) depends on the presence of layers of air of different horizontal velocity. Speed is acquired using gravitational gliding downwind from the higher/faster layers, height is then reacquired by turning into the wind. The pattern is therefore a series of loops inclined to the wind.
Metaphor The structure of groups or individual personalities reflects their mode of movement in relation to public or peer group opinion. Certain groups are able to minimize their degree of exposure to public opinion so as to enable them to 'glide' through society with relatively little effort, from a relatively advantageous position to a less advantageous but distant position, such that the potential loss of advantage is reduced by public expectation. The direction of movement is controlled by adjusting exposure to public opinion. Some suitably adapted groups are able to use a rising current of public opinion to carry them up to a more advantageous position from which they can seek out and glide to some other rising current that will enable them to maintain that advantage. Others glide from position to position, increasing their social momentum as they lose their relative advantage, orienting themselves to the prevailing force of public opinion whenever they need to acquire advantage again despite the associated loss of social momentum.
Special features The elegant manner in which forces in the environment are used to maintain advantage, or to reduce its loss in the most effective manner.
Contrast There is recognition of the ability of individuals or groups to take advantage of a prevailing current of opinion and 'ride with it' especially in relation to political issues and research or other fashion. Public opinion is also tolerant of the misdemeanours of those it favours, cushioning the excesses of media stars, creative geniuses and aristocrats as they manoeuvre through the social system, in a manner which is not possible to those without such relative advantage. But whilst such techniques are vital to maintaining the relative social advantage of many individuals and groups, they have not been explicitly defined.
Further keys Wing span. Sensitivity to air currents. Control of flight. Locating food and recovering height.

♦ MM2060 Driving

Substrate In driving or piloting most vehicles over land or sea, the basic problem is to steer the vehicle towards a particular destination whilst circumventing any intervening obstructions. To maintain direction requires that the driver turn the steering wheel in one direction to correct for deviations towards the other. This tends to result in a deviation in the opposite direction for which a corresponding correction must then be made. Steering therefore requires alternating correction movements of the steering wheel. To circumvent an obstacle however, the driver has to ensure that the vehicle first deviates significantly in one direction and is then brought back 'on course' by deviating in the reverse direction.
Metaphor This reinforces the understanding that alternating movement in opposing policy directions is necessary to ensure that society develops in a controlled manner in the direction towards which it is driven. Problems can only be circumvented by accepting deviations for a period before making corrections for them.
Special features The alternating deviation correction movements tend to be directed at right angles to the resultant direction of movement.
Contrast The driving metaphor is used by leadership to a limited extent. The difficulty is that (like nervous 'back–seat drivers') the passengers in the vehicle on the right–hand side tend to shout 'turn left' to avoid dangers on the right and disagree violently with those on the left–hand side who shout 'turn right' to avoid dangers on the left invisible to those on the right). In the struggle for control of the vehicle, there is little sensitivity to the need to check deviations by alternately moving left and right, or to the need to circumvent obstacles by making extensive deviations in one direction or the other.
Further keys Contrast between expert and novice drivers. Reactions that need to be internalized in 'learning to drive'. Drunken driving. Traffic and rules.

♦ MM2061 Voice and speech

Substrate Speech normally results from the coordination of four operations: voice activation energized by the cycle of respiration; phonating sound generation by transformation of that energy in the larynx; sound moulding into voice patterns in the pharynx; and speech forming articulation, such as by opening or closing the mouth.
Metaphor Respiration may be treated on its own as a metaphor (see Breathing) for various alternating processes (e.g. political shifts between right and left, policy shifts between decentralization and centralization, shifts in awareness between right and left–hemisphere modes). Once psycho–social processes are being energized by such low frequency shifts, the energy of a particular shift may be used to engender shifts of a much higher frequency and of different nature. These can be perceived at a much greater distance. The frequency of these high frequency shifts can be modulated to carry new levels of meaning of much greater subtlety than is possible with the low frequency shifts. Modulation is partly achieved by precise variations in the manner in which collective (or individual) expression is regulated (or deregulated).
Special features The manner in which subtle distinctions are carried on a cycle of much coarser distinctions without which the former could not be expressed.
Contrast Some forms of jargon indicate explicit awareness of the 'sound' a group makes. This awareness does not extend to the manner in which a group 'articulates', its concerns, although the expression is used. People may however be said to act in terms of an implicit awareness of the characteristic meanings conveyed by groups in society.
Further keys Speech disorders. The psychological and physiological distinction between (rhythmic) singing and (arhythmic) speech. Laughter. Transition between vocal registers and their relationship to the size voice types (bass, baritone, and tenor, in the male, in contrast to contralto, mezzo–soprano, and soprano, in the female).

♦ **MM2062 Blindness**
Substrate Inability to see or respond to visual cues.
Metaphor Individuals and groups may be blind to conditions or configurations of events in their environment and thus be unable to adjust their behaviour usefully in relation to them. This can inhibit human and social development. It renders those so afflicted vulnerable to catastrophic accidents.
Special features The total deprivation of a particular type of sensory input. In the case of blindness from birth, this also involves a complete lack of realization of the dimensions revealed by right. It also results in considerably increased emphasis on other forms of sensory input.
Contrast There is little general awareness of the extent to which people and groups may be blind to features of significance for the social development of their environment.
Further keys Transient blindness and its causes. Blindness resulting from infectious or non-infectious diseases, or from diseases of the eye. Congenital blindness. Aids for the blind. Guide dogs.

♦ **MM2063 Traffic regulation**
Substrate The movement of traffic of different kinds, of different densities, at different speeds and with different directions, especially in an urban environment, is (self-) regulated by a range of techniques. These include: basic road rules (driving on right or left), prohibited actions (no entry, speed limit, no stopping, no waiting, no parking), required actions (stop, keep left, yield, turn right), limited access (no cyclists), and special warnings (dangerous crossing, etc). To improve traffic flow, traffic signals may be used permitting an orderly alternation in direction of movement. These may be phased in various ways to improve flow in an area (e.g. green phasing for a group of vehicles moving at constant speed along a route through the area), although area traffic control responsive to a range of traffic conditions must optimize flows by compassing current conditions to models based on past experience. Traffic of different types may also be segregated: pedestrian from vehicles, local from long-distance traffic (e.g. on expressways with merging lanes and cloverleaf junctions).
Metaphor The progress of groups of different kinds, at different rates and with conflicting policies, especially in complex social environments, could be regulated by a range of techniques. A cycle of signals may be used to enable groups with conflicting policies to progress during alternate periods. Actions of groups of different types may be segregated. The progress of groups with conflicting but interrelated policies may be facilitated by devising means for such policies to merge into each other rather than cut across each other.
Special features The precision with which a wide range of potential conflict situations must be handled using a minimum set of rules comprehensible to all concerned. The explicit recognition that priority is given to each according to particular circumstances. The manner in which traffic under many conditions controls itself when each vehicle recognizes under what conditions it has right of way or must give way.
Contrast Under certain conference and air-time rules, groups are given start/stop priority, over each other in succession in order to express their viewpoints. The alternation of political parties in power may be considered in this light; as election being the process through which a decision is taken on the traffic signals. But in general, present policy control in this metaphor can be compared to a procession (or 'progress') in one direction with the support of security forces which ensure that all access roads be blocked off and all opposing traffic suppressed. When the procession has petered out, another such 'convoy' may be organized in another direction to cater for the traffic stream blocked by the first. This corresponds to a very primitive traffic control approach. It takes no account of the sophisticated blend of control and delegation of responsibility to drivers which is characteristic of modern traffic patterns.
Further keys Basic road rule (driving on right or left) and the history of its emergence. Collision avoidance. Co-presence of different developmental stages of traffic management (stop streets, one-way streets, round-abouts, filtering systems, underpasses and cloverleaf intersections). Public and private vehicles. Traffic policemen. Sophisticated area control systems to adjust to congestion, peak periods, and emergencies.

♦ **MM2064 Aerial animal locomotion - flight**
Substrate The structure of animals moving through the air by true flight (in contrast to gliding or soaring) reflects their locomotor habits. True flight is produced by the simultaneous rotation of the left and right wings in a circle or figure-of-eight, which has the appearance of alternating up and down movement. This cycle produces the upward thrust required to overcome gravity and the forward thrust necessary to overcome drag. Lift is produced by the unequal velocities of the air across the upper and lower wing surfaces when the downward and backward phase of the cycle forces the air backward and the body forward. The actual flight pattern and wing movement varies with the different insect, bird or mammal species capable of flight. Guidance and stability are provided by minor alteration in the symmetry of the arched wings.
Metaphor The structure of individual personalities or groups reflects their mode of movement in relation to public or peer group opinion. Certain groups with two (or more) opposing factions are able to ensure that the cycle of actions of each is the reflection of the other. The complementary cycles produce the force necessary to counteract public or peer group opinion, enabling the group to rise to a more advantageous social position or move to a new position. The actual cycle of actions employed, how these are generated, and the resulting pattern of movement varies with the different kinds of group capable of this form of movement. Stability and guidance are provided by continually alternating emphasis on each faction (e.g. the correct adjustment at one time may be 'up' on 'right' and 'down' on 'left', which will tend to cell for the reverse immediately afterwards as a counteracting adjustment).
Special features Simultaneous mirror-image movements of the complementary wings in each pair. Manoeverability is only possible by unbalancing the emphases for an appropriate period of time.
Contrast Extreme factions of a group are often distinguished as right and left 'wings', especially in the case of political parties. In individual modes of thought, right and left brain forms have been distinguished. A group or programme is recognized as having 'taken off' or able 'to fly' if it is able to coordinate its movements appropriately and overcome the resistance of the environment to that initiative. But although terms such as 'wing' are used, the manner in which such wings function together (rather individually or consecutively) to engender social movement, and guide its development, has not been explored. The counteracting controls, or constraints of one on the other, are in fact the subject of acrimonious controversy.
Further keys Controlling imbalance to achieve manoeuvrability. Flight MGder a variety of conditions and with different manoeverability requirements. Aircraft flight, flap movement; control of yawing, pitching and rolling. Controlled turns. Formation flying. Takeoff and landing techniques. Wingless aircraft and propulsive guidance systems.

♦ **MM2065 Composition**
Substrate In many of the arts, but especially in music, the challenge is to hold the perceiver's attention by moving it through a variety of complementary modes such that the pattern of contrasts embodies a new level of significance. In music the composer may for example choose to alternate between different pitches, rhythms, degrees of loudness, or timbres (such as by use of different instruments). Themes may be repeated, moved to an alternate key, contrasted or transformed (retrograde, inversion, or retrograde of inversion). The challenge is to strike a meaningful balance in the alternation between recognizable repetition and introduction of novelty.
Metaphor The ordering of society can be conceived as a problem of interrelating (composing) the characteristic phenomena of social dynamics, especially when the social groups are each perceived as developing musical themes which partially respond to those of others.
Special features The explicit clarification of the elements and variations that are possible. The contrast between technical possibilities and audience appeal. Explicit recognition of the importance of limits.
Contrast Aside from occasional references to 'trumpet blowing' and doing 'the same old number', this metaphor has not been systematically explored.
Further keys Complex possibilities for the development of themes. Multi-part harmony. Need for an interesting balance between harmony and discord. Development of music and music appeal. Alternative tunings. Need for recognizable rules contrasted with the innovator's need to invent new rules (which may not then be recognizable). Lack of contact between those concerned with 'serious' music and those concerned with 'popular' music. Improvisation. The problem of the relationship between composer, interpreter, audience, and intermediaries (studios, manufacturers, distributors).

♦ **MM2066 Ball games**
Substrate Balls may be thrown back and forth collaboratively among a group of individuals, or else control of its movement may be the subject of intense competition between teams of one or more individuals in order to increase the score of one at the expense of the other.
Metaphor An issue may be casually transferred (bounced) back and forth between institutions or groups. These may also form rival coalitions, each competing intensely to place the issue to its own advantage and especially in a position of maximum disadvantage to any opposing coalition.
Special features Ball games are very widespread. The engender patterns of interaction between cooperating and opposing groups which are perceived as fundamental to social behaviour in other settings.
Contrast Whilst there is awareness of the importance of team spirit, of the necessity for a game plan in any social enterprise and of the importance of 'not fumbling a pass', this awareness does not generally extend to team rehearsal of detailed moves or to understanding of how the ball may appear in any such context.
Further keys Game rules, the contextual implications of winning or losing, referees, the process of determining what constitutes winning or losing.

♦ **MM2067 Dance**
Substrate Many dances are characterized by rhythmic movements of two separated partners in response to one being complemented by those of the other. In some dances all the dancers form into patterns with the two sexes facing each other and then moving so that the patterns interweave. The development of the dance may be such that each person of one sex dances successively with each of the other. Or possibly at some stage each of the original couples dances alone for a period, watched by the others who appreciate their relative merits.
Metaphor If the dancers are considered to represent issues or their advocacy groups, matched into opposing pairs, an interestingly stylized interaction between the issues (or their advocates) emerges. Each is given space for its development within the pattern as a whole, possibly ensuring that there is a relationship between each and all the others.
Special features The explicit nature of the alternation between complements and patterns. The place of each within the whole balanced by a counteracting complement to focus and 'contain' its movements. Emphasis on elaborating the variety of possible relationships amongst the dancers. The possibility of switching at the end of each dance to a dance of a different pattern such that the dancers alternate between a range of possible patterns in which their possible relationships are articulated in different ways.
Contrast The metaphor is occasionally used to describe the relationships between social groups (e.g. 'pas de deux') but only in isolation. The ongoing dance of issues and their advocates has not been explored. This may be in part because some relationships are more realistically encoded by 'gutsy' tribal dances, folk dances, and rock, than by the somewhat effete classical ballroom dances. The challenge is to ensure the interrelationship of such patterns which may all be open to the dancers. Dance patterns may be usefully contrasted with military parade formations.
Further keys Varieties of dancers and the patterns they encode. Transitions between dance forms. Development of dance.

♦ **MM2068 Metabolic pathways and cycles**
Substrate The chemical changes that occur in the cells of living organisms may be described in terms of a complex network of metabolic pathways and cycles through which the cells are able to maintain their identity, to grow and to reproduce. In plant-like organisms these pathways include the synthesis of cellular components using sunlight. Such components, when digested by animals through other pathways, permit them to synthesize their own cellular components (anabolism), or to yield energy for life processes through catabolic backdown of the latter. In this extensive network of reactions, one reaction sequence may be repeatedly linked to another through chemical intermediaries that are common to both. Some vitally important reaction pathways form distinct cycles of intermediaries (e.g. Krebs or citric acid cycle, Calvin cycle, glyoxylate cycle, urea cycle), each cycle having particular functions. The network of reactions is maintained in a steady state conferring upon the cell a chemical resilience upon which its survival in a changing environment may depend.
Metaphor The changes that occur in individuals and groups within society may be described in terms of a complex network of processes and cycles through which both are able to maintain their identity, to grow and to reproduce. For some these processes include the formulation of new patterns purely through spontaneous action or creative thinking. Such patterns when absorbed by groups through various processes, enable them to use the patterns as a basis for formulating larger patterns or as a means of yielding energy for action in society by their destructive fragmentation. In this extensive network of processes, one psycho-social process may be repeatedly linked to another through intermediary conditions that are common to both. Some vitally important processes form distinct cycles of intermediary conditions (e.g. production cycle, challenge and response cycle, etc), each cycle having particular functions. The network of processes is maintained in a steady state conferring upon the social organism a resilience upon which its survival in a changing environment may depend.
Special features The ability to interrelate the full range of known metabolic pathways and cycles on one map covering both plant and animal species even though many processes still remain to be identified.
Contrast There is some recognition of the need for a healthy 'social metabolism'. Different disciplines recognize the psycho-social processes and cycles to which they are sensitive. Efforts at interlinking these processes are either highly schematic with few elements or else highly detailed as in specialized economic models.
Further keys Types of processes and cycles. Metabolic diseases.

♦ **MM2069 Juggling**
Substrate The key to the art of juggling lies in alternating the movement of the hands in order to keep a whole set of objects moving through the air in a recurring pattern. Each hand must perform the role of catching and throwing in rapid succession to maintain the pattern.
Metaphor This suggests the need to alternate functions in rapid succession to be able to handle a variety of issues simultaneously with limited resources.
Special features The distinction between mastery of the art and lack of it. Degrees of mastery as measured by the number and variety of objects maintained in the air.
Contrast The art of government, leadership or management is frequently defined in terms of 'juggling factors' suggesting that there is an intuitive understanding of the alternating actions required to maintain the integrity of the social order. The nature of the art has not been explored in these terms, especially with regard to the number and variety of phenomena handled in this way.
Further keys Teamwork to enable an even larger number of objects to be maintained in the air at the same time. Problems of starting, stopping and introducing new objects into an established pattern.

♦ **MM2070 Geography and movement**
Substrate The habitable land surface is the result of a long process of gases cooling into magma, solidification of such magma, and the emergence of lifeforms. This provides a context of land surface, water and air subject to a variety of dynamic processes due to gravitational forces and solar heat. Within this context, and according to the nature of the animal species, movement is hindered or facilitated by mountain barriers, seas and rivers, wind patterns and the nourishment available. Artefacts may be constructed by humans to sail over the sea, to bridge a river or to fly through the air.
Metaphor An ordered social context emerges as the result of a sufficient decrease in intensity of interaction to allow structures to remain stable and processes to fall into patterns providing continuity. Such a context has some domains characterized as solid and enduring, others characterized as fluid and over–changing, and yet others which are insubstantial although capable of exerting an appreciable pressure. Forces external to the context engender processes within it that ensure ongoing interaction between such different domains. Individuals and groups within the context, according to their nature, find their movement within it hindered or facilitated by structural barriers, domains of uncertainty or fluidity, the pressure of prevailing public opinion, or the lack of appropriate support. Social innovation may be devised to traverse domains of uncertainty, to provide a solid structural link between two domains separated by the currents of uncertainty, or to use the rising currents of public opinion to provide the necessary lift to permit movement over both enduring and ever–changing domains.
Special features The range of detailed (but well–known) features and processes and the manner in which they interweave. The relationship between natural features and the range of artefacts which have been developed in response to them.
Contrast Although many features of the natural environment are used as a guide to understanding social change, no attempt has been made to use them together as a means of interpreting the social context. The mediaeval world view may be an exception to this as is the case with the animistic world view.
Further keys Most geogrphical features and phenomena (e.g. storms, tides, mountain formation, deserts, swamps, etc). The characteristics of species associated with different environments.

♦ **MM2071 Project phasing**
Substrate Techniques of management have become so sufficiently explicit that widespread recognition is accorded to the need to organize projects into such phases as: conception, planning, organization, implementation, and evaluation. Management teams (and others) alternate through such phases as well as through a succession of projects organized in that way.
Metaphor In addition to groups, societies can be seen as operating in terms of such phases.
Special features Recognition of the different skills (even personality types) required at each stage. Recognition of the importance of ensuring that the different phases mesh together effectively.
Contrast Although project phasing is a reality where organizations can be managed under appropriate supervision, such phasing is less meaningful in the case of societies as a whole. Even as a metaphor it is much less meaningful when the society is not under some monolithic pattern of control.
Further keys Atlas of Managing Thinking by E de Bono which identifies some 200 phases in the management process.

♦ **MM2072 Magnetic containment of plasma**
Substrate Nuclear fusion can be achieved by confining plasma under certain conditions for sufficient time. A plasma is an electrical conducting medium consisting of positive and negative charges forming a neutrally charged distribution of matter. Plasma, as the fifth state of matter in which 99% of the universe is to be found, is unique in the way it interacts with itself, with electric and magnetic fields, and with its environment. (in contrast with the four other states of matter: reacting elements as in fire, gas, liquid and solid). Its properties depend upon on the collective behaviour of the constituent particles, as distinct from the individual. It is unique in its instability and its tendency to revert to ordinary combinations of matter and energy. In order to generate energy in a fusion reactor, the problem is to find the particular configuration of magnetic fields, values of plasma parameters and means of protecting the plasma from impurities which would quench it. This is achieved by 'bouncing' the plasma around within the configuration of a magnetic cavity (or 'bottle').
Metaphor Individual attention or collective awareness (public opinion) has properties analogous to those of a plasma. The problem of a significant breakthrough in mobilizing the political will to change is comparable to that of designing a fusion reactor. The challenge is to be able to contain and focus individual or collective attention (despite its inherent instability) by a suitable configuration of psycho–social functions that can protect such attention from degenerating into normal modes. Since no one function can be used to this end, the problem is to constrain attention by all of them simultaneously without allowing reliance or dependence on any one of them.
Special features Recognition of the inherent instability of what needs to be contained and the problem of designing a container whose contexts cannot come into contact with its material components.
Contrast The modification of the perception of the environment under incitement to act or rebel suggests an implicit understanding of the need to bend every habitual perception to focus the will to act. Mysticism in both Eastern and Western traditions has stressed the need for an individual to create an inner environment to contain psychic energy (cf. alchemical containers, controlling the movement of 'ch'i', and the circulation of the 'golden fire'). Some mass rallies have design features which a intended to focus collective awareness. In some of these cases are the design considerations as explicit as in the fusion reactor problem.
Further keys Plasma values. Extraordinary temperatures (of the order of 50 million degrees) required for fusion. Extraordinarily short time at that temperature for a viable energy yield to take place. Design of magnetic bottles and toroidal containers.

♦ **MM2073 Animal life cycles**
Substrate The distinction between the phases in the life cycle of some animals is particularly striking in the well–known case of insects such as butterflies (with their caterpillar and pupal phases). An even more striking case is that of the cellular slime molds in which the spores gow independently and then aggregate together into a migration amoebal form. This eventually transforms itself into a 'fruiting body' from which the spores are released to continue the cycle.
Metaphor There is the possibility that some extremely different psycho–social forms can usefully be understood in the above light as constituting stages of transformative cycles.
Special features The radical nature of the tranformation which may involve a complete shift of medium (e.g. from water to air). The contrast in mobility and aesthetic features between the different phases (in the case of butterflies for example).
Contrast It is possible that the radical nature of the transformation prevents this metaphor from being used, despite the exposure of all schoolchildren to such biological cycles in classes on nature.
Further keys Relative lengths of each phase of such a cycle. Change in nature of vulnerability at each portion of the cycle.

♦ **MM2074 Path and journey**
Substrate A path is a way trodden out over time by the feet of humans or animals using it with some frequency or, more generally, the course or route along which any object moves, even if the movement is not repeated (as with a boulder rolling down a hill).
Metaphor The course of action followed by an individual or group may be perceived as a journey over a path. The action may take the form of a time of argument developed through a number of stages. A number of times of argument may intertwine like a network of pathways with varied points of focus.
Special features The surface implicitly defined by any path or journey and the distinction between being on or off the pathway and being on or off that surface.
Contrast Descriptions of arguments make frequent use of journey–related terms (e.g. set out to prove, on the way to solving the problem, getting to the next point, arriving at a conclusion, pointing to the next step) including those related to going 'off–track' (e.g. straying from the path or from the line of argument, missing the point, getting lost in an argument or going around in circles). The paths of arguments are also related to movement over a surface (e.g. covering points or ground, getting on or off the subject, going in circles). In philosophy thinking itself has been considered as a path. The motion of a path, and its stages, is central to many schools of religious and spiritual development.
Further keys That surfaces as contrasted with other surfaces and other geometrics. Closed surfaces. Topological spaces. The Möbius, strip, Klein bottle and the torus. The relationship between the beginning and the end of the path. Topography, geography and cartography. Catastrophe theory.

♦ **MM2075 Good–bad behaviour**
Substrate Whether in controlling one's own behaviour, bringing up a child, commanding a regiment, or managing an enterprise, skill is used in alternating between requiring 'good' behaviour and allowing 'bad' behaviour. In the case of a child, for example, to expect good behaviour all the time stultifies formation of the child's character. It is recognized that 'bad' behaviour is in many cases a healthy expression of a 'free spirit', to be permitted within certain limits. A child always on good behaviour is recognized as lacking some quality of individuality. A similar situation prevails in an army, especially under combat conditions. A good commander knows 'when not to see things'.
Metaphor It is to be expected that healthy groups and societies should also alternate between good and bad behaviour.
Special features The manner in which limits are defined and the struggle to maintain and redefine them. The incompleteness of both extremes. Development takes place by alternation between the limits.
Contrast Whilst there may be some collective tolerance of bad behaviour by groups and societies, this is not understood as a necessary complement to their good behaviour. Societies are supposed to be good all the time. They are not expected to indulge in foolish mistakes.
Further keys Weekly 'booze–ups' and annual carnivals as safety valves to 'let off steam'. 'Sowing wild oats' before 'settling down'.

♦ **MM2076 Battery**
Substrate As a result of the potential chemical reaction between two distinct metals immersed together in an electrolyte (ionized solution of a compound which conducts an electric current), electrons may be released on one (the negative electrode) and allowed to flow to the other (the positive electrode) when the two are connected by an external circuit through which this flow can take place. The process only continues until the circuit is interrupted or the reactants are exhausted by the chemical process. Some forms of battery can be recharged by passing a current through them in the reverse direction.
Metaphor When two distinct individuals or groups are present together in an environment in which they risk a change of condition or status as a result of their relationship, exposing them to each other results in a charged exchange from which one accumulates positive attributes and the other negative attributes. The competitive exchange may be used to activate an appropriate organizational programme. The process only continues whilst the two are exposed to each other or until their condition cannot be further changed by the interaction. In some cases the situation can be regenerated after the participants have refreshed themselves in other social processes.
Special features The method of harnessing energy resulting from difference.
Contrast Phrases such as a 'charged situation' are used to refer to quite distinctive social conditions in which considerable social energy may be creatively or destructively released. Some principles of management or leadership depend upon setting two strong individuals or groups in an (antagonistically) competitive relationship to one another, especially in one form of 'divide and rule'. Some forms of therapy and difficult resolutions are based on discharging the potential energy associated with the relationship between two parties. There does not however seem to be any concern with the design of 'batteries' in order to harness such energy more effectively.
Further keys Distinctive components used in batteries (lead–acid, nickel–cadmium, etc). Connection in series or parallel to give higher voltage or longer capacity respectively. Fuel cells. Solar batteries, thermal batteries and nuclear batteries. Capacity limitations.

♦ **MM2077 Shell games**
Substrate The classic game, much favoured by confidence tricksters at fairs, consists of determining the location of a small object (a pea or a coin) which is moved rapidly beneath one of three cups in such a way as to create the (false) impression that it is obvious under which cup it is to be found. Also known in a card version as 'Find the Lady'.
Metaphor The key to understanding many shifting social situations often creates the impression of being precisely located. Individuals and groups may well take extensive personal, financial or career risks based on their belief of where it is located. Yet the dynamics of the social systems repeatedly shift that key into an unexpected location.

Special features The unpredictability of the alternation as contrasted with the impression of predictability of the location at any one time. The difficulty in 'pinning down' the location.
Contrast There is a widespread confidence amongst groups in their ability to specify where the key to any shifting situation lies at any one time. There is such confidence in this belief that groups may even express considerable hostility towards those who fail to agree with their perception.
Further keys The skill of the shell game artist and how it is developed. The role of his accomplice in misleading the audience. Increasing the number of cups or the number of objects to be found.

♦ MM2079 Vitamins

Substrate Vitamins are organic substances of which small quantities are necessary for normal health and growth in higher forms of animal life. Vitamins are distinct from many other compounds indispensable for animal function in that they cannot be synthesized by the animal in adequate quantities. In the absence of a particular vitamin from the animal's diet, a specific deficiency disease may develop.
Metaphor Individuals and specialized groups experience difficulty in functioning in isolation using only the information they generate themselves. Such groups are dependent upon minimal amounts of different types of information of external origin. In the absence of a particular type of such information from the group's 'information diet', the group may develop a specific kind of dysfunction.
Special features The limited range of vitamins and their specific function in maintaining the health of the animal.
Contrast In the case of both an individual and a group, unrelieved routine is recognized to be unhealthy for development. Importance is attached to challenges and external stimuli. The media are concerned to provide a balanced information diet (news, sport, adventure, romance, music, sex, documentary, how-to-do-it, scandal, who-done-it, religious celebration, etc). There is however little explicit awareness of what basic categories of information make up that diet and which of those kinds the group is unable to generate for itself.
Further keys Sensory deprivation. Types of vitamin and their metabolic function. A balanced diet. Variation in vitamin requirements of different animal species.

♦ MM2080 Climbing

Substrate A special sequence of movements is required when climbing. This is especially evident when climbing up between two smooth parallel walls (a 'chimney' in mountaineering terms). The climber has to ensure that there is alm to sufficient pressure against both walls to enable him to move upward in succession his hands, feet or body.
Metaphor The development of society may be seen as the upward movement between any two constraining extremes (e.g. idealism and materialism) which offer no permanent foothold. Developmental movement may be achieved by ensuring that there is sufficient pressure against both extremes to guarantee a temporarily secure or stable position from which a portion of society may be moved forward.
Special features The sequence of movements required for such a climb to be successfully made, especially when the climber has to rest periodically. The way in which different portions of the climber's anatomy (hands, feet, body) change their function from applying pressure to moving upward. The basic requirement that there always be sufficient counteracting pressure against both wall.
Contrast As with the walking metaphor, the prevailing attitude may be likened to that of a climber attempting to get up a chimney by attempting to cling to the one favoured wall and to avoid touching the other (perceived as anathema). A skilled mountaineer can do this by inserting spikes in the favoured wall. Much less skill is required to climb using pressure on both walls.
Further keys A climber with more extremities. Climbing up a multi-dimensional chimney with N walls and N-1 directions in which to 'fall down'.

♦ MM2081 Getting water

Drawing water — Pumping — Irrigating
Substrate In those situations when water is not available at the turn of a tap, it must be obtained from a river or well, often after a long walk. In the case of a well, some device may be used to get the water out. This may vary in sophistication from a backet on a rope to a pump. If the water is to be used for irrigation, it may be directed along channels through the fields. If it is to be used for washing clothes, this may be done by the well or river.
Metaphor In those social situations governed by well-defined and rigid procedures, special efforts must be made to provide compensatory processes of a fluid and under-defined nature. The social environment may be so over-defined that access to such processes can only be obtained by some special social device. When such processes are plentiful and have established their own context, they may be partially redirected to provide the necessary flexibility in the over-defined environment.
Special features The nature of the work required to compensate for over-definition in social contexts.
Contrast The term 'watering-hole' in occasionally applied to pubs and bars and illustrates the contrast to the over-defined office (or home) environment. Devices such as receptions, night club shows, comedy, house parties, stag parties, sauna parties or even orgies may be used to introduce a necessary informal dimension into social relationships. On the intrapersonal level this may be partially achieved with jokes. On a large scale this may be achieved by regular festivals or carnivals. Some modern management techniques recognize the need to humanize the working environment and therefore encourage, to a limited extent, introduction of compensatory processes into the work-place (e.g. office parties and outings, music, recreation facilities).
Further keys The task of fetching water in containers. Free flowing water in contrast to containerized water. The wide range of water raising devices. Irrigation systems and their dependance on dams. Periodic drying up and flooding of river beds.

♦ MM2082 Dining

Substrate In all societies efforts are made to introduce variety into the food offered at a meal, whether a feast or of the simplest kind. The diner is exposed to contrasting tastes (and textures) between which he shifts in response to the temptations to his palate. The different foods may only appear in one dish. There may however be a succession of dishes each offering different contrasts which he may sample alternately. In addition, whether in a feast or in the succession of daily meals, the style of dish may vary (e.g. breakfast as contrasted with lunch).
Metaphor The quality of life offered in a society may be partially indicated by the variety of experiences offered and the skill with which they are blended to bring out the best in such experiences by contrast or through ensuring their harmonious relationship to other experiences.
Special features The art with which dishes are prepared and presented in an appropriate sequence to enhance harmonies and contrasts.
Contrast There is little official awareness of the need for experiential variety or of the skill with which it can be beneficially prepared and presented. Official recognition is accorded to 'work', 'leisure' and 'rest' with rather crude attempts to manipulate the organization of leisure (as in officially sanctioned sport and similar 'cultural' events). Even for student educational work, the curriculum is usually designed with little awareness of the relationship between the subjects or the overall effect of the combination of experiences.
Further keys Special diets and cultural food preferences. One to four-star restaurant grading. Religious constraints and minimal variety meals. Food snobbery.

♦ MM2083 Stick

Pole — Lever — Club — Stave — Staff
Substrate As one of the first tools, its usefulness emerges when it is distinguished as an object with two distinct ends lying on the principle axis of symmetry. In the simplest case, something is done with the proximate end which imparts an analogous motion or force (or a complementary one) to the distant end (and to whatever the latter is applied). The distance between the ends augments certain effects, especially when the mass at the distant and is increased (as with a club) or when a fulcrum is used between the ends (as with a lever). Sticks and poles are basic elements in the simplest form of construction of shelters or bridges.
Metaphor One of the first conceptual tools emerges when a dimension is distinguished in the light of two distinct but complementary conditions (hot/cold, wet/dry, good/bad, big/little, etc). In the simplest case, value is attached to the favoured condition which imparts an analogous (or complementary) weight to the unfavoured one (and to whatever the latter is applied). The degree of polarization of any such diachotomy increases certain effects when it is used, especially when the charge on the unfavoured end is increased disproportionately or when some intermediate value is used as a standard of comparison against which the favoured value appears in a disproportionately favourable light. Dichotomies, polarities, antinomies and dualities are basic elements in the construction of the simplest social structures.
Special features The simplicity of the device and the extremely wide variety of uses to which it can be put.
Contrast Although stick is used as a metaphor in various forms (e.g. using a 'big stick', not touching someone with a 'large-pole', 'staving someone off', organizational 'staff' (?), and acquiring and using 'leverage'), no attempt is apparently made to distinguish the different ways different kinds of dichotomy may be used as a tool within the social environment. In particular, with the exception of some psychoanalytical approaches, there is no awareness of how several dichotomies may be combined as the necessary basis for a more integrated psycho-social structure.
Further keys Kinds of sticks and range of uses to which they may be put. Finding or making a suitable stick. Methods of tying sticks together.

♦ MM2084 Alluring movement

Substrate Both man and animals in general are fascinated by simple multi-phase bouncing movements (and sounds). These may take the form of: a jiggling toy or rattle in the case of infants; a flashing movement of a fish lure (on a line) or the kind of movements that can be used to attract the attention of a cat or dog; an alluring set of movements in animal courtship ritual, partially reflected (for a man) in fascination with the bouncing movements of a woman's breasts or legs, whether casual or deliberately organized (as in music shows, night clubs or tribal dances). In their most relaxed moments, seated at an open-air cafe, in front of a television, watching waves at the beach or the wind in the trees, people find their attention lured by patterns of movement.
Metaphor In social life also people are fascinated by certain patterns of behaviour that contain a shift between alternatives but are nevertheless to some extent unpredictable, although sufficiently cyclic (and therefore predictable) to be considered non-threatening. Much popular entertainment is based on this property as well as on the 'blow-by-blow' presentation of any good piece of gossip or humour. This fascination may be partially exploited by the continual introduction of new issues into the news and into the political arena.
Special features The almost subliminal quality of the shift between alternatives which makes the movement eternally 'ungraspable' and therefore of continually regenerated fascination.
Contrast Great effort is made to present society and the environment through essentially static and alienating categories when it is precisely the shimmering patterns of movement that bind our attention into experienced reality.
Further keys Brownian movement of particles. Rapid eye movement by which objects are perceived.

♦ MM2085 Orbiting in space

Substrate For a body to maintain its position relative to another body in space, especially a much larger one, it must revolve around it in an orbit. Orbits may be circular in theory but are usually elliptical in practice. Orbits vary greatly in degree of eccentricity.
Metaphor In society smaller groups may maintain an orbit around a larger group to whose activities they are primarily attracted but from which they wish to remain distinct.
Special features Stability is ensured by ensuring that the continuing process of attraction towards the larger centre is displaced around it such that the attraction is renewed and fulfilled thus postponing any final degradation of the orbit.
Contrast In societal development there is little sensitivity to the range of orbits or the range of eccentricities or periodicities. Nor is there much awareness of the conditions required to maintain stable orbits as a bases for action of a different nature.
Further keys Free fall, orbital escape velocity, changing orbit, difficulty of interaction between different orbits, captive orbit; insertion into a selected orbit.

♦ MM2086 Molecular resonance

Substrate Some chemical molecules cannot be satisfactorily described by a single configuration of bonded atoms. The theory of resonance is concerned with the representation of such molecules by a dynamic combination of several structures, rather than by any one of them alone. The molecule is then conceived as 'resonating' among the several conceivable/describable structures and is said to be a 'resonance hybrid' of them. The classic example is the benzene molecule with 6 carbon atoms linked together in a ring. This is one of the basic features of many larger molecules essential to life. Its cyclic form only became credible when it was shown that the structure oscillated between two (and later five) extreme cyclic forms.
Special features The precisely defined nature of the structural extremes in contrast with the lack of precision concerning the ordering of this oscillation between them.
Contrast The concept of resonance is not at present used as a metaphor except occasionally in referring to the relationship between two people (or groups) who are 'in resonance'.
Further keys Typology of molecular resonant structures. Probabilistic nature of chemical resonance. Implication for complementary paradigms of any kind (e.g. wave versus particle explanation).

♦ MM2087 Launching a spacecraft

Substrate For a spacecraft to be placed in orbit around the Earth or to escape its gravitational domination, it must be launched with sufficient acceleration to acquire escape velocity within the constraints of propulsion systems. Three stages of propulsion systems may be ignited successively. If the acceleration is too low, much propellant is required. If it is too high, the stresses on the spacecraft may be too great.
Metaphor Projects, programmes and initiatives are launched, possibly in several stages. For an initiative to 'take off', it must acquire its own dynamic independent of the, setting from which it arises or else it will be absorbed back into the dynamics of that setting.

Special features A successful launch necessitates the design of a complex set of interconnected systems.
Contrast Whilst a few major programmes are launched with attention to the complexity of the interconnected systems involved, in general there is little awareness that such attention to detail is necessary.
Further keys Infra-structure required to assemble the propulsion system in space and time; ground control requirement; amount of energy required to get a significant mass into orbit; discarding used stages; launch windows.

♦ MM2088 Pulsational variable stars

Substrate A significant number of stars are intrinsically variable; that is their total energy output fluctuates over time. Of these the pulsational variables are stars in which light and colour vary as a result of pulsations in the star. Their periods may vary from a few days to nearly a year. Such variability is a characteristic of any star whose evolution carries it to a certain size and luminosity.
Metaphor It can be useful to interpret the energy output of any social group in terms of such variability. Many groups, particularly those with activities limited to large periodic conferences, are distinctly variable if only as evidenced by funds flow, paper output and media coverage.
Special features The detailed investigations of pulsation theory have established important qualitative and quantitative relationships between the cyclic phenomena. In the case of periodic variables such as the Cepheids, the period of variation tends to be proportional to their luminosity so that they are of great importance in the measurement of interstellar and intergalactic distances.
Contrast This metaphor is used in its non-periodic form in connection with the (nova-like) explosion into visibility of some new media 'star'. It is not used in its periodic form.
Further keys Other types of variable star.

♦ MM2090 Musical variations

Substrate Variation is a basic technique in music. A piece of music is changed melodically, harmonically or contrapuntally in such a way as to bring out the different melodies or 'voices' that can be woven around each other in a piece. Variation appears in the music of most cultures and is in fact one of the few universal characteristics of music.
Metaphor An issue in society is explored in a variety of ways by those for whom it is significant. The presentation of the issue is changed by each of them, whether acting in support of it or against it, such that together these 'voices' elaborate all the possibilities of it at that time.
Special features The considerable understanding of variational possibilities whether by scholars in theoretical terms or by practitioners or audiences with 'an ear for music'. The many varieties of variation that have been explored throughout the history of music.
Contrast This metaphor does not appear to have been extensively used except through the phrase 'variations on a theme' where theme is a term common to music bnd to conference programmes.
Further keys Melodic variation (instrumental, figural, Baroque). Harmonic and tonal variation with the introduction of tonal goal orientation, the hierarchical arrangement of keys, the movement to the 'dominant' and back to the 'home' tonic. Ensemble variation. Performance variation by which an organist can transform a liturgical chant into a polyphonic composition – hymn or verse alternating with choir or organ alone. Baroque ornamentation and embellishment. Non-Western musical variation, especially the Indian raga or the Indonesian gamelan in both of which the multi-level variation is conceptually more complex than in Western music. In the gamelan the variations are simultaneous contributions of different members of the orchestra resulting in a highly complex static concept of variation.

♦ MM2092 Mechanical cycles

Substrate Many mechanical and electro-mechanical devices have some cyclic feature basic to the performance of their function. Examples range from the simple pendulum, via the spring or the wheel, to the pump, the motion of cylinders on a crankshaft, motors and dynamos in general, alternators, and even the cyclo-synchroton. In such cycles distinct phases with different functions can usually be distinguished.
Metaphor Certain highly organized psycho-social processes could be perceived in terms of such electro-mechanical metaphors in order to highlight the different phases of any cycle.
Special features The precision with which different portions of any cycle are both distinguished and linked together by designed transitions.
Contrast Although such metaphors are used, for example, 'pumping' money into an operation, the 'dynamo' in an enterprise, or the 'wheel' around which other things turn, the phases, in the cycle are poorly distinguished. The significance of the movement of electro-mechanical energy is not recognized.
Further keys Different current modes (AC or DC). Typology of electro-mechanical cyclic complexity. Associated patterns of stress and strain, torque and vibration.

♦ MM2094 Crop rotation

Substrate In order to maintain the fertility of a field, it has traditionally been the practice to alternate between different crops in some (more or less) regular sequence. It differs from the haphazard change of crops from time to time. Rotations may be of any length, being dependent on soil, climate, and crop. They are commonly 3 to 7 years in duration, but usually with 4 crops (some of which may be grown twice in succession). The different crop rotations on each of the fields of the set making up the farm as a whole, constitute a 'crop rotation system' when integrated optimally.
Metaphor It is possible to perceive the alternation in parties in power in a multi-party system as a form of crop rotation.
Special features The manner in which each crop is deliberately chosen to correct for the weaknesses of the previous crop (e.g. control of pests, soil degradation, soil fertility, etc) and benefit from its enhancement of the soil. Understanding of the merits of particular crops under particular conditions.
Contrast The chaotic switch between policies of the 'right' and the 'left' without any understanding of the merits of each, the need periodically to compensate for the negative consequences ('pests') of the other and to replenish the particular resources which each policy so characteristically depletes.
Further keys Balancing the rotation cycle. Fallow period. Short-term advantages of the use of fertilizers (with dangerous long-term implications). Use of each crop to inhibit the pests encouraged by the previous crop.

Patterns of concepts MP

Rationale

Most conceptual schemes, whether purely theoretical or basic to the practical design of a development programme, are organized into sets of concepts, principles, priorities, or functions. Several such sets may be interrelated in a more elaborate scheme. It is the pattern of such interrelationships which ensures the coherence and integrity of the approach. Patterns therefore constitute a special form of presentation. Such patterns tend to be presented in isolation and often such that only the sub-patterns are explicit. Little is known about them as conceptual patterns. Given the need to interrelate concepts into a coherent pattern, so that they can be effectively communicated without loss of information, it is appropriate to explore the design of conceptual patterns of many elements and interconnections, whether explicit or implicit. Such explorations can contribute to the development of pattern languages through which groups can define and interrelate the non-material features appropriate to the quality of life in their environment.

Contents

The 253 entries are an *editorial experiment* based on a "pattern language" developed by a team led by the environmental designer Christopher Alexander as an aid to designing physical contexts in which quality of life is enhanced. Selected patterns have been used, according to the methods of the previous section, as substrates for metaphors such as to suggest ways in which social, conceptual and intra-personal contexts may also be "designed". A special merit of Alexander's approach is the detailed integration between the component patterns provided by relationships reflecting a profound understanding of the socio-physical environment which is extremely realistic, exceptionally harmonious and unusually sensitive to development potential. The cross-references presented here are metaphorical verions of the relationships indicated by Alexander's group.

Method

The method used is discussed in the comments in Section MZ.

Index

A keyword index to entries is provided in Section MX. The keywords are also incorporated into the index for Volume 2 (Section X)

Bibliography

Bibliographical references, by author, are given in Section MY.

Comment

Detailed comments are given in Section MZ.

Reservations

Although the results of this deliberate editorial experiment are interesting and indicative of further possibilities, the entries raise many questions concerning the appropriateness of the language used. The language is stilted, forced and artificial in order to explore the correspondence between the different metaphorical levels. It highlights the difficulties of making what may be vital distinctions in the social, conceptual and intra-personal realms, compared to the richness and subtlety of the vocabulary and imagery available for similar distinctions at the physical level. This points to the merit of treating the physical level distinctions as a metaphor through which the possibility of equivalent distinctions at other levels may be explored.

Possible future improvements

The selected groups of patterns could be further developed, especially that based on the use of Alexander's work as a substrate. That work was deliberately designed to produce a "pattern language" which people and communities could use to design their own environments. An analogous pattern language could be developed to enable people and groups to design (and redesign) their own conferences and organizations.

ENTRY CONTENT AND ORGANIZATION

Ordering of entries
Entries are in **numeric order** and follow the sequence of Alexander's pattern language on which they are based.

Index access to entries
The location of an entry in this sub-section may be determined from:
- the **Volume Index** (Section Z) on the basis of keywords in the name of the entry or its alternate names
- the **Section Index** (Section MX) on the basis of keywords in the name of the entry

Structure of entries
Entries may be composed of the following descriptive elements:

(a) **Entry number** This number has **no significance**, except that the higher numbers tend to be more specific patterns and the lower numbers more general patterns. It is merely a convenient method of identifying the entry (particularly for indexing and cross-referencing purposes). The last 3 digits of the number are those used by the Alexander group to identify the pattern.

(b) **Qualifier** following entry number. This is a single alphabetic code letter ("a", "b" or "c"). In the scheme from which this was derived these have the significance: a= clearly defined pattern; b= reasonably well-defined pattern; c= poorly defined pattern.

(c) **Entry name** This is printed in bold characters.

(d) **Pattern** Outline of the pattern underlying (e), (f), (g) and (h) below, expressed in abstract terms.

(e) **Physical environment** Brief description of the manifestation of the pattern in the organization of the physical environment as identified by the Alexander group.

(f) **Socio-organizational environment** Brief description of the possible manifestation of the pattern in the organization of the social environment in the light of (d) and (e) above, especially with regard to organizations and networks.

(g) **Conceptual environment** Brief description of the possible manifestation of the pattern in conceptual frameworks and the social organization of knowledge in the light of (d), (e) and (f) above.

(h) **Intra-personal environment** Brief description of the possible manifestation of the pattern in the organization of modes of awareness adopted by a person in the light of (d), (e), (f) and (g) above.

Cross-referencing between entries
At the end of any entry, there may be cross-references to other entries within this range of patterns. There are 2 types of cross-references possible, indicated by a 2-letter code in bold characters:
        Broader = Broader or more general pattern
        Narrower = Narrower or more specific pattern

♦ **MP1001a Independent domains**

Pattern Balance between domains will not be achieved unless each one is small and autonomous enough to be an independent sphere of influence.

Physical environment Metropolitan regions will not come to balance until each one is small and autonomous enough to be an independent sphere of influence. Whenever possible, evolution of such regions should be encouraged; each with its own natural and geographic boundaries; each with its own economy; each one autonomous and self–governing.

Socio–organizational environment Major networks or communities of organizations will not come to functional balance until each one is small and autonomous enough to be an independent sphere of influence.

Conceptual environment Major paradigms, networks of concepts or schools of thought will not come to functional balance until each one is sufficiently well–defined and autonomous to be an independent sphere of insight.

Intra–personal environment Major modes of awareness will not come to functional balance within the individual until each one is sufficiently well–defined and autonomous to be an independent sphere of influence.

Narrower Distribution of organization (#MP1002)
Regenerative resource cultivation areas (#MP1004).

♦ **MP1002c Distribution of organization**

Pattern If a domain is characterized by small clusters of organization to too great an extent, more comprehensive forms of organization cannot emerge. But if a domain is characterized by large clusters of organization to too great an extent, such organization will not be able to ensure the integrity of the domain.

Physical environment If the population of a region is comprised of too many small villages, advanced civilization is unlikely to be generated; but if a region has too many large cities, the surrounding area will go to ruin for there will be no populace there to tend to it. The population needs to be distributed evenly in towns of different sizes, and the towns of the same size need to be distributed evenly throughout the region.

Socio–organizational environment If a functional domain is characterized by smaller groups to too great an extent, larger and more complex forms of organization cannot emerge. But if such a domain is characterized by larger and more complex forms of organization to too great an extent, such forms of organization will prove detrimental to the integrity of the functional domain. Groups of different size and degrees of organization should be distributed evenly throughout the functional domain.

Conceptual environment If a conceptual domain is characterized by small networks or groups of concepts to too great an extent, more integrated and comprehensive forms of organization cannot emerge. But if such a domain is characterized by extensively organized networks of concepts to too great an extent, such a degree of organization will prove detrimental to the integrity of the conceptual domain. Bodies of knowledge of different degrees of organization should be distributed evenly throughout the domain.

Intra–personal environment If a mode of awareness is characterized by too many aspects of limited degree of organization, more integrated and comprehensive forms of awareness cannot emerge. But if such a mode of awareness is characterized by too limited a number of highly organized aspects, then this manner of organization will prove detrimental to the integrity of that mode of awareness.

Broader Independent domains (#MP1001).
Narrower Non–linear organization (#MP1007) Intermediate scale organization (#MP1006)
Regenerative resource cultivation areas (#MP1004)
Interpretation of complementary modes of organization (#MP1003).

♦ **MP1003 Interpretation of complementary modes of organization**

Pattern A continuous pattern of organization and definition denies the existence and emergence of the underdefined and severely diminishes the value of major established patterns of organization. But the degree of integration of such major patterns is also valuable and potent. A compromise can be achieved by ensuring appropriate interpenetration of defined and underdefined modes of organization as complements.

Physical environment Massive urban environments render life mechanized and harsh and detract from enjoyment and health from enjoyment and health. But the size of cities is potent. Interlocking fingers of farmland and urban land are an appropriate compromise, even at the centre of the metropolis.

Socio–organizational environment The proliferation of total patterns of organization and systems of procedures denies the existence and emergence of unorganized activity and severely diminishes the value of major institutions. But the degree of integration of such institutions is also valuable and potent. A compromise can be achieved by ensuring the interpenetration of formal and informal modes of organization as necessary complements.

Conceptual environment The proliferation of totally organized bodies of knowledge and conceptual methods ignores the existence and emergence of underdefined concepts and forms of integration with the consequence that conceptual development is severely inhibited. But the degree of integration of such established bodies of knowledge is also valuable and potent. A compromise can be achieved by ensuring the interpenetration of formal and informal modes of conceptual organization.

Intra–personal environment The proliferation of highly structured modes of awareness ignores the existence and emergence of underdefined modes of awareness and forms of integration, with the consequence that personal development is severely inhibited. But the degree of integration of such established modes of awareness is also valuable and potent. A compromise can be achieved by ensuring the interpenetration of structured and unstructured modes of awareness.

MG M a Intepretation of complementary modes of organization

Broader Distribution of organization (#MP1002).
Narrower Encirclement (#MP1017) Access to intensity (#MP1010)
Non–linear organization (#MP1007) Four–level structural limit (#MP1021)
Network of inter–relationships (#MP1005) Variety of forms and processes (#MP1008)
Local interrelationship domains (#MP1011) Web of general interrelationships (#MP1016)
Regenerative resource cultivation areas (#MP1004).

♦ **MP1004 Regenerative resource cultivation areas**

Pattern Those areas in which resources can best be regenerated are also those most favourable for the construction of frameworks. The availability of such areas is however limited and once denatured by the construction of frameworks, its resource regeneration function cannot easily be recovered. Such areas should therefore be protected.

Physical environment The land which is best for agriculture happens to be best for building too. It is however limited and once the fertility is destroyed or locked up it cannot be regained for centuries. Agricultural valleys should therefore be protected as farmland or as nature reserves, if they are not cultivated.

Socio–organizational environment The functional areas most favourable to regeneration of social resources are also those in which organizations can most easily be established. The availability of such areas is however limited and once denatured by the establishment of organizations, their resource regeneration function cannot easily be recovered. Such functional areas should therefore be protected from attempts at organization.

Conceptual environment The domains most favourable to regeneration of creative resources are also those in which concepts can most easily be ordered into conceptual frameworks. The availability of such areas is however limited and once denatured by the establishment of such frameworks their function in regenerating creative resources cannot easily be recovered. Such creative domains should therefore be protected from attempts at conceptual organization.

Intra–personal environment The modes of awareness most favourable to regeneration of psychic resources are also those which lend themselves most easily to being structured. The availability of such modes is however limited and once denatured by being structured in this way, their regenerative function cannot easily be recovered. Such modes of awareness should therefore be protected from efforts to structure them.

MG M b Regenerative resource cultivation areas

Broader Independent domains (#MP1001) Distribution of organization (#MP1002)
Interpretation of complementary modes of organization (#MP1003).
Narrower Non–linear organization (#MP1007).

♦ **MP1005c Network of inter–relationships**

Pattern There is advantage in relating to centrally organized frameworks as well as to those which are minimally organized. In order to reconcile these contradictory requirements, a network of inter–relationships is necessary which both links the limited number of central positions to the many non–central, minimally– organized areas, and provides a barrier to encroachment on such areas.

Physical environment Many people want to live in the country whilst also living close to a large city. It is however geometrically impossible to have thousands of small farms within a few minutes of a major city. Both are possible by arranging a loose network of country roads around large open squares of countryside or farmland, with houses closely packed along the road, but only one house deep. In these terms, the suburb is an obsolete and contradictory form of human settlement.

Socio–organizational environment There is advantage in participation in major groups and local communities. In order to reconcile these mutually incompatible forms of activity, a loose network of relationships is necessary which links both the limited number of central organizations to the many non–central, minimally organized groups and ensures that the latter provide protection against external encroachment.

Conceptual environment There is advantage in major conceptual frameworks as well as in minimally organized conceptual processes. In order to reconcile these mutually incompatible forms of conceptual organization a loose network of relationships is necessary which links both the limited number of centrally organized frameworks to the minimally organized conceptual processes and ensures that the latter provide protection against progressive formalization.

Intra–personal environment There is advantage in highly ordered modes of awareness as well as in minimally ordered modes of awareness. In order to reconcile these mutually incompatible modes a loose network of associative relationships is necessary which links both the limited number of ordered modes to the minimally ordered modes and ensures that the latter provide protection against loss of the former.

Broader Interpretation of complementary modes of organization (#MP1003).
Narrower Identifiable context (#MP1014) Cluster of frameworks (#MP1037)
Non–linear organization (#MP1007) Four–level structural limit (#MP1021).

♦ **MP1006b Intermediate scale organization**

Pattern More comprehensive forms of organization function as powerful attractors. It is difficult to ensure the viability of intermediate forms of organization in relationship to them. Efforts should be maintained to ensure that such intermediate forms of organization function as attractors in their own right and are not merely dependencies of the more powerful attractors.

Physical environment The big city is a magnet. Small country towns find it difficult to stay alive and healthy in the face of central urban growth. Where they exist, country towns should be preserved. The growth of new self–contained towns should also be encouraged. The region should be collectively concerned to ensure their viability and avoid their development into dormitory towns.

Socio–organizational environment The activities of a large and growing organizational complex attract further participation. It is consequently difficult for smaller organizations to offer processes of equivalent attractiveness to ensure their viability. Where such intermediate bodies exist they should be preserved. The growth of new self–reliant forms of organization should be encouraged. The functional domain should be collectively concerned to ensure their viability and avoid their development into dependent or "front" organizations.

Conceptual environment Major conceptual and ideological frameworks powerfully attract adherents. It is consequently difficult for smaller alternative frameworks to offer knowledge of equivalent interest to ensure their viability. Where such alternatives exist they should be preserved. The growth of new alternatives should be encouraged. There should be collective concern within the conceptual domain to ensure their viability and avoid their development into conceptual outposts of any major framework.

Intra–personal environment Major modes of awareness powerfully attract further involvement. It is consequently difficult for less comprehensive alternative modes of awareness to offer insights of a power sufficient to ensure their viability. Where such alternatives exist they should be preserved. The growth of new alternatives should be encouraged. There should be a general concern to ensure their viability and independence and avoid their development into aspects of any major modes of awareness.

Broader Distribution of organization (#MP1002).
Narrower Functional cycle (#MP1026) Non–linear organization (#MP1007)
Individuality in multiplicity (#MP1012).

♦ **MP1007 Non–linear organization**

Pattern Within each domain, in between the linearly organized areas, there are large areas of non–linear organization whose status and function are crucial to the balance of the domain as a whole.

Physical environment Each region has vast areas of countryside (farmland, parkland, forests, deserts, lakes) which separate towns. The legal and ecological character of this open land is a crucial aspect in the balance of the entire region. parks created in the towns are artificial; farms treated as private property deprive people of their natural rights to enjoy the countryside from whence they came. Every area of open space either is farmed if it is arable or is wardened if it is wild. But the countryside should be open at large to all people, provided they respect the ecological processes going on.

Socio–organizational environment Within each functional domain there are extensive areas of informal organization providing a context for the many formal organizations. The character and status of such informality is crucial to the organizational balance of the domain. Such informality in its many forms should be open to all, provided they respect the special character of social processes associated with it.

Conceptual environment Within each conceptual domain there are extensive areas of non–linear organization of knowledge providing a context for those areas which have been linearly organized. Such non–linear organization is crucial to the balance of knowledge in the domain and as such should be accessible to all, with due respect for the special quality of the conceptual

processes involved.
Intra–personal environment Within each mode of awareness there are many perceptual processes which are non–linearly organized. These non–rational processes provide a context for those which are rationally organized and are crucial to the balance of the mode of awareness.
MG M b Non–linear organization
   Broader Distribution of organization (#MP1002)   Network of inter-relationships (#MP1005)
   Intermediate scale organization (#MP1006)
   Regenerative resource cultivation areas (#MP1004)
   Interpretation of complementary modes of organization (#MP1003)
   Narrower Sub-domain boundary (#MP1013)   Cluster of frameworks (#MP1037)
   Relationship to indeterminacy (#MP1025)
   Hierachy of perspectives favouring the broadest (#MP1114)
   Linear relationships enhanced by non-linear processes (#MP1051).

♦ MP1008 Variety of forms and processes
Pattern Organizations characterized by homogeneity and lack of differentiation inhibit the emergence and growth of variety. By distinguishing an extensive variegated pattern of appropriately juxtaposed sub–areas, the emergence of different forms of organization may be encouraged and protected within each of them.
Physical environment Modern cities are heterogeneous and marked by undifferentiated character which erodes variety of lifestyles and impedes the growth of individuality. By breaking the city, as far as possible, into a vast mosaic of sub–cultures, each with its own spatial territory, the emergence of distinct styles may be encouraged and protected provided that each is stimulated and consolidated by an appropriate degree of contact with the others.
Socio–organizational environment The homogeneous and undifferentiated character of modern organizational complexes and institutional environments inhibits the emergence of alternative life styles and arrests the growth of individual character. By breaking such complexes into a vast mosaic of sub–cultures, each with its own functional territory, the emergence of distinct life styles may be encouraged and protected, provided that each is stimulated and consolidated by an appropriate degree of contact with the others.
Conceptual environment The homogeneous and undifferentiated character of major schools of thought or ideological frameworks inhibits the emergence of alternative conceptual styles deviating from the norm, thus arresting conceptual development. By fragmenting such frameworks into a vast mosaic of sub–cultures, each with its own specialized domain and sharply delineated values, the emergence of distinct intellectual styles may be encouraged and protected, provided that each is stimulated and consolidated by an appropriate degree of contact with the others.
Intra–personal environment The homogeneous and undifferentiated character of a person's principal modes of awareness inhibits the emergence of alternative modes and arrests personal development. By distinguishing a rich mosaic of distinct variants, each with its particular function, the emergence of alternative modes of awareness may be encouraged and protected, provided that each is stimulated and consolidated by an appropriate degree of interaction with the others.
MG M a Variety of forms and processes LK F0 Variety of forms and processes
   Broader Interpretation of complementary modes of organization (#MP1003).
   Narrower Access to intensity (#MP1010)   Sub-domain boundary (#MP1013)
   Identifiable context (#MP1014)   Cluster of frameworks (#MP1037)
   Web of selective interchange (#MP1019)   Individuality in multiplicity (#MP1012)
   Decentralized formal processes (#MP1009)   Local interrelationship domains (#MP1011)
   Cycle of relationship reinforcement (#MP1031)
   Stable density gradient of local relationships (#MP1029).

♦ MP1009a Decentralized formal processes
Pattern The segregation of formal and informal processes and the concentration of each of them within distinct and unrelated areas is artificial and leads to their unhealthy development. Formal processes should be distributed in such a way as to permit interaction with informal processes.
Physical environment The artificial separation of residential and work environments creates rifts in people's inner lives and prevents the emergence of highly differentiated sub– cultures. The concentration and segregation of work leads to dead neighbourhoods. Zoning and other regulations should be used to scatter workplaces throughout the city and to prohibit large concentrations of family life without associated workplaces.
Socio–organizational environment The segregation of formal and informal processes in organizations and groups and the confinement of each to separate and well–defined settings leads to unhealthy social development of a schizophrenic nature. Formal group processes should be distributed throughout the organizational environment so as to facilitate alternations with informal processes.
Conceptual environment The segregation of formal and informal conceptual procedures and methods and the restriction on the use of each to separate and well–defined circumstances leads to unhealthy conceptual development. Formal procedures should be adopted throughout the conceptual domain such as to facilitate alternation with informal, creative processes.
Intra–personal environment The segregation of disciplined and unstructured modes of awareness and the restriction of each to separate and well–defined contexts leads to unhealthy personal development of a schizophrenic nature. Structured modes of awareness should be interrelated with unstructured modes such as to facilitate alternation between them.
   Broader Variety of forms and processes (#MP1008).
   Narrower Interchange (#MP1034)   Complementarity (#MP1027)
   Context boundary (#MP1015)   Sub-domain boundary (#MP1013)
   Accessible non-linearity (#MP1060)   Web of selective interchange (#MP1019)
   Informal context for formal processes (#MP1041)
   Chain of fundamental transformation zones (#MP1042)
   Local opportunities for perspective activity (#MP1157)
   Integrated contexts for perspective dynamics (#MP1080).

♦ MP1010c Access to intensity
Pattern Complexification of structures and processes at centralized locations constitute a powerful attractant but such development progressively limits access to the intensity of relationships in that environment. Access can be increased by limiting the degree of complexification and multiplying the number and variety of such points of focus.
Physical environment City life is experienced as magical because of its intensity. But urban sprawl deprives most people of genuine participation in it because they must live far out from the core. This problem can only be solved by decentralizing the core to form a multitude of smaller cores, each intense and devoted to some special way of life, corresponding to the needs of the region as a whole.
Socio–organizational environment Large organizational complexes are experienced as exciting environments because of the variety and intensity of activity within them. But institutional proliferation deprives most people of genuine participation at the foci where action oriented decisions are taken. This problem can only be solved by decentralization around a multitude of smaller decision–making loci, each exemplifying some special way of life corresponding to the needs of the functional domain as a whole.
Conceptual environment Major schools of thought, conceptual frameworks or ideologies are a powerful attractant because of the intensity and complexity of intellectual activity associated with them. But the proliferation of derivative conceptual developments associated with them progressively deprives most people of genuine participation at those points of excellence where creative development is actually taking place. This problem can only be solved by decentralization around a multitude of smaller innovative areas, each exemplifying some special method or approach corresponding to the needs of the conceptual domain as a whole.
Intra–personal environment The principal modes of awareness exert a powerful influence because of the fascinating richness and intensity of experience they provide. But the proliferation of derivative modes of awareness associated with them progressively inhibits direct creative experience of the fundamental insight governing the further transformation of such a mode. This problem can only be solved by multiplying the number of more limited modes of awareness, each exemplifying some more accessible special insight corresponding to the needs of personal development as a whole.
   Broader Variety of forms and processes (#MP1008)
   Interpretation of complementary modes of organization (#MP1003).
   Narrower Context for disorder (#MP1058)   Underdefined processes (#MP1033)
   Selective interchange axis (#MP1032)   Four-level structural limit (#MP1021)
   Local interrelationship domains (#MP1011)   Web of general interrelationships (#MP1016)
   Cycle of relationship reinforcement (#MP1031)
   Coherent pattern of relationship densities (#MP1028)
   Hospitable contexts for perspectives in transition (#MP1091)
   Cyclic interrelation of complementary perspective in common domains (#MP1063).

♦ MP1011a Local interrelationship domains
Pattern Means of non–local interrelationship, when employed locally, inhibit the articulation of local domains and destroy their integrity. The domain as a whole should be broken down into local interrelationship domains for which the non–local interrelationships help to define a boundary. Appropriate connections should be established between the local and non–local interrelationships.
Physical environment Automobiles provide unlimited opportunities for freedom and access to variety, but at the expense of both the environment and social life. The urban area should be broken down into local transport areas, each surrounded by a ring road. Within such areas, paths and minor local roads (inconvenient for cars) should be built for internal movement, with major roads providing access to the ring roads.
Socio–organizational environment Non–local communication media, when employed locally, inhibit the articulation and social development of community organization and destroy its integrity. The functional domain as a whole should be broken down into local communication areas, each with direct access to non–local communication media. Within such areas, group and interpersonal communication networks should be developed for internal purposes, appropriately connected to (and protected from) the non–local communication media.
Conceptual environment General conceptual relationships, when employed in specialized areas, inhibit the articulation of specialized conceptual domains and destroy their integrity. The domain as a whole should be broken down into domains within which particular local relationships prevail, but appropriately related to a pattern of general relationships.
Intra–personal environment General patterns of insight, when used to order immediate experience, inhibit development of appreciation of the uniqueness of the present and destroy the integrity of such immediacy. The pattern of insight as a whole should be broken down into domains in which the sense of immediacy prevails, with each appropriately related to the contextual pattern.
   Broader Access to intensity (#MP1010)   Variety of forms and processes (#MP1008)
   Interpretation of complementary modes of organization (#MP1003).
   Narrower Interchange (#MP1034)   Encirclement (#MP1017)
   Local relationship loops (#MP1049)   Web of selective interchange (#MP1019)
   Special modes of relationship (#MP1056)   Occupiable temporary site limit (#MP1022)
   Web of general interrelationships (#MP1016)
   Compensating relationships in parallel (#MP1023)
   User-determined specialized communications (#MP1020)
   Linear relationships enhanced by non-linear processes (#MP1051)
   Concealment of necessary monotonous perspective patterns (#MP1097)
   Interfacing vehicles of communication and networks of unmediated relationships (#MP1052).

♦ MP1012 Individuality in multiplicity
Pattern Individual elements, as such, can have no distinguishably unique function in forming a pattern composed of 5,000 to 10,000 such elements. Patterns should therefore be decentralized so as to group not more than approximately 7,000 such elements, thus enabling each element to play a distinct part in the pattern.
Physical environment Territorial communities expression than 5000–10,000 persons inhibit individual voices. City governments should be decentralized in such a manner as to give local control to communities of approximately 7,000 persons; these communities being delineated by geographic and historic boundaries. Each community should be individually responsible for initiating, deciding, and executing those affairs by which it is affected.
Socio–organizational environment Individuals have no effective voice in any non–territorial community of more than 5,000 to 10,000 persons. Decentralize major institutions in a way that gives local control to functional divisions interrelating approximately 7,000 persons. Use cultural and traditional distinctions to reinforce such organizational boundaries whenever possible. Give each organization the power to initiate, decide and execute the affairs that concern it closely.
Conceptual environment Individual concepts can have no distinguishably unique function in constituting a conceptual framework of more than 5,000 to 10,000 such concepts. Conceptual frameworks should therefore be partitioned so as to interrelate not more than 7,000 such concepts, thus enabling each concept to fulfil a distinct role in the pattern.
Intra–personal environment Individual moments of awareness can have no distinguishably unique function in forming a mode of awareness of more than 5,000 to 10,000 such facets. Modes of awareness should therefore be organized so as to interrelate not more than 7,000 such moments, thus enabling each moment to contribute uniquely to the mode as a whole.
MG M b Individuality in multiplicity
   Broader Variety of forms and processes (#MP1008)
   Intermediate scale organization (#MP1006).
   Narrower Activity nodes (#MP1030)   Complementarity (#MP1027)
   Functional cycle (#MP1026)   Context boundary (#MP1015)
   Local focal points (#MP1044)   Sub-domain boundary (#MP1013)
   Identifiable context (#MP1014)   Underdefined processes (#MP1033)
   Points of wider perspective (#MP1062)   Occupiable temporary site limit (#MP1022)
   Cycle of relationship reinforcement (#MP1031)
   Context for emergence of new perspectives (#MP1065)
   Coherent pattern of relationship densities (#MP1028)
   Stable density gradient of local relationships (#MP1029).

♦ MP1013b Sub–domain boundary
Pattern In order to maintain a variety of forms and processes, boundaries are required to insulate from one another the different (at least partially) sub–domains in which each prevails.
Physical environment A mosaic of subcultures requires that hundreds of different cultures live, in their own way, at full intensity, on neighbouring territories. But subcultures have their own

ecology. They can only live at full intensity, unhampered by their territorial neighbours, if they are physically separated by physical boundaries, whether natural or man-made.
Socio-organizational environment A mosaic of subcultures and alternative life styles requires that hundreds of different groups live, in their own way, at full intensity on neighbouring functional territories. But subcultures have their own ecology. They can only live at full intensity, unhampered by their functional neighbours, if they are functionally separated by functional boundaries, whether traditional or designed.
Conceptual environment A mosaic of subcultures and conceptual frameworks requires that hundreds of different schools of thought function, in their own way, at full intensity on neighbouring conceptual domains. But subcultures have their own ecology. They can only live at full intensity, unhampered by their intellectual neighbours, if they are conceptually separated by definitions, whether traditional or designed.
Intra-personal environment A mosaic of distinct modes of awareness requires the co-existence of hundreds of different modes. The full intensity of each mode only emerges, unhampered by its neighbours, if the modes are appropriately separated.
Broader Non-linear organization (#MP1007)
Individuality in multiplicity (#MP1012)
Decentralized formal processes (#MP1009)
Variety of forms and processes (#MP1008).
Narrower Encirclement (#MP1017)
Activity nodes (#MP1030)
Context boundary (#MP1015)
Adaptive interstices (#MP1048)
Identifiable context (#MP1014)
Accessible non-linearity (#MP1060)
Web of selective interchange (#MP1019)
Relationship to indeterminacy (#MP1025)
Positions enabling transcendence (#MP1024)
Access to contained irrationality (#MP1071)
Low-intensity communication pathways (#MP1059)
Informal context for formal processes (#MP1041)
Compensating relationships in parallel (#MP1023)
Chain of fundamental transformation zones (#MP1042)
Access to patterns of active irrationality (#MP1064)
Coherent pattern of relationship densities (#MP1028)
Stable density gradient of local relationships (#MP1029)
Transitional contexts for perspective reorganization (#MP1084)
Concealment of necessary monotonous perspective patterns (#MP1097).

♦ MP1014a Identifiable context

Pattern To establish that a part belongs to a larger whole, the whole needs to be identified with a particular space distinct from other spaces.
Physical environment People need an identifiable spatial unit to belong to (up to 300 yards across; 400-500 inhabitants) distinct from other parts of the urban environment. Today's pattern of development destroys such neighbourhoods. In existing cities, encourage local groups to define physically the neighbourhoods they live in. Give them some degree of autonomy and keep major roads out.
Socio-organizational environment People need an identifiable functional unit to belong to (a sector of social space), distinct from other parts of the social environment. Within existing organizational complexes, encourage groups to define the special activity domains with which they are identified. Give them some degree of autonomy and protect them from the high intensity communication pathways of the external environment.
Conceptual environment People need an identifiable belief system or school of thought to belong to (a sector of conceptual space), distinct from other parts of the conceptual environment. Within existing conceptual frameworks, encourage groups to define the areas of special interest with which they are identified. Give them some degree of autonomy and protect them from being overridden by patterns of general relationships.
Intra-personal environment An individual needs a central focus to his or her personality (a sector of his or her psychic space) as a reference point for a sense of identity or on which to ground any response to the world. Experiential discovery of such a centre should be encouraged. The sense of relative detachment associated with such a centre should be cultivated and protected from the distractions of external patterns of insight.
Broader Sub-domain boundary (#MP1013)
Individuality in multiplicity (#MP1012)
Variety of forms and processes (#MP1008)
Network of inter-relationships (#MP1005).
Narrower Activity nodes (#MP1030)
Complementarity (#MP1027)
Functional cycle (#MP1026)
Context boundary (#MP1015)
Adaptive interstices (#MP1048)
Cluster of frameworks (#MP1037)
Local relationship loops (#MP1049)
Accessible non-linearity (#MP1060)
Principal points of entry (#MP1053)
Occupiable temporary site limit (#MP1022)
Adequate variety of cyclic elements (#MP1035)
Integrating the historical dimension (#MP1040)
Informal context for formal processes (#MP1041)
Compensating relationships in parallel (#MP1023)
Differentiation by relationship density (#MP1036)
Local sources for perspective nourishment (#MP1089)
Context for emergence of new perspectives (#MP1065)
Informal local perspective interface zones (#MP1088)
Bounded common small-scale interaction domains (#MP1061)
Context for acknowledgement of past perspectives (#MP1070).

♦ MP1015b Context boundary

Pattern The strength of the boundary is essential to the maintenance of an identifiable context. An appropriate boundary emerges as the number of relationships to the external environment is limited.
Physical environment The strength of the boundary is essential to the maintenance of the identifiable character of the neighbourhood. The boundary may be formed by closing down some streets crossing the neighbourhood thus limiting access to the neighbourhood.
Socio-organizational environment The strength of the boundary is essential to the maintenance of the identifiable character of a group or functional unit. The boundary may be formed by cutting off some communication pathway through the group and restricting access to the group's activity.
Conceptual environment The strength of the boundary is essential to the maintenance of the identifiable character of a conceptual system. The definition may be articulated by establishing distinctions from the pattern of general relationships within which the system is embedded.
Intra-personal environment The strength of the distinction between modes of awareness is essential to the maintenance of the identifiable character of any focal, ground or "home" mode. The distinction may be reinforced by developing the sense of primacy or groundedness associated with such a mode in contrast to others.
Broader Sub-domain boundary (#MP1013)
Identifiable context (#MP1014)
Individuality in multiplicity (#MP1012)
Decentralized formal processes (#MP1009).
Narrower Activity nodes (#MP1030)
Adaptive interstices (#MP1048)
Accessible non-linearity (#MP1060)
Principal points of entry (#MP1053)
Selective interchange axis (#MP1032)
Low-intensity communication pathways (#MP1059)
Informal context for formal processes (#MP1041)
Common external context for inactivity (#MP1069)
Compensating relationships in parallel (#MP1023)
Access to patterns of active irrationality (#MP1064)
Unstructured context for perspective exchange (#MP1090)
Context for acknowledgement of past perspectives (#MP1070)
Limitation on number of occupiable temporary sites (#MP1103)
Concealment of necessary monotonous perspective patterns (#MP1097)
Competitive interaction opportunities transposed to a concrete level (#MP1072)
Contexts for exploratory relationship formation challenging emerging perspectives (#MP1073).

♦ MP1016b Web of general interrelationships

Pattern A general system of interrelationships can only work if all the parts are well-connected. This tends not to be the case because of the privileged role accorded to the major relationships and the difficulty of linking relationships of different kinds. This difficulty can be reduced by treating the interrelationship nodes as primary and the interrelationship links as secondary.
Physical environment The web of different modes of public transportation can only work if all the parts are well-connected. This tends not to be the case because of the emphasis placed on the high volume modes and the difficulty of coordinating the interchange between different modes, especially to local areas. This difficulty can be reduced by treating the interchange points as primary and the actual transportation lines as secondary. Local communities controlling the interchanges can then require guarantees of a desirable level of local service.
Socio-organizational environment The web of public communication facilities can only work if all social actors are connected by it. This tends not to be the case because of the privileged role accorded to the mass media and the difficulty of translating information between different modes of communication, especially when adapted to particular profiles. This difficulty can be reduced by treating the translation arenas as primary and the actual lines of communication as secondary. Specialized groups controlling the translation arenas can then require guarantees of a desirable level of adaptation to particular profiles.
Conceptual environment The web of general conceptual relationships only acquires its full significance if all concepts are integrated within it. This tends not to be the case because of the fundamental role attributed to intellectual or ideological lines of thought, and the difficulty of integrating concepts based on very different or highly specialized modes of thought. This difficulty can be reduced by treating such interrelationship nexi as primary and the relationships themselves as secondary. Concepts governing such nexi can then require a desirable level of integration of specialized modes of thought.
Intra-personal environment An ordered network associating different modes of awareness only acquires its full significance if all modes are integrated within it. This tends not to be the case because of the primacy accorded to certain preferred modes of awareness and the difficulty of integrating modes based on very different insights or extraordinary experiences. This difficulty can be reduced by recognition of integrative correspondences as primary and the insights themselves as secondary. Modes of awareness integrating such correspondences can then require a desirable level of integration of any specialized modes of awareness.
Broader Access to intensity (#MP1010)
Local interrelationship domains (#MP1011)
Interpretation of complementary modes of organization (#MP1003).
Narrower Interchange (#MP1034)
Encirclement (#MP1017)
User-determined specialized communications (#MP1020).

♦ MP1017c Encirclement

Pattern Fundamental boundary relationships defining an area's relationship to the pattern within which it is embedded are desirable in order to maintain the integrity of that area.
Physical environment High speed ring roads are a necessity which helps to define and generate local transport areas. Care must be taken, however, to insure that the building and placement of the ring roads do not cause extensive damage to communities or the countryside.
Socio-organizational environment High intensity communication pathways are needed whereby an organizational complex deflects such communication around its boundary, thus protecting itself from the disruption that irrelevant communication may cause to the collective life of those involved.
Conceptual environment A clearly defined pattern of concepts is required around a conceptual domain in order to channel away concepts irrelevant to that domain and disruptive of its preoccupations.
Intra-personal environment It is necessary for a person to establish a psychic boundary to unconsciously redirect incoming perceptions which are of no immediate relevance and which would therefore unduly disrupt personal equilibrium.
Broader Sub-domain boundary (#MP1013)
Local interrelationship domains (#MP1011)
Web of general interrelationships (#MP1016)
Interpretation of complementary modes of organization (#MP1003).
Narrower Interchange (#MP1034)
Relationship to indeterminacy (#MP1025)
Compensating relationships in parallel (#MP1023)
Chain of fundamental transformation zones (#MP1042)
Concealment of necessary monotonous perspective patterns (#MP1097).

♦ MP1018b Network of redefinitions

Pattern In a context in which a particular form of order is self-reinforcing at all points, the elements constituting that order do not contribute to the redefinition of it or to the emergence of any new pattern. Such pattern emergences occur when the elements are able progressively to redefine their relationship to each other in the light of their past patterning experience.
Physical environment In a physical environment governed by a fixed plan, the constitutive elements (buildings, traffic, focal points) are perceived as necessarily unchanging in nature. As such they cannot be permitted to contribute to any gradual redefinition of the environment. As such it is difficult to accept that they can contribute to any gradual positive redefinition of the environment through cyclic growth, decay, or adaptation processes. Emergence of new or alternative patterns can only occur when building structures, traffic, etc, adapt to and modify each other's cycles of activity.
Socio-organizational environment In a society or group which emphasizes the inculcation of some particular pattern of behaviour and perception, individuals become passive and unable to think or act for themselves. Creative, active individuals and groups can only emerge in a society which promotes networks of decentralized learning instead of emphasizing structured teaching systems.
Conceptual environment In a conceptual framework which emphasizes the perpetuation of a particular conceptual order or method, the emergence of new viewpoints, challenging and redefining that order, is discouraged. It is through such processes and the emergence of alternative viewpoints that a conceptual pattern is renewed and its development ensured.
Intra-personal environment Adoption of a particular mode of thought prevents the emergence of insights. This inhibits personal development through effectively imposing an experiential straitjacket. Greater personal development is encouraged by the use of a network of alternative modes.
Narrower Protection of emerging foci (#MP1057)
Presentation of new dimensions (#MP1043)
Integrating the historical dimension (#MP1040)
Contexts for care of premature perspectives (#MP1086)
Local opportunities for perspective activity (#MP1157)
Transitional contexts for perspective reorganization (#MP1084)
Perspective imitation contexts for developing perspectives (#MP1085)
Integration of perspective acquisition and perspective maintenance dynamics (#MP1083).

♦ **MP1019 Web of selective interchange**
Pattern Interchange points for similar resources should be distributed evenly throughout the environment for which they are each a process nexus. As such they interact in a self-organizing manner, maintaining a stable nexus web despite changes to it resulting from fluctuating demands upon its parts.
Physical environment Shops and other services rarely locate themselves in those positions which both best serve the needs of those they serve and guarantee their own survival. Similar services can best be evenly distributed throughout the area they serve, filling gaps in the web, whilst different services should be located close to the largest cluster of other services.
Socio-organizational environment Discussion groups, conferences or fairs concerned with similar questions should best be distributed evenly throughout the space-time environment for which they are a process nexus. As such they interact in a self- organizing manner maintaining a stable nexus web despite changes to it resulting from fluctuating demands upon the groupings. Those concerned with different questions can best be located in relationship to the larger groupings in the web.
Conceptual environment Focal concepts through which similar phenomena are ordered should be evenly distributed throughout the conceptual space within which they are each a nexus of deliberative interchange. As such they interact in a self- organizing manner maintaining a stable web of concepts despite changes to it resulting from fluctuating use of its constituent viewpoints.
Intra-personal environment Focal modes of awareness through which perceptions of similar circumstances are processed should be evenly distributed throughout a person's perceptual space. As such they interact in a self-organizing manner maintaining a stable web of modes of awareness despite changes to it resulting from fluctuating use of its constituent modes.
MG M b Web of selective interchange
Broader Sub-domain boundary (#MP1013) Decentralized formal processes (#MP1009)
Variety of forms and processes (#MP1008) Local interrelationship domains (#MP1011).
Narrower Selective interchange axis (#MP1032) Diversified interchange environment (#MP1046)
Local sources for perspective nourishment (#MP1089).

♦ **MP1020b User-determined specialized communications**
Pattern Relationships should be such as to permit contact in an environment between any point and any other point.
Physical environment Public transportation should be such as to take people from any point to any other point within a metropolitan area. This can be accomplished by supplementing high-volume transportation with a mini-bus feeder service over a varying route optimized by computer in response to requirements of individual users.
Socio-organizational environment Communications within a society or an organizational complex should be such as to permit any person to make contact with any other person or group within that community. Computer assistance should be available to enable a person to discover the communications address of the contact appropriate to his needs.
Conceptual environment Relationships within a knowledge- representative system should be such as to permit the relevance of any concept to any other concept to be established, particularly when such a system is computer-enhanced.
Intra-personal environment Associative relationships within any mode of awareness should be such as to enable any perception to enhance any other perception, possibly with the aid of metaphoric transpositions of perspective.
Broader Local interrelationship domains (#MP1011)
Web of general interrelationships (#MP1016).
Narrower Interchange (#MP1034) Hospitable transit points (#MP1092)
Compensating relationships in parallel (#MP1023).

♦ **MP1021a Four-level structural limit**
Pattern Within any framework the number of structural levels varies, with the highest number of levels tending to occur towards the centre. It is desirable that the majority of structures should not have more than four levels because of the confusion that more complex structures tend to engender.
Physical environment Within any urban area the heights of buildings will vary, with the highest tending to be towards the centre. It is desirable that the majority of buildings, especially those destined for human habitation, should not be more than four stories in height. Higher buildings are destructive of the urban environment, offer few genuine advantages, and are psychologically harmful.
Socio-organizational environment Within any society or organization the number of hierarchical levels (or classes) will vary, with the highest number tending to be towards the centre of any such organizational complex. It is desirable that the majority of organizations, especially those associated with the daily life of a community, should not have more than four such levels. Organizations with more hierarchical levels are destructive of social life, offer few genuine advantages, and are psychologically harmful.
Conceptual environment Within any conceptual framework the number of conceptual or category levels will vary, with the highest tending to be towards the centre of the framework. It is desirable that the majority of conceptual structures, especially those destined for frequent use by non-specialists, should be based on not more than four levels. More complex structures tend to confuse the conceptual environment, offer few genuine advantages, and are psychologically disturbing.
Intra-personal environment Within any mode of awareness the number of levels of self-reflectiveness, subtlety or degree of apprehension will vary, with the highest number tending to be associated with the core of the person's being. It is desirable that the majority of perceptual modes, especially those required in normal daily life, should not involve more than four such levels. Modes of awareness based on a greater number of levels tend to confuse, offer few genuine advantages, and are psychologically harmful.
Broader Access to intensity (#MP1010) Network of inter-relationships (#MP1005)
Interpretation of complementary modes of organization (#MP1003).
Narrower Points of wider perspective (#MP1062)
Integrating a new dimension (#MP1039)
Complexification of perspective contexts (#MP1095)
Limitation on number of structural levels (#MP1096)
Coherent pattern of relationship densities (#MP1028)
Stable density gradient of local relationships (#MP1029)
Minimal distance between related operational control contexts (#MP1082).

♦ **MP1022a Occupiable temporary site limit**
Pattern The integrity of a domain depends very much on the limit set on the number of sites available for temporary occupation. These function as attractants for mobile elements, both from within and from outside the domain.
Physical environment The integrity of local communities and neighbourhoods depends very much on the limit set on the number of parking spaces provided. When this exceeds about nine percent the environment becomes unfit for human use because of the number of vehicles attracted to it.
Socio-organizational environment The integrity of groups and organizations depends very much on the limit set on the number of uncommitted participants. The opportunity of such temporary membership attracts excessive and irresponsible involvement which destroys organizational coherence and continuity.
Conceptual environment The integrity of a conceptual domain depends very much on the limit set on the number of unresolved issues within it on which alternative or external explanations are sought. The opportunity to offer such explanations may attract superficial and irresponsible involvement to an excessive degree. This inhibits the development of the domain.
Intra-personal environment The integrity of a mode of awareness depends very much on the limit set on the number of unintegrated perceptions on which alternative insights are sought. Openness to an excessive number of such insights inhibits the development of that mode of awareness.
Broader Identifiable context (#MP1014) Individuality in multiplicity (#MP1012)
Local interrelationship domains (#MP1011).
Narrower Limitation on number of occupiable temporary sites (#MP1103)
Distinct pattern of entry points to complex structures (#MP1102)
Concealment of necessary monotonous perspective patterns (#MP1097).

♦ **MP1023c Compensating relationships in parallel**
Pattern A local domain and its boundaries may be protected from fragmentation by preventing fundamental relationships from traversing it. This however results in the disorderly proliferation of a network of local relationships which itself threatens the integrity of the domain. Such local relationships may however be more effectively integrated by providing for reciprocal relationships, traversing the domain in parallel and linking them to the more fundamental relationships by which the boundary of the domain is defined.
Physical environment Even when ring roads are used to divert through traffic around a local transport area such as a city (or a portion of it), severe congestion results in any network of intersecting streets. This may be avoided by building systems of parallel and alternating one-way roads to carry traffic to and from the ring roads. Cross-streets may be closed to protect neighbourhoods which are defined between the parallel roads if they are an appropriate distance apart (100 to 350 metres).
Socio-organizational environment Even when high intensity, non- local communication media are used to deflect such communication around an organization's boundary, severe local communication overload and underuse may result within any communication networks within the organization or community. This may be avoided by developing systems of parallel and alternating unilateral communication pathways to carry information to and from the bounding non-local communication media. Local and specialized groups can effectively define themselves in relationship to such alternating communication pathways.
Conceptual environment Even when a clearly defined contextual pattern of general concepts is used to filter out irrelevant concepts (which may destroy the integrity of a conceptual domain) confusion may result locally from the proliferation of unordered networks of conceptual relationships thus inhibiting the further development of that domain. This may be avoided by developing alternative systems of assymetric or complementary relationships in parallel as a means of linking locally elaborated concepts into the wider network of general conceptual relationships.
Intra-personal environment Even when a person establishes an adequate psychic boundary to unconsciously redirect non-significant incoming perceptions, undesirable confusion may result from the fragmentation of the person's awareness into a network of ad hoc perceptual modes. This may be avoided by developing parallel systems of complementary modes of perceptions as a means of providing continuity of awareness between immediate perceptions and the general patterns of insight within which they emerge.
Broader Encirclement (#MP1017) Context boundary (#MP1015)
Sub-domain boundary (#MP1013) Identifiable context (#MP1014)
Local interrelationship domains (#MP1011)
User-determined specialized communications (#MP1020).
Narrower Standard frameworks (#MP1038) Local relationship loops (#MP1049)
Selective interchange axis (#MP1032) Protection of emerging foci (#MP1057)
Relationship to indeterminacy (#MP1025) Three-way relationship entrainment (#MP1050)
Protected low intensity relationships (#MP1055)
Intersection of differently paced communications (#MP1054)
Linear relationships enhanced by non-linear processes (#MP1051)
Interfacing vehicles of communication and networks of unmediated relationships (#MP1052).

♦ **MP1024b Positions enabling transcendence**
Pattern In any domain there are points which provide a focus for the relationships between other points. Such a focal point or "centre of gravity" serves an integrative function through which the domain may both itself be integrated as well as being related as a whole to frameworks which transcend the boundary of the domain.
Physical environment In every region, town or locality, there are special places which become symbols embodying the unique characteristics of the area. Unless such sacred sites are protected, whether they are natural or man-made, people cannot maintain their spiritual roots or their sense of historical and cultural identity within a more global framework.
Socio-organizational environment In every society, organization complex or local group, there are regularly occurring special occasions transcending day-to-day preoccupations. These become symbols embodying, through some form of ritual, the unique characteristics of the community in question. Unless such occasions are protected, whether they are traditional, religious or improvised, people cannot maintain their spiritual roots and their sense of historical and cultural identity within a more global community.
Conceptual environment In any conceptual domain there are key, focal, or self-reflective concepts which serve an integrative function whereby the domain may both itself be integrated as well as being related as a whole to more fundamental or meta-frameworks transcending the boundary of the domain. Unless such focal concepts are protected, it becomes difficult to maintain both the cohesion of the domain and its relevance within any larger conceptual framework.
Intra-personal environment Within any mode of awareness there are fundamental or self-reflective insights through which the perceptions in that mode are both integrated and related as a whole to some larger, more fundamental, or transcendent mode of awareness. Unless such key insights are protected, it becomes difficult for the individual to maintain both the coherence of that mode and an understanding of its function within the personality as a whole.
Broader Sub-domain boundary (#MP1013).
Narrower Relationship to indeterminacy (#MP1025)
Low-intensity communication pathways (#MP1059)
Context for transformative experience (#MP1066)
Access to patterns of active irrationality (#MP1064)
Coherent pattern of relationship densities (#MP1028)
Limiting exposure to harmonious perspectives (#MP1134)
Sites for grounding perspectives in non-linearity (#MP1176)
Appropriate relationship of non-linearity to structures (#MP1171).

♦ **MP1025b Relationship to indeterminacy**
Pattern The significance of a domain of defined relationships is enhanced if it includes, or is contiguous with, a domain of indeterminacy in which fixed relationships are partially freed to form other patterns. The boundary between the two domains requires special protection because of the tendency of unharmonious relationships to accumulate there in an uncontrolled manner.
Physical environment Natural bodies of water, whether beaches, lakes or river banks, are of vital

and profound significance to people, if only for recreation or industrial purposes. Measures are required to prevent the accumulation of unsightly structures (buildings, factories, roads, etc) at the water's edge which render it inaccessible and disagreeable.

Socio-organizational environment Informality, whether in the form of parties, festivals or other non-structured occasions providing emotional outlet is of vital and profound significance to people, if only for recreation purposes or to facilitate formal relationships and agreements between organizations. Measures are required to prevent the proliferation of artificial procedures and rituals which render access to informality difficult and disagreeable.

Conceptual environment Indeterminacy, whether in the form of uncertainty, fuzziness, or ignorance, is of vital and profound importance to creative conceptual development, if only to facilitate the emergence of alternative perspectives or to enable beneficial cross-fertilization between conceptual frameworks. Measures are required to prevent the proliferation of artificial methods and procedures which render exposure to uncertainty difficult and conceptually inelegant.

Intra-personal environment Unstructured modes of awareness, whether in the form of insight, intuition, empathy or other non- rational forms, is of vital and profound significance to a person, if only as a catalyst for relaxation or to facilitate integration between more disciplined modes of awareness. Measures are required to prevent the proliferation of artificial attitudes and habits which render exposure to such immediate awareness difficult and painful.

Broader Encirclement (#MP1017) Sub-domain boundary (#MP1013) Non-linear organization (#MP1007) Positions enabling transcendence (#MP1024) Compensating relationships in parallel (#MP1023).
Narrower Access to contained irrationality (#MP1071) Cycle of relationship reinforcement (#MP1031) Access to patterns of active irrationality (#MP1064) Limitation on number of occupiable temporary sites (#MP1103).

♦ MP1026 Functional cycle

Pattern A well-balanced self-organizing domain is characterized by a cycle of interacting phases each of which emphasizes particular functions or processes vital to the integrity of the domain.

Physical environment A well-balanced village or urban community is characterized by the presence of buildings of a range of types (and ages) each of which facilitates a different activity vital to the integrity of the community. The absence of structures of a particular type (or age) may severely endanger the balance and independence of the community as well as restricting the range of experiences a person can experience there.

Socio-organizational environment A well-balanced community, organization, or group is characterized by the presence of a range of distinct activities (some varying correspondingly from the traditional to the innovative), each especially relevant to some phase of the cycle of interacting processes through which the life of a mature group regulates itself and develops. If such activities are not all represented, people can neither fulfil themselves in one phase nor pass successfully on to the next.

Conceptual environment A well-balanced conceptual domain or body of knowledge is characterized by a range of distinct methods or conceptual approaches (some varying correspondingly from the traditional to the innovative), each especially relevant to some phase of the cycle of interacting processes through which the life of a mature school of thought regulates itself and develops. If such approaches are not all represented, ideas cannot be brought to fruition in one phase nor be successfully transformed for development in subsequent phases.

Intra-personal environment A well-balanced mode of awareness is characterized by a range of distinct perceptions (some varying correspondingly from the habitual to the innovative), each especially relevant to some phase of the cycle of interacting processes through which the person's psychic life is integrated and developed. If such perceptions are not all represented, insights cannot be brought to fruition in one phase by the person nor be successfully transformed for development in subsequent phases.

MG M b Functional cycle
Broader Identifiable context (#MP1014) Individuality in multiplicity (#MP1012) Intermediate scale organization (#MP1006).
Narrower Complementarity (#MP1027) Local focal points (#MP1044) Functionality enhancement (#MP1047) Protection of emerging foci (#MP1057) Perspective-adaptable contexts (#MP1079) Adequate variety of cyclic elements (#MP1035) Integrating the historical dimension (#MP1040) Context for transformative experience (#MP1066) Informal context for formal processes (#MP1041) Extended pattern of nuclear interaction (#MP1075) Context for emergence of new perspectives (#MP1065) Contexts for care of premature perspectives (#MP1086) Context for acknowledgement of past perspectives (#MP1070) Semi-autonomous contexts for maturing perspectives (#MP1154) Transitional contexts for perspective reorganization (#MP1084) Opportunities for perspectives of decreasing activity (#MP1156) Perspective imitation contexts for developing perspectives (#MP1085) Semi-autonomous contexts for perspectives of decreasing activity (#MP1155) Structures adaptable to changing number of embodied perspectives (#MP1153) Appropriate configuration for interaction of complementary perspectives (#MP1187) Integration of perspective acquisition and perspective maintenance dynamics (#MP1083).

♦ MP1027c Complementarity

Pattern In order to successfully embody, encompass or reflect potential variety, a domain should be ordered on the basis of two incommensurable but complementary modes of organization.

Physical environment In order to ensure appropriate balance within any urban environment, each building, open space, neighbourhood and work community should be developed with a blend of incommensurable insights (as typified by those of men and women). If structures are developed (whether homes, suburbs, supermarkets or factories), in which either of such insights is repressed, such structures perpetuate and solidify the resulting distortion of reality.

Socio-organizational environment In order to ensure appropriate balance within any organization or community, each procedure and activity should be developed with a blend of incommensurable insights (as typified by those of men and women). If groups and programmes are developed in which either of such insights is repressed, such structures perpetuate and give form to the resulting distortion of reality.

Conceptual environment In order to successfully embody the complexity inherent in any conceptual domain, each framework should be articulated with a combination of incommensurable perspectives (as typified by those of men and women) viewed as a complementary set of descriptions. If a conceptual framework is developed in which one such perspective is ignored, such a framework perpetuates and gives form to the resulting oversimplified representations of reality.

Intra-personal environment In order to successfully comprehend the richness inherent in any mode of awareness, it should be allowed to act in either of two incompatible sub-modes (as typified by those of men and women) understood as permitting complementary perceptions. If a mode of awareness is developed in which one such sub-mode is avoided, the resulting perceptual habits will provide a dangerously oversimplified understanding of reality.

Broader Functional cycle (#MP1026) Identifiable context (#MP1014) Individuality in multiplicity (#MP1012) Decentralized formal processes (#MP1009).
Narrower Local opportunities for perspective activity (#MP1157) Relatively isolated context for each perspective (#MP1141).

♦ MP1028b Coherent pattern of relationship densities

Pattern Within a domain randomness in the density of local relationships fails to reinforce any recognizable pattern on the basis of which the domain as a whole may be identified. A systematic variation in the pattern of relationship densities can provide the necessary coherence.

Physical environment Within an urban environment population densities are higher towards the centre, but there is no recognizable pattern arising from the manner in which this trend is modified by the many component local communities (with their own centres). Such randomness inhibits development of any community identity and creates a chaos in the pattern of land use. Given that the centre of local communities are each located on their community boundary (eccentrically towards the centre of the larger agglomeration), by encouraging them each to bulge both towards the geometric centres of their own local community and in a horseshoe along the neighbouring boundary, a gradient of imbricated horseshoes emerges as an overall pattern supportive of local community life.

Socio-organizational environment Within any organizational complex, network or group, randomness in the local density distribution of formal or informal relationships fails to reinforce any recognizable pattern on the basis of which the organization as a whole may be identified, comprehended and effectively used. A systematic variation in the pattern of relationship densities can provide the necessary coherence.

Conceptual environment Within any conceptual framework or body of knowledge, randomness in the density of relationships between specialized concepts fails to reinforce any recognizable pattern on the basis of which the conceptual framework as a whole may be identified, comprehended and effectively used. A systematic variation in the pattern of specialized relationships can provide the necessary coherence.

Intra-personal environment Within any mode of awareness randomness in the degree of relatedness of sets of perceptions fails to reinforce any recognizable pattern on the basis of which the mode of awareness as a whole may be identified, comprehended and profitably used. A systematic variation in the pattern of sets of perceptions can provide the necessary coherence.

Broader Sub-domain boundary (#MP1013) Access to intensity (#MP1010) Four-level structural limit (#MP1021) Individuality in multiplicity (#MP1012) Positions enabling transcendence (#MP1024).
Narrower Activity nodes (#MP1030) Selective interchange axis (#MP1032) Access to contained irrationality (#MP1071) Cycle of relationship reinforcement (#MP1031) Low-intensity communication pathways (#MP1059) Stable density gradient of local relationships (#MP1029).

♦ MP1029 Stable density gradient of local relationships

Pattern The locus of formation of new local relationships is determined by a balance between proximity to the local centre (as an attraction) and distance from the intensity of processes there (acting as a repellent). The varying ways in which this balance is determined leads to the emergence of rings of different relationship density (a density gradient) around the local centre. This density configuration will be unstable unless measures are adopted to compensate for the instability resulting from continuing pressure to form further relationships at preferred locations.

Physical environment In a local community people want to live close to shops and services for excitement and convenience. They must balance this against the desire to be away from such services in order to experience peace and greenery. The pattern of such choices in a neighbourhood defines density rings around the local centre. Under the continuing pressure of the arrival of new households, such a density gradient becomes unstable (to the disadvantage of the less privileged) unless compensatory measures are adopted.

Socio-organizational environment In a local or specialized organizational complex people want to be close to the centre "where the action is" and where interpersonal interactions are most intense and challenging. But they must balance this against their need for low key unstructured relationships permitting greater freedom of personal expression and growth. The varying ways in which this balance is struck by individuals in an organization or network defines relationship density rings around the nucleus of the organization. Under the continuing pressure of the arrival of new participants, such a density gradient becomes unstable (to the disadvantage of the less privileged) unless compensatory measures are adopted.

Conceptual environment In a specialized conceptual domain specialists experience a need to be close to the intellectual centre of gravity where the latest challenging ideas are being presented and debated. But they must balance this against their need for peaceful, reflective conditions in which their own insights can emerge and be developed. The varying ways in which this balance is struck by individuals within a school of thought or invisible college defines density rings of conceptual relationships around the core of the domain. Under the continuing pressure of the emergence of new thinkers, such a density gradient becomes unstable (to the disadvantage of the less privileged) unless compensatory measures are adopted.

Intra-personal environment In employing a specialized mode of awareness, a balance must be struck between the challenging in- sights to be gained by focusing it in its most intensely disciplined form and the benefit to be derived from allowing it to influence perceptions in a more non-directive manner. The varying ways in which this balance tends to be struck over a period of time by the person using that mode defines density rings of perceptual relationships around the focal awareness. Under the continuing pressure of the emergence of new perceptions, such a gradient becomes unstable (to the disadvantage of some perceptions) unless compensatory measures are adopted.

MG M b Stable density gradient of local relationships
Broader Sub-domain boundary (#MP1013) Four-level structural limit (#MP1021) Individuality in multiplicity (#MP1012) Variety of forms and processes (#MP1008) Coherent pattern of relationship densities (#MP1028).
Narrower Activity nodes (#MP1030) Standard frameworks (#MP1038) Cluster of frameworks (#MP1037) Integrating a new dimension (#MP1039) Cycle of relationship reinforcement (#MP1031) Enhancing function of common domains (#MP1123) Differentiation by relationship density (#MP1036) Bounded common small-scale interaction domains (#MP1061).

♦ MP1030a Activity nodes

Pattern As a framework of relationships is articulated, nodes emerge at the points of convergence of the principal relationships. Such nodes provide loci for processes vital to the self-organizing dynamics of the domain as a whole.

Physical environment As a local community grows, community facilities emerge randomly, thus failing to reinforce each other or the vitality of the community as a whole. This may be remedied by concentrating mutually supportive facilities at activity nodes (such as small public squares distributed evenly throughout the community) on which the network of pathways naturally converge.

Socio-organizational environment As an organizational complex develops, facilitative nodes emerge randomly, thus failing to reinforce each other or the vitality of the group or network as a whole. This may be remedied by concentrating mutually supportive network facilities at activity nodes (such as meetings organized regularly to serve different interest groups) on which the network of relationships naturally converge.

Conceptual environment As a specialized conceptual domain develops, facilitative methods emerge randomly, thus failing to reinforce each other or the interest of the domain as a whole. This may be remedied by deliberately grouping mutually supportive methods in relation to the key concepts (such as those providing a focus for alternative perspectives within the domain) on which the network or conceptual relationships naturally converge.

Intra–personal environment As a specialized mode of awareness develops, facilitative key insights emerge randomly thus failing to reinforce each other or the integrity of that mode. This may be remedied by consciously associating mutually supportive insights in relation to the principal perceptions (such as those providing a focus for alternative perceptions within that mode) on which the network of insights naturally converge.

Broader Context boundary (#MP1015) Sub–domain boundary (#MP1013)
Identifiable context (#MP1014) Individuality in multiplicity (#MP1012)
Coherent pattern of relationship densities (#MP1028)
Stable density gradient of local relationships (#MP1029).
Narrower Interchange (#MP1034) Local focal points (#MP1044)
Underdefined processes (#MP1033) Functionality enhancement (#MP1047)
Presentation of new dimensions (#MP1043) Cycle of relationship reinforcement (#MP1031)
Focus–oriented communication networks (#MP1120)
Informal context for formal processes (#MP1041)
Differentiation by relationship density (#MP1036)
Context for emergence of new perspectives (#MP1065)
Informal local perspective interface zones (#MP1088)
Unstructured context for perspective exchange (#MP1090)
Bounded common small–scale interaction domains (#MP1061)
Hospitable contexts for perspectives in transition (#MP1091)
Exchange contexts controlled by a single perspective (#MP1087)
Transitional contexts for perspective reorganization (#MP1084)
Arrangement of structures ot engender fruitful interfaces (#MP1100)
Perspective imitation contexts for developing perspectives (#MP1085)
Transit point location for sources of perspective nourishment (#MP1093)
Interfacing vehicles of communication and networks of unmediated relationships (#MP1052).

♦ MP1031a Cycle of relationship reinforcement

Pattern Within any domain, linking the activity nodes together through a fundamental cycle reinforces specific relationships as well as the integrity of the domain as a whole.

Physical environment In any urban community the gradual formation of a promenade linking the main activity nodes provides an environment through which people are encouraged to move constantly in order to see and to be seen, especially if the principal nodes are located at each end. Such a promenade provides a focus for the life of the community.

Socio–organizational environment In any organizational complex or group, linking facilitative nodes together in a communication circuit provides an environment through which people are encouraged to move constantly in order to encounter new and alternative activities and the associated relationships. Such a cycle reinforces existing relationships and the integrity of the group as a whole.

Conceptual environment In any conceptual domain or school of thought, linking together the key foci of conceptual activity in a cycle of conceptual processes provides an environment through which ideas are encouraged to move constantly in order to be challenged by new or alternative insights and the associated conceptual relationships. Such a cycle reinforces existing relationships and the integrity of that body of knowledge as a whole.

Intra–personal environment Within any mode of awareness, linking key perceptual processes together within a larger cyclic process provides a context through which attention is encouraged to move constantly in order to encounter new or alternative insights and the patterns of significance within which they are embedded. Such a cycle renews existing relationships and the integrity of the mode of awareness as a whole.

Broader Activity nodes (#MP1030) Access to intensity (#MP1010)
Relationship to indeterminacy (#MP1025) Individuality in multiplicity (#MP1012)
Variety of forms and processes (#MP1008)
Coherent pattern of relationship densities (#MP1028)
Stable density gradient of local relationships (#MP1029).
Narrower Context for disorder (#MP1058) Underdefined processes (#MP1033)
Selective interchange axis (#MP1032) Protection of emerging foci (#MP1057)
Presentation of new dimensions (#MP1043)
Enhancing function of common domains (#MP1123)
Hospitability of communication pathways (#MP1121)
Access to patterns of active irrationality (#MP1064)
Ensuring function of common domain interfaces (#MP1124)
Unstructured context for perspective exchange (#MP1090)
Bounded common small–scale interaction domains (#MP1061)
Hospitable contexts for perspectives in transition (#MP1091)
Arrangement of structures ot engender fruitful interfaces (#MP1100)
Cyclic interrelation of complementary perspective in common domains (#MP1063)
Interfacing vehicles of communication and networks of unmediated relationships (#MP1052).

♦ MP1032 Selective interchange axis

Pattern Effective selective interchange within a domain is achieved by harmonizing the dynamically incompatible requirements of high–intensity non–local relationships and low intensity, local relationships. This can be achieved by arranging that selective interchange axes cut across non–local relationships, providing local relationships between them.

Physical environment The viability of shopping centres depends on an appropriate compromise between their accessibility from major roads and the convenience to pedestrian shoppers. This can be achieved by arranging that pedestrian shopping streets cut across major roads, linking parallel roads, and providing parking space behind the shops.

Socio–organizational environment The viability of environments (such as meetings) for effective exchange between groups depends on an appropriate compromise between their accessibility through structured mass communications pathways and the necessary low key informal interactions between participants. This can be achieved by arranging that low key communication environments cut across major lines of communication, providing a link between them.

Conceptual environment The viability of environments for the cross–fertilization of ideas between different frames of reference or schools of thought (such as in knowledge representation systems) depends on an appropriate compromise between their accessibility through structured high speed information systems and the extended time period required to reflect on new ideas. This can be achieved by connecting local (or personal) information systems enhancing cross–fertilization to different data networks.

Intra–personal environment The viability of conditions for the cross–fertilization of personal insights depends on an appropriate compromise between exposure of the individual to a continuing stream of impressions and the possibility for the individual to filter out irrelevant impressions in order to focus awareness on those which remain. This can be achieved by alternating attention, for appropriate periods of time, between the stream of impressions and a focused awareness, thus maintaining continuity of awareness between the two modes.

MG M b Selective interchange axis

Broader Context boundary (#MP1015) Access to intensity (#MP1010)
Web of selective interchange (#MP1019) Cycle of relationship reinforcement (#MP1031)
Compensating relationships in parallel (#MP1023)
Coherent pattern of relationship densities (#MP1028).
Narrower Flexible interfaces (#MP1244) Adaptive interstices (#MP1048)
Diversified interchange environment (#MP1046)
Hospitability of communication pathways (#MP1121)
Coherent pattern of relationship densities (#MP1028)
Intersection of differently paced communications (#MP1054)
Limitation on number of occupiable temporary sites (#MP1103)
Exchange contexts controlled by a single perspective (#MP1087)
Distinct pattern of entry points to complex structures (#MP1102)
Concealment of necessary monotonous perspective patterns (#MP1097)
Arrangement of structures ot engender fruitful interfaces (#MP1100)
Interfacing vehicles of communication and networks of unmediated relationships (#MP1052).

♦ MP1033b Underdefined processes

Pattern In every domain there are processes that are underdefined or are only evident when the well–defined processes are inactive. Underdefined processes should be clustered to ensure their viability as attractors.

Physical environment In every urban community there is some kind of night life after daytime activities are closed down. Entertainment facilities, bars, discos, restaurants, etc need to be distributed in clusters to provide lively, secure pedestrian environments in order to guarantee their attractiveness and viability.

Socio–organizational environment Associated with every organizational complex or group there are extra–mural activities to which participants are attracted during interruptions in the cycle of normal group activity. These range from drinking together through office recreation clubs, group celebrations, parties, collective participation in carnivals and excursions, to other forms of group entertainment. Such activities need to be clustered in space and/or time if they are to provide lively secure environments in order to guarantee their attractiveness and viability.

Conceptual environment In every conceptual domain there are unconscious, ill–defined processes that are only evident when the well–defined conceptual procedures are inactive or exhausted. Such unconscious processes should be inter–related or associated in order to ensure their viability as catalysts for the emergence of alternative perspectives.

Intra–personal environment In every mode of awareness there are unconscious, ill–defined processes that are only evident when the person's conscious awareness is stilled. Such unconscious processes, including dreams and waking fantasies, should be inter– related in order to ensure their viability as catalysts for the emergence of new insights.

Broader Activity nodes (#MP1030) Access to intensity (#MP1010)
Individuality in multiplicity (#MP1012) Cycle of relationship reinforcement (#MP1031).
Narrower Local focal points (#MP1044) Context for disorder (#MP1058)
Informal local perspective interface zones (#MP1088)
Unstructured context for perspective exchange (#MP1090)
Hospitable contexts for perspectives in transition (#MP1091)
Cyclic interrelation of complementary perspective in common domains (#MP1063).

♦ MP1034 Interchange

Pattern The interconnections between lines of relationships in a domain, namely the nodal points from which alternative relationships originate, play a central role in ensuring the viability and integrity of the overall pattern of relationships.

Physical environment The interchanges in a web of public transportation services play a central role in ensuring the viability and integrity of the transportation system. They should be organized to make them accessible to regular users and to minimize the discontinuity between different modes of transport.

Socio–organizational environment The arenas in which information is translated between different media in the web of public communication services play a central role in ensuring the viability and integrity of the communication system. The arenas should be organized to make them accessible to regular communications from different groups and to minimize the discontinuity between different modes of communication.

Conceptual environment The focal, or interdisciplinary, concepts which provide interconnections between different lines of thought, play a central role in ensuring the viability and integrity of the web of conceptual relationships. Such nexuses of inter–relationships should be organized to make them relevant to ideas emanating from different schools of thought and to minimize the discontinuity between different modes of thought.

Intra–personal environment The integrative modes of awareness, which provide the necessary interconnections between particular modes of awareness, play a central role in ensuring the viability and integrity of any ordered pattern of awareness. The integrative modes should be organized to make them significant to insights emerging from different modes of awareness and to minimize the discontinuity between different modes of awareness.

MG M c Interchange

Broader Encirclement (#MP1017) Activity nodes (#MP1030)
Decentralized formal processes (#MP1009) Local interrelationship domains (#MP1011)
Web of general interrelationships (#MP1016)
User–determined specialized communications (#MP1020).
Narrower Hospitable common domains (#MP1094)
Hospitable transit points (#MP1092)
Integrating a new dimension (#MP1039)
Partially contained interfaces (#MP1119)
Integrating the historical dimension (#MP1040)
Informal context for formal processes (#MP1041)
Provision for temporary perspective inactivity (#MP1150).

♦ MP1035b Adequate variety of cyclic elements

Pattern The coherence and specificity of the cycle of interacting phases, by which a domain is defined, is largely determined by the variety of those cyclic elements.

Physical environment The mix of dwelling types (and ages) in any neighbourhood or urban cluster, namely the presence together of dwellings appropriate for children, single people, working couples, and the old, largely determines the coherence and uniqueness of such a community and its potential for self–renewal. Mixing must be balanced by the need to construct similar dwelling types together.

Socio–organizational environment The mix of people of different ages, namely at different stages in the human life cycle (or the life cycle of a group), is a major factor in determining the degree of coherence and uniqueness of any organizational complex or group. People need support and confirmation from those at other stages in a life cycle, whether older or younger. Mixing different household types must be balanced against the needs for those of the same age group to be together.

Conceptual environment The variety of distinct methods or conceptual approaches (including the classical and the innovative) constituting the learning/discovery cycle of any conceptual domain or school of thought is a major factor in determining its degree of coherence and specificity. The mixture of methods must however be balanced against the need to strengthen the relationships between those of the same kind.

Intra–personal environment The variety of distinct perceptions (including the habitual and the innovative) constituting the cycle of processes of any mode of awareness is a major factor in determining its degree of coherence and uniqueness. The mixture of perceptions must however be balanced against the need to reinforce the relationships between those of the same kind.

Broader Functional cycle (#MP1026) Identifiable context (#MP1014).
Narrower Cluster of frameworks (#MP1037) Integrating the historical dimension (#MP1040)
Informal context for formal processes (#MP1041)
Minimal context for single perspective (#MP1078)
Extended pattern of nuclear interaction (#MP1075)
Minimal context for complementary perspectives (#MP1077)
Minimal context interrelating mature and emerging perspectives (#MP1076)
Spontaneous relationship formation amongst emerging perspectives (#MP1068).

♦ MP1036a Differentiation by relationship density

Pattern A domain is articulated by differentiating zones of high, medium and low relationship densities amongst the domain elements. Such distinctions should be reinforced by the overall clustering of the elements.

Physical environment Within any neighbourhood (or cluster of houses) there are those who wish to live close to central services or the movement of people, there are those who prefer privacy, and there are those who prefer some compromise between such extremes. Such differing preferences for publicness may be satisfied by reinforcing the distinction between busy streets, secluded backstreets or pathways, and streets of an intermediate type.

Socio–organizational environment Within any organization or group there are those who wish to be visibly associated with areas of intense activity or public contact, there are those who prefer seclusion, and there are those who prefer some compromise between those roles involving extreme public relations and personal exposure, those roles requiring privacy, and those involving some mix of the two.

Conceptual environment A conceptual domain is articulated by differentiating areas of intense interaction between mutually challenging and commonplace ideas, isolated areas of exploration and reflection, and areas having some of the characteristics of both. Such distinctions may be usefully maintained by clarifying and reinforcing the pattern of relationships between them.

Intra–personal environment A mode of awareness is articulated by differentiating conditions of intense interaction between perceptions (whether new insights or old), conditions of meditation and withdrawal, and conditions having some of the characteristics of both. Such distinctions may be usefully maintained by clarifying and reinforcing the pattern of transitions between them.

Broader Activity nodes (#MP1030) Identifiable context (#MP1014)
Stable density gradient of local relationships (#MP1029).
Narrower Standard frameworks (#MP1038) Cluster of frameworks (#MP1037)
Integrating a new dimension (#MP1039)
Enhancing function of common domains (#MP1123)
Focus–oriented communication networks (#MP1120)
Protected low intensity relationships (#MP1055)
Hospitability of communication pathways (#MP1121)
Linear relationships enhanced by non–linear processes (#MP1051)
Arrangement of structures ot engender fruitful interfaces (#MP1100).

♦ MP1037a Cluster of frameworks

Pattern The fundamental unit of identifiable local organization within a domain is the cluster of frameworks within which elements are grouped. The limit on the number of frameworks per cluster varies from 8 to 12 depending on the balance between implicit, dynamic and explicit structural organization.

Physical environment The fundamental unit of organization within an identifiable neighbourhood is the cluster of houses. The limit on the number of houses per cluster varies from 8 to 12 depending on the balance between the informality and coherence of the group. The houses should preferably be arranged around some commonly owned land and paths.

Socio–organizational environment The fundamental unit of local organization in any identifiable social span is the (proximity) cluster of groups (or people) with related preoccupations. The limit on the number of groups (or people) per cluster varies from 8 to 12 depending on the balance between the informality of the relationships within the cluster and the coherence of the cluster as a whole. The relationships between the groups (or people) should preferably be such as to define an area of commonly held concern.

Conceptual environment The fundamental unit of local organization in any identifiable conceptual span is the (proximity) cluster of conceptual frameworks with related foci. The limit on the number of frameworks per cluster varies from 8 to 12 depending on the balance between the network of implicit, associative relationships and the degree of explicit, structural organization. The relationships between the frameworks should preferably be such as to define a common focus.

Intra–personal environment The fundamental unit of local organization in any identifiable field of awareness is the cluster of perceptual frameworks with related foci. The limit on the number of frameworks per cluster varies from 8 to 12 depending on the balance between the fluidity of the perceived relationships and the degree of gestalt awareness of the cluster as a whole. The perceptual frameworks should preferably be configured such as to highlight a unifying awareness.

Broader Identifiable context (#MP1014) Non–linear organization (#MP1007)
Variety of forms and processes (#MP1008) Network of inter–relationships (#MP1005)
Adequate variety of cyclic elements (#MP1035) Differentiation by relationship density (#MP1036)
Stable density gradient of local relationships (#MP1029).
Narrower Standard frameworks (#MP1038) Adaptive interstices (#MP1048)
Local relationship loops (#MP1049) Principal points of entry (#MP1053)
Unstructured common domain (#MP1067) Integrating a new dimension (#MP1039)
Perspective–adaptable contexts (#MP1079) Patterning of complex structures (#MP1098)
Access to contained irrationality (#MP1071)
Minimal context for single perspective (#MP1078)
Extended pattern of nuclear interaction (#MP1075)
Local opportunities for perspective activity (#MP1157)
Minimal context for complementary perspectives (#MP1077)
Re–integration of rejected perspective by–products (#MP1178)
Limitation on number of occupiable temporary sites (#MP1103)
Insight capturing non–linear extensions of structures (#MP1175)
Distinct pattern of entry points to complex structures (#MP1102)
Blended integration of formal structure and informal context (#MP1111)
Minimal context interrelating mature and emerging perspectives (#MP1076)
Spontaneous relationship formation amongst emerging perspectives (#MP1068).

♦ MP1038b Standard frameworks

Pattern In certain parts of a domain, clusters of frameworks do not permit a sufficiently high relationship density. Chains of standard frameworks may then be used.

Physical environment In certain parts of an urban community clusters of houses do not permit sufficiently high population densities or degrees of publicness. Row houses are then essential and should preferably be placed along pedestrian paths, running at right angle to local roads, such as to provide common land behind them.

Socio–organizational environment In certain parts of an organizational complex clusters of groups (or people) with related preoccupations do not permit sufficiently high relationship densities or degrees of interaction. Standard groups should then be interlinked in chains, such as to provide low intensity communication pathways between them, connecting into higher intensity communication networks. The groups should be interlinked such as to define a common area of concern.

Conceptual environment In certain parts of a conceptual domain loose clusters of conceptual frameworks do not permit a sufficiently high degree of integration. Standard frameworks (such as matrices of concepts) should then be interlinked by a sequence of procedures such as to encourage non–linear interactions between them, phasing into the linear pattern of more general relationships. The frameworks should be interlinked such as to define a shared common focus.

Intra–personal environment In certain areas of a field of awareness clusters of perceptual frameworks do not provide a sufficiently high degree of integration. Standard perceptual frameworks should then be interlinked together (such as in a rote learning or meditation sequence) such as to engender the implicit relationships between them, phasing into the explicit patterns of overall awareness. The frameworks should be interlinked such as to define a shared core of awareness.

Broader Cluster of frameworks (#MP1037)
Compensating relationships in parallel (#MP1023)
Differentiation by relationship density (#MP1036)
Stable density gradient of local relationships (#MP1029).
Narrower Adaptive interstices (#MP1048) Unstructured common domain (#MP1067)
Perspective–adaptable contexts (#MP1079)
Minimal context for single perspective (#MP1078)
Extended pattern of nuclear interaction (#MP1075)
Complexification of perspective contexts (#MP1095)
Minimal context for complementary perspectives (#MP1077)
Limitation on number of occupiable temporary sites (#MP1103)
Arrangement of structures ot engender fruitful interfaces (#MP1100)
Blended integration of formal structure and informal context (#MP1111)
Minimal context interrelating mature and emerging perspectives (#MP1076)
Organization of structure to enhance autonomy of sub–structures (#MP1109)
Interfacing vehicles of communication and networks of unmediated relationships (#MP1052).

♦ MP1039 Integrating a new dimension

Pattern At the central foci of any domain the density of relationships between elements is such as to require construction of a framework in a new dimension. To avoid a purely mechanistic pattern, each element, whilst acquiring unique characteristics within the framework, should encode its earlier spatial dimensions and the rhythms of its earlier dynamics.

Physical environment In every urban environment there are places so central that high–rise apartments tend to be constructed to accommodate the high population density. To avoid alienating impersonality, each apartment should have a direct connection to the ground, a private garden and the possibility of acquiring a unique identity. This may be accomplished by using stepped terraces on a housing hill.

Socio–organizational environment At the central (or fashionable) foci of any organizational complex the preferred density of relationships is such as to require that many be based on a new mode of action. To avoid alienating impersonality, each group should maintain a direct connection to simpler and more fundamental modes of organizing relationships in time, whilst at the same time acquiring unique characteristics in the new framework.

Conceptual environment At the central foci of any conceptual domain the necessary density of relationships between concepts is such as to require that many be based on a new (or meta) dimension. To avoid arid incomprehensibility, each set of concepts should maintain a direct connection to simpler and more fundamental patterns of relationship, whilst at the same time acquiring unique characteristics within the new framework.

Intra–personal environment At the central foci of any mode of awareness the density of relationships amongst perceptions and insights is such as to require that many be based on a new dimension of understanding. To avoid cold detachment, each set of insights should maintain a direct connection to simpler and more fundamental patterns of insight, whilst at the same time developing unique characteristics in the light of the new dimension.

MG M c Integrating a new dimension

Broader Interchange (#MP1034) Cluster of frameworks (#MP1037)
Four–level structural limit (#MP1021) Differentiation by relationship density (#MP1036)
Stable density gradient of local relationships (#MP1029).
Narrower Unstructured common domain (#MP1067)
Perspective–adaptable contexts (#MP1079)
Extended pattern of nuclear interaction (#MP1075)
External access to higher structural levels (#MP1158)
Local cultivation of sources of perspective nourishment (#MP1177)
Concealment of necessary monotonous perspective patterns (#MP1097)
Arrangement of structures ot engender fruitful interfaces (#MP1100)
Integration of non–linearity into integrative superstructure (#MP1118)
Spontaneous relationship formation amongst emerging perspectives (#MP1068).

♦ MP1040a Integrating the historical dimension

Pattern When a domain is properly formed it encodes the earlier stages in its own development, appropriately relating them to current and emerging stages.

Physical environment When an urban environment is properly designed it preserves structures reflecting the characteristics of its earlier stages of development, appropriately relating them to recent buildings and those planned or in process of construction.

Socio–organizational environment When a social group or organizational complex is properly balanced it integrates within it traditional groups (or elderly people), appropriately relating them to contemporary groups (or adults) as well as to new kinds of groups (or the young). Failure to do so creates dangerous rifts in the sense of historical continuity and development within the community.

Conceptual environment When a conceptual domain is properly formed and balanced it integrates within it the earlier stages in its own development, appropriately relating them to currently accepted concepts and to emerging insights into the probable future development of the domain.

Intra–personal environment When a mode of awareness is properly balanced it integrates within it the earlier learning stages in its own development appropriately relating them to those in which confidence is at present placed, as well as to those emerging insights which are as yet only partially understood.

Broader Interchange (#MP1034) Functional cycle (#MP1026)
Identifiable context (#MP1014) Network of redefinitions (#MP1018)
Adequate variety of cyclic elements (#MP1035).
Narrower Protection of emerging foci (#MP1057)
Extended pattern of nuclear interaction (#MP1075)
Contexts for care of premature perspectives (#MP1086)
Opportunities for perspectives of decreasing activity (#MP1156)
Local cultivation of sources of perspective nourishment (#MP1177)
Semi–autonomous contexts for perspectives of decreasing activity (#MP1155).

♦ MP1041a Informal context for formal processes

Pattern A variety of formal processes should be grouped in clusters to enhance interlinking informal processes that ensure a balanced context for both within each cluster.

Physical environment Workplaces for a variety of employments should be clustered in groups of 10 to 20 around their own courtyards to form an identifiable work community with some collective amenities and nested within a larger community with other services. The work community then provides for a more balanced life outside the house.
Socio-organizational environment Groups or task forces with a variety of preoccupations should be encouraged to form into clusters of 10 to 20 such as to reinforce the informal interaction amongst them. The resulting community or network then provides for a healthier degree of functional balance than is normally possible outside non-directive groups.
Conceptual environment Utilization of a variety of distinct methodologies or conceptual frameworks should be organized into clusters of 10 to 20 such as to reinforce the processes of non-formal interaction amongst them. The resulting network or "invisible college" then provides for a healthier degree of conceptual balance than is normally possible outside contexts of non-deterministic reflection.
Intra-personal environment A variety of distinct perceptual frameworks should be grouped into clusters of 10 to 20 such as to reinforce the processes of unconscious integration amongst them. The resulting gestalt then provides for a more balanced mode of awareness than is normally possible in the absence of "seedless" meditation.
Broader Interchange (#MP1034) Activity nodes (#MP1030) Functional cycle (#MP1026) Context boundary (#MP1015) Sub-domain boundary (#MP1013) Identifiable context (#MP1014) Decentralized formal processes (#MP1009) Adequate variety of cyclic elements (#MP1035).
Narrower Adaptive interstices (#MP1048) Local relationship loops (#MP1049) Accessible non-linearity (#MP1060) Principal points of entry (#MP1053) Unstructured common domain (#MP1067) Protection of emerging foci (#MP1057) Access to contained irrationality (#MP1071) Coordination of perspective nourishment (#MP1147) Chain of fundamental transformation zones (#MP1042) Informal local perspective interface zones (#MP1088) Integrated contexts for perspective dynamics (#MP1080) Bounded common small-scale interaction domains (#MP1061) Limitation on number of occupiable temporary sites (#MP1103) Hospitable contexts for perspectives in transition (#MP1091) Minimally-structured perspective control operations (#MP1081) Insight capturing non-linear extensions of structures (#MP1175) Distinct pattern of entry points to complex structures (#MP1102) Functional integration of unstructured internal domains (#MP1115) Blended integration of formal structure and informal context (#MP1111) Minimal distance between related operational control contexts (#MP1082) Transit point location for sources of perspective nourishment (#MP1093) Competitive interaction opportunities transposed to a concrete level (#MP1072).

♦ MP1042b Chain of fundamental transformation zones

Pattern Zones of fundamental transformations should be situated in chains such as to form boundaries within a domain rather than being isolated within a specialized domain.
Physical environment Industry should be distributed along ribbons such as to form boundaries between communities rather than being completely separated from urban life and thus contributing to the unreality of sheltered residential neighbourhoods and to the sterility of industrial parks.
Socio-organizational environment Managerial and administrative action groups should be situated such as to form boundaries between organizational complexes, rather than being grouped separately from working operations, thus contributing to the unreality of sheltered working environments and to the sterility of bureaucracy.
Conceptual environment Application of theory to concrete problems should be linked such as to form boundaries between conceptual frameworks, rather than being isolated from areas of purely theoretical work, thus contributing to the unreality of such sheltered research environments and to appreciations based on limited insights.
Intra-personal environment Adaptations and use of modes of awareness in real-world situations should be linked such as to form boundaries between fields of awareness, rather than being completely isolated from reflection and meditation and thus contributing to the unreality of such protected activities and to unenlightened action.
Broader Encirclement (#MP1017) Sub-domain boundary (#MP1013) Decentralized formal processes (#MP1009) Informal context for formal processes (#MP1041).
Narrower Integrated contexts for perspective dynamics (#MP1080) Direct relationship between structures and communication pathways (#MP1122) Functional enhancement of domains separating complementary structures (#MP1106).

♦ MP1043c Presentation of new dimensions

Pattern The development of a domain is facilitated by integrating into its structure contexts in which a variety of new dimensions are represented, whether or not they finally provide a basis for such development.
Physical environment The development of an urban environment may be facilitated by providing a range of structures linked by pathways within which people can meet to explore new ideas and projects. These may take the form of a university, a community education centre or a conference centre.
Socio-organizational environment The development of an organizational complex or community may be facilitated by providing a range of interlinked meeting opportunities within which people can choose to receive or offer ideas or instruction, whether in the form of classes, lectures, conferences, or as a marketplace of projects.
Conceptual environment The development of a conceptual domain or a body of knowledge may be facilitated by providing a range of interlinked opportunities for the confrontation of different concepts, whether deriving from different levels of education, from different sources of information, or from different ideology or belief systems.
Intra-personal environment The development of a mode of awareness may be facilitated by ensuring a range of interlinked conditions within which incompatible perceptions are confronted, whether as paradoxes or deriving from different mind-sets.
Broader Activity nodes (#MP1030) Network of redefinitions (#MP1018) Cycle of relationship reinforcement (#MP1031).
Narrower Local action network (#MP1045) Adaptive interstices (#MP1048) Protection of emerging foci (#MP1057) Low-intensity communication pathways (#MP1059) Complexification of perspective contexts (#MP1095) Small-scale perspective interaction contexts (#MP1151) Minimally-structured perspective control operations (#MP1081) Arrangement of structures ot engender fruitful interfaces (#MP1100) Exposure of structural activities to communication pathway (#MP1165) Integration of perspective acquisition and perspective maintenance dynamics (#MP1083).

♦ MP1044b Local focal points

Pattern Local coordination within a domain can only be effectively achieved through the formation of appropriately located local focal points.
Physical environment Local government of communities, with the full participation of inhabitants, can best be ensured by provision of a small town hall for each community of about 7000. This should be located near the busiest intersection in the community. It should provide for a public discussion arena, public services, and community project offices.
Socio-organizational environment Coordination of any specialized group or organizational complex, with the full participation of those involved, can best be ensured by creation of a common service environment. This should be associated with the main nexus of communications amongst participants. It should provide an arena for public discussion, common services, and project management facilities.
Conceptual environment Integration of any specialized conceptual domain or school of thought can best be ensured by promoting the emergence of a common framework, preferably associated with the main nexus of conceptual relationships, to facilitate interrelationships between concepts.
Intra-personal environment Integration of any specialized mode of awareness can best be ensured by encouraging the emergence of a focal insight, preferably associated with the main nexus of perceptual relationships.
Broader Activity nodes (#MP1030) Functional cycle (#MP1026) Underdefined processes (#MP1033) Individuality in multiplicity (#MP1012).
Narrower Local action network (#MP1045) Enhancing function of common domains (#MP1123) Small-scale perspective interaction contexts (#MP1151) Bounded common small-scale interaction domains (#MP1061) Exposure of structural activities to communication pathway (#MP1165).

♦ MP1045c Local action network

Pattern A local focal point only acquires significance in a domain when it is embedded in a local action network whose activities it facilitates.
Physical environment A local town hall functions more effectively when it surrounds itself with shop-size spaces available to community action groups. Availability of low-cost office space is a vital catalyst to the initiatives of effective group action.
Socio-organizational environment A common service environment can only be effective when it is embedded in a network of specialized action projects, each testing its views within the community and mobilizing support and resources. Such a network is basic to the self-regulation of any community.
Conceptual environment An integrative common framework only acquires significance in a specialized conceptual domain when it is embedded in a network of investigative initiatives whose activities it facilitates.
Intra-personal environment A focal insight only acquires significance within any specialized mode of awareness when it is embedded within a network perception whose value it enhances.
Broader Local focal points (#MP1044) Presentation of new dimensions (#MP1043).
Narrower Functionality enhancement (#MP1047) Common external context for inactivity (#MP1069) Minimally-structured perspective control operations (#MP1081) Exchange contexts controlled by a single perspective (#MP1087) Exposure of structural activities to communication pathway (#MP1165) Hospitable interface between structures and external environment (#MP1160) Direct relationship between structures and communication pathways (#MP1122).

♦ MP1046 Diversified interchange environment

Pattern For an interchange environment to be most effective it should be highly diversified within a framework providing the minimum coherence for the variety of styles and modes of operation.
Physical environment A market is most effective and dynamic when it offers a wide variety of products through many shops, each organized according to its own style rather than standardized within a monolithic supermarket. Only a minimum structure of roofing, aisles and common services is required.
Socio-organizational environment A meeting environment is most effective and dynamic when it offers a wide variety of experiences each organized according to its particular style rather than according to a uniform set of procedures. Only a minimum framework defining and interrelating meeting spaces and common services is required.
Conceptual environment A framework for the interchange of ideas is most effective and dynamic when it offers a wide variety of methods, approaches and procedures, rather than being based on a single all-embracing approach. Only a minimum framework defining the relationship between the approaches is required, together with any common facilitative procedures.
Intra-personal environment A mode of awareness ensuring cross- fertilization between perceptions is most effective and significant when it permits a wide variety of perceptions rather than being based on one all-embracing mode of comprehension. Only a minimum comprehension framework is required to clarify the relationship between the different perceptions.
MG M a Diversified interchange environment
Broader Selective interchange axis (#MP1032) Web of selective interchange (#MP1019).
Narrower Flexible interfaces (#MP1244) Local sources for perspective nourishment (#MP1089) Protected low-density communication pathways (#MP1101) Exchange contexts controlled by a single perspective (#MP1087) Arrangement of structures ot engender fruitful interfaces (#MP1100) Exposure of structural activities to communication pathway (#MP1165) Primary inter-level connections at transitions in boundary orientation (#MP1212).

♦ MP1047 Functionality enhancement

Pattern In any domain a significant proportion of the processes tend to be dysfunctional. A network of nodes catalyzing and enhancing functionality is desirable.
Physical environment In any ordinary neighbourhood more than 90 per cent of the people are unhealthy in terms of simple biological criteria. Their health cannot be effectively enhanced by focusing on illnesses. A network of centres is required, with medical facilities, but focusing on the enhancement of physical and mental health through recreational and educational activities.
Socio-organizational environment In any ordinary group or organizational complex a significant proportion of the roles exhibit dysfunctional characteristics in terms of the health of the organization. These cannot be effectively remedied by focusing directly upon them. A network of facilitative environments is required with counselling expertise but providing the opportunity for role health, exploration and development.
Conceptual environment In any ordinary conceptual environment or school of thought a significant proportion of the conceptual portions or processes exhibit dysfunctional characteristics in terms of the integration of the domain as a whole. These cannot be effectively remedied by focusing directly upon them. A network of facilitative conceptual environments is required with consultative expertise but providing the opportunity for creative exercises in the exploration and further development of such positions.
Intra-personal environment Within any ordinary mode of awareness a significant proportion of the perceptions are unhealthy with respect to the holistic integration of that mode. Such perceptions cannot be effectively improved by focusing directly upon them. A network of facilitative processes is required with integrative potential but providing the opportunity for reflective exercises in the exploration and further development of such perceptions.

MG M b Functionality enhancement
Broader Activity nodes (#MP1030) Functional cycle (#MP1026)
Local action network (#MP1045).
Narrower Adaptive interstices (#MP1048) Access to contained irrationality (#MP1071)
Context for emergence of new perspectives (#MP1065)
Local opportunities for perspective activity (#MP1157)
Provision for temporary perspective inactivity (#MP1150)
Minimally-structured perspective control operations (#MP1081)
Local cultivation of sources of perspective nourishment (#MP1177)
Exposure of structural activities to communication pathway (#MP1165)
Competitive interaction opportunities transposed to a concrete level (#MP1072)
Contexts for exploratory relationship formation challenging emerging perspectives (#MP1073).

♦ MP1048a Adaptive interstices

Pattern Wherever there tends to be a sharp boundary between zones that are active throughout the time cycle and those active for only part of it, the presence of sub–zones of the former type scattered throughout those of the latter ensures the necessary maintenance of rhythms in the partially active zones.
Physical environment Wherever there tends to be a sharp separation between residential and non–residential areas, building occasional houses into the fabric of shops, factories and public buildings ensures that the non–residential area is lived–in and does not deteriorate into a slum area.
Socio–organizational environment Wherever there tends to be a sharp separation in an organizational complex between functions or programmes active throughout the time cycle and those active in part of the time cycle only, integrating some communal or family groups into the structure of the latter reduces their vulnerability when they are inactive.
Conceptual environment Wherever there tends to be a sharp distinction made in a conceptual domain between processes active throughout the time cycle and those active for only part of it, integrating some continually active processes into the latter reduces their vulnerability to conditions prevailing when they are inactive.
Intra–personal environment Wherever there tends to be a sharp separation mode within a mode of awareness between perceptual processes intertwining throughout the daily cycle and those active for only part of it, integrating some of the ongoing processes into the latter reduces their vulnerability to impressions engendered when they are inactive.
Broader Context boundary (#MP1015) Standard frameworks (#MP1038)
Sub–domain boundary (#MP1013) Identifiable context (#MP1014)
Cluster of frameworks (#MP1037) Functionality enhancement (#MP1047)
Selective interchange axis (#MP1032) Presentation of new dimensions (#MP1043)
Informal context for formal processes (#MP1041).
Narrower Perspective–adaptable contexts (#MP1079)
Extended pattern of nuclear interaction (#MP1075)
Concealment of necessary monotonous perspective patterns (#MP1097).

♦ MP1049 Local relationship loops

Pattern To ensure the adaptive interactions difficult with general or global relationships, local relationships should be arranged in loops, such that the loops do not facilitate relationships between more distant points.
Physical environment Local roads should be laid out in the form of loops such as to provide access to houses but not to offer any advantages to through traffic seeking a short–cut.
Socio–organizational environment Communication pathways between groups with closely related concerns should be based on circuits structured to facilitate low–key interaction amongst those groups but to offer no advantages to external communications seeking to use them to establish more efficient relationships between distant groups.
Conceptual environment Closely related specialized conceptual domains should be interlinked by self–reinforcing cycles of interaction which are of little significance as a means of establishing more general patterns of relationship between more distant domains.
Intra–personal environment Closely related patterns of perceptions should be interlinked within a specialized mode of awareness by cycles of self–reinforcement which are of little significance as a means of integrating more general modes of awareness.
MG M a Local relationship loops
Broader Identifiable context (#MP1014) Cluster of frameworks (#MP1037)
Local interrelationship domains (#MP1011) Informal context for formal processes (#MP1041)
Compensating relationships in parallel (#MP1023).
Narrower Protection of emerging foci (#MP1057)
Special modes of relationship (#MP1056)
Three–way relationship entrainment (#MP1050)
Hierachy of perspectives favouring the broadest (#MP1114)
Limitation on number of occupiable temporary sites (#MP1103)
Linear relationships enhanced by non–linear processes (#MP1051)
Harmoniously structured entry point for external communication media (#MP1113)
Interfacing vehicles of communication and networks of unmediated relationships (#MP1052).

♦ MP1050b Three–way relationship entrainment

Pattern Discontinuity is reduced when two relationships merge by entrainment at three–way nodes rather than under the conflictual conditions of four–way intersections in which an effort is made to maintain the orientation of each.
Physical environment At intersections where two roads meet, accidents are less frequent when the roads meet in three–way T junctions rather than in four–way junctions involving crossing movements.
Socio–organizational environment The potential for conflict in any inter–organizational environment is reduced when two programmes interact in a mutually entraining manner with a common resultant rather than when an effort is made to maintain the orientation of each toward unrelated outcomes, overriding the frustration and interruption engendered by the interference of the other.
Conceptual environment The potential for disagreement in any conceptual domain is reduced when two lines of research interact in a mutally entraining manner with a common resultant, rather than when each line of thought is pursued, irrespective of the influence exerted by the other and the necessity of overriding the interference and discontinuity it causes.
Intra–personal environment The potential for cognitive dissonance within any mode of awareness is reduced when two distinct perceptual processes merge into a common third process, rather than when each is pursued, irrespective of the influence of the other and the necessity of overriding the interference and discontinuity to which its insights give rise.
Broader Local relationship loops (#MP1049)
Compensating relationships in parallel (#MP1023).
Narrower Linear relationships enhanced by non–linear processes (#MP1051).

♦ MP1051a Linear relationships enhanced by non–linear processes

Pattern Local linear relationships may be enhanced by ensuring the presence of non–linear processes such that the linearity tends to become implicit rather than explicit.
Physical environment Local asphalt roads destroy the micro– climate and the local environment. Substituting paving stones with grass between them alleviates such effects.
Broader Non–linear organization (#MP1007) Local relationship loops (#MP1049)
Local interrelationship domains (#MP1011) Three–way relationship entrainment (#MP1050)
Compensating relationships in parallel (#MP1023)
Differentiation by relationship density (#MP1036).
Narrower Ambiguous boundaries (#MP1243) Unstructured common domain (#MP1067)
Protection of emerging foci (#MP1057) Embedding fixity within variability (#MP1247)
Cultivation of productive non–linearity (#MP1170) Protecting variability to enhance fixity (#MP1245)
Hierachy of perspectives favouring the broadest (#MP1114)
Overview of communication pathway from structure (#MP1164)
Interaction with coherent irrational perspectives (#MP1074)
Limitation on number of occupiable temporary sites (#MP1103)
Distinct pattern of entry points to complex structures (#MP1102)
Spontaneous relationship formation amongst emerging perspectives (#MP1068)
Relative isolation of structural interface with communication pathway (#MP1140)
Interfacing vehicles of communication and networks of unmediated relationships (#MP1052).

♦ MP1052a Interfacing vehicles of communication and networks of unmediated relationships

Pattern Vehicles of communication may endanger the local communication process although significant activities occur at the points at which they interface with zones of unmediated relationships. A protected network of unmediated relationships may be developed distinct from that of the mediated network and orthogonal to it.
Physical environment Automobiles are dangerous to pedestrians and yet social life is enlivened at the points where they meet. A protected network of pedestrian paths may be developed, distinct from the road system, and orthogonal to it.
Broader Activity nodes (#MP1030) Standard frameworks (#MP1038)
Local relationship loops (#MP1049) Selective interchange axis (#MP1032)
Local interrelationship domains (#MP1011) Cycle of relationship reinforcement (#MP1031)
Compensating relationships in parallel (#MP1023)
Linear relationships enhanced by non–linear processes (#MP1051).
Narrower Protection of emerging foci (#MP1057)
Special modes of relationship (#MP1056)
Focus–oriented communication networks (#MP1120)
Protected low intensity relationships (#MP1055)
Hospitability of communication pathways (#MP1121)
Intersection of differently paced communications (#MP1054)
Arrangement of structures ot engender fruitful interfaces (#MP1100).

♦ MP1053a Principal points of entry

Pattern The distinctiveness of any domain is enhanced by the presence of well–identified points of entry through its boundary.
Physical environment The identity of neighbourhoods, housing clusters or other urban precincts is enhanced by the presence of distinctive gateways for the pathways through the boundary.
Broader Context boundary (#MP1015) Identifiable context (#MP1014)
Cluster of frameworks (#MP1037)
Informal context for formal processes (#MP1041).
Narrower Hospitable transit points (#MP1092) Patterning of complex structures (#MP1098)
Context for transformative experience (#MP1066)
Common external context for inactivity (#MP1069)
Complexification of perspective contexts (#MP1095)
Distinctiveness of main entry point to structure (#MP1110)
Limitation on number of occupiable temporary sites (#MP1103)
Distinct pattern of entry points to complex structures (#MP1102)
Transition domain between structure and communication pathway (#MP1112).

♦ MP1054c Intersection of differently paced communications

Pattern Wherever a network of lower intensity relationships is exposed to one of higher intensity, the intersection should be designed such as to protect the integrity of the latter.
Physical environment Where pedestrian pathways cross roads, the crossing should be protected by, for example, providing islands between the lanes and ensuring its recognition by drivers.
Broader Selective interchange axis (#MP1032)
Compensating relationships in parallel (#MP1023)
Interfacing vehicles of communication and networks of unmediated relationships (#MP1052).
Narrower Flexible interfaces (#MP1244) Hospitable transit points (#MP1092)
Protected low intensity relationships (#MP1055)
Bounded common small–scale interaction domains (#MP1061)
Communication pathways enfolded by non–linearity (#MP1174)
Limitation on number of occupiable temporary sites (#MP1103)
Transit point location for sources of perspective nourishment (#MP1093).

♦ MP1055b Protected low intensity relationships

Pattern In order to prevent high intensity relationships from acquiring disproportionate significance, low intensity relationships should be given a prominent position, especially where they have a common boundary with those of high intensity.
Physical environment Pedestrian sidewalks should be raised 0.5 metres above the level of parallel roads carrying fast–moving traffic, with some form of railing to mark the edge, or else pedestrians feel overwhelmed by the vehicles.
Broader Compensating relationships in parallel (#MP1023)
Differentiation by relationship density (#MP1036)
Intersection of differently paced communications (#MP1054)
Interfacing vehicles of communication and networks of unmediated relationships (#MP1052).
Narrower Ambiguous boundaries (#MP1243) Partially contained interfaces (#MP1119)
Hospitability of communication pathways (#MP1121)
Integrating points of perspective into common domains (#MP1125)
Arrangement of structures ot engender fruitful interfaces (#MP1100)
Transit point location for sources of perspective nourishment (#MP1093).

♦ MP1056b Special modes of relationship

Pattern Some beneficial modes of relationship cannot be satisfactorily combined with others and call for the establishment of a distinct network with appropriate adaptation at the interfaces with other modes.
Physical environment Bicycles are endangered by other vehicles on roads and are themselves a danger to pedestrians on pathways, despite being a healthy mode of transport with little negative impact on the environment. This calls for the establishment of a system of appropriately adapted bicycle paths with bike racks located close to common destinations.
Broader Local relationship loops (#MP1049) Local interrelationship domains (#MP1011)
Interfacing vehicles of communication and networks of unmediated relationships (#MP1052).
Narrower Protection of emerging foci (#MP1057)
Partially contained interfaces (#MP1119)
Low–intensity communication pathways (#MP1059)
Distinctiveness of main entry point to structure (#MP1110)
Protecting non–linear contexts from communication pathways (#MP1173)
Contexts for exploratory relationship formation challenging emerging perspectives (#MP1073).

♦ **MP1057c Protection of emerging foci**
Pattern A specially protected network of relationships should be provided to enable new foci to emerge and develop their potential for interaction with the organized domain in which they will become established.
Physical environment A specially protected network of pathways should be provided to enable children to explore the variety of buildings which make up the urban environment in which they have to learn to function as adults.
Broader Functional cycle (#MP1026) Local relationship loops (#MP1049)
Network of redefinitions (#MP1018) Special modes of relationship (#MP1056)
Presentation of new dimensions (#MP1043) Cycle of relationship reinforcement (#MP1031)
Integrating the historical dimension (#MP1040) Informal context for formal processes (#MP1041)
Compensating relationships in parallel (#MP1023)
Linear relationships enhanced by non-linear processes (#MP1051)
Interfacing vehicles of communication and networks of unmediated relationships (#MP1052).
Narrower Contexts for care of premature perspectives (#MP1086)
Overview of communication pathway from structure (#MP1164)
Transitional contexts for perspective reorganization (#MP1084)
Perspective imitation contexts for developing perspectives (#MP1085)
Spontaneous relationship formation amongst emerging perspectives (#MP1068)
Competitive interaction opportunities transposed to a concrete level (#MP1072)
Contexts for exploratory relationship formation challenging emerging perspectives (#MP1073).

♦ **MP1058c Context for disorder**
Pattern A section of any organized domain may be reserved as a context for disorderly processes. These are a necessary complement to order and the stability of ordered structures.
Physical environment Within a town a part of it may be reserved as a place for carnival-type events, whether fairs, displays, street theatre or any kind of unprogrammed happening. These allow the necessary expression of man's unconscious processes and fantasy.
Broader Access to intensity (#MP1010) Underdefined processes (#MP1033)
Cycle of relationship reinforcement (#MP1031).
Narrower Flexible interfaces (#MP1244)
Common external context for inactivity (#MP1069)
Bounded common small-scale interaction domains (#MP1061)
Arrangement of structures ot engender fruitful interfaces (#MP1100)
Transit point location for sources of perspective nourishment (#MP1093)
Cyclic interrelation of complementary perspective in common domains (#MP1063).

♦ **MP1059 Low-intensity communication pathways**
Physical environment Quiet lanes and pathways.
Broader Context boundary (#MP1015) Sub-domain boundary (#MP1013)
Special modes of relationship (#MP1056) Presentation of new dimensions (#MP1043)
Positions enabling transcendence (#MP1024)
Coherent pattern of relationship densities (#MP1028).
Narrower Accessible non-linearity (#MP1060)
Access to patterns of active irrationality (#MP1064)
Context for acknowledgement of past perspectives (#MP1070)
Protecting non-linear contexts from communication pathways (#MP1173)
Relative isolation of structural interface with communication pathway (#MP1140).

♦ **MP1060 Accessible non-linearity**
Physical environment Accessible trees and greenery.
MG M a Accessible non-linearity
Broader Context boundary (#MP1015) Sub-domain boundary (#MP1013)
Identifiable context (#MP1014) Decentralized formal processes (#MP1009)
Low-intensity communication pathways (#MP1059)
Informal context for formal processes (#MP1041).
Narrower Hospitable common domains (#MP1094)
Unstructured common domain (#MP1067)
Context for transformative experience (#MP1066)
Common external context for inactivity (#MP1069)
Hierachy of perspectives favouring the broadest (#MP1114)
Context for acknowledgement of past perspectives (#MP1070)
Interaction with coherent irrational perspectives (#MP1074)
Appropriate relationship of non-linearity to structures (#MP1171)
Perspective imitation contexts for developing perspectives (#MP1085)
Protecting non-linear contexts from communication pathways (#MP1173)
Competitive interaction opportunities transposed to a concrete level (#MP1072)
Functional enhancement of domains separating complementary structures (#MP1106).

♦ **MP1061a Bounded common small-scale interaction domains**
Physical environment Small public squares in urban environments.
Broader Activity nodes (#MP1030) Local focal points (#MP1044)
Context for disorder (#MP1058) Identifiable context (#MP1014)
Points of wider perspective (#MP1062) Cycle of relationship reinforcement (#MP1031)
Informal context for formal processes (#MP1041)
Stable density gradient of local relationships (#MP1029)
Intersection of differently paced communications (#MP1054).
Narrower Hospitable common domains (#MP1094)
Enhancing function of common domains (#MP1123)
Common external context for inactivity (#MP1069)
Informal local perspective interface zones (#MP1088)
Ensuring function of common domain interfaces (#MP1124)
Hierachy of perspectives favouring the broadest (#MP1114)
Overview of communication pathway from structure (#MP1164)
Off-centering the focal point of a common domain (#MP1126)
Integrating points of perspective into common domains (#MP1125)
Transit point location for sources of perspective nourishment (#MP1093)
Direct relationship between structures and communication pathways (#MP1122)
Cyclic interrelation of complementary perspective in common domains (#MP1063)
Functional enhancement of domains separating complementary structures (#MP1106).

♦ **MP1062b Points of wider perspective**
Physical environment High places as vantage points.
Broader Four-level structural limit (#MP1021) Individuality in multiplicity (#MP1012).
Narrower Context for transformative experience (#MP1066)
External access to higher structural levels (#MP1158)
Limiting exposure to harmonious perspectives (#MP1134)
Bounded common small-scale interaction domains (#MP1061)
Off-centering the focal point of a common domain (#MP1126)
Integrating points of perspective into common domains (#MP1125).

♦ **MP1063b Cyclic interrelation of complementary perspective in common domains**
Physical environment Communal dancing in the street.
Broader Access to intensity (#MP1010) Context for disorder (#MP1058)
Underdefined processes (#MP1033) Cycle of relationship reinforcement (#MP1031)
Bounded common small-scale interaction domains (#MP1061).
Narrower Flexible interfaces (#MP1244)
Common external context for inactivity (#MP1069)
Ensuring function of common domain interfaces (#MP1124)
Off-centering the focal point of a common domain (#MP1126)
Transit point location for sources of perspective nourishment (#MP1093).

♦ **MP1064b Access to patterns of active irrationality**
Physical environment Flowing water, streams and pools.
Broader Context boundary (#MP1015) Sub-domain boundary (#MP1013)
Relationship to indeterminacy (#MP1025) Positions enabling transcendence (#MP1024)
Cycle of relationship reinforcement (#MP1031)
Low-intensity communication pathways (#MP1059).
Narrower Partially contained interfaces (#MP1119)
Access to contained irrationality (#MP1071)
Context for transformative experience (#MP1066)
Off-centering the focal point of a common domain (#MP1126)
Arrangement of structures ot engender fruitful interfaces (#MP1100).

♦ **MP1065c Context for emergence of new perspectives**
Physical environment Places for natural childbirth.
Broader Activity nodes (#MP1030) Functional cycle (#MP1026)
Identifiable context (#MP1014) Functionality enhancement (#MP1047)
Individuality in multiplicity (#MP1012).
Narrower Complexification of perspective contexts (#MP1095)
Hospitable domain for perspective nourishment (#MP1139)
Common domain at the focal point of a structure (#MP1129)
Relative isolation of complementary perspectives (#MP1136)
Hospitable contexts for perspectives in transition (#MP1091)
Protecting non-linear contexts from communication pathways (#MP1173)
Blended integration of formal structure and informal context (#MP1111).

♦ **MP1066b Context for transformative experience**
Physical environment Holy ground and sacred places.
Broader Functional cycle (#MP1026) Accessible non-linearity (#MP1060)
Principal points of entry (#MP1053) Points of wider perspective (#MP1062)
Positions enabling transcendence (#MP1024)
Access to patterns of active irrationality (#MP1064).
Narrower Limiting exposure to harmonious perspectives (#MP1134)
Context for acknowledgement of past perspectives (#MP1070)
Appropriate relationship of non-linearity to structures (#MP1171).

♦ **MP1067a Unstructured common domain**
Physical environment Common land in urban areas.
Broader Standard frameworks (#MP1038) Cluster of frameworks (#MP1037)
Accessible non-linearity (#MP1060) Integrating a new dimension (#MP1039)
Informal context for formal processes (#MP1041)
Linear relationships enhanced by non-linear processes (#MP1051).
Narrower Patterning of complex structures (#MP1098)
Access to contained irrationality (#MP1071)
Common external context for inactivity (#MP1069)
Cultivation of productive non-linearity (#MP1170)
Context for acknowledgement of past perspectives (#MP1070)
Hierachy of perspectives favouring the broadest (#MP1114)
Off-centering the focal point of a common domain (#MP1126)
Interaction with coherent irrational perspectives (#MP1074)
Insight capturing non-linear extensions of structures (#MP1175)
Local cultivation of sources of perspective nourishment (#MP1177)
Spontaneous relationship formation amongst emerging perspectives (#MP1068)
Orientation of structures to enhance receptivity to external insight (#MP1105)
Competitive interaction opportunities transposed to a concrete level (#MP1072)
Functional enhancement of domains separating complementary structures (#MP1106)
Contexts for exploratory relationship formation challenging emerging perspectives (#MP1073).

♦ **MP1068b Spontaneous relationship formation amongst emerging perspectives**
Physical environment Connected environment for childrens play.
Broader Cluster of frameworks (#MP1037) Unstructured common domain (#MP1067)
Protection of emerging foci (#MP1057) Integrating a new dimension (#MP1039)
Adequate variety of cyclic elements (#MP1035)
Linear relationships enhanced by non-linear processes (#MP1051).
Narrower Access to contained irrationality (#MP1071)
Domain for developing perspectives (#MP1137)
Contexts for care of premature perspectives (#MP1086)
Interaction with coherent irrational perspectives (#MP1074)
Contexts for exploratory relationship formation challenging emerging perspectives (#MP1073).

♦ **MP1069a Common external context for inactivity**
Physical environment Public outdoor covered place where people can gather.
Broader Context boundary (#MP1015) Context for disorder (#MP1058)
Local action network (#MP1045) Accessible non-linearity (#MP1060)
Principal points of entry (#MP1053) Unstructured common domain (#MP1067)
Bounded common small-scale interaction domains (#MP1061)
Cyclic interrelation of complementary perspective in common domains (#MP1063).
Narrower Inter-level zone (#MP1226) Flexible interfaces (#MP1244)
Hospitable transit points (#MP1092) Attractive temporary positions (#MP1241)
Partially contained interfaces (#MP1119) Access to contained irrationality (#MP1071)
Focus-oriented communication networks (#MP1120)
Hospitability of communication pathways (#MP1121)
Ensuring function of common domain interfaces (#MP1124)
Common domain at the focal point of a structure (#MP1129)
Off-centering the focal point of a common domain (#MP1126)
Hospitable non-linear domain external to structures (#MP1163)
Integrating points of perspective into common domains (#MP1125)
Functional integration of unstructured internal domains (#MP1115)
Arrangement of structures ot engender fruitful interfaces (#MP1100)
Hospitable interface between structures and external environment (#MP1160).

♦ **MP1070 Context for acknowledgement of past perspectives**
Physical environment Grave sites and memorials to the dead.
MG M b Context for acknowledgment of past perspectives
Broader Functional cycle (#MP1026) Context boundary (#MP1015)
Identifiable context (#MP1014) Accessible non-linearity (#MP1060)
Unstructured common domain (#MP1067) Protection of emerging foci (#MP1057)
Low-intensity communication pathways (#MP1059)
Context for transformative experience (#MP1066).
Narrower Attractive temporary positions (#MP1241)
Appropriate relationship of non-linearity to structures (#MP1171).

♦ **MP1071b Access to contained irrationality**
Physical environment Still water, ponds and lakeside areas.

Broader Sub–domain boundary (#MP1013) Cluster of frameworks (#MP1037)
Functionality enhancement (#MP1047) Unstructured common domain (#MP1067)
Relationship to indeterminacy (#MP1025) Informal context for formal processes (#MP1041)
Common external context for inactivity (#MP1069)
Access to patterns of active irrationality (#MP1064)
Coherent pattern of relationship densities (#MP1028)
Spontaneous relationship formation amongst emerging perspectives (#MP1068).
Narrower Ambiguous boundaries (#MP1243)
Contexts of self–organizing non–linearity (#MP1172)
Provision for temporary perspective inactivity (#MP1150)
Off–centering the focal point of a common domain (#MP1126)
Communication pathways enfolded by non–linearity (#MP1174)
Integrating coordinated exposure to irrationality into structures (#MP1144)
Orientation of structures to enhance receptivity to external insight (#MP1105)
Competitive interaction opportunities transposed to a concrete level (#MP1072).

♦ MP1072b Competitive interaction opportunities transposed to a concrete level
Physical environment Local sports facilities.
Broader Context boundary (#MP1015) Accessible non–linearity (#MP1060)
Functionality enhancement (#MP1047) Unstructured common domain (#MP1067)
Protection of emerging foci (#MP1057) Access to contained irrationality (#MP1071)
Informal context for formal processes (#MP1041)
Transitional contexts for perspective reorganization (#MP1084).
Narrower Ambiguous boundaries (#MP1243) Attractive temporary positions (#MP1241)
Complexification of perspective contexts (#MP1095)
Protected low–density communication pathways (#MP1101)
Exposure of structural activities to communication pathway (#MP1165)
Integrating coordinated exposure to irrationality into structures (#MP1144).

♦ MP1073c Contexts for exploratory relationship formation challenging emerging perspectives
Physical environment Adventure playground for children.
Broader Context boundary (#MP1015) Functionality enhancement (#MP1047)
Unstructured common domain (#MP1067) Protection of emerging foci (#MP1057)
Special modes of relationship (#MP1056)
Spontaneous relationship formation amongst emerging perspectives (#MP1068).
Narrower Ambiguous boundaries (#MP1243) Structure–enfolded insight domain (#MP1161)
Contexts of self-organizing non-linearity (#MP1172)
Exclusive spaces for emergent perspectives (#MP1203)
Contexts for care of premature perspectives (#MP1086).

♦ MP1074c Interaction with coherent irrational perspectives
Physical environment Facilities for contact with animals.
Broader Accessible non–linearity (#MP1060) Unstructured common domain (#MP1067)
Linear relationships enhanced by non–linear processes (#MP1051)
Spontaneous relationship formation amongst emerging perspectives (#MP1068).
Narrower Perspective–adaptable contexts (#MP1079)
Contexts for care of premature perspectives (#MP1086)
Re–integration of rejected perspective by–products (#MP1178).

♦ MP1075b Extended pattern of nuclear interaction
Physical environment Communal households.
Broader Functional cycle (#MP1026) Standard frameworks (#MP1038)
Adaptive interstices (#MP1048) Cluster of frameworks (#MP1037)
Integrating a new dimension (#MP1039) Adequate variety of cyclic elements (#MP1035)
Integrating the historical dimension (#MP1040).
Narrower Perspective–adaptable contexts (#MP1079)
Minimal context for single perspective (#MP1078)
Coordination of perspective nourishment (#MP1147)
Complexification of perspective contexts (#MP1095)
Contexts for care of premature perspectives (#MP1086)
Minimal context for complementary perspectives (#MP1077)
Common domain at the focal point of a structure (#MP1129)
Relatively isolated context for each perspective (#MP1141)
Relative isolation of complementary perspectives (#MP1136)
Semi–autonomous contexts for maturing perspectives (#MP1154)
Opportunities for perspectives of decreasing activity (#MP1156)
Minimal context interrelating mature and emerging perspectives (#MP1076)
Semi–autonomous contexts for perspectives of decreasing activity (#MP1155)
Structures adaptable to changing number of embodied perspectives (#MP1153)
Organization of structures to enhance receptivity to external insight (#MP1107).

♦ MP1076b Minimal context interrelating mature and emerging perspectives
Physical environment House for a small family.
Broader Standard frameworks (#MP1038) Cluster of frameworks (#MP1037)
Adequate variety of cyclic elements (#MP1035)
Extended pattern of nuclear interaction (#MP1075).
Narrower Perspective–adaptable contexts (#MP1079)
Domain for developing perspectives (#MP1137)
Minimal context for single perspective (#MP1078)
External access to higher structural levels (#MP1158)
Common domain at the focal point of a structure (#MP1129)
Relatively isolated context for each perspective (#MP1141)
Relative isolation of complementary perspectives (#MP1136)
Semi–autonomous contexts for maturing perspectives (#MP1154)
Domains for non–current elements and those in reserve (#MP1145)
Interrelationship of contexts of developing perspectives (#MP1143)
Organization of structures to enhance receptivity to external insight (#MP1107).

♦ MP1077b Minimal context for complementary perspectives
Physical environment House for a couple.
Broader Standard frameworks (#MP1038) Cluster of frameworks (#MP1037)
Adequate variety of cyclic elements (#MP1035)
Extended pattern of nuclear interaction (#MP1075).
Narrower Perspective–adaptable contexts (#MP1079)
External access to higher structural levels (#MP1158)
Relatively isolated context for each perspective (#MP1141)
Relative isolation of complementary perspectives (#MP1136)
Organization of structures to enhance receptivity to external insight (#MP1107).

♦ MP1078b Minimal context for single perspective
Physical environment House for one person.
Broader Standard frameworks (#MP1038) Cluster of frameworks (#MP1037)
Adequate variety of cyclic elements (#MP1035)
Extended pattern of nuclear interaction (#MP1075)
Minimal context interrelating mature and emerging perspectives (#MP1076).
Narrower Perspective–adaptable contexts (#MP1079)
Partially exposed perspective context (#MP1183)
External access to higher structural levels (#MP1158)
Occupiable sites for perspective inactivity (#MP1188)
Context for forms of perspective presentation (#MP1189)
Occupiable sites exposed to external insight (#MP1180)
Hospitable domain for perspective nourishment (#MP1139)
Semi–autonomous contexts for maturing perspectives (#MP1154)
Semi–autonomous contexts for perspectives of decreasing activity (#MP1155)
Integrating coordinated exposure to irrationality into structures (#MP1144)
Organization of structures to enhance receptivity to external insight (#MP1107).

♦ MP1079a Perspective–adaptable contexts
Physical environment Owning a home that can be modified and adapted.
Broader Functional cycle (#MP1026) Standard frameworks (#MP1038)
Adaptive interstices (#MP1048) Cluster of frameworks (#MP1037)
Integrating a new dimension (#MP1039)
Minimal context for single perspective (#MP1078)
Minimal context for single perspective (#MP1078)
Extended pattern of nuclear interaction (#MP1075)
Minimal context for complementary perspectives (#MP1077)
Minimal context for complementary perspectives (#MP1077)
Interaction with coherent irrational perspectives (#MP1074)
Minimal context interrelating mature and emerging perspectives (#MP1076).
Narrower Complexification of perspective contexts (#MP1095)
Contexts for care of premature perspectives (#MP1086)
Local opportunities for perspective activity (#MP1157)
Blended integration of formal structure and informal context (#MP1111).

♦ MP1080a Integrated contexts for perspective dynamics
Physical environment Self–governing workshops, offices and cooperatives.
Broader Decentralized formal processes (#MP1009)
Informal context for formal processes (#MP1041)
Chain of fundamental transformation zones (#MP1042).
Narrower Flexible domain organization (#MP1146)
Perspective interaction constraints (#MP1148)
Coordination of perspective nourishment (#MP1147)
Complexification of perspective contexts (#MP1095)
External access to higher structural levels (#MP1158)
Minimally–structured perspective control operations (#MP1081)
Domains for non–current elements and those in reserve (#MP1145)
Perspective imitation contexts for developing perspectives (#MP1085)
Hospitable reception of external perspectives by structures (#MP1149)
Minimal distance between related operational control contexts (#MP1082)
Structures adaptable to changing number of embodied perspectives (#MP1153)
Organization of structures to enhance receptivity to external insight (#MP1107)
Integration of perspective acquisition and perspective maintenance dynamics (#MP1083).

♦ MP1081b Minimally–structured perspective control operations
Physical environment Community services without excessive bureaucracy.
Broader Local focal points (#MP1044) Local action network (#MP1045)
Functionality enhancement (#MP1047) Presentation of new dimensions (#MP1043)
Informal context for formal processes (#MP1041)
Integrated contexts for perspective dynamics (#MP1080).
Narrower Flexible domain organization (#MP1146)
Perspective interaction constraints (#MP1148)
Complexification of perspective contexts (#MP1095)
External access to higher structural levels (#MP1158)
Protected low–density communication pathways (#MP1101)
Provision for temporary perspective inactivity (#MP1150)
Transitional contexts for perspective reorganization (#MP1084)
Distinct pattern of entry points to complex structures (#MP1102)
Hospitable reception of external perspectives by structures (#MP1149)
Minimal distance between related operational control contexts (#MP1082)
Structures adaptable to changing number of embodied perspectives (#MP1153)
Organization of structures to enhance receptivity to external insight (#MP1107).

♦ MP1082b Minimal distance between related operational control contexts
Physical environment Connections between offices.
Broader Four–level structural limit (#MP1021) Informal context for formal processes (#MP1041)
Integrated contexts for perspective dynamics (#MP1080)
Minimally–structured perspective control operations (#MP1081).
Narrower Flexible domain organization (#MP1146)
Perspective interaction constraints (#MP1148)
Complexification of perspective contexts (#MP1095)
External access to higher structural levels (#MP1158)
Protected low–density communication pathways (#MP1101)
Provision for temporary perspective inactivity (#MP1150)
Distinct pattern of entry points to complex structures (#MP1102)
Arrangement of structures ot engender fruitful interfaces (#MP1100)
Organization of structures to enhance receptivity to external insight (#MP1107).

♦ MP1083b Integration of perspective acquisition and perspective maintenance dynamics
Physical environment Learning through work in an apprenticeship setting.
Broader Functional cycle (#MP1026) Network of redefinitions (#MP1018)
Presentation of new dimensions (#MP1043)
Integrated contexts for perspective dynamics (#MP1080).
Narrower Partially isolated contexts (#MP1152) Perspective interaction constraints (#MP1148)
Partially exposed perspective context (#MP1183)
Coordination of perspective nourishment (#MP1147)
Small–scale perspective interaction contexts (#MP1151)
Common domain at the focal point of a structure (#MP1129)
Transitional contexts for perspective reorganization (#MP1084)
Organization of structures to enhance receptivity to external insight (#MP1107).

♦ MP1084c Transitional contexts for perspective reorganization
Physical environment Institutional environments for teenagers.
Broader Activity nodes (#MP1030) Functional cycle (#MP1026)
Sub–domain boundary (#MP1013) Network of redefinitions (#MP1018)
Protection of emerging foci (#MP1057)
Minimally–structured perspective control operations (#MP1081)
Integration of perspective acquisition and perspective maintenance dynamics (#MP1083).
Narrower Coordination of perspective nourishment (#MP1147)
Complexification of perspective contexts (#MP1095)
Local opportunities for perspective activity (#MP1157)
Competitive interaction opportunities transposed to a concrete level (#MP1072).

♦ MP1085c Perspective imitation contexts for developing perspectives
Physical environment Shopfront schools.

Broader Activity nodes (#MP1030) Functional cycle (#MP1026)
Accessible non–linearity (#MP1060) Network of redefinitions (#MP1018)
Protection of emerging foci (#MP1057)
Integrated contexts for perspective dynamics (#MP1080).
Narrower Complexification of perspective contexts (#MP1095)
Contexts for care of premature perspectives (#MP1086)
Arrangement of structures ot engender fruitful interfaces (#MP1100)
Exposure of structural activities to communication pathway (#MP1165).

♦ MP1086b Contexts for care of premature perspectives
Physical environment Children's home where they can gather and play.
Broader Functional cycle (#MP1026) Network of redefinitions (#MP1018)
Protection of emerging foci (#MP1057) Perspective–adaptable contexts (#MP1079)
Integrating the historical dimension (#MP1040)
Extended pattern of nuclear interaction (#MP1075)
Interaction with coherent irrational perspectives (#MP1074)
Perspective imitation contexts for developing perspectives (#MP1085)
Spontaneous relationship formation amongst emerging perspectives (#MP1068)
Contexts for exploratory relationship formation challenging emerging perspectives (#MP1073).
Narrower Complexification of perspective contexts (#MP1095)
Exclusive spaces for emergent perspectives (#MP1203)
Protected low–density communication pathways (#MP1101)
Common domain at the focal point of a structure (#MP1129).

♦ MP1087a Exchange contexts controlled by a single perspective
Physical environment Individually owned shops.
Broader Activity nodes (#MP1030) Local action network (#MP1045)
Selective interchange axis (#MP1032) Diversified interchange environment (#MP1046).
Narrower Facilities for perspective adjuncts (#MP1200)
Complexification of perspective contexts (#MP1095)
Local sources for perspective nourishment (#MP1089)
Informal local perspective interface zones (#MP1088)
Substantive distinctions separating domains (#MP1197)
Appropriate proportions of perspective contexts (#MP1191)
Domains for non–current elements and those in reserve (#MP1145)
Concealment of necessary monotonous perspective patterns (#MP1097)
Exposure of structural activities to communication pathway (#MP1165)
Transit point location for sources of perspective nourishment (#MP1093)
Organization of structures to enhance receptivity to external insight (#MP1107).

♦ MP1088a Informal local perspective interface zones
Physical environment Street cafe.
Broader Activity nodes (#MP1030) Identifiable context (#MP1014)
Underdefined processes (#MP1033) Informal context for formal processes (#MP1041)
Bounded common small–scale interaction domains (#MP1061)
Exchange contexts controlled by a single perspective (#MP1087).
Narrower Different settings (#MP1251) Flexible interfaces (#MP1244)
Ambiguous boundaries (#MP1243) Hospitable common domains (#MP1094)
Enhancing function of common domains (#MP1123)
Complexification of perspective contexts (#MP1095)
Ensuring function of common domain interfaces (#MP1124)
Provision for temporary perspective inactivity (#MP1150)
Integrating points of perspective into common domains (#MP1125)
Exposure of structural activities to communication pathway (#MP1165).

♦ MP1089b Local sources for perspective nourishment
Physical environment Corner grocery.
Broader Identifiable context (#MP1014) Web of selective interchange (#MP1019)
Diversified interchange environment (#MP1046)
Exchange contexts controlled by a single perspective (#MP1087).
Narrower Facilities for perspective adjuncts (#MP1200)
Complexification of perspective contexts (#MP1095)
Substantive distinctions separating domains (#MP1197)
Appropriate proportions of perspective contexts (#MP1191)
Distinctiveness of main entry point to structure (#MP1110)
Exposure of structural activities to communication pathway (#MP1165).

♦ MP1090c Unstructured context for perspective exchange
Physical environment Beer hall.
Broader Activity nodes (#MP1030) Context boundary (#MP1015)
Underdefined processes (#MP1033) Cycle of relationship reinforcement (#MP1031).
Narrower Complexification of perspective contexts (#MP1095)
Variation in size of perspective contexts (#MP1190)
Hospitable contexts for perspectives in transition (#MP1091)
Organization of structure to provide occupiable sites (#MP1179)
Maintenance of source of direct insight within structures as a focal point (#MP1181).

♦ MP1091b Hospitable contexts for perspectives in transition
Physical environment Congenial inns for travellers.
Broader Activity nodes (#MP1030) Access to intensity (#MP1010)
Underdefined processes (#MP1033) Cycle of relationship reinforcement (#MP1031)
Informal context for formal processes (#MP1041)
Context for emergence of new perspectives (#MP1065)
Unstructured context for perspective exchange (#MP1090).
Narrower Hospitable common domains (#MP1094)
Coordination of perspective nourishment (#MP1147)
Complexification of perspective contexts (#MP1095)
Common domain at the focal point of a structure (#MP1129)
Appropriate configuration for perspective inactivity (#MP1186).

♦ MP1092b Hospitable transit points
Physical environment Bus stop waiting environments.
Broader Interchange (#MP1034) Principal points of entry (#MP1053)
Common external context for inactivity (#MP1069)
User–determined specialized communications (#MP1020)
Intersection of differently paced communications (#MP1054).
Narrower Attractive temporary positions (#MP1241)
Hospitability of communication pathways (#MP1121)
Provision for temporary perspective inactivity (#MP1150)
Transit point location for sources of perspective nourishment (#MP1093).

♦ MP1093b Transit point location for sources of perspective nourishment
Physical environment Food stands in the street.
Broader Activity nodes (#MP1030) Context for disorder (#MP1058)
Hospitable transit points (#MP1092) Protected low intensity relationships (#MP1055)
Informal context for formal processes (#MP1041)
Bounded common small–scale interaction domains (#MP1061)
Intersection of differently paced communications (#MP1054)
Exchange contexts controlled by a single perspective (#MP1087)
Cyclic interrelation of complementary perspective in common domains (#MP1063).
Narrower Flexible interfaces (#MP1244)
Ensuring function of common domain interfaces (#MP1124)
Integrating coordinated exposure to irrationality into structures (#MP1144).

♦ MP1094c Hospitable common domains
Physical environment Sleeping in public places.
Broader Interchange (#MP1034) Accessible non–linearity (#MP1060)
Informal local perspective interface zones (#MP1088)
Bounded common small–scale interaction domains (#MP1061)
Hospitable contexts for perspectives in transition (#MP1091).
Narrower Attractive temporary positions (#MP1241)
Occupiable sites for perspective inactivity (#MP1188)
Protected low–density communication pathways (#MP1101)
Provision for temporary perspective inactivity (#MP1150)
Arrangement of structures ot engender fruitful interfaces (#MP1100)
Hospitable interface between structures and external environment (#MP1160).

♦ MP1095a Complexification of perspective contexts
Physical environment Building complex.
Broader Standard frameworks (#MP1038) Principal points of entry (#MP1053)
Four–level structural limit (#MP1021) Perspective–adaptable contexts (#MP1079)
Presentation of new dimensions (#MP1043)
Extended pattern of nuclear interaction (#MP1075)
Local sources for perspective nourishment (#MP1089)
Context for emergence of new perspectives (#MP1065)
Informal local perspective interface zones (#MP1088)
Contexts for care of premature perspectives (#MP1086)
Integrated contexts for perspective dynamics (#MP1080)
Unstructured context for perspective exchange (#MP1090)
Hospitable contexts for perspectives in transition (#MP1091)
Minimally–structured perspective control operations (#MP1081)
Exchange contexts controlled by a single perspective (#MP1087)
Transitional contexts for perspective reorganization (#MP1084)
Perspective imitation contexts for developing perspectives (#MP1085)
Minimal distance between related operational control contexts (#MP1082)
Competitive interaction opportunities transposed to a concrete level (#MP1072).
Narrower Interconnected structures (#MP1108) Patterning of complex structures (#MP1098)
Focal centre of a complex structure (#MP1099)
Focus–oriented communication networks (#MP1120)
Patterning integrative superstructure (#MP1116)
Limitation on number of structural levels (#MP1096)
Protected low–density communication pathways (#MP1101)
Distinctiveness of main entry point to structure (#MP1110)
Domains for non–current elements and those in reserve (#MP1145)
Concealment of necessary monotonous perspective patterns (#MP1097)
Arrangement of structures ot engender fruitful interfaces (#MP1100)
Organization of structure to enhance autonomy of sub–structures (#MP1109)
Direct relationship between structures and communication pathways (#MP1122)
Orientation of structures to enhance receptivity to external insight (#MP1105)
Organization of structures to enhance receptivity to external insight (#MP1107)
Structural development designed to counteract deficiencies in pattern harmony (#MP1104)
Congruence between spaces defined by the framework and spaces defined by the processes within it (#MP1205).

♦ MP1096b Limitation on number of structural levels
Physical environment Number of building stories.
Broader Four–level structural limit (#MP1021)
Complexification of perspective contexts (#MP1095).
Narrower Nested levels of accessibility (#MP1127)
Patterning of complex structures (#MP1098)
Focal centre of a complex structure (#MP1099)
Patterning integrative superstructure (#MP1116)
Containment by integrative superstructure (#MP1117)
Distribution of secondary inter–level connections (#MP1213)
Appropriate relationship of non–linearity to structures (#MP1171)
Integration of non–linearity into integrative superstructure (#MP1118)
Orientation of structures to enhance receptivity to external insight (#MP1105)
Structural development designed to counteract deficiencies in pattern harmony (#MP1104)
Congruence between spaces defined by the framework and spaces defined by the processes within it (#MP1205).

♦ MP1097b Concealment of necessary monotonous perspective patterns
Physical environment Shielded parking areas.
Broader Encirclement (#MP1017) Context boundary (#MP1015)
Sub–domain boundary (#MP1013) Adaptive interstices (#MP1048)
Selective interchange axis (#MP1032) Integrating a new dimension (#MP1039)
Occupiable temporary site limit (#MP1022) Local interrelationship domains (#MP1011)
Complexification of perspective contexts (#MP1095)
Exchange contexts controlled by a single perspective (#MP1087).
Narrower Flexible interfaces (#MP1244) Patterning of complex structures (#MP1098)
External access to higher structural levels (#MP1158)
Distinctiveness of main entry point to structure (#MP1110)
Limitation on number of occupiable temporary sites (#MP1103)
Enhancing insight by varying levels of exposure to it (#MP1135)
Distinct pattern of entry points to complex structures (#MP1102)
Overview domains at interfaces of the structure with the external environment (#MP1166)
Congruence between spaces defined by the framework and spaces defined by the processes within it (#MP1205).

♦ MP1098a Patterning of complex structures
Physical environment Recognizable pathways through a building complex.
Broader Cluster of frameworks (#MP1037) Principal points of entry (#MP1053)
Unstructured common domain (#MP1067)
Complexification of perspective contexts (#MP1095)
Limitation on number of structural levels (#MP1096)
Concealment of necessary monotonous perspective patterns (#MP1097).
Narrower Partially contained interfaces (#MP1119)
Focal centre of a complex structure (#MP1099)
Focus–oriented communication networks (#MP1120)
Limiting length of communication pathways (#MP1132)
Protected low–density communication pathways (#MP1101)
Hierachy of perspectives favouring the broadest (#MP1114)
Distinctiveness of main entry point to structure (#MP1110)
Limitation on number of occupiable temporary sites (#MP1103)
Distinct pattern of entry points to complex structures (#MP1102)
Functional integration of unstructured internal domains (#MP1115)
Arrangement of structures ot engender fruitful interfaces (#MP1100)
Structural development designed to counteract deficiencies in pattern harmony (#MP1104).

♦ MP1099b Focal centre of a complex structure
Physical environment Main building as a point of focus.
Broader Patterning of complex structures (#MP1098)
Complexification of perspective contexts (#MP1095)
Limitation on number of structural levels (#MP1096).
Narrower Patterning integrative superstructure (#MP1116)
Common domain at the focal point of a structure (#MP1129)
Congruence between spaces defined by the framework and spaces defined by the processes within it (#MP1205).

♦ MP1100a Arrangement of structures ot engender fruitful interfaces
Physical environment Pedestrian streets and concourses.
Broader Activity nodes (#MP1030)
Standard frameworks (#MP1038)
Context for disorder (#MP1058)
Hospitable common domains (#MP1094)
Selective interchange axis (#MP1032)
Integrating a new dimension (#MP1039)
Presentation of new dimensions (#MP1043)
Patterning of complex structures (#MP1098)
Diversified interchange environment (#MP1046)
Cycle of relationship reinforcement (#MP1031)
Protected low intensity relationships (#MP1055)
Common external context for inactivity (#MP1069)
Differentiation by relationship density (#MP1036)
Complexification of perspective contexts (#MP1095)
Access to patterns of active irrationality (#MP1064)
Perspective imitation contexts for developing perspectives (#MP1085)
Minimal distance between related operational control contexts (#MP1082)
Interfacing vehicles of communication and networks of unmediated relationships (#MP1052).
Narrower Ambiguous boundaries (#MP1243)
Partially contained interfaces (#MP1119)
Enhancing function of common domains (#MP1123)
Hospitability of communication pathways (#MP1121)
External access to higher structural levels (#MP1158)
Protected low–density communication pathways (#MP1101)
Ensuring function of common domain interfaces (#MP1124)
Enfolded overview domains of minimum proportions (#MP1167)
Overview of communication pathway from structure (#MP1164)
Distinct pattern of entry points to complex structures (#MP1102)
Relative isolation of structural interface with communication pathway (#MP1140)
Overview domains at interfaces of the structure with the external environment (#MP1166).

♦ MP1101c Protected low–density communication pathways
Physical environment Building thoroughfares and covered pedestrian concourses.
Broader Hospitable common domains (#MP1094)
Patterning of complex structures (#MP1098)
Diversified interchange environment (#MP1046)
Complexification of perspective contexts (#MP1095)
Contexts for care of premature perspectives (#MP1086)
Minimally–structured perspective control operations (#MP1081)
Arrangement of structures ot engender fruitful interfaces (#MP1100)
Minimal distance between related operational control contexts (#MP1082)
Competitive interaction opportunities transposed to a concrete level (#MP1072).
Narrower Connectedness in isolation (#MP1237)
Partially enclosed internal domains (#MP1193)
Enhancing function of common domains (#MP1123)
Internal connectedness between domains (#MP1194)
Limiting length of communication pathways (#MP1132)
Variation in size of perspective contexts (#MP1190)
External access to higher structural levels (#MP1158)
Occupiable sites exposed to external insight (#MP1180)
Ensuring function of common domain interfaces (#MP1124)
Overview of communication pathway from structure (#MP1164)
Enhancing insight by varying levels of exposure to it (#MP1135)
Distinct pattern of entry points to complex structures (#MP1102)
Exposure of structural activities to communication pathway (#MP1165)
Hospitable reception of external perspectives by structures (#MP1149).

♦ MP1102b Distinct pattern of entry points to complex structures
Physical environment Comprehensible family of entrances to a building complex.
Broader Cluster of frameworks (#MP1037)
Principal points of entry (#MP1053)
Selective interchange axis (#MP1032)
Occupiable temporary site limit (#MP1022)
Patterning of complex structures (#MP1098)
Informal context for formal processes (#MP1041)
Protected low–density communication pathways (#MP1101)
Minimally–structured perspective control operations (#MP1081)
Linear relationships enhanced by non–linear processes (#MP1051)
Concealment of necessary monotonous perspective patterns (#MP1097)
Arrangement of structures ot engender fruitful interfaces (#MP1100)
Minimal distance between related operational control contexts (#MP1082).
Narrower Focus–oriented communication networks (#MP1120)
Emphasizing transitions across boundaries (#MP1224)
External access to higher structural levels (#MP1158)
Distinctiveness of main entry point to structure (#MP1110)
Integrating points of perspective into common domains (#MP1125)
Hospitable reception of external perspectives by structures (#MP1149)
Internal transition spaces enhancing structural entry points (#MP1130)
Transition domain between structure and communication pathway (#MP1112)
Functional enhancement of domains separating complementary structures (#MP1106).

♦ MP1103b Limitation on number of occupiable temporary sites
Physical environment Small parking lots.
Broader Context boundary (#MP1015)
Standard frameworks (#MP1038)
Cluster of frameworks (#MP1037)
Local relationship loops (#MP1049)
Principal points of entry (#MP1053)
Selective interchange axis (#MP1032)
Relationship to indeterminacy (#MP1025)
Occupiable temporary site limit (#MP1022)
Patterning of complex structures (#MP1098)
Informal context for formal processes (#MP1041)
Intersection of differently paced communications (#MP1054)
Linear relationships enhanced by non–linear processes (#MP1051)
Concealment of necessary monotonous perspective patterns (#MP1097).
Narrower Flexible interfaces (#MP1244)
Appropriate relationship of non–linearity to structures (#MP1171)
Protecting non–linear contexts from communication pathways (#MP1173)
Functional enhancement of domains separating complementary structures (#MP1106).

♦ MP1104a Structural development designed to counteract deficiencies in pattern harmony
Physical environment Construction on sites in need of enhancement.
Broader Patterning of complex structures (#MP1098)
Complexification of perspective contexts (#MP1095)
Limitation on number of structural levels (#MP1096).
Narrower Integrative infrastructure (#MP1214)
Contexts of self–organizing non–linearity (#MP1172)
Hierachy of perspectives favouring the broadest (#MP1114)
Distinctiveness of main entry point to structure (#MP1110)
Maintaining distinctions between contextual levels (#MP1169)
Appropriate relationship of non–linearity to structures (#MP1171)
Blended integration of formal structure and informal context (#MP1111)
Organization of structure to enhance autonomy of sub–structures (#MP1109)
Orientation of structures to enhance receptivity to external insight (#MP1105)
Organization of structures to enhance receptivity to external insight (#MP1107)
Functional enhancement of domains separating complementary structures (#MP1106).

♦ MP1105a Orientation of structures to enhance receptivity to external insight
Physical environment Buildings with sheltered sunny spaces.
Broader Unstructured common domain (#MP1067)
Access to contained irrationality (#MP1071)
Complexification of perspective contexts (#MP1095)
Limitation on number of structural levels (#MP1096)
Structural development designed to counteract deficiencies in pattern harmony (#MP1104).
Narrower Structure–enfolded insight domain (#MP1161)
Hierachy of perspectives favouring the broadest (#MP1114)
Distinctiveness of main entry point to structure (#MP1110)
Orientation of domains to receive external insight (#MP1128)
Blended integration of formal structure and informal context (#MP1111)
Organization of structures to enhance receptivity to external insight (#MP1107)
Functional enhancement of domains separating complementary structures (#MP1106)
Organization of integrative superstructure to minimize insight blindspots (#MP1162).

♦ MP1106a Functional enhancement of domains separating complementary structures
Physical environment Enhancing outdoor space.
Broader Accessible non–linearity (#MP1060)
Unstructured common domain (#MP1067)
Chain of fundamental transformation zones (#MP1042)
Bounded common small–scale interaction domains (#MP1061)
Limitation on number of occupiable temporary sites (#MP1103)
Distinct pattern of entry points to complex structures (#MP1102)
Orientation of structures to enhance receptivity to external insight (#MP1105)
Structural development designed to counteract deficiencies in pattern harmony (#MP1104).
Narrower Ambiguous boundaries (#MP1243)
Interconnected structures (#MP1108)
Focus–oriented communication networks (#MP1120)
Hospitability of communication pathways (#MP1121)
Contexts of self–organizing non–linearity (#MP1172)
Appropriate proportions of perspective contexts (#MP1191)
Hierachy of perspectives favouring the broadest (#MP1114)
Communication pathways enfolded by non–linearity (#MP1174)
Hospitable non–linear domain external to structures (#MP1163)
Integrating points of perspective into common domains (#MP1125)
Functional integration of unstructured internal domains (#MP1115)
Protecting non–linear contexts from communication pathways (#MP1173)
Integration of non–linearity into integrative superstructure (#MP1118)
Organization of structure to enhance autonomy of sub–structures (#MP1109)
Hospitable interface between structures and external environment (#MP1160)
Direct relationship between structures and communication pathways (#MP1122)
Organization of structures to permit two sources of external insight (#MP1159)
Organization of structures to enhance receptivity to external insight (#MP1107).

♦ MP1107a Organization of structures to enhance receptivity to external insight
Physical environment Building wings receiving natural light.
Broader Minimal context for single perspective (#MP1078)
Extended pattern of nuclear interaction (#MP1075)
Complexification of perspective contexts (#MP1095)
Integrated contexts for perspective dynamics (#MP1080)
Minimal context for complementary perspectives (#MP1077)
Minimally–structured perspective control operations (#MP1081)
Exchange contexts controlled by a single perspective (#MP1087)
Minimal distance between related operational control contexts (#MP1082)
Minimal context interrelating mature and emerging perspectives (#MP1076)
Orientation of structures to enhance receptivity to external insight (#MP1105)
Functional enhancement of domains separating complementary structures (#MP1106)
Integration of perspective acquisition and perspective maintenance dynamics (#MP1083)
Structural development designed to counteract deficiencies in pattern harmony (#MP1104).
Narrower Interconnected structures (#MP1108)
Nested levels of accessibility (#MP1127)
Partially contained interfaces (#MP1119)
Patterning integrative superstructure (#MP1116)
Containment by integrative superstructure (#MP1117)
Limiting length of communication pathways (#MP1132)
Distinctiveness of main entry point to structure (#MP1110)
Integration of non–linearity into integrative superstructure (#MP1118)
Hospitable interface between structures and external environment (#MP1160)
Direct relationship between structures and communication pathways (#MP1122)
Organization of structures to permit two sources of external insight (#MP1159)
Congruence between spaces defined by the framework and spaces defined by the processes within it (#MP1205).

♦ MP1108b Interconnected structures
Physical environment Connected buildings and connections between buildings.
Broader Complexification of perspective contexts (#MP1095)
Organization of structures to enhance receptivity to external insight (#MP1107)
Functional enhancement of domains separating complementary structures (#MP1106).
Narrower Partially contained interfaces (#MP1119)
Hospitable non–linear domain external to structures (#MP1163)
Functional integration of unstructured internal domains (#MP1115).

♦ MP1109b Organization of structure to enhance autonomy of sub–structures
Physical environment Long thin building to increase privacy.
Broader Standard frameworks (#MP1038)
Complexification of perspective contexts (#MP1095)
Functional enhancement of domains separating complementary structures (#MP1106)
Structural development designed to counteract deficiencies in pattern harmony (#MP1104).
Narrower Nested levels of accessibility (#MP1127)
Patterning integrative superstructure (#MP1116)
Common domain at the focal point of a structure (#MP1129)
Organization of structures to permit two sources of external insight (#MP1159).

♦ MP1110a Distinctiveness of main entry point to structure
Physical environment Prominent main entrance.
Broader Principal points of entry (#MP1053)
Special modes of relationship (#MP1056)
Patterning of complex structures (#MP1098)
Complexification of perspective contexts (#MP1095)
Local sources for perspective nourishment (#MP1089)
Distinct pattern of entry points to complex structures (#MP1102)
Concealment of necessary monotonous perspective patterns (#MP1097)
Orientation of structures to enhance receptivity to external insight (#MP1105)
Organization of structures to enhance receptivity to external insight (#MP1107)
Structural development designed to counteract deficiencies in pattern harmony (#MP1104).
Narrower Intermediate position (#MP1242)
Symbols of integration (#MP1249)

Nested levels of accessibility (#MP1127)
Focus-oriented communication networks (#MP1120)
Emphasizing transitions across boundaries (#MP1224)
Integrating points of perspective into common domains (#MP1125)
Blended integration of formal structure and informal context (#MP1111)
Internal transition spaces enhancing structural entry points (#MP1130)
Transition domain between structure and communication pathway (#MP1112)
Integrating transition pathways between levels into a structure (#MP1133)
Harmoniously structured entry point for external communication media (#MP1113).

♦ MP1111b Blended integration of formal structure and informal context
Physical environment Partially concealed garden.
Broader Standard frameworks (#MP1038) Cluster of frameworks (#MP1037)
Perspective–adaptable contexts (#MP1079) Informal context for formal processes (#MP1041)
Context for emergence of new perspectives (#MP1065)
Distinctiveness of main entry point to structure (#MP1110)
Orientation of structures to enhance receptivity to external insight (#MP1105)
Structural development designed to counteract deficiencies in pattern harmony (#MP1104).
Narrower Ambiguous boundaries (#MP1243) Cultivation of productive non–linearity (#MP1170)
Contexts of self-organizing non-linearity (#MP1172)
Hospitable non–linear domain external to structures (#MP1163)
Local cultivation of sources of perspective nourishment (#MP1177)
Functional integration of unstructured internal domains (#MP1115)
Protecting non–linear contexts from communication pathways (#MP1173)
Integration of non–linearity into integrative superstructure (#MP1118)
Transition domain between structure and communication pathway (#MP1112)
Relative isolation of structural interface with communication pathway (#MP1140).

♦ MP1112a Transition domain between structure and communication pathway
Physical environment Graceful entranceways from the street.
Broader Principal points of entry (#MP1053)
Distinctiveness of main entry point to structure (#MP1110)
Distinct pattern of entry points to complex structures (#MP1102)
Blended integration of formal structure and informal context (#MP1111).
Narrower Nested levels of accessibility (#MP1127)
Limiting exposure to harmonious perspectives (#MP1134)
Communication pathways enfolded by non–linearity (#MP1174)
Enhancing insight by varying levels of exposure to it (#MP1135)
Protecting non–linear contexts from communication pathways (#MP1173)
Internal transition spaces enhancing structural entry points (#MP1130)
Structures adaptable to changing number of embodied perspectives (#MP1153)
Harmoniously structured entry point for external ccmmunication media (#MP1113).

♦ MP1113c Harmoniously structured entry point for external communication media
Physical environment Connection to the car port.
Broader Local relationship loops (#MP1049)
Distinctiveness of main entry point to structure (#MP1110)
Transition domain between structure and communication pathway (#MP1112).
Narrower Nested levels of accessibility (#MP1127)
Partially contained interfaces (#MP1119)
Focus–oriented communication networks (#MP1120)
Protecting variability to enhance fixity (#MP1245)
Common domain at the focal point of a structure (#MP1129)
Hospitable non-linear domain external to structures (#MP1163)
Internal transition spaces enhancing structural entry points (#MP1130)
Organization of integrative superstructure to minimize insight blindspots (#MP1162)
Congruence between spaces defined by the framework and spaces defined by the processes within it (#MP1205).

♦ MP1114b Hierachy of perspectives favouring the broadest
Physical environment Hierarchy of open space with a view of a larger space.
Broader Non–linear organization (#MP1007) Accessible non–linearity (#MP1060)
Local relationship loops (#MP1049) Unstructured common domain (#MP1067)
Patterning of complex structures (#MP1098)
Bounded common small–scale interaction domains (#MP1061)
Linear relationships enhanced by non–linear processes (#MP1051)
Orientation of structures to enhance receptivity to external insight (#MP1105)
Functional enhancement of domains separating complementary structures (#MP1106)
Structural development designed to counteract deficiencies in pattern harmony (#MP1104).
Narrower Ambiguous boundaries (#MP1243)
Ensuring function of common domain interfaces (#MP1124)
Sites for grounding perspectives in non–linearity (#MP1176)
Functional integration of unstructured internal domains (#MP1115)
Protecting non–linear contexts from communication pathways (#MP1173)
Relative isolation of structural interface with communication pathway (#MP1140).

♦ MP1115a Functional integration of unstructured internal domains
Physical environment Lively courtyards.
Broader Interconnected structures (#MP1108) Patterning of complex structures (#MP1098)
Informal context for formal processes (#MP1041)
Common external context for inactivity (#MP1069)
Hierachy of perspectives favouring the broadest (#MP1114)
Blended integration of formal structure and informal context (#MP1111)
Functional enhancement of domains separating complementary structures (#MP1106).
Narrower Partially contained interfaces (#MP1119)
Structure–enfolded insight domain (#MP1161)
Organization of integrative superstructure (#MP1209)
Limiting exposure to harmonious perspectives (#MP1134)
Enfolded overview domains of minimum proportions (#MP1167)
Off–centering the focal point of a common domain (#MP1126)
Hospitable non–linear domain external to structures (#MP1163)
Protecting non–linear contexts from communication pathways (#MP1173)
Overview domains at interfaces of the structure with the external environment (#MP1166).

♦ MP1116b Patterning integrative superstructure
Physical environment Cascade of roofs to integrate a building complex.
Broader Focal centre of a complex structure (#MP1099)
Complexification of perspective contexts (#MP1095)
Limitation on number of structural levels (#MP1096)
Organization of structure to enhance autonomy of sub–structures (#MP1109)
Organization of structures to enhance receptivity to external insight (#MP1107).
Narrower Partially contained interfaces (#MP1119)
Variation in size of perspective contexts (#MP1190)
Containment by integrative superstructure (#MP1117)
Organization of integrative superstructure (#MP1209)
Common domain at the focal point of a structure (#MP1129)
Boundary expansion permitting new level generation (#MP1211)
Integration of non–linearity into integrative superstructure (#MP1118)
Organization of structures to permit two sources of external insight (#MP1159)
Congruence between spaces defined by the framework and spaces defined by the processes within it (#MP1205).

♦ MP1117a Containment by integrative superstructure
Physical environment Sheltering roof.
Broader Patterning integrative superstructure (#MP1116)
Limitation on number of structural levels (#MP1096)
Organization of structures to enhance receptivity to external insight (#MP1107).
Narrower Integration superstructure (#MP1220) Partially contained interfaces (#MP1119)
Organization of integrative superstructure (#MP1209)
Overview sites from integrative superstructure (#MP1231)
Domains for non–current elements and those in reserve (#MP1145)
Integration of non–linearity into integrative superstructure (#MP1118)
Overview domains at interfaces of the structure with the external environment (#MP1166).

♦ MP1118b Integration of non–linearity into integrative superstructure
Physical environment Roof garden.
Broader Integrating a new dimension (#MP1039) Patterning integrative superstructure (#MP1116)
Containment by integrative superstructure (#MP1117)
Limitation on number of structural levels (#MP1096)
Blended integration of formal structure and informal context (#MP1111)
Organization of structures to enhance receptivity to external insight (#MP1107)
Functional enhancement of domains separating complementary structures (#MP1106).
Narrower Flexible interfaces (#MP1244) Integration superstructure (#MP1220)
Integration within context (#MP1246) Structure–enfolded insight domain (#MP1161)
Protecting variability to enhance fixity (#MP1245)
Organization of integrative superstructure (#MP1209)
External access to higher structural levels (#MP1158)
Enfolded overview domains of minimum proportions (#MP1167)
Hospitable non–linear domain external to structures (#MP1163)
Local cultivation of sources of perspective nourishment (#MP1177)
Relative isolation of structural interface with communication pathway (#MP1140)
Overview domains at interfaces of the structure with the external environment (#MP1166).

♦ MP1119a Partially contained interfaces
Physical environment Arcades and covered walkways.
Broader Interchange (#MP1034) Interconnected structures (#MP1108)
Special modes of relationship (#MP1056) Patterning of complex structures (#MP1098)
Patterning integrative superstructure (#MP1116) Protected low intensity relationships (#MP1055)
Common external context for inactivity (#MP1069)
Containment by integrative superstructure (#MP1117)
Access to patterns of active irrationality (#MP1064)
Functional integration of unstructured internal domains (#MP1115)
Arrangement of structures ot engender fruitful interfaces (#MP1100)
Harmoniously structured entry point for external communication media (#MP1113)
Organization of structures to enhance receptivity to external insight (#MP1107).
Narrower Inter–level zone (#MP1226) Flexible interfaces (#MP1244)
Grounded structures (#MP1168) Intermediate position (#MP1242)
Inter–level integrity (#MP1227) Partially enclosed internal domains (#MP1193)
Focus–oriented communication networks (#MP1120)
Hospitability of communication pathways (#MP1121)
Emphasizing transitions across boundaries (#MP1224)
Variation in size of perspective contexts (#MP1190)
Ensuring function of common domain interfaces (#MP1124)
Common domain at the focal point of a structure (#MP1129)
Enfolded overview domains of minimum proportions (#MP1167)
Boundary expansion permitting new level generation (#MP1211)
Hospitable interface between structures and external environment (#MP1160)
Direct relationship between structures and communication pathways (#MP1122)
Overview domains at interfaces of the structure with the external environment (#MP1166)
Congruence between spaces defined by the framework and spaces defined by the processes within it (#MP1205).

♦ MP1120b Focus–oriented communication networks
Physical environment Paths and points of attraction.
Broader Activity nodes (#MP1030) Partially contained interfaces (#MP1119)
Patterning of complex structures (#MP1098)
Common external context for inactivity (#MP1069)
Differentiation by relationship density (#MP1036)
Complexification of perspective contexts (#MP1095)
Distinctiveness of main entry point to structure (#MP1110)
Distinct pattern of entry points to complex structures (#MP1102)
Harmoniously structured entry point for external communication media (#MP1113)
Functional enhancement of domains separating complementary structures (#MP1106)
Interfacing vehicles of communication and networks of unmediated relationships (#MP1052).
Narrower Attractive temporary positions (#MP1241)
Embedding fixity within variability (#MP1247)
Cultivation of productive non–linearity (#MP1170)
Hospitability of communication pathways (#MP1121)
Protecting variability to enhance fixity (#MP1245)
Limiting exposure to harmonious perspectives (#MP1134)
Ensuring function of common domain interfaces (#MP1124)
Communication pathways enfolded by non–linearity (#MP1174)
Off–centering the focal point of a common domain (#MP1126)
Appropriate relationship of non–linearity to structures (#MP1171).

♦ MP1121b Hospitability of communication pathways
Physical environment Path shape to encourage loitering and hanging out.
Broader Hospitable transit points (#MP1092) Selective interchange axis (#MP1032)
Partially contained interfaces (#MP1119) Cycle of relationship reinforcement (#MP1031)
Focus–oriented communication networks (#MP1120)
Protected low intensity relationships (#MP1055)
Common external context for inactivity (#MP1069)
Differentiation by relationship density (#MP1036)
Arrangement of structures ot engender fruitful interfaces (#MP1100)
Functional enhancement of domains separating complementary structures (#MP1106)
Interfacing vehicles of communication and networks of unmediated relationships (#MP1052).
Narrower Ambiguous boundaries (#MP1243) Attractive temporary positions (#MP1241)
Embedding fixity within variability (#MP1247)
Enhancing function of common domains (#MP1123)
Ensuring function of common domain interfaces (#MP1124)
Off–centering the focal point of a common domain (#MP1126)
Overview of communication pathway from structure (#MP1164)
Integrating points of perspective into common domains (#MP1125)
Exposure of structural activities to communication pathway (#MP1165)
Direct relationship between structures and communication pathways (#MP1122).

♦ **MP1122b Direct relationship between structures and communication pathways**
Physical environment Building fronts adapted to the shape of the street.
Broader Local action network (#MP1045) Partially contained interfaces (#MP1119)
Hospitability of communication pathways (#MP1121)
Complexification of perspective contexts (#MP1095)
Chain of fundamental transformation zones (#MP1042)
Bounded common small-scale interaction domains (#MP1061)
Functional enhancement of domains separating complementary structures (#MP1106)
Organization of structures to enhance receptivity to external insight (#MP1107).
Narrower Intermediate position (#MP1242)
Ensuring function of common domain interfaces (#MP1124)
Overview of communication pathway from structure (#MP1164)
Integrating points of perspective into common domains (#MP1125)
Exposure of structural activities to communication pathway (#MP1165)
Hospitable interface between structures and external environment (#MP1160)
Relative isolation of structural interface with communication pathway (#MP1140)
Overview domains at interfaces of the structure with the external environment (#MP1166).

♦ **MP1123b Enhancing function of common domains**
Physical environment Adjusting public spaces for a lively pedestrian density.
Broader Local focal points (#MP1044) Cycle of relationship reinforcement (#MP1031)
Hospitability of communication pathways (#MP1121)
Differentiation by relationship density (#MP1036)
Informal local perspective interface zones (#MP1088)
Protected low–density communication pathways (#MP1101)
Bounded common small–scale interaction domains (#MP1061)
Stable density gradient of local relationships (#MP1029)
Arrangement of structures ot engender fruitful interfaces (#MP1100).
Narrower Ensuring function of common domain interfaces (#MP1124)
Overview of communication pathway from structure (#MP1164)
Integrating points of perspective into common domains (#MP1125)
Exposure of structural activities to communication pathway (#MP1165)
Relative isolation of structural interface with communication pathway (#MP1140)
Overview domains at interfaces of the structure with the external environment (#MP1166).

♦ **MP1124a Ensuring function of common domain interfaces**
Physical environment Pockets of activity around public spaces.
Broader Partially contained interfaces (#MP1119)
Cycle of relationship reinforcement (#MP1031)
Enhancing function of common domains (#MP1123)
Focus–oriented communication networks (#MP1120)
Common external context for inactivity (#MP1069)
Hospitability of communication pathways (#MP1121)
Informal local perspective interface zones (#MP1088)
Protected low–density communication pathways (#MP1101)
Bounded common small–scale interaction domains (#MP1061)
Hierachy of perspectives favouring the broadest (#MP1114)
Arrangement of structures ot engender fruitful interfaces (#MP1100)
Transit point location for sources of perspective nourishment (#MP1093)
Direct relationship between structures and communication pathways (#MP1122)
Cyclic interrelation of complementary perspective in common domains (#MP1063).
Narrower Ambiguous boundaries (#MP1243) Attractive temporary positions (#MP1241)
Provision for temporary perspective inactivity (#MP1150)
Off–centering the focal point of a common domain (#MP1126)
Communication pathways enfolded by non–linearity (#MP1174)
Hospitable non–linear domain external to structures (#MP1163).

♦ **MP1125b Integrating points of perspective into common domains**
Physical environment Seats on stairways in public places.
Broader Points of wider perspective (#MP1062)
Enhancing function of common domains (#MP1123)
Protected low intensity relationships (#MP1055)
Common external context for inactivity (#MP1069)
Hospitability of communication pathways (#MP1121)
Informal local perspective interface zones (#MP1088)
Bounded common small–scale interaction domains (#MP1061)
Distinctiveness of main entry point to structure (#MP1110)
Distinct pattern of entry points to complex structures (#MP1102)
Integrating transition pathways between levels into a structure (#MP1133)
Direct relationship between structures and communication pathways (#MP1122)
Functional enhancement of domains separating complementary structures (#MP1106).
Narrower Time binding (#MP1248) Grounded structures (#MP1168)
Attractive temporary positions (#MP1241) Protecting variability to enhance fixity (#MP1245)
External access to higher structural levels (#MP1158)
Off–centering the focal point of a common domain (#MP1126)
Framework for transition between structural levels (#MP1195)
Hospitable interface between structures and external environment (#MP1160).

♦ **MP1126c Off–centering the focal point of a common domain**
Physical environment Something attractive roughly in the middle of a public square.
Broader Unstructured common domain (#MP1067)
Points of wider perspective (#MP1062)
Access to contained irrationality (#MP1071)
Focus–oriented communication networks (#MP1120)
Common external context for inactivity (#MP1069)
Hospitability of communication pathways (#MP1121)
Access to patterns of active irrationality (#MP1064)
Ensuring function of common domain interfaces (#MP1124)
Bounded common small–scale interaction domains (#MP1061)
Integrating points of perspective into common domains (#MP1125)
Functional integration of unstructured internal domains (#MP1115)
Cyclic interrelation of complementary perspective in common domains (#MP1063).
Narrower Ambiguous boundaries (#MP1243)
Appropriate relationship of non–linearity to structures (#MP1171).

♦ **MP1127a Nested levels of accessibility**
Physical environment Intimacy gradient of spaces in a building.
Broader Limitation on number of structural levels (#MP1096)
Distinctiveness of main entry point to structure (#MP1110)
Transition domain between structure and communication pathway (#MP1112)
Organization of structure to enhance autonomy of sub–structures (#MP1109)
Harmoniously structured entry point for external communication media (#MP1113)
Organization of structures to enhance receptivity to external insight (#MP1107).
Narrower Zoning internal domains (#MP1233) Partially isolated contexts (#MP1152)
Flexible domain organization (#MP1146)
Organization of inter–domain dynamics (#MP1131)
Receptivity to emerging external insight (#MP1138)
Sequence of viewpoint loci within a structure (#MP1142)
Common domain at the focal point of a structure (#MP1129)
Relative isolation of complementary perspectives (#MP1136)
Relatively isolated context for each perspective (#MP1141)
Orientation of domains to receive external insight (#MP1128)
Appropriate configuration for perspective interaction (#MP1185)
Internal transition spaces enhancing structural entry points (#MP1130).

♦ **MP1128b Orientation of domains to receive external insight**
Physical environment Orienting principal rooms for indoor sunlight.
Broader Nested levels of accessibility (#MP1127)
Orientation of structures to enhance receptivity to external insight (#MP1105).
Narrower Displaceable frameworks (#MP1236) Structure–enfolded insight domain (#MP1161)
Partially exposed perspective context (#MP1183)
Receptivity to emerging external insight (#MP1138)
Local opportunities for perspective activity (#MP1157)
Common domain at the focal point of a structure (#MP1129)
Hospitable non–linear domain external to structures (#MP1163)
Exposure of input processing context to external insight (#MP1199)
Organization of integrative superstructure to minimize insight blindspots (#MP1162).

♦ **MP1129a Common domain at the focal point of a structure**
Physical environment Common areas at the heart of any building complex.
Broader Nested levels of accessibility (#MP1127)
Partially contained interfaces (#MP1119)
Focal centre of a complex structure (#MP1099)
Patterning integrative superstructure (#MP1116)
Common external context for inactivity (#MP1069)
Extended pattern of nuclear interaction (#MP1075)
Context for emergence of new perspectives (#MP1065)
Contexts for care of premature perspectives (#MP1086)
Orientation of domains to receive external insight (#MP1128)
Hospitable contexts for perspectives in transition (#MP1091)
Minimal context interrelating mature and emerging perspectives (#MP1076)
Organization of structure to enhance autonomy of sub–structures (#MP1109)
Harmoniously structured entry point for external communication media (#MP1113)
Integration of perspective acquisition and perspective maintenance dynamics (#MP1083).
Narrower Domains of insight (#MP1252) Flexible domain organization (#MP1146)
Domain for developing perspectives (#MP1137)
Organization of inter–domain dynamics (#MP1131)
Coordination of perspective nourishment (#MP1147)
Limiting length of communication pathways (#MP1132)
Sequence of viewpoint loci within a structure (#MP1142)
Hospitable domain for perspective nourishment (#MP1139)
Appropriate proportions of perspective contexts (#MP1191)
Relatively isolated context for each perspective (#MP1141)
Hospitable non–linear domain external to structures (#MP1163)
Appropriate configuration for perspective interaction (#MP1185)
Organization of structure to provide occupiable sites (#MP1179)
Integrating coordinated exposure to irrationality into structures (#MP1144)
Organization of structures to permit two sources of external insight (#MP1159)
Relative isolation of structural interface with communication pathway (#MP1140)
Maintenance of source of direct insight within structures as a focal point (#MP1181).

♦ **MP1130a Internal transition spaces enhancing structural entry points**
Physical environment Entrance room at the main entrance.
Broader Nested levels of accessibility (#MP1127)
Distinctiveness of main entry point to structure (#MP1110)
Distinct pattern of entry points to complex structures (#MP1102)
Transition domain between structure and communication pathway (#MP1112)
Harmoniously structured entry point for external communication media (#MP1113).
Narrower Domains of insight (#MP1252) Grounded structures (#MP1168)
Intermediate position (#MP1242) Connectedness in isolation (#MP1237)
Structure–enfolded occupiable sites (#MP1202)
Organization of inter–domain dynamics (#MP1131)
Internal connectedness between domains (#MP1194)
Occupiable sites exposed to external insight (#MP1180)
Limiting exposure to harmonious perspectives (#MP1134)
Sequence of viewpoint loci within a structure (#MP1142)
Accessible facilities for perspective adjuncts (#MP1201)
Appropriate proportions of perspective contexts (#MP1191)
Enhancing insight by varying levels of exposure to it (#MP1135)
Hospitable reception of external perspectives by structures (#MP1149)
Organization of structures to permit two sources of external insight (#MP1159)
Relative isolation of structural interface with communication pathway (#MP1140)
Overview domains at interfaces of the structure with the external environment (#MP1166).

♦ **MP1131c Organization of inter–domain dynamics**
Physical environment Arrangement of rooms to encourage flow and movement.
Broader Nested levels of accessibility (#MP1127)
Common domain at the focal point of a structure (#MP1129)
Internal transition spaces enhancing structural entry points (#MP1130).
Narrower Partially enclosed internal domains (#MP1193)
Internal connectedness between domains (#MP1194)
Eccentric access to perspective contexts (#MP1196)
Emphasizing transitions across boundaries (#MP1224)
Limiting length of communication pathways (#MP1132)
Limiting exposure to harmonious perspectives (#MP1134)
Enhancing insight by varying levels of exposure to it (#MP1135)
Integrating transition pathways between levels into a structure (#MP1133)
Organization of structures to permit two sources of external insight (#MP1159).

♦ **MP1132b Limiting length of communication pathways**
Physical environment Short room–like passages.
Broader Patterning of complex structures (#MP1098)
Organization of inter–domain dynamics (#MP1131)
Protected low–density communication pathways (#MP1101)
Common domain at the focal point of a structure (#MP1129)
Organization of structures to enhance receptivity to external insight (#MP1107).
Narrower Ground–level visibility (#MP1222) Connectedness in isolation (#MP1237)
Domain for developing perspectives (#MP1137)
Internal connectedness between domains (#MP1194)
Substantive distinctions separating domains (#MP1197)
Limiting exposure to harmonious perspectives (#MP1134)
Occupiable sites exposed to external insight (#MP1180)
Inter–domain contexts for perspective adjuncts (#MP1198)
Appropriate proportions of perspective contexts (#MP1191)
Hospitable non–linear domain external to structures (#MP1163)
Enhancing insight by varying levels of exposure to it (#MP1135)
Organization of structure to provide occupiable sites (#MP1179)

Integrating transition pathways between levels into a structure (#MP1133)
Organization of structures to permit two sources of external insight (#MP1159)
Overview domains at interfaces of the structure with the external environment (#MP1166).

♦ MP1133c Integrating transition pathways between levels into a structure
Physical environment Prominent key staircase as a place of significance.
Broader Organization of inter-domain dynamics (#MP1131)
Limiting length of communication pathways (#MP1132)
Distinctiveness of main entry point to structure (#MP1110).
Narrower Limiting exposure to harmonious perspectives (#MP1134)
Appropriate proportions of perspective contexts (#MP1191)
Framework for transition between structural levels (#MP1195)
Enhancing insight by varying levels of exposure to it (#MP1135)
Integrating points of perspective into common domains (#MP1125)
Appropriate superstructure to contain transitions between levels (#MP1228).

♦ MP1134b Limiting exposure to harmonious perspectives
Physical environment Tantalizing glimpses of a beautiful view.
Broader Points of wider perspective (#MP1062) Positions enabling transcendence (#MP1024)
Organization of inter-domain dynamics (#MP1131)
Context for transformative experience (#MP1066)
Focus-oriented communication networks (#MP1120)
Limiting length of communication pathways (#MP1132)
Functional integration of unstructured internal domains (#MP1115)
Internal transition spaces enhancing structural entry points (#MP1130)
Transition domain between structure and communication pathway (#MP1112)
Integrating transition pathways between levels into a structure (#MP1133).
Narrower Aperture compatibility (#MP1221) Ground-level visibility (#MP1222)
Occupiable sites exposed to external insight (#MP1180)
Sequence of viewpoint loci within a structure (#MP1142)
Enhancing insight by varying levels of exposure to it (#MP1135)
Protecting non-linear contexts from communication pathways (#MP1173).

♦ MP1135b Enhancing insight by varying levels of exposure to it
Physical environment Alternating areas of contrasting light and dark throughout a building.
Broader Organization of inter-domain dynamics (#MP1131)
Limiting length of communication pathways (#MP1132)
Limiting exposure to harmonious perspectives (#MP1134)
Protected low-density communication pathways (#MP1101)
Concealment of necessary monotonous perspective patterns (#MP1097)
Internal transition spaces enhancing structural entry points (#MP1130)
Transition domain between structure and communication pathway (#MP1112)
Integrating transition pathways between levels into a structure (#MP1133).
Narrower Domains of insight (#MP1252) Encouraging emphases (#MP1250)
Connectedness in isolation (#MP1237)
Internal connectedness between domains (#MP1194)
Eccentric access to perspective contexts (#MP1196)
Occupiable sites exposed to external insight (#MP1180)
Hospitable reception of external perspectives by structures (#MP1149).

♦ MP1136b Relative isolation of complementary perspectives
Physical environment Distinct part of the house for the adult couple.
Broader Nested levels of accessibility (#MP1127)
Extended pattern of nuclear interaction (#MP1075)
Context for emergence of new perspectives (#MP1065)
Minimal context for complementary perspectives (#MP1077)
Minimal context interrelating mature and emerging perspectives (#MP1076).
Narrower Domain for developing perspectives (#MP1137)
Receptivity to emerging external insight (#MP1138)
Context for forms of perspective presentation (#MP1189)
Sequence of viewpoint loci within a structure (#MP1142)
Inter-domain contexts for perspective adjuncts (#MP1198)
Appropriate configuration for perspective inactivity (#MP1186)
Appropriate configuration for perspective interaction (#MP1185)
Interrelationship of contexts of developing perspectives (#MP1143)
Integrating coordinated exposure to irrationality into structures (#MP1144)
Organization of structures to permit two sources of external insight (#MP1159)
Appropriate configuration for interaction of complementary perspectives (#MP1187).

♦ MP1137b Domain for developing perspectives
Physical environment Continuous playspace for children in the home.
Broader Limiting length of communication pathways (#MP1132)
Common domain at the focal point of a structure (#MP1129)
Relative isolation of complementary perspectives (#MP1136)
Minimal context interrelating mature and emerging perspectives (#MP1076)
Spontaneous relationship formation amongst emerging perspectives (#MP1068).
Narrower Receptivity to emerging external insight (#MP1138)
Exclusive spaces for emergent perspectives (#MP1203)
Local opportunities for perspective activity (#MP1157)
Hospitable domain for perspective nourishment (#MP1139)
Hospitable non-linear domain external to structures (#MP1163)
Appropriate configuration for perspective inactivity (#MP1186)
Interrelationship of contexts of developing perspectives (#MP1143)
Integrating coordinated exposure to irrationality into structures (#MP1144).

♦ MP1138b Receptivity to emerging external insight
Physical environment Orienting bedrooms towards the rising sun.
Broader Nested levels of accessibility (#MP1127)
Domain for developing perspectives (#MP1137)
Relative isolation of complementary perspectives (#MP1136)
Orientation of domains to receive external insight (#MP1128).
Narrower Filtered insights (#MP1238) Aperture compatibility (#MP1221)
Occupiable sites for perspective inactivity (#MP1188)
Occupiable sites exposed to external insight (#MP1180)
Appropriate configuration for perspective inactivity (#MP1186)
Interrelationship of contexts of developing perspectives (#MP1143)
Integrating coordinated exposure to irrationality into structures (#MP1144)
Appropriate configuration for interaction of complementary perspectives (#MP1187).

♦ MP1139a Hospitable domain for perspective nourishment
Physical environment Kitchen as a family room.
Broader Domain for developing perspectives (#MP1137)
Minimal context for single perspective (#MP1078)
Context for emergence of new perspectives (#MP1065)
Common domain at the focal point of a structure (#MP1129).
Narrower Facilities for perspective adjuncts (#MP1200)
Coordination of perspective nourishment (#MP1147)
Sequence of viewpoint loci within a structure (#MP1142)
Accessible facilities for perspective adjuncts (#MP1201)
Appropriate configuration for input processing (#MP1184)
Appropriate proportions of perspective contexts (#MP1191)
Appropriate conditions for perspective nourishment (#MP1182)
Hospitable non-linear domain external to structures (#MP1163)
Organization of structure to provide occupiable sites (#MP1179)
Exposure of input processing context to external insight (#MP1199)
Organization of structures to permit two sources of external insight (#MP1159).

♦ MP1140a Relative isolation of structural interface with communication pathway
Physical environment Private terrace linking the house to the street.
Broader Enhancing function of common domains (#MP1123)
Low-intensity communication pathways (#MP1059)
Common domain at the focal point of a structure (#MP1129)
Hierachy of perspectives favouring the broadest (#MP1114)
Linear relationships enhanced by non-linear processes (#MP1051)
Arrangement of structures ot engender fruitful interfaces (#MP1100)
Internal transition spaces enhancing structural entry points (#MP1130)
Integration of non-linearity into integrative superstructure (#MP1118)
Blended integration of formal structure and informal context (#MP1111)
Direct relationship between structures and communication pathways (#MP1122).
Narrower Flexible interfaces (#MP1244) Ambiguous boundaries (#MP1243)
Attractive temporary positions (#MP1241) Structure-enfolded insight domain (#MP1161)
Embedding fixity within variability (#MP1247) Partially enclosed internal domains (#MP1193)
Sequence of viewpoint loci within a structure (#MP1142)
Maintaining distinctions between contextual levels (#MP1169)
Hospitable non-linear domain external to structures (#MP1163)
Opportunities for perspectives of decreasing activity (#MP1156)
Protecting non-linear contexts from communication pathways (#MP1173)
Overview domains at interfaces of the structure with the external environment (#MP1166).

♦ MP1141a Relatively isolated context for each perspective
Physical environment One private room for each adult in the home.
Broader Complementarity (#MP1027) Nested levels of accessibility (#MP1127)
Extended pattern of nuclear interaction (#MP1075)
Minimal context for complementary perspectives (#MP1077)
Common domain at the focal point of a structure (#MP1129)
Minimal context interrelating mature and emerging perspectives (#MP1076).
Narrower Partially exposed perspective context (#MP1183)
Meaningful symbols of self-transformation (#MP1253)
Occupiable sites for perspective inactivity (#MP1188)
Context for forms of perspective presentation (#MP1189)
Local opportunities for perspective activity (#MP1157)
Sequence of viewpoint loci within a structure (#MP1142)
Appropriate proportions of perspective contexts (#MP1191)
Semi-autonomous contexts for maturing perspectives (#MP1154)
Opportunities for perspectives of decreasing activity (#MP1156)
Organization of structure to provide occupiable sites (#MP1179)
Semi-autonomous contexts for perspectives of decreasing activity (#MP1155)
Organization of structures to permit two sources of external insight (#MP1159).

♦ MP1142b Sequence of viewpoint loci within a structure
Physical environment Sequence of sitting spaces in a building graded by comfort and degree of enclosure.
Broader Nested levels of accessibility (#MP1127)
Limiting exposure to harmonious perspectives (#MP1134)
Hospitable domain for perspective nourishment (#MP1139)
Common domain at the focal point of a structure (#MP1129)
Relatively isolated context for each perspective (#MP1141)
Relative isolation of complementary perspectives (#MP1136)
Internal transition spaces enhancing structural entry points (#MP1130)
Relative isolation of structural interface with communication pathway (#MP1140).
Narrower Different settings (#MP1251) Partially isolated contexts (#MP1152)
Attractive temporary positions (#MP1241) Structure-enfolded occupiable sites (#MP1202)
Internal connectedness between domains (#MP1194)
Eccentric access to perspective contexts (#MP1196)
Occupiable sites exposed to external insight (#MP1180)
Provision for temporary perspective inactivity (#MP1150)
Hospitable non-linear domain external to structures (#MP1163)
Organization of structure to provide occupiable sites (#MP1179)
Appropriate configuration for perspective interaction (#MP1185)
Maintenance of source of direct insight within structures as a focal point (#MP1181).

♦ MP1143b Interrelationship of contexts of developing perspectives
Physical environment Clustering childrens beds.
Broader Domain for developing perspectives (#MP1137)
Receptivity to emerging external insight (#MP1138)
Relative isolation of complementary perspectives (#MP1136)
Minimal context interrelating mature and emerging perspectives (#MP1076).
Narrower Occupiable sites for perspective inactivity (#MP1188)
Context for forms of perspective presentation (#MP1189)
Inter-domain contexts for perspective adjuncts (#MP1198)
Appropriate proportions of perspective contexts (#MP1191)
Appropriate configuration for perspective inactivity (#MP1186)
Integrating coordinated exposure to irrationality into structures (#MP1144)
Organization of structures to permit two sources of external insight (#MP1159).

♦ MP1144b Integrating coordinated exposure to irrationality into structures
Physical environment Shared bathing facilities in the home.
Broader Nested levels of accessibility (#MP1127)
Access to contained irrationality (#MP1071)
Domain for developing perspectives (#MP1137)
Minimal context for single perspective (#MP1078)
Receptivity to emerging external insight (#MP1138)
Common domain at the focal point of a structure (#MP1129)
Relative isolation of complementary perspectives (#MP1136)
Interrelationship of contexts of developing perspectives (#MP1143)
Transit point location for sources of perspective nourishment (#MP1093)
Competitive interaction opportunities transposed to a concrete level (#MP1072).
Narrower Filtered insights (#MP1238)
Context for forms of perspective presentation (#MP1189)
Appropriate proportions of perspective contexts (#MP1191)
Re-integration of rejected perspective by-products (#MP1178)
Local cultivation of sources of perspective nourishment (#MP1177)
Protecting non-linear contexts from communication pathways (#MP1173)
Organization of structures to permit two sources of external insight (#MP1159).

♦ MP1145c Domains for non-current elements and those in reserve
Physical environment Bulk storage areas.

Broader Complexification of perspective contexts (#MP1095)
Containment by integrative superstructure (#MP1117)
Integrated contexts for perspective dynamics (#MP1080)
Exchange contexts controlled by a single perspective (#MP1087)
Minimal context interrelating mature and emerging perspectives (#MP1076).
Narrower Initial level formation (#MP1215)
Variation in size of perspective contexts (#MP1190)
Maintaining distinctions between contextual levels (#MP1169)
Insight capturing non-linear extensions of structures (#MP1175)
Structures adaptable to changing number of embodied perspectives (#MP1153)
Organization of integrative superstructure to minimize insight blindspots (#MP1162).

♦ MP1146c Flexible domain organization

Physical environment Flexible office space adjustable according to need.
Broader Nested levels of accessibility (#MP1127)
Integrated contexts for perspective dynamics (#MP1080)
Common domain at the focal point of a structure (#MP1129)
Minimally-structured perspective control operations (#MP1081)
Minimal distance between related operational control contexts (#MP1082).
Narrower Inter-level zone (#MP1226)
Domains of insight (#MP1252)
Partially isolated contexts (#MP1152)
Perspective interaction constraints (#MP1148)
Partially exposed perspective context (#MP1183)
Coordination of perspective nourishment (#MP1147)
Variation in size of perspective contexts (#MP1190)
Small-scale perspective interaction contexts (#MP1151)
Organization of structure to provide occupiable sites (#MP1179)
Hospitable reception of external perspectives by structures (#MP1149)
Structures adaptable to changing number of embodied perspectives (#MP1153)
Organization of structures to permit two sources of external insight (#MP1159).

♦ MP1147b Coordination of perspective nourishment

Physical environment Communal eating places for homes and workgroups.
Broader Flexible domain organization (#MP1146)
Informal context for formal processes (#MP1041)
Extended pattern of nuclear interaction (#MP1075)
Integrated contexts for perspective dynamics (#MP1080)
Hospitable domain for perspective nourishment (#MP1139)
Common domain at the focal point of a structure (#MP1129)
Hospitable contexts for perspectives in transition (#MP1091)
Transitional contexts for perspective reorganization (#MP1084)
Integration of perspective acquisition and perspective maintenance dynamics (#MP1083).
Narrower Perspective interaction constraints (#MP1148)
Small-scale perspective interaction contexts (#MP1151)
Appropriate conditions for perspective nourishment (#MP1182).

♦ MP1148a Perspective interaction constraints

Physical environment Shared facilities for small work groups.
Broader Flexible domain organization (#MP1146)
Coordination of perspective nourishment (#MP1147)
Integrated contexts for perspective dynamics (#MP1080)
Minimally-structured perspective control operations (#MP1081)
Minimal distance between related operational control contexts (#MP1082)
Integration of perspective acquisition and perspective maintenance dynamics (#MP1083).
Narrower Partially isolated contexts (#MP1152)
Partially exposed perspective context (#MP1183)
External access to higher structural levels (#MP1158)
Small-scale perspective interaction contexts (#MP1151).

♦ MP1149c Hospitable reception of external perspectives by structures

Physical environment Welcoming reception areas in institutions.
Broader Flexible domain organization (#MP1146)
Protected low-density communication pathways (#MP1101)
Integrated contexts for perspective dynamics (#MP1080)
Minimally-structured perspective control operations (#MP1081)
Enhancing insight by varying levels of exposure to it (#MP1135)
Distinct pattern of entry points to complex structures (#MP1102)
Internal transition spaces enhancing structural entry points (#MP1130).
Narrower Partially exposed perspective context (#MP1183)
Occupiable sites exposed to external insight (#MP1180)
Provision for temporary perspective inactivity (#MP1150)
Appropriate proportions of perspective contexts (#MP1191)
Organization of structure to provide occupiable sites (#MP1179)
Organization of structures to permit two sources of external insight (#MP1159)
Maintenance of source of direct insight within structures as a focal point (#MP1181).

♦ MP1150b Provision for temporary perspective inactivity

Physical environment Facilities to make waiting a positive experience.
Broader Interchange (#MP1034)
Hospitable common domains (#MP1094)
Hospitable transit points (#MP1092)
Functionality enhancement (#MP1047)
Access to contained irrationality (#MP1071)
Informal local perspective interface zones (#MP1088)
Sequence of viewpoint loci within a structure (#MP1142)
Ensuring function of common domain interfaces (#MP1124)
Minimally-structured perspective control operations (#MP1081)
Hospitable reception of external perspectives by structures (#MP1149)
Minimal distance between related operational control contexts (#MP1082).
Narrower Occupiable sites exposed to external insight (#MP1180)
Appropriate proportions of perspective contexts (#MP1191)
Overview of communication pathway from structure (#MP1164)
Sites for grounding perspectives in non-linearity (#MP1176)
Organization of structure to provide occupiable sites (#MP1179)
Exposure of structural activities to communication pathway (#MP1165)
Organization of structures to permit two sources of external insight (#MP1159).

♦ MP1151b Small-scale perspective interaction contexts

Physical environment Small meeting rooms.
Broader Local focal points (#MP1044)
Flexible domain organization (#MP1146)
Presentation of new dimensions (#MP1043)
Perspective interaction constraints (#MP1148)
Coordination of perspective nourishment (#MP1147)
Integration of perspective acquisition and perspective maintenance dynamics (#MP1083).
Narrower Domains of insight (#MP1252)
Different settings (#MP1251)
Appropriate proportions of perspective contexts (#MP1191)
Organization of structure to provide occupiable sites (#MP1179)
Appropriate configuration for perspective interaction (#MP1185)
Organization of structures to permit two sources of external insight (#MP1159).

♦ MP1152c Partially isolated contexts

Physical environment Offices partially open to other workgroups.
Broader Flexible domain organization (#MP1146)
Nested levels of accessibility (#MP1127)
Perspective interaction constraints (#MP1148)
Sequence of viewpoint loci within a structure (#MP1142)
Integration of perspective acquisition and perspective maintenance dynamics (#MP1083).
Narrower Overview of external contexts (#MP1192)
Partially enclosed internal domains (#MP1193)
Partially exposed perspective context (#MP1183)
Appropriate proportions of perspective contexts (#MP1191)
Appropriate configuration for perspective interaction (#MP1185)
Organization of structures to permit two sources of external insight (#MP1159).

♦ MP1153c Structures adaptable to changing number of embodied perspectives

Physical environment Organizing buildings to allow rooms to be rented out whenever not required.
Broader Functional cycle (#MP1026)
Flexible domain organization (#MP1146)
Extended pattern of nuclear interaction (#MP1075)
Integrated contexts for perspective dynamics (#MP1080)
Minimally-structured perspective control operations (#MP1081)
Domains for non-current elements and those in reserve (#MP1145)
Transition domain between structure and communication pathway (#MP1112).
Narrower External access to higher structural levels (#MP1158)
Local opportunities for perspective activity (#MP1157)
Appropriate proportions of perspective contexts (#MP1191)
Semi-autonomous contexts for maturing perspectives (#MP1154)
Semi-autonomous contexts for perspectives of decreasing activity (#MP1155)
Organization of structures to permit two sources of external insight (#MP1159).

♦ MP1154b Semi-autonomous contexts for maturing perspectives

Physical environment Distinct cottage for teenagers associated with the main house.
Broader Functional cycle (#MP1026)
Minimal context for single perspective (#MP1078)
Extended pattern of nuclear interaction (#MP1075)
Relatively isolated context for each perspective (#MP1141)
Minimal context interrelating mature and emerging perspectives (#MP1076)
Structures adaptable to changing number of embodied perspectives (#MP1153).
Narrower Occupiable sites for perspective inactivity (#MP1188)
External access to higher structural levels (#MP1158)
Local opportunities for perspective activity (#MP1157)
Appropriate proportions of perspective contexts (#MP1191)
Appropriate configuration for perspective interaction (#MP1185)
Semi-autonomous contexts for perspectives of decreasing activity (#MP1155)
Congruence between spaces defined by the framework and spaces defined by the processes within it (#MP1205).

♦ MP1155a Semi-autonomous contexts for perspectives of decreasing activity

Physical environment Small cottages for the elderly close to neighbourhood services.
Broader Functional cycle (#MP1026)
Integrating the historical dimension (#MP1040)
Minimal context for single perspective (#MP1078)
Extended pattern of nuclear interaction (#MP1075)
Relatively isolated context for each perspective (#MP1141)
Semi-autonomous contexts for maturing perspectives (#MP1154)
Structures adaptable to changing number of embodied perspectives (#MP1153)
Relative isolation of structural interface with communication pathway (#MP1140).
Narrower Intermediate position (#MP1242)
Appropriate proportions of perspective contexts (#MP1191)
Overview of communication pathway from structure (#MP1164)
Opportunities for perspectives of decreasing activity (#MP1156)
Congruence between spaces defined by the framework and spaces defined by the processes within it (#MP1205).

♦ MP1156b Opportunities for perspectives of decreasing activity

Physical environment Settled productive workplaces, especially for the elderly.
Broader Functional cycle (#MP1026)
Integrating the historical dimension (#MP1040)
Extended pattern of nuclear interaction (#MP1075)
Relatively isolated context for each perspective (#MP1141)
Semi-autonomous contexts for perspectives of decreasing activity (#MP1155)
Relative isolation of structural interface with communication pathway (#MP1140).
Narrower External access to higher structural levels (#MP1158)
Local opportunities for perspective activity (#MP1157)
Exposure of structural activities to communication pathway (#MP1165).

♦ MP1157c Local opportunities for perspective activity

Physical environment Home workshop.
Broader Complementarity (#MP1027)
Cluster of frameworks (#MP1037)
Network of redefinitions (#MP1018)
Functionality enhancement (#MP1047)
Perspective-adaptable contexts (#MP1079)
Decentralized formal processes (#MP1009)
Domain for developing perspectives (#MP1137)
Relatively isolated context for each perspective (#MP1141)
Semi-autonomous contexts for maturing perspectives (#MP1154)
Orientation of domains to receive external insight (#MP1128)
Transitional contexts for perspective reorganization (#MP1084)
Opportunities for perspectives of decreasing activity (#MP1156)
Structures adaptable to changing number of embodied perspectives (#MP1153).
Narrower Overview of external contexts (#MP1192)
Structure-enfolded insight domain (#MP1161)
Partially exposed perspective context (#MP1183)
External access to higher structural levels (#MP1158)
Appropriate proportions of perspective contexts (#MP1191)
Exposure of structural activities to communication pathway (#MP1165)
Organization of structures to permit two sources of external insight (#MP1159).

♦ MP1158b External access to higher structural levels

Physical environment External staircases for institutions.
Broader Points of wider perspective (#MP1062)
Integrating a new dimension (#MP1039)
Perspective interaction constraints (#MP1148)
Minimal context for single perspective (#MP1078)
Local opportunities for perspective activity (#MP1157)
Protected low-density communication pathways (#MP1101)
Integrated contexts for perspective dynamics (#MP1080)
Minimal context for complementary perspectives (#MP1077)
Semi-autonomous contexts for maturing perspectives (#MP1154)
Minimally-structured perspective control operations (#MP1081)
Opportunities for perspectives of decreasing activity (#MP1156)
Integrating points of perspective into common domains (#MP1125)
Distinct pattern of entry points to complex structures (#MP1102)
Concealment of necessary monotonous perspective patterns (#MP1097)
Arrangement of structures ot engender fruitful interfaces (#MP1100)
Integration of non-linearity into integrative superstructure (#MP1118)
Minimal distance between related operational control contexts (#MP1082)

Minimal context interrelating mature and emerging perspectives (#MP1076)
Structures adaptable to changing number of embodied perspectives (#MP1153).
Narrower Structure–enfolded insight domain (#MP1161)
Framework for transition between structural levels (#MP1195).

♦ MP1159a Organization of structures to permit two sources of external insight
Physical environment Light on two sides of every room.
Broader Partially isolated contexts (#MP1152) Flexible domain organization (#MP1146)
Organization of inter–domain dynamics (#MP1131)
Patterning integrative superstructure (#MP1116)
Limiting length of communication pathways (#MP1132)
Local opportunities for perspective activity (#MP1157)
Small–scale perspective interaction contexts (#MP1151)
Hospitable domain for perspective nourishment (#MP1139)
Provision for temporary perspective inactivity (#MP1150)
Common domain at the focal point of a structure (#MP1129)
Relatively isolated context for each perspective (#MP1141)
Relative isolation of complementary perspectives (#MP1136)
Interrelationship of contexts of developing perspectives (#MP1143)
Hospitable reception of external perspectives by structures (#MP1149)
Internal transition spaces enhancing structural entry points (#MP1130)
Organization of structure to enhance autonomy of sub–structures (#MP1109)
Structures adaptable to changing number of embodied perspectives (#MP1153)
Integrating coordinated exposure to irrationality into structures (#MP1144)
Organization of structures to enhance receptivity to external insight (#MP1107)
Functional enhancement of domains separating complementary structures (#MP1106).
Narrower Filtered insights (#MP1238) Aperture compatibility (#MP1221)
Zones of intermediate insight (#MP1223) Overview of external contexts (#MP1192)
Organization of integrative superstructure (#MP1209)
Context for forms of perspective presentation (#MP1189)
Occupiable sites exposed to external insight (#MP1180)
Hospitable interface between structures and external environment (#MP1160)
Organization of integrative superstructure to minimize insight blindspots (#MP1162).

♦ MP1160a Hospitable interface between structures and external environment
Physical environment Inviting spots at the edge of buildings.
Broader Local action network (#MP1045) Hospitable common domains (#MP1094)
Partially contained interfaces (#MP1119)
Common external context for inactivity (#MP1069)
Integrating points of perspective into common domains (#MP1125)
Direct relationship between structures and communication pathways (#MP1122)
Organization of structures to permit two sources of external insight (#MP1159)
Organization of structures to enhance receptivity to external insight (#MP1107)
Functional enhancement of domains separating complementary structures (#MP1106).
Narrower Grounded structures (#MP1168) Intermediate position (#MP1242)
Symbols of integration (#MP1249) Attractive temporary positions (#MP1241)
Structure–enfolded insight domain (#MP1161) Protecting variability to enhance fixity (#MP1245)
Substantive distinctions separating domains (#MP1197)
Enfolded overview domains of minimum proportions (#MP1167)
Overview of communication pathway from structure (#MP1164)
Maintaining distinctions between contextual levels (#MP1169)
Hospitable non–linear domain external to structures (#MP1163)
Organization of integrative superstructure to minimize insight blindspots (#MP1162)
Overview domains at interfaces of the structure with the external environment (#MP1166).

♦ MP1161a Structure–enfolded insight domain
Physical environment Place for basking in the sun.
Broader External access to higher structural levels (#MP1158)
Local opportunities for perspective activity (#MP1157)
Orientation of domains to receive external insight (#MP1128)
Functional integration of unstructured internal domains (#MP1115)
Integration of non–linearity into integrative superstructure (#MP1118)
Hospitable interface between structures and external environment (#MP1160)
Orientation of structures to enhance receptivity to external insight (#MP1105)
Relative isolation of structural interface with communication pathway (#MP1140)
Contexts for exploratory relationship formation challenging emerging perspectives (#MP1073).
Narrower Filtered insights (#MP1238) Flexible interfaces (#MP1244)
Intermediate position (#MP1242) Attractive temporary positions (#MP1241)
Enfolded overview domains of minimum proportions (#MP1167)
Communication pathways enfolded by non–linearity (#MP1174)
Sites for grounding perspectives in non–linearity (#MP1176)
Hospitable non–linear domain external to structures (#MP1163)
Organization of integrative superstructure to minimize insight blindspots (#MP1162).

♦ MP1162c Organization of integrative superstructure to minimize insight blindspots
Physical environment Limiting building shadow.
Broader Structure–enfolded insight domain (#MP1161)
Orientation of domains to receive external insight (#MP1128)
Domains for non–current elements and those in reserve (#MP1145)
Hospitable interface between structures and external environment (#MP1160)
Organization of structures to permit two sources of external insight (#MP1159)
Harmoniously structured entry point for external communication media (#MP1113)
Orientation of structures to enhance receptivity to external insight (#MP1105).
Narrower Inter–domain contexts for perspective adjuncts (#MP1198)
Re–integration of rejected perspective by–products (#MP1178)
Protecting non–linear contexts from communication pathways (#MP1173).

♦ MP1163a Hospitable non–linear domain external to structures
Physical environment Outdoor garden room.
Broader Interconnected structures (#MP1108) Structure–enfolded insight domain (#MP1161)
Domain for developing perspectives (#MP1137)
Common external context for inactivity (#MP1069)
Limiting length of communication pathways (#MP1132)
Sequence of viewpoint loci within a structure (#MP1142)
Hospitable domain for perspective nourishment (#MP1139)
Ensuring function of common domain interfaces (#MP1124)
Common domain at the focal point of a structure (#MP1129)
Orientation of domains to receive external insight (#MP1128)
Functional integration of unstructured internal domains (#MP1115)
Integration of non–linearity into integrative superstructure (#MP1118)
Blended integration of formal structure and informal context (#MP1111)
Hospitable interface between structures and external environment (#MP1160)
Harmoniously structured entry point for external communication media (#MP1113)
Relative isolation of structural interface with communication pathway (#MP1140)
Functional enhancement of domains separating complementary structures (#MP1106).
Narrower Inter–level zone (#MP1226) Flexible interfaces (#MP1244)
Grounded structures (#MP1168) Ambiguous boundaries (#MP1243)
Attractive temporary positions (#MP1241) Embedding fixity within variability (#MP1247)
Appropriate proportions of perspective contexts (#MP1191)

Enfolded overview domains of minimum proportions (#MP1167)
Communication pathways enfolded by non–linearity (#MP1174)
Appropriate relationship of non–linearity to structures (#MP1171)
Exposure of structural activities to communication pathway (#MP1165)
Protecting non–linear contexts from communication pathways (#MP1173)
Overview domains at interfaces of the structure with the external environment (#MP1166)
Congruence between spaces defined by the framework and spaces defined by the processes within it (#MP1205).

♦ MP1164b Overview of communication pathway from structure
Physical environment Window seats and street windows.
Broader Protection of emerging foci (#MP1057)
Enhancing function of common domains (#MP1123)
Hospitability of communication pathways (#MP1121)
Protected low–density communication pathways (#MP1101)
Provision for temporary perspective inactivity (#MP1150)
Bounded common small–scale interaction domains (#MP1061)
Linear relationships enhanced by non–linear processes (#MP1051)
Arrangement of structures ot engender fruitful interfaces (#MP1100)
Hospitable interface between structures and external environment (#MP1160)
Semi–autonomous contexts for perspectives of decreasing activity (#MP1155)
Direct relationship between structures and communication pathways (#MP1122).
Narrower Filtered insights (#MP1238) Aperture compatibility (#MP1221)
Displaceable frameworks (#MP1236) Integration within context (#MP1246)
Occupiable sites exposed to external insight (#MP1180).

♦ MP1165b Exposure of structural activities to communication pathway
Physical environment Opening public places to the street.
Broader Local focal points (#MP1044) Local action network (#MP1045)
Functionality enhancement (#MP1047) Presentation of new dimensions (#MP1043)
Diversified interchange environment (#MP1046)
Enhancing function of common domains (#MP1123)
Hospitability of communication pathways (#MP1121)
Local sources for perspective nourishment (#MP1089)
Informal local perspective interface zones (#MP1088)
Local opportunities for perspective activity (#MP1157)
Protected low–density communication pathways (#MP1101)
Provision for temporary perspective inactivity (#MP1150)
Hospitable non–linear domain external to structures (#MP1163)
Exchange contexts controlled by a single perspective (#MP1087)
Opportunities for perspectives of decreasing activity (#MP1156)
Perspective imitation contexts for developing perspectives (#MP1085)
Direct relationship between structures and communication pathways (#MP1122)
Competitive interaction opportunities transposed to a concrete level (#MP1072).
Narrower Ambiguous boundaries (#MP1243).

♦ MP1166b Overview domains at interfaces of the structure with the external environment
Physical environment Galleries and terraces around buildings.
Broader Partially contained interfaces (#MP1119)
Enhancing function of common domains (#MP1123)
Limiting length of communication pathways (#MP1132)
Containment by integrative superstructure (#MP1117)
Hospitable non–linear domain external to structures (#MP1163)
Functional integration of unstructured internal domains (#MP1115)
Concealment of necessary monotonous perspective patterns (#MP1097)
Arrangement of structures ot engender fruitful interfaces (#MP1100)
Internal transition spaces enhancing structural entry points (#MP1130)
Integration of non–linearity into integrative superstructure (#MP1118)
Hospitable interface between structures and external environment (#MP1160)
Direct relationship between structures and communication pathways (#MP1122)
Relative isolation of structural interface with communication pathway (#MP1140).
Narrower Inter–level zone (#MP1226) Flexible interfaces (#MP1244)
Grounded structures (#MP1168) Ambiguous boundaries (#MP1243)
Inter–level integrity (#MP1227) Partially enclosed internal domains (#MP1193)
Enfolded overview domains of minimum proportions (#MP1167).

♦ MP1167a Enfolded overview domains of minimum proportions
Physical environment Partially–recessed balconies and porches of adequate size.
Broader Partially contained interfaces (#MP1119)
Structure–enfolded insight domain (#MP1161)
Hospitable non–linear domain external to structures (#MP1163)
Functional integration of unstructured internal domains (#MP1115)
Arrangement of structures ot engender fruitful interfaces (#MP1100)
Integration of non–linearity into integrative superstructure (#MP1118)
Hospitable interface between structures and external environment (#MP1160)
Overview domains at interfaces of the structure with the external environment (#MP1166).
Narrower Inter–level zone (#MP1226) Grounded structures (#MP1168)
Ambiguous boundaries (#MP1243) Inter–level integrity (#MP1227)
Partially enclosed internal domains (#MP1193)
Appropriate proportions of perspective contexts (#MP1191).

♦ MP1168a Grounded structures
Physical environment Blurring the boundary between the building and the surrounding earth.
Broader Partially contained interfaces (#MP1119)
Enfolded overview domains of minimum proportions (#MP1167)
Hospitable non–linear domain external to structures (#MP1163)
Integrating points of perspective into common domains (#MP1125)
Internal transition spaces enhancing structural entry points (#MP1130)
Hospitable interface between structures and external environment (#MP1160)
Overview domains at interfaces of the structure with the external environment (#MP1166).
Narrower Time binding (#MP1248) Intermediate position (#MP1242)
Symbols of integration (#MP1249) Initial level formation (#MP1215)
Integrative infrastructure (#MP1214) Attractive temporary positions (#MP1241)
Embedding fixity within variability (#MP1247)
Maintaining distinctions between contextual levels (#MP1169).

♦ MP1169b Maintaining distinctions between contextual levels
Physical environment Terraced slopes following contour lines.
Broader Grounded structures (#MP1168)
Domains for non–current elements and those in reserve (#MP1145)
Hospitable interface between structures and external environment (#MP1160)
Relative isolation of structural interface with communication pathway (#MP1140)
Structural development designed to counteract deficiencies in pattern harmony (#MP1104).
Narrower Ambiguous boundaries (#MP1243) Embedding fixity within variability (#MP1247)
Cultivation of productive non–linearity (#MP1170) Protecting variability to enhance fixity (#MP1245)
Contexts of self–organizing non–linearity (#MP1172)
Local cultivation of sources of perspective nourishment (#MP1177).

♦ MP1170b Cultivation of productive non–linearity
Physical environment Fruit trees and orchards on common land.
Broader Unstructured common domain (#MP1067)
Focus–oriented communication networks (#MP1120)
Maintaining distinctions between contextual levels (#MP1169)
Linear relationships enhanced by non–linear processes (#MP1051)
Blended integration of formal structure and informal context (#MP1111).
Narrower Contexts of self–organizing non–linearity (#MP1172)
Communication pathways enfolded by non–linearity (#MP1174)
Sites for grounding perspectives in non–linearity (#MP1176)
Re–integration of rejected perspective by–products (#MP1178)
Local cultivation of sources of perspective nourishment (#MP1177)
Appropriate relationship of non–linearity to structures (#MP1171).

♦ MP1171a Appropriate relationship of non–linearity to structures
Physical environment Interrelating trees and buildings.
Broader Accessible non–linearity (#MP1060) Positions enabling transcendence (#MP1024)
Focus-oriented communication networks (#MP1120)
Context for transformative experience (#MP1066)
Cultivation of productive non–linearity (#MP1170)
Limitation on number of structural levels (#MP1096)
Context for acknowledgement of past perspectives (#MP1070)
Off–centering the focal point of a common domain (#MP1126)
Limitation on number of occupiable temporary sites (#MP1103)
Hospitable non–linear domain external to structures (#MP1163)
Structural development designed to counteract deficiencies in pattern harmony (#MP1104).
Narrower Ambiguous boundaries (#MP1243) Attractive temporary positions (#MP1241)
Contexts of self–organizing non–linearity (#MP1172)
Communication pathways enfolded by non–linearity (#MP1174)
Sites for grounding perspectives in non–linearity (#MP1176).

♦ MP1172a Contexts of self–organizing non–linearity
Physical environment Garden growing wild.
Broader Access to contained irrationality (#MP1071)
Cultivation of productive non–linearity (#MP1170)
Maintaining distinctions between contextual levels (#MP1169)
Appropriate relationship of non–linearity to structures (#MP1171)
Blended integration of formal structure and informal context (#MP1111)
Functional enhancement of domains separating complementary structures (#MP1106)
Structural development designed to counteract deficiencies in pattern harmony (#MP1104)
Contexts for exploratory relationship formation challenging emerging perspectives (#MP1073).
Narrower Ambiguous boundaries (#MP1243) Protecting variability to enhance fixity (#MP1245)
Sites for grounding perspectives in non–linearity (#MP1176)
Insight capturing non–linear extensions of structures (#MP1175).

♦ MP1173b Protecting non–linear contexts from communication pathways
Physical environment Protective garden wall.
Broader Accessible non–linearity (#MP1060) Special modes of relationship (#MP1056)
Low–intensity communication pathways (#MP1059)
Context for emergence of new perspectives (#MP1065)
Limiting exposure to harmonious perspectives (#MP1134)
Hierachy of perspectives favouring the broadest (#MP1114)
Limitation on number of occupiable temporary sites (#MP1103)
Hospitable non–linear domain external to structures (#MP1163)
Functional integration of unstructured internal domains (#MP1115)
Blended integration of formal structure and informal context (#MP1111)
Transition domain between structure and communication pathway (#MP1112)
Integrating coordinated exposure to irrationality into structures (#MP1144)
Relative isolation of structural interface with communication pathway (#MP1140)
Functional enhancement of domains separating complementary structures (#MP1106)
Organization of integrative superstructure to minimize insight blindspots (#MP1162).
Narrower Symbols of integration (#MP1249) Partially enclosed internal domains (#MP1193)
Protecting variability to enhance fixity (#MP1245)
Communication pathways enfolded by non–linearity (#MP1174).

♦ MP1174a Communication pathways enfolded by non–linearity
Physical environment Trellised walk.
Broader Structure–enfolded insight domain (#MP1161)
Access to contained irrationality (#MP1071)
Focus–oriented communication networks (#MP1120)
Cultivation of productive non–linearity (#MP1170)
Ensuring function of common domain interfaces (#MP1124)
Intersection of differently paced communications (#MP1054)
Hospitable non–linear domain external to structures (#MP1163)
Appropriate relationship of non–linearity to structures (#MP1171)
Protecting non–linear contexts from communication pathways (#MP1173)
Transition domain between structure and communication pathway (#MP1112)
Functional enhancement of domains separating complementary structures (#MP1106).
Narrower Inter–level zone (#MP1226) Filtered insights (#MP1238)
Flexible interfaces (#MP1244) Integration within context (#MP1246)
Attractive temporary positions (#MP1241) Embedding fixity within variability (#MP1247)
Insight capturing non-linear extensions of structures (#MP1175).

♦ MP1175c Insight capturing non–linear extensions of structures
Physical environment Greenhouse as an extension of home or office.
Broader Cluster of frameworks (#MP1037) Unstructured common domain (#MP1067)
Informal context for formal processes (#MP1041)
Contexts of self–organizing non–linearity (#MP1172)
Communication pathways enfolded by non–linearity (#MP1174)
Domains for non–current elements and those in reserve (#MP1145).
Narrower Occupiable sites exposed to external insight (#MP1180)
Accessible facilities for perspective adjuncts (#MP1201)
Sites for grounding perspectives in non–linearity (#MP1176)
Re–integration of rejected perspective by–products (#MP1178)
Local cultivation of sources of perspective nourishment (#MP1177).

♦ MP1176c Sites for grounding perspectives in non–linearity
Physical environment Quiet garden seat.
Broader Positions enabling transcendence (#MP1024)
Structure–enfolded insight domain (#MP1161)
Cultivation of productive non–linearity (#MP1170)
Contexts of self–organizing non–linearity (#MP1172)
Provision for temporary perspective inactivity (#MP1150)
Hierachy of perspectives favouring the broadest (#MP1114)
Insight capturing non–linear extensions of structures (#MP1175)
Appropriate relationship of non–linearity to structures (#MP1171).
Narrower Filtered insights (#MP1238) Attractive temporary positions (#MP1241).

♦ MP1177b Local cultivation of sources of perspective nourishment
Physical environment Vegetable garden on private or fenced common land.
Broader Functionality enhancement (#MP1047) Unstructured common domain (#MP1067)
Integrating a new dimension (#MP1039) Integrating the historical dimension (#MP1040)
Cultivation of productive non–linearity (#MP1170)
Maintaining distinctions between contextual levels (#MP1169)
Insight capturing non–linear extensions of structures (#MP1175)
Integration of non–linearity into integrative superstructure (#MP1118)
Blended integration of formal structure and informal context (#MP1111)
Integrating coordinated exposure to irrationality into structures (#MP1144).
Narrower Re–integration of rejected perspective by–products (#MP1178).

♦ MP1178b Re–integration of rejected perspective by–products
Physical environment Recycling organic garbage and sewage for compost.
Broader Cluster of frameworks (#MP1037) Cultivation of productive non–linearity (#MP1170)
Interaction with coherent irrational perspectives (#MP1074)
Insight capturing non–linear extensions of structures (#MP1175)
Local cultivation of sources of perspective nourishment (#MP1177)
Integrating coordinated exposure to irrationality into structures (#MP1144)
Organization of integrative superstructure to minimize insight blindspots (#MP1162).
Narrower Maintenance of source of direct insight within structures as a focal point (#MP1181).

♦ MP1179a Organization of structure to provide occupiable sites
Physical environment Alcoves at the edge of any common room.
Broader Flexible domain organization (#MP1146)
Limiting length of communication pathways (#MP1132)
Small–scale perspective interaction contexts (#MP1151)
Sequence of viewpoint loci within a structure (#MP1142)
Hospitable domain for perspective nourishment (#MP1139)
Unstructured context for perspective exchange (#MP1090)
Provision for temporary perspective inactivity (#MP1150)
Common domain at the focal point of a structure (#MP1129)
Relatively isolated context for each perspective (#MP1141)
Hospitable reception of external perspectives by structures (#MP1149).
Narrower Inter–level zone (#MP1226) Domains of insight (#MP1252)
Structure–enfolded occupiable sites (#MP1202) Partially enclosed internal domains (#MP1193)
Partially exposed perspective context (#MP1183)
Variation in size of perspective contexts (#MP1190)
Substantive distinctions separating domains (#MP1197)
Occupiable sites exposed to external insight (#MP1180)
Overview sites from integrative superstructure (#MP1231)
Appropriate proportions of perspective contexts (#MP1191)
Boundary expansion permitting new level generation (#MP1211)
Appropriate configuration for perspective inactivity (#MP1186)
Appropriate superstructure to contain transitions between levels (#MP1228).

♦ MP1180a Occupiable sites exposed to external insight
Physical environment Window seats.
Broader Minimal context for single perspective (#MP1078)
Receptivity to emerging external insight (#MP1138)
Limiting length of communication pathways (#MP1132)
Limiting exposure to harmonious perspectives (#MP1134)
Protected low–density communication pathways (#MP1101)
Sequence of viewpoint loci within a structure (#MP1142)
Provision for temporary perspective inactivity (#MP1150)
Overview of communication pathway from structure (#MP1164)
Organization of structure to provide occupiable sites (#MP1179)
Insight capturing non–linear extensions of structures (#MP1175)
Enhancing insight by varying levels of exposure to it (#MP1135)
Hospitable reception of external perspectives by structures (#MP1149)
Internal transition spaces enhancing structural entry points (#MP1130)
Organization of structures to permit two sources of external insight (#MP1159).
Narrower Flexible interfaces (#MP1244) Aperture compatibility (#MP1221)
Displaceable frameworks (#MP1236) Ground–level visibility (#MP1222)
Zones of intermediate insight (#MP1223) Structure–enfolded occupiable sites (#MP1202)
Substantive distinctions separating domains (#MP1197)
Overview sites from integrative superstructure (#MP1231)
Boundary expansion permitting new level generation (#MP1211)
Appropriate configuration for perspective interaction (#MP1185)
Maintenance of source of direct insight within structures as a focal point (#MP1181).

♦ MP1181b Maintenance of source of direct insight within structures as a focal point
Physical environment Fireplace in a common room.
Broader Occupiable sites exposed to external insight (#MP1180)
Sequence of viewpoint loci within a structure (#MP1142)
Unstructured context for perspective exchange (#MP1090)
Common domain at the focal point of a structure (#MP1129)
Re–integration of rejected perspective by–products (#MP1178)
Hospitable reception of external perspectives by structures (#MP1149).
Narrower Appropriate configuration for perspective interaction (#MP1185).

♦ MP1182c Appropriate conditions for perspective nourishment
Physical environment Attractive space for communal eating.
Broader Coordination of perspective nourishment (#MP1147)
Hospitable domain for perspective nourishment (#MP1139).
Narrower Domains of insight (#MP1252) Encouraging emphases (#MP1250)
Structure–enfolded occupiable sites (#MP1202) Facilities for perspective adjuncts (#MP1200)
Accessible facilities for perspective adjuncts (#MP1201).

♦ MP1183a Partially exposed perspective context
Physical environment Workspace enclosure and connectedness.
Broader Partially isolated contexts (#MP1152) Flexible domain organization (#MP1146)
Perspective interaction constraints (#MP1148)
Minimal context for single perspective (#MP1078)
Local opportunities for perspective activity (#MP1157)
Relatively isolated context for each perspective (#MP1141)
Orientation of domains to receive external insight (#MP1128)
Organization of structure to provide occupiable sites (#MP1179)
Hospitable reception of external perspectives by structures (#MP1149)
Integration of perspective acquisition and perspective maintenance dynamics (#MP1083).
Narrower Domains of insight (#MP1252) Overview of external contexts (#MP1192)
Facilities for perspective adjuncts (#MP1200) Partially enclosed internal domains (#MP1193)
Substantive distinctions separating domains (#MP1197)
Inter–domain contexts for perspective adjuncts (#MP1198)
Accessible facilities for perspective adjuncts (#MP1201)
Appropriate proportions of perspective contexts (#MP1191)
Appropriate configuration for perspective interaction (#MP1185).

♦ **MP1184b Appropriate configuration for input processing**
Physical environment Appropriate kitchen layout.
Broader Hospitable domain for perspective nourishment (#MP1139).
Narrower Facilities for perspective adjuncts (#MP1200)
Substantive distinctions separating domains (#MP1197)
Exposure of input processing context to external insight (#MP1199).

♦ **MP1185b Appropriate configuration for perspective interaction**
Physical environment Protected communal seating arrangement.
Broader Partially isolated contexts (#MP1152) Nested levels of accessibility (#MP1127)
Partially exposed perspective context (#MP1183)
Occupiable sites exposed to external insight (#MP1180)
Small-scale perspective interaction contexts (#MP1151)
Sequence of viewpoint loci within a structure (#MP1142)
Common domain at the focal point of a structure (#MP1129)
Relative isolation of complementary perspectives (#MP1136)
Semi-autonomous contexts for maturing perspectives (#MP1154)
Maintenance of source of direct insight within structures as a focal point (#MP1181).
Narrower Different settings (#MP1251) Domains of insight (#MP1252)
Partially enclosed internal domains (#MP1193)
Appropriate proportions of perspective contexts (#MP1191).

♦ **MP1186c Appropriate configuration for perspective inactivity**
Physical environment Communal sleeping area for adults and children.
Broader Domain for developing perspectives (#MP1137)
Receptivity to emerging external insight (#MP1138)
Relative isolation of complementary perspectives (#MP1136)
Hospitable contexts for perspectives in transition (#MP1091)
Organization of structure to provide occupiable sites (#MP1179)
Interrelationship of contexts of developing perspectives (#MP1143).
Narrower Occupiable sites for perspective inactivity (#MP1188)
Context for forms of perspective presentation (#MP1189)
Appropriate configuration for interaction of complementary perspectives (#MP1187).

♦ **MP1187c Appropriate configuration for interaction of complementary perspectives**
Physical environment Partially enclosed marriage bed.
Broader Functional cycle (#MP1026)
Receptivity to emerging external insight (#MP1138)
Relative isolation of complementary perspectives (#MP1136)
Appropriate configuration for perspective inactivity (#MP1186).
Narrower Symbols of integration (#MP1249)
Variation in size of perspective contexts (#MP1190)
Occupiable sites for perspective inactivity (#MP1188)
Context for forms of perspective presentation (#MP1189)
Appropriate proportions of perspective contexts (#MP1191).

♦ **MP1188a Occupiable sites for perspective inactivity**
Physical environment Bed alcove off rooms with other functions.
Broader Hospitable common domains (#MP1094)
Minimal context for single perspective (#MP1078)
Receptivity to emerging external insight (#MP1138)
Relatively isolated context for each perspective (#MP1141)
Semi-autonomous contexts for maturing perspectives (#MP1154)
Appropriate configuration for perspective inactivity (#MP1186)
Interrelationship of contexts of developing perspectives (#MP1143)
Appropriate configuration for interaction of complementary perspectives (#MP1187).
Narrower Aperture compatibility (#MP1221) Facilities for perspective adjuncts (#MP1200)
Partially enclosed internal domains (#MP1193)
Variation in size of perspective contexts (#MP1190)
Substantive distinctions separating domains (#MP1197)
Context for forms of perspective presentation (#MP1189)
Appropriate proportions of perspective contexts (#MP1191).

♦ **MP1189 Context for forms of perspective presentation**
Physical environment Dressing rooms with clothing storage facilities.
tj M b Context for forms of pespective presentation
Broader Minimal context for single perspective (#MP1078)
Occupiable sites for perspective inactivity (#MP1188)
Relatively isolated context for each perspective (#MP1141)
Relative isolation of complementary perspectives (#MP1136)
Appropriate configuration for perspective inactivity (#MP1186)
Interrelationship of contexts of developing perspectives (#MP1143)
Integrating coordinated exposure to irrationality into structures (#MP1144)
Organization of structures to permit two sources of external insight (#MP1159)
Appropriate configuration for interaction of complementary perspectives (#MP1187).
Narrower Facilities for perspective adjuncts (#MP1200)
Substantive distinctions separating domains (#MP1197)
Accessible facilities for perspective adjuncts (#MP1201)
Inter-domain contexts for perspective adjuncts (#MP1198)
Appropriate proportions of perspective contexts (#MP1191).

♦ **MP1190a Variation in size of perspective contexts**
Physical environment Ceiling height variety throughout a building.
Broader Flexible domain organization (#MP1146)
Partially contained interfaces (#MP1119)
Patterning integrative superstructure (#MP1116)
Occupiable sites for perspective inactivity (#MP1188)
Protected low-density communication pathways (#MP1101)
Unstructured context for perspective exchange (#MP1090)
Organization of structure to provide occupiable sites (#MP1179)
Domains for non-current elements and those in reserve (#MP1145)
Appropriate configuration for interaction of complementary perspectives (#MP1187).
Narrower Exclusive contexts (#MP1204) Overview of external contexts (#MP1192)
Level generation of minimum tension (#MP1219)
Substantive distinctions separating domains (#MP1197)
Harmonizing space distribution between levels (#MP1210)
Appropriate proportions of perspective contexts (#MP1191)
Distribution of secondary inter-level connections (#MP1213)
Congruence between spaces defined by the framework and spaces defined by the processes within it (#MP1205).

♦ **MP1191a Appropriate proportions of perspective contexts**
Physical environment Roughly rectangular indoor spaces.
Broader Partially isolated contexts (#MP1152) Partially exposed perspective context (#MP1183)
Variation in size of perspective contexts (#MP1190)
Limiting length of communication pathways (#MP1132)
Local sources for perspective nourishment (#MP1089)
Occupiable sites for perspective inactivity (#MP1188)
Context for forms of perspective presentation (#MP1189)
Local opportunities for perspective activity (#MP1157)
Small-scale perspective interaction contexts (#MP1151)
Hospitable domain for perspective nourishment (#MP1139)
Provision for temporary perspective inactivity (#MP1150)
Common domain at the focal point of a structure (#MP1129)
Enfolded overview domains of minimum proportions (#MP1167)
Relatively isolated context for each perspective (#MP1141)
Semi-autonomous contexts for maturing perspectives (#MP1154)
Hospitable non-linear domain external to structures (#MP1163)
Exchange contexts controlled by a single perspective (#MP1087)
Appropriate configuration for perspective interaction (#MP1185)
Organization of structure to provide occupiable sites (#MP1179)
Interrelationship of contexts of developing perspectives (#MP1143)
Hospitable reception of external perspectives by structures (#MP1149)
Internal transition spaces enhancing structural entry points (#MP1130)
Integrating transition pathways between levels into a structure (#MP1133)
Semi-autonomous contexts for perspectives of decreasing activity (#MP1155)
Structures adaptable to changing number of embodied perspectives (#MP1153)
Integrating coordinated exposure to irrationality into structures (#MP1144)
Functional enhancement of domains separating complementary structures (#MP1106)
Appropriate configuration for interaction of complementary perspectives (#MP1187).
Narrower Overview of external contexts (#MP1192)
Distortion resistant boundaries (#MP1218)
Level generation of minimum tension (#MP1219)
Partially enclosed internal domains (#MP1193)
Substantive distinctions separating domains (#MP1197)
Harmonizing space distribution between levels (#MP1210)
Inter-domain contexts for perspective adjuncts (#MP1198)
Primary inter-level connections at transitions in boundary orientation (#MP1212)
Congruence between spaces defined by the framework and spaces defined by the processes within it (#MP1205).

♦ **MP1192b Overview of external contexts**
Physical environment Rooms with a view overlooking neighbourhood activity.
Broader Partially isolated contexts (#MP1152) Partially exposed perspective context (#MP1183)
Variation in size of perspective contexts (#MP1190)
Local opportunities for perspective activity (#MP1157)
Appropriate proportions of perspective contexts (#MP1191)
Organization of structures to permit two sources of external insight (#MP1159).
Narrower Aperture compatibility (#MP1221) Displaceable frameworks (#MP1236)
Ground-level visibility (#MP1222) Multi-faceted frameworks (#MP1239)
Zones of intermediate insight (#MP1223)
Exposure of input processing context to external insight (#MP1199).

♦ **MP1193b Partially enclosed internal domains**
Physical environment Rooms partially opening into each other.
Broader Partially isolated contexts (#MP1152) Partially contained interfaces (#MP1119)
Partially exposed perspective context (#MP1183)
Organization of inter-domain dynamics (#MP1131)
Occupiable sites for perspective inactivity (#MP1188)
Protected low-density communication pathways (#MP1101)
Appropriate proportions of perspective contexts (#MP1191)
Enfolded overview domains of minimum proportions (#MP1167)
Appropriate configuration for perspective interaction (#MP1185)
Organization of structure to provide occupiable sites (#MP1179)
Protecting non-linear contexts from communication pathways (#MP1173)
Relative isolation of structural interface with communication pathway (#MP1140)
Overview domains at interfaces of the structure with the external environment (#MP1166).
Narrower Inter-level zone (#MP1226) Inter-level integrity (#MP1227)
Symbols of integration (#MP1249) Multi-faceted frameworks (#MP1239)
Internal connectedness between domains (#MP1194)
Primary inter-level connections at transitions in boundary orientation (#MP1212).

♦ **MP1194c Internal connectedness between domains**
Physical environment Connecting windows between interior rooms.
Broader Partially enclosed internal domains (#MP1193)
Organization of inter-domain dynamics (#MP1131)
Limiting length of communication pathways (#MP1132)
Protected low-density communication pathways (#MP1101)
Sequence of viewpoint loci within a structure (#MP1142)
Enhancing insight by varying levels of exposure to it (#MP1135)
Internal transition spaces enhancing structural entry points (#MP1130).
Narrower Multi-faceted frameworks (#MP1239) Connectedness in isolation (#MP1237).

♦ **MP1195b Framework for transition between structural levels**
Physical environment Staircase space requirement.
Broader External access to higher structural levels (#MP1158)
Integrating points of perspective into common domains (#MP1125)
Integrating transition pathways between levels into a structure (#MP1133).
Narrower Exclusive spaces for emergent perspectives (#MP1203)
Appropriate superstructure to contain transitions between levels (#MP1228)
Primary inter-level connections at transitions in boundary orientation (#MP1212).

♦ **MP1196 Eccentric access to perspective contexts**
Physical environment Corner doorways to protect integrity of activities in the room.
TJ M b Excentric access to perspective contexts
Broader Organization of inter-domain dynamics (#MP1131)
Sequence of viewpoint loci within a structure (#MP1142)
Enhancing insight by varying levels of exposure to it (#MP1135).
Narrower Aperture compatibility (#MP1221) Symbols of integration (#MP1249)
Connectedness in isolation (#MP1237) Boundary reinforcement at apertures (#MP1225)
Emphasizing transitions across boundaries (#MP1224)
Inter-domain contexts for perspective adjuncts (#MP1198).

♦ **MP1197a Substantive distinctions separating domains**
Physical environment Thick walls to permit adaptations for alcoves and shelving.
Broader Partially exposed perspective context (#MP1183)
Variation in size of perspective contexts (#MP1190)
Limiting length of communication pathways (#MP1132)
Local sources for perspective nourishment (#MP1089)
Occupiable sites for perspective inactivity (#MP1188)
Context for forms of perspective presentation (#MP1189)
Occupiable sites exposed to external insight (#MP1180)
Appropriate configuration for input processing (#MP1184)
Appropriate proportions of perspective contexts (#MP1191)
Exchange contexts controlled by a single perspective (#MP1087)
Organization of structure to provide occupiable sites (#MP1179)
Hospitable interface between structures and external environment (#MP1160).

Narrower Exclusive contexts (#MP1204) Zones of intermediate insight (#MP1223)
Structure–enfolded occupiable sites (#MP1202) Facilities for perspective adjuncts (#MP1200)
Exclusive spaces for emergent perspectives (#MP1203)
Accessible facilities for perspective adjuncts (#MP1201)
Inter–domain contexts for perspective adjuncts (#MP1198)
Boundary expansion permitting new level generation (#MP1211)
Exposure of input processing context to external insight (#MP1199)
Primary inter–level connections at transitions in boundary orientation (#MP1212).

♦ MP1198b Inter–domain contexts for perspective adjuncts

Physical environment Storage closets on interior walls between rooms.
Broader Partially exposed perspective context (#MP1183)
Eccentric access to perspective contexts (#MP1196)
Limiting length of communication pathways (#MP1132)
Substantive distinctions separating domains (#MP1197)
Context for forms of perspective presentation (#MP1189)
Appropriate proportions of perspective contexts (#MP1191)
Relative isolation of complementary perspectives (#MP1136)
Interrelationship of contexts of developing perspectives (#MP1143)
Organization of integrative superstructure to minimize insight blindspots (#MP1162).
Narrower Exclusive contexts (#MP1204)
Appropriate superstructure to contain transitions between levels (#MP1228).

♦ MP1199b Exposure of input processing context to external insight

Physical environment Kitchen oriented toward the sun.
Broader Overview of external contexts (#MP1192)
Substantive distinctions separating domains (#MP1197)
Hospitable domain for perspective nourishment (#MP1139)
Appropriate configuration for input processing (#MP1184)
Orientation of domains to receive external insight (#MP1128).
Narrower Encouraging emphases (#MP1250) Facilities for perspective adjuncts (#MP1200)
Boundary expansion permitting new level generation (#MP1211).

♦ MP1200b Facilities for perspective adjuncts

Physical environment Open shelving on walls.
Broader Partially exposed perspective context (#MP1183)
Local sources for perspective nourishment (#MP1089)
Substantive distinctions separating domains (#MP1197)
Occupiable sites for perspective inactivity (#MP1188)
Context for forms of perspective presentation (#MP1189)
Hospitable domain for perspective nourishment (#MP1139)
Appropriate configuration for input processing (#MP1184)
Appropriate conditions for perspective nourishment (#MP1182)
Exchange contexts controlled by a single perspective (#MP1087)
Exposure of input processing context to external insight (#MP1199).
Narrower Accessible facilities for perspective adjuncts (#MP1201)
Boundary expansion permitting new level generation (#MP1211).

♦ MP1201c Accessible facilities for perspective adjuncts

Physical environment Waist–high shelving in working and living rooms.
Broader Facilities for perspective adjuncts (#MP1200)
Partially exposed perspective context (#MP1183)
Substantive distinctions separating domains (#MP1197)
Context for forms of perspective presentation (#MP1189)
Hospitable domain for perspective nourishment (#MP1139)
Appropriate conditions for perspective nourishment (#MP1182)
Insight capturing non–linear extensions of structures (#MP1175)
Internal transition spaces enhancing structural entry points (#MP1130).
Narrower Ground–level visibility (#MP1222)
Meaningful symbols of self–transformation (#MP1253)
Boundary expansion permitting new level generation (#MP1211).

♦ MP1202b Structure–enfolded occupiable sites

Physical environment Seats built into walls.
Broader Substantive distinctions separating domains (#MP1197)
Occupiable sites exposed to external insight (#MP1180)
Sequence of viewpoint loci within a structure (#MP1142)
Appropriate conditions for perspective nourishment (#MP1182)
Organization of structure to provide occupiable sites (#MP1179)
Internal transition spaces enhancing structural entry points (#MP1130).
Narrower Different settings (#MP1251) Unmediated supportive emotion (#MP1230)
Boundary expansion permitting new level generation (#MP1211).

♦ MP1203c Exclusive spaces for emergent perspectives

Physical environment Cave–like places for children.
Broader Domain for developing perspectives (#MP1137)
Substantive distinctions separating domains (#MP1197)
Contexts for care of premature perspectives (#MP1086)
Framework for transition between structural levels (#MP1195)
Contexts for exploratory relationship formation challenging emerging perspectives (#MP1073).
Narrower Emphasizing transitions across boundaries (#MP1224)
Boundary expansion permitting new level generation (#MP1211)
Appropriate superstructure to contain transitions between levels (#MP1228).

♦ MP1204c Exclusive contexts

Physical environment Secret places for storage of valuables.
Broader Variation in size of perspective contexts (#MP1190)
Substantive distinctions separating domains (#MP1197)
Inter–domain contexts for perspective adjuncts (#MP1198).
Narrower Level generation of minimum tension (#MP1219)
Boundary expansion permitting new level generation (#MP1211).

♦ MP1205a Congruence between spaces defined by the framework and spaces defined by the processes within it

Pattern No framework is every felt to be appropriate unless the spaces defined within it by boundaries and inter–level connections are congruent with the spaces defined by the pattern of movement within it. Formal construction constraints should not dictate the final form of the framework.
Physical environment No building ever feels right to the people in it unless the physical spaces (defined by columns, walls and ceilings) are congruent with the social spaces (defined by activities and human groups). The social spaces should never be modified to conform to the engineering structure.
Broader Partially contained interfaces (#MP1119)
Focal centre of a complex structure (#MP1099)
Patterning integrative superstructure (#MP1116)
Complexification of perspective contexts (#MP1095)
Variation in size of perspective contexts (#MP1190)
Limitation on number of structural levels (#MP1096)
Appropriate proportions of perspective contexts (#MP1191)
Semi–autonomous contexts for maturing perspectives (#MP1154)
Hospitable non–linear domain external to structures (#MP1163)
Concealment of necessary monotonous perspective patterns (#MP1097)
Semi–autonomous contexts for perspectives of decreasing activity (#MP1155)
Harmoniously structured entry point for external communication media (#MP1113)
Organization of structures to enhance receptivity to external insight (#MP1107).
Narrower Progressive framework definition (#MP1208)
Appropriate construction elements (#MP1207)
Level generation of minimum tension (#MP1219)
Efficient enclosure of spaces with minimal structural distinctions (#MP1206)
Primary inter–level connections at transitions in boundary orientation (#MP1212).

♦ MP1206 Efficient enclosure of spaces with minimal structural distinctions

Pattern The most efficient way of distributing distinctions throughout a framework so as to enclose space, providing appropriate structural integrity with the minimum number of elements, is to conceive of the framework as made from one continuous body of mutually constrained elements.
Physical environment The most efficient way of distributing materials throughout a building, so as to enclose space, strongly and well, with the least amount of expensive, non–depletable material, is to conceive of the building as made from one continuous body of compressive material. In its geometry, conceive it as a three–dimensional system of individually vaulted spaces, most of them roughly rectangular; with thin load–bearing walls, each stiffened by columns at intervals along its length, thickened where walls meet walls, vaults, or openings.
tj M b Efficient enclosure of spaces with minimal structural distinctions
Broader Congruence between spaces defined by the framework and spaces defined by the processes within it (#MP1205).
Narrower Inter–level zone (#MP1226) Perimeter continuity (#MP1217)
Inter–level integrity (#MP1227) Initial level formation (#MP1215)
Inter–level connections (#MP1216) Integration superstructure (#MP1220)
Distortion resistant boundaries (#MP1218) Progressive framework definition (#MP1208)
Appropriate construction elements (#MP1207) Boundary reinforcement at apertures (#MP1225)
Level generation of minimum tension (#MP1219)
Organization of integrative superstructure (#MP1209)
Harmonizing space distribution between levels (#MP1210)
Distribution of secondary inter–level connections (#MP1213)
Provision for pathways for automatic communications (#MP1229)
Appropriate superstructure to contain transitions between levels (#MP1228)
Primary inter–level connections at transitions in boundary orientation (#MP1212).

♦ MP1207a Appropriate construction elements

Pattern Standardized elements used in the construction of any framework tend to prevent the emergence of any organic quality. Appropriate construction elements are required which are small in scale, easy to work with using accessible techniques, easy to adapt, capable of ensuring structural integrity, durable or renewable, not needing specialized or expensive assistance, and universally obtainable at low cost. Isolating such elements from the environment should neither unduly deplete the environment nor prevent their reintegration into it when the framework is no longer required.
Physical environment Mass–produced construction materials tend to destroy the organic quality of natural buildings and to make any such quality impossible. Appropriate materials are required which are small in scale, easy to cut on site, easy to work without sophisticated machinery, easy to vary and adapt, heavy enough to be solid, long–lasting or easy to maintain, and yet easy to build, not needing specialized labour, not expensive in labour, and universally obtainable and cheap. Such materials should be ecologically sound, namely biodegradable, low in energy consumption, and not based on depletable resources.
Broader Maintainable, multi–element external boundaries (#MP1234)
Efficient enclosure of spaces with minimal structural distinctions (#MP1206)
Congruence between spaces defined by the framework and spaces defined by the processes within it (#MP1205).
Narrower Time binding (#MP1248) Encouraging emphases (#MP1250)
Progressive framework definition (#MP1208) Level generation of minimum tension (#MP1219)
Appropriate superstructure to contain transitions between levels (#MP1228).

♦ MP1208a Progressive framework definition

Pattern In order to accommodate subtle design adaptations as the need for them becomes clear, the framework should be conceived from the start as globally complete although tentative in its definition. In subsequent construction phases, as further elements are woven into the framework, it acquires increasing definition up to the final phase at which its full structural integrity is achieved.
Physical environment Buildings should be uniquely adapted to individual needs and sites. The plans should therefore be rather loose and fluid in order to accommodate those subtleties as the need for them emerges. The building should not be thought of as assembled from components but rather as being woven from a structure which starts out globally complete, but flimsy. It is gradually made stiffer, whilst still somewhat flimsy, until it is made completely stiff and strong in the final construction phase. Although the building goes from flimsy to strong, the actual materials added go from the strongest and stiffest to the less stiff, until finally fluid materials are added. In this way each new material is more adaptable, more flexible, more capable of coping with variations (and earlier mistakes) than the last. This can be achieved by constructing a soft skin framework and filling it with a compressive fill.
Broader Appropriate construction elements (#MP1207)
Efficient enclosure of spaces with minimal structural distinctions (#MP1206)
Congruence between spaces defined by the framework and spaces defined by the processes within it (#MP1205).
Narrower Perimeter continuity (#MP1217) Inter–level connections (#MP1216)
Integration superstructure (#MP1220) Distortion resistant boundaries (#MP1218)
Boundary reinforcement at apertures (#MP1225) Level generation of minimum tension (#MP1219)
Overview sites from integrative superstructure (#MP1231)
Primary inter–level connections at transitions in boundary orientation (#MP1212).

♦ MP1209b Organization of integrative superstructure

Pattern Each distinct portion of the integrative superstructure should correspond to an identifiable set of spaces within the framework. The layout will span the largest spaces and it is to them that those spanning lesser spaces should be connected in a self–reinforcing cascade down to those spanning the smallest spaces.
Physical environment Arrange the roofs of a building so that each distinct roof corresponds to an identifiable social span in the building. Place the largest roofs – those which are the highest and have the largest span – over the most important and most communal spaces; build the lesser roofs off these largest and highest roofs; and build the smallest roofs of all off these lesser roofs, in the form of half–vaults and sheds over alcoves and thick walls. The overall arrangement should form a self– buttressing cascade in which each lower roof helps to take up the horizontal thrust generated by the higher roofs.
Broader Patterning integrative superstructure (#MP1116)

Containment by integrative superstructure (#MP1117)
Functional integration of unstructured internal domains (#MP1115)
Integration of non-linearity into integrative superstructure (#MP1118)
Efficient enclosure of spaces with minimal structural distinctions (#MP1206)
Organization of structures to permit two sources of external insight (#MP1159).
Narrower Harmonizing space distribution between levels (#MP1210)
Boundary expansion permitting new level generation (#MP1211)
Primary inter-level connections at transitions in boundary orientation (#MP1212).

♦ MP1210c Harmonizing space distribution between levels

Pattern In order to maintain the integrity of the framework, the boundaries of spaces at a higher level should not be too distant from boundaries to which those forces can be distributed at the lower level beneath it.

Physical environment The floor and ceiling layout must correspond to the important social spaces which will vary from floor to floor. If the spaces are vaulted to permit uses of compression materials only, vaults on different floors do not have to be aligned. To maintain reasonable structural integrity in the system of vaults as a whole, it is sufficient if a vault is placed so that its loads come down in a position from which the forces can spread out downward in a 45 degree cone and be supported by columns of the next vault down.

Broader Variation in size of perspective contexts (#MP1190)
Organization of integrative superstructure (#MP1209)
Appropriate proportions of perspective contexts (#MP1191)
Efficient enclosure of spaces with minimal structural distinctions (#MP1206).
Narrower Perimeter continuity (#MP1217) Zoning internal domains (#MP1233)
Level generation of minimum tension (#MP1219)
Distribution of secondary inter-level connections (#MP1213)
Boundary expansion permitting new level generation (#MP1211)
Primary inter-level connections at transitions in boundary orientation (#MP1212).

♦ MP1211b Boundary expansion permitting new level generation

Pattern Slight expansion of external boundaries adds useful extra sub-space to the framework. They may be integrated into the framework in such a way as to counteract distortion resulting from the addition of a new level within the framework.

Physical environment In a sensibly made building every floor is surrounded, at various places, by small alcoves, window-seats, niches, and counters which thus form low-ceilinged "thick walls" around the outside edge of rooms. In a natural building these thick walls, defined by columns, can work as buttresses to counteract horizontal outward thrust generated by the presence of ceilings or higher floors. In order for such an alcove to work as a buttress, its roof must be built as nearly as possible as a continuation of the curve of the floor vault immediately inside.

Broader Exclusive contexts (#MP1204) Partially contained interfaces (#MP1119)
Structure-enfolded occupiable sites (#MP1202) Facilities for perspective adjuncts (#MP1200)
Patterning integrative superstructure (#MP1116)
Organization of integrative superstructure (#MP1209)
Exclusive spaces for emergent perspectives (#MP1203)
Substantive distinctions separating domains (#MP1197)
Occupiable sites exposed to external insight (#MP1180)
Harmonizing space distribution between levels (#MP1210)
Accessible facilities for perspective adjuncts (#MP1201)
Organization of structure to provide occupiable sites (#MP1179)
Exposure of input processing context to external insight (#MP1199).
Narrower Integration superstructure (#MP1220) Level generation of minimum tension (#MP1219)
Primary inter-level connections at transitions in boundary orientation (#MP1212).

♦ MP1212a Primary inter-level connections at transitions in boundary orientation

Pattern In order to be able to freely adapt the design of a framework during its construction in the light of insights and opportunities emerging during that process, detailed designs elaborated prior to that process should be avoided. The required flexibility can be ensured by first identifying the positions which generate the spaces and their boundaries at the initial level. Those located at transitions in boundary orientation are the site of the primary inter-level connections. These may generate spaces at several other levels of the framework.

Physical environment The common practice of specifying complex building plans in perfect detail cripples the buildings that are generated in this way. The necessities of the drawing itself change the plan, make it more rigid, and limit it to the kind of plan which can be drawn and measured. The rich range of possibilities can only be preserved if the builder is able to generate a living building, with all its slightly uneven lines and imperfect angles. In order to achieve this aim, the building must be generated in an entirely different manner. What must be done, essentially, is to fix those points (representing columns) which generate the space – as few as possible – and then let those points generate the walls, right out on the building site, during the very process of construction.

Broader Progressive framework definition (#MP1208)
Partially enclosed internal domains (#MP1193)
Diversified interchange environment (#MP1046)
Organization of integrative superstructure (#MP1209)
Substantive distinctions separating domains (#MP1197)
Harmonizing space distribution between levels (#MP1210)
Appropriate proportions of perspective contexts (#MP1191)
Boundary expansion permitting new level generation (#MP1211)
Framework for transition between structural levels (#MP1195)
Efficient enclosure of spaces with minimal structural distinctions (#MP1206)
Congruence between spaces defined by the framework and spaces defined by the processes within it (#MP1205).
Narrower Inter-level zone (#MP1226) Perimeter continuity (#MP1217)
Initial level formation (#MP1215) Inter-level connections (#MP1216)
Integration superstructure (#MP1220) Integrative infrastructure (#MP1214)
Level generation of minimum tension (#MP1219)
Overview sites from integrative superstructure (#MP1231)
Distribution of secondary inter-level connections (#MP1213).

♦ MP1213a Distribution of secondary inter-level connections

Pattern Secondary inter-level connections may be required to reinforce the boundary between the primary inter-level connections positioned at transitions in boundary orientation. Such secondary connections should be distributed according to the proportions of the multi-level framework as a whole and of the bounded spaces within any given level.

Physical environment Once spaces have been defined by corner columns, columns are required to stiffen the walls between them. They should be furthest apart on the ground floor (higher walls) and progressively closer together on higher floors (lower walls). On any one floor they should be equally spaced, but closer together along the walls of small rooms and further apart along the walls of large rooms.

Broader Variation in size of perspective contexts (#MP1190)
Limitation on number of structural levels (#MP1096)
Harmonizing space distribution between levels (#MP1210)
Efficient enclosure of spaces with minimal structural distinctions (#MP1206)
Primary inter-level connections at transitions in boundary orientation (#MP1212).
Narrower Inter-level connections (#MP1216) Integrative infrastructure (#MP1214)
Distortion resistant boundaries (#MP1218) Level generation of minimum tension (#MP1219).

♦ MP1214c Integrative infrastructure

Pattern The best form of infrastructure is continuous with the overlying framework whose pattern of inter-level connections is then embedded in the context in a manner entirely integral with it.

Physical environment The best foundations are like those of a tree where the entire structure of the tree simply continues below ground level, creating a system entirely integral with the ground so as to resist tension and horizontal shear as well as compression.

Broader Grounded structures (#MP1168)
Distribution of secondary inter-level connections (#MP1213)
Primary inter-level connections at transitions in boundary orientation (#MP1212)
Primary inter-level connections at transitions in boundary orientation (#MP1212)
Structural development designed to counteract deficiencies in pattern harmony (#MP1104).
Narrower Inter-level connections (#MP1216) Initial level formation (#MP1215)
Level generation of minimum tension (#MP1219).

♦ MP1215c Initial level formation

Pattern The easiest and most natural way to form the initial level of a framework is by enclosing an area of the domain on which it is located and reinforcing it with materials characteristic of that domain that ensure continuity and structural integrity in relation to the inter-level connections.

Physical environment The slab is the easiest, cheapest, and most natural way to lay a ground floor. It is constructed by first building a low perimeter wall around the building, tying it with the column foundations, and then filling it with rubble, gravel and concrete.

Broader Grounded structures (#MP1168) Integrative infrastructure (#MP1214)
Domains for non-current elements and those in reserve (#MP1145)
Efficient enclosure of spaces with minimal structural distinctions (#MP1206)
Primary inter-level connections at transitions in boundary orientation (#MP1212).
Narrower Time binding (#MP1248) Zoning internal domains (#MP1233)
Level generation of minimum tension (#MP1219).

♦ MP1216a Inter-level connections

Pattern Inter-level connections are of prime significance in any framework and should enhance this impression by the manner in which they are constructed and disposed. They should be easy to connect to other elements of the framework in order to establish structural integrity.

Physical environment In all the world's traditional and historic buildings, the columns are expressive, beautiful, and treasured elements. In modern buildings they have become ugly and meaningless because no one knows how to make a column which is at the same time beautiful and structurally efficient. This requires: solidity, especially when using ecologically sound materials of lower strength; inexpensive materials; pleasant surface finish, permitting heavy decoration; concentration of strength towards centre; easy connectibility to other elements; on-site modifiability and subsequent repairability.

Broader Integrative infrastructure (#MP1214) Progressive framework definition (#MP1208)
Distribution of secondary inter-level connections (#MP1213)
Efficient enclosure of spaces with minimal structural distinctions (#MP1206)
Primary inter-level connections at transitions in boundary orientation (#MP1212).
Narrower Inter-level zone (#MP1226) Perimeter continuity (#MP1217)
Inter-level integrity (#MP1227) Aperture compatibility (#MP1221)
Tolerance at level interfaces (#MP1240) Distortion resistant boundaries (#MP1218)
Level generation of minimum tension (#MP1219)
Overview sites from integrative superstructure (#MP1231).

♦ MP1217b Perimeter continuity

Pattern Each new level in a framework, developed by extension of a pattern of inter-level connections, is defined and given form by connecting these together to form a continuous perimeter ring. This helps to resist distortion resulting from the new level partly by redistributing forces from that level to maintain the overall structural integrity.

Physical environment If a room is conceived and built by first placing columns at the corners, and then gradually weaving the walls and ceiling around them, the room needs a perimeter beam around its upper edge. This resists the horizontal thrust of the vault above, spreading the loads from upper stories onto columns, tying the columns together, and functioning as a lintel over openings in the wall.

Broader Inter-level connections (#MP1216) Progressive framework definition (#MP1208)
Harmonizing space distribution between levels (#MP1210)
Efficient enclosure of spaces with minimal structural distinctions (#MP1206)
Primary inter-level connections at transitions in boundary orientation (#MP1212).
Narrower Inter-level zone (#MP1226) Inter-level integrity (#MP1227)
Aperture compatibility (#MP1221) Tolerance at level interfaces (#MP1240)
Level generation of minimum tension (#MP1219)
Overview sites from integrative superstructure (#MP1231).

♦ MP1218b Distortion resistant boundaries

Pattern Boundaries enclosing spaces may contribute to the overall structural integrity of the framework by the manner in which they act to resist its distortion. Such fully integrated boundaries should be contrasted with those which enclose and define spaces but do not contribute to the overall structural integrity.

Physical environment In organic construction, walls, acting as structural membranes, take their share of loads. They work continuously with the structure on all four of their sides; and act to resist shear and bending, and take loads in compression. Such walls are built as a membrane which connects the columns and door frames and window frames and is, at least in part, continuous with them.

Broader Inter-level connections (#MP1216) Progressive framework definition (#MP1208)
Appropriate proportions of perspective contexts (#MP1191)
Distribution of secondary inter-level connections (#MP1213)
Efficient enclosure of spaces with minimal structural distinctions (#MP1206).
Narrower Time binding (#MP1248) Symbols of integration (#MP1249)
Connectedness in isolation (#MP1237) Unmediated supportive emotion (#MP1230)
Unalienating internal boundaries (#MP1235)
Symbolic connection to encompassing domains (#MP1232)
Overview sites from integrative superstructure (#MP1231)
Maintainable, multi-element external boundaries (#MP1234).

♦ MP1219a Level generation of minimum tension

Pattern Generation of a usable new level on a framework calls for a design which will ensure that minimum distortion and tension is engendered by the addition.

Physical environment Elliptical vaults will support a live load on the floor above, whilst forming the ceiling of the room below, and generating minimum bending and tension so that inexpensive compressive materials can be relied on. They can be fitted to rooms of any shape.

Broader Exclusive contexts (#MP1204) Perimeter continuity (#MP1217)
Inter-level connections (#MP1216) Initial level formation (#MP1215)
Integrative infrastructure (#MP1214) Progressive framework definition (#MP1208)
Appropriate construction elements (#MP1207)
Variation in size of perspective contexts (#MP1190)
Harmonizing space distribution between levels (#MP1210)
Appropriate proportions of perspective contexts (#MP1191)

Distribution of secondary inter–level connections (#MP1213)
Boundary expansion permitting new level generation (#MP1211)
Efficient enclosure of spaces with minimal structural distinctions (#MP1206)
Primary inter–level connections at transitions in boundary orientation (#MP1212)
Congruence between spaces defined by the framework and spaces defined by the processes within it (#MP1205).
Narrower Zoning internal domains (#MP1233) Tolerance at level interfaces (#MP1240)
Unmediated supportive emotion (#MP1230) Unalienating internal boundaries (#MP1235)
Overview sites from integrative superstructure (#MP1231)
Provision for pathways for automatic communications (#MP1229)
Appropriate superstructure to contain transitions between levels (#MP1228).

♦ MP1220b Integration superstructure

Pattern The integrative superstructure required to complete the enclosure of any space should be constructed such as to avoid increasing tensions within the framework as a whole. Its design should emerge as the simplest reconciliation of the patterning of forces inherent in the underlying structure. The space enclosed by the superstructure itself should be integrated into the pattern of spaces as a whole rather than simply constituting a superordinate "dead" space.
Physical environment Requirements governing the shape of a roof are: shelter; provision of habitable space within it; roughly rectangular layout; uncontrived, relaxed shape; reduction of bending (unless expensive tension materials are available); and of a steepness appropriate to the nature of any precipitation. Within these constraints a vaulted roof is the most appropriate.
Broader Progressive framework definition (#MP1208)
Level generation of minimum tension (#MP1219)
Containment by integrative superstructure (#MP1117)
Boundary expansion permitting new level generation (#MP1211)
Integration of non–linearity into integrative superstructure (#MP1118)
Efficient enclosure of spaces with minimal structural distinctions (#MP1206)
Primary inter–level connections at transitions in boundary orientation (#MP1212).
Narrower Symbolic connection to encompassing domains (#MP1232)
Overview sites from integrative superstructure (#MP1231)
Maintainable, multi–element external boundaries (#MP1234).

♦ MP1221a Aperture compatibility

Pattern A rigid plan combined with a formal aesthetic results in apertures neither of the appropriate sizes nor compatibly located. Decision on size and placement should be finally determined only when the basic framework has been constructed.
Physical environment Finding the right position for a window or a door is a subtle matter. Minor changes in size or placement make an immense difference. Rather than using standard sizes and pre–planned positions, exact positions and sizes should be determined on site once the framework has been built.
Broader Perimeter continuity (#MP1217) Inter–level connections (#MP1216)
Overview of external contexts (#MP1192)
Eccentric access to perspective contexts (#MP1196)
Receptivity to emerging external insight (#MP1138)
Occupiable sites for perspective inactivity (#MP1188)
Occupiable sites exposed to external insight (#MP1180)
Limiting exposure to harmonious perspectives (#MP1134)
Overview of communication pathway from structure (#MP1164)
Organization of structures to permit two sources of external insight (#MP1159).
Narrower Filtered insights (#MP1238) Displaceable frameworks (#MP1236)
Ground–level visibility (#MP1222) Multi–faceted frameworks (#MP1239)
Zones of intermediate insight (#MP1223) Boundary reinforcement at apertures (#MP1225)
Emphasizing transitions across boundaries (#MP1224).

♦ MP1222c Ground–level visibility

Pattern A primary function of frameworks is to provide a link between a bounded space and the external environment, especially the immediate context in which it is grounded.
Physical environment One of the window's most important functions is to give contact with the outdoors. Low sills provide the most meaningful links because they offer a view of the ground under the window. They should not be too low or the window feels unsafe. Higher sills are desirable on higher floors.
Broader Aperture compatibility (#MP1221) Overview of external contexts (#MP1192)
Limiting length of communication pathways (#MP1132)
Occupiable sites exposed to external insight (#MP1180)
Limiting exposure to harmonious perspectives (#MP1134)
Accessible facilities for perspective adjuncts (#MP1201).
Narrower Displaceable frameworks (#MP1236) Boundary reinforcement at apertures (#MP1225)
Protecting variability to enhance fixity (#MP1245).

♦ MP1223c Zones of intermediate insight

Pattern Well–defined apertures into a bounded space result in strong contrasts in illumination to which it is difficult to adjust. Intermediate levels of contrast are required.
Physical environment Windows with a sharp edge where the frame meets the wall create harsh, blinding glare, and make the rooms they serve uncomfortable. By making the frame a deep, splayed edge, angled to the plane of the window, a gentle gradient of daylight gives a smooth transition between the light of the window and of the dark inner wall.
Broader Aperture compatibility (#MP1221) Overview of external contexts (#MP1192)
Substantive distinctions separating domains (#MP1197)
Occupiable sites exposed to external insight (#MP1180)
Organization of structures to permit two sources of external insight (#MP1159).
Narrower Filtered insights (#MP1238) Integration within context (#MP1246)
Tolerance at level interfaces (#MP1240) Boundary reinforcement at apertures (#MP1225).

♦ MP1224c Emphasizing transitions across boundaries

Pattern Transitions across boundaries may be given special significance by requiring a special position or process through which transition is achieved.
Physical environment Some of the doors in a building play a special role in creating transitions and maintaining privacy. Whilst a high doorway may be simple and convenient, a lower door is often more profound. The act of going through it is a deliberate thoughtful passage from one space to another.
Broader Aperture compatibility (#MP1221) Partially contained interfaces (#MP1119)
Organization of inter–domain dynamics (#MP1131)
Eccentric access to perspective contexts (#MP1196)
Exclusive spaces for emergent perspectives (#MP1203)
Distinctiveness of main entry point to structure (#MP1110)
Distinct pattern of entry points to complex structures (#MP1102).
Narrower Symbols of integration (#MP1249) Boundary reinforcement at apertures (#MP1225).

♦ MP1225a Boundary reinforcement at apertures

Pattern Any homogenous boundary which has holes in it will tend to rupture at the holes, unless the edges of the apertures are reinforced to protect the boundary against the concentrations of stress around them.
Physical environment Door and window frames may be considered as reinforcements of the wall itself rather than as separate rigid structures inserted into holes in the wall. As such they are thickenings of the fabric of the wall itself, protecting it against concentrations of stress.
Broader Aperture compatibility (#MP1221) Ground–level visibility (#MP1222)
Zones of intermediate insight (#MP1223) Progressive framework definition (#MP1208)
Eccentric access to perspective contexts (#MP1196)
Emphasizing transitions across boundaries (#MP1224)
Efficient enclosure of spaces with minimal structural distinctions (#MP1206).
Narrower Inter–level integrity (#MP1227) Symbols of integration (#MP1249)
Displaceable frameworks (#MP1236) Multi–faceted frameworks (#MP1239)
Connectedness in isolation (#MP1237) Tolerance at level interfaces (#MP1240)
Overview sites from integrative superstructure (#MP1231).

♦ MP1226b Inter–level zone

Pattern Isolated inter–level connections in a framework may have useful additional, non–structural functions, but only if they are formed so as to establish a significant zone of influence around them.
Physical environment Free–standing columns play a role in shaping human space. Each marks a point. Two or more define an enclosure. A column thus creates a space for human activity if it permits places to be established around it where people can sit and lean comfortably. A thin column, only fulfilling its structural requirements, does not engender a comfortable environment.
Broader Perimeter continuity (#MP1217) Inter–level connections (#MP1216)
Flexible domain organization (#MP1146) Partially contained interfaces (#MP1119)
Partially enclosed internal domains (#MP1193)
Common external context for inactivity (#MP1069)
Communication pathways enfolded by non–linearity (#MP1174)
Enfolded overview domains of minimum proportions (#MP1167)
Hospitable non–linear domain external to structures (#MP1163)
Organization of structure to provide occupiable sites (#MP1179)
Efficient enclosure of spaces with minimal structural distinctions (#MP1206)
Primary inter–level connections at transitions in boundary orientation (#MP1212)
Overview domains at interfaces of the structure with the external environment (#MP1166).
Narrower Different settings (#MP1251) Ambiguous boundaries (#MP1243)
Inter–level integrity (#MP1227) Symbols of integration (#MP1249)
Protecting variability to enhance fixity (#MP1245).

♦ MP1227a Inter–level integrity

Pattern The integrity of a framework depends on that of the connections between its elements. These are most critical at transitions in orientation, and especially at those where inter– level elements connect with those of a higher level. It is at these positions that some form of reinforcement is required.
Physical environment The strength of a structural framework depends on the strength of its connections. These connections are most critical of all at corners, especially at the corners where the columns meet the beams. Such connections may either be seen as a source of rigidity, which can be strengthened by triangulation using a brace, or as a source of continuity, which helps the forces to flow easily in the process of transferring loads by changing the direction of the force, as in the use of a capital.
Broader Inter–level zone (#MP1226) Perimeter continuity (#MP1217)
Inter–level connections (#MP1216) Partially contained interfaces (#MP1119)
Boundary reinforcement at apertures (#MP1225) Partially enclosed internal domains (#MP1193)
Enfolded overview domains of minimum proportions (#MP1167)
Efficient enclosure of spaces with minimal structural distinctions (#MP1206)
Overview domains at interfaces of the structure with the external environment (#MP1166).
Narrower Symbols of integration (#MP1249).

♦ MP1228b Appropriate superstructure to contain transitions between levels

Physical environment Stair vault.
Broader Appropriate construction elements (#MP1207)
Level generation of minimum tension (#MP1219)
Exclusive spaces for emergent perspectives (#MP1203)
Inter–domain contexts for perspective adjuncts (#MP1198)
Framework for transition between structural levels (#MP1195)
Organization of structure to provide occupiable sites (#MP1179)
Integrating transition pathways between levels into a structure (#MP1133)
Efficient enclosure of spaces with minimal structural distinctions (#MP1206).
Narrower Time binding (#MP1248) Zoning internal domains (#MP1233).

♦ MP1229b Provision for pathways for automatic communications

Physical environment Duct space for pipes and conduits.
Broader Level generation of minimum tension (#MP1219)
Efficient enclosure of spaces with minimal structural distinctions (#MP1206).
Narrower Domains of insight (#MP1252) Unmediated supportive emotion (#MP1230).

♦ MP1230b Unmediated supportive emotion

Physical environment Radiant heating systems.
Broader Distortion resistant boundaries (#MP1218)
Level generation of minimum tension (#MP1219)
Structure–enfolded occupiable sites (#MP1202)
Provision for pathways for automatic communications (#MP1229).

♦ MP1231b Overview sites from integrative superstructure

Physical environment Dormer windows in the roof.
Broader Perimeter continuity (#MP1217) Inter–level connections (#MP1216)
Integration superstructure (#MP1220) Distortion resistant boundaries (#MP1218)
Progressive framework definition (#MP1208) Boundary reinforcement at apertures (#MP1225)
Level generation of minimum tension (#MP1219)
Containment by integrative superstructure (#MP1117)
Occupiable sites exposed to external insight (#MP1180)
Organization of structure to provide occupiable sites (#MP1179)
Primary inter–level connections at transitions in boundary orientation (#MP1212).
Narrower Displaceable frameworks (#MP1236) Multi–faceted frameworks (#MP1239).

♦ MP1232c Symbolic connection to encompassing domains

Pattern The combination of elements in any design calls for some symbolic forms of reduction in completing the highest level, partly in order to identify the design and give it form.
Physical environment There are few cases in traditional architecture where builders have not used some roof detail to cap the building with an ornament. The cap may be structural, but its main function is decorative. It marks the top, the place where the roof penetrates the sky. ("Topping out" ceremonies continue to be a feature of major modern buildings.)
Broader Integration superstructure (#MP1220) Distortion resistant boundaries (#MP1218).
Narrower Symbols of integration (#MP1249).

♦ MP1233a Zoning internal domains

Pattern Some functions associated with internal domains tend to be associated with characteristics of the external environment whereas others are not. Zones may be established within the internal domain to reflect this distinction and to adapt the zones' characteristics accordingly.

Physical environment Zones may be established in a house or building with harder floor materials in the public areas and softer materials in the private or more intimate zones. This avoids the problem arising from the exposure of comfortable but difficult-to-clean materials to foot traffic from the external environment. The distinction between the zones may be emphasized by a step, with different footwear rules on each side of the step.
  Broader Initial level formation (#MP1215)   Nested levels of accessibility (#MP1127)
  Level generation of minimum tension (#MP1219)
  Harmonizing space distribution between levels (#MP1210)
  Appropriate superstructure to contain transitions between levels (#MP1228).
  Narrower Time binding (#MP1248)   Encouraging emphases (#MP1250)
  Symbols of integration (#MP1249).

♦ **MP1234c Maintainable, multi-element external boundaries**
Pattern The purpose of external boundaries is to protect the space from the uncontrollable variability of the external environment. To do so it must be constructed such that the component elements can both be renewed when necessary and locked together to render the boundary impervious.
Physical environment The main function of the outside wall of a building is to keep the weather out. This can best be done with construction elements that lap together but can be repaired individually in patches as necessary. This is not possible with sheets of impervious material which can only be joined together with difficulty.
  Broader Integration superstructure (#MP1220)   Distortion resistant boundaries (#MP1218).
  Narrower Symbols of integration (#MP1249)   Tolerance at level interfaces (#MP1240)
  Appropriate construction elements (#MP1207).

♦ **MP1235b Unalienating internal boundaries**
Pattern Flawless unmalleable internal boundaries provide no basis for further interaction with them. They are difficult to modify in the light of emerging requirements. Malleable and essentially unfinished boundaries encourage continuing interaction throughout their duration.
Physical environment Make inside surfaces warm to the touch with a slight "give" as with soft plaster, or organic materials. A wall which is too hard or too cold or too solid is unpleasant to touch; it makes decoration impossible and creates hollow echoes.
  Broader Distortion resistant boundaries (#MP1218)
  Level generation of minimum tension (#MP1219).
  Narrower Encouraging emphases (#MP1250)   Symbols of integration (#MP1249)
  Tolerance at level interfaces (#MP1240).

♦ **MP1236b Displaceable frameworks**
Pattern Frameworks which can be completely set aside to permit direct interaction with the external world are preferable to fixed or partially movable frameworks.
Physical environment Windows which open wide permit personal contact with the outside environment. They are a place to linger, through which to talk to others outside. Modern buildings tend to be constructed with fixed or partially opening windows which inhibit this degree of contact.
  Broader Aperture compatibility (#MP1221)   Ground-level visibility (#MP1222)
  Overview of external contexts (#MP1192)   Boundary reinforcement at apertures (#MP1225)
  Occupiable sites exposed to external insight (#MP1180)
  Overview sites from integrative superstructure (#MP1231)
  Overview of communication pathway from structure (#MP1164)
  Orientation of domains to receive external insight (#MP1128).
  Narrower Multi-faceted frameworks (#MP1239).

♦ **MP1237c Connectedness in isolation**
Pattern When subdividing a relatively small space, boundaries can be usefully under-emphasized by limiting the number of modes by which contact is maintained across the boundaries. This ensures connectedness throughout the whole space whilst maintaining an adequate degree of isolation for each part of it.
Physical environment In a small building with small rooms, doors with glazing give a sense of visual connection together with the possibility of acoustic isolation. People then feel less isolated.
Socio-organizational environment In a small organization made up of small working units, each unit should receive sufficient general information on the activities of the others without needing to expose the others to the full details of its internal operations. The units then feel connected but appropriately isolated in dealing with their particular responsibilities.
Conceptual environment Within a small conceptual framework with narrow domains of specialization, each specialized framework should be exposed to only some kinds of information from the others in order to maintain the sense of coherence within the larger framework. Other kinds of information, corresponding to the internal preoccupations characterizing each domain, should not need to be exchanged.
Intra-personal environment When further articulating a particular mode of awareness, there should be sufficient interaction between the sub-modes to avoid total fragmentation of the more general mode.
  Broader Distortion resistant boundaries (#MP1218)
  Boundary reinforcement at apertures (#MP1225)
  Internal connectedness between domains (#MP1194)
  Eccentric access to perspective contexts (#MP1196)
  Emphasizing transitions across boundaries (#MP1224)
  Limiting length of communication pathways (#MP1132)
  Protected low-density communication pathways (#MP1101)
  Enhancing insight by varying levels of exposure to it (#MP1135)
  Internal transition spaces enhancing structural entry points (#MP1130).
  Narrower Multi-faceted frameworks (#MP1239).

♦ **MP1238 Filtered insights**
Pattern Unmediated sources of illumination result in strong contrasts, over-emphasizing boundaries and making it difficult to detect detail.
Physical environment Light filtered through a window partially covered by leaves or tracery enhances a sense of well-being in contrast to the dullness associated with uniform lighting. Glare around the window is also reduced by softening the light in this way.
Socio-organizational environment Diffusing the underlying intention of an organization through spontaneously occurring events enhances the organic well-being of the group in contrast to the over-programmed dullness associated with an omnipresent objective. Unnecessary tension is also reduced by indirection of the trend.
Conceptual environment Diluting a fundamental insight by representing it through random phenomena of apparently superficial significance enhances appreciation of its value. Presentation of an insight in undiluted form results in harsh contrasts which make if difficult to accept.
Intra-personal environment (as for conceptual environment)
MG M b Filtered insights
  Broader Aperture compatibility (#MP1221)   Zones of intermediate insight (#MP1223)
  Structure-enfolded insight domain (#MP1161)
  Receptivity to emerging external insight (#MP1138)
  Communication pathways enfolded by non-linearity (#MP1174)
  Overview of communication pathway from structure (#MP1164)
  Sites for grounding perspectives in non-linearity (#MP1176)
  Integrating coordinated exposure to irrationality into structures (#MP1144)
  Organization of structures to permit two sources of external insight (#MP1159).
  Narrower Flexible interfaces (#MP1244)   Encouraging emphases (#MP1250)
  Multi-faceted frameworks (#MP1239)   Integration within context (#MP1246).

♦ **MP1239a Multi-faceted frameworks**
Pattern The smaller the facets offered by a framework, the more intense the connection with that which is framed. The greater the number of facets, the greater the variety of connection to what is framed. The absence of a plurality of facets, however large the framework, engenders alienation from that which is framed.
Physical environment The smaller the windows are, and the smaller the panes are, the more intensely windows help connect us with what is on the other side. Paradoxically large plate glass windows inhibit the relationship to the nature they reveal compared to smaller windows, or smaller-paned windows, which create far more frames through which contact is rendered more intimate. Smaller panes establish a psychologically more acceptable balance between exposure and enclosure.
Socio-organizational environment The smaller the frameworks through which an organization surveys its environment, the more intense the connection with that environment. The greater the number of such frameworks, the greater the variety of connections to the environment. Because of the sense of excessive exposure, larger "windows" on the environment inhibit the organization's sense of contact with it.
Conceptual environment The more specialized the tools with which a conceptual framework maintains contact with its environment, and the more of them, the more adequate its apprehension of that environment is felt to be. Contact maintained through an unspecialized framework of great generality creates uncertainty as to whether an appropriate conceptual distance from the environment has been achieved.
Intra-personal environment The more specialized the modes of awareness through which contact is maintained with the psychic environment, and the more of them, the more intense and intimate that contact is felt to be. Contact maintained through an unspecialized, holistic mode of awareness creates uncertainty as to whether an appropriate distinction is being made between perceiver and perceived.
  Broader Filtered insights (#MP1238)   Aperture compatibility (#MP1221)
  Displaceable frameworks (#MP1236)   Connectedness in isolation (#MP1237)
  Overview of external contexts (#MP1192)   Boundary reinforcement at apertures (#MP1225)
  Partially enclosed internal domains (#MP1193)
  Internal connectedness between domains (#MP1194)
  Overview sites from integrative superstructure (#MP1231).
  Narrower Tolerance at level interfaces (#MP1240).

♦ **MP1240a Tolerance at level interfaces**
Pattern Provision of suitably scaled intermediate positions permits a realistic degree of tolerance at interfaces between frameworks of different levels. Such tolerance is necessary if there is to be freedom in any design and implementation process.
Physical environment A free and natural building cannot be conceived without the possibility of using trim to cover minor variations and mistakes during the construction process. In modern system building the necessary tolerances are reduced by eliminating any possibility of freedom in the building plan. Trim serves the vital additional function of ensuring adequate perceptual continuity between the fine structure of natural materials and the dimensions of the smallest constructional elements. Without such continuity the building is experienced as alienating.
Socio-organizational environment In the natural development of an organization allowances must be made for minor inconsistencies in the working relationships between its constituent groups, especially those at different hierarchical levels. Such inconsistencies can only be avoided by detailed pre-planning intolerant of any spontaneously instituted variations during the organization's development. The manner in which such inconsistencies are integrated into the life of the organization may well contribute to its qualifications as a human organization rather than as an inhuman one.
Conceptual environment In the natural development of a conceptual framework, allowance must be made for minor inconsistencies in the relationships between its constituent modules, especially those at different levels of abstraction. Such inconsistencies can only be avoided by absolute adherence to an all-embracing framework intolerant of any variations which may emerge as desirable during its implementation. The manner in which such inconsistencies are integrated into the conceptual framework may well contribute to its qualifications as realistic rather than as unrealistic.
Intra-personal environment In the natural development of a mode of awareness allowance must be made for minor inconsistencies in the relationships between its various aspects, especially those at different levels. Such inconsistencies can only be avoided by adherence to a single overriding intention intolerant of any variation which may emerge as desirable in practice. The manner in which such inconsistencies are integrated into the mode of awareness may well contribute to its acceptance as acceptable rather than as unacceptable.
  Broader Perimeter continuity (#MP1217)   Inter-level connections (#MP1216)
  Multi-faceted frameworks (#MP1239)   Zones of intermediate insight (#MP1223)
  Unalienating internal boundaries (#MP1235)   Boundary reinforcement at apertures (#MP1225)
  Level generation of minimum tension (#MP1219)
  Maintainable, multi-element external boundaries (#MP1234).
  Narrower Encouraging emphases (#MP1250)   Symbols of integration (#MP1249).

♦ **MP1241a Attractive temporary positions**
Pattern For a temporary position to be attractive, due consideration must be given to the viewpoint it offers, its exposure to agreeable influences, and its proximity to focii of more permanent concern.
Physical environment Outdoor seats, whether public or private, are useless unless they are positioned with due regard for view and climate.
Socio-organizational environment Organizations can usefully permit the emergence of temporary roles which offer participants the opportunity of withdrawing from formal activity, whether to cultivate a sense of perspective, or as a safety valve through which the tensions of organization life may be reduced.
Conceptual environment Conceptual frameworks can usefully recognize the value of temporary viewpoints, not necessarily associated with any particular framework, from which a sense of perspective may be obtained concerning ongoing conceptual activity. Such viewpoints will not however be taken up unless their inherent interest is recognized.
Intra-personal environment Temporary modes of awareness may be usefully adopted as a means of acquiring a sense of perspective on those adopted on a more frequent and regular basis.
  Broader Grounded structures (#MP1168)   Hospitable common domains (#MP1094)
  Hospitable transit points (#MP1092)   Structure-enfolded insight domain (#MP1161)
  Focus-oriented communication networks (#MP1120)
  Common external context for inactivity (#MP1069)

Hospitability of communication pathways (#MP1121)
Sequence of viewpoint loci within a structure (#MP1142)
Ensuring function of common domain interfaces (#MP1124)
Context for acknowledgement of past perspectives (#MP1070)
Communication pathways enfolded by non-linearity (#MP1174)
Sites for grounding perspectives in non-linearity (#MP1176)
Hospitable non-linear domain external to structures (#MP1163)
Integrating points of perspective into common domains (#MP1125)
Appropriate relationship of non-linearity to structures (#MP1171)
Hospitable interface between structures and external environment (#MP1160)
Competitive interaction opportunities transposed to a concrete level (#MP1072)
Relative isolation of structural interface with communication pathway (#MP1140).
Narrower Ambiguous boundaries (#MP1243) Intermediate position (#MP1242).

♦ MP1242b Intermediate position

Pattern Establishing an intermediate position between exposure to a wider range of influences and enclosure within a more controlled environment is strategically advantageous. Such a position is attractive because it facilitates choice of the degree of exposure acceptable at any one time.
Physical environment Front door benches encourage people to linger around the entrance to a building. People, especially the young and the old, like to watch street activity from them because they want to be able to control their degree of involvement, encouraging interaction with passers–by or withdrawing from it.
Broader Grounded structures (#MP1168) Attractive temporary positions (#MP1241)
Partially contained interfaces (#MP1119) Structure–enfolded insight domain (#MP1161)
Distinctiveness of main entry point to structure (#MP1110)
Internal transition spaces enhancing structural entry points (#MP1130)
Hospitable interface between structures and external environment (#MP1160)
Semi–autonomous contexts for perspectives of decreasing activity (#MP1155)
Direct relationship between structures and communication pathways (#MP1122).
Narrower Ambiguous boundaries (#MP1243)
Protecting variability to enhance fixity (#MP1245).

♦ MP1243a Ambiguous boundaries

Pattern There are situations in which the boundaries between spaces are too absolute, but the absence of any boundary would do an injustice to the subtlety of the division between them. Ambiguous boundaries may than be established which both distinguish and bind together.
Physical environment In many places, walls and fences between outdoor spaces are too high, even though some separating boundary is required. In such situations, very low walls may be built so that when sat upon in various ways, or stepped over, the connection between the spaces is emphasized.
Socio–organizational environment In the relationships between groups there are situations in which the boundary between them should be highly permeable, such that participants feel free to shift between actions within one group to those in another, or that there should be an acceptable ambiguity as to which group they are acting for at any one time.
Conceptual environment In the relationship between conceptual frameworks, there are situations in which it is advantageous to be able to shift easily between conceptualization in terms of one to that in terms of the other, accepting a certain ambiguity as to which is appropriate at any one time.
Intra–personal environment In the relationship between different modes of awareness there are situations in which it is advantageous for the person to be able to shift freely between distinct modes such that there is some ambiguity as to which is valid in a given set of circumstances.
Broader Inter–level zone (#MP1226) Intermediate position (#MP1242)
Attractive temporary positions (#MP1241) Access to contained irrationality (#MP1071)
Protected low intensity relationships (#MP1055)
Hospitability of communication pathways (#MP1121)
Contexts of self–organizing non–linearity (#MP1172)
Informal local perspective interface zones (#MP1088)
Ensuring function of common domain interfaces (#MP1124)
Hierachy of perspectives favouring the broadest (#MP1114)
Enfolded overview domains of minimum proportions (#MP1167)
Off–centering the focal point of a common domain (#MP1126)
Maintaining distinctions between contextual levels (#MP1169)
Hospitable non–linear domain external to structures (#MP1163)
Linear relationships enhanced by non–linear processes (#MP1051)
Appropriate relationship of non–linearity to structures (#MP1171)
Arrangement of structures ot engender fruitful interfaces (#MP1100)
Exposure of structural activities to communication pathway (#MP1165)
Blended integration of formal structure and informal context (#MP1111)
Competitive interaction opportunities transposed to a concrete level (#MP1072)
Relative isolation of structural interface with communication pathway (#MP1140)
Functional enhancement of domains separating complementary structures (#MP1106)
Overview domains at interfaces of the structure with the external environment (#MP1166)
Contexts for exploratory relationship formation challenging emerging perspectives (#MP1073).
Narrower Time binding (#MP1248) Symbols of integration (#MP1249)
Protecting variability to enhance fixity (#MP1245).

♦ MP1244b Flexible interfaces

Pattern Flexible interfaces can provide a more appropriate boundary between the fixity of well–defined spaces and the variability of the contexts within which they are embedded.
Physical environment A building using canvas awnings or temporary roofing touches the elements more nearly than when constructed with hard conventional materials only. Canvas has a softness, a suppleness, which is in harmony with wind, light and sun.
Socio–organizational environment A group permitting varying degrees of informal participation responds more naturally to its context than one based on formal membership only. Such flexible informality is in harmony with the shifting currents of opinion in the social environment.
Conceptual environment A conceptual framework able to incorporate informal methods of interacting with its environment is accepted as more realistic than one based on formal methods alone. Such flexibility is more responsive to the problems of representing complex, ill–defined phenomena.
Intra–personal environment A condition of awareness incorporating some subjective modes of understanding can provide a more appropriate interface with the psychic environment than one based on objective modes alone. An interface of such flexibility is more adapted to conditions in which the psychic boundary is ill–defined.
Broader Filtered insights (#MP1238) Context for disorder (#MP1058)
Selective interchange axis (#MP1032) Partially contained interfaces (#MP1119)
Structure–enfolded insight domain (#MP1161) Diversified interchange environment (#MP1046)
Common external context for inactivity (#MP1069)
Informal local perspective interface zones (#MP1088)
Occupiable sites exposed to external insight (#MP1180)
Communication pathways enfolded by non–linearity (#MP1174)
Intersection of differently paced communications (#MP1054)
Limitation on number of occupiable temporary sites (#MP1103)
Hospitable non–linear domain external to structures (#MP1163)
Concealment of necessary monotonous perspective patterns (#MP1097)
Integration of non–linearity into integrative superstructure (#MP1118)
Transit point location for sources of perspective nourishment (#MP1093)
Cyclic interrelation of complementary perspective in common domains (#MP1063)
Relative isolation of structural interface with communication pathway (#MP1140)
Overview domains at interfaces of the structure with the external environment (#MP1166).
Narrower Encouraging emphases (#MP1250) Symbols of integration (#MP1249).

♦ MP1245 Protecting variability to enhance fixity

Pattern Zones of variability require protection to render them accessible so that they enhance zones of fixity effectively.
Physical environment Flowers are beautiful along the edges of paths, buildings, and outdoor rooms where they soften edges. But unless they are protected within a raised bed, they cannot easily survive and are inaccessible to those who would appreciate them.
Socio–organizational environment Spontaneous creativity (and humour) relieves the tedium of organizational rules and procedures thus helping to make them more acceptable. But unless accessible protective contexts are provided for such spontaneity, it cannot easily thrive under normal organizational pressures.
Conceptual environment Imaginative and speculative thinking enlivens intellectual discourses governed by well–defined theories and methods, thus encouraging creative advances. But unless a recognized respectable place is given to such unconstrained speculation it cannot easily survive in disciplined discourse.
Intra–personal environment Imaginative musings and adventures in awareness and modes of being contribute to psychic well–being. But unless they are given an accepted role, it may be difficult for them to fulfil their function adequately.
MG M b Protecting variability to enhance fixity
Broader Inter–level zone (#MP1226) Ambiguous boundaries (#MP1243)
Intermediate position (#MP1242) Ground–level visibility (#MP1222)
Focus–oriented communication networks (#MP1120)
Contexts of self–organizing non–linearity (#MP1172)
Maintaining distinctions between contextual levels (#MP1169)
Integrating points of perspective into common domains (#MP1125)
Linear relationships enhanced by non–linear processes (#MP1051)
Protecting non–linear contexts from communication pathways (#MP1173)
Integration of non–linearity into integrative superstructure (#MP1118)
Hospitable interface between structures and external environment (#MP1160)
Harmoniously structured entry point for external communication media (#MP1113).

♦ MP1246 Integration within context

Pattern A collective framework is successfully integrated within its natural environment when variable processes in that context interact with it as freely as they would if it were natural to that context.
Physical environment A building finally becomes part of its surroundings when plants grow over parts of it as freely as they grow over the ground. In so doing, they effect a smooth transition between the built and the natural, whether in terms of light quality or texture.
Socio–organizational environment An organization becomes a natural part of the community in which it is located when its boundary as a social group offers many points of contact to the informal processes occurring within the community.
Conceptual environment A conceptual framework is successfully integrated into the conceptual domain to which it relates when it is accepted as offering support for the variety of informal creation processes associated with that domain.
Intra–personal environment A psychological construct or mode of awareness is successfully integrated into a person's psychic framework when it supports and reinforces the person's other unstructured modes of awareness.
MG M c Integration within context
Broader Filtered insights (#MP1238) Zones of intermediate insight (#MP1223)
Communication pathways enfolded by non–linearity (#MP1174)
Overview of communication pathway from structure (#MP1164)
Integration of non–linearity into integrative superstructure (#MP1118).

♦ MP1247a Embedding fixity within variability

Pattern When it is necessary to ensure a degree of fixity within a domain of variability, it is an advantage to embed a multiplicity of separated smaller domains of fixity in the space such as to safeguard the contextual variability. This is to be contrasted with the use of a single continuous surface which destroys the characteristics of the variable domain.
Physical environment Use of paving with ample cracks between the stones permits grasses and mosses to grow there, thus preserving a delicate ecology of insect life and allowing rainwater to drain directly into the earth. In contrast to continuous asphalt and concrete surfaces, such paving settles without cracking and is agreeable to walk on.
Socio–organizational environment Any set of rules, regulations, and procedures can usefully be organized in such a way that there is a certain amount of flexibility or "play" between them so as not to totally inhibit the informal life of the group. This is to be contrasted with an alienating set of rules which interlock so completely that everything is prohibited unless it is explicitly permitted. Such a set does not adjust well to the passage of time.
Conceptual environment A set of conceptual tools functions most effectively when each can adapt to the circumstances for which it is most appropriate. This is to be contrasted with a totally integrated set of methods which fails to allow for more complex, unpredictable phenomena to which it is insensitive and for which it is inadequate.
Intra–personal environment In formulating a set of personal rules it is useful to leave a certain amount of "play" between them. Any attempt to subject all processes to such a set of rules inhibits vital processes to which rules are insensitive, possibly to the detriment of psychic health or richness.
Broader Grounded structures (#MP1168)
Focus–oriented communication networks (#MP1120)
Hospitability of communication pathways (#MP1121)
Communication pathways enfolded by non–linearity (#MP1174)
Maintaining distinctions between contextual levels (#MP1169)
Hospitable non–linear domain external to structures (#MP1163)
Linear relationships enhanced by non–linear processes (#MP1051)
Relative isolation of structural interface with communication pathway (#MP1140).
Narrower Time binding (#MP1248).

♦ MP1248c Time binding

Pattern In order to provide a sense of connectedness with a space and its context, it is essential that its boundaries should in some way be sensitive to the passage of time and to the processes that have occurred within that space.
Physical environment Soft tile and brick can be used on ground level surfaces in order that, through the natural processes of wear, they should record the activity of the building as a living entity. Such materials, intermediate in character between the building and the earth, emphasize the connectedness with the earth in a manner impossible for artificial materials that are perceived as impervious and alien.
Socio–organizational environment The transactions of the group, especially with the outside

world, can be conducted in such a way that some meaningful and continuing trace is left of its connections with its historical context and with the manner of its own development. This is to be contrasted with groups who only maintain impersonal records, if any, and have little sense of their own historical continuity.

Conceptual environment In the development of a conceptual space from its relatively unformed beginnings, a sense of continuity and connectedness with that original level of understanding can usefully be cultivated. This tends to correspond to that of the wider conceptual context within which the space is embedded and to which it must continue to relate if its development is not ultimately to be inhibited.

Intra-personal environment In the process of individual development it is useful to maintain a sense of continuity with the earlier states of awareness from which the present forms have emerged. This is to be contrasted with efforts to cut-off or repress any recollection of the past in which present and future development is rooted.
Broader Grounded structures (#MP1168)　　Ambiguous boundaries (#MP1243)
Zoning internal domains (#MP1233)　　Initial level formation (#MP1215)
Distortion resistant boundaries (#MP1218)　　Appropriate construction elements (#MP1207)
Embedding fixity within variability (#MP1247)
Integrating points of perspective into common domains (#MP1125)
Appropriate superstructure to contain transitions between levels (#MP1228).
Narrower Encouraging emphases (#MP1250)　　Symbols of integration (#MP1249).

♦ **MP1249a Symbols of integration**
Pattern Symbols of integration may be used to emphasize the boundaries of a space. At significant transitions between parts of the space, when the connectivity between patterns is weak, they are a means of binding them together to emphasize the larger whole. They function mainly by creating surfaces in which each part is simultaneously figure and boundary and in which the design acts a boundary and figure at several different levels simultaneously.

Physical environment Use of ornamental designs as decoration on buildings where materials meet as a means of providing a seam to knit together such edges to emphasize the space as a whole rather than its constituent parts. Ornamentation is frivolous when there is in fact no lack of connectivity.

Socio-organizational environment Use of logos, banners, totems, rituals, mottoes, or other devices to reinforce recognition of the group as a whole, particularly in contexts in which it is liable to split into factions.

Conceptual environment Use of symbols whose structure or profound significance is congruent with the organization of the conceptual space and therefore serves to bind together the different aspects or dimensions of it.

Intra-personal environment Reflection on symbols which are felt to be keys to the integration of the psyche and the sense of identity. These may emerge from the unconscious in the form of dreams or in certain forms of artistic creation (e.g. sand printing).
Broader Time binding (#MP1248)　　Inter-level zone (#MP1226)
Flexible interfaces (#MP1244)　　Grounded structures (#MP1168)
Ambiguous boundaries (#MP1243)　　Inter-level integrity (#MP1227)
Zoning internal domains (#MP1233)　　Tolerance at level interfaces (#MP1240)
Distortion resistant boundaries (#MP1218)　　Unalienating internal boundaries (#MP1235)
Boundary reinforcement at apertures (#MP1225)　　Partially enclosed internal domains (#MP1193)
Eccentric access to perspective contexts (#MP1196)
Emphasizing transitions across boundaries (#MP1224)
Symbolic connection to encompassing domains (#MP1232)
Occupiable sites exposed to external insight (#MP1180)
Maintainable, multi-element external boundaries (#MP1234)
Distinctiveness of main entry point to structure (#MP1110)
Protecting non-linear contexts from communication pathways (#MP1173)
Hospitable interface between structures and external environment (#MP1160)
Appropriate configuration for interaction of complementary perspectives (#MP1187).
Narrower Encouraging emphases (#MP1250)
Meaningful symbols of self-transformation (#MP1253).

♦ **MP1250a Encouraging emphases**
Pattern The degree to which a space is experienced as congenial depends to a great extent on the manner in which the pervading emphases combine together to define an encouraging environment. This tends to be harder to achieve when some of the emphases are of necessity discouraging.

Physical environment Natural wood, sunlight, and bright colours are warm, in contrast to other colours which tend to be experienced as depressing and cold. In some way this makes a great deal of difference between the comfort and discomfort of a room.

Socio-organizational environment A group is experienced as congenial (or as having "good vibrations") when the sum total of communications and interactions, however contrasted, is felt to be in harmony with its natural development. Unintegrated, destructive communications create the opposite impression.

Conceptual environment A conceptual space is experienced as excitingly meaningful when all the various interrelationships, however contrasted, combine together to suggest the possibility of their further development as an integrated whole. Unintegrated, incompatible sets of relationships are correspondingly unmeaningful and discouraging.

Intra-personal environment An intra-psychic space is experienced as encouraging when all various impressions, however dramatically contrasted, combine together to imply natural possibilities for its further development. Unintegrated, antagonistic impressions are correspondingly depressing and discouraging.

Broader Time binding (#MP1248)　　Filtered insights (#MP1238)
Flexible interfaces (#MP1244)　　Symbols of integration (#MP1249)
Zoning internal domains (#MP1233)　　Tolerance at level interfaces (#MP1240)
Unalienating internal boundaries (#MP1235)　　Appropriate construction elements (#MP1207)
Appropriate conditions for perspective nourishment (#MP1182)
Enhancing insight by varying levels of exposure to it (#MP1135)
Exposure of input processing context to external insight (#MP1199).
Narrower Domains of insight (#MP1252).

♦ **MP1251c Different settings**
Pattern A space can only be effectively used in all its richness if it can be appreciated from a variety of settings according to the felt need of the moment. Efforts to standardize such settings imposes a subtle straightjacket on the manner in which the space is experienced.

Physical environment Provision of a variety of chairs in a space, as opposed to chairs of a single design conceived for an "average person". The latter approach is insensitive to the variety of people, to their sitting habits and to the different needs of any one individual at different times.

Socio-organizational environment Recognition of a variety of roles in a social space, in contrast to trends towards achieving a degree of role uniformity.

Conceptual environment Recognition of a variety of viewpoints in a conceptual domain, in contrast to trends towards achieving a single acceptable viewpoint.

Intra-personal environment Acceptance of a variety of modes of awareness, in contrast to any trend towards achieving a single unvariegated mode.
Broader Inter-level zone (#MP1226)　　Structure-enfolded occupiable sites (#MP1202)
Informal local perspective interface zones (#MP1088)
Small-scale perspective interaction contexts (#MP1151)
Sequence of viewpoint loci within a structure (#MP1142)
Appropriate conditions for perspective nourishment (#MP1182)
Appropriate configuration for perspective interaction (#MP1185).
Narrower Domains of insight (#MP1252).

♦ **MP1252a Domains of insight**
Pattern Space is partly defined by the particular perspectives of those present. Uniform exposure of a space to awareness serves no useful purpose whatsoever. It destroys the social significance of space and leads to a sense of disorientation and unboundedness.

Physical environment Pools of light defining a dappled environment, in contrast to unnatural uniform illumination as typified by many modern offices.

Socio-organizational environment Domains of shared preoccupation, in contrast to a single general concern which undermines the cohesiveness of distinct groups and prevents them from coming into any meaningful form of existence.

Conceptual environment Domains of special insights or foci of attention, in contrast to an overriding general awareness which inhibits the development of a variety of specialized conceptual skills.

Intra-personal environment Modes of personal insights, in contrast to an overriding general awareness or monopolarization such as to inhibit the development of a variety of particular conditions of awareness.
Broader Different settings (#MP1251)　　Encouraging emphases (#MP1250)
Flexible domain organization (#MP1146)　　Partially exposed perspective context (#MP1183)
Small-scale perspective interaction contexts (#MP1151)
Common domain at the focal point of a structure (#MP1129)
Appropriate conditions for perspective nourishment (#MP1182)
Provision for pathways for automatic communications (#MP1229)
Appropriate configuration for perspective interaction (#MP1185)
Appropriate configuration for perspective interaction (#MP1185)
Organization of structure to provide occupiable sites (#MP1179)
Enhancing insight by varying levels of exposure to it (#MP1135)
Internal transition spaces enhancing structural entry points (#MP1130).

♦ **MP1253 Meaningful symbols of self-transformation**
Pattern A space can best be given further definition by associating with it a set of meaningful, self-chosen symbols that have a catalytic power in the continuous process of self-transformation (possibly as an outward counterpart to the unconscious). This function will be inadequately fulfilled by using symbols recommended by external specialists.

Physical environment Things from one's own life (e.g. pictures, objects), as opposed to the recommendations of interior decorators.

Socio-organizational environment Special rituals and behaviour patterns embodying meaningful moments in the history of the integration of the group, as opposed to rituals recommended or imposed by well-intentioned outsiders.

Conceptual environment Special concepts or code words which have helped to define the unique flavour of the particular language or mode of communication used.

Intra-personal environment Personal memories and associations, as opposed to those obtained or imposed via the media.
mg M b Meaningful symbols of self-transformation
Broader Symbols of integration (#MP1249)
Accessible facilities for perspective adjuncts (#MP1201)
Relatively isolated context for each perspective (#MP1141).

Symbols

Rationale

Symbols are a special form of presentation. They are of special importance in embodying significance and giving focus to any campaign or programme and in establishing its identity in relation to other initiatives. Whilst much work has been done on symbols in order to market commercial products or political parties, almost none has been done on their value in communicating the key ideas associated with development strategies. In the case of international campaigns, a common approach is to select the central symbol through an international competition. This totally neglects the psycho-cultural functions of sets of symbols already active in society.

Furthermore, when advocating or imposing the use of particular international sets of values, needs or qualities, it is not recognized that these effectively compete as functional substitutes in traditional societies with other sets of qualities represented by hierarchies of gods, spiritual beings, or natural phenomena perceived as governing those qualities or some of them. Such fundamental sets advocated by the international community are indeed designed to perform many of the regulatory functions previously ascribed to supernatural beings or potencies.

Given the relative rapidity with which such sets are now formulated - compared to the long cultural refinement of a pantheon - it is not surprising if they are viewed as artificial, bloodless and unrelated to the complex pattern of qualities associated with traditional sets of symbols. These are so meaningfully represented (with nested levels of significance) through richly decorated beings and memorable tales exemplifying their relationships, that the qualities and their representation are difficult to distinguish in a particular culture.

The lack of success of public information programmes, using a confusion of unrelated symbolic gimmicks, is therefore understandable. There would therefore seem to be a need to understand the range of symbols that remain active in society and which, as a cultural resource, can be called upon to give focus to international programmes.

Contents

The 103 entries were produced as an *editorial experiment* to identify classes of symbols considered of significance in traditional and modern cultures. As such they offer a means of ordering the large amount of material available on individual symbols.

Method

The information used was obtained from a wide range of specialized reference books.

Index

A keyword index to entries is provided in Section MX. The keywords are also incorporated into the index for Volume 2 (Section X)

Bibliography

Bibliographical references, by author, are given in Section MY.

Comment

Detailed comments are given in Section MZ.

Reservations

The information given here clearly represents the results of a very prelimininary move towards grouping together information on symbols. It is therefore serves mainly as an indication of directions for further exploration.

Possible future improvement

There is a great deal of material available on individual symbols. It should be possible to present entries highlighting the relationships between symbols whilst avoiding the need to reproduce such available information at length. A guideline could well be the need to produce patterns of culturally significant symbols which provide a focus for understanding the relationships between the social processes and phenomena with which they are associated. Of special interest is the possibility of extensively cross-referencing individual symbols to the particular values in Section V with which they are associated, since symbols have traditionally offered a way of ordering and presenting information on values and value complexes.

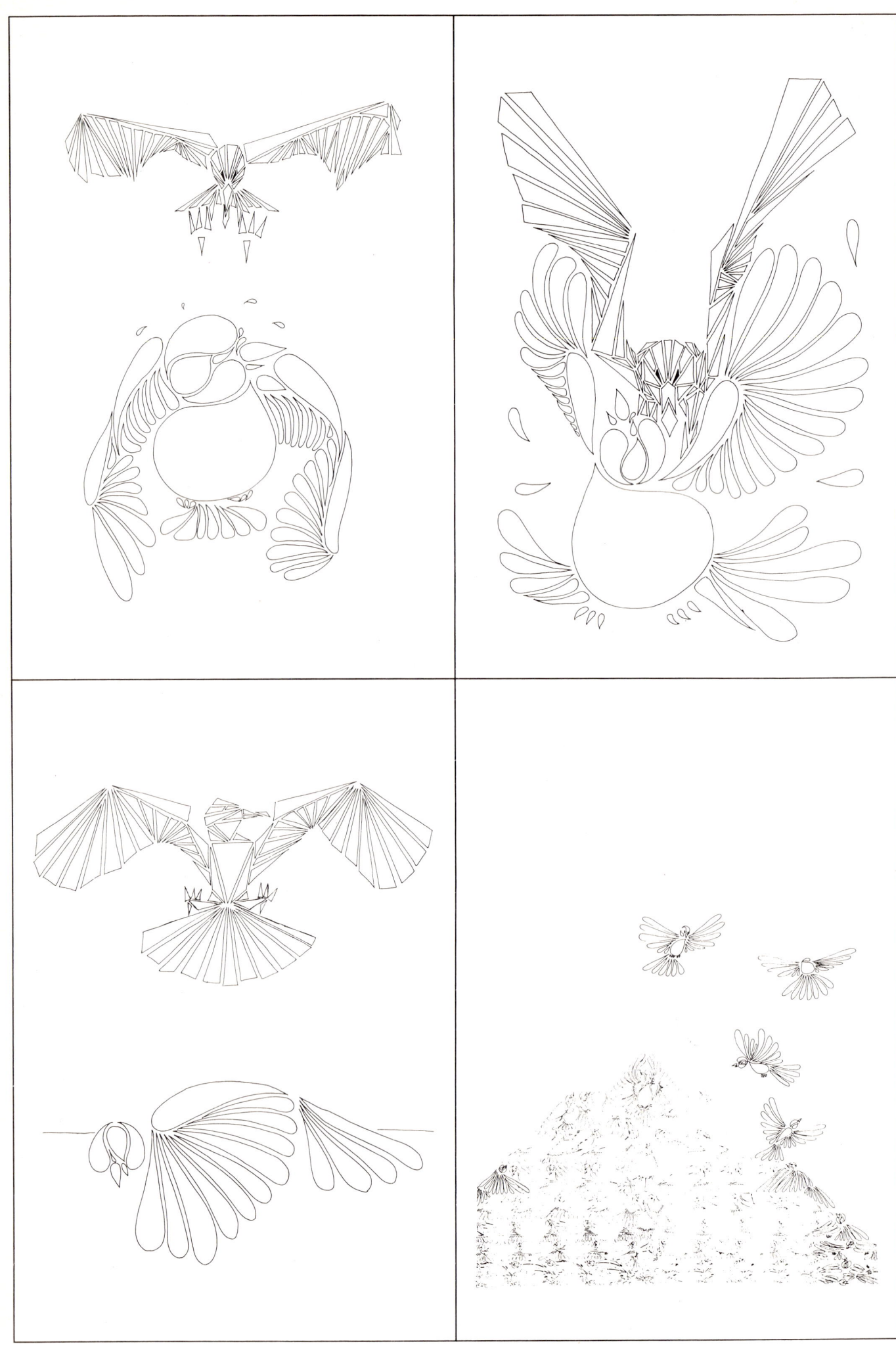

♦ **Craftsman tools** Professional equipment — Domestic devices — Machines and supplies
Description Tools, equipment, machinery and devices provide a vocabulary of symbolism arising from their functionings and purposes. This is a symbolism of technology that includes a vast number of human artifacts ranging from a stone age grinder for masticating food to a modern computer. Symbols that have appeared with frequency are the builder's compasses and rule, the carpenter's nails, the bricklayer's plumbline, and mortar trowel; the smith's furnace, anvil, tongs, and hammer; the farmer's and agricultural labourer's plough, sickle, and scythe; the woodsman's axe; the miner's sieve, the fisherman's net, and the hunter's bow. In another category one might cite the doctor's stethoscope, the surgeon's scalpel, and the apothecary's mortar and pestle. Other trade or profession related symbols are the demi–monde red light, the pawnbroker's tri–sphere, the gambler's roulette wheel, the judge's gavel, the hangman's noose and the undertaker's hearse.
Connotations Some tool symbols have been employed to represent ideologies. The Freemasons use the compasses, the trowel, and the master craftsman's tool apron. Fascist Italy used the double axe and Communist' Russia, the sickle and hammer. Notable among symbolic domestic devices have been the loom, distaff, shuttle, spindle, thimble and needle to indicate what was traditionally women's work, and thus the female sex, but also referring to human generation.
Broader Artifacts (#MS8505).
Narrower Medical instruments (#MS8715).
Related Human gender (#MS8385) Masonic symbols (#MS8843) Scientific symbols (#MS8933) Agricultural symbols (#MS8943) Trades and professions (#MS8343).

♦ **MS8103 Insects**
Bugs — Vermin
Description Insects as symbols include economically valued bees, butterflies appreciated for their beauty, annoyances such as lice and flies, health hazards such as mosquitoes, and crop damaging locusts, weevils and the like.
Connotations In many countries insect names are terms of abuse such as bug, louse, vermin, flea, cockroach. Some insects are at the centre of superstition. The ladybird, for example, must on no account be harmed. Other insects appear in mythology in important roles, as, for example, the ant–people among the Hopi. Ants often symbolically represent industry, tenacity and intelligence. The grasshopper and cicada have been employed as symbols of nature and its rhythms and the June cricket is considered a friendly symbol betokening the halcyon days of summer. The praying manthis is a symbol of voracity. Worker bees (drones) are a symbol of enslavement while the Queen bee represents a female who is the centre of social attention. The scarab or dung–beetle, and the spider to some ancient symbologists, represented the creative force in the universe. The metamorphosis of the caterpillar has always been a symbol of human development endowing man symbolically, with what is perhaps the ultimate archetype: wings.
Broader Animals (#MS8215) Non–human beings (#MS8400).
Related Animal body parts (#MS8133).

♦ **MS8106 Armour and weapons**
Description Weaponry in all its aspects and history is a source of potent symbolism. The ancient weapons still appear in imagery: staff, spear, club, axe, and projectile, whether hurled stone, slingshot or later, bow–flung arrows or poison darts. To these five may be added a weapon of fire: torches first then fiery projectiles. Water also was a weapon. Enemies' water supplies were cut–off, or their walls were undermined by diverting rivers. As 'armour', the watery moat served its purpose for millenia. The Middle Ages saw a further development in body armour, parts of which were used in the symbolic art of heraldry. The modern ages introduced explosives: bullets, bombs, hand grenades, mortars and rockets and their various delivery systems. These have all entered into the stream of symbolic imagery among which the nuclear mushroom–cloud is possibly the most terrifying.
Connotations Weapons and armour symbolize offence and defence; destruction and protection. The meanings may refer also to the deepest levels of the dynamic psyche, where struggle (psychomachia), is also characteristic among those on paths of personal development.
Broader Artifacts (#MS8505).
Narrower Primitive weapons (#MS8525) Nuclear and mass destruction weapons (#MS8723).
Related Death (#MS8603) Jewellery and adornments (#MS8446) Trades and professions (#MS8343).

♦ **MS8113 Symbolic quests and journeys**
Broader Behaviour (#MS8125).
Related Masonic symbols (#MS8843) Religious symbols (#MS8705) Human spiritual phenomena (#MS8373).

♦ **MS8115 Language and literature**
Description Symbol–making, that is, language is the most important class of human behaviour after those that are concerned with sustaining life. The essentially symbolic function of language is illustrated by the instances of language without meaning, that is, unintelligible, indecipherable, obscure or arcane language: for example, the biblical 'handwriting on the wall', the Sagan tablet on the Voyager spare ship which is to communicate enigmatically to extra–terrestrials, Scandinavian runes, petroglyphs, etc. Their meaning firstly is their humaneness, the fact they they are artifacts of intelligent life. Secondly, such language, even as primitive as runes or knots, indicates the purpose of communication which, in a general sense, also indicates the societal or organized nature of human life. Thirdly, the symbolic medium is the message in that there is always a specific content to language that is partly understood by the very circumstances or contexts of its mode, place and time of delivery. Thus two savages from widely separated tribes when initially encountering each other, though having different primary signs for peaceful intentions, by their circumstances alone will give meaning to apparently meaningless peaceful gestures if only simply by indicating a wish to signal or communicate rather than to fight.
Connotations Though built upon animal signalling, human communications took the primal showing of hand and tooth and utterance of whimper, grunt and howl, and developed a universal range of symbols to represent such things as specific objects and beings and classes of these; times of events; relationships of beings to each other and to objects; quantities; qualities, conditions and states; and a certain number of philosophical ideas concerned with orders of reality and extended notions of place and time. Therefore language is both inherently and structurally symbolic. On top of this has been the conscious development of language to be a symbolic meta–language, that is, a language whose function of communicating elusive, difficult, or extremely refined meanings is enhanced by the use of indicators, many of them comparisons and many, categorizations. These indicators may point literally to a comparative object or event or process in the same class or category, or they may point figuratively to a comparison usually in another class. This is the nature of figures of speech, tropes, euphemisms, allusions, images or imagery, personifications, metaphors, similes, etc, among 'well–turned' phrases, and the phrases themselves exist in such symbolic forms of expression as allegories, fables, parables, apologues, parodies and satires. Literature may be symbolic in other senses, for example structurally, as Dante's tripartite Divina Commedia, or as the matrix into which is projected elements from the author's unconscious. It can be said that both language and literature are inherently, essentially or behaviourally symbolic; that they are or may be more or less structurally symbolic; and that they may carry symbolism from conscious or subconscious sources including the archetypes that refer to human development.
Broader Symbolic human activity (#MS8500).
Narrower Stories and fairy–tales (#MS8235) Symbols in sacred literature (#MS8423) Historical, legendary and literary characters (#MS8383).
Related Writing (#MS8245) Behaviour (#MS8125) Race and ethnicity (#MS8395) Beings made by beings (#MS8955) Symbols of civilization (#MS8803).

♦ **MS8117 Deities**
Divinities
Description Primitive thinkers reified the organizing, creative, sustaining, and altering forces and laws of nature (and some of the exemplary objects that reveal them, e.g. the planets) and to the degree that these personifications were perceived as having autonomy in their respective spheres they were considered deities, being equally gods and goddesses, and in some cases, godlings. Some were considered to rule conjointly; some have autonomous functions rather than regions; some have little contact with men, others are presented as having more to do with human affairs.
Connotations Each of the planetary deities, for example, has symbolic meaning. Mainly they correlate to human physical functions, including brain–based cogitation (Saturn). The Sun and Moon Gods are referred to the spiritual and psychic faculties. Other gods symbolize different aspects of human personality and are sometimes represented antithetically or complementarily to one another, viz. Apollo and Dionysos, Anes (Mars) and Aphrodite (Venus). They may also be part of a set, e.g. as the brothers Zeus, Poseidon and Hades, symbolizing the celestial and terrestrial realms and the invisible, other–world; or as the set Oceanos, Ouranos, Chronos, Zeus, Dionysos, symbolizing the ages.
Broader Mythical anthropomorphic beings (#MS8300).
Narrower Classical deities (#MS8613).
Related Infernal beings (#MS8907) Religious symbols (#MS8705) Mythical aerial beings (#MS8357) Mythical aquatic beings (#MS8917) Human spiritual phenomena (#MS8373) Mythical celestial beings (#MS8337) Mythical terrestrial beings (#MS8387) Mythical fire–dwelling beings (#MS8937) Forces, energies and processes (#MS8605).

♦ **MS8123 Sacred calendar**
Broader Religious symbols (#MS8705) Forces, energies and processes (#MS8605).
Related Civil symbols (#MS8223).

♦ **MS8125 Behaviour**
Description Human and animal behaviour provides a vocabulary for symbolism, taken in its entirely. Taken in part, it also exhibits, a high order of symbolic purposes. For example, the image of a man chopping wood can provide a symbol, but a particular man actually chopping wood is normally not communicating symbolically. Animals actually mating, for example, would not be said to be communicating symbolically to one another, but pre–mating or courtship signalling and behaviour may symbolize their instinct or intention to mate. Human behaviour may be symbolic in nature, may provide symbols, or be referred to by symbols. Some classes that are especially significant in these respects are: vital behaviour or functions such as eating, drinking, breathing, resting, sleeping; and social and sexual behaviour and functions which include marriage and mariage rituals such as courtship, betrothal, engagement, wedding and honeymoon; and various phenomena such as banquets, dances and political and other assemblies. Other important classes of notably symbolic behaviour concerns religion, sports, and games. In the religious class there are liturgies, rituals, orgies, saturnalia, mysteries, magic, and rites including ablution, baptism, confirmation, initiation, holy orders, chrism, last communion, bar–mitzvah, circumcision, exorcism, transubstantiation, investiture, and sacrifice or offerings.
Connotations Among sports and games of all types there is a basic symbolism of contest, conflict, struggle, fight or conquest. These activities represent battles and wars, and more fundamentally, the aggressive and competitive instincts. There are solitary games and sports whose purposes are solely individual physical and intellectual development, and a very few philosophical or symbolic games whose purpose is spiritual, aesthetic or ethical development. These last correspond to paths of initiation and their required skills in some traditional religions, and are reflected in the interiorization of some recreational activities, e.g. Zen archery, Tarot, sacred dance, etc. Related to this is the class comprising the symbolism to be found in the major arts of music (e.g. symbolic tones), architecture, (non–monumental) sculpture, painting, drawing, drama, choreography and poetry; and in other arts such as landscaping, flower arrangement, jewellery design and industrial design. Other symbolic behavioural classes are: movement, including running, walking, jumping, journeying, quest, pilgrimage, flight, exile, ascent, descent, passage, sea–crossing, etc; professional, trade and artisanal activities of all kinds; the special class of searching for knowledge; including experimentation, analysis, reasoning, science, occultism, theosophy, meditation, yoga, etc; and the symbologies for the unknown and for the reality or realities that are the objects of the search including the potentials of human development.
Broader Symbolic human activity (#MS8500).
Narrower Animal behaviour (#MS8725) Search for knowledge (#MS8963) Trades and professions (#MS8343) Symbolic quests and journeys (#MS8113).
Related Human gender (#MS8385) Race and ethnicity (#MS8395) Human form and face (#MS8315) Language and literature (#MS8115) Deformed and diseased humans (#MS8375).

♦ **MS8133 Animal body parts**
Broader Animals (#MS8215).
Related Insects (#MS8103).

♦ **MS8200 Mythological reality**
Narrower Mythical artifacts (#MS8915) Beings made by beings (#MS8955).
Related Artifacts (#MS8505) Mythical clothing (#MS8925) Symbolic human activity (#MS8500) Symbols in sacred literature (#MS8423).

♦ **MS8215 Animals**
Creatures
Description Most of the creatures that appear in symbolism can be classified as microbes, joint–legged arthopods (insects, crabs, etc), worms, fish, amphibia, reptilia, birds and mammals, according to their physiology and anatomy. Equally they can be categorized by the habitats they traverse, namely air, water, land, and the sub–classes of these such as sea, lake, river, or mountain, forest, desert, etc. Both the appearance (physical and behavioural) and the habitat provide symbolic connotations. Man's relationship to animals and his assessment of these relationships provide other categories of animals: domestic, pets, transport, economic, wild, dangerous, rare, endangered, exemplary and mysterious. The body parts and secretions of animals also provide symbols as: rhinoceros horn, cock's comb, tentacles, claws, needles, fangs, tail, sting, wings, paw, rattle, antlers, mane, scales, gills, fur, feathers, beak, snout, trunk, jaws, whiskers, shell, venom, egg, musk, pupae, pouch, hump, sac, webfeet, etc. Aggregate creatures provide such symbolic language as herd, flock, drove, swarm, flight, etc; and by their unwelcome presence: plague, infestation, etc.
Connotations From the ancient times specific animals were affiliated to components of the

psyche, taking the well-known forms of totemism and animal heraldic figures. In a deep sense there possibly are at least two animals inside each person, the one that he is born with and the one that he acquires. One represents his inner nature or inclinations; the other (one or two) the circumstances in which these may be expressed, either guarding the person and facilitating his self-expression, or acting otherwise. People's choice of pets man reflect the nature of the inner animal. The fact that people are inordinately and passionately attached to some specific animals is evidence of the fact that as outer beings they correspond to inner realities. The following avid pastimes and pets may be considered from the symbolic viewpoint: keeping monkeys, pigeons, canaries, parrots, tropical birds; falconry and fowling; pearl-diving, crafting, deep-sea fishing, fly-casting, trolling, netting, dolphinarium tending, tropical fish collecting; horseback riding, steeplechasing, race betting; herding, husbandry, hunting; chicken farming, mink ranching, pig farming, cattle ranching; cock-fighting, alligator or bear wrestling; and finally, the keeping of very loved pets of rabbits, turtles, hamsters, mice, and cats. Bee-keeping and butterfly-nettling also appear in symbolic imagery. In a unique place is the dog. He represents the magical, intimate 'familiar' of a person, or the Jungian Shadow, according to some writers. In another psychological sense he represents the 'inferior' functions. Thus his characteristics, e.g. training, are symbolically significant in expressing one aspect of personality development. The colour of the animal asserted in the Middle Ages that the devil was a black dog.

Broader Non-human beings (#MS8400).
Narrower Dog (#MS8853) Fish (#MS8823) Pets (#MS8433)
Birds (#MS8713) Insects (#MS8103) Serpents (#MS8513)
Animal body parts (#MS8133) Mythical theriomorphs (#MS8225).
Related Animal behaviour (#MS8725) Animal artifacts and products (#MS8346)
Objects of animal husbandry and domestication (#MS8336).

♦ MS8223 Civil symbols
Protocol — Perquisites

Description In the affairs of governments a considerable amount of symbolism is involved. Behavioural symbolism is notable in what is termed protocol, affecting, for example, the seating and processional order of persons according to perceived or actual rank, their speaking order, and foods served, clothing and decorations worn, etc. Among the principal civil rituals that are symbol-laden are investiture ceremonies (for monarchs, prime ministers, presidents, mayors, sheriffs, etc,) and the administration of criminal law (trial, sentencing, execution). The day in the year that national sovereignty or independence is celebrated is often characterized by special rituals including parades; particular foods; and revival of former dress, games, occupations and behaviour. Other days may be festal with their own distinctive symbolism, and there may be also a day of mourning or of sombre remembrance, in which symbolic sacrifices are made.
Connotations Civil symbols are closely bound-up with the hierarchical organization of societies and are used to differentiate roles and classes. An almost universal practice, for example, is the assigning to important bureaucrats of an imposing-looking limousine, the automobile being, as in every sphere of human activity, a mobile status-symbol. In general each perquisite obtained by people in civil service or public office is a symbol of their importance.
Broader Symbols of civilization (#MS8803).
Narrower Symbols of office (#MS8913).
Related Automobile (#MS8243) Human roles (#MS8363) Sacred calendar (#MS8123)
Social and hierarchical position (#MS8353).

♦ MS8225 Mythical theriomorphs
Hybrid monsters

Description Animal-formed symbolic beings and partly animal-formed (theriomorphic, zoomorphic), partly human-formed (anthropomorphic) symbolic beings have been known from earliest times. Among the most ancient are winged creatures of this category, and compound creatures with characteristics of several species. These fictional beings could be classified as mythical terrestrial and mythical celestial creatures, and also as malevolent or benevolent. Among them all may be mentioned those that are primarily horse-formed: half-man centaurs, winged horses, and part-horse or part-ass and part other animals; and those that are primarily dog-formed, serpent, fish or bird-formed. Some are recognizably mythic solely by their size and must be included in the mythic-monster category. Well-know examples in this whole class include: wolfman (lycanthrope), unicorn, sphinxes, phoenix, ouroboros, minotaur, medusa, mermaid, leviathan, griffin, gorgon, gargoyle, dragon, cerberus, makara (capricorn goat-fish), and the bird-serpents of Central America (e.g. Quetzalcoatl).
Connotations Some ancient gods took the form of mythic theriomorphs or hybrids, particularly in Egypt and Western Asia. The principle connotation is one of power and in the case of complex composites, of universality.
Broader Animals (#MS8215) Non-human beings (#MS8400).
Narrower Mythical fire-dwelling beings (#MS8937).
Related Infernal beings (#MS8907) Mythical artifacts (#MS8915)
Mythical aerial beings (#MS8357) Mythical aquatic beings (#MS8917)
Mythical celestial beings (#MS8337) Mythical terrestrial beings (#MS8387)
Mythical anthropomorphic beings (#MS8300).

♦ MS8233 Beautiful people
Beauty

Description Beauty in people is often symbolically shown by glowing good physical health evidenced by well-nourished, well-exercised bodies, and joyous or contented facial expressions. Models are chosen normally for their physical beauty.
Connotations Beauty is considered an end in itself, a synonym or symbol for personal fulfilment and happiness, and an image of ultimate personal development. Beautiful people are considered heroes and heroines and their physical qualities are symbolically used by advertisers in association with their products to increase sales. Beautiful people are sometimes used to sell ideas or endorse or become candidates for public office.
Broader Human body (#MS8305).
Related Applied cosmetics (#MS8372) Deformed and diseased humans (#MS8375)
Historical, legendary and literary characters (#MS8383).

♦ MS8235 Stories and fairy-tales
Broader Language and literature (#MS8115).
Related Mythical aerial beings (#MS8357)
Historical, legendary and literary characters (#MS8383).

♦ MS8243 Automobile
Broader Vehicles and parts (#MS8402).
Related Civil symbols (#MS8223).

♦ MS8245 Writing
Broader Cultural objects and devices (#MS8306).
Related Language and literature (#MS8115).

♦ MS8300 Mythical anthropomorphic beings
Narrower Deities (#MS8117) Infernal beings (#MS8907)
Mythical aerial beings (#MS8357) Mythical aquatic beings (#MS8917)
Mythical celestial beings (#MS8337) Mythical terrestrial beings (#MS8387)
Mythical fire-dwelling beings (#MS8937).
Related Non-human beings (#MS8400) Mythical artifacts (#MS8915)
Beings made by beings (#MS8955) Mythical theriomorphs (#MS8225)
Symbols in sacred literature (#MS8423).

♦ MS8303 Signs of human presence
Description Related signs and symbols of human presence include footprints, handprints, bite marks, finger marks, and bodily impressions. Other representations of human presence employ the symbols of man-made objects, particularly various types of dwellings, stone rings and other stone formations, and fire sites.
Connotations Footprints symbolize the path to take, or that someone has gone before. In the East the symbol that is called, Buddha's Footprint, contains, in the arch, the Wheel of the Law, and under the heel, the mystic Lotus. In Sri Lanka, Adam's footprint was located on a mountain. A full hand-print signifies attention and the number five. Finger marks indicate possession. Circular objects may represent the human presence as expressed by the primitive arrangement of habitations or of ring-formed councils.
Related Human body (#MS8305) Human form and face (#MS8315).

♦ MS8304 Boats and ships
Broader Vehicles and parts (#MS8402).

♦ MS8305 Human body
Description The human body and its forms, members, organs and characteristics, appear continually in symbolism.
Connotations This class of symbols is among the most explicitly self-referring when its visual imagery or other expression is conventional. In symbolism, humans appear as one kind of embodied being. Other physical beings present as symbols are the terrestrial fauna; birds, beasts and insects, and types of the aquatic forms of life. However, the range of symbols employing the human body and aspects of its vital physical existence uniquely extends to its economic, and social, and physical life. This includes trades and professions and social or hierarchical positions and their corresponding behaviours.
Broader Body, form and structure (#MS8700).
Narrower Death (#MS8603) Human gender (#MS8385) Human skeleton (#MS8355)
Human plurality (#MS8313) Beautiful people (#MS8233) Human body fluids (#MS8365)
Race and ethnicity (#MS8395) Human form and face (#MS8315)
Human internal organs (#MS8345) Human external organs (#MS8325)
Trades and professions (#MS8343) Human spiritual phenomena (#MS8373)
Human ages and development (#MS8323) Deformed and diseased humans (#MS8375)
Human congenital relationship (#MS8333) Alteration in skin appearance (#MS8335)
Historical, legendary and literary characters (#MS8383).
Related Human roles (#MS8363) Applied cosmetics (#MS8372)
Beings made by beings (#MS8955) Trades and professions (#MS8343)
Signs of human presence (#MS8303) Human spiritual phenomena (#MS8373)
Symbols in sacred literature (#MS8423) Symbols in sacred literature (#MS8423)
Social and hierarchical position (#MS8353)
Historical, legendary and literary characters (#MS8383).

♦ MS8306 Cultural objects and devices
Numeracy

Description A large array of objects that are created by man's intellectual and cultural activities appear in symbolic contexts. These objects include everything to do with writing (pen, pencil, ink, quill, tablet, chalk, blackboard, paper); with reading -book, letter, newspaper, lectern); with verbal or audible communication (telephone, megaphone, loudspeaker, radio, television, telegraph, rostrum); and with pictorial communication and visual fine arts (paintings, easel, palette, brush, statue). Another category of symbolic cultural objects and devices concerns mankind's numeracy. These objects include the abacus, slide rule, adding machine, calculator and computer.
Connotations Writing instruments and written or inscribed surfaces represent the record; factual history and personal deeds. Symbolic records of various types (rolls, ledgers, books) may represent the unalterable past or the forseeable future. In the latter case they can signify a mandate or mission. As a symbol of unalterability, records may appear on stone or metal and be associated with laws and constitutions. Computer records, symbolized by a profusion of paper and other outputs, signify inexhaustible detail, much of it irrelevant; or comprehensive knowledge, when they are represented in scientifically organized information flow charts.
Broader Artifacts (#MS8505).
Narrower Cinema (#MS8371) Writing (#MS8245) Fine arts (#MS8533)
Scientific symbols (#MS8933).
Related Search for knowledge (#MS8963) Symbols of civilization (#MS8803).

♦ MS8313 Human plurality
Description Any number of human beings may be represented symbolically. Symbols may include heads, footprints, houses, automobiles, and anything else that indicates a number of people in a general way. Symbols may also refer to specific groups of people such as workers, soldiers, newborn, elderly, students, etc by reference to one or more attributes.
Connotations Human aggregates can range from a married couple or a family up to and including a nation and all of humanity. Typical aggregates symbolized are military forces, world population and its regional components, the starving, the impoverished, and the world's children.
Broader Human body (#MS8305).
Related Human roles (#MS8363).

♦ MS8315 Human form and face
Hands and feet

Description Related symbols include the whole human body, the head, the arms, legs, feet, hands and other parts such as shoulders, neck and knees. The expression of the facial features and the positions of the fingers and of the hands are the principle bases for symbolic communication along with light and colour, and with sound. These elements taken together, with movement of the head, torso, and limbs, constitute a 'dance' of symbolism in human behaviour and are projected in visual and auditory imagery.
Connotations In the face it is eyes and mouth that convey meanings symbolically. To a secondary extent it is the head. The symbolism is effected by position or attitude and movement. Examples are: upward turned lips (smile of approval, humour, pleasure, etc), nodding head (approval, disapproval) and closed eyes (rejection, ignorance, sleep). The fingers and hands can convey any symbolic message. Special symbolic positions, attitudes and movements of the head, eyes, mouth, hands and fingers have their own names, e.g. nod, stare, frown, fist, thumbs-down.
Broader Human body (#MS8305).
Related Behaviour (#MS8125) Human skeleton (#MS8355)
Human external organs (#MS8325) Signs of human presence (#MS8303)
Alteration in skin appearance (#MS8335).

♦ **MS8316 Musical instruments**
Description Some musical instruments appear very often in symbolic imagery. Typically they are the instruments with the most ancient lineage, namely pipes and horns, lyres, and drums. Among contemporary European instruments, the violin and the piano keyboard have had considerable symbolic employment. Pneumatic instruments such as the organ and the bagipes appear symbolically. Tambourines also appear and even more humble instruments such as castanets, whistles and chimes. Through shape and materials employed (e.g. drums and drumskins) many instruments become symbolic. The symbolic musical instrument presented most often is possibly the bell.
Connotations Musical instruments, as music itself, have a close connection with changes of state or condition. The horns ae representative of the functions of announcement, and sometimes of imperative calls to change, such as to advance or retire. The drums signify a point of charge or a climax, lyres, harps, lutes and similar plucked instruments symbolize harmony, concord, peace and repose. A church organ and its pipes represent spiritual fellowship and heaven. A tolling church bell symbolizes death. A ringing hand–bell signifies sudden good fortune or someone coming. A peal of bells or carillon signifies good news of great proportions. The qualities of tone or tones that an instrument makes is an integral part usually of its symbolism. The birdsong quality of the reed pipe and the thunderous boom of the kettle drum are responsible, for example, for considerable musical symbolism.
Broader Artifacts (#MS8505).
Related Fine arts (#MS8533) Race and ethnicity (#MS8395)
Trades and professions (#MS8343).

♦ **MS8323 Human ages and development**
Description The biological age of symbolic human figures may be indicated approximately or directly. There may also be an indirect representation of a particular stage of physical development, as for example by social role, e.g. an Amerindian brave.
Connotations The developmental age range may include such categories as pre–school, pre–adolescent, adolescent, young adult, mature, advanced mature and senescent. Symbolic figures are often polarized into youth and maiden on the one hand and old man and old lady on the other. The former pair signify vigour, power and hopefulness; the latter, caution, prudence of wisdom, and realism.
LK F0 M Human ages and development
Broader Human body (#MS8305).
Related Human roles (#MS8363) Search for knowledge (#MS8963)
Human congenital relationship (#MS8333).

♦ **MS8325 Human external organs**
Description Related symbols among the sense organs are the skin (the organ of touch) and the ears, eyes, nose and tongue. Organs of action are principally the hands, feet and mouth. The genital organs may be explicitly indicated or suggested by the mons pubis or pelvis. The female breasts and their nipples; the navel and the anus are the remaining organs that have prominent places in visual, verbal or literary symbolism.
Connotations A great deal of the symbolism employing the external organs makes use of the analogy of their functioning to the reference or connotation intended, e.g. smelling (detection), touching (contact), seeing (comprehension), etc.
Broader Human body (#MS8305).
Narrower Alteration in skin appearance (#MS8335).
Related Human form and face (#MS8315) Human internal organs (#MS8345)
Deformed and diseased humans (#MS8375).

♦ **MS8326 Symbolic objects**
Chrematomorphic symbols
Description Many concrete human artifacts have, as their reason for creation, the purpose of serving as symbols for abstract concepts; for example, the flags of nations. In this category are royal crowns, crests, and heraldic devices, seals, sceptres, orbs, and signets. Club emblem jewellery, organization symbol lapel pins, textile insignias (patches), and two–dimensional and three dimensional representation of trade–marks and logos (logotypes) are additional symbolic items. Such objects have as their only, or prime purpose, to serve as symbols. The ecclesiastical crozier and the medical caduceus are ancient examples, as are totem poles and ships' figure-heads. The hood ornaments of automobiles once constituted an array of figures, some simply decorative, others indicative of the imputed qualities of the motor or driver e.g. fleet animals. Some items have a functional as well as symbolic significance such as a royal–throne, a conductor's baton, or a string of prayer beads or rosary. These objects, while at the same time being symbolic, belong to a class of chrematomorphic items. The most common concrete symbol for an abstract concept is money. This takes the form, in most countries, of paper and metal currency, postal and fiscal stamps, and other negotiable financial paper.
Connotations Symbolic objects represent such things as sovereign power, energy, and strength and, more rarely, restraint and wisdom. Symbolic objects can become debased and lose their sign–value. This has happened with currencies in several countries in modern times.
Broader Artifacts (#MS8505).
Narrower Money (#MS8397).
Related Religious symbols (#MS8705) Mythical artifacts (#MS8915)
Symbolic human activity (#MS8500) Man–made forms and structures (#MS8600).

♦ **MS8333 Human congenital relationship**
Kinship — Family
Description Symbolic representations of cogenital and other kinship relationships are usually focused on the nuclear family of parents and children. Symbols include the relationship–defined identities of father, mother, child or siblings (brothers, sisters), and are extended to grandfathers, and grandmothers. A vertical relationship may be depicted by symbols of ancestors or forebears and by a representation of descendants. Diagonally and horizontally, kinship (clan) symbols may encompass fathers' and mothers' brothers and sisters, their spouses and their children, the spouses' parents and the childrens' spouses and the childrens' children. Special relationship symbols are those of twins (though rarely of other multiple human births), of order of birth, and order of spouse (e.g. second wife).
Connotations These symbols are very explicit and on the whole are self–referring. However, the symbol of a father may indicate paternity or authorship; that of a mother, origin; and that a child, development. Ancestors may represent lineage; descendants may indicate long–range implications and effects. Twins may indicate a quantitative doubling, or a qualitative duality or complementarity.
Broader Human body (#MS8305).
Narrower Family (#MS8393).
Related Human roles (#MS8363) Human gender (#MS8385)
Human ages and development (#MS8323).

♦ **MS8335 Alteration in skin appearance**
Description In symbolism involving the human body the skin may be represented as altered. Related symbols include wrinkles lines, scars, brands and wounds. The skin also may be represented as coloured by tattoos, war–paint, grease–paint, cosmetics, camouflage or other agents. The skin may appear spotted, pallid, livid, etc. It may also be represented as hairy, scaly, or blotched. Its condition may be tight or loose.
Connotations The representation of a particular marking, coloration or other condition of the skin may indicate general vitality or morbidity, or particular moral states, personality traits, moods or dispositions to action. The appearance of the skin as a symbol is considered important, for example, in pantomime and theatre.
Broader Human body (#MS8305) Human external organs (#MS8325).
Narrower Applied cosmetics (#MS8372).
Related Human form and face (#MS8315) Deformed and diseased humans (#MS8375).

♦ **MS8336 Objects of animal husbandry and domestication**
Description Among this class of objects employed as symbols are: leash, bridle and reins; saddle, stirrups and yoke; spurs, whip and goad; shepherd's crook, cowboy's lariat and bullfighter's cape; horseshoes, branding iron and bit; milking stool, manger and stall; chicken–coop, dog kennel and pig pen. Other objects are apiary (bee–hive), bird cage and animal trap. A related class of symbolic objects are those connected with fishing, fowling, hunting, slaughtering and dressing game, and animal sports.
Connotations These objects are self–referring symbols in many cases, their function being employed metaphorically. A cowboy's lariat, for example, is a symbol for roping or drawing something in; the shepherd's crook, of guidance; the bullfighter's cape of "seeing red" (becoming incensed). The chicken–coop signifies confinement; the pig pen, disorder and filth; the dog kennel (dog house), disgrace. The bee–hive signifies activity or industry; the trap, imprisonment or being caught; the bird cage, humiliation and repression. Many of these symbols appear in the Bible.
Broader Artifacts (#MS8505) Agricultural symbols (#MS8943).
Related Animals (#MS8215) Food–related objects (#MS8602)
Trades and professions (#MS8343).

♦ **MS8337 Mythical celestial beings**
Angels
Description World symbolism includes classes of beings considered to dwell above mankind in various stations or heavens. Western symbolism knows only the classes of angels, but there are in other traditions men in space, just as there are on earth. These beings, however, differ from earthmen in behaviour, in appearance, in mind and in faculties. Other celestral beings imaged include incorporeal powers and intelligences. In Greek mythology, for example, there are the Furies, the Fates, the Graces, the Harpies, the Hours and the Muses. In Indian mythology also there are Gandharvas, Aswins and others.
Connotations The celestial beings sometimes represent the instrumentalities of divine action. There is also a class of beings who function as messengers, and another who are intermediaries, often intimately associated with the fate of nations, or with individual spiritual life. In the last case there is the reified figure of Wisdom (Sophia). Sophia long has been present in the European consciousness as demonstrated by the mythology of early Gnostic Christianity, and before that of Parmenides' 'Truth'. Sophia reappeared in J Boehme and G Gichtel and their followers in the seventeenth and eighteenth centuries, and is present in the Romantic mysticism of Novalis, Goethe, Steiner, and Jung and their disciples.
Broader Mythical anthropomorphic beings (#MS8300).
Related Deities (#MS8117) Religious symbols (#MS8705)
Mythical theriomorphs (#MS8225).

♦ **MS8338 Aircraft**
Planes — Dirigibles
Broader Vehicles and parts (#MS8402).

♦ **MS8343 Trades and professions**
Description The trades and professions often appear in symbolism. The symbols may vary from an individual clothed in a particular way, or posed according to the reference intended, or at work with or without characteristic tools, equipment or paraphernalia.
Connotations While the symbols employed may be self–referring they may also point to a trade or profession or actions and activities pertaining to them in a metaphorical sense. The indication of a midwife might connote metaphorical birth; that of an undertaker, metaphorical death. A medical doctor may refer to curative or remedial action, and a teacher to the imparting of information. A butcher's work might refer to actions of allotment, and a baker's to those of confection or collection. A judge or lawyer might refer to a test or trial, and so on.
Broader Behaviour (#MS8125) Human body (#MS8305).
Related Money (#MS8397) Clothing (#MS8436) Human body (#MS8305)
Human roles (#MS8363) Craftsman tools (#MS8102) Symbols of office (#MS8913)
Scientific symbols (#MS8933) Vehicles and parts (#MS8402)
Armour and weapons (#MS8106) Medical instruments (#MS8715)
Musical instruments (#MS8316) Agricultural symbols (#MS8943)
Jewellery and adornments (#MS8446)
Objects of animal husbandry and domestication (#MS8336).

♦ **MS8345 Human internal organs**
Description Related symbols are the heart (the foremost symbol among the inner organs), the stomach or belly, the organs of elimination, the brain, the intestines or guts, and the womb, among the more notable.
Connotations A great deal of the symbolism employing the internal organs makes use of the analogy of their functioning to the reference or connotation intended, e.g. digestion, excretion, gestation, etc.
Broader Human body (#MS8305).
Related Human external organs (#MS8325).

♦ **MS8346 Animal artifacts and products**
Description Things that animals make often appear in symbolism. The bird's nest is perhaps the most common symbol in this category, with perhaps the honey–comb or bee–hive second. Other animal created structures include the beaver's dam, the hornets' nest, the tunnel of the mole, the fox–hole, and the spider web. Termite nest, ant colony, dung ball (scarab) and sea–shell also have symbolic values. Quite a special class of symbols derive from creatures' economic products: silk, pearl and coral for ornamentation; milk and honey as human food; musk and ambergris for perfumes; and numerous others including guano and manure, sponge and caviar, and horn and ivory.
Connotations The symbolic meanings of animal artifacts and products are commonplaces. To the above items may be added furs and skins. To the primitive mind, from the Paleolithic era to modern times, the fur and skin of animals has had magical symbolic significance. This increases fur and leather demand, and hence increases the slaughter of wild animals and endangered species.
Broader Artifacts (#MS8505) Forces, energies and processes (#MS8605).
Related Birds (#MS8713) Animals (#MS8215) Human body fluids (#MS8365)
Symbols in sacred literature (#MS8423).

♦ **MS8353 Social and hierarchical position**
Status
Description This class of symbols employs people's positions in the social, economic and other orders. Included are such symbols as the court series of king, queen, prince, princess, counsellor, leader of knights (duke, baron, count, etc), individual sir knights, ambassador or other diplomat, bishop or court priest, herald, guard, jester, gaoler, executioner, court physician, wizard and other posts. In the order or estate of the churches, symbolism may utilize the principal offices, archbishop, Pope, etc, or any lesser offices down to deacons, sextons, etc. Titles of parliamentary or democratic government may be used symbolically such as minister, president, governor, mayor; and military ranks may also be used such as commander, admiral, field marshal, general, or soldier. A popular series of symbols derives from the craft unions which provide apprentice, journeyman, master. High society, on the other hand, gives the figures of the debutante, the matron and the grand dame among the ladies, and among the men, the gentlemen, the 'amateur', and sometimes, the playboy.
Connotations The military, rank and file, is applied symbolically to members in civil organizations. The royal titles may be applied symbolically to anyone preeminent, or as a general superlative. A debutante represents the novice; the master indicates the complete combination of theoretical knowledge and practical skill.
Related Clothing (#MS8436) Human body (#MS8305) Civil symbols (#MS8223) Symbols of office (#MS8913).

♦ **MS8355 Human skeleton**
Bones — Teeth
Description Related symbols are the entire skeletal remains and any of its parts, notably the skull, the ribs and the hands. Teeth in the living face or skull are also present in symbolism.
Connotations The skeleton and the skull symbolize death of some kind. The cross–bones (two bones crossed behind a skull) are an indication of the danger of death, for example, on the 'jolly Roger' pennant of pirates, on containers for toxic substances or near high voltage electrical apparatus. Symbols of death may be metaphorical, referring only to the ideas of cessation or rupture. Teeth barred in a human grimace (or rictus) may symbolize extreme pain. The grimace may indicate enmity, especially with the lower jaw thrust forward in a biting position. The toothy smile reveals that the teeth are not set to bite and is a universal, human symbol of real or artificial friendliness and pleasure.
Broader Human body (#MS8305).
Related Death (#MS8603) Human form and face (#MS8315).

♦ **MS8357 Mythical aerial beings**
Fairies
Description Symbolic non–human beings represented to live in the atmosphere around us or in the air above include such things as fairies, the personified winds, and various spirits.
Connotations Some of these symbols represent phenomena of nature on the one hand. Others stand in close connection with humanity. For example, they may symbolize service or pastoral innocence. Some spirits of the air are also represented as the embodiment of the invisible moral influences, for good and for evil that humanity throws off.
Broader Mythical anthropomorphic beings (#MS8300).
Related Deities (#MS8117) Mythical theriomorphs (#MS8225) Stories and fairy–tales (#MS8235).

♦ **MS8363 Human roles**
Description Role behaviour is used symbolically. Common roles represented include those of husband, wife, friend, lover, neighbour, colleague, superior, subordinate, countryman (compatriot) and peer or equal (by age or interest cohort membership. Less common roles imaged include the always helpful person, saint, hero or heroine, guardian, mediator, and warrior. Other images are of an anti–hero, sleeping beauty, witch, wizard, prophetess or prophet. The roles of a stranger, a wanderer or a pilgrim; or of an individual with a secret or personal mission, or with a public mission (e.g. crusader, evangelist, 'leader', etc) are also represented.
Connotations This class of symbols may be self–referring but also indicates the overriding orientation of the role. For example, the husband's role may be said to be, to husband, that is conserve, preserve or protect, so that it is the functions which may be intended by the role symbolism. The role noun becomes a role verb in order to read the reference: to wive, to befriend, to love and to act neighbourly or collegially, or as superior, inferior, equal, etc. Noun to verb symbols indicate that the anti–hero antagonizes, and the sleeping beauty sleeps and dreams in passive repose, for example. Those who have a nominal role and do not exercise its function are described by a number of symbols such as straw man, lame duck, front, figurehead, etc.
Narrower Human gender (#MS8385).
Related Human body (#MS8305) Civil symbols (#MS8223) Human plurality (#MS8313) Symbols of office (#MS8913) Trades and professions (#MS8343) Symbolic human activity (#MS8500) Human ages and development (#MS8323) Human congenital relationship (#MS8333) Historical, legendary and literary characters (#MS8383).

♦ **MS8365 Human body fluids**
Excreta — Hair and nails
Description Symbols that can be classified as body fluids are many and include blood, sweat, tears and saliva or spittle. Human milk, and urine may also appear symbolically. Other effluxes or emanations from in the body that have a place in visual, verbal or literary symbolism include bile, pus, vomit, and watery excrement, human hair, finger–nails, and faeces can be included in the class of human body fluids and excreta.
Connotations Blood is a symbol of life; sweat of work; tears of sacrifice or sorrow. Spittle and excrement may symbolize rejection. A male urinating may symbolize semen or virility. Long finger–nails may indicate gentility or feminity. Excessive body hair may represent a brutal nature. Luxuriant head hair may indicate energy, mental or physical (e.g. Einstein and Samson).
Broader Human body (#MS8305).
Related Food–related objects (#MS8602) Animal artifacts and products (#MS8346).

♦ **MS8371 Cinema**
Broader Cultural objects and devices (#MS8306).
Related Fine arts (#MS8533).

♦ **MS8372 Applied cosmetics**
Broader Alteration in skin appearance (#MS8335).
Related Artifacts (#MS8505) Human body (#MS8305) Beautiful people (#MS8233).

♦ **MS8373 Human spiritual phenomena**
Description A number of symbols portray alleged phenomena of the spiritual life. Some are of the type of illuministic images of the aureole, the halo, the nimbus, the mandala, and various crowns, and robes as well as emanations of glory from the depicted body. Other imagery may show the glorious emanation as a figure emitted from the body, taking the form of a brilliant winged creature; a butterfly or small angel; or in the form of a person's own likeness. Another phenomenon depicted may be the entry into, or union with, a person's body or soul, by a light–being (e.g., the Iranian Daena angel or the Christian Dove) or by a light–ray (e.g., from the Holy Spirit, Shekmah or Christ).
Connotations These symbols represent spiritual illumination (knowing), spiritual grace (being), sanctification, salvation, or apotheosis.
Broader Human body (#MS8305).
Related Death (#MS8603) Deities (#MS8117) Human body (#MS8305) Religious symbols (#MS8705) Symbolic quests and journeys (#MS8113).

♦ **MS8375 Deformed and diseased humans**
Ugliness
Description Symbols in this class may employ representations of moderately or grossly deformed or diseased humans. A limping, or leg or foot impaired person is a prevalent image. A frail person, or a person in a wheelchair or sick–bed may be represented. A person whose back is deformed (hunchback) or more commonly, whose head or whose facial features are deformed (e.g. lepers) may be imaged. Unusual representations of skin and hair may be included in this category.
Connotations This class of symbols points to distortion, incompletion, and imperfection. It also represents what has already been rejected, or it is intended to evoke rejection. It may also indicate punishment. In the magical world of Grail literature the Loathely Damsel, and in the fairy–tale world, the frog–prince, indicate that ugliness or deformity may conceal beauty. This is somewhat the theme of Cinderella or the Ugly Duckling.
Broader Human body (#MS8305).
Related Death (#MS8603) Behaviour (#MS8125) Health aids (#MS8406) Beautiful people (#MS8233) Religious symbols (#MS8705) Human external organs (#MS8325) Alteration in skin appearance (#MS8335).

♦ **MS8383 Historical, legendary and literary characters**
Description Literary characters, legendary figures, historical and living are used as symbols.
Connotations A number of these symbols, like Caesar and Alexander, are used to indicate military conquest. Lord Kitchener and Uncle Sam on military recruiting posters are twentieth century examples. Rodin's 'Thinker' and the face Albert Einstein represent intellectual activity. The American Santa Claus, Charlie Chaplin's tramp, Napoleon Bonaparte, Superman, Charlemagne, John Bull, Mona Lisa, Hamlet, Harlequin, Joan of Arc, St Francis, Buddha, Don Quixote and countless others are all on the rolls of the world's symbolism, some depicting the virtues, and some the follies of mankind. The most ephemeral symbols are those of feminine beauty for which the public demands new faces and new bodies regularly. Often ephemeral as well, for the length of the publics' span of attention given them, are the more humanly profound symbols of service and sacrifice. The Unknown Soldier, for example, is only officially remembered.
Broader Human body (#MS8305) Language and literature (#MS8115).
Related Human body (#MS8305) Human roles (#MS8363) Beautiful people (#MS8233) Symbolic human activity (#MS8500) Stories and fairy–tales (#MS8235) Symbols in sacred literature (#MS8423).

♦ **MS8385 Human gender**
Sex roles — Masculinity — Femininity
Description The two sexes, male and female are almost constantly differentiated in symbolism. Beyond biological gender, symbolism also employs the behavioural attributes of masculinity and feminity.
Connotations Differentiated genders are used symbolically to indicate the normative 'mutually exclusive' roles of the sexes. They may also refer to reproduction, virility and fecundity. Sexual reversal, that is, males depicted in female dress, or as engaged in feminine activities; and female's shown in male attire or occupations or depicted with moustaches or beards may be associated with confused sex identities either on the personal or the social levels. The range of meanings includes perversion on one end of the scale and implied harmonization of opposites on the other (as also indicated by the imagery of hermaphrodites or bisexuals).
BX M Human gender Sex roles Masculinity Femininity
Broader Human body (#MS8305) Human roles (#MS8363).
Related Behaviour (#MS8125) Craftsman tools (#MS8102) Human congenital relationship (#MS8333).

♦ **MS8387 Mythical terrestrial beings**
Description Mythical anthropomorphic beings (who are non–human) are often represented in symbolism. Those that are represented as ground–dwelling inhabitants of this planet include such creatures as dwarfs, giants, ogres, genies (although the djinn can also take animal shape), banshees (and other mound–dwelling sidh), some classes of fairies, abominable snowmen (yeti), and demons of the deserts and wastelands.
Connotations One of the archetypal symbols is the terrestrial non–human, Pan, uniquely an earth–bound deity. His partly goatish body connects him with the satyrs. Pan and the satyrs symbolize lust and satisfaction of all unbridled appetites in orgiastic abandonment.
Broader Mythical anthropomorphic beings (#MS8300).
Related Deities (#MS8117) Religious symbols (#MS8705) Mythical theriomorphs (#MS8225) Mythical aquatic beings (#MS8917).

♦ **MS8393 Family**
Broader Human congenital relationship (#MS8333).

♦ **MS8395 Race and ethnicity**
Alien — Foreigness
Description Indications of race or ethnic background appear in visual symbolism chiefly as skin colour, physiognomy, and clothing. In auditory symbolism ethnicity or national origin may be indicated by foreign language, foreign accents, and by music. Many objects may symbolize another country or nationality, or particular cultural and ethnic heritage.
Connotations A white or pale human figure to some Asians, or a dark human figure to Europeans may indicate death. A jungle tribesman or American Indian may symbolize the noble savage; the idea of innate purity or simplicity. A cannibal or cave–dweller may indicate depravity.
Broader Human body (#MS8305).
Related Clothing (#MS8436) Behaviour (#MS8125) Artifacts (#MS8505) Musical instruments (#MS8316) Chromatic symbolism (#MS8923) Language and literature (#MS8115).

♦ **MS8397 Money**
Broader Symbolic objects (#MS8326).
Related Precious things (#MS8516) Trades and professions (#MS8343).

♦ **MS8400 Non–human beings**
Narrower Fish (#MS8823) Birds (#MS8713) Insects (#MS8103) Animals (#MS8215) Serpents (#MS8513) Mythical theriomorphs (#MS8225).
Related Symbols in sacred literature (#MS8423) Mythical anthropomorphic beings (#MS8300).

♦ **MS8402 Vehicles and parts**
Transport
Description Vehicles, their parts, and their energy sources are a distinct class of symbols. They include boat, ship, raft, ark, dug–out, galley, life–boat, and parts, such as sail, anchor, oar, rudder, helm or wheel. Specific kinds of sea and waterway vessels have symbolic associations, such as tugs, barges, luxury liners, banana boats, fishing vessels, oil tankers. A number of other symbolic objects and characteristics associated with ships are smokestacks (formerly), foghorns, running lights, signal lights and pennants, masts, and devices such as flare guns, telescopes and sextants, ad charts and logs. Land vehicles that have symbolic connotation include chariots and other horse–drawn conveyances and rigs from farm and Connestoga wagons, to buggies, hansoms, post coaches, sulkies and broughams. Other conventional animal–powered vehicles appearing in Symbolism, are chiefly dog sleds and dog carts. Modern civilian vehicles that have conspicuous symbolism are sport and racing cars, taxis, limousines, hearses, ambulances, fire engines, police patrol cars and motorcyles. Military vehicles, land, sea or air, that often appear in symbolism are fighter planes, tanks and submarines. Some civilian aircraft are also notable symbolically; these are the hot air balloon, the dirigible and the glider. Vehicle parts have symbolic values. Chief among these are steering devices, empowering devices (engines) and tyres and wheels. Power sources (fuels) and batteries are important symbols as well.
Connotations The principle symbolic connotation of vehicles is the notion of empowering. On a simple level of symbolism what is represented by vehicles is that they empower one to travel physically or subjectively. On a psychological level, vehicles are particular embodiments or representations of different kinds of power, all of which are inside a person. Air power, earth power, water power and fire power correspond to the archetypal psychological functions of thought, sensation, emotion and intuition, and the representation of these power, or vehicles indicates the travel direction towards greater individual personality integration and development.
Broader Artifacts (#MS8505).
Narrower Aircraft (#MS8338) Automobile (#MS8243) Boats and ships (#MS8304).
Related Food–related objects (#MS8602) Trades and professions (#MS8343).

♦ **MS8406 Health aids**
Medical preparations
Description This class of symbols includes crutch, casts for broken limbs, bandages, wheelchair, eyepatch, blind person's white stick and dark eye–shades, prescription eyeglasses, hearing aids, false teeth, prosthetic devices (including wooden leg, metal hand, claw or hook), orthopaedic shoe, truss, dental braces, support and figure control devices (corset), iron lung, blood transfusion equipment, x–ray machines, surgical table, surgical instruments, anesthaesia mask, surgeon's mask, surgeon's gloves, jars of medicine, pills, tablets, ointments, bed or hospital bed.
Connotations Objects in this class may symbolically refer to general infirmity or to weakness in the part of the body with which they are associated. Some infirmities are metaphorical, for example, ignorance may be symbolized by blindness, recalcitrance by deafness, apathy by physical immobility, debility by blood transfusion, etc.
Broader Artifacts (#MS8505).
Related Foods and liquids (#MS8606) Medical instruments (#MS8715) Deformed and diseased humans (#MS8375).

♦ **MS8415 Incorporeal and formless symbols**
Ideas — Darkness
Description Symbols that are formless are of two kinds: those of things that never take a form, and those that are of things that are impressible plastic or malleable and can take any form, such as wax and clay. Those things that have body (corporeality or mass) but which are formless include water in its liquid and gaseous state (as well as other liquids and gases, smoke and vapours) earth, mud, and other large aggregates of particles, for example quantities of salt, and flour and, generally, pulverized and powdered substances of all sorts. Any other mass is also in a formless condition, such as lava, or foundry metals and slag. Some things like clouds, which change their shapes often and rapidly, symbolize an incapability to achieve or retain form or organization. Things which are incorporeal in themselves (electro–magnetic phenomena, dreams, thoughts, etc) are classified among energy–related symbols, but symbols exist for such incorporeal realities or concepts as nothingness, abyss, hole, emptiness, vacuum, prime matter, matter, space, sky, heaven, air, nature and existence.
Connotations The archetypal incorporeal and formless symbols are numbers which, according to Greek philosophy, represented each, a specific Idea or Form. A preeminent symbol of incorporeality and formlessness, and thus of origins and beginnings, is darkness. The curdling of milk is a symbol of formation, or form emerging out of formlessness (as in Vedic cosmogenic myth).
Broader Body, form and structure (#MS8700).
Related Scientific symbols (#MS8933) Chromatic symbolism (#MS8923) Man–made forms and structures (#MS8600) Forces, energies and processes (#MS8605).

♦ **MS8416 Home furnishings and amenities**
Description This class of objects which appear in symbolic imagery includes all forms of light devices and fixtures (candle, candelabra, lamp, chandelier, lantern, light bulb, torch, flashlight, floodlight, spotlight, etc) and for classification convenience, industrial and entertainment lighting, including footlights, blacklight, ultra–violet light, and coherent beams including lasers. Home related furnishings that have a universal symbolic presence, in addition to light devices, also include mirrors, and doors and windows and everything pertaining to them: lock, latch, bolt, key, keyhole, chain, bar, window pane, threshhold, doorway, curtain, window shade. A number of symbols are or were kitchen fire connected: oven, fireplace, wood–stove, hearth, woodpile, coal–store, coal pile, fireplace poker and tongs, logs, coals, embers, cinders. In the same category of kitchen–related objects are cutlery (knife, fork, spoon, etc), dishes, and broom, and such structures as pantry, larder, ice–house, smoke–house, woodshed, fountain, well, and water bucket. Of the remaining home–related symbolic objects the bed is most important, along with the cradle. Rocking chair, swing and love–seat or small couch also appear symbolically, as do carpet, and drape. Modern symbolism includes home entertainment, notably the television, and communications via the telephone.
Connotations Lights symbolize knowledge, examination or inspection; keys and locks, the archetypal functions of binding and freeing; kitchen–fire related objects, the libido and its energies and the well represents the inner self and its ancient knowledge. Drapes reveal or conceal, and carpets represent the foundation of a matter, the situation as it exists. On a deeper level, the carpet, as indicated by oriental motifs, may be an archetypal surface representing the reality behind phenomena and the field of interplay of the energies of mind (chitta). The Islamic prayer rug is an intentional symbol of such spiritual reality.
Broader Artifacts (#MS8505).
Related Religious symbols (#MS8705) Chromatic symbolism (#MS8923) Shapes and patterns (#MS8845).

♦ **MS8423 Symbols in sacred literature**
Broader Religious symbols (#MS8705) Language and literature (#MS8115).
Related Fish (#MS8823) Serpents (#MS8513) Artifacts (#MS8505) Human body (#MS8305) Non–human beings (#MS8400) Mythological reality (#MS8200) Beings made by beings (#MS8955) Symbolic human activity (#MS8500) Body, form and structure (#MS8700) Mythical fire–dwelling beings (#MS8937) Man–made forms and structures (#MS8600) Animal artifacts and products (#MS8346) Forces, energies and processes (#MS8605) Mythical anthropomorphic beings (#MS8300) Historical, legendary and literary characters (#MS8383).

♦ **MS8425 Atmosphere and celestial symbols**
Description This class of symbols includes mist, fog, rain, rainbow, cloud, thunder, lightning (thunderbolt) Aurora Borealis, storm, hurricane, tornado, wind, whirlwind, hail, snow, darkness, night, sunlight and day. Celestial objects appearing as symbols include Sun, Moon, Saturn, Morning Star, Evening Star, crescent Moon, Mars, comets, stars, constellations, zodiac, the Milky Way, nova, shooting star. Imaginary cosmic objects that appear in symbolism include the axis of the universe, the cosmic egg (the oviform shape of the universe, or of the universe–to–be), boundaries between dimensions, and the vault of heaven.
Connotations These symbols relate to qualities of action, such as agitated, calm, etc; and to circumstances of position, such as preeminence, transitoriness, etc. The moon, for example, indicates change, and also sometimes represents the mind and the subconscious. The zodiac are symbols representing twelve equal divisions of the sky. One of the symbols, Aquarius, or the Water–power, is said to represent an age of two thousand years which humanity is entering, and which is associated with altruism and service to others.
Broader Body, form and structure (#MS8700).

♦ **MS8426 Container and vessels**
Receptacles and receivers
Description This class of symbols is for objects that receive other things inside them, for storage, concealment, protection, ornamentation, or use. It includes boxes, tins, jars, bottles, cans, baskets, bowls, cups, vases, urns, flower–pots, pitchers, casks, barrels, pots, cauldrons, goblets, chalices, chests, wardrobes, closets, coffins, sarcophagi, mummy cases, reliquaries, pyxes, watch cases, pockets, drawers, pigeon–holes, packages, steamer trunks, bags, suitcases, luggage, sacks, covers, ships' holds, refrigerators, garbage cans and pails, ash trays, quivers, knapsacks, backpacks, kits, carrying cases, cigarette cases, wrappers, sleeping bags, tents, canopies, wallets, purses, hand bags, money bags, safety deposit boxes, mail and post office boxes, safes, vaults, crypts and tombs. Another kind of receptacle includes light bulb sockets, electrical outlets and telephone, headphone or microphone jacks. Receivers include any radio–type or telegraph–type devices and energy–storing batteries.
Connotations Some containers simply symbolize their contents, but passivity is the principle connotations that containers and receptacles of all types suggest. On a deeper level they may represent, a complementarity to the object with which they are associated. They often are what makes that object 'work'; i.e. become operative and dynamic, for example, as matter stands to form which impresses it and shapes it, and as body (formed matter) which, while holding energy, may be moved and vitalized by it.
Broader Artifacts (#MS8505).

♦ **MS8433 Pets**
Broader Animals (#MS8215).
Narrower Dog (#MS8853).
Related Animal behaviour (#MS8725).

♦ **MS8435 Terrestrial natural symbols**
Sub–lunar nature
Description Many of the features and objects of nature found on this planet enter into the psychic life as symbols. One sub–class of symbols is all forms of water: oceans and seas, lakes, ponds, pools, rivers, bays; waves, springs, spouts, whirlpools, currents, tides, ripples, rivulets, drops, deluges, dew; waterfalls, floods, fords, rapids (white–water), tidal waves, frost, snowbanks, ice, icicles, avalanches, etc. Water–related symbols are shoals, reefs, sand bars, breaks, beaches, sea floor, islands, icebergs, driftwood, sea shells, grottoes, sea salt and brine. Symbolic land features are volcanoes, mountain peaks, mountains, hills, valleys, fields, meadows, plains, ravines, gullies, defiles, boulders, rocks, sand, mud, desert, quicksand, dunes, caves and caverns, geysers, glaciers, and earthquakes.
Connotations Some of these symbols refer to the positional relationships of height and depth. Volcanoes and most water symbols refer to energy. Obstacles and barriers are symbolized by reefs, sand bars and icebergs. Psychologically, floods may represent overpowering circumstances; waterfalls and rapids represent danger; quicksand, a trap. Grottoes, caves and caverns may indicate the unconscious; the sea, one's life; sea salt, preservation. Mountain peaks, caves, islands, and rivers are archetypal symbols which comment on the progress of the personality towards psychic integration and personal development.
Broader Body, form and structure (#MS8700).

♦ **MS8436 Clothing**
Apparel
Description Clothing represents two levels of symbolism. In the first instance people's choices of styles and colours of clothing, the amount of clothing they wear, and the conspicuous absence or presence of a particular garment, may be symbolically indicative and constitute a third order of communication after verbal and body languages. The second level of symbolism is the subconscious presentation of articles of clothing to the conscious mind, in dream and waking state imagery, and the conscious use of garment symbols in the arts. In the art of the cinema, for example, just as indispensable as the casting director and the make–up man are to the correct symbolic typology of the performers, so also are the costumers and wardrobe department for the symbolically appropriate garments. Typical examples are the clichés of the black hats and masks, and sometimes totally black costumes, of the hero's or heroine's antagonists. Other clichés in general use are smoking jacket or tuxedo, long fur coat, very low–cut evening dress, man's trench or rain–coat with upturned collar, hoods of various shapes and colours, veils, riding boots, capes, deer–stalker hat, pith helmet, riding breeches, tennis clothes, bathing suit, gun–holster, home–maker's apron, negligee, cowboy hat, the livery of butlers, chauffeurs and maids, women's sweaters, Christmas stockings, Indian feather–head–dress, and primitive peoples' loincloths.
Connotations Symbolic apparel can express the wide range of human behaviour. This includes economic behaviour since many jobs and professions are associated with particular clothing articles and these garments come to symbolize the occupation. Examples are: construction worker's hard–hat, miner's head–lamp, British policeman's helmet, doctor's white coat, judge's robes, (formerly) office–workers' eyeshade, and a number of special hats for bakers, chefs, firemen, military personnel, religious dignitaries or believers, artists (beret), prisoners and plutocrats (homburg). The hat is, in fact, probably the single most important item of apparel from the point of view of what it communicates symbolically. Using hats or head garments as the best example, one can point to other levels of symbolism. For example, the image of a hatless person among others with hats may indicate a lack of identity or of rôle, or an inferior position, socially or psychologically. A person with a higher broader, or more expensive hat, or such a hat itself, may indicate authority or dignity. Generally antithetical are broad brims versus narrow or no brims; stiff upright shapes versus soft, flattened ones, dark colours versus bright, and ornamental versus plain, etc. Hat styles of other nationalities or ethnic groups when worn by a non–member may

indicate respect for that group. This is true also of non–working–class people wearing workers clothing; for example, kerchiefs, jeans, rough sweaters and various sorts of caps. It is a subconscious expression of solidarity with the working masses, often affected by university students.
Broader Artifacts (#MS8505).
Narrower Mythical clothing (#MS8925).
Related Race and ethnicity (#MS8395) Trades and professions (#MS8343)
Jewellery and adornments (#MS8446) Social and hierarchical position (#MS8353).

♦ MS8445 Phytomorphic symbols
Fruits and flowers — Vegetation
Description The organic world or vegetable kingdom is the source of mankind's life as well as its environment, therefore symbology from this realm is profuse. Broadly, this class of symbolic objects includes plants and trees predominantly, and parts of these, notably flowers and fruits. In the case of plants generally, it includes roots, shoots, stalks, leaves, and seed–pods; and of grains, the corn or ear. Grasses, vines and bushes are other vegetation that appears in symbology. Another kind of vegetative symbol is the growth stage: seed, seedling or sprout, pod, mature plant, etc; and another is vegetation in the aggregate: jungle, bush, forest, timber wood, grove, stand, orchard, etc, applying to trees; and crop and harvest applying to farm, vineyard or orchard production. Some forms and quantities of packing are also symbolic, as bales (hay, cotton) and barrels, bushels, etc. The anatomy of flora gives other symbology, for example, peel, core, kernel, nut and stem. The taste qualities; sourness, sweetness and spiciness, are symbolized by some fruits and vegetables, for example, by lemons, melons and peppers. There are said to be languages of symbolism provided by some fifteen to twenty kinds of trees, and by a similar number of flowers.
Broader Body, form and structure (#MS8700).
Related Ornamentation (#MS8816) Foods and liquids (#MS8606)
Agricultural symbols (#MS8943).

♦ MS8446 Jewellery and adornments
Medals, decorations and trophies
Description This class of symbols includes civilian jewellery, badges, awards and adornments; military medals and campaign ribbons; trophies of athletics including fishing and hunting; and souvenirs and booty of war.
Connotations Jewellery and adornments symbolize wealth and social status. Civilian badges and awards (lapel pins, brooches, insignia) are symbols of ostentatious virtue. Sport trophies are symbols of psychological compensation for persons or groups with innate low self–esteem. War booty; the enemy's helmet, insignia, equipment, clothing or weapons; represents the animal satisfaction of brutal conquest.
Broader Artifacts (#MS8505).
Related Clothing (#MS8436) Precious things (#MS8516)
Armour and weapons (#MS8106) Trades and professions (#MS8343).

♦ MS8500 Symbolic human activity
Narrower Behaviour (#MS8125) Language and literature (#MS8115).
Related Artifacts (#MS8505) Human roles (#MS8363) Symbolic objects (#MS8326)
Mythological reality (#MS8200) Beings made by beings (#MS8955)
Symbols in sacred literature (#MS8423)
Historical, legendary and literary characters (#MS8383).

♦ MS8505 Artifacts
Articles
Description Man–made objects, along with the human body and its parts, constitute the bulk of symbols. (This sub–class of symbolic artifacts, for convenience, can exclude those feats of construction and engineering represented by buildings and their architectural elements, bridges, monuments, sculpture, roads, walls, gardens, enclosures, towns and cities). Artifacts, human–sized and smaller, very many hand–held, are represented by such things as armour and weapons, food–associated implements and devices, home furnishings and amenities, clothing, jewellery and precious things, games, craftsmen's tools, musical instruments and many others. There is also a class of objects associated with animal husbandry and domestication, and there is a class of objects that are animal artifacts such as spider webs, beehives, bird nests and beaver dams.
Connotations An unusual class of symbols or of symbolic imagery is beings made by beings. To this class belong Galatea, Pinocchio, Dr Frankenstein's monster, the Kabbalists' golem, the alchemists' homunculus, the cyberneticists' robot, the biologists' clone, and according to some, man himself, the symbol of the macrocosm, as symbolized by Leonardo in the image of the limb–outspread man who measures the universe.
Narrower Clothing (#MS8436) Health aids (#MS8406) Ornamentation (#MS8816)
Precious things (#MS8516) Craftsman tools (#MS8102) Symbolic objects (#MS8326)
Religious symbols (#MS8705) Foods and liquids (#MS8606) Mythical artifacts (#MS8915)
Vehicles and parts (#MS8402) Armour and weapons (#MS8106)
Musical instruments (#MS8316) Amusements and games (#MS8846)
Food–related objects (#MS8602) Container and vessels (#MS8426)
Container and vessels (#MS8426) Jewellery and adornments (#MS8446)
Cultural objects and devices (#MS8306) Animal artifacts and products (#MS8346)
Man–made forms and structures (#MS8600) Home furnishings and amenities (#MS8416)
Objects of animal husbandry and domestication (#MS8336)
Objects of animal husbandry and domestication (#MS8336).
Related Mythical clothing (#MS8925) Applied cosmetics (#MS8372)
Race and ethnicity (#MS8395) Mythological reality (#MS8200)
Beings made by beings (#MS8955) Symbols of civilization (#MS8803)
Symbolic human activity (#MS8500) Symbols in sacred literature (#MS8423)
Man–made forms and structures (#MS8600) Structures and engineering works (#MS8815).

♦ MS8513 Serpents
Snake — Reptile
Description Snakes, lizards, crocodiles, alligators and other dragon–like saurians employed as symbols frequently have a spiritual significance. Zeus is one of many gods who have been imaged in serpent form, and the omphalos stone at Apollo's cult site in Delphi was represented as serpent encircled. The healing staff of Askepius was serpent entwined. Moses showed the image of the brazen serpent during the Exodus, and Christ was represented as a serpent by the Ophite Gnostics.
Connotations These instances (a few among very many) indicate a time when the snake was a very important symbol arising out of the psyche. It was called in one tradition, the agathos deaimon, the good spirit, and was associated with mystical development possibly because of its skin–shedding ability. In India, Shakti as Kundalini in the human body has been visualized as the serpent power and yogic literature speaks of rousing this serpent to effect spiritual realization.
Broader Animals (#MS8215) Non–human beings (#MS8400).
Related Religious symbols (#MS8705) Symbols in sacred literature (#MS8423).

♦ MS8516 Precious things
Personal possessions
Description Objects in this class of symbols include single works of contemporary or ancient art or craftsmanship, rare natural objects and single specimens of exotic fauna or rare minerals. Another kind of precious object is one endowed with spiritual power, a human relic or former possession of a spiritually powerful person; or an object removed from a holy place. Related to these are objects said to have fallen from the sky. Personal possessions that have symbolic connotations (other than purely ornamental, expensive jewellery) are wedding bands, pocket and wrist watches, 'lucky pieces' (coin, rabbit's foot, stone, etc), 'charms', handkerchief, pocket mirror, pocket comb, purse or wallet, walking stick (cane), class ring, jacket buttons, umbrella, 'executive' brief case, and photographs. Any personal possession of a deceased family member has symbolic value as well.
Connotations Some of the foregoing represent 'status symbols'; others reflect magical belief in the power of unique objects. The quintessential expression of the latter superstition is the retention of the scalp or skull of one's enemy, or the burial, beneath one's house, of the bodies of the parents or ancestor, or more currently, the keeping of the urn containing the ashes of a cremated loved one, human or animal. According to the theory of magic one may invest an inanimate object with one's life force, e.g. a tree or a statue, and to some extent at least personal possessions do represent one's life.
Broader Artifacts (#MS8505).
Related Money (#MS8397) Jewellery and adornments (#MS8446).

♦ MS8525 Primitive weapons
Broader Armour and weapons (#MS8106).

♦ MS8533 Fine arts
Broader Cultural objects and devices (#MS8306).
Related Cinema (#MS8371) Musical instruments (#MS8316).

♦ MS8600 Man–made forms and structures
Primary visual elements
Broader Artifacts (#MS8505).
Narrower Chromatic symbolism (#MS8923) Shapes and patterns (#MS8845)
Symbolic structures (#MS8825) Agricultural symbols (#MS8943)
Architectural elements (#MS8835) Structures and engineering works (#MS8815).
Related Artifacts (#MS8505) Symbolic objects (#MS8326) Foods and liquids (#MS8606)
Symbols of civilization (#MS8803) Body, form and structure (#MS8700)
Symbols in sacred literature (#MS8423) Incorporeal and formless symbols (#MS8415).

♦ MS8602 Food–related objects
Description This class of objects includes: food harvesting tools and equipment such as scythes, pitchforks, fish nets and tridents, hunters guns, arrows and traps; food production related objects from a plough, to rolling pins and bread pans, peeling and skinning knives and all preparing, cooking and roasting devices such as spits and barbecues, vats, mixers, slicers, pasteurizers and homogenizers. It includes packaging, canneries and all storage from barns to deep–freeze. Food distribution symbols include wheelbarrows, pushcarts, market stalls, food stores, shopping bags and cash–register tape receipts. Food consumption related objects are those found in a home kitchen, as well as restaurants, their equipment and furnishings and personnel.
Connotations Food–related objects, on one level of symbolism are self–referring, to food or particular items of food or activities related to food. On another level, food is a symbol of knowledge or experience, what the self takes in from its environment (the Pauline Epistles use milk and meat in this sense). Thus food–related objects, psychologically, may represent what is used to handle or process knowledge or experience. Sifting, sorting, distilling, mixing, etc, are operations metaphorically extended to the mental sphere. Some food archetypes, that is milk and blood, are present in the deeply structured imagery of the psyche, evoking the associations of life and death.
Broader Artifacts (#MS8505).
Narrower Agricultural symbols (#MS8943).
Related Foods and liquids (#MS8606) Human body fluids (#MS8365)
Vehicles and parts (#MS8402)
Objects of animal husbandry and domestication (#MS8336).

♦ MS8603 Death
Mortality
Description The symbolism for death ranges from such abstractions as darkness to personifications such as the Grim Reaper, the Daena (Zoroastrian angel), and various other soul–collectors, angelic (such as Gabriel) or daimonic. In between abstractions and personifications are an endless number of objects which symbolize death. Notable among them are an open grave, a coffin, a headstone, a churchyard or cemetery, a hearse, a tolling bell, people in black clothes, a crying person, a veiled person, a boat, a carriage, a black horse, a black dog, a black man, a minister, a skeleton, a skull, an urn, a grotto, cave or cavern, a large hall, a banquet room, a corpse, a ghostly spirit, white–clothed figures, shrouded forms, and human and material instruments of execution: noose, guillotine, firing squad, hangman, judge, police officer, etc.
Connotations The bulk of death symbols are oriented to the biological person and evidence the reactions of his incomprehension of mortality. Thus anxiety, fear, menace and mystery characterize the meaning of most death symbols. Death symbols that are more comprehensible to developed personalities include such things as a wreath of flowers, a torch, a door, a corridor, a pathway, and light, from a gentle glow to a refulgent sun–like blaze.
Broader Human body (#MS8305).
Related Human skeleton (#MS8355) Religious symbols (#MS8705)
Armour and weapons (#MS8106) Symbolic structures (#MS8825)
Human spiritual phenomena (#MS8373) Deformed and diseased humans (#MS8375)
Nuclear and mass destruction weapons (#MS8723).

♦ MS8605 Forces, energies and processes
Light — Number — Time
Description This class of symbols represents the dynamics of the perceived world; the powers and laws that guide everything, as well as the most archetypal processes that they are reflected in. Some may appear both as symbols and as the meaning of symbols. As elements, for example, they include: spark, flame, fire, light, radiance, effulgence, colour, heat, energy, motion, power, sound, noise, vibration, wave, electricity, magnetism, atoms, conservation of energy, relativity, inertia, entropy, time, maya, yang–yin, sephiroth, and other physical and theoretical ingredients of the cosmos at its primal levels. As processes they are reflected in: chaos, creation, cosmogony, theogony, anthropogony, origins, evolution, destruction, disintegration, devolution, catastrophe, metamorphosis, change, cycles, ages, growth, nature, the macrocosm, and all physical phenomena. In addition, they characterize some processes qualitatively as the union of opposites, synchronous events, complementarities and causalities. The symbolic iconography of the gods of time sometimes refers to time, energy and universal dynamics in an integrated way. These gods are the Hindu Kala, the Greek Chronos, the Zoroastrian Zurran and the syncretistic Alexandrian god, Aion, a human figure standing on a globe, with a lion's head and a body enclosed by seven coils of a serpent.
Connotations Concrete expressions of these forces and processes are in man's symbolic orientation to time via the months or seasons, and in his symbolic orientation to space via compass

points or directions. Time and space are symbolically divided. In some systems it is by the number three, and three times three; in others, by the number four, and four times four. Also encountered are the divisions of three plus four and three times four, and three plus two and three times two. The factorial numbers, two, three, four, are the archetypal structuring projected into the ontological and experiential dynamics of the universe. Much of traditional symbolism concerning primal laws and energies as an extrapolation of a still more fundamental arithmological symbolism. This is seen in the symbolic system of Pythagoras who accounted for the universe by assigning it the number 10 and showing that it was a product of the Holy Tetraktys, the first four numbers. In terms of human development, wholeness (number 10) may be symbolized as the outcome of four forces (dynamisms or movements) working to produce four states in each of four functions and or 'bodies', for example, the force of 'fire' working to produce the fourth state of consciousness in the 'mental body'. Any number of symbols of space, time and energy may appear in imagery to reflect this work, such as the directions, east, west, north, south, or the times of the seasons, as seen for example in the I Ching symbology (which is based on 2 x 3, and 2 to the 6th power).
Narrower Sacred calendar (#MS8123) Chromatic symbolism (#MS8923) Animal artifacts and products (#MS8346).
Related Deities (#MS8117) Chromatic symbolism (#MS8923) Shapes and patterns (#MS8845) Body, form and structure (#MS8700) Symbols in sacred literature (#MS8423) Incorporeal and formless symbols (#MS8415).

♦ MS8606 Foods and liquids
Chemicals
Description Foods can be symbols. Items of food that are used symbolically include, in the first instance, bread in all its forms. Sometimes it is as an ear or sheaf of grain, usually wheat, rye or maize (corn). Sometimes it is flour, ground meal or rolled dough, and sometimes baked goods. The latter include breads of all types and forms, such as loaves, rolls, bread slices, crusts, crumbs, crackers, breadsticks, pretzels; and also cookies, patisserie, and cakes, simple or elaborate. Among man–made beverages, wine is most employed symbolically, and, to some extent also, fruit juices, syrups, and particular kinds of alcoholic drinks such as champagne, beer, moonshine, fire–water, cognac, highballs and cocktails. To this may be added particular trade–mark beverages, alcoholic and non–alcoholic, such as the cola drinks. A number of food preparations, ways of cooking and particular recipes are used symbolically. Frying, baking, roasting, cooking and boiling are common metaphors. Soup, stew, ragout, hash and omelette are words used figuratively. Related are mince–meat, hamburger and sausage.
Connotations Some, human types are described by meats such as ham, lamb chops, mutton, beef and chicken. Cream–puff, fruit cake and tart are among many baked goods' metaphors in English. A piece of cake may symbolize luxury and ease, also easiness. A multi–tiered cake is associated with marriage. In itself it represents wealth in the form of the archetypal city. The newly married by removing a slice, take their lot or part, symbolically. A number of chemical products and preparations have figurative and symbolic employment in the imagery of language, visual arts, and the unconscious. They include glue; ink, poison, bleach, starch, soap, dye, wax, paint, varnish, white–wash, solvent, machine oil, petrol (gasoline), fuel oil, polish, hair creme, nail lacquer, mouth wash, deodorant, insecticide, perfume, formaldehyde, DDT, TNT, hair dye and plastic.
Broader Artifacts (#MS8505).
Related Health aids (#MS8406) Agricultural symbols (#MS8943) Phytomorphic symbols (#MS8445) Food–related objects (#MS8602) Man–made forms and structures (#MS8600).

♦ MS8613 Classical deities
Broader Deities (#MS8117).

♦ MS8700 Body, form and structure
Narrower Human body (#MS8305) Phytomorphic symbols (#MS8445) Terrestrial natural symbols (#MS8435) Atmosphere and celestial symbols (#MS8425) Incorporeal and formless symbols (#MS8415).
Related Symbols in sacred literature (#MS8423) Man–made forms and structures (#MS8600) Forces, energies and processes (#MS8605).

♦ MS8705 Religious symbols
Holy scriptures — Icon
Description In religious practice the symbol is said to have the function of representing a reality, truth or dogma, with the nature of disclosing its meaning immediately, or gradually, or only if some condition is met, e.g. the disclosure or apperception of a symbol considered to be previous in a sequence of degrees of importance. Religious symbols may range from an object conceived to be identical physically (in whole or in part) or identical essentially with what is represented; through stages of less and less immediacy passing through the modes of sign and symbol to mere allusive reference. The preeminent behavioural religious symbols are generally the sacred rites and liturgies of divine invocation as they are enacted with symbolic gestures, utterances, steps or dance, music, orientation, and other ritual. In ancient Judaism such symbolism included the Temple service, the Passover circumcision and bar–mitzvah. In the pre–Reformation Church, the Mass and its Eucharist were paramount, but both Judaism and Christianity anciently had a priesthood which observed a sacred year filled with holy days and with symbolic rites of almost limitless variety, for everything from breaking bread to stoning or burning heretics. Mardi gras and carnival, Lent and Easter, Advent and Christmas remain major Christian symbols. Religious symbolic artifacts include holy books (Torah, Gospels, Koran, Vedas, Upanishads, Tantras, Sutras, Talmud, New Testament, Hadith, Logia, Books of the Dead, Corpus Hermitium, Desatir, I Ching, etc); altar implements including daggers and cutting knives, tongs and bells and such utensils as pertain to animal slaughter such as cups to hold blood; and torches or candelabra to give light. Pitchers and other vessels for ritual water, beverages and foods are also represented. All such articles whether functional or not have symbolic connotations. Beside altar objects the second most important class of symbolic religious artifacts are those that are pictorial. These include sculpture and bas–relief but are predominantly two–dimensional. Their expression has been in mural, mosaic and painting, with one of the characteristic forms being the icon. Extending the meaning of icon to any pictorial formalized treatment of a limited number of Christian subjects one might cites as examples of symbolic illustration of the Gospels: the Tree of Jesse (Infancy according to Luke); the Baptism, with the Holy Spirit symbolized as a dove descending (as in John); and the four Evangelists themselves: Matthew symbolized by a man, Mark by a lion, Luke by an ox, and John by an eagle. Other symbolic iconic images show Christ as lamb, pelican or phoenix.
Connotations Most religious symbols are commemorative or didactic. They are addressed to the mind of the worshipper, to cause it to recall the founder of their religion, its saints, and its dogmas of punishment and reward. Some religious symbols are designed to awe, to induce a sense of mystery and in some way, convey the essence of the idea of holiness.
Broader Artifacts (#MS8505) Symbols of civilization (#MS8803).
Narrower Sacred calendar (#MS8123) Symbols in sacred literature (#MS8423).
Related Fish (#MS8823) Death (#MS8603) Deities (#MS8117) Serpents (#MS8513) Infernal beings (#MS8907) Masonic symbols (#MS8843) Symbolic objects (#MS8326) Mythical artifacts (#MS8915) Symbolic structures (#MS8825) Beings made by beings (#MS8955) Architectural elements (#MS8835) Human spiritual phenomena (#MS8373) Mythical celestial beings (#MS8337) Mythical terrestrial beings (#MS8387) Deformed and diseased humans (#MS8375) Symbolic quests and journeys (#MS8113) Mythical fire–dwelling beings (#MS8937) Home furnishings and amenities (#MS8416).

♦ MS8713 Birds
Ornithological symbols
Description The diverse habitats, behaviour, and colours of birds allows them to provide a repertoire of symbols, much as the varieties of flowers do. Birds are still emblems for countries and ethnic groups, as, for example seen by the Gallic cock, the American eagle, and by the peacock among some Asian nations.
Connotations The aggressiveness of the cock and his association with the sun via his crowing at day–break; the solitary, high flight of the eagle; and the splendour of the peacock's colouring illustrate the characteristics which provide symbolism. Some birds have in common the fact that they are raptors, such as hawks, ospreys, eagles, and owls but each has distinctive traits. The hawk or falcon is a symbol of targeting, mission or control; the osprey or sea–hawk of discernment; the owl of knowledge gained in his night–study of nature, etc. Notable bird symbols have been the dove representing love, the Holy Spirit, and peace; and the wild gander representing mystical ecstasy. The black raven, associated with Apollo, has been a symbol of arcane wisdom.
Broader Animals (#MS8215) Non–human beings (#MS8400).
Related Animal artifacts and products (#MS8346).

♦ MS8715 Medical instruments
Broader Craftsman tools (#MS8102).
Related Health aids (#MS8406) Trades and professions (#MS8343).

♦ MS8723 Nuclear and mass destruction weapons
Broader Armour and weapons (#MS8106).
Related Death (#MS8603).

♦ MS8725 Animal behaviour
Broader Behaviour (#MS8125).
Related Pets (#MS8433) Animals (#MS8215).

♦ MS8803 Symbols of civilization
Narrower Civil symbols (#MS8223) Symbols of office (#MS8913) Religious symbols (#MS8705) Scientific symbols (#MS8933) Search for knowledge (#MS8963).
Related Artifacts (#MS8505) Language and literature (#MS8115) Cultural objects and devices (#MS8306) Man–made forms and structures (#MS8600).

♦ MS8815 Structures and engineering works
Broader Man–made forms and structures (#MS8600).
Narrower Architectural elements (#MS8835).
Related Artifacts (#MS8505) Symbolic structures (#MS8825).

♦ MS8816 Ornamentation
Grotesques
Description Architectural and sculptural ornaments and ornamental geometrical forms while themselves intended as symbolic or actual representations of objects or beings, may present a second level of symbolic meaning, arising from their style of execution, position, associated historical circumstances, or general class. One large category of such decorative ornaments are the floral grotesques which include garlands, rosettes, baskets of flowers, vines, sheaves of grain, clusters of grapes, diverse flowers and fruits, shrubs, creepers, vines, vineyards, trees, orchards, and gardens. Inanimate objects among the grotesques includes vases and jars, goblets and other drinking vessels, weapons, the cross tools and implements. Other ornaments in this category represent natural and mythic animals including dragons, winged horses, phoenixes, griffins (eagle–lions) and the well–known Gothic gargoyles.
Connotations Ornaments presented in imagery may have an explicit symbology unless the context suggests otherwise. An Ionic capital, for example, may symbolize Hellenic civilization. A dream of a water–spouting gargoyle may be said to presage a gift or inheritance. Bound sheaves of grain (as used in the FAO emblem) signify abundance, and secondarily unity, while the radical significance is of a peaceful harvest. An olive leaf wreath (as used in the UNO emblem) signifies peace. Many other grotesques symbolize plenty or prosperity and domestic tranquillity.
Broader Artifacts (#MS8505).
Related Shapes and patterns (#MS8845) Phytomorphic symbols (#MS8445).

♦ MS8823 Fish
Piscatological symbols
Description Fishes are among the earliest theriomorphic symbols. Fish was anciently among the sacred foods and this recognition of its symbolic value continued to modern times as the recent fish–on–Friday obligation of Roman Catholics. The sacred fish also appears in the Jewish Passover meal.
Connotations In the natural sphere, fishes are symbols of fecundity. In the cognitive sphere they represent the contents of the deep: knowledge and those that acquire knowledge. In the spiritual sphere Christ has been symbolized as a fish, while in Buddhism the fish–drum board calls people to religious practice. Enormous fishes are among mythical animals which have symbolical associations with the world's creation. The archetypal fish that symbolizes spiritual development is the Salmon, who goes against nature in swimming upwards to its source.
Broader Animals (#MS8215) Non–human beings (#MS8400).
Related Religious symbols (#MS8705) Mythical aquatic beings (#MS8917) Symbols in sacred literature (#MS8423).

♦ MS8825 Symbolic structures
Monuments — Statues
Description A number of large artifacts are constructed as symbols. Some are earthworks, some stone. They include cairns and cromlechs, ashlars and altars, menhirs, markers and memorials (in the forms of plain and adorned tombstones or slabs of symbolic geometric shapes, or of sculpture, or of incised stone testimonials, or of a combination of these). This class includes monuments of every kind: pyramids, arches, etc; and megalith constructions.
Connotations Some of the symbols in this class refer to death. Others signify victory. Large, upright stones with no distinctive feature other than their isolation and massiveness correspond to an archetype in the deep subconscious which is sometimes named, Mystery or Mysterium. Another archetype is represented by the maze and labyrinth which symbolizes difficulty, obstacles and uncertainty. Notable stone symbols include the black stone of the Kazba at Mecca, the Omphaos at Delphi, Egyptian obelisks, Jerusalem's Wailing Wall, France's Arc de Triomphe, Constantine's Arch, Berlin's Brandenburg Gate, the Kremlin wall, Lenin's tomb, the Tomb of the Unknown Soldier, etc.
Broader Man–made forms and structures (#MS8600).
Related Death (#MS8603) Religious symbols (#MS8705) Architectural elements (#MS8835) Structures and engineering works (#MS8815).

♦ **MS8835 Architectural elements**
Description This class of symbols comprises parts of structures. In addition to all portals (gateways, doors, windows) the principal symbols in this class are staircases (or stairs), steps, landings, lifts (or elevators), stories or levels, rooms, pillars, halls, corridors, flows and ceilings. Of especial symbolic note are basements, cellars and wine cellars, and storerooms, lumber rooms, attics, lofts and garrets. Other internal spaces include suites, apartments, sanctums, dens, courts or courtyards, studies, bedrooms and parlours. External architectural elements and attachments that are used symbolically include porticos, porches, balconies, logia, columns, roofs, gables, finicles, cupolas and chimneys. Lightning rods, weathervanes, television antennae, pigeon coops and flagpoles are also associated with building symbology, and at street-level, fences, gates and driveways.
Connotations The most important symbols among architectural elements represent how the 'space' of the psyche is divided up, horizontally and vertically. The corridors represent access to the different functions of the mind, and the lifts, levels, stairs and steps, to the lower or higher realms of consciousness. In sacred buildings, the orientation and the ground plane represent the space of the universe, thus offering a map of the numinous to wayfarers. On a smaller scale mediaeval labyrinth patterns traced in the floor of Gothic churches allowed pilgrims to symbolically arrive at 'Jerusalem'.
Broader Man-made forms and structures (#MS8600)
Structures and engineering works (#MS8815).
Related Religious symbols (#MS8705) Shapes and patterns (#MS8845)
Symbolic structures (#MS8825).

♦ **MS8843 Masonic symbols**
Broader Symbolic objects (#MS8326).
Related Craftsman tools (#MS8102) Religious symbols (#MS8705)
Symbolic quests and journeys (#MS8113).

♦ **MS8845 Shapes and patterns**
Design and figures
Description Symbols that appear mainly as geometrical or abstract designs, patterns, shapes, etc, in two-dimensional representation, fall into some general classes. One class for example, is the geometrical, taking in both one and two-dimensional and projections of multi-dimensional figures. This class includes circles, ellipses and polygons of all kinds, and also the straight and the curved line or arc. Each of the shapes of all the letters of the alphabet do not have more than three or four such geometrical elements combined, whether circle, curve or line. (Only three capital letters in the Roman alphabet, excluding serifs have four elements, the rest have less). Another geometrical symbol is the point. From these fundamental elements of points, lines and surfaces can be built up to such symbolic design components as meanders, swastikas, crescents, discs, loops, spirals, curls, beads, pearls, hatching, crosses, tesselations, chequering etc. Partly and wholly geometrical symbolism is found in mandalas and traditional religious art in general. Magic has also employed geometrical sigils, seals and yantras. masons' marks, potters' signs, and alchemists' and early chemists symbols also have been geometrical.
Connotations Geometric figures and designs often symbolize quantity. The square represents the concept of four; the triangle, of three. Some designs refer to time and its properties (as wave-form or cyclical), such as the meander or swastika. Over-all patterns can refer to the macrocosm and to the world of man with some patterns being archetypes of man's relation to the world around him.
Broader Man-made forms and structures (#MS8600).
Related Ornamentation (#MS8816) Architectural elements (#MS8835)
Forces, energies and processes (#MS8605) Home furnishings and amenities (#MS8416).

♦ **MS8846 Amusements and games**
Sports and contests — Hobbies and pastimes
Description This class of symbols arise from human leisure activities. It includes the symbolic imagery of playing and fortune-telling, cards, chess-pieces, the chequer-board, the roulette wheel, dice, ball games of all sorts, water sports, winter sports, hunting and fishing, races of various kinds, contests physical and mental (such as tug-of-war and quizzes), cross-word puzzles, anagrams, palindromes, carnival, masques, mardi-gras, folk rites of numerous sorts (dancing, mummery, masquerades, bonfires), pastimes such as collecting things (butterflies, bugs stuck on pins, string, buttons, postcards, stamps, antiques, coins, art, jewels, miniature cars, sports cars, books, photographs, music, bottles, beer cans, etc). Other leisure activities appearing in symbolism are travelling and exploring, theatre going, reading, television viewing, protesting, and attending courses, meditation sessions and workshops for greater self-development.
Connotations Some of the leisure time activities (contests, sports) symbolize struggle, others (carnival, masquerades) represent liberation from struggle. Collections are symbolically significant by what is collected, the objects representing unfulfilled wishes. Symbolism of travelling, exploring, or going out to museums, theatres or exhibitions reflects extroversion. Symbolism of self-improvement efforts reflects introversion. Symbolism of social activism may represent inner personal conflicts.
Broader Artifacts (#MS8505).

♦ **MS8853 Dog**
Canine presence
Broader Pets (#MS8433) Animals (#MS8215).

♦ **MS8907 Infernal beings**
Demon
Description Symbolic non-human anthropomorphic beings are represented as dwelling in sub-terrestrial regions, in the earth's hollows or in its centre, and below the earth, i.e. in an inferior other-world or dimension.
Connotations The imagery of an anthropomorphic monster who rises from the bowels of the earth is a symbol produced out of the deep subconsciousness of modern man. It signifies the unleashing of a vast, almost limitless power from the inside of life and nature. It is, on the one hand, the power man has produced by pushing down his animal drives which now threaten to go unchecked. It is also the power arising from the dominations of one segment of mankind by another, and is the evil embodiment of that repression. Since all downward pressure meeting resistance creates energy and heat, the infernal regions have been represented often as hot hells tenanted by powerful devils and evil spirits. At the infernal (as opposed to the supernal) world-centre is said to be enthroned the Prince of Evil, who often represents the extreme of deviant moral energy.
Broader Mythical anthropomorphic beings (#MS8300).
Related Deities (#MS8117) Religious symbols (#MS8705)
Mythical theriomorphs (#MS8225) Mythical fire-dwelling beings (#MS8937).

♦ **MS8913 Symbols of office**
Broader Civil symbols (#MS8223) Symbols of civilization (#MS8803).
Related Human roles (#MS8363) Trades and professions (#MS8343)
Social and hierarchical position (#MS8353).

♦ **MS8915 Mythical artifacts**
Description Symbolic objects that do not exist include implements associated with the gods, for example: Zeus' thunderbolt, Thor's hammer, Diana's bow, Cupid's arrow, Athena's spear; and also various other magical items, such as a magical mirror, shield, and sword, an inexhaustible cornucopia, Pandora's box, Orpheus' lyre, Pan's pipes, Neptune's trident, magic kettles, cauldrons and tripods, magic belts, rings and brooches, flying carpets, magic lamps, the Holy Grail, and flying saucers. Another class of mythic artifacts are human-made objects that never existed. These might include the Tower of Babel, Noah's ark, the Trojan Horse, the wings of Icarus, the automata of Daedalus, the Vimanas (air-ships) of the Hindu legends, the ancient city-state of Atlantis, and the mediaeval alchemists' powders of projection, elixirs or transmuted gold.
Connotations Modern examples of mythic human artifacts are alleged secret weapons and the amount of gold bars in Fort Knox, USA. Almost all mythic artifacts presented as symbols represent a wish for power.
Broader Artifacts (#MS8505) Mythological reality (#MS8200).
Related Symbolic objects (#MS8326) Religious symbols (#MS8705)
Mythical theriomorphs (#MS8225) Mythical anthropomorphic beings (#MS8300).

♦ **MS8917 Mythical aquatic beings**
Description Symbolic non-human anthropomorphic beings are represented as living in water. Such symbolic figures include undines, nymphs, Lorelei, aquatic sirens, various green, man-like beings, as well as Mermen (e.g. Triton) and mermaids.
Connotations Some of these mythic creatures represent wisdom (as for example Oannes, the Babylonian fish-man) since they rise from the deep. Others represent the unconscious world of repressed desires and memories which may appear seductive to the conscious ego.
Broader Mythical anthropomorphic beings (#MS8300).
Related Fish (#MS8823) Deities (#MS8117)
Mythical theriomorphs (#MS8225) Mythical terrestrial beings (#MS8387).

♦ **MS8923 Chromatic symbolism**
Colour
Description Colour is what the visual sense assigns to specific, differentiated ranges of electro-magnetic energy in the visible spectrum. There are, therefore, in the entire spectrum more colours than the human eye sees under normal visual conditions and in normal stages of consciousness. This is why the extremes of visibility: indigo, blackness or ultra-violet darkness, and near infra-red fire-glow, or brilliance, symbolize realities beyond ordinary perception. Red, yellow, blue; orange, green, purple; white, brown and black are basic colours. Grey, tan, mauve, olive, aquamarine, flaxen, rose and pink are among lighter hues of these. Among the more brilliant shades the reds predominate with scarlet, vermilion, carmine, crimson, cerise, etc. The brighter colours also include those symbolized by the metallic-named gold, silver, bronze, copper, platinum, and steel; and jewel-named pearl, coral, ruby, emerald, amethyst and sapphire. Other natural symbol colour names derive from the woods; ebony, mahogony chestnut, hazel; the fruits; cherry, lemon, peach, plum, the plants: lavender, heliotrope; the fur, sable, and the fish, salmon. Colour or light conditions that also appear in symbolism include such terms as sombre, livid; dark, bright; clear, murky; intense, diffused; warm, cold, etc; and also shadowy, chiaroscuro, dappled, dusky, mottled, blotched, dingy, variegated, striped, streaked, chequered, flecked, speckled, stippled, dotted, parti-coloured, piebald, marbled, veined and barred. Polychromatic, Kaleidoscopic and tesselated, are also colour symbolism terms. Related to all the foregoing are such conditions as shining, varnished, glazed, polished, glossy, blazing, shimmery, scintillating, flashing, coruscating and dazzling.
Connotations Attempts at colour symbolism anciently began by using the quality of colour of a thing as its symbol: gold colour for real gold, blood-red for blood; green for most leaves and plants, or for grasses, etc. Also the planets appeared with distinctive colours: Mars, reddish; Saturn, pale yellow; Venus coppery, hence also associated with green verdigris, the two colours therefore suitable for its dual aspects as Morning Star and Evening Star; Jupiter (as rain causing), blush; Sun, fiery gold-red or orange; and Moon, quicksilvery, silvery or white. Since, according to ancient religions' astro-theology, these planet-gods ruled all terrestrial affairs, these matters could be symbolized by the colour of the ruling planet. It was thought that there were colour correspondences, such as between the green grains and grasses and Venus, the 'green' Evening Star; between gold metal and the golden Sun; between bloodshed in battle and red Mars; between Shining water, or cattle-milk (or cheese) and the Moon; etc. From this it was a short step to say that fighting and fighters could be associated with red; herding and herders, with green; water-bearers and milk maids with silver or white. Another step in generalization or abstraction provided such ideas as red symbolizes fighting, therefore aggression, therefore the instincts. Green, then, symbolizes collection (harvesting grain, or herding in the pastures) or cultivation, hence acquired skills, culture, or the noos or intellect. Blue signifies rain and precipitation, therefore heaven-to-earth descent, or spirit. Modern colour symbology employs scientific knowledge of colour, that is, of white light being broken into primary and intermediate colours, and of colour mixing. The metaphor of colour is therefore based often on the position of the colour on the spectrum (horizontally); its degree of more or less brightness (vertically); or its resolution or analysis into component colours. Once blue and yellow are assigned symbolic values, green may be assigned to the outcome of combining blue and yellow, for example. All these considerations explain in good part the confusion within and between cultures as to the symbolic meanings colours bear. The assignment of particular colours to be worn as clothing to distinguish economic and social castes and classes still exists. In some Northern countries, for example, a white or light single-coloured shirt is required for male office workers while very expensive shirts affordable usually only by executives, may be striped, or otherwise evidence more colour. The reverse is true for females. Colourful blouses are often suitable for clerical personnel, but supervisors and managers are expected to dress like male clericals. Factory and construction workers, labourers and miners have been characterized by dark colours or by 'blue collars'. Some professions and roles require specific colours: black for judges, executioners, undertakers and bridegrooms, for example; white for angels, cooks, bakers and brides. Although various religious traditions point to ultimate single colours to represent the final alleged stage of human potential there is no general consensus, reasoning on arbitrary systems of symbolic correspondences, or by accepting revelations, as to which colour represents the pinnacle of mystical, other-worldly achievement. It may be, however, that the highest form of personal development for individuals, in the context of the global society that has emerged in people's awareness, could be symbolized by those conditions of colour which may be called polychrome or rainbow. The universality of the appeal of the musician who called the children to follow him was aptly symbolized by his being called, the Pied Piper. Another universal figure is the Pilgrim-Matto dressed in his bright parti-coloured harlequinade who introduces the procession of the Tarot archetypes. In such symbolism of 'all'-colours there may be the deeper symbolism that 'all'-colours are simply a representation of white light, or the integral reality that co-exists with the reality of coloured differentiation; and that if one of the colours is missing there may be no white light.
Broader Man-made forms and structures (#MS8600)
Forces, energies and processes (#MS8605).
Related Race and ethnicity (#MS8395) Forces, energies and processes (#MS8605)
Home furnishings and amenities (#MS8416) Incorporeal and formless symbols (#MS8415).

♦ **MS8925 Mythical clothing**
Magical garments
Description This class of symbolic artifacts includes items worn by gods and supernatural beings, and by men; but which never existed, or, if they existed, never had magical properties. Magical hats and helmets have made beings invincible or invisible. The wizard's and witch's conical head-gear, in some traditions, invested them with their supernatural powers. Magical footwear includes seven-league boots, red shoes that dance by themselves, shoes that could take anyone anywhere by wishing, and winged sandals to fly through the air. Magical veils, when removed, cause all kinds of havoc. Magical cloaks provide invisibility or invincibility. There have been mythical and magical dresses, aprons, girdles, belts, breeches and trousers, gloves, eyeglasses, and (Mary Poppins et al) magical umbrellas.
Connotations Magical and mythical garments symbolically strengthen the functions of the parts of the body they attire. High hats allow for deep thoughts, long shoes for big steps, delicate gloves for heightening the sensation of touch, small eyeglasses to see finely. The most common magical garment in the European cultures is the long white wedding gown, whose billowing form and extended train symbolize fecundity and progeny.
Broader Clothing (#MS8436).
Related Artifacts (#MS8505)   Mythological reality (#MS8200).

♦ **MS8933 Scientific symbols**
Robot — Computer — Network
Description Symbols in science apply to such things as measurement standards, identification of atomic elements and of molecules, and mathematical notation for formulae. Such symbology has evolved from primitive beginnings in proto-science where things or substances were the primary subjects of symbolization, to an ever more increasing use of symbols for processes as the world-view has changed to a dynamic one with the advance of scientific knowledge. Symbols for processes were not able to be contained in systems that only used letters or numbers or additional unitized signs, but had to move out into graphic displays of linked process flows. Flow-diagrams became imperative in electrical distribution engineering, and later in the early stages of the development of the computer, while Boolean algebra, Venn diagrams and class logic found increasing applications. Network analysis of defence programmes (Critical Path, PERT) and communications switching brought graphic symbology still further. These various developments in mathematics, logic and symbology come together in two of mankind's greatest achievements: computer artificial intelligence and computer modelling. The first leads to robotics and the second towards achieving vastly improved methods of managing the world's human and material resources. The dominant symbols of the imminent future are the robot and the world's interlinked communications-computer networks.
Connotations The robot, the computer and the communications network are presently highly manipulated symbols used either to frighten the public or promise them a utopia, Computer-controlled communications and intelligence satellites may evoke images of "being watched".
BX M Scientific symbols Robot Computer Network
Broader Symbols of civilization (#MS8803)   Cultural objects and devices (#MS8306).
Related Craftsman tools (#MS8102)   Trades and professions (#MS8343)
Incorporeal and formless symbols (#MS8415).

♦ **MS8937 Mythical fire-dwelling beings**
Description Symbolic non-human, anthropomorphic beings are represented as dwelling in the heavens, in the air, and in water, but dwellers in fire in human form are few except for the gods and angels of fire and of the sun. Most pantheistic religions, like those of the American Indians, recognized a spirit in the fire, although the spirit appears to have been usually regarded as amorphous.
Connotations The image of a supernatural being burning, standing in fire, or entering or leaving flames in rarely presented as an animate human-formed creature. Salamanders and phoenixes, on the other hand, are well-known theriomorphic images of living fire. An anthropomorphic being associated with flames is symbolic of human energizing at the highest levesl. Some ancient texts such as the mithraic liturgy speak of dwellers in the fire in connection with their initiation rituals. The symbolism of torch-bearers and other flame handlers may be a substitute for fire-dwellers in the old rites. In Christian religious symbolism, dwellers in a spiritual fire are those surrounded by the burning rings of aureoles, mandalas and halos. In the "Golden Bough" the burning of Hercules-Melkarth at Tyre is noted, and his subsequent apotheosis.
Broader Mythical theriomorphs (#MS8225)   Mythical anthropomorphic beings (#MS8300).
Related Deities (#MS8117)   Infernal beings (#MS8907)   Religious symbols (#MS8705)
Symbols in sacred literature (#MS8423).

♦ **MS8943 Agricultural symbols**
Broader Food-related objects (#MS8602)   Man-made forms and structures (#MS8600).
Narrower Objects of animal husbandry and domestication (#MS8336).
Related Craftsman tools (#MS8102)   Foods and liquids (#MS8606)
Phytomorphic symbols (#MS8445)   Trades and professions (#MS8343).

♦ **MS8955 Beings made by beings**
Man and woman
Description Legends, literature and cinema employ the symbolism of beings created by beings. Among the human-authored creatures are Galatea the living statue, the wooden boy Pinocchio, the alchemists' man in a jar (homunculus), the Kabbalists' drone (golem), Dr Frankenstein's monster, the cyberneticists' robot and the biologists' clone. Another group of symbols are beings created or otherwise caused to exist by non-humans, for example, angels or Adam and Eve.
Connotations Human-authored 'creations' represent the epitome of scientific mastery and also the goals of a magus, Faust or cognitive superman. They symbolize a condition of psychic inflation or megalomania, as well as vain effort. They are also closely connected with illusion and delusion. Their existence is usually represented as unintentionally brief, and their functioning unreliable. As this is true of the angels as well (said by some to be created and to expire daily, and by others to include the rebellions host, led by Lucifer) it is interesting that the being man, is said, to have been created by another being, as a symbol of everything in the universe. Treating man consistently with the other symbols in this class, his existence may also be said to be brief and subject to illusion, and his behaviour unreliable. Man's meaning as a symbol may therefore be metamorphosis or universal change since he is a transitory activity himself, seeking form.
Broader Mythological reality (#MS8200).
Related Artifacts (#MS8505)   Human body (#MS8305)   Religious symbols (#MS8705)
Language and literature (#MS8115)   Symbolic human activity (#MS8500)
Symbols in sacred literature (#MS8423)   Mythical anthropomorphic beings (#MS8300).

♦ **MS8963 Search for knowledge**
Broader Behaviour (#MS8125)   Symbols of civilization (#MS8803).
Related Human ages and development (#MS8323)
Cultural objects and devices (#MS8306).

Index

MX

Metaphors and Patterns

Index scope

This index covers all entries in Section M, namely:

- Metaphors (Section MM)

- Patterns of concepts (Section MP)

- Symbols (Section MS)

Note that **all index entries in the following index are integrated into the Volume Index (Section X)** at the end of this volume.

The main intention of this index is to enable users to obtain a better overview of the individual sub-sections of Section M.

All index entries refer via reference numbers (*eg* MP1234) to the descriptions in the preceding sections. The letters indicate the section (*eg* Section MP).

Index entries

The index entries are of several types:

- Principal name or title of concept;

- Secondary or alternative names of concept (including popular expressions and synonyms)

- Keywords from the principal concept name;

- Keywords from the secondary concept names.

No distinction is made between the different types of entry. Keywords are however recognizable by the presence of a semi-colon in the index entry.

Remarks

To facilitate consultation of the index, and to reduce the space requirement prepositions and articles are normally omitted from the index entries.

Although the index was generated automatically, it has been edited manually to eliminate less helpful keywords resulting from the extraction process. Where there was any doubt, less meaningful words have however been left where they may prove to be of value in locating entries in Sections MM, MP or MS.

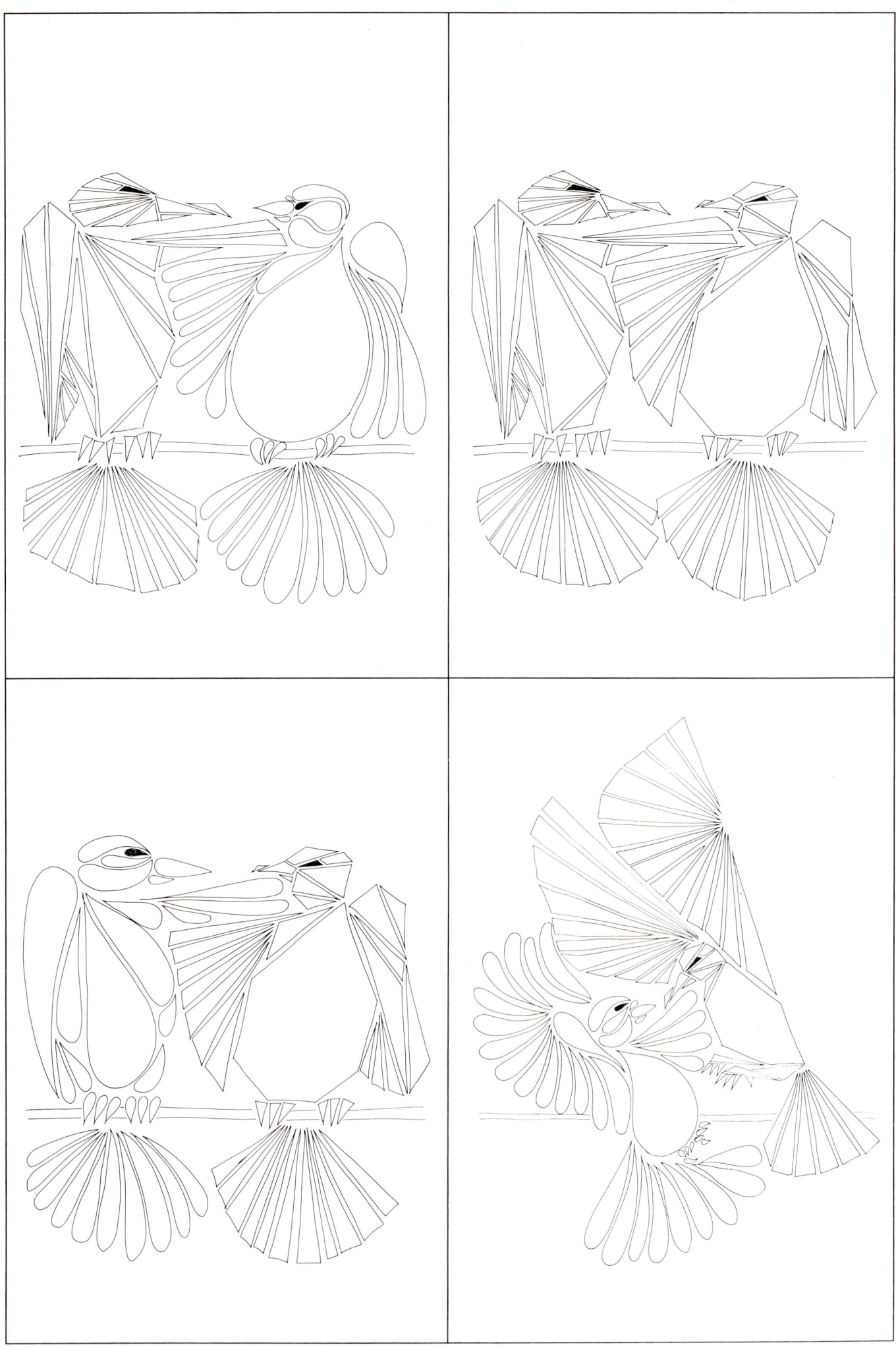

INDEX TO METAPHORS AND PATTERNS

MX

A

MM 2031 Accelerator ; Cyclic resonance
MP 1071 Access contained irrationality
MP 1158 Access higher structural levels ; External
MP 1010 Access intensity
MP 1064 Access patterns active irrationality
MP 1196 Access perspective contexts ; Eccentric
MP 1127 Accessibility ; Nested levels
MP 1201 Accessible facilities perspective adjuncts
MP 1060 Accessible non-linearity
MP 1070 Acknowledgement past perspectives ; Context
MP 1083 Acquisition perspective maintenance dynamics ; Integration perspective
MP 1045 Action network ; Local
MP 1064 Active irrationality ; Access patterns
MP 1165 Activities communication pathway ; Exposure structural
MM 2024 Activities ; Daily round
MP 1157 Activity ; Local opportunities perspective
MP 1030 Activity nodes
MP 1156 Activity ; Opportunities perspectives decreasing
MP 1155 Activity ; Semi autonomous contexts perspectives decreasing
MM 2015 Activity ; Shifting patterns
MS 8500 Activity ; Symbolic human
MP 1153 Adaptable changing number embodied perspectives ; Structures
MP 1079 Adaptable contexts ; Perspective
MP 1048 Adaptive interstices
MP 1035 Adequate variety cyclic elements
MP 1201 Adjuncts ; Accessible facilities perspective
MP 1200 Adjuncts ; Facilities perspective
MP 1198 Adjuncts ; Inter domain contexts perspective
MS 8446 Adornments ; Jewellery
MM 2064 Aerial animal locomotion - flight
MM 2059 Aerial animal locomotion - gliding soaring
MS 8357 Aerial beings ; Mythical
MS 8323 Ages development ; Human
MM 2041 Agricultural development
MS 8943 Agricultural symbols
MS 8338 Aircraft
MS 8395 Alien
MM 2084 Alluring movement
MS 8335 Alteration skin appearance
MP 1243 Ambiguous boundaries
MM 2021 Ambiguous visual illusions
MS 8416 Amenities ; Home furnishings
MS 8846 Amusements games
MS 8337 Angels
MS 8346 Animal artifacts products
MS 8725 Animal behaviour
MS 8133 Animal body parts
MM 2023 Animal-drawn cart
MS 8336 Animal husbandry domestication ; Objects
MM 2073 Animal life cycles
MM 2010 Animal locomotion
MM 2025 Animal locomotion ; Aquatic
MM 2039 Animal locomotion ; Arboreal
MM 2064 Animal locomotion flight ; Aerial
MM 2033 Animal locomotion ; Fossorial
MM 2059 Animal locomotion gliding soaring ; Aerial
MM 2012 Animal locomotion ; Terrestrial
MS 8215 Animals
MS 8300 Anthropomorphic beings ; Mythical
MP 1221 Aperture compatibility
MP 1225 Apertures ; Boundary reinforcement
MS 8436 Apparel
MS 8335 Appearance ; Alteration skin
MS 8372 Applied cosmetics
MP 1182 Appropriate conditions perspective nourishment
MP 1184 Appropriate configuration input processing
MP 1187 Appropriate configuration interaction complementary perspectives
MP 1186 Appropriate configuration perspective inactivity
MP 1185 Appropriate configuration perspective interaction
MP 1207 Appropriate construction elements
MP 1191 Appropriate proportions perspective contexts
MP 1171 Appropriate relationship non-linearity structures
MP 1228 Appropriate superstructure contain transitions levels
MM 2025 Aquatic animal locomotion
MS 8917 Aquatic beings ; Mythical
MM 2039 Arboreal animal locomotion
MS 8835 Architectural elements
MS 8106 Armour weapons
MP 1100 Arrangement structures engender fruitful interfaces
MS 8505 Articles
MS 8505 Artifacts
MS 8915 Artifacts ; Mythical
MS 8346 Artifacts products ; Animal
MS 8533 Arts ; Fine
MS 8425 Atmosphere celestial symbols
MP 1241 Attractive temporary positions
MP 1229 Automatic communications ; Provision pathways
MS 8243 Automobile
MP 1154 Autonomous contexts maturing perspectives ; Semi
MP 1155 Autonomous contexts perspectives decreasing activity ; Semi
MP 1109 Autonomy sub structures ; Organization structure enhance
MP 1032 Axis ; Selective interchange

B

MM 2075 Bad behaviour ; Good
MM 2066 Ball games
MM 2049 Basketry ; Knotting, nets
MM 2076 Battery
MS 8233 Beautiful people
MS 8233 Beauty
MS 8125 Behaviour
MS 8725 Behaviour ; Animal
MM 2075 Behaviour ; Good bad
MS 8955 Beings ; Beings made
MS 8907 Beings ; Infernal
MS 8955 Beings made beings
MS 8357 Beings ; Mythical aerial
MS 8300 Beings ; Mythical anthropomorphic
MS 8917 Beings ; Mythical aquatic
MS 8337 Beings ; Mythical celestial
MS 8937 Beings ; Mythical fire dwelling
MS 8387 Beings ; Mythical terrestrial
MS 8400 Beings ; Non human
MM 2047 Bicycles cycling
MM 2057 Bilingual protocol
MP 1248 Binding ; Time
MS 8713 Birds
MP 1111 Blended integration formal structure informal context
MM 2062 Blindness
MP 1162 Blindspots ; Organization integrative superstructure minimize insight
MS 8304 Boats ships
MS 8365 Body fluids ; Human
MS 8700 Body, form structure
MS 8305 Body ; Human
MS 8133 Body parts ; Animal
MM 2037 Bondage
MS 8355 Bones
MP 1243 Boundaries ; Ambiguous
MP 1218 Boundaries ; Distortion resistant
MP 1224 Boundaries ; Emphasizing transitions across
MP 1234 Boundaries ; Maintainable, multi element external
MP 1235 Boundaries ; Unalienating internal
MM 2014 Boundary ; Container
MP 1015 Boundary ; Context
MP 1211 Boundary expansion permitting new level generation
MP 1212 Boundary orientation ; Primary inter level connections transitions
MP 1225 Boundary reinforcement apertures
MP 1013 Boundary ; Sub domain
MP 1061 Bounded common small-scale interaction domains
MM 2003 Breathing
MP 1114 Broadest ; Hierachy perspectives favouring
MS 8103 Bugs

C

MS 8123 Calendar ; Sacred
MS 8853 Canine presence
MP 1175 Capturing non linear extensions structures ; Insight
MP 1086 Care premature perspectives ; Contexts
MM 2052 Carrot processes ; Stick
MM 2023 Cart ; Animal drawn
MS 8337 Celestial beings ; Mythical
MS 8425 Celestial symbols ; Atmosphere
MP 1126 Centering focal point common domain ; Off
MP 1099 Centre complex structure ; Focal
MM 2026 Ceremony ; Traditional spontaneous
MP 1042 Chain fundamental transformation zones
MM 2013 Chairmanship ; Rotating
MP 1073 Challenging emerging perspectives ; Contexts exploratory relationship formation
MM 2036 Changing fashions
MM 2020 Changing moods individual
MP 1153 Changing number embodied perspectives ; Structures adaptable
MM 2032 Changing physical position
MS 8383 Characters ; Historical, legendary literary
MS 8606 Chemicals
MS 8326 Chrematomorphic symbols
MS 8923 Chromatic symbolism
MS 8371 Cinema
MS 8223 Civil symbols
MS 8803 Civilization ; Symbols
MS 8613 Classical deities
MM 2080 Climbing
MS 8436 Clothing
MS 8925 Clothing ; Mythical
MM 2083 Club
MP 1037 Cluster frameworks
MP 1074 Coherent irrational perspectives ; Interaction
MP 1028 Coherent pattern relationship densities
MS 8923 Colour
MM 2051 Combustion engine ; Internal
MM 2027 Commodities products ; Resources,
MP 1129 Common domain focal point structure
MP 1124 Common domain interfaces ; Ensuring function
MP 1126 Common domain ; Off centering focal point
MP 1067 Common domain ; Unstructured
MP 1063 Common domains ; Cyclic interrelation complementary perspective
MP 1123 Common domains ; Enhancing function
MP 1094 Common domains ; Hospitable
MP 1125 Common domains ; Integrating points perspective
MP 1069 Common external context inactivity
MP 1061 Common small scale interaction domains ; Bounded
MP 1113 Communication media ; Harmoniously structured entry point external
MP 1120 Communication networks ; Focus oriented
MP 1052 Communication networks unmediated relationships ; Interfacing vehicles
MP 1165 Communication pathway ; Exposure structural activities
MP 1140 Communication pathway ; Relative isolation structural interface
MP 1164 Communication pathway structure ; Overview
MP 1112 Communication pathway ; Transition domain structure
MP 1122 Communication pathways ; Direct relationship structures
MP 1174 Communication pathways enfolded non-linearity
MP 1121 Communication pathways ; Hospitability
MP 1132 Communication pathways ; Limiting length
MP 1059 Communication pathways ; Low intensity
MP 1101 Communication pathways ; Protected low density
MP 1173 Communication pathways ; Protecting non linear contexts
MM 2035 Communication ; Radio
MP 1054 Communications ; Intersection differently paced
MP 1229 Communications ; Provision pathways automatic
MP 1020 Communications ; User determined specialized
MP 1221 Compatibility ; Aperture
MP 1023 Compensating relationships parallel
MP 1072 Competitive interaction opportunities transposed concrete level
MP 1027 Complementarity
MP 1003 Complementary modes organization ; Interpretation
MP 1063 Complementary perspective common domains ; Cyclic interrelation
MP 1187 Complementary perspectives ; Appropriate configuration interaction
MP 1077 Complementary perspectives ; Minimal context
MP 1136 Complementary perspectives ; Relative isolation
MP 1106 Complementary structures ; Functional enhancement domains separating
MP 1099 Complex structure ; Focal centre
MP 1102 Complex structures ; Distinct pattern entry points
MP 1098 Complex structures ; Patterning
MP 1095 Complexification perspective contexts
MM 2065 Composition
MS 8933 Computer
MP 1097 Concealment necessary monotonous perspective patterns
MP 1072 Concrete level ; Competitive interaction opportunities transposed
MP 1182 Conditions perspective nourishment ; Appropriate
MM 2016 Conduit
MP 1184 Configuration input processing ; Appropriate
MP 1187 Configuration interaction complementary perspectives ; Appropriate
MP 1186 Configuration perspective inactivity ; Appropriate
MP 1185 Configuration perspective interaction ; Appropriate
MM 2056 Configurations ; Strategic
MS 8333 Congenital relationship ; Human
MP 1205 Congruence spaces defined framework spaces defined processes it
MP 1194 Connectedness domains ; Internal
MP 1237 Connectedness isolation
MP 1232 Connection encompassing domains ; Symbolic
MP 1213 Connections ; Distribution secondary inter level
MP 1216 Connections ; Inter level
MP 1212 Connections transitions boundary orientation ; Primary inter level
MP 1148 Constraints ; Perspective interaction
MP 1207 Construction elements ; Appropriate
MP 1228 Contain transitions levels ; Appropriate superstructure
MP 1119 Contained interfaces ; Partially
MP 1071 Contained irrationality ; Access
MM 2014 Container boundary
MS 8426 Container vessels
MP 1117 Containment integrative superstructure
MM 2072 Containment plasma ; Magnetic
MS 8846 Contests ; Sports
MP 1070 Context acknowledgement past perspectives
MP 1111 Context ; Blended integration formal structure informal
MP 1015 Context boundary
MP 1077 Context complementary perspectives ; Minimal
MP 1058 Context disorder
MP 1141 Context each perspective ; Relatively isolated
MP 1065 Context emergence new perspectives
MP 1199 Context external insight ; Exposure input processing
MP 1041 Context formal processes ; Informal
MP 1189 Context forms perspective presentation
MP 1014 Context ; Identifiable
MP 1069 Context inactivity ; Common external
MP 1246 Context ; Integration
MP 1076 Context interrelating mature emerging perspectives ; Minimal
MP 1183 Context ; Partially exposed perspective
MP 1090 Context perspective exchange ; Unstructured
MP 1078 Context single perspective ; Minimal
MP 1066 Context transformative experience
MP 1191 Contexts ; Appropriate proportions perspective
MP 1086 Contexts care premature perspectives
MP 1173 Contexts communication pathways ; Protecting non linear
MP 1095 Contexts ; Complexification perspective
MP 1087 Contexts controlled single perspective ; Exchange
MP 1143 Contexts developing perspectives ; Interrelationship
MP 1085 Contexts developing perspectives ; Perspective imitation
MP 1196 Contexts ; Eccentric access perspective
MP 1204 Contexts ; Exclusive

MP 1073 Contexts exploratory relationship formation challenging emerging perspectives
MP 1154 Contexts maturing perspectives ; Semi autonomous
MP 1082 Contexts ; Minimal distance related operational control
MP 1192 Contexts ; Overview external
MP 1152 Contexts ; Partially isolated
MP 1079 Contexts ; Perspective adaptable
MP 1198 Contexts perspective adjuncts ; Inter domain
MP 1080 Contexts perspective dynamics ; Integrated
MP 1084 Contexts perspective reorganization ; Transitional
MP 1155 Contexts perspectives decreasing activity ; Semi autonomous
MP 1091 Contexts perspectives transition ; Hospitable
MP 1172 Contexts self-organizing non-linearity
MP 1151 Contexts ; Small scale perspective interaction
MP 1190 Contexts ; Variation size perspective
MP 1169 Contextual levels ; Maintaining distinctions
MP 1217 Continuity ; Perimeter
MM 2050 Contrast significance
MP 1082 Control contexts ; Minimal distance related operational
MP 1081 Control operations ; Minimally structured perspective
MP 1087 Controlled single perspective ; Exchange contexts
MM 2011 Conversation ; Shifting topics
MM 2055 Cooking ; Food, nutrition
MP 1144 Coordinated exposure irrationality structures ; Integrating
MP 1147 Coordination perspective nourishment
MS 8372 Cosmetics ; Applied
MS 8102 Craftsman tools
MS 8215 Creatures
MM 2094 Crop rotation
MP 1004 Cultivation areas ; Regenerative resource
MP 1170 Cultivation productive non-linearity
MP 1177 Cultivation sources perspective nourishment ; Local
MS 8306 Cultural objects devices
MP 1145 Current elements those reserve ; Domains non
MP 1026 Cycle ; Functional
MP 1031 Cycle relationship reinforcement
MM 2073 Cycles ; Animal life
MM 2092 Cycles ; Mechanical
MM 2068 Cycles ; Metabolic pathways
MM 2048 Cycles ; Psycho symbolic
MM 2046 Cycles religious festivals
MP 1035 Cyclic elements ; Adequate variety
MP 1063 Cyclic interrelation complementary perspective common domains
MM 2044 Cyclic migration
MM 2031 Cyclic resonance accelerator
MM 2047 Cycling ; Bicycles

D

MM 2024 Daily round activities
MM 2067 Dance
MS 8415 Darkness
MS 8603 Death
MP 1009 Decentralized formal processes
MS 8446 Decorations trophies ; Medals,
MP 1156 Decreasing activity ; Opportunities perspectives
MP 1155 Decreasing activity ; Semi autonomous contexts perspectives
MP 1104 Deficiencies pattern harmony ; Structural development designed counteract
MP 1205 Defined framework spaces defined processes it ; Congruence spaces
MP 1205 Defined processes it ; Congruence spaces defined framework spaces
MP 1208 Definition ; Progressive framework
MS 8375 Deformed diseased humans
MS 8117 Deities
MS 8613 Deities ; Classical
MS 8907 Demon
MP 1028 Densities ; Coherent pattern relationship
MP 1101 Density communication pathways ; Protected low
MP 1036 Density ; Differentiation relationship
MP 1029 Density gradient local relationships ; Stable
MS 8845 Design figures
MP 1104 Designed counteract deficiencies pattern harmony ; Structural development
MS 8723 Destruction weapons ; Nuclear mass
MP 1137 Developing perspectives ; Domain
MP 1143 Developing perspectives ; Interrelationship contexts
MP 1085 Developing perspectives ; Perspective imitation contexts
MM 2041 Development ; Agricultural
MP 1104 Development designed counteract deficiencies pattern harmony ; Structural
MS 8323 Development ; Human ages
MM 2009 Development ; Interruption thematic
MS 8306 Devices ; Cultural objects
MS 8102 Devices ; Domestic
MM 2022 Diets ; Media
MP 1251 Different settings
MP 1036 Differentiation relationship density
MP 1054 Differently paced communications ; Intersection
MP 1040 Dimension ; Integrating historical
MP 1039 Dimension ; Integrating new
MP 1043 Dimensions ; Presentation new
MM 2082 Dining
MP 1181 Direct insight structures focal point ; Maintenance source
MP 1122 Direct relationship structures communication pathways
MS 8338 Dirigibles
MM 2018 Disease
MS 8375 Diseased humans ; Deformed
MP 1058 Disorder ; Context
MP 1236 Displaceable frameworks
MP 1082 Distance related operational control contexts ; Minimal
MP 1102 Distinct pattern entry points complex structures
MP 1169 Distinctions contextual levels ; Maintaining
MP 1206 Distinctions ; Efficient enclosure spaces minimal structural
MP 1197 Distinctions separating domains ; Substantive
MP 1110 Distinctiveness main entry point structure
MP 1218 Distortion resistant boundaries
MP 1210 Distribution levels ; Harmonizing space
MP 1002 Distribution organization
MP 1213 Distribution secondary inter-level connections
MP 1046 Diversified interchange environment
MS 8117 Divinities
MS 8853 Dog
MP 1013 Domain boundary ; Sub
MP 1198 Domain contexts perspective adjuncts ; Inter
MP 1137 Domain developing perspectives
MP 1131 Domain dynamics ; Organization inter
MP 1163 Domain external structures ; Hospitable non linear
MP 1129 Domain focal point structure ; Common
MP 1124 Domain interfaces ; Ensuring function common
MP 1126 Domain ; Off centering focal point common
MP 1146 Domain organization ; Flexible
MP 1139 Domain perspective nourishment ; Hospitable
MP 1112 Domain structure communication pathway ; Transition
MP 1161 Domain ; Structure enfolded insight
MP 1067 Domain ; Unstructured common
MP 1061 Domains ; Bounded common small scale interaction
MP 1063 Domains ; Cyclic interrelation complementary perspective common
MP 1123 Domains ; Enhancing function common
MP 1115 Domains ; Functional integration unstructured internal
MP 1094 Domains ; Hospitable common
MP 1001 Domains ; Independent
MP 1252 Domains insight
MP 1125 Domains ; Integrating points perspective common
MP 1166 Domains interfaces structure external environment ; Overview
MP 1194 Domains ; Internal connectedness
MP 1011 Domains ; Local interrelationship
MP 1167 Domains minimum proportions ; Enfolded overview
MP 1145 Domains non-current elements those reserve
MP 1193 Domains ; Partially enclosed internal
MP 1128 Domains receive external insight ; Orientation
MP 1106 Domains separating complementary structures ; Functional enhancement
MP 1197 Domains ; Substantive distinctions separating
MP 1232 Domains ; Symbolic connection encompassing
MP 1233 Domains ; Zoning internal
MS 8102 Domestic devices
MS 8336 Domestication ; Objects animal husbandry
MM 2081 Drawing water
MM 2023 Drawn cart ; Animal
MM 2060 Driving
MP 1080 Dynamics ; Integrated contexts perspective
MP 1083 Dynamics ; Integration perspective acquisition perspective maintenance
MP 1131 Dynamics ; Organization inter domain

E

MP 1196 Eccentric access perspective contexts
MP 1206 Efficient enclosure spaces minimal structural distinctions
MM 2043 Electric motors generators
MP 1234 Element external boundaries ; Maintainable, multi
MP 1035 Elements ; Adequate variety cyclic
MP 1207 Elements ; Appropriate construction
MS 8835 Elements ; Architectural
MS 8600 Elements ; Primary visual
MP 1145 Elements those reserve ; Domains non current
MP 1247 Embedding fixity variability
MP 1153 Embodied perspectives ; Structures adaptable changing number
MP 1065 Emergence new perspectives ; Context
MP 1203 Emergent perspectives ; Exclusive spaces
MP 1138 Emerging external insight ; Receptivity
MP 1057 Emerging foci ; Protection
MP 1073 Emerging perspectives ; Contexts exploratory relationship formation challenging
MP 1076 Emerging perspectives ; Minimal context interrelating mature
MP 1068 Emerging perspectives ; Spontaneous relationship formation amongst
MP 1230 Emotion ; Unmediated supportive
MP 1250 Emphases ; Encouraging
MP 1224 Emphasizing transitions across boundaries
MP 1024 Enabling transcendence ; Positions
MP 1017 Encirclement
MP 1193 Enclosed internal domains ; Partially
MP 1206 Enclosure spaces minimal structural distinctions ; Efficient
MP 1232 Encompassing domains ; Symbolic connection
MP 1250 Encouraging emphases
MS 8605 Energies processes ; Forces,
MP 1161 Enfolded insight domain ; Structure
MP 1174 Enfolded non linearity ; Communication pathways
MP 1202 Enfolded occupiable sites ; Structure
MP 1167 Enfolded overview domains minimum proportions
MP 1100 Engender fruitful interfaces ; Arrangement structures
MM 2051 Engine ; Internal combustion
MS 8815 Engineering works ; Structures
MP 1109 Enhance autonomy sub structures ; Organization structure
MP 1245 Enhance fixity ; Protecting variability
MP 1107 Enhance receptivity external insight ; Organization structures
MP 1105 Enhance receptivity external insight ; Orientation structures
MP 1051 Enhanced non linear processes ; Linear relationships
MP 1106 Enhancement domains separating complementary structures ; Functional
MP 1047 Enhancement ; Functionality
MP 1123 Enhancing function common domains
MP 1135 Enhancing insight varying levels exposure it
MP 1130 Enhancing structural entry points ; Internal transition spaces
MP 1124 Ensuring function common domain interfaces
MM 2004 Entities
MP 1050 Entrainment ; Three way relationship
MP 1113 Entry point external communication media ; Harmoniously structured
MP 1110 Entry point structure ; Distinctiveness main
MP 1102 Entry points complex structures ; Distinct pattern
MP 1130 Entry points ; Internal transition spaces enhancing structural
MP 1053 Entry ; Principal points
MP 1046 Environment ; Diversified interchange
MP 1160 Environment ; Hospitable interface structures external
MP 1166 Environment ; Overview domains interfaces structure external
MS 8102 Equipment ; Professional
MS 8395 Ethnicity ; Race
MP 1087 Exchange contexts controlled single perspective
MP 1090 Exchange ; Unstructured context perspective
MP 1204 Exclusive contexts
MP 1203 Exclusive spaces emergent perspectives
MS 8365 Excreta
MM 2005 Exercise rest
MP 1211 Expansion permitting new level generation ; Boundary
MP 1066 Experience ; Context transformative
MP 1073 Exploratory relationship formation challenging emerging perspectives ; Contexts
MP 1180 Exposed external insight ; Occupiable sites
MP 1183 Exposed perspective context ; Partially
MP 1134 Exposure harmonious perspectives ; Limiting
MP 1199 Exposure input processing context external insight
MP 1144 Exposure irrationality structures ; Integrating coordinated
MP 1135 Exposure it ; Enhancing insight varying levels
MP 1165 Exposure structural activities communication pathway
MP 1075 Extended pattern nuclear interaction
MP 1175 Extensions structures ; Insight capturing non linear
MP 1158 External access higher structural levels
MP 1234 External boundaries ; Maintainable, multi element
MP 1113 External communication media ; Harmoniously structured entry point
MP 1069 External context inactivity ; Common
MP 1192 External contexts ; Overview
MP 1160 External environment ; Hospitable interface structures
MP 1166 External environment ; Overview domains interfaces structure
MP 1199 External insight ; Exposure input processing context
MP 1180 External insight ; Occupiable sites exposed
MP 1107 External insight ; Organization structures enhance receptivity
MP 1159 External insight ; Organization structures permit two sources
MP 1128 External insight ; Orientation domains receive
MP 1105 External insight ; Orientation structures enhance receptivity
MP 1138 External insight ; Receptivity emerging
MS 8325 External organs ; Human
MP 1149 External perspectives structures ; Hospitable reception
MP 1163 External structures ; Hospitable non linear domain

F

MS 8315 Face ; Human form
MP 1239 Faceted frameworks ; Multi
MP 1200 Facilities perspective adjuncts
MP 1201 Facilities perspective adjuncts ; Accessible
MS 8357 Fairies
MS 8235 Fairy tales ; Stories
MS 8393 Family
MS 8333 Family
MM 2036 Fashions ; Changing
MP 1114 Favouring broadest ; Hierachy perspectives
MS 8315 Feet ; Hands
MS 8385 Femininity
MM 2046 Festivals ; Cycles religious
MS 8845 Figures ; Design
MP 1238 Filtered insights
MS 8533 Fine arts
MS 8937 Fire dwelling beings ; Mythical
MS 8823 Fish
MP 1245 Fixity ; Protecting variability enhance
MP 1247 Fixity variability ; Embedding
MP 1146 Flexible domain organization
MP 1244 Flexible interfaces
MM 2064 Flight ; Aerial animal locomotion
MS 8445 Flowers ; Fruits
MS 8365 Fluids ; Human body
MP 1099 Focal centre complex structure
MP 1126 Focal point common domain ; Off centering
MP 1181 Focal point ; Maintenance source direct insight structures

INDEX TO METAPHORS AND PATTERNS

MX

MP 1129 Focal point structure ; Common domain
MP 1044 Focal points ; Local
MP 1057 Foci ; Protection emerging
MP 1120 Focus-oriented communication networks
MM 2055 Food, nutrition cooking
MS 8602 Food-related objects
MS 8606 Foods liquids
MM 2037 Forced labour
MS 8605 Forces, energies processes
MS 8395 Foreigness
MS 8315 Form face ; Human
MS 8700 Form structure ; Body,
MP 1009 Formal processes ; Decentralized
MP 1041 Formal processes ; Informal context
MP 1111 Formal structure informal context ; Blended integration
MP 1068 Formation amongst emerging perspectives ; Spontaneous relationship
MP 1073 Formation challenging emerging perspectives ; Contexts exploratory relationship
MP 1215 Formation ; Initial level
MS 8415 Formless symbols ; Incorporeal
MP 1189 Forms perspective presentation ; Context
MP 1008 Forms processes ; Variety
MS 8600 Forms structures ; Man made
MM 2033 Fossorial animal locomotion
MP 1021 Four-level structural limit
MP 1208 Framework definition ; Progressive
MP 1205 Framework spaces defined processes it ; Congruence spaces defined
MP 1195 Framework transition structural levels
MP 1037 Frameworks ; Cluster
MP 1236 Frameworks ; Displaceable
MP 1239 Frameworks ; Multi faceted
MP 1038 Frameworks ; Standard
MP 1100 Fruitful interfaces ; Arrangement structures engender
MS 8445 Fruits flowers
MP 1124 Function common domain interfaces ; Ensuring
MP 1123 Function common domains ; Enhancing
MP 1026 Functional cycle
MP 1106 Functional enhancement domains separating complementary structures
MP 1115 Functional integration unstructured internal domains
MP 1047 Functionality enhancement
MP 1042 Fundamental transformation zones ; Chain
MS 8416 Furnishings amenities ; Home

G

MM 2030 Gaining/losing initiative
MS 8846 Games ; Amusements
MM 2066 Games ; Ball
MM 2077 Games ; Shell
MM 2040 Gamesmanship
MS 8925 Garments ; Magical
MS 8385 Gender ; Human
MP 1016 General interrelationships ; Web
MP 1211 Generation ; Boundary expansion permitting new level
MP 1219 Generation minimum tension ; Level
MM 2043 Generators ; Electric motors
MM 2070 Geography movement
MM 2081 Getting water
MM 2059 Gliding soaring ; Aerial animal locomotion
MM 2075 Good-bad behaviour
MP 1029 Gradient local relationships ; Stable density
MS 8816 Grotesques
MP 1222 Ground-level visibility
MP 1168 Grounded structures
MP 1176 Grounding perspectives non linearity ; Sites

H

MS 8365 Hair nails
MS 8315 Hands feet
MP 1134 Harmonious perspectives ; Limiting exposure
MP 1113 Harmoniously structured entry point external communication media
MP 1210 Harmonizing space distribution levels
MP 1104 Harmony ; Structural development designed counteract deficiencies pattern
MS 8406 Health aids
MP 1114 Hierachy perspectives favouring broadest
MS 8353 Hierarchical position ; Social
MP 1158 Higher structural levels ; External access
MP 1040 Historical dimension ; Integrating
MS 8383 Historical, legendary literary characters
MS 8846 Hobbies pastimes
MS 8705 Holy scriptures
MS 8416 Home furnishings amenities
MP 1121 Hospitability communication pathways
MP 1094 Hospitable common domains
MP 1091 Hospitable contexts perspectives transition
MP 1139 Hospitable domain perspective nourishment
MP 1160 Hospitable interface structures external environment
MP 1163 Hospitable non-linear domain external structures
MP 1149 Hospitable reception external perspectives structures
MP 1092 Hospitable transit points
MS 8500 Human activity ; Symbolic
MS 8323 Human ages development
MS 8400 Human beings ; Non
MS 8305 Human body
MS 8365 Human body fluids
MS 8333 Human congenital relationship
MS 8325 Human external organs
MS 8315 Human form face
MS 8385 Human gender
MS 8345 Human internal organs
MS 8313 Human plurality
MS 8303 Human presence ; Signs
MS 8363 Human roles
MS 8355 Human skeleton
MS 8373 Human spiritual phenomena
MS 8375 Humans ; Deformed diseased
MS 8336 Husbandry domestication ; Objects animal
MS 8225 Hybrid monsters

I

MS 8705 Icon
MS 8415 Ideas
MP 1014 Identifiable context
MM 2021 Illusions ; Ambiguous visual
MP 1085 Imitation contexts developing perspectives ; Perspective
MP 1186 Inactivity ; Appropriate configuration perspective
MP 1069 Inactivity ; Common external context
MP 1188 Inactivity ; Occupiable sites perspective
MP 1150 Inactivity ; Provision temporary perspective
MS 8415 Incorporeal formless symbols
MP 1001 Independent domains
MP 1025 Indeterminacy ; Relationship
MM 2020 Individual ; Changing moods
MP 1012 Individuality multiplicity
MS 8907 Infernal beings
MP 1111 Informal context ; Blended integration formal structure
MP 1041 Informal context formal processes
MP 1088 Informal local perspective interface zones
MP 1214 Infrastructure ; Integrative
MP 1215 Initial level formation
MM 2030 Initiative ; Gaining/losing
MP 1184 Input processing ; Appropriate configuration
MP 1199 Input processing context external insight ; Exposure
MS 8103 Insects
MP 1162 Insight blindspots ; Organization integrative superstructure minimize
MP 1175 Insight capturing non-linear extensions structures
MP 1161 Insight domain ; Structure enfolded
MP 1252 Insight ; Domains
MP 1199 Insight ; Exposure input processing context external
MP 1180 Insight ; Occupiable sites exposed external
MP 1107 Insight ; Organization structures enhance receptivity external
MP 1159 Insight ; Organization structures permit two sources external
MP 1128 Insight ; Orientation domains receive external
MP 1105 Insight ; Orientation structures enhance receptivity external
MP 1138 Insight ; Receptivity emerging external
MP 1181 Insight structures focal point ; Maintenance source direct
MP 1135 Insight varying levels exposure it ; Enhancing
MP 1223 Insight ; Zones intermediate
MP 1238 Insights ; Filtered
MS 8715 Instruments ; Medical
MS 8316 Instruments ; Musical
MP 1080 Integrated contexts perspective dynamics
MP 1144 Integrating coordinated exposure irrationality structures
MP 1040 Integrating historical dimension
MP 1039 Integrating new dimension
MP 1125 Integrating points perspective common domains
MP 1133 Integrating transition pathways levels structure
MP 1246 Integration context
MP 1111 Integration formal structure informal context ; Blended
MP 1118 Integration non-linearity integrative superstructure
MP 1083 Integration perspective acquisition perspective maintenance dynamics
MP 1178 Integration rejected perspective products ; Re
MP 1220 Integration superstructure
MP 1249 Integration ; Symbols
MP 1115 Integration unstructured internal domains ; Functional
MP 1214 Integrative infrastructure
MP 1117 Integrative superstructure ; Containment
MP 1118 Integrative superstructure ; Integration non linearity
MP 1162 Integrative superstructure minimize insight blindspots ; Organization
MP 1209 Integrative superstructure ; Organization
MP 1231 Integrative superstructure ; Overview sites
MP 1116 Integrative superstructure ; Patterning
MP 1227 Integrity ; Inter level
MP 1010 Intensity ; Access
MP 1059 Intensity communication pathways ; Low
MP 1055 Intensity relationships ; Protected low
MP 1198 Inter-domain contexts perspective adjuncts
MP 1131 Inter domain dynamics ; Organization
MP 1216 Inter-level connections
MP 1213 Inter level connections ; Distribution secondary
MP 1212 Inter level connections transitions boundary orientation ; Primary
MP 1227 Inter-level integrity
MP 1226 Inter-level zone
MP 1005 Inter relationships ; Network
MP 1185 Interaction ; Appropriate configuration perspective
MP 1074 Interaction coherent irrational perspectives
MP 1187 Interaction complementary perspectives ; Appropriate configuration
MP 1148 Interaction constraints ; Perspective
MP 1151 Interaction contexts ; Small scale perspective
MP 1061 Interaction domains ; Bounded common small scale
MP 1075 Interaction ; Extended pattern nuclear
MM 2017 Interaction ; Interpersonal
MP 1072 Interaction opportunities transposed concrete level ; Competitive
MP 1034 Interchange
MP 1032 Interchange axis ; Selective
MP 1046 Interchange environment ; Diversified
MP 1019 Interchange ; Web selective
MP 1108 Interconnected structures
MM 2034 Intercourse ; Sexual
MP 1140 Interface communication pathway ; Relative isolation structural
MP 1160 Interface structures external environment ; Hospitable
MP 1088 Interface zones ; Informal local perspective
MP 1100 Interfaces ; Arrangement structures engender fruitful
MP 1124 Interfaces ; Ensuring function common domain
MP 1244 Interfaces ; Flexible
MP 1119 Interfaces ; Partially contained
MP 1166 Interfaces structure external environment ; Overview domains
MP 1240 Interfaces ; Tolerance level
MP 1052 Interfacing vehicles communication networks unmediated relationships
MP 1223 Intermediate insight ; Zones
MP 1242 Intermediate position
MP 1006 Intermediate scale organization
MP 1235 Internal boundaries ; Unalienating
MM 2051 Internal combustion engine
MP 1194 Internal connectedness domains
MP 1115 Internal domains ; Functional integration unstructured
MP 1193 Internal domains ; Partially enclosed
MP 1233 Internal domains ; Zoning
MS 8345 Internal organs ; Human
MP 1130 Internal transition spaces enhancing structural entry points
MM 2017 Interpersonal interaction
MP 1003 Interpretation complementary modes organization
MP 1076 Interrelating mature emerging perspectives ; Minimal context
MP 1063 Interrelation complementary perspective common domains ; Cyclic
MP 1143 Interrelationship contexts developing perspectives
MP 1011 Interrelationship domains ; Local
MP 1016 Interrelationships ; Web general
MM 2009 Interruption thematic development
MP 1054 Intersection differently paced communications
MP 1048 Interstices ; Adaptive
MP 1074 Irrational perspectives ; Interaction coherent
MP 1071 Irrationality ; Access contained
MP 1064 Irrationality ; Access patterns active
MP 1144 Irrationality structures ; Integrating coordinated exposure
MM 2081 Irrigating
MP 1141 Isolated context each perspective ; Relatively
MP 1152 Isolated contexts ; Partially
MP 1136 Isolation complementary perspectives ; Relative
MP 1237 Isolation ; Connectedness
MP 1140 Isolation structural interface communication pathway ; Relative

J

MS 8446 Jewellery adornments
MM 2074 Journey ; Path
MS 8113 Journeys ; Symbolic quests
MM 2069 Juggling

K

MS 8333 Kinship
MM 2049 Knotting, nets basketry
MS 8963 Knowledge ; Search

L

MM 2037 Labour ; Forced
MS 8115 Language literature
MM 2087 Launching spacecraft
MS 8383 Legendary literary characters ; Historical,
MP 1132 Length communication pathways ; Limiting
MP 1072 Level ; Competitive interaction opportunities transposed concrete
MP 1213 Level connections ; Distribution secondary inter
MP 1216 Level connections ; Inter
MP 1212 Level connections transitions boundary orientation ; Primary inter
MP 1215 Level formation ; Initial
MP 1211 Level generation ; Boundary expansion permitting new
MP 1219 Level generation minimum tension
MP 1227 Level integrity ; Inter
MP 1240 Level interfaces ; Tolerance
MP 1021 Level structural limit ; Four
MP 1222 Level visibility ; Ground
MP 1226 Level zone ; Inter
MP 1127 Levels accessibility ; Nested
MP 1228 Levels ; Appropriate superstructure contain transitions
MP 1135 Levels exposure it ; Enhancing insight varying
MP 1158 Levels ; External access higher structural
MP 1195 Levels ; Framework transition structural
MP 1210 Levels ; Harmonizing space distribution
MP 1096 Levels ; Limitation number structural
MP 1169 Levels ; Maintaining distinctions contextual
MP 1133 Levels structure ; Integrating transition pathways
MM 2083 Lever
MM 2073 Life cycles ; Animal
MS 8605 Light
MP 1021 Limit ; Four level structural

MP 1022 Limit ; Occupiable temporary site
MP 1103 Limitation number occupiable temporary sites
MP 1096 Limitation number structural levels
MP 1134 Limiting exposure harmonious perspectives
MP 1132 Limiting length communication pathways
MP 1173 Linear contexts communication pathways ; Protecting non
MP 1163 Linear domain external structures ; Hospitable non
MP 1175 Linear extensions structures ; Insight capturing non
MP 1007 Linear organization ; Non
MP 1051 Linear processes ; Linear relationships enhanced non
MP 1051 Linear relationships enhanced non-linear processes
MP 1060 Linearity ; Accessible non
MP 1174 Linearity ; Communication pathways enfolded non
MP 1172 Linearity ; Contexts self organizing non
MP 1170 Linearity ; Cultivation productive non
MP 1118 Linearity integrative superstructure ; Integration non
MP 1176 Linearity ; Sites grounding perspectives non
MP 1171 Linearity structures ; Appropriate relationship non
MS 8606 Liquids ; Foods
MM 2053 Literacy
MS 8383 Literary characters ; Historical, legendary
MS 8115 Literature ; Language
MS 8423 Literature ; Symbols sacred
MP 1045 Local action network
MP 1177 Local cultivation sources perspective nourishment
MP 1044 Local focal points
MP 1011 Local interrelationship domains
MP 1157 Local opportunities perspective activity
MP 1088 Local perspective interface zones ; Informal
MP 1049 Local relationship loops
MP 1029 Local relationships ; Stable density gradient
MP 1089 Local sources perspective nourishment
MP 1093 Location sources perspective nourishment ; Transit point
MP 1142 Loci structure ; Sequence viewpoint
MM 2010 Locomotion ; Animal
MM 2025 Locomotion ; Aquatic animal
MM 2039 Locomotion ; Arboreal animal
MM 2064 Locomotion flight ; Aerial animal
MM 2033 Locomotion ; Fossorial animal
MM 2059 Locomotion gliding soaring ; Aerial animal
MM 2012 Locomotion ; Terrestrial animal
MP 1049 Loops ; Local relationship
MP 1101 Low density communication pathways ; Protected
MP 1059 Low-intensity communication pathways
MP 1055 Low intensity relationships ; Protected
MS 8435 Lunar nature ; Sub

M

MS 8102 Machines supplies
MS 8925 Magical garments
MM 2072 Magnetic containment plasma
MP 1110 Main entry point structure ; Distinctiveness
MP 1234 Maintainable, multi-element external boundaries
MP 1169 Maintaining distinctions contextual levels
MP 1083 Maintenance dynamics ; Integration perspective acquisition perspective
MP 1181 Maintenance source direct insight structures focal point
MS 8600 Man-made forms structures
MS 8955 Man woman
MS 8385 Masculinity
MS 8843 Masonic symbols
MS 8723 Mass destruction weapons ; Nuclear
MP 1076 Mature emerging perspectives ; Minimal context interrelating
MP 1154 Maturing perspectives ; Semi autonomous contexts
MP 1253 Meaningful symbols self-transformation
MM 2092 Mechanical cycles
MS 8446 Medals, decorations trophies
MM 2022 Media diets
MP 1113 Media ; Harmoniously structured entry point external communication
MS 8715 Medical instruments
MS 8406 Medical preparations
MM 2068 Metabolic pathways cycles
MM 2044 Migration ; Cyclic
MP 1077 Minimal context complementary perspectives
MP 1076 Minimal context interrelating mature emerging perspectives
MP 1078 Minimal context single perspective
MP 1082 Minimal distance related operational control contexts
MP 1206 Minimal structural distinctions ; Efficient enclosure spaces
MP 1081 Minimally-structured perspective control operations
MP 1162 Minimize insight blindspots ; Organization integrative superstructure
MP 1167 Minimum proportions ; Enfolded overview domains
MP 1219 Minimum tension ; Level generation
MP 1003 Modes organization ; Interpretation complementary
MP 1056 Modes relationship ; Special
MM 2086 Molecular resonance
MS 8397 Money
MP 1097 Monotonous perspective patterns ; Concealment necessary
MS 8225 Monsters ; Hybrid
MS 8825 Monuments
MM 2020 Moods individual ; Changing
MS 8603 Mortality
MM 2043 Motors generators ; Electric
MM 2084 Movement ; Alluring
MM 2070 Movement ; Geography
MP 1234 Multi element external boundaries ; Maintainable,
MP 1239 Multi-faceted frameworks
MP 1012 Multiplicity ; Individuality
MS 8316 Musical instruments
MM 2090 Musical variations
MS 8357 Mythical aerial beings
MS 8300 Mythical anthropomorphic beings
MS 8917 Mythical aquatic beings
MS 8915 Mythical artifacts
MS 8337 Mythical celestial beings
MS 8925 Mythical clothing
MS 8937 Mythical fire-dwelling beings
MS 8387 Mythical terrestrial beings
MS 8225 Mythical theriomorphs
MS 8200 Mythological reality

N

MS 8365 Nails ; Hair
MS 8435 Natural symbols ; Terrestrial
MS 8435 Nature ; Sub lunar
MP 1097 Necessary monotonous perspective patterns ; Concealment
MP 1127 Nested levels accessibility
MM 2049 Nets basketry ; Knotting,
MS 8933 Network
MP 1005 Network inter-relationships
MP 1045 Network ; Local action
MP 1018 Network redefinitions
MP 1120 Networks ; Focus oriented communication
MP 1052 Networks unmediated relationships ; Interfacing vehicles communication
MP 1039 New dimension ; Integrating
MP 1043 New dimensions ; Presentation
MP 1211 New level generation ; Boundary expansion permitting
MP 1065 New perspectives ; Context emergence
MP 1030 Nodes ; Activity
MP 1145 Non current elements those reserve ; Domains
MS 8400 Non-human beings
MP 1173 Non linear contexts communication pathways ; Protecting
MP 1163 Non linear domain external structures ; Hospitable
MP 1175 Non linear extensions structures ; Insight capturing
MP 1007 Non-linear organization
MP 1051 Non linear processes ; Linear relationships enhanced
MP 1060 Non linearity ; Accessible
MP 1174 Non linearity ; Communication pathways enfolded
MP 1172 Non linearity ; Contexts self organizing
MP 1170 Non linearity ; Cultivation productive
MP 1118 Non linearity integrative superstructure ; Integration
MP 1176 Non linearity ; Sites grounding perspectives
MP 1171 Non linearity structures ; Appropriate relationship
MP 1182 Nourishment ; Appropriate conditions perspective
MP 1147 Nourishment ; Coordination perspective
MP 1139 Nourishment ; Hospitable domain perspective
MP 1177 Nourishment ; Local cultivation sources perspective
MP 1089 Nourishment ; Local sources perspective
MP 1093 Nourishment ; Transit point location sources perspective
MP 1075 Nuclear interaction ; Extended pattern
MS 8723 Nuclear mass destruction weapons
MS 8605 Number
MP 1153 Number embodied perspectives ; Structures adaptable changing
MP 1103 Number occupiable temporary sites ; Limitation
MP 1096 Number structural levels ; Limitation
MS 8306 Numeracy
MM 2055 Nutrition cooking ; Food,

O

MS 8336 Objects animal husbandry domestication
MS 8306 Objects devices ; Cultural
MS 8602 Objects ; Food related
MS 8326 Objects ; Symbolic
MP 1180 Occupiable sites exposed external insight
MP 1179 Occupiable sites ; Organization structure provide
MP 1188 Occupiable sites perspective inactivity
MP 1202 Occupiable sites ; Structure enfolded
MP 1022 Occupiable temporary site limit
MP 1103 Occupiable temporary sites ; Limitation number
MP 1126 Off-centering focal point common domain
MS 8913 Office ; Symbols
MP 1082 Operational control contexts ; Minimal distance related
MP 1081 Operations ; Minimally structured perspective control
MP 1157 Opportunities perspective activity ; Local
MP 1156 Opportunities perspectives decreasing activity
MP 1072 Opportunities transposed concrete level ; Competitive interaction
MM 2085 Orbiting space
MP 1002 Organization ; Distribution
MP 1146 Organization ; Flexible domain
MP 1209 Organization integrative superstructure
MP 1162 Organization integrative superstructure minimize insight blindspots
MP 1131 Organization inter-domain dynamics
MP 1006 Organization ; Intermediate scale
MP 1003 Organization ; Interpretation complementary modes
MP 1007 Organization ; Non linear
MP 1109 Organization structure enhance autonomy sub-structures
MP 1179 Organization structure provide occupiable sites
MP 1107 Organization structures enhance receptivity external insight
MP 1159 Organization structures permit two sources external insight
MP 1172 Organizing non linearity ; Contexts self
MS 8325 Organs ; Human external
MS 8345 Organs ; Human internal
MP 1128 Orientation domains receive external insight
MP 1212 Orientation ; Primary inter level connections transitions boundary
MM 2029 Orientation ; Spatial
MP 1105 Orientation structures enhance receptivity external insight
MP 1120 Oriented communication networks ; Focus
MS 8816 Ornamentation
MS 8713 Ornithological symbols
MP 1164 Overview communication pathway structure
MP 1166 Overview domains interfaces structure external environment
MP 1167 Overview domains minimum proportions ; Enfolded
MP 1192 Overview external contexts
MP 1231 Overview sites integrative superstructure

P

MP 1054 Paced communications ; Intersection differently
MP 1023 Parallel ; Compensating relationships
MP 1119 Partially contained interfaces
MP 1193 Partially enclosed internal domains
MP 1183 Partially exposed perspective context
MP 1152 Partially isolated contexts
MS 8133 Parts ; Animal body
MS 8402 Parts ; Vehicles
MP 1070 Past perspectives ; Context acknowledgement
MS 8846 Pastimes ; Hobbies
MM 2074 Path journey
MP 1165 Pathway ; Exposure structural activities communication
MP 1140 Pathway ; Relative isolation structural interface communication
MP 1164 Pathway structure ; Overview communication
MP 1112 Pathway ; Transition domain structure communication
MP 1229 Pathways automatic communications ; Provision
MM 2068 Pathways cycles ; Metabolic
MP 1122 Pathways ; Direct relationship structures communication
MP 1174 Pathways enfolded non linearity ; Communication
MP 1121 Pathways ; Hospitability communication
MP 1133 Pathways levels structure ; Integrating transition
MP 1132 Pathways ; Limiting length communication
MP 1059 Pathways ; Low intensity communication
MP 1101 Pathways ; Protected low density communication
MP 1173 Pathways ; Protecting non linear contexts communication
MP 1102 Pattern entry points complex structures ; Distinct
MP 1104 Pattern harmony ; Structural development designed counteract deficiencies
MP 1075 Pattern nuclear interaction ; Extended
MP 1028 Pattern relationship densities ; Coherent
MP 1098 Patterning complex structures
MP 1116 Patterning integrative superstructure
MP 1064 Patterns active irrationality ; Access
MM 2015 Patterns activity ; Shifting
MP 1097 Patterns ; Concealment necessary monotonous perspective
MS 8845 Patterns ; Shapes
MS 8233 People ; Beautiful
MP 1217 Perimeter continuity
MS 8223 Perquisites
MS 8516 Personal possessions
MM 2008 Personification
MP 1083 Perspective acquisition perspective maintenance dynamics ; Integration
MP 1157 Perspective activity ; Local opportunities
MP 1079 Perspective-adaptable contexts
MP 1201 Perspective adjuncts ; Accessible facilities
MP 1200 Perspective adjuncts ; Facilities
MP 1198 Perspective adjuncts ; Inter domain contexts
MP 1063 Perspective common domains ; Cyclic interrelation complementary
MP 1125 Perspective common domains ; Integrating points
MP 1183 Perspective context ; Partially exposed
MP 1191 Perspective contexts ; Appropriate proportions
MP 1095 Perspective contexts ; Complexification
MP 1196 Perspective contexts ; Eccentric access
MP 1190 Perspective contexts ; Variation size
MP 1081 Perspective control operations ; Minimally structured
MP 1080 Perspective dynamics ; Integrated contexts
MP 1087 Perspective ; Exchange contexts controlled single
MP 1090 Perspective exchange ; Unstructured context
MP 1085 Perspective imitation contexts developing perspectives
MP 1186 Perspective inactivity ; Appropriate configuration
MP 1188 Perspective inactivity ; Occupiable sites
MP 1150 Perspective inactivity ; Provision temporary
MP 1185 Perspective interaction ; Appropriate configuration
MP 1148 Perspective interaction constraints
MP 1151 Perspective interaction contexts ; Small scale
MP 1088 Perspective interface zones ; Informal local
MP 1083 Perspective maintenance dynamics ; Integration perspective acquisition
MP 1078 Perspective ; Minimal context single
MP 1182 Perspective nourishment ; Appropriate conditions
MP 1147 Perspective nourishment ; Coordination
MP 1139 Perspective nourishment ; Hospitable domain
MP 1177 Perspective nourishment ; Local cultivation sources
MP 1089 Perspective nourishment ; Local sources
MP 1093 Perspective nourishment ; Transit point location sources
MP 1097 Perspective patterns ; Concealment necessary monotonous
MP 1062 Perspective ; Points wider

INDEX TO METAPHORS AND PATTERNS

MX

MP 1189 Perspective presentation ; Context forms
MP 1178 Perspective products ; Re integration rejected
MP 1141 Perspective ; Relatively isolated context each
MP 1084 Perspective reorganization ; Transitional contexts
MP 1187 Perspectives ; Appropriate configuration interaction complementary
MP 1070 Perspectives ; Context acknowledgement past
MP 1065 Perspectives ; Context emergence new
MP 1086 Perspectives ; Contexts care premature
MP 1073 Perspectives ; Contexts exploratory relationship formation challenging emerging
MP 1156 Perspectives decreasing activity ; Opportunities
MP 1155 Perspectives decreasing activity ; Semi autonomous contexts
MP 1137 Perspectives ; Domain developing
MP 1203 Perspectives ; Exclusive spaces emergent
MP 1114 Perspectives favouring broadest ; Hierachy
MP 1074 Perspectives ; Interaction coherent irrational
MP 1143 Perspectives ; Interrelationship contexts developing
MP 1134 Perspectives ; Limiting exposure harmonious
MP 1077 Perspectives ; Minimal context complementary
MP 1076 Perspectives ; Minimal context interrelating mature emerging
MP 1176 Perspectives non linearity ; Sites grounding
MP 1085 Perspectives ; Perspective imitation contexts developing
MP 1136 Perspectives ; Relative isolation complementary
MP 1154 Perspectives ; Semi autonomous contexts maturing
MP 1068 Perspectives ; Spontaneous relationship formation amongst emerging
MP 1153 Perspectives ; Structures adaptable changing number embodied
MP 1149 Perspectives structures ; Hospitable reception external
MP 1091 Perspectives transition ; Hospitable contexts
MS 8433 Pets
MM 2071 Phasing ; Project
MS 8373 Phenomena ; Human spiritual
MM 2032 Physical position ; Changing
MS 8445 Phytomorphic symbols
MS 8823 Piscatological symbols
MS 8338 Planes
MM 2045 Plants
MM 2072 Plasma ; Magnetic containment
MS 8313 Plurality ; Human
MM 2083 Pole
MM 2032 Position ; Changing physical
MP 1242 Position ; Intermediate
MS 8353 Position ; Social hierarchical
MP 1241 Positions ; Attractive temporary
MP 1024 Positions enabling transcendence
MS 8516 Possessions ; Personal
MM 2007 Practice/performance
MS 8516 Precious things
MP 1086 Premature perspectives ; Contexts care
MS 8406 Preparations ; Medical
MS 8853 Presence ; Canine
MS 8303 Presence ; Signs human
MP 1189 Presentation ; Context forms perspective
MP 1043 Presentation new dimensions
MP 1212 Primary inter-level connections transitions boundary orientation
MS 8600 Primary visual elements
MS 8525 Primitive weapons
MP 1053 Principal points entry
MP 1009 Processes ; Decentralized formal
MS 8605 Processes ; Forces, energies
MP 1041 Processes ; Informal context formal
MP 1205 Processes it ; Congruence spaces defined framework spaces defined
MP 1051 Processes ; Linear relationships enhanced non linear
MM 2052 Processes ; Stick carrot
MP 1033 Processes ; Underdefined
MP 1008 Processes ; Variety forms
MP 1184 Processing ; Appropriate configuration input
MP 1199 Processing context external insight ; Exposure input
MP 1170 Productive non linearity ; Cultivation
MS 8346 Products ; Animal artifacts
MP 1178 Products ; Re integration rejected perspective
MM 2027 Products ; Resources, commodities
MS 8102 Professional equipment
MS 8343 Professions ; Trades
MP 1208 Progressive framework definition
MM 2071 Project phasing
MP 1167 Proportions ; Enfolded overview domains minimum
MP 1191 Proportions perspective contexts ; Appropriate
MP 1101 Protected low-density communication pathways
MP 1055 Protected low intensity relationships
MP 1173 Protecting non-linear contexts communication pathways
MP 1245 Protecting variability enhance fixity
MP 1057 Protection emerging foci
MS 8223 Protocol
MM 2057 Protocol ; Bilingual
MP 1179 Provide occupiable sites ; Organization structure
MP 1229 Provision pathways automatic communications
MP 1150 Provision temporary perspective inactivity
MM 2048 Psycho-symbolic cycles
MM 2088 Pulsational variable stars
MM 2081 Pumping

Q

MS 8113 Quests journeys ; Symbolic

R

MS 8395 Race ethnicity
MM 2035 Radio communication
MP 1178 Re-integration rejected perspective by-products
MS 8200 Reality ; Mythological
MS 8426 Receivers ; Receptacles
MS 8426 Receptacles receivers
MP 1149 Reception external perspectives structures ; Hospitable
MP 1138 Receptivity emerging external insight
MP 1107 Receptivity external insight ; Organization structures enhance
MP 1105 Receptivity external insight ; Orientation structures enhance
MP 1018 Redefinitions ; Network
MP 1004 Regenerative resource cultivation areas
MM 2063 Regulation ; Traffic
MM 2004 Reification
MP 1225 Reinforcement apertures ; Boundary
MP 1031 Reinforcement ; Cycle relationship
MP 1178 Rejected perspective products ; Re integration
MP 1028 Relationship densities ; Coherent pattern
MP 1036 Relationship density ; Differentiation
MP 1050 Relationship entrainment ; Three way
MP 1068 Relationship formation amongst emerging perspectives ; Spontaneous
MP 1073 Relationship formation challenging emerging perspectives ; Contexts exploratory
MS 8333 Relationship ; Human congenital
MP 1025 Relationship indeterminacy
MP 1049 Relationship loops ; Local
MP 1171 Relationship non linearity structures ; Appropriate
MP 1031 Relationship reinforcement ; Cycle
MM 2042 Relationship sharing
MP 1056 Relationship ; Special modes
MP 1122 Relationship structures communication pathways ; Direct
MP 1051 Relationships enhanced non linear processes ; Linear
MP 1052 Relationships ; Interfacing vehicles communication networks unmediated
MP 1005 Relationships ; Network inter
MP 1023 Relationships parallel ; Compensating
MP 1055 Relationships ; Protected low intensity
MP 1029 Relationships ; Stable density gradient local
MP 1136 Relative isolation complementary perspectives
MP 1140 Relative isolation structural interface communication pathway
MP 1141 Relatively isolated context each perspective
MM 2046 Religious festivals ; Cycles
MS 8705 Religious symbols
MP 1084 Reorganization ; Transitional contexts perspective
MS 8513 Reptile
MP 1145 Reserve ; Domains non current elements those
MP 1218 Resistant boundaries ; Distortion
MM 2031 Resonance accelerator ; Cyclic
MM 2086 Resonance ; Molecular
MP 1004 Resource cultivation areas ; Regenerative
MM 2038 Resource sharing
MM 2027 Resources, commodities products
MM 2005 Rest ; Exercise
MS 8933 Robot
MS 8363 Roles ; Human
MS 8385 Roles ; Sex
MM 2013 Rotating chairmanship
MM 2094 Rotation ; Crop

S

MS 8123 Sacred calendar
MS 8423 Sacred literature ; Symbols
MS 8933 Scientific symbols
MS 8705 Scriptures ; Holy
MS 8963 Search knowledge
MM 2054 Seasons weather
MP 1213 Secondary inter level connections ; Distribution
MP 1032 Selective interchange axis
MP 1019 Selective interchange ; Web
MP 1172 Self organizing non linearity ; Contexts
MP 1253 Self transformation ; Meaningful symbols
MP 1154 Semi-autonomous contexts maturing perspectives
MP 1155 Semi-autonomous contexts perspectives decreasing activity
MP 1106 Separating complementary structures ; Functional enhancement domains
MP 1197 Separating domains ; Substantive distinctions
MP 1142 Sequence viewpoint loci structure
MS 8513 Serpents
MP 1251 Settings ; Different
MS 8385 Sex roles
MM 2034 Sexual intercourse
MS 8845 Shapes patterns
MM 2042 Sharing ; Relationship
MM 2038 Sharing ; Resource
MM 2077 Shell games
MM 2015 Shifting patterns activity
MM 2011 Shifting topics conversation
MS 8304 Ships ; Boats
MM 2050 Significance ; Contrast
MS 8303 Signs human presence
MP 1087 Single perspective ; Exchange contexts controlled
MP 1078 Single perspective ; Minimal context
MP 1022 Site limit ; Occupiable temporary
MP 1180 Sites exposed external insight ; Occupiable
MP 1176 Sites grounding perspectives non-linearity
MP 1231 Sites integrative superstructure ; Overview
MP 1103 Sites ; Limitation number occupiable temporary
MP 1179 Sites ; Organization structure provide occupiable
MP 1188 Sites perspective inactivity ; Occupiable
MP 1202 Sites ; Structure enfolded occupiable
MP 1190 Size perspective contexts ; Variation
MS 8355 Skeleton ; Human
MS 8335 Skin appearance ; Alteration
MM 2037 Slavery
MP 1061 Small scale interaction domains ; Bounded common
MP 1151 Small-scale perspective interaction contexts
MS 8513 Snake
MM 2059 Soaring ; Aerial animal locomotion gliding
MS 8353 Social hierarchical position
MP 1181 Source direct insight structures focal point ; Maintenance
MP 1159 Sources external insight ; Organization structures permit two
MP 1089 Sources perspective nourishment ; Local
MP 1177 Sources perspective nourishment ; Local cultivation
MP 1093 Sources perspective nourishment ; Transit point location
MP 1210 Space distribution levels ; Harmonizing
MM 2085 Space ; Orbiting
MM 2087 Spacecraft ; Launching
MP 1205 Spaces defined framework spaces defined processes it ; Congruence
MP 1205 Spaces defined processes it ; Congruence spaces defined framework
MP 1203 Spaces emergent perspectives ; Exclusive
MP 1130 Spaces enhancing structural entry points ; Internal transition
MP 1206 Spaces minimal structural distinctions ; Efficient enclosure
MM 2029 Spatial orientation
MP 1056 Special modes relationship
MP 1020 Specialized communications ; User determined
MM 2061 Speech ; Voice
MS 8373 Spiritual phenomena ; Human
MM 2026 Spontaneous ceremony ; Traditional
MP 1068 Spontaneous relationship formation amongst emerging perspectives
MS 8846 Sports contests
MP 1029 Stable density gradient local relationships
MM 2083 Staff
MP 1038 Standard frameworks
MM 2088 Stars ; Pulsational variable
MS 8825 Statues
MS 8353 Status
MM 2083 Stave
MM 2083 Stick
MM 2052 Stick carrot processes
MS 8235 Stories fairy-tales
MM 2056 Strategic configurations
MP 1165 Structural activities communication pathway ; Exposure
MP 1104 Structural development designed counteract deficiencies pattern memory
MP 1206 Structural distinctions ; Efficient enclosure spaces minimal
MP 1130 Structural entry points ; Internal transition spaces enhancing
MP 1140 Structural interface communication pathway ; Relative isolation
MP 1158 Structural levels ; External access higher
MP 1195 Structural levels ; Framework transition
MP 1096 Structural levels ; Limitation number
MP 1021 Structural limit ; Four level
MS 8700 Structure ; Body, form
MP 1129 Structure ; Common domain focal point
MP 1112 Structure communication pathway ; Transition domain
MP 1110 Structure ; Distinctiveness main entry point
MP 1161 Structure-enfolded insight domain
MP 1202 Structure-enfolded occupiable sites
MP 1109 Structure enhance autonomy sub structures ; Organization
MP 1166 Structure external environment ; Overview domains interfaces
MP 1099 Structure ; Focal centre complex
MP 1111 Structure informal context ; Blended integration formal
MP 1133 Structure ; Integrating transition pathways levels
MP 1164 Structure ; Overview communication pathway
MP 1179 Structure provide occupiable sites ; Organization
MP 1142 Structure ; Sequence viewpoint loci
MP 1113 Structured entry point external communication media ; Harmoniously
MP 1081 Structured perspective control operations ; Minimally
MP 1153 Structures adaptable changing number embodied perspectives
MP 1171 Structures ; Appropriate relationship non linearity
MP 1122 Structures communication pathways ; Direct relationship
MP 1102 Structures ; Distinct pattern entry points complex
MS 8815 Structures engineering works
MP 1107 Structures enhance receptivity external insight ; Organization
MP 1105 Structures enhance receptivity external insight ; Orientation
MP 1160 Structures external environment ; Hospitable interface
MP 1181 Structures focal point ; Maintenance source direct insight
MP 1106 Structures ; Functional enhancement domains separating complementary
MP 1168 Structures ; Grounded
MP 1163 Structures ; Hospitable non linear domain external
MP 1149 Structures ; Hospitable reception external perspectives
MP 1175 Structures ; Insight capturing non linear extensions

MX

MP 1144	Structures ; Integrating coordinated exposure irrationality
MP 1108	Structures ; Interconnected
MS 8600	Structures ; Man made forms
MP 1109	Structures ; Organization structure enhance autonomy sub
MP 1100	Structures engender fruitful interfaces ; Arrangement
MP 1098	Structures ; Patterning complex
MP 1159	Structures permit two sources external insight ; Organization
MS 8825	Structures ; Symbolic
MP 1013	Sub-domain boundary
MS 8435	Sub-lunar nature
MP 1109	Sub structures ; Organization structure enhance autonomy
MM 2004	Substances
MP 1197	Substantive distinctions separating domains
MP 1228	Superstructure contain transitions levels ; Appropriate
MP 1117	Superstructure ; Containment integrative
MP 1220	Superstructure ; Integration
MP 1118	Superstructure ; Integration non linearity integrative
MP 1162	Superstructure minimize insight blindspots ; Organization integrative
MP 1209	Superstructure ; Organization integrative
MP 1231	Superstructure ; Overview sites integrative
MP 1116	Superstructure ; Patterning integrative
MS 8102	Supplies ; Machines
MP 1230	Supportive emotion ; Unmediated
MP 1232	Symbolic connection encompassing domains
MM 2048	Symbolic cycles ; Psycho
MS 8500	Symbolic human activity
MS 8326	Symbolic objects
MS 8113	Symbolic quests journeys
MS 8825	Symbolic structures
MS 8923	Symbolism ; Chromatic
MS 8943	Symbols ; Agricultural
MS 8425	Symbols ; Atmosphere celestial
MS 8326	Symbols ; Chrematomorphic
MS 8223	Symbols ; Civil
MS 8803	Symbols civilization
MS 8415	Symbols ; Incorporeal formless
MP 1249	Symbols integration
MS 8843	Symbols ; Masonic
MS 8913	Symbols office
MS 8713	Symbols ; Ornithological
MS 8445	Symbols ; Phytomorphic
MS 8823	Symbols ; Piscatological
MS 8705	Symbols ; Religious
MS 8423	Symbols sacred literature
MS 8933	Symbols ; Scientific
MP 1253	Symbols self transformation ; Meaningful
MS 8435	Symbols ; Terrestrial natural

T

MM 2058	Tacking
MM 2028	Taking turns
MS 8235	Tales ; Stories fairy
MS 8355	Teeth
MP 1150	Temporary perspective inactivity ; Provision
MP 1241	Temporary positions ; Attractive
MP 1022	Temporary site limit ; Occupiable
MP 1103	Temporary sites ; Limitation number occupiable
MP 1219	Tension ; Level generation minimum
MM 2012	Terrestrial animal locomotion
MS 8387	Terrestrial beings ; Mythical
MS 8435	Terrestrial natural symbols
MM 2009	Thematic development ; Interruption
MS 8225	Theriomorphs ; Mythical
MS 8516	Things ; Precious
MP 1050	Three-way relationship entrainment
MS 8605	Time
MP 1248	Time binding
MP 1240	Tolerance level interfaces
MS 8102	Tools ; Craftsman
MM 2011	Topics conversation ; Shifting
MS 8343	Trades professions
MM 2026	Traditional spontaneous ceremony
MM 2063	Traffic regulation
MP 1024	Transcendence ; Positions enabling
MP 1253	Transformation ; Meaningful symbols self
MP 1042	Transformation zones ; Chain fundamental
MP 1066	Transformative experience ; Context
MP 1093	Transit point location sources perspective nourishment
MP 1092	Transit points ; Hospitable
MP 1112	Transition domain structure communication pathway
MP 1091	Transition ; Hospitable contexts perspectives
MP 1133	Transition pathways levels structure ; Integrating
MP 1130	Transition spaces enhancing structural entry points ; Internal
MP 1195	Transition structural levels ; Framework
MP 1084	Transitional contexts perspective reorganization
MP 1224	Transitions across boundaries ; Emphasizing
MP 1212	Transitions boundary orientation ; Primary inter level connections
MP 1228	Transitions levels ; Appropriate superstructure contain
MS 8402	Transport
MP 1072	Transposed concrete level ; Competitive interaction opportunities
MS 8446	Trophies ; Medals, decorations
MM 2028	Turns ; Taking
MP 1159	Two sources external insight ; Organization structures permit

U

MS 8375	Ugliness
MP 1235	Unalienating internal boundaries
MP 1033	Underdefined processes
MP 1052	Unmediated relationships ; Interfacing vehicles communication networks
MP 1230	Unmediated supportive emotion
MP 1067	Unstructured common domain
MP 1090	Unstructured context perspective exchange
MP 1115	Unstructured internal domains ; Functional integration
MP 1020	User-determined specialized communications

V

MP 1247	Variability ; Embedding fixity
MP 1245	Variability enhance fixity ; Protecting
MM 2088	Variable stars ; Pulsational
MP 1190	Variation size perspective contexts
MM 2090	Variations ; Musical
MP 1035	Variety cyclic elements ; Adequate
MP 1008	Variety forms processes
MP 1135	Varying levels exposure it ; Enhancing insight
MS 8445	Vegetation
MP 1052	Vehicles communication networks unmediated relationships ; Interfacing
MS 8402	Vehicles parts
MS 8103	Vermin
MS 8426	Vessels ; Container
MP 1142	Viewpoint loci structure ; Sequence
MP 1222	Visibility ; Ground level
MS 8600	Visual elements ; Primary
MM 2021	Visual illusions ; Ambiguous
MM 2079	Vitamins
MM 2061	Voice speech

W

MM 2001	Walking
MM 2006	War
MM 2081	Water ; Drawing
MM 2081	Water ; Getting
MS 8106	Weapons ; Armour
MS 8723	Weapons ; Nuclear mass destruction
MS 8525	Weapons ; Primitive
MM 2054	Weather ; Seasons
MP 1016	Web general interrelationships
MP 1019	Web selective interchange
MP 1062	Wider perspective ; Points
MS 8955	Woman ; Man
MS 8815	Works ; Structures engineering
MS 8245	Writing

Z

MP 1226	Zone ; Inter level
MP 1042	Zones ; Chain fundamental transformation
MP 1088	Zones ; Informal local perspective interface
MP 1223	Zones intermediate insight
MP 1233	Zoning internal domains

References

Metaphors and Patterns

Scope of bibliography

The following section constitutes a collection of bibliographical references relevant to the Metaphors and Patterns (Section M).

The entries appear here for any of the following reasons:

(a) Reference is made to them in the individual entries of Section M (where they appear in abridged form, namely with author, title and year of publication).

(b) Reference is made to them in the commentary on Section M (which appears in Section MZ).

(c) The publications are relevant to Section M, even though they have not (yet) been cross-referenced in either of the two ways indicated above.

Bibliographical entries

The entries appear by order of author and, within author, by year.

In certain cases the *International Standard Book Number (ISBN)* is given to facilitate access.

Source of bibliography

The bibliography was derived from a wide variety of sources. These include:

- Information from international organizations, including catalogues of publications and accessions lists

- Information from commercial publishers, especially sales catalogues

- Citations and bibliographies in other publications and reference books

- Systematic compilations of books in print (notably *Books in Print* and *International Books in Print*)

- Book reviews in specialized journals

Extremely comprehensive, and annotated, bibliographies on metaphor are to be found in: W A Shibles (1971), J-P van Noppen (1985), and J-P van Noppen et al. (1990) on which details are given in the bibliography.

Ahsen, A Visuality among other senses and the eidetic process (1981)
in: *Journal of Mental Imagery*, 5, 1, pp 19–24.

Alexander, Christopher The Timeless Way of Building (1979)
New York, Oxford University Press.

Alexander, Christopher, et al The Oregon Experiment (1975)
New York, Oxford University Press. Vol 3.

Angonet, M and Suvin, D L'Implicité du Manifeste: métaphore et imagerie de la démystification dans la Manifeste Communiste
In: *Etudes Françaises* 1980, 16, iii–iv, pp. 43–67.

Arnheim, R Visual Thinking (1969)
Berkeley CA, University of California Press.

Arter, J A; Ortony, A and Reynolds, R E Metaphor: theoretical and empirical research
In: *Psychological Bulletin* 1978, 85, pp. 919–943.

Asch, S The Metaphor: a psychological inquiry (1961)
In: M Henle (Ed), Documents of Gestalt Psychology. Berkeley, University of California Press, pp 323–333.

Asheraft, R Economic Metaphors, Behaviouralism and Political Theory: some observations on the ideological use of language
In: *Western Political Quarterly* 1977, 30, 3, pp 313–328.

Athos, A and Pascale, R The Art of Japanese Management (1981)
New York, Warner Books.

Attali, Jacques Les Trois Mondes: pour une théorie de l'après-crise (1981)
Paris, Fayard.

Avens, Roberts Imagination is Reality: Western Nirvana in Jung, Hillman, Barfield and Cassirer (1980)
Dallas, Spring Publications Inc, 127 p.
ISBN 0–88214–311–5.

Bamberger, J and Schön, D (Eds) The Figural/Formal Transaction: a parable of generative metaphor (1976)
Cambridge MA, MIT.

Bandler, Richard and Grinder, John Reframing: neuro-linguistic programming and the transformation of meaning (1982)
Moab UT, Real People Press.

Barbour, I G Myths, Models and Paradigms: the nature of religious and scientific language (1974)
London, SCM Press.

Barel, Yves Le Paradoxe et le Système: essai sur le fantastique social (1989)
Grenoble, Presses Universitaires de Grenoble, 328 p. rev ed.
ISBN 2–7061–0157–1.

Barfield, Owen Poetic Diction: a study in meaning (1964)
New York, McGraw-Hill. 2nd ed.

Barrett, Frank J and Cooperrider, David L Using Generative Metaphor to Intervene in a System Divided by Turfism and Competition: building a common vision (1988)
In: F Hoy (Ed). Academy of Management Best Paper Proceedings, pp 253–257.

Barwick, D E et al (Ed) Metaphors of Interpretation: essays in honour of W E H Stanner (1985)
Potts Point NSW, Australian National University, 312 p.
ISBN 0–08–029875–3.

Bateson, Gregory Mind and Nature: a necessary unity (1979)
New York, EP Dutton, 238 p. ISBN 0–525–15590–2.

Bateson, Mary Catherine Our Own Metaphor: a personal account of a conference on the effects of conscious purpose on human adaptation (1972)
New York, Knopf.

Baumann, B Imaginative Participation: the career of an organizing concept in a multidisciplinary context (1975)
Dordrecht, Kluwer Academic Publishers Group, 198 p.
ISBN 90–247–1693–4.

Beck, Brenda E F The Metaphor as a Mediator Between Semantic and Analogic Modes of Thought
In: *Current Anthropology* 1978, 19, pp. 83–97.

Berger, Peter The Social Reality of Religion (1969)
London, Faber and Faber.

Berger, Peter L and Luckmann, Thomas The Social Construction of Reality: a treatise in the sociology of knowledge (1972)
New York, Doubleday.

Berggren, Douglas From Myth to Metaphor
In: *The Monist* 1966, 50, pp. 530–552.

Berggren, Douglas The Use and Abuse of Metaphor
In: *Review of Metaphysics* 1963, 3, pp. 450–472.

Betta, J A The Ideology of Rhetoric of Thomas Paine: political justification through metaphor (1975)
New Brunswick NJ, Rutgers University. Ph D Dissertation.

Bettleheim, B The Uses of Enchantment: the meaning of importance of fairy tales (1977)
New York, Vintage.

Billow, R M Metaphor: a review of the psychological literature
In: *Psychological Bulletin* 1977, 84, 1, pp. 81–92.

Billow, R M A Cognitive Developmental Study of Metaphor Comprehension
In: *Developmental Psychology* 1975, II, pp. 415–423.

Binkley, T On the Truth and Probity of Metaphor
In: *Journal of Aesthetics and Art Criticism* 1974, 33, pp. 171–180.

Black, M Models and Metaphors (1962)
Ithaca NY, Cornell University Press.

Bohm, David Wholeness and the Implicate Order (1980)
London, Routledge and Kegan Paul. p.16–25; 6, 1979, 2, p.92–103.

Bolles, Edmund Blair Remembering and Forgetting: an inquiry into the nature of memory (1988)
New York, Walker and Company.

Bosman, Jan Persuasive Effects of Political Metaphors (1987)
Metaphor and Symbolic Activity, 2, 2, 1987, pp 97–113.

Boulding, Elise Futuristics and the imaging capacity of the West (1978)
In: M Maruyama (Ed), Cultures of the Future. The Hague, Mouton, pp 7–31.

Boulding, Elise Building a Global Civic Culture: education for an Interdependent World (1988)
New York, Teachers College Press, Columbia University, 192 p. ISBN 0–8077–2867–5.

Boulding, Elise Using the mind in new ways: new perspectives on educating not found in schooling (1988)
In: Elise Boulding. Building a Global Culture: education for an independent world. New York, Teachers College Press, pp 75–159.

Boulding, Elise Image and action in peace building (1988)
In: *Journal of Social Issues*, 44, 2, pp 7–37.

Boulding, Kenneth E The Image; knowledge in life and society (1956)
Ann Arbor MI, University of Michigan.

Boulding, Kenneth E and Senesh, Lawrence (Eds) The Optimum Utilization of Human Knowledge (1983)
Boulder CO, Westview Press.

Bourgoin, Henry L'Afrique Malade du Management: perspectives 2001 (1984)
Paris, Editions Jean Picollec, 217 p. illus.
ISBN 2–86477–051–2.

Bowers, C A Curriculum as Cultural Reproduction: an examination of metaphor as a carrier of ideology
In: *Teachers College Record* 1980, 82, 2, pp. 267–289.

Boyd, Richard Metaphor and Theory Change: what is "metaphor" a metaphor for? (1979)
In: Andrew Ortony (Ed). Metaphor and Thought. Cambridge, Cambridge University Press, pp 356–408.

Boyden, A Homology and Analogy: a critical review of the meanings and implications of these concepts in biology (1947)
American Midl. Nat. 37, May, pp. 648–669. bibl.

Briggs, John and Monaco, Richard The Logic of Poetry (1974)
New York, McGraw-Hill.

Brooke–Rose, Christine A Grammar of Metaphor (1970)
London, Seeker and Warburg.

Brown, R H Social Theory as Metaphor
In: *Theory and Society* 1976, 3, pp. 169–198.

Brown, R H A Poetic for Sociology (1977)
New York, Cambridge University Press.

Bruner, Jerome Actual Minds, Possible Worlds (1986)
Cambridge MA, Harvard University Press.

Bunge, Mario Analogy, Simulation, Representation (1970)
General Systems Yearbook Society for General Systems Research, 15, p. 27–34.

Burrell, D C Analogy and Philosophical Language (1973)
London, Yale University Press.

Buttimer, Anne Musing on Helicon: root metaphors and geography
In: *Geografiska Annaler* 1982, 64B, 2, pp. 89–96.

Buttimer, Anne Mirrors, Masks and Diverse Milieux
German version published in Münchener Geographische Hefte 1983.

Campbell, Joseph The Inner Reaches of Outer Space: metaphor as myth and as religion

Campbell, Mary B and Rollins, Mark (Eds) Begetting Images: studies in the art and science of symbol production (1989)
New York, Peter Lang Publishing, 328 p. New Connctions: studies in interdisciplinarity. ISBN 0–8204–1045–4.

Carbonnell, J G Metaphor: a key to extensible semantic analysis (1980)
Proceedings of the 18th Annual Meeting of the Association for Computational Linguistics, pp. 17–21. Report on a computer program to paraphrase metaphors based on common formulas in economics and politics.

Casey, Edward S Imagining: a phenomenological study (1976)
Bloomington, Indiana University Press.

Chadwick–Jones, J K Recent Interdisciplinary Exchanges and the Use of Analogy in Social Psychology (1970)
Human Relations 23, August pp. 253–261, bibl.

Chapman, P C and Zashin, E The Uses of Metaphor and Analogy: toward a renewal of political language
In: *Journal of Politics* 1974, 36, 2, pp. 290–336.

Christopher, Alexander, et al A Pattern Language: towns, buildings, construction (1977)
New York, Oxford University Press.

Cirlot, J E A Dictionary of Symbols (1971)
London, Routledge and Kegan Paul.

Click, J, et al The Cultural Context of Learning and Thinking (1971)
New York, Basic Books.

Colin, M The Myth of Metaphor (1970)
University of South Carolina Press.

Cooper, David E Metaphor (1986)
Cambridge MA, Basil Blackwell, 282 p. Aristotelian Society Series: 5. ISBN 0–631–17119–3.

Cooper, J C An Illustrated Encyclopaedia of Traditional Symbols (1978)
London, Thames and Hudson.

Cooper, R C and Gadalla, I E Toward an Epistemology of Management
In: *Social Science Information* 1978, 17, pp. 349–383.

Corbin, Henry Creative Imagination in the Sufism of Ibn 'Arabi (1969)
Princeton NJ, Princeton University Press.

Cowan, James On Totems (1989)
In: *Resurgence*, 138, pp 30–34.

Crocker, J C and Sapir, J D (Eds) The Social Use of Metaphor: essays on the anthropology of rhetoric (1977)
Philadelphia, University of Pennsylvania Press.

De Bono, Edward Atlas of Management Thinking (1983)
New York, Penguin Books, 200 p. illus.
ISBN 0–14–022461–0.

de Bono, Edward Conflicts: a better way to resolve them (1985)
London, Harrap, 207 p. ISBN 0–245–54322–8.

de Laszlo, Violet S (Ed) Psyche and Symbol: a selection from the writings of C G Jung (1958)
Garden City NY, Doubleday/Anchor Books.

Dewey, John Art as Experience (1959)
New York, G P Putnam's Sons.

Dirven, R and Paprotte, W (Eds) The Ubiquity of Metaphor: metaphor in language and thought (1984)
Amsterdam, John Benjamins. CILT: 29.

Douglas, Mary Natural Symbols: explorations in cosmology (1973)
London, Pelikan.

Douglas, Mary Purity and Danger: an analysis of the concepts of pollution and taboo (1984)
New York, Routledge Chapman and Hall, 196 p.
ISBN 0–7448–0011–0.

Dreistadt, Roy An Analysis of the Use of Analogies and Metaphors in Science
In: *Journal of Psychology* 1968, 68, pp. 100.

Duhl, B S From the Inside Out and Other Metaphors: an integrated approach to training in multicentric systems thinking as derived from a family therapy training program (1982)
Boston MA, University of Massachusetts. Ed D Dissertation.

Duncan, H D Symbols in Society (1968)
London, Oxford University Press.

Durka, Gloria and Smith, Joanmarie Aesthetic Dimensions of Religious Education (1979)
New York, Paulist Press.

Ehrenzweig, Anton The Hidden Order of Art: a study in the psychology of artistic imagination (1967)
Los Angeles, University of California Press, London, Weidenfeld and Nicolson.

Eliade, Mircea Images et Symboles (1952)
Paris.

Elliott, Jaques (Ed) Levels of Abstraction in Logic and Human Action
London, Heinemann.

Erickson, M and Rossi, E Hypnotic Realities (1976)
New York, Irvington.

Esnault, Gaston L'Imagination Populaire: métaphors occidentales (1925)
Paris, PUF.

Fairlie, H The Politican Without a Metaphor is a Ship Without Sails: the sense of metaphor
In: *New Republic* 1979, 180, 10, p 10.

Faulkner, Charles Operating Metaphors (1989)
Anchor Point (International Journal for Effective NLP Communicators), 1989, 6 articles in successive monthly issues (from March).

Feinstein, David and Krippner, Stanley Personal Mythology (1988)
San Francisco, J P Tarcher.

Ferguson, E S The Mind's Eye: nonverbal thought in technology
In: *Science* 1977, 197, pp. 827–836.

Forsythe, Kathleen The Secret Garden: learning as the perception of newness (1986)
Unpublished paper.

Forsythe, Kathleen Journeys to the Lands of New: beyond information (1986)
Paper given to the American Society of Cybernetics (Virginia Beach, 1986)

Forsythe, Kathleen Cathedrals in the Mind: the architecture of metaphor in understanding learning (1986)
Dordrecht, R Reidel, 8 p.

Forsythe, Kathleen The Cybernetic Isophor: proceedings of the World Congress on Cybernetics and General Systems (1987)
London.

Forsythe, Kathleen In Search of Unicorns: towards an epistemology of newness (1987)
University of Alberta. Paper 87–2.

Forsythe, Kathleen and Haughey, Margaret Learning Metaphors: the knowledge connectors (1985)
Paper given to the International Council for Distance Education, Melbourne.

Forsythe, Kathleen and Wedder, Candace Designing Arches and Windows: the role of analogical architecture in global communication projects (1988)
Victoria BC, Snowflake Communications, 6 p.

Fox, Sidney and Ho, Mae–Wan (Eds) Process and Metaphors in the Evolutionary Paradigm (1988)
New York, John Wiley and Sons, 350 p.
ISBN 0–471–91801–6.

Franz, Marie–Louise von Alchemical Active Imagination (1979)
Dallas, Spring Publications Inc, p 116.
ISBN 0–88214–114–7.

Friesen, P H and Miller, D Archetypes of Strategy Formulation
In: *Management Science* 1978, 24, pp. 921–933.

Fromm, Erich The Forgotten Language
New York, Grove Press.

Frost, P J; Mitchell, V F and Nord, W R (Eds) Organizational Reality (1982)
Santa Monica CA, Goodyear.

Frost, P; Morgan, G and Pondy, L Organizational Symbolism
In: *L Pondy, et al (Eds) Organizationsl Symbolism* Greenwich CT, JAI Press, 1983.

Fuglesang, Andreas About Understanding: ideas and observations on cross-cultural communication (1982)
Uppsala, Dag Hammerskjld Foundation, 231 p. illus. ISBN 91-85214-09-4.

Galyean, B Mind Sight: learning through imaging (1983)
Long Beach CA, Center for Intergative Learning.

Gardner, H and Winner, E The Development of Metaphoric Competence: implications for humanistic disciplines (1979)
In: S Sacks (Ed), On Metaphor. Chicago, University of Chicago Press, pp 121-139.

Gardner, Howard The Socialization of Human Intelligences Through Symbols
In: *Frames of Mind: the theory of multiple intelligence* 1984, London, Heinemann.

Glaserfeld, Ernst von The Construction of Knowledge: contributions to conceptual semantics (1987)
Seaside CA, Intersystems Publications, 347 p. Systems Inquiry Series. ISBN 0-914105-38-8.

Goldstein, W The Political Metaphors of Environmental Control
In: *Alternations* 1973, 2, 4, pp. 11-17.

Goodman, Nelson Ways of Worldmaking (1978)
Indianapolis IN, Hackett Publishing.

Gordon, David Therapeutic Metaphors: helping others through the looking glass (1976)
Cupertino CA, Meta Publications, 261 p. bibl.

Gordon, W J The Metaphorical Way of Learning and Knowing (1971)
Cambridge, Porpoise Books.

Graves, Robert Seven Days in New Crete (1983)
Oxford, Oxford University Press.

Grote, M D Teaching Metaphors for Educational Theory: a quasi-normative approach (1972)
Pennsylvania, Temple University. Doctoral Dissertation.

Grotjahn, Martin The Voice of the Symbol (1971)
New York, Dell.

Gruber, Edward C Miller Analogy Test: 1400 analogy questions (1967)
New York, Arco Pub, 137 p. 2nd ed.

Gumpel, L Metaphor Reexamined (1984)
Bloomingtðn IN, Indiana University Press.

Hagenbüchle, Roland and Swann, Joseph T (Eds) Poetic Knowledge; Circumference and Centre: papers from the Wuppertal Symposion 1978 (1980)
Bonn, Universitätsbuchhandlung Bouvier, 181 p. ISBN 3-416-01600-9.

Hale, D G The Body Politic: political metaphor in Renaissance English literature (1971)
Den Haag, Mouton de Gruyter. De Proprietatibus Litterarum Series Major: 9.

Halpern, James and Ilsa Projections: our world of imaginary relationships (1983)
New York, Seaview/Putnam.

Handy, Charles Gods of Management: who they are, how they work, and why they fail (1979)
London, Pan Books.

Hansen, B The Complementarity of Science and Magic Before the Scientific Revolution
In: *American Scientist* 1986, 74, March/April, pp. 128-36.

Harman, Willis Participative Wholes and Autnomous Parts: the organism metaphor in science and societal change (1987)
Sausalito CA, Institute of Noetic Sciences, 1987. Paper presented to the Findhorn Foundation conference "From Organization to Organism, Oct 1987.

Haskell, R E Anatomy of Analogy: a new look (1968)
Journal of Humanistic Psychology 8, Fall bibl. (pp. 167-169).

Haskell, R E (Ed) Symbolic, Structures and Cognition: the psychology of metaphoric transformation (1984)
Norwood NJ, Ablex.

Heidegger, Martin Poetry, Language, Thought (1971)
New York, Harper and Row. Trans. by Albert Hofstadter.

Hesse, M B On Defining Analogy (1960)
Journal of Symbolic Logic 25, March pp. 74-75.

Hesse, Mary B The Explanatory Function of Metaphor
In: *Models and Analogies of Science* Notre Dame, University of Notre Dame Press, 1966, pp. 157-177.

Hesse, Mary B Science and the Human Imagination: aspects of the history and logic of physical science (1954)
London, SCM Press.

Hester, Joseph P and Killian, Don R Cartoons for Thinking: issues in ethics and values (1984)
Monroe NY, Trillium Press. illus. ISBN 0-89824-007-7.

Hester, Marcus B The Meaning of Poetic Metaphor (1967)
Den Haag, Mouton.

Hoberman, J M The Body as an Ideological Variable: sportive imagery of leadership and the state
In: *Man and World* 1981, 14, 3, pp. 309-329.

Hoeningswald, Henry M and Wiener, Linda F (Eds) Biological Metaphor and Cladistic Classification: an interdisciplinary approach (1987)
Philadelphia PA, University of Pennsylvania Press, 288 p. illus. ISBN 0-8122-8014-8.

Hoffmann, R R and Honeck, R P (Eds) Cognition and Figurative Language (1980)
Hillsdale NJ, Lawrence Erlbaum Associates.

Howe, J Carrying the Village: Cuna political metaphors
In: *J D Sapir and J C Crocker (Eds) The Social Use of Metaphor* 1977, Philadelphia, University of Philadelphia, pp. 132-163.

Hull, R F C Bibliographical Notes on Active Imagination in the Works of C J Jung (1971)
Spring.

Hunt, H A Cognitive Reinterpretation of Classical Introspectionism: the relation between introspection and altered states of consciousness and their mutual relevance for a cognitive psychology of metaphor and felt meaning, with commentaries (1986)
Annals of Theoretical Psychology, 4, pp 245-313.

Hunt, Harry T and Popham, Coralee Metaphor and States of consciousness: a preliminary correlational study of presentational thinking (1987)
Journal of Mental Imagery, 11, 2, pp 83-100.

Hyde, L The Gift: Imagination and the Erotic Life of Property (1979)
New York, Vintage Books.

Hyman, B The Dominant Metaphors with which Teachers Report they Function in the Classroom Environment (1980)
New York, New York University. Ph D Dissertation.

James, D G Scepticism and Poetry: an essay on the poetic imagination (1937)
London, Allen and Unwin.

Jay, M The Dialectical Imagination (1973)
London, Heinemann.

Jervis, R Perception and Misperception of International Politics (1976)
Princeton NJ, Princeton University Press.

Jimenez, J On Metaphor: a theory of the nature and educational uses of associative thinking (1972)
Boston MA, University of Massachusetts. Ph D Dissertation.

Johnson, J T and Taylor, S E The Effect of Metaphor on Political Attitudes
In: *Basic and Applied Social Psychology* 1981, 2, 1, pp. 305-316.

Johnson, Mark and Lakoff, George Conceptual Metaphor in Everyday Language
In: *The Journal of Philosophy* 1980, 77, pp. 453-486.

Johnson, Mark and Lakoff, George Metaphors We Live By (1980)
Chicago IL, University of Chicago Press, 242 p. ISBN 0-226-46800-3.

Jones, Roger S Physics as Metaphor: a mind-expanding exploration of the human side of science in tradition, of Zen and the art of motorcycle maintenance and the Tao of physics (1983)
Minneapolis MN, University of Minneapolis Press, 254 p. ISBN 0-452-00620-1.

Judge, A J N The Territory Construed as a Map: in search of radical design innovations in the representation of human activities and their relationships
In: *Forms of Presentation and the Future of Comprehension* Brussels, Union of International Associations, 1984, pp. 112-121.

Judge, A J N The Future of Comprehension: conceptual birdcages and functional basketweaving (1982)
Transnational Associations 34, 1982, 6, p.400-404. Also in: Forms of Presentation and the Future of Comprehension (Collection of papers mainly presented to the Forms of Presentation sub-project of the Goals, Processes and Indicators of Development project of the United Nations University, 1978-82). Brussels, Union of International Associations, 1984.

Judge, A J N Comprehension of Appropriateness (1986)
Brussels, Union of International Association. Project on Economic Aspects of Human Development (EAHD) of the Regional and Global Studies Division of the United Nations University: paper submitted informally to Rome workshop, September 1986.

Judge, A J N Governance through Metaphor (1987)
Brussels, Union of International Associations (abridged version in *The USACoR Newletter* (Club of Rome), 12, March 1988, 1). Paper submitted to Geneva workshop, June 1987 of the project on Eonomic Apects of Human Development (EAHD) of the Regional and Global Studies Division of the United Nations University.

Judge, A J N Metaphoric Revolution: in quest of manifesto for governance through metaphor (1988)
Brussels, Union of International Associations. Paper prepared for the 10th World Conference of the World Future Studies Federation (Sept 1988, Beijing) under the auspices of the China Association for Science and Technology, Group 8: Changing political institutions.

Judge, A J N Reordering of Networks of Incommensurable Concepts in Phased Cycles and their Comprehension through Metaphor (1988)
Brussels, Union of International Associations, 36 p. Paper prepared for the International Symposium on Models of Meaning (Bulgaria, September 1988) under the auspices of the Institute of Bulgarian Language of the Bulgarian Academy of Sciences.

Judge, A J N Through Metaphor to a Sustainable Ecology of Development Policies (1989)
In: *Ilze M Gotelli and Thaddeus C Trzyna (Eds) The Power of Convening: collaborative policy forums for sustainable development* Sacramento CA, California Institute of Public Affairs, 1990, pp. 64-81. Paper prepared for an international workshop on collaborative policy for sustainable development convened by the Commission on Sustainable Development of IUCN-World Conservation Union, Claremont CA 1989.

Judge, A J N Recontextualizing Social Problems through Metaphor: transcending the "switch" metaphor (1990)
In: *Revue Belge de Philologie et d'Histoire*, 68, 1990, 3, pp 531-547. Paper prepared for: International Symposium on "How to do things with metaphor" (Brussels, March 1990).

Judge, A J N Aesthetics of Governance in the Year 2490 (1991)
In: *Futures*, 23, 4, March. Paper presented to the 11th World Conference of the World Futures Studies Federation (Budapest, May 1990).

Jung, C G Man and his Symbols (1968)
New York, Dell Books.

Jurgens, H and Saupe, D (Ed) Visualisierung in Mathematik und Naturwissenschaften (1989)
Heidelberg, Springer-Verlag.

Karl, Frederick R and Hamalian, Leo (Eds) The Existential Imagination (1963)
Greenwich CT, Fawcett Publications Inc, 288 p.

Kipp, David Poetic Truth: a theoretical inquiry (1975)
Melbourne VIC, Gold Athena Press. ISBN 0-9598783-1-9.

Kirby, M J L Complexity, Democracy and Governance
In: *The Science and Praxis of Complexity* 1985, Tokyo, United Nations University, pp. 329-337.

Kjärgaard, M S Metaphor and Parable
In: *Acta Theologica Danica* 1986, XIX.

Kliebard, H M Curriculum Theory as Metaphor
In: *Theory Into Practice* 1982, 21, 1, pp. 11-17.

Kosslyn, Stephen M Image and Mind (1981)
Cambridge MA, Harvard University Press.

Kovecses, Zoltan Metaphors of Anger, Pride, and Love: a lexical approach to the structure of concepts (1987)
Philadelphia PA, Benjamins John North America, 147 p. ISBN 1-55619-009-3.

Krishnamurti, J You Are the World (1972)
New York, Harper and Row.

Krueger, D (Ed) The Changing Reality of Modern Man: essays in honor of J H van den Berg (1984)
Capetown, Juta and Company.

Krumhansl, Carol L Concerning the Applicability of Geometric Models
In: *Psychological Review* 1978, 85, pp. 445-463.

Kuhn, Thomas Metaphor in Science (1979)
In: Andrew Ortony (Ed). Metaphor and Thought. Cambridge, Cambridge University Press, pp 409-419.

Lakoff, G Categories and Cognitive Models (1982)
Berkeley CA, University of California. Series A: 96.

Latroff, George and Johnson, Mark Metaphors to Live By (1980)
Chicago, University of Chicago Press.

Leatherdale, W H The Role of Analogy, Model and Metaphor in Science (1974)
Amsterdam, North-Holland, 276 p.

Levin, S R The Semantics of Metaphor (1977)
Baltimore MD, Johns Hopkins University Press.

Levin, Samuel R Metaphoric Worlds: conceptions of a romantic nature (1988)
New Haven, Yale University Press, 251 p. bibl. ISBN 0-300-04172-1.

Levine, Donald N The Flight from Ambiguity: essays in social and cultural theory (1985)
Chicago, University of Chicago Press.

Lightman, Alan P Magic on the Mind: physicists' use of metaphor (1989)
American Scholar, Winter 1989, pp 97-101.

Loewenberg, I Truth and Consequences of Metaphors
In: *Philosophy and Rhetoric* 1973, 6, pp. 30-46.

Low, Albert Zen and Creative Management (1976)
New York, Anchor Books, 255 p. ISBN 0-385-04669-3.

Luck, B T The Influence of Biological Concepts and Metaphors on the Development of the Psychology of Learning (1983)
New York, Columbia University. Dissertation.

Lutz, K A A Bibliography of Research on Imagery and Holistic Learning Strategies (1980)
Iowa City, Iowa University. Visual Scholars Program: 8.

Lynch, Dudley and Kordis, Paul Strategy of the Dolphin; scoring a win in a chaotic world (1988)
New York, Fawcett Columbine.

Mac Cormac, Earl R Religious Metaphors: mediators between biological and cultural evolution that generate transcendent meaning
In: *Zygon* 1983, 18, pp. 45-65.

Mac Cormac, Earl R Scientific Metaphors as Necessary Conceptual Limitations of Science
In: *Nicholas Rescher (Ed) The Limits of Lawfulness* 1983, Pittsburg, University of Pittsburg Center for the Philosophy of Science.

Mac Cormac, Earl R Metaphors and Fuzzy Sets
In: *Fuzzy Sets and Systems* 1982, 7, pp. 243-256.

Mac Cormac, Earl R Metaphor and Myth in Science and Religion (1976)
Durham NC, Duke University Press.

Mac Cormac, Earl R A Cognitive Theory of Metaphor (1985)
Cambridge MA, MIT Press, 254 p. bibl. ISBN 0-262-13212-5.

MacCormac, Earl Scientific Metaphors and Necessary Conceptual Limitations of Science (1983)
In: N Rescher (Ed). The Limits of Lawfulness: studies on the scope and nature of scientific knowledge. Lanham, University Press of America, pp 61-68.

MacCormac, Earl The Cognitive Beauty of Metaphorical Images (1990)
Raleigh NC, Office of the Governor. Paper prepared for an International Symposium "How to do things with metaphor" (Brussels, 1990).

MacDermott, David Meta Metaphor (1974)
Boston MA, Marlborough House, 99 p. illus.

Maloney, P L Metaphor: the poetic mode of knowledge and its implications for education (1980)
Toronto, University of Toronto. Ph D Dissertation.

Man, Paul de The Epistemology of Metaphor (1978)
In: *Critical Enquiry* 5, pp. 13-30.

Manning, P K Metaphors of the Field: varieties of organizational discourse
In: *Administrative Science Quarterly* 1979, 24, pp. 660-671.

Maranda, P The Dialectic of Metaphor: an anthropological essay on hermaneutics
In: *I Crossman and S Sulaiman (Eds) The Reader in the Text* 1980, Princeton, Princeton University Press, pp. 183-204.

Markus, Mario et al Dynamic Pattern Formation in Chemmistry and Mathematics: aesthetics in the sciences (1988)
Dortmund, Max-Planck-Institut für Ernahrungsphysiologie.

Martz, Louis L The Poetry of Meditation (1954)
New Haven, Yale.

Maruyama, Magoroh Mindscapes: the limits to thought (1979)
In: *World Future Society Bulletin*, Sept–Oct, pp 13–23.

Maruyama, Magoroh Mindscapes, social patterns and future development of scientific theory types (1980)
Cybernetica (1980), 23, 1, p. 5–25.

McCarrell, N S and Verbrugge, R R Metaphoric Comprehension: studies in reminding and resembling
In: *Cognitive Psychology* 1977, 9, pp. 494–533.

McCulloch, W S Embodiments of Mind (1965)
Cambridge MA, MIT Press.

McFague, S TeSelle Speaking in Parables: a study in metaphor and theology (1975)
London, SCM Press.

McFague, Sallie Metaphorical Theology: models of God in religious language (1982)
London, SCM Press, 225 p. ISBN 334–00966–0.

McInerny, Ralph M Studies in Analogy (1968)
Den Haag, Martinus Nijhoff, 138 p. bibl., foot.

McKim, Robert Experiences in Visual Thinking (1972)
Brooks Cole, Monterey.

Meyer, A Mechanische und Organische Metaphorik, Politischer Philosophie
In: *Archiv für Begrisffsgeschichte* 1969, 13, 2, pp. 128–199.

Miller, Arthur I Imagery in Scientific Thought: creating 20th-century physics (1984)
Cambridge MA, MIT Press, 355 p. ISBN 0–262–63104–0.

Mills, Wright C Sociological Imagination (1959)

Mitchell, W J T (Ed) The Language of Images (1980)
Chicago IL, University of Chicago Press, 307 p. illus.
ISBN 0–226–53215–1.

Mooij, J J A A Study of Metaphor (1976)
Amsterdam, North–Holland.

Morgan, G Opportunities Arising from Paradigm Diversity
In: *Administration and Society* 1984, 16, pp. 306–327.

Morgan, G The Schismatic Metaphor and Its Implications for Organizational Analysis
In: *Organization Studies* 1981, 2, pp. 23–44.

Morgan, G Paradigms, Metaphors and Puzzle Solving in Organization Theory
In: *Administrative Science Quarterly* 1980, 25, pp. 605–622.

Morgan, G (Ed) Beyond Method: strategies for social research (1983)
Beverly Hills CA, Sage.

Morgan, G and Ramirez, R Action Learning: a holographic metaphor for guiding social change
In: *Human Relations* 1984, 37, pp. 1–28.

Morgan, Gareth Images of Organization (1986)
Beverly Hills CA, SAGE Publications, 421 p. bibl.
ISBN 0–8039–2830–0.

Morrow, Lance Metaphors of the World Unite (1989)
In: *Time*, 16 October 1989, pp 30.
Report on a symposium to find a metaphor for the modern age, on the occasion of the Boston University sesquicentennial (Boston, 1989).

Murry, John Middleton Countries of the Mind (1968)
Freeport NY, Books for Libraries Press.

Nicolas, Antonio de Meditations through the Rg Veda (1978)
Boulder CO, Shambhala Publications.

Oakeshott, M The Voice of Poetry in the Conservation of Mankind (1959)
London, Bowes and Bowes.

Olscamp, Paul How Some Metaphors May be True or False
In: *The Journal of Asthetics and Art Criticism* 1970, 29, pp. 77–86.

Oppenheimer, R Analogy in Science (1956)
American Psychologist 11, pp. 127–35.

Ordway, Samuel The Theory of the Limit to Growth (1953)

Ortony, Alii Theoretical and Methodological Issues in the Empirical Study of Metaphor and Implications and Suggestions for Teachers (1983)
Urbana, Center for the Study of Reading. Reading Education Report: 38.

Ortony, Andrew Why Metaphors Are Necessary and Not Just Nice
In: *Educational Theory* 1975, 25, pp. 45–53.

Ortony, Andrew Metaphor and Thought (1986)
Cambridge, Cambridge University Press, 501 p. bibl.
ISBN 0–521–22727–5.

Paine, R The Political Use of Metaphor and Metonym
In: *Politically Speaking: cross–cultural studies of rhetoric* 1981, Philadelphia PA, Institute for the Study of Human Issues, pp. 87–200.

Pask, Gordon Conversation, Cognition and Learning (1975)
Amsterdam, Elsevier.

Pavel, T G Fictional Worlds (1986)
Cambridge MA, Harvard University Press.

Pepper, S World Hypotheses (1942)
Berkeley CA, University of California Press.

Perrin, S G Metaphor to Mythology: experience as a resonant synthesis of meaning and being (1982)
Boston MA, Boston University.

Petrie, H G Metaphor and Learning
In: *A Ortony (Ed) Metaphor and Thought* Cambridge University Press, 1979, pp. 438–461.

Polak, Fred L The Image of the Future (1973)
Amsterdam, Elsevier, 319 p. Transl. and abridged by Elise Boulding. ISBN 0–444–41053–8.

Polanski, V G A Description of the Spontaneous Production of Similes and Metaphors in the Writing of Students in Grades Four, Eight and Twelve, and the Third Year of College (1981)
Buffalo, SUNY. Ph D Dissertation.

Polanyi, Michael The Creative Imagination
In: *Marjorie Grene (Ed) Toward a Unity of Knowledge* 1969, New York, International Universities Press, pp. 53–70.

Pylshyn, Zenon W Metaphorical Imprecision and the "Top–Down" Research Strategy (1979)
In: Andrew Ortony (Ed). Metaphor and Thought. Cambridge, Cambridge University Press, pp 420–436.

Quina, J Root Metaphor and Interdisciplinary Curriculum: designs for teaching literature in secondary schools
In: *Journal of Mind and Behavior* 1982, 3, 3–4, pp. 345–356.

Raitz, K L An Analysis and Evaluation of Metaphor in Education (1975)
Columbus OH, Ohio State University. Ph D Dissertation.

Reddy, Michael J The Conduit Metaphor: a case of frame conflict in our language about language (1979)
In: Andrew Ortony (Ed). Metaphor and Thought. Cambridge, Cambridge University Press, pp 284–324.

Ricoeur, Paul The Rule of Metaphor: multi–disciplinary studies of the creation of meaning in language (1986)
London, Routledge and Kegan Paul, 384 p. Transl. by Robert Czerny with Katleen McLaughlin and John Costello.
ISBN 0–7100–9329–2.

Rigotti, Francesca Metafore della Politica (1989)
Bologna, Il Mulino.

Rigotti, Francesca La théorie politique et ses métaphores: essai de compréhension (1990)
Revue Belge de Philologie et d'Histoire, 68, 1990, 3. Paper prepared for an International Symposium "How to do things with metaphor" (Brussels, 1990).

Rigotti, Francesca Political Metaphors: real creativity or prestidigitation? (1991)
Transnational Associations, 43, 1.

Romanyshyn, Robert D Metaphors of Experience and Experience as Metaphorical
In: *R S Valle and R von Eckartsberg (Eds) The Metaphors of Consciousness* New York, Plenum Press, 1981, pp. 3–19.

Romanyshyn, Robert D Psychological Life: from science to metaphor (1982)
Austin TX, University of Texas Press.

Romanyshyn, Robert D Technology as Symptom and Dream (1989)
London, Routledge, 254 p. illus. ISBN 0–415–00786–0.

Rosenthal, D C Metaphors, Models and Analogies in Social Science and Public Policy
In: *Political Behaviour* 1982, 4, 3, pp. 283–301.

Ross, J F Analogy and the Resolution of Some Cognitivity Problems (1970)
Journal of Philosophy 67, 22 October pp. 725–746.

Ross, J F Portraying Analogy (1981)
Cambridge, Cambridge University Press.

Saccaro–Battist, G Changing Metaphors of Political Structures
In: *Journal of the History of Ideas* 1983, 44, pp. 31–54.

Sacks, S (Ed) On Metaphor (1979)
Chicago IL, University of Chicago Press.

Samples, Bob The Metaphoric Mind: a celebration of creative consciousness (1976)
Reading MA, Addison–Wesley Publishing, 214 p. illus.
ISBN 0–201–06706–4.

Sartre, Jean–Paul The Psychology of Imagination (1963)
New York, Citadel.

Schneider, Marius El Origen Musical de los Animales–Simbolos en la Mitologia y la Escultura Antiquas (1946)
Barcelona.

Schön, D A The Reflective Practitioner (1983)
New York, Basic Books.

Schon, David The Displacement of Concepts (1963)
London, Tavistock Publications.

Schön, Donald Generative Metaphor: a perspective on problem–setting in social policy (1979)
In: Andrew Ortony (Ed). Metaphor and Thought. Cambridge, Cambridge University Press, pp 254–283.

Schwartz, H S The Usefulness of Myth and the Myth of Usefulness
In: *Journal of Management* 1985, 11, pp. 31–42.

Sewell, E The Human Metaphor (1964)
Notre Dame IN, University of Notre Dame Press.

Sheldrake, Rupert The Presence of the Past (1988)
New Orleans LA, Collins, 390 p.

Shibles, Warren A An Analysis of Metaphor (1971)
Den Haag, Mouton.

Shibles, Warren A Metaphor: an annotated bibliography and history (1971)
Whitewater WI, The Language Press.

Sibbett, David Metaphors, the Managers of Meaning (1984)
Vision Action (Journal of the Bay Area OD Network), 3, March 1984, 3, pp 1–4 (special issue on OD and Mythology).

Solesbury, W Strategic Planning: metaphor or method?
In: *Policy and Polities* 1981, 9, 4, pp. 419–437.

Soskice, J M Metaphor and Religious Language (1987)
Oxford, Clarendon Press.

Srivastva, Suresh and Barrett, Frank J The Transforming Nature of Metaphors in Group Development: a study in group theory (1988)
Human Relations, 41, 1, pp 31–64.

Starhawk Withcraft and Women's Culture
In: *Christ and Plaskow (Eds) Womenspirit Rising*

Steadman, Philip The Evolution of Design: biological analogy in architecture and the applied arts (1979)
Cambridge, Cambridge University Press.

Sternberg, R J; Tourangeau, R and Nigro, G Metaphor, Induction, and Social Policy: the convergence of macroscopic and microscopic views (1979)
In: Andrew Ortony (Ed). Metaphor and Thought. Cambridge, Cambridge University Press, pp 325–353.

Sternberg, Robert J Metaphors of Mind: conceptions of the nature of intelligence (1990)
Cambridge, Cambridge University Press, 344 p. bibl.
ISBN 0–521–35579–6.

Stewart, T C The City as an Image of Man (1970)
London, Latimer Press.

Sticht, T G Educational Uses of Metaphor
In: *A Ortony (Ed) Metaphor and Thought* Cambridge, Cambridge University Press, 1979, pp. 474–485.

Sudhir, Chandra (Ed) Social Transformation and Creative Imagination (1984)
New Delhi, Allied Publishers, 376 p.

Thompson, William Irwin From Nation to Emanation: planetary culture and world governance (1982)
Findhorn, Findhorn Foundation.

Tomkins, Silvan Affect, Imagery, Consciousness
Berlin, Springer–Verlag. 4 Vols.

Tourangeau, E B and Sternberg, R Understanding and Appreciating Metaphors (1982)
Cognition, 11, pp 203–244.

Tracy, David The Analogical Imagination: Christian theology and the culture of pluralism (1981)
New York, SCM Press.

Turbayne, C M The Myth of Metaphor (1962)
New Haven CT, Yale University Press.

Turner, V W Dramas, Fields and Metaphors (1974)
Ithaca NY, Cornell University Press.

Tymieniecka, A T (Ed) Poetics of the Elements in the Human Condition: the sea (1985)
Dordrecht, Kluwer Academic Publishers Group, 532 p.
ISBN 90–277–1906–3.

Tzu, Sun The Art of War (1971)
Oxford, Oxford University Press, 197 p. Chinsese Translations Series of UNESCO. Transl. and introduction by Samuel B Grittith. Foreword by B H Liddell Hart.
ISBN 0–19–501476–6.

United Nations University Science and Praxis of Complexity: contributions held at Montpellier, France, 9–11 May, 1984 (1985)
Tokyo, United Nations University, 384 p.
ISBN 92–808–0560–6.

Valle, Ronald S and Eckartsberg, Rolf von (Eds) The Metaphors of Consciousness (1981)
New York, Plenum Publishing, 521 p. illus.
ISBN 0–306–40520–2.

Van Noppen, Jean–Pierre (Ed) Metaphor and Religion: theolinguistics 2 (1983)
Brussels, Vrije Universiteit Brussel, 290 p. Studiereeks: 12.
ISBN 90–70289–15–6.

Van Noppen, Jean–Pierre and Hols, Edith Metaphor II: a classified bibliography of publications from 1985–1990 (1990)
Amsterdam, John Benjamins Publishing, 342 p.
ISBN 90–272–3746–8.

Van Noppen, Jean–Pierre, et al (Comp) Metaphor: a bibiography of post–1970 publications (1985)
Amsterdam, John Benjamins. Amsterdam Studies in the Theory and History of Linguistic Science.

Varela, Francisco A Calculus for Self–Reference (1975)
International Journal of General Systems 1975, 2, p. 5–24.

Vickers, Brian (Ed) Occult and Scientific Mentalities in the Renaissance (1984)
Cambridge, Cambridge University Press, 408 p.
ISBN 0–521–25879–0.

Warnock, Mary Imagination (1976)
Berkeley CA, University of California Press.

Watkins, Mary Maria Creative Imagination and the Transcendent Function (1972)
Princeton NJ, Princeton University Press, 177 p. unpublished thesis. bibl. (pp. 159–177; works on symbolic experience and its implication for human growth).

Welles, James F Understanding Stupidity: an analysis of the premaladaptive beliefs and behavior of institutions and organizations (1986)
Orient NY, Mount Pleasant Press.

Werner, H and Kaplan B Symbol Formation (1963)
New York, Wiley.

Wheeler, R and Cutsforth T Synaesthesia and Meaning (1922)
American Journal of Psychology, 33, pp 361–384.

Wheelwright, Philip Metaphor and Reality (1968)
Bloomington IN, Indiana University Press.

Whorf, Benjamin Lee Language Thought and Reality: selected writings of B L Whorf (1956)
New York, Wiley. Edited by John B Carroll.

Wilson, P R On the Argument by Analogy (1964)
Philosophy of Science 31, January pp. 34–39 bibl.

Winter, Gibson Liberating Creation: foundations of religious social ethics (1981)
New York, Crossroad.

Yamada, D I Metaphor in Political Thought: an essay in comparison (1975)
Santa Barbara CA, Univerity of California. Ph D Dissertation.

Zdenek, Marilee Inventing the Future: advances in imagery that can change your life (1987)
New York, McGraw–Hill, 196 p. ISBN 0–07–072819–4.

Ziegler, W Mindbook for Imaging/Inventing a World Without Weapons (1987)
Denver CO, Future Invention Association.

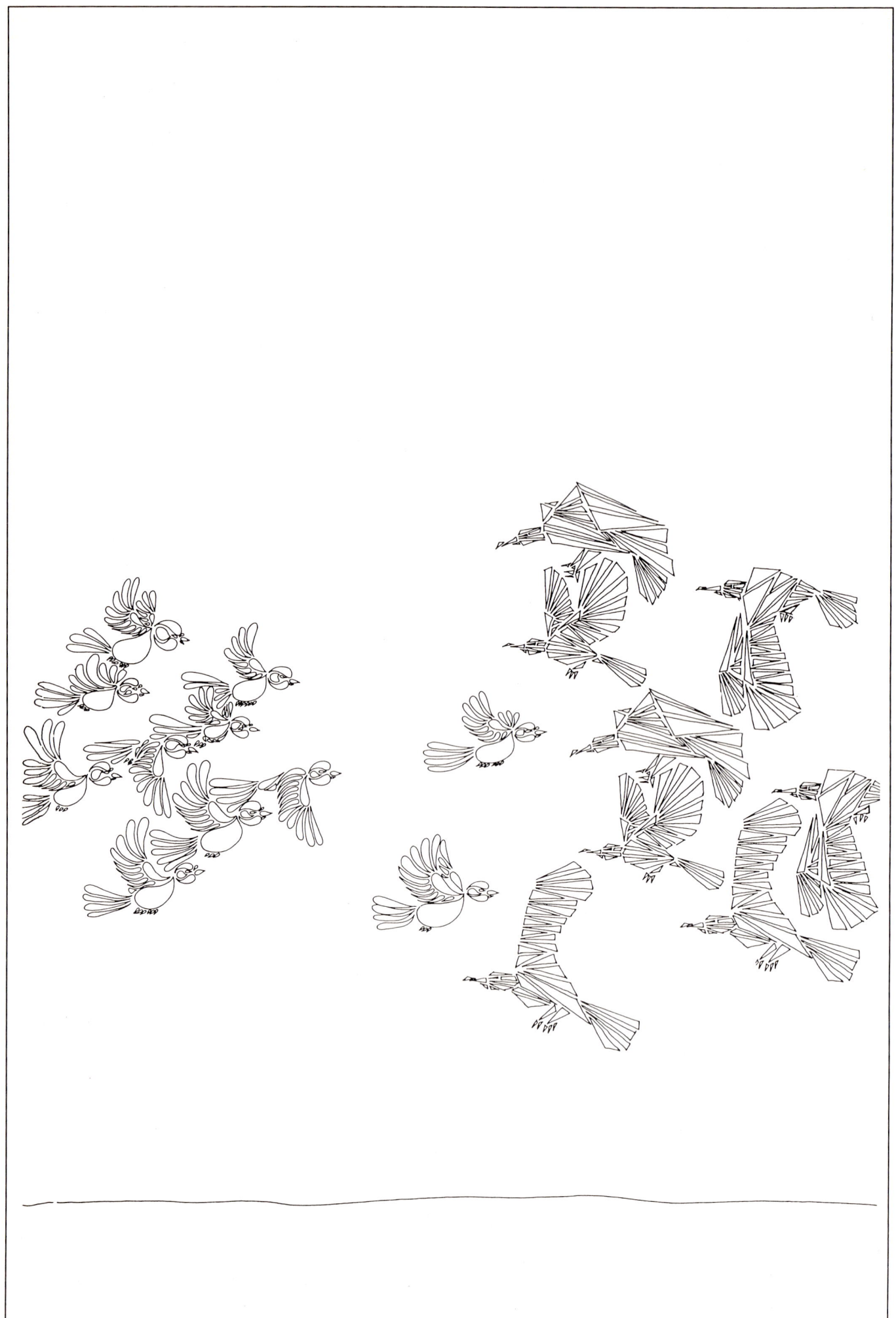

Notes

MZ

Metaphors and Patterns

Significance

Forms of presentation	589
Cognitive role of metaphor	591
Metaphor, analogy, symbol and pattern	593
Root metaphors and social organization	594
Generative metaphor and policy-making	596
Patterns of concepts	597
Communication through symbols	599
Degradation of symbolic forms	601

Challenge

Contemporary crisis of governance	603
Epistemological crisis of governance	605
Unsustainable policies	607
Comprehending complexity and appropriateness	608
Comprehending any new social order	609
Metaphor as an unexplored resource in a time of challenge	610
Transcending the "switch" metaphor	611
Transcending methodological limitations	612
Imaginal deficiency and simplistic metaphors	613
Imaginal deficiency in management and policy-making	614
Distinguishing extended metaphors	615
Designing metaphors and sets of metaphors	616

Opportunity

Programme of metaphoric development	617
From governing metaphors to governance through metaphor	619
Governance through enhancing the movement of meaning	620
Appropriate metaphors of sustainable development	622
Enhancing policy through powerful metaphors	623
Comprehending policy meshing and entrainment	625
Reframing problems as metaphors	626
Reframing the problem of "overpopulation"	627
Reframing cooperation through metaphor	629
Pattern language experiments	631
Metaphoric revolution for the individual	633

Comments

Future possibilities and dangers	635

***   Bibliographical references identified in abridged form in the following section refer to publications detailed, by author, in Section MY, which is the bilbiography for Section M.

Significance: forms of presentation

1. Context
This section is concerned with the forms through which new concepts and insights could be presented or communicated in response to the global problematique. It is based on the recognition that the public information programmes developed by intergovernmental and non-governmental agencies have tended to concentrate on a limited range of forms of presentation, often used such that they do not complement each other or reinforce each others effects.

The United Nations has attached great importance to "mobilizing public opinion", especially in relation to the International Development Decades. But as one report noted in 1978: *"The drive towards world development and a better economic order has made little headway in the present decade. The mobilization of public opinion as an essential effort has not been very successful either. Nevertheless, the approach was right, and a more massive mobilization is indispensable."* (Development Forum, Nov-Dec 1978).

It is not clear that the approach has been right, as was indicated by the conclusions of the UN Secretary-General's report as early as 1973: *"... it is difficult to escape the conclusion that, in spite of governmental efforts and similar programmes by non-governmental organizations, the state of public opinion on matters of development, particularly in the industrialized countries, is generally less favourable today than it has been in the past... It would probably be unfair to conclude that a sudden callousness had overcome public opinion in the developed countries. It is more like a closing of the gates to a pattern of generalizations perceived as out-worn by over-use."* (E/5358, 21 May 1973).

Corresponding concerns exist about the ability of social scientists and change agents to successfully communicate their insights, whether amongst themselves or to wider audiences, especially policy-makers and the general public. These concerns resulted in a special project on forms of presentation within the framework of the Goals, Processes and Indicators of Development project of the Human and Social Development Programme of the United Nations University. At its initial meeting in Geneva in 1979, this group indicated the importance of the following issues (as extracted from the meeting report).

2. Basic questions
(a) What is to be communicated? The importance of communicating what is not evident was stressed. On the one hand a presentation of different problematiques results in excessively balanced and undynamic communication; although on the other hand, the presentation of solutions results in a "missionary" quality which is equally undesirable. Related to this is the question of whether the communication should be didactic or conscientising (not the same thing).

(b) To whom and/or with whom? This is the question of the *Zielpublikum* or "target audience", although these terms reflect an approach to communication which does not conform to the criteria discussed below. They stress communication "to", or "at", rather than "with". Should the concern be primarily with a maximum number of people or should the aim be selective (*eg* decision makers)? What are the advantages of non- mass forms of communication? The disadvantages of mass forms?

(c) To what end? What is the desired effect of such communication? What is expected? Although again these terms imply a certain determining influence and a certain minimum of manipulation of the communication process which restricts the freedom of those involved. Should the intention be simply to interest people in the subject matter, or should they challenged and stimulated to commit themselves to some course of action?

2. Basic constraints
(a) By whom? Whatever institutional or other delivery system is selected to disseminate the chosen form, certain characteristic weaknesses become evident. An intergovernmental organization may thus ensure a very widespread distribution of a publication which then ends up in the private libraries of local officials and is not accessible to those whom it was intended to benefit.

(b) Costs: Communication involves costs which must either be borne or reduced (if not eliminated). Most obvious of these are the financial costs, which of course differ with form of presentation and the audience desired. Environmental costs may be high if the chosen form or method of distribution involves excessive use of natural resources, or energy, or results in pollution of the environment (whether in the form of noise, paper, *etc*). Finally there are social costs associated with a particular form. These include the need for preliminary instruction if the form is to be acceptable and usable, and the social cost of bringing people into the dialogue mode.

3. Means and implementation
(a) Range of alternatives: Emphasis was placed on the importance of broadening the range of acceptable modes of presentation. The alternatives to the written medium, and particularly the traditional academic book or article, were stressed as being complementary to the written mode and not a substitute for it.

(b) Technical constraints: Clearly there is a question of appropriateness of the communication technology associated with a particular form of presentation, in a particular setting. Equipment (TV, video, projectors, *etc*) may involve unacceptable investment, operating, maintenance, or preliminary training costs. This may prevent use of the chosen form at the grass-roots level. Or the equipment may be unsuitable for climatic or cultural reasons.

(c) Technical possibilities: The meeting was somewhat divided as to the desirability of forms of presentation associated with advanced technologies, especially computers but not including TV and video. This issue is dramatised by the rapid spread of such devices (*eg* TV games for children and in cafes, low cost micro-computers, computer conferencing).

4. Criteria for desirable forms
The discussion of the following criteria responded in part to the basic questions noted above, in the sense that the nature of the desirable medium defines the nature of the desirable communication (to paraphrase McLuhan).

(a) Structure/content correspondence: The structure by which communication takes place (whether organizational or technical) should correspond to the content of the communications. Namely undesirable methods of communication should not be used to communicate desirable messages or else the value of the latter is compromised.

(b) Beyond the sender/receiver model: The tendency to perceive communication in terms of a sender/receiver model in which the sender aims to influence the receiver or "to fill a vacuum" in the receiver, should be rejected. It implies a unilateral cause.

(c) Respect for those with whom communication occurs: Those initiating the communication should respect those who become involved in the process. The latter should be understood as capable of introducing dimensions of value to both parties as well as being able to adapt the form to local conditions.

(d) Dialogue: Dialogue should be incorporated into the form of presentation whenever possible to ensure appropriate feedback between both parties. The "dazibao" in China (a wall on which community messages and counter-messages can be hung) was cited as a desirable example. Other forms may permit dialogue to a greater or lesser degree (*eg* letters to editor, books with write-in-option, call-in-radio or TV).

(e) Feedback: The three previous points all imply a form of feedback to those initiating the communication (however, does not the term itself imply an undesirable directionality to the communication process?). The basic point is to avoid alienating or hypnotising those exposed to the chosen forms. Passive communication consumerism should be avoided.

(f) Openness: It was suggested that the chosen form should in some measure leave those exposed to it open to dream, to disorder and to chance, namely free to impose whatever degree of order seemed appropriate at the time. In this sense the whole communication process should be conceived rather like the classic "message in the bottle cast into the sea". A degree of randomness as to how it will be received and by whom and what will be the result.

(g) Reinforcement of the existing good: Communication which reinforces beneficial tendencies and processes is much to be preferred to communication which attempts to introduce alternatives conceived as "good" by outsiders. Aside from the risk that the latter will be rejected the former is more readily understood.

(h) Simplicity: The importance of conceiving forms of communication in terms of the needs of ordinary people was stressed, in contrast to the tendency to develop increasingly complex forms of presentation. It was also argued that even those who have received a specialized education require simplicity in those domains not covered by their speciality. There is therefore a need to communicate with the simple in everybody, including in those in positions of responsibility.

(i) Action possibilities: Exposure to the chosen form should not simply trigger wonder or interest which do not necessarily offer the individual any possibility for action. The content should challenge and indicate action possibilities so that the individual is not left frustrated and alienated by the inability to act.

(j) Schematic approach: Because of the above criteria, and particularly the impossibility of determining the needs of each set of users (especially where a variety of cultures is concerned), there is an advantage in leaving the product somewhat unfinished. It is completed for local distribution, if it is not completed by the user as part of the process.

(k) Complexity: Although there is need for simplicity in communication, as noted above, it is also clear that appropriate forms are required when the relationship pattern is complex (eg ecosystemic notions), often involving non-linear features. The challenge is to find a means of communicating complexity simply and rendering such patterns comprehensible. The conventional tendency to focus on isolated questions (eg children, food, health, disarmament, etc) as though they were independent should be avoided, if only by treating each of the elements in the pattern as part of a pattern. For example, by producing a set (not a "series") of posters on individual topics such that when hung together in a room the pattern between them emerges. It is not the elements which should be stressed but the pattern of relationship between them under different circumstances.

5. Complementarity of forms of presentation

(a) Complementarity: As noted above, stress was placed on the manner in which forms should be conceived as complementating one another. The nature of this complementarity remains to be explored in the light of the complete range of forms. By complementarity is meant that the possibilities, limitations and characteristics of one form may be complemented by use of one or more others - each appealing to groups with different mind-sets and communication preferences.

(b) Transformability: It is desirable that communication through one form should be transformable into communication through another if the latter is likely to be more appropriate. As with complementarity, the limitations on transformability remain to be explored.

(c) Privileged modes: The prevalence of certain forms was noted (eg books, articles). Attention is required to the manner and extent to which alternatives can be used. On the other hand the present degree of commitment to such privileged modes ensures that any substitution can only be limited and may only be possible in areas where the written mode has not yet become habitual. In the final analysis these difficulties are raised by many forms of presentation which have emerged as a consequence of industrialization.

6. Evaluation and research

(a) Successes and failures to date: Much has been done in the way of experiment over the past decades in order to communicate development- related concepts to many different audiences. The work of the different UN Specialized Agencies through their information programmes is one example. Many more experiments have been conducted at the national level. The question is to what extent have these experiments been successful, and to what extent a failure? Is improvement possible and in what areas, in the light of past experiences? What has blocked such improvements in the past?

(b) Evaluation: Attention is required to the method by which different forms of presentation can be evaluated and compared, for example, with regard to the time period over which they effect useful change (e.g immediately or in 15 years) and the number of people which can be involved. There may be various kinds of tradeoff between different advantages and disadvantages.

(c) Forms as an object of research: Beyond the problems raised by evaluation are those concerned with understanding the nature of different forms in terms their strengths and weaknesses in carrying or distorting messages of different degrees of complexity under different conditions.

(d) Influence of new forms on research: It was recognized that innovation in forms of representation had facilitated development of understanding in certain fields (eg chemistry and structural models; geography and maps). Any breakthrough to suitable new forms could therefore have important consequences. New vehicles are required for new concepts.

7. Meta and other questions

(a) Purpose of communication: Whilst the importance of specifying the effect desired on those exposed to the forms was stressed (see above), some attention was given to the question of the purpose of achieving that end. To what extent were the forms not only a means but an end? The goal/process relationship was compared to the zen attitude as exemplified in the art of archery - the two blend into a new unity. The challenge is to improve the objective forming potential. Should attention be on forms or on the process in which they are embedded, or on the process whereby they are generated?

(b) Re-assessment of communicant status: An ideal form places communicants in a situation in which they are encouraged to re-assess their own positions vis--vis each other and the subject matter ("une mise en question"). This process is basic to any approach to non-linear thinking.

(c) Communication vs Communications: Recent years have seen the publication of a major report commissioned by UNESCO through the International Commission on Communication Problems. Aside from the many political consequences of this report, it is much to be regretted that the title and publicity surrounding it implies that it is a report on communication in general. In fact it is a report on the mass media as the composition of the Commission indicates. The attention given to this report as a report on communication has helped to obscure many other dimensions of communication. Matters have not been assisted by what amounts to a counter-programme of the International Telecommunication Union in its organization of an International Communications Year. This focused on telecommunications hardware in such a way as to ignore other forms of communication not based on advanced technology.

(d) Variety of forms: Although the above-mentioned UNU Forms of Presentation project gathered information on a wide variety of forms. This has never been published. The over-emphasis by UNESCO and ITU on particular forms, without any sense of the contextual variety, was therefore a direct stimulus to the collection of the information presented in this section. It can be argued that only through a recognition of the variety of these forms can appropriate methods of communicating insights be developed. Concentration on extensively used, "important" or "effective" forms only raises the question as to why such an approach has been less than successful in "mobilizing public opinion".

Significance: cognitive role of metaphor

1. Resurgence of metaphor
Since the early 1970s there has been a progressive increase in studies of metaphor. Interest in the subject outside the literary world has markedly increased, accompanied by a number of breakthroughs in understanding about the function of metaphor. There is a very extensive literature on metaphor. A bibliography of post-1970 publications on metaphor records 4193 items (J P van Noppen, et al, 1985). A subsequent edition of this bibliography covering the 1980s, contains some 3,500 entries (J P van Noppen, et al, 1990).

Of special interest to many authors in the social and natural sciences is the degree to which concept formation is guided by metaphor or may even be totally based on metaphor. There appears to be increasing recognition of the power of metaphor to facilitate communication in situations where groups are fragmented by disciplinary, language or educational barriers.

2. Conventional use of metaphor
Metaphor is a classic device through which a complex set of elements and relationships can be rendered comprehensible - when any attempt to explain them otherwise could easily be meaningless. It is the peculiar strength of metaphor that it can convey the essential without excessive oversimplification, preserving its complexity by perceiving it through a familiar pattern of equivalent complexity.

A metaphor according to Nelson Goodman, "*typically involves a change not merely of range but also of realm. A label along with others constituting a schema is in effect detached from the home realm of that schema and applied for the sorting and organizing of an alien realm. Partly by thus carrying with it a reorientation of a whole network of labels does a metaphor give clues for its own development and elaboration... A whole set of alternative labels, a whole apparatus of organization takes over a new territory... and the organization they effect in the alien realm is guided by their habitual use in the home realm. A schema may be transported almost anywhere. The choice of territory for invasion is arbitrary; but the operation within that territory is almost never completely so... which elements in the chosen realm are warm, or are warmer than others, is then very largely determinate. Even where a schema is imposed upon a most likely and uncongenial realm, antecedent practice channels the application of the labels.*" (The Language of Art, 1976, p. 72-74)

3. Pervasive cognitive role of metaphor
It is now recognized that metaphors permeate use of both everyday language and the jargons of many disciplines including physics (R Dirven and W Paprotte, *The Ubiquity of Metaphor*, 1984 and R S Jones, *Physics as Metaphor*, 1980).

As George Lakoff and Mark Johnson note: "*Metaphor is for most people a device of the poetic imagination and the rhetorical flourish - a matter of extraordinary rather than ordinary language...most people think they can get along perfectly well without metaphor. We have found, on the contrary, that metaphor is pervasive in everyday life, not just in language but in thought and action. Our ordinary conceptual system, in terms of which we both think and act, is fundamentally metaphorical in nature.*" (Metaphors We Live By, 1980, p.3)

Lakoff and Johnson demonstrate this with many examples which are confirmed in Roger Jones study of *Physics as Metaphor* (1980). The authors conclude that "*If we are right in suggesting that our conceptual system is largely metaphorical, then the way we think, what we experience, and what we do every day is very much a matter of metaphor.*" (Metaphors We Live By, 1980, p.3)

They started their work from a concern that the understanding of meaning as explored by Western philosophy and linguistics had very little to do with what people found meaningful in their lives and quickly discovered that the assumptions of those disciplines precluded them from even raising the kinds of issue they wished to address. "*The problem was not one of extending or patching up some existing theory of meaning but of revising central assumptions in the Western philosophical tradition. In particular, this meant rejecting the possibility of any objective or absolute truth... It also meant supplying an alternative account in which human experience and understanding, rather than objective truth, played the central role.*" (p.x)

4. Beyond objectivism
The authors show how metaphor reveals the limitations of objectivism, namely the assumption that the world is made of distinct objects with inherent properties and fixed relations between them. In a subsequent paper Lakoff takes the investigation a step further with an extensive exploration of classical assumptions about categories and cognitive models.

He concludes: "*Changing our ideas about categories will require changing our ideas about rational thought, the nature of the mind and its relation to the body, and, in the process, changing our conception of man. Rationality, rather than being disembodied, purely mental, asocial, unfeeling and mechanical, is something which essentially involves the body, the senses, the emotions, social structure, interactions with other people, the imagination, and the capacity for idealization and for understanding based on the totality of experience. And the use of many partial models, some of which are inconsistent with each other, to comprehend experience is not irrational, but rather fits the paradigm of human rationality.*" (Categories and Cognitive Models, 1982)

This does not necessarily imply that objectivist categories and models should be abandoned. It does suggest that these constitute only one form of language and that there are others on whose resources society can draw at this critical time.

5. Metaphor as a way of knowing
For Robert Nisbet (*Social Change and History*, 1969): "*Metaphor is a way of knowing -- one of the oldest, most deeply embedded, even indispensable ways of knowing in the history of human consciousness.*" Gibson Winter (1981) argues that artistic process and techno-scientific discourse are bound together on the deepest level through metaphoric power. He quotes Robert Nisbet to the effect that: "*It is easy to dismiss metaphor as "unscientific" or "non-rational", a mere substitute for the hard analysis that rigorous thought requires...but metaphor belongs to philosophy and even to science. It is clear from many studies of the cognitive process generally, and particularly of creative thought that the act of thought in its more intense phases is often inseparable from metaphor -- from that intuitive, iconic, encapsulating grasp of a new entity or process.*"

For Kathleen Forsythe (*Cathedrals in the Mind*, 1986): "*It can be argued that metaphor is the fundamental core of our conceptual system as surely as the logic of form which we use in argument and debate.*" She cites Gregory Bateson: "*...metaphor is not just pretty poetry, it is not either good or bad logic, but it is in fact the logic upon which the biological world gas been built, the main characteristic and organizing glue of this world of mental process...*". Then she argues: "*However, because our conceptual system is not something we are normally aware of, we have failed to account for its metaphorical nature in our discussion of truth and meaning. Yet its pervasiveness suggests a central and basic role in the underlying architecture of thought. Metaphor can create new meaning, create similarities and so define a new insight and new perception of reality.. Such a view has no place in the dominant objectivist picture of the world.*"

Philip Wheelwright (1962) argues: "*What really matters in a metaphor is the psychic depth at which things of the world, whether actual or fancied, are transmuted by the cool heat of the imagination. The transmutative process that is involved may be described as semantic motion: the idea of which is implicit in the very word "metaphor", since the motion (phora) that the word connotes is a semantic motion -- the double' imaginative act of outreaching and combining that essentially marks the metaphoric process.*"

Anne Buttimer (1982) notes that: "*Metaphor, it has been claimed, touches a deeper level of understanding than "paradigm", for it points to the process of learning and discovery - to those analogical leaps from the familiar to the unfamiliar which rally imagination and emotion as well as intellect".* Furthermore she points out that (citing Cassirer and Lange): *"This propensity to make symbolic transformation of reality is the most characteristically human activity of all"*. She quotes E L Doctorow (1977): *"The development of civilizations is essentially a progression of metaphors"*.

Wheelright also argues that there is a release of semantic energy through the tensive fusion of terms. This notion has been elaborated by Paul Ricoeur (*The Rule of Metaphor*, 1977) who argues that a metaphor *"tells us something new about reality"*. In this sense metaphoric utterance is a creative mode of knowledge.

6. Reality construction

According to Andrew Ortony (*Metaphor and Thought*, 1979) one of the dominant presuppositions of our culture is that the description and explanation of physical reality is a respectable and worthwhile enterprise -- an enterprise called "science". Science aims at a precise, unambiguous, literal language. There continues to be considerable faith in such literal language for good reason. A different approach is however possible. *"The central idea of this approach is that cognition is the result of mental construction. Knowledge of reality, whether it is occasioned by perception, language, memory, or anything else, is a result of going beyond the information given."*

The objective world is not directly accessible, but is constructed on the basis of the constraining influences of human knowledge and language. Ortony points out that this constructivist approach seems to entail an important role for metaphors in both language and thought, since the use of language becomes an essentially creative activity. By contrast, the conventional nonconstructivist position has metaphors as rather unimportant, deviant, and parasitic on "normal usage". Both perspectives have advantages and disadvantages. **The advantages of the constructivist use of metaphors have not been adequately explored in relation to the current challenges of society. In this sense metaphors can provide a different way of seeing the world and its problems.**

7. Metaphor and learning

It is interesting, with respect to the challenge of collective learning, that the American Cybernetic Society award for the best paper of the year has recently gone to Kathleen Forsythe, for a paper entitled: *Cathedrals in the Mind; the architecture of metaphor in understanding learning* (1986). In it she notes: *"The pervasiveness of metaphor in our conceptual system suggests a central and basic role in the underlying architecture of thought. Metaphor represents the ability to understand one thing in terms of another as we ascribe an understood pattern to unknown phenomena and perceive their structural integrity within the environment of our experience. We can then begin to perceive the environment of learning as one in which analogical thinking serves as architecture, analytical thinking serves as engineering and the imagination ensures that the interactions which create life and meaning are always being realized anew. The implications for this approach to applied epistemology provide insights into the design and development of learning systems that support the creative nature of learning."*

Forsythe points out, citing David Bohm (*Wholeness and the Implicate Order*, 1980), that the issues of content and process are no longer the key issues in the new ways of thinking about learning. But content and process are now to be seen as two aspects of one whole movement.

For Forsythe: *"The fundamental difference in this new view of learning is to see analogical thinking as the architecture and analytical thinking as the engineering of our mind's view of the world. Thinking and learning become a dynamic "open" geometry (Fuller, Synergetics 2, 1979) characterized by increasing complexity and transformation as a dissipative structure (Prigogine, Order Out of Chaos, 1984) based on a kinetic, relational calculus (Pask, Conservation, Cognition and Learning, 1975). The meta design is not built on inference and syllogism but on analogy and relation thus allowing form to develop from an underlying logic - the morphogenises of an idea. (Sheldrake, A New Science of Life, 1983). Knowledge is seen not as an absolute to be known but always in relation to agreement and disagreement, to coherence and distinction in terms of individual, cultural and social points of view. The language we use to communicate then takes on a heightened importance (Wittgenstein, The Blue and Brown Books, 1972)) whether that be the language of words or the metaphor language of pattern (Alexander, The Timeless Way of Building, 1979)."*

8. Transcendence through metaphor

There is therefore an emerging recognition that univocal language cannot satisfactorily express the degree of complexity or subtlety which has been found necessary to embody the relationships affecting human understanding of man's relation to the universe. There is recognition that not only is human experience metaphorical in nature, but also that metaphor is an essential constituent of the structure of human experience. That is, part of the meaning of any experience is elusive, and it is the use of metaphor that gives form to this elusive meaning and makes available an understandable figure of speech (Robert D Rmanyshyn, *Metaphors of Experience and Experience as Metaphorical*, 1981). Mutual interaction between the metaphor tenor and vehicle has been discovered to be the paragon of all coherent experience in which sensation is able to achieve a state of resonance with some residuum from the experiencer's heritage of remembrance (S G Perrin, *Metaphor to Mythology*, 1982).

It is interesting that the current explorations of the function of metaphor are clarifying its traditional use in conveying subtleties which are denatured by conventional categorization, namely the kind of modes of awareness described in Section HM. J P van Noppen (Metaphor and Religion, 1983, p.4) points out that *"while it is becoming clear that metaphor is not a panacea providing the final answer to all questions raised by human attempts at framing a transcendent mode of being in man-centred language, the present evolution traces paths of thought and investigation which deserve to be pursued and which are...being trodden with a great deal of enthusiasm."* This has been stimulated by explorations of the mechanism whereby man's words could be "stretched" beyond the usual limits of this worldly reference.

For Gibson Winter (1981): *"...if the present age faces a crisis of root metaphors, a shift in metaphors may open new vistas of human possibilities. Metaphor is, in this sense, a vehicle for transcendence and freedom."* And again: *"Metaphoric insights in science, philosophy, religion, and morality open vistas of humanization which had not been imagined.. Here transcendence means and inner, historical distancing from the determining forces that encompass human dwelling."* And: *"the deeper reality of human dwelling is that it thrives upon creative, metaphoric disclosures and decays when such powers degenerate into mechanical repetition."*

Van Noppen stresses however that exponents of metaphor have not been blinded to the limitations of the medium. The contributors to the reader edited by him repeatedly emphasize that the metaphor *"should not be taken beyond its point, ie should remain subordinate to the insight it was coined to express, and perhaps even be adapted when the actual insight is blurred or swamped by secondary associations."* (p.4)

Significance: metaphor, analogy, symbol and pattern

In the humanities subtle distinctions have long been made between metaphor, analogy, allegory, synecdoche, metonymy, parable, symbol, and the like. The assumption made here is that these may usefully be considered as forming a continuum which may indeed be segmented in a number of ways. But it would seem to be the case that the long and intense debate on the appropriateness of some particular pattern of segmentation has obscured the possibility of vital and unexplored uses of metaphor in response to the crisis of the times.

1. Models, analogies and metaphors
There is an ongoing debate in the philosophy of science concerning the status of models and analogies in relation to scientific "explanations" (Mary B Hesse, *The Explanatory Function of Metaphor*, 1966). Sharp distinctions have been made by some between models, analogies, metaphors and isomorphs in this debate. As a technical term, analogy has been used to designate a kind of predication midway between univocation and equivocation.

For many years it was therefore of no interest to scientists who perceived themselves as dealing only in univocal terms. But research in fundamental physics involving phenomena only describable by complex equations has given increasing legitimacy to analogy and metaphor as tools with which to create, comprehend and communicate complex intangibles. Analogy and metaphor, however they may be distinguished, have thus come to serve the same function as they have traditionally had in theology where metaphor has been used to focus the mind on the dynamic real.

Gibson Winter (1981) provides one possible clarification of the relationship between metaphor and analogy. *"Analogy is an abstract way of talking about a working metaphor, since analogy draws upon a conjunction in things which is discerned in reflection. However, analogy draws out the similarity in the metaphoric imagery, letting the dissimilarity fall into the background. In this sense, analogy functions as a weakened metaphor."*

But he argues that: *"Analogy and metaphor have the common element of drawing upon similarities for understanding the less known from the better known. However, tensive metaphors are much more than this, since they conjoin similar and dissimilar realities in an explosive disclosure of insight. This is the paradoxical quality of the tensive metaphor. A metaphor thus resolves contradictions in an unexpected way which generates new understanding."*

Kathleen Forsythe (1986) in a paper to a meeting of cyberneticians argues: *"Analogy and its poetic expression, metaphor, may be the "meta-forms" necessary to understanding those aspects of our mind that make connections, often in non-verbal and implicit fashion, that allow us to understand the world in a whole way."*

2. Isophors
Forsythe uses the term isophors for isomorphisms experienced in the use of language. Isophors are distinct from metaphors in that they are experienced directly. With the isophor there is no separation between thought and action, between feeling and experience. The experience itself is evoked through the relation. She suggests that the experience of one thing in terms of another, the isophor, is the means by which we map domain to domain and that our consciousness of this meta-action, when we observe ourselves experiencing this, lies at the heart of cognition. She has postulated the development of an epistemology of newness in which learning is the perception of newness and cognition depends on a disposition for wonder leading to this domain of conception-perception interactions.

She argues that the notion of metaphor is commonly understood to mean the description of one thing in terms of another. This notion presupposes an objective reality. This objectivity may be questioned and if, as suggested by Maturana, (objectivity) is placed in parentheses, *"we can begin to appreciate clearly the role we play in the construction of our own perception of reality. for this reason, the notion of the experience of one thing in terms of another, the isophor, suggests that it is this dynamic constructing ability that involves conception and perception -- unfolding and enfolding, that this gives rise to the coordination of actions in recursion which we know as language."*

3. Symbols
Winter (1981) points out that: *"Metaphors do not settle everything, but they are guides to the rich possibilities of life and nature."*

"Metaphors furnish clues to transformation, but they are not the powers that resist or engender such new realities...Symbols are the powers that resist change or open the way to creative change when it is needed. Root metaphors are interpretations of these founding symbols...Symbols are, in fact, metaphoric events that arise through the poetic powers of the human species."

He calls for caution regarding symbols: *"The symbol is ambiguous and, indeed, conceals as well as it discloses the powers and rhythms of life and cosmos which it mediates. To identify the symbol wholly with the higher order reality it mediates is to reify that which transcends...When the mediation of higher order realities supplants the reality, symbolization becomes idolatry...This reification of the higher order realities that are mediated in symbolization is the peculiar idolatry of the modern age."*

He then raises a fundamental issue: *"The question remains as to how one determines an authentic rather than inauthentic symbolization of life. With all their promise, Western symbols have revealed their corruption. To transcend the crisis of Western spirituality and the imminent destruction of non-Western peoples, at least on the level of understanding and ideology, a new paradigm and spiritual foundation are required."*

Pattern language
Forsythe (1986) stress the relationship between metaphor and pattern language: *"The architecture of how we structure the reality of our imagination is metaphoric. Metaphors are bridges that order the nature of our collective and individual humanity. Metaphor provides the reality to the pattern language of thought for it is the mechanism of ordering newness. Language only lives when each person has his or her own version that must constantly be re-created in each person's mind as he or she interacts with others in the environment. It is only through understanding these inner patterns that we can begin to consciously bring the outer pattern of our lives into harmony."*

Significance: root metaphors and social organization

1. Root metaphors
There is of course a multitude of metaphors on which politicians draw to increase the power of their communication and these have been extensively studied (see Van Noppen, 1985), including metaphors implicit in the Communist Manifesto (Angonet et al, 1980). It has even been said that *"The politician without metaphor is a ship without sails"* (Fairlie, 1979). But such uses of metaphor tend to correspond to the communicative or illustrative function.

There seem to be very few studies of the use of metaphor in economics (Asheraft, 1977), public policy-making (Rosenthal, 1982), strategic planning (Solesbury, 1981) or decision-making (Maranda, 1980), namely the internal, initiatory processes of governance. The challenge for governance is to discover whether there are not some key "root" metaphors, each especially applicable under certain circumstances. Of even greater value would be the existence of any systematization of them relevant to the problems of governance. The existence of such root metaphors has been reviewed in work by the geographer Anne Buttimer (1982, 1983).

Buttimer examines geographic thought and practice in terms of a contemporary trend away from observation (of reality) to participation (in reality). This trend is not so much a linear or chronological progression as a conceptual (even ideological) transformation, involving epistemological, dialectical and hermeneutical phases. Her basic approach is inspired largely by Stephen Pepper's theory of world hypotheses (1942) by which four distinct world views are claimed to have stood the test of time in Western intellectual history: formism, organicism, mechanism and contextualism.

Each of these hypotheses about the nature of world reality grounds its claims to truth, and its categories of analysis, on a "root metaphor". As an example, in geography she sees a reflection of these macro world views in the root metaphors of "map", "organism", "mechanism" and "arena".

Thus: formism grounds itself on the common sense experience of similarity and a correspondence theory of truth, expressed in the case of geography in a preoccupation with mapping. Mechanism, based on a causal adjustment theory of truth, takes the machine as root metaphor, resulting in a preoccupation with special systems and functional mechanisms in the case of geography. Organicism, based on a coherence theory of truth, regards every event as a more or less concealed process within an organic whole. Finally, contextualism, based on an operational theory of truth, sees the world as an arena of unique events.

Buttimer points out that creative thinkers move quite freely between the different styles of these root metaphors. Each of them *"spells a distinct design for the physical and functional arrangements of space, time, and activities on the ground; their often incompatible demands eventually becoming legible in the texts and textures of urban life. Which metaphor, or combination of metaphors, will endure or dominate at any particular will, of course, be a function of economic and political power interests..."* (1982).

2. Metaphoric networks
The concept of a root metaphor has acquired importance since the 1970s. Thus according to Gibson Winter (1981): *"The human species abides in its world through the meanings borne in thought and discourse. This is not to suggest that the world is merely a mental phenomenon, for by the way people treat things and others, by the way they organize work and family life, people express their understandings and feelings. At the same time, ways of working, acting, loving, and fearing shape ways of understanding and interpreting. To this extent, the comprehensive metaphors that give clues to the coherence of things serve to shape human activities even as they are reshaped by patterns of life and work."*

Root metaphors may be related to the notion of a metaphoric network of which Paul Ricoeur (1976) states: *"Metaphorical functioning would be completely inadequate as a way of expressing the different temporality of symbols, what we might call their insistence, if metaphors did not save themselves from complete evanescence by means of a who array of intersignifications. One metaphor, in effect, calls for another and each one stays alive by conserving its power to evoke the whole network. the network engenders what we can call root metaphors."*

3. Societal coherence and the war of worlds
Gibson Winter (1981) points out: *"However, root metaphors furnish important clues to the institutional; struggles and symbolic clashes.. This is a war of worlds, for it is a contention between total views of life and foundational symbolizations of the world. Root metaphors furnish important clues to such totalities. In fact, it is only though such clues tat one can gain a sense of the coherence of the total world that encompasses thought and life. There is no way to subsume one's total world under a concept. Only a comprehensive metaphor can guide thought toward that totality. In this sense the root metaphor is the first step toward foundational understanding."* Following Pepper (World Hypotheses, 1961), he argues that *"certain metaphoric networks become dominant in a total society, shaping modes of thought, action, decision, and life."*

He continues: *"comprehension of the clash of worlds invites retrieval of a more encompassing symbolic heritage and its conversion to an emerging,; more human world. The artistic vision opens the way to such a retrieval and creative transformation."*

On this basis he further argues that *"if one assumes that comprehensive metaphors furnish coherence in a people's world, there is no way to reduce such metaphors to rational formulae or systems of thought. a people may shift its organizing metaphor over time, but it cannot dispense with a comprehensive metaphor without losing a sense of coherence.."*

4. Influence of natural symbols on social organization
Such an approach is not as incongruous as might be suspected. Mary Douglas, an anthropologist, has argued that the organic system provides an analogy of the social system which, other things being equal, is used in the same way and understood in the same way all over the world. The human body is capable of furnishing a natural system of symbols, but the problem is to identify the elements in the social dimension which are reflected in views of how the body should function or how its waste-products should be judged (*Purity and Danger*, 1966). In a more recent study she points out that according to the "purity rule": *"the more the social situation exerts pressure on persons involved in it, the more the social demand for conformity tends to be expressed by a demand for physical control. Bodily processes are more ignored and more firmly set outside the social discourse, the more the latter is important. A natural way of investing a social occasion with dignity is to hide organic processes."* (*Natural Symbols*, 1973, p.12)

But such dignity, despite its value, is essentially static and conservative, denying the dynamics of development, decay and renewal - more effectively contained by the essentially human folk rituals of carnival, etc. It is then easier to understand how oversimplified and "inhuman" our highest ideals become when they reject such bodily functions as digestion, excretion and intercourse. Douglas points out how uncomfortable some religions are with the association of such processes with a deity and consequently the difficulty they have in dealing with whatever they reject. Similarly in society's major institutions, there is no explicit conceptual link with that of themselves which they reject. The attitude towards bodily waste products is indicative of the degree of creative acceptance of the "loss" portion of any learning cycles.

As might be expected from arguments in Section KD, she identifies **four** distinctive systems of natural symbols, namely **social systems in which the image of the body is used in different ways to reflect and enhance each person's experience of society:**

(a) Body conceived as an organ of communication: *"The major preoccupations will be with its functioning effectively; the relation of head to subordinate members will be a model of the central control system, the favourite metaphors of statecraft will harp upon the flow of blood in the arteries, sustenance and restoration of strength."*

(b) Body seen as a vehicle of life: As such "*it will be vulnerable in different ways. The dangers to it will come...from failure to control the quality of what it absorbs through the orifices; fear of poisoning, protection of boundaries, aversion to bodily waste products, and medical theory that enjoins frequent purging.*"

(c) Practical concern with possible uses of bodily rejects: As such it will be "*very cool about recycling waste matter and about the pay-off from such practices...In the control areas of this society controversies about spirit and matter will scarcely arise.*"

(d) Life seen as purely spiritual, and the body as irrelevant matter: "*In these types of social experience, a person feels that his personal relations, so inexplicably unprofitable, are in the sinister grip of a social system. It follows that the body tends to serve as a symbol of evil, as a structured system contrasted with pure spirit which by its nature is free and undifferentiated. The millennialist...believes in a Utopian world in which goodness of heart can prevail without institutional devices.*" (p.16-17)

Clearly such distinct attitudes can well determine the kinds of political tendencies discussed earlier. It is unfortunate that Douglas did not broaden the scope of her study to include sexual behaviour. For although she recognizes its fundamental importance (p.93), she confines her concern to the significance of attitudes to the waste-products (of a single body) in determining behaviour within family systems. An equivalent focus on sexual behaviour would provide insight into the ways in which attitudes to alternation are similarly encoded and into the possibility of employing courtship and sexual symbols to enrich understanding of alternation processes in society.

Another bodily activity which encodes alternation is of course respiration, a favourite preoccupation in eastern philosophies. Again, however, this is focused on a single body and is therefore far less controversial and "seductive" as a form of presentation. This is the price of being less rich as a substrate for the generative dynamics of the relationship between opposites.

5. Family structure and social organization
Emmanuel Todd has explored the hypothesis that family relations constitute a model for the socio-political relations in each society. He points out that until recently this old hypothesis has proved quite useless due to the embryonic state of social anthropology. He argues that any such comparisons have lacked significance because of the narrowly eurocentric (*cf* Herb Addo, *Beyond Eurocentricity*, 1985) concept of valid socio-political forms: "*Est-il difficile d'admettre que la répartition mondiale des idéologies politiques et religieuses ne définit pas une structure dichotomique mais un ensemble multi-polaire et dont tous les poles - communistes, libéraux, catholiques, sociaux-démocrates, hindous, musulumans, bouddhistes - sont également normaux, légitimes et dignes d'analyse.*" (Emmanuel Todd, *La Troisième Planète*, 1983, p.12)

For Todd the family structure is an infralogical mechanism governing the reproduction of specific human values. This leads him to question the "grande illusion" that politics make society rather than the converse. Each culture, founded on a specific anthropological base, then engenders an ideological form of its own family values (p.24).

6. Images of social organization
In a much-lauded book *Images of Organization* (1986), Gareth Morgan considers the challenge to managers, administrators, organizational consultants, and others concerned with the effective functioning of organizations. The more proficient, he argues, develop a skill in the art of "reading" the situations that they are attempting to organize or manage. This skill usually develops as an intuitive process, learned through experience and natural ability. Such skilled readers develop the knack of reading situations with various scenarios in mind, and of forging actions that seem appropriate to the readings thus obtained. By contrast, less effective managers and problem solvers seem to interpret everything from a fixed standpoint with the consequence that their actions and behaviours are often rigid and inflexible and a source of conflict.

The basic premise of Morgan's approach is that "*our theories and explanations of organizational life are based on metaphors that lead us to see and understand organizations in distinctive yet partial ways. Metaphor is often just regarded as a device for embellishing discourse, but its significance is much greater than this. For the use of metaphor implies a way of thinking and a way of seeing that pervade how we understand our world generally.*" He points out, for example, that organizations are frequently discussed as if they were machines, designed to achieve predetermined goals and objectives, and which should operate smoothly. As a result of this way of thinking, attempts are often made to organize and manage them in a mechanistic way, forcing human qualities into the background.

His book explores and develops the art of reading and understanding organizations: "*First it seeks to show how many of our conventional ideas about organization and management build on a small number of taken-for-granted images, especially mechanical and biological ones. Second, by exploring these and a number of alternative images, it seeks to show how we can create new ways of thinking about organization. Third, it seeks to show how this general method of analysis can be used as a practical tool for diagnosing organizational problems, and for the management and design of organizations more generally.*" So, by using different metaphors to understand the complex and paradoxical character of organization life, it becomes possible manage and design organizations in ways that may not have appeared possible before. Morgan devotes separate chapters to the understanding of organizations in the light of the following metaphors:

(a) Organizations as organisms: From this perspective different types of organization are seen as belonging to different species, each suited for coping with the demands of different environments. This encourages the exploration of the relations between such species of organization as well as the evolutionary patterns in inter-organizational ecology.

(b) Organizations as brains: From this perspective organizations are seen in terms of information processing, learning, and intelligence. It provides a frame of reference through which such qualities can be enhanced. Some of the different metaphors already used for understanding the brain may also then be used, such as the brain as a computer and the brain as a hologram. The latter indicates principles of self-organization.

(c) Organizations as cultures: From this perspective organization is seen to reside in the ideas, values, norms, rituals and beliefs that sustain organizations as socially constructed realities. It focuses on the patterns of shared meaning that guide organizational life.

(d) Organizations as political systems: From this perspective, using a political metaphor, it possible to focus on the different sets of interests, conflicts and powerplays that shape organizational activities. Organizations can then be explored as systems of government based on various political principles through which different kinds of rule are legitimized.

(e) Organizations as psychic prisons: From this perspective organizations can be explored as places where people become trapped by their own thoughts, ideas, and beliefs, or by preoccupations originating in the unconscious mind. This offers insights into the psychodynamic and ideological aspects of the organization.

(e) Organization as flux and transformation: From this perspective the organization is understood in terms of the different logics of change shaping social life. Examples given are: organizations as self-producing systems creating themselves in their own image; organizations as produced through the results of circular and negative feedback; and organizations as the product of a dialectical logic through which every phenomenon tends to generate its opposite.

(f) Organizations as instruments of domination: From this perspective the potentially exploitative aspects of organization can be explored, especially in order to facilitate understanding of the ways in which management-labour relations have been radicalized and of the attitudes of exploited groups.

Significance: generative metaphor and policy-making

1. Societal problem setting
In a key paper, Donald Schön (1979) argues that *"the essential difficulties in social policy have more to do with problem setting than with problem solving, more to do with ways in which we frame the purposes to be achieved than with the selection of optimal means for achieving them."* For Schön *"the framing of problems often depends upon metaphors underlying the stories which generate problem setting and set the direction of problem solving."*

As an example he explores the case of slum housing. If the underlying metaphor is a slum is a "blight" or "disease", then this encourages an approach governed by the corresponding medical remedies, including the surgery whereby the blight is removed. On the other hand, if the underlying metaphor is that the slum is a "natural community", then this orients any response in terms of enhancing the life of that community. The two perceptions and approaches are quite distinct and have quite different consequences in practice.

In this sense problems are not given. For Schön, they are constructed by human beings in their attempts to make sense of complex and troubling situations. Ways of describing problems move into and out of good currency." Furthermore, new understanding of problems does not necessarily result as a logical consequence of the success or failure of instrumental responses to problems as previously defined. Rather attempts at solutions, based on partial understanding, generate other kinds of problem situations, without necessarily resolving those that existed initially. For Schön, the *"social situations confronting us have turned out to be far more complex than we had supposed, and it becomes increasingly doubtful that in the domain of social policy, we can make accurate temporal predictions, design models which converge upon a true description of reality, and carry out experiments which yield unambiguous results. Moreover, the unexpected problems created by our search for acceptable means to the ends we have chosen reveals...a stubborn conflict of ends traceable to the problem setting itself."*

2. Metaphoric "spells"
There is a marked tendency to be unaware of the metaphors that shape perceptions and understanding of social situations. Although metaphors are at once tacit and often outside of explicit experience, they represent a special way of seeing, a way of selecting and naming "facts" that can filter or distort what is taken in. For Schön, if the metaphor remains tacit, it effectively exerts a "spell" that conditions the way in which a situation is perceived as problematic. The consequence of such tacit metaphors can be dangerously counter-productive. Metaphors can constrain and sometimes dangerously control the way in which people and groups construct the world in which they act.

If, however, the metaphor no longer remains tacit, because an effort is made to "spell it out", other possibilities emerge. The assumptions which flow from the metaphor can be elaborated in terms of the appropriateness of the metaphor to the circumstances in question.

3. Conflicting frames
From this perspective, the difficulties of collective response to societal problem situations are severely compounded by different perceptions of the nature of the problems. Society's ills receive conflicting descriptions, often couched as metaphors. For Schön, *"Such a multiplicity of conflicting stories about the situation makes it dramatically apparent that we are dealing not with "reality" but with various ways of making sense of a reality."* Some of these may be based on an inappropriate or simplistic understanding of the situation. Inadequate metaphors carry with them implicitly, and often insidiously, natural "solutions".

The disagreements may be more fruitfully understood as frame conflicts. Conflicts of frames, according to Schön, cannot be resolved by any appeal to the facts, because all the "relevant" facts have already been selected, filtered, and embedded in the metaphors through which the situation is variously perceived. The dilemma of such frame conflicts may of course be avoided, through a form of conceptual surgery, by deliberately omitting consideration of any inconvenient values that give rise to the dilemma.

4. Generative metaphor
A generative metaphor in Schön's terms is characterized by the carrying over of frames or perspectives from one domain to another. It allows for frame re-structuring when frame conflict exists. The notion of a generative metaphor may be extended to include a proposed metaphor that frames socially constructed reality in a new, more complex way (S Srivastva and F Barrett, 1988).

Not all metaphors are generative. Some merely capitalize upon existing ways of seeing things. In the case of generative metaphor, however, there is an actual generation of new perceptions, explanations and inventions. The sense of the obviousness of what is wrong in a problematic situation, and what needs fixing, is the hallmark of a generative metaphor in Schön's sense. A generative metaphor *"derives its normative force from certain purposes and values, certain normative images, which have long been powerful in our culture."*

For Schön, the notion of generative metaphor *"then becomes an interpretive tool for the critical analysis of social policy"*. He argues that *"it is not that we ought to think metaphorically about social problems, but that we do already think about them in terms of certain pervasive, tacit generative metaphors; and that we ought to become critically aware of these generative metaphors, to increase the rigour and precision of our analysis of social policy problems by examining the analogies and "disanalogies" between the familiar descriptions...and the actual problematic situations that confront us."*

There is therefore reason to be wary of generative metaphors, especially when they carry their own solutions to problems. More often than not they will fail to present an objective characterization of the problem situation.

5. Frame restructuring
In the presence of generative metaphors governing perception of a problem situation, Schön advocates a process of frame restructuring. This involves the design of a new problem-setting story or metaphor. Under the new metaphor, an attempt is made to integrate conflicting frames by including features and relations drawn from earlier metaphors, yet without sacrificing internal coherence or the degree of simplicity required for action. He gives a concrete example involving a squatter settlement in a developing country.

He stresses that conflicting descriptions cannot be effectively mapped onto one another by matching corresponding elements in each. Rather it is restructured descriptions of a situation from different perspectives which are coordinated with one another: *"some pairs of restructured elements now match one another, and others are juxtaposed in the new description as components of larger elements."* The new description is also not a compromise or average of the values implicit in earlier descriptions. Rather *"there is a shift in the meanings of these terms, and along with this, a shift in the distribution of the redescribed functions of initiative and control."*

6. Transforming nature of metaphors in group development
Frame restructuring can be understood as co-inquiry. Experimental procedures have been developed by the Department of Organizational Behaviour of Case Western Reserve University (Cleveland, USA) to explore group development (S Srivastva and F Barrett, 1988; F Barrett and D Cooperrider, 1988). They put forward propositions concerning metaphor and group process that suggest how: paying attention to metaphors are indicators of a group's phase of development; metaphor facilitates learning and overcomes resistance to otherwise difficult subjects; metaphor facilitates growth and development of the group; metaphor enables the group to construct its own social reality.

They are currently extending this work to international associations (P Johnson and D Cooperrider, 1991).

Significance: patterns of concepts

1. Context
With the explosion of information in a multiplicity of fields, it is natural that many concepts have emerged in different domains. Whether in presenting new concepts for communication amongst peers, or for the education of a wider audience, or in organizing shared concepts in policy making, there is a tendency to group concepts into sets.

An agenda, in the forms of a list, is the simplest example, possibly resulting in a series of items in a resolution or an agreement. A matrix of issues, factors or categories, is a more complex example frequently used in academic papers. Although in the latter case the tendency is to employ simple matrices *eg* 2x2, 2x3, 3x3, *etc*. Much more complex is a structure such as the periodic "table" of chemical elements, for which better forms of presentation continue to be sought. In the religious traditions of many cultures there are also examples of lists, tables and concentric presentations of categories, especially qualities (vices or virtues), energies, and deities.

Such devices clearly serve as a form of presentation enabling people to obtain an overview of the relationship between the categories so ordered. They have an important integrative function. Faced as society is with a multiplicity of relevant concepts, it is appropriate to ask whether sufficient attention has been given to the patterns within which such concepts are ordered. This is especially the case when faced with inconsistent or incompatible concepts from different perspectives, as in usually the case when dealing with the global problematique.

The question of how incompatible concepts should be interrelated was brought to the fore during explorations of the possibility of integrating the insights of the Goals, Processes and Indicators of Development project (1978-82). This project of the United Nations University's Human and Social Development Programme was unique in the variety of ideologies, cultures and disciplinary backgrounds, that the participating individuals and research institutes represented. How indeed could the results be presented if they were to be of significance as more than an administratively ordered collection of contributions? How is the pattern of insights to be conveyed, reflected or constrained by the patterning of the form chosen to embody them?

It is of course possible to use film, theatre, poetry, posters, and many other methods. But, because of their nature, these do not lend themselves to a **systematic** precise elaboration of the concerns of a complex project. They can only pick out **portions** of the total pattern for special emphasis. And even if a series of them is used, it is not clear that any such "sequences" would protect the essentially **non-linear** characteristics of a pattern involving conceptual discontinuities between perspectives.

When it is required to convey carefully a complex pattern, it is of course possible to do so using conventional text (even with appropriate illustrations). But the more carefully and precisely this is done, the longer will be such a document. And the longer it is, the more it will tend to defeat its purpose as a medium of communication.

This problem is not new. For example, in the case of the major *Capacity Study of the United Nations Development System* (the "Jackson Report", 1969), the results were presented in two volumes, the first of which was a short summary volume, itself introduced by a summary. But in order to be "readable", the underlying pattern of concepts had to be distorted and abridged. This method is primarily useful for making a single point (*eg* "something must be done") and not for presenting the pattern of concepts which need to be taken into consideration in ensuring that appropriate balance is maintained.

Another widely-used approach is to produce a list of points, as in the International Development Strategy or the Universal Declaration of Human Rights. Such lists may be enriched by annotations. The lists are structured somewhat like conference "resolutions" or "recommendations". Whilst they are convenient as checklists, they imply by their structure a simplistic **linear** "pattern" of relationships between the points in the list - even if some of the points have sub-points. But unfortunately such lists, as they are used in practice, strongly de-emphasize any relationships between the items (including the linear ones), even if they were recognized when the list was elaborated.

How then is the underlying "master" pattern to be expressed, protected from "erosion", and rendered widely communicable, given the constraints noted above? For whilst parts of the pattern may be significant in themselves, their special significance if any lies in the attempt to harmonize the parts, arising from complementary concerns, into a coherent whole.

2. Investigation of precedents
If, as noted above, this is not a new problem, then there is much to be gained by reviewing a wide range of concept schemes to see whether there are lessons to be drawn which could be of value. In selecting such examples three major considerations need to be borne in mind:

(a) Examples are required which cover, in each case, more than one **single** set of concepts (like the single set of 31 articles of the Universal Declaration of Human Rights). The examples should have **multiple** sets of concepts in which there is some relationship, or pattern of relationships, to be maintained between such sets. This is necessary because sub-projects will each tend to generate concept sets in response to the few sets which govern any project's conception as a whole.

(b) Some of the examples should be from domains in which the prime emphasis is **not** intellectual but rather expressive (as in the case of music, art, dance, *etc*). This consideration had already been accepted by the group in question and gave rise to the exploration documented in Section CF.

(c) Some of the examples should be from the **non-western** cultures which characterize many of the peoples to whose condition such a project must necessarily respond - especially in a "dialogue" situation.

In addition to the above considerations, it would also be valuable to examine:

(d) **Traditional** concept schemes which have retained their significance and communicability (above all at the village level), despite the passage of considerable periods of time since they were elaborated.

(e) Elaborate **modern** concept schemes which attempt to respond either to the complexity of the social condition using the full benefit of the range of western disciplines, or to a complex body of data.

(f) Concept schemes elaborated in the light of problems of comprehensibility and communicability.

Clearly the schemes selected should preferably be as independent of one another as possible.

3. Presentation of patterns
In Section CP, three completely different approaches to such concept patterns are presented.

(a) **Pattern language:** This uses the pattern language for environmental design, elaborated by Christopher Alexander and his colleagues, as a set of 253 interrelated metaphors from which equivalent patterns were then developed as an experiment for the socio-organizational, conceptual and intra-personal environments. In this case the patterning derives from Alexander's art as a designer constrained by his practicality and sensitivity as an architect.

(b) **Pattern of changes:** This uses the highly developed pattern of the Chinese classic, the *Book of Changes*, based on 64 interlinked conditions and adapts this, as an experiment, into western organizational jargon. It is therefore an exploration of the possibility of making further use of a traditional pattern of concepts which is still highly valued.

(c) Pattern of disagreement: This is an exercise in designing a multi-set concept scheme. The intention was to "internalize" the maximum amount of disagreement within the scheme as a guarantee of its relevance to a society in which disagreement is rife, whether constructive or destructive. The exercise resulted in an ordered series of 210 mutually- incompatible, transformation-oriented statements. It may be considered as an initial step towards more realistic organization of psycho-social development, cured of the tendency to "disagreement phobia" and of the desperate compensatory pursuit of agreement-promoting processes.

4. Ordering concept scheme material for comparison

Any investigation, such as that suggested above, can easily run the risk of being overly ambitious and falling victim to the problem of "presentation" it is designed to clarify. A "filium ariadnis" is required as a guide through the conceptual labyrinths to be investigated. The following procedure was therefore used in support of the above experiment in designing a multi-set concept scheme.

An earlier investigation considered the role of number in the formulation of **complete** sets, and the associated problems of their representation, comprehension and communication (A J N Judge, *Representation, Comprehension and Communication of Sets*, 1978). Since the concept schemes it is proposed to investigate are supposedly made up of complete sets, it could well be that the number of elements in any such set would provide a relatively simple and unambiguous way of ordering the material. That investigation in fact reinforced the arguments of other authors in favour of such an approach.

Examples of concept sets were collected from 20 integrated multi-set concept schemes of the most diverse nature and compared with the organization of the sets of concepts on which the UN University's project were based (A J N Judge, *Patterns of N-foldness*, 1978). To facilitate comparison they are ordered by number of set elements. The paper itself discusses some tentative approaches to the process of comparison. A special merit of the approach outlined lies in the concern with protecting the concept patterns as a whole.

5. Comment

A great danger with many exciting "sensitizing" forms of presentation is that they "excite" people's awareness very successfully to levels at which there is no coherent pattern which can ensure the permanent stabilization of their awareness at that level. Such **temporary** increases in "attention potential" are associated with high "forgetability" (witness the track record of various UN public information programmes). Conceived in this way the challenge is to find ways of stabilizing such patterns of nested sets and of facilitating shifts of attention between them. This has been compared to the problem of a "conceptual gearbox" in a separate paper (A J N Judge, *The Future of Comprehension*, 1984). One problem is to find ways of using "higher gears" to mesh more effectively with the rate of social change. This is the question of how to develop more operationally significant concept sets.

The more excitingly communicative, audio-visual forms are based on the axiom of minimum number of explicit messages per presentation, thus fragmenting the whole pattern. It would therefore be important to establish what forms are useful to communicate what parts of that pattern. it is important also for people to be able to transfer, from the sets and associated symbols of a concept scheme with which they are familiar (and have faith in), to whatever new ones can be presented, with a sense of **continuity** - otherwise the latter will be perceived as of ersatz quality. This is a reason for establishing the relationship between a variety of concept schemes.

Significance: communication through symbols

1. Ubiquity of symbols
The international community is faced with the twofold challenge of information overload and fragmentation of relevant patterns of information. Many of the possibilities of conceptual integration are reviewed in Section K, but it is not clear whether enough can be achieved in time nor whether such conceptual integration will meet the communication need, especially in relation to the wider public. The possibilities of using metaphors and their relationship to patterns of concepts are reviewed in Section MM and MP. It would however be a mistake to neglect the role of symbols, especially to the extent that they can play a more fundamental integrative role.

Symbols are widely used by the international community. Attempts are made to give an identity to most major international programmes through the use of a symbol. Most international organizations make use of a distinctive symbol on their letterheads and other documents. More informally the mass media are constantly in the process of searching for symbols which will attract the attention of their audiences, whether it be Mère Thérése, a Live Aid happening, Baby Doc Duvalier, the Challenger space shuttle, child ambassadors (exchanged between the USA and the USSR), or some media star.

Whilst the role of symbols is scorned in academic and administrative responses to the global problematique, symbols are heavily exploited by the advertising media in helping people to define their self-image and personal development in terms of the acquisition of consumer products. Symbols are also used, under the guidance of media consultants, in helping to define the identity of national political parties. Such manipulation of contemporary symbols is only possible because of public sensitivity to a large body of symbolism which has become traditional over the ages and constituting an international language transcending the normal limits of communication.

The question is whether some form of symbolism can provide any assistance in new responses to the global problematique.

2. Definition
A distinction must be made, following Erich Fromm (*The Forgotten Language*, 1952), between three kinds of symbol: (a) the conventional, such as the signs used in industry, in mathematics and in other fields; (b) the accidental, resulting from temporary association whereby one thing symbolizes another; (c) the universal, where there is an intrinsic relation between the symbol and what it represents. It is the latter which is the concern of this section.

According to J E Cirlot, definitions and analyses of symbols and symbolism are all too frequent (*A Dictionary of Symbols*, 1971). In many ways it is the excesses and mutual antagonism of those sensitive to the symbolic dimension which has prevented the symbolic heritage of humanity from being explored as a resource relevant to an integrative response to the global problematique.

Because of their integrative function, symbols lend themselves to many definitions and interpretations. it is therefore doubtful whether it is possible or useful to seek any precise definition. Cirlot adopts the approach of pointing to useful insights into their nature, rather than reducing them to some formula. Thus for Lin Yu-tang symbolism has the virtue of containing within a few conventional lines the thought of the ages and the dreams of a race. It kindles our imaginational and leads us to realms of wordless thought. Commenting on this J C Cooper states that: *"This thought is not that of the individual ego; the symbol cannot be created artificially or invented for some purely personal interpretation or whim: it goes beyond the individual to the universal and is innate in the life of the spirit. It is the external, or lower, expression of the higher truth which is symbolized, and is a means of communicating realities which might otherwise be either obscured by the limitations of language or too complex for adequate expression."* (*An Illustrated Encyclopaedia of Traditional Symbols*, 1978, p.7) This certainly suggest that symbols have something to offer to the challenge of communicating paradoxical or mutually antagonistic insights as reviewed in Section KD. This is confirmed by Cooper's subsequent observation that: *"Much of symbolism directly concerns the dramatic interplay and interaction of the opposing forces in the dualistic world of manifestation, their conflicting but also complementary and compensating characteristics, and their final union, symbolized by the androgyne or sacred marriage. These are expressions of the unity of life which is the central point of all traditional symbolism."* (p.8).

3. Function of symbols
Cirlot summarizes the unique function of symbolism as a language of images and emotions based upon a precise and crystallized means of expression, revealing transcendent truths, both in the cosmic order external to man and within him. Furthermore it possesses a quality which increases its dynamism and gives it a truly dramatic character. This quality, the essence of the symbol, is its ability to express simultaneously the various aspects, namely both thesis and antithesis, of the idea it represents. The symbolic function may therefore be said to appear or become active when a state of tension is set up between opposites which the normal conscious mind cannot resolve by itself. The symbolic function is thus effectively centred within the unconscious and does not recognize the inherent distinctions of contraposition (J E Cirlot, 1971, p.xxxi).

Mircea Eliade perceives the function of the symbol as going beyond the limitations of man and his particular concerns as "fragments" and of integrating such fragments into entities of wider scope such as society, culture or the universe. But an object transmuted into a symbol is not simply dissolved into totality, for the symbol does not restrict movement or circulation between levels of reality or interpretation, rather it enables it, thus performing an integrative function. What the symbol makes manifest restates point by point what the totality manifests. (*Images et Symboles*, 1952, p.17)

The analogy between two planes of reality is founded upon recognition in both of a common rhythm according to Schneider (El origen musical de los animales-simbolos en la mitologia y la escultura antiguas, 1946), namely a *"coherent, determinate and dynamic factor which a character or figure possesses and which is transmitted to the object over which it presides or form which it emanates. This rhythm is fundamentally a movement resulting from a certain vitality or from a given "number"... Rhythm may be understood as a grouping of distances, of quantitative values, but also as a formal pattern determined by rhythmic numbers, that is, as spatial, formal and positional similitude."* (J E Cirlot, 1971, p.xxxii)

Rhythms and modes thus allow relationships to be perceived between different planes of reality in a manner similar to the approach of general systems theory in its preoccupation with isomorphy. But while natural science establishes relationships only between "horizontal" groups of beings following the classificatory approach of Linnaeus, symbols construct "vertical" relationships between objects having the same rhythm, namely objects which correspond or are analogous to others on a different plane. Symbols thus function in terms of a kind of magnetic force that draws together for comprehension phenomena having the same rhythm. (p.xxxiii). Symbolism thus reveals that the profound meaning behind all series of symbolic objects is the very cause of their repeated appearance in the world of phenomena. This can be related to David Bohm's explorations of the manner in which objects are unfolded to perception and re-enfolded into the wholeness of the implicate order (*Wholeness and the Implicate Order*, 1980). Such "objects" can include man's image of himself, which reinforces Donadeo's insight that: *"It could perhaps even be said that rather than having created symbols, man has himself been created by his capacity to form symbols."* Jung uses the word "archetype" to designate those universal symbols which possess the greatest constancy and efficiency, the greatest potentiality for psychic evolution, and which point away from the inferior to the superior. In discussing psychic energy he specifically states: *"The psychological mechanism that transforms energy is the symbol."* (*Symbols of Transformation*, 1956)

For Martin Grotjahn: *"The word symbol implies throwing together, bringing together, integrating. The process of symbol formation is a continuous process of bringing together and integrating the internal with the external world, the subject with the object, and the earlier emotional experiences with later experience."* (*The Voice of the Symbol*, 1971, p.178)

In the face of the fragmentation resulting from the explosion of information and specialized knowledge, analogy, as a unifying and ordering process, appears continuously in art, myth and poetry. It functions to unite what has been dispersed, exploring the unknown and, paradoxically, communicating the incommunicable (J E Cirlot, 1971, p.xli). Given the increasing impossibility for an individual to absorb more than a small fraction of the knowledge of experts, reliance must necessarily be placed by educators, media documentaries, and policy presentations, on the use of analogy to convey complex insights. Such analogies are more easily remembered and have greater impact if they are drawn from patterns of symbols which reinforce one another.

4. Symbolic patterning and syntax

As Cirlot remarks in discussing "symbolic syntax", symbols, in whatever form they appear, are not usually isolated. They appear in clusters, giving rise to symbolic compositions which may be evolved over time (as in the case of story-telling), in space (works of art, graphic designs), or in both space and time (drama and dreams). Combinations of symbols thus evidence a cumulative meaning (*A Dictionary of Symbols*, 1971).

There is thus a rich network of relations between symbols in the form of correspondences and analogies. For Mary Douglas: *"A symbol only has meaning from its relation to other symbols in a pattern. The pattern gives the meaning. Therefore no one item in the pattern can carry meaning by itself isolated from the rest."* (*Natural Symbols*, 1973)

Unfortunately little effort is made by those concerned with symbols to map out such networks. The tendency in the literature is to focus on a few relationships at a time, explored in lengthy texts presupposing a sense of perspective which is not offered in any readily accessible form. There is a marked disinclination to map. Exceptions to this may be seen in some eastern systems using forms of mandala in which different symbols are associated with particular parts of the mandala and are thus interrelated. Perhaps the richest map of this kind, and the most explicitly documented, is that constituted by the network of explicit relations between the 64 hexagrams of the Chinese *Book of Changes* (see Section TP).

The pattern of concepts developed from Alexander's pattern language for building design (see Section MP) can also be usefully viewed in symbolic terms. Cirlot notes that the symbolism of architecture is complex and wide-ranging. *"It is founded upon "correspondences" between various patterns of spatial organization, consequent upon the relationships, on the abstract plane, between architectural structures and the organized pattern of space."* (p.15) He gives as example a Romanesque cloister with its central fountain, which although it corresponds exactly to the concept of a sacred precinct (temenos) with a silver thread (sutratma) linking from the centre to its origin, *"does not invalidate or even modify the architectural and utilitarian reality of this cloister; it enriches its significance by identifying it with an "inner form"."* Of special interest in Alexander's initiative is his effort to map the network of relations between some 253 design patterns which, to the extent that they are in his terms "archetypal", must necessarily also be understood in symbolic terms. His deliberate effort to embody the "quality without a name" may indeed by successful to the extent that such patterns reflect such an "inner form".

The exercise in Section MP to identify parallel pattern languages at the socio-organizational, conceptual and intra-personal level, may consequently be given further legitimacy on the basis of symbolic considerations. Cirlot stresses that there is an *"immense weight of testimony offered by human faith and wisdom"* proving that intangible order, whether conceptual, psychological or spiritual, is analogous in form to the material order." This is evident in the saying of Plato: *"What is perceptible to the senses is the reflection of what is intelligible to the mind."* Or, more recently, in studies on the calculus of self-reference, Francisco Varela concludes: *"In contrast with what is commonly assumed, a description, when carefully inspected reveals the properties of the observer. We, observers, distinguish ourselves precisely by distinguishing what we apparently are not, the world."* (*A Calculus of Self-Reference*, 1975). In this sense the exercise in Section MP is an effort to describe verbally a network of forms or patterns at various levels, which can most probably be much better described through the use of symbols. Alexander's architectural patterns may be interpreted in their symbolic sense to this end. Other symbols most probably exist for many of the intra-personal patterns. The network is however unique in its explicitness and merits further exploration in terms of its possible symbolic significance.

5. Resistance to the symbolic dimension

The widespread resistance to recognition of the special integrative function of symbols is in large part a backlash against the excesses of the pre-scientific period. The scientific method can even be said to have emerged in reaction to such excesses.

Cirlot points out that: *"The error of symbolist artists and writers has always been precisely this: that they sought to turn the entire sphere of reality into a vehicle for impalpable "correspondences", into an obsessive conjunction of analogics, without being aware that the symbolic is opposed to the existential and instrumental and without realizing that the laws of symbolism hold good only within it own particular sphere."* (*A Dictionary of Symbols*, 1971, p.xii)

It has been with great reluctance, and largely as a result of Freud and Jung, that any significance has been attached in the West to symbols in relation to the processes of human development. Such acceptance is however limited to the psychoanalytical disciplines, which are themselves viewed as of doubtful significance by most of the natural and social sciences. The only exception, as noted above, being those applied social sciences dealing with the use of symbols in advertising and propaganda or engaging in descriptive exercises on the use of symbols in different cultures. Outside the sciences however there is not such resistance, whether in the arts, in religion or in relation to the status symbols of an individual, a corporation or a country. This is especially the case in non-western cultures.

The resistance is however justified because of the continuing tendency, despite the emergence of scientific disciplines, towards sloppy and undisciplined thinking in relation to symbolism. It has proved difficult to distinguish between uncritical, undisciplined attitudes towards symbols and inspired, creative explorations of symbols. As pointed out above, symbols encourage freedom of association but in the absence of any countervailing discipline, the whole approach is easily associated with primitive, non-scientific thinking or purely aesthetic exercises.

Jung, who contributed so much to clarifying the transformative function of symbols noted that: *"For the modern mind, analogies - even when they are analogics with he most unexpected symbolic meanings - are noting but self-evident absurdities. This worthy judgement does not, however, in any way alter the fact that such affinities of thought do exist and that they have been playing an important role for centuries. Psychology has a duty to recognize these facts; it should leave it to the profane to denigrate them as absurdities or obscurantism."* (*Symbols of Transformation*, 1956)

Despite the wealth of work by psychoanalysts, and possibly because of the archetypal nature of their disagreements, psychologists and sociologists are still far from according any significance to symbolism. This resistance is reviewed by H D Duncan (*Symbols in Society*, 1968) who notes that systematic explanations of human relationships are determined by communication of significant symbols have not been much in vogue in America. In a major review of contemporary sociology, under the auspices of the American Sociological Association, "symbol" did not appear in the index. As sociology becomes more quantified it provides less insight into the nature of symbols. In discussing the confusion over the social function of symbols in sociology Duncan points to factors such as: confusion of the symbolic with the subjective, failure to study symbolic forms, the "trained incapacity" of sociologists in the use of non-mechanistic models, the definition of a symbol itself, and the fundamental ambiguity of symbols. In discussing the latter he suggest that it may indeed be the ambiguity of symbols which makes them so useful in society as a kind of bridge allowing people to alternate between different interpretations (namely the possibility reviewed in Section KD).

Symbols: degradation of symbolic forms

1. Central role of "symbol"
The resistance to symbols as criticized by Duncan has recently been reduced as a result of the Harvard Project on Human Potential (see Section H). In the first product of this project Howard Gardner writes at length concerning the central role of symbols: *"It is through symbols and symbol systems that our present framework, rooted in the psychology of intelligences, can be effectively linked with the concerns of culture, including the rearing of children and their ultimate placement in niches of responsibility and competence. Symbols pave the royal route from raw intelligences to finished cultures."* (*The Socialization of Human Intelligence Through Symbols*, 1984, p.300)

Gardner points to the vital bridging role of symbols between incommensurable domains such as biology (nervous system structure and function) an anthropology (cultural roles and activities): *"So far as I can see, there is no ready way to build a bridge between these two bodies of information: their vocabularies, their frames of reference are too disparate. It is as if one were asked to build a connecting link between the structure of a harpsichord and the sound of Bach's music: these entities are incommensurate."* (p.300) There is even a whole section on "symbolic development", based on the Harvard Project Zero. Unfortunately none of these references to symbols in any way distinguishes between the use of symbols as signs and the presence of some intrinsic relation between the symbol and what it represents. Gardner is quite clear about this: *"I conceive of a symbol as any entity (material or abstract) that can denote or refer to any other entity."* (p.301).

2. Symbols vs. Sign
Despite discussion of stages in symbolic development during an individual's development, this totally ignores the symbolic dimensions discussed in the previous note and presumably is totally unable to relate to them. Signs cannot apparently be distinguished from symbols in current theories of education. Indeed the index of Gardner's book contains no reference to "signs".

To the extent that people identify with symbols (often to the point of being enthraled by them), rather than with signs, the resulting forms of education can only enable people to manipulate signs in the *"niches of responsibility and competence"* in which they are placed within the economic system. This leaves people vulnerable to other experiences in the psychosocial system through which they may be able to learn the significance of symbols, as opposed to signs. Religions and other sects may be considered a relatively benign source of such learning, as are certain forms of advertising, and media events. However, the vulnerability of teutonic cultures to symbolic manipulation during the Nazi period indicates the possibility of more extreme forms of societal learning for which modern theories of education are unprepared.

3. Confusion about symbolism
Cirlot notes that much scepticism about symbolism, especially among psychologists, arises because of the confusion of two quite different aspects of symbolism: the manifestation of the true meaning of the symbolic object as against the manifestation of a distorted meaning superimposed by an individual mind prejudiced by circumstantial or psychological factors.

He points to the Freudians as offering one example of prejudiced interpretation of all forms as unveiling universal sexuality. He contrasts the Freudian approach with several classical symbol systems based on the essential polarity of the world of phenomena according to generic principles which in each case include the sexual division, but not to the exclusion of all other interpretations. Such distortion of the true meaning of symbols arises from an over-restriction of their function, as well as from over-identification with the psychological mechanism which construes it. (*A Dictionary of Symbols*, 1971, p.xlvii-xlviii)

Symbolic interpretations may thus be confined within the narrow limitations of allegory, restricting it to some particular level. Cirlot points out that such degraded meaning may not only affect the interpretation the symbol receives, but also the symbol itself. Other forms of degradation he notes include: trivial vulgarization; over-particularized interpretation, leading to lengthy and arbitrary descriptions; over-intellectualized, allegorical interpretations; and identifications through so-called analogy (p.li).

This has in no way discouraged interest in symbols in many sectors of society. But it is the disagreement between symbolists, itself symbolized by the relationship between Freud and Jung (and perpetuated by their interpreters), which one might expect to be contained and given significance by the symbolic domain. Symbolists, like other specialists, have however proved unable to heal their own domain.

4. Avoidance of symbols by international organizations
This environment, its encouragement of mystification, and the neglect by sociologists, has reinforced the tendency to ignore symbols in the conception of any organization or programme. As pointed out above, symbols are consequently of little interest to international institutions except for public relations purposes, as political tokens in campaigns to preserve symbols of a country's cultural heritage, or, ironically as indicators of the status of the institutions and those who work there.

Rather than explore the integrative function of symbols, the international community limits itself to the use of symbols solely for public relations purposes in a manner totally divorced from the challenges of interrelating the substantive problems under discussion. Symbols are used to paper over the substantive cracks and institutional rivalry. This is perhaps typified by the widespread use of the NASA photograph of the Earth taken from the Moon. This has been treated as an extremely powerful and meaningful symbol by many, especially by those concerned with the environment ("One Earth").

But it has not encouraged any further exploration of symbols as symbols. It is a symbolical irony that the symbols (or "logos" in their terms) of the various United Nations Specialized Agencies, for example, are invariably presented in isolation even though it might be assumed that they represent the different parts of a symbolic composition. There is no call to explore ways of representing their functional complementarity and it is therefore not surprising that their relationships should be primarily characterized by petty bureaucratic territorial squabbles. There is no attempt to capture and communicate the "cumulative meaning" of the functional pattern they represent.

The irony is all the greater in that it is such Specialized Agencies, or their corresponding ministries at the national level, which are in many ways (in the public eye) the modern equivalent of the different temples to the gods in ancient civilizations such as Rome, Greece or Egypt. This is in part confirmed by the symbols associated with such gods as their own (*eg* FAO, UNESCO, WHO). But whereas the relationships between the gods had been extensively dramatized in legends and folk tales, the relationships between the functions represented by such agencies is buried in the proceedings of a maze of coordinating bodies.

5. Signs of quantity vs. Symbols of quality
There is a strange historical symmetry to the fact that such ancient civilizations were so heavily influenced by qualitative symbols in all their decision-making but relatively indifferent to quantitative indicators. Modern institutions are however heavily influenced by quantitative "symbols" in the form of statistics and indicators and almost totally incapable of identifying qualitative indicators. This is especially clear in efforts to assess quality of life.

This irony is perhaps matched by the harmony in their cosmologies between macrocosm and microcosm as being reflections of each other, in contrast to the complete lack of ability to provide a framework to relate individual (micro) with global (macro) concerns in modern development programmes as evidenced by the United Nations University's Goals, Processes and Indicators of Development project which fragmented on this issue. It was symbols which provided the bridge, and in many ways continue to do so in non-western cultures.

6. Symbolic pattern implied by international institutions
For those who would see in such institutions as the United Nations a symbol of hope, it would seem to be vital to devote care to the nature of the symbolic pattern they constitute. What is the symbolic pattern constituted by the international community of institutions as a whole? To the extent that such institutions are foci of societal learning, Gregory Bateson's basic point is again relevant: once the pattern which connects learning foci is broken, all quality is necessarily destroyed (*Mind and Nature*, 1979, p.8)

J C Cooper notes: "*Symbolism is basic to the human mind; to ignore it is to suffer a serious deficiency; it is fundamental to thinking, and the perfect symbol should satisfy every aspect of man - his spirit, intellect and emotions.*" (*An Illustrated Encyclopaedia of Traditional Symbols*, 1978, p.8) For Dean Inge: "*Indifference to them is not, as many have supposed, a sign of enlightenment and spirituality. It is, in fact, an unhealthy symptom.*"

7. Beyond symbol degradation: symbolism for the future
In addition to the important function symbolism can play in integrating understanding of the range of opposing factors so characteristic of the global problematique, symbolism also has a role to play in defining man's relation to his future (to the extent that this may be dissociated from the problematique).

For Martin Grotjahn the understanding of symbol gives the clue to understanding people and it also provides a key to the future. "*We cannot hope to predict or plan developments without analyzing the trends in unconscious motivation. To have neglected this has been the failure, in the past, of proper planning for the future... When we understand the symbol and when we related ourselves to problems of the future as if the future itself were a symbol of our unconscious, then we will develop that kind of creative integration which we need in order to understand the past, to master the present and to predict the future.*" (*The Voice of the Symbol*, 1971, p.189)

Elsewhere he suggests that: "*The next step in human development will be a creative and courageous integration of our knowledge. Human life is a symbol of existence in this world. A return to naive, primitive symbolization is unacceptable. If tried it world lead to psychosis.*" (p.181). For him the creative integration of symbolic vision would combine respect for incompatibles such as respect for individual rights with duties to the community. Such creative integration does not imply a return to pre-logical thinking in the future. It combines symbolic thinking with logical, rational, and scientific thinking. "*Once, at the beginning of human development, it was no longer enough to react to signs and signals as an animal does; now it is no longer enough to form symbols in the form of rituals or myths.*" (p.183-4)

It is perhaps a tragedy, and a necessary consequence of the human condition, that those most consciously involved in the design of motivating symbols are those with the greatest commitment to entrap society in the unthinking consumerism which is aggravating the ecological and resource problems so characteristic of the global problematique.

8. Method
Recognizing the importance of symbolism as an unexplored resource, Section MS was designed as a way of providing a brief overview of the range of symbols in order to get some sense of the symbolic patterns. Given the overwhelming amount of information available on symbols, the approach was to group symbols into classes and to indicate tentative relationships between such classes. This first step in many ways denatures the symbols to which reference is made since, as Cirlot points out, it is difficult to classify symbols with exactitude.

By placing symbols in this context however the stage is set for opening entries on specific groups of symbols which can be interrelated by other types of cross-reference.

In particular the challenge is to find ways of reflecting to some degree the patterns of symbols discussed above. Of great interest also is the possibility of relating specific symbols to entries in other sections, especially values, modes of awareness and even, possibly, to problems and to the organizations responding to them.

Challenge: contemporary crisis of governance

1. Scope of governance

The experience of the past decades in designing and implementing international development-related strategies, and governing the process through which they become possible, is not especially encouraging. Major disaster has been averted but the early hopes are far from being fulfilled. The situation has become worse for many and the risks of major disaster have increased for everyone. Particularly tragic is the recognition that the international system of institutions is defective in its management of the development process, riddled with inefficiencies and lacking in credibility, especially in the eye of public opinion. This situation has recently been officially documented for the first time for the United Nations system by Maurice Bertrand of the Joint Inspection Unit (*Some Reflections on Reform of the United Nations*, 1985). It is within the constraints of this context that the sustainability of development advocated in the Brundtland Report needs to be considered.

This paper follows earlier work on the challenges of collective comprehension of appropriateness and the special constraints it imposes on the design and implementation of any development initiative (A J N Judge, *Comprehension of Appropriateness*, 1986). The paper addressed the resulting challenges for "governance". This term has been resuscitated by John Fobes, former Deputy Director-General of UNESCO, in order to promote a reconceptualization of the commonly used terms "governing" and "government". In recent remarks to a Club of Rome conference he states:

"The concept of governance emphasizes that order in society is created and maintained by a spectrum of institutions, only one of which is known as government. By examining that spectrum at all levels of society, we can obtain a broader sense of "governability" as it is exercised in policy-making, in providing services and the application of law. Order is certainly part of governance. But I believe that one should also consider governance, at least at the international level, as a global learning exercise. By so doing, politicians, practitioners, activists and academies may expand their thinking beyond the traditional concepts of government, of international organizations and of the exercise of sovereignty". (*Next Steps in World Governance*, 1985)

Of special value in Fobes' remarks is his creative response to the complexities of the situation. He recognizes that the processes of governance have become increasingly complex and are no longer strictly limited to governments. He points out that the fact that so many individuals and groups, whether NGO's or IGO's, at all levels, want to "get into the act" of learning, if not governing, is both hopeful and chaotic. It is for this reason that he points to the need to re-examine attitudes to different "learning modes". *"Learning, and learning to "govern", or to participate in governance, on the part of citizens and their civic and special interest groups, have become part of the survival skills for nations and for humanity as a whole."* (John Fobes, 1985)

The dimension of the challenge is indicated, if only within the international community of organizations, by that the latest edition of the *Yearbook of International Organizations* (UIA, 1990/1991). It identifies over 25,800 international governmental and nongovernmental bodies, acting in 3,000 subject areas, on the world problems described in Section P.

The focus in this paper on the use of metaphor in governance is one response to the recognition articulated by Fobes that: *"The stresses from social change that require a broader sense of governance have called into play Ashby's "law of requisite variety"* (which may be interpreted as stating that *"the regulators or governors of a system must reflect the variety in that system in order to be of service to it"*). This applies as much to the government of a country, as of a small group, or even an individual's endeavours to govern his or her own behaviour in a turbulent social environment.

The question explored here is that of the need to provide a sufficiently rich medium for the communication of complex insights in a world in which the possibilities of governance are constrained by the explanations and proposals that can be made meaningful to public opinion. The complexity of econometric and global models in their present form makes it improbable that they can be of any significance to those who must justify their actions to public opinion and receive their mandates from an informed electorate.

2. Clusters of dilemmas

This section endeavours to order the principal factors contributing to the contemporary crisis of governance and of bringing about any form of sustainable development. Such factors may be clustered of course in different ways. The number of such clusters it is useful to select is partially determined by constraints explored in earlier papers (A J N Judge, 1987).

In order therefore to maximize the number of explicit factors identified as contributing to the crisis of governance the following eight clusters are proposed:

(a) Simplicity: Governance, to be feasible, requires that the number of factors or issues on which a mandate is sought, or for which policies must be developed, should be limited in number and defined simply enough to be meaningful. They should be interesting rather than boring. Failing this the preoccupations of governance lose their focus, and the governing body becomes vulnerable to loss of its mandate in favour of some other coalition whose focus is appropriately simple. Conventional strategies in response to this dilemma include:

- only focusing on those issues which through their identification can conveniently come to be perceived as important as the result of a self-fulfilling process;
- only focusing on a few macro-issues which lend themselves to a multiplicity of simple descriptions, whilst failing to encompass their inherent complexity.

(b) Complexity: Governance, to be practical, must necessarily deal with the complexities and crises of the real world, whether or not they lend themselves to any meaningful ordering or pattern of mandates for specialized agencies. Failing this governance is overwhelmed by the many pressures of the moment and becomes vulnerable to loss of its mandate in favour of some other coalition that can deal with them. Conventional strategies in response to complexity and the associated information overload include:

- elaboration of an array of administrative procedures, plus filtering and delaying mechanisms for every conceivable circumstance;

- displacement of new issues and pressures by other issues and pressures for which procedural responses already exist.

(c) Requisite variety: Governance, in order to be able to exert some long-term degree of control over the dynamics of society, must itself be sufficiently varied in its policy-making capacity to respond to the variety of issues which may emerge. Failing this the governing body is caught off-balance by the dynamics of the society and is vulnerable to loss of its mandate in favour of some appropriately dynamic coalition. Conventional strategies in response to this challenge include:

- emphasis on short-term issues and programmes to disguise any lack of ability to handle long-term trends;

- emphasis on publicizing long-term projects, whilst disguising the degree to which they themselves will aggravate other problems for which no remedy has been envisaged.

(d) Operational relevance: Governance, in order to be credible to those mandating it, must be able to formulate its policies in a form which is readily implementable, especially in response to issues which call for immediate action. Failing this the governing body is perceived as irrelevant to the solution of pressing issues and is vulnerable to loss of its mandate in favour of some more practical coalition. Conventional strategies in response to this requirement include:

- emphasis on short-term remedial programmes, irrespective of whether these effectively respond to the problem which evoked their creation;

- focusing attention away from the more obvious solution onto the necessity for some alternative programme of effective remedial action (for which an appropriate mandate may not be obtainable).

(e) Complementarity: Governance, in order to attract support from a plurality of unrelated (or even mutually hostile) sectors, must be able to configure those sectors into a pattern such that they appear as complementary to one another. Failure of the governing body to establish such a context, or community of interest, leads to fragmentation and erosion of its support, rendering it vulnerable to any coalition of wider appeal. Conventional strategies in response to this requirement include:

- promotion of superficial consensus in such a way as to disguise irreconcilable differences between sectors;

- cultivation of distinct communications with each sector, concealing any contradictions between the undertakings made.

(f) Difference: Governance, in order to respond effectively to disagreement, critical opposition and alternative insights, must develop some means of dealing with incommensurable positions. Failure of the governing body to develop such skills makes any form of co-existence with its opponents unstable and renders it highly vulnerable to attack. Conventional strategies in response to such differences include:

- disparagement, neutralization or suppression of any dissidence (possibly through judicious manipulation of information), implicitly denying any merit in such viewpoints;

- efforts to persuade the dissident group to modify its position or to coopt its members.

(g) Containment: Governance, to be able to maintain its domain of influence, must reinforce a certain order within definable boundaries. Failure of the governing body to do so results in an open system vulnerable to the effects of uncontrollable variations in external influences. Conventional strategies in response to this requirement include:

- strengthening of boundaries and gate-keeping functions, justified by the necessity of excluding "undesirable" influences;

- limiting freedom of action in order to facilitate the maintenance of the favoured order.

(h) Empowerment: Governance, to be able to encourage the growth and development expected by those who mandate it, must be able to empower people and groups to undertake and sustain new initiatives of their own accord. Failure of the governing body to do so results in stagnation and disaffection rendering it vulnerable to replacement by a coalition encouraging such initiative. Conventional strategies in response to this requirement include:

- mobilization of people and groups in support of some defined programme, irrespective of the initiatives they would otherwise choose to take;

- manipulation, subversion or cooptation of initiatives if they achieve any degree of social significance.

3. Fourfold principle of uncertainty in governance

As argued elsewhere (A J N Judge, *Beyond Method*, 1981), especially in the light of epistemological problems in the social sciences which suggest that a generalized Heizenberg principle operates in the social sciences (Garrison Sposito, *Does a Generalized Heisenberg Principle Operate in the Social Sciences?*, 1969), the dilemmas of the previous section could well be summarized in a four-fold principle of uncertainty as follows:

(a) A governing mode in which it is easy to say "no" overtly, makes it very difficult to say "yes" except covertly, whereas one in which it is easy to say "yes" overtly makes it very difficult to say "no" except covertly.

(b) A governing mode which encourages overt declarations of consensus has great difficulty in accepting fundamental differences in practice except covertly, whereas one in which differences are realistically accepted has great difficulty in establishing consensus except covertly.

(c) A governing mode of requisite variety for long-term continuity has great difficulty in elaborating appropriate short-term programmes except covertly, whereas one in which operationally relevant short-term programmes are easily elaborated has great difficulty in ensuring any policy of long-term significance except covertly.

(d) A governing mode which can be made meaningful and inspiring has great difficulty in taking into account the full complexity of a practical situation except covertly, whereas one which takes into account that complexity in all its operational detail cannot be meaningful and inspiring except covertly.

Use of the terms "overt" and "covert" could be considered as unnecessarily value-loaded. Alternatives might be "formal" and "informal" or else "public" and "private".

The merit of using "covert" is that it emphasizes the potential for procedural abuse and manipulative processes in certain situations, namely insidious corruption. These points are perhaps well illustrated by the difference between the overt processes in international organizations and those occurring behind the scenes (and covered by security clauses in employment contacts).

Whilst there is much overt discussion of the efficiencies in the overt processes (as in the recent reviews of the United Nations and UNESCO), the dysfunctional features of the covert processes are only discussed in corridor gossip and newsworthy exposés. There has never been any overt study by an international body of corruption in governance at all levels, and especially of corruption in such international bodies. Yet "corruption" is frequently cited in informal reports as a cause of inefficiencies in the implementation of programmes.

This paper is not about corruption but about the inability to fully encompass conceptually the processes of governance in an adequate model or set of models. This results in grey areas in which dysfunctional processes proliferate, however carefully the overt processes are defined. These are the shadow side of governance. Any attempt to envisage new approaches to governance that neglects this dimension, or fails to come to terms with it, must necessarily fall victim to the ways in which it undermines effectiveness.

Challenge: epistemological crisis of governance

Sustainable development is usually conceived as a problem of instrumentality - namely deploying the available organizational and conceptual resources to achieve what seems appropriate. An earlier paper (A J N Judge, *Comprehension of Appropriateness*, 1986) argues that this approach fails completely to recognize the inherent difficulties in comprehending the instrumental design which is appropriate - and of communicating that comprehension, with all its nuances through the processes of governance.

1. Questionable assumptions of governance
The following hidden assumptions concerning any advocated new mode of governance were listed to illustrate this failure:

(a) New mode as an absolute improvement: That the mode is inherently better in some absolute sense and that, - conversely, the old mode must necessarily be permanently abandoned as historically outmoded;

- the defects in the new mode will not eventually prove to be as significant as those under the old mode.

(b) New mode as universally appropriate: That the new mode is equally appropriate to all societies and to all sub-cultures within those societies, especially if adapted to local contexts and requirements.

(c) Requisite complexity of comprehensible new mode: That, if it can be comprehended, represented and discussed within one frame of reference, the mode can nevertheless be of sufficient complexity to respond to the concerns perceived by constituencies preferring other frames of reference.

(d) Completely articulable new mode: That an appropriate new mode can be readily articulated in its entirety, rather than necessarily provoking a set of partial comprehensions which people, of whatever level of competence, experience considerable difficulty in integrating/reconciling, even if they are motivated to do so.

(e) Implementability of new mode through pre-planned actions: That an appropriate mode can be readily implemented by a consistent pattern of actions, rather than requiring set of seemingly inconsistent and incompatible actions, each favoured or condemned by some different configuration of constituencies.

(f) Hierarchically-derived integrity of new mode: That the coherence and integrity of an appropriate mode derives from a hierarchical relationship between its components, as opposed to other possibilities with characteristics such as:

- configurations of incommensurable conceptual or organizational groupings in which the hierarchical dimension, if any, is secondary or implicit;

- cyclic phases of emphasis over time;

- alternation between seemingly opposed or contradictory policy modes.

(g) Recognition of hard realities implied by new mode: That credible articulations of a seemingly attractive approach do not effectively obscure hard realities to which the advocating group may be insensitive (or anxious to avoid discussing in order to further some hidden agenda).

(h) Non-provocation and non-evocation of counter-initiatives: That any readily devised approach will not necessarily provoke counter-strategies or strategies which exploit the situation created by the implementation of the new approach, undermining it and eventually rendering it ineffective.

(i) Avoidance of polarized assessments of the new mode: That, during the implementation of the appropriate new mode, it is possible for any given constituency to avoid being trapped into recognizing any necessary practical strategy in either a "positive" of a "negative" light, and consequently to be entrained to further or oppose that partial strategy, without consideration of whether such effort is excessive in the light of the contextual mode to which it contributes.

(j) Avoidance of ambiguity in the new mode: That the essence of being human, and of human development, involves processes free from ambiguity, paradox and counter-intuitive phases, permitting an appropriate new mode to be articulated in an manner free of such non-rational characteristics.

2. Epistemological landscapes
The remainder of that paper*** considered the probability that the appropriate global socio-economic mode of organization is necessarily more complex than can be recognized or comprehended within any particular frame of reference - whether conceptual or organizational. The question here is how to describe and handle this epistemological challenge for governance.

The question has been helpfully highlighted by the recent study prepared by Development Alternatives (New Delhi) on "A transcultural view of sustainable development; the landscape of design" as a contribution to the final deliberations of the World Commission on Environment and Development (*A Transcultural View of Sustainable Development*, 1986). The study outlines "transform grammar of design" based on a "phase space" model using a n-dimensional space to show the evolution of a system (where n is the number of degrees of freedom, or independent variables, needed to describe the system at the level of recursion or aggregation of the model under study). The work draws on recent theoretical advances, including those of Shannon (1962), Ashby (1956), Beer (1979), Prigogine (1985), Zadeh (1965) and de Laet (1985).

It is apparently necessary to "freeze" any such "epistemological landscape" into a well-defined model in order to navigate over the landscape. And within the short time scales (and electoral periods) characteristic of the majority of the problems of governance (and the budgetary periods of international organizations) such a landscape may legitimately be considered to be unchanging. Governance can then endeavour to move the social system over the landscape.

3. Competing epistemological landscapes
The epistemological problem lies in the fact that different constituencies are sensitive to different dimensions of the "n-dimensional phase space" out of which the model is extracted or abstracted. Consequently the epistemological landscape perceived by one group may be very different from that which is meaningful to another - such that each may be the basis for the strategies and programmes of a different intergovernmental agency. This has the further consequence between agencies of reinforcing incompatibilities, contradictions, competition for resources and even the undermining of one strategy by another - as has been noted on many occasions, and recently by Maurice Bertrand (*Some Reflections on Reform of the United Nations*, 1985).

It is therefore less fruitful to focus initially on any particular way of viewing the n-dimensional phase space. Rather it would seem more appropriate to consider the epistemological challenge of how to open up any "window of comprehension" onto such complexity - and how to perceive the relationship between such windows, whether used simultaneously (by different groups) or consecutively.

4. Navigating through complexity
Before taking the argument further it is necessary to avoid the trap of using the phase space notion itself as a fundamental window. It is a powerful tool but not necessarily convenient for all. "Complexity" has itself recently attracted attention in its own right (United Nations University, Complexity). "Chaos" is now a key descriptor for some interesting breakthroughs in mathematics (H O Peitgen and P H Richter, *The Beauty of Fractals*, 1986 and B B Mandelbrot, *The Fractal Geometry of Nature*, 1983). Although it would be incompatible with the theme of this section to favour any one such description, it is

important to recognize the range of attempts to indicate the epistemological attributes at this level of abstraction.

It is somewhat ironic that the earlier Greek philosophers made use of the Greek term "hyle" (matter) and viewed such matter as fundamentally alive, either in itself or by its participation in the operation of a world soul or some similar principle. Characteristically they did not distinguish between kinds of matter, forces and qualities nor between physical and emotional qualities, making any such distinction with an important degree of ambiguity.

The contemporary epistemological challenge remains one of dealing with a form of "conceptual hyle" or "mindstuff" within which the variety of possible models and concepts is implicit and from which they may be explicated, as described by David Bohm (*Wholeness and the Implicate Order*, 1980). This is not to suggest that the "hyle" is purely conceptual. As contemporary studies of this intimate relationship between consciousness and fundamental understanding in physics are clarifying, there is a matter-consciousness continuum of perhaps greater significance than the space-time continuum. Relevant insights from Eastern philosophies are also increasingly (G Zukav, *The Dancing Wu-Li Masters*, 1979 and Capra, *The Tao of Physics*, 1975) noted. The comprehension of features explicated from the "hyle" is as much constrained by the realities dear to materialists as it is by individual (or collective) ability to formulate appropriate models of requisite variety and to communicate them.

The challenge of governance is to enable society to navigate through the "hyle", avoiding catastrophic disasters in a manner such as to sustain a process of "development" over the long-term - whatever "development" is understood to mean in the short-term under different circumstances, within different cultures and at different stages of that process. But since governance is above all constrained by daily practicalities, there is a dramatic problem of ensuring some kind of meaningful epistemological bridge between the multidimensional fluidity or ambiguity of the "hyle" - with all the innovative potential that implies - and the concrete socio-political realities to which it must respond effectively or be called into question.

5. Schizophrenic practices
As noted above, extensive use of metaphor is made by politicians and statesmen in endeavouring to communicate policy options and positions. It is a characteristic of political discourse. However, metaphor is seldom if ever consciously used in policy documents and in the documents of experts legitimating such policies. Such documents are characterized by bureaucratic jargon and the supposedly metaphor-free language of experts appropriate to the objective discussion of scenarios and theoretical models.

It is not the purpose here to query these two modes of discourse. Rather it is to question the epistemological nature of the "conceptual bridge" which integrates them. What in fact is the current link between these two functions ? In practice, if the policy model emerges first, then public relations consultants are engaged to discover means of "packaging" it for communication to wider constituencies. If the concept emerges as a politicians insight from the cut-and-thrust of the political arena, then experts are called upon to dust off some model which can give theoretical credibility to it. Those associated with each mode of discourse have little respect for the contributions of the other. No scholar has any appreciation of the constraints of public relations, just as no media consultant has any respect for the niceties of scholarly methods. Policy-makers navigate in an essentially schizophrenic domain of discourse.

In a very real sense governance essentially takes place in an epistemological "war zone" where the battle between metaphoric and modelling modes takes place.

The challenge is to move beyond the limitations of a discussion in which either (a) metaphors are claimed to be purely figurative and of no cognitive significance, or (b) models are claimed to be fundamentally metaphoric in nature. There is presumably some truth and some exaggeration to both claims. The question is how this epistemological battle affects the problem of governance in any effort to pursue future policies of "sustainable development".

6. Metaphor/model hybrids: an epistemological quest
It is important to stress that the focus on the metaphoric dimensions does not in any way deny the importance of the modelling function when conceived non-metaphorically as a purely conceptual device (e.g. as in econometrics, global modelling, etc.) The point is rather that in order to present and explain such models successfully to those preoccupied with the many dimensions of governance, they must anyway be imbued with metaphoric dimensions - however distasteful this may be to modelling purists. But for those concerned with governance, it is precisely through imbuing the models with metaphoric dimensions that they become meaningful and can be related, through the political insight and experience of the governors to concrete realities which models, as abstractions, do not fully take into account. It is the ability of the governors to project themselves into the metaphor which enables them to find ways of fitting the model to the decision-making realities of the world they are dealing with and to the mindsets of the governed. Both model and metaphor are epistemological crutches - one facilitating left-hemisphere information processing and the other right-hemisphere processing. As Jeremy Cambell says: *"Another kind of context supplied by the right brain comes from its superior grasp of metaphor"* (*Grammatical Man*, 1982)

Expressed in these terms, it becomes clearer that many of the inadequacies of modelling for governance are precisely due to the lack of attention to the need to imbue them with metaphoric dimensions. Equally many of the inadequacies of metaphors for governance are due to the lack of attention to the need to imbue them with modelling dimensions. In metaphorical terms, the former furnish clothes of appropriate strength, but which are so uncomfortable and ugly, that nobody is inclined to try them on or be seen wearing them. Whereas the latter furnish clothes which are a delight to try on, but cannot be taken more seriously than fancy dress, because they are not appropriate to the varieties of weather conditions which they must withstand. This is a problem of design.

7. Donning and doffing metaphors
For both the governors and the governed it is a question of the extent to which they are able to "get into" the "metaphor-model". In relation to this question of "getting into", Anne Buttimer notes the most profound transformation in twentieth century knowledge as being the movement from observation (of reality) to participation (in reality) (*Musing on Helicon*, 1982 and *Mirrors, Masks and Diverse Milieux*, 1983) - a theme explored by Michael Polanyi (Personal Knowledge). What degrees of "epistemological participation" does a "metaphor-model" offer? Are there more powerful forms of participation, or at least forms more powerful in different circumstances? These need not be trivial questions for governance, because in a sense epistemological participation can be more powerfully attractive than participation limited to political processes, which it effectively underlies.

Challenge: unsustainable policies

1. Policy fashions
Policies and issues move into and out of fashion according to the vagaries of the political process and the priorities of the moment. This is true even within the international community, as noted by Johan Galtung. There is a "flavour-of-the-month" quality to policy-making, however serious the long-term issues may appear to be. Governance suffers in consequence through lack of any conceptual continuity. Past policy flavours within the international community, according to Galtung, include basic needs, self-reliance, new international economic order, appropriate technology, health-for-all, community participation, primary health care and common heritage of mankind. The current flavour is sustainable development. It is useful to ask how sustainable is the concept of sustainable development, and what dimensions does it fail to take into account.

2. Sustainability of "sustainable development"
In the light of the series of integrative foci of the past decades, "sustainable development" can be considered humanity's best and latest effort to reconceptualize "the good, the true and the beautiful" for the international community. Given the responses to past efforts, notably the Brandt Commission, it is fruitful to ask how sustainable over time is the concept of "sustainable development". Already there is evidence of multiple interpretations (Pezzey, 1989), some of them quite incompatible, just as has been the case with "development" alone. In any policy forum, such differences are immediately apparent through the factions and coalitions to which they give rise. As with past foci, there are those who perceive it to be totally legitimate to "milk" a concept to their own benefit whilst it still has "mileage" left in it. Johan Galtung has described the life cycle of such concepts in relation to the international community. The position in the life cycle determines how themes are handled by policy agendas.

3. Unsustainable policy assumptions
It is not fruitful to view such concept cycles cynically, although exposure to them can justify this. The challenge is to identify how the development of any such insight can be sustained, especially in a policy forum. The difficulty lies in assumptions made by those actively involved in promoting or clarifying such an insight. These include:

(a) Monopoly of validity: A tendency to consider it the only valid integrative concept that has ever been formulated. This ignores the history of previous concepts which have created the context for the emergence of this latest one. It also ignores what happened to the previous ones and the nature of the relationships they established with other competing policy concepts.

(b) Non-emergence of more valid concepts: A tendency to consider that no further valid integrative concepts will emerge to replace the current one. This structures reflection on the concept to preclude the future emergence of more appropriate concepts. It engenders dogmatism and identification of heresies. (Do the advocates of sustainable development have the right conceptual posture to respond appropriately to the policy insight which will succeed it -- or will there be no such innovation?)

(c) Inherent credibility and attractiveness: A tendency to believe that the concept is inherently credible and desirable to those who have not been involved in its formulation. The next step is to assume that they should be persuaded to that conviction if they are not.

(d) Integrable within a single policy framework: A tendency to believe that policy insights of requisite variety can be adequately embodied within a single policy framework.

(e) Universal acceptance of legitimating information: A tendency to fail to recognize that groups are sensitive to quite different forms of information in relation to any issue, and frequently consider other forms as having marginal significance, if any.

As more people and groups are touched by the insight, they reinterpret it to better reflect their own understanding. This leads to factionalism and multiple interpretations which may be highly critical of each other, even to the point of subverting each others initiatives in competition for resources. "Sustainable development" has to survive in this environment. To be of any significance, policy forums must respond effectively to such factionalism -- whether or not they are effectively represented at any forum.

4. Sustaining policy through imagery and metaphor
There is little need to remake points concerning the role of the media in politics or the problems of information overload. In such a context a vitally important issue for policy acceptance is the process whereby policy proposals are communicated for clarification and approbation. The constraints and opportunities are most evident in the case of politicians and political parties concerned to "get a message across". The same may be said concerning the communication of any proposal in a policy forum (Majone, 1989).

Much has been written about the deliberate cultivation of an image by politicians and their increasing investment in media consultants and image makers, following the example of corporations. It has been argued that image is becoming as important as content in politics, if not more important. The need for visionary leadership is stressed (Dror, 1988a). Given the intimate relationship between policies and the politicians presenting them, it is appropriate to ask to what degree policy-making is now "image-led" as opposed to "content-led". For whilst it is possible to formulate policies based on the most appropriate scientific models and the greatest of expertise, it is increasingly recognized that if such policies do not communicate well they have little chance of being either understood or approved.

These points are made, not in order to denigrate sophisticated models and conscientiously articulated policies, but in order to suggest that the leading edge of the policy approval process is now the image through which the policy is envisioned and presented. But although these are clear examples of the extensive use of metaphor in relation to governance or in support of it, the question remains whether the role of metaphor is limited to a public relations function, namely, to the communication function noted earlier. Metaphors may affect the way people think about the governance of complex issues (e.g. references to Reagan's "John Wayne"/"Rambo" approach to governance), but do they affect the processes of governance itself and the way choices are made? The literature cited above provides ample evidence of the use of metaphor by politicians in parliamentary debate to clarify an issue or attack the position of the opposition. Criticism of Thatcher's policy of privatization was recently given a very sharp focus through former Prime Minister Harold Macmillan's phrase "selling off the family silver". Thatcher's subsequent reply in justifying and reiterating that policy was "selling the family silver back to the family". This is a good example of policy clarification at the metaphoric level.

But such examples, even though significant in parliamentary debate, do not indicate whether metaphors are used in the initiation, elaboration or long-term guidance policy. This is the difference between cabinet level debate and parliamentary debate. But even cabinet level debate is about policies already outlined in draft form through the services of secretariat professionals who normally make great effort to avoid metaphor in an effort to present texts "professionally". However such secretariats are supposedly the infrastructure of governance, so the question remains how actively metaphor is used in the "smoke filled rooms" where policy options are conceived and mulled over, prior to being formulated on paper and in feasibility studies. What are the conceptual processes through which policy options are conceived by those taking new initiatives in governance? And what part does metaphor play in such processes?

In the case of President Reagan, it has frequently been pointed out that he preferred to receive information in the form of video films and imagery in general, rather then through briefing documents. Is it possible that the kinds of policy which he supported were limited by the kinds of metaphors to which he was sensitive? Would more appropriate policies have become credible if their conception could have been supported by richer metaphors?

Challenge: comprehending complexity and appropriateness

The appropriate global socio-economic mode of organization must necessarily be more complex than is accepted within any particular frame of reference. It is therefore more than probable that it cannot be fully comprehended within any single frame of reference.

1. Limitations of axiomatically defined systems

The probability is increased in the light of the classic study of axiomatic systems by Kurt Gödel (1931). Prior to this a climate of opinion existed among mathematicians in which it was tacitly assumed that each sector of mathematical thought can be supplied with a set of axioms sufficient for developing systematically the endless totality of true propositions about a given area of inquiry. Gödel demonstrated that no system can be comprehensible without being self-contradictory. In doing so he showed that it is impossible to establish the internal logical consistency of a very large class of deductive systems, unless principles of reasoning are adopted which are so complex that their internal consistency is as open to doubt as that of the systems themselves. (E Nagel et al. *Gödels Proof*, 1958)

Care should be taken in dismissing the relevance of such insights to the comprehension of social systems. A recent discussion of their relevance in the *Financial Times* concludes that although *"he was concerned specifically with systems of symbols such as mathematics and other languages... experience indicates that the principle applies to organisational systems too. Its implications are particularly destructive for bureaucratic attempts at management. Their tendency is to lay down systematic rules intended to cover every eventuality and, when they don't, to lay down more rules supposed to close the loophole. Whilst Gödel's principle suggests that any such process is necessarily self-frustrating, almost all bureaucracies seem determined to believe otherwise."* (Michael Dixon. *The Common Laws of Organizational Stupidity*, 4 Sept 1986).

This also suggests the merit of reflecting on the relationship between the political axioms in terms of which attempts are made to govern countries and groups of countries, especially to the extent that they are embodied in political slogans reflecting values which are "axiomatic". Such slogans presumably preclude consideration of modes of organization which are not built directly upon such axioms. But Gödel also showed that there is an endless number of true arithmetical statements which cannot be formally deduced from any given set of axioms by a closed set of rules of inference. The dramatic implications of this has just recently been demonstrated in the world of chess, traditionally referred to as the "Game of Kings" because of the manner in which it simulated the strategic problems of a leader. There are many axioms governing the different possibilities of winning in a chess endgame situation. To the considerable astonishment of the chess community, a very recent computer analysis of endgames has however demonstrated that there are many other ways of winning, unforeseen by such axioms, and in some cases inconsistent with them (K L Thompson, 1986).

2. Comprehension of appropriateness

Comprehending appropriateness might be illustrated by the problem of comprehending the nature of an n-dimensional object (e.g. a hypercube) whose elements represent factors in a mathematical model of a socio-economic system. Portions of the representation can be comprehended when they are represented as 2 or 3-dimensional diagrams of cross-sections of the n-dimensional structure. Integrating a set of such representations tends to be a task beyond the current abilities of the human mind.

It could be argued that it is not vital that the n-dimensional object be comprehensible in its entirety, provided the mathematical representation can be proved to be satisfactory. This is the case with the current use of hypercubes as the basis for the organization of new, and more efficient, forms of computer memory. Provided the product finally works, people do not feel that it is necessary to understand it in its entirety. The wiring diagram can be represented, even though it is meaningless to those making the sequence of connections between the parts. A similar argument is made concerning the lack of need for a driver to understand the engine.

This attitude is not however acceptable in the case of the presentation of some new mode of socio-economic organization, whether at the macro or the micro level. It is one thing for people to have confidence in leaders (or experts) who can claim to comprehend such a mode in its entirety, even though their followers do not. Most social innovations in the past have been implemented on this basis. It is quite another thing when the appropriate mode of organization cannot be fully comprehended by any leader (or expert), especially, as is the case at present, when the motivations of such elites are increasingly considered questionable.

3. Constraints on learning

The situation is further complicated by the learning dimension. If the appropriate mode was fully comprehensible, it would then exclude the possibility of a learning dimension. It would permit learning within that mode, but it could not render explicit (and would therefore probably preclude) learning beyond the framework imposed by that mode. It could not permit learning by which the framework itself would be challenged. It would thus be consistent with human development within a framework implemented at a particular historical moment, but opposed to human development arising from insights emerging subsequently. Such a mode would therefore be consistent with human development in a "minor" key but not with human development in a "major" key, namely development requiring paradigm shifts.

4. Constraints on social organization

A response to this situation is not to expect or require that the appropriate mode be comprehensible in its entirety to any one person or group. It could be expected that different people or groups would be capable of comprehending different features or processes of that mode and would then act to ensure their implementation. But, necessarily, such people or groups would then not comprehend the justification for the activities of other people or groups concerned with other features or processes.

5. Partial comprehension of contextual appropriateness

The socio-political situation would then be one in which:

- Constituency A would support strategic components P, Q and R, whilst opposing (violently) strategic components I, J and K;
- Constituency B would support strategic components E, F and K, whilst opposing Q, J and T;
- Constituency C would support strategic components I, T and R, whilst opposing F, G and P.

The difficulty is that, in such a context of partial comprehension, no group (e.g. Group A) would be in a position to distinguish between:

(a) A condition in which some other group (e.g. Group C) was acting dangerously, inadequately or irresponsibly in terms of its contribution to the contextual mode, and should therefore be considered as inconsistent with the successful implementation of that mode.

(b) A condition in which its own group (Group A) was fulfilling its function, in relation to the contextual mode, by acting in opposition to some other group (Group C), even though that group was itself fulfilling its function in relation to the contextual mode in an appropriate manner.

In such a situation, for the contextual mode to function appropriately, the groups would act in support or opposition to each other to provide a system of checks and balances that would permit human development to occur in the optimum manner. No group could effectively take a position within this context in support of the contextual mode. It would, because of the necessary partiality of its comprehension, quite validly be perceived as acting to further certain interests consistent with that partiality.

The challenge then is to explore ways of improving comprehension of fruitful patterns of interaction between groups and perspectives which of necessity must function in shifting coalitions in support and opposition to one another.

Challenge: comprehending any new social order

If a more appropriate mode of socio-economic organization is advocated, the question is how it is to be comprehended in relation to those which preceded it. Acting on the belief in continual linear forward progress, its advocates may hope that it will completely replace preceding modes, since their functions are supposedly more satisfactorily performed by the new mode. Advocates of other modes, relegated to the status of historical curiosities, will not of course see things in that light. In which case the new mode must enter into competition and struggle with the older modes. In the dynamics of the social system it is one more mode, which seeks to improve its "market share" of public opinion.

2. Innovation as element or as patterning of elements

It may however be argued that the appropriate new mode should not be compared to a new species (a more evolved mutation) entering into an ecosystem and thereby modifying the pattern of relationships amongst the species present. Rather it should be compared to the pattern of relationships between the species, namely to the pattern of interdependence itself. In this light the appropriate new mode is a new pattern of interdependence between contrasting modes of socio-economic organization. This corresponds to Gregory Bateson's central thesis: *"The pattern which connects is a metapattern. It is a pattern of patterns. It is that metapattern which defines the vast generalization that, indeed, it is patterns which connect."*(*Mind and Nature*, 1979, p. 11). And it is in this connection that he warns: *"Break the pattern which connects the items of learning and you necessarily destroy all quality."* (p. 8)

The difficulty, as stressed earlier, is that given that such patterns of interdependence cannot be comprehended in their entirety, any new alternative mode of organization comes to be perceived not at the ecosystemic level but simply as another species. And as such it fails to respond to the need for an alternative of global significance, however much success there is in imposing the alternative as the dominant species (of organization).

3. Understanding relationships between competing alternatives

In such a situation, there is merit in exploring how the relationships between the existing alternatives are to be understood. Whether or not a new species is introduced, there would seem to be a need to understand what function each of the existing modes performs, under what conditions, and with what characteristic negative effects (which have to be remedied by some other mode). For it is the current spastic alternation between these existing modes which somehow ensures the relative viability of the existing system, not the sole contribution of any one of them (handicapped by the others, as its advocates would assume).

It is clearly easier to deny this probability by arguing that some modes are clearly "useful", up to a certain period of historical development, but they, and others, are completely inappropriate beyond that point. The dangers of such an argument become much clearer if the problem is compared to one of determining which species are useful and which should now be "phased out". It is not clear that man has developed sufficient understanding of nature to eliminate species which may be of unrecognized future importance (e.g. the case of medicinal plants) or in some way vital to the maintenance of food chains on which man and others species depend.

In terms of this metaphor, man is still operating on the basis that any species which is in some way a nuisance or a danger should be eliminated. This would lead to the elimination of most carnivores, other animals and plants which endanger man's food supply, together with most insects and smaller species which do not directly serve man's immediate needs. More enlightened understanding of the environment has established that even the most undesirable species (e.g. crocodiles, wolves, spiders, snakes) have important functions to perform in particular environmental niches.

4. Caring for the ecosystem of alternatives

To the extent that it is accepted that any new mode will not be met with universal support, and may well provoke the emergence of other modes to exploit or counteract it, then it can be argued that a major opportunity for significant advance lies in understanding how the dynamics of this ecosystem may be most beneficially "cared for".

In this sense the dilemma of man in discovering the most fruitful relationship to the species in the natural environment reflects man's dilemma in discovering the most fruitful relationship to the variety of co-existing modes of socio-economic organization. (This is not a dilemma which the "greens" have solved, or properly addressed, since they have apparently been unable to develop any coherent understanding of either their relationship to competing viewpoints, or of the appropriate "stewardship function" in relation to the pattern of different schools of thought within the green movement itself.)

5. Healthy alternation between alternatives

The challenge of appropriateness may well be less a question of replacing the existing condition as of finding ways of shifting between its sub-conditions in a healthy manner. In arguing for a heterogeneity of epistemologies, Magoroh Maruyama offers a beautiful metaphor in response to the (homogenistic) question "but which one is correct?"

He suggests that in binocular vision it is irrelevant to raise the question as to which eye is correct and which wrong. *"Binocular vision works, not because two eyes see different sides of the same object, but because the differential between the two images enables the brain to compute the invisible dimension"* (*Paradigmatology and its Applications*, 1974, p. 84). The brain computes a third dimension which cannot be directly perceived and if we live in a multidimensional space even more epistemological "eyes" are required (*Heterogenistics*, 1977, p. 269-272). Reducing such vision to the parts in common provides much less than monocular vision. Each "eye" has its inherent limitations and strengths, and the homogenistic "eye" presumably also has its own vital contribution to make to the process of encompassing (or responding to) the complexity of our collective condition. His work, together with that of J O Harvey's (*Experience, Structure and Adaptability*, 1966), demonstrates that a minimum of four such "eyes" are required to describe the variety of perceptions of our collective reality.

Challenge: metaphor as an unexplored resource in a time of challenge

1. Context
As noted earlier, each form of presentation has both strengths and weaknesses, depending on a number of factors but especially on the subtlety or complexity of what needs to be communicated and to whom. The question is whether it is possible to devise some means of by-passing the desperately slow learning cycle associated with research, education, policy formulation-implementation in a world in which the education gap is increasing rapidly. If the current crisis is to be taken seriously, people need to acquire access to an appropriate response by some other means.

2. Prevalence of mechanical metaphors
The unfortunate characteristic of answer propagation in response to the global problematique, as currently practised with all the skills of media specialists, is that it is conceived in terms of mechanical metaphors such as "hitting" a "target" audience and achieving "impact". This is the approach used both by the public information programmes of the United Nations family of organizations and by many grass-roots initiatives. This could be described as a "particle" approach acting to achieve the **displacement** of people from one mind-set to another.

3. Existence of complementary non-mechanical metaphors
Arguments such as those in Section KD suggest the need for a complementary "wave" approach acting to achieve the **entrainment** of people in terms of their current mind-sets. Propagating an answer by resonance may prove to be a more appropriate mode in dealing with the "field" of world opinion. Particle propagation tends to be considerably slower than wave propagation, as well as being easily blocked or deflected.

The challenge is to make available something simple enough to be comprehensible and yet "seductive" enough to retain peoples involvement. On the other hand, if it is to be of any value at this time, it must also be sufficiently complex and coherent to encompass the complexity of a social reality in crisis, and yet empower people to act together to contain the crisis in such a way as to be transformed by the unique learning opportunity it constitutes. This is a tall order, far beyond the capability or ambition of conventional international programmes.

4. Cultural "carriers"
Under the circumstances it is appropriate to look at unconventional possibilities. One approach is through existing processes, penetrating all levels of society, which already hold most peoples attention, transform their awareness, and govern their actions. The challenge would then be whether it was possible to "code" onto these, as a kind of "carrier", a second level of meaning. The "double meaning" should then offer a totally new set of insights suggesting new patterns of action.

5. Construing the territory as the map
Some possibilities for this approach are: popular music and dance, spectator competitive sports, strip cartoons, rumour and scandal, humour, astrology and divination, myths and legends, fables and parables, sex, courtship and family life, nature and weather patterns. The merit of the last possibilities is that they effectively involve **coding the world problematique back onto the world and onto human beings,** which would seem to be a conceptually elegant response to the problem of self-reference (D R Hofstadter, *Gödel, Escher, Bach: an eternal golden braid*, 1979). There is a remarkable possibility of "re-reading" the world, as articulated by the sciences, for the metaphoric insights it offers.

There is also merit in relating a conscious pattern of significance to a substrate by which people are usually governed unconsciously. In Jungian terms this is an appropriate and fruitful form of marriage between conscious and unconscious elements. Humanity's inability to relate creatively to aspects of these unconscious elements (*eg* the environment and the reproductive instinct) severely aggravates the problematique (*eg* environmental degradation and the population explosion).

6. Metaphors as a short-cut
The approach advocated therefore involves the simple pleasure of activating new metaphors which can enchant, empower, explain and orient approaches to the problematique through the user's own comprehension of each metaphor's significance. But such metaphors is are only new in that they have not been widely used before, despite the fact that everyone has access to them.

In Boulding's words: *"Our consciousness of the unity of the self in the middle of a vast complexity of images or material structures is at least a suitable metaphor for the unity of a group, organization, department, discipline, or science. If personification is only a metaphor, let us not despise metaphors - we might be one ourselves."* (*Ecodynamics*, 1978).

Or, as the poet John Keats puts it: *"A man's life is a continual allegory - and very few eyes can see the mystery of his life - a life like the scriptures, figurative."*

The charm of it, as Gregory Bateson stated in concluding a conference on the effects of conscious purpose on human adaptation, is that: *"We are our own metaphor."* (Catherine Bateson. *Our Own Metaphor*, 1972, p.304). Unfortunately we have over-identified with the metaphor and have been unable to see ourselves in perspective. The lack of such self-reflexiveness could well prove to be an important contributory factor to the current uncontrolled attitude to procreation which is at the root of many current problems.

Metaphors are much used in every culture by people of every kind as vital short cuts to the communication of nuance and complexity. There is a desperate need for any such short cuts at a time when new intellectual and other insights are virtually inaccessible to most people unfamiliar with the professional jargons in which they are formulated.

7. Accessibility of metaphors
Metaphors have the tremendous advantage of being grounded in what is familiar, often at a gut level. As such, not only do they facilitate rapid comprehension, but they often suggest new dimensions to what is being conveyed through them. These unforeseen dimensions can provide subtle poetic linkages between isolated mechanistic concepts, as well as totally new insights to be explored.

The natural environment, for example, gives perceptible, concrete, three-dimensional illustrations of the kinds of subtle distinctions which the mind is capable of making. Metaphors based upon any such phenomena therefore firm up intuitions of relationships between non-physical phenomena - rather than reducing them to simplistic, mechanistic forms (as tends, to happen when the natural environment is destroyed, impoverished or inaccessible). They thus offer insights into the management of differences.

8. Reframing and re-enchantment through metaphor
Metaphors have a unique ability to enchant people and capture their imagination - at a time when alienation and cynicism are the rule. This in fact is what has made them extremely suspect in the eyes of professional intellectuals. As with any tool, however, the issue is really one of learning when and how to use it, and to what purpose. Comprehension of problems and their possible solutions, for example, may lie in understanding how their metaphorical equivalents may be interrelated.

Challenge: transcending the "switch" metaphor

1. Limitations of dualistic thinking
Much has been said in recent years about the inappropriateness of conventional western mind-sets in responding to the complexities of the environment. Particular criticism has focused on "dualistic" and "linear" thinking. "Holistic" approaches are advocated as more desirable alternatives, but unfortunately without any insights into the practicalities of their implementation. Metaphor may be helpful in this respect.

2. Implicit "switch" metaphor
Consider the implicit switch metaphor which governs much of our thinking concerning major problems of society:

- unemployment: an individual has a job, or does not have a job.

- ignorance: an individual is educated, or is uneducated.

- violence: an individual is subject to violence, or is not.

- illegality and criminality: an individual is acting illegally, or is not.

- illness: an individual is healthy, or is not.

- corruption: an individual is corrupt, or is not.

- uncleanliness: an individual is unhygienic, or is not.

- discrimination: an individual is subject to discrimination, or is not.

- environmental exploitation: an individual wastes resources and degrades the environment, or does not.

- substance abuse: an individual is addicted to drugs (over-eating, smoking, alcohol, etc), or is not.

3. Policy-making and the "switch "metaphor
Many advocated policies are explicitly designed to "switch" individuals from one condition to the other in each case (e.g. from "on" to "off") -- from an undesirable condition to a desirable one. And once such a transition has been accomplished, the object is to prevent backsliding into the undesirable condition. The switch metaphor is a simple device through which ambiguity can be avoided (D N Levine, *The Flight from Ambiguity*, 1985).

Ironically this switch metaphor is also implicit in the thinking of those who identify most closely with a holistic, non-linear, appropriate and sustainable alternative. For them it is a question of how to switch from the inappropriate to the appropriate -- and stay there.

It would be a mistake to consider that this metaphor is "just a way of thinking" without any concrete implications. Much legislation is designed around whether a person is in Condition A or Condition B of some such switch, with immediate consequences in terms of social security benefits, various forms of aid, and varieties of social sanction. An extreme example, the apartheid policy in South Africa distinguishing between "white" and "non-white", became administratively feasible following a seemingly innocent census in which people were requested to identify their "racial group".

4. Ambiguity and limitations of the "switch" metaphor
It is important to recognize the extent to which this switch metaphor is natural to western modes of thinking. It is debateable how meaningful such polarities are in other cultures, or within many sub-cultures of western societies (Maruyama 1980, Hofstede 1980). Indications of this are to be found in the ambiguity of attitudes towards corruption in non-western society -- and even in western society. If comprehension of the issue is more complex than that implied by the switch metaphor, and if the dynamics associated with each problem dimension call for a more complex description, then unquestioning use of the switch metaphor constitutes a real danger at this time (Judge, 1986).

Individuals and groups escape into ambiguity to capture the wider reality on which the options of the switch metaphor have been imposed. There are obviously more degrees of freedom than are implied by the switch metaphor. People have direct experience of those opportunities even though they may be poorly articulated into sets of categories.

5. "Smoking" as a metaphor of experiential ambiguity
The issue of smoking is an extremely valuable illustration of many dimensions of individual and collective response to the challenges of these times. It is a neat metaphor of the experiential ambiguities in discovering a more appropriate relationship to these challenges. It is especially valuable because it offers us a framework within which to discuss much more charged or controversial issues such as overpopulation and environmental degradation. Consider:

- as an illustration of switch thinking in public policy -- legislation on smoking vs. not-smoking (and ways of circumventing such restrictions in restaurants and the workplace)

- as a major source of tax revenue for governments -- can governments afford to recommend against it

- the struggle of the individual -- whether to smoke, how often to smoke, whether to "stop"

- the fashionable image of smoking -- macho, cool, sophisticated, a shared experience, low-key bonding, etc

- the health aspect -- the risk of lung cancer against the challenge of gaining weight

- as a stimulant and tranquillizer -- what alternatives are available for mood adjustment

- as a means of self-assertion -- imposing a style and subjecting others to its waste products; revolting against parental and other authorities

Using this metaphor, it is much easier -- especially for smokers -- to understand the ambiguity of governments and industrialists in restraining their exploitation of the environment. For industrialists "sustainable development" then lends itself to other interpretations far from those of conservationists, for example "sustainable competitive advantage". It is not simply a matter of the inherent logic of switching from unsustainable policies to sustainable ones. The assumption that industrialists will willingly espouse environmentally-friendly sustainable policies seems extremely naive in this light, even when all the arguments are clearly evident -- as the partial results of "health warnings" to smokers illustrate. Can any smoker genuinely expect industry to stop air pollution through smokestacks ?

6. Transcending the "switch" metaphor
In the light of the above arguments it should be possible to look anew at many of the conventional problems with which people are obliged to deal personally. This process should be legitimated by the probability of detecting forms of response by individuals which are not captured by the categories that the switch metaphor reinforces. The existence of additional categories, however confusedly they are currently understood, would then call for richer, and less mechanistic, metaphors to capture the relationship between them.

The issue is, as Mark Twain succinctly put it, *"If your only tool is a hammer, then all problems look like nails"*. The principal tool of the international community would appear to be the switch.

Challenge: transcending methodological limitations

1. Availability of imagery as a policy constraint
Many studies contributing to policy proposals continue to be made totally independently of any consideration of the imagery through which they may ultimately need to be presented. Many disciplines have a strong bias against imagery of any kind as well as against any consideration of the process whereby insights are communicated.

Such biases are inappropriate if only because of the recognized importance of metaphor and imagery in creative thinking (Van Noppen at al., 1985) even in the hardest of sciences such as fundamental physics (Miller, 1986). It is clear however that within any disciplinary framework or jargon there is little need for imagery because the practitioners share a common imaginal framework. There are terms for everything that needs to be communicated.

The situation is quite different when dealing with policy proposals emanating from different disciplinary, political, cultural and ideological contexts. In such settings each faction tends to view the methods and explanations of others with suspicion or contempt. The language and concepts used communicate increasingly poorly according to the conceptual distance between them (Feyerabend, 1987). In parliamentary debate this is frequently signalled by the use of "absurd", "irrelevant", "naive", "irresponsible", "incomprehensible" and "ridiculous" in referring to proposals from opposing factions.

2. Prosthetics and scaffolding for interdisciplinarity
"Interdisciplinary method" is at this point a contradiction in terms. A discipline is characterized by its methods. Despite three decades of general systems, no interdisciplinary method appropriate to the complex challenge of the times has achieved any degree of acceptance. Where such "methods" have been used in very specific situations, they take the form of administrative procedures for ensuring that a succession of experts comment on or discuss issues, but without any pretence at conceptual integration in the final report. Integration is left to the end-user, as exemplified by a term in German translating as "book-binding synthesis".

Since this situation has prevailed through several development decades, during which "interdisciplinarity" and "integration" have been favoured buzz words, it is worth asking whether a more radical approach could not be fruitfully explored. Is it possible that the functionality which "interdisciplinarity" and "integration" endeavour to denote is to be found at a different level, and in a different form, than that at which the methodological and other differences are so evident?

Specifically, are there comprehensible images or metaphors, of requisite complexity, onto which the insights of different constituencies of expertise can be mapped so as to establish the dynamics and boundaries of their relationships without eroding or destroying their identity? This possibility, explored by Bateson (1987), appears to call for much comment and detailed explanation in the light of this or that methodology. But it could be argued that any such explanation would merely be a further contribution to the existing communication problem. A more fruitful route forward would be to consider ways of identifying, designing and testing such metaphors in practice.

3. Marrying metaphor and interactive graphics
This proposal is not as radical as it might appear. The most advanced thinking in many disciplines is expressed in terms of objects and surfaces in a complex space. In some cases computer techniques are used to assist visualization of such spaces as a guide to further theoretical development. The suggestion is that some effort be devoted to "marrying" such uses of imagery with those developed by animators or with those based on features of the environment with which people have a familiar relationship.

4. Extra-paradigmatic dimensions: beyond the binary
Before considering the implications in response to real problems, it is appropriate to note the constructive criticism by Kinhide Mushakoji of what he calls "binary" approaches in science and disciplines affected by its methodology. *"By the very nature of scientific logic which is binary, intellectuals tend to form bi-polar structures with two opposed camps rallied under two paradigmatic banners. The polarization often takes place even within each of the two poles which then divide themselves into sub-poles, and so on...An inter-paradigmatic process should be able to break the bi-polarity of the intellectual community by introducing a third pole in the dialogical process... The role of such a pole is to introduce extra-paradigmatic considerations (into the discussion) and to break the dichotomic argumentation thus bringing into the discussion innovative ideas."* (Mushakoji, 1978). Edward de Bono has advocated the use of a special term "po" to accomplish precisely this (De Bono, 1973).

5. Fourfold grasp of reality
But Mushakoji goes on to draw attention to the "logico-real" problem of the relationship between the logical and the reality levels. He suggests that catastrophe theory can help to shed light on the different logical positions in the morphogenetic space by relating the continuous reality (i.e. *signifié*) to the discrete set of concepts (i.e. *significant*).

This leads him to advocate a four-fold non-formal logic model to provide a logical basis for inter-paradigmatic dialogues. Such a logic emerges from another Japanese scholar, Tokuryu Yamauchi (Yamauchi, 1974) who interrelates oriental thinking based on "lemmas" with occidental thinking based on "logos". Lemma concerns the modalities according to which the human mind grasps reality, rather than how human intellect reasons about it. Mushakoji sees the lemmic approach as offering a breakthrough in response to the static ontology of the West.

The tetralemmic model Mushakoji describes stipulates the existence of four lemmas: (a) affirmation, (b) negation, non-affirmation and non-negation, (c) affirmation and negation. Here (a) and (b) both belong to formal logic, whereas (c) and (d) are unacceptable to it, although they are necessary in theoretical physics. *"Only an acceptance of the third and fourth lemmas can allow a full representation of the contemporary world problematique in its totality since contemporary world reality is full of cases where a mere affirmation or negation does not make sense"*.

6. Reframing reality
It is unfortunate that Mushakoji has limited his concern to representing or grasping reality for the purposes of revolution in thinking. This does not respond to the problem of how to intervene in that reality on the basis of any such conceptual revolution -- the vital preoccupation in furthering human and social development. And yet the four lemmas lend themselves to such an action-oriented interpretation as the basis for a more general "action logic" discussed elsewhere (UIA, 1986).

Following Mushakoji's lead concerning catastrophe theory, essentially what could usefully be explored is the possibility of enabling people to recognize how they redefine the morphogenetic surface on which they function. The switch metaphor is associated with a surface with two focal positions (attractors or wells) separated by a "coll" and surrounded by impracticable "mountains". The challenge is to modify that topography to offer a multiplicity of alternatives -- including the original positions.

In this light it is interesting to note that Denis Postle (*Catastrophe Theory; predict and avoid personal disasters*, 1980) has demonstrated how catastrophe theory may be used by an individual to map his own behaviour and the critical areas on that map at which stress and breakdown are possible.

Challenge: imaginal deficiency and simplistic metaphors

1. Imaginal deficiency
In the light of earlier arguments concerning the "switch metaphor", it is useful to explore further on the assumption that the inadequacies of existing strategies are partly due to poverty of the imagination -- namely to imaginal deficiency at the policy level. The question to be asked is whether there is some pattern to current thinking -- such as reliance on the switch metaphor -- which effectively limits the complexity of the policy options which tend to emerge, especially at the international level.

The concern is that, perhaps more crucially, the question should be asked whether such imaginal deficiency is not a prime handicap for those most vulnerable to the problems of the times -- unemployment, illness, etc. It is well-recognized that rich use of imagination is made by those in underprivileged circumstances, whether in the form of visual imagery or metaphor, and irrespective of educational background. So it is not imagination that is lacking. The question may be rather:

- whether that imagination is appropriately harnessed in response to the challenges

- whether the imaginative form is appropriate to any new response to those challenges

- whether external forces (educators, experts, officials, priests, media, etc) do not actively de-legitimize the effective use of that imagination.

2. Prevalence of simplistic metaphors
Unfortunately those assuming responsibility for advocating and implementing new policies in response to the social problematique do not consider the social system to involve problems of comprehension as complex as those encountered in fundamental physics and theology. They continue to believe that policies of requisite complexity can be envisaged using non-metaphoric language, with the consequence that such policies tend to be simplistic, inadequate, incoherent (except to those involved) and incapable of arousing the enthusiasm of broad constituencies.

When it is a matter of practical politics, however, politicians make considerable use of metaphor in communicating their positions to the masses (*eg* "nuclear umbrella", "Star Wars", "green revolution"), as the studies cited by van Noppen show (1985). Instead of policy-makers using richer metaphors to envisage more imaginative policies of an inherently more integrative nature, policy-making is done using a sterile administrative jargon (legitimated by its academic equivalent). The simplistic product is then discussed and communicated to a wider public using metaphors which fail to conceal the poverty of the imagination on which the policy is based (*eg* "umbrella").

3. Examples of simplistic metaphors

(a) Geometrical metaphor: As a first and very basic illustration, consider the language in which policy arguments are made. In any policy debate, much reference is made to the "points" made and occasionally to the "line" of argument. Agendas, declarations, policy documents and organization charters are structured in terms of "points". It is important to recognize that metaphoric use is being made of the most primitive geometric elements -- points and lines.

It is true that within geometry much can be constructed with points and lines, because of their fundamental nature. But it is also true that any such construction depends on clear recognition of intermediary structures such as surfaces and volumes of various well-established forms (polygons, conics, polyhedrons, etc). In the policy world occasional vague reference is made to "areas" (of specialized activity) and to "spheres" (of influence). But it clear that the kind of understanding required in architecture to move from such basic geometric insights to the construction of the simplest arch required for the most basic forms of building is totally lacking at the policy level -- except in the intuitive understanding of the need for "checks and balances". Any discussion of "bridge building" policies to link two opposing factions therefore tends to lack the conceptual scaffolding through which effective bridges could be constructed.

This may help to explain why the principal means of describing and designing organizations is the organization chart which is normally a hierarchical tree structure -- quite primitive in geometric terms.

Is it not appropriate to ask why no exploration has been made of the potential implications of more complex geometrical objects as a scaffolding for new forms of policy ? What might emerge from endeavouring to structure an agenda, a declaration or a set of resolutions into a polyhedral form of some appropriate degree of complexity ? Through such a form, relationships between "points" can be defined more explicitly. "Areas" can be identified with precision. But potentially of the greatest significance, the assembly of points, lines and areas to form a volume. In this way conceptual scaffolding is provided for an integrated whole, which is not the case with the normal jumble of policy recommendations. It is much to be regretted that the implications of Buckminster Fuller's work, especially on tensegrity structures, have not been explored in the design of more appropriate conceptual, policy and institutional frameworks.

(b) Positional metaphors: Much current policy discussion is based on identification of the "position" taken by each party, to the point that each is called upon "to make his position clear". It is assumed that a reasonable debate can only take place if parties each take up a "stance" on some position. Shifting "position" is perceived as recognition of the weakness of that policy position. Such static metaphors reinforce the static quality of policies and prevent the emergence of more dynamic policies which might be of requisite complexity to contain dynamic problem situations.

In life, only plants occupy fixed positions. To survive animals need to shift position and take up different stances according to the changing challenges of their environment. One insightful book on policy by Geoffrey Vickers, is entitled "*Freedom in a Rocking Boat*". On rocking boats, fixity of position is associated with instability. What kinds of policy might emerge if policies were debated in terms of dynamic positional metaphors such as "walking" or "dancing" ?

(c) Tool metaphors: It is quite amazing to note the way in which unimaginative use is made of the simplest tools to illustrate what are supposedly the most critical and sensitive policies. The saddest examples relate to military strategies as defined by nuclear "umbrellas" and desert "shields" and "swords". Such thinking is reflected in the naming of military programs and defense systems, for example "trident".

Such metaphors raise the question as to whether there are not more complex tools which could provide the conceptual scaffolding to understand richer policy options.

(d) Domestic metaphors: The power of the metaphors used by politicians may effectively distract attention from the poverty of imagination. This is best illustrated by Harold Macmillan (former British Prime Minister) in attacking Margaret Thatcher's privatization policy as being a case of "selling the family silver". This powerful indictment of a complex policy was deflected by Thatcher's subsequent acknowledgement that she was indeed "selling the family silver", but that she was "selling it back to the family". Again are there not more complex features of domestic life which could be used to open up a wider range of policy options ?

Challenge: imaginal deficiency in management and policy making

It is intriguing to note the kinds of policy-making environment envisaged in science fiction for the distant future. But even million of years hence, there is an unfortunate similarity to the dynamics of board meetings today, just as they themselves appear to bear a strong resemblance to those in Roman times. Even in imagining such distant futures, galactic councils are envisaged as operating under some variant of Robert's *Rules of Order*.

1. Use of metaphors in business

In the corporate world, very extensive use is made of metaphor to communicate the essence of policies and strategies and responses to competing initiatives. Such metaphors are almost entirely based on military or sporting situations: "zapping the competition", "target audiences", "advertising ammunition", "keeping the ball in play", "scoring points", etc. Politicians have recently taken to declaring "war" on problems (cf "war on want", "war on drugs"). Such terms, and especially "mobilization", have been taken over by intergovernmental agencies, even when their aims are ostensibly and peaceful.

It is interesting to note that in the West, the favoured metaphors are derived from ball games (football, cricket, baseball, etc) and military combat. The sports call for a more mechanistic understanding. Whereas the Japanese and Chinese make use of a more non-linear, organic or poetic understanding of the sports they use as metaphors. A standard Japanese management text is concerned with the art and strategy of swordsmanship (Musashi, 1982). Note that a study has explored how the USA forces were defeated in Vietnam because of their dependence on military strategies modelled on chess in comparison to the Vietnamese strategies modelled on go.

It is appropriate to ask whether the use of richer metaphors is not a major factor in the continuing success of Japanese business strategies. Conversely the relative economic weakness of some societies may in part be due to the inappropriateness of the metaphors through which their entrepreneurial initiatives are contextualized or to their metaphoric impoverishment.

2. Initiative

One interesting example in a management context is the tendency to describe an initiative as "taking off", whether it then "flies" or "crashes", and whether it was flown "by the seat of the pants". It would be interesting to investigate whether the richness and value of this metaphor could be further developed by imbuing it with modelling dimensions. Whether as an airplane or a bird, further dimensions could be added to increase the match between the concrete actions, controlling initiatives and feedback requirements, and the way in which these are integrated within the flying metaphor. Developing such metaphors could offer a whole new approach to educating people in the art of taking initiative and entrepreneurship. Other examples include "in the field" and "cultivating" contacts.

3. Team work

The question might be asked as to whether the success of team work was determined in part by the sophistication of the metaphor through which it was perceived. A limited amount of research has been undertaken on the personality characteristics associated with the distinct roles required to build a succesful team. Typically much of the earlier work focussed on building crews for strategic bomber planes. The more interesting recent work on management teams has identified eight roles or functions (Belbin, 1981).

The question requiring investigation is whether managers performing such distinct roles can further increase their effectiveness by their use of more powerful metaphors, both as a personal creativity aid and as an aid to communication between members of the team. One of the advantages of such metaphors is that they can be used to encode both the positive qualities of the role and the characteristic weakness. An example of this is the "resource investigator" role involving a capacity for contacting people and exploring anything new, especially in response to a challenge. This is a well-known informal role in an army platoon during war-time (the person who can "obtain" a camera inside a prisoner of war camp.) Many metaphorical terms are used to describe it. The point is whether the such metaphors used are complementary or mutually undermining.

At the community level the use of different animal totems within an Indian tribe as symbolizing a special function could provide insight of relevance to governance of contemporary society. At the international level is there any significance for governance in the relationship between the "eagle" (USA) and the "bear" (USSR)? What sort of "menagerie" or "ecosystem" is the United Nations in these terms?

4. Management styles

Although metaphors are frequently used in management, the problem remains of establishing their use to the policy-making processes of governance. At this level distinct styles of management may be active or available as opportunities.

One recent study by Charles Handy of the London Graduate School of Business Studies (*The Gods of Management; who they are, how they work and why they fail*, 1979) uses four Greek deities to characterize the different styles of management. The four gods (and the associated organizational styles) are: Zeus (club), Apollo (role), Athena (task) and Dionysus (existential). He notes: "*Each of the four gods gives its name to a cult or philosophy of management and to an organizational culture. Each of these cultures has also got a formal, more technical name, as well as a diagrammatic picture. The names, picture and Greek God each carries its own overtones, and these overtones combine to build up the concept I am trying to convey. They also help to keep the ideas in one's memory. These names and signs and Gods do not amount to definitions, for the cultures cannot be precisely defined, only recognized when you see them... It is important to realize that each of these cultures, or ways of running things is good - for something.*"

As Handy stresses, the problem is to know how to choose which god for which circumstances. It is the constraints and opportunities of the process of choosing that need to be embodied in metaphor and which call for further investigation.

5. Management texts

Metaphor is much used in selling new approaches to management and policy-making. Thus the editor of the Harvard Business Review has authored a book entitled *When Giants Learn to Dance* (R Kanter, 1989). Another by Dudley Lynch and Paul Kordis is entitled *Strategy of the Dolphin; scoring a win in a chaotic world* (1988). Here management is urged to think like a "dolphin", rather than a "shark", in order to keep on top of the "carps". A reviewer in a management journal greeted it as "*a welcome respite from other management books that urge us to think like samurais, Attila the Hun, or members of the Prussian General Staff.*"

The point to be made, however, is whether a more systematic approach is required to discover what metaphors are beneficial to management thinking under what circumstances. It may indeed be useful to think like a shark or like a carp under certain circumstances.

6. Neuro-linguistic programming

It is important to note that aspects of the work on language and metaphor of Gregory Bateson, and of George Lakoff and Mark Johnson, have been developed by Richard Bandler and John Grinder (students of Bateson), and their followers, into a training programme for communication and therapy known as neuro-linguistic programming. This has given rise to an extensive literature, some of it specifically oriented to achieving more successful communication in business. Institutes and courses have been created in a number of countries, with provision for assessment of the competence of practitioners. Much specific attention is devoted to the use of metaphor. The degree of effort applied to this particular approach, and the quantity of material (whether published or in the form of course material), makes it difficult to determine whether the special style of neuro-linguistic programming does not effectively obscure the possibility of other applications of metaphor which have so far failed to attract significant support.

Challenge: distinguishing extended metaphors

It is neither possible nor appropriate to review here the literature on the many dimensions of metaphor relevant to this section. Figure 1 distinguishes different concerns.

1. Functions (rows in Fgure 1)
(a) Initiatory functions: The importance of metaphor in relationship to creativity, whether in the arts or the sciences, has been frequently noted. Through exploration of "lateral thinking", for example, this has been extended to management (E de Bono, *Lateral Thinking*). In such cases metaphor is the vehicle of insight and provides the first ordering of a previously inchoate set of possibilities and constraints. It is thus a vital tool for concept design.

Through a metaphor the earlier confusion is seen in a new way. Once this is possible, other tools may build on this foundation. In the case of governance, this may mean the formulation of a strategy, a slogan, a model, etc. Any such formulation may well make no reference to the triggering metaphor. It can be argued that exposure to the rigours of team sport and military training (or combat) ensures an unconscious formative influence on the categories people use to act and to comprehend social dynamics.

(b) Communicative functions: Once a concept has been formulated, it usually has to be communicated to people and groups who are unfamiliar with the specialized jargon in which it is embodied -- and are quite possibly completely disinclined to learn it (even if they have the background to do so). In such a situation, metaphor can be called upon to convey the essentials of the concept. In the case of governance, this may mean the presentation of a model or a strategy. Such presentation may, or may not, use the same metaphor as that through which such a strategy was conceived.

Despite such extensive use in creativity and communication, ironically metaphor has a very "bad press". The fact that metaphor may be used with great elegance in literature in no way compensates for the fact that in the "real world" of governance, technology and social problems, metaphors are usually perceived as a nuisance and a sign of sloppy thinking. A goal in computerized information is to ensure metaphor-free communication to avoid ambiguity and confusion. But at the same time, those concerned with such real world issues find themselves obliged to make use of metaphors to explain their concerns in seeking resources for them. They therefore tend to associate their use with methods of public relations which may be necessary but cannot be taken seriously. The best examples of this are the extensive use of metaphors by politicians, whether speaking to their constituencies or in parliamentary debate. But no policy document would be taken seriously if its language was based on metaphor.

2. Duration and intensity of use (columns in Figure 1)
This section aims to distinguish two approaches to metaphor as indicated by the columns of Figure 1.

(a) Ephemeral uses: Metaphor is most frequently used as a literary device to illustrate some ideas for rhetorical purposes. The charm and beauty of much literature is based on this effect. For communication and education purposes, metaphors may be briefly used to help people to understand an idea. The aim being that people should discard the metaphor once they have got the idea. The cognitive value of the metaphor is as a temporary piece of conceptual scaffolding. Similarly in any creative endeavour, insight may come through use of metaphor to help give form to an idea. The metaphor may then be discarded -- many ideas in fundamental physics are reported to arise in this way, for example.

(b) Extended use: A metaphoric framework may be used over an extended period of time as a vehicle for communication in practical situations. A clear example of this is the ball-and-stick models of complex chemical molecules, although this is known to be an unrealistic representation. Another is in business management where military metaphor is extensively used when describing the dynamics of relations with the competition and the markets "targeted", and sporting metaphors are used to describe the dynamics of teamwork within the corporation. It can be argued that the metaphor here performs an important cognitive function in giving form to the complex action oriented dynamics in a manner which is useful to the pursuit of business and is significant in maintaining a viable pattern of communications within a group (often of very diverse backgrounds).

3. Conscious vs Unconscious uses (diagonals of Figure 1).
(a) Unconscious uses: Despite all that may be taught about metaphor in any literary education, people tend not to be conscious of using such devices or of their cognitive implications -- just as people tend not to be conscious of using particular jargons or grammatical constructions. People tend to be trapped in root metaphors of which they are unconscious. The use of military and sporting metaphor in management tends to be unconscious and its cognitive consequences are implicit, as in the use of any language.

(b) Conscious uses: Metaphors may of course be deliberately and consciously selected in order to present phenomena in a particular light. An author or poet may do so quite intentionally, as may a politician, an educator, a psychotherapist, a salesman -- or any skilled negotiator.

This section is primarily concerned with the use of metaphor in an extended manner, whether for creative or communication purposes, but especially by design, when used consciously. The focus is thus on the use of metaphor for "reframing" understanding of the social environment as a basic for a conceptually sustainable basis for sustainable development in the future.

4. Illustrative metaphors (cells of the Figure 1)
(a) Ephemeral use for communication functions (cell A): In this case metaphors are selected and used in passing for purely rhetorical purposes. Using a clothing metaphor, it is rather like making a point by flashing a handkerchief.

(b) Ephemeral use for creative functions (cell B): Here metaphors are used as part of a brainstorming, creative process in which briefly the metaphor forms and carries the integrative insight. Using a clothing metaphor, it is rather like briefly trying on an outlandish garb as an expression of an unexplored aspect of one's prsonality.

(c) Extended use for creative functions (cell C): In this case a metaphor is selected for extended use, rather in the management use of military and sporting metaphors to provide a language through which to discuss teamwork. Using a clothing metaphor, it is like wearing a favourite jacket to make a statement.

(d) Extended use for communication functions (cell D): Here a metaphor is used as a way of articulating (all) relationships with the external environment. The metaphor provides a complete language. Using a clothing metaphor, it is rather like wearing a diving suit in order to work underwater, or a spacesuit to work in space.

Figure 1.	Ephemeral use	Extended use
Initiatory + Creative functions	Conscious * **** B ******* Unconscious ************** ******************	Conscious * **** C ******* Unconscious ************** ******************
Illustrative + Communicative functions	Conscious * **** A ******* Unconscious ************** ******************	Conscious * **** D ******* Unconscious ************** ******************

Challenge: designing metaphors and sets of metaphors

1. Construction of metaphor
It is strange that of the 4193 items recorded in the post-1970 bibliography of metaphor (J P van Noppen, 1985), only one is concerned with the **construction** of metaphors - and that only secondarily. The bibliography has a special "*Index of tenors, vehicles, and semantic fields*" which indicates 500 items covering all the substrates reviewed by **studies** of metaphor. They include everything from automobile through birth, building, dance, oven, spaceship, theatre to zoological.

It would appear that no deliberate attempts have been made to **design** metaphors which could provide new insight into man's relation to the global problematique. Nor has there been any attempt to collect together metaphors relevant to understanding of the global problematique.

There seems to have been little systematic search for root metaphors by which social organization is dominated. The challenge would seem to move beyond such domination to discover how it is possible to move creatively between such metaphors. Clues are required to the projection of such fundamental metaphors into decision-making contexts of organizations and governance. According to Doctorow (1977), as cited earlier: "*The development of civilization is essentially a progression of metaphors.*" This descriptive approach needs to be converted into the operational challenge of governance: *How is the progression of metaphors to be designed in order to develop civilization?.*

It is of course poets who are most sensitive and skilled in the design of metaphors. Poetry might be called the art of giving form to the invisible through metaphor. This is however no reason for other disciplines to avoid the challenge of clarifying the processes involved as an aid to more appropriate forms of social action.

2. Metaphors in education
Surprisingly there are a number of studies (see Van Noppen, 1985, 1990) of the metaphors governing theories of education, learning, growth, curriculum design and the use of metaphor by teachers. It is even argued that some forms of education oblige students to learn an imposed metaphor. The use of particular metaphors tends to create a symbolic metalanguage that conditions research and the kinds of theories that emerge. The metaphor implicit in the curriculum may perpetuate a culture or ideology. But except in the case of younger children, educators have not chosen to incorporate metaphor into the process of formal education as a means of offering short- cuts to the acquisition of learning. Given the choice between educating the few to mastery of formal skills for well-defined jobs (which increasingly do not exist) and educating the many to enable them to use metaphors by which they may be empowered conceptually to act in unforeseen ways, education systems favour the former.

3. Editorial experiment on metaphors of alternation
This was therefore the context for the editorial experiment in Section MM in which 88 metaphors have been deliberately constructed. The origin of this exercise was the recognition emerging from the many authors reviewed in Section KD that each development policy may be considered as a particular "answer" to the social problematique.

No such answer appears to be free from weaknesses. A shift to an alternative policy becomes progressively more necessary as the effects of these weaknesses accumulate. Since each such policy is a "language" or mindset whereby a world-view is organized, no adequate "logical" framework can exist to facilitate comprehension of the nature of such a shift or the transition between alternatives. Many familiar metaphors of alternation exist through which the characteristics of such a shift may however be understood. A wide range of metaphors were therefore adapted or constructed using a standard format. In considering them it is appropriate to reflect on those which have a developmental feature and on the distinctions between alternation, oscillation, vacillation or variation.

A further group of metaphors were subsequently elaborated to illustrate other possibilities of understanding the development problematique. In choosing the substrates for all the metaphors part of the intention was to demonstrate the variety of natural and technological phenomena which are available as a rich source of insight.

4. Resonance between partial models
Lakoff and Johnson (1980), refer to the need to reject "*the possibility of any objective or absolute truth*". Lakoff again, as quoted above, calls for acceptance of "*the use of many partial models, some of which are inconsistent with each other*".

The authors reviewed in Section KD tend to confirm these positions, but with the implication that a more fundamental truth can be embodied in a set of complementary partial models or languages, in which each model reflects one or more aspects of that truth. The truth is then embodied in the resonance between the various partial models, none of which is capable of encoding that truth in isolation (especially since some of the models may be essentially incommensurable or basically opposed to one another). The set of models thus constitutes a **resonance hybrid,** namely the minimum set required to communicate the various dimensions of the theory, paradigm, policy or insight which cannot be adequately embodied in any one of the models.

5. Embedding complementary metaphors in a set
The challenge is therefore not simply one of developing a "library" of individual metaphors - although this would be an extremely valuable communication tool in its own right - but rather of identifying complementary metaphors which can be usefully grouped as a set, but especially within a pattern. And presumably the richer the pattern, the more subtle or fundamental the insight which it is then able to carry.

6. Patterns which connect
Gregory Bateson's central thesis is: "*The pattern which connects is a metapattern. It is a pattern of patterns. It is that metapattern which defines the vast generalization that, indeed, it is patterns which connect.*" (*Mind and Nature*, 1979, p.11)

It is for this reason that conservation of the natural environment can usefully be seen in a new light. It is after all one of the richest integrated patterns to which man has access, even in resource-poor developing countries. As such it is a storehouse of metaphors, interrelated into patterns of whatever degree of complexity is required (A J N Judge, *The Territory Construed as a Map*, 1984). Destruction of that environment sets up a vicious circle that progressively endangers man's ability to acquire the necessary familiarity with patterns of a degree of complexity appropriate to his stewardship role. Such pattern erosion limits his ability to the simplistic requirements of a "hydroponic planet", whether or not this is viable in the longer term. As a biologist, Bateson asks: "*The pattern which connects: why do schools teach almost nothing of the pattern which connects? .... What is the pattern which connects all living creatures?*" (*Mind and Nature*, 1979, p.8).

The identification of patterns of concepts in Section MP may be seen in this light. From the manner of their construction, they are sets of metaphors using a "pattern language" as a substrate.

Opportunity: programme of metaphoric development

1. Empowerment through metaphor
There is a widely held academic view that use of metaphor may be valid for rhetorical and illustrative purposes but that its use for any serious purpose is to be considered highly suspect. This is not the place to repeat earlier arguments demonstrating how basic metaphor is to concept formation in the most respectable disciplines. Nor will any attempt be made to justify the well-established use of metaphor as an aid to creativity and social innovation. Nor is it useful to focus on the long-term debate concerning the appropriate distinctions between metaphor, analogy, allegory, parable, symbol, and the like, as literary devices, especially since such explorations do not seem to have empowered people to make more effective use of metaphor (however defined) in policy-related situations.

The key question requiring further exploration is whether metaphor can in reality be effectively used to enhance innovative policy-making for global management. The question is whether metaphors, used non-rhetorically, can provide the conceptual scaffolding for new policies and the structures resulting from them. Can metaphors offer new way for surviving in, and navigating through, a complex environment?

2. Speculative scenario for navigating through complexity
An imaginative stimulus for such investigation is provided by a science fiction scenario explored by a number of writers. It focuses on the challenge of comprehending high degrees of complexity calling for decision-making under operational conditions (as is the case in global management).

The problem envisaged is that of piloting or navigating a spacecraft through "hyperspace" or "sub-space", as imagined in the light of recent advances in theoretical physics and mathematics. Because of the inherent complexity of such environments, writers have explored the possibility that pilots and navigators might choose appropriate metaphors through which to perceive and order their task in relation to qualitative features of that complexity - for example, flying like a bird, windsurfing, swimming like a fish, tunnelling like a mole, etc.

The mass of data input derived from various arrays of sensors, and otherwise completely unmanageable, is then channelled to the pilot in the form of appropriate sensory inputs to the nerve synapses corresponding to his "wings" or his "fins". Perception through the chosen metaphor is assisted by artificial intelligence software and appropriate graphic displays.

The pilot switches between metaphors according to the nature of the hyperspace terrain. Such speculations do at least stimulate imagination concerning a possible marriage between metaphor and artificial intelligence in relation to governance.

3. Areas for investigation
Although it could be argued that successful entrepreneurs and leaders may in effect use some such metaphoric device already, as a way of ordering their strategic perceptions (even without the high-tech devices), the question as to whether more fundamental use can be made of metaphor in policy contexts can only be effectively answered by further work along the following lines:

(a) Design: Investigations are required into the way extended metaphors can be designed as an aid to governance, but above all to enable people at all levels of society to reconfigure their environment to open up the possibility of new initiatives and to provide the insights to ensure that they are viable and sustainable.

(b) Education: Educational techniques on the practical use of metaphor should be documented. People are not helped to learn how to select and use appropriate metaphors, nor how to order perceived complexity through metaphor. This failure deprives those with limited access to formal education of many possibilities, especially for self-education. A high proportion of the 120 studies of metaphor in education cited by van Noppen (1985) focus on metaphor as a problem for younger children or for students, especially when developing reading skills or learning a new language. Many studies however show the importance of figurative language for developing comprehension and recall, even on a long-term basis. But a study by H Polio et al. shows that even in the language arts, figures of speech are typically dealt with in minor subsections under the more general heading of poetry. Figurative language as such is never explained or developed as an important topic in its own right. But the same authors demonstrate that, in mundane contexts, people in their culture produce an average of about 1.5 novel and 3.4 cliche figures of speech per 100 words spoken. Van Noppen cites 31 studies of figurative language production, some of which focus on the learning environment. Although educational material to encourage use of metaphors could easily be developed to enrich peoples response to their social environment, such material does not appear to exist. The only exception appears to be the workbooks explicitly designed by W J Gordon to increase figurative language use in creative thinking (1968, 1971). He and others have described the significance of metaphors for learning.

(c) Development of metaphoric indicators: Irrespective of whether enhanced use of metaphors is encouraged, there is a need to develop aids to the recognition of what might be considered metaphoric "aggression" or "entrapment". Understanding is required of the conditions under which the use of metaphor is excessive, inadequate or inappropriate in some other way.

(d) Engaging and disengaging from a metaphor: Effort should be made to articulate the skills required to "take-on" an extended metaphor to guide understanding of complex issues, whether individually or in a group -- especially where concrete action is called for. Corresponding effort is required to develop the skills which make it possible to "take-off" the metaphor, possibly in order to move to a more appropriate metaphor. These are the processes of "donning" and "doffing" metaphors in response to different environments. Of special concern is developing the ability, when working within one metaphor, to determine that it is no longer as appropriate to the circumstances as some other metaphor might be.

(e) Empowerment in relation to social problems: The metaphors to which people have access should be examined to determine to what extent these are empowering or disempowering. The question is whether it is possible to design, or bring about the emergence of, empowering metaphors. There appear to have been no studies of the relevance of metaphor for societal learning (a major concern of the United Nations University). And there is no detectable effort to assist people in learning metaphoric skills as a means of responding more effectively to their social problems and opportunities and to the global problematique in general.

If, as Fritjof Capra claims, the contemporary crisis is a crisis of perception, there is no effort to provide people with metaphoric tools with which they can re-imagine the evolutionary challenge with which they are faced. The nature of this opportunity for innovative practical use of metaphor is unfortunately obscured by the plethora of figurative material available and the kind of attention devoted to it in literary studies undertaken for their own sake. The relevance of such studies to the social problematique has not been effectively established. It is therefore difficult for people to transfer figurative skills from that context to the non-literary world.

As a trivial, but interesting example, Lakoff and Johnson (*Metaphors We Live By*, 1980, p.143-4) cite the case of a foreign student of theirs who, on encountering the expression "solution of your problems", assumed that this was a well-recognized chemical metaphor. Through it he had immediately obtained an understanding of the set of problems as being made up of some dissolved into a solution whilst others had been precipitated out (perhaps later to be redissolved again).

In this light problems never completely disappear, some are perceptible, whereas others have been temporarily "solved". Any attempt to solve some problems may quite probably precipitate out others. As Lakoff and Johnson say: *"To live by the chemical metaphor...rather than direct your energies towards solving your problems once and for all, you would direct your energies toward finding out what catalysts will dissolve your most pressing ones for the longest time without precipitating out worse ones."*

(f) Identification of metaphors of specialized agencies: It is not recognized, when advocating or imposing the use of particular sets of values, needs or programmes, that these effectively compete as functional substitutes in traditional societies with other sets of qualities and modes of action symbolized by hierarchies of gods or spiritual beings governing those qualities. The fundamental sets society now attempts to implant, whether embodied in the Specialized Agencies of the United Nations or the equivalent government ministries, are indeed designed to perform many of the regulatory functions previously ascribed to supernatural beings or potencies. Given the ersatz quality of the academic and administrative approaches to legitimating such initiatives, in contrast with the cultural richness popularly associated in the past with pantheons or Camelot, for example, it is not surprising that public information programmes have relatively little success in arousing enthusiasm and generating "a political will to change".

The question is therefore how such agencies could make creative use of the metaphoric and symbolic dimensions to counteract their superficial and "bloodless" images, and give credibility to their initiatives. Given the criticisms of inefficiency and fragmentation, such investigations could uncover ways in which the metaphors governing agency action could be seen as components of a self-organizing organic pattern of fundamental significance - even to the governance of the planet as a whole. Such investigations could highlight the necessary functional complementarity between the metaphors in any such pattern.

(g) Investigation of problems as metaphors: It is seldom realized that a societal problem, as such, is a problem (at least to some degree) precisely because it escapes any attempt to encompass it within any conventional set of categories. Such problems cannot be "defined" in any scientific way. Of special interest are the metaphors through which the global problematique may be perceived.

(h) Investigation of metaphors implicit in development action: It is seldom realized that a significant proportion of organization vocabulary results from innovations made by the Cistercian Order of monks after the 12th century in an early form of transnational organization. The notions of "assembly", "commission", "constitution", "agenda" and "ballot", for example, derive from that context.

(i) Relevance of therapeutic metaphors to development action: Metaphors traditionally have played an important role in therapeutic situations, both in the case of individuals and for communities as a whole. David Gordon's study of *Therapeutic Metaphors* (1978) represents an extremely valuable articulation of the therapeutic possibilities which are highly suggestive of new approaches to development and societal learning. Gordon's work is part of the general approach of neuro-linguistic programming which deliberately makes extensive use of metaphor, perhaps the most extensive conscious use by a professional group. The wider significance of this calls for investigation.

(j) Investigation of policy cycles: It has been argued earlier that seemingly incommensurable theoretical positions or social policies could be fruitfully explored as "frozen" portions of social learning cycles. In this light such particular positions are each naturally valid (i.e. appropriate) for a part of the cycle, but are inappropriate under conditions to which positions in other parts of the cycle respond.

Well-articulated positions or policies, taken in isolation, may thus be judged as attractive by those sensitive to the range of conditions which they address, namely by those in the same portion of the learning cycle. But such positions are essentially "sub-cyclic". Thus policy-making today, with its short-term focus, becomes a victim of cycles whose temporal scope it is unable to encompass. Any such policy naturally engenders what is perceived as "opposition", once it fails to respond to emerging conditions in the learning cycle. An interesting feature of this approach is the recognition that a position or policy rejected as inappropriate today may well re-emerge as appropriate some time in the future -- when the cycle repeats. Typical examples of this are alternation between phases of "centralization" and "decentralization".

This raises the question of how to design a cycle of "incompatible" but complementary policies, and how to select or design a metaphor through which to comprehend its phases (each of which may itself need to be communicated in metaphoric form). One intriguing example along these lines is the Chinese classic the *I Ching* (or Book of Changes) -- a traditional policy guide to the Emperor. This involves transitions between 64 contrasting conditions in a cyclic sequence, each described in metaphoric terms. A version of this has been interpreted into Western management jargon and applied to the clarification of sustainable policy cycles (see Section TP).

The argument for a shift to a cyclic focus needs to be based on further theoretical understanding of cycles in relation to social phenomena. Kinhide Mushakoji is exploring the effects of the introduction of cyclic assumptions into understanding of nature/society interactions, which may result in a proposal for a quasi-Buddhist understanding of transient reality based on an underlying non-aristotelian logic.

(k) Adapting insights from the arts fiction, poetry and music: It is one of the recognized functions of the arts to give form to visions of new ways of organizing perceptions of the world. The arts are therefore an important resource in exploring new visions of social organization and visions of the future. As such it might be expected that they would suggest new approaches to governance.

In explaining why "we are our own metaphor", biologist Gregory Bateson (1972) pointed out to a conference on the effects of conscious purpose on human adaptation that

"One reason why poetry is important for finding out about the world is because in poetry a set of relationships get mapped onto a level of diversity in us that we don't ordinarily have access to. We bring it out in poetry. We can give to each other in poetry the access to a set of relationships in the other person and in the world that we're not usually conscious of in ourselves. So we need poetry as knowledge about the world and about ourselves, because of this mapping from complexity to complexity."

Bateson is thus pointing to the advantages of poetry in providing access to a level of complexity in people of which they are not normally aware. This could well be of significance for the governance of social processes characterized by patterns of relationships normally too complex for the mind to grasp. Of special interest in comprehending non-linear cyclic processes in relation to linear thinking, are the potential insights arising from the relation of rhythm to metre in poetry. In this sense the current "spastic" development of society, as a victim of economic cycles, may be seen as resulting from an a-rhythmic approach to governance.

The challenge in exploring the rich range of metaphor available from the arts is to develop a method of culling materials, identifying "useful" metaphors and storing them in some appropriately designed database through which their wider significance could be explored.

Opportunity: from governing metaphors to governance through metaphor

1. Governing metaphors
At a time when there is much discussion of new paradigms, quantum leaps, breakthroughs and imaginative alternatives, it could be useful to explore collective and individual behaviour in search of the implicit metaphors by which they may be governed -- or govern themselves (Judge, 1987). Such exploration tends to take the form of identifying the "belief" or "value" systems within which people operate. And in these terms there has been concern in the international community as to ways of communicating more appropriate value systems -- especially those enshrined in human rights conventions. There are also many constituencies actively promoting particular belief systems.

Whilst promotion of belief and value systems opens opportunities for some, the track record of this approach does not suggest that it will make a difference in the time available. They also tend to be presented in relatively diffuse texts that call for special education processes before the full benefit can be derived from them. At the other extreme are the slogans favoured by political groups. In this case the difference made, if any, tends not to reflect the complexities of the situation -- thus engendering further difficulties.

There have been suggestions concerning the existence of "root metaphors" governing particular world views. Such root metaphors have also been noted in relation to images of social organization (Morgan, 1986). There is currently much emphasis, in the case of particular corporations, of identifying or designing an appropriate "corporate culture". In the past at least, great emphasis has been placed on family mottoes (at least amongst the western aristocracy). Such mottoes were also developed by guilds. In some non-western cultures totems have played an even more powerful role in providing a metaphoric view of the world (Cowan, 1990). In various religious traditions, phrases based on particular metaphors are used to guide personal transformation, often through meditation.

2. Enhancing conscious use of metaphor
In the light of the recognized cognitive function of metaphor (MacCormac, 1985), these examples suggest the possibility of encouraging more active use of metaphor by individuals in order to creatively "redesign" their cognitive environments so that new opportunities become apparent and acquire legitimacy. The role of metaphor in scientific and artistic innovation suggests that equivalent uses of metaphor are possible in the realm of social innovation.

It should be quickly noted that there are clearly limitations to any metaphor and that it is easy to get trapped in an inappropriate metaphor -- or rather in a metaphor which is inappropriate to the circumstances. Current entrapment by the switch metaphor (discussed earlier) might be an example. The challenge is therefore to provide contextual metaphors which enable people to shift around within a set of metaphors, where each is appropriate to different conditions (Judge, 1989b). This is especially important when it is becoming increasingly apparent that no one explanation, theory, model or paradigm can encompass the complexity within which people have to navigate. It would therefore be a mistake to imply that any particular metaphor can encompass more than an aspect of the reality with which people have to deal.

Given the increasing problems of the educational system, typified by the increasing number of functionally illiterate adults, it is necessary to look to other means of disseminating such metaphors. Of greater interest than such "dissemination from the centre" is the desirability of finding ways to encourage people to select or design their own metaphors using material natural to their own (sub)culture. It is more a question of enabling people to harness the social innovation potential of metaphors with which they are already familiar.

3. Governance through metaphor
Metaphor is widely used by politicians to communicate policy options -- both amongst themselves and to their constituencies. However it is used simplistically and in a rhetorical manner divorced from the written articulation of the policy and its implementation in practice. The metaphors currently favoured do not reflect the exigencies of sustainable development or the dynamics between the advocates of competing policy alternatives. (Earl MacCormac (cited above), who is both Science Advisor to the Office of the Governor of the State of North Carolina, and author of a book on *A Cognitive Theory of Metaphor* (1985) is deeply concerned with policy issues on the frontier between science and government. It is therefore strange that his book, and other studies, makes no reference to the possible relevance of metaphor for new approaches to development.)

The argument here is that governance could be more effectively based on processes facilitating:

(a) the emergence and movement of policy relevant metaphors;
(b) their relationship (as comprehensible meaning complexes) to more conventional forms of information; and
(c) their reflection in organizational form.

The merit of this vision of governance -- whether of a society, a group, a family, or as "self-governance" -- is that it does not call for an improbable, radical transformation of institutions and programmes. Rather it calls for a change in the way of thinking about what is circulated through society's information systems as the triggering force for any action (Judge, 1987b).

Resources can be usefully devoted to identifying, selecting, designing disseminating and employing more appropriate metaphors in policy contexts. Such a shift in focus should open up new ways of reflecting collectively on the more complex, cyclic and incommensurable perspectives currently lost in the savage interactions between factions. It is such complex perspectives which constitute the real policy challenge.

This suggests that the design of a desirable policy forum would focus attention on the emergence and movement of policy-relevant metaphors, their relationship (as comprehensible meaning complexes) to more conventional forms of information, and their reflection in organizational form. Stewardship in the governance of a forum opens possibilities in the governance of society as a whole.

4. Integrative function of metaphor for governance
The use of metaphors for communicative purposes clearly has an important integrative function in relating the governors and the governed. But it is the initiatory function which is of prime importance to the internal processes of governance. In a sense metaphor here has a "keystone" function as the ordering pattern or matrix through which strategies, models and programmes take form. It provides the implicit bridge between the disparate tools of governance.

Governance, especially when faced with the complex challenge of sustaining development, makes use of metaphor (whether explicitly or implicitly) in ordering its priorities and strategies. It is such fundamental metaphors imposed upon the "hyle", which give form and stability to a "landscape" on which the hazards and opportunities of governance are mapped. A major attribute of governance is the skill required to traverse such a terrain, possibly whilst under attack from hostile or destabilizing forces. But of equal importance, especially in the long-term, is the ability to switch to a new metaphor through which the epistemological domain is ordered. For, given the inherent complexity of the "hyle", no one metaphor can adequately encompass the dimensions to which governance must respond.

To fulfil its functions governance must be able to orient itself in terms of a succession of more appropriate "landscapes". It is possible for a single root metaphor to last the duration of a period of government (and electoral period) and engender a variety of needed strategies. But in a highly turbulent socio-political context, such a single metaphor is more then likely to prove inadequate. Governance then requires the skill to move between a set of metaphors each capable of rendering comprehensible certain sets of dimensions of the hyle. For this skill to become communicable it must itself be embodied in a metaphor

Opportunity: governance through enhancing the movement of meaning

1. Beyond the movement of information
Much has been made in recent years of the emergence of the "information society". Enthusiasts have envisaged this resulting in a "global village", given the facility of information access and transfer. Great care should however be taken in building on such hopes in envisioning new forms of governance.

In order to identify the opportunity for the emergence of a form of governance which responds with requisite variety to the issues identified in this paper, it is useful to distinguish three major ways of exploring the potential of the information society (Judge, *Reflections on Associative Constraints and Possibilities in an Information Society*, 1987). These are:

-- an Adaptive Approach, anchored in existing mind-sets;

-- an Innovative Approach, committed to change (but without disrupting fundamental patterns); and

-- a Transformative Approach, specifically concerned with the ability to switch between fundamental patterns.

Most effort and attention concerning the information society focuses on the Adaptive Approach. Some effort is devoted to the Innovative Approach, whilst very little is devoted to the Transformative Approach.

For there to be a real breakthrough in processes of governance, there has to be a real breakthrough in the movement of meaning in society. The mere movement of information (as represented by the Adaptive Approach) will not suffice, even if its is described as the "dissemination of knowledge". It leads to information overload and information underuse (a project of the United Nations University).

2. Concept life-cycles
It is at the level of the Innovative Approach that new key concepts emerge and, in the case of the international community, result in new programmes and institutions with new emphases. The manner in which this occurs at the moment is inadequate to the challenge. One useful way to envision the governance of the future is in contrast to Johan Galtung's insightful but disillusioned analysis of "concept careers" within the UN system, meaning both how innovative concepts undergo a career of stages or phases, a life-cycle in other words, and how concepts may move from one organization to another. Thus, as to their life-cycle at present, he notes:

(a) a fresh concept is co-opted into the system from the outside (almost never from the inside because the inside is not creative enough for the reasons mentioned). The concept is broad, unspecified, full of promises because of its (as yet) virgin character, capable of instilling some enthusiasm in people who do not suffer too much from a feeling of déjà-vu having been through a number of concept life cycles already. Examples: basic needs, self-reliance, new international economic order, appropriate technology, health for all, community participation, primary health care, inner/outer limits, common heritage of mankind;

(b) the organization receives the concept and it is built into preambles of resolutions; drafters and secretaries get dexterity in handling it. The demand then arises to make it more precise so that it can reappear in the operational part of a resolution. A number of studies are commissioned, very carefully avoiding too close contact with people and groups behind the more original formulations as "they do not need to be convinced." The concept thus moves from birth via adolescence to maturity, meaning that it has been changed sufficiently to become structure and culture compatible (it will not threaten states except states singled out by the majority to be threatened); the idiom will be that of the saxonic intellectual style, rich in documentation and poor in theory and insights; very precise but limited in connotations and emotive overtones; "politically adequate" meaning that it can be used to build consensus or dissent; depending on what is wanted where and when.

(c) From maturity to senescence and death is but a short step: the concept thus emasculated can no longer serve the purpose of renewal as what was new has largely been taken a away and what was old has been added in its place - except, possibly, the term itself. Even the word will then, after a period of grace, tend to disappear, those who believe in it now no longer identify with it; those who did not get tired of saying "we knew it would not work, it did not stand the test of reality". In this phase outside originators of the concept may be called in for last ditch efforts of resuscitation, usually in vain. There is no official funeral ceremony as the concept will linger on in some resolutions, but there will be a feeling of a void, of bereavement. Consequently, the search will be on, by concept scouts, for new concepts to kindle frustrated and sluggish consciences. And as a result-

(d) a fresh concept is co-opted into the system from the outside, e.g. one that has already been through its life cycle in another part of the UN system. For the rest read the story once more.

Nevertheless, each concept leaves some trace behind, more than its denigrators would like to believe, less than the protagonists might have hoped for. If this were not the case the cognitive framework for the system would have undergone no change during the 35 years of its existence". (Johan Galtung, *Processes in the UN System*, 1980)

3. Encompassing conceptual renewal
In the light of the arguments of this paper, the weakness of the system highlighted by Galtung is that it is focused on concepts as they move into and out of fashion, rather then on the metaphor-models through which concepts emerge and may be associated. Effort is made to create the impression that such "concepts" as self-reliance should be understood as meaningful in their own right, as the product of academic, political and administrative expertise. At the same time, in order to communicate their significance and ensure support for them, they form the subject of public information programmes, documentaries, propaganda and sloganneering. Through this process they also become metaphors (as well as symbols of an approach which others attack).

The problem is that as such these metaphor-models are not very rich. As conceptual models, they may be, but those dimensions are not well reflected in the metaphorical presentation that migrates through the field of public opinion - they were not designed to be. They do not excite the imagination of many as metaphors can.

How are current preoccupations with the concept of "sustainable development" to be understood in this temporal context ? How long will the concept be able to sustain its "career" ? What factors will contribute to the emergence of a new concept ?

4. Sustaining the movement of meaning
It is a truism that development is an essentially dynamic process. It is however less evident that the modes of thought enabling that process need to be equally dynamic if that process is to be sustained (cf Ashby's Law). The dilemma is that any concrete action tends to have to be designed in terms of specific goals, models and institutions which must necessarily be characterized by a certain static quality in conflict with such dynamic flexibility. Loss of dynamism appears to be the price of concreteness. The argument here is that loss of specifity is the price of sustainability.

Earlier points make it possible to suggest that a desirable form of **governance should focus its attention on the emergence and movement of policy-relevant metaphor-models in society**. Instead of regretting or resisting the life-cycle that Galtung identifies, many possibilities lie in enhancing and ordering that movement, which is better conceived as the life-blood of the international community. The challenge lies in bonding metaphors to concepts to provide vehicles for the latter to move effectively through information and institutional systems - as motivating concepts rather than solely as part of the streams of information processed.

Governance is then fundamentally the process of ensuring the emergence and movement of such "guiding" metaphor-models

through an information society, as well as their embodiment in organizational form. Such stewardship also requires sensitively to the progressive devaluation of any metaphor-model (at the end of its current cycle) and the need to adapt institutions accordingly. The stewardship required of the metaphor-model "gene pool" is analogous to that currently called for in the care of tropical forest ecosystems - as the richest pools of species and as vital to the condition of the atmosphere. There are conceptual "rain forests" which are currently endangered.

5. Reframing the movement of information
The merit of this vision of governance, as noted earlier, is that it does not call for a radical transformation of institutions. Rather it calls for a change in the way of thinking about what is circulated through society's information systems as the triggering force for any action. At present governance in the international community is haunted by a form of collective schizophrenia - a left-brain preoccupation with "serious" academic models and administrative programmes, and a right-brain preoccupation with the proclivities of public opinion avid for "meaningful" action (even if "sensational"). This schizophrenic battle between models and metaphors could be resolved by legitimating the metaphoric dimensions already so vital to any motivation of public opinion as a vehicle for the models. There needs to be a two-away flow however from model-to-metaphor and from metaphor-to-model, as in any interesting learning process.

In a sense this proposal is only radical in that it advocates the legitimation and improvement of processes which already occur - if only in the sterile and demotivating manner highlighted by Galtung. New metaphors are constantly emerging in the arts and sciences. They are used by politicians. Presumably some of them are used in the existing policy-making processes of governance. But the ecosystem of metaphor-models is an impoverished one. It is totally divorced from the cultural heritage of the world. There is a need to shift the level of analysis to the Transformative Approach. This shift is consistent with the analyses of the "post-modern" predicament (Hilary Lawson, *Reflexivity*, 1985).

6. Configurations of options: the contrast to relativism
This section is not simply an argument for relativism. The current approach to governance focuses on the attempt to achieve consensus on a single policy which is appropriate -- usually for an unspecified long-time. This does not correspond to the challenge of sustainable development in a learning society.

The next stage of development could well be described in terms of a configuration of metaphors -- each one of which is a distinct mode of perceiving the development process. Such metaphors are to be perceived as complementary -- appropriate to different conditions or different actors -- and implying the necessity to shift between them according to circumstance. No one of them reveals the whole truth, but the pattern of alternation between them provides a best approximation to appropriate action.

7. Resonance between alternative policies
The set of alternative structures, between which alternation takes place in any learning cycle, may be more clearly understood in the light of the theory of resonance. Johan Galtung first explored the possibility of using the organization of chemical molecules to clarify the description of social organization. He dealt with fixed structures and not with the transition between alternatives. The theory of resonance in chemistry is concerned with the representation of the actual normal state of molecules by a combination of several alternative "reasonable" structures, rather than by a single valence-bond structure. The molecule is then conceived as resonating among the several valence-bond structures, or rather to have a structure that is a resonance hybrid of these structures.

The best illustration of this is a resonance hybrid which is the form that best describes the dynamic structure of the benzene molecule -- basic to most organic substances.

8. Countervailing initiatives
In such a light, it would be important in policy-making to design a set of complementary policies which would be successively activated or deactivated according to circumstances. Some might be implemented simultaneously by organizations whose countervailing preoccupations would ensure an appropriate pattern of checks and balances. To ensure sustainability of development the focus would be on building complementarity into the design of the set. Such complementarity might necessarily entail that certain policies be opposed to one another or have opposite effects in order to respond appropriately to turbulent conditions in the social environment.

It could be argued that such a set would be increasingly appropriate to sustaining development to the extent that the number of complementary policies composing it -- namely the diversity of policies -- increased. The challenge is to design large coherent policy sets but in which the coherence is defined dynamically rather than statically. The dilemma is that the larger and the more diverse such a set becomes, the more difficult it is to comprehend, and the more difficult it is to understand the dynamics through which its coherence can be recognized. Hence the importance of metaphors to providing insight into coherent patterns of alternation between policies within the set.

9. Cycles of policies: illustrative metaphors
The difficulty in exploring patterns of alternation between modes of organization is the seeming lack of concrete (as opposed to abstract) examples by which the credibility of such patterns in practice may become apparent. In an effort to clarify the nature of such alternation, some 88 metaphors have been explored in Section MM.

In searching for appropriate metaphors to illustrate the need for cycles of policies there is a certain appropriateness to using a process which has traditionally been considered basic to sustaining the productivity of the land, namely crop rotation (see later note). The rotation of agricultural crops is an interesting "earthy" practice to explore in the light of the mind-set which it has required of farmers for several thousand years.

A cycle of seemingly incompatible practices, such as crop rotation, appears coherent to a large extent because of the establishment of a rhythm. Recognition of the complementarity over time is reinforced by the rhythm. Particular practices are eventually recognized to be appropriate to particular phases in the cycle. This sense has been lost in high-tech agriculture, just as it has in contemporary policy-making. Policy-making today, with its short-term focus, may be said to be essentially "sub-cyclic". As such it becomes the victim of cycles whose temporal scope it is unable to encompass. Focusing on the design of individual policies, rather than a cycle of policies, effectively builds inappropriateness and nonsustainability into the policy.

As an example, a recent newspaper article (entitled *"Scandals grow when a party has been too long at the trough"*) states: *"What the Reaganites face...is a growing national feeling that the Republican occupants of Washington's executive branch have been at the trough too long. They have become too caught up in self-interest to pay attention to the public interest"*. The same article notes a similar situation with respect to the Democrats in 1952.

10. Sustaining "sustainable development"
Sustainable development can only be sustained by a sustainable cycle of policies. This must necessarily encompass the necessary changes of particular policies associated with changes of government. Whilst government needs to be free to change policies in the short-term, a different attitude is required to cultivating and enriching the metaphors which give coherence to the cycle of such policies.

In a very real sense those metaphors symbolize for people the significance of the process of human development in which they are engaged -- irrespective of the phases of abundance or austerity by which the policy cycle may be characterized. Without such metaphors, abundance and austerity are simply associated with the promises and failures of different policies rather than as characteristics of a cycle of policies through which the society develops.

Opportunity: appropriate metaphors of sustainable development

1. Reframing the challenge of sustainable development
In response to the challenge of sustainable development, the potential of metaphor may be used to redefine that challenge. In both conceptual and policy terms it may be conceived as being one of designing metaphors to give form to a "sustainable ecology of development policies"(Judge, 1989).

In relation to the issues raised by Srivastva and Barrett (1988) and Barrett and Cooperrider (1989), it could prove appropriate to use richer metaphors to integrate (and render comprehensible as sets), the individual metaphors which govern the action of groups over time, or which govern opposing factions during a given period.

2. Levels of relationship between competing policies
A simplistic metaphor of the relationship between "environment" and "development" is that of "having one's cake and eating it too". It makes a critical difference what metaphor is used, whether implicitly or explicitly, to view the relationship between competing policy concepts:

(a) From a particular concept: From any given policy concept other concepts can only be viewed as threatening since that concept provides no sense of context, other than itself. "Enemy" is then an appropriate metaphor. Such defensive postures are not uncommon in policy forums. "Sustainable development" can be perceived in these terms with any other policy perspective as the enemy.

(b) As a group of competing concepts: Here context is provided by the sense of a "marketplace of ideas" in which the most appropriate products survive, if the market mechanism works satisfactorily. A more powerful metaphor is that of the "gladiatorial arena", in which one concept strives to emerge triumphant at the expense of the others, possibly learning from them in order to do so. Metaphors of this type, including those based on competitive sports, are widely used as noted above. "Sustainable development" can then be perceived as a set of competing concepts from which the most appropriate will emerge triumphant -- as the ideal result of a policy forum.

(c) As a homeostatic ecology of concepts: The two previous perspectives can however be perceived as subsystems or processes within an "ecology" of policy concepts. Here there are a variety of relationships between alternative policies (including "predation", "parasitism", and "commensalism"), but these function such as to maintain a balance between the different "species" of policy within the ecology (see following note). "Sustainable development" can then be perceived as a stewardship function of ensuring the stability of an ecology of policy concepts in which each fulfils particular developmental functions under particular conditions and there is a niche for developmental policies of all sizes and orientations.

(d) As an evolving ecology of concepts: Of greater interest is the possibility of perceiving "sustainable development" as an evolving ecology of developmental policies. Here there is a maintenance dimension corresponding to a homeostatic ecology as well as a longer-term evolutionary dimension as the various species adapt and evolve to emerging conditions, with new species emerging as the creative result of mutation processes.

3. Reframing the relationship between competing policies
If "sustainable development" is associated with metaphors of the first two kinds, its long-term value is questionable. If it can be perceived through metaphors of richness equivalent to the last two kinds, it can perform the integrative function necessary to incorporate both the policy priorities of "development" (in its many forms) and of "environment" (in its many forms). Note that only the last kind encompasses the continuing proliferation of alternative interpretations through a recognition of "speciation" processes.

There is an attractive conceptual elegance in endeavouring to use the natural environment as a metaphoric map to provide conceptual handles on the many policy dimensions of sustainable development. It suggests the need for a certain isomorphism between the pattern of development policies and the structure of the natural environment within which (and in response to which) they are implemented (Judge, 1984c). The ecological metaphor is explored in a following note.

4. Implications for policy forums
In any policy forum the question may then be asked as to the nature of the metaphor used to sustain the relationship between the range of policy perspectives represented. If that metaphor is not of requisite variety any result of such a forum can only be of value limited in time and space.

The insight of "sustainable development" cannot be satisfactorily embodied in a single policy or set of policies if no coherent context is provided for those who have to understand or approve it. Whatever the multiple, alternative or competing articulations of "sustainable development" at the conceptual or policy level, the insight integrating their dynamic relationships can only be adequately communicated at the metaphoric level.

If new approaches cannot be effectively implemented so as transform the functioning of international meetings, then they are also of little significance outside that arena. It is for this reason that challenge of transformative conferencing (see Section TC) constitutes an acid test for new proposals.

5. Implications for human development
In the light of the challenge of sustainable development, the question might well be asked as to how many metaphors people need to ensure their survival -- and especially their psychological survival? Is there a problem of metaphor impoverishment and deprivation associated with both ineffectual policies and individual alienation?

Is it possible that a metaphoric measure is necessary as a complement to the questionable value of current social indicators and the questionable educational role played by the exclusive use of the IQ measure of intelligence? To the extent that we ourselves are metaphors, do we need to develop richer metaphors through which to experience and express our self-image?

6. Implications for social development
If individual learning is governed by metaphors (as a number of studies indicate), how is it that metaphors governing societal learning and development have not been studied? In the light of Andreas Fuglesang's severe criticism of western assumptions concerning communication in developing countries (1982), would it not be more useful to conceive of different cultures as operating within different root metaphors?

Is it possible that social transformation is essentially a question of offering people (and empowering them to discover from their own traditions) richer and more meaningful metaphors through which to live, act and empower themselves?

Opportunity: enhancing policy through powerful metaphors

To a large extent the patterns of understanding appropriate to social innovations for global management cannot be effectively presented in the conventional linear mode (of which this text is an example). It is indeed possible to present a highly articulated argument, but the exercise bears some resemblance to the classic attempt to describe a spiral staircase verbally. The description, although exact, is not meaningful.

To be consistent with the argument, the kinds of insights to be gained from metaphors are presented using selected metaphors as examples of relevance to global management. Three such metaphors are presented below. A fourth, based on traffic, is presented in the following note. Section MP represents a sophisticated pattern of metaphor(s based on building design. Section TP may also be considered as an exercise in metaphor development to clarify issues of sustainable policy cycles.

1. Natural ecosystems as a source of policy insights

It is unfortunate that those claiming to be most sensitive to the need for sustainable development, namely the "greens", are unable to organize policy forums which reconcile policy differences in a significantly new way. It is worth exploring whether they at least could make creative use of an ecological metaphor to integrate their factions in a manner from which others could benefit. However, it is especially ironic that they seem to have felt no need to apply insights from their extremely valuable ecological thinking to new understanding of their own policy processes.

(a) Interacting species: Any policy forum constitutes a social system. Such a social system can be likened to an ecosystem with a range of interacting species. Each policy faction can be perceived as a species with varying numbers of members. The relationship between such factions can be observed in the light of the ways in which species may interact (symbiosis, commensalism, parasitism, allopathy, synnecrosis, amensalism, predation, allotropy, or none). It is clear that some policies are "predatory" and that complementary policies may be perceived as "symbiotic". Such relationships may effectively vary over time. Predation only takes place when the predator needs to eat. At other times the relationship is "peaceful".

(b) Food webs: A major ecological insight is that every species is some other species "lunch". It is not useful to think in terms of "good" species and "bad" species -- although members of any given species are obliged to perceive those that threaten them as "bad" to give focus to their fight for survival. Nor is it helpful to aim naively to have only symbiotic relationships between species -- eliminating the carnivores. Species are woven together in food chains. It is not helpful to focus attention solely on the top of any food chain (however magnificent the species there may appear) -- it is the food chain as a whole that needs to be understood. An endangered species is an important indicator of dangers to the ecosystem as a whole. On the other hand the cyclic rise and fall in numbers of particular species under different environmental conditions is a dynamic to which a resilient ecosystem responds appropriately -- any such rise or fall may be neither "good" nor "bad".

(c) Endangered ecosystems: In this light, a policy is naturally experienced as "bad" by other policies to which it is a threat. It is in turn experienced as "bad" and "good" by other policies. This level of perception does not help to understand the dynamics of the ecosystem of policies. At the ecosystemic level the issue is whether the numbers and dynamics of the species are destabilizing the ecosystem irreversibly and in what way. Excessive proliferation of any species, "swarming", endangers the ecosystem. This suggests that the predominance of any particular policy might have disastrous consequences. The health of the ecosystem lies in the healthy relationship between the species, even though this involves many predatory relationships.

(d) Shifting the level of debate: Through this metaphor, the level of debate is shifted. The natural tendency of any species to proliferate must be constrained by other species. The necessary "consumption" of some "innocent" policy by "predatory" interests needs to be explored in this light, as with the "regrettable decline" of other "predatory" policies for lack of resources. Any short-sighted effort to prevent the "nice" herbivores from being so "cruelly" consumed by the "nasty" carnivores invokes the need to "cull" their numbers periodically or prevent them breeding.

(e) Enriching the ecosystem: In these terms it is possible to shift the debate from consideration of species to consideration of whether the ecosystem could be usefully enriched: which ecosystems are "unhealthy", when should swamps be drained and arid zones "irrigated"? The difficulty here is that with the prevailing emphasis on monoculture, there is little shared understanding of how to diversify an ecosystem in ways such as those recommended by the Permaculture Movement (**). It is no wonder that many policy initiatives amount to a form of policy monoculture, fertilized by inappropriate use of resources and leading to pollution of the food chain.

(f) Designing systems of policies: An ecosystem calling for enrichment might be one which had been degraded by excesses of the past. The system of policies currently prevailing there would need to be redesigned. But note that it is the system of policies that needs to be redesigned, which does not imply that some single policy should prevail -- and the design needs to be an organic rather than a mechanistic one. It may mean that new "predators" should be introduced and that some population of "herbivores" should be cut back. Enrichment may involve introduction of many smaller species -- a reminder that the answer does not necessarily lie in mega-policies at the top of the food chain.

(g) Constraining species of policy: It should be noted that this metaphor does not suggest a form of policy relativism -- a tolerance of all policies. It suggests that any policy is dangerous in excess and needs counter-acting policies to contain. It suggests that whether a policy has a function depends on the ecosystem and that many policies may have a function within a policy ecosystem of a variety necessary to make it sustainable. This may mean that some policies can be usefully perceived as "prehistoric" but it does not deny that some prehistoric species (such as sharks) may still have a function, perhaps only in certain special niches.

Within this metaphor the many development policies are represented by species, each contributing to the health of the ecosystem. That ecosystem can be enriched by introducing new species to improve its sustainability. But members of those species, in the form of particular programmes and proposals may have a "life and death" relationship to one another -- reflected in such common phrases as "they killed our programme" or "they got our budget allocation".

Both in a policy forum and in the organized initiatives to which it gives rise, the information system needs to be designed to facilitate initiatives which sustain the ecosystem as a whole and which contribute to its redesign. In this sense the system of development policies should have a self-organizing dimension. Such an information system is in many ways a reflection of the food chain. Through it meaning is passed to nourish initiatives at different levels.

2. Crop rotation as a metaphor for a sustainable cycle of policies

Linear thinking encourages adoption of policies without thought to the nature of the policies which will have to follow them to remedy the havoc they cause, however incidentally. Given the many cycles essential to the coherence of the natural environment, a non-linear approach would suggest the exploration of "policy cycles" -- within which any "linear" policies are perceived as phases.

In searching for appropriate metaphors to illustrate the need for cycles of policies, there is a certain appropriateness to using a process which has traditionally been considered basic to sustaining the productivity of the land, namely crop rotation. The rotation of agricultural crops is an interesting "earthy" practice to explore in the light of the mind-set which it has required of farmers for several thousand years.

essential to the coherence of the natural environment, a non-linear approach would suggest the exploration of "policy cycles" -- within which any "linear" policies are perceived as phases.

In searching for appropriate metaphors to illustrate the need for cycles of policies, there is a certain appropriateness to using a process which has traditionally been considered basic to sustaining the productivity of the land, namely crop rotation. The rotation of agricultural crops is an interesting "earthy" practice to explore in the light of the mind-set which it has required of farmers for several thousand years.

Crop rotation is the alternation of different crops in the same field in some (more or less) regular sequence. It differs from the haphazard change of crops from time to time, in that a deliberately chosen set of crops is grown in succession in cycles over a period of years. Rotations may be of any length, being dependent on soil, climate, and crop. They are commonly of 3 to 7 years duration, usually with 4 crops (some of which may be grown twice in succession). The different crop rotations on each of the fields of the set making up the farm as a whole constitute a "crop rotation system" when integrated optimally. Long before crop rotation became a science, practice demonstrated that crop yields decline if the same crop is grown continuously in the same place. There are therefore many benefits, both direct and indirect to be obtained from good rotational cycles.

(a) Control of pests: With each crop grown the emergence of characteristic weeds, insects and diseases is facilitated. Changing to another crop inhibits the spread of such pests which would otherwise become uncontrollable.

(b) Maintenance of organic matter: Some crops deplete the organic matter in the soil, other increase it.

(c) Maintenance of soil nitrogen supply: No single cropping system will ordinarily maintain the nitrogen supply unless leguminous crops are alternated with others.

(d) Economy of labour: Several crops may be grown in succession with only one soil preparation (ploughing). For example: the land is ploughed for maize, the maize stubble is disked for wheat, then grass and clover are seeded in the wheat.

(e) Protection of soil: It was once believed necessary to leave land fallow for part of the cycle. Now it is known that a proper rotation of crops, with due attention to maintaining the balance of nutrients, is more successful than leaving the land bare and exposed to leaching and erosion.

(f) Complete use of soil: By alternation between deep and shallow-rooted crops the soil may be utilized more completely.

(g) Balanced use of plant nutrients: When appropriately alternated, crops reduce the different nutrient materials of the soil in more desirable proportions.

(h) Orderly farming: Work is more evenly distributed throughout the year. The farm layout is usually simplified and costs of production are reduced. The rushed work characteristic of haphazard cropping is avoided.

(i) Risk reduction: Risks are distributed among several crops as a guarantee against complete failure.

There is a striking parallel between the rotation of crops and the succession of (governmental) policies applied in a society. The contrast is also striking because of the essentially haphazard switch between "right" and "left" policies. There is little explicit awareness of the need for any rotation to correct for negative consequences ("pests") encouraged by each and to replenish the resources of society ("nutrients", "soil structure") which each policy so characteristically depletes.

There is no awareness, for example, of the number of distinct policies or modes of organization through which it is useful to rotate. Nor is it known how many such distinct cycles are necessary for an optimally integrated world society in which the temporary failure of one paradigm or mode of organization, due to adverse circumstances (disaster), is compensated by the success of others. It is also interesting that during a period of increasing complaints regarding cultural homogenization ("monoculture"), voters are either confronted with single-party systems or are frustrated by the lack of real choice between the alternatives offered.

There is something to be learnt from the mind-sets and social organizations associated with the stages in the history of crop rotation which evolved, beyond the slash-and-burn stage, through a 2-year crop-fallow rotation, to more complex 3 and 4-year rotations. Given the widespread sense of increasing impoverishment of the quality-of-life, consideration of crop rotation may clarify ways of thinking about what is being depleted, how to counteract this process, and the nature of the resources that are so vainly (and expensively) used as "fertilizer" and "pesticide" to keep the system going in the short-term. The "yield" to be maximized is presumably human and social development. The concern is whether current approaches are a dangerous "policy monoculture" trap.

3. Configuration of incommensurable policies as a resonance hybrid

In a world community characterized by distinct and often opposing views, the possibility of interrelating them so as to form the basis for an overarching structure, without denying the distinctness of those structures, can be usefully illustrated by a concept from chemistry. Chemical resonance hybrids are in fact basic to the molecular structures characteristic of living organisms.

Some chemical molecules cannot be satisfactorily described by a single configuration of bonded atoms. The theory of resonance is concerned with the representation of such molecules by a dynamic combination of several alternative structures, rather than by any one alone. The molecule is then conceived as "resonating" among the several structures and is said to be a "resonance hybrid" of them. The classic example is the benzene molecule (see Figure 2) with 6 carbon atoms. This is one of the basic components of many larger molecules essential to life. Its cyclic form only became credible when Kekulé showed that it oscillated between structures A and B. Linus Pauling later showed that it in fact alternates between all five forms above (and as such requires less energy than for any one of them).

This insight could be used in designing, describing or operating organizations, especially fragile coalitions or volatile meetings, and in giving form to complex agreements and policy configurations which would otherwise not exist. It may provide a key to the successful "marriage" between hierarchies and networks, or between centrally-planned and market-economies. It could also be used to interrelate alternative definitions (or theories, paradigms, policies, etc), especially where none of them is completely satisfactory in isolation. The underlying significance -- in contrast to the essentially unsustainable significance of each in isolation -- emerges through the resonance between the set of alternatives.

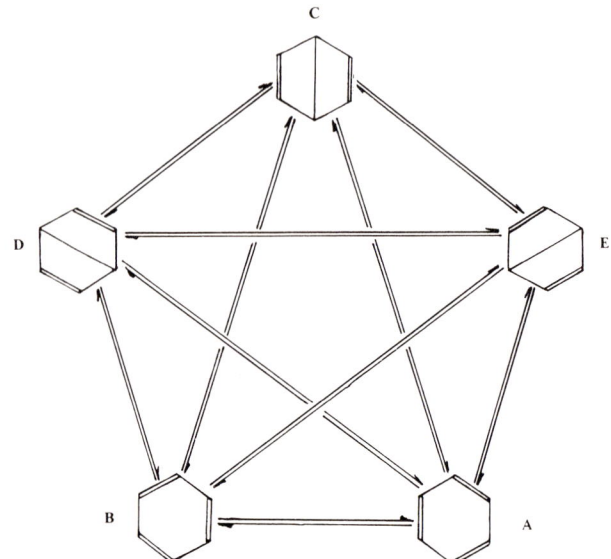

Figure 2: Resonance hybrid: alternation between distinct structures

Opportunity: comprehending policy meshing and entrainment

1. Policy cycles
In order to be able to base some appropriate new form of socio-economic organization on a new pattern of insights, it is clearly necessary that collective comprehension of that pattern should persist in a coherent manner over a period of time. The length of time required must clearly be at least of the same order as that of any major cycle of processes through which the well-being of that new mode is ensured. By a major cycle is meant one which encompasses the shifts between the alternative paradigms or modes of operation required to correct for deviations or the accumulation of characteristics impeding the long-term development of that society.

Since most policy-making is tied to electoral cycles of from 1 to 7 years, and several electoral cycles may be required to compensate for each others excesses, it is probable that collective comprehension of such larger cycles is inadequate, to the extent that it exists at all.

2. Vehicle traffic as a metaphor
It is interesting that the dimensions of this collective comprehension problem of time cycles can be beautifully illustrated by a metaphor whose features are very familiar to all, namely the circulation of traffic. The movement of traffic of different kinds, of different densities, at different speeds and with different directions, especially in an urban environment, is (self-) regulated by a range of techniques.

These include:
- basic road rules (driving on right or left);
- prohibited actions (no entry, speed limit, no stopping, no waiting, no parking);
- required actions (stop, keep left, yield, turn right);
- limited access (no cyclists); and
- special warnings (dangerous crossing, etc).

Whilst drivers may bend and break such rules occasionally, they recognize the wisdom of them in most situations -- for their own continued survival, if not for that of others.

To improve traffic flow, traffic signals may be used permitting an orderly alternation in direction of movement (e.g. from the right or from the left). These may be phased in various ways to improve flow in an area (e.g. the green phasing for a group of vehicles moving at constant speed along a route through the area). Area traffic control responsive to a range of traffic conditions may optimize flows by comparing current conditions to models based on past experience. Traffic of different types may also be segregated: pedestrians from vehicles, local from long-distance (e.g. on expressways with merging lanes and cloverleaf junctions).

3. Collective comprehension span
Movement of traffic under such conditions is only possible because the collective comprehension span exceeds that of:

- traffic signal cycles;
- waiting periods at stop or giveaway junctions;
- delays associated with travelling in peak period traffic;
- delays associated with multiple accidents and road blockages.

People are prepared to tolerate the priority given to others, knowing that priority will be given to the group of vehicles with which they are currently associated in accordance with the priority of traffic flows at that point. They are prepared to tolerate traffic moving in directions other than their own, knowing that at some time they may find it necessary to be moving in those same directions, perhaps on a return journey.

However, if the time span comprehended was reduced to the same order as that of the traffic signal cycles or less, a very different situation would prevail. A stoplight would be perceived as an unjust deprivation of rights. The traffic able to move at that time would be perceived as having acquired undue privileges, since it is composed of vehicles moving in other (and therefore irrelevant) directions, and especially if relatively few vehicles were travelling in that direction at that time. The sense of pattern would be completely lost. Only the direction of one's current journey would be of any significance. Traffic moving in the opposite direction could quite legitimately be forbidden.

4. Social implications
If the appropriate mode of socio-economic organization is to be comprehended in this light, the implication from the earlier arguments is that it is highly probable that the prevailing collective comprehension span is less than that of the cycles by which that mode is to be sustained. Constituencies advocating opposing policies are perceived like streams of traffic coming "from the right" or "from the left" or even "from the opposite direction". Since the pattern of the appropriate mode cannot be comprehended in its entirety, such alternative policies can only be considered as a misguided or dangerous use of resources. The possibility that implementation of the policy favoured by one's group might be temporarily interrupted, to ease the build up in the pressure in favour of another, could only be considered unreasonable.

5. Enriching the interplay of priorities
Use of this metaphor is not intended as a means of rendering existing injustices acceptable. Even in terms of this metaphor, there are some groups that have been waiting at a traffic junction a very long time for a green light to give them some degree of priority. The purpose of the metaphor is to provide a more accessible framework within which the interplay of priorities can be discussed. When should the flow of traffic on an expressway be interrupted to allow traffic from a side road to cross or enter the mainstream? Where is it appropriate to construct merging lane ("cloverleaf") junctions or underpasses? Where is it more appropriate for low frequency traffic to detour to an entry/crossover point? What are the major routes on the road map?

7. Phasing progress
The metaphor indicates a way of thinking about how the progress of groups of different kinds, developing at different rates in complex social environments, could order their conflicting policies with minimal mutual interference. Some form of cycle of signals might be used to enable groups with conflicting policies to progress during alternate periods. Actions of groups of different types may be segregated. The progress of groups with conflicting but interrelated policies may be facilitated by devising means for such policies to filter through each other (as at traffic "roundabouts") rather than cut across each other.

8. Limitations of present thinking
The prevailing approach may be seen, in the light of this metaphor, as one in which groups promoting different policies are given start/stop priority over each other in succession, in order to express their viewpoints. In the case of the alternation of political parties in power, an election is a process through which a decision is taken on the traffic signals. But in general, present policy control in this metaphor can be compared to a procession (or "progress") in one direction with the support of security forces which ensure that all access roads be blocked off and all opposing traffic suppressed. When the procession has petered out, another such "convoy" may be organized, by another coalition of forces, in another direction to cater for the traffic stream blocked by the first. This corresponds to a very primitive traffic control approach. It takes no account of the sophisticated blend of control and delegation of responsibility to drivers which is characteristic of modern traffic patterns.

Opportunity: reframing problems as metaphors

It is useful to challenge the thinking trap of "problem-solving". The approach to problems may then be reframed by asking what a problem is "trying to tell us" -- or, better still, is the problem as understood in effect a metaphor for something we would prefer not to understand? From this perspective "institutionalized" problems may in effect be a sort of metaphorical euphemism -- a package which it is better not to unwrap. Problems are not only nasty in themselves, they are also nasty in what they imply about ourselves -- however saintly we might wish to appear as disinterested change agents, victims or innocent bystanders. Consider the following:

1. Substance abuse (including drugs and alcohol)
Is it too trite to suggest that substance abuse is signalling a desperate need for different modes of thinking, feeling and experience than those sanctioned by a society governed by antiquated thinking patterns which have been only too effectively institutionalized in "acceptable" modes of work and leisure? Again, since many in key positions in such institutions also use drugs or alcohol "to relax", what should be learnt from the level of stress -- and schizophrenia -- at which the prevailing mode of thought is requiring them to function? Is substance abuse not effectively offering a remedy for the imaginal deficiency and mechanistic patterning characteristic of "acceptable" individual and collective behaviour? And consequently would not substance abuse become less necessary if society acknowledged more imaginative opportunities? What is the incidence of substance abuse in cultures whose languages make very extensive use of metaphor? Too what extent is it useful to perceive our relation to the prevailing thinking pattern as a form of "addiction" -- a habit that we do not know how to kick?

2. Unemployment (including underemployment and absenteeism)
It is no longer fruitful to argue that a significant proportion of unemployment is simply due to laziness, reluctance to learn new skills, lack of initiative or lack of opportunities. Is it possible that the prevailing mode of thinking is inhibiting peoples ability to imagine new forms of action of value to others, encouraging people to perceive existing employment opportunities as worthless both to themselves and to others, as well as impoverishing the manner in which people consider what to do with their lives? Is unemployment telling us that much of the work on offer is not worth doing -- and that much which is done is pointless? This would certainly be consistent with many criticisms of the consumer society and of industrial exploitation of the environment. Perhaps it is also saying that what we value doing, or are obliged to do, is not appropriately valued (as "work") in an economic system governed by an inadequate mode of thinking. This would certainly be consistent with the debate about the economic value of housework. Contrasting employment with recreation (as opposed to unemployment) is somewhat ironic in that unimaginative leisure opportunities are increasingly incapable of offering "re-creation". Is the level of unemployment also indicating that we really do not know to what society could usefully devote its human resources? Worse still, is it indicating that we have dissociated the challenges to human society from opportunities for "work" because of the way such challenges are perceived within the prevailing pattern of thinking?

3. Ignorance (including functional illiteracy)
Is the level of ignorance, even in industrialized countries, telling us that much of the knowledge on which that judgement is based is not worth learning? This concern has certainly been expressed in debates about existing curricula. Is it suggesting that for their psychic survival people are educating themselves along pathways which are not considered meaningful, or indicative of intelligence, within the prevailing pattern of thinking? This is suggested by the immense resources devoted to music and to "alternative" therapies and belief systems. Is it suggesting that people feel deprived of an imaginal education, faced with the formal (even rote) learning so frequently considered most appropriate (especially "to the needs of industry")? This is suggested by the enthusiasm for graphics, cartoon books, science fiction, fantasy and the archetypal portrayal of cult figures in music. Is our concern with the ignorance of many concealing the fact that those with most expertise and power are really quite ignorant about how to navigate through current and future crises?

4. Homelessness
Is the lack of appropriate shelter, even in industrialized countries, indicating that with our current pattern of thinking we are ineffective in our ability to provide, construct, or acquire cognitive and affective frameworks to shelter us appropriately from the turbulence of the times? This would be consistent with concerns about alienation in modern society. It would also follow from the recognition that many traditional frameworks and belief systems have been torn down or discredited. Even where people are well sheltered, it is often in houses or apartments which reflect an impoverishment of architectural imagination as reinforced by unimaginative building regulations and construction economics. Are our imaginative lives so impoverished by the media that the ability to provide a hospitable "interior decoration" for our psyches has been degraded?

5. Illness
Disease as a metaphor has been explored, especially by Susan Sontag. Nevertheless the preoccupation of the World Health Organization with "Heath for all by the Year 2000" fails to address the increasing prevalence of stress, neurosis and personality disorder -- especially in industrialized countries. Just as the range of individual diseases provides admirable metaphors for a taxonomic study of the world problematique, so it might also be used to explore the diseases of the imagination and of imaginal deficiency.

6. Hunger (including malnutrition)
At the time of writing some 4 million people are threatened with death by starvation in Ethiopia alone. Is this problem not signalling the existence of a subtler and more widespread form of deprivation -- a malnutrition of the psyche and a spiritual hunger which we are even less capable of addressing? This would be consistent with concern about the artificiality and superficiality of experience offered in the emerging "information society" or "global village" -- and with the desperate attempts to increase the level of "realism" by increasing the quantity and degrading quality of violence portrayed in the media. To what extent are our imaginations appropriately nourished at this time -- despite the surfeit of imaginative material available.?

7. Wastage (including environmental degradation)
Is our insensitivity to the processes of wastage and pollution, for which we are individually responsible, signalling the existence of an indifference to the "salubrity" of our imaginative lives? This would be consistent with the concern expressed by some non-western cultures and constituencies at the indifference to "spiritual purity". There is little consensus on what is or is not healthy for the psyche -- just as we are no longer clear, with the increasing scope of pollution, to what extent which foodstuffs are safe. The depletion of natural resources associated with wastage calls for reflection on the possibility that western-inspired culture is depleting its psychic resources in ways that we have yet to understand? Can the imaginative resources of a culture be depleted to a point of "bankruptcy" and how can such resources be conserved and "recycled"? Do empires fall through imaginative failure?

8. Corruption (including crime)
A major criticism of the development aid process is that the resources are diverted away from those most in need, despite agreements to prevent this. Various forms of bribery or "commission" are a common feature, even in industrialized countries. In any position (including intergovernmental agencies), people endeavour to obtain perks and privileges for themselves, for relatives or for friends -- whether this is limited to pilferage of office supplies, extended into the imposition of a "socially acceptable taxation" (or "sweetner") on any transactions which they control, or developed into a full-blown criminal activity. What can be learnt from this degree of self-interest and the associated rule-breaking propensity? Is this an indication that people cannot survive within the mechanistic regulations which emerge from the current pattern of thinking -- or at least choose not to do so, and feel free not to do so when possible? This would be consistent with the admiration for people who can get things done despite the rules, because they are capable of imagining more subtle opportunities. To what extent is corruption associated with a more creative world view -- as reflected in the term "creative accounting"?

Opportunity: reframing the problem of "overpopulation"

The previous note has indicated how problems may be seen in a new light by exploring the implications for the sustainable development of the individual -- through the individual's eyes. This provides an integrative focus which is absent when such problems are projected onto the global level, where mutually exclusive perspectives retain some measure of credibility. But, however valuable, it is not sufficient just to see such problems in a new light. The key question is whether they enable some new approach to them. The ultimate test is the case of "overpopulation", which many would argue to be at the origin of the problems outlined above.

The following constitutes an exploratory exercise to determine whether there are more fruitful ways of comprehending the issue of overpopulation.

1. Difficulties

There are many well-known difficulties in approaching this problem (as opposed to those in the previous note):

(a) Overpopulation as a non-problem: There are strong constituencies which do not consider overpopulation to be a problem in the first place. These include religious groups, such as the Catholic Church, with a vested interest in increasing the number of people of that faith, as well as countries which believe that their own population does not reflect their desired importance, either in absolute terms, or because of declining birthrate, or because of the need for expanding markets, or because of a threatened reduction in the influence of one ethnic group due to the high reproductive rate of some other ethnic group.

(b) Opposition to birthrate reduction: There are strong constituencies which view with suspicion any suggestion, especially by outsiders, that their birthrate should be restrained or reduced. This may be seen as a violation of their rights, as catering to the interests of the outsider group, or as an effort to deprive them of the socio-economic benefit of children in the form of labour, income, and social security in old age.

(c) Opposition to contraception: There are strong constituencies which view prevention of conception and/or termination of pregnancy with repugnance (the abortion issue, etc).

(d) Opposition to discussion of sexual relationships: There is a very strong constituency which views any discussion touching on the intimacies of sexual relationship as improper and to be avoided.

(e) Political evasion of the issue: The above factors reinforce the tendency of politicians to avoid such a controversial issue or to dramatically de-emphasize it -- as evidenced by the fate of international population programmes under the influence of President Reagan.

2. Family planning and sexless euphemisms

Because of all the above factors, even discussion of the "overpopulation" issue takes place through euphemism and indirection. This is compounded by the tendency, reinforced by intergovernmental agencies, to discuss problems through terms denoting programmes to solve them or through the values enhanced by solving them. Examples include: the "peace" and "youth" problems, or literacy (instead of illiteracy). In this case we have "family planning", "demography", "fertility", "population dynamics", etc. Programme agencies favour this approach because it lends itself to upbeat reporting concerning their programmes, irrespective of the impact of such programmes on the problems. Academics favour it because it emphasizes theoretical and methodological issues, irrespective of their relevance to any substantive problem.

Use of such sanitized terms to refer to an extremely charged issue may undoubtedly be appropriate under many conditions -- just as any reference to sexual relations tends to be avoided in the presence of children. It is questionable whether the discussion can be confined in this way in attempting to respond to the problem in a more innovative way. The sanitized terms, which are in effect euphemisms when they are not deliberate avoidance mechanisms, can be viewed as a sort of "metaphorical dissociation". The term used provides an uncharged metaphor through which to view an uncontroversial aspect of the problem. In this sense metaphors are being used to distort perception of the problem.

In discussion of the population issue there is in fact remarkably little discussion of sexual relations (especially the individual's perception thereof) -- as though one had almost nothing to do with the other. This is all the more remarkable given the importance of sex in the media and especially in advertising -- and, much more explicitly, in worldwide warning campaigns concerning AIDS. Advertising has made an art form out of metaphoric references to sex -- reinforced by visual metaphors in product design and packaging. In contrast, and paradoxically, there is a sexless quality to "family planning" which impedes any imaginative response to the issues involved -- especially to those aspects exacerbated by advertising, given the natural interest in sex.

This sexless quality is rendered even more unrealistic to the popular imagination given the widespread and extensive use of sexual metaphors in informal discussion. This is most remarkable in many job situations, including those at the highest level -- as President Richard Nixon demonstrated in his choice of expletives. Such metaphors are a basic characteristic of management dialogue in most corporations, as well as on the shop floor where things get done. Some people make use of them in every sentence. For example, other than its use as a simple expletive, "fucking" ("screwing", etc) is widely used to articulate an attitude to a group (another department, clients, competitors, etc) to whom one is doing something or by whom one is being manipulated.

The question which merits exploration is how the tremendous amount of psychic energy articulated (however inappropriately) by this metaphor is related to the "fertility" issue. Is it that the use of this basic metaphor for "doing" or "being done to" channels -- as a form of sublimation -- some of the energy which would otherwise go into sexual intercourse? The contexts in which the metaphor is used must certainly feedback onto the perception of sexual intercourse.

The challenge of "family planning" and "contraception" is that these are essentially processes of "not-doing" and as such do not excite the imagination -- except to those inclined to philosophies of inaction (and action through inaction). It is questionable whether the metaphors for such processes are sufficiently meaningful in competition with the richer sexual metaphors. It could be argued that the "contraception" issue involves only the prevention of contraception, not the prevention of sexual intercourse, and therefore does not detract from the energy of sexual relations. This brings into focus the core issue of whether "contraception" calls for any modification in the attitude to sexual relations in order to be successfully implemented by those gripped by a variety of powerful metaphors of sexual relations. Specifically is there a possibility of discovering metaphors which would enable people to articulate their attitudes to sexual relations in a manner consistent with the objectives of "family planning"?

3. Sexual intercourse as a metaphor

It would be presumptuous to hope to explore here this central theme in literature, psychoanalysis and advertising techniques. But some questions can be touched upon as they relate to the "overpopulation" issue.

In social conditions widely characterized by turbulence, insecurity, savage competition for resources, deprivation and the like, the privacy and intimacy of sexual intercourse creates a world in which individuals can experience a sense of security, caring and personal value, whilst at the same time offering them opportunities for imaginative self-expression and enjoyment away from the censure of wider society. It is a world in which they have a real opportunity to fulfil their desires, to experience a sense of personal integrity and to repair the psychic damage suffered in daily life.

Morris Berman (*Coming to Our Senses*, 1989) cites Denis de Rougemont's *Love in the Western World* to argue that *"falling in love is literally the one ecstatic or mystical experience left open, and it serves a haven from the culture of repression and control... It is the one tiny portion of their lives in which they can be truly kinaesthetic rather than visual; in which they can (theoretically, at least) live true selves as opposed to false ones...it remains, for millions, the only real counterculture they can enjoy, the only expression of inwardness left in the modern world."* Sexual intercourse is the embodiment of this process, as well as being a substitute for it when the early intensity of love no longer hold sway.

But, as a metaphor, this private "world" also encodes many of the problems and dilemmas of the "sustainable development" of wider society -- the macrocosm mirrored in microcosm. It is a world in which one partner may seek to dominate and subjugate the other, a world of resources which may be exploited until they are depleted beyond any measures of conservation, and a world in which the frustrations of wider society may be given a new, and often crueller, focus -- often without any court of appeal. The shared intimacy may decay into alienation.

It is into this world that "family planning" endeavours to insert "contraception". But little is said concerning the implications for the psychic life of the individual. The matter tends to be discussed and described using plumbing metaphors, in "practical, down-to-earth" terms. And undoubtedly this may be totally appropriate for those of unimaginative temperament who believe that problems can be "fixed" with an appropriate device -- or for those who are so desperate that they will use anything provided it works in practice.

It is not clear that this approach is appropriate to others, and especially to those for whom contraception acquires some symbolic significance. In this sense it is useful to consider how contraception can function as a metaphor. Understanding its potential, and limitations, as a metaphor may suggest other approaches. Lines of investigation might include:

(a) Preventing, or aborting, completion of a process: With the increasing mechanization of society, and the increasing fragmentation of fields of activity, there are few integrative processes of which people are personally aware. The processes to which people are exposed in society are increasingly embedded in bureaucratic procedures, manufacturing cycles or information systems. Most processes are subject to "production deadlines" -- including academic research. There are few opportunities for process completion, which it could be argued are vital to the psychic integration of the individual. Even in manufacturing there is increasing recognition of the merit of allowing people to personally complete a process (eg assembly of a car). For those for whom the significance of the sexual process is not limited to the act, but includes the socio-biological consequence (and possibly religious implications), to what extent does contraception become a metaphor for a restraint which is increasingly intolerable in an alienating society?

How is contraceptive technology to be understood in the light of Heisenberg's observation that the purpose of technology is to arrange the world so that people do not have to experience it?

(b) Reproduction and impotence: Reproduction is a basic purpose of sexual intercourse. But in how many ways are people currently able to reproduce themselves, and how many of these are increasingly frustrated in modern society? People can, for example, feel that they are reproducing themselves by impregnating an individual or group with a pattern of ideas -- possibly to be passed on to future generations. To a lesser degree, but with more immediate impact, people "produce" themselves before an audience. Indeed professional performers, especially before mass audiences, describe this process in explicitly sexual terms. Such forms of reproduction (harmonics to the sexual process, are increasingly unavailable or of decreasing significance -- to the point that people experience a sense of impotence. To what extent then does this place an unconstrainable burden on the physical process which can compensate for this inadequacy in some measure?

(c) Rechannelling sexual energy: Does contraception have no effect on sexual energy and the way it is channelled? It can easily be argued that it is a liberator of sexual expression. It is less clear how it affects the quality of that expression. It is possible that exploring the limitations of the "channel" metaphor (see Lakoff and Johnson, 1980 on the conduit metaphor), and that of contraception in relation to it, might suggest a more appropriate approach. The use of such mechanistic metaphors, of the same class as the switch metaphor discussed earlier, inhibits recognition of less polarized insights into the movement of sexual energy (such as through diffusion or resonance processes, for example). The standard argument that access to a television at home reduces fertility needs to be reviewed in this light. What function is the television exploring and how is it affecting the imaginative life in relation to the need to fulfil sexual desires? Is the television an example of a wider class of opportunities for the movement of sexual energy that obviates the need for sexual intercourse? What is the function of dance and partying in this respect? Do these suggest the existence of processes which are metaphorically equivalent to intercourse, but diffused beyond the confines of the switch metaphor (making-it, or not)?

(d) The "developer mentality" of family planning: The community of international development agencies seldom accords attention to the "developer's" view of development -- by which land, for example, is "developed" when it is "cleared" of unproductive trees and wildlife, drained of unnecessary surface water, and segmented by access roads permitting construction of any required buildings. In arguing for greater literacy for women, UNICEF indicates that four years of schooling enables women to plan smaller families, to space the children for the better welfare of all, and to make use of preventive health care. It is quite unclear what effect the rationality of such procedures has on the imaginative life of those who accept them, and whether any resistance to them arises from a repugnance analogous to that of "romantic" conservationists towards the initiatives of developers. To what extent is psychoanalytical expertise used in population programmes?

4. Alternative metaphors

The above possibilities raise the question of whether other kinds of metaphor might prove more appropriate as a way of articulating the imaginative life of sexually active people. Or rather, whether they have access to other metaphors which conflict with those implicit in their perception of contraception and "family planning". And how a greater proportion of sexual energy might be expressed through activities which are metaphors for sexual intercourse -- a concern close to the interests of advertising agencies endeavouring to market products which effect this transfer. Can the objectives of "fertility reduction" be served by a transfer of this kind. In this light something as simple as more cafes (social "intercourse") and dances might achieve more than strenuous attempts to extend family planning programmes -- however ridiculous this might appear to those seeking a technical fix.

As with the other problems, discussed in the previous note, it might then be possible to move on to the metaphoric implications of the global dimensions of "overpopulation". Somehow the proliferation of the species has become an absolute good. The action of any inhibitory feedback mechanism has itself been inhibited. The same phenomenon may be seen with the proliferation of information and products -- and any form of creativity. All such processes could be explored as metaphors of a human attitude in which withholding or holding back is inhibited. This suggests the need to discover attractive metaphors for "withholding", but without becoming trapped in the switch metaphor. An early, and perhaps inadequate, example is the increasing fashionability of "soft" sex (as opposed to penetrative sex), as a result of the rising threat of AIDS.

The creative value of exploring such metaphors of relationship is well illustrate by a study of metaphorical theology, namely of possible new metaphors of a person's relationship to God (McFague, 1983). The justification is a similar one.

Opportunity: reframing cooperation through metaphor

The following set of metaphors endeavours to highlight the range of mind-sets through which cooperation has been so enthusiastically pursued in the last 30 years -- with questionable success. Fundamental problems associated with each are briefly noted. The metaphors provide contrasting windows through which the imagination can explore the ways in which people, groups, factions and governments organize meetings, projects and long-term cooperation to improve the condition of the world.

1. Networking and Teleconferencing
Cooperation may be understood as networking -- the sending and receiving of messages amongst a network of people, groups and institutions within the "global village". This bypasses the conventional difficulties of communicating through and between different levels of organizational hierarchies and opens the doors to new opportunities for cooperation. **Problem:** Despite initial enthusiasm, such exchanges tend to evolve either into chatting, soliloquies or ("under strong leadership") narrow technical exchanges. They rely on mutual appreciation -- there is only limited capacity for management of conflict. Tension and negative feedback are designed out -- networks become incestuous and ineffectual. If the non-viability of a network is recognized, it decays into token exchanges or is abandoned -- possibly to give rise to another. There is little provision for collective learning -- insight and wisdom are not accumulated.

2. Revolution
Cooperation arises when we "bury our differences" in a revolutionary struggle to bury some common enemy, usually a group of people responsible for an iniquitous social structure or for an erroneous belief system. Self-interest, normally the principal obstacle to successful cooperation is transmuted into self-righteousness in a "holy war". **Problem:** In order to mobilize successfully for such a war, systems of restraint have to be abandoned. Once abandoned, there is no check on extreme violence (which may be non-physical) or exploitation by those able to manipulate the situation to their own ends -- in the name of the common cause. There is almost no capacity to distinguish what should be kept from what should be abandoned. Collective learning results only after collective revulsion at the pain and bloodshed and after recognition of the true colours of those who thrive on their necessity.

3. Trade and Development
Cooperation, especially for some French-speaking governments, is equivalent to development -- or the policies and procedures through which it is brought about. In practice this means evolving terms of trade -- "let's trade" -- perceived as mutually beneficial, whatever the constraints and recognized inequities under the agreement. **Problem:** As in the simplest deal, there is considerable scope within the terms of the agreement, for poor quality, unserviceable, obsolescent or hazardous goods. Purchases on credit may be such as to place the purchaser in semi-permanent bondage. The seller may over-sell, ensuring the placement of essentially inappropriate products which create more difficulties than they resolve. The weaker party may be persuaded by a skilled negotiator to part with assets of considerable value, especially when there is pressure to sacrifice long-term benefits to short-term relief. Those dealing on both sides may be more interested in how they benefit personally (kick-backs, career advancement), irrespective of the longer-term consequences to those whose interests they are supposed to represent. Collective learning only results when it is recognized who benefits systematically from such deals and who is systematically impoverished by them.

4. Sexual intercourse
At its best "making love" is one of the principal examples of effective cooperation between people -- "make love, not war". It calls for sensitivity, initiative and receptivity, and enhances mutual respect. Ideally it ranges from the reassuring to the transforming, and through such dynamics a new generation is conceived. **Problem:** As has been well-publicized, there are many far from ideal ways in which people engage in sex, from brutal domination by one partner through various exploitative sexual games -- not to mention the implications of prostitution, pornography or what some choose to perceive as perversion. It is questionable how often partners are mutually satisfied by such cooperation. Considerable emphasis is placed on preliminary techniques for arousing interest, on short-term "performance", on the level of personal "pay-off" (such as the quality and quantity of orgasms), and on avoiding any long-term consequences. "Safe-sex" is advocated to avoid mutual infection and contraception is practised to avoid the conception of any product from the union -- except amongst those without the means to care for such issue. In the unfortunate event of effective conception, considerable means are deployed to ensure abortion or disposal of the issue by other means. Every effort is made by the majority to avoid any tangible consequences of such acts of cooperation -- whilst a minority goes to great lengths to rectify infertility -- through artificial insemination and the use of surrogates.

5. Environmental ecosystems
The ecosystems interlinking flora and fauna are a valued example of how different species can cooperate -- the ideal of symbiosis is a much favoured model. The Gaia Hypothesis is explored as a model for cooperation at the global scale. Such insights are fundamental to the "green" movement. **Problem:** In the less challenging interpretation, humankind is to be seen as a single species whose members should cooperate as peacefully as those of any other species. This loses sight of the hierarchical "pecking order" obtaining within most such species and the dominance of one or other sex. It loses sight of the competition for territory and exclusion from herds. And, except as the dominant species, it loses sight of the consequences of being part of a food chain. In a more challenging interpretation, humankind forms a multiplicity of species -- not so much by race as by vocation, specialization, ideology or culture. In this case food chains, if only in the form of information, raise many questions -- such as why the factions of the green movement are unsuccessful in functioning symbiotically, and instead draw attention to the other (seven), less symbiotic, forms of interaction between species. In perhaps the most challenging interpretation, each person constitutes an ecosystem of roles and mind-sets, which interweave amongst themselves and with others -- raising questions about who (or what) it is that cooperates.

6. Drama and Opera
A dramatic work can be construed as a design for cooperation -- in which the actors cooperate in exploring themes and dramatic moments which play off each other to bring out certain qualities and insights. For the integrity of the work there is necessarily a deep commitment to ensuring the effectiveness of such cooperation. **Problem:** Even within a dramatic work, a distinction is made between those having minor roles and the stars who have a fundamental need to be set above the others. And, despite increasing exploration into ways of reducing the separation between actors and "audience", whether within a theatre, on television or in the street, there remains a basic distinction between the dramatic reality and that no longer governed by a particular work -- from which the audience is drawn. In effect actors play at cooperating -- as do many who pretend to cooperate -- in contrast to the often less than cooperative relationships obtaining between them off the stage. What is to be said of the basic commitment to "seduce" the audience, who are paying in order to be captured, at least temporarily, by the reality presented. There is also the question as to whether effective cooperation must necessarily be scripted or directed, or at least to what degree actors can improvise. If "all the world's a stage", are there many scripts, and what does that say about cooperation and the need for its direction?

7. Sharing in spirit
When spiritual values predominate, whether in an established religious tradition, a sect, a charismatic movement, or a religious community, then self-interest as an inhibitor of cooperation is bypassed. Cooperation becomes a sharing in spirit -- in the name of such as Christ, Allah, Buddha, Gaia, or of their enlightened representatives. People are "born again" into a new mode of interaction. **Problem:** Difficulties arise when the priorities are not clear and different factions emerge favouring distinct strategies. Everything then depends on the manner in which the spiritual values

are interpreted and articulated. Groups become vulnerable to skilled operators who can successfully manipulate peoples' relationship to their evolving understanding of spiritual values. It becomes difficult to distinguish between skilled "supervision" for the good of the whole (as part of a spiritual journey) and skilled manipulation at the expense of those who accept the process -- for the benefit of the "disciples" who lead it. An important device used in this process is the stress on some external threat, its insidious influence on those within the group, and the need to maintain a strong "non-cooperative" relationship with those who can be named as vehicles of it -- especially when they follow other practices.

8. Building

Cooperation may be seen as "building together". Emphasis is placed on the tangible, if not on construction in its most concrete sense, whether houses, barns, schools, clinics or community amenities. It may take the form of major projects (joint ventures) such as dams, aircraft, defence systems or satellites. Or it may take the form of building communication networks or distribution networks. Differences are necessarily resolved in the practicalities of ensuring the viability of whatever is constructed -- the process may even be facilitated by common membership in some group such as the freemasons for whom building and architecture are fundamental symbols. **Problem:** Difficulties arise from the easy association with the economic, financial and political interests which approve or underwrite such projects and are involved in their subsequent exploitation. Once their interest has been aroused, it becomes difficult to dissociate such vested interests from any larger purpose for which the cooperative project was conceived. Such interests are totally insensitive (except under legislative constraint) to such issues as the inappropriateness of the project, wastage of scarce resources, or any negative social impact -- which are denied or viewed as unfortunate necessities. Each project is viewed in isolation (often ignoring the resources needed for its upkeep), irrespective of its unfortunate impact on other projects -- thus corrupting the purpose of the original concept. This leads to a legacy of silting dams, uninhabitable buildings, inappropriate monoculture, inoperable factories, obsolete weaponry and abandoned community projects.

9. Games and Teamwork

Games necessarily involve significant cooperation between the players, whether the games take the form of board games, competitive or team sports, or war games. In team games, cooperation operates in one way amongst those of the same team and in another in relationship to the opposing team(s). Successful business and military strategy is developed through a strong awareness of the importance of teamwork in relation to opposing teams. Within a team, explicit recognition is given to the role of each and the manner in which they should be able to support and substitute for each other in the event of crisis. Special attention is given by each to "marking the opposite number" in the opposing team. Each must endeavour to know the games his opponents (and his team mates) endeavour to play. **Problem:** In their least challenging form it is questionable whether games are a useful model of all but the most sterile form of cooperation -- as when two people hit a ball over a net purely for entertainment. Teams are built in order to win a continuing series of games -- not just a single game. As a result both teams and their members shift their focus increasingly to the way in which their status is measured in series or league tables. Increasing those measurements becomes the objective of the game -- whether it be the statistics of ball players or teams, the number of police convictions, bodycounts from military operations, or financial indicators of corporations. Gamesmanship, and questionable devices for increasing convictions and bodycounts, become the rule. The decay of the Olympic spirit, under the influence of politicization, commercialism, medal counting, and the pressure to improve performance with drugs, bears witness to the vulnerability of this approach to cooperation.

10. Celebration

People cooperate through gathering together in some ceremonial, for a celebration, or for a "happening". This form of cooperation may be extended through media events such as Live Aid, Hands Across America, or a World Run. It may take the form of celebrating achievements such as the 40th Anniversary of the United Nations, or the annual celebration of "days", such as One Earth. It may also fulfil a psychologically important ritual or liturgical function within the life of a group -- rekindling enthusiasm and commitment, and reinforcing a sense of community. **Problem:** The great attention aroused by such events, particularly through the media, easily creates the impression that some lasting cooperation has been achieved -- bypassing the obstacles confronted by conventional initiatives. Such events legitimately build hopes and create visions of what might be, but they delude when presented as cooperation of other than the most ephemeral kind. In contrast to the sacrifices normally demanded by any significant cooperation, participants have little to lose by being seen to attend or contribute briefly to a happening. Such events salve consciences, draining resources away from longer-term projects. Symbols of achievement parade as realities, disguising healthy responses to non-achievement.

11. Rule of law

The elaboration of agreements and networks of regulations binding the relationship between social actors is cooperation in one of its most lasting forms. Much effort is devoted to formulating resolutions, declarations of shared principles, and multilateral treaties -- as a means of evolving the framework of law, whether national or international. The stream of regulations from the EEC is a prime example. **Problem:** Much of the effort devoted to articulating such instruments is in response to the need for visible symbols of achievement at the time they were voted or signed -- whether for public relations purposes or to justify participation in a meeting. Many such instruments remain dead letters -- and indeed many are only produced for valid short-term effect, as reminders of what ought to be done. Treaties either fail to enter into force for lack of ratifications, or only govern the behaviour of a minority of potential parties, or are systematically violated, in the letter or in the spirit. Little provision is made for enforcement of obligations. Little is learnt from the lack of commitment to last year's resolutions in the throes of articulating those for this year.

12. Conspiracy of elites

Real cooperation may be seen as associated with the unpublicized, long-term working relationships between elites of whatever kind. This may range from a group of community "elders", through "old boy networks" or "nomenklatura", through academic "invisible colleges", to semi-secret societies such as the freemasons and Opus Dei. It may be cultivated in closed meetings (Trilateral Commission, Bilderberg Group) and by secret diplomacy. It may be articulated in secret agreements, whether between governments, classified research establishments, intelligence agencies, corporations, crime syndicates or revolutionary groups. It may take a seemingly innocent form in conspiracies of the spiritually "initiated" or of like-minded social change agents (the "Aquarian Conspiracy"). **Problem:** The successes of this form enhance complacency amongst the elites -- the belief that their power and insight provide adequate social guidance -- as well as encouraging non-elites in this same belief, thus disempowering them. The difficulty with such forms is that there are no checks ensuring that the self-selected participants act in the interest of the wider community rather than their own -- as with cartels and organized crime. Consequently groups such as the freemasons and Opus Dei must check each others excesses in continuing battles hidden from the public eye. Invisible colleges must engage in primitive skirmishes to deprive each other of larger shares of scarce resources. Such groups are often poorly equipped to regulate the excesses of their members, as the publicized excesses of the insider traders, the freemasons, and irresponsible researchers make clear.

The challenge for the 1990s may involve not so much abandoning any one of these mind-sets but rather of learning how to avoid being trapped within any such metaphor as providing "the one solution". In each case there is a need to see through the veils of opportunistic reporting and media hype establishing claims of successful cooperation. The danger is one of being deluded by semblances of cooperation and symbols portrayed as achievements. Their current status constitutes a re-emergence of idolatry -- the perfection and worship of new forms of "golden calf". Such idols of cooperation should not disguise the questionable value of efficient rearrangement of the deck-chairs on the Titanic or of effective use of a tea cup in bailing out a life-boat being swamped in heavy seas.

Is the bitter lesson to be learnt from the last 30 years that: **Until we understand how we -- "the enlightened cooperators" -- are part of the problem, we cannot understand the nature of the solution required?**

Opportunity: pattern language experiments

1. The Alexander initiative
Christopher Alexander and his colleagues at the Centre for Environmental Structure (Berkeley) have published a 3-volume study to *"lay the basis for an entirely new approach to architecture, building and planning."* (*The Timeless Way of Building*, 1979; *A Pattern Language*, 1977 and *The Oregon Experiment*, 1975) The approach is based on the idea that people should design for themselves their own houses, streets and communities. This seemingly radical idea is based on their documented observation that most of the places in the world where it is attractive to be were designed not by professional architects but by ordinary people.

(a) Unnameable quality: They have a delightfully elegant way (consistent with the arguments of Section KD) of deliberately not naming the central quality which they believe should be engendered by any development process. *"There is a central quality which is the root criterion of life and spirit in a man, a town, a building, or a wilderness. This quality is objective and precise, but it cannot be named. The search we make for this quality, in our own lives, is the central search of any person, and the crux of any individual person's story. It is the search for those moments when we are most alive."* (*The Timeless Way of Building*, p.x)

(b) Living patterns: They show how that in order to embody this quality in buildings and in communities, it is necessary to recognize that every place acquires its character by certain "patterns" of events that keep on happening therse. These patterns are interlocked with certain geometric patterns in the space in question and it is out of these patterns that our environment is effectively constructed. The vital point however is that such patterns may be alive or dead. *"To the extent they are alive, they let our inner forces loose, and set us free; but when they are dead, they keep us locked in inner conflict. The more living patterns there are in a place... the more it comes to life as an entirety, the more it glows, the more it has that self-maintaining fire which is the quality without a name."* (p.x)

(c) Reliance on pattern language: At the core of their approach is the point that in designing their environments people always rely on certain "languages", which, like the languages we speak, allow them to articulate and communicate an infinite variety of designs within a formal system which gives them coherence. *"A pattern language gives each person who uses it the power to create an infinite variety of new and unique buildings, just as his ordinary language gives him the power to create an infinite variety of sentences."* (p.xi)

Alexander's team spent over ten years in formulating what he carefully points out is merely one possible pattern language. It is made up of 253 interrelated patterns. Each consists of a problem statement, a discussion of the design problem with an illustration and a solution. Many of the problems are, in his terms, archetypal in that they are so deeply rooted in the nature of the environment that it is to be expected that they will be as much a part of human nature and action now as in five hundred years time.

(d) Personal pattern language: They perceive the role of their proposed language as the first step in a society-wide process by which people will gradually become conscious of their own pattern languages adding to or adapting those that his team has formulated. They believe that in modern society the design languages people use have broken down. *"Since they are no longer shared, the processes which keep them deep have broken down; and it is therefore virtually impossible for anybody, in our time, to make a building live."* (p.xii) There is therefore a need to discover and improve collectively the patterns capable of generating and maintaining the living quality.

(e) Complete patterns of forces: In contrast to many fragmented approaches, Alexander shows that the structure of a pattern language is created by the fact that individual patterns are not isolated. *"Each pattern depends both on the smaller patterns it contains and on the larger patterns within which it is contained... Each pattern sits at the centre of a network of connections which connect it to certain other patterns that help to complete it... And it is the network of these connections between patterns which creates the language... In this network, the links between the patterns are almost as much a part of the language as the patterns themselves."* (p.312-4) He points out that the language is a good one, capable of making something whole, only when it is morphologically and functionally complete. It is morphologically complete when it produces clearly defined designs and it is functionally complete *"when the system of patterns it defines is fully capable of allowing all its inner forces to resolve themselves."* This requires that the individual patterns should themselves be complete.

2. Possibility of parallel languages
The Alexander initiative is explicitly concerned with the design of buildings and physical environments. And yet so many of his examples respond to social needs. The philosophy and thinking underlying the approach contain many features which are just as relevant to the design of the psycho-social environment as they are to the design of the physical environment. Indeed for the Alexander team any such distinction may be artificial and meaningless.

(a) Design of social structures: There are those however for whom building design is not an option in resolving the psycho-social problems they are facing, as is the case for the many bodies responding to the global problematique. For them the question is whether they could use the principles so admirably formulated in Alexander's *The Timeless Way of Building* (1979) to assist in building social structures as opposed to physical structures. Is it possible to envisage a pattern language which would enable groups, organizations and institutional complexes to be designed by those concerned, using patterns that are "alive" rather than "dead", as is so often the case? And would a group or organization come to life and acquire that *"self-maintaining fire which is the quality without a name"*, the more living patterns there are active in it?

(b) Design of conceptual structures: Given this possibility, and the frequently deplored fragmentation of the conceptual environment, is it possible that an equivalent pattern language might also he developed to enable people to improve the relationship between conceptual frameworks and bodies of knowledge? Is it possible that a network-based pattern language could render more fruitful and humane the structuring of systems of ideas, especially in the light of the way people are increasingly able to interact with patterns of ideas through interactive information systems? It is extremely probable, both in the socio-organizational and in the conceptual case, that many patterns already exist and simply need to be reconsidered as part of a pattern language which each is able to explore and develop in his own way.

(c) Design of psychic structures: There is also a final possibility that the more basic patterns are of such a fundamental or archetypal nature in relation to man that an equivalent pattern language could be usefully recognized as a way for individuals to order the relationship between their own modes of awareness (Section HM).

(d) Isomorphism between pattern languages: If indeed the configurational attributes making for a fundamental pattern derive as much from inherent characteristics in man's pattern recognizing ability, then it is to be expected that there would be a fairly high degree of isomorphism between the patterns recognized in any domain, whether physical, socio-organizational, conceptual or intra-personal. Evidence pointing towards this conclusion includes that from general systems and especially J G Miller (*Living Systems*, 1978), that on parallel developments in isolated cultures, and Jung's extensively documented emergence of archetypes in a person with no possible experience of the culture from which the form derived (*The Collected Works*, 1953-71).

3. Experimental elaboration of parallel languages
Alexander's initiative is unique in the manner in which it concretizes subtle relationships amongst a comprehensive range of phenomena oriented around the needs of man. It does not involve the imposition of a closed conceptual framework and yet a measure of coherence

is provided. The referents for his pattern language have the merit of being comprehensible to all in their own physical environments.

(a) Physical patterns as templates: As a first step in exploring the elaboration of parallel languages, it is therefore appropriate to attempt to use Alexander's pattern language as a "template" or substrate to provide guidance in the identification and interrelationship of patterns in such parallel languages.

The procedure adopted was to endeavour to formulate an abstract pattern from Alexander's example in the physical environment (see Section MP). This was used with the physical pattern to guide the elaboration of a socio-organizational equivalent. All three were then used to guide the formulation of an equivalent for the conceptual domain. Finally all of them were used to clarify a possible pattern within the intra-personal domain. The full procedure was applied to some 60 of Alexander's 253 patterns. An abstract pattern name was however specified for all of them. This is presented with an indication of Alexander's physical referent. The structure of the network of cross-references presented corresponds to the network identified by Alexander between smaller patterns and the larger patterns which they constitute. The names given are however the abstract pattern names derived as indicated above.

(b) Physical patterns as metaphors: In terms of the earlier discussion (relating to Section MP) on metaphors, each set of patterns, whether at the socio-organizational, conceptual or intra-personal level, bears a metaphorical relationship to Alexander's physical set of patterns. But although the procedure involved a search for a parallel to the physical set in other domains, it is the physical set which can be more usefully considered as the metaphor through which the patterns and relationships at the other levels can be clarified and comprehended. In a number of cases, however, the obvious attractiveness of the parallel in one of the other domains may offer insight into the merits of the physical pattern identified by Alexander. Others may raise questions about the completeness or appropriateness of the pattern at any level, especially in those cases where the Alexander team have flagged the pattern as being less than satisfactory as formulated.

(c) Questionable generality: Another question is whether the patterns identified by Alexander are of general relevance rather than being largely determined by his own context. The work does contain many examples from outside the USA. As to whether any such pattern language is of a more arbitrary nature than a spoken language, he states: *"The language, and the processes which stem from it, merely release the fundamental order which is native to us. They do not teach us, they only remind us of what we know already, and of what we shall discover time and time again, when we give up our ideas and opinions, and do exactly what emerges from ourselves."* (*The Timeless Way of Building,* 1979, p.xv)

(c) Questionable significance: The basic question this exercise raises however is whether the jargonistic language conveys any new meaning or whether the whole exercise is totally artificial. It is clear that the vocabulary used to identify the parallel patterns is barbaric compared to the poetry of Alexander's elegant presentation. In part this is due to the nature of the exercise and to editorial inadequacies. But it is in part due to the lack of an attractive vocabulary with which such a rich set of distinctions can be made. Are there is fact equivalent distinctions to be made?

(d) Quality enhancing distinctions: The whole exercise was undertaken on the assumption that new responses are desperately needed to the global problematique. These new responses can benefit from recognizing the kinds of quality-enhancing distinctions which can be usefully and fairly unambiguously made with regard to the organization of the physical environment. It is possible that it may be many years, if not centuries, before a patterning of such detailed distinctions is called for in the domains of the parallel languages. But even if the exercise is only a stimulus to further exploration of the relevance of Alexander's approach in other domains, much will have been accomplished.

Opportunity: metaphoric revolution for the individual

1. Individual opportunity
The complexities of society and the global problematique are such that the shift in focus advocated in the preceding notes may well only occur in isolated groups, corporations and countries, if at all. Although it can be demonstrated that such a shift is a natural evolution beyond the current situation, and that it is pre-figured in many ways by current uses of metaphor in government, the pressures in favour of short-term political crisis management will in all probability prevail.

The opportunities for the individual and for affinity groups are entirely different. Individuals may relatively easily choose to make much more extensive use of metaphor to provide themselves with quite different ways of restructuring their perceptual and cognitive environments. This may be done, as it is to some degree at present, quite superficially and primarily for rhetorical or illustrative purposes. There is however little to prevent individuals and groups from selecting or designing metaphors to be used over an extended period of time to structure their perceptions and their communications. Such use is evident in the implicit use of military and sporting jargon amongst management groups already. The same be said of the language of students and youth gangs.

Such use of metaphor may become "revolutionary" in the following two ways, as consciously cultivated cognitive dissonance, and through a rhythmic change of cognitive framework.

2. Consciously cultivated cognitive dissonance
Individuals alienated by mind-sets and policies prevailing in society may choose metaphors which enable them to totally reinterpret social dynamics, attributing value according to a very different pattern. They may associate with others sharing that metaphor. Strongly bonded associations may well depend a great deal on the implicit or explicit metaphors which their members share.

The key question is whether this is in any way different from the current freedom of individuals to hold (or convert to) certain beliefs or work with certain paradigms. In many ways it is not, except perhaps in the greater recognition that individuals are free to do so, and can personally respond to the creative opportunity. The shift becomes more radical and revolutionary to the extent that individuals choose metaphors which provide them with insights into dynamic relationships about which they can communicate amongst those who share the metaphor but who are then totally unable to communicate meaningfully with those who do not. This too is already a characteristic of those using specialized jargons, whether academics or gang members. The question is how would society be if the number of active specialized jargons increased by several orders of magnitude -- if individuals effectively felt empowered to develop their own specialized languages and cognitive systems.

It is one thing for such specialized jargons to emerge from scholarly or technological preoccupations legitimized by establishment institutions. It is quite another when people are actively developing uses of metaphors which effectively ignore or devalue such structures and the cognitive systems on which they are based.

None of this is especially improbable, as can be seen in the development and seductiveness of the cognitive systems associated with cults. In this sense the metaphoric revolution opens the gates to a new cognitive frontier, a set of parallel conceptual universes, possibly richer and more challenging, in which people can develop new relationships to their available resources.

2. Rhythmic change of cognitive framework
(a) Cultivating a cognitive dance: If people are enriched by having a range of metaphors amongst which they can select and move in creative response to pressures from the social environment (and especially information overload), how should they govern their choice of metaphor? Rather than clinging to any one metaphor (with the false sense of security that gives), or shifting reactively from one to another in spastic response to external pressures, the real challenge is to enable people to cultivate a rhythm of changes amongst a set of metaphors -- to evolve a cognitive dance with their environment.

(b) Beyond spastic shifts of perspective: Again such a transition is not improbable in that it is prefigured in many ways by the manner in which people switch cognitive frameworks in switching from home to work to café to leisure activity, or in their dealings with people in different roles (e.g. as spouse, as helpmate, as lover, as companion, etc). But people are offered little insight as to how such switches are to be governed and consequently tend to live them spastically unless they can evolve some sense of pattern and rhythm for themselves.

(c) Defining a new psychic centre of gravity: In this sense the metaphoric revolution is one of revolving through a cycle of cognitive frameworks such that the revolution itself defines a new psychic centre of gravity for the individual immersed in a socially turbulent environment.

A very sophisticated version of this is to be found in the Chinese classic the *I Ching* (or Book of Changes), which involves transitions between 64 conditions, each described in metaphoric terms. This has been interpreted into Western management jargon in (Section TP. The fact that it is traditionally recommended for the over-60s is an indication that simpler cycles could usefully be developed and explored.

(d) "The answer" barrier: The basic point of this section is the individual and collective need to respond creatively to the apparently fragmented reality of society, whether within or between cultures. In the light of recent historical trends it is very difficult to sustain the prevailing assumption that people and groups can (or should) all be persuaded -- within the foreseeable future -- to subscribe to any one particular paradigm, belief system or form of sustainable development (or the institutions and policies they engender). Rather than placing all hope in the possibility of finding this one magical "mega-answer", through which all ills are to be finally dispelled, a radical alternative can be usefully explored.

Conventional approaches to social transformation tend to be based on changes to material conditions (as well as to social and attitudinal structures) recommended as necessary and desirable by some group in power in the light of advice by some elite group of experts. Such "mono-perspective" approaches tend to respect the views and needs of the majority in any territory, possibly with compromises to take account of minorities. It is extremely difficult for such changes to be implemented so as fully to meet the perceived needs of all on a socially and culturally diverse planet. This is a major reason for the fragmentation of conceptual and belief systems and their associated institutions.

(e) Epistemological diaspora: A contrasting approach would be one in which such epistemological divergence was encouraged -- moving with the process of fragmentation rather than attempting vainly to oppose it. This is in accord with a fundamental principle of Eastern martial arts. The integration and consensus so desperately sought is then achieved in a more subtle and elegant manner.

The "epistemological diaspora" advocated here is already a reality of increasing significance -- although it may be said to have commenced with the diversification of man's first reflections on the universe. The use of metaphor as advocated here could however result in a metaphoric revolution which would dramatically encourage such epistemological divergence in the interests of those who engage in it.

Such a revolution would encourage and enable people and groups to select, adapt or design their own conceptual frameworks and manner of perceiving their environment as well as their own way of comprehending and communicating about their action on it. Whilst they might at any one time use frameworks favoured or advocated by others, they would in no way feel obliged to continue to use them.

3. Implications

(a) De-linking from authoritative explanations: The emphasis following such a revolution would shift from the present situation of dependence on specialists, experts and political leaders putting forward "ultimate" explanations, models and developmental policy recommendations. The implication that such explanations should be accepted in preference to all previous ones would then become questionable. Earlier explanations, no longer need necessarily be rejected as reflecting various levels of misunderstanding or downright stupidity -- irrespective of any fundamental disagreement amongst the elites responsible for them.

Such a shift in emphasis honours the complexity and variety of peoples needs and the increasing difficulty for the average person to even remotely comprehend the justification of such explanations. These they are therefore expected to take on trust -- but which they often simply ignore.

In a condition of continuous metaphoric revolution an explanation loses its character of permanence as the authoritative pattern of reference. Rather people select between alternative explanations according to their circumstances and immediate needs -- shifting to other explanations as the circumstances change. This does not preclude the possibility of staying permanently with one explanation -- but continuously shifting between explanations becomes a meaningful alternative.

Metaphors are required through which to balance the harmonies and discords engendered by mind-sets of particular authorities, whether taken together or in succession. Perhaps the skill called for is the creation of "melodies" or "dances" through which to give place, duration and significance to each such form, whether in turn or through new combinations and permutations. In practical terms, this suggests the need to transcend the extremes of "hierarchy vs. network" or "majority vs. consensus voting" in seeking more viable forms of cooperation. The route forward could be through ordering patterns of differences in which different perspectives and strategies are interrelated....perhaps as phases in policy cycles....or like the tracery of struts in a dome....a cathedral of interdependent insights?

(b) Cognitive empowerment: Under such circumstances the value of an explanation to the user comes as much from the consciousness of having chosen it -- however temporarily -- as from its intrinsic merits. This is equivalent to the value attached by a climber to the particular branches of a tree or ledges on a mountain -- they are of value as part of the climbing process in providing temporary security and a foundation for further progress. But equally, staying on any one ledge may offer a satisfactory view of the world which reduces any need to continue climbing.

It might be considered strange that in a rapidly changing world, considerable effort should be made to incarcerate comprehension of society in particular explanations. In a context of planned obsolescence, changing priorities and shifting fashions, such explanations do not last long. It would seem to be more appropriate to open up the possibility of shifting explanations, thus freeing people to explore the many dimensions of comprehension and the opportunities to which they give rise.

4. Response to reservations

(a) Loss of permanence: The major objection to the acceptance of such "epistemological chaos" is the seeming loss of permanence and order which have been the object of so much effort in the past -- and what of the various "bodies of knowledge" so painfully built up? How could society function under such circumstances? Can development be sustained in such a turbulent epistemological context? The argument of this section is that to a large extent this is already the case, but by attempting to avoid such seeming chaos, policies and institutions are designed which are inadequate to the real challenge of sustainable development.

(b) Cultivating the gift of pattern: Citing Gregory Bateson's concern with the "pattern that connects" and Christopher Alexander's proposal for a living pattern language, Kathleen Forsythe (1986) argues: *"It is time to acknowledge that we are all born with a gift of pattern. It is as evident in us as the pattern of the whole corn plant is potential in the kernel. We have a living language, a metaphoric language of analogy and relation, that lies at the heart of our conceptual system and probably does not take place in the brain! Each time we move, we seek to complete the pattern, we speak, engage in conversation and each time the conversation is new. We understand, when we perceive the distinction that makes the pattern new.*

Our wholeness derives from having at the core of our being, one pattern, one rule that derives from our kernel and orders our movements in each of our consensual domains whether these be the domain of physiological relations or the consensual domains of information relations, structural relations or conversational relations. When we are true to our one rule then our decisions, our actions and our behaviour have integrity that is the essence of harmony. To do this we must, however, move beyond information, to be able to self-reference ourself in all the consensual domains at once. Such consciousness transcends goals and allows us a direct experience of being and becoming simultaneously.

We may formulate our one rule differently -- it is our purpose our meaning. But the nature of our rule will determine, as we move through the cosmic web of new beginnings, whether our interactions empower and free us or ensnare and entrap us. When we chain the real in worlds that follow one on one, we enslave the kinetic pattern from the dance towards wholeness, clouding our eyes from the vision as we move."

Forsythe also cites Christopher Alexander (1979): *"Imagine that one day millions of people are using pattern languages and making them again. Won't it impress itself then, as extraordinary, that these poems which they exchange, this giant tapestry of images, which they create, is coming alive before their eyes. Will it be possible then, for people to say stonily, that poems are not real, and that patterns are nothing but images; when in fact, the world of images controls the world of matter...a pattern language...shows each person his connection to the world in terms so powerful that he can re-affirm it daily by using it to create new life in all the places round about him. And in this sense, finally, as we shall see, the living language is a gate."*

(c) Acknowledging virtual centres of reference: Development of the ability to dance through cognitive frameworks could encourage recognition of the underlying perspective from which the decision to shift is made, the emptiness at the centre of change, or the silent place to which mystics respond.

Comments: future possibilities and dangers

Despite the number of studies of metaphor and the extent of its use, especially in some non-western languages, it can be considered a largely unexplored resource with its attendant dangers.

1. Complement to models

Part of the alienating nature of modern society derives from the extent to which everything perceptible is governed by packaged "explanations" provided by authority figures, experts and the media. Such "models" are essentially non-participative and elitist, whereas metaphors facilitate individual interpretation, each according to his ability. An excess of explanations is experienced as disempowering. Metaphor offers a possibility of empowering the individual in that he has greater control over what metaphor is to be used when. He is not obliged to communicate with acquaintances using an explanatory language effectively imposed by distant third parties.

There is in fact some merit in perceiving conventional "explanations", whether scientific, political, religious, legal or administrative, as needing to be complemented by appropriate metaphors to enhance the quality of communication. Although some already argue that theories may themselves be usefully considered as metaphors.

2. Metaphoric aids to development

In development terms, the question might well be asked as to how many metaphors people need for psychological survival? Is there a problem of metaphor deprivation associated with alienation? Is it possible that a metaphoric measure is necessary as a complement to counterbalance the questionable educational role played by the IQ measure of intelligence?

To the extent that we are ourselves metaphors, do we need to develop richer metaphors through which to experience and express our self-image? If individual learning is governed by metaphors (as some studies indicate), how is it that the metaphors governing societal learning and development have not been studied?

In the light of Andreas Fuglesang's (1982) severe criticism of western assumptions concerning communication in developing countries, would it not be more useful to conceive of different cultures as operating with different root metaphors? Is it possible that social transformation is essentially a question of offering people richer and more meaningful root metaphors through which to live, act and empower themselves?

Are people enriched by having at their disposal a pattern of metaphors within which they can select and move in response to pressures from the social environment, like information overload?

Metaphors may even be seen as opening the gates to a new conceptual frontier, a set of parallel universes, possibly richer or more challenging, in which people can have a different relationship to the available resources (as is implied by Fuglesang's comments on African cultures). And to the extent that the drug problem is the consequence of a search for new ways of perceiving the world, development of metaphoric skills may offer a more meaningful alternative than unrealistic medical attempts to simply "get people off drugs" and legalistic attempts to "stamp out drug-taking".

3. Reframing problems

If indeed use of metaphors allows people to perceive problems in new ways in order to develop a new relationship to them, this leads to the question as to when a particular social phenomena should be interpreted by any particular metaphor. Authors of science fiction have explored highly complex situations in which those involved had to deliberately choose a metaphor convenient to them personally as the only means of ordering a chaotic range of information input through which the situation could be controlled. Are more appropriate metaphors required with which groups can navigate through the global problematique and contain its expansion?

4. Skills in the use of metaphors

What are the beneficial ways to shift metaphors? Are there useful pathways to be discovered between complementary metaphors in a set? For what period of time is it useful to perceive a particular phenomenon through a particular metaphor?

When should particular metaphors, or any metaphor, not be used? How can people or groups using different metaphors interrelate their insights in the control of some collective enterprise such as a research project or a community development programme. Would it be helpful to perceive the international community of organizations (and especially the United Nations system) through new metaphors?

In this light perhaps the most basic question is to what extent is some particular phenomena not a metaphor of some other phenomena? Does there always exist some level of comprehension at which a given substrate is significant for understanding some aspect of social, conceptual or intra-personal dynamics? Of what characteristic of modern society is a particular historical event, natural phenomenon, or man-made artifact an appropriate metaphor?

5. Reservations and dangers

It is possible to elaborate a vision of desirable form of governance based on the meaningful movement of metaphor-models as discussed earlier. It was emphasized that the radicalness lay essentially in the approach to information rather than in institutional change. The apparent similarity to the prevailing system, with all its defects, was also stressed.

This suggests the possibility of the emergence, or the existence already in some form, of highly undesirable forms of governance based on the deliberate design and manipulation of metaphor-models and the control of their movement through the international community. At the time of writing, the Gulf War could usefully be reframed as a conflict at the metaphoric level -- and it as at that level that peace might usefully have been sought.

Because of the prevailing focus on conceptual models, whether for academic or ideological reasons, little attention is paid to the manner in which they lend themselves to manipulation as metaphor. If, as has been argued here, metaphors can exert a more powerful influence than paradigmus, then the international community is highly vulnerable to manipulation at the metaphoric level.

These dangers are vaguely perceived in disguised form in concerns with "cultural imperialism" and various forms of "disinformation". But current processes of governance are more or less impotent in responding to them because the metaphoric level cannot be taken seriously at this time. It could be argued that "Dallas" and "Coca Cola" have a more powerful impact on the collective imagination than all the programmes of the international agencies combined. This suggests the value of exploring the vulnerability of the world community to a modern equivalent of the metaphoric manipulation so successfully carried out in Nazi Germany. The dividing line between "manipulation" and "challenging approach to human development" is difficult to establish using conceptual tools alone. "Dallas" and "Coca Cola" are perceived by many as desirable symbols of the American Way of Life.

As the many analyses of the cult phenomenon in the West have shown, people are powerfully attracted by the metaphors embodied in such cults, more so than by sterile establishment concepts. Presumably the way is wide open for groups, with somewhat greater skills in manipulating metaphors, to mobilize society in totally unforeseeable ways and to ends quite in contrast with those currently conceived as desirable.

The problem is that any positive shift in the approach to governance will create awareness of the manipulative opportunities. So whether such a shift is initiated to counteract such manipulation or not, such manipulation will become increasingly evident. Many would argue that in current approaches to government, considerable attention is devoted to such covert manipulation, whatever the concepts or models overtly debated and implemented.

World Problems P

Reminder

Section P is located in Volume 1 of this Encyclopedia, because of its size and importance.

The index to the world problems in Section P is also located in Volume 1 (see Section PX)

Transformative approaches   T

Scope

The purpose of this section is to provide a context for the presentation of accessible techniques which offer possibilities of making an immediate difference to the manner in which resources are mobilized in response to the global problematique.

Sub-sections

The section contains 304 entries. It is divided into two parts: Section TC and Section TP. The first contains 240 entries with descriptions on new ways of conceiving meetings and meeting processes. The second contains 64 entries is an editorial experiment towards understanding the nature of transformative policy cycles, in the light of a western management interpretation of the Chinese *Book of Changes*.

Method

The procedures used in preparing this section are discussed in detail in Section TX.

Overview

Detailed discussion of is given in the Notes of Section TZ. Meetings, and especially international meetings, are a vital feature of social processes and the initiation of change. They are a principal means whereby different perspectives are "assembled". Through such occasions resources are brought to bear upon questions of common concern. They may also provide the environment in which supposedly unrelated topics can emerge and be juxtaposed. But despite the assistance of professionals and the increasing number of such events, there is rising concern that many do not fulfil the expectations of participants, nor of those whose future may depend upon the outcome. This is particularly true of events most concerned with social transformation. Current meeting procedures, despite efforts at innovation, on such questions tend to give rise to little more than short-term public relations impact and in this form can themselves constitute an important obstacle to social change. In a very real sense meetings model collective (in)ability to act and the ineffectiveness of collective action. The challenge in Section TC is therefore to provoke reflection on a new attitude or conceptual framework through which meeting dynamics may be perceived and organized in order that they may fulfil their potential role in response to the global problematique.

Given the marked tendency towards the polarization of issues, especially in meetings, and the unfruitful consequences, Section TP, composed of 64 entries is an *editorial experiment* based on the pattern of concepts implicit in the much-publicized Chinese classic, the *Book of Changes*. These are transposed into a language which highlights the significance of such a complex pattern of transformations in any organizational or meeting environment. Its special merit is the explicit recognition of the need to shift from condition to condition in order to ensure both healthy development and the ability to respond to a turbulent environment. One traditionally recognized set of 384 transformational pathways between the conditions is indicated.

Index

A keyword index to entries is provided in Section TX. The keywords are also incorporated into the index for Volume 2 (Section X)

Context

The contents of this section may be considered as complementing the other sections in ways such as the following:

Human development: By the importance of taking into consideration different modes of human development in designing techniques to be of relevance to people with different needs and modes of response.

Integrative knowledge: By the integrative characteristics required of innovative techniques.

World problems: By the importance of such techniques in offering some new leverage in responding to the global problematique whilst avoiding the creation of more complex problems than they resolve.

Metaphors and patterns: By the evolution of communication techniques (especially in meetings), and by the need to communicate innovative techniques and the essentially integrative function of meetings.

Human values: By the values expressed or enhanced by innovative techniques.

Transformative conferencing TC

Rationale

Although meetings, and especially international meetings, are recognized as performing a vital role in the initiation of change, few attempts are made to rethink procedures which have changed remarkably little over the past century. It is widely assumed that no special skills or insights are required for fruitful meeting participation. Those with such expertise rarely use it to advance the interests of the meeting as a whole.

Considerable efforts have however been made to use new forms of communication hardware, especially in order to handle the logistics of meeting organization and administration. Many experiments have been made with small group processes and facilitators, but it has not proved possible to adapt such innovations to major international meetings where they are often perceived as gimmicky or culture-specific, and for good reason.

The fundamental problem seems to be disguised by the apparent success with which agenda items, documents and participants are processed through well-tried procedures. To the extent that meetings arouse the expectation of bringing a diversity of implicit and explicit viewpoints into focus in order to give birth to new possibilities for social charge, current procedures are however deeply disappointing in that they are primarily concerned with implementation of a pre-established schedule of speakers, document distribution and receptions. They are in no way designed to facilitate the conceptual dynamics through which significant new possibilities can emerge at the meeting itself.

Communication hardware is designed to help presenters to communicate to other participants. It is not designed to interrelate, even in computer conferencing environments. As a result, meetings satisfy the lower shared expectation rather than rendering explicit the subtle pattern of integrative insights to which the diversity of participants can give form. They reinforce fragmentation thinly disguised behind an array of resolutions for collective action to which participants are only minimally committed, except when others are called upon to act. Most striking is the inability to make use of available technologies to enable meeting programmes to be redesigned in response to new insights emerging during the meeting.

Paradoxically meetings are at present models of rigid design in which people gather to discuss social change in the external environment (over which they have minimal control). It would be difficult to design a system which could hinder more effectively the process of giving form to subtle patterns of integrative insights and empowering people to organize for action in terms of them.

The challenge would therefore appear to be to elaborate new conceptual frameworks within which a meeting may be perceived. This should highlight the implicit dimensions of the problem and point to more appropriate options. For unless a new attitude to the meeting process can be elaborated, it seems highly probable that concealed inherent weaknesses will continue to undermine and erode the value for social change of any meeting outcome.

Content

The section contains 240 entries each describing some aspect of meeting organization. The selection of entries and the descriptive content endeavour to highlight neglected perspectives. The entries point to possibilities, some of which have been explored in practice. Others call for much more extensive investigation but nevertheless provide a conceptual context for a more innovative approach to meetings. Such could be built into a pattern language on which new approaches to meeting organization could be based.

Method

The information used was obtained from a range of studies on meeting organization, some of which derive from a long-term programme of the Union of International Associations, convenors of the International Congress on Congress Organization series.

Index

A keyword index to entries is provided in Section TX. The keywords are also incorporated into the index for Volume 2 (Section X)

Comment

Detailed comments are given in Section TZ.

Reservations

This section is not concerned with those types of meeting which are already fulfilling the expectation of both organizers and participants. Many meetings use well-tried procedures to achieve well-defined results and there is no call to enhance the ability of participants to interact or to interrelate emerging concepts. This section is primarily addressed to those larger international conferences where collective objective has to emerge within complex constraints in order, namely where the conference design is itself symptomatic of the external social challenge being faced by the conference to mobilize people and groups to act in new ways.

ENTRY CONTENT AND ORGANIZATION

Ordering of entries
Entries are in **numeric order**. Entry numbers have been **allocated randomly**; they have no significance other than as a permanent point of reference to facilitate indexing, cross- referencing, and updating between editions.

Index access to entries
The location of an entry in this sub-section may be determined from:
- the **Volume Index** (Section X) on the basis of keywords in the name of the entry or its alternate names
- the **Section Index** (Section TX) on the basis of keywords in the name of the entry or its alternate names

Structure of entries
Entries may be composed of the following descriptive elements:

(a) **Entry number** This number has **no significance**, except as a convenient method of identifying the entry (particularly for indexing purposes), of filing information on it, and as an identifier to which cross-references from other entries (possibly in other sections) may refer in this and future editions. The first letter of the entry number refers to the section of this volume in which the sub-section, denoted by the second letter, is located.

(b) **Entry name** This is printed in bold characters. It may be followed by alternative names.

(c) **Description** Brief description of the conferencing feature with, where appropriate, an indication of possible innovative developments and its transformative potential.

(d) **Advantages** Brief indication of the implied advantages.

(e) **Disadvantages** Brief indication of the implied disadvantages.

♦ **TC1001 Image development**

Description Image creation attempts to create for an organization as a whole, as well as for the body gathered in a conference, shared jargon, experiences, mental pictures and assumptions. This "common memory" serves as a basis for continuing dialogue and communication.

Image creation aims to develop for any group a "common history". This simply means that the group has, in addition to their varied experiences of the development of the effort in which they are engaged, some threads of common understanding of the origin and intent of this effort.

Image creation steers between several tensions. First, images are both based in real facts but they are also a motivating interpretation of the facts. Avoid trying to create motivation on the basis of something other than the objective data about the situation. Even the most dreary state of things can be articulated in a way that allows the group to grasp it and move on into the future.

Secondly, images must be both applicable to the whole group, but also specific. Beware of both the universal abstractions which mean equally little to anyone, and the glorification of one local example at the cost of everything else.

Thirdly, images need to be complex enough to hold the reality which people experience: simple black–and–white divisions have little impact in the present period. At the same time, they must be simple enough to understand easily.

Image creation is presently being explored widely in advertising, but is often strangely unconscious in the process of conference design. Conferencing is however an image–transmission process for which values of simplicity, brevity and clarity are most important to honour.

♦ **TC1002 Conceptual framework development**

Conceptual screen development

Description Conceptual screens are developed out of a dialogue—whether internal to an individual or amongst members of a screen creating team. The first aspect of this dialogue is that it occurs between a Platonic or deductive approach to ideation and an Aristotelian or inductive approach. Neither are seen as preferable, but both are needed to develop an effective conceptual framework.

The Platonic approach asks what is true about this situation generally; what does it have to do with life universally. It is out to create a model which can be applied everywhere. It is often helpful here to begin with a basis of outside research. The Aristotelian approach begins with the particular encounters that the group has with an issue. Research is done into the specific dynamics of known cases. Case–study type research is helpful here. It then seeks to look at the idea from the inside, and see what things are consistently a part of it and what things are apparently incidental. Out of this review of the data, an integration emerges.

Screen creation is doing both of these things with the same material, up to such a point that the screen seems to emerge. It then gets checked back against its inductive and deductive origins, and gets looked at for its own sake from as many perspectives as possible. The next step in screen development is to create occasions, such as pre–conference local gatherings, or publications, which force the originators to articulate the screen to a larger audience and to alter it or not in response to feedback.

For the most part the development of screens for a conference is done before the event but for those events that create screens during the programme the decision making process is the key point of feedback. The tension between the Platonic and Aristotelian approaches is most evident in this feedback. Usually the participants best play the role of the grounded, Aristotelian pole of the tension, and the facilitators and procedures team play the Platonic pole of the tension.

The participants hold the local experiential pole of the tension. They create out of their own experience the practical aspect of the model. They can, through brainstorming and organizing data and experience, give a sense of reality to the screen. As the screen develops in the process of the conference they test the reality of the screen.

The facilitators and the procedures group test the development of the screen over against the ideal. They might use a series of sets of extremely abstract images like knowing, doing and being or foundational, social and universal for a theoretical model, or the past, the future, the inclusive and the profound for a social model as partial tests for the screen that is emerging. The choice of these abstract images depends upon the abstract rationale that is being used to develop the screen. These abstract images must be in actual dialogue with the experience of the participants and not dictate the meaning of the experience if a creative screen is to emerge.

One of the most interesting points of feedback is between the conference spirit and the screens. When these are in creative dialogue then quite unexpected break throughs can result. There have been break throughs in individual and group creativity when the limits of social care and interior awareness are being explored at the same time.

One of the merits of the development of screens is that they are usable by many people. This does require simplicity, profundity, and realism. These screens, while not necessarily acceptable in some circles, are very much what is required by development efforts across the world.

Another merit in the development of screens is developing complex and lasting screens during the process of the conference itself. Most screens developed during a conference are usable only for the event and are very simple. This process would require new technologies of making decisions and of developing a corporate understanding that are beyond current capacities.

♦ **TC1005 Integrative failure**

Description Although integrative skills may be successfully applied to a situation, their elusive nature can be partially defined by the ways in which such skills may fail or be used to conceal abuse:

1. Reduction in variety: A simple way to ease the integrative problem is to reduce the diversity of elements present in the situation using an argument for standardization and against any "hodge podge" mixture of elements. This of course eliminates some minority interests. In the extreme case of destructive or "meltdown" synthesis, all variety is eliminated.

2. Reduction in quantity: By eliminating a significant number of the elements, the problem may also be eased. The argument that can be used is that they are well–represented by the variety of elements that remain and that any "proliferation" of elements is disorderly. In practice this results in the absorption of some elements by others, such as in the case of minority groups.

3. Simplification: Subtleties and nuances, possibly defended by specific minority groups, may be ignored. Interconnecting webs of relations can be ignored.

4. Tokenism: Emphasis may be placed on the image or desirability of synthesis in order to conceal inability to achieve any steps towards it.

5. Temporary synthesis: In a dynamic situation it may be possible to achieve some measure of integration in the short–term by ignoring factors temporarily absent or only emerging over longer time cycles.

6. Coloured synthesis: A significant degree of synthesis may be achieved, but from a particular viewpoint or in terms of a particular mode, approach or strategy. The narrowness of such a synthesis, coloured by the perspective of those who achieve it, may be difficult to communicate within the framework established by that synthesis.

7. Enforced synthesis: In some instances, as with a dynamic set of minority interests, a form of integration may be imposed by constraining the dynamics (although without reducing the number or variety of the elements).

8. Dogmatic synthesis: An impression of synthesis may be achieved by stating frequently and forcefully that it has been achieved and thus eroding expectation that a greater degree of synthesis is possible.

9. Laissez faire synthesis: By reinterpreting the nature of synthesis or integration, it may be deemed to exist under any circumstances as the pattern of interaction amongst the elements. No intervention is required, although if undertaken it would merely add to the pattern of interaction.

10. Agglomerative synthesis: Appropriate integration may be assumed to have been achieved simply by ensuring the juxtaposition of the various elements or viewpoints. This corresponds to the use of the prefix "multi" (eg in multidisciplinary). In books reflecting such a multidisciplinary synthesis, it is the binding which provides the synthesis, given the absence of any relationship between the constituent disciplinary chapters.

11. Comparative or cross–referential synthesis: Integration may be asumed to have been achieved by recording comparisons between the perspectives or elements. This often corresponds to the use of the prefix "cross" (eg in cross–cultural).

12. Cross–impact synthesis: Integration may be assumed to have been achieved by taking into account the constraints and feedback loops emerging from other disciplinary perspectives. This may correspond to use of the prefix "inter–" (eg in interdisciplinary). Note however that it is only with the emergence of a new level of order that a synthesis breakthrough may be said to have occurred (this may correspond to the use of the prefix "trans–" as in trans–disciplinary).

Related Interdisciplinarity (#TC1208)
Meeting focal processes (#TC1351)
Losing focus in meetings (#TC1432)
Meeting energy dissipation (#TC1699)
Varieties of meeting focus (#TC1285)
Alienation of committed activists (#TC1962).

♦ **TC1011 Consensus building**

Agreement — Public opinion — Synthesis — Compromise — Synergy

Description The general agreement of the people concerned. The oft–cited aim of conferencing and the oft–cited justification for decisions and action; the synthesis of ideas rather than simply the recording of opinion; creativity rather than votes. New technology is introducing new and more streamlined methods to assist the consensus building process and there are some fairly sophisticated approaches that are using the latest insights of group processes and structure.

Advantages When a genuinely creative consensus is arrived at, it can generate a great sense of achievement and motivation in all those involved to implement the decisions. Emphasis on the contribution and the importance of each person's point of view and perspective ensures that all the participants "own" the consensus and feel a certain responsibility for it.

Disadvantages It requires a high level of involvement, time and commitment. There is a tendency towards endless circular discussion if the focus of the group is lost. Consensus is often claimed when there is no other justification for decisions. There is a tendency towards the lowest common denominator rather than the highest common factor; and opinion polls are cited as consensus when the questions and alternatives clearly affect the results and even in the most "objective" polls the answers reflect existing prejudice rather than a consensus that has been built in the creative tension of the workshop. The mood of the group can be interpreted as the consensus by those with emotive skill.

♦ **TC1056 Event planning methods**

Description The most effective design of an event that will allow people to develop new understandings of a topic is one that is consistent with the experience of the topic. Planning may be experienced in communities in the following sequence. First, any community of people have an operating image of what the future needs to be. This operating image is commonly held but largely unconscious. Second, this operating image is blocked from being realized by real social forms and internalized interpretations of social dynamics. These blocks can be discerned and analyzed. Third, the blocks are capable of providing insights to broad social changes required to reach the operating vision. These broad social changes are articulated as proposals for transforming the blocks into possibilities. Finally, the proposals may be broken down into specific small easily accomplished, independent activities or tactics. Tactics may be directly aimed at doing the proposal or be tangential but supportive of the proposal. The planning process follows the thinking of any planning process of a group of people. This consistency between experience and the flow of the planning event enhances the capacity of people to participate creatively.

♦ **TC1061 Conference tone**

Description The conference tone is the combination of the corporate body's emotional mood and the group's will to be creative. It is the spirit and mood of the group. Each of these often neglected aspects needs to be seriously thought through.

♦ **TC1062 Integrative skills**

Description Although during meetings there is much discussion of "integration" and there are many attempts at producing a "synthesis", the skills called upon seem to be poorly understood, hard to communicate, and very difficult to put into practice. It is therefore useful to note very different domains where integrative skills are practised successfully, even if it is not immediately clear what can be learnt from them for use in a meeting environment.

1. Design and composition: This is the process through which creative intuition influences the selection of elements and the manner and proportion in which they are to be balanced – what is to be put together and how. In each of the following the configuration of elements tends to relate to an emergent focal point: Composing music; Painting a picture; Flower arrangement (Ikebana); Landscaping; Building and community design; Interior decoration; Designing a meal (or menu); Putting together a group a team, or an evening party; Writing a novel; Casting for a film or play.

2. Managing dynamic situations: This is the process whereby the relationships between a complex set of given elements is kept in focus. Examples are: Juggling; Leadership of a group (including use of charisma); Production of a show; Conducting a military campaign; Controlling a chemical plant; Scheduling railways, deliveries, etc; Making a party "go" (hosting); Conducting an orchestra; Gardening.

3. Analyzing complex situations: This is clearly oriented to understanding whatever can be analyzed irrespective of whether this leads to broader synthesis. Examples are: Operations research; Systems research; Cybernetics; Management research; Political analysis.

4. Communicating synthesis: This is the process whereby a sense of wholeness or unity among diverse parts is imparted to others, even if only as a symbol or token of what may later be achieved in practice: Environmental appreciation ("One Earth"); Art education; Art of speaking; Political commentator.

5. Embodying synthesis: Whereas each of the above is in some way a manipulation of synthesis, however necessary, there seem to be instances where a person acts as the focal point for synthesis and is so perceived by those whose interests are reinterpreted and focused in this way. Examples are perhaps: Spiritual leaders (including saints, gurus, and charismatic evangelists); Political heroes (including statesmen, military and revolutionary leaders); Cultural heroes (including pop–stars, film–stars).

Related Interdisciplinarity (#TC1208)
Varieties of meeting focus (#TC1285).

♦ **TC1082 Conference**

Description Meetings that bring together people with the intent of (a) pooling their experience, expertise, skills, information and problems; (b) juxtaposing their different disciplines, and polarities;

(c) sharing their differing perceptions, concepts, opinions and futures; and (d) channelling their motivations and enthusiasms. Purposes vary but often include: (a) information exchange, (b) planning and decision-making, (c) inspiration, (d) problem-identification, (e) problem-solving, (f) fact-finding, etc. Aims, though these are sometimes unconscious, are to do with (a) development, change or transformation of situations, (b) attitudes, thinking, (c) strategy, direction and/or relationships, (d) beliefs, etc.

Advantages Conferencing is an opportunity for many things including: (a) recontexting, reframing, rehearsing one's perceptions and interpretations; (b) participating, involvement in creativity, decision-making and change processes; (c) networking, connecting, interchanging, coordinating, exploring different issues and initiatives, (d) personal growth, challenge and stimulation, renewal, remotivation, myth-working consciousness-raising, inter-personal dialogue and trust-working; (e) deciding about one's own life, changing direction, new commitments, becoming something new.

Disadvantages Although some participants do take advantage of the opportunities of conferencing, very few conferences are designed to provide these opportunities but in their structure and process reflect the inability of the group to risk genuine openness and the uncertainty of transformation. Many conferences are an exercise in organization-centred consumer-oriented, low risk events, structured so as to restrict any possbility for them to "get out of control" and perhaps become dangerously transformative. The predictable and "safe" events are very expensive in terms of the finance, organization and time invested and in terms of the drain on motivation, spirit and enthusiasm of those who participate.

♦ TC1123 Non-linear agendas

Beyond linear thinking

Description There is increasing expression of regret at the prevalence of "linear thinking". By this is meant any ordering of concepts which is sequential between (or within) subdivisions but contains no loops linking non-proximate elements in the sequence. Such linearity constitutes a method of ordering experience which is recognized as crude in relationship to the complexity of the environment.

Linear thinking is reinforced by many of the conventional responses to constraints on presentation of information: (a) The necessarily linear sequence of: words in sentences, paragraphs, sections, and chapters in documents. (This is only slightly modified by the device of parallel columns of text); (b) The linear schemes for numbering subdivisions of any structured document or thesaurus; (c) The sequential ordering of words of a speaker at a meeting.

The agenda of a meeting conforms to this pattern of linearity in the sequence of agenda items. Even the use of parallel sessions or sequences of sessions maintains the linearity. There are no particularly satisfactory procedures to ensure cross-fertilization between sessions and convergence on new levels of significance or synthesis.

An interesting alternative to the conventional representation of an agenda by items in a linear sequence having a beginning an end, is to treat the sequence as circular, so that the end joins the beginning. The agenda items are then associated with points on the circumference, through which the meeting may progress sequentially. This raises questions such as: (a) Should the subdivision of the circumference into agenda items constitute a complete set as implied by this approach - thus "exhausting" the topic ? And does it, if only by an "other matters" item ?; (b) Should the last element in any such sequence link back to the first - "closing the loop" ? Or is the relationship between the beginning and the end unclear and, if so, why ?

For more complex agendas, with distinct themes considered to be complementary or in some way related, one circular sequence may be subdivided for each such principal theme. But rather than separating the circles, they may then be represented as overlapping, such that the related agenda items in different thematic circles are at the points of overlap. Since such circles necessarily overlap at two points, one can indicate the priority of theme A over B, and the other the priority of theme B over A - necessary conditions for functional interweaving.

In order to move beyond this simple representation of non-linear interconnectedness, the communication links between non-adjacent items, necessary to preserve the topology of the representation, may then be inserted. This permits the agenda to be represented as a 3-dimensional configuration of functionally related items in which the necessary relationships to maintain the integrity of the configuration are explicitly indicated.

This procedure has the advantage of challenging any simplistic comprehension of the verbal description normally used to identify individual agenda items. Then the meaning to be associated with such descriptors emerges to a greater extent from the position of the items within the configuration. The latter also raises useful questions about the relative importance of agenda items possibly leading to the combination or subdivision of some of them.

Clarifying the non-linear relationships between the agenda items can guide conceptualization and action concerning the relationship between meeting sub-division (into groups, commissions, etc) and any attempt at synthesis in plenary. Configurations of the kind described may also be considered as representing functional subdivision through the subdivision of a spherical surface area rather than a line. From this point of view, the implications of subdivision by triangulation (the basis of topographical survey techniques), rather than by linear subdivision, should be considered. The former respects relationships, the latter ignores them.

♦ TC1129 Marginal participant type

Description Participants left out of the main group and who are outside the system and on the fringe.

Advantages Can represent creative alternatives and unusual perspectives. They can constitute the only place in some conferences where real transformative potential is present.

Disadvantages Can be irrelevant and completely unrelated to the conference process and serve only as a distraction and a drain on the energy in the process.

♦ TC1133 Converger participant type

Description Participants whose dominant abilities in the conference situation are abstract conceptualization and active experimentation. In terms of those abilities, they are at the opposite pole to the divergers.

Advantages The practical application of ideas executed in a relatively objective un-emotional way.

Disadvantages Tend to prefer to deal with things rather than people. Narrow interests; tend to specialize.

♦ TC1135 Conferencing as meditation

Description Just as individuals perceive meditation as a process to introduce increasing harmony into their jumble of thoughts, emotions and activities, conferencing can be considered as a process of collective meditation. During the initial stages of meditation, central themes emerge and are blended into an even larger theme. This is "attunement" or "alignment". If an aim of conferencing can be described as introducing harmony into the jumble of issues, attitudes, and behaviour of the different groups and coalitions, the process can be recognized as being surprisingly similar to those of personal attunement. There are five main concerns:

(a) The physical well being, its health, balance and posture. (There are many parallels between the health of an individual and the health of a conference).

(b) Psychological well-being of the conference. Its "breathing", flexibility and suppleness; its internal relationships and awareness of itself and the polarities within it.

(c) Control and expression of emotions. The flowing and the channelling of emotional energy.

(d) Organization of concepts. The contradictions and inconsistencies and the process for integration and synthesis. The relationship of the conference to its larger environment.

(e) Once these concerns are being attended to, the next concern has to do with the focus on the conference and its nature as a whole.

The whole jumble of events of a conference can be viewed as the surface manifestation of a collective entity struggling to be. The efforts of the parts and the factions to control the whole, according to their limited perspectives, is the cause of confusion and turmoil, and yet is necessary as the parts develop their own self-awareness, prior to relating harmoniously to the other parts, in order to be able to express the whole. The real challenge is to maintain a sufficiently strong vision of the nature of the whole in order to ensure that this preliminary chaos of attunement does not completely absorb the attention of the participants, and prevent the whole from coming to maturity in the time available.

In its maturity, as a conscious meditation, the congress constitutes a chalice into which energies can be focused and through which they can flow. This is not just a beautiful image of the difficulty in reaching and attaining this level of consciously integrated focus. (Such synthesis is to analysis, just as fission is to fusion - and people do not yet have access to fusion energy, despite much research on the required configuration to bring it about).

♦ TC1142 Team promoter role

Description An affinity-role played by one or more members of a group or team. They are usually sensitive to others, sociable but unobtrusive. They support and build on the suggestions of others while covering for the weaknessess in others.

Advantages They can improve the level of communication within a group. They have an ability to respond to people and to situations that weaves the group together into a whole, and they foster team spirit.

Disadvantages Indecisiveness at moments of crisis.

♦ TC1152 Meetings as games and contests

Description A gathering may be described as a pattern of interlocking games, whether recreational, therapeutic or "serious" in intent. An underlying objective may be the emergence of qualitatively superior games (eg in the style of Hesse's Glass Bead Game). The following may be used as metaphors to explore alternative ways of understanding meetings:

1. Mediaeval tournament: Participants may be viewed as knights gathered for a tournament. Each bearing a heraldic coat of arms representing his qualities and territorial origins to be defended at all costs. Contests are ritualized under an elaborate code of honour.

2. Miss Universe contest: Issues are paraded before eminent panelists who discuss their qualities before ranking them and selecting the "issue of the year". The whole process being immersed in a sea of public relations and other interests.

3. Martial art: The struggle between issues or their representatives may be viewed in the light of the "holds" and "throws" of Eastern martial arts (aikido, judo, etc). In these the supreme achievement is to use the enemy's energy to defeat him, and ultimately to see the enemy as but a reflection of oneself.

4. Market-place: The production, exchange and consumption of perspectives may be seen in terms of the dynamics of the market and the economic laws governing supply, demand and marketing considerations.

Broader Metaphors of meetings (#TC1603).

♦ TC1155 Socio-structural meeting configurations

Description The following are useful metaphors through which the structure of meetings may be understood in new ways:

1. Orchestra: With stationary groups of musicians/instruments, usually forming an incomplete circle around the conductor.

2. Auditorium: With seats ordered by row, aisle and tier, usually in a semi-circle facing a stage area, but occasionally surrounding the stage (eg colosseum, sports arena, circus) or inter-penetrating the stage (eg some avant-garde theatres).

3. Parliament: With seats arranged and allocated in terms of the parties and a perception of their relationship to each other (eg facing each other) and to the government (eg facing the podium).

4. Temple or cathedral: With participants arranged in relation to a symbolic focal point before which one or more intermediaries may officiate; minor chapels may be located within the temple or disposed around it in an appropriate configuration. Special significance may be attached to location and orientation.

5. Fortress or castle: With elements appropriately arranged to ensure successive lines of defence in order to maximize the protection of what is most valued. Importance is attached to the strategic location and the relationship to the surrounding terrain.

6. City: When planned as a whole from the start may be specially divided into zones appropriately (often symmetrically) arranged in relation to each other according to their function and the lines of communication required. usually located in relation to natural resources or a transport nexus.

7. Battle plan: Whereby opposing generals locate the different functional units of their respective forces both in relation to one another and to the opposing force, in order to favour respective strategies. Special attention is given to terrain, logistics and the training and morale of participants.

8. Table design and seating: Whereby an attempt is made to reflect the status of the participating parties (eg in negotiations or mediation) or at functions requiring careful attention to protocol.

9. Ritual, dramatic or dance movements: In which participants continually modify their relationship to each other, possibly to bring about a sequence of changes in the overall pattern they constitute. Some forms are completely pre-determined, others are partially or completely improvised. In some forms all participate all the time, in others they may be absent for the part of the sequence in which their role is most stressed, or when it is stressed by their absence.

Broader Metaphors of meetings (#TC1603).

♦ TC1164 Accommodator participant type

Description Participants whose dominant abilities in the conference situation are "concrete experience" and "active experimentation". In terms of abilities they represent the opposite pole to the assimilators.

Advantages Doing things, carrying out plans and getting involved in experiments. Involving themselves in new experiences. They are risk-takers. They adapt themselves to their immediate circumstances and favour intuition, trial and error problem-solving.

Disadvantages Rely heavily on people for information and feedback and less on their own understanding of logic.

♦ TC1188 Dynamic design methods

Description The dynamic of designing methods for a conference is the dialogue between the organization of the participants into teams and time blocks. It involves determination what products will result, the steps to achieve the products with the group in the time design and

according to the intent of the conference.
The conference designers work over the six months before the may event on several aspects. Any work done on the overall design, flow, intent, existential aim or rational objective could be done jointly, as well as regular reports on progress, breakthroughs and blocks to progress. This provides common images of the programme and coordinates efforts between the groups. Points at which a major decision is required by a subgroup would result in joint meetings. When the input or reflections of a larger group is required to move things along, this also could be done in joint meetings. Joint meetings might happen as often as daily for one hour, or weekly half or all day meetings, depending on the proximity of the event.
In addition, the process of creating an intellectual framework for the design group is important. This is in part creating a balance between clarifying ideas and maintaining openness to change. Quite often a partially completed model will be set aside for a time while work on an entirely different arena is continued. This allows a basic model to emerge with a minimum of intellectual and psychological commitment to it at the early, changeable stages. At the same time it enables further steps to be developed in other arenas. Finally, it provides time for the group's unconscious mind to work on solutions for the design.
The ongoing definition of the intent is in fact the process of refining and redefining what is expected by the event. Initially the intent is just an image of what could be done. By writing out a prose description of that image, the refinement begins. Dialogue with and presentations to others further refines it. Writing out the rational objective and the existential aims gives additional clarity. This initial clarification may then be set aside before it becomes too fixed in people's minds.
The next step is to look at where the participants will be when they arrive. A conversation or workshop about who is coming and what context they will be operating out of when they arrive begins this process. As this analysis begins to emerge it too is set aside before it becomes too fixed in the designers' minds. This sets the beginning and ending limits to the conference. How to get from the beginning to the end is the next question. This involves deciding what are the two, three or four major time blocks over the course of the conference. If the event is over a month, the weeks may be used. For shorter time periods, days may be used. During the course of these time blocks what needs to be accomplished ? What concrete products will be the result of each of these blocks ? What materials will be required to develop each of these products ? In the course of the discussions about products, materials and time blocks, insights may emerge about the intent and participants. It is helpful to keep note of these insights.
When the larger blocks have been worked on for an image, as well as for materials and products for each of them, then contextual statements can be be written about each time block. This context would include an operating image, the major steps of the process, the materials, rational objectives and existential aim for each block.
This overall description of the flow of the event would be set aside for the time being.
If there are conceptual frameworks to be developed this could be the time to intensify work on them. This development of conceptual screens would further clarify the intent. Other work on things like conference spirit, decor, etc. will benefit work on the intent. After a few weeks of work on these arenas the group can back to the intent and the detailed process of the conference itself.
Each of the major blocks may then be broken down into the next smaller blocks of time, for example, days. The products for each day would be described. Contextual statements would be written, and materials and so forth named. In the process of refinement a new existential aim and rational objective of the conference would be written. At the point where blocks of three or four hours each are defined and worked through, the overall intent of the conference should be clear if the conference is two or more weeks long. At this point a decision by the conference designers is required whether: (a) to do the more refined procedures, that is, the minute by minute directions for each workshop leader before the conference begins or (b) to write these procedures in the midst of the event itself. Partly this depends on how different the group's approach to the topics will be in the midst of the event. If the writing is done during the event, there are two approaches. One is for a procedure writing group that continues to write on a day-by-day basis the procedures on behalf of the whole. The second approach is for the facilitators for each group to do the writing on a day-to-day basis.
The advantage of the first is that the whole conference is kept somewhat in sync and somewhat in the same set of images as far as procedure goes. It allows the whole conference to be coordinated from a central perspective. This type of procedure development is necessary when a topic is dealt with that has not been approached in this way before. The procedures group is developing the methods, context, etc. on a day-by-day basis. Major shifts in the whole conference are easier. The disadvantages in this approach are that the procedures writing group has little or no direct contact with participants. The mechanics of writing, reproducing and distributing/training people during the procedures requires that they stay a day or two ahead of the group in terms of what is required next. When a set of procedures leads to a blind alley the whole conference may be stopped (methodologically, of course other things can be done) while the procedures group thinks through a new direction.
The advantages of the second approach is that there is a great deal of sensitivity toward the needs of the individual groups, and topics can develop at the speed at which the group can work. This approach also allows a greater sense of responsibility for the final products to be placed in the hands of the smaller groups. This approach is most effective when there are several topics requiring several different methods of treatment, when there is little or no need for feedback between groups and when one group's products are not dependent upon those of another group. The disadvantage of this approach is that the facilitation team of any given group could spend a great deal of non-session time designing procedures. The products and direction of any given group can end up as quite divergent from the rest of the conference.
If the procedure writing is done before the event, then a quite extensive set of procedures may be written. They could be a script type word by word description of what the facilitator says from the beginning of the day to the end. This is not to be slavishly followed but to give a detailed idea of how the workshop or session might be run. They could be simply a set of steps with an accompanying description of the operating context for the sessions, suggestions of what is needed for each part and a list of hints on how to do it.
The process of writing procedures begins with the intent of this session, the group's context from the previous sessions and the time frame in which the session is to operate. Basically a session would include creating a question in the participants minds, developing or focusing a context, and the model building, decision making or study session itself. Generally a conversation may be used to raise a question, a talk may be used to create a context and a seminar may be used for a study and a workshop may be used to crated a model or make a decision. It is possible in a highly structured and centrally controlled event that each 3 or 4 hour session would have all three. Usually during a more loosely structured conference that is developing new methods only the contextual and workshop part of the above would be used. It is then important for the conference spirit to be very well thought through so that the participants are in an open frame of mind.

♦ TC1191 Conference viscidity

Description The degree to which the participants in a conference function as a unit. It is reflected by the lack of dissension and personal conflict among participants, by the absence of vested interest, by the ability of the conference to resist disrupting forces, and by the belief of the participants that the conference is unified and does act as a unit.

♦ TC1204 Participant personality needs

Description Wherever individuals groups or institutions work to remedy social problems, there is an inability of all concerned to admit openly the psychosocial needs of the individuals and groups involved. It is only in informal discussion, and in the absence of the concerned individual, that there is frank discussion of how to confer a sense of prestige by suitable juggling of organizational procedures and positions, appropriate use of flattery, etc. The facilitation of individual "ego trips", for example, is often an absolutely essential condition for their further support of a meeting or project. Even when two organizations or initiatives should be merged in the light of all available information, this will be opposed (behind-the- scenes), by the personalities involved, unless their status needs can be fulfilled. Such concerns, whether for a person individually, or for a group as represented by an individual, are basic to all social action. When they are not even recognized in behind-the-scenes planning, they are recognized tacitly in the dynamics of interaction with the person in question.
The inability to handle these matters in open debate severely inhibits the manner in which organizations or meetings can function. Even in crisis situations, discussion of action to be taken during a meeting will not occur until these other matters have been satisfactorily resolved through behind-the-scenes manoeuvering. Frequently it is questionable, even in a crisis situation, whether a given individual is not more interested in the recognition accorded to himself or his group than in any substantive matter which may be discussed.
Organizational action of any kind (and even in response to crises) may be perceived primarily as providing a legitimate opportunity for appropriate conference and organizational ritual to satisfy the psychosocial needs of the individuals and groups involved. The situation is particularly serious when the personality needs are neurotic or border on the psychopathic. There are many well-documented examples of this amongst national leadership, and in the leadership of groups represented in conferences or having responsibilities during them. Such matters cannot currently be discussed in open debate.
Clearly the priority accorded to these needs, and the inability to give explicit recognition to them in organizational documents or debate, despite their fundamental importance to organized action emerging during conferences, constitute a constraint upon the full realization of human potential. This is the case both because it distorts the manner by which a person develops through action within an organization or meeting, and because it distorts the manner by which an organization or meeting is able to act.
Related Participant pre-logical biases (#TC1965) Participant strategic preferences (#TC1828) Participant interaction preferences (#TC1845).

♦ TC1208 Interdisciplinarity

Description The need in meetings to interrelate the approaches of different disciplines, in order to understand a social problem situation and to be able to recommend appropriate remedial programmes, is now increasingly recognized. The "inter-disciplinary" approach in now in fashion and an essential element in many requests for programme funds. On closer examination, however it is possible to discover that this requirement, far from constituting any form of progress, is only the symptom of the pathological state of knowledge at this time. The specialization without limit of scientific disciplines has resulted in an increasing fragmentation of the epistemological horizon.
Specialists cannot be asked to testify in meetings with regard to the unification of the sciences, or an "integrated" action programme, insofar as these specialists by their vocation and training are ignorant of, or deny, this very unity. Even those who profess to stand for the unification of the sciences cannot always be trusted, for each one of them would be satisfied in defining his familiar point of view, and more or less justifying his own individual presuppositions.
Teaching and research institutions reinforce the above separation through administrative procedures which tend to eliminate communications with institutions associated with other disciplines. This is reflected in conference programme events sponsored in parallel by such bodies. The division of intellectual space into smaller and smaller compartments, and the multiplication of institutions which assume the management of each such territory, results in the formation of a feudal system which governs the majority of scientific teaching and research enterprises and is clearly reflected in the organization of meetings.
When an "interdisciplinary" approach is used in a meeting, it most often consists in bringing together specialists from different disciplines, in the simplistic belief that such an assembly would suffice to bring about a common ground and a common language between individuals who have nothing else in common. The reports or results of such meetings neither achieve, nor attempt to achieve, any synthesis - other than the purely spatial juxtaposition of viewpoints and constraints, and subsequently, a judiciously worded editorial overview for the published proceedings.
Few of the societal problems which give rise to large conferences at this time can adequately be handled within any one discipline. Such problems result from the interaction of social, economic, technological, political, religious, psychological, biological and other factors. Understanding requires an integration of the relevant disciplinary perspectives. Such integration however must be much more than the synthesis of results obtained by independent unidisciplinary studies conducted prior to the meeting. The synthesis, to be useful, must come before the unidisciplinary commitments have been made and the conclusions frozen, without having been tempered by exposure to other constraints. This should be the true function of an interdisciplinary "meeting" - to act as a "transformative crucible" from which a new perspective emerges and is tempered in a number of stages. If the result is merely an agglomeration, then no transformation has taken place and the process has failed.
Where such interdisciplinary synthesis does take place, however, it is most successful between two closely related disciplines. Such integration is decreasingly successful as the number of disciplines involved increases. This is matched by a rapid decrease in the sophistication of the synthesis and a reduction in expectation of its benefits by those involved. A "synthesis" of results in itself dangerous in a meeting if it is superficial, but nevertheless succeeds in removing the stimulus to greater collective effort. The difficulties are increased when the disciplines are of a different nature, have fundamentally different methodologies, or focus on very different subject matter. As the variety of disciplinary perspectives increases, so does the tendency of each subgroup to perceive the activity of others as being of marginal relevance or importance.
The challenge in meetings is to face up to the failures of the past (particularly those disguised as successes) and to find new ways of interrelating the intellectual resources available in order to guide significant change.
Related Integrative skills (#TC1062) Integrative failure (#TC1005) Metaphors of meetings (#TC1603) Losing focus in meetings (#TC1432) Varieties of meeting focus (#TC1285).

♦ TC1218 Video teleconferencing

Electronic meeting — Freeze frame — Slowscan
Description Meetings that involve participants in a number of geographically separated groups through the use of video systems. Freeze frame (or slowscan) techniques enable transmitted images to be projected onto a screen line by line, and these can be refreshed every 30 seconds.
Advantages Currently the most economical medium particularly where there is a large amount of document manipulation to be done. Relatively low- cost equipment and a narrower band width is used.

Disadvantages Can detract from the intensity and rhythm of the meeting if it is not well managed.

♦ **TC1225 Diverger participant type**
Description Participants whose dominant abilities in a conference situation are "concrete experience" and "reflective observation". In terms of these abilities, they are at the opposite pole from the "convergers".
Advantages Imaginative ability and the ability to view real situations from many perspectives and to organise many relationships into a meaningful gestalt. They perform well in situations requiring idea generation, brainstorming, etc. They are interested in people; imaginative, emotional, and have broad cultural interests.
Disadvantages Their broad range of interests generally precludes specific specialist knowledge or expertise.

♦ **TC1229 Alienative participant involvement**
Description Conference participation by those having an intensely negative orientation, a hostile polarity against the organizers. This may be partly due to an actual or perceived use of coercive power by the organizing group or sponsors.
Advantages Though dangerous, the force of the suppressed energy in an alienative group can be highly creative -and if nothing else, its presence demands that some response needs to be made.
Disadvantages Explosive, volatile, potentially destructive force, extremely difficult to control, channel or even to communicate with, except in the most crude and coercive way.

♦ **TC1244 Conference authority**
Description Can be derived from three main sources: 1. Institutional or formal authority based essentially on the power of patronage or sponsorship -"those who pay the piper call the tune".
2. Personal, "authentic" authority based on knowledge, and experience. The authority of the expert consultant, and equally the "local person".
3. Group or corporate authority based on the actual consensus of the participants in the whole conference.
4. Heterogenic, "absolute" authority, based on reference to super-mundane or divine principles or personages.
5. Hierarchical, aristocratic authority based on longevity and earned or granted status within the group.

♦ **TC1264 Team grounder role**
Description An affinity-role played by one or more members of a small group or team. They are usually loyal, steady, and considered to be well balanced. Turns concepts and plans into practical working procedures and carries out plans systematically and efficiently.
Advantages They are a source of organizing ability, common sense and hardworking self-discipline. They recognize the relevance of proposals to the particular situation and can inject a pragmatic realism into the group process.
Disadvantages Can tend to be inflexible and unresponsive towards new and interesting ideas, and generally rather predictable and resistant to change.

♦ **TC1269 Creative care groups**
Special taskforce
Description Small groups (3-6 people) with a goal, clearly assigned by the conference who have the brief to work through until they are ready for more feedback or have achieved the goal. They define their own project procedures and are generally most effective when there is an unchallenged leadership, with 3-5 in close dialogue. They can be long-term groups or very temporary assignments -four or five groups could be assigned to work on specific points during a plenary session.
Advantages Can produce high quality work and at the same time allow the conferencing to continue with other issues.
Disadvantages They have a considerable responsibility and may misunderstand or misuse their assignment to express their own bias. This can be corrected by feedback in the plenary but it can be distracting.

♦ **TC1275 Associative networks**
Relevance maps — Relevance networks — Thesaurus network
Description Documentation systems or hierarchical data bases that associate information and concepts across the conventional boundaries of categorization (based on the kind of word-association that occurs in thesaurus entries).
Advantages Word-associations are fairly direct mappings of the organizational structure of peoples minds and are direct indicators of the degrees of relevance between the concepts. They provide a whole new range of analyses of value to the user. Not only do they indicate which elements need to be thought about in contiguity with each other, but they also indicate the degree of cohesion existing between them. Documentation and data bases based on these associative or relevance networks can highlight and facilitate possibilities for integrative approaches.
Disadvantages Although the production of relevance maps would be a major aid to international document users, the challenging requirements for comprehension and innovative learning already make such an advance inadequate. The problem is that such maps are too complex and disorganized to facilitate contextual memory and comprehension. They are difficult to memorize as a whole.

♦ **TC1277 Meetings as aesthetic symbols**
Description The following may be used as metaphors to explore alternative ways of understanding meetings:
1. Abstract forms: Some may wish to see the gathering as energies patterned onto more abstract forms: spiral, hierarchy, network, tensegrity, mandala, matrix, torus, polyhedron, knot.
2. Symbol systems: Some may be attracted by seeing the interweaving energies at the gathering in terms of a particular symbol system such as astrology, the I Ching, any pantheon, etc. These could even be used to identify imbalance in the energies represented, blockages in the evolution of the event, or threshold tests and challenges.
3. Imagery and dance: Such a gathering can also lend itself to comprehension as a pattern of aesthetic images, or as a dance of energies.

♦ **TC1285 Varieties of meeting focus**
Description In a discussion an individual may be rebuked for not "keeping to the point". In a meeting this may refer to relevance to a point on the agenda. It is the agenda which is used to focus the meeting process, although when there is a programme, focus may only be achieved through the agendas of individual sessions or possibly through a concluding plenary session. What is focus in a meeting and what is its significance, especially in relation to the aim or objective of the meeting ?
1. Imposed focus: A meeting may be convened to focus on a particular concern decided in advance. In such a case those present, and the points raised, will be clearly related to that concern, although perhaps not in the view of all present. Focus is thus a question of establishing and maintaining the relationship of a variety of subsidiary concerns to one central concern, even though the proponents of particular subsidiary concerns may not recognize each others relevance to that central concern.
2. Emergent focus: A meeting may be convened in the hope that a point of common focus will emerge as a basis for interlinking a variety of partially (or un-) related concerns. The problem is then to facilitate its identification and emergence.
3. Multiple focus: Whether imposed or emergent, it may be a question of a multiple focus, rather than a single one. There may be no intention, desire or ability to relate the multiple points of fFcessocus to one another or to a single underlying concern. This may be reflected in a variety of unrelated points in an agenda or meeting programme.
4. Degree of focus: Whether a matter of ability or intention, the meeting may resist any classification or sharpening of focus in preference to a diffuse focus or none at all. An unfocused meeting may be viewed as more creative or effective under certain conditions, or perhaps all that is feasible. Note that focus may be achieved without any verbal acknowledgement of its nature.
5. Aims, objectives and goals: Although it is possible to make useful distinctions between these, it is their difference from focus which should be noted. Each of them is in one way or another an intention or desire as opposed to the definite achievement characteristic of focus. But focus is also a precondition for them, in that it interrelates the relevant elements necessary for their achievement, whether any subsequent action is taken or not. In this way a meeting can focus on its objective, for example, or may fail to do so because its ability to focus is inadequate.
6. Focus and transformation: To achieve whatever transformation it intends, a meeting must bring the resources it has assembled to bear, bringing them appropriately into focus. This establishes the critical quantity or variety of factors necessary to the transformation. Focus ensures that the configuration of factors assembled will direct the energy of the meeting participants appropriately, rather than allowing it to dissipate ineffectually. Individual actions are then mutually reinforcing rather than nullifying. Depending on the nature of the meeting, focus may also be required to disseminate or contain the energy released by the transformative process.
7. Strategy and process: Focus may be brought about, from the prior unfocused condition, by an appropriate strategy for a process - a focusing procedure. Such strategy may even be considered the time dimension of focus.
8. Structure and focus: One method of ensuring focus is through the conventional hierarchical structure of executive and other programme committees and officers, culminating in the meeting president. The weakness of this approach results from the limitations of the simple hierarchy as a means of appropriately channelling and interrelating the information flows associated with interrelated topics. This is especially true when the hierarchy also has to perform protocol and other non-substantive functions which prevent either the executive director or the president from ensuring a substantive synthesis, even if they were able.
9. Focus and configuration: Where hierarchical ordering of the meeting programme or lines of responsibility no longer suffices to contain the complexity of the subject matter, a programme matrix may be used. When this is inadequate more complex configurations are required (eg critical path and network diagrams). There is however a major constraint in that focus is no longer possible if the complexity exceeds the ability of participants to comprehend. And in order to maintain comprehensibility the configuration of issues must contain elements of symmetry and pattern to reinforce memorability and communicability. Whilst it is not necessary for all participants to comprehend the whole configuration, there must be sufficient overlap both to maintain connectedness and to prevent loss of confidence in the chain of overlaps linking the most distant parts of the configuration.
10. Focus and the individual: The adequacy of the configuration depends on the quality of the participants and the extent to which its features engage their attention and energy. The greater the variety reflected in the configuration, the greater the potential, but also the greater the risk that participants will only be engaged partially or superficially and that the focus will be trivial. Powerful focus is achieved when the meeting configuration matches to a significant degree the psychic configuration of the participants. Participants respond to finding their own condition reflected in the meeting configuration, and the meeting reflected within themselves - it is this resonance which energizes the meeting. Any action through the meeting is then directly consistent with the individual's own development and calls upon all the participant is able to contribute because of the manner in which that contribution results in personal growth through the meeting. The meeting configuration thus reinforces connections which enable focus and transformation at a new level of significance, both collective and individual.
Related Integrative skills (#TC1062) Interdisciplinarity (#TC1208) Integrative failure (#TC1005) Meeting focal processes (#TC1351).

♦ **TC1304 Meeting space**
Description Meeting space must coincide with the size of the group and the intent of the conference. Ideally, each one has his or her place. But if it it necessary to estimate, it is often better to err on the side of slightly too small a space than very much too large a space. Few factors are more dampening to conference participation than a few people hunched together at the end of a huge, empty space, or speaking loudly to one another across vast distances: it already feels as if the event is a flop. On the other hand, moving in a few extra chairs at the last minute stresses people's interest and involvement in the subject matter.
Small groups may have separate small meeting rooms if there are long sessions of small group work. For short group conversations, however, it is stimulating to form small groups informally right in the large meeting space. Groups can see and feel the progress of other groups around them, and much time is saved in transit.

♦ **TC1306 Huddle groups**
Phillips 66 — Discussion 66
Description A device for breaking a large assembly into small units to facilitate discussion. Sometimes conceived as being groups of six people discussing a subject for six minutes. (Hence the "66"). The process can be used to obtain information from the groups, regarding their interests, needs, problems, desires and suggestions, to be used in the planning of programmes, activities, evaluation procedures, and policies. It is important to clearly explain the procedures for forming the groups, the procedures to be followed by the groups, the aims and objectives of the exercise, and to carefully prepare the questions. To facilitate dividing into the groups, prior seating arrangements that allow a rapid reorganization with the minimum of movement ensures that the process is more informal and can be used with more spontaneity. "Counting off" into sixes has the advantage of breaking up cliques, but it takes longer and can be confusing in large groups. It is helpful to have an identified chairperson and note-keeper in each of the groups.
Advantages An informal atmosphere that can provide a welcome break from meeting fatigue and allow participation by all those present. If the mechanics of re-organization are handled well, it can be an extremely rapid way to tap the total resources of the assembly, to obtain a consensus, or to determine if a consensus exists. Huddle groups make it difficult for discussions and recommendations to be controlled by authoritarian leaders or vociferous minorities. They help to free individuals of their inhibitions against participation by allowing them to share their ideas

... with.
... to be overused. The mechanical problems of re–organization ... Huddle groups do not often produce above the level of knowledge and ...nce possessed by the individuals. If the results of the group work are not used or followed up on, those who worked hard to produce them can get feelings of futility and cynicism about the whole conference process. Unless the huddle groups are clearly seen to be an integral part of the whole process they tend to be undervalued with the resulting loss of commitment and a drop in the value of the contribution.

♦ **TC1311 Buzz group**
Dyads — Pairs
Description A two–person group. Can be used in a general session by having each participant turn to a neighbour to discuss a presentation. Usually of short duration. It can also be used very effectively to "warm–up" a large group by having them meet several people through several dyads which can be formed on the basis of various arbitrary criteria.
Advantages Extremely informal and simple to set up. Buzz groups virtually guarantee total participation and have great potential for individual involvement. They provide a measure of support to individuals preparing to risk a contribution to the whole assembly. Many separate aspects of the issue at hand can be considered simultaneously and there is the widest possibility of expression of the different characteristics of the members with respect to background, knowledge, or point of view.
Disadvantages A large number of people talking causes an increasingly high noise level. The informality can get out of control. An extremely dominant member of the dyad can submerge any contribution that the other might make, although time allotted to each member can help to avoid this. Turning to your neighbour may mean turning to your colleague with little prospect of a new discussion; or it may mean turning to someone with whom it is rather uncomfortable to communicate, for personal, linguistic, or ideological reasons. There may be wide variations in the time required to cover a subject by the different dyads which can lead to mild attacks of boredom or frustration.

♦ **TC1313 Managerial–political conferencing**
Description Meetings that are focused on the coordination and control of people and resources and on the adjudication between groups with competing interests. Often used by government agencies and arbitration groups but also by pressure groups, labour unions, and special interest organizations.

♦ **TC1315 Amorphous conferencing**
Description Meetings that are comparatively unstructured, undirected and involve free movement of participants. The most common forms are: exhibitions –allowing exposure to a wide variety of different experiences and information on exhibition stands according to their special interests, social occasions, receptions and parties involving self selected interaction between participants. Open meetings with a relatively free access to the public address system and minimal direction.
Advantages These include considerable opportunity for participants to make contact with one another on the basis of their special interests and to choose the manner in which those interests should be developed (whether by holding a small meeting immediately, or planning some collaborative enterprise for some later date).
Disadvantages These include a considerable restriction on general coordination and consensus formation verging in some cases on a general state of disorder.

♦ **TC1323 Images**
Description Images need first of all to be consistent with the intent of the conference. Groups involved in supporting self–development efforts in developing countries wisely use individual portraits of people in their projects, often working. The stress is on empathy with individuals, rather than on the extent of their problems. Community service agencies show pictures of worthy facilities in which they have invested the taxpayers' money or of workers interacting with people served. This need for consistency is equally true for phrases, language use, illustrations and kinds of humour. It is important to decide what visual and verbal impressions participants need to create the appropriate milieu for the dialogue which is intended.
Secondly, images need to be consistent with the conference spirit. Stress needs to be placed upon those images which are in keeping with the kind of significance the conference is intended to create. Sometimes a sales conference is out to stress faith in new products; at other times to develop the solidarity of sales teams or to create corporate concern over critical problems.
Visual images take time and people to think through. Diverse media communicate differently—works of art, symbolic diagrams and logos, photographs, displays of actual products and equipment, films or videos, slides, etc. Consider what media creates the relationship required to the subject matter. Then consider the subject matter. Having the conference facility in mind, mockups of different combinations of media and content can be tried until a decision is made.
Verbal images often emerge as the planning and registration process proceeds. Some of these come out of the dialogue over the direction of the conference. Correspondence with potential participants can be combed for helpful articulation that the whole conference can use later. The process of soliciting and circulating position papers, seeking and reviewing bibliography, interviewing people in related fields and showing films and documentaries develops such a storehouse of imagery.
Conference imagery creates an underlying base of commonality across a conference which intentionally seeks to enhance the decision–making process. As decision–making proceeds in certain directions, imagery can be adapted accordingly. Often the working charts, phrases, rhetoric or diagrams used to arrive at decisions come to symbolize for the group the process they went through and their own relationship to the decision or decisions they have made. This sort of opportunity should be watched for and maximized wherever possible.

♦ **TC1332 Transformative event**
Description The creation of a situation through which a shared concern or emotion is rehearsed and released. These events function as socio–dramas. There are dramas of unity within a group; rehearsals of support or solidarity with some cause or group. They may be liturgical or secular rites marking significant events, like memorials or rites of passage, like inaugurations. They may be simply opportunities to interact with others, like reunions. A peace rally or an arms negotiation conference might be such an event.
Most conferences may in reality be more events of transformation than either decision–making occasions or opportunities to share information. Drama is the important factor for a transforming event. Participation is important to the decision making type of conference. Understanding is crucial to the information sharing type. Naturally very few conferences are purely one or the other.

♦ **TC1334 Mood and motivation assumptions**
Description The mood and motivation of a conference arises from the creation of an environment which reflects the significance of the conference, encourages appropriate participation and reinforces the existential aim and rational objectives of the event. Every event, no matter how dry, has some mood: this is not merely a matter of peaks and valleys. Sometimes it might be called "quiet resolve", "angry contention" or "unanticipated delight". The object is not to mark high points and low points, but rather to be aware of the interior relationship that the group as a whole is taking to the ongoing work. There is in many offices a "Friday afternoon" or "Saturday morning" mood, when anticipation of free time colours relationships. It is recognized that that is not the time to open a new agenda. In fact it is not the mood for starting a new agenda. Conference mood, like weekend mood, can be influenced, but it is not created. It is in the group, it is what the people have to work with.
The mood of the conference and the motivation of the participants are influenced in three ways: the environment, the time design and the relations among individuals and groups. Deciding about conference mood and motivation has two phases. One occurs long before the event, as basic decisions are made about the shape and drama of the conference and how it will be dramatized. The other occurs in the midst of the conference itself, as practical decisions are made about specific sessions, and shifts made to adapt to the group which actually attends.
The mood of the group responds to many things, but the decision–making process is certainly a major one. A very negative, contentious mood in a group can often be a response to a decision–making process deemed offensively inappropriate. In such cases, mood cannot be dealt with directly; adaptation is needed in many arenas at the same time. On the other hand, authentic struggle in a group over issues which they must confront, or extreme jubilation over some victory can also result in moods difficult to manage, but which must be allowed to achieve the conference ends.
Mood manipulation of any kind is seen by some to be an invasion of privacy. People are however continually subjected to attempts at such manipulation in daily life. It is increasingly important both to be self–conscious of the conference mood and to address it in some way. At the same time, this activity is no secret. There are times when it is perfectly appropriate to describe to a group what is sensed the mood to be. On the whole, what is intended is enablement, not clandestine manipulation, in order to adequately honour the investment of participants in the event.

♦ **TC1339 Conferencing evaluation**
Description Dimensions for evaluating conferences include:
(a) Nature of relationship of the conference to individual interests: directly related, indirectly related, unrelated
(b) Nature of relationship to community and to the societal structure: integrated, integrative, alienating.
(c) Nature of the conference focus: instrumental (on behalf of others), expressive (on behalf of themselves).
(d) Nature of the power of the conference over the members: high coercion, low coercion.
(e) Nature of association with an institution: part of an institution, related to an institution, independent of any institution.
(f) Nature of motivation: high voluntary enthusiasm, low voluntary enthusiasm.

♦ **TC1343 Networking conferencing**
Description This is an emergent form of meeting organization characterized by the following:
(a) Flexibility: rapid conversion, in the light of emerging consensus during the course of the meeting, to and from the other forms of meeting organization.
(b) Emergent issues: identification of emergent issues and formation of subgroups to clarify them rapidly so as to maintain the momentum of the meeting.
(c) Alternative sessions: Organization of alternative sessions not originally envisaged in the programme or room allocation, where significant numbers of participants find that they have more in common on subjects not scheduled in the pre–established meeting programme.
Advantages These include a much greater response to the needs of participants present rather than the imposition upon them of a programme which may not reflect their pre–occupations or the areas in which they consider interaction to be both possible and useful.
Disadvantages These include a considerable strain on the ability of the conference organizers to maintain the coherence of the meeting without having it endangered by emerging issues and desires for programme restructuring.

♦ **TC1345 Moral participant involvement**
Commitment — Loyalty — Devotion
Description High intensity positive orientation towards the organizing group and sponsors. High level of commitment and loyalty and even devotion marks this kind of involvement.
Advantages A powerful force that tends to trust the organisers and the procedures.
Disadvantages A tendency towards blind unquestioning acceptance that does not hold a creative tension and can lead to sterility.

♦ **TC1347 Conference procedure feedback**
Description When there is a procedures writing group it is very important to have regular feedback between the procedures group and the facilitators on a daily basis. This can be done through daily check–signals meetings. These meetings usually include reports on the progress of each sub–group, reflection on the state of the participants and what the next steps in the procedures are.
This is the time for reflection on conceptual frameworks and their refinement. It is also important for the procedures group to encourage input from participants and facilitators about the methods of the conference. This is not so much the overall flow as the specific enabling steps and processes. It is beneficial to the whole for the meetings to include a section on special issues like translation difficulties or participation difficulties.
When necessary, a sub–group may be visited by a representative of the procedures group to assess progress. The facilitators of all sub–groups may be called in for a more detailed report. These meetings are for the purpose of gaining greater sensitivity to all of the participants. Occasionally it is necessary for the procedures group to walk through a set of procedures for the sub–group as a whole or to make suggestions about direction or content from the perspective of the whole conference.
The procedures group can also sit in on the on the reflections on the conference spirit with the facilitators. This usually is done on a daily basis. When necessary to clarify a question or work though a change of direction the conference spirit group and the procedures group might meet jointly with the symbolic leadership of the conference and the practices people. These meetings can be done on an ad hoc basis and may not be needed if things are going well.
All of this feedback about methods is for the purpose of altering the direction of the conference when and if required. With this approach a conference need not be locked into a fixed set of steps months ahead of time but may be radically altered during the event itself.

♦ **TC1351 Meeting focal processes**
Description Below are listed, in no particular order, different aspects of focus or processes which tend to occur when a meeting is in focus:
1. Category transformation: A condition of focus should permit a reordering of the categories governing the meeting (or the organization of its subject matter) into a less procrustean pattern corresponding more appropriately to the reality encoded.
2. Organizational transformation: In a condition of focus the organizational units or sub–

divisions whereby it has been brought about can be reformed into a pattern more appropriate to the functional categories.
3. Problem sensitivity (resolving power): A condition of focus permits problems (otherwise considered identical) to be appropriately distinguished.
4. Problem subtlety: Certain all–pervading subtle problems can only be detected in a condition of low "noise–level" characteristic of focus.
5. Stabilized overview: Focus is a necessary condition for a stable overview of the meeting's domain (possibly as a meta–dimension) otherwise viewed as a multi–faceted image.
6. Contribution of the seemingly irrelevant: Only in a condition of focus can the contribution of otherwise "irrelevant" resources to the balance of the whole be understood.
7. Hospitable to divergent perspectives: A condition of focus is hospitable to overwise "divergent" perspectives.
8. Sensitivity to new options: The reduction in "noise–level" associated with a condition of focus permits new options and directions to emerge.
9. Transformation of collective self–awareness: The condition of focus facilitates the emergence of a collective sense of identity at a new level of integration and immediacy.
10. Transformation of personal awareness: A condition of focus enhances the processes of personal transformation in each participant and in relation to the here–and–now.
11. Energy containment and release: A focused configuration is able to contain and anchor the synergy normally dissipated during a meeting (possibly as a temporary state of enthusiasm or euphoria).
12. Emergence of simplifying perspectives: A condition of focus enables simpler descriptions of complex conditions to emerge, possibly as appropriate metaphors.
Related Meeting magic (#TC1567) Integrative failure (#TC1005)
Varieties of meeting focus (#TC1285).

♦ TC1354 Facilitative techniques
Description The extensive range of traditional and modern methods used to facilitate the conferencing process, including audio–visual presentation and recording techniques, group organization and procedural techniques, computerized networking techniques, and personal and inter–personal development techniques.
Advantages The range of techniques makes conferencing potentially innovative and a transformative dynamic within the ongoing social process.

♦ TC1358 Expertist participant type
Blind belief type
Description Conference participants who judge the value of a meeting by the number and the qualifications of the experts present tend to believe and trust in expert opinion and mistrust contributions from other participants including their own.
Strengths A respectful relationship to speakers and presenters, if they are experts or specialists. Become aware of dimensions to an issue that non–specialists may not have been exposed to. A useful balance to grassrooter types.
Weaknesses Mistrust of "grassroot opinion" including their own leads to a passive consumer type participation. Tends to accept "expertise" without question. Can become cynical when disillusionment sets in.

♦ TC1359 Image development
Description Image creation attempts to create for an organization as a whole, as well as for the body gathered in a conference, shared jargon, experiences, mental pictures and assumptions. This "common memory" serves as a basis for continuing dialogue and communication.
Image creation aims to develop for any group a "common history". This simply means that the group has, in addition to their varied experiences of the development of the effort in which they are engaged, some threads of common understanding of the origin and intent of this effort.
Image creation steers between several tensions. First, images are both based in real facts but they are also a motivating interpretation of the facts. Avoid trying to create motivation on the basis of something other than the objective data about the situation. Even the most dreary state of things can be articulated in a way that allows the group to grasp it and move on into the future.
Secondly, images must be both applicable to the whole group, but also specific. Beware of both the universal abstractions which mean equally little to anyone, and the glorification of one local example at the cost of everything else.
Thirdly, images need to be complex enough to hold the reality which people experience: simple black–and–white divisions have little impact in the present period. At the same time, they must be simple enough to understand easily.
Image creation is presently being explored widely in advertising, but is often strangely unconscious in the process of conference design. Conferencing is however an image–transmission process for which values of simplicity, brevity and clarity are most important to honour.

♦ TC1361 Team driver role
Team shaper
Description An affinity–role of one or more members of a small group or team. They are often outgoing and dynamic people with a readiness to challenge inertia, ineffectiveness and any signs of complacency or self–delusion. They concentrate on shaping the direction and the process of the group work, calling for the articulation of objectives and priorities.
Advantages Their drive is a source of momentum and stamina for the team.
Disadvantages Their desire to introduce or impose some shape or pattern on the group process can be threatening. They are prone to impatience, irritation and provocative behaviour.

♦ TC1362 Conference potency
Description The degree to which the conference has a primary significance for the participants. It is reflected the kinds of needs the conference is satisfying or has the potentiality of satisfying. By the extent of re–adjustment which would be required of participants should the conference fail, and by the degree to which the conference has meaning to the participants with reference to their central values.

♦ TC1366 Group conversations
Description Establishing a context sets parameters for the group and holding a workshop welds individual opinions into a group statement of some kind, but conversations in a group allow people to share their reflections. Conversations are best held at a small group level, although they can be done, if carefully orchestrated, with as many as 100 people. Generally, however, 5–30 people make the best conversational unit. If possible, conversations should be held in no more than two languages at the same time.
The function of a conversation is to raise questions about a topic in the minds of the participants in such a way that the topic of the conference is opened up to them in some new way or from some perspective that they might not have considered previously. The facilitator thinks through precisely what questions to ask. Movement should be from the simple to the more complex, from the more obvious to the more interpretive. The facilitator considers ways of phrasing questions that bridge obvious conflicts and untouchable subjects for the group. The facilitator avoids rhetorical questions, any question which has a right or w... facilitator cannot answer.

♦ TC1367 Meetings as physical processes
Description The following may be used as metaphors to explore alternative ways of understanding meetings:
1. Thermodynamics: The social processes in the meeting may be viewed in terms of the relationships between "pressure", "volume", "temperature" and various measures of energy stored and released.
2. Magnetothermohydrodynamics: The challenge of assembling the different participant orientations into a coherent configuration, generating and focusing the associated energies, and reaching a new level of significance, may be seen in the light of a fusion approach to plasma in a magnetic bottle.
3. Meteorology: The condition of a meeting may be viewed in terms of meteorological phenomena: wind, fog, heat, cold, visibility, precipitation (rain, snow), clouds, warm/cold fronts, wind patterns, etc.
4. Geology/topography: Participants and their interests may be viewed as geographical features (continents, islands, mountains) isolated or linked by seas, rivers, rifts, etc.
Broader Metaphors of meetings (#TC1603).

♦ TC1370 Programme modification
Exceeded time allotments
Description Sometimes sessions necessarily exceed the allotted time. This is no problem in itself. The problem which must immediately be resolved is the flow of the rest of the conference. One overtime session means something, somewhere, in the whole of the rest of the event. Procedures should be altered accordingly. Sometimes the entire conference direction shifts. This is a time to reconsider the flow of the rest of the event in the light of the intent of the event. What are other ways, in the light of the new focus of this event, that the intent can be accomplished ? What is the form of the new kind of product called for and how can the group produce it in the time now available ? This is where the creativity of facilitators and participants is required.
The only real error that can be made here is to begin to lose time and think that somehow it will take care of itself. The time will in fact take care of itself, and it will almost surely change the intent of the conference in the process. Although it is always a favourite complaint for a participant to say that there wasn't enough time to deal with the issue of the conference, this complaint is justified when corrections are not made immediately for time lost. The easiest way to make these corrections is to have in mind before the conference starts the points at which time is scarce, and alternative ways to proceed after that session or day.

♦ TC1373 Distortion
Description One of the three universal elements of "model" or image creation, distortion refers to the way that the model of a system differs from the system it represents. It has survival value because distortion is often necessary in the process of interpreting and communicating the complexity of "reality". However, distortion can also produce an "impoverished" model which has a reduced structure of choice. The most common examples of distortion used in conference presentations and other forms of communication are:
1. The representation of a process by an event (nominalization);
2. Use of unidentified presuppositions;
3. Distorted analysis of cause and effect.

♦ TC1374 Expressive conferencing
Description Meetings that tend to be ends in themselves are oriented exclusively towards the participants, and include activities designed to provide gratification immediately. Meetings can be classified along an instrumental–expressive continuum.

♦ TC1375 Study group
Description A group that gathers to conduct some particular coordinated line of research. This could be, for example, into the thought of an author, a situation, or a subject. There is generally a combination of "individual work" and group reflection that leads towards a conclusion that indicates the next step that is required for the group.
Advantages Intensity of focus, the richness that arises from the pooling of minds and insights stimulated by a disciplined structure. There is a strong sense of personal creativity that enables a much deeper than usual dialogue between each individual and the material or the subject at hand.
Disadvantages Can tend to become academically sterile and divorced from interaction in the wider social context.

♦ TC1377 Workshop facilitation
Description Although establishing a common context comes first, in fact workshops are the most important sphere of the facilitators' responsibility. It is necessary in any workshop to guide the group from one place to another place in their thinking: from an assortment of divergent individual opinions to a corporate statement. A workshop is not an informal exchange of views, in which views are exchanged, but each maintains his own. Views in a workshop are somehow related together and reorganized so that they become the domain of the whole group. There are many ways of doing this, but the group is helped if they know the approach they are using.
First, it is wise to delimit the question for workshopping in such a way that rapid responses can be given. Asking for information in some modular form, like a phrase or a few words, demands that the individual process his own data for the sake of the group. Procedures which call for recording input from a group on a chart, cards or blackboard are facilitated by short answers.
Secondly, the initial commitment of an individual to a group is the simple hearing of his or her own voice as part of the group. Workshop questions are designed to ensure that all members can participate. The first few questions, answerable by all, might be asked right around the group. Avoid allowing a whole meeting to go by in which there are participants who say nothing at all, or there dissidents may emerge later on.
Thirdly, use in every workshop a visual means of showing the group where their information is going. Information on cards can be grouped and groups titled. Information on lists can be checked off. Charts showing quantifiable data can be manipulated and tested for different results. Statements written by the group can be immediately copied and distributed to everyone, or even written with markers on wall charts to share first drafts. This visual movement of information allows a workshop to make tangible the decisions which are being made, encouraging participation and clarifying for everyone the direction the group is taking. They also make reports by a subgroup to a larger group immediately simpler to give and to understand. Where there is great linguistic diversity in a group, these visual tools permit a commonizing of images which language cannot perform.
Finally, a facilitator needs to consider workshop composition. Sometimes having people volunteer for workshops is appropriate. At other times, the subject matter is such that expertise determines who works in what group. Whenever possible, it is well to hold the principle that a workshop group is not the same group of people who work together locally, day–to–day. They have already determined a posture out of their own situation, and have little to gain in coming to a conference

to meet together again. Wherever diversity can be promoted in the workshop assignments themselves, this aids the capacity of the larger group to make decisions. Wherever expertise can be mixed in a workshop group, this aids the capacity of the larger group to understand the report or recommendation, as the group itself must clarify technical jargon.

♦ TC1378 Verbal interaction preferences

Description Participants may have distinct preferences amongst the following primarily verbal modes:
1. Fact-oriented: The stress is on stating information (often quantitative) considered to be factual, querying such facts, comparing them, and extrapolating from them to domains about which fewer facts are known by those present.
2. Affect-oriented: The stress is on the expression of emotional opinion concerning different experiences and facts. Participants may be emotionally aroused by the repeated reinforcement of certain opinions.
3. Concept-oriented: In this mode, categories of fact and experience are compared, criticized, re-ordered, possibly with only incidental reference to the referents.
4. Doctrine-oriented: A set of beliefs shared by participants may give rise to statements reaffirming and justifying them, as well as extending their application to new domains. This includes interaction about legal and procedural matters.
5. Value-oriented: Statements stressing the qualitative importance of particular approaches to any of the above.
6. Action prescriptive: Here the stress is on what should be done, usually in the light of any of the above.
Broader Participant interaction preferences (#TC1845).

♦ TC1381 Co-counsel conferencing

Description The meeting in a conference situation of two groups with the aim of mutual help and guidance. Derived from the "self-help therapy" groups where individuals in a group situation gather in pairs and take turns to be "client" and "counsellor". In the conference situation the groups pair off and adopt similar roles of "client" and "counsellor". The "counsellor" group listens to the "client" group; and through a questioning process that can be more or less structured allows the "client" to "talk-through" its situation and gather strength from its internal resources. The two groups then reverse roles and the other group has the chance to "talk-through" the issues or the crisis it is facing. The groups can belong to the same organization or be completely different organizations. The keys to the process include: (a) the active listening which probes for the hidden distress within the organization and encourages its expression; (b) the positive feedback that emphasizes the positive qualities exhibited by the group, which also tends to encourage the release and expression of feelings; and (c) the proposal and rehearsing of alternatives to the rigid patterns of behaviour they had become locked into.
Advantages Rapid creation of a therapeutic situation. The groups know why they are there and the process is understood. The reversing of the counsellor/client roles enables the establishment of trust between the groups. Re-activation of the stress and tension in the group allows for a breakthrough in the rigid patterns of behaviour that were adopted to cope with tension. This gives the possibility to widen the group system giving a more open structure of choice for assessment and action on their situation.
Disadvantages There is the danger that the "distress" of the "client" group can stimulate the experience of similar tension in the "counsellor" to the extent that it can no longer listen, but becomes caught up with its own internal issues. The process depends on the sensitivity of the members of the counsellor group to allow the expression of the distress of the "client" group and not to simply attempt to impose its patterns in place of the old patterns.

♦ TC1383 Lecture

Presentation — Dissertation — Monologue — Paper — Address
Description A carefully prepared one-way presentation by an individual resource person. It is a means to communicate ideas to large numbers of people, who then process the ideas on their own or in groups, and who then may have the chance to respond to the speaker or to the whole conference.
Advantages The lecture is a chance for a coherent set of ideas and concepts to be presented in a carefully structured form. It allows for the formation of a common context on which can be based the future interaction of the members of the audience. The value of the lecture method can be greatly enhanced by active listening techniques that increase the involvement of the audience. These include audience reaction teams, observation teams, and various forms of the participant interaction messaging process. The use of visual displays, variation in style and format, increase the percentage of information actually retained by the audience, and the use of metaphor and images that relate key concepts to the experience of the participants are known to retain their power for years.
Disadvantages The lecture process tends to emphasize the consumer role of the audience. Participants do not want to merely consume what the speakers produce. Members of the conference can exert little control over the content of the presentation which may be interesting but unrelated to the specified goals of the group. The lecture can be used to present distorted information or to make emotional appeals. It is an over-used technique. Its familiarity reduces its effectiveness and alternative ways to achieve the same objective should be considered. Unless a group definitely wishes to learn, it will react poorly to being lectured.

♦ TC1384 Doctrinaire participant type

Description Participants whose adherence to doctrine forms the basis for all their judgements, actions and decision-making. This can include political, religious or in-group doctrines.
Advantages Commitment and dedication.
Disadvantages Rigidity and an inability to consider other options or alternatives.

♦ TC1385 Decision-making conferencing

Description Meetings where it is intended that important decisions are to be made as opposed to contacts, plans, exchanges, presentations, etc. (though these may be part of a decision-making process). The most basic decision is: "Who shall decide ?" There are many decision-making processes for individuals and groups -the greater the number of participants in the process the longer the cost in time, energy, and other resources. Yet the smaller the decision-making group is within any organization, the smaller the chance for apparent and actual representation which can lead to frustration and a lack of cooperation with the decisions.
Advantages The conference situation can bring together many different people and different perspectives in order to focus on the decisions to be made. People experience themselves as involved and can commit themselves to the plan. The limited time-frame can catalyse intense concentration on the issues in order to be able to make the decision in the time available. The change from the routine can also allow participants to get "distance" from their immediate issues and to see them in a broader context.
Disadvantages The limited time-frame can prove to be inadequate and can be exploited by groups who seek to delay the process. The conference process is usually costly and the actual decisions that are made may not appear to justify the costs of the exercise.

♦ TC1389 Generalization

Description One of three universal elements of model or image creation, generalization refers to the way that the model of a system assumes more universality or commonality than exists in the system it represents. It has survival value because generalization enables deciding and choosing to go on, based on previous experiences selected from the complexity of "reality". However, generalization can also produce an impoverished model which has a reduced structure of choice. The most common examples of generalization are:
1. The use of unspecified collective nouns;
2. The use of "always", "never" and collective adjectives such as "every", "all", "none", "some" and "any";
3. The use of incompletely specified verbs or process words;
4. The use of assumed and unidentified presuppositions.

♦ TC1391 Self-reflexer participant type

Description Conference participants who recognize and examine how they are "part of the problem". They emphasize the having of personal experience and involvement and seek to find ways so that they can be part of the solution.
Advantages Problems are "earthed" in the real experience of these participants. They can examine how they block understanding of the issues and action that they could do in response to the issues. Useful balance to detached-operator types.
Disadvantages Can ignore issues that are perceived as outside their experience or they can be so overwhelmed with the complexity and dismay at their own involvement in the problem that they don't have the detachment to be free to act.

♦ TC1392 Conceptual screens

Conceptual frameworks — Conceptual models
Description Developing screens for a conference operating models that help people to stand in some perspective and to view some reality. It is a way of interpreting reality, and although limited, it may be the only way. Screens provide contexts out of which to make decisions.
A helpful screen is relevant, internally consistent, simple, elegant and beautiful. In order for a screen to be of use it must be consistent with the experience of people using it. To be readily understandable it needs to be built on a rationale which is consistently applied throughout the model and its component parts. Multiple rationales may be used but their application must be consistent in order for the screen to be effective. Simplicity enhances the ability of any design to be communicated and to be used with flexibility. Elegance increases its effectiveness as a screen. Beauty motivates people to use it.
A screen is developed out of the synthesis of an abstract conceptual framework and practical experience. An ideal is created of the phenomenon for which the screen is being developed. This ideal includes a number of components, the rationale for that number and the rationale for each part of the number. For instance polarities might be used to describe conflicting concepts in which parts are seen in the rationale."is and is not". A triangular rationale might be used to describe stable tensional relationships in which components are seen as "limiting, creating and sustaining" or in which dynamics are either "foundational, communal or significating". This ideal is placed in dialogue with experience. Using the rationale forces new relations to be brought to consciousness as more levels of the screen are developed. As the screen is created the experience of the group doing the development, writings about the topic and other peoples' insights are used to test the screen. As the dialogue goes on, the overall rationale may change, the specific components may be reinterpreted to fit the rationale or experience may be understood in new ways in light of the rationale.
A certain dogmatism and a certain openness to change are required for the screen to be developed. The dogmatism lies in maintaining the existing rationale until there is overwhelming evidence of the need for a different one. It is also including every insight or experience in the screen.
The openness lies in being willing to reinterpret the data, rework the rationales and rethink the dynamics of the screen. Naturally, the more developed the screen becomes, the more it is tested against experience and the more resistant to major change it becomes. This resistance is the result of increasing confidence in the viability of the screen and in its capacity to interpret data in helpful ways.
The most effective way of presenting screens to participants in a conference is for them to actually develop them in the process of preparing for the event. This of course is not always desirable or possible. The next best way is for the participants to use the screens in a variety of ways.

♦ TC1394 Conference maturation

Description An important question in conferencing is how to mature the power of a meeting to: (a) reflect the complexity of the external environment in an ordered manner (representation), to reflect about that environment (conceptual processes), and to reflect about itself (self-reference or self- reflexiveness); (b) focus the variety of perspectives represented, without destroying it in some simplistic formula of superficial consensus; (c) transform the issue presented, and the organizational groups which take responsibility for them, into new configurations of operational significance; (d) act or empower those represented to act, in the light of the level of understanding achieved during the meeting. The task is therefore to discover the nature of the "complete meeting" of the future, through which a new order may be brought into being.

♦ TC1399 Conference flexibility

Description The degree to which the conference activities, process and structure are marked by informal procedures rather than by adherence to established procedures. It is not reflected in the extent to which the participation of members is free from prescribed ways of behaving through customs, traditions, written or unwritten rules, regulations or codes of procedure.

♦ TC1410 Coordination conferencing

Description Meetings aimed to link together groups and individuals for short or long term action projects.
Advantages Can be events where there is an exciting discovery of complementarity, common concerns and exchangeable skills and information. The sense of potential for change can increase and give courage to the participating groups and individuals to explore new avenues of their own involvement and action.
Disadvantages Coordinating bodies are short-lived (or out-live their usefulness) and are too frequently "letterhead" or "talkshop" devices. Networks are frequently exercises in optimism which fail to work or attract commitment for any length of time.

♦ TC1418 Audio-graphic teleconferencing

Electronic meeting
Description Meetings that involve participants in a number of different groups in different geographical locations with a visual element that might include facsimile electronic blackboards, interactive drawing with a light pencil on a T.V. screen, digi-pad (a stylus with a writing tablet which shows the written print in a video memory).
Advantages The ability to see images greatly enhances the quality of communication, reinforcing the participants' sense of "presence" and confidence in the process.

Disadvantages Requires a lot of equipment and a very tight discipline and a certain degree of adaptation to this meeting process. There are constraints on the informal contacts that can be made at conferences.

♦ **TC1420 Seminar method**
Description A discussion group session in which all members of the group have something to offer. The process of the seminar generally includes a context or reference point around which the study is focused. These are usually papers selected or prepared by a faculty or organizing team. Whole papers or even books can be studied.
Advantages Useful for serious study of seminar papers in a conference situation. Non-confrontational approach that begins by seriously examining the content, relevance and validity of a person's thought by comparing it with personal experience. It can authentically promote changes in attitude and understanding which may lead to subsequent changes in practice. It tends to encourage maturity of thought through the interchange of ideas within the group. The responsibility remains with the participants for the contents, conclusions, and implications of the seminar.
Disadvantages This method is not useful for introducing new ideas that are unrelated to the experience of the members, since each person is expected to have something worthwhile to contribute. The effectiveness of the seminar depends heavily on the skill of the faculty in involving all the members of the group and in preparing the flow and focus of the questions that will catalyse creative discussion. The effectiveness is also dependent on the quality and appropriateness of the paper that is being studied.

♦ **TC1421 Majoral conferencing**
Description Conferences made up of representation from the higher levels of the political, social and economic establishment.
Advantages The high level of the delegates in the societal power structure gives a high functional significance to any products of the conference deliberations.
Disadvantages The high functional significance of any deliberations tends to encourage low risk procedures and participation and therefore low risk conservative proposals.

♦ **TC1423 Deletion**
Description One of the three universal elements of model or image creation, deletion refers to the way that the model of a system conveys less information than the system it represents. It has survival value because it allows for a focus on the "important" elements in a system, by cutting down on the distraction of the less important elements in the complexity of "reality". In this way it enhances the structure of choice. However, deletion can also produce an "impoverished" model which has a reduced structure of choice. The most common examples of deletion used in presentations and other forms of verbal communication are:
1. Missing information about specific referential indices;
2. Use of incomplete comparatives;
3. Use of unrelated superlatives;
4. Use of unrelated adverbs;
5. Use of implicit but unspecific necessity;
6. Use of implicit but unspecified impossibility.

♦ **TC1424 Conference stability**
Description The degree to which a conference persists over a period of time with essentially the same characteristics. It is reflected by the rate of participant turnover, by the frequency of reorganization and by the constancy of group size.

♦ **TC1430 Grassrooter participant type**
Description Conference participants who have a mistrustful relationship towards experts and specialists and who judge the value of a meeting by the participation of the grassroot constituency. Believes that the grassroot participants are in close touch with the "real" situation, and tend to trust action rather than theory.
Advantages Can make non-specialist participants feel they have valuable contributions to make. Can be sensitive to local and specific issues. Tend to encourage practical action. A useful balance to expertist types.
Disadvantages Can be naive and paternalistic about the contribution of grassroot participants. Can ignore dimensions of local issues that are more universal. Can be trapped by obvious and popular solutions.

♦ **TC1432 Losing focus in meetings**
Description The nature of focus may be partially understood from the various ways in which it may be lost during a meeting. These are the processes which may be guarded against although they are not necessarily independent:
1. **Loss of immediacy:** Participants may lose any sense of immediacy and allow discussion to focus on questions which erode their sense of urgency and responsibility. The assumption that necessary action can be taken on some other occasion, possibly by others, gradually holds sway.
2. **Attention absorption:** Topics become a focus for attention for different participants to the exclusion of any understanding of the context from which they emerge and by which they are linked.
3. **Attention span:** The complexity of the topic is such that participants do not have the patience to attend to any discussion of its intricacies and thus fail to comprehend it. This situation may be aggravated by poor verbal presentation, particularly when an audio-visual presentation would be clearer and quicker.
4. **Topic change too rapid:** When the meeting is switching between supposedly related topics, this may be done too rapidly for the participants to retain any permanent understanding of their connection.
5. **Topic change too slow:** Time spent by participants in treating one topic may be too great to retain adequate understanding of the previous topic. In this way they lose sight of the whole and may in fact become bored with excessive detail if they are not unnecessarily fascinated by it.
6. **Loss of connectedness:** Participants, for any of the above reasons, may lose understanding of the web of relevance interlinking the different topics under discussion. Conceptual fragmentation holds sway and most topics appear irrelevant to the participants major interest.
7. **"Topic twigging":** Topics may be explored with such enthusiasm that issues are broken up into sub-issues, sub-sub-issues, etc without any control over how to maintain the connection between such "twigs" or branches and the trunk of the "tree" from which they spring.
8. **Games and traps:** Discussion of topics may become enmeshed in various games and traps from which participants find it impossible to extract themselves. Such "sub-routines" may divert all energy from the fundamental or underlying issues.
9. **Superficiality:** The focus of the meeting may be trivialized by unnecessary enthusiastic interventions which do not take participants forward.
10. **Disruption:** The "noise-level" of the meeting may be such that no focus may be shared amongst participants.
11. **Polarization:** Discussion of the focus may provoke some participants to advocate a counter-focus, thus dividing the meeting.
12. **Energy drain:** The structure and processes of the meeting may be such as to drain participant energy rather than enhancing it. This weakens any focus which is still possible.
Related Interdisciplinarity (#TC1208)　　Integrative failure (#TC1005)
Meeting energy dissipation (#TC1699)　　Alienation of committed activists (#TC1962).

♦ **TC1439 Dread-filled conferencing**
Description Meetings convened and conducted in an atmosphere of dread and fear that the crisis or conflict at hand cannot be resolved or dissolved.
Advantages Apt to be extremely realistic, often a time when many hitherto hidden issues are brought to light and can be examined openly by all. The crisis can throw all other issues into another perspective and can, at the moment when all appears lost, catalyse a transformation.
Disadvantages If conflict and crises are treated as though they cannot be resolved or dissolved they tend to behave as if this were the case.

♦ **TC1441 Dynamic tensions**
Description Several levels of dynamic tensions are to be considered. The first level might be called the conceptual framework of the conference. This level has several parts: the method of the conference and the operating image of organizers and participants of the topic being discussed. If the operating image of organizers and participants is that the conference topic can only be conceived as a polarity i.e. good and bad, yes or no, etc., then diversity results in direct conflict and there is little that can be done to avoid it. If the conference topic can be conceived as a variety of alternative perspectives coming to a best of possible solutions, in which all sides win or at least lose somewhat equally, then there are greater possibilities for diversity within a unifying image and task.
The third level of the dynamic tension has to do with organization. If the organization of the conference encourages interchange among different perspectives and kinds of expertise, then unifying decisions are more likely. If within small groups all major perspectives are represented, then the decisions of the small group will more likely represent the decisions of the whole body. If the organization encourages isolation of divergent views from each other or direct conflict among irreconcilable perspectives without the possibility of change, then the process will likely break down.
The method and organization of the conference determine to a large extent the capacity of the participants to create unifying images and to maintain diversity. If the methods encourage dialogue from multiple perspectives, enable decisions that unify and honour diversity within a common framework then there are more likely to be common directions. If the methods polarize discussions, emphasize disagreement and maximize disunity then there is more likely to be a breakdown in the process of the event.

♦ **TC1449 Workshop methods**
Description The workshop method at its most basic level is consistent with the art form method. The major points of difference are the point of reference and the intent of the method. The art form conversation uses the object or situation being discussed and the participants' relationship to it as points of reference. It uses the objectivity of the art form to maintain focus to the conversation and the structure of the conversation to deepen the participants' relationship to the issue being raised indirectly by the art form. The workshop method uses the focus question and the experience and ideas of the participants as the points of reference. The structure of the sequence of the workshop process is as follows.
The first process in the workshop is to set the contextual framework for the discussion. This process sets the parameters for the thinking of the participants and gives a focus to the conversation.
The second process is to have the participants "brainstorm" ideas about the focus question. This is best done individually first in order to broaden the potential input from the group. A large number of items in a limited amount of time maximizes the use of the right brain. Intuitions that are unreflected are often quite useful in creating new ideas. The group may then be asked to share their insights one item at a time. This process is getting out people's lists; not reflecting on them. It is most helpful that these items be put up without comment about their relevance or irrelevance. This encourages participants to participate and creates an atmosphere of openness to ideas.
The third process is to organize the data, to create a "gestalt". Because of the large number of ways in which any big body of data can be organized, it is important to reach agreement on what the organizing principle will be. This organizing principle will help form the questions for the brainstorm and help focus the brainstormed ideas into helpful insights. As the pattern of the organization begins to emerge new insights may emerge; participants should feel free to add them.
The fourth process is to give names to the groups, sub-groups, and sub-sub-groups, etc. into which the data has been organized. These names should be as concrete as possible while being inclusive of all the data in the group. Some attention should be paid to the language used in naming. The type of language used should reflect the organizing principle. For example, if specific steps to implement a plan are being organized then action verbs and verb modifiers should be used for the names.
The final process is to write prose descriptions of the groups and sub-groups. Where possible the data should be used for this purpose along with additional insights as the data is organized, named and written. This written material is the basis of agreement for the workshop. It is the product of the workshop. It is at this point that the group may be asked if that is their decision. While the group has been making decisions all along here is where the final decision is articulated and symbolized.
While it is possible to have workshops in size from an individual to over 1000 people participating, the optimum size is from 15 to 30 people. For groups larger than this it is best to divide into several groups dealing with the same questions and to gestalt the results of these workshops in a plenary session. In this case each group would be treated as an individual in the smaller workshops.
In the dynamics of the workshop the objective level is the context and the brainstorm. The reflective level is the gestalting process. The naming process is the level of interpretation. The last step is the decisional level.

♦ **TC1451 Creative ambience**
Facilities — Space design
Description The atmosphere that the physical environment is able to create around the participation at a meeting. The "magic" that is stirred by the "rightness" of, among other things, the seating arrangement, the lighting, the decor, the sound quality, plants, etc. An elusive quality relying more on intuition than reason and highly specific in relationship to the content structure and process of each different meeting.
Advantages While extremely creative, conferencing can take place in the most sterile environment. There is no doubt that effort put into the ambience focuses and reinforces participation. The physical environment can be modelled on the social environment, reflecting the degree of structure, seriousness, and openness, and the contextual framework within which the conferencing is taking place.
Disadvantages It is possible to overemphasize the ambience and to over- invest in gimmicky, distracting and artificial environments, which by their elaborateness communicate a message of inflexible, prescribed procedures. The ambience, effective for one conference, may not hold the

same magic for another –an issue apparently overlooked by the standardized "five star" environments of many conference centres.

♦ TC1454 Debate

Description A formal and highly structured presentation by two teams holding different points of view about an identified issue. Each presenter speaks in turn without immediate response, discussion or interruption, while the audience (described as "the floor") listens. Can be used as a stimulus for Huddle groups, Buzz Groups, and Group Discussion, or as an alternative to a speaker.
Debating procedures: (a) Listen to the expression of the others' views until it is possible to formulate their argument in a way that is acceptable to them. (b) When both parties can do this they then formulate the conditions on which the others' views would be valid. (c) These "dissolving conditions", which tend to dissolve the argument, are then taken by each participant who formulates a concept of how these conditions could actually be determined. (d) The opponents attempt to determine what these conditions actually are.
(e) A matrix showing the different positions and against them, the agreed justifying conditions.
(f) Examination of the errors of judgement that are related and subsequent choice of path.
Advantages This process can be a welcomed relief from the strain that synergistic consensus building can impose on a group. It allows and encourages the statement of extremely polarized even polemic arguments in a kind of "dance of disagreement". There is an emphasis on the exposure of weakness in the position of the other side. It can be quite persuasive though those most easily persuaded are usually those who already hold the position that is being advocated. It can help to strengthen the "converted" by increasing their arsenal of effective arguments and developing their skill in presenting their point of view.
Disadvantages The focus on "differences" and analytical polarization, can lead to emotionally powerful divisive forces within the group. Form tends to be more important than content and rhetoric tends to preside over reason. It can be very time–consuming, distracting, and conspicuously unproductive. While exhilarating for skilled speakers, it can leave less articulate people intimidated and frustrated, perhaps losing valuable contradictions.

♦ TC1465 Meetings as energy–processing configurations

Description The following are useful metaphors through which energy processing in meetings may be understood in new ways:
1. Antenna: For which the constituent elements are precisely located in relation to one another to constitute a configuration (often parabolic) to focus incoming electromagnetic radiation for subsequent processing. The orientation of the configuration as a whole is vital to its operation, as in the case of the microwave receiver or the radio–telescope.
2. Magnetic "bottle": Whereby a configuration of precisely located magnets is used to contain plasma in such a way that its temperature may be maintained at over one million degrees for a period of seconds in order that fusion can take place (as an alternative source of energy to nuclear fission). It is only through the use of magnetic forces that the plasma may be kept from destroying any material container.
3. Reactor: In which particular attention is given to the configuration of heating, cooling, agitating, input and output elements, in order that an optimum transformation of materials should take place. This applies as much to the simple crucible, although the precision and symmetry of the configuration is most evident in nuclear reactor design.
4. Mirror configurations: As used for focusing sunlight in certain solar power furnaces, or alternatively for directing light, as in search–lights and lighthouses.
5. Optical systems of lenses: As used in telescopes and microscopes.
6. Acoustical configurations of walls and baffles: As used in an auditorium, required to ensure the balanced distribution of sound and the elimination of unwanted echoes.
7. Electric motor or generator: In which electricity is used or generated by the controlled movement of one configuration of elements in relation to another due to the effects of polarized forces operating in phase.
8. Factory complex: Usually designed with special attention to the transfer of energy and materials to processing locations which are therefore appropriately arranged in relation to each other. Usually located in relation to natural resources, a transport nexus or associated factories.
Broader Metaphors of meetings (#TC1603).

♦ TC1472 Overhead projector

Description Versatile display system that projects images from transparent sheets onto a large screen behind the presenter. Can be used with screens prepared in advance, with sheets that progressively overlay the original image and can be made up, altered and added to during the presentation itself.
Advantages The presenter remains facing the group while drawing on the sheets. The sheets can be specially printed and copied and transported with comparative ease.
Disadvantages Can be a distracting element.

♦ TC1475 Meetings as ecological and natural processes

Description The following may be used as metaphors to explore alternative ways of understanding meetings:
1. Chemistry: The "chemistry" of a meeting may be explored as the sequence or pattern of reactions taking place at a certain rate, possibly in the presence of catalysts. A meeting may also be seen as a "chemical soup" within which new varieties of complex molecules may emerge under certain conditions.
2. Biochemical and metabolic processes: The range of possible meeting processes may be seen as constituting a map of pathways whereby various kinds of essential transformation take place with the assistance of specific enzymes.
3. Environmental genetics: The viewpoints represented and emerging at a meeting may be seen in terms of species and gene pools linked and isolated by food webs and ecological niches, but subject to genetic drift and mutation. Such environments may be poor, vulnerable, or in process of enrichment. Meetings may be seen as ruled by the "law of the jungle".
4. Ecosystem: The various perspectives and processes may be best mapped by some onto an image of some environmental system with different species interacting, procreating and developing somewhat at the mercy of the elements.
5. Energy sources and sinks: Some may choose to see the event in terms of sources of different qualities to be cultivated, energy receptacles to be created and maintained, and energy sinks or traps to be avoided. The whole event may be seen in terms of gathering and using ch'i energy.
Broader Metaphors of meetings (#TC1603).

♦ TC1478 Interstitial conferencing

Description Filling of gaps in the social structure to enable the operation of a subsystem or to enable the integration of subsystems. The conference tends to focus on analysis and planning of the structural needs, configurations and logistical systems, that can alleviate or fill the gap in social structure.

♦ TC1481 Conference mythology

Description The process of creating a "mythology" involves several factors. As used here, "mythology" means the story, based on experience, which relates that experience to some larger social significance. The rationale created by the organizers about the intent of the conference, their recruitment brochures and materials and the "rumours" about it help create this story. The history of the event enhances the story. Who organizes and sponsors the event become components of the mythology. All of these things and more create a swirl of emotional, intellectual and decisional processes that place the event in the world.

♦ TC1484 Conference conviviality

Description The degree to which conference participation is accompanied in a general feeling of pleasantness or agreeableness. It is reflected by the frequency of laughter, pleasant anticipation of meetings and by the absence of gripping and complaining.

♦ TC1485 Workgroup

Taskforce
Description A group of about 15 people with a specifically assigned task. The particular issues that are handled by the group are identified and worked on relying on the interest, enthusiasm, ingenuity, and creativity of the members. The process of the workshop is usually structured in a logical progression that works towards the goal to be achieved.
Advantages The high level of involvement elicits a high level of responsibility from the group. The high level of responsibility creates a high level of relevance and appropriateness of the content for the members of the group. The short term nature of the assignment enables an intensive commitment to the task and its completion.
Disadvantages In a conference situation the various workshop groups can become isolated and disrelated from each other. Can be sterile and frustrating if they are overstructured and they can be confusing and wasteful if they are unplanned.

♦ TC1487 Chalkboard

Blackboard — Whiteboard
Description A surface that is written and drawn upon with chalk or markers used during presentations and discussions to temporarily record and display diagrams, models, lists and texts. Generally black or green and used with various colours of chalk or marker. Also used to display messages and even to allow inter–participant dialogues to take place.
Advantages Failproof familiar simplicity. Low cost, and available to participants at any level of training or experience.
Disadvantages Needs to be erased frequently to accommodate new data. Sometimes hard to read. Markers tend to dry out extremely quickly especially when waved about without being capped during presentations or discussions.

♦ TC1488 Meetings through analytical frameworks

Description
1. Structure: The gathering may be "objectified" in terms of any of the following: Agenda, critical pathway, system diagram, Programme matrix, event timetable, programme or topic "tracks".
2. Risk: Participants may prefer to assess their participation in terms of "risk tracks". Some may be entirely conventional low–risk lecture/discussion type events. Others may be designed to make the participant take or defend a position as a person. Others may involve the participant in some personal transformation process; and some may be high–risk experiments which may fail, as experiments do, providing lessons for the future.
3. Socio–political analysis: The gathering will lend itself to description and interpretation in terms of power politics and societal dynamics.
4. Psycho–cultural analysis: The forms and expressions of the gathering can be seen in psychoanalytical terms with necessary archetypal confrontations.
5. Special theories: The transitions in the event may be best understood by some in the light of the mathematics of catastrophe or chaos theory.
Broader Metaphors of meetings (#TC1603).

♦ TC1490 Conference participation

Description The degree to which individuals and groups apply time and effort to conference activities. It is reflected in the members who contribute in discussion, the different kinds of contributions, in voluntary assumption of non–assigned duties, and in the amount of time spent in small group activities.

♦ TC1491 Initiator participant type

Contributors — Genius
Description Meeting participants who suggest or propose new ideas, different ways to perceive problems or goals or new forms of procedure or organization for the group to follow. They tend to be serious–minded but individualistic and unorthodox.
Advantages The introduction of new imagination, intellect, knowledge and even at times, genius.
Disadvantages Inclined to disregard practical details and protocol, can be impatient with those slow to grasp what they are communicating. Liable to lose interest quickly once the proposal is adopted, leaving it to others to follow through.

♦ TC1493 Participant consensus buttons

Description A voting device that can instantly and continuously give feedback on various opinions in the audience. Participants can, for example, indicate any of the following: agreement or disagreement with the speaker, agreement or disagreement with the proposal under discussion, desire to move onto the next point on the agenda, desire for clarification of the point being made, desire to adjourn the session, desire to break into small group discussion sessions, or similar points. The device given to each participant consists of a set of 6 (or more) switches corresponding to each of the above points. The switches are linked to a counting device such that when 27 participants press the first switch a counter visible to all participants (including the speaker and the chairman) indicates "27". The total for each other point can also be indicated at the same time.
Advantages In this way, at a glance, all participants in the meeting session can determine with greater accuracy the sense of the meeting and how it should be continued. This helps to avoid meandering. The great advantage of the device is that it helps to change the pattern of communication. Instead of all communications being mediated by the chairman or speaker, participants are able to indicate to one another their assessment of the meeting in a way which prevents the chairman from manipulating the meeting on the basis of his own interpretation of the desires of participants. The use of such a device introduces much more immediacy into debates since at every moment, in effect, a continuing vote is being made on a number of features of the meeting. (If recorded, as is technically feasible, this would be extremely valuable data for the evaluation of meeting effectiveness).
Disadvantages Can be distracting, misused and misunderstood and most conference speakers and participants are unfamiliar and uncomfortable with such a change in communication patterns.

♦ **TC1494 Dissidence**
Description The manner in which dissidence is handled is crucial for the effectiveness of a conference. This process begins with recruitment. It is the process of creating understanding of what the programme is about before participants arrive. Familiarity with the methods used helps participants function effectively. The development of conceptual frameworks or screens by all or most of the participants before the event, helps create commitment to the success of the event, common understanding of the perspective being suggested and camaraderie among participants.
Dissidence arising in the midst of the event itself has to be carefully evaluated. First of all, much of the dissidence is invisible. People who disagree with majority opinions most often do not participate or just leave early. The more vocal type is more noticeable. Once organizers have noticed dissident opinions, they must decide if objections are valid or not. Most often vocal dissent is helpful because it can be analyzed and responded to in some way. If the objection has to do with a minor issue, then often an announcement of change suffices to resolve the issue. If the objection is of such magnitude that the intent of the conference is being challenged then it may be that in a small conference everyone reconsiders the direction in a plenary session. If the conference involves several hundred people, then the organizers and dissidents may need to rethink the intent as an ad hoc committee. It is important that the process of altering the intent of the conference be done rapidly enough to permit the conference to reach some conclusion in the time available.
When the dissidence is toward creative resolution of issues that are valid within the scope of the programme, then it is incumbent upon the organizers and dissidents to create a resolution of the conflict. If, however, the dissidence is simply disruptive, then the organizers must find a way of minimizing its impact for the sake of the larger group. Simply permitting the event to be disrupted for the sake of a few does not serve the group as a whole. Nor is it helpful to ignore authentic dissidence simply for the sake of continuing the event.

♦ **TC1498 Audio–tape recording**
Description There are many sophisticated ways to tape–record meetings using reel to reel cassette or mini–cassette recorders; these can be transcribed into printed material, stored, or copied onto other tapes for distribution. Speech activated microphones and other new technology is greatly improving this recording system.
Advantages A low–profile, unobtrusive way to make a record of everything that is said. It can be encouraging and reassuring to participants that the insights and breakthroughs that occur in lively interaction are on record for later processing and recall.
Disadvantages Some people find it disconcerting to be recorded and find their expression cramped. Quiet speakers and speakers competing with each other or with other extraneous noises like coughing or laughter may not be audible unless there are well placed and well adjusted microphones.

♦ **TC1501 Public relations**
Description Although strong criticism can be made of the conventional use of "public relations" techniques in meetings, especially when crudely done for simplistic purposes, skilled practitioners are sensitive to dimensions otherwise ignored. It is this sensitivity which can contribute considerably to the "magic" of whatever occurs in a condition of focus. The major problem of public relations as applied in meetings is that it is conceived in terms of the priorities of the meeting sponsor or organizer. A major concern then is to stress at all cost the qualities and significance: of those responsible for the event, of the event itself, of the participants, and of whatever is achieved. The techniques are necessarily so pervasive in their application that they cloak every facet of the event in a concealing garb of seeming glamour and significance. This of course serves to "paper over any cracks" in the arrangement, effectively turning each moment of the meeting into a piece of theatre, however flimsy the sets. Participant awareness of the reality, as it contrasts with the image, generates cynicism and is counter–productive.
One feature of this problem is the tendency to reinforce the status quo and to conceal weaknesses and conflicts which can provoke and justify healthy change. Existing categories are effectively treated like icons requiring appropriate praise and decoration. Another feature of this problem is dependence upon the "showmanship" strengths of PR techniques to provide "attention grabbers" to absorb the time of participants. These may extend from glossy audio–visuals through sumptuous feasts to tourist attractions. This leads to a simplistic conception of meetings, and a total disrespect for participants and the issues on which they supposedly hope to act. Sad to say, many meeting sponsors are evaluated by their peers in terms of "how good a show they put on" and the meeting market is such that it is unlikely that they would fail by underestimating the level of sophistication of participants.
Another feature of this problem is the stress on the impact on participants of "messages" fired at them as "targets" in the marketing "communications" approach which has given birth to most public relations techniques. Despite these present defects, the practitioners are nevertheless especially sensitive to configuration, place, timing, non–verbal stimuli and their effect on image. The question is whether these skills can be employed in the interest of participants and their concerns, rather than as a manipulation of them.
The question is how can meeting participants themselves engender collective sensitivity to these dimensions, correcting continually for any excesses. The process of building up and focusing significance collectively is one known through the rituals of less artificial cultures with a more organic response to a happening. It would appear that the "civilized" conscious emphasis on rational discourse in meetings has left them exposed to manipulation of any unconscious emotional needs which would otherwise provide a healthy equilibrium. How can the power of any such emotional arousal of the imagination be consciously evoked by participants to meld their perspectives together more effectively to "get their act together" and get the meeting into focus ? The "primitive" approach, the "PR communicator" approach, and the "small groups process" approach are extremes, each with important clues and dangerous traps.
The clues suggest the importance of articulating, "feeding" and reinforcing images (which well up and breed within the meeting) by an almost rhythmic alternation between different information modes, sensed to be in some comprehensible configuration. If harmoniously balanced through a timed progression of intensity, the meeting will then "take off". If not, the significance leaks away, leaving an empty, brittle shell of programme elements.

♦ **TC1506 Conference control**
Description The degree to which the conference directs and regulates the behaviour of individuals and groups as they participate as conference members. It is reflected in the modifications which the conference participation imposes upon the complete freedom of individual behaviour.

♦ **TC1511 Discussion group**
Description This can be a long–term or short–term group for an exchange of ideas between the participants. There is usually a leader or leaders and rotational leadership is especially effective in maintaining an equal level of "group responsibility" amongst the participants. There is usually a focus to the discussion, and generally some kind of report or synthesis is required. In the conference situation the feedback that comes from the group can be in the form of new questions, new data, problems, proposals, etc.
Advantages Equality of participation, and a chance for everyone to contribute. Input towards the conference process receives its first refinement, before being passed on to the whole assembly. The exchange encourages the airing of different perspectives, anticipations and needs and works towards the development of common understanding.
Disadvantages In a conference situation the use of discussion groups can be undervalued by the participants and organizers and become a "chewing– over" of the ideas that have been presented by the previous speaker, in which case participants experience themselves as mere consumers.

♦ **TC1524 Educative conferencing**
Contextual conferencing
Description Meetings designed to help participants to achieve some new level of understanding of the complex of issues with which the world and their own situation are confronted.
Advantages A common context or understanding is often the starting point for further involvement with others who share the context. As part of a larger conference this can provide a common reference point for all further creativity.
Disadvantages Can absorb much programme time. The question is whether the focus on learning is not a disguise for inability to focus collectively on the fact that there is a marked inability to act even when everyone has learnt the same lessons.

♦ **TC1525 Minoral conferencing**
Description Conferences made up of representation from the "minority interest" groups which tend to be marginal, single–issue, and disestablishment. Concern is with the relatively non–powerful subgroups within the population.
Advantages High risk, high potential, conferencing is possible with the maximum chance of transformative eventfulness. Grassroots awareness of injustice. Rights and special group interests lends fuel to the creative process. High tension deliberations.
Disadvantages Relatively non–powerful. Prone to fragmentation rather than integration of all the single–issue perspectives into common action, it often takes the identification of a common enemy to justify the high tension and brittle integrative alliances between the different minority groups.

♦ **TC1526 Team completer role**
Team finisher
Description An affinity–role played by one or more members of a small group or team. They tend to be anxious, orderly, conscientious people with a capacity for follow through and perfectionism. They actively search for aspects and details that require, and are not receiving, greater attention.
Advantages Can protect the team from mistakes and omissions. They have the stamina to carry out a project through to its completion and a willingness to work on the details. They maintain a sense of urgency. Can be the key to actual products resulting from teamwork.
Disadvantages Their tendency to worry about small issues can distract the group. They can hold up the team by their reluctance to "let go" of an issue. Their anxiety can become a source of distraction.

♦ **TC1528 Team roles**
Affinity roles
Description Team–role is a term used to describe the behaviour pattern characteristic of an individual participant in the interaction that takes place in a team situation, particularly with regard to their positive contribution to the effectiveness of the team. There are various screens that distinguish useful types which are generally based on personality types that can be measured by different testing procedures. Most individuals appear to have a primary team–role to which they have the greatest affinity and a back–up or secondary team–role. There is also evidence that roles can be chosen or even assigned on a rotational basis.
Advantages Understanding the gifts that different roles can bring to team creativity allows for a greater use of human resources. The complementary role played by people who would not immediately be considered as compatible can dramatically enhance team performance. Rotating the roles played by different members has many advantages; in particular 1) it increases the repertoire of each member; 2) it continuously reframes the structure and interpersonal system of the team; 3) it gives members the chance to experience different perceptions of each other and of the team process; 4) it puts members into a situation of equality; 5) it generates an ongoing training component; 6) it distributes the burden of responsibility onto each member.
Disadvantages Delineating roles within a team can constrain creative innovation and reduce the structure of choice when selecting team members. Some criteria are reductionistic. For example, selecting members for specific technical demands is most often done on the basis of their past experience and qualifications, and although they may fit this "functional role" there is little consideration given to personal characteristics or aptitudes that will influence the effectiveness of the team as a creative whole. Role delineation is useful if it is used to encourage flexible structures and response patterns and problematic if it becomes rigid and constricting.

♦ **TC1529 Meetings as agriculture and food processing**
Description The following may be used as metaphors to explore alternative ways of understanding meetings:
1. Horticulture and gardening: A meeting may be seen as a garden of flowers, vegetables and other species (with "a hundred flowers blooming"). The challenge is to care appropriately for these species: to water, to cover, to prune, to weed, to encourage or reduce certain insects, etc.
2. Cooking: A meeting may be viewed as a menu of dishes amongst which participants select. Balance is important both in selecting the dishes an individual consumes (the art of the gourmet) and in combining the ingredients whereby a dish is prepared (the culinary art).
3. Diet: A meeting may be viewed in terms of the dietary regime appropriate to participant nourishment, namely the quantity of carbohydrates, protein, and vitamins, interpreted as various kinds of information. The question of "calories", "exercise" and "obsesity" may also be raised.
Broader Metaphors of meetings (#TC1603).

♦ **TC1531 Panel discussion**
Description A group of three to six people who have a discussion in front of a larger audience. It is not a series of speeches but a purposeful discussion usually utilizing a moderator. It is most often used in the following ways:
1. To create an informal atmosphere for communication within the group or groups in the audience.
2. To identify problems or issues and to begin to explore them.
3. To give the audience an understanding of the component parts of the problem.
4. To get different facts and points of view into a discussion framework.
5. To weigh the advantages and disadvantages of a course or courses of action.
6. As a stimulus for small group discussion.
Advantages Advantages include:
1. It can expose and focus on different points of view, different facts, and different attitudes, on a subject.

2. It allows for maximum interaction and interstimulation between panel members.
3. Maintains interest of an audience because of the dramatic presentation and the differences of opinion.
4. It is useful for defining points of agreement, areas of disagreement, and for working towards consensus.
5. It divides the responsibility for comprehensive coverage of a particular issue between the panel members who can be asked to prepare specific components of the whole presentation.
Disadvantages The panel discussion depends heavily on the members and their willingness and ability to communicate with each other and the audience. Weak members' point of view can be lost. Strong members can dominate and monopolize the discussion, make speeches etc. It is difficult for the panel members to avoid making speeches on the subject they have prepared for and they may be unprepared for the line of questioning that arises out of the discussion.

♦ TC1532 Conference practics
Description The practical aspects of conferencing or practics are the care of the materials, facilities, communication and accommodation of a group gathering. The practics dimension must be carefully coordinated with the intent of the conference in order to ensure that it is invisible. Like most services, practics in a conference are usually only visible, only commented on, when something goes wrong. It is wise to assume that this dimension needs a full time coordinator just as much as does the programme itself. The progress of practics needs to be watched throughout the conference just as carefully as does any other dimension.
These "small" decisions should not be placed in the hands of professionals without very clear direction as to the tone and intent of the gathering. An effective conference facility can only embody as much of the conference image as they are given to work with. Getting participants "inside the picture" as early as possible and coordinating with them at every point allows outside services to really serve, rather than have to guess at images of an event, which in the end may not coincide with what was intended.

♦ TC1535 Colloquy
Description A discussion between two teams representing different points of view in the presence of an audience.
Advantages Tends to enable a thorough airing of divergent points of view. High interest level for the audience. Particularly useful when there are observation teams listening for specific arenas of agreement or complementarity between the different views expressed by members of the teams.
Disadvantages Tends to polarize the audience who can fall back into the debating mode in their small group discussions, rather than seeking for the synergistic and creative way forward. As in many of the conferencing modes that emphasize the passive audience, there is an overemphasis on the importance of the individual performance of those on the platform, with the risk of missing the opportunity to tap the creativity of all the participants.

♦ TC1538 Simulation exercises
Games — Structured experiences
Description Activities performed or played within the framework of specific rules and objectives which are designed to create situations similar to real life in order to raise awareness of the issues and to prepare the participants to deal with them. Often involving role playing.
Advantages Can be good to involve the participants who can even be motivated to additional voluntary work outside formal discussions. Actual decisions need to be made in most simulation games and the consequences become clearly revealed; and immediate feedback is possible. Critical examination and learning from the experience of playing can prove very powerful in exploring and influencing attitudes. The enjoyment of playing can create a favourable atmosphere at the conference.
Disadvantages Some participants may not wish to participate for fear of making fools of themselves. There are often different degrees of seriousness in the group. Some games are under-prepared, poorly administrated and/or inappropriate or irrelevant to the conference. Can be counter-productive by distracting or irritating participants. Very heavy on time consumption and can dominate conversation long after the game is over.

♦ TC1544 Conclave
Description A gathering that is generally focused on the process of selecting a person or persons to take up a position of special responsibility, leadership and influence. It may also be called when there is a crisis that involves and extends beyond the leadership.
Advantages Can be a very intense process of leadership selection that goes beyond the mere mechanics of a voting system. The process generally inspires confidence in the new leadership and a commitment from those who participated in the process.
Disadvantages It can be a lengthy and costly process. There can be an overemphasis on voting and the repeated recording of opinions can overshadow the importance of ideas and creative selection of leadership that meets and balances the criteria of all participants.

♦ TC1546 Joker
Fool — Jester — Clown — Buffoon
Description A role that may be played during the conferencing process. It can be an assigned role or a role adopted by one of the participants masked or who may be dressed in an appropriate and distinguishing costume traditionally including a cape and bells. Court jesters, clowns, fools or buffoons are mythic figures representing the inversion of the powers of leaders (as the possessor of supreme powers) –or as their alter ego. They are therefore often the victims chosen in folklore as the substitute or foil for the leader in rites whereby the people respond frankly and unceremoniously to such powers. They were often masters of song and dance, and could be a dramatic foil to pomp, superficiality and falsehood of any kind.
Advantages Ideally they are a powerful reminder of the distortion of the human condition –more immediate than the photographs disseminated via the media. Additionally, due to the freedom from censure and responsibility for their actions which they are accorded, they are able to mirror, parody and mimic court situations in such a way as to bring out truths which would otherwise be collectively and carefully ignored. As an ambiguous and often androgynous figure, the jester can function as a powerful social catalyst – for good or for ill, depending upon the response of those by whom he is surrounded. The fool is an enigmatic symbol of the point of crisis when the normal or conscious appears to become the irrational, the unconscious and the abnormal. As such, the fools are to be found on the fringe of all orders and systems, outside all conventional categories, processes and social rules. They are the bridge between the conscious and the unconscious (and between the attributes of the right and left hemispheres of the brain) –a reminder that, after having failed in our effort to order and understand the universe in the light of our intellect and instinct there nevertheless remains another way. Eliminating the jester from the conference is as risky as allowing him to play this role. For, if "foolishness" is not given a channel through which to express itself, it seeks its own channel anyway. Summit meetings and international assemblies, particularly those in which each is conscious of the high purpose and seriousness of his role, run a considerable risk of incorporating distortion into their proceedings and results because of an inability to accept what a jester would reveal. (Political cartoons offer a partial remedy, but they lack the significance of being accepted as part of the proceedings and thus have little effect on them.)
Disadvantages It requires greater maturity on the part of all participants, especially the chairperson and principal speakers, to play their parts in the face of such instant feedback. In the absence of children at international assemblies, who can say whether international emperors wear any clothes ?

♦ TC1547 Structural dimensions of conferencing
Description In describing and comparing different kinds of conferences, these four structural dimensions are considered to be basic: (a) The "size" expressed in terms of numbers of participants, duration and investment.
(b) The "territoriality"; the field of concern, the geography, the dicipline, the constituency.
(c) The degree of "formality"; in style, structure and process organization.
(d) The degree of "complexity". Differentiation between meetings of a group with a relatively simple, concrete purpose and meetings made up of many groups with relatively abstract aims, which are themselves compounded from the aims of several sub–groups.

♦ TC1548 Conference rituals
Description Every event has a set of rituals. At minimum there are rituals that mark as significant the beginning and ending of the event. The welcoming and closing banquets may be the simplest and most common of these. In addition there are usually literally hundreds of rites that are a part of the life of a conference. They start the day, open a meeting, introduce a speaker, get people to meals, create relations, and catalyze other acts. Most are associated with the larger culture but some may be created for the event or the organization holding the event.

♦ TC1550 Team genius role
Imagineer
Description An affinity–role played by one or more members of a small group or team. Typically, individualistic serious–minded, yet unpredictable and unorthodox. One who contributes new ideas, concepts and strategies, and who looks for alternative approaches that the group might take in regard to the major issues.
Advantages Imagination, intellect, knowledge, can provide the creative input that will be the grist of the meeting; in this sense providing the raw material for group discussion, decision making and new strategies.
Disadvantages "Up in the clouds" often oblivious to practical details, and prone to ignore procedure and structural guidelines. If there are two or more playing this role in one group, the group's creativity is reduced to the same level as a group with no one playing this role. Ideas and creativity on their own are not effective and require the good processing and implementation skills of others in the team.

♦ TC1553 Conferencing as risk
Description An investment of time, energy, money, human and material resources; an investment of enthusiasm, hope, passion; an investment of knowledge and experience. The stakes are high. Risk is essential to pioneering. It is practically synonymous with innovation, transformation and change. Conferencing is risking; risking the security of your own conceptual framework against the challenge of exposure to other frameworks; risking the fragility of personal idealism and enthusiasm in the heat of the interchange; risking the challenge to change, to become transformed, risking the danger that accompanies the opportunity of every crisis.
Risk is assymetrical. Possible loss can be weighed against possible gain. Generally, however, transformative conferencing is high–risk as the potential gain and the potential loss of change are both very high. Adopting traditional well–tried methods and approaches to conferencing will give more or less predictable results. This is a low risk "safe" approach. It is controllable, there are fewer surprises. Organizers and sponsors tend to want to control the process and structure of conferencing. In avoiding the risk of surrendering their control they may miss the tremendous transformative potential of the gathering. The conferencing process that is adopted by an organization, is a metaphor of the (in)ability of that organization to respond effectively. Transformative conferencing is a risk that is worth encouraging and entering into. It is a risk that cannot be avoided by organizations who wish to demonstrate their readiness to become part of a solution; to become part of a future; to become part of the change.

♦ TC1555 Conference size
Description The number of participants in the conference. Size is relative to previous gatherings, other similar meetings, anticipated numbers, and the extent of the constituency represented.

♦ TC1556 Conference time design
Description The time design of the conference creates a balance between the agenda and the time it will require and the celebrational life, free time for personal contacts and other non–agenda time. Each conference has its own rhythm. Some schedules must be loose because the important factor of the event is personal contact among delegates. Others are a focused flow of plenary sessions, in which participants expect to work as a whole group.
In relation to the time designing of a conference, an appropriate balance should be sought between the "continuity" of similar uses of time and the "discontinuity" of different kinds of time. This is in fact as true within a single session as in the overall time design of the event. Think through the expectations of the group, and then look over the whole flow from the perspective of mood. If there are extreme imbalances of kinds of time, like two heavy working days and two very open days, it is best to even them out. Then also, ensure that the time of important plenary or decision–making sessions is free enough before and after so that attendance will be ensured and the group as prepared for it as possible. Do not, for example, expect to start the major reporting session early in the morning the day after the big night out on the town. Many time issues are resolved ahead of time by careful thinking through of the experience of participants going through the time allotted.

♦ TC1558 Unity and diversity
Description In order for a conference to achieve anything, to understand anything, or to experience anything, some level of unity is required. At the same time, for a conference to be creative, to go anywhere, or to make contributions to the larger society, some level of disunity is required. The task of the organizers is to create a balance between the two. An overly unified set of images, life styles, expectations and experiences among the participants will lead to little or no change in action, thought or experience. A totally disunited conference (if one could exist) would result in a series of individual experiences but little corporate direction or decision.
The conception of any conference takes into consideration some degree of unity and some degree of diversity. Most conferences are unified either through a limited topic, which rules out the breadth of possible discussion, or through a limited makeup of participants. The more focused the subject matter and more homogeneous the group, the easier it is to hold a unified conference and to arrive at consensus. Once made, however the consensus may be rather superficial and limited. The more diversified the group and broad the topics for discussion, the more difficult it is to unify the conference and to arrive at shared decisions; but such decisions can be well worth the effort.

No matter the degree of diversity, however, holding the conference requires creating a perspective about the topic and the participants which maximizes the possibility of unity. All aspects of unity are created rather than "natural". The "earthrise" image, for example, is one which holds great uniqueness and richness of diversity in a unitary image. This perspective is expressed in how the conference is described, its story. This commonizing imagery is not intended to manipulate a group into homogeneity. It is the basis or the boundaries within which diversity can be encouraged to flourish.
This balance between unity and disunity, between homogeneity and diversity is created from the very concept of the conference. The process of recruiting participants is the early key in this dynamic tension.

♦ TC1559 Facilitation
Description Facilitators are enable the group to make decisions; they do not make decisions for the group. Facilitators are not only responsible for the care of the decision–making dynamics, but also for guarding and, where needed, promoting, the capacity of the group to transform its fundamental assumptions. Thus, those with vested interests in an existing perspective or in some idealized opposition perspective are inappropriate facilitators, as neither can readily gain the group's trust. For this reason the facilitator is necessarily possibility–oriented, but from a practical rather than a theoretical perspective.
Facilitation is done by a team of people, whose job it is to translate into day–to–day activities the intents and methods of the conference. Facilitation is also performed by those participants and organizers who take it upon themselves to promote wherever possible values and perspectives which affirm the group's struggles and decisions. They care enough to comprehend the intellectual framework and give permission for the group to move positively. This care about the group and its task is an important basis of conferencing that works.

♦ TC1562 Conscious conferencing
Conscious congress — Conscious groups
Description It is useful to distinguish between (a) the awareness individuals may have of the group of other individuals with whom they interact, namely, "group consciousness" and (b) an awareness by a group as a whole of itself and its activities, namely a "conscious group". The first is necessary to enable individuals to respond appropriately to each other within a group. The second arises when the individuals are collectively and simultaneously aware of the pattern of those interactions between the group members. A sign of the emergence of a conscious group –from the point of view of anyone involved– is that each is moved to act in the right way at the right time, although there does not appear to be any central coordinating agent or any explicit design. The actions of the whole are very much greater than can be comprehended from the individual actions. How each awareness interpenetrates the others is not yet clear. The "eyes" do not understand how they are related to the "feet" or the "hands", and the right and left "feet" do not understand how their movements harmonize through their opposition to each other (a yin–yang cycle) to move the body forward. A similar situation arises early in the growth of a child.
Advantages The prime characteristic of conscious group is its awareness of itself and its place and rhythm in the scheme of things. Within itself it mirrors an awareness of how its environment is organized. Each action on the environment is paralleled by an equivalent displacement of energies within itself. There is a "magical sympathy" between the outer and the inner worlds. It is through this inner/outer attunement that the group is able to increase considerably the amount and range of energies that it can handle and focus in order to transform itself and its environment as it evolves into a greater identity.
When learning to ride a bicycle, a person has to deliberately correct excessive responses in order to maintain balance –until such correctional moves are made instinctively in a conscious group, excessive responses resulting in energy disequilibrium are also smoothly corrected by an integrated response within the group– whereas this would normally only be achieved through a series of sporadic procedures, characterized by a heated mix of rational and irrational argument and expression, leading to changes of an almost spastic quality.
Disadvantages The insecurity and high risk factor involved in such conferencing –which are of course essential elements– represent a high threshold for conference organisers and coordinating committees to overcome before they would be prepared to try a process of this kind.

♦ TC1564 Sustenance conferencing
Maintenance conferencing
Description Consideration is given to the sustenance of the membership, coping with strain, frustration, stress, and any inabilities to deal with the situation. The conferencing tends to focus on reflection, problem solving, training, re–motivation and inspiration of the membership.

♦ TC1566 Key–note speakers
Super stars — Conference heavies — Guest speakers
Description Speakers at a conference who are invited to attract participation to publicize and give significance to its programme. They are well known, preferably experienced and qualified in the subject or theme of the conference and are generally expected to be sympathetic towards the aims of the conference; though these latter qualifications are not always considered necessary.
Advantages Publicity and weight for the conference. Entertainment and inspiration –a form of blessing from people of high esteem. What they have to say is generally listened to and heard and reported –it is worth discussing this with them in advance. They can have the charisma to generate high levels of enthusiasm.
Disadvantages Such people may be encouraged to take up much programme time, even though there is little to say that is not already available (e.g. in their books or in reports of previous meetings). It may be quite impossible to prevent them from abusively exceeding the speaking time allotted to them. Little attention is given to what people can do once they have been stimulated to an appropriate level of enthusiasm. Little attention is given to the lack of integration (if not conflict) between those providing such inspiration. Any overemphasis on inspiration necessarily presupposes that participants are not already adequately inspired or motivated and thus alienates those who hope to use the occasion to move collectively on to the next step.

♦ TC1567 Meeting magic
Meeting serendipity — Meeting take off
Description Occasionally, perhaps under special circumstances, meetings "come together" and "take off" as if by magic. It might be called serendipity. There is very little indication of why this comes about or how it is to be described objectively. It can happen when every care has been put into arranging the meeting and selecting the participants, or it can happen under extremely non–ideal circumstances. The following notes indicate some possible directions for further reflection on the question: **1. Indirection:** In such a case there seems to be a strength in defining the central point of focus by discussions which use it as an unspoken reference point. The totality of tangential dialogues is then facilitated by this approach, whereas "going to the heart of the matter", and efforts to render it explicit, effectively only introduce perturbation and fragmentation. (Note that non–directiveness, being the non–imposition of a line of discussion, is only loosely related to indirection in this sense).
2. Paradox: There usually seems to be a strong element of paradox in such cases, or at least a tolerance of it and a suspension of judgement. (The meeting could almost be considered a collective reflection on a Zen koan).
3. Incompatibility: Associated with paradox is a context which permits incompatible perspectives to be "bracketed" and held in complementary juxtaposition. It is the shared attitude underlying this contextual awareness which provides a subtle interface between the perspectives.
4. Attunement: The magic tends to occur when participants are attuned to each other or empathize with each other, possibly stimulated by a quota of antipathy which provokes a search for a more fundamental level of harmony (cf the use of this concept in certain group meditation techniques).
5. "Chemistry": As in the previous point, when the right mix of participants is present, they react in unpredictable ways to produce interesting transformation for all concerned. (The "recipe" analogy may also be used).
6. Aesthetic elegance: There seems to be a special economy and proportion of structure and process which can only be described in aesthetic terms.
7. Drama: Relating to the previous point, there is often a sense of evolving and mounting drama, engendering appropriate events at each stage. There is a collective awareness of how each event is charged with significance.
8. "Invisible hand": Relating to the previous point, at certain moments events seem to be guided by an unseen hand, so well do they emerge spontaneously and fall into place unplanned. There is a strange "rightness" to the flow of events.
9. Non–action: During the course of such meetings, deliberate actions usually tend to be of less significance or else their significance emerges totally transformed in relation to the original intent. The more participants can approximate to the Taoist attitude of non–action, the better the event for all concerned (cf the adage: "Don't push the river. Guide the canoe").
10. Non–conscious: Relating to the previous point, participant appreciation of the event depends on ability to "let go" and "flow with the stream of things". This seems to call upon instinctual and intuitive aspects of personality, appropriately blended by the participant (cf the Japanese concept of hara). It should perhaps be contrasted with unconsciousness and "stream of consciousness" monologue.
11. Humorously quixotic: In contrast to the heavy quality of conventional meetings, such events have an underlying thread of humour strangely blended with wisdom (cf the Sufi tales of Nasruddin). This also serves as a very powerful and rapid means of conveying an explanation.
12. Innocence: The flow of such events tends to evoke a childlike innocence and sense of wonder in participants, which is to be contrasted in conventional meetings with the defensive attitude towards ignorance, a pervasive cynicism, and childishness under certain circumstances.
13. Magical shifts of perspective: Characteristically in such meetings, apparently insignificant events brought about in an unforeseen manner can trigger major shifts of perspective (cf the Zen tales concerning achievement of satori).
Related Meeting focal processes (#TC1351).

♦ TC1568 Conference papers
Description Written contributions, which can be read, presented, passed out to each participant or simply made available.
Advantages Written work can generally be clearly researched, thought through and presented. It can stimulate discussion and generate other "papers" in reply. They can be used very well in the seminar process using the structure mapping method which enables very effective small group study.
Disadvantages Can be a substitute for discussion and tend to bury valuable insights under a mass of words. When read out they tend to be boring and since they have been prepared in advance they can tend to be out of touch with the flow of the conference. When not read out they are often stored or discarded unread.

♦ TC1574 Judicial conferencing
Description Exposure and consciousness raising about injustice and the denial or suppression of legitimate rights form the main part of this form of conferencing. Focus is given to the identification of injustice and new definitions of rights, often taking the form of a published declaration.

♦ TC1581 Nonchalent participant type
Playboy
Description Conference participants who make a display of their lack of involvement through open cynicism, nonchalance, horseplay, and other kinds of studied inappropriate behaviour.
Advantages A catalytic challenge to all the other members of the groups they are in; they are a reminder of the real world.
Disadvantages They can distract a lot of attention in their direction which rarely has any effect on their level of involvement and can frustrate the others.

♦ TC1587 Coercive power
Description Based on the application or threat of application of punitive sanctions upon the participants: such as the infliction of pain, deformity or death; restriction of movement, food, comfort, etc.
Advantages Very effective in controlling and directing behaviour in the short term. Particularly useful with "alienative" hostile participants.
Disadvantages Very high investment of time and energy. Least effective in controlling thought –it encourages alienative hostile involvement. Discourages creative input, adaptation, initiative.

♦ TC1588 Conference intent
Description Central to the process of conference design is deciding what the event is about. Not only must the organizers decide what the point or points of the programme are, they must convey that understanding to potential participants. While this may seem extremely basic it is the cause of great problems at many events. Participants and organizers have expectations about what will happen at the event and what they will experience. When these expectations are not met departures, dissatisfaction or rebellion are likely to result.
When the organizers have thought through the intent of the programme, they then can modify the process of the conference in the midst of it. When the intent is clearly understood by the organizers and the participants, then self–conscious decisions can be made. Changing the intent is possible if necessary.
The intent of a conference is two fold: the rational objective and the existential aim.
As a programme is developed, the organizers may write several versions of the existential aim and rational objective. Usually this is in fact refining the initial concept. Occasionally quite dramatic changes in understanding can occur that are unpredictable. When a dramatic change occurs the whole design of the event should be reviewed in the light of change. This may change the flow, content, and timing of events during the programme. The targeted audience of the conference may be expanded or changed.
The organizers most directly designing the event begin by creating for themselves a reason for the event. They then write a prospectus of the event which includes an overview of the intent, the process, the audience, and the intended products and experience of participants. It is shared

with the staff of the organizing group that are not designing the event, with sponsors and potential participants. In the process of making these presentations, new and more significant reasons become clear. While feedback from the groups addressed is quite useful, more important is the clarity gained in presenting it.
As the process of designing the intent of the conference is being done, potential participants are considered. The organizers hold a workshop about who may come and what they might expect at the event. What will their mood be ? What will they struggle over in the event ? What issues will they bring to the event ? Which of these issues are relevant to the programme and which are not helpful to the programme ? How can the relevant issues be rendered explicit the table and how can the unhelpful ones be avoided ?
There is the possibility of the intent of the conference changing during the event itself. If this is to occur with some creativity, then it is best for all the participants and organizers to be a part of the decision. This type of change is difficult to predict and even more difficult to make. Often this type of change begins with a dissident individual or group that creates a movement to restructure the direction of the programme. The organizers need to be extremely sensitive to this possibility. The disruption caused by such an attempt during the programme can result in the group abandoning the original intent while replacing it with no new one. There is also the danger of not actually changing the intent but simply broadening it to include a greater diversity of intents. This is usually redefining the intent at a higher level of abstraction. A highly abstract rational objective and existential aim may render the conference meaningless and create a majority of dissatisfied participants.

♦ TC1593 Special interest groups

Description Groups that meet in a conference situation that are made up of individuals who have a specific interest or position, which may but probably will not have any direct connections with the flow and focus of the conference as a whole.
Advantages Conferences are rarely homogeneous. As part of the integrating process, special interest groups can be planned into the design of the conference and their deliberations included in the overall process. Special interest groups that meet after, or toward the end of a conference can include in their deliberations the input of the conference, as they relate to their specific situation.
Disadvantages Where the special groups are not formalized the mechanism for integrating their input into the conference may be difficult. There is the danger of splinter groups emerging that schedule meetings that conflict with the other events in the programme. Allocating time for the groups to meet can disrupt the flow of the conference, and there will be some participants who are not part of any of the special groups. Different groups will need different amounts of time to complete their business. If they are not well integrated into the conference they can be distracting and disruptive.

♦ TC1599 Repetition of learning cycles

Description In many social domains, reflected in conferences, time and a variety of collective experiences have created amongst those concerned an awareness of which actions are feasible, viable and useful and which are not. Such collective learning is difficult to transfer to others in such a manner as to enable them to understand the (usually relatively sophisticated) dynamics which limit the value of seemingly obvious positive actions. Since there is a certain turnover of organizations, groups and individuals concerned with the problem in that domain, and represented at relevant conferences, those entering a meeting for the first time tend to initiate proposals, recommendations and programmes which past experience has shown to be a waste of resources or of otherwise limited value. They will however have difficulty in recognizing this and will attribute past failure to ineffectiveness of those involved at that time.
The consequence is that any group (possibly of institutions) with experience extending over several "programme generations" always has latecomers who are drawn together at a meeting in support of projects which constitute the repetition of a learning cycle. Such cycles must play themselves out in order that the latecomers may acquire the understanding as to why those particular actions are of limited effectiveness. They will however then be repeated when the number of newcomers again becomes great enough to make it difficult to redirect their attention during a meeting from such seemingly obvious courses of action, particularly when the obvious courses attract good press coverage with its immediate pay-off.
This repeated fragementation of groups and the use of resources in support of ineffective programmes clearly limits the ability of meetings to respond adequately to any problem situation. It is also discouraging to those who have already acquired, through such learning cycles, the necessary knowledge base from which more effective programmes could be designed. However, it is also the desire of the latecomers to apply their creative energies without regard for past experience which leads to the acquisition of new knowledge. The situation is such that it is seldom possible to blend both forms of knowledge in a meeting in an effective response to the problem situation.
Related Alienation of committed activists (#TC1962).

♦ TC1602 Self-reflective conferencing

Group-counsel conferencing — Inter-group therapy — Affinity group conferencing
Description A meeting or a part of a meeting when the different groupings that are identifiable at the conference attempt to explore and to deal creatively with the different expectations, frustrations that each group has in relationship to the others. The process can take the form of a group therapy session, except that instead of individual participants speaking for themselves, the various groupings congregate in different places around the hall and make their contributions as a group. There are many kinds of questions which can open up "therapeutic dialogue" between the groups, for example: (a) What are our expectations for this conference ? (b) How do we perceive the qualities and the role of our grouping furthering the conference process ? (c) How de we perceive the qualities and the roles of the other groupings in furthering the process ? (d) How do we perceive other groupings hindering the process ? (e) How can we move towards a resolution of the conflicts and the realization of all our expectations ? Responses to these and similar questions can be worked on by the groupings and then shared with each other in the open assembly. The conference participants can be grouped in different ways and participants can be left free to identify themselves with any of the groupings -often having to choose between two and three appropriate options. The groupings may be formal categories or informal, affinity groups, for example; (a) Original organizing group (b) Structure-oriented group (i.e. favouring adherence to a predetermined programme, with emphasis on lectures and workshops by key resource people) (c) Process-oriented group (i.e. favouring flexibility, with emphasis on all participants as resource people) (d) "Super-class" resource people (i.e. those who participated with the intention of giving a lecture) (e) Lecture attenders (i.e. those specially in favour of lectures by key resource people) (f) Workshop attenders (i.e. those specially in favour of workshops) (g) Detached observers (i.e. those uncommitted to the ends of the conference) (h) Floaters (i.e. those drawn to a variety of experiences) (i) Visionary instigators (i.e. the group concerned to ensure that something new and significant emerges from the conference) (j) Intellectual modellers (i.e. those intent on the possibilities and fruits of conceptual synthesis) (k) "Action-now" group (i.e. those wanting to act immediately and to stop talking) (l) "Here-and-now" group (i.e. those impressed by the immediacy and "rightness" of the present and the lack of pressure to act) (m) Artists and visualizers (n) Educator group.

Advantages High-risk manoeuvre but one that has enormous potential for individual organizational transformation. Participants feel themselves involved in an unprecedented way. Groups become aware of themselves and of others in a way that increases mutual understanding and trust. The burden for the "event" of the conference is taken on by all the participants within their affinity groups.
Disadvantages Takes an unpredictable period of time without a guarantee of a creative outcome since it is completely dependent on the participants and the interaction between them. Can be threatening and frustrating to participants who prefer the role of consumers at a conference.

♦ TC1603 Metaphors of meetings

Pattern language for participants — Configurative models of meetings — Meeting analogies
Description In a complex gathering people need to have some image through which to make sense of the event as a whole and of where it is going, and to help them to decide on how to participate in it. Whatever the images used they are needed to give a sense of continuity and context. Different people prefer one or more different images. It is easy to get locked into a conventional pattern of reflection about meetings. This blocks the opportunity offered by many metaphors and analogies to highlight alternative or complementary perspectives. These can be useful in suggesting more fruitful approaches, if only under special circumstances. Focus emerges as a consequence of an appropriate configuration of perspectives, people or groups within a meeting. To assist the exploration of the possibilities associated with configuration, it is also appropriate to note different kinds of configuration in use in other domains in the hope that they may offer clues to its significance in meetings.
It is useful to look at these as a kind of pattern language for participants. "Pattern" is a suggestive general term to describe any particular (and usually familiar) way of organizing the flow of energies in a gathering. Patterns can be combined into a network within a "pattern language". Some of the resulting arrangements are "better" than others, and the challenge is to find arrangements which enhance the hidden quality which makes them "feel right" in a given set of circumstances.
1. Macro-patterns include: Conference, fair, market/bazaar, agora/forum, symposium, workshop, demonstration, drama show, reception, exhibition, court, festival, lecture, pilgrimage, passion play, ceremony/ritual, panel session, sharing, brainstorming, songfest, games, holiday camp, contest, public blessing, celebration, discussion, group meditation, carnival, show/music hall, majlis, dance, happening, procession, retreat, audio-visual.
2. Micro-patterns include: Talking to speaker, speaking to group, sharing with another, protesting, learning, coffee table discussion, swapping information, lobbying/persuading, having fun, changing, distributing papers, receiving documents, show and tell, meeting new people, non-verbal experience.
3. Role patterns: Many of the above patterns are "activated" only by the presence of people playing appropriate roles. People may take up these roles irrespective of the formal reason for their participation in the gathering and their performance may be more significant for the gathering than their concerns. These roles may in fact be considered as sub-patterns in their own right. They include: Speaker, listener, jester, facilitator, writer, therapist, devil's advocate, priest, sympathizer, strategist, rapporteur, interpreter, musician, creative artist, performer, "accompanying person", game organizer, child, ego stroker, agent provocateur, improviser, note-taker, critic, organizer, lobbyist, caterer, adviser, old person, fixer, presenter, animator, super-star, wise person, networker, mediator, handicapped, fan, appreciator, material arranger, discussant, ritualist, chairperson, security person, helper.
4. Pattern concerns: People participate in events because of "concerns" which they wish in some way to advance or promote. These concerns colour the energy content of the patterns through which they are expressed:
(a) Theoretical concerns, as represented by the intellectual disciplines of which, ungrouped, there are some 1,800.
(b) Substantive concerns, namely societal problems and conditions, typically including: population, inflation, unemployment, refugees, energy, environment, illiteracy, human rights.
(c) Aesthetic concerns, especially their expression and involving others in that expression: music, song, poetry, art, theatre, dance, textures, perfumes.
(d) Intangible experiential concerns: prayer, meditation, power, humour, risk, renewal, ego trip, other negative values, other positive values.
Narrower Meetings as social activities (#TC1789)
Meetings as physical processes (#TC1367)
Meetings as games and contests (#TC1152)
Meetings as physical constructs (#TC1689)
Meetings as symbolic configurations (#TC1909)
Meetings as psycho-physical processes (#TC1922)
Meetings through analytical frameworks (#TC1488)
Meetings as agriculture and food processing (#TC1529)
Meetings as ecological and natural processes (#TC1475)
Meetings as energy-processing configurations (#TC1465).
Related Interdisciplinarity (#TC1208) Meetings as models of reality (#TC1727).

♦ TC1613 Participant expectations

Description People attend conferences for many reasons that have very little to do with the aims and objectives of the organisers. Those personal reasons are important to the participants and it is worthwhile bearing them in mind.
There are practical benefits, including the possibility of making contacts or meeting people who could further a career; they may be able to link into a new network and may be able to advance a project to "sell" themselves, their ideas and their concepts.
There are also psychological benefits. Conferences can cultivate a sense of belonging. Participants can be part of a group that others are not part of, and this can lead to a rise in prestige, as well as admiration and respect from those who were not there. They can feel acknowledged, having had their voice heard, having been noticed.
There are sociological benefits; a conference is a chance to assume and to demonstrate authority, power and responsibility.
Other benefits include the interest generated by the possibility of travel, meeting new people with similar interests who may become friends; catching up on the latest information and gossip; or simply the opportunity of a refreshing interruption to daily usual routine.

♦ TC1615 Conferencing inertia

Description Inertia is the force of inaction. It is the tendency of a group of individuals or organizations to avoid change and to continue on with more of the same. It is the difficulty of a large system to re-act to new stimuli, new ideas and new challenges.
Advantages Protection against opportunistic and rash activity. Inertia is a force that can demand the utmost creativity from the change-agent who must seek out and exert pressure on the sensitive nodes in the system.
Disadvantages It appears to make evolutionary organic change impossible, demanding rather, that an equal force and mass be exerted in a wasteful destructive and cataclysmic clash. If the sensitive pressure point cannot be reached, inertia can be very costly in terms of discouragement - without being anyone's fault.

♦ **TC1616 In-conference computer contact system**
Description Computerized registration systems that can put participants in contact with the conference environment and with what is going on, and with the organizers and other participants. Forms can be filled in that record relevant items of identification and preference which can then be used as a basis for distributing messages of particular relevance to particular groups of people, as well as a basis for arranging for meetings with people of similar interests. Computer terminals strategically located in all parts of the building can be used to update and maintain continuous communication. Messages can be exchanged generally or specifically addressed to other participants and "pigeon-holed" for collection.
Advantages The possibility of the exchange of messages of "serious" or "frivolous" nature increases the fluidity of any occasion allowing for more spontaneity and flexibility. Up to date reports of proceedings in different groups within the conference can be available, and allow participants to choose where they get involved and to communicate something when they are not able to attend in person. Those who type in an "identifier" can receive messages and thereby speed up the process of making contact with people who share their interests.
Disadvantages Cost (though this can be covered by participants paying for facilities they use) can become a distracting fad, appearing to be gimmicky. Some participants are put off by high technology gadgets and the system is very dependent on the style of assistants and the level of enthusiasm shown for putting in material. The system is open to abuse either casual or imaginatively destructive.

♦ **TC1623 Conference autonomy**
Description The degree of independence in which a conference is held. It is reflected in the degree to which the conference can determine its own activities, process and structure, and in its absence of allegiance, deference and/or dependence relative to any particular groups.

♦ **TC1627 Decision-making process**
Description Decision making is the process of the participant or the organization the participant represents actually altering their behaviour as a result of the process of the conference. Proposals may be made, plans developed, resolutions passed or groups mobilized. Actions may be initiated, organized or coordinated.
The decision-making process of a conference is the means by which a gathering of people develop and select a common direction which implies some action for them. Some conferences are called purely for the purpose of decision-making, while others make decisions only passively. No matter how weak or how strong the emphasis, however, there is in every gathering a decisional aspect which cannot be ignored. The decision-making dynamic is holding a creative tension between participation and facilitation. The tension which occurs in decision-making is a tension between conference input and output. New input is from participants of diverse perspectives and frames of reference, while output is required of the facilitators in a coherent, unified body. When this tension loses its force, there may be false peace on the side of the participants or on the side of the facilitators, but it is then probable that nothing new will be created. When tension is removed by dominance of the participation dynamic, a conference becomes an airing of views without conclusion or focus, which is difficult to convey to non-participants and does little to change the thinking of anyone. When tension is removed by dominance of the facilitation dynamic, a conference is aimed toward a direction—perhaps adopted prior to the event itself—which participants may agree to or not.
In contrast can be a conference an occasion in which struggle takes place. Subtly or loudly there is a conflict the articulation of directions, suggestions, reports and proposals on the one hand, and the need to develop as detailed and focused a report as possible. It is always possible to determine the long-range effectiveness of a conference's decision-making by seeing how thoroughly the decisions made by the conference are implemented.

♦ **TC1630 Task force**
Description A small group with a specifically assigned task, usually of a practical nature and related to the "enablement" of the conference.
Advantages Useful for the many practical tasks associated with the organization of a conference. Preparing displays, re-organizing seating arrangements, collating papers and reports. The intense short-term and discontinuous nature of the assignment is good for the development of team spirit.
Disadvantages Too structured for those who believe in spontaneous voluntary responses to things that need doing.

♦ **TC1632 Hierarchical conferencing**
Description Highly structured meetings with clear, distinct hierarchy that is usually demonstrated by the mass seating arrangements, probably in rows, and elevated podium and an stage for speakers. Generally used in the following modes: a. Protocol and policy: These tend to involve a speech by an eminent person which participants must listen to, either as a gesture of respect, or for reasons of protocol, or as a matter of good public relations, or because it may outline new policies for the first time.
b. Exhortative: These tend to involve a speech by a respected person exhorting participants to some new effort, namely a speech by a skilled orator conceived as a means of arousing enthusiasm or of changing beliefs in support of some new action.
c. Information: These tend to involve a speech by some technically competent person in which new facts are presented, or the results of programmes, or a detailed outline of new programmes.
d. Administration: These tend to involve the presentation of annual or financial reports, election of officers, etc.
Advantages These include the absence of restriction on the number of participants; the ability for those organizing the meeting to inform large numbers of some current situation; and the ability of participants to hear the views of individuals who would otherwise be inaccessible to them.
Disadvantages These include the restriction on participant expression; the suppression of viewpoints not in accord with those of the organizers of the meeting, or at least not envisaged within the programme framework; and the channeling of participant expression via the podium rather than directly between participants.

♦ **TC1636 Conference environment**
Description The environment of a conference has to do with the size, coloring, acoustics, arrangement and decor of meeting spaces. There is not one "ideal" environment, nor is this ultimately answered by providing higher quality appointments and more beautiful space. The environment needs to be considered in relation to the group and the existential and rational intents of the event. Consider first what kind of environment is needed for the event to be created. Then consider the space available and devise how to bridge the gaps. Decor is important. It should not be automatically assume that hunting prints or whatever is already on the walls is going to be helpful to the event. Consideration should be given to what photographs, brochures, symbols, banners or even displays of work in progress could help conference participants to make tangible the conversations they are having ?

♦ **TC1637 Accident tactics**
Description The lights fail -there has been a power cut and the generator isn't working. A miscalculation means that there isn't enough food for everyone, and there aren't enough chairs. The special visit has to be cancelled at the last minute. The keynote speaker is ill. The conference centre is suddenly completely shut off by snow, floods, a terrorist action or... Accidents, major or minor, precipitate crisis within the conference and crisis generate a response from the participants. It is always revealing to observe the ways that different people respond in crisis. Some people simply wait for the organizing committee to deal with the problem. Others complain and grow increasingly frustrated and angry with the incompetence of those who are trying to do something. Others drift off and begin doing something else like playing cards. Others act as though they are panicking, rushing around and causing confusion with frantic and contradictory immediate solutions. Others calmly appraise the situation and formulate plans that will deal with it. Some people organize the waiting, anxious and angry participants into alternative and temporary structures that allow a sense of order within the chaos and enable an effective flow of information to and from the participants. Some people simply observe the ways that different people respond in crisis. While, by definition, accidents, chaos and crisis are unexpected and unplanned, it is sometimes tempting to arrange for crisis, because of the catalytic effect that it has on all the participants.
Advantages Crisis and chaos produce a highrisk situation. Shared risk generates trust. And greater trust encourages further risking. Even a small crisis can be recognized as a metaphor of our global historical situation and reflecting on our ability to handle it reveals insights about our ability to handle chaos as a planet. The immediate chaos shared crisis brings people together. Strategically used, an accident can catalyse individual and group creativity and transformation.
Disadvantages Unpredictable and necessarily risky. Can catalyse a negative reaction. Participants may experience the "accident" as negligence and proof of the worthlessness of the conference, or even as a manipulative gimmick. Although the anger and frustration are often important dynamics they can be overwhelming and simply destructive. Can provide an excuse for returning to "safe" controlled conferencing.
References C0380 Failure tactics

♦ **TC1640 Non-verbal interaction preferences**
Description Participants may have distinct preferences amongst the following primarily non-verbal modes:
1. Physical sharing: Feasting/drinking, dance, physical games, group exercises.
2. Emotional sharing: Drama, song, music, group empathy exercises.
3. Intellectual sharing: Conceptual "resonance" of participants ("on the same wavelength"), usually stimulated by occasional words; drama, music.
4. Status affirming: Actions which reinforce the importance of a participant and of those who articulate the beliefs or doctrines he shares.
5. Communal celebration: Partially ritualized collective affirmation of values, and renewal of participant belief therein.
6. Action: Shared work, whether constructive or destructive.
Broader Participant interaction preferences (#TC1845).

♦ **TC1642 Team evaluator role**
Hard-head — Monitor
Description An affinity-role played by one or more members of a small group or team. Typically, rather sober, prudent and unemotional. One who analyses problems, evaluates ideas, suggestions, and proposals, so that the team can formulate more balanced decisions.
Advantages Judgement, discretion, hard headed, acts as a check and even a challenge to the input of others. Can in some cases intuitively recognize whether the contributor's enthusiasm is behind the suggestion or proposal.
Disadvantages Can be a demotivating role, which does not inspire or motivate the group in its task.

♦ **TC1644 Conference symbols**
Description The symbols of an event are important and can be self-consciously developed. Symbols are either re-created or more difficultly created during the development and implementation of a conference. The development of new symbols requires a great deal of sensitivity and artistic ability. Most successfully utilized symbols are old ones that have been reinvested with meaning; that is, a new understanding of their significance has been created and interiorized by the organizers and participants.

♦ **TC1650 Zetetic panel**
Interrogator panel — Responsive panel — Interrogation-discussion — Interchange panel — Investigative panel
Description A group of three to six resource persons who are questioned by one or more "enquirers" in the presence of a moderator and observed by an audience. The interchange is conversational but the answers are intended to be precise and to the point. Zetetic literally means "proceeding by enquiry" and this process is best used to obtain specific details and particular information from those who have special knowledge and experience.
Advantages This is a variation on the panel discussion and is a most effective use of "experts" and resource people. It is often a preferable alternative to a series of lectures. Advantages include:
1) Many questions can be covered in a short period.
2) The interaction between the panel and enquirers should lead to a full development of the subject.
3) A high interest level is maintained.
4) Specificity and appropriateness of questions and answers is at a high level.
5) Many facets of a problem can be explored simultaneously.
6) There is a high probability of the emergence of a synthesis of ideas and concepts.
7) The process is particularly valuable in conjunction with other techniques that involve the total audience in the formation of the questions and in the contributions. (see Buzz groups)
Disadvantages The process is time consuming. The process depends heavily on the approach of the "enquirers" who can misinterpret their role, and set logical traps or seek to clinch debating points, make speeches, or simply distract the panel by erratic lines of enquiry. The panelists may find themselves under pressure and become flustered by the rapid interchange of ideas. Not all people "think on their feet" with the same alacrity and weak members of the panel can become overshadowed and valuable contributions lost which might have emerged in another format.

♦ **TC1651 Followers participant type**
Sheep — Conference fodder — Party-liners — Sycophants — "Yes" person
Description Conference participants who more or less passively go along with the movement of the group, accepting the ideas of others.
Advantages They serve as an audience for group discussion and decision- making.
Disadvantages They rarely risk creative contributions of their own, when obliged to contribute they will rephrase a previous offering or produce something "immune" from contradiction.

♦ TC1657 Process-oriented participant type
Description Conference participants who mistrust structures and procedures, and believe that the process of the conference can and should emerge spontaneously and be the focus of the event.
Advantages Openness to change and to personal and group transformation, they don't simply attend the conference but become the conference. Unexpected juxtaposition of people and concepts can generate unexpected insights and creativity. A useful balance to structurists who are threatened by the apparent chaos of unstructured meetings.
Disadvantages Can be diffuse and unable to focus. A large high risk investment of time and energy in a project that while it may be something significant tends not to achieve significant results.

♦ TC1663 Non-conformist participant type
Idiosyncratics — Antipathetics — Disagreers
Description Participants who display individuality when the individual view is considered of less value than that of the group, or established doctrine. They are neither part of the trend or of the establishment. They are unpredictable and reasons for their behaviour are unknown and suspected of being entirely personal. They tend to be unpopular, to have little status and to have competing loyalties and affiliations. They show a marked ability to take the opposite point of view and to be apparently unable to understand or sympathize with concepts or feelings of others.
Advantages Can play an important role in stimulating unifying group responses to their idiosyncratic presence. Can perhaps fulfil the function of the random mutation element that is critical for evolution.
Disadvantages Can be seriously distracting and consume an inordinate amount of time.

♦ TC1666 Computer teleconferencing
Description Meetings involving geographically separated participants linked through computer terminals in a telephone or data network.
Advantages Flexibility. The storage and retrieval capacity of most computer terminals allows for intermittent and uncoordinated participation as well as simultaneous interaction when appropriate.
Disadvantages There are still relatively few people who are "at home" at a computer keyboard and the lack of standardization and limits of compatibility put constraints on this form of communication as a medium for conferencing.

♦ TC1683 Conference homogeneity
Description The degree to which conference participants are similar with respect to socially relevant characteristics. It is reflected in the relative uniformity of members with respect to age, sex, race, socio-economic status, interests, attitudes and habits.

♦ TC1684 Problem solving conferencing
Description Meetings that concentrate on the solving of certain problems, rather than pursuing new objectives.
Advantages Problems tend to be real issues and in response to reality many creative proposals can be formulated which are in fact new technological or social inventions.
Disadvantages Focusing on problems can tend to obscure new objectives by simply attempting to deal with issues that hinder realization of old objectives. Many of the "problems" groups or individuals are perceived to have, are actually solutions to other unseen problems that they experience. There is always the danger of "solving" surface symptoms, which tend to escalate if the underlying issue is left untouched. It is often harder to solve a problem deliberately created by others than it is for them to find a way round the problem created by the solution.

♦ TC1689 Meetings as physical constructs
Description The following may be used as metaphors to explore alternative ways of understanding meetings:
1. **Architecture:** The structural and functional divisions of a meeting may be viewed in terms of architectural analogues, from the simple one-room hut to the complex cathedral, fortress or palace. This raises questions of design and practicality of layout.
2. **Tensegrity:** This recent advance in architectural possibilities (and the basis of the geodesic dome) suggests new ways of balancing configurations of opposing forces in a meeting.
3. **Circuits:** The variety of components in electric, electronic and fluidic circuits suggest ways of combining well differentiated modes of participant information processing.
4. **Information processing device:** The whole gathering may be interpreted as a complex bio-mechanical computer processing different types of information, storing it, and forming it into various images of the whole possibly with some final output.
Broader Metaphors of meetings (#TC1603).

♦ TC1691 Flipchart
Noteboard
Description An easel that holds a "pad" of large sheets of paper that can be drawn on with markers, and folded back or torn off and displayed somewhere else in the room; diagrams and text can also be prepared in advance and "flipped" over at the appropriate points in the presentation or discussion.
Advantages Low-cost, low technology equipment that is very available to all users. Diagrams and drawings made during a discussion can be easily retained, displayed and conveyed to other meetings.
Disadvantages Dependent on the skill and neatness of the user. If used during presentations, it can enhance the level of communication but may distract both speaker and listeners if they become too involved in the drawings and neglect the direct communication.

♦ TC1696 Service conferencing
Description Prime beneficiaries are those who require and are offered the services. It is characterized by the high level of participation and input by client-participants and the high level of receptivity of organizers and conference leadership in response to this input.
Advantages Useful mode for service organizations, social work agencies, hospitals, schools, legal aid societies, mental health clinics, etc. which have a particular client group.
Disadvantages Encourages a dependency relationship as clients voice their needs and anticipate a response. The problem of the gap between the "felt needs" and proposals of the client group and the actual possibility of the service group practically meeting those needs. The problem of the different perception of actual needs leading to mistrust and frustration.

♦ TC1699 Meeting energy dissipation
Unconstrained meeting configurations
Description In organizing a meeting there is concern that it should be sufficiently "stimulating" to attract and maintain the interest of participants. There is however also a concern that any "controversy" should not exceed what can be contained by the meeting structure and processes. A low risk meeting therefore runs the risk of being boring and without significance. The question is whether this dilemma can be understood in a new light in order to be able to organize interesting and significant meetings, whilst minimizing the risk of their being torn apart.

As one extreme case, what needs to be done to avoid all controversy. The relationships between the participants, the topics or the meeting sessions need to be such that only supportive, reinforcing information is exchanged between them but none which challenges, denies, accuses, limits or questions assumptions. If any such challenges are effectively transferred to the relationship between the meeting and the external world, the meeting can maintain its positive harmonious nature. This could be called exporting or projecting problems, inconsistencies or contradictions. For this to be possible however, no effective link should be established between those participants, topics or meeting sessions which would draw attention to such contradictions by the nature of their interaction. This can best be illustrated by a grid, reflecting (according to its size) the variety of participants, topics or meeting sessions. In it supportive information of one kind is transferred from point to point along grid lines. Only by confronting information from distant points which is avoided in the meeting, would the challenge they constitute be evident. In the meeting the challenge between them (at any particular grid location) is minimized. Expressed differently, every effort is made to ensure that feed back loops are not completed. Or alternatively, the meeting is perceived as a grid on an infinite plane.

This approach ensures that energy is effectively drained into or absorbed by the meeting environment. There it merely goes to reinforce any positive or negative images of external problems or organizations. It does not enhance the ability of the meeting to get to grips with such problems or its own. The meeting is essentially escapist, dumping its own problems on the environment. A grid configuration is a de-motivating, energy-dissipating pattern, not an energy conserving pattern. For this reason care should be taken when basing meetings on linear agendas, coding or classification schemes.

At the opposite extreme in which conflict is internalized and challenge is accepted as an integral feature of the meeting. If the meeting is not to be torn apart, the opposing participants, viewpoints or meeting sessions need to be held in relation to one other by a configuration which distributes the stress evenly throughout it. This calls for the completion of all feedback circuits and the juxtaposition of integrative (harmonizing) and dissipative (challenging) forces at every point throughout the configuration. Such configurations are not constrained by the environment, as in the previous case. They are selfconstraining. Energy is not dissipated; it is conserved as synergy. It is such self-constraining patterns of curvature which provide the focus which is absent in a "planar" meeting. The question is how to "foldup" a grid into an appropriate configuration.
Related Integrative failure (#TC1005)   Communication patterns (#TC1943)
Losing focus in meetings (#TC1432).

♦ TC1704 Rational objective
Description The rational objective is first, the objective intent of the programme. It is a description of what is to be accomplished. It is deciding whether the conference will be basically decision making, experiential or informative. It is deciding the breadth or focus of the topics being covered.
It is, second, deciding what the end product of the event will be. Is the product a resolution, a more understanding group, or a position paper ? Is it renewed friendships or motivated members of a movement ? The more concrete the product is for organizers and participants the greater the possibility of actually reaching it.
The third part of creating the rational objective is deciding what broad steps will accomplish the intent. These broad steps of the event are thought through and articulated.
When no rational objective is decided either before or during the first few hours of the event, there are at least two potential problems. The participants and organizers alike may experience a great deal of anxiety about the overall direction. The lack of a framework for what the conference will and will not cover encourages irrelevant diversions in the discussions.

♦ TC1717 Conference structure
Description While the decision-making process and the intellectual momentum of a conference give it meaning, the structure gives it life. The methods used by the organizers to run the conference make the difference between the possibility of transformational events and the dull rehearsal of dull ideas. The practical aspects of the conference mean the difference between constant distraction over irritants and focused attention on creativity. Leadership organization means the difference between facilitated participation and wasted time.
Many conferences are accused of having too much structure but in fact the most common problem is not too much structure, but misdirected or dysfunctional structure or an inappropriate style of facilitation. Few people would accuse a football game of having too much organization. Participation, creativity and passion are all possible. Nevertheless, it is an intricately structured event.
If the conference's structure does not move the group toward what they believe to be the intent of the event then it may be seen as overly structured. This is especially true if the content is appropriate but there is no perceived progress toward the goals. There is a growing value in people participating in the decision making process of the event. If participants are frustrated by the process and feel that little can be done to effect the course of the programme, then it may be accused of excessive structure. Many participants and the organizers have multiple reasons for being involved in the conference and if these are not allowed for, there may be complaints.

♦ TC1718 Cost clock
Description A special type of clock that records the cost of a meeting in terms of travel and other expenses, and person-hours, as it mounts minute by minute.
Advantages A silent reminder to the participants of the investment involved which raises the question of whether the time is being used in ways that justify the expenditure. This can increase the pressure on the participants to take responsibility for the proceedings and to become involved in controlling and guiding them. Best when not taken too seriously.
Disadvantages Increased pressure of costs and time can be counterproductive when sensitive and difficult issues require unhurried deliberations and some participants become moralistic about meetings taking too long.

♦ TC1721 Trust working
Trust building activities
Description Exercises, games and techniques that are used in a conference situation to raise the level of interpersonal trust amongst the participants. There are an enormous range of trust working methods, most of them based on the circular relationship between trust and risk. Trust-building activities emphasize the development of a supportive climate and norms; exploring perceived similarities in attitude and experiences; the development of a shared identity that implies substantial direction by the participants themselves; and a gradual release of undisclosed information that is easily available.
Advantages Can create an atmosphere that encourages participants to risk their contribution. Generally the processes are gradual and controlled by the individuals in the group and have less of a "casualty rate" than the riskworking activities.
Disadvantages Time-consuming and if not introduced and given an appropriate context, they can appear to be irrelevant and confusing. Less of a break in routine, they can fail to make a significant impression.

TC1722

♦ TC1722 Encounter group conferencing
Description An intensive experiential learning process participated in by groups that are seeking enrichment, positive learning and growth, (rather than recovery from dysfunction, or to establish intergroup contracts or networks). It is generally a temporary collection of groups, with a planned time framework that stresses inter-group and intra-group enquiry. The process enables the expression and exploration of motivations, perceptions and interpretations, and a reflection on the difficulties, processes, effects and meaning of such sharing. The encounter can occur on many levels including the experience of: (a) the sharing between the groups and what is communicated; (b) the process of exchange and how communication happens; (c) the meaning of the experience of connectedness; and (d) the awarenesss of values and qualities of the atmosphere within the encounter.

The organization of an encounter group conference varies with the number of groups involved. It is valuable to allow each group time to go through the process with each of the other groups in pairs or triads. Triads are best guided to choose two groups who will dialogue while observed by the third, which will join the interchange at some point with an intervention formulated from its perspective as an observer. Thus for a conference of two groups, each group would take its time as presenter and listener. With three groups, each group would play the observer in turn for the other two with a fourth session for reflecting on the event as a whole group. For eight groups which is about the maximum, the following combination can be used with groups labelled A,B,C,D,E,F,G,H.

Session I ABD/GCH/FE
Session II GFB/CDE/HA
Session III CAF/BEH/GD
Session IV HFD/GAE/BC

The typical process has three phases although it is without formal agenda or any predictable content. Phase I, engagement, is characterized by testing and trial, and in the opening of the dialogue much that is said is negative, even hostile and generally referring to some other time and place. Yet this very hostility indicates that the linking relationship has been initiated. Phase II, trust development and active exploration, is characterized by exchange that is a deeper exploration of themselves and each others group and can be described as the healing phase. Phase III, encounter and change. If Phase I is hostility, Phase II healing, Phase III has to do with helping. The exchange shifts in content and texture (and is recognized to have done so) towards deeper personal sharing –greater self-searching within the groups and noticeable change is seen in the attitudes, perceptions, interactions and behaviour of the groups and there is the emergence of a warm supportive and encouraging relationship between the groups.

Advantages It is a place to break new ground and it is almost impossible for a group to avoid trying out new and unfamiliar perspectives and linkages. Each group must explore in a kind of self-expressive inward search, stimulated and assisted by others. Other effects that can be anticipated are:
(a) an increased repertoire in communication skills, greater expressiveness, openness, empathy and sensitivity in relation to other groups.
(b) an increased sense of group "wholeness" and unity.
(c) a more differentiated and reflexive awareness within the group which will allow it to learn to move from subsequent encounters and become more adaptive in crisis or stress.
(d) the groups will be able to increase their connections with others.
(e) a change in the self-image of the group due to the feedback received and the experience of new behaviour. This tends to be an increased understanding of group-worth and self-esteem.
(f) a change in the process by which the group forms images of itself towards "discovering" itself in experience and away from depending on the established image which has to this point regulated awareness and action.
(g) the "ripple" effect, that can begin to link small organizations and groups in mutual awareness and intuitively complementary action and interwoven systems of change.

♦ TC1724 Multi-meetings
Description Conferences arranged by several related organizations with parallel or concurrent meetings to allow for interchange and joint sessions when appropriate, but otherwise to continue with their own separate agendas.
Advantages Participants with interests in other organizations can attend sessions within the different programme frameworks, saving on time, travel expenses, etc. Formative interchange can take place between the different organizations during the time they most need and are most open to input.
Disadvantages Can be distracting and confusing to have two or more parallel conferences and the particular focus of each of them can be lost.

♦ TC1727 Meetings as models of reality
Description It is normally assumed that meetings are either concerned with issues in wider society (the external world) or constitute an environment or vehicle for interaction between persons or viewpoints (or possibly a mixture of both). Both perspectives fail, in an important respect, to focus on the meeting itself. They treat the meeting as a vehicle or device but fail to consider the significance of the structures and processes constituting the vehicle, whether as a result of forces emerging within the meeting or during its planning stages.

Meetings may usefully be viewed as models of the reality of the forces and perspectives in the wider external society as comprehended within the meeting. This is only partly acknowledged in concern for the representivity of the meeting. This concern only reflects an awareness, from a particular perspective, of who or what should be represented at the meeting. The meeting structures and processes reflect more than the simple list of participants or themes, they reflect their possible relationship in the light of the constraints imposed on the meeting. As such they constitute a map of the external reality, significant in its own right and especially because of any detectable limitations. In a different sense a meeting also provides a convenient "surface" onto which concerns may be projected. As such, some meetings may be treated as new opportunities to redefine and concretize "the good, the true, and the beautiful", following the failures of previous attempts. The problems of the external world are also reflected in the decisions and compromises required to organize the meeting. Clear examples arise from policies (or their absence) on: handicapped participants, interpretation budgets, travel budgets, privileges, space and time constraints, use of recycled paper, etc.

Aside from such technical problems, the more fundamental societal problems can also emerge to some degree in embryonic form in the meeting environment, if only as analogues. Examples are: limitations on the human rights of participants; alienation, structural violence; problems arising from the multitude of participants each concerned o populate society with their particular perspectives; intellectual or emotional undernourishment of participants in the meeting process; problems associated with the different levels of education/experience of participants, and the constraints imposed by ever present ignorance; overconsumption and privileged use of resources. In each case the forces contributing to the problem may be observed.

Given the central role of meetings in society, they may also be seen as the focal point from which arise programmes, organizations, information systems (including periodicals, bibliographies, etc), and recognized problems. Such societal artifacts emerge, "peel off" and acquire separate identity, partly because of insensitivity to the significance of the meeting and avoidance of the issues it raises. In this sense such artifacts are an escape from the immediacy of the issues raised by the meeting and a delegation of action to others beyond the here-and-now. A loss of vitality and information content goes with this loss of immediacy.

Meetings also usefully model the capacity of those assembled to interweave their perspectives and skills within a viable whole – a whole capable of encompassing creatively the problems to which those same perspectives give rise. In this sense failure to bring about a new level of significance within the meeting is a strong indication of the limited relevance of the assembly to wider society.

Following from the previous point, meetings can be used by participants as a social micrococosm within which the significance of emergent insights can be tested. As such they are extremely valuable laboratories which have the immense advantage of being immediately accessible to those participating.

Related Metaphors of meetings (#TC1603).

♦ TC1733 Instrumental conferencing
Description Meetings that tend to be means to external ends, are oriented towards persons other than the participants, and are made up of activities that are designed to take effect at some later time. Meetings can be classified along an instrumental-expressive continuum.

♦ TC1735 Declaration conferencing
Description Declarations that are intended to collect and focus the ideas and consensus of the participants, are the end products of many conferences. These are carefully worded, (in advance, during or after the conference) by a small committee using actual or anticipated input from all those present.
Advantages A visible product to mark the event and to publicize the views arrived at. A device that focuses the creativity of the conference and gives a sense of purpose. The participants tend to "own" the declaration and it gives fuel to their enthusiasm and missionary zeal after the conference is over.
Disadvantages The preparation and discussion of such a document tends to consume considerable time. Little is heard of the declaration after the event and it seldom provides the platform for the collective action originally intended. The same is true for the written report or audio-visual record of such events. They seldom have the same power for others as they have for those who participated in their creation.

♦ TC1736 Detached-operator type
Description Conference participants who act on problems as though they are in some way white coated and hygienically detached from them. They tend to prefer statistical and theoretical analysis to self analysis.
Advantages Some detachment is necessary to acquire the freedom to act – and it can give enough distance to take an overview that forms a context for specific action.
Disadvantages Overdependence on theoretical and statistical analysis – dependent as it is on the particular bias of the research structure– can obscure the insights of intuitive analysis based on the recognition of their own involvement in the problems.

♦ TC1738 Conference insulation
Description The degree to which participants in a meeting are insulated from their communicative environment. Insulation does not imply isolation but rather that the channels of communication are interrupted by the introduction of something in between that slows down all or particular elements of the communication.
Advantages It is sometimes necessary to "distance" conference participants from the immediacy of an extremely pressing or urgent issue or a situation where the messages are very emotive or provocative, in order to allow them to deal with the issue with a sense of perspective and without distraction.
Disadvantages The risk of insulation is the decrease in sensitivity.

♦ TC1739 Less-haste participant type
Description Conference participants who point out that change requires time and who distrust immediate and impulsive action. They often claim that there has not been enough time to deal seriously with the issues and that more time must be given.
Advantages This perspective is a useful counterbalance to actionists who are frustrated by the delay in the implementation of action plans. They tend to be aware of the dangers and risks of impetuous action and tend to favour a more organic cultivation of change.
Disadvantages Can be advocates of procrastination and delay simply as an excuse for their own lack of results. They can seriously damage momentum and motivation that could have achieved much. They can be afraid of the pace of change and the risk involved.

♦ TC1742 Observation teams
Listening teams
Description Groupings within an audience that are assigned to watch out for special elements in the presentation and/or to assume a particular perspective in acting out their role as audience. Each group can receive different instructions; to record, for example, questions for clarification (like the audience reaction teams), points of disagreement they may have personally or in their assigned role, points that are new, surprising, particularly relevant, points that synthesize with other presentations in the conference, or points about the style and the means of presentation. Observations made by individuals in each of the teams can be written down and collected at the end of the presentation, briefly processed by a few members of each group and reported to the chairperson. Alternatively the observation teams can be divided into "buzz-groups" to synthesize their observations and to report them to the total meeting at the appropriate point in the discussion.
Advantages Particularly applicable to very large gatherings. They encourage audience involvement and "active listening", and help to ensure constructive input during the discussion period. Individual contributions undergo an initial selection and refinement process before being presented to the whole group, and the points that are made can cover a wide spectrum. If the conference is to divide up into smaller meeting spaces for the discussion, members of the different observation teams can be spread through different meetings after they have collected and processed their findings as teams. Clarity and appropriateness of the instructions is very important. The process can be very constructive and work well towards a synthesis between the different presentations at a conference, and will probably be welcomed by those who prefer clearly structured procedures which enable and encourage participation.
Disadvantages People who do not prefer structured processes could find this method manipulative. The special instructions and assignments could get in the way of their direct appreciation of the presentation, and they may feel that they were being programmed to receive the message in a prescribed way. Concentration on particular points may mean that other valuable points are missed.

♦ TC1743 Travelling microphone
Description A device used in meeting situations where there are no individual microphones for members of the audience, that enables the rapid delivery of a microphone to whoever wishes to speak. Methods include: carriers who take the microphone to the waiting participant, a micro-

phone carried on an overhead track framework which can reach all parts of the room. Hanging mikes in each section of the room which can be lowered and passed to the speaker.
Advantages Less time and continuity is lost if participants can speak from the floor and not from the front platform. Suspended microphones can be delivered more quickly and with less trouble. A relatively simple system for suspending and delivering the mikes can be set up before most meetings.
Disadvantages The delivery of the microphone almost always causes delay and distraction.

♦ **TC1748 Builder type**
Description Conference participants who feel that the overall product of the conference is of an importance that overides the necessity for any recognition of the importance of their own particular contribution. They work to build the consensus as though building a cathedral using as bricks the contributions of all the other participants, heedless of those who wish to keep their concepts clearly distinguished from the others'.
Advantages A very synergistic force within a conference that equalizes and yet values all contributions. An important counterbalance to bricksigners who are keen that their specific domain of insight be distinguished and given special recognition.
Disadvantages They can sometimes perhaps be building a consensus or a cathedral from inadequate materials and the construction is unstable and can rapidly deteriorate. The building depends on the contributions that are made towards its construction.

♦ **TC1749 Conference spirit**
Description The spirit of an event is not something that can be directed or controlled, and yet it is something that can be influenced and should be respected and worked on by the organizers.
The development of an effective conference spirit first requires some understanding of the process of creativity. Most people require some form of internally or externally imposed limits within which to work. Deadlines enhance output and quality of work for many. For example, in writing poetry the structure of a sonnet or an ode gives focus and direction to thinking. The process of developing a conference spirit requires paying attention to the development of limits which enhance creativity.
Secondly, conference orchestration requires conveying to the participants the fact that there is the possibility of creativity in every moment. This is best conveyed by the facilitators, and is a three-fold process; the first part is discerning the real situation. This is extremely difficult because most people prefer their own biases. The second is to discern the possible alternatives within that situation. The third is choosing the better alternative. This alternative may be the best of extremely bad alternatives or extremely good alternatives.
The third factor required to develop an effective conference spirit is that while people are always free to make choices most people most of the time would rather have someone impose decisions on them. Many people would rather someone make the decisions without seeming to impose. The decision-making component of a conference calls for clear contexting and careful procedures, because of this unconscious rejection of freedom and its responsibilities.
The conference spirit is the interrelated process of creating, symbolizing and interiorizing the story of significance of the conference, its topics and the role of the participants. Both the participants and the organizers place the event in the largest possible historical and social context. This context must be consistent with the experience of the event.
The creation of the spirit of the conference begins with the development of the intent of the conference. First, it is the articulation of the existential aim and rational objectives for the whole and each part. Second, the flow of each part and of the whole are tested against the phenomenology of the topic being discussed. Does the internal sequence of the event parallel people's experience of the topic ? Third, the spirit images of the event are developed. The spirit images include images of the opening talk, and any talks during the event given by the organizers. The themes of reflective conversations, exercises and celebrations are developed. Fourth, the motifs of the symbolic life (rituals, symbols and decor) are decided. Last, language and art forms are developed or found which are consistent with the existential aim.
Most dimensions of the spirit of conferencing presently operate on an unconscious level. Human potential movements and sales and advertising organizations are more attentive to them than other groups, but often out of a very narrow perspective. It is well to assume that there is always some mythology, some ritual and some symbol taking place in any human group. In coordinating the conference, the organizer's responsibility is to see what these factors are and to honour and use them for the sake of the intent of the conference as a whole.

♦ **TC1750 Information seeking participants**
Description Conference participants who tend to ask for more information, clarification of suggestions made, authoritative information, facts and figures.
Advantages Useful at the onset of discussion to bring out the objective data for examination by the group.
Disadvantages Can unnecessarily slow-up discussion by insistence on objective data. They tend to be rather unwilling to share their own personal experience.

♦ **TC1753 In-conference computer conferencing**
Description Miniconferences with up to 50 people on a special subject that take place on computer in parallel with the main sessions. Participants can register with the focal persons and can then obtain a lists of other participants, the current "agenda", and the text of any statements already made on this agenda. The participants are able to contribute generally or to specific people or to propose changes and additions to the agenda. An "editor" can periodically select and publish statement bulletins. While it is helpful, it is not strictly necessary for each participant to have his own terminal. As long as there is fairly easy access to terminals the miniconference can continue during and even after the main conference is over. In a conference where computer terminals are used for many different activities and where there may be more than one computer mini-conference going on, it is useful to designate certain terminals for the use of the "mini-conference" participants.
Advantages A very intensive medium of interchange can provide for a focus on issues of minority concern. The medium itself contributes, to the form and content of the messages exchanged, a new style which can be an experience of a new language with new metaphors and images. The potential for intercultural exchange is enormous.
Disadvantages Prone to misuse and abuse; expensive to set up and still rather exclusive of those who are suspicious or ignorant of computers. The lack of face to face communication introduces some constraints to trusting and risking.

♦ **TC1760 Conference relationships**
Description Conference relationships are not only the responsibility of the participants themselves. Consider in light of the intent of the event what procedure will be used for introducing people. Design an approach that fits the group and its work, but do not simply pass it over. Consider also the times and dynamics appropriate for people to make personal contacts. If only social occasions are provided for this, the assumption may be implicit that participants' relationships are basically social. Sometimes a time of interchange of work can be valuable for allowing people the opportunity not simply to know one another as individuals but to gain appreciation of one another's work.
Relationships are often enhanced when small groups of divergent interests work intensely on the resolution of a problem. Further, when one small group works as part of a team of such small groups, motivation can be intensified. Within the setting of the small group or team, structured conversations about rather tangential topics can deepen the human relationships within the group.

♦ **TC1765 Action-oriented type**
Description Conference types who tend to subscribe to the statement "action will remove the doubt that theory cannot solve". They are impatient of delay, discussion and caution.
Advantages They are a useful counterbalance to analysts, theoreticians and those who attend the conference simply for their personal, emotional or intellectual growth. By using the momentum and motivation generated by the conference they can encourage constructive action and achieve much in a short time. They tend to remind the conference of the implications of the deliberations.
Disadvantages Prone to exhaustion and "burn out". Prone to dramatic but naive and short term achievements. Prone to simple solutions and may prove distracting in a meeting trying to uncover the complexity of the issues at hand.

♦ **TC1772 Assimilator participant type**
Description Participants whose dominant abilities in a conference situation are "abstract conceptualization" and "reflective observation". In terms of these abilities, they represent the opposite pole from the accommodators.
Advantages Creating theoretical models, inductive reasoning, assimilating observations into integrated explanations. Useful in a research and development arena.
Disadvantages Less interested in people and the practical application theories. It is more important that the theory be logically sound and precise.

♦ **TC1775 Mediator participant type**
Conciliators — Harmonizers
Description Conference participants who attempt as "third parties" to resolve conflict between polarized groups or individuals. This is sometimes done publicly but more often occurs "behind the scenes" in a number of visits between the different parties.
Advantages Mediation can sometimes resolve conflict and disputes in a creative manner that produces a constructive settlement. It can allow for cooperation in moving onto the next step in a process.
Disadvantages Mediation is time consuming and can fail to resolve conflict. Negotiations conducted through a third party are prone to misinterpretation. Conflict resolved through mediation can in fact be merely hidden. Some conflict can be mediated away at a point when it is most creative, only to resurface when it is destructive.

♦ **TC1776 Meeting fatigue**
Prattle fatigue — Secretariat fatigue — Participant burn-out — Conference stamina
Description The well known phenomenon of tiredness and exhaustion at conferences, participants sleeping or absent during presentations, the post conference "low". The issue is sometimes underestimated and sometimes overestimated during planning conference. It is important to guard against people going beyond the limit to which they are productive and useful, and important to remember that people become more prone to manipulation when fatigued. The causes of fatigue are varied and complex. Some obvious causes include: the strain of working in a foreign language, or in a "foreign" jargon or conceptual system; the stress of frustration caused by discomfort with the procedures (or lack of them); the strain of maintaining communication in a new group of social relationships; overwork and lack of sleep (possibly due to jet lag).
Proposals Built-in flexibility, time for rest and relaxation, a varied rhythm of activities, varied style of presentations, a focused environment with variation of the meeting spaces, appropriate celebrations, are all obvious and yet valuable ways to maintain the stamina of conference participants. The most important factor appears to be the level of comment, questions and contributions which increases participants' involvement, alertness and feeling of fulfilment. It appears that more energy is dissipated in the inhibition of comments and contributions than in the expression of them.
Counter-proposals Where the fatigue factor is over emphasized there can be too much time set aside for rest and relaxation which results in some of the participants who may have invested heavily in the conference, feeling frustrated at the wastage of time. Some attempts to "enliven" a conference are gimmicky and distracting, or may appear manipulative. Super-effective communication techniques may simply produce "message saturation" more quickly. Although simple "involvement" does seem to be the key to alert participation there are some participants who do not wish to be involved and can obstruct processes that do not allow them to opt out.

♦ **TC1777 Risk working**
Risk triggering activities
Description Exercises, games and techniques that are used in a conference situation to raise the level of interpersonal trust and the ability and willingness to risk personally their own commitment and contribution to the creative process. Risk triggering activities emphasize an immediate and even sudden involvement under the guidance of a charismatic facilitator. The costs of the deep interventions need to be rapidly balanced by the sense of deep benefits from the exercise which can include the uncovering of hidden personal responses which may even be below conscious thresholds.
Advantages The exercises are, by definition, unusual. They are a definite break in any routine and can enable a breakthrough in interpersonal sharing and creativity. They are a challenge or a rewarding ordeal which can rapidly unify a group which experiences it together.
Disadvantages Can be very offensive and disturbing to some participants, especially when the leader of the group applies pressure on them to participate against their will. It is very dependent on the skill and sensitivity of the facilitator and when it is not introduced and presented in an appropriate context it can appear to be irrelevant and gimmicky.

♦ **TC1782 Conference tensions**
Description Any conference operates within a set of polar tensions of which five seem important in the design of the event: amount of structure, diversity of participants, open endedness of product, focus of topic and style of decision making. The amount of structure has ranges from no structure like a sensitivity training group to extremely structured, like a liturgy or drama. The diversity of participants ranges from a homogeneous group to extreme disparity. The open endedness of the product runs from predetermined results to results being designed by the participants in the process of the conference. The topic may be as focused as a single, narrow, concrete subject or as broad as dealing with any concern the participants may wish to discuss. The style of decision making runs from consensus to confrontation. Most conferences operate toward the middle of these tensions. Organizers of any event need to decide what the emphasis is within these polarities.
The tensions and aims of a conference have to be taken into consideration by the organizers as the design is developed. A conference is not so much an isolated event as an unfolding process

of development which effectively runs for several months or even years. The focal point of the process is the meeting itself, or perhaps a series of related meetings. In some cases, however, other points along the journey, like rulings on conference participation or the local followup events, can be more important. Above all, the image of conferencing as a process of gradually developing an event, allows the necessary reflection and adaptation.

♦ **TC1789 Meetings as social activities**
Description The following may be used as metaphors to explore alternative ways of understanding meetings:
1. **Orchestra:** The challenge of interrelating participant view points to produce a new balance between harmony and dissonance may be seem in terms of an orchestra.
2. **Theatre:** The possibilities of drama, dramatic tension, the roles of actors, and the audience relationship have often been used to describe meetings.
3. **Dance:** The rhythmic interweaving of dancers may also be used to describe the rhythm of meeting processes and participant interaction.
4. **Temple ritual:** The meeting as a ceremony of celebration of the values to which the participants subscribe may be seem in terms of temple processes with extremes of sacrifice and communion accompanied by ritual chants.
5. **Ceremonial and celebration:** The gathering may be decided as a grouping of sub–ceremonies culminating or constituting some macro– event. This may involve, or be seen as, the high point of a pilgrimage with associated festival activity.
6. **Community:** Some may prefer to experience the event as an "instant community", enriched by the presence of children, old people, the handicapped, etc.
7. **Group formation:** For some there will be ways of using information which could make of the whole gathering a gigantic experiment in forming and reforming groups until the most mature groups emerge suitably empowered and able to relate appropriately to other groups emphasizing other energies.
8. **Drama:** The gathering should be dramatic, and some may want to participate in it in such a way as to heighten the dramatic effects and the significance of the event as a whole.
9. **Learning pathways:** To those oriented towards education, the gathering may best be understood as a complex set of interweaving learning pathways.
10. **Quest:** The gathering may be attractive to some when interpreted as a mystical quest or an exercise in collective alchemical marriage.
Broader Metaphors of meetings (#TC1603).

♦ **TC1798 Intercultural consultants**
Intercultural interpreter
Description Appointed consultants for each cultural group at conferences who ensure that images, metaphors, idioms, gestures and other communicative behaviour are interpreted into their own culture as they were intended by the presenting culture.
Advantages Bridging cultural gaps and creating an awareness of cultural sensibilities which might otherwise be ignored creating offence or otherwise hindering the establishment of good communications between participants.
Disadvantages Interpreters tend to de–emphasize the responsibility for effective communication which rests on whoever is presenting. Too much can be dumped onto the interpreter while presenters can continue to be blind and unaware of the intercultural issues thereby impoverishing both their understanding of the audience and their ability to communicate effectively with it.

♦ **TC1800 Flannel board**
Velcro–board — Magnetic board — Sticky board
Description A versatile vertical display apparatus on which prepared items can be easily placed, moved, and removed. Originally made from stretched flannel to which could be "stuck" items backed with sandpaper. There are now many variations of this, including the use of velcro (hook and loop), special adhesive boards, magnetic boards, and other adhesion processes.
Advantages Elaborate, visual presentations, which can be built up progressively, and changed quickly during a talk and/or discussion. A "failproof" display tool that doesn't depend on electronics. Useful in situations where good diagrams are essential and cannot be presented in other ways.
Disadvantages Inflexible to the extent that items which are prepared in advance are difficult to change on the spot. In the world of hi–tech audio visual aids, this kind of display may in itself have a low–interest rating.

♦ **TC1808 Conference accessibility**
Description The degree to which a conference permits ready access to membership. It is reflected the absence of entrance requirements of any kind, and by the degree to which participation is solicited.

♦ **TC1814 Team explorer role**
Resource investigator
Description An affinity–role played by one or more members of a small group or team. Typically rather curious and communicative, they are those who explore and report on ideas, developments and resources outside the group. They have an interest and a capacity for making contacts, exploring anything new, and responding to challenge.
Advantages A source of enthusiasm within the team. The contacts and "findings" from outside the group enrich and inform the group–process.
Disadvantages Once their initial fascination with something new is over they are liable to lose interest in it. Can be distracting and uncoordinated with the group.

♦ **TC1815 Conferencing as an energy sink**
Conferencing as dissipation
Description A great deal of energy in various forms and from a variety of sources goes into conferencing. Many participants, and most organizers, leave a conference drained and exhausted in spite of their exhilaration. It can be felt that this energy is wastefully dissipated and that it should be employed more usefully. Yet the dissipation of energy is not always a wasteful exercise. Particularly in times of conflict and hostility transformative conferencing can provide a way to dissipate anger, aggression, and the heat of the friction as two opposing forces meet. In every conference there is dissipation of energy: the dissipation of accumulated knowledge, concepts, and perspectives; all the effort that goes into the preparation and distribution of papers and other means of communication most of which appear to be largely futile; the anxiousness to be able to contribute and get one's voice heard. Even this kind of energy dissipation, as wasteful as it may appear, can have the effect of clearing the ground; providing room for preparation to begin on the next project and the development of fresh perspectives. Energy dissipation of this form is rather like the leaves falling from trees in autumn. The trees may appear to be dead and indeed some may be dying but for most it is a natural cyclical process providing the way for new growth.

♦ **TC1816 Electronic communications unit**
Description Terminal given to each participant at a conference which can be used in all meeting and accommodation rooms. Messages can be stored and transferred either from the organizers to all (or selected) participants or between participants themselves.
Advantages Used in conjunction with computer assisted contact formation, this process can dramatically decrease paper shuffling and increase the creative potential of the conferencing interchange.
Disadvantages This is still an expensive investment to make and the problems of system compatibility are not yet resolved so that notes stored on this system may not transfer easily to another.

♦ **TC1824 Slide projection**
Description Projection of photographic images onto a screen. Becoming a very sophisticated art, using coordinated tape recorded "sound tracks", several screens, and several projectors simultaneously.
Advantages Surprisingly effective process for illustrating a talk or making a point. Relatively low costs involved in making up the "show" and in presenting it. Photography is still a powerful medium for communicating atmosphere and ideas inspite of video, films and other cinematic processes.
Disadvantages Can tend to define rather than illustrate a talk.

♦ **TC1828 Participant strategic preferences**
Participant change preferences
Description People tend to move or drift through the social system into those groups and organizations which are engaged in the change processes most congenial to them. As individuals develop they may reach stages when a given change process and its organizational support seems unfruitful or unsuited to their desire for self–expression. The individual needs fresh fields to conquer, a new life–style or a new mode of work. The development of the individual implies life–style mobility and organizational and social change. Social change and development requires development of the individual to adapt to new challenges. This is also the case in meetings. The difficulty is that society currently sanctions movement within organizational and career systems but not between them. The individual is therefore forced into one particular mode of self–expression for his whole working life unless he wishes to run the risk of being labelled a grasshopper or dilettante, or of being viewed as an ignorant outsider (a "foreigner") in the systems into which he attempts to move. Participants are faced with this difficulty in conferences which have groups emphasizing distinct modes. Within one system an individual can of course develop other modes of self– expression but only as secondary modes within the constant and overriding primary mode (eg an executive in the business system, an individual can move from a high technology corporation to a commercial art corporation; the switch from science to art is then contained within the unchanging management framework).
The problem in conferences is therefore whether it is possible to provide an organizational setting in which an individual can develop secondary modes of expression and allow any of them to become primary for any desired length of time in response to the flow of the meeting. The problem is complicated by the very radical nature of the differences between approaches to change advocated or undertaken in meetings, as well as between the corresponding modes of expression of the individual engaged in them. There does not appear to be any systematic listing of change strategies but the following list is an indication of the variety: 1. political action
2. scientific and technological development
3. economic and financial development
4. education, training
5. art, music
6. architectural and machine design, urban planning
7. religious faith, prayer
8. social engineering, social development
9. philosophical or esoteric understanding
10. behavioural and perceptual modifications by drugs
11. public information, media, propaganda
12. community development
13. drama, theatre
14. organizational development
15. legislative action
16. military or police action
17. direct action, violent civilian protest
18. military or police action
19. self–exploration, meditation
20. mediation, negotiation
21. manual labour.
Ironically, the proponents of a particular form of change tend to perceive it as the only viable or significant form (eg to a political activist everything of any significance is political). They are unable to detect the manner in which their action is counter–balanced, checked, contained or even undermined by the other forms of change. Similarly it is not possible to determine how such different strategies can be blended harmoniously together into a mix which can ensure appropriate change. No body has a mandate to attempt this, and no integrative discipline exists to legitimate such an approach.
Related Participant personality needs (#TC1204)   Participant pre–logical biases (#TC1965)
Participant interaction preferences (#TC1845).

♦ **TC1831 Remunerative power**
Utilitarian
Description Based on control over material resources and rewards: such as salaries, commissions, fringe benefits, services and commodities.
Advantages Clearly recognisable benefits that influence behaviour directly especially of "calculative" participants.
Disadvantages An inflationary process, can lead to a tendency towards the need to escalate the rewards continually to maintain power. The problem of correctly identifying the appropriate reward for each group. The short life of the motivation that is generated on this contractual basis.

♦ **TC1835 Art form method**
Description This technique of leading a conversation is based on the assumption that people perceive reality in a sequence of steps. As an individual enters a new situation the senses initially perceive raw data. This data is related within itself and with conceptual frameworks and with experiences from memory. From these relationships and data the individual interprets the significance for himself, decides the appropriate responses to the situation and acts on those decisions. This process is quite unconscious and occurs extremely quickly.
The art form method follows this same sequence in the structure of the sequence of its questions. There are four levels of conversation; objective, reflective, interpretive and decisional. The first level asks the group to simply state what is perceived, not their opinions about what is perceived but just what is there. Next are questions of reflection; what are associations that can

be made. What emotions are related to the data ? What are other experiences that may be related ? Next is the level of interpretation. What is the meaning of the data ? What would you call this experience ? What is a story about the experience ? The final level is that of asking for a decision. What would you do about this ?

♦ TC1843 Adaptive conferencing
Research conferencing — Participatory research
Description Intended to work on and create new knowledge, innovative solutions to problems, etc.
Advantages When breakthroughs are achieved there is a very high return on the investment of time and energy.
Disadvantages High risk conferencing in the sense that return on the investment cannot be ensured.

♦ TC1845 Participant interaction preferences
Description Meetings as a whole, or groups of participants within a meeting, may give preference to one or more modes of interaction possibly at different stages of the meeting. This effectively determines the styles of the meeting and may either attract or alienate certain participants. Preferences may be for primarily verbal or primarily non-verbal interaction.
Narrower Verbal interaction preferences (#TC1378)
Non-verbal interaction preferences (#TC1640).
Related Participant personality needs (#TC1204)   Participant pre-logical biases (#TC1965)
Participant strategic preferences (#TC1828).

♦ TC1847 Conference as procreation
Description The challenge of transformative conferencing can be described by the analogy to the process of sexual congress and human birth. Conferencing as currently envisaged may involve titillation leading, if successful, merely to mild participant arousal. Or it can result in some form of effective intercourse, possibly leading to orgasms of collective enthusiasm with little outcome. Additional factors of major importance are however required to ensure any quickening of a new departure as a result of effective conception through cross-fertilization of ideas. And even then the result may be a miscarriage, a still-birth, or even malformed.
Unfortunately it would seem from past events that excessive attention has been given to the processes engendering enthusiasm, with little concern for the adequacy of the receptacle (a womb within which the results of the exercise can be contained and brought to fruition to say nothing of the question of midwifery and post-natal care). There is a problem of collective, responsible parenthood, as opposed to the short-term, and frequently irresponsible, concerns of casual intercourse, however pleasurable.
It is only very recently that sex education has made any inroads on the, often quaint, misinformation and furtive "dirty" secrets by which the physical process has been surrounded – at the price of untold misery and disillusionment to those initially caught in the compensating glamour. Is it not possible that many conferencing processes are in effect couched in terms which reinforce an analogously quaint misunderstanding of what is really involved in the "gutsy" process of engendering new psycho-social forms ?

♦ TC1848 Commonweal conferencing
Description Prime beneficiary is the general public. It is characterized by analysis of the social process and goal formation. The aims of the conference transcend personal or group needs.
Advantages Useful mode for government departments, police and fire services, etc. and for organizations concerned with maintaining or changing the established social system. Useful for building inclusive plans from an objective perspective.
Disadvantages Danger of a paternal and detached view that loses touch with the "beneficiaries", the "general public". The basic problems of maintaining the external democratic influence and finding a way to influence authentically, involve, and evaluate creative input of the "general public".

♦ TC1850 Conference impact
Description There are three levels of impact on the participants in a conference. The first is the immediate impact. The participants during and immediately following the event say that they have had a good time: "It was a significant event". This is largely due to the style of the facilitators of the event; they are friendly, open, engaging, funny, etc. They convey a sense of listening and they elicit participation.
The second level is the medium term impact which is reflected in the participants' memory of the event up to six months after the program. This is largely a function of the content of the event. The topics were interesting, engaging or challenging. Materials were presented in understandable ways. They were personally relevant to the participants or perceived as relevant for those they represented.
The third level is the long term impact of the occasion on the participants. The life and thinking of participants has been altered for one to five years after the event. Such an impact is largely a function of the structure of the event. The sequence in which material was presented, the design of the daily activities and the use of space potentially modify the perception and response of participants. The structure of the conference consistent with the experience the organizers intended and consistent with the experience of the issues being dealt with.
Naturally a "good" conference has short, medium and long term impact on participants and organizers alike.

♦ TC1852 Interpersonal trust
Description The willingness to risk significant communication and creativity with others; accepting the possibility of rejection, pain and uncertainty; being reliant on others and acting from a position of confidence and optimistic expectation.
Advantages People in a trusting relationship experience more flexibility as well as commitment to each other and to the task at hand. They are less defensive and more receptive to new and challenging information. They tend to be more search-oriented, self-determining and more capable of interdependent relationships with others.
Disadvantages Extreme forms of trust can become almost pathological gullibility or naivety resulting in suspension, surrender or distortion of judgement. Trusting behaviour is sometimes confused with despairing acceptance, social conformity, innocence, impulsiveness, masochism, blind faith, and gambling.

♦ TC1853 Value-monger participant type
Description Conference participants who believe in and adhere to certain values and who feel the need to strongly encourage others to adopt these values.
Advantages Commitment to their own beliefs can become a strong force within a group that appears to accept the same values. They are useful in maintaining a position in a polarity that can generate genuinely creative tension.
Disadvantages Commitment to their own beliefs can become a divisive force within a group that does not wholeheartedly appear to endorse their values. This can be particularly difficult when decisions have to be made in moments of ambiguity and crisis, when conflicting values need to be balanced.

♦ TC1856 Peer-mediated-learning group
Description A specialized type of small group session where the leader is one of the group. To be effective the leader needs specially prepared materials, and provision for feedback to the coordinator.
Advantages Useful with small groups with specified and limited learning objectives. Good for structurally spreading the responsibility for the whole process.
Disadvantages There is often a certain amount of confusion which can waste time as the group gets clear on its objectives. A lot depends on the make-up of the group.

♦ TC1858 Normative power
Description Based on the use of symbolic reward or deprivation such as appointment to leadership, media coverage, prestige and status symbols, administration of rituals, and influence of "acceptance" and "positive" responses.
Advantages Effective over the long term especially with "moral", positively loyal participants in influencing behaviour and thought.
Disadvantages Ineffective with "alienatives" or "calculatives". It is prone to sudden and extreme disillusionment.

♦ TC1860 Structure-oriented participant type
Description Conference participants who feel that valuable work is done, when time, space and relationships are clearly structured. Prefers procedures and clear agendas. Mistrusts and is frustrated by unstructured, dis-ordered, and spontaneous forms of interchange. Tends to favour the standard structure of lecture presentation and small group-workshops.
Advantages Prepared to trust procedures structured by the organizers and to get on and work through them as productively as possible. This allow a greater focus on the aims of the conference and less distracting questioning of the structure itself. A useful balance to processists who are threatened by and question all structural guidelines.
Disadvantages Unquestioning acceptance of the structures can lead to constraints on creativity and to a sterile conference. Unwillingness to be exposed to new and unexpected forms of inter-change reflects an unwillingness to risk unexpected directions of change.

♦ TC1862 Medial conferencing
Description Conferences made up of representation from medial level associations who are relatively un-integrated into the major structural processes of society.
Advantages These groups have formed and come together because of the gaps in the major structural processes of society and therefore bring a perspective that reveals points of strain and stress in the system.
Disadvantages The lack of the functional significance of the "majoral associations" and often more or less dependent on them tends to encourage low risk compromise proposals.

♦ TC1866 Audience reaction teams
Conceptual interpretation teams — Inter-cultural communication teams — Interruption teams
Description A group, or groups, made up of members of the audience, who are specially assigned to interrupt a presentation when they feel that clarification is required. They are to ensure that the presenter and audience remain in touch with each other. Where there are different groups in the audience, their role is to ensure that these groups remain in communication with each other during the discussions, and that the speaker correctly "hears" the points and questions from the floor. They may ask for clarification and/or rephrase the point themselves in order to check "have we understood correctly...?" It is important that those selected for the teams are familiar with the groups they represent.
Advantages If an audience is made up of groups from different cultural backgrounds, the idioms, terms and concepts used by the presenters or by representatives of these different groups may be unfamiliar and confusing. This process helps to get the points cleared up immediately before they lead to further confusion, wastage of time, and unhelpful conflict. This process can be used to assist the process of interpretation, not only between various national languages but also between different jargon languages, pattern languages, and ideologies.
Disadvantages Overzealous teams can interrupt unnecessarily. The audience can experience the team and the process as patronizing. The speaker can be irritated and distracted by the interruptions. The process can be misused by a team that is anxious to indicate its disagreement with the speaker, and use their privilege of interruption to debate each point rather than assist in the clarification. Teams that do disagree with the point of view of the speaker may, in extreme cases, find it hard to act out their role for fear that they may be seen as agreeing with, and even arguing for, the points they are trying to clarify. Some concepts do not translate into their cultures, and some groups and speakers are unwilling to accept reworded interpretations which threaten to defuse or distort their message.

♦ TC1870 Telephone conference
Description Use of the telephone and public address systems to allow resource people or presenters to speak at a meeting when they are unable to attend in person. Can be coordinated with a slide show or other forms of presentation and can enable two-way communication.
Advantages The inclusion of people into the conferencing process who would not otherwise be able to communicate directly. "Live" communication commands more attention than written or recorded messages.
Disadvantages The difficulties imposed by the absence of visual contact.

♦ TC1875 Bricksigner type
Recognition seekers
Description Conference participants who wish that their particular domain of insight and expertise be properly recognized and distinguished like those who would sign the bricks which can be used for building a cathedral. They struggle to avoid being placed in an "inferior" position and tend to spend time reporting on personal achievements.
Advantages They tend to put a lot of care into the shaping, presentation, and preservation of their insight – their "brick". This can lead to high quality material for building a consensus. Thus they are useful with the balancing synergistic force of builder types.
Disadvantages The care they put in their own presentations tends to limit their potential to let their insight go, and to limit their awareness of other complementary insights. Unwillingness to allow synergy often results in conference reports which are a collection of uncoordinated papers which lie like a pile of beautifully constructed bricks that have not been designed for the same building.

♦ TC1876 Full motion video teleconferencing
Electronic meeting
Description Meetings involving participants in geographically separated groups connected by full audio and visual circuits. Recent technical development has produced advances in full-motion

video compression and large screen video projection units are becoming readily available. Used with high resolution graphics or facsimile facilities, together with an adaptive bit-sharing multiplexer which allows simultaneous transmission of audio and video data.

Advantages These systems may eliminate the need for many of the meetings that consume so much time and energy. The opportunity for simultaneous international interactive communication has enormous implications for international relations.

Disadvantages Costly. The lack (as yet) of standardization of the systems. Many factors have not yet been assessed; such as fatigue, distraction and other user reactions. The feeling that greater communications may lead to greater efficiency may be an illusion; communications may be used as a substitute for delegating responsibility thus placing the participants in a constraining network where they feel obliged to "cable back for instructions".

♦ TC1877 Participation assumptions

Description The participation dynamic of conferencing is the point of input. This is not simply information, but also outlook, mood and morale about the subject matter and intentions for the organization or organizations involved. Different types of conferences maximize different kinds of input. Effective conferencing demands only that participants, facilitators and conference organizers each consider carefully what sorts of input they are responsible for generating.

The attitude of conference organizers toward the participants is an important factor in organizing participation. Considering whether participants are more benevolent and worthy of participating, or whether they are more troublesome and needing to be kept in line is a trap to be avoided. As conference organizers it is necessary to assume, at least in public, that the intent of the conference is sufficiently shared by participants so that they can be expected to cooperate.

♦ TC1879 Maximization conferencing

Description Major concern is given to the question of how to "maximize the interests" of the membership for example through political pressure to obtain favourable legislation or efforts to promote new awareness or beliefs. The conferencing tends to focus on the analysis of problems and constraints faced by the membership and proposals for action to influence those in a position to change or remove these constraints and problems.

♦ TC1880 Existential aim

Description The existential aim is what the organizers envision that the participants to experience during the conference. It is projecting what the internal mood of the individuals will be as they participate in the event. It is also an initial attempt to anticipate the spirit of the event. Is the mood intended to be reflective, joyful or decisive ? Are the participants to experience themselves as a part of a great movement of people, or to experience their will to make a personal sacrifice for a cause ? Are the participants to experience a new understanding of material being presented ? Whatever the intent of the programme, a self-conscious decision in this arena does not limit the participants' experience but allows the organizers to operate out of a decision that can be changed.

The existential aim is also orienting the journey of the participants, or the internal experience of the event. This description may be as refined as a session-by-session analysis of the programme or as simple as anticipating shifts in mood during blocks of sessions.

The existential aim is a tool. The spirit of the programme and the social environment of the programme are both based on this understanding. The setting of the event, the decor and even the type of meals are informed by the existential aim. It is also a self-conscious guide for testing how the conference is going in the midst of the event. This description of the journey helps the organizers to decide the timing of events. An understanding of shifts in mood might determine which speaker or topic is introduced before another. The danger of not having an existential aim self-consciously described is that the mood of the conference may be turned over to the most dominant or charismatic personality.

♦ TC1886 Contextual facilitation

Description Facilitators are responsible for maintaining the contextual framework of the conference. The parameters of the conference are set so that participants are in the same universe of discourse insofar as possible from the beginning. The context establishes what is admissible. The context needs to be simple, motivating and common among all of the facilitators of the conference.

The context given in the conference itself can build upon backup material or pre-conference gatherings. Some conferences use reports from pre-conference meetings to begin, so that the whole group is on a common footing to start. Others study together one or several of the key contextual papers at the beginning of the conference to achieve the same end.

♦ TC1888 Conceptual framework development

Conceptual screen development

Description Conceptual screens are developed out of a dialogue—whether internal to an individual or amongst members of a screen creating team. The first aspect of this dialogue is that it occurs between a Platonic or deductive approach to ideation and an Aristotelian or inductive approach. Neither are seen as preferable, but both are needed to develop an effective conceptual framework.

The Platonic approach asks what is true about this situation generally; what does it have to do with life universally. It is out to create a model which can be applied everywhere. It is often helpful here to begin with a basis of outside research. The Aristotelian approach begins with the particular encounters that the group has with an issue. Research is done into the specific dynamics of known cases. Case-study type research is helpful here. It then seeks to look at the idea from the inside, and see what things are consistently a part of it and what things are apparently incidental. Out of this review of the data, an integration emerges.

Screen creation is doing both of these things with the same material, up to such a point that the screen seems to emerge. It then gets checked back against its inductive and deductive origins, and gets looked at for its own sake from as many perspectives as possible. The next step in screen development is to create occasions, such as pre-conference local gatherings, or publications, which force the originators to articulate the screen to a larger audience and to alter it or not in response to feedback.

For the most part the development of screens for a conference is done before the event but for those events that create screens during the programme the decision making process is the key point of feedback. The tension between the Platonic and Aristotelian approaches is most evident in this feedback. Usually the participants best play the role of the grounded, Aristotelian pole of the tension, and the facilitators and procedures team play the Platonic pole of the tension.

The participants hold the local experiential pole of the tension. They create out of their own experience the practical aspect of the model. They can, through brainstorming and organizing data and experience, give a sense of reality to the screen. As the screen develops in the process of the conference they test the reality of the screen.

The facilitators and the procedures group test the development of the screen over against the ideal. They might use a series of sets of extremely abstract images like knowing, doing and being or foundational, social and universal for a theoretical model, or the past, the future, the inclusive and the profound for a social model as partial tests for the screen that is emerging. The choice of these abstract images depends upon the abstract rationale that is being used to develop the screen. These abstract images must be in actual dialogue with the experience of the participants and not dictate the meaning of the experience if a creative screen is to emerge.

One of the most interesting points of feedback is between the conference spirit and the screens. When these are in creative dialogue then quite unexpected break throughs can result. There have been break throughs in individual and group creativity when the limits of social care and interior awareness are being explored at the same time.

One of the merits of the development of screens is that they are usable by many people. This does require simplicity, profundity, and realism. These screens, while not necessarily acceptable in some circles, are very much what is required by development efforts across the world.

Another merit in the development of screens is developing complex and lasting screens during the process of the conference itself. Most screens developed during a conference are usable only for the event and are very simple. This process would require new technologies of making decisions and of developing a corporate understanding that are beyond current capacities.

♦ TC1894 Language constraints

Description During the conference itself, document production facilities are needed by smaller working groups as well as by the group as a whole. The bottlenecks occur most often prior to plenary sessions, when many reports are needed simultaneously. The extent to which materials need to be translated increases the complexity of the job. It is worthwhile to devise a flow chart of anticipated materials needs in advance and staff and equip the materials production team accordingly.

Language, of course, is a major issue in the mutual honouring of conference participants, although it does assume larger than necessary significance when other cultural variables are passed over. In a single language conference, it is essential to watch carefully the development of a common vocabulary by the conference. Different delegations come with different jargons. Much of the launching of any meeting is allowing opportunities for the commonizing of jargon among the group. Even when formal conferencing calls for very precise translations of vocabulary, the jargon dimension must be watched and brought to self-consciousness to enable genuine communication.

Where translation is used, just as in the use of any other outside services, the individuals involved need to be offered the opportunity to share the conference intents as early on and as deeply as possible. They are greatly assisted by opportunities to meet as a team to work out ongoing vocabulary issues, and to work as needed with individual participants. It is not always inappropriate by any means to engage the linguistic skills of participants in the translation enterprise, allowing more face-to-face communication among participants.

♦ TC1895 Calculative participant involvement

Description Low intensity orientation towards the organizers either vaguely positive or roughly negative, compliance on a contractual basis. A retail/consumer relationship that depends on the maintenance of interest and does not demand creative contributions or feedback from the consumer.

Advantages This is generally a long term relationship of involvement based on mutual benefits. The majority of conference goers would probably make up this category.

Disadvantages Self-interest is the main criterion for evaluation and participation rather than an analysis of the more general worth of the conference.

♦ TC1897 Video tape

Description An audio-visual recording - playback process that can be monitored, edited, transmitted, stored, copied and distributed.

Advantages Captures visual as well as audio signals. Very useful for recording sessions which can be repeated when the resource people are not available, and for broadcasting to a wider audience, and edited into a distributable film of the event for those unable to attend.

Disadvantages Can be distracting and disconcerting to some people, especially if special lighting and obstrusive equipment is used. A high investment process that can tend to take control of the proceedings. The camera is a selective eye and its apparent objectivity can misrepresent and manipulate an event depending on the person controlling the camera; close- ups tend to expose weakness in the subject and "respected" people are rarely subjected to this.

♦ TC1900 Sharing information

Description The process wherein a participant or the organization that he represents gains new knowledge from and gives new knowledge to other participants. Data is interchanged, issues are identified, implications are explored and research shared. A scientific symposium or mathematics teachers' conference might be this kind of event.

♦ TC1902 Interpersonal suspicion

Description The unwillingness to risk significant communication and creativity with others; lack of trust for people and events; being independent of others and acting from a position of scepticism, and pessimistic expectations.

Advantages A useful balance to gullibility and naivety. Stimulates questioning and thorough examination of presentations, models, reports and proposals. A useful "foil" in group discussion that draws out all the issues into the focus of the group.

Disadvantages May develop into defensiveness and even paranoia. Tends towards structural rigidity, tight control systems, and the inhibition of uncertainty and change. There can be a tendency to protect against intimacy and personal challenge and growth using camoflage and masks. The unwillingness to risk can inhibit the conferencing process.

♦ TC1903 Conference polarization

Description The degree to which a conference is split into rigidly defined groupings. It is reflected the absence of consensus, polemic, rhetoric, increasing fragmentation and a breakdown in communication between the groupings.

♦ TC1908 Weakwilled conferencing

Description Meetings are "weakwilled" when the participants can be said to allow the circumstances to shape their intentions. Often this results in avoiding proposals and plans that are "impossible in the circumstances". The effect of the conference is not very visible.

Advantages Tends to be sensitive and realistic about the "real situation" and to follow a path that evolves according to the circumstances.

Disadvantages Good intentions and sensible proposals can be structured not only by "force of circumstances" but even by fear and timidity.

♦ TC1909 Meetings as symbolic configurations

Meetings as aesthetic symbols

Description The following are useful symbols through which meeting structure may be understood in new ways:

1. Monument or memorial: Whereby architectural or decorative elements are disposed in relation to some central focal point. Such elements often reflect aspects of the central theme of the monument. In the larger memorials a considerable degree of symmetry is usually to be found.

2. Memory devices: Whereby items to be remembered are associated with or "impressed upon" some easily remembered configuration such as the elements of a memorial, the features of an ornamental garden, a suitable pantheon, etc. The items may then be "recovered" by progressing through the configuration in whatever order is appropriate.

3. Mandalas: Whereby a complete set of complementary figures are disposed symmetrically in relation to one another around a central focal point in order to indicate both a succession of possible experiences and a progression to more or less fundamental levels of experience. Each such experience is understood as essential to the harmony and evolution of the whole. Mandalas, or their equivalents, are used as attention focusing devices in different cultures.

4. Symbols: Such as a crown, a chalice or a stupa, which may be viewed as a configuration of elements constituting a "receptacle" for energies, qualities or attributes thus held in balance.

5. Abstract forms: Some may wish to see the gathering as energies patterned onto more abstract forms: spiral, hierarchy, network, tensegrity, mandala, matrix, torus, polyhedron, knot.

6. Symbol systems: Some may be attracted by seeing the interweaving energies at the gathering in terms of a particular symbol system such as astrology, the I Ching, any pantheon, etc. These could even be used to identify imbalance in the energies represented, blockages in the evolution of the event, or threshold tests and challenges.

7. Imagery and dance: Such a gathering can also lend itself to comprehension as a pattern of aesthetic images, or as a dance of energies.

Broader Metaphors of meetings (#TC1603).

♦ **TC1910 Rule-governed behaviour**
Patterned behaviour
Description Behaviour that can be consistently recognized, predicted, anticipated, and missed by "intuition" can be described as following or corresponding to norms or rules that are accepted by the system in which the behaviour occurs. This does not imply that the behaviour is determined or fixed by these rules but that it "makes sense" only when it corresponds to the rules.
Advantages It has survival value because it forms a basis for interaction between actors within the system. The most obvious example of rule governed behaviour is the use of language. Native speakers of a language can immediately detect ill-formed sentences, inconsistent structure, and illogical semantic relations, even without being aware of the rules that govern well formedness, consistent structure and semantic logic.
Disadvantages Certain models of the rule system are impoverished representations of the system they represent leading to a reduced structure of choice.

♦ **TC1913 Creative listening process**
Systematic listening process
Description A set of questions that probe beneath the "surface structure" of the speaker's words to the "deep structure" which is their personal representation or model of "reality".
Advantages Words, phrases, messages and whole presentations at conferences can be grammatically ill-formed, either in syntax (the order and patterning of words and phrases) or semantically (their portrayal of meaning). This can be unintentional or deliberate, and when it occurs, elements of the "deep structure" of the speaker's model can be obscured. The process develops the intuitional recognition of ill-formed communication patterns and provides a way to recover, challenge, and re-connect them. Not only does it reveal more of the "deep structure" but can also "enrich" it, by calling for the inclusion of more "reality" and experience. The process is based on the insights of transformational grammar.
Disadvantages Sometimes the questions and interventions can be confusing and distracting when posed to speakers who are unaware of the way in which they have obscured their "deep structure" and indeed may have no understanding of their thought in this way.

♦ **TC1915 Prayer meeting**
Description A meeting of people, generally focused on a particular concern, who articulate their concerns, their fears and resolves and gratitude, in a form of address towards the God in which they all, to greater or lesser extent, hold belief. Participants especially look for a sense of guidance or reassurance which is experienced either as a sign from outside or some deep state of being within them.
Advantages Can be an effective mode for making corporate decisions. The sense of commitment and seriousness can be very strong and enduring. The external focus lifts the group above and beyond the limitations and irritations of human interaction.
Disadvantages Can be misused, abused, or simply distracting. There can be a surrendering of responsibility which can be simply a giving-up of power or more seriously can be a handing-over of power into the hands of an elite.

♦ **TC1916 Rap group**
Description Small discussion group to explore and compare the personal experience of each of the members in order to arrive at political conclusions. Typically the leadership is rotated and a disciplined structure ensures that each person is able to share and to be listened to on a basis of equality. Contributions are generally accepted without comment until all have spoken in turn and then reflections and conclusions are discussed, each person having time allocated for their participation.
Advantages Personal experience is discovered to be shared by others and this consciousness of affinity both releases them from what they thought a solitary burden and engages them in common effort to respond creatively to their situation. A very effective way to initiate a conference as it builds interpersonal solidarity, warmth and trust, and creates a participant-generated context for the process. The time allocations to allow each person time to be listened to, encourages both speaking and listening skills.
Disadvantages The disciplined structure can be disconcerting to people who are unaccustomed to being listened to without interruption. Although much valuable material is contributed, little is recorded, and what is a very significant experience can be forgotten because there are no written or even verbal reports or products.

♦ **TC1919 Conferencing failure**
Description Most conferences measure failure in terms that can be described as a lack of transformation of those involved from one state to another more effective, appropriate, or enlightened state. Transformation implies a certain permanence and a recognisable and positive change that has involved the participant in some new awareness and decision-making. Most conferences are designed to ensure that this does not occur. In order to increase the chances of failing consistently, the following list, while not comprehensive, will assist even unskilled conference organisers to decrease the likelihood of any positive transformation occurring amongst participants in their care.
Steps
(a) Ensure that the issues and problems experienced and raised by the participants, in other words "real" issues, are considered of little or no importance to the conference. Instead appoint eloquent keynote speakers who name the issues that the participants should discuss in order to occupy their time and prevent distracting conversation about their real problems.
(b) Ensure that small group discussion is controlled and remains focussed on the remarks made by the expert speaker.
(c) In the case of real issues or problems inadvertently being aired, insist that they cannot be handled directly but that extensive research, committee work and specialist or expert analysis is needed and will be presented at a later date. Or point out that this point is to be covered by another speaker or is in fact irrelevant to the current discussion and theme of the conference.
(d) Encourage the spread of anxiety that the attempt to deal with the "real" issue will in fact lead to something frighteningly worse, thereby developing a fear of a real breakthrough and enlisting the support of the participants themselves in maintaining the controlled and safe structures with the minimum of risk.
(e) Take care to ensure that diagnosis of any situation is concluded in language that makes it impossible to think of creative proposals and interventions.
(f) Emphasize a single approach, as narrow as possible, to all issues, no matter how diverse. Problems and situations that fail to respond to this approach should be defined chronic, untreatable and then abandoned. Once an approach has proven to be consistently ineffective it should never be given up. Anyone who suggest or attempts variations and alternatives should be sharply condemned as inexperienced, naive, untrained, and ignorant of the real nature of the situation.
(g) Ensure that every problem is treated in isolation and avoid any attempts to confuse the issue by exploring any context, patterns or links to other situations.
(h) Ensure that proposals are vague and ambiguous in case anyone might attempt to follow them and actually achieve something.
(i) Insist that conferencing is a process of seeing what is wrong and finding out how it got that way.
(j) Make sure that any goals set are visionary, unmeasurable, and cannot be investigated, so that no one can judge if they have ever been reached. If possible make sure they cannot be reached.
(k) In the case of untrained and inexperienced participants who insist on something that will actually occasion change, and who fail to be shamed into acquiescence, it might be necessary to offer some kind of ambiguous, general idea which is untestable but which ensures that they have something to do. This may require insisting that they work to interpret the problem and its solution in an ideological framework. Transfer the issue from one hypothetical entity to another for as long as possible.
(l) In the case of conferencing, that in spite of all this is fruitful and effective, it is sometimes possible to limit the extent and possible consequences of the transformation by persuading participants that they haven't really changed and that further conferencing is necessary. Point out that the feeling of exhilaration is really illusory and escapist.
(m) Continually rehearse the dangers associated with transformative conferencing. People and situations might actually change causing no end of chaos and re-arranging. Participants may re-value their own creativity and find it difficult to remain quiet in constricting structures and processes. They may become uncomfortable and perhaps intolerable, increasing stress and pushing them towards alcoholism or a career change. Emphasize the importance of the well-tried, safe and predictable and consistently ineffective methods to avoid these problems and others like them.

♦ **TC1920 Conferencing intentions**
Conferencing objectives
Description There are, of course, many varied and particular objectives and intentions that are sought through conferencing. There are some which are generally accepted, including: maximization of beneficial contact between participants with complementary interests and commitments; maximization of participant ability to initiate new action, inform and involve other interested participants, form groups and formalize group action (to the degree necessary). Such objectives are often unspoken, neglected and ignored in the procedures. In fact it appears, at many conferences, that these objectives are deliberately avoided and thwarted by restrictions, built into the agenda, the structure and the process.

♦ **TC1922 Meetings as psycho-physical processes**
Description The following may be used as metaphors to explore alternative ways of understanding meetings:
1. Respiration: The meeting may be viewed as composed of cycles of inbreathing and outbreathing of information in the light of yoga attitudes towards the ultimate significance of such processes.
2. Meditation: The meeting may be viewed as an exercise in collective meditation and group consciousness, with all the consequent problems of physical, emotional and mental alignment.
3. Alchemy: The various alchemical processes explored by psychoanalysts may be used to model the progressive transmutation of the crude (material) perspectives initially present in the meeting.
4. Group healing exercise: The gathering may been seen as a body to be healed and rendered whole.
Broader Metaphors of meetings (#TC1603).

♦ **TC1924 Symposium**
Description A group of talks, presented by several individuals on various aspects of a single subject. A moderator coordinates the subject matter and timing. Each presentation takes about twenty minutes and the whole symposium can be limited to one hour.
Advantages Comparatively easy to organize, it allows for systematic and relatively complete expression of ideas, without interruption. Complex issues can be divided up and dealt with by separate speakers in logical component parts. Careful preparation with the speakers allows for close control of subject, matter, timing and style of presentations.
Disadvantages Dependent on the empathy and skill of the moderator and speakers. Very little interaction with and between members of the audience. Tendency towards duplication and confusing semantic differences between speakers. There is a difficulty of ensuring the appropriateness of the presentations to the conference objectives. The formality and compartmentalization of the issues limit the possibility of integration of concepts. Symposia, in spite of the best moderators, tend to run on and on.

♦ **TC1927 Participation methods**
Description The means by which participation is solicited in a conference very much influence the degree of participation. Methods must be considered to evoke the degree and kind of participation suitable to the task at hand. There is no such thing as a "participatory" or a "non-participatory" conference. Eliciting participation requires thinking through, either as an organizer or as a conference goer, the points at which the group's input is most urgent, and focusing efforts there. It is also necessary to consider how to prevent a flood of impractical "good ideas" and "free advice" from the group.
Voting on a measure offered from someone else is a very restrained form of participation. Creating such measures is much more involved and requires considerably more group involvement and responsibility. Small group discussions, plenary reporting, the giving of papers and reports must not be automatically assumed, but thought through.

♦ **TC1928 Participation dynamics**

Description The trend toward increased participation of all kinds is growing. There is no reason to believe that this trend is a temporary one. The group which will attentively sit in rows and be told things becomes rarer and rarer. Planners which envision, however subtly, such docile groups will simply find their conferencing difficulties on the rise. This trend requires of conference planners an unprecedented dedication to devising ways of utilizing unexpected input in conferencing.

It is also necessary to focus more of conference planning on the outside parameters, the limits, of the conference than on the ideal. Stating what is outside the scope of the discussion allows participants to stay within that scope far more helpfully than does indicating what their participation should be like.

Professional consultation in conferencing can be especially helpful in providing ways to orchestrate participation which advances rather than obstructs the intended direction of the conference. The art of group facilitation in this context can only grow in its importance.

Participants, the most important actors in the conference, can only do as much as they are given the opportunity to prepare for. Material which acquaints the participant as specifically as possible with the subject matter of the conference is urgent, whether this means working papers, a questionnaire designed to let the participant consider the issues which the conference will face, or pre–conference informational meetings and videos that allow participants to share the conference context as it is developed. Participation is limited by the use of passive media for preparation—long documents of background data to absorb, involved questionnaires to submit and so on. This is because a great deal of training and discipline is required by this kind of media if it is to be useful. Participation is enhanced by active media: setting the parameters of the conference itself, writing position statements, preparing displays and presentations, attending and setting up preparatory meetings.

Many consider the more passive media to be the more appropriately dignified mode for an intellectual or business endeavour. While use of media must take serious consideration of the group's habitual expectations and carefully honour them, it is often found that active media, appropriately tailored to the group being prepared for the conference, are outstandingly effective.

Preparation is basically of two types. Most commonly thought of is preparation of the group for the conceptual aspect of the conference. It is quite important, however, to prepare the group also for the methods which the conference will use. If consensus–making is to be used, participants need "field" or "hands–on" experience with the process of creating a consensus. Consensus–making on, for instance, a local position or the appropriate components of a subgroup's display can be held before the conference. The content of the activity is quite secondary to the need of participants having the opportunity to prepare themselves for the approach to be used. This is as true for the conference organizer as it is for the group member who wishes to participate meaningfully.

♦ **TC1929 Aggressive participant type**

Antagonist — Devil's advocate — Disagreer

Description Conference participants who tend to take aggressive attitudes towards other members of their group. They use various methods, deflating status of others, strongly expressed disapproval or rejection of others' values, acts or contributions; attacking the very reason for the meeting.

Advantages They can be a catalytic challenge to the other members of the group and can stimulate a creative and powerful response. Can be useful to stir up an apathetic or passive group.

Disadvantages Can be very distracting and time consuming and they rarely relinquish their role even when they get their own way. They can dominate timid members of the group and have the effect of making some people "clam– up" and say nothing.

♦ **TC1932 Key resource persons**

Description Conference participants who have access to or who are channels for restricted information in the sense that they have answers and data that others need and do not have.

Advantages They can be an effective resource for the opening up of information and the unlocking of answers. Identified in a conference situation they can be sought by those who require access to the information they represent.

Disadvantages They can, like keys, be used not so much to open doors, as to ensure that they remain closed to others.

♦ **TC1940 Vision conferencing**

Design conferencing — Visionary conferencing

Description Meetings aspiring to envisage, design and plan for the future.

Advantages Visions of the future throw present reality into new perspectives. Creating idealized future scenarios allow participants to recognize elements of the vision as they are encountered in emerging trends and be able to make decisions about which initiatives and movements to support. They can also reveal the underlying obstacles, upon which, realistic action can be focused.

Disadvantages The vision, if any, tends to remain at the level of enthusiasm and is not confronted with constraints essential to the design process from which action blueprints can emerge. If designs do emerge, they tend to be simplistic and to reflect the views of the self–selected group which worked on them.

♦ **TC1943 Communication patterns**

Constrained meeting configurations

Description To understand how different meeting configurations may emerge, it is useful to look at the variety of communication patterns which may be characteristic of a meeting:. At one extreme, for example, participants, viewpoints or conditions interact more or less randomly with no detectable pattern or order. This is characteristic of "idea fairs".

One approach to the possible patterns in a large group is to review the well–studied communication patterns in small groups of 3–6 participants. These are described as the circle, the star, the Y, the line, and the starred circle. Although these patterns have been examined in groups of participants, they may also be characteristic of groups of groups, or groups of themes, or the relationship between meeting sessions. For example, the star pattern emerges when all participant groups (or themes), except one, are related to that one but not to each other. Each pattern has well–recognized advantages under certain conditions.

Another approach is to see the large group as a complex, but reasonably stable, network of relationships between participants, themes or meeting sessions. Social networks are studied by the discipline of social network analysis.

Another approach is to imagine all participants, viewpoints or meeting sessions as being represented as points on the surface of a simple stretched rubber sheet. If they are located, such that lines of communication drawn between them cross to a minimal extent, then the meeting pattern as a whole may be viewed as a particular deformation of a regular grid. For example, if stretched in one way, many of the lines of communication might converge on one point as in the star pattern (above). Or several such points of relative convergence might emerge. Alternatively, by stretching the sheet so that a single space emerged in the centre, all the communication lines between points would be pushed into an approximate circle around it, creating the circle pattern.

The grid deformation approach briefly outlined may also be associated with a possible application of catastrophe theory to an analysis of meeting events, structural stability and morphogenesis. But even if a highly ordered communication pattern is achieved, this does not necessarily mean that it can focus usefully in terms of any objectives. It may simply be an efficient way of dissipating energy generated at the central point (see elsewhere). Supposing however that, in this case, a "counter–grid" or "counter– pattern" is delineated in the same way for the problems with which the elements of the communication pattern are confronted. This problem pattern could simply be the reflection of the other (effectively generated by "reflection" within it on the perceived problems and how they are linked). But if the one pattern "comes to grips" with the other and is "constrained" by it, a balanced configuration could emerge in one of two forms.

Such forms may correspond to those of convex or concave lenses. The resulting optical analogy draws attention to the significance of the fact that, constrained into curvature in this way, the focal point, previously "at infinity", is brought closer to the lens according to the degree to which the configuration is constrained into curvature (being at the centre of a spherical configuration, as the limiting case).

Related Meeting energy dissipation (#TC1699).

♦ **TC1944 Conference leadership**

Description Leadership and/or various working groups within a conference do not control but simply influence the dynamics of the event. Each group, as well as the team of leaders, facilitators, etc. has influence over their given specific arena which might greatly influence the direction of the event. Usually this influence on the direction of the event is temporary and is somewhat arbitrary, depending on the composition, mood, topic, etc. of the conference.

The conference leadership is the team of people whose common discipline and internal organization are designed to allow the fullest possible responsibility for the achievement of the intent or intents of the conference. Their concern is both their arena of responsibility and the whole conference. The larger the event, the more minds it takes together to catch all of the details of the life of the conference.

The overall leadership team needs members with to have overlapping responsibilities. The tension among two or three individuals dealing together with an arena provides the necessary creativity to respond to unusual needs should they arise. One person in charge of one arena can lead to territoriality.

Conference leadership does more preparation than anything else. They prepare for the conference, and then during the conference prepare for every session, every workshop and every plenary. It is not altogether necessary that they all participate in any but the most highly symbolic sessions of the conference unless they have unique individual roles to fill. Rather, they are the ones who reflect on the results of each session, consider new alternative directions as they are needed, and watch the conference spirit. They are necessarily in evidence "around the edges" of the conference, and make themselves available to participants who have special concerns about the conference direction.

Leadership of a conference may consist of groups dealing with (a) the overall leadership of the event, (b) conference procedures, (c) spirit of the , meeting, (d) facilitation and (e) practic.

The overall leadership devises meetings with facilitators, and the coordinators of the programme, of the conference life and of the practics, to allow everyone to share their learnings and to hear arenas for new directions, ideas for how to implement them, and assignments to teams and individuals to carry them out. When a crisis occurs in the direction of the conference, the leadership team has the time to locate the contradiction underlying the issue, and to work out a number of approaches to dealing with it. For this sort of purpose, they may well ask the counsel of participants who are appropriately placed to be helpful.

The overall leadership is very much concerned with polity. They try to devise new strategies for reconciling opposing viewpoints. They look for ways to ensure that decisions made are as inclusive as possible. They function occasionally as a sounding board for issues and individuals who feel themselves left out of the process, and for tempering issues and individuals who may be dominating the process. They watch the intent of the conference above all, and seek to ensure that it is achieved both through the conference process and beyond the process, through individual conversations and negotiations as needed.

The overall leadership serves as the centre of ongoing conference coordination. They are the place to go for underlying issues and concerns. They delegate the ongoing work of the conference in order to be available for rethinking and recoordinating as it becomes necessary. In a way they are the eyes and ears of the conference. This team, whether a full time group in session or a daily check–signals meeting, symbolizes the legitimacy and the responsiveness of the conferencing process. They symbolically and quite substantially exist to make it clear to anyone that the conference is not a system set in motion which now runs on driverless according to programme, but is a living creation to be improved upon and adapted as the content and the group need arises. The very existence of such a group is often enough not only to ensure the sensitivity of conference activities to participant input, but also to convince dissidents that their outlook is being taken into consideration, in whatever form that is taking place.

One of the most effective ways of designing the organization of leadership for a conference is to create several autonomous groups with some overlapping responsibilities and with ample opportunity to dialogue. Several levels of leadership can be used. The overall leaders (possibly a team of four or five) are responsible for many of the master of ceremonies roles in the conference. They are "In Charge" of the event. This is the role of the first person who speaks in the meeting. "Good evening, ladies and gentlemen," is the first line of this role. It establishes that someone is in charge so the group can relax. Whereas a round table without a leader creates anxiety about what is required of whom, a more formal style paradoxically creates confidence that someone is in charge to whom a participant may go if displeased. At the same time this role has to appear to be quiet, nonchalant and natural and at the same time requires meticulous staging and orchestration. "The best leader is the one who, when the war is won, the people say, 'We did it ourselves.'" (Sun Tsu) The mark of a successful conferencing event is one where the leadership is not intrusive.

The second level of leadership are the facilitators. They are the up front people during the workshops and the studies. They have the closest contact with the participants. They need to be capable of leading their sub–group to a product and yet enable all of the participants to actually contribute to the product. They, stylistically, need to be open and flexible and yet understand where they are going and how to get there.

The third level of leadership are the support forces. These include the practics people and the procedures group. Their role is that of seeing to it that the participants and facilitators are not distracted from the point of the conference, that they are not bothered by food or accommodations or foggy directions. They coordinate translation, production, food, lodgings, decor, procedures and any of the other practical aspect of making a conference successful. They translate the existential aim and rational objective of the event and of each session into practical steps and environments.

The leadership team is concerned with consensus, within whatever decision– making process is being applied. If the form of polity of the conference is dictate, they are after the acceptance and creative response of the body to the dictates of the authorities that be. If, for example, the method is majority rule, they seek a reconciliation among opposing forces by which they can move with whatever the majority opinion turns out to be, rather than withdrawing support into a camp of one of the opposing forces.

This is a concern not only for the process of the conference itself, but more importantly for the

quality, the depth and the long–term effectiveness of the decisions arrived at. This role clearly is one which requires insiders to the organization, but also a strong sense of team collaboration and considerable trust by the conference delegates. In effect, the leadership team functions as a microcosm of the conference, in order to work through its living dynamics and to affirm its struggles.
The leadership team is basically a symbolic dynamic, which serves as a "dumping ground" or a "lightning rod" for feedback of all kinds. Its role is to suggest direction, tone and approaches to all of the other arenas of conference life.
It is certainly clear that non–political power roles play a very important part in the conferencing dynamic, but this arena is relatively unclear. A good source for this sort of research is the type of conference which exists purely for symbolic purposes, in which protocol is more important than decisions, products or the intellectual life. This is an arena in which much consideration and new models are badly needed.

♦ TC1947 Opaque projector
Description A machine that projects written material directly onto a screen. Used simply by placing a book, or diagram in position under the high wattage camp and it is enlarged and projected onto a screen.
Advantages Mechanically simple and thus perhaps less liable to go wrong or to be misused.
Disadvantages Tends to be heavy, cumbersome and slow to use. It needs a darkened room which can restrict note taking by participants.

♦ TC1952 Conference intimacy
Description The degree to which participants in a conference are mutually acquainted with one another and are familiar with the most personal details of each other's lives. It is reflected in the nature of topics discussed, modes of greeting, forms of address and in interactions which presuppose a knowledge of the probable reaction of others under widely differing circumstances, as well as in the extent and type of knowledge each member has about other members of the conference.

♦ TC1955 Conference stratification
Description The degree to which a conference is ordered into status hierarchies. It is reflected by differential distribution of power, privileges, duties, and by assymetrical patterns of differential behaviour among members.

♦ TC1962 Alienation of committed activists
Description Each generation produces a number of well–qualified individuals concerned with one or more social problems and prepared to commit themselves, and possibly their careers, in an effort to achieve a significant impact upon them. Such people frequently instigate or function actively in meetings. As in any occupation, some years are spent learning the dimensions of the problem and the possibilities for action, especially in a conference environment. Thereafter, however, many of them find themselves forced into positions of compromise. In an effort to stick to their original values, they come into conflict with conference structures and resource realities which often prevent anything more than token action. They are encouraged to be patient and find that patience changes little. They find that those activists who have preceded them, and continue to attend meetings, lapse easily into cynicism or are satisfied with minimal change. They find that those who are similarly inspired, and who should be their allies, are frequently hostile and suspicious of any form of cooperation of more than a token nature.
Some become aware that even when their recommendations are fully accepted in a conference, and implemented by some organizational system with apparent success, the system in effect nullifies such achievements by adjusting itself so that other different problems emerge. There is then no end to such a chain of displaced problems, many of which are as much internal to the system of meetings and organizations as they are external foci of the action of a meeting or an organization. These situations finally lead to a withdrawal (or "loss of faith") of many of the committed activists.
This withdrawal takes place without transfer of acquired experience and insight to others who might later be able to overcome the dynamics of entrapment. There is no accumulation of learning. Those who know about the dynamics are often unable to speak about them, or have lost the desire to do so. Those who do speak about them are frequently ill–informed, self– interested and merely provoke a repetition of learning cycles. This withdrawal may well take the form of a refusal to participate in meetings in which their insight would be invaluable. They may argue that "large conferences are a useless waste of time". Such conferences then become meetings of the uninitiated with all that implies for their outcome.
Some withdraw partially and are willing to attend conferences if they are given some significant role in their organization, or as speakers. As such they may be totally indifferent to the impact on participants of the conflicting views disseminated by themselves and their colleagues of the same frame of mind. Other eminent individuals attend conferences but remain silent in order to allow time for the uninitiated participants to interact and learn from the experience. Again this may prevent their experience from being appropriately reflected in the outcome.
Related Integrative failure (#TC1005) Losing focus in meetings (#TC1432) Repetition of learning cycles (#TC1599).

♦ TC1965 Participant pre–logical biases
Description At the basis of the personality of every person or group active in a meeting, it is useful to recognize a set of pre–rational temperamental biases which are reflected in the aesthetic, theoretical, value and action preferences and in the preferred mode of discussion. The preferred mode of each individual or group may be positioned somewhere along axes of bias such as the following:
1. Order vs disorder: Namely the range between a preference for fluidity, muddle, chaos, etc and a preference for system, structure, conceptual clarity, etc.
2. Static vs dynamic: Namely the range between a preference for the changeless, eternal, etc and a preference for movement, for explanation in genetic and process terms, etc.
3. Continuity vs discrete: Namely the range between a preference for wholeness, unity, etc and a preference for discreteness, plurality, diversity, etc.
4. Inner vs outer: Namely the range between a preference for being able to project oneself into the objects of one's experience (to experience them as one experiences oneself), and a preference for a relatively external, objective relation to them.
5. Sharp focus vs soft focus: Namely the range between a preference for clear, direct experience and a preference for threshold experiences which are felt to be saturated with more meaning than is immediately present.
6. This world vs other world: Namely the range between a preference for belief in the spatio–temporal world as self–explanatory and a preference for belief that it is not self–explanatory (but can only be comprehended in the light of other factors and frames of reference).
7. Spontaneity vs process: Namely the range between a preference for chance, freedom, accident, etc and a preference for explanations subject to laws and definable processes.
Such pre–logical biases may be at the base of choice of life–style, discipline, policy, mode of action, mode of presentation of information, etc. To the extent that people have very different profiles in terms of these axes, every particular meeting position, viewpoint or programme will have only limited appeal. The challenge is to design meetings which are hospitable to all these biases in order that any outcome will be significant to those in the outside world who share them. Meetings are usually "successful" when participants share a set of biases and are a "frustrating failure" when a wider spectrum is represented. In neither case is the problem of overcoming such differences recognized, nor is the validity of any bias respected if it is represented by a minority. It is not surprising therefore that the results of "successful" meetings have little impact on those whose biases were not blended into their outcomes.
Related Participant personality needs (#TC1204) Participant strategic preferences (#TC1828) Participant interaction preferences (#TC1845).

♦ TC1969 Team focalizer role
Chairperson — Facilitator — Moderator
Description An affinity–role played by one or more members of a group or team. In style they tend to be calm, confident, and controlled. They have the ability to encourage potential contributors and to input without prejudice.
Advantages They can recognize the strengths and weaknesses within the team and help to ensure that best use is made of each person's potential. A strong sense of objectives enables them to maintain the focus of the group's deliberations.
Disadvantages Although they can recognize talent in others, they tend not to have more than an ordinary intellect or creative ability. Focalizers or chairpersons have perhaps the greatest tendency to forget that they are one of many roles that make up a creative team and to fall into the trap of assuming they play the key role which can stifle and frustrate the participation of others.

♦ TC1971 Intellectual momentum
Description The intellectual momentum of a conference begins long before the event itself. The process develops and refines the thinking of the organizers and the participants alike. Both intuitive and rational aspects are utilized. Screens and charts used as models of reality are developed using both approaches, but are directed at the more rational aspects of the conference. Images used as symbols of more complex and profound relationships are developed using both, but are more directed toward the intuitive aspects of the event.
People operate out of internalized pictures of the world, themselves and their role in the world. These operating–images are created as solutions to problems. They provide people with ways of appropriating and understanding the world. They sort information into relevant and irrelevant data at an unconscious level. They make sense out of the world for people. They can be invented by the individual but by far the bulk are given by society. In fact, society provides us with solutions to problems of living and to problems that do not yet exit. Operating–images can change and when an operating–image changes individual behaviour changes. Operating–images may be added to or minor adjustments may be made. Operating–images may be dramatically altered to the point that a whole new world view becomes operative.
It is not possible for an individual to change his own operating–image but he may place himself in a situation where his operating–image may change and he may be open to the possibility of change. Messages from the environment add to, confirm or change them. Operating–images resist change, by denying the validity of the message or the source of the message or by screening out the message altogether.
The intellectual momentum of the conference depends on intelligent evaluation of the practical needs of the group, more so the larger the group is. At the same time, it is up to the intellectual momentum of the conference to take into consideration the actual capabilities for production and translation, to ensure that they are envisioning a flow of thinking and learning that is within the realm of possibility.

♦ TC1972 Audio teleconferencing
Speakerphone meeting — Conference line — Loud speaker telephone
Description A meeting involving a number of participants in different locations through the telephone system, whereby all the participants can hear all the contributions made. Recent developments include customized rooms with special accoustic equipment.
Advantages Participation by many people in different parts of the world in small groups, without the expense and time wastage of travel and special accommodation.
Disadvantages Without visual contact and with limited possibility of transferring visual data immediately, the audio teleconferencing cannot always command the confidence of the groups involved. Its decision–making and consensus building function is therefore constrained.

♦ TC1974 Consultation conferencing
Diagnostic conferencing
Description A meeting between a local action group and other regional, national or international agencies representing various fields of expertise. The aim of the conference is to diagnose the situation confronting the local group and to assist in the creation of a plan of action for them to appropriately deal with their situation. The "external" agencies contribute their experience and expertise, the local group its commitment, local knowledge, local wisdom and experience. The "external" perspective introduces a new perspective and is outside the local system.
Advantages The importance of this kind of multisector coalition is more than the pooling of material experience and expertise. The presence of people who are outside the local system and are without direct vested interest and involvement is an intervention which allows the local system itself to change and this can release new energy into the problem–solving process.
Disadvantages The power, influence and large material resources that appear to be in the control of the external agencies seen from the local perspective, can distract the local group from recognizing its own potential and in turn affect their participation in the consultation towards manipulating the external consultants into solving their problem, thus tending to "learn" dependency.

♦ TC1989 Strongwilled conferencing
Description Meetings are strongwilled when the participants can be said to shape the circumstances to suit their intentions, and the proposals and plans can be seen to be carried out.
Advantages Effectiveness, changing circumstances don't affect the intentions. Successes and mistakes are dramatic and have a definite effect on the group's environment.
Disadvantages Strongwill is overrated because strongwilled failures are not identified with strongwill but with stubbornness. Lacking in sensitivity, there is a tendency towards rigidity, ruthlessness, and callousness.

♦ TC1990 Encourager participant type
Enabler
Description Conference participants who openly recognize, accept and encourage the contributions of others. They indicate warmth and solidarity in their attitude to the group and in various ways, indicate understanding and acceptance of other points of view, ideas and suggestions.
Advantages Tends to enable the discussion process by encouraging participation by all members of the group.
Disadvantages Can appear to be and in fact be rather insincere and obsequious; the group may distrust them.

TC1996
Business concern conferencing
Productive conferencing — Economic conferencing
Description Prime beneficiaries are those who have something to sell. Their ideas, insights, products, time, services, are offered to the consumer audience. Characterized by the high level of activity by presenters and the high level of passivity by the participants.
Advantages Useful mode for commercial concerns: industrial firms, mail order houses, wholesale and retail stores, banks, insurance companies, etc. but also for other groups with a product or a message to "sell", and a group to sell to.
Disadvantages The basic problem of operating efficiency –balancing the investment and returns. The problem of communication overload in a field of intense competition. The "consumers" who either "buy" or reject the product, and have very little opportunity for other forms of creative feedback or input.

Transformative policy cycles TP

Rationale

Most conceptual schemes, whether purely theoretical or basic to the practical design of a development programme, are organized into sets of concepts, principles, priorities, or functions. Several such sets may be interrelated in a more elaborate scheme. It is the pattern of such interrelationships which ensures the coherence and integrity of the approach.

Policies as conceptual schemes are notorious for their inability to respond to longer-term cycles of change associated with sustainable development. There is therefore merit in exploring ways of thinking about cycles of policies which are deliberately designed to allow for the emergence of the negative features of any particular policy phase, and to correct them in subsequent phases (that will in turn engender their own negative consequences).

Contents

The 64 entries constitute an *editorial experiment* based on the pattern of concepts implicit in the much-publicized Chinese classic, the *Book of Changes*. These are transposed into a language which highlights the significance of such a complex pattern of transformations in any organizational or meeting environment. Its special merit is the explicit recognition of the need to shift from condition to condition in order to ensure both healthy development and the ability to respond to a turbulent environment. One traditionally recognized set of 384 transformational pathways between the conditions is indicated.

Method

The method used is discussed in the comments in Section TZ.

Index

A keyword index to entries is provided in Section TX. THe keywords are also incorporated into the index for Volume 2 (Section X)

Comment

Detailed comments are given in Section TZ.

Reservations

Although the results of this deliberate editorial experiment are interesting and indicative of further possibilities, the entries raise many questions concerning the appropriateness of the language used. The language as been deliberately forced into western organizational jargon which can be claimed to be contrary to the poetic spirit of the original. It is interesting to note that these entries reflect a "top-down" approach to policies and management in that they emphasize the subtle and abstract perspectives through which continuity can be maintained through change. In doing so they sacrifice the kinds of precision which are normally required in "practical" managment over the short-term.

Possible future improvements

Further thought needs to be given to ways of linking such abstract patterns of change to those of mode immediate concern to policy-makers.

ENTRY CONTENT AND ORGANIZATION

Ordering of entries
Entries are in **numeric order**.

Index access to entries
The location of an entry in this sub-section may be determined from:
- the **Volume Index** (Section X) on the basis of keywords in the name of the entry or its alternate names
- the **Section Index** (Section TX) on the basis of the name of the entry

Structure of entries
Entries may be composed of the following descriptive elements:

(a) **Entry number** The last two digits are the number traditionally used to sequence the entries in the *Book of Changes*. Here it is merely a convenient method of identifying the entry (particularly for indexing and cross-reference purposes). The first letter of the entry number refers to the section of this volume in which the sub-section, denoted by the second letter, is located.

(b) **Qualifier** following entry number. There are **no qualifiers** for this group.

(c) **Entry name** This is printed in bold characters.

(d) **Condition** General indication of the condition as applied to a sustainable policy cycle.

(e) **Subconditions** Indication of 6 possible variants of the basic condition through which it may be transformed into an another condition (cross-referenced in papentheses).

(f) **Transformation sequence** Indication of a basic constraint on the prolongation of the condition and the manner in which it may be transformed into the subsequent condition in the basic sequence (eventually repeating the cycle, since TP0064 transforms back to TP0001).

Cross-referencing between entries
These are of two forms and are explained in points (e) and (f) immediately above.

♦ **TP0001 Creativity** (Heaven)
Strong action
Condition Creative energy and inspiration may engender new patterns as a result of unrestrained action by a sustainable policy cycle.
Subconditions:
1. Patient caution by the sustainable policy cycle may be required to avoid premature action. (To: TP0044)
2. Exertion of a recognized positive influence by the sustainable policy cycle may be required prior to action. (To: TP0013)
3. Attraction of mass support by the sustainable policy cycle can lead to the temptations of over-ambition. (To: TP0010)
4. The sustainable policy cycle may be faced with the choice between internal development and external social action. (To: TP0009)
5. Widespread recognition of the effects of the action of the sustainable policy cycle on society may result in long-term positive consequences. (To: TP0014)
6. Catastrophe may result when the sustainable policy cycle indulges in aspirations exceeding its capacity. (To: TP0043)
Transformation sequence In order to bear fruit, creativity eventually requires the existence of a receptive environment.

♦ **TP0002 Receptivity** (Earth)
Acquiescence — Responsive service — Passivity
Condition A sustainable policy cycle may respond to the actions and opportunities of its environment through which it may then bring about change.
Subconditions:
1. The sustainable policy cycle may take heed of the first signs of deterioration in its environment. (To: TP0024)
2. The sustainable policy cycle may respond naturally to its environment. (To: TP0007)
3. Care should be taken to avoid premature public attention to the maturing work of the sustainable policy cycle. (To: TP0015)
4. The strictest reticence is required by the sustainable policy cycle to avoid both the enmity of antagonists and the dangers of misplaced acclaim. (To: TP0016)
5. The sustainable policy cycle may express its qualities indirectly and discreetly as its actions emerge into prominence. (To: TP0008)
6. The sustainable policy cycle may make an inappropriate attempt to take the leading role, thus causing a struggle destructive to all concerned. (To: TP0023)
Transformation sequence Initiatives emerging in a receptive environment first experience difficulties.

♦ **TP0003 Initial difficulty**
Difficulty at the beginning — Gathering support — Organizational growth pains
Condition Due to the profusion of changes being brought about, confusing obstacles to the growth of sustainable policy cycle action occur, calling for a cooperative response to bring order out of chaos.
Subconditions:
1. The sustainable policy cycle may proceed cautiously in pursuit of its goal by attracting suitable assistance. (To: TP0008)
2. The sustainable policy cycle should avoid the obligations entailed by early acceptance of assistance from unexpected sources in times of difficulties. (To: TP0060)
3. The sustainable policy cycle should renounce immediate objectives that have proved unfruitful due to premature action without adequate guidance. (To: TP0063)
4. Necessary sustainable policy cycle action may be rendered successful, despite inadequate resources, by obtaining appropriate assistance (To: TP0017)
5. Its position in society may cause direct action by the sustainable policy cycle to be distorted, necessitating cautious indirect action to overcome the obstacles arising from such misinterpretations. (To: TP0024)
6. The sustainable policy cycle may cease its struggle due to an acceptance of the initial difficulties as overwhelming. (To: TP0042)
Transformation sequence When first launched, initiatives tend to be handicapped by inexperience.

♦ **TP0004 Inexperience** (Young shoot)
Youthful folly — Acquiring experience — Immaturity — Uncultivated growth — Darkness
Condition Aided by enthusiasm, sustainable policy cycle action may succeed despite inexperience, provided appropriate guidance is sought with the right attitude.
Subconditions:
1. Discipline is a necessary counterweight to dissipative carelessness in sustainable policy cycle action, although excessive discipline has itself a crippling effect on the development of sustainable policy cycle potential. (To: TP0041)
2. Tolerance of shortcomings is a prerequisite for assumption of social responsibility by the sustainable policy cycle. (To: TP0023)
3. The inexperienced sustainable policy cycle does not develop by applying itself to simple problems which readily offer themselves for solution. (To: TP0018)
4. The sustainable policy cycle entangled in action fantasies of imagined significance can often only free itself by experiencing the humiliation which finally results. (To: TP0064)
5. The inexperienced sustainable policy cycle that seeks guidance in an unassuming manner may develop successfully. (To: TP0059)
6. Constraints should be applied to the sustainable policy cycle that persists in careless action, but only to prevent unjustified excesses. (To: TP0007)
Transformation sequence After overcoming problems of inexperience, initiatives await further support.

♦ **TP0005 Waiting** (Getting wet)
Biding one's time — Calculated inaction
Condition A sustainable policy cycle can only derive the strength to confront crises by being able to wait, however long is necessary for opportunities to emerge, rather than being panicked into action by immediate dangers.
Subconditions:
1. Before the sensed crisis takes form, the sustainable policy cycle should continue as long as possible to engage in the longterm processes by which its strength is renewed. (To: TP0048)
2. As the crisis takes form, disagreements may emerge and escalate dangerously if the sustainable policy cycle is unable to maintain its tranquillity. (To: TP0063)
3. Premature response to the crisis leaves the sustainable policy cycle waiting in an exposed and vulnerable position which calls for serious reassessment. (To: TP0060)
4. In the midst of danger any action may aggravate the situation; the composure of the sustainable policy cycle is the only guarantee of survival as events take their course. (To: TP0043)
5. Despite the crisis there are moments of calm which the sustainable policy cycle should use to fortify itself for renewed struggle. (To: TP0011)
6. When the crisis strikes, the sustainable policy cycle must yield to the inevitable whilst being ready to respond to the potential of unforeseen developments. (To: TP0009)
Transformation sequence Initiatives awaiting support engender conflict over the allocation of available resources.

♦ **TP0006 Conflict**
Grievance — Contention — Strife
Condition When a sustainable policy cycle encounters opposition in pursuing a course of action it considers appropriate, conflict arises which can only be usefully resolved by coming to terms with the opponent.
Subconditions:
1. In its incipient stage, especially when the opposition is strong, it may be best for the sustainable policy cycle to drop an issue rather than risk open conflict. (To: TP0010)
2. If the opposition is of superior strength, timely withdrawal by the sustainable policy cycle may prevent undesirable consequences for the community as a whole. (To: TP0012)
3. If the sustainable policy cycle subordinates itself to a strong ally, conflict can be avoided by not acting to acquire prestige. (To: TP0044)
4. If the opposition is of weaker strength, the sustainable policy cycle will have difficulty in justifying the success of any conflict and can best achieve its ends by redefining its goals. (To: TP0059)
5. If it is in the right, the sustainable policy cycle can derive great benefit if the conflict takes place under the auspices of a powerful and just arbiter. (To: TP0064)
6. If the sustainable policy cycle carries the conflict successfully to the bitter end, it will find its success short lived and constantly exposed to further attack. (To: TP0047)
Transformation sequence When there is conflict a controlled threat eventually emerges to regulate it.

♦ **TP0007 Controlled threat** (Army)
Condition For a sustainable policy cycle to struggle successfully, discipline must be instilled in the community by arousing enthusiasm, sustaining the people and eliciting confidence in the value of its actions.
Subconditions:
1. For an entreprise to be successful when it is initiated, order must prevail within the sustainable policy cycle. (To: TP0019)
2. The sustainable policy cycle should remain in close touch with the community, sharing its condition and receiving recognition on its behalf, in order to be able to meet the demands made on it. (To: TP0002)
3. If the community takes the initiative from the sustainable policy cycle, misfortune will ensue. (To: TP0046)
4. If the sustainable policy cycle is faced with superior opposition, orderly withdrawal avoids the risk of disintegration. (To: TP0040)
5. Energetic struggle by the sustainable policy cycle is required to counteract any attacks, but a disorderly response may become counterproductive. (To: TP0029)
6. Following successful action by the sustainable policy cycle, it is important that power should not be given to those who would abuse it. (To: TP0004)
Transformation sequence The emergence of a controlled threat eventually promotes solidarity.

♦ **TP0008 Solidarity**
Coordination — Leadership — Alliance — Holding together — Accord
Condition Complementarity of action by the different parts of a sustainable policy cycle requires that they should be held together by a central symbol whose significance reinforces each in his understanding of his role within the action of the whole.
Subconditions:
1. Only a fundamental sincerity provides the basic cohesive power through which appropriate sustainable policy cycle relationships can be formed. (To: TP0003)
2. To avoid losing the dignity and intrinsic clarity of its relationships, the sustainable policy cycle should persevere in the appropriate response to any summons to action. (To: TP0029)
3. The sustainable policy cycle should avoid engendering relationships based on false intimacy - formed when differences of habit are underestimated - which subsequently prevents the formation of more genuine and appropriate bonds. (To: TP0039)
4. When the sustainable policy cycle relationships with the central rallying point are well established, they should be acknowledged openly. (To: TP0045)
5. The cultivated quality of the central symbol recognized within the sustainable policy cycle should be such as to engender the voluntary dependence of those who hold to it, leaving others free to go their own way. (To: TP0002)
6. Coherence within a sustainable policy cycle calls for right timing which if miscalculated through hesitancy may be a cause for regret. (To: TP0020)
Transformation sequence Solidarity ensures a subtle restraint.

♦ **TP0009 Subtle restraint**
Restraint by the weak — Taming power of the small — Nurturance by the small — Small obstructions
Condition When the influence of a sustainable policy cycle is as yet unable to produce great or lasting effects, it is best that it should act in a restraining or subduing manner in anticipation of ultimate success.
Subconditions:
1. Avoidance of forceful action by the sustainable policy cycle leaves it free to advance or retreat when obstructions are encountered. (To: TP0057)
2. The sustainable policy cycle benefits by retreating with others of similar orientation when obstructions are too great. (To: TP0037)
3. Forceful action by the sustainable policy cycle is bound to fail when circumstances combine to enhance the powers of seemingly minor hindrances. (To: TP0061)
4. Disinterested restraining action by the sustainable policy cycle to ensure that the right prevails will eventually succeed, despite the real dangers with which such action is threatened. (To: TP0001)
5. Relationships based on loyalty and trust reinforce the effectiveness of the sustainable policy cycle and of the complementary roles of those involved. (To: TP0026)
6. When the sustainable policy cycle has achieved limited success from a position of weakness, it is dangerous to pursue the advantage any further until circumstances are more favourable. (To: TP0005)
Transformation sequence Subtle restraints give rise to careful conduct.

♦ **TP0010 Careful conduct** (Treading)
Stepping carefully — Tranquillity
Condition A sustainable policy cycle can best relate to those who are strong and intractable by conducting itself with due respect, reinforcing recognition of inner worth whenever it is reflected in external rank.

Subconditions:
1. The sustainable policy cycle can make progress when in an inferior position through simple unassuming actions, since it is not yet bound by social obligations. (To: TP0006)
2. The sustainable policy cycle may act successfully in isolation, free of conventional entanglements and the enticements of conventional goals. (To: TP0025)
3. When handicapped, reckless action beyond its capabilities invites disaster, which is only justified when the sustainable policy cycle is struggling for a higher cause. (To: TP0001)
4. The sustainable policy cycle may succeed in a dangerous entreprise if it acts with caution, conscious of its inner resources. (To: TP0061)
5. For a resolute action to succeed, the sustainable policy cycle should be aware of the dangers of resoluteness. (To: TP0038)
6. The success of the sustainable policy cycle can only be assessed in terms of effects of its actions; these then determine the consequences for the sustainable policy cycle. (To: TP0058)
Transformation sequence Careful conduct ensures that peaceful relationships prevail.

♦ **TP0011 Peace**
Flowing
Condition When harmony prevails in a sustainable policy cycle and in its relationship with society a period of fruitful action is assured. To benefit from this condition processes must be ordered and adjusted to increase their natural yield.
Subconditions:
1. Under such favourable circumstances the sustainable policy cycle attracts together those with similar preoccupations and is encouraged to extend its activities. (To: TP0046)
2. The sustainable policy cycle can overcome the danger of becoming slack in times of harmony by working with the imperfect, risking dangerous undertakings, taking the seemingly insignificant into account, and avoiding factionalism. (To: TP0036)
3. The sustainable policy cycle can avoid succumbing to the illusion that the period of harmony will never end by recognizing how its inner strengths are independent of external circumstances. (To: TP0019)
4. In times of mutual confidence, the sustainable policy cycle can spontaneously establish close contact with the alienated, by emphasizing an inner bond rather than external inequality. (To: TP0034)
5. The sustainable policy cycle may successfully achieve its ends by uniting with others who have fewer advantages. (To: TP0005)
6. When the period of peace comes to an end the only recourse is to accept the transition and attempt to maintain a measure of harmony within the sustainable policy cycle itself. (To: TP0026)
Transformation sequence If peaceful relationships continue to prevail, stagnation eventually results.

♦ **TP0012 Stagnation**
Blocked — Obstruction — Standstill
Condition Disharmony prevailing in a sustainable policy cycle and in its relationships with society ensures an uncreative period of confusion and disorder. As the exertion of effective influence is impossible, a sustainable policy cycle can best remain faithful to its principles by withdrawing into seclusion rather than by accepting the temptation of public action.
Subconditions:
1. Under such unfavourable circumstances, the sustainable policy cycle may best protect its values by retiring into seclusion with other of similar preoccupation. (To: TP0025)
2. The sustainable policy cycle should not interact with those of inferior values, even though they might welcome such action as a way of reducing their disorder, any consequent suffering to the sustainable policy cycle is a guarantee of its ultimate success. (To: TP0006)
3. Those of inferior values, who have illegitimately acquired power within the sustainable policy cycle or in society, eventually recognize their lack of ability. (To: TP0033)
4. Those seeking to restore order within the sustainable policy cycle or in society should feel capable of responding to the challenge in collaboration with others rather than risk acting in the light of their own limited perceptions. (To: TP0020)
5. Once the sustainable policy cycle has emerged into a position from which order can be restored, success is only assured through the greatest attention to the possibilities of failure. (To: TP0035)
6. The sustainable policy cycle must act deliberately and creatively to end the condition of stagnation and disintegration. (To: TP0045)
Transformation sequence Stagnation cannot persist indefinitely and therefore fellowship finally emerges.

♦ **TP0013 Fellowship**
Universal brotherhood — Identification with others
Condition True fellowship can be brought about within and by the sustainable policy cycle through the emergence of clear, convincing, and inspiring aims. These should be based upon a concern that is universal and be accompanied by the strength to carry them out. To ensure the appropriate functional relationships amongst diverse elements, an organic mode of organization is required.
Subconditions:
1. The sustainable policy cycle should ensure that the fundamental principles upon which any union is based are equally accessible to all those involved. (To: TP0033)
2. The emergence of exclusive factions based upon self–interest and the rejection of others is a danger to the sustainable policy cycle and to the achievement of its aims. (To: TP0001)
3. Mistrust and reservation within the sustainable policy cycle undermine fellowship, leading to strategies based on guile which engender further alienation. (To: TP0025)
4. Confrontation may reach a point at which the opposing parties are no longer able to act against each other, and in this way the situation of the sustainable policy cycle is usefully clarified. (To: TP0037)
5. Its position in society may cause the sustainable policy cycle to be able to relate effectively to others who share its fundamental preoccupations only after a long struggle to overcome the obstacles unfortunately separating them. (To: TP0030)
6. The sustainable policy cycle may be able to engage with others only in a limited alliance based upon mutual interest rather than on a shared approach to universal concerns. (To: TP0049)
Transformation sequence Through fellowship values emerge, leading to acquisition of wealth.

♦ **TP0014 Wealth**
Possession in great measure
Condition A sustainable policy cycle may acquire a position of power in relation to the strong by acting disinterestedly with a low profile. In this way wealth is appropriately administered in a graceful and controlled manner.
Subconditions:
1. The sustainable policy cycle can avoid the temptation of wealth only by developing an awareness of the many difficulties to be overcome and of the possibilities of mistakes in its use. (To: TP0050)
2. The sustainable policy cycle should delegate responsibility in order to ensure that the resources at its disposal are used most effectively in new undertakings. (To: TP0030)
3. The sustainable policy cycle is most successful when it seeks to place itself and the wealth it has acquired at the service of a higher cause, or of society as a whole, rather than vainly attempting to maintain a hold on it for itself. (To: TP0038)
4. The sustainable policy cycle should carefully distinguish its own position from that of the strong with whom it is in contact, in order to avoid the dangers of vying with them and thus jeopardizing the very basis of its power. (To: TP0026)
5. Even when the benevolent action of the sustainable policy cycle succeeds in attracting support based solely on unaffected sincerity, the tendency for insolence to emerge must be kept in check through the strength of dignity. (To: TP0001)
6. When at the height of its power, the sustainable policy cycle can best enhance the value of its position by cultivating an unassuming attitude and honouring values which transcend the mundane affairs of society. (To: TP0034)
Transformation sequence To retain valuable possessions, the amount should be modest and the attitude unpretentious.

♦ **TP0015 Unpretentiousness**
Modesty — Humility
Condition A sustainable policy cycle prospers best by acting in an unassuming manner, whether in a position of influence or not. This principle also favours its efforts to establish order by reducing those extremes and inequalities which are the source of social discontent.
Subconditions:
1. The sustainable policy cycle may successfully undertake dangerous enterprises if the situation is not confused by unnecessary claims and by the resistance of others resulting from such claims. (To: TP0036)
2. When an unassuming manner is natural to the sustainable policy cycle's mode of action, the possibilities of exerting a lasting influence emerge of their own accord. (To: TP0046)
3. If the sustainable policy cycle responds immodestly to widespread recognition of its achievements, criticism develops, preventing the work from being carried through to its final fruition. (To: TP0002)
4. The sustainable policy cycle should guard against the danger that an unassuming manner of action may become an effective disguise for irresponsibility and inaction. (To: TP0062)
5. Circumstances may call for energetic corrective action by the sustainable policy cycle, which should not interpret the merit of an unassuming mode as an excuse for letting events take an inappropriate course. (To: TP0039)
6. The sustainable policy cycle should act vigorously to defend itself and order its environment, especially when the root of the problem lies in weaknesses of own. (To: TP0052)
Transformation sequence Valued possessions and unpretentiousness together engender enthusiasm.

♦ **TP0016 Enthusiasm**
Contentment — Repose
Condition A sustainable policy cycle can arouse enthusiasm by acting in harmony with the needs of the time and coopting assistance for the completion of an undertaking. Such enthusiasm releases people from the grip of mundane tensions and allows them to express the hidden potentials of their society.
Subconditions:
1. When in a position of weakness, the sustainable policy cycle invites misfortune if it makes enthusiastic claims about its own connection with those in positions of power. (To: TP0051)
2. The sustainable policy cycle should not allow itself to be misled by illusory manifestations of enthusiasm, but should be sensitive to the emerging tendencies of the time, acting self–reliantly in response to those in positions of strength or weakness. (To: TP0040)
3. The sustainable policy cycle must choose the right moment to act, for otherwise either the opportunity will be lost or else it will become unnecessarily dependent on an external leader capable of engendering enthusiasm. (To: TP0062)
4. The sustainable policy cycle can arouse enthusiasm and cooperation through its own self–confidence and the sincere support it gives to those who collaborate with it. (To: TP0002)
5. The sustainable policy cycle can be obstructed in its ability to engender enthusiasm, but this may usefully prolong its existence by preventing it from depleting its energies. (To: TP0045)
6. Being misled by false enthusiasm may constitute a valuable learning experience for the sustainable policy cycle, provided it is subsequently capable of further development. (To: TP0035)
Transformation sequence Where enthusiasm persists, a following emerges.

♦ **TP0017 Following**
Entrainment — Acquiring followers — Hunting — Attunement
Condition In order to be capable of inducing people voluntarily to follow its lead without resistance, the sustainable policy cycle must first adapt itself to their circumstances to be able to serve them.
Subconditions:
1. In order for the sustainable policy cycle to identify how it should adapt to its environment, it must open itself to contact with a wide range of different views. (To: TP0045)
2. In developing its pattern of contacts, the sustainable policy cycle should take care to avoid those holding inferior values or risk losing those holding superior values by which its action can be benefitted. (To: TP0058)
3. The development of its contacts with those holding superior values will lead the sustainable policy cycle to obtain what it needs for its own development, despite the loss of stimulating distractions with those holding inferior values. (To: TP0049)
4. Once it is successful in its influence, the sustainable policy cycle should develop the ability to distinguish insincere supporters, attracted for their own advantage, who must be kept at a distance if success is not to be jeopardized. (To: TP0003)
5. The sustainable policy cycle must itself follow something which guides, legitimates and empowers its initiatives. (To: TP0051)
6. Having developed its activities to the point of detachment from mundane affairs, the sustainable policy cycle may be confronted with a persistent following which once more draws it back into a guiding role. (To: TP0025)
Transformation sequence Following others leads to undertakings and remedial action.

♦ **TP0018 Remedial action**
Arresting decay — Responding to illness — Degeneration
Condition Inertia, indifference and the abuse of human freedom lead to deterioration of the sustainable policy cycle or society and call for decisive, energetic action, if regeneration is to occur.
Subconditions:
1. Deterioration due simply to rigid adherence to conventional patterns of action may easily be

remedied, provided that the sustainable policy cycle is conscious of the dangers associated with any such reform. (To: TP0026)
2. The deterioration may be the result of inherent weakness, in which case the sustainable policy cycle should avoid drastic action so as not to further aggravate the situation. (To: TP0052)
3. If the sustainable policy cycle proceeds somewhat too energetically in rectifying the mistakes of the past, difficulties will arise, but this is preferable to the results of insufficiently vigorous action. (To: TP0004)
4. Misfortune will result if the sustainable policy cycle is itself too weak to take action against progressive deterioration resulting from past mistakes. (To: TP0050)
5. Even though it is inadequate to the challenge of past neglect and corruption, the sustainable policy cycle may achieve partial success with the assistance of others. (To: TP0057)
6. The development of the sustainable policy cycle may be such that it is unnecessary for it to engage in any remedial action, provided that, in its withdrawal from mundane affairs, it engenders new values for the future. (To: TP0046)
Transformation sequence Where there is scope for remedial action, there is growth through initiative.

♦ TP0019 Initiative
Leadership — Approach — Getting ahead — Overseeing
Condition When conditions are appropriate for a sustainable policy cycle to initiate action on mundane affairs, this should be done with determination and perseverance, bearing in mind the need to prepare for unfavourable conditions which in their turn will later prevail.
Subconditions:
1. When superior values find a response in influential circles, the sustainable policy cycle could well associate itself with this trend, provided this does not distract it from its own line of action. (To: TP0007)
2. When the initiative originates in the light of superior values, the sustainable policy cycle should not hesitate to apply its own resources to the task, for such action must necessarily contribute to ultimate success. (To: TP0024)
3. When the sustainable policy cycle is succeeding in its initiative, there is a danger that lack of vigilance may lead to careless mistakes, which may however be remedied by responsible action. (To: TP0011)
4. The action is benefitted when the sustainable policy cycle is open-minded in its approach to those of ability who are attracted by its initiative. (To: TP0054)
5. The sustainable policy cycle should act with self-restraint in order to attract those of quality capable of undertaking all that is required by the initiative without interference. (To: TP0060)
6. A sustainable policy cycle which has withdrawn from mundane affairs may under certain circumstances initiate new action for the benefit of those it attracts. (To: TP0041)
Transformation sequence The results of initiative call for recognition.

♦ TP0020 Recognition
Contemplation — Watching — Observing
Condition Through the effort it devotes to comprehending the significance underlying external events, a sustainable policy cycle acquires the power to apply that understanding to influence events. This power can be recognized by others, who may in turn be influenced by it to take the actions of the sustainable policy cycle as a model for their own.
Subconditions:
1. Whilst it is to be expected that some can only be superficially affected by a profound understanding of events, it is to be regretted when the sustainable policy cycle of superior values contents itself with a shallow, disconnected view of the forces prevailing in society as a whole. (To: TP0042)
2. Whilst for some it is sufficient to view the world from a subjectively limited standpoint, this narrowness is harmful in the case of the sustainable policy cycle which must take an active part in the affairs of the world. (To: TP0059)
3. When it focuses on recognition of its own nature and the effects it creates, this may be a basis for the sustainable policy cycle to determine whether or not it is developing. (To: TP0053)
4. The sustainable policy cycle should facilitate independent action by those who understand how it can be made to flourish. (To: TP0012)
5. Self-evaluation by the sustainable policy cycle of superior values will only bring satisfaction when its effects are beneficial and free of mistakes. (To: TP0023)
6. The sustainable policy cycle detached from mundane affairs will most benefit society when exploration of psycho-social processes brings recognition of how it may avoid being responsible for generating negative effects. (To: TP0008)
Transformation sequence Recognition of the relationship between results engenders decisive integrative action.

♦ TP0021 Decisive action (Biting through)
Punitive action
Condition When faced with deliberate hindrance to integrative development, a sustainable policy cycle must take a just measure of decisive action against those responsible. Such hindrances increase when norms are unclear and there is negligence in ensuring that they are respected.
Subconditions:
1. If the sustainable policy cycle responds mildly to any initial departure from norms this should constitute sufficient warning against repetition of the infringement (To: TP0035)
2. If the response of the sustainable policy cycle to frequent infringement of norms is excessive, this should not be regretted since the results are merited. (To: TP0038)
3. Although no other course is possible, if the sustainable policy cycle lacks the power and authority to back up its censure when norms have been infringed, strong negative feelings will be engendered against it, placing it in a somewhat humiliating position. (To: TP0030)
4. If those infringing norms are powerful, the sustainable policy cycle can only succeed in censuring them by acting with great clarity and force. (To: TP0027)
5. In order to respond impartially, the sustainable policy cycle should be constantly aware of the dangers associated with the responsibility it has assumed in censuring infringement of norms. (To: TP0025)
6. If those infringing norms fail to respond to censorship by the sustainable policy cycle, misfortune inevitably results. (To: TP0051)
Transformation sequence Rather than acting crudely, decisive integrative action calls for a graceful style.

♦ TP0022 Style
Elegance — Image — Adornment
Condition A sustainable policy cycle may succeed in matters of lesser importance by gracefully respecting the sensitivities of those concerned. Fundamental or controversial issues cannot however be resolved by cultivating an appropriate image in this way.
Subconditions:
1. When the sustainable policy cycle is in a subordinate role, the gracefulness of self-reliance leads to greater success than the surreptitious acceptance of assistance. (To: TP0052)
2. The sustainable policy cycle risks deluding itself if it attaches greater importance to the form of its actions than to their substance. (To: TP0026)
3. The gracious style of the sustainable policy cycle may prove so enchanting to all concerned that the vigilance necessary for the success of its action is lost. (To: TP0027)
4. The sustainable policy cycle may find that more significant relationships are possible by acting simply than by depending on the trappings associated with a gracious mode of response. (To: TP0030)
5. Once the sustainable policy cycle has chosen to abandon dependence on a gracious mode of action, it will at first be embarrassed when attempting to relate to those of superior values who only attach significance to the substantive contributions it has to make (To: TP0037)
6. In the final stage of the sustainable policy cycle's development, the form of its action no longer disguises the substance but rather expresses its value to the full. (To: TP0036)
Transformation sequence Excessive emphasis on style leads to deterioration

♦ TP0023 Deterioration (Stripping away)
Destruction — Intrigue
Condition Under certain conditions of society inferior values may predominate. A sustainable policy cycle of superior values is wise to accept this phase of events calmly rather than vainly attempting to counteract it.
Subconditions:
1. Those of inferior values may initiate schemes to undermine the position of the sustainable policy cycle by intriguing against its supporters. (To: TP0027)
2. The sustainable policy cycle, isolated by the initiatives of those of inferior values, may be destroyed unless it can rapidly adjust its position. (To: TP0004)
3. Provided it is able to enhance the expression of its superior values, the sustainable policy cycle may disassociate itself from those of inferior values, who will then oppose it actively. (To: TP0052)
4. Events can deteriorate to the point at which the sustainable policy cycle is unable to avoid misfortune. (To: TP0035)
5. Those of inferior values may be attracted by the superior values of the sustainable policy cycle and voluntarily accept its guidance. (To: TP0020)
6. As support for the sustainable policy cycle increases, the strategies of those of inferior values become progressively more self-destructive. (To: TP0002)
Transformation sequence Deterioration cannot continue indefinitely, thus recovery finally commences.

♦ TP0024 Recovery (Turning point)
Return
Condition A sustainable policy cycle may recover spontaneously from adverse conditions, with the old patterns being transformed naturally into the new. This process of renewal should not be disturbed by acting prematurely.
Subconditions:
1. Occasionally the sustainable policy cycle will not be able to avoid adopting inferior values, at least to some degree; such errors should not be regretted if they are rectified promptly. (To: TP0002)
2. Renewal calls for a positive decision by the sustainable policy cycle to confirm the stability of the new order; this is best done in a supportive environment. (To: TP0019)
3. Renewal is not impossible, even if the sustainable policy cycle is so unstable as to be repeatedly attracted to inferior values, only to renounce them after each such deviation. (To: TP0036)
4. Although in an environment dominated by inferior values, the sustainable policy cycle may renew itself in isolation by responding to superior values. (To: TP0051)
5. If the time is appropriate for renewal, the sustainable policy cycle should publicly recognize any errors in its old pattern of actions, rather than reinforcing them with trivial arguments. (To: TP0003)
6. If the sustainable policy cycle does not take advantage of an appropriate occasion for renewal, it is condemned, by its own attitude, to an extended period of unfortunate conflictual relationships with its environment. (To: TP0027)
Transformation sequence Recovery lifts the weight of the past leading to innocent spontaneity.

♦ TP0025 Spontaneity
Innocence — Instinctive goodness — Integrity — Without Expectation — Error free — Fidelity
Condition A sustainable policy cycle is most successful when it acts spontaneously in response to emerging events rather than on the basis of some pre-defined programme. However, the guidance of such instinctive certainty leads to misfortune unless it is correctly rooted in superior values.
Subconditions:
1. The sustainable policy cycle can be confident of success when it acts on impulses involving no expectation of gain. (To: TP0012)
2. The sustainable policy cycle's activity can succeed if each phase is carried out for its own sake and irrespective of any possible result. (To: TP0010)
3. Even though the sustainable policy cycle acts without expectation of gain, it should be prepared to adjust to the possibility of misfortunes arising from external events. (To: TP0013)
4. No catastrophe can deprive the sustainable policy cycle of its inherent qualities, provided it continues to uphold them. (To: TP0042)
5. In the event of catastrophe arising from external causes, the sustainable policy cycle should take time to heal itself rather than call on external assistance. (To: TP0021)
6. When the time is not appropriate, any spontaneous response by the sustainable policy cycle is likely to be counterproductive. (To: TP0017)
Transformation sequence The excesses of spontaneity are contained through conservation measures.

♦ TP0026 Conservation
Consolidation — Taming power of the great — Restraint by the strong
Condition A sustainable policy cycle may be called upon to bind together, restrain, and care for valued features of society. Such an intimate relationship with the products of past initiatives is in itself valuable to a sustainable policy cycle's development.
Subconditions:
1. Vigorous action by the sustainable policy cycle may be so obstructed that, to avoid misfortune, further efforts are best restrained. (To: TP0018)
2. The forces restraining sustainable policy cycle action may be so superior that energy is best conserved in anticipation of a later opportunity. (To: TP0022)
3. When there is an opportunity for action, the sustainable policy cycle should move forward with others sharing its intent, meanwhile preparing its defences against unforeseen problems. (To: TP0041)
4. Prompt action by the sustainable policy cycle is necessary to forestall initiatives which are not self-restraining. (To: TP0014)
5. The sustainable policy cycle may best counteract unruly action by changing its nature or diverting it into appropriate channels. (To: TP0009)

6. The sustainable policy cycle may achieve a position in which its influence prevails because its action is no longer inhibited by opposing forces. (To: TP0011)
Transformation sequence Conservation measures ensure that support is provided where necessary.

♦ TP0027 Support
Providing nourishment — Wise counsel
Condition A sustainable policy cycle should be attentive to the manner in which it supports both its own activities and those of others able to contribute to the development of society.
Subconditions:
1. Misfortune results when the sustainable policy cycle's self-reliance is undermined in aspiring to the apparent advantages of others. (To: TP0023)
2. Misfortune results when the sustainable policy cycle fails to become self-reliant and persists in depending on others. (To: TP0041)
3. The sustainable policy cycle cannot be successful if it seeks advantages for their own sake and thus becomes dependent upon them. (To: TP0022)
4. When the sustainable policy cycle is in a position of influence, it should seek out others of the right quality to assist in achieving advances for society as a whole. (To: TP0021)
5. If deficiences in its mode of action prevent it from contributing effectively to the development of society, the sustainable policy cycle should seek the advice of those of superior values. (To: TP0042)
6. When the sustainable policy cycle becomes a source of influence in sustaining society, it can best continue in this role by being aware of the dangers of such responsibility. (To: TP0024)
Transformation sequence Continual build-up of support leads to importance.

♦ TP0028 Importance
Preponderance of the great — Excesses of the great — Great gains — Passing of greatness
Condition Circumstances may be such that a sustainable policy cycle of superior values experiences a period of great potential influence. This condition is necessarily unstable and the possible transition to other conditions should be carefully explored, whatever sacrifices these may then demand.
Subconditions:
1. In undertaking any new initiative under favourable conditions, the sustainable policy cycle should take extreme care in its preparations. (To: TP0043)
2. Under favourable conditions, the establishing of a relationship with those of inferior values may offer the possibility of renewal to the sustainable policy cycle. (To: TP0031)
3. If the sustainable policy cycle is reckless in its initiatives and ignores advice, favourable conditions are destabilized and catastrophe may result. (To: TP0047)
4. The situation may be stabilized with the assistance of those of inferior values, but the result will be unfortunate if the sustainable policy cycle achieves this out of self-interest. (To: TP0048)
5. If the sustainable policy cycle abandons its contacts with those of inferior values, the cultivation of its contacts with those of superior values will further destabilize the situation rather than leading to its renewal. (To: TP0032)
6. Under exceptional conditions the task faced can be so dangerous that the sustainable policy cycle may have to accept that it may accomplish its aim only by sacrificing its very existence for the values in question. (To: TP0044)
Transformation sequence Excessive importance is underminded by persistence.

♦ TP0029 Persistence (Abyss)
Danger — Mastering pitfalls — Multiple dangers
Condition A sustainable policy cycle may succeed through persisting in its course of action, responding appropriately to difficulties as they emerge. In this way the difficulties may subsequently be used as a form of protection.
Subconditions:
1. The sustainable policy cycle should avoid adapting permanently to dangers for this may prevent it from functioning appropriately in a normal environment. (To: TP0060)
2. When faced with danger, the sustainable policy cycle should assess the situation and act with caution. (To: TP0008)
3. In certain dangerous circumstances, inaction is preferable to action which may aggravate the situation for the sustainable policy cycle. (To: TP0048)
4. In times of danger, the sustainable policy cycle can usefully base its relationships on simplicity of substance rather than on complex forms of protocol. (To: TP0047)
5. Danger will be increased if the sustainable policy cycle has ambitions beyond its capacities and the opportunities of the moment. (To: TP0007)
6. If the sustainable policy cycle becomes confused in its strategy in a highly dangerous situation, it is unlikely that this can be immediately remedied. (To: TP0059)
Transformation sequence Persistence is only effective if there are normative constraints.

♦ TP0030 Normative constraint (Fire)
Cosmic mean — Model of elegance — Shining light
Condition By its nature a sustainable policy cycle is conditioned and unable to act freely. It may best achieve success by recognizing the beneficial limitations on which it can usefully depend. Through such voluntary compliance, a sustainable policy cycle develops the clarity of perception required for effective action.
Subconditions:
1. The sustainable policy cycle should maintain its composure in the midst of the confusion of society, so it may concentrate attention on the initial phases of any new action. (To: TP0056)
2. The sustainable policy cycle acts with greatest skill when striking a fruitful balance between extreme strategies. (To: TP0014)
3. Recognition of the sustainable policy cycle's transitory nature should induce comprehension of how it is fulfilled by playing its role in historical processes, rather than encouraging despair or efforts to avoid despair. (To: TP0021)
4. If the sustainable policy cycle is overactive it may rapidly deplete its internal resources, without achieving any lasting effect. (To: TP0022)
5. At the peak of its activity the sustainable policy cycle may derive long-term benefit by recognizing the transitory nature of its preoccupations. (To: TP0013)
6. In inducing discipline within itself or society, the sustainable policy cycle should eradicate the promoters of inferior values, whilst tolerating the weakness of those persuaded to follow them. (To: TP0055)
Transformation sequence Normative constraints operate through mutual influence.

♦ TP0031 Influence
Movement — Sensing — Sensitivity
Condition Success results from mutual attraction. This may be induced by a sustainable policy cycle of superior values whose openness to counsel is a fruitful influence on such relationships.
Subconditions:
1. Until the intention of the sustainable policy cycle has a visible effect it has no positive or negative influence on society. (To: TP0049)
2. The sustainable policy cycle runs the risk of misfortune if it acts before being impelled to do so by a genuine influence. (To: TP0028)
3. To avoid humiliation, the sustainable policy cycle should cultivate restraint in selecting the influences to which it responds and should exercise control on the response itself. (To: TP0045)
4. The influence of the sustainable policy cycle is most successful and widespread when it results from an appreciation of its intentions rather than from a deliberate effort to manipulate some target group. (To: TP0039)
5. If the sustainable policy cycle's influence is primarily focused on its own actions, such closure to outside influence in turn limits its influence upon society. (To: TP0062)
6. Any attempt by the sustainable policy cycle to influence society through words alone is necessarily insignificant and without consequence. (To: TP0033)
Transformation sequence Influence can only be effective if it endures.

♦ TP0032 Endurance
Constancy — Duration
Condition A sustainable policy cycle may be characterized by a self-renewing movement acting alternately on itself and on society. For a sustainable policy cycle of superior values this ensures a flexibility in response to the environment which is grounded on an inner directive that governs all its actions.
Subconditions:
1. The sustainable policy cycle can only ensure enduring effects through careful action over a long period that precludes any form of precipitate action. (To: TP0034)
2. If the strength of the sustainable policy cycle is greater than its material resources, successful control may avoid an inappropriate response. (To: TP0062)
3. Inconsistency on the part of the sustainable policy cycle, in response to external events, leads to unexpected forms of humiliation. (To: TP0040)
4. For the sustainable policy cycle to achieve success through persistence, it is necessary that the action should be appropriate. (To: TP0046)
5. If the sustainable policy cycle undertakes an active role, it should remain flexible in adapting to circumstances in the light of its enduring values; whereas in a passive role, it should be consistent in conforming to external guidelines. (To: TP0028)
6. If the sustainable policy cycle is permanently agitated, any attempt to produce enduring effects is undermined. (To: TP0050)
Transformation sequence Endurance cannot continue indefinitely, therefore withdrawal takes place.

♦ TP0033 Withdrawal
Retreat — Yielding
Condition A sustainable policy cycle may usefully withdraw when faced with opposing forces favoured by the current circumstances of society. For the retreat to be constructive it should be carried out with acts of resistance which prepare the way for later counter-movement.
Subconditions:
1. The retreating sustainable policy cycle should not take any initiative if it is in immediate contact with the opposing forces. (To: TP0013)
2. Those of inferior values may maintain such close contact with the sustainable policy cycle that they are successful in achieving superior goals. (To: TP0044)
3. The sustainable policy cycle may only achieve the freedom to retreat by taking responsibility for those who would otherwise prevent it, but this course carries its own risks. (To: TP0012)
4. The sustainable policy cycle of superior values adapts easily and harmoniously to the process of retreat from those of inferior values who degenerate when deprived of such guidance. (To: TP0053)
5. The sustainable policy cycle must judge the time for retreat correctly, and act firmly, or else run the risk of unpleasant discussion of irrelevant matters. (To: TP0056)
6. Once the sustainable policy cycle has ceased to identify with the prevailing conditions it acquires the ability to act fully in following the most appropriate line of retreat. (To: TP0031)
Transformation sequence Withdrawal cannot continue indefinitely, hence power becomes evident.

♦ TP0034 Power
Power of the great — Great vigour
Condition A sustainable policy cycle of superior values may acquire great strength and run the risk of depending upon that strength alone. True power is only exhibited when that strength is used in the service of a higher cause.
Subconditions:
1. If the sustainable policy cycle attempts to use its strength from an inferior position it courts disaster. (To: TP0032)
2. As resistance breaks down, the sustainable policy cycle may easily become self-confident and lose the advantage of balanced use of its force. (To: TP0055).
3. The sustainable policy cycle should avoid displays of power for their own sake, especially because of the complications to which they lead. (To: TP0054)
4. When all resistance disappears, the sustainable policy cycle is free to use all its powers, although the less this is apparent the greater its effectiveness. (To: TP0011)
5. When all resistance has disappeared, it is no longer desirable for the sustainable policy cycle to act forcefully and decisively. (To: TP0043)
6. The sustainable policy cycle should discontinue its initiative if, having proceeded too far in its actions, it encounters complications which hinder any further action. (To: TP0014)
Transformation sequence Power cannot be restrained indefinitely, hence progressive expansion occurs.

♦ TP0035 Progress
Advancement
Condition A sustainable policy cycle may achieve great progress when it is able to influence others to collaborate in the light of superior values. Progress may be accompanied by expansion.
Subconditions:
1. The sustainable policy cycle's initiative may fail to meet with a positive response from those calling for progress and it should not run the risk of making mistakes through being perturbed by this. (To: TP0021)
2. The sustainable policy cycle should continue in its efforts, even though progress is blocked and inspiration lost, for the latter will return when it can be based on fundamental principles not centred on the narrow preoccupations of the sustainable policy cycle. (To: TP0064)
3. The sustainable policy cycle may be encouraged by the support of others, even though it is unable to succeed without their assistance. (To: TP0056)
4. The sustainable policy cycle should avoid the temptation of using its position to accumulate advantages, especially since such abuse tends to be discovered in times of progress. (To:

TP0023)
5. The sustainable policy cycle should appreciate the values of its influential position in promoting the progress of society, rather than regretting lost opportunities in which its own narrower interests could have been advanced. (To: TP0012)
6. The sustainable policy cycle may act aggressively to rectify conditions opposing progress among its own contacts but should be aware of the dangers of such a procedure, particularly when extended to others. (To: TP0016)
Transformation sequence Progressive expansion eventually encounters resistance leading to decline.

♦ **TP0036 Decline** (Darkening of the light)
Unappreciated intelligence — Injury — Concealment of illumination — Adulteration of insight
Condition In adverse circumstances a sustainable policy cycle should not reveal the values it holds and thus provoke opposition. Rather it should appear to accept the prevailing standards and mode of behaviour whenever this is necessary.
Subconditions:
1. Faced with opposition, the sustainable policy cycle may limit its objectives but will nevertheless face continuing opposition if it remains true to its principles. (To: TP0015)
2. Although handicapped by opposing forces, the sustainable policy cycle may concentrate beneficially on assisting others who are also at risk. (To: TP0011)
3. In the process of establishing a new order, the sustainable policy cycle may contain the initiator of the opposition, but premature consolidation of such a victory should be carried out with caution if the habits of the old order have become too well–entrenched. (To: TP0024)
4. The sustainable policy cycle may be able to avoid being drawn into disaster by being well informed concerning the intentions of the initiator of the opposition. (To: TP0055)
5. If the sustainable policy cycle is obliged to remain under the influence of the opposing forces, it can only survive intact through the exercise of dissimulation and considerable caution. (To: TP0063)
6. The forces in opposition to the sustainable policy cycle of superior values turn upon themselves at the height of their power and cause their own destruction. (To: TP0022)
Transformation sequence Decline eventually necessitates a withdrawal into a community context.

♦ **TP0037 Community** (Family)
Clan — Household
Condition A sustainable policy cycle can only influence others effectively when its external initiatives are consistent with its own internal mode of organization. A community context is most favourable to this.
Subconditions:
1. Within the sustainable policy cycle a measure of discipline is necessary in order that each member learns to fulfil his or her own function to enable the sustainable policy cycle to undertake external initiatives successfully. (To: TP0053)
2. The sustainable policy cycle should concentrate on 'keeping its own house in order' rather than undertaking initiatives based on force. (To: TP0009)
3. In disciplining itself the sustainable policy cycle should seek a careful mean between the excesses of indulgence and severity, although under exceptional conditions the latter may be necessary. (To: TP0042)
4. In manifesting its principles in a role of stewardship, the sustainable policy cycle contributes significantly to the well–being of society. (To: TP0013)
5. The character of the principles governing the sustainable policy cycle may be such that no disciplinary action is required to achieve the necessary effects (To: TP0022)
6. The quality of the achievements engendered by the principles of the sustainable policy cycle is the fundamental force holding it together. (To: TP0063)
Transformation sequence When the community context proves inadequate, misunderstandings and opposition arise.

♦ **TP0038 Opposition**
Disharmony — Alienation — Estrangement
Condition The preservation of the individuality of a sustainable policy cycle of superior values can only be achieved through creative opposition to those of inferior values. It on this basis that order is engendered. Faced with opposition and misunderstandings a sustainable policy cycle should concentrate on minor initiatives.
Subconditions:
1. The sustainable policy cycle should avoid the consequences of attempting to ensure unity through forceful action since the temporarily estranged will re–establish contact of their own accord and those who impose themselves will eventually drift away. (To: TP0064)
2. Informal contacts may suffice when misunderstandings prevent the sustainable policy cycle from establishing formal relationships with its natural partners. (To: TP0021)
3. Despite opposition and discouragement, the sustainable policy cycle can succeed if it discovers a trustworthy partner of complementary nature. (To: TP0014)
4. Although isolated in opposition to others, the sustainable policy cycle will eventually succeed through maintaining contact with a natural partner. (To: TP0041)
5. The opposition faced by the sustainable policy cycle may initially prevent recognition of a trustworthy partner with whom it is beneficial for it to work. (To: TP0010)
6. The isolation experienced by the sustainable policy cycle may be due to opposition based upon misunderstanding which once clarified permits fruitful collaboration. (To: TP0054)
Transformation sequence Through misunderstandings and opposition, difficulties and obstructions are created.

♦ **TP0039 Obstruction**
Stumbling — Trouble
Condition When faced with difficulties and obstacles to the achievement of its intentions, a sustainable policy cycle of superior values searches for errors in the assumptions underlying its initiative, thus creating the opportunity for its own further development.
Subconditions:
1. The sustainable policy cycle, when faced with obstacles, should retreat temporarily in anticipation of a more appropriate occasion for action. (To: TP0063)
2. When its obligations so dictate, the sustainable policy cycle should attack the obstacle directly rather than seeking ways to circumvent it. (To: TP0048)
3. If the sustainable policy cycle has others dependent upon it, whose existence would be endangered by its failure, it is preferable for it to avoid tackling the obstacle. (To: TP0008)
4. It is preferable for the sustainable policy cycle to avoid hasty action against an obstacle in order to gather support and make adequate preparations. (To: TP0031)
5. Despite the importance of the obstruction, if the sustainable policy cycle is totally committed to the task it will attract collaborators with whom success may be achieved. (To: TP0015)
6. If a sustainable policy cycle no longer concerned with mundane affairs is faced with obstructions, it can through its experience and insight bring about a solution of special significance, rather than vainly attempting to avoid the issue. (To: TP0053)
Transformation sequence Obstructions cannot persist indefinitely, thus eventually liberation is achieved.

♦ **TP0040 Liberation**
Release — Deliverance — Getting free — Elimination of obstacles — Resolution
Condition When a sustainable policy cycle is liberated from the obstacles which have hindered its initiatives, any remaining problems should be cleared up rapidly without dwelling unduly on the misdeeds of those responsible for the obstacles.
Subconditions:
1. After liberation has been achieved, the sustainable policy cycle should recuperate in peace and refrain from immediate action. (To: TP0054)
2. Full commitment by the sustainable policy cycle to just action undermines the efforts of those who seek to prevent liberation by influencing those in power. (To: TP0016)
3. After liberation, the sustainable policy cycle should take care not to flaunt its successes and thus attract those capable of appropriating them. (To: TP0032)
4. In anticipation of liberation, the sustainable policy cycle should disassociate itself from casual collaborators who are liable to discourage offers of support from more deeply committed potential partners. (To: TP0007)
5. To achieve liberation successfully, the sustainable policy cycle of superior values must avoid affirming any contribution from those of inferior values, thus encouraging them to withdraw. (To: TP0047)
6. The sustainable policy cycle will achieve liberation successfully it acts forcefully to remove those of inferior values who have reached key positions of power. (To: TP0064)
Transformation sequence The process of liberation necessarily results in some loss and deficiency.

♦ **TP0041 Deficiency**
Reduction of excesses — Decrease
Condition A sustainable policy cycle may experience a decrease in the external resources available. This situation may be used to clarify and strengthen the inner resources on which it can draw for future undertakings.
Subconditions:
1. In accepting assistance, the sustainable policy cycle of superior values should take care not to overstrain the abilities of those offering it, nor should it exploit them. (To: TP0004)
2. In order for the sustainable policy cycle to be able to provide assistance of enduring significance to others, it must take care not to overstrain its own resources. (To: TP0027)
3. Collaborating with two partners in an undertaking leads to an unstable situation for the sustainable policy cycle resulting in the alienation of one of them, whereas a partner will be found if the sustainable policy cycle undertakes an initiative alone. (To: TP0026)
4. If the sustainable policy cycle is able to reduce its defects this encourages those who are well–disposed to collaborate. (To: TP0038)
5. If those in power appoint the sustainable policy cycle to perform a key role there is nothing to prevent its success. (To: TP0061)
6. Every increase in the power of the sustainable policy cycle may be such as to benefit others, rather than to deprive others of benefits. (To: TP0019)
Transformation sequence If deficiency persists it eventually evokes assistance.

♦ **TP0042 Assistance**
Increase
Condition By sacrificing its own interests in favour of the development of others, a sustainable policy cycle may temporarily create conditions in which great progress can be made. The development of the sustainable policy cycle itself may be brought about by adopting the positive attributes of others and eliminating its own defects.
Subconditions:
1. The sustainable policy cycle should use any major assistance it receives from those in power for initiatives which it would not otherwise be able to undertake. (To: TP0020)
2. The sustainable policy cycle brings about its own development by producing in itself the necessary conditions and identifying with the progressive spirit of the times. (To: TP0061)
3. In times of great progress even unfortunate events may be advantageous, releasing the sustainable policy cycle from error and enabling it to act with authority according to the needs of the moment. (To: TP0037)
4. The sustainable policy cycle may usefully function, especially with respect to major undertakings, as an intermediary between those in power and those who should benefit from such progress. (To: TP0025)
5. The beneficial influence of the sustainable policy cycle's assistance results from its expression of an inner necessity for which recognition is sufficient reward. (To: TP0027)
6. If the sustainable policy cycle fails to assist in the progress of others it isolates itself and exposes itself to attack. (To: TP0003)
Transformation sequence If assistance continues long enough a new resolution emerges.

♦ **TP0043 Resolution** (Flight)
Breakthrough — Parting
Condition As any struggle against opposing forces begins to bear fruit, a sustainable policy cycle should ensure that the resolution of the conflict is based on an amicable union from a position of strength, free from compromise or any concealment of the sustainable policy cycle's own defects. It should avoid the use of force and concentrate on the redistribution of the advantages it accumulates.
Subconditions:
1. Whilst resistance is still strong, the sustainable policy cycle should avoid any hasty implementation of its resolution that could result in major setbacks. (To: TP0028)
2. If the sustainable policy cycle is vigilant it engenders attitudes appropriate to cautious resolution of the conflict. (To: TP0049)
3. If the sustainable policy cycle is obliged to maintain relationships with those of inferior values to avoid jeopardizing resolution of the conflict, it will have to endure a period of condemnation by those actively combatting such values. (To: TP0058)
4. The sustainable policy cycle may identify so closely with action to resolve the conflict that its obstinacy raises obstacles and prevents it from benefiting from advice. (To: TP0005)
5. The efforts of the sustainable policy cycle against those of inferior values in positions of power must constantly be renewed, especially because of the obligation to maintain a relationship with them. (To: TP0034)
6. When the objective appears to have been achieved, the sustainable policy cycle should act resolutely to eliminate any remaining vestiges of the old order, especially those rooted in its own attitudes. (To: TP0001)
Transformation sequence Resolution and the associated action lead to new encounters.

♦ **TP0044 Encounter**
Contact — Infiltration of the inferior — Meeting — Subjugation
Condition A sustainable policy cycle may find itself attracted by initiatives made by those of

inferior values. Although apparently harmless, according them recognition allows them to develop, possibly leading to a dangerous condition, unless they are free from ulterior motives.
Subconditions:
1. If a measure of acceptance has been accorded to those of inferior values by the sustainable policy cycle, they must be constantly held in check to prevent undesirable developments. (To: TP0001)
2. If those of inferior values have been successfully contained by the sustainable policy cycle, care must be taken not to allow them to develop their influence through contact with others unable to maintain such control. (To: TP0033)
3. If the sustainable policy cycle is tempted to collaborate with those of inferior values, but is prevented by circumstances from doing so, the errors that are liable to result from such indecision may be avoided by becoming aware of the dangers. (To: TP0006)
4. If the sustainable policy cycle is unable to tolerate those of inferior values, they cannot be called upon for assistance in time of need. (To: TP0057)
5. The sustainable policy cycle may tolerate and protect collaborators of inferior values, relying solely on the superior qualities of its influence as a means of successfully controlling them. (To: TP0050)
6. A sustainable policy cycle no longer concerned with mundane affairs may be able to tolerate the reproaches of those of inferior values with whom it refuses to associate. (To: TP0028)
Transformation sequence A multiplicity of encounters leads to congregation.

♦ **TP0045 Congregation**
Assembly — Gathering together
Condition When circumstances promote congregation in society, a sustainable policy cycle of superior values, to be capable of focusing this process, should ensure that it is itself well integrated. It should also be prepared to counteract uncontrolled consequences.
Subconditions:
1. The sustainable policy cycle can facilitate formation of a group by encouraging individually those that have not yet committed themselves. (To: TP0017)
2. The process of congregation is assisted by a recognition of mutual complementarity with which the sustainable policy cycle should work, rather than acting on the basis of arbitrary decisions. (To: TP0047)
3. If a group has already formed from which the sustainable policy cycle is isolated, it can best succeed by allying itself with some of those at the centre, despite the initial humiliation. (To: TP0031)
4. The sustainable policy cycle is successful when it acts as a focus for a group united for a higher cause. (To: TP0008)
5. In acting as a focus for a group, the sustainable policy cycle attracts some people only because of the influence it acquires in the process; this necessitates special efforts to gain their confidence. (To: TP0016)
6. If the desire of some to group together is misunderstood, their expression of regret can usefully enable the sustainable policy cycle to revise its views and bring about the alliance. (To: TP0012)
Transformation sequence Congregation creates an environment permitting advancement.

♦ **TP0046 Advancement**
Promotion — Rising
Condition A sustainable policy cycle may benefit from circumstances to rise to a position of influence through unrelenting effort in circumventing obstacles to its progress.
Subconditions:
1. In a position of obscurity, the sustainable policy cycle can derive strength and encouragement for its progress from those in position of power, who also benefit thereby from such a link to their origins. (To: TP0011)
2. Even though the sustainable policy cycle is unsubtle in its relationships with others, it may succeed in advancing because of recognition of the strength of its inherent qualities. (To: TP0015)
3. In the absence of obstacles, the sustainable policy cycle should take advantage of the opportunity to advance, rather than being preoccupied prematurely with how long such advance will be possible. (To: TP0007)
4. In attaining its goal, the sustainable policy cycle becomes accepted by those in positions of influence and thus achieves enduring significance. (To: TP0032)
5. The sustainable policy cycle should progress steadily rather than becoming overconfident and impatient. (To: TP0048)
6. The sustainable policy cycle should take care to avoid committing itself to advancement for its own sake, and thus become unable to retreat when necessary. (To: TP0018)
Transformation sequence Continual advancement eventually leads to adversity and exhaustion.

♦ **TP0047 Adversity**
Oppression — Exhaustion — Burdened
Condition Under adverse circumstances a sustainable policy cycle should accept restraint, whilst remaining true to its principles in anticipation of future opportunities. Such restraint may be due to oppression or to the exhaustion of its own resources.
Subconditions:
1. Faced with adversity, the sustainable policy cycle should overcome its own negative response to the situation which otherwise will undermine its ablity to act. (To: TP0058)
2. Even though external circumstances are satisfactory, the sustainable policy cycle must concentrate on overcoming inner restraints in order to be able to respond to opportunities offered by those in power. (To: TP0045)
3. The sustainable policy cycle should avoid being oppressed by restraints which are engendered solely by its own indecisive mode of action. (To: TP0028)
4. If endowed with resources it wishes to use for the benefit of others, the sustainable policy cycle may find itself temporarily impeded by its own uncertainty and the distractions of its peers. (To: TP0029)
5. Although intent on initiatives for the general well-being, the sustainable policy cycle may find itself obstructed by those in power, in which case progress comes slowly provided it does not lose its equanimity. (To: TP0040)
6. The sustainable policy cycle may be restrained principally by the assumption that any action is fruitless, in which case a change of attitude should enable it to break free. (To: TP0006)
Transformation sequence Extremes of adversity necessitate a concentration on basic needs.

♦ **TP0048 Basic need** (Well)
Fulfilled potentialities
Condition In order to engender appropriate order in society, a sustainable policy cycle must ensure that this fulfils the basic needs of humanity, rather than those defined by convention. In doing so care is required and excesses should be avoided.
Subconditions:
1. If the sustainable policy cycle dissipates its energies on trivia, it loses all significance for others and will be ignored. (To: TP0005)
2. If the sustainable policy cycle neglects to make use of its positive qualities and associates with those of inferior values, it will deteriorate and be unable to accomplish anything of significance. (To: TP0039)
3. Under unfortunate circumstances, the sustainable policy cycle of superior quality may not be known to those in power nor made use of by others. (To: TP0029)
4. It may be of greater long-term benefit for the sustainable policy cycle to reorganize itself, even though it is temporarily unable to act. (To: TP0028)
5. Despite the value of the sustainable policy cycle as a catalyst for social renewal, it is useless unless this potential is translated into practice. (To: TP0046)
6. The sustainable policy cycle is of greatest value when as a result of the demands made upon it, it becomes a self-renewing source of inspiration and assistance to all in need. (To: TP0057)
Transformation sequence Persisting inequalities in access to basic needs eventually engender revolution.

♦ **TP0049 Revolution** (Moulting)
Change
Condition A sustainable policy cycle having the confidence of others may be obliged to respond to emerging crisis conditions by promoting social transformation in order to meet the needs of the underprivileged.
Subconditions:
1. The sustainable policy cycle should refrain from initiating radical change until it is absolutely necessary. (To: TP0031)
2. When all other initiatives have failed, revolution may be initiated by the sustainable policy cycle after careful preparation and bearing in mind the condition to be brought about. (To: TP0043)
3. After the need for social transformation has been repeatedly expressed, action may be undertaken by the sustainable policy cycle; this should avoid the errors of ruthless haste and hesitant conservatism. (To: TP0017)
4. To be successful, the sustainable policy cycle should ensure that the radical change undertaken is based on the superior values it embodies which people will support as being instinctively just. (To: TP0063)
5. The authority of the sustainable policy cycle of superior values will be acknowledged during social transformation if clear principles of organization are formulated in a manner all can understand. (To: TP0055)
6. Once the social transformation is underway, those of inferior values will adapt in the light of their own interests and the sustainable policy cycle should not expect more of them than conditions permit. (To: TP0013)
Transformation sequence The most transformative revolution is that available through the cultural heritage.

♦ **TP0050 Cultural heritage** (Cauldron)
Rejuvenation — — Sacrificial vessel
Condition Society is nourished by its cultural heritage, a vehicle through which human values are consecrated. A sustainable policy cycle embodying this heritage can succeed by ensuring an appropriate relationship between its existence and its sense of destiny.
Subconditions:
1. Irrespective of its humble origins, the sustainable policy cycle may succeed if it is prepared to refine and develop its mode of action. (To: TP0014)
2. The sustainable policy cycle should undertake significant initiatives, for even though these may expose it to criticism, the later will not prevent success. (To: TP0056)
3. If its effectiveness is severely handicapped by lack of recognition, associating the sustainable policy cycle with superior values will ensure success when the opportunities emerge. (To: TP0064)
4. If the sustainable policy cycle is inadequate to the challenge it faces, and associates with those of inferior values, its initiatives will probably fail. (To: TP0018)
5. If the sustainable policy cycle is in a position of influence, and attracts able assistance through the quality of its action, it should then take care not to modify its style. (To: TP0044)
6. The sustainable policy cycle is most successful if the power and severity of its counsel is expressed in a form which others find agreeable. (To: TP0032)
Transformation sequence The protection of the cultural heritage necessitates crisis preparedness.

♦ **TP0051 Crisis preparedness** (Thunderbolts)
Vigilance — Shock — Arousal
Condition In order to fulfil a leadership role, a sustainable policy cycle should be capable of accepting any external shock and recognizing the nature of the response required by it.
Subconditions:
1. The relief experienced following a shock may place the sustainable policy cycle at an advantage compared to others who did not experience it. (To: TP0016)
2. If the sustainable policy cycle is endangered as a result of severe shock, it should withdraw until the crisis is over rather than acting vainly to recover its losses before the appropriate opportunity. (To: TP0054)
3. The sustainable policy cycle is liable to be overwhelmed by an external shock unless it can learn from it and respond to the opportunities it presents. (To: TP0055)
4. The response of the sustainable policy cycle will be dangerously handicapped if the shock offers no opportunity for it to act. (To: TP0024)
5. In the event of a multiplicity of shocks, the sustainable policy cycle can best survive by moving with the flow of events. (To: TP0017)
6. If the shock is confusing others, so that they are unable to respond effectively, the sustainable policy cycle can best prepare to respond by withdrawing from the situation, even though this may invite disapproval. (To: TP0021)
Transformation sequence Crises cannot continue to emerge if inaction is cultivated.

♦ **TP0052 Inaction** (Mountain)
Keeping still — Desisting — Resting
Condition The effectiveness of action initiated by a sustainable policy cycle is largely dependent on the equanimity with which it is able to assess what is required. A sustainable policy cycle should be able to pause before action is required.
Subconditions:
1. By pausing before action has been initiated, the sustainable policy cycle avoids mistakes although it runs the risk of irresoluteness. (To: TP0022)
2. By ceasing to act, the sustainable policy cycle may avoid a disaster into which those it supports are drawn, without however being able to assist them. (To: TP0018)
3. Enforced inaction will not induce in the sustainable policy cycle the tranquillity required to envisage appropriate initiatives. (To: TP0023)
4. The ability of the sustainable policy cycle to restrain its impulsive responses is valuable, even

through this does not prevent it from being perturbed by doubts and restlessness. (To: TP0056)
5. In contrast to occasional well-formulated statements, injudicious pronouncements by the sustainable policy cycle can have regrettable consequences. (To: TP0053)
6. The sustainable policy cycle may achieve a continuing quality of tranquillity from which it can respond appropriately to all demands made upon it. (To: TP0015)
Transformation sequence Inaction cannot continue indefinitely, thus at some stage development commences.

♦ **TP0053 Development**
Gradual progress — Growth — Gradual advance
In order to engender lasting development, a sustainable policy cycle should act slowly over an extended period of time, both to establish cooperative relationships and to increase its own influence so that its initiatives carry weight
Subconditions:
1. Difficulties and criticism experienced by the sustainable policy cycle in the early phases of its action may ensure successful development by preventing excessive haste. (To: TP0037)
2. The initial success of the sustainable policy cycle provides it with a sense of security and encouragement as a basis for further action and collaboration. (To: TP0057)
3. The sustainable policy cycle will succeed if it avoids provoking conflict and concentrates on developing and protecting its own position. (To: TP0020)
4. If the process of development places the sustainable policy cycle in an awkward or dangerous position, it should be adaptable in order to locate a secure position from which to continue its action. (To: TP0033)
5. Due to action by those of inferior values, once it is in a position of influence the sustainable policy cycle may become isolated and be temporarily misjudged by those on whom it is most dependent. (To: TP0052)
6. Once the work of a sustainable policy cycle of superior values is completed it can become a striking example for those who may follow in its stead. (To: TP0039)
Transformation sequence Development permits the establishment of elective affinities.

♦ **TP0054 Elective affinity** (Marrying maiden)
Propriety
Condition As a complement to its formal relationships, a sustainable policy cycle may beneficially engage in integrative initiatives, based on spontaneously emergent sympathetic relations with others, provided that these are conducted with reserve and mutual respect.
Subconditions:
1. The sustainable policy cycle may successfully wield influence through informal relationship with those in power, even though this relationship cannot be formally recognized. (To: TP0040)
2. The sustainable policy cycle can benefit by maintaining its loyalty to those in power, even though they no longer acknowledge some pre-existing informal relationships with them. (To: TP0051)
3. Frustrated by the lack of success of its formal initiatives, the sustainable policy cycle may enter into a constraining informal relationship which is not compatible with its assessment of its own value. (To: TP0034)
4. Out of respect for its principles, the sustainable policy cycle may beneficially delay establishing relationships until the appropriate opportunity occurs. (To: TP0019)
5. If the sustainable policy cycle originated in an influential context, it may enter into a beneficial relationship with others of more humble origins, provided it places itself at their service and does not draw attention to any fortuitous differences. (To: TP0008)
6. The sustainable policy cycle will not benefit from entering into a superficial relationship with others, especially if it is not based on shared respect for superior values. (To: TP0038)
Transformation sequence The establishment of elective affinities creates an environment favourable to general prosperity.

♦ **TP0055 Prosperity**
Abundance — Fullness — Richness
Condition Because of the probability of subsequent decline, only a sustainable policy cycle that acts optimistically without regret can effectively sustain a time of general prosperity. In so doing careful attention should be paid to the enforcement of agreed rules.
Subconditions:
1. To engender prosperity through a relationship with those in power, the sustainable policy cycle needs to temper the qualities of wisdom with those of energetic action for an adequate period, in order to achieve an acknowledged influence. (To: TP0062)
2. If the sustainable policy cycle undertakes initiatives, when its relationship with those in power has been distorted by mistrust due to usurpers, action becomes impossible; it is then preferable for it to uphold its principles as an indirect influence that can ultimately have the necessary effect. (To: TP0034)
3. When in a role as immediate assistant to those in power, the sustainable policy cycle does not merit censure if usurpers prevent it from acting. (To: TP0051)
4. The sustainable policy cycle can expect success when it is able to temper the qualities of energy with those of wisdom in a favourable relationship with those in power. (To: TP0036)
5. If in a position of power the sustainable policy cycle is open to counsel from those of ability, it will accumulate useful proposals resulting in benefit to all. (To: TP0049)
6. The sustainable policy cycle may achieve a position of power and affluence for itself, but only at the cost of alienating all those depending on it. (To: TP0030)
Transformation sequence When prosperity declines from its own excesses, estrangement and marginality result from the destruction of relationships.

♦ **TP0056 Marginality** (Wanderer)
Newcomer — Traveller
Condition When a sustainable policy cycle has no established position or relationships in society, it succeeds best by engaging in short-term activities with those of superior values such that it is not drawn into conflict situations.
Subconditions:
1. It is counterproductive for the isolated sustainable policy cycle to undertake trivial initiatives in order to achieve favour in the eyes of others. (To: TP0030)
2. The isolated sustainable policy cycle viewed favourably by others eventually finds a foothold in society and attracts permanent support. (To: TP0050)
3. By acting discourteously and interfering in the affairs of others, the isolated sustainable policy cycle may well lose any foothold it has in society and alienate its support, thus placing it in a very vulnerable position. (To: TP0035)
4. Disguising its aspirations, the isolated sustainable policy cycle may obtain a provisional position in society by limiting its publicly voiced requirements, but any resources it can the accumulate will have to be constantly protected, leaving it with a permanent sense of insecurity. (To: TP0052)
5. In order to establish a relationship with those in power and develop contacts with others, the isolated sustainable policy cycle can best succeed through a demonstration of its qualities, on the basis of which it can then be recommended. (To: TP0033)
6. If, having established a position in society, the sustainable policy cycle acts imprudently, forgetting its marginal status as a newcomer, it may be rejected, losing all it was in the process of building up. (To: TP0062)
Transformation sequence Marginality cannot be absorbed or controlled by the environment and thus gives rise to penetrating clarity of perception.

♦ **TP0057 Penetrating clarity** (Wind)
Gentleness — Willing submission
Condition A sustainable policy cycle of limited resources may best achieve lasting success by acting gradually and persistently towards a clearly defined goal in association with others in a position of power. Its influence results from penetrating clarity of judgement that disempowers those with ulterior motives.
Subconditions:
1. The clarity of understanding of the sustainable policy cycle may promote indecision when resolute action is to be preferred. (To: TP0009)
2. Strenuous effort should be devoted by the sustainable policy cycle to tracing and eliminating any elusive negative influences by which initiatives are being distorted. (To: TP0003)
3. Excessive reflection by the sustainable policy cycle on a possible initiative undermines its credibility. (To: TP0059)
4. The sustainable policy cycle is assured of success if, in a position of responsibility, it combines the qualities of experience, unpretentiousness and energetic action. (To: TP0044)
5. When reforms can be gradually introduced by the sustainable policy cycle, this is best done on the basis of careful preliminary study, with corresponding follow-up evaluations of the appropriateness of the action. (To: TP0018)
6. If the sustainable policy cycle has the ability to trace negative influences to the instigating body, but no longer has the strength to combat it, then such action is best avoided. (To: TP0048)
Transformation sequence Use of penetrating clarity leads to a sense of vitality.

♦ **TP0058 Vitality** (Lake)
Joy — Delight — Standing straight
Condition A sustainable policy cycle can best ensure the injection of vitality into its undertakings by engaging in stimulating interaction with others so as to provide a multi-faceted optimistic basis for its initiatives.
Subconditions:
1. The sustainable policy cycle may benefit from a self-sustaining sense of vitality and optimism which is not dependent upon reinforcement by others. (To: TP0047)
2. The sustainable policy cycle can avoid regrettable consequences by not indulging in the interaction proposed as stimulating by those of inferior values. (To: TP0017)
3. The vitality of the sustainable policy cycle should be engendered by it in light of its own values rather than deriving from participation in external distractions in which it may become dangerously absorbed. (To: TP0043)
4. So long as the sustainable policy cycle has difficulty in choosing between the stimulation of dynamics based on superior and inferior values, it remains subject to inner conflicts. (To: TP0060)
5. The sustainable policy cycle should protect itself from association with disintegrative influences, however stimulating, because of the harmful effects they may gradually engender. (To: TP0054)
6. The sustainable policy cycle may become so involved in the stimulation of external distractions that it no longer retains any effective control over it own actions. (To: TP0010)
Transformation sequence Vitality in action leads to the dissolution of barriers.

♦ **TP0059 Barrier dissolution** (Flooding)
Dispersion — Scattering — Disintegration — Overcoming dissension
Condition A sustainable policy cycle can best dissolve divisive barriers preventing collaboration by promoting awareness of underlying unity and solidarity in a manner which engages emotions engendered by superior values.
Subconditions:
1. The sustainable policy cycle should act vigorously to counteract divisive misunderstanding before it has fully taken form. (To: TP0061)
2. When the sustainable policy cycle recognizes a tendency on its own part to establish barriers against others, it should make deliberate efforts to remedy the situation through its supporters. (To: TP0020)
3. Circumstances may be such that the sustainable policy cycle can act best by dissolving all barriers distinguishing itself from others, in order to marshal resources for an initiative in the interest of all. (To: TP0057)
4. When acting in the general interest, the sustainable policy cycle can only have a lasting effect if it ceases attaching special importance to its immediate relationships and supporters. (To: TP0006)
5. When society is fragmented by many barriers, the sustainable policy cycle may provide a powerful idea to dispel misunderstandings, as a focus for the emergence of a new order. (To: TP0004)
6. Faced with extreme divisiveness, the sustainable policy cycle may usefully reduce the danger to itself and to its immediate contacts by dispersing in order to re-assemble on another occasion. (To: TP0029)
Transformation sequence The elimination of barriers cannot continue indefinitely, thus the need for limitation emerges.

♦ **TP0060 Limitation**
Discipline — Restraint — Regulation
Condition In order that its freedom of action may acquire significance, a sustainable policy cycle should operate under constraints that distinguish its activities from those of others. Limits should however be set upon limitation of this kind to prevent such discrimination from becoming unbearable to the sustainable policy cycle itself or to others.
Subconditions:
1. Faced with insurmountable limitations, the sustainable policy cycle should forego action until an appropriate opportunity arises for a forceful initiative. (To: TP0029)
2. When the moment for action arises, the sustainable policy cycle should not hesitate in seizing the opportunity. (To: TP0003)
3. If the sustainable policy cycle acts only in its self-interests, it may easily fail to recognize the need for the limits and restraints without which it will make regrettable mistakes. (To: TP0005)
4. The sustainable policy cycle avoids waste of its resources, and may achieve success, through working with limitations rather than against them. (To: TP0058)
5. If a sustainable policy cycle in a position of influence first imposes limitations upon its own action, its achievements under these conditions constitute an example to others who will then accept similar restrictions more readily. (To: TP0019)
6. Although imposition of excessive limitations may prove unbearable to the sustainable policy cycle and to others, such ruthlessness applied to itself may under certain circumstances be the only means for the sustainable policy cycle to uphold its principles. (To: TP0061)
Transformation sequence Through limitation, dependence on essential quality is assured.

♦ TP0061 Essential quality
Inner truth — Sincerity at the centre — Inner confidence — Wholehearted allegiance — Faithfulness at the centre
Condition A sustainable policy cycle may succeed by influencing the most intractable, if it is able to identify with their condition, sincerely affirming the importance of essential qualities that it shares with them. These may take the form of fundamental principles.
Subconditions:
1. The power of fundamental principles upheld by the sustainable policy cycle is progressively undermined to the extent that it loses its self-reliance through dependence on secret agreements with others. (To: TP0059)
2. Through the fundamental principles which it upholds, the sustainable policy cycle of superior values may exert a far-reaching influence that attracts others without any intent to achieve this end. (To: TP0042)
3. The vitality of the sustainable policy cycle may be dependent, for better or for worse, on the vagaries of the fundamental nature of its relationship with others. (To: TP0009)
4. To increase the power of its fundamental principles, the sustainable policy cycle should concentrate on deepening its understanding of superior values governing action beyond the domain of factionalism. (To: TP0010)
5. The sustainable policy cycle in a position of power may succeed in linking others together in a non-superficial manner through the fundamental quality of the multi-faceted influences engendered by its action. (To: TP0041)
6. The sustainable policy cycle should be aware of relying on accepted formulas to awaken a shared sense of fundamental solidarity, for such standard appeals may fail when they are most needed. (To: TP0060)

Transformation sequence Limitation and dependence on essential quality enable actions to be undertaken conscientiously.

♦ TP0062 Conscientiousness
Preponderance of the small — Small excesses
Condition Faced with a challenge for which it is not fully competent, a sustainable policy cycle of superior values can best succeed by acting with extreme prudence and attention to detail, especially in support of those not in a position of influence.
Subconditions:
1. The sustainable policy cycle should employ conventional measures whenever possible, avoiding the depletion of resources and the risk of failure associated with extraordinary measures. (To: TP0055)
2. Under exceptional circumstances the sustainable policy cycle may best succeed through extreme restraint and conscientious fulfilment of its obligations. (To: TP0032)
3. Exceptional circumstances require that the sustainable policy cycle pay attention to details through which it may learn of dangers that otherwise it would be unable to avoid. (To: TP0016)
4. Under certain circumstances the sustainable policy cycle should refrain from action until a more opportune moment, rather than render itself vulnerable to dangers through persisting in its own initiatives. (To: TP0015)
5. In exceptional circumstances the isolated sustainable policy cycle, able to bring about order in society, should seek assistance from others on the basis of their genuine achievements rather than their claims to fame. (To: TP0031)
6. If the sustainable policy cycle fails to exercise self-discipline at a time when attention to detail is required, it is unlikely to be successful in its initiative. (To: TP0056)

Transformation sequence Conscientiousness ensures the accomplishment of initiatives.

♦ TP0063 Accomplishment
Settlement — After completion
Condition Once the transition from the old to the new order has been accomplished with the exception of details, a sustainable policy cycle should take considerable care to ensure that the harmony with which events evolve during their final phases does not encourage negligence that enables the seeds of an inferior order to take root once again.
Subconditions:
1. A successful transition immediately encourages widespread pressure for further development, which the sustainable policy cycle should resist in order to avoid the dangers of enthusiastically overshooting the goal and jeopardizing all that has been achieved. (To: TP0039)
2. Following a successful transition, those acquiring power tend to become arrogant and neglect to offer roles to those wishing to participate, thus encouraging discreditable position-seeking manoeuvres; these the sustainable policy cycle should avoid in the expectation that appropriate opportunities will emerge in due time. (To: TP0005)
3. Following successful transition, the sustainable policy cycle together with others will tend to struggle to expand the new order into neighbouring domains functioning under an old order; any success should not then be undermined by using those of inferior values to control such domains. (To: TP0003)
4. Following successful transition, the scandals that may come to light, and which are readily forgotten by others, should be treated by the sustainable policy cycle as important indicators of possible future difficulties. (To: TP0049)
5. Following successful transition the sustainable policy cycle should take care to continue the sincere affirmation of superior values, since this tends to evolve under the new order into an elaborate ritual from which significance easily disappears. (To: TP0036)
6. Fascination for the old order, from which a successful transition has been made, may prevent the sustainable policy cycle from appropriately consolidating what has been achieved. (To: TP0037)

Transformation sequence Accomplishment cannot exhaust the potential for further transformation.

♦ TP0064 Transformation threshold
Unsettlement — Before completion — Tasks remaining
Condition When all has been prepared for transition to a new order that can transform a fragmented condition of society, a sustainable policy cycle should act with deliberation and caution to determine how the available resources can best be applied to achieve the desired effect.
Subconditions:
1. As a response to the prevailing lack of order, the sustainable policy cycle may act prematurely in order to achieve something tangible, thus increasing the risk of failure. (To: TP0038)
2. The sustainable policy cycle should develop its own resources so that they are adequate to the task, but should refrain from using them until the time is ripe. (To: TP0035)
3. At the moment for transition, the sustainable policy cycle may lack the resources to complete the task as required, in which case qualified assistance should be obtained. (To: TP0050)
4. During the struggle to bring about the transition and overthrow the old order, the sustainable policy cycle should avoid doubt and lay the foundation for the future. (To: TP0004)
5. Justifying its efforts, the sustainable policy cycle may succeed in its struggle so that superior values become explicit in the envisaged order and the influence it has on society, especially in contrast to that which preceded it. (To: TP0006)
6. On the threshold of the new era convivial celebration is appropriate, but the sustainable policy cycle should take care not to lose its self-control and thus jeopardize what could be achieved. (To: TP0040)

Transformation sequence Further transformation calls for creativity.

Index

TX

Transformative Approaches

Index scope

This index covers all entries in Section T, namely:

- Transformative conferencing (Section TC)

- Transformative policy cycles (Section TP)

Note that **all index entries in the following index are integrated into the Volume Index (Section X)** at the end of this volume.

The main intention of this index is to enable users to obtain a better overview of the individual sub-sections of Section T.

All index entries refer via reference numbers (eg TC1234) to the descriptions in the preceding sections. The letters indicate the section (eg Section TC).

Index entries

The index entries are of several types:

- Principal name or title of concept;

- Secondary or alternative names of concept (including popular expressions and synonyms)

- Keywords from the principal concept name;

- Keywords from the secondary concept names.

No distinction is made between the different types of entry. Keywords are however recognizable by the presence of a semi-colon in the index entry.

Remarks

To facilitate consultation of the index, and to reduce the space requirement prepositions and articles are normally omitted from the index entries.

Although the index was generated automatically, it has been edited manually to eliminate less helpful keywords resulting from the extraction process. Where there was any doubt, less meaningful words have however been left where they may prove to be of value in locating entries in Section TC and TP.

INDEX TO TRANSFORMATIVE APPROACHES   TX

A

TP 0055 Abundance
TP 0029 Abyss Danger
TP 0029 Abyss Mastering pitfalls
TP 0029 Abyss Multiple dangers
TP 0029 Abyss Persistence
TC 1808 Accessibility ; Conference
TC 1637 Accident tactics
TC 1164 Accommodator participant type
TP 0063 Accomplishment
TP 0008 Accord
TP 0002 Acquiescence (Earth)
TP 0004 Acquiring experience (Young shoot)
TP 0017 Acquiring followers
TP 0021 Action ; Decisive (Biting through)
TC 1765 Action-oriented type
TP 0021 Action ; Punitive (Biting through)
TP 0018 Action ; Remedial
TP 0001 Action ; Strong (Heaven)
TC 1962 Activists ; Alienation committed
TC 1789 Activities ; Meetings social
TC 1777 Activities ; Risk triggering
TC 1721 Activities ; Trust building
TC 1843 Adaptive conferencing
TP 1383 Address
TP 0022 Adornment
TP 0036 Adulteration insight (Darkening of the light)
TP 0053 Advance ; Gradual
TP 0035 Advancement
TP 0046 Advancement
TP 0047 Adversity
TC 1929 Advocate ; Devil's
TC 1909 Aesthetic symbols ; Meetings
TP 0054 Affinity ; Elective (Marrying maiden)
TC 1602 Affinity group conferencing
TC 1528 Affinity roles
TP 0063 After completion
TC 1123 Agendas ; Non linear
TC 1929 Aggressive participant type
TP 1011 Agreement
TC 1529 Agriculture food processing ; Meetings
TP 0019 Ahead ; Getting
TC 1880 Aim ; Existential
TP 0038 Alienation
TC 1962 Alienation committed activists
TC 1229 Alienative participant involvement
TP 0061 Allegiance ; Wholehearted
TP 0008 Alliance
TC 1370 Allotments ; Exceeded time
TC 1451 Ambience ; Creative
TC 1315 Amorphous conferencing
TC 1603 Analogies ; Meeting
TC 1488 Analytical frameworks ; Meetings
TC 1929 Antagonist
TC 1663 Antipathetics
TP 0019 Approach
TP 0007 Army Controlled threat
TP 0051 Arousal (Thunderbolts)
TP 0018 Arresting decay
TC 1835 Art form method
TP 0045 Assembly
TC 1772 Assimilator participant type
TP 0042 Assistance
TC 1275 Associative networks
TC 1334 Assumptions ; Mood motivation
TC 1877 Assumptions ; Participation
TP 0017 Attunement
TC 1866 Audience reaction teams
TC 1418 Audio-graphic teleconferencing
TC 1498 Audio-tape recording
TC 1972 Audio teleconferencing
TC 1244 Authority ; Conference
TC 1623 Autonomy ; Conference

B

TP 0059 Barrier dissolution (Flooding)
TP 0048 Basic need (Well)
TP 0064 Before completion
TP 0003 Beginning ; Difficulty
TC 1910 Behaviour ; Patterned
TC 1910 Behaviour ; Rule governed
TC 1358 Belief type ; Blind
TC 1123 Beyond linear thinking
TC 1965 Biases ; Participant pre logical
TP 0005 Biding one's time (Getting wet)
TP 0021 Biting through Decisive action
TP 0021 Biting through Punitive action
TC 1487 Blackboard
TC 1358 Blind belief type
TP 0012 Blocked
TC 1800 Board ; Flannel
TC 1800 Board ; Magnetic
TC 1800 Board ; Sticky
TC 1800 Board ; Velcro
TP 0043 Breakthrough (Flight)
TC 1875 Bricksigner type
TP 0013 Brotherhood ; Universal
TC 1546 Buffoon
TC 1748 Builder type
TC 1721 Building activities ; Trust
TP 1011 Building ; Consensus
TP 0047 Burdened
TC 1776 Burn out ; Participant

TC 1996 Business concern conferencing
TC 1493 Buttons ; Participant consensus
TC 1311 Buzz group

C

TP 0005 Calculated inaction (Getting wet)
TC 1895 Calculative participant involvement
TC 1269 Care groups ; Creative
TP 0010 Careful conduct (Treading)
TP 0010 Carefully ; Stepping (Treading)
TP 0050 Cauldron Cultural heritage
TP 0050 Cauldron Rejuvenation
TP 0050 Cauldron Sacrificial vessel
TP 0061 Centre ; Faithfulness
TP 0061 Centre ; Sincerity
TC 1969 Chairperson
TC 1487 Chalkboard
TP 0049 Change (Moulting)
TC 1828 Change preferences ; Participant
TP 0037 Clan (Family)
TP 0057 Clarity ; Penetrating (Wind)
TC 1718 Clock ; Cost
TC 1546 Clown
TC 1381 Co-counsel conferencing
TC 1587 Coercive power
TC 1535 Colloquy
TC 1345 Commitment
TC 1962 Committed activists ; Alienation
TC 1848 Commonweal conferencing
TC 1943 Communication patterns
TC 1866 Communication teams ; Inter cultural
TC 1816 Communications unit ; Electronic
TP 0037 Community (Family)
TC 1526 Completer role ; Team
TP 0063 Completion ; After
TP 0064 Completion ; Before
TP 1011 Compromise
TC 1753 Computer conferencing ; In conference
TC 1616 Computer contact system ; In conference
TC 1666 Computer teleconferencing
TP 0036 Concealment illumination (Darkening of the light)
TC 1888 Conceptual framework development
TC 1002 Conceptual framework development
TC 1392 Conceptual frameworks
TC 1866 Conceptual interpretation teams
TC 1392 Conceptual models
TC 1888 Conceptual screen development
TC 1002 Conceptual screen development
TC 1392 Conceptual screens
TC 1996 Concern conferencing ; Business
TC 1775 Conciliators
TC 1544 Conclave
TP 0010 Conduct ; Careful (Treading)
TC 1082 Conference
TC 1808 Conference accessibility
TC 1244 Conference authority
TC 1623 Conference autonomy
TC 1753 Conference computer conferencing ; In
TC 1616 Conference computer contact system ; In
TC 1506 Conference control
TC 1484 Conference conviviality
TC 1636 Conference environment
TC 1399 Conference flexibility
TC 1651 Conference fodder
TC 1566 Conference heavies
TC 1683 Conference homogeneity
TC 1850 Conference impact
TC 1738 Conference insulation
TC 1588 Conference intent
TC 1952 Conference intimacy
TC 1944 Conference leadership
TC 1972 Conference line
TC 1394 Conference maturation
TC 1481 Conference mythology
TC 1568 Conference papers
TC 1490 Conference participation
TC 1903 Conference polarization
TC 1362 Conference potency
TC 1532 Conference practics
TC 1347 Conference procedure feedback
TC 1847 Conference procreation
TC 1760 Conference relationships
TC 1548 Conference rituals
TC 1555 Conference size
TC 1749 Conference spirit
TC 1424 Conference stability
TC 1776 Conference stamina
TC 1955 Conference stratification
TC 1717 Conference structure
TC 1644 Conference symbols
TC 1870 Conference ; Telephone
TC 1782 Conference tensions
TC 1556 Conference time design
TC 1061 Conference tone
TC 1191 Conference viscidity
TC 1843 Conferencing ; Adaptive
TC 1602 Conferencing ; Affinity group
TC 1315 Conferencing ; Amorphous
TC 1996 Conferencing ; Business concern
TC 1381 Conferencing ; Co counsel
TC 1848 Conferencing ; Commonweal
TC 1562 Conferencing ; Conscious
TC 1974 Conferencing ; Consultation
TC 1524 Conferencing ; Contextual

TC 1410 Conferencing ; Coordination
TC 1385 Conferencing ; Decision making
TC 1735 Conferencing ; Declaration
TC 1940 Conferencing ; Design
TC 1974 Conferencing ; Diagnostic
TC 1815 Conferencing dissipation
TC 1439 Conferencing ; Dread filled
TC 1996 Conferencing ; Economic
TC 1524 Conferencing ; Educative
TC 1722 Conferencing ; Encounter group
TC 1815 Conferencing energy sink
TC 1339 Conferencing evaluation
TC 1374 Conferencing ; Expressive
TC 1919 Conferencing failure
TC 1602 Conferencing ; Group counsel
TC 1632 Conferencing ; Hierarchical
TC 1753 Conferencing ; In conference computer
TC 1615 Conferencing inertia
TC 1733 Conferencing ; Instrumental
TC 1920 Conferencing intentions
TC 1478 Conferencing ; Interstitial
TC 1574 Conferencing ; Judicial
TC 1564 Conferencing ; Maintenance
TC 1421 Conferencing ; Majoral
TC 1313 Conferencing ; Managerial political
TC 1879 Conferencing ; Maximization
TC 1862 Conferencing ; Medial
TC 1135 Conferencing meditation
TC 1525 Conferencing ; Minoral
TC 1343 Conferencing ; Networking
TC 1920 Conferencing objectives
TC 1684 Conferencing ; Problem solving
TC 1996 Conferencing ; Productive
TC 1843 Conferencing ; Research
TC 1553 Conferencing risk
TC 1602 Conferencing ; Self reflective
TC 1696 Conferencing ; Service
TC 1989 Conferencing ; Strongwilled
TC 1547 Conferencing ; Structural dimensions
TC 1564 Conferencing ; Sustenance
TC 1940 Conferencing ; Vision
TC 1940 Conferencing ; Visionary
TC 1908 Conferencing ; Weakwilled
TP 0061 Confidence ; Inner
TC 1943 Configurations ; Constrained meeting
TC 1465 Configurations ; Meetings energy processing
TC 1909 Configurations ; Meetings symbolic
TC 1155 Configurations ; Socio structural meeting
TC 1699 Configurations ; Unconstrained meeting
TC 1603 Configurative models meetings
TP 0006 Conflict
TC 1663 Conformist participant type ; Non
TP 0045 Congregation
TC 1562 Congress ; Conscious
TP 0062 Conscientiousness
TC 1562 Conscious conferencing
TC 1562 Conscious congress
TC 1562 Conscious groups
TP 1011 Consensus building
TC 1493 Consensus buttons ; Participant
TP 0026 Conservation
TP 0026 Consolidation
TP 0032 Constancy
TC 1943 Constrained meeting configurations
TP 0030 Constraint ; Normative (Fire)
TC 1894 Constraints ; Language
TC 1689 Constructs ; Meetings physical
TC 1798 Consultants ; Intercultural
TC 1974 Consultation conferencing
TP 0044 Contact
TC 1616 Contact system ; In conference computer
TP 0020 Contemplation
TP 0006 Contention
TP 0016 Contentment
TC 1152 Contests ; Meetings games
TC 1524 Contextual conferencing
TC 1886 Contextual facilitation
TC 1491 Contributors
TC 1506 Control ; Conference
TP 0007 Controlled threat (Army)
TC 1133 Converger participant type
TC 1366 Conversations ; Group
TC 1484 Conviviality ; Conference
TP 0008 Coordination
TC 1410 Coordination conferencing
TP 0030 Cosmic mean (Fire)
TC 1718 Cost clock
TC 1381 Counsel conferencing ; Co
TC 1602 Counsel conferencing ; Group
TP 0027 Counsel ; Wise
TC 1451 Creative ambience
TC 1269 Creative care groups
TC 1913 Creative listening process
TP 0001 Creativity (Heaven)
TP 0051 Crisis preparedness (Thunderbolts)
TC 1866 Cultural communication teams ; Inter
TP 0050 Cultural heritage (Cauldron)
TC 1599 Cycles ; Repetition learning

D

TP 0029 Danger (Abyss)
TP 0029 Dangers ; Multiple (Abyss)
TP 0036 Darkening of the light Adulteration insight
TP 0036 Darkening of the light Concealment illumination

-679-

TX

TP 0036	Darkening of the light Decline
TP 0036	Darkening of the light Injury
TP 0036	Darkening of the light Unappreciated intelligence
TP 0004	Darkness (Young shoot)
TC 1454	Debate
TP 0018	Decay ; Arresting
TC 1385	Decision-making conferencing
TC 1627	Decision-making process
TP 0021	Decisive action (Biting through)
TC 1735	Declaration conferencing
TP 0036	Decline (Darkening of the light)
TP 0041	Decrease
TP 0041	Deficiency
TP 0018	Degeneration
TC 1423	Deletion
TP 0058	Delight (Lake)
TP 0040	Deliverance
TC 1556	Design ; Conference time
TC 1940	Design conferencing
TC 1188	Design methods ; Dynamic
TC 1451	Design ; Space
TP 0052	Desisting (Mountain)
TP 0023	Destruction (Stripping away)
TC 1736	Detached-operator type
TP 0023	Deterioration (Stripping away)
TP 0053	Development
TC 1888	Development ; Conceptual framework
TC 1002	Development ; Conceptual framework
TC 1888	Development ; Conceptual screen
TC 1002	Development ; Conceptual screen
TC 1359	Development ; Image
TC 1001	Development ; Image
TC 1929	Devil's advocate
TC 1345	Devotion
TC 1974	Diagnostic conferencing
TP 0003	Difficulty beginning
TP 0003	Difficulty ; Initial
TC 1547	Dimensions conferencing ; Structural
TC 1929	Disagreer
TC 1663	Disagreers
TP 0060	Discipline
TC 1306	Discussion 66
TC 1511	Discussion group
TC 1650	Discussion ; Interrogation
TC 1531	Discussion ; Panel
TP 0038	Disharmony
TP 0059	Disintegration (Flooding)
TP 0059	Dispersion (Flooding)
TP 0059	Dissension ; Overcoming (Flooding)
TC 1383	Dissertation
TC 1494	Dissidence
TC 1815	Dissipation ; Conferencing
TC 1699	Dissipation ; Meeting energy
TP 0059	Dissolution ; Barrier (Flooding)
TC 1373	Distortion
TC 1225	Diverger participant type
TC 1558	Diversity ; Unity
TC 1384	Doctrinaire participant type
TC 1439	Dread-filled conferencing
TC 1361	Driver role ; Team
TP 0032	Duration
TC 1311	Dyads
TC 1188	Dynamic design methods
TC 1441	Dynamic tensions
TC 1928	Dynamics ; Participation

E

TP 0002	Earth Acquiescence
TP 0002	Earth Passivity
TP 0002	Earth Receptivity
TP 0002	Earth Responsive service
TC 1475	Ecological natural processes ; Meetings
TC 1996	Economic conferencing
TC 1524	Educative conferencing
TP 0054	Elective affinity (Marrying maiden)
TC 1816	Electronic communications unit
TC 1418	Electronic meeting
TC 1218	Electronic meeting
TC 1876	Electronic meeting
TP 0022	Elegance
TP 0030	Elegance ; Model (Fire)
TP 0040	Elimination obstacles
TC 1990	Enabler
TP 0044	Encounter
TC 1722	Encounter group conferencing
TC 1990	Encourager participant type
TP 0032	Endurance
TC 1699	Energy dissipation ; Meeting
TC 1465	Energy processing configurations ; Meetings
TC 1815	Energy sink ; Conferencing
TP 0016	Enthusiasm
TP 0017	Entrainment
TC 1636	Environment ; Conference
TP 0025	Error free
TP 0061	Essential quality
TP 0038	Estrangement
TC 1339	Evaluation ; Conferencing
TC 1642	Evaluator role ; Team
TC 1056	Event planning methods
TC 1332	Event ; Transformative
TC 1370	Exceeded time allotments
TP 0028	Excesses great
TP 0041	Excesses ; Reduction
TP 0062	Excesses ; Small

TC 1538	Exercises ; Simulation
TP 0047	Exhaustion
TC 1880	Existential aim
TP 0025	Expectation ; Without
TC 1613	Expectations ; Participant
TP 0004	Experience ; Acquiring (Young shoot)
TC 1538	Experiences ; Structured
TC 1358	Expertist participant type
TC 1814	Explorer role ; Team
TC 1374	Expressive conferencing

F

TC 1559	Facilitation
TC 1886	Facilitation ; Contextual
TC 1377	Facilitation ; Workshop
TC 1354	Facilitative techniques
TC 1969	Facilitator
TC 1451	Facilities
TC 1919	Failure ; Conferencing
TC 1005	Failure ; Integrative
TP 0061	Faithfulness centre
TP 0037	Family Clan
TP 0037	Family Community
TP 0037	Family Household
TC 1776	Fatigue ; Meeting
TC 1776	Fatigue ; Prattle
TC 1776	Fatigue ; Secretariat
TC 1347	Feedback ; Conference procedure
TP 0013	Fellowship
TP 0025	Fidelity
TC 1439	Filled conferencing ; Dread
TC 1526	Finisher ; Team
TP 0030	Fire Cosmic mean
TP 0030	Fire Model elegance
TP 0030	Fire Normative constraint
TP 0030	Fire Shining light
TC 1800	Flannel board
TC 1399	Flexibility ; Conference
TP 0043	Flight Breakthrough
TP 0043	Flight Parting
TP 0043	Flight Resolution
TC 1691	Flipchart
TP 0059	Flooding Barrier dissolution
TP 0059	Flooding Disintegration
TP 0059	Flooding Dispersion
TP 0059	Flooding Overcoming dissension
TP 0059	Flooding Scattering
TP 0011	Flowing
TC 1351	Focal processes ; Meeting
TC 1969	Focalizer role ; Team
TC 1432	Focus meetings ; Losing
TC 1285	Focus ; Varieties meeting
TC 1651	Fodder ; Conference
TP 0017	Followers ; Acquiring
TC 1651	Followers participant type
TP 0017	Following
TP 0004	Folly ; Youthful (Young shoot)
TC 1529	Food processing ; Meetings agriculture
TC 1546	Fool
TC 1630	Force ; Task
TC 1218	Frame ; Freeze
TC 1888	Framework development ; Conceptual
TC 1002	Framework development ; Conceptual
TC 1392	Frameworks ; Conceptual
TC 1488	Frameworks ; Meetings analytical
TP 0040	Free ; Getting
TC 1218	Freeze frame
TP 0048	Fulfilled potentialities (Well)
TC 1876	Full motion video teleconferencing
TP 0055	Fullness

G

TP 0028	Gains ; Great
TC 1538	Games
TC 1152	Games contests ; Meetings
TP 0003	Gathering support
TP 0045	Gathering together
TC 1389	Generalization
TC 1491	Genius
TC 1550	Genius role ; Team
TP 0057	Gentleness (Wind)
TP 0019	Getting ahead
TP 0040	Getting free
TP 0005	Getting wet Biding one's time
TP 0005	Getting wet Calculated inaction
TP 0005	Getting wet Waiting
TP 0025	Goodness ; Instinctive
TP 0053	Gradual advance
TP 0053	Gradual progress
TC 1418	Graphic teleconferencing ; Audio
TC 1430	Grassrooter participant type
TP 0028	Great ; Excesses
TP 0028	Great gains
TP 0014	Great measure ; Possession
TP 0034	Great ; Power
TP 0028	Great ; Preponderance
TP 0026	Great ; Taming power
TP 0034	Great vigour
TP 0028	Greatness ; Passing
TP 0006	Grievance
TC 1264	Grounder role ; Team
TC 1311	Group ; Buzz
TC 1602	Group conferencing ; Affinity

TC 1722	Group conferencing ; Encounter
TC 1366	Group conversations
TC 1602	Group-counsel conferencing
TC 1511	Group ; Discussion
TC 1856	Group ; Peer mediated learning
TC 1916	Group ; Rap
TC 1375	Group ; Study
TC 1602	Group therapy ; Inter
TC 1562	Groups ; Conscious
TC 1269	Groups ; Creative care
TC 1306	Groups ; Huddle
TC 1593	Groups ; Special interest
TP 0053	Growth
TP 0003	Growth pains ; Organizational
TP 0004	Growth ; Uncultivated (Young shoot)
TC 1566	Guest speakers

H

TC 1642	Hard-head
TC 1775	Harmonizers
TC 1739	Haste participant type ; Less
TC 1642	Head ; Hard
TP 0001	Heaven Creativity
TP 0001	Heaven Strong action
TC 1566	Heavies ; Conference
TP 0050	Heritage ; Cultural (Cauldron)
TC 1632	Hierarchical conferencing
TP 0008	Holding together
TC 1683	Homogeneity ; Conference
TP 0037	Household (Family)
TC 1306	Huddle groups
TP 0015	Humility
TP 0017	Hunting

I

TP 0013	Identification others
TC 1663	Idiosyncratics
TP 0018	Illness ; Responding
TP 0036	Illumination ; Concealment (Darkening of the light)
TP 0022	Image
TC 1359	Image development
TC 1001	Image development
TC 1323	Images
TC 1550	Imagineer
TP 0004	Immaturity (Young shoot)
TC 1850	Impact ; Conference
TP 0028	Importance
TC 1753	In-conference computer conferencing
TC 1616	In-conference computer contact system
TP 0005	Inaction ; Calculated (Getting wet)
TP 0052	Inaction (Mountain)
TP 0042	Increase
TC 1615	Inertia ; Conferencing
TP 0004	Inexperience (Young shoot)
TP 0044	Inferior ; Infiltration
TP 0044	Infiltration inferior
TP 0031	Influence
TC 1750	Information seeking participants
TC 1900	Information ; Sharing
TP 0003	Initial difficulty
TP 0019	Initiative
TC 1491	Initiator participant type
TP 0036	Injury (Darkening of the light)
TP 0061	Inner confidence
TP 0061	Inner truth
TP 0025	Innocence
TP 0036	Insight ; Adulteration (Darkening of the light)
TP 0025	Instinctive goodness
TC 1733	Instrumental conferencing
TC 1738	Insulation ; Conference
TC 1005	Integrative failure
TC 1062	Integrative skills
TP 0025	Integrity
TC 1971	Intellectual momentum
TP 0036	Intelligence ; Unappreciated (Darkening of the light)
TC 1588	Intent ; Conference
TC 1920	Intentions ; Conferencing
TC 1866	Inter-cultural communication teams
TC 1602	Inter-group therapy
TC 1640	Interaction preferences ; Non verbal
TC 1845	Interaction preferences ; Participant
TC 1378	Interaction preferences ; Verbal
TC 1650	Interchange panel
TC 1798	Intercultural consultants
TC 1798	Intercultural interpreter
TC 1208	Interdisciplinarity
TC 1593	Interest groups ; Special
TC 1902	Interpersonal suspicion
TC 1852	Interpersonal trust
TC 1866	Interpretation teams ; Conceptual
TC 1798	Interpreter ; Intercultural
TC 1650	Interrogation-discussion
TC 1650	Interrogator panel
TC 1478	Interstitial conferencing
TC 1952	Intimacy ; Conference
TP 0023	Intrigue (Stripping away)
TC 1650	Investigative panel
TC 1814	Investigator ; Resource
TC 1229	Involvement ; Alienative participant
TC 1895	Involvement ; Calculative participant
TC 1345	Involvement ; Moral participant

-680-

INDEX TO TRANSFORMATIVE APPROACHES

TX

J

TC 1546	Jester
TC 1546	Joker
TP 0058	Joy (Lake)
TC 1574	Judicial conferencing

K

TP 0052	Keeping still (Mountain)
TC 1566	Key-note speakers
TC 1932	Key resource persons

L

TP 0058	Lake Delight
TP 0058	Lake Joy
TP 0058	Lake Standing straight
TP 0058	Lake Vitality
TC 1894	Language constraints
TC 1603	Language participants ; Pattern
TP 0019	Leadership
TP 0008	Leadership
TC 1944	Leadership ; Conference
TC 1599	Learning cycles ; Repetition
TC 1856	Learning group ; Peer mediated
TC 1383	Lecture
TC 1739	Less-haste participant type
TP 0040	Liberation
TP 0030	Light ; Shining (Fire)
TP 0060	Limitation
TC 1972	Line ; Conference
TC 1123	Linear agendas ; Non
TC 1123	Linear thinking ; Beyond
TC 1651	Liners ; Party
TC 1913	Listening process ; Creative
TC 1913	Listening process ; Systematic
TC 1742	Listening teams
TC 1965	Logical biases ; Participant pre
TC 1432	Losing focus meetings
TC 1972	Loud speaker telephone
TC 1345	Loyalty

M

TC 1567	Magic ; Meeting
TC 1800	Magnetic board
TC 1564	Maintenance conferencing
TC 1421	Majoral conferencing
TC 1313	Managerial-political conferencing
TC 1275	Maps ; Relevance
TC 1129	Marginal participant type
TP 0056	Marginality (Wanderer)
TP 0054	Marrying maiden Elective affinity
TP 0054	Marrying maiden Propriety
TP 0029	Mastering pitfalls (Abyss)
TC 1394	Maturation ; Conference
TC 1879	Maximization conferencing
TP 0030	Mean ; Cosmic (Fire)
TP 0014	Measure ; Possession great
TC 1862	Medial conferencing
TC 1856	Mediated learning group ; Peer
TC 1775	Mediator participant type
TC 1135	Meditation ; Conferencing
TP 0044	Meeting
TC 1603	Meeting analogies
TC 1943	Meeting configurations ; Constrained
TC 1155	Meeting configurations ; Socio structural
TC 1699	Meeting configurations ; Unconstrained
TC 1418	Meeting ; Electronic
TC 1218	Meeting ; Electronic
TC 1876	Meeting ; Electronic
TC 1699	Meeting energy dissipation
TC 1776	Meeting fatigue
TC 1351	Meeting focal processes
TC 1285	Meeting focus ; Varieties
TC 1567	Meeting magic
TC 1915	Meeting ; Prayer
TC 1567	Meeting serendipity
TC 1304	Meeting space
TC 1972	Meeting ; Speakerphone
TC 1567	Meeting take off
TC 1909	Meetings aesthetic symbols
TC 1529	Meetings agriculture food processing
TC 1488	Meetings analytical frameworks
TC 1603	Meetings ; Configurative models
TC 1475	Meetings ecological natural processes
TC 1465	Meetings energy-processing configurations
TC 1152	Meetings games contests
TC 1432	Meetings ; Losing focus
TC 1603	Meetings ; Metaphors
TC 1727	Meetings models reality
TC 1724	Meetings ; Multi
TC 1689	Meetings physical constructs
TC 1367	Meetings physical processes
TC 1922	Meetings psycho-physical processes
TC 1789	Meetings social activities
TC 1909	Meetings symbolic configurations
TC 1603	Metaphors meetings
TC 1835	Method ; Art form
TC 1420	Method ; Seminar
TC 1188	Methods ; Dynamic design
TC 1056	Methods ; Event planning
TC 1927	Methods ; Participation
TC 1449	Methods ; Workshop
TC 1743	Microphone ; Travelling
TC 1525	Minoral conferencing
TP 0030	Model elegance (Fire)
TC 1392	Models ; Conceptual
TC 1603	Models meetings ; Configurative
TC 1727	Models reality ; Meetings
TC 1969	Moderator
TP 0015	Modesty
TC 1370	Modification ; Programme
TC 1971	Momentum ; Intellectual
TC 1853	Monger participant type ; Value
TC 1642	Monitor
TC 1383	Monologue
TC 1334	Mood motivation assumptions
TC 1345	Moral participant involvement
TC 1876	Motion video teleconferencing ; Full
TC 1334	Motivation assumptions ; Mood
TP 0049	Moulting Change
TP 0049	Moulting Revolution
TP 0052	Mountain Desisting
TP 0052	Mountain Inaction
TP 0052	Mountain Keeping still
TP 0052	Mountain Resting
TP 0031	Movement
TC 1724	Multi-meetings
TP 0029	Multiple dangers (Abyss)
TC 1481	Mythology ; Conference

N

TC 1475	Natural processes ; Meetings ecological
TP 0048	Need ; Basic (Well)
TC 1204	Needs ; Participant personality
TC 1275	Network ; Thesaurus
TC 1343	Networking conferencing
TC 1275	Networks ; Associative
TC 1275	Networks ; Relevance
TP 0056	Newcomer (Wanderer)
TC 1663	Non-conformist participant type
TC 1123	Non-linear agendas
TC 1640	Non-verbal interaction preferences
TC 1581	Nonchalent participant type
TP 0030	Normative constraint (Fire)
TC 1858	Normative power
TC 1691	Noteboard
TP 0027	Nourishment ; Providing
TP 0009	Nurturance small

O

TC 1704	Objective ; Rational
TC 1920	Objectives ; Conferencing
TC 1742	Observation teams
TP 0020	Observing
TP 0040	Obstacles ; Elimination
TP 0039	Obstruction
TP 0012	Obstruction
TP 0009	Obstructions ; Small
TP 0005	One's time ; Biding (Getting wet)
TC 1947	Opaque projector
TC 1736	Operator type ; Detached
TC 1011	Opinion ; Public
TP 0038	Opposition
TP 0047	Oppression
TP 0003	Organizational growth pains
TC 1657	Oriented participant type ; Process
TC 1860	Oriented participant type ; Structure
TC 1765	Oriented type ; Action
TP 0013	Others ; Identification
TP 0059	Overcoming dissension (Flooding)
TC 1472	Overhead projector
TP 0019	Overseeing

P

TP 0003	Pains ; Organizational growth
TC 1311	Pairs
TC 1531	Panel discussion
TC 1650	Panel ; Interchange
TC 1650	Panel ; Interrogator
TC 1650	Panel ; Investigative
TC 1650	Panel ; Responsive
TC 1650	Panel ; Zetetic
TC 1383	Paper
TC 1568	Papers ; Conference
TC 1776	Participant burn-out
TC 1828	Participant change preferences
TC 1493	Participant consensus buttons
TC 1613	Participant expectations
TC 1845	Participant interaction preferences
TC 1229	Participant involvement ; Alienative
TC 1895	Participant involvement ; Calculative
TC 1345	Participant involvement ; Moral
TC 1204	Participant personality needs
TC 1965	Participant pre-logical biases
TC 1828	Participant strategic preferences
TC 1164	Participant type ; Accommodator
TC 1929	Participant type ; Aggressive
TC 1772	Participant type ; Assimilator
TC 1133	Participant type ; Converger
TC 1225	Participant type ; Diverger
TC 1384	Participant type ; Doctrinaire
TC 1990	Participant type ; Encourager
TC 1358	Participant type ; Expertist
TC 1651	Participant type ; Followers
TC 1430	Participant type ; Grassrooter
TC 1491	Participant type ; Initiator
TC 1739	Participant type ; Less haste
TC 1129	Participant type ; Marginal
TC 1775	Participant type ; Mediator
TC 1663	Participant type ; Non conformist
TC 1581	Participant type ; Nonchalent
TC 1657	Participant type ; Process oriented
TC 1391	Participant type ; Self reflexer
TC 1860	Participant type ; Structure oriented
TC 1853	Participant type ; Value monger
TC 1750	Participants ; Information seeking
TC 1603	Participants ; Pattern language
TC 1877	Participation assumptions
TC 1490	Participation ; Conference
TC 1928	Participation dynamics
TC 1927	Participation methods
TC 1843	Participatory research
TP 0043	Parting (Flight)
TC 1651	Party-liners
TP 0028	Passing greatness
TP 0002	Passivity (Earth)
TC 1603	Pattern language participants
TC 1910	Patterned behaviour
TC 1943	Patterns ; Communication
TP 0011	Peace
TC 1856	Peer-mediated-learning group
TP 0057	Penetrating clarity (Wind)
TP 0029	Persistence (Abyss)
TC 1651	Person ; Yes
TC 1204	Personality needs ; Participant
TC 1932	Persons ; Key resource
TC 1306	Phillips 66
TC 1689	Physical constructs ; Meetings
TC 1367	Physical processes ; Meetings
TC 1922	Physical processes ; Meetings psycho
TP 0029	Pitfalls ; Mastering (Abyss)
TC 1056	Planning methods ; Event
TC 1581	Playboy
TC 1903	Polarization ; Conference
TC 1313	Political conferencing ; Managerial
TP 0014	Possession great measure
TC 1362	Potency ; Conference
TP 0048	Potentialities ; Fulfilled (Well)
TP 0034	Power
TC 1587	Power ; Coercive
TP 0034	Power great
TP 0026	Power great ; Taming
TC 1858	Power ; Normative
TC 1831	Power ; Remunerative
TP 0009	Power small ; Taming
TC 1532	Practics ; Conference
TC 1776	Prattle fatigue
TC 1915	Prayer meeting
TC 1965	Pre logical biases ; Participant
TC 1640	Preferences ; Non verbal interaction
TC 1828	Preferences ; Participant change
TC 1845	Preferences ; Participant interaction
TC 1828	Preferences ; Participant strategic
TC 1378	Preferences ; Verbal interaction
TP 0051	Preparedness ; Crisis (Thunderbolts)
TP 0028	Preponderance great
TP 0062	Preponderance small
TC 1383	Presentation
TC 1684	Problem solving conferencing
TC 1347	Procedure feedback ; Conference
TC 1913	Process ; Creative listening
TC 1627	Process ; Decision making
TC 1657	Process-oriented participant type
TC 1913	Process ; Systematic listening
TC 1351	Processes ; Meeting focal
TC 1475	Processes ; Meetings ecological natural
TC 1367	Processes ; Meetings physical
TC 1922	Processes ; Meetings psycho physical
TC 1465	Processing configurations ; Meetings energy
TC 1529	Processing ; Meetings agriculture food
TC 1847	Procreation ; Conference
TC 1996	Productive conferencing
TC 1370	Programme modification
TP 0035	Progress
TP 0053	Progress ; Gradual
TC 1824	Projection ; Slide
TC 1947	Projector ; Opaque
TC 1472	Projector ; Overhead
TC 1142	Promoter role ; Team
TP 0046	Promotion
TP 0054	Propriety (Marrying maiden)
TP 0055	Prosperity
TP 0027	Providing nourishment
TC 1922	Psycho physical processes ; Meetings
TC 1011	Public opinion
TC 1501	Public relations
TP 0021	Punitive action (Biting through)

Q

TP 0061	Quality ; Essential

R

TC 1916	Rap group
TC 1704	Rational objective
TC 1866	Reaction teams ; Audience
TC 1727	Reality ; Meetings models
TP 0002	Receptivity (Earth)

-681-

TX

TP 0020	Recognition	TP 0012	Stagnation	TP 0010	Treading Tranquillity
TC 1875	Recognition seekers	TC 1776	Stamina ; Conference	TP 0039	Trouble
TC 1498	Recording ; Audio tape	TP 0058	Standing straight (Lake)	TC 1721	Trust building activities
TP 0024	Recovery (Turning point)	TP 0012	Standstill	TC 1852	Trust ; Interpersonal
TP 0041	Reduction excesses	TC 1566	Stars ; Super	TC 1721	Trust working
TC 1602	Reflective conferencing ; Self	TP 0010	Stepping carefully (Treading)	TP 0061	Truth ; Inner
TC 1391	Reflexer participant type ; Self	TP 1800	Sticky board	TP 0024	Turning point Recovery
TP 0060	Regulation	TP 0052	Still ; Keeping (Mountain)	TP 0024	Turning point Return
TP 0050	Rejuvenation (Cauldron)	TP 0058	Straight ; Standing (Lake)	TC 1164	Type ; Accommodator participant
TC 1501	Relations ; Public	TC 1828	Strategic preferences ; Participant	TC 1765	Type ; Action oriented
TC 1760	Relationships ; Conference	TC 1955	Stratification ; Conference	TC 1929	Type ; Aggressive participant
TP 0040	Release	TP 0006	Strife	TC 1772	Type ; Assimilator participant
TC 1275	Relevance maps	TP 0023	Stripping away Destruction	TC 1358	Type ; Blind belief
TC 1275	Relevance networks	TP 0023	Stripping away Deterioration	TC 1875	Type ; Bricksigner
TP 0064	Remaining ; Tasks	TP 0023	Stripping away Intrigue	TC 1748	Type ; Builder
TP 0018	Remedial action	TP 0001	Strong action (Heaven)	TC 1133	Type ; Converger participant
TC 1831	Remunerative power	TP 0026	Strong ; Restraint	TC 1736	Type ; Detached operator
TC 1599	Repetition learning cycles	TC 1989	Strongwilled conferencing	TC 1225	Type ; Diverger participant
TP 0016	Repose	TC 1547	Structural dimensions conferencing	TC 1384	Type ; Doctrinaire participant
TC 1843	Research conferencing	TC 1155	Structural meeting configurations ; Socio	TC 1990	Type ; Encourager participant
TC 1843	Research ; Participatory	TC 1717	Structure ; Conference	TC 1358	Type ; Expertist participant
TP 0040	Resolution	TC 1860	Structure-oriented participant type	TC 1651	Type ; Followers participant
TP 0043	Resolution (Flight)	TC 1538	Structured experiences	TC 1430	Type ; Grassrooter participant
TC 1814	Resource investigator	TC 1375	Study group	TC 1491	Type ; Initiator participant
TC 1932	Resource persons ; Key	TP 0039	Stumbling	TC 1739	Type ; Less haste participant
TP 0018	Responding illness	TP 0022	Style	TC 1129	Type ; Marginal participant
TC 1650	Responsive panel	TP 0044	Subjugation	TC 1775	Type ; Mediator participant
TP 0002	Responsive service (Earth)	TP 0057	Submission ; Willing (Wind)	TC 1663	Type ; Non conformist participant
TP 0052	Resting (Mountain)	TP 0009	Subtle restraint	TC 1581	Type ; Nonchalent participant
TP 0060	Restraint	TC 1566	Super stars	TC 1657	Type ; Process oriented participant
TP 0026	Restraint strong	TP 0027	Support	TC 1391	Type ; Self reflexer participant
TP 0009	Restraint ; Subtle	TP 0003	Support ; Gathering	TC 1860	Type ; Structure oriented participant
TP 0009	Restraint weak	TC 1902	Suspicion ; Interpersonal	TC 1853	Type ; Value monger participant
TP 0033	Retreat	TC 1564	Sustenance conferencing		
TP 0024	Return (Turning point)	TC 1651	Sycophants		**U**
TP 0049	Revolution (Moulting)	TC 1909	Symbolic configurations ; Meetings		
TP 0055	Richness	TC 1644	Symbols ; Conference	TP 0036	Unappreciated intelligence (Darkening of the light)
TP 0046	Rising	TC 1909	Symbols ; Meetings aesthetic	TC 1699	Unconstrained meeting configurations
TC 1553	Risk ; Conferencing	TC 1924	Symposium	TP 0004	Uncultivated growth (Young shoot)
TC 1777	Risk triggering activities	TC 1011	Synergy	TC 1816	Unit ; Electronic communications
TC 1777	Risk working	TC 1011	Synthesis	TC 1558	Unity diversity
TC 1548	Rituals ; Conference	TC 1616	System ; In conference computer contact	TP 0013	Universal brotherhood
TC 1526	Role ; Team completer	TC 1913	Systematic listening process	TP 0015	Unpretentiousness
TC 1361	Role ; Team driver			TP 0064	Unsettlement
TC 1642	Role ; Team evaluator		**T**	TC 1831	Utilitarian
TC 1814	Role ; Team explorer				
TC 1969	Role ; Team focalizer	TC 1637	Tactics ; Accident		**V**
TC 1550	Role ; Team genius	TC 1567	Take off ; Meeting		
TC 1264	Role ; Team grounder	TP 0026	Taming power great	TC 1853	Value-monger participant type
TC 1142	Role ; Team promoter	TP 0009	Taming power small	TC 1285	Varieties meeting focus
TC 1528	Roles ; Affinity	TC 1498	Tape recording ; Audio	TP 1800	Velcro-board
TC 1528	Roles ; Team	TC 1897	Tape ; Video	TC 1378	Verbal interaction preferences
TC 1910	Rule-governed behaviour	TC 1630	Task force	TC 1640	Verbal interaction preferences ; Non
		TC 1485	Taskforce	TP 0050	Vessel ; Sacrificial (Cauldron)
	S	TC 1269	Taskforce ; Special	TC 1897	Video tape
		TP 0064	Tasks remaining	TC 1218	Video teleconferencing
TP 0050	Sacrificial vessel (Cauldron)	TC 1526	Team completer role	TC 1876	Video teleconferencing ; Full motion
TP 0059	Scattering (Flooding)	TC 1361	Team driver role	TP 0051	Vigilance (Thunderbolts)
TC 1888	Screen development ; Conceptual	TC 1642	Team evaluator role	TP 0034	Vigour ; Great
TC 1002	Screen development ; Conceptual	TC 1814	Team explorer role	TC 1191	Viscidity ; Conference
TC 1392	Screens ; Conceptual	TC 1526	Team finisher	TC 1940	Vision conferencing
TC 1776	Secretariat fatigue	TC 1969	Team focalizer role	TC 1940	Visionary conferencing
TC 1875	Seekers ; Recognition	TC 1550	Team genius role	TP 0058	Vitality (Lake)
TC 1750	Seeking participants ; Information	TC 1264	Team grounder role		
TC 1602	Self-reflective conferencing	TC 1142	Team promoter role		**W**
TC 1391	Self-reflexer participant type	TC 1528	Team roles		
TC 1420	Seminar method	TC 1361	Team shaper	TP 0005	Waiting (Getting wet)
TP 0031	Sensing	TC 1866	Teams ; Audience reaction	TP 0056	Wanderer Marginality
TP 0031	Sensitivity	TC 1866	Teams ; Conceptual interpretation	TP 0056	Wanderer Newcomer
TC 1567	Serendipity ; Meeting	TC 1866	Teams ; Inter cultural communication	TP 0056	Wanderer Traveller
TC 1696	Service conferencing	TC 1866	Teams ; Interruption	TP 0020	Watching
TP 0002	Service ; Responsive (Earth)	TC 1742	Teams ; Listening	TP 0009	Weak ; Restraint
TP 0063	Settlement	TC 1742	Teams ; Observation	TC 1908	Weakwilled conferencing
TC 1361	Shaper ; Team	TC 1354	Techniques ; Facilitative	TP 0014	Wealth
TC 1900	Sharing information	TC 1972	Teleconferencing ; Audio	TP 0048	Well Basic need
TC 1651	Sheep	TC 1418	Teleconferencing ; Audio graphic	TP 0048	Well Fulfilled potentialities
TP 0030	Shining light (Fire)	TC 1666	Teleconferencing ; Computer	TC 1487	Whiteboard
TP 0051	Shock (Thunderbolts)	TC 1876	Teleconferencing ; Full motion video	TP 0061	Wholehearted allegiance
TC 1538	Simulation exercises	TC 1218	Teleconferencing ; Video	TP 0057	Willing submission (Wind)
TP 0061	Sincerity centre	TC 1870	Telephone conference	TP 0057	Wind Gentleness
TC 1815	Sink ; Conferencing energy	TC 1972	Telephone ; Loud speaker	TP 0057	Wind Penetrating clarity
TC 1555	Size ; Conference	TC 1782	Tensions ; Conference	TP 0057	Wind Willing submission
TC 1062	Skills ; Integrative	TC 1441	Tensions ; Dynamic	TP 0027	Wise counsel
TC 1824	Slide projection	TC 1602	Therapy ; Inter group	TP 0033	Withdrawal
TC 1218	Slowscan	TC 1275	Thesaurus network	TP 0025	Without Expectation
TP 0062	Small excesses	TC 1123	Thinking ; Beyond linear	TC 1485	Workgroup
TP 0009	Small ; Nurturance	TP 0007	Threat ; Controlled (Army)	TC 1777	Working ; Risk
TP 0009	Small obstructions	TP 0064	Threshold ; Transformation	TC 1721	Working ; Trust
TP 0062	Small ; Preponderance	TP 0051	Thunderbolts Arousal	TC 1377	Workshop facilitation
TP 0009	Small ; Taming power	TP 0051	Thunderbolts Crisis preparedness	TC 1449	Workshop methods
TC 1789	Social activities ; Meetings	TP 0051	Thunderbolts Shock		
TC 1155	Socio-structural meeting configurations	TP 0051	Thunderbolts Vigilance		**Y**
TP 0008	Solidarity	TC 1370	Time allotments ; Exceeded		
TC 1684	Solving conferencing ; Problem	TP 0005	Time ; Biding one's (Getting wet)	TC 1651	Yes person
TC 1451	Space design	TC 1556	Time design ; Conference	TP 0033	Yielding
TC 1304	Space ; Meeting	TP 0045	Together ; Gathering	TP 0004	Young shoot Acquiring experience
TC 1972	Speaker telephone ; Loud	TP 0003	Together ; Holding	TP 0004	Young shoot Darkness
TC 1972	Speakerphone meeting	TC 1061	Tone ; Conference	TP 0004	Young shoot Immaturity
TC 1566	Speakers ; Guest	TP 0010	Tranquillity (Treading)	TP 0004	Young shoot Inexperience
TC 1566	Speakers ; Key note	TP 0064	Transformation threshold	TP 0004	Young shoot Uncultivated growth
TC 1593	Special interest groups	TC 1332	Transformative event	TP 0004	Young shoot Youthful folly
TC 1269	Special taskforce	TP 0056	Traveller (Wanderer)	TP 0004	Youthful folly (Young shoot)
TC 1749	Spirit ; Conference	TC 1743	Travelling microphone		
TP 0025	Spontaneity	TP 0010	Treading Careful conduct		
TC 1424	Stability ; Conference	TP 0010	Treading Stepping carefully		

Z

TC 1650 Zetetic panel

NUMBERS

TC 1306 66 ; Discussion
TC 1306 66 ; Phillips

Notes

TZ

Transformative Approaches

Significance

    Transformative conferencing   687

Challenge

    Conceptual challenge   688
    Conceptual weaknesses of conferencing   689

Opportunity

    Marrying modes of information   690
    Conceptual scaffolding and prosthetics   691
    Policy forums as metaphors   693
    Metaphors of transformation in conferences   694

Envisioning conferencing

    Aesthetics of governance   695
    Insights from music   697
    Insights from poetry and painting   698
    Insights from drama and dance   699
    Insights from architecture   700
    Current possibilities of implementation   701

Transformative policy cycles

    Patterning cycles   702
    Modelling of sustainable development   704
    Alternation between complementary conditions   706

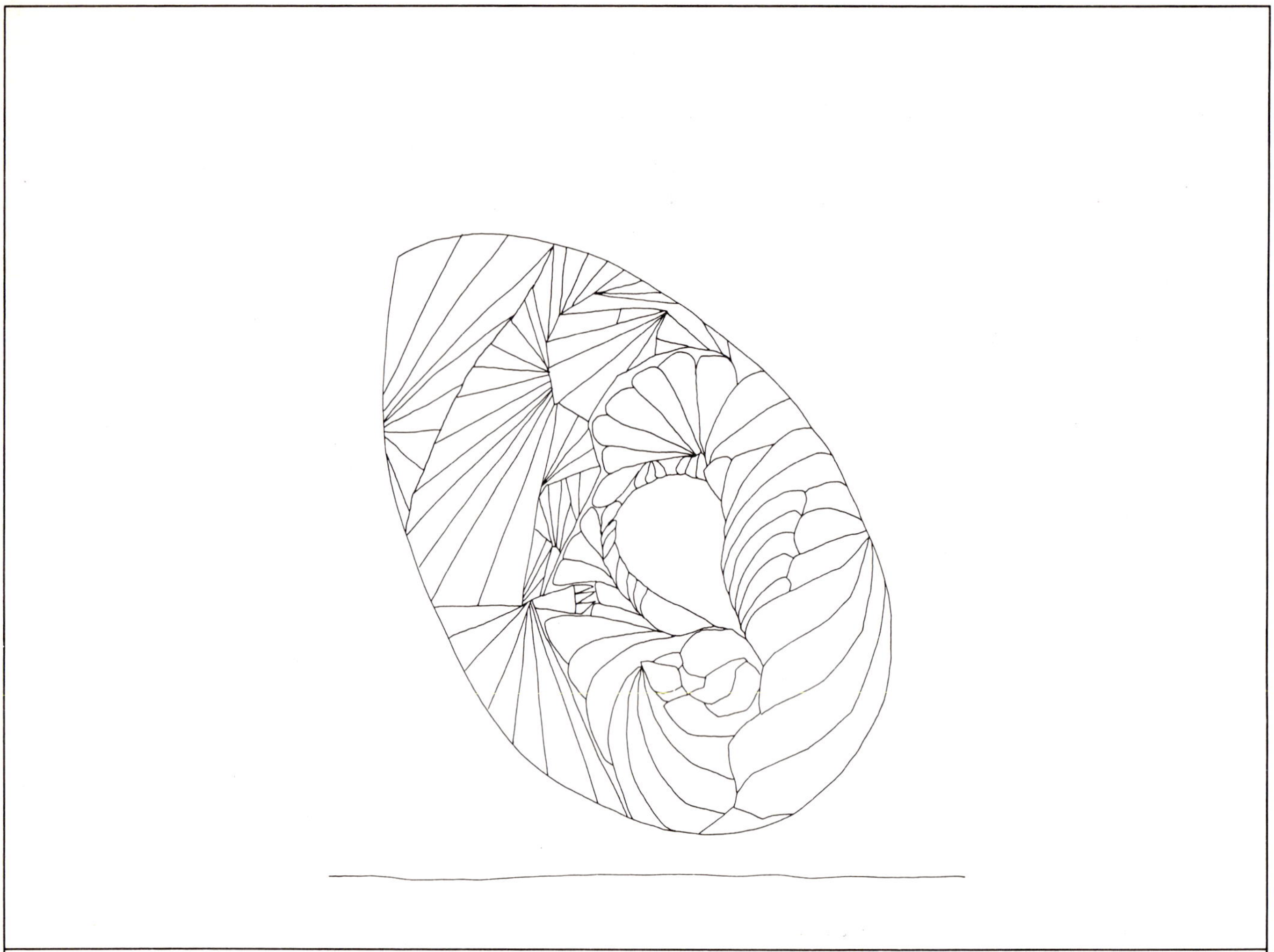

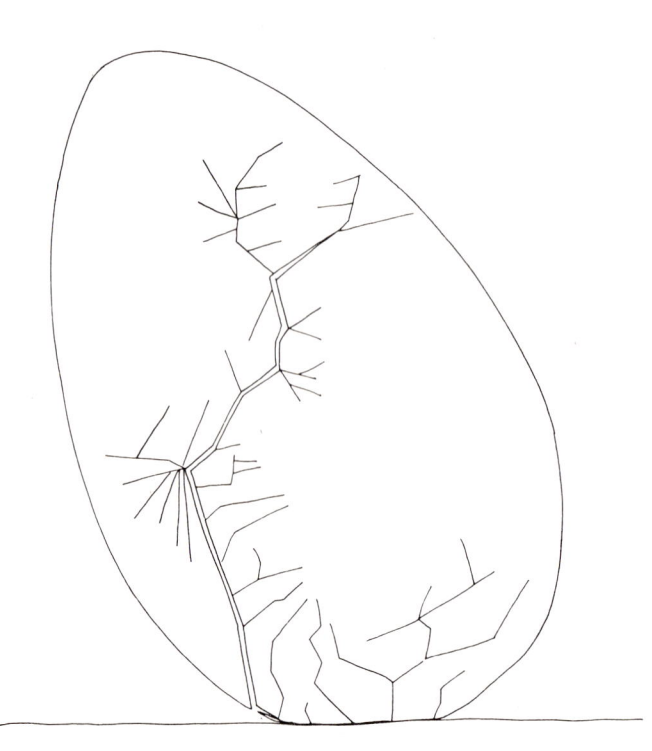

Significance: transformative conferencing

1. Reservations
The concern in this section is with those large-group meetings or conferences which are **not**:

(a) Organized according to procedures considered reasonably satisfactory by most of those directly involved, possibly on the basis of experience of previous meetings in the same series;

(b) Deliberately structured by the instigators to achieve a certain objective, irrespective of the individual preoccupations of those who choose to participate under such circumstances;

(c) Conceived around a pre-defined set of topics, irrespective of any other topics which may emerge during the meeting as common to a number of participants present;

(d) Deliberately unstructured as an environment for spontaneous exchange between participants, but without any concern that such exchanges should lead to the emergence of some larger pattern.

The main concern is with highlighting problems and possibilities relevant to the organization of more mature meetings on the new frontier of high- risk gatherings in response to social development issues and the global problematique. Attention is only given to the "mechanics" of meeting organization (covered in the many books available on such matters) in so far as they directly affect the psycho-social dynamic of the meeting.

2. Meeting maturation
The topics are therefore oriented around the possibility of maturing the power of a larger meeting to:

(a) reflect the complexity of the external environment in an ordered manner (representation), to reflect about that environment (conceptual processes), and to reflect about itself (self-reference or self- reflexiveness);

(b) focus the variety of perspectives represented, without destroying that variety in some simplistic formula of superficial consensus;

(c) transform the issues presented, and the organizational groups which take responsibility for them, into new configurations of operational significance;

(d) act, or empower those represented to act, in the light of the level of understanding achieved during the meeting.

In line with the general theme of this project, there is a concern that meeting innovation is being severely hindered by the limited vocabularly by which meeting processes and structures are defined: programme, session, speaker, participant, topic, organizer, *etc*. This is especially the case in that most of this vocabularly focuses on the logistics and administration of the meeting. The challenge is to find ways of enriching understanding of the range of meeting processes, including "conceptual logistics", moving beyond the limitations of that vocabularly, clarifying new distinctions and reinforcing those new distinctions by a new vocabularly.

3. Envisioning the perfect meeting
In recent years many people have deplored the inadequacies of the visions of society in the future. It is argued that credible visions offer a vital guideline to long-term policy. Clarifying such visions is a useful focus for debate. As a central process in society, meetings also merit this form of concern. Indeed if the problems inherent in meetings cannot be solved, is it possible to move toward any better society? What could constitute a perfect meeting in the future? Adequate images of such ideal meetings can guide reflection on present inadequacies and on how they may be overcome. The following points identify aspects which can be usefully borne in mind.

(a) Inter-weaving resources: Rather than the present emphasis on isolated participant contributions, the emphasis will be on interrelating contributions to form a pattern whose form evokes further contributions thus bringing about an appropriate balance of perspectives. Representatives of each discipline or approach will strive for better ways to evoke that pattern. Lengthy contributions (in time or on paper) will become secondary to the contribution of specific ideas, values, facts, problems or relationships. Those which significantly improve the emerging pattern will be valued most.

(b) Pace: Rather than the present hectic exercises in maximizing "communication", many meetings or sessions will bear a greater resemblance to a public game of chess or go. Periods of silence will be interspersed with brief contributions to the emerging pattern on whose evolution all are reflecting.

(c) Status and reward: Rather than status being accorded or acknowledged by protocol and "prime time" privileges, it will be self-evident from the record of the relative significance of the contributions made to the emerging pattern. This will be the prime source of personal satisfaction.

(d) Process: Rather than the simplistic overt processes of present meetings (made possible by a complex of covert processes), the range of processes will be understood to interweave as they do in a complex but healthy ecosystem - of which there are many types.

(e) Maturity: Rather than the present possibility of immaturity in a meeting of the most eminent, the maturity level of the meeting will be a matter of explicit concern and many will have skills to evolve the meeting beyond the characteristic traps of the present.

(f) Roles: Rather than the limited range of roles in present meetings, those of the future will be characterized by a rich variety of supporting, guiding, informing, facilitating roles. The potential of a meeting may well be judged by the "participant/supporting role" ratio (*cf* the teacher/pupil ratio in schools) as well as the number of "jargons" between which "interpretation" is provided.

(g) Modes: Rather than the limited range of modes now permissable in a given meeting, it will be possible for a meeting to move flexibly between many modes according to the energy requirements of the participants - and without losing a sense of coherence.

(h) Conceptual environment: Rather than the crude (lack of) awareness of meeting conceptual dynamics, participants will be much more conscious of the "species" of each contribution made, the effect it can have on the evolution of the conceptual environment, and the constraint on its viability.

(i) Physical environment: To those ivolved in such perfect meetings, the negative effects of the many subtle and less subtle design factors in present conference centres will be obvious. Conference environment design will focus on enabling the many aspects of conceptual pattern formation rather than "processing" participants and inhibiting synthesis. Flexible settings will adapt to the changing conceptual environment.

(j) Technology: Aside from the already evident move towards "electronic meetings" between distant participants, much greater use will be made of technology to enable spontaneous comunication between participants (rather than at them), to represent grahically the pattern emerging from the contributions made, and to facilitate synthesis whilst protecting variety.

(k) New challenges: Because the environment will enable collective reflection on much more subtle questions than at present, new challenges will emerge - possibly to be recognized as of greater (or more fundamental) significance than the often simplistic preoccupations of present meetings.

Challenge: conceptual challenge

1. Promise of the 1970s
The original excitement of the conceptual implications of computers in the early 1970s was inspired by statements such as the following:

"*Concepts can be viewed as manifolds in the multidimensional variate space spanned by the parameters describing the situation. If a correspondence is established that represents our incomplete knowledge by altitude functions, we can seek the terrae incognitae, plateaus, enclaves of knowledge, cusps, peaks, and saddles by a conceptual photogrammetry. Exploring the face of a new concept would be comparable to exploring the topography of the back of the moon. Commonly heard remarks such as 'Now I'm beginning to get the picture' are perhaps an indication that these processes already play an unsuspected role in conceptualization...*" (Dean Brown and Joan Lewis)

"*Unfortunately, my abstract model tends to fade out when I get a circuit that is a little bit too complex. I can't remember what is happening in one place long enough to see what is going to happen somewhere else. My model evaporates. If I could somehow represent that abstract model in the computer to see a circuit in animation, my abstraction wouldn't evaporate. I could take the vague notion that 'fades out at the edges' and solidify it. I could analyze bigger circuits. In all fields there are such abstractions.*" (Ivan Sutherland)

"*Concepts seem to be structurable, in that a new concept can be composed of an organization of established concepts... A given structure of concepts can be represented by any of an infinite number of different symbol structures, some of which would be much better than others for enabling the human perceptual and cognitive apparatus to search out and comprehend the conceptual matter of significance...Besides the forms of symbol structures that can be constructed and portrayed, we are very much concerned with the speed and flexibility with which one form can be transformed into another, and with which new material can be located and portrayed...With a computer manipulating our symbols and generating their portrayals to us on a display, we no longer need think of our looking at the symbol structure which is stored as we think of looking at the structures stored in notebooks, memos and books...In fact, this structuring has immensely greater potential for accurately mapping a complex concept structure than does a structure an individual would find it practical to construct or use on paper. The computer can transform back and forth between the two-dimensional portrayal on the screen, of some limited view of the total structure, and the aspect of the n-dimensional internal image that represents this 'view'.*" (Douglas Engelbart)

It will be possible to use computer devices as a sort of "*electronic vehicle with which one could drive around with extraordinary freedom through the information domain. Imagine driving a car through a landscape which instead of buildings, roads and trees, had groves of facts, structures of ideas, and so on, relevant to your professional interests. But this information landscape is a remarkably organized one; not only can you drive around a grove of certain arranged facts and look at it from many aspects, you have the capability of totally reorganizing that grove almost instantaneously.*" (Nilo Lundgren)

2. Disappointment of the 1980s
Re-reading these early texts in the light of the achievements of 20 years, it is possible to argue that we have access to these features if they are understood simplistically. Much manipulation is indeed possible. A user can indeed "drive around" an electronic network, dipping into and out of conferences and examining arrays of ideas.

But it is also possible to view current electronic networking as quite disappointing in the light of those early aspirations. Consider the following:

(a) Concept representation: In most networking environments concepts are represented by "message item numbers". The message may have a "title". To this may be attached a questionable selection of "keywords". There is no question of representation by an infinite variety of "symbols". In fact the graphic dimension is totally lacking.

(b) Concept organization: In most networking environments concepts are organized as messages by "conference" and/or clustered by "keyword". Within any message there may be a reference to an earlier message which can then be accessed, but it may be more difficult (if not impossible) to locate any subsequent message referring to it. In a highly organized conference, the theme may be structured into "agenda points". The addition and removal of agenda points is resisted in order to give stability to the conference. Just as the addition and removal of conferences may be resisted to isolate zones of stability -- although there may be other constraints.

(c) Concept access: In most networking environments permitting message keywording, the thesaurus structure through which to explore the pattern of keywords is poorly developed. As in many library systems, it is hierarchical and simplistic. Keywords are used too broadly or in such unusual ways that they are unreliable as a retrieval method. They do not provide a meaningful conceptual overview.

3. An important distinction
It should be quickly said that the existing software is obviously satisfactory to many users of electronic networking. Distinctions can usefully be made between:

(a) "Focused networking": In which the focus is provided by:

- the interaction process itself, namely the bonding process with particular participants with whom dialogue is experienced as fruitful;

- information exchange (including schedule matching) in which the content at any one time is what provides the focus.

(b) "Unfocused networking": In which the commitment of participants is to discover and articulate a shared domain of concern, amongst a network of participants, for which a network of concepts can be articulated and brought into focus.

It would seem to be the case that user satisfaction is due to the predominance of focused networking. Where users work within specific conferences or maintain contact with specific people, the question of conceptual organization is implicit in the text of the messages exchanged. But in the case of unfocused networking, the question is whether the networking environment responds to the conceptual challenge of the focusing process, namely interrelating complex networks of organizations and problems, articulating agendas, identifying conference participants. The view taken here is that the broader conceptual challenge is to provide an electronic networking system which facilitates the emergence and articulation of conceptual clusters as a prelude to focused networking, if required.

4. Challenge of innovation under complexity and turbulence
Focused networking may be adequate to well-defined stable issues. However, to respond proactively to a complex, turbulent environment requires some form of unfocused networking.

It is unfocused networking which facilitates the emergence of different perspectives and alternative agendas -- different foci. Such a context can be used during the creative, tentative period when agendas and coalitions are being formed. It is able to respond rapidly to crises, surprises and creative opportunities when new patterns of categories are required.

Focused networking is vital, but it needs to emerge from a context of unfocused networking, if it is not to become a trap from which those involved cannot detach themselves.

Challenge: conceptual weaknesses of conferencing

It is useful to consider the following conceptual weaknesses in both electronic networking and in conventional face-to-face conferences:

1. Conceptual amnesia
The tendency for a network of participants to forget, or repress, points made in earlier time frames. Participants become addicted to novelty and devalue concepts articulated earlier. The ability to build a complex conceptual structure over time is therefore constantly undermined. Such amnesia is in effect a process of conceptual resource destruction.

2. Conceptual fade out
The tendency for a complex conceptual structure to fade at the edges, so that the scope of any emerging conceptual structure is constantly being eroded by limitations of conceptual span, whether for an individual or for the network of participants collectively. Just as there is a need for screen refreshment, so there is a need for systematic concept refreshment.

3. Conceptual burial
The tendency for concepts to be buried in a mass of text, whether explanatory, anecdotal or otherwise. Little effort is made to distinguish concepts from contextual material which may not be essential to their subsequent use in articulating a complex conceptual network. Many environments are designed to bury concepts almost as quickly as they are generated. The conditions ensure a high concept mortality rate.

4. Conceptual haze
The tendency for a multiplicity of concepts to be simultaneously present in a diffuse haze through which participants wander (or blunder) with little more than a confused sense of orientation. Everything is relevant to everything, but little can be effectively distinguished. People enter and leave the conference environment confused.

5. Conceptual swamping
The well-documented phenomenon of information overload. The amount of information inhibits creativity.

6. Conceptual mouse-trapping
The tendency of a conference to premature conceptual closure. Given the conceptual confusion which tends to prevail, any ordering which emerges tends to be seized upon and used to impose order before alternative perspectives acquire sufficient weight to call for their integration in a more complex conceptual structure. This is associated with conceptual big-game hunting, namely the tendency to focus on the most obvious and dominant concepts and to ignore other aspects of the conceptual ecology represented within the conference -- and possibly vital to the healthy growth of the conceptual ecosystem.

7. Conceptual collapse
The reductionist tendency to blur subtle distinctions, collapsing them into a simpler concept. The complexity of a soap bubble would thus be reduced to that of a blob of water on a two-dimensional surface. Unusual, counter-intuitive or paradoxical structures are thus not adequately protected in the normal conference environment.

8. Conceptual stasis
The tendency to define concepts in static terms, when a dynamic definition might be more appropriate in response to an evolving, turbulent social environment. The concepts needed at this time may only be representable in dynamic terms (as with resonance hybrids in chemistry).

9. Conceptual consensus-mania (or disagreement-phobia)
The tendency of a conference to avoid disagreement and seek consensus, when more realistic conceptual articulations might be based on appropriate configurations of complementary, but opposing perspectives. Within a network this tends to result in the effective exclusion of those who disagree -- leading to a form of conceptual incest or inbreeding. The conference environment is not designed for conceptual variety, unless variants are screened off in their own sub-conferences.

10. Conceptual contraception
The tendency for conferences to be designed in a sanitized, "safe-sex" mode to avoid conception and the collective creation of viable new conceptual configurations. There is a strong emphasis on conceptual foreplay and titillation, with success being associated with a form of conceptual orgasm, hopefully to be repeated on subsequent occasions. But conceptual progeny are as unwelcome as the risk of being infected by dangerous ideas.

In response to such assessments, it is usually argued that these difficulties can be avoided if the conference is appropriately organized with a "strong chairperson" or "moderator" to "keep things in focus". This is in effect a betrayal of the original non-hierarchical inspiration of networking.

The further suggestion that the conference should have a "clear agenda" tends to imply that the agenda is decided in advance, thus inhibiting the creative, self-organizing process whereby responsible people redesign the framework through which they interact in response to new insights emerging from that interaction.

Most of the burning international issues call for conference environments in which the agenda is constantly redesigned as an evolving conceptual framework. A frozen agenda precludes creativity and implies a frozen, still-borne outcome. The formation of only the most probable coalitions is possible at a time when only the less probable are appropriate to the task. The emergence of more imaginative coalitions is not facilitated.

Is there no way that responsible individuals can get there act together without a "policeperson" or a conceptual straitjacket ? It can be argued that much more could be done with networking software to facilitate conceptual activity.

Opportunity: marrying modes of information

One way to explore future possibilities for conferencing is to consider the implications of various possible marriages between modes of information:

1. Text and Data

The classic separation between text processing and data processing has severely impeded the evolution of conferencing. A fruitful marriage would allow users their current freedom of expression but would also enable them to navigate more effectively through the maze of messages. Various approaches could be taken:

(a) an outline facility would structure lengthy communications so that users could explore them to different depths using an onion-skin approach. Of particular interest would be to code such levels to indicate their relevance to the core message (e.g. background or context, argument or justification, precedents, counter-arguments, action implications, explanatory or learning mode material, anecdotal illustrations, etc). Archiving could then be done selectively, gradually reducing to the core concept only.

(b) a hypertext facility would obviously empower users in new and interesting ways.

The issue in both cases is how to code levels of the text and embed hypertext links in the text as part of the message generation process. This is an extension of the classic problem of how to motivate authors to provide abstracts. The long-term solution is to shift the focus of attention from the text to the representation of the knowledge implied by the text. A transitional solution is to develop what might be called a "text compressor" or "concept processor" based on artificial intelligence procedures.

As has been repeatedly noted, the desk-top publishing revolution and its conferencing parallel will more than overwhelm a saturated readership. Desk-top readers do not accomplish what we would like their name to imply. They do not help us to filter and comprehend the content. Some form of text analysis and restructuring by a concept processor is required to mine the conceptual ore from what needs to be dumped or filed at a lower priority level. The most practical approach would to provide users with a minimum facility which they could adapt and tune to their personal idiosyncrasies. Users could of course view and edit the structured product generated from their own outgoing communications. Such a processor might usefully be related to the need for machine-assisted translation.

2. Data and Graphics

Much has been accomplished with respect to this marriage in the form of representing data in graphical form (business graphics, statistical graph plotting). But this quantitative challenge for conferencing is possibly of much less interest than the non-quantitative one of how to represent graphically the concept networks being articulated within a conference. Possibilities include:

(a) mind mapping, whereby concepts are indicated as interlinked nodes on a network. Some existing packages permit users to do this as an individual exercise. One variant of this is the use of arrow diagrams in certain areas of documentation.

(b) social network mapping, whereby the relationship between participants, in the light of their profiles or patterns of communication, can be viewed when this is considered desirable. One variant of this can be seen in the citation analysis graphs.

There are two challenges here:

(a) enabling a group of users to address the emerging articulation on a shared map (possibly with personal overlays, etc);

(b) escaping the conceptual straitjacket of packages based on a directed graph or tree structure in order to use an associative structure (on which alternative tree structures can be temporarily imposed).

It is worth noting that a heroic attempt was made to do just this by Stafford Beer and Gordon Pask at the first international conference of the Society for General Systems Research (London, 1979) before the PC era. Both concept maps and participant network maps were produced and used to orient discussion. Such experiments would be infinitely easier now and many refinements could be incorporated.

Comment

The absence of such tools is an indication of the priorities of conferencing at this time. Questions such as the following need to be asked:

(a) Why is it that participants in a conference have experienced no need to represent the conceptual structure which they are collectively attempting to articulate?

(b) Is it that participants are satisfied with the schematic representation in an agenda or programme? Or is it that they prefer a discursive mode in which the structure is implied or left ambiguous?

(c) Why is it that in the academic analysis of social networks almost no attention has been devoted to the graphical problems of representing complex networks -- despite the extensive manipulation of data on them.

(d) Why is it that in the current enthusiasm with hypertext, no effort is made to provide the user with a map of the hypertext pathways between the set of frames? It is almost as though a hypertext stack was designed like a rat maze, which the user has to explore like the rat, without any sense of perspective. Learning is the process whereby the rat builds up its own mental model. The map of the relations in a relational database is not considered as valuable information to orient new forms of inquiry or modification of the pattern of relations. It can be argued that it is that map which constitutes knowledge, in contrast to information.

There is every possibility that users have different preferred cognitive modes (possibly under different circumstances) and that it remains important to cater flexibly for those who feel constrained by explicit structures.

One possible reason for the relative lack of interest in conferencing systems in the international community is that in the present form they do not reflect the dynamics of factional interaction. The action is perceived as being elsewhere. Even the texts produced can be viewed as conceptual shells discarded by a dynamic beast that has moved elsewhere. The consensus-mania pervading explicit conferences forces the real, tension-filled, business of factional wheeling and dealing into other arenas -- if only the corridors and bars outside meeting rooms or in one-to-one messaging. This clearly suggests the need for handling the public-private interface more flexibly, veiling and unveiling explicit structure when appropriate. The conferencing of the future may yet prove to be a conceptual dance of the seven veils !

Opportunity: conceptual scaffolding and prosthetics

1. Scaffolding

The possibilities of the preceding note point to quite concrete features which could provide a major new facility for conferences, whether electronic or otherwise. These possibilities are basically concerned with the whole issue of what might be called "conceptual scaffolding". In the process of constructing a building scaffolding is necessary, especially to hold structures in position until appropriate permanent building elements can be inserted to lock them into place. Much can be learnt from architecture in considering the challenges of developing more powerful and appropriate forms of conceptual architecture.

Structurally an agenda or a conference programme, even a multi-track program, is rather simple -- even simplistic -- especially when considered in relation to the complex ecology of problems and organizations which are supposedly to be interrelated effectively through it. Is it any wonder that conferences are relatively ineffective at coming to grips with complex issues? What is being attempted is in defiance of Ashby's Law of Requisite Variety.

The issue is therefore how to enable users to collectively design more complex forms of conceptual scaffolding to hold in place embryonic or unstable concepts until other concepts can be fitted into the pattern to lock them into place. Ideally, of course, it is the conferencing software which should provide such scaffolding. And, like the scaffolding for buildings, it should be adjustable to different structural configurations as the building grows.

A typical function of scaffolding in a conference is to provide a framework within which complementary perspectives can be articulated, especially when there is a major tension between them. When Concept A is formulated, the scaffolding holds a space for Concept B to counter-balance it. Such scaffolding is even more essential when more than two concepts have to be held in balance. As with buildings, the scaffolding provides a protection against disruptive forces in the conference process. A typical disruptive force in a contemporary conference might focus narrowly on "industry is exploitative", when the larger issue is to provide a sustainable framework in which to balance the exploitative characteristics of industry against the socio-economic benefits that it provides in the light of environmental constraints. The more complex the balance, the more vulnerable is the conference to disruptive forces.

Four forms of scaffolding are especially interesting:

(a) Symmetrical structures: Geometry supplies a vast repertoire of geometrical patterns which can be used to interrelate concepts. Of special interest are the symmetrical polygons in 2-dimensions and polyhedra in 3-dimensions. Symmetry has the merit of being in some way associated with global or integrative comprehensibility. To the extent that opposing perspectives can be mapped onto such structures, there is greater possibility of collective recognition of the distinct functions they perform in relation to one another. It is also possible that the more complex the structure, the greater its stability. Eastern religions have made extensive use of such conceptual patterns in the form of mandalas. These hold the complex relationship between a multiplicity of complementary insights, whilst maintaining an integrative focus on the whole. The software issue here is how to massage an associative network of concepts into the pattern (or a range of alternative patterns) which can give the most appropriate overall order to it. Maybe there is a place for marrying networking concepts to those of sacred geometry.

(b) Tensegrity structures: A feature missing from such geometrical structures is any explicit recognition of the dynamics between the elements and of how they contribute to the dynamic integrity of the whole. Again architecture points to the importance of appropriately interrelating tensional and comprehension elements. In conferences the art is to creatively interrelate perspectives that are in sympathy and in opposition to each other. Buckminster Fuller pointed to the existence of a whole family of tensegrity structures which underlie the structure of his well-known geodesic domes. Tensegrity (or tensional integrity) has many suggestive implications for more effective conferencing:

-- such structures make explicit the value of having discontinuous (antagonistic) relations between concepts (or their advocates) embedded in a continuous (mutually supportive) network of relationships. Both have a role to play.

-- such structures make clear how an appropriate combination of appropriately positioned elements can give rise to a totally unsuspected structure of unsuspected stability. Whilst it is relatively easy to comprehend the logic of such a structure in 3-dimensions, the process of constructing it is much less clear. This suggests that the conceptual elements and dynamics characteristic of today's conferences could lend themselves to structural patterning of a totally new kind.

-- such structures make clear that facilitating communication between all parties (all to all) is not the only way forward, even if it were feasible in practice. They suggest that much may be accomplished by ensuring a supportive relationship with neighbouring nodes, provided that position is "challenged" by an appropriate opposing node. This is a step beyond all the work done on social networks. It implies that software could be used to configure communication pathways (opening some, closing others) to bring about much more healthy (non-flabby) networking.

-- of special interest is that such structures have empty centres so that every point is visible from every other. The centre is a virtual one rather than being occupied by some dominant individual or concept.

-- as will be seen below, such structures also imply a range of global transformations through which the conference can grow to encompass greater variety.

It is clear that only with the use of appropriate software could tensegrity-based conferences be explored. The scaffolding problem is an ideal computer challenge. It opens the door to a totally new way of representing agendas non-hierarchically.

(c) Resonance hybrids: There is a certain class of chemical molecules whose structure cannot be meaningfully defined by a single pattern of atoms. Thus the benzene ring, present in most organic compounds, is best understood as oscillating between 5 distinct patterns of bonds between its constituent atoms. The resulting resonance hybrid is much more stable than any of the 5 individual patterns -- even though that stability is dynamic. This suggests the possibility that there may be conceptual and organizational structures which can only come into existence by allowing them to alternate between essentially unstable (or unsustainable) extremes. The challenge of an appropriate response to the issues of sustainable development may depend on the ability to discover such structures. Computer conferencing may be absolutely essential in providing the conceptual scaffolding through which they can emerge. It is even possible that the legal and accounting structures to maintain institutions based on them could only be managed through some such environment. (Just as the newest aircraft can only be flown with computer assistance, it is possible that the most advanced organizations need to be conceived in the same light.)

(d) Embedding data in images: It has long been recognized that some of the most complex problems of process control, call for a totally new way of presenting hard data to the human brain. Instead of a multiplicity of dials and graphs, use is made of the full range of visual images (landscapes, animals, imaginary objects) as vehicles onto which to project or hang complex patterns of data so that they can be more readily comprehended. Thus when the wind agitates a tree on a landscape image, a particular control action is called for. Very large amounts of data can be compressed into such images. Recalling Douglas Engelbart's vision, this suggests the need to explore how conference participants can embed their insights into comprehensible images. In particular it suggests the possibility that the collective task of a conference might also be perceived in terms of sculpting such an image -- with every conceptual contribution leading to a modification or articulation of it. This calls for a very special marriage between conceptual contributions and image processing. Of special interest is the possibility that the insights of some conferences could only be effectively carried by dynamic imagery, and especially by imagery governed by other rules than those of the physical world (as is the case with some computer generated imagery). It is clear that computer image manipulation skills are well developed, but much needs to be done to determine how to hang data on them such that changes to the data modify the image, and changes to the image modify the data.»

2. Conceptual transformation

The need for conceptual scaffolding is clear given the kinds of complexity with which society has to work. The challenge of making the more complex structures comprehensible is also clear -- those most appropriate to the challenge of sustainable development may be beyond the ability of any single human mind to grasp. But any form of development implies structural transformation. Whilst transforming simplistic structures like conference agendas and organization charts may pose little challenge, the transformation of the complex structures described earlier are quite another matter.

The process of conceptual or social transformation appears to call for a form of dynamic scaffolding which provides some form of continuity -- from stage to stage -- through the transformation process. What we are looking for is a form of scaffolding onto which the conference's insights can be mapped at Stage I. The relationships in this mapping would then be stretched or changed in the transformation to Stage II, which might be some very different kind of structure -- suggesting new kinds of relationships between the concepts so bound (and between their proponents in the conference).

There are few examples of this kind of structure:

(a) **Image transformation:** The skills of image-transformation on computer suggest many possibilities. The challenge is to find ways of relating real-world issues and challenges to such images so as to benefit from this facility. Of special interest is the way in which development is to be understood or encoded in such image transformation. If the many details of the global problematique could be encoded onto one (or more) archetypal animals, suitably animated, this would be of major conceptual and symbolic significance -- especially when the animation can be used to represent a transformation process. The media advantages are obvious.

(b) **Vector equilibrium:** Buckminster Fuller drew attention to a very unusual symmetrical polyhedron, the vector equilibrium (normally known as the cuboctahedron) as the common denominator of the tetrahedron, octahedron and cube. It is unusual in that it lies on a transformational pathway to a variety of other structures. An appropriately jointed model can be transformed into an icosahedron and from there to an octahedron and on to a tetrahedron. The merit of this model, aside from the many claims made by Fuller himself, is that it provides a way of understanding the structural transformation process. The challenge in a conferencing environment is not to focus on this particular structure, but rather to use it as an example to persuade topologists to locate other transformational systems of this kind so as to build up a library of possibilities on which to draw.

Presumably it will only be through such explorations that conferences can anchor their transformative insights so that people can recognize and have confidence in the structural continuity of appropriate change, rather than being threatened by change of any kind -- and therefore resistant to it.»

There is a dearth of imaginative ideas to respond to the challenge of sustainable development in this period of crisis and crisis-management thinking. An immediate challenge for the West is how to respond to the radical transformation in the Eastern European countries. Given the scarcity of resources, what can they be given to catalyze the fruitful reorganization of their societies? And even more challenging, how can advantage be taken of the very high level of education achieved by a high proportion of the younger generation? Rather than thinking in terms of how such societies can make use of various Western styles of organization, which have resulted in many significant failures in other societies to which they have been exported, is there an alternative? Is it possible to provide some communication package, to run on stand-alones or small networks, which could provide them with the conceptual scaffolding that would enhance their ability to apply their own imaginative insight to their own problems?

The challenge of the Eastern bloc is in effect a metaphor of the challenge that the world as a whole faces with respect to sustainable development. The economists will continue to be given every opportunity to apply their unimaginative insights to the task, whatever suffering their austerity measures imply. This will not change. And the degree of alienation of the population, and especially the young, will continue to increase. But in the many creative interstices, there is a receptive audience for devices which open up opportunities for more complex, and more fruitful, modes of thinking and organizing. With appropriate imagination, limited resources can be applied in new ways. Computers can provide an environment to assist that process.»

Opportunity: policy forums as metaphors

1. Constraints of meeting design
The organization of a meeting and its processes in fact provide a remarkable metaphor of wider society and the challenge of using resources more appropriately. To use Gregory Bateson's insight, "we are our own metaphor" (Bateson, 1972). The challenge of formulating more appropriate policies is highlighted by the difficulties in meeting design:

(a) Building constraints: The constraints of the building and the regulations governing how it is to be used, including simple questions like the ability to reconfigure the seating arrangement in the light of the emerging processes of the meeting. These are a reminder of the constraints imposed by existing physical structures and regulations in wider society.

(b) Protocol constraints: The protocol constraints necessitating special focus around certain individuals. These are a reminder of the constraints imposed by existing social structures, whether relevant to social change or not.

(c) Meeting procedures: Meeting procedures based on rules of order (Robert, 1985) which have not changed to any significant degree throughout the 20th century and which fail to take into account the best thinking on self-organizing systems. These are a reminder that "plus ça change...", as those with revolutionary inclinations delight to point out.

(d) Meeting agendas: Meeting agendas designed months (or years) before the event, thus to a large degree pre-programming the process and outcome and blocking any unplanned initiatives in the light of emerging opportunities. This is a reminder of the dead weight of prior commitments under which policy-makers operate. The structure of such agendas also tends to reinforce linear thinking and fails to reflect the non-linear relationships between the items -- a reminder of the clumsiness with which we endeavour to respond to the cyclic complexities of the social and natural environment.

(e) Hidden symbolism: Much more controversially, except for those acknowledging the implications of Freudian symbols and sexual politics, is the body-language of speakers, especially in relationship to the microphone and any proscenium, and that of the audience seated in expectation of stimulation. This is a reminder that unconscious factors may play a determining role in meeting processes.

(f) Misuse of attention time: Use and abuse of one of the principal resources in meetings, namely time, especially in the form of the attention time of a captive audience. This is a reminder of how policy-makers tend to exploit their position in relation to the resources of captive constituencies and markets, whether this takes the form of "cartel formation", "asset stripping", "environmental degradation" or "resource depletion".

(g) Limitation on forms of presentation: Obligation of the audience to accept the form of presentation favoured by the speaker, with little recognition of the need to translate the content into other modes (except in the extreme case of language interpretation, but not including that between disciplinary languages). This is a reminder of the widespread assumption that people are all naturally capable of processing a complete spectrum of information forms, unless they are of reduced mental competence.

Again it is not the purpose of this paper to explore such intractable issues further. They must be circumvented by other means if there is to be any hope for timely breakthroughs.

2. Facilitating integrative breakthroughs
In any policy forum, integrative breakthroughs are facilitated by:

(a) Imagery within the conference: recognizing the implicit or explicit metaphors favoured by the factions represented, namely what imagery they use to communicate within their group and with their constituencies;

(b) Imagery for external communication: recognizing the implicit or explicit metaphors of the policy forum as a whole, namely what imagery is acceptable and how that may relate to that of any subsequent public relations campaign;

(c) Design of new imagery and metaphors: encouraging the deliberate selection and design of more powerful metaphors to encode the dynamics of the relations between incompatible perspectives and especially between the factions represented. For if one faction perceives the other as "sharks", and are perceived by the latter as "sheep", no amount of rational discussion will overcome the "ecosystemic" constraints on their harmonious relationship. The same may be said of "hawks" and "doves"; both know who "eats" whom.

Metaphor is widely used to communicate policy options. However it is used simplistically and in a rhetorical manner divorced from the actual written articulation of policy. The metaphors currently favoured do not reflect the exigencies of sustainable development or the dynamics between the advocates of competing policy alternatives. Resources can be usefully devoted to identifying, selecting, designing disseminating and employing more appropriate metaphors in policy contexts. Such a shift in focus should open up new ways of reflecting collectively on the more complex, cyclic and incommensurable perspectives currently lost in the savage interactions between factions. It is such complex perspectives which constitute the real policy challenge.

This suggests that a desirable policy forum design would focus attention on the emergence and movement of policy-relevant metaphors, their relationship (as comprehensible meaning complexes) to more conventional forms of information, and their reflection in organizational form. Such stewardship in the governance of a forum opens up new possibilities in the governance of society as a whole.

Opportunity: metaphors of transformation in conferences

1. Imagination barrier
Earlier notes have stressed the special need for software capable of facilitating more complex forms of conceptual communication in a conferencing environment. This argument is based on the assumption that just as aircraft were faced with the technological challenge of the sound barrier, software faces the challenge of the imagination barrier. The sub-sonic conferencing problems have been largely solved. But we do not yet know how to ensuring the stability and integrity of conferences functioning at a high imaginative level. The conventional organizational and conceptual structures tend to get shaken apart.

2. Restricted access to significant innovation
There would seem to be a number of fruitful steps that can be taken, as indicated in earlier notes. When it is recognized what strategic advantage they offer to the networks that use them, it is probable that resources will be devoted to their development. It is very probable that such software will be restricted to those major corporations for whom strategic advantage is a vital consideration. It is also probable that versions of such software will be developed by certain alternative groups. It seems less likely that the core of the electronic networking and conferencing constituencies will have access to such facilities or perceive the need to do so. Unfortunately this is also likely to be the case with users in the international community of organizations.

3. Short-termism
The difficulty is that it is always possible to argue that concrete, short-term, simple procedures are sufficient in a crisis-management environment. Much of what passes for international projects and programmes is in effect reactive, crisis management. Upbeat reporting of their successes is always possible. But in strategic terms it is rather like a chess novice playing a grand master. The novice can be allowed to delude himself by many short-term gains as he progressively sinks into a more and more disadvantages strategic situation from which recovery is hopeless. This is the dilemma of sustainable development.

4. Countering deficiency of imagination
The real challenge for conferencing in relation to the crises of our times is to provide people with tools to counter the imaginal deficiency from which we collectively suffer when dealing with complexity. The texty, linear-environment of massaging and documents has a poor track record. Eminent experts, with suitable budgetary encouragement, can now be found to negate the importance of any problem, whether over-population, acid rain, low-level radiation exposure, or smoking. Their "facts" are no longer a reliable basis for action.

5. Insight through metaphor
In this context, there is a most intriguing unexplored resource. That is the use of metaphor as a guide to the elaboration of more complex conceptual frameworks and organizational structures. In effect the arguments already made with respect to tensegrity, resonance hybrids and imagery rely to some extent on the power of metaphor, especially visual metaphor. Metaphor is renowned as a key to creative thinking and innovation. Information systems have traditionally been ruthless in eliminating the ambiguity of metaphor from the communications they support. But the classical triangle of text, data and graphics processing is only 2-dimensional. Imaginative insight can be usefully placed at the apex of the (tetrahedral) pyramid based on that triangle. Metaphor is the prime vehicle for such insight.

6. Carrying complex insights by metaphor
How then can we marry metaphor processing into conferencing environments as a way of breaking through the imaginative barrier? There are clearly possibilities for doing so with computer assistance. But it seems doubtful that advances will be made fast enough on these fronts. However one great advantage of metaphor is that, like rumour and humour, it travels rapidly through any network, whether computer-assisted or not.

Consider the fashionable focus for the international community at this time, namely sustainable development. How is this complex notion to be carried and addressed in the imagination, and especially in the media. Metaphor can be used to highlight our collective difficulty in developing strategies to bring it about. Metaphors such as "global village" or "gaia" do not give focus to the strategic dilemma and the operational opportunities. Due to our imaginal deficiency, sustainable development is best understood at this time through the metaphor "having our cake and eating it too". This corresponds to the corporate interpretation of "sustainable competitive advantage". Both are tragic examples of poverty of imagination in a complex environment.

7. Shifting the conceptual centre of gravity of conferences
Imagine a conferencing environment in which text (including speech), data and graphics were treated as infrastructure "plumbing" and the conceptual centre of gravity shifted to an imaginative level sustained and disciplined by the computer-assisted use of metaphor. A major concern in the conference would be to ensure the circulation of meaning through metaphor. Complex notions would be expressed briefly by metaphor. The challenge would not be who could dominate the discussion in quantitative air-time terms or resolutions passed. Rather it would be a question of who could produce the most seductive metaphor to capture the strategic complexities and the opportunities for the formation of hitherto impossible coalitions of bodies.

Computer could do much to assist the management of such creative environments. Essentially they have three tasks. Firstly to render a repertoire of metaphors appropriately accessible, in the light of their structural and patterning characteristics. Secondly, to provide a disciplined communication framework to channel forces hindering the emergence of imaginative new patterns, and providing a protective framework (a "matrix") for such patterns in their embryonic stages. Thirdly, to give stability to the stages between the imaginative level and the organizational and operational implications (a sort of "Jacob's ladder" or "gearing down" facility), which need to be articulated at the "plumbing" level.

Envisioning conferencing: aesthetics of governance

1. Context

One of the difficulties of these interesting times is the vast outpouring of information, insightful and otherwise. Even the most creative people with many helping hands have large piles of documents and periodicals in their offices labelled "To Read" -- where many remain unread. In an era of "desktop-publishing" the "desktop-reader" does not accomplish for us what its name implies. It is a mark of eminence for a person to be able to claim lack of time to read all the relevant documents in his or her field. This has serious implications for those with policy-making responsibilities and for the insightfulness of the innovations to which they subscribe. Our society seems to be decreasingly capable of channelling its best insights to the places where decisions are taken and interrelating them in such a way as to empower those capable of acting in terms of new paradigms -- although upbeat reporting might lead us to believe otherwise.

Information specialists delight in describing what computers will be able to do for us to resolve such difficulties with new gadgets and fancy software. But they focus on fact shuffling -- at a time when many "facts" have become questionable. The question of how creative, integrative insights emerge, are comprehended and rendered appealing to a wider audience is not addressed. How do we collectively sense and grasp a fragile new gestalt that is an emerging paradigm in embryonic form?

2. Imaging exercise: conferencing in the Year 2490

What follows, in this note and the following, is an exercise in imagining how the creative imagination might be used some time in the future, possibly 500 years in the future -- unfettered and unconstrained by the obvious difficulties arising from our present priorities and understanding. The focus is on the contribution of the arts to more appropriate forms of policy-making and to the design of more appropriate forms of social and conceptual structure.

One stimulus for this exercise is the poverty of imagination associated with fictional and dramatic scenarios of how executive councils function in the distant future -- as reflected in science fiction films and books. Even when entities gather from "the 100 galaxies", thousands of years hence, their encounter (even through "holographic projections") still seems to be modelled on the United Nations Security Council or its unfortunate imitations. This organizational archetype is no challenge to our imagination, especially when other styles might be more appropriate. The degree of innovation in such policy councils since classical Greek or Roman times is laughable compared to that in any technology. High tech Pentagon-style "war rooms" and corporate "situation rooms" do not empower participants to interweave value-laden views that differ and cross-pollinate in realms beyond the quantifiable. It is sad indeed to see this same archetype impoverishing the gatherings of spiritual leaders of different faiths.

A second stimulus is the failure of artists to nourish our imaginations with better insights into the technicalities of governing our world -- and specifically the failure of the poet Robert Graves in endeavouring to describe a country ruled by poets (in "*Seven Days in New Crete*") and of the author Doris Lessing in her, otherwise remarkable, "*Canopus*" series.

3. Clarification

The concern in this exercise is not with new forms of information technology, nor with new kinds of business graphics "for the decision-maker", nor with the communication of words. The focus is also not on the design of conference rooms and the associated communication technology, nor is it on group dynamics and the manner in which such meetings might then be facilitated. The group processes and interpersonal dynamics of that time, and their relationship to the personal growth of participants are not the concern.

The focus is on meaningful insight, its communication and its comprehension -- and especially when participants hold quite incompatible views. The concern is with the embodiment of new patterns of meaning -- whatever media are used to carry those meanings.

Although the preoccupation is with how more appropriate forms of policy will emerge at that time, the theme is the contribution of the arts to that process. What might be the interface between the arts and the most creative aspects of policy-making as a "high art" in its own right? And let it quickly be emphasized that the issue to be explored here is not whether symphony concerts should be held "on the occasion" of any such assembly, or whether the walls should be monopolized by the mural of some distinguished artist.

In pursuing this theme it is assumed that by that time it may be possible to distinguish more effectively and creatively between insights and their expression and between personal concerns and those of the collective. Put bluntly the challenge seems to lie in making policies more seductive and enthralling to the individual, on the one hand, and in finding ways to permit the arts to be more an evolving expression of collective insight rather than a series of isolated works associated with the personalities (and idiosyncrasies) of their individual creators. Our tragedy at this time is that the longer-term policies to move us beyond the crises of our times, and the processes by which they are formulated, are inherently boring to the vast majority of the population. But the creative expressions to which we are all attracted, whatever the form (music, poster art, TV drama, etc), do not offer us a means of articulating the frameworks for collective action -- however well they may express our aspirations. Live Aid can raise consciousness, enthusiasm and money, but as a process it cannot articulate and ensure its appropriate use.

What seems to be called for is a form of marriage between Beauty and the Beast, in which both need to compromise in ways quite foreign to their natures. The Beast needs to be more sensitive to the harmonies through which its force could be more appropriately expressed and Beauty needs to be less narcissistic in order to respond to the earthly priorities of the collective and the way work can be done collectively.

4. Movement of meaning

In that far and distant time a gathering of the wise may best be imagined as blending the characteristics of policy councils as we now know them with those of an art workshop, a poetry reading, a classical music concert, a theatre, a folk song-fest and a dance, together with other dimensions we would have difficulty recognizing -- and might find awe-inspiring, if not personally quite threatening. It is difficult to imagine how these seemingly distinct forms of activity blend in this way, but that is because we have difficulty in understanding how the same meaning can be taken up, articulated and developed through different forms. We see this most clearly today in music, where different instruments develop the same musical theme, responding to each other's contribution. In that future setting policy-related themes are developed across artistic forms, much as happens in the relation between song and music, or in relation to a dramatic setting as in opera.

But it is less the form and appearance of the occasion which is the concern, for to explore those would keep us trapped in their meanings for us at this time. And it is obvious that the arts will have evolved in ways we can clearly not suspect. At this point it suffices to note the presence of a spectrum of arts. Of much greater relevance is the manner in which they open up and develop themes essential to the policy process. For lack of a better word, "meaning" will be used to refer to the emotion-mind-intuitive "stuff" with which the gathering is working and on which its attention is focused. What are they trying to do with it and what opposing and complementary forces are brought to bear upon it?

Those attending the gathering each bring to it their own contributions. These may be quite distinct, whether compatible with others or not. The participants are there because the meanings they bring are those which others wish them to articulate. The process of the gathering allows these meanings to play off against each other. Through what conceptual or other frameworks do participants (and external observers) comprehend these movements of meanings? This is what we can endeavour to explore.

Before engaging in that exploration, it may be useful to clarify the relation between meaning and policy. Put briefly, policy is that which the collective concludes that it is most meaningful to undertake. Not all meaning is directed towards action. Some may articulate the context for action (or inaction). Some levels of policy may indeed be more concerned with maintaining a context within which other policies may be pursued. It may be argued that the highest and most appropriate form of governance would be that which ensured the generation and circulation of meaning within a society, whilst intervening minimally.

5. Artistic vehicles for meaning

Participants at the gathering therefore make use of different artistic vehicles at different times to introduce new meanings and to sustain the movement of meaning as a whole. We need not be too concerned about how they do this is practice. Clearly extensive use might be made of electronic devices (or their successors) to run video or audio sequences extracted from a library of the world's cultural heritage. Perhaps participants might call upon artistic "staff" support to endeavour to articulate a theme for which no cultural referents were known. The artist might use some visual or other sequences from the library, manipulating them in the light of his or her own insight (and in response to feedback from the participant) -- much as is done by computer graphics enthusiasts, by computer-enhanced music synthesizers, and by experts in special film effects. Individual participants might choose to use a poetic form, music or song, depending on their skills, in order to supplement any statements in prose form. Clearly the future will hold many possibilities of this kind -- but that is not the point.

What we would have difficulty grasping in following this process would be the connection between one "intervention" and the next if the sequence moved through different artistic forms. We can begin to understand when we think of our response to the normal musical accompaniment to the drama evolving in a film. But in that time instead of simply reinforcing the meaning, the music may also carry the meaning to a new level of insight. Any words which then followed might be considered by us as a non sequitur -- we would have missed the link carried meaningfully by the music. We are more used to this process when a lecturer introduces slides and other graphics to make points which cannot be effectively made in verbal form. Such graphics seldom, if ever, appear in the proceedings of policy bodies. But how could we grasp what was being articulated when the gathering shifted from a univocal to a polyphonic mode, where the "voices" might take visual as well as audio form? Again we can begin to understand when we think of how "voices" interplay in a choir or in symphonic music. There is a logic to the relationships -- a harmonic logic -- to which we respond both instinctively and intuitively. There is a "rightness" to the harmonic integration so achieved. In our meetings today, people speaking simultaneously are seen as disorderly and various procedures are used to inhibit or prevent it. To the extent that such interventions represent the "voices" of distinct factions, we are deprived of the richness of any polyphonic integration -- one "voice" is expected to drown out all the others (as in majority voting), or all are expected to "speak with the same voice" (as in consensus procedures). Where there are many speakers who can only speak in succession, it now takes much experience to be able to follow the emerging pattern and to integrate the threads of the discussion at a higher level of real significance to participants. And effective integration of any current debate tends in our era to be more tokenistic -- its meaning lies mainly in its value for public relations, whatever the policy implications.

What could we understand from the arts today that would help us to understand how they in the future could work collectively in this way?

6. Artistic discipline

One key to understanding how such gatherings work is their preoccupation with a well recognized concern of anyone in the arts, namely "finding the appropriate medium". The emphasis is on the insights to be expressed. The challenge is to find one or more vehicles through which to express any such insight. The dilemma is that many of the most complex and valued insights often cannot be adequately expressed through a single medium or even in a single moment of time. The insight can then only be carried by an interplay of forms over a period of time. The concern therefore shifts to the "design", "orchestration" or "choreography" of that interplay.

But of what relevance are these concerns to the articulation of policy? A major handicap for policy-makers of our day is that their insights must invariably take their final expression as words in prose form. Much has been written about the turgidity of that prose, especially in its extreme legal form. The prose is usually structured into a nested hierarchy of "points" -- which emerge from policy meetings governed by agendas of similar form. It is difficult to imagine a less creative way of expressing insights, however carefully the document is "crafted". Its great merit lies in the fact that each point in principle corresponds to a course of action for which some person, group or institution may be made responsible. Unfortunately this leads to the creation of institutions which mirror the structural poverty of the policy document. And, although there is much benefit in the stability of static structures like agendas, policy-documents and organization charts, they are almost totally inappropriate to the ambiguous, fluid, cyclic or evolving conditions, so characteristic of a real world full of "surprises" -- and such forms have proven to be incredibly difficult to change in response to the surprises.

The crises of our times, and of those to come, forced future generations to embody the temporal dimension into their design of conceptual, policy and institutional structures. The dangers of embalming such structures as monuments to the insights of a particular moment -- and then allowing subsequent actions to be governed by the self-serving priesthoods which accumulated around them -- became only too obvious. As will be seen, incorporating the temporal dimension involves more than producing a "Five Year Plan" which is totally insensitive to insights emerging either after its adoption or as feedback from the phases of its implementation.

But the only way that they could take this major design step, comprehend the complexity of the outcome, and (above all) engage the interest, participation and understanding of the population as a whole, was through the use of artistic disciplines. Indeed integrating what had seemed so totally irrelevant to the policy-makers of our times was seen as an essential healing process for the collective ("two cultures") schizophrenia which had engendered the contradictions at the root of so many of our problems. This healing demanded as much radical rethinking of policy-making as it did of the social role of artistic expression.

Let us now look at some of the disciplines and insights from the arts and see how they were woven by our descendants into the high art of policy-making. We must of course remember that from our perspective the reality of that integration would appear quite magically incomprehensible to us -- all we can do is note certain threads and principles which were significant to that magic.

Part of our difficulty in comprehending their achievement is that this healing involved more than a simplistic putting together of policy and artistic skills. The integration was based on a paradoxical (and uncomfortable) level of insight (with associated skills) in order to transcend the easy duality by which we now find it convenient to separate them (and many other things). The beginnings of this insight are only now becoming familiar to us in the discussions of the relationship between physics and consciousness and with related insights from the East.

Envisioning conferencing: insights from music

This note continues the exercise of envisioning the cognitive contributions of the arts to conferencing in the Year 2490.

1. Harmony and the language of music
One of the mysteries to that future era was our reluctance, in our constantly declared search for social "harmony", to draw upon the articulation of harmony in music. Our excuse, in the midst of factional squabbles over concrete urgent problems, is that no serious person could imagine that music had anything to offer other than some pleasant distraction before or after the reception on the occasion of some such gathering. And yet music could be called the science of harmony. An immense amount of effort has been devoted over the past centuries to exploring the nature of harmony in music.

Music scholars and philosophers of our era have long disputed whether or not music actually "means" anything. They recognized that all composers of tonal music have used the same "language" of melodic phrases, harmonies and rhythms to evoke the same patterns of significance (whether intellectual or emotional). They developed and used a pattern language of musical idioms to carry dynamically complex insights. These were especially powerful in interweaving shorter and longer developmental cycles.

2. Comprehending complexity through harmony
Where we had vainly sought for the keys to controlling our environment through systems science and cybernetics, they married such explorations to the science of harmony as articulated in music. In our era much has been written about the relationship of music and time -- music as time made audible. We have seen the efforts of systems scientists and "world modellers" to represent complex systems dynamics using equations, flow charts and sophisticated graphics -- denying comprehension by most of us. Our descendants projected such dynamics into musical relationships which could be played. The "business graphics" of that time had musical variants. People could hear the various harmonies which provided integration to any policy represented, and they could hear the dissonances which challenged that harmony -- whether as a stimulus to social growth or as a potential crisis. The only equivalent we have to this is the ability of any motor mechanic to listen to an engine as a means of diagnosing its state of health. One great advantage is that everyone could listen to such musical representations, irrespective of the sophistication with which they understood it. The major integrating features were obvious to all, however little they understood the detailed harmonic organization.

3. Enriching patterns of insight through harmony
Such representations of systems insights were not just public relations devices. By listening to the musical representation it became possible to identify and discuss features which could be changed and improved, in the light of musical insights, into richer or more challenging patterns of harmony. The musical perspective highlighted features which made a policy boring -- namely "monotonous" to their ears -- and thus uninspiring to those in whose interest it was being elaborated. We can get some understanding of this process from the way jazz and pop groups collectively develop a piece of music until it sounds right.

4. Policy implications
Space limitations here preclude detailed explorations of the policy significance that they were able to attach to all the many attributes of musical organization. But, for example, where today international development agencies have a range of programmatic approaches on which they rely, in that era such approaches would be recognizable by what are effectively melodic signatures. Such signatures became a way of communicating complex programmatic proposals. And whilst there were many "old favourites", there was greater sensitivity to those which had been superseded, and to the emergence of new melodies which addressed issues in a more interesting way. This clarified the relationship between the fashionable programmatic melodies of the moment and those of more enduring quality.

Of special interest is their use of insights from the temporal organization of music as it impacted on the programme and budgetary cycles which are the skeletal structure of any concrete action programme. A major concern in administering an organization is to ensure financial discipline. They resolved this problem by using a musical discipline of far greater flexibility and more subtle articulation. The cyclic aspect of organizational life acquired whole new dimensions, for in music there can be many cycles of different length and involving different instruments. They also made intriguing use of rhythm and tempo -- partly as a way of dealing with urgency and the need for an appropriately timed response.

5. Encoding competing perspectives
But perhaps of most interest to us are the insights they gained from musical notation and the harmonic relationship between different chords and instrumental qualities. They took the typically politicized factional spectrum around any issue in our time (which undermines any appropriate response) and effectively coded the spectral elements into musical notation. Interventions in any discussion were thus comprehended within a musical framework, whether as isolated notes or chords, but above all in terms of their relationship to the emerging theme. The art of debate thus became one of contributing to the emergence of better music -- recognizing the role and limitations of the particular contribution one could make. The characteristic intervention of our time -- the frequent repetition of a single note, louder than those preceding it -- was an obvious musical disaster (although see below). In this context, "note taking" acquired a whole new meaning in recording the proceedings of the gathering.

6. Proactive response to dissonance
We would however be completely misunderstanding their achievement if it were taken to be a simplistic exploration of harmonies. Their society, like ours, was constantly challenged by deep divisions of perspective. But, whereas we resolve these in the organizational equivalent of a gladiatorial arena, they reinterpreted such dissonance in musical terms. To our ears the music they played would at different times have such qualities as: gothic immensity; the challenging intensity and immediacy of hard rock; the supportive, solidarity of folk tunes; the intellectual intricacies of computer generated music; as well as many others. They had a tool to work effectively with differences and to use those differences to enhance the dimensions of their policies.

Envisioning conferencing: insights from poetry and painting

This note continues the exercise of envisioning the cognitive contributions of the arts to conferencing in the Year 2490.

1. Poetry

(a) Beyond turgid prose: The written and spoken word during the policy-making process has lost those attributes of language which focus our aspirations and inspire us to collective action. There is a divorce between the "rousing speech" of a leader and the articulation of the policy by men in grey suits -- a divorce concealed by the use of public relations consultants to sooth our concern. We are governed through turgid prose embodied in the bureaucratic procedures we love to hate.

In that far time their deliberations and conclusions depended heavily on insights into poetic form. They rediscovered the merits of poetry as "the other way of using language". It should not be forgotten that poetry was first used in rituals to regulate the life of a community and ensure a good harvest, only later to become an important means of giving form to the life of the spirit. It should quickly be emphasized that it was the discipline of the poetic form through which they worked, not some convoluted effort to translate policy conclusions into a poetic "press communiqué" for public consumption.

(b) Mapping complexity onto subtly known patterns: Is there even the faintest recognition in our times of the need to make use of poetic disciplines in response to the challenges we face? Surprisingly there is. The recognition comes from those who recognize the limitations of scientific disciplines in dealing with the complexity of the problematique -- and specifically with the limitations of the human mind, or any particular language, in comprehending and encompassing the subtle dimensions amongst which a dynamic balance needs to be maintained. For example, the biologist Gregory Bateson, in explaining why "we are our own metaphor", pointed out to a conference on the effects of conscious purpose on human adaptation that:

"One reason why poetry is important for finding out about the world is because in poetry a set of relationships get mapped onto a level of diversity in us that we don't ordinarily have access to. We bring it out in poetry. We can give to each other in poetry the access to a set of relationships in the other person and in the world that we are not usually conscious of in ourselves. So we need poetry as knowledge about the world and about ourselves, because of this mapping from complexity to complexity." (1972, p. 288-9)

(c) Poetically moving agendas: Again space precludes no more than brief references to how the poetic form was used in the future. For example, where we make extensive use of point filled agendas, charters and declarations to express and order our policies, they recognized the inappropriateness of these forms to the subtleties of what they were called upon to govern. To our times it would appear as though they had simply expressed their agendas as poems, of which declarations and charters were more complex elaborations.

This perception would be to misunderstand their achievement. For the poetic form allowed them to interrelate insights and challenges which we only link mechanically or as "budget items". Their use of such forms immediately engaged the attention -- people were not only "moved" by them, but the articulation focused understanding of how "being moved" could be translated into implementation and what complex environmental relationships they needed to be sensitive to during that process.

(d) Poetry of the collective: Such approaches appear totally impractical to us, locked as we are into our schizophrenically dissociated roles. For us a poem is the work of an individual (often marginalized) making few concessions to the collective -- it is a voice crying in the wilderness. For them their highest achievements were poems designed by groups (of inspired individuals) representing the aspirations of the collective -- faced with its own shadow. We can only laugh at such possibilities because we perceive in it various echoes of totalitarian art (just as we would question the collective function of martial music). Group creativity is the rare exception in the arts -- and then only in pop music, experimental theatre, and group murals, none of which are held to be of great long-term value. But from their perspective our charters and declarations could only be understood as aesthetic abominations whose form, distorted the spirit of collective action and ensured the reinforcement of precisely those problems which we deplore. For them, ironically, such forms were conceptual totalitarianism par excellence.

2. Painting

(a) Challenge of colour: It would be a mistake to believe that policy-makers in our day are blind to any distinctions of colour. On the contrary their task is bedeviled by problems of colour, especially at the national level. For policy-making is highly politicized and the factions are usually strongly associated with a particular colour: from the socialist reds, to the conservative blues, with of course the ecologist greens. Some distinguish fascist blacks and technocratic yellows. And there are even efforts to distinguish shadings: pink to dark red, light to dark green, etc. Such use of colours goes back to early military needs to be able to identify soldiers on a battlefield and rally them around a distinctive banner to be loyally defended "to the last".

(b) Aesthetics of colour coded policies: In that time to come, the objective of policy had shifted from explicit attempts to ensure that any particular colour triumphed to the suppression and exclusion of all others. The significant contributions and dramatic weaknesses of each such approach had become only too apparent.
Treating each policy as a sort of action vector, they were able to represent the range of possible action vectors through a complex classification of the complete range of several thousand colours distinguishable to the human eye. One policy-making tool they then used was the art of combining colours from this "palette" into a meaningful painting, whether in two or three dimensions (or more, by cycling through pattern sequences). Some representations also took the form of "light sculptures". Others bore more resemblance to tapestries.

(c) Policy balance and appeal: Discussion about policy thus shifted from the implicit objective of maximizing blue or green, to the challenge of how to combine many such colours on a complex surface. In this light our efforts at global policy-making were primitive in the extreme, without any sense of form, diversity or balance. It makes clear how little respect technocratic policy-makers now have for the complex issues of balance and appeal to which aesthetics devotes so much attention.
They were able to use colours to encode the policy dimensions which needed to be held in balance in a complex social ecology. A credible policy was therefore designed and represented by some form of painting with a strong aesthetic appeal -- with the colours and shapes indicative of details necessary to the pattern of the whole. Indeed, their technology permitted such paintings (on computer-enhanced screens) to be used as "control panels" through which the health of a society could be assessed by all. The elements of the painting became indicators (so by-passing the statistical difficulties of the innumerate). Such pictures were truly worth a million words.

(d) Aesthetics of sustainable development: We can speculate on how they would represent to us on some such painting the appropriate policy mix to respond to the challenges of "sustainable development" in the 1990s. Obviously there would be some green, but how much of each shade. How would it be related to the conservative blue? And what of the shades of red? And how would the colours be disposed and interwoven? What would justify the exclusion of any particular range of colours?

To respond to concerns at both global and local level, the painting would have to be very large indeed -- and beyond our current imaginings. But this would allow many policy variants at the detailed level, where different constituencies experimented with different policy combinations. The merit would lie in the ability to discuss the aesthetics of the painting as a whole -- would the detail form a meaningful global pattern?

Envisioning governance: insights from drama and dance

This note continues the exercise of envisioning the cognitive contributions of the arts to conferencing in the Year 2490.

1. Drama

(a) Beyond scripted conferences: There is a true story of the visit of the President of one country to another not so long ago. Each had his principal public speech carefully crafted by a speech writer to appropriately stress the policy issues in question from his position. They delivered their speeches to each other before a large audience and all were content. Unknowingly they had used the services of the same speech writer.

There are many tales of conference conclusions having been prepared, before the gathering, in "draft" form for approval on the occasion. Together with the above tale, this suggests that all policy gatherings are to some extent scripted, possibly during the course of preliminary meetings. A "dry run" is common for critical business meetings. Academic meetings may be almost totally scripted, given that papers may have to be submitted months in advance to determine whether they can be accepted in the programme and "read" at the meeting. Whatever the degree of pre-scripting, some time is usually given for "free discussion" or "questions from the floor" -- this may also be scripted by the use of appropriate "plants".

(b) Policy dramatics: In that future era they approached these matters as dramatic opportunities. A policy gathering was also designed and assessed by the criteria of the dramatic arts. As such this view of a gathering is not too strange to us. We talk of the "main players" and are sensitive to "dramatic moments". The media are especially sensitive to such aspects, to the point of placing pressure on organizers to structure the event so that there are such moments. Events are "staged" because of the media opportunities they offer. The key speakers prepare themselves accordingly -- even to the point of being appropriately dressed, if not made-up and bewigged.

(c) Casting factional representation: But our efforts in this direction make rather uninteresting drama, except for the participants. In the future the challenge was to ensure that the different policy factions were represented by a cast of characters capable of giving adequate dramatic emphasis to the complex issues that needed to be aired, interrelated and resolved. Even today the organizers of conferences are sensitive to the question of "casting" -- who can most appropriately represent a particular perspective. But our descendants made this into a high art. Thus if some hard decision had to be made, the tragic dimensions were appropriately drawn out so that all were aware of what opportunities had to be sacrificed and the suffering that would cause. If there were ridiculous inconsistencies under discussion, their potential as comedy was fully explored (as it is today, outside the gathering, by cartoonists and political comedians).

(d) Improvization and psychodrama: But the special merit of their dramatic approach was that they had skilled techniques for blending scripting and improvisation. In contrast with our programmed gatherings, the outcome was not necessarily predetermined. The inherent logic of the drama as it unfolded through the unscripted interventions of the participants could move the drama to some unsuspected conclusion. In our time we understand this best in psychodrama and indeed their gatherings were to a high degree sophisticated psychodramas in which participants took the role of factions or constituencies rather than personalities.

(e) Orchestrating dramatic moments to "make a point": The dramatic dimension to their gatherings provided ways of giving form to otherwise "bloodless" debates in which policy implications could take only a purely abstract form. Faced with a complex of challenges and opportunities (which could only be represented on an essentially incomprehensible complex mathematical "surface"), the drama articulated the tensions between values such as joy and despair associated with different policy dimensions, however the gathering finally resolved them. "Points" were made through dramatic moments (as at the origin of the phrase, "to make a point", in the century theatre). In our day policy-makers do not weep at the suffering caused by the decisions they may be forced to make. In that era, such dramatizations provided every justification for weeping when appropriate. The emotional implications of policies were thus fully explored during the policy-making process.

2. Dance

(a) Embodiment of aspirations in dance: In the closing years of the 20th century much is made of the of the increasing proportion of young people in the world, despite the reverse situation in western countries. Much is also made of rising levels of functional illiteracy -- even in western countries. And it is clear that the young in general have relatively little interest in the organization of society that is being thrust on them by their elders. Although they are deeply concerned by some of the issues, it is fair to say that a high proportion of young people have their core aspirations articulated through music and its embodiment in dance. In a world of many languages, it is one of the few that is shared worldwide.

One of the shocking features of our era, to those of the future, was that those involved in policy-making had lost the art of dancing. Formal dances, where they are held, have atrophied into a formal shuffle of little significance. Disco dancing on the occasion of any gathering is provided solely as a means of relaxing and cultivating relationships. Complex dances from our cultural heritage are executed as entertainment with little insight into their significance.

(b) Exploration of patterns of dualities: Our descendants developed the use of dance during such gatherings into a way of exploring the pattern of dualities by which our policy debates are variously polarized beyond any logical reconciliation. Such dualities and factional differences could be encoded in music as described earlier. But dance offered the possibility of acting out those tensions so that they acquired a felt reality -- and the sequence of the dance allowed particular polarities to be transcended in the pattern of the dance.

Such dances bore some resemblance to ceremonial dances, and masked dances, of earlier times and cultures. They also borrowed heavily from insights into self-organizing systems. Many formal patterns existed, but the dance itself, through choices made by participants, might stabilize temporarily in one, before switching into another, or through a cycle of patterns.

(c) Interfacing the personal and the collective: Dances of this kind allowed participants to explore the boundary between their personal preferences, those of others, and the organization of the whole. They offered insights into patterns of organization in which sacrifices made to others under certain conditions, could be compensated by benefits under other conditions. They illustrated the art of "winning" and "losing". People were able to feel out where they could take initiative, leading some part of the dance, and where they could more appropriately respond to the initiatives of others.

(d) Giving form to larger patterns: Of most importance, such dances gave felt reality to complex patterns which could be used to interweave polarizing tendencies in social organization. They provided a means of understanding the "temporal logic" of combining opposing (factional) policies as phases in policy cycles -- themselves interwoven in more complex patterns (reminiscent of Buckminster Fuller's geodesic patterns). In effect, the whirl of the dance kept opposing elements within the larger pattern. This creative use of time was their key to the use of more appropriate and sustainable styles of organization. We can hear a faint echo of this insight in the peasant farmer's traditional use of crop rotation to sustain the productivity of his fields and in our current approaches to traffic circulation.

Envisioning conferencing: insights from architecture

This note continues the exercise of envisioning the cognitive contributions of the arts to conferencing in the Year 2490.

1. Beyond "top-down" conference communication

Vast sums are invested these days in the design and construction of prestigious conference centres, as one of the principal environments in which policy is articulated and approved. Although much is made of the advanced "communications technology" installed there, no attention is paid to the fact that none of it is designed to facilitate unmediated communication between participants. Such centres are fundamentally totalitarian in concept. All is controlled and articulated from the top and feedback from the floor is severely controlled or impossible.

In many cases this extends to the pattern of seating -- unmovably bolted to the floor for maximum exposure to messages from the podium. Not only does this mirror our social organization, it is also reflects the way in which meaning is communicated in such policy environments. It reinforces the patterns by which we tend to organize knowledge and insight -- and how we subsequently impose them on others.

2. Conference architecture

In the time to come, the principles of architecture were basic to the organization of policy insights and their implementation. The point here is not the way in which such principles were used in the physical design of conference environments, rather it is the way they informed the conceptual organization -- however that might reflect on the physical layout and communications technology.

One of their insights was that in conceptual terms gatherings had to be "constructed". A meaningful policy conference was one which provided appropriate conceptual spaces for different purposes -- and ensured communication between them. In part the task of the conference was to build anew, on each occasion, such a pattern of spaces. To some degree this already happens in our time through the design of the programme.

They made "conference architecture" into an art form at the conceptual level. But the conference had to be designed and built by the participants -- the viability of the resulting "building" was a measure of their success in policy design.

3. Designing conceptual spaces for factions

This is not the place to discuss their approach to the "foundations" or to many other features of the conceptual construct. Most striking perhaps was their use of space. Each faction found it reasonably easy to design a space for itself and its own "wares" -- somewhat as do major exhibitors in designing their stands at an exhibition associated with a conference.

The first real challenge was to be able to design with others a conceptual context in which participants with similar priorities and values could successfully explore their relationships. In this phase, corresponding to the meeting of sub-plenary groups, the design views of participants were constrained and inspired by their immediate peers.

Then followed the challenge of relating that space to those of other groups with other priorities, so that participants could move from space to space. At this design stage, each group had to take into account requirements of other groups -- compromises had to be made.

4. Construction of collective conceptual spaces

The most challenging phase was the construction of the collective conceptual space in which all viewpoints were interrelated, providing integrity to the whole, namely the equivalent of a plenary conference room. A central architectural insight lay in the means of constructing an arch -- or a series of arches which could be roofed over to protect the space.

In effect, even for the smaller spaces, participants were often obliged to retrace the history of architectural principles and techniques. The challenge was to use opposing conceptual elements as columns and to use various ways of bridging between them to create the desired space -- whatever scaffolding was temporarily acquired to install keystones or their equivalent. For the smaller spaces this tended to call upon principles from the very early history of architecture.

To create a space for all views -- the conference in plenary form -- required a much more sophisticated understanding because of the wide expanse that had to be covered with minimum intervening supports.

5. Transcending duality

Their achievement was to use opposition between policy perspectives as "compression elements" and to use mutually supportive perspectives as "tension elements". Their skill, inspired by physical buildings, lay in finding ways of using the dynamic interplay between two types of element to create structures which would be impossible with either of them alone.

They effectively used the elements of a duality so that the 2-dimensional stresses between them -- which normally render any conceptual construction impossible -- could only be resolved by engendering a space in 3-dimensions. In some cases this resulted in "gothic" structures -- "cathedrals of the mind" -- in others it resulted in what we might understand through Buckminster Fuller's tensegrity structures (basic to his geodesic domes).

Envisioning conferencing: current possibilities of implementation

This note continues the exercise of envisioning the cognitive contributions of the arts to conferencing in the Year 2490.

1. Problem-solving illusions
Speculation about approaches to problems in the distant future is most useful when it sharpens our understanding of new possibilities in the present. Our difficulty today is that few problems are insoluble, rather most of the solutions are themselves perceived as problems. Success is claimed, through upbeat reporting, at the elimination of a problem in one domain, only by carefully avoiding recognition of its displacement into some other form or jurisdiction. Effective action on problems continually eludes us -- its always associated with some other opportunity we have been unable to take.

2. Habitual constraints
One difficulty seems to be that we are trapped by habitual conceptual and procedural approaches to problems -- and their reflection in institutions and programmes. It suits most of us to point a finger at seemingly isolated problems like the ozone layer because our degree of accountability for them is limited. And when it comes to allocating resources to solve problems, no matter how severe, the process is most characterized by cynical tradeoffs for the short-term advantage of constituencies already privileged -- whatever media packaging is offered to make such solutions appear desirable.

3. Rechannelling "negative" forces
What we are looking for is a way of working with large complexes of problems, perceptions and organizational networks that would provide a more fruitful context for the healthy features of political horsetrading. But to be of any value it also needs to rechannel and refocus what currently manifests in institutional operations as mutual accusation, suspicion, deception, manipulation, alienation, corruption, subversion and sabotage -- dynamics seldom discussed by enthusiastic problem solvers surprised at the ways in which their efforts get undermined in the real world. Whilst much may be accomplished in the long-term by exploring processes through which people can "come to know each other", "reach consensus on values", "love one another", and "identify with humanity as a whole" or with Gaia, it is useful to question whether these "positive" initiatives do not effectively serve as a rather beautiful avoidance mechanism -- at least in their present form.

4. Harmonies and discords of conference design
An alternative approach could make extensive use of aesthetic insights into the discipline of harmony and into the role of dissonance in enriching that harmony, especially as articulated in music. Such an approach would recognize the place of easy harmonies, their limitations, and the role of more complex harmonies brought out by effective response to more challenging discords. But note that the "discords" are not the nasty problems, but rather other groups opposing the "harmonious" way favoured by my group in solving a problem -- our policy "theme song" to whose irritating limitations we are totally insensitive. Until we can work within contexts allowing each participating group to be recognized as part of the problem, we cannot collectively determine the nature of the solution that would be appropriate or sustainable.

5. Conceptual scaffolding through aesthetics
In any gathering the aim would be to use aesthetic devices (music, colour, drama, etc) to register the different perspectives represented (and their associated dynamics), to provide a conceptual scaffolding to hold their relationships as they developed during the event, and to suggest directions through which richer harmonies could be explored. In contrast with the present preoccupation with a majority or consensus vote, the outcome would be expressed by a pattern or tapestry of views. Superficial or token unity would be replaced by a more complex, and more dynamic, set of relationships, reflecting the reality of the deepfelt differences between those represented within it -- as well as being both comprehensible and challenging to those in the outside world investing hope in the outcome of such gatherings.

6. Incorporation of requisite variety
The acid test would be the manner in which such dynamic patterns were reflected in the design of programmes, budgets, institutions and information systems. The key feature here would be the way in which policies ensured that opposing perspectives were brought into play at appropriate times to correct for programmatic weaknesses resulting from the excesses of any one insight or set of priorities. It is through a more disciplined use of time that it becomes possible to overcome the apparent impracticality of ensuring that a configuration of non-consensual insights guides policies of requisite variety. In this light budgetary cycles at present can only be perceived as crude and clumsy, completely failing to take advantage of the flexibility and responsiveness that current computer software techniques could permit (perhaps best seen in the rapid reallocation of resources through worldwide exchange and money market operations, despite their weaknesses). However it is the aesthetic insights that are needed to give form to appropriate patterns of complementarity.

7. Immediate implementation?
(a) Availability of resources: And is any of this really possible in the immediate future ? The tragedy is that we are already using the software techniques and technology needed -- but not in response to the dilemma of our time. Similarly many of the aesthetic, scientific and policy disciplines, whose insights would be beneficial, are locked into expediently self-serving activities rendering them insensitive to external constraints. Those with a mandate to fund exploration of social innovations avoid criticism by accepting advice resulting in more of the same.

(b) Entrapment: So yes it is possible, but it is not probable. We are stuck in a vicious circle such that gatherings of the wise, for the purpose of improving such gatherings, are rendered ineffective by the processes which they aspire to rechannel -- disguising their collective impotence under expressions of appreciation at their achievements, however minimal. We are very much our own metaphor.

(c) Positive indications: For those locked into bureaucratic procedures, academic or artistic traditions, or into the prevailing conventions of policy-making, that future will appear fantastic indeed. But at a time when actors and playwrights become presidents, when policy is articulated through carefully staged photo opportunities, when major policies are communicated and discussed through their metaphoric wrappings, and when policy successes at the global level seem few and far between, then more open-ended approaches merit exploration.

(d) Possible first steps: It is a nice challenge to ask ourselves why the possibilities mentioned above could not be explored now rather than in the year 2490 -- if only for smaller groups and communities. The first step would require a clear distinction between such initiatives and those characterized by enthusiastic attempts to add on to a conference yet another performance of "The Ode to Joy" or to "celebrate" once again (while the world is literally burning). What would it take to determine what might be feasible ? To represent Beauty, it would be necessary to have those with artistic skills of course -- but it would be vital that they not be locked into the need for a platform for themselves and their own work, rather than for collective concerns. To represent the Beast, much could be accomplished with accountants, lawyers and those from the organizational development world, in addition to those with policy skills -- but it would be vital that they not be locked into a narrow conception of their role. When they gather together it would be vital to recognize that the personal needs of facilitators, with their favourite "processes", are also part of the problem.

We need to disillusion ourselves that the task just involves bringing appropriately skilled people together -- as in so many delightful gatherings and task forces of little consequence. It calls for long-term commitment by many -- perhaps equivalent to the Apollo programme -- in order to escape from the conceptual gravity well in which we are stuck.

Transformative policy cycles: patterning cycles

1. Inadequacy of particular policies
In the light of the previous sections, and especially Sections KD and MZ, it is useful to ask whether the characteristics of the appropriate new mode of socio-economic organization are such that it can only be sustained by a cycle of policies, or even a pattern of such cycles. If this were the case then whilst particular policies, such as those of the "left" or the "right" or of other political hues, are necessary during particular phases of such cycles, they are not however sufficient individually to sustain the mode most appropriate to long-term human development.

2. Concealed dilemma of democratic voting
This question can be related to the dramatic problem, central to social organization, of whether a system of voting can be devised that is at the same time rational, decisive and egalitarian. In the classic analysis of this problem, Kenneth J Arrow (*Social Choice and Individual Values*, 1983) advanced five intuitively appealing axioms (including unanimity and universal scope) that any procedure for combining or aggregating the preferences of individuals into collective judgements should satisfy.

Treating "non-dictatorship" as a sixth axiom, Arrow demonstrated that no constitution can exist which will obey all six simultaneously. What happens is that when three or more alternatives are faced, majority rule gives rise to voting cycles in which: Alternative A defeats Alternative B, B defeats C, C defeats D, D defeats E and E defeats A, as noted in a recent discussion of Arrow's "impossibility theorem" by D Blair and R Pollak (*Rational Collective Choice*, 1983). For them: *"Thus the designer of voting procedures for legislatures, committees and clubs who accepts these conditions as necessary properties of constitutions is simply out of luck... If society foregoes collective rationality, thereby accepting the necessary arbitrariness and manipulation of irrational procedures, majority rule is likely to be the choice because it attains the remaining goals. If society insists on retaining a degree of collective rationality, it can achieve equality by adopting the rule of consensus, but only at the price of extreme indecisiveness. Society can increase decisiveness by concentrating veto power in progressively fewer hands; the most decisive rule, dictatorship, is also the least egalitarian."*

It is worth noting here that Gheorghe Paun (*An Impossibility Theorem for Indicators Aggregation*, 1983) has explored an aspect of this dilemma using fuzzy set theory to demonstrate the impossibility of aggregating a small set of good social indicators to fulfil three natural conditions of a good indicator, namely sensitivity, anticastrophism and noncompensation. This establishes theoretically the noncomparability of certain social issues, which must somehow be "managed" in an appropriate new mode of socio-economic organization.

Blair and Pollak explore the difficulty of designing **acyclic constitutions for organizations** which would avoid such voting cycles. The Eastern insights from the *I Ching* (see Section TP and following notes) suggest that it might be more valuable to look for ways of designing **cyclic constitutions** to permit an organization to alternate through such a network of alternatives, each of which exerts **a dominant influence for a period of the cycle**, before in turn being overthrown or undermined by a succeeding alternative in that cycle.

3. Embodying cycles into social design
Blair and Pollak explore the possibility of designing acyclic constitutions which would avoid such voting cycles. The arguments of this paper indicate the value of exploring ways of designing "constitutions" which embody such, seemingly unavoidable, cyclic phenomena, especially since they are evident in the necessary policy changes required to remedy the inadequacies of particular policies. The question is how to initiate such a design process, given the nature of the design required.

In such a context, the process whereby any such particular policy comes into favour, and is subsequently displaced, is an integral part of such a policy cycle. The emphasis on such a cycle is in marked contrast to the prevailing emphasis on the dominance of a particular policy and the desirability of its continuing dominance for the long-term well-being of the society in question. However, by its very nature (as discussed above), no such policy cycle can be planned or programmed, for this would make of it merely another policy competing with other policies in the cycle. It is here that the core of the challenge lies. It is the paradoxical problem of organizing self-organization.

Section M suggests the merit of metaphors in catalyzing the emergence of an awareness of the necessity of policy cycles. It points to the lack of understanding of the nature of policy cycles and patterns of such cycles, especially as they might function in different cultures, resulting in the entrainment, and synchronization, of such cycles between cultures. This is surprising given the considerable research on economic cycles, which presumably call for some understanding of a corresponding cycle of policies to respond appropriately to the changing circumstances. This lack is probably due to to the fact that current policies are of such short-term scope that longer-term cyclicity appears irrelevant. Things may be changing however. *The Wall Street Journal* (4 Sept 1986) recently reported on work being undertaken at the prestigious Japan Economic Research Center by Yuji Shimanaka demonstrating the relationship of economic cycles, technological innovation and periods of social conflict to 11-year and 55-year solar cycles; the latter corresponding to Nikolai Kondratieff's long-term economic cycles.

As an illustration, consider four contrasting policies currently competing savagely with each other for a larger "market share" of public opinion support. The arguments of this paper suggest that this savage competition contributes to the emergence of an appropriate new mode only to the extent that it ensures successive dominance phases amongst the four policies according to a periodicity or rhythmn to which there is, as yet, little collective sensitivity.

Each policy acquires dominance in a cycle at some stage because of the need to correct for deficiencies resulting from the (necessary) imperfections and excesses of the preceding policy, only to be displaced in its turn. The appropriateness to human development results, ultimately, not from any particular policy but from the extent to which the pattern of policies and the rhythmn of their phasing becomes increasingly self-organizing.

4. Cyclic operations and the nature of transformation
Understanding how such cycles of contrasting phases accomplish effective transformative work in society may be facilitated by a thermodynamic metaphor. The Carnot cycle of heat and work, basic to the operation of any heat engine, itself involves four successive and contrasting operations (expansion at constant temperature, expansion without change in amount of heat, compression at constant temperature, and compression without change in amount of heat). Any attempt to isolate and prolong unduly the most effective work phase simply jeopardizes the ability of the engine to continue operating. It is then quite inappropriate to view the non-work phases as "inefficient". The operation of a task force (or meeting) of individuals with distinct functions may also be interpreted as involving a cycle of phases in which each function enters and leaves the limelight in turn. This is best illustrated by the results of research by R Meredith Belbin (*Management Teams*, 1981) into the roles required for good teamwork. These have been labelled as: chairman, company worker, completer-finisher, monitor-evaluator, plant, resource investigator, shaper and team worker. A preponderance of any one role type, especially the "most productive", jeopardizes both the appropriateness of the group's work and its ability to renew itself and continue functioning.

5. Distinct levels of attention
The different levels of attention required in discussing the relationship of distinct policies to policy cycles may be illustrated by the metaphors of walking and dancing. In walking the right and left foot are moved forward alternately, shifting the weight of the body from one to the other. Although in places of difficulty attention may be focussed on one foot to the exclusion of the other, the body can be more satisfactorily moved forward by focussing on the process of walking, namely on the alternation between the two contrasting positions. In a 2-party political process however, there is a necessary struggle between the "right" and the "left", with no institutionalized

awareness of what is achieved by the process of alternation between them. There is little recognition of when it is appropriate to relinquish a policy in favour of an alternative and then renew it to fulfil a new role. This may perhaps be more accurately compared to the preoccupation of a drunkard, or a spastic, with the forward movement of one leg (temporarily forgetting the need for the other).

5. Levels of appropriateness
Appropriateness of the 1st order may be compared to movement of a foot, whereas 2nd order appropriateness may be compared to the process of walking. Higher orders of appropriateness may be compared to dancing and to a cycle of dances. It is the movement between the steps, and the manner in which they are ordered, which renders the dance meaningful. Focusing attention exclusively on any individual step prevents the rhythmn from emerging and thus obscures the meaning of the dance. It is the rhythmn which guides the self-organization of a dance, based on the execution of the individual steps, whose importance can in no way be neglected. The test of the appropriateness of any new mode is whether it embodies a more "seductive" pattern in Attali's sense (24). In terms of 2nd order appropriateness current policy initiatives may be compared to a drunkard's walk, a monotonous dance or, more dangerously, a lock-step march.

6. Impotence of appropriateness: the dilemma of nth order modes
Cyclic patterns of policies clarify the essential dilemma of any appropriate mode. In any concrete socio-economic context, it is only possible to mobilize people in support of a basically short-term policy in response to the deficiencies of any policies currently dominant in the short-term. And this is indeed what is required to remedy those deficiencies. Such a "new" policy can easily acquire an inherent moral rectitude, implying that any other policy is a dangerous aberration. The difficulty is that such moral rectitude continues to be associated with the policy long after it ceases to be appropriate to that particular cycle of policies.

Policies contributing to a policy cycle may be considered to constitute a lst order degree of appropriateness. The policy cycle itself may be considered a 2nd order degree of appropriateness. Higher orders of appropriateness, cycles of policy cycles, may in fact be what is required for viable long-term human development. As the arguments of earlier sections have indicated, such higher order forms of appropriateness are increasingly difficult to comprehend. They cannot therefore inspire a sense of moral rectitude and consequently would appear to be necessarily associated with political impotence. Political power is concerned with struggle within the cycle, not with the movement of the cycle - and yet it is from cyclic movement that enduring social development emerges.

7. Constraints under democracy
Such difficulties are further aggravated by the constraints of democratic systems in which education and minimal levels of literacy are a continuing problem. This obtains both in industrialized countries, but especially in those developing countries characterized by population explosions. Grass-roots political wisdom, as well as the experience of sophisticated organizers of political campaigns, requires that issues be kept simple and comprehensible. Paradoxically in such circumstances "ignorance is right", at least in political terms. "Information is power" only in the sense of the power of elites to manipulate. But any such manipulation is itself constrained by the political necessity of communicating it in terms comprehensible to the largest constituencies. "I do not understand, therefore I will not vote for you (because I question your motives)", is the ultimate constraint in a democracy.

In complex social systems, such ignorance may also be the result of cultural preferences and background, even amongst the educated, whereby particular policies are viewed as inherently "bad" or "evil". Much political "mileage" may be guaranteed by the process of reinforcing such views and cultivating suitable political scapegoats. But this too is an inherent feature of policy cycles. Each policy acquires dominance to the extent that it can successfully cast other policies in a "negative" light, such that its own "positive" features are enhanced by contrast. The characteristic of any particular mode of organization are such that others must necssarily appear in a "negative" light from that perspective. It is this positive/negative polarization which drives the cycle through a succession of inadequate perspectives which compensate for each others distortions. It also prevents any "purely objective" discussion of which policy is appropriate at which time.

In traditional societies this dilemma was partially resolved by considerable use of metaphors, parables, myths and legends to render comprehensible the need for 2nd and higher order policy neglected in industrialized societies in favour of economic models which are essentially incomprehensible to all but the very few. And it is valid to question whether those who claim to understand their significance are adequately informed about the dimensions of society they choose to exclude from such models.

Transformative policy cycles: modelling sustainable development

1. Introduction
This exercise is concerned with identifying and representing patterns of change and with the development of better ways of responding to its possibilities in various forms of socially organized activity, such that developmental momentum is conserved within the pattern rather than being dissipated unusefully.

The focus is on moving beyond the inadequacies of single policies in an effort to provide an adequate policy framework for sustainable development. The challenge, as explored in Section MZ, is to avoid single-policy weaknesses such as:

(a) single policies create the impression of being viable and successfull by filtering out conflict and opposition. They are thus ill-equipped to interrelate a diversity of perspectives, many of which may involve fundamental disagreements (sometimes manageable by hierarchies in an "objectionable" manner);

(b) single policies may be used as temporary vehicles for enthusiastic response to problems, only to be abandoned as soon as unpleasant realities have to be faced;

(c) single policies are often geared to "positive thinking", negating the possibility of criticism and especially self- criticism, thus hindering collective learning.

The question is then whether there are any clues to ways of "tensing" policies to correct such tendencies). What can be done to prevent the energy from draining out of policies? One approach has been discussed under the heading of "tensegrity organization" as a hybrid "marriage" between networks and hierarchies (A J N Judge. *Implementing Principles by Balancing Configurations of Functions,* 1979).

A related approach is to assume that policies fail to contain problems because they are effectively out-manoevered by the dynamics of such problems. As in the martial arts, a policy must swiftly re-order conceptual and organizational resources to keep up with shape-shifting and hydra-like transformations of the problematique. The policy may need to alternate between several modes of action and conception in order to repond effectively (Judge, *Policy Alternation for Development,* 1984). If this is the case how can we come to recognize the pattern of transformation pathways to which a cycle of policies needs to respond?

2. Chinese insights
It is debatable whether Western-style organization has reached the limits of its ability to improve its "effectiveness". Even if this is not the case, it is possible that new insights can be derived from non-Western approaches, as is indicated by the current Western concern with the art of Japanese management. These would have the merit of breaking out of the currently criticized constraints of "eurocentric" modes of thought.

For example, the above challenge can be usefully clarified by an exercice in adapting the insights of the *Book of Changes,* otherwise known as the *I Ching*. This has been a major influence on Chinese thinking for 3,000 years, providing a common source for both Confucian and Taoist philosophy. As noted by R G H Siu (1968): *"For centuries, the* I Ching *has served as a principal guide in China on how to govern a country, organize an enterprise, deal with people, conduct oneself under difficult conditions, and contemplate the future. It has been studied carefully by philosophers like Confucius and men of the world like Mao Tse-tung."* For this reason the popularity of its (ab)use as an oracle should not be confused with the philosophy and insight embodied in its structure. With the benediction of C G Jung, it has achieved wide popularity in the West over the past decades.

Part of the merit of the book, as its title indicated, is that it purports to indicate complete patterns of changes, one of which has 384 **pathways** between 64 **conditions** that are recognizable both in an individual and in society. These insights have hitherto been interpreted in terms of the needs of the individual (of whatever degree of influence in society). Although basically they are addressed to the condition of any social entity, they have not been applied to organizations as such. Thus even though R G H Siu, cited above as one of the commentators on the *I Ching*, has managerial interest in addition to his research role as a biochemist at the Massachusetts Institute of Technology (MIT), his commentary is addressed to the individual. It is interesting to note that not only did MIT publish his commentary, it also published a study by Siu on the nature of "Ch'i" (1974). This is the psychic energy that an individual can accumulate according to neo-taoist philosophy. It may also be useful to conceive of it as the kind of "energy" which leaks out of policy frameworks when they fail to respond appropriately to the dynamics of change and development.

3. Interpretative exercise
The structure of the *I Ching* is based on 64 **conditions** (dynamic situations, perspectives, challenges, phases, or modes of action or conception) with which an entity may be faced. The underlying scheme is based on sets of 2 or 8 more fundamental conditions. The series could be expanded geometrically to 128, 256, 512 or more conditions. But as Siu (1968) notes: *"The originators of the* I Ching *judiciously stopped at the practical limit of sixty-four. This number constitutes a classification sufficiently fine so as to provide useful types of situations, against which specific cases can be matched. Yet the subdivisions are not so numerous as to be too cumbersome for a single scheme."* For each of the 64 conditions there are six possible **sub-conditions** (behavioural responses) on which statements are also provided.

The text of the *Book of Changes* is often written in a notoriously subtle and poetic style. This in no way precludes an interpretation of its significance for organizations or, more specifically, for the policies of organizations. Such an interpretation has therefore been undertaken as an exercise in Section TP. By making the interpretation specific to a sustainable policy cycle, there is clearly a loss of generality, but this is compensated by a reduction in ambiguity. Subsequent evaluation will show whether this constitutes an unfortunate degree of distortion of the original insights.

The interpretation given is as faithful to the spirit of the texts of the Richard Wilhelm translation (1950) as seemed feasible. Some of the condition names have been adapted from those suggested by Siu and others. Hopefully this exercise will encourage others to produce a more helpful interpretation. No extraneous insights have been introduced. In elaborating each statement the basic constraint was that it should be **briefly** formulated with respect to a "sustainable policy cycle" and that any terms used should be credible in a policy context. A similar exrcise was performed in the 1986 edition of this Encyclopedia where the focus was placed on "network".

The formulation of the statements here can be criticized because the orientation is not always consistent. In some cases they are formulated as injunctions as to what the policy "should" do. In other cases they are formulated in terms of explanations as to the probable consequences of the policy acting in a certain manner. Or else they are expressed in terms of what the policy cycle "could" or "might" do. The original texts place the burden of choosing between such interpretations on the reader.

It is important to recognize that the original text permits a complex of interpretations, encouraged by the nature of the Chinese language. For each condition the central meaning is underdefined, although clearly delimited by a complex of connotations based on terms that "alternate" subtlety in meaning between emphasis on: abstract or concrete; operator or operand; noun or verb; action or actor; problem or opportunity. Any word can often be beneficially replaced by a synonym or an alternative grammatical form. Quite distinct conditions may acquire apparent similarity as a result of the specificity of the words finally chosen - a choice that amounts to a "frozen" distortion of the connotation dynamics by which the underlying meaning is embodied.

The (undeterministic) significance in fact emerges through alternation of attention between the possible (deterministic) interpretations - in sympathy with the theme of this commentary. Note also the concept of chemical molecules which can only exist as a "resonance hybrid", namely a dynamic combination of several alternative structures, when none of them individually is stable.

An exercise of this kind is therefore rather like attempting to "tune" a "semantic piano" in order to distinguish meanings effectively, even though no one tuning system can satisfactorily bring out all the possible relationships between the connotations. Valuable insights into the nature of this semantic problem, given the possibilities of alternative tuning systems, can be found in the works of E G McClain (1978). An earlier experiment focussed on "tuning" interrelated cross-cultural concept sets having from 2 to 20 statements each (see Section KP). Longer interpretations may offer greater clarity, as in those of Wilhelm (1950) or Siu (1968). Needless to say, as an exercise by one person, the results given here for policy cycles call for further "tuning" and should therefore be viewed with reservation. Furthermore, it should be noted that the presentation given here does not do justice to the more sophisticated relationships embedded in the structure of the *I Ching*.

4. Tranformation pathways
It is the network of 384 transformation pathways between the 64 conditions into which an entity can supposedly get "trapped" that is perhaps the most interesting feature of this exercise.

In Section TP each of the 64 numbered conditions is briefly described, accompanied in each case by descriptions of 6 possible transformation pathways from that condition. These may also be understood as the possible "levels" of skill with which that condition can be faced. **The number following each transformation possibility indicates the new condition with which the policy cycle is then purportedly forced.** It should be emphasized however that these are merely the high probability transformation pathways. Another set of pathways given is that of the actual sequence of the numbered conditions. The "a-causal" reason for each such transformation is given at the end of each condition (as the "Transformation sequence") on the basis of one of the classic commentaries on the sequence. Read separately, **the transformation sequence constitutes an interesting acausal cycle**, with many links of immediately comprehensible relevance to current world conditions (eg progress-decline-community, adversity-basic need-revolution, or liberation-deficiency-aid).

If in a particular condition the policy cycle engages in lower probability multiple transformations the result is not apparent here, although the *Book of Changes* does employ a binary coding system from which this can be determined without ambiguity. Leibniz is reported to have been influenced in the 17th century by the binary code of the *I Ching*, which could therefore be said to have influenced the design of modern computers. The striking relationship to the genetic coding system has also been explored (Martin Schönberger, 1977).

The range of possible transformation pathways encoded in this way is of great value in the light of contemporary efforts to grasp the nature of change in relation to human and social development.

5. Contrasting exercises
As a work of political philosophy, it is useful to contrast interpretations of the *I Ching* with an early Western equivalent, namely Machiavelli's *The Prince*. Both provide recommendations to rulers, but the *I Ching* also adapts its recommendations to the initiatives of the ruled. *The Prince* has been severely criticized (often inappropriately, given the instabilities of its historical context), because of the distinctly undemocratic values of the princes for whom it was designed. In contrast, built into the *I Ching* is the progressive discovery of "superior values", however these are to be understood by the user. As with Machiavelli's advice, the policy cycle precepts from the *I Ching* could prove as valuable to the "ill-intentioned" as to the "well- intentioned".

It is worth noting that another set of 394 Chinese precepts, in Sun Tzu's classic *The Art of War*, has received considerable attention in modern military academies. It is based on the principle that it is the supreme art of war to subdue the enemy without fighting.

Contemporary students of organizational life have also benefitted from an adaptation of Machiavelli's insights by Antony Jay (1968) to the management of corporations.

Organization sociologists do not appear to have had the ambition (or the presumption) to attempt such a transformation map. Although in 1958 March and Simon published a study, now a classic, tracing parts of what might have become such a map. This does not appear to have been followed up. Literature reviews have since resulted in the production of "inventories" of concepts for organization effectiveness, as in that of J L Price (1968) with 31 propositions, or more recently in that of D H and B L Smith with approximately 400 concrete suggestions, especially for voluntary associations (1979).

Of special interest is the exercise of Edward de Bono who has produced an *Atlas of Management Thinking* (1983). This identifies 200 functions or "complex situations" which bear a striking resemblance to those derived from the *I Ching*. The Western managerial sciences have given rise to many treatises on problem solving in organizations. One of the originators of systems science, Russell Ackoff, has condensed his understanding of the art of problem solving into 34 "fables" (1978). Semi-humorous insights have also emerged in the form of numerous "laws" (Parson, Peter, etc), culminating in their synthesis in John Gall's 32 "axioms" in *Systemantics* (1978). Another semi-humorous approach, inspired by the holds and positions in the martial arts, is that of Thierry Gaudin (1977) who has identified 21 institutional "katas". It is appropriate to note that the control of "ch'i", mentioned earlier, is basic to the Eastern martial arts.

Western efforts to provide (world) systems models of the interrelationships between socio-political conditions of societies (as opposed to socio-economic conditions) have been modest and of limited success. For a recent general review, see J M Richardson jr (1981), reporting in a special issue on "Models - tools for shaping reality", (as well as reference 36), compared to the preferences for lengthy textual discourses of which Machiavelli's is an early form.

It is surprising to note that in the East a number of societies have produced religiously inspired board games with squares denoting value-based psychosocial conditions, linked by a variety of transformation pathways, in a manner similar to systems flow charts. Precepts (possibly embodied in chants) are associated with the definition of each condition and the developmental challenge it constitutes. Examples are: a Tibetan game (72 conditions) with a Bhutanese version (64 + 13 conditions) and a Nepalese version (25); a Korean game (169 conditions) and a Hindu equivalent (72 conditions), supposedly the prototype of Western "snakes and ladders". It has been argued by Stewart Culin (Games of the Orient, 1985) that the similarity between such games provides *"the most perfect existing evidence of the underlying foundation of mythic concepts upon which so much of the fabric of our culture is built."*

Transformative policy cycles: alternation between complementary conditions

1. Alternation

The vital point that emerges from this Chinese perspective is that it is not sufficient to conceive of organizational conditions in isolation, as is the prevalent tendency among Western networkers. The processes of change in which a policy cycle is embedded, or to which it responds, require that the policy cycle consider itself in a state of transience within a set of potential conditions. It courts disaster if it attempts to "stick" to one condition such as "peace". If the dynamics of problem networks are not being contained by present strategies, as would appear to be the case, then organizational self-satisfaction is a recipe for the disaster-prone or the ineffectual. It creates a false sense of security. **Any condition may be right temporarily, none is right permanently**. The situation is somewhat analogous to many team ball games where if a player tries to retain the ball it will be taken from him by the opposing side, or else the team is penalized. Furthermore policy cycles opposing the "team" of world problems find themselves having to deal with an opponent which handles the ball with a dynamism such as that of the Haarlem Globetrotters or a shell-game con-artist. The focus shifts continually and is often where it is least to be expected in order to take advantage of weaknesses.

A policy cycle must continually "alternate" its stance within the network of transformation pathways in order to "keep on the ball" and "keep its act together". As with a surfer, a wind sailor, or a sailor on a rocking boat, if it fails to change its stance it will be destabilized, according to the *I Ching*, by one of 64 changing conditions through which it is forced to move in a turbulent environment.

The developmental goal can then be conceived as somehow lying "through" the exit of this labyrinth of traps for the unwary. More satisfactorily, it is perhaps "in" the art of moving through these conditions as progressively clarifying the locus of a common point of reference undefined by any of them (cf, the Sanskrit phrase "Neti Neti", roughly translated as "not this, not that"). It is this art which is extolled in describing the use of the *I Ching* or of Eastern board games. A similar notion has recently emerged from theoretical physics through the work of David Bohm (1980). He stresses the nature of an underlying "holomovement" from which particularities are successively "unfolded" once again. The significance is more readily apparent in the case of "resonance hybrids" mentioned earlier.

The problem for a policy cycle, an organization, an intentional community, a meeting, or even an individual, is then **how to "network the alternation pathways together"** and **how to "alternate through a transformative policy cycle"**. Given that understanding of alternation seems only to be well-developed at the instinctual or sub-conscious level (eg walking, breathing, sex, dancing), the nature of alternation processes is explored in Section MZ. Extending the earlier metaphor of the "semantic piano" however, the challenge for policy cycles is then not simply to try to activate people by monotonous playing of single notes (eg "peace", "liberation", "development"), as presently tends to be the case. It is rather to acquire a perspective enabling them to collaborate in improvising exciting, rippling tunes with such notes (each of which might be *I Ching* condition) in order to bring out all the musical possibilities of alternation as explored in harmony, counterpoint, discord and rhythm.

In this sense **the true potential of "policy cycling" lies in the transformational possibilities of "playing"** on such instruments. Such an approach could perhaps provide the "requisite variety" by which the world problematique may be tamed, without breaking the spirit it embodies. A related challenge is then how to represent or map these transformation pathways in a memorable manner so that the range of possibilities becomes clear. In the *Book of Changes* a mnemonic system for the 64 conditions is given on the basis of 8 natural features of which people have both a instinctive and a poetic understanding. The features used as metaphors include: mountain, lake, wind, thunder, fight, ravine, earth and sky. Arguments in favour of some such topographically based mnemonic system are given in an earlier paper: *"The territory construed as a map"* (Judge, 1983). Such features contribute significantly to dissemination of understanding about relationships between such conditions in contrast to the restriction of interest in such matters in the West to scientific elites. The Eastern board games mentioned above are deliberately used for educational purposes, whereas very few in the West have access to the computer simulation exercises with an equivalent orientation.

The following remarks, and those in the following notes, indicate some possibilities for producing an adequate general map of the transformation pathways are discussed.

2. Challenge of representation

The challenge for any organization is then to learn how to "alternate" through such a policy cycle rather than get trapped in any particular condition. To facilitate the response to this challenge, ways must be found to map this set of transformation pathways so that it becomes comprehensible as a whole that can be consciously negotiated. Some mapping possibilities are discussed below.

3. Elaboration of a circular sequence

Helmut Wilhelm reports that in the Sung period (960-1127) of Confucianism the scholar Shao Yung produced a tabular representation of the *I Ching* elements. This "table" was also represented as a circle which he reproduces. It was Shao Yung's scheme which so excited Leibniz in the course of his reflections on the binary system.

In this traditional representation the transformation pathways are implicit except for the circular sequence itself. It is however possible to render them explicit by simple adding them to the representation. One way of doing this results in a diagram such as Figure 1. The only lines added are for the six "high probability" transformation pathways associated with the six sub-conditions of each of the 64 conditions, as described in Section TP.

Before commenting further on Figure 1, some basic points must be made about the traditional circular sequence. It is made up of 64 distinct "hexagrams". The hexagram is the traditional Chinese way of representing a change condition by a binary code of 6 broken or unbroken lines (which can be considered identical to the binary bit-code used in modern computers). But there are at least two fundamental points about any such code, as pointed out in the case of computers by Xavier Sallantin (1975):

- there must be agreement as to what represents "broken" (or "on"), as opposed to "unbroken" (or "off"), or else the code may be mis-read as its own "negative";

- there must be agreement as to how the hexagram (or computer bit sequence) should be read, whether up-to-down (or right-to-left) or down-to-up (or left-to-right), or else the code may be mis-read in an "inverted" form. The traditional circular sequence does not distinguish between these two possibilities.

The second point as applied to Figure 1 means that in relating the 64 condition names to their traditional hexagram representations a decision has to be taken as to the direction in which a hexagram is to be read. In Figure 1 the decision has been made to read the hexagrams with the "top" of each towards the centre and the numbered conditions have been allocated accordingly. This means that there is an alternative interpretation, Figure 2, in which the bottom of each is towards the centre. Note that the order of the numbered conditions is then quite different. The pattern of transformation pathways remains the same, although the sub-conditions to which they relate are now different. The 3 transformation pathways for each hexagram that were originally indicated inside the circle in Figure 1 are indicated by the lines outside the circle in Figure 2.

4. Interpretation problems

The diagrams give rise to three problems:

(a) First problem: Either Figure 1 or Figure 2 can thus be considered as a very compact map of the 384 high probability transformation pathways. But the existence of two different and seemingly conflicting maps is obviously cause for reflection.

With regard to this problem, the existence of two interpretations can be explained as due to the manner in which the *I Ching* perspective is grounded on **alternation between perspectives** rather than being tied arbitrarily to one perspective. If two interpretations are possible there is necessarily an alternation between them according to the Chinese perspective. What then could the alternation between such contrasting interpretations signify? From the significance traditionally attached to the top and bottom of the *I Ching* hexagrams, it could be argued that in the case of organizations the two contrasting interpretations could relate to an inward global worldview alternating with an outward local worldview. The top-in perspective (Figure 1) would then correspond to a map of consciously interrelated contrasting perspectives on the wholeness in which they are embedded, signalled to some extent by the process whereby leaders of a group "put their heads together" and "share their views". The "enemy" is recognized as being within the group ("he is us"). The alternative top-out perspective would then correspond to a map of unexplicated solidarity in response to the challenges of the immediately perceived external environment, signalled to some extent by the process whereby group members "stand back-to-back" to face an external "enemy" as he manifests differently to each. To survive the group must to some extent alternate between these contextual and particular worldviews, rather as an individual alternates between right and left-brain perspectives. Lama Govinda (1981) notes that hexagrams are read from bottom-to-top to represent the sub-conditions of individual life, in contrast to the top-to-bottom direction for more fundamental or universal transformation.

(b) Second problem: Also of concern in their non-evident relation to the numbered sequence of conditions, which itself constitutes a single transformation cycle. This lack of relationship is especially evident when lines are traced between the conditions in that traditional sequence, as in the case of Figure 3 (using the Figure 1 order) or Figure 4 (using the Figure 2 order).

With regard to this problem, using Figure 3 or 4, inspection will show that the continuing alternation between "global inwardness" and "local outwardness" forces every second hexagram in the numbered sequence into its opposite form (*eg* 3 in Figure 1 becomes 4 in Figure 2; 5 becomes 6; *etc*) and back again. Only the hexagrams 1, 2, 27, 28, 29, 30, 61 and 62 are not "driven" through the numbered sequence by this alternation process (which here acts in a manner reminiscent of the effects of current alternation in the coil windings of an electric motor). The map is a map of alternation **dynamics** and cannot be appropriately understood as a conventional map of **static** structural elements.

(c) Third problem: In addition, other than the striking elegance of the pattern, it is not obvious why either the order of Figure 1 or 2 should be the basis for an appropriate map.

With regard to this problem, the "logic" of the circular representation is that every condition denoted by a hexagram is conterbalanced by its "opposite" across the circle. In effect the broken lines are converted into unbroken lines and vice versa (thus partially containing the variations in significance of broken and unbroken lines noted above). In addition to the six high probability transformations from (and to) each condition, there is therefore a seventh transformation through the numbered sequence (by inversion of the code reading direction) and an eighth transformation into its opposite (through "negative" code bits of a hexagram acquiring a "positive" connotation and vice versa).

Given the striking relationship already noted by Schönberger between the *I Ching* 64-hexagram code and the genetic 64-codon code, the fundamental nature of the circular representation may also be illustrated by using it to map the 20 amino acids basic to biological organization. In Figure 1 these are denoted completely by the set of (long) transformation lines linking quarters of the circle. For example, according to Schnberger, asparagine is denoted by (the transformation between) the hexagram pair 34-43, the more complex amino acid threonin is denoted by (the symmetrically balanced transformation lines) 11:5:26:9, and the "stop" codes amber and ochre are denoted by the individual hexagrams 56 and 33 respectively. In the Figure 2 map the hexagrams denoting each amino acid, rather than being equidistant, are brought **together** side-by-side, as is illustrated around the circumference of Figure 4. Whether this suggests that certain well-defined transformation processes are as essential for the life of an organization or policy cycle as those 20 amino acids are for biological organization, is a question for further investigation.

5. Transformation cycles
A striking feature of Figure 1 (or 2) is the manner in which the transformation pathways of different types differentiate the circle so clearly into:

      (a) 2 halves of 32
      (b) 4 quarters of 16
      (c) 8 groups of 8
      (d) 16 groups of 4
      (e) 32 groups of 2
      (f) 64 groups of 1

In the light of current interest in the distinct functions of right and left brain perspectives, group (a) can be considered an interesting representation of the limited number of pathways linking such halves and the manner in which the halves are each separately integrated. In the light of Jungian investigation of the four basic psychological functions (sensation, feeling, intellect, intuition), group (b) can be considered an interesting representation of the transformation pathways by which these are linked and separately integrated as semi-independent functions. The 4 masculine and 4 feminine archetypal versions of these functions distinguised by psychoanalysts can in turn perhaps usefully be represented by group (c).

The question that now emerges is whether it is possible to elaborate some kind of typology of transformation "cycles" for organizations or policy cycles. Such a typology would clarify the different kinds of way that, for example, the two functional halves, or the four functional quarters are interlinked. For it is highly probable that organizations or policy cycles can "survive" by using the simplest possible transformation cycles that enable them to renew themselves, but that richer and more effective policy cycleing is only possible when more complex transformation pathway cycles are used. It is therefore to be expected that some organizations only manage a 4-transformation cycle linking four functional quarters but are quite incapable of handling the subtler functional transformations. Many organizations probably get stuck in cyclic "traps" because they cannot enrich the transformative cycles on which they depend. In addition to what has been termed the "high probability" transformations, based on the modification of a single line in a hexagram denoting a policy cycle condition, some other transformations of lower probability are shown in Figure 5. These too may form part of transformation cycles.

6. Circular representation: inner structure
A different approach to circular representation forms part of the conclusion of an extensive study by the renowned Buddhist scholar Lama Anagarika Govinda in a recent book entitled: *The Inner Structure of the I Ching: The Book of Transformations* (1981). His preference for "transformation" in the title is to be compared with the conventional translation as "change".

The special interest of this study, in contrast to the many studies of *I Ching* commentaries, is that it focuses on the structure of the *I Ching* itself as a system of signs in which *"two values were alternated and finally combined into eight symbols, which by replication yielded sixty-four hexagrams."*

Lama Govinda concentrates on the problem of the relationship between two traditional representations of the set of transformations. The first is the "abstract order" of Fu Hi which essentially determines the order of balanced polarities from which Figures 1 and 2 were derived. The second is the "temporal order" of King Wen which emphasizes the developmental sequence of phenomena. In order to make the movements from one condition to another graphically visible the author concludes that it only seems possible to find a unifying principle in the Fu Hi system.

His detailed investigations lead him to propose Figure 6. This shows the position of all 64 *I Ching* conditions projected onto a circular diagram. A unique feature of his focus on the "inner structure" is that this diagram results from the interplay between the 8 fundamental conditions from which the 64 are derived. The 8 are each denoted by a half-hexagram, namely a trigram. Depending on the order in which any given pair of trigrams is read, one of two hexagrams is thus defined. It is the condition numbers of these alternatives which are indicated on the straight lines within the circle. Each line thus represents two transformative movements. The eight conditions around the circumference represent those cases when the two trigrams are identical. Thus the straight lines denote transformations governed by the relationship between the 8 fundamental conditions denoted by each doubled trigram on the circumference.

What then is the relationship between Figure 6 and Figures 1 to 5? As noted above, in Figures 1 to 5 the circle of hexagrams may be split into eight parts in each of which the trigram on the inside is identical. One of the hexagrams in each part also has the outside trigram equal to the inside one. It is these eight (1, 2, 29, 30, 51, 52, 57 and 58) that are positioned around the circumference in the "top-out" order of Figures 2 and 4. Comparison with these Figures will show that the transformations from any numbered condition are here indicated by the lines (or points) to which it is connected through these fundamental positions, whether one or more hexagram lines are modified. In this sense Figure 6 is a much more compact representation than Figure 2 and 5. There is an intriguing resemblance between some of Lama Govinda's other diagrams of transformation between trigrams (represented by "curves" and "lines") and aspects of the structure of Figures 1 and 2. In graph theory terms, Figure 6 is a "dual" of Figures 2 and 5 combined, in that the transformation lines in the latter correspond to the transformation points in the former. Even in this representational convention there is advantage in alternating between both forms.

Also of great interest is Lama Govinda's very detailed investigation of sub-patterns of transformation connecting groups of 8 conditions traditionally called "houses". These patterns provide an important basis for any further investigation of the typology of transformation cycles called for above. It also enables him to clarify the relationship between the numerical sequence and the abstract order of Figure 6 by determining in Figure 7 the four symmetrical sub-patterns from which Figure 6 is constituted.

7. Elaboration of a spherical map

One interesting approach to this is to consider how Figure 6 would be transformed if it were to correspond to the alternative "top-in" order of Figures 1 and 3, instead of the "top-out" order of Figure 2. In effect the square formed by conditions 51, 52, 57, 58 in Figure 6 is simply rotated about the axis of conditions 1, 2; Conditions 1, 2, 29 and 30 do not move. The new sequence around the circumference is then 1, 58, 29, 51, 2, 52, 30, 57, as in Figures 1 and 3. If conditions 1 and 2 are considered as fixed "poles", a continuous rotation between the fixed positions 29 and 30 may be seen as transforming the circular representation into a spheric one. This dynamic model would need to be interpreted in terms of lines of force, as in the analysis of an electric motor or dynamo.

For reasons discussed in earlier papers, there are advantages in seeking a representation whose completeness is highlighted by basing it on an approximation to a spheric surface. The question then becomes how to cut up that surface into 64 units which will be assumed firstly to take the form of regular areas and secondly to be of identical form. (Other approaches are of course worth exploring.) Since the 64 phases (hexagrams) result from a conceptual system based on an eightfold complexification of 8 fundamental phases of change (trigrams), the problem can initially be reduced to one of representing the latter on a spherical approximation. The simplest such polyhedral approximation is the regular octahedron with eight triangular facets (see Figure 8). In allocating the 8 phases to these facets it would obviously be advantageous to do so such that their three high probability transformation pathways are highlighted.

Returning to the 64 phases, the problem can now be defined as one of how to divide up each of the triangular facets of the octahedron into eight equal parts so that eight phases can be represented within each such triangle. This can be done as shown in Figure 9. In this way the 64 phases can each be given a unique location on a polyhedral structure which can be easily projected onto the surface of a sphere.

There remains the problem of how to order the eight phases within each facet in Figure 8 so that within the completed figure the six high probability transformation pathways of the 64 phases are highlighted. It would seem, as with the standard problem of geographical map projections onto a two-dimensional surface, that there are a number of approaches to be explored. Each would be based on a different convention and would lead to a different arrangement with different advantages.

8. Comment

The *Book of Changes* is recognized as striking a remarkable balance between logical, structural (left-brain) precision and intuitive, contextual (right-brain) nuances of comprehension. For 3,000 years it has proved to be a unique achievement in relating the qualitative to the quantitative in a manner which is both practical and poetically appealing -- qualities for any blueprint for a new world order.

In the exercise for Section TP, most of the poetic appeal has been sacrificed. It does demonstrate that it is possible to interpret the insights of an Eastern classic into the jargon of Western management, however much of a "profanation" this may appear to those who know the original. An important consequence of the elimination of metaphor (despite the argument of Section MZ) is the loss of vital mnemonic keys with which the original is replete with good reason. Much of value has therefore been lost, as in any interpretation, despite the seeming advantages to be gained from the precision of the alternative presentation. Clearly some of the distortion is due to the alternative framework, whilst much is due to the limitations of the interpreter. Other interpretations could strike a more graceful balance between jargon and insight.

The acid test is of course whether this interpretation is useful to the formulation of sustainable policy cycles. Is it possible to relate the conditions described to the practical issues to be encountered? Can policy-makers use or adapt the maps of transformation pathways reproduced here? The answers are for the future. But the precision of the framework of the *Book of Changes*, linking such contemporary topics as "development", "liberation", "peace", "revolution", with what have here been termed "basic need", "deficiency" and "cultural heritage", offers an intriguing challenge to reflection and comprehension. The topics recall many of the concerns of the Goals, Processes and Indicators of Development project (1978-82) of the United Nations University.

With regard to the important problem of representation, it is appropriate to note that schematic diagrams of similar form have already been produced in combining Eastern insights and a Western management emphasis. A striking example is that of Figure 10, from *Zen and Creative Management* by Albert Low (1976). Erich Jantsch (1980), in his wide-ranging synthesis of self-organizing systems and their implications for policy-making and human development, draws attention to metabolic transformation cycles such as the carbon cycle shown in Figure 11. Indeed, given the fundamental nature of the representation system and its relationship to the basic amino acids, it is worth investigating to what extent the set of interconnected metabolic cycles and pathways does not illustrate the kinds of transformation pathways which need to be identified for organizations. The map of metabolic pathways could prove to be a provocative challenge to organizational sociologists of the future.

It is also tempting to see the 6 (+1) basic transformations from each condition (in Figures 1 and 2) in terms of catastrophe theory, as qualitative equivalents to the 7 characteristics kinds of catastrophe to which natural conditions are subject. The containment of plasma in fusion research suggest other insights concerning the containment of energy and the avoidance of "quenching".

This commentary began with a concern with how to reduce the drain of "energy" and significance from policies, organizations and meetings to which some of the transformation conditions respond. Is there not some possibility, like the search for the Holy Grail, that the challenge of giving form to sustainable policy cycles may be of equivalent complexity and form as that of containing plasma energy?

Numbers in this figure are abridged versions of entry numbers: eg Stagnation = 12 = TP 0012

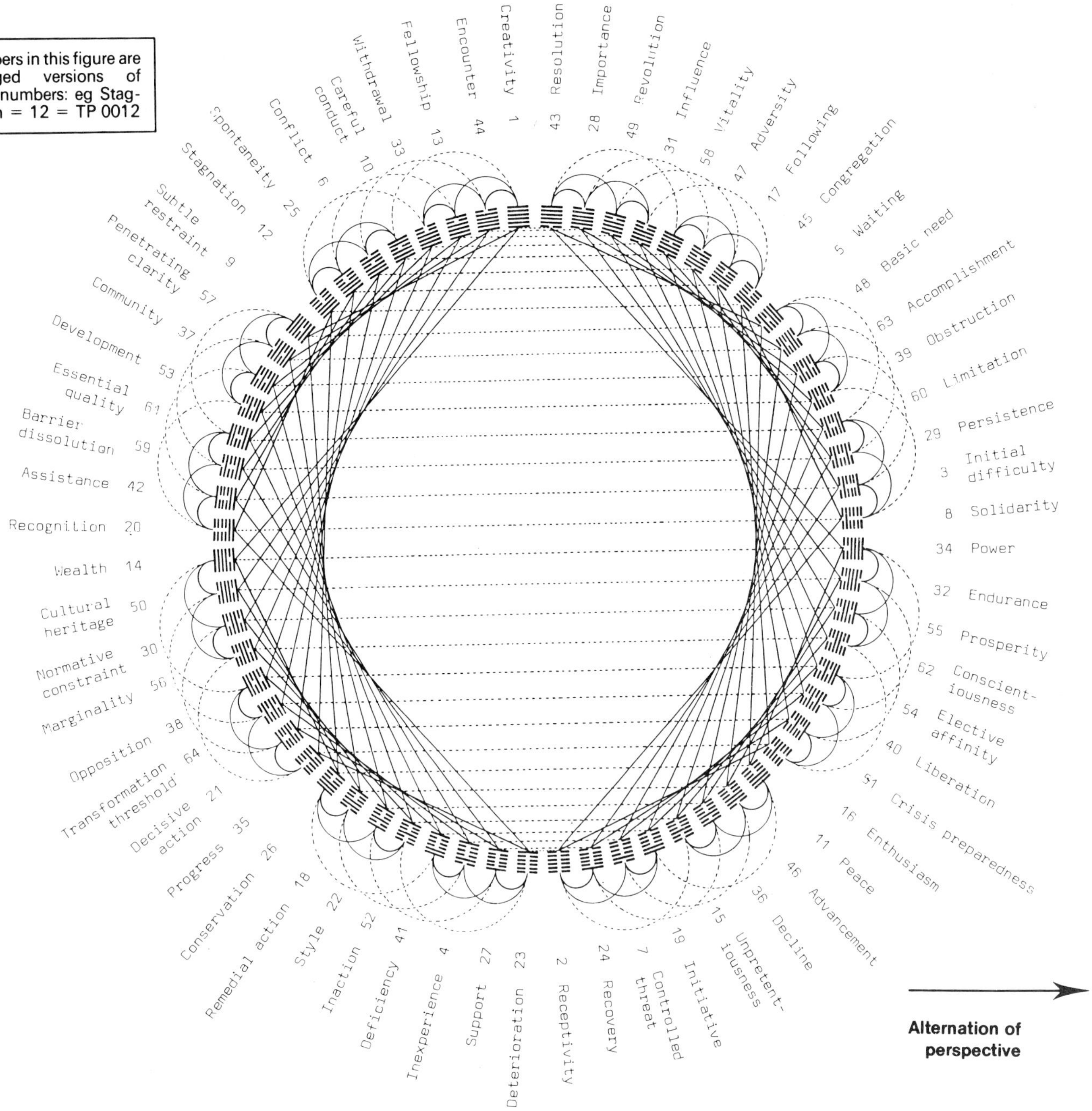

Alternation of perspective

Transformations (curves)
(- - -) 3rd sub-condition
(long) 2nd sub-condition
(short) 1st sub-condition

Transformations (straight lines)
(- - -) 6th sub-condition
(long) 5th sub-condition
(short) 4th sub-condition

Figure 1 — Map of transformations between global, 'heads-together' policy cycle conditions ('top-in')

The conditions are denoted by hexagrams in a traditional circular order (each facing its negative image). The 6 transformations shown interlinking these conditions are those described in the accompanying text (in which only one line of each hexagram code is modified; see Figure 5 for multiple line modifications).
The hexagram code is read here with the **top line closest to the centre** (in contrast to Figure 2), thus determining the condition numbers added. Note that a 7th transformation from each condition is that to its negative across the circle; an 8th is to its inversion, in the equivalent position in Figure 2.

Figure 3 — Transformation sequence through conditions in numerical order using Figure 1 hexagram positions
Odd-to-even transformations indicated by unbroken lines.

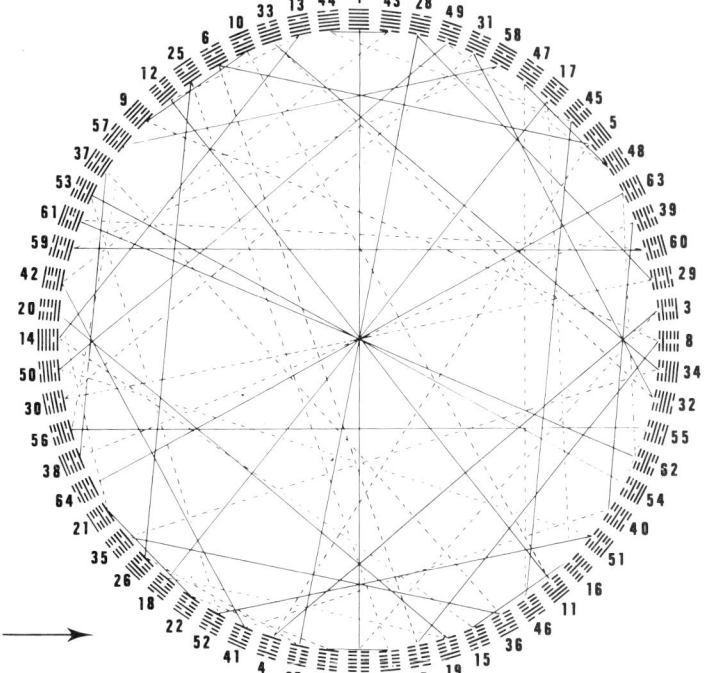

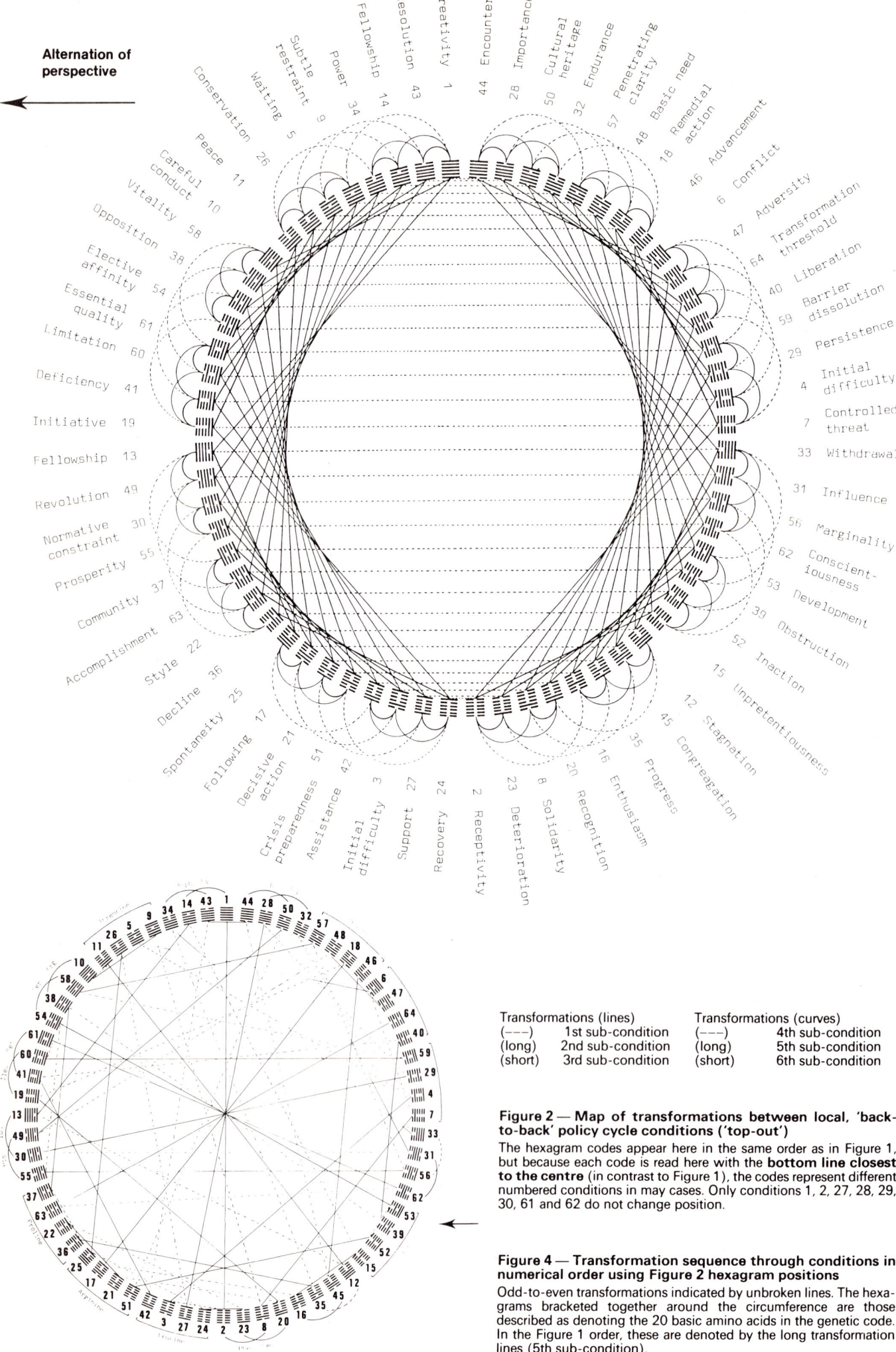

Transformations (lines)
(---) 1st sub-condition
(long) 2nd sub-condition
(short) 3rd sub-condition

Transformations (curves)
(---) 4th sub-condition
(long) 5th sub-condition
(short) 6th sub-condition

Figure 2 — Map of transformations between local, 'back-to-back' policy cycle conditions ('top-out')

The hexagram codes appear here in the same order as in Figure 1, but because each code is read here with the **bottom line closest to the centre** (in contrast to Figure 1), the codes represent different numbered conditions in may cases. Only conditions 1, 2, 27, 28, 29, 30, 61 and 62 do not change position.

Figure 4 — Transformation sequence through conditions in numerical order using Figure 2 hexagram positions

Odd-to-even transformations indicated by unbroken lines. The hexagrams bracketed together around the circumference are those described as denoting the 20 basic amino acids in the genetic code. In the Figure 1 order, these are denoted by the long transformation lines (5th sub-condition).

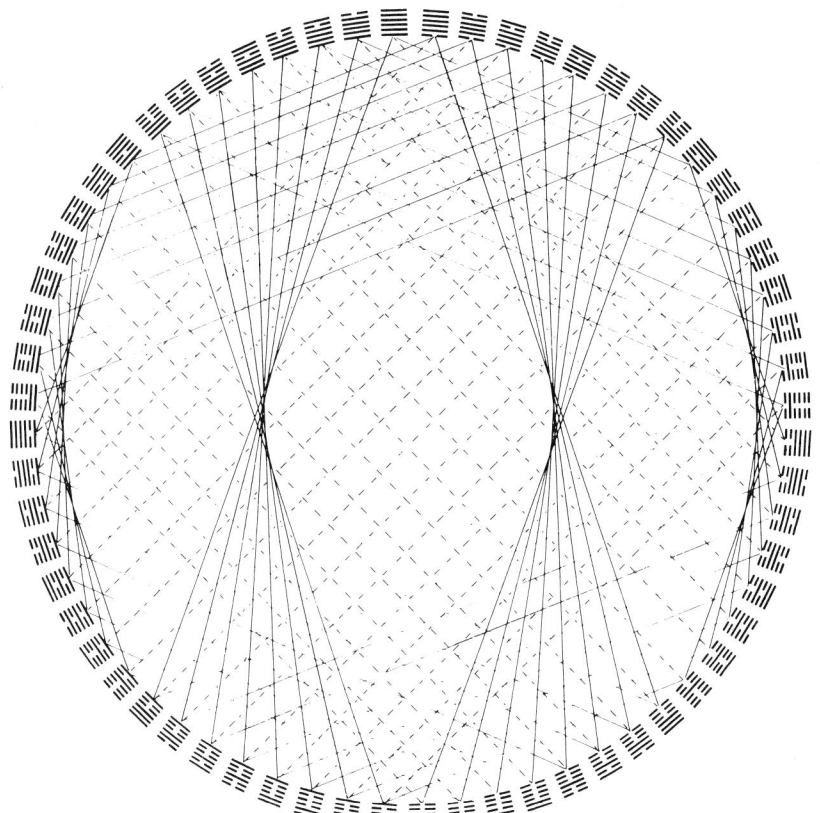

Figure 5 - Map of selected complex transformations between network conditions

Using the same circular order as for Figures 1 to 4, transformations are indicated between hexagrams for cases where **two** lines of the hexagram code are modified (see Figures 1 and 2 for single line transformations). The transformations selected are for different combinations of the **inner** three lines of each code (since those for the outer three link neighbouring hexagrams in a pattern similar to that around the circumference of Figures 1 and 2). Other combinations do not appear to result in significantly different patterns. The hexagram codes may be read either in terms of the Figure 1 (« top-in ») or the Figure 2 (« top-out ») orders from which the corresponding numbered conditions may be obtained.

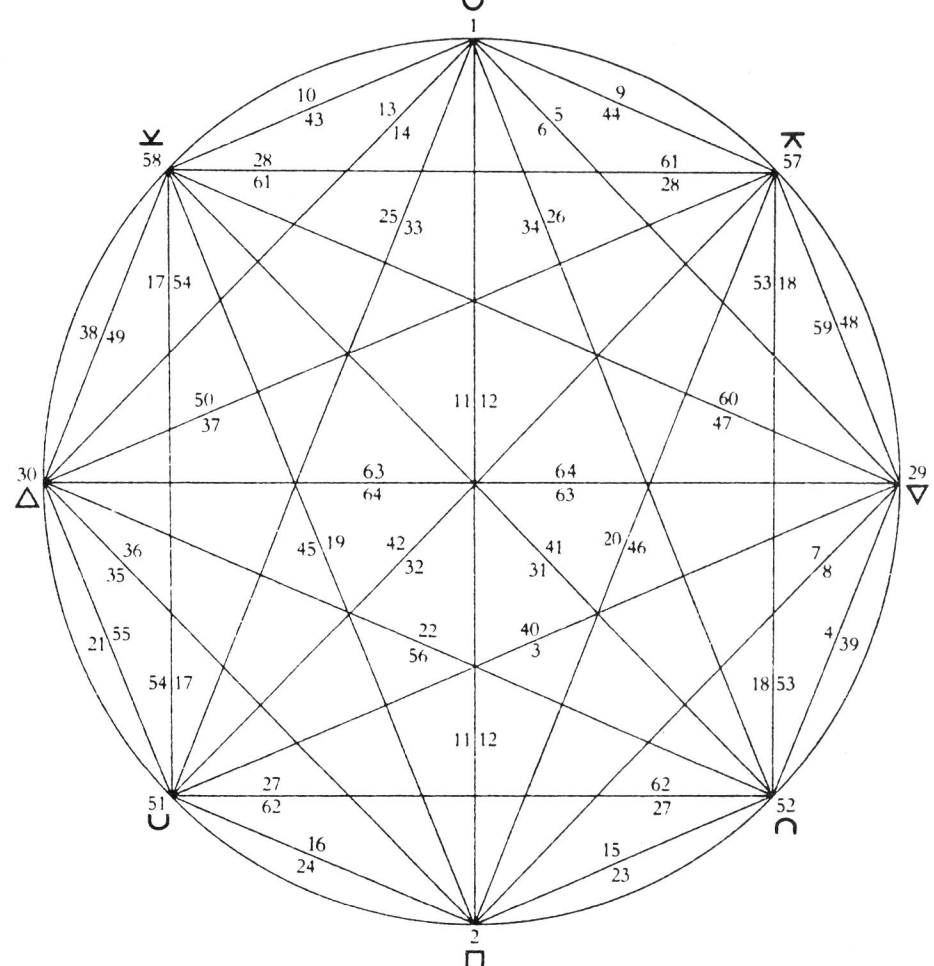

Figure 6 - Projection of all conditions (hexagrams) onto a circle
(Reproduced with the kind permission of Lama Anagarika Govinda, author of the **Inner Structure of the I Ching; the Book of Transformations**.

In Figures 1 to 5 the transformations between conditions are indicated by lines and curves (whether broken or unbroken). In Figure 6 those transformations are all represented as occurring within the 8 points around the circumference, whereas the lines represent the dynamic conditions denoted by the individual hexagrams positioned in a circle in Figures 1 to 5. Each line in Figure 6 indicates two possible conditions of change (just as each line in Figures 1 to 5 indicates two possible directions of transformation). The order of the 8 points around the circumference of Figure 6 corresponds to the order of the same points around the circumference of Figure 2 (« top-out » interpretation).

Figure 7 — Sub-patterns of policy cycle conditions extracted from Figure 6 (Adapted from diagrams of Lama Anagarika Govinda).

The numbered sequence of 64 conditions is split into 4 groups in numerical order. The patterns for each group are shown in the relevant diagram as a part of Figure 6. This establishes a relationship between the numerical sequence and an abstract order (which is the basis for Figures 1 to 5).

Note that the reconstruction of this arrangement is only possible as a result of recognition, from internal structural evidence, of the error noted below.

N.B. In producing Figure 6 from the elements of Figure 7, Lama Govinda concludes (pp 145-147) with Richard Wilhelm, that the traditional numerical order of the hexagrams in current works is slightly in **error**: 35 and 36 should replace 3 and 4; 21 and 22 should replace 35 and 36; and 3 and 4 should be inserted between 56 and 57.

This **does not affect** the patterns in Figures 1 to 5, with the exception of the broken lines in Figures 3 and 4. It **does affect** the 'logic' of the italic sequence of text linking the conditions. The explanation given for the error is that the Chinese original was on loose-leaf pages of which some were misplaced.

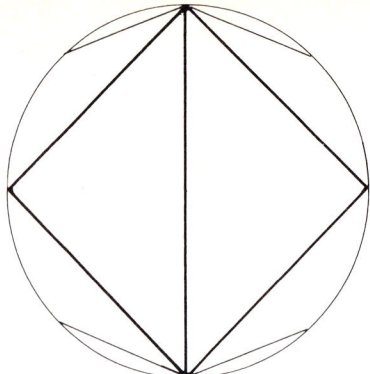

Policy cycle conditions 1 to 16

Policy cycle conditions 17 to 32

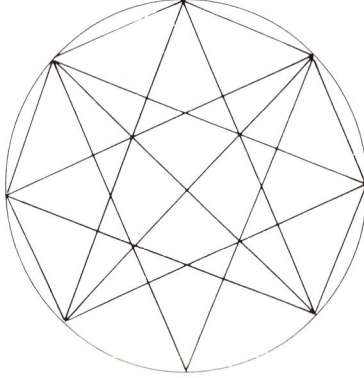

Policy cycle conditions 33 to 48

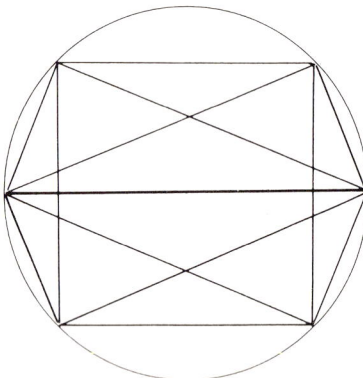

Policy cycle conditions 49 to 64

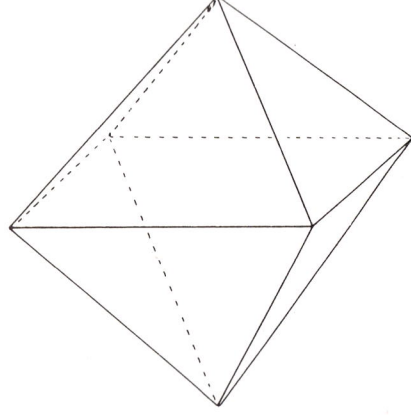

Figure 8 — Octahedron as basis for mapping 8 fundamental policy cycle conditions onto a sphere

The 64 policy cycle conditions are derived from 8 fundamental conditions (represented by the doubled hexagrams indicated on the circumference of Figure 6). Each of the 8 may be denoted by one triangular facet of the octahedron. The allocation of the conditions, and the transformational relationships between them, can then be mapped onto the geometry of the octahedron (as one of the simplest polyhedral approximations to a sphere).

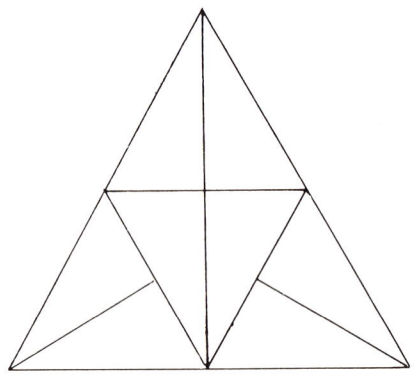

Figure 9 — Eightfold subdivision of the triangular facet of an octahedron.

In order to represent all 64 policy cycle conditions on an octahedron (Figure 8), each triangular face can be sub-divided into 8 equal areas as shown.

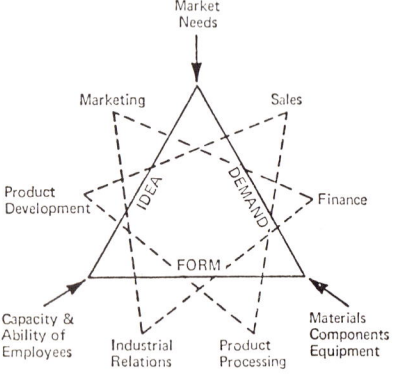

Figure 10 - Interrelationship of economic functions in management systems.

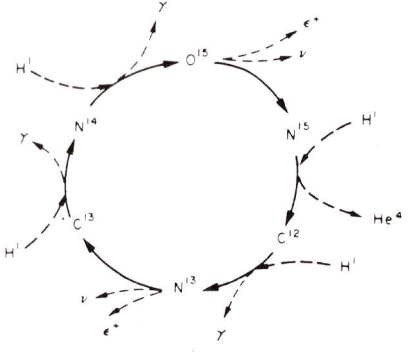

Figure 11 - Carbon cycle as a detail of metabolic pathways.

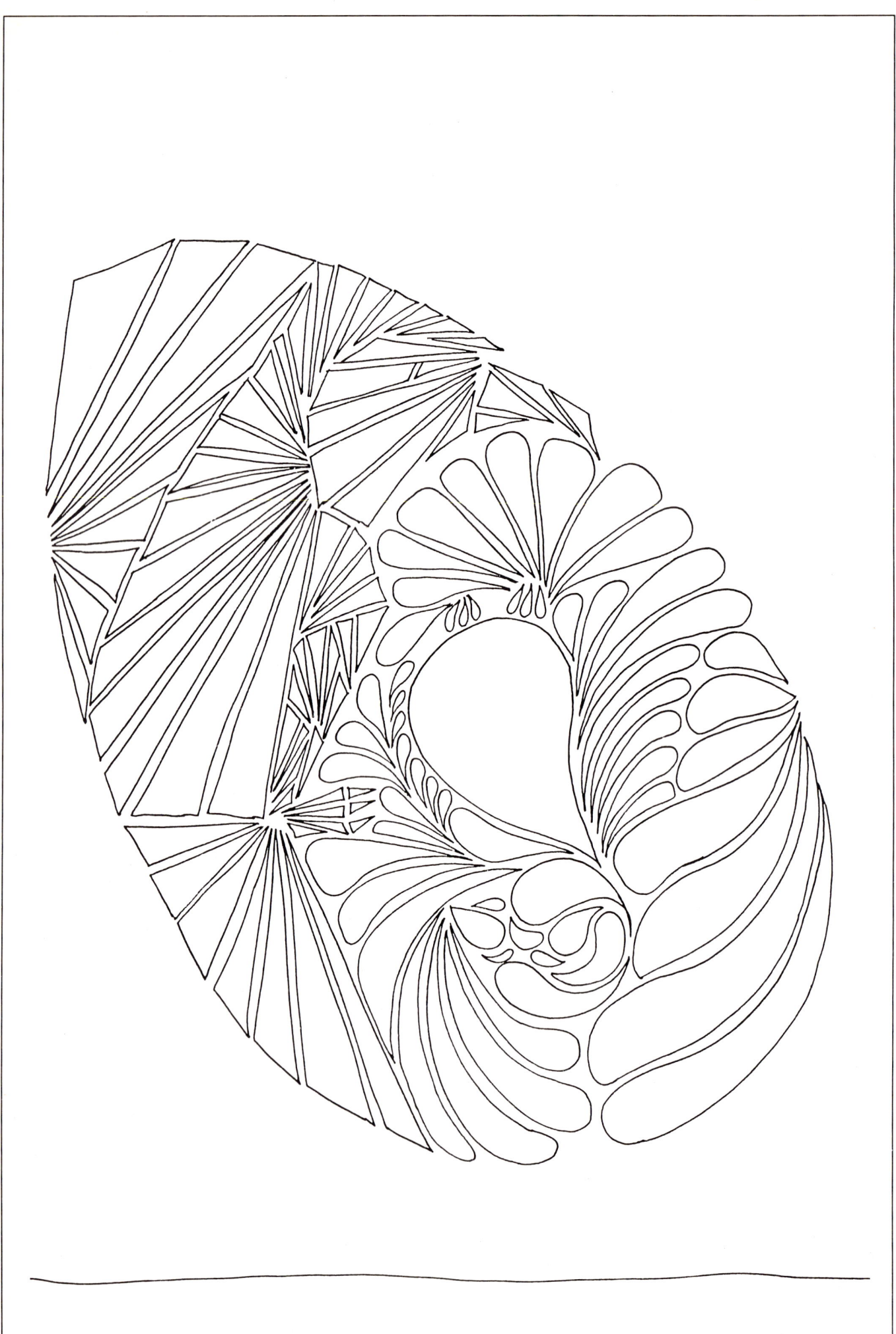

Human Values and Wisdom — V

Scope
The importance of values is frequently cited in relation to the global problematique, whether it be in debates in international assemblies, in studies criticizing "value-free" approaches to research, or in discussion of quality of life and individual fulfillment. Values are deemed especially important in questions of cultural development and are central to concern for the preservation of cultural heritage.

The purpose of this section is to register a comprehensive range of values with which people identify, to which they are attracted or which they reject as abhorrent. Whilst it had been hoped to develop such lists from documents of international bodies, no adequate lists of values were located, even within the intergovernmental agencies (such as UNESCO) specifically concerned with human values, and despite numerous reports and meetings on "values" in recent years. The values referred to are very seldom named, although the commonest may be cited as examples. The list presented here has therefore been elaborated by the editors as an experiment based on the selection and interrelationship of constructive and destructive value words.

The appreciation of the subtlest values, and especially value dilemmas, is intimately related to "wisdom", however it is to be understood. There is therefore a concern to explore the possibility of a framework in which explicit links between "values" and "wisdom" could be established.

Sub-sections
The section contains 2,270 entries linked by 14,463 cross-references. It is divided into four parts:

-- Section VC contains 960 constructive value words (*eg* peace, harmony, beauty);

-- Section VD contains 1,040 destructive value words (*eg* conflict, depravity, ugliness).

-- Section VP contains 225 entries on value-polarities (*eg* agreement-disagreement, freedom- restraint, pleasure-displeasure) derived from the organization of *Roget's Thesaurus*. These group and link the entries of Sections VC and VD.

-- Section VT consists of 45 "value types" or "value complexes" which are used to group the value-polarities of Section VP.

Method
The procedures used in preparing this section are discussed in detail in Section VX. They are based on use of the standard reference work *Roget's Thesaurus* as representing a much-used example of the way one international language is used.

Overview
Detailed comments are provided in Section VZ. None of the entries contain "descriptions" of the value(s) implied. In most cases this would be superfluous. The words in Section VC reflect values which tend to be accepted without questioning. Those in Section VD reflect values which tend to be rejected without questioning. The emphasis is placed on using the cross-references to indicate the range of connotations of particular value words. The entries on value polarities, Section VP, do however list proverbs, aphorisms or quotations selected to illustrate the dynamic counter-intuitive relationship between constructive and destructive values. They endeavour to draw on popular wisdom or insight to demonstrate the negative consequences and limitations of blind adherence to constructive values or to demonstrate the positive consequences and creative opportunity of judicious action in the light of destructive values. They point to the existence of a more fundamental and challenging dynamic than that implied, for example, by peace-at-all-costs and total rejection of conflict.

This exploration of values is of special interest in relation to the world problems in Section P. Many problems are named in international debate using a destructive value word (*eg* insufficient, unrealistic, unjust, inappropriate). Problems defined in this way imply the existence of some corresponding value whose expression is infringed by the problem. Such values may or may not be noted in defining the purposes underlying remedial action in response to the problem, although often they form part of the wording of any rallying slogan in support of some international strategy.

The set of constructive and destructive value words does indicate a way of coming to grips with the range of problems which the existing language renders perceiveable and nameable. Such values also indicate possible dimensions of human development. This section is of course limited at this stage by the biases inherent in *Roget's Thesaurus* and the English language. It does however create a framework which could enable these limitations to be transcended.

Context
The contents of this section may be considered as complementing the other sections in ways such as the following:

Human development: By the manner in which values acquire their significance through the pursuit of different modes of human development and through the association of many specific modes of awareness with the experience of particular values.

Integrative knowledge: By the challenge of providing integrative frameworks to interrelate seemingly unrelated values and by the inherently integrative nature of value perspectives.

World problems: By the direct correspondence between disvalues and problems, and by the manner in which problems only become perceptible in the light of the values upon which they infringe.

Metaphors and patterns: By the manner in which human values are communicated, and through the intrinsic value of communication in maintaining the fabric of global society.

Transformative approaches: By the values expressed or enhanced by innovative techniques.

Constructive values

Rationale

Widespread recognition is currently given to the importance of identifying constructive values as a guide to formulation of policy and action programmes in response to problems. Indeed it may be argued that such problems are only perceptible in the light of the values they infringe. Specific values are frequently cited in political discourse as a rallying focus around which people may be incited to action.

It is therefore appropriate to determine the range of "concepts" that can, and possibly should, be cited in this way. Yet despite the frequent references to "values", it is not clear in most cases to what values reference is being made. There exists a core group of values (*eg* peace, justice) on which there is a very extensive literature. Social scientists responding to the needs of market researchers have identified value complexes with which different consumer groups identify. Religions have traditionally identified lists of virtues. But it would appear that no attempt has been made to list the values which are identifiable in contemporary society. And yet it is supposedly in the light of such values that people and groups guide the development processes in which they engage, whether consciously or unconsciously.

Contents

This section lists in alphabetic order 960 words which can be considered as reflecting "constructive" values. Because of the ambiguous connotations of many such words, they are each cross-referenced to a number of entries in Section VP. The entries in Section VP are each value categories or dimensions denoted by a pair of "constructive" and "destructive" value words in opposition, namely a "value polarity". This polar relationship sharpens the meaning that can be associated with a particular interpretation of the constructive value word identified in this section.

Associating value words with value polarities also responds significantly to potential differences of opinion as to whether a particular value should be considered "constructive" as opposed to "destructive" (as could well be the case under certain circumstances). It is for this reason that the more usual terms "positive" and "negative" were not used. It is more understandable that "destructive" action may be necessary to clear the way for "constructive" action, and that "constructive" action may reach the point at which "destructive" action is necessary to initiate some new phase.

Method

The method is described in Section VZ. It is based on selecting and interrelating words from *Roget's Thesaurus* which was considered a comprehensive reference work providing a useful framework for this exploratory exercise.

Index

A keyword index to entries is provided in Section VX. THe keywords are also incorporated into the index for Volume 2 (Section X)

Bibliography

Bibliographical references, by author, are given in Section VY.

Comment

Detailed comments are given in Section VZ.

Reservations

Distinguishing value words as "constructive" rather than "destructive" raises many useful questions. Although this may often correspond to the reality of a conventional, first-order response, the reverse may be true under other more complex circumstances. Such difficulties have been partially resolved by stressing the value dimension, indicated by the value polarities cross-referenced in Section VP, rather than the individual value words.

Sections VC and VD could therefore have been merged. As an exploratory exercise the results must necessarily be considered as preliminary, and limited by dependence on the Roget framework and the English language. To the extent that some values (*eg* quality of life) are associated with phrases rather than single words (*eg* quality, life), this single-word approach can only be considered a preliminary exercise. In conformity with the general editorial policy, borderline value words have been included if they raise interesting questions concerning the criteria governing their inclusion.

Possible future improvements

In addition to refining and extending the set of words and cross-references, ways of incorporating equivalent words from other languages need to be considered.

ENTRY CONTENT AND ORGANIZATION

Ordering of entries
Entries are numbered in **alphabetic order**. Entry numbers have no significance other than as a permanent point of reference to facilitate indexing, cross-referencing, and updating between editions.

Index access to entries
This sub-section is self-indexing since the entries are in alphabetic order. The location of an entry in this sub-section may also be determined from:
- the **Volume Index** (Section X)
- the **Section Index** (Section VX) on the basis of keywords in the name of the entry or its alternate names

Structure of entries
Entries may be composed of the following descriptive elements:

(a) **Entry number** This number has **no significance**, except as a convenient method of identifying the entry (particularly for indexing purposes), of filing information on it, and as an identifier to which cross-references from other entries (possibly in other sections) may refer in this and future editions. The first letter of the entry number refers to the section of this volume in which the sub-section, denoted by the second letter, is located.

(b) **Entry name**   This is printed in bold characters

Cross-referencing of entries
At the end of any entry, there may be cross-references to other entries. These indicate the number and name of the cross- referenced entry, whether within this Section or in other Sections. In this sub-section there is a single type of cross-reference, indicated by a 2-letter code in bold characters:
  Integrative polarities = Broader value or value polarity

CONSTRUCTIVE VALUES

♦ **VC0003 Ability**
 Integrative polarities Power–Impotence (#VP5157)
 Skilfulness–Unskilfulness (#VP5733)
 Preparedness–Unpreparedness (#VP5720).

♦ **VC0005 Abnegation**
 Integrative polarities Energy–Moderation (#VP5161)
 Temperance–Intemperance (#VP5992).

♦ **VC0007 Absolution**
 Integrative polarities Forgiveness–Vengeance (#VP5947).

♦ **VC0009 Absoluteness**
 Integrative polarities Truth–Error (#VP5516)　Certainty–Uncertainty (#VP5513)
 Simplicity–Complexity (#VP5045)　Perfection–Imperfection (#VP5677)
 Completeness–Incompleteness (#VP5054).

♦ **VC0011 Abstinence**
 Integrative polarities Energy–Moderation (#VP5161)
 Chastity–Unchastity (#VP5987)
 Temperance–Intemperance (#VP5992).

♦ **VC0013 Abstraction**
 Integrative polarities Attention–Inattention (#VP5530)
 Thought–Thoughtlessness (#VP5478).

♦ **VC0015 Abundance**
 Integrative polarities Numerousness–Fewness (#VP5101)
 Conciseness–Diffuseness (#VP5592)
 Oversufficiency–Insufficiency (#VP5660)
 Productiveness–Unproductiveness (#VP5165).

♦ **VC0017 Acceptance**
 Integrative polarities Assent–Dissent (#VP5521)
 Consent–Refusal (#VP5775)
 Belief–Unbelief (#VP5501)
 Freedom–Restraint (#VP5762)
 Patience–Impatience (#VP5861)
 Leniency–Compulsion (#VP5756)
 Approval–Disapproval (#VP5968)
 Contentment–Discontentment (#VP5868).

♦ **VC0019 Accessibility**
 Integrative polarities Possession–Loss (#VP5808)
 Presence–Absence (#VP5186)
 Increase–Decrease (#VP5038)
 Approach–Recession (#VP5296)
 Possibility–Impossibility (#VP5509)
 Influence–Influencelessness (#VP5172)
 Appropriateness–Inappropriateness (#VP5665)
 Communicativeness–Uncommunicativeness (#VP5554).

♦ **VC0021 Acclaim**
 Integrative polarities Assent–Dissent (#VP5521)
 Approval–Disapproval (#VP5968).

♦ **VC0023 Accomplishment**
 Integrative polarities Evolution–Revolution (#VP5147)
 Production–Reproduction (#VP5167)
 Skilfulness–Unskilfulness (#VP5733)
 Completeness–Incompleteness (#VP5054)
 Accomplishment–Nonaccomplishment (#VP5722).

♦ **VC0025 Accord**
 Integrative polarities Assent–Dissent (#VP5521)
 Consent–Refusal (#VP5775)
 Harmony–Discord (#VP5462)
 Accord–Disaccord (#VP5794)
 Sharing–Appropriation (#VP5815)
 Agreement–Disagreement (#VP5026)
 Conformity–Nonconformity (#VP5082)
 Concurrence–Counteraction (#VP5177).

♦ **VC0027 Accumulation**
 Integrative polarities Possession–Loss (#VP5808)
 Increase–Decrease (#VP5038)
 Concurrence–Counteraction (#VP5177).

♦ **VC0029 Accuracy**
 Integrative polarities Truth–Error (#VP5516)　Carefulness–Neglect (#VP5533)
 Discrimination–Indiscrimination (#VP5492).

♦ **VC0031 Achievement**
 Integrative polarities Courage–Fear (#VP5890)　Action–Inaction (#VP5705)
 Production–Reproduction (#VP5167)
 Accomplishment–Nonaccomplishment (#VP5722).

♦ **VC0033 Acknowledgement**
 Integrative polarities Assent–Dissent (#VP5521)
 Approval–Disapproval (#VP5968)
 Gratitude–Ingratitude (#VP5949)
 Attributability–Chance (#VP5155)
 Provability–Unprovability (#VP5505).

♦ **VC0035 Acquiescence**
 Integrative polarities Assent–Dissent (#VP5521)
 Belief–Unbelief (#VP5501)
 Freedom–Restraint (#VP5762)
 Patience–Impatience (#VP5861)
 Obedience–Disobedience (#VP5766)
 Conformity–Nonconformity (#VP5082)
 Willingness–Unwillingness (#VP5621).

♦ **VC0037 Acquisition**
 Integrative polarities Knowledge–Ignorance (#VP5475)
 Sharing–Appropriation (#VP5815).

♦ **VC0039 Action**
 Integrative polarities Action–Inaction (#VP5705)　Amusement–Boredom (#VP5878)
 Motion–Quiescence (#VP5267)　Energy–Moderation (#VP5161)
 Expedience–Inexpedience (#VP5670)　Appropriateness–Inappropriateness (#VP5665).

♦ **VC0041 Activity**
 Integrative polarities Action–Inaction (#VP5705)　Motion–Quiescence (#VP5267)
 Energy–Moderation (#VP5161).

♦ **VC0043 Actuality**
 Integrative polarities Truth–Error (#VP5516)　Futurity–Antiquity (#VP5119)
 Certainty–Uncertainty (#VP5513)　Existence–Nonexistence (#VP5001).

♦ **VC0045 Acuity**
 Integrative polarities Carefulness–Neglect (#VP5533)
 Intelligence–Unintelligence (#VP5467).

♦ **VC0047 Acumen**
 Integrative polarities Sensation–Insensibility (#VP5422)
 Intelligence–Unintelligence (#VP5467)
 Discrimination–Indiscrimination (#VP5492).

♦ **VC0049 Adaptability**
 Integrative polarities Harmony–Discord (#VP5462)
 Change–Permanence (#VP5139)
 Elasticity–Toughness (#VP5358)
 Evolution–Revolution (#VP5147)
 Agreement–Disagreement (#VP5026)
 Stability–Changeableness (#VP5141)
 Conformity–Nonconformity (#VP5082)
 Skilfulness–Unskilfulness (#VP5733)
 Conventionality–Unconventionality (#VP5642).

♦ **VC0051 Adequacy**
 Integrative polarities Power–Impotence (#VP5157)
 Contentment–Discontentment (#VP5868)
 Oversufficiency–Insufficiency (#VP5660).

♦ **VC0053 Adherence**
 Integrative polarities Probity–Improbity (#VP5974)
 Approval–Disapproval (#VP5968)
 Cohesion–Disintegration (#VP5050).

♦ **VC0055 Adjustment**
 Integrative polarities Change–Permanence (#VP5139)
 Equality–Inequality (#VP5030)
 Neutrality–Compromise (#VP5806)
 Agreement–Disagreement (#VP5026)
 Skilfulness–Unskilfulness (#VP5733).

♦ **VC0057 Admiration**
 Integrative polarities Love–Hate (#VP5930)　Respect–Disrespect (#VP5964)
 Approval–Disapproval (#VP5968).

♦ **VC0059 Adoration**
 Integrative polarities Love–Hate (#VP5930)　Piety–Impiety (#VP6028)
 Respect–Disrespect (#VP5964)　Pleasure–Displeasure (#VP5865).

♦ **VC0061 Adornment**
 Integrative polarities Beauty–Ugliness (#VP5899).

♦ **VC0063 Advancement**
 Integrative polarities Goodness–Badness (#VP5674)
 Increase–Decrease (#VP5038)
 Support–Opposition (#VP5785)
 Prosperity–Adversity (#VP5728)
 Evolution–Revolution (#VP5147)
 Improvement–Impairment (#VP5691)
 Accomplishment–Nonaccomplishment (#VP5722).

♦ **VC0065 Advantage**
 Integrative polarities Possession–Loss (#VP5808)
 Goodness–Badness (#VP5674)
 Support–Opposition (#VP5785)
 Superiority–Inferiority (#VP5036)
 Appropriateness–Inappropriateness (#VP5665).

♦ **VC0067 Adventure**
 Integrative polarities Courage–Fear (#VP5890)　Action–Inaction (#VP5705).

♦ **VC0069 Advocacy**
 Integrative polarities Attack–Defence (#VP5798)
 Friendship–Enmity (#VP5927)
 Approval–Disapproval (#VP5968)
 Motivation–Dissuasion (#VP5648)
 Vindication–Condemnation (#VP6005).

♦ **VC0071 Aesthetics**
 Integrative polarities Beauty–Ugliness (#VP5899)
 Taste–Vulgarity (#VP5896).

♦ **VC0073 Affability**
 Integrative polarities Kindness–Unkindness (#VP5938)
 Leniency–Compulsion (#VP5756)
 Courtesy–Discourtesy (#VP5936)
 Sociability–Unsociability (#VP5922)
 Pleasantness–Unpleasantness (#VP5863).

♦ **VC0075 Affection**
 Integrative polarities Love–Hate (#VP5930)　Desire–Avoidance (#VP5631)
 Kindness–Unkindness (#VP5938)　Feeling–Unfeelingness (#VP5855).

♦ **VC0077 Affiliation**
 Integrative polarities Ancestry–Posterity (#VP5170).

♦ **VC0079 Affinity**
Integrative polarities Accord–Disaccord (#VP5794)
Friendship–Enmity (#VP5927)
Attraction–Repulsion (#VP5288)
Agreement–Disagreement (#VP5026).

♦ **VC0081 Affirmation**
Integrative polarities Consent–Refusal (#VP5775)
Affirmation–Denial (#VP5523)
Provability–Unprovability (#VP5505).

♦ **VC0083 Affluence**
Integrative polarities Prosperity–Adversity (#VP5728)
Oversufficiency–Insufficiency (#VP5660).

♦ **VC0087 Aggrandizement**
Integrative polarities Repute–Disrepute (#VP5914)
Expansion–Contraction (#VP5197)
Representation–Misrepresentation (#VP5573).

♦ **VC0089 Agility**
Integrative polarities Action–Inaction (#VP5705)   Carefulness–Neglect (#VP5533)
Skilfulness–Unskilfulness (#VP5733).

♦ **VC0091 Agreeableness**
Integrative polarities Beauty–Ugliness (#VP5899)
Consent–Refusal (#VP5775)
Accord–Disaccord (#VP5794)
Friendship–Enmity (#VP5927)
Freedom–Restraint (#VP5762)
Kindness–Unkindness (#VP5938)
Courtesy–Discourtesy (#VP5936)
Agreement–Disagreement (#VP5026)
Hospitality–Inhospitality (#VP5925)
Willingness–Unwillingness (#VP5621)
Contentment–Discontentment (#VP5868)
Pleasantness–Unpleasantness (#VP5863).

♦ **VC0093 Agreement**
Integrative polarities Assent–Dissent (#VP5521)
Consent–Refusal (#VP5775)
Accord–Disaccord (#VP5794)
Agreement–Disagreement (#VP5026)
Concurrence–Counteraction (#VP5177).

♦ **VC0095 Aid**
Integrative polarities Support–Opposition (#VP5785)
Facility–Difficulty (#VP5732)
Motivation–Dissuasion (#VP5648).

♦ **VC0097 Alacrity**
Integrative polarities Action–Inaction (#VP5705)   Desire–Avoidance (#VP5631)
Timeliness–Untimeliness (#VP5129)   Willingness–Unwillingness (#VP5621).

♦ **VC0099 Alertness**
Integrative polarities Carefulness–Neglect (#VP5533)
Attention–Inattention (#VP5530)
Timeliness–Untimeliness (#VP5129)
Preparedness–Unpreparedness (#VP5720)
Intelligence–Unintelligence (#VP5467).

♦ **VC0101 Aliveness**
Integrative polarities Life–Death (#VP5407)   Health–Disease (#VP5685)
Action–Inaction (#VP5705)   Carefulness–Neglect (#VP5533)
Intelligence–Unintelligence (#VP5467).

♦ **VC0103 Allegiance**
Integrative polarities Probity–Improbity (#VP5974)
Obedience–Disobedience (#VP5766).

♦ **VC0105 Alliance**
Integrative polarities Support–Opposition (#VP5785)
Conjunction–Separation (#VP5047)
Concurrence–Counteraction (#VP5177)
Relatedness–Unrelatedness (#VP5009).

♦ **VC0107 Allurement**
Integrative polarities Love–Hate (#VP5930)   Attraction–Repulsion (#VP5288)
Motivation–Dissuasion (#VP5648).

♦ **VC0109 Almightiness**
Integrative polarities Power–Impotence (#VP5157)
Godliness–Ungodliness (#VP6013).

♦ **VC0111 Alternation**
Integrative polarities Identity–Difference (#VP5013).

♦ **VC0113 Altruism**
Integrative polarities Kindness–Unkindness (#VP5938)
Unselfishness–Selfishness (#VP5978).

♦ **VC0115 Ambition**
Integrative polarities Action–Inaction (#VP5705)   Desire–Avoidance (#VP5631).

♦ **VC0117 Amelioration**
Integrative polarities Change–Permanence (#VP5139)
Improvement–Impairment (#VP5691).

♦ **VC0119 Amenability**
Integrative polarities Dueness–Undueness (#VP5960)
Patience–Impatience (#VP5861)
Willingness–Unwillingness (#VP5621).

♦ **VC0121 Amiability**
Integrative polarities Friendship–Enmity (#VP5927)
Kindness–Unkindness (#VP5938)
Hospitality–Inhospitality (#VP5925)
Sociability–Unsociability (#VP5922)
Pleasantness–Unpleasantness (#VP5863).

♦ **VC0123 Amicability**
Integrative polarities Accord–Disaccord (#VP5794)
Friendship–Enmity (#VP5927)
Support–Opposition (#VP5785)
Pleasantness–Unpleasantness (#VP5863).

♦ **VC0125 Amity**
Integrative polarities Accord–Disaccord (#VP5794)
Friendship–Enmity (#VP5927).

♦ **VC0127 Amorousness**
Integrative polarities Love–Hate (#VP5930).

♦ **VC0129 Amplification**
Integrative polarities Increase–Decrease (#VP5038)
Evolution–Revolution (#VP5147)
Expansion–Contraction (#VP5197).

♦ **VC0131 Amusement**
Integrative polarities Amusement–Boredom (#VP5878)
Cheerfulness–Solemnity (#VP5870)
Contentment–Discontentment (#VP5868).

♦ **VC0133 Ancestry**
Integrative polarities Ancestry–Posterity (#VP5170).

♦ **VC0135 Animation**
Integrative polarities Life–Death (#VP5407)   Action–Inaction (#VP5705)
Desire–Avoidance (#VP5631)   Energy–Moderation (#VP5161)
Motivation–Dissuasion (#VP5648)   Cheerfulness–Solemnity (#VP5870).

♦ **VC0137 Anonymity**
Integrative polarities Communicativeness–Uncommunicativeness (#VP5554).

♦ **VC0139 Anticipation**
Integrative polarities Intuition–Reason (#VP5481)
Carefulness–Neglect (#VP5533)
Timeliness–Untimeliness (#VP5129)
Expectation–Inexpectation (#VP5539).

♦ **VC0141 Apology**
Integrative polarities Exultation–Lamentation (#VP5873).

♦ **VC0143 Appeal**
Integrative polarities Love–Hate (#VP5930)   Motivation–Dissuasion (#VP5648)
Pleasantness–Unpleasantness (#VP5863).

♦ **VC0145 Appearance**
Integrative polarities Form–Formlessness (#VP5246)
Appearance–Disappearance (#VP5446).

♦ **VC0147 Appeasement**
Integrative polarities Accord–Disaccord (#VP5794)
Energy–Moderation (#VP5161)
Comfort–Aggravation (#VP5885).

♦ **VC0149 Appetite**
Integrative polarities Desire–Avoidance (#VP5631)
Chastity–Unchastity (#VP5987)
Willingness–Unwillingness (#VP5621).

♦ **VC0151 Application**
Integrative polarities Action–Inaction (#VP5705)   Dueness–Undueness (#VP5960)
Resolution–Irresolution (#VP5624).

♦ **VC0153 Appreciation**
Integrative polarities Assent–Dissent (#VP5521)
Goodness–Badness (#VP5674)
Increase–Decrease (#VP5038)
Respect–Disrespect (#VP5964)
Knowledge–Ignorance (#VP5475)
Approval–Disapproval (#VP5968)
Pleasure–Displeasure (#VP5865)
Gratitude–Ingratitude (#VP5949)
Judgement–Misjudgement (#VP5494)
Discrimination–Indiscrimination (#VP5492)
Intelligibility–Unintelligibility (#VP5548).

♦ **VC0155 Approachability**
Integrative polarities Friendship–Enmity (#VP5927)
Approach–Recession (#VP5296)
Convergence–Divergence (#VP5298)
Possibility–Impossibility (#VP5509)
Communicativeness–Uncommunicativeness (#VP5554).

♦ **VC0157 Approbation**
Integrative polarities Consent–Refusal (#VP5775)
Repute–Disrepute (#VP5914)
Respect–Disrespect (#VP5964)
Approval–Disapproval (#VP5968).

♦ **VC0159 Appropriateness**
Integrative polarities Taste–Vulgarity (#VP5896)   Dueness–Undueness (#VP5960)
Rightness–Wrongness (#VP5957)   Elegance–Inelegance (#VP5589)
Expedience–Inexpedience (#VP5670)   Timeliness–Untimeliness (#VP5129)
Appropriateness–Inappropriateness (#VP5665).

CONSTRUCTIVE VALUES

VC0251

♦ **VC0161 Approval**
 Integrative polarities Consent–Refusal (#VP5775)
 Repute–Disrepute (#VP5914)
 Choice–Necessity (#VP5637)
 Respect–Disrespect (#VP5964)
 Approval–Disapproval (#VP5968).

♦ **VC0163 Aptitude**
 Integrative polarities Education–Miseducation (#VP5562)
 Agreement–Disagreement (#VP5026)
 Skilfulness–Unskilfulness (#VP5733)
 Intelligence–Unintelligence (#VP5467).

♦ **VC0165 Ardour**
 Integrative polarities Love–Hate (#VP5930)   Action–Inaction (#VP5705)
 Desire–Avoidance (#VP5631)   Energy–Moderation (#VP5161)
 Willingness–Unwillingness (#VP5621).

♦ **VC0167 Aristocracy**
 Integrative polarities Pride–Humility (#VP5905)   Repute–Disrepute (#VP5914)
 Authority–Lawlessness (#VP5739).

♦ **VC0169 Art**
 Integrative polarities Taste–Vulgarity (#VP5896)   Skilfulness–Unskilfulness (#VP5733).

♦ **VC0171 Artfulness**
 Integrative polarities Skilfulness–Unskilfulness (#VP5733).

♦ **VC0173 Artlessness**
 Integrative polarities Truth–Error (#VP5516)   Probity–Improbity (#VP5974)
 Skilfulness–Unskilfulness (#VP5733).

♦ **VC0175 Ascendancy**
 Integrative polarities Victory–Defeat (#VP5726)   Breadth–Narrowness (#VP5204)
 Elevation–Depression (#VP5317)   Influence–Influencelessness (#VP5172).

♦ **VC0177 Asceticism**
 Integrative polarities Chastity–Unchastity (#VP5987)
 Atonement–Punishment (#VP6010)
 Sociability–Unsociability (#VP5922).

♦ **VC0179 Aspiration**
 Integrative polarities Hope–Hopelessness (#VP5888)
 Elevation–Depression (#VP5317)
 Conventionality–Unconventionality (#VP5642).

♦ **VC0181 Assembly**
 Integrative polarities Completeness–Incompleteness (#VP5054).

♦ **VC0183 Assiduity**
 Integrative polarities Action–Inaction (#VP5705)   Carefulness–Neglect (#VP5533)
 Attention–Inattention (#VP5530)   Resolution–Irresolution (#VP5624).

♦ **VC0185 Assistance**
 Integrative polarities Support–Opposition (#VP5785).

♦ **VC0187 Association**
 Integrative polarities Support–Opposition (#VP5785)
 Sharing–Appropriation (#VP5815)
 Thought–Thoughtlessness (#VP5478)
 Sociability–Unsociability (#VP5922)
 Concurrence–Counteraction (#VP5177).

♦ **VC0189 Assurance**
 Integrative polarities Courage–Fear (#VP5890)   Safety–Danger (#VP5697)
 Belief–Unbelief (#VP5501)   Hope–Hopelessness (#VP5888)
 Comfort–Aggravation (#VP5885)   Certainty–Uncertainty (#VP5513)
 Excitement–Inexcitability (#VP5857)
 Accomplishment-Nonaccomplishment (#VP5722).

♦ **VC0193 Attachment**
 Integrative polarities Love–Hate (#VP5930)   Probity–Improbity (#VP5974)
 Attributability–Chance (#VP5155)   Conjunction–Separation (#VP5047).

♦ **VC0195 Attainment**
 Integrative polarities Possession–Loss (#VP5808)
 Knowledge–Ignorance (#VP5475)
 Skilfulness–Unskilfulness (#VP5733)
 Accomplishment-Nonaccomplishment (#VP5722).

♦ **VC0197 Attention**
 Integrative polarities Respect–Disrespect (#VP5964)
 Kindness–Unkindness (#VP5938)
 Carefulness–Neglect (#VP5533)
 Courtesy–Discourtesy (#VP5936)
 Attention–Inattention (#VP5530).

♦ **VC0199 Attraction**
 Integrative polarities Beauty–Ugliness (#VP5899)
 Attraction–Repulsion (#VP5288)
 Motivation–Dissuasion (#VP5648)
 Attention–Inattention (#VP5530)
 Pleasantness–Unpleasantness (#VP5863).

♦ **VC0201 Audacity**
 Integrative polarities Courage–Fear (#VP5890).

♦ **VC0203 Authenticity**
 Integrative polarities Truth–Error (#VP5516)   Certainty–Uncertainty (#VP5513)
 Originality–Imitation (#VP5023)   Existence–Nonexistence (#VP5001)
 Provability–Unprovability (#VP5505).

♦ **VC0205 Authority**
 Integrative polarities Truth–Error (#VP5516)   Power–Impotence (#VP5157)
 Authority–Lawlessness (#VP5739)   Skilfulness–Unskilfulness (#VP5733)
 Intelligence–Unintelligence (#VP5467)   Influence–Influencelessness (#VP5172).

♦ **VC0207 Autonomy**
 Integrative polarities Freedom–Restraint (#VP5762)
 Willingness–Unwillingness (#VP5621).

♦ **VC0209 Availability**
 Integrative polarities Possession–Loss (#VP5808)
 Presence–Absence (#VP5186)
 Possibility–Impossibility (#VP5509)
 Appropriateness–Inappropriateness (#VP5665).

♦ **VC0211 Awareness**
 Integrative polarities Knowledge–Ignorance (#VP5475)
 Attention–Inattention (#VP5530)
 Sensation–Insensibility (#VP5422).

♦ **VC0213 Awe**
 Integrative polarities Courage–Fear (#VP5890)   Respect–Disrespect (#VP5964)
 Sanctity–Unsanctity (#VP6026)   Wonder–Unastonishment (#VP5920)
 Naturalness–Affectation (#VP5900)   Importance–Unimportance (#VP5672).

♦ **VC0215 Balance**
 Integrative polarities Sanity–Insanity (#VP5472)   Justice–Injustice (#VP5976)
 Elegance–Inelegance (#VP5589)   Symmetry–Distortion (#VP5248)
 Equality–Inequality (#VP5030)   Stability–Changeableness (#VP5141)
 Excitement–Inexcitability (#VP5857)   Intelligence–Unintelligence (#VP5467).

♦ **VC0217 Beatification**
 Integrative polarities Repute–Disrepute (#VP5914)
 Sanctity–Unsanctity (#VP6026)
 Pleasure–Displeasure (#VP5865).

♦ **VC0219 Beatitude**
 Integrative polarities Sanctity–Unsanctity (#VP6026)
 Pleasure–Displeasure (#VP5865).

♦ **VC0221 Beauty**
 Integrative polarities Beauty–Ugliness (#VP5899)
 Improvement–Impairment (#VP5691).

♦ **VC0223 Behaviour**
 Integrative polarities Action–Inaction (#VP5705)   Behaviour–Misbehaviour (#VP5737).

♦ **VC0225 Being**
 Integrative polarities Futurity–Antiquity (#VP5119)
 Existence–Nonexistence (#VP5001).

♦ **VC0227 Belief**
 Integrative polarities Piety–Impiety (#VP6028)   Belief–Unbelief (#VP5501)
 Certainty–Uncertainty (#VP5513).

♦ **VC0229 Belongingness**
 Integrative polarities Location–Dislocation (#VP5185).

♦ **VC0231 Benediction**
 Integrative polarities Gratitude–Ingratitude (#VP5949).

♦ **VC0233 Beneficence**
 Integrative polarities Support–Opposition (#VP5785)
 Kindness–Unkindness (#VP5938).

♦ **VC0235 Benevolence**
 Integrative polarities Goodness–Badness (#VP5674)
 Support–Opposition (#VP5785)
 Kindness–Unkindness (#VP5938)
 Forgiveness–Vengeance (#VP5947)
 Compassion–Pitilessness (#VP5944).

♦ **VC0237 Benignity**
 Integrative polarities Goodness–Badness (#VP5674)
 Support–Opposition (#VP5785)
 Kindness–Unkindness (#VP5938)
 Healthfulness–Unhealthfulness (#VP5683).

♦ **VC0239 Betterment**
 Integrative polarities Change–Permanence (#VP5139)
 Improvement–Impairment (#VP5691).

♦ **VC0241 Bigness**
 Integrative polarities Repute–Disrepute (#VP5914)
 Bigness–Littleness (#VP5195)
 Importance–Unimportance (#VP5672)
 Unselfishness–Selfishness (#VP5978).

♦ **VC0243 Blamelessness**
 Integrative polarities Innocence–Guilt (#VP5983)
 Probity–Improbity (#VP5974).

♦ **VC0247 Blessedness**
 Integrative polarities Safety–Danger (#VP5697)   Sanctity–Unsanctity (#VP6026)
 Approval–Disapproval (#VP5968)   Pleasure–Displeasure (#VP5865)
 Prosperity–Adversity (#VP5728)   Gratitude–Ingratitude (#VP5949).

♦ **VC0249 Bliss**
 Integrative polarities Pleasure–Displeasure (#VP5865)
 Pleasantness–Unpleasantness (#VP5863).

♦ **VC0251 Boldness**
 Integrative polarities Courage–Fear (#VP5890)   Support–Opposition (#VP5785)
 Chastity–Unchastity (#VP5987).

VC0253

♦ **VC0253 Boundlessness**
Integrative polarities Greatness–Smallness (#VP5034)
Numerousness–Fewness (#VP5101)
Godliness–Ungodliness (#VP6013).

♦ **VC0255 Bountifulness**
Integrative polarities Economy–Prodigality (#VP5851)
Oversufficiency–Insufficiency (#VP5660)
Productiveness–Unproductiveness (#VP5165).

♦ **VC0257 Bravery**
Integrative polarities Courage–Fear (#VP5890)   Patience–Impatience (#VP5861).

♦ **VC0259 Breathtaking**
Integrative polarities Wonder–Unastonishment (#VP5920)
Excitement–Inexcitability (#VP5857).

♦ **VC0261 Breeding**
Integrative polarities Courtesy–Discourtesy (#VP5936).

♦ **VC0263 Brevity**
Integrative polarities Conciseness–Diffuseness (#VP5592).

♦ **VC0265 Brightness**
Integrative polarities Light–Darkness (#VP5335)
Beauty–Ugliness (#VP5899)
Hope–Hopelessness (#VP5888)
Carefulness–Neglect (#VP5533)
Cleanness–Uncleanness (#VP5681)
Cheerfulness–Solemnity (#VP5870)
Education–Miseducation (#VP5562)
Pleasantness–Unpleasantness (#VP5863)
Intelligence–Unintelligence (#VP5467).

♦ **VC0267 Brilliance**
Integrative polarities Light–Darkness (#VP5335)
Beauty–Ugliness (#VP5899)
Repute–Disrepute (#VP5914)
Amusement–Boredom (#VP5878)
Naturalness–Affectation (#VP5900)
Skilfulness–Unskilfulness (#VP5733)
Intelligence–Unintelligence (#VP5467).

♦ **VC0269 Broadmindedness**
Integrative polarities Freedom–Restraint (#VP5762)
Intelligence–Unintelligence (#VP5467)
Broadmindedness–Narrowmindedness (#VP5526).

♦ **VC0271 Broadness**
Integrative polarities Broadmindedness–Narrowmindedness (#VP5526).

♦ **VC0273 Brotherhood**
Integrative polarities Love–Hate (#VP5930)   Friendship–Enmity (#VP5927)
Kindness–Unkindness (#VP5938).

♦ **VC0275 Brotherliness**
Integrative polarities Friendship–Enmity (#VP5927)
Kindness–Unkindness (#VP5938).

♦ **VC0277 Buoyancy**
Integrative polarities Weight–Lightness (#VP5352)
Elasticity–Toughness (#VP5358)
Cheerfulness–Solemnity (#VP5870).

♦ **VC0279 Business**
Integrative polarities Order–Disorder (#VP5059)   Action–Inaction (#VP5705).

♦ **VC0281 Calmness**
Integrative polarities Accord–Disaccord (#VP5794)
Motion–Quiescence (#VP5267)
Energy–Moderation (#VP5161)
Excitement–Inexcitability (#VP5857).

♦ **VC0283 Candor**
Integrative polarities Justice–Injustice (#VP5976)
Probity–Improbity (#VP5974)
Skilfulness–Unskilfulness (#VP5733)
Communicativeness–Uncommunicativeness (#VP5554).

♦ **VC0285 Canniness**
Integrative polarities Caution–Rashness (#VP5894)
Economy–Prodigality (#VP5851).

♦ **VC0287 Capability**
Integrative polarities Power–Impotence (#VP5157)
Skilfulness–Unskilfulness (#VP5733)
Preparedness–Unpreparedness (#VP5720).

♦ **VC0289 Capacity**
Integrative polarities Power–Impotence (#VP5157)
Skilfulness–Unskilfulness (#VP5733)
Intelligence–Unintelligence (#VP5467).

♦ **VC0291 Care**
Integrative polarities Safety–Danger (#VP5697)   Caution–Rashness (#VP5894)
Support–Opposition (#VP5785)   Economy–Prodigality (#VP5851)
Carefulness–Neglect (#VP5533)   Attention–Inattention (#VP5530).

♦ **VC0293 Carefulness**
Integrative polarities Safety–Danger (#VP5697)   Caution–Rashness (#VP5894)
Probity–Improbity (#VP5974)   Support–Opposition (#VP5785)
Economy–Prodigality (#VP5851)   Carefulness–Neglect (#VP5533)
Attention–Inattention (#VP5530)   Intelligence–Unintelligence (#VP5467).

♦ **VC0295 Caution**
Integrative polarities Caution–Rashness (#VP5894)
Expectation–Inexpectation (#VP5539).

♦ **VC0297 Cautiousness**
Integrative polarities Caution–Rashness (#VP5894)
Carefulness–Neglect (#VP5533).

♦ **VC0299 Celebration**
Integrative polarities Approval–Disapproval (#VP5968)
Exultation–Lamentation (#VP5873).

♦ **VC0301 Certainty**
Integrative polarities Truth–Error (#VP5516)   Belief–Unbelief (#VP5501)
Certainty–Uncertainty (#VP5513)
Accomplishment–Nonaccomplishment (#VP5722).

♦ **VC0303 Challenge**
Integrative polarities Support–Opposition (#VP5785)
Motivation–Dissuasion (#VP5648).

♦ **VC0305 Championship**
Integrative polarities Safety–Danger (#VP5697)   Attack–Defence (#VP5798)
Victory–Defeat (#VP5726)   Superiority–Inferiority (#VP5036)
Vindication–Condemnation (#VP6005).

♦ **VC0307 Chance**
Integrative polarities Timeliness–Untimeliness (#VP5129)
Possibility–Impossibility (#VP5509).

♦ **VC0309 Change**
Integrative polarities Change–Permanence (#VP5139).

♦ **VC0311 Character**
Integrative polarities Repute–Disrepute (#VP5914)
Probity–Improbity (#VP5974)
Approval–Disapproval (#VP5968).

♦ **VC0313 Charisma**
Integrative polarities Power–Impotence (#VP5157)
Repute–Disrepute (#VP5914)
Motivation–Dissuasion (#VP5648)
Influence–Influencelessness (#VP5172).

♦ **VC0315 Charity**
Integrative polarities Love–Hate (#VP5930)   Virtue–Vice (#VP5980)
Accord–Disaccord (#VP5794)   Kindness–Unkindness (#VP5938)
Sharing–Appropriation (#VP5815)   Compassion–Pitilessness (#VP5944)
Broadmindedness–Narrowmindedness (#VP5526).

♦ **VC0317 Charm**
Integrative polarities Love–Hate (#VP5930)   Beauty–Ugliness (#VP5899)
Pleasure–Displeasure (#VP5865)   Motivation–Dissuasion (#VP5648)
Pleasantness–Unpleasantness (#VP5863)   Influence–Influencelessness (#VP5172).

♦ **VC0319 Chastity**
Integrative polarities Virtue–Vice (#VP5980)   Taste–Vulgarity (#VP5896)
Chastity–Unchastity (#VP5987)   Elegance–Inelegance (#VP5589)
Simplicity–Complexity (#VP5045)   Temperance–Intemperance (#VP5992)
Naturalness–Affectation (#VP5900)   Perfection–Imperfection (#VP5677).

♦ **VC0321 Cheerfulness**
Integrative polarities Hope–Hopelessness (#VP5888)
Pleasure–Displeasure (#VP5865)
Cheerfulness–Solemnity (#VP5870)
Sociability–Unsociability (#VP5922)
Pleasantness–Unpleasantness (#VP5863).

♦ **VC0323 Chivalry**
Integrative polarities Courage–Fear (#VP5890)   Repute–Disrepute (#VP5914)
Courtesy–Discourtesy (#VP5936)   Unselfishness–Selfishness (#VP5978).

♦ **VC0325 Choice**
Integrative polarities Taste–Vulgarity (#VP5896)   Goodness–Badness (#VP5674)
Freedom–Restraint (#VP5762)   Judgement–Misjudgement (#VP5494)
Willingness–Unwillingness (#VP5621).

♦ **VC0327 Circumspection**
Integrative polarities Caution–Rashness (#VP5894)
Carefulness–Neglect (#VP5533)
Intelligence–Unintelligence (#VP5467).

♦ **VC0329 Civil rights**
Integrative polarities Freedom–Restraint (#VP5762).

♦ **VC0331 Civility**
Integrative polarities Taste–Vulgarity (#VP5896)   Courtesy–Discourtesy (#VP5936)
Formality–Informality (#VP5646)   Improvement–Impairment (#VP5691)
Sociability–Unsociability (#VP5922).

♦ **VC0333 Civilization**
Integrative polarities Improvement–Impairment (#VP5691)
Conventionality–Unconventionality (#VP5642).

♦ **VC0335 Clarity**
Integrative polarities Facility–Difficulty (#VP5732)
Visibility–Invisibility (#VP5444).

♦ **VC0337 Cleanliness**
Integrative polarities Innocence–Guilt (#VP5983)
Probity–Improbity (#VP5974)
Chastity–Unchastity (#VP5987)
Cleanness–Uncleanness (#VP5681).

CONSTRUCTIVE VALUES VC0413

♦ **VC0339 Clearness**
Integrative polarities Light–Darkness (#VP5335)
Innocence–Guilt (#VP5983)
Opening–Closure (#VP5265)
Freedom–Restraint (#VP5762)
Facility–Difficulty (#VP5732)
Prosperity–Adversity (#VP5728)
Certainty–Uncertainty (#VP5513)
Visibility–Invisibility (#VP5444)
Completeness–Incompleteness (#VP5054)
Intelligibility–Unintelligibility (#VP5548).

♦ **VC0341 Clemency**
Integrative polarities Leniency–Compulsion (#VP5756)
Compassion–Pitilessness (#VP5944).

♦ **VC0343 Cleverness**
Integrative polarities Amusement–Boredom (#VP5878)
Education–Miseducation (#VP5562)
Skilfulness–Unskilfulness (#VP5733)
Intelligence–Unintelligence (#VP5467).

♦ **VC0345 Clout**
Integrative polarities Impact–Reaction (#VP5283)
Power–Impotence (#VP5157)
Authority–Lawlessness (#VP5739)
Causation–Culmination (#VP5153)
Influence–Influencelessness (#VP5172).

♦ **VC0347 Coalition**
Integrative polarities Support–Opposition (#VP5785).

♦ **VC0349 Cogency**
Integrative polarities Truth–Error (#VP5516)   Power–Impotence (#VP5157)
Goodness–Badness (#VP5674)   Intuition–Reason (#VP5481)
Intelligence–Unintelligence (#VP5467).

♦ **VC0351 Coherence**
Integrative polarities Weight–Lightness (#VP5352)
Agreement–Disagreement (#VP5026)
Cohesion–Disintegration (#VP5050)
Intelligibility–Unintelligibility (#VP5548).

♦ **VC0353 Collaboration**
Integrative polarities Support–Opposition (#VP5785)
Concurrence–Counteraction (#VP5177).

♦ **VC0355 Comfort**
Integrative polarities Support–Opposition (#VP5785)
Comfort–Aggravation (#VP5885)
Pleasure–Displeasure (#VP5865)
Prosperity–Adversity (#VP5728)
Compassion–Pitilessness (#VP5944)
Contentment–Discontentment (#VP5868).

♦ **VC0357 Comfortableness**
Integrative polarities Comfort–Aggravation (#VP5885)
Pleasure–Displeasure (#VP5865)
Prosperity–Adversity (#VP5728)
Contentment–Discontentment (#VP5868).

♦ **VC0359 Commendability**
Integrative polarities Goodness–Badness (#VP5674)
Approval–Disapproval (#VP5968)
Sharing–Appropriation (#VP5815).

♦ **VC0361 Commitment**
Integrative polarities Desire–Avoidance (#VP5631)
Friendship–Enmity (#VP5927)
Neutrality–Compromise (#VP5806)
Resolution–Irresolution (#VP5624)
Unselfishness–Selfishness (#VP5978).

♦ **VC0363 Common sense**
Integrative polarities Intuition–Reason (#VP5481)
Intelligence–Unintelligence (#VP5467).

♦ **VC0365 Commonweal**
Integrative polarities Kindness–Unkindness (#VP5938).

♦ **VC0367 Communication**
Integrative polarities Sharing–Appropriation (#VP5815)
Communicativeness–Uncommunicativeness (#VP5554).

♦ **VC0369 Communicativeness**
Integrative polarities Sociability–Unsociability (#VP5922).

♦ **VC0371 Communion**
Integrative polarities Accord–Disaccord (#VP5794)
Sharing–Appropriation (#VP5815)
Communicativeness–Uncommunicativeness (#VP5554).

♦ **VC0373 Community**
Integrative polarities Accord–Disaccord (#VP5794)
Support–Opposition (#VP5785)
Sharing–Appropriation (#VP5815)
Similarity–Dissimilarity (#VP5020)
Sociability–Unsociability (#VP5922).

♦ **VC0375 Companionship**
Integrative polarities Friendship–Enmity (#VP5927)
Sociability–Unsociability (#VP5922).

♦ **VC0377 Compassion**
Integrative polarities Kindness–Unkindness (#VP5938)
Leniency–Compulsion (#VP5756)
Compassion–Pitilessness (#VP5944).

♦ **VC0379 Compatibility**
Integrative polarities Accord–Disaccord (#VP5794)
Agreement–Disagreement (#VP5026)
Sociability–Unsociability (#VP5922)
Pleasantness–Unpleasantness (#VP5863).

♦ **VC0381 Compensation**
Integrative polarities Sharing–Appropriation (#VP5815).

♦ **VC0383 Competence**
Integrative polarities Power–Impotence (#VP5157)
Authority–Lawlessness (#VP5739)
Skilfulness–Unskilfulness (#VP5733)
Preparedness–Unpreparedness (#VP5720)
Oversufficiency–Insufficiency (#VP5660).

♦ **VC0385 Competition**
Integrative polarities Accord–Disaccord (#VP5794)
Support–Opposition (#VP5785).

♦ **VC0387 Completion**
Integrative polarities Causation–Culmination (#VP5153)
Completeness–Incompleteness (#VP5054)
Accomplishment–Nonaccomplishment (#VP5722).

♦ **VC0389 Compliance**
Integrative polarities Assent–Dissent (#VP5521)
Consent–Refusal (#VP5775)
Freedom–Restraint (#VP5762)
Hardness–Softness (#VP5356)
Patience–Impatience (#VP5861)
Obedience–Disobedience (#VP5766)
Willingness–Unwillingness (#VP5621).

♦ **VC0391 Composure**
Integrative polarities Motion–Quiescence (#VP5267)
Resolution–Irresolution (#VP5624)
Excitement–Inexcitability (#VP5857)
Contentment–Discontentment (#VP5868).

♦ **VC0393 Comprehension**
Integrative polarities Knowledge–Ignorance (#VP5475)
Inclusion–Exclusion (#VP5076)
Intelligence–Unintelligence (#VP5467).

♦ **VC0395 Comprehensiveness**
Integrative polarities Inclusion–Exclusion (#VP5076)
Greatness–Smallness (#VP5034)
Completeness–Incompleteness (#VP5054).

♦ **VC0397 Compromise**
Integrative polarities Accord–Disaccord (#VP5794)
Neutrality–Compromise (#VP5806).

♦ **VC0399 Compunction**
Integrative polarities Willingness–Unwillingness (#VP5621).

♦ **VC0401 Comradeship**
Integrative polarities Friendship–Enmity (#VP5927)
Support–Opposition (#VP5785)
Sociability–Unsociability (#VP5922).

♦ **VC0403 Concentration**
Integrative polarities Action–Inaction (#VP5705)   Attention–Inattention (#VP5530)
Convergence–Divergence (#VP5298)   Centrality–Environment (#VP5226)
Resolution–Irresolution (#VP5624)   Thought–Thoughtlessness (#VP5478).

♦ **VC0405 Conception**
Integrative polarities Belief–Unbelief (#VP5501)   Knowledge–Ignorance (#VP5475)
Thought–Thoughtlessness (#VP5478)   Production–Reproduction (#VP5167)
Intelligence–Unintelligence (#VP5467)   Imaginativeness–Unimaginativeness (#VP5535).

♦ **VC0407 Concern**
Integrative polarities Kindness–Unkindness (#VP5938)
Carefulness–Neglect (#VP5533)
Attention–Inattention (#VP5530)
Importance–Unimportance (#VP5672)
Feeling–Unfeelinglessness (#VP5855)
Relatedness–Unrelatedness (#VP5009).

♦ **VC0409 Conciliation**
Integrative polarities Accord–Disaccord (#VP5794)
Forgiveness–Vengeance (#VP5947).

♦ **VC0411 Concord**
Integrative polarities Assent–Dissent (#VP5521)
Order–Disorder (#VP5059)
Accord–Disaccord (#VP5794)
Support–Opposition (#VP5785)
Agreement–Disagreement (#VP5026)
Concurrence–Counteraction (#VP5177).

♦ **VC0413 Concurrence**
Integrative polarities Assent–Dissent (#VP5521)
Support–Opposition (#VP5785)
Identity–Difference (#VP5013)
Convergence–Divergence (#VP5298)
Concurrence–Counteraction (#VP5177).

VC0415

♦ **VC0415 Condolence**
Integrative polarities Comfort–Aggravation (#VP5885)
Compassion–Pitilessness (#VP5944).

♦ **VC0417 Conduct**
Integrative polarities Behaviour–Misbehaviour (#VP5737).

♦ **VC0419 Confidence**
Integrative polarities Courage–Fear (#VP5890)   Belief–Unbelief (#VP5501)
Certainty–Uncertainty (#VP5513)   Excitement–Inexcitability (#VP5857).

♦ **VC0421 Confirmation**
Integrative polarities Truth–Error (#VP5516)   Certainty–Uncertainty (#VP5513)
Provability–Unprovability (#VP5505).

♦ **VC0423 Conformation**
Integrative polarities Assent–Dissent (#VP5521)
Agreement–Disagreement (#VP5026)
Conformity–Nonconformity (#VP5082).

♦ **VC0425 Conformity**
Integrative polarities Piety–Impiety (#VP6028)   Symmetry–Distortion (#VP5248)
Obedience–Disobedience (#VP5766)   Agreement–Disagreement (#VP5026)
Conformity–Nonconformity (#VP5082).

♦ **VC0427 Congeniality**
Integrative polarities Accord–Discaccord (#VP5794)
Friendship–Enmity (#VP5927)
Agreement–Disagreement (#VP5026)
Sociability–Unsociability (#VP5922)
Pleasantness–Unpleasantness (#VP5863).

♦ **VC0429 Congruity**
Integrative polarities Symmetry–Distortion (#VP5248)
Agreement–Disagreement (#VP5026)
Conformity–Nonconformity (#VP5082).

♦ **VC0431 Connectedness**
Integrative polarities Cohesion–Disintegration (#VP5050)
Continuity–Discontinuity (#VP5071)
Accomplishment–Nonaccomplishment (#VP5722)
Intelligibility–Unintelligibility (#VP5548).

♦ **VC0433 Connection**
Integrative polarities Attributability–Chance (#VP5155)
Cohesion–Disintegration (#VP5050)
Relatedness–Unrelatedness (#VP5009)
Communicativeness–Uncommunicativeness (#VP5554).

♦ **VC0435 Conquest**
Integrative polarities Victory–Defeat (#VP5726).

♦ **VC0437 Consanguinity**
Integrative polarities Ancestry–Posterity (#VP5170).

♦ **VC0439 Conscience**
Integrative polarities Rightness–Wrongness (#VP5957)
Selfactualization–Neurosis (#VP5690).

♦ **VC0441 Conscientiousness**
Integrative polarities Taste–Vulgarity (#VP5896)   Probity–Improbity (#VP5974)
Dueness–Undueness (#VP5960)   Carefulness–Neglect (#VP5533).

♦ **VC0443 Consciousness**
Integrative polarities Life–Death (#VP5407)   Knowledge–Ignorance (#VP5475)
Attention–Inattention (#VP5530)   Sensation–Insensibility (#VP5422)
Intelligence–Unintelligence (#VP5467).

♦ **VC0445 Conservation**
Integrative polarities Safety–Danger (#VP5697)   Change–Permanence (#VP5139).

♦ **VC0447 Conservativeness**
Integrative polarities Energy–Moderation (#VP5161).

♦ **VC0449 Consideration**
Integrative polarities Belief–Unbelief (#VP5501)   Repute–Disrepute (#VP5914)
Respect–Disrespect (#VP5964)   Kindness–Unkindness (#VP5938)
Carefulness–Neglect (#VP5533)   Courtesy–Discourtesy (#VP5936)
Attention–Inattention (#VP5530)   Judgement–Misjudgement (#VP5494)
Importance–Unimportance (#VP5672)   Thought–Thoughtlessness (#VP5478)
Intelligence–Unintelligence (#VP5467).

♦ **VC0451 Consistency**
Integrative polarities Truth–Error (#VP5516)   Probity–Improbity (#VP5974)
Symmetry–Distortion (#VP5248)   Agreement–Disagreement (#VP5026)
Cohesion–Disintegration (#VP5050)   Conformity–Nonconformity (#VP5082)
Intelligibility–Unintelligibility (#VP5548).

♦ **VC0453 Consonance**
Integrative polarities Agreement–Disagreement (#VP5026).

♦ **VC0455 Constancy**
Integrative polarities Probity–Improbity (#VP5974)
Friendship–Enmity (#VP5927)
Change–Permanence (#VP5139)
Durability–Transience (#VP5110)
Resolution–Irresolution (#VP5624)
Stability–Changeableness (#VP5141)
Continuity–Discontinuity (#VP5071)
Perpetuity–Instantaneousness (#VP5112).

♦ **VC0457 Constructiveness**
Integrative polarities Support–Opposition (#VP5785)
Production–Reproduction (#VP5167).

♦ **VC0459 Consummation**
Integrative polarities Causation–Culmination (#VP5153)
Perfection–Imperfection (#VP5677)
Completeness–Incompleteness (#VP5054)
Accomplishment–Nonaccomplishment (#VP5722).

♦ **VC0461 Consumption**
Integrative polarities Possession–Loss (#VP5808)
Increase–Decrease (#VP5038)
Improvement–Impairment (#VP5691)
Restoration–Destruction (#VP5693)
Appropriateness–Inappropriateness (#VP5665).

♦ **VC0463 Contact**
Integrative polarities Nearness–Distance (#VP5200)
Communicativeness–Uncommunicativeness (#VP5554).

♦ **VC0465 Content**
Integrative polarities Container–Content (#VP5193).

♦ **VC0467 Contentment**
Integrative polarities Assent–Dissent (#VP5521)
Consent–Refusal (#VP5775)
Pleasure–Displeasure (#VP5865)
Willingness–Unwillingness (#VP5621)
Contentment–Discontentment (#VP5868).

♦ **VC0469 Continence**
Integrative polarities Energy–Moderation (#VP5161)
Chastity–Unchastity (#VP5987)
Temperance–Intemperance (#VP5992).

♦ **VC0471 Continuity**
Integrative polarities Continuance–Cessation (#VP5143)
Frequency–Infrequency (#VP5135)
Durability–Transience (#VP5110)
Resolution–Irresolution (#VP5624)
Cohesion–Disintegration (#VP5050).

♦ **VC0473 Convenience**
Integrative polarities Support–Opposition (#VP5785)
Comfort–Aggravation (#VP5885)
Facility–Difficulty (#VP5732)
Appropriateness–Inappropriateness (#VP5665).

♦ **VC0475 Conversion**
Integrative polarities Change–Permanence (#VP5139)
Conversion–Reversion (#VP5145).

♦ **VC0477 Convertibility**
Integrative polarities Equality–Inequality (#VP5030)
Conversion–Reversion (#VP5145).

♦ **VC0479 Conviction**
Integrative polarities Belief–Unbelief (#VP5501)   Hope–Hopelessness (#VP5888)
Certainty–Uncertainty (#VP5513).

♦ **VC0481 Cooperation**
Integrative polarities Support–Opposition (#VP5785)
Identity–Difference (#VP5013)
Sharing–Appropriation (#VP5815)
Agreement–Disagreement (#VP5026)
Sociability–Unsociability (#VP5922)
Concurrence–Counteraction (#VP5177).

♦ **VC0483 Cooperativeness**
Integrative polarities Support–Opposition (#VP5785)
Identity–Difference (#VP5013)
Sharing–Appropriation (#VP5815)
Agreement–Disagreement (#VP5026)
Sociability–Unsociability (#VP5922)
Willingness–Unwillingness (#VP5621)
Concurrence–Counteraction (#VP5177).

♦ **VC0485 Coordination**
Integrative polarities Symmetry–Distortion (#VP5248)
Equality–Inequality (#VP5030)
Agreement–Disagreement (#VP5026)
Skilfulness–Unskilfulness (#VP5733).

♦ **VC0487 Cordiality**
Integrative polarities Friendship–Enmity (#VP5927)
Kindness–Unkindness (#VP5938)
Refreshment–Relapse (#VP5695)
Formality–Informality (#VP5646)
Hospitality–Inhospitality (#VP5925)
Feeling–Unfeelinglessness (#VP5855)
Pleasantness–Unpleasantness (#VP5863).

♦ **VC0489 Correctness**
Integrative polarities Truth–Error (#VP5516)   Elegance–Inelegance (#VP5589)
Carefulness–Neglect (#VP5533)   Courtesy–Discourtesy (#VP5936)
Restoration–Destruction (#VP5693)   Conformity–Nonconformity (#VP5082).

♦ **VC0491 Correspondence**
Integrative polarities Accord–Discaccord (#VP5794)
Symmetry–Distortion (#VP5248)
Equality–Inequality (#VP5030)
Agreement–Disagreement (#VP5026)
Similarity–Dissimilarity (#VP5020)
Concurrence–Counteraction (#VP5177)
Communicativeness–Uncommunicativeness (#VP5554).

♦ **VC0493 Cosmopolitan**
Integrative polarities Humanity–Nonhumanity (#VP5417)

CONSTRUCTIVE VALUES VC0571

Skilfulness–Unskilfulness (#VP5733)
Broadmindedness–Narrowmindedness (#VP5526)
Conventionality–Unconventionality (#VP5642).

♦ **VC0495 Courage**
Integrative polarities Courage–Fear (#VP5890)　Certainty–Uncertainty (#VP5513)
Resolution–Irresolution (#VP5624).

♦ **VC0497 Courtesy**
Integrative polarities Respect–Disrespect (#VP5964)
Kindness–Unkindness (#VP5938)
Courtesy–Discourtesy (#VP5936)
Behaviour–Misbehaviour (#VP5737)
Sociability–Unsociability (#VP5922).

♦ **VC0499 Creation**
Integrative polarities Form–Formlessness (#VP5246)
Godliness–Ungodliness (#VP6013)
Originality–Imitation (#VP5023)
Production–Reproduction (#VP5167)
Productiveness–Unproductiveness (#VP5165)
Imaginativeness–Unimaginativeness (#VP5535).

♦ **VC0501 Creativity**
Integrative polarities Originality–Imitation (#VP5023)
Intelligence–Unintelligence (#VP5467)
Imaginativeness–Unimaginativeness (#VP5535).

♦ **VC0503 Credibility**
Integrative polarities Belief–Unbelief (#VP5501)　Intuition–Reason (#VP5481)
Probity–Improbity (#VP5974).

♦ **VC0505 Culmination**
Integrative polarities Height–Lowness (#VP5207)
Causation–Culmination (#VP5153)
Perfection–Imperfection (#VP5677)
Completeness–Incompleteness (#VP5054)
Accomplishment–Nonaccomplishment (#VP5722).

♦ **VC0507 Culture**
Integrative polarities Taste–Vulgarity (#VP5896)　Knowledge–Ignorance (#VP5475)
Courtesy–Discourtesy (#VP5936)　Improvement–Impairment (#VP5691).

♦ **VC0509 Curiosity**
Integrative polarities Wonder–Unastonishment (#VP5920)
Attention–Inattention (#VP5530)
Curiosity–Incuriosity (#VP5528).

♦ **VC0511 Custom**
Integrative polarities Newness–Oldness (#VP5122)
Behaviour–Misbehaviour (#VP5737).

♦ **VC0513 Daring**
Integrative polarities Courage–Fear (#VP5890)　Action–Inaction (#VP5705)
Support–Opposition (#VP5785).

♦ **VC0515 Dauntlessness**
Integrative polarities Courage–Fear (#VP5890)　Resolution–Irresolution (#VP5624).

♦ **VC0517 Decency**
Integrative polarities Taste–Vulgarity (#VP5896)　Probity–Improbity (#VP5974)
Chastity–Unchastity (#VP5987)　Rightness–Wrongness (#VP5957)
Kindness–Unkindness (#VP5938)　Formality–Informality (#VP5646).

♦ **VC0519 Decisiveness**
Integrative polarities Certainty–Uncertainty (#VP5513)
Judgement–Misjudgement (#VP5494)
Resolution–Irresolution (#VP5624)
Timeliness–Untimeliness (#VP5129)
Willingness–Unwillingness (#VP5621).

♦ **VC0521 Decoration**
Integrative polarities Beauty–Ugliness (#VP5899)
Repute–Disrepute (#VP5914)
Naturalness–Affectation (#VP5900).

♦ **VC0523 Decorum**
Integrative polarities Taste–Vulgarity (#VP5896)　Chastity–Unchastity (#VP5987)
Rightness–Wrongness (#VP5957)　Formality–Informality (#VP5646)
Cheerfulness–Solemnity (#VP5870).

♦ **VC0525 Dedication**
Integrative polarities Desire–Avoidance (#VP5631)
Dueness–Undueness (#VP5960)
Friendship–Enmity (#VP5927)
Sharing–Appropriation (#VP5815)
Resolution–Irresolution (#VP5624)
Unselfishness–Selfishness (#VP5978).

♦ **VC0527 Deference**
Integrative polarities Dueness–Undueness (#VP5960)
Freedom–Restraint (#VP5762)
Courtesy–Discourtesy (#VP5936).

♦ **VC0529 Deification**
Integrative polarities Repute–Disrepute (#VP5914)
Respect–Disrespect (#VP5964)
Approval–Disapproval (#VP5968)
Elevation–Depression (#VP5317).

♦ **VC0531 Delectableness**
Integrative polarities Pleasantness–Unpleasantness (#VP5863).

♦ **VC0533 Deliberateness**
Integrative polarities Exertion–Rest (#VP5709)　Caution–Rashness (#VP5894)
Thought–Thoughtlessness (#VP5478).

♦ **VC0535 Delicacy**
Integrative polarities Truth–Error (#VP5516)　Beauty–Ugliness (#VP5899)
Taste–Vulgarity (#VP5896)　Chastity–Unchastity (#VP5987)
Kindness–Unkindness (#VP5938)　Carefulness–Neglect (#VP5533)
Sensation–Insensibility (#VP5422).

♦ **VC0537 Delight**
Integrative polarities Pleasure–Displeasure (#VP5865)
Exultation–Lamentation (#VP5873)
Pleasantness–Unpleasantness (#VP5863).

♦ **VC0539 Deliverance**
Integrative polarities Safety–Danger (#VP5697)　Freedom–Restraint (#VP5762)
Sharing–Appropriation (#VP5815)　Judgement–Misjudgement (#VP5494).

♦ **VC0541 Democracy**
Integrative polarities Sharing–Appropriation (#VP5815).

♦ **VC0543 Desirableness**
Integrative polarities Love–Hate (#VP5930)　Choice–Necessity (#VP5637)
Desire–Avoidance (#VP5631)　Hope–Hopelessness (#VP5888)
Expedience–Inexpedience (#VP5670)　Hospitality–Inhospitality (#VP5925)
Willingness–Unwillingness (#VP5621)　Pleasantness–Unpleasantness (#VP5863).

♦ **VC0545 Detachment**
Integrative polarities Unity–Duality (#VP5089)　Desire–Avoidance (#VP5631)
Justice–Injustice (#VP5976)　Curiosity–Incuriosity (#VP5528)
Conjunction–Separation (#VP5047)　Sociability–Unsociability (#VP5922)
Feeling–Unfeelinglessness (#VP5855)　Broadmindedness–Narrowmindedness (#VP5526)
Communicativeness–Uncommunicativeness (#VP5554).

♦ **VC0547 Determination**
Integrative polarities Action–Inaction (#VP5705)　Research–Discovery (#VP5485)
Certainty–Uncertainty (#VP5513)　Judgement–Misjudgement (#VP5494)
Resolution–Irresolution (#VP5624)　Willingness–Unwillingness (#VP5621)
Provability–Unprovability (#VP5505)　Circumscription–Intrusion (#VP5234).

♦ **VC0549 Development**
Integrative polarities Youth–Age (#VP5124)　Increase–Decrease (#VP5038)
Evolution–Revolution (#VP5147)　Expansion–Contraction (#VP5197)
Causation–Culmination (#VP5153)　Improvement–Impairment (#VP5691)
Production–Reproduction (#VP5167).

♦ **VC0551 Devotion**
Integrative polarities Love–Hate (#VP5930)　Piety–Impiety (#VP6028)
Desire–Avoidance (#VP5631)　Probity–Improbity (#VP5974)
Dueness–Undueness (#VP5960)　Friendship–Enmity (#VP5927)
Sharing–Appropriation (#VP5815)　Resolution–Irresolution (#VP5624)
Unselfishness–Selfishness (#VP5978).

♦ **VC0553 Dexterity**
Integrative polarities Skilfulness–Unskilfulness (#VP5733)
Intelligence–Unintelligence (#VP5467).

♦ **VC0555 Differentiation**
Integrative polarities Identity–Difference (#VP5013)
Uniformity–Nonuniformity (#VP5017).

♦ **VC0557 Dignity**
Integrative polarities Pride–Humility (#VP5905)　Repute–Disrepute (#VP5914)
Eloquence–Uneloquence (#VP5600)　Importance–Unimportance (#VP5672).

♦ **VC0559 Diligence**
Integrative polarities Action–Inaction (#VP5705)　Carefulness–Neglect (#VP5533)
Attention–Inattention (#VP5530)　Education–Miseducation (#VP5562)
Resolution–Irresolution (#VP5624).

♦ **VC0561 Direction**
Integrative polarities Direction–Deviation (#VP5290)
Motivation–Dissuasion (#VP5648).

♦ **VC0563 Directness**
Integrative polarities Truth–Error (#VP5516)　Probity–Improbity (#VP5974)
Direction–Deviation (#VP5290)　Naturalness–Affectation (#VP5900)
Timeliness–Untimeliness (#VP5129)　Skilfulness–Unskilfulness (#VP5733)
Intelligibility–Unintelligibility (#VP5548).

♦ **VC0565 Discernment**
Integrative polarities Knowledge–Ignorance (#VP5475)
Intelligence–Unintelligence (#VP5467)
Discrimination–Indiscrimination (#VP5492).

♦ **VC0567 Discipline**
Integrative polarities Order–Disorder (#VP5059)　Patience–Impatience (#VP5861)
Leniency–Compulsion (#VP5756)　Education–Miseducation (#VP5562)
Resolution–Irresolution (#VP5624).

♦ **VC0569 Discovery**
Integrative polarities Change–Permanence (#VP5139)
Research–Discovery (#VP5485).

♦ **VC0571 Discretion**
Integrative polarities Caution–Rashness (#VP5894)
Freedom–Restraint (#VP5762)
Willingness–Unwillingness (#VP5621)
Expectation–Inexpectation (#VP5539)
Intelligence–Unintelligence (#VP5467).

VC0573

♦ **VC0573 Discrimination**
 Integrative polarities Taste–Vulgarity (#VP5896)   Choice–Necessity (#VP5637)
 Identity–Difference (#VP5013)   Intelligence–Unintelligence (#VP5467)
 Discrimination–Indiscrimination (#VP5492).

♦ **VC0575 Dispassion**
 Integrative polarities Justice–Injustice (#VP5976)
 Frequency–Infrequency (#VP5135)
 Excitement–Inexcitability (#VP5857)
 Broadmindedness–Narrowmindedness (#VP5526).

♦ **VC0577 Distinction**
 Integrative polarities Repute–Disrepute (#VP5914)
 Greatness–Smallness (#VP5034)
 Identity–Difference (#VP5013)
 Importance–Unimportance (#VP5672)
 Discrimination–Indiscrimination (#VP5492)
 Intelligibility–Unintelligibility (#VP5548).

♦ **VC0579 Diversity**
 Integrative polarities Change–Permanence (#VP5139)
 Similarity–Dissimilarity (#VP5020)
 Uniformity–Nonuniformity (#VP5017).

♦ **VC0581 Docility**
 Integrative polarities Freedom–Restraint (#VP5762)
 Education–Miseducation (#VP5562)
 Willingness–Unwillingness (#VP5621).

♦ **VC0587 Dominance**
 Integrative polarities Victory–Defeat (#VP5726)   Authority–Lawlessness (#VP5739)
 Influence–Influencelessness (#VP5172).

♦ **VC0589 Durability**
 Integrative polarities Change–Permanence (#VP5139)
 Durability–Transience (#VP5110)
 Substantiality–Unsubstantiality (#VP5003).

♦ **VC0591 Duration**
 Integrative polarities Change–Permanence (#VP5139)
 Durability–Transience (#VP5110).

♦ **VC0593 Duty**
 Integrative polarities Piety–Impiety (#VP6028)   Dueness–Undueness (#VP5960)
 Respect–Disrespect (#VP5964)   Obedience–Disobedience (#VP5766).

♦ **VC0595 Dynamism**
 Integrative polarities Action–Inaction (#VP5705)   Energy–Moderation (#VP5161).

♦ **VC0597 Eagerness**
 Integrative polarities Consent–Refusal (#VP5775)
 Desire–Avoidance (#VP5631)
 Willingness–Unwillingness (#VP5621).

♦ **VC0599 Earnestness**
 Integrative polarities Desire–Avoidance (#VP5631)
 Attention–Inattention (#VP5530)
 Resolution–Irresolution (#VP5624).

♦ **VC0601 Ease**
 Integrative polarities Comfort–Aggravation (#VP5885)
 Facility–Difficulty (#VP5732)
 Pleasure–Displeasure (#VP5865)
 Prosperity–Adversity (#VP5728)
 Eloquence–Uneloquence (#VP5600)
 Contentment–Discontentment (#VP5868).

♦ **VC0603 Eclecticism**
 Integrative polarities Choice–Necessity (#VP5637)
 Increase–Decrease (#VP5038).

♦ **VC0605 Economy**
 Integrative polarities Economy–Prodigality (#VP5851).

♦ **VC0607 Ecstasy**
 Integrative polarities Love–Hate (#VP5930)   Pleasure–Displeasure (#VP5865)
 Excitement–Inexcitability (#VP5857)   Feeling–Unfeelinglessness (#VP5855).

♦ **VC0609 Edification**
 Integrative polarities Knowledge–Ignorance (#VP5475)
 Education–Miseducation (#VP5562).

♦ **VC0611 Education**
 Integrative polarities Knowledge–Ignorance (#VP5475)
 Improvement–Impairment (#VP5691)
 Education–Miseducation (#VP5562).

♦ **VC0613 Effectiveness**
 Integrative polarities Power–Impotence (#VP5157)
 Eloquence–Uneloquence (#VP5600)
 Expedience–Inexpedience (#VP5670)
 Influence–Influencelessness (#VP5172)
 Appropriateness–Inappropriateness (#VP5665).

♦ **VC0615 Efficiency**
 Integrative polarities Power–Impotence (#VP5157)
 Expedience–Inexpedience (#VP5670)
 Skilfulness–Unskilfulness (#VP5733)
 Appropriateness–Inappropriateness (#VP5665).

♦ **VC0617 Effortlessness**
 Integrative polarities Facility–Difficulty (#VP5732).

♦ **VC0619 Elan**
 Integrative polarities Desire–Avoidance (#VP5631)
 Energy–Moderation (#VP5161)
 Cheerfulness–Solemnity (#VP5870).

♦ **VC0621 Elation**
 Integrative polarities Pleasure–Displeasure (#VP5865)
 Exultation–Lamentation (#VP5873)
 Cheerfulness–Solemnity (#VP5870).

♦ **VC0623 Elegance**
 Integrative polarities Beauty–Ugliness (#VP5899)
 Taste–Vulgarity (#VP5896)
 Chastity–Unchastity (#VP5987)
 Elegance–Inelegance (#VP5589)
 Courtesy–Discourtesy (#VP5936)
 Formality–Informality (#VP5646)
 Eloquence–Uneloquence (#VP5600)
 Naturalness–Affectation (#VP5900).

♦ **VC0625 Elevation**
 Integrative polarities Repute–Disrepute (#VP5914)
 Increase–Decrease (#VP5038)
 Elevation–Depression (#VP5317)
 Eloquence–Uneloquence (#VP5600)
 Unselfishness–Selfishness (#VP5978).

♦ **VC0627 Eloquence**
 Integrative polarities Eloquence–Uneloquence (#VP5600).

♦ **VC0629 Emancipation**
 Integrative polarities Freedom–Restraint (#VP5762).

♦ **VC0631 Eminence**
 Integrative polarities Repute–Disrepute (#VP5914)
 Greatness–Smallness (#VP5034)
 Authority–Lawlessness (#VP5739)
 Importance–Unimportance (#VP5672)
 Influence–Influencelessness (#VP5172).

♦ **VC0633 Emotion**
 Integrative polarities Excitement–Inexcitability (#VP5857)
 Feeling–Unfeelinglessness (#VP5855).

♦ **VC0635 Empathy**
 Integrative polarities Accord–Disaccord (#VP5794)
 Sensation–Insensibility (#VP5422)
 Feeling–Unfeelinglessness (#VP5855).

♦ **VC0637 Enchantment**
 Integrative polarities Love–Hate (#VP5930)   Pleasure–Displeasure (#VP5865)
 Motivation–Dissuasion (#VP5648)   Pleasantness–Unpleasantness (#VP5863)
 Influence–Influencelessness (#VP5172).

♦ **VC0639 Encouragement**
 Integrative polarities Courage–Fear (#VP5890)   Hope–Hopelessness (#VP5888)
 Comfort–Aggravation (#VP5885)   Motivation–Dissuasion (#VP5648)
 Cheerfulness–Solemnity (#VP5870).

♦ **VC0641 Endearment**
 Integrative polarities Love–Hate (#VP5930)   Motivation–Dissuasion (#VP5648).

♦ **VC0643 Endowment**
 Integrative polarities Power–Impotence (#VP5157)
 Sharing–Appropriation (#VP5815)
 Exultation–Lamentation (#VP5873)
 Skilfulness–Unskilfulness (#VP5733).

♦ **VC0645 Endurance**
 Integrative polarities Life–Death (#VP5407)   Strength–Weakness (#VP5159)
 Change–Permanence (#VP5139)   Patience–Impatience (#VP5861)
 Continuance–Cessation (#VP5143)   Durability–Transience (#VP5110)
 Resolution–Irresolution (#VP5624).

♦ **VC0647 Energy**
 Integrative polarities Action–Inaction (#VP5705)   Power–Impotence (#VP5157)
 Energy–Moderation (#VP5161)   Strength–Weakness (#VP5159).

♦ **VC0649 Enhancement**
 Integrative polarities Increase–Decrease (#VP5038)
 Improvement–Impairment (#VP5691).

♦ **VC0651 Enjoyment**
 Integrative polarities Pleasure–Displeasure (#VP5865).

♦ **VC0653 Enlightenment**
 Integrative polarities Light–Darkness (#VP5335)
 Knowledge–Ignorance (#VP5475)
 Improvement–Impairment (#VP5691)
 Education–Miseducation (#VP5562)
 Intelligence–Unintelligence (#VP5467)
 Interpretability–Misinterpretability (#VP5552)
 Communicativeness–Uncommunicativeness (#VP5554).

♦ **VC0655 Enrichment**
 Integrative polarities Improvement–Impairment (#VP5691).

♦ **VC0657 Entente**
 Integrative polarities Agreement–Disagreement (#VP5026).

♦ **VC0659 Enterprise**
 Integrative polarities Courage–Fear (#VP5890)   Action–Inaction (#VP5705)
 Energy–Moderation (#VP5161).

CONSTRUCTIVE VALUES　　　　　　　　　　　　　　　　　　　　　　　　　　　　　　　　　　　　　　VC0735

♦ **VC0661 Entertainment**
Integrative polarities Amusement–Boredom (#VP5878)
Pleasure–Displeasure (#VP5865).

♦ **VC0663 Enthusiasm**
Integrative polarities Desire–Avoidance (#VP5631)
Energy–Moderation (#VP5161)
Eloquence–Uneloquence (#VP5600)
Attention–Inattention (#VP5530)
Willingness–Unwillingness (#VP5621).

♦ **VC0665 Enumeration**
Integrative polarities Numbered–Unnumbered (#VP5086).

♦ **VC0667 Equality**
Integrative polarities Symmetry–Distortion (#VP5248)
Equality–Inequality (#VP5030)
Identity–Difference (#VP5013).

♦ **VC0669 Equanimity**
Integrative polarities Excitement–Inexcitability (#VP5857).

♦ **VC0671 Equilibrium**
Integrative polarities Symmetry–Distortion (#VP5248)
Equality–Inequality (#VP5030)
Identity–Difference (#VP5013)
Stability–Changeableness (#VP5141)
Excitement–Inexcitability (#VP5857).

♦ **VC0673 Equity**
Integrative polarities Justice–Injustice (#VP5976)
Legality–Illegality (#VP5998)
Equality–Inequality (#VP5030).

♦ **VC0675 Equivalence**
Integrative polarities Equality–Inequality (#VP5030)
Identity–Difference (#VP5013)
Agreement–Disagreement (#VP5026).

♦ **VC0677 Erudition**
Integrative polarities Knowledge–Ignorance (#VP5475)
Education–Miseducation (#VP5562)
Intelligence–Unintelligence (#VP5467).

♦ **VC0679 Esprit**
Integrative polarities Accord–Disaccord (#VP5794)
Amusement–Boredom (#VP5878)
Support–Opposition (#VP5785)
Cheerfulness–Solemnity (#VP5870)
Feeling–Unfeelinglessness (#VP5855)
Intelligence–Unintelligence (#VP5467).

♦ **VC0681 Essence**
Integrative polarities Simplicity–Complexity (#VP5045)
Existence–Nonexistence (#VP5001)
Importance–Unimportance (#VP5672)
Meaning–Meaninglessness (#VP5545)
Intrinsicality–Extrinsicality (#VP5005).

♦ **VC0683 Establishment**
Integrative polarities Location–Dislocation (#VP5185).

♦ **VC0685 Euphony**
Integrative polarities Harmony–Discord (#VP5462)
Elegance–Inelegance (#VP5589).

♦ **VC0687 Evenness**
Integrative polarities Energy–Moderation (#VP5161)
Symmetry–Distortion (#VP5248)
Equality–Inequality (#VP5030)
Regularity–Irregularity (#VP5137).

♦ **VC0689 Eventuation**
Integrative polarities Eventuation–Imminence (#VP5151).

♦ **VC0691 Evolution**
Integrative polarities Evolution–Revolution (#VP5147)
Improvement–Impairment (#VP5691).

♦ **VC0693 Exactness**
Integrative polarities Truth–Error (#VP5516)　　Carefulness–Neglect (#VP5533).

♦ **VC0695 Exaltation**
Integrative polarities Repute–Disrepute (#VP5914)
Sanctity–Unsanctity (#VP6026)
Greatness–Smallness (#VP5034)
Approval–Disapproval (#VP5968)
Pleasure–Displeasure (#VP5865)
Elevation–Depression (#VP5317)
Cheerfulness–Solemnity (#VP5870)
Unselfishness–Selfishness (#VP5978).

♦ **VC0697 Excellence**
Integrative polarities Taste–Vulgarity (#VP5896)　　Repute–Disrepute (#VP5914)
Goodness–Badness (#VP5674)　　Superiority–Inferiority (#VP5036)
Skilfulness–Unskilfulness (#VP5733).

♦ **VC0699 Excitement**
Integrative polarities Desire–Avoidance (#VP5631)
Motivation–Dissuasion (#VP5648)
Eloquence–Uneloquence (#VP5600)
Attention–Inattention (#VP5530)
Excitement–Inexcitability (#VP5857)
Feeling–Unfeelinglessness (#VP5855).

♦ **VC0701 Exclusiveness**
Integrative polarities Taste–Vulgarity (#VP5896)　　Respect–Disrespect (#VP5964)
Inclusion–Exclusion (#VP5076)　　Sociability–Unsociability (#VP5922).

♦ **VC0703 Exhilaration**
Integrative polarities Energy–Moderation (#VP5161)
Refreshment–Relapse (#VP5695)
Pleasure–Displeasure (#VP5865)
Motivation–Dissuasion (#VP5648)
Cheerfulness–Solemnity (#VP5870)
Excitement–Inexcitability (#VP5857).

♦ **VC0705 Existence**
Integrative polarities Life–Death (#VP5407)　　Presence–Absence (#VP5186)
Existence–Nonexistence (#VP5001).

♦ **VC0707 Expansion**
Integrative polarities Increase–Decrease (#VP5038)
Breadth–Narrowness (#VP5204)
Evolution–Revolution (#VP5147)
Expansion–Contraction (#VP5197).

♦ **VC0709 Expectation**
Integrative polarities Belief–Unbelief (#VP5501)　　Hope–Hopelessness (#VP5888)
Expectation–Inexpectation (#VP5539).

♦ **VC0711 Expedience**
Integrative polarities Goodness–Badness (#VP5674)
Expedience–Inexpedience (#VP5670)
Timeliness–Untimeliness (#VP5129)
Appropriateness–Inappropriateness (#VP5665).

♦ **VC0713 Expeditious**
Integrative polarities Action–Inaction (#VP5705)　　Timeliness–Untimeliness (#VP5129).

♦ **VC0715 Experience**
Integrative polarities Knowledge–Ignorance (#VP5475)
Skilfulness–Unskilfulness (#VP5733).

♦ **VC0717 Expertise**
Integrative polarities Taste–Vulgarity (#VP5896)　　Knowledge–Ignorance (#VP5475)
Skilfulness–Unskilfulness (#VP5733).

♦ **VC0719 Explanation**
Integrative polarities Research–Discovery (#VP5485)
Facility–Difficulty (#VP5732)
Causation–Culmination (#VP5153)
Vindication–Condemnation (#VP6005)
Intelligibility–Unintelligibility (#VP5548).

♦ **VC0721 Exquisiteness**
Integrative polarities Beauty–Ugliness (#VP5899)
Taste–Vulgarity (#VP5896)
Goodness–Badness (#VP5674)
Carefulness–Neglect (#VP5533)
Sensation–Insensibility (#VP5422)
Pleasantness–Unpleasantness (#VP5863).

♦ **VC0723 Exuberance**
Integrative polarities Pleasure–Displeasure (#VP5865)
Prosperity–Adversity (#VP5728)
Cheerfulness–Solemnity (#VP5870)
Oversufficiency–Insufficiency (#VP5660)
Productiveness–Unproductiveness (#VP5165).

♦ **VC0725 Facility**
Integrative polarities Power–Impotence (#VP5157)
Freedom–Restraint (#VP5762)
Support–Opposition (#VP5785)
Facility–Difficulty (#VP5732)
Eloquence–Uneloquence (#VP5600)
Education–Miseducation (#VP5562)
Skilfulness–Unskilfulness (#VP5733).

♦ **VC0727 Fairness**
Integrative polarities Beauty–Ugliness (#VP5899)
Goodness–Badness (#VP5674)
Justice–Injustice (#VP5976)
Probity–Improbity (#VP5974)
Broadmindedness–Narrowmindedness (#VP5526).

♦ **VC0729 Faith**
Integrative polarities Virtue–Vice (#VP5980)　　Piety–Impiety (#VP6028)
Belief–Unbelief (#VP5501)　　Desire–Avoidance (#VP5631)
Probity–Improbity (#VP5974)　　Hope–Hopelessness (#VP5888)
Certainty–Uncertainty (#VP5513)　　Obedience–Disobedience (#VP5766).

♦ **VC0731 Faithfulness**
Integrative polarities Love–Hate (#VP5930)　　Truth–Error (#VP5516)
Piety–Impiety (#VP6028)　　Belief–Unbelief (#VP5501)
Desire–Avoidance (#VP5631)　　Probity–Improbity (#VP5974)
Friendship–Enmity (#VP5927)　　Certainty–Uncertainty (#VP5513)
Obedience–Disobedience (#VP5766)　　Resolution–Irresolution (#VP5624).

♦ **VC0733 Fame**
Integrative polarities Repute–Disrepute (#VP5914)
Greatness–Smallness (#VP5034)
Importance–Unimportance (#VP5672).

♦ **VC0735 Familiarity**
Integrative polarities Friendship–Enmity (#VP5927)
Knowledge–Ignorance (#VP5475)
Sociability–Unsociability (#VP5922).

VC0737

- **VC0737 Fascination**
  Integrative polarities Desire–Avoidance (#VP5631)
  Wonder–Unastonishment (#VP5920)
  Motivation–Dissuasion (#VP5648)
  Attention–Inattention (#VP5530)
  Pleasantness–Unpleasantness (#VP5863).

- **VC0739 Fatherhood**
  Integrative polarities Ancestry–Posterity (#VP5170).

- **VC0741 Faultlessness**
  Integrative polarities Truth–Error (#VP5516)   Innocence–Guilt (#VP5983)
  Perfection–Imperfection (#VP5677).

- **VC0743 Favour**
  Integrative polarities Repute–Disrepute (#VP5914)
  Choice–Necessity (#VP5637)
  Friendship–Enmity (#VP5927)
  Respect–Disrespect (#VP5964)
  Support–Opposition (#VP5785)
  Rightness–Wrongness (#VP5957)
  Kindness–Unkindness (#VP5938)
  Approval–Disapproval (#VP5968)
  Courtesy–Discourtesy (#VP5936)
  Compassion–Pitilessness (#VP5944)
  Influence–Influencelessness (#VP5172).

- **VC0745 Fearlessness**
  Integrative polarities Courage–Fear (#VP5890).

- **VC0747 Feasibility**
  Integrative polarities Facility–Difficulty (#VP5732)
  Expedience–Inexpedience (#VP5670).

- **VC0749 Fecundity**
  Integrative polarities Conciseness–Diffuseness (#VP5592)
  Productiveness–Unproductiveness (#VP5165)
  Imaginativeness–Unimaginativeness (#VP5535).

- **VC0751 Feeling**
  Integrative polarities Belief–Unbelief (#VP5501)   Intuition–Reason (#VP5481)
  Research–Discovery (#VP5485)   Compassion–Pitilessness (#VP5944).

- **VC0753 Felicity**
  Integrative polarities Taste–Vulgarity (#VP5896)   Elegance–Inelegance (#VP5589)
  Pleasure–Displeasure (#VP5865)   Prosperity–Adversity (#VP5728)
  Eloquence–Uneloquence (#VP5600)   Agreement–Disagreement (#VP5026)
  Skilfulness–Unskilfulness (#VP5733).

- **VC0755 Fellowship**
  Integrative polarities Accord–Disaccord (#VP5794)
  Friendship–Enmity (#VP5927)
  Support–Opposition (#VP5785)
  Sociability–Unsociability (#VP5922).

- **VC0757 Feminity**
  Integrative polarities Masculinity–Femininity (#VP5420).

- **VC0759 Fertility**
  Integrative polarities Conciseness–Diffuseness (#VP5592)
  Oversufficiency–Insufficiency (#VP5660)
  Productiveness–Unproductiveness (#VP5165)
  Imaginativeness–Unimaginativeness (#VP5535).

- **VC0761 Fervour**
  Integrative polarities Love–Hate (#VP5930)   Action–Inaction (#VP5705)
  Desire–Avoidance (#VP5631)   Eloquence–Uneloquence (#VP5600)
  Feeling–Unfeelinglessness (#VP5855).

- **VC0763 Fidelity**
  Integrative polarities Truth–Error (#VP5516)   Desire–Avoidance (#VP5631)
  Probity–Improbity (#VP5974)   Resolution–Irresolution (#VP5624).

- **VC0765 Finesse**
  Integrative polarities Taste–Vulgarity (#VP5896)   Skilfulness–Unskilfulness (#VP5733)
  Discrimination–Indiscrimination (#VP5492).

- **VC0767 Fitness**
  Integrative polarities Health–Disease (#VP5685)
  Taste–Vulgarity (#VP5896)
  Power–Impotence (#VP5157)
  Choice–Necessity (#VP5637)
  Rightness–Wrongness (#VP5957)
  Agreement–Disagreement (#VP5026)
  Expedience–Inexpedience (#VP5670)
  Timeliness–Untimeliness (#VP5129)
  Preparedness–Unpreparedness (#VP5720).

- **VC0769 Flawlessness**
  Integrative polarities Truth–Error (#VP5516)   Perfection–Imperfection (#VP5677).

- **VC0771 Flexibility**
  Integrative polarities Freedom–Restraint (#VP5762)
  Hardness–Softness (#VP5356)
  Leniency–Compulsion (#VP5756)
  Facility–Difficulty (#VP5732)
  Elasticity–Toughness (#VP5358)
  Stability–Changeableness (#VP5141)
  Skilfulness–Unskilfulness (#VP5733).

- **VC0773 Flourish**
  Integrative polarities Health–Disease (#VP5685)
  Harmony–Discord (#VP5462)
  Energy–Moderation (#VP5161)
  Prosperity–Adversity (#VP5728)
  Expansion–Contraction (#VP5197)
  Naturalness–Affectation (#VP5900)
  Accomplishment–Nonaccomplishment (#VP5722).

- **VC0775 Fluency**
  Integrative polarities Elegance–Inelegance (#VP5589)
  Eloquence–Uneloquence (#VP5600)
  Conciseness–Diffuseness (#VP5592).

- **VC0777 Fondness**
  Integrative polarities Love–Hate (#VP5930)   Possession–Loss (#VP5808)
  Belief–Unbelief (#VP5501)   Desire–Avoidance (#VP5631)
  Support–Opposition (#VP5785)   Feeling–Unfeelinglessness (#VP5855).

- **VC0779 Forbearance**
  Integrative polarities Patience–Impatience (#VP5861)
  Leniency–Compulsion (#VP5756)
  Forgiveness–Vengeance (#VP5947)
  Temperance–Intemperance (#VP5992)
  Compassion–Pitilessness (#VP5944)
  Broadmindedness–Narrowmindedness (#VP5526).

- **VC0781 Forcefulness**
  Integrative polarities Truth–Error (#VP5516)   Action–Inaction (#VP5705)
  Power–Impotence (#VP5157)   Energy–Moderation (#VP5161)
  Strength–Weakness (#VP5159)   Motivation–Dissuasion (#VP5648)
  Importance–Unimportance (#VP5672).

- **VC0783 Foresight**
  Integrative polarities Caution–Rashness (#VP5894)
  Intelligence–Unintelligence (#VP5467).

- **VC0785 Forethought**
  Integrative polarities Caution–Rashness (#VP5894)
  Carefulness–Neglect (#VP5533)
  Expectation–Inexpectation (#VP5539)
  Intelligence–Unintelligence (#VP5467).

- **VC0787 Forgiveness**
  Integrative polarities Accord–Disaccord (#VP5794)
  Leniency–Compulsion (#VP5756)
  Forgiveness–Vengeance (#VP5947)
  Compassion–Pitilessness (#VP5944)
  Remembrance–Forgetfulness (#VP5537).

- **VC0789 Forthrightness**
  Integrative polarities Probity–Improbity (#VP5974)
  Direction–Deviation (#VP5290).

- **VC0791 Fortune**
  Integrative polarities Wealth–Poverty (#VP5837)
  Prosperity–Adversity (#VP5728)
  Attributability–Chance (#VP5155)
  Accomplishment–Nonaccomplishment (#VP5722).

- **VC0793 Frankness**
  Integrative polarities Probity–Improbity (#VP5974)
  Elegance–Inelegance (#VP5589)
  Communicativeness–Uncommunicativeness (#VP5554).

- **VC0795 Fraternity**
  Integrative polarities Friendship–Enmity (#VP5927)
  Support–Opposition (#VP5785)
  Sociability–Unsociability (#VP5922).

- **VC0797 Freedom**
  Integrative polarities Exertion–Rest (#VP5709)   Probity–Improbity (#VP5974)
  Freedom–Restraint (#VP5762)   Rightness–Wrongness (#VP5957)
  Economy–Prodigality (#VP5851)   Facility–Difficulty (#VP5732)
  Willingness–Unwillingness (#VP5621).

- **VC0799 Freshness**
  Integrative polarities Heat–Cold (#VP5328)   Health–Disease (#VP5685)
  Newness–Oldness (#VP5122)   Futurity–Antiquity (#VP5119)
  Refreshment–Relapse (#VP5695)   Cleanness–Uncleanness (#VP5681)
  Originality–Imitation (#VP5023)   Perfection–Imperfection (#VP5677)
  Appropriateness–Inappropriateness (#VP5665).

- **VC0801 Frictionlessness**
  Integrative polarities Accord–Disaccord (#VP5794).

- **VC0803 Friendliness**
  Integrative polarities Friendship–Enmity (#VP5927)
  Support–Opposition (#VP5785)
  Comfort–Aggravation (#VP5885)
  Hospitality–Inhospitality (#VP5925)
  Sociability–Unsociability (#VP5922).

- **VC0805 Friendship**
  Integrative polarities Friendship–Enmity (#VP5927).

- **VC0807 Frugality**
  Integrative polarities Economy–Prodigality (#VP5851)
  Temperance–Intemperance (#VP5992).

- **VC0809 Fulfilment**
  Integrative polarities Contentment–Discontentment (#VP5868)
  Selfactualization–Neurosis (#VP5690)
  Completeness–Incompleteness (#VP5054)
  Oversufficiency–Insufficiency (#VP5660)
  Accomplishment–Nonaccomplishment (#VP5722).

- **VC0811 Fullness**
  Integrative polarities Breadth–Narrowness (#VP5204)
  Greatness–Smallness (#VP5034)

CONSTRUCTIVE VALUES   **VC0887**

Perfection–Imperfection (#VP5677)
Continuity–Discontinuity (#VP5071)
Completeness–Incompleteness (#VP5054)
Oversufficiency–Insufficiency (#VP5660).

♦ **VC0813 Fun**
Integrative polarities Amusement–Boredom (#VP5878)
Pleasure–Displeasure (#VP5865)
Cheerfulness–Solemnity (#VP5870).

♦ **VC0815 Gaiety**
Integrative polarities Amusement–Boredom (#VP5878)
Pleasure–Displeasure (#VP5865)
Cheerfulness–Solemnity (#VP5870)
Sociability–Unsociability (#VP5922).

♦ **VC0817 Gain**
Integrative polarities Victory–Defeat (#VP5726)   Possession–Loss (#VP5808)
Goodness–Badness (#VP5674)   Increase–Decrease (#VP5038)
Swiftness–Slowness (#VP5269)   Motivation–Dissuasion (#VP5648)
Improvement–Impairment (#VP5691).

♦ **VC0819 Gallantry**
Integrative polarities Courage–Fear (#VP5890)   Courtesy–Discourtesy (#VP5936).

♦ **VC0821 Generosity**
Integrative polarities Kindness–Unkindness (#VP5938)
Economy–Prodigality (#VP5851)
Greatness–Smallness (#VP5034)
Forgiveness–Vengeance (#VP5947)
Unselfishness–Selfishness (#VP5978)
Hospitality–Inhospitality (#VP5925)
Oversufficiency–Insufficiency (#VP5660).

♦ **VC0823 Geniality**
Integrative polarities Friendship–Enmity (#VP5927)
Kindness–Unkindness (#VP5938)
Cheerfulness–Solemnity (#VP5870)
Hospitality–Inhospitality (#VP5925)
Sociability–Unsociability (#VP5922)
Pleasantness–Unpleasantness (#VP5863).

♦ **VC0825 Genius**
Integrative polarities Power–Impotence (#VP5157)
Motivation–Dissuasion (#VP5648)
Skilfulness–Unskilfulness (#VP5733)
Intelligence–Unintelligence (#VP5467)
Imaginativeness–Unimaginativeness (#VP5535).

♦ **VC0827 Gentility**
Integrative polarities Taste–Vulgarity (#VP5896)   Repute–Disrepute (#VP5914)
Courtesy–Discourtesy (#VP5936).

♦ **VC0829 Gentleness**
Integrative polarities Freedom–Restraint (#VP5762)
Energy–Moderation (#VP5161)
Kindness–Unkindness (#VP5938)
Courtesy–Discourtesy (#VP5936)
Compassion–Pitilessness (#VP5944)
Excitement–Inexcitability (#VP5857).

♦ **VC0831 Genuineness**
Integrative polarities Truth–Error (#VP5516)   Probity–Improbity (#VP5974)
Existence–Nonexistence (#VP5001)   Skilfulness–Unskilfulness (#VP5733).

♦ **VC0833 Giftedness**
Integrative polarities Skilfulness–Unskilfulness (#VP5733)
Intelligence–Unintelligence (#VP5467).

♦ **VC0835 Gladness**
Integrative polarities Pleasure–Displeasure (#VP5865)
Cheerfulness–Solemnity (#VP5870).

♦ **VC0837 Glamour**
Integrative polarities Beauty–Ugliness (#VP5899)
Repute–Disrepute (#VP5914)
Motivation–Dissuasion (#VP5648)
Pleasantness–Unpleasantness (#VP5863).

♦ **VC0839 Globality**
Integrative polarities Inclusion–Exclusion (#VP5076)
Completeness–Incompleteness (#VP5054).

♦ **VC0841 Glory**
Integrative polarities Light–Darkness (#VP5335)
Beauty–Ugliness (#VP5899)
Repute–Disrepute (#VP5914)
Goodness–Badness (#VP5674)
Greatness–Smallness (#VP5034)
Approval–Disapproval (#VP5968)
Godliness–Ungodliness (#VP6013)
Exultation–Lamentation (#VP5873)
Naturalness–Affectation (#VP5900)
Importance–Unimportance (#VP5672).

♦ **VC0843 Godliness**
Integrative polarities Virtue–Vice (#VP5980)   Piety–Impiety (#VP6028)
Sanctity–Unsanctity (#VP6026)   Godliness–Ungodliness (#VP6013).

♦ **VC0845 Good**
Integrative polarities Virtue–Vice (#VP5980)   Truth–Error (#VP5516)
Piety–Impiety (#VP6028)   Goodness–Badness (#VP5674)
Justice–Injustice (#VP5976)   Probity–Improbity (#VP5974)
Rightness–Wrongness (#VP5957)   Kindness–Unkindness (#VP5938)
Godliness–Ungodliness (#VP6013)   Expedience–Inexpedience (#VP5670)

Skilfulness–Unskilfulness (#VP5733)   Expectation–Inexpectation (#VP5539)
Savouriness–Unsavouriness (#VP5427)   Pleasantness–Unpleasantness (#VP5863)
Healthfulness–Unhealthfulness (#VP5683)   Oversufficiency–Insufficiency (#VP5660).

♦ **VC0847 Goodness**
Integrative polarities Virtue–Vice (#VP5980)   Piety–Impiety (#VP6028)
Goodness–Badness (#VP5674)   Probity–Improbity (#VP5974)
Rightness–Wrongness (#VP5957)   Kindness–Unkindness (#VP5938)
Pleasantness–Unpleasantness (#VP5863)   Healthfulness–Unhealthfulness (#VP5683).

♦ **VC0849 Goodwill**
Integrative polarities Friendship–Enmity (#VP5927)
Support–Opposition (#VP5785)
Kindness–Unkindness (#VP5938)
Willingness–Unwillingness (#VP5621).

♦ **VC0851 Gorgeousness**
Integrative polarities Beauty–Ugliness (#VP5899).

♦ **VC0853 Government**
Integrative polarities Safety–Danger (#VP5697)   Authority–Lawlessness (#VP5739).

♦ **VC0855 Grace**
Integrative polarities Beauty–Ugliness (#VP5899)
Sanctity–Unsanctity (#VP6026)
Forgiveness–Vengeance (#VP5947)
Compassion–Pitilessness (#VP5944)
Vindication–Condemnation (#VP6005).

♦ **VC0857 Gracefulness**
Integrative polarities Beauty–Ugliness (#VP5899)
Taste–Vulgarity (#VP5896)
Courtesy–Discourtesy (#VP5936)
Eloquence–Uneloquence (#VP5600)
Skilfulness–Unskilfulness (#VP5733)
Pleasantness–Unpleasantness (#VP5863).

♦ **VC0859 Graciousness**
Integrative polarities Kindness–Unkindness (#VP5938)
Economy–Prodigality (#VP5851)
Leniency–Compulsion (#VP5756)
Courtesy–Discourtesy (#VP5936)
Hospitality–Inhospitality (#VP5925)
Pleasantness–Unpleasantness (#VP5863).

♦ **VC0861 Grandeur**
Integrative polarities Pride–Humility (#VP5905)   Repute–Disrepute (#VP5914)
Greatness–Smallness (#VP5034)   Naturalness–Affectation (#VP5900).

♦ **VC0863 Gratification**
Integrative polarities Leniency–Compulsion (#VP5756)
Pleasure–Displeasure (#VP5865)
Motivation–Dissuasion (#VP5648).

♦ **VC0865 Gratitude**
Integrative polarities Gratitude–Ingratitude (#VP5949).

♦ **VC0867 Greatness**
Integrative polarities Greatness–Smallness (#VP5034)
Authority–Lawlessness (#VP5739)
Importance–Unimportance (#VP5672)
Unselfishness–Selfishness (#VP5978).

♦ **VC0869 Gregariousness**
Integrative polarities Sociability–Unsociability (#VP5922).

♦ **VC0871 Growth**
Integrative polarities Increase–Decrease (#VP5038)
Evolution–Revolution (#VP5147)
Expansion–Contraction (#VP5197).

♦ **VC0873 Guiltlessness**
Integrative polarities Virtue–Vice (#VP5980)   Innocence–Guilt (#VP5983).

♦ **VC0879 Handsomeness**
Integrative polarities Beauty–Ugliness (#VP5899)
Economy–Prodigality (#VP5851)
Unselfishness–Selfishness (#VP5978).

♦ **VC0881 Happiness**
Integrative polarities Taste–Vulgarity (#VP5896)   Pleasure–Displeasure (#VP5865)
Prosperity–Adversity (#VP5728)   Cheerfulness–Solemnity (#VP5870)
Contentment–Discontentment (#VP5868).

♦ **VC0883 Hardiness**
Integrative polarities Courage–Fear (#VP5890)   Health–Disease (#VP5685)
Weight–Lightness (#VP5352)   Strength–Weakness (#VP5159)
Elasticity–Toughness (#VP5358)   Durability–Transience (#VP5110)
Substantiality–Unsubstantiality (#VP5003).

♦ **VC0885 Harmlessness**
Integrative polarities Safety–Danger (#VP5697)   Goodness–Badness (#VP5674)
Perfection–Imperfection (#VP5677).

♦ **VC0887 Harmony**
Integrative polarities Assent–Dissent (#VP5521)
Order–Disorder (#VP5059)
Harmony–Discord (#VP5462)
Accord–Discord (#VP5794)
Friendship–Enmity (#VP5927)
Support–Opposition (#VP5785)
Elegance–Inelegance (#VP5589)
Symmetry–Distortion (#VP5248)
Agreement–Disagreement (#VP5026)
Concurrence–Counteraction (#VP5177)
Pleasantness–Unpleasantness (#VP5863).

-729-

VC0889

♦ **VC0889 Health**
Integrative polarities Health–Disease (#VP5685)
Conformity–Nonconformity (#VP5082).

♦ **VC0891 Healthiness**
Integrative polarities Health–Disease (#VP5685)
Goodness–Badness (#VP5674)
Conformity–Nonconformity (#VP5082)
Healthfulness–Unhealthfulness (#VP5683).

♦ **VC0893 Helpfulness**
Integrative polarities Goodness–Badness (#VP5674)
Support–Opposition (#VP5785)
Kindness–Unkindness (#VP5938)
Appropriateness–Inappropriateness (#VP5665).

♦ **VC0895 Heritage**
Integrative polarities Ancestry–Posterity (#VP5170).

♦ **VC0897 Heroism**
Integrative polarities Courage–Fear (#VP5890)   Greatness–Smallness (#VP5034)
Originality–Imitation (#VP5023)   Unselfishness–Selfishness (#VP5978).

♦ **VC0899 High-mindedness**
Integrative polarities Pride–Humility (#VP5905)   Probity–Improbity (#VP5974)
Unselfishness–Selfishness (#VP5978).

♦ **VC0901 High-spiritedness**
Integrative polarities Pleasure–Displeasure (#VP5865)
Cheerfulness–Solemnity (#VP5870)
Excitement–Inexcitability (#VP5857).

♦ **VC0903 Hilarity**
Integrative polarities Amusement–Boredom (#VP5878)
Exultation–Lamentation (#VP5873)
Cheerfulness–Solemnity (#VP5870).

♦ **VC0905 Holiness**
Integrative polarities Piety–Impiety (#VP6028)   Sanctity–Unsanctity (#VP6026)
Godliness–Ungodliness (#VP6013).

♦ **VC0907 Holism**
Integrative polarities Completeness–Incompleteness (#VP5054).

♦ **VC0909 Homage**
Integrative polarities Freedom–Restraint (#VP5762)
Respect–Disrespect (#VP5964)
Approval–Disapproval (#VP5968)
Forgiveness–Vengeance (#VP5947)
Obedience–Disobedience (#VP5766).

♦ **VC0913 Honesty**
Integrative polarities Virtue–Vice (#VP5980)   Truth–Error (#VP5516)
Probity–Improbity (#VP5974)   Naturalness–Affectation (#VP5900).

♦ **VC0915 Honour**
Integrative polarities Probity–Improbity (#VP5974)
Respect–Disrespect (#VP5964)
Chastity–Unchastity (#VP5987)
Approval–Disapproval (#VP5968)
Exultation–Lamentation (#VP5873)
Naturalness–Affectation (#VP5900)
Importance–Unimportance (#VP5672).

♦ **VC0917 Hope**
Integrative polarities Virtue–Vice (#VP5980)   Safety–Danger (#VP5697)
Belief–Unbelief (#VP5501)   Hope–Hopelessness (#VP5888)
Cheerfulness–Solemnity (#VP5870).

♦ **VC0919 Hospitality**
Integrative polarities Friendship–Enmity (#VP5927)
Comfort–Aggravation (#VP5885)
Economy–Prodigality (#VP5851)
Hospitality–Inhospitality (#VP5925)
Sociability–Unsociability (#VP5922).

♦ **VC0921 Humaneness**
Integrative polarities Kindness–Unkindness (#VP5938)
Leniency–Compulsion (#VP5756)
Compassion–Pitilessness (#VP5944).

♦ **VC0923 Humanism**
Integrative polarities Piety–Impiety (#VP6028)   Knowledge–Ignorance (#VP5475)
Humanity–Nonhumanity (#VP5417).

♦ **VC0925 Humanitarianism**
Integrative polarities Kindness–Unkindness (#VP5938).

♦ **VC0927 Humanity**
Integrative polarities Kindness–Unkindness (#VP5938)
Leniency–Compulsion (#VP5756)
Humanity–Nonhumanity (#VP5417)
Compassion–Pitilessness (#VP5944).

♦ **VC0929 Humility**
Integrative polarities Modesty–Vanity (#VP5908)
Pride–Humility (#VP5905)
Freedom–Restraint (#VP5762)
Patience–Impatience (#VP5861)
Unselfishness–Selfishness (#VP5978).

♦ **VC0931 Humour**
Integrative polarities Amusement–Boredom (#VP5878).

♦ **VC0933 Hygiene**
Integrative polarities Cleanness–Uncleanness (#VP5681)
Healthfulness–Unhealthfulness (#VP5683).

♦ **VC0935 Idealism**
Integrative polarities Desire–Avoidance (#VP5631)
Perfection–Imperfection (#VP5677)
Thought–Thoughtlessness (#VP5478)
Unselfishness–Selfishness (#VP5978).

♦ **VC0937 Illumination**
Integrative polarities Light–Darkness (#VP5335)
Knowledge–Ignorance (#VP5475)
Education–Miseducation (#VP5562)
Interpretability–Misinterpretability (#VP5552).

♦ **VC0939 Imagination**
Integrative polarities Illusion–Disillusionment (#VP5519)
Imaginativeness–Unimaginativeness (#VP5535).

♦ **VC0941 Immediacy**
Integrative polarities Presence–Absence (#VP5186)
Nearness–Distance (#VP5200)
Futurity–Antiquity (#VP5119)
Direction–Deviation (#VP5290)
Continuity–Discontinuity (#VP5071)
Perpetuity–Instantaneousness (#VP5112).

♦ **VC0943 Immensity**
Integrative polarities Goodness–Badness (#VP5674)
Greatness–Smallness (#VP5034).

♦ **VC0945 Imminence**
Integrative polarities Expectation–Inexpectation (#VP5539).

♦ **VC0947 Immortality**
Integrative polarities Life–Death (#VP5407)   Repute–Disrepute (#VP5914)
Stability–Changeableness (#VP5141)   Perpetuity–Instantaneousness (#VP5112).

♦ **VC0949 Immunity**
Integrative polarities Safety–Danger (#VP5697)   Health–Disease (#VP5685)
Freedom–Restraint (#VP5762)   Rightness–Wrongness (#VP5957)
Forgiveness–Vengeance (#VP5947).

♦ **VC0951 Immutability**
Integrative polarities Change–Permanence (#VP5139)
Godliness–Ungodliness (#VP6013)
Durability–Transience (#VP5110)
Resolution–Irresolution (#VP5624)
Stability–Changeableness (#VP5141).

♦ **VC0953 Impartiality**
Integrative polarities Justice–Injustice (#VP5976)
Energy–Moderation (#VP5161)
Neutrality–Compromise (#VP5806)
Broadmindedness–Narrowmindedness (#VP5526).

♦ **VC0955 Implementation**
Integrative polarities Action–Inaction (#VP5705)
Accomplishment–Nonaccomplishment (#VP5722).

♦ **VC0957 Importance**
Integrative polarities Repute–Disrepute (#VP5914)
Authority–Lawlessness (#VP5739)
Importance–Unimportance (#VP5672)
Influence–Influencelessness (#VP5172).

♦ **VC0959 Impressivensss**
Integrative polarities Belief–Unbelief (#VP5501)   Eloquence–Uneloquence (#VP5600)
Naturalness–Affectation (#VP5900)   Sensation–Insensibility (#VP5422)
Excitement–Inexcitability (#VP5857).

♦ **VC0961 Improvement**
Integrative polarities Change–Permanence (#VP5139)
Conversion–Reversion (#VP5145)
Improvement–Impairment (#VP5691)
Education–Miseducation (#VP5562)
Restoration–Destruction (#VP5693).

♦ **VC0963 Inclusion**
Integrative polarities Support–Opposition (#VP5785)
Inclusion–Exclusion (#VP5076)
Conjunction–Separation (#VP5047).

♦ **VC0965 Increase**
Integrative polarities Increase–Decrease (#VP5038)
Comfort–Aggravation (#VP5885)
Elevation–Depression (#VP5317)
Expansion–Contraction (#VP5197).

♦ **VC0967 Indefatigableness**
Integrative polarities Action–Inaction (#VP5705)   Continuance–Cessation (#VP5143)
Resolution–Irresolution (#VP5624).

♦ **VC0969 Independence**
Integrative polarities Pride–Humility (#VP5905)   Wealth–Poverty (#VP5837)
Freedom–Restraint (#VP5762)   Support–Opposition (#VP5785)
Neutrality–Compromise (#VP5806)   Resolution–Irresolution (#VP5624)
Willingness–Unwillingness (#VP5621).

♦ **VC0971 Individuality**
Integrative polarities Unity–Duality (#VP5089).

CONSTRUCTIVE VALUES VC1047

♦ **VC0973 Indubitableness**
Integrative polarities Greatness–Smallness (#VP5034)
Certainty–Uncertainty (#VP5513).

♦ **VC0975 Industriousness**
Integrative polarities Exertion–Rest (#VP5709) Action–Inaction (#VP5705)
Carefulness–Neglect (#VP5533) Resolution–Irresolution (#VP5624).

♦ **VC0977 Infallibility**
Integrative polarities Truth–Error (#VP5516) Certainty–Uncertainty (#VP5513)
Perfection–Imperfection (#VP5677).

♦ **VC0979 Influence**
Integrative polarities Power–Impotence (#VP5157)
Authority–Lawlessness (#VP5739)
Motivation–Dissuasion (#VP5648)
Causation–Culmination (#VP5153)
Influence–Influencelessness (#VP5172).

♦ **VC0981 Ingenuity**
Integrative polarities Skilfulness–Unskilfulness (#VP5733)
Imaginativeness–Unimaginativeness (#VP5535).

♦ **VC0983 Ingenuousness**
Integrative polarities Probity–Improbity (#VP5974)
Skilfulness–Unskilfulness (#VP5733).

♦ **VC0985 Initiative**
Integrative polarities Action–Inaction (#VP5705) Energy–Moderation (#VP5161).

♦ **VC0987 Innocence**
Integrative polarities Virtue–Vice (#VP5980) Innocence–Guilt (#VP5983)
Belief–Unbelief (#VP5501) Goodness–Badness (#VP5674)
Chastity–Unchastity (#VP5987) Skilfulness–Unskilfulness (#VP5733).

♦ **VC0988 Inoffensiveness**
Integrative polarities Goodness–Badness (#VP5674)
Vindication–Condemnation (#VP6005).

♦ **VC0991 Insight**
Integrative polarities Intuition–Reason (#VP5481)
Knowledge–Ignorance (#VP5475)
Intelligence–Unintelligence (#VP5467)
Discrimination–Indiscrimination (#VP5492).

♦ **VC0993 Inspiration**
Integrative polarities Courage–Fear (#VP5890) Hope–Hopelessness (#VP5888)
Motivation–Dissuasion (#VP5648) Cheerfulness–Solemnity (#VP5870)
Thought–Thoughtlessness (#VP5478) Intelligence–Unintelligence (#VP5467)
Imaginativeness–Unimaginativeness (#VP5535).

♦ **VC0995 Integrality**
Integrative polarities Unity–Duality (#VP5089) Completeness–Incompleteness (#VP5054).

♦ **VC0997 Integrity**
Integrative polarities Unity–Duality (#VP5089) Probity–Improbity (#VP5974)
Simplicity–Complexity (#VP5045) Perfection–Imperfection (#VP5677)
Completeness–Incompleteness (#VP5054).

♦ **VC0999 Intellectuality**
Integrative polarities Intuition–Reason (#VP5481)
Knowledge–Ignorance (#VP5475)
Thought–Thoughtlessness (#VP5478)
Intelligence–Unintelligence (#VP5467).

♦ **VC1001 Intelligence**
Integrative polarities Knowledge–Ignorance (#VP5475)
Intelligence–Unintelligence (#VP5467).

♦ **VC1003 Intelligibility**
Integrative polarities Facility–Difficulty (#VP5732)
Meaning–Meaninglessness (#VP5545)
Intelligibility–Unintelligibility (#VP5548).

♦ **VC1005 Intensity**
Integrative polarities Desire–Avoidance (#VP5631)
Energy–Moderation (#VP5161)
Greatness–Smallness (#VP5034)
Feeling–Unfeelinglessness (#VP5855).

♦ **VC1007 Intention**
Integrative polarities Motivation–Dissuasion (#VP5648)
Willingness–Unwillingness (#VP5621).

♦ **VC1009 Intentness**
Integrative polarities Desire–Avoidance (#VP5631)
Attention–Inattention (#VP5530).

♦ **VC1011 Interest**
Integrative polarities Possession–Loss (#VP5808)
Goodness–Badness (#VP5674)
Intuition–Reason (#VP5481)
Justice–Injustice (#VP5976)
Probity–Improbity (#VP5974)
Support–Opposition (#VP5785)
Rightness–Wrongness (#VP5957)
Motivation–Dissuasion (#VP5648)
Attention–Inattention (#VP5530)
Curiosity–Incuriosity (#VP5528)
Causation–Culmination (#VP5153)
Eventuation–Imminence (#VP5151)
Importance–Unimportance (#VP5672)
Unselfishness–Selfishness (#VP5978)
Relatedness–Unrelatedness (#VP5009)
Influence–Influencelessness (#VP5172)
Appropriateness–Inappropriateness (#VP5665).

♦ **VC1013 Intimacy**
Integrative polarities Friendship–Enmity (#VP5927)
Knowledge–Ignorance (#VP5475)
Sociability–Unsociability (#VP5922)
Relatedness–Unrelatedness (#VP5009).

♦ **VC1015 Intricacy**
Integrative polarities Prosperity–Adversity (#VP5728)
Simplicity–Complexity (#VP5045)
Intelligibility–Unintelligibility (#VP5548).

♦ **VC1017 Intuitiveness**
Integrative polarities Intuition–Reason (#VP5481).

♦ **VC1019 Inventiveness**
Integrative polarities Change–Permanence (#VP5139)
Research–Discovery (#VP5485)
Originality–Imitation (#VP5023)
Production–Reproduction (#VP5167)
Imaginativeness–Unimaginativeness (#VP5535).

♦ **VC1021 Invincibility**
Integrative polarities Certainty–Uncertainty (#VP5513)
Resolution–Irresolution (#VP5624)
Superiority–Inferiority (#VP5036)
Stability–Changeableness (#VP5141).

♦ **VC1023 Involvement**
Integrative polarities Innocence–Guilt (#VP5983)
Justice–Injustice (#VP5976)
Sharing–Appropriation (#VP5815)
Attention–Inattention (#VP5530)
Simplicity–Complexity (#VP5045)
Vindication–Condemnation (#VP6005)
Feeling–Unfeelinglessness (#VP5855).

♦ **VC1025 Invulnerability**
Integrative polarities Safety–Danger (#VP5697) Stability–Changeableness (#VP5141).

♦ **VC1027 Irrepressibility**
Integrative polarities Cheerfulness–Solemnity (#VP5870).

♦ **VC1029 Irreproachability**
Integrative polarities Innocence–Guilt (#VP5983)
Probity–Improbity (#VP5974)
Perfection–Imperfection (#VP5677).

♦ **VC1031 Irresistability**
Integrative polarities Power–Impotence (#VP5157)
Greatness–Smallness (#VP5034)
Motivation–Dissuasion (#VP5648)
Pleasantness–Unpleasantness (#VP5863).

♦ **VC1033 Joy**
Integrative polarities Pleasure–Displeasure (#VP5865)
Exultation–Lamentation (#VP5873)
Cheerfulness–Solemnity (#VP5870).

♦ **VC1035 Joyfulness**
Integrative polarities Pleasure–Displeasure (#VP5865)
Cheerfulness–Solemnity (#VP5870).

♦ **VC1037 Judgement**
Integrative polarities Belief–Unbelief (#VP5501) Judgement–Misjudgement (#VP5494)
Vindication–Condemnation (#VP6005) Intelligence–Unintelligence (#VP5467)
Discrimination–Indiscrimination (#VP5492).

♦ **VC1039 Judiciousness**
Integrative polarities Caution–Rashness (#VP5894)
Energy–Moderation (#VP5161)
Intelligence–Unintelligence (#VP5467)
Discrimination–Indiscrimination (#VP5492).

♦ **VC1041 Justice**
Integrative polarities Virtue–Vice (#VP5980) Truth–Error (#VP5516)
Intuition–Reason (#VP5481) Justice–Injustice (#VP5976)
Probity–Improbity (#VP5974) Dueness–Undueness (#VP5960)
Equality–Inequality (#VP5030) Judgement–Misjudgement (#VP5494)
Broadmindedness–Narrowmindedness (#VP5526).

♦ **VC1043 Keenness**
Integrative polarities Goodness–Badness (#VP5674)
Desire–Avoidance (#VP5631)
Amusement–Boredom (#VP5878)
Energy–Moderation (#VP5161)
Carefulness–Neglect (#VP5533)
Sensation–Insensibility (#VP5422)
Feeling–Unfeelinglessness (#VP5855)
Intelligence–Unintelligence (#VP5467).

♦ **VC1045 Kindness**
Integrative polarities Goodness–Badness (#VP5674)
Friendship–Enmity (#VP5927)
Support–Opposition (#VP5785)
Kindness–Unkindness (#VP5938)
Forgiveness–Vengeance (#VP5947)
Compassion–Pitilessness (#VP5944).

♦ **VC1047 Knowledge**
Integrative polarities Knowledge–Ignorance (#VP5475)
Education–Miseducation (#VP5562)
Intelligence–Unintelligence (#VP5467)
Communicativeness–Uncommunicativeness (#VP5554).

VC1049

♦ **VC1049 Largeness**
Integrative polarities Economy–Prodigality (#VP5851)
Greatness–Smallness (#VP5034).

♦ **VC1051 Lasting**
Integrative polarities Change–Permanence (#VP5139)
Durability–Transience (#VP5110)
Resolution–Irresolution (#VP5624)
Stability–Changeableness (#VP5141).

♦ **VC1053 Laughter**
Integrative polarities Exultation–Lamentation (#VP5873)
Cheerfulness–Solemnity (#VP5870).

♦ **VC1055 Lawfulness**
Integrative polarities Justice–Injustice (#VP5976).

♦ **VC1057 Leadership**
Integrative polarities Authority–Lawlessness (#VP5739)
Superiority–Inferiority (#VP5036)
Influence–Influencelessness (#VP5172).

♦ **VC1059 Learning**
Integrative polarities Knowledge–Ignorance (#VP5475)
Education–Miseducation (#VP5562)
Intelligence–Unintelligence (#VP5467).

♦ **VC1061 Legality**
Integrative polarities Consent–Refusal (#VP5775)
Justice–Injustice (#VP5976)
Legality–Illegality (#VP5998).

♦ **VC1063 Legitimacy**
Integrative polarities Truth–Error (#VP5516) Intuition–Reason (#VP5481)
Authority–Lawlessness (#VP5739) Vindication–Condemnation (#VP6005).

♦ **VC1065 Leisure**
Integrative polarities Exertion–Rest (#VP5709) Action–Inaction (#VP5705).

♦ **VC1067 Leniency**
Integrative polarities Energy–Moderation (#VP5161)
Kindness–Unkindness (#VP5938)
Patience–Impatience (#VP5861)
Leniency–Compulsion (#VP5756)
Compassion–Pitilessness (#VP5944)
Broadmindedness–Narrowmindedness (#VP5526).

♦ **VC1069 Liberality**
Integrative polarities Economy–Prodigality (#VP5851)
Education–Miseducation (#VP5562)
Unselfishness–Selfishness (#VP5978)
Hospitality–Inhospitality (#VP5925)
Oversufficiency–Insufficiency (#VP5660).

♦ **VC1071 Liberation**
Integrative polarities Safety–Danger (#VP5697) Desire–Avoidance (#VP5631)
Freedom–Restraint (#VP5762) Sharing–Appropriation (#VP5815).

♦ **VC1073 Liberty**
Integrative polarities Freedom–Restraint (#VP5762)
Rightness–Wrongness (#VP5957)
Timeliness–Untimeliness (#VP5129).

♦ **VC1075 Life**
Integrative polarities Life–Death (#VP5407) Action–Inaction (#VP5705)
Desire–Avoidance (#VP5631) Energy–Moderation (#VP5161)
Cheerfulness–Solemnity (#VP5870) Substantiality–Unsubstantiality (#VP5003).

♦ **VC1077 Light**
Integrative polarities Light–Darkness (#VP5335)
Weight–Lightness (#VP5352)
Facility–Difficulty (#VP5732)
Cheerfulness–Solemnity (#VP5870)
Skilfulness–Unskilfulness (#VP5733).

♦ **VC1079 Lightness**
Integrative polarities Light–Darkness (#VP5335)
Weight–Lightness (#VP5352)
Cheerfulness–Solemnity (#VP5870)
Skilfulness–Unskilfulness (#VP5733).

♦ **VC1081 Limitlessness**
Integrative polarities Freedom–Restraint (#VP5762)
Numerousness–Fewness (#VP5101)
Godliness–Ungodliness (#VP6013).

♦ **VC1083 Lineage**
Integrative polarities Ancestry–Posterity (#VP5170).

♦ **VC1085 Liveliness**
Integrative polarities Life–Death (#VP5407) Action–Inaction (#VP5705)
Desire–Avoidance (#VP5631) Energy–Moderation (#VP5161)
Attention–Inattention (#VP5530) Cheerfulness–Solemnity (#VP5870)
Feeling–Unfeelinglessness (#VP5855).

♦ **VC1087 Location**
Integrative polarities Location–Dislocation (#VP5185).

♦ **VC1089 Longanimity**
Integrative polarities Patience–Impatience (#VP5861)
Forgiveness–Vengeance (#VP5947).

♦ **VC1091 Longevity**
Integrative polarities Life–Death (#VP5407) Health–Disease (#VP5685)
Durability–Transience (#VP5110).

♦ **VC1093 Love**
Integrative polarities Love–Hate (#VP5930) Virtue–Vice (#VP5980)
Accord–Discaccord (#VP5794) Desire–Avoidance (#VP5631)
Friendship–Enmity (#VP5927) Kindness–Unkindness (#VP5938)
Sexiness–Unsexiness (#VP5419) Pleasure–Displeasure (#VP5865).

♦ **VC1095 Loyalty**
Integrative polarities Desire–Avoidance (#VP5631)
Probity–Improbity (#VP5974)
Dueness–Undueness (#VP5960)
Obedience–Disobedience (#VP5766)
Resolution–Irresolution (#VP5624).

♦ **VC1097 Lucidity**
Integrative polarities Light–Darkness (#VP5335)
Sanity–Insanity (#VP5472)
Intelligibility–Unintelligibility (#VP5548).

♦ **VC1099 Luck**
Integrative polarities Prosperity–Adversity (#VP5728)
Expectation–Inexpectation (#VP5539).

♦ **VC1101 Luxury**
Integrative polarities Comfort–Aggravation (#VP5885)
Pleasure–Displeasure (#VP5865)
Prosperity–Adversity (#VP5728)
Naturalness–Affectation (#VP5900)
Pleasantness–Unpleasantness (#VP5863).

♦ **VC1103 Machismo**
Integrative polarities Masculinity–Femininity (#VP5420).

♦ **VC1105 Magnanimity**
Integrative polarities Economy–Prodigality (#VP5851)
Greatness–Smallness (#VP5034)
Forgiveness–Vengeance (#VP5947)
Unselfishness–Selfishness (#VP5978)
Broadmindedness–Narrowmindedness (#VP5526).

♦ **VC1107 Magnificence**
Integrative polarities Goodness–Badness (#VP5674)
Naturalness–Affectation (#VP5900).

♦ **VC1109 Majesty**
Integrative polarities Pride–Humility (#VP5905) Greatness–Smallness (#VP5034)
Godliness–Ungodliness (#VP6013) Authority–Lawlessness (#VP5739)
Eloquence–Uneloquence (#VP5600) Naturalness–Affectation (#VP5900).

♦ **VC1111 Manageability**
Integrative polarities Freedom–Restraint (#VP5762)
Energy–Moderation (#VP5161)
Facility–Difficulty (#VP5732)
Expensiveness–Cheapness (#VP5848)
Accomplishment–Nonaccomplishment (#VP5722).

♦ **VC1113 Manoeuvrability**
Integrative polarities Energy–Moderation (#VP5161)
Facility–Difficulty (#VP5732).

♦ **VC1115 Manliness**
Integrative polarities Courage–Fear (#VP5890) Probity–Improbity (#VP5974)
Masculinity–Femininity (#VP5420).

♦ **VC1117 Manners**
Integrative polarities Courtesy–Discourtesy (#VP5936)
Formality–Informality (#VP5646)
Behaviour–Misbehaviour (#VP5737)
Conventionality–Unconventionality (#VP5642).

♦ **VC1121 Mastery**
Integrative polarities Victory–Defeat (#VP5726) Knowledge–Ignorance (#VP5475)
Authority–Lawlessness (#VP5739) Skilfulness–Unskilfulness (#VP5733)
Preparedness–Unpreparedness (#VP5720) Influence–Influencelessness (#VP5172).

♦ **VC1123 Materialization**
Integrative polarities Eventuation–Imminence (#VP5151).

♦ **VC1125 Maternity**
Integrative polarities Ancestry–Posterity (#VP5170).

♦ **VC1127 Maturity**
Integrative polarities Youth–Age (#VP5124) Skilfulness–Unskilfulness (#VP5733)
Preparedness–Unpreparedness (#VP5720) Completeness–Incompleteness (#VP5054)
Accomplishment–Nonaccomplishment (#VP5722).

♦ **VC1129 Meaningfulness**
Integrative polarities Eloquence–Uneloquence (#VP5600)
Meaning–Meaninglessness (#VP5545).

♦ **VC1131 Mediation**
Integrative polarities Accord–Discaccord (#VP5794).

♦ **VC1133 Meekness**
Integrative polarities Modesty–Vanity (#VP5908)
Pride–Humility (#VP5905)
Freedom–Restraint (#VP5762)
Patience–Impatience (#VP5861).

♦ **VC1135 Mellowness**
Integrative polarities Youth–Age (#VP5124) Pleasantness–Unpleasantness (#VP5863).

♦ **VC1137 Mercifulness**
Integrative polarities Leniency–Compulsion (#VP5756)
Compassion–Pitilessness (#VP5944).

CONSTRUCTIVE VALUES   **VC1225**

♦ **VC1139 Merit**
Integrative polarities Repute–Disrepute (#VP5914)
Goodness–Badness (#VP5674)
Approval–Disapproval (#VP5968)
Importance–Unimportance (#VP5672).

♦ **VC1141 Merriment**
Integrative polarities Amusement–Boredom (#VP5878)
Exultation–Lamentation (#VP5873)
Cheerfulness–Solemnity (#VP5870)
Sociability–Unsociability (#VP5922).

♦ **VC1143 Meticulousness**
Integrative polarities Truth–Error (#VP5516)   Taste–Vulgarity (#VP5896)
Probity–Improbity (#VP5974)   Carefulness–Neglect (#VP5533)
Attention–Inattention (#VP5530).

♦ **VC1145 Mettlesomeness**
Integrative polarities Courage–Fear (#VP5890)   Energy–Moderation (#VP5161)
Resolution–Irresolution (#VP5624).

♦ **VC1147 Might**
Integrative polarities Power–Impotence (#VP5157)
Strength–Weakness (#VP5159)
Greatness–Smallness (#VP5034)
Authority–Lawlessness (#VP5739).

♦ **VC1149 Mildness**
Integrative polarities Freedom–Restraint (#VP5762)
Energy–Moderation (#VP5161)
Kindness–Unkindness (#VP5938)
Leniency–Compulsion (#VP5756)
Savouriness–Unsavouriness (#VP5427).

♦ **VC1151 Mindfulness**
Integrative polarities Caution–Rashness (#VP5894)
Kindness–Unkindness (#VP5938)
Carefulness–Neglect (#VP5533)
Knowledge–Ignorance (#VP5475)
Attention–Inattention (#VP5530).

♦ **VC1153 Mitigation**
Integrative polarities Energy–Moderation (#VP5161)
Change–Permanence (#VP5139)
Comfort–Aggravation (#VP5885)
Compassion–Pitilessness (#VP5944).

♦ **VC1155 Mobility**
Integrative polarities Motion–Quiescence (#VP5267)
Stability–Changeableness (#VP5141).

♦ **VC1157 Moderation**
Integrative polarities Goodness–Badness (#VP5674)
Energy–Moderation (#VP5161)
Neutrality–Compromise (#VP5806)
Temperance–Intemperance (#VP5992)
Excitement–Inexcitability (#VP5857).

♦ **VC1159 Modesty**
Integrative polarities Modesty–Vanity (#VP5908)
Pride–Humility (#VP5905)
Chastity–Unchastity (#VP5987)
Unselfishness–Selfishness (#VP5978)
Willingness–Unwillingness (#VP5621).

♦ **VC1161 Momentousness**
Integrative polarities Authority–Lawlessness (#VP5739)
Eventuation–Imminence (#VP5151)
Importance–Unimportance (#VP5672)
Influence–Influencelessness (#VP5172).

♦ **VC1163 Morality**
Integrative polarities Virtue–Vice (#VP5980)   Rightness–Wrongness (#VP5957).

♦ **VC1165 Motherhood**
Integrative polarities Ancestry–Posterity (#VP5170).

♦ **VC1167 Motherliness**
Integrative polarities Ancestry–Posterity (#VP5170).

♦ **VC1169 Motivation**
Integrative polarities Motivation–Dissuasion (#VP5648)
Education–Miseducation (#VP5562).

♦ **VC1171 Motive**
Integrative polarities Impact–Reaction (#VP5283)
Motivation–Dissuasion (#VP5648).

♦ **VC1173 Munificence**
Integrative polarities Economy–Prodigality (#VP5851).

♦ **VC1175 Mutuality**
Integrative polarities Accord–Discord (#VP5794)
Support–Opposition (#VP5785)
Identity–Difference (#VP5013).

♦ **VC1179 Naturalness**
Integrative polarities Truth–Error (#VP5516)   Elegance–Inelegance (#VP5589)
Naturalness–Affectation (#VP5900)   Conformity–Nonconformity (#VP5082)
Skilfulness–Unskilfulness (#VP5733)   Preparedness–Unpreparedness (#VP5720).

♦ **VC1181 Neatness**
Integrative polarities Order–Disorder (#VP5059)   Simplicity–Complexity (#VP5045).

♦ **VC1183 Necessity**
Integrative polarities Wealth–Poverty (#VP5837)
Choice–Necessity (#VP5637)
Leniency–Compulsion (#VP5756)
Certainty–Uncertainty (#VP5513).

♦ **VC1185 Neighbourliness**
Integrative polarities Friendship–Enmity (#VP5927)
Support–Opposition (#VP5785)
Hospitality–Inhospitality (#VP5925).

♦ **VC1187 Neutrality**
Integrative polarities Justice–Injustice (#VP5976)
Freedom–Restraint (#VP5762)
Energy–Moderation (#VP5161)
Neutrality–Compromise (#VP5806)
Broadmindedness–Narrowmindedness (#VP5526).

♦ **VC1189 Newness**
Integrative polarities Newness–Oldness (#VP5122)
Futurity–Antiquity (#VP5119)
Originality–Imitation (#VP5023).

♦ **VC1191 Nimbleness**
Integrative polarities Action–Inaction (#VP5705)   Carefulness–Neglect (#VP5533)
Skilfulness–Unskilfulness (#VP5733)   Intelligence–Unintelligence (#VP5467).

♦ **VC1193 Nobility**
Integrative polarities Pride–Humility (#VP5905)   Repute–Disrepute (#VP5914)
Probity–Improbity (#VP5974)   Greatness–Smallness (#VP5034)
Authority–Lawlessness (#VP5739)   Eloquence–Uneloquence (#VP5600)
Naturalness–Affectation (#VP5900)   Importance–Unimportance (#VP5672)
Unselfishness–Selfishness (#VP5978)

♦ **VC1195 Nonagressivity**
Integrative polarities Accord–Discord (#VP5794).

♦ **VC1197 Nonalignment**
Integrative polarities Freedom–Restraint (#VP5762)
Neutrality–Compromise (#VP5806).

♦ **VC1199 Nonviolence**
Integrative polarities Accord–Discord (#VP5794)
Energy–Moderation (#VP5161).

♦ **VC1201 Normality**
Integrative polarities Sanity–Insanity (#VP5472)   Rightness–Wrongness (#VP5957).

♦ **VC1203 Notability**
Integrative polarities Repute–Disrepute (#VP5914)
Greatness–Smallness (#VP5034)
Importance–Unimportance (#VP5672).

♦ **VC1205 Novelty**
Integrative polarities Newness–Oldness (#VP5122)
Originality–Imitation (#VP5023).

♦ **VC1207 Obedience**
Integrative polarities Dueness–Undueness (#VP5960)
Freedom–Restraint (#VP5762)
Patience–Impatience (#VP5861)
Obedience–Disobedience (#VP5766).

♦ **VC1209 Objectivity**
Integrative polarities Broadmindedness–Narrowmindedness (#VP5526).

♦ **VC1211 Observance**
Integrative polarities Piety–Impiety (#VP6028)   Dueness–Undueness (#VP5960)
Carefulness–Neglect (#VP5533)   Attention–Inattention (#VP5530)
Obedience–Disobedience (#VP5766)   Observance–Nonobservance (#VP5768).

♦ **VC1213 Occurrence**
Integrative polarities Eventuation–Imminence (#VP5151).

♦ **VC1215 Omnipotence**
Integrative polarities Power–Impotence (#VP5157)
Godliness–Ungodliness (#VP6013).

♦ **VC1217 Omnipresence**
Integrative polarities Presence–Absence (#VP5186)
Godliness–Ungodliness (#VP6013)
Completeness–Incompleteness (#VP5054).

♦ **VC1219 Omniscience**
Integrative polarities Knowledge–Ignorance (#VP5475)
Godliness–Ungodliness (#VP6013).

♦ **VC1221 Openheartedness**
Integrative polarities Probity–Improbity (#VP5974)
Economy–Prodigality (#VP5851)
Hospitality–Inhospitality (#VP5925)
Skilfulness–Unskilfulness (#VP5733).

♦ **VC1223 Openmindedness**
Integrative polarities Freedom–Restraint (#VP5762)
Influence–Influencelessness (#VP5172)
Broadmindedness–Narrowmindedness (#VP5526).

♦ **VC1225 Openness**
Integrative polarities Opening–Closure (#VP5265)
Probity–Improbity (#VP5974)
Skilfulness–Unskilfulness (#VP5733)
Influence–Influencelessness (#VP5172)
Broadmindedness–Narrowmindedness (#VP5526)
Communicativeness–Uncommunicativeness (#VP5554).

VC1227

♦ **VC1227 Opportunity**
Integrative polarities Timeliness–Untimeliness (#VP5129).

♦ **VC1229 Optimism**
Integrative polarities Hope–Hopelessness (#VP5888)
Cheerfulness–Solemnity (#VP5870).

♦ **VC1231 Order**
Integrative polarities Order–Disorder (#VP5059)   Accord–Disaccord (#VP5794)
Education–Miseducation (#VP5562)   Judgement–Misjudgement (#VP5494)
Conformity–Nonconformity (#VP5082)   Influence–Influencelessness (#VP5172).

♦ **VC1233 Orderliness**
Integrative polarities Order–Disorder (#VP5059)   Accord–Disaccord (#VP5794)
Elegance–Inelegance (#VP5589)   Regularity–Irregularity (#VP5137)
Conformity–Nonconformity (#VP5082).

♦ **VC1235 Ordinariness**
Integrative polarities Taste–Vulgarity (#VP5896)   Frequency–Infrequency (#VP5135)
Naturalness–Affectation (#VP5900)   Perfection–Imperfection (#VP5677)
Superiority–Inferiority (#VP5036)   Conformity–Nonconformity (#VP5082)
Conventionality–Unconventionality (#VP5642).

♦ **VC1237 Organization**
Integrative polarities Order–Disorder (#VP5059)   Production–Reproduction (#VP5167).

♦ **VC1239 Originality**
Integrative polarities Newness–Oldness (#VP5122)
Originality–Imitation (#VP5023)
Imaginativeness–Unimaginativeness (#VP5535).

♦ **VC1241 Orthodoxy**
Integrative polarities Leniency–Compulsion (#VP5756)
Orthodoxy–Unorthodoxy (#VP6024)
Conformity–Nonconformity (#VP5082).

♦ **VC1243 Pacifism**
Integrative polarities Accord–Disaccord (#VP5794)
Energy–Moderation (#VP5161).

♦ **VC1245 Painstakingness**
Integrative polarities Carefulness–Neglect (#VP5533).

♦ **VC1247 Pardon**
Integrative polarities Forgiveness–Vengeance (#VP5947)
Compassion–Pitilessness (#VP5944)
Vindication–Condemnation (#VP6005).

♦ **VC1249 Parenthood**
Integrative polarities Ancestry–Posterity (#VP5170).

♦ **VC1251 Parsimony**
Integrative polarities Economy–Prodigality (#VP5851).

♦ **VC1253 Participation**
Integrative polarities Inclusion–Exclusion (#VP5076)
Sharing–Appropriation (#VP5815)
Sociability–Unsociability (#VP5922).

♦ **VC1255 Partnership**
Integrative polarities Support–Opposition (#VP5785)
Sharing–Appropriation (#VP5815).

♦ **VC1257 Passion**
Integrative polarities Love–Hate (#VP5930)   Desire–Avoidance (#VP5631)
Eloquence–Uneloquence (#VP5600)   Attention–Inattention (#VP5530)
Feeling–Unfeelinglessness (#VP5855)   Willingness–Unwillingness (#VP5621).

♦ **VC1259 Paternity**
Integrative polarities Ancestry–Posterity (#VP5170).

♦ **VC1261 Patience**
Integrative polarities Patience–Impatience (#VP5861)
Leniency–Compulsion (#VP5756)
Forgiveness–Vengeance (#VP5947)
Resolution–Irresolution (#VP5624)
Broadmindedness–Narrowmindedness (#VP5526).

♦ **VC1265 Peace**
Integrative polarities Order–Disorder (#VP5059)   Accord–Disaccord (#VP5794)
Comfort–Aggravation (#VP5885)   Agreement–Disagreement (#VP5026).

♦ **VC1267 Peacefulness**
Integrative polarities Order–Disorder (#VP5059)   Accord–Disaccord (#VP5794)
Comfort–Aggravation (#VP5885)   Excitement–Inexcitability (#VP5857).

♦ **VC1271 Percipience**
Integrative polarities Knowledge–Ignorance (#VP5475)
Intelligence–Unintelligence (#VP5467).

♦ **VC1273 Perfection**
Integrative polarities Truth–Error (#VP5516)   Improvement–Impairment (#VP5691)
Perfection–Imperfection (#VP5677)   Completeness–Incompleteness (#VP5054)
Accomplishment–Nonaccomplishment (#VP5722).

♦ **VC1275 Permanence**
Integrative polarities Change–Permanence (#VP5139)
Godliness–Ungodliness (#VP6013)
Durability–Transience (#VP5110)
Resolution–Irresolution (#VP5624)
Stability–Changeableness (#VP5141)
Perpetuity–Instantaneousness (#VP5112).

♦ **VC1277 Perpetuation**
Integrative polarities Continuance–Cessation (#VP5143)
Frequency–Infrequency (#VP5135)
Durability–Transience (#VP5110)
Perpetuity–Instantaneousness (#VP5112).

♦ **VC1279 Perseverance**
Integrative polarities Patience–Impatience (#VP5861)
Resolution–Irresolution (#VP5624)
Accomplishment–Nonaccomplishment (#VP5722).

♦ **VC1281 Persistence**
Integrative polarities Change–Permanence (#VP5139)
Durability–Transience (#VP5110)
Resolution–Irresolution (#VP5624)
Cohesion–Disintegration (#VP5050).

♦ **VC1283 Perspective**
Integrative polarities Nearness–Distance (#VP5200)
Location–Dislocation (#VP5185)
Visibility–Invisibility (#VP5444)
Appearance–Disappearance (#VP5446).

♦ **VC1285 Perspicacity**
Integrative polarities Knowledge–Ignorance (#VP5475)
Intelligence–Unintelligence (#VP5467)
Discrimination–Indiscrimination (#VP5492).

♦ **VC1287 Perspicuity**
Integrative polarities Intelligence–Unintelligence (#VP5467)
Intelligibility–Unintelligibility (#VP5548).

♦ **VC1289 Persuasiveness**
Integrative polarities Belief–Unbelief (#VP5501)   Influence–Influencelessness (#VP5172).

♦ **VC1291 Pertinacity**
Integrative polarities Courage–Fear (#VP5890)   Resolution–Irresolution (#VP5624).

♦ **VC1293 Philosophy**
Integrative polarities Intuition–Reason (#VP5481)
Thought–Thoughtlessness (#VP5478)
Excitement–Inexcitability (#VP5857).

♦ **VC1295 Piety**
Integrative polarities Piety–Impiety (#VP6028).

♦ **VC1297 Piousness**
Integrative polarities Piety–Impiety (#VP6028).

♦ **VC1299 Pity**
Integrative polarities Leniency–Compulsion (#VP5756)
Compassion–Pitilessness (#VP5944).

♦ **VC1301 Plainness**
Integrative polarities Pride–Humility (#VP5905)   Probity–Improbity (#VP5974)
Simplicity–Complexity (#VP5045)   Naturalness–Affectation (#VP5900)
Skilfulness–Unskilfulness (#VP5733)   Intelligibility–Unintelligibility (#VP5548)
Communicativeness–Uncommunicativeness (#VP5554).

♦ **VC1303 Plausibility**
Integrative polarities Belief–Unbelief (#VP5501)   Savouriness–Unsavouriness (#VP5427).

♦ **VC1305 Playfulness**
Integrative polarities Amusement–Boredom (#VP5878)
Cheerfulness–Solemnity (#VP5870).

♦ **VC1307 Pleasantness**
Integrative polarities Goodness–Badness (#VP5674)
Friendship–Enmity (#VP5927)
Cheerfulness–Solemnity (#VP5870)
Pleasantness–Unpleasantness (#VP5863).

♦ **VC1309 Pleasure**
Integrative polarities Choice–Necessity (#VP5637)
Desire–Avoidance (#VP5631)
Amusement–Boredom (#VP5878)
Pleasure–Displeasure (#VP5865)
Expensiveness–Cheapness (#VP5848)
Willingness–Unwillingness (#VP5621)
Pleasantness–Unpleasantness (#VP5863).

♦ **VC1311 Plenitude**
Integrative polarities Greatness–Smallness (#VP5034)
Perfection–Imperfection (#VP5677)
Continuity–Discontinuity (#VP5071)
Completeness–Incompleteness (#VP5054)
Oversufficiency–Insufficiency (#VP5660).

♦ **VC1313 Plenty**
Integrative polarities Oversufficiency–Insufficiency (#VP5660)
Productiveness–Unproductiveness (#VP5165).

♦ **VC1315 Pluck**
Integrative polarities Courage–Fear (#VP5890)   Resolution–Irresolution (#VP5624).

♦ **VC1317 Politeness**
Integrative polarities Courtesy–Discourtesy (#VP5936)
Formality–Informality (#VP5646).

♦ **VC1319 Popularity**
Integrative polarities Love–Hate (#VP5930)   Repute–Disrepute (#VP5914)
Approval–Disapproval (#VP5968).

CONSTRUCTIVE VALUES

VC1401

♦ **VC1321 Position**
Integrative polarities Location–Dislocation (#VP5185).

♦ **VC1323 Positiveness**
Integrative polarities Belief–Unbelief (#VP5501)   Support–Opposition (#VP5785)
Certainty–Uncertainty (#VP5513)   Agreement–Disagreement (#VP5026)
Existence–Nonexistence (#VP5001).

♦ **VC1325 Possession**
Integrative polarities Possession–Loss (#VP5808)
Sanity–Insanity (#VP5472)
Sharing–Appropriation (#VP5815)
Resolution–Irresolution (#VP5624)
Excitement–Inexcitability (#VP5857).

♦ **VC1327 Potency**
Integrative polarities Power–Impotence (#VP5157)
Energy–Moderation (#VP5161)
Strength–Weakness (#VP5159)
Authority–Lawlessness (#VP5739)
Masculinity–Femininity (#VP5420)
Influence–Influencelessness (#VP5172).

♦ **VC1329 Potentiality**
Integrative polarities Power–Impotence (#VP5157)
Skilfulness–Unskilfulness (#VP5733).

♦ **VC1331 Power**
Integrative polarities Power–Impotence (#VP5157)
Energy–Moderation (#VP5161)
Strength–Weakness (#VP5159)
Greatness–Smallness (#VP5034)
Authority–Lawlessness (#VP5739)
Resolution–Irresolution (#VP5624)
Skilfulness–Unskilfulness (#VP5733)
Influence–Influencelessness (#VP5172).

♦ **VC1333 Practicability**
Integrative polarities Energy–Moderation (#VP5161)
Appropriateness–Inappropriateness (#VP5665).

♦ **VC1335 Practicality**
Integrative polarities Facility–Difficulty (#VP5732)
Intelligence–Unintelligence (#VP5467)
Appropriateness–Inappropriateness (#VP5665)
Imaginativeness–Unimaginativeness (#VP5535).

♦ **VC1337 Praise**
Integrative polarities Repute–Disrepute (#VP5914)
Approval–Disapproval (#VP5968)
Gratitude–Ingratitude (#VP5949).

♦ **VC1339 Precedence**
Integrative polarities Authority–Lawlessness (#VP5739)
Originality–Imitation (#VP5023)
Importance–Unimportance (#VP5672)
Superiority–Inferiority (#VP5036).

♦ **VC1341 Precision**
Integrative polarities Truth–Error (#VP5516)   Taste–Vulgarity (#VP5896)
Carefulness–Neglect (#VP5533)   Discrimination–Indiscrimination (#VP5492).

♦ **VC1343 Preparedness**
Integrative polarities Carefulness–Neglect (#VP5533)
Preparedness–Unpreparedness (#VP5720).

♦ **VC1345 Presence**
Integrative polarities Presence–Absence (#VP5186)
Behaviour–Misbehaviour (#VP5737)
Existence–Nonexistence (#VP5001).

♦ **VC1347 Preservation**
Integrative polarities Safety–Danger (#VP5697)   Possession–Loss (#VP5808)
Change–Permanence (#VP5139)   Godliness–Ungodliness (#VP6013)
Perpetuity–Instantaneousness (#VP5112).

♦ **VC1349 Prestige**
Integrative polarities Repute–Disrepute (#VP5914)
Respect–Disrespect (#VP5964)
Authority–Lawlessness (#VP5739)
Importance–Unimportance (#VP5672)
Influence–Influencelessness (#VP5172).

♦ **VC1351 Prettiness**
Integrative polarities Beauty–Ugliness (#VP5899).

♦ **VC1353 Principle**
Integrative polarities Belief–Unbelief (#VP5501)   Motivation–Dissuasion (#VP5648)
Causation–Culmination (#VP5153)   Conformity–Nonconformity (#VP5082)
Intrinsicality–Extrinsicality (#VP5005).

♦ **VC1355 Priority**
Integrative polarities Leading–Following (#VP5292)
Authority–Lawlessness (#VP5739)
Importance–Unimportance (#VP5672).

♦ **VC1357 Privacy**
Integrative polarities Unity–Duality (#VP5089)   Intrinsicality–Extrinsicality (#VP5005).

♦ **VC1359 Privilege**
Integrative polarities Consent–Refusal (#VP5775)
Rightness–Wrongness (#VP5957).

♦ **VC1361 Probity**
Integrative polarities Virtue–Vice (#VP5980)   Probity–Improbity (#VP5974).

♦ **VC1363 Prodigiousness**
Integrative polarities Greatness–Smallness (#VP5034)
Wonder–Unastonishment (#VP5920)
Conformity–Nonconformity (#VP5082).

♦ **VC1365 Productiveness**
Integrative polarities Possession–Loss (#VP5808)
Power–Impotence (#VP5157)
Increase–Decrease (#VP5038)
Production–Reproduction (#VP5167)
Oversufficiency–Insufficiency (#VP5660)
Productiveness–Unproductiveness (#VP5165)
Imaginativeness–Unimaginativeness (#VP5535).

♦ **VC1367 Proficiency**
Integrative polarities Power–Impotence (#VP5157)
Knowledge–Ignorance (#VP5475)
Perfection–Imperfection (#VP5677)
Skilfulness–Unskilfulness (#VP5733)
Preparedness–Unpreparedness (#VP5720).

♦ **VC1369 Profitability**
Integrative polarities Possession–Loss (#VP5808)
Goodness–Badness (#VP5674)
Appropriateness–Inappropriateness (#VP5665).

♦ **VC1371 Profundity**
Integrative polarities Depth–Shallowness (#VP5209)
Intelligence–Unintelligence (#VP5467).

♦ **VC1373 Progress**
Integrative polarities Prosperity–Adversity (#VP5728)
Evolution–Revolution (#VP5147)
Continuance–Cessation (#VP5143)
Improvement–Impairment (#VP5691)
Progression–Regression (#VP5294).

♦ **VC1375 Promise**
Integrative polarities Hope–Hopelessness (#VP5888)
Expectation–Inexpectation (#VP5539).

♦ **VC1377 Promptness**
Integrative polarities Exertion–Rest (#VP5709)   Consent–Refusal (#VP5775)
Action–Inaction (#VP5705)   Desire–Avoidance (#VP5631)
Carefulness–Neglect (#VP5533)   Willingness–Unwillingness (#VP5621).

♦ **VC1379 Propriety**
Integrative polarities Taste–Vulgarity (#VP5896)   Justice–Injustice (#VP5976)
Chastity–Unchastity (#VP5987)   Rightness–Wrongness (#VP5957)
Timeliness–Untimeliness (#VP5129)   Conformity–Nonconformity (#VP5082).

♦ **VC1381 Prosperity**
Integrative polarities Wealth–Poverty (#VP5837)
Prosperity–Adversity (#VP5728)
Accomplishment–Nonaccomplishment (#VP5722).

♦ **VC1383 Protection**
Integrative polarities Safety–Danger (#VP5697)   Support–Opposition (#VP5785).

♦ **VC1385 Proximity**
Integrative polarities Nearness–Distance (#VP5200)
Relatedness–Unrelatedness (#VP5009).

♦ **VC1387 Prudence**
Integrative polarities Virtue–Vice (#VP5980)   Caution–Rashness (#VP5894)
Energy–Moderation (#VP5161)   Economy–Prodigality (#VP5851)
Carefulness–Neglect (#VP5533)   Intelligence–Unintelligence (#VP5467).

♦ **VC1389 Punctuality**
Integrative polarities Probity–Improbity (#VP5974)
Carefulness–Neglect (#VP5533)
Regularity–Irregularity (#VP5137)
Timeliness–Untimeliness (#VP5129).

♦ **VC1391 Purity**
Integrative polarities Virtue–Vice (#VP5980)   Truth–Error (#VP5516)
Piety–Impiety (#VP6028)   Unity–Duality (#VP5089)
Innocence–Guilt (#VP5983)   Taste–Vulgarity (#VP5896)
Probity–Improbity (#VP5974)   Chastity–Unchastity (#VP5987)
Cleanness–Uncleanness (#VP5681)   Simplicity–Complexity (#VP5045)
Naturalness–Affectation (#VP5900)   Perfection–Imperfection (#VP5677).

♦ **VC1393 Purposefulness**
Integrative polarities Resolution–Irresolution (#VP5624).

♦ **VC1395 Qualification**
Integrative polarities Power–Impotence (#VP5157)
Agreement–Disagreement (#VP5026)
Skilfulness–Unskilfulness (#VP5733)
Preparedness–Unpreparedness (#VP5720)
Accomplishment–Nonaccomplishment (#VP5722).

♦ **VC1397 Quality**
Integrative polarities Taste–Vulgarity (#VP5896)   Repute–Disrepute (#VP5914)
Goodness–Badness (#VP5674)   Intrinsicality–Extrinsicality (#VP5005).

♦ **VC1399 Quickness**
Integrative polarities Life–Death (#VP5407)   Action–Inaction (#VP5705)
Desire–Avoidance (#VP5631)   Carefulness–Neglect (#VP5533)
Education–Miseducation (#VP5562)   Timeliness–Untimeliness (#VP5129)
Skilfulness–Unskilfulness (#VP5733)   Intelligence–Unintelligence (#VP5467).

♦ **VC1401 Quiescence**
Integrative polarities Motion–Quiescence (#VP5267).

VC1403

♦ **VC1403 Quietness**
Integrative polarities Exertion–Rest (#VP5709)   Modesty–Vanity (#VP5908)
Order–Disorder (#VP5059)   Taste–Vulgarity (#VP5896)
Accord–Disaccord (#VP5794)   Silence–Loudness (#VP5451)
Motion–Quiescence (#VP5267)   Energy–Moderation (#VP5161)
Excitement–Inexcitability (#VP5857)
Communicativeness–Uncommunicativeness (#VP5554).

♦ **VC1405 Quietude**
Integrative polarities Order–Disorder (#VP5059)   Accord–Disaccord (#VP5794)
Motion–Quiescence (#VP5267)   Excitement–Inexcitability (#VP5857).

♦ **VC1407 Quintessence**
Integrative polarities Goodness–Badness (#VP5674)
Perfection–Imperfection (#VP5677)
Intrinsicality–Extrinsicality (#VP5005).

♦ **VC1409 Radiance**
Integrative polarities Light–Darkness (#VP5335)
Beauty–Ugliness (#VP5899)
Repute–Disrepute (#VP5914)
Pleasure–Displeasure (#VP5865)
Godliness–Ungodliness (#VP6013)
Cheerfulness–Solemnity (#VP5870).

♦ **VC1411 Rapture**
Integrative polarities Love–Hate (#VP5930)   Pleasure–Displeasure (#VP5865).

♦ **VC1413 Rareness**
Integrative polarities Numerousness–Fewness (#VP5101)
Wonder–Unastonishment (#VP5920)
Importance–Unimportance (#VP5672).

♦ **VC1415 Rationality**
Integrative polarities Sanity–Insanity (#VP5472)   Intuition–Reason (#VP5481)
Intelligence–Unintelligence (#VP5467)   Imaginativeness–Unimaginativeness (#VP5535).

♦ **VC1417 Readiness**
Integrative polarities Consent–Refusal (#VP5775)
Action–Inaction (#VP5705)
Desire–Avoidance (#VP5631)
Carefulness–Neglect (#VP5533)
Timeliness–Untimeliness (#VP5129)
Skilfulness–Unskilfulness (#VP5733)
Willingness–Unwillingness (#VP5621)
Preparedness–Unpreparedness (#VP5720).

♦ **VC1419 Reality**
Integrative polarities Truth–Error (#VP5516)   Certainty–Uncertainty (#VP5513)
Existence–Nonexistence (#VP5001).

♦ **VC1421 Realization**
Integrative polarities Knowledge–Ignorance (#VP5475)
Causation–Culmination (#VP5153)
Production–Reproduction (#VP5167)
Remembrance–Forgetfulness (#VP5537)
Completeness–Incompleteness (#VP5054)
Accomplishment–Nonaccomplishment (#VP5722).

♦ **VC1423 Reason**
Integrative polarities Intuition–Reason (#VP5481)
Motivation–Dissuasion (#VP5648)
Causation–Culmination (#VP5153)
Judgement–Misjudgement (#VP5494)
Thought–Thoughtlessness (#VP5478)
Vindication–Condemnation (#VP6005)
Intelligence–Unintelligence (#VP5467)
Interpretability–Misinterpretability (#VP5552).

♦ **VC1425 Reasonableness**
Integrative polarities Sanity–Insanity (#VP5472)   Intuition–Reason (#VP5481)
Energy–Moderation (#VP5161)   Vindication–Condemnation (#VP6005)
Intelligence–Unintelligence (#VP5467)   Imaginativeness–Unimaginativeness (#VP5535).

♦ **VC1427 Recognition**
Integrative polarities Assent–Dissent (#VP5521)
Repute–Disrepute (#VP5914)
Knowledge–Ignorance (#VP5475)
Approval–Disapproval (#VP5968)
Gratitude–Ingratitude (#VP5949).

♦ **VC1429 Recollection**
Integrative polarities Futurity–Antiquity (#VP5119).

♦ **VC1431 Recommendation**
Integrative polarities Approval–Disapproval (#VP5968)
Motivation–Dissuasion (#VP5648).

♦ **VC1433 Reconciliation**
Integrative polarities Accord–Disaccord (#VP5794)
Patience–Impatience (#VP5861)
Agreement–Disagreement (#VP5026)
Contentment–Discontentment (#VP5868).

♦ **VC1435 Recreation**
Integrative polarities Amusement–Boredom (#VP5878).

♦ **VC1437 Rectitude**
Integrative polarities Virtue–Vice (#VP5980)   Probity–Improbity (#VP5974).

♦ **VC1439 Redemption**
Integrative polarities Safety–Danger (#VP5697)   Sanctity–Unsanctity (#VP6026)
Atonement–Punishment (#VP6010)   Forgiveness–Vengeance (#VP5947)
Sharing–Appropriation (#VP5815).

♦ **VC1441 Refinement**
Integrative polarities Truth–Error (#VP5516)   Taste–Vulgarity (#VP5896)
Carefulness–Neglect (#VP5533)   Courtesy–Discourtesy (#VP5936)
Cleanness–Uncleanness (#VP5681)   Simplicity–Complexity (#VP5045)
Improvement–Impairment (#VP5691)   Discrimination–Indiscrimination (#VP5492).

♦ **VC1443 Reform**
Integrative polarities Conversion–Reversion (#VP5145)
Restoration–Destruction (#VP5693)
Production–Reproduction (#VP5167).

♦ **VC1445 Regularity**
Integrative polarities Order–Disorder (#VP5059)   Symmetry–Distortion (#VP5248)
Frequency–Infrequency (#VP5135)   Regularity–Irregularity (#VP5137)
Conformity–Nonconformity (#VP5082)   Uniformity–Nonuniformity (#VP5017).

♦ **VC1447 Regulation**
Integrative polarities Legality–Illegality (#VP5998)
Agreement–Disagreement (#VP5026).

♦ **VC1449 Rehabilitation**
Integrative polarities Restoration–Destruction (#VP5693)
Vindication–Condemnation (#VP6005).

♦ **VC1451 Reinforcement**
Integrative polarities Strength–Weakness (#VP5159)
Support–Opposition (#VP5785)
Provability–Unprovability (#VP5505).

♦ **VC1453 Relation**
Integrative polarities Meaning–Meaninglessness (#VP5545)
Relatedness–Unrelatedness (#VP5009).

♦ **VC1455 Relaxation**
Integrative polarities Exertion–Rest (#VP5709)   Amusement–Boredom (#VP5878)
Energy–Moderation (#VP5161)   Comfort–Aggravation (#VP5885)
Leniency–Compulsion (#VP5756).

♦ **VC1457 Relevance**
Integrative polarities Agreement–Disagreement (#VP5026)
Meaning–Meaninglessness (#VP5545)
Relatedness–Unrelatedness (#VP5009).

♦ **VC1459 Reliability**
Integrative polarities Belief–Unbelief (#VP5501)   Probity–Improbity (#VP5974)
Certainty–Uncertainty (#VP5513)   Stability–Changeableness (#VP5141).

♦ **VC1461 Relief**
Integrative polarities Support–Opposition (#VP5785)
Comfort–Aggravation (#VP5885)
Compassion–Pitilessness (#VP5944).

♦ **VC1463 Remission**
Integrative polarities Energy–Moderation (#VP5161)
Forgiveness–Vengeance (#VP5947)
Sharing–Appropriation (#VP5815)
Vindication–Condemnation (#VP6005).

♦ **VC1465 Renewableness**
Integrative polarities Change–Permanence (#VP5139)
Restoration–Destruction (#VP5693).

♦ **VC1467 Renunciation**
Integrative polarities Temperance–Intemperance (#VP5992).

♦ **VC1469 Repentance**
Integrative polarities Atonement–Punishment (#VP6010)
Exultation–Lamentation (#VP5873).

♦ **VC1471 Reputation**
Integrative polarities Repute–Disrepute (#VP5914)
Importance–Unimportance (#VP5672).

♦ **VC1473 Reservedness**
Integrative polarities Safety–Danger (#VP5697)   Modesty–Vanity (#VP5908)
Conciseness–Diffuseness (#VP5592).

♦ **VC1475 Resilience**
Integrative polarities Elasticity–Toughness (#VP5358)
Restoration–Destruction (#VP5693)
Stability–Changeableness (#VP5141).

♦ **VC1477 Resistance**
Integrative polarities Health–Disease (#VP5685)
Hardness–Softness (#VP5356)
Support–Opposition (#VP5785)
Prosperity–Adversity (#VP5728)
Elasticity–Toughness (#VP5358)
Resolution–Irresolution (#VP5624)
Willingness–Unwillingness (#VP5621)
Concurrence–Counteraction (#VP5177)
Selfactualization–Neurosis (#VP5690).

♦ **VC1479 Resolution**
Integrative polarities Courage–Fear (#VP5890)   Action–Inaction (#VP5705)
Harmony–Discord (#VP5462)   Accord–Disaccord (#VP5794)
Desire–Avoidance (#VP5631)   Research–Discovery (#VP5485)
Judgement–Misjudgement (#VP5494)   Resolution–Irresolution (#VP5624)
Willingness–Unwillingness (#VP5621).

♦ **VC1481 Resourcefulness**
Integrative polarities Skilfulness–Unskilfulness (#VP5733).

CONSTRUCTIVE VALUES　　　VC1573

♦ **VC1483 Respectability**
　Integrative polarities Repute–Disrepute (#VP5914)
　Goodness–Badness (#VP5674)
　Probity–Improbity (#VP5974).

♦ **VC1485 Respectfulness**
　Integrative polarities Repute–Disrepute (#VP5914)
　Kindness–Unkindness (#VP5938)
　Approval–Disapproval (#VP5968)
　Courtesy–Discourtesy (#VP5936)
　Attention–Inattention (#VP5530).

♦ **VC1487 Resplendence**
　Integrative polarities Light–Darkness (#VP5335)
　Beauty–Ugliness (#VP5899)
　Repute–Disrepute (#VP5914)
　Naturalness–Affectation (#VP5900).

♦ **VC1489 Responsibility**
　Integrative polarities Probity–Improbity (#VP5974)
　Dueness–Undueness (#VP5960).

♦ **VC1491 Responsiveness**
　Integrative polarities Hardness–Softness (#VP5356)
　Research–Discovery (#VP5485)
　Elasticity–Toughness (#VP5358)
　Willingness–Unwillingness (#VP5621)
　Influence–Influencelessness (#VP5172)
　Communicativeness–Uncommunicativeness (#VP5554).

♦ **VC1493 Restfulness**
　Integrative polarities Accord–Disaccord (#VP5794)
　Energy–Moderation (#VP5161)
　Comfort–Aggravation (#VP5885).

♦ **VC1495 Restraint**
　Integrative polarities Modesty–Vanity (#VP5908)
　Taste–Vulgarity (#VP5896)
　Energy–Moderation (#VP5161)
　Temperance–Intemperance (#VP5992)
　Resolution–Irresolution (#VP5624)
　Excitement–Inexcitability (#VP5857).

♦ **VC1497 Revelation**
　Integrative polarities Research–Discovery (#VP5485)
　Appearance–Disappearance (#VP5446).

♦ **VC1499 Reverence**
　Integrative polarities Piety–Impiety (#VP6028)　　Respect–Disrespect (#VP5964).

♦ **VC1501 Revival**
　Integrative polarities Energy–Moderation (#VP5161)
　Sharing–Appropriation (#VP5815)
　Improvement–Impairment (#VP5691)
　Restoration–Destruction (#VP5693).

♦ **VC1503 Richness**
　Integrative polarities Harmony–Discord (#VP5462)
　Amusement–Boredom (#VP5878)
　Attention–Inattention (#VP5530)
　Expensiveness–Cheapness (#VP5848)
　Oversufficiency–Insufficiency (#VP5660)
　Productiveness–Unproductiveness (#VP5165).

♦ **VC1505 Righteousness**
　Integrative polarities Virtue–Vice (#VP5980)　　Piety–Impiety (#VP6028)
　Probity–Improbity (#VP5974)　　Rightness–Wrongness (#VP5957).

♦ **VC1507 Rightness**
　Integrative polarities Truth–Error (#VP5516)　　Sanity–Insanity (#VP5472)
　Justice–Injustice (#VP5976)　　Probity–Improbity (#VP5974)
　Freedom–Restraint (#VP5762)　　Rightness–Wrongness (#VP5957)
　Authority–Lawlessness (#VP5739)　　Restoration–Destruction (#VP5693).

♦ **VC1509 Rigorousness**
　Integrative polarities Truth–Error (#VP5516)　　Chastity–Unchastity (#VP5987)
　Leniency–Compulsion (#VP5756)　　Carefulness–Neglect (#VP5533).

♦ **VC1511 Romance**
　Integrative polarities Love–Hate (#VP5930)　　Harmony–Discord (#VP5462)
　Representation–Misrepresentation (#VP5573)　　Imaginativeness–Unimaginativeness (#VP5535).

♦ **VC1513 Royalty**
　Integrative polarities Repute–Disrepute (#VP5914)
　Authority–Lawlessness (#VP5739).

♦ **VC1515 Sacredness**
　Integrative polarities Harmony–Discord (#VP5462)
　Sanctity–Unsanctity (#VP6026)
　Godliness–Ungodliness (#VP6013).

♦ **VC1517 Safety**
　Integrative polarities Safety–Danger (#VP5697).

♦ **VC1519 Sagacity**
　Integrative polarities Knowledge–Ignorance (#VP5475)
　Skilfulness–Unskilfulness (#VP5733)
　Intelligence–Unintelligence (#VP5467).

♦ **VC1521 Saintliness**
　Integrative polarities Virtue–Vice (#VP5980)　　Piety–Impiety (#VP6028)
　Pleasantness–Unpleasantness (#VP5863).

♦ **VC1523 Salubrity**
　Integrative polarities Healthfulness–Unhealthfulness (#VP5683).

♦ **VC1525 Salvation**
　Integrative polarities Safety–Danger (#VP5697)　　Sanctity–Unsanctity (#VP6026)
　Restoration–Destruction (#VP5693).

♦ **VC1527 Sanctification**
　Integrative polarities Repute–Disrepute (#VP5914)
　Sanctity–Unsanctity (#VP6026).

♦ **VC1529 Sanctity**
　Integrative polarities Piety–Impiety (#VP6028)　　Sanctity–Unsanctity (#VP6026).

♦ **VC1531 Sanity**
　Integrative polarities Sanity–Insanity (#VP5472)　　Intuition–Reason (#VP5481)
　Intelligence–Unintelligence (#VP5467).

♦ **VC1533 Satisfaction**
　Integrative polarities Pleasure–Displeasure (#VP5865)
　Contentment–Discontentment (#VP5868)
　Oversufficiency–Insufficiency (#VP5660).

♦ **VC1535 Scrupulousness**
　Integrative polarities Taste–Vulgarity (#VP5896)　　Probity–Improbity (#VP5974)
　Dueness–Undueness (#VP5960)　　Carefulness–Neglect (#VP5533)
　Formality–Informality (#VP5646).

♦ **VC1537 Security**
　Integrative polarities Safety–Danger (#VP5697)　　Hope–Hopelessness (#VP5888)
　Prosperity–Adversity (#VP5728)　　Certainty–Uncertainty (#VP5513)
　Stability–Changeableness (#VP5141).

♦ **VC1539 Self-actualization**
　Integrative polarities Accomplishment–Nonaccomplishment (#VP5722).

♦ **VC1541 Self-advancement**
　Integrative polarities Support–Opposition (#VP5785).

♦ **VC1543 Self-assurance**
　Integrative polarities Certainty–Uncertainty (#VP5513)
　Excitement–Inexcitability (#VP5857).

♦ **VC1545 Self-confidence**
　Integrative polarities Pride–Humility (#VP5905)　　Certainty–Uncertainty (#VP5513)
　Excitement–Inexcitability (#VP5857).

♦ **VC1547 Self-containedness**
　Integrative polarities Freedom–Restraint (#VP5762).

♦ **VC1549 Self-control**
　Integrative polarities Energy–Moderation (#VP5161)
　Patience–Impatience (#VP5861)
　Leniency–Compulsion (#VP5756)
　Resolution–Irresolution (#VP5624)
　Sociability–Unsociability (#VP5922)
　Excitement–Inexcitability (#VP5857).

♦ **VC1551 Self-denial**
　Integrative polarities Energy–Moderation (#VP5161)
　Chastity–Unchastity (#VP5987)
　Temperance–Intemperance (#VP5992)
　Resolution–Irresolution (#VP5624)
　Unselfishness–Selfishness (#VP5978).

♦ **VC1553 Self-determination**
　Integrative polarities Freedom–Restraint (#VP5762)
　Willingness–Unwillingness (#VP5621).

♦ **VC1555 Self-direction**
　Integrative polarities Freedom–Restraint (#VP5762).

♦ **VC1557 Self-discipline**
　Integrative polarities Resolution–Irresolution (#VP5624)
　Sociability–Unsociability (#VP5922).

♦ **VC1559 Self-effacement**
　Integrative polarities Modesty–Vanity (#VP5908)
　Unselfishness–Selfishness (#VP5978).

♦ **VC1561 Self-expression**
　Integrative polarities Accomplishment–Nonaccomplishment (#VP5722).

♦ **VC1563 Self-fulfilment**
　Integrative polarities Accomplishment–Nonaccomplishment (#VP5722).

♦ **VC1565 Self-government**
　Integrative polarities Freedom–Restraint (#VP5762)
　Resolution–Irresolution (#VP5624).

♦ **VC1567 Self-help**
　Integrative polarities Support–Opposition (#VP5785).

♦ **VC1569 Self-interest**
　Integrative polarities Modesty–Vanity (#VP5908)
　Unselfishness–Selfishness (#VP5978).

♦ **VC1571 Self-possession**
　Integrative polarities Resolution–Irresolution (#VP5624)
　Excitement–Inexcitability (#VP5857).

♦ **VC1573 Self-preservation**
　Integrative polarities Attack–Defence (#VP5798).

VC1577

♦ **VC1577 Self-reliance**
Integrative polarities Pride–Humility (#VP5905)   Freedom–Restraint (#VP5762)
Certainty–Uncertainty (#VP5513).

♦ **VC1579 Self-respect**
Integrative polarities Pride–Humility (#VP5905).

♦ **VC1581 Self-restraint**
Integrative polarities Energy–Moderation (#VP5161)
Temperance–Intemperance (#VP5992)
Resolution–Irresolution (#VP5624)
Excitement–Inexcitability (#VP5857).

♦ **VC1583 Self-sacrifice**
Integrative polarities Unselfishness–Selfishness (#VP5978).

♦ **VC1585 Self-sufficiency**
Integrative polarities Pride–Humility (#VP5905)   Freedom–Restraint (#VP5762).

♦ **VC1587 Selflessness**
Integrative polarities Justice–Injustice (#VP5976)
Unselfishness–Selfishness (#VP5978).

♦ **VC1589 Sensation**
Integrative polarities Wonder–Unastonishment (#VP5920)
Sensation–Insensibility (#VP5422)
Excitement–Inexcitability (#VP5857)
Feeling–Unfeelinglessness (#VP5855)
Intelligence–Unintelligence (#VP5467)
Accomplishment–Nonaccomplishment (#VP5722).

♦ **VC1591 Sensibility**
Integrative polarities Hardness–Softness (#VP5356)
Knowledge–Ignorance (#VP5475)
Sensation–Insensibility (#VP5422)
Intelligence–Unintelligence (#VP5467)
Discrimination–Indiscrimination (#VP5492).

♦ **VC1593 Sensibleness**
Integrative polarities Sanity–Insanity (#VP5472)   Intuition–Reason (#VP5481)
Knowledge–Ignorance (#VP5475)   Gratitude–Ingratitude (#VP5949)
Intelligence–Unintelligence (#VP5467)   Imaginativeness–Unimaginativeness (#VP5535).

♦ **VC1595 Sensitivity**
Integrative polarities Taste–Vulgarity (#VP5896)   Sensation–Insensibility (#VP5422)
Discrimination–Indiscrimination (#VP5492).

♦ **VC1597 Sentiment**
Integrative polarities Love–Hate (#VP5930)   Belief–Unbelief (#VP5501)
Feeling–Unfeelinglessness (#VP5855).

♦ **VC1599 Serendipity**
Integrative polarities Attributability–Chance (#VP5155).

♦ **VC1601 Serenity**
Integrative polarities Accord–Disaccord (#VP5794)
Motion–Quiescence (#VP5267)
Energy–Moderation (#VP5161)
Excitement–Inexcitability (#VP5857).

♦ **VC1603 Seriousness**
Integrative polarities Desire–Avoidance (#VP5631)
Cheerfulness–Solemnity (#VP5870)
Importance–Unimportance (#VP5672)
Resolution–Irresolution (#VP5624)
Thought–Thoughtlessness (#VP5478)
Excitement–Inexcitability (#VP5857).

♦ **VC1605 Service**
Integrative polarities Freedom–Restraint (#VP5762)
Support–Opposition (#VP5785)
Kindness–Unkindness (#VP5938)
Formality–Informality (#VP5646)
Obedience–Disobedience (#VP5766)
Restoration–Destruction (#VP5693)
Appropriateness–Inappropriateness (#VP5665).

♦ **VC1607 Settlement**
Integrative polarities Location–Dislocation (#VP5185).

♦ **VC1609 Sex appeal**
Integrative polarities Sexiness–Unsexiness (#VP5419)
Motivation–Dissuasion (#VP5648).

♦ **VC1611 Shelter**
Integrative polarities Safety–Danger (#VP5697).

♦ **VC1613 Significance**
Integrative polarities Repute–Disrepute (#VP5914)
Importance–Unimportance (#VP5672)
Meaning–Meaninglessness (#VP5545).

♦ **VC1615 Similarity**
Integrative polarities Similarity–Dissimilarity (#VP5020)
Relatedness–Unrelatedness (#VP5009).

♦ **VC1617 Simplicity**
Integrative polarities Truth–Error (#VP5516)   Unity–Duality (#VP5089)
Taste–Vulgarity (#VP5896)   Facility–Difficulty (#VP5732)
Simplicity–Complexity (#VP5045)   Naturalness–Affectation (#VP5900)
Skilfulness–Unskilfulness (#VP5733)   Intelligibility–Unintelligibility (#VP5548).

♦ **VC1619 Sincerity**
Integrative polarities Truth–Error (#VP5516)   Desire–Avoidance (#VP5631)
Probity–Improbity (#VP5974)   Resolution–Irresolution (#VP5624)
Skilfulness–Unskilfulness (#VP5733).

♦ **VC1621 Skill**
Integrative polarities Skilfulness–Unskilfulness (#VP5733).

♦ **VC1623 Sobriety**
Integrative polarities Pride–Humility (#VP5905)   Sanity–Insanity (#VP5472)
Energy–Moderation (#VP5161)   Cheerfulness–Solemnity (#VP5870)
Temperance–Intemperance (#VP5992)   Excitement–Inexcitability (#VP5857)
Intelligence–Unintelligence (#VP5467).

♦ **VC1625 Sociability**
Integrative polarities Friendship–Enmity (#VP5927)
Behaviour–Misbehaviour (#VP5737)
Sociability–Unsociability (#VP5922)
Communicativeness–Uncommunicativeness (#VP5554).

♦ **VC1627 Solemnity**
Integrative polarities Pride–Humility (#VP5905)   Respect–Disrespect (#VP5964)
Eloquence–Uneloquence (#VP5600)   Importance–Unimportance (#VP5672)
Excitement–Inexcitability (#VP5857).

♦ **VC1629 Solicitude**
Integrative polarities Caution–Rashness (#VP5894)
Kindness–Unkindness (#VP5938)
Carefulness–Neglect (#VP5533)
Courtesy–Discourtesy (#VP5936).

♦ **VC1631 Solidarity**
Integrative polarities Unity–Duality (#VP5089)   Accord–Disaccord (#VP5794)
Support–Opposition (#VP5785)   Completeness–Incompleteness (#VP5054).

♦ **VC1633 Solidity**
Integrative polarities Truth–Error (#VP5516)   Unity–Duality (#VP5089)
Strength–Weakness (#VP5159)   Change–Permanence (#VP5139)
Certainty–Uncertainty (#VP5513)   Completeness–Incompleteness (#VP5054).

♦ **VC1635 Solitude**
Integrative polarities Unity–Duality (#VP5089).

♦ **VC1637 Soundness**
Integrative polarities Truth–Error (#VP5516)   Health–Disease (#VP5685)
Sanity–Insanity (#VP5472)   Goodness–Badness (#VP5674)
Intuition–Reason (#VP5481)   Strength–Weakness (#VP5159)
Certainty–Uncertainty (#VP5513)   Perfection–Imperfection (#VP5677)
Stability–Changeableness (#VP5141)   Intelligence–Unintelligence (#VP5467).

♦ **VC1639 Sovereignty**
Integrative polarities Possession–Loss (#VP5808)
Authority–Lawlessness (#VP5739).

♦ **VC1641 Spirituality**
Integrative polarities Piety–Impiety (#VP6028)   Godliness–Ungodliness (#VP6013)
Materiality–Immateriality (#VP5376).

♦ **VC1643 Splendour**
Integrative polarities Light–Darkness (#VP5335)
Beauty–Ugliness (#VP5899)
Repute–Disrepute (#VP5914)
Naturalness–Affectation (#VP5900).

♦ **VC1645 Spontaneity**
Integrative polarities Resolution–Irresolution (#VP5624)
Willingness–Unwillingness (#VP5621).

♦ **VC1647 Sportsmanship**
Integrative polarities Justice–Injustice (#VP5976).

♦ **VC1649 Spotlessness**
Integrative polarities Innocence–Guilt (#VP5983)
Probity–Improbity (#VP5974)
Chastity–Unchastity (#VP5987)
Cleanness–Uncleanness (#VP5681)
Perfection–Imperfection (#VP5677).

♦ **VC1651 Stability**
Integrative polarities Energy–Moderation (#VP5161)
Strength–Weakness (#VP5159)
Change–Permanence (#VP5139)
Certainty–Uncertainty (#VP5513)
Durability–Transience (#VP5110)
Resolution–Irresolution (#VP5624)
Stability–Changeableness (#VP5141)
Continuity–Discontinuity (#VP5071)
Substantiality–Unsubstantiality (#VP5003).

♦ **VC1653 Stainlessness**
Integrative polarities Innocence–Guilt (#VP5983)
Probity–Improbity (#VP5974)
Chastity–Unchastity (#VP5987)
Cleanness–Uncleanness (#VP5681)
Perfection–Imperfection (#VP5677).

♦ **VC1655 Stamina**
Integrative polarities Courage–Fear (#VP5890)   Strength–Weakness (#VP5159)
Resolution–Irresolution (#VP5624).

♦ **VC1657 Standing**
Integrative polarities Location–Dislocation (#VP5185).

CONSTRUCTIVE VALUES

♦ **VC1659 Stateliness**
Integrative polarities Pride–Humility (#VP5905)
Eloquence–Uneloquence (#VP5600)
Formality–Informality (#VP5646)
Naturalness–Affectation (#VP5900).

♦ **VC1663 Status**
Integrative polarities Repute–Disrepute (#VP5914).

♦ **VC1665 Staunchness**
Integrative polarities Opening–Closure (#VP5265)
Probity–Improbity (#VP5974)
Friendship–Enmity (#VP5927)
Strength–Weakness (#VP5159)
Certainty–Uncertainty (#VP5513)
Resolution–Irresolution (#VP5624).

♦ **VC1667 Steadfastness**
Integrative polarities Probity–Improbity (#VP5974)
Friendship–Enmity (#VP5927)
Change–Permanence (#VP5139)
Certainty–Uncertainty (#VP5513)
Durability–Transience (#VP5110)
Resolution–Irresolution (#VP5624)
Stability–Changeableness (#VP5141).

♦ **VC1669 Steadiness**
Integrative polarities Safety–Danger (#VP5697)
Energy–Moderation (#VP5161)
Resolution–Irresolution (#VP5624)
Probity–Improbity (#VP5974)
Frequency–Infrequency (#VP5135)
Regularity–Irregularity (#VP5137).

♦ **VC1671 Stewardship**
Integrative polarities Safety–Danger (#VP5697)
Appropriateness–Inappropriateness (#VP5665).
Carefulness–Neglect (#VP5533)

♦ **VC1673 Stillness**
Integrative polarities Silence–Loudness (#VP5451)
Motion–Quiescence (#VP5267).

♦ **VC1675 Stimulation**
Integrative polarities Energy–Moderation (#VP5161)
Refreshment–Relapse (#VP5695)
Motivation–Dissuasion (#VP5648).

♦ **VC1677 Stintlessness**
Integrative polarities Economy–Prodigality (#VP5851).

♦ **VC1679 Straightforwardness**
Integrative polarities Probity–Improbity (#VP5974)
Facility–Difficulty (#VP5732)
Elegance–Inelegance (#VP5589)
Direction–Deviation (#VP5290)
Simplicity–Complexity (#VP5045)
Naturalness–Affectation (#VP5900)
Intelligibility–Unintelligibility (#VP5548).

♦ **VC1681 Strength**
Integrative polarities Health–Disease (#VP5685)
Power–Impotence (#VP5157)
Energy–Moderation (#VP5161)
Strength–Weakness (#VP5159)
Greatness–Smallness (#VP5034)
Authority–Lawlessness (#VP5739)
Eloquence–Uneloquence (#VP5600)
Resolution–Irresolution (#VP5624)
Substantiality–Unsubstantiality (#VP5003).

♦ **VC1683 Strictness**
Integrative polarities Truth–Error (#VP5516)
Carefulness–Neglect (#VP5533).
Taste–Vulgarity (#VP5896)

♦ **VC1685 Sturdiness**
Integrative polarities Health–Disease (#VP5685)
Strength–Weakness (#VP5159)
Substantiality–Unsubstantiality (#VP5003).

♦ **VC1687 Style**
Integrative polarities Behaviour–Misbehaviour (#VP5737)
Skilfulness–Unskilfulness (#VP5733).

♦ **VC1689 Submission**
Integrative polarities Pride–Humility (#VP5905)
Consent–Refusal (#VP5775)
Patience–Impatience (#VP5861)
Assent–Dissent (#VP5521)
Freedom–Restraint (#VP5762)
Obedience–Disobedience (#VP5766).

♦ **VC1691 Subservience**
Integrative polarities Freedom–Restraint (#VP5762)
Support–Opposition (#VP5785).

♦ **VC1693 Substantiality**
Integrative polarities Truth–Error (#VP5516)
Existence–Nonexistence (#VP5001)
Oversufficiency–Insufficiency (#VP5660)
Certainty–Uncertainty (#VP5513)
Stability–Changeableness (#VP5141)
Substantiality–Unsubstantiality (#VP5003).

♦ **VC1695 Subtlety**
Integrative polarities Truth–Error (#VP5516)
Carefulness–Neglect (#VP5533)
Discrimination–Indiscrimination (#VP5492).
Taste–Vulgarity (#VP5896)
Identity–Difference (#VP5013).

♦ **VC1697 Success**
Integrative polarities Victory–Defeat (#VP5726)
Accomplishment–Nonaccomplishment (#VP5722).
Prosperity–Adversity (#VP5728)

♦ **VC1699 Sufficiency**
Integrative polarities Truth–Error (#VP5516)
Contentment–Discontentment (#VP5868)
Power–Impotence (#VP5157)
Oversufficiency–Insufficiency (#VP5660).

♦ **VC1701 Suitability**
Integrative polarities Taste–Vulgarity (#VP5896)
Rightness–Wrongness (#VP5957)
Timeliness–Untimeliness (#VP5129)
Choice–Necessity (#VP5637)
Agreement–Disagreement (#VP5026)
Preparedness–Unpreparedness (#VP5720).

♦ **VC1703 Superabundance**
Integrative polarities Conciseness–Diffuseness (#VP5592)
Oversufficiency–Insufficiency (#VP5660)
Productiveness–Unproductiveness (#VP5165).

♦ **VC1705 Superfluity**
Integrative polarities Greatness–Smallness (#VP5034)
Conciseness–Diffuseness (#VP5592).

♦ **VC1707 Superhumanness**
Integrative polarities Godliness–Ungodliness (#VP6013).

♦ **VC1709 Superiority**
Integrative polarities Power–Impotence (#VP5157)
Goodness–Badness (#VP5674)
Authority–Lawlessness (#VP5739)
Importance–Unimportance (#VP5672)
Superiority–Inferiority (#VP5036).

♦ **VC1711 Support**
Integrative polarities Courage–Fear (#VP5890)
Assent–Dissent (#VP5521)
Strength–Weakness (#VP5159)
Comfort–Aggravation (#VP5885)
Approval–Disapproval (#VP5968)
Vindication–Condemnation (#VP6005)
Safety–Danger (#VP5697)
Hope–Hopelessness (#VP5888)
Support–Opposition (#VP5785)
Patience–Impatience (#VP5861)
Sharing–Appropriation (#VP5815)
Provability–Unprovability (#VP5505).

♦ **VC1713 Supremacy**
Integrative polarities Goodness–Badness (#VP5674)
Authority–Lawlessness (#VP5739)
Importance–Unimportance (#VP5672)
Superiority–Inferiority (#VP5036)
Influence–Influencelessness (#VP5172).

♦ **VC1715 Sureness**
Integrative polarities Belief–Unbelief (#VP5501)
Certainty–Uncertainty (#VP5513)
Probity–Improbity (#VP5974)
Expectation–Inexpectation (#VP5539).

♦ **VC1717 Surety**
Integrative polarities Safety–Danger (#VP5697)
Certainty–Uncertainty (#VP5513).
Belief–Unbelief (#VP5501)

♦ **VC1719 Surprise**
Integrative polarities Attack–Defence (#VP5798)
Wonder–Unastonishment (#VP5920)
Expectation–Inexpectation (#VP5539).

♦ **VC1721 Survival**
Integrative polarities Durability–Transience (#VP5110).

♦ **VC1723 Sustenance**
Integrative polarities Support–Opposition (#VP5785)
Continuance–Cessation (#VP5143).

♦ **VC1725 Sweetness**
Integrative polarities Love–Hate (#VP5930)
Kindness–Unkindness (#VP5938)
Cleanness–Uncleanness (#VP5681)
Pleasantness–Unpleasantness (#VP5863).
Fragrance–Stench (#VP5435)
Elegance–Inelegance (#VP5589)
Savouriness–Unsavouriness (#VP5427)

♦ **VC1727 Swiftness**
Integrative polarities Action–Inaction (#VP5705)
Timeliness–Untimeliness (#VP5129).
Swiftness–Slowness (#VP5269)

♦ **VC1729 Symmetry**
Integrative polarities Order–Disorder (#VP5059)
Symmetry–Distortion (#VP5248)
Identity–Difference (#VP5013)
Elegance–Inelegance (#VP5589)
Equality–Inequality (#VP5030)
Agreement–Disagreement (#VP5026).

♦ **VC1731 Sympathy**
Integrative polarities Accord–Disaccord (#VP5794)
Friendship–Enmity (#VP5927)
Kindness–Unkindness (#VP5938)
Comfort–Aggravation (#VP5885)
Compassion–Pitilessness (#VP5944)
Sensation–Insensibility (#VP5422)
Feeling–Unfeelinglessness (#VP5855)
Relatedness–Unrelatedness (#VP5009).

♦ **VC1733 Synergy**
Integrative polarities Exultation–Lamentation (#VP5873)
Concurrence–Counteraction (#VP5177).

♦ **VC1735 Synthesis**
Integrative polarities Intuition–Reason (#VP5481)
Identity–Difference (#VP5013).

♦ **VC1737 Tact**
Integrative polarities Kindness–Unkindness (#VP5938)
Courtesy–Discourtesy (#VP5936)
Sensation–Insensibility (#VP5422)
Skilfulness–Unskilfulness (#VP5733)
Discrimination–Indiscrimination (#VP5492).

♦ **VC1739 Talent**
Integrative polarities Power–Impotence (#VP5157)
Skilfulness–Unskilfulness (#VP5733)
Intelligence–Unintelligence (#VP5467).

♦ **VC1741 Tameness**
Integrative polarities Freedom–Restraint (#VP5762)
Energy–Moderation (#VP5161)
Leniency–Compulsion (#VP5756).

♦ **VC1743 Tangibility**
Integrative polarities Tangibility–Intangibility (#VP5425)
Substantiality–Unsubstantiality (#VP5003).

♦ **VC1745 Tastefulness**
Integrative polarities Taste–Vulgarity (#VP5896) Choice–Necessity (#VP5637)
Desire–Avoidance (#VP5631) Discrimination–Indiscrimination (#VP5492).

♦ **VC1747 Temperance**
Integrative polarities Virtue–Vice (#VP5980) Temperance–Intemperance (#VP5992)
Excitement–Inexcitability (#VP5857).

♦ **VC1749 Tenacity**
Integrative polarities Courage–Fear (#VP5890) Resolution–Irresolution (#VP5624)
Cohesion–Disintegration (#VP5050).

♦ **VC1751 Tenderness**
Integrative polarities Weight–Lightness (#VP5352)
Leniency–Compulsion (#VP5756)
Compassion–Pitilessness (#VP5944)
Sensation–Insensibility (#VP5422).

♦ **VC1753 Thoroughness**
Integrative polarities Caution–Rashness (#VP5894)
Carefulness–Neglect (#VP5533)
Perfection–Imperfection (#VP5677)
Completeness–Incompleteness (#VP5054).

♦ **VC1755 Thoughtfulness**
Integrative polarities Kindness–Unkindness (#VP5938)
Carefulness–Neglect (#VP5533)
Courtesy–Discourtesy (#VP5936)
Cheerfulness–Solemnity (#VP5870)
Thought–Thoughtlessness (#VP5478)
Intelligence–Unintelligence (#VP5467).

♦ **VC1757 Thriftiness**
Integrative polarities Economy–Prodigality (#VP5851).

♦ **VC1759 Timelessness**
Integrative polarities Godliness–Ungodliness (#VP6013)
Perpetuity–Instantaneousness (#VP5112).

♦ **VC1761 Timeliness**
Integrative polarities Timeliness–Untimeliness (#VP5129).

♦ **VC1763 Togetherness**
Integrative polarities Assent–Dissent (#VP5521)
Support–Opposition (#VP5785)
Conjunction–Separation (#VP5047)
Concurrence–Counteraction (#VP5177).

♦ **VC1765 Tolerance**
Integrative polarities Consent–Refusal (#VP5775)
Freedom–Restraint (#VP5762)
Patience–Impatience (#VP5861)
Leniency–Compulsion (#VP5756)
Inclusion–Exclusion (#VP5076)
Forgiveness–Vengeance (#VP5947)
Broadmindedness–Narrowmindedness (#VP5526).

♦ **VC1767 Toughness**
Integrative polarities Courage–Fear (#VP5890) Strength–Weakness (#VP5159)
Leniency–Compulsion (#VP5756) Elasticity–Toughness (#VP5358)
Resolution–Irresolution (#VP5624) Cohesion–Disintegration (#VP5050)
Substantiality–Unsubstantiality (#VP5003).

♦ **VC1769 Tractability**
Integrative polarities Freedom–Restraint (#VP5762)
Hardness–Softness (#VP5356)
Willingness–Unwillingness (#VP5621).

♦ **VC1771 Tradition**
Integrative polarities Newness–Oldness (#VP5122).

♦ **VC1773 Tranquillity**
Integrative polarities Exertion–Rest (#VP5709) Order–Disorder (#VP5059)
Accord–Discaccord (#VP5794) Silence–Loudness (#VP5451)
Motion–Quiescence (#VP5267) Energy–Moderation (#VP5161)
Excitement–Inexcitability (#VP5857).

♦ **VC1775 Transcendence**
Integrative polarities Godliness–Ungodliness (#VP6013)
Superiority–Inferiority (#VP5036).

♦ **VC1777 Transcience**
Integrative polarities Life–Death (#VP5407) Stability–Changeableness (#VP5141).

♦ **VC1779 Transformation**
Integrative polarities Change–Permanence (#VP5139)
Improvement–Impairment (#VP5691).

♦ **VC1781 Triumph**
Integrative polarities Victory–Defeat (#VP5726) Exultation–Lamentation (#VP5873)
Accomplishment–Nonaccomplishment (#VP5722).

♦ **VC1783 Trust**
Integrative polarities Belief–Unbelief (#VP5501) Hope–Hopelessness (#VP5888)
Certainty–Uncertainty (#VP5513).

♦ **VC1785 Trustworthiness**
Integrative polarities Safety–Danger (#VP5697) Belief–Unbelief (#VP5501)
Probity–Improbity (#VP5974) Certainty–Uncertainty (#VP5513).

♦ **VC1787 Truth**
Integrative polarities Truth–Error (#VP5516) Probity–Improbity (#VP5974)
Certainty–Uncertainty (#VP5513) Existence–Nonexistence (#VP5001).

♦ **VC1789 Truthfulness**
Integrative polarities Truth–Error (#VP5516) Probity–Improbity (#VP5974).

♦ **VC1791 Ubiquity**
Integrative polarities Presence–Absence (#VP5186)
Godliness–Ungodliness (#VP6013)
Completeness–Incompleteness (#VP5054).

♦ **VC1793 Unadornment**
Integrative polarities Elegance–Inelegance (#VP5589)
Simplicity–Complexity (#VP5045)
Skilfulness–Unskilfulness (#VP5733).

♦ **VC1795 Unadulteration**
Integrative polarities Truth–Error (#VP5516) Cleanness–Uncleanness (#VP5681)
Simplicity–Complexity (#VP5045) Naturalness–Affectation (#VP5900)
Perfection–Imperfection (#VP5677).

♦ **VC1797 Unambiguity**
Integrative polarities Certainty–Uncertainty (#VP5513)
Intelligibility–Unintelligibility (#VP5548).

♦ **VC1799 Unanimity**
Integrative polarities Assent–Dissent (#VP5521)
Agreement–Disagreement (#VP5026).

♦ **VC1801 Unassuming**
Integrative polarities Truth–Error (#VP5516) Modesty–Vanity (#VP5908)
Probity–Improbity (#VP5974) Formality–Informality (#VP5646)
Naturalness–Affectation (#VP5900) Skilfulness–Unskilfulness (#VP5733).

♦ **VC1803 Unbigoted**
Integrative polarities Freedom–Restraint (#VP5762)
Broadmindedness–Narrowmindedness (#VP5526).

♦ **VC1805 Unblemished**
Integrative polarities Innocence–Guilt (#VP5983)
Probity–Improbity (#VP5974)
Chastity–Unchastity (#VP5987)
Cleanness–Uncleanness (#VP5681)
Perfection–Imperfection (#VP5677).

♦ **VC1807 Unchangeable**
Integrative polarities Change–Permanence (#VP5139)
Frequency–Infrequency (#VP5135)
Stability–Changeableness (#VP5141).

♦ **VC1809 Uncorrupted**
Integrative polarities Virtue–Vice (#VP5980) Innocence–Guilt (#VP5983)
Probity–Improbity (#VP5974) Simplicity–Complexity (#VP5045).

♦ **VC1811 Understanding**
Integrative polarities Assent–Dissent (#VP5521)
Accord–Discaccord (#VP5794)
Patience–Impatience (#VP5861)
Knowledge–Ignorance (#VP5475)
Compassion–Pitilessness (#VP5944)
Intelligence–Unintelligence (#VP5467)
Intelligibility–Unintelligibility (#VP5548).

♦ **VC1813 Undivided**
Integrative polarities Unity–Duality (#VP5089) Completeness–Incompleteness (#VP5054).

♦ **VC1815 Unequivocalness**
Integrative polarities Belief–Unbelief (#VP5501) Probity–Improbity (#VP5974)
Freedom–Restraint (#VP5762) Greatness–Smallness (#VP5034)
Certainty–Uncertainty (#VP5513) Limitation–Unlimitedness (#VP5507)
Intelligibility–Unintelligibility (#VP5548).

♦ **VC1817 Unerring**
Integrative polarities Virtue–Vice (#VP5980) Truth–Error (#VP5516)
Certainty–Uncertainty (#VP5513).

♦ **VC1819 Unfailing**
Integrative polarities Probity–Improbity (#VP5974)
Certainty–Uncertainty (#VP5513)
Resolution–Irresolution (#VP5624).

♦ **VC1821 Uniformity**
Integrative polarities Unity–Duality (#VP5089) Order–Disorder (#VP5059)
Symmetry–Distortion (#VP5248) Agreement–Disagreement (#VP5026)
Regularity–Irregularity (#VP5137) Stability–Changeableness (#VP5141)
Continuity–Discontinuity (#VP5071) Uniformity–Nonuniformity (#VP5017).

♦ **VC1823 Uninfluence**
Integrative polarities Justice–Injustice (#VP5976)
Willingness–Unwillingness (#VP5621)
Broadmindedness–Narrowmindedness (#VP5526).

♦ **VC1825 Union**
Integrative polarities Accord–Discaccord (#VP5794)
Nearness–Distance (#VP5200)
Support–Opposition (#VP5785)
Conjunction–Separation (#VP5047)
Agreement–Disagreement (#VP5026)
Concurrence–Counteraction (#VP5177)
Relatedness–Unrelatedness (#VP5009).

CONSTRUCTIVE VALUES VC1911

♦ **VC1827 Uniqueness**
  Integrative polarities Unity–Duality (#VP5089)
  Identity–Difference (#VP5013)
  Frequency–Infrequency (#VP5135)
  Superiority–Inferiority (#VP5036).
  Newness–Oldness (#VP5122)
  Wonder–Unastonishment (#VP5920)
  Originality–Imitation (#VP5023)

♦ **VC1829 Unity**
  Integrative polarities Unity–Duality (#VP5089)
  Weight–Lightness (#VP5352)
  Simplicity–Complexity (#VP5045)
  Accord–Disaccord (#VP5794)
  Godliness–Ungodliness (#VP6013)
  Completeness–Incompleteness (#VP5054).

♦ **VC1831 Universality**
  Integrative polarities Inclusion–Exclusion (#VP5076)
  Numerousness–Fewness (#VP5101)
  Materiality–Immateriality (#VP5376)
  Completeness–Incompleteness (#VP5054).

♦ **VC1833 Unlimited**
  Integrative polarities Power–Impotence (#VP5157)
  Freedom–Restraint (#VP5762)
  Numerousness–Fewness (#VP5101)
  Godliness–Ungodliness (#VP6013).

♦ **VC1835 Unobtrusiveness**
  Integrative polarities Modesty–Vanity (#VP5908)
  Taste–Vulgarity (#VP5896)
  Communicativeness–Uncommunicativeness (#VP5554).

♦ **VC1837 Unpretention**
  Integrative polarities Modesty–Vanity (#VP5908)
  Pride–Humility (#VP5905)
  Naturalness–Affectation (#VP5900)
  Unselfishness–Selfishness (#VP5978)
  Skilfulness–Unskilfulness (#VP5733).

♦ **VC1839 Unselfishness**
  Integrative polarities Justice–Injustice (#VP5976)
  Economy–Prodigality (#VP5851)
  Unselfishness–Selfishness (#VP5978).

♦ **VC1841 Unspotted**
  Integrative polarities Innocence–Guilt (#VP5983)
  Probity–Improbity (#VP5974)
  Chastity–Unchastity (#VP5987)
  Cleanness–Uncleanness (#VP5681)
  Perfection–Imperfection (#VP5677).

♦ **VC1843 Uplift**
  Integrative polarities Repute–Disrepute (#VP5914)
  Cheerfulness–Solemnity (#VP5870)
  Improvement–Impairment (#VP5691).

♦ **VC1845 Uprightness**
  Integrative polarities Virtue–Vice (#VP5980)   Probity–Improbity (#VP5974).

♦ **VC1847 Usefulness**
  Integrative polarities Goodness–Badness (#VP5674)
  Support–Opposition (#VP5785)
  Appropriateness–Inappropriateness (#VP5665).

♦ **VC1849 Utility**
  Integrative polarities Support–Opposition (#VP5785)
  Appropriateness–Inappropriateness (#VP5665).

♦ **VC1851 Valour**
  Integrative polarities Courage–Fear (#VP5890).

♦ **VC1853 Validity**
  Integrative polarities Truth–Error (#VP5516)
  Power–Impotence (#VP5157)
  Legality–Illegality (#VP5998)
  Consent–Refusal (#VP5775)
  Goodness–Badness (#VP5674)
  Certainty–Uncertainty (#VP5513).

♦ **VC1855 Value**
  Integrative polarities Goodness–Badness (#VP5674)
  Respect–Disrespect (#VP5964)
  Judgement–Misjudgement (#VP5494)
  Expensiveness–Cheapness (#VP5848)
  Importance–Unimportance (#VP5672)
  Meaning–Meaninglessness (#VP5545)
  Appropriateness–Inappropriateness (#VP5665).

♦ **VC1857 Vantage**
  Integrative polarities Superiority–Inferiority (#VP5036)
  Influence–Influencelessness (#VP5172).

♦ **VC1859 Variety**
  Integrative polarities Change–Permanence (#VP5139)
  Numerousness–Fewness (#VP5101)
  Uniformity–Nonuniformity (#VP5017).

♦ **VC1861 Vastness**
  Integrative polarities Originality–Imitation (#VP5023).

♦ **VC1863 Venerableness**
  Integrative polarities Pride–Humility (#VP5905)
  Repute–Disrepute (#VP5914)
  Sanctity–Unsanctity (#VP6026).
  Newness–Oldness (#VP5122)
  Respect–Disrespect (#VP5964)

♦ **VC1865 Veneration**
  Integrative polarities Piety–Impiety (#VP6028)   Respect–Disrespect (#VP5964).

♦ **VC1867 Venturesomeness**
  Integrative polarities Courage–Fear (#VP5890)   Action–Inaction (#VP5705).

♦ **VC1869 Veracity**
  Integrative polarities Truth–Error (#VP5516)   Probity–Improbity (#VP5974).

♦ **VC1871 Veritableness**
  Integrative polarities Truth–Error (#VP5516)   Probity–Improbity (#VP5974)
  Existence–Nonexistence (#VP5001).

♦ **VC1873 Versatility**
  Integrative polarities Appropriateness–Inappropriateness (#VP5665).

♦ **VC1875 Viability**
  Integrative polarities Life–Death (#VP5407)   Energy–Moderation (#VP5161)
  Contentment–Discontentment (#VP5868).

♦ **VC1877 Victory**
  Integrative polarities Victory–Defeat (#VP5726)
  Accomplishment–Nonaccomplishment (#VP5722).

♦ **VC1879 Vigilance**
  Integrative polarities Safety–Danger (#VP5697)   Carefulness–Neglect (#VP5533)
  Preparedness–Unpreparedness (#VP5720).

♦ **VC1881 Vigour**
  Integrative polarities Health–Disease (#VP5685)
  Power–Impotence (#VP5157)
  Energy–Moderation (#VP5161)
  Strength–Weakness (#VP5159)
  Prosperity–Adversity (#VP5728)
  Eloquence–Uneloquence (#VP5600)
  Cheerfulness–Solemnity (#VP5870).

♦ **VC1883 Virginity**
  Integrative polarities Newness–Oldness (#VP5122)
  Chastity–Unchastity (#VP5987)
  Preparedness–Unpreparedness (#VP5720).

♦ **VC1885 Virility**
  Integrative polarities Youth–Age (#VP5124)   Power–Impotence (#VP5157).

♦ **VC1887 Virtue**
  Integrative polarities Virtue–Vice (#VP5980)   Courage–Fear (#VP5890)
  Power–Impotence (#VP5157)   Goodness–Badness (#VP5674)
  Probity–Improbity (#VP5974)   Chastity–Unchastity (#VP5987)
  Rightness–Wrongness (#VP5957).

♦ **VC1889 Vision**
  Integrative polarities Beauty–Ugliness (#VP5899)
  Imaginativeness–Unimaginativeness (#VP5535).

♦ **VC1891 Vitality**
  Integrative polarities Life–Death (#VP5407)   Health–Disease (#VP5685)
  Power–Impotence (#VP5157)   Desire–Avoidance (#VP5631)
  Energy–Moderation (#VP5161)   Strength–Weakness (#VP5159)
  Eloquence–Uneloquence (#VP5600)   Cheerfulness–Solemnity (#VP5870).

♦ **VC1893 Vivacity**
  Integrative polarities Life–Death (#VP5407)   Action–Inaction (#VP5705)
  Desire–Avoidance (#VP5631)   Eloquence–Uneloquence (#VP5600)
  Cheerfulness–Solemnity (#VP5870).

♦ **VC1895 Voluptuousness**
  Integrative polarities Chastity–Unchastity (#VP5987)
  Pleasantness–Unpleasantness (#VP5863).

♦ **VC1897 Warmheartedness**
  Integrative polarities Friendship–Enmity (#VP5927)
  Kindness–Unkindness (#VP5938)
  Compassion–Pitilessness (#VP5944)
  Hospitality–Inhospitality (#VP5925).

♦ **VC1899 Warmth**
  Integrative polarities Heat–Cold (#VP5328)   Desire–Avoidance (#VP5631)
  Friendship–Enmity (#VP5927)   Kindness–Unkindness (#VP5938)
  Eloquence–Uneloquence (#VP5600)   Hospitality–Inhospitality (#VP5925).

♦ **VC1901 Wealth**
  Integrative polarities Wealth–Poverty (#VP5837)
  Prosperity–Adversity (#VP5728)
  Oversufficiency–Insufficiency (#VP5660).

♦ **VC1903 Welfare**
  Integrative polarities Goodness–Badness (#VP5674)
  Kindness–Unkindness (#VP5938)
  Prosperity–Adversity (#VP5728).

♦ **VC1905 Well-disposed**
  Integrative polarities Friendship–Enmity (#VP5927)
  Support–Opposition (#VP5785)
  Kindness–Unkindness (#VP5938)
  Approval–Disapproval (#VP5968)
  Willingness–Unwillingness (#VP5621).

♦ **VC1907 Well-grounded**
  Integrative polarities Truth–Error (#VP5516)   Intuition–Reason (#VP5481)
  Knowledge–Ignorance (#VP5475)   Certainty–Uncertainty (#VP5513)
  Substantiality–Unsubstantiality (#VP5003).

♦ **VC1909 Well-informed**
  Integrative polarities Knowledge–Ignorance (#VP5475).

♦ **VC1911 Well-made**
  Integrative polarities Beauty–Ugliness (#VP5899)
  Symmetry–Distortion (#VP5248)
  Substantiality–Unsubstantiality (#VP5003).

♦ **VC1913 Well-mannered**
Integrative polarities Courtesy–Discourtesy (#VP5936)
Formality–Informality (#VP5646).

♦ **VC1915 Well-meaning**
Integrative polarities Friendship–Enmity (#VP5927)
Support–Opposition (#VP5785)
Kindness–Unkindness (#VP5938).

♦ **VC1917 Well-provided**
Integrative polarities Oversufficiency–Insufficiency (#VP5660).

♦ **VC1919 Wholeness**
Integrative polarities Unity–Duality (#VP5089)   Health–Disease (#VP5685)
Goodness–Badness (#VP5674)   Perfection–Imperfection (#VP5677)
Completeness–Incompleteness (#VP5054).

♦ **VC1921 Wholesomeness**
Integrative polarities Health–Disease (#VP5685)
Sanity–Insanity (#VP5472)
Intuition–Reason (#VP5481)
Healthfulness–Unhealthfulness (#VP5683).

♦ **VC1923 Will**
Integrative polarities Choice–Necessity (#VP5637)
Resolution–Irresolution (#VP5624)
Willingness–Unwillingness (#VP5621).

♦ **VC1925 Willingness**
Integrative polarities Consent–Refusal (#VP5775)
Action–Inaction (#VP5705)
Obedience–Disobedience (#VP5766)
Education–Miseducation (#VP5562)
Willingness–Unwillingness (#VP5621).

♦ **VC1927 Wisdom**
Integrative polarities Belief–Unbelief (#VP5501)   Knowledge–Ignorance (#VP5475)
Intelligence–Unintelligence (#VP5467).

♦ **VC1929 Wittiness**
Integrative polarities Amusement–Boredom (#VP5878).

♦ **VC1931 Womanliness**
Integrative polarities Masculinity–Femininity (#VP5420).

♦ **VC1933 Wonder**
Integrative polarities Greatness–Smallness (#VP5034)
Wonder–Unastonishment (#VP5920).

♦ **VC1935 Wonderfulness**
Integrative polarities Goodness–Badness (#VP5674).

♦ **VC1937 Work**
Integrative polarities Exertion–Rest (#VP5709)   Action–Inaction (#VP5705)
Research–Discovery (#VP5485)   Causation–Culmination (#VP5153)
Production–Reproduction (#VP5167)   Influence–Influencelessness (#VP5172)
Oversufficiency–Insufficiency (#VP5660)
Accomplishment–Nonaccomplishment (#VP5722).

♦ **VC1939 Worth**
Integrative polarities Repute–Disrepute (#VP5914)
Goodness–Badness (#VP5674)
Importance–Unimportance (#VP5672)
Appropriateness–Inappropriateness (#VP5665).

♦ **VC1941 Worthiness**
Integrative polarities Pride–Humility (#VP5905)   Repute–Disrepute (#VP5914)
Choice–Necessity (#VP5637)   Probity–Improbity (#VP5974)
Approval–Disapproval (#VP5968)   Expensiveness–Cheapness (#VP5848)
Skilfulness–Unskilfulness (#VP5733).

♦ **VC1943 Youth**
Integrative polarities Youth–Age (#VP5124)   Health–Disease (#VP5685)
Newness–Oldness (#VP5122).

♦ **VC1945 Youthfulness**
Integrative polarities Youth–Age (#VP5124)   Health–Disease (#VP5685).

♦ **VC1947 Zeal**
Integrative polarities Courage–Fear (#VP5890)   Piety–Impiety (#VP6028)
Desire–Avoidance (#VP5631)   Feeling–Unfeelinglessness (#VP5855)
Willingness–Unwillingness (#VP5621).

♦ **VC1949 Zest**
Integrative polarities Desire–Avoidance (#VP5631)
Energy–Moderation (#VP5161)
Pleasure–Displeasure (#VP5865)
Cheerfulness–Solemnity (#VP5870).

Destructive values

Rationale

Most of the world problems in this volume are identified by names incorporating a word implying some form of destructive value (*eg* injustice, imbalance, inadequate, endangered). Indeed it may be argued that it is the availability of such words as operators which enable certain conditions to be defined and perceived as problematic. For example, the word "endangered" as a value operator, may be applied to many physical and social phenomena, thus raising the question as to whether they constitute recognized potential problems (*eg* endangered whales, endangered minorities, endangered cultures).

Given the considerable difficulty in distinguishing between problems identified by similar value operators of this type, it is important to explore the way available destructive value words lead to perception of problems.

Contents

This section lists in alphabetic order 1,040 words which can be considered as reflecting "destructive" values. Because of the ambiguous connotations of many such words, they are each cross-referenced to a number of entries in Section VP. The entries in Section VP are each value categories or dimensions denoted by a pair of "constructive" and "destructive" value words in opposition, namely a "value polarity".

This polar relationship sharpens the meaning that can be associated with a particular interpretation of the constructive value word identified in this section. Associating value words with value polarities also responds significantly to potential differences of opinion as to whether a particular value should be considered "constructive" as opposed to "destructive" (as could well be the case under certain circumstances). It is for this reason that the more usual terms "positive" and "negative" were not used. It is more understandable that "destructive" action may be necessary to clear the way for "constructive" action, and that "constructive" action may reach the point at which "destructive" action is necessary to initiate some new phase.

Method

The method is described in Section VZ. It is based on selecting and interrelating words from *Roget's Thesaurus*, which was considered a comprehensive reference work providing a useful framework for this exploratory excercise.

Index

A keyword index to entries is provided in Section VX. The keywords are also incorporated into the index for Volume 2 (Section X)

Bibliography

Bibliographical references, by author, are given in Section VY.

Comment

Detail comments are given in Section VZ.

Reservations

As an exploratory exercise the results must necessarily be considered as preliminary, and limited by dependence on the Roget framework and the English language. To the extent that some destructive values are associated with phrases rather than single words (*eg* intolerant misunderstanding), this survey at present only covers "elementary" values (*eg* intolerance, misunderstanding). In conformity with the general editorial policy, borderline value words have been included if they raise interesting questions concerning the criteria governing their inclusion.

Possible future improvements

In addition to refining and extending the set of words and cross-references, ways of incorporating equivalent words from other languages need to be considered.

ENTRY CONTENT AND ORGANIZATION

Ordering of entries
Entries are numbered in **alphabetic order**. Entry numbers have no significance other than as a permanent point of reference to facilitate indexing, cross-referencing, and updating between editions.

Index access to entries
This sub-section is self-indexing since the entries are in alphabetic order. The location of an entry in this sub-section may also be determined from:
- the **Volume Index** (Section X)
- the **Section Index** (Section VX) on the basis of keywords in the name of the entry of its alternate names

Structure of entries
Entries may be composed of the following descriptive elements:

(a) **Entry number** This number has **no significance**, except as a convenient method of identifying the entry (particularly for indexing purposes), of filing information on it, and as an identifier to which cross-references from other entries (possibly in other sections) may refer in this and future editions. The first letter of the entry number refers to the section of this volume in which the sub-section, denoted by the second letter, is located.

(b) **Entry name** This is printed in bold characters

Cross-referencing of entries
At the end of any entry, there may be cross-references to other entries. These indicate the number and name of the cross-referenced entry, whether within this Section or in other Sections. In this sub-section there is a single type of cross-reference, indicated by a 2-letter code in bold characters:
    Integrative polarities = Broader value or value polarity

DESTRUCTIVE VALUES

♦ **VD2004 Abandonment**
Integrative polarities Virtue–Vice (#VP5980)
Presence–Absence (#VP5186)
Chastity–Unchastity (#VP5987)
Hospitality–Inhospitality (#VP5925)
Desire–Avoidance (#VP5631)
Freedom–Restraint (#VP5762)
Carefulness–Neglect (#VP5533)
Oversufficiency–Insufficiency (#VP5660).

♦ **VD2006 Abasement**
Integrative polarities Pride–Humility (#VP5905).

♦ **VD2008 Aberration**
Integrative polarities Truth–Error (#VP5516)
Direction–Deviation (#VP5290)
Sanity–Insanity (#VP5472)
Conformity–Nonconformity (#VP5082).

♦ **VD2010 Abhorrence**
Integrative polarities Love–Hate (#VP5930)
Pleasantness–Unpleasantness (#VP5863).

♦ **VD2012 Abjection**
Integrative polarities Virtue–Vice (#VP5980)
Repute–Disrepute (#VP5914).
Pride–Humility (#VP5905)

♦ **VD2014 Abnormality**
Integrative polarities Health–Disease (#VP5685)
Sanity–Insanity (#VP5472)
Rightness–Wrongness (#VP5957)
Agreement–Disagreement (#VP5026).

♦ **VD2016 Abomination**
Integrative polarities Love–Hate (#VP5930)
Goodness–Badness (#VP5674)
Cleanness–Uncleanness (#VP5681)
Virtue–Vice (#VP5980)
Rightness–Wrongness (#VP5957)
Pleasantness–Unpleasantness (#VP5863).

♦ **VD2018 Absence**
Integrative polarities Presence–Absence (#VP5186)
Oversufficiency–Insufficiency (#VP5660).

♦ **VD2020 Absolutism**
Integrative polarities Freedom–Restraint (#VP5762).

♦ **VD2022 Absurdity**
Integrative polarities Meaning–Meaninglessness (#VP5545)
Intelligence–Unintelligence (#VP5467)
Appropriateness–Inappropriateness (#VP5665).

♦ **VD2024 Abuse**
Integrative polarities Goodness–Badness (#VP5674)
Appropriateness–Inappropriateness (#VP5665).

♦ **VD2026 Abusiveness**
Integrative polarities Respect–Disrespect (#VP5964)
Approval–Disapproval (#VP5968).

♦ **VD2028 Accident**
Integrative polarities Prosperity–Adversity (#VP5728).

♦ **VD2030 Acquisitiveness**
Integrative polarities Desire–Avoidance (#VP5631)
Unselfishness–Selfishness (#VP5978).

♦ **VD2032 Acrimony**
Integrative polarities Friendship–Enmity (#VP5927)
Energy–Moderation (#VP5161)
Congratulation–Envy (#VP5948).

♦ **VD2034 Affectation**
Integrative polarities Naturalness–Affectation (#VP5900)
Communicativeness–Uncommunicativeness (#VP5554).

♦ **VD2036 Affliction**
Integrative polarities Health–Disease (#VP5685)
Goodness–Badness (#VP5674)
Prosperity–Adversity (#VP5728)
Pleasantness–Unpleasantness (#VP5863).

♦ **VD2038 Affrontery**
Integrative polarities Respect–Disrespect (#VP5964)
Support–Opposition (#VP5785)
Congratulation–Envy (#VP5948).

♦ **VD2040 Age**
Integrative polarities Youth–Age (#VP5124)
Newness–Oldness (#VP5122).

♦ **VD2042 Aggravation**
Integrative polarities Congratulation–Envy (#VP5948)
Comfort–Aggravation (#VP5885)
Prosperity–Adversity (#VP5728)
Pleasantness–Unpleasantness (#VP5863).

♦ **VD2044 Aggression**
Integrative polarities Attack–Defence (#VP5798)
Accord–Discord (#VP5794)
Courtesy–Discourtesy (#VP5936).

♦ **VD2046 Agitation**
Integrative polarities Courage–Fear (#VP5890)
Oscillation–Agitation (#VP5323)
Healthfulness–Unhealthfulness (#VP5683).
Energy–Moderation (#VP5161)
Excitement–Inexcitability (#VP5857)

♦ **VD2048 Aimlessness**
Integrative polarities Order–Disorder (#VP5059)
Meaning–Meaninglessness (#VP5545)
Attributability–Chance (#VP5155)
Appropriateness–Inappropriateness (#VP5665).

♦ **VD2050 Alarmism**
Integrative polarities Courage–Fear (#VP5890)
Safety–Danger (#VP5697).

♦ **VD2052 Alienation**
Integrative polarities Unity–Duality (#VP5089)
Sanity–Insanity (#VP5472)
Support–Opposition (#VP5785)
Assent–Dissent (#VP5521)
Friendship–Enmity (#VP5927)
Conversion–Reversion (#VP5145).

♦ **VD2054 Ambiguity**
Integrative polarities Identity–Difference (#VP5013)
Certainty–Uncertainty (#VP5513)
Agreement–Disagreement (#VP5026)
Intelligibility–Unintelligibility (#VP5548).

♦ **VD2056 Ambivalence**
Integrative polarities Identity–Difference (#VP5013)
Agreement–Disagreement (#VP5026)
Resolution–Irresolution (#VP5624)
Selfactualization–Neurosis (#VP5690).

♦ **VD2058 Amorality**
Integrative polarities Virtue–Vice (#VP5980)
Rightness–Wrongness (#VP5957).

♦ **VD2060 Anarchism**
Integrative polarities Order–Disorder (#VP5059)
Legality–Illegality (#VP5998)
Authority–Lawlessness (#VP5739).
Energy–Moderation (#VP5161)
Evolution–Revolution (#VP5147)

♦ **VD2062 Anarchy**
Integrative polarities Order–Disorder (#VP5059)
Authority–Lawlessness (#VP5739)
Legality–Illegality (#VP5998)
Cohesion–Disintegration (#VP5050).

♦ **VD2064 Anathema**
Integrative polarities Approval–Disapproval (#VP5968).

♦ **VD2066 Anger**
Integrative polarities Virtue–Vice (#VP5980)
Congratulation–Envy (#VP5948).

♦ **VD2068 Anguish**
Integrative polarities Cheerfulness–Solemnity (#VP5870)
Sensation–Insensibility (#VP5422)
Pleasantness–Unpleasantness (#VP5863).

♦ **VD2070 Animality**
Integrative polarities Taste–Vulgarity (#VP5896)
Chastity–Unchastity (#VP5987)
Energy–Moderation (#VP5161)
Kindness–Unkindness (#VP5938).

♦ **VD2072 Animosity**
Integrative polarities Love–Hate (#VP5930)
Congratulation–Envy (#VP5948)
Friendship–Enmity (#VP5927)
Feeling–Unfeelinglessness (#VP5855).

♦ **VD2074 Annihilation**
Integrative polarities Life–Death (#VP5407)
Restoration–Destruction (#VP5693).
Existence–Nonexistence (#VP5001)

♦ **VD2076 Annoyance**
Integrative polarities Congratulation–Envy (#VP5948)
Comfort–Aggravation (#VP5885)
Facility–Difficulty (#VP5732)
Prosperity–Adversity (#VP5728)
Pleasantness–Unpleasantness (#VP5863).

♦ **VD2078 Annulment**
Integrative polarities Assent–Dissent (#VP5521)
Conjugality–Celibacy (#VP5931)
Restoration–Destruction (#VP5693).

♦ **VD2080 Antagonism**
Integrative polarities Accord–Discord (#VP5794)
Friendship–Enmity (#VP5927)
Support–Opposition (#VP5785)
Identity–Difference (#VP5013)
Agreement–Disagreement (#VP5026)
Pleasantness–Unpleasantness (#VP5863).

♦ **VD2082 Antediluvian**
Integrative polarities Newness–Oldness (#VP5122).

♦ **VD2084 Antipathy**
Integrative polarities Love–Hate (#VP5930)
Support–Opposition (#VP5785)
Willingness–Unwillingness (#VP5621)
Friendship–Enmity (#VP5927)
Identity–Difference (#VP5013)
Pleasantness–Unpleasantness (#VP5863).

♦ **VD2086 Anxiety**
Integrative polarities Courage–Fear (#VP5890)
Facility–Difficulty (#VP5732)
Patience–Impatience (#VP5861)
Pleasantness–Unpleasantness (#VP5863).

♦ **VD2088 Apartheid**
Integrative polarities Broadmindedness–Narrowmindedness (#VP5526).

♦ **VD2090 Apathy**
Integrative polarities Desire–Avoidance (#VP5631)
Hope–Hopelessness (#VP5888)
Motion–Quiescence (#VP5267)
Curiosity–Incuriosity (#VP5528)
Feeling–Unfeelinglessness (#VP5855).

♦ **VD2092 Apprehension**
Integrative polarities Courage–Fear (#VP5890)
Belief–Unbelief (#VP5501).

♦ **VD2094 Archaism**
Integrative polarities Newness–Oldness (#VP5122).

♦ **VD2096 Argumentativeness**
Integrative polarities Accord–Discord (#VP5794)
Intuition–Reason (#VP5481)
Congratulation–Envy (#VP5948).

VD2098

◆ **VD2098 Arrant**
Integrative polarities Virtue–Vice (#VP5980)   Repute–Disrepute (#VP5914)
Goodness–Badness (#VP5674).

◆ **VD2100 Arrested development**
Integrative polarities Selfactualization–Neurosis (#VP5690)
Intelligence–Unintelligence (#VP5467).

◆ **VD2102 Arrogance**
Integrative polarities Modesty–Vanity (#VP5908)
Pride–Humility (#VP5905)
Respect–Disrespect (#VP5964)
Support–Opposition (#VP5785).

◆ **VD2104 Artifice**
Integrative polarities Skilfulness–Unskilfulness (#VP5733)
Communicativeness–Uncommunicativeness (#VP5554).

◆ **VD2106 Artificiality**
Integrative polarities Naturalness–Affectation (#VP5900)
Communicativeness–Uncommunicativeness (#VP5554).

◆ **VD2108 Asperity**
Integrative polarities Congratulation–Envy (#VP5948)
Kindness–Unkindness (#VP5938).

◆ **VD2110 Asymmetry**
Integrative polarities Symmetry–Distortion (#VP5248)
Equality–Inequality (#VP5030)
Agreement–Disagreement (#VP5026).

◆ **VD2112 Atheism**
Integrative polarities Piety–Impiety (#VP6028)   Attention–Inattention (#VP5530).

◆ **VD2114 Atrocity**
Integrative polarities Virtue–Vice (#VP5980)   Repute–Disrepute (#VP5914)
Goodness–Badness (#VP5674)   Energy–Moderation (#VP5161)
Respect–Disrespect (#VP5964)   Rightness–Wrongness (#VP5957)
Kindness–Unkindness (#VP5938)   Sensation–Insensibility (#VP5422)
Pleasantness–Unpleasantness (#VP5863).

◆ **VD2116 Austerity**
Integrative polarities Chastity–Unchastity (#VP5987)
Kindness–Unkindness (#VP5938)
Leniency–Compulsion (#VP5756)
Oversufficiency–Insufficiency (#VP5660).

◆ **VD2118 Authoritarianism**
Integrative polarities Leniency–Compulsion (#VP5756)
Authority–Lawlessness (#VP5739)
Broadmindedness–Narrowmindedness (#VP5526).

◆ **VD2120 Autism**
Integrative polarities Truth–Error (#VP5516)   Unselfishness–Selfishness (#VP5978)
Sociability–Unsociability (#VP5922)   Feeling–Unfeelinglessness (#VP5855)
Imaginativeness–Unimaginativeness (#VP5535).

◆ **VD2122 Avarice**
Integrative polarities Virtue–Vice (#VP5980)   Desire–Avoidance (#VP5631)
Expensiveness–Cheapness (#VP5848).

◆ **VD2124 Aversion**
Integrative polarities Love–Hate (#VP5930)   Willingness–Unwillingness (#VP5621)
Pleasantness–Unpleasantness (#VP5863).

◆ **VD2126 Avoidance**
Integrative polarities Desire–Avoidance (#VP5631)
Temperance–Intemperance (#VP5992).

◆ **VD2128 Awfulness**
Integrative polarities Courage–Fear (#VP5890)   Beauty–Ugliness (#VP5899)
Goodness–Badness (#VP5674)   Pleasantness–Unpleasantness (#VP5863).

◆ **VD2130 Awkwardness**
Integrative polarities Facility–Difficulty (#VP5732)
Elegance–Inelegance (#VP5589)
Knowledge–Ignorance (#VP5475)
Expedience–Inexpedience (#VP5670)
Skilfulness–Unskilfulness (#VP5733)
Pleasantness–Unpleasantness (#VP5863).

◆ **VD2132 Backlash**
Integrative polarities Causation–Culmination (#VP5153).

◆ **VD2134 Badness**
Integrative polarities Virtue–Vice (#VP5980)   Safety–Danger (#VP5697)
Health–Disease (#VP5685)   Goodness–Badness (#VP5674)
Fragrance–Stench (#VP5435)   Behaviour–Misbehaviour (#VP5737)
Healthfulness–Unhealthfulness (#VP5683).

◆ **VD2136 Bafflement**
Integrative polarities Victory–Defeat (#VP5726)   Facility–Difficulty (#VP5732)
Certainty–Uncertainty (#VP5513)   Expectation–Inexpectation (#VP5539).

◆ **VD2138 Banality**
Integrative polarities Amusement–Boredom (#VP5878).

◆ **VD2140 Banefulness**
Integrative polarities Life–Death (#VP5407)   Goodness–Badness (#VP5674)
Restoration–Destruction (#VP5693)   Pleasantness–Unpleasantness (#VP5863).

◆ **VD2142 Barbarism**
Integrative polarities Taste–Vulgarity (#VP5896)   Knowledge–Ignorance (#VP5475).

◆ **VD2144 Barbarity**
Integrative polarities Taste–Vulgarity (#VP5896)   Energy–Moderation (#VP5161)
Kindness–Unkindness (#VP5938).

◆ **VD2146 Barrenness**
Integrative polarities Power–Impotence (#VP5157)
Amusement–Boredom (#VP5878)
Productiveness–Unproductiveness (#VP5165)
Appropriateness–Inappropriateness (#VP5665)
Imaginativeness–Unimaginativeness (#VP5535).

◆ **VD2148 Barrier**
Integrative polarities Opening–Closure (#VP5265)
Freedom–Restraint (#VP5762)
Facility–Difficulty (#VP5732).

◆ **VD2150 Baseness**
Integrative polarities Virtue–Vice (#VP5980)   Courage–Fear (#VP5890)
Pride–Humility (#VP5905)   Taste–Vulgarity (#VP5896)
Repute–Disrepute (#VP5914)   Goodness–Badness (#VP5674)
Probity–Improbity (#VP5974)   Perfection–Imperfection (#VP5677)
Superiority–Inferiority (#VP5036)   Pleasantness–Unpleasantness (#VP5863).

◆ **VD2152 Bastardy**
Integrative polarities Legality–Illegality (#VP5998).

◆ **VD2154 Battle**
Integrative polarities Accord–Disaccord (#VP5794).

◆ **VD2156 Bawdiness**
Integrative polarities Chastity–Unchastity (#VP5987).

◆ **VD2158 Beastliness**
Integrative polarities Goodness–Badness (#VP5674)
Chastity–Unchastity (#VP5987)
Kindness–Unkindness (#VP5938)
Cleanness–Uncleanness (#VP5681)
Pleasantness–Unpleasantness (#VP5863).

◆ **VD2160 Bedevilment**
Integrative polarities Sanity–Insanity (#VP5472)   Godliness–Ungodliness (#VP6013)
Pleasantness–Unpleasantness (#VP5863).

◆ **VD2162 Bedraggled**
Integrative polarities Order–Disorder (#VP5059)   Cleanness–Uncleanness (#VP5681).

◆ **VD2164 Belligerence**
Integrative polarities Accord–Disaccord (#VP5794)
Friendship–Enmity (#VP5927)
Congratulation–Envy (#VP5948).

◆ **VD2166 Bereavement**
Integrative polarities Life–Death (#VP5407)   Possession–Loss (#VP5808)
Conjugality–Celibacy (#VP5931)   Sharing–Appropriation (#VP5815).

◆ **VD2168 Bestiality**
Integrative polarities Taste–Vulgarity (#VP5896)   Goodness–Badness (#VP5674)
Chastity–Unchastity (#VP5987)   Kindness–Unkindness (#VP5938).

◆ **VD2170 Betrayal**
Integrative polarities Desire–Avoidance (#VP5631)
Probity–Improbity (#VP5974)
Resolution–Irresolution (#VP5624).

◆ **VD2172 Bewilderment**
Integrative polarities Facility–Difficulty (#VP5732)
Certainty–Uncertainty (#VP5513).

◆ **VD2174 Bias**
Integrative polarities Justice–Injustice (#VP5976)
Symmetry–Distortion (#VP5248)
Broadmindedness–Narrowmindedness (#VP5526).

◆ **VD2176 Bigotry**
Integrative polarities Love–Hate (#VP5930)   Sanity–Insanity (#VP5472)
Certainty–Uncertainty (#VP5513)   Resolution–Irresolution (#VP5624)
Broadmindedness–Narrowmindedness (#VP5526).

◆ **VD2178 Bitterness**
Integrative polarities Love–Hate (#VP5930)   Friendship–Enmity (#VP5927)
Energy–Moderation (#VP5161)   Congratulation–Envy (#VP5948)
Kindness–Unkindness (#VP5938)   Pleasantness–Unpleasantness (#VP5863).

◆ **VD2180 Blame**
Integrative polarities Virtue–Vice (#VP5980)   Goodness–Badness (#VP5674)
Approval–Disapproval (#VP5968)   Vindication–Condemnation (#VP6005).

◆ **VD2181 Blasphemy**
Integrative polarities Piety–Impiety (#VP6028).

◆ **VD2182 Blatancy**
Integrative polarities Naturalness–Affectation (#VP5900).

◆ **VD2184 Blemish**
Integrative polarities Beauty–Ugliness (#VP5899)
Symmetry–Distortion (#VP5248)
Inclusion–Exclusion (#VP5076)
Improvement–Impairment (#VP5691)
Perfection–Imperfection (#VP5677).

◆ **VD2186 Blight**
Integrative polarities Improvement–Impairment (#VP5691)
Restoration–Destruction (#VP5693)
Expectation–Inexpectation (#VP5539).

DESTRUCTIVE VALUES   **VD2272**

♦ **VD2188 Bloody-mindedness**
   Integrative polarities Life–Death (#VP5407)   Accord–Disaccord (#VP5794)
   Kindness–Unkindness (#VP5938).

♦ **VD2190 Boastfulness**
   Integrative polarities Modesty–Vanity (#VP5908)
   Pride–Humility (#VP5905).

♦ **VD2192 Boredom**
   Integrative polarities Action–Inaction (#VP5705)   Amusement–Boredom (#VP5878)
   Curiosity–Incuriosity (#VP5528)   Uniformity–Nonuniformity (#VP5017)
   Pleasantness–Unpleasantness (#VP5863).

♦ **VD2194 Boringness**
   Integrative polarities Action–Inaction (#VP5705)   Amusement–Boredom (#VP5878)
   Curiosity–Incuriosity (#VP5528)   Pleasantness–Unpleasantness (#VP5863).

♦ **VD2196 Bothersomeness**
   Integrative polarities Courage–Fear (#VP5890)   Facility–Difficulty (#VP5732)
   Pleasantness–Unpleasantness (#VP5863).

♦ **VD2198 Brainwash**
   Integrative polarities Conversion–Reversion (#VP5145)
   Education–Miseducation (#VP5562).

♦ **VD2200 Brokenness**
   Integrative polarities Victory–Defeat (#VP5726)   Obedience–Disobedience (#VP5766)
   Improvement–Impairment (#VP5691)   Conjunction–Separation (#VP5047)
   Restoration–Destruction (#VP5693)   Cohesion–Disintegration (#VP5050)
   Continuity–Discontinuity (#VP5071).

♦ **VD2202 Brutality**
   Integrative polarities Taste–Vulgarity (#VP5896)   Goodness–Badness (#VP5674)
   Energy–Moderation (#VP5161)   Chastity–Unchastity (#VP5987)
   Kindness–Unkindness (#VP5938).

♦ **VD2204 Bumptiousness**
   Integrative polarities Modesty–Vanity (#VP5908)
   Support–Opposition (#VP5785).

♦ **VD2206 Burdensomeness**
   Integrative polarities Action–Inaction (#VP5705)   Weight–Lightness (#VP5352)
   Facility–Difficulty (#VP5732)   Pleasantness–Unpleasantness (#VP5863).

♦ **VD2208 Calamity**
   Integrative polarities Life–Death (#VP5407)   Goodness–Badness (#VP5674)
   Prosperity–Adversity (#VP5728)   Restoration–Destruction (#VP5693).

♦ **VD2210 Callousness**
   Integrative polarities Virtue–Vice (#VP5980)   Kindness–Unkindness (#VP5938)
   Exultation–Lamentation (#VP5873)   Tangibility–Intangibility (#VP5425)
   Feeling–Unfeelinglessness (#VP5855).

♦ **VD2212 Capriciousness**
   Integrative polarities Order–Disorder (#VP5059)   Durability–Transience (#VP5110)
   Resolution–Irresolution (#VP5624)   Regularity–Irregularity (#VP5137)
   Stability–Changeableness (#VP5141)   Uniformity–Nonuniformity (#VP5017).

♦ **VD2214 Carelessness**
   Integrative polarities Order–Disorder (#VP5059)   Caution–Rashness (#VP5894)
   Desire–Avoidance (#VP5631)   Leniency–Compulsion (#VP5756)
   Carefulness–Neglect (#VP5533)   Attention–Inattention (#VP5530)
   Curiosity–Incuriosity (#VP5528)   Resolution–Irresolution (#VP5624)
   Skilfulness–Unskilfulness (#VP5733).

♦ **VD2216 Censoriousness**
   Integrative polarities Approval–Disapproval (#VP5968)
   Naturalness–Affectation (#VP5900)
   Vindication–Condemnation (#VP6005).

♦ **VD2218 Censure**
   Integrative polarities Repute–Disrepute (#VP5914)
   Approval–Disapproval (#VP5968)
   Judgement–Misjudgement (#VP5494)
   Vindication–Condemnation (#VP6005).

♦ **VD2220 Chagrin**
   Integrative polarities Pride–Humility (#VP5905)   Pleasantness–Unpleasantness (#VP5863).

♦ **VD2222 Changeableness**
   Integrative polarities Strength–Weakness (#VP5159)
   Certainty–Uncertainty (#VP5513)
   Durability–Transience (#VP5110)
   Resolution–Irresolution (#VP5624)
   Stability–Changeableness (#VP5141)
   Uniformity–Nonuniformity (#VP5017).

♦ **VD2224 Chaos**
   Integrative polarities Order–Disorder (#VP5059)   Form–Formlessness (#VP5246)
   Energy–Moderation (#VP5161)   Authority–Lawlessness (#VP5739)
   Attention–Inattention (#VP5530)   Cohesion–Disintegration (#VP5050).

♦ **VD2226 Characterlessness**
   Integrative polarities Presence–Absence (#VP5186)
   Amusement–Boredom (#VP5878)
   Form–Formlessness (#VP5246).

♦ **VD2228 Chauvinism**
   Integrative polarities Accord–Disaccord (#VP5794)
   Broadmindedness–Narrowmindedness (#VP5526).

♦ **VD2230 Cheapness**
   Integrative polarities Repute–Disrepute (#VP5914)
   Expensiveness–Cheapness (#VP5848)
   Perfection–Imperfection (#VP5677)
   Importance–Unimportance (#VP5672)
   Appropriateness–Inappropriateness (#VP5665).

♦ **VD2232 Cheerlessness**
   Integrative polarities Hope–Hopelessness (#VP5888)
   Cheerfulness–Solemnity (#VP5870)
   Pleasantness–Unpleasantness (#VP5863).

♦ **VD2234 Childishness**
   Integrative polarities Intelligence–Unintelligence (#VP5467).

♦ **VD2236 Closeness**
   Integrative polarities Heat–Cold (#VP5328)   Motion–Quiescence (#VP5267)
   Breadth–Narrowness (#VP5204)   Expensiveness–Cheapness (#VP5848)
   Communicativeness–Uncommunicativeness (#VP5554).

♦ **VD2238 Clumsiness**
   Integrative polarities Beauty–Ugliness (#VP5899)
   Goodness–Badness (#VP5674)
   Elegance–Inelegance (#VP5589)
   Carefulness–Neglect (#VP5533)
   Skilfulness–Unskilfulness (#VP5733).

♦ **VD2240 Coarseness**
   Integrative polarities Taste–Vulgarity (#VP5896)   Chastity–Unchastity (#VP5987)
   Courtesy–Discourtesy (#VP5936)   Perfection–Imperfection (#VP5677)
   Preparedness–Unpreparedness (#VP5720).

♦ **VD2242 Coldness**
   Integrative polarities Heat–Cold (#VP5328)   Desire–Avoidance (#VP5631)
   Friendship–Enmity (#VP5927)   Kindness–Unkindness (#VP5938)
   Sociability–Unsociability (#VP5922)   Feeling–Unfeelinglessness (#VP5855).

♦ **VD2244 Collusion**
   Integrative polarities Communicativeness–Uncommunicativeness (#VP5554).

♦ **VD2246 Combativeness**
   Integrative polarities Attack–Defence (#VP5798)
   Accord–Disaccord (#VP5794)
   Intuition–Reason (#VP5481).

♦ **VD2248 Commonness**
   Integrative polarities Taste–Vulgarity (#VP5896)   Amusement–Boredom (#VP5878)
   Perfection–Imperfection (#VP5677).

♦ **VD2250 Complacency**
   Integrative polarities Modesty–Vanity (#VP5908)
   Contentment–Discontentment (#VP5868).

♦ **VD2252 Complexity**
   Integrative polarities Intelligibility–Unintelligibility (#VP5548).

♦ **VD2254 Complication**
   Integrative polarities Health–Disease (#VP5685)
   Facility–Difficulty (#VP5732)
   Simplicity–Complexity (#VP5045)
   Intelligibility–Unintelligibility (#VP5548).

♦ **VD2256 Complicity**
   Integrative polarities Innocence–Guilt (#VP5983)
   Support–Opposition (#VP5785).

♦ **VD2258 Compulsiveness**
   Integrative polarities Sanity–Insanity (#VP5472).

♦ **VD2260 Conceit**
   Integrative polarities Modesty–Vanity (#VP5908)
   Pride–Humility (#VP5905)
   Naturalness–Affectation (#VP5900).

♦ **VD2262 Concupiscence**
   Integrative polarities Chastity–Unchastity (#VP5987).

♦ **VD2264 Condemnation**
   Integrative polarities Approval–Disapproval (#VP5968)
   Vindication–Condemnation (#VP6005).

♦ **VD2266 Condescension**
   Integrative polarities Modesty–Vanity (#VP5908)
   Pride–Humility (#VP5905).

♦ **VD2268 Confinement**
   Integrative polarities Freedom–Restraint (#VP5762)
   Breadth–Narrowness (#VP5204)
   Atonement–Punishment (#VP6010)
   Centrality–Environment (#VP5226).

♦ **VD2270 Conflict**
   Integrative polarities Accord–Disaccord (#VP5794)
   Friendship–Enmity (#VP5927)
   Support–Opposition (#VP5785)
   Identity–Difference (#VP5013)
   Agreement–Disagreement (#VP5026)
   Selfactualization–Neurosis (#VP5690).

♦ **VD2272 Confusion**
   Integrative polarities Modesty–Vanity (#VP5908)
   Victory–Defeat (#VP5726)
   Order–Disorder (#VP5059)
   Form–Formlessness (#VP5246)
   Authority–Lawlessness (#VP5739)
   Attention–Inattention (#VP5530)

VD2272

Certainty–Uncertainty (#VP5513)
Cohesion–Disintegration (#VP5050)
Pleasantness–Unpleasantness (#VP5863).

♦ **VD2274 Congestion**
Integrative polarities Opening–Closure (#VP5265)
Weight–Lightness (#VP5352)
Oversufficiency–Insufficiency (#VP5660).

♦ **VD2276 Conspiracy**
Integrative polarities Accomplishment–Nonaccomplishment (#VP5722)
Communicativeness–Uncommunicativeness (#VP5554).

♦ **VD2278 Constraint**
Integrative polarities Freedom–Restraint (#VP5762)
Leniency–Compulsion (#VP5756).

♦ **VD2280 Contemptuousness**
Integrative polarities Modesty–Vanity (#VP5908)
Choice–Necessity (#VP5637)
Respect–Disrespect (#VP5964)
Support–Opposition (#VP5785)
Approval–Disapproval (#VP5968).

♦ **VD2282 Contentiousness**
Integrative polarities Accord–Discaccord (#VP5794)
Intuition–Reason (#VP5481)
Congratulation–Envy (#VP5948)
Comfort–Aggravation (#VP5885).

♦ **VD2284 Contradiction**
Integrative polarities Assent–Dissent (#VP5521)
Intuition–Reason (#VP5481)
Support–Opposition (#VP5785)
Identity–Difference (#VP5013)
Agreement–Disagreement (#VP5026)
Provability–Unprovability (#VP5505).

♦ **VD2286 Contrariety**
Integrative polarities Support–Opposition (#VP5785)
Equality–Inequality (#VP5030)
Pleasantness–Unpleasantness (#VP5863).

♦ **VD2288 Controversy**
Integrative polarities Accord–Discaccord (#VP5794)
Intuition–Reason (#VP5481)
Agreement–Disagreement (#VP5026).

♦ **VD2290 Contumaciousness**
Integrative polarities Obedience–Disobedience (#VP5766)
Resolution–Irresolution (#VP5624).

♦ **VD2292 Contumeliousness**
Integrative polarities Modesty–Vanity (#VP5908)
Respect–Disrespect (#VP5964)
Approval–Disapproval (#VP5968).

♦ **VD2294 Corruption**
Integrative polarities Virtue–Vice (#VP5980)   Goodness–Badness (#VP5674)
Probity–Improbity (#VP5974)   Symmetry–Distortion (#VP5248)
Conversion–Reversion (#VP5145)   Cleanness–Uncleanness (#VP5681)
Motivation–Dissuasion (#VP5648)   Simplicity–Complexity (#VP5045)
Improvement–Impairment (#VP5691)   Education–Miseducation (#VP5562).

♦ **VD2296 Counterproductivity**
Integrative polarities Power–Impotence (#VP5157)
Goodness–Badness (#VP5674)
Facility–Difficulty (#VP5732).

♦ **VD2298 Cowardice**
Integrative polarities Courage–Fear (#VP5890)   Strength–Weakness (#VP5159)
Resolution–Irresolution (#VP5624).

♦ **VD2300 Credulousness**
Integrative polarities Intelligence–Unintelligence (#VP5467).

♦ **VD2302 Crisis**
Integrative polarities Safety–Danger (#VP5697).

♦ **VD2304 Cruelty**
Integrative polarities Kindness–Unkindness (#VP5938)
Compassion–Pitilessness (#VP5944).

♦ **VD2306 Culpability**
Integrative polarities Innocence–Guilt (#VP5983)
Approval–Disapproval (#VP5968).

♦ **VD2308 Cunning**
Integrative polarities Skilfulness–Unskilfulness (#VP5733)
Communicativeness–Uncommunicativeness (#VP5554).

♦ **VD2310 Curtness**
Integrative polarities Courtesy–Discourtesy (#VP5936)
Communicativeness–Uncommunicativeness (#VP5554).

♦ **VD2312 Cynicism**
Integrative polarities Hope–Hopelessness (#VP5888)
Respect–Disrespect (#VP5964)
Kindness–Unkindness (#VP5938).

♦ **VD2314 Damage**
Integrative polarities Possession–Loss (#VP5808)
Goodness–Badness (#VP5674)
Improvement–Impairment (#VP5691)
Expedience–Inexpedience (#VP5670).

♦ **VD2316 Damnation**
Integrative polarities Kindness–Unkindness (#VP5938)
Approval–Disapproval (#VP5968)
Restoration–Destruction (#VP5693)
Vindication–Condemnation (#VP6005).

♦ **VD2318 Dangerousness**
Integrative polarities Safety–Danger (#VP5697)   Certainty–Uncertainty (#VP5513).

♦ **VD2320 Darkness**
Integrative polarities Light–Darkness (#VP5335)
Vision–Blindness (#VP5439)
Knowledge–Ignorance (#VP5475)
Cheerfulness–Solemnity (#VP5870)
Visibility–Invisibility (#VP5444)
Intelligibility–Unintelligibility (#VP5548).

♦ **VD2322 Deadness**
Integrative polarities Light–Darkness (#VP5335)
Amusement–Boredom (#VP5878)
Sensation–Insensibility (#VP5422)
Savouriness–Unsavouriness (#VP5427).

♦ **VD2324 Death**
Integrative polarities Life–Death (#VP5407)   Restoration–Destruction (#VP5693).

♦ **VD2326 Debauchery**
Integrative polarities Virtue–Vice (#VP5980)   Chastity–Unchastity (#VP5987)
Temperance–Intemperance (#VP5992).

♦ **VD2328 Debility**
Integrative polarities Health–Disease (#VP5685)
Action–Inaction (#VP5705)
Newness–Oldness (#VP5122)
Strength–Weakness (#VP5159).

♦ **VD2330 Decadence**
Integrative polarities Virtue–Vice (#VP5980)   Improvement–Impairment (#VP5691).

♦ **VD2332 Deceit**
Integrative polarities Probity–Improbity (#VP5974)
Skilfulness–Unskilfulness (#VP5733)
Communicativeness–Uncommunicativeness (#VP5554).

♦ **VD2334 Deception**
Integrative polarities Truth–Error (#VP5516)
Communicativeness–Uncommunicativeness (#VP5554).

♦ **VD2336 Decline**
Integrative polarities Consent–Refusal (#VP5775)
Choice–Necessity (#VP5637)
Strength–Weakness (#VP5159)
Prosperity–Adversity (#VP5728)
Improvement–Impairment (#VP5691)
Expensiveness–Cheapness (#VP5848)
Overrunning–Shortcoming (#VP5313)
Intelligence–Unintelligence (#VP5467).

♦ **VD2338 Decomposition**
Integrative polarities Improvement–Impairment (#VP5691)
Cohesion–Disintegration (#VP5050).

♦ **VD2340 Decrepitness**
Integrative polarities Newness–Oldness (#VP5122)
Strength–Weakness (#VP5159)
Improvement–Impairment (#VP5691)
Intelligence–Unintelligence (#VP5467).

♦ **VD2342 Defamation**
Integrative polarities Repute–Disrepute (#VP5914)
Approval–Disapproval (#VP5968).

♦ **VD2344 Default**
Integrative polarities Wealth–Poverty (#VP5837)
Presence–Absence (#VP5186)
Carefulness–Neglect (#VP5533)
Overrunning–Shortcoming (#VP5313).

♦ **VD2346 Defeat**
Integrative polarities Victory–Defeat (#VP5726)   Facility–Difficulty (#VP5732)
Restoration–Destruction (#VP5693)   Expectation–Inexpectation (#VP5539)
Accomplishment–Nonaccomplishment (#VP5722).

♦ **VD2348 Defection**
Integrative polarities Desire–Avoidance (#VP5631)
Perfection–Imperfection (#VP5677)
Resolution–Irresolution (#VP5624).

♦ **VD2350 Defectiveness**
Integrative polarities Truth–Error (#VP5516)   Health–Disease (#VP5685)
Perfection–Imperfection (#VP5677)   Intelligence–Unintelligence (#VP5467)
Completeness–Incompleteness (#VP5054)   Oversufficiency–Insufficiency (#VP5660).

♦ **VD2352 Defensiveness**
Integrative polarities Attack–Defence (#VP5798).

♦ **VD2354 Defiance**
Integrative polarities Accord–Discaccord (#VP5794)
Support–Opposition (#VP5785)
Obedience–Disobedience (#VP5766)
Resolution–Irresolution (#VP5624).

♦ **VD2356 Deficiency**
Integrative polarities Perfection–Imperfection (#VP5677)

DESTRUCTIVE VALUES **VD2446**

Superiority–Inferiority (#VP5036)
Completeness–Incompleteness (#VP5054)
Oversufficiency–Insufficiency (#VP5660).

♦ **VD2358 Deformation**
Integrative polarities Beauty–Ugliness (#VP5899)
Symmetry–Distortion (#VP5248)
Perfection–Imperfection (#VP5677)
Conformity–Nonconformity (#VP5082).

♦ **VD2360 Degeneration**
Integrative polarities Virtue–Vice (#VP5980)   Improvement–Impairment (#VP5691).

♦ **VD2362 Degradation**
Integrative polarities Virtue–Vice (#VP5980)   Repute–Disrepute (#VP5914)
Elevation–Depression (#VP5317)   Improvement–Impairment (#VP5691)
Cohesion–Disintegration (#VP5050).

♦ **VD2364 Dehumanization**
Integrative polarities Kindness–Unkindness (#VP5938).

♦ **VD2366 Dejection**
Integrative polarities Congratulation–Envy (#VP5948)
Cheerfulness–Solemnity (#VP5870)
Selfactualization–Neurosis (#VP5690).

♦ **VD2368 Delerium**
Integrative polarities Health–Disease (#VP5685)
Sanity–Insanity (#VP5472)
Selfactualization–Neurosis (#VP5690).

♦ **VD2370 Deleteriousness**
Integrative polarities Goodness–Badness (#VP5674)
Expedience–Inexpedience (#VP5670).

♦ **VD2372 Delusion**
Integrative polarities Truth–Error (#VP5516)   Selfactualization–Neurosis (#VP5690)
Communicativeness–Uncommunicativeness (#VP5554).

♦ **VD2374 Demeaning**
Integrative polarities Superiority–Inferiority (#VP5036)
Communicativeness–Uncommunicativeness (#VP5554).

♦ **VD2376 Denial**
Integrative polarities Assent–Dissent (#VP5521)
Possession–Loss (#VP5808)
Consent–Refusal (#VP5775)
Belief–Unbelief (#VP5501)
Choice–Necessity (#VP5637)
Support–Opposition (#VP5785)
Affirmation–Denial (#VP5523)
Provability–Unprovability (#VP5505).

♦ **VD2378 Denigration**
Integrative polarities Approval–Disapproval (#VP5968).

♦ **VD2380 Denunciation**
Integrative polarities Approval–Disapproval (#VP5968)
Vindication–Condemnation (#VP6005).

♦ **VD2382 Dependence**
Integrative polarities Freedom–Restraint (#VP5762)
Conventionality–Unconventionality (#VP5642).

♦ **VD2384 Depersonalization**
Integrative polarities Selfactualization–Neurosis (#VP5690).

♦ **VD2386 Depletion**
Integrative polarities Possession–Loss (#VP5808)
Appropriateness–Inappropriateness (#VP5665).

♦ **VD2388 Depradation**
integrative polarities Sharing–Appropriation (#VP5815)
Restoration–Destruction (#VP5693).

♦ **VD2390 Depravation**
Integrative polarities Virtue–Vice (#VP5980)   Improvement–Impairment (#VP5691).

♦ **VD2392 Depravity**
Integrative polarities Virtue–Vice (#VP5980)   Repute–Disrepute (#VP5914).

♦ **VD2394 Depreciation**
Integrative polarities Possession–Loss (#VP5808)
Approval–Disapproval (#VP5968)
Improvement–Impairment (#VP5691)
Judgement–Misjudgement (#VP5494)
Expensiveness–Cheapness (#VP5848).

♦ **VD2396 Depression**
Integrative polarities Elevation–Depression (#VP5317)
Cheerfulness–Solemnity (#VP5870)
Improvement–Impairment (#VP5691)
Pleasantness–Unpleasantness (#VP5863).

♦ **VD2398 Deprivation**
Integrative polarities Wealth–Poverty (#VP5837)
Possession–Loss (#VP5808)
Presence–Absence (#VP5186)
Existence–Nonexistence (#VP5001)
Oversufficiency–Insufficiency (#VP5660).

♦ **VD2400 Dereliction**
Integrative polarities Virtue–Vice (#VP5980)   Desire–Avoidance (#VP5631)
Probity–Improbity (#VP5974)   Carefulness–Neglect (#VP5533).

♦ **VD2402 Derogation**
Integrative polarities Approval–Disapproval (#VP5968)
Improvement–Impairment (#VP5691).

♦ **VD2404 Desecration**
Integrative polarities Piety–Impiety (#VP6028)   Rightness–Wrongness (#VP5957)
Appropriateness–Inappropriateness (#VP5665).

♦ **VD2406 Desertion**
Integrative polarities Piety–Impiety (#VP6028)   Desire–Avoidance (#VP5631)
Conversion–Reversion (#VP5145)   Resolution–Irresolution (#VP5624)
Sociability–Unsociability (#VP5922).

♦ **VD2408 Desolation**
Integrative polarities Cheerfulness–Solemnity (#VP5870)
Restoration–Destruction (#VP5693)
Sociability–Unsociability (#VP5922)
Pleasantness–Unpleasantness (#VP5863)
Productiveness–Unproductiveness (#VP5165).

♦ **VD2410 Despair**
Integrative polarities Hope–Hopelessness (#VP5888)
Cheerfulness–Solemnity (#VP5870)
Pleasantness–Unpleasantness (#VP5863).

♦ **VD2412 Desperation**
Integrative polarities Beauty–Ugliness (#VP5899).

♦ **VD2414 Despicableness**
Integrative polarities Repute–Disrepute (#VP5914)
Goodness–Badness (#VP5674)
Importance–Unimportance (#VP5672)
Pleasantness–Unpleasantness (#VP5863).

♦ **VD2416 Despoliation**
Integrative polarities Goodness–Badness (#VP5674)
Sharing–Appropriation (#VP5815)
Restoration–Destruction (#VP5693).

♦ **VD2418 Despondency**
Integrative polarities Hope–Hopelessness (#VP5888)
Cheerfulness–Solemnity (#VP5870).

♦ **VD2420 Despotism**
Integrative polarities Authority–Lawlessness (#VP5739).

♦ **VD2422 Destructiveness**
Integrative polarities Life–Death (#VP5407)   Prosperity–Adversity (#VP5728)
Restoration–Destruction (#VP5693)   Healthfulness–Unhealthfulness (#VP5683).

♦ **VD2424 Deterioration**
Integrative polarities Comfort–Aggravation (#VP5885)
Improvement–Impairment (#VP5691).

♦ **VD2426 Devastation**
Integrative polarities Restoration–Destruction (#VP5693).

♦ **VD2428 Deviation**
Integrative polarities Truth–Error (#VP5516)   Symmetry–Distortion (#VP5248)
Conformity–Nonconformity (#VP5082)   Uniformity–Nonuniformity (#VP5017).

♦ **VD2430 Devilishness**
Integrative polarities Virtue–Vice (#VP5980)   Goodness–Badness (#VP5674)
Kindness–Unkindness (#VP5938).

♦ **VD2432 Deviousness**
Integrative polarities Probity–Improbity (#VP5974)
Direction–Deviation (#VP5290)
Simplicity–Complexity (#VP5045)
Intelligence–Unintelligence (#VP5467).

♦ **VD2434 Devitalized**
Integrative polarities Power–Impotence (#VP5157)
Strength–Weakness (#VP5159).

♦ **VD2436 Diabolic**
Integrative polarities Virtue–Vice (#VP5980)   Goodness–Badness (#VP5674)
Kindness–Unkindness (#VP5938).

♦ **VD2438 Dictatorship**
Integrative polarities Modesty–Vanity (#VP5908)
Authority–Lawlessness (#VP5739).

♦ **VD2440 Difficulty**
Integrative polarities Accord–Disaccord (#VP5794)
Facility–Difficulty (#VP5732)
Prosperity–Adversity (#VP5728)
Pleasantness–Unpleasantness (#VP5863)
Intelligibility–Unintelligibility (#VP5548).

♦ **VD2442 Dilapidation**
Integrative polarities Improvement–Impairment (#VP5691)
Cohesion–Disintegration (#VP5050).

♦ **VD2444 Dirtiness**
Integrative polarities Virtue–Vice (#VP5980)   Goodness–Badness (#VP5674)
Chastity–Unchastity (#VP5987)   Approval–Disapproval (#VP5968)
Cleanness–Uncleanness (#VP5681)   Perfection–Imperfection (#VP5677).

♦ **VD2446 Disability**
Integrative polarities Health–Disease (#VP5685)
Power–Impotence (#VP5157)
Strength–Weakness (#VP5159)
Improvement–Impairment (#VP5691).

-749-

VD2448

♦ **VD2448 Disaccord**
Integrative polarities Assent–Dissent (#VP5521)
Accord–Disaccord (#VP5794)
Friendship–Enmity (#VP5927)
Support–Opposition (#VP5785)
Identity–Difference (#VP5013)
Agreement–Disagreement (#VP5026).

♦ **VD2450 Disadvantage**
Integrative polarities Goodness–Badness (#VP5674)
Facility–Difficulty (#VP5732)
Expedience–Inexpedience (#VP5670).

♦ **VD2452 Disaffection**
Integrative polarities Accord–Disaccord (#VP5794)
Probity–Improbity (#VP5974)
Pleasantness–Unpleasantness (#VP5863).

♦ **VD2454 Disaffinity**
Integrative polarities Accord–Disaccord (#VP5794)
Friendship–Enmity (#VP5927)
Attraction–Repulsion (#VP5288)
Feeling–Unfeelinglessness (#VP5855)
Pleasantness–Unpleasantness (#VP5863).

♦ **VD2456 Disagreeableness**
Integrative polarities Congratulation–Envy (#VP5948)
Kindness–Unkindness (#VP5938)
Agreement–Disagreement (#VP5026)
Willingness–Unwillingness (#VP5621)
Pleasantness–Unpleasantness (#VP5863).

♦ **VD2458 Disappearance**
Integrative polarities Desire–Avoidance (#VP5631)
Presence–Absence (#VP5186)
Appearance–Disappearance (#VP5446).

♦ **VD2460 Disappointment**
Integrative polarities Truth–Error (#VP5516)   Hope–Hopelessness (#VP5888)
Approval–Disapproval (#VP5968)   Expectation–Inexpectation (#VP5539)
Contentment–Discontentment (#VP5868).

♦ **VD2462 Disapproval**
Integrative polarities Assent–Dissent (#VP5521)
Choice–Necessity (#VP5637)
Congratulation–Envy (#VP5948)
Approval–Disapproval (#VP5968)
Pleasantness–Unpleasantness (#VP5863).

♦ **VD2464 Disarrangement**
Integrative polarities Order–Disorder (#VP5059)   Location–Dislocation (#VP5185).

♦ **VD2466 Disaster**
Integrative polarities Life–Death (#VP5407)   Prosperity–Adversity (#VP5728)
Restoration–Destruction (#VP5693).

♦ **VD2468 Disbelief**
Integrative polarities Belief–Unbelief (#VP5501)   Certainty–Uncertainty (#VP5513).

♦ **VD2470 Discomfort**
Integrative polarities Sensation–Insensibility (#VP5422)
Pleasantness–Unpleasantness (#VP5863).

♦ **VD2472 Discomposure**
Integrative polarities Order–Disorder (#VP5059)   Attention–Inattention (#VP5530)
Certainty–Uncertainty (#VP5513)   Oscillation–Agitation (#VP5323)
Pleasantness–Unpleasantness (#VP5863).

♦ **VD2474 Disconcertion**
Integrative polarities Courage–Fear (#VP5890)   Facility–Difficulty (#VP5732)
Attention–Inattention (#VP5530)   Certainty–Uncertainty (#VP5513)
Pleasantness–Unpleasantness (#VP5863).

♦ **VD2476 Discontentment**
Integrative polarities Congratulation–Envy (#VP5948)
Approval–Disapproval (#VP5968)
Cheerfulness–Solemnity (#VP5870)
Contentment–Discontentment (#VP5868)
Pleasantness–Unpleasantness (#VP5863).

♦ **VD2478 Discontinuity**
Integrative polarities Location–Dislocation (#VP5185)
Conjunction–Separation (#VP5047)
Regularity–Irregularity (#VP5137)
Continuity–Discontinuity (#VP5071).

♦ **VD2480 Discord**
Integrative polarities Harmony–Discord (#VP5462)
Accord–Disaccord (#VP5794)
Silence–Loudness (#VP5451)
Agreement–Disagreement (#VP5026).

♦ **VD2482 Discouragement**
Integrative polarities Courage–Fear (#VP5890)   Motivation–Dissuasion (#VP5648)
Cheerfulness–Solemnity (#VP5870).

♦ **VD2484 Discourtesy**
Integrative polarities Respect–Disrespect (#VP5964)
Courtesy–Discourtesy (#VP5936)
Behaviour–Misbehaviour (#VP5737).

♦ **VD2486 Discredit**
Integrative polarities Belief–Unbelief (#VP5501)   Repute–Disrepute (#VP5914)
Approval–Disapproval (#VP5968)   Provability–Unprovability (#VP5505).

♦ **VD2488 Discrimination**
Integrative polarities Justice–Injustice (#VP5976)
Broadmindedness–Narrowmindedness (#VP5526).

♦ **VD2490 Disdain**
Integrative polarities Modesty–Vanity (#VP5908)
Choice–Necessity (#VP5637)
Respect–Disrespect (#VP5964)
Support–Opposition (#VP5785).

♦ **VD2492 Disenchantment**
Integrative polarities Truth–Error (#VP5516)   Approval–Disapproval (#VP5968)
Conversion–Reversion (#VP5145).

♦ **VD2494 Disesteem**
Integrative polarities Repute–Disrepute (#VP5914)
Respect–Disrespect (#VP5964)
Approval–Disapproval (#VP5968).

♦ **VD2496 Disfavour**
Integrative polarities Repute–Disrepute (#VP5914)
Accord–Disaccord (#VP5794)
Approval–Disapproval (#VP5968)
Pleasantness–Unpleasantness (#VP5863).

♦ **VD2498 Disfigurement**
Integrative polarities Beauty–Ugliness (#VP5899)
Symmetry–Distortion (#VP5248)
Perfection–Imperfection (#VP5677).

♦ **VD2500 Disgrace**
Integrative polarities Virtue–Vice (#VP5980)   Pride–Humility (#VP5905)
Repute–Disrepute (#VP5914)   Rightness–Wrongness (#VP5957)
Approval–Disapproval (#VP5968).

♦ **VD2502 Disgust**
Integrative polarities Savouriness–Unsavouriness (#VP5427)
Pleasantness–Unpleasantness (#VP5863).

♦ **VD2504 Disharmony**
Integrative polarities Order–Disorder (#VP5059)   Harmony–Discord (#VP5462)
Accord–Disaccord (#VP5794)   Agreement–Disagreement (#VP5026).

♦ **VD2506 Dishonesty**
Integrative polarities Probity–Improbity (#VP5974)
Communicativeness–Uncommunicativeness (#VP5554).

♦ **VD2508 Dishonour**
Integrative polarities Piety–Impiety (#VP6028)   Repute–Disrepute (#VP5914)
Probity–Improbity (#VP5974)   Respect–Disrespect (#VP5964).

♦ **VD2510 Disillusionment**
Integrative polarities Truth–Error (#VP5516)   Approval–Disapproval (#VP5968)
Illusion–Disillusionment (#VP5519)   Expectation–Inexpectation (#VP5539).

♦ **VD2512 Disingenuousness**
Integrative polarities Intuition–Reason (#VP5481)
Probity–Improbity (#VP5974)
Communicativeness–Uncommunicativeness (#VP5554).

♦ **VD2514 Disintegration**
Integrative polarities Order–Disorder (#VP5059)   Improvement–Impairment (#VP5691)
Conjunction–Separation (#VP5047)   Restoration–Destruction (#VP5693)
Cohesion–Disintegration (#VP5050).

♦ **VD2516 Disinterest**
Integrative polarities Desire–Avoidance (#VP5631)
Curiosity–Incuriosity (#VP5528)
Feeling–Unfeelinglessness (#VP5855).

♦ **VD2518 Dislike**
Integrative polarities Love–Hate (#VP5930)   Pleasantness–Unpleasantness (#VP5863).

♦ **VD2520 Disloyalty**
Integrative polarities Probity–Improbity (#VP5974)
Resolution–Irresolution (#VP5624).

♦ **VD2522 Dismay**
Integrative polarities Courage–Fear (#VP5890)   Certainty–Uncertainty (#VP5513)
Pleasantness–Unpleasantness (#VP5863).

♦ **VD2524 Disobedience**
Integrative polarities Consent–Refusal (#VP5775)
Authority–Lawlessness (#VP5739)
Obedience–Disobedience (#VP5766)
Willingness–Unwillingness (#VP5621).

♦ **VD2526 Disorder**
Integrative polarities Health–Disease (#VP5685)
Order–Disorder (#VP5059)
Form–Formlessness (#VP5246)
Authority–Lawlessness (#VP5739)
Attention–Inattention (#VP5530)
Certainty–Uncertainty (#VP5513)
Oscillation–Agitation (#VP5323)
Behaviour–Misbehaviour (#VP5737).

♦ **VD2528 Disorganization**
Integrative polarities Order–Disorder (#VP5059)   Authority–Lawlessness (#VP5739)
Attention–Inattention (#VP5530)   Improvement–Impairment (#VP5691)
Restoration–Destruction (#VP5693)   Cohesion–Disintegration (#VP5050).

♦ **VD2530 Disorientation**
Integrative polarities Sanity–Insanity (#VP5472)   Attention–Inattention (#VP5530)
Certainty–Uncertainty (#VP5513)   Selfactualization–Neurosis (#VP5690).

DESTRUCTIVE VALUES

VD2610

◆ **VD2532 Disparagement**
Integrative polarities Repute–Disrepute (#VP5914)
Respect–Disrespect (#VP5964)
Approval–Disapproval (#VP5968)
Judgement–Misjudgement (#VP5494).

◆ **VD2534 Disparity**
Integrative polarities Assent–Dissent (#VP5521)
Accord–Disaccord (#VP5794)
Equality–Inequality (#VP5030)
Identity–Difference (#VP5013)
Agreement–Disagreement (#VP5026).

◆ **VD2536 Displeasure**
Integrative polarities Congratulation–Envy (#VP5948)
Approval–Disapproval (#VP5968)
Pleasure–Displeasure (#VP5865)
Cheerfulness–Solemnity (#VP5870)
Pleasantness–Unpleasantness (#VP5863).

◆ **VD2538 Disputatiousness**
Integrative polarities Accord–Disaccord (#VP5794)
Intuition–Reason (#VP5481)
Support–Opposition (#VP5785)
Congratulation–Envy (#VP5948).

◆ **VD2540 Disquiet**
Integrative polarities Courage–Fear (#VP5890)   Oscillation–Agitation (#VP5323)
Pleasantness–Unpleasantness (#VP5863).

◆ **VD2542 Disregard**
Integrative polarities Choice–Necessity (#VP5637)
Desire–Avoidance (#VP5631)
Respect–Disrespect (#VP5964)
Support–Opposition (#VP5785)
Carefulness–Neglect (#VP5533)
Attention–Inattention (#VP5530)
Obedience–Disobedience (#VP5766).

◆ **VD2544 Disrespect**
Integrative polarities Modesty–Vanity (#VP5908)
Respect–Disrespect (#VP5964)
Approval–Disapproval (#VP5968).

◆ **VD2546 Dissatisfaction**
Integrative polarities Assent–Dissent (#VP5521)
Congratulation–Envy (#VP5948)
Approval–Disapproval (#VP5968)
Expectation–Inexpectation (#VP5539)
Contentment–Discontentment (#VP5868)
Pleasantness–Unpleasantness (#VP5863).

◆ **VD2548 Dissension**
Integrative polarities Assent–Dissent (#VP5521)
Accord–Disaccord (#VP5794)
Support–Opposition (#VP5785)
Agreement–Disagreement (#VP5026).

◆ **VD2550 Dissidence**
Integrative polarities Assent–Dissent (#VP5521)
Accord–Disaccord (#VP5794)
Agreement–Disagreement (#VP5026).

◆ **VD2552 Dissipation**
Integrative polarities Possession–Loss (#VP5808)
Temperance–Intemperance (#VP5992)
Appearance–Disappearance (#VP5446).

◆ **VD2554 Dissociation**
Integrative polarities Cohesion–Disintegration (#VP5050)
Selfactualization–Neurosis (#VP5690).

◆ **VD2556 Dissolution**
Integrative polarities Life–Death (#VP5407)   Improvement–Impairment (#VP5691)
Conjunction–Separation (#VP5047)   Restoration–Destruction (#VP5693)
Cohesion–Disintegration (#VP5050)   Appearance–Disappearance (#VP5446).

◆ **VD2558 Dissonance**
Integrative polarities Harmony–Discord (#VP5462)
Identity–Difference (#VP5013)
Agreement–Disagreement (#VP5026).

◆ **VD2560 Distaste**
Integrative polarities Approval–Disapproval (#VP5968)
Willingness–Unwillingness (#VP5621)
Pleasantness–Unpleasantness (#VP5863).

◆ **VD2562 Distortion**
Integrative polarities Truth–Error (#VP5516)   Symmetry–Distortion (#VP5248)
Perfection–Imperfection (#VP5677)   Representation–Misrepresentation (#VP5573)
Interpretability–Misinterpretability (#VP5552)
Communicativeness–Uncommunicativeness (#VP5554).

◆ **VD2564 Distraction**
Integrative polarities Sanity–Insanity (#VP5472)   Attention–Inattention (#VP5530).

◆ **VD2566 Distress**
Integrative polarities Courage–Fear (#VP5890)   Wealth–Poverty (#VP5837)
Goodness–Badness (#VP5674)   Facility–Difficulty (#VP5732)
Sharing–Appropriation (#VP5815)   Sensation–Insensibility (#VP5422)
Pleasantness–Unpleasantness (#VP5863).

◆ **VD2568 Distrust**
Integrative polarities Belief–Unbelief (#VP5501)   Caution–Rashness (#VP5894)
Congratulation–Envy (#VP5948).

◆ **VD2570 Disturbance**
Integrative polarities Courage–Fear (#VP5890)   Order–Disorder (#VP5059)
Energy–Moderation (#VP5161)   Attention–Inattention (#VP5530)
Certainty–Uncertainty (#VP5513)   Oscillation–Agitation (#VP5323)
Excitement–Inexcitability (#VP5857)   Pleasantness–Unpleasantness (#VP5863).

◆ **VD2572 Disunity**
Integrative polarities Friendship–Enmity (#VP5927)
Conjunction–Separation (#VP5047)
Agreement–Disagreement (#VP5026).

◆ **VD2574 Disuse**
Integrative polarities Newness–Oldness (#VP5122)
Desire–Avoidance (#VP5631)
Appropriateness–Inappropriateness (#VP5665).

◆ **VD2576 Divergence**
Integrative polarities Accord–Disaccord (#VP5794)
Direction–Deviation (#VP5290)
Convergence–Divergence (#VP5298)
Agreement–Disagreement (#VP5026)
Conformity–Nonconformity (#VP5082)
Uniformity–Nonuniformity (#VP5017).

◆ **VD2578 Divisiveness**
Integrative polarities Accord–Disaccord (#VP5794).

◆ **VD2580 Division**
Integrative polarities Accord–Disaccord (#VP5794)
Inclusion–Exclusion (#VP5076)
Convergence–Divergence (#VP5298)
Conjunction–Separation (#VP5047)
Discrimination–Indiscrimination (#VP5492).

◆ **VD2582 Dolorousness**
Integrative polarities Cheerfulness–Solemnity (#VP5870)
Pleasantness–Unpleasantness (#VP5863).

◆ **VD2584 Doubtfulness**
Integrative polarities Safety–Danger (#VP5697)   Belief–Unbelief (#VP5501)
Probity–Improbity (#VP5974)   Certainty–Uncertainty (#VP5513).

◆ **VD2586 Dubiousness**
Integrative polarities Piety–Impiety (#VP6028)   Safety–Danger (#VP5697)
Belief–Unbelief (#VP5501)   Probity–Improbity (#VP5974)
Certainty–Uncertainty (#VP5513)   Resolution–Irresolution (#VP5624)
Communicativeness–Uncommunicativeness (#VP5554).

◆ **VD2588 Dullness**
Integrative polarities Action–Inaction (#VP5705)   Amusement–Boredom (#VP5878)
Strength–Weakness (#VP5159)   Perfection–Imperfection (#VP5677)
Sensation–Insensibility (#VP5422)   Feeling–Unfeelinglessness (#VP5855)
Pleasantness–Unpleasantness (#VP5863)   Intelligence–Unintelligence (#VP5467)
Imaginativeness–Unimaginativeness (#VP5535).

◆ **VD2590 Duplicity**
Integrative polarities Probity–Improbity (#VP5974)
Communicativeness–Uncommunicativeness (#VP5554).

◆ **VD2592 Duress**
Integrative polarities Freedom–Restraint (#VP5762)
Leniency–Compulsion (#VP5756).

◆ **VD2594 Eccentricity**
Integrative polarities Stability–Changeableness (#VP5141)
Conformity–Nonconformity (#VP5082)
Intelligence–Unintelligence (#VP5467).

◆ **VD2596 Eeriness**
Integrative polarities Life–Death (#VP5407)   Courage–Fear (#VP5890).

◆ **VD2598 Effetism**
Integrative polarities Power–Impotence (#VP5157)
Amusement–Boredom (#VP5878)
Strength–Weakness (#VP5159)
Improvement–Impairment (#VP5691)
Appropriateness–Inappropriateness (#VP5665).

◆ **VD2600 Elaborateness**
Integrative polarities Beauty–Ugliness (#VP5899)
Naturalness–Affectation (#VP5900).

◆ **VD2602 Embarrassment**
Integrative polarities Pride–Humility (#VP5905)   Wealth–Poverty (#VP5837)
Facility–Difficulty (#VP5732)   Attention–Inattention (#VP5530)
Certainty–Uncertainty (#VP5513)   Pleasantness–Unpleasantness (#VP5863).

◆ **VD2604 Encroachment**
Integrative polarities Dueness–Undueness (#VP5960)
Overrunning–Shortcoming (#VP5313)
Circumscription–Intrusion (#VP5234).

◆ **VD2606 Encumbrance**
Integrative polarities Weight–Lightness (#VP5352)
Facility–Difficulty (#VP5732)
Pleasantness–Unpleasantness (#VP5863).

◆ **VD2608 Enforcement**
Integrative polarities Leniency–Compulsion (#VP5756).

◆ **VD2610 Enmity**
Integrative polarities Love–Hate (#VP5930)   Accord–Disaccord (#VP5794)
Friendship–Enmity (#VP5927)   Support–Opposition (#VP5785)
Pleasantness–Unpleasantness (#VP5863).

-751-

♦ **VD2612 Ennui**
Integrative polarities Action–Inaction (#VP5705)   Amusement–Boredom (#VP5878)
Pleasantness–Unpleasantness (#VP5863).

♦ **VD2614 Envy**
Integrative polarities Virtue–Vice (#VP5980)   Congratulation–Envy (#VP5948)
Contentment–Discontentment (#VP5868).

♦ **VD2616 Equivocation**
Integrative polarities Desire–Avoidance (#VP5631)
Intuition–Reason (#VP5481)
Identity–Difference (#VP5013)
Resolution–Irresolution (#VP5624)
Intelligibility–Unintelligibility (#VP5548)
Communicativeness–Uncommunicativeness (#VP5554).

♦ **VD2618 Erosion**
Integrative polarities Improvement–Impairment (#VP5691)
Cohesion–Disintegration (#VP5050)
Appropriateness–Inappropriateness (#VP5665).

♦ **VD2620 Erroneousness**
Integrative polarities Truth–Error (#VP5516)   Perfection–Imperfection (#VP5677)
Communicativeness–Uncommunicativeness (#VP5554).

♦ **VD2622 Evasion**
Integrative polarities Desire–Avoidance (#VP5631)
Intuition–Reason (#VP5481)
Probity–Improbity (#VP5974)
Communicativeness–Uncommunicativeness (#VP5554).

♦ **VD2624 Evil**
Integrative polarities Virtue–Vice (#VP5980)   Goodness–Badness (#VP5674)
Rightness–Wrongness (#VP5957)   Facility–Difficulty (#VP5732).

♦ **VD2626 Exaggeration**
Integrative polarities Judgement–Misjudgement (#VP5494)
Representation–Misrepresentation (#VP5573)
Communicativeness–Uncommunicativeness (#VP5554).

♦ **VD2628 Exasperation**
Integrative polarities Congratulation–Envy (#VP5948)
Comfort–Aggravation (#VP5885)
Pleasantness–Unpleasantness (#VP5863).

♦ **VD2630 Excess**
Integrative polarities Dueness–Undueness (#VP5960)
Temperance–Intemperance (#VP5992)
Expensiveness–Cheapness (#VP5848)
Oversufficiency–Insufficiency (#VP5660)
Communicativeness–Uncommunicativeness (#VP5554).

♦ **VD2632 Excessiveness**
Integrative polarities Sanity–Insanity (#VP5472)   Temperance–Intemperance (#VP5992)
Expensiveness–Cheapness (#VP5848)   Oversufficiency–Insufficiency (#VP5660).

♦ **VD2634 Exclusion**
Integrative polarities Consent–Refusal (#VP5775)
Choice–Necessity (#VP5637)
Inclusion–Exclusion (#VP5076)
Approval–Disapproval (#VP5968).

♦ **VD2636 Exclusiveness**
Integrative polarities Taste–Vulgarity (#VP5896)   Respect–Disrespect (#VP5964)
Support–Opposition (#VP5785)   Sociability–Unsociability (#VP5922).

♦ **VD2638 Execration**
Integrative polarities Love–Hate (#VP5930)   Approval–Disapproval (#VP5968).

♦ **VD2640 Exhaustion**
Integrative polarities Health–Disease (#VP5685)
Possession–Loss (#VP5808)
Action–Inaction (#VP5705)
Appropriateness–Inappropriateness (#VP5665).

♦ **VD2642 Exiguity**
Integrative polarities Bigness–Littleness (#VP5195)
Greatness–Smallness (#VP5034)
Oversufficiency–Insufficiency (#VP5660).

♦ **VD2644 Exploitation**
Integrative polarities Expensiveness–Cheapness (#VP5848).

♦ **VD2646 Expulsion**
Integrative polarities Inclusion–Exclusion (#VP5076).

♦ **VD2648 Extermination**
Integrative polarities Life–Death (#VP5407)   Restoration–Destruction (#VP5693).

♦ **VD2650 Extinction**
Integrative polarities Heat–Cold (#VP5328)   Life–Death (#VP5407)
Restoration–Destruction (#VP5693)   Appearance–Disappearance (#VP5446).

♦ **VD2652 Extortion**
Integrative polarities Sharing–Appropriation (#VP5815)
Expensiveness–Cheapness (#VP5848).

♦ **VD2654 Extravagance**
Integrative polarities Sanity–Insanity (#VP5472)   Temperance–Intemperance (#VP5992)
Naturalness–Affectation (#VP5900)   Expensiveness–Cheapness (#VP5848)
Conciseness–Diffuseness (#VP5592)   Oversufficiency–Insufficiency (#VP5660)
Communicativeness–Uncommunicativeness (#VP5554).

♦ **VD2656 Extremism**
Integrative polarities Sanity–Insanity (#VP5472)   Obedience–Disobedience (#VP5766)
Oversufficiency–Insufficiency (#VP5660).

♦ **VD2658 Fabrication**
Integrative polarities Communicativeness–Uncommunicativeness (#VP5554).

♦ **VD2660 Failure**
Integrative polarities Virtue–Vice (#VP5980)   Truth–Error (#VP5516)
Victory–Defeat (#VP5726)   Power–Impotence (#VP5157)
Improvement–Impairment (#VP5691)   Perfection–Imperfection (#VP5677)
Overrunning–Shortcoming (#VP5313)   Superiority–Inferiority (#VP5036)
Expectation–Inexpectation (#VP5539)
Accomplishment–Nonaccomplishment (#VP5722).

♦ **VD2662 Faithlessness**
Integrative polarities Belief–Unbelief (#VP5501)   Probity–Improbity (#VP5974)
Resolution–Irresolution (#VP5624)
Communicativeness–Uncommunicativeness (#VP5554).

♦ **VD2664 Fakery**
Integrative polarities Naturalness–Affectation (#VP5900)
Communicativeness–Uncommunicativeness (#VP5554).

♦ **VD2666 Fallacy**
Integrative polarities Truth–Error (#VP5516)   Intuition–Reason (#VP5481)
Communicativeness–Uncommunicativeness (#VP5554).

♦ **VD2668 Fallibility**
Integrative polarities Certainty–Uncertainty (#VP5513)
Perfection–Imperfection (#VP5677).

♦ **VD2670 Falsehood**
Integrative polarities Communicativeness–Uncommunicativeness (#VP5554).

♦ **VD2672 Falseness**
Integrative polarities Truth–Error (#VP5516)   Probity–Improbity (#VP5974)
Communicativeness–Uncommunicativeness (#VP5554).

♦ **VD2674 Fantasy**
Integrative polarities Resolution–Irresolution (#VP5624).

♦ **VD2676 Fanaticism**
Integrative polarities Sanity–Insanity (#VP5472)   Desire–Avoidance (#VP5631)
Energy–Moderation (#VP5161)   Resolution–Irresolution (#VP5624)
Broadmindedness–Narrowmindedness (#VP5526).

♦ **VD2678 Fatigue**
Integrative polarities Health–Disease (#VP5685)
Action–Inaction (#VP5705)
Strength–Weakness (#VP5159).

♦ **VD2680 Fatuity**
Integrative polarities Power–Impotence (#VP5157)
Thought–Thoughtlessness (#VP5478)
Intelligence–Unintelligence (#VP5467)
Appropriateness–Inappropriateness (#VP5665).

♦ **VD2682 Faultiness**
Integrative polarities Truth–Error (#VP5516)   Innocence–Guilt (#VP5983)
Intuition–Reason (#VP5481)   Perfection–Imperfection (#VP5677).

♦ **VD2684 Fear**
Integrative polarities Courage–Fear (#VP5890)   Resolution–Irresolution (#VP5624)
Excitement–Inexcitability (#VP5857).

♦ **VD2686 Featurelessness**
Integrative polarities Presence–Absence (#VP5186)
Form–Formlessness (#VP5246).

♦ **VD2688 Fecklessness**
Integrative polarities Power–Impotence (#VP5157)
Preparedness–Unpreparedness (#VP5720)
Appropriateness–Inappropriateness (#VP5665).

♦ **VD2690 Feebleness**
Integrative polarities Health–Disease (#VP5685)
Power–Impotence (#VP5157)
Newness–Oldness (#VP5122)
Intuition–Reason (#VP5481)
Strength–Weakness (#VP5159)
Resolution–Irresolution (#VP5624)
Visibility–Invisibility (#VP5444)
Intelligence–Unintelligence (#VP5467).

♦ **VD2692 Feloniousness**
Integrative polarities Virtue–Vice (#VP5980)   Probity–Improbity (#VP5974)
Legality–Illegality (#VP5998).

♦ **VD2694 Ferocity**
Integrative polarities Accord–Disaccord (#VP5794)
Kindness–Unkindness (#VP5938).

♦ **VD2696 Fiendish**
Integrative polarities Virtue–Vice (#VP5980)   Goodness–Badness (#VP5674)
Kindness–Unkindness (#VP5938)   Godliness–Ungodliness (#VP6013).

♦ **VD2698 Fierceness**
Integrative polarities Accord–Disaccord (#VP5794)
Energy–Moderation (#VP5161)
Congratulation–Envy (#VP5948)
Kindness–Unkindness (#VP5938)
Excitement–Inexcitability (#VP5857).

DESTRUCTIVE VALUES

♦ **VD2700 Filth**
Integrative polarities Goodness–Badness (#VP5674)
Chastity–Unchastity (#VP5987)
Approval–Disapproval (#VP5968)
Cleanness–Uncleanness (#VP5681).

♦ **VD2702 Fixation**
Integrative polarities Sanity–Insanity (#VP5472)   Facility–Difficulty (#VP5732)
Selfactualization–Neurosis (#VP5690).

♦ **VD2704 Flaw**
Integrative polarities Truth–Error (#VP5516)   Intuition–Reason (#VP5481)
Legality–Illegality (#VP5998)   Perfection–Imperfection (#VP5677).

♦ **VD2706 Forcelessness**
Integrative polarities Power–Impotence (#VP5157)
Influence–Influencelessness (#VP5172).

♦ **VD2708 Forgetfulness**
Integrative polarities Kindness–Unkindness (#VP5938)
Carefulness–Neglect (#VP5533)
Remembrance–Forgetfulness (#VP5537).

♦ **VD2710 Foulness**
Integrative polarities Virtue–Vice (#VP5980)   Beauty–Ugliness (#VP5899)
Opening–Closure (#VP5265)   Repute–Disrepute (#VP5914)
Goodness–Badness (#VP5674)   Fragrance–Stench (#VP5435)
Justice–Injustice (#VP5976)   Chastity–Unchastity (#VP5987)
Approval–Disapproval (#VP5968)   Cleanness–Uncleanness (#VP5681)
Restoration–Destruction (#VP5693)   Savouriness–Unsavouriness (#VP5427)
Pleasantness–Unpleasantness (#VP5863).

♦ **VD2712 Fragility**
Integrative polarities Health–Disease (#VP5685)
Strength–Weakness (#VP5159)
Elasticity–Toughness (#VP5358).

♦ **VD2714 Fraudulence**
Integrative polarities Probity–Improbity (#VP5974)
Sharing–Appropriation (#VP5815)
Communicativeness–Uncommunicativeness (#VP5554).

♦ **VD2716 Frenzy**
Integrative polarities Health–Disease (#VP5685)
Sanity–Insanity (#VP5472)
Desire–Avoidance (#VP5631)
Energy–Moderation (#VP5161)
Attention–Inattention (#VP5530).

♦ **VD2718 Fretfulness**
Integrative polarities Congratulation–Envy (#VP5948)
Patience–Impatience (#VP5861).

♦ **VD2720 Friction**
Integrative polarities Accord–Disaccord (#VP5794)
Friendship–Enmity (#VP5927)
Support–Opposition (#VP5785).

♦ **VD2722 Frightfulness**
Integrative polarities Courage–Fear (#VP5890)   Beauty–Ugliness (#VP5899).

♦ **VD2724 Frivolity**
Integrative polarities Order–Disorder (#VP5059)   Attention–Inattention (#VP5530)
Importance–Unimportance (#VP5672)   Intelligence–Unintelligence (#VP5467).

♦ **VD2726 Frustration**
Integrative polarities Victory–Defeat (#VP5726)   Facility–Difficulty (#VP5732)
Expectation–Inexpectation (#VP5539)   Selfactualization–Neurosis (#VP5690).

♦ **VD2728 Fulsomeness**
Integrative polarities Repute–Disrepute (#VP5914)
Goodness–Badness (#VP5674)
Fragrance–Stench (#VP5435)
Chastity–Unchastity (#VP5987)
Pleasantness–Unpleasantness (#VP5863).

♦ **VD2730 Furtiveness**
Integrative polarities Communicativeness–Uncommunicativeness (#VP5554).

♦ **VD2732 Fury**
Integrative polarities Sanity–Insanity (#VP5472)   Desire–Avoidance (#VP5631)
Energy–Moderation (#VP5161)   Congratulation–Envy (#VP5948).

♦ **VD2734 Fuss**
Integrative polarities Modesty–Vanity (#VP5908)
Order–Disorder (#VP5059)
Accord–Disaccord (#VP5794)
Energy–Moderation (#VP5161)
Patience–Impatience (#VP5861)
Approval–Disapproval (#VP5968)
Attention–Inattention (#VP5530)
Oscillation–Agitation (#VP5323)
Exultation–Lamentation (#VP5873)
Excitement–Inexcitability (#VP5857).

♦ **VD2736 Futility**
Integrative polarities Power–Impotence (#VP5157)
Hope–Hopelessness (#VP5888)
Importance–Unimportance (#VP5672)
Expedience–Inexpedience (#VP5670)
Meaning–Meaninglessness (#VP5545)
Accomplishment–Nonaccomplishment (#VP5722)
Appropriateness–Inappropriateness (#VP5665).

♦ **VD2738 Gaudiness**
Integrative polarities Taste–Vulgarity (#VP5896)   Colour–Colourlessness (#VP5362).

♦ **VD2740 Glibness**
Integrative polarities Courtesy–Discourtesy (#VP5936).

♦ **VD2742 Godlessness**
Integrative polarities Piety–Impiety (#VP6028).

♦ **VD2744 Gracelessness**
Integrative polarities Beauty–Ugliness (#VP5899)
Elegance–Inelegance (#VP5589)
Skilfulness–Unskilfulness (#VP5733).

♦ **VD2746 Greed**
Integrative polarities Virtue–Vice (#VP5980)   Desire–Avoidance (#VP5631)
Temperance–Intemperance (#VP5992)   Unselfishness–Selfishness (#VP5978).

♦ **VD2748 Grief**
Integrative polarities Prosperity–Adversity (#VP5728)
Cheerfulness–Solemnity (#VP5870)
Pleasantness–Unpleasantness (#VP5863).

♦ **VD2750 Grievance**
Integrative polarities Assent–Dissent (#VP5521)
Goodness–Badness (#VP5674)
Justice–Injustice (#VP5976)
Exultation–Lamentation (#VP5873)
Pleasantness–Unpleasantness (#VP5863).

♦ **VD2752 Grimness**
Integrative polarities Courage–Fear (#VP5890)   Beauty–Ugliness (#VP5899)
Goodness–Badness (#VP5674)   Hope–Hopelessness (#VP5888)
Congratulation–Envy (#VP5948)   Kindness–Unkindness (#VP5938)
Cheerfulness–Solemnity (#VP5870)   Resolution–Irresolution (#VP5624)
Pleasantness–Unpleasantness (#VP5863).

♦ **VD2754 Grossness**
Integrative polarities Taste–Vulgarity (#VP5896)   Repute–Disrepute (#VP5914)
Goodness–Badness (#VP5674)   Chastity–Unchastity (#VP5987)
Pleasantness–Unpleasantness (#VP5863)   Intelligence–Unintelligence (#VP5467).

♦ **VD2756 Gruesomeness**
Integrative polarities Life–Death (#VP5407)   Courage–Fear (#VP5890)
Beauty–Ugliness (#VP5899).

♦ **VD2758 Guile**
Integrative polarities Skilfulness–Unskilfulness (#VP5733)
Communicativeness–Uncommunicativeness (#VP5554).

♦ **VD2760 Guilt**
Integrative polarities Innocence–Guilt (#VP5983).

♦ **VD2762 Gullibleness**
Integrative polarities Belief–Unbelief (#VP5501)   Intelligence–Unintelligence (#VP5467).

♦ **VD2764 Hallucination**
Integrative polarities Truth–Error (#VP5516)   Selfactualization–Neurosis (#VP5690)
Communicativeness–Uncommunicativeness (#VP5554).

♦ **VD2766 Haphazardness**
Integrative polarities Order–Disorder (#VP5059)   Carefulness–Neglect (#VP5533)
Attributability–Chance (#VP5155).

♦ **VD2768 Harrassment**
Integrative polarities Courage–Fear (#VP5890)   Pleasantness–Unpleasantness (#VP5863)
Appropriateness–Inappropriateness (#VP5665).

♦ **VD2770 Hardness**
Integrative polarities Virtue–Vice (#VP5980)   Strength–Weakness (#VP5159)
Kindness–Unkindness (#VP5938)   Facility–Difficulty (#VP5732)
Elasticity–Toughness (#VP5358)   Exultation–Lamentation (#VP5873)
Compassion–Pitilessness (#VP5944)   Feeling–Unfeelingness (#VP5855)
Intelligibility–Unintelligibility (#VP5548).

♦ **VD2772 Harmfulness**
Integrative polarities Goodness–Badness (#VP5674)
Kindness–Unkindness (#VP5938)
Healthfulness–Unhealthfulness (#VP5683).

♦ **VD2774 Harshness**
Integrative polarities Harmony–Discord (#VP5462)
Energy–Moderation (#VP5161)
Kindness–Unkindness (#VP5938)
Courtesy–Discourtesy (#VP5936)
Compassion–Pitilessness (#VP5944)
Pleasantness–Unpleasantness (#VP5863).

♦ **VD2776 Haste**
Integrative polarities Caution–Rashness (#VP5894)
Patience–Impatience (#VP5861)
Timeliness–Untimeliness (#VP5129).

♦ **VD2778 Hastiness**
Integrative polarities Caution–Rashness (#VP5894)
Congratulation–Envy (#VP5948)
Patience–Impatience (#VP5861)
Carefulness–Neglect (#VP5533)
Timeliness–Untimeliness (#VP5129).

♦ **VD2780 Hatred**
Integrative polarities Love–Hate (#VP5930)   Friendship–Enmity (#VP5927)
Pleasantness–Unpleasantness (#VP5863).

♦ **VD2782 Hazardousness**
Integrative polarities Safety–Danger (#VP5697)   Certainty–Uncertainty (#VP5513).

VD2784

◆ **VD2784 Heartlessness**
Integrative polarities Virtue–Vice (#VP5980)
Kindness–Unkindness (#VP5938)
Compassion–Pitilessness (#VP5944)
Courage–Fear (#VP5890)
Cheerfulness–Solemnity (#VP5870)
Feeling–Unfeelinglessness (#VP5855).

◆ **VD2786 Heedlessness**
Integrative polarities Desire–Avoidance (#VP5631)
Kindness–Unkindness (#VP5938)
Carefulness–Neglect (#VP5533)
Attention–Inattention (#VP5530)
Curiosity–Incuriosity (#VP5528)
Remembrance–Forgetfulness (#VP5537).

◆ **VD2788 Helplessness**
Integrative polarities Power–Impotence (#VP5157)
Temperance–Intemperance (#VP5992).

◆ **VD2790 Heresy**
Integrative polarities Truth–Error (#VP5516)
Agreement–Disagreement (#VP5026).
Belief–Unbelief (#VP5501)

◆ **VD2792 Hesitation**
Integrative polarities Caution–Rashness (#VP5894)
Certainty–Uncertainty (#VP5513)
Resolution–Irresolution (#VP5624)
Timeliness–Untimeliness (#VP5129).

◆ **VD2794 Hideousness**
Integrative polarities Courage–Fear (#VP5890)
Pleasantness–Unpleasantness (#VP5863).
Beauty–Ugliness (#VP5899)

◆ **VD2796 Homelessness**
Integrative polarities Unity–Duality (#VP5089)
Location–Dislocation (#VP5185)
Wealth–Poverty (#VP5837)
Sociability–Unsociability (#VP5922).

◆ **VD2798 Hopelessness**
Integrative polarities Hope–Hopelessness (#VP5888)
Cheerfulness–Solemnity (#VP5870)
Attributability–Chance (#VP5155)
Feeling–Unfeelinglessness (#VP5855)
Possibility–Impossibility (#VP5509).

◆ **VD2800 Horribleness**
Integrative polarities Courage–Fear (#VP5890)
Goodness–Badness (#VP5674)
Beauty–Ugliness (#VP5899)
Pleasantness–Unpleasantness (#VP5863).

◆ **VD2802 Hostility**
Integrative polarities Accord–Disaccord (#VP5794)
Friendship–Enmity (#VP5927)
Support–Opposition (#VP5785)
Identity–Difference (#VP5013)
Feeling–Unfeelinglessness (#VP5855)
Pleasantness–Unpleasantness (#VP5863).

◆ **VD2804 Humiliation**
Integrative polarities Pride–Humility (#VP5905)
Respect–Disrespect (#VP5964)
Repute–Disrepute (#VP5914)
Pleasantness–Unpleasantness (#VP5863).

◆ **VD2806 Hurtfulness**
Integrative polarities Goodness–Badness (#VP5674)
Sensation–Insensibility (#VP5422).

◆ **VD2808 Hypocrisy**
Integrative polarities Naturalness–Affectation (#VP5900)
Communicativeness–Uncommunicativeness (#VP5554).

◆ **VD2810 Idleness**
Integrative polarities Action–Inaction (#VP5705)   Importance–Unimportance (#VP5672).

◆ **VD2812 Ignobility**
Integrative polarities Taste–Vulgarity (#VP5896)   Repute–Disrepute (#VP5914)
Pleasantness–Unpleasantness (#VP5863).

◆ **VD2814 Ignominiousness**
Integrative polarities Repute–Disrepute (#VP5914).

◆ **VD2816 Ignominy**
Integrative polarities Rightness–Wrongness (#VP5957).

◆ **VD2818 Ignorance**
Integrative polarities Knowledge–Ignorance (#VP5475)
Skilfulness–Unskilfulness (#VP5733)
Intelligence–Unintelligence (#VP5467).

◆ **VD2820 Illegality**
Integrative polarities Consent–Refusal (#VP5775)
Legality–Illegality (#VP5998)
Chastity–Unchastity (#VP5987).

◆ **VD2822 Illegitimate**
Integrative polarities Legality–Illegality (#VP5998)
Communicativeness–Uncommunicativeness (#VP5554).

◆ **VD2824 Illiberalism**
Integrative polarities Expensiveness–Cheapness (#VP5848)
Unselfishness–Selfishness (#VP5978)
Broadmindedness–Narrowmindedness (#VP5526).

◆ **VD2826 Illness**
Integrative polarities Health–Disease (#VP5685)
Goodness–Badness (#VP5674)
Kindness–Unkindness (#VP5938)
Expedience–Inexpedience (#VP5670).

◆ **VD2828 Illogic**
Integrative polarities Truth–Error (#VP5516)   Intuition–Reason (#VP5481).

◆ **VD2830 Imbecility**
Integrative polarities Power–Impotence (#VP5157)
Intelligence–Unintelligence (#VP5467).

◆ **VD2832 Immaturity**
Integrative polarities Perfection–Imperfection (#VP5677)
Skilfulness–Unskilfulness (#VP5733)
Preparedness–Unpreparedness (#VP5720)
Intelligence–Unintelligence (#VP5467)
Completeness–Incompleteness (#VP5054).

◆ **VD2834 Immodesty**
Integrative polarities Modesty–Vanity (#VP5908)
Chastity–Unchastity (#VP5987).

◆ **VD2836 Immorality**
Integrative polarities Virtue–Vice (#VP5980)   Probity–Improbity (#VP5974).

◆ **VD2838 Impairment**
Integrative polarities Improvement–Impairment (#VP5691)
Perfection–Imperfection (#VP5677)
Expedience–Inexpedience (#VP5670).

◆ **VD2840 Impatience**
Integrative polarities Patience–Impatience (#VP5861)
Resolution–Irresolution (#VP5624).

◆ **VD2842 Imperfection**
Integrative polarities Virtue–Vice (#VP5980)   Perfection–Imperfection (#VP5677)
Overrunning–Shortcoming (#VP5313)   Superiority–Inferiority (#VP5036)
Oversufficiency–Insufficiency (#VP5660).

◆ **VD2844 Impertinence**
Integrative polarities Modesty–Vanity (#VP5908)
Support–Opposition (#VP5785)
Circumscription–Intrusion (#VP5234).

◆ **VD2846 Imperviousness**
Integrative polarities Strength–Weakness (#VP5159)
Feeling–Unfeelinglessness (#VP5855).

◆ **VD2848 Impetuousity**
Integrative polarities Action–Inaction (#VP5705)   Caution–Rashness (#VP5894)
Patience–Impatience (#VP5861)   Durability–Transience (#VP5110)
Resolution–Irresolution (#VP5624)   Stability–Changeableness (#VP5141).

◆ **VD2850 Impiety**
Integrative polarities Piety–Impiety (#VP6028).

◆ **VD2852 Implausibility**
Integrative polarities Belief–Unbelief (#VP5501).

◆ **VD2854 Impoliteness**
Integrative polarities Courtesy–Discourtesy (#VP5936).

◆ **VD2856 Impossibility**
Integrative polarities Hope–Hopelessness (#VP5888)
Attributability–Chance (#VP5155)
Possibility–Impossibility (#VP5509).

◆ **VD2858 Impotence**
Integrative polarities Power–Impotence (#VP5157)
Strength–Weakness (#VP5159)
Influence–Influencelessness (#VP5172)
Productiveness–Unproductiveness (#VP5165)
Appropriateness–Inappropriateness (#VP5665).

◆ **VD2860 Impoverishment**
Integrative polarities Wealth–Poverty (#VP5837)
Sharing–Appropriation (#VP5815)
Appropriateness–Inappropriateness (#VP5665).

◆ **VD2862 Impracticality**
Integrative polarities Facility–Difficulty (#VP5732)
Possibility–Impossibility (#VP5509).

◆ **VD2864 Imprecision**
Integrative polarities Truth–Error (#VP5516)   Certainty–Uncertainty (#VP5513)
Perfection–Imperfection (#VP5677).

◆ **VD2866 Impressionability**
Integrative polarities Influence–Influencelessness (#VP5172).

◆ **VD2868 Improbity**
Integrative polarities Probity–Improbity (#VP5974)
Communicativeness–Uncommunicativeness (#VP5554).

◆ **VD2870 Impropriety**
Integrative polarities Virtue–Vice (#VP5980)   Taste–Vulgarity (#VP5896)
Justice–Injustice (#VP5976)   Dueness–Undueness (#VP5960)
Chastity–Unchastity (#VP5987)   Rightness–Wrongness (#VP5957)
Behaviour–Misbehaviour (#VP5737)   Agreement–Disagreement (#VP5026)
Timeliness–Untimeliness (#VP5129).

◆ **VD2872 Improvidence**
Integrative polarities Caution–Rashness (#VP5894)
Preparedness–Unpreparedness (#VP5720).

DESTRUCTIVE VALUES

♦ **VD2874 Imprudence**
Integrative polarities Caution–Rashness (#VP5894)
Intelligence–Unintelligence (#VP5467)
Discrimination–Indiscrimination (#VP5492).

♦ **VD2876 Impudence**
Integrative polarities Modesty–Vanity (#VP5908)
Caution–Rashness (#VP5894)
Respect–Disrespect (#VP5964)
Support–Opposition (#VP5785).

♦ **VD2878 Impulsiveness**
Integrative polarities Action–Inaction (#VP5705)   Resolution–Irresolution (#VP5624)
Stability–Changeableness (#VP5141).

♦ **VD2880 Impurity**
Integrative polarities Virtue–Vice (#VP5980)   Chastity–Unchastity (#VP5987)
Cleanness–Uncleanness (#VP5681)   Perfection–Imperfection (#VP5677).

♦ **VD2882 Inability**
Integrative polarities Power–Impotence (#VP5157)
Skilfulness–Unskilfulness (#VP5733).

♦ **VD2884 Inaccessibility**
Integrative polarities Sociability–Unsociability (#VP5922)
Communicativeness–Uncommunicativeness (#VP5554).

♦ **VD2886 Inaccuracy**
Integrative polarities Truth–Error (#VP5516)   Carefulness–Neglect (#VP5533)
Certainty–Uncertainty (#VP5513)   Perfection–Imperfection (#VP5677)
Representation–Misrepresentation (#VP5573).

♦ **VD2888 Inactivity**
Integrative polarities Action–Inaction (#VP5705).

♦ **VD2890 Inadequacy**
Integrative polarities Power–Impotence (#VP5157)
Equality–Inequality (#VP5030)
Perfection–Imperfection (#VP5677)
Overrunning–Shortcoming (#VP5313)
Superiority–Inferiority (#VP5036)
Skilfulness–Unskilfulness (#VP5733)
Contentment–Discontentment (#VP5868)
Completeness–Incompleteness (#VP5054)
Oversufficiency–Insufficiency (#VP5660).

♦ **VD2892 Inadvertence**
Integrative polarities Truth–Error (#VP5516)   Carefulness–Neglect (#VP5533)
Attention–Inattention (#VP5530).

♦ **VD2894 Inadvisability**
Integrative polarities Expedience–Inexpedience (#VP5670)
Intelligence–Unintelligence (#VP5467).

♦ **VD2896 Inanity**
Integrative polarities Power–Impotence (#VP5157)
Amusement–Boredom (#VP5878)
Knowledge–Ignorance (#VP5475)
Importance–Unimportance (#VP5672)
Meaning–Meaninglessness (#VP5545)
Thought–Thoughtlessness (#VP5478)
Savouriness–Unsavouriness (#VP5427)
Intelligence–Unintelligence (#VP5467)
Appropriateness–Inappropriateness (#VP5665).

♦ **VD2898 Inapplicability**
Integrative polarities Agreement–Disagreement (#VP5026)
Appropriateness–Inappropriateness (#VP5665).

♦ **VD2900 Inappropriateness**
Integrative polarities Taste–Vulgarity (#VP5896)   Dueness–Undueness (#VP5960)
Chastity–Unchastity (#VP5987)   Rightness–Wrongness (#VP5957)
Agreement–Disagreement (#VP5026)   Expedience–Inexpedience (#VP5670)
Appropriateness–Inappropriateness (#VP5665).

♦ **VD2902 Inaptitude**
Integrative polarities Agreement–Disagreement (#VP5026)
Expedience–Inexpedience (#VP5670)
Skilfulness–Unskilfulness (#VP5733).

♦ **VD2904 Inarticulation**
Integrative polarities Modesty–Vanity (#VP5908)
Intelligibility–Unintelligibility (#VP5548).

♦ **VD2906 Inattention**
Integrative polarities Desire–Avoidance (#VP5631)
Carefulness–Neglect (#VP5533)
Attention–Inattention (#VP5530)
Intelligence–Unintelligence (#VP5467).

♦ **VD2908 Incapability**
Integrative polarities Power–Impotence (#VP5157)
Skilfulness–Unskilfulness (#VP5733)
Preparedness–Unpreparedness (#VP5720).

♦ **VD2910 Incapacity**
Integrative polarities Power–Impotence (#VP5157)
Strength–Weakness (#VP5159)
Skilfulness–Unskilfulness (#VP5733)
Intelligence–Unintelligence (#VP5467).

♦ **VD2912 Incivility**
Integrative polarities Taste–Vulgarity (#VP5896)   Courtesy–Discourtesy (#VP5936).

♦ **VD2914 Inclemency**
Integrative polarities Energy–Moderation (#VP5161)
Kindness–Unkindness (#VP5938)
Compassion–Pitilessness (#VP5944).

♦ **VD2916 Incoherence**
Integrative polarities Sanity–Insanity (#VP5472)   Certainty–Uncertainty (#VP5513)
Agreement–Disagreement (#VP5026)   Cohesion–Disintegration (#VP5050)
Intelligibility–Unintelligibility (#VP5548).

♦ **VD2918 Incompatibility**
Integrative polarities Accord–Discord (#VP5794)
Friendship–Enmity (#VP5927)
Agreement–Disagreement (#VP5026)
Sociability–Unsociability (#VP5922).

♦ **VD2920 Incompetence**
Integrative polarities Power–Impotence (#VP5157)
Superiority–Inferiority (#VP5036)
Skilfulness–Unskilfulness (#VP5733)
Preparedness–Unpreparedness (#VP5720)
Oversufficiency–Insufficiency (#VP5660).

♦ **VD2922 Incompleteness**
Integrative polarities Power–Impotence (#VP5157)
Perfection–Imperfection (#VP5677)
Completeness–Incompleteness (#VP5054)
Oversufficiency–Insufficiency (#VP5660).

♦ **VD2924 Incomprehensibility**
Integrative polarities Knowledge–Ignorance (#VP5475)
Intelligence–Unintelligence (#VP5467)
Intelligibility–Unintelligibility (#VP5548).

♦ **VD2926 Incongruity**
Integrative polarities Intuition–Reason (#VP5481)
Agreement–Disagreement (#VP5026)
Expedience–Inexpedience (#VP5670).

♦ **VD2928 Inconsequence**
Integrative polarities Intuition–Reason (#VP5481)
Greatness–Smallness (#VP5034)
Importance–Unimportance (#VP5672).

♦ **VD2930 Inconsiderateness**
Integrative polarities Kindness–Unkindness (#VP5938)
Carefulness–Neglect (#VP5533)
Courtesy–Discourtesy (#VP5936)
Resolution–Irresolution (#VP5624)
Intelligence–Unintelligence (#VP5467).

♦ **VD2932 Inconsistency**
Integrative polarities Intuition–Reason (#VP5481)
Identity–Difference (#VP5013)
Agreement–Disagreement (#VP5026)
Cohesion–Disintegration (#VP5050)
Stability–Changeableness (#VP5141).

♦ **VD2934 Inconstancy**
Integrative polarities Probity–Improbity (#VP5974)
Resolution–Irresolution (#VP5624)
Stability–Changeableness (#VP5141).

♦ **VD2936 Incontinence**
Integrative polarities Desire–Avoidance (#VP5631)
Freedom–Restraint (#VP5762)
Chastity–Unchastity (#VP5987)
Temperance–Intemperance (#VP5992)
Oversufficiency–Insufficiency (#VP5660).

♦ **VD2938 Inconvenience**
Integrative polarities Dueness–Undueness (#VP5960)
Facility–Difficulty (#VP5732)
Expedience–Inexpedience (#VP5670)
Timeliness–Untimeliness (#VP5129).

♦ **VD2940 Incorrectness**
Integrative polarities Truth–Error (#VP5516)   Rightness–Wrongness (#VP5957).

♦ **VD2942 Incorrigibility**
Integrative polarities Virtue–Vice (#VP5980)   Hope–Hopelessness (#VP5888)
Resolution–Irresolution (#VP5624).

♦ **VD2944 Incredulity**
Integrative polarities Belief–Unbelief (#VP5501).

♦ **VD2946 Incuriousity**
Integrative polarities Desire–Avoidance (#VP5631)
Attention–Inattention (#VP5530)
Curiosity–Incuriosity (#VP5528).

♦ **VD2948 Indecency**
Integrative polarities Taste–Vulgarity (#VP5896)   Chastity–Unchastity (#VP5987).

♦ **VD2950 Indecisiveness**
Integrative polarities Form–Formlessness (#VP5246)
Strength–Weakness (#VP5159)
Certainty–Uncertainty (#VP5513)
Resolution–Irresolution (#VP5624).

♦ **VD2952 Indecorum**
Integrative polarities Taste–Vulgarity (#VP5896)   Chastity–Unchastity (#VP5987)
Rightness–Wrongness (#VP5957).

VD2954

♦ **VD2954 Indefensibility**
Integrative polarities Justice–Injustice (#VP5976)
Contentment–Discontentment (#VP5868).

♦ **VD2956 Indelicacy**
Integrative polarities Taste–Vulgarity (#VP5896) Chastity–Unchastity (#VP5987).

♦ **VD2958 Indifference**
Integrative polarities Action–Inaction (#VP5705) Desire–Avoidance (#VP5631)
Motion–Quiescence (#VP5267) Carefulness–Neglect (#VP5533)
Attention–Inattention (#VP5530) Curiosity–Incuriosity (#VP5528)
Perfection–Imperfection (#VP5677) Importance–Unimportance (#VP5672)
Feeling–Unfeelinglessness (#VP5855).

♦ **VD2960 Indignation**
Integrative polarities Congratulation–Envy (#VP5948)
Approval–Disapproval (#VP5968).

♦ **VD2962 Indirection**
Integrative polarities Probity–Improbity (#VP5974)
Direction–Deviation (#VP5290)
Conciseness–Diffuseness (#VP5592)
Communicativeness–Uncommunicativeness (#VP5554).

♦ **VD2964 Indiscretion**
Integrative polarities Virtue–Vice (#VP5980) Truth–Error (#VP5516)
Caution–Rashness (#VP5894) Chastity–Unchastity (#VP5987)
Intelligence–Unintelligence (#VP5467) Discrimination–Indiscrimination (#VP5492).

♦ **VD2966 Indiscrimination**
Integrative polarities Choice–Necessity (#VP5637)
Desire–Avoidance (#VP5631)
Discrimination–Indiscrimination (#VP5492).

♦ **VD2968 Indisposition**
Integrative polarities Health–Disease (#VP5685)
Willingness–Unwillingness (#VP5621).

♦ **VD2970 Indocility**
Integrative polarities Obedience–Disobedience (#VP5766)
Resolution–Irresolution (#VP5624)
Willingness–Unwillingness (#VP5621).

♦ **VD2972 Indolence**
Integrative polarities Action–Inaction (#VP5705) Motion–Quiescence (#VP5267).

♦ **VD2974 Indulgence**
Integrative polarities Consent–Refusal (#VP5775)
Temperance–Intemperance (#VP5992).

♦ **VD2976 Ineffectiveness**
Integrative polarities Power–Impotence (#VP5157)
Skilfulness–Unskilfulness (#VP5733)
Influence–Influencelessness (#VP5172)
Appropriateness–Inappropriateness (#VP5665).

♦ **VD2978 Ineffectuality**
Integrative polarities Power–Impotence (#VP5157)
Importance–Unimportance (#VP5672)
Skilfulness–Unskilfulness (#VP5733)
Influence–Influencelessness (#VP5172)
Appropriateness–Inappropriateness (#VP5665).

♦ **VD2980 Inefficiency**
Integrative polarities Power–Impotence (#VP5157)
Skilfulness–Unskilfulness (#VP5733).

♦ **VD2982 Inelegance**
Integrative polarities Beauty–Ugliness (#VP5899)
Taste–Vulgarity (#VP5896)
Chastity–Unchastity (#VP5987)
Elegance–Inelegance (#VP5589)
Skilfulness–Unskilfulness (#VP5733).

♦ **VD2984 Ineptitude**
Integrative polarities Power–Impotence (#VP5157)
Expedience–Inexpedience (#VP5670)
Intelligence–Unintelligence (#VP5467).

♦ **VD2986 Inequality**
Integrative polarities Justice–Injustice (#VP5976)
Equality–Inequality (#VP5030)
Agreement–Disagreement (#VP5026).

♦ **VD2988 Inequity**
Integrative polarities Justice–Injustice (#VP5976)
Equality–Inequality (#VP5030).

♦ **VD2990 Inertia**
Integrative polarities Action–Inaction (#VP5705) Motion–Quiescence (#VP5267).

♦ **VD2992 Inexpedience**
Integrative polarities Goodness–Badness (#VP5674)
Expedience–Inexpedience (#VP5670)
Timeliness–Untimeliness (#VP5129)
Skilfulness–Unskilfulness (#VP5733)
Intelligence–Unintelligence (#VP5467).

♦ **VD2994 Inexperience**
Integrative polarities Newness–Oldness (#VP5122)
Knowledge–Ignorance (#VP5475)
Skilfulness–Unskilfulness (#VP5733).

♦ **VD2996 Infamy**
Integrative polarities Virtue–Vice (#VP5980) Repute–Disrepute (#VP5914)
Rightness–Wrongness (#VP5957).

♦ **VD2998 Infatuation**
Integrative polarities Belief–Unbelief (#VP5501) Sanity–Insanity (#VP5472)
Desire–Avoidance (#VP5631) Intelligence–Unintelligence (#VP5467).

♦ **VD3000 Infelicity**
Integrative polarities Cheerfulness–Solemnity (#VP5870)
Agreement–Disagreement (#VP5026)
Expedience–Inexpedience (#VP5670)
Timeliness–Untimeliness (#VP5129)
Pleasantness–Unpleasantness (#VP5863).

♦ **VD3002 Inferiority**
Integrative polarities Power–Impotence (#VP5157)
Goodness–Badness (#VP5674)
Freedom–Restraint (#VP5762)
Perfection–Imperfection (#VP5677)
Importance–Unimportance (#VP5672)
Overrunning–Shortcoming (#VP5313)
Superiority–Inferiority (#VP5036)
Conformity–Nonconformity (#VP5082).

♦ **VD3004 Infertility**
Integrative polarities Productiveness–Unproductiveness (#VP5165)
Imaginativeness–Unimaginativeness (#VP5535).

♦ **VD3006 Infidelity**
Integrative polarities Belief–Unbelief (#VP5501) Probity–Improbity (#VP5974).

♦ **VD3008 Inflexibility**
Integrative polarities Hardness–Softness (#VP5356)
Leniency–Compulsion (#VP5756)
Resolution–Irresolution (#VP5624).

♦ **VD3010 Ingloriousness**
Integrative polarities Pride–Humility (#VP5905) Repute–Disrepute (#VP5914).

♦ **VD3012 Ingratitude**
Integrative polarities Gratitude–Ingratitude (#VP5949).

♦ **VD3014 Inharmonious**
Integrative polarities Accord–Disaccord (#VP5794)
Agreement–Disagreement (#VP5026).

♦ **VD3016 Inhibition**
Integrative polarities Consent–Refusal (#VP5775)
Freedom–Restraint (#VP5762)
Facility–Difficulty (#VP5732).

♦ **VD3018 Inhospitality**
Integrative polarities Friendship–Enmity (#VP5927)
Kindness–Unkindness (#VP5938)
Hospitality–Inhospitality (#VP5925).

♦ **VD3020 Inhumanity**
Integrative polarities Energy–Moderation (#VP5161)
Kindness–Unkindness (#VP5938).

♦ **VD3022 Iniquity**
Integrative polarities Virtue–Vice (#VP5980) Justice–Injustice (#VP5976).

♦ **VD3024 Injury**
Integrative polarities Virtue–Vice (#VP5980) Possession–Loss (#VP5808)
Goodness–Badness (#VP5674) Justice–Injustice (#VP5976)
Respect–Disrespect (#VP5964) Improvement–Impairment (#VP5691)
Expedience–Inexpedience (#VP5670) Pleasantness–Unpleasantness (#VP5863)
Appropriateness–Inappropriateness (#VP5665).

♦ **VD3026 Injustice**
Integrative polarities Virtue–Vice (#VP5980) Justice–Injustice (#VP5976)
Equality–Inequality (#VP5030) Representation–Misrepresentation (#VP5573).

♦ **VD3028 Inobservance**
Integrative polarities Attention–Inattention (#VP5530)
Obedience–Disobedience (#VP5766).

♦ **VD3030 Inquietude**
Integrative polarities Courage–Fear (#VP5890) Oscillation–Agitation (#VP5323)
Pleasantness–Unpleasantness (#VP5863).

♦ **VD3032 Inquisition**
Integrative polarities Research–Discovery (#VP5485).

♦ **VD3034 Insanity**
Integrative polarities Sanity–Insanity (#VP5472) Selfactualization–Neurosis (#VP5690)
Intelligence–Unintelligence (#VP5467).

♦ **VD3036 Insecurity**
Integrative polarities Safety–Danger (#VP5697) Certainty–Uncertainty (#VP5513).

♦ **VD3038 Insensibility**
Integrative polarities Knowledge–Ignorance (#VP5475)
Sensation–Insensibility (#VP5422)
Selfactualization–Neurosis (#VP5690)
Intelligence–Unintelligence (#VP5467)
Discrimination–Indiscrimination (#VP5492).

♦ **VD3040 Insensitivity**
Integrative polarities Kindness–Unkindness (#VP5938)
Courtesy–Discourtesy (#VP5936)

DESTRUCTIVE VALUES

Sensation–Insensibility (#VP5422)
Feeling–Unfeelinglessness (#VP5855)
Discrimination–Indiscrimination (#VP5492).

♦ **VD3042 Insignificance**
Integrative polarities Importance–Unimportance (#VP5672)
Meaning–Meaninglessness (#VP5545).

♦ **VD3044 Insincerity**
Integrative polarities Intuition–Reason (#VP5481)
Probity–Improbity (#VP5974)
Naturalness–Affectation (#VP5900)
Communicativeness–Uncommunicativeness (#VP5554).

♦ **VD3046 Insinuation**
Integrative polarities Pride–Humility (#VP5905)   Approval–Disapproval (#VP5968)
Vindication–Condemnation (#VP6005)   Circumscription–Intrusion (#VP5234).

♦ **VD3048 Insobriety**
Integrative polarities Temperance–Intemperance (#VP5992).

♦ **VD3050 Insolence**
Integrative polarities Modesty–Vanity (#VP5908)
Caution–Rashness (#VP5894)
Respect–Disrespect (#VP5964)
Support–Opposition (#VP5785)
Courtesy–Discourtesy (#VP5936)
Exultation–Lamentation (#VP5873).

♦ **VD3052 Insouciance**
Integrative polarities Desire–Avoidance (#VP5631)
Carefulness–Neglect (#VP5533)
Curiosity–Incuriosity (#VP5528)
Feeling–Unfeelinglessness (#VP5855).

♦ **VD3054 Instability**
Integrative polarities Safety–Danger (#VP5697)   Strength–Weakness (#VP5159)
Certainty–Uncertainty (#VP5513)   Durability–Transience (#VP5110)
Resolution–Irresolution (#VP5624)   Stability–Changeableness (#VP5141).

♦ **VD3056 Insubordination**
Integrative polarities Authority–Lawlessness (#VP5739)
Obedience–Disobedience (#VP5766).

♦ **VD3058 Insubstantiality**
Integrative polarities Certainty–Uncertainty (#VP5513).

♦ **VD3060 Insufficiency**
Integrative polarities Power–Impotence (#VP5157)
Greatness–Smallness (#VP5034)
Equality–Inequality (#VP5030)
Overrunning–Shortcoming (#VP5313)
Superiority–Inferiority (#VP5036)
Contentment–Discontentment (#VP5868)
Oversufficiency–Insufficiency (#VP5660).

♦ **VD3062 Insularity**
Integrative polarities Inclusion–Exclusion (#VP5076)
Broadmindedness–Narrowmindedness (#VP5526).

♦ **VD3064 Insult**
Integrative polarities Respect–Disrespect (#VP5964).

♦ **VD3066 Intemperence**
Integrative polarities Freedom–Restraint (#VP5762)
Chastity–Unchastity (#VP5987)
Temperance–Intemperance (#VP5992)
Oversufficiency–Insufficiency (#VP5660).

♦ **VD3068 Interference**
Integrative polarities Facility–Difficulty (#VP5732)
Circumscription–Intrusion (#VP5234).

♦ **VD3070 Intervention**
Integrative polarities Circumscription–Intrusion (#VP5234).

♦ **VD3072 Intimidation**
Integrative polarities Courage–Fear (#VP5890)   Modesty–Vanity (#VP5908)
Leniency–Compulsion (#VP5756)   Approval–Disapproval (#VP5968)
Motivation–Dissuasion (#VP5648).

♦ **VD3074 Intolerance**
Integrative polarities Patience–Impatience (#VP5861)
Resolution–Irresolution (#VP5624)
Broadmindedness–Narrowmindedness (#VP5526).

♦ **VD3076 Intoxication**
Integrative polarities Health–Disease (#VP5685)
Temperance–Intemperance (#VP5992).

♦ **VD3078 Intractability**
Integrative polarities Hardness–Softness (#VP5356)
Obedience–Disobedience (#VP5766)
Resolution–Irresolution (#VP5624).

♦ **VD3080 Intransigence**
Integrative polarities Support–Opposition (#VP5785)
Resolution–Irresolution (#VP5624).

♦ **VD3082 Intricacy**
Integrative polarities Facility–Difficulty (#VP5732)
Intelligibility–Unintelligibility (#VP5548).

♦ **VD3084 Intrigue**
Integrative polarities Probity–Improbity (#VP5974).

♦ **VD3086 Intrusiveness**
Integrative polarities Action–Inaction (#VP5705)   Inclusion–Exclusion (#VP5076)
Timeliness–Untimeliness (#VP5129)   Circumscription–Intrusion (#VP5234).

♦ **VD3088 Invalidity**
Integrative polarities Health–Disease (#VP5685)
Power–Impotence (#VP5157)
Intuition–Reason (#VP5481).

♦ **VD3090 Invasion**
Integrative polarities Attack–Defence (#VP5798)
Circumscription–Intrusion (#VP5234).

♦ **VD3092 Involution**
Integrative polarities Simplicity–Complexity (#VP5045)
Improvement–Impairment (#VP5691).

♦ **VD3094 Irascibility**
Integrative polarities Accord–Disaccord (#VP5794)
Congratulation–Envy (#VP5948)
Resolution–Irresolution (#VP5624).

♦ **VD3096 Irksomeness**
Integrative polarities Amusement–Boredom (#VP5878)
Facility–Difficulty (#VP5732)
Pleasantness–Unpleasantness (#VP5863).

♦ **VD3098 Irrationality**
Integrative polarities Sanity–Insanity (#VP5472)   Intuition–Reason (#VP5481)
Intelligence–Unintelligence (#VP5467).

♦ **VD3100 Irreconcilability**
Integrative polarities Friendship–Enmity (#VP5927)
Support–Opposition (#VP5785)
Forgiveness–Vengeance (#VP5947)
Agreement–Disagreement (#VP5026)
Resolution–Irresolution (#VP5624).

♦ **VD3102 Irredeemability**
Integrative polarities Virtue–Vice (#VP5980)   Hope–Hopelessness (#VP5888).

♦ **VD3104 Irreformability**
Integrative polarities Virtue–Vice (#VP5980)   Hope–Hopelessness (#VP5888).

♦ **VD3106 Irrelevance**
Integrative polarities Agreement–Disagreement (#VP5026)
Importance–Unimportance (#VP5672)
Timeliness–Untimeliness (#VP5129).

♦ **VD3108 Irreligiousness**
Integrative polarities Piety–Impiety (#VP6028).

♦ **VD3110 Irremediability**
Integrative polarities Hope–Hopelessness (#VP5888)
Restoration–Destruction (#VP5693).

♦ **VD3112 Irresolution**
Integrative polarities Strength–Weakness (#VP5159)
Certainty–Uncertainty (#VP5513)
Resolution–Irresolution (#VP5624).

♦ **VD3114 Irresponsibility**
Integrative polarities Probity–Improbity (#VP5974)
Authority–Lawlessness (#VP5739)
Stability–Changeableness (#VP5141).

♦ **VD3116 Irreverence**
Integrative polarities Piety–Impiety (#VP6028)   Respect–Disrespect (#VP5964).

♦ **VD3118 Irritability**
Integrative polarities Probity–Improbity (#VP5974)
Congratulation–Envy (#VP5948)
Sensation–Insensibility (#VP5422).

♦ **VD3120 Isolation**
Integrative polarities Unity–Duality (#VP5089)   Freedom–Restraint (#VP5762)
Inclusion–Exclusion (#VP5076).

♦ **VD3122 Jealousy**
Integrative polarities Congratulation–Envy (#VP5948).

♦ **VD3124 Joylessness**
Integrative polarities Cheerfulness–Solemnity (#VP5870)
Pleasantness–Unpleasantness (#VP5863).

♦ **VD3126 Lack**
Integrative polarities Wealth–Poverty (#VP5837)
Presence–Absence (#VP5186)
Perfection–Imperfection (#VP5677)
Completeness–Incompleteness (#VP5054)
Oversufficiency–Insufficiency (#VP5660).

♦ **VD3128 Languor**
Integrative polarities Action–Inaction (#VP5705)   Motion–Quiescence (#VP5267)
Strength–Weakness (#VP5159).

♦ **VD3130 Lapse**
Integrative polarities Virtue–Vice (#VP5980)   Truth–Error (#VP5516)
Courage–Fear (#VP5890)   Piety–Impiety (#VP6028)
Refreshment–Relapse (#VP5695)   Carefulness–Neglect (#VP5533)
Improvement–Impairment (#VP5691)   Progression–Regression (#VP5294).

♦ **VD3132 Lateness**
Integrative polarities Timeliness–Untimeliness (#VP5129).

VD3134

◆ **VD3134 Lavishness**
Integrative polarities Oversufficiency–Insufficiency (#VP5660).

◆ **VD3136 Lawlessness**
Integrative polarities Dueness–Undueness (#VP5960)
Legality–Illegality (#VP5998)
Authority–Lawlessness (#VP5739)
Obedience–Disobedience (#VP5766).

◆ **VD3138 Laxity**
Integrative polarities Truth–Error (#VP5516)　　Chastity–Unchastity (#VP5987)
Carefulness–Neglect (#VP5533)　　Certainty–Uncertainty (#VP5513).

◆ **VD3140 Laxness**
Integrative polarities Freedom–Restraint (#VP5762)
Leniency–Compulsion (#VP5756)
Carefulness–Neglect (#VP5533)
Timeliness–Untimeliness (#VP5129).

◆ **VD3142 Laziness**
Integrative polarities Action–Inaction (#VP5705)　　Carefulness–Neglect (#VP5533)
Timeliness–Untimeliness (#VP5129).

◆ **VD3144 Lethargy**
Integrative polarities Action–Inaction (#VP5705)　　Feeling–Unfeelinglessness (#VP5855)
Intelligence–Unintelligence (#VP5467).

◆ **VD3146 Levity**
Integrative polarities Respect–Disrespect (#VP5964)
Attention–Inattention (#VP5530)
Importance–Unimportance (#VP5672)
Resolution–Irresolution (#VP5624).

◆ **VD3148 Liability**
Integrative polarities Safety–Danger (#VP5697)　　Wealth–Poverty (#VP5837)
Expedience–Inexpedience (#VP5670).

◆ **VD3150 Licentiousness**
Integrative polarities Dueness–Undueness (#VP5960)
Freedom–Restraint (#VP5762)
Chastity–Unchastity (#VP5987)
Authority–Lawlessness (#VP5739)
Temperance–Intemperance (#VP5992).

◆ **VD3152 Lifelessness**
Integrative polarities Life–Death (#VP5407)　　Light–Darkness (#VP5335)
Action–Inaction (#VP5705)　　Amusement–Boredom (#VP5878)
Motion–Quiescence (#VP5267).

◆ **VD3154 Limitedness**
Integrative polarities Centrality–Environment (#VP5226)
Oversufficiency–Insufficiency (#VP5660).

◆ **VD3156 Listlessness**
Integrative polarities Action–Inaction (#VP5705)　　Desire–Avoidance (#VP5631)
Strength–Weakness (#VP5159)　　Curiosity–Incuriosity (#VP5528)
Feeling–Unfeelinglessness (#VP5855).

◆ **VD3158 Littleness**
Integrative polarities Repute–Disrepute (#VP5914)
Bigness–Littleness (#VP5195)
Importance–Unimportance (#VP5672)
Superiority–Inferiority (#VP5036)
Unselfishness–Selfishness (#VP5978)
Broadmindedness–Narrowmindedness (#VP5526).

◆ **VD3160 Loathsomeness**
Integrative polarities Love–Hate (#VP5930)　　Beauty–Ugliness (#VP5899)
Goodness–Badness (#VP5674)　　Pleasantness–Unpleasantness (#VP5863).

◆ **VD3162 Loneliness**
Integrative polarities Unity–Duality (#VP5089)　　Sociability–Unsociability (#VP5922).

◆ **VD3164 Looseness**
Integrative polarities Truth–Error (#VP5516)　　Order–Disorder (#VP5059)
Chastity–Unchastity (#VP5987)　　Leniency–Compulsion (#VP5756)
Carefulness–Neglect (#VP5533)　　Certainty–Uncertainty (#VP5513)
Cohesion–Disintegration (#VP5050).

◆ **VD3166 Lordliness**
Integrative polarities Modesty–Vanity (#VP5908)
Pride–Humility (#VP5905).

◆ **VD3168 Loudness**
Integrative polarities Taste–Vulgarity (#VP5896)　　Silence–Loudness (#VP5451)
Colour–Colourlessness (#VP5362).

◆ **VD3170 Lovelessness**
Integrative polarities Desire–Avoidance (#VP5631)
Pleasantness–Unpleasantness (#VP5863).

◆ **VD3172 Lowness**
Integrative polarities Height–Lowness (#VP5207)
Taste–Vulgarity (#VP5896)
Repute–Disrepute (#VP5914)
Cheerfulness–Solemnity (#VP5870)
Superiority–Inferiority (#VP5036).

◆ **VD3174 Lucklessness**
Integrative polarities Prosperity–Adversity (#VP5728).

◆ **VD3176 Lust**
Integrative polarities Virtue–Vice (#VP5980)　　Desire–Avoidance (#VP5631)
Chastity–Unchastity (#VP5987).

◆ **VD3178 Madness**
Integrative polarities Health–Disease (#VP5685)
Sanity–Insanity (#VP5472)
Intelligence–Unintelligence (#VP5467).

◆ **VD3180 Maladjustment**
Integrative polarities Agreement–Disagreement (#VP5026)
Skilfulness–Unskilfulness (#VP5733)
Selfactualization–Neurosis (#VP5690).

◆ **VD3182 Maladroit**
Integrative polarities Superiority–Inferiority (#VP5036)
Skilfulness–Unskilfulness (#VP5733).

◆ **VD3184 Malaise**
Integrative polarities Courage–Fear (#VP5890)　　Health–Disease (#VP5685)
Oscillation–Agitation (#VP5323)　　Cheerfulness–Solemnity (#VP5870)
Sensation–Insensibility (#VP5422)　　Pleasantness–Unpleasantness (#VP5863).

◆ **VD3186 Malevolence**
Integrative polarities Love–Hate (#VP5930)　　Goodness–Badness (#VP5674)
Friendship–Enmity (#VP5927)　　Kindness–Unkindness (#VP5938).

◆ **VD3188 Malfeasance**
Integrative polarities Virtue–Vice (#VP5980)　　Conjugality–Celibacy (#VP5931)
Appropriateness–Inappropriateness (#VP5665).

◆ **VD3190 Malformation**
Integrative polarities Beauty–Ugliness (#VP5899)
Symmetry–Distortion (#VP5248)
Conformity–Nonconformity (#VP5082).

◆ **VD3192 Malice**
Integrative polarities Love–Hate (#VP5930)　　Friendship–Enmity (#VP5927)
Kindness–Unkindness (#VP5938).

◆ **VD3194 Malignancy**
Integrative polarities Life–Death (#VP5407)　　Goodness–Badness (#VP5674)
Kindness–Unkindness (#VP5938).

◆ **VD3196 Malingering**
Integrative polarities Desire–Avoidance (#VP5631)
Carefulness–Neglect (#VP5533)
Communicativeness–Uncommunicativeness (#VP5554).

◆ **VD3198 Malpractice**
Integrative polarities Virtue–Vice (#VP5980)　　Skilfulness–Unskilfulness (#VP5733)
Appropriateness–Inappropriateness (#VP5665).

◆ **VD3200 Maltreatment**
Integrative polarities Appropriateness–Inappropriateness (#VP5665).

◆ **VD3204 Meanness**
Integrative polarities Pride–Humility (#VP5905)　　Taste–Vulgarity (#VP5896)
Repute–Disrepute (#VP5914)　　Congratulation–Envy (#VP5948)
Kindness–Unkindness (#VP5938)　　Expensiveness–Cheapness (#VP5848)
Perfection–Imperfection (#VP5677)　　Importance–Unimportance (#VP5672)
Superiority–Inferiority (#VP5036)　　Unselfishness–Selfishness (#VP5978)
Oversufficiency–Insufficiency (#VP5660)
Broadmindedness–Narrowmindedness (#VP5526).

◆ **VD3206 Mediocrity**
Integrative polarities Perfection–Imperfection (#VP5677)
Importance–Unimportance (#VP5672)
Superiority–Inferiority (#VP5036)
Skilfulness–Unskilfulness (#VP5733).

◆ **VD3208 Melancholy**
Integrative polarities Amusement–Boredom (#VP5878)
Congratulation–Envy (#VP5948)
Cheerfulness–Solemnity (#VP5870)
Pleasantness–Unpleasantness (#VP5863).

◆ **VD3210 Menace**
Integrative polarities Safety–Danger (#VP5697)　　Goodness–Badness (#VP5674)
Approval–Disapproval (#VP5968).

◆ **VD3212 Mendacity**
Integrative polarities Communicativeness–Uncommunicativeness (#VP5554).

◆ **VD3214 Mendicancy**
Integrative polarities Wealth–Poverty (#VP5837).

◆ **VD3216 Mercilessness**
Integrative polarities Energy–Moderation (#VP5161)
Congratulation–Envy (#VP5948).

◆ **VD3218 Meritlessness**
Integrative polarities Importance–Unimportance (#VP5672).

◆ **VD3220 Messiness**
Integrative polarities Order–Disorder (#VP5059)　　Form–Formlessness (#VP5246)
Carefulness–Neglect (#VP5533)　　Cleanness–Uncleanness (#VP5681).

◆ **VD3222 Miasma**
Integrative polarities Goodness–Badness (#VP5674)
Fragrance–Stench (#VP5435)
Pleasantness–Unpleasantness (#VP5863).

◆ **VD3224 Mischievousness**
Integrative polarities Goodness–Badness (#VP5674)
Godliness–Ungodliness (#VP6013)
Behaviour–Misbehaviour (#VP5737).

DESTRUCTIVE VALUES VD3316

♦ **VD3226 Militancy**
Integrative polarities Accord–Disaccord (#VP5794).

♦ **VD3228 Mindlessness**
Integrative polarities Desire–Avoidance (#VP5631)
Energy–Moderation (#VP5161)
Knowledge–Ignorance (#VP5475)
Intelligence–Unintelligence (#VP5467).

♦ **VD3230 Misalliance**
Integrative polarities Agreement–Disagreement (#VP5026)
Relatedness–Unrelatedness (#VP5009).

♦ **VD3232 Misapplication**
Integrative polarities Truth–Error (#VP5516)   Intuition–Reason (#VP5481)
Relatedness–Unrelatedness (#VP5009)   Appropriateness–Inappropriateness (#VP5665)
Interpretability–Misinterpretability (#VP5552).

♦ **VD3234 Misapprehension**
Integrative polarities Truth–Error (#VP5516)   Interpretability–Misinterpretability (#VP5552).

♦ **VD3236 Misbelief**
Integrative polarities Truth–Error (#VP5516)   Belief–Unbelief (#VP5501)
Orthodoxy–Unorthodoxy (#VP6024).

♦ **VD3238 Miscalculation**
Integrative polarities Truth–Error (#VP5516)   Judgement–Misjudgement (#VP5494).

♦ **VD3240 Miscarriage**
Integrative polarities Truth–Error (#VP5516)   Justice–Injustice (#VP5976)
Skilfulness–Unskilfulness (#VP5733).

♦ **VD3242 Mischief**
Integrative polarities Accord–Disaccord (#VP5794)
Goodness–Badness (#VP5674)
Behaviour–Misbehaviour (#VP5737)
Improvement–Impairment (#VP5691)
Expedience–Inexpedience (#VP5670).

♦ **VD3244 Misconception**
Integrative polarities Truth–Error (#VP5516)   Interpretability–Misinterpretability (#VP5552).

♦ **VD3246 Misconduct**
Integrative polarities Virtue–Vice (#VP5980)   Truth–Error (#VP5516)
Behaviour–Misbehaviour (#VP5737)   Skilfulness–Unskilfulness (#VP5733)
Appropriateness–Inappropriateness (#VP5665).

♦ **VD3248 Misconstruction**
Integrative polarities Truth–Error (#VP5516)   Symmetry–Distortion (#VP5248)
Judgement–Misjudgement (#VP5494)   Interpretability–Misinterpretability (#VP5552)
Communicativeness–Uncommunicativeness (#VP5554).

♦ **VD3250 Misdemeanor**
Integrative polarities Virtue–Vice (#VP5980)   Legality–Illegality (#VP5998)
Behaviour–Misbehaviour (#VP5737).

♦ **VD3252 Misdirection**
Integrative polarities Symmetry–Distortion (#VP5248)
Education–Miseducation (#VP5562)
Skilfulness–Unskilfulness (#VP5733)
Communicativeness–Uncommunicativeness (#VP5554).

♦ **VD3254 Misery**
Integrative polarities Cheerfulness–Solemnity (#VP5870)
Sensation–Insensibility (#VP5422)
Pleasantness–Unpleasantness (#VP5863).

♦ **VD3256 Misfeasance**
Integrative polarities Virtue–Vice (#VP5980)   Truth–Error (#VP5516)
Skilfulness–Unskilfulness (#VP5733)   Appropriateness–Inappropriateness (#VP5665).

♦ **VD3258 Misfortune**
Integrative polarities Prosperity–Adversity (#VP5728).

♦ **VD3260 Misguidance**
Integrative polarities Education–Miseducation (#VP5562)
Skilfulness–Unskilfulness (#VP5733)
Communicativeness–Uncommunicativeness (#VP5554).

♦ **VD3262 Misinformation**
Integrative polarities Knowledge–Ignorance (#VP5475)
Education–Miseducation (#VP5562)
Communicativeness–Uncommunicativeness (#VP5554).

♦ **VD3264 Misinterpretation**
Integrative polarities Truth–Error (#VP5516)   Symmetry–Distortion (#VP5248)
Judgement–Misjudgement (#VP5494)   Interpretability–Misinterpretability (#VP5552).

♦ **VD3266 Misjudgement**
Integrative polarities Truth–Error (#VP5516)   Judgement–Misjudgement (#VP5494)
Interpretability–Misinterpretability (#VP5552).

♦ **VD3268 Mismanagement**
Integrative polarities Skilfulness–Unskilfulness (#VP5733)
Appropriateness–Inappropriateness (#VP5665).

♦ **VD3270 Misrepresentation**
Integrative polarities Symmetry–Distortion (#VP5248)
Representation–Misrepresentation (#VP5573)
Communicativeness–Uncommunicativeness (#VP5554).

♦ **VD3272 Misrule**
Integrative polarities Order–Disorder (#VP5059)   Authority–Lawlessness (#VP5739)
Skilfulness–Unskilfulness (#VP5733).

♦ **VD3274 Misstatement**
Integrative polarities Truth–Error (#VP5516)   Representation–Misrepresentation (#VP5573)
Communicativeness–Uncommunicativeness (#VP5554).

♦ **VD3276 Mistake**
Integrative polarities Truth–Error (#VP5516)   Skilfulness–Unskilfulness (#VP5733)
Accomplishment–Nonaccomplishment (#VP5722)   Interpretability–Misinterpretability (#VP5552).

♦ **VD3278 Mistrust**
Integrative polarities Belief–Unbelief (#VP5501)   Caution–Rashness (#VP5894)
Congratulation–Envy (#VP5948).

♦ **VD3280 Misunderstanding**
Integrative polarities Truth–Error (#VP5516)   Accord–Disaccord (#VP5794)
Interpretability–Misinterpretability (#VP5552).

♦ **VD3282 Misusage**
Integrative polarities Appropriateness–Inappropriateness (#VP5665).

♦ **VD3284 Misuse**
Integrative polarities Truth–Error (#VP5516)   Symmetry–Distortion (#VP5248)
Improvement–Impairment (#VP5691)   Appropriateness–Inappropriateness (#VP5665).

♦ **VD3286 Morbidity**
Integrative polarities Courage–Fear (#VP5890)   Health–Disease (#VP5685).

♦ **VD3288 Mortification**
Integrative polarities Pride–Humility (#VP5905)   Health–Disease (#VP5685)
Chastity–Unchastity (#VP5987)   Improvement–Impairment (#VP5691)
Pleasantness–Unpleasantness (#VP5863).

♦ **VD3290 Mournfulness**
Integrative polarities Exultation–Lamentation (#VP5873)
Cheerfulness–Solemnity (#VP5870)
Pleasantness–Unpleasantness (#VP5863).

♦ **VD3292 Muddle**
Integrative polarities Order–Disorder (#VP5059)   Form–Formlessness (#VP5246)
Attention–Inattention (#VP5530)   Certainty–Uncertainty (#VP5513)
Simplicity–Complexity (#VP5045)   Skilfulness–Unskilfulness (#VP5733)
Discrimination–Indiscrimination (#VP5492)
Accomplishment–Nonaccomplishment (#VP5722).

♦ **VD3294 Mustiness**
Integrative polarities Fragrance–Stench (#VP5435)
Amusement–Boredom (#VP5878).

♦ **VD3296 Mutability**
Integrative polarities Durability–Transience (#VP5110)
Stability–Changeableness (#VP5141)
Uniformity–Nonuniformity (#VP5017).

♦ **VD3298 Muteness**
Integrative polarities Silence–Loudness (#VP5451)
Eloquence–Uneloquence (#VP5600)
Communicativeness–Uncommunicativeness (#VP5554).

♦ **VD3300 Mutilation**
Integrative polarities Symmetry–Distortion (#VP5248)
Improvement–Impairment (#VP5691).

♦ **VD3302 Mystification**
Integrative polarities Intuition–Reason (#VP5481)
Education–Miseducation (#VP5562)
Intelligibility–Unintelligibility (#VP5548)
Communicativeness–Uncommunicativeness (#VP5554).

♦ **VD3304 Naïvety**
Integrative polarities Belief–Unbelief (#VP5501)   Newness–Oldness (#VP5122)
Knowledge–Ignorance (#VP5475)   Skilfulness–Unskilfulness (#VP5733)
Intelligence–Unintelligence (#VP5467)
Accomplishment–Nonaccomplishment (#VP5722).

♦ **VD3306 Negation**
Integrative polarities Assent–Dissent (#VP5521)
Support–Opposition (#VP5785)
Agreement–Disagreement (#VP5026)
Restoration–Destruction (#VP5693).

♦ **VD3308 Negativity**
Integrative polarities Assent–Dissent (#VP5521)
Consent–Refusal (#VP5775)
Hope–Hopelessness (#VP5888)
Support–Opposition (#VP5785)
Agreement–Disagreement (#VP5026).

♦ **VD3310 Neglect**
Integrative polarities Respect–Disrespect (#VP5964)
Carefulness–Neglect (#VP5533)
Skilfulness–Unskilfulness (#VP5733)
Accomplishment–Nonaccomplishment (#VP5722).

♦ **VD3312 Negligence**
Integrative polarities Truth–Error (#VP5516)   Order–Disorder (#VP5059)
Desire–Avoidance (#VP5631)   Carefulness–Neglect (#VP5533)
Attention–Inattention (#VP5530)   Skilfulness–Unskilfulness (#VP5733)
Preparedness–Unpreparedness (#VP5720).

♦ **VD3314 Nervousness**
Integrative polarities Courage–Fear (#VP5890)   Oscillation–Agitation (#VP5323)
Excitement–Inexcitability (#VP5857).

♦ **VD3316 Noisome**
Integrative polarities Goodness–Badness (#VP5674)

VD3316

Savouriness–Unsavouriness (#VP5427)
Pleasantness–Unpleasantness (#VP5863)
Healthfulness–Unhealthfulness (#VP5683).

♦ **VD3318 Nonacceptance**
Integrative polarities Consent–Refusal (#VP5775)
Choice–Necessity (#VP5637).

♦ **VD3320 Nonaccomplishment**
Integrative polarities Accomplishment–Nonaccomplishment (#VP5722).

♦ **VD3322 Nonadherence**
Integrative polarities Obedience–Disobedience (#VP5766)
Cohesion–Disintegration (#VP5050).

♦ **VD3324 Nonconformity**
Integrative polarities Assent–Dissent (#VP5521)
Inclusion–Exclusion (#VP5076)
Obedience–Disobedience (#VP5766)
Agreement–Disagreement (#VP5026)
Conformity–Nonconformity (#VP5082)
Concurrence–Counteraction (#VP5177).

♦ **VD3326 Nonfeasance**
Integrative polarities Virtue–Vice (#VP5980)   Carefulness–Neglect (#VP5533)
Skilfulness–Unskilfulness (#VP5733)
Accomplishment–Nonaccomplishment (#VP5722).

♦ **VD3328 Nonfulfilment**
Integrative polarities Oversufficiency–Insufficiency (#VP5660)
Accomplishment–Nonaccomplishment (#VP5722).

♦ **VD3330 Nonobservance**
Integrative polarities Consent–Refusal (#VP5775)
Attention–Inattention (#VP5530)
Obedience–Disobedience (#VP5766).

♦ **VD3332 Nonsense**
Integrative polarities Meaning–Meaninglessness (#VP5545)
Intelligence–Unintelligence (#VP5467).

♦ **VD3334 Nonuniformity**
Integrative polarities Order–Disorder (#VP5059)   Equality–Inequality (#VP5030)
Regularity–Irregularity (#VP5137)   Stability–Changeableness (#VP5141)
Continuity–Discontinuity (#VP5071)   Uniformity–Nonuniformity (#VP5017).

♦ **VD3336 Nothingness**
Integrative polarities Presence–Absence (#VP5186)
Existence–Nonexistence (#VP5001).

♦ **VD3338 Notoriety**
Integrative polarities Repute–Disrepute (#VP5914)
Goodness–Badness (#VP5674)
Probity–Improbity (#VP5974)
Chastity–Unchastity (#VP5987).

♦ **VD3340 Nuisance**
Integrative polarities Amusement–Boredom (#VP5878)
Pleasantness–Unpleasantness (#VP5863).

♦ **VD3342 Nullity**
Integrative polarities Existence–Nonexistence (#VP5001)
Importance–Unimportance (#VP5672)
Meaning–Meaninglessness (#VP5545).

♦ **VD3344 Oblivion**
Integrative polarities Carefulness–Neglect (#VP5533)
Thought–Thoughtlessness (#VP5478)
Feeling–Unfeelinglessness (#VP5855)
Remembrance–Forgetfulness (#VP5537).

♦ **VD3346 Obscurity**
Integrative polarities Form–Formlessness (#VP5246)
Certainty–Uncertainty (#VP5513)
Importance–Unimportance (#VP5672)
Visibility–Invisibility (#VP5444)
Intelligibility–Unintelligibility (#VP5548).

♦ **VD3348 Obsession**
Integrative polarities Sanity–Insanity (#VP5472)   Selfactualization–Neurosis (#VP5690).

♦ **VD3350 Obsolescence**
Integrative polarities Newness–Oldness (#VP5122)
Appropriateness–Inappropriateness (#VP5665).

♦ **VD3352 Obstacle**
Integrative polarities Opening–Closure (#VP5265)
Facility–Difficulty (#VP5732).

♦ **VD3354 Obstinacy**
Integrative polarities Support–Opposition (#VP5785)
Obedience–Disobedience (#VP5766)
Resolution–Irresolution (#VP5624)
Willingness–Unwillingness (#VP5621).

♦ **VD3356 Obstruction**
Integrative polarities Opening–Closure (#VP5265)
Facility–Difficulty (#VP5732)
Timeliness–Untimeliness (#VP5129).

♦ **VD3358 Offence**
Integrative polarities Virtue–Vice (#VP5980)   Attack–Defence (#VP5798)
Respect–Disrespect (#VP5964)   Legality–Illegality (#VP5998)
Congratulation–Envy (#VP5948)   Obedience–Disobedience (#VP5766).

♦ **VD3360 Omission**
Integrative polarities Virtue–Vice (#VP5980)   Truth–Error (#VP5516)
Carefulness–Neglect (#VP5533)   Skilfulness–Unskilfulness (#VP5733)
Completeness–Incompleteness (#VP5054)   Oversufficiency–Insufficiency (#VP5660)
Accomplishment–Nonaccomplishment (#VP5722).

♦ **VD3362 Opacity**
Integrative polarities Intelligence–Unintelligence (#VP5467)
Intelligibility–Unintelligibility (#VP5548).

♦ **VD3364 Opposition**
Integrative polarities Assent–Dissent (#VP5521)
Support–Opposition (#VP5785)
Facility–Difficulty (#VP5732)
Identity–Difference (#VP5013)
Approval–Disapproval (#VP5968)
Agreement–Disagreement (#VP5026)
Willingness–Unwillingness (#VP5621).

♦ **VD3366 Oppression**
Integrative polarities Cheerfulness–Solemnity (#VP5870)
Pleasantness–Unpleasantness (#VP5863)
Appropriateness–Inappropriateness (#VP5665).

♦ **VD3368 Opprobrium**
Integrative polarities Repute–Disrepute (#VP5914)
Approval–Disapproval (#VP5968).

♦ **VD3370 Orderlessness**
Integrative polarities Order–Disorder (#VP5059)   Form–Formlessness (#VP5246)
Certainty–Uncertainty (#VP5513).

♦ **VD3372 Ostentation**
Integrative polarities Naturalness–Affectation (#VP5900)
Communicativeness–Uncommunicativeness (#VP5554).

♦ **VD3374 Outrage**
Integrative polarities Virtue–Vice (#VP5980)   Goodness–Badness (#VP5674)
Justice–Injustice (#VP5976)   Respect–Disrespect (#VP5964)
Congratulation–Envy (#VP5948)   Appropriateness–Inappropriateness (#VP5665).

♦ **VD3376 Overabundance**
Integrative polarities Oversufficiency–Insufficiency (#VP5660).

♦ **VD3378 Overactivity**
Integrative polarities Action–Inaction (#VP5705).

♦ **VD3380 Overcompensation**
Integrative polarities Selfactualization–Neurosis (#VP5690).

♦ **VD3382 Overconfidence**
Integrative polarities Caution–Rashness (#VP5894).

♦ **VD3384 Overconscientiousnes**
Integrative polarities Taste–Vulgarity (#VP5896).

♦ **VD3386 Overdeveloped**
Integrative polarities Bigness–Littleness (#VP5195)
Oversufficiency–Insufficiency (#VP5660).

♦ **VD3388 Overemphasis**
Integrative polarities Oversufficiency–Insufficiency (#VP5660)
Communicativeness–Uncommunicativeness (#VP5554).

♦ **VD3390 Overestimation**
Integrative polarities Judgement–Misjudgement (#VP5494)
Communicativeness–Uncommunicativeness (#VP5554).

♦ **VD3392 Overexpansion**
Integrative polarities Oversufficiency–Insufficiency (#VP5660).

♦ **VD3394 Overextension**
Integrative polarities Action–Inaction (#VP5705)   Oversufficiency–Insufficiency (#VP5660).

♦ **VD3396 Overload**
Integrative polarities Weight–Lightness (#VP5352)
Prosperity–Adversity (#VP5728)
Oversufficiency–Insufficiency (#VP5660).

♦ **VD3398 Overreligiousness**
Integrative polarities Sanity–Insanity (#VP5472).

♦ **VD3400 Oversensitiveness**
Integrative polarities Taste–Vulgarity (#VP5896)   Congratulation–Envy (#VP5948).

♦ **VD3402 Oversight**
Integrative polarities Truth–Error (#VP5516)   Carefulness–Neglect (#VP5533).

♦ **VD3404 Oversimplification**
Integrative polarities Simplicity–Complexity (#VP5045)
Preparedness–Unpreparedness (#VP5720).

♦ **VD3406 Overstrain**
Integrative polarities Action–Inaction (#VP5705)   Oversufficiency–Insufficiency (#VP5660).

♦ **VD3408 Oversupply**
Integrative polarities Oversufficiency–Insufficiency (#VP5660).

♦ **VD3410 Overtax**
Integrative polarities Weight–Lightness (#VP5352)
Expensiveness–Cheapness (#VP5848)
Oversufficiency–Insufficiency (#VP5660).

DESTRUCTIVE VALUES

♦ **VD3412 Overturn**
  Integrative polarities Victory–Defeat (#VP5726)   Elevation–Depression (#VP5317)
  Evolution–Revolution (#VP5147)   Restoration–Destruction (#VP5693).

♦ **VD3414 Overweight**
  Integrative polarities Weight–Lightness (#VP5352)
  Bigness–Littleness (#VP5195)
  Oversufficiency–Insufficiency (#VP5660).

♦ **VD3416 Overwork**
  Integrative polarities Oversufficiency–Insufficiency (#VP5660).

♦ **VD3418 Overzealousness**
  Integrative polarities Sanity–Insanity (#VP5472)   Caution–Rashness (#VP5894)
  Desire–Avoidance (#VP5631)   Resolution–Irresolution (#VP5624).

♦ **VD3420 Paganism**
  Integrative polarities Piety–Impiety (#VP6028)   Knowledge–Ignorance (#VP5475).

♦ **VD3422 Pain**
  Integrative polarities Health–Disease (#VP5685)
  Sensation–Insensibility (#VP5422)
  Pleasantness–Unpleasantness (#VP5863).

♦ **VD3424 Panic**
  Integrative polarities Courage–Fear (#VP5890)   Victory–Defeat (#VP5726)
  Excitement–Inexcitability (#VP5857).

♦ **VD3426 Paralysis**
  Integrative polarities Health–Disease (#VP5685)
  Action–Inaction (#VP5705).

♦ **VD3428 Passionlessness**
  Integrative polarities Desire–Avoidance (#VP5631)
  Feeling–Unfeelinglessness (#VP5855).

♦ **VD3430 Passivity**
  Integrative polarities Action–Inaction (#VP5705)   Motion–Quiescence (#VP5267)
  Patience–Impatience (#VP5861)   Obedience–Disobedience (#VP5766)
  Feeling–Unfeelinglessness (#VP5855).

♦ **VD3432 Pedantry**
  Integrative polarities Formality–Informality (#VP5646)
  Naturalness–Affectation (#VP5900).

♦ **VD3434 Perdition**
  Integrative polarities Godliness–Ungodliness (#VP6013)
  Restoration–Destruction (#VP5693).

♦ **VD3436 Perfidiousness**
  Integrative polarities Probity–Improbity (#VP5974)
  Communicativeness–Uncommunicativeness (#VP5554).

♦ **VD3438 Permissiveness**
  Integrative polarities Consent–Refusal (#VP5775)
  Leniency–Compulsion (#VP5756)
  Carefulness–Neglect (#VP5533).

♦ **VD3440 Perplexity**
  Integrative polarities Facility–Difficulty (#VP5732)
  Attention–Inattention (#VP5530)
  Certainty–Uncertainty (#VP5513)
  Intelligibility–Unintelligibility (#VP5548).

♦ **VD3442 Perturbation**
  Integrative polarities Courage–Fear (#VP5890)   Order–Disorder (#VP5059)
  Attention–Inattention (#VP5530)   Certainty–Uncertainty (#VP5513)
  Oscillation–Agitation (#VP5323).

♦ **VD3444 Perversion**
  Integrative polarities Truth–Error (#VP5516)   Intuition–Reason (#VP5481)
  Symmetry–Distortion (#VP5248)   Improvement–Impairment (#VP5691)
  Education–Miseducation (#VP5562)   Representation–Misrepresentation (#VP5573)
  Appropriateness–Inappropriateness (#VP5665)   Interpretability–Misinterpretability (#VP5552)
  Communicativeness–Uncommunicativeness (#VP5554).

♦ **VD3446 Perversity**
  Integrative polarities Congratulation–Envy (#VP5948)
  Identity–Difference (#VP5013)
  Resolution–Irresolution (#VP5624).

♦ **VD3448 Pessimism**
  Integrative polarities Hope–Hopelessness (#VP5888)
  Cheerfulness–Solemnity (#VP5870).

♦ **VD3450 Pestiferousness**
  Integrative polarities Health–Disease (#VP5685)
  Pleasantness–Unpleasantness (#VP5863)
  Healthfulness–Unhealthfulness (#VP5683).

♦ **VD3452 Pettiness**
  Integrative polarities Repute–Disrepute (#VP5914)
  Importance–Unimportance (#VP5672)
  Superiority–Inferiority (#VP5036)
  Unselfishness–Selfishness (#VP5978)
  Broadmindedness–Narrowmindedness (#VP5526).

♦ **VD3454 Petulance**
  Integrative polarities Congratulation–Envy (#VP5948)
  Exultation–Lamentation (#VP5873)
  Resolution–Irresolution (#VP5624)
  Contentment–Discontentment (#VP5868).

♦ **VD3456 Pitilessness**
  Integrative polarities Energy–Moderation (#VP5161)
  Kindness–Unkindness (#VP5938)
  Compassion–Pitilessness (#VP5944).

♦ **VD3458 Pleasurelessness**
  Integrative polarities Cheerfulness–Solemnity (#VP5870)
  Pleasantness–Unpleasantness (#VP5863).

♦ **VD3460 Pointlessness**
  Integrative polarities Amusement–Boredom (#VP5878)
  Appropriateness–Inappropriateness (#VP5665).

♦ **VD3462 Poisonousness**
  Integrative polarities Goodness–Badness (#VP5674)
  Improvement–Impairment (#VP5691)
  Healthfulness–Unhealthfulness (#VP5683).

♦ **VD3464 Pollution**
  Integrative polarities Goodness–Badness (#VP5674)
  Cleanness–Uncleanness (#VP5681)
  Simplicity–Complexity (#VP5045)
  Improvement–Impairment (#VP5691)
  Healthfulness–Unhealthfulness (#VP5683)
  Appropriateness–Inappropriateness (#VP5665).

♦ **VD3466 Pomposity**
  Integrative polarities Naturalness–Affectation (#VP5900).

♦ **VD3468 Ponderousness**
  Integrative polarities Amusement–Boredom (#VP5878)
  Facility–Difficulty (#VP5732)
  Skilfulness–Unskilfulness (#VP5733).

♦ **VD3470 Poverty**
  Integrative polarities Wealth–Poverty (#VP5837)
  Oversufficiency–Insufficiency (#VP5660).

♦ **VD3472 Powerlessness**
  Integrative polarities Power–Impotence (#VP5157)
  Strength–Weakness (#VP5159)
  Influence–Influencelessness (#VP5172).

♦ **VD3474 Prejudice**
  Integrative polarities Goodness–Badness (#VP5674)
  Judgement–Misjudgement (#VP5494)
  Expedience–Inexpedience (#VP5670)
  Broadmindedness–Narrowmindedness (#VP5526).

♦ **VD3476 Prematurity**
  Integrative polarities Timeliness–Untimeliness (#VP5129).

♦ **VD3478 Presumption**
  Integrative polarities Modesty–Vanity (#VP5908)
  Caution–Rashness (#VP5894)
  Dueness–Undueness (#VP5960)
  Judgement–Misjudgement (#VP5494)
  Circumscription–Intrusion (#VP5234).

♦ **VD3480 Pretentiousness**
  Integrative polarities Modesty–Vanity (#VP5908)
  Naturalness–Affectation (#VP5900).

♦ **VD3482 Prevarication**
  Integrative polarities Intuition–Reason (#VP5481)
  Communicativeness–Uncommunicativeness (#VP5554).

♦ **VD3484 Privation**
  Integrative polarities Wealth–Poverty (#VP5837)
  Possession–Loss (#VP5808).

♦ **VD3486 Problem**
  Integrative polarities Facility–Difficulty (#VP5732)
  Certainty–Uncertainty (#VP5513)
  Perfection–Imperfection (#VP5677)
  Pleasantness–Unpleasantness (#VP5863)
  Intelligibility–Unintelligibility (#VP5548).

♦ **VD3488 Prodigality**
  Integrative polarities Virtue–Vice (#VP5980)   Economy–Prodigality (#VP5851)
  Temperance–Intemperance (#VP5992)   Expensiveness–Cheapness (#VP5848)
  Oversufficiency–Insufficiency (#VP5660)
  Communicativeness–Uncommunicativeness (#VP5554).

♦ **VD3490 Profanation**
  Integrative polarities Piety–Impiety (#VP6028)   Rightness–Wrongness (#VP5957)
  Appropriateness–Inappropriateness (#VP5665).

♦ **VD3492 Profligacy**
  Integrative polarities Virtue–Vice (#VP5980)   Taste–Vulgarity (#VP5896)
  Expensiveness–Cheapness (#VP5848).

♦ **VD3494 Prohibition**
  Integrative polarities Consent–Refusal (#VP5775)
  Freedom–Restraint (#VP5762)
  Facility–Difficulty (#VP5732)
  Inclusion–Exclusion (#VP5076).

♦ **VD3496 Proliferation**
  Integrative polarities Increase–Decrease (#VP5038).

♦ **VD3498 Prolixity**
  Integrative polarities Amusement–Boredom (#VP5878)
  Conciseness–Diffuseness (#VP5592)
  Oversufficiency–Insufficiency (#VP5660).

VD3500

- **VD3500 Promiscuity**
  Integrative polarities Order–Disorder (#VP5059)   Chastity–Unchastity (#VP5987)
  Discrimination–Indiscrimination (#VP5492).

- **VD3502 Prostitution**
  Integrative polarities Chastity–Unchastity (#VP5987)
  Improvement–Impairment (#VP5691)
  Appropriateness–Inappropriateness (#VP5665).

- **VD3504 Provocation**
  Integrative polarities Congratulation–Envy (#VP5948)
  Comfort–Aggravation (#VP5885)
  Pleasantness–Unpleasantness (#VP5863).

- **VD3506 Puerility**
  Integrative polarities Newness–Oldness (#VP5122)
  Intelligence–Unintelligence (#VP5467).

- **VD3508 Purposelessness**
  Integrative polarities Attributability–Chance (#VP5155)
  Meaning–Meaninglessness (#VP5545)
  Appropriateness–Inappropriateness (#VP5665).

- **VD3510 Putrefacton**
  Integrative polarities Improvement–Impairment (#VP5691).

- **VD3512 Quarrelsomeness**
  Integrative polarities Accord–Discaccord (#VP5794)
  Intuition–Reason (#VP5481)
  Friendship–Enmity (#VP5927)
  Congratulation–Envy (#VP5948).

- **VD3514 Randomness**
  Integrative polarities Order–Disorder (#VP5059)   Certainty–Uncertainty (#VP5513)
  Attributability–Chance (#VP5155).

- **VD3516 Rapacity**
  Integrative polarities Desire–Avoidance (#VP5631)
  Sharing–Appropriation (#VP5815)
  Temperance–Intemperance (#VP5992).

- **VD3518 Ravishment**
  Integrative polarities Chastity–Unchastity (#VP5987)
  Sharing–Appropriation (#VP5815).

- **VD3520 Recalcitrance**
  Integrative polarities Support–Opposition (#VP5785)
  Obedience–Disobedience (#VP5766)
  Resolution–Irresolution (#VP5624)
  Willingness–Unwillingness (#VP5621).

- **VD3522 Recrimination**
  Integrative polarities Vindication–Condemnation (#VP6005).

- **VD3524 Regression**
  Integrative polarities Refreshment–Relapse (#VP5695)
  Conversion–Reversion (#VP5145)
  Improvement–Impairment (#VP5691)
  Progression–Regression (#VP5294).

- **VD3526 Rejection**
  Integrative polarities Assent–Dissent (#VP5521)
  Consent–Refusal (#VP5775)
  Belief–Unbelief (#VP5501)
  Choice–Necessity (#VP5637)
  Support–Opposition (#VP5785)
  Inclusion–Exclusion (#VP5076)
  Approval–Disapproval (#VP5968)
  Appropriateness–Inappropriateness (#VP5665).

- **VD3528 Relentlessness**
  Integrative polarities Leniency–Compulsion (#VP5756)
  Compassion–Pitilessness (#VP5944)
  Resolution–Irresolution (#VP5624).

- **VD3530 Remorselessness**
  Integrative polarities Exultation–Lamentation (#VP5873)
  Compassion–Pitilessness (#VP5944).

- **VD3532 Repression**
  Integrative polarities Possession–Loss (#VP5808)
  Consent–Refusal (#VP5775)
  Freedom–Restraint (#VP5762)
  Facility–Difficulty (#VP5732)
  Remembrance–Forgetfulness (#VP5537).

- **VD3534 Reproach**
  Integrative polarities Repute–Disrepute (#VP5914)
  Approval–Disapproval (#VP5968)
  Vindication–Condemnation (#VP6005).

- **VD3536 Repudiation**
  Integrative polarities Assent–Dissent (#VP5521)
  Consent–Refusal (#VP5775)
  Choice–Necessity (#VP5637)
  Inclusion–Exclusion (#VP5076)
  Resolution–Irresolution (#VP5624).

- **VD3538 Repugnance**
  Integrative polarities Love–Hate (#VP5930)   Friendship–Enmity (#VP5927)
  Identity–Difference (#VP5013)   Agreement–Disagreement (#VP5026)
  Willingness–Unwillingness (#VP5621)   Pleasantness–Unpleasantness (#VP5863).

- **VD3540 Repulsion**
  Integrative polarities Goodness–Badness (#VP5674)
  Support–Opposition (#VP5785)
  Attraction–Repulsion (#VP5288)
  Cleanness–Uncleanness (#VP5681)
  Pleasantness–Unpleasantness (#VP5863).

- **VD3542 Resentment**
  Integrative polarities Congratulation–Envy (#VP5948)
  Contentment–Discontentment (#VP5868).

- **VD3544 Restlessness**
  Integrative polarities Patience–Impatience (#VP5861)
  Oscillation–Agitation (#VP5323)
  Stability–Changeableness (#VP5141)
  Contentment–Discontentment (#VP5868).

- **VD3546 Restriction**
  Integrative polarities Freedom–Restraint (#VP5762)
  Facility–Difficulty (#VP5732)
  Inclusion–Exclusion (#VP5076)
  Centrality–Environment (#VP5226).

- **VD3548 Retaliation**
  Integrative polarities Forgiveness–Vengeance (#VP5947).

- **VD3550 Retardation**
  Integrative polarities Freedom–Restraint (#VP5762)
  Facility–Difficulty (#VP5732)
  Timeliness–Untimeliness (#VP5129)
  Intelligence–Unintelligence (#VP5467).

- **VD3552 Retribution**
  Integrative polarities Atonement–Punishment (#VP6010)
  Forgiveness–Vengeance (#VP5947).

- **VD3554 Retrogression**
  Integrative polarities Conversion–Reversion (#VP5145)
  Improvement–Impairment (#VP5691)
  Progression–Regression (#VP5294).

- **VD3556 Revenge**
  Integrative polarities Forgiveness–Vengeance (#VP5947).

- **VD3558 Reversion**
  Integrative polarities Refreshment–Relapse (#VP5695)
  Conversion–Reversion (#VP5145)
  Progression–Regression (#VP5294).

- **VD3560 Revilement**
  Integrative polarities Approval–Disapproval (#VP5968).

- **VD3562 Revolution**
  Integrative polarities Evolution–Revolution (#VP5147)
  Authority–Lawlessness (#VP5739)
  Obedience–Disobedience (#VP5766)
  Conjunction–Separation (#VP5047).

- **VD3564 Revulsion**
  Integrative polarities Evolution–Revolution (#VP5147)
  Conversion–Reversion (#VP5145).

- **VD3566 Ridicule**
  Integrative polarities Modesty–Vanity (#VP5908)
  Respect–Disrespect (#VP5964)
  Approval–Disapproval (#VP5968).

- **VD3568 Rigidity**
  Integrative polarities Change–Permanence (#VP5139)
  Leniency–Compulsion (#VP5756)
  Resolution–Irresolution (#VP5624).

- **VD3570 Risk**
  Integrative polarities Safety–Danger (#VP5697)   Certainty–Uncertainty (#VP5513).

- **VD3572 Riskiness**
  Integrative polarities Respect–Disrespect (#VP5964)
  Certainty–Uncertainty (#VP5513).

- **VD3574 Rivalry**
  Integrative polarities Accord–Discaccord (#VP5794)
  Support–Opposition (#VP5785)
  Congratulation–Envy (#VP5948).

- **VD3576 Roguery**
  Integrative polarities Probity–Improbity (#VP5974)
  Behaviour–Misbehaviour (#VP5737).

- **VD3578 Rot**
  Integrative polarities Health–Disease (#VP5685)
  Goodness–Badness (#VP5674)
  Cleanness–Uncleanness (#VP5681)
  Improvement–Impairment (#VP5691)
  Meaning–Meaninglessness (#VP5545).

- **VD3580 Roughness**
  Integrative polarities Taste–Vulgarity (#VP5896)   Energy–Moderation (#VP5161)
  Kindness–Unkindness (#VP5938)   Preparedness–Unpreparedness (#VP5720).

- **VD3582 Rubbish**
  Integrative polarities Meaning–Meaninglessness (#VP5545)
  Appropriateness–Inappropriateness (#VP5665).

DESTRUCTIVE VALUES

♦ **VD3584 Rudeness**
Integrative polarities Modesty–Vanity (#VP5908)
Taste–Vulgarity (#VP5896)
Courtesy–Discourtesy (#VP5936)
Preparedness–Unpreparedness (#VP5720).

♦ **VD3586 Ruin**
Integrative polarities Victory–Defeat (#VP5726)   Possession–Loss (#VP5808)
Facility–Difficulty (#VP5732)   Improvement–Impairment (#VP5691)
Restoration–Destruction (#VP5693).

♦ **VD3588 Ruthlessness**
Integrative polarities Kindness–Unkindness (#VP5938)
Compassion–Pitilessness (#VP5944).

♦ **VD3590 Sabotage**
Integrative polarities Power–Impotence (#VP5157)
Facility–Difficulty (#VP5732)
Improvement–Impairment (#VP5691)
Restoration–Destruction (#VP5693).

♦ **VD3592 Sacrifice**
Integrative polarities Life–Death (#VP5407)   Possession–Loss (#VP5808).

♦ **VD3594 Sacrilege**
Integrative polarities Piety–Impiety (#VP6028)   Rightness–Wrongness (#VP5957).

♦ **VD3596 Sadism**
Integrative polarities Kindness–Unkindness (#VP5938)
Pleasantness–Unpleasantness (#VP5863).

♦ **VD3598 Sadness**
Integrative polarities Cheerfulness–Solemnity (#VP5870)
Importance–Unimportance (#VP5672)
Pleasantness–Unpleasantness (#VP5863).

♦ **VD3600 Sameness**
Integrative polarities Amusement–Boredom (#VP5878).

♦ **VD3602 Sanctimony**
Integrative polarities Naturalness–Affectation (#VP5900)
Communicativeness–Uncommunicativeness (#VP5554).

♦ **VD3604 Sarcasm**
Integrative polarities Respect–Disrespect (#VP5964).

♦ **VD3606 Satire**
Integrative polarities Respect–Disrespect (#VP5964)
Approval–Disapproval (#VP5968).

♦ **VD3608 Savagery**
Integrative polarities Taste–Vulgarity (#VP5896)   Energy–Moderation (#VP5161)
Kindness–Unkindness (#VP5938)   Knowledge–Ignorance (#VP5475).

♦ **VD3610 Scandal**
Integrative polarities Virtue–Vice (#VP5980)   Repute–Disrepute (#VP5914)
Rightness–Wrongness (#VP5957)   Approval–Disapproval (#VP5968).

♦ **VD3612 Scantiness**
Integrative polarities Bigness–Littleness (#VP5195)
Greatness–Smallness (#VP5034)
Oversufficiency–Insufficiency (#VP5660).

♦ **VD3614 Scarcity**
Integrative polarities Oversufficiency–Insufficiency (#VP5660).

♦ **VD3616 Seclusion**
Integrative polarities Unity–Duality (#VP5089)   Freedom–Restraint (#VP5762)
Inclusion–Exclusion (#VP5076)   Sociability–Unsociability (#VP5922).

♦ **VD3618 Secrecy**
Integrative polarities Sociability–Unsociability (#VP5922)
Communicativeness–Uncommunicativeness (#VP5554).

♦ **VD3620 Sedition**
Integrative polarities Probity–Improbity (#VP5974)
Obedience–Disobedience (#VP5766).

♦ **VD3622 Seduction**
Integrative polarities Chastity–Unchastity (#VP5987).

♦ **VD3624 Segregation**
Integrative polarities Freedom–Restraint (#VP5762)
Inclusion–Exclusion (#VP5076)
Sociability–Unsociability (#VP5922)
Discrimination–Indiscrimination (#VP5492)
Broadmindedness–Narrowmindedness (#VP5526).

♦ **VD3626 Self-absorption**
Integrative polarities Probity–Improbity (#VP5974)
Feeling–Unfeelingness (#VP5855).

♦ **VD3628 Self-admiration**
Integrative polarities Modesty–Vanity (#VP5908)
Unselfishness–Selfishness (#VP5978).

♦ **VD3630 Self-centredness**
Integrative polarities Modesty–Vanity (#VP5908)
Unselfishness–Selfishness (#VP5978).

♦ **VD3632 Self-contradiciton**
Integrative polarities Truth–Error (#VP5516)   Identity–Difference (#VP5013)
Agreement–Disagreement (#VP5026).

♦ **VD3634 Self-destruction**
Integrative polarities Life–Death (#VP5407).

♦ **VD3636 Self-esteem**
Integrative polarities Modesty–Vanity (#VP5908)
Pride–Humility (#VP5905)
Unselfishness–Selfishness (#VP5978).

♦ **VD3638 Self-importance**
Integrative polarities Modesty–Vanity (#VP5908)
Naturalness–Affectation (#VP5900).

♦ **VD3640 Self-indulgence**
Integrative polarities Temperance–Intemperance (#VP5992)
Unselfishness–Selfishness (#VP5978).

♦ **VD3642 Self-righteousness**
Integrative polarities Piety–Impiety (#VP6028).

♦ **VD3644 Self-satisfaction**
Integrative polarities Modesty–Vanity (#VP5908)
Contentment–Discontentment (#VP5868).

♦ **VD3646 Selfishness**
Integrative polarities Modesty–Vanity (#VP5908)
Unselfishness–Selfishness (#VP5978).

♦ **VD3648 Senselessness**
Integrative polarities Order–Disorder (#VP5059)   Sanity–Insanity (#VP5472)
Intuition–Reason (#VP5481)   Meaning–Meaninglessness (#VP5545)
Intelligence–Unintelligence (#VP5467).

♦ **VD3650 Sensuality**
Integrative polarities Chastity–Unchastity (#VP5987).

♦ **VD3652 Sentimentality**
Integrative polarities Feeling–Unfeelinglessness (#VP5855)
Intelligence–Unintelligence (#VP5467).

♦ **VD3654 Separateness**
Integrative polarities Life–Death (#VP5407)   Unity–Duality (#VP5089)
Conjunction–Separation (#VP5047)   Cohesion–Disintegration (#VP5050)
Relatedness–Unrelatedness (#VP5009).

♦ **VD3656 Servility**
Integrative polarities Pride–Humility (#VP5905)   Freedom–Restraint (#VP5762)
Respect–Disrespect (#VP5964)   Obedience–Disobedience (#VP5766)
Superiority–Inferiority (#VP5036).

♦ **VD3658 Severance**
Integrative polarities Inclusion–Exclusion (#VP5076)
Conjunction–Separation (#VP5047).

♦ **VD3660 Severity**
Integrative polarities Energy–Moderation (#VP5161)
Kindness–Unkindness (#VP5938)
Leniency–Compulsion (#VP5756)
Courtesy–Discourtesy (#VP5936)
Naturalness–Affectation (#VP5900).

♦ **VD3662 Sexuality**
Integrative polarities Chastity–Unchastity (#VP5987)
Sexiness–Unsexiness (#VP5419).

♦ **VD3664 Shallowness**
Integrative polarities Knowledge–Ignorance (#VP5475)
Attention–Inattention (#VP5530)
Importance–Unimportance (#VP5672)
Intelligence–Unintelligence (#VP5467).

♦ **VD3666 Shamelessness**
Integrative polarities Virtue–Vice (#VP5980)   Modesty–Vanity (#VP5908)
Probity–Improbity (#VP5974)   Chastity–Unchastity (#VP5987)
Rightness–Wrongness (#VP5957)   Exultation–Lamentation (#VP5873)
Naturalness–Affectation (#VP5900).

♦ **VD3668 Shapelessness**
Integrative polarities Order–Disorder (#VP5059)   Beauty–Ugliness (#VP5899)
Form–Formlessness (#VP5246)   Certainty–Uncertainty (#VP5513)
Stability–Changeableness (#VP5141)   Conformity–Nonconformity (#VP5082)
Intelligibility–Unintelligibility (#VP5548).

♦ **VD3670 Shiftlessness**
Integrative polarities Action–Inaction (#VP5705)   Preparedness–Unpreparedness (#VP5720).

♦ **VD3672 Shock**
Integrative polarities Courage–Fear (#VP5890)   Health–Disease (#VP5685)
Prosperity–Adversity (#VP5728)   Pleasantness–Unpleasantness (#VP5863).

♦ **VD3674 Shortage**
Integrative polarities Perfection–Imperfection (#VP5677)
Overrunning–Shortcoming (#VP5313)
Completeness–Incompleteness (#VP5054)
Oversufficiency–Insufficiency (#VP5660).

♦ **VD3676 Sickness**
Integrative polarities Health–Disease (#VP5685)
Sanity–Insanity (#VP5472).

♦ **VD3678 Slant**
Integrative polarities Direction–Deviation (#VP5290)
Symmetry–Distortion (#VP5248)
Representation–Misrepresentation (#VP5573)
Communicativeness–Uncommunicativeness (#VP5554).

VD3680

♦ **VD3680 Slothfulness**
Integrative polarities Virtue–Vice (#VP5980)   Action–Inaction (#VP5705)
Desire–Avoidance (#VP5631)   Hope–Hopelessness (#VP5888)
Feeling–Unfeelinglessness (#VP5855).

♦ **VD3682 Slowness**
Integrative polarities Action–Inaction (#VP5705)   Amusement–Boredom (#VP5878)
Swiftness–Slowness (#VP5269)   Timeliness–Untimeliness (#VP5129)
Intelligence–Unintelligence (#VP5467).

♦ **VD3684 Smallness**
Integrative polarities Pride–Humility (#VP5905)   Repute–Disrepute (#VP5914)
Bigness–Littleness (#VP5195)   Greatness–Smallness (#VP5034)
Numerousness–Fewness (#VP5101)   Importance–Unimportance (#VP5672)
Superiority–Inferiority (#VP5036)   Unselfishness–Selfishness (#VP5978)
Oversufficiency–Insufficiency (#VP5660)
Broadmindedness–Narrowmindedness (#VP5526).

♦ **VD3686 Snobbery**
Integrative polarities Modesty–Vanity (#VP5908)
Taste–Vulgarity (#VP5896)
Respect–Disrespect (#VP5964)
Inclusion–Exclusion (#VP5076).

♦ **VD3688 Softness**
Integrative polarities Courage–Fear (#VP5890)   Belief–Unbelief (#VP5501)
Power–Impotence (#VP5157)   Strength–Weakness (#VP5159)
Intelligence–Unintelligence (#VP5467).

♦ **VD3690 Sophistry**
Integrative polarities Intuition–Reason (#VP5481)
Education–Miseducation (#VP5562)
Skilfulness–Unskilfulness (#VP5733)
Communicativeness–Uncommunicativeness (#VP5554).

♦ **VD3692 Sordidness**
Integrative polarities Order–Disorder (#VP5059)   Repute–Disrepute (#VP5914)
Goodness–Badness (#VP5674)   Desire–Avoidance (#VP5631)
Cleanness–Uncleanness (#VP5681)   Expensiveness–Cheapness (#VP5848).

♦ **VD3694 Soreness**
Integrative polarities Friendship–Enmity (#VP5927)
Congratulation–Envy (#VP5948)
Sensation–Insensibility (#VP5422).

♦ **VD3696 Sorrow**
Integrative polarities Exultation–Lamentation (#VP5873)
Cheerfulness–Solemnity (#VP5870)
Pleasantness–Unpleasantness (#VP5863).

♦ **VD3698 Soullessness**
Integrative polarities Feeling–Unfeelinglessness (#VP5855).

♦ **VD3700 Sourness**
Integrative polarities Friendship–Enmity (#VP5927)
Congratulation–Envy (#VP5948)
Contentment–Discontentment (#VP5868).

♦ **VD3702 Sparsity**
Integrative polarities Oversufficiency–Insufficiency (#VP5660).

♦ **VD3704 Spasmodicness**
Integrative polarities Order–Disorder (#VP5059)   Energy–Moderation (#VP5161)
Oscillation–Agitation (#VP5323)   Sensation–Insensibility (#VP5422)
Continuity–Discontinuity (#VP5071).

♦ **VD3706 Spasticity**
Integrative polarities Health–Disease (#VP5685)
Energy–Moderation (#VP5161)
Oscillation–Agitation (#VP5323).

♦ **VD3708 Speciousness**
Integrative polarities Intuition–Reason (#VP5481)
Appearance–Disappearance (#VP5446)
Communicativeness–Uncommunicativeness (#VP5554).

♦ **VD3710 Spiritlessness**
Integrative polarities Courage–Fear (#VP5890)   Amusement–Boredom (#VP5878)
Cheerfulness–Solemnity (#VP5870)   Feeling–Unfeelinglessness (#VP5855).

♦ **VD3712 Spoilage**
Integrative polarities Improvement–Impairment (#VP5691).

♦ **VD3714 Spoliation**
Integrative polarities Possession–Loss (#VP5808)
Sharing–Appropriation (#VP5815)
Restoration–Destruction (#VP5693).

♦ **VD3716 Spunklessness**
Integrative polarities Courage–Fear (#VP5890)   Feeling–Unfeelinglessness (#VP5855).

♦ **VD3718 Spurious**
Integrative polarities Truth–Error (#VP5516)   Legality–Illegality (#VP5998)
Naturalness–Affectation (#VP5900)
Communicativeness–Uncommunicativeness (#VP5554).

♦ **VD3720 Squalor**
Integrative polarities Order–Disorder (#VP5059)   Repute–Disrepute (#VP5914)
Goodness–Badness (#VP5674)   Cleanness–Uncleanness (#VP5681).

♦ **VD3722 Stagnation**
Integrative polarities Action–Inaction (#VP5705)   Motion–Quiescence (#VP5267).

♦ **VD3724 Stain**
Integrative polarities Repute–Disrepute (#VP5914)
Cleanness–Uncleanness (#VP5681)
Perfection–Imperfection (#VP5677).

♦ **VD3726 Stalemate**
Integrative polarities Facility–Difficulty (#VP5732).

♦ **VD3728 Sterility**
Integrative polarities Amusement–Boredom (#VP5878)
Productiveness–Unproductiveness (#VP5165).

♦ **VD3730 Stiffness**
Integrative polarities Amusement–Boredom (#VP5878)
Leniency–Compulsion (#VP5756)
Elasticity–Toughness (#VP5358)
Formality–Informality (#VP5646)
Resolution–Irresolution (#VP5624).

♦ **VD3732 Strain**
Integrative polarities Courage–Fear (#VP5890)   Action–Inaction (#VP5705)
Friendship–Enmity (#VP5927)   Improvement–Impairment (#VP5691)
Resolution–Irresolution (#VP5624)   Excitement–Inexcitability (#VP5857)
Oversufficiency–Insufficiency (#VP5660)
Communicativeness–Uncommunicativeness (#VP5554).

♦ **VD3734 Strangeness**
Integrative polarities Sanity–Insanity (#VP5472)   Conformity–Nonconformity (#VP5082).

♦ **VD3736 Stridency**
Integrative polarities Energy–Moderation (#VP5161).

♦ **VD3738 Stubbornness**
Integrative polarities Hardness–Softness (#VP5356)
Leniency–Compulsion (#VP5756)
Obedience–Disobedience (#VP5766)
Resolution–Irresolution (#VP5624)
Willingness–Unwillingness (#VP5621).

♦ **VD3740 Stuffiness**
Integrative polarities Fragrance–Stench (#VP5435)
Amusement–Boredom (#VP5878)
Motion–Quiescence (#VP5267)
Naturalness–Affectation (#VP5900)
Resolution–Irresolution (#VP5624)
Intelligence–Unintelligence (#VP5467)
Broadmindedness–Narrowmindedness (#VP5526)
Imaginativeness–Unimaginativeness (#VP5535).

♦ **VD3742 Stupidity**
Integrative polarities Truth–Error (#VP5516)   Intelligence–Unintelligence (#VP5467).

♦ **VD3744 Stupor**
Integrative polarities Action–Inaction (#VP5705)   Feeling–Unfeelinglessness (#VP5855)
Selfactualization–Neurosis (#VP5690).

♦ **VD3746 Subjection**
Integrative polarities Freedom–Restraint (#VP5762)
Obedience–Disobedience (#VP5766)
Superiority–Inferiority (#VP5036).

♦ **VD3748 Subjugation**
Integrative polarities Victory–Defeat (#VP5726)   Freedom–Restraint (#VP5762)
Sharing–Appropriation (#VP5815).

♦ **VD3750 Subordination**
Integrative polarities Freedom–Restraint (#VP5762)
Superiority–Inferiority (#VP5036).

♦ **VD3752 Subterfuge**
Integrative polarities Intuition–Reason (#VP5481)
Skilfulness–Unskilfulness (#VP5733)
Communicativeness–Uncommunicativeness (#VP5554).

♦ **VD3754 Subversion**
Integrative polarities Evolution–Revolution (#VP5147)
Conversion–Reversion (#VP5145)
Restoration–Destruction (#VP5693).

♦ **VD3756 Suddenness**
Integrative polarities Action–Inaction (#VP5705)   Resolution–Irresolution (#VP5624).

♦ **VD3758 Sufferance**
Integrative polarities Sensation–Insensibility (#VP5422)
Pleasantness–Unpleasantness (#VP5863).

♦ **VD3760 Suffocation**
Integrative polarities Life–Death (#VP5407)   Restoration–Destruction (#VP5693).

♦ **VD3762 Sumptuousness**
Integrative polarities Naturalness–Affectation (#VP5900).

♦ **VD3764 Superciliousness**
Integrative polarities Modesty–Vanity (#VP5908)
Respect–Disrespect (#VP5964).

♦ **VD3766 Superficiality**
Integrative polarities Amusement–Boredom (#VP5878)
Knowledge–Ignorance (#VP5475)
Attention–Inattention (#VP5530)
Importance–Unimportance (#VP5672)
Appearance–Disappearance (#VP5446)
Intelligence–Unintelligence (#VP5467).

DESTRUCTIVE VALUES

♦ **VD3768 Superstition**
Integrative polarities Belief–Unbelief (#VP5501).

♦ **VD3770 Suppression**
Integrative polarities Possession–Loss (#VP5808)
Consent–Refusal (#VP5775)
Freedom–Restraint (#VP5762)
Facility–Difficulty (#VP5732)
Restoration–Destruction (#VP5693).

♦ **VD3772 Surfeit**
Integrative polarities Oversufficiency–Insufficiency (#VP5660).

♦ **VD3774 Surliness**
Integrative polarities Congratulation–Envy (#VP5948)
Courtesy–Discourtesy (#VP5936).

♦ **VD3776 Surplus**
Integrative polarities Oversufficiency–Insufficiency (#VP5660).

♦ **VD3778 Surreptitiousness**
Integrative polarities Communicativeness–Uncommunicativeness (#VP5554).

♦ **VD3780 Susceptibility**
Integrative polarities Safety–Danger (#VP5697)   Influence–Influencelessness (#VP5172).

♦ **VD3782 Suspense**
Integrative polarities Courage–Fear (#VP5890)   Motion–Quiescence (#VP5267)
Certainty–Uncertainty (#VP5513).

♦ **VD3784 Suspension**
Integrative polarities Action–Inaction (#VP5705)   Inclusion–Exclusion (#VP5076)
Timeliness–Untimeliness (#VP5129)   Appropriateness–Inappropriateness (#VP5665).

♦ **VD3786 Suspicion**
Integrative polarities Belief–Unbelief (#VP5501).

♦ **VD3788 Tactlessness**
Integrative polarities Carefulness–Neglect (#VP5533)
Courtesy–Discourtesy (#VP5936)
Discrimination–Indiscrimination (#VP5492).

♦ **VD3790 Tardiness**
Integrative polarities Timeliness–Untimeliness (#VP5129).

♦ **VD3792 Tastelessness**
Integrative polarities Taste–Vulgarity (#VP5896)   Amusement–Boredom (#VP5878)
Strength–Weakness (#VP5159).

♦ **VD3794 Tedium**
Integrative polarities Amusement–Boredom (#VP5878)
Perfection–Imperfection (#VP5677)
Uniformity–Nonuniformity (#VP5017)
Pleasantness–Unpleasantness (#VP5863).

♦ **VD3796 Temper**
Integrative polarities Congratulation–Envy (#VP5948).

♦ **VD3798 Tempestuousness**
Integrative polarities Energy–Moderation (#VP5161).

♦ **VD3800 Temptation**
Integrative polarities Motivation–Dissuasion (#VP5648).

♦ **VD3802 Tension**
Integrative polarities Courage–Fear (#VP5890)   Accord–Disaccord (#VP5794)
Friendship–Enmity (#VP5927)   Oversufficiency–Insufficiency (#VP5660).

♦ **VD3804 Tenuousness**
Integrative polarities Bigness–Littleness (#VP5195)
Cohesion–Disintegration (#VP5050).

♦ **VD3806 Terror**
Integrative polarities Courage–Fear (#VP5890)   Energy–Moderation (#VP5161).

♦ **VD3808 Thinness**
Integrative polarities Numerousness–Fewness (#VP5101)
Savouriness–Unsavouriness (#VP5427)
Intelligence–Unintelligence (#VP5467)
Oversufficiency–Insufficiency (#VP5660).

♦ **VD3810 Thoughtlessness**
Integrative polarities Kindness–Unkindness (#VP5938)
Carefulness–Neglect (#VP5533)
Attention–Inattention (#VP5530)
Skilfulness–Unskilfulness (#VP5733)
Intelligence–Unintelligence (#VP5467).

♦ **VD3812 Threat**
Integrative polarities Safety–Danger (#VP5697)   Approval–Disapproval (#VP5968).

♦ **VD3814 Thriftlessness**
Integrative polarities Preparedness–Unpreparedness (#VP5720).

♦ **VD3816 Tightness**
Integrative polarities Inclusion–Exclusion (#VP5076)
Numerousness–Fewness (#VP5101)
Expensiveness–Cheapness (#VP5848).

♦ **VD3818 Timidity**
Integrative polarities Courage–Fear (#VP5890).

VD3866

♦ **VD3820 Tiresomeness**
Integrative polarities Action–Inaction (#VP5705)   Amusement–Boredom (#VP5878)
Pleasantness–Unpleasantness (#VP5863).

♦ **VD3822 Toilsomeness**
Integrative polarities Action–Inaction (#VP5705)   Facility–Difficulty (#VP5732).

♦ **VD3824 Torment**
Integrative polarities Courage–Fear (#VP5890)   Goodness–Badness (#VP5674)
Facility–Difficulty (#VP5732)   Sensation–Insensibility (#VP5422)
Pleasantness–Unpleasantness (#VP5863)   Appropriateness–Inappropriateness (#VP5665).

♦ **VD3826 Torpor**
Integrative polarities Action–Inaction (#VP5705)   Motion–Quiescence (#VP5267)
Feeling–Unfeelinglessness (#VP5855).

♦ **VD3828 Torture**
Integrative polarities Goodness–Badness (#VP5674)
Symmetry–Distortion (#VP5248)
Sensation–Insensibility (#VP5422)
Pleasantness–Unpleasantness (#VP5863).

♦ **VD3830 Tortuousness**
Integrative polarities Sensation–Insensibility (#VP5422)
Pleasantness–Unpleasantness (#VP5863).

♦ **VD3832 Toxicity**
Integrative polarities Goodness–Badness (#VP5674)
Healthfulness–Unhealthfulness (#VP5683).

♦ **VD3834 Traitorousness**
Integrative polarities Probity–Improbity (#VP5974)
Conversion–Reversion (#VP5145)
Obedience–Disobedience (#VP5766)
Resolution–Irresolution (#VP5624).

♦ **VD3836 Transgression**
Integrative polarities Virtue–Vice (#VP5980)   Legality–Illegality (#VP5998)
Obedience–Disobedience (#VP5766).

♦ **VD3838 Transience**
Integrative polarities Durability–Transience (#VP5110)
Stability–Changeableness (#VP5141).

♦ **VD3840 Travesty**
Integrative polarities Respect–Disrespect (#VP5964)
Representation–Misrepresentation (#VP5573)
Communicativeness–Uncommunicativeness (#VP5554).

♦ **VD3842 Treachery**
Integrative polarities Probity–Improbity (#VP5974)
Certainty–Uncertainty (#VP5513)
Communicativeness–Uncommunicativeness (#VP5554).

♦ **VD3844 Treason**
Integrative polarities Probity–Improbity (#VP5974)
Conversion–Reversion (#VP5145)
Resolution–Irresolution (#VP5624).

♦ **VD3846 Tremulousness**
Integrative polarities Courage–Fear (#VP5890).

♦ **VD3848 Trepidation**
Integrative polarities Courage–Fear (#VP5890)   Oscillation–Agitation (#VP5323)
Excitement–Inexcitability (#VP5857).

♦ **VD3850 Trespass**
Integrative polarities Virtue–Vice (#VP5980)   Dueness–Undueness (#VP5960)
Legality–Illegality (#VP5998)   Obedience–Disobedience (#VP5766)
Circumscription–Intrusion (#VP5234).

♦ **VD3852 Tribulation**
Integrative polarities Prosperity–Adversity (#VP5728)
Pleasantness–Unpleasantness (#VP5863).

♦ **VD3854 Trickery**
Integrative polarities Skilfulness–Unskilfulness (#VP5733)
Communicativeness–Uncommunicativeness (#VP5554).

♦ **VD3856 Triviality**
Integrative polarities Importance–Unimportance (#VP5672)
Superiority–Inferiority (#VP5036)
Intelligence–Unintelligence (#VP5467)
Appropriateness–Inappropriateness (#VP5665).

♦ **VD3858 Trouble**
Integrative polarities Courage–Fear (#VP5890)   Dueness–Undueness (#VP5960)
Facility–Difficulty (#VP5732)   Prosperity–Adversity (#VP5728)
Expedience–Inexpedience (#VP5670)   Pleasantness–Unpleasantness (#VP5863).

♦ **VD3860 Truculence**
Integrative polarities Accord–Disaccord (#VP5794)
Kindness–Unkindness (#VP5938)
Courtesy–Discourtesy (#VP5936).

♦ **VD3862 Truthlessness**
Integrative polarities Communicativeness–Uncommunicativeness (#VP5554).

♦ **VD3864 Turbidity**
Integrative polarities Oscillation–Agitation (#VP5323).

♦ **VD3866 Turbulence**
Integrative polarities Order–Disorder (#VP5059)   Energy–Moderation (#VP5161)
Oscillation–Agitation (#VP5323).

VD3868

♦ **VD3868 Turpitude**
  Integrative polarities Virtue–Vice (#VP5980)   Probity–Improbity (#VP5974).

♦ **VD3870 Tyranny**
  Integrative polarities Freedom–Restraint (#VP5762).

♦ **VD3872 Ugliness**
  Integrative polarities Beauty–Ugliness (#VP5899)
  Congratulation–Envy (#VP5948)
  Pleasantness–Unpleasantness (#VP5863).

♦ **VD3874 Unadvisedness**
  Integrative polarities Resolution–Irresolution (#VP5624)
  Intelligence–Unintelligence (#VP5467).

♦ **VD3876 Unauthenticity**
  Integrative polarities Truth–Error (#VP5516)   Intuition–Reason (#VP5481)
  Certainty–Uncertainty (#VP5513)
  Communicativeness–Uncommunicativeness (#VP5554).

♦ **VD3878 Unbearableness**
  Integrative polarities Pleasantness–Unpleasantness (#VP5863).

♦ **VD3880 Uncertainty**
  Integrative polarities Safety–Danger (#VP5697)   Belief–Unbelief (#VP5501)
  Certainty–Uncertainty (#VP5513)   Resolution–Irresolution (#VP5624)
  Visibility–Invisibility (#VP5444)   Stability–Changeableness (#VP5141)
  Intelligibility–Unintelligibility (#VP5548).

♦ **VD3882 Unchastity**
  Integrative polarities Love–Hate (#VP5930)   Virtue–Vice (#VP5980)
  Chastity–Unchastity (#VP5987).

♦ **VD3884 Uncleanliness**
  Integrative polarities Virtue–Vice (#VP5980)   Goodness–Badness (#VP5674)
  Chastity–Unchastity (#VP5987)   Cleanness–Uncleanness (#VP5681).

♦ **VD3886 Uncomfortableness**
  Integrative polarities Pleasantness–Unpleasantness (#VP5863).

♦ **VD3888 Uncommunicativeness**
  Integrative polarities Caution–Rashness (#VP5894)
  Sociability–Unsociability (#VP5922)
  Communicativeness–Uncommunicativeness (#VP5554).

♦ **VD3890 Unconscionableness**
  Integrative polarities Justice–Injustice (#VP5976)
  Probity–Improbity (#VP5974)
  Energy–Moderation (#VP5161)
  Expensiveness–Cheapness (#VP5848)
  Oversufficiency–Insufficiency (#VP5660).

♦ **VD3892 Unctuousness**
  Integrative polarities Piety–Impiety (#VP6028)   Approval–Disapproval (#VP5968)
  Communicativeness–Uncommunicativeness (#VP5554).

♦ **VD3894 Undependability**
  Integrative polarities Safety–Danger (#VP5697)   Probity–Improbity (#VP5974)
  Certainty–Uncertainty (#VP5513)   Resolution–Irresolution (#VP5624)
  Stability–Changeableness (#VP5141).

♦ **VD3896 Underdevelopment**
  Integrative polarities Preparedness–Unpreparedness (#VP5720)
  Completeness–Incompleteness (#VP5054).

♦ **VD3898 Underestimation**
  Integrative polarities Judgement–Misjudgement (#VP5494).

♦ **VD3900 Undesirableness**
  Integrative polarities Expedience–Inexpedience (#VP5670)
  Hospitality–Inhospitality (#VP5925)
  Contentment–Discontentment (#VP5868)
  Pleasantness–Unpleasantness (#VP5863).

♦ **VD3902 Uneasiness**
  Integrative polarities Courage–Fear (#VP5890)   Patience–Impatience (#VP5861)
  Excitement–Inexcitability (#VP5857)   Contentment–Discontentment (#VP5868)
  Pleasantness–Unpleasantness (#VP5863).

♦ **VD3904 Unevenness**
  Integrative polarities Equality–Inequality (#VP5030)
  Perfection–Imperfection (#VP5677).

♦ **VD3906 Unfaithfulness**
  Integrative polarities Probity–Improbity (#VP5974).

♦ **VD3908 Unfamiliarity**
  Integrative polarities Knowledge–Ignorance (#VP5475)
  Skilfulness–Unskilfulness (#VP5733).

♦ **VD3910 Unfriendliness**
  Integrative polarities Accord–Disaccord (#VP5794)
  Friendship–Enmity (#VP5927)
  Courtesy–Discourtesy (#VP5936)
  Sociability–Unsociability (#VP5922).

♦ **VD3912 Ungodliness**
  Integrative polarities Virtue–Vice (#VP5980).

♦ **VD3914 Ungraciousness**
  Integrative polarities Kindness–Unkindness (#VP5938)
  Courtesy–Discourtesy (#VP5936)
  Hospitality–Inhospitality (#VP5925).

♦ **VD3916 Unhappiness**
  Integrative polarities Approval–Disapproval (#VP5968)
  Cheerfulness–Solemnity (#VP5870)
  Contentment–Discontentment (#VP5868)
  Pleasantness–Unpleasantness (#VP5863).

♦ **VD3918 Unimaginativeness**
  Integrative polarities Amusement–Boredom (#VP5878)
  Imaginativeness–Unimaginativeness (#VP5535).

♦ **VD3920 Unimportance**
  Integrative polarities Pride–Humility (#VP5905)   Greatness–Smallness (#VP5034)
  Importance–Unimportance (#VP5672).

♦ **VD3922 Unintelligence**
  Integrative polarities Knowledge–Ignorance (#VP5475)
  Thought–Thoughtlessness (#VP5478)
  Skilfulness–Unskilfulness (#VP5733)
  Intelligence–Unintelligence (#VP5467).

♦ **VD3924 Unkindness**
  Integrative polarities Kindness–Unkindness (#VP5938).

♦ **VD3926 Unlawfulness**
  Integrative polarities Justice–Injustice (#VP5976)
  Legality–Illegality (#VP5998)
  Rightness–Wrongness (#VP5957).

♦ **VD3928 Unmanageability**
  Integrative polarities Facility–Difficulty (#VP5732)
  Resolution–Irresolution (#VP5624)
  Skilfulness–Unskilfulness (#VP5733).

♦ **VD3930 Unnaturalness**
  Integrative polarities Sanity–Insanity (#VP5472)   Kindness–Unkindness (#VP5938)
  Elegance–Inelegance (#VP5589)   Naturalness–Affectation (#VP5900)
  Conformity–Nonconformity (#VP5082)
  Communicativeness–Uncommunicativeness (#VP5554).

♦ **VD3932 Unneighbourliness**
  Integrative polarities Hospitality–Inhospitality (#VP5925).

♦ **VD3934 Unpeacefulness**
  Integrative polarities Accord–Disaccord (#VP5794)
  Oscillation–Agitation (#VP5323).

♦ **VD3936 Unpleasantness**
  Integrative polarities Accord–Disaccord (#VP5794)
  Goodness–Badness (#VP5674)
  Pleasantness–Unpleasantness (#VP5863).

♦ **VD3938 Unpredictability**
  Integrative polarities Safety–Danger (#VP5697)   Certainty–Uncertainty (#VP5513)
  Resolution–Irresolution (#VP5624)   Stability–Changeableness (#VP5141).

♦ **VD3940 Unpreparedness**
  Integrative polarities Carefulness–Neglect (#VP5533)
  Timeliness–Untimeliness (#VP5129)
  Preparedness–Unpreparedness (#VP5720).

♦ **VD3942 Unproductiveness**
  Integrative polarities Productiveness–Unproductiveness (#VP5165)
  Appropriateness–Inappropriateness (#VP5665).

♦ **VD3944 Unreality**
  Integrative polarities Truth–Error (#VP5516)   Existence–Nonexistence (#VP5001)
  Substantiality–Unsubstantiality (#VP5003)   Imaginativeness–Unimaginativeness (#VP5535).

♦ **VD3946 Unreasonableness**
  Integrative polarities Intuition–Reason (#VP5481)
  Expensiveness–Cheapness (#VP5848)
  Intelligence–Unintelligence (#VP5467)
  Oversufficiency–Insufficiency (#VP5660).

♦ **VD3948 Unreliability**
  Integrative polarities Safety–Danger (#VP5697)   Belief–Unbelief (#VP5501)
  Probity–Improbity (#VP5974)   Certainty–Uncertainty (#VP5513)
  Resolution–Irresolution (#VP5624)   Stability–Changeableness (#VP5141).

♦ **VD3950 Unruliness**
  Integrative polarities Freedom–Restraint (#VP5762)
  Energy–Moderation (#VP5161)
  Authority–Lawlessness (#VP5739)
  Obedience–Disobedience (#VP5766)
  Resolution–Irresolution (#VP5624).

♦ **VD3952 Unsophistication**
  Integrative polarities Belief–Unbelief (#VP5501)   Skilfulness–Unskilfulness (#VP5733).

♦ **VD3954 Unsuitability**
  Integrative polarities Taste–Vulgarity (#VP5896)   Rightness–Wrongness (#VP5957)
  Agreement–Disagreement (#VP5026)   Expedience–Inexpedience (#VP5670)
  Timeliness–Untimeliness (#VP5129)   Contentment–Discontentment (#VP5868)
  Preparedness–Unpreparedness (#VP5720)   Oversufficiency–Insufficiency (#VP5660)
  Appropriateness–Inappropriateness (#VP5665).

♦ **VD3956 Unthoughtfulness**
  Integrative polarities Kindness–Unkindness (#VP5938)
  Resolution–Irresolution (#VP5624)
  Intelligence–Unintelligence (#VP5467).

♦ **VD3958 Untrustworthiness**
  Integrative polarities Safety–Danger (#VP5697)   Probity–Improbity (#VP5974)
  Certainty–Uncertainty (#VP5513).

DESTRUCTIVE VALUES VD4046

♦ **VD3960 Unwholesomeness**
  Integrative polarities Health–Disease (#VP5685)
  Healthfulness–Unhealthfulness (#VP5683).

♦ **VD3962 Upheaval**
  Integrative polarities Energy–Moderation (#VP5161)
  Restoration–Destruction (#VP5693)
  Excitement–Inexcitability (#VP5857).

♦ **VD3964 Uproar**
  Integrative polarities Order–Disorder (#VP5059)   Energy–Moderation (#VP5161).

♦ **VD3966 Upset**
  Integrative polarities Courage–Fear (#VP5890)   Order–Disorder (#VP5059)
  Energy–Moderation (#VP5161)   Facility–Difficulty (#VP5732)
  Evolution–Revolution (#VP5147)   Attention–Inattention (#VP5530)
  Certainty–Uncertainty (#VP5513)   Oscillation–Agitation (#VP5323)
  Restoration–Destruction (#VP5693).

♦ **VD3968 Urgency**
  Integrative polarities Choice–Necessity (#VP5637).

♦ **VD3970 Uselessness**
  Integrative polarities Power–Impotence (#VP5157)
  Expedience–Inexpedience (#VP5670)
  Accomplishment–Nonaccomplishment (#VP5722)
  Appropriateness–Inappropriateness (#VP5665).

♦ **VD3972 Usurpation**
  Integrative polarities Dueness–Undueness (#VP5960)
  Sharing–Appropriation (#VP5815).

♦ **VD3974 Vacillation**
  Integrative polarities Certainty–Uncertainty (#VP5513)
  Resolution–Irresolution (#VP5624)
  Stability–Changeableness (#VP5141).

♦ **VD3976 Vagrancy**
  Integrative polarities Action–Inaction (#VP5705).

♦ **VD3978 Vagueness**
  Integrative polarities Form–Formlessness (#VP5246)
  Certainty–Uncertainty (#VP5513)
  Visibility–Invisibility (#VP5444)
  Substantiality–Unsubstantiality (#VP5003)
  Intelligibility–Unintelligibility (#VP5548).

♦ **VD3980 Vainness**
  Integrative polarities Modesty–Vanity (#VP5908)
  Power–Impotence (#VP5157)
  Intuition–Reason (#VP5481)
  Hope–Hopelessness (#VP5888)
  Importance–Unimportance (#VP5672)
  Appropriateness–Inappropriateness (#VP5665).

♦ **VD3982 Valueless**
  Integrative polarities Importance–Unimportance (#VP5672)
  Appropriateness–Inappropriateness (#VP5665).

♦ **VD3984 Vandalism**
  Integrative polarities Energy–Moderation (#VP5161)
  Kindness–Unkindness (#VP5938)
  Behaviour–Misbehaviour (#VP5737)
  Restoration–Destruction (#VP5693).

♦ **VD3986 Vanity**
  Integrative polarities Modesty–Vanity (#VP5908)
  Pride–Humility (#VP5905)
  Importance–Unimportance (#VP5672)
  Appropriateness–Inappropriateness (#VP5665).

♦ **VD3988 Vanquishment**
  Integrative polarities Victory–Defeat (#VP5726).

♦ **VD3990 Vapidity**
  Integrative polarities Amusement–Boredom (#VP5878)
  Strength–Weakness (#VP5159)
  Perfection–Imperfection (#VP5677)
  Importance–Unimportance (#VP5672)
  Savouriness–Unsavouriness (#VP5427)
  Intelligence–Unintelligence (#VP5467).

♦ **VD3992 Variance**
  Integrative polarities Assent–Dissent (#VP5521)
  Accord–Discord (#VP5794)
  Agreement–Disagreement (#VP5026).

♦ **VD3994 Variation**
  Integrative polarities Direction–Deviation (#VP5290)
  Identity–Difference (#VP5013)
  Stability–Changeableness (#VP5141)
  Uniformity–Nonuniformity (#VP5017).

♦ **VD3996 Vehemence**
  Integrative polarities Energy–Moderation (#VP5161)
  Congratulation–Envy (#VP5948).

♦ **VD3998 Venality**
  Integrative polarities Probity–Improbity (#VP5974).

♦ **VD4000 Venial**
  Integrative polarities Vindication–Condemnation (#VP6005).

♦ **VD4002 Venom**
  Integrative polarities Goodness–Badness (#VP5674)
  Friendship–Enmity (#VP5927)
  Energy–Moderation (#VP5161)
  Kindness–Unkindness (#VP5938)
  Healthfulness–Unhealthfulness (#VP5683).

♦ **VD4004 Verbosity**
  Integrative polarities Oversufficiency–Insufficiency (#VP5660).

♦ **VD4006 Vexation**
  Integrative polarities Courage–Fear (#VP5890)   Goodness–Badness (#VP5674)
  Congratulation–Envy (#VP5948)   Pleasantness–Unpleasantness (#VP5863).

♦ **VD4008 Viciousness**
  Integrative polarities Virtue–Vice (#VP5980)   Goodness–Badness (#VP5674)
  Energy–Moderation (#VP5161)   Kindness–Unkindness (#VP5938).

♦ **VD4010 Vicissitude**
  Integrative polarities Prosperity–Adversity (#VP5728)
  Stability–Changeableness (#VP5141).

♦ **VD4012 Victimization**
  Integrative polarities Appropriateness–Inappropriateness (#VP5665)
  Communicativeness–Uncommunicativeness (#VP5554).

♦ **VD4014 Vileness**
  Integrative polarities Virtue–Vice (#VP5980)   Taste–Vulgarity (#VP5896)
  Repute–Disrepute (#VP5914)   Goodness–Badness (#VP5674)
  Probity–Improbity (#VP5974)   Chastity–Unchastity (#VP5987)
  Approval–Disapproval (#VP5968)   Cleanness–Uncleanness (#VP5681)
  Importance–Unimportance (#VP5672)   Pleasantness–Unpleasantness (#VP5863).

♦ **VD4016 Villainy**
  Integrative polarities Virtue–Vice (#VP5980)   Probity–Improbity (#VP5974).

♦ **VD4018 Vindictiveness**
  Integrative polarities Forgiveness–Vengeance (#VP5947).

♦ **VD4020 Violation**
  Integrative polarities Energy–Moderation (#VP5161)
  Legality–Illegality (#VP5998)
  Rightness–Wrongness (#VP5957)
  Obedience–Disobedience (#VP5766)
  Appropriateness–Inappropriateness (#VP5665).

♦ **VD4022 Violence**
  Integrative polarities Energy–Moderation (#VP5161)
  Congratulation–Envy (#VP5948)
  Kindness–Unkindness (#VP5938)
  Leniency–Compulsion (#VP5756)
  Appropriateness–Inappropriateness (#VP5665).

♦ **VD4024 Virulence**
  Integrative polarities Life–Death (#VP5407)   Goodness–Badness (#VP5674)
  Friendship–Enmity (#VP5927)   Energy–Moderation (#VP5161)
  Congratulation–Envy (#VP5948)   Kindness–Unkindness (#VP5938)
  Healthfulness–Unhealthfulness (#VP5683).

♦ **VD4026 Vitiation**
  Integrative polarities Virtue–Vice (#VP5980)   Improvement–Impairment (#VP5691).

♦ **VD4028 Vituperation**
  Integrative polarities Approval–Disapproval (#VP5968).

♦ **VD4030 Volatility**
  Integrative polarities Durability–Transience (#VP5110)
  Resolution–Irresolution (#VP5624)
  Intelligence–Unintelligence (#VP5467).

♦ **VD4032 Voracity**
  Integrative polarities Desire–Avoidance (#VP5631)
  Temperance–Intemperance (#VP5992).

♦ **VD4034 Vulgarity**
  Integrative polarities Taste–Vulgarity (#VP5896)   Chastity–Unchastity (#VP5987)
  Courtesy–Discourtesy (#VP5936)   Superiority–Inferiority (#VP5036).

♦ **VD4036 Vulnerability**
  Integrative polarities Elasticity–Toughness (#VP5358)
  Preparedness–Unpreparedness (#VP5720).

♦ **VD4038 Wantonness**
  Integrative polarities Virtue–Vice (#VP5980)   Chastity–Unchastity (#VP5987)
  Resolution–Irresolution (#VP5624).

♦ **VD4040 War**
  Integrative polarities Accord–Discord (#VP5794).

♦ **VD4042 Warlike**
  Integrative polarities Accord–Discord (#VP5794).

♦ **VD4044 Warpedness**
  Integrative polarities Virtue–Vice (#VP5980)   Justice–Injustice (#VP5976)
  Symmetry–Distortion (#VP5248)   Perfection–Imperfection (#VP5677)
  Broadmindedness–Narrowmindedness (#VP5526)
  Communicativeness–Uncommunicativeness (#VP5554).

♦ **VD4046 Waste**
  Integrative polarities Health–Disease (#VP5685)
  Possession–Loss (#VP5808)
  Improvement–Impairment (#VP5691)

VD4046

Expensiveness–Cheapness (#VP5848)
Restoration–Destruction (#VP5693)
Appropriateness–Inappropriateness (#VP5665).

♦ **VD4048 Wastefulness**
Integrative polarities Expensiveness–Cheapness (#VP5848)
Restoration–Destruction (#VP5693).

♦ **VD4050 Weak–mindedness**
Integrative polarities Resolution–Irresolution (#VP5624)
Intelligence–Unintelligence (#VP5467).

♦ **VD4052 Weakness**
Integrative polarities Virtue–Vice (#VP5980)   Courage–Fear (#VP5890)
Safety–Danger (#VP5697)   Action–Inaction (#VP5705)
Belief–Unbelief (#VP5501)   Power–Impotence (#VP5157)
Strength–Weakness (#VP5159)   Perfection–Imperfection (#VP5677)
Resolution–Irresolution (#VP5624)   Savouriness–Unsavouriness (#VP5427)
Intelligence–Unintelligence (#VP5467)   Influence–Influencelessness (#VP5172).

♦ **VD4054 Wear**
Integrative polarities Possession–Loss (#VP5808)
Amusement–Boredom (#VP5878)
Improvement–Impairment (#VP5691)
Cohesion–Disintegration (#VP5050).

♦ **VD4056 Weariness**
Integrative polarities Action–Inaction (#VP5705)   Amusement–Boredom (#VP5878)
Strength–Weakness (#VP5159).

♦ **VD4058 Wearisomeness**
Integrative polarities Action–Inaction (#VP5705)   Amusement–Boredom (#VP5878)
Cheerfulness–Solemnity (#VP5870)   Pleasantness–Unpleasantness (#VP5863).

♦ **VD4060 Willfulness**
Integrative polarities Authority–Lawlessness (#VP5739)
Obedience–Disobedience (#VP5766)
Resolution–Irresolution (#VP5624).

♦ **VD4062 Withdrawal**
Integrative polarities Unity–Duality (#VP5089)   Assent–Dissent (#VP5521)
Desire–Avoidance (#VP5631)   Inclusion–Exclusion (#VP5076)
Curiosity–Incuriosity (#VP5528)   Progression–Regression (#VP5294)
Resolution–Irresolution (#VP5624)   Feeling–Unfeelinglessness (#VP5855).

♦ **VD4064 Withering**
Integrative polarities Respect–Disrespect (#VP5964)

Kindness–Unkindness (#VP5938)
Improvement–Impairment (#VP5691)
Restoration–Destruction (#VP5693).

♦ **VD4066 Witlessness**
Integrative polarities Sanity–Insanity (#VP5472)   Knowledge–Ignorance (#VP5475)
Attention–Inattention (#VP5530)   Intelligence–Unintelligence (#VP5467).

♦ **VD4068 Woe**
Integrative polarities Goodness–Badness (#VP5674)
Cheerfulness–Solemnity (#VP5870)
Pleasantness–Unpleasantness (#VP5863).

♦ **VD4070 Worry**
Integrative polarities Courage–Fear (#VP5890)   Facility–Difficulty (#VP5732)
Pleasantness–Unpleasantness (#VP5863).

♦ **VD4072 Worthlessness**
Integrative polarities Goodness–Badness (#VP5674)
Importance–Unimportance (#VP5672)
Expedience–Inexpedience (#VP5670)
Appropriateness–Inappropriateness (#VP5665).

♦ **VD4074 Wound**
Integrative polarities Health–Disease (#VP5685)
Goodness–Badness (#VP5674)
Congratulation–Envy (#VP5948)
Improvement–Impairment (#VP5691)
Sensation–Insensibility (#VP5422)
Pleasantness–Unpleasantness (#VP5863).

♦ **VD4076 Wrangle**
Integrative polarities Accord–Disaccord (#VP5794)
Intuition–Reason (#VP5481).

♦ **VD4078 Wrongness**
Integrative polarities Virtue–Vice (#VP5980)   Truth–Error (#VP5516)
Goodness–Badness (#VP5674)   Justice–Injustice (#VP5976)
Legality–Illegality (#VP5998)   Rightness–Wrongness (#VP5957)
Timeliness–Untimeliness (#VP5129).

♦ **VD4080 Xenophobia**
Integrative polarities Love–Hate (#VP5930)   Courage–Fear (#VP5890)
Inclusion–Exclusion (#VP5076)
Broadmindedness–Narrowmindedness (#VP5526).

♦ **VD4082 Zealotry**
Integrative polarities Sanity–Insanity (#VP5472)   Desire–Avoidance (#VP5631).

Value polarities   VP

Rationale

The words used to indicate values tend to be those most subject to misinterpretation and misunderstanding. Indeed, part of their strength in evoking an integrative response from people lies in the way that people can project their different aspirations onto the same word, as can be seen in the case of "peace". Such words therefore lend themselves to the production of lengthy treatises clarifying the ways each may be interpreted, or ought to be.

This approach has not resulted in any well-defined map of the value ecology that is presumably a governing factor in social dynamics. It is therefore appropriate to explore alternative approaches which reflect the multiple interpretations, sharpen the meanings, but avoid the trap of lengthy explication of nuances on which there is little consensus. The objective is to relate the resulting network of values to the extensive networks of problems, strategies and organizations at a much geater level of detail than has hitherto been possible.

Contents

This section lists 225 value polarities identified from *Roget's Thesaurus*. They are in a numerical sequence adapted from the order in which they appear in Roget. The entries cross-reference synonymous "constructive" or "destructive" value words in Section VC and VD respectively.

Rather than attempt the possibly sterile exercise of producing descriptive texts on each value polarity, the entries include selected proverbs, aphorisms and quotations. This has the merit of highlighting the significance of each value dimension in a succinct and pithy manner linking it both to cultural lore and to the insights of those to whom wisdom is attributed in contemporary society. This material has been deliberately selected to highlight the dynamic relationship between values guiding constructive and destructive action.

Method

The method is described in detail in Section VZ.

Index

A keyword index to entries is provided in Section VX. The keywords are also incorporated into the index for Volume 2 (Section X)

Bibliography

Bibliographical references, by author, are given in Section VY.

Comment

Detailed comments on the approach used in identifying value words, grouping them into polar categories and further organizing the result are presented in Section VZ at the end of this volume. It is interesting that insightful proverbs and quotations tend to indicate the limitation of the obvious, first-order meanings of value words.

Constructive values become destructive if pursued obsessively; destructive values become essential in clearing the way for developmental breakthroughs. The inclusion of proverbs and quotations has the practical merit to of grounding understanding of value dimensions. It also has the merit of linking the volume to the world of literature with all that it represents for the clarification and reinforcement of values.

Since people identify more readily with fictional representations than with global problems (or perhaps via fictional representations with global issues), such a link merits further exploration by the social sciences and those concerned with mobilizing resources in response to the global problematique.

Reservations

As an exploratory exercise the results must necessarily be considered as preliminary and limited by dependence on the Roget framework and the English language. Selecting a word as having primarily "constructive" or "destructive" value connotations raises many questions. Some words selected as "destructive" may under certain circumstances appear "constructive", and vice versa. Sections VC and VD may therefore be seen as attempts to identify first-order responses to value words. The limitations of this approach are partially resolved by grouping such words within the value polarity categories of this section.

Possible future improvements

In addition to refining the existing entries and introducing other polarities, explicit cross-references could be included to the world problems and strategies sections, and possibly also to international organizations promoting particular values.

ENTRY CONTENT AND ORGANIZATION

Ordering of entries
Entries are in **numeric order**. Entry numbers are based on an adaptation of the numeric ordering of concepts in Roget's Theasurus. They serve primarily as a permanent point of reference to facilitate indexing, cross-referencing, and updating between editions.

Index access to entries
The location of an entry in this sub-section may be determined from:
- the **Volume Index** (Section X) on the basis of keywords in the name of the entry or its alternate names
- the **Section Index** (see Section VX) on the basis of keywords in the name of the entry or its alternate names
- Section VT functions as a classified index to the entries in this sub-section

Structure of entries
Entries may be composed of the following descriptive elements:

(a) **Entry number** The numeric portion has been designed to correspond approximately to the code number for the equivalent concepts in the edition of Roget's Thesaurus used to identify polarities. This correspondance may however be ignored for some purposes. The number may then be considered a convenient method of identifying the entry (particularly for indexing purposes), of filing information on it, and as an identifier to which cross-references from other entries (possibly in other sections) may refer in this and future editions. The first letter of the entry number refers to the section of this volume in which the sub-section, denoted by the second letter, is located.

(b) **Qualifier** following the entry number. A single alphabetic character ("a", "b" or "c") may follow the entry number. When regrouping the polarity entry with others within an entry of Section VT, this letter indicates possible clusters of related polarities.

(c) **Entry name** The name of the polarity is printed in bold characters. The two words used are only indicative of the nature of the value polarity; they are not intended to completely define its nature. The polarity name may be followed by alternative words indicating aspects of the value polarity.

(d) **Dynamics** Quotations and aphorisms may be included to indicate the dilemma associated with the value polarity, especially negative aspects of constructive values and positive aspects of destructive values

Cross-referencing of entries
At the end of any entry, there may be cross-references to other entries. These indicate the number and name the cross- referenced entry, whether within this Section or in other Sections. In this sub-section there are 2 types of cross-references, indicated by a 2-letter code in bold characters:
    Integrative complex = Broader value complex
    Constituent values = Narrower or more specific values

♦ VP5001a Existence–Nonexistence

Dynamics There is no reality except the one contained within us. That is why so many people live such an unreal life. They take the images outside them for reality and never allow the world within to assert itself. (Hermann Hesse)
Man is the only animal for whom his own existence is a problem which he has to solve. (Erich Fromm)
We are prudent people. We are afraid to let go of our petty reality in order to grasp at a great shadow. (Saint-Exupéry)
The basic fact about human existence is not that it is a tragedy, but that it is a bore. It is not so much a war as an endless standing in line. (H L Mencken)
He who confronts the paradoxical exposes himself to reality. (Friedrich Durrenmatt)

Integrative complex Existence (#VT8011).

Constituent values Truth (#VC1787)	Being (#VC0225)
Reality (#VC1419)	Essence (#VC0681)
Presence (#VC1345)	Existence (#VC0705)
Actuality (#VC0043)	Genuineness (#VC0831)
Positiveness (#VC1323)	Authenticity (#VC0203)
Veritableness (#VC1871)	Substantiality (#VC1693).
Constituent values Nullity (#VD3342)	Unreality (#VD3944)
Nothingness (#VD3336)	Deprivation (#VD2398)
Annihilation (#VD2074).	

♦ VP5003b Substantiality–Unsubstantiality

Dynamics Beware lest you lose the substance by grasping at the shadow. (Aesop's Fables)

Integrative complex Existence (#VT8011).

Constituent values Life (#VC1075)	Strength (#VC1681)
Well-made (#VC1911)	Toughness (#VC1767)
Stability (#VC1651)	Hardiness (#VC0883)
Sturdiness (#VC1685)	Durability (#VC0589)
Tangibility (#VC1743)	Well-grounded (#VC1907)
Substantiality (#VC1693).	
Constituent values Vagueness (#VD3978)	Unreality (#VD3944).

♦ VP5005c Intrinsicality–Extrinsicality

Dynamics The man who acts the least, upbraids the most. (Homer)

Integrative complex Existence (#VT8011).

Constituent values Quality (#VC1397)	Privacy (#VC1357)
Essence (#VC0681)	Principle (#VC1353)
Quintessence (#VC1407).	

♦ VP5009a Relatedness–Unrelatedness

Dynamics Almost all of our relationships begin and most of them continue as forms of mutual exploitation, a mental or physical barter, to be terminated when one or both parties run out of goods. (W H Auden)
Man is a knot, a web, a mesh into which relationships are tied. Only those relationships matter. (Saint-Exupéry)

Integrative complex Relationship (#VT8012).

Constituent values Union (#VC1825)	Concern (#VC0407)
Sympathy (#VC1731)	Relation (#VC1453)
Intimacy (#VC1013)	Interest (#VC1011)
Alliance (#VC0105)	Relevance (#VC1457)
Proximity (#VC1385)	Similarity (#VC1615)
Connection (#VC0433).	
Constituent values Misalliance (#VD3230)	Separateness (#VD3654)
Misapplication (#VD3232).	

♦ VP5013a Identity–Difference

Dynamics We must resemble each other a little in order to understand each other, but we must be a little different to love each other. (Paul Géraldy)
Anybody who is any good is different from anybody else. (Felix Frankfurter)
The indwelling ideal lends all the gods their divinity. (George Santayana)
There is as much difference between us and ourselves as between us and others. (Montaigne)

Integrative complex Relationship (#VT8012).

Constituent values Symmetry (#VC1729)	Subtlety (#VC1695)
Equality (#VC0667)	Synthesis (#VC1735)
Mutuality (#VC1175)	Uniqueness (#VC1827)
Equivalence (#VC0675)	Equilibrium (#VC0671)
Distinction (#VC0577)	Cooperation (#VC0481)
Concurrence (#VC0413)	Alternation (#VC0111)
Discrimination (#VC0573)	Differentiation (#VC0555)
Cooperativeness (#VC0483).	
Constituent values Conflict (#VD2270)	Variation (#VD3994)
Hostility (#VD2802)	Disparity (#VD2534)
Disaccord (#VD2448)	Antipathy (#VD2084)
Ambiguity (#VD2054)	Repugnance (#VD3538)
Perversity (#VD3446)	Opposition (#VD3364)
Dissonance (#VD2558)	Antagonism (#VD2080)
Ambivalence (#VD2056)	Equivocation (#VD2616)
Inconsistency (#VD2932)	Contradiction (#VD2284)
Self-contradiciton (#VD3632).	

♦ VP5017a Uniformity–Nonuniformity

Dynamics Every central government worships uniformity: uniformity relieves it from inquiry into an infinity of details, which must be attended to if rules have to be adapted to different men, instead of indiscriminately subjecting all men to the same rule. (Alexis de Tocqueville)

Integrative complex Relationship (#VT8012).

Constituent values Variety (#VC1859)	Diversity (#VC0579)
Uniformity (#VC1821)	Regularity (#VC1445)
Differentiation (#VC0555).	
Constituent values Tedium (#VD3794)	Boredom (#VD2192)
Variation (#VD3994)	Deviation (#VD2428)
Mutability (#VD3296)	Divergence (#VD2576)
Nonuniformity (#VD3334)	Changeableness (#VD2222)
Capriciousness (#VD2212).	

♦ VP5020a Similarity–Dissimilarity

Dynamics There never were, in the world, two opinions alike, no more than two hairs, or two grains; the most universal quality is diversity. (Montaigne)
Men are created different; they lose their social freedom and their individual autonomy in seeking to become like each other. (David Riesman)
It is by disease that health is pleasant; by evil that good is pleasant; by hunger, satiety; by weariness, rest. (Heraclitus)

Integrative complex Relationship (#VT8012).

Constituent values Diversity (#VC0579)	Community (#VC0373)
Similarity (#VC1615)	Correspondence (#VC0491).

♦ VP5023b Originality–Imitation

Dynamics The merit of originality is not novelty; it is sincerity. The believing man is the original man; whatsoever he believes, he believes it for himself, not for another. (Thomas Carlyle)
All good things which exist are the fruits of originality. (John Stuart Mill)
When people are free to do as they please, they usually imitate each other. Originality is deliberate and forced, and partakes of the nature of a protest. (Eric Hoffer)
He is great who is what he is from nature and who never reminds us of others. (Emerson)

Integrative complex Relationship (#VT8012).

Constituent values Novelty (#VC1205)	Newness (#VC1189)
Heroism (#VC0897)	Vastness (#VC1861)
Creation (#VC0499)	Freshness (#VC0799)
Uniqueness (#VC1827)	Precedence (#VC1339)
Creativity (#VC0501)	Originality (#VC1239)
Authenticity (#VC0203)	Inventiveness (#VC1019).

♦ VP5026b Agreement–Disagreement

Dynamics We cannot remain consistent with the world save by growing inconsistent with our past selves. (Havelock Ellis)
It is by universal misunderstanding that all agree. For if, by ill luck, people understood each other, they would never agree. (Charles Baudelaire)
We are more inclined to hate one another for points on which we differ, than to love one another for points on which we agree. (Charles Caleb Colton)
Agreement is made more precious by disagreement. (Publilius Syrus)
Conformity is the ape of harmony. (Emerson)
There is nothing more likely to start disagreement among people or countries than an agreement. (E B White)
It were not best that we should all think alike; it is difference of opinion that makes horse races. (Mark Twain)

Integrative complex Relationship (#VT8012).

Constituent values Union (#VC1825)	Peace (#VC1265)
Accord (#VC0025)	Harmony (#VC0887)
Fitness (#VC0767)	Entente (#VC0657)
Concord (#VC0411)	Symmetry (#VC1729)
Felicity (#VC0753)	Aptitude (#VC0163)
Affinity (#VC0079)	Unanimity (#VC1799)
Relevance (#VC1457)	Congruity (#VC0429)
Coherence (#VC0351)	Agreement (#VC0093)
Uniformity (#VC1821)	Regulation (#VC1447)
Consonance (#VC0453)	Conformity (#VC0425)
Adjustment (#VC0055)	Suitability (#VC1701)
Equivalence (#VC0675)	Cooperation (#VC0481)
Consistency (#VC0451)	Positiveness (#VC1323)
Coordination (#VC0485)	Congeniality (#VC0427)
Conformation (#VC0423)	Adaptability (#VC0049)
Qualification (#VC1395)	Compatibility (#VC0379)
Agreeableness (#VC0091)	Reconciliation (#VC1433)
Correspondence (#VC0491)	Cooperativeness (#VC0483).
Constituent values Heresy (#VD2790)	Discord (#VD2480)
Variance (#VD3992)	Negation (#VD3306)
Disunity (#VD2572)	Conflict (#VD2270)
Disparity (#VD2534)	Disaccord (#VD2448)
Asymmetry (#VD2110)	Ambiguity (#VD2054)
Repugnance (#VD3538)	Opposition (#VD3364)
Negativity (#VD3308)	Infelicity (#VD3000)
Inequality (#VD2986)	Inaptitude (#VD2902)
Divergence (#VD2576)	Dissonance (#VD2558)
Dissidence (#VD2550)	Dissension (#VD2548)
Disharmony (#VD2504)	Antagonism (#VD2080)
Misalliance (#VD3230)	Irrelevance (#VD3106)
Incongruity (#VD2926)	Incoherence (#VD2916)
Impropriety (#VD2870)	Controversy (#VD2288)
Ambivalence (#VD2056)	Abnormality (#VD2014)
Unsuitability (#VD3954)	Nonconformity (#VD3324)
Maladjustment (#VD3180)	Inharmonious (#VD3014)
Inconsistency (#VD2932)	Contradiction (#VD2284)
Incompatibility (#VD2918)	Inapplicability (#VD2898)
Disagreeableness (#VD2456)	Irreconcilability (#VD3100)
Inappropriateness (#VD2900)	Self-contradiciton (#VD3632).

♦ VP5030c Equality–Inequality

Dynamics We clamour for equality chiefly in matters in which we ourselves cannot hope to obtain excellence. (Eric Hoffer)
Whatever may be the general endeavour of a community to render its members equal and alike, the personal pride of individuals will always seek to rise above the line, and to form somewhere an inequality to their own advantage. (Alexis de Tocqueville)

Integrative complex Relationship (#VT8012).

Constituent values Equity (#VC0673)	Justice (#VC1041)
Balance (#VC0215)	Symmetry (#VC1729)
Evenness (#VC0687)	Equality (#VC0667)
Adjustment (#VC0055)	Equivalence (#VC0675)
Equilibrium (#VC0671)	Coordination (#VC0485)
Correspondence (#VC0491)	Convertibility (#VC0477).
Constituent values Inequity (#VD2988)	Injustice (#VD3026)
Disparity (#VD2534)	Asymmetry (#VD2110)
Unevenness (#VD3904)	Inequality (#VD2986)
Inadequacy (#VD2890)	Contrariety (#VD2286)
Nonuniformity (#VD3334)	Insufficiency (#VD3060).

♦ VP5034a Greatness–Smallness

Dynamics The essence of greatness is the perception that virtue is enough. (Emerson)
Magnanimity in politics is not seldom the truest wisdom; and a great empire and little minds go ill together. (Edmund Burke)
All the glory of greatness has no lustre for people who are in search of understanding. (Pascal)
Of all virtues, magnanimity is the rarest. There are a hundred persons of merit for one who willingly acknowledges it in another. (William Hazlitt)
Desire of greatness is a godlike sin. (John Dryden)
Small is beautiful. (Schumacher)

Integrative complex Quantity (#VT8013).

Constituent values Fame (#VC0733)	Power (#VC1331)
Might (#VC1147)	Glory (#VC0841)
Wonder (#VC1933)	Majesty (#VC1109)
Heroism (#VC0897)	Strength (#VC1681)
Nobility (#VC1193)	Grandeur (#VC0861)
Fullness (#VC0811)	Eminence (#VC0631)
Plenitude (#VC1311)	Largeness (#VC1049)

VP5034

Intensity (#VC1005)
Greatness (#VC0867)
Generosity (#VC0821)
Superfluity (#VC1705)
Distinction (#VC0577)
Prodigiousness (#VC1363)
Irresistability (#VC1031)
Comprehensiveness (#VC0395).
Constituent values Exiguity (#VD2642)
Scantiness (#VD3612)
Insufficiency (#VD3060)

Immensity (#VC0943)
Notability (#VC1203)
Exaltation (#VC0695)
Magnanimity (#VC1105)
Boundlessness (#VC0253)
Unequivocalness (#VC1815)
Indubitableness (#VC0973)

Smallness (#VD3684)
Unimportance (#VD3920)
Inconsequence (#VD2928).

♦ **VP5036a Superiority–Inferiority**
Dynamics In our society to admit inferiority is to be a fool, and to admit superiority is to be an outcast. Those who are in reality superior in intelligence can be accepted by their fellows only if they pretend they are not. (Mannes)
Superiority and inferiority are individual, not racial or national. (Wylie)
Racism is the dogma that one ethnic group is condemned by nature to congenital inferiority and another group is destined to congenital superiority. (Benedict)
The surrender of life is nothing to sinking down into acknowledgment of inferiority. (Calhoun)
Integrative complex Quantity (#VT8013).
Constituent values Vantage (#VC1857)
Advantage (#VC0065)
Precedence (#VC1339)
Excellence (#VC0697)
Ordinariness (#VC1235)
Transcendence (#VC1775)
Constituent values Lowness (#VD3172)
Meanness (#VD3204)
Vulgarity (#VD4034)
Servility (#VD3656)
Maladroit (#VD3182)
Triviality (#VD3856)
Mediocrity (#VD3206)
Inadequacy (#VD2890)
Inferiority (#VD3002)
Imperfection (#VD2842)
Insufficiency (#VD3060).

Supremacy (#VC1713)
Uniqueness (#VC1827)
Leadership (#VC1057)
Superiority (#VC1709)
Championship (#VC0305)
Invincibility (#VC1021).
Failure (#VD2660)
Baseness (#VD2150)
Smallness (#VD3684)
Pettiness (#VD3452)
Demeaning (#VD2374)
Subjection (#VD3746)
Littleness (#VD3158)
Deficiency (#VD2356)
Incompetence (#VD2920)
Subordination (#VD3750)

♦ **VP5038a Increase–Decrease**
Dynamics Growth itself contains the germ of happiness. (Pearl S Buck)
Thus the sum of things is ever being renewed, and mortals live dependent one upon another. Some nations increase, others diminish, and in a short space the generations of living creatures are changed and like runners pass on the torch of life. (Lucretius)
Absence diminishes mediocre passions and increases great ones, as the wind blows out candles and fans fire. (La Rochefoucauld)
In much wisdom is much grief: and he that increaseth knowledge increaseth sorrow. (Bible)
Integrative complex Quantity (#VT8013).
Constituent values Gain (#VC0817)
Increase (#VC0965)
Elevation (#VC0625)
Eclecticism (#VC0603)
Consumption (#VC0461)
Appreciation (#VC0153)
Amplification (#VC0129)
Productiveness (#VC1365)
Constituent values Proliferation (#VD3496).

Growth (#VC0871)
Expansion (#VC0707)
Enhancement (#VC0649)
Development (#VC0549)
Advancement (#VC0063)
Accumulation (#VC0027)
Accessibility (#VC0019)

♦ **VP5045b Simplicity–Complexity**
Dynamics Beauty of style and harmony and grace and good rhythm depend on simplicity. (Plato)
One who understands much displays a greater simplicity of character than one who understands little. (Alexander Chase)
Power, whether exercised over matter or over man, is partial to simplification. (Eric Hoffer)
Integrative complex Quantity (#VT8013).
Constituent values Unity (#VC1829)
Essence (#VC0681)
Chastity (#VC0319)
Intricacy (#VC1015)
Simplicity (#VC1617)
Uncorrupted (#VC1809)
Involvement (#VC1023)
Unadulteration (#VC1795)
Constituent values Muddle (#VD3292)
Involution (#VD3092)
Deviousness (#VD2432)
Oversimplification (#VD3404).

Purity (#VC1391)
Neatness (#VC1181)
Plainness (#VC1301)
Integrity (#VC0997)
Refinement (#VC1441)
Unadornment (#VC1793)
Absoluteness (#VC0009)
Straightforwardness (#VC1679)
Pollution (#VD3464)
Corruption (#VD2294)
Complication (#VD2254)

♦ **VP5047c Conjunction–Separation**
Dynamics The only solid and lasting peace between a man and his wife is, doubtless, a separation. (Lord Chesterfield)
Ever has it been that love knows not its own depth until the hour of separation. (Gibran)
The return makes one love the farewell. (de Musset)
Singularity is dangerous in everything. (Fénelon)
Integrative complex Quantity (#VT8013).
Constituent values Union (#VC1825)
Inclusion (#VC0963)
Attachment (#VC0193)
Constituent values Division (#VD2580)
Severance (#VD3658)
Brokenness (#VD2200)
Separateness (#VD3654)
Disintegration (#VD2514).

Alliance (#VC0105)
Detachment (#VC0545)
Togetherness (#VC1763).
Disunity (#VD2572)
Revolution (#VD3562)
Dissolution (#VD2556)
Discontinuity (#VD2478)

♦ **VP5050c Cohesion–Disintegration**
Dynamics The trouble with the age we live in is that it lacks consistence. People start reasoning things out, and as soon as they begin to get anywhere they drop everything and run. (Ugo Betti)
Integrative complex Quantity (#VT8013).
Constituent values Tenacity (#VC1749)
Coherence (#VC0351)
Continuity (#VC0471)
Persistence (#VC1281)
Connectedness (#VC0431).
Constituent values Wear (#VD4054)
Erosion (#VD2618)

Toughness (#VC1767)
Adherence (#VC0053)
Connection (#VC0433)
Consistency (#VC0451)

Chaos (#VD2224)
Anarchy (#VD2062)

Looseness (#VD3164)
Brokenness (#VD2200)
Incoherence (#VD2916)
Degradation (#VD2362)
Dissociation (#VD2554)
Nonadherence (#VD3322)
Decomposition (#VD2338)
Disorganization (#VD2528).

Confusion (#VD2272)
Tenuousness (#VD3804)
Dissolution (#VD2556)
Separateness (#VD3654)
Dilapidation (#VD2442)
Inconsistency (#VD2932)
Disintegration (#VD2514)

♦ **VP5054c Completeness–Incompleteness**
Dynamics When we rejoice in our fullness, then we can part with our fruits with joy. (Rabindranath Tagore)
It is the highest creatures who take the longest to mature, and are the most helpless during their immaturity. (George Bernard Shaw)
Integrative complex Quantity (#VT8013).
Constituent values Unity (#VC1829)
Ubiquity (#VC1791)
Maturity (#VC1127)
Assembly (#VC0181)
Undivided (#VC1813)
Integrity (#VC0997)
Clearness (#VC0339)
Perfection (#VC1273)
Completion (#VC0387)
Integrality (#VC0995)
Universality (#VC1831)
Omnipresence (#VC1217)
Absoluteness (#VC0009)
Comprehensiveness (#VC0395).
Constituent values Lack (#VD3126)
Omission (#VD3360)
Immaturity (#VD2832)
Defectiveness (#VD2350)
Underdevelopment (#VD3896).

Holism (#VC0907)
Solidity (#VC1633)
Fullness (#VC0811)
Wholeness (#VC1919)
Plenitude (#VC1311)
Globality (#VC0839)
Solidarity (#VC1631)
Fulfilment (#VC0809)
Realization (#VC1421)
Culmination (#VC0505)
Thoroughness (#VC1753)
Consummation (#VC0459)
Accomplishment (#VC0023)

Shortage (#VD3674)
Inadequacy (#VD2890)
Deficiency (#VD2356)
Incompleteness (#VD2922)

♦ **VP5059a Order–Disorder**
Dynamics Good order is the foundation of all things. (Edmund Burke)
It is best to do things systematically, since we are only human, and disorder is our worst enemy. (Hesiod)
Chaos often breeds life, when order breeds habit. (Henry Adams)
Integrative complex Order (#VT8014).
Constituent values Peace (#VC1265)
Harmony (#VC0887)
Symmetry (#VC1729)
Neatness (#VC1181)
Quietness (#VC1403)
Regularity (#VC1445)
Orderliness (#VC1233)
Peacefulness (#VC1267)
Constituent values Fuss (#VD2734)
Chaos (#VD2224)
Muddle (#VD3292)
Misrule (#VD3272)
Disorder (#VD2526)
Looseness (#VD3164)
Confusion (#VD2272)
Turbulence (#VD3866)
Randomness (#VD3514)
Disharmony (#VD2504)
Promiscuity (#VD3500)
Aimlessness (#VD2048)
Discomposure (#VD2472)
Spasmodicness (#VD3704)
Senselessness (#VD3648)
Nonuniformity (#VD3334)
Disintegration (#VD2514)
Capriciousness (#VD2212)

Order (#VC1231)
Concord (#VC0411)
Quietude (#VC1405)
Business (#VC0279)
Uniformity (#VC1821)
Discipline (#VC0567)
Tranquillity (#VC1773)
Organization (#VC1237).
Upset (#VD3966)
Uproar (#VD3964)
Squalor (#VD3720)
Anarchy (#VD2062)
Messiness (#VD3220)
Frivolity (#VD2724)
Anarchism (#VD2060)
Sordidness (#VD3692)
Negligence (#VD3312)
Bedraggled (#VD2162)
Disturbance (#VD2570)
Perturbation (#VD3442)
Carelessness (#VD2214)
Shapelessness (#VD3668)
Orderlessness (#VD3370)
Haphazardness (#VD2766)
Disarrangement (#VD2464)
Disorganization (#VD2528).

♦ **VP5071a Continuity–Discontinuity**
Dynamics The most active lives have so much routine as to preclude progress almost equally with the most inactive. (Emerson)
Integrative complex Order (#VT8014).
Constituent values Fullness (#VC0811)
Plenitude (#VC1311)
Constancy (#VC0455)
Connectedness (#VC0431).
Constituent values Brokenness (#VD2200)
Nonuniformity (#VD3334)

Stability (#VC1651)
Immediacy (#VC0941)
Uniformity (#VC1821)

Spasmodicness (#VD3704)
Discontinuity (#VD2478)

♦ **VP5076b Inclusion–Exclusion**
Dynamics Generalization is necessary to the advancement of knowledge; but particularity is indispensable to the creations of the imagination. (Thomas Babington Macaulay)
Integrative complex Order (#VT8014).
Constituent values Tolerance (#VC1765)
Globality (#VC0839)
Participation (#VC1253)
Comprehension (#VC0393)
Constituent values Blemish (#VD2184)
Division (#VD2580)
Severance (#VD3658)
Rejection (#VD3526)
Expulsion (#VD2646)
Xenophobia (#VD4080)
Suspension (#VD3784)
Segregation (#VD3624)
Repudiation (#VD3536)
Nonconformity (#VD3324)

Inclusion (#VC0963)
Universality (#VC1831)
Exclusiveness (#VC0701)
Comprehensiveness (#VC0395).
Snobbery (#VD3686)
Tightness (#VD3816)
Seclusion (#VD3616)
Isolation (#VD3120)
Exclusion (#VD2634)
Withdrawal (#VD4062)
Insularity (#VD3062)
Restriction (#VD3546)
Prohibition (#VD3494)
Intrusiveness (#VD3086).

♦ **VP5082c Conformity–Nonconformity**
Dynamics Eccentricity has always abounded when and where strength of character has abounded; and the amount of eccentricity in a society has generally been proportional to the amount of genius, mental vigour, and moral courage which it contained. (John Stuart Mill)
Adapt or perish, now as ever, is Nature's inexorable imperative. (H G Wells)
One of the saddest things about conformity is the ghastly sort of non-conformity it breeds; the noisy protesting, the aggressive rebelliousness, the rigid counterfetishism. (Louis Kronenberger)
How protean are the devices available to human intelligence when it lends itself to the persistence of the conformist error. (Robert Lindner)

VALUE POLARITIES

Integrative complex Order (#VT8014).
Constituent values Order (#VC1231)
Accord (#VC0025)
Principle (#VC1353)
Congruity (#VC0429)
Conformity (#VC0425)
Naturalness (#VC1179)
Correctness (#VC0489)
Ordinariness (#VC1235)
Adaptability (#VC0049)
Prodigiousness (#VC1363).
Health (#VC0889)
Propriety (#VC1379)
Orthodoxy (#VC1241)
Regularity (#VC1445)
Orderliness (#VC1233)
Healthiness (#VC0891)
Consistency (#VC0451)
Conformation (#VC0423)
Acquiescence (#VC0035)
Constituent values Deviation (#VD2428)
Aberration (#VD2008)
Inferiority (#VD3002)
Malformation (#VD3190)
Unnaturalness (#VD3930)
Nonconformity (#VD3324).
Divergence (#VD2576)
Strangeness (#VD3734)
Deformation (#VD2358)
Eccentricity (#VD2594)
Shapelessness (#VD3668)

♦ **VP5086a Numbered–Unnumbered**
Dynamics Are not two sparrows sold for a farthing? and not one of them shall fall on the ground without your Father. But the very hairs of your head are all numbered. (Bible)
I have seen all the works that are done under the sun; and behold, all is vanity and vexation of spirit. That which is crooked cannot be made straight; and that which is wanting cannot be numbered. (Bible)
Integrative complex Number (#VT8015).
Constituent values Enumeration (#VC0665).

♦ **VP5089b Unity–Duality**
Dynamics Plurality which is not reduced to unity is confusion; unity which does not depend on plurality is tyranny. (Pascal)
In necessary things, unity; in doubtful things, liberty; in all things, charity. (Richard Baxter)
When spider webs unite, they can tie up a lion. (Ethiopian proverb)
It is always possible to bind together a considerable number of people in love, so long as there are other people left over to receive the manifestations of their aggressiveness. (Freud)
There are only two forces that unite men – fear and interest. (Napoleon)
Integrative complex Number (#VT8015).
Constituent values Unity (#VC1829)
Privacy (#VC1357)
Solidity (#VC1633)
Undivided (#VC1813)
Uniqueness (#VC1827)
Solidarity (#VC1631)
Detachment (#VC0545)
Individuality (#VC0971).
Purity (#VC1391)
Solitude (#VC1635)
Wholeness (#VC1919)
Integrity (#VC0997)
Uniformity (#VC1821)
Simplicity (#VC1617)
Integrality (#VC0995)
Constituent values Seclusion (#VD3616)
Withdrawal (#VD4062)
Alienation (#VD2052)
Homelessness (#VD2796).
Isolation (#VD3120)
Loneliness (#VD3162)
Separateness (#VD3654)

♦ **VP5101c Numerousness–Fewness**
Dynamics As writers become more numerous, it is natural for readers to become more indolent. (Goldsmith)
Nations are not truly great solely because the individuals composing them are numerous, free, and active; but they are great when these numbers, this freedom, and this activity are employed in the service of an ideal higher than that of an ordinary man, taken by himself. (Arnold)
And for the few that only lend their ear, that few is all the world. (Samuel Daniel)
Never in the field of human conflict was so much owed by so many to so few. (Churchill)
Let thy speech be short, comprehending much in few words. (Bible)
Integrative complex Number (#VT8015).
Constituent values Variety (#VC1859)
Unlimited (#VC1833)
Universality (#VC1831)
Limitlessness (#VC1081).
Rareness (#VC1413)
Abundance (#VC0015)
Boundlessness (#VC0253)
Constituent values Thinness (#VD3808)
Smallness (#VD3684).
Tightness (#VD3816)

♦ **VP5110a Durability–Transience**
Dynamics People are too durable, that's their main trouble. They can do too much to themselves, they last too long. (Brecht)
To endure what is unendurable is true endurance. (Japanese proverb)
When two people are under the influence of the most violent, most insane, most delusive and most transient of passions, they are required to swear that they will remain in that excited, abnormal, and exhausting condition continuously until death do them part. (Shaw)
Art is long, life short; judgement difficult, opportunity transient. (Goethe)
Integrative complex Time (#VT8021).
Constituent values Lasting (#VC1051)
Duration (#VC0591)
Longevity (#VC1091)
Endurance (#VC0645)
Permanence (#VC1275)
Continuity (#VC0471)
Perpetuation (#VC1277)
Steadfastness (#VC1667).
Survival (#VC1721)
Stability (#VC1651)
Hardiness (#VC0883)
Constancy (#VC0455)
Durability (#VC0589)
Persistence (#VC1281)
Immutability (#VC0951)
Constituent values Volatility (#VD4030)
Mutability (#VD3296)
Impetuousity (#VD2848)
Capriciousness (#VD2212).
Transience (#VD3838)
Instability (#VD3054)
Changeableness (#VD2222)

♦ **VP5112a Perpetuity–Instantaneousness**
Dynamics The iniquity of oblivion blindly scattereth her poppy, and deals with the memory of men without distinction to merit of perpetuity. (Broune)
Be instant in season, out of season. (Bible)
The instant, trivial as it is, is all we have unless – unless things the imagination feeds upon, the scent of the rose, startle us anew. (William Carlos Williams)
Integrative complex Time (#VT8021).
Constituent values Immediacy (#VC0941)
Permanence (#VC1275)
Timelessness (#VC1759)
Perpetuation (#VC1277)
Constancy (#VC0455)
Immortality (#VC0947)
Preservation (#VC1347)

♦ **VP5119b Futurity–Antiquity**
Dynamics It seems to be the fate of man to seek all his consolations in futurity. The time present is seldom able to fill desire or imagination with immediate enjoyment, and we are forced to supply its deficiencies by recollection or anticipation. (Samuel Johnson)
What is actual is actual only for one time, and only for one place. (T S Eliot)

The future is the only transcendental value for men without God. (Albert Camus)
A preoccupation with the future not only prevents us from seeing the present as it is but often prompts us to rearrange the past. (Eric Hoffer)
I believe the future is only the past again, entered through another gate. (Sir Arthur Wing Pinero)
Respect the past in the full measure of its desserts, but do not make the mistake of confusing it with the present nor seek in it the ideas of the future. (José Ingenieros)
Antiquity is full of the praises of another antiquity still more remote. (Voltaire)
Integrative complex Time (#VT8021).
Constituent values Being (#VC0225)
Immediacy (#VC0941)
Actuality (#VC0043)
Newness (#VC1189)
Freshness (#VC0799)
Recollection (#VC1429).

♦ **VP5122b Newness–Oldness**
Dynamics As soon as we are shown the existence of something old in a new thing, we are pacified. (Nietzsche)
Everywhere the basis of principle is tradition. (Oliver Wendell Holmes)
They that reverence too much old times are but a scorn to the new. (Francis Bacon)
The new always carries with it the sense of violation, of sacrilege. What is dead is sacred; what is new, that is, different, is evil, dangerous, or subversive. (Henry Miller)
When a new book appears, read an old one. (Proverb)
Who leaves the old way for the new will find himself deceived. (Proverb)
Preserve the old, but know the new. (Chinese proverb)
Integrative complex Time (#VT8021).
Constituent values Youth (#VC1943)
Novelty (#VC1205)
Virginity (#VC1883)
Freshness (#VC0799)
Originality (#VC1239)
Custom (#VC0511)
Newness (#VC1189)
Tradition (#VC1771)
Uniqueness (#VC1827)
Venerableness (#VC1863).
Constituent values Age (#VD2040)
Naïvety (#VD3304)
Archaism (#VD2094)
Feebleness (#VD2690)
Inexperience (#VD2994)
Antediluvian (#VD2082).
Disuse (#VD2574)
Debility (#VD2328)
Puerility (#VD3506)
Obsolescence (#VD3350)
Decrepitness (#VD2340)

♦ **VP5124b Youth–Age**
Dynamics Believe me, all evil comes from the old. They grow fat on ideas and young men die of them. (Jean Anouilh)
In youth, it is common to measure right and wrong by the opinion of the world, and in age, to act without any measure but interest, and to lose shame without substituting virtue. (Samuel Johnson)
For the complete life, the perfect pattern includes old age as well as youth and maturity. (W Somerset Maugham)
With the ancient is wisdom; and in length of days understanding. (Bible)
All sorts of allowances are made for the illusions of youth; and none, or almost none, for the disenchantments of age. (Robert Louis Stevenson)
Rashness is the error of youth, timid caution of age. Manhood is...the ripe and fertile season of action, when alone we can hope to find the head to contrive, united with the hand to execute. (Charles Caleb)
Age is opportunity no less than youth itself, though in another dress, and as the evening twilight fades away, the sky is filled with stars, invisible by day. (Longfellow)
Age in a virtuous person, of either sex, carries in it an authority which makes it preferable to all the pleasures of youth. (Richard Steele)
Integrative complex Time (#VT8021).
Constituent values Youth (#VC1943)
Maturity (#VC1127)
Development (#VC0549)
Virility (#VC1885)
Mellowness (#VC1135)
Youthfulness (#VC1945).
Constituent values Age (#VD2040).

♦ **VP5129c Timeliness–Untimeliness**
Dynamics Timeliness is best in all matters. (Hesiod)
There is a slowness in affairs which ripens them, and a slowness which rots them. (Joseph Roux)
There is no security on this earth; there is only opportunity. (Douglas MacArthur)
Whatever is produced in haste goes hastily to waste. (Sa'di)
Decisiveness is often the art of timely cruelty. (Henri Becque)
Integrative complex Time (#VT8021).
Constituent values Chance (#VC0307)
Fitness (#VC0767)
Swiftness (#VC1727)
Quickness (#VC1399)
Alertness (#VC0099)
Expedience (#VC0711)
Suitability (#VC1701)
Opportunity (#VC1227)
Decisiveness (#VC0519)
Appropriateness (#VC0159).
Liberty (#VC1073)
Alacrity (#VC0097)
Readiness (#VC1417)
Propriety (#VC1379)
Timeliness (#VC1761)
Directness (#VC0563)
Punctuality (#VC1389)
Expeditious (#VC0713)
Anticipation (#VC0139)
Constituent values Haste (#VD2776)
Slowness (#VD3682)
Lateness (#VD3132)
Tardiness (#VD3790)
Suspension (#VD3784)
Hesitation (#VD2792)
Prematurity (#VD3476)
Irrelevance (#VD3106)
Inexpedience (#VD2992)
Intrusiveness (#VD3086)
Unpreparedness (#VD3940).
Laxness (#VD3140)
Laziness (#VD3142)
Wrongness (#VD4078)
Hastiness (#VD2778)
Infelicity (#VD3000)
Retardation (#VD3550)
Obstruction (#VD3356)
Impropriety (#VD2870)
Unsuitability (#VD3954)
Inconvenience (#VD2938)

♦ **VP5135c Frequency–Infrequency**
Dynamics Hit hard, hit fast, hit often. (Halsey)
The art of war is simple enough. Find out where your enemy is. Get at him as soon as you can. Strike at him as hard as you can and as often as you can, and keep moving on. (Grant)
Integrative complex Time (#VT8021).
Constituent values Uniqueness (#VC1827)
Regularity (#VC1445)
Continuity (#VC0471)
Perpetuation (#VC1277)
Steadiness (#VC1669)
Dispassion (#VC0575)
Unchangeable (#VC1807)
Ordinariness (#VC1235).

♦ **VP5137c Regularity–Irregularity**
Dynamics Our whole knowledge of the world hangs on this very slender thread: the regularity of our experiences. (Luigi Pirandello)
Integrative complex Time (#VT8021).

VP5137

Constituent values Evenness (#VC0687)
Steadiness (#VC1669)
Punctuality (#VC1389)
Constituent values Nonuniformity (#VD3334)
Capriciousness (#VD2212).
Uniformity (#VC1821)
Regularity (#VC1445)
Orderliness (#VC1233).
Discontinuity (#VD2478)

♦ **VP5139a Change–Permanence**
Dynamics All is change; all yields its place and goes. (Euripides)
Nothing is permanent but change. (Heraclitus)
The more things change, the more they remain the same. (Alphonse Karr)
We used to think that revolutions are the cause of change. Actually it is the other revolution. (Eric Hoffer)
Happiness is never really so welcome as changelessness. (Graham Greene)
Seven changes of address are a guarantee of poverty. (Malay proverb)
Weep not that the world changes – did it keep a stable, changeless state, 't were cause indeed to weep. (Bryant)
Change as change is mere flux and lapse; it insults intelligence. Genuinely to know is to grasp a permanent end that realizes itself through changes. (Dewey)
Life is not a static thing. The only people who do not change their minds are incompetents in asylums, who can't, and those in cemetaries. (Dirksen)
Integrative complex Change (#VT8022).
Constituent values Change (#VC0309)
Lasting (#VC1051)
Duration (#VC0591)
Endurance (#VC0645)
Discovery (#VC0569)
Permanence (#VC1275)
Durability (#VC0589)
Betterment (#VC0239)
Persistence (#VC1281)
Unchangeable (#VC1807)
Immutability (#VC0951)
Amelioration (#VC0117)
Steadfastness (#VC1667)
Inventiveness (#VC1019)
Constituent values Rigidity (#VD3568).
Variety (#VC1859)
Solidity (#VC1633)
Stability (#VC1651)
Diversity (#VC0579)
Constancy (#VC0455)
Mitigation (#VC1153)
Conversion (#VC0475)
Adjustment (#VC0055)
Improvement (#VC0961)
Preservation (#VC1347)
Conservation (#VC0445)
Adaptability (#VC0049)
Renewableness (#VC1465)
Transformation (#VC1779).

♦ **VP5141a Stability–Changeableness**
Dynamics There are those who would misteach us that to stick in a rut is consistency and a virtue, and that to climb out of the rut is inconsistency and a vice. (Mark Twain)
Integrative complex Change (#VT8022).
Constituent values Lasting (#VC1051)
Security (#VC1537)
Stability (#VC1651)
Constancy (#VC0455)
Resilience (#VC1475)
Transcience (#VC1777)
Immortality (#VC0947)
Equilibrium (#VC0671)
Immutability (#VC0951)
Steadfastness (#VC1667)
Substantiality (#VC1693)
Constituent values Variation (#VD3994)
Mutability (#VD3296)
Vacillation (#VD3974)
Instability (#VD3054)
Restlessness (#VD3544)
Eccentricity (#VD2594)
Shapelessness (#VD3668)
Inconsistency (#VD2932)
Changeableness (#VD2222)
Undependability (#VD3894)
Irresponsibility (#VD3114).
Balance (#VC0215)
Mobility (#VC1155)
Soundness (#VC1637)
Uniformity (#VC1821)
Permanence (#VC1275)
Reliability (#VC1459)
Flexibility (#VC0771)
Unchangeable (#VC1807)
Adaptability (#VC0049)
Invincibility (#VC1021)
Invulnerability (#VC1025).
Transience (#VD3838)
Vicissitude (#VD4010)
Uncertainty (#VD3880)
Inconstancy (#VD2934)
Impetuousity (#VD2848)
Unreliability (#VD3948)
Nonuniformity (#VD3334)
Impulsiveness (#VD2878)
Capriciousness (#VD2212)
Unpredictability (#VD3938)

♦ **VP5143b Continuance–Cessation**
Dynamics Continuity in everything is unpleasant. Cold is agreeable, that we may get warm. (Pascal)
Integrative complex Change (#VT8022).
Constituent values Progress (#VC1373)
Sustenance (#VC1723)
Perpetuation (#VC1277)
Endurance (#VC0645)
Continuity (#VC0471)
Indefatigableness (#VC0967).

♦ **VP5145b Conversion–Reversion**
Dynamics Except ye be converted, and become as little children, ye shall not enter into the kingdom of heaven. (Bible)
You have not converted a man because you have silenced him. (Viscount Morley)
Tao invariably takes no action, and yet there is nothing left undone. Reversion is the action of Tao. Weakness is the function of Tao. All things in the world come from being. And being comes from non-being. (Lao-tzu)
Integrative complex Change (#VT8022).
Constituent values Reform (#VC1443)
Improvement (#VC0961)
Constituent values Treason (#VD3844)
Reversion (#VD3558)
Brainwash (#VD2198)
Regression (#VD3524)
Alienation (#VD2052)
Traitorousness (#VD3834)
Conversion (#VC0475)
Convertibility (#VC0477).
Revulsion (#VD3564)
Desertion (#VD2406)
Subversion (#VD3754)
Corruption (#VD2294)
Retrogression (#VD3554)
Disenchantment (#VD2492).

♦ **VP5147c Evolution–Revolution**
Dynamics In nothing is there more evolution than the American mind. (Whitman)
Women is perpetual revolution, and is that element in the world which continually destroys and re-creates. (Warner)
Every revolution was first a thought in one man's mind; and when the same thought occurs to another man, it is the key to that era. (Emerson)
The Revolution must take place in men before it can be manifest in things. (Graffito)
Not actual suffering but the hope of better things incites people to revolt. (Hoffer)
Integrative complex Change (#VT8022).
Constituent values Growth (#VC0871)
Expansion (#VC0707)
Development (#VC0549)
Adaptability (#VC0049)
Accomplishment (#VC0023)
Constituent values Upset (#VD3966)
Revulsion (#VD3564)
Subversion (#VD3754)
Progress (#VC1373)
Evolution (#VC0691)
Advancement (#VC0063)
Amplification (#VC0129)
Overturn (#VD3412)
Anarchism (#VD2060)
Revolution (#VD3562).

♦ **VP5151a Eventuation–Imminence**
Dynamics What may be done at any time will be done at no time. (Fuller)
Integrative complex Causation (#VT8023).
Constituent values Interest (#VC1011)
Eventuation (#VC0689)
Materialization (#VC1123).
Occurrence (#VC1213)
Momentousness (#VC1161)

♦ **VP5153b Causation–Culmination**
Dynamics Take away the cause, and the effect ceases; what the eye ne'er sees, the heart ne'er rues. (Cervantes)
Nothing comes from nothing. (Lucretius)
Every why hath a wherefore. (Shakespeare)
Integrative complex Causation (#VT8023).
Constituent values Work (#VC1937)
Reason (#VC1423)
Principle (#VC1353)
Completion (#VC0387)
Explanation (#VC0719)
Culmination (#VC0505)
Constituent values Backlash (#VD2132).
Clout (#VC0345)
Interest (#VC1011)
Influence (#VC0979)
Realization (#VC1421)
Development (#VC0549)
Consummation (#VC0459).

♦ **VP5155c Attributability–Chance**
Dynamics What is the use of working out chances ? There are no chances against God. (Bernanos)
We cannot bear to regard ourselves simply as playthings of blind chance; we cannot admit to feeling ourselves abandoned. (Ugo Betti)
Do we, holding that the gods exist, deceive ourselves with unsubstantial dreams and lies, while random careless chance and change alone control the world ? (Euripides)
We do not what we ought; what we ought not, we do, and lean upon the thought that chance will bring us through. But our own acts, for good or ill, are mightier powers. (Matthew Arnold)
Integrative complex Causation (#VT8023).
Constituent values Fortune (#VC0791)
Attachment (#VC0193)
Acknowledgement (#VC0033).
Constituent values Randomness (#VD3514)
Hopelessness (#VD2798)
Haphazardness (#VD2766)
Connection (#VC0433)
Serendipity (#VC1599)
Aimlessness (#VD2048)
Impossibility (#VD2856)
Purposelessness (#VD3508).

♦ **VP5157a Power–Impotence**
Dynamics Time goes by: reputation increases, ability declines. (Dag Hammarskjold)
Powerlessness frustrates; absolute powerlessness frustrates absolutely. Absolute frustration is a dangerous emotion to run a world with. (Russell Baker)
Where might is master, justice is servant. (German Proverb)
Where there is no might, right loses itself. (Portuguese Proverb)
There is nothing so imperious as feebleness which feels itself supported by force. (Napoleon I)
Power tends to corrupt and absolute power corrupts absolutely. (Lord Acton)
Life is a search after power; and this is an element with which the world is so saturated, there is no chink or crevice in which it is not lodged, that no honest seeking goes unrewarded. (Emerson)
There is a universal need to exercise some kind of power, or to create for one's self the appearance of some power, if only temporarily, in the form of intoxication. (Nietzsche)
Man is as full of potentiality as he is of impotence. (George Santayana)
Integrative complex Power (#VT8024).
Constituent values Power (#VC1331)
Clout (#VC0345)
Vigour (#VC1881)
Genius (#VC0825)
Potency (#VC1327)
Cogency (#VC0349)
Vitality (#VC1891)
Validity (#VC1853)
Facility (#VC0725)
Capacity (#VC0289)
Unlimited (#VC1833)
Endowment (#VC0643)
Efficiency (#VC0615)
Capability (#VC0287)
Sufficiency (#VC1699)
Omnipotence (#VC1215)
Forcefulness (#VC0781)
Qualification (#VC1395)
Productiveness (#VC1365)
Constituent values Inanity (#VD2896)
Failure (#VD2660)
Vainness (#VD3980)
Sabotage (#VD3590)
Effetism (#VD2598)
Impotence (#VD2858)
Ineptitude (#VD2984)
Inadequacy (#VD2890)
Feebleness (#VD2690)
Barrenness (#VD2146)
Inferiority (#VD3002)
Inefficiency (#VD2980)
Incapability (#VD2908)
Fecklessness (#VD2688)
Insufficiency (#VD3060)
Ineffectuality (#VD2978)
Ineffectiveness (#VD2976)
Might (#VC1147)
Virtue (#VC1887)
Talent (#VC1739)
Energy (#VC0647)
Fitness (#VC0767)
Ability (#VC0003)
Virility (#VC1885)
Strength (#VC1681)
Charisma (#VC0313)
Adequacy (#VC0051)
Influence (#VC0979)
Authority (#VC0205)
Competence (#VC0383)
Superiority (#VC1709)
Proficiency (#VC1367)
Potentiality (#VC1329)
Almightiness (#VC0109)
Effectiveness (#VC0613)
Irresistability (#VC1031).
Fatuity (#VD2680)
Weakness (#VD4052)
Softness (#VD3688)
Futility (#VD2736)
Inability (#VD2882)
Invalidity (#VD3088)
Incapacity (#VD2910)
Imbecility (#VD2830)
Disability (#VD2446)
Uselessness (#VD3970)
Devitalized (#VD2434)
Incompetence (#VD2920)
Helplessness (#VD2788)
Powerlessness (#VD3472)
Forcelessness (#VD2706)
Incompleteness (#VD2922)
Counterproductivity (#VD2296).

♦ **VP5159a Strength–Weakness**
Dynamics The turning point in the process of growing up is when you discover the core of strength within you that survives all hurt. (Max Lerner)
Strength and strength's will are the supreme ethic. All else are dreams from hospital beds, the sly, crawling goodness of sneaking souls. (Elbert Hubbard)
Integrative complex Power (#VT8024).
Constituent values Power (#VC1331)
Vigour (#VC1881)
Support (#VC1711)
Potency (#VC1327)
Strength (#VC1681)
Toughness (#VC1767)
Soundness (#VC1637)
Endurance (#VC0645)
Staunchness (#VC1665)
Reinforcement (#VC1451).
Might (#VC1147)
Energy (#VC0647)
Stamina (#VC1655)
Vitality (#VC1891)
Solidity (#VC1633)
Stability (#VC1651)
Hardiness (#VC0883)
Sturdiness (#VC1685)
Forcefulness (#VC0781)

VALUE POLARITIES   VP5197

Constituent values Languor (#VD3128)
Decline (#VD2336)
Vapidity (#VD3990)
Hardness (#VD2770)
Dullness (#VD2588)
Weariness (#VD4056)
Fragility (#VD2712)
Incapacity (#VD2910)
Disability (#VD2446)
Devitalized (#VD2434)
Irresolution (#VD3112)
Tastelessness (#VD3792)
Indecisiveness (#VD2950)
Changeableness (#VD2222).
Fatigue (#VD2678)
Weakness (#VD4052)
Softness (#VD3688)
Effetism (#VD2598)
Debility (#VD2328)
Impotence (#VD2858)
Cowardice (#VD2298)
Feebleness (#VD2690)
Instability (#VD3054)
Listlessness (#VD3156)
Decrepitness (#VD2340)
Powerlessness (#VD3472)
Imperviousness (#VD2846)

♦ VP5161a Energy–Moderation
Dynamics Moderation has been created a virtue to limit the ambition of great men, and to console undistinguished people for their want of fortune and their lack of merit. (La Rochefoucauld)
The great mind knows the power of gentleness; only tries force, because persuasion fails. (Robert Browning)
Forcible ways make not an end of evil, but leave hatred and malice behind them. (Sir Thomas Browne)
Fanaticism consists in redoubling your effort when you have forgotten your aim. (George Santayana)
Extremism in the pursuit of the Presidency is an unpardonable vice. Moderation in the affairs of the nation is the highest virtue. (Lyndon B Johnson)
Integrative complex Power (#VT8024).
Constituent values Zest (#VC1949)
Elan (#VC0619)
Vigour (#VC1881)
Action (#VC0039)
Potency (#VC1327)
Vitality (#VC1891)
Strength (#VC1681)
Serenity (#VC1601)
Pacifism (#VC1243)
Leniency (#VC1067)
Flourish (#VC0773)
Dynamism (#VC0595)
Activity (#VC0041)
Stability (#VC1651)
Remission (#VC1463)
Intensity (#VC1005)
Steadiness (#VC1669)
Neutrality (#VC1187)
Mitigation (#VC1153)
Initiative (#VC0985)
Enthusiasm (#VC0663)
Continence (#VC0469)
Abnegation (#VC0005)
Self-denial (#VC1551)
Nonviolence (#VC1199)
Tranquillity (#VC1773)
Impartiality (#VC0953)
Exhilaration (#VC0703)
Judiciousness (#VC1039)
Reasonableness (#VC1425)
Mettlesomeness (#VC1145)
Conservativeness (#VC0447).
Life (#VC1075)
Power (#VC1331)
Energy (#VC0647)
Revival (#VC1501)
Ardour (#VC0165)
Tameness (#VC1741)
Sobriety (#VC1623)
Prudence (#VC1387)
Mildness (#VC1149)
Keenness (#VC1043)
Evenness (#VC0687)
Calmness (#VC0281)
Viability (#VC1875)
Restraint (#VC1495)
Quietness (#VC1403)
Animation (#VC0135)
Relaxation (#VC1455)
Moderation (#VC1157)
Liveliness (#VC1085)
Gentleness (#VC0829)
Enterprise (#VC0659)
Abstinence (#VC0011)
Stimulation (#VC1675)
Restfulness (#VC1493)
Appeasement (#VC0147)
Self-control (#VC1549)
Forcefulness (#VC0781)
Manageability (#VC1111)
Self-restraint (#VC1581)
Practicability (#VC1333)
Manoeuvrability (#VC1113)
Constituent values Fuss (#VD2734)
Venom (#VD4002)
Chaos (#VD2224)
Terror (#VD3806)
Violence (#VD4022)
Severity (#VD3660)
Atrocity (#VD2114)
Virulence (#VD4024)
Vehemence (#VD3996)
Stridency (#VD3736)
Harshness (#VD2774)
Barbarity (#VD2144)
Anarchism (#VD2060)
Unruliness (#VD3950)
Spasticity (#VD3706)
Inclemency (#VD2914)
Fanaticism (#VD2676)
Viciousness (#VD4008)
Pitilessness (#VD3456)
Spasmodicness (#VD3704)
Tempestuousness (#VD3798)
Fury (#VD2732)
Upset (#VD3966)
Uproar (#VD3964)
Frenzy (#VD2716)
Upheaval (#VD3962)
Savagery (#VD3608)
Acrimony (#VD2032)
Violation (#VD4020)
Vandalism (#VD3984)
Roughness (#VD3580)
Brutality (#VD2202)
Animality (#VD2070)
Agitation (#VD2046)
Turbulence (#VD3866)
Inhumanity (#VD3020)
Fierceness (#VD2698)
Bitterness (#VD2178)
Disturbance (#VD2570)
Mindlessness (#VD3228)
Mercilessness (#VD3216)
Unconscionableness (#VD3890).

♦ VP5165b Productiveness–Unproductiveness
Dynamics The raging productivity of the Victorians shattered nerves and punctured stomachs, but it was a thing noble, glorious, awesome in itself. (Hardwick)
Integrative complex Power (#VT8024).
Constituent values Plenty (#VC1313)
Creation (#VC0499)
Fecundity (#VC0749)
Exuberance (#VC0723)
Superabundance (#VC1703)
Constituent values Sterility (#VD3728)
Barrenness (#VD2146)
Desolation (#VD2408)
Richness (#VC1503)
Fertility (#VC0759)
Abundance (#VC0015)
Bountifulness (#VC0255)
Productiveness (#VC1365).
Impotence (#VD2858)
Infertility (#VD3004)
Unproductiveness (#VD3942).

♦ VP5167b Production–Reproduction
Dynamics Capitalist production begets, with the inexorability of a law of nature, its own negation. (Marx)
All the acquisitions or losses wrought by nature in individuals are preserved by reproduction to the new individuals which arise. (Lamarck)
The flower is the poetry of reproduction. It is an example of the eternal seductiveness of life. (Giraudoux)
Integrative complex Power (#VT8024).
Constituent values Work (#VC1937)
Creation (#VC0499)
Realization (#VC1421)
Achievement (#VC0031)
Inventiveness (#VC1019)
Accomplishment (#VC0023)
Reform (#VC1443)
Conception (#VC0405)
Development (#VC0549)
Organization (#VC1237)
Productiveness (#VC1365)
Constructiveness (#VC0457).

♦ VP5170b Ancestry–Posterity
Dynamics To forget one's ancestors is to be a brook without a source, a tree without a root. (Chinese proverb)
Good birth is a fine thing, but the merit is our ancestors'. (Plutarch)
The behaviour of an individual is determined not by his racial affiliation, but by the character of his ancestry and his cultural environment. (Franz Boas)
Integrative complex Power (#VT8024).
Constituent values Lineage (#VC1083)
Ancestry (#VC0133)
Maternity (#VC1125)
Motherhood (#VC1165)
Affiliation (#VC0077)
Consanguinity (#VC0437).
Heritage (#VC0895)
Paternity (#VC1259)
Parenthood (#VC1249)
Fatherhood (#VC0739)
Motherliness (#VC1167)

♦ VP5172c Influence–Influencelessness
Dynamics A cock has great influence on his own dunghill. (Publius Syrus)
Influence is neither good nor bad in an absolute manner, but only in relation to the one who experiences it. (Gide)
A teacher affects eternity; he can never tell where his influence stops. (Henry Brooks Adams)
Men make situations, and not situations men. (Italian proverb)
Integrative complex Power (#VT8024).
Constituent values Work (#VC1937)
Order (#VC1231)
Charm (#VC0317)
Vantage (#VC1857)
Mastery (#VC1121)
Openness (#VC1225)
Eminence (#VC0631)
Supremacy (#VC1713)
Dominance (#VC0587)
Leadership (#VC1057)
Ascendancy (#VC0175)
Momentousness (#VC1161)
Accessibility (#VC0019)
Persuasiveness (#VC1289)
Power (#VC1331)
Clout (#VC0345)
Favour (#VC0743)
Potency (#VC1327)
Prestige (#VC1349)
Interest (#VC1011)
Charisma (#VC0313)
Influence (#VC0979)
Authority (#VC0205)
Importance (#VC0957)
Enchantment (#VC0637)
Effectiveness (#VC0613)
Responsiveness (#VC1491)
Openmindedness (#VC1223)
Constituent values Weakness (#VD4052)
Powerlessness (#VD3472)
Susceptibility (#VD3780)
Ineffectiveness (#VD2976)
Impotence (#VD2858)
Forcelessness (#VD2706)
Ineffectuality (#VD2978)
Impressionability (#VD2866).

♦ VP5177c Concurrence–Counteraction
Dynamics One man may hit the ark, another blunder; but heed not these distinctions. Only from the alliance of the one, working with and through the other, are great things born. (Saint-Exupéry)
Integrative complex Power (#VT8024).
Constituent values Union (#VC1825)
Synergy (#VC1733)
Concord (#VC0411)
Agreement (#VC0093)
Cooperation (#VC0481)
Association (#VC0187)
Accumulation (#VC0027)
Correspondence (#VC0491)
Accord (#VC0025)
Harmony (#VC0887)
Alliance (#VC0105)
Resistance (#VC1477)
Concurrence (#VC0413)
Togetherness (#VC1763)
Collaboration (#VC0353)
Cooperativeness (#VC0483).
Constituent values Nonconformity (#VD3324).

♦ VP5185a Location–Dislocation
Integrative complex Space (#VT8025).
Constituent values Standing (#VC1657)
Location (#VC1087)
Perspective (#VC1283)
Belongingness (#VC0229).
Position (#VC1321)
Settlement (#VC1607)
Establishment (#VC0683)
Constituent values Homelessness (#VD2796)
Disarrangement (#VD2464).
Discontinuity (#VD2478)

♦ VP5186b Presence–Absence
Dynamics Those who are absent are always wrong. (English proverb)
When you part from your friend, you grieve not; for that which you love most in him may be clearer in his absence, as the mountain to the climber is clearer from the plain. (Kahlil Gibran)
Absence lessens ordinary passions and augments great ones, as the wind blows out a candle and makes a fire blaze. (La Rochefoucauld)
Greater things are believed of those who are absent. (Tacitus)
Integrative complex Space (#VT8025).
Constituent values Ubiquity (#VC1791)
Immediacy (#VC0941)
Omnipresence (#VC1217)
Accessibility (#VC0019).
Presence (#VC1345)
Existence (#VC0705)
Availability (#VC0209)
Constituent values Lack (#VD3126)
Absence (#VD2018)
Deprivation (#VD2398)
Disappearance (#VD2458)
Characterlessness (#VD2226).
Default (#VD2344)
Nothingness (#VD3336)
Abandonment (#VD2004)
Featurelessness (#VD2686)

♦ VP5193c Container–Content
Dynamics (His) mind had reached the calm of water which receives and reflects images without absorbing them; it contained nothing but itself. (Henry Brook Adams)
The cistern contains: the fountain overflows. (Blake)
I knew a woman, lovely in her bones, when small birds sighed, she would sigh back at them; Ah, when she moved, she moved more ways than one: The shapes a bright container can contain. (Roethke)
Integrative complex Space (#VT8025).
Constituent values Content (#VC0465).

♦ VP5195a Bigness–Littleness
Dynamics Looking at small advantages prevents great affairs from being accomplished. (Confucius)
In life's small things be resolute and great to keep thy muscle trained. (James Russell Lowell)
Integrative complex Dimension (#VT8031).
Constituent values Bigness (#VC0241).
Constituent values Exiguity (#VC2642)
Scantiness (#VD3612)
Littleness (#VD3158)
Overdeveloped (#VD3386).
Smallness (#VD3684)
Overweight (#VD3414)
Tenuousness (#VD3804)

♦ VP5197a Expansion–Contraction
Dynamics Life shrinks or expands in proportion to one's courage. (Nin)
Aging people should know that their lives are not mounting and unfolding but that an inexorable

-775-

VP5197

inner process forces the contraction of life. For a young person it is almost a sin – and certainly a danger – to be much too occupied with himself; but for the aging person it is a duty and a necessity to give serious attention to himself. (Jung)
Integrative complex Dimension (#VT8031).
Constituent values Growth (#VC0871)
Flourish (#VC0773)
Development (#VC0549)
Aggrandizement (#VC0087).
Increase (#VC0965)
Expansion (#VC0707)
Amplification (#VC0129)

◆ **VP5200b Nearness–Distance**
Dynamics Better is a neighbour that is near than a brother far off. (Bible)
The path of duty lies in what is near, and man seeks for it in what is remote. (Mencius)
Slight not what's near through aiming at what's far. (Euripides)
Good government obtains when those who are near are made happy, and those who are far off are attracted. (Confucius)
Integrative complex Dimension (#VT8031).
Constituent values Union (#VC1825)
Proximity (#VC1385)
Perspective (#VC1283).
Contact (#VC0463)
Immediacy (#VC0941)

◆ **VP5204c Breadth–Narrowness**
Dynamics In narrow streets watch out for the dagger. (Chinese Proverb)
Narrowly gathered, widely spent. (Proverb)
The most fatal illusion is the settled point of view. Since life is growth and motion, a fixed point of view kills anyone who has one. (Atkinson)
It is with narrow-souled people as with narrow-necked bottles: the less they have in them, the more noise they make in pouring it out. (Pope)
Where there is an open mind, there will always be a frontier. (Kettering)
Integrative complex Dimension (#VT8031).
Constituent values Fullness (#VC0811)
Ascendancy (#VC0175).
Constituent values Closeness (#VD2236)
Expansion (#VC0707)
Confinement (#VD2268).

◆ **VP5207c Height–Lowness**
Dynamics Detestation of the high is the involuntary homage of the low. (Dickens)
The heights by great men reached and kept were not attained by sudden flight, but they, while their companions slept, were toiling upward in the night. (Longfellow)
None can usurp this height....but those to whom the miseries of the world are misery, and will not let them rest. (Keats)
Measure your mind's height by the shade it casts. (Robert Browning)
How do I lqve thee ? Let me count the ways, I love thee to the depth and breadth and height my soul can reach, when feeling out of sight for the ends of Being and ideal Grace. (Elizebeth Browning)
Integrative complex Dimension (#VT8031).
Constituent values Culmination (#VC0505)
Constituent values Lowness (#VD3172).

◆ **VP5209c Depth–Shallowness**
Dynamics The profound thinker always suspects that he is superficial. (Benjamin Disraeli)
Integrative complex Dimension (#VT8031).
Constituent values Profundity (#VC1371).

◆ **VP5224a Exteriority–Interiority**
Dynamics A fair exterior is a silent recommendation. (Publilius Syrus)
Happiness depends, as Nature shows, less on exterior things than most suppose (Conper)
Man was formed by his struggle with exterior forces and it is only easy for him to discern things which are outside of himself. (José Ortega y Gasset)
Integrative complex Contextuality (#VT8032).

◆ **VP5226b Centrality–Environment**
Dynamics Adapt yourself to the environment in which your lot has been cast, and show true love to the fellow-mortals with whom destiny has surrounded you. (Marcus Aurelius)
Never have people been more the masters of their environment. Yet never has a people felt more deceived and disappointed. For never has a people expected so much more than the world could offer. (Boorstin)
Many people live in ugly wastelands, but in the absence of imaginative standards, most of them do not even know it. (C Wright Mills)
Integrative complex Contextuality (#VT8032).
Constituent values Concentration (#VC0403)
Constituent values Restriction (#VD3546)
Confinement (#VD2268).
Limitedness (#VD3154)

◆ **VP5234c Circumscription–Intrusion**
Dynamics Eyes, you know, are the great intruders. (Erving Goffman)
Hell hath no limits, nor is circumscribed in one safe place; for where we are is hell, and where hell is there must we ever be. (Shakespeare)
Personal space refers to an area with invisible boundaries surrounding a person's body into which intruders may not come. (Robert Sommer)
Integrative complex Contextuality (#VT8032).
Constituent values Determination (#VC0547).
Constituent values Trespass (#VD3850)
Presumption (#VD3478)
Intervention (#VD3070)
Impertinence (#VD2844)
Intrusiveness (#VD3086).
Invasion (#VD3090)
Insinuation (#VD3046)
Interference (#VD3068)
Encroachment (#VD2604)

◆ **VP5246a Form–Formlessness**
Dynamics Fine words and an insinuating appearance are seldom associated with true virtue. (Confucius)
It is only shallow people who do not judge by appearances. The true mystery of the world is the visible, not the invisible. (Oscar Wilde)
Integrative complex Structure (#VT8033).
Constituent values Creation (#VC0499)
Constituent values Chaos (#VD2224)
Disorder (#VD2526)
Obscurity (#VD3346)
Confusion (#VD2272)
Orderlessness (#VD3370)
Featurelessness (#VD2686)
Appearance (#VC0145).
Muddle (#VD3292)
Vagueness (#VD3978)
Messiness (#VD3220)
Shapelessness (#VD3668)
Indecisiveness (#VD2950)
Characterlessness (#VD2226).

◆ **VP5248b Symmetry–Distortion**
Dynamics The history of our race, and each individual's experience, are sown thick with evidence that a truth it is not hard to kill and that a lie told well is immortal. (Mark Twain)
Symmetry is ennui, and ennui is the very essence of grief and melancholy. Despair yawns. (Victor Hugo)
Integrative complex Structure (#VT8033).
Constituent values Harmony (#VC0887)
Symmetry (#VC1729)
Equality (#VC0667)
Congruity (#VC0429)
Regularity (#VC1445)
Equilibrium (#VC0671)
Coordination (#VC0485)
Constituent values Bias (#VD2174)
Misuse (#VD3284)
Blemish (#VD2184)
Asymmetry (#VD2110)
Perversion (#VD3444)
Distortion (#VD2562)
Deformation (#VD2358)
Malformation (#VD3190)
Misconstruction (#VD3248)
Misinterpretation (#VD3264)
Balance (#VC0215)
Evenness (#VC0687)
Well-made (#VC1911)
Uniformity (#VC1821)
Conformity (#VC0425)
Consistency (#VC0451)
Correspondence (#VC0491).
Slant (#VD3678)
Torture (#VD3828)
Deviation (#VD2428)
Warpedness (#VD4044)
Mutilation (#VD3300)
Corruption (#VD2294)
Misdirection (#VD3252)
Disfigurement (#VD2498)
Misrepresentation (#VD3270)

◆ **VP5265c Opening–Closure**
Dynamics Strategy requires sensitive alternation of opening and closing according to advantages perceived. Therefore, openness or closedness is not a fixed structural feature of a system but a changing life strategy of organisms and groups. (Klapp)
Have an open face, but conceal your thoughts. (Italian proverb)
It is nothing won to admit men with an open door, and to receive them with a shut and reserved countenance. (Bacon)
Integrative complex Structure (#VT8033).
Constituent values Openness (#VC1225)
Staunchness (#VC1665).
Constituent values Barrier (#VD2148)
Foulness (#VD2710)
Obstruction (#VD3356).
Clearness (#VC0339)
Obstacle (#VD3352)
Congestion (#VD2274)

◆ **VP5267a Motion–Quiescence**
Dynamics Certainty generally is illusion, and repose is not the destiny of man. (Oliver Wendell Holmes, Jr)
What a day may bring a day may take away. (Thomas Fuller)
Integrative complex Motion (#VT8034).
Constituent values Action (#VC0039)
Quietude (#VC1405)
Calmness (#VC0281)
Stillness (#VC1673)
Composure (#VC0391)
Tranquillity (#VC1773)
Constituent values Torpor (#VD3826)
Languor (#VD3128)
Suspense (#VD3782)
Indolence (#VD2972)
Stuffiness (#VD3740)
Lifelessness (#VD3152)
Serenity (#VC1601)
Mobility (#VC1155)
Activity (#VC0041)
Quietness (#VC1403)
Quiescence (#VC1401)
Apathy (#VD2090)
Inertia (#VD2990)
Passivity (#VD3430)
Closeness (#VD2236)
Stagnation (#VD3722)
Indifference (#VD2958).

◆ **VP5269a Swiftness–Slowness**
Dynamics It is not strength, but art, obtains the prize, and to be swift is less than to be wise. (Homer)
Too swift arrives as tardy as too slow. (Shakespeare)
Wisely, and slow. They stumble that run fast. (Shakespeare)
Slowness of movement in a quick-thinking person makes you feel some complication of thought or feeling behind anything that is done. (Bowen)
There is a slowness in affairs which ripens them, and a slowness which rots them. (Roux)
Integrative complex Motion (#VT8034).
Constituent values Gain (#VC0817)
Constituent values Slowness (#VD3682).
Swiftness (#VC1727).

◆ **VP5283b Impact–Reaction**
Dynamics Attack is the reaction; I never think I have hit hard unless it rebounds. (Johnson)
The meeting of two personalities is like the contact of two chemical substances: if there is any reaction, both are transformed. (Jung)
Religion is a man's total reaction upon life. (James)
To every action there is always opposed an equal reaction: or, the mutual actions of two bodies upon each other are always equal, and directed to contrary parts. (Newton)
Integrative complex Motion (#VT8034).
Constituent values Clout (#VC0345)
Motive (#VC1171).

◆ **VP5288c Attraction–Repulsion**
Dynamics Young man, there is America – which at this day serves for little more than to amuse you with stories of savage men and uncouth manners; yet shall, before you taste of death, show itself equal to the whole of that commerce which now attracts the envy of the world. (Burke)
Like attracts like. (Anonymous)
Integrative complex Motion (#VT8034).
Constituent values Affinity (#VC0079)
Allurement (#VC0107).
Constituent values Repulsion (#VD3540)
Attraction (#VC0199)
Disaffinity (#VD2454).

◆ **VP5290a Direction–Deviation**
Dynamics Nothing is beneath you if it is in the direction of your life. (Emerson)
For all the compasses in the world, there's only one direction, and time is its only measure. (Tom Stoppard)
But we, we have no sense of direction; impetus is all we have. We do not proceed, we only roll down the mountain, like disbalanced boulders, crushing before us many delicate springing things, whose plan it was to grow. (Millay)
If all pulled in one direction, the world would keel over. (Yiddish proverbs)
Integrative complex Relative motion (#VT8035).
Constituent values Immediacy (#VC0941)
Directness (#VC0563)
Straightforwardness (#VC1679)
Constituent values Slant (#VD3678)
Divergence (#VD2576)
Indirection (#VD2962)
Direction (#VC0561)
Forthrightness (#VC0789)
Variation (#VD3994)
Aberration (#VD2008)
Deviousness (#VD2432)

VALUE POLARITIES

♦ **VP5292a Leading–Following**
Dynamics To get others to come into our ways of thinking, we must go over to theirs; and it is necessary to follow, in order to lead. (Hazlitt)
He who has never learned to obey cannot be a good commander. (Aristotle)
For the most part our leaders are merely following out in front; they do but marshal us the way that we are going. (Bergen Evans)
The weaknesses of the many make the leader possible. (Elvert Hubbard)
No man is great enough or wise enough for any of us to surrender our destiny to. The only way in which any one can lead us is to restore to us the belief in our own guidance. (Henry Miller)
The real leader has no need to lead – he is content to point the way. (Henry Miller)
  Integrative complex Relative motion (#VT8035).
  Constituent values Priority (#VC1355).

♦ **VP5294b Progression–Regression**
Dynamics A thousand things advance; nine hundred and ninety-nine retreat: that is progress. (Amiel)
All that is human must retrograde if it do not advance. (Gibbon)
Is it progress if a cannibal uses knife and fork ? (Stanislaw Lec)
The desire to understand the world and the desire to reform it are the two great engines of progress, without which human society would stand still or retrogress. (Bertrand Russell)
The fatal metaphor of progress which means leaving things behind us, has utterly obscured the real idea of growth, which means leaving things inside us. (G K Chesterton)
  Integrative complex Relative motion (#VT8035).
  Constituent values Progress (#VC1373).
  Constituent values Lapse (#VD3130)   Reversion (#VD3558)
  Withdrawal (#VD4062)   Regression (#VD3524)
  Retrogression (#VD3554).

♦ **VP5296b Approach–Recession**
Dynamics In bed we laugh, in bed we cry; and, born in bed, in bed we die. The near approach a bed may show of human bliss to human woe. (de Benserade)
  Integrative complex Relative motion (#VT8035).
  Constituent values Accessibility (#VC0019)   Approachability (#VC0155).

♦ **VP5298b Convergence–Divergence**
Dynamics I shall be telling this with a sigh, somewhere ages and ages hence: two roads diverged in a wood, and I – I took the one less travelled by, and that has made all the difference. (Frost)
Infinitely often it is clear that we appreciate, even respect – not a multitude – but ten people gathered in a room, each of whom, taken by himself, we consider of no account. (Leopardi)
  Integrative complex Relative motion (#VT8035).
  Constituent values Concurrence (#VC0413)   Concentration (#VC0403)
  Approachability (#VC0155).
  Constituent values Division (#VD2580)   Divergence (#VD2576).

♦ **VP5313b Overrunning–Shortcoming**
Dynamics Men fall from great fortune because of the same short comings that led to their rise. (La Bruyère)
Our shortcomings are the eyes with which we see the ideal. (Nietzsche)
  Integrative complex Relative motion (#VT8035).
  Constituent values Failure (#VD2660)   Default (#VD2344)
  Decline (#VD2336)   Shortage (#VD3674)
  Inadequacy (#VD2890)   Inferiority (#VD3002)
  Imperfection (#VD2842)   Encroachment (#VD2604)
  Insufficiency (#VD3060).

♦ **VP5317c Elevation–Depression**
Dynamics I know of no more encouraging fact than the unquestionable ability of man to elevate his life by a conscious endeavour. (Thoreau)
If the heart of a man is depressed with cares, the mist is dispelled when a woman appears. (Gay)
If we didn't live venturously, plucking the wild goat by the beard, and trembling over precipices, we should never be depressed, I've no doubt; but already should be faded, fatalistic, and aged. (Virginia Woolf)
Noble deeds and hot baths are the best cure for depression. (Smith)
And whosoever shall exalt himself shall be abased; and he that shall humble himself shall be exalted. (Bible)
  Integrative complex Relative motion (#VT8035).
  Constituent values Increase (#VC0965)   Elevation (#VC0625)
  Exaltation (#VC0695)   Aspiration (#VC0179)
  Ascendancy (#VC0175)   Deification (#VC0529).
  Constituent values Overturn (#VD3412)   Depression (#VD2396)
  Degradation (#VD2362).

♦ **VP5323c Oscillation–Agitation**
Dynamics The whole history of the progress of human liberty shows that all concesssions yet made to her august claims have been born of earnest struggle... If there is no struggle, there is no progress. Those who profess to favour freedom, and yet deprecate agitation, are men who want crops without plowing up the ground, they want rain without thunder and lightning. They want the ocean without the awful roar of its many waters. (Frederick Douglass)
  Integrative complex Relative motion (#VT8035).
  Constituent values Fuss (#VD2734)   Upset (#VD3966)
  Malaise (#VD3184)   Disquiet (#VD2540)
  Disorder (#VD2526)   Turbidity (#VD3864)
  Agitation (#VD2046)   Turbulence (#VD3866)
  Spasticity (#VD3706)   Inquietude (#VD3030)
  Trepidation (#VD3848)   Nervousness (#VD3314)
  Disturbance (#VD2570)   Restlessness (#VD3544)
  Perturbation (#VD3442)   Discomposure (#VD2472)
  Spasmodicness (#VD3704)   Unpeacefulness (#VD3934).

♦ **VP5328a Heat–Cold**
Dynamics Continuity in everything is unpleasant. Cold is agreeable that we may get warm. (Pascal)
For all who move in the mortal sun know halfway warm is better than freezing, as half a love is better than none. (McGinley)
Everyone feels the heat, but not everyone feels the cold. (Chinese proverb)
  Integrative complex Absolute properties (#VT8041).
  Constituent values Warmth (#VC1899)   Freshness (#VC0799)
  Constituent values Coldness (#VD2242)   Closeness (#VD2236)
  Extinction (#VD2650).

♦ **VP5335b Light–Darkness**
Dynamics It is better to light a candle than to curse the darkness. (Chinese proverb)
The best way to see divine light is to put out thine own candle. (Fuller)
Sadness flies on the wings of the morning and out of the heart of darkness comes the light. (Giraudoux)
The darkness of night, like pain, is dumb; the darkness of dawn, like peace, is silent. (Rabindranath Tagore)
  Integrative complex Absolute properties (#VT8041).
  Constituent values Light (#VC1077)   Glory (#VC0841)
  Radiance (#VC1409)   Lucidity (#VC1097)
  Splendour (#VC1643)   Lightness (#VC1079)
  Clearness (#VC0339)   Brilliance (#VC0267)
  Brightness (#VC0265)   Resplendence (#VC1487)
  Illumination (#VC0937)   Enlightenment (#VC0653).
  Constituent values Deadness (#VD2322)   Darkness (#VD2320)
  Lifelessness (#VD3152).

♦ **VP5339b Transparency–Opaqueness**
Dynamics I wish that every human life might be pure transparent freedom. (de Beauvoir)
It is the bright, the bold, the transparent who are cleverest among those who are silent: their ground is down so deep that even the brightest water does not betray it. (Nietzsche)
There's no one so transparent as the person who thinks he's devilish deep. (Maugham)
  Integrative complex Absolute properties (#VT8041).

♦ **VP5352c Weight–Lightness**
Dynamics Wearing all that weight of learning lightly like a flower. (Tennyson)
None knows the weight of another's burden. (Herbert)
A simple child that lightly draws its breath, and feels its life in every limb; what should it know of death ? (Wordsworth)
May the earth rest lightly on you. (Anonymous)
  Integrative complex Absolute properties (#VT8041).
  Constituent values Unity (#VC1829)   Light (#VC1077)
  Buoyancy (#VC0277)   Lightness (#VC1079)
  Hardiness (#VC0883)   Coherence (#VC0351)
  Tenderness (#VC1751).
  Constituent values Overtax (#VC3410)   Overload (#VD3396)
  Overweight (#VD3414)   Congestion (#VD2274)
  Encumbrance (#VD2606)   Burdensomeness (#VD2206).

♦ **VP5356a Hardness–Softness**
Dynamics There is a homely adage which runs: 'Speak softly and carry a big stick; you will go far'. (Theodore Roosevelt)
The bitter and the sweet come from the outside, the hard from within, from one's own efforts. (Einstein)
What was hard to bear is sweet to remember. (Portuguese proverb)
Harsh words can be a medicine; gentle words encourage illness. (Chinese proverb)
  Integrative complex Relative properties (#VT8042).
  Constituent values Resistance (#VC1477)   Compliance (#VC0389)
  Sensibility (#VC1591)   Flexibility (#VC0771)
  Tractability (#VC1769)   Responsiveness (#VC1491).
  Constituent values Stubbornness (#VD3738)   Inflexibility (#VD3008)
  Intractability (#VD3078).

♦ **VP5358b Elasticity–Toughness**
Dynamics To him whose elastic and vigorous thought keeps pace with the sun, the day is a perpetual morning. (Thoreau)
The time which we have at our disposal every day is elastic; the passions that we feel expand it, those that we inspire contract it, and habit fills up what remains. (Proust)
Life's a tough proposition, and the first hundred years are the hardest. (Mizner)
  Integrative complex Relative properties (#VT8042).
  Constituent values Buoyancy (#VC0277)   Toughness (#VC1767)
  Hardiness (#VC0883)   Resistance (#VC1477)
  Resilience (#VC1475)   Flexibility (#VC0771)
  Adaptability (#VC0049)   Responsiveness (#VC1491).
  Constituent values Hardness (#VD2770)   Stiffness (#VD3730)
  Fragility (#VD2712)   Vulnerability (#VD4036).

♦ **VP5362c Colour–Colourlessness**
Dynamics Colour speaks all languages. (Addison)
Grey is a colour that always seems on the eve of changing to some other colour. (Chesterton)
The purest and mot thoughtful minds are those which love colour the most. (Ruskin)
The Lord so constituted everybody that no matter what colour you are you require the same amount of nourishment. (Will Rogers)
Naked I came into the world, naked I shall go out of it. And a very good thing too, for it reminds me that I am naked under my shirt, whatever its colour. (Forster)
  Integrative complex Relative properties (#VT8042).
  Constituent values Loudness (#VD3168)   Gaudiness (#VD2738).

♦ **VP5376a Materiality–Immateriality**
Dynamics High thinking is inconsistent with complicated material life based on high speed imposed on us by Mammon worship. (Gandhi)
What is at the heart of all our national problems ? It is that we have seen the hand of material interest sometimes about to close upon our dearest rights and possessions. (Woodrow Wilson)
A tough but nervous, tenacious but restless race the Yankees; materially ambitious, yet prone to introspection, and subject to waves of religious emotion... A race whose typical member is eternally torn between a passion for righteousness and a desire to get on in the world. (Morison)
  Integrative complex Life (#VT8043).
  Constituent values Universality (#VC1831)   Spirituality (#VC1641).

♦ **VP5407a Life–Death**
Dynamics Death is the supreme festival on the road to freedom. (Dietrich Bonhoeffer)
You never know what life, means till you die; even throughout life, 'tis death that makes life live, gives it whatever the significance. (Robert Browning)
Death holds no horrors. It is simply the ultimate horror of life. (Jean Giraudoux)
Death takes away the commonplace of life. (Alexander Smith)
Life and fame and wealth – all these must, I say, be defended by fighting. Death in battle is the most glorious for men. Who lives under the sway of his foe – it is he that is dead. (Panchatantra)
Some people seem to think that death is the only reality in life. Others, happier and rightlier minded,

VP5407

see and feel that life is the true reality in death. (Julius Charels Hare and Augustus William Hare)
Integrative complex Life (#VT8043).
Constituent values Life (#VC1075)
Vitality (#VC1891)
Quickness (#VC1399)
Existence (#VC0705)
Animation (#VC0135)
Liveliness (#VC1085)
Immortality (#VC0947)
Vivacity (#VC1893)
Viability (#VC1875)
Longevity (#VC1091)
Endurance (#VC0645)
Aliveness (#VC0101)
Transcience (#VC1777)
Consciousness (#VC0443).
Constituent values Death (#VD2324)
Disaster (#VD2466)
Virulence (#VD4024)
Malignancy (#VD3194)
Suffocation (#VD3760)
Bereavement (#VD2166)
Lifelessness (#VD3152)
Annihilation (#VD2074)
Destructiveness (#VD2422)
Bloody-mindedness (#VD2188).
Eeriness (#VD2596)
Calamity (#VD2208)
Sacrifice (#VD3592)
Extinction (#VD2650)
Dissolution (#VD2556)
Banefulness (#VD2140)
Gruesomeness (#VD2756)
Extermination (#VD2648)
Self-destruction (#VD3634)

♦ **VP5417b Humanity–Nonhumanity**
Dynamics Be ashamed to die until you have won some victory for humanity. (Horace Mann)
After all there is but one race – humanity. (Moore)
The life of humanity upon this planet may yet come to an end, and a very terrible end. But I would have you notice that this end is threatened in our time not by anything that the universe may do to us, but only by what man may do to himself. (Holmes)
I have thought some of nature's journeymen had made men and not made them well they imitated humanity so abominably. (Shakespeare)
The worst sin towards our fellow creatures is not to hate them but to be indifferent to them: that's the essence of inhumanity. (Shaw)
Man's inhumanity to man makes countless thousands mourn. (Burns)
Integrative complex Life (#VT8043).
Constituent values Humanity (#VC0927)
Cosmopolitan (#VC0493).
Humanism (#VC0923)

♦ **VP5419c Sexiness–Unsexiness**
Dynamics The degree and kind of a man's sexuality reach up into the ultimate pinnacle of his spirit. (Nietzsche)
Civilized people cannot fully satisfy their sexual instinct without love. (Bertrand Russell)
Integrative complex Life (#VT8043).
Constituent values Love (#VC1093)
Sexuality (#VD3662).
Sex appeal (#VC1609).

♦ **VP5420c Masculinity–Femininity**
Dynamics A woman's guess is much more accurate than a man's certainty. (Kipling)
How men hate waiting while their wives shop for clothes and trinkets; how women hate waiting, often for much of their lives, while their husbands shop for fame and glory. (Szasz)
A women who strives to be like a man lacks ambition. (Graffito)
The great question which I have not been able to answer, despite my thirty years of research into the feminine soul, is 'What does a woman want?' (Freud)
When she stopped conforming to the conventional picture of feminity she finally began to enjoy being a woman. (Friedan)
The human race, in its intellectual life, is organized like the bees: the masculine soul is a worker, sexually atrophied, and essentially dedicated to impersonal and universal arts; the feminine is a queen, infinitely fertile, omnipresent in its brooding industry, but passive and abounding in intuitions without method and passions without justice. (Santayana)
Integrative complex Life (#VT8043).
Constituent values Potency (#VC1327)
Feminity (#VC0757)
Womanliness (#VC1931).
Machismo (#VC1103)
Manliness (#VC1115)

♦ **VP5422a Sensation–Insensibility**
Dynamics There are times when fear is good. It must keep its watchful place at the heart's controls. There is advantage in the wisdom won from pain. (Aeschylus)
Suffering is the sole origin of consciousness. (Dostoevsky)
It is axiomatic that we should all think of ourselves as being more sensitive than other people because, when we are insensitive in our dealings with others, we cannot be aware of it at the time: conscious insensitivity is a self-contradiction. (W H Auden)
When people fall in true distress, their native sense departs. (Sophocles)
A deep distress hath humanised my Soul. (William Wordsworth)
To the dull mind all nature is leaden. To the illumined mind the whole world burns and sparkles with light. (Emerson)
Integrative complex Sense (#VT8044).
Constituent values Tact (#VC1737)
Empathy (#VC0635)
Keenness (#VC1043)
Sensation (#VC1589)
Tenderness (#VC1751)
Sensibility (#VC1591)
Consciousness (#VC0443)
Constituent values Pain (#VD3422)
Misery (#VD3254)
Torment (#VD3824)
Anguish (#VD2068)
Dullness (#VD2588)
Deadness (#VD2322)
Sufferance (#VD3758)
Hurtfulness (#VD2806)
Irritability (#VD3118)
Insensitivity (#VD3040)
Acumen (#VC0047)
Sympathy (#VC1731)
Delicacy (#VC0535)
Awareness (#VC0211)
Sensitivity (#VC1595)
Exquisiteness (#VC0721)
Impressivenss (#VC0959).
Wound (#VD4074)
Torture (#VD3828)
Malaise (#VD3184)
Soreness (#VD3694)
Distress (#VD2566)
Atrocity (#VD2114)
Discomfort (#VD2470)
Tortuousness (#VD3830)
Spasmodicness (#VD3704)
Insensibility (#VD3038).

♦ **VP5425b Tangibility–Intangibility**
Dynamics If we suppose that physical events can be reduced to spatial motions of material particles we impose on the creations of thought the limitations of the visible and tangible. (Ernest Mach)
The Tartar saw a pudding in his dream, but he didn't have a spoon; he took a spoon to bed – he didn't dream about the pudding. (Russian Proverb)
A person who says that a meeting will not be held because he is absent, deceives himself. (Nigerian Proverb)
Integrative complex Sense (#VT8044).
Constituent values Tangibility (#VC1743)
Constituent values Callousness (#VD2210).

♦ **VP5427b Savouriness–Unsavouriness**
Dynamics You are the salt of the earth: but if the salt have lost his savour, wherewith shall it be salted ? (Bible)
Sweet are the thoughts that savour of content; the quiet mind is richer than a crown. (Robert Greene)
Wisdom and goodness to the vile seem vile; filths savour but themselves. (Shakespeare)
Integrative complex Sense (#VT8044).
Constituent values Good (#VC0845)
Sweetness (#VC1725)
Constituent values Noisome (#VD3316)
Disgust (#VD2502)
Vapidity (#VD3990)
Foulness (#VD2710)
Mildness (#VC1149)
Plausibility (#VC1303).
Inanity (#VD2896)
Weakness (#VD4052)
Thinness (#VD3808)
Deadness (#VD2322).

♦ **VP5435b Fragrance–Stench**
Dynamics The smell of garlic takes away the smell of onions. (Proverb)
Just as the first taste of inferior tea has its fragrance even a devil may be pretty at eighteen. (Japanese Proverb)
Nothing awakens a reminiscence like an odour. (Victor Hugo)
A rose too often smelled loses its fragrance. (Spanish Proverb)
A piece of incense may be as large as the knee but, unless burnt, emits no fragrance. (Malay Proverb)
What's in a name ? That which we call a rose by any other name would smell as sweet. (Shakespeare)
Integrative complex Sense (#VT8044).
Constituent values Sweetness (#VC1725).
Constituent values Miasma (#VD3222)
Foulness (#VD2710)
Stuffiness (#VD3740)
Badness (#VD2134)
Mustiness (#VD3294)
Fulsomeness (#VD2728).

♦ **VP5439b Vision–Blindness**
Dynamics Genius seems to consist merely in trueness of sight, in using such words as show that the man was an eye-witness, and not a repeater of what was told. (Emerson)
Integrative complex Sense (#VT8044).
Constituent values Darkness (#VD2320).

♦ **VP5444b Visibility–Invisibility**
Dynamics A matter that becomes clear ceases to concern us. (Nietzsche)
Obscurity often brings safety. (Aesop)
Obscurity is the refuge of incompetence. (Robert A Heinlein)
Integrative complex Sense (#VT8044).
Constituent values Clarity (#VC0335)
Perspective (#VC1283).
Constituent values Darkness (#VD2320)
Obscurity (#VD3346)
Uncertainty (#VD3880).
Clearness (#VC0339)
Vagueness (#VD3978)
Feebleness (#VD2690)

♦ **VP5446b Appearance–Disappearance**
Dynamics Anyone who has looked deeply into the world may guess how much wisdom lies in the superficiality of men. The instinct that reserves them teaches them to be flighty, light, and false. (Nietzsche)
Integrative complex Sense (#VT8044).
Constituent values Revelation (#VC1497)
Perspective (#VC1283).
Constituent values Extinction (#VD2650)
Dissipation (#VD2552)
Disappearance (#VD2458)
Appearance (#VC0145)
Dissolution (#VD2556)
Speciousness (#VD3708)
Superficiality (#VD3766).

♦ **VP5448b Audibility–Inaudibility**
Integrative complex Sense (#VT8044).

♦ **VP5451b Silence–Loudness**
Dynamics Liberty is the right to silence. (Graffito written during French student revolt)
There is the silence of age, too full of wisdom for the tongue to utter it in words intelligible to those who have not lived the great range of life. (Edgard Lee Masters)
In silence alone does a man's truth bind itself together and strike root. (Saint-Exupéry)
Integrative complex Sense (#VT8044).
Constituent values Stillness (#VC1673)
Tranquillity (#VC1773).
Constituent values Discord (#VD2480)
Loudness (#VD3168).
Quietness (#VC1403)
Muteness (#VD3298)

♦ **VP5462c Harmony–Discord**
Dynamics You can achieve victory better by deliberation than by wrath. (Proverb)
Never cut what you can untie. (Portuguese Proverb)
All your strength is in your union. All your danger is in discord; Therefore be at peace hence forward, And as brothers live together. (Longfellow)
We will not think noble because we are not noble. We will not live in beautiful harmony because there is no such thing in this world, nor should there be. We promise only to do our best and live out our lives. Dear God, that's all we can promise in truth. (Hellman)
Integrative complex Sense (#VT8044).
Constituent values Accord (#VC0025)
Euphony (#VC0685)
Flourish (#VC0773)
Resolution (#VC1479)
Constituent values Discord (#VD2480)
Dissonance (#VD2558)
Romance (#VC1511)
Richness (#VC1503)
Sacredness (#VC1515)
Adaptability (#VC0049).
Harshness (#VD2774)
Disharmony (#VD2504).

♦ **VP5467a Intelligence–Unintelligence**
Dynamics The public is wonderfully tolerant. It forgives everything except genius. (Oscar Wilde)
Our credulity is greatest concerning the things we know least about. And since we know least about ourselves, we are ready to believe all that is said about us. Hence the mysterious power of both flattery and calumny. (Eric Hoffer)
Stupidity often saves a man from going mad. (Oliver Wendell)
Nothing in all the world is more dangerous than sincere ignorance and conscientious stupidity. (Martin Luther King)
All that is necessary to raise imbecility into what the mob regards as profundity is to lift it off the floor and put it on a platform. (George Jean Nathan)
It is a profitable thing, if one is wise, to seem foolish. (Aeschylus)
The bold are helpless without cleverness. (Euripides)
Here's a good rule of thumb: too clever is dumb. (Ogden Nash)
The wisdom of providence is as much revealed in the rarity of genius, as in the circumstance that

VALUE POLARITIES

not everyone is deaf or blind. (Georg Christoph Lichtenberg)
The test of a first rate intelligence is the ability to hold two opposed ideas in the mind at the same time, and still retain the ability to function. (F Scott Fitzgerald)
The greatest intelligence is precisely the one that suffers most from its own limitations. (André Gide)
Intelligence is silence, truth being invisible. But what a racket I make in declaring this. (Ned Rorem)
Integrative complex Intellectual faculties (#VT8045).
Constituent values Wisdom (#VC1927)
Sanity (#VC1531)
Genius (#VC0825)
Acumen (#VC0047)
Insight (#VC0991)
Balance (#VC0215)
Sagacity (#VC1519)
Learning (#VC1059)
Capacity (#VC0289)
Soundness (#VC1637)
Quickness (#VC1399)
Judgement (#VC1037)
Erudition (#VC0677)
Authority (#VC0205)
Alertness (#VC0099)
Nimbleness (#VC1191)
Discretion (#VC0571)
Conception (#VC0405)
Brilliance (#VC0267)
Sensibility (#VC1591)
Perspicuity (#VC1287)
Inspiration (#VC0993)
Discernment (#VC0565)
Sensibleness (#VC1593)
Perspicacity (#VC1285)
Common sense (#VC0363)
Judiciousness (#VC1039)
Consideration (#VC0449)
Comprehension (#VC0393)
Reasonableness (#VC1425)
Circumspection (#VC0327)
Broadmindedness (#VC0269).
Talent (#VC1739)
Reason (#VC1423)
Esprit (#VC0679)
Acuity (#VC0045)
Cogency (#VC0349)
Sobriety (#VC1623)
Prudence (#VC1387)
Keenness (#VC1043)
Aptitude (#VC0163)
Sensation (#VC1589)
Knowledge (#VC1047)
Foresight (#VC0783)
Dexterity (#VC0553)
Aliveness (#VC0101)
Profundity (#VC1371)
Giftedness (#VC0833)
Creativity (#VC0501)
Cleverness (#VC0343)
Brightness (#VC0265)
Rationality (#VC1415)
Percipience (#VC1271)
Forethought (#VC0785)
Carefulness (#VC0293)
Practicality (#VC1335)
Intelligence (#VC1001)
Understanding (#VC1811)
Enlightenment (#VC0653)
Consciousness (#VC0443)
Thoughtfulness (#VC1755)
Discrimination (#VC0573)
Intellectuality (#VC0999).
Constituent values Opacity (#VD3362)
Inanity (#VD2896)
Decline (#VD2336)
Vapidity (#VD3990)
Softness (#VD3688)
Nonsense (#VD3332)
Lethargy (#VD3144)
Dullness (#VD2588)
Puerility (#VD3506)
Grossness (#VD2754)
Absurdity (#VD2022)
Triviality (#VD3856)
Ineptitude (#VD2984)
Imprudence (#VD2874)
Imbecility (#VD2830)
Witlessness (#VD4066)
Retardation (#VD3550)
Inattention (#VD2906)
Mindlessness (#VD3228)
Indiscretion (#VD2964)
Eccentricity (#VD2594)
Childishness (#VD2234)
Senselessness (#VD3648)
Insensibility (#VD3038)
Credulousness (#VD2300)
Superficiality (#VD3766)
Inadvisability (#VD2894)
Thoughtlessness (#VD3810)
Unreasonableness (#VD3946)
Incomprehensibility (#VD2924)
Madness (#VD3178)
Fatuity (#VD2680)
Weakness (#VD4052)
Thinness (#VD3808)
Slowness (#VD3682)
Naïvety (#VD3304)
Insanity (#VD3034)
Stupidity (#VD3742)
Ignorance (#VD2818)
Frivolity (#VD2724)
Volatility (#VD4030)
Stuffiness (#VD3740)
Incapacity (#VD2910)
Immaturity (#VD2832)
Feebleness (#VD2690)
Shallowness (#VD3664)
Infatuation (#VD2998)
Deviousness (#VD2432)
Inexpedience (#VD2992)
Gullibleness (#VD2762)
Decrepitness (#VD2340)
Unadvisedness (#VD3874)
Irrationality (#VD3098)
Defectiveness (#VD2350)
Unintelligence (#VD3922)
Sentimentality (#VD3652)
Weak-mindedness (#VD4050)
Unthoughtfulness (#VD3956)
Inconsiderateness (#VD2930)
Arrested development (#VD2100).

♦ **VP5472a Sanity–Insanity**
Dynamics 'Mad' is a term we use to describe a man who is obsessed with one idea and nothing else. (Ugo Betti)
When we remember that we are all mad, the mysteries disappear and life stands explained. (Mark Twain)
There are times when sense may be unreasonable, as well as truth. (William Congreve)
That which is virtue in season is madness out of season, as when an old man makes love. (George Santayana)
Alienation ends where yours begins. (Graffito written during French student revolt)
Integrative complex Intellectual faculties (#VT8045).
Constituent values Sanity (#VC1531)
Sobriety (#VC1623)
Soundness (#VC1637)
Normality (#VC1201)
Rationality (#VC1415)
Wholesomeness (#VC1921)
Balance (#VC0215)
Lucidity (#VC1097)
Rightness (#VC1507)
Possession (#VC1325)
Sensibleness (#VC1593)
Reasonableness (#VC1425).
Constituent values Fury (#VD2732)
Madness (#VD3178)
Zealotry (#VD4082)
Insanity (#VD3034)
Delerium (#VD2368)
Extremism (#VD2656)
Alienation (#VD2052)
Witlessness (#VD4066)
Infatuation (#VD2998)
Distraction (#VD2564)
Abnormality (#VD2014)
Unnaturalness (#VD3930)
Irrationality (#VD3098)
Disorientation (#VD2530)
Overzealousness (#VD3418)
Frenzy (#VD2716)
Bigotry (#VD2176)
Sickness (#VD3676)
Fixation (#VD2702)
Obsession (#VD3348)
Fanaticism (#VD2676)
Aberration (#VD2008)
Strangeness (#VD3734)
Incoherence (#VD2916)
Bedevilment (#VD2160)
Extravagance (#VD2654)
Senselessness (#VD3648)
Excessiveness (#VD2632)
Compulsiveness (#VD2258)
Overreligiousness (#VD3398).

♦ **VP5475b Knowledge–Ignorance**
Dynamics If others had not been foolish, we should be so. (William Blake)
It is fortunate that each generation does not comprehend its own ignorance. We are thus enabled to call our ancestors barbarous. (Charles Dudley Warner)
It is better to have wisdom without learning, than to have learning without wisdom; just as it is better to be rich without being the possessor of a mine, than to be the possessor of a mine without being rich. (Charles Caleb Colton)
Our experience is composed rather of illusions lost than of wisdom acquired. (Joseph Roux)
A man who is so dull that he can learn only by personal experience is too dull to learn anything important by experience. (Donn Marquis)
To be conscious that you are ignorant is a great step to knowledge. (Benjamin Disraeli)
From ignorance our comfort flows, the only wretched are the wise. (Matthew Prior)
Knowing what thou knowest not is in a sense omniscience. (Piet Hein)
A desire of knowledge is the natural feeling of mankind; and every human being, whose mind is not debauched, will be willing to give all that he has to get knowledge. (Samuel Johnson)
The greater our knowledge increases, the greater our ignorance unfolds. (John F Kennedy)
Perfect understanding will sometimes almost extinguish pleasure. (A E Housman)
Men are wise in proportion, not to their experience, but to their capacity for experience. (George Bernard Shaw)
There's nothing in the world so unfair as a man who has no experience of life; he thinks nothing is done right except what he's doing himself. (Terence)
What we call education and culture is for the most part nothing but the substitution of reading for experience, of literature for life, of the obsolete fictitious for the contemporary real. (George Bernard Shaw)
There is much more learning than knowing in the world. (Thomas Fuller)
A learned fool is sillier than an ignorant one. (Molière)
Knowledge is recognition of something absent; it is a salutation, not an embrace. (George Santayana)
Integrative complex Intellectual faculties (#VT8045).
Constituent values Wisdom (#VC1927)
Insight (#VC0991)
Sagacity (#VC1519)
Intimacy (#VC1013)
Knowledge (#VC1047)
Erudition (#VC0677)
Awareness (#VC0211)
Conception (#VC0405)
Sensibility (#VC1591)
Realization (#VC1421)
Percipience (#VC1271)
Mindfulness (#VC1151)
Edification (#VC0609)
Acquisition (#VC0037)
Perspicacity (#VC1285)
Illumination (#VC0937)
Well-informed (#VC1909)
Understanding (#VC1811)
Consciousness (#VC0443)
Intellectuality (#VC0999).
Mastery (#VC1121)
Culture (#VC0507)
Learning (#VC1059)
Humanism (#VC0923)
Expertise (#VC0717)
Education (#VC0611)
Experience (#VC0715)
Attainment (#VC0195)
Recognition (#VC1427)
Proficiency (#VC1367)
Omniscience (#VC1219)
Familiarity (#VC0735)
Discernment (#VC0565)
Sensibleness (#VC1593)
Intelligence (#VC1001)
Appreciation (#VC0153)
Well-grounded (#VC1907)
Enlightenment (#VC0653)
Comprehension (#VC0393)
Constituent values Inanity (#VD2896)
Paganism (#VD3420)
Darkness (#VD2320)
Barbarism (#VD2142)
Shallowness (#VD3664)
Mindlessness (#VD3228)
Unfamiliarity (#VD3908)
Unintelligence (#VD3922)
Misinformation (#VD3262)
Savagery (#VD3608)
Naïvety (#VD3304)
Ignorance (#VD2818)
Witlessness (#VD4066)
Awkwardness (#VD2130)
Inexperience (#VD2994)
Insensibility (#VD3038)
Superficiality (#VD3766)
Incomprehensibility (#VD2924).

♦ **VP5478c Thought–Thoughtlessness**
Dynamics Men use thought only as authority for their injustice, and employ speech only to conceal their thoughts. (Voltaire)
The quality of the thought differences the Egyptian and the Roman, the Austrian and the American. (Emerson)
Thought is essentially practical in the sense that but for thought no motion would be an action, no change a progress. (George Santayana)
Reason is man's faculty for grasping the world by thought, in contradiction to intelligence, which is man's ability to manipulate the world with the help of thought. (Eric Fromm)
Integrative complex Intellectual faculties (#VT8045).
Constituent values Reason (#VC1423)
Philosophy (#VC1293)
Seriousness (#VC1603)
Association (#VC0187)
Consideration (#VC0449)
Thoughtfulness (#VC1755)
Intellectuality (#VC0999).
Idealism (#VC0935)
Conception (#VC0405)
Inspiration (#VC0993)
Abstraction (#VC0013)
Concentration (#VC0403)
Deliberateness (#VC0533)
Constituent values Inanity (#VD2896)
Oblivion (#VD3344)
Fatuity (#VD2680)
Unintelligence (#VD3922).

♦ **VP5481c Intuition–Reason**
Dynamics In what we really understand, we reason but little. (William Hazlitt)
A moment's insight is sometimes worth a life's experience. (Oliver Wendell)
Peace rules the day, where reason rules the mind. (William Collins)
All that is vital is irrational, and all that is rational is anti-vital, for reason is essentially skeptical. (Miguel de Unamuno)
Reason never acts in vain. (Chinese Proverb)
The reasons of the poor weigh not. (Proverb)
Integrative complex Intellectual faculties (#VT8045).
Constituent values Sanity (#VC1531)
Justice (#VC1041)
Feeling (#VC0751)
Interest (#VC1011)
Soundness (#VC1637)
Legitimacy (#VC1063)
Credibility (#VC0503)
Common sense (#VC0363)
Wholesomeness (#VC1921)
Intuitiveness (#VC1017)
Intellectuality (#VC0999).
Reason (#VC1423)
Insight (#VC0991)
Cogency (#VC0349)
Synthesis (#VC1735)
Philosophy (#VC1293)
Rationality (#VC1415)
Sensibleness (#VC1593)
Anticipation (#VC0139)
Well-grounded (#VC1907)
Reasonableness (#VC1425)
Constituent values Flaw (#VD2704)
Illogic (#VD2828)
Evasion (#VD2622)
Sophistry (#VD3690)
Perversion (#VD3444)
Feebleness (#VD2690)
Insincerity (#VD3044)
Controversy (#VD2288)
Equivocation (#VD2616)
Prevarication (#VD3482)
Irrationality (#VD3098)
Inconsequence (#VD2928)
Combativeness (#VD2246)
Wrangle (#VD4076)
Fallacy (#VD2666)
Vainness (#VD3980)
Subterfuge (#VD3752)
Invalidity (#VD3088)
Faultiness (#VD2682)
Incongruity (#VD2926)
Speciousness (#VD3708)
Senselessness (#VD3648)
Mystification (#VD3302)
Inconsistency (#VD2932)
Contradiction (#VD2284)
Unauthenticity (#VD3876)

VP5481

Misapplication (#VD3232)
Contentiousness (#VD2282)
Disputatiousness (#VD2538)
Argumentativeness (#VD2096).

Quarrelsomeness (#VD3512)
Unreasonableness (#VD3946)
Disingenuousness (#VD2512)

Assurance (#VC0189)
Confidence (#VC0419)
Acceptance (#VC0017)
Expectation (#VC0709)
Positiveness (#VC1323)
Faithfulness (#VC0731)
Consideration (#VC0449)
Impressivensss (#VC0959)
Trustworthiness (#VC1785).
Constituent values Heresy (#VD2790)
Weakness (#VD4052)
Naïvety (#VD3304)
Distrust (#VD2568)
Rejection (#VD3526)
Discredit (#VD2486)
Infidelity (#VD3006)
Infatuation (#VD2998)
Dubiousness (#VD2586)
Gullibleness (#VD2762)
Apprehension (#VD2092)
Faithlessness (#VD2662)
Unsophistication (#VD3952).

Conviction (#VC0479)
Conception (#VC0405)
Reliability (#VC1459)
Credibility (#VC0503)
Plausibility (#VC1303)
Acquiescence (#VC0035)
Persuasiveness (#VC1289)
Unequivocalness (#VC1815)

Denial (#VD2376)
Softness (#VD3688)
Mistrust (#VD3278)
Suspicion (#VD3786)
Misbelief (#VD3236)
Disbelief (#VD2468)
Uncertainty (#VD3880)
Incredulity (#VD2976)
Superstition (#VD3768)
Doubtfulness (#VD2584)
Unreliability (#VD3948)
Implausibility (#VD2852)

♦ **VP5485a Research–Discovery**
Dynamics One doesn't discover new lands without consenting to lose sight of the shore for a very long time. (André Gide)
A man of genius makes no mistakes. His errors are volitional and are the portals of discovery. (Joyce)
In completing one discovery we never fail to get an imperfect knowledge of others of which we could have no idea before, so that we cannot solve one doubt without creating several new ones. (Priestly)
When you steal from one author, it's plagiarism; if you steal from many, it's research. (Mizner)
Integrative complex Evaluation (#VT8051).
Constituent values Work (#VC1937)
Discovery (#VC0569)
Resolution (#VC1479)
Inventiveness (#VC1019)
Responsiveness (#VC1491)
Constituent values Inquisition (#VD3032).

Feeling (#VC0751)
Revelation (#VC1497)
Explanation (#VC0719)
Determination (#VC0547)

♦ **VP5492b Discrimination–Indiscrimination**
Dynamics Those who would administer wisely must, indeed, be wise, for one of the serious obstacles to the improvement of our race is indiscriminate charity. (Carnegie)
Integrative complex Evaluation (#VT8051).
Constituent values Tact (#VC1737)
Insight (#VC0991)
Subtlety (#VC1695)
Precision (#VC1341)
Refinement (#VC1441)
Sensibility (#VC1591)
Discernment (#VC0565)
Perspicacity (#VC1285)
Judiciousness (#VC1039)
Constituent values Muddle (#VD3292)
Imprudence (#VD2874)
Promiscuity (#VD3500)
Indiscretion (#VD2964)
Insensibility (#VD3038)

Acumen (#VC0047)
Finesse (#VC0765)
Accuracy (#VC0029)
Judgement (#VC1037)
Sensitivity (#VC1595)
Distinction (#VC0577)
Tastefulness (#VC1745)
Appreciation (#VC0153)
Discrimination (#VC0573).
Division (#VD2580)
Segregation (#VD3624)
Tactlessness (#VD3788)
Insensitivity (#VD3040)
Indiscrimination (#VD2966).

♦ **VP5494b Judgement–Misjudgement**
Dynamics Exaggeration is a prodigality of the judgement which shows the narrowness of one's knowledge or one's taste. (Baltasar Gracian)
He has a good judgement that relies not wholly on his own. (Proverb)
The number of those who undergo the fatigue of judging for themselves is very small indeed. (Sheridan)
Next to sound judgement, diamonds and pearls are the rarest things in the world. (La Bruyère)
Rightness of judgement is bitterness to the heart. (Euripides)
A mistake in judgement isn't fatal, but too much anxiety about judgement is. (Pauline Kael)
We are mistaken in believing the mind and judgement two separate things; judgement is only the extent of the mind's illumination. (La Rochefoucauld)
A hasty judgement is the first step to recantation. (Publius Syrus)
Most people suspend their judgement till somebody else has expressed his own and then they repeat it. (Earnest Dimnet).
Integrative complex Evaluation (#VT8051).
Constituent values Value (#VC1855)
Reason (#VC1423)
Justice (#VC1041)
Resolution (#VC1479)
Decisiveness (#VC0519)
Determination (#VC0547)
Constituent values Censure (#VD2218)
Presumption (#VD3478)
Exaggeration (#VD2626)
Disparagement (#VD2532)
Miscalculation (#VD3238)
Misconstruction (#VD3248)

Order (#VC1231)
Choice (#VC0325)
Judgement (#VC1037)
Deliverance (#VC0539)
Appreciation (#VC0153)
Consideration (#VC0449).
Prejudice (#VD3474)
Misjudgement (#VD3266)
Depreciation (#VD2394)
Overestimation (#VD3390)
Underestimation (#VD3898)
Misinterpretation (#VD3264).

♦ **VP5497c Overestimation–Underestimation**
Dynamics No one ever went broke underestimating the intelligence (or taste) of the American people (Mencken)
Integrative complex Evaluation (#VT8051).

♦ **VP5501a Belief–Unbelief**
Dynamics When faith is lost, when honour dies, the man is dead. (John Greenleaf)
If faith can move mountains, disbelief can deny their existence. And faith is impotent against such impotence. (Arnold Schoenberg)
A man consists of the faith that is in him. Whatever his faith is, he is. (Bhagavadgita)
Faith embraces itself and the doubt itself. (Paul Tillich)
Belief in a Divine mission is one of the many forms of certainty that have afflicted the human race. (Bertrand Russell)
Doubt is faith in the main: but faith, on the whole, is doubt: we cannot believe by proof: but could we believe without? (Algernon Charles Swinburne)
First there is a time when we believe everything without reasons, then for a little while we believe with discrimination, then we believe nothing whatever, and then we believe everything again – and, moreover, give reasons why we believe everything. (Georg Christoph Lichtenberg)
You're not free until you've been made captive by supreme belief. (Marianne Moore)
Our doubts are traitors and make us lose the good we oft might win by fearing to attempt. (Shakespeare)
The fact that an opinion has been widely held is no evidence whatever that it is not utterly absurd; indeed in view of the silliness of the majority of mankind, a widespread belief is more likely too foolish than sensible. (Bertrand Russell)
To believe with certainty we must begin with doubting. (Stanislaus)
Integrative complex Credibility (#VT8052).
Constituent values Hope (#VC0917)
Faith (#VC0729)
Surety (#VC1717)
Feeling (#VC0751)
Fondness (#VC0777)
Principle (#VC1353)
Innocence (#VC0987)

Trust (#VC1783)
Wisdom (#VC1927)
Belief (#VC0227)
Sureness (#VC1715)
Sentiment (#VC1597)
Judgement (#VC1037)
Certainty (#VC0301)

♦ **VP5505b Provability–Unprovability**
Dynamics You can prove almost anything with the evidence of a small enough segment of time. How often, in any search for truth, the answer of the minute is positive, the answer of the hour qualified, the answers of the year contradictory. (Teale)
We can prove whatever we want to, and the real difficulty is to know what we want to prove. (Alain)
You cannot demonstrate an emotion or prove an aspiration. (Morley)
Integrative complex Credibility (#VT8052).
Constituent values Support (#VC1711)
Confirmation (#VC0421)
Reinforcement (#VC1451)
Acknowledgement (#VC0033).
Constituent values Denial (#VD2376)
Contemptuousness (#VD2280).

Affirmation (#VC0081)
Authenticity (#VC0203)
Determination (#VC0547)

Discredit (#VD2486)

♦ **VP5507b Limitation–Unlimitedness**
Dynamics The high strength of men / know no content with limitation. (Aeschylus)
Freedom is the supreme good – freedom from self-imposed limitation. (Elbert Hubbard)
Never exceed your rights, and they will soon become unlimited. (Rousseau)
Men cease to interest us when we find their limitations. (Emerson)
Integrative complex Credibility (#VT8052).
Constituent values Unequivocalness (#VC1815).

♦ **VP5509b Possibility–Impossibility**
Dynamics They, believe me, who await no gifts from chance, have conquered fate. (Matthew Arnold)
Anyone prepared to survive on cabbage-stalks can accomplish a hundred possibilities. (Chinese Proverb)
Possibilities are infinite. (Proverb)
Nothing is impossible to a willing heart. (Proverb)
The difficult is done at once; the impossible takes a little longer. (Proverb)
All things are possible until they are proved impossible – and even the impossible may only be so, as of now. (Pearl Buck)
Integrative complex Credibility (#VT8052).
Constituent values Chance (#VC0307)
Accessibility (#VC0019)
Constituent values Hopelessness (#VD2798)
Impracticality (#VD2862).

Availability (#VC0209)
Approachability (#VC0155).
Impossibility (#VD2856)

♦ **VP5513c Certainty–Uncertainty**
Dynamics A man who has humility will have acquired in the last reaches of his beliefs the saving doubt of his own certainty. (Walter Lippmann)
If life were eternal all interest and anticipation would vanish. It is uncertainty which lends its fascination. (Yoshida Kenko)
There is one thing certain, namely, that we can have nothing certain; therefore it is not certain that we can have nothing certain. (Samuel Butler)
We can be absolutely certain only about things we do not understand. (Eric Hoffer)
Diffidence is the better part of knowledge. (Charles Caleb Colton)
The quest for certainty blocks the search for meaning. Uncertainty is the very condition to impel man to unfold his powers. (Erich Fromm)
Convictions are more dangerous enemies of truth than lies. (Nietzsche)
In this unbelievable universe in which we live, there are no absolutes. Even parallel lines, reaching into infinity, meet somewhere yonder. (Pearl S Buck)
Integrative complex Credibility (#VT8052).
Constituent values Truth (#VC1787)
Faith (#VC0729)
Belief (#VC0227)
Courage (#VC0495)
Unerring (#VC1817)
Solidity (#VC1633)
Unfailing (#VC1819)
Soundness (#VC1637)
Clearness (#VC0339)
Assurance (#VC0189)
Conviction (#VC0479)
Unambiguity (#VC1797)
Reliability (#VC1459)
Faithfulness (#VC0731)
Confirmation (#VC0421)
Well-grounded (#VC1907)
Self-reliance (#VC1577)
Infallibility (#VC0977)
Absoluteness (#VC0009)
Unequivocalness (#VC1815)
Self-confidence (#VC1545)
Indubitableness (#VC0973).
Constituent values Risk (#VD3570)
Muddle (#VD3292)
Dismay (#VD2522)
Bigotry (#VD2176)
Disorder (#VD2526)
Treachery (#VD3842)
Obscurity (#VD3346)

Trust (#VC1783)
Surety (#VC1717)
Reality (#VC1419)
Validity (#VC1853)
Sureness (#VC1715)
Security (#VC1537)
Stability (#VC1651)
Necessity (#VC1183)
Certainty (#VC0301)
Actuality (#VC0043)
Confidence (#VC0419)
Staunchness (#VC1665)
Positiveness (#VC1323)
Decisiveness (#VC0519)
Authenticity (#VC0203)
Steadfastness (#VC1667)
Invincibility (#VC1021)
Determination (#VC0547)
Substantiality (#VC1693)
Trustworthiness (#VC1785)
Self-assurance (#VC1543)

Upset (#VD3966)
Laxity (#VD3138)
Problem (#VD3486)
Suspense (#VD3782)
Vagueness (#VD3978)
Riskiness (#VD3572)
Looseness (#VD3164)

VALUE POLARITIES

Disbelief (#VD2468)
Ambiguity (#VD2054)
Perplexity (#VD3440)
Inaccuracy (#VD2886)
Bafflement (#VD2136)
Uncertainty (#VD3880)
Incoherence (#VD2916)
Fallibility (#VD2668)
Disturbance (#VD2570)
Irresolution (#VD3112)
Discomposure (#VD2472)
Unreliability (#VD3948)
Orderlessness (#VD3370)
Embarrassment (#VD2602)
Dangerousness (#VD2318)
Indecisiveness (#VD2950)
Changeableness (#VD2222)
Unpredictability (#VD3938)
Untrustworthiness (#VD3958).

Confusion (#VD2272)
Randomness (#VD3514)
Insecurity (#VD3036)
Hesitation (#VD2792)
Vacillation (#VD3974)
Instability (#VD3054)
Imprecision (#VD2864)
Dubiousness (#VD2586)
Perturbation (#VD3442)
Doubtfulness (#VD2584)
Bewilderment (#VD2172)
Shapelessness (#VD3668)
Hazardousness (#VD2782)
Disconcertion (#VD2474)
Unauthenticity (#VD3876)
Disorientation (#VD2530)
Undependability (#VD3894)
Insubstantiality (#VD3058)

♦ **VP5516c Truth–Error**
Dynamics All error, not merely verbal, is a strong way of stating that the current truth is incomplete. (Robert Louis Stevenson)
Not Truth, but Faith, it is That keeps the world alive. (Edna St Vincent Millay)
Faith embraces many truths which seem to contradict each other. (Pascal)
The heresy of one age becomes the orthodoxy of the next. (Helen Keller)
People who honestly mean to be true really contradict themselves much more rarely than those who try to be 'consistent'. (Oliver Wezndell Holmes Sr)
Life is a paradox. Every truth has its counterpart which contradicts it; and every philosopher supplies the logic for his own undoing. (Elbert Hubbard)
The greatest triumphs of propaganda have been accomplished, not by doing something, but by refraining from doing. Great is truth, but still greater, from a practical point of view, is silence about truth. (Aldous Huxley)
Even under the most favourable circumstances no mortal can be asked to seize the truth in its wholeness or at its centre. (George Santayana)
If a man fasten his attention on a single aspect of truth and apply himself to that alone for a long time, the truth becomes distorted and not itself but falsehood. (Emerson)
No doubt about it: error is the rule, truth is the accident of error. (Georges Duhamel)
Integrative complex Credibility (#VT8052).
Constituent values Good (#VC0845)
Purity (#VC1391)
Justice (#VC1041)
Cogency (#VC0349)
Validity (#VC1853)
Subtlety (#VC1695)
Fidelity (#VC0763)
Accuracy (#VC0029)
Sincerity (#VC1619)
Precision (#VC1341)
Certainty (#VC0301)
Actuality (#VC0043)
Strictness (#VC1683)
Refinement (#VC1441)
Legitimacy (#VC1063)
Sufficiency (#VC1699)
Genuineness (#VC0831)
Consistency (#VC0451)
Truthfulness (#VC1789)
Forcefulness (#VC0781)
Faithfulness (#VC0731)
Authenticity (#VC0203)
Veritableness (#VC1871)
Faultlessness (#VC0741)
Unadulteration (#VC1795)
Meticulousness (#VC1143).

Truth (#VC1787)
Reality (#VC1419)
Honesty (#VC0913)
Veracity (#VC1869)
Unerring (#VC1817)
Solidity (#VC1633)
Delicacy (#VC0535)
Soundness (#VC1637)
Rightness (#VC1507)
Exactness (#VC0693)
Authority (#VC0205)
Unassuming (#VC1801)
Simplicity (#VC1617)
Perfection (#VC1273)
Directness (#VC0563)
Naturalness (#VC1179)
Correctness (#VC0489)
Artlessness (#VC0173)
Rigorousness (#VC1509)
Flawlessness (#VC0769)
Confirmation (#VC0421)
Well-grounded (#VC1907)
Infallibility (#VC0977)
Absoluteness (#VC0009)
Substantiality (#VC1693)

Constituent values Flaw (#VD2704)
Misuse (#VD3284)
Heresy (#VD2790)
Mistake (#VD3276)
Fallacy (#VD2666)
Spurious (#VD3718)
Delusion (#VD2372)
Unreality (#VD3944)
Oversight (#VD3402)
Looseness (#VD3164)
Deviation (#VD2428)
Perversion (#VD3444)
Misconduct (#VD3246)
Faultiness (#VD2682)
Aberration (#VD2008)
Miscarriage (#VD3240)
Misstatement (#VD3274)
Indiscretion (#VD2964)
Misconception (#VD3244)
Hallucination (#VD2764)
Defectiveness (#VD2350)
Miscalculation (#VD3238)
Disenchantment (#VD2492)
Misconstruction (#VD3248)
Disillusionment (#VD2510)
Misinterpretation (#VD3264)

Lapse (#VD3130)
Laxity (#VD3138)
Autism (#VD2120)
Illogic (#VD2828)
Failure (#VD2660)
Omission (#VD3360)
Wrongness (#VD4078)
Stupidity (#VD3742)
Misbelief (#VD3236)
Falseness (#VD2672)
Deception (#VD2334)
Negligence (#VD3312)
Inaccuracy (#VD2886)
Distortion (#VD2562)
Misfeasance (#VD3256)
Imprecision (#VD2864)
Misjudgement (#VD3266)
Inadvertence (#VD2892)
Incorrectness (#VD2940)
Erroneousness (#VD2620)
Unauthenticity (#VD3876)
Misapplication (#VD3232)
Disappointment (#VD2460)
Misapprehension (#VD3234)
Misunderstanding (#VD3280)
Self-contradiciton (#VD3632).

♦ **VP5519a Illusion–Disillusionment**
Dynamics Anyone who can handle a needle convincingly can make us see a thread which is not there. (Gombrich)
Nature rejoices in illusion. If a man destroys the power of illusion, either in himself or in others, she punishes him like the harshest tyrant. (Goethe)
Clemenceau had one illusion – France; and one disillusion – mankind, including Frenchmen. (Keynes)
The usual masculine disillusionment in discovering that a woman has a brain. (Mitchell)
The most dangerous of our calculations are those we call illusions. (Bernanos)
The illusion that times that were were better than those that are, has probably pervaded all ages. (Horace Greeley)
Integrative complex Truth (#VT8053).
Constituent values Imagination (#VC0939)
Constituent values Disillusionment (#VD2510).

♦ **VP5521b Assent–Dissent**
Dynamics No man can sit down and withhold his hands from the warfare against wrong and get peace from his acquiescence. (Woodrow Wilson)
Dissension, well directed, may divide even the true-hearted, as a mighty stream of waters divides mountains of solid rock. (Panchatantra)
In formal logic, a contradiction is the signal of a defeat: but in the evolution of real knowledge it marks the first step in reason for the utmost toleration of variety of opinion. (Alfred North Whitehead)
Integrative complex Truth (#VT8053).
Constituent values Accord (#VC0025)
Harmony (#VC0887)
Acclaim (#VC0021)
Agreement (#VC0093)
Compliance (#VC0389)
Recognition (#VC1427)
Concurrence (#VC0413)
Conformation (#VC0423)
Acquiescence (#VC0035)
Acknowledgement (#VC0033).

Support (#VC1711)
Concord (#VC0411)
Unanimity (#VC1799)
Submission (#VC1689)
Acceptance (#VC0017)
Contentment (#VC0467)
Togetherness (#VC1763)
Appreciation (#VC0153)
Understanding (#VC1811)

Constituent values Denial (#VD2376)
Negation (#VD3306)
Grievance (#VD2750)
Disaccord (#VD2448)
Withdrawal (#VD4062)
Negativity (#VD3308)
Dissension (#VD2548)
Repudiation (#VD3536)
Nonconformity (#VD3324)
Dissatisfaction (#VD2546).

Variance (#VD3992)
Rejection (#VD3526)
Disparity (#VD2534)
Annulment (#VD2078)
Opposition (#VD3364)
Dissidence (#VD2550)
Alienation (#VD2052)
Disapproval (#VD2462)
Contradiction (#VD2284)

♦ **VP5523c Affirmation–Denial**
Dynamics All our affirmations are mere matters of chronology; and even our bad taste is nothing more than the bad taste of the age we live in. (Logan Pearsall Smith)
A civil denial is better than a rude grant. (Fuller)
Better a friendly denial than unwilling compliance. (German proverb)
Integrative complex Truth (#VT8053).
Constituent values Affirmation (#VC0081).
Constituent values Denial (#VD2376).

♦ **VP5526c Broadmindedness–Narrowmindedness**
Dynamics All good men are international. Nearly all bad men are cosmopolitan. If we are to be international we must be national. (G K Chesterton)
Those wearing Tolerance for a label call other views intolerant. (Phyllis McGinley)
As in political so in literary action a man wins friends for himself mostly by the passion of his prejudices and by the consistent narrowness of his outlook. (Joseph Conrad)
Knowledge humanizes mankind, and reason inclines to mildness; but prejudices eradicate every tender disposition. (Montaigne)
No loss by flood and lightning, no destruction of cities and temples by the hostile forces of nature, has deprived man of so many noble lives and impulses as those which his intolerance has destroyed. (Helen Keller)
It is as though nature must needs make men narrow in order to give them force. (W E B du Bois)
World peace, like community peace, does not require that each man love his neighbour – it requires only that they live together with mutual tolerance, submitting their disputes to a just and peaceful settlement. (John F Kennedy)
Integrative complex Attitude (#VT8054).
Constituent values Justice (#VC1041)
Patience (#VC1261)
Leniency (#VC1067)
Unbigoted (#VC1803)
Broadness (#VC0271)
Dispassion (#VC0575)
Uninfluence (#VC1823)
Magnanimity (#VC1105)
Impartiality (#VC0953)
Openmindedness (#VC1223)

Charity (#VC0315)
Openness (#VC1225)
Fairness (#VC0727)
Tolerance (#VC1765)
Neutrality (#VC1187)
Detachment (#VC0545)
Objectivity (#VC1209)
Forbearance (#VC0779)
Cosmopolitan (#VC0493)
Broadmindedness (#VC0269).

Constituent values Bias (#VD2174)
Meanness (#VD3204)
Prejudice (#VD3474)
Apartheid (#VD2088)
Warpedness (#VD4044)
Littleness (#VD3158)
Fanaticism (#VD2676)
Segregation (#VD3624)
Illiberalism (#VD2824)
Authoritarianism (#VD2118).

Bigotry (#VD2176)
Smallness (#VD3684)
Pettiness (#VD3452)
Xenophobia (#VD4080)
Stuffiness (#VD3740)
Insularity (#VD3062)
Chauvinism (#VD2228)
Intolerance (#VD3074)
Discrimination (#VD2488)

♦ **VP5528b Curiosity–Incuriosity**
Dynamics Glory and curiosity are the two scourges of the soul; the last prompts us to thrust our noses into everything, the other forbids us to leave anything doubtful and undecided. (Montaigne)
A man should live if only to satisfy his curiosity. (Yiddish Proverbs)
Curiosity is ill manners in another house. (Proverb)
Ignorance and incuriosity are two very soft pillows. (French proverb)
Creatures whose mainspring is curiosity will enjoy the accumulating of facts, far more than the pausing at times to reflect on those facts. (Clarence Day)
Integrative complex Attitude (#VT8054).
Constituent values Interest (#VC1011)
Detachment (#VC0545).
Constituent values Apathy (#VD2090)
Withdrawal (#VD4062)
Insouciance (#VD3052)
Listlessness (#VD3156)
Incuriosity (#VD2946)
Carelessness (#VD2214).

Curiosity (#VC0509)

Boredom (#VD2192)
Boringness (#VD2194)
Disinterest (#VD2516)
Indifference (#VD2958)
Heedlessness (#VD2786)

♦ **VP5530c Attention–Inattention**
Dynamics Attention is like a narrow-mouthed vessel; pour into it what you have to say cautiously, and, as it were, drop by drop. (Joseph Joubert)
The tongues of dying men enforce attention like deep harmony. (Shakespeare)
Calling attention isn't the same thing as explaining. (Ashbery)
Women are systematically degraded by receiving the trivial attentions which men think it manly to say to the sex, when, in fact, men are insultingly supporting their own superiority. (Wollstonecraft)
Integrative complex Attitude (#VT8054).

VP5530

Constituent values Care (#VC0291)
Concern (#VC0407)
Interest (#VC1011)
Curiosity (#VC0509)
Attention (#VC0197)
Alertness (#VC0099)
Liveliness (#VC1085)
Excitement (#VC0699)
Attraction (#VC0199)
Involvement (#VC1023)
Earnestness (#VC0599)
Abstraction (#VC0013)
Consciousness (#VC0443)
Respectfulness (#VC1485)
Constituent values Fuss (#VD2734)
Chaos (#VD2224)
Levity (#VD3146)
Atheism (#VD2112)
Frivolity (#VD2724)
Confusion (#VD2272)
Negligence (#VD3312)
Shallowness (#VD3664)
Disturbance (#VD2570)
Perturbation (#VD3442)
Indifference (#VD2958)
Inadvertence (#VD2892)
Discomposure (#VD2472)
Nonobservance (#VD3330)
Disconcertion (#VD2474)
Disorientation (#VD2530)
Disorganization (#VD2528)
Passion (#VC1257)
Richness (#VC1503)
Diligence (#VC0559)
Awareness (#VC0211)
Assiduity (#VC0183)
Observance (#VC1211)
Intentness (#VC1009)
Enthusiasm (#VC0663)
Mindfulness (#VC1151)
Fascination (#VC0737)
Carefulness (#VC0293)
Consideration (#VC0449)
Concentration (#VC0403)
Meticulousness (#VC1143).
Upset (#VD3890)
Muddle (#VD3292)
Frenzy (#VD2716)
Disorder (#VD2526)
Disregard (#VD2542)
Perplexity (#VD3440)
Witlessness (#VD4066)
Inattention (#VD2906)
Distraction (#VD2564)
Inobservance (#VD3028)
Incuriousity (#VD2946)
Heedlessness (#VD2786)
Carelessness (#VD2214)
Embarrassment (#VD2602)
Superficiality (#VD3766)
Thoughtlessness (#VD3810).

♦ **VP5533c Carefulness–Neglect**
Dynamics A pound of care will not pay an ounce of debt. (Proverb)
Neglect will kill an injury sooner than revenge. (Proverb)
Perpetual devotion to what man calls his business, is only to be sustained by perpetual neglect of many other things. (R L Stevenson)
It will all be the same a hundred years hence. (Proverb)
Integrative complex Attitude (#VT8054).
Constituent values Care (#VC0291)
Concern (#VC0407)
Subtlety (#VC1695)
Keenness (#VC1043)
Accuracy (#VC0029)
Readiness (#VC1417)
Precision (#VC1341)
Diligence (#VC0559)
Assiduity (#VC0183)
Alertness (#VC0099)
Solicitude (#VC1629)
Promptness (#VC1377)
Nimbleness (#VC1191)
Stewardship (#VC1671)
Mindfulness (#VC1151)
Correctness (#VC0489)
Thoroughness (#VC1753)
Preparedness (#VC1343)
Anticipation (#VC0139)
Consideration (#VC0449)
Scrupulousness (#VC1535)
Circumspection (#VC0327)
Industriousness (#VC0975)
Constituent values Lapse (#VD3130)
Neglect (#VD3310)
Default (#VD2344)
Oblivion (#VD3344)
Oversight (#VD3402)
Looseness (#VD3164)
Disregard (#VD2542)
Inaccuracy (#VD2886)
Nonfeasance (#VD3326)
Insouciance (#VD3052)
Dereliction (#VD2400)
Tactlessness (#VD3788)
Inadvertence (#VD2892)
Carelessness (#VD2214)
Forgetfulness (#VD2708)
Permissiveness (#VD3438)
Inconsiderateness (#VD2930).
Acuity (#VC0045)
Agility (#VC0089)
Prudence (#VC1387)
Delicacy (#VC0535)
Vigilance (#VC1879)
Quickness (#VC1399)
Exactness (#VC0693)
Attention (#VC0197)
Aliveness (#VC0101)
Strictness (#VC1683)
Refinement (#VC1441)
Observance (#VC1211)
Brightness (#VC0265)
Punctuality (#VC1389)
Forethought (#VC0785)
Carefulness (#VC0293)
Rigorousness (#VC1509)
Cautiousness (#VC0297)
Exquisiteness (#VC0721)
Thoughtfulness (#VC1755)
Meticulousness (#VC1143)
Painstakingness (#VC1245)
Conscientiousness (#VC0441).
Laxity (#VD3138)
Laxness (#VD3140)
Omission (#VD3360)
Laziness (#VD3142)
Messiness (#VD3220)
Hastiness (#VD2778)
Negligence (#VD3312)
Clumsiness (#VD2238)
Malingering (#VD3196)
Inattention (#VD2906)
Abandonment (#VD2004)
Indifference (#VD2958)
Heedlessness (#VD2786)
Haphazardness (#VD2766)
Unpreparedness (#VD3940)
Thoughtlessness (#VD3810).

♦ **VP5535b Imaginativeness–Unimaginativeness**
Dynamics Vision is the art of seeing things invisible. (Jonathan Swift)
The right honourable gentleman is indebted to his memory for his jests, and to his imagination for his facts. (Sheridan)
The problems of the world cannot possibly be solved by sceptics or cynics whose horizons are limited by the obvious realities. We need men who can dream of things that never were. (J F Kennedy)
I imagine, therefore I belong and am free. (Lawrence Durrell)
The imagination may be compared to Adam's dream – he awoke and found it truth. (Keats)
Integrative complex Attitude (#VT8054).
Constituent values Vision (#VC1889)
Romance (#VC1511)
Ingenuity (#VC0981)
Fecundity (#VC0749)
Conception (#VC0405)
Originality (#VC1239)
Imagination (#VC0939)
Practicality (#VC1335)
Reasonableness (#VC1425)
Constituent values Autism (#VD2120)
Unreality (#VD3944)
Barrenness (#VD2146)
Unimaginativeness (#VD3918).
Genius (#VC0825)
Creation (#VC0499)
Fertility (#VC0759)
Creativity (#VC0501)
Rationality (#VC1415)
Inspiration (#VC0993)
Sensibleness (#VC1593)
Inventiveness (#VC1019)
Productiveness (#VC1365).
Dullness (#VD2588)
Stuffiness (#VD3740)
Infertility (#VD3004)

♦ **VP5537c Remembrance–Forgetfulness**
Dynamics Not the power to remember, but its very opposite, the power to forget, is a necessary condition for our existence. (Sholem Asch)
A retentive memory may be a good thing, but the ability to forget is the true token of greatness. (Elbert Hubbard)
The richness of life lies in memories we have forgotten. (Cesare Pavese)
To endeavour to forget anyone is a certain way of thinking of nothing else. (La Bruyère)
The things we remember best are those better forgotten. (Gracian)
Integrative complex Attitude (#VT8054).
Constituent values Realization (#VC1421)
Constituent values Oblivion (#VD3344)
Heedlessness (#VD2786)
Forgiveness (#VC0787)
Repression (#VD3532)
Forgetfulness (#VD2708).

♦ **VP5539b Expectation–Inexpectation**
Dynamics Oft expectation fails and most oft there where most it promises, and oft it hits, where hope is coldest and despair most fits. (Shakespeare)
The wise man avoids evil by anticipating it. (Publilius Syrus)
Uncertainty and expectation are the joys of life. Security is an insipid thing, and the overtaking and possessing of a wish, discovers the folly of the chase. (William Congreve)
Blessed is he who expects nothing, for he shall never be disappointed. (Pope)
For people who live on expectations, to face up to their realisation is something of an ordeal. (Elizabeth Bowen)
Nothing is so good as it seems beforehand. (George Eliot)
Integrative complex Attitude (#VT8054).
Constituent values Luck (#VC1099)
Promise (#VC1375)
Surprise (#VC1719)
Imminence (#VC0945)
Forethought (#VC0785)
Anticipation (#VC0139).
Constituent values Defeat (#VD2346)
Failure (#VD2660)
Frustration (#VD2726)
Dissatisfaction (#VD2546)
Good (#VC0845)
Caution (#VC0295)
Sureness (#VC1715)
Discretion (#VC0571)
Expectation (#VC0709)
Blight (#VD2186)
Bafflement (#VD2136)
Disappointment (#VD2460)
Disillusionment (#VD2510)

♦ **VP5545a Meaning–Meaninglessness**
Dynamics We accept every person in the world as that for which he gives himself out, only he must give himself out for something. We can put up with the unpleasant more easily than we can endure the insignificant. (Goethe)
The grand thing about the human mind is that it can turn its own tables and see meaninglessness as ultimate meaning. (John Cage)
The absurd is born of the confrontation between the human call and the unreasonable silence of the world. (Albert Camus)
Every true man, sir, who is a little above the level of the beasts and plants does not live for the sake of living, without knowing how to live; but he lives so as to give a meaning and a value of his own to life. (Luigi Pirandello)
Nonsense is good only because common sense is so limited. (George Santayana)
The privilege of absurdity; to which no living creature is subject but man only. (Thomas Hobbes)
Heaven knows what seeming nonsense may not tomorrow be demonstrated truth. (Alfred North Whitehead)
Integrative complex Meaning (#VT8055).
Constituent values Value (#VC1855)
Relation (#VC1453)
Significance (#VC1613)
Intelligibility (#VC1003).
Constituent values Rot (#VD3578)
Nullity (#VD3342)
Nonsense (#VD3332)
Absurdity (#VD2022)
Senselessness (#VD3648)
Purposelessness (#VD3508).
Essence (#VC0681)
Relevance (#VC1457)
Meaningfulness (#VC1129)
Rubbish (#VD3582)
Inanity (#VD2896)
Futility (#VD2736)
Aimlessness (#VD2048)
Insignificance (#VD3042)

♦ **VP5548b Intelligibility–Unintelligibility**
Dynamics If clearness about things produces a fundamental despair, a fundamental despair in turn produces a remarkable clearness or even playfulness about ordinary matters. (George Santayana)
There is a poignancy in all things clear, In the stare of the deer, in the ring of a hammer in the morning. Seeing a bucket of perfectly lucid water we fall to imagining prodigious honesties. (Richard Wilbur)
The mystic (is) too full of God to speak intelligibly to the world. (Symons)
Integrative complex Meaning (#VT8055).
Constituent values Lucidity (#VC1097)
Intricacy (#VC1015)
Clearness (#VC0339)
Directness (#VC0563)
Perspicuity (#VC1287)
Distinction (#VC0577)
Appreciation (#VC0153)
Connectedness (#VC0431)
Intelligibility (#VC1003)
Constituent values Problem (#VD3486)
Hardness (#VD2770)
Vagueness (#VD3978)
Intricacy (#VD3082)
Perplexity (#VD3440)
Complexity (#VD2252)
Incoherence (#VD2916)
Complication (#VD2254)
Mystification (#VD3302)
Incomprehensibility (#VD2924).
Plainness (#VC1301)
Coherence (#VC0351)
Simplicity (#VC1617)
Unambiguity (#VC1797)
Explanation (#VC0719)
Consistency (#VC0451)
Understanding (#VC1811)
Unequivocalness (#VC1815)
Straightforwardness (#VC1679).
Opacity (#VD3362)
Darkness (#VD2320)
Obscurity (#VD3346)
Ambiguity (#VD2054)
Difficulty (#VD2440)
Uncertainty (#VD3880)
Equivocation (#VD2616)
Shapelessness (#VD3668)
Inarticulation (#VD2904).

♦ **VP5552c Interpretability–Misinterpretability**
Dynamics It is more of a job to interpret the interpretations than to interpret the things, and there are more books about other books than about anyother subject: we do nothing but write glosses about each other. (Montaigne)
In every house of marriage there's room for an interpreter. (Stanley Kunitz)
Conscience is the perfect interpreter of life. (Karl Barth)
A great interpreter of life ought not himself to need interpretation. (Morley)
Integrative complex Meaning (#VT8055).
Constituent values Reason (#VC1423)
Enlightenment (#VC0653).
Constituent values Mistake (#VD3276)
Distortion (#VD2562)
Misconception (#VD3244)
Misconstruction (#VD3248)
Misunderstanding (#VD3280)
Illumination (#VC0937)
Perversion (#VD3444)
Misjudgement (#VD3266)
Misapplication (#VD3232)
Misapprehension (#VD3234)
Misinterpretation (#VD3264)

♦ VP5554a Communicativeness–Uncommunicativeness
Dynamics We are least open to precise knowledge concerning the things we are most vehement about. (Eric Hoffer)
Knowledge can be communicated, but not wisdom. One can find it, live it, be fortified by it, do wonders through it, but one cannot communicate and teach it. (Hermann Hesse)
Whoever interrupts the conversation of others to make a display of his fund of knowledge, makes notorious his own stock of ignorance. (Sa'di)
Wisdom sets bounds even to knowledge. (Nietzsche)
Often, the surest way to convey misinformation is to tell the strict truth. (Mark Twain)
Human society is founded on mutual deceit; few friendships would endure if each knew what his friend said of him in his absence. (Pascal)
All cruel people describe themselves as paragons of frankness. (Tennessee Williams)
Integrative complex Communication (#VT8061).
Constituent values Candor (#VC0283)
Openness (#VC1225)
Plainness (#VC1301)
Frankness (#VC0793)
Anonymity (#VC0137)
Connection (#VC0433)
Enlightenment (#VC0653)
Accessibility (#VC0019)
Correspondence (#VC0491)
Approachability (#VC0155).
Contact (#VC0463)
Quietness (#VC1403)
Knowledge (#VC1047)
Communion (#VC0371)
Detachment (#VC0545)
Sociability (#VC1625)
Communication (#VC0367)
Responsiveness (#VC1491)
Unobtrusiveness (#VC1835)
Constituent values Slant (#VD3678)
Strain (#VD3732)
Excess (#VD2630)
Secrecy (#VD3618)
Evasion (#VD2622)
Trickery (#VD3854)
Spurious (#VD3718)
Impurity (#VD2880)
Curtness (#VD2310)
Treachery (#VD3842)
Mendacity (#VD3212)
Hypocrisy (#VD2808)
Falsehood (#VD2670)
Demeaning (#VD2374)
Collusion (#VD2244)
Warpedness (#VD4044)
Sanctimony (#VD3602)
Distortion (#VD2562)
Conspiracy (#VD2276)
Ostentation (#VD3372)
Malingering (#VD3196)
Indirection (#VD2962)
Fraudulence (#VD2714)
Dubiousness (#VD2586)
Unctuousness (#VD3892)
Overemphasis (#VD3388)
Misdirection (#VD3252)
Extravagance (#VD2654)
Equivocation (#VD2616)
Unnaturalness (#VD3930)
Prevarication (#VD3482)
Hallucination (#VD2764)
Erroneousness (#VD2620)
Unauthenticity (#VD3876)
Overestimation (#VD3390)
Misconstruction (#VD3248)
Surreptitiousness (#VD3778)
Uncommunicativeness (#VD3888).
Guile (#VD2758)
Fakery (#VD2664)
Deceit (#VD2332)
Fallacy (#VD2666)
Cunning (#VD2308)
Travesty (#VD3840)
Muteness (#VD3298)
Delusion (#VD2372)
Artifice (#VD2104)
Sophistry (#VD3690)
Improbity (#VD2868)
Falseness (#VD2672)
Duplicity (#VD2590)
Deception (#VD2334)
Closeness (#VD2236)
Subterfuge (#VD3752)
Perversion (#VD3444)
Dishonesty (#VD2506)
Prodigality (#VD3488)
Misguidance (#VD3260)
Insincerity (#VD3044)
Furtiveness (#VD2730)
Fabrication (#VD2658)
Affectation (#VD2034)
Speciousness (#VD3708)
Misstatement (#VD3274)
Illegitimate (#VD2822)
Exaggeration (#VD2626)
Victimization (#VD4012)
Truthlessness (#VD3862)
Mystification (#VD3302)
Faithlessness (#VD2662)
Artificiality (#VD2106)
Perfidiousness (#VD3436)
Misinformation (#VD3262)
Disingenuousness (#VD2512)
Misrepresentation (#VD3270)

♦ VP5562b Education–Miseducation
Dynamics Docility is the observable half of reason. (George Santayana)
Education makes a people easy to lead, possible to enslave. (Lord Brougham)
Learning, the destroyer of arrogance, begets arrogance in fools; even as light, that illumines the eye, makes owls blind. (Panchatantra)
How is it that little children are so intelligent and men so stupid? It must be education that does it. (Alexandre Dumas)
Education then, beyond all other devices of human origin, is a great equalizer of the conditions of men, the balance wheel of the social machinery. (Horace Mann)
Education consists mainly in what we have unlearned. (Mark Twain)
Integrative complex Communication (#VT8061).
Constituent values Order (#VC1231)
Facility (#VC0725)
Aptitude (#VC0163)
Knowledge (#VC1047)
Education (#VC0611)
Motivation (#VC1169)
Discipline (#VC0567)
Brightness (#VC0265)
Improvement (#VC0961)
Illumination (#VC0937)
Learning (#VC1059)
Docility (#VC0581)
Quickness (#VC1399)
Erudition (#VC0677)
Diligence (#VC0559)
Liberality (#VC1069)
Cleverness (#VC0343)
Willingness (#VC1925)
Edification (#VC0609)
Enlightenment (#VC0653).
Constituent values Sophistry (#VD3690)
Perversion (#VD3444)
Misguidance (#VD3260)
Mystification (#VD3302)
Brainwash (#VD2198)
Corruption (#VD2294)
Misdirection (#VD3252)
Misinformation (#VD3262)

♦ VP5573b Representation–Misrepresentation
Dynamics Taxation without representation is tyranny. (James Otis)
We ought not to extract pernicious honey from poison-blossoms of misrepresentation and mendacious half-truth, to pamper the coarse appetite of bigotry and self-love. (Samuel Taylor Coleridge)
Integrative complex Communication (#VT8061).
Constituent values Romance (#VC1511)
Constituent values Slant (#VD3678)
Injustice (#VD3026)
Inaccuracy (#VD2886)
Misstatement (#VD3274)
Misrepresentation (#VD3270).
Aggrandizement (#VC0087).
Travesty (#VD3840)
Perversion (#VD3444)
Distortion (#VD2562)
Exaggeration (#VD2626)

♦ VP5589c Elegance–Inelegance
Dynamics Superior men know how to combine elegance with simplicity (Chinese Proverb)
Integrative complex Communication (#VT8061).
Constituent values Harmony (#VC0887)
Euphony (#VC0685)
Symmetry (#VC1729)
Fluency (#VC0775)
Balance (#VC0215)
Felicity (#VC0753)
Elegance (#VC0623)
Sweetness (#VC1725)
Unadornment (#VC1793)
Naturalness (#VC1179)
Appropriateness (#VC0159)
Constituent values Inelegance (#VD2982)
Awkwardness (#VD2130)
Gracelessness (#VD2744).
Chastity (#VC0319)
Frankness (#VC0793)
Orderliness (#VC1233)
Correctness (#VC0489)
Straightforwardness (#VC1679)
Clumsiness (#VD2238)
Unnaturalness (#VD3930)

♦ VP5592c Conciseness–Diffuseness
Dynamics Brevity is the soul of wit. (Shakespeare)
Diffused knowledge immortalizes itself. (Sir James Mackintosh)
In all pointed sentences, some degree of accuracy must be sacrificed to conciseness. (Samuel Johnson)
Integrative complex Communication (#VT8061).
Constituent values Fluency (#VC0775)
Fertility (#VC0759)
Abundance (#VC0015)
Reservedness (#VC1473)
Constituent values Prolixity (#VD3498)
Extravagance (#VD2654).
Brevity (#VC0263)
Fecundity (#VC0749)
Superfluity (#VC1705)
Superabundance (#VC1703).
Indirection (#VD2962)

♦ VP5600c Eloquence–Uneloquence
Dynamics Continuous eloquence wearies. (Pascal)
Well-timed silence hath more eloquence than speech. (Martin Tupper)
Talking and eloquence are not the same; to speak, and to speak well, are two things. A fool may talk, but a wise man speaks. (Ben Jonson)
When the cause is lost words are useless. (Italian Proverb)
What you are doing, Sir, speaks so loudly I cannot hear what you are saying. (American proverb)
Integrative complex Communication (#VT8061).
Constituent values Ease (#VC0601)
Vigour (#VC1881)
Majesty (#VC1109)
Dignity (#VC0557)
Vitality (#VC1891)
Nobility (#VC1193)
Felicity (#VC0753)
Elegance (#VC0623)
Eloquence (#VC0627)
Excitement (#VC0699)
Stateliness (#VC1659)
Effectiveness (#VC0613)
Impressivensss (#VC0959).
Constituent values Muteness (#VD3298).
Warmth (#VC1899)
Passion (#VC1257)
Fluency (#VC0775)
Vivacity (#VC1893)
Strength (#VC1681)
Fervour (#VC0761)
Facility (#VC0725)
Solemnity (#VC1627)
Elevation (#VC0625)
Enthusiasm (#VC0663)
Gracefulness (#VC0857)
Meaningfulness (#VC1129)

♦ VP5621a Willingness–Unwillingness
Dynamics Smallness of mind is the cause of stubbornness, and we do not credit readily what is beyond our view. (La Rochefoucauld)
Be there a will, and wisdom finds a way. (George Crabbe)
The measure of an enthusiasm must be taken between interesting events. It is between bites that the lukewarm angler loses heart. (Edwin Way Teale)
An education which does not cultivate the will is an education that depraves the mind. (Anatole France)
When a man's willing and eager, God joins in. (Aeschylus)
Nothing is easy to the unwilling. (Thomas Fuller)
Integrative complex Choice (#VT8062).
Constituent values Zeal (#VC1947)
Choice (#VC0325)
Modesty (#VC1159)
Ardour (#VC0165)
Goodwill (#VC0849)
Autonomy (#VC0207)
Alacrity (#VC0097)
Intention (#VC1007)
Resolution (#VC1479)
Promptness (#VC1377)
Discretion (#VC0571)
Willingness (#VC1925)
Spontaneity (#VC1645)
Compunction (#VC0399)
Tractability (#VC1769)
Decisiveness (#VC0519)
Well-disposed (#VC1905)
Desirableness (#VC0543)
Responsiveness (#VC1491)
Self-determination (#VC1553).
Constituent values Distaste (#VD2560)
Obstinacy (#VD3354)
Repugnance (#VD3538)
Indocility (#VD2970)
Disobedience (#VD2524)
Indisposition (#VD2968)
Will (#VC1923)
Passion (#VC1257)
Freedom (#VC0797)
Pleasure (#VC1309)
Docility (#VC0581)
Appetite (#VC0149)
Readiness (#VC1417)
Eagerness (#VC0597)
Resistance (#VC1477)
Enthusiasm (#VC0663)
Compliance (#VC0389)
Uninfluence (#VC1823)
Contentment (#VC0467)
Amenability (#VC0119)
Independence (#VC0969)
Acquiescence (#VC0035)
Determination (#VC0547)
Agreeableness (#VC0091)
Cooperativeness (#VC0483)
Aversion (#VD2124)
Antipathy (#VD2084)
Opposition (#VD3364)
Stubbornness (#VD3738)
Recalcitrance (#VD3520)
Disagreeableness (#VD2456).

♦ VP5624b Resolution–Irresolution
Dynamics Nothing that is not a real crime makes a man appear so contemptible and little in the eyes of the world as inconstancy. (Joseph Addison)
No man is good for anything who has not some particle of obstinacy to use upon occasion. (Henry Ward Beecher)
Often we are firm from weakness, and audacious from timidity. (La Rochefoucauld)
There is no more miserable human being than one in whom nothing is habitual but indecision. (William James)
Intellect, without firmness, is craft and chicanery; and firmness, without intellect, perverseness and obstinacy. (Sa'di)
Integrative complex Choice (#VT8062).
Constituent values Will (#VC1923)
Pluck (#VC1315)
Loyalty (#VC1095)
Courage (#VC0495)
Strength (#VC1681)
Fidelity (#VC0763)
Unfailing (#VC1819)
Stability (#VC1651)
Restraint (#VC1495)
Diligence (#VC0559)
Composure (#VC0391)
Power (#VC1331)
Stamina (#VC1655)
Lasting (#VC1051)
Tenacity (#VC1749)
Patience (#VC1261)
Devotion (#VC0551)
Toughness (#VC1767)
Sincerity (#VC1619)
Endurance (#VC0645)
Constancy (#VC0455)
Assiduity (#VC0183)

VP5624

Steadiness (#VC1669)
Resistance (#VC1477)
Permanence (#VC1275)
Dedication (#VC0525)
Commitment (#VC0361)
Spontaneity (#VC1645)
Self-denial (#VC1551)
Persistence (#VC1281)
Application (#VC0151)
Perseverance (#VC1279)
Immutability (#VC0951)
Decisiveness (#VC0519)
Invincibility (#VC1021)
Dauntlessness (#VC0515)
Self-restraint (#VC1581)
Mettlesomeness (#VC1145)
Self-government (#VC1565)
Industriousness (#VC0975)
Constituent values Fear (#VD2684)
Levity (#VD3146)
Bigotry (#VD2176)
Rigidity (#VD3568)
Fantasy (#VD2674)
Betrayal (#VD2170)
Petulance (#VD3454)
Desertion (#VD2406)
Cowardice (#VD2298)
Wantonness (#VD4038)
Unruliness (#VD3950)
Stuffiness (#VD3740)
Indocility (#VD2970)
Hesitation (#VD2792)
Fanaticism (#VD2676)
Willfulness (#VD4060)
Uncertainty (#VD3880)
Intolerance (#VD3074)
Inconstancy (#VD2934)
Ambivalence (#VD2056)
Irresolution (#VD3112)
Impetuousity (#VD2848)
Carelessness (#VD2214)
Unadvisedness (#VD3874)
Intransigence (#VD3080)
Impulsiveness (#VD2878)
Traitorousness (#VD3834)
Intractability (#VD3078)
Changeableness (#VD2222)
Weak-mindedness (#VD4050)
Undependability (#VD3894)
Incorrigibility (#VD2942)
Unpredictability (#VD3938)
Irreconcilability (#VD3100)

Resolution (#VC1479)
Possession (#VC1325)
Discipline (#VC0567)
Continuity (#VC0471)
Staunchness (#VC1665)
Seriousness (#VC1603)
Pertinacity (#VC1291)
Earnestness (#VC0599)
Self-control (#VC1549)
Independence (#VC0969)
Faithfulness (#VC0731)
Steadfastness (#VC1667)
Determination (#VC0547)
Concentration (#VC0403)
Purposefulness (#VC1393)
Self-possession (#VC1571)
Self-discipline (#VC1557)
Indefatigableness (#VC0967).
Strain (#VD3732)
Treason (#VD3844)
Weakness (#VD4052)
Grimness (#VD2752)
Defiance (#VD2354)
Stiffness (#VD3730)
Obstinacy (#VD3354)
Defection (#VD2348)
Withdrawal (#VD4062)
Volatility (#VD4030)
Suddenness (#VD3756)
Perversity (#VD3446)
Impatience (#VD2840)
Feebleness (#VD2690)
Disloyalty (#VD2520)
Vacillation (#VD3974)
Repudiation (#VD3536)
Instability (#VD3054)
Dubiousness (#VD2586)
Stubbornness (#VD3738)
Irascibility (#VD3094)
Equivocation (#VD2616)
Unreliability (#VD3948)
Recalcitrance (#VD3520)
Inflexibility (#VD3008)
Faithlessness (#VD2662)
Relentlessness (#VD3528)
Indecisiveness (#VD2950)
Capriciousness (#VD2212)
Unmanageability (#VD3928)
Overzealousness (#VD3418)
Unthoughtfulness (#VD3956)
Contumaciousness (#VD2290)
Inconsiderateness (#VD2930).

Disappearance (#VD2458)
Overzealousness (#VD3418)
Indiscrimination (#VD2966)

Passionlessness (#VD3428)
Acquisitiveness (#VD2030)

◆ **VP5637c Choice–Necessity**
Dynamics Necessity reconciles and brings men together; and this accidental connection afterward forms itself into laws. (Montaigne)
Hold it wise...to make a virtue of necessity. (Chaucer)
Necessity relieves us from the embarrassment of choice. (Vauvenargues)
  Integrative complex Choice (#VT8062).
  Constituent values Will (#VC1923)
  Fitness (#VC0767)
  Approval (#VC0161)
  Worthiness (#VC1941)
  Eclecticism (#VC0603)
  Desirableness (#VC0543)
  Constituent values Denial (#VD2376)
  Disdain (#VD2490)
  Rejection (#VD3526)
  Disregard (#VD2542)
  Disapproval (#VD2462)
  Indiscrimination (#VD2966)

  Favour (#VC0743)
  Pleasure (#VC1309)
  Necessity (#VC1183)
  Suitability (#VC1701)
  Tastefulness (#VC1745)
  Discrimination (#VC0573).
  Urgency (#VD3968)
  Decline (#VD2336)
  Exclusion (#VD2634)
  Repudiation (#VD3536)
  Nonacceptance (#VD3318)
  Contemptuousness (#VD2280)

◆ **VP5642a Conventionality–Unconventionality**
Dynamics It is not difficult to be unconventional in the eyes of the world when your unconventionality is but the convention of your set. (W Somerset Maugham)
Ascend above the restrictions and conventions of the world, but not so high as to lose sight of them. (Richard Garnett)
Conventionality is not morality. Self-righteousness is not religion. To attack the first is not to assail the last. (Charlotte Brontë)
When she stopped conforming to the conventional picture of feminity she finally began to enjoy being a woman. (Betty Friedan)
What men call civilization is the condition of present customs; what they call barbarism, the condition of past ones. (Anatole France)
  Integrative complex Motivation (#VT8063).
  Constituent values Manners (#VC1117)
  Ordinariness (#VC1235)
  Civilization (#VC0333)
  Constituent values Dependence (#VD2382).

  Aspiration (#VC0179)
  Cosmopolitan (#VC0493)
  Adaptability (#VC0049).

◆ **VP5646b Formality–Informality**
Dynamics In statesmanship get the formalities right, never mind about the moralities. (Mark Twain)
It is superstition to put one's hope in formalities; but it is pride to be unwilling to submit to them. (Pascal)
  Integrative complex Motivation (#VT8063).
  Constituent values Service (#VC1605)
  Decorum (#VC0523)
  Elegance (#VC0623)
  Unassuming (#VC1801)
  Cordiality (#VC0487)
  Well-mannered (#VC1913)
  Constituent values Pedantry (#VD3432)

  Manners (#VC1117)
  Decency (#VC0517)
  Civility (#VC0331)
  Politeness (#VC1317)
  Stateliness (#VC1659)
  Scrupulousness (#VC1535).
  Stiffness (#VD3730).

◆ **VP5648c Motivation–Dissuasion**
Dynamics Right is its own defence. (Bertolt Brecht)
Religion is a great force – the only real motive force in the world; but what you fellows don't understand is that you must get at a man through his own religion and not through yours. (Shaw)
With the exception of the instinct of self-preservation, the propensity for emulation is probably the strongest and most alert and persistent of the economic motives proper. (Veblen)
Good and evil, reward and punishment, are the only motives to a rational creature: these are the spur and reins whereby all mankind are set on work, and guided. (Locke)
We should often feel ashamed of our best actions if the world could see all of the motives which produced them. (La Rochefoucauld)
It is good to be without vices, but it is not good to be without temptations. (Walter Bagehot)
  Integrative complex Motivation (#VT8063).
  Constituent values Aid (#VC0095)
  Charm (#VC0317)
  Motive (#VC1171)
  Appeal (#VC0143)
  Interest (#VC1011)
  Advocacy (#VC0069)
  Intention (#VC1007)
  Direction (#VC0561)
  Animation (#VC0135)
  Motivation (#VC1169)
  Endearment (#VC0641)
  Allurement (#VC0107)
  Inspiration (#VC0993)
  Enchantment (#VC0637)
  Exhilaration (#VC0703)
  Encouragement (#VC0639)
  Irresistability (#VC1031).
  Constituent values Temptation (#VD3800)
  Intimidation (#VD3072)

  Gain (#VC0817)
  Reason (#VC1423)
  Genius (#VC0825)
  Glamour (#VC0837)
  Charisma (#VC0313)
  Principle (#VC1353)
  Influence (#VC0979)
  Challenge (#VC0303)
  Sex appeal (#VC1609)
  Excitement (#VC0699)
  Attraction (#VC0199)
  Stimulation (#VC1675)
  Fascination (#VC0737)
  Forcefulness (#VC0781)
  Gratification (#VC0863)
  Recommendation (#VC1431)

  Corruption (#VD2294)
  Discouragement (#VD2482).

◆ **VP5631b Desire–Avoidance**
Dynamics A wise man is cured of ambition by ambition itself; his aim is so exalted that riches, office, fortune, and favour cannot satisfy him. (La Bruyère)
Though ambition may be a fault in itself, it is often the mother of virtues. (Quintilian)
Nothing in the world is so incontinent as a man's accursed appetite. (Homer)
When he has no lust, no hatred, a man walks safely among the things of lust and hatred. (Bhagavadgita)
From the satisfaction of desire there may arise, accompanying joy and as it were sheltering behind it, something not unlike despair. (André Gide)
The desire for imaginary benefits often involves the loss of present blessings. (Aesop)
Some desire is necessary to keep life in motion, and he whose real wants are supplied must admit those of fancy. (Samuel Johnson)
Nothing great was ever achieved without enthusiasm. (Emerson)
Science is the great antidote to the poison of enthusiasm and superstition. (Adam Smith)
  Integrative complex Choice (#VT8062).
  Constituent values Zest (#VC1949)
  Love (#VC1093)
  Elan (#VC0619)
  Warmth (#VC1899)
  Loyalty (#VC1095)
  Vivacity (#VC1893)
  Pleasure (#VC1309)
  Idealism (#VC0935)
  Fidelity (#VC0763)
  Devotion (#VC0551)
  Ambition (#VC0115)
  Sincerity (#VC1619)
  Quickness (#VC1399)
  Eagerness (#VC0597)
  Affection (#VC0075)
  Promptness (#VC1377)
  Liberation (#VC1071)
  Excitement (#VC0699)
  Detachment (#VC0545)
  Commitment (#VC0361)
  Fascination (#VC0737)
  Tastefulness (#VC1745)
  Desirableness (#VC0543).
  Constituent values Lust (#VD3176)
  Greed (#VD2746)
  Disuse (#VD2574)
  Evasion (#VD2622)
  Zealotry (#VD4082)
  Rapacity (#VD3516)
  Betrayal (#VD2170)
  Desertion (#VD2406)
  Avoidance (#VD2126)
  Sordidness (#VD3692)
  Fanaticism (#VD2676)
  Insouciance (#VD3052)
  Inattention (#VD2906)
  Dereliction (#VD2400)
  Slothfulness (#VD3680)
  Lovelessness (#VD3170)
  Indifference (#VD2958)
  Incontinence (#VD2936)
  Equivocation (#VD2616)

  Zeal (#VC1947)
  Life (#VC1075)
  Faith (#VC0729)
  Passion (#VC1257)
  Ardour (#VC0165)
  Vitality (#VC1891)
  Keenness (#VC1043)
  Fondness (#VC0777)
  Fervour (#VC0761)
  Appetite (#VC0149)
  Alacrity (#VC0097)
  Readiness (#VC1417)
  Intensity (#VC1005)
  Animation (#VC0135)
  Resolution (#VC1479)
  Liveliness (#VC1085)
  Intentness (#VC1009)
  Enthusiasm (#VC0663)
  Dedication (#VC0525)
  Seriousness (#VC1603)
  Earnestness (#VC0599)
  Faithfulness (#VC0731)

  Fury (#VD2732)
  Frenzy (#VD2716)
  Apathy (#VD2090)
  Avarice (#VD2122)
  Voracity (#VD4032)
  Coldness (#VD2242)
  Disregard (#VD2542)
  Defection (#VD2348)
  Withdrawal (#VD4062)
  Negligence (#VD3312)
  Malingering (#VD3196)
  Infatuation (#VD2998)
  Disinterest (#VD2516)
  Abandonment (#VD2004)
  Mindlessness (#VD3228)
  Listlessness (#VD3156)
  Incuriosity (#VD2946)
  Heedlessness (#VD2786)
  Carelessness (#VD2214)

◆ **VP5660a Oversufficiency–Insufficiency**
Dynamics Elephants suffer from too much patience. Their exhibitions of it may seem superb, – such power and such restraint, combined, are noble – but a quality carried to excess defeats itself. (Clarence Day)
The sovereign source of melancholy is repletion. Need and struggle are what excite and inspire us; our hour of triumph is what brings the void. (William James)
Good and evil...are not what vulgar opinion accounts them; many who seem to be struggling with adversity are happy; many, amid great affluence, are utterly miserable. (Tacitus)
  Integrative complex Adaptation (#VT8064).
  Constituent values Work (#VC1937)
  Wealth (#VC1901)
  Richness (#VC1503)
  Adequacy (#VC0051)
  Fertility (#VC0759)
  Abundance (#VC0015)
  Generosity (#VC0821)
  Exuberance (#VC0723)
  Sufficiency (#VC1699)

  Good (#VC0845)
  Plenty (#VC1313)
  Fullness (#VC0811)
  Plenitude (#VC1311)
  Affluence (#VC0083)
  Liberality (#VC1069)
  Fulfilment (#VC0809)
  Competence (#VC0383)
  Satisfaction (#VC1533)

VALUE POLARITIES — VP5674

Well-provided (#VC1917)
Superabundance (#VC1703)
Productiveness (#VC1365).
Constituent values Lack (#VD3126)
Excess (#VD2630)
Surplus (#VD3776)
Poverty (#VD3470)
Absence (#VD2018)
Sparsity (#VD3702)
Scarcity (#VD3614)
Overload (#VD3396)
Meanness (#VD3204)
Verbosity (#VD4004)
Prolixity (#VD3498)
Austerity (#VD2116)
Overweight (#VD3414)
Overstrain (#VD3406)
Inadequacy (#VD2890)
Congestion (#VD2274)
Limitedness (#VD3154)
Abandonment (#VD2004)
Intemperence (#VD3066)
Incompetence (#VD2920)
Extravagance (#VD2654)
Overextension (#VD3394)
Overdeveloped (#VD3386)
Nonfulfilment (#VD3328)
Excessiveness (#VD2632)
Incompleteness (#VD2922)
Unconscionableness (#VD3890).

Bountifulness (#VC0255)
Substantiality (#VC1693)
Strain (#VD3732)
Tension (#VD3802)
Surfeit (#VD3772)
Overtax (#VD3410)
Thinness (#VD3808)
Shortage (#VD3674)
Overwork (#VD3416)
Omission (#VD3360)
Exiguity (#VD2642)
Smallness (#VD3684)
Extremism (#VD2656)
Scantiness (#VD3612)
Oversupply (#VD3408)
Lavishness (#VD3134)
Deficiency (#VD2356)
Prodigality (#VD3488)
Deprivation (#VD2398)
Overemphasis (#VD3388)
Incontinence (#VD2936)
Imperfection (#VD2842)
Unsuitability (#VD3954)
Overexpansion (#VD3392)
Overabundance (#VD3376)
Insufficiency (#VD3060)
Defectiveness (#VD2350)
Unreasonableness (#VD3946).

♦ **VP5665b Appropriateness–Inappropriateness**
Dynamics Send not for a hatchet to break open an egg with. (Thomas Fuller)
With our mortal minds we should seek from the gods that which becomes us. (Pindar)
Every beauty, when out of its place, is a beauty no longer. (Voltaire)
Integrative complex Adaptation (#VT8064).
Constituent values Worth (#VC1939)
Action (#VC0039)
Service (#VC1605)
Freshness (#VC0799)
Usefulness (#VC1847)
Efficiency (#VC0615)
Stewardship (#VC1671)
Convenience (#VC0473)
Practicality (#VC1335)
Profitability (#VC1369)
Accessibility (#VC0019)
Appropriateness (#VC0159).
Constituent values Waste (#VD4046)
Vanity (#VD3986)
Injury (#VD3024)
Torment (#VD3824)
Outrage (#VD3374)
Fatuity (#VD2680)
Violence (#VD4022)
Misusage (#VD3282)
Effetism (#VD2598)
Valueless (#VD3982)
Pollution (#VD3464)
Depletion (#VD2386)
Absurdity (#VD2022)
Suspension (#VD3784)
Oppression (#VD3366)
Exhaustion (#VD2640)
Uselessness (#VD3970)
Misfeasance (#VD3256)
Malfeasance (#VD3188)
Desecration (#VD2404)
Prostitution (#VD3502)
Maltreatment (#VD3200)
Worthlessness (#VD4072)
Unsuitability (#VD3954)
Mismanagement (#VD3268)
Ineffectuality (#VD2978)
Purposelessness (#VD3508)
Inapplicability (#VD2898)
Inappropriateness (#VD2900).

Value (#VC1855)
Utility (#VC1849)
Interest (#VC1011)
Advantage (#VC0065)
Expedience (#VC0711)
Versatility (#VC1873)
Helpfulness (#VC0893)
Consumption (#VC0461)
Availability (#VC0209)
Effectiveness (#VC0613)
Practicability (#VC1333)

Abuse (#VD2024)
Misuse (#VD3284)
Disuse (#VD2574)
Rubbish (#VD3582)
Inanity (#VD2896)
Erosion (#VD2618)
Vainness (#VD3980)
Futility (#VD2736)
Violation (#VD4020)
Rejection (#VD3526)
Impotence (#VD2858)
Cheapness (#VD2230)
Triviality (#VD3856)
Perversion (#VD3444)
Misconduct (#VD3246)
Barrenness (#VD2146)
Profanation (#VD3490)
Malpractice (#VD3198)
Harrassment (#VD2768)
Aimlessness (#VD2048)
Obsolescence (#VD3350)
Fecklessness (#VD2688)
Victimization (#VD4012)
Pointlessness (#VD3460)
Misapplication (#VD3232)
Impoverishment (#VD2860)
Ineffectiveness (#VD2976)
Unproductiveness (#VD3942)

♦ **VP5670b Expedience–Inexpedience**
Dynamics Effective action is always unjust. (Jean Anouilh)
No man is justified in doing evil on the ground of expediency. (Theodore Roosevelt)
Decision by majorities is as much an expedient as lighting by gas. (Gladstone)
Free trade is not a principal, it is an expedient. (Disraeli)
It is expedient for us, that one man should die for the people. (Bible)
Integrative complex Adaptation (#VT8064).
Constituent values Good (#VC0845)
Fitness (#VC0767)
Efficiency (#VC0615)
Effectiveness (#VC0613)
Appropriateness (#VC0159).
Constituent values Injury (#VD3024)
Trouble (#VD3858)
Mischief (#VD3242)
Prejudice (#VD3474)
Infelicity (#VD3000)
Inaptitude (#VD2902)
Uselessness (#VD3970)
Awkwardness (#VD2130)
Disadvantage (#VD2450)
Unsuitability (#VD3954)
Inadvisability (#VD2894)
Deleteriousness (#VD2370)

Action (#VC0039)
Expedience (#VC0711)
Feasibility (#VC0747)
Desirableness (#VC0543)

Damage (#VD2314)
Illness (#VD2826)
Futility (#VD2736)
Liability (#VD3148)
Ineptitude (#VD2984)
Impairment (#VD2838)
Incongruity (#VD2926)
Inexpedience (#VD2992)
Worthlessness (#VD4072)
Inconvenience (#VD2938)
Undesirableness (#VD3900)
Inappropriateness (#VD2900).

♦ **VP5672a Importance–Unimportance**
Dynamics No human thing is of serious importance. (Plato)
It is completely unimportant. That is why it is so interesting. (Agatha Christie)
We do not succeed in changing things according to our desire, but gradually our desire changes. The situation that we hoped to change because it was intolerable becomes unimportant. We have not managed to surmount the obstacle, as we were resolutely determined to do, but life has taken us round it, led us past it, and then if we turn round to gaze at the remote past, we can barely catch sight of it, so imperceptible has it become. (Proust)
Where I was born and where and how I have lived is unimportant. It is what I have done with where I have been that should be of interest. (Georgia O'Keeffe)
Integrative complex Adaptation (#VT8064).
Constituent values Awe (#VC0213)
Worth (#VC1939)
Merit (#VC1139)
Honour (#VC0915)
Dignity (#VC0557)
Bigness (#VC0241)
Priority (#VC1355)
Nobility (#VC1193)
Eminence (#VC0631)
Solemnity (#VC1627)
Reputation (#VC1471)
Notability (#VC1203)
Superiority (#VC1709)
Distinction (#VC0577)
Forcefulness (#VC0781)
Consideration (#VC0449).
Constituent values Vanity (#VD3986)
Sadness (#VD3598)
Inanity (#VD2896)
Vapidity (#VD3990)
Meanness (#VD3204)
Futility (#VD2736)
Smallness (#VD3684)
Obscurity (#VD3346)
Cheapness (#VD2230)
Mediocrity (#VD3206)
Shallowness (#VD3664)
Inferiority (#VD3002)
Indifference (#VD2958)
Meritlessness (#VD3218)
Superficiality (#VD3766)
Ineffectuality (#VD2978)

Fame (#VC0733)
Value (#VC1855)
Glory (#VC0841)
Essence (#VC0681)
Concern (#VC0407)
Rareness (#VC1413)
Prestige (#VC1349)
Interest (#VC1011)
Supremacy (#VC1713)
Greatness (#VC0867)
Precedence (#VC1339)
Importance (#VC0957)
Seriousness (#VC1603)
Significance (#VC1613)
Momentousness (#VC1161)

Levity (#VD3146)
Nullity (#VD3342)
Vileness (#VD4014)
Vainness (#VD3980)
Idleness (#VD2810)
Valueless (#VD3982)
Pettiness (#VD3452)
Frivolity (#VD2724)
Triviality (#VD3856)
Littleness (#VD3158)
Irrelevance (#VD3106)
Unimportance (#VD3920)
Worthlessness (#VD4072)
Inconsequence (#VD2928)
Insignificance (#VD3042)
Despicableness (#VD2414).

♦ **VP5674c Goodness–Badness**
Dynamics Whatever harm the evil may do, the harm done by the good is the most harmful harm. (Nietzsche)
Yes, that is what good is: to forgive evil. There is no other good. (Antonio Porchia)
I am afraid that good people do a great deal of harm in this world. Certainly the greatest harm they do is that they make badness of such extraordinary importance. (Oscar Wilde)
Every sweet hath its sour; every evil its good. (Emerson)
To be nobly wrong is more manly than to be meanly right. (Thomas Paine)
There is no odour so bad as that which arises from goodness tainted. (Thoreau)
A good is never productive of evil but when it is carried to a culpable excess, in which case it completely ceases to be a good. (Voltaire)
It's the bad that's in the best of us leaves the saint so like the rest of us. (Arthur Stringer)
A certain alloy of expediency improves the gold of morality and makes it wear all the longer. (Don Marquis)
All nature is but art, unknown to thee; all chance, direction, which thou canst not see; all discord, harmony not understood; all partial evil, universal good. (Alexander Pope)
Those who are fond of setting things to rights, have no great objection to seeing them wrong. (William Hazlitt)
Let no man presume to think that he can devise any plan of extensive good, unalloyed and unadulterated with evil. (Charles Caleb Colton)
Good and bad may not be dissevered; there is, as there should be, a commingling. (Euripides)
He that can't endure the bad, will not live to see the good. (Yiddish Proverb)
We cannot freely and wisely choose the right way for ourselves unless we know both good and evil. (Helen Keller)
The betrothed of good is evil, the betrothed of life is death, the betrothed of love is divorce. (Malay Proverb)
If we could see all the evil that may spring from good, what should we do ? (Luigi Pirandello)
The omission of good is no less reprehensible than the commission of evil. (Plutarch)
A good thing which prevents us from enjoying a greater good is in truth an evil. (Spinoza)
When the bad imitate the good, there is no knowing what mischief is intended. (Publilius Syrus)
Integrative complex Adaptation (#VT8064).
Constituent values Good (#VC0845)
Worth (#VC1939)
Merit (#VC1139)
Virtue (#VC1887)
Welfare (#VC1903)
Cogency (#VC0349)
Kindness (#VC1045)
Interest (#VC1011)
Fairness (#VC0727)
Supremacy (#VC1713)
Innocence (#VC0987)
Benignity (#VC0237)
Usefulness (#VC1847)
Expedience (#VC0711)
Superiority (#VC1709)
Healthiness (#VC0891)
Advancement (#VC0063)
Pleasantness (#VC1307)
Harmlessness (#VC0885)
Wonderfulness (#VC1935)
Exquisiteness (#VC0721)
Respectability (#VC1483)
Constituent values Woe (#VD4068)
Evil (#VD2624)
Venom (#VD4002)
Blame (#VD2180)
Miasma (#VD3222)
Injury (#VD3024)
Arrant (#VD2098)
Torment (#VD3824)
Outrage (#VD3374)
Illness (#VD2826)
Vileness (#VD4014)
Toxicity (#VD3832)
Grimness (#VD2752)
Fiendish (#VD2696)

Gain (#VC0817)
Value (#VC1855)
Glory (#VC0841)
Choice (#VC0325)
Quality (#VC1397)
Validity (#VC1853)
Keenness (#VC1043)
Goodness (#VC0847)
Wholeness (#VC1919)
Soundness (#VC1637)
Immensity (#VC0943)
Advantage (#VC0065)
Moderation (#VC1157)
Excellence (#VC0697)
Helpfulness (#VC0893)
Benevolence (#VC0235)
Quintessence (#VC1407)
Magnificence (#VC1107)
Appreciation (#VC0153)
Profitability (#VC1369)
Commendability (#VC0359)
Inoffensiveness (#VC0988).

Rot (#VD3578)
Wound (#VD4074)
Filth (#VD2700)
Abuse (#VD2024)
Menace (#VD3210)
Damage (#VD2314)
Torture (#VD3828)
Squalor (#VD3720)
Noisome (#VD3316)
Badness (#VD2134)
Vexation (#VD4006)
Mischief (#VD3242)
Foulness (#VD2710)
Distress (#VD2566)

VP5674

Diabolic (#VD2436)
Baseness (#VD2150)
Wrongness (#VD4078)
Repulsion (#VD3540)
Pollution (#VD3464)
Grievance (#VD2750)
Brutality (#VD2202)
Sordidness (#VD3692)
Malignancy (#VD3194)
Clumsiness (#VD2238)
Affliction (#VD2036)
Malevolence (#VD3186)
Hurtfulness (#VD2806)
Fulsomeness (#VD2728)
Banefulness (#VD2140)
Inexpedience (#VD2992)
Disadvantage (#VD2450)
Despoliation (#VD2416)
Uncleanliness (#VD3884)
Loathsomeness (#VD3160)
Despicableness (#VD2414)
Deleteriousness (#VD2370)

Calamity (#VD2208)
Atrocity (#VD2114)
Virulence (#VD4024)
Prejudice (#VD3474)
Grossness (#VD2754)
Dirtiness (#VD2444)
Awfulness (#VD2128)
Notoriety (#VD3338)
Corruption (#VD2294)
Bestiality (#VD2168)
Viciousness (#VD4008)
Inferiority (#VD3002)
Harmfulness (#VD2772)
Beastliness (#VD2158)
Abomination (#VD2016)
Horribleness (#VD2800)
Devilishness (#VD2430)
Worthlessness (#VD4072)
Poisonousness (#VD3462)
Unpleasantness (#VD3936)
Mischievousness (#VD3224)
Counterproductivity (#VD2296).

♦ VP5677c Perfection–Imperfection

Dynamics If you shut your door to all errors truth will be shut out. (Rabindranath Tagore)
It seems to be the fate of idealists to obtain what they have struggled for in a form which destroys their ideals. (Bertrand Russell)
Mud-pies gratify one of our first and best instincts. So long as we are dirty, we are pure. (Charles Dudley Warner)
There is a tragedy in perfection, because the universe in which perfection arises is itself imperfect. (George Santayana)
The indefatigable pursuit of an unattainable Perfection, even though it consist in nothing more than the pounding of an old piano, alone gives a meaning to our life on this unavailing star. (Logan Pearsall)
When we get sick, we want an uncommon doctor. If we have a construction job, we want an uncommon engineer. When we get into a war, we dreadfully want an uncommon admiral and an uncommon general. Only when we get into politics are we content with the common man. (Herbert Hoover)
Every age confutes old errors and begets new. (Thomas Fuller)
Style is the perfection of a point of view. (Richard Eberhart)
It takes a certain courage and a certain greatness even to be truly base. (Jean Anouilh)
The greatest of faults, I should say, is to be conscious of none. (Thomas Carlyle)
He is lifeless that is faultless. (English Proverb)
There is no such source of error as the pursuit of absolute truth. (Samuel Butler)
Faults shared are comfortable as bedroom slippers and as easy to slip into. (Phyllis McGinley)
Integrative complex Adaptation (#VT8064).
Constituent values Purity (#VC1391)
Fullness (#VC0811)
Wholeness (#VC1919)
Soundness (#VC1637)
Integrity (#VC0997)
Perfection (#VC1273)
Proficiency (#VC1367)
Thoroughness (#VC1753)
Quintessence (#VC1407)
Harmlessness (#VC0885)
Consummation (#VC0459)
Infallibility (#VC0977)
Absoluteness (#VC0009)
Irreproachability (#VC1029).
Constituent values Lack (#VD3126)
Stain (#VD3724)
Problem (#VD3486)
Blemish (#VD2184)
Vapidity (#VD3990)
Meanness (#VD3204)
Dullness (#VD2588)
Dirtiness (#VD2444)
Cheapness (#VD2230)
Unevenness (#VD3904)
Inadequacy (#VD2890)
Impairment (#VD2838)
Faultiness (#VD2682)
Deficiency (#VD2356)
Coarseness (#VD2240)
Imprecision (#VD2864)
Deformation (#VD2358)
Imperfection (#VD2842)
Disfigurement (#VD2498)
Incompleteness (#VD2922).

Idealism (#VC0935)
Chastity (#VC0319)
Unspotted (#VC1841)
Plenitude (#VC1311)
Freshness (#VC0799)
Unblemished (#VC1805)
Culmination (#VC0505)
Spotlessness (#VC1649)
Ordinariness (#VC1235)
Flawlessness (#VC0769)
Stainlessness (#VC1653)
Faultlessness (#VC0741)
Unadulteration (#VC1795).

Flaw (#VD2704)
Tedium (#VD3794)
Failure (#VD2660)
Weakness (#VD4052)
Shortage (#VD3674)
Impurity (#VD2880)
Baseness (#VD2150)
Defection (#VD2348)
Warpedness (#VD4044)
Mediocrity (#VD3206)
Inaccuracy (#VD2886)
Immaturity (#VD2832)
Distortion (#VD2562)
Commonness (#VD2248)
Inferiority (#VD3002)
Fallibility (#VD2668)
Indifference (#VD2958)
Erroneousness (#VD2620)
Defectiveness (#VD2350)

♦ VP5681a Cleanness–Uncleanness

Dynamics What separates two people most profoundly is a different sense and degree of cleanliness. (Nietzsche)
The secret thoughts of a man run over all things, holy, profane, clean, obscene, grave, and light, without shame or blame. (Thomas Hobbes)
Soldiers are dreamers; when the guns begin they think of firelit homes, clean beds, and wives. (Siegfried Sasson)
Will all great Neptune's ocean wash this blood clean from my hand ? No, this my hand will rather the multitudinous seas incarnadine, making the green one red. (Shakespeare)
Mud-pies gratify one of our first and best instincts. So long as we are dirty, we are pure. (Charles Dudley Warner)
Integrative complex Integrity (#VT8065).
Constituent values Purity (#VC1391)
Unspotted (#VC1841)
Freshness (#VC0799)
Brightness (#VC0265)
Cleanliness (#VC0337)
Stainlessness (#VC1653)
Constituent values Rot (#VD3578)
Filth (#VD2700)
Vileness (#VD4014)
Foulness (#VD2710)
Pollution (#VD3464)
Dirtiness (#VD2444)
Corruption (#VD2294)
Beastliness (#VD2158)
Uncleanliness (#VD3884)

Hygiene (#VC0933)
Sweetness (#VC1725)
Refinement (#VC1441)
Unblemished (#VC1805)
Spotlessness (#VC1649)
Unadulteration (#VC1795).
Stain (#VD3724)
Squalor (#VD3720)
Impurity (#VD2880)
Repulsion (#VD3540)
Messiness (#VD3220)
Sordidness (#VD3692)
Bedraggled (#VD2162)
Abomination (#VD2016)

♦ VP5683a Healthfulness–Unhealthfulness

Dynamics No man was ever made more healthful by a dangerous sickness, or came home better from a long voyage. (Proverb)
The healthful man can give counsel to the sick. (Proverb)
Integrative complex Integrity (#VT8065).
Constituent values Good (#VC0845)
Goodness (#VC0847)
Benignity (#VC0237)
Wholesomeness (#VC1921).
Constituent values Venom (#VD4002)
Badness (#VD2134)
Virulence (#VD4024)
Agitation (#VD2046)
Poisonousness (#VD3462)
Pestiferousness (#VD3450)

Hygiene (#VC0933)
Salubrity (#VC1523)
Healthiness (#VC0891)

Noisome (#VD3316)
Toxicity (#VD3832)
Pollution (#VD3464)
Harmfulness (#VD2772)
Unwholesomeness (#VD3960)
Destructiveness (#VD2422).

♦ VP5685a Health–Disease

Dynamics Old age and sickness bring out the essential characteristics of a man. (Felix Frankfurter)
There is a certain state of health that does not allow us to understand everything; and perhaps illness shuts us off from certain truths: but health shuts us off just as effectively from others. (André Gide)
What some call health, if purchased by perpetual anxiety about diet, isn't much better than tedious disease. (George Dennison Prentice)
How sickness enlarges the dimensions of a man's self to himself He is his own exclusive object. Supreme selfishness is inculcated upon him as his only duty. (Charles Lamb)
Integrative complex Integrity (#VT8065).
Constituent values Youth (#VC1943)
Health (#VC0889)
Vitality (#VC1891)
Immunity (#VC0949)
Wholeness (#VC1919)
Longevity (#VC1091)
Freshness (#VC0799)
Sturdiness (#VC1685)
Healthiness (#VC0891)
Wholesomeness (#VC1921).
Constituent values Rot (#VD3578)
Wound (#VD4074)
Shock (#VD3672)
Malaise (#VD3184)
Illness (#VD2826)
Badness (#VD2134)
Disorder (#VD2526)
Debility (#VD2328)
Morbidity (#VD3286)
Spasticity (#VD3706)
Feebleness (#VD2690)
Disability (#VD2446)
Abnormality (#VD2014)
Complication (#VD2254)
Indisposition (#VD2968)
Unwholesomeness (#VD3960)

Vigour (#VC1881)
Fitness (#VC0767)
Strength (#VC1681)
Flourish (#VC0773)
Soundness (#VC1637)
Hardiness (#VC0883)
Aliveness (#VC0101)
Resistance (#VC1477)
Youthfulness (#VC1945)

Pain (#VD3422)
Waste (#VD4046)
Frenzy (#VD2716)
Madness (#VD3178)
Fatigue (#VD2678)
Sickness (#VD3676)
Delerium (#VD2368)
Paralysis (#VD3426)
Fragility (#VD2712)
Invalidity (#VD3088)
Exhaustion (#VD2640)
Affliction (#VD2036)
Intoxication (#VD3076)
Mortification (#VD3288)
Defectiveness (#VD2350)
Pestiferousness (#VD3450).

♦ VP5690a Selfactualization–Neurosis

Dynamics Conscience is the frame of character, and love is the covering for it. (Henry Ward Beecher)
We believe that humanness consists in what we call conscience, in that courage, if you wish, which we have shown on one single occasion rather than in the cowardice which on many occasions has counselled prudence. (Luigi Pirandello)
We live in a hemisphere whose one revolution has given birth to the most powerful force of the modern age – the search for the freedom and self-fulfilment of man. (John F Kennedy)
The moment one is on the side of life 'peace and security' drop out of consciousness. The only peace, the only security, is in fulfilment. (Henry Miller)
Human salvation lies in the hands of the creatively maladjusted. (Martin Luther King Jr)
Until we know what motivates the hearts and minds of men we can understand nothing outside ourselves, nor will we ever reach fulfilment as that greatest miracle of all, the human being. (Marya Mannes)
It is in self-limitation that a master first shows himself. (Goethe)
The test of a civilized person is first self-awareness, and then depth after depth of sincerity in self-confrontation. (Clarence Day)
Assurance is contemptible and fatal unless it is self-knowledge (George Santayana)
Men can starve from a lack of self-realization as much as they can from a lack of bread. (Richard Wright)
Integrative complex Integrity (#VT8065).
Constituent values Resistance (#VC1477)
Conscience (#VC0439).
Constituent values Stupor (#VD3744)
Fixation (#VD2702)
Delerium (#VD2368)
Obsession (#VD3348)
Frustration (#VD2726)
Dissociation (#VD2554)
Insensibility (#VD3038)
Disorientation (#VD2530)
Depersonalization (#VD2384)

Fulfilment (#VC0809)

Insanity (#VD3034)
Delusion (#VD2372)
Conflict (#VD2270)
Dejection (#VD2366)
Ambivalence (#VD2056)
Maladjustment (#VD3180)
Hallucination (#VD2764)
Overcompensation (#VD3380)
Arrested development (#VD2100).

♦ VP5691b Improvement–Impairment

Dynamics The civilized man is a larger mind but a more imperfect nature than the savage. (Margaret Fuller)
Not a having and a resting, but a growing and a becoming, is the character of perfection as culture conceives it. (Matthew Arnold)
It is an error to imagine that evolution signifies a constant tendency to increased perfection. That process undoubtedly involves a constant remodelling of the organism in adaptation to new conditions; but it depends on the nature of those conditions whether the direction of the modifications effected shall be upward or downward. (Thomas Henry Huxley)
The crimes of extreme civilization are probably worse than those of extreme barbarism, because of their refinement, the corruption they presuppose, and their superior degree of intellectuality. (Jules Barbey)
Integrative complex Integrity (#VT8065).
Constituent values Gain (#VC0817)
Beauty (#VC0221)
Culture (#VC0507)
Civility (#VC0331)

Uplift (#VC1843)
Revival (#VC1501)
Progress (#VC1373)
Evolution (#VC0691)

-786-

VALUE POLARITIES VP5720

Education (#VC0611)
Perfection (#VC1273)
Betterment (#VC0239)
Enhancement (#VC0649)
Consumption (#VC0461)
Civilization (#VC0333)
Enlightenment (#VC0653)
Constituent values Rot (#VD3578)
Ruin (#VD3586)
Waste (#VD4046)
Strain (#VD3732)
Injury (#VD3024)
Blight (#VD2186)
Erosion (#VD2618)
Blemish (#VD2184)
Sabotage (#VD3590)
Effetism (#VD2598)
Vitiation (#VD4026)
Decadence (#VD2330)
Perversion (#VD3444)
Involution (#VD3092)
Disability (#VD2446)
Depression (#VD2396)
Brokenness (#VD2200)
Dissolution (#VD2556)
Degradation (#VD2362)
Dilapidation (#VD2442)
Degeneration (#VD2360)
Retrogression (#VD3554)
Mortification (#VD3288)
Decomposition (#VD2338)
Disorganization (#VD2528).
Refinement (#VC1441)
Enrichment (#VC0655)
Improvement (#VC0961)
Development (#VC0549)
Advancement (#VC0063)
Amelioration (#VC0117)
Transformation (#VC1779).
Wear (#VD4054)
Wound (#VD4074)
Lapse (#VD3130)
Misuse (#VD3284)
Damage (#VD2314)
Failure (#VD2660)
Decline (#VD2336)
Spoilage (#VD3712)
Mischief (#VD3242)
Withering (#VD4064)
Pollution (#VD3464)
Regression (#VD3524)
Mutilation (#VD3300)
Impairment (#VD2838)
Derogation (#VD2402)
Corruption (#VD2294)
Putrefaction (#VD3510)
Depravation (#VD2390)
Prostitution (#VD3502)
Depreciation (#VD2394)
Decrepitness (#VD2340)
Poisonousness (#VD3462)
Deterioration (#VD2424)
Disintegration (#VD2514).

♦ **VP5693b Restoration–Destruction**
Dynamics All destruction, by violent revolution or however it be, is but new creation on a wider scale. (Thomas Carlyle)
Only in growth, reform, and change, paradoxically enough, is true security to be found. (Anne Morrow Lindbergh)
Public calamity is a mighty leveller. (Edmund Burke)
If anything ail a man, so that he does not perform his functions, if he have a pain in his bowels even, – for that is the seat of sympathy, – he forthwith sets about reforming the world. (Thoreau)
Integrative complex Integrity (#VT8065).
Constituent values Reform (#VC1443)
Revival (#VC1501)
Rightness (#VC1507)
Improvement (#VC0961)
Consumption (#VC0461)
Rehabilitation (#VC1449).
Constituent values Ruin (#VD3586)
Upset (#VD3966)
Defeat (#VD2346)
Upheaval (#VD3962)
Overturn (#VD3412)
Foulness (#VD2710)
Calamity (#VD2208)
Vandalism (#VD3984)
Damnation (#VD2316)
Subversion (#VD3754)
Extinction (#VD2650)
Suppression (#VD3770)
Dissolution (#VD2556)
Desolation (#VD2408)
Banefulness (#VD2140)
Despoliation (#VD2416)
Extermination (#VD2648)
Irremediability (#VD3110)
Destructiveness (#VD2422).
Service (#VC1605)
Salvation (#VC1525)
Resilience (#VC1475)
Correctness (#VC0489)
Renewableness (#VC1465)
Waste (#VD4046)
Death (#VD2324)
Blight (#VD2186)
Sabotage (#VD3590)
Negation (#VD3306)
Disaster (#VD2466)
Withering (#VD4064)
Perdition (#VD3434)
Annulment (#VD2078)
Spoilation (#VD3714)
Brokenness (#VD2200)
Suffocation (#VD3760)
Devastation (#VD2426)
Depradation (#VD2388)
Wastefulness (#VD4048)
Annihilation (#VD2074)
Disintegration (#VD2514)
Disorganization (#VD2528).

♦ **VP5695b Refreshment–Relapse**
Dynamics Cato said the best way to keep good acts in memory was to refresh them with new. (Bacon)
What country before ever existed a century and a half without a rebellion ?... The tree of liberty must be refreshed from time to time with the blood of patriots and tyrants. It is its natural manure. (Thomas Jefferson)
Talk not of wasted affection. Affection never was wasted. If it enrich not the heart of another, its waters, returning back to their springs, like the rain, shall fill them full of refreshment. That which the fountain sends forth returns again to the fountain (Longfellow)
Integrative complex Integrity (#VT8065).
Constituent values Freshness (#VC0799)
Stimulation (#VC1675)
Constituent values Lapse (#VD3130)
Regression (#VD3524).
Cordiality (#VC0487)
Exhilaration (#VC0703).
Reversion (#VD3558)

♦ **VP5697c Safety–Danger**
Dynamics Everything is sweetened by risk. (Alexander Smith)
A common danger unites even the bitterest enemies. (Aristotle)
He that is too secure is not safe. (Thomas Fuller)
The path is smooth that leadeth on to danger. (Shakespeare)
Most people want security in this world, not liberty. (H L Mencken)
When written in Chinese, the word 'crisis' is composed of two characters – one represents danger and the other represents opportunity. (John F Kennedy)
Our insignificance is often the cause of our safety. (Aesop)
It may well be that we shall by a process of sublime irony have reached a state in this story where safety will be the sturdy child of terror, and survival the twin brother of annihilation. (Sir Winston Churchill)
Integrative complex Integrity (#VT8065).
Constituent values Hope (#VC0917)
Surety (#VC1717)
Support (#VC1711)
Security (#VC1537)
Vigilance (#VC1879)
Assurance (#VC0189)
Redemption (#VC1439)
Liberation (#VC1071)
Stewardship (#VC1671)
Carefulness (#VC0293)
Care (#VC0291)
Safety (#VC1517)
Shelter (#VC1611)
Immunity (#VC0949)
Salvation (#VC1525)
Steadiness (#VC1669)
Protection (#VC1383)
Government (#VC0853)
Deliverance (#VC0539)
Blessedness (#VC0247)
Reservedness (#VC1473)
Harmlessness (#VC0885)
Championship (#VC0305)
Invulnerability (#VC1025).
Constituent values Risk (#VD3570)
Menace (#VD3210)
Badness (#VD2134)
Alarmism (#VD2050)
Insecurity (#VD3036)
Instability (#VD3054)
Doubtfulness (#VD2584)
Hazardousness (#VD2782)
Susceptibility (#VD3780)
Unpredictability (#VD3938)
Preservation (#VC1347)
Conservation (#VC0445)
Trustworthiness (#VC1785)
Threat (#VD3812)
Crisis (#VD2302)
Weakness (#VD4052)
Liability (#VD3148)
Uncertainty (#VD3880)
Dubiousness (#VD2586)
Unreliability (#VD3948)
Dangerousness (#VD2318)
Undependability (#VD3894)
Untrustworthiness (#VD3958).

♦ **VP5705a Action–Inaction**
Dynamics There is nothing more explosive than a skilled population condemned to inaction. Such a population is likely to become a hotbed of extremism and intolerance, and be receptive to any proselytizing ideology, however absurd and vicious, which promises vast action. (Eric Hoffer)
It is the just doom of laziness and gluttony to be inactive without ease and drowsy without tranquillity. (Samuel Johnson)
What is this self inside us, this silent observer, severe and speechless critic, who can terrorize us and urge us on to futile activity, and in the end, judge us still more severely for the errors into which his own reproaches drove us? (T S Eliot)
The lazy are always wanting to do something. (Vauvenargues)
The majority prove their worth by keeping busy. A busy life is the nearest thing to a purposeful life. (Eric Hoffer)
The highest pleasure to be got out of freedom, and having nothing to do, is labour. (Mark Twain)
Perpetual devotion to what a man calls his business, is only to be sustained by perpetual neglect of many other things. (Robert Louis Stevenson)
There is no conceivable human action which custom has not at one time justified and at another condemned. (Joseph Wood Krutch)
The reason men oppose progress is not that they hate progress, but that they love inertia. (Elbert Hubbard)
Generally among intelligent people are found nothing but paralytics and among men of action nothing but fools. (André Gide)
Action is thought tempered by illusion. (Elbert Hubbard)
The great end of life is not knowledge but action. (Thomas Henry Huxley)
Action (is) the great business of mankind, and the whole matter about which all laws are conversant. (John Locke)
Integrative complex Action (#VT8071).
Constituent values Work (#VC1937)
Energy (#VC0647)
Action (#VC0039)
Ardour (#VC0165)
Vivacity (#VC1893)
Dynamism (#VC0595)
Ambition (#VC0115)
Activity (#VC0041)
Readiness (#VC1417)
Diligence (#VC0559)
Assiduity (#VC0183)
Aliveness (#VC0101)
Resolution (#VC1479)
Nimbleness (#VC1191)
Initiative (#VC0985)
Willingness (#VC1925)
Application (#VC0151)
Forcefulness (#VC0781)
Concentration (#VC0403)
Venturesomeness (#VC1867)
Indefatigableness (#VC0967).
Constituent values Ennui (#VD2612)
Stupor (#VD3744)
Languor (#VD3128)
Fatigue (#VD2678)
Weakness (#VD4052)
Slowness (#VD3682)
Laziness (#VD3142)
Dullness (#VD2588)
Weariness (#VD4056)
Paralysis (#VD3426)
Suspension (#VD3784)
Stagnation (#VD3722)
Inactivity (#VD2888)
Boringness (#VD2194)
Tiresomeness (#VD3820)
Overactivity (#VD3378)
Lifelessness (#VD3152)
Impetuousity (#VD2848)
Shiftlessness (#VD3670)
Intrusiveness (#VD3086)
Burdensomeness (#VD2206).
Life (#VC1075)
Daring (#VC0513)
Leisure (#VC1065)
Agility (#VC0089)
Fervour (#VC0761)
Business (#VC0279)
Alacrity (#VC0097)
Swiftness (#VC1727)
Quickness (#VC1399)
Behaviour (#VC0223)
Animation (#VC0135)
Adventure (#VC0067)
Promptness (#VC1377)
Liveliness (#VC1085)
Enterprise (#VC0659)
Expeditious (#VC0713)
Achievement (#VC0031)
Determination (#VC0547)
Implementation (#VC0955)
Industriousness (#VC0975).
Torpor (#VD3826)
Strain (#VD3732)
Inertia (#VD2990)
Boredom (#VD2192)
Vagrancy (#VD3976)
Lethargy (#VD3144)
Idleness (#VD2810)
Debility (#VD2328)
Passivity (#VD3430)
Indolence (#VD2972)
Suddenness (#VD3756)
Overstrain (#VD3406)
Exhaustion (#VD2640)
Toilsomeness (#VD3822)
Slothfulness (#VD3680)
Listlessness (#VD3156)
Indifference (#VD2958)
Wearisomeness (#VD4058)
Overextension (#VD3394)
Impulsiveness (#VD2878).

♦ **VP5709b Exertion–Rest**
Dynamics One cannot rest except after steady practice. (George Ade)
Restfulness is a quality for cattle; the virtues are all active, life is alert. (Robert Louis Stevenson)
Even in the hottest weather a restful mind will keep you cool. (Chinese Proverb)
How can a rational being be ennobled by anything that is not obtained by its own exertions ? (Mary Wollstonecraft)
Integrative complex Action (#VT8071).
Constituent values Work (#VC1937)
Freedom (#VC0797)
Relaxation (#VC1455)
Tranquillity (#VC1773)
Industriousness (#VC0975).
Leisure (#VC1065)
Quietness (#VC1403)
Promptness (#VC1377)
Deliberateness (#VC0533)

♦ **VP5720c Preparedness–Unpreparedness**
Dynamics Eternal vigilance is the price of liberty. (John Philpot Curran)
The price of eternal vigilance is indifference. (Marshall McLuhan)
Magnificently unprepared for the long littleness of life. (Rupert Brooke)
Ready money is Aladdin's lamp. (Byron)
You'll find us rough, sir, but you'll find us ready. (Dickens)
Integrative complex Action (#VT8071).

VP5720

Constituent values Mastery (#VC1121)
Ability (#VC0003)
Virginity (#VC1883)
Readiness (#VC1417)
Competence (#VC0383)
Suitability (#VC1701)
Naturalness (#VC1179)
Qualification (#VC1395).
Constituent values Rudeness (#VD3584)
Negligence (#VD3312)
Coarseness (#VD2240)
Incapability (#VD2908)
Fecklessness (#VD2688)
Unsuitability (#VD3954)
Unpreparedness (#VD3940)
Underdevelopment (#VD3896)

Fitness (#VC0767)
Maturity (#VC1127)
Vigilance (#VC1879)
Alertness (#VC0099)
Capability (#VC0287)
Proficiency (#VC1367)
Preparedness (#VC1343)
Roughness (#VD3580)
Immaturity (#VD2832)
Incompetence (#VD2920)
Improvidence (#VD2872)
Vulnerability (#VD4036)
Shiftlessness (#VD3670)
Thriftlessness (#VD3814)
Oversimplification (#VD3404).

Constituent values Luck (#VC1099)
Wealth (#VC1901)
Luxury (#VC1101)
Success (#VC1697)
Comfort (#VC0355)
Progress (#VC1373)
Felicity (#VC0753)
Happiness (#VC0881)
Affluence (#VC0083)
Prosperity (#VC1381)
Blessedness (#VC0247)
Comfortableness (#VC0357).
Constituent values Shock (#VD3672)
Trouble (#VD3858)
Overload (#VD3396)
Calamity (#VD2208)
Annoyance (#VD2076)
Difficulty (#VD2440)
Vicissitude (#VD4010)
Aggravation (#VD2042)
Destructiveness (#VD2422).

Ease (#VC0601)
Vigour (#VC1881)
Welfare (#VC1903)
Fortune (#VC0791)
Security (#VC1537)
Flourish (#VC0773)
Intricacy (#VC1015)
Clearness (#VC0339)
Resistance (#VC1477)
Exuberance (#VC0723)
Advancement (#VC0063)

Grief (#VD2748)
Decline (#VD2336)
Disaster (#VD2466)
Accident (#VD2028)
Misfortune (#VD3258)
Affliction (#VD2036)
Tribulation (#VD3852)
Lucklessness (#VD3174)

♦ **VP5722a Accomplishment–Nonaccomplishment**
Dynamics Is there anything in life so disenchanting as attainment ? (Robert Louis Stevenson)
Knowledge may give weight, but accomplishments give lustre, and many more people see than weigh. (Lord Chesterfield)
There is no beauty like that which was spoiled by an accident; no accomplishments and graces are so to be envied as those that circumstances rudely hindered the development of. (Charles Dudley)
Achievement, n. The death of endeavour and the birth of disgust. (Ambrose Bierce)
There is much to be said for failure. It is more interesting than success. (Max Beerbohm)
There is not a fiercer hell than the failure in a great object. (John Keats)
Constant success shows us but one side of the world. For as it surrounds us with friends who will tell us only our merits, so it silences those enemies from whom alone we can learn our defects. (Charles Caleb Colton)
Success makes men rigid and they tend to exalt stability over all the other virtues; tired of the effort of willing they become fanatics about conservatism. (Walter Lippmann)
Nothing fails like success; nothing is so defeated as yesterday's triumphant Cause. (Phyllis McGinley)
Integrative complex Achievement (#VT8072).
Constituent values Work (#VC1937)
Triumph (#VC1781)
Fortune (#VC0791)
Flourish (#VC0773)
Certainty (#VC0301)
Prosperity (#VC1381)
Fulfilment (#VC0809)
Attainment (#VC0195)
Culmination (#VC0505)
Achievement (#VC0031)
Consummation (#VC0459)
Manageability (#VC1111)
Implementation (#VC0955)
Self-fulfilment (#VC1563)
Self-actualization (#VC1539).
Constituent values Muddle (#VD3292)
Neglect (#VD3310)
Failure (#VD2660)
Futility (#VD2736)
Uselessness (#VD3970)
Nonfulfilment (#VD3328)

Victory (#VC1877)
Success (#VC1697)
Maturity (#VC1127)
Sensation (#VC1589)
Assurance (#VC0189)
Perfection (#VC1273)
Completion (#VC0387)
Realization (#VC1421)
Advancement (#VC0063)
Perseverance (#VC1279)
Qualification (#VC1395)
Connectedness (#VC0431)
Accomplishment (#VC0023)
Self-expression (#VC1561)

Defeat (#VD2346)
Mistake (#VD3276)
Omission (#VD3360)
Conspiracy (#VD2276)
Nonfeasance (#VD3326)
Nonaccomplishment (#VD3320).

♦ **VP5726a Victory–Defeat**
Dynamics Defeat is a school in which truth always grows strong. (Henry Ward Beecher)
There are defeats more triumphant than victories. (Montaigne)
Far better it is to dare mighty things, to win glorious triumphs, even though checkered by failure, than to take rank with those poor spirits who neither enjoy much nor suffer much, because they live in the grey twilight that knows not victory nor defeat. (Theodore Roosevelt)
Any coward can fight a battle when he's sure of winning; but give me the man who has pluck to fight when he's sure of losing. That's my way, sir; and there are many victories worse than a defeat. (George Eliot)
Another such victory over the Romans, and we are undone. (Pyrrhus)
Victory at all costs, victory in spite of all terror, victory however long and hard the road may be; for without victory there is no survival. (Churchill)
He got the better of himself, and that's the best kind of victory one can wish for. (Cervantes)
Death is swallowed up in victory. O death, where is thy sting ? O grave, where is thy victory ? (Bible)
We fight for lost causes because we know that our defeat and dismay may be the preface to our successors' victory, though that victory itself will be temporary; we fight rather to keep something alive than in the expectation that anything will triumph. (T S Eliot)
Integrative complex Achievement (#VT8072).
Constituent values Gain (#VC0817)
Triumph (#VC1781)
Mastery (#VC1121)
Dominance (#VC0587)
Championship (#VC0305).
Constituent values Ruin (#VD3586)
Defeat (#VD2346)
Overturn (#VD3412)
Brokenness (#VD2200)
Subjugation (#VD3748)
Vanquishment (#VD3988).

Victory (#VC1877)
Success (#VC1697)
Conquest (#VC0435)
Ascendancy (#VC0175)

Panic (#VD3424)
Failure (#VD2660)
Confusion (#VD2272)
Bafflement (#VD2136)
Frustration (#VD2726)

♦ **VP5728b Prosperity–Adversity**
Dynamics There is no success without hardship. (Sophocles)
There is in the worst of fortune the best of chances for a happy change. (Euripides)
One likes people much better when they're battered down by a prodigious siege of misfortune than when they triumph. (Virginia Woolf)
He that has never suffered extreme adversity, knows not the full extent of his own depravation. (Charles Caleb Colton)
Every calamity is a spur and valuable hint. (Emerson)
He knows not his own strength that hath not met adversity. (Ben Jonson)
One who was abhorred by all in prosperity is adored by all in adversity. (Baltasar Gracian)
Prosperity is a great teacher; adversity is a greater. (William Hazlitt)
Not even a collapsing world looks dark to a man who is about to make his fortune. (E B White)
Luxury either comes of riches or makes them necessary; it corrupts at once rich and poor, the rich by possession and the poor by covetousness. (Rousseau)
Integrative complex Achievement (#VT8072).

♦ **VP5732c Facility–Difficulty**
Dynamics He who accounts all things easy will have many difficulties. (Lao Tse)
In the difficult are the friendly forces, the hands that work on us. (Rainer Maria Rilke)
For easy things, that may be got at will, most sorts of men do set but little store. (Edmund Spenser)
Integrative complex Achievement (#VT8072).
Constituent values Aid (#VC0095)
Light (#VC1077)
Clarity (#VC0335)
Clearness (#VC0339)
Flexibility (#VC0771)
Explanation (#VC0719)
Practicality (#VC1335)
Effortlessness (#VC0617)
Intelligibility (#VC1003)
Constituent values Ruin (#VD3586)
Worry (#VD4070)
Defeat (#VD2346)
Torment (#VD3824)
Barrier (#VD2148)
Sabotage (#VD3590)
Hardness (#VD2770)
Distress (#VD2566)
Intricacy (#VD3082)
Repression (#VD3532)
Opposition (#VD3364)
Difficulty (#VD2440)
Suppression (#VD3770)
Restriction (#VD3546)
Obstruction (#VD3356)
Frustration (#VD2726)
Awkwardness (#VD2130)
Interference (#VD3068)
Complication (#VD2254)
Ponderousness (#VD3468)
Embarrassment (#VD2602)
Impracticality (#VD2862)
Bothersomeness (#VD2196)
Counterproductivity (#VD2296).

Ease (#VC0601)
Freedom (#VC0797)
Facility (#VC0725)
Simplicity (#VC1617)
Feasibility (#VC0747)
Convenience (#VC0473)
Manageability (#VC1111)
Manoeuvrability (#VC1113)
Straightforwardness (#VC1679).
Evil (#VD2624)
Upset (#VD3966)
Trouble (#VD3858)
Problem (#VD3486)
Anxiety (#VD2086)
Obstacle (#VD3352)
Fixation (#VD2702)
Stalemate (#VD3726)
Annoyance (#VD2076)
Perplexity (#VD3440)
Inhibition (#VD3016)
Bafflement (#VD2136)
Retardation (#VD3550)
Prohibition (#VD3494)
Irksomeness (#VD3096)
Encumbrance (#VD2606)
Toilsomeness (#VD3822)
Disadvantage (#VD2450)
Bewilderment (#VD2172)
Inconvenience (#VD2938)
Disconcertion (#VD2474)
Burdensomeness (#VD2206)
Unmanageability (#VD3928)

♦ **VP5733c Skilfulness–Unskilfulness**
Dynamics A brand new mediocrity is thought more of than accustomed excellence. (Baltasar Gracian)
There is great ability in knowing how to conceal one's ability. (La Rochefoucauld)
There is one art, no more, no less: to do all things with art-lessness. (Piet Hein)
The greatest cunning is to have none at all. (Carl Sandburg)
To do easily what is difficult for others is the mark of talent. To do what is impossible for talent is the mark of genius. (Henri Frédéric Amiel)
Natural abilities are like natural plants, that need pruning by study; and studies themselves do give forth directions too much at large, except they be bounded in by experience. (Francis Bacon)
Nothing doth more hurt in a state than that cunning men pass for wise. (Francis Bacon)
What a man is begins to betray itself when his talent decreases – when he stops showing what he can do. Talent, too, is finery; finery, too, is a hiding place. (Nietzsche)
The awareness of the ambiguity of one's highest achievements (as well as one's deepest failures) is a definite symptom of maturity. (Paul Tillich)
Integrative complex Achievement (#VT8072).
Constituent values Art (#VC0169)
Good (#VC0845)
Skill (#VC1621)
Light (#VC1077)
Genius (#VC0825)
Mastery (#VC1121)
Agility (#VC0089)
Sagacity (#VC1519)
Maturity (#VC1127)
Facility (#VC0725)
Aptitude (#VC0163)
Readiness (#VC1417)
Plainness (#VC1301)
Innocence (#VC0987)
Expertise (#VC0717)
Dexterity (#VC0553)
Worthiness (#VC1941)
Simplicity (#VC1617)
Giftedness (#VC0833)
Excellence (#VC0697)
Directness (#VC0563)
Cleverness (#VC0343)
Brilliance (#VC0267)
Artfulness (#VC0171)
Unadornment (#VC1793)
Naturalness (#VC1179)
Flexibility (#VC0771)
Unpretention (#VC1837)
Gracefulness (#VC0857)
Coordination (#VC0485)
Qualification (#VC1395)

Tact (#VC1737)
Style (#VC1687)
Power (#VC1331)
Talent (#VC1739)
Candor (#VC0283)
Finesse (#VC0765)
Ability (#VC0003)
Openness (#VC1225)
Felicity (#VC0753)
Capacity (#VC0289)
Sincerity (#VC1619)
Quickness (#VC1399)
Lightness (#VC1079)
Ingenuity (#VC0981)
Endowment (#VC0643)
Authority (#VC0205)
Unassuming (#VC1801)
Nimbleness (#VC1191)
Experience (#VC0715)
Efficiency (#VC0615)
Competence (#VC0383)
Capability (#VC0287)
Attainment (#VC0195)
Adjustment (#VC0055)
Proficiency (#VC1367)
Genuineness (#VC0831)
Artlessness (#VC0173)
Potentiality (#VC1329)
Cosmopolitan (#VC0493)
Adaptability (#VC0049)
Ingenuousness (#VC0983)

VALUE POLARITIES   VP5766

Accomplishment (#VC0023).
Openheartedness (#VC1221).
Constituent values Guile (#VD2758)
Deceit (#VD2332)
Mistake (#VD3276)
Cunning (#VD2308)
Omission (#VD3360)
Artifice (#VD2104)
Maladroit (#VD3182)
Ignorance (#VD2818)
Negligence (#VD3312)
Mediocrity (#VD3206)
Incapacity (#VD2910)
Inadequacy (#VD2890)
Clumsiness (#VD2238)
Misguidance (#VD3260)
Miscarriage (#VD3240)
Awkwardness (#VD2130)
Inexperience (#VD2994)
Inefficiency (#VD2980)
Incapability (#VD2908)
Unfamiliarity (#VD3908)
Mismanagement (#VD3268)
Gracelessness (#VD2744)
Ineffectuality (#VD2978)
Thoughtlessness (#VD3810)
Unsophistication (#VD3952).

Resourcefulness (#VC1481)
Muddle (#VD3292)
Neglect (#VD3310)
Misrule (#VD3272)
Trickery (#VD3854)
Naïvety (#VD3304)
Sophistry (#VD3690)
Inability (#VD2882)
Subterfuge (#VD3752)
Misconduct (#VD3246)
Inelegance (#VD2982)
Inaptitude (#VD2902)
Immaturity (#VD2832)
Nonfeasance (#VD3326)
Misfeasance (#VD3256)
Malpractice (#VD3198)
Misdirection (#VD3252)
Inexpedience (#VD2992)
Incompetence (#VD2920)
Carelessness (#VD2214)
Ponderousness (#VD3468)
Maladjustment (#VD3180)
Unintelligence (#VD3922)
Unmanageability (#VD3928)
Ineffectiveness (#VD2976)

♦ **VP5737c Behaviour–Misbehaviour**
Dynamics Conduct is three-fourths of our life and its largest concern. (Matthew Arnold)
Customs represent the experience of mankind. (Henry Ward Beecher)
What humanity abhors, custom reconciles and recommends to us. (John Locke)
Anything awful makes me laugh. I misbehaved once at a funeral. (Charles Lamb)
The test of a man or woman's breeding is how they behave in a quarrel. (Shaw)
We should behave to our friends as we would wish our friends to behave to us. (Aristotle)
Integrative complex Achievement (#VT8072).
Constituent values Style (#VC1687)
Manners (#VC1117)
Presence (#VC1345)
Behaviour (#VC0223)
Constituent values Roguery (#VD3576)
Mischief (#VD3242)
Vandalism (#VD3984)
Misdemeanor (#VD3250)
Discourtesy (#VD2484)

Custom (#VC0511)
Conduct (#VC0417)
Courtesy (#VC0497)
Sociability (#VC1625).
Badness (#VD2134)
Disorder (#VD2526)
Misconduct (#VD3246)
Impropriety (#VD2870)
Mischievousness (#VD3224).

♦ **VP5739a Authority–Lawlessness**
Dynamics Government is a contrivance of human wisdom to provide for human wants. (Edmund Burke)
Chaos and ineptitude are anti-human but so too is a superlatively efficient government, equipped with all the products of a highly developed technology. (Aldous Huxley)
A good government remains the greatest of human blessings, and no nation has ever enjoyed it. (William Ralph Inge)
Most men, after a little freedom, have preferred authority with the consoling assurances and the economy of effort which it brings. (Walter Lippmann)
Government and co-operation are in all things the laws of life; anarchy and competition the laws of death. (John Ruskin)
Whenever you have an efficient government you have a dictatorship. (Harry S Truman)
Anarchy is the stepping stone to absolute power. (Napoleon I)
The worst thing in this world, next to anarchy, is government. (Henry Ward Beecher)
Integrative complex Compliance (#VT8073).
Constituent values Power (#VC1331)
Clout (#VC0345)
Potency (#VC1327)
Majesty (#VC1109)
Priority (#VC1355)
Nobility (#VC1193)
Supremacy (#VC1713)
Influence (#VC0979)
Dominance (#VC0587)
Precedence (#VC1339)
Leadership (#VC1057)
Government (#VC0853)
Superiority (#VC1709)
Aristocracy (#VC0167)
Constituent values Chaos (#VD2224)
Anarchy (#VD2062)
Despotism (#VD2420)
Anarchism (#VD2060)
Revolution (#VD3562)
Lawlessness (#VD3136)
Dictatorship (#VD2438)
Insubordination (#VD3056)
Irresponsibility (#VD3114)

Might (#VC1147)
Royalty (#VC1513)
Mastery (#VC1121)
Strength (#VC1681)
Prestige (#VC1349)
Eminence (#VC0631)
Rightness (#VC1507)
Greatness (#VC0867)
Authority (#VC0205)
Legitimacy (#VC1063)
Importance (#VC0957)
Competence (#VC0383)
Sovereignty (#VC1639)
Momentousness (#VC1161).
Misrule (#VD3272)
Disorder (#VD2526)
Confusion (#VD2272)
Unruliness (#VD3950)
Willfulness (#VD4060)
Disobedience (#VD2524)
Licentiousness (#VD3150)
Disorganization (#VD2528)
Authoritarianism (#VD2118).

♦ **VP5756b Leniency–Compulsion**
Dynamics A great deal may be done by severity, more by love, but most by clear discernment and impartial justice. (Goethe)
Clemency is the support of justice. (Russian Proverb)
Sometimes clemency is cruelty, and cruelty clemency. (Proverb)
Bodily exercise, when compulsory, does no harm to the body; but knowledge which is acquired under compulsion obtains no hold on the mind. (Plato)
He that spares the bad injures the good. (Thomas Fuller)
Pardon one offence, and you encourage the commission of many. (Publilius Syrus)
I have with me two gods, Persuasion and Compulsion. (Heraclitus)
Integrative complex Compliance (#VT8073).
Constituent values Pity (#VC1299)
Patience (#VC1261)
Leniency (#VC1067)
Clemency (#VC0341)
Tolerance (#VC1765)
Necessity (#VC1183)
Relaxation (#VC1455)
Discipline (#VC0567)
Affability (#VC0073)
Forgiveness (#VC0787)
Flexibility (#VC0771)

Tameness (#VC1741)
Mildness (#VC1149)
Humanity (#VC0927)
Toughness (#VC1767)
Orthodoxy (#VC1241)
Tenderness (#VC1751)
Humaneness (#VC0921)
Compassion (#VC0377)
Acceptance (#VC0017)
Forbearance (#VC0779)
Self-control (#VC1549)

Rigorousness (#VC1509)
Graciousness (#VC0859)
Constituent values Duress (#VD2592)
Violence (#VD4022)
Rigidity (#VD3568)
Looseness (#VD3164)
Constraint (#VD2278)
Stubbornness (#VD3738)
Carelessness (#VD2214)
Relentlessness (#VD3528)
Authoritarianism (#VD2118).

Mercifulness (#VC1137)
Gratification (#VC0863).
Laxness (#VD3140)
Severity (#VD3660)
Stiffness (#VD3730)
Austerity (#VD2116)
Enforcement (#VD2608)
Intimidation (#VD3072)
Inflexibility (#VD3008)
Permissiveness (#VD3438)

♦ **VP5762b Freedom–Restraint**
Dynamics Atheism. There is not a single exalting and emancipating influence that does not in turn become inhibitory. (André Gide)
What country can preserve its liberties, if its rulers are not warned from time to time, that this people preserve the spirit of resistance ? (Thomas Jefferson)
Liberation is not deliverance. (Victor Hugo)
If the self-discipline of the free cannot match the iron discipline of the mailed fist, in economic, political, scientific, and all the other kinds of struggles, as well as the military, then the evil to freedom will continue to rise. (John F Kennedy)
Tyranny is always better organized than freedom. (Charles Péguy)
The bonds that unite another person to ourself exist only in our mind. (Marcel Proust)
What more oft in nations grown corrupt, and by their vices brought to servitude, than to love bondage more than liberty, bondage with ease than strenuous liberty. (Milton)
If the world knew how to use freedom without abusing it, tyranny would not exist. (Tehyi Hsieh)
Liberty exists in proportion to wholesome restraint. (Daniel Webster)
It seems that it is madder never to abandon one's self than often to be infatuated; better to be wounded, a captive and a slave, than always to walk in armour. (Margaret Fuller)
The spirit of improvement is not always a spirit of liberty, for it may aim at forcing improvements on an unwilling people. (John Stuart Mill)
Self-restraint may be alien to the human temperament, but humanity without restraint will dig its own grave. (Marya Mannes)
Freedom and constraint are two aspects of the same necessity, the necessity of being the man you are and not another. You are free to be that man, but not free to be another. (Saint-Exupéry)
Integrative complex Compliance (#VT8073).
Constituent values Homage (#VC0909)
Service (#VC1605)
Freedom (#VC0797)
Mildness (#VC1149)
Immunity (#VC0949)
Facility (#VC0725)
Autonomy (#VC0207)
Unbigoted (#VC1803)
Rightness (#VC1507)
Deference (#VC0527)
Submission (#VC1689)
Liberation (#VC1071)
Discretion (#VC0571)
Acceptance (#VC0017)
Deliverance (#VC0539)
Subservience (#VC1691)
Independence (#VC0969)
Civil rights (#VC0329)
Self-reliance (#VC1577)
Agreeableness (#VC0091)
Openmindedness (#VC1223)
Unequivocalness (#VC1815)
Broadmindedness (#VC0269)
Self-determination (#VC1553)
Constituent values Duress (#VD2592)
Laxness (#VD3140)
Servility (#VD3656)
Isolation (#VD3120)
Subjection (#VD3746)
Inhibition (#VD3016)
Constraint (#VD2278)
Suppression (#VD3770)
Segregation (#VD3624)
Restriction (#VD3546)
Inferiority (#VD3002)
Abandonment (#VD2004)
Incontinence (#VD2936)
Licentiousness (#VD3150).

Choice (#VC0325)
Liberty (#VC1073)
Tameness (#VC1741)
Meekness (#VC1133)
Humility (#VC0929)
Docility (#VC0581)
Unlimited (#VC1833)
Tolerance (#VC1765)
Obedience (#VC1207)
Clearness (#VC0339)
Neutrality (#VC1187)
Gentleness (#VC0829)
Compliance (#VC0389)
Flexibility (#VC0771)
Tractability (#VC1769)
Nonalignment (#VC1197)
Emancipation (#VC0629)
Acquiescence (#VC0035)
Manageability (#VC1111)
Self-direction (#VC1555)
Limitlessness (#VC1081)
Self-government (#VC1565)
Self-sufficiency (#VC1585)
Self-containedness (#VC1547).
Tyranny (#VD3870)
Barrier (#VD2148)
Seclusion (#VD3616)
Unruliness (#VD3950)
Repression (#VD3532)
Dependence (#VD2382)
Absolutism (#VD2020)
Subjugation (#VD3748)
Retardation (#VD3550)
Prohibition (#VD3494)
Confinement (#VD2268)
Intemperence (#VD3066)
Subordination (#VD3750)

♦ **VP5766c Obedience–Disobedience**
Dynamics Disobedience, in the eyes of anyone who has read history, is man's original virtue. It is through disobedience that progress has been made, through disobedience and through rebellion. (Oscar Wilde)
Forgive us for bypassing political duties; for condemning civil disobedience when we will not obey You; for reducing Your holy law to average virtues, by trying to be no better or worse than most men. (Litany for Holy Communion – United Presbyterian Church)
Obedience, bane of all genius, virtue, freedom, truth. Makes slaves of men, and, of the human frame, a mechanized automation. (Shelley).
Integrative complex Compliance (#VT8073).
Constituent values Duty (#VC0593)
Homage (#VC0909)
Loyalty (#VC1095)
Submission (#VC1689)
Conformity (#VC0425)
Allegiance (#VC0103)
Faithfulness (#VC0731)
Constituent values Trespass (#VD3850)
Offence (#VD3358)
Violation (#VD4020)
Passivity (#VD3430)
Extremism (#VD2656)
Unruliness (#VD3950)
Revolution (#VD3562)
Brokenness (#VD2200)
Lawlessness (#VD3136)
Inobservance (#VD3028)
Transgression (#VD3836)
Nonobservance (#VD3330)

Faith (#VC0729)
Service (#VC1605)
Obedience (#VC1207)
Observance (#VC1211)
Compliance (#VC0389)
Willingness (#VC1925)
Acquiescence (#VC0035).
Sedition (#VD3620)
Defiance (#VD2354)
Servility (#VD3656)
Obstinacy (#VD3354)
Disregard (#VD2542)
Subjection (#VD3746)
Indocility (#VD2970)
Willfulness (#VD4060)
Stubbornness (#VD3738)
Disobedience (#VD2524)
Recalcitrance (#VD3520)
Nonconformity (#VD3324)

VP5766

Nonadherence (#VD3322)
Intractability (#VD3078)
Contumaciousness (#VD2290).

Traitorousness (#VD3834)
Insubordination (#VD3056)

◆ **VP5768c Observance–Nonobservance**
Integrative complex Compliance (#VT8073).
Constituent values Observance (#VC1211).

◆ **VP5775c Consent–Refusal**
Dynamics The art of acceptance is the art of making someone who has just done you a small favour wish that he might have done you a greater one. (Russell Lynes)
To refuse and to give tardily is all the same. (Proverb)
No one can make you feel inferior without your consent. (Eleanor Roosevelt)
Silence gives consent. (Oliver Goldsmith)
A little still she strove, and much repented, and whispering 'I will ne'er consent' – consented. (Byron)
My poverty, but not my will, consents. (Shakespeare)
I shall never ask, never refuse, nor ever resign an office. (Benjamin Franklin)
Integrative complex Compliance (#VT8073).
Constituent values Accord (#VC0025)
Legality (#VC1061)
Tolerance (#VC1765)
Privilege (#VC1359)
Agreement (#VC0093)
Promptness (#VC1377)
Acceptance (#VC0017)
Contentment (#VC0467)
Affirmation (#VC0081)
Validity (#VC1853)
Approval (#VC0161)
Readiness (#VC1417)
Eagerness (#VC0597)
Submission (#VC1689)
Compliance (#VC0389)
Willingness (#VC1925)
Approbation (#VC0157)
Agreeableness (#VC0091).
Constituent values Denial (#VD2376)
Rejection (#VD3526)
Repression (#VD3532)
Inhibition (#VD3016)
Illegality (#VD2820)
Repudiation (#VD3536)
Disobedience (#VD2524)
Nonacceptance (#VD3318)
Decline (#VD2336)
Exclusion (#VD2634)
Negativity (#VD3308)
Indulgence (#VD2974)
Suppression (#VD3770)
Prohibition (#VD3494)
Nonobservance (#VD3330)
Permissiveness (#VD3438).

◆ **VP5785a Support–Opposition**
Dynamics In comradeship is danger countered best. (Goethe)
Human service is the highest form of self-interest for the person who serves. (Elbert Hubbard)
No government can be long secure without formidable opposition. (Benjamin Disraeli)
There is in the human race some dark spirit of recalcitrance, always pulling us in the direction contrary to that in which we are reasonably expected to go. (Max Beerbohm)
As soon as public service ceases to be the chief business of the citizens, and they would rather serve with their money than with their persons, the State is not far from its fall. (Rousseau)
Integrative complex Interaction (#VT8074).
Constituent values Aid (#VC0095)
Union (#VC1825)
Favour (#VC0743)
Daring (#VC0513)
Support (#VC1711)
Harmony (#VC0887)
Comfort (#VC0355)
Interest (#VC1011)
Fondness (#VC0777)
Boldness (#VC0251)
Self-help (#VC1567)
Inclusion (#VC0963)
Coalition (#VC0347)
Benignity (#VC0237)
Usefulness (#VC1847)
Solidarity (#VC1631)
Protection (#VC1383)
Fellowship (#VC0755)
Partnership (#VC1255)
Cooperation (#VC0481)
Concurrence (#VC0413)
Competition (#VC0385)
Benevolence (#VC0235)
Association (#VC0187)
Advancement (#VC0063)
Togetherness (#VC1763)
Positiveness (#VC1323)
Friendliness (#VC0803)
Reinforcement (#VC1451)
Neighbourliness (#VC1185)
Self-advancement (#VC1541)
Care (#VC0291)
Relief (#VC1461)
Esprit (#VC0679)
Utility (#VC1849)
Service (#VC1605)
Concord (#VC0411)
Kindness (#VC1045)
Goodwill (#VC0849)
Facility (#VC0725)
Alliance (#VC0105)
Mutuality (#VC1175)
Community (#VC0373)
Challenge (#VC0303)
Advantage (#VC0065)
Sustenance (#VC1723)
Resistance (#VC1477)
Fraternity (#VC0795)
Assistance (#VC0185)
Helpfulness (#VC0893)
Convenience (#VC0473)
Comradeship (#VC0401)
Carefulness (#VC0293)
Beneficence (#VC0233)
Amicability (#VC0123)
Well-meaning (#VC1915)
Subservience (#VC1691)
Independence (#VC0969)
Well-disposed (#VC1905)
Collaboration (#VC0353)
Cooperativeness (#VC0483)
Constructiveness (#VC0457).
Constituent values Enmity (#VD2610)
Rivalry (#VD3574)
Negation (#VD3306)
Defiance (#VD2354)
Repulsion (#VD3540)
Obstinacy (#VD3354)
Impudence (#VD2876)
Disregard (#VD2542)
Arrogance (#VD2102)
Opposition (#VD3364)
Dissension (#VD2548)
Antagonism (#VD2080)
Affrontery (#VD2038)
Impertinence (#VD2844)
Intransigence (#VD3080)
Contradiction (#VD2284)
Disputatiousness (#VD2538)
Irreconcilability (#VD3100).
Denial (#VD2376)
Disdain (#VD2490)
Friction (#VD2720)
Conflict (#VD2270)
Rejection (#VD3526)
Insolence (#VD3050)
Hostility (#VD2802)
Disaccord (#VD2448)
Antipathy (#VD2084)
Negativity (#VD3308)
Complicity (#VD2256)
Alienation (#VD2052)
Contrariety (#VD2286)
Recalcitrance (#VD3520)
Exclusiveness (#VD2636)
Bumptiousness (#VD2204)
Contemptuousness (#VD2280)

◆ **VP5794a Accord–Disaccord**
Dynamics Mankind has grown strong in eternal struggles and it will only perish through eternal peace. (Adolf Hitler)
It is better to have a war for justice than peace in injustice. (Charles Péguy)
Peace hath higher tests of manhood than battle ever knew. (John Greenleaf)
It is always possible to bind together a considerable number of people in love, so long as there are other people left over to receive the manifestations of their aggressiveness. (Sigmund Freud)
Harmony would lose its attractiveness if it did not have a background of discord. (Tehyi Hsieh)

As peace is of all goodness, so war is an emblem, a hieroglyphic, of all misery. (John Donne)
Integrative complex Interaction (#VT8074).
Constituent values Love (#VC1093)
Union (#VC1825)
Order (#VC1231)
Esprit (#VC0679)
Harmony (#VC0887)
Concord (#VC0411)
Sympathy (#VC1731)
Quietude (#VC1405)
Calmness (#VC0281)
Quietness (#VC1403)
Mediation (#VC1131)
Communion (#VC0371)
Solidarity (#VC1631)
Fellowship (#VC0755)
Restfulness (#VC1493)
Nonviolence (#VC1199)
Competition (#VC0385)
Amicability (#VC0123)
Peacefulness (#VC1267)
Conciliation (#VC0409)
Compatibility (#VC0379)
Reconciliation (#VC1433)
Correspondence (#VC0491)
Unity (#VC1829)
Peace (#VC1265)
Amity (#VC0125)
Accord (#VC0025)
Empathy (#VC0635)
Charity (#VC0315)
Serenity (#VC1601)
Pacifism (#VC1243)
Affinity (#VC0079)
Mutuality (#VC1175)
Community (#VC0373)
Agreement (#VC0093)
Resolution (#VC1471)
Compromise (#VC0397)
Orderliness (#VC1233)
Forgiveness (#VC0787)
Appeasement (#VC0147)
Tranquillity (#VC1773)
Congeniality (#VC0427)
Understanding (#VC1811)
Agreeableness (#VC0091)
Nonagressivity (#VC1195)
Frictionlessness (#VC0801).
Constituent values War (#VD4040)
Enmity (#VD2610)
Wrangle (#VD4076)
Tension (#VD3802)
Discord (#VD2480)
Mischief (#VD3242)
Ferocity (#VD2694)
Defiance (#VD2634)
Militancy (#VD3226)
Disparity (#VD2534)
Disaccord (#VD2448)
Fierceness (#VD2698)
Dissidence (#VD2550)
Disharmony (#VD2504)
Chauvinism (#VD2228)
Aggression (#VD2044)
Controversy (#VD2288)
Divisiveness (#VD2578)
Belligerence (#VD2164)
Combativeness (#VD2246)
Unpeacefulness (#VD3934)
Quarrelsomeness (#VD3512)
Contentiousness (#VD2282)
Disputatiousness (#VD2538)
Argumentativeness (#VD2096).
Fuss (#VD2734)
Battle (#VD2154)
Warlike (#VD4042)
Rivalry (#VD3574)
Variance (#VD3992)
Friction (#VD2720)
Division (#VD2580)
Conflict (#VD2270)
Hostility (#VD2802)
Disfavour (#VD2496)
Truculence (#VD3860)
Divergence (#VD2576)
Dissension (#VD2548)
Difficulty (#VD2440)
Antagonism (#VD2080)
Disaffinity (#VD2454)
Irascibility (#VD3094)
Disaffection (#VD2452)
Inharmonious (#VD3014)
Unpleasantness (#VD3936)
Unfriendliness (#VD3910)
Incompatibility (#VD2918)
Misunderstanding (#VD3280)
Bloody-mindedness (#VD2188).

◆ **VP5798b Attack–Defence**
Dynamics It is an unfortunate fact that we can secure peace only by preparing for war. (J F Kennedy)
Whatever needs to be maintained by force is doomed. (Henry Miller)
No nation has ever had an army large enough to guarantee it against attack in time of peace or insure it victory in time of war. (Calvin Coolidge)
We are advocates of the abolition of war, we do not want war; but war can only be abolished through war... (Mao Tse-Tung)
Armies kept for three years are deployed on a single morning. (Chinese proverb)
Attack is the best form of defence. (Proverb)
We are so outnumbered there's only one thing to do. We must attack. (Sir Andrew B Cunningham)
Strength lies not in defence but in attack. (Hitler)
Integrative complex Interaction (#VT8074).
Constituent values Surprise (#VC1719)
Championship (#VC0305)
Constituent values Offence (#VD3358)
Aggression (#VD2044)
Combativeness (#VD2246).
Advocacy (#VC0069)
Self-preservation (#VC1573).
Invasion (#VD3090)
Defensiveness (#VD2352)

◆ **VP5806c Neutrality–Compromise**
Dynamics Adjustment, that synonym for conformity that comes more easily to the modern tongue, is the theme of our swan song, the piper's tune to which we dance on the brink of the abyss, the siren's melody that destroys our senses and paralyzes our wills. (Robert Lindner)
All government, – indeed, every human benefit and enjoyment, every virtue and every prudent act, – is founded on compromise and barter. (Edmund Burke)
Compromise, if not the spice of life, is its solidity. (Phyllis McGinley)
What people call impartiality may simply mean indifference, and what people call partiality may simply mean mental activity. (G K Chesterton)
That expression 'positive neutrality' is a contradiction in terms. There can be no more positive neutrality than there can be a vegetarian tiger. (V K Krishna Menon)
Integrative complex Interaction (#VT8074).
Constituent values Neutrality (#VC1187)
Compromise (#VC0397)
Adjustment (#VC0055)
Independence (#VC0969)
Moderation (#VC1157)
Commitment (#VC0361)
Nonalignment (#VC1197)
Impartiality (#VC0953).

◆ **VP5808a Possession–Loss**
Dynamics Conspicuous consumption of valuable goods is a means of reputability to the gentleman of leisure. (Thorstein Veblen)
The trouble with the profit system has always been that it was highly unprofitable to most people. (E B White)
Loss is nothing else but change, and change is Nature's delight. (Marcus Aurelius)
There are occasions when it is undoubtedly better to incur loss than to make gain. (Plautus)
Integrative complex Possession (#VT8075).
Constituent values Gain (#VC0817)
Fondness (#VC0777)
Possession (#VC1325)
Sovereignty (#VC1639)
Preservation (#VC1347)
Accumulation (#VC0027)
Accessibility (#VC0019)
Interest (#VC1011)
Advantage (#VC0065)
Attainment (#VC0195)
Consumption (#VC0461)
Availability (#VC0209)
Profitability (#VC1369)
Productiveness (#VC1365)
Constituent values Wear (#VD4054)
Waste (#VD4046)
Denial (#VD2376)
Ruin (#VD3586)
Injury (#VD3024)
Damage (#VD2314)

VALUE POLARITIES **VP5861**

Sacrifice (#VD3592)
Depletion (#VD2386)
Repression (#VD3532)
Suppression (#VD3770)
Deprivation (#VD2398)
Depreciation (#VD2394).

Privation (#VD3484)
Spoliation (#VD3714)
Exhaustion (#VD2640)
Dissipation (#VD2552)
Bereavement (#VD2166)

♦ VP5815b Sharing–Appropriation

Dynamics The Holy Supper is kept, indeed, in whatso we share with another's need; not what we give, but what we share. For the gift without the giver is bare. (James Russell Lowell)
In faith and hope the world will disagree, But all mankind's concern is charity. (Alexander Pope)
The want of a thing is perplexing enough, but the possession of it is intolerable. (Sir John Vanbrugh)
Too many have dispensed with generosity to practice charity. (Albert Camus)
Mankind, why do ye set your hearts on things that, of necessity, may not be shared ? (Dante)
Integrative complex Possession (#VT8075).
Constituent values Accord (#VC0025)
Revival (#VC1501)
Devotion (#VC0551)
Endowment (#VC0643)
Community (#VC0373)
Redemption (#VC1439)
Liberation (#VC1071)
Partnership (#VC1255)
Deliverance (#VC0539)
Association (#VC0187)
Compensation (#VC0381)
Communication (#VC0367)
Cooperativeness (#VC0483).
Support (#VC1711)
Charity (#VC0315)
Remission (#VC1463)
Democracy (#VC0541)
Communion (#VC0371)
Possession (#VC1325)
Dedication (#VC0525)
Involvement (#VC1023)
Cooperation (#VC0481)
Acquisition (#VC0037)
Participation (#VC1253)
Commendability (#VC0359)
Constituent values Rapacity (#VD3516)
Extortion (#VD2652)
Spoliation (#VD3714)
Subjugation (#VD3748)
Depradation (#VD2388)
Despoliation (#VD2416)
Distress (#VD2566)
Usurpation (#VD3972)
Ravishment (#VD3518)
Fraudulence (#VD2714)
Bereavement (#VD2166)
Impoverishment (#VD2860).

♦ VP5837c Wealth–Poverty

Dynamics Fortune always will confer an aura of worth, unworthily; and in this world the lucky person passes for a genius. (Euripides)
A great fortune is a great slavery. (Seneca)
For every talent that poverty has stimulated it has blighted a hundred. (John W Gardner)
Wealth...and poverty: the one is the parent of luxury and indolence, and the other of meanness and viciousness, and both of discontent. (Plato)
Immoderate power, like other intemperence, leaves the progeny weaker and weaker, until Nature, as in compassion, covers it with her mantle and it is seen no more. (Walter Savage Landor)
Integrative complex Possession (#VT8075).
Constituent values Wealth (#VC1901)
Necessity (#VC1183)
Independence (#VC0969).
Fortune (#VC0791)
Prosperity (#VC1381)
Constituent values Lack (#VD3126)
Default (#VD2344)
Privation (#VD3484)
Mendicancy (#VD3214)
Homelessness (#VD2796)
Impoverishment (#VD2860).
Poverty (#VD3470)
Distress (#VD2566)
Liability (#VD3148)
Deprivation (#VD2398)
Embarrassment (#VD2602)

♦ VP5848c Expensiveness–Cheapness

Dynamics What we obtain too cheap, we esteem too lightly; it is dearness (expensiveness) only that gives everything its value. (Thomas Paine)
The superior gratification derived from the use and contemplation of costly and supposedly beautiful products is, commonly, in great measure a gratification of our sense of costliness masquerading under the name of beauty. (Thorstein Veblen)
It is cheap enough to say, 'God help you'. (Proverb)
Never cheapen unless you mean to buy. (Proverb)
Never buy what you do not want because it is cheap; it will be costly to you. (Thomas Jefferson)
Prosperity has everything cheap. (Thomas Fuller)
Integrative complex Possession (#VT8075).
Constituent values Value (#VC1855)
Pleasure (#VC1309)
Manageability (#VC1111).
Richness (#VC1503)
Worthiness (#VC1941)
Constituent values Waste (#VD4046)
Overtax (#VD3410)
Avarice (#VD2122)
Tightness (#VD3816)
Closeness (#VD2236)
Sordidness (#VD3692)
Prodigality (#VD3488)
Illiberalism (#VD2824)
Exploitation (#VD2644)
Excessiveness (#VD2632)
Unconscionableness (#VD3890)
Excess (#VD2630)
Decline (#VD2336)
Meanness (#VD3204)
Extortion (#VD2652)
Cheapness (#VD2230)
Profligacy (#VD3492)
Wastefulness (#VD4048)
Extravagance (#VD2654)
Depreciation (#VD2394)
Unreasonableness (#VD3946)

♦ VP5851c Economy–Prodigality

Dynamics If a man is prodigal, he cannot be truly generous. (James Boswell)
Lavishness is not generosity. (Thomas Browne)
Frugality is the sure guardian of our virtues. (Brahman Proverb)
Expense, and great expense, may be an essential part of true economy. (Edmund Burke)
The dignity of the individual demands that he be not reduced to vassalage by the largesse of others. (Saint-Exupéry)
Integrative complex Possession (#VT8075).
Constituent values Care (#VC0291)
Economy (#VC0605)
Parsimony (#VC1251)
Frugality (#VC0807)
Liberality (#VC1069)
Thriftiness (#VC1757)
Magnanimity (#VC1105)
Carefulness (#VC0293)
Graciousness (#VC0859)
Stintlessness (#VC1677)
Openheartedness (#VC1221).
Freedom (#VC0797)
Prudence (#VC1387)
Largeness (#VC1049)
Canniness (#VC0285)
Generosity (#VC0821)
Munificence (#VC1173)
Hospitality (#VC0919)
Handsomeness (#VC0879)
Unselfishness (#VC1839)
Bountifulness (#VC0255)
Constituent values Prodigality (#VD3488).

♦ VP5855a Feeling–Unfeelinglessness

Dynamics Before I die, I shall leave a will, because if you want something done, sentimentality is effective. (John Cage)
Great eaters and great sleepers are incapable of anything else that is great. (Henry IV of France)
The ultimate foundation of free society is the binding tie of cohesive sentiment. (Felix Frankfurter)
To spare oneself from grief at all cost can be achieved only at the price of total detachment, which excludes the ability to experience happiness. (Erich Fromm)
Wisdom must go with Sympathy, else the emotions will become maudlin and pity may be wasted on a poodle instead of a child – on a field-mouse instead of a human soul. (Elbert Hubbard)
Acquisition means life to miserable mortals. (Hesiod)
Without passion man is a mere latent force and possibility, like the flint which awaits the shock of the iron before it can give forth its spark. (Henri Frédéric)
Nothing great in the world has been accomplished without passion. (Hegel)
Seeing's believing, but feeling's the truth. (Thomas Fuller)
The important thing is being capable of emotions, but to experience only one's own would be a sorry limitation. (André Gide)
Life is the enjoyment of emotion, derived from the past and aimed at the future. (Alfred North Whitehead)
Integrative complex Feeling (#VT8081).
Constituent values Zeal (#VC1947)
Passion (#VC1257)
Emotion (#VC0633)
Concern (#VC0407)
Keenness (#VC1043)
Fervour (#VC0761)
Sensation (#VC1589)
Affection (#VC0075)
Excitement (#VC0699)
Cordiality (#VC0487)
Esprit (#VC0679)
Empathy (#VC0635)
Ecstasy (#VC0607)
Sympathy (#VC1731)
Fondness (#VC0777)
Sentiment (#VC1597)
Intensity (#VC1005)
Liveliness (#VC1085)
Detachment (#VC0545)
Involvement (#VC1023).
Constituent values Torpor (#VD3826)
Autism (#VD2120)
Oblivion (#VD3344)
Hardness (#VD2770)
Coldness (#VD2242)
Hostility (#VD2802)
Withdrawal (#VD4062)
Disinterest (#VD2516)
Callousness (#VD2210)
Slothfulness (#VD3680)
Indifference (#VD2958)
Spunklessness (#VD3716)
Heartlessness (#VD2784)
Sentimentality (#VD3652)
Self-absorption (#VD3626)
Stupor (#VD3744)
Apathy (#VD2090)
Lethargy (#VD3144)
Dullness (#VD2588)
Passivity (#VD3430)
Animosity (#VD2072)
Insouciance (#VD3052)
Disaffinity (#VD2454)
Soullessness (#VD3698)
Listlessness (#VD3156)
Hopelessness (#VD2798)
Insensitivity (#VD3040)
Spiritlessness (#VD3710)
Imperviousness (#VD2846)
Passionlessness (#VD3428).

♦ VP5857a Excitement–Inexcitability

Dynamics Ah, men do not know how much strength is in poise, that he goes the farthest who goes far enough. (James Russell Lowell)
It is not merely cruelty that leads men to love war, it is excitement. (Henry Ward Beecher)
Dullness is the coming of age of seriousness. (Oscar Wilde)
Take all away from me, but leave me Ecstasy, and I am richer then than all my Fellow Men. (Emily Dickinson)
Integrative complex Feeling (#VT8081).
Constituent values Emotion (#VC0633)
Balance (#VC0215)
Serenity (#VC1601)
Calmness (#VC0281)
Sensation (#VC1589)
Quietness (#VC1403)
Assurance (#VC0189)
Possession (#VC1325)
Moderation (#VC1157)
Excitement (#VC0699)
Dispassion (#VC0575)
Seriousness (#VC1603)
Tranquillity (#VC1773)
Peacefulness (#VC1267)
Breathtaking (#VC0259)
Impressivenss (#VC0959)
Self-confidence (#VC1545)
High-spiritedness (#VC0901).
Ecstasy (#VC0607)
Sobriety (#VC1623)
Quietude (#VC1405)
Solemnity (#VC1627)
Restraint (#VC1495)
Composure (#VC0391)
Temperance (#VC1747)
Philosophy (#VC1293)
Gentleness (#VC0829)
Equanimity (#VC0669)
Confidence (#VC0419)
Equilibrium (#VC0671)
Self-control (#VC1549)
Exhilaration (#VC0703)
Self-restraint (#VC1581)
Self-possession (#VC1571)
Self-assurance (#VC1543)
Constituent values Fuss (#VD2734)
Panic (#VD3424)
Upheaval (#VD3962)
Uneasiness (#VD3902)
Trepidation (#VD3848)
Disturbance (#VD2570).
Fear (#VD2684)
Strain (#VD3732)
Agitation (#VD2046)
Fierceness (#VD2698)
Nervousness (#VD3314)

♦ VP5861a Patience–Impatience

Dynamics He who has patience may accomplish anything. (Rabelais)
It is not he who gains the exact point in dispute who scores most in controversy, but he who has shown the most forbearance and the better temper. (Samuel Butler)
A wise man will not dispute with one that is hasty. (Sa'di)
The man who knows when not to act is wise. To my mind, bravery is forethought. (Euripides)
Integrative complex Feeling (#VT8081).
Constituent values Support (#VC1711)
Patience (#VC1261)
Leniency (#VC1067)
Tolerance (#VC1765)
Endurance (#VC0645)
Discipline (#VC0567)
Acceptance (#VC0017)
Forbearance (#VC0779)
Self-control (#VC1549)
Acquiescence (#VC0035)
Reconciliation (#VC1433).
Bravery (#VC0257)
Meekness (#VC1133)
Humility (#VC0929)
Obedience (#VC1207)
Submission (#VC1689)
Compliance (#VC0389)
Longanimity (#VC1089)
Amenability (#VC0119)
Perseverance (#VC1279)
Understanding (#VC1811)
Constituent values Fuss (#VD2734)
Anxiety (#VD2086)
Hastiness (#VD2778)
Impatience (#VD2840)
Fretfulness (#VD2718)
Impetuousity (#VD2848).
Haste (#VD2776)
Passivity (#VD3430)
Uneasiness (#VD3902)
Intolerance (#VD3074)
Restlessness (#VD3544)

♦ VP5863b Pleasantness–Unpleasantness

Dynamics One moment may with bliss repay unnumbered hours of pain. (Thomas Campbell)

It seldom happens that any felicity comes so pure as not to be tempered and allayed by some mixture of sorrow. (Cervantes)

The beauty of the world which is so soon to perish, has two edges, one of laughter, one of anguish, cutting the heart asunder. (Virginia Woolf)

It is better to drink of deep griefs than to taste shallow pleasures. (William Hazlitt)

To a great extent, suffering is a sort of need felt by the organism to make itself familiar with a new state, which makes it uneasy, to adapt its sensibility to that state. (Marcel Proust)

Affability contains no hatred of men, but for that very reason too much contempt for men. (Nietzsche)

The most intolerable pain is produced by prolonging the keenest pleasure. (George Bernard Shaw)

All fits of pleasure are balanced by an equal degree of pain or langour; 'tis like spending this year part of the next year's revenue. (Jonathan Swift)

Custom determines what is agreeable. (Pascal)

They must know but little of mankind who can imagine that, after they have been once seduced by luxury, they can ever renounce it. (Rousseau)

Integrative complex Feeling (#VT8081).
Constituent values Good (#VC0845)
Bliss (#VC0249)
Appeal (#VC0143)
Glamour (#VC0837)
Pleasure (#VC1309)
Sweetness (#VC1725)
Mellowness (#VC1135)
Brightness (#VC0265)
Amiability (#VC0121)
Saintliness (#VC1521)
Enchantment (#VC0637)
Pleasantness (#VC1307)
Gracefulness (#VC0857)
Cheerfulness (#VC0321)
Desirableness (#VC0543)
Agreeableness (#VC0091)
Delectableness (#VC0531)
Charm (#VC0317)
Luxury (#VC1101)
Harmony (#VC0887)
Delight (#VC0537)
Goodness (#VC0847)
Geniality (#VC0823)
Cordiality (#VC0487)
Attraction (#VC0199)
Affability (#VC0073)
Fascination (#VC0737)
Amicability (#VC0123)
Graciousness (#VC0859)
Congeniality (#VC0427)
Exquisiteness (#VC0721)
Compatibility (#VC0379)
Voluptuousness (#VC1895)
Irresistibility (#VC1031).

Constituent values Woe (#VD4068)
Wound (#VD4074)
Shock (#VD3672)
Ennui (#VD2612)
Sorrow (#VD3696)
Misery (#VD3254)
Injury (#VD3024)
Enmity (#VD2610)
Trouble (#VD3858)
Torment (#VD3824)
Problem (#VD3486)
Malaise (#VD3184)
Disgust (#VD2502)
Chagrin (#VD2220)
Anxiety (#VD2086)
Vileness (#VD4014)
Ugliness (#VD3872)
Grimness (#VD2752)
Dullness (#VD2588)
Distaste (#VD2560)
Baseness (#VD2150)
Atrocity (#VD2114)
Hostility (#VD2802)
Grossness (#VD2754)
Disfavour (#VD2496)
Awfulness (#VD2128)
Annoyance (#VD2076)
Sufferance (#VD3758)
Oppression (#VD3366)
Inquietude (#VD3030)
Ignobility (#VD2812)
Difficulty (#VD2440)
Boringness (#VD2194)
Antagonism (#VD2080)
Abhorrence (#VD2010)
Tribulation (#VD3852)
Joylessness (#VD3124)
Humiliation (#VD2804)
Harrassment (#VD2768)
Encumbrance (#VD2606)
Displeasure (#VD2536)
Disaffinity (#VD2454)
Contrariety (#VD2286)
Beastliness (#VD2158)
Awkwardness (#VD2130)
Abomination (#VD2016)
Tiresomeness (#VD3820)
Lovelessness (#VD3170)
Exasperation (#VD2628)
Discomposure (#VD2472)
Wearisomeness (#VD4058)
Loathsomeness (#VD3160)
Disconcertion (#VD2474)
Unpleasantness (#VD3936)
Discontentment (#VD2476)
Burdensomeness (#VD2206)
Undesirableness (#VD3900)
Dissatisfaction (#VD2546)
Disagreeableness (#VD2456)
Pain (#VD3422)
Worry (#VD4070)
Grief (#VD2748)
Tedium (#VD3794)
Sadism (#VD3596)
Miasma (#VD3222)
Hatred (#VD2780)
Dismay (#VD2522)
Torture (#VD3828)
Sadness (#VD3598)
Noisome (#VD3316)
Dislike (#VD2518)
Despair (#VD2410)
Boredom (#VD2192)
Anguish (#VD2068)
Vexation (#VD4006)
Nuisance (#VD3340)
Foulness (#VD2710)
Distress (#VD2566)
Disquiet (#VD2540)
Aversion (#VD2124)
Repulsion (#VD3540)
Harshness (#VD2774)
Grievance (#VD2750)
Confusion (#VD2272)
Antipathy (#VD2084)
Uneasiness (#VD3902)
Repugnance (#VD3538)
Melancholy (#VD3208)
Infelicity (#VD3000)
Discomfort (#VD2470)
Depression (#VD2396)
Bitterness (#VD2178)
Affliction (#VD2036)
Unhappiness (#VD3916)
Provocation (#VD3504)
Irksomeness (#VD3096)
Hideousness (#VD2794)
Fulsomeness (#VD2728)
Disturbance (#VD2570)
Disapproval (#VD2462)
Desolation (#VD2408)
Bedevilment (#VD2160)
Banefulness (#VD2140)
Aggravation (#VD2042)
Tortuousness (#VD3830)
Mournfulness (#VD3290)
Horribleness (#VD2800)
Dolorousness (#VD2582)
Disaffection (#VD2452)
Mortification (#VD3288)
Embarrassment (#VD2602)
Cheerlessness (#VD2232)
Unbearableness (#VD3878)
Despicableness (#VD2414)
Bothersomeness (#VD2196)
Pestiferousness (#VD3450)
Pleasurelessness (#VD3458)
Uncomfortableness (#VD3886).

♦ VP5865b Pleasure–Displeasure

Dynamics Can you learn to live ? Yes, if you are not happy. There is no virtue in felicity. (Colette)

Not joy but joylessness is the mother of debauchery. (Nietzsche)

The greatest happiness you can have is knowing that you do not necessarily require happiness. (William Saroyan)

Nature, with her customary beneficence, has ordained that man shall not learn how to live until the reasons for living are stolen from him, that he shall find no enjoyment until he has become incapable of vivid pleasure. (Giacomo Leopardi)

Life is a progress from want to want, not from enjoyment to enjoyment. (Samuel Johnson)

The greatest happiness of the greatest number is the foundation of morals and legislation. (Jeremy Bentham)

Integrative complex Feeling (#VT8081).
Constituent values Joy (#VC1033)
Zest (#VC1949)
Ease (#VC0601)
Bliss (#VC0249)
Gaiety (#VC0815)
Elation (#VC0621)
Delight (#VC0537)
Radiance (#VC1409)
Gladness (#VC0835)
Happiness (#VC0881)
Beatitude (#VC0219)
Joyfulness (#VC1035)
Exaltation (#VC0695)
Contentment (#VC0467)
Satisfaction (#VC1533)
Cheerfulness (#VC0321)
Gratification (#VC0863)
Beatification (#VC0217)
High-spiritedness (#VC0901).
Fun (#VC0813)
Love (#VC1093)
Charm (#VC0317)
Luxury (#VC1101)
Rapture (#VC1411)
Ecstasy (#VC0607)
Comfort (#VC0355)
Pleasure (#VC1309)
Felicity (#VC0753)
Enjoyment (#VC0651)
Adoration (#VC0059)
Exuberance (#VC0723)
Enchantment (#VC0637)
Blessedness (#VC0247)
Exhilaration (#VC0703)
Appreciation (#VC0153)
Entertainment (#VC0661)
Comfortableness (#VC0357).

Constituent values Displeasure (#VD2536).

♦ VP5868b Contentment–Discontentment

Dynamics The search for happiness is one of the chief sources of unhappiness. (Eric Hoffer)

True contentment is a thing as active as agriculture. It is the power of getting out of any situation all that there is in it. It is arduous and it is rare. (G K Chesterton)

Every man's happiness is built on the unhappiness of another. (Ivan Turgenev)

Discontent is the first step in the progress of a man or a nation. (Oscar Wilde)

Just as a cautious businessman avoids tying up all his capital in one concern, so, perhaps, worldly wisdom will advise us not to look for the whole of our satisfaction from a single aspiration. (Sigmund Freud)

Back of tranquillity lies always conquered unhappiness. (David Grayson)

That action is best which procures the greatest happiness for the greatest numbers. (Francis Hutcheson)

Integrative complex Feeling (#VT8081).
Constituent values Ease (#VC0601)
Adequacy (#VC0051)
Happiness (#VC0881)
Amusement (#VC0131)
Acceptance (#VC0017)
Contentment (#VC0467)
Agreeableness (#VC0091)
Comfortableness (#VC0357).
Comfort (#VC0355)
Viability (#VC1875)
Composure (#VC0391)
Fulfilment (#VC0809)
Sufficiency (#VC1699)
Satisfaction (#VC1533)
Reconciliation (#VC1433)

Constituent values Envy (#VD2614)
Petulance (#VD3454)
Resentment (#VD3542)
Unhappiness (#VD3916)
Restlessness (#VD3544)
Insufficiency (#VD3060)
Disappointment (#VD2460)
Indefensibility (#VD2954)
Self-satisfaction (#VD3644).
Sourness (#VD3700)
Uneasiness (#VD3902)
Inadequacy (#VD2890)
Complacency (#VD2250)
Unsuitability (#VD3954)
Discontentment (#VD2476)
Undesirableness (#VD3900)
Dissatisfaction (#VD2546)

♦ VP5870b Cheerfulness–Solemnity

Dynamics The value of ourselves is but the value of our melancholy and our disquiet. (Maurice Maeterlinck)

The gaiety of life, like the beauty and the moral worth of life, is a saving grace, which to ignore is folly, and to destroy is crime. There is no more than we need, – there is barely enough to go round. (Agnes Repplier)

Happiness is beneficial for the body, but it is grief that develops the powers of the mind. (Marcel Proust)

In heaven above, and earth below, they best can serve true gladness who meet most feelingly the calls of sadness. (William Wordsworth)

Solemnity is the shield of idiots. (Montesquieu)

You have to have a serious streak in you or you can't see the funny side in the other fellow. (Will Rogers)

Gladness of heart is the life of man, and the rejoicing of a man is length of days. (Apocrypha)

Integrative complex Feeling (#VT8081).
Constituent values Joy (#VC1033)
Zest (#VC1949)
Hope (#VC0917)
Light (#VC1077)
Uplift (#VC1843)
Esprit (#VC0679)
Decorum (#VC0523)
Vitality (#VC1891)
Radiance (#VC1409)
Laughter (#VC1053)
Gladness (#VC0835)
Merriment (#VC1141)
Happiness (#VC0881)
Animation (#VC0135)
Liveliness (#VC1085)
Exuberance (#VC0723)
Brightness (#VC0265)
Playfulness (#VC1305)
Pleasantness (#VC1307)
Cheerfulness (#VC0321)
Thoughtfulness (#VC1755)
High-spiritedness (#VC0901).
Fun (#VC0813)
Life (#VC1075)
Elan (#VC0619)
Vigour (#VC1881)
Gaiety (#VC0815)
Elation (#VC0621)
Vivacity (#VC1893)
Sobriety (#VC1623)
Optimism (#VC1229)
Hilarity (#VC0903)
Buoyancy (#VC0277)
Lightness (#VC1079)
Geniality (#VC0823)
Amusement (#VC0131)
Joyfulness (#VC1035)
Exaltation (#VC0695)
Seriousness (#VC1603)
Inspiration (#VC0993)
Exhilaration (#VC0703)
Encouragement (#VC0639)
Irrepressibility (#VC1027)

Constituent values Woe (#VD4068)
Sorrow (#VD3696)
Sadness (#VD3598)
Lowness (#VD3172)
Anguish (#VD2068)
Darkness (#VD2320)
Dejection (#VD2366)
Melancholy (#VD3208)
Depression (#VD2396)
Joylessness (#VD3124)
Despondency (#VD2418)
Mournfulness (#VD3290)
Dolorousness (#VD2582)
Heartlessness (#VD2784)
Spiritlessness (#VD3710)
Discontentment (#VD2476)
Grief (#VD2748)
Misery (#VD3254)
Malaise (#VD3184)
Despair (#VD2410)
Grimness (#VD2752)
Pessimism (#VD3448)
Oppression (#VD3366)
Infelicity (#VD3000)
Unhappiness (#VD3916)
Displeasure (#VD2536)
Desolation (#VD2408)
Hopelessness (#VD2798)
Wearisomeness (#VD4058)
Cheerlessness (#VD2232)
Discouragement (#VD2482)
Pleasurelessness (#VD3458).

VALUE POLARITIES

♦ VP5873b Exultation–Lamentation
Dynamics He who binds to himself a joy does the winged life destroy; but he who kisses the joy as it flies lives in eternity's sun rise. (William Blake)
Through loyalty to the past, our mind refuses to realize that tomorrow's joy is possible only if today's makes way for it; that each wave owes the beauty of its line only to the withdrawal of the preceding one. (André Gide)
Grief should be the instructor of the wise; sorrow is Knowledge. (Byron)
Ay, in the very temple of Delight, Veiled Melancholy has her sovran shrine. (John Keats)
Integrative complex Feeling (#VT8081).
Constituent values Joy (#VC1033)
Honour (#VC0915)
Synergy (#VC1733)
Delight (#VC0537)
Laughter (#VC1053)
Merriment (#VC1141)
Repentance (#VC1469)
Glory (#VC0841)
Triumph (#VC1781)
Elation (#VC0621)
Apology (#VC0141)
Hilarity (#VC0903)
Endowment (#VC0643)
Celebration (#VC0299).
Constituent values Fuss (#VD2734)
Hardness (#VD2770)
Insolence (#VD3050)
Callousness (#VD2210)
Shamelessness (#VD3666)
Sorrow (#VD3696)
Petulance (#VD3454)
Grievance (#VD2750)
Mournfulness (#VD3290)
Remorselessness (#VD3530).

♦ VP5878b Amusement–Boredom
Dynamics One can be bored until boredom becomes the most sublime of all emotions. (Logan Pearsall Smith)
By his very success in inventing labour-saving devices, modern man has manufactured an abyss of boredom that only the privileged classes in earlier civilizations have ever fathomed. (Lewis Mumford)
One must choose in life between boredom and torment. (Mme de Stael)
Great minds tend toward banality. It is the noblest effort of individualism. But it implies a sort of modesty, which is so rare that it is scarcely found except in the greatest, or in beggars. (André Gide)
The man who suspects his own tediousness is yet to be born. (Thomas Bailey Aldrich)
When men are rightly occupied, their amusement grows out of their work, as the colour-petals out of a fruitful flower. (John Ruskin)
Nothing is so aggravating as calmness. There is something positively brutal about the good temper of most modern men. (Oscar Wilde)
Integrative complex Feeling (#VT8081).
Constituent values Fun (#VC0813)
Gaiety (#VC0815)
Action (#VC0039)
Pleasure (#VC1309)
Hilarity (#VC0903)
Merriment (#VC1141)
Relaxation (#VC1455)
Cleverness (#VC0343)
Playfulness (#VC1305)
Humour (#VC0931)
Esprit (#VC0679)
Richness (#VC1503)
Keenness (#VC1043)
Wittiness (#VC1929)
Amusement (#VC0131)
Recreation (#VC1435)
Brilliance (#VC0267)
Entertainment (#VC0661).
Constituent values Wear (#VD4054)
Tedium (#VD3794)
Boredom (#VD2192)
Slowness (#VD3682)
Nuisance (#VD3340)
Dullness (#VD2588)
Banality (#VD2138)
Stiffness (#VD3730)
Prolixity (#VD3498)
Stuffiness (#VD3740)
Commonness (#VD2248)
Barrenness (#VD2146)
Tiresomeness (#VD3820)
Wearisomeness (#VD4058)
Ponderousness (#VD3468)
Superficiality (#VD3766)
Unimaginativeness (#VD3918)
Ennui (#VD2612)
Inanity (#VD2896)
Vapidity (#VD3990)
Sameness (#VD3600)
Effetism (#VD2598)
Deadness (#VD2322)
Weariness (#VD4056)
Sterility (#VD3728)
Mustiness (#VD3294)
Melancholy (#VD3208)
Boringness (#VD2194)
Irksomeness (#VD3096)
Lifelessness (#VD3152)
Tastelessness (#VD3792)
Pointlessness (#VD3460)
Spiritlessness (#VD3710)
Characterlessness (#VD2226).

♦ VP5885c Comfort–Aggravation
Dynamics We seldom break our leg so long as life continues a toilsome upward climb. The danger comes when we begin to take things easily and choose the convenient paths. (Nietzsche)
The seed-bed of life is sorrow and misfortune; the seed-bed of death is comfort and pleasure. (Chinese proverb)
Comfort, n. A state of mind produced by contemplation of a neighbour's uneasiness. (Bierce)
Integrative complex Feeling (#VT8081).
Constituent values Ease (#VC0601)
Relief (#VC1461)
Support (#VC1711)
Sympathy (#VC1731)
Assurance (#VC0189)
Mitigation (#VC1153)
Restfulness (#VC1493)
Convenience (#VC0473)
Peacefulness (#VC1267)
Encouragement (#VC0639)
Peace (#VC1265)
Luxury (#VC1101)
Comfort (#VC0355)
Increase (#VC0965)
Relaxation (#VC1455)
Condolence (#VC0415)
Hospitality (#VC0919)
Appeasement (#VC0147)
Friendliness (#VC0803)
Comfortableness (#VC0357).
Constituent values Annoyance (#VD2076)
Aggravation (#VD2042)
Deterioration (#VD2424)
Provocation (#VD3504)
Exasperation (#VD2628)
Contentiousness (#VD2282).

♦ VP5888a Hope–Hopelessness
Dynamics He who has never hoped can never despair. (George Bernard Shaw)
Not actual suffering but the hope of better things incites people to revolt. (Eric Hoffer)
It takes a certain level of aspiration before one can take advantage of opportunities that are clearly offered. (Michael Harrington)
An aspiration is a joy for ever, a possession as solid as a landed estate, a fortune which we can never exhaust and which gives us year by year a revenue of pleasurable activity. (Robert Louis Stevenson)
Hope is necessary in every condition. The miseries of poverty, sickness, of captivity, would, without this comfort, be insupportable. (Samuel Johnson)
Hope in reality is the worst of all evils, because it prolongs the torments of man. (Nietzsche)
Just as dumb creatures are snared by food, human beings would not be caught unless they had a nibble of hope. (Petronius)
Hope springs eternal in the human breast: man never is, but always to be blest. (Alexander Pope)
The mind which renounces, once and for ever, a futile hope, has its compensation in ever-growing calm. (George Gising)
A calm despair, without angry convulsions or reproaches directed at heaven, is the essence of wisdom. (Alfred de Vigny)
That glittering hope is immemorial and beckons many men to their undoing. (Euripides)
Integrative complex Anticipation (#VT8082).
Constituent values Hope (#VC0917)
Faith (#VC0729)
Promise (#VC1375)
Optimism (#VC1229)
Conviction (#VC0479)
Aspiration (#VC0179)
Expectation (#VC0709)
Encouragement (#VC0639)
Trust (#VC1783)
Support (#VC1711)
Security (#VC1537)
Assurance (#VC0189)
Brightness (#VC0265)
Inspiration (#VC0993)
Cheerfulness (#VC0321)
Desirableness (#VC0543).
Constituent values Apathy (#VD2090)
Vainness (#VD3980)
Futility (#VD2736)
Pessimism (#VD3448)
Despondency (#VD2418)
Hopelessness (#VD2798)
Cheerlessness (#VD2232)
Irremediability (#VD3110)
Irredeemability (#VD3102)
Despair (#VD2410)
Grimness (#VD2752)
Cynicism (#VD2312)
Negativity (#VD3308)
Slothfulness (#VD3680)
Impossibility (#VD2856)
Disappointment (#VD2460)
Irreformability (#VD3104)
Incorrigibility (#VD2942).

♦ VP5890b Courage–Fear
Dynamics Fools rush in through the door; for folly is always bold. (Baltasar Gracian)
Between cowardice and despair, valour is gendered. (John Donne)
Character begins to form at the first pinch of anxiety about ourselves. (Yevgeny Yevtushenko)
If you knew how cowardly your enemy is, you would slap him. Bravery is knowledge of the cowardice in the enemy. (Edgar Watson Howe)
Everyone becomes brave when he observes one who despairs. (Nietzsche)
That man is not truly brave who is afraid either to seem to be, or to be, when it suits him, a coward. (Edgar Allan Poe)
Whatever there be of progress in life comes not through adaptation but through daring, through obeying the blind urge. (Henry Miller)
Courage is resistance to fear, mastery of fear of absence of fear. Except a creature be part coward it is not a compliment to say it is brave. (Mark Twain)
Perfect simplicity is unconsciously audacious. (George Meredith)
Just as courage imperils life, fear protects it. (Leonardo da Vinci)
To live with fear and not be afraid is the final test of maturity. (Edward Weeks)
In difficult situations when hope seems feeble, the boldest plans are safest. (Livy)
With audacity one can undertake anything, but not do everything. (Napoleon I)
Integrative complex Anticipation (#VT8082).
Constituent values Awe (#VC0213)
Pluck (#VC1315)
Daring (#VC0513)
Support (#VC1711)
Heroism (#VC0897)
Bravery (#VC0257)
Chivalry (#VC0323)
Audacity (#VC0201)
Manliness (#VC1115)
Gallantry (#VC0819)
Adventure (#VC0067)
Enterprise (#VC0659)
Pertinacity (#VC1291)
Achievement (#VC0031)
Encouragement (#VC0639)
Mettlesomeness (#VC1145)
Zeal (#VC1947)
Virtue (#VC1887)
Valour (#VC1851)
Stamina (#VC1655)
Courage (#VC0495)
Tenacity (#VC1749)
Boldness (#VC0251)
Toughness (#VC1767)
Hardiness (#VC0883)
Assurance (#VC0189)
Resolution (#VC1479)
Confidence (#VC0419)
Inspiration (#VC0993)
Fearlessness (#VC0745)
Dauntlessness (#VC0515)
Venturesomeness (#VC1867).
Constituent values Fear (#VD2684)
Upset (#VD3966)
Panic (#VD3424)
Terror (#VD3806)
Dismay (#VD2522)
Torment (#VD3824)
Malaise (#VD3184)
Weakness (#VD4052)
Timidity (#VD3818)
Softness (#VD3688)
Eeriness (#VD2596)
Disquiet (#VD2540)
Alarmism (#VD2050)
Cowardice (#VD2298)
Agitation (#VD2046)
Uneasiness (#VD3902)
Trepidation (#VD3848)
Hideousness (#VD2794)
Disturbance (#VD2570)
Intimidation (#VD3072)
Gruesomeness (#VD2756)
Tremulousness (#VD3846)
Heartlessness (#VD2784)
Disconcertion (#VD2474)
Discouragement (#VD2482)
Worry (#VD4070)
Shock (#VD3672)
Lapse (#VD3130)
Strain (#VD3732)
Trouble (#VD3858)
Tension (#VD3802)
Anxiety (#VD2086)
Vexation (#VD4006)
Suspense (#VD3782)
Grimness (#VD2752)
Distress (#VD2566)
Baseness (#VD2150)
Morbidity (#VD3286)
Awfulness (#VD2128)
Xenophobia (#VD4080)
Inquietude (#VD3030)
Nervousness (#VD3314)
Harrassment (#VD2768)
Perturbation (#VD3442)
Horribleness (#VD2800)
Apprehension (#VD2092)
Spunklessness (#VD3716)
Frightfulness (#VD2722)
Spiritlessness (#VD3710)
Bothersomeness (#VD2196).

♦ VP5894c Caution–Rashness
Dynamics When a man feels the difficulty of doing, can he be other than cautious and slow in speaking? (Confucius)
A dram of discretion is worth a pound of wisdom. (German Proverb)
Folly often goes beyond her bonds, but impudence knows none. (Ben Jonson)
The better part of valour is discretion. (Shakespeare)
Prudence is sometimes stretched too far, until it blocks the road of progress. (Tehyi Hsieh)
Rashness succeeds often, still more often fails. (Napoleon I)
Integrative complex Anticipation (#VT8082).
Constituent values Care (#VC0291)
Prudence (#VC1387)
Canniness (#VC0285)
Discretion (#VC0571)
Forethought (#VC0785)
Thoroughness (#VC1753)
Judiciousness (#VC1039)
Circumspection (#VC0327).
Caution (#VC0295)
Foresight (#VC0783)
Solicitude (#VC1629)
Mindfulness (#VC1151)
Carefulness (#VC0293)
Cautiousness (#VC0297)
Deliberateness (#VC0533)
Constituent values Haste (#VD2776)
Distrust (#VD2568)
Impudence (#VD2876)
Imprudence (#VD2874)
Presumption (#VD3478)
Improvidence (#VD2872)
Carelessness (#VD2214)
Overzealousness (#VD3418)
Mistrust (#VD3278)
Insolence (#VD3050)
Hastiness (#VD2778)
Hesitation (#VD2792)
Indiscretion (#VD2964)
Impetuousity (#VD2848)
Overconfidence (#VD3382)
Uncommunicativeness (#VD3888).

VP5896

♦ VP5896a Taste–Vulgarity
Dynamics Without art, the crudeness of reality would make the world unbearable. (George Bernard Shaw)
Happiness is in the taste, and not in the things. (La Rochefoucauld)
A certain crudity makes people interesting just as much as a certain cultivation; in many ways, the next best thing to what Harvard represents is what the school of hard knocks does. (Louis Kronenberger)
Barbarism is the absence of standards to which appeal can be made. (José Ortega y Gasset)
Good taste and humour are a contradiction in terms, like a chaste whore. (Malcolm Muggeridge)
A good style should show no sign of effort. What is written should seem a happy accident. (W Somerset Maugham)
Elegance is good taste plus a dash of daring. (Carmel Snow)
People care more about being thought to have taste than about being thought either good, clever, or amiable. (Samuel Butler)
Integrative complex Discriminative affection (#VT8083).
Constituent values Art (#VC0169)
Choice (#VC0325)
Fitness (#VC0767)
Decorum (#VC0523)
Culture (#VC0507)
Felicity (#VC0753)
Delicacy (#VC0535)
Chastity (#VC0319)
Quietness (#VC1403)
Precision (#VC1341)
Gentility (#VC0827)
Strictness (#VC1683)
Refinement (#VC1441)
Aesthetics (#VC0071)
Sensitivity (#VC1595)
Ordinariness (#VC1235)
Exquisiteness (#VC0721)
Scrupulousness (#VC1535)
Discrimination (#VC0573)
Appropriateness (#VC0159)
Purity (#VC1391)
Quality (#VC1397)
Finesse (#VC0765)
Decency (#VC0517)
Subtlety (#VC1695)
Elegance (#VC0623)
Civility (#VC0331)
Restraint (#VC1495)
Propriety (#VC1379)
Happiness (#VC0881)
Expertise (#VC0717)
Simplicity (#VC1617)
Excellence (#VC0697)
Suitability (#VC1701)
Tastefulness (#VC1745)
Gracefulness (#VC0857)
Exclusiveness (#VC0701)
Meticulousness (#VC1143)
Unobtrusiveness (#VC1835)
Conscientiousness (#VC0441).
Constituent values Lowness (#VD3172)
Snobbery (#VD3686)
Rudeness (#VD3584)
Loudness (#VD3168)
Vulgarity (#VD4034)
Indecorum (#VD2952)
Grossness (#VD2754)
Brutality (#VD2202)
Barbarism (#VD2142)
Profligacy (#VD3492)
Indelicacy (#VD2956)
Ignobility (#VD2812)
Coarseness (#VD2240)
Impropriety (#VD2870)
Tastelessness (#VD3792)
Oversensitiveness (#VD3400)
Overconscientiousnes (#VD3384).
Vileness (#VD4014)
Savagery (#VD3608)
Meanness (#VD3204)
Baseness (#VD2150)
Roughness (#VD3580)
Indecency (#VD2948)
Gaudiness (#VD2738)
Barbarity (#VD2144)
Animality (#VD2070)
Inelegance (#VD2982)
Incivility (#VD2912)
Commonness (#VD2248)
Bestiality (#VD2168)
Unsuitability (#VD3954)
Exclusiveness (#VD2636)
Inappropriateness (#VD2900)

♦ VP5899a Beauty–Ugliness
Dynamics Beauty is truth, truth beauty, that is all ye know on earth, and all ye need to know. (John Keats)
Remember that the most beautiful things in the world are the most useless: peacocks and lilies, for instance. (John Ruskin)
Outstanding beauty, like outstanding gifts of any kind, tends to get in the way of normal emotional development, and thus of that particular success in life which we call happiness. (Milton R Sapirstein)
Joy and sorrow, beauty and deformity, equally pass away. (Sa'di)
Beauty is feared more than death. (William Carlos Williams)
Without grace beauty is an unbaited hook. (French Proverb)
The habit of looking for beauty in everything makes us notice the shortcomings of things; our sense, hungry for complete satisfaction, misses the perfection it demands. (George Santayana)
We fly to Beauty as an asylum from the terrors of finite nature. (Emerson)
The beautiful rests on the foundations of the necessary. (Emerson)
Oh, what a vileness human beauty is, corroding, corrupting everything it touches (Euripides)
Integrative complex Discriminative affection (#VT8083).
Constituent values Grace (#VC0855)
Charm (#VC0317)
Beauty (#VC0221)
Radiance (#VC1409)
Elegance (#VC0623)
Well-made (#VC1911)
Adornment (#VC0061)
Decoration (#VC0521)
Brightness (#VC0265)
Aesthetics (#VC0071)
Handsomeness (#VC0879)
Gorgeousness (#VC0851)
Agreeableness (#VC0091).
Glory (#VC0841)
Vision (#VC1889)
Glamour (#VC0837)
Fairness (#VC0727)
Delicacy (#VC0535)
Splendour (#VC1643)
Prettiness (#VC1351)
Brilliance (#VC0267)
Attraction (#VC0199)
Resplendence (#VC1487)
Gracefulness (#VC0857)
Exquisiteness (#VC0721)
Constituent values Blemish (#VD2184)
Grimness (#VD2752)
Awfulness (#VD2128)
Clumsiness (#VD2238)
Desperation (#VD2412)
Malformation (#VD3190)
Gruesomeness (#VD2756)
Loathsomeness (#VD3160)
Frightfulness (#VD2722)
Disfigurement (#VD2498).
Ugliness (#VD3872)
Foulness (#VD2710)
Inelegance (#VD2982)
Hideousness (#VD2794)
Deformation (#VD2358)
Horribleness (#VD2800)
Shapelessness (#VD3668)
Gracelessness (#VD2744)
Elaborateness (#VD2600)

♦ VP5900b Naturalness–Affectation
Dynamics The first duty in life is to be as artificial as possible. What the second duty is no one has yet discovered. (Oscar Wilde)
Fine conduct is always spontaneous. (Seneca)
I am certainly convinced that it is one of the greatest impulses of mankind to arrive at something higher than a natural state. (James Baldwin)
Affected simplicity is an elegant imposture. (La Rochefoucauld)
Hypocrisy is the homage which vice pays to virtue. (La Rochefoucauld)
Our greatest pretenses are built up not to hide the evil and the ugly in us, but our emptiness. The hardest thing to hide is something that is not there. (Eric Hoffer)
To be natural means to dare to be as immoral as Nature is. (Nietzsche)
Some degree of affectation is as necesary to the mind as dress is to the body; we must overact our part in some measure, in order to produce any effect at all. (William Hazlitt)
Affection is an awkward and forced imitation of what should be genuine and easy, wanting the beauty that accompanies what is natural. (John Locke)
Nature, in denying us perennial youth, has at least invited us to become unselfish and noble. (George Santayana)
The mark of the man of the world is absence of pretension. (Emerson)
Integrative complex Discriminative affection (#VT8083).
Constituent values Awe (#VC0213)
Purity (#VC1391)
Honour (#VC0915)
Honesty (#VC0913)
Grandeur (#VC0861)
Elegance (#VC0623)
Splendour (#VC1643)
Unassuming (#VC1801)
Directness (#VC0563)
Brilliance (#VC0267)
Naturalness (#VC1179)
Resplendence (#VC1487)
Magnificence (#VC1107)
Impressivensss (#VC0959)
Glory (#VC0841)
Luxury (#VC1101)
Majesty (#VC1109)
Nobility (#VC1193)
Flourish (#VC0773)
Chastity (#VC0319)
Plainness (#VC1301)
Simplicity (#VC1617)
Decoration (#VC0521)
Stateliness (#VC1659)
Unpretention (#VC1837)
Ordinariness (#VC1235)
Unadulteration (#VC1795)
Straightforwardness (#VC1679).
Constituent values Fakery (#VD2664)
Spurious (#VD3718)
Pedantry (#VD3432)
Hypocrisy (#VD2808)
Stuffiness (#VD3740)
Ostentation (#VD3372)
Affectation (#VD2034)
Unnaturalness (#VD3930)
Shamelessness (#VD3666)
Artificiality (#VD2106)
Self-importance (#VD3638)
Conceit (#VD2260)
Severity (#VD3660)
Pomposity (#VD3466)
Blatancy (#VD2182)
Sanctimony (#VD3602)
Insincerity (#VD3044)
Extravagance (#VD2654)
Sumptuousness (#VD3762)
Elaborateness (#VD2600)
Censoriousness (#VD2216)
Pretentiousness (#VD3480).

♦ VP5905b Pride–Humility
Dynamics One may be humble out of pride. (Montaigne)
The first test of a truly great man is his humility. (John Ruskin)
There is a price which is too great to pay for peace, and that price can be put in one word. One cannot pay the price of self-respect. (Woodrow Wilson)
Self respect will keep a man from being abject when he is in the power of enemies, and will enable him to feel that he may be in the right when the world is against him. (Bertrand Russell)
To know a man, observe how he wins his object, rather than how he loses it; for when we fail our pride supports us, when we succeed, it betrays us. (Charles Caleb Colton)
Pride, perceiving humility honourable, often borrows her cloak. (Thomas Fuller)
It takes a kind of shabby arrogance to survive in our time, and a fairly romantic nature to want to. (Edgar Z Friedenberg)
Integrative complex Discriminative affection (#VT8083).
Constituent values Modesty (#VC1159)
Dignity (#VC0557)
Nobility (#VC1193)
Humility (#VC0929)
Solemnity (#VC1627)
Worthiness (#VC1941)
Stateliness (#VC1659)
Unpretention (#VC1837)
Independence (#VC0969)
Self-reliance (#VC1577)
High-mindedness (#VC0899)
Majesty (#VC1109)
Sobriety (#VC1623)
Meekness (#VC1133)
Grandeur (#VC0861)
Plainness (#VC1301)
Submission (#VC1689)
Aristocracy (#VC0167)
Self-respect (#VC1579)
Venerableness (#VC1863)
Self-confidence (#VC1545)
Self-sufficiency (#VC1585).
Constituent values Vanity (#VD3986)
Chagrin (#VD2220)
Disgrace (#VD2500)
Smallness (#VD3684)
Arrogance (#VD2102)
Abasement (#VD2006)
Self-esteem (#VD3636)
Humiliation (#VD2804)
Boastfulness (#VD2190)
Embarrassment (#VD2602)
Ingloriousness (#VD3010).
Conceit (#VD2260)
Meanness (#VD3204)
Baseness (#VD2150)
Servility (#VD3656)
Abjection (#VD2012)
Lordliness (#VD3166)
Insinuation (#VD3046)
Unimportance (#VD3920)
Mortification (#VD3288)
Condescension (#VD2266)

♦ VP5908b Modesty–Vanity
Dynamics Virtue brings honour, and honour vanity. (Thomas Fuller)
Pride that dines on vanity sups on contempt. (Benjamin Franklin)
Nothing is more amiable than true modesty, and nothing more contemptible than the false. The one guards virtue, the other betrays it. (Joseph Addison)
Vanity is truly the motive-power that moves humanity, and it is flattery that greases the wheels. (Jerome K Jerome)
Integrative complex Discriminative affection (#VT8083).
Constituent values Modesty (#VC1159)
Humility (#VC0929)
Quietness (#VC1403)
Unpretention (#VC1837)
Self-interest (#VC1569)
Self-effacement (#VC1559).
Meekness (#VC1133)
Restraint (#VC1495)
Unassuming (#VC1801)
Reservedness (#VC1473)
Unobtrusiveness (#VC1835)
Constituent values Fuss (#VD2734)
Disdain (#VD2490)
Vainness (#VD3980)
Rudeness (#VD3584)
Insolence (#VD3050)
Immodesty (#VD2834)
Arrogance (#VD2102)
Disrespect (#VD2544)
Self-esteem (#VD3636)
Complacency (#VD2250)
Impertinence (#VD2844)
Boastfulness (#VD2190)
Condescension (#VD2266)
Inarticulation (#VD2904)
Self-admiration (#VD3628)
Superciliousness (#VD3764)
Contumeliousness (#VD2292)
Self-satisfaction (#VD3644).
Vanity (#VD3986)
Conceit (#VD2260)
Snobbery (#VD3686)
Ridicule (#VD3566)
Impudence (#VD2876)
Confusion (#VD2272)
Lordliness (#VD3166)
Selfishness (#VD3646)
Presumption (#VD3478)
Intimidation (#VD3072)
Dictatorship (#VD2438)
Shamelessness (#VD3666)
Bumptiousness (#VD2204)
Self-importance (#VD3638)
Pretentiousness (#VD3480)
Self-centredness (#VD3630)
Contemptuousness (#VD2280)

♦ VP5914c Repute–Disrepute
Dynamics Worldly renown is but a breath of wind; coming first from here and then from there, and changes name because it changes quarter. (Dante)
All fame is dangerous: good bringeth envy; bad, shame. (Thomas Fuller)

VALUE POLARITIES   VP5930

When glory comes, memory departs. (French Proverb)
The purest treasure mortal times afford is spotless reputation. (Shakespeare)
One crowded hour of glorious life is worth an age without a name. (Sir Walter Scott)
The desire of glory is the last infirmity cast off even by the wise. (Tacitus)
Even workhouses have their aristocracy. (English Proverb)
There is hardly that person to be found who is not more concerned for the reputation of wit and sense, than honesty and virtue. (Richard Steele)
He who practices right, but in the hope of acquiring great renown, is very near to vice. (Napoleon I)
The great despisers are the great reverers. (Nietzsche)
Whoever rises above those who once pleased themselves with equality, will have many malevolent gazers at his eminence. (Samuel Johnson)
Integrative complex Discriminative affection (#VT8083).
Constituent values Fame (#VC0733)

Merit (#VC1139)	Worth (#VC1939)
Uplift (#VC1843)	Glory (#VC0841)
Praise (#VC1337)	Status (#VC1663)
Royalty (#VC1513)	Favour (#VC0743)
Glamour (#VC0837)	Quality (#VC1397)
Bigness (#VC0241)	Dignity (#VC0557)
Prestige (#VC1349)	Radiance (#VC1409)
Grandeur (#VC0861)	Nobility (#VC1193)
Chivalry (#VC0323)	Eminence (#VC0631)
Approval (#VC0161)	Charisma (#VC0313)
Gentility (#VC0827)	Splendour (#VC1643)
Character (#VC0311)	Elevation (#VC0625)
Reputation (#VC1471)	Worthiness (#VC1941)
Notability (#VC1203)	Popularity (#VC1319)
Excellence (#VC0697)	Importance (#VC0957)
Decoration (#VC0521)	Exaltation (#VC0695)
Recognition (#VC1427)	Brilliance (#VC0267)
Distinction (#VC0577)	Immortality (#VC0947)
Aristocracy (#VC0167)	Deification (#VC0529)
Significance (#VC1613)	Approbation (#VC0157)
Venerableness (#VC1863)	Resplendence (#VC1487)
Beatification (#VC0217)	Consideration (#VC0449)
Respectfulness (#VC1485)	Sanctification (#VC1527)
Respectability (#VC1483).	Aggrandizement (#VC0087)

Constituent values Stain (#VD3724)

Arrant (#VD2098)	Infamy (#VD2996)
Scandal (#VD3610)	Squalor (#VD3720)
Censure (#VD2218)	Lowness (#VD3172)
Reproach (#VD3534)	Vileness (#VD4014)
Foulness (#VD2710)	Meanness (#VD3204)
Baseness (#VD2150)	Disgrace (#VD2500)
Smallness (#VD3684)	Atrocity (#VD2114)
Grossness (#VD2754)	Pettiness (#VD3452)
Disfavour (#VD2496)	Dishonour (#VD2508)
Discredit (#VD2486)	Disesteem (#VD2494)
Cheapness (#VD2230)	Depravity (#VD2392)
Sordidness (#VD3692)	Abjection (#VD2012)
Notoriety (#VD3338)	Opprobrium (#VD3368)
Ignobility (#VD2812)	Littleness (#VD3158)
Humiliation (#VD2804)	Defamation (#VD2342)
Degradation (#VD2362)	Fulsomeness (#VD2728)
Ingloriousness (#VD3010)	Disparagement (#VD2532)
Ignominiousness (#VD2814).	Despicableness (#VD2414)

♦ **VP5920c Wonder–Unastonishment**
Dynamics Mystery magnifies danger as the fog the sun. (Charles Caleb Colton)
Surprise is the greatest gift which life can grant us. (Boris Pasternak)
To be surprised, to wonder, is to begin to understand. (José Ortega y Gasset)
The ultimate gift of conscious life is a sense of the mystery that encompasses it. (Lewis Mumford)
Integrative complex Discriminative affection (#VT8083).
Constituent values

Awe (#VC0213)	Wonder (#VC1933)
Surprise (#VC1719)	Rareness (#VC1413)
Sensation (#VC1589)	Curiosity (#VC0479)
Uniqueness (#VC1827)	Fascination (#VC0737)
Breathtaking (#VC0259)	Prodigiousness (#VC1363).

♦ **VP5922a Sociability–Unsociability**
Dynamics To dare to live alone is the rarest courage; since there are many who had rather meet their bitterest enemy in the field, than their own hearts in their closet. (Charles Caleb Colton)
Togetherness is a substitute sense of community, a counterfeit communion. (Gabriel Vahanian)
To every man it is decreed: Thou shalt live alone. Happy they who imagine that they have escaped the common lot; happy, whilst they imagine it. (George Gissing)
In solitude it is possible to love mankind; in the world, for one who knows the world, there can be nothing but secret or open war. (George Santayana)
We seek pitifully to convey to others the treasures of our heart, but they have not the power to accept them, and so we go lonely, side by side but not together, unable to know our fellows and unknown by them. (W Somerset Maugham)
The world is always curious, and people become valuable merely for their inaccessibility. (F Scott Fitzgerald)
He who must needs have company, must needs have sometimes bad company. (Sir Thomas Browne)
The kind of relatedness to the world may be noble or trivial, but even being related to the basest kind of pattern is immensely preferable to being alone. (Erich Fromm)
When is man strong until he feels alone ? (Robert Browning)
At bottom, and just in the deepest and most important things, we are unutterably alone, and for one person to be able to advise or even help another, a lot must happen, a lot must go well, a whole constellation of things must come right in order once to succeed. (Rainer Maria Rilke)
The human animal needs a freedom seldom mentioned, freedom from intrusion. He needs a little privacy quite as much as he wants understanding or vitamins or exercise or praise. (Phyllis McGinley)
Integrative complex Socialization (#VT8084).
Constituent values Gaiety (#VC0815)

Courtesy (#VC0497)	Intimacy (#VC1013)
Merriment (#VC1141)	Civility (#VC0331)
Community (#VC0373)	Geniality (#VC0823)
Fellowship (#VC0755)	Fraternity (#VC0795)
Asceticism (#VC0177)	Detachment (#VC0545)
Affability (#VC0073)	Amiability (#VC0121)
Hospitality (#VC0919)	Sociability (#VC1625)
	Familiarity (#VC0735)

Cooperation (#VC0481)	Comradeship (#VC0401)
Association (#VC0187)	Self-control (#VC1549)
Friendliness (#VC0803)	Congeniality (#VC0427)
Cheerfulness (#VC0321)	Participation (#VC1253)
Exclusiveness (#VC0701)	Compatibility (#VC0379)
Companionship (#VC0375)	Gregariousness (#VC0869)
Self-discipline (#VC1557)	Cooperativeness (#VC0483)
Communicativeness (#VC0369).	

Constituent values Autism (#VD2120)

Coldness (#VD2242)	Secrecy (#VD3618)
Desertion (#VD2406)	Seclusion (#VD3616)
Segregation (#VD3624)	Loneliness (#VD3162)
Homelessness (#VD2796)	Desolation (#VD2408)
Unfriendliness (#VD3910)	Exclusiveness (#VD2636)
Inaccessibility (#VD2884)	Incompatibility (#VD2918)
	Uncommunicativeness (#VD3888).

♦ **VP5925a Hospitality–Inhospitality**
Dynamics When hospitality becomes an art, it loses its very soul. (Max Beerbohm)
Drinking together in the evening we are human. When dawn comes, animals, we rise up against each other. (Antimedon)
So saying, with dispatchful looks in haste. She turns, on hospitable thoughts intent. (Milton)
It is nothing won to admit men with an open door, and to receive them with a shut and reserved countenance. (Francis Bacon)
Fish and guests smell at three days old. (Danish proverb)
My evening visitors, if they cannot see the clock should find the time in my face. (Emerson)
Integrative complex Socialization (#VT8084).
Constituent values Warmth (#VC1899)

Liberality (#VC1069)	Geniality (#VC0823)
Cordiality (#VC0487)	Generosity (#VC0821)
Hospitality (#VC0919)	Amiability (#VC0121)
Friendliness (#VC0803)	Graciousness (#VC0859)
Agreeableness (#VC0091)	Desirableness (#VC0543)
Openheartedness (#VC1221)	Warmheartedness (#VC1897)
	Neighbourliness (#VC1185).

Constituent values Abandonment (#VD2004)

Ungraciousness (#VD3914)	Inhospitality (#VD3018)
Unneighbourliness (#VD3932).	Undesirableness (#VD3900)

♦ **VP5927b Friendship–Enmity**
Dynamics Amity itself can only be maintained by reciprocal respect, and true friends are punctilious equals. (Herman Melville)
Greater love hath no man than this, that a man lay down his life for his friends. (Bible)
The firmest friendships have been formed in mutual adversity, as iron is most strongly united by the fiercest flame. (Charles Caleb Colton)
A wise man gets more use from his enemies than a fool from his friends. (Baltasar Gracian)
The real enemy can always be met and conquered, or won over. Real antagonism is based on love, a love which has not recognized itself. (Henry Miller)
He who lives by fighting with an enemy has an interest in the preservation of the enemy's life. (Nietzsche)
The condition which high friendship demands is ability to do without it. (Emerson)
You want to hate somebody, if you can, just to keep your powers of discrimination bright, and to save yourself from becoming a mere mush of good–nature. (Charles Dudley Warner)
Couldn't we even argue that it is because men are unequal that they have that much more need to be brothers ? (Charles du Bos)
At the heart of our friendly or purely social relations, there lurks a hostility momentarily cured but recurring by fits and starts. (Marcel Proust)
The response man has the greatest difficulty in tolerating is pity, especially when he warrants it. Hatred is a tonic, it makes one live, it inspires vengeance, but pity kills, it makes our weakness weaker. (Balzac)
A low capacity for getting along with those near us often goes hand in hand with a high receptivity to the idea of the brotherhood of men. (Eric Hoffer)
Integrative complex Socialization (#VT8084).
Constituent values Love (#VC1093)

Warmth (#VC1899)	Amity (#VC0125)
Harmony (#VC0887)	Favour (#VC0743)
Kindness (#VC1045)	Sympathy (#VC1731)
Goodwill (#VC0849)	Intimacy (#VC1013)
Affinity (#VC0079)	Devotion (#VC0551)
Geniality (#VC0823)	Advocacy (#VC0069)
Friendship (#VC0805)	Constancy (#VC0455)
Fellowship (#VC0755)	Fraternity (#VC0795)
Cordiality (#VC0487)	Dedication (#VC0525)
Amiability (#VC0121)	Commitment (#VC0361)
Sociability (#VC1625)	Staunchness (#VC1665)
Familiarity (#VC0735)	Hospitality (#VC0919)
Brotherhood (#VC0273)	Comradeship (#VC0401)
Well-meaning (#VC1915)	Amicability (#VC0123)
Friendliness (#VC0803)	Pleasantness (#VC1307)
Congeniality (#VC0427)	Faithfulness (#VC0731)
Steadfastness (#VC1667)	Well-disposed (#VC1905)
Brotherliness (#VC0275)	Companionship (#VC0375)
Warmheartedness (#VC1897)	Agreeableness (#VC0091)
Approachability (#VC0155)	Neighbourliness (#VC1185)

Constituent values Venom (#VD4002)

Malice (#VD3192)	Strain (#VD3732)
Enmity (#VD2610)	Hatred (#VD2780)
Sourness (#VD3700)	Tension (#VD3802)
Friction (#VD2720)	Soreness (#VD3694)
Conflict (#VD2270)	Disunity (#VD2572)
Acrimony (#VD2032)	Coldness (#VD2242)
Hostility (#VD2802)	Virulence (#VD4024)
Antipathy (#VD2084)	Discord (#VD2448)
Repugnance (#VD3538)	Animosity (#VD2072)
Antagonism (#VD2080)	Bitterness (#VD2178)
Malevolence (#VD3186)	Alienation (#VD2052)
Belligerence (#VD2164)	Disaffinity (#VD2454)
Unfriendliness (#VD3910)	Inhospitality (#VD3018)
Incompatibility (#VD2918)	Quarrelsomeness (#VD3512)
	Irreconcilability (#VD3100).

♦ **VP5930b Love–Hate**
Dynamics Praise is well, compliment is well, but affection – that is the last and final and most precious reward that any man can win, whether by character or achievement. (Mark Twain)
That trial is not fair where affection is judge. (Thomas Fuller)
It is impossible to love and be wise. (Francis Bacon)
If you hate a person, you hate something in him that is part of yourself. What isn't part of ourselves doesn't disturb us. (Hermann Hesse)
Hatred is a feeling which leads to the extinction of values. (José Ortega y Gasset)

VP5930

To the eye of enmity virtue appears the ugliest blemish. (Sa'di)
As the best wine doth make the sharpest vinegar, so the deepest love turneth to the deadliest hate. (John Lyly)
Even as love crowns you so shall he crucify you. Even as he is for your growth so is he for your pruning. (Kahlil Gibran)
Love is more afraid of change than destruction. (Nietzsche)
I am of this mind, that both might and malice, deceit and treachery, all perjury, any impiety may lawfully be committed in love, which is lawless. (John Lyly)
What dire offence from amorous causes springs What mighty contests rise from trivial things (Alexander Pope)
In abstract love of humanity one almost always only loves oneself. (Dostoevsky)
Whatever is done for love always occurs beyond good and evil. (Nietzsche)

Integrative complex Socialization (#VT8084).
Constituent values Love (#VC1093)
- Appeal (#VC0143)
- Rapture (#VC1411)
- Ecstasy (#VC0607)
- Ardour (#VC0165)
- Fervour (#VC0761)
- Sweetness (#VC1725)
- Affection (#VC0075)
- Popularity (#VC1319)
- Attachment (#VC0193)
- Admiration (#VC0057)
- Brotherhood (#VC0273)
- Faithfulness (#VC0731)
- Charm (#VC0317)
- Romance (#VC1511)
- Passion (#VC1257)
- Charity (#VC0315)
- Fondness (#VC0777)
- Devotion (#VC0551)
- Sentiment (#VC1597)
- Adoration (#VC0059)
- Endearment (#VC0641)
- Allurement (#VC0107)
- Enchantment (#VC0637)
- Amorousness (#VC0127)
- Desirableness (#VC0543).

Constituent values Malice (#VD3192)
- Enmity (#VD2610)
- Bigotry (#VD2176)
- Antipathy (#VD2084)
- Xenophobia (#VD4080)
- Repugnance (#VD3538)
- Bitterness (#VD2178)
- Malevolence (#VD3186)
- Loathsomeness (#VD3160).
- Hatred (#VD2780)
- Dislike (#VD2518)
- Aversion (#VD2124)
- Animosity (#VD2072)
- Unchastity (#VD3882)
- Execration (#VD2638)
- Abhorrence (#VD2010)
- Abomination (#VD2016)

♦ **VP5931c Conjugality–Celibacy**
Dynamics Marriage has many pains, but celibacy has no pleasures. (Samuel Johnson)
Every day men sleep with women whom they do not love, and do not sleep with women whom they love. (Diderot)
Few romantic intrigues can be kept secret; many women are as well known by the names of their lovers as they are by those of their husbands. (La Bruyère)
When a widow finds a new husband, she should consummate the marriage immediately. (Chinese proverb)

Integrative complex Socialization (#VT8084).
Constituent values Annulment (#VD2078)   Malfeasance (#VD3188)
Bereavement (#VD2166).

♦ **VP5936a Courtesy–Discourtesy**
Dynamics It is good breeding alone that can prepossess people in your favour at first sight, more time being necessary to discover greater talents. (Lord Chesterfield)
The whole essence of true gentle-breeding (one does not like to say gentility) lies in the wish and the art to be agreeable. Good-breeding is surface-Christianity. (Oliver Wendell Holmes, Sr)
Good breeding consists in concealing how much we think of ourselves and how little we think of other persons. (Mark Twain)
All doors open to courtesy. (Thomas Fuller)
There can be no defence like elaborate courtesy. (E V Lucas)
Politeness is to human nature what warmth is to wax. (Schopenhauer)
Manners are the hypocrisy of a nation. (Balzac)

Integrative complex Benevolence (#VT8085).
Constituent values Tact (#VC1737)
- Manners (#VC1117)
- Elegance (#VC0623)
- Civility (#VC0331)
- Breeding (#VC0261)
- Gallantry (#VC0819)
- Attention (#VC0197)
- Refinement (#VC1441)
- Gentleness (#VC0829)
- Correctness (#VC0489)
- Gracefulness (#VC0857)
- Consideration (#VC0449)
- Thoughtfulness (#VC1755)
- Favour (#VC0743)
- Culture (#VC0507)
- Courtesy (#VC0497)
- Chivalry (#VC0323)
- Gentility (#VC0827)
- Deference (#VC0527)
- Solicitude (#VC1629)
- Politeness (#VC1317)
- Affability (#VC0073)
- Graciousness (#VC0859)
- Well-mannered (#VC1913)
- Agreeableness (#VC0091)
- Respectfulness (#VC1485).

Constituent values Severity (#VD3660)
- Glibness (#VD2740)
- Vulgarity (#VD4034)
- Insolence (#VD3050)
- Truculence (#VD3860)
- Coarseness (#VD2240)
- Discourtesy (#VD2484)
- Impoliteness (#VD2854)
- Ungraciousness (#VD3914)
- Inconsiderateness (#VD2930).
- Rudeness (#VD3584)
- Curtness (#VD2310)
- Surliness (#VD3774)
- Harshness (#VD2774)
- Incivility (#VD2912)
- Aggression (#VD2044)
- Tactlessness (#VD3788)
- Insensitivity (#VD3040)
- Unfriendliness (#VD3910)

♦ **VP5938a Kindness–Unkindness**
Dynamics Man's inhumanity to man / Makes countless thousands mourn. (Robert Burns)
Cruelty is, perhaps, the worst kind of sin. Intellectual cruelty is certainly the worst kind of cruelty. (G K Chesterton)
The ignorance of the world leaves one at the mercy of its malice. (William Hazlitt)
Loving-kindness is the better part of goodness. It lends grace to the sterner qualities of which this consists. (W Somerset Maugham)
I must be cruel, only to be kind. (Shakespeare)
Love compels cruelty to those who do not understand love. (T S Eliot)
Savagery is necessary every four or five hundred years in order to bring the world back to life. Otherwise the world would die of civilization. (Edmound and Jules de Goncourt)
Kindness effects more than severity. (Aesop)
The bird thinks it is an act of kindness to give the fish a lift in the air. (Rabindranath Tagore)

Integrative complex Benevolence (#VT8085).
Constituent values Tact (#VC1737)
- Good (#VC0845)
- Favour (#VC0743)
- Service (#VC1605)
- Concern (#VC0407)
- Sympathy (#VC1731)
- Leniency (#VC1067)
- Love (#VC1093)
- Warmth (#VC1899)
- Welfare (#VC1903)
- Decency (#VC0517)
- Charity (#VC0315)
- Mildness (#VC1149)
- Kindness (#VC1045)
- Humanity (#VC0927)
- Goodness (#VC0847)
- Courtesy (#VC0497)
- Sweetness (#VC1725)
- Benignity (#VC0237)
- Affection (#VC0075)
- Humaneness (#VC0921)
- Generosity (#VC0821)
- Compassion (#VC0377)
- Amiability (#VC0121)
- Mindfulness (#VC1151)
- Brotherhood (#VC0273)
- Beneficence (#VC0233)
- Graciousness (#VC0859)
- Consideration (#VC0449)
- Agreeableness (#VC0091)
- Respectfulness (#VC1485)
- Humanitarianism (#VC0925).
- Goodwill (#VC0849)
- Delicacy (#VC0535)
- Altruism (#VC0113)
- Geniality (#VC0823)
- Attention (#VC0197)
- Solicitude (#VC1629)
- Gentleness (#VC0829)
- Cordiality (#VC0487)
- Commonweal (#VC0365)
- Affability (#VC0073)
- Helpfulness (#VC0893)
- Benevolence (#VC0235)
- Well-meaning (#VC1915)
- Well-disposed (#VC1905)
- Brotherliness (#VC0275)
- Thoughtfulness (#VC1755)
- Warmheartedness (#VC1897)

Constituent values Venom (#VD4002)
- Malice (#VD3192)
- Cruelty (#VD2304)
- Severity (#VD3660)
- Meanness (#VD3204)
- Grimness (#VD2752)
- Ferocity (#VD2694)
- Cynicism (#VD2312)
- Atrocity (#VD2114)
- Withering (#VD4064)
- Vandalism (#VD3984)
- Harshness (#VD2774)
- Brutality (#VD2202)
- Austerity (#VD2116)
- Unkindness (#VD3924)
- Malignancy (#VD3194)
- Inclemency (#VD2914)
- Bitterness (#VD2178)
- Viciousness (#VD4008)
- Harmfulness (#VD2772)
- Beastliness (#VD2158)
- Pitilessness (#VD3456)
- Devilishness (#VD2430)
- Insensitivity (#VD3040)
- Heartlessness (#VD2784)
- Ungraciousness (#VD3914)
- Thoughtlessness (#VD3810)
- Disagreeableness (#VD2456)
- Bloody-mindedness (#VD2188).
- Sadism (#VD3596)
- Illness (#VD2826)
- Violence (#VD4022)
- Savagery (#VD3608)
- Hardness (#VD2770)
- Fiendish (#VD2696)
- Diabolic (#VD2436)
- Coldness (#VD2242)
- Asperity (#VD2108)
- Virulence (#VD4024)
- Roughness (#VD3580)
- Damnation (#VD2316)
- Barbarity (#VD2144)
- Animality (#VD2070)
- Truculence (#VD3860)
- Inhumanity (#VD3020)
- Fierceness (#VD2698)
- Bestiality (#VD2168)
- Malevolence (#VD3186)
- Callousness (#VD2210)
- Ruthlessness (#VD3588)
- Heedlessness (#VD2786)
- Unnaturalness (#VD3930)
- Inhospitality (#VD3018)
- Forgetfulness (#VD2708)
- Dehumanization (#VD2364)
- Unthoughtfulness (#VD3956)
- Inconsiderateness (#VD2930)

♦ **VP5944a Compassion–Pitilessness**
Dynamics Of all cruelties those are the most intolerable that come under the name of condolence and consolation. (Walter Savage)
Nothing emboldens sin so much as mercy. (Shakespeare)
'Pity for all' would be hardness and tyranny toward you, my dear neighbour. (Nietzsche)
The entire world would perish, if pity were not to limit anger. (Seneca)
Compassion is the property of the privileged classes. When the pitier lowers himself to give to a beggar, he throbs with contempt. (Peter Weiss)
Compassionate understanding too often buys a long-range with a small-change gesture – like giving a quarter to a beggar. (Ross Wetzsteon)

Integrative complex Benevolence (#VT8085).
Constituent values Pity (#VC1299)
- Relief (#VC1461)
- Favour (#VC0743)
- Comfort (#VC0355)
- Sympathy (#VC1731)
- Kindness (#VC1045)
- Clemency (#VC0341)
- Mitigation (#VC1153)
- Gentleness (#VC0829)
- Compassion (#VC0377)
- Forbearance (#VC0779)
- Mercifulness (#VC1137)
- Warmheartedness (#VC1897).
- Grace (#VC0855)
- Pardon (#VC1247)
- Feeling (#VC0751)
- Charity (#VC0315)
- Leniency (#VC1067)
- Humanity (#VC0927)
- Tenderness (#VC1751)
- Humaneness (#VC0921)
- Condolence (#VC0415)
- Forgiveness (#VC0787)
- Benevolence (#VC0235)
- Understanding (#VC1811)

Constituent values Cruelty (#VD2304)
- Harshness (#VD2774)
- Ruthlessness (#VD3588)
- Heartlessness (#VD2784)
- Remorselessness (#VD3530).
- Hardness (#VD2770)
- Inclemency (#VD2914)
- Pitilessness (#VD3456)
- Relentlessness (#VD3528)

♦ **VP5947b Forgiveness–Vengeance**
Dynamics Not by a radiant jewel, not by the sun nor by fire, but by conciliation alone is dispelled the darkness born of enmity. (Panchatantra)
There is no forgiveness in nature. (Ugo Betti)
An act by which we make one friend and one enemy is a losing game; because revenge is a much stronger principle than gratitude. (Charles Caleb Colton)
Just vengeance does not call for punishment. (Corneille)
No revenge is more honourable than the one not taken. (Spanish Proverb)
Pardon one offence, and you encourage the commission of many. (Publilius Syrus)
Forgive all but thyself. (Proverb)
To err is human; to forgive, divine. (Alexander Pope)
Forgiving the unrepentant is like making pictures on water. (Japanese proverb)
There is no austerity like forgiveness. (Indian proverb)

Integrative complex Benevolence (#VT8085).
Constituent values Grace (#VC0855)
- Homage (#VC0909)
- Kindness (#VC1045)
- Tolerance (#VC1765)
- Redemption (#VC1439)
- Absolution (#VC0007)
- Longanimity (#VC1089)
- Forbearance (#VC0779)
- Conciliation (#VC0409)
- Pardon (#VC1247)
- Patience (#VC1261)
- Immunity (#VC0949)
- Remission (#VC1463)
- Generosity (#VC0821)
- Magnanimity (#VC1105)
- Forgiveness (#VC0787)
- Benevolence (#VC0235)

Constituent values Revenge (#VD3556)
- Retaliation (#VD3548)
- Irreconcilability (#VD3100).
- Retribution (#VD3552)
- Vindictiveness (#VD4018)

VALUE POLARITIES   VP5968

♦ **VP5948c Congratulation–Envy**
Dynamics Nothing mortal is enduring, and there is nothing sweet which does not presently end in bitterness. (Petrarch)
Where there is no jealousy there is no love. (German Proverb)
Jealousy is the greatest of all evils, and the one which arouses the least pity in the person who causes it. (La Rochefoucauld)
If envy were a fever, all the world would be ill. (Danish Proverb)
Envy is the basis of democracy. (Bertrand Russell)
Integrative complex Benevolence (#VT8085).
Constituent values Fury (#VD2732)
Wound (#VD4074)
Temper (#VD3796)
Outrage (#VD3374)
Vexation (#VD4006)
Sourness (#VD3700)
Offence (#VD3358)
Meanness (#VD3204)
Grimness (#VD2752)
Asperity (#VD2108)
Virulence (#VD4024)
Surliness (#VD3774)
Hastiness (#VD2778)
Annoyance (#VD2076)
Resentment (#VD3542)
Melancholy (#VD3208)
Bitterness (#VD2178)
Provocation (#VD3504)
Fretfulness (#VD2718)
Disapproval (#VD2462)
Irritability (#VD3118)
Exasperation (#VD2628)
Mercilessness (#VD3216)
Quarrelsomeness (#VD3512)
Contentiousness (#VD2282)
Disagreeableness (#VD2456)
Argumentativeness (#VD2096).
Envy (#VD2614)
Anger (#VD2066)
Rivalry (#VD3574)
Violence (#VD4022)
Ugliness (#VD3872)
Soreness (#VD3694)
Mistrust (#VD3278)
Jealousy (#VD3122)
Distrust (#VD2568)
Acrimony (#VD2032)
Vehemence (#VD3996)
Petulance (#VD3454)
Dejection (#VD2366)
Animosity (#VD2072)
Perversity (#VD3446)
Fierceness (#VD2698)
Affrontery (#VD2038)
Indignation (#VD2960)
Displeasure (#VD2536)
Aggravation (#VD2042)
Irascibility (#VD3094)
Belligerence (#VD2164)
Discontentment (#VD2476)
Dissatisfaction (#VD2546)
Disputatiousness (#VD2538)
Oversensitiveness (#VD3400)

♦ **VP5949c Gratitude–Ingratitude**
Dynamics Next to ingratitude, the most painful thing to bear is gratitude. (Henry Ward Beecher)
Gratitude is the most exquisite form of courtesy. (Jacques Maritain)
Gratitude preserves old friendships, and procures new. (Proverb)
Gratitude is the least of virtues, but ingratitude is the worst of vices. (Proverb)
Too great haste in paying off an obligation is a kind of ingratitude. (La Rochefoucauld)
Blow, blow, thou winter wind. Thou art not so unkind as man's ingratitude. (Shakespeare)
The gratitude of most men is merely a secret desire to receive greater benefits. (La Rochefoucauld)
Gratitude is a fruit of great cultivation; you do not find it among gross people. (Samuel Johnson)
What soon grows old ? Gratitude (Aristotle)
Integrative complex Benevolence (#VT8085).
Constituent values Praise (#VC1337)
Recognition (#VC1427)
Benediction (#VC0231)
Appreciation (#VC0153)
Constituent values Ingratitude (#VD3012).
Gratitude (#VC0865)
Blessedness (#VC0247)
Sensibleness (#VC1593)
Acknowledgement (#VC0033).

♦ **VP5957a Rightness–Wrongness**
Dynamics It is not man, but the world that has become abnormal. (Graffito written during French student revolt)
The State not seldom tolerates a comparatively great evil to keep out millions of lesser ills and inconveniences which otherwise would be inevitable and without remedy. (La Bruyère)
From a worldly point of view there is no mistake so great as that of being always right. (Samuel Butler)
Not conforming to standard. In matters of thought and conduct, to be independent is to be abnormal, to be abnormal is to be detested. (Ambrose Bierce)
To do a great right, do a little wrong. (Shakespeare)
The right is more precious than peace. (Woodrow Wilson)
Integrative complex Appropriateness (#VT8091).
Constituent values Good (#VC0845)
Favour (#VC0743)
Freedom (#VC0797)
Decorum (#VC0523)
Morality (#VC1163)
Immunity (#VC0949)
Rightness (#VC1507)
Privilege (#VC1359)
Conscience (#VC0439)
Righteousness (#VC1505)
Constituent values Evil (#VD2624)
Scandal (#VD3610)
Disgrace (#VD2500)
Wrongness (#VD4078)
Sacrilege (#VD3594)
Amorality (#VD2058)
Impropriety (#VD2870)
Abomination (#VD2016)
Unlawfulness (#VD3926)
Shamelessness (#VD3666)
Inappropriateness (#VD2900).
Virtue (#VC1887)
Liberty (#VC1073)
Fitness (#VC0767)
Decency (#VC0517)
Interest (#VC1011)
Goodness (#VC0847)
Propriety (#VC1379)
Normality (#VC1201)
Suitability (#VC1701)
Appropriateness (#VC0159).
Infamy (#VD2996)
Ignominy (#VD2816)
Atrocity (#VD2114)
Violation (#VD4020)
Indecorum (#VD2952)
Profanation (#VD3490)
Desecration (#VD2404)
Abnormality (#VD2014)
Unsuitability (#VD3954)
Incorrectness (#VD2940)

♦ **VP5960b Dueness–Undueness**
Dynamics Without duty, life is soft and boneless; it cannot hold itself together. (Joseph Joubert)
Duty then is the sublimest word in our language. Do your duty in all things. You cannot do more. You should never wish to do less. (Robert E Lee)
Liberty means responsibility. That is why most men dread it. (George Bernard Shaw)
Responsibility is to oneself and the highest form of it is irresponsibility to oneself which is to say the calm acceptance of whatever responsibility to others and things comes along. (John Cage)
Integrative complex Appropriateness (#VT8091).
Constituent values Duty (#VC0593)
Justice (#VC1041)
Obedience (#VC1207)
Observance (#VC1211)
Loyalty (#VC1095)
Devotion (#VC0551)
Deference (#VC0527)
Dedication (#VC0525)
Application (#VC0151)
Scrupulousness (#VC1535)
Appropriateness (#VC0159)
Constituent values Excess (#VD2630)
Trespass (#VD3850)
Presumption (#VD3478)
Impropriety (#VD2870)
Inconvenience (#VD2938)
Inappropriateness (#VD2900).
Amenability (#VC0119)
Responsibility (#VC1489)
Conscientiousness (#VC0441).
Trouble (#VD3858)
Usurpation (#VD3972)
Lawlessness (#VD3136)
Encroachment (#VD2604)
Licentiousness (#VD3150)

♦ **VP5964c Respect–Disrespect**
Dynamics Always we like those who admire us, but we do not always like those whom we admire. (La Rochefoucauld)
Irreverence is the champion of liberty and its only sure defence. (Mark Twain)
'Tis much to gain universal admiration; more, universal love. (Baltasar Gracian)
Wealth gives a man the respect due to someone thirty years his senior. (Chinese proverb)
Respect a man, he will do the more. (Proverb)
Respect is greater from a distance. (Proverb)
One word is too often profaned for me to profane it; one feeling too falsely disdained for thee to disdain it. (Shelley)
We confide in our strength, without boasting of it; we respect that of others, without fearing it. (Thomas Jefferson)
Integrative complex Appropriateness (#VT8091).
Constituent values Awe (#VC0213)
Value (#VC1855)
Homage (#VC0909)
Prestige (#VC1349)
Approval (#VC0161)
Reverence (#VC1499)
Adoration (#VC0059)
Admiration (#VC0057)
Approbation (#VC0157)
Venerableness (#VC1863)
Consideration (#VC0449).
Constituent values Satire (#VD3606)
Insult (#VD3064)
Sarcasm (#VD3604)
Neglect (#VD3310)
Travesty (#VD3840)
Ridicule (#VD3566)
Cynicism (#VD2312)
Withering (#VD4064)
Riskiness (#VD3572)
Impudence (#VD2876)
Dishonour (#VD2508)
Arrogance (#VD2102)
Affrontery (#VD2038)
Humiliation (#VD2804)
Abusiveness (#VD2026)
Disparagement (#VD2532)
Contumeliousness (#VD2292)
Duty (#VC0593)
Honour (#VC0915)
Favour (#VC0743)
Courtesy (#VC0497)
Solemnity (#VC1627)
Attention (#VC0197)
Veneration (#VC1865)
Deification (#VC0529)
Appreciation (#VC0153)
Exclusiveness (#VC0701)
Levity (#VD3146)
Injury (#VD3024)
Outrage (#VD3374)
Disdain (#VD2490)
Snobbery (#VD3686)
Offence (#VD3358)
Atrocity (#VD2114)
Servility (#VD3656)
Insolence (#VD3050)
Disregard (#VD2542)
Disesteem (#VD2494)
Disrespect (#VD2544)
Irreverence (#VD3116)
Discourtesy (#VD2484)
Exclusiveness (#VD2636)
Superciliousness (#VD3764)
Contemptuousness (#VD2280).

♦ **VP5968a Approval–Disapproval**
Dynamics Praises from an enemy imply real merit. (Thomas Fuller)
Do not expect to be acknowledged for what you are, much less for what you would be; since no one can well measure a great man but upon the bier. (Walter Savage)
Our entire life, with our fine moral code and our precious freedom, consists ultimately in accepting ourselves as we are. (Jean Anouilh)
We are all motivated by a keen desire for praise, and the better a man is, the more he is inspired by glory. The very philosophers themselves, even in those books which they write on contempt of glory, inscribe their names. (Cicero)
There is no reward so delightful, no pleasure so exquisite, as having one's work known and acclaimed by those whose applause confers honour. (Molière)
The greatest humiliation in life, is to work hard on something from which you expect great appreciation, and then fail to get it. (Edgar Watson Howe)
It is difficult to like those whom we do not esteem; but it is no less so to like those whom we esteem more than ourselves. (La Rochefoucauld)
The most effective way of attacking vice is to expose it to public ridicule. People can put up with rebukes but they cannot bear being laughed at: they are prepared to be wicked but they dislike appearing ridiculous. (Molière)
Lean too much upon the approval of people, and it becomes a bed of thorns. (Tehyi Hsieh)
Integrative complex Judgement (#VT8092).
Constituent values Merit (#VC1139)
Praise (#VC1337)
Homage (#VC0909)
Support (#VC1711)
Approval (#VC0161)
Character (#VC0311)
Worthiness (#VC1941)
Exaltation (#VC0695)
Acceptance (#VC0017)
Deification (#VC0529)
Blessedness (#VC0247)
Appreciation (#VC0153)
Respectfulness (#VC1485)
Recommendation (#VC1431)
Constituent values Fuss (#VD2734)
Blame (#VD2180)
Satire (#VD3606)
Scandal (#VD3610)
Vileness (#VD4014)
Reproach (#VD3534)
Distaste (#VD2560)
Anathema (#VD2064)
Exclusion (#VD2634)
Disesteem (#VD2494)
Dirtiness (#VD2444)
Revilement (#VD3560)
Opposition (#VD3364)
Disrespect (#VD2544)
Defamation (#VD2342)
Insinuation (#VD3046)
Displeasure (#VD2536)
Denigration (#VD2378)
Abusiveness (#VD2026)
Unctuousness (#VD3892)
Depreciation (#VD2394)
Glory (#VC0841)
Honour (#VC0915)
Favour (#VC0743)
Acclaim (#VC0021)
Advocacy (#VC0069)
Adherence (#VC0053)
Popularity (#VC1319)
Admiration (#VC0057)
Recognition (#VC1427)
Celebration (#VC0299)
Approbation (#VC0157)
Well-disposed (#VC1905)
Commendability (#VC0359)
Acknowledgement (#VC0033).
Filth (#VD2700)
Threat (#VD3812)
Menace (#VD3210)
Censure (#VD2218)
Ridicule (#VD3566)
Foulness (#VD2710)
Disgrace (#VD2500)
Rejection (#VD3526)
Disfavour (#VD2496)
Discredit (#VD2486)
Damnation (#VD2316)
Opprobrium (#VD3368)
Execration (#VD2638)
Derogation (#VD2402)
Unhappiness (#VD3916)
Indignation (#VD2960)
Disapproval (#VD2462)
Culpability (#VD2306)
Vituperation (#VD4028)
Intimidation (#VD3072)
Denunciation (#VD2380)

VP5968

Condemnation (#VD2264)
Disenchantment (#VD2492)
Disappointment (#VD2460)
Dissatisfaction (#VD2546)
Contumeliousness (#VD2292)
Disparagement (#VD2532)
Discontentment (#VD2476)
Censoriousness (#VD2216)
Disillusionment (#VD2510)
Contemptuousness (#VD2280).

♦ VP5974b Probity–Improbity
Dynamics Morality regulates the acts of man as a private individual; honour, his acts as a public man. (Esteban Echeverria)
Nothing is at last sacred but the integrity of our own mind. Absolve you to yourself, and you shall have the suffrage of the world. (Emerson)
Happiness is not the end of life: character is. (Henry Ward Beecher)
Integrity without knowledge is weak and useless, and knowledge without integrity is dangerous and dreadful. (Samuel Johnson)
I could not love thee, dear, so much, loved I not honour more. (Richard Lovelace)
The truth is, hardly any of us have ethical energy enough for more than one really inflexible point of honour. (George Bernard Shaw)
Even honour and virtue make enemies, condemning, as they do, their opposites by too close a contrast. (Tacitus)
The corruption of every government begins nearly always with that of principles. (Montesquieu)
The value of culture is its effect on character. It avails nothing unless it ennobles and strengthens that. Its use is for life. Its aim is not beauty but goodness. (W Somerset Maugham)
A world of vested interests is not a world which welcomes the disruptive force of candour. (Agnes Repplier)
Honesty is a good thing but it is not profitable to its possessor unless it is kept under control. If you are not honest at all everybody hates you, and if you are absolutely honest, you get martyred. (Don Marquis)
All faults may be forgiven of him who has perfect candour. (Walt Whitman)
Integrative complex Judgement (#VT8092).
Constituent values Good (#VC0845)
Faith (#VC0729)
Purity (#VC1391)
Candor (#VC0283)
Loyalty (#VC1095)
Honesty (#VC0913)
Decency (#VC0517)
Sureness (#VC1715)
Nobility (#VC1193)
Goodness (#VC0847)
Fairness (#VC0727)
Unspotted (#VC1841)
Sincerity (#VC1619)
Rectitude (#VC1437)
Manliness (#VC1115)
Frankness (#VC0793)
Character (#VC0311)
Worthiness (#VC1941)
Steadiness (#VC1669)
Attachment (#VC0193)
Uprightness (#VC1845)
Unblemished (#VC1805)
Reliability (#VC1459)
Genuineness (#VC0831)
Consistency (#VC0451)
Carefulness (#VC0293)
Truthfulness (#VC1789)
Faithfulness (#VC0731)
Steadfastness (#VC1667)
Righteousness (#VC1505)
Blamelessness (#VC0243)
Responsibility (#VC1489)
Forthrightness (#VC0789)
Trustworthiness (#VC1785)
Openheartedness (#VC1221)
Irreproachability (#VC1029)
Straightforwardness (#VC1679).
Truth (#VC1787)
Virtue (#VC1887)
Honour (#VC0915)
Probity (#VC1361)
Justice (#VC1041)
Freedom (#VC0797)
Veracity (#VC1869)
Openness (#VC1225)
Interest (#VC1011)
Fidelity (#VC0763)
Devotion (#VC0551)
Unfailing (#VC1819)
Rightness (#VC1507)
Plainness (#VC1301)
Integrity (#VC0997)
Constancy (#VC0455)
Adherence (#VC0053)
Unassuming (#VC1801)
Directness (#VC0563)
Allegiance (#VC0103)
Uncorrupted (#VC1809)
Staunchness (#VC1665)
Punctuality (#VC1389)
Credibility (#VC0503)
Cleanliness (#VC0337)
Artlessness (#VC0173)
Spotlessness (#VC1649)
Veritableness (#VC1871)
Stainlessness (#VC1653)
Ingenuousness (#VC0983)
Scrupulousness (#VC1535)
Meticulousness (#VC1143)
Unequivocalness (#VC1815)
Respectability (#VC1483)
High-mindedness (#VC0899)
Conscientiousness (#VC0441).
Constituent values Deceit (#VD2332)
Roguery (#VD3576)
Villainy (#VD4016)
Venality (#VD3998)
Intrigue (#VD3084)
Baseness (#VD2150)
Treachery (#VD3842)
Falseness (#VD2672)
Dishonour (#VD2508)
Infidelity (#VD3006)
Disloyalty (#VD2520)
Corruption (#VD2294)
Indirection (#VD2962)
Fraudulence (#VD2714)
Deviousness (#VD2432)
Irritability (#VD3118)
Disaffection (#VD2452)
Shamelessness (#VD3666)
Faithlessness (#VD2662)
Traitorousness (#VD3834)
Undependability (#VD3894)
Irresponsibility (#VD3114)
Untrustworthiness (#VD3958)
Treason (#VD3844)
Evasion (#VD2622)
Vileness (#VD4014)
Sedition (#VD3620)
Betrayal (#VD2170)
Turpitude (#VD3868)
Improbity (#VD2868)
Duplicity (#VD2590)
Notoriety (#VD3338)
Immorality (#VD2836)
Dishonesty (#VD2506)
Insincerity (#VD3044)
Inconstancy (#VD2934)
Dubiousness (#VD2586)
Dereliction (#VD2400)
Doubtfulness (#VD2584)
Unreliability (#VD3948)
Feloniousness (#VD2692)
Unfaithfulness (#VD3906)
Perfidiousness (#VD3436)
Self-absorption (#VD3626)
Disingenuousness (#VD2512)
Unconscionableness (#VD3890).

♦ VP5976c Justice–Injustice
Dynamics Justice Custodian of the world But since the world errs, justice must be custodian of the world's errors. (Ugo Betti)
Rigid justice is the greatest injustice. (Thomas Fuller)
Even the laws of justice themselves cannot subsist without mixture of injustice. (Montaigne)
The firm basis of government is justice, not pity. (Woodrow Wilson)
A man who deals in fairness with his own, he can make manifest justice in the state. (Sophocles)
Absolute freedom mocks at justice. Absolute justice denies freedom. (Albert Camus)
Justice is reason enough for anything ugly. It balances the beauty in the world. (Diane Wakoski)
To have a grievance is to have a purpose in life. (Eric Hoffer)
Propriety is the least of all laws, and the most observed. (La Rochefoucauld)
Integrative complex Judgement (#VT8092).
Constituent values Good (#VC0845)
Equity (#VC0673)
Candor (#VC0283)
Balance (#VC0215)
Interest (#VC1011)
Rightness (#VC1507)
Neutrality (#VC1187)
Dispassion (#VC0575)
Uninfluence (#VC1823)
Selflessness (#VC1587)
Unselfishness (#VC1839)
Constituent values Bias (#VD2174)
Outrage (#VD3374)
Inequity (#VD2988)
Wrongness (#VD4078)
Grievance (#VD2750)
Inequality (#VD2986)
Impropriety (#VD2870)
Discrimination (#VD2488)
Unconscionableness (#VD3890).
Justice (#VC1041)
Legality (#VC1061)
Fairness (#VC0727)
Propriety (#VC1379)
Lawfulness (#VC1055)
Detachment (#VC0545)
Involvement (#VC1023)
Impartiality (#VC0953)
Sportsmanship (#VC1647).
Injury (#VD3024)
Iniquity (#VD3022)
Foulness (#VD2710)
Injustice (#VD3026)
Warpedness (#VD4044)
Miscarriage (#VD3240)
Unlawfulness (#VD3926)
Indefensibility (#VD2954).

♦ VP5978a Unselfishness–Selfishness
Dynamics Though peace be made, yet it is interest that keeps peace. (Oliver Cromwell)
The very act of sacrifice magnifies the one who sacrifices himself to the point where his sacrifice is much more costly to humanity than would have been the loss of those for whom he is sacrificing himself. But in his abnegation lies the secret of his grandeur. (André Gide)
Sacrifice may be a flower that virtue will pluck on its road, but it was not to gather this flower that virtue set forth on its travels. (Maurice Maeterlinck)
What appears to be generosity is often only ambition disguised, which despises small interests to pursue great ones. (La Rochefoucauld)
Integrative complex Morality (#VT8093).
Constituent values Modesty (#VC1159)
Bigness (#VC0241)
Interest (#VC1011)
Humility (#VC0929)
Chivalry (#VC0323)
Greatness (#VC0867)
Liberality (#VC1069)
Exaltation (#VC0635)
Commitment (#VC0361)
Magnanimity (#VC1105)
Selflessness (#VC1587)
Unselfishness (#VC1839)
Self-sacrifice (#VC1583)
High-mindedness (#VC0899).
Heroism (#VC0897)
Nobility (#VC1193)
Idealism (#VC0935)
Devotion (#VC0551)
Altruism (#VC0113)
Elevation (#VC0625)
Generosity (#VC0821)
Dedication (#VC0525)
Self-denial (#VC1551)
Unpretention (#VC1837)
Handsomeness (#VC0879)
Self-interest (#VC1569)
Self-effacement (#VC1559)
Constituent values Greed (#VD2746)
Meanness (#VD3204)
Pettiness (#VD3452)
Selfishness (#VD3646)
Illiberalism (#VD2824)
Self-admiration (#VD3628)
Self-centredness (#VD3630).
Autism (#VD2120)
Smallness (#VD3684)
Littleness (#VD3158)
Self-esteem (#VD3636)
Self-indulgence (#VD3640)
Acquisitiveness (#VD2030)

♦ VP5980a Virtue–Vice
Dynamics The highest virtue is always against the law. (Emerson)
It is the function of vice to keep virtue within reasonable bounds. (Samuel Butler)
One unable to dance blames the unevenness of the floor. (Malay Proverb)
When evil acts in the world it always manages to find instruments who believe that what they do is not evil but honourable. (Max Lerner)
The great epochs of our life come when we gain the courage to rechristen our evil as what is best in us. (Nietzsche)
Our greatest evils flow from ourselves. (Rousseau)
If a man has no vices, he's in great danger of making vices about his virtues, and there's a spectacle. (Thornton Wilde)
It is convention and arbitrary rewards which make all the merit and demerit of what we call vice and virtue. (Julien Offroy de la Mettrie)
Some rise by sin, and some by virtue fall. (Shakespeare)
The only absolute morality is absolute stagnation. (Samuel Butler)
True eloquence makes light of eloquence, true morality makes light of morality. (Pascal)
The world is shocked, or amused, by the sight of saintly old people hindering in the name of morality the removal of obvious brutalities from a legal system. (Alfred North Whitehead)
Sin is a dangerous toy in the hands of the virtuous. It should be left to the congenitally sinful, who know when to play with it and when to let it alone. (H L Mencken)
Without civic morality communities perish; without personal morality their survival has no value. (Bertrand Russell)
We have, in fact, two kinds of morality side by side; one which we preach but do not practise, and another which we practise but seldom preach. (Bertrand Russell)
The best moral virtues are those of which the vulgar are perhaps the best judges. (Lord Chesterfield)
Moral men have many children. (Japanese proverb)
Integrative complex Morality (#VT8093).
Constituent values Love (#VC1093)
Good (#VC0845)
Virtue (#VC1887)
Probity (#VC1361)
Honesty (#VC0913)
Unerring (#VC1817)
Morality (#VC1163)
Chastity (#VC0319)
Innocence (#VC0987)
Temperance (#VC1747)
Uncorrupted (#VC1809)
Righteousness (#VC1505)
Constituent values Lust (#VD3176)
Envy (#VD2614)
Greed (#VD2746)
Anger (#VD2066)
Infamy (#VD2996)
Scandal (#VD3610)
Failure (#VD2660)
Avarice (#VD2122)
Villainy (#VD4016)
Trespass (#VD3850)
Offence (#VD3358)
Impurity (#VD2880)
Foulness (#VD2710)
Disgrace (#VD2500)
Baseness (#VD2150)
Hope (#VC0917)
Faith (#VC0729)
Purity (#VC1391)
Justice (#VC1041)
Charity (#VC0315)
Prudence (#VC1387)
Goodness (#VC0847)
Rectitude (#VC1437)
Godliness (#VC0843)
Uprightness (#VC1845)
Saintliness (#VC1521)
Guiltlessness (#VC0873).
Evil (#VD2624)
Lapse (#VD3130)
Blame (#VD2180)
Injury (#VD3024)
Arrant (#VD2098)
Outrage (#VD3374)
Badness (#VD2134)
Weakness (#VD4052)
Vileness (#VD4014)
Omission (#VD3360)
Iniquity (#VD3022)
Hardness (#VD2770)
Fiendish (#VD2696)
Diabolic (#VD2436)
Atrocity (#VD2114)

VALUE POLARITIES VP6024

Wrongness (#VD4078)
Turpitude (#VD3868)
Dirtiness (#VD2444)
Decadence (#VD2330)
Abjection (#VD2012)
Wantonness (#VD4038)
Profligacy (#VD3492)
Immorality (#VD2836)
Corruption (#VD2294)
Ungodliness (#VD3912)
Nonfeasance (#VD3326)
Misdemeanor (#VD3250)
Malfeasance (#VD3188)
Dereliction (#VD2400)
Degradation (#VD2362)
Abomination (#VD2016)
Slothfulness (#VD3680)
Imperfection (#VD2842)
Degeneration (#VD2360)
Transgression (#VD3836)
Heartlessness (#VD2784)
Irreformability (#VD3104)
Incorrigibility (#VD2942).
Vitiation (#VD4026)
Injustice (#VD3026)
Depravity (#VD2392)
Amorality (#VD2058)
Warpedness (#VD4044)
Unchastity (#VD3882)
Misconduct (#VD3246)
Debauchery (#VD2326)
Viciousness (#VD4008)
Prodigality (#VD3488)
Misfeasance (#VD3256)
Malpractice (#VD3198)
Impropriety (#VD2870)
Depravation (#VD2390)
Callousness (#VD2210)
Abandonment (#VD2004)
Indiscretion (#VD2964)
Devilishness (#VD2430)
Uncleanliness (#VD3884)
Shamelessness (#VD3666)
Feloniousness (#VD2692)
Irredeemability (#VD3102)

♦ VP5983b Innocence–Guilt
Dynamics There may be guilt when there is too much virtue. (Racine)
We have no choice but to be guilty. God is unthinkable if we are innocent. (Archibald MacLeish)
Without guilt what is a man? An animal, isn't he? A wolf forgiven at his meat, a beetle innocent in his copulation. (Archibald MacLeish)
When we quarrel, how we wish we had been blameless. (Emerson)
Bad company is as instructive as debauchery: one is indemnified for the loss of innocence by the loss of prejudice. (Denis Diderot)
Through our own recovered innocence we discern the innocence of our neighbours. (Thoreau)
After the great destruction, everyone will prove that he was innocent. (Gunter Eich)
Integrative complex Morality (#VT8093).
Constituent values Purity (#VC1391)
Innocence (#VC0987)
Uncorrupted (#VC1809)
Involvement (#VC1023)
Spotlessness (#VC1649)
Guiltlessness (#VC0873)
Blamelessness (#VC0243)
Constituent values Guilt (#VD2760)
Complicity (#VD2256)
Unspotted (#VC1841)
Clearness (#VC0339)
Unblemished (#VC1805)
Cleanliness (#VC0337)
Stainlessness (#VC1653)
Faultlessness (#VC0741)
Irreproachability (#VC1029).
Faultiness (#VD2682)
Culpability (#VD2306).

♦ VP5987c Chastity–Unchastity
Dynamics Intemperance is the plague of sensuality, and temperance is not its bane but its seasoning. (Montaigne)
The essence of chastity is not the suppression of lust, but the total orientation of one's life towards a goal. (Dietrich Bonhoeffer)
Chastity more rarely follows fear, or a resolution, or a vow, than it is the mere effect of lack of appetite and, sometimes even, of distaste. (André Gide)
People who lack a coarse streak, I have discovered, almost always possess a cruel one. (Louis Kronenberger)
Of all sexual aberrations, chastity is the strangest. (Anatole France)
Give me chastity and continence, but not just now. (St Augustine)
The only chaste woman is the one who has not been asked. (Spanish proverb)
Integrative complex Morality (#VT8093).
Constituent values Virtue (#VC1887)
Honour (#VC0915)
Decorum (#VC0523)
Elegance (#VC0623)
Chastity (#VC0319)
Appetite (#VC0149)
Unspotted (#VC1841)
Innocence (#VC0987)
Asceticism (#VC0177)
Unblemished (#VC1805)
Cleanliness (#VC0337)
Rigorousness (#VC1509)
Voluptuousness (#VC1895).
Constituent values Lust (#VD3176)
Laxity (#VD3138)
Impurity (#VD2880)
Vulgarity (#VD4034)
Seduction (#VD3622)
Indecorum (#VD2952)
Immodesty (#VD2834)
Dirtiness (#VD2444)
Bawdiness (#VD2156)
Animality (#VD2070)
Unchastity (#VD3882)
Ravishment (#VD3518)
Inelegance (#VD2982)
Illegality (#VD2820)
Coarseness (#VD2240)
Promiscuity (#VD3500)
Fulsomeness (#VD2728)
Abandonment (#VD2004)
Intemperance (#VD3066)
Incontinence (#VD2936)
Shamelessness (#VD3666)
Concupiscence (#VD2262)
Inappropriateness (#VD2900).
Purity (#VC1391)
Modesty (#VC1159)
Decency (#VC0517)
Delicacy (#VC0535)
Boldness (#VC0251)
Virginity (#VC1883)
Propriety (#VC1379)
Continence (#VC0469)
Abstinence (#VC0011)
Self–denial (#VC1551)
Spotlessness (#VC1649)
Stainlessness (#VC1653)
Filth (#VD2700)
Vileness (#VD4014)
Foulness (#VD2710)
Sexuality (#VD3662)
Looseness (#VD3164)
Indecency (#VD2948)
Grossness (#VD2754)
Brutality (#VD2202)
Austerity (#VD2116)
Wantonness (#VD4038)
Sensuality (#VD3650)
Notoriety (#VD3338)
Indelicacy (#VD2956)
Debauchery (#VD2326)
Bestiality (#VD2168)
Impropriety (#VD2870)
Beastliness (#VD2158)
Prostitution (#VD3502)
Indiscretion (#VD2964)
Uncleanliness (#VD3884)
Mortification (#VD3288)
Licentiousness (#VD3150)

♦ VP5992c Temperance–Intemperance
Dynamics Half the vices which the world condemns most loudly have seeds of good in them and require moderate use rather than total abstinence. (Samuel Butler)
Excess on occasion is exhilarating. It prevents moderation from acquiring the deadening effect of a habit. (W Somerset)
Temperance is the acknowledged ruler of the pleasures and desires, and no pleasure ever masters Love; he is their master and they are his servants; and if he conquers them he must be temperate indeed. (Plato)
Intemperance is the only vulgarity. (Emerson)
Integrative complex Morality (#VT8093).
Constituent values Sobriety (#VC1623)
Restraint (#VC1495)
Temperance (#VC1747)
Continence (#VC0469)
Abnegation (#VC0005)
Forbearance (#VC0779)
Self–restraint (#VC1581).
Constituent values Greed (#VD2746)
Voracity (#VD4032)
Avoidance (#VD2126)
Indulgence (#VD2974)
Prodigality (#VD3488)
Intoxication (#VD3076)
Incontinence (#VD2936)
Extravagance (#VD2654)
Licentiousness (#VD3150)
Chastity (#VC0319)
Frugality (#VC0807)
Moderation (#VC1157)
Abstinence (#VC0011)
Self–denial (#VC1551)
Renunciation (#VC1467)
Excess (#VD2630)
Rapacity (#VD3516)
Insobriety (#VD3048)
Debauchery (#VD2326)
Dissipation (#VD2552)
Intemperance (#VD3066)
Helplessness (#VD2788)
Excessiveness (#VD2632)
Self–indulgence (#VD3640).

♦ VP5998a Legality–Illegality
Dynamics For lawless joys a bitter ending waits. (Pindar)
The law is the true embodiment of everything that's excellent. It has no kind of fault or flaw, and I, my Lords, embody the Law. (W S Gilbert)
Integrative complex Retribution (#VT8094).
Constituent values Equity (#VC0673)
Legality (#VC1061)
Constituent values Flaw (#VD2704)
Trespass (#VD3850)
Offence (#VD3358)
Wrongness (#VD4078)
Anarchism (#VD2060)
Misdemeanor (#VD3250)
Unlawfulness (#VD3926)
Transgression (#VD3836)
Validity (#VC1853)
Regulation (#VC1447).
Anarchy (#VD2062)
Spurious (#VD3718)
Bastardy (#VD2152)
Violation (#VD4020)
Illegality (#VD2820)
Lawlessness (#VD3136)
Illegitimate (#VD2822)
Feloniousness (#VD2692).

♦ VP6005b Vindication–Condemnation
Dynamics A man is sometimes extolled to the skies for the very thing which occasioned his misfortune. (Anonymous)
Do you want to injure someone's reputation? Don't speak ill of him, speak too well. (André Siefried)
To accuse requires less eloquence (such is man's nature) than to excuse. (Hobbes)
Integrative complex Retribution (#VT8094).
Constituent values Grace (#VC0855)
Pardon (#VC1247)
Advocacy (#VC0069)
Judgement (#VC1037)
Involvement (#VC1023)
Championship (#VC0305)
Reasonableness (#VC1425)
Constituent values Blame (#VD2180)
Censure (#VD2218)
Damnation (#VD2316)
Denunciation (#VD2380)
Recrimination (#VD3522)
Reason (#VC1423)
Support (#VC1711)
Remission (#VC1463)
Legitimacy (#VC1063)
Explanation (#VC0719)
Rehabilitation (#VC1449)
Inoffensiveness (#VC0988).
Venial (#VD4000)
Reproach (#VD3534)
Insinuation (#VD3046)
Condemnation (#VD2264)
Censoriousness (#VD2216).

♦ VP6010c Atonement–Punishment
Dynamics He that spareth the rod hateth his son. (Bible)
Beat your child once a day. If you don't know why, he does. (Chinese proverb)
He only may chastise who loves. (Rabindranath Tagore)
Happiness comes to the good man as a reward, to the bad man as a punishment. (Chinese proverb)
You should punish your appetites rather than allow yourself to be punished by them. (Epictetus)
Remorse is impotence; it will sin again. Only repentance is strong; it can end everything. (Henry Miller)
Integrative complex Retribution (#VT8094).
Constituent values Repentance (#VC1469)
Asceticism (#VC0177).
Constituent values Retribution (#VD3552)
Redemption (#VC1439)
Confinement (#VD2268).

♦ VP6013a Godliness–Ungodliness
Dynamics The indwelling ideal lends all the gods their divinity. (George Santayana)
Bodily exercise profiteth little; but godliness is profitable unto all things. (Bible)
Cleanliness is, indeed, next to godliness. (Wesley)
Integrative complex Redemption (#VT8095).
Constituent values Good (#VC0845)
Glory (#VC0841)
Ubiquity (#VC1791)
Holiness (#VC0905)
Unlimited (#VC1833)
Sacredness (#VC1515)
Omniscience (#VC1219)
Timelessness (#VC1759)
Preservation (#VC1347)
Immutability (#VC0951)
Transcendence (#VC1775)
Superhumanness (#VC1707)
Constituent values Fiendish (#VD2696)
Bedevilment (#VD2160)
Unity (#VC1829)
Majesty (#VC1109)
Radiance (#VC1409)
Creation (#VC0499)
Godliness (#VC0843)
Permanence (#VC1275)
Omnipotence (#VC1215)
Spirituality (#VC1641)
Omnipresence (#VC1217)
Almightiness (#VC0109)
Boundlessness (#VC0253)
Limitlessness (#VC1081).
Perdition (#VD3434)
Mischievousness (#VD3224).

♦ VP6024b Orthodoxy–Unorthodoxy
Dynamics The most universal quality is diversity. (Montaigne)
No bird soars too high, if he soars with his own wings. (Blake)
Tradition by itself is not enough; it must be perpetually criticized and brought up–to–date under the supervision of what I call orthodoxy. (T S Eliot)
That community is already in the process of dissolution where each man begins to eye his neighbour as a possible enemy; where non-conformity with the accepted creed, political as well as religious, is a mark of disaffection; where denunciation, without specification or backing, takes the place of evidence; where orthodoxy chokes freedom of dissent; where faith in the eventual supremacy of reason has become so timid that we dare not enter our convictions in the open lists, to win or lose. (Learned Hand)
Most people are other people. Their thoughts are someone else's opinions, their lives a mimicry, their passions a quotation. (Oscar Wilde)
Orthodoxy: That peculiar condition where the patient can neither eliminate an old idea nor absorb a new one. (Elbert Hubbard)
Integrative complex Redemption (#VT8095).
Constituent values Orthodoxy (#VC1241).
Constituent values Misbelief (#VD3236).

VP6026

♦ VP6026c Sanctity–Unsanctity
Dynamics In our era, the road to holiness necessarily passes through the world of action. (Dag Hammarskjold)
Many of the insights of the saint stem from his experience as a sinner. (Eric Hoffer)
No doubt alcohol, tobacco, and so forth, are things that a saint must avoid, but sainthood is also a thing that human beings must avoid. (George Orwell)
Integrative complex Redemption (#VT8095).
Constituent values Awe (#VC0213)
Sanctity (#VC1529)
Salvation (#VC1525)
Beatitude (#VC0219)
Redemption (#VC1439)
Blessedness (#VC0247)
Beatification (#VC0217)
Grace (#VC0855)
Holiness (#VC0905)
Godliness (#VC0843)
Sacredness (#VC1515)
Exaltation (#VC0695)
Venerableness (#VC1863)
Sanctification (#VC1527).

♦ VP6028c Piety–Impiety
Dynamics A state of temperance, sobriety and justice without devotion is a cold, lifeless, insipid condition of virtue, and is rather to be styled philosophy than religion. (Joseph Adison)
Philanthropic and religious bodies do not commonly make their executive officers out of saints. (Emerson)
Beware of the community in which blasphemy does not exist; underneath, atheism runs rampant. (Antonio Machado)
All great truths begin as blasphemies. (George Bernard Shaw)
Saintliness is also a temptation. (Jean Anouilh)
The saints indulge in subtleties in order to think themselves criminals, and to impeach their better actions. (Pascal)
Integrative complex Redemption (#VT8095).
Constituent values Zeal (#VC1947)
Duty (#VC0593)
Faith (#VC0729)
Belief (#VC0227)
Humanism (#VC0923)
Goodness (#VC0847)
Reverence (#VC1499)
Godliness (#VC0843)
Veneration (#VC1865)
Conformity (#VC0425)
Spirituality (#VC1641)
Righteousness (#VC1505).
Good (#VC0845)
Piety (#VC1295)
Purity (#VC1391)
Sanctity (#VC1529)
Holiness (#VC0905)
Devotion (#VC0551)
Piousness (#VC1297)
Adoration (#VC0059)
Observance (#VC1211)
Saintliness (#VC1521)
Faithfulness (#VC0731)
Constituent values Lapse (#VD3130)
Atheism (#VD2112)
Sacrilege (#VD3594)
Desertion (#VD2406)
Profanation (#VD3490)
Godlessness (#VD2742)
Desecration (#VD2404)
Irreligiousness (#VD3108)
Impiety (#VD2850)
Paganism (#VD3420)
Dishonour (#VD2508)
Blasphemy (#VD2181)
Irreverence (#VD3116)
Dubiousness (#VD2586)
Unctuousness (#VD3892)
Self-righteousness (#VD3642).

Value types

Rationale

The range of value dimensions tentatively identified in Section VP suggests the need to group the information into an even smaller number of categories to obtain a clearer overview of the clusters it implies. This is a further step in exploring the ordering of value information so that some sense of the ecology of values emerges. In taking such a step in this context there is a clear advantage in also exploring the relationships to the strategies which supposedly underlie such values.

Content

This section is composed of 45 entries. The entries do not have descriptions but they do contain cross-references to the value polarities in Section VP. The entries may therefore be considered as clusters of value dimensions (as implied by appending "*complex" to each name). These compensate for any overlap between the value polarity entries. The names given to each entry are simply approximate labels pointing to the nature of the cluster. They are not intended to be value-type names.

Method

The 45 entries were established as an extension of the approach whereby the Section VP strategic polarity entries were defined. This is discussed in detail in Section VZ at the end of the volume.

Index

A keyword index to entries is provided in Section VX. The keywords are also incorporated into the index for Volume 2 (Section X)

Bibliography

Bibliographical references, by author, are given in Section VY.

Comment

Detailed comments are given in Section VZ.

Reservations

Although these entries successfully define a comprehensive set of value domains, further work is required to establish whether this ordering device has any wider significance, especially through the deliberate parallelism with the set of strategy domains in Section VT.

Possible future improvements

As with Section VP, many opportunities exist for developing this approach and extending the pattern of cross-references to clarify the nature of the ecology of values that are operating in society.

ENTRY CONTENT AND ORGANIZATION

Ordering of entries
Entries are in **numeric order**. Entry numbers are based on an adaptation of the numeric ordering of concepts in Roget's Thesaurus as used for Section VP (and Section SP). They serve primarily as a permanent point of reference to facilitate indexing, cross-referencing, and updating between editions.

Index access to entries
The location of an entry in this sub-section may be determined from:
- **Volume Index** (Section X) on the basis of keywords in the name of the entry or its alternate names

Structure of entries
Entries may be composed of the following descriptive elements:

(a) **Entry number** This number has **no significance**, although it does follow the sequence of entries in Sections VP (and Section SP). It is simply a convenient method of identifying the entry (particularly for indexing purposes), of filing information on it, and as an identifier to which cross-references from other entries (possibly in other sections) may refer in this and future editions. The first letter of the entry number refers to the section of this volume in which the sub-section, denoted by the second letter, is located.

(b) **Qualifier** following the entry number. There are **no qualifiers** in this sub-sections.

(c) **Entry name** This is printed in bold characters.

Cross-referencing of entries
At the end of any entry, there may be cross-references to other entries. These indicate the number and name of the cross-referenced entry, whether within this Section or in other Sections. In this sub-section there is a single type of cross-reference, indicated by a 2-letter code in bold characters:
    Narrower = Narrower or more specific value, namely the value polarities grouped by the entry
    This sub-section functions as a classified index to the value polarities in Section VP.

VALUE TYPES

VT8011 Existence∗complex
Constituent polarities Existence–Nonexistence (#VP5001)
Intrinsicality–Extrinsicality (#VP5005)
Substantiality–Unsubstantiality (#VP5003).

VT8012 Relationship∗complex
Constituent polarities Equality–Inequality (#VP5030)
Identity–Difference (#VP5013)
Originality–Imitation (#VP5023)
Agreement–Disagreement (#VP5026)
Similarity–Dissimilarity (#VP5020)
Uniformity–Nonuniformity (#VP5017)
Relatedness–Unrelatedness (#VP5009).

VT8013 Quantity∗complex
Constituent polarities Increase–Decrease (#VP5038)
Greatness–Smallness (#VP5034)
Simplicity–Complexity (#VP5045)
Conjunction–Separation (#VP5047)
Cohesion–Disintegration (#VP5050)
Superiority–Inferiority (#VP5036)
Completeness–Incompleteness (#VP5054).

VT8014 Order∗complex
Constituent polarities Order–Disorder (#VP5059)
Inclusion–Exclusion (#VP5076)
Conformity–Nonconformity (#VP5082)
Continuity–Discontinuity (#VP5071).

VT8015 Number∗complex
Constituent polarities Unity–Duality (#VP5089)   Numbered–Unnumbered (#VP5086)
Numerousness–Fewness (#VP5101).

VT8021 Time∗complex
Constituent polarities Youth–Age (#VP5124)   Newness–Oldness (#VP5122)
Futurity–Antiquity (#VP5119)   Frequency–Infrequency (#VP5135)
Durability–Transience (#VP5110)   Regularity–Irregularity (#VP5137)
Timeliness–Untimeliness (#VP5129)   Perpetuity–Instantaneousness (#VP5112).

VT8022 Change∗complex
Constituent polarities Change–Permanence (#VP5139)
Evolution–Revolution (#VP5147)
Conversion–Reversion (#VP5145)
Continuance–Cessation (#VP5143)
Stability–Changeableness (#VP5141).

VT8023 Causation∗complex
Constituent polarities Causation–Culmination (#VP5153)
Eventuation–Imminence (#VP5151)
Attributability–Chance (#VP5155).

VT8024 Power∗complex
Constituent polarities Power–Impotence (#VP5157)
Energy–Moderation (#VP5161)
Strength–Weakness (#VP5159)
Ancestry–Posterity (#VP5170)
Production–Reproduction (#VP5167)
Concurrence–Counteraction (#VP5177)
Influence–Influencelessness (#VP5172)
Productiveness–Unproductiveness (#VP5165).

VT8025 Space∗complex
Constituent polarities Presence–Absence (#VP5186)
Container–Content (#VP5193)
Location–Dislocation (#VP5185).

VT8031 Dimension∗complex
Constituent polarities Height–Lowness (#VP5207)
Depth–Shallowness (#VP5209)
Nearness–Distance (#VP5200)
Breadth–Narrowness (#VP5204)
Bigness–Littleness (#VP5195)
Expansion–Contraction (#VP5197).

VT8032 Contextuality∗complex
Constituent polarities Centrality–Environment (#VP5226)
Circumscription–Intrusion (#VP5234).

VT8033 Structure∗complex
Constituent polarities Opening–Closure (#VP5265)
Form–Formlessness (#VP5246)
Symmetry–Distortion (#VP5248).

VT8034 Motion∗complex
Constituent polarities Impact–Reaction (#VP5283)
Motion–Quiescence (#VP5267)
Swiftness–Slowness (#VP5269)
Attraction–Repulsion (#VP5288).

VT8035 Relative motion∗complex
Constituent polarities Leading–Following (#VP5292)
Approach–Recession (#VP5296)
Direction–Deviation (#VP5290)
Elevation–Depression (#VP5317)
Oscillation–Agitation (#VP5323)
Convergence–Divergence (#VP5298)
Progression–Regression (#VP5294)
Overrunning–Shortcoming (#VP5313).

VT8041 Absolute properties∗complex
Constituent polarities Heat–Cold (#VP5328)   Light–Darkness (#VP5335)
Weight–Lightness (#VP5352)   Transparency–Opaqueness (#VP5339).

VT8042 Relative properties∗complex
Constituent polarities Hardness–Softness (#VP5356)
Elasticity–Toughness (#VP5358)
Colour–Colourlessness (#VP5362).

VT8043 Life∗complex
Constituent polarities Life–Death (#VP5407)   Sexiness–Unsexiness (#VP5419)
Humanity–Nonhumanity (#VP5417)   Masculinity–Femininity (#VP5420)
Materiality–Immateriality (#VP5376).

VT8044 Sense∗complex
Constituent polarities Harmony–Discord (#VP5462)
Silence–Loudness (#VP5451)
Vision–Blindness (#VP5439)
Fragrance–Stench (#VP5435)
Audibility–Inaudibility (#VP5448)
Visibility–Invisibility (#VP5444)
Sensation–Insensibility (#VP5422)
Appearance–Disappearance (#VP5446)
Savouriness–Unsavouriness (#VP5427)
Tangibility–Intangibility (#VP5425).

VT8045 Intellectual faculties∗complex
Constituent polarities Sanity–Insanity (#VP5472)
Intuition–Reason (#VP5481)
Knowledge–Ignorance (#VP5475)
Thought–Thoughtlessness (#VP5478)
Intelligence–Unintelligence (#VP5467).

VT8051 Evaluation∗complex
Constituent polarities Research–Discovery (#VP5485)
Judgement–Misjudgement (#VP5494)
Overestimation–Underestimation (#VP5497)
Discrimination–Indiscrimination (#VP5492).

VT8052 Credibility∗complex
Constituent polarities Truth–Error (#VP5516)   Belief–Unbelief (#VP5501)
Certainty–Uncertainty (#VP5513)   Limitation–Unlimitedness (#VP5507)
Possibility–Impossibility (#VP5509)   Provability–Unprovability (#VP5505).

VT8053 Truth∗complex
Constituent polarities Assent–Dissent (#VP5521)
Affirmation–Denial (#VP5523)
Illusion–Disillusionment (#VP5519).

VT8054 Attitude∗complex
Constituent polarities Carefulness–Neglect (#VP5533)
Attention–Inattention (#VP5530)
Curiosity–Incuriosity (#VP5528)
Expectation–Inexpectation (#VP5539)
Remembrance–Forgetfulness (#VP5537)
Broadmindedness–Narrowmindedness (#VP5526)
Imaginativeness–Unimaginativeness (#VP5535).

VT8055 Meaning∗complex
Constituent polarities Meaning–Meaninglessness (#VP5545)
Intelligibility–Unintelligibility (#VP5548)
Interpretability–Misinterpretability (#VP5552).

VT8061 Communication∗complex
Constituent polarities Elegance–Inelegance (#VP5589)
Eloquence–Uneloquence (#VP5600)
Education–Miseducation (#VP5562)
Conciseness–Diffuseness (#VP5592)
Representation–Misrepresentation (#VP5573)
Communicativeness–Uncommunicativeness (#VP5554).

VT8062 Choice∗complex
Constituent polarities Choice–Necessity (#VP5637)
Desire–Avoidance (#VP5631)
Resolution–Irresolution (#VP5624)
Willingness–Unwillingness (#VP5621).

VT8063 Motivation∗complex
Constituent polarities Motivation–Dissuasion (#VP5648)
Formality–Informality (#VP5646)
Conventionality–Unconventionality (#VP5642).

VT8064 Adaptation∗complex
Constituent polarities Goodness–Badness (#VP5674)
Perfection–Imperfection (#VP5677)
Importance–Unimportance (#VP5672)
Expedience–Inexpedience (#VP5670)
Oversufficiency–Insufficiency (#VP5660)
Appropriateness–Inappropriateness (#VP5665).

VT8065 Integrity∗complex
Constituent polarities Safety–Danger (#VP5697)
Health–Disease (#VP5685)
Refreshment–Relapse (#VP5695)
Cleanness–Uncleanness (#VP5681)
Improvement–Impairment (#VP5691)
Restoration–Destruction (#VP5693)
Selfactualization–Neurosis (#VP5690)
Healthfulness–Unhealthfulness (#VP5683).

VT8071 Action∗complex
Constituent polarities Exertion–Rest (#VP5709)   Action–Inaction (#VP5705)
Preparedness–Unpreparedness (#VP5720).

VT8072 Achievement∗complex
Constituent polarities Victory–Defeat (#VP5726)
Facility–Difficulty (#VP5732)
Prosperity–Adversity (#VP5728)
Behaviour–Misbehaviour (#VP5737)
Skilfulness–Unskilfulness (#VP5733)
Accomplishment–Nonaccomplishment (#VP5722).

VT8073 Compliance∗complex
Constituent polarities Consent–Refusal (#VP5775)
Freedom–Restraint (#VP5762)

VT8073

Leniency–Compulsion (#VP5756)
Authority–Lawlessness (#VP5739)
Obedience–Disobedience (#VP5766)
Observance–Nonobservance (#VP5768).

♦ **VT8074 Interaction∗complex**
  Constituent polarities Attack–Defence (#VP5798)
  Accord–Disaccord (#VP5794)
  Support–Opposition (#VP5785)
  Neutrality–Compromise (#VP5806).

♦ **VT8075 Possession∗complex**
  Constituent polarities Wealth–Poverty (#VP5837)
  Possession–Loss (#VP5808)
  Economy–Prodigality (#VP5851)
  Sharing–Appropriation (#VP5815)
  Expensiveness–Cheapness (#VP5848).

♦ **VT8081 Feeling∗complex**
  Constituent polarities Amusement–Boredom (#VP5878)
  Comfort–Aggravation (#VP5885)
  Patience–Impatience (#VP5861)
  Pleasure–Displeasure (#VP5865)
  Exultation–Lamentation (#VP5873)
  Cheerfulness–Solemnity (#VP5870)
  Excitement–Inexcitability (#VP5857)
  Feeling–Unfeelinglessness (#VP5855)
  Contentment–Discontentment (#VP5868)
  Pleasantness–Unpleasantness (#VP5863).

♦ **VT8082 Anticipation∗complex**
  Constituent polarities Courage–Fear (#VP5890)   Caution–Rashness (#VP5894)
  Hope–Hopelessness (#VP5888).

♦ **VT8083 Discriminative affection∗complex**
  Constituent polarities Modesty–Vanity (#VP5908)
  Pride–Humility (#VP5905)
  Beauty–Ugliness (#VP5899)
  Taste–Vulgarity (#VP5896)

Repute–Disrepute (#VP5914)
Wonder–Unastonishment (#VP5920)
Naturalness–Affectation (#VP5900).

♦ **VT8084 Socialization∗complex**
  Constituent polarities Love–Hate (#VP5930)   Friendship–Enmity (#VP5927)
  Conjugality–Celibacy (#VP5931)   Hospitality–Inhospitality (#VP5925)
  Sociability–Unsociability (#VP5922).

♦ **VT8085 Benevolence∗complex**
  Constituent polarities Congratulation–Envy (#VP5948)
  Kindness–Unkindness (#VP5938)
  Courtesy–Discourtesy (#VP5936)
  Gratitude–Ingratitude (#VP5949)
  Forgiveness–Vengeance (#VP5947)
  Compassion–Pitilessness (#VP5944).

♦ **VT8091 Appropriateness∗complex**
  Constituent polarities Dueness–Undueness (#VP5960)
  Respect–Disrespect (#VP5964)
  Rightness–Wrongness (#VP5957).

♦ **VT8092 Judgement∗complex**
  Constituent polarities Justice–Injustice (#VP5976)
  Probity–Improbity (#VP5974)
  Approval–Disapproval (#VP5968).

♦ **VT8093 Morality∗complex**
  Constituent polarities Virtue–Vice (#VP5980)   Innocence–Guilt (#VP5983)
  Chastity–Unchastity (#VP5987)   Temperance–Intemperance (#VP5992)
  Unselfishness–Selfishness (#VP5978).

♦ **VT8094 Retribution∗complex**
  Constituent polarities Legality–Illegality (#VP5998)
  Atonement–Punishment (#VP6010)
  Vindication–Condemnation (#VP6005).

♦ **VT8095 Redemption∗complex**
  Constituent polarities Piety–Impiety (#VP6028)   Sanctity–Unsanctity (#VP6026)
  Orthodoxy–Unorthodoxy (#VP6024)   Godliness–Ungodliness (#VP6013).

Index

Human Values and Wisdom

Index scope

This index covers all entries in Section V, namely:

- Constructive values (Section VC)

- Destructive values (Section VD)

- Value poliarities (Section VP)

- Value types (Section VT)

Note that **all index entries in the following index are integrated into the Volume Index (Section X)** at the end of this volume.

The main intention of this index is to enable users to obtain a better overview of the individual sub-sections of Section V.

All index entries refer via reference numbers (eg VC1234) to the descriptions in the preceding sections. The letters indicate the section (eg Section VC).

Index entries

The index entries are of several types:

- Principal name or title of concept;

- Secondary or alternative names of concept (including popular expressions and synonyms)

- Keywords from the principal concept name;

- Keywords from the secondary concept names.

No distinction is made between the different types of entry. Keywords are however recognizable by the presence of a semi-colon in the index entry.

Remarks

To facilitate consultation of the index, and to reduce the space requirement prepositions and articles are normally omitted from the index entries.

Although the index was generated automatically, it has been edited manually to eliminate less helpful keywords resulting from the extraction process. Where there was any doubt, less meaningful words have however been left where they may prove to be of value in locating entries in Sections VC, VD, VP or VT.

Note that the value word entries in Section VC and VD are in **alphabetic order** and therefore are self-indexing. The value polarity entries in Section VP are cross-referenced from the value word entries in Section VC and VD which therefore contribute indexes to Section VP. Section VT constitutes a classified index to the value polarities in Section VP.

INDEX TO HUMAN VALUES VX

A

VD 2004	Abandonment
VD 2006	Abasement
VD 2008	Aberration
VD 2010	Abhorrence
VC 0003	Ability
VD 2012	Abjection
VC 0005	Abnegation
VD 2014	Abnormality
VD 2016	Abomination
VD 2018	Absence
VP 5186	Absence ; Presence
VT 8041	Absolute properties∗complex
VC 0009	Absoluteness
VC 0007	Absolution
VD 2020	Absolutism
VD 3626	Absorption ; Self
VC 0011	Abstinence
VC 0013	Abstraction
VD 2022	Absurdity
VC 0015	Abundance
VD 2024	Abuse
VD 2026	Abusiveness
VC 0017	Acceptance
VC 0019	Accessibility
VD 2028	Accident
VC 0021	Acclaim
VC 0023	Accomplishment
VP 5722	Accomplishment-Nonaccomplishment
VC 0025	Accord
VP 5794	Accord-Disaccord
VC 0027	Accumulation
VC 0029	Accuracy
VC 0031	Achievement
VT 8072	Achievement∗complex
VC 0033	Acknowledgement
VC 0035	Acquiescence
VC 0037	Acquisition
VD 2030	Acquisitiveness
VD 2032	Acrimony
VC 0039	Action
VT 8071	Action∗complex
VP 5705	Action-Inaction
VC 0041	Activity
VC 0043	Actuality
VC 1539	Actualization ; Self
VC 0045	Acuity
VC 0047	Acumen
VC 0049	Adaptability
VT 8064	Adaptation∗complex
VC 0051	Adequacy
VC 0053	Adherence
VC 0055	Adjustment
VC 0057	Admiration
VD 3628	Admiration ; Self
VC 0059	Adoration
VC 0061	Adornment
VC 0063	Advancement
VC 1541	Advancement ; Self
VC 0065	Advantage
VC 0067	Adventure
VP 5728	Adversity ; Prosperity
VC 0069	Advocacy
VC 0071	Aesthetics
VC 0073	Affability
VD 2034	Affectation
VP 5900	Affectation ; Naturalness
VC 0075	Affection
VT 8083	Affection∗complex ; Discriminative
VC 0077	Affiliation
VC 0079	Affinity
VC 0081	Affirmation
VP 5523	Affirmation-Denial
VD 2036	Affliction
VC 0083	Affluence
VD 2038	Affrontery
VD 2040	Age
VP 5124	Age ; Youth
VC 0087	Aggrandizement
VD 2042	Aggravation
VP 5885	Aggravation ; Comfort
VD 2044	Aggression
VC 0089	Agility
VD 2046	Agitation
VP 5323	Agitation ; Oscillation
VC 0091	Agreeableness
VC 0093	Agreement
VP 5026	Agreement-Disagreement
VC 0095	Aid
VD 2048	Aimlessness
VC 0097	Alacrity
VD 2050	Alarmism
VC 0099	Alertness
VD 2052	Alienation
VC 0101	Aliveness
VC 0103	Allegiance
VC 0105	Alliance
VC 0107	Allurement
VC 0109	Almightiness
VC 0111	Alternation
VC 0113	Altruism
VD 2054	Ambiguity
VC 0115	Ambition
VD 2056	Ambivalence
VC 0117	Amelioration
VC 0119	Amenability
VC 0121	Amiability
VC 0123	Amicability
VC 0125	Amity
VD 2058	Amorality
VC 0127	Amorousness
VC 0129	Amplification
VC 0131	Amusement
VP 5878	Amusement-Boredom
VD 2060	Anarchism
VD 2062	Anarchy
VD 2064	Anathema
VC 0133	Ancestry
VP 5170	Ancestry-Posterity
VD 2066	Anger
VD 2068	Anguish
VD 2070	Animality
VC 0135	Animation
VD 2072	Animosity
VD 2074	Annihilation
VD 2076	Annoyance
VD 2078	Annulment
VC 0137	Anonymity
VD 2080	Antagonism
VD 2082	Antediluvian
VC 0139	Anticipation
VT 8082	Anticipation∗complex
VD 2084	Antipathy
VP 5119	Antiquity ; Futurity
VD 2086	Anxiety
VD 2088	Apartheid
VD 2090	Apathy
VC 0141	Apology
VC 0143	Appeal
VC 1609	Appeal ; Sex
VC 0145	Appearance
VP 5446	Appearance-Disappearance
VC 0147	Appeasement
VC 0149	Appetite
VC 0151	Application
VC 0153	Appreciation
VD 2092	Apprehension
VP 5296	Approach-Recession
VC 0155	Approachability
VC 0157	Approbation
VC 0159	Appropriateness
VT 8091	Appropriateness∗complex
VP 5665	Appropriateness-Inappropriateness
VP 5815	Appropriation ; Sharing
VC 0161	Approval
VP 5968	Approval-Disapproval
VC 0163	Aptitude
VD 2094	Archaism
VC 0165	Ardour
VD 2096	Argumentativeness
VC 0167	Aristocracy
VD 2098	Arrant
VD 2100	Arrested development
VD 2102	Arrogance
VC 0169	Art
VC 0171	Artfulness
VD 2104	Artifice
VD 2106	Artificiality
VC 0173	Artlessness
VC 0175	Ascendancy
VC 0177	Asceticism
VD 2108	Asperity
VC 0179	Aspiration
VC 0181	Assembly
VP 5521	Assent-Dissent
VC 0183	Assiduity
VC 0185	Assistance
VC 0187	Association
VC 0189	Assurance
VC 1543	Assurance ; Self
VD 2110	Asymmetry
VD 2112	Atheism
VP 6010	Atonement-Punishment
VD 2114	Atrocity
VC 0193	Attachment
VP 5798	Attack-Defence
VC 0195	Attainment
VC 0197	Attention
VP 5530	Attention-Inattention
VT 8054	Attitude∗complex
VC 0199	Attraction
VP 5288	Attraction-Repulsion
VP 5155	Attributability-Chance
VC 0201	Audacity
VP 5448	Audibility-Inaudibility
VD 2116	Austerity
VC 0203	Authenticity
VD 2118	Authoritarianism
VC 0205	Authority
VP 5739	Authority-Lawlessness
VD 2120	Autism
VC 0207	Autonomy
VC 0209	Availability
VD 2122	Avarice
VD 2124	Aversion
VD 2126	Avoidance
VP 5631	Avoidance ; Desire
VC 0211	Awareness
VC 0213	Awe
VD 2128	Awfulness
VD 2130	Awkwardness

B

VD 2132	Backlash
VD 2134	Badness
VP 5674	Badness ; Goodness
VD 2136	Bafflement
VC 0215	Balance
VD 2138	Banality
VD 2140	Banefulness
VD 2142	Barbarism
VD 2144	Barbarity
VD 2146	Barrenness
VD 2148	Barrier
VD 2150	Baseness
VD 2152	Bastardy
VD 2154	Battle
VD 2156	Bawdiness
VD 2158	Beastliness
VC 0217	Beatification
VC 0219	Beatitude
VC 0221	Beauty
VP 5899	Beauty-Ugliness
VD 2160	Bedevilment
VD 2162	Bedraggled
VC 0223	Behaviour
VP 5737	Behaviour-Misbehaviour
VC 0225	Being
VC 0227	Belief
VP 5501	Belief-Unbelief
VD 2164	Belligerence
VC 0229	Belongingness
VC 0231	Benediction
VC 0233	Beneficence
VC 0235	Benevolence
VT 8085	Benevolence∗complex
VC 0237	Benignity
VD 2166	Bereavement
VD 2168	Bestiality
VD 2170	Betrayal
VC 0239	Betterment
VD 2172	Bewilderment
VD 2174	Bias
VC 0241	Bigness
VP 5195	Bigness-Littleness
VD 2176	Bigotry
VD 2178	Bitterness
VD 2180	Blame
VC 0243	Blamelessness
VD 2181	Blasphemy
VD 2182	Blatancy
VD 2184	Blemish
VC 0247	Blessedness
VD 2186	Blight
VP 5439	Blindness ; Vision
VC 0249	Bliss
VD 2188	Bloody-mindedness
VD 2190	Boastfulness
VC 0251	Boldness
VD 2192	Boredom
VP 5878	Boredom ; Amusement
VD 2194	Boringness
VD 2196	Bothersomeness
VC 0253	Boundlessness
VC 0255	Bountifulness
VD 2198	Brainwash
VC 0257	Bravery
VP 5204	Breadth-Narrowness
VC 0259	Breathtaking
VC 0261	Breeding
VC 0263	Brevity
VC 0265	Brightness
VC 0267	Brilliance
VC 0269	Broadmindedness
VP 5526	Broadmindedness-Narrowmindedness
VC 0271	Broadness
VD 2200	Brokenness
VC 0273	Brotherhood
VC 0275	Brotherliness
VD 2202	Brutality
VD 2204	Bumptiousness
VC 0277	Buoyancy
VD 2206	Burdensomeness
VC 0279	Business

C

VD 2208	Calamity
VD 2210	Callousness
VC 0281	Calmness
VC 0283	Candor
VC 0285	Canniness
VC 0287	Capability
VC 0289	Capacity
VD 2212	Capriciousness
VC 0291	Care
VC 0293	Carefulness
VP 5533	Carefulness-Neglect
VD 2214	Carelessness
VT 8023	Causation∗complex
VP 5153	Causation-Culmination
VC 0295	Caution
VP 5894	Caution-Rashness
VC 0297	Cautiousness
VC 0299	Celebration
VP 5931	Celibacy ; Conjugality

VX

VD 2216	Censoriousness	VC 0409	Conciliation
VD 2218	Censure	VP 5592	Conciseness-Diffuseness
VP 5226	Centrality-Environment	VC 0411	Concord
VD 3630	Centredness ; Self	VD 2262	Concupiscence
VC 0301	Certainty	VC 0413	Concurrence
VP 5513	Certainty-Uncertainty	VP 5177	Concurrence-Counteraction
VP 5143	Cessation ; Continuance	VD 2264	Condemnation
VD 2220	Chagrin	VP 6005	Condemnation ; Vindication
VC 0303	Challenge	VD 2266	Condescension
VC 0305	Championship	VC 0415	Condolence
VC 0307	Chance	VC 0417	Conduct
VP 5155	Chance ; Attributability	VC 0419	Confidence
VC 0309	Change	VC 1545	Confidence ; Self
VT 8022	Change∗complex	VD 2268	Confinement
VP 5139	Change-Permanence	VC 0421	Confirmation
VD 2222	Changeableness	VD 2270	Conflict
VP 5141	Changeableness ; Stability	VC 0423	Conformation
VD 2224	Chaos	VC 0425	Conformity
VC 0311	Character	VP 5082	Conformity-Nonconformity
VD 2226	Characterlessness	VD 2272	Confusion
VC 0313	Charisma	VC 0427	Congeniality
VC 0315	Charity	VD 2274	Congestion
VC 0317	Charm	VP 5948	Congratulation-Envy
VC 0319	Chastity	VC 0429	Congruity
VP 5987	Chastity-Unchastity	VP 5931	Conjugality-Celibacy
VD 2228	Chauvinism	VP 5047	Conjunction-Separation
VD 2230	Cheapness	VC 0431	Connectedness
VP 5848	Cheapness ; Expensiveness	VC 0433	Connection
VC 0321	Cheerfulness	VC 0435	Conquest
VP 5870	Cheerfulness-Solemnity	VC 0437	Consanguinity
VD 2232	Cheerlessness	VC 0439	Conscience
VD 2234	Childishness	VC 0441	Conscientiousness
VC 0323	Chivalry	VC 0443	Consciousness
VC 0325	Choice	VP 5775	Consent-Refusal
VT 8062	Choice∗complex	VC 0445	Conservation
VP 5637	Choice-Necessity	VC 0447	Conservativeness
VP 5234	Circumscription-Intrusion	VC 0449	Consideration
VC 0327	Circumspection	VC 0451	Consistency
VC 0329	Civil rights	VC 0453	Consonance
VC 0331	Civility	VD 2276	Conspiracy
VC 0333	Civilization	VC 0455	Constancy
VC 0335	Clarity	VD 2278	Constraint
VC 0337	Cleanliness	VC 0457	Constructiveness
VP 5681	Cleanness-Uncleanness	VC 0459	Consummation
VC 0339	Clearness	VC 0461	Consumption
VC 0341	Clemency	VC 0463	Contact
VC 0343	Cleverness	VC 1547	Containedness ; Self
VD 2236	Closeness	VP 5193	Container-Content
VP 5265	Closure ; Opening	VD 2280	Contemptuousness
VC 0345	Clout	VC 0465	Content
VD 2238	Clumsiness	VP 5193	Content ; Container
VC 0347	Coalition	VD 2282	Contentiousness
VD 2240	Coarseness	VC 0467	Contentment
VC 0349	Cogency	VP 5868	Contentment-Discontentment
VC 0351	Coherence	VT 8032	Contextuality∗complex
VP 5050	Cohesion-Disintegration	VC 0469	Continence
VP 5328	Cold ; Heat	VP 5143	Continuance-Cessation
VD 2242	Coldness	VC 0471	Continuity
VC 0353	Collaboration	VP 5071	Continuity-Discontinuity
VD 2244	Collusion	VP 5197	Contraction ; Expansion
VP 5362	Colour-Colourlessness	VD 3632	Contradiciton ; Self
VP 5362	Colourlessness ; Colour	VD 2284	Contradiction
VD 2246	Combativeness	VD 2286	Contrariety
VC 0355	Comfort	VC 1549	Control ; Self
VP 5885	Comfort-Aggravation	VD 2288	Controversy
VC 0357	Comfortableness	VD 2290	Contumaciousness
VC 0359	Commendability	VD 2292	Contumeliousness
VC 0361	Commitment	VC 0473	Convenience
VC 0363	Common sense	VP 5642	Conventionality-Unconventionality
VD 2248	Commonness	VP 5298	Convergence-Divergence
VC 0365	Commonweal	VC 0475	Conversion
VC 0367	Communication	VP 5145	Conversion-Reversion
VT 8061	Communication∗complex	VC 0477	Convertibility
VC 0369	Communicativeness	VC 0479	Conviction
VP 5554	Communicativeness-Uncommunicativeness	VC 0481	Cooperation
VC 0371	Communion	VC 0483	Cooperativeness
VC 0373	Community	VC 0485	Coordination
VC 0375	Companionship	VC 0487	Cordiality
VC 0377	Compassion	VC 0489	Correctness
VP 5944	Compassion-Pitilessness	VC 0491	Correspondence
VC 0379	Compatibility	VD 2294	Corruption
VC 0381	Compensation	VC 0493	Cosmopolitan
VC 0383	Competence	VP 5177	Counteraction ; Concurrence
VC 0385	Competition	VD 2296	Counterproductivity
VD 2250	Complacency	VC 0495	Courage
VP 5054	Completeness-Incompleteness	VP 5890	Courage-Fear
VC 0387	Completion	VC 0497	Courtesy
VD 2252	Complexity	VP 5936	Courtesy-Discourtesy
VP 5045	Complexity ; Simplicity	VD 2298	Cowardice
VC 0389	Compliance	VC 0499	Creation
VT 8073	Compliance∗complex	VC 0501	Creativity
VD 2254	Complication	VC 0503	Credibility
VD 2256	Complicity	VT 8052	Credibility∗complex
VC 0391	Composure	VD 2300	Credulousness
VC 0393	Comprehension	VD 2302	Crisis
VC 0395	Comprehensiveness	VD 2304	Cruelty
VC 0397	Compromise	VC 0505	Culmination
VP 5806	Compromise ; Neutrality	VP 5153	Culmination ; Causation
VP 5756	Compulsion ; Leniency	VD 2306	Culpability
VD 2258	Compulsiveness	VC 0507	Culture
VC 0399	Compunction	VD 2308	Cunning
VC 0401	Comradeship	VC 0509	Curiosity
VD 2260	Conceit	VP 5528	Curiosity-Incuriosity
VC 0403	Concentration	VD 2310	Curtness
VC 0405	Conception	VC 0511	Custom
VC 0407	Concern	VD 2312	Cynicism

D

VD 2314	Damage
VD 2316	Damnation
VP 5697	Danger ; Safety
VD 2318	Dangerousness
VC 0513	Daring
VD 2320	Darkness
VP 5335	Darkness ; Light
VC 0515	Dauntlessness
VD 2322	Deadness
VD 2324	Death
VP 5407	Death ; Life
VD 2326	Debauchery
VD 2328	Debility
VD 2330	Decadence
VD 2332	Deceit
VC 0517	Decency
VD 2334	Deception
VC 0519	Decisiveness
VD 2336	Decline
VD 2338	Decomposition
VC 0521	Decoration
VC 0523	Decorum
VP 5038	Decrease ; Increase
VD 2340	Decrepitness
VC 0525	Dedication
VD 2342	Defamation
VD 2344	Default
VD 2346	Defeat
VP 5726	Defeat ; Victory
VD 2348	Defection
VD 2350	Defectiveness
VP 5798	Defence ; Attack
VD 2352	Defensiveness
VC 0527	Deference
VD 2354	Defiance
VD 2356	Deficiency
VD 2358	Deformation
VD 2360	Degeneration
VD 2362	Degradation
VD 2364	Dehumanization
VC 0529	Deification
VD 2366	Dejection
VC 0531	Delectableness
VD 2368	Delerium
VD 2370	Deleteriousness
VC 0533	Deliberateness
VC 0535	Delicacy
VC 0537	Delight
VC 0539	Deliverance
VD 2372	Delusion
VD 2374	Demeaning
VC 0541	Democracy
VD 2376	Denial
VP 5523	Denial ; Affirmation
VC 1551	Denial ; Self
VD 2378	Denigration
VD 2380	Denunciation
VD 2382	Dependence
VD 2384	Depersonalization
VD 2386	Depletion
VD 2388	Depradation
VD 2390	Depravation
VD 2392	Depravity
VD 2394	Depreciation
VD 2396	Depression
VP 5317	Depression ; Elevation
VD 2398	Deprivation
VP 5209	Depth-Shallowness
VD 2400	Dereliction
VD 2402	Derogation
VD 2404	Desecration
VD 2406	Desertion
VC 0543	Desirableness
VP 5631	Desire-Avoidance
VD 2408	Desolation
VD 2410	Despair
VD 2412	Desperation
VD 2414	Despicableness
VD 2416	Despoliation
VD 2418	Despondency
VD 2420	Despotism
VP 5693	Destruction ; Restoration
VD 3634	Destruction ; Self
VD 2422	Destructiveness
VC 0545	Detachment
VD 2424	Deterioration
VC 0547	Determination
VC 1553	Determination ; Self
VD 2426	Devastation
VC 0549	Development
VD 2100	Development ; Arrested
VD 2428	Deviation
VP 5290	Deviation ; Direction
VD 2430	Devilishness
VD 2432	Deviousness
VD 2434	Devitalized
VC 0551	Devotion
VC 0553	Dexterity
VD 2436	Diabolic
VD 2438	Dictatorship
VP 5013	Difference ; Identity
VC 0555	Differentiation
VD 2440	Difficulty
VP 5732	Difficulty ; Facility

INDEX TO HUMAN VALUES

VX

VP 5592	Diffuseness ; Conciseness	
VC 0557	Dignity	
VD 2442	Dilapidation	
VC 0559	Diligence	
VT 8031	Dimension∗complex	
VC 0561	Direction	
VP 5290	Direction-Deviation	
VC 1555	Direction ; Self	
VC 0563	Directness	
VD 2444	Dirtiness	
VD 2446	Disability	
VD 2448	Disaccord	
VP 5794	Disaccord ; Accord	
VD 2450	Disadvantage	
VD 2452	Disaffection	
VD 2454	Disaffinity	
VD 2456	Disagreeableness	
VP 5026	Disagreement ; Agreement	
VD 2458	Disappearance	
VP 5446	Disappearance ; Appearance	
VD 2460	Disappointment	
VD 2462	Disapproval	
VP 5968	Disapproval ; Approval	
VD 2464	Disarrangement	
VD 2466	Disaster	
VD 2468	Disbelief	
VC 0565	Discernment	
VC 0567	Discipline	
VC 1557	Discipline ; Self	
VD 2470	Discomfort	
VD 2472	Discomposure	
VD 2474	Disconcertion	
VD 2476	Discontentment	
VP 5868	Discontentment ; Contentment	
VD 2478	Discontinuity	
VP 5071	Discontinuity ; Continuity	
VD 2480	Discord	
VP 5462	Discord ; Harmony	
VD 2482	Discouragement	
VD 2484	Discourtesy	
VP 5936	Discourtesy ; Courtesy	
VC 0569	Discovery	
VP 5485	Discovery ; Research	
VD 2486	Discredit	
VC 0571	Discretion	
VC 0573	Discrimination	
VD 2488	Discrimination	
VP 5492	Discrimination-Indiscrimination	
VT 8083	Discriminative affection∗complex	
VD 2490	Disdain	
VP 5685	Disease ; Health	
VD 2492	Disenchantment	
VD 2494	Disesteem	
VD 2496	Disfavour	
VD 2498	Disfigurement	
VD 2500	Disgrace	
VD 2502	Disgust	
VD 2504	Disharmony	
VD 2506	Dishonesty	
VD 2508	Dishonour	
VD 2510	Disillusionment	
VP 5519	Disillusionment ; Illusion	
VD 2512	Disingenuousness	
VD 2514	Disintegration	
VP 5050	Disintegration ; Cohesion	
VD 2516	Disinterest	
VD 2518	Dislike	
VP 5185	Dislocation ; Location	
VD 2520	Disloyalty	
VD 2522	Dismay	
VD 2524	Disobedience	
VP 5766	Disobedience ; Obedience	
VD 2526	Disorder	
VP 5059	Disorder ; Order	
VD 2528	Disorganization	
VD 2530	Disorientation	
VD 2532	Disparagement	
VD 2534	Disparity	
VC 0575	Dispassion	
VD 2536	Displeasure	
VP 5865	Displeasure ; Pleasure	
VC 1905	Disposed ; Well	
VD 2538	Disputatiousness	
VD 2540	Disquiet	
VD 2542	Disregard	
VP 5914	Disrepute ; Repute	
VD 2544	Disrespect	
VP 5964	Disrespect ; Respect	
VD 2546	Dissatisfaction	
VD 2548	Dissension	
VP 5521	Dissent ; Assent	
VD 2550	Dissidence	
VP 5020	Dissimilarity ; Similarity	
VD 2552	Dissipation	
VD 2554	Dissociation	
VD 2556	Dissolution	
VD 2558	Dissonance	
VP 5648	Dissuasion ; Motivation	
VP 5200	Distance ; Nearness	
VD 2560	Distaste	
VC 0577	Distinction	
VD 2562	Distortion	
VP 5248	Distortion ; Symmetry	
VD 2564	Distraction	
VD 2566	Distress	
VD 2568	Distrust	
VD 2570	Disturbance	
VD 2572	Disunity	
VD 2574	Disuse	
VD 2576	Divergence	
VP 5298	Divergence ; Convergence	
VC 0579	Diversity	
VD 2580	Division	
VD 2578	Divisiveness	
VC 0581	Docility	
VD 2582	Dolorousness	
VC 0587	Dominance	
VD 2584	Doubtfulness	
VP 5089	Duality ; Unity	
VD 2586	Dubiousness	
VP 5960	Dueness-Undueness	
VD 2588	Dullness	
VD 2590	Duplicity	
VC 0589	Durability	
VP 5110	Durability-Transience	
VC 0591	Duration	
VD 2592	Duress	
VC 0593	Duty	
VC 0595	Dynamism	

E

VC 0597	Eagerness	
VC 0599	Earnestness	
VC 0601	Ease	
VD 2594	Eccentricity	
VC 0603	Eclecticism	
VC 0605	Economy	
VP 5851	Economy-Prodigality	
VC 0607	Ecstasy	
VC 0609	Edification	
VC 0611	Education	
VP 5562	Education-Miseducation	
VD 2596	Eeriness	
VC 1559	Effacement ; Self	
VC 0613	Effectiveness	
VD 2598	Effetism	
VC 0615	Efficiency	
VC 0617	Effortlessness	
VD 2600	Elaborateness	
VC 0619	Elan	
VP 5358	Elasticity-Toughness	
VC 0621	Elation	
VC 0623	Elegance	
VP 5589	Elegance-Inelegance	
VC 0625	Elevation	
VP 5317	Elevation-Depression	
VC 0627	Eloquence	
VP 5600	Eloquence-Uneloquence	
VC 0629	Emancipation	
VD 2602	Embarrassment	
VC 0631	Eminence	
VC 0633	Emotion	
VC 0635	Empathy	
VC 0637	Enchantment	
VC 0639	Encouragement	
VD 2604	Encroachment	
VD 2606	Encumbrance	
VC 0641	Endearment	
VC 0643	Endowment	
VC 0645	Endurance	
VC 0647	Energy	
VP 5161	Energy-Moderation	
VD 2608	Enforcement	
VC 0649	Enhancement	
VC 0651	Enjoyment	
VC 0653	Enlightenment	
VD 2610	Enmity	
VP 5927	Enmity ; Friendship	
VD 2612	Ennui	
VC 0655	Enrichment	
VC 0657	Entente	
VC 0659	Enterprise	
VC 0661	Entertainment	
VC 0663	Enthusiasm	
VC 0665	Enumeration	
VP 5226	Environment ; Centrality	
VD 2614	Envy	
VP 5948	Envy ; Congratulation	
VC 0667	Equality	
VP 5030	Equality-Inequality	
VC 0669	Equanimity	
VC 0671	Equilibrium	
VC 0673	Equity	
VC 0675	Equivalence	
VD 2616	Equivocation	
VD 2618	Erosion	
VD 2620	Erroneousness	
VP 5516	Error ; Truth	
VC 0677	Erudition	
VC 0679	Esprit	
VC 0681	Essence	
VC 0683	Establishment	
VD 3636	Esteem ; Self	
VC 0685	Euphony	
VT 8051	Evaluation∗complex	
VD 2622	Evasion	
VC 0687	Evenness	
VC 0689	Eventuation	
VP 5151	Eventuation-Imminence	
VD 2624	Evil	
VC 0691	Evolution	
VP 5147	Evolution-Revolution	
VC 0693	Exactness	
VD 2626	Exaggeration	
VC 0695	Exaltation	
VD 2628	Exasperation	
VC 0697	Excellence	
VD 2630	Excess	
VD 2632	Excessiveness	
VC 0699	Excitement	
VP 5857	Excitement-Inexcitability	
VD 2634	Exclusion	
VP 5076	Exclusion ; Inclusion	
VD 2636	Exclusiveness	
VC 0701	Exclusiveness	
VD 2638	Execration	
VP 5709	Exertion-Rest	
VD 2640	Exhaustion	
VC 0703	Exhilaration	
VD 2642	Exiguity	
VC 0705	Existence	
VT 8011	Existence∗complex	
VP 5001	Existence-Nonexistence	
VC 0707	Expansion	
VP 5197	Expansion-Contraction	
VC 0709	Expectation	
VP 5539	Expectation-Inexpectation	
VC 0711	Expedience	
VP 5670	Expedience-Inexpedience	
VC 0713	Expeditious	
VP 5848	Expensiveness-Cheapness	
VC 0715	Experience	
VC 0717	Expertise	
VC 0719	Explanation	
VD 2644	Exploitation	
VC 1561	Expression ; Self	
VD 2646	Expulsion	
VC 0721	Exquisiteness	
VP 5224	Exteriority-Interiority	
VD 2648	Extermination	
VD 2650	Extinction	
VD 2652	Extortion	
VD 2654	Extravagance	
VD 2656	Extremism	
VP 5005	Extrinsicality ; Intrinsicality	
VC 0723	Exuberance	
VP 5873	Exultation-Lamentation	

F

VD 2658	Fabrication	
VC 0725	Facility	
VP 5732	Facility-Difficulty	
VT 8045	Faculties∗complex ; Intellectual	
VD 2660	Failure	
VC 0727	Fairness	
VC 0729	Faith	
VC 0731	Faithfulness	
VD 2662	Faithlessness	
VD 2664	Fakery	
VD 2666	Fallacy	
VD 2668	Fallibility	
VD 2670	Falsehood	
VD 2672	Falseness	
VC 0733	Fame	
VC 0735	Familiarity	
VD 2676	Fanaticism	
VD 2674	Fantasy	
VC 0737	Fascination	
VC 0739	Fatherhood	
VD 2678	Fatigue	
VD 2680	Fatuity	
VD 2682	Faultiness	
VC 0741	Faultlessness	
VC 0743	Favour	
VD 2684	Fear	
VP 5890	Fear ; Courage	
VC 0745	Fearlessness	
VC 0747	Feasibility	
VD 2686	Featurelessness	
VD 2688	Fecklessness	
VC 0749	Fecundity	
VD 2690	Feebleness	
VC 0751	Feeling	
VT 8081	Feeling∗complex	
VP 5855	Feeling-Unfeelinglessness	
VC 0753	Felicity	
VC 0755	Fellowship	
VD 2692	Feloniousness	
VP 5420	Femininity ; Masculinity	
VC 0757	Feminity	
VD 2694	Ferocity	
VC 0759	Fertility	
VC 0761	Fervour	
VP 5101	Fewness ; Numerousness	
VC 0763	Fidelity	
VD 2696	Fiendish	
VD 2698	Fierceness	
VD 2700	Filth	
VC 0765	Finesse	
VC 0767	Fitness	
VD 2702	Fixation	
VD 2704	Flaw	
VC 0769	Flawlessness	
VC 0771	Flexibility	

VX

VC 0773 Flourish		VD 2842 Imperfection
VC 0775 Fluency	**H**	VP 5677 Imperfection ; Perfection
VP 5292 Following ; Leading	VD 2764 Hallucination	VD 2844 Impertinence
VC 0777 Fondness	VC 0879 Handsomeness	VD 2846 Imperviousness
VC 0779 Forbearance	VD 2766 Haphazardness	VD 2848 Impetuousity
VC 0781 Forcefulness	VC 0881 Happiness	VD 2850 Impiety
VD 2706 Forcelessness	VC 0883 Hardiness	VP 6028 Impiety ; Piety
VC 0783 Foresight	VD 2770 Hardness	VD 2852 Implausibility
VC 0785 Forethought	VP 5356 Hardness-Softness	VC 0955 Implementation
VD 2708 Forgetfulness	VD 2772 Harmfulness	VD 2854 Impoliteness
VP 5537 Forgetfulness ; Remembrance	VC 0885 Harmlessness	VC 0957 Importance
VC 0787 Forgiveness	VC 0887 Harmony	VD 3638 Importance ; Self
VP 5947 Forgiveness-Vengeance	VP 5462 Harmony-Discord	VP 5672 Importance-Unimportance
VP 5246 Form-Formlessness	VD 2768 Harrassment	VD 2856 Impossibility
VP 5646 Formality-Informality	VD 2774 Harshness	VP 5509 Impossibility ; Possibility
VP 5246 Formlessness ; Form	VD 2776 Haste	VD 2858 Impotence
VC 0789 Forthrightness	VD 2778 Hastiness	VP 5157 Impotence ; Power
VC 0791 Fortune	VP 5930 Hate ; Love	VD 2860 Impoverishment
VD 2710 Foulness	VD 2780 Hatred	VD 2862 Impracticality
VD 2712 Fragility	VD 2782 Hazardousness	VD 2864 Imprecision
VP 5435 Fragrance-Stench	VC 0889 Health	VD 2866 Impressionability
VC 0793 Frankness	VP 5685 Health-Disease	VC 0959 Impressivensss
VC 0795 Fraternity	VP 5683 Healthfulness-Unhealthfulness	VD 2868 Improbity
VD 2714 Fraudulence	VC 0891 Healthiness	VP 5974 Improbity ; Probity
VC 0797 Freedom	VD 2784 Heartlessness	VD 2870 Impropriety
VP 5762 Freedom-Restraint	VP 5328 Heat-Cold	VC 0961 Improvement
VD 2716 Frenzy	VD 2786 Heedlessness	VP 5691 Improvement-Impairment
VP 5135 Frequency-Infrequency	VP 5207 Height-Lowness	VD 2872 Improvidence
VC 0799 Freshness	VC 1567 Help ; Self	VD 2874 Imprudence
VD 2718 Fretfulness	VC 0893 Helpfulness	VD 2876 Impudence
VD 2720 Friction	VD 2788 Helplessness	VD 2878 Impulsiveness
VC 0801 Frictionlessness	VD 2790 Heresy	VD 2880 Impurity
VC 0803 Friendliness	VC 0895 Heritage	VD 2882 Inability
VC 0805 Friendship	VC 0897 Heroism	VD 2884 Inaccessibility
VP 5927 Friendship-Enmity	VD 2792 Hesitation	VD 2886 Inaccuracy
VD 2722 Frightfulness	VD 2794 Hideousness	VP 5705 Inaction ; Action
VD 2724 Frivolity	VC 0899 High-mindedness	VD 2888 Inactivity
VC 0807 Frugality	VC 0901 High-spiritedness	VD 2890 Inadequacy
VD 2726 Frustration	VC 0903 Hilarity	VD 2892 Inadvertence
VC 0809 Fulfilment	VC 0905 Holiness	VD 2894 Inadvisability
VC 1563 Fulfilment ; Self	VC 0907 Holism	VD 2896 Inanity
VC 0811 Fullness	VC 0909 Homage	VD 2898 Inapplicability
VD 2728 Fulsomeness	VD 2796 Homelessness	VD 2900 Inappropriateness
VC 0813 Fun	VC 0913 Honesty	VP 5665 Inappropriateness ; Appropriateness
VD 2730 Furtiveness	VC 0915 Honour	VD 2902 Inaptitude
VD 2732 Fury	VC 0917 Hope	VD 2904 Inarticulation
VD 2734 Fuss	VP 5888 Hope-Hopelessness	VD 2906 Inattention
VD 2736 Futility	VD 2798 Hopelessness	VP 5530 Inattention ; Attention
VP 5119 Futurity-Antiquity	VP 5888 Hopelessness ; Hope	VP 5448 Inaudibility ; Audibility
	VD 2800 Horribleness	VD 2908 Incapability
G	VC 0919 Hospitality	VD 2910 Incapacity
	VP 5925 Hospitality-Inhospitality	VD 2912 Incivility
VC 0815 Gaiety	VD 2802 Hostility	VD 2914 Inclemency
VC 0817 Gain	VC 0921 Humaneness	VC 0963 Inclusion
VC 0819 Gallantry	VC 0923 Humanism	VP 5076 Inclusion-Exclusion
VD 2738 Gaudiness	VC 0925 Humanitarianism	VD 2916 Incoherence
VC 0821 Generosity	VC 0927 Humanity	VD 2918 Incompatibility
VC 0823 Geniality	VP 5417 Humanity-Nonhumanity	VD 2920 Incompetence
VC 0825 Genius	VD 2804 Humiliation	VD 2922 Incompleteness
VC 0827 Gentility	VC 0929 Humility	VP 5054 Incompleteness ; Completeness
VC 0829 Gentleness	VP 5905 Humility ; Pride	VD 2924 Incomprehensibility
VC 0831 Genuineness	VC 0931 Humour	VD 2926 Incongruity
VC 0833 Giftedness	VD 2806 Hurtfulness	VD 2928 Inconsequence
VC 0835 Gladness	VC 0933 Hygiene	VD 2930 Inconsiderateness
VC 0837 Glamour	VD 2808 Hypocrisy	VD 2932 Inconsistency
VD 2740 Glibness		VD 2934 Inconstancy
VC 0839 Globality	**I**	VD 2936 Incontinence
VC 0841 Glory		VD 2938 Inconvenience
VD 2742 Godlessness	VC 0935 Idealism	VD 2940 Incorrectness
VC 0843 Godliness	VP 5013 Identity-Difference	VD 2942 Incorrigibility
VP 6013 Godliness-Ungodliness	VD 2810 Idleness	VC 0965 Increase
VC 0845 Good	VD 2812 Ignobility	VP 5038 Increase-Decrease
VC 0847 Goodness	VD 2814 Ignominiousness	VD 2944 Incredulity
VP 5674 Goodness-Badness	VD 2816 Ignominy	VP 5528 Incuriosity ; Curiosity
VC 0849 Goodwill	VD 2818 Ignorance	VD 2946 Incuriousness
VC 0851 Gorgeousness	VP 5475 Ignorance ; Knowledge	VD 2948 Indecency
VC 0853 Government	VD 2820 Illegality	VD 2950 Indecisiveness
VC 1565 Government ; Self	VP 5998 Illegality ; Legality	VD 2952 Indecorum
VC 0855 Grace	VD 2822 Illegitimate	VC 0967 Indefatigableness
VC 0857 Gracefulness	VD 2824 Illiberalism	VD 2954 Indefensibility
VD 2744 Gracelessness	VD 2826 Illness	VD 2956 Indelicacy
VC 0859 Graciousness	VD 2828 Illogic	VC 0969 Independence
VC 0861 Grandeur	VC 0937 Illumination	VD 2958 Indifference
VC 0863 Gratification	VP 5519 Illusion-Disillusionment	VD 2960 Indignation
VC 0865 Gratitude	VC 0939 Imagination	VD 2962 Indirection
VP 5949 Gratitude-Ingratitude	VP 5535 Imaginativeness-Unimaginativeness	VD 2964 Indiscretion
VC 0867 Greatness	VD 2830 Imbecility	VD 2966 Indiscrimination
VP 5034 Greatness-Smallness	VP 5023 Imitation ; Originality	VP 5492 Indiscrimination ; Discrimination
VD 2746 Greed	VP 5376 Immateriality ; Materiality	VD 2968 Indisposition
VC 0869 Gregariousness	VD 2832 Immaturity	VC 0971 Individuality
VD 2748 Grief	VC 0941 Immediacy	VD 2970 Indocility
VD 2750 Grievance	VC 0943 Immensity	VD 2972 Indolence
VD 2752 Grimness	VC 0945 Imminence	VC 0973 Indubitableness
VD 2754 Grossness	VP 5151 Imminence ; Eventuation	VD 2974 Indulgence
VC 1907 Grounded ; Well	VD 2834 Immodesty	VD 3640 Indulgence ; Self
VC 0871 Growth	VD 2836 Immorality	VC 0975 Industriousness
VD 2756 Gruesomeness	VC 0947 Immortality	VD 2976 Ineffectiveness
VD 2758 Guile	VC 0949 Immunity	VD 2978 Ineffectuality
VD 2760 Guilt	VC 0951 Immutability	VD 2980 Inefficiency
VP 5983 Guilt ; Innocence	VP 5283 Impact-Reaction	VD 2982 Inelegance
VC 0873 Guiltlessness	VD 2838 Impairment	VP 5589 Inelegance ; Elegance
VD 2762 Gullibleness	VP 5691 Impairment ; Improvement	VD 2984 Ineptitude
	VC 0953 Impartiality	VD 2986 Inequality
	VD 2840 Impatience	VP 5030 Inequality ; Equality
	VP 5861 Impatience ; Patience	VD 2988 Inequity
		VD 2990 Inertia

INDEX TO HUMAN VALUES

VX

VP 5857	Inexcitability ; Excitement	VC 1025	Invulnerability	VD 3172	Lowness		
VP 5539	Inexpectation ; Expectation	VD 3094	Irascibility	VP 5207	Lowness ; Height		
VD 2992	Inexpedience	VD 3096	Irksomeness	VC 1095	Loyalty		
VP 5670	Inexpedience ; Expedience	VD 3098	Irrationality	VC 1097	Lucidity		
VD 2994	Inexperience	VD 3100	Irreconcilability	VC 1099	Luck		
VC 0977	Infallibility	VD 3102	Irredeemability	VD 3174	Lucklessness		
VC 2996	Infamy	VD 3104	Irreformability	VD 3176	Lust		
VD 2998	Infatuation	VP 5137	Irregularity ; Regularity	VC 1101	Luxury		
VD 3000	Infelicity	VD 3106	Irrelevance				
VD 3002	Inferiority	VD 3108	Irreligiousness		**M**		
VP 5036	Inferiority ; Superiority	VD 3110	Irremediability				
VD 3004	Infertility	VC 1027	Irrepressibility	VC 1103	Machismo		
VD 3006	Infidelity	VC 1029	Irreproachability	VD 3178	Madness		
VD 3008	Inflexibility	VC 1031	Irresistability	VC 1105	Magnanimity		
VC 0979	Influence	VD 3112	Irresolution	VC 1107	Magnificence		
VP 5172	Influence-Influencelessness	VP 5624	Irresolution ; Resolution	VC 1109	Majesty		
VP 5172	Influencelessness ; Influence	VD 3114	Irresponsibility	VD 3180	Maladjustment		
VP 5646	Informality ; Formality	VD 3116	Irreverence	VD 3182	Maladroit		
VC 1909	Informed ; Well	VD 3118	Irritability	VD 3184	Malaise		
VP 5135	Infrequency ; Frequency	VD 3120	Isolation	VD 3186	Malevolence		
VC 0981	Ingenuity			VD 3188	Malfeasance		
VC 0983	Ingenuousness		**J**	VD 3190	Malformation		
VD 3010	Ingloriousness			VD 3192	Malice		
VD 3012	Ingratitude	VD 3122	Jealousy	VD 3194	Malignancy		
VP 5949	Ingratitude ; Gratitude	VC 1033	Joy	VD 3196	Malingering		
VD 3014	Inharmonious	VC 1035	Joyfulness	VD 3198	Malpractice		
VD 3016	Inhibition	VD 3124	Joylessness	VD 3200	Maltreatment		
VD 3018	Inhospitality	VC 1037	Judgement	VC 1111	Manageability		
VP 5925	Inhospitality ; Hospitality	VT 8092	Judgement*complex	VC 1115	Manliness		
VD 3020	Inhumanity	VP 5494	Judgement-Misjudgement	VC 1913	Mannered ; Well		
VD 3022	Iniquity	VC 1039	Judiciousness	VC 1117	Manners		
VC 0985	Initiative	VC 1041	Justice	VC 1113	Manoeuvrability		
VD 3024	Injury	VP 5976	Justice-Injustice	VP 5420	Masculinity-Femininity		
VD 3026	Injustice			VC 1121	Mastery		
VP 5976	Injustice ; Justice		**K**	VP 5376	Materiality-Immateriality		
VC 0987	Innocence			VC 1123	Materialization		
VP 5983	Innocence-Guilt	VC 1043	Keenness	VC 1125	Maternity		
VD 3028	Inobservance	VC 1045	Kindness	VC 1127	Maturity		
VC 0988	Inoffensiveness	VP 5938	Kindness-Unkindness	VT 8055	Meaning*complex		
VD 3030	Inquietude	VC 1047	Knowledge	VP 5545	Meaning-Meaninglessness		
VD 3032	Inquisition	VP 5475	Knowledge-Ignorance	VC 1915	Meaning ; Well		
VD 3034	Insanity			VC 1129	Meaningfulness		
VP 5472	Insanity ; Sanity		**L**	VP 5545	Meaninglessness ; Meaning		
VD 3036	Insecurity			VD 3204	Meanness		
VD 3038	Insensibility	VD 3126	Lack	VC 1131	Mediation		
VP 5422	Insensibility ; Sensation	VP 5873	Lamentation ; Exultation	VD 3206	Mediocrity		
VD 3040	Insensitivity	VD 3128	Languor	VC 1133	Meekness		
VC 0991	Insight	VD 3130	Lapse	VD 3208	Melancholy		
VD 3042	Insignificance	VC 1049	Largeness	VC 1135	Mellowness		
VD 3044	Insincerity	VC 1051	Lasting	VD 3210	Menace		
VD 3046	Insinuation	VD 3132	Lateness	VD 3212	Mendacity		
VD 3048	Insobriety	VC 1053	Laughter	VD 3214	Mendicancy		
VD 3050	Insolence	VD 3134	Lavishness	VC 1137	Mercifulness		
VD 3052	Insouciance	VC 1055	Lawfulness	VD 3216	Mercilessness		
VC 0993	Inspiration	VD 3136	Lawlessness	VC 1139	Merit		
VD 3054	Instability	VP 5739	Lawlessness ; Authority	VD 3218	Meritlessness		
VP 5112	Instantaneousness ; Perpetuity	VD 3138	Laxity	VC 1141	Merriment		
VD 3056	Insubordination	VD 3140	Laxness	VD 3220	Messiness		
VD 3058	Insubstantiality	VD 3142	Laziness	VD 3222	Miasma		
VD 3060	Insufficiency	VC 1057	Leadership	VC 1143	Meticulousness		
VP 5660	Insufficiency ; Oversufficiency	VP 5292	Leading-Following	VC 1145	Mettlesomeness		
VD 3062	Insularity	VC 1059	Learning	VC 1147	Might		
VD 3064	Insult	VC 1061	Legality	VC 1149	Mildness		
VP 5425	Intangibility ; Tangibility	VP 5998	Legality-Illegality	VD 3226	Militancy		
VC 0995	Integrality	VC 1063	Legitimacy	VD 2188	Mindedness ; Bloody		
VC 0997	Integrity	VC 1065	Leisure	VC 0899	Mindedness ; High		
VT 8065	Integrity*complex	VC 1067	Leniency	VD 4050	Mindedness ; Weak		
VT 8045	Intellectual faculties*complex	VP 5756	Leniency-Compulsion	VC 1151	Mindfulness		
VC 0999	Intellectuality	VD 3144	Lethargy	VD 3228	Mindlessness		
VC 1001	Intelligence	VD 3146	Levity	VD 3230	Misalliance		
VP 5467	Intelligence-Unintelligence	VD 3148	Liability	VD 3232	Misapplication		
VC 1003	Intelligibility	VC 1069	Liberality	VD 3234	Misapprehension		
VP 5548	Intelligibility-Unintelligibility	VC 1071	Liberation	VP 5737	Misbehaviour ; Behaviour		
VP 5992	Intemperance ; Temperance	VC 1073	Liberty	VD 3236	Misbelief		
VD 3066	Intemperence	VD 3150	Licentiousness	VD 3238	Miscalculation		
VC 1005	Intensity	VC 1075	Life	VD 3240	Miscarriage		
VC 1007	Intention	VT 8043	Life*complex	VD 3242	Mischief		
VC 1009	Intentness	VP 5407	Life-Death	VD 3224	Mischievousness		
VT 8074	Interaction*complex	VD 3152	Lifelessness	VD 3244	Misconception		
VC 1011	Interest	VC 1077	Light	VD 3246	Misconduct		
VC 1569	Interest ; Self	VP 5335	Light-Darkness	VD 3248	Misconstruction		
VD 3068	Interference	VC 1079	Lightness	VD 3250	Misdemeanor		
VP 5224	Interiority ; Exteriority	VP 5352	Lightness ; Weight	VD 3252	Misdirection		
VP 5552	Interpretability-Misinterpretability	VP 5507	Limitation-Unlimitedness	VP 5562	Miseducation ; Education		
VD 3070	Intervention	VD 3154	Limitedness	VD 3254	Misery		
VC 1013	Intimacy	VC 1081	Limitlessness	VD 3256	Misfeasance		
VD 3072	Intimidation	VC 1083	Lineage	VD 3258	Misfortune		
VD 3074	Intolerance	VD 3156	Listlessness	VD 3260	Misguidance		
VD 3076	Intoxication	VD 3158	Littleness	VD 3262	Misinformation		
VD 3078	Intractability	VP 5195	Littleness ; Bigness	VP 5552	Misinterpretability ; Interpretability		
VD 3080	Intransigence	VC 1085	Liveliness	VD 3264	Misinterpretation		
VD 3082	Intricacy	VD 3160	Loathsomeness	VD 3266	Misjudgement		
VC 1015	Intricacy	VC 1087	Location	VP 5494	Misjudgement ; Judgement		
VD 3084	Intrigue	VP 5185	Location-Dislocation	VD 3268	Mismanagement		
VP 5005	Intrinsicality-Extrinsicality	VD 3162	Loneliness	VD 3270	Misrepresentation		
VP 5234	Intrusion ; Circumscription	VC 1089	Longanimity	VP 5573	Misrepresentation ; Representation		
VD 3086	Intrusiveness	VC 1091	Longevity	VD 3272	Misrule		
VP 5481	Intuition-Reason	VD 3164	Looseness	VD 3274	Misstatement		
VC 1017	Intuitiveness	VD 3166	Lordliness	VD 3276	Mistake		
VD 3088	Invalidity	VP 5808	Loss ; Possession	VD 3278	Mistrust		
VD 3090	Invasion	VD 3168	Loudness	VD 3280	Misunderstanding		
VC 1019	Inventiveness	VP 5451	Loudness ; Silence	VD 3282	Misusage		
VC 1021	Invincibility	VC 1093	Love	VD 3284	Misuse		
VP 5444	Invisibility ; Visibility	VP 5930	Love-Hate	VC 1153	Mitigation		
VD 3092	Involution	VD 3170	Lovelessness	VC 1155	Mobility		
VC 1023	Involvement						

VC 1157 Moderation	VP 5339 Opaqueness ; Transparency	VC 1301 Plainness
VP 5161 Moderation ; Energy	VC 1221 Openheartedness	VC 1303 Plausibility
VC 1159 Modesty	VP 5265 Opening-Closure	VC 1305 Playfulness
VP 5908 Modesty-Vanity	VC 1223 Openmindedness	VC 1307 Pleasantness
VC 1161 Momentousness	VC 1225 Openness	VP 5863 Pleasantness-Unpleasantness
VC 1163 Morality	VC 1227 Opportunity	VC 1309 Pleasure
VT 8093 Morality*complex	VD 3364 Opposition	VP 5865 Pleasure-Displeasure
VD 3286 Morbidity	VP 5785 Opposition ; Support	VD 3458 Pleasurelessness
VD 3288 Mortification	VD 3366 Oppression	VC 1311 Plenitude
VC 1165 Motherhood	VD 3368 Opprobrium	VC 1313 Plenty
VC 1167 Motherliness	VC 1229 Optimism	VC 1315 Pluck
VT 8034 Motion*complex	VC 1231 Order	VD 3460 Pointlessness
VT 8035 Motion*complex ; Relative	VT 8014 Order*complex	VD 3462 Poisonousness
VP 5267 Motion-Quiescence	VP 5059 Order-Disorder	VC 1317 Politeness
VC 1169 Motivation	VD 3370 Orderlessness	VD 3464 Pollution
VT 8063 Motivation*complex	VC 1233 Orderliness	VD 3466 Pomposity
VP 5648 Motivation-Dissuasion	VC 1235 Ordinariness	VD 3468 Ponderousness
VC 1171 Motive	VC 1237 Organization	VC 1319 Popularity
VD 3290 Mournfulness	VC 1239 Originality	VC 1321 Position
VD 3292 Muddle	VP 5023 Originality-Imitation	VC 1323 Positiveness
VC 1173 Munificence	VC 1241 Orthodoxy	VC 1325 Possession
VD 3294 Mustiness	VP 6024 Orthodoxy-Unorthodoxy	VT 8075 Possession*complex
VD 3296 Mutability	VP 5323 Oscillation-Agitation	VP 5808 Possession-Loss
VD 3298 Muteness	VD 3372 Ostentation	VC 1571 Possession ; Self
VD 3300 Mutilation	VD 3374 Outrage	VP 5509 Possibility-Impossibility
VC 1175 Mutuality	VD 3376 Overabundance	VP 5170 Posterity ; Ancestry
VD 3302 Mystification	VD 3378 Overactivity	VC 1327 Potency
	VD 3380 Overcompensation	VC 1329 Potentiality
N	VD 3382 Overconfidence	VD 3470 Poverty
	VD 3384 Overconscientiousness	VP 5837 Poverty ; Wealth
VD 3304 Naïvety	VD 3386 Overdeveloped	VC 1331 Power*
VP 5526 Narrowmindedness ; Broadmindedness	VD 3388 Overemphasis	VT 8024 Power*complex
VP 5204 Narrowness ; Breadth	VD 3390 Overestimation	VP 5157 Power-Impotence
VC 1179 Naturalness	VP 5497 Overestimation-Underestimation	VD 3472 Powerlessness
VP 5900 Naturalness-Affectation	VD 3392 Overexpansion	VC 1333 Practicability
VP 5200 Nearness-Distance	VD 3394 Overextension	VC 1335 Practicality
VC 1181 Neatness	VD 3396 Overload	VC 1337 Praise
VC 1183 Necessity	VD 3398 Overreligiousness	VC 1339 Precedence
VP 5637 Necessity ; Choice	VP 5313 Overrunning-Shortcoming	VC 1341 Precision
VD 3306 Negation	VD 3400 Oversensitiveness	VD 3474 Prejudice
VD 3308 Negativity	VD 3402 Oversight	VD 3476 Prematurity
VD 3310 Neglect	VD 3404 Oversimplification	VC 1343 Preparedness
VP 5533 Neglect ; Carefulness	VD 3406 Overstrain	VP 5720 Preparedness-Unpreparedness
VD 3312 Negligence	VP 5660 Oversufficiency-Insufficiency	VC 1345 Presence
VC 1185 Neighbourliness	VD 3408 Oversupply	VP 5186 Presence-Absence
VD 3314 Nervousness	VD 3410 Overtax	VC 1347 Preservation
VP 5690 Neurosis ; Selfactualization	VD 3412 Overturn	VC 1573 Preservation ; Self
VC 1187 Neutrality	VD 3414 Overweight	VC 1349 Prestige
VP 5806 Neutrality-Compromise	VD 3416 Overwork	VD 3478 Presumption
VC 1189 Newness	VD 3418 Overzealousness	VD 3480 Pretentiousness
VP 5122 Newness-Oldness		VC 1351 Prettiness
VC 1191 Nimbleness	**P**	VD 3482 Prevarication
VC 1193 Nobility		VP 5905 Pride-Humility
VD 3316 Noisome	VC 1243 Pacifism	VC 1353 Principle
VD 3318 Nonacceptance	VD 3420 Paganism	VC 1355 Priority
VD 3320 Nonaccomplishment	VD 3422 Pain	VC 1357 Privacy
VP 5722 Nonaccomplishment ; Accomplishment	VC 1245 Painstakingness	VD 3484 Privation
VD 3322 Nonadherence	VD 3424 Panic	VC 1359 Privilege
VC 1195 Nonagressivity	VD 3426 Paralysis	VC 1361 Probity
VC 1197 Nonalignment	VC 1247 Pardon	VP 5974 Probity-Improbity
VD 3324 Nonconformity	VC 1249 Parenthood	VD 3486 Problem
VP 5082 Nonconformity ; Conformity	VC 1251 Parsimony	VD 3488 Prodigality
VP 5001 Nonexistence ; Existence	VC 1253 Participation	VP 5851 Prodigality ; Economy
VD 3326 Nonfeasance	VC 1255 Partnership	VC 1363 Prodigiousness
VD 3328 Nonfulfilment	VC 1257 Passion	VP 5167 Production-Reproduction
VP 5417 Nonhumanity ; Humanity	VD 3428 Passionlessness	VC 1365 Productiveness
VD 3330 Nonobservance	VD 3430 Passivity	VP 5165 Productiveness-Unproductiveness
VP 5768 Nonobservance ; Observance	VC 1259 Paternity	VD 3490 Profanation
VD 3332 Nonsense	VC 1261 Patience	VC 1367 Proficiency
VD 3334 Nonuniformity	VP 5861 Patience-Impatience	VC 1369 Profitability
VP 5017 Nonuniformity ; Uniformity	VC 1265 Peace	VD 3492 Profligacy
VC 1199 Nonviolence	VC 1267 Peacefulness	VC 1371 Profundity
VC 1201 Normality	VD 3432 Pedantry	VC 1373 Progress
VC 1203 Notability	VC 1271 Percipience	VP 5294 Progression-Regression
VD 3336 Nothingness	VD 3434 Perdition	VD 3494 Prohibition
VD 3338 Notoriety	VC 1273 Perfection	VD 3496 Proliferation
VC 1205 Novelty	VP 5677 Perfection-Imperfection	VD 3498 Prolixity
VD 3340 Nuisance	VD 3436 Perfidiousness	VD 3500 Promiscuity
VD 3342 Nullity	VC 1275 Permanence	VC 1375 Promise
VT 8015 Number*complex	VP 5139 Permanence ; Change	VC 1377 Promptness
VP 5086 Numbered-Unnumbered	VD 3438 Permissiveness	VT 8041 Properties*complex ; Absolute
VP 5101 Numerousness-Fewness	VC 1277 Perpetuation	VT 8042 Properties*complex ; Relative
	VP 5112 Perpetuity-Instantaneousness	VC 1379 Propriety
O	VD 3440 Perplexity	VC 1381 Prosperity
	VC 1279 Perseverance	VP 5728 Prosperity-Adversity
VC 1207 Obedience	VC 1281 Persistence	VD 3502 Prostitution
VP 5766 Obedience-Disobedience	VC 1285 Perspective	VC 1383 Protection
VC 1209 Objectivity	VC 1285 Perspicacity	VP 5505 Provability-Unprovability
VD 3344 Oblivion	VC 1287 Perspicuity	VC 1917 Provided ; Well
VD 3346 Obscurity	VC 1289 Persuasiveness	VD 3504 Provocation
VC 1211 Observance	VC 1291 Pertinacity	VC 1385 Proximity
VP 5768 Observance-Nonobservance	VD 3442 Perturbation	VC 1387 Prudence
VD 3348 Obsession	VD 3444 Perversion	VD 3506 Puerility
VD 3350 Obsolescence	VD 3446 Perversity	VC 1389 Punctuality
VD 3352 Obstacle	VD 3448 Pessimism	VP 6010 Punishment ; Atonement
VD 3354 Obstinacy	VD 3450 Pestiferousness	VC 1391 Purity
VD 3356 Obstruction	VD 3452 Pettiness	VC 1393 Purposefulness
VC 1213 Occurrence	VD 3454 Petulance	VD 3508 Purposelessness
VD 3358 Offence	VC 1293 Philosophy	VD 3510 Putrefaction
VP 5122 Oldness ; Newness	VC 1295 Piety	
VD 3360 Omission	VP 6028 Piety-Impiety	**Q**
VC 1215 Omnipotence	VC 1297 Piousness	
VC 1217 Omnipresence	VD 3456 Pitilessness	VC 1395 Qualification
VC 1219 Omniscience	VP 5944 Pitilessness ; Compassion	VC 1397 Quality
VD 3362 Opacity	VC 1299 Pity	VT 8013 Quantity*complex

INDEX TO HUMAN VALUES

VD 3512	Quarrelsomeness
VC 1399	Quickness
VC 1401	Quiescence
VP 5267	Quiescence ; Motion
VC 1403	Quietness
VC 1405	Quietude
VC 1407	Quintessence

R

VC 1409	Radiance
VD 3514	Randomness
VD 3516	Rapacity
VC 1411	Rapture
VC 1413	Rareness
VP 5894	Rashness ; Caution
VC 1415	Rationality
VD 3518	Ravishment
VP 5283	Reaction ; Impact
VC 1417	Readiness
VC 1419	Reality
VC 1421	Realization
VC 1423	Reason
VP 5481	Reason ; Intuition
VC 1425	Reasonableness
VD 3520	Recalcitrance
VP 5296	Recession ; Approach
VC 1427	Recognition
VC 1429	Recollection
VC 1431	Recommendation
VC 1433	Reconciliation
VC 1435	Recreation
VD 3522	Recrimination
VC 1437	Rectitude
VC 1439	Redemption
VT 8095	Redemption∗complex
VC 1441	Refinement
VC 1443	Reform
VP 5695	Refreshment-Relapse
VP 5775	Refusal ; Consent
VD 3524	Regression
VP 5294	Regression ; Progression
VC 1445	Regularity
VP 5137	Regularity-Irregularity
VC 1447	Regulation
VC 1449	Rehabilitation
VC 1451	Reinforcement
VD 3526	Rejection
VP 5695	Relapse ; Refreshment
VP 5009	Relatedness-Unrelatedness
VC 1453	Relation
VT 8012	Relationship∗complex
VT 8035	Relative motion∗complex
VT 8042	Relative properties∗complex
VC 1455	Relaxation
VD 3528	Relentlessness
VC 1457	Relevance
VC 1459	Reliability
VC 1577	Reliance ; Self
VC 1461	Relief
VP 5537	Remembrance-Forgetfulness
VC 1463	Remission
VD 3530	Remorselessness
VC 1465	Renewableness
VC 1467	Renunciation
VC 1469	Repentance
VP 5573	Representation-Misrepresentation
VD 3532	Repression
VD 3534	Reproach
VP 5167	Reproduction ; Production
VD 3536	Repudiation
VD 3538	Repugnance
VD 3540	Repulsion
VP 5288	Repulsion ; Attraction
VC 1471	Reputation
VP 5914	Repute-Disrepute
VP 5485	Research-Discovery
VD 3542	Resentment
VC 1473	Reservedness
VC 1475	Resilience
VC 1477	Resistance
VC 1479	Resolution
VP 5624	Resolution-Irresolution
VC 1481	Resourcefulness
VP 5964	Respect-Disrespect
VC 1579	Respect ; Self
VC 1483	Respectability
VC 1485	Respectfulness
VC 1487	Resplendence
VC 1489	Responsibility
VC 1491	Responsiveness
VP 5709	Rest ; Exertion
VC 1493	Restfulness
VD 3544	Restlessness
VP 5693	Restoration-Destruction
VC 1495	Restraint
VP 5762	Restraint ; Freedom
VC 1581	Restraint ; Self
VD 3546	Restriction
VD 3548	Retaliation
VD 3550	Retardation
VD 3552	Retribution
VT 8094	Retribution∗complex
VD 3554	Retrogression
VC 1497	Revelation
VD 3556	Revenge
VC 1499	Reverence
VD 3558	Reversion
VP 5145	Reversion ; Conversion
VD 3560	Revilement
VC 1501	Revival
VD 3562	Revolution
VP 5147	Revolution ; Evolution
VD 3564	Revulsion
VC 1503	Richness
VD 3566	Ridicule
VC 1505	Righteousness
VD 3642	Righteousness ; Self
VC 1507	Rightness
VP 5957	Rightness-Wrongness
VC 0329	Rights ; Civil
VD 3568	Rigidity
VC 1509	Rigorousness
VD 3570	Risk
VD 3572	Riskiness
VD 3574	Rivalry
VD 3576	Roguery
VC 1511	Romance
VD 3578	Rot
VD 3580	Roughness
VC 1513	Royalty
VD 3582	Rubbish
VD 3584	Rudeness
VD 3586	Ruin
VD 3588	Ruthlessness

S

VD 3590	Sabotage
VC 1515	Sacredness
VD 3592	Sacrifice
VC 1583	Sacrifice ; Self
VD 3594	Sacrilege
VD 3596	Sadism
VD 3598	Sadness
VC 1517	Safety
VP 5697	Safety-Danger
VC 1519	Sagacity
VC 1521	Saintliness
VC 1523	Salubrity
VC 1525	Salvation
VD 3600	Sameness
VC 1527	Sanctification
VD 3602	Sanctimony
VC 1529	Sanctity
VP 6026	Sanctity-Unsanctity
VC 1531	Sanity
VP 5472	Sanity-Insanity
VD 3604	Sarcasm
VD 3606	Satire
VC 1533	Satisfaction
VD 3644	Satisfaction ; Self
VD 3608	Savagery
VP 5427	Savouriness-Unsavouriness
VD 3610	Scandal
VD 3612	Scantiness
VD 3614	Scarcity
VC 1535	Scrupulousness
VD 3616	Seclusion
VD 3618	Secrecy
VC 1537	Security
VD 3620	Sedition
VD 3622	Seduction
VD 3624	Segregation
VD 3626	Self-absorption
VC 1539	Self-actualization
VD 3628	Self-admiration
VC 1541	Self-advancement
VC 1543	Self-assurance
VD 3630	Self-centredness
VC 1545	Self-confidence
VC 1547	Self-containedness
VD 3632	Self-contradiciton
VC 1549	Self-control
VC 1551	Self-denial
VD 3634	Self-destruction
VC 1553	Self-determination
VC 1555	Self-direction
VC 1557	Self-discipline
VC 1559	Self-effacement
VD 3636	Self-esteem
VC 1561	Self-expression
VC 1563	Self-fulfilment
VC 1565	Self-government
VC 1567	Self-help
VD 3638	Self-importance
VD 3640	Self-indulgence
VC 1569	Self-interest
VC 1571	Self-possession
VC 1573	Self-preservation
VC 1577	Self-reliance
VC 1579	Self-respect
VC 1581	Self-restraint
VD 3642	Self-righteousness
VC 1583	Self-sacrifice
VD 3644	Self-satisfaction
VD 3585	Self-sufficiency
VP 5690	Selfactualization-Neurosis
VD 3646	Selfishness
VP 5978	Selfishness ; Unselfishness
VC 1587	Selflessness
VC 1589	Sensation
VP 5422	Sensation-Insensibility
VC 0363	Sense ; Common
VT 8044	Sense∗complex
VD 3648	Senselessness
VC 1591	Sensibility
VC 1593	Sensibleness
VC 1595	Sensitivity
VD 3650	Sensuality
VC 1597	Sentiment
VD 3652	Sentimentality
VD 3654	Separateness
VP 5047	Separation ; Conjunction
VC 1599	Serendipity
VC 1601	Serenity
VC 1603	Seriousness
VC 1605	Service
VD 3656	Servility
VC 1607	Settlement
VD 3658	Severance
VD 3660	Severity
VC 1609	Sex appeal
VP 5419	Sexiness-Unsexiness
VD 3662	Sexuality
VD 3664	Shallowness
VP 5209	Shallowness ; Depth
VD 3666	Shamelessness
VD 3668	Shapelessness
VP 5815	Sharing-Appropriation
VC 1611	Shelter
VD 3670	Shiftlessness
VD 3672	Shock
VD 3674	Shortage
VP 5313	Shortcoming ; Overrunning
VD 3676	Sickness
VC 1613	Significance
VP 5451	Silence-Loudness
VC 1615	Similarity
VP 5020	Similarity-Dissimilarity
VC 1617	Simplicity
VP 5045	Simplicity-Complexity
VC 1619	Sincerity
VP 5733	Skilfulness-Unskilfulness
VC 1621	Skill
VD 3678	Slant
VD 3680	Slothfulness
VD 3682	Slowness
VP 5269	Slowness ; Swiftness
VD 3684	Smallness
VP 5034	Smallness ; Greatness
VD 3686	Snobbery
VC 1623	Sobriety
VC 1625	Sociability
VP 5922	Sociability-Unsociability
VT 8084	Socialization∗complex
VD 3688	Softness
VP 5356	Softness ; Hardness
VC 1627	Solemnity
VP 5870	Solemnity ; Cheerfulness
VC 1629	Solicitude
VC 1631	Solidarity
VC 1633	Solidity
VC 1635	Solitude
VD 3690	Sophistry
VD 3692	Sordidness
VD 3694	Soreness
VD 3696	Sorrow
VD 3698	Soullessness
VC 1637	Soundness
VD 3700	Sourness
VC 1639	Sovereignty
VT 8025	Space∗complex
VD 3702	Sparsity
VD 3704	Spasmodicness
VD 3706	Spasticity
VD 3708	Speciousness
VC 0901	Spiritedness ; High
VD 3710	Spiritlessness
VC 1641	Spirituality
VC 1643	Splendour
VD 3712	Spoilage
VD 3714	Spoilation
VC 1645	Spontaneity
VC 1647	Sportsmanship
VC 1649	Spotlessness
VD 3716	Spunklessness
VD 3718	Spurious
VD 3720	Squalor
VC 1651	Stability
VP 5141	Stability-Changeableness
VD 3722	Stagnation
VD 3724	Stain
VC 1653	Stainlessness
VD 3726	Stalemate
VC 1655	Stamina
VC 1657	Standing
VC 1659	Stateliness
VC 1663	Status
VC 1665	Staunchness
VC 1667	Steadfastness
VC 1669	Steadiness
VP 5435	Stench ; Fragrance
VD 3728	Sterility
VC 1671	Stewardship
VD 3730	Stiffness

-813-

VX

VC 1673 Stillness	VP 5129 Timeliness-Untimeliness	VP 5535 Unimaginativeness ; Imaginativeness
VC 1675 Stimulation	VD 3818 Timidity	VD 3920 Unimportance
VC 1677 Stintlessness	VD 3820 Tiresomeness	VP 5672 Unimportance ; Importance
VC 1679 Straightforwardness	VC 1763 Togetherness	VC 1823 Uninfluence
VD 3732 Strain	VD 3822 Toilsomeness	VD 3922 Unintelligence
VD 3734 Strangeness	VC 1765 Tolerance	VP 5467 Unintelligence ; Intelligence
VC 1681 Strength	VD 3824 Torment	VP 5548 Unintelligibility ; Intelligibility
VP 5159 Strength-Weakness	VD 3826 Torpor	VC 1825 Union
VC 1683 Strictness	VD 3830 Tortuousness	VC 1827 Uniqueness
VD 3736 Stridency	VD 3828 Torture	VC 1829 Unity
VT 8033 Structure∗complex	VC 1767 Toughness	VP 5089 Unity-Duality
VD 3738 Stubbornness	VP 5358 Toughness ; Elasticity	VC 1831 Universality
VD 3740 Stuffiness	VD 3832 Toxicity	VD 3924 Unkindness
VD 3742 Stupidity	VC 1769 Tractability	VP 5938 Unkindness ; Kindness
VD 3744 Stupor	VC 1771 Tradition	VD 3926 Unlawfulness
VC 1685 Sturdiness	VD 3834 Traitorousness	VC 1833 Unlimited
VC 1687 Style	VC 1773 Tranquillity	VP 5507 Unlimitedness ; Limitation
VD 3746 Subjection	VC 1775 Transcendence	VD 3928 Unmanageability
VD 3748 Subjugation	VC 1777 Transcience	VD 3930 Unnaturalness
VC 1689 Submission	VC 1779 Transformation	VD 3932 Unneighbourliness
VD 3750 Subordination	VD 3836 Transgression	VP 5086 Unnumbered ; Numbered
VC 1691 Subservience	VD 3838 Transience	VC 1835 Unobtrusiveness
VC 1693 Substantiality	VP 5110 Transience ; Durability	VP 6024 Unorthodoxy ; Orthodoxy
VP 5003 Substantiality-Unsubstantiality	VP 5339 Transparency-Opaqueness	VD 3934 Unpeacefulness
VD 3752 Subterfuge	VD 3840 Travesty	VD 3936 Unpleasantness
VC 1695 Subtlety	VD 3842 Treachery	VP 5863 Unpleasantness ; Pleasantness
VD 3754 Subversion	VD 3844 Treason	VD 3938 Unpredictability
VC 1697 Success	VD 3846 Tremulousness	VD 3940 Unpreparedness
VD 3756 Suddenness	VD 3848 Trepidation	VP 5720 Unpreparedness ; Preparedness
VD 3758 Sufferance	VD 3850 Trespass	VC 1837 Unpretention
VC 1699 Sufficiency	VD 3852 Tribulation	VD 3942 Unproductiveness
VC 1585 Sufficiency ; Self	VD 3854 Trickery	VP 5165 Unproductiveness ; Productiveness
VD 3760 Suffocation	VC 1781 Triumph	VP 5505 Unprovability ; Provability
VC 1701 Suitability	VD 3856 Triviality	VD 3944 Unreality
VD 3762 Sumptuousness	VD 3858 Trouble	VD 3946 Unreasonableness
VC 1703 Superabundance	VD 3860 Truculence	VP 5009 Unrelatedness ; Relatedness
VD 3764 Superciliousness	VC 1783 Trust	VD 3948 Unreliability
VD 3766 Superficiality	VC 1785 Trustworthiness	VD 3950 Unruliness
VC 1705 Superfluity	VC 1787 Truth	VP 6026 Unsanctity ; Sanctity
VC 1707 Superhumanness	VT 8053 Truth∗complex	VP 5427 Unsavouriness ; Savouriness
VC 1709 Superiority	VP 5516 Truth-Error	VC 1839 Unselfishness
VP 5036 Superiority-Inferiority	VC 1789 Truthfulness	VP 5978 Unselfishness-Selfishness
VD 3768 Superstition	VD 3862 Truthlessness	VP 5419 Unsexiness ; Sexiness
VC 1711 Support	VD 3864 Turbidity	VP 5733 Unskilfulness ; Skilfulness
VP 5785 Support-Opposition	VD 3866 Turbulence	VP 5922 Unsociability ; Sociability
VD 3770 Suppression	VD 3868 Turpitude	VD 3952 Unsophistication
VC 1713 Supremacy	VD 3870 Tyranny	VC 1841 Unspotted
VC 1715 Sureness		VP 5003 Unsubstantiality ; Substantiality
VC 1717 Surety	**U**	VD 3954 Unsuitability
VD 3772 Surfeit		VD 3956 Unthoughtfulness
VD 3774 Surliness	VC 1791 Ubiquity	VP 5129 Untimeliness ; Timeliness
VD 3776 Surplus	VD 3872 Ugliness	VD 3958 Untrustworthiness
VC 1719 Surprise	VP 5899 Ugliness ; Beauty	VD 3960 Unwholesomeness
VD 3778 Surreptitiousness	VC 1793 Unadornment	VP 5621 Unwillingness ; Willingness
VC 1721 Survival	VC 1795 Unadulteration	VD 3962 Upheaval
VD 3780 Susceptibility	VD 3874 Unadvisedness	VC 1843 Uplift
VD 3782 Suspense	VC 1797 Unambiguity	VC 1845 Uprightness
VD 3784 Suspension	VC 1799 Unanimity	VD 3964 Uproar
VD 3786 Suspicion	VC 1801 Unassuming	VD 3966 Upset
VC 1723 Sustenance	VP 5920 Unastonishment ; Wonder	VD 3968 Urgency
VC 1725 Sweetness	VD 3876 Unauthenticity	VC 1847 Usefulness
VC 1727 Swiftness	VD 3878 Unbearableness	VD 3970 Uselessness
VP 5269 Swiftness-Slowness	VP 5501 Unbelief ; Belief	VD 3972 Usurpation
VC 1729 Symmetry	VC 1803 Unbigoted	VC 1849 Utility
VP 5248 Symmetry-Distortion	VC 1805 Unblemished	
VC 1731 Sympathy	VD 3880 Uncertainty	**V**
VC 1733 Synergy	VP 5513 Uncertainty ; Certainty	
VC 1735 Synthesis	VC 1807 Unchangeable	VD 3974 Vacillation
	VD 3882 Unchastity	VD 3976 Vagrancy
T	VP 5987 Unchastity ; Chastity	VD 3978 Vagueness
	VD 3884 Uncleanliness	VD 3980 Vainness
VC 1737 Tact	VP 5681 Uncleanness ; Cleanness	VC 1853 Validity
VD 3788 Tactlessness	VD 3886 Uncomfortableness	VC 1851 Valour
VC 1739 Talent	VD 3888 Uncommunicativeness	VC 1855 Value
VC 1741 Tameness	VP 5554 Uncommunicativeness ; Communicativeness	VD 3982 Valueless
VC 1743 Tangibility	VD 3890 Unconscionableness	VD 3984 Vandalism
VP 5425 Tangibility-Intangibility	VP 5642 Unconventionality ; Conventionality	VD 3986 Vanity
VD 3790 Tardiness	VC 1809 Uncorrupted	VP 5908 Vanity ; Modesty
VP 5896 Taste-Vulgarity	VD 3892 Unctuousness	VD 3988 Vanquishment
VC 1745 Tastefulness	VD 3894 Undependability	VC 1857 Vantage
VD 3792 Tastelessness	VD 3896 Underdevelopment	VD 3990 Vapidity
VD 3794 Tedium	VD 3898 Underestimation	VD 3992 Variance
VD 3796 Temper	VP 5497 Underestimation ; Overestimation	VD 3994 Variation
VC 1747 Temperance	VC 1811 Understanding	VC 1859 Variety
VP 5992 Temperance-Intemperance	VD 3900 Undesirableness	VC 1861 Vastness
VD 3798 Tempestuousness	VC 1813 Undivided	VD 3996 Vehemence
VD 3800 Temptation	VP 5960 Undueness ; Dueness	VD 3998 Venality
VC 1749 Tenacity	VD 3902 Uneasiness	VC 1863 Venerableness
VC 1751 Tenderness	VP 5600 Uneloquence ; Eloquence	VC 1865 Veneration
VD 3802 Tension	VC 1815 Unequivocalness	VP 5947 Vengeance ; Forgiveness
VD 3804 Tenuousness	VC 1817 Unerring	VD 4000 Venial
VD 3806 Terror	VD 3904 Unevenness	VD 4002 Venom
VD 3808 Thinness	VC 1819 Unfailing	VC 1867 Venturesomeness
VC 1753 Thoroughness	VD 3906 Unfaithfulness	VC 1869 Veracity
VP 5478 Thought-Thoughtlessness	VD 3908 Unfamiliarity	VD 4004 Verbosity
VC 1755 Thoughtfulness	VP 5855 Unfeelinglessness ; Feeling	VC 1871 Veritableness
VD 3810 Thoughtlessness	VD 3910 Unfriendliness	VC 1873 Versatility
VP 5478 Thoughtlessness ; Thought	VD 3912 Ungodliness	VD 4006 Vexation
VD 3812 Threat	VP 6013 Ungodliness ; Godliness	VC 1875 Viability
VC 1757 Thriftiness	VD 3914 Ungraciousness	VP 5980 Vice ; Virtue
VD 3814 Thriftlessness	VD 3916 Unhappiness	VD 4008 Viciousness
VD 3816 Tightness	VP 5683 Unhealthfulness ; Healthfulness	VD 4010 Vicissitude
VT 8021 Time∗complex	VC 1821 Uniformity	VD 4012 Victimization
VC 1759 Timelessness	VP 5017 Uniformity-Nonuniformity	VC 1877 Victory
VC 1761 Timeliness	VD 3918 Unimaginativeness	VP 5726 Victory-Defeat

INDEX TO HUMAN VALUES

V (cont.)

- VC 1879 Vigilance
- VC 1881 Vigour
- VD 4014 Vileness
- VD 4016 Villainy
- VP 6005 Vindication-Condemnation
- VD 4018 Vindictiveness
- VD 4020 Violation
- VD 4022 Violence
- VC 1883 Virginity
- VC 1885 Virility
- VC 1887 Virtue
- VP 5980 Virtue-Vice
- VD 4024 Virulence
- VP 5444 Visibility-Invisibility
- VC 1889 Vision
- VP 5439 Vision-Blindness
- VC 1891 Vitality
- VD 4026 Vitiation
- VD 4028 Vituperation
- VC 1893 Vivacity
- VD 4030 Volatility
- VC 1895 Voluptuousness
- VD 4032 Voracity
- VD 4034 Vulgarity
- VP 5896 Vulgarity ; Taste
- VD 4036 Vulnerability

W

- VD 4038 Wantonness
- VD 4040 War
- VD 4042 Warlike
- VC 1897 Warmheartedness
- VC 1899 Warmth
- VD 4044 Warpedness
- VD 4046 Waste
- VD 4048 Wastefulness
- VD 4050 Weak-mindedness
- VD 4052 Weakness
- VP 5159 Weakness ; Strength
- VC 1901 Wealth
- VP 5837 Wealth-Poverty
- VD 4054 Wear
- VD 4056 Weariness
- VD 4058 Wearisomeness
- VP 5352 Weight-Lightness
- VC 1903 Welfare
- VC 1905 Well-disposed
- VC 1907 Well-grounded
- VC 1909 Well-informed
- VC 1911 Well-made
- VC 1913 Well-mannered
- VC 1915 Well-meaning
- VC 1917 Well-provided
- VC 1919 Wholeness
- VC 1921 Wholesomeness
- VC 1923 Will
- VD 4060 Willfulness
- VC 1925 Willingness
- VP 5621 Willingness-Unwillingness
- VC 1927 Wisdom
- VD 4062 Withdrawal
- VD 4064 Withering
- VD 4066 Witlessness
- VC 1929 Wittiness
- VD 4068 Woe
- VC 1931 Womanliness
- VC 1933 Wonder
- VP 5920 Wonder-Unastonishment
- VC 1935 Wonderfulness
- VC 1937 Work
- VD 4070 Worry
- VC 1939 Worth
- VC 1941 Worthiness
- VD 4072 Worthlessness
- VD 4074 Wound
- VD 4076 Wrangle
- VD 4078 Wrongness
- VP 5957 Wrongness ; Rightness

X

- VD 4080 Xenophobia

Y

- VC 1943 Youth
- VP 5124 Youth-Age
- VC 1945 Youthfulness

Z

- VC 1947 Zeal
- VD 4082 Zealotry
- VC 1949 Zest

References

Human Values and Wisdom

Scope of bibliography

The following section constitutes a collection of bibliographical references relevant to the Human Values (Section V).

The entries appear here for any of the following reasons:

(a) Reference is made to them in the individual entries of Section V (where they appear in abridged form, namely with author, title and year of publication).

(b) Reference is made to them in the commentary on Section V (which appears in Section VZ).

(c) The publications are relevant to Section V, even though they have not (yet) been cross-referenced in either of the two ways indicated above.

Bibliographical entries

The entries appear by order of author and, within author, by year.

In certain cases the *International Standard Book Number (ISBN)* is given to facilitate access.

Source of bibliography

The bibliography was derived from a wide variety of sources. These include:

- Information from international organizations, including catalogues of publications and accessions lists

- Information from commercial publishers, especially sales catalogues

- Citations and bibliographies in other publications and reference books

- Systematic compilations of books in print (notably *Books in Print* and *International Books in Print*)

- Book reviews in specialized journals

Abbott, Ann A Professional Choices: values at work (1988)
Silver Spring MD, National Association of Social Workers, 258 p.

Abita, E A Los valores de los mexicanos Mexico: entre la tradicion y la modernidad (1986)
Mexico, Famento Cultural Banamex.

Abrahams, M; Gerard, D and Timms, N (Eds) Values and social change in Britain (1985)
London, Macmillan.

Adels, Jill The Wisdom of the Saints: an anthology of voices (1989)
Oxford, Oxford University Press, 256 p.
ISBN 0–19–505915–8.

Agricola, Johannes Die Sprichwoertersammlungen (1971)
Berlin, Walter de Gruyter, 989 p. 2 Vols. Ausgaben Deutscher Literatur des XV bis XVIII Jahrhunderts.
ISBN 3–11–003710–6.

Al-Arabi, Muhyiddin The Seals of Wisdom (1983)
Santa Barbara CA, Concord Grove Press. illus. Sacred Texts Series. ISBN 0–88695–010–4.

Albert, Ethel M The Classification of Values: a method and illustration
In: *American Anthropologist* 1956, 58, pp. 221–248.

Albert, Ethel M and Clyde, Kluckhohn A Selected Bibliography on Values, Ethics and Esthetics in the Behavioral Sciences and Philosophy: 1920 –1958 (1959)
Bloomington, Glencoe.

Alexander, Kern and Mc Keown, Mary P (Eds) Values in Conflict: funding priorities for higher education (1987)
Cambridge MA, Ballinger, 360 p. ISBN 0–88730–102–9.

Alger, Chadwick F Values in Global Issues: the global dalectic in value clarification (1980)
Columbus OH, Ohio State University.

Alisjahbana, Sutan T Values as Integrating Forces in Personality, Society and Culture: essay of a new anthropology (1966)
London, Oxford University Press, 248 p. bibl. foot.

Allport, Gordon W; Gardner Lindsey and Philip, E Vernon A Study of Values (1951)
Boston MA. rev ed.

Almond, Brenda and Wilson, Bryan (Eds) Values: a symposium (1988)
Atlantic Highlands NJ, 300 p. ISBN 0–391–03368–9.

American Institute for Psychological Research The Psychological Wisdom from the Sanskrit (1985)
American Institute for Psychological Research, 301 p. 2 Vols.
ISBN 0–89920–080–X.

Anon Encyclopedia of Arcane Wisdom
Falls Church VA, Arcane Order Studio of Contemplation, 133 p. ISBN 0–318–13345–8.

Anon Shaping the Future: biological research and human values (1989)
Washington DC, National Academy Press, 112 p.
ISBN 0–309–03944–4.

Aron, Raymond; Dyson, Freeman and Robinson, Joan Values at War: selected Tanner lectures on the nuclear crisis (1983)
Salt Lake City UT, University of Utah Press, 130 p.
ISBN 0–87480–226–1.

Arora, Ramesh K and Raghavulu, C V (Eds) Values in Administration (1990)
New York, Advent Books.

Arrow, Kenneth J Social Choice and Individual Values (1963)
New York. 2nd ed.

Attig, Thomas; Callen, Donald and Sumner, Wayne L (Eds) Values and Moral Standing (1987)
Bowling Green OH, Bowling Green State University, 167 p. Studies in Applied Technology: VIII. ISBN 0–935756–09–4.

Bagley, Ayers (Ed) An Investigation to Wisdom and Schooling (1985)
Edinburg TX, Society of Professors of Education. SPE Monograph Series. ISBN 0–933669–35–6.

Bahadur, K P The Wisdom of Yoga (1989)
New York, Apt Books. ISBN 81–207–0330–8.

Bahadur, K P The Wisdom of Upanisads (1989)
New York, Apt Books, 400 p. ISBN 81–207–0896–2.

Bahm, Orchie J Axiology: the science of values (1984)
Albuquerque NM, Archie J Bahm, 84 p. abbreviated ed.
ISBN 0–911714–14–6.

Baier, Kurt The Concept of Value
In: *E Laszlo and J B Wilbur (Eds) Value Theory in Philosophy and Social Science* New York, Gordon and Breach, 1973, pp. 1–11.

Baier, Kurt What is Value?: an analysis of the concept
In: *K Baier and N Rescher (Eds) Values and the Future* New York, Free Press, 1969, pp. 33–67.

Baier, Kurt and Rescher, Nicholas (Eds) Values and the Future: the impact of technological change on American values (1969)
New York, The Free Press, 527 p. bibl.

Baird, Mary K and Mitchell, Arnold American Values (1969)
SRI International.

Bancroft, Anne Weavers of Wisdom (1990)
New York, Penguin Books, 192 p. ISBN 0–14–019193–3.

Bankowski, Z and Bryant, J H (Eds) Health Policy, Ethics and Human Values: European and North American perspectives: conference highlights, papers and conclusions, XXIst CIOMS Conference (1988)
Geneva, WHO, 223 p. ISBN 92–9036–034–8.

Bawa Muhaiyaddeen, M R Wisdom of Man: selected discourses (1980)
Philadelphia PA, Fellowship Press, 168 p. illus. Transl. from Tamil. ISBN 0–914390–16–3.

Baz, Petros D (Ed) A Dictionary of Proverbs (1984)
New York, Philosophical Library. ISBN 0–8022–0086–9.

Belcher Humanities: life styles and human values (1983)
Dubuque IA, Kendall/Hunt Publishing, 240 p. vol 3.
ISBN 0–8403–2924–5.

Bence, Kathy Developing Christian Values (1990)
Cincinnati OH, Standard Publishing, 80 p. illus. Discipling Young Teens Series. ISBN 0–87403–640–2.

Benson, Larry D and Wenzel, Siegfried (Eds) The Wisdom of Poetry: essays in early English literature in honor of Morton W Bloomfield (1982)
Kalamazoo MI, Medieval Institute Publications, 315 p.
ISBN 0–918720–15–X.

Berenda, C W World Visions and the Image of Man: cosmologies as reflections of man (1965)
New York, Vantage Press.

Berg, Ivar and Murphey, Murray G Values and Value Theory in Twentieth Century America: essays in honor of Elizabeth Flower (1988)
Philadelphia PA, Temple University Press, 308 p.
ISBN 0–87722–557–5.

Bingham, Golin Wit and Wisdom: a public affairs miscellany (1982)
Portland OR, International Specialized Book Services, 368 p.
ISBN 0–522–84241–0.

Black Antony State, Community and Human Desire: a group-centured account of political values (1988)
New York, Saint Martin's Press, 232 p.
ISBN 0–312–01984–X.

Blofeld, John Gateway to Wisdom: Taoist and Buddhist contemplative and healing yogas adapted for Western students of the way

Boadt, Lawrence E Introduction to Wisdom Literature Proverbs (1986)
Collegeville MN, Liturgical Press, 104 p. Collegeville Bible Commentary, Old Testament Series: 18.
ISBN 0–8146–1475–2.

Bohn, Henry G Handbook of Proverbs
New York, AMS Press. Repr of 1855. Bohn's Antiquarian Library Series. ISBN 0–404–50003–X.

Bohn, Henry G Polyglot of Foreign Proverbs
New York, AMS Press. Repr of 1857. Bohn's Antiquarian Library Series. ISBN 0–404–50004–8.

Boulding, Kenneth E Ecodynamics: a new theory of societal evolution (1978)
London, Sage.

Bredo, Eric and Feinberg, Walter (Eds) Knowledge and Values in Social and Educational Research (1982)
Philadelphia PA, Temple University Press, 456 p.
ISBN 0–87722–242–8.

Brestin, Dee and Brestin, Steve Proverbs and Parables: God's wisdom for living (1975)
Wheaton IL, Harold Shaw, 75 p. ISBN 0–87788–694–6.

Bronowski, J Science and Human Values (1990)
New York, Harper and Row, 128 p. Repr of 1956.
ISBN 0–06–097281–5.

Brown, B The Essence of Chinese Wisdom (1986)
Albuquerque NM, Foundation for Classical Reprints, 227 p.
ISBN 0–89901–279–5.

Brown, Brian The Wisdom of the Chinese: their philosophy in sayings and proverbs
New York, Krishna Press. ISBN 0–87968–138–1.

Brown, Hanbury The Wisdom of Science: its relevance to culture and religion (1986)
Cambridge, Cambridge University Press, 194 p. illus.
ISBN 0–521–31448–8.

Brown, J and Comber, M Family values – a secondary analysis of EVSSG data
Guilford, University of Surrey.

Browne, Lewis Wisdom of the Jewish People (1988)
Northvale NJ, Aronson Jason, 773 p.
ISBN 0–87668–985–3.

Brunton, Paul The Wisdom of the Overself (1970)
New York, Samuel Weiser.

Buchanan, Daniel C (Ed) Japanese Proverbs and Sayings (1987)
Norman OK, University of Oklahoma Press, 296 p.
ISBN 0–8061–1082–1.

Bühler, Charlotte Values in Psychotherapy (1962)
New York, Free Press.

Bukovskaia, M V A Dictionary of English Proverbs in Modern Use (1985)
New York, State Mutual Book and Periodical Service, 232 p.

Buono, Anthony F and Nichols, Lawrence T Corporate Policy, Values and Social Responsibility (1985)
New York, Praeger Publishers, 240 p.
ISBN 0–275–90068–1.

Burckhardt, J L Arabic Proverbs (1988)
New York, State Mutual Book and Periodical Service, 296 p.
ISBN 0–85077–183–9.

Burton, Richard F Wit and Wisdom from West Africa: a book of proverbial philosophy, idioms, enigmas and laconisms (1969)
Cheshire CT, Biblo and Tannen Booksellers and Publishers. Repr of 1865. ISBN 0–8196–0243–4.

Calvaruso, C and Abbruzzese, S Indagine sui valori in Italia: dai post-materialismi ala ricerca di senso (1985)
Torino, SEI.

Camp, Claudia Wisdom and the Feminine in the Book of Proverbs
Bible and Literature Series: 11.

Cannon, W B The Wisdom of the Body (1963)
New York, WW Norton, 333 p.

Capps, Donald Deadly and Sins Saving Virtues (1987)
Philadelphia PA, Fortress Press, 162 p.
ISBN 0–8006–1948–X.

Capra, Fritjof Uncommon Wisdom (1989)
New York, Bantam Books, 224 p. New Age Series.
ISBN 0–553–34610–5.

Carballo de Cilley, M Qué pensamos los argentinos? Los valores de los argentinos de nuestro tiempo (1987)
Buenos Aires, Sadei.

Carr, Anne E A Search for Wisdom and Spirit: Thomas Merton's theology of the self (1989)
Notre Dame IN, University of Notre Dame Press, 182 p.
ISBN 0–268–01735–2.

Caws, Peter Science and the Theory of Value (1967)
New York.

Cecil, Andrew R, et al Traditional Moral Values in the Age of Technology (1987)
Austin TX, University of Texas Press, 210 p. Andrew R Cecil Lectures on Moral Values in a Free Society: VIII.
ISBN 0–292–78098–2.

Center for the Study of Social Policy Changing Images of Man (1974)
Menlo Park, Stanford Research Institute, 319 p. Policy Research Report: 4.

Chapman, Robert L (Ed) Roget's International Thesaurus
New York, Thomas Y Crowell, 1979, 4th ed.

Chatterji, J C Wisdom of the Vedas (1980)
Wheaton IL, Theosophical Publishing House, 100 p.
ISBN 0–8356–0538–8.

Chevlot, Andrew (Ed) Proverbs, Proverbial Expressions and Popular Rhymes of Scotland (1990)
Maple Shade NJ, Omnigraphics. Repr of 1896.
ISBN 1–55888–177–8.

Churchman, C West Thought and Wisdom (1982)
Salinas CA, Intersystems Publications, 150 p.
ISBN 0–914105–03–5.

Cirillo and Wapner (Eds) Value Presuppositions in Theories of Human Development
Hillsdale NJ, Lawrense Erlbaum. ISBN 0–89859–753–6.

Clements, Ronald E Wisdom for a Changing World: wisdom in Old Testament theology (1990)
Berkeley CA, Bibal Press, 80 p. Ed by Duane L Christensen.
ISBN 0–941037–13–4.

Clinton, Stephen, et al Values and Public Policy (1988)
Washington DC, Family Research Council of America.
ISBN 1–55872–000–6.

Cochran, Thomas C Challenges to American Values: society, business and religion (1985)
Oxford, Oxford University Press. ISBN 0–19–503534–8.

Colin, Jean Pierre La Société Multidimensionnelle: d'une crise de la conscience mondiale à de nouvelles formes de solidarité (1988)
Paris, UNESCO, 95 p. Grand Programme I: Réflexion sur les problèmes mondiaux et études prospectives: BEP/GPI/20.

Coll, Alberto R The Wisdom of Statecraft: Sir Butterfield and the philosphy of international politics (1985)
Durham NC, Duke University Press, 173 p. Intro by Adam Watson. ISBN 0–8223–0607–7.

Collier (Ed) Values and Moral Development in Higher Education
New York, Routledge, Chapman and Hall.
ISBN 0–85664–171–5.

Conservation Foundation Conservation and Values: the Conservation Foundation's Thirtieth Anniversary Symposium (1979)
Washington DC, Conservation Foundation. illus.
ISBN 0–89164–053–3.

Conze, Edward (Transl) The Perfection of Wisdom in Eight Thousand Lines and Its Verse Summary (1973)
San Francisco CA, Four Seasons Foundation, 348 p. Wheel Series: 1. Transl. from Sanskrit. ISBN 0–87704–048–6.

Conze, Edward (Ed) The Large Sutra on Perfect Wisdom: with the divisions of the Abhisamayālankāra (1984)
Berkeley LA, University of California Press, 679 p.
ISBN 0–520–05312–4.

Conze, Edward Buddhist Wisdom Books (1988)
Winchester MA, Unwin Hyman, 132 p.
ISBN 0–04–440259–7.

Crenshaw, Floyd D and Flanders, John A (Eds) Christian Values and the Academic Disciplines (1985)
Lanham MD, University Press of America, 224 p.
ISBN 0–8191–4306–5.

Csikszentmihalyi, M and Rathunde, K The Psychology of Wisdom: an evolutionary interpretation (1989)
In: *R J Sternberg (Ed) The Psychology of Wisdom* New York, Cambridge University Press.

Cummings, W D and Tomoda, Y (Eds) The Revival of Values Education in Asia and the West (1988)
Elmsford NY, Pergamon Press, 192 p. illus. Comparative and International Education Series: 7. ISBN 0–08–035854–3.

Cunningham, Lawrence and Reich, John Culture and Values (1985)
New York, Holt, Rinehart and Winston, 504 p. alternate ed.
ISBN 0–03–063511–X.

Dalai Lama Ocean of Wisdom: guidelines for living by the Dalai Lama of Tibet (1989)
Weehawken NJ, Clear Light Publications.
ISBN 0–940666–09–X.

Das, Mitra and Kolack, Shirley Technology, Values and Society: social forces in technological change (1989)
New York, Peter Lang Publishing, 200 p. American University Studies, Anthropology and Sociology: 27.
ISBN 0–8204–0824–7.

Dashiell, J Frederick An Introductory Bibliography on Value
In: *The Journal of Philosophy* 1913, 10, pp. 472–476.

de Guinzbourg, V S M Wit and Wisdom of the United Nations/Esprit et Sagesse des Nations Unies: proverbs and apothegms on diplomacy (1961)
New York, Paroemiological Society.

de Vries, Joop, et al Values and Strategic Planning (1980)
SRI International.

Deane, Darshani Bliss and Common Sense: secrets of self-transformation (1989)
Wheaton IL, Theosophical Publishing House, 251 p.
ISBN 0–8356–0644–9.

Delamotte, Yves and Takezawa, Shin–ichi Quality of Working Life in International Perspective (1984)
Geneva, ILO, 89 p. ISBN 92–2–103402–X.

Delooz, P Une enquête européenne sur les valeurs
In: *La Revue Nouvelle*, Bruxelles, Janvier 1984.

Delooz, P Who believes in what?
In: *Lumen Vitae* 38, 1983/4, pp 367–380.

Derber, Milton Competing Values in America Industrial Relations (1987)
Honolulu HI, University of Hawaii at Manoa, 16 p. Occasional Publication: 160. ISBN 0–318–23505–6.

Dollen, Charles The Book of Catholic Wisdom (1986)
Huntington IN, Our Sunday Visitor, 205 p. ISBN 0–87973–535–X.

Donnelly, John (Ed) Reflective Wisdom: Richard Taylor on issues that matter (1989)
Buffalo NJ, Prometheus Books, 300 p. ISBN 0–87975–522–9.

Dukes, Willliam Psychological Studies of Values
In: *The Psychological Bulletin* 1955, 52, pp. 24–50.

Dunn (Ed) Values, Ethics and the Practice of Policy Analysis
New York, Lexington Books. ISBN 0–669–05707–X.

Erasmus, Desiderius Proverbs or Adages (1977)
Delmar NY, Scholars' Facsimiles and Reprints. ISBN 0–8201–1232–1.

Evans, Stephen C Wisdom and Humanness in Psychology: prospects for a Christian approach (1989)
Grand Rapids MI, Baker Book House, 144 p. ISBN 0–8010–3449–3.

Ewin, R E Cooperation and Human Values: a study of moral reasoning (1981)
New York, Saint Martin's Press. ISBN 0–312–16956–6.

Feinberg, Gerald The Prometheus Project: mankind's search for long–range goals (1969)
New York, Doubleday, 264 p.

Findlay, J N Values and Intentions (1961)
London.

Fischer, Frank and Forester, John (Eds) Confronting Values in Policy Analysis: the politics of criteria (1987)
San Mateo CA, Dage Publications, 320 p. Sage Yearbooks in Politics and Public Policy: 14. ISBN 0–8039–2616–2.

Fogarty, M; Ryan, L and Lee, J Irish values and attitudes: the Irish report of the European value systems study (1984)
Dublin, Dominican Publications.

Forrester, D B Beliefs, Values and Policies: conviction politics in a secular age: Hensley Henson lectures in theology 1987–1988 (1990)
Oxford, Oxford University Press, 120 p. ISBN 0–19–826194–2.

Fox, Douglas A The Heart of Buddhist Wisdom: a translation of the Heart Sutra with historical introduction and commentary (1986)
Lewistone NY, Mellen Edwin, 195 p. Studies in Asian Thought and Religion: 3. ISBN 0–88946–053–1.

Francis, Dave and Woodcock, Mike Unblocking Organizational Values (1989)
Glenview IL, Scott, Foresman and Company. ISBN 0–673–38917–0.

Frick, Dieter (Ed) Quality of Urban Life: social, psychological and physical conditions (1986)
Berlin, Walter de Gruyter, 262 p. ISBN 3–11–010577–2.

Frost, Peter J; Mitchell, Vance Fand Nord, Walter R Managerial Reality: balancing technique, practice and values (1990)
Glenview IL, Scott, Foresman and Company. ISBN 0–673–38600–7.

Gabriel, Marcel The Decline of Wisdom (1955)
New York, Allied Publications.

Gage, N L Review of Allport–Vernon Literature
In: *Fifth Mental Measurements Yearbook* New Jersey, 1959, pp. 199–202.

Gallup, George, et al Human Needs and Satisfactions: a global survey (1977)
Princeton NJ, Gallup International Research Institutes.

Galston, Orthur W Green Wisdom (1983)
Putnam Publishing Group, 240 p. illus. ISBN 0–399–50713–2.

Gangrade, K D Crisis of Values: a study in generation gap (1975)
New Delhi, Chetana Publications, 295 p.

Ghose, Sri A Wisdom of the Upanishads (1988)
Wilmot WI, Lotus Light Publications, 134 p. ISBN 0–941524–43–4.

Girard, A and Stoetzel, J Les Français et les valeurs du temps présent
In: *Revue Française de sociologie* XXVI, 1985, pp 3–31.

Global Research Institute Live One Hundred Happy Years: your personal health guide based on the wisdom of the long–living people of the world (1987)
Santa Barbara CA, Global Research Institute, 112 p. ISBN 0–9619630–1–8.

Gluski, J (Ed) Proverbs: a comparative book of English, French, German, Italian, Spanish and Russian proverbs with a Latin appendix (1989)
New York, Elsevier, 448 p. ISBN 0–444–87350–3.

Goddard, Dwight (Comp) Self–Realization of Noble Wisdom (The Lankavatara Sutra)
Compiled based on D T Suzuki's rendering.

Goldberg, David T Ethics and Social Issues (1989)
New York, Holt, Rinehart and Winston, 480 p. ISBN 0–03–014194–X.

Goldstein, Joseph and Kornfield, Jack Seeking the Heart of Wisdom: the path of insight meditation (1987)
Boston MA, Shambhala Publications, 195 p. Dragon Editions Series. ISBN 0–87773–327–9.

Grasberg, Lynn and Most, Stephen (Eds) Broken Circle: a search for wisdom in the nuclear age (1980)
Palo Alto CA, Consulting Psychologists Press, 160 p. ISBN 0–89106–038–3.

Graves, C W Levels of Existence: an open system theory of values
In: *J Humanistic Psychology* 1970, 10, pp. 131–155.

Greene, Carol Proverbs–Important Things to Know (1980)
Saint Louis, MO, Concordia Publishing House. ISBN 0–570–06140–7.

Grimm, George Buddhist Wisdom (1989)
Columbia MO, South Asia Books. ISBN 81–208–0510–0.

Grof, Stanislav (Ed) Ancient Wisdom and Modern Science (1984)
Albany NY, State University of New York Press, 285 p. ISBN 0–87395–848–9.

Grof, Stanislav East and West: ancient wisdom and modern science (1985)
Mill Valley CA, Robert Briggs Associates, 30 p. Broadside Series. ISBN 0–931191–00–9.

Gross, Feliks Ideologies, Goals and Values (1985)
Westport CT, Greenwood Publishing Group, 343 p. illus. Contributions in Sociology Series: 52. ISBN 0–8371–6377–3.

Gross, John The Oxford Book of Aphorisms (1983)
Oxford, Oxford University Press, 383 p. ISBN 0–19–214111–2.

Haddad, Yvonne Y and Lummis, Adair T Islamic Values in the United States: a comparative study (1987)
Oxford, Oxford University Press, 196 p. illus. ISBN 0–19–504112–7.

Hagenaars, J A and Halman L C Searching for idealtypes: the potentialities of latent class analysis
In: *European Sociological Review* vol 5, no 1, 1989, pp 1–16.

Halberstam, Joshua Virtues and Values: an introduction to ethics (1988)
Englewood Cliffs NJ, Prentice Hall, 384 p. ISBN 0–13–942202–1.

Halman, L Values in East and West: some tentative results and the question of comparability (1988)
Tilburg, Tilburg University, Department of Sociology, Working Paper Series no 34. Paper prepared for the International Values Conference Hungarian Academy of Sciences, Budapest, April 6–8, 1988.

Halman, L C J M Ethostypen in de Nederlandse samenleving
In: *Sociale Wetenschappen* 30, 1987, pp 117–143.

Halman, L; Heunks, F; Moor, R de and Zanders, H Traditie, secularisatie en individualisering: een studie naar de waarden van de Nederlanders in een Europese context (1987)
Tilburg, Tilburg University Press.

Hamilton, Arthur W Malay Proverbs
New York, AMS Press. Repr of 1957. ISBN 0–404–16825–6.

Hans, James S The Question of Value: thinking through Nietzsche, Heidegger and Freud (1989)
Carbondale IL, Southern Illinois University Press, 224 p. ISBN 0–8093–1506–8.

Harding, S Unraveling the moral code: towards an understanding of cross–national differences in moral outlook
Papoer prepared for the International Values Conference, Hungarian Academy of Sciences, Budapest, April 6–8 1988.

Harding, S and Walley, L How do people express their views on moral issues? Towards a conceptual framework of moral judgement, based on secondary analysis of SCPR British Social Attitudes and EVSSG European Values Surveys (1986)
Northampton, nene College, Faculty of Education and Social Science, Department of Psychology.

Harding, S D Contrasting values in western Europe: some methodological issues arising from the EVSSG European Values Project
Paper presented at the third cross–national research seminar Language and Culture in Cross–National Research at Aston University, Birmingham, October 1986.

Harding, S D Political values change and the social context of post–materialism: results from the EVSSG European Values Survey
Paper presented at the fifth conference of Europeanists on changing consciousness, values and culture in advanced industrial societies, Washington DC, 1985.

Harding, S; Phillips, D and Fogarty, M Contrasting values in Western Europe: unity, diversity and change (1986)
London, Macmillan.

Hardon, John A Treasury of Catholic Wisdom (1987)
New York, Doubleday, 768 p. ISBN 0–385–23079–6.

Hart, Gordon M Values Classification for Counselors: how counselors, social workers, psychologists and other human service workers can use available techniques (1978)
Springfield IL, C C Thomas, 104 p. illus. ISBN 0–398–03847–3.

Hartley, Elda Perennial Wisdom (1986)
Warwick NY, Amity House, 80 p. Chrysalis Books. Illus. by Nall Page. ISBN 0–916349–09–8.

Hartman, Robert S The Hartman Value Inventory (1966)
Boston MA, Miller Associates.

Hartman, Robert S The Structure of Value: foundations of scientific axiology (1967)
Carbondale IL, Illinois University Press.

Hartono, Prio Inner Wisdom (1988)
Warwick NY, Amity House, 144 p. Chrysalis Books. ISBN 0–916349–21–7.

Hartshorne, Charles Wisdom as Moderation: a philosophy of the middle way (1987)
Albany NY, State University of New York Press, 157 p. SUNY Series in Philosophy. ISBN 0–88706–472–8.

Hayward, Jeremy W Perceiving Ordinary Magic: science and intuitive wisdom (1984)
London, New Science Library, 323 p. bibl. ISBN 0–87773–297–3.

Healy, Eloise K Ordinary Wisdom (1981)
Santa Monica CA, Paradise Press. illus. ISBN 0–940806–00–2.

Heard, Gerry C Basic Values and Ethical Decisions (1990)
Melbourne FL, Robert E Krieger. ISBN 0–89464–431–9.

Helmer, O A Use of Simuliation for the Study of Future Values (1966)
Santa Monica CA, Rand Corporation.

Heunks, F J The values by which we live in Europe
Paper prepared for presentation at the European Confernce of PAx Romana – MIIC, Innsbruck, 1985.

Heunks, Felix Individualism and beyond: social and political values in the nineties
Paper prepared for the International Values Conference, Hungarian Academy of Sciences, Budapest, April 6–8 1988.

Hilliard, Albert Leroy The Forms of Value: the extension of a hedonistic axiology (1950)
New York.

Hodson Concealed Wisdom in World Mythology
Wheaton IL, Theosophical Publishing House. ISBN 0–8356–7556–4.

Hofstede, Geert Cultures Consequences: international differences in work–related values (1980)
London, Sage.

Holliday, St G Wisdom: explorations in adult competence (1986)
New York, Karger S, 102 p. illus. Contributions to Human Development Series: 17. Ed by M J Chandler. ISBN 3–8055–4283–6.

Horne, Michael Values in Social Work (1988)
Brookfield VT, Gower Publishing, 100 p. Community Care Practice Handbooks: 26. ISBN 0–7045–0581–9.

Howe, Leland W; Kirschenbaum, Howard and Simon, Sidney B Values Clarification: a handbook of practical strategies for teachers and students (1985)
New York, Dodd, Mead and Company, 400 p. ISBN 0–396–08470–2.

Hulme, F E Proverb Lore
New York, Gordon Press Publishers. ISBN 0–8490–0909–X.

Humphreys, Christmas (Ed) The Wisdom of Buddhism (1987)
Atlantic Highlands NJ, Humanities Press International, 280 p. ISBN 0–391–03464–2.

Hunt, Donald Pondering the Proverbs (1974)
Joplin MO, College Press Publishing. illus. The Bible Study Textbook Series. ISBN 0–89900–018–5.

Huntsman, B W (Comp) Wisdom is One (1985)
Rutland VT, Charles E Tuttle, 175 p. ISBN 0–8048–1434–1.

Hutterer, Karl L; Lovelace, George and Rambo, A Terry (Eds) Cultural Values and Human Ecology in Southeast Asia (1985)
Ann Arbor MI, University of Michigan, 416 p. illus. Michigan Papers on South and Southeast Asia: 27. ISBN 0–89148–039–0.

Huxley, Aldous The Wisdom of the Ages (1989)
Albuquerque NM, The Foundation for Classical Reprints, 365 p. illus. ISBN 0–89901–404–6.

Hyman, Eric L and Stiftel, Bruce Combining Facts and Values in Environmental Impact Assessment: theories and techniques (1988)
Boulder CO, Westview Press, 300 p. ISBN 0–8133–7162–7.

Idries, Shah The Hundred Tales of Wisdom (1978)
Cambridge MA, Institute for the Study of Human Knowledge. ISBN 0–900860–60–X.

Ikeda, Diasaku and Wilson, Bryan Human Values in a Changing World (1987)
Carol Publishing Group, 384 p. ISBN 0–8184–0427–2.

Inglehart, R Culture Shift in advanced industrial society (1990)
Princeton University Press, pp 484.

Jack, Dana C and Jack, Rand Moral Vision and Professional Decisions: the changing values of women and men lawyers (1989)
Cambridge MA, Cambridge University Press. illus. ISBN 0–521–37161–9.

Jameson, Kenneth P and Wilber, Charles K (Eds) Religious Values and Development (1981)
Sausalito CA, Pergamon Press, 154 p. illus. ISBN 0–08–026107–8.

Jaspers, Karl Way to Wisdom: an introduction to philosophy (1954)
New Haven CT, Yale University Press. ISBN 0–300–00603–9.

Jensen, Kenneth L Wisdom, the Principal Thing
Lake City Way NE, Pacific Meridian Publishing. ISBN 0–685–25806–2.

Jenson, Herman A Classified Collection of Tamil Proverbs (1986)
Columbia MO, South Asia Books, 449 p. Repr. of 1897. ISBN 0–8364–1683–X.

Johnson, Pamela and Cooperrider, David L The Global Integrity Ethic: defining global social change organizations and the organizing principles which make transnational organizing possible (1991)
Transnational Associations.

Jones, W T The Romantic Syndrome: toward a new method in cultural anthropology and history of ideas (1961)
The Hague, Martinus Nijhoff, 255 p.

Judge, A J N Needs Communication: viable needs patterns and their identification (1980)
In: *Katrin Lederer (Ed) Human Needs; a contribution to the current debate* Konigstein, Verlag Anton Hain, 1980, p.279–312. Also in: Forms of Presentation and the Future of

Comprehension (Collection of papers mainly presented to the Forms of Presentation sub-project of the Goals, Processes and Indicators of Development project of the United Nations University, 1978–82). Brussels, Union of International Associations, 1984.

Kahle, Lynn R (Ed) Social Values and Social Change: adaptations to life in America (1983)
New York, Praeger Publishers, 346 p.
ISBN 0-275-91018-0.

Kao, S R and Sinha, Durganand (Eds) Social Values and Development: Asian perspectives (1989)
San Mateo CA, Sage Publications, 320 p.
ISBN 0-8039-9568-7.

Kaplan, Abraham In Pursuit of Wisdom: the scope of philosophy (1988)
Lanham MD, University Press of America, 664 p. repr. of 1977.
ISBN 0-8191-6749-5.

Kaviratna, Harischandra (Transl) Dhammapada, Wisdom of the Buddha (1980)
Wheaton IL, Theosophical University Press.
ISBN 0-911500-40-5.

Kealey, Robert J Everyday Issues Related to Justice and Other Gospel Values (1984)
Washington DC, National Catholic Educational Association, 80 p.
ISBN 0-318-17779-X.

Keleti, Peter From Action to Interaction: values, methods and goals in philosophy, culture and education (1989)
New York, Peter Lang Publishing, 300 p. American University Studies, Series V, Philosophy: 52.
ISBN 0-8204-0641-4.

Kelly, Eugene and Navia, Luis E (Eds) Ethics and the Search for Values: a comprehensive anthology placing man's perennial search for ethical values in historical perspective (1980)
Buffalo NY, Prometheus Books, 530 p. bibl.
ISBN 0-87975-139-8.

Kenny, Anthony The Heritage of Wisdom: essays in the history of philosophy (1987)
New York, Blackwell Basil, 250 p. ISBN 0-631-15269-5.

Kerkhofs, J Les valeurs du temps présent: une enquête européenne
In: *Chosir* no 289, Janvier 1984, pp 13–18.

Kerkhofs, J and Rezsohazy, R De stille ommekeer: oude en nieuwe waarden in het België van de jaaren tachtig (1984)
Lannoo, Tielt en Weesp.

Keyes, Charles D Foundations for an Ethic of Dignity: a study in the degradation of the good (1989)
Lewiston NY, Edwin Mellen, 280 p. Toronto Studies in Theology: 36.
ISBN 0-88946-757-9.

Khawam, Rene R (Transl) The Subtle Ruse: the book of Arabic wisdom and guile (1982)
New York, State Mutual Book and Periodical Service, 400 p.
ISBN 0-85692-035-5.

Kidner, Derek The Wisdom of Proverbs, Job and Ecclesiastes (1985)
Downers Grove IL, Inter-Varsity Press, 176 p.
ISBN 0-87784-405-4.

Kimmel, Allan J (Ed) Ethics and Values in Applied Social Research (1988)
San Mateo CA, Sage Publications, 160 p. Applied Social Research Methods Series: 12. ISBN 0-8039-2631-6.

Kluckhohn, Clyde K Universal Values and Anthropological Relativism (1952)
In: *Modern Education and Human Values* Pillesburgh, University Press.

Kluckhohn, Florence R and Strodbeck, F L Variations in Value Orientations (1961)
Evanston.

Kochen, Manfred (Ed) Information for Action: from knowledge to wisdom (1975)
New York, Academic Press, 248 p. Library and Information Science Series.

Kogane, Yoshihiro The Impact of Changing Values on Economic Structure (1982)
New York, Columbia University Press, 228 p.
ISBN 0-86008-317-9.

Kohlenberg, Philip and Ogilvy, James Values and Religion (1981)
SRI International.

Köhler, Wolfgang The Place of Value in a World of Fact (1938)
New York.

Kolak, Daniel and Martin, Raymond Wisdom Without Answers: a guide to the experience of philosophy (1989)
Belmont CA, Wadsworth Publishing, 104 p.
ISBN 0-534-10236-0.

Lang, Bernard Wisdom and the Book of Proverbs (1985)
New York, Pilgrim Press/United Church Press, 192 p.
ISBN 0-8298-0568-0.

Lasswell (Ed) Values and Development
Cambridge MA, MIT Press. ISBN 0-262-12074-7.

Laszlo, Ervin (Ed) Goals in a Global Community: studies on the conceptual foundations (1977)
Elmsford NY, Pergamon. 2 vols.

Laszlo, Ervin The Inner Limits of Mankind: heretical reflections on today's values, culture and politics (1989)
London, Oneworld Publications, 143 p.
ISBN 1-85168-015-2.

Laszlo, Ervin and Wilbur, James B (Eds) Human Values and Natural Science (1970)
New York, Gordon and Breach, 306 p. Current Topics of Contemporary Thought Series: 4. ISBN 0-677-13960-8.

Laszlo, Ervin and Wilbur, James B (Eds) Value Theory in Philosophy and Social Science (1973)
New York, Gordon and Breach Science Publishers, 154 p. Current Topics of Contemporary Thought: 2.
ISBN 0-677-14160-2.

Laurence Urdang Associates The Penguin Dictionary of Proverbs (1983)
New York, Penguin Books, 256 p. Ed by Rosalind Ferguson.
ISBN 0-14-051118-0.

Lenz, Elinor and Myerhoff, Barbara The Feminization of America: how women's values are changing our public and private lives (1986)
Los Angeles CA, J P Tarcher, 288 p.
ISBN 0-87477-415-2.

Lepley, Roy (Ed) Value: a cooperative inquiry (1949)
New York.

Lewis, Hunter Question of Values: six ways we make the personal choices that shape our lives (1990)
New York, Harper and Row. ISBN 0-06-250521-1.

Lifton, Robert J Adaptation and Value Development: self-process in protean man (1968)
In: *The Development and Acquisition of Values* report of a Conference, National Institute of Child Health and Human Development, Washington, D C, 15–17 May.

Listhaug, O; Jenssen, A T and Mysen, H T Values in Norway: study description and codebook (1983)
Universitetet i Trondheim, ISS-rapport no 11.

Litvak, Stuart Seeking Wisdom: the Sufi path (1985)
York Beach ME, Samuel Weiser, 128 p.
ISBN 0-87728-543-8.

Lowe, Sigmund Seventy Steps Towards Wisdom (1981)
Marina del Rey CA, DeVorss and Company, 95 p. 2nd ed.
ISBN 0-87516-050-6.

Lowrance, William W Modern Science and Human Values (1985)
Oxford, Oxford University Press, 250 p.
ISBN 0-19-503605-0.

Lunde, Paul and Wintle, Justin A Dictionary of Arabic and Islamic Proverbs (1984)
New York, Routledge, Chapman and Hall, 200 p.
ISBN 0-7102-0179-6.

Lyttlens, Lorentz Of Human Discipline: social control and long-term shifts in values (1989)
Philadelphia PA, Coronet Books, 260 p. Transl. from Swedish by Roger Tanner. ISBN 91-22-00861-6.

Mace, C A Homeostasis, Needs and Values (1953)
British Journal Psychology 44, pp. 200–10.

MacGregor, Geddes The Gospels as a Mandala of Wisdom (1982)
Wheaton IL, Theosophical Publishing House, 224 p.
ISBN 0-8356-0554-X.

MacLean, Douglas (Ed) Values at Risk (1986)
Totowa JN, Rowman and Littlefield, 192 p. Maryland Studies in Public Philosophy. ISBN 0-8476-7414-2.

MacNulty, Christine Values and Lifestyles in Western Europe (1981)
SRI International.

Maheu, René Comptes Rendus des Débats: 15e conférence générale (1968)
Paris, UNESCO.

Mannheim, Karl Ideology and Utopia (1936)
New York, Harcourt, Brace and Company. Trans. by Edward Shils and Louis Wirth.

Marcel, Gabriel Tragic Wisdom and Beyond (1973)
Evanston IL, Northwestern University Press, 250 p. Studies in Phenomenology and Existential Philosophy.
ISBN 0-8101-0414-8.

Markley, O W, et al Changing Images of Man (1973)
Menlo Park: Stanford Research Institute, 347 p. Policy Research Report: 3.

Maslow, A H Hierarchy of Human Needs
In: *Motivation and Personality* 1954, New York, Harper and Row.

Maslow, Abraham H New Knowledge in Human Values (1959)
New York, Harper and Row.

Maslow, Abraham H A Theory of Metamotivation: the biological rooting of the value-life (1967)
Journal of Humanistic Psychology 7, 1967 pp. 93–127.

Maslow, Abraham H Religious Values and Peak Experiences (1970)
New York, Viking Compass Book.

Matthews, Bruce and Nagata, Judith (Eds) Religion, Values and Development in Southeast Asia (1986)
Brookfield VT, Gower Publishing, 168 p.
ISBN 9971-988-20-8.

Maxwell, Nicholas From Knowledge to Wisdom: a revolution in the aims and methods of science (1987)
New York, Basil Blackwell, 298 p. ISBN 0-631-13602-9.

Maziarz, E Value and Values in Evolution: a symposium (1979)
New York, Gordon and Breach Science Publishers, 208 p. Current Topics of Contempary Thought Series.
ISBN 0-677-15240-X.

Maziarz, Edward A (Ed) Evolution of Man's Values
New York, Gordon and Breach.

McClain, Ernest G The Myth of Invariance (1978)
Boulder CO, Shambhala Publications.

McClelland, David C (Ed) Education for Values (1984)
Mercersburg PA, Irvington Publishers, 220 p.
ISBN 0-8290-1557-4.

McCoy, Charles S Management of Values: the ethical difference in corporate policy and performance (1985)
Cambridge MA, Ballinger Publishing, 394 p. Business and Public Policy Series. ISBN 0-273-01988-0.

McDonald, Julie J (Ed) Scandinavian Proverbs (1985)
Ioawa City IA, Penfield Press, 32 p. ISBN 0-941016-11-0.

McElhinney, Thomas K Human Values Teaching Programs for Health Professionals (1981)
Ardmore PA, Whitmore Publishing, 200 p. Foreword by Edmund D Pelligrino. ISBN 0-87426-051-5.

McGann, Jerome J Social Values and Poetic Acts: the historical judgment of literary work (1988)
Cambridge MA, Harvard University Press, 296 p.
ISBN 0-674-81495-9.

McLean, George F and Pegoraro, Olinto (Eds) The Social Context and Values: perspectives of the Americas (1989)
Lanham MD, University Press of America. Cultural Heritage and Contemporary Life Series: V. ISBN 0-8191-7354-1.

Meider, Wolfgang The Prentice Hall Encyclopedia of World Proverbs (1986)
Englewood Cliffs NJ, Prentice Hall, 582 p.
ISBN 0-13-695586-X.

Mendlovitz, Saul H and Weiss, T G Towards Consensus: the world order models project of the Institute for World Order
In: *Grenville Clark and Louis B Sohn* (Eds) Introduction to Peace through World Law Chicago IL, World Without War Publications, 1973, pp. 74–97.

Messina, James J (Ed) Personal Values Analysis Handbook (1982)
Tampa FL, Advanced Development Systems, 26 p. Professional Handbook Series. ISBN 0-931975-16-6.

Michael, Donald N The New Competence: the organization as a learning system (1980)
SRI International.

Midgley, Mary Wisdom, Information and Wonder: what is knowledge for? (1989)
New York, Routledge, Chapman and Hall, 275 p.

Mieder, Wolfgang International Bibliography of Explanatory Essays on Individual Proverbs and Proverbial Expressions: German language and literature (1977)
New York, Peter Lang, 146 p. European University Studies Series: 1, 191. ISBN 3-261-02932-3.

Mieder, Wolfgang International Proverb Scholarship: an annotated bibliography (1982)
New York, Garland Publishing, 633 p.
ISBN 0-8240-9262-7.

Miller, James Measures of Wisdom: the cosmic dance in classical and Christian antiquity (1986)
Cheektowage NY, University of Toronto Press, 672 p.
ISBN 0-8020-2553-6.

Miller, Lynn H Global Order: values and power in international politics (1990)
Boulder CO, Westview Press, 288 p. 2nd ed.
ISBN 0-8133-0931-X.

Millgate, Irvine H and Millgate, Rachel W (Eds) The Common Language of Values and Ethics (1987)
Cutler ME, Six Lights, 209 p. illus. rev ed.
ISBN 0-938919-01-6.

Mills, Wright C Sociological Imagination (1959)

Mitchell, Arnold Consumer Values: a trypology (1978)
SRI International.

Mitchell, Arnold Social Change: implications of trends in values and lifestyles (1979)
SRI International.

Mitchell, Arnold Values Scenarios for the 1980s (1981)
SRI International.

Mitchell, Arnold and Royce, William S Stakeholder Values and Corporate Success (1980)
SRI International.

Mitchell, Arnold, et al Attitudinal and Other Correlations of Values (1979)
SRI International.

Mitchell, Arnold, et al Trends in Values: 1973–1978 (1979)
SRI International.

Monier-Williams, Monier Indian Wisdom (1978)
Livingston NJ, Orient Book Distributors, 575 p.
ISBN 0-89684-105-7.

Moor, R A de Valeurs du temps présent: une étude de l'Europe occidental
In: Bruggen, L van der; Ladrière, J et Morren, L *Ethique, science at foi chrétienne*, Presses de Louvain-la-Neuve, 1985, pp 85–107.

Morris, Charles Varieties of Human Value (1956)
Chicago IL.

Moustakas, Clark E Personal Growth: the struggle for identity and human values (1969)
East Dennis MA, Howard A Doyle Publishing.

Mukherjee, R The Quality of Life: valuation in social research (1989)
London, Sage.

Murphy, Roland E Wisdom Literature: Ruth, Esther, job, proverbs, ecclesiastes, canticles (1981)
Grand Rapids MI, William B Eerdmans. The Forms of the Old Testament Literature Series. ISBN 0-8028-1877-3.

Murphy, Roland E (Ed) Medieval Exegesis of Wisdom Literature: essays by Beryl Smalley (1986)
Brookfield VT, Scholars Press. Scholars Press Reprints and Translations Series. ISBN 1-55540-026-4.

Neihardt, John G The Divine Enchantment: a mystical poem and poetic values: their reality and our need of them (1989)
Lincoln NE, University of Nebraska Press, 200 p.
ISBN 0-8032-3319-1.

Nicholson, Shirley Ancient Wisdom: modern insight (1985)
Wheaton IL, Theosophical Publishing House, 198 p. illus.
ISBN 0-8356-0595-7.

Norrick, Neal R How Proverbs Mean: semantic studies in English proverbs (1985)
Hawthorne NY, Mouton de Gruyter, 228 p. Trends in Linguistics: 27. ISBN 0-89925-037-8.

Nyembezi, C L Zulu Proverbs (1963)
IBC Limited. rev ed.
ISBN 0-85494-051-0.

O'Connor, Kathleen Wisdom Literature (1988)
Wilmington DE, Glazier Michael. Message Biblical Spirituality Series: 5. ISBN 0-89453-571-4.

Ogilvy, James and Schwartz, Peter The Emergent Paradigm: changing patterns of thought and belief (1979)
SRI International.

Orizo, F A España, entre la apatia y el cambio social: una encuesta sobre el sistema europeo de valores: el caso español (1983)
Madrid, Mapfre SA.

Otten, C Michael Power, Values and Society: an introduction to sociology (1981)
New York, Random House. ISBN 0–394–33295–4.

Owens, Claire M The Mystical Experience: facts and values (1967)
Main Currents in Modern Thought 23, 4, March/April.

Oxenham, John (Ed) Education and Values in Developing Nations (1989)
New York, Paragon House, 170 p. ISBN 0–89226–050–5.

Palmer, Earl The Search for Values (1990)
Waco TX, Word.

Panin, I M Alyosha Russian Proverbs (1989)
Chicago IL, Imported Publications, 110 p. illus. ISBN 5–200–00411–X.

Parsons, Howard L Value and Mental Health in the Thought of Marx (1964)
Philosophy and Phenomenological Research 24, pp. 355–65.

Peck, M Scott Road Less Traveled: a new psychology of love, traditional values and spiritual growth (1980)
New York, Simon and Schuster. ISBN 0–671–25067–1.

Penfield, Joyce Communicating with Quotes: the Igbo case (1983)
Greenwich CT, Greenwich Publishing Group, 138 p. illus. Contributions in Intercultural and Comparative Studies: 8. ISBN 0–313–23767–0.

Pepper, Stephen C A Digest of Purposive Values (1947)
Berkeley.

Peterson, Grethe B, et al (Eds) The Tanner Lectures on Human Values
Salt Lake City UT, University of Utah Press. 11 Vols. Tanner Lectures.

Pollack, Rachel Seventy-Eight Degrees of Wisdom: the minor arcana and readings, no II (1986)
San Bernardino CA, Borgo Press, 176 p. ISBN 0–8095–7026–2.

Presno, C and Presno, V The Value Realms: activities for helping students develop values (1980)
New York, Columbia University. ISBN 0–8077–2584–6.

Quigley, Pat Unconventional Wisdom (1988)
Evergreen CO, Cordillera Press, 160 p. ISBN 0–917895–21–5.

Quintas, Alfonso L The Knowledge of Values: a methodological introduction (1989)
Landham MD, University Press of America, 132 p. Cultural Heritage and Contemporary Life Series I, Culture and Values: 2. ISBN 0–8191–7418–1.

Rajaee, Farhang Islamic Values and World View: Khomeiny on man, the state and international politics (1984)
Lanham MD, University Press of America, 162 p. American Values Projected Abroad: Exxon Series: XIII. ISBN 0–8191–3578–X.

Rajneesh, S Book of Wisdom (1983)
Boulder CO, Chidvilas, 420 p. Atisha Series. ISBN 0–88050–530–3.

Reid, T E H (Ed) Values in Conflict (1963)
Toronto.

Rescher, Nicholas What is Value Change?: a framework for research
In: *K Baier and N Rescher (Eds) Values and the Future* New York, Free Press, 1969, pp. 68–109.

Rescher, Nicholas Introduction to Value Theory (1968)
Englewood Cliffs. bibl.

Rezsohazy, R and Kerkhofs, J L'univers des Belges, valeurs anciennes et valeurs dans les années 80 (1984)
Louvan-la-Neuve, Ciaco sc.

Riggio, Milla Cozart (Ed) The Wisdom Symposium: papers from the Trinity College Medieval Festival (1986)
New York, AMS Press. illus. Studies in the Middle Ages: 11. ISBN 0–404–61441–8.

Riley, Sue S How to Generate Values in Young Children: integrity, honesty, individuality, self-confidence and wisdom (1984)
Washington DC, National Association for the Education of Young Children, 94 p. ISBN 0–912674–88–1.

Rinpoche, Lama Z and Yeshe, Lama T Wisdom Energy (1984)
Newburyport MA, Wisdom Publications, 152 p. illus. 2 Vols. Wisdom Basic Books: Orange Series.

Robert, L Humphrey J D, et al Teach the Universal Values (1984)
Coronado CA, Life Values Press, 100 p. illus. ISBN 0–915761–00–9.

Rogers, Michael (Ed) Contradictory Quotations (1983)
Essex, Longman Group, 224 p. ISBN 0–582–55698–8.

Rohr, J A Ethics for Bureaucrats: an essay on law and values (1988)
New York, Marcel Dekker. 2nd ed. Public Administration Public Policy Series: 36. ISBN 0–8247–8032–9.

Rokeach, Milton Understanding Human Values: individual and societal (1979)
New York, Free Press. illus. ISBN 0–02–926760–9.

Rosenbaum, Max (Ed) Ethics and Values in Psychotherapy: a guidebook (1981)
New York, Free Press, 480 p. ISBN 0–02–927090–1.

Rosenberg, Morris Occupations and Values (1980)
Salem NH, Ayer Company. Repr of 1957. Edited by Robert K Merton and Harriet Zuckerman. Dissertations on Sociology Series. ISBN 0–405–12989–0.

Rosener, Lynn and Spengler, Marie Business Uses of Values and Lifestyles (1979)
SRI International.

Ross, Nancy W Three Ways of Asian Wisdom: Hinduism, Buddhism, Zen (1978)
New York, Simon and Schuster. illus. ISBN 0–671–24230–X.

Ruhela, S P (Ed) Human Values and Education (1986)
New York, Apt Books, 243 p. ISBN 81–207–0152–6.

Rushworth, Kidder M Reinventing the Future: global goals for the 21st century (1989)
Cambridge MA, MIT Press, 194 p.

Russell, Bertrand Wisdom of the West (1989)
New York, Outlet Book Company. ISBN 0–517–69041–1.

Sakharov, Alexander Human Values Have to Save the World: Alexander Sakharov Speaks About US–Soviet Relations (1985)
Berkeley CA, Fine Line Productions, 40 p. illus. Ed by Stephen Most. Intro by Gloria Duffy. ISBN 0–936413–01–8.

Salk, Jonas The Survival of the Wisest (1973)
New York, Harper and Row, 124 p.

Sampson, Charles Values, Bureaucracy and Public Policy (1983)
Lanham MD, Univesity Press of America, 276 p. illus. ISBN 0–8191–3482–1.

Saraydarian, Torkom The Ageless Wisdom (1989)
TSG Enterprises Publications and Communications, 250 p. ISBN 0–929874–13–7.

Savage, Robert C Pocket Wisdom: seven hundred seventy seven golden nuggets (1984)
Wheaton IL, Tyndale House Publishers, 160 p. ISBN 0–8423–4905–7.

Schloegl, Irmgard (Ed) The Wisdom of the Zen Masters (1976)
New York, New Directions Publishing Corporation (96 p. The Wisdom Books. ISBN 0–8112–0610–6.

Schnall, Maxine Limits: a search for new values (1981)
New York, Clarkson N Potter, 339 p. ISBN 0–517–541432.

Schneider, Bertrand In Search of a Wisdom for the World: the role of ethical values in education: a collective investigation of the Club of Rome (February - October 1986) (1987)
Paris, UNESCO, 44 p. Major Programme I, Reflection on World Problems and Future-Oriented Studies: BEP/GPI/7.

Schuon, Frithjof In the Face of the Absolute: the library of traditional wisdom (1989)
Bloomington IN, World Wisdom Books, 249 p. Transl. from French. Repr of 1980. ISBN 0–941532–07–0.

Schwartz, Benjamin China's Cultural Values (1985)
Tempe AZ, Arizona State University Center for Asian Studies. Arizona State University Center for Asian Studies Occasional Paper: 18. ISBN 0–939252–14–7.

Scott, R B Way of Wisdom (1972)
New York, Macmillan Publishing Company. ISBN 0–02–089280–2.

Seals, Thomas L Proverbs: wisdom for all ages
Lakewood OH, Quality Publications. ISBN 0–89137–529–5.

Segraves, Daniel L Ancient Wisdom for Today's World: a commentary on the book of proverbs (1990)
Hazelwood MO, World Aflame Press, 350 p. Ed by David K Bernard. ISBN 0–932581–60–9.

Sen, A Resources, Values and Development (1984)
Oxford, Oxford University Press, 547 p.

Shah, Idries Wisdom of the Idiots (1971)
New York, E P Dutton. ISBN 0–525–47307–6.

Shephard, Gerald T (Ed) Wisdom as a Hermeneutical Construct (1980)
Berlin, Walter de Gruyter, 178 p. ISBN 3–1100–7504–0.

Sheppard, Gerald T Wisdom as a Hermeneutical Construct: a study in the Sapientalizing of the Old Testament (1979)
Berlin, Walter de Gruyter. Beihefte zur Zeitschrift für die Alttestamentiche Wissenschaft: 151. ISBN 3–1100–7504–0.

Simpson, J A (Ed) The Concise Oxford Dictionary of Proverbs (1983)
Oxford, Oxford University Press. ISBN 0–19–866131–2.

Siu, R G H The Tao of Science: an essay on western knowledge and eastern wisdom (1957)
Cambridge MA, MIT Press. illus. ISBN 0–262–69004–7.

Slack, Kenneth The Seven Deadly Sins (1985)
London, SCM Press, 104 p. ISBN 0–334–01503–0.

Sloan, Douglas Education and Values (1980)
New York, Columbia University. ISBN 0–8077–2574–9.

Smith, D Howard The Wisdom of Taoists (1980)
New York, New Directions Publishing Corporation, 96 p. Wisdom Series. ISBN 0–8112–777–3.

Smith, F LaGard Insights for Today: the wisdom of the proverbs (1985)
Eugene OR, Harvest House Pubishers. ISBN 0–89081–499–6.

Smith, M Brewster Social Psychology and Human Values (1988)
New York, Irvington Publishers, 440 p. ISBN 0–8290–0744–X.

Smith, Philip L The Problem of Values in Educational Thought (1982)
Ames IA, Iowa State University Press, 92 p. ISBN 0–8138–1853–2.

Smith, Steve (Ed) Ways of Wisdom: readings on good life (1983)
Lanham MD, University Press of America, 312 p. ISBN 0–8191–3387–6.

Somers, Adele (Ed) Wisdom Through the Ages: ethics for the living experience (1989)
Century City CA, World Relations Press, 80 p. illus. Illus. by Stanley Somers. ISBN 0–9615032–2–X.

Srivastva, Suresh, et al Executive Integrity: the search for high human values in organizational life (1988)
San Francisco CA, Jossey-Bass, 376 p. Management Series. ISBN 1–55542–085–0.

Stang, Sondra and Wiltenberg, Robert Collective Wisdom (1987)
New York, McGraw-Hill Publishing, 352 p. ISBN 0–07–554961–1.

Stephens, Thomas A Proverb Literature: a bibliography of works relating to proverbs
Millwood NY, Kraus Reprint and Periodicals. Repr of 1928. Folk-Lore Society of London Monographs: 89. ISBN 0–8115–0535–9.

Stephenson, R H Goethe's Wisdom Literature: a study in aesthetic transmutation (1983)
New York, Peter Lang, 274 p. British and Irish Studies in German Language and Literature: 6. ISBN 3–261–05025–X.

Sternberg, Robert J Wisdom: its nature, origins and development (1990)
Cambridge, Cambridge University Press, 375 p. illus. ISBN 0–521–36718–2.

Sternberg, Robert J and Wagner, Richard K (Eds) Practical Intelligence: origins of competence in the everyday world (1986)
Cambridge, Cambridge University Press, 240 p. illus. ISBN 0–521–30253–6.

Stevenson, Burton The Macmillan Book of Proverbs, Maxim's and Famous Phrases (1987)
New York, Macmillan Publishing, 2976 p. ISBN 0–02–614500–6.

Stevenson, C L Twentieth Century Philosophy: the analytic tradition (1966)
Ivan Illich Descooling Society.

Stock, Irvin Fiction as Wisdom: from Goethe to Bellow (1980)
Pennsylvania PA, Pennsylvania State Univeristy Press.

Stocker, Michael Plural and Conflicting Values (1990)
Oxford, Oxford University Press, 376 p. ISBN 0–19–824447–9.

Stoetzel, Jean Les Valeurs du Temps Present: une enquête européenne (1983)
Paris, PUF.

Telushkin, Joseph Uncommon Sense: the world's fullest compendium of wisdom (1987)
New York, Shapolsky Publishers, 238 p. ISBN 0–933503–48–2.

The Journal of Value Inquiry

The Theosophy Science Study Group Holistic Science and Human Values
Journal.

Thirring, Hans The Step from Knowledge to Wisdom (1956)
American Scientist 46, Oct. pp. 445–56.

Thompson, John M The Form and Function of Proverbs in Ancient Israel (1974)
Hawthorn NY, Mouton de Gruyter, 156 p. ISBN 90–2792–675–1.

Thompson, William Irwin From Nation to Emanation: planetary culture and world governance (1982)
Findhorn, Findhorn Foundation.

Tingley, Katherine The Wisdom of the Heart: Katherine Tinley speaks (1978)
San Diego CA, Point Loma Publications. ISBN 0–913004–33–2.

Tishby, Isaiah (Ed) The Wisdom of the Zohar (1989)
Oxford, Oxford University Press, 1596 p. 3 Vols. The Litman Library of Jewish Civilization. Transl. by David Goldstein. ISBN 0–19–710043–0.

Tripp, Rhoda Thomas (Comp) The International Thesaurus of Quotations: a companion to Roget's international thesaurus (1970)
New York, Thomas Y Crowell, approx 1500 p. ISBN 0–690–44584–9.

Tropman, John American Values and Social Welfare (1988)
Englewood Cliffs NJ, Prentice Hall, 256 p. ISBN 0–13–031675–X.

Turner, Fr C The varieties of democratic values: religion, patriotism, and equality
Paper written for inclusion in a book on the pluralist democracies, edited by Mattei Dogan, University of Connecticut, undated.

Tyberg, Judith M Sanskrit Keys to the Wisdom-Religion (1976)
San Diego CA, Point Loma Publications, 180 p. ISBN 0–913004–29–4.

Umbaugh, Duane C Social Values: index of modern information with bibliography (1988)
Annandale VA, ABBA Publishers Association, 150 p. ISBN 0–88164–894–9.

UNESCO Regional Office for Education in Asia and the Pacific Quality of Life: an orientation to population education (1981)
Bangkok, Population Education Clearing House.

Unger, Peter Identity, Consciousness and value (1990)
Oxford, Oxford University Press, 400 p. ISBN 0–19–505401–6.

Upton, Cyril What is Wisdom?: the world's oldest question posed in the light of contemporary perplexity
An Arbor MI, Books on Demand. Repr of 1959. ISBN 0–317–08824–6.

Upton, Cyril What is Wisdom? (1959)
Woodstock NY, Beekman Publishers, 148 p. ISBN 0–8464–0967–4.

Vaill, P B Process Wisdom for a New Age
In: *ReVision* 1984, 7, pp. 39–49.

Vallance, Theodore R Values and Ethics in Human Development Professions (1984)
Dubuque IA, Kendall/Hunt Publishing, 512 p. ISBN 0–8403–3465–6.

Vanderbilt, Vito C Proverbs Twisted with Wit and Humor for Laughs or Tumor (1988)
Annandale VA, ABBE Publishers Association of Washington, 150 p. ISBN 0–88164–872–8.

VanEtten, Teresa Ways of Indian Wisdom (1987)
Santa Fe NM, Sunstone Press, 117 p. ISBN 0–86534–090–0.

Vawter, Bruce The Path of Wisdom: biblical investigations (1986)
Wilmington DE, Glazier Michael. Background Books: 3. ISBN 0–89453–466–1.

Vickers, G Value Systems and Social Process: choosing, planning, controlling, revaluing, appreciating, learning, surviving (1970)
New York, Pelican Books, 221 p.

Vickers, Geoffrey Values, Norms and Policies
In: *Policy Sciences* March 1973, 4, 1, pp. 103–111.

Vickers, Geoffrey Freedom in a Rocking Boat: changing values in an unstable society (1970)
London, Penguin Books, 215 p. ISBN 0-14-021205-1.

Vico, Giambattista On the Most Ancient Wisdom of the Italians: unearthed from the origins of the Latin language (1988)
Conel University Press, 208 p. Transl. by L Palmer.
ISBN 0-8014-1280-3.

Vojcanin, Sava A (Ed) Law, Culture and Values (1989)
New Brunswick NJ, Transaction Books, 212 p.
ISBN 0-88738-305-X.

Von Der Heydt, Vera Psychology and the Care of Souls and Standards and Values (1985)
New York, State Mutual Book and Periodical Service.
ISBN 0-317-62295-1.

von Mering, Otto A Grammar of Human Values (1961)
Pittsburgh. bibl.

von Wright, Georg Henrik The Varieties of Goodness (1963)
London.

Voorwinde, Stephen Wisdom for Today's Issues: a topical arrangement of the proverbs (1981)
Phillipsburg NJ, Presbyterian and Reformed Publishing Company. ISBN 0-87552-472-9.

Waddington, C H Values, Life Styles and the Future of Technological Society
In: *Anticipation* Geneva, World Council of Churches, May 1974, 17, pp. 36–45.

Walchars, John Resurrection of Values (1986)
New York, Crossroad Publishing, 176 p.
ISBN 0-8245-0746-0.

Walker, Barbara The Crone: women of age, wisdom and power (1985)
San Francisco CA, Harper Religious Books, 160 p. illus.
ISBN 0-06-250928-4.

Wallace, B Alan (Ed) Transcendent Wisdom: a commentary on the ninth chapter of Shantideva's guide to the Bodhisattva way of life (1988)
Ithaca NY, Snow Lion Pubications, 150 p.
ISBN 0-937938-65-3.

Walsh, James (Ed) The Pursuit of Wisdom and Other Works (1988)
Mahwah NJ, Paulist Press, 384 p. Classics of Western Spirituality Series: 58. ISBN 0-8091-2972-8.

Watts, A W The Wisdom of Insecurity (1968)
New York, Vintage Books.

Weber, Rhiannon Signposts from Proverbs: an introduction to proverbs (1988)
Carlisle PA, Banner of Truth, 128 p. illus.
ISBN 0-85151-517-7.

Weinberg, Daniela Peasant Wisdom: cultural adaptation in a Swiss village (1975)
Berkeley CA, University of California Press, 226 p. illus.
ISBN 0-520-02789-2.

White, John K The New Politics of Old Values (1990)
Hanover NH, University Press of New England, 224 p. 2nd ed. ISBN 0-87451-508-4.

Whiting, Bartlett J Modern Proverbs and Proverbial Sayings (1989)
Cambridge MA, Oxford Univeristy Press, 752 p.
ISBN 0-674-58053-2.

Wiggins, David Values, Deliberation and Truth: essays in the philosophy of value (1986)
New York, Basil Blackwell, 240 p. Aristotelian Series.
ISBN 0-631-14044-1.

Wigner, Eugene (Ed) Physical Science and Human Values (1947)
Princeton.

Wilson, Frank P Oxford Dictionary of English Proverbs (1970)
Oxford, Oxford University Press. 3rd ed. Intro by J Wilson.
ISBN 0-19-869118-1.

Winter, Gibson Liberating Creation: foundations of religious social ethics (1981)
New York, Crossroad.

Wright, David Wisdom as a Lifestyle: building biblical life-codes (1987)
Grand Rapids MI, Zondervan Publishing House.
ISBN 0-310-44311-3.

Yatsushiro, Toshio Politics and Cultural Values (1979)
Salem NH, Ayer Company Publishers. Asian Experience in North America Series. ISBN 0-405-11299-8.

Yetiv, Isaac One Thousand Proverbs from Tunesia, Bouraoui, Hedi (1987)
Washington DC, Three Continents Press, 170 p.
ISBN 0-89410-615-5.

Ying-Ming, Hung The Roots of Wisdom (1985)
New York, Harper and Row/Kodansha International, 136 p. illus. ISBN 0-87011-701-7.

Yoo, Young H Wisdom of the Far East: a dictionary of proverbs, maxims and famous classical phrases of the Chinesen Japanese and Korean (1972)
Washington DC, Far Eastern Research and Publications Center. Dictionary Series: 5. ISBN 0-912580-00-3.

Notes

VZ

Human Values and Wisdom

Significance	827
Previous, parallel and related initiatives	828
Definitions	829
Method	831
Comments	
Interrelating values	837
Wisdom and requisite variety	839

***   Bibliographical references identified in abridged form in the following section refer to publications detailed, by author, in Section VY, which is the bilbiography for Section V.

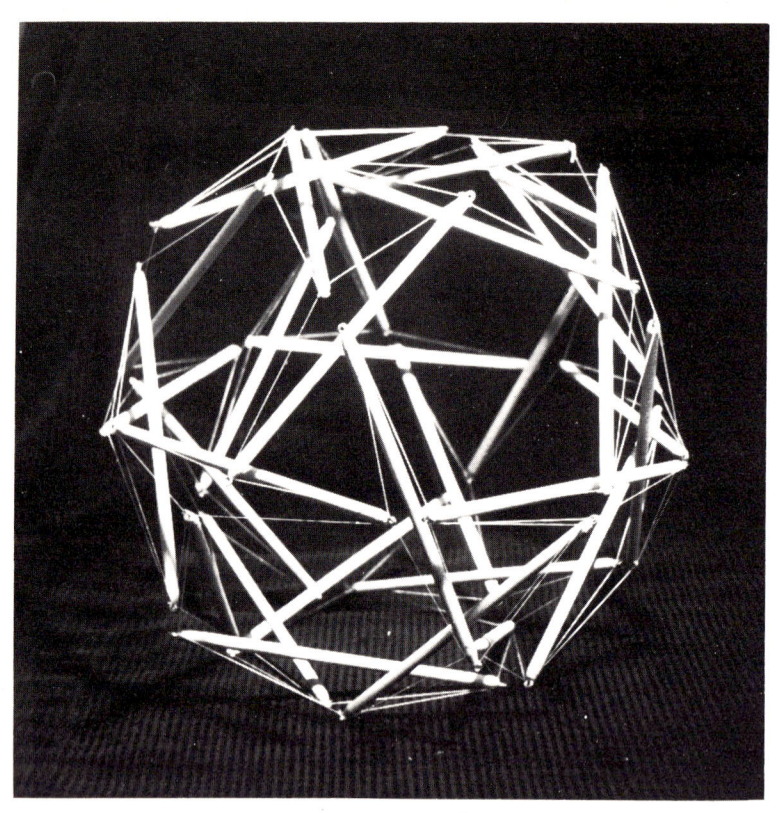

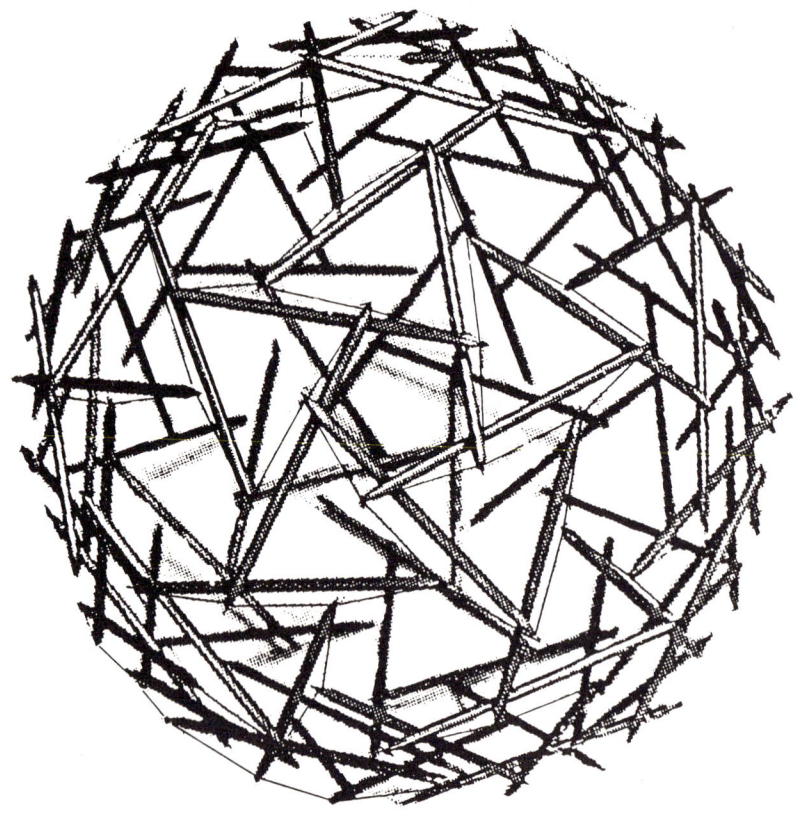

Spherical tensional integrity (or 'tensegrity') structure. Composed of non-touching struts which, through their resistance to compression within a continuous network of cords in tension, collectively form a coherent structure in three dimensions.
Such a structure indicates new possibilities of thinking about the relationship between polar opposites and the coherent structures they form if they are appropriately balanced.

Significance

1. Intent

(a) Identify a wide range of concepts which may be termed human values as a preliminary to determining their relationship to one another, to entries in other sections of this publications, but specially to the world problems which their recognition makes evident.

(b) Provide sufficient contextual material on each value to provide an understanding of the conceptual domain with which it is associated.

(c) Provide a context for values which are cited in essentially different and frequently non-interacting sectors of society, without excluding those values which are not normally accorded recognition in public debate.

(d) Clarify any distinction between a value, an attribute, a quality, a need, and a human right.

(e) Identify the specific world problems which become evident in the light of recognition of the importance of a particular value as a preliminary, firstly, to obtaining some understanding of the degree of mismatch between the network of recognized values and the network of world problems and, secondly, to predicting the emergence of value-related problems.

(f) Identify relationships between the values included and entries in other sections of this publication as a preliminary to predicting the emergence of new values.

2. Significance

(a) Appeals to values: The debate on social policy at the local, national or world level makes many references to concepts such as equality, justice and liberty. These are abstract concepts of great ambiguity and imprecision. In part their power and value is due to this, since each generation is then obliged to redefine the content to be associated with such terms. But values and norms are currently terms of unusable vagueness, not because they cannot be usefully defined, but because they have not yet been sufficiently analyzed (Vickers, 1973). Although some go as far as to hold that value judgements in general, and moral judgments in particular, do not state any facts and may hence be said to be descriptively meaningless (Stevenson, 1973).

(b) Recognition of importance of value change: The vagueness attached to the notion of values in the formulation of social policy has led to a multiplicity of definitions and a vigorous ongoing debate. There is widespread recognition that the rate of value change is increasing to a point at which it is no longer possible to predict with any accuracy the values of the next generation. Major shifts in the value system of a society become apparent within the span of a single lifetime or within even shorter periods, shattering the presumed identity between one generation and the next. This acceleration of change is considered by some to be one of the most dramatic developments in the entire cultural history of the human race. In attempting to formulate social policy for the future, values must however be fed into the decision- making process. The utility of any such policy depends therefore on an understanding of the complex and shifting architecture of values that regulates human behaviour. What are values, how do they relate to one another, and how do they change? How do they relate to the problems with which society is confronted? Knowledge of these matters remains primitive relative to the needs of the time.

(c) Recognition of deceptive uniformity of terminology: Despite this ignorance, there is an increasing uniformity of terminology relating to values which may be noticed in international meetings for which it undoubtedly facilitates formal communication and apparent agreement. This should not, according to René Maheu (past Director-General of UNESCO) lead to any belief that such agreements are solidly grounded. The diverse, even contradictory, interpretations, motivations and utilizations, are an indication of fundamental divisions concerning values (René Maheu, 1968)

(d) Institutionalization of values: There is also concern that the institutionalization of values leads inevitably to physical pollution, social polarization and physiological impotence: three dimensions in a process of global degradation and modernized misery. Most of the research now going on about the future may then be considered as advocating further increases in the institutionalization of values, when conditions are required which would permit precisely the contrary to happen and ensure the continual emergence of values which cannot be substantially controlled by technocrats (Baier, 1969).

(e) Necessity for re-examination of value systems: At the same time, throughout the whole developed and developing world, there is a widespread feeling that the systems of value by which man has guided his actions in the last few generations, require re-examination and almost certainly should be altered in many important respects. Hardly any of the older ethical assumptions remain unquestioned. The basis for this is partly a moral revulsion against some of the old value systems, which are thought to be unworthy in various ways; but the need to change the ethical bases on which society is organized and people act is highlighted by the practical demands of world problems. A solution to the problem of population may only be found under the guidance of a new attitude to the sanctity of life; and the issues of urbanisation, transport, increased leisure, safeguarding of the environment, and so on, which at first sight may appear to be simple material questions, turn out on deeper inspection to involve motivations and aspirations and the value systems on which these are based (Waddington, 1974)

(f) Desperate search for common values: Under the heading "A Desperate Cause?", a UNESCO-sponsored Club of Rome report (UNESCO, 1987) notes: *"There has been an endless succession throughout history of speeches, discussions and writings on ethical values and education -- but have they had any significant impact? Given such an avalanche of words, one may rightly wonder why all the values and principles thus enumerated, voted upon and proclaimed have so little effect on the behaviour of people and nations and why there exists such a gulf between words and real life."* Noting the negative and cynical reactions to double standards, the report continues *"Beyond such reactions, however, there exists a deeper search, which is often difficult to discern because it has few means of self-expression. In societies threatened with breaking up, human beings, rootless and pulled in all directions, are searching for common values and compatible visions of the future."*

(g) Values implicit in problem perception: In the final analysis, no problem can be recognized, or adequately formulated, unless the values involved, and the apparent threat to them, are stated. These values and their imperilment constitute the terms of the problem itself. The values that have been the thread of classic social analysis are freedom and reason. In any formulation of problems, it must be made clear what values are really threatened in the troubles and issues involved, who accepts them as values, and by whom or by what they are threatened (C Wright Mills, 1959).

Many world problems can be specifically associated with the values which they threaten or violate in some way. Some values can be more closely associated with entities in other sections of this Encyclopedia, and only indirectly with specific world problems.

Previous, parallel and related initiatives

1. Previous editions of this Encyclopedia
For the values section of the first edition of this book, prepared during the period 1972-76, efforts were made to trace comprehensive listings of values. Those that were found were either very short or explicitly oriented to one country. This situation had not changed at the time of preparation of the second edition (1983-86).

2. World Order Models Project
This world wide project, initiated in the late 1960s, has been specifically concerned with identifying values appropriate to a more desirable world order. Four such values, or value clusters, were identified. These are discussed in a following note.

3. Goals for Mankind Project
The Club of Rome undertook a study of the goals for mankind envisaged in different regions of the world. The project was directed by Ervin Laszlo and resulted in several volumes (Laszlo, 1977).

4. Values and Lifestyles (VALS) Project
This project initiated in the USA in 1978 has specifically focused on the identification of values in American society, notably in relation to the segmentation of consumer markets. The project has generated many publications and stimulated the production of many others. Because of the methodology of the project, no specific lists of values appear to have been produced. The emphasis appears to have been on the manner in which differences in values (and related concepts) can be used to identify a limited number of market segments. The resulting clusters, or value complexes, are discussed in a following note.

4. European Value Systems Study Group
In 1980 an international group of researchers coordinated by pastoral-theologian Jan Kerkofs of the University of Louvain and sociologist Ruud de Moor of Tilburg University initiated a major study of European values. This was done through the European Value Systems Study Group set up for that purpose. The group undertook, with collaborating bodies, a survey involving 12,463 responses from ten countries. The results of the study were summarized in a publication by Jean Stoetzel (1983).

The survey covered the EEC countries and Spain. The collaborating bodies were Gallup International, in cooperation with DATA SA (Spain) and Institut für Demoskopie (Germany) and Faits et Opinions (France). The questionnaire contained several value domains (religion, ethics, work, politics, family, etc), with approximately 1,200 interviews in each country. The initial intention was to produce the single comparative study noted above. The data proved to be so rich that it was decided to produce national reports on the study by researchers in each country.

Researchers all over the world showed interest in the questionnaire which was subsequently used in more than 30 countries. This proved to be a unique initiative not only because of the number of countries involved but because of the range of values. The result is a large database on values permitting comparative studies. The data for Western European countries has been deposited with the Economic and Social Research Council (ESRC) Data Archive at the University of Essex.

In 1988 plans were made for a second questionnaire to permit study of value change over time. Where possible the same questions were to be used, modifying those which had been subject to methodological criticism. New questions were added around the core domains of religion, family and politics. It was intended that the field work would be undertaken in 1990.

When the plans were discussed at the Budapest conference on values in 1988, a proposal was made to establish a World Values Working Group to coordinate the extension of the study to non-European countries.

Prior to the 1981 survey the focus of much contemporary social research had been on economic and political indicators and on measures of quality of life. There had been little international comparative research into people's values and value systems. Attitudes towards specific issues had often been surveyed but there was a lack of systematically collected data on underlying values and beliefs.

The 1981 survey stimulated a rich academic literature, totalling over 72 books and papers. Numerous other publications made use of the findings. The results of the survey have been widely commented in the media.

The methodology of the study is such that there is some difficulty in identifying values in the manner intended here. The study does not seem to permit questions of the kind "to how many values are Europeans sensitive ?" It seems to be designed to respond to questions of the kind "is job satisfaction important to you ?" The results of the study are therefore highly dependent on the value domains selected and the values selected for comment in the questionnaire. The range of values is therefore predefined by the researchers to a large degree.

6. Ethical Values for the 21st Century
For one of its programme activities for 1986-87, UNESCO requested the Club of Rome to undertake an international investigation into the ethical values of 21st century man in the light of the rapid and sometimes violent changes that have affected society in recent decades. The purpose of the study (UNESCO, 1987), based on 41 contributions, is to guide curriculum developers, educators and educational administrators through the problems of ethical values and to encourage further investigation. The study does not endeavour to identify systematically the range of ethical values in existence or desirable.

7. International Future Survey
With the sponsorship of UNESCO, the organization Futuribles International conducted a worldwide survey of eminent persons from varying geopolitical, socio-cultural and ideological backgrounds in the world of culture of science. On section of the questionnaire asked respondents to identify the *"twenty individual and collective values to which people in your region attach the greatest importance."* A report was produced in 1987 on the analysis of the 180 replies received (UNESCO, 1987).

Definitions

1. Multiplicity of definitions

A Club of Rome report for UNESCO (1987) noted that: "*The concept of value refers to two contrasting ideas. At one extreme we speak of economic values based on products, wealth, prices -- on highly material things. In another context, however, the word value acquires and abstract, intangible and non-measurable meaning. Among such spiritual values are freedom, peace, justice, equity. A value system is a group of interconnected values that form a system and reinforce each other. They are anchored in religion or in humanist traditions. To be precise, it is necessary to distinguish clearly between the values themselves and the means of attaining them. In many cases there is broad agreement over ethical goals, but there are differences of opinion over rules of conduct...In any society, therefore, you will now find different systems of values co-existing -- but not peacefully -- side by side.*"

There is considerable confusion surrounding the definition of values. Kurt Baier (1969) notes that sociologists employ a bewildering profusion of terms, ranging from what a person wants, desires, needs, enjoys, prefers, through what he thinks desirable, preferable, rewarding, obligatory, to what the community enjoins, sanctions, or enforces. He cites the following more popular definitions to show the great variety and looseness of the terms employed:

1. "*A thing has or is a value if and when people behave toward it so as to retain or increase their possession of it.*" (George Lundberg)

2. "*Anything capable of being appreciated (wished for) is a value.*" (Robert Part and E W Burgess)

3. "*Values are the obverse of motives...the object, quality, or condition that satisfies the motivation.*" (Richard T LaPiere)

4. "*Values are any object of any need.*" (Howard Becker)

5. "*A desideratum or anything desired or chosen by someone, at sometime - operationally: what the respondent says he wants.*" (Stuart C Dodd)

6. "*By a social value we understand any datum having an empirical content accessible to the members of some social group and a meaning with regard to which it is or may be an object of activity.*" (Znaniecki)

7. "*(A value is) a conception, explicit or implicit, distinctive of an individual or characteristic of a group, of the desirable which influences the selection from available means and ends of action.*" (Clyde Kluckholn)

8. "*Values are the desirable end states which act as a guide to human endeavour or the most general statements of legitimate ends which guide social action.*" (Neil J Smelser)

9. "*The noun "value" has usually been used to imply some code or standard which persists through time and provides a criterion by which people order their intensities of desiring various desiderata. To the extent that people are able to place objects, actions, ways of life, and so on, on a continuum of approval-disapproval with some reliability, it appears that their responses to a particular desideratum are functions of culturally acquired values.*" (William R Catton, Jr)

10. "*Values are normative standards by which human beings are influenced in their choice among the alternative courses of action which they perceive.*" (Philip E Jacob and James J Flink)

11. "*What we properly call a value in life is an organic mixture of need, interest, feeling, purpose and goal..the production and conservation of values is one of the main concerns of human existence.*" (Lewis Mumford)

12. "*A value is anything of interest to a human subject.*" (Perry)

13. "*Values may refer to interests, pleasures, likes, preferences duties, moral obligations, desires, wants, needs, aversions and attractions, and many other modalities of selective orientation.*" (Stephen C Pepper)

14. "*I find it confusing to give the word "values" any narrower meaning than will comprehend interests and expectations, as well as standards of judgement.*" (G Vickers)

15. "*A value is a belief upon which a man acts by preference.*" (Gordon W Allport)

2. Related concepts

Understanding of the nature of human values may be so intimately associated with what might otherwise be considered to be distinct concepts that they cannot be effectively separated from some perspectives:

(a) Economic value: The concept of the value of a thing is central to traditional economic value theory for which value is the so-called exchange or market value of a commodity. Economists distinguish between value in this sense and the values of individuals or societies which in welfare economics mean much the same as preferences or tastes. Such values may then be realized by the appropriate allocation of resources.

(b) Value assessments and imputations: Baier (1973) distinguishes between value assessments and value imputations. Value assessments are assertions to the effect that something did, will or would favourably affect the life of someone. Value imputations are assertions to the effect that someone or some group has, holds, or subscribes to some value (eg achievement, work, altruism, comfort, equality, thrift, friendship), or that some such thing is one of his values. The word value then means different things in these two contexts. Assessed values then become measures of the capacities of various kinds of entities, including persons, to confer benefits, whereas imputed values are measures of tendencies of persons to promote certain ends, for certain reasons.

(c) Instrumental and intrinsic values: A distinction may also be made between instrumental values, which are the means to something else, and intrinsic values, which are those desired for themselves (such as goodness, truth, and beauty).

(d) Attitudes and opinions: Many surveys of the "values" held by people do not find it useful to distinguish between attitudes or opinions held by people and the values that they hold. A survey of values then becomes a survey of attitudes and opinions. Presumably some attitudes may be considered as relating to values, but the distinction is then difficult to establish in that context. It is difficult to identify "values" from such survey data.

(e) Lifestyles: Increasingly clusters of attitudes and opinions arising from survey data are used to identify distinct lifestyles. Each such lifestyle is then seen as reflecting a cluster of values, although these are usually considered implicit.

3. Elusive nature of values

Despite extensive discussion of values, and the importance given to "values" in the abstract, it seems to be quite difficult to identify specific values. Although many international organizations claim a strong interest in values, and would claim to be acting to enhance certain values, it is quite difficult to determine exactly what the range of possible values is.

It might for example be assumed that some of the major international conventions or declarations, such as the Universal Declaration of Human Rights, would constitute a prime source from which values could be identified unambiguously. This does not appear to be the case. It would be difficult to respond to the question "How many values are specified in the Declaration?".

Such difficulties are disguised by the ease with which obvious values are named: peace, justice, health, security, and the like. These might perhaps be treated as "first order" values. Typically, in many studies and list, the number of such values is of the order 5 to 20.

These values can usefully be seen as elements in a much larger set of values. This would then lead to several questions:

(a) How large is that set and what other elements does it contain?

(b) Is it useful to look at sets of values composed of 500, 1,000 or 10,000 elements?

(c) How might such a set be clustered or ordered?

(d) Do particular values subsume other values in the set?

(e) How do the first order values emerge from such larger sets?

(f) Which values tend to be neglected or ignored in focusing on first order values?

(g) To what extent do the identified values in the set overlap one another, namely to what extent is the set artificially enlarged by the presence of synonyms?

(h) Where distinct values terms may be judged by some to be synonymous, is the distinction meaningful to others, and valued by them?

4. Working definition

From the above confusion of definitions, anything may be taken to be a value. For the purposes of this exercise, human values are understood in the very broad sense covered by Nicholas Rescher (1969) in the following. *"Sometimes "human value" is restricted to the area of personal values (of character and personality). But we take it to include not only what the individual may prize in himself and his associates, but also what he prizes in his society, his nation, his culture, his fellowmen in general, and his environment. We thus view this idea extended over a very broad domain - ranging from individual to social and universal values."* In his own tentative list of (American) values, Rescher notes *"We deal here with overtly espoused and publicly appealed to values to the exclusion of (a) unconscious motives..and (b) traits of national character...The factors included in the register are such that explicit or overt appeal to them can well be expected from publicly recognized spokesmen for values: newspaper editorialists, graduation exercise speakers, religio-moral sermonizers, and political orators."*

5. Value complexes and "baskets" of values

Many surveys of values use questionnaires which identify value clusters or domains using phrases composed of what amount to "code words". It is often difficult to determine what values are in relation to such domains, if this question can be considered to be a meaningful one. The intention with such surveys is often to identify which "baskets" of values are considered important, without having to deal with the nature or contents of the basket. The survey is then free to focus on the preferences of people for different baskets in the light of their projection of significance onto the labels attached to the baskets. Conclusions concerning life-style preferences and the like may then be made without looking any further into what specific values are involved. Efforts may also be made to impose specific baskets of values as being universal, or required, for a desirable world society. Some examples of such value complexes are given below:

(a) **World Order Models Project:** A fourfold universal value framework was first suggested by a transnational group of scholars representing all major regions of the world and participating in the World Order Models Project (Saul Mendlovitz). The value complexes identified were: "peace without national military arsenals", "economic well-being for all inhabitants on the earth", "universal rights and social justice", and "ecological balance". Presumably the values in question are the key qualitative operators in each phrase, for example: peace, balance, universality, *etc*.

(b) **US Foreign Policy:** In a study of the values implicit in US foreign policy (Robert C Johansen, The National Interest and the Human Interest, 1980) identified the following professed values: "to serve US security interests", "to protect or improve economic benefits from international trade and investment", and "to express humanitarian concern for helping people in need". The study then compared the second of these with the four values of "global humanism" identified by the World Order Models Project.

(c) **Futuribles International Survey:** With UNESCO support the organization Futuribles International surveyed eminent persons, world wide. One portion of the 180 questionnaires received focused on social values, prioritized by continental region. The values included a mix of single-term and multi-term values:
- single-term values included: freedom, health, money, friendship, peace, education, leisure, honesty, responsibility, power, fulfilment, solidarity, honour;
- multi-term values included: social equality, cultural tolerance, humanization of work, sexual satisfaction, frugal way of life, human rights, environmental protection, law and order, political independence, material possessions, social status.

(d) **VALS Project:** A major study of values in the USA has been made through the VALS project at SRI International since 1978. It has given rise to many papers. For them: *"By the term "values" we mean the entire constellation of a person's attitudes, beliefs, opinions, hopes, fears, prejudices, needs, desires, and aspirations that, taken together, govern how one behaves. One's interior set of values -- numerous, complex, overlapping, and contradictory though they are -- finds holistic expression in a lifestyle."* In their work they have looked at *"well over 800 facets of people and find that different lifestyle groups have unique patterns in almost every area. We now have powerful evidence that the classification of an individual on the basis of a few dozen attitudes and demographics tells us a good deal about what to expect of that person in hundreds of other domains."* (Arnold Mitchell, 1983).

The VALS typology comprises four comprehensive groups that are subdivided into nine lifestyles, *"each intended to describe a unique way of life defined by its distinctive array of values, drives, beliefs, needs, dreams, and special points of view."* These are:
- Need-driven groups (survivor lifestyle; sustainer lifestyle);
- Outer-directed groups (belonger lifestyle; emulator lifestyle; achiever lifestyle);
- Inner-directed groups (I-am-Me lifestyle; experiential lifestyle; societally conscious lifestyle);
- Combined outer- and inner-directed group (integrated lifestyle).

(e) **Western values:**
Using the general structure of the VALS typology, Joseph Plummer (Changing Values, *The Futurist*, 1989, Jan-Feb, pp 8-13) compares traditional and new values. The pairs identified are: self-denial ethic/self-fulfilment ethic; higher standard of living/better quality of life; traditional sex roles/blurring of sex roles; accepted definition of success/individualized definition of success; traditional family life/alternative families; faith in industry, institutions/self-reliance; live to work/work to live; hero worship/love of ideas; expansionism/pluralism; patriotism/less nationalistic; unparalleled growth/growing sense of limits; industrial growth/information and service growth; receptivity to technology/technology orientation.

6. Exclusions
For the purpose of this exercise, the following are specifically excluded from the values identified:

(a) **Tangible objects:** Specifically excluded from this exercise however are tangible objects which an individual "may prize", for example: automobile, house, stereo- equipment, pet, children, land, *etc*. The first is best illustrated by the advertisement "Buy a Buick; something to believe in."

(b) **Value complexes:** Because of the preliminary nature of this exercise, and the effort to identify the components of more complex values, no effort is made to include value complexes in the scope of this study.

Method

1. Methodological bias
As has been frequently noted, there is a very extensive literature on values, with many books on single values such as justice, love, or peace. Such studies will continue to be produced in quantity. Despite the possible resulting distortion, the bias here is to focus on sets and systems of values, endeavouring to identify the individual values which might form part of any comprehensive collection of values.

The bias is therefore against wordy attempts to (re)define this or that value yet again. And it also against the many attempts to limit the range of useful values to 5 to 20 commonly recognized values. Such approaches may well fulfil a vital function, but in terms of this approach they assume the existence of information on the full range of values when this information seems to be totally lacking. There is no information on the value context from which selected values are chosen for detailed study or as prime values necessary for healthy social order.

The bias here is therefore to be simplistic, and even naive, in attempting to locate and interrelate a rich and extensive range of candidates for inclusion in the universe of values. As a preliminary exercise, this is also biased by the focus on the values carried by words in English.

These biases follow from the kinds of assumptions detailed in the discussion of the method used in collecting information on world problems (Section PZ) and on modes of awareness (Section HZ).

2. Challenge of language and artificial distinctions
As discussed in the introduction to the Encyclopedia and in relation to the identification of world problems (Section PZ) and of modes of awareness (Section HZ), the values that can be distinguished are intimately related to the kinds of distinctions established by the vocabulary used. The question is whether these distinctions are real and meaningful, for whom and under what circumstances.

Little need be said concerning the rich pattern of confusion obscured by any verbal consensus on the meaning and importance of "peace". Such words have many different connotations in practice. It is not very helpful to limit attention to such degrees of consensus.

Words which carry values are especially characterized by multiple connotations and overlapping patterns of meaning. It would seem that there is little hope of exploring value systems unless some attempt is made to encode this realm of fuzziness. That has been the orientation explored for this section.

3. Previous initiatives
When this project was initiated in 1972, efforts were made to trace comprehensive listings of values. Those that were found were either very short or explicitly oriented to one country. This situation had not changed when work was done on the second edition over the period (1983-86), despite the various projects identified above. For this reason, it was considered appropriate to generate, as an experimental exercise, a set of value words which could (at some stage) be cross-referenced to the world problems in this Encyclopedia.

A number of partial lists were used at that time to build up a preliminary comprehensive list. This was still inadequate in terms of the objective. There was difficulty in finding a satisfactory means of defining, distinguishing or regrouping values which could be considered synonyms. But the realization that value antonyms were reasonably indicative of the (world) problems to which the value word could be cross-referenced, suggested the use of synonym/antonym dictionaries both as a means of obtaining further values and as a means of identifying "values" in terms of a clusters of synonyms and antonyms. This resulted, for the 1976 edition, in a section containing 614 entries, each composed of the synonyms and antonyms of the particular value word.

In considering how to update this section for the 1986 edition, a further search was made for value lists without any success, despite the considerable interest in "values" in many sectors. After considering various approaches to updating the section, it was finally decided to abandon the entries from the previous edition and to use an entirely new approach which would build on the synonym/antonym pair.

The decision was made to take advantage of the well-established structure of *Roget's Thesaurus* (R L Chapman, 1979), which through its many editions can arguably be said to reflect a reasonably acceptable stable pattern of relationships between words in terms of their various connotations. It has stood the test of time. It is of course a very pragmatic approach to the organization of words and could be subject to many kinds of criticism, especially since it only orders words in English. Nevertheless, as a first step, it provides a better framework than is otherwise available.

4. Value words
The procedure adopted was to scan through the index to the thesaurus in two phases. In the first phase "constructive" value words were identified. In a second phase the same approach was used for "destructive" value words. Clearly these two phases could raise many difficulties. What is to be considered a "value word" and what can be usefully meant by "constructive" or "destructive" ?

The value words included are those which imply some non-material quality which people might commonly be expected to value. An effort was made to select words in the noun form, although the quality might be more usually associated with the adjective form. The presence of words denoting physical phenomena, but also used metaphorically to denote non-material qualities was considered acceptable in this exercise. The tendency, common to many sections of this volume, was to include words if they raised interesting questions, rather than excluding them because no immediate answer was available to the difficulties they raised.

In distinguishing "constructive" from "destructive" value words, the initial approach was to consider them as indicative of "positive" and "negative" values respectively. This quickly raised difficulties in the case of perhaps 5-10% of the words which could be associated with either positive or negative values, or be associated with values which some would consider positive and others negative. This difficulty is to some extent resolved by a subsequent phase, but the difficulty with such exceptions was further reduced by accepting the possibility of a first-order and a second- order response to a value word.
Thus, for example, a first-order response to "peace" (especially in its absence) is positive and peace is perceived as constructive. Whereas a second-order response (especially faced with the total absence of any form of conflict) is negative and peace is perceived as stultifying, monotonous and uninspiring. In this sense the perception of peace metamorphoses into one of stagnation. Similarly a first-order response to "danger" is negative, whereas a second-order response (especially in a risk-free context) is positive because of the excitement and challenge.

5. Value connotations
The value words were then coded by the Roget number indicative of their connotations. On average each word was linked to three to four Roget categories by such codes. At this stage it remained unclear how the information could be usefully presented. Much micro-computer time was used in exploring ways of grouping the material so coded. This in itself produced some interesting results as indicated by Figures 1 and 2, which list the constructive (positive) and destructive (negative) value words with the largest number of cross-references to distinct Roget categories.

6. Value polarities
The next stage was undertaken in order to transcend the somewhat simplistic implications of grouping value words by "first order" responses, neglecting what amounted to subtle qualities associated with certain value words. The difficulty is that a given value is not necessarily valued in the same way under different conditions. A seemingly positive value may become counter-productive or

destructive, whilst a seemingly negative value may become productive or constructive.

To avoid reinforcing simplistic interpretations, it was decided to group the Roget categories (by which the words were coded) into opposing pairs or value polarities. Such polarities then became indicative of a value dimension. People might identify more strongly with one pole, rather than the other, as suggested by the study by W T Jones (1961) on the axes of bias. The value appropriate to certain circumstances might be perceived as shifting between the two extremes, possibly in some cyclic manner, as in the process of enantiodromia discussed by William Irwin Thompson (1982). Many decisions might involve a counter-intuitive compromise between the two poles.

In determining the value polarities, it was again decided to use Roget as a framework. The categories there are in the majority of cases more or less grouped by pairs, providing a synonym/antonym relationship. The Roget sequence of potential pairs was therefore used, suitably adapted with the omission of polarities that were not required because no value words were coded to them. The pairs were named by a word pair indicative of the value dimension. Where appropriate alternative words to those used by Roget were used to label the word pair. Roget was therefore used as a major guideline, but the framework was adapted wherever appropriate in response to the value orientation. Some seemingly artificial or farfetched polar categories were left in, more because they raised questions about the nature of values that might be associated with them, than because they constituted immediately significant categories for the value words coded. This was also true for the categories which in Roget are explicitly used for material phenomena. As noted earlier, in practice these are often used as a substrate for metaphorical connotations. In this sense the Roget framework was completely abandoned, because all such material categories are interpreted here in their metaphorical sense. The majority of the value polarities are presented in Figure 3 in order of the total number of cross-references to value words in Section VC and VD.

7. Word coding results
The results of these stages are presented in the first three parts of Section V. As a first-order presentation, the constructive, positive value words appear in Section VC. The corresponding, destructive, negative value words appear in Section VD. Entries in both sections are cross-referenced to the value polarities in Section VP.

8. Value dynamic
In order to clarify understanding of the value dynamic associated with the polarity, a range of proverbs, quotations and aphorisms were selected for inclusion in the entry. These were deliberately chosen to highlight the fundamental importance of one or other pole, the limiting conditions under which it was necessary or wise to re-align with the other, the nature of this reorientation process (its enantiodromic characteristics) and relevance to individual or societal development. This material therefore points to insights into undervalued values or the limitations of conventional "first-order" responses, whether these are derived from folk wisdom in the form of proverbs or as quotations from those whose wisdom or insight is considered by society to be of sufficient value to be quoted. Of particular relevance in this connection is the effort by V S M Guinzbourg (1961) to publish one of the most extensive multi-cultural collections of proverbs *The Wit and Wisdom of the United Nations*. This was largely collected from delegates and permanent representative to the United Nations as offering insight into the diplomatic processes in which they were engaged.

9. Classification
The next stage in the procedure was to explore ways of further regrouping the information to bring out a more meaningful and comprehensible pattern. The question was how to reorder the 220 polarities of Section V into a smaller number of categories. Having ordered the information from Sections VC and VD into polarities (namely using a factor of two) in the absence of any other guidelines, it was decided to further divide the data set by factors of three and five. This was based on studies of patterning which indicate that new patterns only emerge when a new prime number factor is used to divide up the data (McClain, 1976).

In the course of imposing these constraints, the opportunity was taken to associate polarities which (in Section VP) might be considered too similar to be usefully distinguished. This exercise resulted in 45 three-fold categories, the 224 polarities having been "condensed" to 135 by associating those that were similar. The possibility of using this condensed set of polarities as the basis for Section VP to group the entries in Sections VC and VD was considered and rejected in favour of using the extended set which corresponded more closely to the original Roget pattern. The 45 triplet categories are presented in Section VT with cross-references to the polarities in Section VP which they group. Where several polarities cross-referenced within an entry in Section VT bear the same qualifier code "a", "b" or "c", these could be combined in any condensed set of polarities.

The names given to the entries in Section VT are merely indicative of a "transcendent" complex of value polarities. The significance of a word used to label such an entry should not be confused with the significance of the same word in Section VP or Section VC.

10. Tabular presentation
In an attempt to explore the results of further ordering the information the factor five constraint was imposed on the 45 categories. The resulting categories are presented in a 5 x 9 matrix as Figure 4. This gives an overview of the Section VP polarities grouped into triplets (by matrix cell (with the elements coded "a", "b" or "c"), with each triplet cell forming part of a five-fold group (a column) and a nine-fold group (a row). Totals for each portion of Figure 4 are given in Figure 5. The 45 value types are listed in Figure 6 in order of the total number of cross-references (via Section VP) to Sections VC and VD.

The arbitrary guidelines governing this classification experiment would have required that a factor seven constraint be used instead of allowing the data to form nine-fold groups. This possibility was explored, but it was decided that the nine-fold group raised the possibility of comparison with a more complex classification system currently being used experimentally to group international organizations and the world problems (from this volume). That is the code matrix used for the companion series the *Yearbook of International Organizations* (Vol 3: Global Action Networks).

11. Global patterning
In other studies, exploring the possibility of more comprehensible classification systems highlighting global patterns of significance within sets of data, the importance of seeking ways to project the data onto new patterns has been stressed (Judge, 1980). In particular the need to bypass the misleading limitations of a matrix have been emphasized in order to reinforce non-linear relationships within patterns. One possibility considered was that of projecting the data onto a tensegrity structure (of which an example is given on page 926). Such structures have the unusual property of existing in three dimensions as a result of a dynamic balance between tension elements (cords) and compression elements (struts). What makes such structures unique is that the struts do not touch one another and there are no privileged struts, especially at the centre (where there are in fact no structural elements at all). It is such properties which usefully encode the possibility of non-linear relationships between the value categories. Of special interest is the possibility of projecting value polarities onto struts.

The particular model presented to highlight these possibilities for further investigation has been selected because it has 90 struts. Speculating that it may be fruitful to distinguish between a conscious (explicit) and an unconscious (implicit) form of the 45 value dimensions, the two sets of 45 could be projected onto such a model.

The distinction between two such sets is itself modelled by the impossibility of viewing more than half such a global structure from any one perspective. Other properties of interest are the three-fold and five- fold groupings of polar elements within the model as well as many symmetry effects which contribute to the integrity of the model. It is the use of such properties to stabilize and render comprehensible interpretation of the complex relationships between values that merits attention.

Notes \ Method **VZ**

VC No	Value word		VC No	Value word		VC No	Value word		VC No	Value word	
VC1011	Interest	17	VC1399	Quickness	8	VC1605	Service	7	VC0829	Gentleness	6
VC0845	Good	16	VC1417	Readiness	8	VC1623	Sobriety	7	VC0857	Gracefulness	6
VC0091	Agreeableness	12	VC1423	Reason	8	VC1667	Steadfastness	7	VC0859	Graciousness	6
VC1391	Purity	12	VC1441	Refinement	8	VC1679	Straightforwardness	7	VC0941	Immediacy	6
VC1711	Support	12	VC1507	Rightness	8	VC1725	Sweetness	7	VC0987	Innocence	6
VC0153	Appreciation	11	VC1617	Simplicity	8	VC1765	Tolerance	7	VC1045	Kindness	6
VC0449	Consideration	11	VC1731	Sympathy	8	VC1767	Toughness	7	VC1067	Leniency	6
VC0743	Favour	11	VC1821	Uniformity	8	VC1773	Tranquillity	7	VC1075	Life	6
VC0887	Harmony	11	VC1891	Vitality	8	VC1811	Understanding	7	VC1109	Majesty	6
VC0339	Clearness	10	VC1937	Work	8	VC1815	Unequivocalness	7	VC1121	Mastery	6
VC0731	Faithfulness	10	VC0035	Acquiescence	7	VC1825	Union	7	VC1179	Naturalness	6
VC0841	Glory	10	VC0063	Advancement	7	VC1827	Uniqueness	7	VC1211	Observance	6
VC1403	Quietness	10	VC0159	Appropriateness	7	VC1855	Value	7	VC1225	Openness	6
VC1637	Soundness	10	VC0267	Brilliance	7	VC1881	Vigour	7	VC1231	Order	6
VC0049	Adaptability	9	VC0315	Charity	7	VC1887	Virtue	7	VC1257	Passion	6
VC0265	Brightness	9	VC0389	Compliance	7	VC1941	Worthiness	6	VC1275	Permanence	6
VC0545	Detachment	9	VC0451	Consistency	7	VC0039	Action	6	VC1327	Potency	6
VC0551	Devotion	9	VC0483	Cooperativeness	7	VC0135	Animation	6	VC1377	Promptness	6
VC0767	Fitness	9	VC0487	Cordiality	7	VC0205	Authority	6	VC1379	Propriety	6
VC0799	Freshness	9	VC0491	Correspondence	7	VC0213	Awe	6	VC1387	Prudence	6
VC1041	Justice	9	VC0535	Delicacy	7	VC0247	Blessedness	6	VC1409	Radiance	6
VC1193	Nobility	9	VC0549	Development	7	VC0291	Care	6	VC1421	Realization	6
VC1477	Resistance	9	VC0563	Directness	7	VC0317	Charm	6	VC1425	Reasonableness	6
VC1479	Resolution	9	VC0645	Endurance	7	VC0355	Comfort	6	VC1445	Regularity	6
VC1651	Stability	9	VC0653	Enlightenment	7	VC0403	Concentration	6	VC1491	Responsiveness	6
VC1681	Strength	9	VC0725	Facility	7	VC0405	Conception	6	VC1495	Restraint	6
VC0017	Acceptance	8	VC0735	Familiarity	7	VC0407	Concern	6	VC1503	Richness	6
VC0019	Accessibility	8	VC0771	Flexibility	7	VC0411	Concord	6	VC1549	Self-control	6
VC0025	Accord	8	VC0773	Flourish	7	VC0481	Cooperation	6	VC1589	Sensation	6
VC0189	Assurance	8	VC0781	Forcefulness	7	VC0489	Correctness	6	VC1593	Sensibleness	6
VC0215	Balance	8	VC0797	Freedom	7	VC0499	Creation	6	VC1603	Seriousness	6
VC0293	Carefulness	8	VC0817	Gain	7	VC0517	Decency	6	VC1633	Solidity	6
VC0319	Chastity	8	VC0821	Generosity	7	VC0525	Dedication	6	VC1665	Staunchness	6
VC0455	Constancy	8	VC0883	Hardiness	7	VC0577	Distinction	6	VC1669	Steadiness	6
VC0543	Desirableness	8	VC0915	Honour	7	VC0601	Ease	6	VC1689	Submission	6
VC0547	Determination	8	VC0969	Independence	7	VC0679	Esprit	6	VC1693	Substantiality	6
VC0623	Elegance	8	VC0993	Inspiration	7	VC0699	Excitement	6	VC1701	Suitability	6
VC0695	Exaltation	8	VC1023	Involvement	7	VC0703	Exhilaration	6	VC1729	Symmetry	6
VC0729	Faith	8	VC1085	Liveliness	7	VC0721	Exquisiteness	6	VC1755	Thoughtfulness	6
VC0847	Goodness	8	VC1235	Ordinariness	7	VC0777	Fondness	6	VC1801	Unassuming	6
VC1043	Keenness	8	VC1301	Plainness	7	VC0779	Forbearance	6	VC1829	Unity	6
VC1093	Love	8	VC1309	Pleasure	7	VC0811	Fullness	6	VC1853	Validity	6
VC1331	Power	8	VC1365	Productiveness	7	VC0823	Geniality	6	VC1899	Warmth	6

Figure 1. Constructive value words (Section VC)

VD No	Value word		VD No	Value word		VD No	Value word		VD No	Value word	
VD2710	Foulness	13	VD2570	Disturbance	8	VD2016	Abomination	6	VD2802	Hostility	6
VD3204	Meanness	12	VD2690	Feebleness	8	VD2052	Alienation	6	VD2848	Impetuousity	6
VD4052	Weakness	12	VD3002	Inferiority	8	VD2080	Antagonism	6	VD2964	Indiscretion	6
VD2150	Baseness	10	VD3130	Lapse	8	VD2084	Antipathy	6	VD3050	Insolence	6
VD2294	Corruption	10	VD3292	Muddle	8	VD2130	Awkwardness	6	VD3054	Instability	6
VD2660	Failure	10	VD3526	Rejection	8	VD2178	Bitterness	6	VD3158	Littleness	6
VD2734	Fuss	10	VD3732	Strain	8	VD2212	Capriciousness	6	VD3184	Malaise	6
VD3684	Smallness	10	VD3740	Stuffiness	8	VD2222	Changeableness	6	VD3304	Naïvety	6
VD4014	Vileness	10	VD4062	Withdrawal	8	VD2224	Chaos	6	VD3324	Nonconformity	6
VD2114	Atrocity	9	VD2134	Badness	7	VD2242	Coldness	6	VD3334	Nonuniformity	6
VD2214	Carelessness	9	VD2200	Brokenness	7	VD2270	Conflict	6	VD3358	Offence	6
VD2272	Confusion	9	VD2542	Disregard	7	VD2284	Contradiction	6	VD3374	Outrage	6
VD2588	Dullness	9	VD2566	Distress	7	VD2320	Darkness	6	VD3464	Pollution	6
VD2752	Grimness	9	VD2586	Dubiousness	7	VD2350	Defectiveness	6	VD3488	Prodigality	6
VD2770	Hardness	9	VD2654	Extravagance	7	VD2444	Dirtiness	6	VD3538	Repugnance	6
VD2870	Impropriety	9	VD2736	Futility	7	VD2448	Disaccord	6	VD3692	Sordidness	6
VD2890	Inadequacy	9	VD2900	Inappropriateness	7	VD2528	Disorganization	6	VD3766	Superficiality	6
VD2896	Inanity	9	VD3060	Insufficiency	7	VD2546	Dissatisfaction	6	VD3824	Torment	6
VD2958	Indifference	9	VD3164	Looseness	7	VD2556	Dissolution	6	VD3858	Trouble	6
VD3024	Injury	9	VD3312	Negligence	7	VD2562	Distortion	6	VD3930	Unnaturalness	6
VD3444	Perversion	9	VD3360	Omission	7	VD2576	Divergence	6	VD3948	Unreliability	6
VD3954	Unsuitability	9	VD3364	Opposition	7	VD2602	Embarrassment	6	VD3980	Vainness	6
VD3966	Upset	9	VD3366	Shamelessness	7	VD2616	Equivocation	6	VD3990	Vapidity	6
VD2004	Abandonment	8	VD3668	Shapelessness	7	VD2754	Grossness	6	VD4044	Warpedness	6
VD2336	Decline	8	VD3880	Uncertainty	7	VD2784	Harshness	6	VD4046	Waste	6
VD2376	Denial	8	VD4024	Virulence	7	VD2786	Heartlessness	6	VD4074	Wound	6
VD2526	Disorder	8	VD4078	Wrongness	7	VD2786	Heedlessness	6			

Figure 2. Destructive value words (Section VD)

Figures 1 and 2 list the value words with the greatest number of cross-references to distinct value polarities (Section VP) listed in Figure 3. Such cross-references correspond to distinct connotations of a given word. The value words with the largest number of distinct connotations could be considered as indicative of the value concepts which are most fundamental (for English speakers). Value concepts could also be identifies with the polarities as value dimensions.

VP No	VALUE POLARITY	VC	VD	TOT.	%	VP No	VALUE POLARITY	VC	VD	TOT.	%
VP5863	Pleasantness-Unpleasantness	34	118	152	2.1	VP5076	Inclusion-Exclusion	8	20	28	0.3
VP5624	Resolution-Irresolution	58	68	126	1.7	VP5492	Discrimination-Indiscrimination	18	10	28	0.3
VP5467	Intelligence-Unintelligence	63	60	123	1.7	VP5562	Education-Miseducation	20	8	28	0.3
VP5974	Probity-Improbity	73	46	119	1.6	VP6013	Godliness-Ungodliness	24	4	28	0.3
VP5674	Goodness-Badness	44	72	116	1.6	VP5600	Eloquence-Uneloquence	25	1	26	0.3
VP5733	Skilfulness-Unskilfulness	65	49	114	1.6	VP5848	Expensiveness-Cheapness	5	21	26	0.3
VP5938	Kindness-Unkindness	49	57	106	1.5	VP5885	Comfort-Aggravation	20	6	26	0.3
VP5161	Energy-Moderation	63	42	105	1.4	VP5535	Imaginativeness-Unimaginativeness	18	7	25	0.3
VP5516	Truth-Error	51	52	103	1.4	VP5960	Dueness-Undueness	14	11	25	0.3
VP5980	Virtue-Vice	24	75	99	1.4	VP5494	Judgement-Misjudgement	12	12	24	0.3
VP5785	Support-Opposition	62	35	97	1.3	VP5637	Choice-Necessity	12	12	24	0.3
VP5794	Accord-Disaccord	46	49	95	1.3	VP5873	Exultation-Lamentation	14	10	24	0.3
VP5513	Certainty-Uncertainty	43	51	94	1.3	VP6005	Vindication-Condemnation	14	10	24	0.3
VP5554	Communicativeness-Uncommunicativeness	19	75	94	1.3	VP5045	Simplicity-Complexity	16	7	23	0.3
VP5631	Desire-Avoidance	45	43	88	1.2	VP5267	Motion-Quiescence	11	12	23	0.3
VP5914	Repute-Disrepute	49	37	86	1.2	VP5030	Equality-Inequality	12	10	22	0.3
VP5705	Action-Inaction	41	41	82	1.1	VP5089	Unity-Duality	15	7	22	0.3
VP5890	Courage-Fear	32	50	82	1.1	VP5110	Durability-Transience	15	7	22	0.3
VP5665	Appropriateness-Inappropriateness	23	57	80	1.1	VP5851	Economy-Prodigality	21	1	22	0.3
VP5968	Approval-Disapproval	28	52	80	1.1	VP5947	Forgiveness-Vengeance	17	5	22	0.3
VP5533	Carefulness-Neglect	46	33	79	1.1	VP5122	Newness-Oldness	10	11	21	0.2
VP5026	Agreement-Disagreement	36	42	78	1.1	VP5589	Elegance-Inelegance	16	5	21	0.2
VP5660	Oversufficiency-Insufficiency	23	53	76	1.0	VP5690	Selfactualization-Neurosis	3	18	21	0.2
VP5762	Freedom-Restraint	48	27	75	1.0	VP5726	Victory-Defeat	9	11	20	0.2
VP5870	Cheerfulness-Solemnity	43	32	75	1.0	VP5998	Legality-Illegality	4	16	20	0.2
VP5896	Taste-Vulgarity	40	33	73	1.0	VP5539	Expectation-Inexpectation	11	8	19	0.2
VP5157	Power-Impotence	38	34	72	1.0	VP5683	Healthfulness-Unhealthfulness	7	12	19	0.2
VP5691	Improvement-Impairment	22	49	71	1.0	VP5925	Hospitality-Inhospitality	14	5	19	0.2
VP5927	Friendship-Enmity	41	30	71	1.0	VP5323	Oscillation-Agitation	0	18	18	0.2
VP5987	Chastity-Unchastity	25	45	70	0.9	VP5545	Meaning-Meaninglessness	7	11	18	0.2
VP5677	Perfection-Imperfection	27	39	66	0.9	VP5737	Behaviour-Misbehaviour	8	10	18	0.2
VP5732	Facility-Difficulty	18	47	65	0.9	VP5983	Innocence-Guilt	14	4	18	0.2
VP5672	Importance-Unimportance	31	32	63	0.8	VP5001	Existence-Nonexistence	12	5	17	0.2
VP5530	Attention-Inattention	28	33	61	0.8	VP5177	Concurrence-Counteraction	16	1	17	0.2
VP5475	Knowledge-Ignorance	39	18	57	0.8	VP5478	Thought-Thoughtlessness	13	4	17	0.2
VP5501	Belief-Unbelief	31	25	56	0.7	VP5038	Increase-Decrease	15	1	16	0.2
VP5964	Respect-Disrespect	21	34	55	0.7	VP5145	Conversion-Reversion	4	12	16	0.2
VP5481	Intuition-Reason	21	33	54	0.7	VP5165	Productiveness-Unproductiveness	10	6	16	0.2
VP5948	Congratulation-Envy	0	53	53	0.7	VP5186	Presence-Absence	7	9	16	0.2
VP5059	Order-Disorder	16	36	52	0.7	VP5837	Wealth-Poverty	5	11	16	0.2
VP5878	Amusement-Boredom	18	34	52	0.7	VP5047	Conjunction-Separation	6	9	15	0.2
VP5621	Willingness-Unwillingness	39	12	51	0.7	VP5147	Evolution-Revolution	9	6	15	0.2
VP5685	Health-Disease	19	32	51	0.7	VP5335	Light-Darkness	12	3	15	0.2
VP5855	Feeling-Unfeelinglessness	20	30	50	0.7	VP5009	Relatedness-Unrelatedness	11	3	14	0.1
VP5900	Naturalness-Affectation	28	22	50	0.7	VP5017	Uniformity-Nonuniformity	5	9	14	0.1
VP5693	Restoration-Destruction	11	37	48	0.6	VP5246	Form-Formlessness	2	12	14	0.1
VP5697	Safety-Danger	27	20	47	0.6	VP5528	Curiosity-Incuriosity	3	11	14	0.1
VP5159	Strength-Weakness	19	27	46	0.6	VP5646	Formality-Informality	12	2	1	0.1
VP5739	Authority-Lawlessness	28	18	46	0.6	VP6026	Sanctity-Unsanctity	14	0	14	0.1
VP5857	Excitement-Inexcitability	35	11	46	0.6	VP5003	Substantiality-Unsubstantiality	11	2	13	0.1
VP5908	Modesty-Vanity	11	35	46	0.6	VP5153	Causation-Culmination	12	1	13	0.1
VP5922	Sociability-Unsociability	31	14	45	0.6	VP5352	Weight-Lightness	7	6	13	0.1
VP5936	Courtesy-Discourtesy	26	19	45	0.6	VP5462	Harmony-Discord	8	4	12	0.1
VP5899	Beauty-Ugliness	25	19	44	0.6	VP5552	Interpretability-Misinterpretability	3	10	13	0.1
VP5141	Stability-Changeableness	22	21	43	0.6	VP5023	Originality-Imitation	12	0	12	0.1
VP5756	Leniency-Compulsion	26	17	43	0.6	VP5167	Production-Reproduction	12	0	12	0.1
VP5766	Obedience-Disobedience	14	29	43	0.6	VP5358	Elasticity-Toughness	8	4	12	0.1
VP5905	Pride-Humility	22	21	43	0.6	VP5427	Savouriness-Unsavouriness	4	8	12	0.1
VP5930	Love-Hate	26	17	43	0.6	VP5071	Continuity-Discontinuity	7	4	11	0.1
VP5472	Sanity-Insanity	12	30	42	0.5	VP5155	Attributability-Chance	5	6	11	0.1
VP5722	Accomplishment-Nonaccomplishment	29	12	41	0.5	VP5170	Ancestry-Posterity	11	0	11	0.1
VP5957	Rightness-Wrongness	20	21	41	0.5	VP5290	Direction-Deviation	5	6	11	0.1
VP5129	Timeliness-Untimeliness	19	21	40	0.5	VP5573	Representation-Misrepresentation	2	9	11	0.1
VP5728	Prosperity-Adversity	23	17	40	0.5	VP5592	Conciseness-Diffuseness	8	3	11	0.1
VP5978	Unselfishness-Selfishness	27	13	40	0.5	VP5101	Numerousness-Fewness	7	3	10	0.1
VP5526	Broadmindedness-Narrowmindedness	20	19	39	0.5	VP5185	Location-Dislocation	7	3	10	0.1
VP6028	Piety-Impiety	23	16	39	0.5	VP5234	Circumscription-Intrusion	1	9	10	0.1
VP5521	Assent-Dissent	19	19	38	0.5	VP5485	Research-Discovery	9	1	10	0.1
VP5865	Pleasure-Displeasure	37	1	38	0.5	VP5505	Provability-Unprovability	7	3	10	0.1
VP5548	Intelligibility-Unintelligibility	18	19	37	0.5	VP5920	Wonder-Unastonishment	10	0	10	0.1
VP5648	Motivation-Dissuasion	33	4	37	0.5	VP5137	Regularity-Irregularity	6	3	9	0.1
VP5815	Sharing-Appropriation	25	12	37	0.5	VP5313	Overrunning-Shortcoming	0	9	9	0.1
VP5976	Justice-Injustice	20	17	37	0.5	VP5317	Elevation-Depression	6	3	9	0.1
VP5054	Completeness-Incompleteness	27	9	36	0.5	VP5356	Hardness-Softness	6	3	9	0.1
VP5172	Influence-Influencelessness	28	8	36	0.5	VP5446	Appearance-Disappearance	3	6	9	0.1
VP5034	Greatness-Smallness	29	6	35	0.4	VP5709	Exertion-Rest	9	0	9	0.1
VP5422	Sensation-Insensibility	14	20	34	0.4	VP5798	Attack-Defence	4	5	9	0.1
VP5775	Consent-Refusal	18	16	34	0.4	VP5949	Gratitude-Ingratitude	8	1	9	0.1
VP5888	Hope-Hopelessness	16	18	34	0.4	VP5135	Frequency-Infrequency	8	0	8	0.1
VP5944	Compassion-Pitilessness	25	9	34	0.4	VP5195	Bigness-Littleness	1	7	8	0.1
VP5036	Superiority-Inferiority	12	21	33	0.4	VP5265	Opening-Closure	3	5	8	0.1
VP5248	Symmetry-Distortion	14	19	33	0.4	VP5444	Visibility-Invisibility	3	5	8	0.1
VP5407	Life-Death	14	19	33	0.4	VP5806	Neutrality-Compromise	8	0	8	0.1
VP5670	Expedience-Inexpedience	9	24	33	0.4	VP5112	Perpetuity-Instantaneousness	7	0	7	0.0
VP5013	Identity-Difference	15	17	32	0.4	VP5124	Youth-Age	6	1	7	0.0
VP5861	Patience-Impatience	21	11	32	0.4	VP5197	Expansion-Contraction	7	0	7	0.0
VP5868	Contentment-Discontentment	15	17	32	0.4	VP5435	Fragrance-Stench	1	6	7	0.0
VP5720	Preparedness-Unpreparedness	15	16	31	0.4	VP5509	Possibility-Impossibility	4	3	7	0.0
VP5808	Possession-Loss	14	17	31	0.4	VP5642	Conventionality-Unconventionality	6	1	7	0.0
VP5894	Caution-Rashness	15	16	31	0.4	VP5695	Refreshment-Relapse	4	3	7	0.0
VP5992	Temperance-Intemperance	13	18	31	0.4	VP5119	Futurity-Antiquity	6	0	6	0.0
VP5082	Conformity-Nonconformity	19	11	30	0.4	VP5143	Continuance-Cessation	6	0	6	0.0
VP5139	Change-Permanence	28	1	29	0.4	VP5294	Progression-Regression	1	5	6	0.0
VP5681	Cleanness-Uncleanness	12	17	29	0.4	VP5451	Silence-Loudness	3	3	6	0.0
VP5050	Cohesion-Disintegration	9	19	28	0.3						

Figure 3. Value polarities (Section VP). Presented in order of the number of cross-references to constructive and destructive value words (Sections VC and VD). Low order polarities have been omitted. Polarity names are only indicative.

Notes \ Method **VZ**

	FOCUS IN CONTEXT	CERTAINTY	INTRINSIC CONSTRAINT	NECESSITY	EXTERNAL CONSTRAINT
O R D E R	**VT8011 Existence*complex** VP5001 Existence-Nonexistence VP5005 Intrinsicality-Extrinsicality VP5003 Substantiality-Unsubstant.	**VT8012 Relationship*complex** VP5009 Relatedness-Unrelated. VP5017 Uniformity-Nonuniformity VP5023 Originality-Imitation VP5030 Equality-Inequality VP5013 Identity-Difference VP5020 Similarity-Dissimilarity VP5026 Agreement-Disagreement	**VT8013 Quantity*complex** VP5034 Greatness-Smallness VP5038 Increase-Decrease VP5047 Conjunction-Separation VP5054 Completeness-Incomplet. VP5036 Superiority-Inferiority VP5045 Simplicity-Complexity VP5050 Cohesion-Disintegration	**VT8014 Order*complex** VP5059 Order-Disorder VP5076 Inclusion-Exclusion VP5071 Continuity-Discontinuity VP5082 Conformity-Nonconformity	**VT8015 Number*complex** VP5086 Numbered-Unnumber. VP5101 Numerousness-Fewness VP5089 Unity-Duality
C H A N G E	**VT8021 Time*complex** VP5110 Durability-Transience VP5119 Futurity-Antiquity VP5124 Youth-Age VP5135 Frequency-Infrequency VP5112 Perpetuity-Instantaneous. VP5122 Newness-Oldness VP5129 Timeliness-Untimeliness VP5137 Regularity-Irregularity	**VT8022 Change*complex** VP5139 Change-Permanence VP5143 Continuance-Cessation VP5147 Evolution-Revolution VP5141 Stability-Changeableness VP5145 Conversion-Reversion	**VT8023 Causation*complex** VP5151 Eventuation-Imminence VP5155 Attributability-Chance VP5153 Causation-Culmination	**VT8024 Power*complex** VP5157 Power-Impotence VP5161 Energy-Moderation VP5167 Production-Reproduction VP5172 Influence-Influenceless. VP5159 Strength-Weakness VP5165 Productiveness-Unproduct. VP5170 Ancestry-Posterity VP5177 Concurrence-Counterac.	**VT8025 Space*complex** VP5185 Location-Dislocation VP5193 Container-Content VP5186 Presence-Absence
F O R M	**VT8031 Dimension*complex** VP5195 Bigness-Littleness VP5200 Nearness-Distance VP5207 Height-Lowness VP5197 Expansion-Contraction VP5204 Breadth-Narrowness VP5209 Depth-Shallowness	**VT8032 Contextuality*complex** VP5226 Centrality-Environment VP5234 Circumscription-Intrusion	**VT8033 Structure*complex** VP5246 Form-Formlessness VP5265 Opening-Closure VP5248 Symmetry-Distortion	**VT8034 Motion*complex** VP5267 Motion-Quiescence VP5283 Impact-Reaction VP5269 Swiftness-Slowness VP5288 Attraction-Repulsion	**VT8035 Relativemotion*compl.** VP5290 Direction-Deviation VP5294 Progression-Regression VP5298 Convergence-Divergen. VP5317 Elevation-Depression VP5292 Leading-Following VP5296 Approach-Recession VP5313 Overrunning-Shortcom. VP5323 Oscillation-Agitation
Q U A L I T Y	**VT8041 Absolute properties*complex** VP5328 Heat-Cold VP5339 Transparency-Opaque. VP5335 Light-Darkness VP5352 Weight-Lightness	**VT8042 Relative properties*comp.** VP5356 Hardness-Softness VP5362 Colour-Colourlessness VP5358 Elasticity-Toughness	**VT8043 Life*complex** VP5376 Materiality-Immateriality VP5417 Humanity-Nonhumanity VP5420 Masculinity-Femininity VP5407 Life-Death VP5419 Sexiness-Unsexiness	**VT8044 Sense*complex** VP5422 Sensation-Insensibility VP5427 Savouriness-Unsavour. VP5439 Vision-Blindness VP5446 Appearance-Disappear. VP5451 Silence-Loudness VP5425 Tangibility-Intangibility VP5435 Fragrance-Stench VP5444 Visibility-Invisibility VP5448 Audibility-Inaudibility VP5462 Harmony-Discord	**T8045 Intellectual faculties*.** VP5467 Intelligence-Unintell. VP5475 Knowledge-Ignorance VP5481 Intuition-Reason VP5472 Sanity-Insanity VP5478 Thought-Thoughtless.
S I G N I F I C.	**VT8051 Evaluation*complex** VP5485 Research-Discovery VP5494 Judgement-Misjudgement VP5492 Discrimination-Indiscrim. VP5497 Overestimation-Underest.	**VT8052 Credibility*complex** VP5501 Belief-Unbelief VP5507 Limitation-Unlimitedness VP5513 Certainty-Uncertainty VP5505 Provability-Unprovability VP5509 Possibility-Impossibility VP5516 Truth-Error	**VT8053 Truth*complex** VP5519 Illusion-Disillusionment VP5523 Affirmation-Denial VP5521 Assent-Dissent	**VT8054 Attitude*complex** VP5528 Curiosity-Incuriosity VP5533 Carefulness-Neglect VP5537 Remembrance-Forgetful. VP5526 Broadmindedness-Narrow. WP5530 Attention-Inattention VP5535 Imaginativeness-Unimag. VP5539 Expectation-Inexpectation	**VT8055 Meaning*complex** VP5545 Meaning-Meaninglessn. VP5552 Interpretability-Misinterp. VP5548 Intelligibility-Unintellig.
I N I T I A T I V E	**VT8061 Communication*complex** VP5562 Education-Miseducation VP5589 Elegance-Inelegance VP5600 Eloquence-Uneloquence VP5554 Communicativeness-Unco. VP5573 Representation-Misrepres. VP5592 Conciseness-Diffuseness	**VT8062 Choice*complex** VP5621 Willingness-Unwillingness VP5631 Desire-Avoidance VP5624 Resolution-Irresolution VP5637 Choice-Necessity	**VT8063 Motivation*complex** VP5642 Conventionality-Unconvent. VP5648 Motivation-Dissuasion VP5646 Formality-Informality	**VT8064 Adaptation*complex** VP5660 Oversufficiency-Insuffic. VP5670 Expedience-Inexpedience VP5674 Goodness-Badness VP5665 Appropriateness-Inappropr. VP5672 Importance-Unimportance VP5677 Perfection-Imperfection	**VT8065 Integrity*complex** VP5681 Cleanness-Uncleanness VP5685 Health-Disease VP5691 Improvement-Impairm. VP5695 Refreshment-Relapse VP5683 Healthfulness-Unhealthf. VP5690 Selfactualization-Neuros. VP5693 Restoration-Destruction VP5697 Safety-Danger
A C H I E V E M E N T	**VT8071 Action*complex** VP5705 Action-Inaction VP5720 Preparedness-Unprepar. VP5709 Exertion-Rest	**VT8072 Achievement*complex** VP5726 Victory-Defeat VP5732 Facility-Difficulty VP5737 Behaviour-Misbehaviour VP5722 Accomplishment-Nonacco. VP5728 Prosperity-Adversity VP5733 Skilfulness-Unskilfulness	**VT8073 Compliance*complex** VP5739 Authority-Lawlessness VP5762 Freedom-Restraint VP5768 Observance-Nonobserv. VP5756 Leniency-Compulsion VP5766 Obedience-Disobedience VP5775 Consent-Refusal	**VT8074 Interaction*complex** VP5785 Support-Opposition VP5798 Attack-Defence VP5794 Accord-Disaccord VP5806 Neutrality-Compromise	**VT8075 Possession*complex** VP5808 Possession-Loss VP5837 Wealth-Poverty VP5851 Economy-Prodigality VP5815 Sharing-Appropriation VP5848 Expensiveness-Cheap.
C O N S E Q U E N C E	**VT8081 Feeling*complex** VP5855 Feeling-Unfeelinglessness VP5861 Patience-Impatience VP5865 Pleasure-Displeasure VP5870 Cheerfulness-Solemnity VP5878 Amusement-Boredom VP5857 Excitement-Inexcitability VP5863 Pleasantness-Unpleasant. VP5868 Contentment-Discontent. VP5873 Exultation-Lamentation VP5885 Comfort-Aggravation	**VT8082 Anticipation*complex** VP5888 Hope-Hopelessness VP5894 Caution-Rashness VP5890 Courage-Fear	**VT8083 Discriminative affection*complex** VP5896 Taste-Vulgarity VP5900 Naturalness-Affectation VP5908 Modesty-Vanity VP5920 Wonder-Unastonishment VP5899 Beauty-Ugliness VP5905 Pride-Humility VP5914 Repute-Disrepute	**VT8084 Socialization*complex** VP5922 Sociability-Unsociability VP5927 Friendship-Enmity VP5931 Conjugality-Celibacy VP5925 Hospitality-Inhospitality VP5930 Love-Hate	**VT8085 Benevolence*complex** VP5936 Courtesy-Discourtesy VP5944 Compassion-Pitilessness VP5948 Congratulation-Envy VP5938 Kindness-Unkindness VP5947 Forgiveness-Vengeance VP5949 Gratitude-Ingratitude
R E A D A P T.	**VT8091 Appropriateness*complex** VP5957 Rightness-Wrongness VP5964 Respect-Disrespect VP5960 Dueness-Undueness	**VT8092 Judgement*complex** VP5968 Approval-Disapproval VP5976 Justice-Injustice VP5974 Probity-Improbity	**VT8093 Morality*complex** VP5978 Unselfishness-Selfishness VP5983 Innocence-Guilt VP5992 Temperance-Intemper. VP5980 Virtue-Vice VP5987 Chastity-Unchastity	**VT8094 Retribution*complex** VP5998 Legality-Illegality VP6010 Atonement-Punishment VP6005 Vindication-Condemnation	**VT8095 Redemption*complex** VP6013 Godliness-Ungodliness VP6026 Sanctity-Unsanctity VP6024 Orthodoxy-Unorthodoxy VP6028 Piety-Impiety

Figure 4. Value types (Section VT) Clusters of value polarities. Cluster type names are only indicative.

Notes \ Method **VZ**

	Focus in context	Certainty	Intrin. constraint	Necessity	External constraint	TOTAL
Order	VT8011: VC: 28 VD: 7 Tot: 35 %: 0.5	VT8012: VC: 95 VD: 81 Tot: 176 %: 2.5	VT8013: VC: 114 VD: 72 Tot: 186 %: 2.7	VT8014: VC: 50 VD: 74 Tot: 121 %: 1.7	VT8015: VC: 23 VD: 10 Tot: 33 %: 0.5	310 241 551 %7.9
Change	VT8021: VC: 77 VD: 43 Tot: 120 %: 1.7	VT8022: VC: 69 VD: 40 Tot:109 %: 1.6	VT8023: VC: 22 VD:7 Tot: 29 %: 0.4	VT8024: VC: 197 VD:118 Tot: 315 %: 4.5	VT8025: VC: 15 VD: 12 Tot: 27 %: 0.4	380 220 600 %8.6
Form	VT8031: VC: 18 VD: 10 Tot: 28 %: 0.4	VT8032: VC: 2 VD: 12 Tot: 14 %: 0.2	VT8033: VC: 19 VD: 36 Tot: 55 %: 0.8	VT8034: VC: 18 VD: 15 Tot: 33 %: 0.5	VT8035: VC: 18 VD: 43 Tot: 61 %: 0.9	75 116 191 %2.7
Quality	VT8041: VC: 21 VD: 12 Tot: 33 %: 0.5	VT8042: VC: 14 VD: 9 Tot: 23 %: 0.3	VT8043: VC: 26 VD: 20 Tot: 46 %: 0.7	VT8044: VC: 38 VD: 54 Tot: 92 %: 1.3	VT8045: VC: 148 VD:145 Tot: 293 %: 4.2	247 240 487 %7.0
Significance	VT8051: VC: 39 VD: 23 Tot: 62 %: 0.9	VT8052: VC: 137 VD: 134 Tot: 271 %: 3.9	VT8053: VC: 21 VD: 21 Tot: 42 %: 0.6	VT8054: VC: 128 VD:115 Tot: 243 %: 3.5	VT8055: VC: 28 VD: 40 Tot: 68 %: 1.0	353 333 686 %9.8
Initiative	VT8061: VC: 90 VD: 101 Tot: 191 %: 2.7	VT8062: VC: 154 VD: 135 Tot: 289 %: 4.1	VT8063: VC: 51 VD: 7 Tot: 58 %: 0.8	VT8064: VC: 157 VD: 277 Tot: 434 %: 6.2	VT8065: VC: 105 VD:188 Tot: 293 %: 4.2	557 708 1265 %18.1
Achievement	VT8071: VC: 65 VD: 57 Tot: 122 %: 1.7	VT8072: VC: 152 VD: 146 Tot: 298 %: 4.3	VT8073: VC: 135 VD: 107 Tot: 242 %: 3.5	VY8074: VC: 120 VD: 89 Tot: 209 %: 3.0	VT8075: VC: 70 VD: 62 Tot: 132 %: 1.9	542 461 1003 %14.3
Consequence	VT8081: VC: 257 VD 270 Tot: 527 %: 7.5	VT8082: VC: 63 VD: 84 Tot: 147 %: 2.1	VT8083: VC: 185 VD: 167 Tot: 352 %: 5.0	VT8084: VC: 112 VD: 69 Tot: 181 %: 2.6	VT8085: VC: 125 VD:144 Tot: 269 %: 3.8	742 734 1476 %21.1
Readjustment	VT8091: VC: 55 VD: 66 Tot: 121 %: 1.7	VT8092: VC: 121 VD: 115 Tot: 236 %: 3.4	VT8093: VC: 103 VD: 155 Tot: 258 %: 3.7	VT8094: VC: 21 VD: 28 Tot: 49 %: 0.7	VT8095: VC: 62 VD: 21 Tot: 83 %: 1.2	362 385 747 %10.7
TOTALS	VC: 650 VD: 589 Tot:1239 %: 17.7	VC: 807 VD: 756 Tot:1563 %: 22.3	VC: 676 VD: 592 Tot:1268 %: 18.1	VC: 841 VD: 836 Tot:1677 %: 23.9	VC: 594 VD:665 Tot:1259 %: 18.0	3568 3438 7006 %100

Figure 5. Value types data. Corresponds in presentation to Figure 4. Indicates totals by value type for the constructive and destructive value word cross-references (Sections VC and VD) via the value polarities (Section VP)

VT No	Value type	Constructive values	Destructive values	TOTAL	Percentage
VT8081	Feeling	257	270	527	7.52
VT8064	Adaptation	157	277	434	6.19
VT8083	Discriminative affection	185	167	352	5.02
VT8024	Power	197	118	315	4.49
VT8072	Achievement	152	146	298	4.25
VT8045	Intellectual faculties	148	145	293	4.18
VT8065	Integrity	105	188	293	4.18
VT8062	Choice	154	135	289	4.12
VT8052	Credibility	137	134	271	3.86
VT8085	Benevolence	125	144	269	3.83
VT8093	Morality	103	155	258	3.68
VT8054	Attitude	128	115	243	3.46
VT8073	Compliance	135	107	242	3.45
VT8092	Judgement	121	115	236	3.36
VT8074	Interaction	120	89	209	2.98
VT8061	Communication	90	101	191	2.72
VT8013	Quantity	114	72	186	2.65
VT8084	Socialization	112	69	181	2.58
VT8012	Relationship	95	81	176	2.51
VT8082	Anticipation	63	84	147	2.09
VT8075	Possession	70	62	132	1.88
VT8071	Action	65	57	122	1.74
VT8014	Order	50	71	121	1.72
VT8091	Appropriateness	55	66	121	1.72
VT8021	Time	77	43	120	1.71
VT8022	Change	69	40	109	1.55
VT8044	Sense	38	54	92	1.31
VT8095	Redemption	62	21	83	1.18
VT8055	Meaning	28	40	68	0.97
VT8051	Evaluation	39	23	62	0.88
VT8035	Relative motion	18	43	61	0.87
VT8063	Motivation	51	7	58	0.82
VT8033	Structure	19	36	55	0.78
VT8094	Retribution	21	28	49	0.69
VT8043	Life	26	20	46	0.65
VT8053	Truth	21	21	42	0.59
VT8011	Existence	28	7	35	0.49
VT8015	Number	23	10	33	0.47
VT8034	Motion	18	15	33	0.47
VT8041	Absolute properties	21	12	33	0.47
VT8023	Causation	22	7	29	0.41
VT8031	Dimension	18	10	28	0.39
VT8025	Space	15	12	27	0.38
VT8042	Relative properties	14	9	23	0.32
VT8032	Contextuality	2	12	14	0.19
		3568	3438	7006	99.77

Figure 6. Value types (Section VT) Presented in order of the total number of cross-references to constructive and destructive value words (Sections VC and VD)

Comments: interrelating values

1. Universal set of values
There is a continuing belief that a universal set of values can be formulated for the global community, possibly elaborated in the form of a hierarchy. This view is echoed in a recent study of the Club of Rome for UNESCO (1987): *"It seems strange that we should be obliged today to proclaim, at the most acute stage in the transition towards a new civilization, the obvious fact that no culture is possible without agreement on a foundation or solid base of common ethical values. This search for fundamental guidelines has been a constant feature of our historical and probably prehistoric past....The most recent form of this need to have a consistent set of mental images is represented by "human rights". It is interesting to note the similar concerns expressed by the sacred texts of the Vedas, the Gita, the Bible and the Koran....If there is no basic system of reference, there is no possibility of consensus and not even a possibility of challenge or rejection: discussion necessarily takes place in a vacuum. It would therefore be a tragic backward step to lose such basic values or not to replace them by more effective ones....The human person is nothing in itself, from the practical or even from the moral point of view, if it shuts itself off in isolation. Its value lies in its relations with others. It only acts, it is only capable of acting, in "communication" and this communication is impossible unless there exists a broad base of values common to the individuals involved. That is why, whether we like it or not, we must give special prominence as the fundamental data base, to a "system of ethical and moral values". However, the sharing of a common base is not yet enough. We must add the availability of a set of models representing the rules that will enable communication to take place..."*. The report argues that: *"Values such as collective human survival, the primacy and protection of human life, the preservation of nature and the dignity of mankind, justice, freedom and equity, already form the nucleus of universally accepted values upon which a real consensus has formed among peoples, but not among governments."*

2. Conflicting systems of values
However, the same report of the Club of Rome to UNESCO continues: *"Meanwhile, in many societies we find a growing antagonism between some of these new values, conveyed by the mass media, and the traditional values inherited from the past."*
The history of the relations between traditional value systems (exemplified by those of the major religions), as well as the present stresses between fundamentalist interpretations of them, is not encouraging for those believing that a universal set of values is emerging. The report is obliged to acknowledge the continuing presence of conflicting values: *"The emergence of certain universal values such as human rights or respect for nature does not mean the end of ancestral values. They may contradict each other...in addition, individual values may sometimes conflict with collective values or one value with another....Certain changes and contradictions in the hierarchy of values are taking a dramatic turn for individuals or communities when they are intensified by agitation aimed at arousing the emotions needed to achieve political ends."*

The report accepts that the harmonious co-existence of very different values is nothing new and asks the question of how co-existing and contradictory value systems are to be reconciled. It suggests, hopefully, that: *"We should not attach too much importance to the problem as value systems do not function in the same way as logical systems. The human mind is capable of absorbing systems which include traditional elements and other more modern or future-oriented elements as well as individual and collective criteria."* On this point it concludes that: *"the interesting and important point is that different systems of values do in fact co-exist even though their co-existence is sometimes coloured by opposition and mistrust. Indeed it is not so much a question of the co-existence of contradictory value systems as of the values being interpreted in different terms. When all is said, the factor that makes such co-existence, the plurality of interpretations and the society of uncertainty, possible is the capacity for dialogue."*

3. Beyond "credos" to requisite variety
There is little recognition in the discussion of universal values that society is likely to remain complex and that this complexity is necessary to the long-term viability of society. The impression is frequently created that an updated set of "ten commandments" could be based on worldwide consensus on ten universal values. There is even concern as to how such values should be "implemented". This raises the question of the treatment to be accorded to those who do not subscribe with equal enthusiasm to that same set of values. It also raises the question of how those values are to be refined if simplistic efforts to institutionalize them are not challenged.

A great deal of hope is placed in the possibility that everyone naturally accepts that "peace", "love", and "justice", for example, are unquestionable "goods", or that people can be educated into this understanding. The arguments of Section KD, for example, suggest that, whilst efforts in this direction are necessary, they are of local rather than global significance. There is a strong possibility that efforts to reduce the complexity of the value system to a limited number of values may be quite unhealthy. Using an ecological metaphor, it may be equivalent to the naive belief that only a few species are required in the natural environment to sustain human life.

The issue is one of requisite variety. How much variety and diversity is required in the value system(s) to ensure the sustainable development of human society. The question raised by Sections VC and VD is therefore which of the 960 constructive or 1040 destructive "values" implied can the psycho-social ecosystem afford to lose. What others need to be reflected there from other cultures and languages?

There is a great difference between: (a) designing a limited universal set of values in the light of questionable assumptions and (b) identifying the range of extant (or potential) values which might have some function in a universal system of values. It is this second strategy which has been tentatively explored in this section.

4. Challenge to comprehension
The problem is partly one of comprehension. Such value terms are understood very differently in different cultures and languages. They also lend themselves to every variety of (mis)interpretation. Most of the world's problems can be said to result from actions guided by differing interpretations of "peace", "love" and "justice" - the other person's interpretation always being perceived as at fault.

As noted by the Director-General of UNESCO, René Maheu: *"Behind the misty wall of words, the diverse, even contradictory, interpretations, motivations and utilizations are an indication of fundamental divisions concerning values. In particular, the most basic human rights are more frequently invoked as a weapon of attack or defence against some party, rather than recognized and practised as the royal road to a positive relationship between individuals and groups in an objective form of fraternity."* (15th General Conference, Paris, UNESCO, 1968).

The arguments of this section suggest that it is somewhat simplistic to expect the word "justice", for example, to carry the full set of connotations necessary for it to fulfil all the functions expected of it. As noted in Section KD, global consensus on any of these terms, or any set of them, can best be conceived as being characterized by an inherent uncertainty. This uncertainty is not to be regretted for it is that which is a guarantee of the dynamics through which a more profound understanding of values emerges. A neat definition of any value can only be of significance to a necessarily limited local group prepared to be bound in that way for some period of time, until its members are once again transformed by the global dynamics.

It may be argued that a significant number of words in Sections VC or VD should not be considered as carrying values. As with the possibility of "problem-less" problem names in Section P, these would therefore be "value-less" value names. But, as with the "name-less" problems discussed in Section PZ, there is also the extremely interesting question of "name-less" values. Clearly it is more than probable that the present vocabulary does not have words to denote many values which will be distinguished in the future, and which some may already distinguish at this time (as implied by Section HM). The challenge of "qualities without a name" implies the need for an organization of value information which will facilitate their emergence and recognition -- somewhat as with the role of the periodic table of chemical elements.

5. Nature of the requisite dynamic framework
This said it is not simply a question of accepting value relativism. As Kenneth Boulding (1978) points out: *"There is not, of course, a single set of human values and each human being has his or her own set. There are however processes in the ecological interaction or society by which these differing values, though not reduced to a single set, are at least coordinated in an ongoing evolutionary process."*

The question is then how the holding of any particular value fits into some such dynamic framework through which it is transformed by learning processes. Particular understandings are then better conceived as local way stations on learning cycles composed of complementary value sets. What is as yet far from clear is how such cycles are interlinked and how the transition to cycles encoding greater uncertainty can be accomplished. This question have been explored in relation to sets of human needs (Judge, 1980).

It is quite amazing that efforts to articulate value sets or systems seem to limit themselves to the simplicity of checklists or simple hierarchies. Whilst structural reductionism of this kind may be possible by collapsing rich structural detail, it is totally inadequate to the task of demonstrating how the ecosystem of values is woven together. Such simplistic structures totally fail to incorporate any degree of dynamic challenge to counteract the naturally tendency towards simplistic interpretations of values such as "peace", "love", etc. It is for this reason that a structure based on the natural challenge of value polarities opens up richer possibilities of embodying greater variety.

In such polarities there is truth and falsehood in both extremes. The polarity itself reflects the lived dilemma of acting in terms of the values represented. The polarities signal the existence of learning arenas in which individuals and societies respond to the tensions of the dilemma. Through experience of the dilemma comes understanding of the hidden weakness in "constructive" values and of the hidden truth in "destructive" values. There is then recognition that "constructive" initiatives are not always appropriate. As many religious traditions and mythologies recognize, appropriate renewal may well need to be preceded by "destruction".

6. Local vs Global
The key question then remains by what norms should action be guided. Clearly people can only be adequately motivated by the values they fully understand. Local values necessarily avoid the uncertainty inherent in global values (to which local communities may have an equivalent of the body's immune response reaction). Until such local values are acknowledged, respected and given a place within any global value framework, it is not to be expected that local communities will respond, other than in token form, to global values. This response is effectively a built-in safeguard. Local "shoulds" are a response to local conditions. Global "shoulds", as society is currently able to define them, are insensitive to the variety of local demands and are therefore effectively disempowered. They would engender a highly vulnerable society if expressed locally in their present form, aside from the possibilities of abuse.

At present the need is therefore for different local groups to act in terms of the different local values they perceive as meaningful. "Local" includes the "peace" movement(s), the "human rights" movement(s), the "green" movement(s), the "development" movement(s), *etc*, whose fundamental differences are an indication of the non-global nature of their specialized preoccupations. The spastic or paralysing global consequences of such differences may be overcome when values can be embodied as phases in learning cycles, with a local/global dimension, rather than perceived as static categories invoking territorial dynamics.

7. Human development as the resolution of value dilemmas
Erik Erikson (*Childhood and Society*, 1963) has explored the possibility that, at eight stages of the development of an individual through the life-cycle, distinct value conflicts are encountered. The implications of this sort of approach are outlined in the Introduction on *Phases of human development through challenging problems*.

8. Interrelating values through tensional integrity
In order to interrelate such competing value systems, frameworks such as the tensional integrity (tensegrity) structure can usefully be explored (see page 826). For such structures are brought into existence, and acquire their globality, precisely because of the countervailing dynamics between the local systems of which they are composed. Their integrity as a whole is engendered by the dynamic relationship, whether supportive or antagonistic, between the parts.

A truly universal system of values is thus better modelled by a tensional integrity structure which offers distinct niches for competing value systems, whilst ensuring a dynamic relationship between them. It is brought into existence by using the tensions between competing value perspectives -- using the energy of opposing perspectives to engender a larger structure. The complexity of such structures can be scaled up or down, by collapsing particular features, to the degree required for particular levels of comprehension.

Conventional presentations of value systems, as lists or hierarchies, reinforce the unfortunate illusion that the subtleties to which many are sensitive are adequately embodied in them. In spherical tensional integrities, the focal point around which the individual elements are structured is unoccupied by any concrete element. As such the structure leaves open the nature of that focal point, in sympathy with many of the arguments and intuitions outlined in Section H of this Encyclopedia. The centre is empty, and it is through the emptiness of that centre that the parts specified on the periphery acquire their meaning in relation to one another.

Given the possibility of such elegant representations of the universal system of values, honouring their rich complexity, it is difficult to justify the simplistic patterns of values which are favoured conventionally. Why is it that such simplistic representations are so rarely challenged?

9. The dance of positive and negative
This scheme suggests a much healthier approach to "positivity" as a slogan and "negativity" as an anathema in society today. The more responsible approach is well-illustrated by the following: *"What does it mean, to be whole? It means that we must be willing to conceive of, to contain within ourselves, whatever is "other than" any limited idea. It means knowing that when we emphasize a positive, we are at the same time creating a negative. When we choose an ideal of knowledge, then we must deal with the ignorance that is other than the knowledge. When we emphasize an ideal of holiness, then we must live with the sin that is its companion, and accept our responsibility for having awareness... If we allow that ugliness is always within us, then we are free to create beauty. If we know that stupidity is always within us, then we are free to emphasize this intelligence."* (Thaddeus Golas, 1971)

10. Values as strange attractors
In the spirit of the previous paragraphs, it is useful to speculate on the possible relevance of the most recent leap in understanding of the principles by which many natural and social phenomena are governed, namely chaos theory. It is widely accepted that the importance of values lies in the manner in which they govern the seemingly chaotic behaviour of individuals and groups. It is quite clear that the greatest difficulty is experienced in determining what values are and what form their existence takes.

Chaos theory has been faced with a similar challenge in determining what governs and constrains seemingly chaotic patterns of behaviour, especially in natural phenomena. Over time the behaviour of a complex system becomes constrained so that any action is increasingly influenced by such invisible attractors. It might be argued that values perform exactly that function in relation to the behaviour of individuals and groups. Their nature is as difficult to visualize as that of the strange attractors of chaos theory, which in no way denies the importance of the function that they perform.

The search for a universal system of values, and an adequate way of representing them, might be usefully explored in this light. How might it be possible to map the axiological landscape on which such attractors are located? How might the competing attractions of "constructive" and "destructive" values emerge?

Comments: wisdom and requisite variety

1. Challenge of communication and the search for wisdom

In a collective investigation for UNESCO entitled *In Search of a Wisdom for the World: the role of ethical values in education* (1987), the Club of Rome concludes: *"Successful development is very closely bound up with society's capacity to learn.... The role of communication and the revolution it is bringing about in the transmission of ideas may radically transform the problem of ethical values -- but the whole question needs careful thought and the will to succeed.*

Such thinking must be a collective enterprise, associating men and women from all countries and all fields of study, since it is an immense undertaking, a grand design that is at stake. Much patience and tolerance will be needed because not everyone gives the same importance to each fact.

Nor is the objective equally obvious to everyone. With the modern world as it is, the search for wisdom will not necessarily strike people as a priority issue and many will be sceptical and ironical. Nevertheless, all are invited to lay the foundations for a new humanism that will enable the peoples of tomorrow to live together harmoniously."

It is intriguing that a report by such authorities should endeavour so explicitly to marry the "search for wisdom" with the role ethical values in education (as signalled by the title of the report).

2. Images of wisdom

Times of crisis are signalled not only by the severity of the problems faced but also by the failure of standard solutions and by the loss of confidence in the paradigms through which new responses are recommended. At such times people search for "wisdom" to guide them through the crisis. It is therefore valuable to explore where people look for such wisdom and what forms it is perceived to take. A prime requirement would seem to be that it should transcend habitual forms of thought whilst reaffirming in a regenerative way the integrity of society. Achieving this would seem paradoxically to call for both a negation of received modes of thought and an integration of them into a larger or more firmly grounded context, perhaps affirming values recognized as fundamental or eternal. Wisdom is thus signalled by the perception of a more challenging opportunity opened up by the crisis itself.

The following are some of the sources of wisdom acknowledged by different constituencies, roughly grouped.

3. Sources of wisdom: genius

(a) Nobel laureates: The ultimate accolade in the advancement of understanding is the Nobel Prize. It is the peak achievement sought in many careers. Pronouncements of individual laureates are viewed as a form of wisdom. Meetings of laureates are one of the acknowledged means of benefitting from the best insights that modern society has to offer. And yet the prizes are given for very specific achievements, especially in the case of the sciences. It is not clear how such specialization is any guarantee of a wisdom that must transcend specific biases. Meetings of laureates are renowned for the expectations they arouse but not for the insights to which they give rise.

(b) Super-intelligence: In any society there are those who are gifted with a higher order of intelligence. They are recognized in concern for the super-gifted as a unique resource or through the associations of high-IQ people that are created. For some, such levels of intelligence offer the insights desired from wisdom. And yet the impact of the super-gifted on the current crisis has been limited. Often the intelligence demonstrated is narrow in focus and unable to handle the fuzzy complexity of a turbulent environment. Gatherings of such people are not renowned for the insights that they generate.

4. Sources of wisdom: expertise and experience

(a) Universities: Academics continue to cultivate the belief that university faculties provide the environment in which wisdom is to be found, if it exists. This follows from the traditional image of such centres of academic excellence. And yet it has become increasingly clear, from internal and external critics, that the competition between faculties (in the desperate search for resources and career advancement) has severely eroded the role of the university as a source of wisdom relevant to the challenge of the times.

(b) Think tanks: A key source of insight for policy-makers and others in this time of crisis are the policy institutes and think tanks, some of which have emerged as a reaction to the inappropriate information emerging from traditional universities. Such centres are designed as centres of excellence to focus conceptual resources on the global problematique and to envisage appropriate responses. And yet such centres seem constantly to be caught unprepared for the new form the crisis takes and seem to have little to offer that is meaningful within the political constraints under which their proposals have to be approved.

(c) General systems and cybernetics
The phenomena of society and nature can be described and interrelated through the discipline of general systems. This deliberately integrative discipline, and the cybernetic insights with which it is associated, in theory embodies all the principles necessary to control a society navigating through turbulent times. And yet the practitioners of the discipline, and the computer-based models they build, have not proved capable of embodying the factors which oppose the wider implementation of such insights. They have proved unable to deal with competing viewpoints and seem insensitive to shifting values and to the human scale, thus limiting the application of their wisdom to compartmentalized situations.

5. Sources of wisdom: advisors

(a) Counsellors and consultants: Under whatever name, organizations continue to call upon counsellors, consultants or experts, as a source of wisdom and expertise in designing more appropriate responses to the crisis of the times. This function continues to develop, supported by the increasing professionalization of management and policy skills. And yet, despite such investment, it is not clear whether the result is merely an adaptation to immediate challenges or is empowering society as a whole to respond to the larger dimensions of the crisis that exceed the mandates under which such consultants are required to operate.

(b) Community of peers: There is a recognition that although no one individual in a community may be able to provide the insights to respond to a complex crisis, the wisdom to guide an appropriate response is however available within a community of peers (such as an "invisible college" of academics). And yet although such groups of peers have accomplished much in sensitizing the world to different dimensions of the crisis and have explored many alternatives, they have remained unable to contain the crisis as it continues to evolve.

(c) Friendships: Wisdom may also be sought from a respected friend, not necessarily in the expectation of profundity, but because a caring friend (possibly a "soul friend") can provide a sensitive alternative perspective. Such a relationship can evoke wisdom, irrespective of the level of insight of either party individually. And yet, whatever the value to the individual, this form of wisdom is difficult to render useful to the condition of society as a whole.

6. Sources of wisdom: strategists

(a) Statesmen (and women): Wisdom is often associated with statesmanship, whether in the person of an emperor, a president or an "éminence grise". Such people are seen as rising above the concerns of petty politics and national preoccupations in the service of wider interests. In modern times such wisdom might be assumed to be institutionalized in the directorship of major intergovernmental organizations. And yet it would seem that the degree of ambition and political infighting required to achieve and maintain such positions inhibits the demonstration of a level of wisdom adequate to the present crisis. Too often nationalistic priorities are all too evident.

(b) **Policy scientists:** In an effort to provide a system for the subtle insights of those concerned with policy-making and the development of strategies, the policy sciences have emerged as a discipline. This appears to have been most appreciated as a source of insight to the extent that it focused either on the more immediate priorities of those in power concerned for the short-term survival of their initiatives or else on narrower ranges of concerns. In this respect it has become a tool of technocrats despite its wider potential in responding to the dimensions faced by those with broader concerns.

(c) **Military generals:** From earliest times the strategic sense developed by leaders of successful military campaigns as been considered as source of insight beyond purely military concerns. This is partly due to the human insight required of generals in responding to the non-technical factors governing the success of their actions. The classic example is that of Sun Tzu (*The Art of War*).

7. Hidden and revealed knowledge

(a) **Religious authorities:** For many, the process whereby religious authorities are elected or appointed is a guarantee that the wisdom available to that community is maximized in the person chosen, whether it is a pope, a priest or a spiritual "elder". Such focal figures perform important functions in this respect. And yet the disaffection with organized religion has progressively undermined the relevance of the insights of such people to the point at which their positions are even perceived by others as exacerbating the crisis rather than alleviating it.

(b) **Esoteric societies:** Wisdom continues to be sought within esoteric societies, often of a secret nature although "schools of wisdom" continue to be founded, inspired by those of the antiquity. Such societies are usually structured so that participants pass through a series of stages associated with increasing levels of wisdom that cannot be effectively communicated to the uninitiated without danger. At the highest level it is not unusual for a relationship to be claimed to hidden masters as a font of wisdom. And yet, despite rumours of the activities of such societies at the highest levels of society, it remains unclear whether the wisdom to which they claim access can be considered relevant to the challenge of the times.

(c) **Sacred books**: For many, wisdom is embodied in particular sacred books, and no further wisdom is called for in responding to any crisis. It is argued that if humanity were to abide by the principles indicated therein, society could rise above its problems with ease. And yet society has been torn apart by the constituencies inspired by different sacred books. And even those inspired by the same book have proved unable to reconcile their differences. Somehow the wisdom of such works is insufficient to the task of reconciling the many ways in which it may be interpreted.

8. Sources of wisdom: traditional

(a) **Proverbial folk wisdom:** There is a large reservoir of confidence in folk wisdom, whether expressed in proverbs or aphorisms or in the person of particular individuals seen to be "close to the earth" and its rhythms. Such wisdom is especially important to those in rural areas. And yet this wisdom has proved vulnerable to novel ideas associated with modernization and has largely lost its attractiveness to the young.

(b) **Witchcraft:** In some forms witchcraft is held to be a source of earth wisdom, traditionally associated with powerful feminine insight. The value of this insight is believed by some to have been distorted as a result of periods of religious repression, especially since its threat to the dominance of male approaches encouraged deliberate misunderstanding of its nature. Past decades have seen a marked resurgence of witchcraft, partly encouraged by certain trends in feminism. Some distinguish the spiritual orientation of the "magician" from the material orientation of the "witch" (with one displacing the other under different circumstances). And yet, despite the ways in which it compensates for the sterility of organized society, such earth wisdom permeates with difficulty into wider society and is frequently corrupted by the dubious initiatives of other forms of witchcraft.

(c) **Divination:** Many cultures have devices or procedures through which divination is possible without the need for any intermediary. This may involve "casting the bones", casting a horoscope, use of runes, or other such procedures. The results are treated as a form of wisdom by which many continue to live on a daily basis.

(d) **Seers:** One of the oldest recognized sources of wisdom is the seer, epitomized by the oracle at Delphi. Their authority may come from their skill in divination. Although it might be assumed to be obsolete, this function is now actively filled by a wide spectrum of astrologers, psychics, channels, shaman and others. The number of such people exceeds that of the established clergy in many countries and the amounts invested in consulting them, whether by presidents or paupers, exceed the turnover of major industries. "Channelled" insights are widely sought on a variety of media. And yet, despite the hopes they raise, it is difficult to point to wisdom from such sources that enables society to respond in more appropriate ways, whatever the perceived value for individuals in adjusting to the stresses to which they are exposed.

9. Sources of wisdom: spiritual insight

(a) **Spiritual leaders and gurus:** Many are drawn to people of special spiritual insight. Such spiritual leaders exert a powerful influence on people in gatherings, through their writings, or in the communities they create. They are often viewed as wisdom incarnate. And yet when they appear together, if they are willing to do so, their interaction does not appear to reconcile their constituencies in a manner which enables a larger wisdom to prevail in initiatives for the future. Where the interaction among them is not indicative of conflicts that their wisdom is unable to resolve, their particular wisdom emerges as lacking in novelty and immediate relevance (often to the point of being banal).

(b) **Psychotherapists:** Recent decades have seen a burgeoning interest in various forms of psychotherapy perceived as a way for individuals to obtain insight into a more appropriate organization of their lives. As one modern form of the seer or shaman, psychotherapy is recognized as a valued source of wisdom by those who use it, especially when the practitioner is of the status of Freud or Jung.

(c) **Wise people:** In any period or society, it would seem that certain individuals, usually of some prominence for other reasons, are widely recognized as wise. While being valued for their particular expertise, it is because of this quality that they become a symbol in society. Examples might include: Bateson, Krishnamurti, Schumacher, Jung, Einstein, Bohm, Jantsch. And yet, although their insights continue to inspire and inform many current initiatives, the question remains as to whether such wisdom is adequate to the challenge of the times.

10. Sources of wisdom: inner wisdom

There are many who would argue that ultimately the most meaningful source of wisdom is that derived by the individual from some inner source, however that is experienced or understood.

Index

Human Potential (Volume 2)

Index scope

This index covers **all sections in this volume**. It does **not** include information on entries in Volume 1.

All index entries refer via reference numbers (eg HH1234) to the descriptions in the preceding sections. The letters indicate the section (eg Section HH).

Index entries

The index entries are of several types:

- Principal name or title of concept;

- Secondary or alternative names of concept (including popular expressions and synonyms)

- Keywords from the principal concept name;

- Keywords from the secondary concept names.

Since the concepts do not necessarily have formally defined names, it was not considered useful to distinguish typographically between the different types of index entry. Keywords are however recognizable by the presence of a semi-colon in the index entry.

Remarks

To facilitate consultation of the index, and to reduce the space requirement:

- Prepositions and articles are normally omitted from the index entries;

- Sub-headings in bold are generated either when several entries commence with a minimum of three common words, or when the second word (of a group) starts a new alphabetic sequence.

Although the index was generated automatically, it has been edited manually to eliminate less helpful keywords resulting from the extraction process. Where there was any doubt, less meaningful words have however been left where they may prove to be of value in locating problems.

Users may find it more convenient to consult the individual indexes for each section, if the purpose is to get an overview of the contents of a section. They are Sections HX, KX, MX, TX and VX. (Note that Section HX is however a special index by numbered sets of concepts.)

A

HM 2212 Abaddon
HM 3615 Abaissement niveau mental (Jung)
HM 4348 Abandoning ; Full understanding (Buddhism)
HM 3057 Abandonings ; Thorough (Buddhism)
VD 2004 Abandonment
HH 1223 Abandonment God (Christianity)
HH 0868 Abandonment ; Spiritual self
VD 2006 Abasement
HM 2209 Abasement ; Abounding (ICA)
HM 3587 Abasement ; Self
VD 2008 Aberration
HM 1534 Abgeschiedenheit (Christianity)
HM 3208 Abhavashunyata (Buddhism)
HM 2448 Abhavasvabhavashunyata (Buddhism)
HH 1765 Abhidhamma (Pali)
HH 1765 Abhidharma (Buddhism)
HM 3563 Abhijjha (Pali)
HH 1809 Abhijna (Buddhism)
HM 2385 Abhimukhi (Buddhism)
HM 3432 Abhinivesa (Pali)
HH 1809 Abhinna (Pali)
HM 3871 Abhinvesa (Yoga)
HM 1687 Abhippasada (Pali)
HM 2337 Abhirati (Buddhism)
HM 5134 Abhirati ; Sunnagare (Buddhism)
HM 9845 Abhisneha (Hinduism)
HM 9845 Abhisvanga (Hinduism)
VD 2010 Abhorrence
HH 2772 Abhyantara vritti yoga (Yoga)
HH 0544 Abhyasa
HM 2424 Abi'l Khayr ; Stations consciousness ibn (Sufism)
HM 0815 'Abid (Sufism)
HM 2147 Abiding ; Calm (Buddhism)
HM 2277 Abiding equipoise ; Mental (Buddhism)
HM 3485 Abiding ; Non (Zen)
Abiding states
HM 2181 — Disciplining mental (Buddhism)
HM 2729 — Full pacification mental (Buddhism)
HM 2753 — One pointedness mental (Buddhism)
HM 2205 — Pacifying mental (Buddhism)
HM 0330 Abiding (Sufism)
HH 3534 Abidings meditation subjects ; Divine (Buddhism)
HM 2145 Abidings ; Meditative states mental (Buddhism)
VC 0003 Ability
HH 0481 Ability ; Problem solving
HH 2334 Ability ; Psychic
HM 0979 Ability ; Working (Buddhism)
HM 2429 Abject helplessness (ICA)
VD 2012 Abjection
VC 0005 Abnegation
VD 2014 Abnormality
HH 0100 Abnormality ; Psychic
HM 6093 Abolition passion (Buddhism)
VD 2016 Abomination
HH 1301 Aboriginal dream-time (Australian)
HM 0672 Aboulia
HM 2209 Abounding abasement (ICA)
HM 3974 Abraham one's being (Sufism)
HH 0640 Abreaction
HM 3857 Abruptness (Buddhism)
VD 2018 Absence
HM 2670 Absence consciousness
HM 2695 Absence delusion (Buddhism)
HM 2695 Absence dullness (Buddhism)
HM 7344 Absence love (Hinduism)
VP 5186 Absence ; Presence
HH 1333 Absent healing
HM 4976 Absent-mindedness
HM 4976 Absentmindedness
HM 2397 Absolute-body awareness (Buddhism)
HM 3228 Absolute conditional ; Reciprocal relation (Sufism)
HM 0634 Absolute (Confucianism)
HH 0893 Absolute idealism
HM 3243 Absolute intelligence ; Path (Judaism)
HM 4326 Absolute plane (Leela)
VT 8041 Absolute properties*complex
HM 5265 Absolutely open nothingness (Taoism)
VC 0009 Absoluteness
HH 0472 Absolution
VC 0007 Absolution
HM 2370 Absolution ; Personal (ICA)
VD 2020 Absolutism
HM 1618 Absorption ; Appearance (Buddhism)
HM 0450 Absorption (Buddhism)
HM 0311 Absorption concentration (Buddhism)
HM 2027 Absorption ; Consciousness nothingness (Buddhism)
HM 0137 Absorption (Hinduism)
Absorption immaterial sphere
HM 2110 — First (Buddhism)
HM 2051 — Fourth (Buddhism)
HM 3043 — Second (Buddhism)
HM 2027 — Third (Buddhism)
HM 3043 Absorption ; Limitless consciousness (Buddhism)
HM 2110 Absorption ; Limitless space (Buddhism)
HM 1690 Absorption ; Meditative (Yoga)
HM 3047 Absorption nothingness (ICA)
HH 0827 Absorption ; Path meditative (Yoga)
HM 2051 Absorption ; Peak cyclic existence (Buddhism)
VD 3626 Absorption ; Self
HH 4777 Absorption ; Skill (Buddhism)
HM 3150 Absorptions ; Serial (Buddhism)
HM 2669 Absorptions ; Tibetan meditative states formless (Buddhism)
HM 2122 Absorptions ; Trances mental (Buddhism)
HH 0361 Abstaining misappropriation
Abstaining refraining
HM 1781 — four deviations speech ; Leaving off, abstaining, totally (Buddhism)
HM 1133 — three deviations body ; Leaving off, abstaining, totally (Buddhism)
HM 0252 — wrong livelihood ; Leaving off, abstaining, totally (Buddhism)
Abstaining totally abstaining
HM 1781 — refraining four deviations speech ; Leaving off, (Buddhism)
HM 1133 — refraining three deviations body ; Leaving off, (Buddhism)
HM 0252 — refraining wrong livelihood ; Leaving off, (Buddhism)
HH 0978 Abstention craving sensual enjoyment
HH 1008 Abstention sleep
Abstinence
HH 0070 — [Abstinence]
HH 0600 — [Abstinence]
HH 0011 — [Abstinence]
VC 0011 — [Abstinence]
HM 5600 Abstinence bodily misconduct (Buddhism)
HM 5600 Abstinence misconduct kaya (Buddhism)
HH 0045 Abstinence ; Sexual
HM 4450 Abstinence (Sufism)
HM 7171 Abstinence verbal misconduct (Buddhism)
HM 7887 Abstinence wrong livelihood (Buddhism)
HM 0286 Abstinence wrongdoing (Sufism)
HM 3213 Abstract dancing
KC 0386 Abstract mathematical space
KC 0129 Abstract objects
KC 0627 Abstract systems
HH 0348 Abstract thinking
Abstraction
KC 0271 — [Abstraction]
HH 0120 — [Abstraction]
VC 0013 — [Abstraction]
HH 0829 Abstraction (Yoga)
HM 2331 Absurd existence (ICA)
VD 2022 Absurdity
HM 0672 Abulia
VC 0015 Abundance
TP 0055 Abundance
VD 2024 Abuse
HH 5110 Abuse ; Alternatives drug
VD 2026 Abusiveness
HM 1734 Abyapada (Pali)
TP 0029 Abyss Danger
TP 0029 Abyss Mastering pitfalls
TP 0029 Abyss Multiple dangers
TP 0029 Abyss Persistence
HM 2816 Acala (Buddhism)
HH 4511 Accelerated learning
MM 2031 Accelerator ; Cyclic resonance
VC 0017 Acceptance
HM 1875 Acceptance
HH 3663 Acceptance female principle
HM 3912 Acceptance ; Self (Jung)
HM 2460 Acceptance ; Terrifying (ICA)
HH 0204 Accepting shadow
HM 4999 Access concentration (Buddhism)
MP 1071 Access contained irrationality
MP 1158 Access higher structural levels ; External
MP 1010 Access intensity
MP 1064 Access patterns active irrationality
MP 1196 Access perspective contexts ; Eccentric
VC 0019 Accessibility
TC 1808 Accessibility ; Conference
MP 1127 Accessibility ; Nested levels
MP 1201 Accessible facilities perspective adjuncts
MP 1060 Accessible non-linearity
VD 2028 Accident
TC 1637 Accident tactics
HM 4033 Accidental psychic attack (Psychism)
HM 3259 Accidie
VC 0021 Acclaim
TC 1164 Accommodator participant type
HM 4191 Accompanied bliss ; Concentration (Buddhism)
HM 1454 Accompanied bliss ; Concentration (Buddhism)
HM 4191 Accompanied ease ; Concentration (Buddhism)
HM 1454 Accompanied ease ; Concentration (Buddhism)
Accompanied equanimity
HM 5008 — Concentration (Buddhism)
HM 8021 — Concentration (Buddhism)
HM 7525 — Understanding (Buddhism)
Accompanied even mindedness
HM 5008 — Concentration (Buddhism)
HM 8021 — Concentration (Buddhism)
HM 7525 — Understanding (Buddhism)
HM 4433 Accompanied happiness ; Concentration (Buddhism)
HM 5008 Accompanied indifference ; Concentration (Buddhism)
HM 8021 Accompanied indifference ; Concentration (Buddhism)
HM 6245 Accompanied joy ; Understanding (Buddhism)
HM 4433 Accompanied rapture ; Concentration (Buddhism)
HM 4320 Accomplishing wisdom ; All
VC 0023 Accomplishment
TP 0063 Accomplishment
VP 5722 Accomplishment-Nonaccomplishment
HM 1342 Accomplishment ; Ultimate (Taoism)
VC 0025 Accord
TP 0008 Accord
HM 1405 Accord ; Anonymous
VP 5794 Accord-Disaccord
HM 5309 Accordance righteousness ; Meeting God both resting working (Christianity)
HM 2008 Accountability ; Called (ICA)
HM 2309 Accountability ; Destinal (ICA)
HM 3206 Accountability ; Final (ICA)
HH 0783 Acculturation
KC 0198 Acculturation
VC 0027 Accumulation
HM 2962 Accumulation application ascension stages game ; Sutra paths (Buddhism)
HM 2848 Accumulation application ascension stages game ; Tantric paths (Buddhism)
KD 2215 Accumulation learning society ; Pattern
KD 2007 Accumulation significance ; Answer production
KD 2009 Accumulation significance ; Development processes
VC 0029 Accuracy
HH 0584 Accusation ; Self
HM 3259 Acedia
HM 3259 Acedy
HM 4390 Acheta (Japanese)
HM 0298 Achieve concentration ; Inability (Yoga)
VC 0031 Achievement
HH 3188 Achievement
VT 8072 Achievement*complex
HH 6018 Achievement excellence ; Education early
HH 0456 Achievement motivation
VC 0033 Acknowledgement
HM 2463 Acknowledgement ; Painful (ICA)
MP 1070 Acknowledgement past perspectives ; Context
VC 0035 Acquiescence
TP 0002 Acquiescence (Earth)
HM 4190 Acquiescence (Sufism)
TP 0004 Acquiring experience (Young shoot)
TP 0017 Acquiring followers
VC 0037 Acquisition
MP 1083 Acquisition perspective maintenance dynamics ; Integration perspective
Acquisition sinful unwholesome
HM 0552 — dharmas ; Being ashamed (Buddhism)
HM 1208 — dharmas ; Feeling remorse (Buddhism)
HM 3257 — dharmas ; Not ashamed (Buddhism)
HM 3353 — dharmas ; Not feeling remorse (Buddhism)
VD 2030 Acquisitiveness
VD 2032 Acrimony
MP 1224 Across boundaries ; Emphasizing transitions
HM 4487 Act ; Wish (Buddhism)
HH 0567 Action
VC 0039 Action
VT 8071 Action*complex
HM 0439 Action ; Correct (Buddhism)
HM 0687 Action ; Correct (Buddhism)
Action [Decisive...]
TP 0021 — Decisive (Biting through)
HM 2491 — desire (Hinduism)
HH 5487 — desire (Hinduism)
KD 2353 Action formulation ; Alternation: implications
VP 5705 Action-Inaction
HM 2466 Action irrelevant (ICA)
HH 0315 Action learning
Action [Naturalness...]
HM 3189 — Naturalness bodily
MP 1045 — network ; Local
HH 1073 — Non
TC 1765 Action-oriented type
Action [Path...]
HM 2021 — Path intelligence spiritual (Judaism)
HM 2417 — principle conscious states ; Least (Physical sciences)
TP 0021 — Punitive (Biting through)
Action [Remedial...]
TP 0018 — Remedial
HM 0198 — Right aims (Buddhism)
HM 0198 — Right (Buddhism)
HM 3462 Action ; Saviors God: (Christianity)
TP 0001 Action ; Strong (Heaven)
HM 2864 Action tantra: kriya (Buddhism)
HH 1211 Action ; Way (Hinduism)
HH 0372 Action ; Yoga (Yoga)
HM 3080 Actional existence (ICA)
HM 6534 Activating mind Tao (Taoism)
HM 6204 Active desire ; Meeting God (Christianity)
HM 1430 Active emotional identification
Active [imagination...]
HM 0867 — imagination
HM 2218 — intelligence ; Path (Judaism)
HM 1997 — intelligence ; Way (Esotericism)
MP 1064 — irrationality ; Access patterns
HH 4008 Active meditation
HM 1905 Active remembrance God (Sufism)
HH 0091 Active therapy
HM 0442 Active trance (Psychism)
TC 1962 Activists ; Alienation committed
HM 2050 Activities ; Aggregate mental (Buddhism)
MP 1165 Activities communication pathway ; Exposure structural
MM 2024 Activities ; Daily round
TC 1789 Activities ; Meetings social
TC 1777 Activities ; Risk triggering
TC 1721 Activities ; Trust building
VC 0041 Activity
HH 0492 Activity group therapy
MP 1157 Activity ; Local opportunities perspective
MP 1030 Activity nodes
MP 1156 Activity ; Opportunities perspectives decreasing
Activity [Semi...]
MP 1155 — Semi autonomous contexts perspectives decreasing
MM 2015 — Shifting patterns
MS 8500 — Symbolic human
HM 0198 Acts ; Right (Buddhism)
HH 6548 Actual grace (Christianity)

Actualism

HH 0406	Actualism (Yoga)
VC 0043	Actuality
HH 0764	Actualization
VC 1539	Actualization ; Self
HH 0412	Actualization ; Self
HH 0971	Actualizing personality ; Self
VC 0045	Acuity
VC 0047	Acumen
HH 0728	Acupuncture
HM 3151	Acute inadequacy (ICA)
HM 4542	Ad-Darr (Sufism)
HM 3425	Adam one's being (Sufism)
HM 2028	Adamantine-hell awareness (Buddhism)
HM 2393	Adapability ; God consciousness
VC 0049	Adaptability
MP 1153	Adaptable changing number embodied perspectives ; Structures
MP 1079	Adaptable contexts ; Perspective
KC 0817	Adaptation
HH 1233	Adaptation
VT 8064	Adaptation∗complex
TC 1843	Adaptive conferencing
KC 0133	Adaptive cultural integration
MP 1048	Adaptive interstices
HM 1510	Adassana (Pali)
HH 0325	Addiction group therapy
HM 0110	Addiction objects (Yoga)
TC 1383	Address
HM 1643	Addukkhamasukha vedana ; Cetosamphassaja (Pali)
HH 0805	Adept
HH 1552	Adept
VC 0051	Adequacy
MP 1035	Adequate variety cyclic elements
HM 3697	Adha-loka (Jainism)
HM 0670	Adharma (Leela)
VC 0053	Adherence
HH 2387	Adhicitta sikkha (Buddhism)
HM 2933	Adhimoksha (Buddhism)
HM 2719	Adhimokshikamanaskara (Buddhism)
HM 2744	Adhosa (Pali)
HM 3198	Adhyasa
HM 2794	Adhyatmabahirdhashunyata (Buddhism)
HM 2671	Adhyatmashunyata (Buddhism)
HM 0822	Adi plane (Psychism)
MP 1201	Adjuncts ; Accessible facilities perspective
MP 1200	Adjuncts ; Facilities perspective
MP 1198	Adjuncts ; Inter domain contexts perspective
VC 0055	Adjustment
HH 4009	Adjustment human face ; Structural (United Nations)
HM 3807	'Adl ; Al (Sufism)
KC 0021	Administration
HM 2095	Administrative intelligence ; Path (Judaism)
HM 3013	Admirable intelligence ; Path (Judaism)
HM 3701	Admiration
VC 0057	Admiration
HM 0453	Admiration (Islam)
VD 3628	Admiration ; Self
HH 4012	Admission sin
HH 0021	Adolescence ; Rites
HM 4637	Adoptive knowledge (Buddhism)
HM 2412	Adoration
VC 0059	Adoration
HM 2111	Adoration ; Singular (ICA)
HH 3424	Adore
VC 0061	Adornment
TP 0022	Adornment
HH 0555	Adornment ; Self
MS 8446	Adornments ; Jewellery
HM 2744	Adosa (Buddhism)
HM 1399	Adosa (Pali)
HM 3560	Adoso kusala mulam (Pali)
HM 1889	Adrenergia (Psychism)
HM 5118	'adudi ; Wuiquf (Sufism)
HM 1352	Adukkhamasukham vedayitam ; Cetosamphassajam (Pali)
TP 0036	Adulteration insight (Darkening of the light)
HM 1557	Adussana (Pali)
HM 1597	Adussitattam (Pali)
HH 0518	Advaita vedanta (Hinduism)
TP 0053	Advance ; Gradual
HH 0521	Advanced training
	Advancement
VC 0063	— [Advancement]
TP 0035	— [Advancement]
TP 0046	— [Advancement]
VC 1541	Advancement ; Self
VC 0065	Advantage
VC 0067	Adventure
TP 0047	Adversity
VP 5728	Adversity ; Prosperity
HM 8336	Adverting (Buddhism)
HM 8336	Adverting ; Five door (Buddhism)
HM 8336	Adverting ; Mind door (Buddhism)
HM 8336	Adverting ; Six door (Buddhism)
HM 2744	Advesha (Buddhism)
VC 0069	Advocacy
TC 1929	Advocate ; Devil's
HM 2784	Advocate ; Heavenly (ICA)
HH 2002	Aeons dissolution evolution (Buddhism)
MM 2064	Aerial animal locomotion - flight
MM 2059	Aerial animal locomotion - gliding soaring
MS 8357	Aerial beings ; Mythical
HM 3387	Aesthetic attitude (Japanese)
HM 4350	Aesthetic contrasts ; Harmony (Japanese)
HH 0029	Aesthetic development
HH 0029	Aesthetic education
	Aesthetic emotion
HM 8762	— anger (Hinduism)
HM 0966	— disgust (Hinduism)
HM 4085	— energy (Hinduism)
HM 1644	— fear (Hinduism)
HM 0399	— grief (Hinduism)
HM 0205	— (Hinduism)
HM 1780	— love (Hinduism)
HM 0645	— mirth (Hinduism)
HM 0291	— wonder (Hinduism)
HM 7219	Aesthetic enjoyment (Hinduism)
HM 2762	Aesthetic feeling (Japanese)
HM 2542	Aesthetic LSD experience
HM 0218	Aesthetic perception (Psychism)
	Aesthetic [silence...]
HM 1664	— silence
HM 1365	— simplicity (Japanese)
HM 3416	— state (Psychism)
HH 0993	— surgery
TC 1909	— symbols ; Meetings
HM 3597	Aesthetic understatement (Japanese)
VC 0071	Aesthetics
KD 2065	Aesthetics ; Containing discontinuity
KD 2005	Aesthetics ; Forms truth: uniformity versus
VC 0073	Affability
HH 0819	Affect
HM 6087	Affect ; Dissociation
VD 2034	Affectation
VP 5900	Affectation ; Naturalness
	Affection
HM 5409	— [Affection]
VC 0075	— [Affection]
HH 0258	— [Affection]
HH 0422	— [Affection]
VT 8083	Affection∗complex ; Discriminative
HM 9845	Affection (Hinduism)
HM 8000	Affection ; Non (Hinduism)
HM 3590	Affection (Psychism)
HM 3498	Affectional bonding
HM 0973	Affectionate bonding
HH 1890	Affections ; Gracious
HH 1890	Affections ; Religious
HH 1890	Affections ; Spiritual
HM 2891	Affective dementia
HH 0260	Affective transformation
HM 7132	Affects
HM 1748	Affects
VC 0077	Affiliation
HH 2098	Affinities
VC 0079	Affinity
TP 0054	Affinity ; Elective (Marrying maiden)
TC 1602	Affinity group conferencing
HM 5304	Affinity (Islam)
TC 1528	Affinity roles
VC 0081	Affirmation
VP 5523	Affirmation-Denial
HM 2199	Affirmation (ICA)
HH 1453	Affirmations
HH 0458	Affirmative healing
HM 1564	Afflatus
HM 3611	Afflicted mother ; Jesus meets his (Christianity)
HM 2996	Afflicted views (Buddhism)
VD 2036	Affliction
HH 5007	Afflictions hindrances (Buddhism)
HM 2470	Afflictions ; Mansions favours (Christianity)
HH 0270	Afflictions ; Root (Buddhism)
HH 0781	Afflictions ; Secondary (Buddhism)
HH 1047	Afflictions (Yoga)
VC 0083	Affluence
VD 2038	Affrontery
HH 5224	African spirituality
TP 0063	After completion
HM 0665	After death dream state (Psychism)
HM 0794	After-images
HH 0399	Afterlife
HM 5543	'Afuw ; Al (Sufism)
HH 0576	Again ; Born (Christianity)
HH 0258	Agape
HM 3105	Agape appreciation (ICA)
HM 2774	Agape compassion (ICA)
HM 2828	Agape motivity (ICA)
HH 5908	Agape philo ; Journeying transcendence
HM 3217	Agape responsibility (ICA)
VD 2040	Age
HM 2895	Age Father (Christianity)
HH 4572	Age ; Golden
HM 3499	Age Holy Spirit (Christianity)
HH 0239	Age ; Iron
HM 3172	Age movement ; New
HH 5223	Age ; Old
HM 3087	Age Son (Christianity)
VP 5124	Age ; Youth
HH 0887	Ageing
TC 1123	Agendas ; Non linear
MS 8323	Ages development ; Human
HH 0397	Ages ; Doctrine three (Christianity)
VC 0087	Aggrandizement
HM 4671	Aggrandizing love ; Self
VD 2042	Aggravation
VP 5885	Aggravation ; Comfort
	Aggregate cognition aggregate
HM 1377	— synergies ; Composedness aggregate sensation, (Buddhism)
HM 3397	— synergies ; Flexibility aggregate sensation, (Buddhism)
HM 0349	— synergies ; Good quality aggregate sensation, (Buddhism)
HM 1435	— synergies ; Lightness aggregate sensation, (Buddhism)
HM 0409	— synergies ; Straightness aggregate sensation, (Buddhism)
HM 1381	— synergies ; Workability aggregate sensation, (Buddhism)
	Aggregate [consciousness...]
HM 1741	— consciousness ; Good quality (Buddhism)
HM 2098	— consciousness (Buddhism)
HM 2098	— consciousness (Buddhism)
	Aggregate consciousness
HM 4605	— Composedness (Buddhism)
HM 1796	— Flexibility (Buddhism)
HM 3403	— Lightness (Buddhism)
HM 1338	— Straightness (Buddhism)
HM 4605	— Workability (Buddhism)
HM 4983	Aggregate ; Feeling (Buddhism)
HM 2050	Aggregate ; Formations (Buddhism)
HM 2692	Aggregate (Hinduism)
	Aggregate [Materiality...]
HM 2108	— Materiality (Buddhism)
HM 2108	— matter (Buddhism)
HM 2050	— mental activities (Buddhism)
HM 4143	Aggregate ; Perception (Buddhism)
	Aggregate sensation aggregate
HM 1377	— cognition, aggregate synergies ; Composedness (Buddhism)
HM 3397	— cognition, aggregate synergies ; Flexibility (Buddhism)
HM 0349	— cognition, aggregate synergies ; Good quality (Buddhism)
HM 1435	— cognition, aggregate synergies ; Lightness (Buddhism)
HM 0409	— cognition, aggregate synergies ; Straightness (Buddhism)
HM 1381	— cognition, aggregate synergies ; Workability (Buddhism)
	Aggregate synergies
HM 1377	— Composedness aggregate sensation, aggregate cognition, (Buddhism)
HM 3397	— Flexibility aggregate sensation, aggregate cognition, (Buddhism)
HM 0349	— Good quality aggregate sensation, aggregate cognition, (Buddhism)
HM 1435	— Lightness aggregate sensation, aggregate cognition, (Buddhism)
HM 0409	— Straightness aggregate sensation, aggregate cognition, (Buddhism)
HM 1381	— Workability aggregate sensation, aggregate cognition, (Buddhism)
HH 3321	Aggregates clinging (Buddhism)
HH 3321	Aggregates ; Five (Buddhism)
VD 2044	Aggression
TC 1929	Aggressive participant type
VC 0089	Agility
VD 2046	Agitation
HM 3154	Agitation (Buddhism)
HM 0436	Agitation (Buddhism)
HH 0021	Agitation ; Non
VP 5323	Agitation ; Oscillation
HH 0406	Agni yoga (Yoga)
HH 0859	Agnih-loka (Leela)
HM 0859	Agnih ; Plane (Leela)
HM 8932	Agnosia
HM 6539	Agnosia ; Auditory
HM 8079	Agnosia ; Tactile
HM 7019	Agnosia ; Visual
HH 0904	Agnosticism
KC 0556	Agonemmetry
HM 2889	Agonizing prediction (ICA)
VC 0091	Agreeableness
VC 0093	Agreement
TC 1011	Agreement
KD 2350	Agreement consensus ; Alternation: implications
VP 5026	Agreement-Disagreement
HM 7981	Agreement (Sufism)
MM 2041	Agricultural development
KD 2285	Agricultural key crop rotation ; Patterns alternation:
MS 8943	Agricultural symbols
TC 1529	Agriculture food processing ; Meetings
HM 6117	Ahad ; Al (Sufism)
HM 2059	Ahamkara (Hinduism)
HM 1726	Ahamkara (Leela)
HM 2590	Ahamkara ; Lower manas (Hinduism)
HM 2059	Ahankara (Hinduism)
TP 0019	Ahead ; Getting
HH 0088	Ahimsa
HM 2986	Ahirika (Buddhism)
HM 1724	Ahirikabala (Pali)
HM 2986	Ahrikya (Buddhism)
HM 2365	Ahwal (Sufism)
HM 3371	Ahwal (Sufism)
HM 0573	Aia (Psychism)
VC 0095	Aid
MS 8406	Aids ; Health
HH 0252	Aikido
KC 0097	Aim
TC 1880	Aim ; Existential
VD 2048	Aimlessness
HM 0198	Aims action ; Right (Buddhism)
HM 1142	Aims ; Right (Buddhism)
KC 0717	Aims ; Short term
HM 2955	Air ; Triplicity (Astrology)

MS 8338	Aircraft		Al [Musawwir...] cont'd	MS 8335	Alteration skin appearance

MS 8338	Aircraft		Al [Musawwir...] cont'd	MS 8335	Alteration skin appearance
HM 2955	Airy awareness (Astrology)	HM 5786	— Musawwir (Sufism)	HM 4120	Altered ego states
HM 1054	Ajimhata (Pali)	HM 8019	— Muta'ali (Sufism)	HM 0062	Altered perceptions sport
HM 0252	Ajiva arati virati pativirati veramani ; Miccha (Pali)	HM 3850	— Mutakabbir (Sufism)	HM 2553	Altered state consciousness ; Death
	Ajiva [Samma...]	HM 4510	— Muzill (Sufism)		**Altered states**
HM 0549	— Samma (Pali)		Al [Qabid...]	HM 2391	— awareness
HM 1763	— Samma (Pali)	HM 2767	— Qabid (Sufism)	HH 0739	— Causative factor
HM 1341	— Samyag (Buddhism)	HM 1500	— Qadir (Sufism)	HM 2391	— consciousness
HM 2144	Ajna (Yoga)	HM 1318	— Qahhar (Sufism)	HM 0461	— consciousness induction device (ASCID)
HM 1028	Akakkhalata (Pali)	HM 4988	— Qawi (Sufism)	HM 1537	Alternate state consciousness (ASC)
HM 2813	Akanistha (Buddhism)	HM 5006	— Qayyum (Sufism)	VC 0111	Alternation
HM 3375	Akaranam (Pali)	HM 0235	— Quddus (Sufism)	KD 2285	Alternation: agricultural key crop rotation ; Patterns
HM 2110	Akasanancayatana (Pali)		Al [Rafi...]	KD 2090	Alternation discontinuous learning ; Opening closing:
HM 2110	Akasanantyayatana (Buddhism)	HM 4040	— Rafi (Sufism)	KD 2300	Alternation energetic expansion mentalistic reduction
HM 3014	Akathinata (Pali)	HM 4043	— Rahim (Sufism)	KC 0383	Alternation-fluctuation
HM 7112	Akhir ; Al (Sufism)	HM 6098	— Raqib (Sufism)	KD 2353	Alternation: implications action formulation
HM 2027	Akimcanyayatana (Buddhism)	HM 7021	— Razzaq (Sufism)	KD 2350	Alternation: implications agreement consensus
HM 2027	Akincannaayatana (Pali)	HM 4495	— Al-Sabur (Sufism)	KD 2280	Alternation: musical key political philosopher ; Patterns
HM 3118	Akirame (Japanese)	HM 0419	— Al-Sami' (Sufism)	KD 2100	Alternation ; Revolutionary cycles
HM 3429	Akiriya (Pali)	HM 4334	— Al-Tawwab (Sufism)	KD 2095	Alternation ; Third perspective container
HM 6898	Akrodha (Hinduism)		Al [Wadud...]	HH 4510	Alternative futures ; Imagining
HM 0123	Akshara consciousness (Psychism)	HM 7712	— Wadud (Sufism)	HH 0477	Alternative lifestyles
HM 1615	Akuhaiamio (Psychism)	HM 0069	— Wahhab (Sufism)	HM 2473	Alternative mode awareness while unconscious
HM 1954	Akusalamulam ; Lobho (Pali)	HM 5506	— Wahid (Sufism)	HH 5110	Alternative pursuits
HM 1439	Akusalamulam ; Moho (Pali)	HM 0743	— Wajid (Sufism)	HH 0477	Alternative ways life (AWL)
	Akusalanam dhammanam samapattiya	HM 1382	— Wakil (Sufism)	HM 1561	Alternatives ; Being doubt two (Buddhism)
HM 0552	— Hiriyati papakanam (Pali)	HM 0525	— Wali (Sufism)	HH 5110	Alternatives drug abuse
HM 3257	— Na kiriyati papakanam (Pali)	HM 1633	— Wali (Sufism)	KD 2262	Alternatives: resonance hybrids ; Interwoven
HM 3353	— Na ottapati papakanam (Pali)	HM 0113	— Warith (Sufism)	KD 2260	Alternatives: tensegrity organization ; Interwoven
HM 1208	— Ottappati papakanam (Pali)	HM 4169	— Wasi' (Sufism)	HM 0265	Alternobaric vertigo
HM 1008	Akutilata (Pali)	VC 0097	Alacrity	HH 0081	Altruism
	Al [Adl...]	HH 6282	Alam-i-amr (Sufism)	VC 0113	Altruism
HM 3807	— 'Adl (Sufism)	HH 6282	Alam-i-khalq (Sufism)	HM 3947	Alubbhana (Pali)
HM 5543	— 'Afuw (Sufism)	VD 2050	Alarmism	HM 1825	Alubbhitattam (Pali)
HM 6117	— Ahad (Sufism)	HM 2730	Alaya-vijnana (Buddhism)	HH 2877	Am ; I
HM 7112	— Akhir (Sufism)	HM 6328	Albedo (Esotericism)	HM 2937	Am this ; I (Yoga)
HM 4162	— 'Ali (Sufism)	HM 5345	Alchemical blackening (Esotericism)	HM 7332	Amarsa (Hinduism)
HM 4010	— 'Alim (Sufism)	HM 2696	Alchemical-flask (Buddhism)	TC 1451	Ambience ; Creative
HM 6139	— Awwal (Sufism)	HM 1409	Alchemical imagination (Esotericism)	VD 2054	Ambiguity
HM 5198	— 'Azim (Sufism)	HM 3631	Alchemical reddening (Esotericism)	MP 1243	Ambiguous boundaries
HM 4562	— 'Aziz (Sufism)	HM 6328	Alchemical whitening (Esotericism)	MM 2021	Ambiguous visual illusions
	Al [Baith...]	KC 0579	Alchemy	VC 0115	Ambition
HM 4017	— Ba'ith (Sufism)	HH 0221	Alchemy (Esotericism)	HH 6089	Ambition
HM 4299	— Badi (Sufism)	HM 5887	Alchemy (Taoism)	VD 2056	Ambivalence
HM 6786	— Baqi (Sufism)	HM 0134	Alcohol consumption ; Modes awareness associated	HM 3468	Ambivalence (Jung)
HM 3636	— Bari' (Sufism)	HM 0035	Alert immediacy	VC 0117	Amelioration
HM 7022	— Barr (Sufism)	HM 3959	Alert passivity	VC 0119	Amenability
HM 6512	— Basir (Sufism)	HM 4469	Alert passivity	MS 8416	Amenities ; Home furnishings
HM 0115	— Basit (Sufism)	HM 4967	Alert thinking	HM 2069	Amerindian Monadic awareness
HM 1209	— Batin (Sufism)	HM 3476	Alert wakefulness	HM 2036	Amerindian Nagual state awareness
HM 0002	Al-Fattah (Sufism)	HM 3003	Alertness	HH 4746	Amerindian Spirituality
HM 1264	Al-fikr fi 'l-manzur (Islam)	VC 0099	Alertness	HM 2066	Amerindian Tonal state awareness
	Al [Ghaffar...]	HM 2792	Alertness heart (Sufism)	HH 8219	Amerindian Way warrior
HM 1257	— Ghaffar (Sufism)	HM 1461	Alertness ; Pure (Psychism)	VC 0121	Amiability
HM 6255	— Ghafur (Sufism)	HM 7512	Alertness ; Relaxed (Japanese)	VC 0123	Amicability
HM 3822	— Ghani (Sufism)	HM 0996	Alertness ; Restful	VC 0125	Amity
	Al [Hadi...]	HH 0352	Alexander technique	HM 4897	Amnesia
HM 1287	— Hadi (Sufism)	HM 4162	'Ali ; Al (Sufism)	HM 7608	Amnesia ; Dissociative
HM 7099	— Hafiz (Sufism)	MS 8395	Alien	HM 7608	Amnesia ; Hysterical
HM 5014	— Hakam (Sufism)	HM 2329	Alien consciousness	HM 2406	Amnesic behaviour
HM 4799	— Hakim (Sufism)		**Alienation**	HM 4065	Amodana (Pali)
HM 4519	— Halim (Sufism)	HM 3030	— [Alienation]	HM 2783	Amoghasiddhi-karma awareness (Buddhism)
HM 0392	— Hamid (Sufism)	VD 2052	— [Alienation]	HM 2695	Amoha (Pali)
HM 3945	— Haqq (Sufism)	HH 0145	— [Alienation]	HM 3594	Amok
HM 0741	— haqq (Sufism)	TP 0038	— [Alienation]	MP 1068	Amongst emerging perspectives ; Spontaneous relationship formation
HM 1467	— Hasib (Sufism)	TC 1962	Alienation committed activists		
HM 4236	— Hayy (Sufism)	HM 3030	Alienation ; Self	VD 2058	Amorality
HM 2920	Al-insan al-kamil (Sufism)	TC 1229	Alienative participant involvement	HM 6003	Amorous regard (Islam)
	Al [Jabbar...]	HM 6508	Alientation thought	VC 0127	Amorousness
HM 0215	— Jabbar (Sufism)	HM 4010	'Alim ; Al (Sufism)	TC 1315	Amorphous conferencing
HM 6771	— Jalil (Sufism)	HM 3509	Alinga stage gunas (Yoga)	HM 0803	Amphetamines similar drugs ; Modes awareness associated use
HM 0009	— Jame' (Sufism)	VC 0101	Aliveness		
HM 0009	— Jami' (Sufism)	HM 4320	All-accomplishing wisdom	HH 3228	Amping
	Al [Kabir...]	HM 3097	All-being-in-myself (ICA)	VC 0129	Amplification
HM 4437	— Kabir (Sufism)	HM 2730	All-conserving mind (Buddhism)	KC 0427	Amplification ; Deviation
HM 5289	— Karim (Sufism)	HM 0476	All-pervading knowledge (Sufism)	HH 6282	Amr ; Alam (Sufism)
HM 4400	— Khabir (Sufism)	HM 3088	All-that-ever-was (ICA)	VC 0131	Amusement
HM 3644	— Khafid (Sufism)	HM 3083	All-that's-yet-to-be (ICA)	VP 5878	Amusement-Boredom
HM 0034	— Khaliq (Sufism)	HM 2561	Allah ; Ninety nine names (Sufism)	MS 8846	Amusements games
HM 1626	— khatim (Sufism)	HM 3496	Allah ; Presence (Sufism)	HM 6539	Amusia ; Sensory
HM 0308	Al-Latif (Sufism)	HM 4500	Allah (Sufism)	HM 1655	An ; Experience (Sufism)
	Al [mahabba...]	HM 3864	Allah ; Unification (Sufism)	HM 6350	An-Nafi' (Sufism)
HM 4712	— mahabba fi 'llah (Islam)	VC 0103	Allegiance	HM 1354	An-Nur (Sufism)
HM 5580	— mahabba ma'a 'llah (Islam)	HM 2073	Allegiance ; Awareness (ICA)	HM 4449	An (Sufism)
HM 4526	— Majid (Sufism)	TP 0061	Allegiance ; Wholehearted	HM 0684	Anabhijjha (Pali)
HM 5446	— Majid (Sufism)	VC 0105	Alliance	HM 4090	Anabhisamaya (Pali)
HM 4061	— Malik (Sufism)	TP 0008	Alliance	HM 7344	Anabhisneha (Hinduism)
HM 7220	— Mani (Sufism)	HH 1249	Allocentric attribution	HM 8000	Anabhisvanga (Hinduism)
HM 5633	— Matin (Sufism)	KC 0476	Allometry ; Principle	HM 3898	Anaesthesia
HM 4052	— Mu'akhkhir (Sufism)	KC 0066	Allopoietic system	HM 3242	Anahata (Yoga)
HM 4093	— Mu'id (Sufism)	TC 1370	Allotments ; Exceeded time	HM 3413	Anajjhapatti (Pali)
HM 1116	— Mu'izz (Sufism)	VC 0107	Allurement	KC 0196	Analog computer language
HM 1245	— Mu'min (Sufism)	MM 2084	Alluring movement	TC 1603	Analogies ; Meeting
HM 0622	— Mubdi (Sufism)	VC 0109	Almightiness	KC 0731	Analogy
HM 4510	— Mudhill (Sufism)	HH 5643	Almsgiving (Islam)	HM 1538	Analogy ; Language divine
HM 3633	— Mughni (Sufism)	HM 1799	Alobha (Pali)	HM 1538	Analogy ; Sacred
HM 0526	— Muhaymin (Sufism)	HM 2128	Alobha (Pali)	HM 0485	Analyses ; Four (Buddhism)
HM 4544	— Muhsi (Sufism)	HM 3752	Alobhokusala mulam (Pali)	HH 0652	Analysis ; Bioenergetic
HM 1007	— Muhyi (Sufism)	HM 2232	Aloba-vriddhi samadhi (Buddhism)	HM 5089	Analysis (Buddhism)
HM 0504	— Mujib (Sufism)	HM 2604	Alone awareness ; Ox forgotten leaving man (Zen)		**Analysis [Comparative...]**
HM 3868	— Mumit (Sufism)	HH 0992	Alone ; Presence	KC 0647	— Comparative
HM 4513	— Muntaqim (Sufism)	HM 6762	Aloneness (Islam)	KC 0016	— Cost benefit
HM 5331	— Muqaddim (Sufism)	HM 3869	Aloneness (Yoga)	KC 0754	— Cross impact
HM 3171	— Muqit (Sufism)	HM 2345	Alpha wave consciousness	HH 0695	Analysis ; Dasein
HM 8001	— Muqsit (Sufism)	HM 0763	Alter personality	HH 0695	Analysis ; Existence
HM 3576	— Muqtadir (Sufism)	HM 0768	Alteration	HH 0416	Analysis ; Existential

Analysis

KC 0344	Analysis ; Factor
KC 0665	Analysis ; Input/Output
KC 0545	Analysis ; Integrational
HM 0169	Analysis ; Knowledge (Buddhism)
KC 0964	Analysis ; Morphological
KC 0036	Analysis ; Multivariate
KC 0842	Analysis ; Network
KC 0696	Analysis ; Policy
KC 0696	Analysis ; Public policy
	Analysis [Self...]
HH 0906	— Self
KC 0567	— Sensitivity
KC 0365	— Structural functional
KC 0718	— Synergetic
KC 0610	— Systems
KC 0327	Analysis ; Transaction
HH 0487	Analysis ; Transactional
KC 0256	Analysis ; Value
HH 0507	Analytic group therapy
	Analytical [cessations...]
HM 3310	— cessations (Buddhism)
HM 3310	— cessations ; Non (Buddhism)
HM 2415	— consciousness
TC 1488	Analytical frameworks ; Meetings
HM 3345	Analytical investigation (Buddhism)
HH 1420	Analytical psychology
HM 2659	Analytical subtle contemplation ; Critical (Buddhism)
HH 0507	Analytical therapy
HH 7123	Analytical trilogy
HM 1872	Anamnesis
HM 6901	Anand-loka (Leela)
HM 3227	Ananda (Hinduism)
	Ananda [Sat...]
HM 0592	— Sat cit
HM 3227	— Sat cit (Hinduism)
HM 3212	— stage (Yoga)
HM 3212	Anandamaya kosha (Yoga)
HM 2015	Anantaryamarga (Buddhism)
HM 0403	Ananubodha (Pali)
HM 3203	Anapatrapya (Buddhism)
VD 2060	Anarchism
VD 2062	Anarchy
VD 2064	Anathema
HM 3435	Anatikkama ; Vela (Buddhism)
HM 3435	Anatikkamo ; Vela (Pali)
HH 0241	Anatma (Buddhism)
HH 0241	Anatta (Buddhism)
HM 3949	Anattamanata cittassa (Pali)
HM 1915	Anattappabala (Pali)
HM 3072	Anavakarashunyata (Buddhism)
HM 3032	Anavaragrashunyata (Buddhism)
HM 3224	Ancestor ; Primordial (ICA)
HM 3122	Ancestral obligation (ICA)
VC 0133	Ancestry
VP 5170	Ancestry-Posterity
HH 0958	Anchoress
HH 0958	Anchorite
HH 4508	Ancient spirituality
HM 0212	Androgyne ; Soul
HM 0212	Androgynous love
HM 0212	Androgyny
HM 3001	Android consciousness
HM 0870	Anekamasaggaha (Pali)
HM 3146	Angel death
HM 3130	Angel ; Guardian (ICA)
	Angelic awareness
HM 2222	— Gabriel
HM 2248	— Logos
HM 2198	— Michael
HM 2188	— Orders (Judaism)
HM 2162	— Raphael
HM 3123	— Uriel
HM 3022	— Zacharael
HM 2282	— Zoharariel
HM 2212	Angelic evil ; Awareness
HM 2150	Angelic frame awareness
HM 3146	Angelic transformation ; Awareness
MS 8337	Angels
HM 7089	Angels ; Saintship (Sufism)
VD 2066	Anger
HM 8762	Anger ; Aesthetic emotion (Hinduism)
HM 8665	Anger (Hinduism)
HM 1433	Anger (Leela)
HM 6898	Anger ; Non (Hinduism)
HM 0936	Anger ; Spiritual (Christianity)
HM 2959	Anger ; State (Buddhism)
HM 0362	Angst
VD 2068	Anguish
HM 9021	Anicca (Buddhism)
HM 4131	Anikkhittachandata (Pali)
HM 1403	Anikkhittadharata (Pali)
KC 0302	Anima mundi
MS 8346	Animal artifacts products
MS 8725	Animal behaviour
MS 8133	Animal body parts
MM 2023	Animal-drawn cart
	Animal [hell...]
HM 2636	— hell awareness (Buddhism)
HM 2007	— hell awareness ; Divine (Buddhism)
MS 8336	— husbandry domestication ; Objects
MM 2073	Animal life cycles
	Animal locomotion
MM 2010	— [Animal locomotion]
MM 2025	— Aquatic
MM 2039	— Arboreal
MM 2064	— flight ; Aerial
	Animal locomotion cont'd
MM 2033	— Fossorial
MM 2059	— gliding soaring ; Aerial
MM 2012	— Terrestrial
VD 2070	Animality
HM 0847	Animality ; State (Buddhism)
MS 8215	Animals
VC 0135	Animation
HM 1182	Animation ; Suspended
HH 7219	Animistic yoga (Yoga)
VD 2072	Animosity
HH 0910	Aniyata (Buddhism)
HM 4987	Anjuman ; Khilwat dar (Sufism)
HM 1394	Annana (Pali)
VD 2074	Annihilation
HM 2330	Annihilation (Buddhism)
HM 0052	Annihilation ; Emptiness what is free permanence (Buddhism)
HM 7622	Annihilation heart (Sufism)
HM 3358	Annihilation personal sensation (Sufism)
HM 7626	Annihilation (Sufism)
HM 1270	Annihilation ; Non (Sufism)
HM 5775	Annihilation ; Total (Sufism)
VD 2076	Annoyance
VD 2078	Annulment
HM 8186	Anomalies awareness one's self
HM 1122	Anomalies consistency consciousness
HM 2548	Anomalies experience reality one's self environment
	Anomalies experience self
HM 8186	— [Anomalies experience self]
HM 4754	— distinct outside world
HM 1336	— recognized personal performance
HM 9132	Anomalies experience unity self
HM 4312	Anomalies flexibility associations
HM 4521	Anomalies integrity consciousness
HM 8105	Anomalous emphasis
HM 2947	Anomie
VC 0137	Anonymity
HM 1405	Anonymous accord
HM 7298	Another's thoughts ; Knowledge (Hinduism)
HM 4049	Another world ; Being (Balinese)
HM 0188	Anottappa (Pali)
HM 2317	Ansari ; Stations consciousness (Sufism)
KD 2045	Answer ; Constraints meta
KD 2006	Answer: gladiatorial arena ; Prevailing meta
KD 2060	Answer patterning ; Meta
KD 2007	Answer production accumulation significance
KD 2003	Answers ; Assumptions concerning appropriate
KD 2001	Answers: monopolarization ; Questionable
VD 2080	Antagonism
TC 1929	Antagonist
HM 2738	Antar-vritti (Yoga)
	Antarabhava consciousness
HM 2407	— Fifth (Buddhism)
HM 2335	— Fourth (Buddhism)
HM 2371	— Second (Buddhism)
HM 2431	— Sixth (Buddhism)
HM 2311	— Third (Buddhism)
HM 4512	Antariksha (Leela)
VD 2082	Antediluvian
HH 3329	Anthropocentrism
MS 8300	Anthropomorphic beings ; Mythical
HH 1449	Anthropomorphism
HH 0018	Anthroposophical system
HH 0018	Anthroposophy
HM 2901	Anthroposophy Imagination
HM 2901	Anthroposophy Imaginative consciousness
VC 0139	Anticipation
VT 8082	Anticipation∗complex
HH 2099	Antinomianism
KD 2030	Antinomies ; Paradoxes
TC 1663	Antipathetics
HM 2481	Antipathy
VD 2084	Antipathy
VP 5119	Antiquity ; Futurity
HM 2763	Anujna (Buddhism)
HM 4665	Anupadisesanibbana (Pali)
HM 2501	Anupalambhashunyata (Buddhism)
HM 1930	Anupekkhanata (Pali)
HM 0405	Anussati (Pali)
HM 3263	Anuttara samyak sambodhi
HM 2864	Anuttarayoga ; Highest yoga tantra: (Buddhism)
HM 0649	Anuttassa (Pali)
HM 1982	Anuvicara (Pali)
	Anxiety
HM 2465	— [Anxiety]
HM 3094	— [Anxiety]
HM 3530	— [Anxiety]
VD 2086	— [Anxiety]
HM 1746	Apacakkhakamma (Pali)
HH 0604	Apacayana (Pali)
HM 1902	Apaccavekkhana (Pali)
HM 0145	Apana-loka (Leela)
HM 2519	Apara-godaniya (Pali)
HH 1268	Aparigraha
HM 5665	Aparisphutamanovijnana (Buddhism)
HM 4228	Apariyobahana (Pali)
HM 0427	Apariyogahana (Pali)
VD 2088	Apartheid
VD 2090	Apathy
HM 2210	Apatrapya (Buddhism)
MP 1221	Aperture compatibility
MP 1225	Apertures ; Boundary reinforcement
HM 1689	Apilapanata (Pali)
HH 6071	Apokatastasis
HM 2390	Apollo awareness
VC 0141	Apology
HM 4604	Apophany
HM 1356	Apoplexy (Hinduism)
HM 1617	Appana (Pali)
HM 2586	Appana samadhi (Pali)
MS 8436	Apparel
HM 0850	Apparition ; Crisis (Psychism)
HM 1637	Appativedha (Pali)
VC 0143	Appeal
VC 1609	Appeal ; Sex
VC 0145	Appearance
HM 1618	Appearance absorption (Buddhism)
MS 8335	Appearance ; Alteration skin
VP 5446	Appearance-Disappearance
HH 0993	Appearance ; Physical
HM 7114	Appearance terror ; Knowledge (Buddhism)
VC 0147	Appeasement
HM 1961	Apperception
HM 7268	Apperception (Buddhism)
HM 2610	Apperceptive awareness
VC 0149	Appetite
HM 0129	Appetite ; Concupiscible (Islam)
HH 0544	Application
VC 0151	Application
	Application [ascension...]
HM 2962	— ascension stages game ; Sutra paths accumulation (Buddhism)
HM 2848	— ascension stages game ; Tantric paths accumulation (Buddhism)
HM 2716	— awareness ; Discipleship (Buddhism)
	Application awareness Mahayana
HM 2232	— climax (Buddhism)
HM 2208	— heat (Buddhism)
HM 2271	— highest teachings (Buddhism)
HM 2247	— receptivity (Buddhism)
HM 3020	Application awareness ; Pratyeka Buddha (Buddhism)
	Application awareness Tantra
HM 2220	— climax (Buddhism)
HM 2696	— heat (Buddhism)
HM 2756	— receptivity (Buddhism)
HM 2280	— union learning (Buddhism)
HM 1982	Application reflection (Buddhism)
MS 8372	Applied cosmetics
HM 1177	Applied thought (Buddhism)
	Applied thought sustained
HM 4908	— thought ; Concentration (Buddhism)
HM 5199	— thought ; Concentration (Buddhism)
HM 7543	— thought ; Concentration (Buddhism)
VC 0153	Appreciation
HM 3105	Appreciation ; Agape (ICA)
KC 0527	Appreciative system
VD 2092	Apprehension
TP 0019	Approach
HH 0649	Approach ; Holistic
VP 5296	Approach-Recession
HH 1763	Approach transcendence ; Psychological
VC 0155	Approachability
HM 2446	Approaches consciousness ; Indian (Hinduism)
HM 1647	Approaching reflection (Buddhism)
VC 0157	Approbation
KD 2003	Appropriate answers ; Assumptions concerning
MP 1182	Appropriate conditions perspective nourishment
	Appropriate configuration
MP 1184	— input processing
MP 1187	— interaction complementary perspectives
MP 1186	— perspective inactivity
MP 1185	— perspective interaction
MP 1207	Appropriate construction elements
HM 2292	Appropriate intellect (Buddhism)
MP 1191	Appropriate proportions perspective contexts
MP 1171	Appropriate relationship non-linearity structures
MP 1228	Appropriate superstructure contain transitions levels
HM 3054	Appropriated passion (ICA)
VC 0159	Appropriateness
VT 8091	Appropriateness∗complex
VP 5665	Appropriateness-Inappropriateness
VP 5815	Appropriation ; Sharing
VC 0161	Approval
VP 5968	Approval-Disapproval
HM 3220	Aramada (Buddhism)
HM 2981	Apt religion (Leela)
VC 0163	Aptitude
HM 2973	Aquarius-consciousness (Astrology)
MM 2025	Aquatic animal locomotion
MS 8917	Aquatic beings ; Mythical
	Ar [Rauf...]
HM 0206	— Ra'uf (Sufism)
HM 3539	— Rahman (Sufism)
HM 5234	— Rashid (Sufism)
HH 0233	Arahant (Buddhism)
HM 7055	Arahantship ; Knowledge path (Buddhism)
	Arati virati pativirati
HM 1781	— veramani ; Catuhi vaciduccaritehi (Pali)
HM 0252	— veramani ; Miccha ajiva (Pali)
HM 1133	— veramani ; Tihi kayaduccaritehi (Pali)
MM 2039	Arboreal animal locomotion
KC 0908	Arboreal organization
KC 0908	Arborization
HM 2097	Arcana conscious states ; Tarot (Tarot)
HM 2189	Archaic consciousness
HM 0216	Archaic ego states
HH 0748	Archaic-paralogical thinking
HH 4508	Archaic spirituality
HH 0748	Archaic thought
VD 2094	Archaism

GENERAL INDEX TO VOLUME 2
Attraction

HM 3007 Archaism ; Awareness (ICA)
HM 3112 Archetypal humanness (ICA)
Archetypal image
HM 3028 — Betrothal initiation (Tarot)
HM 3045 — Charioteer (Tarot)
HM 2285 — Female sovereignty (Tarot)
HM 2260 — Hierophant (Tarot)
HM 2237 — Magician (Tarot)
HM 2202 — Old wise man (Tarot)
HM 2225 — Pilgrim (Tarot)
HM 2261 — Wisdom (Tarot)
HH 5119 Archetypal psychology
KC 0487 Archetype
HH 0019 Archetype
HH 3020 Archetypes ; Guiding
HM 2201 Archetypes psyche ; Emblem (Tarot)
MS 8835 Architectural elements
HM 2191 Arcis-mati (Buddhism)
KC 0440 Arcology
VC 0165 Ardour
HM 3357 Ardour (Buddhism)
KC 0074 Area studies
MP 1004 Areas ; Regenerative resource cultivation
KD 2006 Arena ; Prevailing meta answer: gladiatorial
HM 3012 Arete
VD 2096 Argumentativeness
HM 2776 Arhat awareness ; Pratyeka Buddha (Buddhism)
HH 0233 Arhat (Buddhism)
HM 3024 Arhat sanctity awareness (Buddhism)
HH 0024 Arica training
HM 2282 Ariel
HM 2665 Aries-consciousness (Astrology)
HM 1045 'Arif (Sufism)
HM 7655 Arising ; Twenty four causes consciousness (Buddhism)
VC 0167 Aristocracy
HH 0523 Ariya (Buddhism)
HH 0307 Armament ; Moral re
MS 8106 Armour weapons
TP 0007 Army Controlled threat
HH 5080 Aromatherapy
TP 0051 Arousal (Thunderbolts)
MP 1100 Arrangement structures engender fruitful interfaces
VD 2098 Arrant
HM 2860 Arrested consciousness ; Dreamy
VD 2100 Arrested development
TP 0018 Arresting decay
HH 4880 Arriving reality (Taoism)
VD 2102 Arrogance
HM 2553 Ars moriendi
HM 9124 Arsajnana (Hinduism)
Art
HH 0570 — [Art]
KC 0751 — [Art]
VC 0169 — [Art]
TC 1835 Art form method
HM 1094 Art ; Inspirational (Psychism)
HH 0775 Art therapy
VC 0171 Artfulness
HH 0353 Artha (Hinduism)
MS 8505 Articles
MS 8505 Artifacts
MS 8915 Artifacts ; Mythical
MS 8346 Artifacts products ; Animal
VD 2104 Artifice
HH 2189 Artificial intelligence
KC 0510 Artificial organisms
VD 2106 Artificiality
HH 0029 Artistic education
KC 0847 Artistic expression
KC 0847 Artistic images
KC 0115 Artistic life
HH 0754 Artistic life (Hinduism)
HM 3387 Artistic refinement (Japanese)
KC 0083 Artistic synthesis
KC 0847 Artistic vision
VC 0173 Artlessness
MS 8533 Arts ; Fine
HH 0545 Arts ; Mantic
HH 0085 Arts ; Martial
HM 0220 Arts training ; One pointedness martial
HM 3144 Arupa-dhatu (Buddhism)
HM 2281 Arupadhatu (Buddhism)
HM 2012 Arupaloka consciousness (Buddhism)
HM 2669 Arupayasamapatti (Buddhism)
HM 1969 Arupinam dhammanam ayu (Pali)
HH 3198 Aruppa-niddesa (Pali)
HM 2339 Aryastangamarga (Buddhism)
HM 1221 As-Salam (Sufism)
HM 4980 As-Samad (Sufism)
HM 0451 Asagahana (Pali)
HM 5091 Asakti (Hinduism)
HM 1965 Asamapekkhana (Pali)
HM 1362 Asambodha (Pali)
HM 5332 Asamjnisamapatti (Buddhism)
HM 6504 Asamkhata (Pali)
HM 1705 Asammusanata (Pali)
HM 1725 Asampajanna (Pali)
HM 2063 Asamprajanya (Buddhism)
HM 3041 Asamprajnata samadhi (Hinduism)
HM 6504 Asamskrita (Buddhism)
HM 2089 Asamskrtashunyata (Buddhism)
HH 0669 Asana (Hinduism)
HM 6153 Asanga (Hinduism)
HM 3394 Asappana (Pali)
HM 1688 Asaraga (Pali)
HM 3183 Asarajjana (Pali)
HM 4081 Asarajjitattam (Pali)
HM 1090 Asata vedana ; Cetosamphassaja (Pali)
HM 0234 Asatam ; Manovinnanadhatusampassa jam cetasikam (Pali)
HM 1537 ASC Alternate state consciousness
VC 0175 Ascendancy
HM 2327 Ascension awareness (Sufism)
HM 8634 Ascension meditation
Ascension stages game
HH 4000 — Consciousness (Buddhism)
HM 3066 — Masters wisdom (Buddhism)
HM 3341 — Religious traditions (Buddhism)
HM 3214 — Supreme heavens (Buddhism)
HM 2962 — Sutra paths accumulation application (Buddhism)
HM 2848 — Tantric paths accumulation application (Buddhism)
HM 5563 Ascertaining truth (Sufism)
HH 4298 Ascetic practices (Buddhism)
HM 5354 Ascetic states (Buddhism)
HH 3652 Ascetical theology (Christianity)
HH 0556 Asceticism
VC 0177 Asceticism
HM 6122 Asceticism ; Erotic
HM 4450 Asceticism (Sufism)
HM 0461 ASCID Altered states consciousness induction device
HM 3198 Ascription
HM 0375 Ash-Shahid (Sufism)
HM 1934 Ash-Shakur (Sufism)
HM 0552 Ashamed acquisition sinful unwholesome dharmas ; Being (Buddhism)
HM 3257 Ashamed acquisition sinful unwholesome dharmas ; Not (Buddhism)
HM 3828 Ashamed ; Being ashamed what one ought be (Buddhism)
HM 1023 Ashamed ; Not ashamed ought be (Buddhism)
HM 1023 Ashamed ought be ashamed ; Not (Buddhism)
HM 3828 Ashamed what one ought be ashamed ; Being (Buddhism)
HM 2461 Ashraddhya (Buddhism)
HH 2563 Ashramas (Hinduism)
HH 0862 Ashtanga yoga (Yoga)
HM 3261 Asithilaparakkamata (Pali)
HM 2312 Asiyah ; Olam (Judaism)
HM 2079 Asleep trance ; Double awake body
HM 2640 Asleep trance ; Mind awake body (Psychism)
HM 3509 Asmita stage (Yoga)
HM 2937 Asmita (Yoga)
HM 0019 Asomatic experience (Psychism)
HH 3904 Asparsa yoga (Yoga)
HH 0108 Aspects psychic health ; Spiritual
VD 2108 Asperity
VC 0179 Aspiration
HM 2578 Aspiration (Buddhism)
HH 0656 Aspiration level
HM 2726 Aspiration ; Spiritual (Astrology)
HM 2512 Aspiration unity (Astrology)
HM 3002 Assault identity (Brainwashing)
HM 4338 Assembling five elements (Taoism)
KC 0177 Assembling resources
VC 0181 Assembly
TP 0045 Assembly
VP 5521 Assent-Dissent
HM 3587 Assertion ; Self
HH 0685 Assertion-structured therapy
HH 5563 Assertion truth (Sufism)
HM 2376 Assertive self-consciousness (Astrology)
HH 0994 Assertiveness training
HM 3542 Assessment mystic experience ; Psychological (Sufism)
HH 0314 Assessment ; Self
KC 0943 Assessment ; Technology
VC 0183 Assiduity
HH 0802 Assignment therapy
HH 0783 Assimilation
HH 0783 Assimilation ; Cultural
TC 1772 Assimilator participant type
VC 0185 Assistance
TP 0042 Assistance
KC 0019 Association
VC 0187 Association
KC 0508 Association ; Data
HM 4312 Associations ; Anomalies flexibility
TC 1275 Associative networks
HH 0644 Associative thinking
HM 4550 Assumption forms (Psychism)
KD 2003 Assumptions concerning appropriate answers
TC 1334 Assumptions ; Mood motivation
TC 1877 Assumptions ; Participation
KP 3016 Assumptions ; Values
VC 0189 Assurance
HM 1687 Assurance (Buddhism)
VC 1543 Assurance ; Self
HH 0779 Astanga yoga (Yoga)
HH 0361 Asteya
HM 3197 Astonishment
HH 0206 Astral body
HM 0183 Astral consciousness (Psychism)
HM 1887 Astral flight (Psychism)
Astral [plane...]
HM 3651 — plane (Leela)
HM 4600 — plane (Leela)
HM 1887 — projection (Psychism)
HH 0552 Astrology
KC 0490 Astrology
HM 2955 Astrology Airy awareness
HM 2973 Astrology Aquarius-consciousness
HM 2665 Astrology Aries-consciousness
HM 2512 Astrology Aspiration unity
HM 2376 Astrology Assertive self-consciousness
HM 2239 Astrology Cancer-consciousness
HM 2822 Astrology Capricorn-consciousness
HM 2259 Astrology Cardinal awareness
HM 3092 Astrology Common cross
HM 2822 Astrology Conscious wisdom
HM 2259 Astrology Cross transcendence
HM 2554 Astrology Cross transmutation
HM 3235 Astrology Earth triplicity
HM 3235 Astrology Earthy awareness
HM 2924 Astrology Energetic conceptualization
HM 2124 Astrology Fiery awareness
HM 2554 Astrology Fixed awareness
HM 2924 Astrology Gemini-consciousness
HM 2973 Astrology Illumined effort
HM 2239 Astrology Individualized awareness
HM 2439 Astrology Intuitive consciousness
HM 2376 Astrology Leo-consciousness
HM 2512 Astrology Libra-consciousness
HM 3092 Astrology Mutable awareness
HM 2856 Astrology Pisces-consciousness
HM 2726 Astrology Sagittarius-consciousness
HM 2615 Astrology Scorpio-consciousness
HM 2615 Astrology Self-sacrifice
HM 2726 Astrology Spiritual aspiration
HM 2815 Astrology Taurus-consciousness
HM 2955 Astrology Triplicity air
HM 2124 Astrology Triplicity fire
HM 2384 Astrology Triplicity water
HM 2856 Astrology Ultimate union
HM 2665 Astrology Undifferentiated potentiality
HM 2815 Astrology Undifferentiated receptivity
HM 2439 Astrology Virgo-consciousness
HM 2384 Astrology Watery awareness
HM 2713 Astrology Zodiacal forms awareness
HM 2579 Asura-world awareness (Buddhism)
HM 3857 Asuropa (Pali)
VD 2110 Asymmetry
HM 2242 Asyndesis
HM 0021 Ataraxia
HM 0401 Atavism ; Psychic (Psychism)
VD 2112 Atheism
HH 4223 Atheism ; Human development
HH 0938 Atheistic education
HM 2103 Atma awareness (Hinduism)
HM 3509 Atma (Yoga)
HH 0774 Atma yoga (Yoga)
HM 1195 Atman ; Being (Hinduism)
HH 0103 Atman (Hinduism)
HM 2153 Atman project
HM 1195 Atmanubhava (Hinduism)
MS 8425 Atmosphere celestial symbols
HH 0267 Atonement
HM 3837 Atonement (Leela)
VP 6010 Atonement-Punishment
VD 2114 Atrocity
VC 0193 Attachment
HM 6144 Attachment due insight (Buddhism)
HM 3914 Attachment (Hinduism)
HM 0697 Attachment (Leela)
HH 0868 Attachment ; Non
HM 2128 Attachment ; Non (Buddhism)
HM 2001 Attachment ; Total freedom (Hinduism)
HM 3119 Attachment (Yoga)
HM 4033 Attack ; Accidental psychic (Psychism)
VP 5798 Attack-Defence
HM 0837 Attack ; Psychic (Psychism)
HH 0325 Attack therapy ; Verbal
HM 3291 Attaining balance ; Fourth chakra: (Leela)
VC 0195 Attainment
HM 0159 Attainment (Buddhism)
HM 5438 Attainment cessation (Buddhism)
HM 6346 Attainment cessation (Buddhism)
HM 2110 Attainment ; First formless (Buddhism)
HM 2051 Attainment ; Fourth formless (Buddhism)
HM 0922 Attainment ; Perfect (Taoism)
HM 1342 Attainment ; Perfect (Taoism)
HM 3043 Attainment ; Second formless (Buddhism)
HM 0140 Attamanata cittassa (Pali)
KP 3018 Attempts ; Inadequate transformation
HH 0383 Attending needs superiors (Buddhism)
Attention
HH 0756 — [Attention]
VC 0197 — [Attention]
HM 1817 — [Attention]
HM 3237 Attention (Buddhism)
HM 2305 Attention ; Constant (Sufism)
VP 5530 Attention-Inattention
HM 1817 Attention ; Inward
HM 2423 Attention ; Loss
HM 2580 Attention ; Sated
HM 4309 Attention ; Wise (Buddhism)
HM 3378 Attentiveness ; Induced (Buddhism)
HH 0463 Attentiveness nature (Zen)
HM 3378 Attentiveness ; Other induced (Buddhism)
HM 4383 Attentiveness ; Quiet
HM 3378 Attentiveness ; Self induced (Buddhism)
HM 7143 Attentiveness ; Spiritual (Christianity)
HM 1663 Attha-patisambhida (Pali)
HM 3387 Attitude ; Aesthetic (Japanese)
HH 0856 Attitude change
VT 8054 Attitude*complex
HM 0610 Attitude ; Failure hold meditative (Yoga)
HH 0676 Attitude therapy
KC 0497 Attitudinal integration ; International
VC 0199 Attraction

Attraction

HM 5123	Attraction consciousness (Psychism)		HM 3829	Avijjapariyutthana (Pali)		HM 2393	Awareness ; Dionysos
HM 2907	Attraction divine grace ; Liberation spirit (Sufism)		HM 4470	Avijjayoga (Pali)			Awareness Discipleship
HM 1026	Attraction ; Divine (Sufism)		HM 3145	Avijjogha (Pali)		HM 2716	— application (Buddhism)
VP 5288	Attraction-Repulsion		HM 4572	Avikkhepa (Pali)		HM 2192	— karma (Buddhism)
MP 1241	Attractive temporary positions		HM 0449	Avisahara (Pali)		HM 2240	— vision (Buddhism)
VP 5155	Attributability-Chance		HM 0715	Avisahatamanasata (Pali)			Awareness [Divine...]
KC 0307	Attribute space		HM 3374	Avitthanata (Pali)		HM 2007	— Divine animal hell (Buddhism)
HM 2775	Attributes Buddha ; Eighteen unshared (Buddhism)		HM 3460	Avivesa stage gunas (Yoga)		HM 2401	— Divine love (Sufism)
HH 2561	Attributes ; Divine (Sufism)		VD 2126	Avoidance		HM 2416	— divine presences (Sufism)
HM 3260	Attributes their essence ; Identity (Sufism)		VP 5631	Avoidance ; Desire		HM 2371	— Dream state bardo (Buddhism)
HH 1249	Attribution ; Allocentric		HM 0436	Avupasama (Pali)		HM 2930	— Dyadic
HM 1336	Attribution ; Loss personal		HM 4565	Avyapajja (Pali)			Awareness [Earthy...]
HH 4112	Attrition		HM 2079	Awake body asleep trance ; Double		HM 3235	— Earthy (Astrology)
HM 3574	Attunement		HM 2640	Awake body asleep trance ; Mind (Psychism)		HH 1267	— Ecological
TP 0017	Attunement		HM 2180	Awakening ; Great (Buddhism)		HM 2059	— Ego (Hinduism)
HM 3574	Attunement ; Psychic (Psychism)		HM 2972	Awakening ; Sensory		HM 2816	— Eighth scriptural bodhisattva (Buddhism)
HM 0052	Atyantashunyata (Buddhism)		HM 3518	Awakening stage life ; Heart		HM 2301	— Eighth vajra master (Buddhism)
HM 2408	Atziluth ; Olam (Judaism)		HM 3833	Aware ; Mono no (Japanese)		HM 2340	— Emanation body (Buddhism)
VC 0201	Audacity			Awareness		HM 2873	— Enjoyment body (Buddhism)
HM 2534	Auddhatya (Buddhism)		HH 0756	— [Awareness]		HM 3068	— Entering city bliss bestowing hands (Zen)
VP 5448	Audibility-Inaudibility		VC 0211	— [Awareness]		HM 2006	— Evolutionary
HM 1317	Audible thought		HM 2096	— [Awareness]		HM 2126	— expansion
TC 1866	Audience reaction teams			Awareness [Absolute...]		HM 2637	— externality (ICA)
TC 1418	Audio-graphic teleconferencing		HM 2397	— Absolute body (Buddhism)			Awareness [Fatigued...]
TC 1498	Audio-tape recording		HM 2028	— Adamantine hell (Buddhism)		HM 2423	— Fatigued conscious
TC 1972	Audio teleconferencing		HM 2955	— Airy (Astrology)		HM 4983	— feeling-group conscious existence (Buddhism)
HM 7212	Audiogravic illusions		HM 2073	— allegiance (ICA)		HM 2124	— Fiery (Astrology)
HM 7212	Audiogyral illusions		HM 2391	— Altered states		HM 2909	— Fifth scriptural bodhisattva (Buddhism)
HH 1076	Auditing		HM 2783	— Amoghasiddhi karma (Buddhism)		HM 2763	— Fifth vajra master (Buddhism)
HM 6539	Auditory agnosia		HM 2212	— angelic evil		HM 2849	— Final vajra master (Buddhism)
HM 6704	Auditory hallucination		HM 2150	— Angelic frame		HM 2155	— First scriptural bodhisattva (Buddhism)
KC 0206	Augmentation intellect		HM 3146	— angelic-transformation		HM 2554	— Fixed (Astrology)
KC 0206	Augmentation ; Team		HM 2636	— Animal hell (Buddhism)		HM 2142	— Form realm (Buddhism)
HH 2267	Aura ; Human		HM 2390	— Apollo		HM 3144	— Formless realm (Buddhism)
HM 0679	Auric clairvoyance (Psychism)		HM 2610	— Apperceptive		HM 0650	— Foundation
HH 2552	Auric healing		HM 3007	— archaism (ICA)		HM 2082	— Four great kings heaven (Buddhism)
HM 2749	Auspicious moment		HM 3024	— Arhat sanctity (Buddhism)		HM 2191	— Fourth scriptural bodhisattva (Buddhism)
HH 1248	Austerities (Hinduism)		HM 2327	— Ascension (Sufism)		HM 2251	— Fourth vajra master (Buddhism)
VD 2116	Austerity			Awareness associated		HM 2364	— Fragrance (Buddhism)
HM 7703	Austerity (Buddhism)		HM 0134	— alcohol consumption ; Modes		HM 2316	— Freedom (ICA)
HM 4412	Austerity ; Plane (Leela)		HM 4546	— barbiturates ; Modes		HM 2609	— futurity (ICA)
HM 5244	Austerity ; Plane (Leela)		HM 0584	— psychoactive substances ; Modes			Awareness [Gabriel...]
HM 4450	Austerity (Sufism)		HM 2313	— schizophrenia ; Modes		HM 2222	— Gabriel ; Angelic
HM 6690	Austerity (Sufism)			Awareness associated use		HM 2118	— Great black lord (Buddhism)
HH 1301	Australian Aboriginal dream-time		HM 0803	— amphetamines similar drugs ; Modes		HM 2656	— Great path tantra (Buddhism)
HH 1301	Australian Dreaming		HM 1368	— caffeine ; Modes			Awareness Great vehicle
HM 2316	Authentic relation (ICA)		HM 3691	— cannabis ; Modes		HM 3048	— higher path (Buddhism)
VC 0203	Authenticity		HM 1594	— cocaine ; Modes		HM 3268	— lower path (Buddhism)
HH 1357	Authenticity		HM 0801	— hallucinogens ; Modes		HM 2160	— middle path (Buddhism)
VD 2118	Authoritarianism		HM 3743	— inhalants ; Modes		HM 2699	Awareness ; Guhya samaja urgyan (Buddhism)
HM 2144	Authority ; Awareness (Yoga)		HM 4277	— khat ; Modes			Awareness [Hatred...]
KC 0652	Authority ; Collegiate		HM 1881	— nicotine ; Modes		HM 4596	— Hatred
TC 1244	Authority ; Conference		HM 4060	— opium similar drugs ; Modes		HM 2169	— Hearing (Buddhism)
VP 5739	Authority-Lawlessness		HM 4254	— phencyclidine similar drugs ; Modes		HM 2565	— Heaven (Psychism)
VD 2120	Autism		HM 4546	— sedatives ; Modes		HM 2010	— Heavenly highway (Buddhism)
HM 4058	Autism (Yoga)			Awareness [Asura...]		HM 3174	— Her favourite resort (Yoga)
HH 1235	Autocontrol consciousness		HM 2579	— Asura world (Buddhism)		HM 2560	— Herding ox (Zen)
HM 0707	Autocracy (Systematics)		HM 2103	— Atma (Hinduism)		HM 2136	— Hermes
HH 0072	Autogenic imagogy		HM 2144	— authority (Yoga)		HM 2115	— Hindu states (Buddhism)
HH 2338	Autogenic relaxation sequence		HM 3906	— Automaton		HM 2723	— Hindu wisdom holder (Buddhism)
HH 0506	Autogenic training			Awareness [Back...]		HM 2100	— Howling hells (Buddhism)
MP 1229	Automatic communications ; Provision pathways		HM 3251	— Back source (Zen)		HM 2112	— Hungry ghosts hell (Buddhism)
KC 0345	Automatic systems ; Integrated		HM 2091	— Barbarian state (Buddhism)		HM 2337	— Hyper bliss realm (Buddhism)
HM 3906	Automatization		HM 3744	— biofeedback training ; States			Awareness [I...]
HM 8105	Automatization ; Breakdown		HM 2603	— black freedom ; Rudra (Buddhism)		HM 3225	— I Thou
HM 3906	Automaton awareness		HM 2220	— Bodhicitta (Buddhism)		HM 2201	— Image (Tarot)
MS 8243	Automobile		HM 2972	— Body		HM 0035	— Immediate boundary
MP 1154	Autonomous contexts maturing perspectives ; Semi		HM 2639	— Bon practitioner state (Buddhism)		HM 2620	— Inactive
MP 1155	Autonomous contexts perspectives decreasing activity ; Semi		HM 2663	— Bon wisdom (Buddhism)		HM 2239	— Individualized (Astrology)
HH 0328	Autonomous discipline		HM 2099	— Buddhi (Hinduism)		HM 0934	— infinity (Judaism)
VC 0207	Autonomy			Awareness [Cardinal...]		HH 3321	— inter-dependency conscious existence phenomena (Buddhism)
TC 1623	Autonomy ; Conference		HM 2259	— Cardinal (Astrology)		HM 3225	— Interhuman
MP 1109	Autonomy sub structures ; Organization structure enhance		HM 2979	— Catching ox (Zen)		HM 2052	— Interminable hell (Buddhism)
KD 2085	Autopoiesis ; Organization self renewal:		HM 2519	— cattle ; Western continent (Buddhism)		HM 1347	— Internal (Psychism)
KC 0956	Autopoietic system		HM 4426	— certitude (Sufism)		HM 1044	— Internalization
HM 7342	Autoscopy		HM 2172	— Cessation (Buddhism)		HM 2850	— Islamic transformation
HH 0766	Autosuggestion		HH 4221	— Change			Awareness [Jewelled...]
HM 1410	Autotelic experience		HM 4469	— Choiceless		HM 2275	— Jewelled peaks realm (Buddhism)
HH 7004	Auxiliary methods Taoism ; Sidetracks (Taoism)		HM 2880	— Christian stations (Christianity)		HM 2167	— Joy land (Buddhism)
HH 0141	Avadhoot		HM 3063	— City shining jewel (Yoga)		HM 2022	— Joyful heaven (Buddhism)
VC 0209	Availability		HM 0087	— coding		HM 2106	— Joyous
HM 4037	Avankata (Pali)		HM 2040	— Cold hells (Buddhism)			Awareness [Kalacakra...]
VD 2122	Avarice		HM 3153	— Coming home ox's back (Zen)		HM 2151	— Kalacakra tantra shambhala (Buddhism)
HM 0227	Avarice (Buddhism)		HM 2067	— community ; Northern continent (Buddhism)		HM 2048	— Karma (Buddhism)
HM 0987	Avarice (Leela)		HM 0634	— Complete (Buddhism)		HM 0563	— Kingdom heaven
HM 0642	Avarice ; Spiritual (Christianity)		HM 2233	— Concrete		HM 2558	— Kriya tantra (Buddhism)
HM 2265	Avasthapana (Buddhism)		HM 2098	— consciousness-group conscious existence (Buddhism)			Awareness [Levels...]
HM 2032	Avasthas (Hinduism)		HM 2556	— consciousness-group conscious existence - senses (Buddhism)		HM 2391	— Levels
HH 0079	Avatar		HM 2108	— corporeality-group conscious existence (Buddhism)		HM 2347	— Life state bardo (Buddhism)
HM 2921	Avatar ; Scorching (ICA)		HM 3609	— Critical		HM 3381	— light ; Near death (NDE)
HM 6006	Avataric level musical inspiration		HM 3142	— Crushing hells (Buddhism)		HM 2320	— Logical metalogical
HM 3471	Avatthiti ; Cittasa (Pali)			Awareness Death bardo		HM 2248	— Logos ; Angelic
VD 2124	Aversion		HM 2335	— clear light (Buddhism)		HM 2088	— Lord death (Buddhism)
HH 0256	Aversion therapy		HM 2407	— heaven reality (Buddhism)		HM 0750	— Lucid (Psychism)
HM 3504	Aversions ; Emotional (Buddhism)		HM 2431	— rebirth seeking (Buddhism)		HM 2679	Awareness ; Magical forces (Buddhism)
HM 2052	Avici-hell (Buddhism)			Awareness [Delusional...]			Awareness Mahayana
HM 3196	Avidya (Buddhism)		HM 5003	— Delusional		HM 2232	— climax application (Buddhism)
HM 0310	Avidya (Leela)		HM 2055	— Demon island (Buddhism)		HM 2208	— heat application (Buddhism)
HM 2608	Avihimsa (Buddhism)		HM 2447	— depth (ICA)		HM 2271	— highest teachings application (Buddhism)
HM 3035	Avijja (Pali)		HM 2044	— Devachon (Psychism)		HM 2247	— receptivity application (Buddhism)
HM 3196	Avijja (Pali)			Awareness Diminished			Awareness [Manas...]
HM 1013	Avijjalanga (Pali)		HM 6201	— clarity		HM 2902	— Manas (Hinduism)
HM 1375	Avijjanusaya (Pali)		HM 1353	— focus		HM 2311	— Meditation state bardo (Buddhism)
			HM 1353	— range		HM 2127	— men ; Southern continent (Buddhism)

-848-

GENERAL INDEX TO VOLUME 2

Being

Awareness [mental...] cont'd
- HM 2050 — mental-formation group conscious existence (Buddhism)
- HM 2198 — Michael ; Angelic
- HM 3026 — Middle path tantra (Buddhism)
- HM 2069 — Monadic (Amerindian)
- HM 3005 — Multi level (Psychism)
- HM 0652 — Mundane (Buddhism)
- HM 6534 — mundane ; Restoration celestial (Taoism)
- HM 3092 — Mutable (Astrology)
- HM 2900 — mystic journey

Awareness [Naga...]
- HM 2031 — Naga world (Buddhism)
- HM 2036 — Nagual state (Amerindian)
- HM 2377 — Nearness (Sufism)
- HM 2292 — Ninth scriptural bodhisattva (Buddhism)
- HM 2325 — Ninth vajra master (Buddhism)
- HM 2543 — noble figures ; Eastern continent (Buddhism)
- HM 3242 — Not hit (Yoga)
- HM 5118 — Numerical (Sufism)

Awareness [Objective...]
- HM 2203 — Objective
- HM 2598 — Objective (ICA)
- HM 8186 — one's self ; Anomalies
- HM 2188 — Orders angelic (Judaism)
- HM 2266 — Ordinary
- HM 2390 — Orpheus
- HM 5123 — Overall (Psychism)
- HM 0827 — Oversoul (Psychism)
- HM 2604 — Ox forgotten leaving man alone (Zen)
- HM 2492 — Ox man both out sight (Zen)

Awareness [Parallel...]
- HM 5856 — Parallel
- HM 4143 — perception-group conscious existence (Buddhism)
- HM 1344 — Perfection discriminating (Buddhism)
- HM 2551 — Phenomena (Buddhism)
- HM 1911 — Philosophical
- HM 2393 — Phoenix
- HM 2175 — Potala island (Buddhism)

Awareness Pratyeka Buddha
- HM 3020 — application (Buddhism)
- HM 2776 — arhat (Buddhism)
- HM 2180 — (Buddhism)
- HM 2252 — cultivation (Buddhism)
- HM 2228 — vision path (Buddhism)

Awareness [Primordial...]
- HM 2163 — Primordial (Zen)
- HM 2760 — Prometheus
- HM 2075 — Psycho physical faculties (Buddhism)
- HM 2168 — Pure lands (Buddhism)
- HM 3461 — Purification (Yoga)

Awareness [Raphael...]
- HM 2162 — Raphael ; Angelic
- HM 4493 — reality ; Pure (Yoga)
- HM 2032 — relative reality (Hinduism)
- HM 2516 — Reviving hell (Buddhism)
- HM 2893 — Root base (Yoga)

Awareness [sacred...]
- HM 0876 — sacred (Christianity)
- HM 2409 — Satisfaction
- HM 3036 — Searching ox (Zen)
- HM 2739 — Second scriptural bodhisattva (Buddhism)
- HM 2703 — Second vajra master (Buddhism)
- HM 2755 — Seeing ox (Zen)
- HM 3302 — Seeing traces (Zen)

Awareness Self
- HM 2486 — [Awareness ; Self]
- HM 1874 — (Buddhism)
- HM 2436 — (Psychism)

Awareness [self...]
- HM 2984 — self spiritual intelligence (Yoga)
- HM 2972 — Sensory
- HM 2664 — Sensual (Buddhism)
- HM 2361 — Seventh scriptural bodhisattva (Buddhism)
- HM 2789 — Seventh vajra master (Buddhism)
- HM 2374 — Sexual
- HM 2002 — Shekinah (Judaism)
- HM 2385 — Sixth scriptural bodhisattva (Buddhism)
- HM 2287 — Sixth vajra master (Buddhism)
- HM 2176 — Spatial
- HM 3232 — spiritual identity (Yoga)
- HM 2286 — spiritual poverty (ICA)
- HM 2305 — State (Sufism)
- HM 2596 — State universal cessation (Buddhism)
- HM 2476 — Sufi infused (Sufism)
- HM 1067 — Supramundane (Buddhism)
- HM 2813 — Supreme heaven (Buddhism)

Awareness Tantra
- HM 2452 — beginner (Buddhism)
- HM 2220 — climax application (Buddhism)
- HM 2696 — heat application (Buddhism)
- HM 2235 — master form realm (Buddhism)
- HM 2211 — master sense realm (Buddhism)
- HM 2756 — receptivity application (Buddhism)
- HM 2280 — union learning application (Buddhism)

Awareness [Taste...]
- HM 2263 — Taste (Buddhism)
- HM 2454 — Temporary hells (Buddhism)
- HM 2421 — Tenth scriptural bodhisattva (Buddhism)
- HM 2215 — Third scriptural bodhisattva (Buddhism)
- HM 2727 — Third vajra master (Buddhism)
- HM 2606 — Thirty three god heaven (Buddhism)
- HM 2321 — Threshold
- HM 2033 — Tibetan definitions (Buddhism)
- HM 2066 — Tonal state (Amerindian)
- HM 4341 — Totemic

Awareness [Touch...] cont'd
- HM 2562 — Touch (Buddhism)
- HM 2395 — Transcendental (Hinduism)
- HM 0768 — Transpersonal
- HM 2772 — Triadic
- HM 2395 — Turiya (Hinduism)

Awareness [Ultimate...]
- HM 2388 — Ultimate (ICA)
- HM 1456 — ultimate reality
- HM 2392 — Upanishadic stages (Hinduism)
- HM 3123 — Uriel ; Angelic
- HM 2625 — Utter (ICA)
- HM 2576 — Awareness ; Very hot hells (Buddhism)
- HM 2074 — Awareness ; Visual (Buddhism)

Awareness [Watery...]
- HM 2384 — Watery (Astrology)
- HM 2759 — Wheel turner king (Buddhism)
- HM 2058 — Wheel turning king (Buddhism)
- HM 2473 — while unconscious ; Alternative mode
- HM 2179 — Wish dominated

Awareness [Zacharael...]
- HM 3022 — Zacharael ; Angelic
- HM 2760 — Zeus
- HM 2713 — Zodiacal forms (Astrology)
- HM 2282 — Zoharariel ; Angelic
- HM 1087 — Away ; Passing (Islam)

Away Passing Sufism
- HM 3799 — [Away ; Passing (Sufism)]
- HM 5775 — [Away ; Passing (Sufism)]
- HM 1270 — [Away ; Passing (Sufism)]
- HM 0748 — Away reappearance beings ; Knowledge passing (Buddhism)
- HM 2076 — Awayness ; Passing (ICA)
- HM 2592 — Awe
- VC 0213 — Awe
- HM 2302 — Awe ; Consciousness
- HM 2619 — Aweful encounter (ICA)
- VD 2128 — Awfulness
- VD 2130 — Awkwardness
- HH 0477 — AWL Alternative ways life
- HM 6139 — Awwal ; Al (Sufism)
- HH 0512 — Axiodrama
- MP 1032 — Axis ; Selective interchange
- HM 1969 — Ayu ; Arupinam dhammanam (Pali)
- HM 4042 — Az-Zahir (Sufism)
- HM 2212 — Azazel
- HM 5198 — 'Azim ; Al (Sufism)
- HM 4562 — 'Aziz ; Al (Sufism)

B

- HM 2474 B-cognition
- HM 1732 B beginning ; Serene
- HH 0871 B-values
- HH 0578 Ba ; Dge (Buddhism)
- HM 4017 Ba'ith ; Al (Sufism)
- HM 2177 Ba ; Khor (Buddhism)
- KC 0313 Babel syndrome
- HM 3153 Back awareness ; Coming home ox's (Zen)
- HM 3251 Back-to-the-source awareness (Zen)
- VD 2132 Backlash
- HH 2321 Backsliding
- MM 2075 Bad behaviour ; Good
- HM 0764 Bad company (Leela)
- HM 0818 Bad trip
- HM 4299 Badi ; Al (Sufism)
- VD 2134 Badness
- VP 5674 Badness ; Goodness
- VD 2136 Bafflement
- HM 2314 Bag-med-pa (Tibetan)
- HM 3220 Bag-yod-pa (Tibetan)
- HM 2435 Baha'ism Human development
- HM 2581 Bahir-vritti (Yoga)
- HM 2522 Bahirdhashunyata (Buddhism)
- HH 0053 Bakarah (Islam)
- KC 0267 Balance
- VC 0215 Balance
- HM 3291 Balance ; Fourth chakra: attaining (Leela)
- KC 0968 Balance nature
- HM 3291 Balance ; Plane (Leela)
- HM 1351 Balance ; Plane (Leela)
- HH 4997 Balancing yin yang (Taoism)
- HM 4049 Balinese Being another world
- HM 4049 Balinese Nadi
- MM 2066 Ball games
- HM 1864 Balya (Pali)
- VD 2138 Banality
- HH 8007 Bandhas (Yoga)
- HM 2530 Bands ; Transpersonal
- VD 2140 Banefulness
- HH 0035 Baptism
- HM 1267 Baptism ; Initiation (Esotericism)
- HM 0330 Baqa (Sufism)
- HM 6786 Baqi ; Al (Sufism)
- HM 4594 Baqi ; Al (Sufism)
- HM 0101 Bar-do (Tibetan)
- HM 1882 Baraka (Islam)
- HM 2091 Barbarian-state awareness (Buddhism)
- VD 2142 Barbarism
- VD 2144 Barbarity
- HM 4546 Barbiturates ; Modes awareness associated

Bardo awareness
- HM 2371 — Dream state (Buddhism)
- HM 2347 — Life state (Buddhism)
- HM 2311 — Meditation state (Buddhism)

- HM 2335 Bardo clear light awareness ; Death (Buddhism)
- HM 0698 Bardo consciousness (Buddhism)
- HM 2407 Bardo heaven reality awareness ; Death (Buddhism)
- HM 2431 Bardo rebirth seeking awareness ; Death (Buddhism)
- HM 2098 Bare cognition (Buddhism)
- HM 4651 Bare perception (Buddhism)
- HM 3265 Bare perception ; Seemingly (Buddhism)
- HM 3636 Bari' ; Al (Sufism)
- HM 7022 Barr ; Al (Sufism)
- VD 2146 Barrenness
- VD 2148 Barrier
- TP 0059 Barrier dissolution (Flooding)
- HM 8621 Barriers ; Consciousness rid (Buddhism)
- HM 2893 Base awareness ; Root (Yoga)
- HM 1736 Base mind (Buddhism)
- KD 2327 Based consensus games ; Resonance
- VD 2150 Baseness
- HM 3274 Bashriyya ; Sifat (Sufism)
- HH 0708 Basic conflict
- HM 3209 Basic faith (Buddhism)
- HM 2799 Basic fear ; Breaking point (Brainwashing)
- TP 0048 Basic need (Well)
- HM 6512 Basir ; Al (Sufism)
- HM 0476 Basir ; 'Ilm (Sufism)
- HM 0115 Basit ; Al (Sufism)
- MM 2049 Basketry ; Knotting, nets
- HM 2106 Bast (Sufism)
- VD 2152 Bastardy
- HH 3664 Bathing (Taoism)
- HM 1209 Batin ; Al (Sufism)
- MM 2076 Battery
- VD 2154 Battle
- VD 2156 Bawdiness
- HM 2245 Bden-pa-bzhi (Tibetan)
- HM 3500 Bear cross ; Jesus made (Christianity)
- VD 2158 Beastliness
- HM 3347 Beatific vision
- HH 0608 Beatification
- VC 0217 Beatification
- VC 0219 Beatitude
- HH 2561 Beautiful names ; Most (Sufism)
- MS 8233 Beautiful people

Beauty
- HH 0488 — [Beauty]
- VC 0221 — [Beauty]
- MS 8233 — [Beauty]
- HM 3031 Beauty ; Sphere (Kabbalah)
- VP 5899 Beauty-Ugliness
- HM 0763 Beauty ; Way (Esotericism)
- HH 0902 Become ; Capability
- HH 0405 Become ; Capacity
- HH 0171 Become ; Propensity
- HM 0933 Becomes himself ; Fifth chakra: man (Leela)
- HM 5909 Becoming (Buddhism)
- HH 5328 Becoming Christ
- HH 6211 Becoming Christian (Christianity)
- HM 7672 Becoming many ; One (Buddhism)
- HM 2717 Becoming ; Perpetual (ICA)
- HM 5909 Becoming ; Wheel (Buddhism)
- VD 2160 Bedevilment
- VD 2162 Bedraggled
- TP 0064 Before completion
- HM 2452 Beginner awareness ; Tantra (Buddhism)
- HM 0102 Beginner spiritual life (Christianity)
- HM 2171 Beginner subtle contemplation ; Only (Buddhism)
- TP 0003 Beginning ; Difficulty
- VC 0223 Behaviour
- MS 8125 Behaviour
- HM 2406 Behaviour ; Amnesic
- MS 8725 Behaviour ; Animal
- HH 0447 Behaviour development ; Human
- HH 6109 Behaviour ; Ethological interpretation
- MM 2075 Behaviour ; Good bad
- VP 5737 Behaviour-Misbehaviour
- HH 0771 Behaviour modification
- TC 1910 Behaviour ; Patterned
- HM 0198 Behaviour ; Right (Buddhism)
- TC 1910 Behaviour ; Rule governed
- HH 0795 Behaviour therapy
- HM 3335 Behavioural transformation ; Will (Sufism)
- HH 1150 Behaviourism
- VC 0225 Being

Being [Abraham...]
- HM 3974 — Abraham one's (Sufism)
- HM 3425 — Adam one's (Sufism)
- HM 4049 — another world (Balinese)
- HM 0552 — ashamed acquisition sinful unwholesome dharmas (Buddhism)
- HM 3828 — ashamed what one ought be ashamed (Buddhism)
- HM 1195 — Atman (Hinduism)

Being [Causally...]
- HH 1099 — Causally continuous doctrine (Buddhism)
- HM 2474 — cognition
- HM 4514 — conscious (Buddhism)
- HM 2099 — consciousness (Hinduism)
- HM 0836 — contacted (Buddhism)
- HM 0684 — covetous ; Not (Buddhism)
- HM 4570 — ; David one's (Sufism)
- HM 1561 Being doubt two alternatives (Buddhism)

Being [Emblems...]
- HM 2422 — Emblems well (Tarot)
- HM 6111 — Essential unity (Christianity)
- HH 0602 — Existential unity
- HM 2409 Being ; Feeling well
- HM 4103 Being ; First chakra: fundamentals (Leela)

-849-

Being

	Being [good...]	HM 3323	Bhavanga-mano (Buddhism)
HM 0872	— good quality (Buddhism)	HM 2989	Bhavashunyata (Buddhism)
HM 1442	— greedy (Buddhism)	HM 1194	Bhaya (Hinduism)
HM 3947	— greedy ; State not (Buddhism)	HM 1024	Bhu-loka (Leela)
	Being [history...]	HM 4637	Bhu nanam ; Gotra (Pali)
HM 3075	— history (ICA)	HM 2155	Bhumi (Buddhism)
HM 1809	— hooked	HM 1623	Bhuri (Pali)
HH 1542	— Human	HM 7549	Bhuta jnana darsana ; Yatha (Buddhism)
	Being [ICA...]	HM 7549	Bhuta nana dassana ; Yatha (Pali)
HM 2983	— (ICA)	HM 4600	Bhuvar-loka (Leela)
HM 2632	— in-the-world	VD 2174	Bias
HM 1683	— infatuated (Buddhism)	HM 3685	Bias ; Sectarian (Buddhism)
HM 3183	— infatuated ; State not (Buddhism)	TC 1965	Biases ; Participant pre logical
HM 1412	— intended (Buddhism)	MM 2047	Bicycles cycling
HM 2246	Being ; Jesus one's (Sufism)	TP 0005	Biding one's time (Getting wet)
HM 3229	— Life (ICA)	VC 0241	Bigness
	Being [Mohammad...]	VP 5195	Bigness-Littleness
HM 3513	— Mohammad one's (Sufism)	VD 2176	Bigotry
HM 3616	— Moses one's (Sufism)	MM 2057	Bilingual protocol
HM 3097	— myself ; All (ICA)	HM 2372	Binah sephira (Kabbalah)
HM 2154	— myself (ICA)	KC 0870	Binary recoding
	Being [Nature...]	HH 6112	Bind ; Therapeutic double
HH 0343	— Nature (Zen)		**Binding Time**
HM 3286	— Noah one's (Sufism)	HH 0585	— [Binding ; Time]
HH 0587	— Non temporal dimension	HH 0815	— [Binding ; Time]
HM 1596	— Own (Buddhism)	MP 1248	— [Binding ; Time]
	Being [plane...]	HH 1267	Bio-globalism
HM 4128	— plane culture ; Understanding (Buddhism)	HM 2262	Bio-information (ESP)
HM 6192	— plane discernment ; Understanding (Buddhism)	HH 1267	Bio-revolution
HH 0718	— psychology	HH 2315	Biocentric equality
HM 0889	— revitalized (Thai)	KC 0061	Biocybernetics
	Being [Solitary...]	HH 0652	Bioenergetic analysis
HM 2404	— Solitary (ICA)	HH 0765	Biofeedback training
HH 0269	— Spiritual	HM 3744	Biofeedback training ; States awareness
HM 1210	— sync	HH 0475	Biological rhythms
HM 4341	Being ; Totemic	HH 0806	Biology ; Humanistic
	Being [Unchanging...]	HH 0962	Biomagnetic therapy
HM 3316	— Unchanging reality (Sufism)	KC 0935	Bionics
HM 2702	— Unity	HH 0475	Biorhythm
HM 1965	— unobservant (Buddhism)	HM 0044	Biosocial consciousness
HM 2611	— Unveiled (ICA)	KC 0494	Biosphere
HM 3716	Being workable (Buddhism)	HH 0616	Biotechnology
MS 8955	Beings ; Beings made	HH 0765	Biotelemetry
HH 1306	Beings ; Communication supernatural	MS 8713	Birds
MS 8907	Beings ; Infernal	HM 1529	Birth consciousness
HM 0748	Beings ; Knowledge passing away reappearance (Buddhism)	HM 6522	Birth God soul's ground (Christianity)
		HM 1337	Birth ; Initiation (Esotericism)
MS 8955	Beings made beings	HM 0978	Birth man (Leela)
	Beings Mythical	HH 4098	Birth ; Second (Christianity)
MS 8357	— aerial	HM 3175	Birth ; Second (ICA)
MS 8300	— anthropomorphic	HM 2666	Birth ; Virgin (ICA)
MS 8917	— aquatic	HM 2368	Bisociated consciousness ; Creative
MS 8337	— celestial	TP 0021	Biting through Decisive action
MS 8937	— fire dwelling	TP 0021	Biting through Punitive action
MS 8387	— terrestrial	VD 2178	Bitterness
MS 8400	Beings ; Non human	HM 1598	Bitterness (Japanese)
VC 0227	Belief	HM 2603	Black freedom ; Rudra awareness (Buddhism)
HM 2719	Belief-arising subtle contemplation (Buddhism)	HM 2118	Black lord awareness ; Great (Buddhism)
HM 2933	Belief (Buddhism)	HM 3142	Black rope hell (Buddhism)
TC 1358	Belief type ; Blind	HM 2213	Black void ; Near death (NDE)
VP 5501	Belief-Unbelief	TC 1487	Blackboard
HH 0974	Beliefs ; Right (Buddhism)	HM 5345	Blackening ; Alchemical (Esotericism)
VD 2164	Belligerence	VD 2180	Blame
HM 2264	Belligerence (Buddhism)	HM 8112	Blame ; Dread (Buddhism)
VC 0229	Belongingness	HM 0649	Blame ; Fearlessness (Buddhism)
VC 0231	Benediction	HM 4280	Blame (Sufism)
VC 0233	Beneficence	VC 0243	Blamelessness
KC 0016	Benefit analysis ; Cost	HM 4902	Blankness ; Mind
VC 0235	Benevolence	VD 2181	Blasphemy
HH 0989	Benevolence	VD 2182	Blatancy
VT 8085	Benevolence*complex	VD 2184	Blemish
HH 4066	Benevolent desires soul ; Innermost	MP 1111	Blended integration formal structure informal context
VC 0237	Benignity	HH 4997	Blending celestial consciousness earthly consciousness (Taoism)
VD 2166	Bereavement		
HM 2379	Besetting sin (ICA)	HM 2741	Blessed
VD 2168	Bestiality	HM 2741	Blessedness
HM 3476	Beta wave consciousness	VC 0247	Blessedness
VD 2170	Betrayal	HM 2958	Blessedness ; Final (ICA)
HM 2229	Betrayal ; Self (Brainwashing)	HM 1882	Blessedness ; Sense (Islam)
HM 3028	Betrothal initiation archetypal image (Tarot)	HH 0039	Blessing
VC 0239	Betterment	HM 2229	Blessings ; Manifold (ICA)
HM 1600	Bettschwere (German)	VD 2186	Blight
VD 2172	Bewilderment	TC 1358	Blind belief type
HM 3522	Bewilderment (Japanese)	MM 2062	Blindness
HM 0819	Bewitched (Psychism)	HM 7019	Blindness ; Mind
HM 1141	Beyond chakras: gods themselves (Leela)	HH 1354	Blindness sight ; Journeying transcendence
HM 4570	Beyond form ; World (Sufism)	VP 5439	Blindness ; Vision
HM 2681	Beyond individual distinctions (Sufism)	MP 1162	Blindspots ; Organization integrative superstructure minimize insight
TC 1123	Beyond linear thinking		
KD 2040	Beyond method	HM 4048	Bliss
HM 3049	Beyond morality (ICA)	VC 0249	Bliss
HM 2702	Beyond number		**Bliss [bestowing...]**
HM 2941	Beyond success (ICA)	HM 3068	— bestowing hands awareness ; Entering city (Zen)
HM 2653	Beyond-the-third-realm consciousness (Buddhism)	HM 4298	— born seclusion ; Jhana happiness (Buddhism)
HM 1475	Bhakti-loka (Leela)	HM 0747	— (Buddhism)
HH 0628	Bhakti marga (Hinduism)	HM 4191	Bliss ; Concentration accompanied (Buddhism)
HH 0337	Bhakti yoga (Yoga)	HM 1454	Bliss ; Concentration accompanied (Buddhism)
HM 0400	Bhantattam cittassa (Pali)	HM 5088	Bliss due insight (Buddhism)
HM 3619	Bhava samadhi	HM 3227	Bliss (Japanese)
HM 2051	Bhavagra (Buddhism)	HM 3436	Bliss mystical seduction ; Surrender (Sufism)
HH 0551	Bhavana (Buddhism)	HM 6901	Bliss ; Plane (Leela)
HH 5973	Bhavana ; Citta (Buddhism)	HM 2337	Bliss realm awareness ; Hyper (Buddhism)
HH 0796	Bhavana ; Magga (Buddhism)	HM 8772	Blissful love (Christianity)
HH 8231	Bhavana ; Samadhi (Buddhism)	HM 3148	Blissful seizure (ICA)
HH 0710	Bhavana ; Samatha	HM 1089	Blissing out
HH 0680	Bhavana ; Vipassana (Buddhism)	KC 0706	Block diagram

KC 0706	Block language
TP 0012	Blocked
VD 2188	Bloody-mindedness
HM 4344	Blues
HM 4754	Blurring ego boundaries
TC 1800	Board ; Flannel
TC 1800	Board ; Magnetic
TC 1800	Board ; Sticky
TC 1800	Board ; Velcro
VD 2190	Boastfulness
MS 8304	Boats ships
HM 2029	Bodhi
HH 4019	Bodhi (Buddhism)
HH 5645	Bodhi-marga (Buddhism)
HM 2793	Bodhi-pakkhiya dhamma (Buddhism)
HM 2793	Bodhi-pakshika dharma (Buddhism)
HM 2220	Bodhicitta awareness (Buddhism)
	Bodhisattva awareness
HM 2816	— Eighth scriptural (Buddhism)
HM 2909	— Fifth scriptural (Buddhism)
HM 2155	— First scriptural (Buddhism)
HM 2191	— Fourth scriptural (Buddhism)
HM 2292	— Ninth scriptural (Buddhism)
HM 2739	— Second scriptural (Buddhism)
HM 2361	— Seventh scriptural (Buddhism)
HM 2385	— Sixth scriptural (Buddhism)
HM 2421	— Tenth scriptural (Buddhism)
HM 2215	— Third scriptural (Buddhism)
HM 1380	Bodhisattva (Buddhism)
HM 1225	Bodhisattva ; State (Buddhism)
HM 2769	Bodhisattvas ; Meditation way (Buddhism)
HM 3189	Bodily action ; Naturalness
HM 1134	Bodily disability (Yoga)
HH 0662	Bodily-kinesthetic intelligence
HM 5600	Bodily misconduct ; Abstinence (Buddhism)
HM 2183	Bodily perfection (Hinduism)
HM 4332	Bodily relaxation
	Body [asleep...]
HM 2079	— asleep trance ; Double awake
HM 2640	— asleep trance ; Mind awake (Psychism)
HH 0206	— Astral
	Body awareness
HM 2972	— [Body awareness]
HM 2397	— Absolute (Buddhism)
HM 2340	— Emanation (Buddhism)
HM 2873	— Enjoyment (Buddhism)
	Body Buddha
HM 3019	— emanation (Buddhism)
HM 3113	— enjoyment (Buddhism)
HM 3019	— form (Buddhism)
HM 3113	— form (Buddhism)
HM 2039	— nature (Buddhism)
HM 2834	— truth (Buddhism)
HM 2039	— truth (Buddhism)
HM 2834	— wisdom (Buddhism)
HM 1867	Body ; Composedness (Buddhism)
HM 2562	Body consciousness (Buddhism)
HH 0299	Body dualism ; Mind
HH 0206	Body ; Etheric
	Body experience Out
HM 6120	— [Body experience ; Out]
HM 0777	— (NDE)
HM 5534	— (Psychism)
HM 0332	Body experience ; Spontaneous out
	Body [Fitness...]
HM 1455	— Fitness mental (Buddhism)
HM 3370	— Flexibility (Buddhism)
MS 8365	— fluids ; Human
MS 8700	— form structure
HM 4185	Body glory (Psychism)
MS 8305	Body ; Human
HM 1766	Body image
HM 2858	Body knowledge
KC 0528	Body knowledge
	Body [language...]
HH 0392	— language
HM 1133	— Leaving off, abstaining, totally abstaining refraining three deviations (Buddhism)
HM 4395	— Lightness (Buddhism)
HM 7636	— Lightness mental (Buddhism)
HM 3696	— Body ; Malleability mental (Buddhism)
HH 0172	Body movement therapy
MS 8133	Body parts ; Animal
HM 1766	Body perception
HM 5402	Body ; Rectitude mental (Buddhism)
	Body [Second...]
HH 0206	— Second
HM 3349	— separation ; Near death (NDE)
HH 0249	— space
HM 1424	— Straightness (Buddhism)
HM 4704	Body ; Tranquillity mental (Buddhism)
HM 7969	Body ; Wieldiness mental (Buddhism)
HM 0739	Body ; Workability (Buddhism)
HM 1222	Body ; Yoga illusory (Buddhism)
VC 0251	Boldness
HM 6533	Bombardment ; Rhythmic sensory
HH 0218	Bombardment therapy ; Rhythmic sensory
HH 0785	Bombu Zen (Zen)
HM 2639	Bon-practitioner state awareness (Buddhism)
HM 2663	Bon-wisdom awareness (Buddhism)
MM 2037	Bondage
HH 2661	Bondage freedom ; Journeying transcendence (Christianity)
HH 3498	Bonding ; Affectional
HH 0973	Bonding ; Affectionate
HH 3498	Bonding ; Human

-850-

GENERAL INDEX TO VOLUME 2

Buddhism

HH 3498 Bonding ; Pair
HH 3498 Bonding ; Parent infant
HH 3498 Bonding ; Sexual
HH 3498 Bonding ; Social
MS 8355 Bones
HH 0004 Book Changes (Taoism)
HH 0382 Book Thoth (Tarot)
Boredom
HM 2423 — [Boredom]
HM 2620 — [Boredom]
VD 2192 — [Boredom]
VP 5878 Boredom ; Amusement
VD 2194 Boringness
HH 0576 Born-again (Christianity)
HM 8273 Born materiality ; Consciousness (Buddhism)
HM 4298 Born seclusion ; Jhana happiness bliss (Buddhism)
HH 0715 Born ; Twice
VD 2196 Bothersomeness
HM 2872 Bottomless centre (ICA)
MP 1243 Boundaries ; Ambiguous
HM 4754 Boundaries ; Blurring ego
MP 1218 Boundaries ; Distortion resistant
MP 1224 Boundaries ; Emphasizing transitions across
MP 1234 Boundaries ; Maintainable, multi element external
MP 1235 Boundaries ; Unalienating internal
KC 0806 Boundary
HM 0035 Boundary awareness ; Immediate
MM 2014 Boundary ; Container
MP 1015 Boundary ; Context
MP 1211 Boundary expansion permitting new level generation
MP 1212 Boundary orientation ; Primary inter level connections transitions
MP 1225 Boundary reinforcement apertures
MP 1013 Boundary ; Sub domain
MP 1061 Bounded common small-scale interaction domains
VC 0253 Boundlessness
VC 0255 Bountifulness
HM 2041 Brahma consciousness (Hinduism)
HM 4326 Brahma-loka (Leela)
HH 0978 Brahmacarya
HH 1987 Brahmacharyashrama (Hinduism)
HH 1226 Brahman
HM 4331 Brahman ; Nirguna
HM 0576 Brahman ; Saguna
HH 3534 Brahmavihara-niddesa (Buddhism)
HM 2295 Brain-based consciousness
HH 0772 Brain building
Brain consciousness
HM 2415 — Left
HM 2355 — Mammalian limbic human
HM 2427 — Neomammalian neocortex human
HM 2307 — Reptilian human
HM 2367 — Right
HH 0995 Brain ; Electrical stimulation
HH 3129 Brain waves
KC 0357 Brain ; World
HH 0772 Brainpower enhancement
VD 2198 Brainwash
HM 3002 Brainwashing Assault identity
HH 0865 Brainwashing Brainwashing
HH 0865 Brainwashing (Brainwashing)
HM 2799 Brainwashing Breaking point basic fear
HM 2354 Brainwashing Breaking point total conflict
HM 3204 Brainwashing Channelling guilt
HM 2745 Brainwashing Compulsion confess
HH 1020 Brainwashing cults
HM 3002 Brainwashing Deprivation selfhood
HM 3017 Brainwashing Desperate gratitude
HM 3156 Brainwashing Final confession
HM 3017 Brainwashing Grasping straws
HM 2843 Brainwashing Identification
HM 2843 Brainwashing Ideological rebirth
HH 0865 Brainwashing Ideological reform
HM 2564 Brainwashing Logical dishonouring
HH 0865 Brainwashing Menticide
HM 2923 Brainwashing Recognition guilt
HM 2229 Brainwashing Self-betrayal
HM 2149 Brainwashing Sense harmony progress
HH 0865 Brainwashing Sensory deprivation
HM 3156 Brainwashing Summing-up
HH 0865 Brainwashing Thought reform
HM 2934 Brainwashing Transitional limbo
VC 0257 Bravery
HH 0677 Bravery (Zen)
VP 5204 Breadth-Narrowness
HM 1623 Breadth wisdom (Buddhism)
HM 8105 Breakdown automatization
HM 2799 Breaking point basic fear (Brainwashing)
HM 2354 Breaking point total conflict (Brainwashing)
TP 0043 Breakthrough (Flight)
HH 0687 Breath ; Concentration flow
HH 2763 Breath control
HH 2997 Breath ; Healing
MM 2003 Breathing
HH 0752 Breathing therapy
VC 0259 Breathtaking
VC 0261 Breeding
VC 0263 Brevity
HM 3038 Briah ; Olam (Judaism)
TC 1875 Bricksigner type
HM 1114 Bridge consciousness (Psychism)
VC 0265 Brightness
VC 0267 Brilliance
HM 3237 Bringing-to-mind (Buddhism)
HM 3053 Brjed-nges-pa (Tibetan)
HM 9384 Broadcasting ; Thought
MP 1114 Broadest ; Hierachy perspectives favouring
VC 0269 Broadmindedness
VP 5526 Broadmindedness-Narrowmindedness
VC 0271 Broadness
VD 2200 Brokenness
HM 4312 Brooding
HM 2814 Brother ; Ever present (ICA)
VC 0273 Brotherhood
HM 2432 Brotherhood ; Global (ICA)
TP 0013 Brotherhood ; Universal
VC 0275 Brotherliness
HM 2389 Brtson-'grus (Tibetan)
VD 2202 Brutality
Bsam gtan
HM 2586 — bzhi pa (Buddhism)
HM 2450 — dang-po (Buddhism)
HM 2038 — gnyis-pa (Buddhism)
HM 2062 — gsum-pa (Tibetan)
HM 2693 — (Tibetan)
HM 7769 Btang-snyoms (Tibetan)
Buddha [application...]
HM 3020 — application awareness ; Pratyeka (Buddhism)
HM 2776 — arhat awareness ; Pratyeka (Buddhism)
HM 2180 — awareness ; Pratyeka (Buddhism)
HM 2735 Buddha-consciousness (Buddhism)
HM 2252 Buddha cultivation awareness ; Pratyeka (Buddhism)
Buddha [Eighteen...]
HM 2775 — Eighteen unshared attributes (Buddhism)
HM 3019 — emanation body (Buddhism)
HM 3113 — enjoyment body (Buddhism)
HM 3019 Buddha form body (Buddhism)
HM 3113 Buddha form body (Buddhism)
HM 3263 Buddha illumination (Buddhism)
HM 2039 Buddha nature body (Buddhism)
HM 2834 Buddha truth body (Buddhism)
HM 2039 Buddha truth body (Buddhism)
HM 2228 Buddha vision path awareness ; Pratyeka (Buddhism)
HM 2834 Buddha wisdom body (Buddhism)
HH 3514 Buddhahood (Buddhism)
HM 1873 Buddhahood ; State (Buddhism)
HM 2185 Buddhas ; Way (Buddhism)
HM 2099 Buddhi awareness (Hinduism)
HH 0484 Buddhi yoga (Yoga)
HM 3208 Buddhism Abhavashunyata
HM 2448 Buddhism Abhavasvabhavashunyata
HH 1765 Buddhism Abhidhamma (Pali)
HH 1765 Buddhism Abhidharma
HM 3563 Buddhism Abhijjha (Pali)
HH 1809 Buddhism Abhijna
HM 2385 Buddhism Abhimukhi
HM 3432 Buddhism Abhinivesa (Pali)
HH 1809 Buddhism Abhinna (Pali)
HM 1687 Buddhism Abhippasada (Pali)
HM 2337 Buddhism Abhirati
HM 6093 Buddhism Abolition passion
HM 3857 Buddhism Abruptness
HM 2695 Buddhism Absence delusion
HM 2695 Buddhism Absence dullness
HM 2397 Buddhism Absolute-body awareness
HM 0634 Buddhism Absolute (Confucianism)
HM 0137 Buddhism Absorption
HM 0450 Buddhism Absorption
HM 0311 Buddhism Absorption concentration
HM 5600 Buddhism Abstinence bodily misconduct
HM 5600 Buddhism Abstinence misconduct kaya
HM 7171 Buddhism Abstinence verbal misconduct
HM 7887 Buddhism Abstinence wrong livelihood
HM 1734 Buddhism Abyapada (Pali)
HM 2816 Buddhism Acala
HM 4999 Buddhism Access concentration
HM 2864 Buddhism Action tantra: kriya
HM 2028 Buddhism Adamantine-hell awareness
HM 1510 Buddhism Adassana (Pali)
HH 2387 Buddhism Adhicitta sikkha
HM 2933 Buddhism Adhimoksha
HM 2719 Buddhism Adhimokshikamanaskara
HM 2744 Buddhism Adhosa (Pali)
HM 2794 Buddhism Adhyatmabahirdhashunyata
HM 2671 Buddhism Adhyatmashunyata
HM 4637 Buddhism Adoptive knowledge
HM 2744 Buddhism Adosa
HM 1399 Buddhism Adosa (Pali)
HM 3560 Buddhism Adoso kusala mulam (Pali)
HM 1557 Buddhism Adussana (Pali)
HM 1597 Buddhism Adussitattam (Pali)
HM 8336 Buddhism Adverting
HM 2744 Buddhism Advesha
HM 2002 Buddhism Aeons dissolution evolution
HM 2996 Buddhism Afflicted views
HH 5007 Buddhism Afflictions hindrances
HM 2098 Buddhism Aggregate consciousness
HM 2108 Buddhism Aggregate matter
HM 2050 Buddhism Aggregate mental activities
HM 3321 Buddhism Aggregates clinging
HM 3154 Buddhism Agitation
HM 0436 Buddhism Agitation
HM 2986 Buddhism Ahirika
HM 1724 Buddhism Ahirikabala (Pali)
HM 2986 Buddhism Ahrikya
HM 1054 Buddhism Ajimhata (Pali)
HM 1028 Buddhism Akakkhalata (Pali)
HM 2813 Buddhism Akanistha
HM 3375 Buddhism Akaranam (Pali)
HM 2110 Buddhism Akasanancayatana (Pali)
HM 2110 Buddhism Akasanantyayatana
HM 3014 Buddhism Akathinata (Pali)
HM 2027 Buddhism Akimcanyayatana
HM 2027 Buddhism Akincannaayatana (Pali)
HM 3429 Buddhism Akiriya (Pali)
HM 1008 Buddhism Akutilata (Pali)
HM 2730 Buddhism Alaya-vijnana
HM 2696 Buddhism Alchemical-flask
HM 2730 Buddhism All-conserving mind
HM 1799 Buddhism Alobha (Pali)
HM 2128 Buddhism Alobha (Pali)
HM 3752 Buddhism Alobhokusala mulam (Pali)
HM 2232 Buddhism Aloka-vriddhi samadhi
HM 3947 Buddhism Alubbhana (Pali)
HM 1825 Buddhism Alubbhitattam (Pali)
HM 4065 Buddhism Amodana (Pali)
HM 2783 Buddhism Amoghasiddhi-karma awareness
HM 2695 Buddhism Amoha (Pali)
HM 0684 Buddhism Anabhijjha (Pali)
HM 4090 Buddhism Anabhisamaya (Pali)
HM 3413 Buddhism Anajjhapatti (Pali)
HM 5089 Buddhism Analysis
HM 3310 Buddhism Analytical cessations
HM 3345 Buddhism Analytical investigation
HM 2015 Buddhism Anantaryamarga
HM 0403 Buddhism Ananubodha (Pali)
HM 3203 Buddhism Anapatrapya
HH 0241 Buddhism Anatma
HH 0241 Buddhism Anatta
HM 3949 Buddhism Anattamanata cittassa (Pali)
HM 1915 Buddhism Anattappabala (Pali)
HM 3072 Buddhism Anavakarashunyata
HM 3032 Buddhism Anavaragrashunyata
HM 0870 Buddhism Anekamasaggaha (Pali)
HM 9021 Buddhism Anicca
HM 4131 Buddhism Anikkhittacchandata (Pali)
HM 1403 Buddhism Anikkhittadharata (Pali)
HM 2636 Buddhism Animal-hell awareness
HH 0910 Buddhism Aniyata
HM 1394 Buddhism Annana (Pali)
HM 2330 Buddhism Annihilation
HM 0188 Buddhism Anottappa (Pali)
HM 2763 Buddhism Anujna
HM 4665 Buddhism Anupadisesanibbana (Pali)
HM 2501 Buddhism Anupalambhashunyata
HM 1930 Buddhism Anupekkhanata (Pali)
HM 0405 Buddhism Anussati (Pali)
HM 0649 Buddhism Anuttassa (Pali)
HM 1982 Buddhism Anuvicara (Pali)
HM 1746 Buddhism Apacakkhakamma (Pali)
HH 0604 Buddhism Apacayana
HM 1902 Buddhism Apaccavekkhana (Pali)
HM 2519 Buddhism Apara-godaniya
HM 5665 Buddhism Aparisphutamanovijnana
HM 4228 Buddhism Apariyobahana (Pali)
HM 0427 Buddhism Apariyogahana (Pali)
HM 2210 Buddhism Apatrapya
HM 1689 Buddhism Apilapanata (Pali)
HM 1617 Buddhism Appana (Pali)
HM 2586 Buddhism Appana samadhi (Pali)
HM 1637 Buddhism Appativedha (Pali)
HM 1618 Buddhism Appearance absorption
HM 7268 Buddhism Apperception
HM 1982 Buddhism Application reflection
HM 1177 Buddhism Applied thought
HM 1647 Buddhism Approaching reflection
HM 2292 Buddhism Appropriate intellect
HM 3220 Buddhism Apramada
HH 0233 Buddhism Arahant
HM 2191 Buddhism Arcis-mati
HM 3357 Buddhism Ardour
HH 0233 Buddhism Arhat
HM 3024 Buddhism Arhat sanctity awareness
HH 0523 Buddhism Ariya
HM 3144 Buddhism Arupa-dhatu
HM 2281 Buddhism Arupadhatu
HM 2012 Buddhism Arupaloka consciousness
HM 2669 Buddhism Arupayasamapatti
HM 1969 Buddhism Arupinam dhammanam ayu (Pali)
HH 3198 Buddhism Aruppa-niddesa (Pali)
HM 2339 Buddhism Aryastangamarga
HM 0451 Buddhism Asagahana (Pali)
HM 1965 Buddhism Asamapekkhana (Pali)
HM 1362 Buddhism Asambodha (Pali)
HM 5332 Buddhism Asamjnisamapatti
HM 6504 Buddhism Asamkhata (Pali)
HM 1705 Buddhism Asammusanata (Pali)
HM 1725 Buddhism Asampajanna (Pali)
HM 2063 Buddhism Asamprajanya
HM 6504 Buddhism Asamskrita
HM 2089 Buddhism Asamskrtashunyata
HM 3394 Buddhism Asappana (Pali)
HM 1688 Buddhism Asaraga (Pali)
HM 3183 Buddhism Asarajjana (Pali)
HM 4081 Buddhism Asarajjitattam (Pali)
HH 4298 Buddhism Ascetic practices
HM 5354 Buddhism Ascetic states
HM 2461 Buddhism Ashraddhya
HM 3261 Buddhism Asithilaparakkamata (Pali)
HM 2578 Buddhism Aspiration
HM 1687 Buddhism Assurance
HM 2579 Buddhism Asura-world awareness
HM 3857 Buddhism Asuropa (Pali)
HM 6144 Buddhism Attachment due insight
HM 0159 Buddhism Attainment
HM 5438 Buddhism Attainment cessation
HM 6346 Buddhism Attainment cessation
HM 0140 Buddhism Attamanata cittassa (Pali)

HH 0383 Buddhism Attending needs superiors
HM 3237 Buddhism Attention
HM 1663 Buddhism Attha-patisambhida (Pali)
HM 0052 Buddhism Atyantashunyata
HM 2534 Buddhism Auddhatya
HM 7703 Buddhism Austerity
HM 4037 Buddhism Avankata (Pali)
HM 0227 Buddhism Avarice
HM 2265 Buddhism Avasthapana
HM 2052 Buddhism Avici-hell
HM 3196 Buddhism Avidya
HM 2608 Buddhism Avihimsa
HM 3035 Buddhism Avijja (Pali)
HM 3196 Buddhism Avijja (Pali)
HM 1013 Buddhism Avijjalanga (Pali)
HM 1375 Buddhism Avijjanusaya (Pali)
HM 3829 Buddhism Avijjapariyutthana (Pali)
HM 4470 Buddhism Avijjayoga (Pali)
HM 3145 Buddhism Avijjogha (Pali)
HM 4572 Buddhism Avikkhepa (Pali)
HM 0449 Buddhism Avisahara (Pali)
HM 0715 Buddhism Avisahatamanasata (Pali)
HM 3374 Buddhism Avitthanata (Pali)
HM 0436 Buddhism Avupasama (Pali)
HM 4565 Buddhism Avyapajja (Pali)
HM 2098 Buddhism Awareness consciousness-group conscious existence
HM 2556 Buddhism Awareness consciousness-group conscious existence - senses mind
HM 2108 Buddhism Awareness corporeality-group conscious existence
HM 4983 Buddhism Awareness feeling-group conscious existence
HH 3321 Buddhism Awareness inter-dependency conscious existence phenomena
HM 2050 Buddhism Awareness mental-formation group conscious existence
HM 4143 Buddhism Awareness perception-group conscious existence
HM 2314 Buddhism Bag-med-pa (Tibetan)
HM 3220 Buddhism Bag-yod-pa (Tibetan)
HM 2522 Buddhism Bahirdhashunyata
HM 1864 Buddhism Balya (Pali)
HM 0101 Buddhism Bar-do (Tibetan)
HM 2091 Buddhism Barbarian-state awareness
HM 0698 Buddhism Bardo consciousness
HM 2098 Buddhism Bare cognition
HM 4651 Buddhism Bare perception
HM 1736 Buddhism Base mind
HM 3209 Buddhism Basic faith
HM 2245 Buddhism Bden-pa-bzhi (Tibetan)
HM 5909 Buddhism Becoming
HM 0552 Buddhism Being ashamed acquisition sinful unwholesome dharmas
HM 3828 Buddhism Being ashamed what one ought be ashamed
HM 4514 Buddhism Being conscious
HM 0836 Buddhism Being contacted
HM 1561 Buddhism Being doubt two alternatives
HM 0872 Buddhism Being good quality
HM 1442 Buddhism Being greedy
HM 1683 Buddhism Being infatuated
HM 1412 Buddhism Being intended
HM 1965 Buddhism Being unobservant
HM 3716 Buddhism Being workable
HM 2933 Buddhism Belief
HM 2719 Buddhism Belief-arising subtle contemplation
HM 2264 Buddhism Belligerence
HM 2653 Buddhism Beyond-the-third-realm consciousness
HM 0400 Buddhism Bhantattam cittassa (Pali)
HM 2051 Buddhism Bhavagra
HH 0551 Buddhism Bhavana
HM 3323 Buddhism Bhavanga-mano
HM 2989 Buddhism Bhavashunyata
HM 2155 Buddhism Bhumi
HM 1623 Buddhism Bhuri (Pali)
HM 3142 Buddhism Black rope hell
HM 0747 Buddhism Bliss
HM 5088 Buddhism Bliss due insight
HH 4019 Buddhism Bodhi
HH 5645 Buddhism Bodhi-marga
HM 2793 Buddhism Bodhi-pakkhiya dhamma
HM 2793 Buddhism Bodhi-pakshika dharma
HM 2220 Buddhism Bodhicitta awareness
HH 1380 Buddhism Bodhisattva
HM 2562 Buddhism Body consciousness
HM 2639 Buddhism Bon-practitioner state awareness
HM 2663 Buddhism Bon-wisdom awareness
HH 3534 Buddhism Brahmavihara-niddesa
HM 1623 Buddhism Breadth wisdom
HM 3237 Buddhism Bringing-to-mind
HM 3053 Buddhism Brjed-nges-pa (Tibetan)
HM 2389 Buddhism Brtson-'grus (Tibetan)
HM 2586 Buddhism Bsam gtan bzhi pa
HM 2450 Buddhism Bsam-gtan-dang-po
HM 2038 Buddhism Bsam-gtan-gnyis-pa
HM 2062 Buddhism Bsam-gtan-gsum-pa (Tibetan)
HM 2693 Buddhism Bsam gtan (Tibetan)
HM 7769 Buddhism Btang-snyoms (Tibetan)
HM 2735 Buddhism Buddha-consciousness
HM 3019 Buddhism Buddha emanation body
HM 3113 Buddhism Buddha enjoyment body
HM 3019 Buddhism Buddha form body
HM 3113 Buddhism Buddha form body
HM 3263 Buddhism Buddha illumination
HM 2039 Buddhism Buddha nature body
HM 2834 Buddhism Buddha truth body
HM 2039 Buddhism Buddha truth body
HM 2834 Buddhism Buddha wisdom body
HM 3514 Buddhism Buddhahood
HH 0551 Buddhism Buddhist meditation
HM 6033 Buddhism Buoyancy citta (Pali)
HM 7636 Buddhism Buoyancy kaya (Pali)
HM 7636 Buddhism Buoyancy mental factors
HM 6033 Buddhism Buoyancy mind
HH 4019 Buddhism Byan-chub (Tibetan)
HM 2769 Buddhism Byang-chub-sems-dpa (Tibetan)
HM 1111 Buddhism Byapada
HM 1591 Buddhism Byapajjana (Pali)
HM 3210 Buddhism Byapajjitatta (Pali)
HM 1111 Buddhism Byapatti (Pali)
HM 1271 Buddhism Cakkhuvinnanadhatusamphassaja sanna (Pali)
HM 0972 Buddhism Cakkhuvinnanadhatusamphassajam cetasikam neva satam nasatam (Pali)
HM 2058 Buddhism Cakravartin
HM 2147 Buddhism Calm abiding
HM 0634 Buddhism Calm stability
HM 1761 Buddhism Calming down
HM 0226 Buddhism Calming mind
HM 0945 Buddhism Candikka (Pali)
HM 3808 Buddhism Capacity easy transformation
HM 1781 Buddhism Catuhi vaciduccaritehi arati virati pativirati veramani (Pali)
HH 1099 Buddhism Causally continuous doctrine being
HM 1591 Buddhism Causing harm
HM 2330 Buddhism Cessation
HM 2172 Buddhism Cessation-awareness
HM 5676 Buddhism Cessation hatred
HH 2119 Buddhism Cessation ill
HH 2119 Buddhism Cessation suffering
HM 3310 Buddhism Cessations
HM 2589 Buddhism Cetana (Pali)
HM 0805 Buddhism Cetasikam dukkham (Pali)
HM 1862 Buddhism Cetasikam sukham (Pali)
HM 1263 Buddhism Cetasiko viriyarambho (Pali)
HM 4425 Buddhism Cetaso vikkhepo (Pali)
HM 1412 Buddhism Cetayitattam (Pali)
HM 5587 Buddhism Ceto-samadhi
HM 0709 Buddhism Ceto samatha
HM 0831 Buddhism Ceto vimutti
HM 1643 Buddhism Cetosamphassaja addukkhamasukha vedana (Pali)
HM 1090 Buddhism Cetosamphassaja asata vedana (Pali)
HM 4076 Buddhism Cetosamphassaja dukkha vedana (Pali)
HM 0309 Buddhism Cetosamphassaja sata vedana (Pali)
HM 1352 Buddhism Cetosamphassajam adukkhamasukham vedayitam (Pali)
HM 1031 Buddhism Cetosamphassajam dukkam vedayitam (Pali)
HM 0901 Buddhism Cetosamphassajamasatam vedayitam (Pali)
HM 2605 Buddhism 'Chab-pa (Tibetan)
HM 2074 Buddhism Chakshurvinjana
HH 0910 Buddhism Changeable mental factors
HH 0470 Buddhism Character defects
HH 1234 Buddhism Charity
HM 2586 Buddhism Chaturthadhyana
HM 2245 Buddhism Chatvari satvani
HM 2589 Buddhism Chetana
HM 2578 Buddhism Chhanda
HM 2335 Buddhism Chikhai-bardo (Tibetan)
HM 1864 Buddhism Childishness
HM 2217 Buddhism Chittasthapana
HM 2407 Buddhism Chonyid bardo (Tibetan)
HM 2407 Buddhism Chos-nid
HM 0945 Buddhism Churlishness
HM 3966 Buddhism Cinta (Pali)
HM 2730 Buddhism Citta
HM 3529 Buddhism Citta
HH 5973 Buddhism Citta-bhavana
HM 1810 Buddhism Citta-pagunnata (Pali)
HM 6149 Buddhism Citta visuddhi
HM 1584 Buddhism Cittakammannata (Pali)
HM 1830 Buddhism Cittalahuta (Pali)
HM 1805 Buddhism Cittamuduta (Pali)
HM 1490 Buddhism Cittapassaddhi (Pali)
HM 3471 Buddhism Cittasa avatthiti (Pali)
HM 6663 Buddhism Cittassa ekaggata
HM 0223 Buddhism Cittassa manovilekha (Pali)
HM 1066 Buddhism Cittassa santhiti (Pali)
HM 0051 Buddhism Cittassa thiti (Pali)
HM 0975 Buddhism Cittassa uddhacca (Pali)
HM 6663 Buddhism Cittassekaggata (Pali) One-directedness thought
HM 0253 Buddhism Cittujukata (Pali)
HM 2157 Buddhism Close setting mind
HM 1389 Buddhism Cognition
HM 1271 Buddhism Cognition born contact element eye-consciousness
HM 1501 Buddhism Cognition born contact element mind-consciousness
HM 3419 Buddhism Cognizing
HM 2040 Buddhism Cold-hells awareness
HM 4999 Buddhism Collectedness mind
HM 0311 Buddhism Collectedness mind
HM 6663 Buddhism Collectedness moral thought
HH 3246 Buddhism Colour kasina
HM 5634 Buddhism Compassion
HM 0513 Buddhism Compassion
HM 0634 Buddhism Complete awareness
HM 1956 Buddhism Complete delusion
HM 3411 Buddhism Complete tranquillization
HM 1528 Buddhism Completely tranquillized
HM 4605 Buddhism Composedness aggregate consciousness
HM 1377 Buddhism Composedness aggregate sensation, aggregate cognition, aggregate synergies
HM 1867 Buddhism Composedness body
HM 1490 Buddhism Composedness thought
HM 1738 Buddhism Composure
HM 3490 Buddhism Comprehension known
HM 4348 Buddhism Comprehension rejection
HM 4552 Buddhism Comprehension scrutiny
HM 2605 Buddhism Concealment
HM 2440 Buddhism Concentration
HM 6663 Buddhism Concentration
HM 4191 Buddhism Concentration accompanied bliss
HM 1454 Buddhism Concentration accompanied bliss
HM 4191 Buddhism Concentration accompanied ease
HM 1454 Buddhism Concentration accompanied ease
HM 5008 Buddhism Concentration accompanied equanimity
HM 8021 Buddhism Concentration accompanied equanimity
HM 5008 Buddhism Concentration accompanied even-mindedness
HM 8021 Buddhism Concentration accompanied even-mindedness
HM 4433 Buddhism Concentration accompanied happiness
HM 5008 Buddhism Concentration accompanied indifference
HM 8021 Buddhism Concentration accompanied indifference
HM 4433 Buddhism Concentration accompanied rapture
HM 4908 Buddhism Concentration applied thought sustained thought
HM 5199 Buddhism Concentration applied thought sustained thought
HM 7543 Buddhism Concentration applied thought sustained thought
HM 5992 Buddhism Concentration difficult progress sluggish direct-knowledge
HM 7859 Buddhism Concentration difficult progress swift direct-knowledge
HM 8365 Buddhism Concentration due energy
HM 7434 Buddhism Concentration due inquiry
HM 7961 Buddhism Concentration due natural purity consciousness
HM 4969 Buddhism Concentration due zeal
HM 4977 Buddhism Concentration easy progress quick intuition
HM 1010 Buddhism Concentration easy progress sluggish direct-knowledge
HM 1010 Buddhism Concentration easy progress sluggish intuition
HM 4977 Buddhism Concentration easy progress swift direct-knowledge
HM 5143 Buddhism Concentration fifth jhana five
HM 0297 Buddhism Concentration first jhana five
HM 4456 Buddhism Concentration first jhana four
HM 4882 Buddhism Concentration fourth jhana five
HM 7202 Buddhism Concentration fourth jhana four
HM 5767 Buddhism Concentration happiness
HM 0730 Buddhism Concentration happiness
HM 7859 Buddhism Concentration painful progress quick intuition
HM 5992 Buddhism Concentration painful progress sluggish intuition
HM 8002 Buddhism Concentration partaking diminution
HM 7363 Buddhism Concentration partaking distinction
HM 5930 Buddhism Concentration partaking penetration
HM 6956 Buddhism Concentration partaking stability
HM 6956 Buddhism Concentration partaking stagnation
HM 8002 Buddhism Concentration partaking worsening
HM 5767 Buddhism Concentration rapture
HM 0730 Buddhism Concentration rapture
HM 4575 Buddhism Concentration second jhana five
HM 4380 Buddhism Concentration second jhana four
HM 8532 Buddhism Concentration third jhana five
HM 2284 Buddhism Concentration third jhana four
HM 2220 Buddhism Conception
HM 0713 Buddhism Conceptual cognition
HM 7655 Buddhism Conditions consciousness
HM 1998 Buddhism Confidence
HM 5403 Buddhism Conformity
HM 1590 Buddhism Conscience
HM 4394 Buddhism Consciencelessness
HM 3220 Buddhism Conscientiousness
HM 1590 Buddhism Conscientiousness
HM 6652 Buddhism Conscious duration
HM 3617 Buddhism Consciousness
HM 5321 Buddhism Consciousness
HM 2098 Buddhism Consciousness aggregate
HH 4000 Buddhism Consciousness ascension stages game
HM 8273 Buddhism Consciousness-born materiality
HM 2200 Buddhism Consciousness Buddhism
HM 2200 Buddhism ; Consciousness (Buddhism)
HM 2027 Buddhism Consciousness nothingness absorption
HM 5588 Buddhism Consciousness reinforced concentration
HM 4396 Buddhism Consciousness reinforced energy
HM 7902 Buddhism Consciousness reinforced faith
HM 5499 Buddhism Consciousness reinforced mindfulness
HM 5901 Buddhism Consciousness reinforced understanding
HM 8621 Buddhism Consciousness rid barriers
HM 2177 Buddhism Consciousness states cyclic existence
HM 2730 Buddhism Consciousness storehouse
HM 1508 Buddhism Conservation
HM 1066 Buddhism Constancy thought
HM 2708 Buddhism Contact
HM 1380 Buddhism Contact
HM 0137 Buddhism Contemplation
HM 3630 Buddhism Contemplations
HM 4563 Buddhism Contentment
HM 2468 Buddhism Continuance
HM 2241 Buddhism Continuous setting mind
HM 7209 Buddhism Contrition
HM 0439 Buddhism Correct action

HM 0687 Buddhism Correct action
HM 1735 Buddhism Correct concentration
HM 1451 Buddhism Correct construing
HM 1657 Buddhism Correct endeavour
HM 0549 Buddhism Correct livelihood
HM 1763 Buddhism Correct livelihood
HM 0180 Buddhism Correct recollection
HM 4608 Buddhism Correct speech
HM 1821 Buddhism Correct speech
HM 3252 Buddhism Correct view
HM 9722 Buddhism Corruption insight
HM 3563 Buddhism Covetousness
HM 2659 Buddhism Critical-analytical subtle contemplation
HM 3142 Buddhism Crushing-hells awareness
HH 8231 Buddhism Cultivation concentration
HH 0680 Buddhism Cultivation experiential insight
HM 2054 Buddhism Current views opinions
HM 3209 Buddhism Dad-pa (Tibetan)
HM 2181 Buddhism Damana
HH 1234 Buddhism Dana
HM 2931 Buddhism Darsana
HH 0446 Buddhism Dasakusalakamma
HM 8003 Buddhism De-kho-na (Tibetan)
HM 3932 Buddhism Death
HM 2335 Buddhism Death-bardo clear-light awareness
HM 2407 Buddhism Death-bardo heaven-reality awareness
HM 2431 Buddhism Death-bardo rebirth-seeking awareness
HM 3932 Buddhism Decease
HM 3246 Buddhism Deceit
HM 7632 Buddhism Deciding
HM 1689 Buddhism Deep penetration memory
HH 2718 Buddhism Defining mentality-materiality
HM 4065 Buddhism Delight
HM 1398 Buddhism Delightful
HM 2546 Buddhism Delightful-emanation-heaven consciousness
HM 4246 Buddhism Delightfulness
HH 2718 Buddhism Delimitation formations
HM 3196 Buddhism Delusion
HM 0184 Buddhism Delusion
HM 0918 Buddhism Delusion
HM 1119 Buddhism Delusion consciousness
HM 3152 Buddhism Delusion personal self
HM 1439 Buddhism Delusion unwholesome root
HM 2055 Buddhism Demon-island awareness
HM 0509 Buddhism Dependence ceremonies
HM 3949 Buddhism Depression thought
HM 4487 Buddhism Desire
HM 4969 Buddhism Desire-concentration
HM 6144 Buddhism Desire due insight
HM 1261 Buddhism Desire formless life
HM 3583 Buddhism Desire life form
HM 2733 Buddhism Desire-realm consciousness
HM 4487 Buddhism Desire-to-do
HM 1098 Buddhism Destroying cause
HM 2128 Buddhism Detachment
HM 7632 Buddhism Determining
HH 0170 Buddhism Determining mental factors
HM 5982 Buddhism Deva hearing
HM 0748 Buddhism Deva sight
HH 5973 Buddhism Development mind
HH 0578 Buddhism Dge-ba
HM 4726 Buddhism Dhamma-patisambhida (Pali)
HH 1022 Buddhism Dhammadesana
HM 2551 Buddhism Dhammas (Pali)
HM 1715 Buddhism Dhammavicaya (Pali)
HM 1423 Buddhism Dharanata (Pali)
HM 2397 Buddhism Dharma-kaya
HM 2421 Buddhism Dharma-negha
HM 2208 Buddhism Dharmaloka-labdha-samadhi
HM 2551 Buddhism Dhatu
HM 1974 Buddhism Dhiti (Pali)
HM 1081 Buddhism Dhurasampaggaha (Pali)
HM 2693 Buddhism Dhyana
HM 0137 Buddhism Dhyana
HH 0309 Buddhism Diamond vehicle Buddhism
HH 0309 Buddhism ; Diamond vehicle (Buddhism)
HM 0249 Buddhism Differentiation
HM 5532 Buddhism Direct intuition
HM 4297 Buddhism Direct knowledge
HM 5232 Buddhism Direct knowledge
HM 5982 Buddhism Direct knowledge
HM 7672 Buddhism Direct knowledge
HH 5652 Buddhism Direct knowledge
HM 0748 Buddhism Direct knowledge
HM 2033 Buddhism Direct non-conceptual cognition emptiness
HM 3630 Buddhism Discernment
HM 1143 Buddhism Discernment
HM 2716 Buddhism Discipleship-application awareness
HM 2192 Buddhism Discipleship-karma awareness
HM 2240 Buddhism Discipleship-vision awareness
HH 1376 Buddhism Discipline
HM 2181 Buddhism Disciplining mental abiding states
HM 2399 Buddhism Discrimination
HM 1799 Buddhism Disinterestedness
HM 2128 Buddhism Disinterestedness
HM 3752 Buddhism Disinterestedness wholesome root
HM 2098 Buddhism Dispositions consciousness
HM 3093 Buddhism Dissimulation
HM 8092 Buddhism Dissociated mind
HM 2668 Buddhism Distorted cognition
HM 1407 Buddhism Distortion views
HM 3154 Buddhism Distraction
HM 1191 Buddhism Ditthi (Pali)
HH 2718 Buddhism Ditthi-visuddhi-niddesa (Pali)
HM 1613 Buddhism Ditthigahana (Pali)
HM 2054 Buddhism Ditthigata (Pali)
HM 0974 Buddhism Ditthijjukamma
HM 4548 Buddhism Ditthikantara (Pali)
HM 1511 Buddhism Ditthisannojana (Pali)
HM 2722 Buddhism Ditthivipphandita (Pali)
HM 1407 Buddhism Ditthivisukayika (Pali)
HH 3534 Buddhism Divine abidings meditation subjects
HM 2007 Buddhism Divine-animal-hell awareness
HH 3534 Buddhism Divine states
HM 2110 Buddhism Dngos gzhii snyoms jug (Tibetan)
HM 2433 Buddhism 'Dod-chags
HM 2733 Buddhism Dod-khams (Tibetan)
HM 4502 Buddhism Dosa (Pali)
HM 0607 Buddhism Dosa (Pali)
HM 5676 Buddhism Dosakkhaya (Pali)
HM 5676 Buddhism Dosaksaya
HM 3306 Buddhism Doubt
HM 4347 Buddhism Doubt
HM 4467 Buddhism Doubt
HM 1294 Buddhism Doubt efficacy good life
HM 0177 Buddhism Doubting
HM 5089 Buddhism Dpyod-pa (Tibetan)
HM 2659 Buddhism Dpyod-pa-yid-byed (Tibetan)
HM 2847 Buddhism Dran-pa (Tibetan)
HM 8112 Buddhism Dread blame
HM 2371 Buddhism Dream-state bardo awareness
HH 4087 Buddhism Dream yoga
HM 2996 Buddhism Drishti
HM 2399 Buddhism 'Du-shes (Tibetan)
HM 3934 Buddhism Dubiety
HM 2574 Buddhism Duhkha
HH 0523 Buddhism Duhkha
HM 2574 Buddhism Dukkha (Pali)
HH 0523 Buddhism Dukkha (Pali)
HM 3556 Buddhism Dukkhindriya (Pali)
HM 2181 Buddhism Dul-bar-byed-pa (Tibetan)
HM 3196 Buddhism Dullness
HM 1857 Buddhism Dummajjha (Pali)
HM 2578 Buddhism 'Dun pa (Tibetan)
HM 2361 Buddhism Duramgama
HM 1969 Buddhism Duration formless dharmas
HM 4323 Buddhism Dussana (Pali)
HM 1967 Buddhism Dussitattam (Pali)
HM 1286 Buddhism Dvedhapatha (Pali)
HM 1561 Buddhism Dvelhaka (Pali)
HM 2038 Buddhism Dvitiyadhyana
HM 5643 Buddhism Dwelling equanimity
HM 4298 Buddhism Dwelling first jhana
HM 6553 Buddhism Dwelling fivefold jhana
HM 8087 Buddhism Dwelling fourth jhana
HM 7121 Buddhism Dwelling second jhana
HM 6553 Buddhism Dwelling second jhana fivefold system
HM 5643 Buddhism Dwelling third jhana
HM 2127 Buddhism Dzam-bu-gling (Tibetan)
HM 2169 Buddhism Ear consciousness
HH 3246 Buddhism Earth kasina
HM 2543 Buddhism Eastern-continent awareness noble figures
HM 0311 Buddhism Ecstatic concentration
HM 7703 Buddhism Effacement
HM 2389 Buddhism Effort
HH 0241 Buddhism Egolessness
HM 2635 Buddhism Eight liberations
HM 2679 Buddhism Eight siddhis
HM 2775 Buddhism Eighteen unshared attributes Buddha
HM 2339 Buddhism Eightfold way
HM 2816 Buddhism Eighth scriptural bodhisattva awareness
HM 2301 Buddhism Eighth vajra-master awareness
HM 2753 Buddhism Ekotikarana
HM 0425 Buddhism Elation
HM 1834 Buddhism Element mind-consciousness
HM 2340 Buddhism Emanation-body awareness
HM 8241 Buddhism Emancipated mind
HM 2210 Buddhism Embarrassment
HM 3504 Buddhism Emotional aversions
HM 3301 Buddhism Emotional desires
HM 2193 Buddhism Emptiness
HM 3032 Buddhism Emptiness cyclic existence
HM 2946 Buddhism Emptiness definitions
HM 2671 Buddhism Emptiness five senses
HM 3072 Buddhism Emptiness indestructible Mahayana
HM 2448 Buddhism Emptiness inherent existence non-things
HM 2794 Buddhism Emptiness loci senses
HM 3205 Buddhism Emptiness nature
HM 2899 Buddhism Emptiness nature phenomena
HM 2089 Buddhism Emptiness non-products
HM 2522 Buddhism Emptiness objects sense mental consciousness
HM 3000 Buddhism Emptiness phenomena
HM 2414 Buddhism Emptiness products
HM 2995 Buddhism Emptiness ten directions
HM 2989 Buddhism Emptiness things
HM 3208 Buddhism Emptiness things
HM 2294 Buddhism Emptiness ultimate nirvana
HM 2501 Buddhism Emptiness unapprehendable
HM 0052 Buddhism Emptiness what is free permanence annihilation
HM 2944 Buddhism Emptinesses paths view Buddhism
HM 2944 Buddhism ; Emptinesses paths view (Buddhism)
HM 0229 Buddhism Endeavour
HM 1333 Buddhism Energy
HM 0450 Buddhism Engaku (Japanese)
HM 2873 Buddhism Enjoyment-body awareness
HM 6336 Buddhism Enlightenment factors
HM 4239 Buddhism Enmity
HM 2638 Buddhism Envy (Pali)
HM 7002 Buddhism Equanimity
HM 7769 Buddhism Equanimity
HM 8566 Buddhism Equanimity due insight
HM 1912 Buddhism Erroneous way
HM 1499 Buddhism Erudition
HM 6544 Buddhism Establishment due insight
HM 2997 Buddhism Establishments mindfulness
HM 3394 Buddhism Evasion
HM 7769 Buddhism Even-mindedness
HH 0320 Buddhism Ever-recurring mental factors
HM 4036 Buddhism Exalted concentration
HM 7324 Buddhism Examining
HM 2534 Buddhism Excitement
HM 1469 Buddhism Excitement
HM 0975 Buddhism Excitement thought
HM 1496 Buddhism Exercise
HM 3469 Buddhism Exertion
HM 5986 Buddhism Exertion due insight
HM 1632 Buddhism Experiential insight
HM 3390 Buddhism Expertise
HM 6124 Buddhism Extinction illusion
HH 1809 Buddhism Extraordinary knowledge
HM 2074 Buddhism Eye consciousness
HM 3454 Buddhism Faculty concentration
HM 1470 Buddhism Faculty energy
HM 1829 Buddhism Faculty energy
HM 0066 Buddhism Faculty faith
HM 3803 Buddhism Faculty indifference
HM 0261 Buddhism Faculty life
HM 3404 Buddhism Faculty living
HM 1649 Buddhism Faculty mental gladness
HM 1369 Buddhism Faculty mind
HM 4248 Buddhism Faculty recollection
HM 0047 Buddhism Faculty recollection
HM 3556 Buddhism Faculty suffering
HM 3233 Buddhism Faculty wisdom
HM 1840 Buddhism Faculty wisdom
HM 2933 Buddhism Faith
HM 0649 Buddhism Fearlessness blame
HM 2270 Buddhism Feeling
HM 4983 Buddhism Feeling aggregate
HM 1513 Buddhism Feeling greed
HM 4363 Buddhism Feeling infatuation
HM 1208 Buddhism Feeling remorse acquisition sinful unwholesome dharmas
HM 1757 Buddhism Feeling remorse what one ought be remorseful
HM 1545 Buddhism Felicity
HM 4470 Buddhism Fetters ignorance
HM 1511 Buddhism Fetters views
HM 0460 Buddhism Fewness wishes
HM 2407 Buddhism Fifth antarabhava consciousness
HM 2909 Buddhism Fifth scriptural bodhisattva awareness
HM 2763 Buddhism Fifth vajra-master awareness
HM 2683 Buddhism Final-training subtle contemplation
HM 2849 Buddhism Final vajra-master awareness
HM 2536 Buddhism Fine-material sphere
HM 4265 Buddhism Fine-material-sphere concentration
HM 1974 Buddhism Firmness
HM 2110 Buddhism First absorption immaterial sphere
HM 2347 Buddhism First antarabhava consciousness (Tibetan)
HM 2450 Buddhism First form-realm concentration
HM 2110 Buddhism First formless attainment
HM 2155 Buddhism First scriptural bodhisattva awareness
HM 2450 Buddhism First trance fine-material sphere
HM 2187 Buddhism First vajra-master stage
HM 1810 Buddhism Fitness consciousness
HM 1455 Buddhism Fitness mental body
HM 1810 Buddhism Fitness thought
HH 3321 Buddhism Five aggregates
HM 2841 Buddhism Five clairvoyances
HH 6773 Buddhism Five covers
HM 8336 Buddhism Five-door adverting
HH 1010 Buddhism Five eyes
HM 2948 Buddhism Five forces
HH 6773 Buddhism Five hindrances
HH 1010 Buddhism Five levels understanding
HM 3238 Buddhism Five powers
HM 0747 Buddhism Fivefold happiness
HM 1617 Buddhism Fixation
HM 2191 Buddhism Flaming enlightenment
HM 1796 Buddhism Flexibility aggregate consciousness
HM 3397 Buddhism Flexibility aggregate sensation, aggregate cognition, aggregate synergies
HM 3370 Buddhism Flexibility body
HM 1805 Buddhism Flexibility thought
HM 3145 Buddhism Flood ignorance
HM 0487 Buddhism Focusing
HM 0847 Buddhism Foolishness
HM 3053 Buddhism Forgetfulness
HM 2142 Buddhism Form-realm awareness
HM 4265 Buddhism Form-realm concentration
HM 2257 Buddhism Form-realm consciousness
HH 3226 Buddhism Formation demerit
HH 5543 Buddhism Formation imperturbable
HH 5122 Buddhism Formation merit
HM 2050 Buddhism Formations aggregate
HM 3144 Buddhism Formless-realm awareness
HM 0696 Buddhism Formless-realm concentration
HM 2281 Buddhism Formless-realm consciousness
HH 3198 Buddhism Formless realms
HM 2886 Buddhism Forms enlightenment
HM 0485 Buddhism Four analyses
HM 0485 Buddhism Four discriminations
HM 2064 Buddhism Four doors retention
HM 1252 Buddhism Four evil paths
HM 3044 Buddhism Four fearlessnesses
HM 2082 Buddhism Four-great-kings-heaven awareness

Buddhism

HM 2507 Buddhism Four immeasurables
HM 4026 Buddhism Four noble paths
HM 0523 Buddhism Four noble truths
HM 3216 Buddhism Four sciences
HM 2051 Buddhism Fourth absorption immaterial sphere
HM 2335 Buddhism Fourth antarabhava consciousness
HM 2586 Buddhism Fourth form-realm concentration
HM 2051 Buddhism Fourth formless attainment
HM 2191 Buddhism Fourth scriptural bodhisattva awareness
HM 2586 Buddhism Fourth trance fine-material sphere
HM 2251 Buddhism Fourth vajra-master awareness
HM 2364 Buddhism Fragrance awareness
HM 2743 Buddhism Full-isolation subtle contemplation
HM 2729 Buddhism Full pacification mental abiding states
HM 4348 Buddhism Full understanding abandoning
HM 4552 Buddhism Full understanding investigation
HM 3490 Buddhism Full understanding known
HM 3093 Buddhism G'yo (Tibetan)
HM 0700 Buddhism Gaha (Pali)
HH 1765 Buddhism General knowledge
HM 2364 Buddhism Ghranavijnana
HH 1234 Buddhism Giving
HM 5224 Buddhism Gladness
HM 7921 Buddhism Gnyid (Tibetan)
HH 6773 Buddhism Gogai (Zen)
HM 1691 Buddhism Going
HM 0634 Buddhism Gold pill (Taoism)
HM 3424 Buddhism Good quality
HM 1741 Buddhism Good quality aggregate consciousness
HM 0349 Buddhism Good quality aggregate sensation, aggregate cognition, aggregate synergies
HM 4637 Buddhism Gotra-bhu-nanam (Pali)
HM 0881 Buddhism Grasp
HM 1751 Buddhism Grasping inverted views
HM 0700 Buddhism Grasping view
HM 2180 Buddhism Great awakening
HM 2118 Buddhism Great-black-lord awareness
HM 2807 Buddhism Great compassion
HM 3630 Buddhism Great insights
HM 2929 Buddhism Great love
HM 2656 Buddhism Great-path tantra awareness
HH 0900 Buddhism Great vehicle Buddhism
HH 0900 Buddhism ; Great vehicle (Buddhism)
HM 3048 Buddhism Great-vehicle higher-path awareness
HM 2268 Buddhism Great-vehicle lower-path awareness
HM 2160 Buddhism Great-vehicle middle-path awareness
HM 3283 Buddhism Greed
HM 0150 Buddhism Greed
HM 0802 Buddhism Greed
HM 1954 Buddhism Greed unwholesome root
HM 7120 Buddhism Grief
HM 2695 Buddhism Gti-mug-med-pa (Tibetan)
HM 3863 Buddhism Gtum-mo (Tibetan)
HM 2699 Buddhism Guhya-samaja urgyan awareness
HM 7209 Buddhism 'Gyod-pa (Tibetan)
HH 0910 Buddhism Gzhan-'gyur (Tibetan)
HM 2257 Buddhism Gzugs-khams (Tibetan)
HM 2669 Buddhism Gzugs-med-kyi-snyoms-jug (Tibetan)
HM 2281 Buddhism Gzugs-sku (Tibetan)
HM 2866 Buddhism Happiness
HM 0747 Buddhism Happiness
HM 3210 Buddhism Harmfulness
HM 4565 Buddhism Harmless
HH 0603 Buddhism Harmonies enlightenment
HM 1421 Buddhism Hasa (Pali)
HM 4502 Buddhism Hate
HM 4323 Buddhism Hating
HM 4502 Buddhism Hatred
HM 0607 Buddhism Hatred
HM 2528 Buddhism Haughtiness
HM 3158 Buddhism Having faith
HM 2169 Buddhism Hearing awareness
HM 1973 Buddhism Heaven
HM 2130 Buddhism Heaven-without-fighting consciousness
HM 2010 Buddhism Heavenly-highway awareness
HM 4297 Buddhism Higher knowledge
HM 5232 Buddhism Higher knowledge
HM 5982 Buddhism Higher knowledge
HM 7672 Buddhism Higher knowledge
HH 5652 Buddhism Higher knowledge
HM 0748 Buddhism Higher knowledge
HM 3529 Buddhism Higher mind
HM 2864 Buddhism Highest yoga tantra: anuttarayoga
HH 0845 Buddhism Hinayana Buddhism
HH 0845 Buddhism ; Hinayana (Buddhism)
HM 2115 Buddhism Hindu-states awareness
HM 2723 Buddhism Hindu-wisdom-holder awareness
HM 2881 Buddhism Hiri (Pali)
HM 1590 Buddhism Hiri (Pali)
HM 0352 Buddhism Hiribala (Pali)
HM 3828 Buddhism Hiriyati hiriyitabbena (Pali)
HM 0552 Buddhism Hiriyati papakanam akusalanam dhammanam samapattiya (Pali)
HM 3690 Buddhism Hod-gsal (Tibetan)
HM 1992 Buddhism Holding paramount one's view
HM 0719 Buddhism Hostility
HM 2100 Buddhism Howling-hells awareness
HM 5122 Buddhism Hpho-ba (Tibetan)
HM 2881 Buddhism Hri
HH 0650 Buddhism Human development
HM 2112 Buddhism Hungry-ghosts-hell awareness
HM 2337 Buddhism Hyper-bliss-realm awareness
HM 5498 Buddhism Idamatthita
HH 5652 Buddhism Iddhi (Pali)
HM 3196 Buddhism Ignorance
HM 3035 Buddhism Ignorance dependent origination formula
HM 3493 Buddhism Ignorance nature essential self
HM 6208 Buddhism Illuminated consciousness
HM 6208 Buddhism Illuminated mind
HM 2215 Buddhism Illumination
HM 4337 Buddhism Illumination due insight
HM 0696 Buddhism Immaterial-sphere concentration
HH 3198 Buddhism Immaterial states meditation subjects
HM 9722 Buddhism Imperfection insight
HM 9021 Buddhism Impermanence
HH 0241 Buddhism Impersonality
HM 7268 Buddhism Impulsion
HM 0427 Buddhism Inability compare
HM 1746 Buddhism Inability demonstrate
HM 3348 Buddhism Inattentive perception
HM 3432 Buddhism Inclination towards view
HM 0870 Buddhism Indecision
HM 3074 Buddhism Indecisive wavering
HM 3745 Buddhism Indecisiveness
HM 6243 Buddhism Independent consciousness
HM 6243 Buddhism Independent mind
HM 4761 Buddhism Indeterminate consciousness fine-material sphere - functional
HM 0594 Buddhism Indeterminate consciousness fine-material sphere - resultant
HM 4761 Buddhism Indeterminate consciousness form-realm - inoperative
HM 0594 Buddhism Indeterminate consciousness form-realm - resultant
HM 0282 Buddhism Indeterminate consciousness formless realm - inoperative
HM 4982 Buddhism Indeterminate consciousness formless realm - resultant
HM 0282 Buddhism Indeterminate consciousness immaterial sphere - functional
HM 4982 Buddhism Indeterminate consciousness immaterial sphere - resultant
HM 3852 Buddhism Indeterminate consciousness sense sphere - functional
HM 3852 Buddhism Indeterminate consciousness sense sphere - inoperative
HM 5721 Buddhism Indeterminate consciousness sense sphere - resultant
HM 5129 Buddhism Indeterminate consciousness supramundane plane - resultant
HM 5129 Buddhism Indeterminate consciousness transcendental plane - resultant
HM 7769 Buddhism Indifference
HM 8566 Buddhism Indifference due insight
HM 7769 Buddhism Indifferent feeling
HM 2195 Buddhism Individual knowledge character
HM 2075 Buddhism Indriya phenomena (Pali)
HM 3378 Buddhism Induced attentiveness
HM 3250 Buddhism Infatuated
HM 5562 Buddhism Inferior concentration
HM 0496 Buddhism Infinite concentration
HM 5214 Buddhism Infinite concentration infinite object
HM 4653 Buddhism Infinite concentration limited object
HM 6521 Buddhism Inflexible mind
HM 1596 Buddhism Inherent existence
HM 6502 Buddhism Insight
HM 2574 Buddhism Insufficiency
HM 5532 Buddhism Intelligible intuition
HM 2589 Buddhism Intention
HM 0705 Buddhism Intention
HM 2052 Buddhism Interminable-hell awareness
HM 7109 Buddhism Introspection
HM 1632 Buddhism Intuition
HM 1344 Buddhism Intuitional insight
HM 7324 Buddhism Investigating
HM 1177 Buddhism Investigation
HM 7434 Buddhism Investigation-concentration
HM 1972 Buddhism Iriyana (Pali)
HM 2638 Buddhism Irshya
HM 2127 Buddhism Jambudvipa
HM 2638 Buddhism Jealousy
HM 1486 Buddhism Jewel wisdom
HM 2275 Buddhism Jewelled-peaks-realm awareness
HM 7193 Buddhism Jhana
HM 4298 Buddhism Jhana happiness bliss born seclusion
HM 2536 Buddhism Jhana (Pali)
HM 0137 Buddhism Jhana (Pali)
HM 2122 Buddhism Jhana (Pali)
HM 4282 Buddhism Jigoku (Japanese)
HM 0853 Buddhism Jihi
HM 2263 Buddhism Jihvavijnana
HM 3379 Buddhism Jivita (Pali)
HM 3404 Buddhism Jivitindriya (Pali)
HM 0261 Buddhism Jivitindriya (Pali)
HM 5321 Buddhism Jna
HM 8737 Buddhism Joy
HM 0747 Buddhism Joy
HM 2167 Buddhism Joy-land awareness
HM 8737 Buddhism Joyful feeling
HM 2022 Buddhism Joyful-heaven awareness
HH 1187 Buddhism Junior stream-winner
HM 2151 Buddhism Kalacakra-tantra shambhala awareness
HM 2211 Buddhism Kamadera-vidyadhara
HM 2733 Buddhism Kamadhatu
HM 2060 Buddhism Kamaloka consciousness
HM 2048 Buddhism Kamma (Pali)
HM 3716 Buddhism Kammannabhava (Pali)
HM 0979 Buddhism Kammannattam (Pali)
HM 4347 Buddhism Kankha (Pali)
HH 1187 Buddhism Kankhavitarana-visuddhi-niddesa (Pali) Lesser stream-enterer
HM 0177 Buddhism Kankhayana (Pali)
HM 3934 Buddhism Kankhayitattam (Pali)
HM 2048 Buddhism Karma awareness
HM 2783 Buddhism Karma-paripurana
HM 0130 Buddhism Karma-result consciousness
HM 5634 Buddhism Karuna
HH 3246 Buddhism Kasina meditation
HM 7209 Buddhism Kaukritya
HM 3163 Buddhism Kausidya
HM 1455 Buddhism Kaya-pagunnata (Pali)
HM 5600 Buddhism Kayaduccaritavirati (Pali)
HM 0739 Buddhism Kayakammannata (Pali)
HM 4395 Buddhism Kayalahuta (Pali)
HM 3370 Buddhism Kayamuduta (Pali)
HM 1867 Buddhism Kayapassaddhi (Pali)
HM 2562 Buddhism Kayavijnana
HM 1424 Buddhism Kayujukata (Pali)
HM 3327 Buddhism Keeping going
HM 2177 Buddhism Khams-gsum (Tibetan)
HH 3321 Buddhism Khandha (Pali)
HH 3321 Buddhism Khandhas
HM 2134 Buddhism 'Khon-'dzin (Tibetan)
HM 2959 Buddhism Khong-khro (Tibetan)
HM 2177 Buddhism Khor-ba
HM 3203 Buddhism Khrel-med-pa (Tibetan)
HM 2210 Buddhism Khrel-yod-pa (Tibetan)
HM 2264 Buddhism Khro-ba (Tibetan)
HM 7345 Buddhism Kilesappahana
HM 2655 Buddhism Knowledge
HH 0538 Buddhism Knowledge
HM 5498 Buddhism Knowledge
HM 6502 Buddhism Knowledge
HH 1971 Buddhism Knowledge
HM 0169 Buddhism Knowledge analysis
HM 7114 Buddhism Knowledge appearance terror
HM 4637 Buddhism Knowledge change lineage
HM 5403 Buddhism Knowledge conformity truth
HM 7297 Buddhism Knowledge contemplation danger
HM 5364 Buddhism Knowledge contemplation dispassion
HM 3385 Buddhism Knowledge contemplation dissolution
HM 0169 Buddhism Knowledge contemplation reflection
HM 3723 Buddhism Knowledge contemplation rise fall
HM 0766 Buddhism Knowledge desire deliverance
HM 0766 Buddhism Knowledge desire release
HM 5982 Buddhism Knowledge divine ear element
HM 0748 Buddhism Knowledge divine sight
HM 5221 Buddhism Knowledge due insight
HM 0424 Buddhism Knowledge equanimity formations
HM 3527 Buddhism Knowledge ideas
HM 0424 Buddhism Knowledge indifference complexes
HM 5232 Buddhism Knowledge others' thoughts
HM 0748 Buddhism Knowledge passing away reappearance beings
HM 7055 Buddhism Knowledge path arahantship
HM 6920 Buddhism Knowledge path non-return
HM 7563 Buddhism Knowledge path once-return
HM 1088 Buddhism Knowledge path stream entry
HM 3527 Buddhism Knowledge penetration
HM 5232 Buddhism Knowledge penetration minds
HM 7114 Buddhism Knowledge presence fear
HM 4297 Buddhism Knowledge recollection past life
HM 4297 Buddhism Knowledge recollection previous existence
HM 5364 Buddhism Knowledge repulsion
HM 7672 Buddhism Knowledge supernormal powers
HM 7297 Buddhism Knowledge tribulation
HM 1554 Buddhism Kosalla (Pali)
HM 2558 Buddhism Kriya-tantra awareness
HM 4535 Buddhism Ksana
HM 2247 Buddhism Kshanti
HM 0581 Buddhism Ku
HM 9101 Buddhism Kukkucca (Pali)
HM 1912 Buddhism Kummagga (Pali)
HH 0320 Buddhism Kun-'gro
HM 2067 Buddhism Kuru
HH 0578 Buddhism Kushula
HM 4228 Buddhism Lack real grasping
HM 3808 Buddhism Lahuparinamata (Pali)
HM 2946 Buddhism Lakshanashunyata
HM 1846 Buddhism Lamp wisdom
HM 2171 Buddhism Las-dang-po-pa-tsam-kyi-yid-byed (Tibetan)
HM 2271 Buddhism Laukikagra-dharma
HH 3398 Buddhism Laws
HM 3163 Buddhism Laziness
HM 2263 Buddhism Lce'i rnam shes pa (Tibetan)
HM 3163 Buddhism Le-lo (Tibetan)
HM 1781 Buddhism Leaving off, abstaining, totally abstaining refraining four deviations speech
HM 1133 Buddhism Leaving off, abstaining, totally abstaining refraining three deviations body
HM 0252 Buddhism Leaving off, abstaining, totally abstaining refraining wrong livelihood
HM 3429 Buddhism Leaving undone
HH 1187 Buddhism Lesser stream enterer
HH 0845 Buddhism Lesser vehicle Buddhism
HH 0845 Buddhism ; Lesser vehicle (Buddhism)
HM 2926 Buddhism Lethargy
HM 2623 Buddhism Lhag mthong manaskara
HM 2207 Buddhism Lhag-mthong (Tibetan)
HM 8241 Buddhism Liberated consciousness
HM 5977 Buddhism Liberation
HM 0201 Buddhism Life
HM 6221 Buddhism Life-continuum
HM 3404 Buddhism Life principle
HM 2347 Buddhism Life-state bardo awareness
HM 3403 Buddhism Lightness aggregate consciousness
HM 1435 Buddhism Lightness aggregate sensation, aggregate cognition, aggregate synergies

HM 4395 Buddhism Lightness body
HM 6033 Buddhism Lightness consciousness
HM 7636 Buddhism Lightness mental body
HM 1830 Buddhism Lightness thought
HM 1073 Buddhism Limited concentration
HM 5547 Buddhism Limited concentration infinite object
HM 1175 Buddhism Limited concentration limited object
HM 5547 Buddhism Limited concentration measureless object
HM 3043 Buddhism Limitless consciousness absorption
HM 2110 Buddhism Limitless space absorption
HH 1183 Buddhism Listening dharma
HM 3379 Buddhism Living
HH 0873 Buddhism Lo-rig (Tibetan)
HM 3283 Buddhism Lobha (Pali)
HM 0802 Buddhism Lobha (Pali)
HM 1954 Buddhism Lobho akusalamulam (Pali)
HM 2072 Buddhism Loka consciousness
HM 2120 Buddhism Lokuttara consciousness
HM 2088 Buddhism Lord-of-death awareness
HH 3443 Buddhism Lotus sutra
HM 7607 Buddhism Love
HH 1234 Buddhism Love
HM 7607 Buddhism Lovingkindness
HM 6342 Buddhism Lower fetters
HM 2996 Buddhism Lta-ba-nyon-mongs-can (Tibetan)
HM 1442 Buddhism Lubbhana (Pali)
HM 1513 Buddhism Lubbhitattam (Pali)
HH 4087 Buddhism Lucid dreams
HM 2562 Buddhism Lus-kyi-rnam-par-shes-pa (Tibetan)
HM 1547 Buddhism Lustre wisdom
HM 2128 Buddhism Ma-chags-pa (Tibetan)
HM 2461 Buddhism Ma-dad-pa (Tibetan)
HM 3196 Buddhism Ma-rig-pa (Tibetan)
HM 2528 Buddhism Mada
HM 0191 Buddhism Maddavata (Pali)
HH 1038 Buddhism Madhyamaka
HH 1010 Buddhism Madhyamaka buddhism
HH 1010 Buddhism ; Madhyamaka (Buddhism)
HM 0796 Buddhism Magga-bhavana
HH 4007 Buddhism Maggamagga-nanadassana-visuddhi-niddesa (Pali)
HM 2679 Buddhism Magical-forces awareness
HM 2639 Buddhism Magical-shamanic state
HM 2118 Buddhism Mahakala
HM 2656 Buddhism Mahamudra
HM 2995 Buddhism Mahashunyata
HH 0900 Buddhism Mahayana Buddhism
HH 0900 Buddhism ; Mahayana (Buddhism)
HM 2232 Buddhism Mahayana climax-application awareness
HM 2208 Buddhism Mahayana heat-application awareness
HM 2271 Buddhism Mahayana highest-teachings-application awareness
HM 2268 Buddhism Mahayana-path
HM 3048 Buddhism Mahayana path
HM 2160 Buddhism Mahayana path
HM 2247 Buddhism Mahayana receptivity-application awareness
HH 0306 Buddhism Maithuna
HM 1111 Buddhism Maliciousness
HM 1155 Buddhism Malleability consciousness
HM 3696 Buddhism Malleability mental body
HM 0251 Buddhism Manasa (Pali)
HM 6644 Buddhism Manasa-pratyaksa
HM 3237 Buddhism Manasi-kara (Pali)
HM 3237 Buddhism Manaskara
HM 2171 Buddhism Manaskaradhikarmika
HM 1736 Buddhism Manayatana (Pali)
HM 1369 Buddhism Manindriya (Pali)
HM 1743 Buddhism Mano (Pali)
HM 2838 Buddhism Manovijnana
HM 1834 Buddhism Manovinnanadhatu (Pali)
HM 1921 Buddhism Manovinnanadhatu samphasajam cetasikam satam (Pali)
HM 0404 Buddhism Manovinnanadhatu samphassaja cetana (Pali)
HM 1501 Buddhism Manovinnanadhatu samphassaja sanna (Pali)
HM 0234 Buddhism Manovinnanadhatusampassa jam cetasikam asatam (Pali)
HM 3314 Buddhism Manovinnanadhatusamphassajam cetasikam neva satam nasatam (Pali)
HM 2838 Buddhism Manovynana
HM 8673 Buddhism Mark calm
HM 3066 Buddhism Masters wisdom ascension stages game
HM 2108 Buddhism Materiality aggregate
HM 2964 Buddhism Matsarya
HM 2110 Buddhism Maulasamapatti
HM 3246 Buddhism Maya
HM 3306 Buddhism Mayoi (Zen)
HM 0227 Buddhism Meanness
HM 0496 Buddhism Measureless concentration
HM 4653 Buddhism Measureless concentration limited object
HM 5214 Buddhism Measureless concentration measureless object
HM 1750 Buddhism Medha (Pali)
HH 2918 Buddhism Meditating kasinas
HM 0137 Buddhism Meditation
HM 2311 Buddhism Meditation-state bardo awareness
HM 2769 Buddhism Meditation way bodhisattvas
HM 2245 Buddhism Meditation way four truths
HM 2161 Buddhism Meditation way hearers
HM 2709 Buddhism Meditation way solitary realizers
HM 2968 Buddhism Meditative stabilizations
HM 2145 Buddhism Meditative states mental abidings
HM 6089 Buddhism Medium concentration
HM 0251 Buddhism Mental
HM 2277 Buddhism Mental abiding equipoise
HM 2838 Buddhism Mental consciousness
HM 3237 Buddhism Mental engagement
HM 0223 Buddhism Mental perturbation thought
HM 6149 Buddhism Mental purity
HM 0709 Buddhism Mental quiescence
HM 0831 Buddhism Mental release
HM 6644 Buddhism Mental sensation
HM 2655 Buddhism Mental wisdom
HM 1061 Buddhism Merriment
HH 1765 Buddhism Metaphysics
HM 3306 Buddhism Mi
HM 0252 Buddhism Miccha ajiva arati virati pativirati veramani (Pali)
HM 1710 Buddhism Micchaditthi (Pali)
HM 0863 Buddhism Micchapatha (Pali)
HM 1802 Buddhism Micchasamadhi (Pali)
HM 3356 Buddhism Micchatta (Pali)
HM 1585 Buddhism Micchavayama (Pali)
HM 7921 Buddhism Middha
HM 0264 Buddhism Middha (Pali)
HM 3026 Buddhism Middle-path tantra awareness
HH 1010 Buddhism Middle way
HH 1038 Buddhism Middle way
HM 6089 Buddhism Middling concentration
HM 2074 Buddhism Mig-gi-rnam-par-shes-pa (Tibetan)
HM 0191 Buddhism Mildness
HM 2659 Buddhism Mimamsamanaskara
HM 1743 Buddhism Mind
HM 7961 Buddhism Mind-concentration
HM 3323 Buddhism Mind consciousness
HM 6173 Buddhism Mind-consciousness element
HM 8336 Buddhism Mind-door adverting
HM 2838 Buddhism Mind-element consciousness
HM 5588 Buddhism Mind upheld concentration
HM 4396 Buddhism Mind upheld energy
HM 7902 Buddhism Mind upheld faith
HM 5499 Buddhism Mind upheld mindfulness
HM 5901 Buddhism Mind upheld understanding
HM 2847 Buddhism Mindfulness
HH 6221 Buddhism Mindfulness
HM 1423 Buddhism Mindfulness
HM 0847 Buddhism Mindlessness
HM 1421 Buddhism Mirth
HM 2964 Buddhism Miserliness
HH 1809 Buddhism Mngon shes pa (Tibetan)
HM 2277 Buddhism Mnyam-par-jog-pa (Tibetan)
HM 6720 Buddhism Modes occurrence consciousness
HM 3035 Buddhism Moha
HM 0184 Buddhism Moha (Pali)
HM 0918 Buddhism Moha (Pali)
HM 6124 Buddhism Mohakkhaya (Pali)
HM 6124 Buddhism Mohaksaya
HM 1439 Buddhism Moho akusalamulam (Pali)
HM 2603 Buddhism Mokshakala
HM 5338 Buddhism Moral consciousness form-realm
HM 4701 Buddhism Moral consciousness formless realm
HM 4930 Buddhism Moral consciousness transcendental plane
HM 2719 Buddhism Mos-pa-las-byung-bai-yid-byed (Tibetan)
HM 2933 Buddhism Mos-pa (Tibetan)
HM 2605 Buddhism Mraksha
HH 0270 Buddhism Mulaklesha
HM 0652 Buddhism Mundane awareness
HM 7234 Buddhism Mundane concentration
HM 0130 Buddhism Mundane resultant consciousness
HM 7628 Buddhism Mundane understanding
HM 3053 Buddhism Mushitasmrtita
HH 2657 Buddhism Mutual possession ten worlds
HM 1023 Buddhism Na kiriyati hiriyitabbena (Pali)
HM 3257 Buddhism Na kiriyati papakanam akusalanam dhammanam samapattiya (Pali)
HM 1673 Buddhism Na ottappati ottappitabbena (Pali)
HM 3353 Buddhism Na ottappati papakanam akusalanam dhammanam samapattiya (Pali)
HM 2031 Buddhism Naga-world awareness
HM 2051 Buddhism Naivasamjnanasamjnayatana
HH 3443 Buddhism Nam-myoho-renge-kyo
HH 0680 Buddhism Nana-dassana
HM 6502 Buddhism Nanadassana
HH 3025 Buddhism Nanadassana-visuddhi-niddesa (Pali)
HH 3246 Buddhism Nature kasina
HM 2145 Buddhism Navakara chittasthiti
HH 0581 Buddhism Neither existence nor non-existence
HM 3314 Buddhism Neither physical ease nor unease born contact element mind-consciousness
HM 0972 Buddhism Neither psychical ease nor unease born contact element eye-consciousness
HM 1352 Buddhism Neither suffering nor pleasure experienced born contact psychical
HM 3390 Buddhism Nepunna (Pali)
HM 1394 Buddhism Nescience
HM 2051 Buddhism Nevasanna-n'asannayatana (Pali)
HM 2823 Buddhism Nga-rgyal
HM 2986 Buddhism Ngo-tsha-med-pa (Tibetan)
HM 2881 Buddhism Ngo-tsha-shes-pa (Tibetan)
HM 2330 Buddhism Nibbana (Pali)
HH 3443 Buddhism Nichiren shoshu buddhism
HH 3443 Buddhism ; Nichiren shoshu (Buddhism)
HM 1496 Buddhism Nikkama (Pali)
HM 2292 Buddhism Ninth scriptural bodhisattva awareness
HM 2325 Buddhism Ninth vajra-master awareness
HM 2340 Buddhism Nirmana-kaya
HM 2172 Buddhism Nirodha
HM 6346 Buddhism Nirodhasamapatti
HM 5026 Buddhism Nirutti-patisambhida (Pali)
HM 2330 Buddhism Nirvana
HM 2546 Buddhism Nirvana-rati
HM 4665 Buddhism Nirvana remainder
HM 2339 Buddhism Noble eightfold path
HH 1099 Buddhism Noble system
HM 3310 Buddhism Non-analytical cessations
HM 2128 Buddhism Non-attachment
HM 2314 Buddhism Non-conscientiousness
HM 1902 Buddhism Non-consideration
HM 2695 Buddhism Non-delusion
HM 4382 Buddhism Non-distraction
HM 2070 Buddhism Non-emanating consciousness
HM 3203 Buddhism Non-embarrassment
HM 2461 Buddhism Non-faith
HM 2847 Buddhism Non-forgetfulness
HM 0451 Buddhism Non-grasping
HM 2128 Buddhism Non-greed
HM 2608 Buddhism Non-harmfulness
HM 2744 Buddhism Non-hate
HM 1557 Buddhism Non-hating
HM 2744 Buddhism Non-hatred
HM 1399 Buddhism Non-hatred
HM 3560 Buddhism Non-hatred wholesome root
HM 2695 Buddhism Non-ignorance
HM 3374 Buddhism Non-inertness
HM 2063 Buddhism Non-introspection
HM 1637 Buddhism Non-penetration
HM 1028 Buddhism Non-rigidity
HM 6644 Buddhism Non-sensuous feeling
HM 2986 Buddhism Non-shame
HM 3014 Buddhism Non-stiffness
HM 2067 Buddhism Northern-continent awareness community
HM 2364 Buddhism Nose consciousness
HM 3257 Buddhism Not ashamed acquisition sinful unwholesome dharmas
HM 1023 Buddhism Not ashamed ought be ashamed
HM 0684 Buddhism Not being covetous
HM 3353 Buddhism Not feeling remorse acquisition sinful unwholesome dharmas
HM 1673 Buddhism Not feeling remorse what one ought be remorseful
HM 3413 Buddhism Not incurring guilt
HM 1688 Buddhism Not infatuated
HM 1510 Buddhism Not-seeing
HH 0241 Buddhism Not-self
HM 3435 Buddhism Not trespassing limit
HM 0403 Buddhism Not understanding
HM 4382 Buddhism Not-wavering
HM 3612 Buddhism Notion one's self entity
HM 2663 Buddhism Nyamsrtsal
HM 2161 Buddhism Nyan thos (Tibetan)
HM 2157 Buddhism Nye bar jog pa (Tibetan)
HM 2729 Buddhism Nye-bar-zhi-bar-byed-pa (Tibetan)
HH 0781 Buddhism Nye-nyon
HM 2623 Buddhism Nyer bsdogs (Tibetan)
HH 0278 Buddhism Observing moral precepts
HM 3829 Buddhism Obsession ignorance
HM 0425 Buddhism Odagya (Pali)
HM 1998 Buddhism Okappana (Pali)
HH 0320 Buddhism Omnipresent mental factors
HM 7672 Buddhism One becoming many
HM 7843 Buddhism One-centred mind
HM 6663 Buddhism One directedness thought
HM 6663 Buddhism One-pointedness citta
HM 2753 Buddhism One-pointedness mental abiding states
HM 2171 Buddhism Only-a-beginner subtle contemplation
HM 3378 Buddhism Other-induced attentiveness
HM 8112 Buddhism Ottappa (Pali)
HM 2538 Buddhism Ottappabala (Pali)
HM 1757 Buddhism Ottappati ottappitabbena (Pali)
HM 1208 Buddhism Ottappati papakanam akusalanam dhammanam samapattiya (Pali)
HM 2655 Buddhism Overcoming doubt
HM 1596 Buddhism Own-being
HM 2690 Buddhism ; Oxherding pictures Zen (Zen)
HM 2599 Buddhism Paccaya
HH 1099 Buddhism Paccayakara
HM 3955 Buddhism Paccekabodhi
HM 0249 Buddhism Paccupalakkhana (Pali)
HM 2205 Buddhism Pacifying mental abiding states
HM 0881 Buddhism Paggaha (Pali)
HM 0872 Buddhism Pagunabhava (Pali)
HM 3424 Buddhism Pagunattam (Pali)
HM 1061 Buddhism Pahasa (Pali)
HM 8010 Buddhism Pain
HM 8010 Buddhism Painful feeling
HM 0158 Buddhism Pajanana (Pali)
HM 2508 Buddhism Palace wisdom
HM 1508 Buddhism Palana (Pali)
HM 4246 Buddhism Pamodana (Pali)
HM 1371 Buddhism Pamoha (Pali)
HM 1398 Buddhism Pamojja (Pali)
HH 0278 Buddhism Panca sila
HM 1499 Buddhism Pandicca (Pali)
HM 4523 Buddhism Panna (Pali)
HM 0082 Buddhism Panna (Pali)
HM 1547 Buddhism Pannaabhasa (Pali)
HM 1846 Buddhism Pannaaloka (Pali)
HM 1792 Buddhism Pannabala (Pali)
HM 0192 Buddhism Pannapajjota (Pali)
HM 2508 Buddhism Pannapasada (Pali)
HM 1486 Buddhism Pannaratana (Pali)
HM 1645 Buddhism Pannasattha (Pali)
HM 3233 Buddhism Pannindriya (Pali)
HM 1840 Buddhism Pannindriya (Pali)
HM 3469 Buddhism Parakkama (Pali)
HM 1067 Buddhism Paramartha
HM 2294 Buddhism Paramarthashunyata
HM 1992 Buddhism Paramasa (Pali)

Buddhism

HM 2070 Buddhism Paranirmita-vasavartin
HM 2330 Buddhism Pari-nirvana
HM 4665 Buddhism Pari-nirvana
HM 4358 Buddhism Parinayika (Pali)
HM 3745 Buddhism Parisappana (Pali)
HM 1761 Buddhism Passambhana (Pali)
HH 5645 Buddhism Path enlightenment
HH 3875 Buddhism Path purification
HH 3875 Buddhism Path purity
HM 2987 Buddhism Paths calming
HM 3194 Buddhism Paths insight
HM 3079 Buddhism Paths special qualities
HM 2240 Buddhism Paths vision cultivation
HM 4958 Buddhism Patibhana-patisambhida (Pali)
HH 1099 Buddhism Paticca samuppada
HM 3942 Buddhism Patiggaha (Pali)
HH 3550 Buddhism Patipada-nanadassana-niddesa (Pali)
HM 1738 Buddhism Patipassaddhi (Pali)
HM 3411 Buddhism Patipassambhana (Pali)
HM 1528 Buddhism Patipassambhatattam (Pali)
HM 3806 Buddhism Patissati (Pali)
HM 0719 Buddhism Pativirodha (Pali)
HM 0633 Buddhism Patoda (Pali)
HH 0932 Buddhism Pattanumodana (Pali)
HH 1266 Buddhism Pattanumodana (Pali)
HM 1602 Buddhism Pavicaya (Pali)
HM 2051 Buddhism Peak cyclic existence absorption
HM 2931 Buddhism Perception
HM 3617 Buddhism Perception
HM 4143 Buddhism Perception aggregate
HM 7844 Buddhism Perfect wisdom
HM 1344 Buddhism Perfection discriminating awareness
HM 2864 Buddhism Performance tantra: charya
HM 0529 Buddhism Perplexity
HM 0812 Buddhism Perplexity
HM 0932 Buddhism Phala sacchikiriya
HM 0831 Buddhism Phala samadhi
HM 2599 Buddhism Phase-conditions consciousness
HM 2708 Buddhism Phassa (Pali)
HM 4212 Buddhism Phassa (Pali)
HM 2551 Buddhism Phenomena awareness
HM 2638 Buddhism Phrag-dog (Tibetan)
HM 1959 Buddhism Phusana (Pali)
HM 1263 Buddhism Physical inception energy
HM 8737 Buddhism Piti
HM 0747 Buddhism Piti (Pali)
HM 2038 Buddhism Piti (Pali)
HM 0513 Buddhism Pity
HM 6722 Buddhism Pleasant feeling
HM 2866 Buddhism Pleasure
HM 6722 Buddhism Pleasure
HM 5088 Buddhism Pleasure due insight
HM 3162 Buddhism Pliancy
HM 1155 Buddhism Pliancy citta (Pali)
HM 3696 Buddhism Pliancy kaya (Pali)
HM 3696 Buddhism Pliancy mental factors
HM 1155 Buddhism Pliancy mind
HM 4488 Buddhism Pondering
HM 2175 Buddhism Potala-island awareness
HM 1759 Buddhism Power concentration
HM 1943 Buddhism Power energy
HM 4404 Buddhism Power faith
HM 3450 Buddhism Power recollection
HM 2538 Buddhism Power remorse
HM 0352 Buddhism Power shame
HM 1724 Buddhism Power shamelessness
HM 1915 Buddhism Power unremorsefulness
HM 1792 Buddhism Power wisdom
HM 2215 Buddhism Prabha-kari
HH 4006 Buddhism Practice presence mindfulness
HM 2778 Buddhism Pradasha
HM 2655 Buddhism Prajna
HM 4523 Buddhism Prajna
HM 2756 Buddhism Prajna-jnana
HM 1344 Buddhism Prajnaparamita
HM 3205 Buddhism Prakrtishunyata
HM 2314 Buddhism Pramada
HH 4388 Buddhism Pramana
HM 2155 Buddhism Pramudita
HM 3162 Buddhism Prasrabdhi
HM 2450 Buddhism Prathamadhyana
HM 2959 Buddhism Pratigha
HM 2709 Buddhism Pratyckabuddha
HM 3020 Buddhism Pratyeka Buddha application awareness
HM 2776 Buddhism Pratyeka Buddha arhat awareness
HM 2180 Buddhism Pratyeka Buddha awareness
HM 2252 Buddhism Pratyeka Buddha cultivation awareness
HM 2228 Buddhism Pratyeka Buddha vision-path awareness
HM 2743 Buddhism Pravivekyamanaskara
HM 2683 Buddhism Prayoganishthamanaskara
HH 1022 Buddhism Preaching dharma
HH 5543 Buddhism Preparation stationariness
HM 3170 Buddhism Prerequisites manifestation
HM 6544 Buddhism Presentation due insight
HM 4032 Buddhism Presumption
HM 2112 Buddhism Pretas
HM 3630 Buddhism Principal insights
HM 1554 Buddhism Proficiency
HM 1455 Buddhism Proficiency mental factors
HM 1810 Buddhism Proficiency mind
HM 5338 Buddhism Profitable consciousness fine-material sphere
HM 4701 Buddhism Profitable consciousness immaterial sphere
HM 4447 Buddhism Profitable consciousness sense sphere
HM 4930 Buddhism Profitable consciousness supramundane plane
HM 1972 Buddhism Progression
HH 5652 Buddhism Psychic powers
HM 1921 Buddhism Psychical ease born contact element mind-consciousness
HM 4425 Buddhism Psychical perplexity
HM 0234 Buddhism Psychical unease born contact element mind-consciousness
HM 1862 Buddhism Psychically pleasant
HM 0805 Buddhism Psychically unpleasant
HM 2075 Buddhism Psycho-physical faculties awareness
HM 2168 Buddhism Pure-lands awareness
HH 2332 Buddhism Purification consciousness
HH 3025 Buddhism Purification knowledge vision
HH 3550 Buddhism Purification knowledge vision way
HH 4007 Buddhism Purification knowledge vision what is path what is not path
HH 1187 Buddhism Purification overcoming doubt
HH 2718 Buddhism Purification view
HH 1865 Buddhism Purification virtue
HH 3875 Buddhism Purifications
HH 3875 Buddhism Purities
HH 3025 Buddhism Purity knowledge discernment
HH 3550 Buddhism Purity knowledge discernment middle way
HH 4007 Buddhism Purity knowledge discernment right path wrong path
HH 2332 Buddhism Purity mind
HH 1865 Buddhism Purity morality
HH 1187 Buddhism Purity transcending doubt
HH 2718 Buddhism Purity views
HM 2543 Buddhism Purva-videha
HM 4008 Buddhism Puzzlement
HM 1498 Buddhism Quietude
HM 2743 Buddhism Rab-tu-dben-pai-yid-byed (Tibetan)
HM 6093 Buddhism Ragakkhaya (Pali)
HM 6093 Buddhism Ragaksaya
HM 2055 Buddhism Rakshasas
HM 2709 Buddhism Rang-sang-rgyas (Tibetan)
HM 0747 Buddhism Rapture
HM 4886 Buddhism Rapturous happiness due insight
HM 2267 Buddhism Ratisamgrahakamanaskara
HM 2275 Buddhism Ratna-kuta
HM 2028 Buddhism Rdo-rje (Tibetan)
HM 1177 Buddhism Reasoning
HM 8266 Buddhism Rebirth-linking
HM 3806 Buddhism Recalling
HM 7092 Buddhism Receiving
HM 1563 Buddhism Recollection
HH 6221 Buddhism Recollections meditation subjects
HM 1563 Buddhism Recolleection
HM 8266 Buddhism Reconception
HM 4250 Buddhism Rectitude
HM 9001 Buddhism Rectitude citta (Pali)
HM 9001 Buddhism Rectitude consciousness
HM 5402 Buddhism Rectitude kaya (Pali)
HM 5402 Buddhism Rectitude mental body
HM 5402 Buddhism Rectitude mental factors
HM 9001 Buddhism Rectitude mind
HM 5089 Buddhism Reflection
HM 2708 Buddhism Reg-pa (Tibetan)
HM 1646 Buddhism Registering
HM 1646 Buddhism Registration
HH 0932 Buddhism Rejoicing merit others
HM 3341 Buddhism Religious traditions ascension stages game
HM 0405 Buddhism Remembering
HM 1484 Buddhism Remembrance
HM 8112 Buddhism Remorse
HM 2128 Buddhism Renunciation
HM 0226 Buddhism Repose citta (Pali)
HM 7115 Buddhism Repose due insight
HM 4704 Buddhism Repose kaya (Pali)
HM 4704 Buddhism Repose mental factors
HM 1602 Buddhism Research
HM 2134 Buddhism Resentment
HM 2265 Buddhism Resetting mind
HM 3800 Buddhism Resolution
HM 5432 Buddhism Resolution faith due insight
HM 3800 Buddhism Resolve
HH 0604 Buddhism Respect superiors
HM 2516 Buddhism Reviving-hell awareness
HM 2534 Buddhism Rgod-pa (Tibetan)
HM 2528 Buddhism Rgyags-pa (Tibetan)
HM 2241 Buddhism Rgyun-du-jog-pa (Tibetan)
HM 0198 Buddhism Right action
HM 0198 Buddhism Right acts
HM 1142 Buddhism Right aims
HM 0198 Buddhism Right aims action
HM 0198 Buddhism Right behaviour
HH 0974 Buddhism Right beliefs
HM 0931 Buddhism Right concentration
HM 1735 Buddhism Right concentration
HM 1295 Buddhism Right effort
HM 1295 Buddhism Right endeavour
HM 1341 Buddhism Right livelihood
HM 0931 Buddhism Right meditative stabilization
HM 0704 Buddhism Right mindfulness
HM 1280 Buddhism Right outlook
HM 0931 Buddhism Right rapture concentration
HM 1142 Buddhism Right realization
HM 1142 Buddhism Right resolves
HM 1157 Buddhism Right speech
HM 1142 Buddhism Right thinking
HM 1142 Buddhism Right thought
HM 1280 Buddhism Right understanding
HM 0931 Buddhism Right unification
HM 1280 Buddhism Right view
HM 0600 Buddhism Rmi-lam (Tibetan)
HM 2926 Buddhism Rmugs-pa (Tibetan)
HM 2169 Buddhism Rna-ba'i-rnam-par-shes-pa (Tibetan)
HM 3154 Buddhism Rnam-par-g'yeng-ba (Tibetan)
HM 2608 Buddhism Rnam-par-mi-'tshe-ba (Tibetan)
HM 3210 Buddhism Rnam-par-'tshe-ba (Tibetan)
HM 3617 Buddhism Rnam shes (Tibetan)
HH 0270 Buddhism Root afflictions
HM 1177 Buddhism Rtog-pa (Tibetan)
HH 0270 Buddhism Rtsa-nyon
HM 2753 Buddhism Rtse-geig-tu-byed-pa (Tibetan)
HM 2603 Buddhism Rudra awareness black freedom
HM 2142 Buddhism Rupa-dhatu
HM 2235 Buddhism Rupa-dhatu-vidyadhara
HM 2108 Buddhism Rupa-khanda (Pali)
HM 2257 Buddhism Rupadhatu
HM 2536 Buddhism Rupaloka consciousness
HH 0538 Buddhism Sabda
HM 5403 Buddhism Saccanulomikanan (Pali)
HM 7120 Buddhism Sad feeling
HM 3158 Buddhism Saddahana (Pali)
HM 2933 Buddhism Saddha (Pali)
HM 4404 Buddhism Saddhabala (Pali)
HM 0066 Buddhism Saddhindriya (Pali)
HM 2292 Buddhism Sadhu-mati
HM 1750 Buddhism Sagacity
HM 1143 Buddhism Sallakkhana (Pali)
HM 2277 Buddhism Samadhana
HM 2440 Buddhism Samadhi
HM 6663 Buddhism Samadhi
HH 8231 Buddhism Samadhi-bhavana
HM 1759 Buddhism Samadhibala (Pali)
HM 3454 Buddhism Samadhindriya (Pali)
HM 2623 Buddhism Samantaka
HM 4514 Buddhism Samapajanna (Pali)
HM 0159 Buddhism Samapatti
HM 8673 Buddhism Samatha-nimitta
HM 1498 Buddhism Samatha (Pali)
HM 2147 Buddhism Samatha (Pali)
HM 2873 Buddhism Sambhoga-kaya
HM 7195 Buddhism Sambodhi
HM 2399 Buddhism Samjna
HM 5438 Buddhism Samjnavedayitanirodha
HM 0549 Buddhism Samma ajiva (Pali)
HM 1763 Buddhism Samma ajiva (Pali)
HM 1735 Buddhism Samma-samadhi (Pali)
HM 7091 Buddhism Samma-sambodhi
HM 3252 Buddhism Sammaditthi (Pali)
HM 0439 Buddhism Sammakammanta (Pali)
HM 0687 Buddhism Sammakammanta (Pali)
HM 1735 Buddhism Sammasamadhi
HM 1451 Buddhism Sammasankappa (Pali)
HM 0180 Buddhism Sammasati (Pali)
HM 4608 Buddhism Sammavaca (Pali)
HM 1821 Buddhism Sammavaca (Pali)
HM 1657 Buddhism Sammavayama (Pali)
HM 1956 Buddhism Sammoha (Pali)
HM 1874 Buddhism Sampajanna (Pali)
HM 1380 Buddhism Samphusana (Pali)
HM 0836 Buddhism Samphusitattam (Pali)
HM 2177 Buddhism Samsara
HM 0808 Buddhism Samsaya (Pali)
HM 2414 Buddhism Samskrtashunyata
HM 2241 Buddhism Samsthapara
HM 6221 Buddhism Samtana
HM 2596 Buddhism Samvatta-kappa (Pali)
HM 0652 Buddhism Samvriti
HM 1341 Buddhism Samyag-ajiva
HM 1280 Buddhism Samyag-drishti
HM 1142 Buddhism Samyag-samkalpa
HM 1157 Buddhism Samyag-vaca
HM 1295 Buddhism Samyag-vyayama
HM 0198 Buddhism Samyak-karmanta
HM 0931 Buddhism Samyak-samadhi
HM 0704 Buddhism Samyak-smriti
HM 0705 Buddhism Sancetana (Pali)
HM 2551 Buddhism Sankhara
HM 2050 Buddhism Sankhara-khanda (Pali)
HM 4143 Buddhism Sanna-khanda (Pali)
HM 1389 Buddhism Sanna (Pali)
HM 3419 Buddhism Sannanana (Pali)
HM 3250 Buddhism Saraga (Pali)
HM 4363 Buddhism Sarajitattam (Pali)
HM 1683 Buddhism Sarajjana (Pali)
HM 1484 Buddhism Saranata (Pali)
HM 3000 Buddhism Sarvadharmashunyata
HM 5334 Buddhism Sarvajnata
HH 0320 Buddhism Sarvatraga
HM 2847 Buddhism Sati (Pali)
HM 1563 Buddhism Sati (Pali)
HM 3450 Buddhism Satibala (Pali)
HM 4248 Buddhism Satindriya (Pali)
HM 0047 Buddhism Satindriya (Pali)
HH 4006 Buddhism Satipatthana (Pali)
HH 4019 Buddhism Satori
HM 0680 Buddhism Savakabodhi
HM 2683 Buddhism Sbyor-mthai-yid-byed (Tibetan)
HM 2722 Buddhism Scuffle views
HM 3408 Buddhism Search
HM 1715 Buddhism Search dharma
HM 6312 Buddhism Seclusion
HM 3043 Buddhism Second absorption immaterial sphere
HM 2371 Buddhism Second antarabhava consciousness
HM 2038 Buddhism Second form-realm concentration
HM 3043 Buddhism Second formless attainment
HM 2739 Buddhism Second scriptural bodhisattva awareness
HM 2038 Buddhism Second trance fine-material sphere

HM 2703 Buddhism Second vajra-master awareness
HH 0781 Buddhism Secondary afflictions
HH 1348 Buddhism Secondary mental factors
HM 3685 Buddhism Sectarian bias
HM 3265 Buddhism Seemingly bare perception
HM 1874 Buddhism Self-awareness
HM 2200 Buddhism Self-consciousness
HM 2016 Buddhism Self-emptiness
HM 3378 Buddhism Self-induced attentiveness
HM 2016 Buddhism Selflessness
HM 2145 Buddhism Sems gnas dgu (Tibetan)
HM 2217 Buddhism Sems-jogpa (Tibetan)
HM 2589 Buddhism Sems-pa (Tibetan)
HM 2270 Buddhism Sensation
HM 0309 Buddhism Sensation ease born contact psychical
HM 1643 Buddhism Sensation neither suffering nor pleasure born contact psychical
HM 1090 Buddhism Sensation unease born contact psychical
HM 4076 Buddhism Sensation unpleasant born contact psychical
HM 2664 Buddhism Sense consciousness
HM 4389 Buddhism Sense mode consciousness occurrence
HM 2881 Buddhism Sense shame
HM 1097 Buddhism Sense-sphere concentration
HM 2664 Buddhism Sensual awareness
HM 2964 Buddhism Ser-sna (Tibetan)
HM 2147 Buddhism Serenity
HM 0226 Buddhism Serenity mind
HM 3150 Buddhism Serial absorptions
HM 2217 Buddhism Setting mind
HM 1098 Buddhism Setughata (Pali)
HM 2361 Buddhism Seventh-scriptural bodhisattva awareness
HM 2789 Buddhism Seventh vajra-master awareness
HH 1108 Buddhism Seventy-five dharmas
HM 1222 Buddhism Sgyu-lus (Tibetan)
HM 3246 Buddhism Sgyu (Tibetan)
HH 0538 Buddhism Shabda
HM 2205 Buddhism Shamana
HM 2147 Buddhism Shamatha
HM 2881 Buddhism Shame
HM 8112 Buddhism Shame
HM 1590 Buddhism Shame
HM 2986 Buddhism Shamelessness
HM 0649 Buddhism Shamelessness
HM 3093 Buddhism Shathya
HM 2663 Buddhism Shen-siddhi
HM 2063 Buddhism Shes-bzhin-ma-yin-pa (Tibetan)
HM 5321 Buddhism Shes pa (Tibetan)
HM 2655 Buddhism Shes-rab (Tibetan)
HM 3162 Buddhism Shin-tu-sbyangs-pa (Tibetan)
HM 2169 Buddhism Shotravijnana
HM 2051 Buddhism Shpere neither perception nor non-perception
HM 3209 Buddhism Shraddha
HM 2161 Buddhism Shravaka
HM 2193 Buddhism Shunyata
HM 2899 Buddhism Shunyatashunyata
HH 0380 Buddhism Siddhis
HH 0278 Buddhism Sila
HM 8336 Buddhism Six-door adverting
HM 1914 Buddhism Six paths
HM 2431 Buddhism Sixth antarabhava consciousness
HM 2385 Buddhism Sixth scriptural bodhisattva awareness
HM 2287 Buddhism Sixth vajra-master awareness
HM 2692 Buddhism Skandha
HH 3321 Buddhism Skandha
HH 4777 Buddhism Skill absorption
HH 4777 Buddhism Skill ecstasy
HM 2265 Buddhism Slan-te-jog-pa (Tibetan)
HM 7921 Buddhism Sleep
HM 5667 Buddhism Sloth
HH 0845 Buddhism Small vehicle Buddhism
HM 2847 Buddhism Smriti
HM 2364 Buddhism Sna'i-rnam-par-shes-pa (Tibetan)
HM 1013 Buddhism Snares ignorance
HM 5634 Buddhism Snying re (Tibetan)
HM 3514 Buddhism Sokushin jobutsu
HM 6312 Buddhism Solitude
HM 1649 Buddhism Somanassindriya (Pali)
HM 3246 Buddhism Sound kasina
HH 0845 Buddhism Southern Buddhism
HM 2127 Buddhism Southern-continent awareness men
HM 3246 Buddhism Space kasina
HM 2708 Buddhism Sparsa
HM 2708 Buddhism Sparsha
HM 2623 Buddhism Special insight preparations states
HM 7002 Buddhism Specific neutrality
HM 3043 Buddhism Sphere infinite consciousness
HM 2110 Buddhism Sphere infinite space
HM 2051 Buddhism Sphere neither cognition nor non-cognition
HM 2027 Buddhism Sphere nothing
HM 2852 Buddhism Spiritual pride
HH 3907 Buddhism Spirituality
HM 2778 Buddhism Spite
HM 0192 Buddhism Splendour wisdom
HM 2240 Buddhism Sravaka-darsana-marga
HM 2716 Buddhism Sravaka prayoga-marga
HM 2192 Buddhism Sravaka-sambhara-marga
HM 2051 Buddhism Srid-rtse (Tibetan)
HM 2431 Buddhism Sridpa bardo (Tibetan)
HM 2440 Buddhism Stabilization
HM 1930 Buddhism Stabilization thought
HM 1749 Buddhism Stamina
HM 1286 Buddhism Standing crossroads
HM 1418 Buddhism Stasis
HM 0051 Buddhism Stasis thought
HM 2959 Buddhism State anger
HM 0847 Buddhism State animality
HM 1225 Buddhism State bodhisattva
HM 1873 Buddhism State buddhahood
HM 1967 Buddhism State feeling hatred
HM 4282 Buddhism State hell
HM 0150 Buddhism State hunger
HM 0581 Buddhism State latency
HM 3662 Buddhism State learning
HM 3947 Buddhism State not being greedy
HM 3183 Buddhism State not being infatuated
HM 1825 Buddhism State not feeling greed
HM 1597 Buddhism State not feeling hatred
HM 4081 Buddhism State not feeling infatuation
HM 1973 Buddhism State rapture
HM 0450 Buddhism State realization
HM 3492 Buddhism State tranquillity
HM 4131 Buddhism State unabated desire
HM 1403 Buddhism State unabated endurance
HM 3261 Buddhism State unfaltering exertion
HM 2596 Buddhism State universal cessation awareness
HM 2207 Buddhism States special insight
HM 3471 Buddhism Steadfastness thought
HM 6652 Buddhism Steadiness consciousness
HH 4019 Buddhism Steps enlightenment
HM 3942 Buddhism Sticking strongly
HM 5667 Buddhism Stiffness
HM 2730 Buddhism Stored consciousness
HM 1338 Buddhism Straightness aggregate consciousness
HM 0409 Buddhism Straightness aggregate sensation, aggregate cognition, aggregate synergies
HM 1424 Buddhism Straightness body
HM 0253 Buddhism Straightness thought
HM 1436 Buddhism Strive
HM 1081 Buddhism Strong grip burden
HM 1857 Buddhism Stupidity
HM 2926 Buddhism Styana
HM 6221 Buddhism Subconsciousness
HH 3987 Buddhism Subjects meditation
HM 4036 Buddhism Sublime concentration
HM 3508 Buddhism Subsequent cognition
HM 2195 Buddhism Subtle contemplation
HM 2267 Buddhism Subtle contemplation withdrawal joy
HM 0808 Buddhism Succumbing hesitation
HM 2435 Buddhism Suchness
HM 2168 Buddhism Suddhavasa
HM 2909 Buddhism Sudurjaya
HM 2574 Buddhism Suffering
HH 0523 Buddhism Suffering
HM 2866 Buddhism Sukha
HM 2167 Buddhism Sukhavati
HM 5134 Buddhism Sunnagare abhirati
HM 2193 Buddhism Sunyata
HH 9764 Buddhism Superhuman qualities
HM 6327 Buddhism Superior concentration
HH 5652 Buddhism Supernormal powers
HM 1067 Buddhism Supramundane awareness
HM 4243 Buddhism Supramundane concentration
HM 8838 Buddhism Supramundane understanding
HM 2813 Buddhism Supreme-heaven awareness
HM 3214 Buddhism Supreme heavens ascension stages game
HM 5089 Buddhism Sustained thought
HM 2962 Buddhism Sutra paths accumulation application ascension stages game
HM 7109 Buddhism Sva-samvedana
HM 1596 Buddhism Svabhava
HM 1645 Buddhism Sword wisdom
HM 5224 Buddhism Sympathy
HH 0306 Buddhism Tantra
HM 2452 Buddhism Tantra-beginner awareness
HM 2220 Buddhism Tantra climax-application awareness
HM 2696 Buddhism Tantra heat-application awareness
HM 2235 Buddhism Tantra master form-realm awareness
HM 2211 Buddhism Tantra master-in-sense-realm awareness
HM 2864 Buddhism Tantra meditation emptiness
HM 2756 Buddhism Tantra receptivity-application awareness
HM 2280 Buddhism Tantra union-in-learning-application awareness
HM 4603 Buddhism Tantric formless path
HM 2848 Buddhism Tantric paths accumulation application ascension-stages-game
HH 0306 Buddhism Tantric yoga
HM 2263 Buddhism Taste awareness
HM 2435 Buddhism Tathata
HM 7002 Buddhism Tatra-majjhattata (Pali)
HM 8003 Buddhism Tattva
HM 2454 Buddhism Temporary-hells awareness
HH 0446 Buddhism Ten meritorious deeds
HM 3159 Buddhism Ten powers
HM 2657 Buddhism Ten worlds
HM 1375 Buddhism Tendency towards ignorance
HH 2918 Buddhism Tenfold powers
HM 2421 Buddhism Tenth scriptural bodhisattva awareness
HM 1749 Buddhism Thama (Pali)
HM 0077 Buddhism Thambhatattam (Pali)
HM 8003 Buddhism Thatness
HH 0845 Buddhism Theravada
HH 1613 Buddhism Thicket views
HM 5667 Buddhism Thina (Pali)
HM 3966 Buddhism Thinking
HM 2027 Buddhism Third absorption immaterial sphere
HM 2311 Buddhism Third antarabhava consciousness
HM 2062 Buddhism Third form-realm concentration
HM 2027 Buddhism Third formless attainment (Tibetan)
HM 2215 Buddhism Third scriptural bodhisattva awareness
HM 2062 Buddhism Third trance fine-material sphere
HM 2727 Buddhism Third vajra-master awareness
HM 2606 Buddhism Thirty-three-god-heaven awareness
HM 1418 Buddhism Thiti (Pali)
HM 3057 Buddhism Thorough abandonings
HM 3529 Buddhism Thought
HM 0847 Buddhism Thoughtlessness
HM 0923 Buddhism Three evil paths
HM 2033 Buddhism Tibetan definitions awareness
HM 2693 Buddhism Tibetan meditative states form concentrations
HM 2669 Buddhism Tibetan meditative states formless absorptions
HH 0873 Buddhism Tibetan studies
HM 1133 Buddhism Tihi kayaduccaritehi arati virati pativirati veramani (Pali)
HM 2440 Buddhism Ting-nge-'dzin (Tibetan)
HM 3685 Buddhism Titthayatana (Pali)
HM 2263 Buddhism Tongue consciousness
HM 0264 Buddhism Torpor
HM 2708 Buddhism Touch
HM 4212 Buddhism Touch
HM 2562 Buddhism Touch awareness
HM 1959 Buddhism Touching
HH 2387 Buddhism Training higher mind
HM 2122 Buddhism Trances mental absorptions
HM 0226 Buddhism Tranquillity consciousness
HM 7115 Buddhism Tranquillity due insight
HM 4704 Buddhism Tranquillity mental body
HM 0709 Buddhism Tranquillity thoughts
HM 4243 Buddhism Transcendental concentration
HM 8838 Buddhism Transcendental understanding
HH 1266 Buddhism Transferring merit
HM 9021 Buddhism Transitoriness
HM 2606 Buddhism Trayatrimsa
HM 2177 Buddhism Tridhatu
HM 3707 Buddhism Trisna
HM 2062 Buddhism Tritiyadhyana
HM 2778 Buddhism 'Tshig-pa (Tibetan)
HM 2270 Buddhism Tshor-ba (Tibetan)
HM 2340 Buddhism Tulku (Tibetan)
HM 0400 Buddhism Turmoil thought
HM 2022 Buddhism Tushita
HM 7655 Buddhism Twenty-four causes consciousness arising
HM 3527 Buddhism Two-fold knowledge truths
HM 3662 Buddhism Two vehicles
HM 0450 Buddhism Two vehicles
HM 3026 Buddhism Ubhaya-carya-tantra
HM 1469 Buddhism Uddhacca (Pali)
HM 4250 Buddhism Ujuta (Pali)
HM 4090 Buddhism Un-comprehension
HM 3375 Buddhism Unaffected
HM 8092 Buddhism Unassociated consciousness
HM 5335 Buddhism Unattracted consciousness
HM 1362 Buddhism Unawakened
HM 1725 Buddhism Unawareness
HM 5335 Buddhism Unbending mind
HM 0812 Buddhism Uncertainty
HM 0870 Buddhism Uncertainty
HM 6504 Buddhism Uncompounded existence
HM 6504 Buddhism Unconditioned existence
HM 8621 Buddhism Unconfined mind
HM 4394 Buddhism Unconscientiousness
HM 4037 Buddhism Uncrookedness
HM 1054 Buddhism Undeflectedness
HM 6521 Buddhism Undejected consciousness
HM 3617 Buddhism Understanding
HM 4523 Buddhism Understanding
HM 0158 Buddhism Understanding
HM 7525 Buddhism Understanding accompanied equanimity
HM 7525 Buddhism Understanding accompanied even-mindedness
HM 6245 Buddhism Understanding accompanied joy
HM 4128 Buddhism Understanding being plane culture
HM 6192 Buddhism Understanding being plane discernment
HM 7826 Buddhism Understanding consisting development
HM 5298 Buddhism Understanding consisting what is learned
HM 7154 Buddhism Understanding consisting what is reasoned
HM 7826 Buddhism Understanding culture
HM 4968 Buddhism Understanding defining materiality
HM 5627 Buddhism Understanding defining mentality
HM 4958 Buddhism Understanding discrimination perspicuity
HM 0756 Buddhism Understanding free cankers
HM 3296 Buddhism Understanding having exalted object
HM 6681 Buddhism Understanding having infinite object
HM 2495 Buddhism Understanding having limited object
HM 6681 Buddhism Understanding having measureless object
HM 3296 Buddhism Understanding having sublime object
HM 1128 Buddhism Understanding interpreting external
HM 0956 Buddhism Understanding interpreting internal
HM 4490 Buddhism Understanding interpreting internal external
HM 8163 Buddhism Understanding knowledge cessation suffering
HM 4726 Buddhism Understanding knowledge doctrine
HM 5026 Buddhism Understanding knowledge interpretation
HM 4958 Buddhism Understanding knowledge kinds knowledge
HM 5026 Buddhism Understanding knowledge language
HM 4726 Buddhism Understanding knowledge law
HM 1663 Buddhism Understanding knowledge meaning
HM 7290 Buddhism Understanding knowledge origin suffering
HM 6870 Buddhism Understanding knowledge suffering
HM 7645 Buddhism Understanding knowledge way leading cessation suffering
HM 4128 Buddhism Understanding plane development
HM 6192 Buddhism Understanding plane seeing
HM 8163 Buddhism Understanding reference truth cessation ill
HM 6870 Buddhism Understanding reference truth ill
HM 7290 Buddhism Understanding reference truth origin ill

HM 7645 Buddhism Understanding reference truth path leading cessation ill
HM 7602 Buddhism Understanding skill detriment
HM 8290 Buddhism Understanding skill improvement
HM 8290 Buddhism Understanding skill increase
HM 7602 Buddhism Understanding skill loss
HM 6601 Buddhism Understanding skill means
HM 8290 Buddhism Understanding skill profit
HM 4194 Buddhism Understanding subject cankers
HM 4968 Buddhism Understanding way fixing material qualities
HM 5627 Buddhism Understanding way fixing mental qualities
HM 7154 Buddhism Understanding way imagination
HM 5298 Buddhism Understanding way tradition
HM 7154 Buddhism Understanding what is thought out
HM 0901 Buddhism Unease experienced born contact psychical
HM 5623 Buddhism Unelated consciousness
HM 5623 Buddhism Unelated mind
HM 6791 Buddhism Unfettered mind
HM 1705 Buddhism Unforgetfulness
HM 3020 Buddhism Unicorn
HM 6663 Buddhism Unification mind
HM 7843 Buddhism Unified consciousness
HM 5768 Buddhism Unincluded concentration
HM 2015 Buddhism Uninterrupted release paths
HM 1734 Buddhism Unmaliciousness
HM 5665 Buddhism Unmanifest thinking consciousness
HM 5908 Buddhism Unoffended mind
HM 4572 Buddhism Unperplexity thought
HM 0715 Buddhism Unperturbed mindedness
HM 4572 Buddhism Unperturbedness
HM 0449 Buddhism Unperturbedness thought
HM 1031 Buddhism Unpleasant experienced born contact psychical
HM 8375 Buddhism Unprofitable consciousness sense sphere
HM 0188 Buddhism Unremorsefulness
HM 5908 Buddhism Unrepelled consciousness
HM 2574 Buddhism Unsatisfactoriness
HM 6791 Buddhism Untrammelled consciousness
HM 1008 Buddhism Untwistedness
HM 4999 Buddhism Upacara samadhi
HH 0781 Buddhism Upaklesha
HM 2399 Buddhism Upalakkhana
HM 2134 Buddhism Upanaha
HM 3345 Buddhism Upaparikkha (Pali)
HM 2157 Buddhism Upasthapura
HM 1647 Buddhism Upavicara (Pali)
HM 7769 Buddhism Upekkha (Pali)
HM 2062 Buddhism Upekkha (Pali)
HM 3803 Buddhism Upekkhindriya (Pali)
HM 7769 Buddhism Upeksha
HM 5986 Buddhism Uplift due insight
HM 0140 Buddhism Uprisedness thought
HM 4487 Buddhism Ussaha
HM 3357 Buddhism Ussolhi (Pali)
HH 9764 Buddhism Uttarimanussa
HM 1371 Buddhism Utter delusion
HM 1436 Buddhism Uyyama (Pali)
HM 0077 Buddhism Vacillation
HH 1108 Buddhism Vaibhasika system
HM 2028 Buddhism Vajra-hell
HM 2187 Buddhism Vajracarya
HH 0309 Buddhism Vajrayana Buddhism
HH 0309 Buddhism ; Vajrayana (Buddhism)
HH 4388 Buddhism Validities
HM 2468 Buddhism Vattana (Pali)
HM 0229 Buddhism Vayama (Pali)
HM 4488 Buddhism Vebhabya (Pali)
HM 2270 Buddhism Vedana
HM 4983 Buddhism Vedana-khanda
HM 1381 Buddhism Vedanakkhandhassa sannakkhandhassa sankharakkhandhassa kammannata (Pali)
HM 1435 Buddhism Vedanakkhandhassa sannakkhandhassa sankharakkhandhassa labuta (Pali)
HM 3397 Buddhism Vedanakkhandhassa sannakkhandhassa sankharakkhandhassa muduta (Pali)
HM 0349 Buddhism Vedanakkhandhassa sannakkhandhassa sankharakkhandhassa (Pali)
HM 1377 Buddhism Vedanakkhandhassa sannakkhandhassa sankharakkhandhassa passaddhi (Pali)
HM 0409 Buddhism Vedanakkhandhassa sannakkhandhassa sankharakkhandhassa ujukata (Pali)
HM 3435 Buddhism Vela anatikkama
HM 3435 Buddhism Vela anatikkamo (Pali)
HM 2576 Buddhism Very-hot-hells awareness
HM 5089 Buddhism Vicara (Pali)
HM 3408 Buddhism Vicaya (Pali)
HM 5089 Buddhism Vichara
HM 4467 Buddhism Vichikitsa - The-tshom (Tibetan)
HM 0529 Buddhism Vicikiccha (Pali)
HM 0812 Buddhism Vicikiccha (Pali)
HM 2723 Buddhism Vidyadhara
HM 2759 Buddhism Vidyadhara-emperor
HM 3066 Buddhism Vidyahara
HM 1191 Buddhism View
HM 3210 Buddhism Vihimsa
HM 3617 Buddhism Vijna
HM 3617 Buddhism Vijnana
HM 3043 Buddhism Vijnananantyayatana
HM 3306 Buddhism Vikalpa
HM 3154 Buddhism Vikshepa
HM 2739 Buddhism Vimala
HM 4008 Buddhism Vimati (Pali)
HM 5977 Buddhism Vimokkha (Pali)
HM 5977 Buddhism Vimoksa
HM 2015 Buddhism Vimuktimarga
HH 1376 Buddhism Vinaya
HH 3398 Buddhism Vinaya
HM 0927 Buddhism Vinivaranata
HH 0170 Buddhism Viniyata
HM 2098 Buddhism Vinnana-khanda development (Pali)
HM 2556 Buddhism Vinnana-khanda (Pali)
HM 2599 Buddhism Vinnana-kicca (Pali)
HM 3617 Buddhism Vinnana (Pali)
HM 3043 Buddhism Vinnanacayatana (Pali)
HM 3403 Buddhism Vinnanakkhandhassa lahuta (Pali)
HM 1796 Buddhism Vinnanakkhandhassa muduta (Pali)
HM 1741 Buddhism Vinnanakkhandhassa pagunata (Pali)
HM 4605 Buddhism Vinnanakkhandhassa (Pali)
HM 1338 Buddhism Vinnanakkhandhassa ujukata (Pali)
HM 2098 Buddhism Vinnananakkhandha
HM 1751 Buddhism Vipariyasaggaha (Pali)
HM 2207 Buddhism Vipashyana
HM 2623 Buddhism Vipashyana
HH 0680 Buddhism Vipassana-bhavana
HM 1632 Buddhism Vipassana (Pali)
HM 1333 Buddhism Viriya (Pali)
HM 1943 Buddhism Viriyabala (Pali)
HM 1470 Buddhism Viriyindriya (Pali)
HM 1829 Buddhism Viriyindriya (Pali)
HM 4239 Buddhism Virodha (Pali)
HH 0578 Buddhism Virtuous mental factors
HM 2389 Buddhism Virya
HM 2074 Buddhism Visual awareness
HH 3875 Buddhism Visuddhimagga (Pali)
HM 1177 Buddhism Vitakka (Pali)
HM 2450 Buddhism Vitakka-vicara (Pali)
HM 3404 Buddhism Vitality
HM 1177 Buddhism Vitarka
HM 1545 Buddhism Vitti (Pali)
HM 2193 Buddhism Voidness
HM 2589 Buddhism Volition
HM 0404 Buddhism Volition born contact element mind consciousness
HM 0487 Buddhism Vyappana (Pali)
HM 2729 Buddhism Vyvpashamana
HH 3246 Buddhism Water kasina
HM 2185 Buddhism Way Buddhas
HM 3090 Buddhism Way vajra masters
HH 0873 Buddhism Ways knowing
HM 6652 Buddhism Weak concentration
HM 2519 Buddhism Western-continent awareness cattle
HM 5909 Buddhism Wheel becoming
HM 2759 Buddhism Wheel-turner-king awareness
HM 2058 Buddhism Wheel-turning-king awareness
HM 6556 Buddhism Wieldiness citta (Pali)
HM 6556 Buddhism Wieldiness consciousness
HM 7969 Buddhism Wieldiness kaya (Pali)
HM 7969 Buddhism Wieldiness mental body
HM 7969 Buddhism Wieldiness mental factors
HM 6556 Buddhism Wieldiness mind
HM 4548 Buddhism Wilderness views
HM 4523 Buddhism Wisdom
HM 0082 Buddhism Wisdom
HM 6336 Buddhism Wisdom factors
HM 0633 Buddhism Wisdom goad
HM 4358 Buddhism Wisdom guide
HM 4309 Buddhism Wise attention
HM 4487 Buddhism Wish act
HM 4605 Buddhism Workability aggregate consciousness
HM 1381 Buddhism Workability aggregate sensation, aggregate cognition, aggregate synergies
HM 0739 Buddhism Workability body
HM 1584 Buddhism Workability thought
HM 0979 Buddhism Working ability
HH 2002 Buddhism World cycles
HM 1973 Buddhism World desires
HM 1973 Buddhism World form
HM 1973 Buddhism World formlessness
HM 3492 Buddhism World humanity
HM 3214 Buddhism World-transcending spheres
HM 7234 Buddhism Worldly concentration
HM 7628 Buddhism Worldly understanding
HM 2072 Buddhism Worlds conscious existence
HM 9101 Buddhism Worry
HM 1802 Buddhism Wrong concentration
HM 1585 Buddhism Wrong endeavour
HM 0863 Buddhism Wrong path
HM 5324 Buddhism Wrong view
HM 1710 Buddhism Wrong view
HM 3356 Buddhism Wrongness
HH 4019 Buddhism Wu (Zen)
HM 2130 Buddhism Yama-devas
HM 2088 Buddhism Yama state
HM 2435 Buddhism Yan-dag-pa-ji-lta-ba-bzin-du (Tibetan)
HM 3327 Buddhism Yapana (Pali)
HM 1691 Buddhism Yapana (Pali)
HM 7549 Buddhism Yatha-bhuta-jnana-darsana
HM 7549 Buddhism Yatha-bhuta-nana-dassana (Pali)
HM 2435 Buddhism Yathabhuta
HM 2623 Buddhism Yid byed pa
HM 2838 Buddhism Yid-kyi-rnam-par-shes-pa (Tibetan)
HM 3237 Buddhism Yid-la-byed-pa (Tibetan)
HM 3690 Buddhism Yoga clear light
HM 5122 Buddhism Yoga consciousness transference
HM 0600 Buddhism Yoga dream state
HM 1222 Buddhism Yoga illusory body
HM 3863 Buddhism Yoga inner fire
HM 0101 Buddhism Yoga intermediary state
HM 2864 Buddhism Yoga tantra
HM 5532 Buddhism Yogi-pratyaksa
HH 0603 Buddhism Yogic paths
HM 2849 Buddhism Yugbanaddha
HH 0170 Buddhism Yul-nges
HM 4487 Buddhism Zeal
HM 0747 Buddhism Zest
HM 2744 Buddhism Zhe-sdang-med-pa (Tibetan)
HM 2205 Buddhism Zhi-bar-byed-pa (Tibetan)
HM 2147 Buddhism Zhi gnas (Tibetan)
HH 0551 Buddhist meditation (Buddhism)
KC 0967 Budgeting
KC 0566 Budgeting System ; Planning Programming (PPBS)
HH 0085 Budo
TC 1546 Buffoon
HM 4094 Bug ; Cocaine
MS 8103 Bugs
TC 1748 Builder type
TC 1721 Building activities ; Trust
HH 0772 Building ; Brain
HH 0895 Building ; Character
TC 1011 Building ; Consensus
HH 0717 Building group , Team
VD 2204 Bumptiousness
VC 0277 Buoyancy
HM 6033 Buoyancy citta (Pali)
HM 2136 Buoyancy ; God consciousness
HM 7636 Buoyancy kaya (Pali)
HM 7636 Buoyancy mental factors (Buddhism)
HM 6033 Buoyancy mind (Buddhism)
HM 1081 Burden ; Strong grip (Buddhism)
TP 0047 Burdened
VD 2206 Burdensomeness
HH 3783 Burmese meditation
TC 1776 Burn out ; Participant
VC 0279 Business
TC 1996 Business concern conferencing
KC 0913 Business cycles
TC 1493 Buttons ; Participant consensus
TC 1311 Buzz group
HH 4019 Byan-chub (Tibetan)
HM 2769 Byang-chub-sems-dpa (Tibetan)
HM 1111 Byapada (Buddhism)
HM 1591 Byapajjana (Pali)
HM 3210 Byapajjitatta (Pali)
HM 1111 Byapatti (Pali)
HM 2623 Byed pa ; Yid (Buddhism)
HM 2586 Bzhi pa ; Bsam gtan (Buddhism)

C

HM 4344 Cafard
HM 1368 Caffeine ; Modes awareness associated use
HM 1271 Cakkhuvinnanadhatusamphassaja sanna (Pali)
HM 0972 Cakkhuvinnanadhatusamphassajam cetasikam neva satam nasatam (Pali)
HM 2058 Cakravartin (Buddhism)
VD 2208 Calamity
TP 0005 Calculated inaction (Getting wet)
TC 1895 Calculative participant involvement
HM 2283 Calculative thinking ; Rational,
MS 8123 Calendar ; Sacred
HH 3352 Call mission ; Journeying transcendence (Christianity)
HM 2008 Called accountability (ICA)
HH 0619 Calling (Christianity)
HM 2346 Calling ; Surrender one's (ICA)
VD 2210 Callousness
HM 2147 Calm abiding (Buddhism)
HM 8673 Calm ; Mark (Buddhism)
HM 0634 Calm stability (Buddhism)
HM 1761 Calming down (Buddhism)
HM 0226 Calming mind (Buddhism)
HM 2987 Calming ; Paths (Buddhism)
VC 0281 Calmness
HM 2239 Cancer-consciousness (Astrology)
HM 0945 Candikka (Pali)
VC 0283 Candor
MS 8853 Canine presence
HM 0756 Cankers ; Understanding free (Buddhism)
HM 4194 Cankers ; Understanding subject (Buddhism)
HM 3691 Cannabis ; Modes awareness associated use
VC 0285 Canniness
HM 1167 Canonical hours (Christianity)
HH 0331 Canonization
HH 2806 Capabilities ; Enhancement human
VC 0287 Capability
HH 0902 Capability become
HM 6133 Capability ; Negative
VC 0289 Capacity
HH 0405 Capacity become
KC 0906 Capacity ; Communication channel
HM 3808 Capacity easy transformation (Buddhism)
HH 0937 Capital formation ; Human
HH 6209 Capitalism ; Humanistic
HH 6122 Capitalist human development ; Liberal
VD 2212 Capriciousness
HM 2822 Capricorn-consciousness (Astrology)
HM 3190 Captive ; Divine (ICA)
MP 1175 Capturing non linear extensions structures ; Insight
HM 2259 Cardinal awareness (Astrology)
HH 0712 Cardinal Virtues
VC 0291 Care
HH 0042 Care
TC 1269 Care groups ; Creative
HM 2170 Care ; Mountain (ICA)
HH 0581 Care movement ; Holistic health
MP 1086 Care premature perspectives ; Contexts
HH 1199 Care ; Soul (Christianity)
HH 0117 Career development ; Personal

TP 0010	Careful conduct (Treading)		**Cessation [Attainment...] cont'd**	HH 2178	Change ; Structural
TP 0010	Carefully ; Stepping (Treading)	HM 6346	— Attainment (Buddhism)	HH 4561	Change (Yoga)
VC 0293	Carefulness	HM 2172	— awareness (Buddhism)	HH 0910	Changeable mental factors (Buddhism)
VP 5533	Carefulness-Neglect	HM 2596	— awareness ; State universal (Buddhism)	VD 2222	Changeableness
VD 2214	Carelessness	HM 2330	Cessation (Buddhism)	VP 5141	Changeableness ; Stability
HM 0361	Carelessness (Yoga)	VP 5143	Cessation ; Continuance	HH 0004	Changes ; Book (Taoism)
HH 1523	Cargo cults	HM 5676	Cessation hatred (Buddhism)	MM 2036	Changing fashions
KC 0926	Carlo simulation technique ; Monte		**Cessation ill**	MM 2020	Changing moods individual
HM 9406	Carnal love	HH 2119	— (Buddhism)	MP 1153	Changing number embodied perspectives ; Structures adaptable
HM 7336	Carnal soul ; Opposition (Sufism)	HM 8163	— Understanding reference truth (Buddhism)		
MM 2052	Carrot processes ; Stick	HM 7645	— Understanding reference truth path leading (Buddhism)	MM 2032	Changing physical position
MM 2023	Cart ; Animal drawn			KC 0906	Channel capacity ; Communication
HM 3344	Carry cross ; Jesus helped (Christianity)	HM 3354	Cessation objectivity (Hinduism)	HH 0878	Channelling
HM 7219	Carvana (Hinduism)		**Cessation suffering**	HM 3204	Channelling guilt (Brainwashing)
HM 3026	Carya tantra ; Ubhaya (Buddhism)	HH 2119	— (Buddhism)	HM 3456	Channelling (Psychism)
HH 1409	Castle Saint Teresa ; Interior (Christianity)	HM 8163	— Understanding knowledge (Buddhism)	HM 3456	Channelling ; Trance (Psychism)
HM 4319	Catalepsy (Psychism)	HM 7645	— Understanding knowledge way leading (Buddhism)	HH 1257	Chanting
HM 4319	Cataleptic trance (Psychism)	HM 3310	Cessations ; Analytical (Buddhism)	HH 0496	Chants
HM 1955	Cataplexy	HM 3310	Cessations (Buddhism)	VD 2224	Chaos
KC 0462	Catastrophe theory	HM 3310	Cessations ; Non analytical (Buddhism)	KC 0222	Chaos ; Order
HM 2979	Catching-the-ox awareness (Zen)	HM 0404	Cetana ; Manovinnanadhatu samphassaja (Pali)	VC 0311	Character
HH 0348	Categoried thinking	HM 2589	Cetana (Pali)	HH 0895	Character
KC 0779	Categories	HM 0234	Cetasikam asatam ; Manovinnanadhatusampassa jam (Pali)	HH 0895	Character building
KC 0879	Categories ; Part whole			HH 0470	Character defects (Buddhism)
KD 2146	Categorization poly ocular vision ; De	HM 0805	Cetasikam dukkham (Pali)	HH 0895	Character development
HH 0560	Catharsis	HM 0972	Cetasikam neva satam nasatam ; Cakkhuvinnanadhatusamphassajam (Pali)	HH 0895	Character education
HM 7621	Catharsis (Sufism)			HM 2195	Character ; Individual knowledge (Buddhism)
HM 2519	Cattle ; Western continent awareness (Buddhism)	HM 3314	Cetasikam neva satam nasatam ; Manovinnanadhatusamphassajam (Pali)	KC 0636	Character recognition
HM 1781	Catuhi vaciduccaritehi arati virati pativirati veramani (Pali)				**Character [Social...]**
		HM 1921	Cetasikam satam ; Manovinnanadhatu samphasajam (Pali)	HH 0251	— Social
TP 0050	Cauldron Cultural heritage			HH 0797	— Strength
TP 0050	Cauldron Rejuvenation	HM 1862	Cetasikam sukham (Pali)	HM 3759	— (Sufism)
TP 0050	Cauldron Sacrificial vessel	HM 1263	Cetasiko viriyarambho (Pali)	VD 2226	Characterlessness
HH 1099	Causally continuous doctrine being (Buddhism)	HM 4425	Cetaso vikkhepo (Pali)	MS 8383	Characters ; Historical, legendary literary
VT 8023	Causation*complex	HM 1412	Cetayitattam (Pali)	HM 3045	Chariot (Tarot)
VP 5153	Causation-Culmination	HM 5587	Ceto-samadhi (Buddhism)	HM 3045	Charioteer archetypal image (Tarot)
HH 0739	Causative factor altered states	HM 0709	Ceto samatha (Buddhism)	VC 0313	Charisma
HM 1098	Cause ; Destroying (Buddhism)	HM 0831	Ceto vimutti (Buddhism)	HH 0053	Charisma
HM 7655	Causes consciousness arising ; Twenty four (Buddhism)	HM 1643	Cetosamphassaja addukkhamasukha vedana (Pali)	HH 3124	Charismatic renewal (Christianity)
		HM 1090	Cetosamphassaja asata vedana (Pali)		**Charity**
HH 1047	Causes misery (Yoga)	HM 4076	Cetosamphassaja dukkha vedana (Pali)	VC 0315	— [Charity]
HM 1591	Causing harm (Buddhism)	HM 0309	Cetosamphassaja sata vedana (Pali)	HH 1206	— [Charity]
VC 0295	Caution	HM 1352	Cetosamphassajam adukkhamasukham vedayitam (Pali)	HH 0258	— [Charity]
VP 5894	Caution-Rashness			HH 1234	Charity (Buddhism)
VC 0297	Cautiousness	HM 1031	Cetosamphassajam dukkam vedayitam (Pali)	HM 3056	Charity ; Embodying (ICA)
HH 1932	Cave meditation	HM 0901	Cetosamphassajamasatam vedayitam (Pali)	HM 4387	Charity (Leela)
HM 1169	Cave phenomenon	HM 0125	Cetovimutti (Pali)	VC 0317	Charm
HM 3295	Ceasing thought	HH 0589	Ch'an (Zen)	KC 0816	Chart ; Flow
VC 0299	Celebration	HH 0620	Ch'i	HM 2864	Charya ; Performance tantra: (Buddhism)
HH 0956	Celebration ; Discipline (Christianity)	HH 0282	Ch'uan ; T'ai Chi	VC 0319	Chastity
HM 6534	Celestial awareness mundane ; Restoration (Taoism)	HM 2605	'Chab-pa (Tibetan)	HH 0978	Chastity
MS 8337	Celestial beings ; Mythical	VD 2220	Chagrin	HH 0583	Chastity (Christianity)
HH 4997	Celestial consciousness earthly consciousness ; Blending (Taoism)	HM 2433	Chags ; 'Dod (Buddhism)	HM 2506	Chastity ; Life (ICA)
		KC 0771	Chain ; Food	VP 5987	Chastity-Unchastity
HM 5265	Celestial immortality (Taoism)	MP 1042	Chain fundamental transformation zones	HM 2586	Chaturthadhyana (Buddhism)
HM 4403	Celestial musicians (Leela)	MM 2013	Chairmanship ; Rotating	HM 2245	Chatvari satvani (Buddhism)
HM 0717	Celestial plane (Leela)	TC 1969	Chairperson	HM 2278	Chaupud ; Gyan
HM 1052	Celestial plane (Leela)	HM 3291	Chakra: attaining balance ; Fourth (Leela)	VD 2228	Chauvinism
MS 8425	Celestial symbols ; Atmosphere	HM 2219	Chakra centres consciousness	VD 2230	Cheapness
HH 0045	Celibacy		**Chakra [Fifth...]**	VP 5848	Cheapness ; Expensiveness
VP 5931	Celibacy ; Conjugality	HM 3461	— Fifth (Yoga)	VC 0321	Cheerfulness
HH 1987	Celibate stage life (Hinduism)	HM 2893	— First (Yoga)	VP 5870	Cheerfulness-Solemnity
VD 2216	Censoriousness	HM 3242	— Fourth (Yoga)	VD 2232	Cheerlessness
VD 2218	Censure	HM 4103	Chakra: fundamentals being ; First (Leela)	HH 2552	Cheirothesy
HH 0527	Centering	HM 0933	Chakra: man becomes himself ; Fifth (Leela)	KC 0135	Chemical synthesis
MP 1126	Centering focal point common domain ; Off	HM 3398	Chakra ; Niralambapuri (Yoga)	MS 8606	Chemicals
HH 1994	Centering prayer (Christianity)	HM 0754	Chakra: plane reality ; Seventh (Leela)	HM 0917	Chen-hsing (Zen)
KC 0220	Central place theory	HM 3651	Chakra: realm fantasy ; Second (Leela)	HM 2420	Chesed sephira (Kabbalah)
KC 0120	Centrality		**Chakra [Second...]**	HM 2589	Chetana (Buddhism)
VP 5226	Centrality-Environment	HM 3174	— Second (Yoga)	HM 2578	Chhanda (Buddhism)
KC 0122	Centralization	HM 3398	— Seventh (Yoga)	HH 0282	Chi Ch'uan ; T'ai
KC 0122	Centralized system	HM 3398	— Shunya (Yoga)	HH 6129	Chi ; Lin (Zen)
HM 2872	Centre ; Bottomless (ICA)	HM 2144	— Sixth (Yoga)	HM 2335	Chikhai-bardo (Tibetan)
	Centre [Certitude...]	HM 0176	— Soma (Yoga)	HH 0500	Child creativity
HM 2273	— Certitude (ICA)	HM 0717	Chakra: theatre karma ; Third (Leela)	HH 0850	Child guidance
MP 1099	— complex structure ; Focal	HM 3063	Chakra ; Third (Yoga)	HH 0894	Child guidance therapy
HM 2529	— Contentment (ICA)	HM 4412	Chakra: time penance ; Sixth (Leela)	HM 2158	Childhood ; Spiritual (Christianity)
HM 2437	Centre ; Everlastingness (ICA)	HM 1141	Chakras: gods themselves ; Beyond (Leela)	VD 2234	Childishness
TP 0061	Centre ; Faithfulness	HM 2074	Chakshurvinjana (Buddhism)	HM 1864	Childishness (Buddhism)
HM 3015	Centre ; Problemlessness (ICA)	TC 1487	Chalkboard	HH 0611	Children ; Exceptionally gifted
TP 0061	Centre ; Sincerity	VC 0303	Challenge	HH 2901	Chinese Feng-shui
HM 1233	Centre (Taoism)	MP 1073	Challenging emerging perspectives ; Contexts exploratory relationship formation	HH 2901	Chinese Geomancy
HH 3607	Centred development ; Human			HM 1700	Chinese Takuan
HM 7843	Centred mind ; One (Buddhism)	VC 0305	Championship	HM 3011	Chitta-dependent consciousness (Yoga)
HH 0095	Centred therapy ; Client	VC 0307	Chance	HM 2217	Chitasthapana (Buddhism)
HH 0607	Centred therapy ; Group	VP 5155	Chance ; Attributability	HM 2145	Chittasthiti ; Navakara (Buddhism)
HM 2345	Centredness ; Present		**Change**	VC 0323	Chivalry
VD 3630	Centredness ; Self	KC 0645	— [Change]	HH 0329	Chivalry
HM 2219	Centres consciousness ; Chakra	VC 0309	— [Change]	VC 0325	Choice
HH 0562	Centres ; Growth	HH 1116	— [Change]	VT 8062	Choice*complex
HH 5120	Ceremonial magic (Magic)	HM 0856	Change ; Attitude	HH 4712	Choice ; Evolutionary
HH 0423	Ceremonies	HH 4221	Change awareness	HH 0789	Choice ; Freedom
HM 0509	Ceremonies ; Dependence (Buddhism)	VT 8022	Change*complex	VP 5637	Choice-Necessity
MM 2026	Ceremony ; Traditional spontaneous		**Change [Constraints...]**	HM 4469	Choiceless awareness
HM 0930	Certain knowledge God (Sufism)	KP 3008	— Constraints	HM 2348	Chokmah sephira (Kabbalah)
HM 4556	Certain truth (Sufism)	HH 0476	— Constructive personality	HM 2407	Chonyid bardo (Tibetan)
KC 0371	Certainty	HH 0476	— Constructive psychotherapeutic	KC 0395	Choreography
VC 0301	Certainty	HM 4637	Change lineage ; Knowledge (Buddhism)	HM 2407	Chos-nid (Buddhism)
HM 0930	Certainty ; Science (Sufism)	HM 3065	Change ; Luminous (ICA)	MS 8326	Chrematomorphic symbols
HM 4426	Certainty (Sufism)		**Change [Mathematics...]**	HH 5328	Christ ; Becoming
HM 4556	Certainty ; Truth (Sufism)	KC 0462	— Mathematics discontinuous		**Christ consciousness**
VP 5513	Certainty-Uncertainty	KP 3007	— Modes	HM 2013	— (Christianity)
HM 4426	Certitude ; Awareness (Sufism)	TP 0049	— (Moulting)	HM 1402	— dying-rising (Christianity)
HM 2273	Certitude centre (ICA)	VP 5139	Change-Permanence	HM 0778	— empty mind (Christianity)
	Cessation [Attainment...]	TC 1828	Change preferences ; Participant	HM 1728	— forgiving mind (Christianity)
HM 5438	— Attainment (Buddhism)	HM 3421	Change ; Resistance (Physical sciences)	HM 0976	— loving mind (Christianity)

Christ

Christ consciousness cont'd
HM 2880 — Mystical (Christianity)
Christ [Imitation...]
HH 0810 — Imitation (Christianity)
HM 3140 — Imitation (ICA)
HM 2480 — Incarnate (ICA)
HH 1167 Christ ; Names (Christianity)
HM 2013 Christ ; Presence (Christianity)
HM 2206 Christ ; Presence Jesus (ICA)
HM 3140 Christ ; Replication (ICA)
HM 2644 Christ ; Universal (ICA)
HH 6211 Christian ; Becoming (Christianity)
HM 2013 Christian consciousness (Christianity)
HH 5023 Christian meditation (Christianity)
HM 2077 Christian recollection (Christianity)
Christian [self...]
HH 6732 — self-love (Christianity)
HM 2880 — stations awareness (Christianity)
HH 3121 — stewardship (Christianity)
HH 1223 Christianity Abandonment God
HM 1534 Christianity Abgeschiedenheit
HH 6548 Christianity Actual grace
HM 2895 Christianity Age Father
HM 3499 Christianity Age Holy Spirit
HM 3087 Christianity Age Son
HH 3652 Christianity Ascetical theology
HM 0876 Christianity Awareness sacred
HH 6211 Christianity Becoming Christian
HM 0102 Christianity Beginner spiritual life
HM 6522 Christianity Birth God soul's ground
HM 8772 Christianity Blissful love
HH 0576 Christianity Born-again
HH 0619 Christianity Calling
HM 1167 Christianity Canonical hours
HH 1994 Christianity Centering prayer
HH 3124 Christianity Charismatic renewal
HH 0583 Christianity Chastity
HM 2013 Christianity Christ consciousness
HM 1402 Christianity Christ consciousness dying-rising
HM 0778 Christianity Christ consciousness empty mind
HM 1728 Christianity Christ consciousness forgiving mind
HM 0976 Christianity Christ consciousness loving mind
HM 2013 Christianity Christian consciousness
HH 5023 Christianity Christian meditation
HM 2077 Christianity Christian recollection
HH 6732 Christianity Christian self-love
HM 2880 Christianity Christian stations awareness
HH 3121 Christianity Christian stewardship
HH 4201 Christianity Christotherapy
HM 4321 Christianity Compline
HH 7634 Christianity Compromise
HM 4251 Christianity Compunction
HM 2823 Christianity Conceit
HM 3502 Christianity Consolation
HH 2145 Christianity Contemplative life
HH 2768 Christianity Corporate worship
HH 1098 Christianity Crusade evangelism
HM 0059 Christianity Cry
HM 1714 Christianity Dark night
HM 1714 Christianity Dark night Saint John Cross
HM 1727 Christianity Dark night senses
HM 3941 Christianity Dark night soul
HH 3012 Christianity Degrees love
HM 5119 Christianity Desolation
HM 1534 Christianity Detachment
HH 0349 Christianity Devotion
HM 0810 Christianity Diaconia
HM 0810 Christianity Diakonia
HH 0956 Christianity Discipline celebration
HH 0825 Christianity Discipline confession
HH 1244 Christianity Discipline guidance
HH 1688 Christianity Discipline meditation
HH 1193 Christianity Discipline prayer
HH 0232 Christianity Discipline service
HH 0333 Christianity Discipline solitude
HH 1323 Christianity Discipline study
HH 0225 Christianity Discipline submission
HH 3776 Christianity Discipline worship
HH 0298 Christianity Discipline worship
HH 3900 Christianity Discretion
HH 0197 Christianity Discursive meditation
HM 6997 Christianity Discursive reasoning
HH 4001 Christianity Disruptive thoughts
HM 4340 Christianity Divine contemplation
HH 0158 Christianity Divine reading
HH 0397 Christianity Doctrine three ages
HM 3311 Christianity Dread
HH 0099 Christianity Ecumenicism
HH 0134 Christianity Elevation man
HM 3098 Christianity Envy
HM 6111 Christianity Essential unity being
HH 3798 Christianity Evangelism
HM 0284 Christianity Experience miracles
HH 1801 Christianity Family life
HM 3311 Christianity Fear
HM 0673 Christianity Fear God
HM 6989 Christianity First degree prayer
HM 2224 Christianity First duty
HM 3748 Christianity First step: ego
HH 6019 Christianity Following Jesus
HH 0899 Christianity Forgiveness
HM 3570 Christianity Fourth step: earth
HH 4020 Christianity Freedom
HM 4497 Christianity Fund individual
HH 1608 Christianity Glossolalia
HH 4202 Christianity Gratitude
HH 0205 Christianity Hesychasm
HM 7554 Christianity Higher knowing
HM 0876 Christianity Holy
HM 0673 Christianity Holy fear
HM 5776 Christianity Holy indifference
HM 8022 Christianity Holy resignation
HH 0199 Christianity Hope
HM 4321 Christianity Hours compline
HM 0894 Christianity Hours lauds
HM 0160 Christianity Hours matins
HM 0466 Christianity Hours none
HM 1904 Christianity Hours prime
HM 3725 Christianity Hours sext
HM 2965 Christianity Hours terce
HM 1468 Christianity Hours vespers
HH 4533 Christianity Hubris
HH 2198 Christianity Human development
HM 0446 Christianity Human holiness
HH 0861 Christianity Humility
HM 4533 Christianity Hybris
HH 0816 Christianity Imageless prayer
HH 0810 Christianity Imitation
HH 0810 Christianity Imitation Christ
HH 6211 Christianity Initiation
HM 3412 Christianity Innocence
HH 1409 Christianity Interior castle Saint Teresa
HH 3008 Christianity Interior life
HM 3098 Christianity Jealousy
HM 2535 Christianity Jesus condemned death
HM 3719 Christianity Jesus dies cross
HM 4009 Christianity Jesus falls first time His cross
HM 3824 Christianity Jesus falls second time cross
HM 2289 Christianity Jesus falls third time cross
HM 3344 Christianity Jesus helped carry cross
HM 3500 Christianity Jesus made bear cross
HM 3611 Christianity Jesus meets his afflicted mother
HM 4136 Christianity Jesus nailed cross
HM 3536 Christianity Jesus placed sepulchre
HH 0205 Christianity Jesus prayer
HM 2737 Christianity Jesus speaks daughters Jerusalem
HM 3166 Christianity Jesus stripped his garments
HM 3410 Christianity Jesus taken down cross
HH 6505 Christianity Journeying transcendence
HH 2661 Christianity Journeying transcendence - bondage freedom
HH 3352 Christianity Journeying transcendence - call mission
HH 4110 Christianity Journeying transcendence - extraordinary ordinary
HH 6012 Christianity Journeying transcendence - served servant
HM 1608 Christianity Jubilatio
HH 0956 Christianity Jubilee spirit
HH 0341 Christianity Justification
HM 4256 Christianity Kerygma
HM 0144 Christianity Koinonia
HM 0894 Christianity Lauds
HH 2023 Christianity Laying hands
HH 4552 Christianity Liberation theology
HH 4209 Christianity Liturgical prayer
HH 5232 Christianity Love God
HM 6997 Christianity Lower knowing
HM 2823 Christianity Mana
HH 3007 Christianity Manifestation conscience
HM 2969 Christianity Mansions exemplary life
HM 2470 Christianity Mansions favours afflictions
HM 3382 Christianity Mansions humility
HM 4110 Christianity Mansions incipient union
HM 3218 Christianity Mansions practice prayer
HH 1409 Christianity Mansions soul
HM 3443 Christianity Mansions spiritual consolations
HM 3971 Christianity Mansions spiritual marriage
HH 1098 Christianity Mass evangelism
HM 0160 Christianity Matins
HM 0541 Christianity Meeting God
HM 6204 Christianity Meeting God active desire
HM 5309 Christianity Meeting God both resting working accordance righteousness
HM 8772 Christianity Meeting God emptiness
HM 7384 Christianity Melting
HH 8672 Christianity Mental prayer
HH 0888 Christianity Metanoia
HH 3798 Christianity Mission
HM 6997 Christianity Mundane knowledge
HM 3889 Christianity Mystic union God
HM 2880 Christianity Mystical Christ-consciousness
HM 0402 Christianity Mystical journey
HM 3941 Christianity Mystical ladder divine love
HH 5217 Christianity Mystical theology
HH 1167 Christianity Names Christ
HM 5119 Christianity Negative feeling towards God
HM 2823 Christianity Nga-rgyal (Buddhism, Tibetan)
HM 1727 Christianity Night correction
HM 9330 Christianity Non-desire
HM 0466 Christianity None
HM 8123 Christianity Passive state
HH 3124 Christianity Pentecostal movement
HH 7343 Christianity Perfection soul
HH 3652 Christianity Perfection spiritual life
HM 6335 Christianity Personal salvation
HM 1265 Christianity Phenomenon presence
HM 3502 Christianity Positive feeling towards God
HH 3007 Christianity Practice confidence
HM 0238 Christianity Prayer repose
HM 0238 Christianity Prayer silence
HM 0238 Christianity Prayer simplicity
HM 1608 Christianity Praying tongues
HH 0563 Christianity Predestination
HM 2013 Christianity Presence Christ
HM 2823 Christianity Pride
HM 1904 Christianity Prime
HH 6434 Christianity Progress
HH 2909 Christianity Psychospiritual growth
HH 4201 Christianity Psychospiritual therapy
HM 0074 Christianity Pure love
HM 1727 Christianity Purification senses
HH 4022 Christianity Quaker meditation
HH 4477 Christianity Quietism
HH 0164 Christianity Reconciliation
HH 0167 Christianity Redemption
HH 0618 Christianity Regeneration
HM 3595 Christianity Relationship God man
HM 2766 Christianity Relationship man man
HM 3488 Christianity Relationship man nature
HH 5022 Christianity Religious psychology
HM 4111 Christianity Religious states
HM 8123 Christianity Repose God
HM 9213 Christianity Resting God
HM 8004 Christianity Resting Spirit
HM 0560 Christianity Sainthood
HH 5116 Christianity Sanctifying grace
HM 0560 Christianity Sanctity
HM 3420 Christianity Saviors God
HM 3462 Christianity Saviors God: action
HM 3439 Christianity Saviors God: march
HM 3264 Christianity Saviors God: preparation
HM 3977 Christianity Saviors God: silence
HM 2855 Christianity Saviors God: vision
HH 4098 Christianity Second birth
HM 0238 Christianity Second degree prayer
HM 2861 Christianity Second duty
HM 2953 Christianity Second step: race
HH 3710 Christianity Self-culture
HH 2121 Christianity Self-denial
HH 3710 Christianity Self-development
HH 2121 Christianity Self-discipline
HM 2823 Christianity Self-regard
HH 1223 Christianity Self-renunciation
HH 4477 Christianity Selfless love
HM 6989 Christianity Sense presence God
HM 0284 Christianity Sense wonder
HM 3725 Christianity Sext
HH 1199 Christianity Soul care
HH 1199 Christianity Soul friendship
HM 7384 Christianity Speaking forth
HM 1608 Christianity Speaking tongues
HM 6522 Christianity Speaking Word soul
HM 0936 Christianity Spiritual anger
HM 7143 Christianity Spiritual attentiveness
HM 0642 Christianity Spiritual avarice
HM 2158 Christianity Spiritual childhood
HH 3900 Christianity Spiritual discernment
HM 3818 Christianity Spiritual envy
HM 3420 Christianity Spiritual exercises Nikos Kazantzakis
HH 9760 Christianity Spiritual exercises Saint Ignatius
HH 1199 Christianity Spiritual friendship
HM 0507 Christianity Spiritual gluttony
HM 0642 Christianity Spiritual greed
HH 1199 Christianity Spiritual guidance
HM 0102 Christianity Spiritual immaturity
HM 7554 Christianity Spiritual knowledge
HM 8977 Christianity Spiritual love
HM 4180 Christianity Spiritual lust
HH 0465 Christianity Spiritual marriage
HM 6226 Christianity Spiritual meekness
HM 4533 Christianity Spiritual pride
HH 4098 Christianity Spiritual rebirth
HH 0576 Christianity Spiritual renewal
HH 3124 Christianity Spiritual renewal
HM 6552 Christianity Spiritual repose
HH 4522 Christianity Spiritual self-denial
HH 4522 Christianity Spiritual self restraint
HM 0491 Christianity Spiritual sloth
HH 3652 Christianity Spiritual theology
HH 5217 Christianity Spiritual theology
HH 0465 Christianity Spiritual union
HH 0792 Christianity Spirituality
HM 6445 Christianity State grace
HH 1723 Christianity State holiness
HM 7656 Christianity State hope
HM 0074 Christianity State perfect love
HM 4340 Christianity State purity
HM 3771 Christianity State quies
HM 8772 Christianity State rest
HM 5467 Christianity State self-surrender
HM 0074 Christianity State transformation
HM 0357 Christianity States prayer
HM 2880 Christianity Stations cross
HM 3516 Christianity Stations cross
HH 2207 Christianity Suffering
HM 4340 Christianity Superessential contemplation
HH 1098 Christianity Television evangelism
HM 2965 Christianity Terce
HM 4279 Christianity Third duty
HM 3501 Christianity Third step: mankind
HH 0631 Christianity Three spiritual ways
HH 1173 Christianity Traditionalism
HH 0631 Christianity Triple way
HH 5767 Christianity True knowledge
HH 1471 Christianity Uncertainty
HM 3889 Christianity Unio mystica
HH 8206 Christianity Unity personality
HM 3401 Christianity Veronica wipes face Jesus
HM 1468 Christianity Vespers

GENERAL INDEX TO VOLUME 2
Communion

HH 6030 Christianity Via illuminativa
HH 4090 Christianity Via purgativa
HH 4100 Christianity Via unitiva
HM 0357 Christianity Visions Trinity
HH 0619 Christianity Vocation
HM 6129 Christianity Waiting God
HM 3516 Christianity Way cross
HH 3335 Christianity Way paradox
HH 6566 Christianity Way transcendence
HH 1616 Christianity Ways knowing
HH 5022 Christianity We-psychology
HM 1608 Christianity Xenoglossis
HH 4201 Christotherapy (Christianity)
MS 8923 Chromatic symbolism
HH 0007 Chromotherapy
HM 2472 Chronic irritability
HM 3485 Chu ; Wu (Zen)
KC 0870 Chunking
HM 0945 Churlishness (Buddhism)
MS 8371 Cinema
HM 3966 Cinta (Pali)
HM 1397 Circadian rhythm
HH 3487 Circle ; Home
HH 0350 Circles enlightenment (Zen)
KC 0600 Circuit ; Integrated
KC 0069 Circuit language ; Energy
HH 6452 Circulation light (Taoism)
VP 5234 Circumscription-Intrusion
VC 0327 Circumspection
HM 0592 Cit ananda ; Sat
HM 3227 Cit ananda ; Sat (Hinduism)
KC 0067 Citizenship ; International
KC 0067 Citizenship ; Planetary
KC 0067 Citizenship ; World
HM 3631 Citrinistas (Esotericism)
Citta [bhavana...]
HH 5973 — bhavana (Buddhism)
HM 2730 — (Buddhism)
HM 6033 — Buoyancy (Pali)
HM 3529 Citta (Hinduism)
HM 6663 Citta ; One pointedness (Buddhism)
HM 1810 Citta-pagunnata (Pali)
HM 1155 Citta ; Pliancy (Pali)
HM 9001 Citta ; Rectitude (Pali)
HM 0226 Citta ; Repose (Pali)
HM 6149 Citta visuddhi (Buddhism)
HM 6556 Citta ; Wieldiness (Pali)
HM 1584 Cittakammannata (Pali)
HM 1830 Cittalahuta (Pali)
HM 1805 Cittamuduta (Pali)
HM 1490 Cittapassaddhi (Pali)
HM 3471 Cittasa avatthiti (Pali)
HM 3949 Cittassa ; Anattamanata (Pali)
HM 0140 Cittassa ; Attamanata (Pali)
HM 0400 Cittassa ; Bhantattam (Pali)
HM 6663 Cittassa ekaggata (Buddhism)
HM 0223 Cittassa manovilekha (Pali)
HM 1066 Cittassa santhiti (Pali)
HM 0051 Cittassa thiti (Pali)
HM 0975 Cittassa uddhacca (Pali)
HM 6663 Cittassekaggata (Pali)
HM 0253 Cittujukata (Pali)
HM 3068 City bliss bestowing hands awareness ; Entering (Zen)
HM 3063 City shining jewel awareness (Yoga)
HH 0759 Civil disobedience
VC 0329 Civil rights
MS 8223 Civil symbols
VC 0331 Civility
KC 0678 Civilization
VC 0333 Civilization
MS 8803 Civilization ; Symbols
KC 0410 Cladistics
HM 4245 Clairaudience ; Emotional (Psychism)
HM 4333 Clairaudience (Psychism)
HM 4516 Clairempathy (Psychism)
HM 0412 Clairgustance (Psychism)
HM 0318 Clairscent (Psychism)
HM 0679 Clairvoyance ; Auric (Psychism)
HM 1118 Clairvoyance ; Emotional (Psychism)
HM 2262 Clairvoyance (ESP)
HM 4071 Clairvoyance ; Fourth dimensional (Psychism)
HM 3498 Clairvoyance (Psychism)
HM 2841 Clairvoyances ; Five (Buddhism)
HM 4875 Clairvoyant intuition (Hinduism)
TP 0037 Clan (Family)
VC 0335 Clarity
HM 6201 Clarity awareness ; Diminished
HM 0455 Clarity consciousness (Leela)
TP 0057 Clarity ; Penetrating (Wind)
KC 0768 Class
KC 0671 Classic
MS 8613 Classical deities
KC 0333 Classical field theory
HM 2757 Classical story (ICA)
KC 0045 Classification
KC 0622 Classification
KC 0944 Classification schedule ; Thesaurus
HH 3664 Cleaning (Taoism)
VC 0337 Cleanliness
VP 5681 Cleanness-Uncleanness
HM 6932 Cleansing heart (Sufism)
HH 0401 Cleansing ; Ritual
HM 5353 Cleansing sirr (Sufism)
HM 2335 Clear light awareness ; Death bardo (Buddhism)
HM 3690 Clear light ; Yoga (Buddhism)
HM 2428 Clear state consciousness ; Theta (Scientology)
HH 1076 Clearing
VC 0339 Clearness
VC 0341 Clemency
VC 0343 Cleverness
HH 0095 Client-centred therapy
HM 2232 Climax application awareness ; Mahayana (Buddhism)
HM 2220 Climax application awareness ; Tantra (Buddhism)
MM 2080 Climbing
HH 3321 Clinging ; Aggregates (Buddhism)
HM 6153 Clinging (Hinduism)
TC 1718 Clock ; Cost
HM 2157 Close setting mind (Buddhism)
KC 0646 Closed system
VD 2236 Closeness
HM 2377 Closeness God (Sufism)
KD 2090 Closing: alternation discontinuous learning ; Opening
VP 5265 Closure ; Opening
MS 8436 Clothing
MS 8925 Clothing ; Mythical
HM 5754 Clouding consciousness
VC 0345 Clout
TC 1546 Clown
MM 2083 Club
VD 2238 Clumsiness
MP 1037 Cluster frameworks
Co [consciousness...]
HH 0763 — consciousness
HM 2094 — consciousness
TC 1381 — counsel conferencing
HH 0342 — counselling
HH 0763 Co-presence
VC 0347 Coalition
VD 2240 Coarseness
HM 4094 Cocaine bug
HM 1594 Cocaine ; Modes awareness associated use
HH 0183 Code ; Holiness
HM 0087 Coding ; Awareness
HM 1406 Coding ; Self conscious
HH 0766 Coéism
TC 1587 Coercive power
KC 0533 Coevolution
VC 0349 Cogency
Cognition aggregate synergies
HM 1377 — Composedness aggregate sensation, aggregate (Buddhism)
HM 3397 — Flexibility aggregate sensation, aggregate (Buddhism)
HM 0349 — Good quality aggregate sensation, aggregate (Buddhism)
HM 1435 — Lightness aggregate sensation, aggregate (Buddhism)
HM 0409 — Straightness aggregate sensation, aggregate (Buddhism)
HM 1381 — Workability aggregate sensation, aggregate (Buddhism)
Cognition [B...]
HM 2474 — B
HM 2098 — Bare (Buddhism)
HM 2474 — Being
HM 1271 — born contact element eye-consciousness (Buddhism)
HM 1501 — born contact element mind-consciousness (Buddhism)
HM 1389 — (Buddhism)
HM 0713 Cognition ; Conceptual (Buddhism)
HM 3157 Cognition ; Deficiency
HM 2668 Cognition ; Distorted (Buddhism)
HM 2033 Cognition emptiness ; Direct non conceptual (Buddhism)
HM 0888 Cognition ; Innocent
HM 3104 Cognition ; Intuitional (Japanese)
HM 2272 Cognition ; Mystical
HM 2051 Cognition nor non cognition ; Sphere neither (Buddhism)
HM 3182 Cognition ; Obstacles soul (Yoga)
HM 2262 Cognition ; Paranormal (ESP)
HM 1732 Cognition ; Plateau
Cognition [Self...]
HM 3604 — Self (Sufism)
HM 1732 — Serene B
HM 2051 — Sphere neither cognition nor non (Buddhism)
HM 2829 — Spiritual intuitive
HM 3508 — Subsequent (Buddhism)
HH 8204 Cognitive dissonance
HH 0146 Cognitive evolution
HH 0146 Cognitive growth development
KC 0108 Cognitive mobilization
KC 0783 Cognitive system
KD 2225 Cognitive systematization
HM 3419 Cognizing (Buddhism)
KC 0866 Coherence
VC 0351 Coherence
KP 3006 Coherence renewal
MP 1074 Coherent irrational perspectives ; Interaction
MP 1028 Coherent pattern relationship densities
VP 5050 Cohesion-Disintegration
HM 3102 Cohesive intelligence ; Path (Judaism)
KC 0458 Coincidence
KC 0026 Coincidence
VP 5328 Cold ; Heat
HM 2040 Cold-hells awareness (Buddhism)
VD 2242 Coldness
VC 0353 Collaboration
Collectedness [mind...]
HM 4999 — mind (Buddhism)
HM 0311 — mind (Buddhism)
Collectedness [moral...] cont'd
HM 6663 — moral thought (Buddhism)
HM 2811 Collective conscience (Psychosynthesis)
HH 3764 Collective consciousness ; Theory
HM 3046 Collective intelligence ; Path (Judaism)
HH 0388 Collective liberation
HM 2602 Collective mob mind
KC 0662 Collective ; Socialist
HM 0085 Collective unconscious (Jung)
HM 2811 Collective unconscious (Psychosynthesis)
KC 0057 Collectivization mankind
HM 2812 Collegiality ; Disinterested (ICA)
HM 2333 Collegiality ; Inclusive (ICA)
KC 0652 Collegiate authority
TC 1535 Colloquy
HM 2352 Colloquy ; Primordial (ICA)
VD 2244 Collusion
MS 8923 Colour
VP 5362 Colour-Colourlessness
HH 3246 Colour kasina (Buddhism)
HH 0007 Colour therapy
VP 5362 Colourlessness ; Colour
HM 2670 Coma
HH 0975 Coma ; Insulin
VD 2246 Combativeness
MM 2051 Combustion engine ; Internal
VC 0355 Comfort
VP 5885 Comfort-Aggravation
VC 0357 Comfortableness
HM 0645 Comic sentiment (Hinduism)
HM 3153 Coming-home-on-the-ox's-back awareness (Zen)
VC 0359 Commendability
KC 0663 Commensalism
KC 0274 Commensurability
Commitment
HH 0558 — [Commitment]
VC 0361 — [Commitment]
TC 1345 — [Commitment]
HH 0571 Commitment ; Ideological
HM 3230 Commitment ; Unlimited (ICA)
TC 1962 Committed activists ; Alienation
HM 3179 Committed teacher (ICA)
MM 2027 Commodities products ; Resources,
HM 3092 Common cross (Astrology)
Common domain
MP 1129 — focal point structure
MP 1124 — interfaces ; Ensuring function
MP 1126 — Off centering focal point
MP 1067 — Unstructured
Common domains
MP 1063 — Cyclic interrelation complementary perspective
MP 1123 — Enhancing function
MP 1094 — Hospitable
MP 1125 — Integrating points perspective
HM 2023 Common earth (ICA)
MP 1069 Common external context inactivity
KD 2272 Common focus ; Undefined
Common sense
HH 0510 — [Common sense]
KC 0735 — [Common sense]
VC 0363 — [Common sense]
MP 1061 Common small scale interaction domains ; Bounded
VD 2248 Commonness
VC 0365 Commonweal
TC 1848 Commonweal conferencing
HH 5522 Communalist human development
HH 0058 Commune
KC 0905 Commune
KD 2207 Communicable insights: geometry connectivity
KC 0265 Communication
VC 0367 Communication
KC 0906 Communication channel capacity
VT 8061 Communication*complex
KD 2202 Communication ; Encompassing varieties
HH 0588 Communication ; Mass
MP 1113 Communication media ; Harmoniously structured entry point external
MP 1120 Communication networks ; Focus oriented
MP 1052 Communication networks unmediated relationships ; Interfacing vehicles
Communication pathway
MP 1165 — Exposure structural activities
MP 1140 — Relative isolation structural interface
MP 1164 — structure ; Overview
MP 1112 — Transition domain structure
Communication pathways
MP 1122 — Direct relationship structures
MP 1174 — enfolded non-linearity
MP 1121 — Hospitability
MP 1132 — Limiting length
MP 1059 — Low intensity
MP 1101 — Protected low density
MP 1173 — Protecting non linear contexts
TC 1943 Communication patterns
MM 2035 Communication ; Radio
HH 1306 Communication supernatural beings
TC 1866 Communication teams ; Inter cultural
MP 1054 Communications ; Intersection differently paced
MP 1229 Communications ; Provision pathways automatic
TC 1816 Communications unit ; Electronic
MP 1020 Communications ; User determined specialized
KC 0140 Communicative social integration
VC 0369 Communicativeness
VP 5554 Communicativeness-Uncommunicativeness
KC 0425 Communion
VC 0371 Communion

Communion

HM 2675 Communion deity
HM 2675 Communion God
HM 0915 Communion nature
HM 2338 Communion saints (ICA)
Community
KC 0152 — [Community]
VC 0373 — [Community]
HH 0058 — [Community]
HM 2777 Community ; Everlasting (ICA)
TP 0037 Community (Family)
KC 0786 Community ; International
HM 2067 Community ; Northern continent awareness (Buddhism)
HH 0058 Community ; Religious
HH 0336 Community ; Therapeutic
HM 1404 Comoción (Spanish)
HM 0973 Companionate love
VC 0375 Companionship
HM 0764 Company ; Bad (Leela)
HM 1355 Company ; Good (Leela)
KC 0647 Comparative analysis
KC 0647 Comparative studies
HM 0427 Compare ; Inability (Buddhism)
HM 2919 Compassion
VC 0377 Compassion
HM 2774 Compassion ; Agape (ICA)
HM 5634 Compassion (Buddhism)
HM 0513 Compassion (Buddhism)
HM 2807 Compassion ; Great (Buddhism)
VP 5944 Compassion-Pitilessness
HM 2734 Compassion ; Universal (ICA)
KC 0434 Compatibility
VC 0379 Compatibility
MP 1221 Compatibility ; Aperture
MP 1023 Compensating relationships parallel
VC 0381 Compensation
VC 0383 Competence
KC 0556 Competition
VC 0385 Competition
MP 1072 Competitive interaction opportunities transposed concrete level
VD 2250 Complacency
KC 0722 Complementarity
MP 1027 Complementarity
KD 2083 Complementarity determinism fluctuation
KD 2070 Complementarity ; Observer entrapment micro macro
KD 2154 Complementary languages
KD 2342 Complementary metaphors ; Patterns
MP 1003 Complementary modes organization ; Interpretation
MP 1063 Complementary perspective common domains ; Cyclic interrelation
Complementary perspectives
MP 1187 — Appropriate configuration interaction
MP 1077 — Minimal context
MP 1136 — Relative isolation
MP 1106 Complementary structures ; Functional enhancement domains separating
KC 0349 Complementation
KC 0174 Complementology
HM 0634 Complete awareness (Buddhism)
HM 1956 Complete delusion (Buddhism)
HH 0727 Complete man
HH 3998 Complete personality
HM 3411 Complete tranquillization (Buddhism)
HM 1528 Completely tranquillized (Buddhism)
Completeness
HM 2725 — [Completeness]
HM 3070 — [Completeness]
KC 0834 — [Completeness]
KC 0142 Completeness equipment
VP 5054 Completeness-Incompleteness
TC 1526 Completer role ; Team
VC 0387 Completion
TP 0063 Completion ; After
TP 0064 Completion ; Before
KC 0010 Complex ; Mathematical
Complex [social...]
KC 0907 — social systems
MP 1099 — structure ; Focal centre
MP 1102 — structures ; Distinct pattern entry points
MP 1098 — structures ; Patterning
KC 0336 — systems
HM 0424 Complexes ; Knowledge indifference (Buddhism)
KC 0628 Complexification
MP 1095 Complexification perspective contexts
KC 0800 Complexity
VD 2252 Complexity
KC 0604 Complexity ; Management
KC 0336 Complexity ; Organized
VP 5045 Complexity ; Simplicity
KC 0183 Complexity ; System
VC 0389 Compliance
VT 8073 Compliance*complex
HM 2917 Compliance (Japanese)
VD 2254 Complication
VD 2256 Complicity
HM 4321 Compline (Christianity)
HM 4321 Compline ; Hours (Christianity)
HM 4605 Composedness aggregate consciousness (Buddhism)
HM 1377 Composedness aggregate sensation, aggregate cognition, aggregate synergies (Buddhism)
HM 1867 Composedness body (Buddhism)
HM 1490 Composedness thought (Buddhism)
MM 2065 Composition
HM 1406 Composition
VC 0391 Composure
HM 1738 Composure (Buddhism)
KD 2110 Comprehensibility: fourfold minimal system ; Threshold
KC 0187 Comprehensibility systems
HM 2660 Comprehension
VC 0393 Comprehension
KD 2205 Comprehension holes ; Non
Comprehension [identity...]
KD 2325 — identity transformation ; Game
HM 2256 — Inclusive (ICA)
KD 2210 — internalization ; Discontinuity:
HM 3490 Comprehension known (Buddhism)
HM 4348 Comprehension rejection (Buddhism)
HM 4552 Comprehension scrutiny (Buddhism)
KD 2270 Comprehension structuring phenomenon learning society ; Non
HM 4090 Comprehension ; Un (Buddhism)
KC 0106 Comprehensive thinking
KC 0708 Comprehensiveness
VC 0395 Comprehensiveness
VC 0397 Compromise
TC 1011 Compromise
HH 7634 Compromise (Christianity)
VP 5806 Compromise ; Neutrality
HH 0345 Compulsion
HM 2745 Compulsion confess (Brainwashing)
VP 5756 Compulsion ; Leniency
VD 2258 Compulsiveness
VC 0399 Compunction
HM 4251 Compunction (Christianity)
MS 8933 Computer
TC 1753 Computer conferencing ; In conference
TC 1616 Computer contact system ; In conference
KC 0196 Computer language ; Analog
KC 0836 Computer program
TC 1666 Computer teleconferencing
VC 0401 Comradeship
HM 2976 Comradeship ; Missional (ICA)
HM 2605 Concealment (Buddhism)
TP 0036 Concealment illumination (Darkening of the light)
MP 1097 Concealment necessary monotonous perspective patterns
VD 2260 Conceit
HM 2823 Conceit (Christianity)
HM 0098 Conceit (Leela)
Concentration
HM 3169 — [Concentration]
HM 4226 — [Concentration]
VC 0403 — [Concentration]
HM 0311 Concentration ; Absorption (Buddhism)
HM 4999 Concentration ; Access (Buddhism)
Concentration accompanied
HM 4191 — bliss (Buddhism)
HM 1454 — bliss (Buddhism)
HM 4191 — ease (Buddhism)
HM 1454 — ease (Buddhism)
HM 5008 — equanimity (Buddhism)
HM 8021 — equanimity (Buddhism)
HM 5008 — even-mindedness (Buddhism)
HM 8021 — even-mindedness (Buddhism)
HM 4433 — happiness (Buddhism)
HM 5008 — indifference (Buddhism)
HM 8021 — indifference (Buddhism)
HM 4433 — rapture (Buddhism)
Concentration applied thought
HM 4908 — sustained thought (Buddhism)
HM 5199 — sustained thought (Buddhism)
HM 7543 — sustained thought (Buddhism)
HM 2440 Concentration (Buddhism)
HM 6663 Concentration (Buddhism)
Concentration [Consciousness...]
HM 5588 — Consciousness reinforced (Buddhism)
HM 1735 — Correct (Buddhism)
HH 8231 — Cultivation (Buddhism)
Concentration [Desire...]
HM 4969 — Desire (Buddhism)
HM 5992 — difficult progress sluggish direct-knowledge (Buddhism)
HM 7859 — difficult progress swift direct-knowledge (Buddhism)
Concentration due
HM 8365 — energy (Buddhism)
HM 7434 — inquiry (Buddhism)
HM 7961 — natural purity consciousness (Buddhism)
HM 4969 — zeal (Buddhism)
Concentration easy progress
HM 4977 — quick intuition (Buddhism)
HM 1010 — sluggish direct-knowledge (Buddhism)
HM 1010 — sluggish intuition (Buddhism)
HM 4977 — swift direct-knowledge (Buddhism)
HM 0311 Concentration ; Ecstatic (Buddhism)
HM 4036 Concentration ; Exalted (Buddhism)
Concentration [Faculty...]
HM 3454 — Faculty (Buddhism)
HM 5143 — fifth jhana five (Buddhism)
HM 4265 — Fine material sphere (Buddhism)
HM 2450 — First form realm (Buddhism)
HM 0297 — first jhana five (Buddhism)
HM 4456 — first jhana four (Buddhism)
HH 0687 — flow breath
HM 4265 — Form realm (Buddhism)
HM 0696 — Formless realm (Buddhism)
HM 2586 — Fourth form realm (Buddhism)
HM 4882 — fourth jhana five (Buddhism)
HM 7202 — fourth jhana four (Buddhism)
Concentration [happiness...]
HM 5767 — happiness (Buddhism)
HM 0730 — happiness (Buddhism)
HM 2566 — (Hinduism)
Concentration [Immaterial...]
HM 0696 — Immaterial sphere (Buddhism)
HM 0298 — Inability achieve (Yoga)
HM 5562 — Inferior (Buddhism)
HM 0496 — Infinite (Buddhism)
HM 5214 — infinite object ; Infinite (Buddhism)
HM 5547 — infinite object ; Limited (Buddhism)
HM 7434 — Investigation (Buddhism)
HM 4006 Concentration ; Lack
HM 1073 Concentration ; Limited (Buddhism)
Concentration limited object
HM 4653 — Infinite (Buddhism)
HM 1175 — Limited (Buddhism)
HM 4653 — Measureless (Buddhism)
Concentration [Measureless...]
HM 0496 — Measureless (Buddhism)
HM 5547 — measureless object ; Limited (Buddhism)
HM 5214 — measureless object ; Measureless (Buddhism)
HM 6089 — Medium (Buddhism)
HM 6089 — Middling (Buddhism)
HM 7961 — Mind (Buddhism)
HM 5588 — Mind upheld (Buddhism)
HM 7234 — Mundane (Buddhism)
HM 7859 Concentration painful progress quick intuition (Buddhism)
HM 5992 Concentration painful progress sluggish intuition (Buddhism)
Concentration partaking
HM 8002 — diminution (Buddhism)
HM 7363 — distinction (Buddhism)
HM 5930 — penetration (Buddhism)
HM 6956 — stability (Buddhism)
HM 6956 — stagnation (Buddhism)
HM 8002 — worsening (Buddhism)
HM 0824 Concentration ; Perceptual
HM 1759 Concentration ; Power (Buddhism)
HM 5767 Concentration rapture (Buddhism)
HM 0730 Concentration rapture (Buddhism)
Concentration Right
HM 0931 — (Buddhism)
HM 1735 — (Buddhism)
HM 0931 — rapture (Buddhism)
Concentration [Second...]
HM 2038 — Second form realm (Buddhism)
HM 4575 — second jhana five (Buddhism)
HM 4380 — second jhana four (Buddhism)
HM 1097 — Sense sphere (Buddhism)
HM 0220 — Single hearted
HM 4036 — Sublime (Buddhism)
HM 6327 — Superior (Buddhism)
HM 4243 — Supramundane (Buddhism)
Concentration [Third...]
HM 2062 — Third form realm (Buddhism)
HM 8532 — third jhana five (Buddhism)
HM 2284 — third jhana four (Buddhism)
HM 4243 — Transcendental (Buddhism)
HM 5768 Concentration ; Unincluded (Buddhism)
Concentration [Weak...]
HM 6652 — Weak (Buddhism)
HM 7234 — Worldly (Buddhism)
HM 1802 — Wrong (Buddhism)
HM 2693 Concentrations ; Tibetan meditative states form (Buddhism)
KC 0630 Concept
HH 1222 Concept development
HH 0471 Concept ; Ego
HH 1222 Concept formation
HH 0351 Concept man
HH 0297 Concept ; Polytheism
HH 0166 Concept progress
HH 0641 Concept ; Sarvodaya
HH 0471 Concept ; Self
HH 2189 Concept ; Three world
VC 0405 Conception
HM 2220 Conception (Buddhism)
HH 3663 Conception social responsibility ; Maternal
KC 0922 Concepts ; Ecology
KC 0961 Concepts integrated science ; Teaching
HM 4000 Concepts ; Ninth order perceptions control systems
HM 0713 Conceptual cognition (Buddhism)
HM 2033 Conceptual cognition emptiness ; Direct non (Buddhism)
Conceptual [framework...]
TC 1888 — framework development
TC 1002 — framework development
TC 1392 — frameworks
KC 0686 Conceptual integration
TC 1866 Conceptual interpretation teams
TC 1392 Conceptual models
Conceptual [screen...]
TC 1888 — screen development
TC 1002 — screen development
TC 1392 — screens
KC 0627 — systems
HM 2924 Conceptualization ; Energetic (Astrology)
VC 0407 Concern
TC 1996 Concern conferencing ; Business
HM 2939 Concern ; Particular (ICA)
HM 3128 Concern ; Passionate (ICA)
HM 2774 Concern ; Universal (ICA)
HM 2582 Concerned judge (ICA)
KD 2003 Concerning appropriate answers ; Assumptions
HH 3211 Conciliarity
VC 0409 Conciliation
HM 2105 Conciliation ; Path intelligence (Judaism)
TC 1775 Conciliators

Consciousness

GENERAL INDEX TO VOLUME 2

VP 5592	Conciseness-Diffuseness
TC 1544	Conclave
VC 0411	Concord
HM 2233	Concrete awareness
MP 1072	Concrete level ; Competitive interaction opportunities transposed
HH 0348	Concrete thinking
HH 4302	Concretization
VD 2262	Concupiscence
HM 0129	Concupiscible appetite (Islam)
VC 0413	Concurrence
VP 5177	Concurrence-Counteraction
VD 2264	Condemnation
VP 6005	Condemnation ; Vindication
HM 2535	Condemned death ; Jesus (Christianity)
VD 2266	Condescension
HM 3455	Condition ; Intellectual (Hinduism)
HM 2336	Condition ; Material (Hinduism)
HM 3040	Condition ; Mental (Hinduism)
HM 3480	Condition ; Paradisal (Psychism)
HM 2032	Conditional consciousness states (Hinduism)
HM 3228	Conditional ; Reciprocal relation absolute (Sufism)
HM 3060	Conditional self-realization
	Conditioned [reflex...]
HH 0266	— reflex
HH 0985	— reflex therapy
HH 0266	— response
HH 0266	Conditioning
HM 7655	Conditions consciousness (Buddhism)
HM 2599	Conditions consciousness ; Phase (Buddhism)
MP 1182	Conditions perspective nourishment ; Appropriate
HM 0891	Conditions self (Hinduism)
VC 0415	Condolence
VC 0417	Conduct
TP 0010	Conduct ; Careful (Treading)
HH 0429	Conduct ; Good (Zen)
HM 4648	Conduct (Sufism)
MM 2016	Conduit
KC 0803	Confederation
TC 1082	Conference
	Conference [accessibility...]
TC 1808	— accessibility
TC 1244	— authority
TC 1623	— autonomy
	Conference [computer...]
TC 1753	— computer conferencing ; In
TC 1616	— computer contact system ; In
TC 1506	— control
TC 1484	— conviviality
TC 1636	Conference environment
TC 1399	Conference flexibility
TC 1651	Conference fodder
TC 1566	Conference heavies
TC 1683	Conference homogeneity
	Conference [impact...]
TC 1850	— impact
TC 1738	— insulation
TC 1588	— intent
TC 1952	— intimacy
TC 1944	Conference leadership
TC 1972	Conference line
TC 1394	Conference maturation
TC 1481	Conference mythology
	Conference [papers...]
TC 1568	— papers
TC 1490	— participation
TC 1903	— polarization
TC 1362	— potency
TC 1532	— practics
TC 1347	— procedure feedback
TC 1847	— procreation
TC 1760	Conference relationships
TC 1548	Conference rituals
	Conference [size...]
TC 1555	— size
TC 1749	— spirit
TC 1424	— stability
TC 1776	— stamina
TC 1955	— stratification
TC 1717	— structure
TC 1644	— symbols
	Conference [Telephone...]
TC 1870	— Telephone
TC 1782	— tensions
TC 1556	— time design
TC 1061	— tone
TC 1191	Conference viscidity
	Conferencing [Adaptive...]
TC 1843	— Adaptive
TC 1602	— Affinity group
TC 1315	— Amorphous
TC 1996	Conferencing ; Business concern
	Conferencing [Co...]
TC 1381	— Co counsel
TC 1848	— Commonweal
TC 1562	— Conscious
TC 1974	— Consultation
TC 1524	— Contextual
TC 1410	— Coordination
	Conferencing [Decision...]
TC 1385	— Decision making
TC 1735	— Declaration
TC 1940	— Design
TC 1974	— Diagnostic
TC 1815	— dissipation
TC 1439	— Dread filled
	Conferencing [Economic...]
TC 1996	— Economic
TC 1524	— Educative
TC 1722	— Encounter group
TC 1815	— energy sink
TC 1339	— evaluation
TC 1374	— Expressive
TC 1919	Conferencing failure
TC 1602	Conferencing ; Group counsel
TC 1632	Conferencing ; Hierarchical
	Conferencing [In...]
TC 1753	— In conference computer
TC 1615	— inertia
TC 1733	— Instrumental
TC 1920	— intentions
TC 1478	— Interstitial
TC 1574	Conferencing ; Judicial
	Conferencing [Maintenance...]
TC 1564	— Maintenance
TC 1421	— Majoral
TC 1313	— Managerial political
TC 1879	— Maximization
TC 1862	— Medial
TC 1135	— meditation
TC 1525	— Minoral
TC 1343	Conferencing ; Networking
TC 1920	Conferencing objectives
TC 1684	Conferencing ; Problem solving
TC 1996	Conferencing ; Productive
TC 1843	Conferencing ; Research
TC 1553	Conferencing risk
	Conferencing [Self...]
TC 1602	— Self reflective
TC 1696	— Service
TC 1989	— Strongwilled
TC 1547	— Structural dimensions
TC 1564	— Sustenance
TC 1940	Conferencing ; Vision
TC 1940	Conferencing ; Visionary
TC 1908	Conferencing ; Weakwilled
HM 2745	Confess ; Compulsion (Brainwashing)
HM 2718	Confess ; Struggle (ICA)
HH 0825	Confession ; Discipline (Christianity)
HM 3156	Confession ; Final (Brainwashing)
VC 0419	Confidence
HM 1998	Confidence (Buddhism)
TP 0061	Confidence ; Inner
HH 4233	Confidence ; Mutual
HH 3007	Confidence ; Practice (Christianity)
VC 1545	Confidence ; Self
HM 0807	Confidence (Sufism)
MP 1184	Configuration input processing ; Appropriate
MP 1187	Configuration interaction complementary perspectives ; Appropriate
MP 1186	Configuration perspective inactivity ; Appropriate
MP 1185	Configuration perspective interaction ; Appropriate
KC 0380	Configurational cultural integration
KC 0753	Configurations
TC 1943	Configurations ; Constrained meeting
TC 1465	Configurations ; Meetings energy processing
TC 1909	Configurations ; Meetings symbolic
TC 1155	Configurations ; Socio structural meeting
MM 2056	Configurations ; Strategic
HM 0312	Configurations ; Third order perceptions
TC 1699	Configurations ; Unconstrained meeting
TC 1603	Configurative models meetings
KC 0406	Configurative thinking
VD 2268	Confinement
VC 0421	Confirmation
HH 0433	Confirmation
HM 3225	Confirmation otherness
VD 2270	Conflict
TP 0006	Conflict
HH 0708	Conflict ; Basic
HM 2354	Conflict ; Breaking point total (Brainwashing)
HM 0044	Conflict ; Consciousness ego
HH 0421	Conflict ; Management
KD 2170	Conflicting styles ; Modes managing: embracing
VC 0423	Conformation
TC 1663	Conformist participant type ; Non
VC 0425	Conformity
HM 5403	Conformity (Buddhism)
VP 5082	Conformity-Nonconformity
HM 5403	Conformity truth ; Knowledge (Buddhism)
HH 5213	Confucianism
HM 0634	Confucianism Absolute
HH 4301	Confucianism Spirituality
VD 2272	Confusion
VC 0427	Congeniality
MS 8333	Congenital relationship ; Human
VD 2274	Congestion
VP 5948	Congratulation-Envy
TP 0045	Congregation
TC 1562	Congress ; Conscious
MP 1205	Congruence spaces defined framework spaces defined processes it
VC 0429	Congruity
HH 1125	Coniunctio
HH 0362	Conjoint family therapy
VP 5931	Conjugality-Celibacy
VP 5047	Conjunction-Separation
HM 3359	Conjunctive faith
KC 0837	Connectedness
VC 0431	Connectedness
MP 1194	Connectedness domains ; Internal
MP 1237	Connectedness isolation
KC 0871	Connection
VC 0433	Connection
MP 1232	Connection encompassing domains ; Symbolic
MP 1213	Connections ; Distribution secondary inter level
MP 1216	Connections ; Inter level
MP 1212	Connections transitions boundary orientation ; Primary inter level
KC 0360	Connective cultural integration
KD 2207	Connectivity ; Communicable insights: geometry
VC 0435	Conquest
VC 0437	Consanguinity
HH 0466	Conscience
VC 0439	Conscience
HM 1590	Conscience (Buddhism)
HM 2811	Conscience ; Collective (Psychosynthesis)
HH 2344	Conscience ; Freedom
HM 2892	Conscience ; Intentional (ICA)
HM 2249	Conscience ; Keeping (ICA)
HM 1601	Conscience (Leela)
HH 3007	Conscience ; Manifestation (Christianity)
HM 4394	Consciencelessness (Buddhism)
HH 1219	Conscientious objection
VC 0441	Conscientiousness
TP 0062	Conscientiousness
HM 3220	Conscientiousness (Buddhism)
HM 1590	Conscientiousness (Buddhism)
HM 2314	Conscientiousness ; Non (Buddhism)
HH 0027	Conscientization
HM 1741	Consciousness ; Good quality aggregate (Buddhism)
HM 2423	Conscious awareness ; Fatigued
HM 4514	Conscious ; Being (Buddhism)
	Conscious [coding...]
HM 1406	— coding ; Self
TC 1562	— conferencing
TC 1562	— congress
HM 2610	— consciousness ; Self
HM 6652	Conscious duration (Buddhism)
	Conscious energy
HM 2087	— Ida (Yoga)
HM 2871	— Kundalini (Yoga)
HM 2890	— Pingala (Yoga)
	Conscious existence Awareness
HM 2098	— consciousness group (Buddhism)
HM 2108	— corporeality group (Buddhism)
HM 4983	— feeling group (Buddhism)
HM 2050	— mental formation group (Buddhism)
HM 4143	— perception group (Buddhism)
	Conscious existence [phenomena...]
HH 3321	— phenomena ; Awareness inter dependency (Buddhism)
HM 2556	— senses mind ; Awareness consciousness group (Buddhism)
HM 2072	— Worlds (Buddhism)
HM 3321	Conscious groups
TC 1562	Conscious groups
HM 7622	Conscious heart (Sufism)
HM 0168	Conscious learning
HM 2000	Conscious love
HM 2788	Conscious pure intelligence ; Pre
	Conscious [self...]
HM 2123	— self (Psychosynthesis)
HM 0467	— self-relatedness (Yoga)
HM 2417	— states ; Least action principle (Physical sciences)
	Conscious states Physical
HM 2357	— conservation (Physical sciences)
HM 2381	— duality (Physical sciences)
HM 2322	— relativity (Physical sciences)
HM 2097	Conscious states ; Tarot arcana (Tarot)
HM 1084	Conscious thinking ; Pre
HM 2822	Conscious wisdom (Astrology)
HM 2426	Consciousness
VC 0443	Consciousness
	Consciousness [Absence...]
HM 2670	— Absence
HM 3043	— absorption ; Limitless (Buddhism)
HM 2393	— adapability ; God
HM 2098	— aggregate (Buddhism)
HM 2098	— Aggregate (Buddhism)
HM 0123	— Akshara (Psychism)
HM 2329	— Alien
HM 2345	— Alpha wave
HM 2391	— Altered states
HM 1537	— Alternate state (ASC)
HM 2415	— Analytical
HM 3001	— Android
HM 1122	— Anomalies consistency
HM 4521	— Anomalies integrity
HM 2317	— Ansari ; Stations (Sufism)
HM 2973	— Aquarius (Astrology)
HM 2189	— Archaic
HM 2665	— Aries (Astrology)
HM 7655	— arising ; Twenty four causes (Buddhism)
HM 2012	— Arupaloka
HH 4000	— ascension stages game (Buddhism)
HM 2376	— Assertive self (Astrology)
HM 0183	— Astral (Psychism)
HM 5123	— Attraction (Psychism)
HH 1235	— Autocontrol
HM 2302	— awe
	Consciousness [Bardo...]
HM 0698	— Bardo (Buddhism)
HM 2099	— Being (Hinduism)
HM 3476	— Beta wave
HM 2653	— Beyond third realm (Buddhism)
HM 0044	— Biosocial
HM 1529	— Birth

Consciousness

	Consciousness [Blending...] cont'd
HH 4997	— Blending celestial consciousness earthly (Taoism)
HM 2562	— Body (Buddhism)
HM 8273	— born materiality (Buddhism)
HM 2041	— Brahma (Hinduism)
HM 2295	— Brain based
HM 1114	— Bridge (Psychism)
HM 2735	— Buddha (Buddhism)
	Consciousness Buddhism
HM 3617	— [Consciousness (Buddhism)]
HM 5321	— [Consciousness (Buddhism)]
HM 2200	— (Buddhism)
HM 2136	Consciousness buoyancy ; God
	Consciousness [Cancer...]
HM 2239	— Cancer (Astrology)
HM 2822	— Capricorn (Astrology)
HM 2219	— Chakra centres
HM 3011	— Chitta dependent (Yoga)
HM 2013	— Christ (Christianity)
HM 2013	— Christian (Christianity)
HM 0455	— Clarity (Leela)
HM 5754	— Clouding
HH 0763	— Co
HM 2094	— Co
HM 1271	— Cognition born contact element eye (Buddhism)
HM 1501	— Cognition born contact element mind (Buddhism)
HM 4605	— Composedness aggregate (Buddhism)
HM 7961	— Concentration due natural purity (Buddhism)
HM 7655	— Conditions (Buddhism)
HM 2380	— Contextual
	Consciousness Cosmic
HM 2291	— [Consciousness ; Cosmic]
HM 2405	— [Consciousness ; Cosmic]
HM 2156	— [Consciousness ; Cosmic]
	Consciousness [Crazy...]
HM 2590	— Crazy monkey ego (Hinduism)
HM 2368	— Creative bisociated
HM 0376	— Creature
HM 2602	— Crowd
HH 4908	— Crystal
HM 2751	— cyclical time
	Consciousness [Daily...]
HM 1397	— Daily variations
HM 0921	— Dawning (Psychism)
HM 2553	— Death altered state
HM 1966	— Death (Psychism)
HM 0891	— Degrees (Hinduism)
HM 2546	— Delightful emanation heaven (Buddhism)
HM 1785	— Delta wave
HM 1119	— Delusion (Buddhism)
HM 2600	— Delusive
HM 2173	— Depth (Jung)
HM 2733	— Desire realm (Buddhism)
HM 2298	— Discrete states (Physical sciences)
HM 2098	— Dispositions (Buddhism)
HM 0563	— Divine
HM 2980	— Dormant
HM 2860	— Dreamy arrested
HM 1402	— dying rising ; Christ (Christianity)
	Consciousness [Ear...]
HM 2169	— Ear (Buddhism)
HH 4997	— earthly consciousness ; Blending celestial (Taoism)
HM 2006	— Ecological
HH 1267	— Ecological planetary
HM 0570	— Ego
HM 0044	— ego conflict
HM 6173	— element ; Mind (Buddhism)
HM 1834	— Element mind (Buddhism)
HM 2499	— Embodied
HM 3590	— Emotional (Psychism)
HM 2522	— Emptiness objects sense mental (Buddhism)
HM 0778	— empty mind ; Christ (Christianity)
HM 2620	— Enervated
HM 3169	— Enstatic
HM 4024	— Esoteric divine (Psychism)
HM 2140	— Evolution
HM 3336	— Existential
HM 0801	— expanding drugs
HM 2126	— expansion
HM 2074	— Eye (Buddhism)
	Consciousness [Fashion...]
HH 0555	— Fashion
HM 3590	— Feeling (Psychism)
HM 2768	— Feminine
HM 2990	— Field
HM 2383	— field ; Personal (Psychosynthesis)
HM 2407	— Fifth antarabhava (Buddhism)
HM 0750	— Fifth state (Psychism)
	Consciousness fine material
HM 4761	— sphere functional ; Indeterminate (Buddhism)
HM 5338	— sphere ; Profitable (Buddhism)
HM 0594	— sphere resultant ; Indeterminate (Buddhism)
	Consciousness [Fitness...]
HM 1810	— Fitness (Buddhism)
HM 1796	— Flexibility aggregate (Buddhism)
HM 1728	— forgiving mind ; Christ (Christianity)
HM 2257	— Form realm (Buddhism)
	Consciousness form realm
HM 4761	— inoperative ; Indeterminate (Buddhism)
HM 5338	— Moral (Buddhism)
HM 0594	— resultant ; Indeterminate (Buddhism)
HM 2281	Consciousness ; Formless realm (Buddhism)
	Consciousness formless realm
HM 0282	-- inoperative ; Indeterminate (Buddhism)
HM 4701	— Moral (Buddhism)
HM 4982	— resultant ; Indeterminate (Buddhism)
HM 2335	Consciousness ; Fourth antarabhava (Buddhism)
HM 3667	Consciousness ; Fringe (Psychism)
	Consciousness [Gemini...]
HM 2924	— Gemini (Astrology)
HM 2006	— Global
HM 0563	— God
HM 2166	— God
HM 3109	— Group
HM 2098	— group conscious existence ; Awareness (Buddhism)
HM 2556	— group conscious existence senses mind ; Awareness (Buddhism)
HM 2102	— Growth
HM 2130	Consciousness ; Heaven fighting (Buddhism)
	Consciousness Higher
HM 0866	— state (Psychism)
HM 0935	— states
HM 2365	— states (Sufism)
	Consciousness [Holistic...]
HM 2367	— Holistic
HM 3012	— human service
HM 3511	— Hypnagogic (Psychism)
HM 3688	— Hypnopompic (Psychism)
HM 2133	— Hypnotic states
	Consciousness [ibn...]
HM 2424	— ibn Abi'l Khayr ; Stations (Sufism)
HM 6208	— Illuminated (Buddhism)
HM 2901	— Imaginative (Anthroposophy)
	Consciousness immaterial sphere
HM 0282	— functional ; Indeterminate (Buddhism)
HM 4701	— Profitable (Buddhism)
HM 4982	— resultant ; Indeterminate (Buddhism)
	Consciousness [Independent...]
HM 6243	— Independent (Buddhism)
HM 3869	— independent vehicle (Yoga)
HM 2446	— Indian approaches (Hinduism)
HM 2692	— Individual tendencies (Hinduism)
HM 0461	— induction device ; Altered states (ASCID)
HM 3125	— Inspired
HM 2152	— Integral
HM 0099	— International (Yoga)
HM 2439	— Intuitive (Astrology)
HM 2738	— Inward flowing (Yoga)
HM 2651	— irreversible direction time
HM 2913	— Ishvara (Yoga)
HM 2141	Consciousness ; Jagrat state waking (Yoga)
	Consciousness [Kamaloka...]
HM 2060	— Kamaloka (Buddhism)
HM 0130	— Karma result (Buddhism)
HM 2300	— Krishna
	Consciousness [Left...]
HM 2415	— Left brain
KC 0060	— Legal
HM 2376	— Leo (Astrology)
HM 8241	— Liberated (Buddhism)
HM 2196	— Liberation (Hinduism)
HM 2512	— Libra (Astrology)
HM 3403	— Lightness aggregate (Buddhism)
HM 6033	— Lightness (Buddhism)
HM 2072	— Loka (Buddhism)
HM 2120	— Lokuttara (Buddhism)
HM 2000	— Love
HM 0976	— loving mind ; Christ (Christianity)
HM 1219	— Lowered state (Psychism)
	Consciousness [Magical...]
HM 2090	— Magical
HM 1155	— Malleability (Buddhism)
HM 2355	— Mammalian limbic human brain
HM 2464	— Mantra (Yoga)
HH 8903	— Maps
HM 2319	— Mental
HM 2838	— Mental (Buddhism)
HM 3323	— Mind
HM 2838	— Mind element (Buddhism)
HM 5123	— Mineral (Psychism)
HM 6720	— Modes occurrence (Buddhism)
HM 2780	— Multi dimensional
HM 0130	— Mundane resultant (Buddhism)
HM 2880	— Mystical Christ (Christianity)
HM 3241	— mystical poverty
HM 0043	— Mythical
HM 2078	— Mythical
	Consciousness [National...]
HM 1370	— National (Yoga)
HM 3314	— Neither physical ease nor unease born contact element mind (Buddhism)
HM 0972	— Neither psychical ease nor unease born contact element eye (Buddhism)
HM 2427	— Neomammalian neocortex human brain
HM 2070	— Non emanating (Buddhism)
HM 2364	— Nose (Buddhism)
HM 2027	-- nothingness absorption (Buddhism)
KD 2130	Consciousness: number time ; Quaternary
	Consciousness [Obfuscation...]
HM 5754	— Obfuscation
HM 2203	— Objective
HM 2402	— Objective fourth state
HM 4389	— occurrence ; Sense mode (Buddhism)
HM 1993	— Onlooker
HM 1105	— Ontogenetic model human
HM 2581	— Outward flowing (Yoga)
	Consciousness [Perfect...]
HM 2920	— Perfect man (Sufism)
HM 2599	— Phase conditions (Buddhism)
HM 2856	— Pisces (Astrology)
HM 1141	— Plane cosmic (Leela)
HM 1301	— Plane cosmic (Leela)
	Consciousness [Planetary...] cont'd
HM 4621	— Planetary
HM 2006	— Planetary
HM 2006	— Planetization
HM 2760	— power ; God
HM 2870	— Preoccupied
HM 1919	— Psyche (Psychism)
HM 0885	— Psychic field (FC)
HM 1919	— Psychic (Psychism)
HM 1921	— Psychical ease born contact element mind (Buddhism)
HM 0234	— Psychical unease born contact element mind (Buddhism)
HM 3455	— Pure (Hinduism)
HM 0521	— Pure (Psychism)
HM 2332	— Purification (Buddhism)
HM 0820	Consciousness ; Quickened (Psychism)
	Consciousness [Racial...]
HM 1070	— Racial (Yoga)
HH 0427	— raising
HM 2390	— rapture ; God
HM 9001	— Rectitude (Buddhism)
HM 3222	— Refined cosmic
	Consciousness reinforced
HM 5588	— concentration (Buddhism)
HM 4396	— energy (Buddhism)
HM 7902	— faith (Buddhism)
HM 5499	— mindfulness (Buddhism)
HM 5901	— understanding (Buddhism)
	Consciousness [Religious...]
HM 3008	— Religious
HM 2307	— Reptilian human brain
HM 2701	— reversible direction time
HM 8621	— rid barriers (Buddhism)
HM 2367	— Right brain
HM 2993	— River (ICA)
HM 2536	— Rupaloka (Buddhism)
	Consciousness [Sagittarius...]
HM 2726	— Sagittarius (Astrology)
HM 0345	— Scientific
HM 2615	— Scorpio (Astrology)
HM 2371	— Second antarabhava (Buddhism)
HM 2486	— Self
HM 2571	— self
	Consciousness Self
HM 2571	— [Consciousness ; Self]
HM 2200	— (Buddhism)
HM 2610	— conscious
HM 0467	— (Yoga)
HM 2664	Consciousness ; Sense (Buddhism)
	Consciousness sense sphere
HM 3852	— functional ; Indeterminate (Buddhism)
HM 3852	— inoperative ; Indeterminate (Buddhism)
HM 4447	— Profitable (Buddhism)
HM 5721	— resultant ; Indeterminate (Buddhism)
HM 8375	— Unprofitable (Buddhism)
	Consciousness [shift...]
HM 1793	— shift
HM 2341	— Simnami ; Stations (Sufism)
HM 2431	— Sixth antarabhava (Buddhism)
HM 8232	— Sound
HM 2530	— spectrum
HM 3043	— Sphere infinite (Buddhism)
HM 1330	— Spirit open (Taoism)
HM 2032	— states ; Conditional (Hinduism)
HM 2177	— states cyclic existence (Buddhism)
HM 2938	— States (Physical sciences)
HM 6652	— Steadiness (Buddhism)
HH 3220	— Steps cosmic
HM 2730	— Stored (Buddhism)
HM 2730	— storehouse (Buddhism)
HM 1338	— Straightness aggregate (Buddhism)
HM 2026	— Subjective domains (Yoga)
HM 2800	— Sufi contracted (Sufism)
HM 0935	— Superior
HM 2156	— Supra
HM 4930	— supramundane plane ; Profitable (Buddhism)
HM 5129	— supramundane plane resultant ; Indeterminate (Buddhism)
HM 3193	— supreme ; God
HM 2781	— Swapna state dream (Yoga)
	Consciousness [Tacit...]
HM 2297	— Tacit knowledge
HM 2815	— Taurus (Astrology)
HM 2129	— Telepathic
HH 3764	— Theory collective
HM 2428	— Theta clear state (Scientology)
HM 2321	— Theta wave
HM 2311	— Third antarabhava (Buddhism)
HM 0768	— three
HM 2601	— Time
HM 2263	— Tongue (Buddhism)
HM 1483	— Torpid
HM 0226	— Tranquillity (Buddhism)
HM 2020	— Transcendental
HM 2405	— Transcendental cosmic
HM 4930	— transcendental plane ; Moral (Buddhism)
HM 5129	— transcendental plane resultant ; Indeterminate (Buddhism)
HM 3193	— Transcendental unity
HM 5122	— transference ; Yoga (Buddhism)
HM 4423	— Transformation (Yoga)
HM 0411	— Tribal (Yoga)
HM 2196	— Truth (Hinduism)
HM 2406	— Twilight
HM 3336	— two

Consciousness [Ultimate...]
- HM 2318 — Ultimate Sufi state (Sufism)
- HM 2156 — Ultra
- HM 8092 — Unassociated (Buddhism)
- HM 5335 — Unattracted (Buddhism)
- HM 6521 — Undejected (Buddhism)
- HM 5623 — Unelated (Buddhism)
- HM 7843 — Unified (Buddhism)
- HM 2702 — Unitary
- HM 3193 — Unity
- HM 2156 — Universal
- HM 5665 — Unmanifest thinking (Buddhism)
- HM 5908 — Unrepelled (Buddhism)
- HM 6791 — Untrammelled (Buddhism)
- HM 1999 — Usual state (USC)

Consciousness [Virgo...]
- HM 2439 — Virgo (Astrology)
- HM 2176 — Visual
- HM 0404 — Volition born contact element mind (Buddhism)
- HM 2502 — Voodoo trance

Consciousness [Wakening...]
- HM 3021 — Wakening
- HM 3021 — Waking
- HM 6556 — Wieldiness (Buddhism)
- HM 2396 — Witness (Yoga)
- HM 4605 — Workability aggregate (Buddhism)
- HH 0238 Consecration
- HH 0064 Consecration ; Personal
- KC 0532 Consensus
- KD 2350 Consensus ; Alternation: implications agreement
- TC 1011 Consensus building
- TC 1493 Consensus buttons ; Participant
- KD 2327 Consensus games ; Resonance based
- KC 0525 Consensus politics
- VP 5775 Consent-Refusal
- VC 0445 Conservation
- TP 0026 Conservation
- HM 1508 Conservation (Buddhism)
- HM 2357 Conservation conscious states ; Physical (Physical sciences)
- KC 0569 Conservation laws
- VC 0447 Conservativeness
- HM 2730 Conserving mind ; All (Buddhism)
- VC 0449 Consideration
- HM 1902 Consideration ; Non (Buddhism)
- KC 0934 Consistency
- VC 0451 Consistency
- HM 1122 Consistency consciousness ; Anomalies
- HM 7826 Consisting development ; Understanding (Buddhism)
- HM 5298 Consisting what is learned ; Understanding (Buddhism)
- HM 7154 Consisting what is reasoned ; Understanding (Buddhism)
- HM 3502 Consolation (Christianity)
- HM 3443 Consolations ; Mansions spiritual (Christianity)
- KC 0180 Consolidated
- KC 0890 Consolidated
- TP 0026 Consolidation
- VC 0453 Consonance
- VD 2276 Conspiracy
- VC 0455 Constancy
- TP 0032 Constancy
- HM 1066 Constancy thought (Buddhism)
- HM 2305 Constant attention (Sufism)
- HM 2117 Constituting intelligence ; Path (Judaism)
- TC 1943 Constrained meeting configurations
- KP 3020 Constraining forms ; Significance mutually
- VD 2278 Constraint
- TP 0030 Constraint ; Normative (Fire)
- KP 3008 Constraints change
- KP 3005 Constraints existence
- TC 1894 Constraints ; Language
- KD 2045 Constraints meta-answer
- MP 1148 Constraints ; Perspective interaction
- KP 3015 Construction development form
- MP 1207 Construction elements ; Appropriate
- HH 2076 Constructive imagination
- HH 0476 Constructive personality change
- HH 0476 Constructive psychotherapeutic change
- VC 0457 Constructiveness
- TC 1689 Constructs ; Meetings physical
- HM 1451 Construing ; Correct (Buddhism)
- TC 1798 Consultants ; Intercultural
- TC 1974 Consultation conferencing
- VC 0459 Consummation
- VC 0461 Consumption
- HM 0134 Consumption ; Modes awareness associated alcohol
- VC 0463 Contact
- TP 0044 Contact
- HM 2708 Contact (Buddhism)
- HM 1380 Contact (Buddhism)
- HM 1271 Contact element eye consciousness ; Cognition born (Buddhism)
- HM 0972 Contact element eye consciousness ; Neither psychical ease nor unease born (Buddhism)

Contact element mind
- HM 1501 — consciousness ; Cognition born (Buddhism)
- HM 3314 — consciousness ; Neither physical ease nor unease born (Buddhism)
- HM 1921 — consciousness ; Psychical ease born (Buddhism)
- HM 0234 — consciousness ; Psychical unease born (Buddhism)
- HM 0404 — consciousness ; Volition born (Buddhism)
- HM 2552 Contact healing
- HM 1352 Contact psychical ; Neither suffering nor pleasure experienced born (Buddhism)

Contact psychical Sensation
- HM 0309 — ease born (Buddhism)

Contact psychical Sensation cont'd
- HM 1643 — neither suffering nor pleasure born (Buddhism)
- HM 1090 — unease born (Buddhism)
- HM 4076 — unpleasant born (Buddhism)
- HM 0901 Contact psychical ; Unease experienced born (Buddhism)
- HM 1031 Contact psychical ; Unpleasant experienced born (Buddhism)
- TC 1616 Contact system ; In conference computer
- HM 0836 Contacted ; Being (Buddhism)
- MP 1228 Contain transitions levels ; Appropriate superstructure
- MP 1119 Contained interfaces ; Partially
- MP 1071 Contained irrationality ; Access
- VC 1547 Containedness ; Self
- KD 2095 Container alternation ; Third perspective
- MM 2014 Container boundary
- VP 5193 Container-Content
- MS 8426 Container vessels
- KD 2065 Containing discontinuity aesthetics
- MP 1117 Containment integrative superstructure
- MM 2072 Containment plasma ; Magnetic
- HM 2952 Contemplation
- TP 0020 Contemplation
- HM 2719 Contemplation ; Belief arising subtle (Buddhism)
- HM 2659 Contemplation ; Critical analytical subtle (Buddhism)

Contemplation [danger...]
- HM 7297 — danger ; Knowledge (Buddhism)
- HM 5364 — dispassion ; Knowledge (Buddhism)
- HM 3385 — dissolution ; Knowledge (Buddhism)
- HM 4340 — Divine (Christianity)
- HM 2683 Contemplation ; Final training subtle (Buddhism)
- HM 2743 Contemplation ; Full isolation subtle (Buddhism)
- HM 0137 Contemplation (Hinduism)
- HM 2710 Contemplation ; Infused
- HM 2109 Contemplation ; Life (ICA)
- HM 2710 Contemplation ; Mystical
- HH 0580 Contemplation nature
- HM 1264 Contemplation object (Islam)
- HM 2171 Contemplation ; Only beginner subtle (Buddhism)
- HH 0169 Contemplation reflection ; Knowledge (Buddhism)
- HM 3723 Contemplation rise fall ; Knowledge (Buddhism)

Contemplation [Simple...]
- HH 0816 — Simple
- HM 2195 — Subtle (Buddhism)
- HM 4340 — Superessential (Christianity)
- HM 3521 Contemplation unitary experience (Sufism)
- HM 2267 Contemplation withdrawal joy ; Subtle (Buddhism)
- HM 3630 Contemplations (Buddhism)

Contemplative [indifference...]
- HM 1405 — indifference
- HH 0816 — intuitive meditation
- HH 0683 — invocation (Japanese)
- HM 2254 Contemplative knowledge (Sufism)
- HH 2145 Contemplative life (Christianity)
- HH 0816 Contemplative prayer

Contemplative state
- HM 3521 — Islamic (Sufism)
- HM 2818 — Super (Yoga)
- HM 3232 — Super (Yoga)
- VD 2280 Contemptuousness
- VC 0465 Content
- VP 5193 Content ; Container
- TP 0006 Contention
- VD 2282 Contentiousness
- HM 3106 Contentless transformation (ICA)
- HM 2373 Contentless word (ICA)

Contentment
- HM 2409 — [Contentment]
- HM 3729 — [Contentment]
- VC 0467 — [Contentment]
- TP 0016 — [Contentment]
- HM 4563 Contentment (Buddhism)
- HM 2529 Contentment centre (ICA)
- VP 5868 Contentment-Discontentment
- HM 4190 Contentment (Sufism)
- HM 7024 Contentment (Sufism)
- HM 2898 Contentment (Yoga)
- HM 5000 Contento ; Cuor (Italian)
- TC 1152 Contests ; Meetings games
- MS 8846 Contests ; Sports
- KC 0456 Context
- MP 1070 Context acknowledgement past perspectives
- MP 1111 Context ; Blended integration formal structure informal
- MP 1015 Context boundary
- MP 1077 Context complementary perspectives ; Minimal
- MP 1058 Context disorder

Context [each...]
- MP 1141 — each perspective ; Relatively isolated
- MP 1065 — emergence new perspectives
- MP 1199 — external insight ; Exposure input processing
- MP 1041 Context formal processes ; Informal
- MP 1189 Context forms perspective presentation

Context [Identifiable...]
- MP 1014 — Identifiable
- MP 1069 — inactivity ; Common external
- MP 1246 — Integration
- MP 1076 — interrelating mature emerging perspectives ; Minimal
- MP 1183 Context ; Partially exposed perspective
- MP 1090 Context perspectives exchange ; Unstructured
- MP 1078 Context single perspective ; Minimal
- HM 3176 Context ; Symbolizing eternal (ICA)
- MP 1066 Context transformative experience
- MP 1191 Contexts ; Appropriate proportions perspective

Contexts [care...]
- MP 1086 — care premature perspectives
- MP 1173 — communication pathways ; Protecting non linear

Contexts [Complexification...] cont'd
- MP 1095 — Complexification perspective
- MP 1087 — controlled single perspective ; Exchange
- MP 1143 Contexts developing perspectives ; Interrelationship
- MP 1085 Contexts developing perspectives ; Perspective imitation

Contexts [Eccentric...]
- MP 1196 — Eccentric access perspective
- MP 1204 — Exclusive
- MP 1073 — exploratory relationship formation challenging emerging perspectives
- MP 1154 Contexts maturing perspectives ; Semi autonomous
- MP 1082 Contexts ; Minimal distance related operational control
- MP 1192 Contexts ; Overview external
- MP 1152 Contexts ; Partially isolated
- MP 1079 Contexts ; Perspective adaptable

Contexts perspective
- MP 1198 — adjuncts ; Inter domain
- MP 1080 — dynamics ; Integrated
- MP 1084 — reorganization ; Transitional
- MP 1155 Contexts perspectives decreasing activity ; Semi autonomous
- MP 1091 Contexts perspectives transition ; Hospitable
- MP 1172 Contexts self-organizing non-linearity
- MP 1151 Contexts ; Small scale perspective interaction
- MP 1190 Contexts ; Variation size perspective
- TC 1524 Contextual conferencing
- HM 2380 Contextual consciousness
- TC 1886 Contextual facilitation
- MP 1169 Contextual levels ; Maintaining distinctions
- KC 0059 Contextual thinking ; Trans
- HM 2842 Contextual world-view (ICA)
- KC 0059 Contextuality
- VT 8032 Contextuality*complex
- HH 0070 Continence
- VC 0469 Continence

Continent awareness
- HM 2519 — cattle ; Western (Buddhism)
- HM 2067 — community ; Northern (Buddhism)
- HM 2127 — men ; Southern (Buddhism)
- HM 2543 — noble figures ; Eastern (Buddhism)
- HH 2992 Contingency
- HM 2104 Contingency ; Human (ICA)
- HM 2477 Contingency ; Radical (ICA)
- HM 2456 Contingent eternality (ICA)
- HM 2468 Continuance (Buddhism)
- VP 5143 Continuance-Cessation
- HH 0955 Continuing education
- VC 0471 Continuity
- KC 0008 Continuity discontinuity
- VP 5071 Continuity-Discontinuity
- MP 1217 Continuity ; Perimeter
- HM 4788 Continuous all-pervading knowledge (Sufism)
- HH 1099 Continuous doctrine being ; Causally (Buddhism)
- HM 2241 Continuous setting mind (Buddhism)
- KC 0702 Continuum
- HM 6221 Continuum ; Life (Buddhism)
- HM 2800 Contracted consciousness ; Sufi (Sufism)
- VP 5197 Contraction ; Expansion
- VD 3632 Contradiciton ; Self
- VD 2284 Contradiction
- HM 2951 Contradiction ; Exclusive (ICA)
- HH 0708 Contradictions ; Inner
- VD 2286 Contrariety
- MM 2050 Contrast significance
- HM 4350 Contrasts ; Harmony aesthetic (Japanese)
- TC 1491 Contributors
- HH 4112 Contrition
- HM 7209 Contrition (Buddhism)
- KC 0930 Control
- HH 2763 Control ; Breath
- TC 1506 Control ; Conference
- MP 1082 Control contexts ; Minimal distance related operational

Control [Hierarchial...]
- KC 0795 — Hierarchial principle
- HH 7637 — human development ; Mythical
- HH 7637 — human development ; Psychic
- HH 1988 Control involuntary functions
- MP 1081 Control operations ; Minimally structured perspective
- HM 1325 Control principles ; Eighth order perceptions
- HM 0103 Control relationships ; Sixth order perceptions

Control Self
- HH 0600 — [Control ; Self]
- HH 0778 — [Control ; Self]
- VC 1549 — [Control ; Self]

Control [sequence...]
- HM 0772 — sequence ; Fifth order perceptions
- HM 1001 — Seventh order perceptions programme
- HH 3635 — Silva mind
- KC 0607 — system
- HM 4000 — systems concepts ; Ninth order perceptions
- HH 7637 — systems ; Spiritual
- HM 1516 Control transitions ; Fourth order perceptions
- KC 0256 Control; Value engineering ; Value
- KP 3012 Controlled relationship ; Harmoniously transformative
- MP 1087 Controlled single perspective ; Exchange contexts
- HH 0418 Controlled spontaneity
- TP 0007 Controlled threat (Army)
- VD 2288 Controversy
- VD 2290 Contumaciousness
- VD 2292 Contumeliousness
- VC 0473 Convenience
- HM 1814 Conventional faith ; Synthetic
- VP 5642 Conventionality-Unconventionality
- KC 0113 Convergence
- VP 5298 Convergence-Divergence

Converger

TC 1133	Converger participant type
MM 2011	Conversation ; Shifting topics
TC 1366	Conversations ; Group
HH 0077	Conversion
VC 0475	Conversion
HH 0757	Conversion ; Ideological
VP 5145	Conversion-Reversion
HM 3484	Conversion (Sufism)
VC 0477	Convertibility
VC 0479	Conviction
HM 1160	Conviction sin
HM 2916	Conviction (Sufism)
TC 1484	Conviviality ; Conference

Convulsive therapy
HH 0566	— [Convulsive therapy]
HH 0975	— [Convulsive therapy]
HH 0866	— Electric (ECT)
MM 2055	Cooking ; Food, nutrition
KC 0377	Cooperation
VC 0481	Cooperation
KC 0384	Cooperation ; Cultural
KC 0555	Cooperation ; Intellectual
KC 0850	Cooperation ; International
KC 0897	Cooperation ; Regional international
VC 0483	Cooperativeness
KC 0303	Coordinate indexing
MP 1144	Coordinated exposure irrationality structures ; Integrating

Coordination
KC 0924	— [Coordination]
VC 0485	— [Coordination]
TP 0008	— [Coordination]
TC 1410	Coordination conferencing
MP 1147	Coordination perspective nourishment
VC 0487	Cordiality
HM 0777	Core-experience (NDE)
HM 2080	Core-religious experiences
HM 4664	Coriolis illusion
HM 4664	Coriolis reaction ; Vestibular
KC 0550	Corporate database
HM 2694	Corporate duty (ICA)
HH 2768	Corporate worship (Christianity)
HH 0080	Corporateness
HM 3149	Corporeal intelligence ; Path (Judaism)
HM 2108	Corporeality group conscious existence ; Awareness (Buddhism)
KC 0528	Corpus knowledge
HM 0439	Correct action (Buddhism)
HM 0687	Correct action (Buddhism)
HM 1735	Correct concentration (Buddhism)
HM 1451	Correct construing (Buddhism)
HM 1657	Correct endeavour (Buddhism)
HM 0549	Correct livelihood (Buddhism)
HM 1763	Correct livelihood (Buddhism)
HM 0180	Correct recollection (Buddhism)
HM 4608	Correct speech (Buddhism)
HM 1821	Correct speech (Buddhism)
HM 3252	Correct view (Buddhism)
HM 1727	Correction ; Night (Christianity)
VC 0489	Correctness
KC 0473	Correlation
VC 0491	Correspondence
VD 2294	Corruption
HM 9722	Corruption insight (Buddhism)
HH 0993	Cosmetic surgery
MS 8372	Cosmetics ; Applied

Cosmic consciousness
HM 2291	— [Cosmic consciousness]
HM 2405	— [Cosmic consciousness]
HM 2156	— [Cosmic consciousness]

Cosmic consciousness [Plane...]
HM 1141	— Plane (Leela)
HM 1301	— Plane (Leela)
HM 3222	— Refined
HH 3220	— Steps
HM 2405	— Transcendental
HH 0629	Cosmic creativity
HM 4343	Cosmic good ; Plane (Leela)
HH 0599	Cosmic identification
HH 0808	Cosmic integration
HM 2278	Cosmic love
TP 0030	Cosmic mean (Fire)
HM 2945	Cosmic sanctions (ICA)
HM 1635	Cosmological mysticism
KC 0502	Cosmology
VC 0493	Cosmopolitan
KC 0278	Cosmos
KC 0016	Cost-benefit analysis
TC 1718	Cost clock
TC 1381	Counsel conferencing ; Co
TC 1602	Counsel conferencing ; Group
TP 0027	Counsel ; Wise
HH 0227	Counseling ; Group
HH 0842	Counselling
HH 0342	Counselling
HH 0342	Counselling ; Co
HH 0357	Counselling ; Genetic
HH 0095	Counselling ; Non directive
HH 0342	Counselling ; Re evaluation
KC 0611	Counter-intuitive
HH 0235	Counter-transference
MP 1104	Counteract deficiencies pattern harmony ; Structural development designed
VP 5177	Counteraction ; Concurrence
VD 2296	Counterproductivity

	Courage
VC 0495	— [Courage]
HH 0929	— [Courage]
HH 5640	— [Courage]
VP 5890	Courage-Fear
VC 0497	Courtesy
HH 0305	Courtesy
VP 5936	Courtesy-Discourtesy
HM 6122	Courtly love
HH 6773	Covers ; Five (Buddhism)
HM 0684	Covetous ; Not being (Buddhism)
HM 3563	Covetousness (Buddhism)
HM 2690	Cow-herding simile (Zen)
VD 2298	Cowardice
HM 1594	Crack ; Use
HH 1909	Craft ; Wicca
HH 1909	Craft work
MS 8102	Craftsman tools
HM 3707	Craving (Pali)
HH 0978	Craving sensual enjoyment ; Abstention
HM 2590	Crazy-monkey ego consciousness (Hinduism)
VC 0499	Creation
HM 2017	Creation ; Meaning (ICA)
HM 3101	Creation ; Sheer re (ICA)
HM 3038	Creation ; World (Judaism)
TC 1451	Creative ambience
HM 2368	Creative bisociated consciousness
TC 1269	Creative care groups
HM 3618	Creative dreaming
HH 0513	Creative existence
HM 2894	Creative existence (ICA)
HM 2493	Creative futility (ICA)
HM 0786	Creative imagination
HH 0679	Creative intelligence
TC 1913	Creative listening process
HM 4449	Creative moment (Sufism)
HM 1655	Creative moment (Sufism)
HH 4099	Creative psychology
KP 3013	Creative renewal
HH 0896	Creative synthesis
HH 0703	Creative thinking
	Creativity
HH 0481	— [Creativity]
KC 0323	— [Creativity]
VC 0501	— [Creativity]
HH 0500	Creativity ; Child
HH 0629	Creativity ; Cosmic
HM 1804	Creativity ; Experience (Systematics)
HH 0917	Creativity group workshop
TP 0001	Creativity (Heaven)
HM 2037	Creativity ; Spiritual (ICA)
HM 0376	Creature consciousness
MS 8215	Creatures
VC 0503	Credibility
VT 8052	Credibility∗complex
VD 2300	Credulousness
KC 0750	Crises ; World
VD 2302	Crisis
HM 0850	Crisis apparition (Psychism)
KD 2295	Crisis learning response ; Entropic
KC 0293	Crisis management
TP 0051	Crisis preparedness (Thunderbolts)
HH 0021	Crisis rites ; Life
HM 2659	Critical-analytical subtle contemplation (Buddhism)
HM 3609	Critical awareness
KC 0387	Critical mass
KC 0736	Critical path method
KC 0387	Critical quantity
MM 2094	Crop rotation
KD 2285	Crop rotation ; Patterns alternation: agricultural key
HM 3092	Cross ; Common (Astrology)
KC 0411	Cross cultural study
HM 1714	Cross ; Dark night Saint John (Christianity)
KC 0754	Cross-impact analysis
HM 3719	Cross ; Jesus dies (Christianity)
	Cross Jesus falls
HM 4009	— first time His (Christianity)
HM 3824	— second time (Christianity)
HM 2289	— third time (Christianity)
	Cross Jesus [helped...]
HM 3344	— helped carry (Christianity)
HM 3500	— made bear (Christianity)
HM 4136	— nailed (Christianity)
HM 3410	— taken down (Christianity)
HM 2880	Cross ; Stations (Christianity)
HM 3516	Cross ; Stations (Christianity)
HM 2259	Cross transcendence (Astrology)
HM 2554	Cross transmutation (Astrology)
HM 3516	Cross ; Way (Christianity)
KC 0896	Crossdisciplinarity
HM 1286	Crossroads ; Standing (Buddhism)
HM 2602	Crowd consciousness
HM 1404	Crowd emotion (Spanish)
HM 3132	Crown ; Sphere supreme (Kabbalah)
HM 0210	Crucifixion ; Initiation (Esotericism)
HM 2303	Cruciform exaltation (ICA)
VD 2304	Cruelty
HH 1098	Crusade evangelism (Christianity)
HM 5991	Crushes
HM 3142	Crushing-hells awareness (Buddhism)
HM 0059	Cry (Christianity)
HM 2854	Cryptic disclosure (ICA)
HM 8235	Cryptomnesia
HH 4908	Crystal consciousness
HH 4908	Crystal healing
HH 4908	Crystals ; Meditation

VC 0505	Culmination
VP 5153	Culmination ; Causation
VD 2306	Culpability
MP 1004	Cultivation areas ; Regenerative resource
HM 2252	Cultivation awareness ; Pratyeka Buddha (Buddhism)
HH 8231	Cultivation concentration (Buddhism)
HH 0680	Cultivation experiential insight (Buddhism)
HM 2240	Cultivation ; Paths vision (Buddhism)
MP 1170	Cultivation productive non-linearity
MP 1177	Cultivation sources perspective nourishment ; Local
HH 0710	Cultivation tranquillity
HH 0423	Cults
HH 1020	Cults ; Brainwashing
HH 1523	Cults ; Cargo
HH 0783	Cultural assimilation
TC 1866	Cultural communication teams ; Inter
KC 0384	Cultural cooperation
HH 0737	Cultural development
KC 0470	Cultural evolution
HH 0926	Cultural evolution
TP 0050	Cultural heritage (Cauldron)
HH 1929	Cultural identity
	Cultural integration
KC 0243	— [Cultural integration]
KC 0133	— Adaptive
KC 0380	— Configurational
KC 0360	— Connective
KC 0133	— Functional
KC 0460	— Logical
KC 0143	— Regulative
KC 0145	— Stylistic
KC 0380	— Thematic
HH 0754	Cultural life (Hinduism)
MS 8306	Cultural objects devices
	Cultural [personality...]
HH 0724	— personality
HH 0955	— potential ; Development personal
HH 1778	— primitivism
KC 0076	Cultural relativity
HH 0837	Cultural revolution
HH 0090	Cultural science psychology
KC 0411	Cultural study ; Cross
HH 3120	Cultural transformation
KC 0422	Cultural unity
	Culture
KC 0051	— [Culture]
VC 0507	— [Culture]
HH 1754	— [Culture]
KC 0609	Culture ; Mass
HH 0730	Culture ; Physical
KC 0057	Culture ; Planetary
HH 3710	Culture ; Self (Christianity)
	Culture [Understanding...]
HM 4128	— Understanding being plane (Buddhism)
HM 7826	— Understanding (Buddhism)
KC 0422	— Universal
KC 0422	Culture ; World
VD 2308	Cunning
HM 5000	Cuor contento (Italian)
HH 1208	Curative education
HH 1264	Cure
HH 1772	Cure ; Mind
HH 1264	Cure souls
VC 0509	Curiosity
VP 5528	Curiosity-Incuriosity
MP 1145	Current elements those reserve ; Domains non
HM 2054	Current views opinions (Buddhism)
KC 0428	Curricula
KC 0961	Curricula ; Integrated science
KC 0961	Curricula ; Interdisciplinary science
KC 0961	Curricula ; Unified science
KC 0976	Curriculum integration
HH 0594	Curse ; Official
HH 0594	Cursing
VD 2310	Curtness
VC 0511	Custom
HM 3167	Cut-off unknownness (ICA)
KC 0116	Cybernation
KC 0116	Cybernetic management
KC 0230	Cybernetics
KC 0963	Cybernetics ; Economic
KC 0910	Cybernetics ; Engineering
KD 2160	Cybernetics ; Nonlinear
HH 0681	Cybernetics ; Psycho
HM 1772	Cyberspace
KP 3014	Cycle development processes
MP 1026	Cycle ; Functional
KC 0965	Cycle ; Life
MP 1031	Cycle relationship reinforcement
KC 0322	Cycles
KD 2100	Cycles alternation ; Revolutionary
MM 2073	Cycles ; Animal life
KC 0913	Cycles ; Business
KD 2275	Cycles ; Learning
MM 2092	Cycles ; Mechanical
MM 2068	Cycles ; Metabolic pathways
KD 2120	Cycles ; Omnitriangulation: interlocking
MM 2048	Cycles ; Psycho symbolic
MM 2046	Cycles religious festivals
TC 1599	Cycles ; Repetition learning
HH 2002	Cycles ; World (Buddhism)
MP 1035	Cyclic elements ; Adequate variety
	Cyclic existence
HM 2051	— absorption ; Peak (Buddhism)
HM 2177	— Consciousness states (Buddhism)
HM 3032	— Emptiness (Buddhism)

-866-

GENERAL INDEX TO VOLUME 2   Denial

HH 0933	Cyclic history		**Death [Dying...] cont'd**		**Defining [mentality...] cont'd**
MP 1063	Cyclic interrelation complementary perspective common domains	HM 3099	— Dying (ICA)	HH 2718	— mentality-materiality (Buddhism)
			Death [Ego...]	HM 5627	— mentality ; Understanding (Buddhism)
HM 5187	Cyclic knowledge (Systematics)	HM 3060	— Ego	MP 1208	Definition ; Progressive framework
MM 2044	Cyclic migration	HH 6921	— Ego	HM 2033	Definitions awareness ; Tibetan (Buddhism)
KD 2329	Cyclic phase ; Learning loss	HM 3121	— emblems (Tarot)	HM 2946	Definitions ; Emptiness (Buddhism)
MM 2031	Cyclic resonance accelerator	HM 3059	— encounter higher self ; Near (NDE)	HM 2796	Definitive effectivity (ICA)
KD 2180	Cyclic self-organization requirements	HM 3294	— entering darkness ; Near (NDE)	HM 2131	Definitive predestination (ICA)
KC 0394	Cyclical theory history	HM 3180	— entering light ; Near	VD 2358	Deformation
HM 2751	Cyclical time ; Consciousness	HM 3422	— evil force ; Near	MS 8375	Deformed diseased humans
MM 2047	Cycling ; Bicycles	HM 0777	— experience ; Near (NDE)	HH 0565	Degeneracy
HH 0321	Cyclothyme	HM 3177	Death fear ; Near (NDE)	VD 2360	Degeneration
HM 1616	Cyclothymia	HM 3165	Death hell ; Near (NDE)	TP 0018	Degeneration
VD 2312	Cynicism	HM 6705	Death (Hinduism)	VD 2362	Degradation
		HM 2535	Death ; Jesus condemned (Christianity)	HM 6989	Degree prayer ; First (Christianity)
	D		**Death [Life...]**	HM 0238	Degree prayer ; Second (Christianity)
		VP 5407	— Life	HM 0891	Degrees consciousness (Hinduism)
HM 4387	Daan (Leela)	HH 2987	— life ; Journeying transcendence	KC 0137	Degrees freedom
HM 0126	Daath-sephira (Kabbalah)	HM 2808	— Living (ICA)	HH 3012	Degrees love (Christianity)
HM 3209	Dad-pa (Tibetan)	HM 3115	Death negative detachment ; Near (NDE)	HH 0659	Degrees ; Masonic (Freemasonry)
MM 2024	Daily round activities	HM 2494	Death peace ; Near (NDE)	HM 3030	Dehumanization
HM 1397	Daily variations consciousness	HM 2888	Death review life ; Near (NDE)	VD 2364	Dehumanization
VD 2314	Damage	HM 3294	Death transition ; Near (NDE)	HH 3221	Dehumanizing process
HM 2181	Damana (Buddhism)	HM 3200	Death ; Unmitigated (ICA)	VC 0529	Deification
VD 2316	Damnation	HH 0574	Death-urge	HH 1049	Deification
HH 1234	Dana (Buddhism)	HH 8105	Deautomatization	HM 0946	Deification (Sufism)
MM 2067	Dance	HH 2331	Deautomatization	MS 8117	Deities
HH 0445	Dance	HM 4398	Deautomatization mystic experience	MS 8613	Deities ; Classical
HM 3213	Dance-induced experience	TC 1454	Debate	HM 2675	Deity ; Communion
HM 0303	Dance performance ; Spiritual	VD 2326	Debauchery	HM 1240	Déjà entendu
HH 0445	Dance therapy	VD 2328	Debility	HM 1240	Déjà éprouvé
HM 3213	Dancing ; Abstract	VD 2330	Decadence	HM 1240	Déjà fait
HM 2450	Dang ; Bsam gtan (Buddhism)	HM 1600	Decadent sleepiness (German)	HM 1240	Déjà pensé
HM 1363	Danger	KC 0217	Decay	HM 1240	Déjà raconté
TP 0029	Danger (Abyss)	TP 0018	Decay ; Arresting	HM 1240	Déjà voulu
HM 7297	Danger ; Knowledge contemplation (Buddhism)	HM 3932	Decease (Buddhism)	HM 1240	Déjà vu
VP 5697	Danger ; Safety	VD 2332	Deceit	VD 2366	Dejection
HM 2449	Dangerous intrusion (ICA)	HM 3246	Deceit (Buddhism)	HM 0883	Dejection (Hinduism)
VD 2318	Dangerousness	HH 3421	Deceit ; Heavenly	VC 0531	Delectableness
TP 0029	Dangers ; Multiple (Abyss)	VC 0517	Decency	VD 2368	Delerium
HM 4987	Dar anjuman ; Khilwat (Sufism)	MP 1009	Decentralized formal processes	VD 2370	Deleteriousness
VC 0513	Daring	VD 2334	Deception	TC 1423	Deletion
HM 2009	Daring embracement (ICA)	HM 7632	Deciding (Buddhism)	VC 0533	Deliberateness
	Dark night	HM 2721	Decision ; Freedom (ICA)	VC 0535	Delicacy
HM 1714	— (Christianity)	HM 0322	Decision ; Initiation (Esotericism)	VC 0537	Delight
HM 1714	— Saint John Cross (Christianity)	HH 3366	Decision ; Journeying transcendence discussion	HM 4065	Delight (Buddhism)
HM 1727	— senses (Christianity)		**Decision making**	TP 0058	Delight (Lake)
HM 3941	— soul (Christianity)	KC 0340	— [Decision-making]	HM 1398	Delightful (Buddhism)
TP 0036	Darkening of the light Adulteration insight	TC 1385	— conferencing	HM 2546	Delightful-emanation-heaven consciousness (Buddhism)
TP 0036	Darkening of the light Concealment illumination	TC 1627	— process	HM 4246	Delightfulness (Buddhism)
TP 0036	Darkening of the light Decline	KC 0312	Decision theory procedure	HH 2718	Delimitation formations (Buddhism)
TP 0036	Darkening of the light Injury	HM 3009	Decisional nothingness (ICA)	HH 0565	Delinquency ; Moral
TP 0036	Darkening of the light Unappreciated intelligence	TP 0021	Decisive action (Biting through)	HM 7186	Delirium
VD 2320	Darkness	VC 0519	Decisiveness	HM 7805	Delirium tremens
MS 8415	Darkness	HH 0110	Declaration	VC 0539	Deliverance
VP 5335	Darkness ; Light	TC 1735	Declaration conferencing	TP 0040	Deliverance
HM 3294	Darkness ; Near death entering (NDE)	VD 2336	Decline	HM 0766	Deliverance ; Knowledge desire (Buddhism)
HM 0317	Darkness (Taoism)	TP 0036	Decline (Darkening of the light)	HM 1785	Delta wave consciousness
TP 0004	Darkness (Young shoot)	VD 2338	Decomposition	HM 2600	Delusion
HM 4542	Darr ; Ad (Sufism)	VC 0521	Decoration	VD 2372	Delusion
HM 2931	Darsana (Buddhism)	MS 8446	Decorations trophies ; Medals,	HM 2695	Delusion ; Absence (Buddhism)
HM 2240	Darsana marga ; Sravaka (Buddhism)	VC 0523	Decorum		**Delusion Buddhism**
HM 7549	Darsana ; Yatha bhuta jnana (Buddhism)	TP 0041	Decrease	HM 3196	— [Delusion (Buddhism)]
HH 0446	Dasakusalakamma (Buddhism)	VP 5038	Decrease ; Increase	HM 0184	— [Delusion (Buddhism)]
HM 2632	Dasein	MP 1156	Decreasing activity ; Opportunities perspectives	HM 0918	— [Delusion (Buddhism)]
HH 0695	Dasein analysis	MP 1155	Decreasing activity ; Semi autonomous contexts perspectives	HM 1956	Delusion ; Complete (Buddhism)
HM 2380	Dasein ; Existential			HM 1119	Delusion consciousness (Buddhism)
HH 0680	Dassana ; Nana (Buddhism)	VD 2340	Decrepitness	HM 0697	Delusion (Leela)
HM 7549	Dassana ; Yatha bhuta nana (Pali)	VC 0525	Dedication	HM 2695	Delusion ; Non (Buddhism)
KC 0508	Data association	HH 0238	Dedication	HM 3152	Delusion personal self (Buddhism)
KC 0526	Data processing system	HH 0446	Deeds ; Ten meritorious (Buddhism)	HM 1439	Delusion unwholesome root (Buddhism)
KC 0118	Data structure	HH 2315	Deep ecology	HM 1371	Delusion ; Utter (Buddhism)
KC 0218	Data structures ; Network	HM 3554	Deep feeling ; Lack (Japanese)	HM 3592	Delusion ; Veils (Hinduism)
KC 0550	Database ; Corporate	HM 1689	Deep penetration memory (Buddhism)	HM 5003	Delusional awareness
HM 2737	Daughters Jerusalem ; Jesus speaks (Christianity)	HM 3664	Deep psychophysiological relaxation (DPR)	HM 2795	Delusional ideas
VC 0515	Dauntlessness	HM 2135	Deep relaxation	HM 4504	Delusional ideas
HM 4570	David one's being (Sufism)		**Deep [sleep...]**	HM 4504	Delusional memories
HM 0921	Dawning consciousness (Psychism)	HM 6307	— sleep	HM 0464	Delusional perception
HM 3434	Daya (Leela)	HM 2957	— sleep (Hinduism)	HM 2795	Delusional systematization
HM 2138	Daydream	HM 1226	— state hypnosis	HM 7208	Delusional zoopathy
HM 2138	Daydreaming	KC 0361	— structure	HM 9002	Delusions grandeur
HM 0248	Daydreaming ; Escapist	HM 1886	Deep trance (Psychism)	HM 6036	Delusions guilt
KD 2146	De-categorization poly-ocular vision	HM 2591	Deeps ; Defender (ICA)	HM 1440	Delusions ill-health
HM 8003	De-kho-na (Tibetan)	VD 2342	Defamation	HM 6993	Delusions jealousy
VD 2322	Deadness	VD 2344	Default	HM 6276	Delusions love
HM 6539	Deafness ; Word	VD 2346	Defeat		**Delusions [passivity...]**
VD 2324	Death	VP 5726	Defeat ; Victory	HM 7203	— passivity
MS 8603	Death	VD 2348	Defection	HM 7760	— persecution
	Death [altered...]	VD 2350	Defectiveness	HM 9701	— poverty
HM 2553	— altered state consciousness	HH 0470	Defects ; Character (Buddhism)	HM 4604	— Primary
HM 3146	— Angel	VP 5798	Defence ; Attack	HM 2795	Delusions ; Secondary
HM 3381	— awareness light ; Near (NDE)	HM 2591	Defender deeps (ICA)	HM 4604	Delusions ; True
HM 2088	— awareness ; Lord (Buddhism)	VD 2352	Defensiveness	HM 2600	Delusive consciousness
	Death bardo	VC 0527	Deference	VD 2374	Demeaning
HM 2335	— clear-light awareness (Buddhism)	VD 2354	Defiance	HM 2891	Dementia ; Affective
HM 2407	— heaven-reality awareness (Buddhism)	MP 1104	Deficiencies pattern harmony ; Structural development designed counteract	HH 3226	Demerit ; Formation (Buddhism)
HM 2431	— rebirth-seeking awareness (Buddhism)			VC 0541	Democracy
	Death [black...]	VD 2356	Deficiency	MS 8907	Demon
HM 2213	— black void ; Near (NDE)	TP 0041	Deficiency	HM 2055	Demon-island awareness (Buddhism)
HM 3349	— body separation ; Near (NDE)	HM 3157	Deficiency cognition	HM 1746	Demonstrate ; Inability (Buddhism)
HM 3932	— (Buddhism)	MP 1205	Defined framework spaces defined processes it ; Congruence spaces	HH 0839	Demoralization ; Personal
HM 1966	Death consciousness (Psychism)			VD 2376	Denial
	Death [detachment...]	MP 1205	Defined processes it ; Congruence spaces defined framework spaces	VP 5523	Denial ; Affirmation
HM 3349	— detachment ; Near (NDE)				**Denial Self**
HM 0665	— dream state ; After (Psychism)		**Defining [materiality...]**	VC 1551	— [Denial ; Self]
		HM 4968	— materiality ; Understanding (Buddhism)	HH 0964	— [Denial ; Self]

-867-

Denial

Denial Self cont'd
HH 2121 — (Christianity)
HM 2174 Denial ; Spiritual (ICA)
HH 4522 Denial ; Spiritual self (Christianity)
VD 2378 Denigration
MP 1028 Densities ; Coherent pattern relationship
MP 1101 Density communication pathways ; Protected low
MP 1036 Density ; Differentiation relationship
MP 1029 Density gradient local relationships ; Stable
VD 2380 Denunciation
VD 2382 Dependence
HM 0509 Dependence ceremonies (Buddhism)
HH 3321 Dependency conscious existence phenomena ; Awareness inter (Buddhism)
HM 3011 Dependent consciousness ; Chitta (Yoga)
HM 3035 Dependent origination formula ; Ignorance (Buddhism)
Depersonalization
VD 2384 — [Depersonalization]
HH 0281 — [Depersonalization]
HM 1248 — [Depersonalization]
VD 2386 Depletion
VD 2388 Depradation
VD 2390 Depravation
HM 2840 Depravity
VD 2392 Depravity
VD 2394 Depreciation
HM 2563 Depression
VD 2396 Depression
VP 5317 Depression ; Elevation
HM 1616 Depression ; Manic
HM 3949 Depression thought (Buddhism)
VD 2398 Deprivation
Deprivation [selfhood...]
HM 3002 — selfhood (Brainwashing)
HH 1478 — Sensory
HH 0865 — Sensory (Brainwashing)
HH 4102 Deprogramming
HM 2447 Depth ; Awareness (ICA)
HM 2173 Depth consciousness (Jung)
HH 1420 Depth psychology
VP 5209 Depth-Shallowness
HM 5128 Derealization
HM 1248 Derealization
HM 1248 Dereism
VD 2400 Dereliction
VD 2402 Derogation
HH 0741 Dervish (Islam)
HM 3427 Dervish (Sufism)
HH 0741 Derwish (Islam)
HM 2903 Descendant ; Expectant (ICA)
HM 3592 Descending steps ignorance (Hinduism)
VD 2404 Desecration
KC 0992 Desegregation
VD 2406 Desertion
KC 0150 Design
TC 1556 Design ; Conference time
TC 1940 Design conferencing
MS 8845 Design figures
KC 0529 Design ; Industrial
KC 0408 Design ; Information system
TC 1188 Design methods ; Dynamic
TC 1451 Design ; Space
MP 1104 Designed counteract deficiencies pattern harmony ; Structural development
VC 0543 Desirableness
HM 2433 Desire
Desire [Action...]
HM 2491 — Action (Hinduism)
HH 5487 — Action (Hinduism)
VP 5631 — Avoidance
HM 4487 Desire (Buddhism)
HM 4969 Desire-concentration (Buddhism)
HM 0766 Desire deliverance ; Knowledge (Buddhism)
HM 6144 Desire due insight (Buddhism)
HM 1261 Desire formless life (Buddhism)
HM 3554 Desire ; Freedom (Japanese)
HH 0129 Desire (Hinduism)
HM 0877 Desire (Islam)
Desire [Leela...]
HM 3627 — (Leela)
HM 3583 — life form (Buddhism)
HM 3871 — life (Yoga)
HM 6204 Desire ; Meeting God active (Christianity)
HM 9330 Desire ; Non (Christianity)
HM 4772 Desire possess (Islam)
HM 2733 Desire-realm consciousness (Buddhism)
HM 0766 Desire release ; Knowledge (Buddhism)
HM 4131 Desire ; State unabated (Buddhism)
HM 4487 Desire-to-do (Buddhism)
HM 3231 Desire (Yoga)
HM 2765 Desiredness ; Single (Sufism)
HM 4406 Desirelessness (Hinduism)
HM 0575 Desirelessness (Sufism)
HH 3301 Desires ; Emotional (Buddhism)
HH 4066 Desires soul ; Innermost benevolent
HM 1973 Desires ; World (Buddhism)
TP 0052 Desisting (Mountain)
VD 2408 Desolation
HM 5119 Desolation (Christianity)
HM 3405 Despair
VD 2410 Despair
HM 3017 Desperate gratitude (Brainwashing)
VD 2412 Desperation
VD 2414 Despicableness
VD 2416 Despoliation

VD 2418 Despondency
VD 2420 Despotism
HM 2309 Destinal accountability (ICA)
HM 3027 Destinal elector (ICA)
HH 1362 Destiny
HM 1098 Destroying cause (Buddhism)
HM 0655 Destruction ; Ego
VP 5693 Destruction ; Restoration
VD 3634 Destruction ; Self
TP 0023 Destruction (Stripping away)
MS 8723 Destruction weapons ; Nuclear mass
HH 1020 Destructive manipulation
VD 2422 Destructiveness
TC 1736 Detached-operator type
HH 0229 Detached self-observation meditation
VC 0545 Detachment
HM 2128 Detachment (Buddhism)
HM 1534 Detachment (Christianity)
HM 5091 Detachment (Hinduism)
HM 0210 Detachment ; Initiation (Esotericism)
HM 2867 Detachment ; Mystical (Sufism)
HM 3349 Detachment ; Near death (NDE)
HM 3115 Detachment ; Near death negative (NDE)
HM 4450 Detachment (Sufism)
HM 7621 Detachment (Sufism)
HM 2451 Detachment ; Worldly (ICA)
VD 2424 Deterioration
TP 0023 Deterioration (Stripping away)
KC 0216 Determinate system
VC 1553 Determination
HH 0317 Determination ; Self
MP 1020 Determined specialized communications ; User
KC 0827 Determined system ; State
HM 7632 Determining (Buddhism)
HM 2682 Determining history (ICA)
HH 0170 Determining mental factors (Buddhism)
HM 3076 Determining ; Self (ICA)
KD 2083 Determinism fluctuation ; Complementarity
HH 2009 Deterministic psychology ; Non
HH 7602 Detriment ; Understanding skill (Buddhism)
HH 0704 Deutero learning
HM 5982 Deva hearing (Buddhism)
HH 0748 Deva sight (Buddhism)
HM 2044 Devachon awareness (Psychism)
HM 2130 Devas ; Yama (Buddhism)
VD 2426 Devastation
HM 0377 Devekuth (Judaism)
HH 0971 Developed personality ; Fully
Developing perspectives
MP 1137 — Domain
MP 1143 — Interrelationship contexts
MP 1085 — Perspective imitation contexts
Development
KC 0363 — [Development]
VC 0549 — [Development]
TP 0053 — [Development]
HH 0437 — [Development]
Development [Aesthetic...]
HH 0029 — Aesthetic
MM 2041 — Agricultural
VD 2100 — Arrested
HH 4223 — atheism ; Human
Development [Character...]
HH 0895 — Character
HH 0146 — Cognitive growth
HH 5522 — Communalist human
HH 1222 — Concept
Development Conceptual
TC 1888 — framework
TC 1002 — framework
TC 1888 — screen
TC 1002 — screen
HH 0737 Development ; Cultural
Development [designed...]
MP 1104 — designed counteract deficiencies pattern harmony ; Structural
HH 7324 — diet ; Human
KC 0596 — Discipline
Development [Economic...]
HH 0437 — Economic social
HH 0955 — Educational self
HH 0371 — Ego
HH 0936 — Emotional growth
HH 6902 — Esoteric
KP 3015 Development form ; Construction
Development [General...]
HH 1127 — General system theory human
HH 0617 — group ; Leadership
HH 0617 — group ; Organizational
HH 0347 Development ; Healthy human growth
Development Human
MS 8323 — ages
HH 2435 — Baha'ism)
HH 0447 — behaviour
HH 0650 — (Buddhism)
HH 3607 — centred
HH 2198 — (Christianity)
HH 3098 — (Existentialism)
HH 0330 — (Hinduism)
HH 1799 — (Islam)
HH 0622 — (Jainism)
HH 3029 — (Judaism)
HH 0447 — psychological
HH 0745 — resources (United Nations)
HH 2651 — scale

Development Human cont'd
HH 6292 — (Sikhism)
HH 0436 — (Sufism)
HH 0689 — (Taoism)
HH 3290 — (United Nations)
HH 1003 — (Zen)
HH 1903 — (Zoroastrianism)
HH 5343 Development humour ; Human
Development [Ideological...]
HH 2187 — Ideological
TC 1359 — Image
TC 1001 — Image
HH 0984 — Inappropriate patterns
HH 5101 — index ; Human (United Nations)
HH 0624 — Indicators
HH 1543 — Individual
HH 0630 — infants ; Psychological
HH 0146 — Intellectual growth
MM 2009 — Interruption thematic
Development [Language...]
HH 0485 — Language
HH 0516 — Leadership
HH 6122 — Liberal capitalist human
HH 4033 — Liberal humanist human
HH 0986 — Libido
Development [Magical...]
HH 0720 — Magical
HH 0745 — Manpower resource (United Nations)
HH 0146 — Mental growth
HH 5973 — mind (Buddhism)
HH 0565 — Moral
HH 0565 — moral values
HH 0055 — Motivational
HH 3003 — music ; Human
HH 7637 — Mythical control human
HH 1523 Development new religious movements ; Human
Development [panentheism...]
HH 1908 — panentheism ; Human
HH 5190 — pantheism ; Human
HH 0655 — Perceptual
HH 0117 — Personal career
HH 0955 — personal cultural potential
HH 0117 — Personal self
HH 0281 — Personality
HH 0447 — Personality
HH 0836 — Physical
HH 1902 — primal religion ; Human
KD 2009 — processes accumulation significance
KP 3014 — processes ; Cycle
HH 7637 — Psychic control human
HH 0855 — Psychic (Psychism)
HH 0407 — Psychosexual
HH 3117 — psychosomatic power
HH 0647 Development quality human life
HH 1198 Development religion ; Human
Development [Self...]
HH 0651 — Self
HH 3710 — Self (Christianity)
HH 0195 — Sensorimotor
HH 0046 — Social
HH 0046 — Socio economic
HH 0017 — Spiritual
HH 0855 — Spiritual (Psychism)
HH 0285 — Stages personality
HH 3997 — State socialist human
HH 0247 — strategy ; Human (United Nations)
HH 3181 — Sustainable
HH 3662 Development ; Technological
HH 4663 Development theism ; Human
HM 7826 Development ; Understanding consisting (Buddhism)
HM 4128 Development ; Understanding plane (Buddhism)
HM 2098 Development ; Vinnana khanda (Pali)
HH 1976 Development women ; Psychological
KP 3004 Developmental interaction
HH 3698 Developmental psychology
HH 0697 Deviance
VD 2428 Deviation
KC 0427 Deviation-amplification
VP 5290 Deviation ; Direction
HM 1133 Deviations body ; Leaving off, abstaining, totally abstaining refraining three (Buddhism)
HM 1781 Deviations speech ; Leaving off, abstaining, totally abstaining refraining four (Buddhism)
HM 0461 Device ; Altered states consciousness induction (ASCID)
MS 8306 Devices ; Cultural objects
MS 8102 Devices ; Domestic
TC 1929 Devil's advocate
HM 2378 Devil (Tarot)
VD 2430 Devilishness
VD 2432 Deviousness
VD 2434 Devitalized
VC 0551 Devotion
TC 1345 Devotion
HH 0349 Devotion (Christianity)
HM 8776 Devotion God (Sufism)
HH 1016 Devotion lord
HM 1475 Devotion ; Spiritual (Leela)
HM 6554 Devotion ; Way (Esotericism)
HM 0561 Devotion worship ; Religious experience personal
VC 0553 Dexterity
HH 0578 Dge-ba (Buddhism)
HH 0093 Dhamma
HM 2793 Dhamma ; Bodhi pakkhiya (Buddhism)
HM 4726 Dhamma-patisambhida (Pali)
HH 1022 Dhammadesana (Buddhism)

-868-

HM 1969 Dhammanam ayu ; Arupinam (Pali)
Dhammanam samapattiya
HM 0552 — Hiriyati papakanam akusalanam (Pali)
HM 3257 — Na kiriyati papakanam akusalanam (Pali)
HM 3353 — Na ottappati papakanam akusalanam (Pali)
HM 1208 — Ottappati papakanam akusalanam (Pali)
HM 2551 Dhammas (Pali)
HM 1715 Dhammavicaya (Pali)
HM 2566 Dharana (Hinduism)
HM 1423 Dharanata (Pali)
HH 0093 Dharma
HM 2793 Dharma ; Bodhi pakshika (Buddhism)
HM 2397 Dharma-kaya (Buddhism)
Dharma [Laukikagra...]
HM 2271 — Laukikagra (Buddhism)
HH 1183 — Listening (Buddhism)
HM 0481 — loka (Leela)
HM 0194 Dharma megha samadhi (Yoga)
HM 2421 Dharma-negha (Buddhism)
HM 0481 Dharma ; Plane (Leela)
HH 1022 Dharma ; Preaching (Buddhism)
Dharma [Sadharana...]
HH 0093 — Sadharana
HH 0093 — Sanatana
HM 1715 — Search (Buddhism)
HM 2208 Dharmaloka-labdha-samadhi (Buddhism)
HM 0552 Dharmas ; Being ashamed acquisition sinful unwholesome (Buddhism)
HM 1969 Dharmas ; Duration formless (Buddhism)
HM 1208 Dharmas ; Feeling remorse acquisition sinful unwholesome (Buddhism)
HM 3257 Dharmas ; Not ashamed acquisition sinful unwholesome (Buddhism)
HM 3353 Dharmas ; Not feeling remorse acquisition sinful unwholesome (Buddhism)
HH 1108 Dharmas ; Seventy five (Buddhism)
HM 3144 Dhatu ; Arupa (Buddhism)
HM 2551 Dhatu (Buddhism)
HM 2142 Dhatu ; Rupa (Buddhism)
HM 2235 Dhatu vidyadhara ; Rupa (Buddhism)
HM 2792 Dhikr-i-Qalbi (Sufism)
HM 2351 Dhikr (Islam)
HM 6562 Dhikr ; Practice (Sufism)
HM 8330 Dhikr ; Sultan (Sufism)
HM 1974 Dhiti (Pali)
HM 6719 Dhul-Jalal-wal-Ikram (Sufism)
HM 6719 Dhul-Jalali wal-Ikram (Sufism)
HM 1081 Dhurasampaggaha (Pali)
HM 2693 Dhyana (Buddhism)
HM 0137 Dhyana (Hinduism)
HH 0827 Dhyana yoga (Yoga)
HH 0589 Dhyana (Zen)
VD 2436 Diabolic
HM 0810 Diaconia (Christianity)
TC 1974 Diagnostic conferencing
KC 0706 Diagram ; Block
KC 0816 Diagram ; Flow
HM 0810 Diakonia (Christianity)
KC 0282 Dialectic
KP 3003 Dialectic synthesis
HH 0529 Dialectical materialism
KC 0457 Dialectical-materialist synthesis
HM 3225 Dialogue ; Fundamental
HM 1762 Dialogue ; Fundamental
HH 2434 Dialogue ; Inter religious
HH 2434 Dialogue ; Living
KD 2140 Dialogue ; Logos lemma interparadigmatic
HH 0309 Diamond vehicle Buddhism (Buddhism)
HH 0106 Dianetics
HM 2927 Diaphanous intuition (ICA)
VD 2438 Dictatorship
KC 0944 Dictionary
HM 3719 Dies cross ; Jesus (Christianity)
HH 7324 Diet ; Human development
MM 2022 Diets ; Media
VP 5013 Difference ; Identity
MP 1251 Different settings
HH 3772 Differentiated non-duality (Hinduism)
KC 0276 Differentiation
VC 0555 Differentiation
HM 0249 Differentiation (Buddhism)
MP 1036 Differentiation relationship density
MP 1054 Differently paced communications ; Intersection
HM 5992 Difficult progress sluggish direct knowledge ; Concentration (Buddhism)
HM 7859 Difficult progress swift direct knowledge ; Concentration (Buddhism)
VD 2440 Difficulty
TP 0003 Difficulty beginning
VP 5732 Difficulty ; Facility
TP 0003 Difficulty ; Initial
VP 5592 Diffuseness ; Conciseness
VC 0557 Dignity
HH 3179 Dignity
VD 2442 Dilapidation
VC 0559 Diligence
HH 0587 Dimension being ; Non temporal
VT 8031 Dimension∗complex
MP 1040 Dimension ; Integrating historical
MP 1039 Dimension ; Integrating new
HM 4071 Dimensional clairvoyance ; Fourth (Psychism)
HM 2780 Dimensional consciousness ; Multi
TC 1547 Dimensions conferencing ; Structural
MP 1043 Dimensions ; Presentation new
HM 6201 Diminished clarity awareness
HM 1353 Diminished focus awareness

HM 1353 Diminished range awareness
HM 8002 Diminution ; Concentration partaking (Buddhism)
HM 2290 Din sephira ; Geburah (Kabbalah)
MM 2082 Dining
HM 2393 Dionysos awareness
Direct [insight...]
MP 1181 — insight structures focal point ; Maintenance source
HM 5532 — intuition (Buddhism)
HM 2099 — intuitive knowledge (Hinduism)
Direct knowledge Buddhism
HM 4297 — [Direct knowledge (Buddhism)]
HM 5232 — [Direct knowledge (Buddhism)]
HM 5982 — [Direct knowledge (Buddhism)]
HM 7672 — [Direct knowledge (Buddhism)]
HH 5652 — [Direct knowledge (Buddhism)]
HM 0748 — [Direct knowledge (Buddhism)]
Direct knowledge Concentration
HM 5992 — difficult progress sluggish (Buddhism)
HM 7859 — difficult progress swift (Buddhism)
HM 1010 — easy progress sluggish (Buddhism)
HM 4977 — easy progress swift (Buddhism)
HM 2033 Direct non-conceptual cognition emptiness (Buddhism)
HM 4469 Direct perception
MP 1122 Direct relationship structures communication pathways
HH 0072 Directed fantasy
HM 0034 Directedness ; Goal
HM 6663 Directedness thought ; One (Buddhism)
VC 0561 Direction
VP 5290 Direction-Deviation
HH 0442 Direction ; Educative spiritual
VC 1555 Direction ; Self
HH 0442 Direction ; Spiritual
HM 2651 Direction time ; Consciousness irreversible
HM 2701 Direction time ; Consciousness reversible
HM 2995 Directions ; Emptiness ten (Buddhism)
HH 0095 Directive counselling ; Non
HH 0034 Directiveness ; Potential
VC 0563 Directness
MS 8338 Dirigibles
VD 2444 Dirtiness
VD 2446 Disability
HM 1134 Disability ; Bodily (Yoga)
VD 2448 Disaccord
VP 5794 Disaccord ; Accord
VD 2450 Disadvantage
VD 2452 Disaffection
VD 2454 Disaffinity
VD 2456 Disagreeableness
VP 5026 Disagreement ; Agreement
TC 1929 Disagreer
TC 1663 Disagreers
VD 2458 Disappearance
VP 5446 Disappearance ; Appearance
VD 2460 Disappointment
VD 2462 Disapproval
VP 5968 Disapproval ; Approval
VD 2464 Disarrangement
VD 2466 Disaster
HM 2350 Disaster ; Emblems (Tarot)
VD 2468 Disbelief
VC 0565 Discernment
HM 3630 Discernment (Buddhism)
HM 1143 Discernment (Buddhism)
HM 8303 Discernment ; Loss (Islam)
HH 3550 Discernment middle way ; Purity knowledge (Buddhism)
HH 3025 Discernment ; Purity knowledge (Buddhism)
HH 4007 Discernment right path wrong path ; Purity knowledge (Buddhism)
HH 3900 Discernment ; Spiritual (Christianity)
HH 6192 Discernment ; Understanding being plane (Buddhism)
HM 2998 Discernment (Yoga)
HH 2716 Discipleship-application awareness (Buddhism)
HM 2192 Discipleship-karma awareness (Buddhism)
HM 8901 Discipleship (Sufism)
HM 2240 Discipleship-vision awareness (Buddhism)
KC 0077 Disciplinarity
Discipline
KC 0077 — [Discipline]
VC 0567 — [Discipline]
HH 0163 — [Discipline]
TP 0060 — [Discipline]
HH 0328 Discipline ; Autonomous
HH 1376 Discipline (Buddhism)
HH 0956 Discipline celebration (Christianity)
HH 0825 Discipline confession (Christianity)
KC 0596 Discipline development
HH 5282 Discipline ; Ecstatic
HH 1086 Discipline ; Ethical
HH 1244 Discipline guidance (Christianity)
HH 0925 Discipline ; Heteronomous
HH 0740 Discipline ; Interactive
HM 2851 Discipline ; Interior (ICA)
HH 1688 Discipline meditation (Christianity)
HH 1086 Discipline ; Moral
HH 1193 Discipline prayer (Christianity)
HM 6122 Discipline ; Romance spiritual
HH 0959 Discipline secret
Discipline Self
HH 0877 — [Discipline ; Self]
VC 1557 — [Discipline ; Self]
HM 2121 — (Christianity)
HM 3280 — (Hinduism)
Discipline [service...]
HH 0232 — service (Christianity)
HH 0333 — solitude (Christianity)

Discipline [Spiritual...] cont'd
HH 1021 — Spiritual
HH 1323 — study (Christianity)
HH 0225 — submission (Christianity)
HH 3776 Discipline worship (Christianity)
HH 0298 Discipline worship (Christianity)
HH 0707 Disciplines ; Spiritual
HM 2181 Disciplining mental abiding states (Buddhism)
HM 2824 Disclosure ; Cryptic (ICA)
VD 2470 Discomfort
VD 2472 Discompose
VD 2474 Disconcertion
VD 2476 Discontentment
VP 5868 Discontentment ; Contentment
VD 2478 Discontinuity
KD 2065 Discontinuity aesthetics ; Containing
KD 2210 Discontinuity: Comprehension internalization
KC 0008 Discontinuity ; Continuity
VP 5071 Discontinuity ; Continuity
KC 0462 Discontinuous change ; Mathematics
KD 2090 Discontinuous learning ; Opening closing: alternation
VD 2480 Discord
VP 5462 Discord ; Harmony
HM 2498 Discouragement
VD 2482 Discourteousness
KC 0797 Discourse ; Universe
VD 2484 Discourtesy
VP 5936 Discourtesy ; Courtesy
VC 0569 Discovery
HM 6633 Discovery natural mind (Taoism)
VP 5485 Discovery ; Research
HH 1126 Discovery ; Self
VD 2486 Discredit
HM 2298 Discrete states consciousness (Physical sciences)
VC 0571 Discretion
HH 3900 Discretion (Christianity)
HM 1344 Discriminating awareness ; Perfection (Buddhism)
VC 0573 Discrimination
VD 2488 Discrimination
HM 2399 Discrimination (Buddhism)
HM 2099 Discrimination (Hinduism)
VP 5492 Discrimination-Indiscrimination
HM 4958 Discrimination perspicuity ; Understanding (Buddhism)
HM 2998 Discrimination (Yoga)
HM 0485 Discriminations ; Four (Buddhism)
VT 8083 Discriminative affection∗complex
HM 4002 Discriminative knowledge ; Non (Systematics)
HM 1234 Discriminative knowledge (Systematics)
HH 0197 Discursive meditation (Christianity)
HH 0084 Discursive power
HM 6997 Discursive reasoning (Christianity)
TC 1306 Discussion 66
HH 3366 Discussion decision ; Journeying transcendence
TC 1511 Discussion group
TC 1650 Discussion ; Interrogation
TC 1531 Discussion ; Panel
VD 2490 Disdain
HM 6001 Disease
MM 2018 Disease
VP 5685 Disease ; Health
MS 8375 Diseased humans ; Deformed
VD 2492 Disenchantment
HM 2204 Disenchantment
HM 2552 Disengagement work (ICA)
VD 2494 Disesteem
VD 2496 Disfavour
VD 2498 Disfigurement
VD 2500 Disgrace
VD 2502 Disgust
HM 0966 Disgust ; Aesthetic emotion (Hinduism)
VD 2504 Disharmony
TP 0038 Disharmony
VD 2506 Dishonesty
VD 2508 Dishonour
HM 2564 Dishonouring ; Logical (Brainwashing)
HM 6174 Disillusionment
VD 2510 Disillusionment
VP 5519 Disillusionment ; Illusion
VD 2512 Disingenuousness
HH 0561 Disintegration
VD 2514 Disintegration
VP 5050 Disintegration ; Cohesion
TP 0059 Disintegration (Flooding)
VD 2516 Disinterest
HM 2547 Disinterest ; Passionate (ICA)
HM 2812 Disinterested collegiality (ICA)
HM 1799 Disinterestedness (Buddhism)
HM 2128 Disinterestedness (Buddhism)
HM 3752 Disinterestedness wholesome root (Buddhism)
VD 2518 Dislike
VP 5185 Dislocation ; Location
VD 2520 Disloyalty
VD 2522 Dismay
VD 2524 Disobedience
HH 0759 Disobedience ; Civil
VP 5766 Disobedience ; Obedience
VD 2526 Disorder
MP 1058 Disorder ; Context
HH 0763 Disorder ; Multiple personality
VP 5059 Disorder ; Order
VD 2528 Disorganization
VD 2530 Disorientation
VD 2532 Disparagement
VD 2534 Disparity
VC 0575 Dispassion
HM 5364 Dispassion ; Knowledge contemplation (Buddhism)

Dispassion

HM 0110 Dispassion ; Lack (Yoga)
HM 0903 Dispassion (Yoga)
TP 0059 Dispersion (Flooding)
MP 1236 Displaceable frameworks
VD 2536 Displeasure
VP 5865 Displeasure ; Pleasure
VC 1905 Disposed ; Well
HM 2569 Disposing intelligence ; Path (Judaism)
HM 2098 Dispositions consciousness (Buddhism)
VD 2538 Disputatiousness
VD 2540 Disquiet
VD 2542 Disregard
VP 5914 Disrepute ; Repute
VD 2544 Disrespect
VP 5964 Disrespect ; Respect
HH 3356 Disruption lifestyle
HH 4001 Disruptive thoughts (Christianity)
VD 2546 Dissatisfaction
HM 5448 Dissatisfaction (Hinduism)
VD 2548 Dissension
TP 0059 Dissension ; Overcoming (Flooding)
VP 5521 Dissent ; Assent
TC 1383 Dissertation
VD 2550 Dissidence
TC 1494 Dissidence
VP 5020 Dissimilarity ; Similarity
HM 3093 Dissimulation (Buddhism)
VD 2552 Dissipation
TC 1815 Dissipation ; Conferencing
TC 1699 Dissipation ; Meeting energy
KD 2081 Dissipative structures ; Order fluctuation:
HM 8092 Dissociated mind (Buddhism)
VD 2554 Dissociation
HH 1294 Dissociation
HM 6087 Dissociation affect
HM 7608 Dissociative amnesia
HH 0763 Dissociative states
VD 2556 Dissolution
TP 0059 Dissolution ; Barrier (Flooding)
HH 2002 Dissolution evolution ; Aeons (Buddhism)
HM 3385 Dissolution ; Knowledge contemplation (Buddhism)
HM 6534 Dissolving human mind (Taoism)
VD 2558 Dissonance
HH 8204 Dissonance ; Cognitive
KD 2250 Dissonant harmony holistic resonance
VP 5648 Dissuasion ; Motivation
VP 5200 Distance ; Nearness
MP 1082 Distance related operational control contexts ; Minimal
VD 2560 Distaste
HM 4754 Distinct outside world ; Anomalies experience self
MP 1102 Distinct pattern entry points complex structures
VC 0577 Distinction
HM 7363 Distinction ; Concentration partaking (Buddhism)
HM 2681 Distinctions ; Beyond individual (Sufism)
MP 1169 Distinctions contextual levels ; Maintaining
MP 1206 Distinctions ; Efficient enclosure spaces minimal structural
MP 1197 Distinctions separating domains ; Substantive
MP 1110 Distinctiveness main entry point structure
HM 2668 Distorted cognition (Buddhism)
VD 2562 Distortion
TC 1373 Distortion
MP 1218 Distortion resistant boundaries
VP 5248 Distortion ; Symmetry
HM 1407 Distortion views (Buddhism)
VD 2564 Distraction
HM 3154 Distraction (Buddhism)
HM 4382 Distraction ; Non (Buddhism)
VD 2566 Distress
KC 0518 Distributed information system
MP 1210 Distribution levels ; Harmonizing space
MP 1002 Distribution organization
MP 1213 Distribution secondary inter-level connections
HH 0896 Distributive synthesis
VD 2568 Distrust
VD 2570 Disturbance
VD 2572 Disunity
VD 2574 Disuse
HM 1191 Ditthi (Pali)
HH 2718 Ditthi-visuddhi-niddesa (Pali)
HM 1613 Ditthigahana (Pali)
HM 2054 Ditthigata (Pali)
HH 0974 Ditthijjukamma (Buddhism)
HM 4548 Ditthikantara (Pali)
HM 1511 Ditthisannojana (Pali)
HM 2722 Ditthivipphandita (Pali)
HM 1407 Ditthivisukayika (Pali)
VD 2576 Divergence
VP 5298 Divergence ; Convergence
HM 2368 Divergent thinking
TC 1225 Diverger participant type
KC 0231 Diversification
MP 1046 Diversified interchange environment
HH 0596 Diversional therapy
VC 0579 Diversity
KC 0002 Diversity ; Unity
TC 1558 Diversity ; Unity
HH 0158 Divina ; Lectio
HH 0545 Divination
Divine [abidings...]
HH 3534 — abidings meditation subjects (Buddhism)
HM 1538 — analogy ; Language
HM 2007 — animal-hell awareness (Buddhism)
HM 1026 — attraction (Sufism)
HH 2561 — attributes (Sufism)
Divine [captive...]
HM 3190 — captive (ICA)
HM 0563 — consciousness
HM 4024 — consciousness ; Esoteric (Psychism)
HM 4340 — contemplation (Christianity)
HM 5982 Divine ear element ; Knowledge (Buddhism)
HM 3513 Divine essence (Sufism)
HM 2907 Divine grace ; Liberation spirit attraction (Sufism)
HM 2859 Divine hosts (ICA)
HH 0814 Divine indwelling
HM 2401 Divine-love awareness (Sufism)
HM 3941 Divine love ; Mystical ladder (Christianity)
HM 1452 Divine mysteries ; Unveiling (Sufism)
Divine [names...]
HH 2561 — names (Sufism)
HH 0700 — nature
HM 2246 — nature (Sufism)
HM 3042 — nothingness (ICA)
HM 3371 Divine path (Sufism)
HM 2416 Divine presences ; Awareness (Sufism)
HH 0158 Divine reading (Christianity)
HM 0748 Divine sight ; Knowledge (Buddhism)
HH 3534 Divine states (Buddhism)
HM 0762 Divine unity ; Intense longing (Sufism)
HM 1027 Divini ; Furor
MS 8117 Divinities
VD 2580 Division
VD 2578 Divisiveness
HM 2110 Dngos gzhii snyoms jug (Tibetan)
VC 0581 Docility
HH 0973 Doctor ; Witch
TC 1384 Doctrinaire participant type
HH 1099 Doctrine being ; Causally continuous (Buddhism)
HH 0589 Doctrine ; Mind (Zen)
HH 0397 Doctrine three ages (Christianity)
HM 4726 Doctrine ; Understanding knowledge (Buddhism)
HM 2433 'Dod-chags (Buddhism)
HM 2733 Dod-khams (Tibetan)
MS 8853 Dog
HM 3018 Doing ; Life (ICA)
HM 2143 Doing mystery (ICA)
HH 0367 Doing ; Not
VD 2582 Dolorousness
MP 1013 Domain boundary ; Sub
MP 1198 Domain contexts perspective adjuncts ; Inter
MP 1137 Domain developing perspectives
MP 1131 Domain dynamics ; Organization inter
MP 1163 Domain external structures ; Hospitable non linear
MP 1129 Domain focal point structure ; Common
MP 1124 Domain interfaces ; Ensuring function common
MP 1126 Domain ; Off centering focal point common
MP 1146 Domain organization ; Flexible
MP 1139 Domain perspective nourishment ; Hospitable
MP 1112 Domain structure communication pathway ; Transition
MP 1161 Domain ; Structure enfolded insight
MP 1067 Domain ; Unstructured common
MP 1061 Domains ; Bounded common small scale interaction
HM 2026 Domains consciousness ; Subjective (Yoga)
MP 1063 Domains ; Cyclic interrelation complementary perspective common
MP 1123 Domains ; Enhancing function common
MP 1115 Domains ; Functional integration unstructured internal
MP 1094 Domains ; Hospitable common
Domains [Independent...]
MP 1001 — Independent
MP 1252 — insight
MP 1125 — Integrating points perspective common
MP 1166 — interfaces structure external environment ; Overview
MP 1194 — Internal connectedness
MP 1011 Domains ; Local interrelationship
MP 1167 Domains minimum proportions ; Enfolded overview
MP 1145 Domains non-current elements those reserve
MP 1193 Domains ; Partially enclosed internal
MP 1128 Domains receive external insight ; Orientation
Domains [separating...]
MP 1106 — separating complementary structures ; Functional enhancement
MP 1197 — Substantive distinctions separating
MP 1232 — Symbolic connection encompassing
MP 1233 Domains ; Zoning internal
MS 8102 Domestic devices
MS 8336 Domestication ; Objects animal husbandry
VC 0587 Dominance
HM 1466 Dominant ideas
HH 0477 Dominant ways life (DWL)
HM 2179 Dominated awareness ; Wish
HM 0065 Domination ; Experience (Systematics)
HM 3451 Doors (Sufism)
HM 7342 Doppelgänger
HM 2980 Dormant consciousness
HM 4502 Dosa (Pali)
HM 0607 Dosa (Pali)
HM 5676 Dosakkhaya (Pali)
HM 5676 Dosaksaya (Buddhism)
HM 2079 Double-awake body-asleep trance
HH 6112 Double bind ; Therapeutic
HM 3362 Double mindedness
HM 2490 Doubt
Doubt Buddhism
HM 3306 — [Doubt (Buddhism)]
HM 4347 — [Doubt (Buddhism)]
HM 4467 — [Doubt (Buddhism)]
HM 1294 Doubt efficacy good life (Buddhism)
HM 2655 Doubt ; Overcoming (Buddhism)
HH 1187 Doubt ; Purification overcoming (Buddhism)
HH 1187 Doubt ; Purity transcending (Buddhism)
HM 1561 Doubt two alternatives ; Being (Buddhism)
HM 1080 Doubt (Yoga)
VD 2584 Doubtfulness
HM 0177 Doubting (Buddhism)
HM 3664 DPR Deep psychophysiological relaxation
HM 5089 Dpyod-pa (Tibetan)
HM 2659 Dpyod-pa-yid-byed (Tibetan)
KC 0194 Dramatic unity
HM 2847 Dran-pa (Tibetan)
MM 2081 Drawing water
MM 2023 Drawn cart ; Animal
HM 8112 Dread blame (Buddhism)
HM 3311 Dread (Christianity)
TC 1439 Dread-filled conferencing
HM 2781 Dream consciousness ; Swapna state (Yoga)
HM 0147 Dream ; High (Psychism)
HM 3147 Dream (Hinduism)
Dream state
HM 0665 — After death (Psychism)
HM 2371 — bardo awareness (Buddhism)
HM 0600 — Yoga (Buddhism)
HH 2116 Dream therapy
HH 1301 Dream time ; Aboriginal (Australian)
HM 2014 Dream ; Wakeful (Hinduism)
HM 3510 Dream wakefulness (Hinduism)
HM 4087 Dream yoga (Buddhism)
HH 1301 Dreaming (Australian)
HM 3618 Dreaming ; Creative
HM 3908 Dreaming experience
HM 3618 Dreaming ; Lucid
HM 3173 Dreaming ; Mastery
HM 2950 Dreams
HM 4087 Dreams ; Lucid (Buddhism)
HM 2085 Dreams ; Threatening
HM 1376 Dreams ; Waking
HM 2860 Dreamy-arrested consciousness
HH 0555 Dress
HM 4006 Drift ; Mind
HM 2996 Drishti (Buddhism)
HM 1280 Drishti ; Samyag (Buddhism)
TC 1361 Driver role ; Team
MM 2060 Driving
HM 6231 Drowsiness
HH 5110 Drug abuse ; Alternatives
HH 0075 Drug-induced experience
HH 1123 Drug use transcendent experience
HM 0801 Drugs ; Consciousness expanding
HH 0075 Drugs ; Hallucinogenic
HH 0075 Drugs ; Mind expanding
Drugs Modes awareness
HM 0803 — associated use amphetamines similar
HM 4060 — associated use opium similar
HM 4254 — associated use phencyclidine similar
HH 0075 Drugs ; Psychedelic
HH 0075 Drugs ; Psychotomimetic
HH 0788 Drugs ; Sacred
HM 7174 Drunk ; Punch
HM 6097 Drunkenness (Sufism)
HM 7805 DTs
HM 2399 'Du-shes (Tibetan)
HM 7131 Dual personality
HH 0299 Dualism ; Mind body
HH 4455 Dualism ; Psychospiritual
HH 0872 Duality
HM 2381 Duality conscious states ; Physical (Physical sciences)
HM 3772 Duality ; Differentiated non (Hinduism)
HH 0518 Duality ; Non (Hinduism)
VP 5089 Duality ; Unity
HM 3934 Dubiety (Buddhism)
HM 3135 Dubiety ; Essential (ICA)
VD 2586 Dubiousness
VP 5960 Dueness-Undueness
HM 2574 Duhkha (Buddhism)
HH 0523 Duhkha (Buddhism)
HM 1782 Dukh (Leela)
HM 1031 Dukkam vedayitam ; Cetosamphassajam (Pali)
HM 2574 Dukkha (Pali)
HH 0523 Dukkha (Pali)
HM 4076 Dukkha vedana ; Cetosamphassaja (Pali)
HM 0805 Dukkham ; Cetasikam (Pali)
HM 3556 Dukkhindriya (Pali)
HM 2181 Dul-bar-byed-pa (Tibetan)
VD 2588 Dullness
HM 2695 Dullness ; Absence (Buddhism)
HM 3196 Dullness (Buddhism)
HM 1857 Dummajjha (Pali)
HM 2578 'Dun pa (Tibetan)
VD 2590 Duplicity
VC 0589 Durability
VP 5110 Durability-Transience
HM 2361 Duramgama (Buddhism)
VC 0591 Duration
TP 0032 Duration
HM 6652 Duration ; Conscious (Buddhism)
HM 1969 Duration formless dharmas (Buddhism)
HM 3623 Durbuddhi (Leela)
VD 2592 Duress
HM 4323 Dussana (Pali)
HM 1967 Dussitattam (Pali)
VC 0593 Duty
HM 2694 Duty ; Corporate (ICA)
HM 2224 Duty ; First (Christianity)
HM 4327 Duty ; Religious experience realization
HM 2861 Duty ; Second (Christianity)
HM 4279 Duty ; Third (Christianity)
HM 1286 Dvedhapatha (Pali)

HM 1561 Dvelhaka (Pali)
HM 3620 Dvesa (Yoga)
HH 0715 Dvija (Hinduism)
HM 2038 Dvitiyadhyana (Buddhism)
MS 8937 Dwelling beings ; Mythical fire
HM 5643 Dwelling equanimity (Buddhism)
Dwelling [first...]
HM 4298 — first jhana (Buddhism)
HM 6553 — fivefold jhana (Buddhism)
HM 8087 — fourth jhana (Buddhism)
HM 7121 Dwelling second jhana (Buddhism)
HM 6553 Dwelling second jhana fivefold system (Buddhism)
HM 5643 Dwelling third jhana (Buddhism)
HM 0554 Dwesh (Leela)
HH 0477 DWL Dominant ways life
HM 2930 Dyadic awareness
TC 1311 Dyads
HM 3099 Dying death (ICA)
HH 1407 Dying ; Instructing
HM 2553 Dying person ; Psychological growth
HM 1402 Dying rising ; Christ consciousness (Christianity)
TC 1188 Dynamic design methods
HH 0787 Dynamic self-opening meditation
HM 3091 Dynamic selfhood (ICA)
TC 1441 Dynamic tensions
HH 0629 Dynamic universal
KC 0147 Dynamical similarity
KC 0954 Dynamics ; Group
Dynamics [Industrial...]
KC 0426 — Industrial
MP 1080 — Integrated contexts perspective
MP 1083 — Integration perspective acquisition perspective maintenance
KD 2192 Dynamics: learning ; Encompassing system
MP 1131 Dynamics ; Organization inter domain
TC 1928 Dynamics ; Participation
KD 2190 Dynamics: sixfold restraint ; Encompassing system
KC 0426 Dynamics ; System
KC 0426 Dynamics ; Urban
VC 0595 Dynamism
HH 0357 Dysgenics
HM 3407 Dysphoria
HM 2127 Dzam-bu-gling (Tibetan)

E

MP 1141 Each perspective ; Relatively isolated context
VC 0597 Eagerness
HM 2169 Ear consciousness (Buddhism)
HM 5982 Ear element ; Knowledge divine (Buddhism)
HH 6018 Early achievement excellence ; Education
VC 0599 Earnestness
TP 0002 Earth Acquiescence
HM 2023 Earth ; Common (ICA)
HM 3570 Earth ; Fourth step: (Christianity)
HH 3246 Earth kasina (Buddhism)
HM 6112 Earth (Leela)
KC 0179 Earth ; One
TP 0002 Earth Passivity
TP 0002 Earth Receptivity
TP 0002 Earth Responsive service
HM 3235 Earth triplicity (Astrology)
HH 4997 Earthly consciousness ; Blending celestial consciousness (Taoism)
HM 3235 Earthy awareness (Astrology)
VC 0601 Ease
HM 1921 Ease born contact element mind consciousness ; Psychical (Buddhism)
HM 0309 Ease born contact psychical ; Sensation (Buddhism)
HM 4191 Ease ; Concentration accompanied (Buddhism)
HM 1454 Ease ; Concentration accompanied (Buddhism)
HM 0972 Ease nor unease born contact element eye consciousness ; Neither psychical (Buddhism)
HM 3314 Ease nor unease born contact element mind consciousness ; Neither physical (Buddhism)
HM 2543 Eastern-continent awareness noble figures (Buddhism)
HH 9219 Eastern mysticism
Easy progress
HM 4977 — quick intuition ; Concentration (Buddhism)
HM 1010 — sluggish direct knowledge ; Concentration (Buddhism)
HM 1010 — sluggish intuition ; Concentration (Buddhism)
HM 4977 — swift direct knowledge ; Concentration (Buddhism)
HM 3808 Easy transformation ; Capacity (Buddhism)
MP 1196 Eccentric access perspective contexts
VD 2594 Eccentricity
HH 3435 Eccentricity
HM 6704 Echo pensées
KC 0581 Eclecticism
VC 0603 Eclecticism
HM 0800 Eco-spirituality
KD 2335 Ecodynamics societal evolution
HH 1267 Ecological awareness
HM 2006 Ecological consciousness
HH 4965 Ecological humanism
TC 1475 Ecological natural processes ; Meetings
HH 1267 Ecological planetary consciousness
HH 2315 Ecological wisdom
KC 0950 Ecology
KC 0922 Ecology concepts
HH 2315 Ecology ; Deep
KC 0922 Ecology knowledge
HH 2315 Ecology movement
KC 0415 Ecology ; Systems
TC 1996 Economic conferencing

KC 0963 Economic cybernetics
HH 0046 Economic development ; Socio
KC 0792 Economic integration ; International
KC 0651 Economic integration ; Voluntary
KC 0277 Economic order ; International
KC 0080 Economic policy
Economic [social...]
HH 0437 — social development
KC 0162 — structure ; Socio
KC 0530 — system
VC 0605 Economy
VP 5851 Economy-Prodigality
KC 0821 Economy ; World
HH 1778 Ecosophy
HH 2315 Ecosophy
KC 0447 Ecosphere
KC 0138 Ecosystem
VC 0607 Ecstasy
HM 2046 Ecstasy
HM 1400 Ecstasy ; Hypnotic
HM 4805 Ecstasy ; Musical
HM 3521 Ecstasy ; Mystical (Sufism)
Ecstasy [Sexual...]
HM 0829 — Sexual
HH 4777 — Skill (Buddhism)
HM 3650 — (Sufism)
HM 4501 Ecstasy ; Talmudic (Judaism)
HM 0311 Ecstatic concentration (Buddhism)
HH 5282 Ecstatic discipline
HM 4501 Ecstatic experience God (Judaism)
HM 7100 Ecstatic perception (Hinduism)
HH 0866 ECT Electric convulsive therapy
HH 0099 Ecumenicism (Christianity)
HH 0099 Ecumenicism (Christianity)
KC 0543 Ecumenism
VC 0609 Edification
HH 0945 Education
VC 0611 Education
Education [Aesthetic...]
HH 0029 — Aesthetic
HH 0029 — Artistic
HH 0938 — Atheistic
Education [Character...]
HH 0895 — Character
HH 0955 — Continuing
HH 1208 — Curative
HH 6018 Education early achievement excellence
Education [Ideological...]
HH 0996 — Ideological re
KC 0976 — Integrated
KC 0711 — Integrated science
KC 0290 — Integrative
HH 6451 — international understanding
HH 0955 Education ; Lifelong
VP 5562 Education-Miseducation
HH 0565 Education ; Moral
HH 0575 Education ; Physical
KC 0690 Education ; Racially ethnically integrated
HH 0709 Education ; Religious
Education [Seishin...]
HH 1109 — Seishin (Japanese)
HH 0417 — Self
HH 0547 — Sex
HH 1109 — Spiritual (Japanese)
KC 0920 — Systems
HH 0178 Educational guidance
HH 5786 Educational guidance
KC 0976 Educational integration
HH 0924 Educational mobilization
HH 0924 Educational motivation
HH 0955 Educational self-development
KC 0720 Educational system
HH 0010 Educational transformation human personality ; Lifelong
TC 1524 Educative conferencing
HH 0442 Educative spiritual direction
VD 2596 Eeriness
HM 7703 Effacement (Buddhism)
HM 1559 Effacement ; Self
VC 0613 Effectiveness
HM 2796 Effectivity ; Definitive (ICA)
HM 8080 Effectual knowledge (Systematics)
VD 2598 Effetism
HM 1294 Efficacy good life ; Doubt (Buddhism)
VC 0615 Efficiency
MP 1206 Efficient enclosure spaces minimal structural distinctions
HH 0544 Effort
HM 2389 Effort (Buddhism)
HM 2973 Effort ; Illumined (Astrology)
HM 1295 Effort ; Right (Buddhism)
HM 6099 Effort (Sufism)
VC 0617 Effortlessness
HH 2188 Effortlessness ; Path
HM 2521 Effulgence
HH 0636 Ego
HM 2059 Ego awareness (Hinduism)
HM 4754 Ego boundaries ; Blurring
Ego [concept...]
HM 0471 — concept
HM 0044 — conflict ; Consciousness
HM 0570 — consciousness
HM 2590 — consciousness ; Crazy monkey (Hinduism)
Ego [death...]
HM 3060 — death
HM 6921 — death
HM 0655 — destruction

Ego [development...] cont'd
HH 0371 — development
HM 3748 Ego ; First step: (Christianity)
HH 0875 Ego identity
HH 0634 Ego level
HM 3094 Ego pain
Ego [Spiritual...]
HH 0698 — Spiritual rebirth
HM 6138 — splitting
HH 0217 — Stability
HM 4120 — states ; Altered
HM 0216 — states ; Archaic
HH 0217 — strength
HM 1907 — strength
HM 2230 Ego transcendence (Jung)
HM 0203 Egoic musical inspiration
HH 0130 Egoism
HM 3477 Egoism (Yoga)
HH 0241 Egolessness (Buddhism)
HM 1726 Egotism (Leela)
HH 7234 Egyptian spirituality
KC 0527 Eiconic system
HM 0344 Eight-fold knowledge (Systematics)
HM 2635 Eight liberations (Buddhism)
HM 2679 Eight siddhis (Buddhism)
HM 2775 Eighteen unshared attributes Buddha (Buddhism)
HM 2339 Eightfold path ; Noble (Buddhism)
HH 0779 Eightfold path yoga (Yoga)
HM 2339 Eightfold way (Buddhism)
HM 1325 Eighth order perceptions - control principles
HM 2816 Eighth scriptural bodhisattva awareness (Buddhism)
HM 2318 Eighth stage progress (Sufism)
HM 2301 Eighth vajra-master awareness (Buddhism)
HM 0832 Eirsha (Leela)
HM 6663 Ekaggata ; Cittassa (Buddhism)
HM 5734 Ekagra (Yoga)
HM 3337 Ekagrata parinama (Yoga)
KC 0225 Ekistics
HM 2753 Ekotikarana (Buddhism)
VD 2600 Elaborateness
VC 0619 Elan
VP 5358 Elasticity-Toughness
HM 4899 Elation
VC 0621 Elation
HM 0425 Elation (Buddhism)
TP 0054 Elective affinity (Marrying maiden)
HM 3027 Elector ; Destinal (ICA)
HH 0866 Electric convulsive therapy (ECT)
MM 2043 Electric motors generators
HH 0995 Electrical stimulation brain
HH 0866 Electroconvulsive therapy
TC 1816 Electronic communications unit
HM 1772 Electronic high
HH 1100 Electronic meditation
Electronic meeting
TC 1418 — [Electronic meeting]
TC 1218 — [Electronic meeting]
TC 1876 — [Electronic meeting]
KC 0782 Electronics ; Integrated
Electroshock therapy
HH 0866 — [Electroshock therapy]
HH 0866 — Regressive
HH 0866 — Sleep
VC 0623 Elegance
TP 0022 Elegance
KC 0586 Elegance ; Formal
VP 5589 Elegance-Inelegance
KC 0586 Elegance ; Mathematical
TP 0030 Elegance ; Model (Fire)
HM 2838 Element consciousness ; Mind (Buddhism)
Element [external...]
MP 1234 — external boundaries ; Maintainable, multi
HM 1271 — eye consciousness ; Cognition born contact (Buddhism)
HM 0972 — eye consciousness ; Neither psychical ease nor unease born contact (Buddhism)
HM 5982 Element ; Knowledge divine ear (Buddhism)
HM 6173 Element ; Mind consciousness (Buddhism)
Element mind consciousness
HM 1834 — (Buddhism)
HM 1501 — Cognition born contact (Buddhism)
HM 3314 — Neither physical ease nor unease born contact (Buddhism)
HM 1921 — Psychical ease born contact (Buddhism)
HM 0234 — Psychical unease born contact (Buddhism)
HM 0404 — Volition born contact (Buddhism)
Elements [Adequate...]
MP 1035 — Adequate variety cyclic
MP 1207 — Appropriate construction
MS 8835 — Architectural
HM 4338 — Assembling five (Taoism)
MS 8600 Elements ; Primary visual
MP 1145 Elements those reserve ; Domains non current
VC 0625 Elevation
VP 5317 Elevation-Depression
HH 0134 Elevation man (Christianity)
HM 0351 Elevator illusion
HM 0065 Eleven-fold knowledge (Systematics)
TP 0040 Elimination obstacles
HM 1330 Elixir ; Formation gold (Taoism)
VC 0627 Eloquence
VP 5600 Eloquence-Uneloquence
HM 2070 Emanating consciousness ; Non (Buddhism)
HM 2340 Emanation-body awareness (Buddhism)
HM 3019 Emanation body ; Buddha (Buddhism)

Emanation

HM 2546 Emanation heaven consciousness ; Delightful (Buddhism)
HM 2408 Emanation ; World (Judaism)
HM 8241 Emancipated mind (Buddhism)
VC 0629 Emancipation
HH 0953 Emancipation
KD 2150 Emancipation particular languages
HH 0907 Emancipation self
VD 2602 Embarrassment
HM 2210 Embarrassment (Buddhism)
HM 3203 Embarrassment ; Non (Buddhism)
MP 1247 Embedding fixity variability
HM 2201 Emblem archetypes psyche (Tarot)
HM 3121 Emblems ; Death (Tarot)
HM 2350 Emblems disaster (Tarot)
Emblems [Measuring...]
HM 2178 — Measuring (Tarot)
HM 2250 — Mutability (Tarot)
HM 2398 — mysterious (Tarot)
HM 3127 Emblems ; Natural forces (Tarot)
HM 2378 Emblems personified evil (Tarot)
HM 2315 Emblems renewal (Tarot)
HM 3137 Emblems star ; Life fluid (Tarot)
HM 2411 Emblems supremacy (Tarot)
HM 3289 Emblems temperance ; Life fluid (Tarot)
HM 2387 Emblems totality (Tarot)
HM 2190 Emblems ; Victimization (Tarot)
HM 2422 Emblems well-being (Tarot)
HM 2499 Embodied consciousness
MP 1153 Embodied perspectives ; Structures adaptable changing number
HM 3409 Embodiment
HH 0686 Embodiment ; Re
HM 3056 Embodying charity (ICA)
HM 2442 Embodying equity (ICA)
HM 2876 Embodying peace (ICA)
HM 3199 Embodying service (ICA)
HM 2009 Embracement ; Daring (ICA)
KD 2170 Embracing conflicting styles ; Modes managing:
HM 4762 Embryo ; Formation spiritual (Taoism)
HM 5265 Embryo ; Incubation spiritual (Taoism)
MP 1065 Emergence new perspectives ; Context
MP 1203 Emergent perspectives ; Exclusive spaces
KC 0017 Emergent systems
MP 1138 Emerging external insight ; Receptivity
MP 1057 Emerging foci ; Protection
Emerging perspectives
MP 1073 — Contexts exploratory relationship formation challenging
MP 1076 — Minimal context interrelating mature
MP 1068 — Spontaneous relationship formation amongst
VC 0631 Eminence
HH 0819 Emotion
VC 0633 Emotion
HM 0205 Emotion ; Aesthetic (Hinduism)
HM 8762 Emotion anger ; Aesthetic (Hinduism)
HM 1404 Emotion ; Crowd (Spanish)
HM 0966 Emotion disgust ; Aesthetic (Hinduism)
HM 4085 Emotion energy ; Aesthetic (Hinduism)
HM 1644 Emotion fear ; Aesthetic (Hinduism)
HM 0399 Emotion grief ; Aesthetic (Hinduism)
HM 1780 Emotion love ; Aesthetic (Hinduism)
HM 0645 Emotion mirth ; Aesthetic (Hinduism)
MP 1230 Emotion ; Unmediated supportive
HM 0291 Emotion wonder ; Aesthetic (Hinduism)
HM 3504 Emotional aversions (Buddhism)
Emotional [clairaudience...]
HM 4245 — clairaudience (Psychism)
HM 1118 — clairvoyance (Psychism)
HM 3590 — consciousness (Psychism)
HM 3301 Emotional desires (Buddhism)
HH 0936 Emotional growth development
HM 1400 Emotional high
HM 1430 Emotional identification ; Active
HM 1400 Emotional peak experience
HM 2954 Emotional-sexual stage life
HM 0889 Emotional spiritual thirst ; Slaking (Thai)
VC 0635 Empathy
HH 0145 Empathy
HM 2411 Emperor (Tarot)
HM 2759 Emperor ; Vidyadhara (Buddhism)
MP 1250 Emphases ; Encouraging
HM 8105 Emphasis ; Anomalous
MP 1224 Emphasizing transitions across boundaries
KP 3011 Empowerment importance form
HM 2285 Empress (Tarot)
HM 0672 Emptiness
HM 2193 Emptiness (Buddhism)
HM 3032 Emptiness cyclic existence (Buddhism)
HM 2946 Emptiness definitions (Buddhism)
HM 2033 Emptiness ; Direct non conceptual cognition (Buddhism)
HM 2671 Emptiness five senses (Buddhism)
HM 3072 Emptiness indestructible Mahayana (Buddhism)
HM 2448 Emptiness inherent existence non-things (Buddhism)
HM 2794 Emptiness loci senses (Buddhism)
HM 8772 Emptiness ; Meeting God (Christianity)
Emptiness [nature...]
HM 3205 — nature (Buddhism)
HM 2899 — nature phenomena (Buddhism)
HM 2089 — non-products (Buddhism)
HM 2522 Emptiness objects sense mental consciousness (Buddhism)
HM 3000 Emptiness phenomena (Buddhism)
HM 2414 Emptiness products (Buddhism)
HM 2016 Emptiness ; Self (Buddhism)
HM 0672 Emptiness ; Spiritual
Emptiness [Tantra...]
HM 2864 — Tantra meditation (Buddhism)
HM 2995 — ten directions (Buddhism)
HM 2989 — things (Buddhism)
HM 3208 — things (Buddhism)
HM 2294 Emptiness ultimate nirvana (Buddhism)
HM 2501 Emptiness unapprehendable (Buddhism)
HM 0052 Emptiness what is free permanence annihilation (Buddhism)
HM 2944 Emptinesses paths view Buddhism (Buddhism)
HM 0882 Empty-field myopia
HM 0778 Empty mind ; Christ consciousness (Christianity)
HM 5353 Emptying sirr (Sufism)
HM 0934 En-sof (Judaism)
TC 1990 Enabler
MP 1024 Enabling transcendence ; Positions
KD 2320 Enantiomorphic policy ; Toward
VC 0637 Enchantment
HM 2204 Enchantment
MP 1017 Encirclement
MP 1193 Enclosed internal domains ; Partially
MP 1206 Enclosure spaces minimal structural distinctions ; Efficient
MP 1232 Encompassing domains ; Symbolic connection
KD 2192 Encompassing system dynamics: learning
KD 2190 Encompassing system dynamics: sixfold restraint
KD 2202 Encompassing varieties communication
KD 2200 Encompassing varieties form
TP 0044 Encounter
HM 2619 Encounter ; Aweful (ICA)
HH 0162 Encounter group
TC 1722 Encounter group conferencing
HM 3059 Encounter higher self ; Near death (NDE)
HH 0817 Encounter marathon
VC 0639 Encouragement
TC 1990 Encourager participant type
MP 1250 Encouraging emphases
VD 2604 Encroachment
HH 0954 Enculturation
VD 2606 Encumbrance
KC 0931 Encyclopaedia
KC 0357 Encyclopaedia ; World
KC 0730 End
VC 0641 Endearment
HH 1281 Endeavour
HM 0229 Endeavour (Buddhism)
HM 1657 Endeavour ; Correct (Buddhism)
HM 1295 Endeavour ; Right (Buddhism)
HM 6099 Endeavour (Sufism)
HM 1585 Endeavour ; Wrong (Buddhism)
HM 2437 Endless life (ICA)
VC 0643 Endowment
HH 0718 Ends ; Psychology
Endurance
HH 0729 — [Endurance]
VC 0645 — [Endurance]
TP 0032 — [Endurance]
KP 3010 Endurance form
HM 3418 Endurance ; Patient (Sufism)
HM 1403 Endurance ; State unabated (Buddhism)
HM 2821 Enemy ; Everlasting (ICA)
HM 2924 Energetic conceptualization (Astrology)
KD 2300 Energetic expansion mentalistic reduction ; Alternation
MS 8605 Energies processes ; Forces,
VC 0647 Energy
HM 4085 Energy ; Aesthetic emotion (Hinduism)
HM 1333 Energy (Buddhism)
Energy [circuit...]
KC 0069 — circuit language
HM 8365 — Concentration due (Buddhism)
HM 4396 — Consciousness reinforced (Buddhism)
TC 1699 Energy dissipation ; Meeting
HM 1470 Energy ; Faculty (Buddhism)
HM 1829 Energy ; Faculty (Buddhism)
HM 2087 Energy ; Ida conscious (Yoga)
HH 0620 Energy ; Ki
HM 2871 Energy ; Kundalini conscious (Yoga)
HM 4396 Energy ; Mind upheld (Buddhism)
VP 5161 Energy-Moderation
Energy [Physical...]
HM 1263 — Physical inception (Buddhism)
HM 2890 — Pingala conscious (Yoga)
HM 1943 — Power (Buddhism)
TC 1465 — processing configurations ; Meetings
HH 0734 — Psychic
HH 0750 — Psychic
TC 1815 Energy sink ; Conferencing
HM 1882 Energy ; Spiritual (Islam)
HM 4762 Energy ; Unification (Taoism)
HH 0731 Energy ; Universal life
HM 2620 Enervated consciousness
MP 1161 Enfolded insight domain ; Structure
MP 1174 Enfolded non linearity ; Communication pathways
MP 1202 Enfolded occupiable sites ; Structure
MP 1167 Enfolded overview domains minimum proportions
VD 2608 Enforcement
HM 2588 Engagement history (ICA)
HM 3237 Engagement ; Mental (Buddhism)
HM 2343 Engagement ; Missional (ICA)
HM 2736 Engagement ; Transparent (ICA)
HM 0450 Engaku (Japanese)
MP 1100 Engender fruitful interfaces ; Arrangement structures
MM 2051 Engine ; Internal combustion
KC 0910 Engineering cybernetics
HH 0357 Engineering ; Genetic
Engineering Human
KC 0707 — [Engineering ; Human]
HH 0276 — [Engineering ; Human]
HH 0276 — factors
KC 0807 Engineering ; Industrial
KC 0167 Engineering ; Social
KC 0710 Engineering ; Systems
KC 0256 Engineering ; Value control; Value
MS 8815 Engineering works ; Structures
HH 0514 Engrams
MP 1109 Enhance autonomy sub structures ; Organization structure
MP 1245 Enhance fixity ; Protecting variability
MP 1107 Enhance receptivity external insight ; Organization structures
MP 1105 Enhance receptivity external insight ; Orientation structures
MP 1051 Enhanced non linear processes ; Linear relationships
VC 0649 Enhancement
HH 0772 Enhancement ; Brainpower
MP 1106 Enhancement domains separating complementary structures ; Functional
MP 1047 Enhancement ; Functionality
HH 2806 Enhancement human capabilities
HH 6131 Enhancement ; Intelligence
MP 1123 Enhancing function common domains
MP 1135 Enhancing insight varying levels exposure it
MP 1130 Enhancing structural entry points ; Internal transition spaces
HM 1756 Enjoying senses
HM 0829 Enjoying sex
HM 0666 Enjoying thought
HM 4877 Enjoying work
HM 2883 Enjoyment
VC 0651 Enjoyment
HH 0978 Enjoyment ; Abstention craving sensual
HM 7219 Enjoyment ; Aesthetic (Hinduism)
HM 2873 Enjoyment-body awareness (Buddhism)
HM 3113 Enjoyment body ; Buddha (Buddhism)
HM 4987 Enjoyment solitude (Sufism)
VC 0653 Enlightenment
HM 2029 Enlightenment
HH 0350 Enlightenment ; Circles (Zen)
Enlightenment [factors...]
HM 6336 — factors (Buddhism)
HM 2191 — Flaming (Buddhism)
HM 2886 — Forms (Buddhism)
HH 0603 Enlightenment ; Harmonies (Buddhism)
HH 0215 Enlightenment humanism
HH 5645 Enlightenment ; Path (Buddhism)
HH 0591 Enlightenment ; Philosophy
Enlightenment [Spiritual...]
HM 2029 — Spiritual
HH 4019 — Steps (Buddhism)
HH 8487 — Synergic
HH 0495 Enlightenment ; Way (Hinduism)
HM 3869 Enlightenment (Yoga)
VD 2610 Enmity
HM 4239 Enmity (Buddhism)
VP 5927 Enmity ; Friendship
HH 3945 Enneagram patterning (Sufism)
HH 3945 Enneagram personality types (Sufism)
VD 2612 Ennui
HM 2443 Enquiry (Hinduism)
VC 0655 Enrichment
HH 6131 Enrichment ; Instrumental
KC 0270 Ensemble
HM 5876 Enslavement (Islam)
HM 3169 Enstasy
HM 2226 Enstasy (Hinduism)
HM 3041 Enstasy (Hinduism)
HM 0148 Enstasy ; Unconscious (Yoga)
HM 3169 Enstatic consciousness
MP 1124 Ensuring function common domain interfaces
HM 1240 Entendu ; Déjà
VC 0657 Entente
HH 1187 Enterer ; Lesser stream (Buddhism)
HM 3068 Entering-city-with-bliss-bestowing-hands awareness (Zen)
HM 3294 Entering darkness ; Near death (NDE)
HM 3180 Entering light ; Near death
VC 0659 Enterprise
VC 0661 Entertainment
HM 4403 Entertainment (Leela)
Enthusiasm
VC 0663 — [Enthusiasm]
TP 0016 — [Enthusiasm]
HM 2146 — [Enthusiasm]
HM 0875 Enthusiasm ; Poetic
HM 2146 Enthusiasm ; Religious
MM 2004 Entities
HH 6115 Entitlements ; Expression human
HM 3612 Entity ; Notion one's self (Buddhism)
TP 0017 Entrainment
HH 3466 Entrainment
KC 0337 Entrainment frequencies
MP 1050 Entrainment ; Three way relationship
HM 3236 Entrancement
KD 2070 Entrapment micro macro complementarity ; Observer
KD 2295 Entropic crisis learning response
KC 0136 Entropy
KC 0003 Entropy ; Negative
HM 1088 Entry ; Knowledge path stream (Buddhism)
Entry [point...]
MP 1113 — point external communication media ; Harmoniously structured

Existence*complex

	Entry [point...] cont'd
MP 1110	— point structure ; Distinctiveness main
MP 1102	— points complex structures ; Distinct pattern
MP 1130	— points ; Internal transition spaces enhancing structural
MP 1053	— Principal points
VC 0665	Enumeration
HM 3096	Enveloped mystery (ICA)
HH 0663	Environment
KC 0416	Environment
HM 2548	Environment ; Anomalies experience reality one's self
VP 5226	Environment ; Centrality
TC 1636	Environment ; Conference
MP 1046	Environment ; Diversified interchange
MP 1160	Environment ; Hospitable interface structures external
MP 1166	Environment ; Overview domains interfaces structure external
HM 0800	Environmental mysticism
VD 2614	Envy
HM 3098	Envy (Christianity)
VP 5948	Envy ; Congratulation
HM 0832	Envy (Leela)
HM 2638	Envy (Pali)
HM 3818	Envy ; Spiritual (Christianity)
HH 0760	Epigenesis
KC 0516	Epigenetic landscape
HM 2595	Epiphany ; Personal (ICA)
KD 2145	Epistemological mindscapes
KC 0479	Epistemology
KC 0212	Epistemology ; Genetic
HM 2272	Epithalamian mysticism
HM 1240	Eprouvé ; Déjà
VC 0667	Equality
HH 2315	Equality ; Biocentric
VP 5030	Equality-Inequality
HM 2731	Equanimity
VC 0669	Equanimity
HM 7002	Equanimity (Buddhism)
HM 7769	Equanimity (Buddhism)
HM 5008	Equanimity ; Concentration accompanied (Buddhism)
HM 8021	Equanimity ; Concentration accompanied (Buddhism)
HM 8566	Equanimity due insight (Buddhism)
HM 5643	Equanimity ; Dwelling (Buddhism)
HM 0424	Equanimity formations ; Knowledge (Buddhism)
HM 7525	Equanimity ; Understanding accompanied (Buddhism)
HM 1586	Equanimity (Yoga)
KC 0616	Equifinality
KC 0846	Equilibrium
VC 0671	Equilibrium
HM 2630	Equilibrium ; Narcissistic
KD 2080	Equilibrium structures ; Non
KC 0142	Equipment ; Completeness
MS 8102	Equipment ; Professional
HM 2277	Equipoise ; Mental abiding (Buddhism)
VC 0673	Equity
HM 2442	Equity ; Embodying (ICA)
VC 0675	Equivalence
KC 0932	Equivalent network
VD 2616	Equivocation
HH 0958	Eremetical life
HH 0436	Erfan (Sufism)
HH 0276	Ergonomics
HH 0115	Erhard seminars training
	Eros
HH 0574	— [Eros]
HH 7231	— [Eros]
HH 0258	— [Eros]
VD 2618	Erosion
HM 6122	Erotic asceticism
HM 0036	Erotic frenzy
HM 1780	Erotic sentiment (Hinduism)
HM 7045	Erotic transference
HM 6276	Erotomania
HM 0036	Erotomania
HM 1077	Erroneous perception (Yoga)
HM 1912	Erroneous way (Buddhism)
VD 2620	Erroneousness
TP 0025	Error free
VP 5516	Error ; Truth
VC 0677	Erudition
HM 1499	Erudition (Buddhism)
HM 0248	Escapist daydreaming
HH 6902	Esoteric development
HM 4024	Esoteric divine consciousness (Psychism)
HH 4032	Esoteric exoteric practice
HM 6328	Esotericism Albedo
HM 5345	Esotericism Alchemical blackening
HM 1409	Esotericism Alchemical imagination
HM 3631	Esotericism Alchemical reddening
HM 6328	Esotericism Alchemical whitening
HM 0221	Esotericism Alchemy
HM 3631	Esotericism Citrinistas
HM 1417	Esotericism Group spiritual initiation
HM 0221	Esotericism Hermetic philosophy
HM 1267	Esotericism Initiation baptism
HM 1337	Esotericism Initiation birth
HM 0210	Esotericism Initiation crucifixion
HM 0322	Esotericism Initiation decision
HM 0210	Esotericism Initiation detachment
HM 1020	Esotericism Initiation refusal
HM 0210	Esotericism Initiation renunciation
HM 1153	Esotericism Initiation resurrection
HM 0181	Esotericism Initiation revelation
HM 0428	Esotericism Initiation transfiguration
HM 1258	Esotericism Initiation transition
HM 6328	Esotericism Leucosis

HM 5345	Esotericism Melanosis
HM 5345	Esotericism Nigredo
HM 3631	Esotericism Rubedo
HM 2665	Esotericism Seven rays
HH 3390	Esotericism Spiritual initiation
HM 5234	Esotericism Spirituality
HM 1997	Esotericism Way active intelligence
HM 0763	Esotericism Way beauty
HM 6554	Esotericism Way devotion
HM 0763	Esotericism Way harmony
HM 6554	Esotericism Way idealism
HM 1201	Esotericism Way knowledge
HM 1507	Esotericism Way love-wisdom
HM 1605	Esotericism Way organization
HM 0454	Esotericism Way power
HM 1605	Esotericism Way ritual
HM 1201	Esotericism Way science
HM 0454	Esotericism Way will
HH 2665	Esotericism Ways spiritual realization
HM 3631	Esotericism Xantosis
HM 2262	ESP Bio-information
HM 2262	ESP Clairvoyance
HM 2262	ESP Extrasensory perception
HM 2262	ESP Paranormal cognition
HM 2262	ESP Psychotronics
HM 2262	ESP Telepathy
VC 0679	Esprit
VC 0681	Essence
HM 3513	Essence ; Divine (Sufism)
HM 3614	Essence ; First individuation (Sufism)
	Essence [Identity...]
HM 3260	— Identity attributes their (Sufism)
HM 2113	— Indivisibility (Sufism)
HM 3138	— Inventing (ICA)
HM 3340	Essence ; Unification life (Sufism)
HM 2642	Essence veils ; Manifestation (Sufism)
HM 3135	Essential dubiety (ICA)
TP 0061	Essential quality
HM 3493	Essential self ; Ignorance nature (Buddhism)
HM 6111	Essential unity being (Christianity)
HH 0107	Essential wisdom
HH 0115	est
VC 0683	Establishment
HM 6544	Establishment due insight (Buddhism)
HM 3006	Establishment truth (Hinduism)
HM 2997	Establishments mindfulness (Buddhism)
	Esteem Self
HM 2750	— [Esteem ; Self]
HH 0634	— [Esteem ; Self]
VD 3636	— [Esteem ; Self]
TP 0038	Estrangement
HM 3176	Eternal context ; Symbolizing (ICA)
HM 3095	Eternal friends (ICA)
	Eternal [identification...]
HM 2541	— identification (ICA)
HM 2517	— insecurity (ICA)
HM 2593	— intelligence ; Path (Judaism)
HM 2786	Eternal moment (ICA)
HM 0933	Eternal return
HM 2253	Eternal saviour (ICA)
HM 2994	Eternal void (ICA)
HM 2456	Eternality ; Contingent (ICA)
HH 0206	Etheric Body
HH 0206	Etheric pathology
KC 0467	Ethic ; Self realization
HH 1086	Ethical discipline
HM 2661	Ethical existence (ICA)
HH 0893	Ethical idealism
KC 0992	Ethnic integration ; Racial
KC 0690	Ethnically integrated education ; Racially
MS 8395	Ethnicity ; Race
HH 0639	Ethnocentricity
HH 0639	Ethnocentrism
HH 0439	Ethnopathology
HM 3594	Ethnopsychosis
HH 6109	Ethological interpretation behaviour
HH 0725	Ethos
HM 2675	Eucharist
HH 0357	Eugenics
HH 0747	Euphenics
VC 0685	Euphony
HM 3763	Euphoria
HH 0830	Eurhythmics
VT 8051	Evaluation*complex
TC 1339	Evaluation ; Conferencing
HH 0342	Evaluation counselling ; Re
KC 0736	Evaluation review technique ; Programme
HH 0922	Evaluation ; Self
TC 1642	Evaluator role ; Team
HM 3798	Evangelism (Christianity)
HM 1098	Evangelism ; Crusade (Christianity)
HM 1098	Evangelism ; Mass (Christianity)
HM 1098	Evangelism ; Television (Christianity)
VD 2622	Evasion
HM 3394	Evasion (Buddhism)
	Even mindedness
HM 7769	— (Buddhism)
HM 5008	— Concentration accompanied (Buddhism)
HM 8021	— Concentration accompanied (Buddhism)
HM 7525	— Understanding accompanied (Buddhism)
VC 0687	Evenness
TC 1056	Event planning methods
TC 1332	Event ; Transformative
VC 0689	Eventuation
VP 5151	Eventuation-Imminence
HM 2814	Ever-present brother (ICA)

HH 0320	Ever-recurring mental factors (Buddhism)
HM 3088	Ever was ; All that (ICA)
HM 2777	Everlasting community (ICA)
HM 2821	Everlasting enemy (ICA)
HM 2791	Everlasting inescapability (ICA)
HM 2437	Everlastingness centre (ICA)
HM 2991	Every situation (ICA)
HM 2266	Everyday thinking
VD 2624	Evil
HH 1034	Evil
HM 2212	Evil ; Awareness angelic
HM 2378	Evil ; Emblems personified (Tarot)
HM 3422	Evil force ; Near death
HM 1252	Evil paths ; Four (Buddhism)
HM 0923	Evil paths ; Three (Buddhism)
HM 0062	Evocative sports ; Spiritually
	Evolution
KC 0801	— [Evolution]
VC 0691	— [Evolution]
HH 0425	— [Evolution]
HH 2002	Evolution ; Aeons dissolution (Buddhism)
	Evolution [Cognitive...]
HH 0146	— Cognitive
HM 2140	— consciousness
KC 0470	— Cultural
HH 0926	— Cultural
KD 2335	Evolution ; Ecodynamics societal
HH 0246	Evolution human mind
HH 1673	Evolution ; Organic
HH 3662	Evolution prosthesis
VP 5147	Evolution-Revolution
HH 2653	Evolution ; Social
HH 0615	Evolution ; Technological implications human
HM 2006	Evolutionary awareness
HH 4712	Evolutionary choice
VC 0693	Exactness
VD 2626	Exaggeration
VC 0695	Exaltation
HM 2303	Exaltation ; Cruciform (ICA)
HM 4152	Exaltation ; Extreme (Psychism)
HH 2329	Exaltation ; Journeying transcendence humiliation
HM 4036	Exalted concentration (Buddhism)
HM 3296	Exalted object ; Understanding having (Buddhism)
HH 0922	Examination ; Psychotherapeutic self
HH 0314	Examination ; Self
HM 7324	Examining (Buddhism)
HM 3188	Example ; Human (ICA)
VD 2628	Exasperation
TC 1370	Exceeded time allotments
HH 0749	Excellence
VC 0697	Excellence
HH 6018	Excellence ; Education early achievement
HH 0611	Exceptionally gifted children
VD 2630	Excess
TP 0028	Excesses great
TP 0041	Excesses ; Reduction
TP 0062	Excesses ; Small
VD 2632	Excessiveness
MP 1087	Exchange contexts controlled single perspective
MP 1090	Exchange ; Unstructured context perspective
VC 0699	Excitement
HM 2534	Excitement (Buddhism)
HM 1469	Excitement (Buddhism)
VP 5857	Excitement-Inexcitability
HM 0975	Excitement thought (Buddhism)
HM 2218	Exciting intelligence ; Path (Judaism)
VD 2634	Exclusion
VP 5076	Exclusion ; Inclusion
MP 1204	Exclusive contexts
HM 2951	Exclusive contradiction (ICA)
MP 1203	Exclusive spaces emergent perspectives
VD 2636	Exclusiveness
VC 0701	Exclusiveness
MS 8365	Excreta
VD 2638	Execration
HM 2969	Exemplary life ; Mansions (Christianity)
HM 1496	Exercise (Buddhism)
MM 2005	Exercise rest
HH 6452	Exercise ; Waterwheel (Taoism)
HM 3420	Exercises Nikos Kazantzakis ; Spiritual (Christianity)
	Exercises [Saint...]
HH 9760	— Saint Ignatius ; Spiritual (Christianity)
TC 1538	— Simulation
HH 0707	— Spiritual
HM 3469	Exertion (Buddhism)
HM 5986	Exertion due insight (Buddhism)
VP 5709	Exertion-Rest
HM 3261	Exertion ; State unfaltering (Buddhism)
VD 2640	Exhaustion
TP 0047	Exhaustion
VC 0703	Exhilaration
HM 0063	Exhilaration
VD 2642	Exiguity
VC 0705	Existence
	Existence [absorption...]
HM 2051	— absorption ; Peak cyclic (Buddhism)
HM 2331	— Absurd (ICA)
HM 3080	— Actional (ICA)
HH 0695	— analysis
	Existence Awareness
HM 2098	— consciousness group conscious (Buddhism)
HM 2108	— corporeality group conscious (Buddhism)
HM 4983	— feeling group conscious (Buddhism)
HM 2050	— mental formation group conscious (Buddhism)
HM 4143	— perception group conscious (Buddhism)
VT 8011	Existence*complex

Existence

Existence [Consciousness...]
- HM 2177 — Consciousness states cyclic (Buddhism)
- KP 3005 — Constraints
- HH 0513 — Creative
- HM 2894 — Creative (ICA)
- HM 3032 Existence ; Emptiness cyclic (Buddhism)
- HM 2661 Existence ; Ethical (ICA)
- HM 1596 Existence ; Inherent (Buddhism)
- HM 4297 Existence ; Knowledge recollection previous (Buddhism)

Existence [Neither...]
- HM 0581 — Neither existence nor non (Buddhism)
- HM 2448 — non things ; Emptiness inherent (Buddhism)
- VP 5001 — Nonexistence
- HM 0581 — nor non existence ; Neither (Buddhism)
- HM 3457 Existence ontological reality (Sufism)

Existence [phenomena...]
- HH 3321 — phenomena ; Awareness inter dependency conscious (Buddhism)
- HH 1339 — Pre
- HH 1339 — Prior
- HM 2557 Existence ; Representational (ICA)
- HM 2631 Existence ; Resurrectional (ICA)

Existence [Self...]
- HH 0743 — Self
- HM 2556 — senses mind ; Awareness consciousness group conscious (Buddhism)
- HM 2904 — Soteriological (ICA)
- HM 2862 Existence ; Transformed (ICA)
- HM 3067 Existence ; Transparent (ICA)
- HM 6504 Existence ; Uncompounded (Buddhism)
- HM 6504 Existence ; Unconditioned (Buddhism)
- HM 2072 Existence ; Worlds conscious (Buddhism)
- TC 1880 Existential aim
- HH 0416 Existential analysis
- HM 3336 Existential consciousness
- HM 2380 Existential dasein
- HM 2114 Existential guidance (ICA)
- HH 3098 Existential imagination (Existentialism)
- HH 0832 Existential integration
- HM 0334 Existential love
- HH 0602 Existential unity being
- HM 2430 Existential vacuum
- HH 3098 Existentialism Existential imagination
- HH 3098 Existentialism Human development
- HH 4032 Exoteric practice ; Esoteric
- HM 0801 Expanding drugs ; Consciousness
- HH 0075 Expanding drugs ; Mind
- VC 0707 Expansion
- HM 2126 Expansion ; Awareness
- HM 2126 Expansion ; Consciousness
- VP 5197 Expansion-Contraction
- KD 2300 Expansion mentalistic reduction ; Alternation energetic
- HM 2126 Expansion ; Mind
- HM 7007 Expansion ; Perceptual
- MP 1211 Expansion permitting new level generation ; Boundary
- HM 7098 Expansion (Sufism)
- HM 2903 Expectant descendant (ICA)
- VC 0709 Expectation
- VP 5539 Expectation-Inexpectation
- TP 0025 Expectation ; Without
- TC 1613 Expectations ; Participant
- VC 0711 Expedience
- VP 5670 Expedience-Inexpedience
- VC 0713 Expeditious
- VP 5848 Expensiveness-Cheapness
- VC 0715 Experience
- HH 0182 Experience

Experience [Acquiring...]
- TP 0004 — Acquiring (Young shoot)
- HM 2542 — Aesthetic LSD
- HM 1655 — An (Sufism)
- HM 0019 — Asomatic (Psychism)
- HM 1410 — Autotelic

Experience [Contemplation...]
- HM 3521 — Contemplation unitary (Sufism)
- MP 1066 — Context transformative
- HM 0777 — Core (NDE)
- HM 1804 — creativity (Systematics)

Experience [Dance...]
- HM 3213 — Dance induced
- HM 4398 — Deautomatization mystic
- HM 0065 — domination (Systematics)
- HM 3908 — Dreaming
- HH 0075 — Drug induced
- HH 1123 — Drug use transcendent
- HM 1400 — Experience ; Emotional peak
- HM 2344 Experience ; Flow
- HM 4100 Experience ; God
- HM 4501 Experience God ; Ecstatic (Judaism)

Experience [individuality...]
- HM 0344 — individuality (Systematics)
- HM 3221 — Induced religious
- HM 0899 — Introvertive mystical
- HM 3559 Experience ; Light induced

Experience [Mantra...]
- HM 2464 — Mantra induced (Yoga)
- HM 1593 — meditational insight ; Religious
- HM 0666 — mind ; Flow
- HM 0284 — miracles (Christianity)
- HM 2272 — Mystic
- HM 3445 — Mystical
- HM 4349 — mystical knowledge (Sufism)
- HM 0530 — Mystical (Sufism)
- HM 0777 — Near death (NDE)
- HM 3811 Experience ; Numinous (Jung)
- HM 5329 Experience ; Obsessional
- HM 1655 Experience original moments (Sufism)

Experience Out body
- HM 6120 — [Experience ; Out body]
- HM 0777 — (NDE)
- HM 5534 — (Psychism)

Experience [pattern...]
- HM 1408 — pattern (Systematics)
- HM 0561 — personal devotion worship ; Religious
- HM 1234 — polarity (Systematics)
- HM 8080 — potentiality (Systematics)
- HM 0801 — Psychedelic
- HM 3271 — Psychodynamic LSD
- HM 3542 — Psychological assessment mystic (Sufism)

Experience [real...]
- HM 3605 — real ; Inward (Sufism)
- HM 2548 — reality ; Loss
- HM 2548 — reality one's self environment ; Anomalies
- HM 4327 — realization duty ; Religious
- HM 0811 — relatedness (Systematics)
- HM 3445 — Religious
- HM 5187 — repetition (Systematics)
- HM 3538 — revelation
- HM 6687 Experience ; Selective filtering

Experience self
- HM 8186 — Anomalies
- HM 4754 — distinct outside world ; Anomalies
- HM 1336 — recognized personal performance ; Anomalies

Experience [senses...]
- HM 1756 — senses ; Flow
- HM 0829 — Sex flow
- HM 0829 — Sexual
- HM 1766 — Somatic
- HM 0332 — Spontaneous out body
- HM 2624 — Spontaneous religious
- HM 0776 — structure (Systematics)
- HM 1991 — subsistence (Systematics)

Experience [Task...]
- HH 0716 — Task oriented group
- HM 0666 — thought ; Flow
- HM 0502 — Time gap
- HM 6193 — Tip tongue
- HM 9022 — Trans sensate
- HM 3445 — Transcendent
- HM 2712 — Transcendental
- HM 2332 Experience unconscious
- HM 9132 Experience unity self ; Anomalies
- HM 1766 Experience ; Visceral
- HM 9218 Experience ; Visionary
- HM 4002 Experience wholeness (Systematics)
- HM 4877 Experience work ; Flow
- HM 3169 Experience ; Yogic (Yoga)

Experienced born contact
- HM 1352 — psychical ; Neither suffering nor pleasure (Buddhism)
- HM 0901 — psychical ; Unease (Buddhism)
- HM 1031 — psychical ; Unpleasant (Buddhism)
- HM 2080 Experiences ; Core religious
- HM 7203 Experiences ; Passivity
- HM 2080 Experiences ; Peak
- HM 3542 Experiences ; Spiritual (Sufism)
- TC 1538 Experiences ; Structured
- HM 2080 Experiences ; Transcendent
- HM 1632 Experiential insight
- HH 0680 Experiential insight ; Cultivation (Buddhism)
- HH 0885 Experiential therapy
- HM 2272 Experiential wisdom
- VC 0717 Expertise
- HM 3390 Expertise (Buddhism)
- TC 1358 Expertist participant type
- VC 0719 Explanation
- VD 2644 Exploitation
- MP 1073 Exploratory relationship formation challenging emerging perspectives ; Contexts
- TC 1814 Explorer role ; Team
- MP 1180 Exposed external insight ; Occupiable sites
- MP 1183 Exposed perspective context ; Partially
- MP 1134 Exposure harmonious perspectives ; Limiting

Exposure [input...]
- MP 1199 — input processing context external insight
- MP 1144 — irrationality structures ; Integrating coordinated
- MP 1135 — it ; Enhancing insight varying levels
- MP 1165 Exposure structural activities communication pathway
- HM 2764 Exposure ; Total (ICA)
- KC 0847 Expression ; Artistic
- HH 6115 Expression human entitlements
- HH 0791 Expression ; Self
- VC 1561 Expression ; Self
- TC 1374 Expressive conferencing
- HH 0149 Expressive therapy
- VD 2646 Expulsion
- HH 0948 Expurgation
- VC 0721 Exquisiteness
- MP 1075 Extended pattern nuclear interaction
- MP 1175 Extensions structures ; Insight capturing non linear
- VP 5224 Exteriority-Interiority
- VD 2648 Extermination
- MP 1158 External access higher structural levels
- MP 1234 External boundaries ; Maintainable, multi element

External [communication...]
- MP 1113 — communication media ; Harmoniously structured entry point
- MP 1069 — context inactivity ; Common
- MP 1192 — contexts ; Overview
- MP 1160 External environment ; Hospitable interface structures
- MP 1166 External environment ; Overview domains interfaces structure

External insight
- MP 1199 — Exposure input processing context
- MP 1180 — Occupiable sites exposed
- MP 1107 — Organization structures enhance receptivity
- MP 1159 — Organization structures permit two sources
- MP 1128 — Orientation domains receive
- MP 1105 — Orientation structures enhance receptivity
- MP 1138 — Receptivity emerging
- MS 8325 External organs ; Human
- MP 1149 External perspectives structures ; Hospitable reception
- HM 2471 External relation (ICA)
- MP 1163 External structures ; Hospitable non linear domain
- HM 1128 External ; Understanding interpreting (Buddhism)
- HM 4490 External ; Understanding interpreting internal (Buddhism)
- HM 2637 Externality ; Awareness (ICA)
- VD 2650 Extinction
- HM 6124 Extinction illusion (Buddhism)
- VD 2652 Extortion
- HM 2329 Extra-terrestrial minds
- HH 1809 Extraordinary knowledge (Buddhism)
- HH 4110 Extraordinary ordinary ; Journeying transcendence (Christianity)
- HM 2262 Extrasensory perception (ESP)
- VD 2654 Extravagance
- HM 4152 Extreme exaltation (Psychism)
- VD 2656 Extremism
- VP 5005 Extrinsicality ; Intrinsicality
- VC 0723 Exuberance
- VP 5873 Exultation-Lamentation

Eye consciousness
- HM 2074 — (Buddhism)
- HM 1271 — Cognition born contact element (Buddhism)
- HM 0972 — Neither psychical ease nor unease born contact element (Buddhism)
- HM 5974 Eye movement ; Rapid
- HH 1010 Eyes ; Five (Buddhism)

F

- HH 0370 Fables - Parables
- VD 2658 Fabrication
- MS 8315 Face ; Human form
- HM 3401 Face Jesus ; Veronica wipes (Christianity)
- HH 4009 Face ; Structural adjustment human (United Nations)
- MP 1239 Faceted frameworks ; Multi
- KC 0877 Facilitation
- TC 1559 Facilitation
- TC 1886 Facilitation ; Contextual
- HH 0161 Facilitation ; Personal growth
- TC 1377 Facilitation ; Workshop
- TC 1354 Facilitative techniques
- TC 1969 Facilitator
- TC 1451 Facilities
- MP 1200 Facilities perspective adjuncts
- MP 1201 Facilities perspective adjuncts ; Accessible
- VC 0725 Facility
- VP 5732 Facility-Difficulty
- HH 0739 Factor altered states ; Causative
- KC 0344 Factor analysis
- HH 0739 Factor ; Trigger
- HM 3003 Factor-X
- HM 7636 Factors ; Buoyancy mental (Buddhism)
- HH 0910 Factors ; Changeable mental (Buddhism)
- HH 0170 Factors ; Determining mental (Buddhism)

Factors [engineering...]
- HH 0276 — engineering ; Human
- HM 6336 — Enlightenment (Buddhism)
- HH 0320 — Ever recurring mental (Buddhism)
- HH 0320 — Factors ; Omnipresent mental (Buddhism)
- HM 3696 Factors ; Pliancy mental (Buddhism)
- HM 1455 Factors ; Proficiency mental (Buddhism)
- HM 5402 Factors ; Rectitude mental (Buddhism)
- HM 4704 Factors ; Repose mental (Buddhism)
- HH 1348 Factors ; Secondary mental (Buddhism)
- HH 0578 Factors ; Virtuous mental (Buddhism)
- HM 7969 Factors ; Wieldiness mental (Buddhism)
- HM 6336 Factors ; Wisdom (Buddhism)
- HM 2075 Faculties awareness ; Psycho physical (Buddhism)
- VT 8045 Faculties∗complex ; Intellectual
- HH 6282 Faculties ; Subtle (Sufism)
- HM 3454 Faculty concentration (Buddhism)
- HM 1470 Faculty energy (Buddhism)
- HM 1829 Faculty energy (Buddhism)
- HM 0066 Faculty faith (Buddhism)
- HM 3803 Faculty indifference (Buddhism)
- HM 0261 Faculty life (Buddhism)
- HM 3404 Faculty living (Buddhism)
- HM 1649 Faculty mental gladness (Buddhism)
- HM 1369 Faculty mind (Buddhism)
- HM 4248 Faculty recollection (Buddhism)
- HM 0047 Faculty recollection (Buddhism)
- HM 3556 Faculty suffering (Buddhism)
- HM 3233 Faculty wisdom (Buddhism)
- HM 1840 Faculty wisdom (Buddhism)
- VD 2660 Failure
- HH 1433 Failure
- TC 1919 Failure ; Conferencing
- HM 0610 Failure hold meditative attitude (Yoga)
- TC 1005 Failure ; Integrative
- HM 2728 Failure ; Social (ICA)
- MS 8357 Fairies
- VC 0727 Fairness

MS 8235 Fairy tales ; Stories
HM 1240 Fait ; Déjà
HH 0694 Faith
VC 0729 Faith
HM 3209 Faith ; Basic (Buddhism)
HM 2933 Faith (Buddhism)
HM 3359 Faith ; Conjunctive
HM 7902 Faith ; Consciousness reinforced (Buddhism)
HM 5432 Faith due insight ; Resolution (Buddhism)
HM 0066 Faith ; Faculty (Buddhism)
HM 3158 Faith ; Having (Buddhism)
HH 0458 Faith healing
HM 1665 Faith ; Individuative reflective
HM 1425 Faith ; Intuitive projective
HM 7902 Faith ; Mind upheld (Buddhism)
HM 1525 Faith ; Mythic literal
HM 2461 Faith ; Non (Buddhism)
Faith [Passive...]
HH 3412 — Passive way
HM 4404 — Power (Buddhism)
HM 8672 — Pre stage
HH 2341 — Profession (Islam)
Faith Stage
HM 3359 — five
HM 1665 — four
HM 1425 — one
HM 1465 — six
HM 1814 — three
HM 1525 — two
HH 2097 Faith ; Stages
HM 1814 Faith ; Synthetic conventional
Faith [Undifferentiated...]
HM 8672 — Undifferentiated
KC 0378 — Unity
HM 1465 — Universalizing
HH 0628 Faith ; Way loving (Hinduism)
HM 2081 Faithful intelligence ; Path (Judaism)
KC 0164 Faithful ; Unity
VC 0731 Faithfulness
TP 0061 Faithfulness centre
VD 2662 Faithlessness
VD 2664 Fakery
HM 3723 Fall ; Knowledge contemplation rise (Buddhism)
HH 1778 Fall man
HH 3567 Fall man
VD 2666 Fallacy
VD 2668 Fallibility
HM 5087 Falling love
HM 4009 Falls first time His cross ; Jesus (Christianity)
HM 3824 Falls second time cross ; Jesus (Christianity)
HM 2289 Falls third time cross ; Jesus (Christianity)
HH 3013 False paths (Taoism)
VD 2670 Falsehood
VD 2672 Falseness
VC 0733 Fame
VC 0735 Familiarity
Family
KC 0273 — [Family]
MS 8393 — [Family]
MS 8333 — [Family]
TP 0037 Family Clan
TP 0037 Family Community
HH 0362 Family group therapy
TP 0037 Family Household
HH 3543 Family life
HH 1801 Family life (Christianity)
HH 0362 Family psychotherapy
Family therapy
HH 0362 — [Family therapy]
HH 0362 — Conjoint
HH 0362 — Structural
HM 7622 Fana-i-qalb (Sufism)
HM 1087 Fana (Islam)
Fana Sufism
HM 3241 — [Fana (Sufism)]
HM 3799 — [Fana (Sufism)]
HM 1270 — [Fana (Sufism)]
VD 2676 Fanaticism
HM 4064 Fani (Sufism)
HH 3020 Fantasies ; Guiding
VD 2674 Fantasy
HM 0892 Fantasy
HH 0072 Fantasy ; Directed
HM 1011 Fantasy ; Free floating
HH 0627 Fantasy ; Guided
HM 1485 Fantasy ; Plane (Leela)
HH 0939 Fantasy ; Schizophrenic
HM 3651 Fantasy ; Second chakra: realm (Leela)
HM 2138 Fantasying
HM 3427 Faqir (Sufism)
HM 3427 Faqr (Sufism)
HM 2632 Fascinans
VC 0737 Fascination
HH 0555 Fashion consciousness
MM 2036 Fashions ; Changing
HH 0011 Fasting
HH 4216 Fasting (Islam)
HH 1362 Fate
HM 2687 Fate ; Universal (ICA)
HM 3052 Fate ; Wonder filled (ICA)
HM 2223 Fatefulness ; Individual (ICA)
HM 2895 Father ; Age (Christianity)
HH 1773 Father ; Journeying transcendence son
HM 2418 Father ; Universal (ICA)
VC 0739 Fatherhood
VD 2678 Fatigue
TC 1776 Fatigue ; Meeting
TC 1776 Fatigue ; Prattle
TC 1776 Fatigue ; Secretariat
HM 2423 Fatigued conscious awareness
HM 0002 Fattah ; Al (Sufism)
VD 2680 Fatuity
VD 2682 Faultiness
VC 0741 Faultlessness
VC 0743 Favour
MP 1114 Favouring broadest ; Hierachy perspectives
HM 3174 Favourite resort awareness ; Her (Yoga)
HM 2470 Favours afflictions ; Mansions (Christianity)
HM 0568 Fayd (Sufism)
HM 0885 FC Psychic field consciousness
VD 2684 Fear
HM 1644 Fear ; Aesthetic emotion (Hinduism)
HM 2799 Fear ; Breaking point basic (Brainwashing)
HM 3311 Fear (Christianity)
VP 5890 Fear ; Courage
HM 0673 Fear God (Christianity)
HM 1047 Fear God (Sufism)
HM 1194 Fear (Hinduism)
HM 0673 Fear ; Holy (Christianity)
HM 7114 Fear ; Knowledge presence (Buddhism)
HM 3177 Fear ; Near death (NDE)
HM 2465 Fear ; Station
HM 3553 Fearlessness
VC 0745 Fearlessness
HM 0649 Fearlessness blame (Buddhism)
HM 3044 Fearlessnesses ; Four (Buddhism)
VC 0747 Feasibility
HH 0288 Feasting
HH 4398 Feats ; Yogic
VD 2686 Featurelessness
VD 2688 Fecklessness
VC 0749 Fecundity
KC 0153 Federal government ; World
VD 2690 Feebleness
KC 0830 Feedback
TC 1347 Feedback ; Conference procedure
KC 0427 Feedback ; Positive
HH 0819 Feeling
VC 0751 Feeling
HM 2762 Feeling ; Aesthetic (Japanese)
HM 4983 Feeling aggregate (Buddhism)
HM 2270 Feeling (Buddhism)
VT 8081 Feeling*complex
HM 3590 Feeling consciousness (Psychism)
Feeling [greed...]
HM 1513 — greed (Buddhism)
HM 1825 — greed ; State not (Buddhism)
HM 2559 — Group
HM 4983 — group conscious existence ; Awareness (Buddhism)
HM 1967 Feeling hatred ; State (Buddhism)
HM 1597 Feeling hatred ; State not (Buddhism)
Feeling [I...]
HH 0840 — I
HM 7769 — Indifferent (Buddhism)
HM 4363 — infatuation (Buddhism)
HM 4081 — infatuation ; State not (Buddhism)
HM 8737 Feeling ; Joyful (Buddhism)
HM 3554 Feeling ; Lack deep (Japanese)
HM 6644 Feeling ; Non sensuous (Buddhism)
HM 2750 Feeling ; Oceanic
HM 8010 Feeling ; Painful (Buddhism)
HM 6722 Feeling ; Pleasant (Buddhism)
HM 4335 Feeling ; Religious
Feeling remorse
HM 1208 — acquisition sinful unwholesome dharmas (Buddhism)
HM 3353 — acquisition sinful unwholesome dharmas ; Not (Buddhism)
HM 1757 — what one ought be remorseful (Buddhism)
HM 1673 — what one ought be remorseful ; Not (Buddhism)
HM 7120 Feeling ; Sad (Buddhism)
HH 1384 Feeling ; Self
HM 5119 Feeling towards God ; Negative (Christianity)
HM 3502 Feeling towards God ; Positive (Christianity)
VP 5855 Feeling-Unfeelinglessness
HM 2409 Feeling well-being
HH 0643 Feelings ; Hostile
HH 0912 Feelings ; Natural (Japanese)
MS 8315 Feet ; Hands
HM 1798 Feierabend (German)
VC 0753 Felicity
HM 1545 Felicity (Buddhism)
VC 0755 Fellowship
TP 0013 Fellowship
VD 2692 Feloniousness
HH 6432 Female heroism
HH 3663 Female principle ; Acceptance
HM 2285 Female sovereignty archetypal image (Tarot)
HM 2768 Feminine consciousness
HH 4003 Feminine spirituality
MS 8385 Femininity
VP 5420 Femininity ; Masculinity
VC 0757 Feminity
HH 2901 Feng-shui (Chinese)
VD 2694 Ferocity
VC 0759 Fertility
VC 0761 Fervour
MM 2046 Festivals ; Cycles religious
HM 4470 Fetters ignorance (Buddhism)
HM 6342 Fetters ; Lower (Buddhism)
HM 1511 Fetters views (Buddhism)
VP 5101 Fewness ; Numerousness
HM 0460 Fewness wishes (Buddhism)
HM 1264 Fi 'l manzur ; Al fikr (Islam)
HM 4712 Fi 'llah ; Al mahabba (Islam)
KC 0032 Fibonacci numbers
VC 0763 Fidelity
TP 0025 Fidelity
KC 0970 Field
HM 2990 Field consciousness
HM 0885 Field consciousness ; Psychic (FC)
HM 0882 Field myopia ; Empty
HM 2383 Field ; Personal consciousness (Psychosynthesis)
KC 0723 Field ; Semantic
Field theory
KC 0333 — Classical
KC 0407 — Social
KC 0833 — Unified
VD 2696 Fiendish
VD 2698 Fierceness
HM 2124 Fiery awareness (Astrology)
HM 2523 Fiery intelligence ; Path (Judaism)
HM 3975 Fifteen ; Focus (Psychism)
HM 2407 Fifth antarabhava consciousness (Buddhism)
HM 0933 Fifth chakra: man becomes himself (Leela)
HM 3461 Fifth chakra (Yoga)
HM 5143 Fifth jhana five ; Concentration (Buddhism)
HM 0772 Fifth order perceptions - control sequence
HM 2001 Fifth plane wisdom (Hinduism)
Fifth [scriptural...]
HM 2909 — scriptural bodhisattva awareness (Buddhism)
HM 3191 — stage life
HM 0750 — state consciousness (Psychism)
HM 2405 — state (Yoga)
HM 2763 Fifth vajra-master awareness (Buddhism)
HM 2130 Fighting consciousness ; Heaven (Buddhism)
MS 8845 Figures ; Design
HM 2543 Figures ; Eastern continent awareness noble (Buddhism)
HM 1264 Fikr fi 'l manzur ; Al (Islam)
TC 1439 Filled conferencing ; Dread
MP 1238 Filtered insights
HM 6687 Filtering experience ; Selective
VD 2700 Filth
HM 3206 Final accountability (ICA)
HM 2958 Final blessedness (ICA)
HM 3156 Final confession (Brainwashing)
HH 0157 Final integration personality
HM 2674 Final limits (ICA)
HM 2119 Final situation (ICA)
HM 2683 Final-training subtle contemplation (Buddhism)
HM 2849 Final vajra-master awareness (Buddhism)
KC 0642 Finality
MS 8533 Fine arts
Fine material sphere
HM 2536 — (Buddhism)
HM 4265 — concentration (Buddhism)
HM 2450 — First trance (Buddhism)
HM 2586 — Fourth trance (Buddhism)
HM 4761 — functional ; Indeterminate consciousness (Buddhism)
HM 5338 — Profitable consciousness (Buddhism)
HM 0594 — resultant ; Indeterminate consciousness (Buddhism)
HM 2038 — Second trance (Buddhism)
HM 2062 — Third trance (Buddhism)
VC 0765 Finesse
TC 1526 Finisher ; Team
HM 4112 Firaq (Sufism)
TP 0030 Fire Cosmic mean
MS 8937 Fire dwelling beings ; Mythical
TP 0030 Fire Model elegance
TP 0030 Fire Normative constraint
HM 0859 Fire ; Plane (Leela)
TP 0030 Fire Shining light
HM 2124 Fire ; Triplicity (Astrology)
HM 3863 Fire ; Yoga inner (Buddhism)
HM 1974 Firmness (Buddhism)
HM 2110 First absorption immaterial sphere (Buddhism)
HM 2347 First antarabhava consciousness (Tibetan)
HM 4103 First chakra: fundamentals being (Leela)
HM 2893 First chakra (Yoga)
HM 6989 First degree prayer (Christianity)
HM 2224 First duty (Christianity)
HM 2450 First form-realm concentration (Buddhism)
HM 2110 First formless attainment (Buddhism)
HM 3614 First individuation essence (Sufism)
First jhana
HM 4298 — Dwelling (Buddhism)
HM 0297 — five ; Concentration (Buddhism)
HM 4456 — four ; Concentration (Buddhism)
HM 0543 First order perceptions - intensity
HM 3100 First plane wisdom (Hinduism)
First [scriptural...]
HM 2155 — scriptural bodhisattva awareness (Buddhism)
HM 3320 — stage life
HM 3748 — step: ego (Christianity)
HM 4009 First time His cross ; Jesus falls (Christianity)
HM 2450 First trance fine-material sphere (Buddhism)
HM 2187 First vajra-master stage (Buddhism)
MS 8823 Fish
VC 0767 Fitness
HM 1810 Fitness consciousness (Buddhism)
HM 1455 Fitness mental body (Buddhism)
HM 1810 Fitness thought (Buddhism)
HH 0625 Fitness ; Total
HH 0758 Fitness work
HH 3321 Five aggregates (Buddhism)
HM 2841 Five clairvoyances (Buddhism)
Five Concentration
HM 5143 — fifth jhana (Buddhism)

Five

Five Concentration cont'd
- HM 0297 — first jhana (Buddhism)
- HM 4882 — fourth jhana (Buddhism)
- HM 4575 — second jhana (Buddhism)
- HM 8532 — third jhana (Buddhism)
- HH 6773 Five covers (Buddhism)
- HH 1108 Five dharmas ; Seventy (Buddhism)
- HM 8336 Five-door adverting (Buddhism)
- HM 4338 Five elements ; Assembling (Taoism)
- HH 1010 Five eyes (Buddhism)

Five [faith...]
- HM 3359 — faith ; Stage
- HM 8080 — fold knowledge (Systematics)
- HM 2948 — forces (Buddhism)
- HH 6773 Five hindrances (Buddhism)
- HH 0390 Five kleshas (Yoga)
- HH 1010 Five levels understanding (Buddhism)
- HM 3238 Five powers (Buddhism)
- HM 2671 Five senses ; Emptiness (Buddhism)
- HM 0747 Fivefold happiness (Buddhism)
- HM 6553 Fivefold jhana ; Dwelling (Buddhism)
- HH 4536 Fivefold moral law
- HM 6553 Fivefold system ; Dwelling second jhana (Buddhism)
- VD 2702 Fixation
- HM 1617 Fixation (Buddhism)
- HM 2554 Fixed awareness (Astrology)
- HM 4968 Fixing material qualities ; Understanding way (Buddhism)
- HM 5627 Fixing mental qualities ; Understanding way (Buddhism)
- MP 1245 Fixity ; Protecting variability enhance
- MP 1247 Fixity variability ; Embedding
- HH 0151 Flagellation
- HM 2191 Flaming enlightenment (Buddhism)
- TC 1800 Flannel board
- HM 4875 Flash intuition (Hinduism)
- HM 2696 Flask ; Alchemical (Buddhism)
- VD 2704 Flaw
- VC 0769 Flawlessness
- VC 0771 Flexibility

Flexibility [aggregate...]
- HM 1796 — aggregate consciousness (Buddhism)
- HM 3397 — aggregate sensation, aggregate cognition, aggregate synergies (Buddhism)
- HM 4312 — associations ; Anomalies
- HM 3370 Flexibility body (Buddhism)
- TC 1399 Flexibility ; Conference
- HM 1805 Flexibility thought (Buddhism)
- MP 1146 Flexible domain organization
- MP 1244 Flexible interfaces
- MM 2064 Flight ; Aerial animal locomotion
- HM 1887 Flight ; Astral (Psychism)
- TP 0043 Flight Breakthrough
- TP 0043 Flight Parting
- TP 0043 Flight Resolution
- HM 6120 Flight ; Soul
- HH 3442 Flight world
- TC 1691 Flipchart
- HM 3903 Floating
- HM 1011 Floating fantasy ; Free
- HM 3145 Flood ignorance (Buddhism)
- TP 0059 Flooding Barrier dissolution
- TP 0059 Flooding Disintegration
- TP 0059 Flooding Dispersion
- TP 0059 Flooding Overcoming dissension
- TP 0059 Flooding Scattering
- VC 0773 Flourish
- HH 0687 Flow breath ; Concentration
- KC 0816 Flow chart
- KC 0816 Flow diagram

Flow experience
- HM 2344 — [Flow experience]
- HM 0666 — mind
- HM 1756 — senses
- HM 0829 — Sex
- HM 0666 — thought
- HM 4877 — work
- KC 0816 Flow graph
- MS 8445 Flowers ; Fruits
- TP 0011 Flowing
- HM 2738 Flowing consciousness ; Inward (Yoga)
- HM 2581 Flowing consciousness ; Outward (Yoga)
- KC 0383 Fluctuation ; Alternation
- KD 2083 Fluctuation ; Complementarity determinism
- KD 2081 Fluctuation: dissipative structures ; Order
- VC 0775 Fluency
- HM 3137 Fluid emblems star ; Life (Tarot)
- HM 3289 Fluid emblems temperance ; Life (Tarot)
- MS 8365 Fluids ; Human body
- HM 8634 Flying ; Yogic
- MP 1099 Focal centre complex structure

Focal point
- MP 1126 — common domain ; Off centering
- MP 1181 — Maintenance source direct insight structures
- MP 1129 — structure ; Common domain
- MP 1044 Focal points ; Local
- TC 1351 Focal processes ; Meeting
- TC 1969 Focalizer role ; Team
- MP 1057 Foci ; Protection emerging
- HM 1353 Focus awareness ; Diminished
- HM 3975 Focus fifteen (Psychism)
- TC 1432 Focus meetings ; Losing
- HM 0629 Focus one (Psychism)
- MP 1120 Focus-oriented communication networks
- HM 1136 Focus ten (Psychism)
- HM 0267 Focus twelve (Psychism)
- KD 2272 Focus ; Undefined common

- KD 2272 Focus variable geometries global order
- TC 1285 Focus ; Varieties meeting
- HM 3003 Focusing
- HM 0487 Focusing (Buddhism)
- TC 1651 Fodder ; Conference
- HH 0248 Folk healing
- HM 3321 Folk soul
- TP 0017 Followers ; Acquiring
- TC 1651 Followers participant type
- TP 0017 Following
- HH 6019 Following Jesus (Christianity)
- VP 5292 Following ; Leading
- HM 1771 Following Tao (Taoism)
- TP 0004 Folly ; Youthful (Young shoot)
- VC 0777 Fondness
- KC 0771 Food chain
- MM 2055 Food, nutrition cooking
- TC 1529 Food processing ; Meetings agriculture
- MS 8602 Food-related objects
- MS 8606 Foods liquids
- TC 1546 Fool
- HM 2225 Fool (Tarot)
- HM 0847 Foolishness (Buddhism)
- VC 0779 Forbearance
- HM 3422 Force ; Near death evil
- HH 0213 Force ; Regulation vital (Hinduism)
- HH 0138 Force ; Soul
- TC 1630 Force ; Task
- MM 2037 Forced labour
- VC 0781 Forcefulness
- VD 2706 Forcelessness
- HM 2679 Forces awareness ; Magical (Buddhism)
- HM 3127 Forces emblems ; Natural (Tarot)
- MS 8605 Forces, energies processes
- HM 2948 Forces ; Five (Buddhism)
- HM 2820 Forces ; Negative intrapsychic (Psychism)
- KC 0547 Forecasting
- KC 0974 Forecasting ; Social
- KC 0687 Forecasting ; Technological
- MS 8395 Foreignness
- VC 0783 Foresight
- HM 3606 Foresight (Psychism)
- VC 0785 Forethought
- VD 2708 Forgetfulness
- HM 3053 Forgetfulness (Buddhism)
- HM 7383 Forgetfulness God (Sufism)
- HH 5219 Forgetfulness memory ; Journeying transcendence
- HM 2847 Forgetfulness ; Non (Buddhism)
- VP 5537 Forgetfulness ; Remembrance
- VC 0787 Forgiveness
- HH 0899 Forgiveness (Christianity)
- VP 5947 Forgiveness-Vengeance
- HM 1728 Forgiving mind ; Christ consciousness (Christianity)
- HM 2604 Forgotten leaving man alone awareness ; Ox (Zen)
- KC 0156 Form
- HM 3019 Form body ; Buddha (Buddhism)
- HM 3113 Form body ; Buddha (Buddhism)
- HM 2693 Form concentrations ; Tibetan meditative states (Buddhism)
- KP 3015 Form ; Construction development
- HM 3583 Form ; Desire life (Buddhism)

Form [Empowerment...]
- KP 3011 — Empowerment importance
- KD 2200 — Encompassing varieties
- KP 3010 — Endurance
- MS 8315 Form face ; Human
- VP 5246 Form-Formlessness
- KC 0093 Form language
- TC 1835 Form method ; Art
- KC 0203 Form ; Musical
- KC 0721 Form ; Organic
- HM 1690 Form ; Path (Yoga)
- HM 3300 Form ; Perception (Yoga)
- HM 2142 Form-realm awareness (Buddhism)
- HM 2235 Form realm awareness ; Tantra master (Buddhism)

Form realm concentration
- HM 4265 — (Buddhism)
- HM 2450 — First (Buddhism)
- HM 2586 — Fourth (Buddhism)
- HM 2038 — Second (Buddhism)
- HM 2062 — Third (Buddhism)

Form realm [consciousness...]
- HM 2257 — consciousness (Buddhism)
- HM 4761 — inoperative ; Indeterminate consciousness (Buddhism)
- HM 5338 — Moral consciousness (Buddhism)
- HM 0594 — resultant ; Indeterminate consciousness (Buddhism)
- KP 3017 Form ; Relationship potential
- MS 8700 Form structure ; Body,
- HM 4570 Form ; World beyond (Sufism)
- HM 1973 Form ; World (Buddhism)
- KC 0586 Formal elegance
- MP 1009 Formal processes ; Decentralized
- MP 1041 Formal processes ; Informal context
- MP 1111 Formal structure informal context ; Blended integration
- KC 0261 Formalism
- VP 5646 Formality-Informality
- MP 1068 Formation amongst emerging perspectives ; Spontaneous relationship
- MP 1073 Formation challenging emerging perspectives ; Contexts exploratory relationship
- HH 1222 Formation ; Concept
- HH 3226 Formation demerit (Buddhism)

Formation [gold...]
- HM 1330 — gold elixir (Taoism)
- HM 2559 — Group

Formation [group...] cont'd
- HM 2050 — group conscious existence ; Awareness mental (Buddhism)
- KC 0908 Formation ; Hierarchy
- HH 0937 Formation ; Human capital
- HH 5543 Formation imperturbable (Buddhism)
- MP 1215 Formation ; Initial level
- HH 5122 Formation merit (Buddhism)
- HM 4762 Formation spiritual embryo (Taoism)
- HM 2360 Formation ; World (Judaism)
- HM 2050 Formations aggregate (Buddhism)
- HH 2718 Formations ; Delimitation (Buddhism)
- HM 0424 Formations ; Knowledge equanimity (Buddhism)
- HH 7403 Formative spirituality
- HM 4094 Formication
- HM 2669 Formless absorptions ; Tibetan meditative states (Buddhism)

Formless attainment
- HM 2110 — First (Buddhism)
- HM 2051 — Fourth (Buddhism)
- HM 3043 — Second (Buddhism)
- HM 1969 Formless dharmas ; Duration (Buddhism)
- HM 1261 Formless life ; Desire (Buddhism)
- HM 4603 Formless path ; Tantric (Buddhism)

Formless realm
- HM 3144 — awareness (Buddhism)
- HM 0696 — concentration (Buddhism)
- HM 2281 — consciousness (Buddhism)
- HM 0282 — inoperative ; Indeterminate consciousness (Buddhism)
- HM 4701 — Moral consciousness (Buddhism)
- HM 4982 — resultant ; Indeterminate consciousness (Buddhism)
- HH 3198 Formless realms (Buddhism)
- MS 8415 Formless symbols ; Incorporeal
- VP 5246 Formlessness ; Form
- HM 1973 Formlessness ; World (Buddhism)
- HM 0910 Formlessness (Zen)
- HM 4550 Forms ; Assumption (Psychism)
- HM 2713 Forms awareness ; Zodiacal (Astrology)
- HM 2886 Forms enlightenment (Buddhism)
- MP 1189 Forms perspective presentation ; Context
- MP 1008 Forms processes ; Variety
- KP 3020 Forms ; Significance mutually constraining
- MS 8600 Forms structures ; Man made
- KD 2005 Forms truth: uniformity versus aesthetics
- HM 3286 Forms ; World (Sufism)
- HM 3035 Formula ; Ignorance dependent origination (Buddhism)
- KD 2353 Formulation ; Alternation: implications action
- KP 3001 Formulations ; Inadequacy
- VC 0789 Forthrightness

Fortitude
- HH 0729 — [Fortitude]
- HH 0929 — [Fortitude]
- HH 5640 — [Fortitude]
- VC 0791 Fortune
- HM 2250 Fortune-wheel (Tarot)
- HM 2424 Forty ; Maqamat (Sufism)
- MM 2033 Fossorial animal locomotion
- VD 2710 Foulness
- HM 0650 Foundation awareness
- HM 4007 Foundation ; Setting up (Taoism)
- HM 2410 Foundation ; Sphere (Kabbalah)
- HM 3039 Fountains light (Sufism)
- HM 0485 Four analyses (Buddhism)
- HM 7655 Four causes consciousness arising ; Twenty (Buddhism)

Four Concentration
- HM 4456 — first jhana (Buddhism)
- HM 7202 — fourth jhana (Buddhism)
- HM 4380 — second jhana (Buddhism)
- HM 2284 — third jhana (Buddhism)

Four [deviations...]
- HM 1781 — deviations speech ; Leaving off, abstaining, totally abstaining refraining (Buddhism)
- HM 0485 — discriminations (Buddhism)
- HM 2064 — doors retention (Buddhism)
- HM 1252 Four evil paths (Buddhism)

Four [faith...]
- HM 1665 — faith ; Stage
- HM 3044 — fearlessnesses (Buddhism)
- HM 1991 — fold knowledge (Systematics)
- HM 2082 Four-great-kings-heaven awareness (Buddhism)
- HM 2507 Four immeasurables (Buddhism)
- MP 1021 Four-level structural limit
- HM 4026 Four noble paths (Buddhism)
- HH 0523 Four noble truths (Buddhism)

Four [sciences...]
- HM 3216 — sciences (Buddhism)
- HM 6307 — sleep ; Stage
- HH 2563 — stages life (Hinduism)
- HM 6754 — stations mindfulness (Pali)
- HM 2245 Four truths ; Meditation way (Buddhism)
- KD 2110 Fourfold minimal system ; Threshold comprehensibility:
- HM 2051 Fourth absorption immaterial sphere (Buddhism)
- HM 2335 Fourth antarabhava consciousness (Buddhism)
- HM 2391 Fourth chakra: attaining balance (Leela)
- HM 3242 Fourth chakra (Yoga)
- HM 4071 Fourth dimensional clairvoyance (Psychism)
- HM 2586 Fourth form-realm concentration (Buddhism)
- HM 2051 Fourth formless attainment (Buddhism)

Fourth jhana
- HM 8087 — Dwelling (Buddhism)
- HM 4882 — five ; Concentration (Buddhism)
- HM 7202 — four ; Concentration (Buddhism)
- HM 1516 Fourth order perceptions - control transitions
- HM 3006 Fourth plane wisdom (Hinduism)

Global

Fourth [scriptural...]
HM 2191 — scriptural bodhisattva awareness (Buddhism)
HM 3518 — stage
HM 2402 — state consciousness ; Objective
HM 2020 — state (Yoga)
HM 3570 — step: earth (Christianity)
HM 2586 Fourth trance fine-material sphere (Buddhism)
HM 2251 Fourth vajra-master awareness (Buddhism)
VD 2712 Fragility
HM 1212 Fragmentation thinking
HM 2364 Fragrance awareness (Buddhism)
HM 0008 Fragrance ; Plane (Leela)
VP 5435 Fragrance-Stench
HM 2150 Frame awareness ; Angelic
TC 1218 Frame ; Freeze
KC 0437 Frame reference
Framework [definition...]
MP 1208 — definition ; Progressive
TC 1888 — development ; Conceptual
TC 1002 — development ; Conceptual
MP 1205 Framework spaces defined processes it ; Congruence spaces defined
MP 1195 Framework transition structural levels
MP 1037 Frameworks ; Cluster
TC 1392 Frameworks ; Conceptual
MP 1236 Frameworks ; Displaceable
TC 1488 Frameworks ; Meetings analytical
MP 1239 Frameworks ; Multi faceted
MP 1038 Frameworks ; Standard
VC 0793 Frankness
VC 0795 Fraternity
VD 2714 Fraudulence
HM 0756 Free cankers ; Understanding (Buddhism)
TP 0025 Free ; Error
HM 1011 Free-floating fantasy
TP 0040 Free ; Getting
HM 0052 Free permanence annihilation ; Emptiness what is (Buddhism)
HH 0508 Free self-inquiry meditation
HH 0632 Free-thinking meditation
HH 0920 Free will ; Good will
VC 0797 Freedom
HM 2001 Freedom attachment ; Total (Hinduism)
HM 2316 Freedom awareness (ICA)
Freedom [choice...]
HH 0789 — choice
HH 4020 — (Christianity)
HH 2344 — conscience
Freedom [decision...]
HM 2721 — decision (ICA)
KC 0137 — Degrees
HM 3554 — desire (Japanese)
HM 2894 Freedom inventiveness (ICA)
HH 2661 Freedom ; Journeying transcendence bondage (Christianity)
HM 3274 Freedom lower qualities (Sufism)
HM 3206 Freedom obligation (ICA)
VP 5762 Freedom-Restraint
HM 2603 Freedom ; Rudra awareness black (Buddhism)
HH 0659 Freemasonry Initiation
HH 0659 Freemasonry Masonic degrees
TC 1218 Freeze frame
HM 3594 Frenzy
VD 2716 Frenzy
HM 0036 Frenzy ; Erotic
KC 0337 Frequencies ; Entrainment
HH 5328 Frequency ; God
VP 5135 Frequency-Infrequency
VC 0799 Freshness
HM 2472 Fretfulness
VD 2718 Fretfulness
VD 2720 Friction
VC 0801 Frictionlessness
HM 3124 Friend ; Immutable (ICA)
HM 2139 Friend ; Persistent (ICA)
VC 0803 Friendliness
HM 3095 Friends ; Eternal (ICA)
Friendship
VC 0805 — [Friendship]
HH 0156 — [Friendship]
HH 0258 — [Friendship]
VP 5927 Friendship-Enmity
Friendship [Sacrificial...]
HM 2754 — Sacrificial (ICA)
HH 1199 — Soul (Christianity)
HH 1199 — Spiritual (Christianity)
HM 4902 Fright ; Stage
HM 3117 Frightful possibility (ICA)
VD 2722 Frightfulness
HM 3667 Fringe consciousness (Psychism)
HM 3770 Fringe space-time (Psychism)
VD 2724 Frivolity
KC 0799 Front
VC 0807 Frugality
MP 1100 Fruitful interfaces ; Arrangement structures engender
MS 8445 Fruits flowers
HM 3084 Frustration
VD 2726 Frustration
HM 7131 Fugue state
HH 0223 Fulfilled living ; Immediacy
TP 0048 Fulfilled potentialities (Well)
Fulfilment
HM 2521 — [Fulfilment]
VC 0809 — [Fulfilment]
HM 0728 — [Fulfilment]
VC 1563 Fulfilment ; Self

HH 0412 Fulfilment ; Self
HM 2743 Full-isolation subtle contemplation (Buddhism)
TC 1876 Full motion video teleconferencing
HM 2729 Full pacification mental abiding states (Buddhism)
HM 1886 Full trance (Psychism)
Full understanding
HM 4348 — abandoning (Buddhism)
HM 4552 — investigation (Buddhism)
HM 3490 — known (Buddhism)
VC 0811 Fullness
TP 0055 Fullness
HH 0971 Fully developed personality
VD 2728 Fulsomeness
VC 0813 Fun
KC 0738 Function
MP 1124 Function common domain interfaces ; Ensuring
MP 1123 Function common domains ; Enhancing
KD 2305 Function ignorance ; Uncertainty
HM 2577 Function ; Religious (ICA)
HH 0324 Function ; Transcendent
KC 0365 Functional analysis ; Structural
KC 0133 Functional cultural integration
MP 1026 Functional cycle
MP 1106 Functional enhancement domains separating complementary structures
HM 5504 Functional hallucination
Functional Indeterminate consciousness
HM 4761 — fine material sphere (Buddhism)
HM 0282 — immaterial sphere (Buddhism)
HM 3852 — sense sphere (Buddhism)
MP 1115 Functional integration unstructured internal domains
HM 0196 Functional learning
KC 0169 Functional multiplicity ; Organic
KC 0620 Functional social integration
KC 0232 Functionalism
MP 1047 Functionality enhancement
HH 1988 Functions ; Control involuntary
HM 4497 Fund individual (Christianity)
HM 3225 Fundamental dialogue
HM 1762 Fundamental dialogue
MP 1042 Fundamental transformation zones ; Chain
KC 0634 Fundamentalism
HM 4103 Fundamentals being ; First chakra: (Leela)
HM 8762 Furious sentiment (Hinduism)
MS 8416 Furnishings amenities ; Home
HM 3594 Furor
HM 1027 Furor divini
VD 2730 Furtiveness
VD 2732 Fury
HM 3387 Furyu (Japanese)
HH 0368 Fusion
VD 2734 Fuss
VD 2736 Futility
HM 2493 Futility ; Creative (ICA)
HH 4510 Future ; Inventing
HH 4510 Future ; Social imaging
HH 4510 Futures ; Imagining alternative
KC 0450 Futures research
KC 0450 Futuribles
HM 3016 Futuric responsibility (ICA)
KC 0450 Futurism
KC 0450 Futuristics
VP 5119 Futurity-Antiquity
HM 2609 Futurity ; Awareness (ICA)
KC 0450 Futurology
HM 3012 Futuwwa (Sufism)
KC 0758 Fuzzy logic
KC 0332 Fuzzy sets

G

HM 3093 G'yo (Tibetan)
HM 2222 Gabriel ; Angelic awareness
HM 0700 Gaha (Pali)
VC 0815 Gaiety
VC 0817 Gain
MM 2030 Gaining/losing initiative
TP 0028 Gains ; Great
VC 0819 Gallantry
KD 2325 Game comprehension identity transformation
HH 4000 Game ; Consciousness ascension stages (Buddhism)
HM 2278 Game knowledge
HM 2278 Game life
HM 3066 Game ; Masters wisdom ascension stages (Buddhism)
HM 3341 Game ; Religious traditions ascension stages (Buddhism)
HM 3214 Game ; Supreme heavens ascension stages (Buddhism)
HM 2962 Game ; Sutra paths accumulation application ascension stages (Buddhism)
Game [Tantric...]
HM 2848 — Tantric paths accumulation application ascension stages (Buddhism)
KC 0916 — theory
HH 5634 — Transformation
HH 0705 Games
TC 1538 Games
MS 8846 Games ; Amusements
MM 2066 Games ; Ball
TC 1152 Games contests ; Meetings
HH 0175 Games ; Growth
KC 0637 Games ; Meta
HH 0692 Games ; Organized
KD 2327 Games ; Resonance based consensus
MM 2077 Games ; Shell

HH 0582 Games ; Theatre
HH 0175 Games ; Transformative
MM 2040 Gamesmanship
KC 0840 Gaming ; Operational
HM 0008 Gandha-loka (Leela)
HM 4403 Gandharvas (Leela)
HM 3779 Ganga (Leela)
HM 3779 Ganges (Leela)
HM 0502 Gap experience ; Time
HM 3166 Garments ; Jesus stripped his (Christianity)
MS 8925 Garments ; Magical
HM 6434 Gaseous plane (Leela)
HM 3226 Gateway (Sufism)
HM 2318 Gathering-separation (Sufism)
TP 0003 Gathering support
TP 0045 Gathering together
HM 5197 Gati ; Uttam (Leela)
VD 2738 Gaudiness
HM 6003 Gazing (Islam)
HM 2034 Ge ; Ji ji mu (Zen)
HM 2290 Geburah-din sephira (Kabbalah)
HH 0972 Gedatsu therapy
HM 2924 Gemini-consciousness (Astrology)
MS 8385 Gender ; Human
HH 1334 Gene therapy
MP 1016 General interrelationships ; Web
HH 1765 General knowledge (Buddhism)
KC 0171 General plan
General [semantics...]
HH 0585 — semantics
KC 0330 — semantics
HH 1127 — system theory human development
KD 2220 — systems holonomy
KC 1000 — systems theory
KC 0432 Generalist
KC 0477 Generalization
TC 1389 Generalization
HH 0692 Generalized
MP 1211 Generation ; Boundary expansion permitting new level
MP 1219 Generation minimum tension ; Level
HH 2764 Generative man
MM 2043 Generators ; Electric motors
VC 0821 Generosity
HM 4474 Genesis (Leela)
HH 0357 Genetic counselling
HH 0357 Genetic engineering
KC 0212 Genetic epistemology
HH 0357 Genetic improvement ; Human
HH 0357 Genetic selection
VC 0823 Geniality
HH 0407 Genitality ; Maturity
Genius
VC 0825 — [Genius]
TC 1491 — [Genius]
HH 0295 — [Genius]
TC 1550 Genius role ; Team
VC 0827 Gentility
VC 0829 Gentleness
TP 0057 Gentleness (Wind)
VC 0831 Genuineness
MM 2070 Geography movement
HH 2901 Geomancy (Chinese)
KD 2272 Geometries global order ; Focus variable
KD 2207 Geometry connectivity ; Communicable insights:
KC 0621 Geometry ; Projective
KC 0621 Geometry ; Synthetic
HM 1600 German Bettschwere
HM 1600 German Decadent sleepiness
HM 1798 German Feierabend
HM 7221 German Passion miracles
HM 7221 German Wundersucht
HH 0805 Geront
KC 0781 Gestalt
HH 0751 Gestalt therapy
TP 0019 Getting ahead
TP 0040 Getting free
HM 4477 Getting nearer ; Hope (Sufism)
MM 2081 Getting water
TP 0005 Getting wet Biding one's time
TP 0005 Getting wet Calculated inaction
TP 0005 Getting wet Waiting
HM 1257 Ghaffar ; Al (Sufism)
HM 7383 Ghaflah (Sufism)
HM 6255 Ghafur ; Al (Sufism)
HM 2243 Ghaiba (Sufism)
HM 3822 Ghani ; Al (Sufism)
HM 2961 Ghost ; Presence Holy
HM 2112 Ghosts hell awareness ; Hungry (Buddhism)
HM 2364 Ghranavijnana (Buddhism)
HH 0611 Gifted children ; Exceptionally
VC 0833 Giftedness
HH 0295 Giftedness
HH 0300 Giri (Japanese)
HM 2101 Givenness ; Stark (ICA)
HH 1234 Giving (Buddhism)
KD 2006 Gladiatorial arena ; Prevailing meta answer:
VC 0835 Gladness
HM 5224 Gladness (Buddhism)
HM 1649 Gladness ; Faculty mental (Buddhism)
VC 0837 Glamour
HM 6003 Glancing (Islam)
VD 2740 Glibness
MM 2059 Gliding soaring ; Aerial animal locomotion
KC 0004 Global
HM 2432 Global brotherhood (ICA)
HM 2006 Global consciousness

Global

HM 2817	Global guardianship (ICA)	
KD 2272	Global order ; Focus variable geometries	
KD 2082	Global relations ; Unexpected	

Global [society...]
- KC 0506 — society
- HH 1676 — spirituality
- KC 0805 — system
- KC 0997 — system ; Regional subsystem
- HH 1267 Globalism ; Bio
- VC 0839 Globality
- HM 2042 Glorious mystery (ICA)
- VC 0841 Glory
- HM 4185 Glory ; Body (Psychism)
- HM 2238 Glory ; Sphere (Kabbalah)
- HM 1608 Glossolalia (Christianity)
- HM 3743 Glue-sniffing
- HM 0507 Gluttony ; Spiritual (Christianity)
- HH 0927 Gnani yoga (Yoga)
- HM 3060 Gnosis
- HM 0413 Gnosis
- HM 8667 Gnosis kardias (Sufism)
- HM 2254 Gnosis (Sufism)
- HM 0413 Gnostic knowledge
- HM 1045 Gnostic (Sufism)
- KC 0632 Gnosticism
- HH 3546 Gnosticism
- HM 7921 Gnyid (Tibetan)
- HM 0633 Goad ; Wisdom (Buddhism)
- HH 0034 Goal-directedness
- HM 3377 Goal ; Supreme (Sufism)
- KC 0717 Goals
- KC 0977 Goals mankind ; Long range
- HH 1223 God ; Abandonment (Christianity)
- HM 3462 God: action ; Saviors (Christianity)
- HM 6204 God active desire ; Meeting (Christianity)
- HM 1905 God ; Active remembrance (Sufism)
- HM 5309 God both resting working accordance righteousness ; Meeting (Christianity)

God [Certain...]
- HM 0930 — Certain knowledge (Sufism)
- HM 2377 — Closeness (Sufism)
- HM 2675 — Communion

God consciousness
- HM 0563 — [God consciousness]
- HM 2166 — [God-consciousness]
- HM 2393 — adapability
- HM 2136 — buoyancy
- HM 2760 — power
- HM 2390 — rapture
- HM 3193 — supreme
- HM 8776 God ; Devotion (Sufism)

God [Ecstatic...]
- HM 4501 — Ecstatic experience (Judaism)
- HM 8772 — emptiness ; Meeting (Christianity)
- HM 4100 — experience

God [Fear...]
- HM 0673 — Fear (Christianity)
- HM 1047 — Fear (Sufism)
- HM 7383 — Forgetfulness (Sufism)
- HH 5328 — frequency
- HM 2606 God heaven awareness ; Thirty three (Buddhism)
- HM 4393 God ; Hope (Sufism)

God [Identification...]
- HM 3520 — Identification (Sufism)
- HM 0946 — Identification (Sufism)
- HM 2214 — Identity (Sufism)
- HM 1957 — Intimacy (Sufism)

God [Kingdom...]
- HM 1606 — Kingdom
- HM 4501 — Kiss (Judaism)
- HM 2254 — Knowledge (Sufism)
- HH 5232 God ; Love (Christianity)

God Love Islam
- HM 4712 — [God ; Love (Islam)]
- HM 5116 — [God ; Love (Islam)]
- HM 5580 — [God ; Love (Islam)]
- HM 4273 God ; Love (Sufism)
- HM 3595 God man ; Relationship (Christianity)
- HM 3439 God: march ; Saviors (Christianity)

God [Meditation...]
- HM 6227 — Meditation (Sufism)
- HM 0541 — Meeting (Christianity)
- HH 1306 — Messengers
- HM 3889 — Mystic union (Christianity)
- HM 2561 God ; Names (Sufism)
- HM 5119 God ; Negative feeling towards (Christianity)

God [People...]
- HM 2484 — People (ICA)
- HM 3502 — Positive feeling towards (Christianity)
- HH 0992 — Practice presence
- HM 3264 God: preparation ; Saviors (Christianity)
- HM 2961 God ; Presence
- HM 2377 God-proximity (Sufism)

God [Remembrance...]
- HM 6562 — Remembrance (Sufism)
- HM 8123 — Repose (Christianity)
- HH 1016 — Resignation will
- HM 9213 — Resting (Christianity)
- HM 3420 God ; Saviors (Christianity)
- HM 6989 God ; Sense presence (Christianity)
- HM 3977 God: silence ; Saviors (Christianity)

God [souls...]
- HM 6522 — soul's ground ; Birth (Christianity)
- HM 3521 — Spiritual vision (Sufism)
- HM 0330 — Subsistence (Sufism)
- HM 1799 — Surrender will (Islam)
- HM 0807 God ; Trust (Sufism)
- HM 4062 God ; Turning (Sufism)

God [Unification...]
- HM 3438 — Unification (Sufism)
- HM 3864 — Unification (Sufism)
- HM 7119 — Union (Sufism)
- HM 2855 God: vision ; Saviors (Christianity)

God [Waiting...]
- HM 6129 — Waiting (Christianity)
- HM 4501 — wholly (Judaism)
- HM 4291 — wholly same (Judaism)
- HM 0128 — Worship (Sufism)
- HM 2519 Godaniya ; Apara (Buddhism)
- HH 1909 Goddess ; Way
- KC 0444 Gödel's Theorem
- VD 2742 Godlessness
- VC 0843 Godliness
- VP 6013 Godliness-Ungodliness
- HH 0364 Godly ; Qualities
- HH 0364 Godly ; Treasures
- HM 2700 Godo ; Kensho (Japanese)
- HM 1141 Gods themselves ; Beyond chakras: (Leela)
- HH 6773 Gogai (Zen)
- HM 1691 Going (Buddhism)
- HM 3327 Going ; Keeping (Buddhism)
- HH 1330 Gold elixir ; Formation (Taoism)
- HH 4880 Gold pill ; Refining (Taoism)
- HM 0634 Gold pill (Taoism)
- HM 4572 Golden age
- HH 3862 Gong ; Qi
- VC 0845 Good
- MM 2075 Good-bad behaviour
- HM 1355 Good company (Leela)
- HH 0429 Good conduct (Zen)
- HM 1294 Good life ; Doubt efficacy (Buddhism)
- HH 2764 Good man
- HM 4343 Good ; Plane cosmic (Leela)

Good quality
- HM 1741 — aggregate consciousness (Buddhism)
- HM 0349 — aggregate sensation, aggregate cognition, aggregate synergies (Buddhism)
- HM 0872 — Being (Buddhism)
- HM 3424 — (Buddhism)
- HM 3143 Good stewardship (ICA)
- HM 5197 Good tendencies (Leela)
- HH 0920 Good will - Free will
- HH 0568 Goodness
- VC 0847 Goodness
- VP 5674 Goodness-Badness
- TP 0025 Goodness ; Instinctive
- VC 0849 Goodwill
- VC 0851 Gorgeousness
- HM 4637 Gotra-bhu-nanam (Pali)
- TC 1910 Governed behaviour ; Rule
- KC 0021 Government
- VC 0853 Government
- VC 1565 Government ; Self
- KC 0153 Government ; World
- KC 0153 Government ; World federal
- VC 0855 Grace
- HH 0169 Grace
- HM 6548 Grace ; Actual (Christianity)
- HH 4330 Grace (Hinduism)
- HM 2907 Grace ; Liberation spirit attraction divine (Sufism)
- HH 5116 Grace ; Sanctifying (Christianity)
- HM 6445 Grace ; State (Christianity)
- HH 0809 Gracefulness
- VC 0857 Gracefulness
- VD 2744 Gracelessness
- HH 1890 Gracious affections
- VC 0859 Graciousness
- HH 2003 Gradations knowledge
- HH 7734 Grade ; Higher upper (Taoism)

Grade [low...]
- HH 3667 — low paths ; Middle (Taoism)
- HH 4289 — low paths ; Upper (Taoism)
- HH 3296 — Lower middle (Taoism)
- HH 4711 — Lower upper (Taoism)
- HH 5207 Grade ; Middle middle (Taoism)
- HH 6304 Grade ; Middle upper (Taoism)
- HH 2633 Grade ; Upper middle (Taoism)
- MP 1029 Gradient local relationships ; Stable density
- TP 0053 Gradual advance
- HH 5342 Gradual method ; Three vehicles (Taoism)
- TP 0053 Gradual progress
- HH 2343 Grahastashrama (Hinduism)
- HH 1179 Grail quest
- KC 0262 Grammar ; Universal
- VC 0861 Grandeur
- HM 9002 Grandeur ; Delusions
- KC 0816 Graph ; Flow
- KC 0584 Graph theory
- TC 1418 Graphic teleconferencing ; Audio
- HM 0881 Grasp (Buddhism)
- HM 1751 Grasping inverted views (Buddhism)
- HM 4228 Grasping ; Lack real (Buddhism)
- HM 0451 Grasping ; Non (Buddhism)
- HM 3017 Grasping straws (Brainwashing)
- HM 0700 Grasping view (Buddhism)
- HM 4312 Grass-hopper mind
- TC 1430 Grassrooter participant type
- VC 0863 Gratification
- VC 0865 Gratitude
- HH 4202 Gratitude (Christianity)
- HM 3017 Gratitude ; Desperate (Brainwashing)
- VP 5949 Gratitude-Ingratitude
- HM 3105 Gratitude ; Original (ICA)
- HM 3025 Gratitude ; Spontaneous (ICA)
- HM 4154 Gratitude (Sufism)
- HM 2180 Great awakening (Buddhism)
- HM 2118 Great-black-lord awareness (Buddhism)
- HM 2807 Great compassion (Buddhism)
- TP 0028 Great ; Excesses
- TP 0028 Great gains
- HM 3630 Great insights (Buddhism)
- HM 2082 Great kings heaven awareness ; Four (Buddhism)
- HM 2929 Great love (Buddhism)
- TP 0014 Great measure ; Possession

Great [path...]
- HM 2656 — path tantra awareness (Buddhism)
- HH 0550 — perfection (Zen)
- TP 0034 — Power
- TP 0028 — Preponderance
- TP 0026 Great ; Taming power

Great vehicle
- HH 0900 — Buddhism (Buddhism)
- HM 3048 — higher-path awareness (Buddhism)
- HM 2268 — lower-path awareness (Buddhism)
- HM 2160 — middle-path awareness (Buddhism)
- TP 0034 Great vigour
- HM 3055 Great wakefulness (Hinduism)
- VC 0867 Greatness
- TP 0028 Greatness ; Passing
- VP 5034 Greatness-Smallness
- HM 2420 Greatness ; Sphere mercy (Kabbalah)
- VD 2746 Greed

Greed Buddhism
- HM 3283 — [Greed (Buddhism)]
- HM 0150 — [Greed (Buddhism)]
- HM 0802 — [Greed (Buddhism)]
- HM 1513 Greed ; Feeling (Buddhism)
- HM 1931 Greed (Leela)
- HM 2128 Greed ; Non (Buddhism)
- HM 0642 Greed ; Spiritual (Christianity)
- HM 1825 Greed ; State not feeling (Buddhism)
- HM 1954 Greed unwholesome root (Buddhism)
- HM 1442 Greedy ; Being (Buddhism)
- HM 3947 Greedy ; State not being (Buddhism)
- HH 4865 Greek spirituality
- HH 2315 Green movement
- HM 0800 Green spirituality
- VC 0869 Gregariousness
- HM 2685 Grief
- VD 2748 Grief
- HM 0399 Grief ; Aesthetic emotion (Hinduism)
- HM 7120 Grief (Buddhism)
- HM 9392 Grief (Hinduism)
- VD 2748 Grievance
- TP 0006 Grievance
- HM 3121 Grim reaper (Tarot)
- VD 2752 Grimness
- HM 1081 Grip burden ; Strong (Buddhism)
- HH 0320 'gro ; Kun (Buddhism)
- VD 2754 Grossness
- MS 8816 Grotesques
- HM 6522 Ground ; Birth God soul's (Christianity)
- MP 1222 Ground-level visibility
- HM 2721 Ground ; Moral (ICA)
- MP 1168 Grounded structures
- VC 1907 Grounded ; Well
- TC 1264 Grounder role ; Team
- HH 0318 Grounding
- MP 1176 Grounding perspectives non linearity ; Sites
- KC 0928 Group
- TC 1311 Group ; Buzz

Group [centred...]
- HH 0607 — centred therapy
- TC 1602 — conferencing ; Affinity
- TC 1722 — conferencing ; Encounter

Group conscious existence
- HM 2098 — Awareness consciousness (Buddhism)
- HM 2108 — Awareness corporeality (Buddhism)
- HM 4983 — Awareness feeling (Buddhism)
- HM 2050 — Awareness mental formation (Buddhism)
- HM 4143 — Awareness perception (Buddhism)
- HM 2556 — senses mind ; Awareness consciousness (Buddhism)

Group [consciousness...]
- HM 3109 — consciousness
- TC 1366 — conversations
- TC 1602 — counsel conferencing
- HH 0227 — counseling
- TC 1511 Group ; Discussion
- KC 0954 Group dynamics
- HH 0162 Group ; Encounter
- HH 0716 Group experience ; Task oriented
- HM 2559 Group feeling
- HM 2559 Group formation
- HM 4001 Group harmony (Japanese)
- HH 4502 Group initiation ; Peer
- HH 4502 Group initiation ; Work
- HH 0617 Group ; Leadership development

Group [Marathon...]
- HH 0817 — Marathon
- KC 0054 — Mathematical
- HM 3321 — mind
- HM 6119 — mindset
- HH 0617 Group ; Organizational development
- TC 1856 Group ; Peer mediated learning
- HH 0851 Group psychotherapy
- TC 1916 Group ; Rap

Group [Social...]
- KC 0915 — Social
- HM 3321 — soul
- HM 1417 — spiritual initiation (Esotericism)
- TC 1375 — Study
- HH 0717 Group ; Team building

Group therapy
- HH 0851 — [Group therapy]
- HH 0492 — Activity
- HH 0325 — Addiction
- HH 0507 — Analytic
- HH 0362 — Family
- TC 1602 — Inter
- HH 0482 — Nude
- HM 6119 Group think
- HM 2602 Group unconsciousness
- HH 0801 Group work ; Social
- HH 0917 Group workshop ; Creativity

Groups [Conscious...]
- HM 3321 — Conscious
- TC 1562 — Conscious
- TC 1269 — Creative care
- TC 1306 Groups ; Huddle
- HH 0672 Groups ; Human relations
- TC 1593 Groups ; Special interest
- HH 0672 Groups ; T
- KC 0372 Groups ; World mind

Growth
- KC 0376 — [Growth]
- VC 0871 — [Growth]
- TP 0053 — [Growth]
- HH 0562 Growth centres
- HM 2102 Growth consciousness

Growth development
- HH 0146 — Cognitive
- HH 0936 — Emotional
- HH 0347 — Healthy human
- HH 0146 — Intellectual
- HH 0146 — Mental
- HM 2553 Growth dying person ; Psychological
- HH 0161 Growth facilitation ; Personal
- HH 0175 Growth games

Growth [Maturity...]
- HH 5434 — Maturity
- HH 0565 — Moral
- HH 1321 — Moral

Growth [pains...]
- TP 0003 — pains ; Organizational
- HH 0836 — Physical
- HH 6669 — Primary
- HH 3117 — Psychic
- HH 5434 — Psychic
- HH 2909 — Psychospiritual (Christianity)
- HH 1321 Growth ; Religious
- HH 0835 Growth spirals
- HH 5009 Growth ; Spiritual
- TP 0004 Growth ; Uncultivated (Young shoot)
- VD 2756 Gruesomeness
- HM 2586 Gtan bzhi pa ; Bsam (Buddhism)
- HM 2450 Gtan dang ; Bsam (Buddhism)
- HM 2038 Gtan gnyis pa ; Bsam (Buddhism)
- HM 2695 Gti-mug-med-pa (Tibetan)
- HM 3863 Gtum-mo (Tibetan)
- HM 3130 Guardian angel (ICA)
- HM 2817 Guardianship ; Global (ICA)
- TC 1566 Guest speakers
- HM 2699 Guhya-samaja urgyan awareness (Buddhism)
- HH 0178 Guidance
- HH 0850 Guidance ; Child
- HH 1244 Guidance ; Discipline (Christianity)

Guidance [Educational...]
- HH 0178 — Educational
- HH 5786 — Educational
- HM 2114 — Existential (ICA)

Guidance Spiritual
- HH 0878 — [Guidance ; Spiritual]
- HH 0442 — [Guidance ; Spiritual]
- HH 1199 — (Christianity)
- HH 0894 Guidance therapy ; Child
- HH 0178 Guidance ; Vocational
- HH 8005 Guidance ; Vocational
- HH 0442 Guide ; Spiritual
- HM 4358 Guide ; Wisdom (Buddhism)
- HH 0627 Guided fantasy
- HH 0962 Guided-hypnosis
- HH 0627 Guided imagery
- HH 2304 Guided meditation
- HH 6053 Guided visualization (Magic)
- HH 3020 Guiding archetypes
- HH 3020 Guiding fantasies
- HH 3020 Guiding images
- HH 3020 Guiding metaphors
- HH 3020 Guiding myths
- VD 2758 Guile
- HM 3391 Guilt
- VD 2760 Guilt
- HM 3204 Guilt ; Channelling (Brainwashing)
- HM 6036 Guilt ; Delusions
- VP 5983 Guilt ; Innocence
- HM 3413 Guilt ; Not incurring (Buddhism)
- HM 2923 Guilt ; Recognition (Brainwashing)
- VC 0873 Guiltlessness
- VD 2762 Gullibleness
- HM 3509 Gunas ; Alinga stage (Yoga)
- HM 3460 Gunas ; Avivesa stage (Yoga)
- HM 3212 Gunas ; Linga stage (Yoga)
- HM 2805 Gunas ; Stages (Yoga)
- HH 0413 Gunas ; Three
- HM 2912 Gunas ; Vivesa stage (Yoga)
- HH 0805 Guru
- HM 2019 Guru ; Radiant (ICA)
- HM 3058 Guru ; Transcendent (ICA)
- HM 6995 Gustatory hallucination
- HM 2278 Gyan chaupad
- HM 3846 Gyana (Leela)
- HM 7209 'Gyod-pa (Tibetan)
- HH 0910 Gzhan-'gyur (Tibetan)
- HM 2257 Gzugs-khams (Tibetan)
- HM 2669 Gzugs-med-kyi-snyoms-jug (Tibetan)
- HM 2281 Gzugs-sku (Tibetan)

H

- HH 3665 Habit
- HH 0638 Habit training
- HH 0200 Habits
- HM 2416 Hadarat (Sufism)
- HM 1287 Hadi ; Al (Sufism)
- HM 2416 Hadrat (Sufism)
- HM 7099 Hafiz ; Al (Sufism)
- MS 8365 Hair nails
- HH 1265 Hajj (Islam)
- HM 5014 Hakam ; Al (Sufism)
- HM 4799 Hakim ; Al (Sufism)
- HM 2365 Hal (Sufism)
- HM 4519 Halim ; Al (Sufism)
- HM 2680 Hallowed honour (ICA)
- HM 4580 Hallucination
- VD 2764 Hallucination
- HM 6704 Hallucination ; Auditory
- HM 5504 Hallucination ; Functional
- HM 6995 Hallucination ; Gustatory
- HM 6704 Hallucination ; Imperative
- HM 5904 Hallucination ; Negative
- HM 4559 Hallucination ; Olfactory
- HM 4336 Hallucination ; Pseudo
- HM 7208 Hallucination ; Somatic
- HM 4094 Hallucination ; Tactile
- HM 6118 Hallucination ; Tactile
- HM 0092 Hallucination ; Visual
- HM 1317 Hallucinations ; Psychic
- HH 0075 Hallucinogenic drugs
- HM 0801 Hallucinogens ; Modes awareness associated use
- HM 0392 Hamid ; Al (Sufism)
- HM 3068 Hands awareness ; Entering city bliss bestowing (Zen)
- MS 8315 Hands feet
- HH 2552 Hands ; Laying
- HH 2023 Hands ; Laying (Christianity)
- VC 0879 Handsomeness
- HM 2190 Hanged man (Tarot)
- HM 0134 Hangover
- VD 2766 Haphazardness
- HM 2409 Happiness
- VC 0881 Happiness

Happiness [bliss...]
- HM 4298 — bliss born seclusion ; Jhana (Buddhism)
- HM 2866 — (Buddhism)
- HM 0747 — (Buddhism)

Happiness Concentration
- HM 4433 — accompanied (Buddhism)
- HM 5767 — (Buddhism)
- HM 0730 — (Buddhism)
- HM 4886 Happiness due insight ; Rapturous (Buddhism)
- HM 0747 Happiness ; Fivefold (Buddhism)
- HM 5099 Happiness (Leela)
- HM 5000 Happy-heartedness (Italian)
- HM 0124 Haqiqa (Sufism)

Haqq [Al...]
- HM 3945 — Al (Sufism)
- HM 0741 — Al (Sufism)
- HM 4556 — al-yaqin (Sufism)
- HM 2662 Hara (Japanese)
- TC 1642 Hard-head
- VC 0883 Hardness
- VD 2770 Hardness
- VP 5356 Hardness-Softness
- HH 0031 Hare kegare cycle ; Ke (Japanese)
- HM 1591 Harm ; Causing (Buddhism)
- VD 2772 Harmfulness
- HM 3210 Harmfulness (Buddhism)
- HM 2608 Harmfulness ; Non (Buddhism)
- HM 4565 Harmless (Buddhism)
- VC 0885 Harmlessness
- KC 0110 Harmonic series
- HH 0603 Harmonies enlightenment (Buddhism)
- MP 1134 Harmonious perspectives ; Limiting exposure
- MP 1113 Harmoniously structured entry point external communication media
- KP 3012 Harmoniously transformative controlled relationship
- TC 1775 Harmonizers
- HM 0486 Harmonizing illumination (Taoism)
- MP 1210 Harmonizing space distribution levels

Harmony
- KC 0434 — [Harmony]
- KC 0424 — [Harmony]
- VC 0887 — [Harmony]
- HM 0974 — [Harmony]
- HM 4350 Harmony aesthetic contrasts (Japanese)
- VP 5462 Harmony-Discord
- HM 4001 Harmony ; Group (Japanese)
- KD 2250 Harmony holistic resonance ; Dissonant
- HM 2149 Harmony progress ; Sense (Brainwashing)

Harmony [Self...]
- HH 0778 — Self
- HM 4001 — Social (Japanese)
- MP 1104 — Structural development designed counteract deficiencies pattern
- HM 0763 Harmony ; Way (Esotericism)
- HM 3070 Harmony whole
- HH 0429 Harmony (Zen)
- VD 2768 Harrassment
- HM 8098 Harsa (Hinduism)
- HM 4376 Harsha-loka (Leela)
- VD 2774 Harshness
- HM 1421 Hasa (Pali)
- HM 1467 Hasib ; Al (Sufism)
- HH 0597 Hasidism (Judaism)
- VD 2776 Haste
- TC 1739 Haste participant type ; Less
- VD 2778 Hastiness
- HM 4596 Hate
- HM 4502 Hate (Buddhism)
- VP 5930 Hate ; Love
- HM 2744 Hate ; Non (Buddhism)
- HH 0862 Hatha yoga (Yoga)
- HM 4323 Hating (Buddhism)
- HM 1557 Hating ; Non (Buddhism)
- VD 2780 Hatred
- HM 4596 Hatred awareness
- HM 4502 Hatred (Buddhism)
- HM 0607 Hatred (Buddhism)
- HM 5676 Hatred ; Cessation (Buddhism)
- HM 2744 Hatred ; Non (Buddhism)
- HM 1399 Hatred ; Non (Buddhism)
- HM 1967 Hatred ; State feeling (Buddhism)
- HM 1597 Hatred ; State not feeling (Buddhism)
- HM 3560 Hatred wholesome root ; Non (Buddhism)
- HM 2528 Haughtiness (Buddhism)
- HM 3158 Having faith (Buddhism)
- HM 2762 Having heart (Japanese)
- HM 0120 Hawa (Islam)
- HM 4236 Hayy ; Al (Sufism)
- VD 2782 Hazardousness
- TC 1642 Head ; Hard
- HH 4974 Head heart ; Journeying transcendence
- HH 0458 Healer ; Psychic
- HH 0216 Healing

Healing [Absent...]
- HH 1333 — Absent
- HH 0458 — Affirmative
- HH 2552 — Auric
- HH 2997 Healing breath
- HH 2552 Healing ; Contact
- HH 4908 Healing ; Crystal
- HH 0458 Healing ; Faith
- HH 0248 Healing ; Folk
- HM 3925 Healing ; Ikia
- HH 2552 Healing ; Magnetic
- HH 0458 Healing ; Mental
- HH 2314 Healing ; Primal
- HH 0458 Healing ; Psychic
- HH 2997 Healing ; Spirit
- HH 0458 Healing ; Spiritual
- HH 2552 Healing ; Touch
- HH 0458 Healing ; Trance
- HH 0509 Health
- VC 0889 Health
- MS 8406 Health aids
- HH 0581 Health care movement ; Holistic
- HM 1440 Health ; Delusions ill
- VP 5685 Health-Disease
- HH 0296 Health ; Mental
- HH 0675 Health ; Positive mental

Health [Sound...]
- HH 2543 — Sound
- KD 2240 — space-time
- HH 0108 — Spiritual aspects psychic
- HH 0109 — spirituality ; Psychic
- HH 4908 Health ; Vibrational
- VP 5683 Healthfulness-Unhealthfulness
- VC 0891 Healthiness
- HH 0347 Healthy human growth development
- HH 0971 Healthy personality
- HH 0257 Healthy personality
- HM 2161 Hearers ; Meditation way (Buddhism)
- HM 2169 Hearing awareness (Buddhism)
- HM 5982 Hearing ; Deva (Buddhism)
- HH 1973 Hearing music spheres

Heart [Alertness...]
- HM 2792 — Alertness (Sufism)
- HM 7622 — Annihilation (Sufism)
- HM 3518 — awakening stage life
- HM 6932 Heart ; Cleansing (Sufism)
- HM 7622 Heart ; Conscious (Sufism)
- HM 2762 Heart ; Having (Japanese)
- HH 4974 Heart ; Journeying transcendence head
- HM 3244 Heart ; Liberation (Sufism)
- HH 5124 Heart ; Listening
- HM 1612 Heart ; Purity
- HM 1434 Heart ; Soundness
- HM 3438 Heart ; Trans substantiation (Sufism)
- HM 0935 Heart ; Wisdom
- HM 8667 Heart ; Wisdom (Sufism)
- HM 0220 Hearted concentration ; Single
- HM 5000 Heartedness ; Happy (Italian)
- HM 3085 Heartedness ; Single (Sufism)
- VD 2784 Heartlessness

Heat

HM 2208	Heat application awareness ; Mahayana (Buddhism)	VC 0899	High-mindedness	HM 2590	Hinduism Crazy-monkey ego consciousness
HM 2696	Heat application awareness ; Tantra (Buddhism)	HM 2260	High priest (Tarot)	HH 0754	Hinduism Cultural life
VP 5328	Heat-Cold	HM 2261	High priestess (Tarot)	HM 6705	Hinduism Death
HM 4302	Heaven	VC 0901	High-spiritedness	HM 2957	Hinduism Deep sleep
HH 3564	Heaven	HM 7550	Higher knowing (Christianity)	HM 0891	Hinduism Degrees consciousness
	Heaven awareness		**Higher knowledge Buddhism**	HM 0883	Hinduism Dejection
HM 2082	— Four great kings (Buddhism)	HM 4297	— [Higher knowledge (Buddhism)]	HM 3592	Hinduism Descending steps ignorance
HM 2022	— Joyful (Buddhism)	HM 5232	— [Higher knowledge (Buddhism)]	HM 7800	Hinduism Desire
HM 0563	— Kingdom	HM 5982	— [Higher knowledge (Buddhism)]	HM 4406	Hinduism Desirelessness
HM 2565	— (Psychism)	HM 7672	— [Higher knowledge (Buddhism)]	HM 5091	Hinduism Detachment
HM 2813	— Supreme (Buddhism)	HH 5652	— [Higher knowledge (Buddhism)]	HM 2566	Hinduism Dharana
HM 2606	— Thirty three god (Buddhism)	HM 0748	— [Higher knowledge (Buddhism)]	HM 0137	Hinduism Dhyana
HM 1973	Heaven (Buddhism)	HM 3529	Higher mind (Hinduism)	HH 3772	Hinduism Differentiated non-duality
HM 2546	Heaven consciousness ; Delightful emanation (Buddhism)	HH 2387	Higher mind ; Training (Buddhism)	HM 2099	Hinduism Direct intuitive knowledge
TP 0001	Heaven Creativity	HM 3048	Higher path awareness ; Great vehicle (Buddhism)	HM 2099	Hinduism Discrimination
HM 2407	Heaven reality awareness ; Death bardo (Buddhism)		**Higher [self...]**	HM 5448	Hinduism Dissatisfaction
TP 0001	Heaven Strong action	HM 3059	— self ; Near death encounter (NDE)	HM 3147	Hinduism Dream
HM 2130	Heaven-without-fighting consciousness (Buddhism)	HM 2970	— self (Psychosynthesis)	HM 3510	Hinduism Dream wakefulness
HM 2784	Heavenly advocate (ICA)	HM 0866	— state consciousness (Psychism)	HM 2574	Hinduism Duhkha
HH 3421	Heavenly deceit	HM 0935	— states consciousness	HM 2574	Hinduism Dukkha (Pali)
HM 2010	Heavenly-highway awareness (Buddhism)	HM 2365	— states consciousness (Sufism)	HH 0715	Hinduism Dvija
HM 2918	Heavenly secret (ICA)	MP 1158	— structural levels ; External access	HM 7100	Hinduism Ecstatic perception
HM 2216	Heavenly sorrow (ICA)		**Higher [unconscious...]**	HM 2059	Hinduism Ego awareness
HM 3214	Heavens ascension stages game ; Supreme (Buddhism)	HM 2960	— unconscious	HM 2443	Hinduism Enquiry
		HM 2057	— unconscious (Psychosynthesis)	HM 2226	Hinduism Enstasy
TC 1566	Heavies ; Conference	HH 7734	— upper grade (Taoism)	HM 3041	Hinduism Enstasy
HM 2891	Hebetude	HH 7386	Higher vehicle path (Taoism)	HM 1780	Hinduism Erotic sentiment
HH 0130	Hedonism	HM 2271	Highest teachings application awareness ; Mahayana (Buddhism)	HM 3006	Hinduism Establishment truth
VD 2786	Heedlessness			HM 1194	Hinduism Fear
VP 5207	Height-Lowness	HH 4742	Highest vehicle (Taoism)	HM 2001	Hinduism Fifth plane wisdom
HH 2007	Hell	HM 2864	Highest yoga tantra: anuttarayoga (Buddhism)	HM 3100	Hinduism First plane wisdom
HM 2052	Hell ; Avici (Buddhism)	HM 2010	Highway awareness ; Heavenly (Buddhism)	HM 4875	Hinduism Flash intuition
	Hell awareness	VC 0903	Hilarity	HH 2563	Hinduism Four stages life
HM 2028	— Adamantine (Buddhism)	HM 8667	Himma (Sufism)	HM 3006	Hinduism Fourth plane wisdom
HM 2636	— Animal (Buddhism)	HM 1276	Himsa-loka (Leela)	HM 8762	Hinduism Furious sentiment
HM 2007	— Divine animal (Buddhism)	HM 0933	Himself ; Fifth chakra: man becomes (Leela)	HH 4330	Hinduism Grace
HM 2112	— Hungry ghosts (Buddhism)	HH 0845	Hinayana Buddhism (Buddhism)	HM 2343	Hinduism Grahastashrama
HM 2052	— Interminable (Buddhism)	HH 5007	Hindrances ; Afflictions (Buddhism)	HM 3055	Hinduism Great wakefulness
HM 2516	— Reviving (Buddhism)	HH 6773	Hindrances ; Five (Buddhism)	HM 9392	Hinduism Grief
HM 3142	Hell ; Black rope (Buddhism)	HM 2115	Hindu-states awareness (Buddhism)	HM 8098	Hinduism Harsa
HM 3165	Hell ; Near death (NDE)	HM 2723	Hindu-wisdom-holder awareness (Buddhism)	HM 4085	Hinduism Heroic sentiment
HM 4282	Hell ; State (Buddhism)	HM 9845	Hinduism Abhisneha	HM 3529	Hinduism Higher mind
HM 2028	Hell ; Vajra (Buddhism)	HM 9845	Hinduism Abhisvanga	HH 2343	Hinduism Householder stage life
	Hells awareness	HM 7344	Hinduism Absence love	HH 0330	Hinduism Human development
HM 2040	— Cold (Buddhism)	HM 0137	Hinduism Absorption	HH 1004	Hinduism Illusion
HM 3142	— Crushing (Buddhism)	HM 2491	Hinduism Action desire	HM 2446	Hinduism Indian approaches consciousness
HM 2100	— Howling (Buddhism)	HM 5487	Hinduism Action desire	HM 2692	Hinduism Individual tendencies consciousness
HM 2454	— Temporary (Buddhism)	HH 0518	Hinduism Advaita vedanta	HM 2574	Hinduism Insufficiency
HM 2576	— Very hot (Buddhism)	HM 0205	Hinduism Aesthetic emotion	HM 3455	Hinduism Intellectual condition
VC 1567	Help ; Self	HM 8762	Hinduism Aesthetic emotion anger	HM 0992	Hinduism Intoxication
HM 3344	Helped carry cross ; Jesus (Christianity)	HM 0966	Hinduism Aesthetic emotion disgust	HM 1256	Hinduism Intuition
VC 0893	Helpfulness	HM 4085	Hinduism Aesthetic emotion energy	HM 9124	Hinduism Intuition sages
VD 2788	Helplessness	HM 1644	Hinduism Aesthetic emotion fear	HH 1016	Hinduism Isvara pranidhana
HM 2429	Helplessness ; Abject (ICA)	HM 0399	Hinduism Aesthetic emotion grief	HM 0137	Hinduism Jhana (Pali)
HM 2182	Helplessness ; Psychic	HM 1780	Hinduism Aesthetic emotion love	HM 3890	Hinduism Jivan-mukti
HM 3174	Her favourite resort awareness (Yoga)	HM 0645	Hinduism Aesthetic emotion mirth	HM 3890	Hinduism Jivanmukti
HM 2559	Herd instinct	HM 0291	Hinduism Aesthetic emotion wonder	HH 0610	Hinduism Jnana
HM 2690	Herding simile ; Cow (Zen)	HM 7219	Hinduism Aesthetic enjoyment	HH 0495	Hinduism Jnana kanda
HM 2560	Herding-the-ox awareness (Zen)	HM 9845	Hinduism Affection	HH 0495	Hinduism Jnana marga
VD 2790	Heresy	HH 1047	Hinduism Afflictions	HM 3060	Hinduism Jnana samadhi
VC 0895	Heritage	HM 2692	Hinduism Aggregate	HM 8098	Hinduism Joy
TP 0050	Heritage ; Cultural (Cauldron)	HM 2059	Hinduism Ahamkara	HH 0754	Hinduism Kama
KC 0125	Hermeneutics	HM 2059	Hinduism Ahankara	HM 7800	Hinduism Kama
KC 0545	Hermeneutics ; Structural	HM 6898	Hinduism Akrodha	HM 4666	Hinduism Kamachanda
HM 2136	Hermes awareness	HM 7332	Hinduism Amarsa	HH 1211	Hinduism Karma marga
HH 0221	Hermetic philosophy (Esotericism)	HM 7344	Hinduism Anabhisneha	HM 2692	Hinduism Khandha (Pali)
HH 0958	Hermit	HM 8000	Hinduism Anabhisvanga	HM 2390	Hinduism Kirtan
HM 2202	Hermit (Tarot)	HM 3227	Hinduism Ananda	HH 1047	Hinduism Klesas
HM 2654	Hero ; Revered (ICA)	HM 8665	Hinduism Anger	HH 0610	Hinduism Knowledge
HH 0653	Hero worship	HM 1356	Hinduism Apoplexy	HH 0353	Hinduism Knowledge
HM 4085	Heroic sentiment (Hinduism)	HM 9124	Hinduism Arsajnana	HM 7298	Hinduism Knowledge another's thoughts
HM 4060	Heroin ; Use	HH 0353	Hinduism Artha	HM 6558	Hinduism Knowledge liberated souls
VC 0897	Heroism	HH 0754	Hinduism Artistic life	HH 4309	Hinduism Kriya sakti
HH 0929	Heroism	HM 5091	Hinduism Asakti	HM 8665	Hinduism Krodha
HH 6432	Heroism ; Female	HM 3041	Hinduism Asamprajnata samadhi	HM 2196	Hinduism Liberation consciousness
VD 2792	Hesitation	HH 0669	Hinduism Asana	HM 9845	Hinduism Love
HM 0808	Hesitation ; Succumbing (Buddhism)	HM 6153	Hinduism Asanga	HM 2590	Hinduism Lower manas ahamkara
HM 0205	Hesychasm (Christianity)	HH 2563	Hinduism Ashramas	HM 4666	Hinduism Lust
KC 0606	Heterogeneity	HM 2103	Hinduism Atma awareness	HM 7298	Hinduism Manahparyayajnana
KC 0231	Heterogenization	HH 0103	Hinduism Atman	HM 2902	Hinduism Manas awareness
HH 0962	Heterohypnotherapy	HM 1195	Hinduism Atmanubhava	HM 6705	Hinduism Marana
HH 0925	Heteronomous discipline	HM 3914	Hinduism Attachment	HM 0291	Hinduism Marvellous sentiment
KC 0018	Heterostasis	HM 1248	Hinduism Austerities	HM 2336	Hinduism Material condition
KC 0126	Heuristic	HM 2032	Hinduism Avasthas	HH 1004	Hinduism Maya
HH 3873	Hexing	HM 2032	Hinduism Awareness relative reality	HM 0137	Hinduism Meditation
HM 2444	Hidden intelligence ; Path (Judaism)	HM 1195	Hinduism Being Atman	HH 2056	Hinduism Mendicant stage life
HM 3013	Hidden intelligence ; Path (Judaism)	HM 2099	Hinduism Being consciousness	HM 3040	Hinduism Mental condition
HM 2531	Hidden reality ; Perception (Sufism)	HH 0628	Hinduism Bhakti marga	HM 0992	Hinduism Moha
VD 2794	Hideousness	HM 1194	Hinduism Bhaya	HM 2196	Hinduism Moksa
MP 1114	Hierachy perspectives favouring broadest	HM 3227	Hinduism Bliss	HM 2196	Hinduism Moksha
KC 0795	Hierarchial principle control	HM 2183	Hinduism Bodily perfection	HM 6558	Hinduism Muktajnana
TC 1632	Hierarchical conferencing	HM 2041	Hinduism Brahma consciousness	HM 1159	Hinduism Murccha
MS 8353	Hierarchical position ; Social	HH 1987	Hinduism Brahmacharyashrama	HM 2957	Hinduism Nidra
KC 0047	Hierarchical restructuring	HM 2099	Hinduism Buddhi awareness	HM 4406	Hinduism Nirasih
KC 0400	Hierarchical system	HM 7219	Hinduism Carvana	HM 2125	Hinduism Nirbija samadhi
KC 0400	Hierarchy	HH 1047	Hinduism Causes misery	HM 2226	Hinduism Nirvikalpa samadhi
KC 0908	Hierarchy formation	HH 1987	Hinduism Celibate stage life	HM 2061	Hinduism Nirvikalpa samadhi
HH 0913	Hierarchy ; Needs	HM 3354	Hinduism Cessation objectivity	HM 2491	Hinduism Niskama karma
HM 3028	Hierogamy (Tarot)	HM 3529	Hinduism Citta	HH 5487	Hinduism Niskama karma
HM 2260	Hierophant archetypal image (Tarot)	HM 4875	Hinduism Clairvoyant intuition	HM 3280	Hinduism Niyama
HM 3763	High	HM 6153	Hinduism Clinging	HM 8000	Hinduism Non-affection
HM 0147	High dream (Psychism)	HM 0645	Hinduism Comic sentiment	HM 6898	Hinduism Non-anger
HM 1772	High ; Electronic	HM 2566	Hinduism Concentration	HH 0518	Hinduism Non-duality
HM 1400	High ; Emotional	HM 2032	Hinduism Conditional consciousness states	HM 0239	Hinduism Nonecstatic perception
HH 0720	High magic	HM 0891	Hinduism Conditions self	HM 1285	Hinduism Occult perception
		HM 0137	Hinduism Contemplation	HM 0966	Hinduism Odious sentiment

HM 6558	Hinduism Omniscience	
HH 0103	Hinduism Paramatman	
HM 0399	Hinduism Pathetic sentiment	
HM 2183	Hinduism Perfect man	
HM 3298	Hinduism Planes wisdom	
HM 3455	Hinduism Prajna	
HH 0213	Hinduism Pranayama	
HH 4330	Hinduism Prasada	
HM 4875	Hinduism Pratibhajnana	
HM 4875	Hinduism Prophetic intuition	
HM 3455	Hinduism Pure consciousness	
HM 3100	Hinduism Pure intention	
HM 3914	Hinduism Raga	
HM 4666	Hinduism Raga	
HM 0205	Hinduism Rasa	
HM 4085	Hinduism Rasa resoluteness	
HM 0966	Hinduism Rasa revulsion	
HH 2782	Hinduism Recluse stage life	
HH 0213	Hinduism Regulation vital force	
HM 6558	Hinduism Released soul's perception	
HM 0966	Hinduism Repulsive sentiment	
HM 7332	Hinduism Resentment	
HM 7298	Hinduism Rijumati	
HM 1780	Hinduism Romantic rasa	
HM 2308	Hinduism Sabija samadhi	
HM 3227	Hinduism Saccidananda	
HM 2043	Hinduism Sahaja samadhi	
HM 2226	Hinduism Samadhi	
HM 0155	Hinduism Samkhya	
HM 1356	Hinduism Samnyasa	
HM 1006	Hinduism Samsara	
HH 8116	Hinduism Santosa	
HH 2056	Hinduism Sanyasashrama	
HM 3227	Hinduism Sat-cit-ananda	
HH 8116	Hinduism Satisfaction	
HM 3329	Hinduism Satsang	
HM 2226	Hinduism Savikalpa samadhi	
HM 2650	Hinduism Savikalpa samadhi	
HM 3526	Hinduism Savitarka samadhi	
HM 2443	Hinduism Second plane wisdom	
HM 2159	Hinduism Seed state wakefulness	
HH 0103	Hinduism Self	
HM 3280	Hinduism Self-discipline	
HM 1004	Hinduism Sheaths masking reality	
HM 2053	Hinduism Shushumna	
HM 1285	Hinduism Siddhadarsana	
HM 3354	Hinduism Sixth plane wisdom	
HM 2692	Hinduism Skandha (Buddhism)	
HM 2853	Hinduism Sleep	
HM 0992	Hinduism Slight unconsciousness	
HM 9392	Hinduism Soka	
HM 9392	Hinduism Sorrow	
HM 0399	Hinduism Sorrowful rasa	
HH 3000	Hinduism Spirituality	
HM 2395	Hinduism Subramania	
HM 3312	Hinduism Subtlety mind	
HM 2574	Hinduism Suffering	
HM 2957	Hinduism Sushupti	
HM 2957	Hinduism Susupti state unconsciousness	
HH 0957	Hinduism Svadhyaya	
HM 1159	Hinduism Swoon	
HM 3040	Hinduism Taijasa	
HM 1248	Hinduism Tapas	
HM 1644	Hinduism Terrible sentiment	
HM 1644	Hinduism Terrifying rasa	
HM 3312	Hinduism Third plane wisdom	
HM 7308	Hinduism Thirst	
HM 3529	Hinduism Thought	
HM 2001	Hinduism Total freedom attachment	
HM 3139	Hinduism Total union	
HM 2395	Hinduism Transcendental awareness	
HM 7308	Hinduism Trisna	
HM 2196	Hinduism Truth consciousness	
HM 2395	Hinduism Turiya awareness	
HM 3139	Hinduism Turiyatita	
HM 3139	Hinduism Turyatita	
HM 2957	Hinduism Unconscious	
HM 2099	Hinduism Understanding	
HM 2574	Hinduism Unsatisfactoriness	
HM 2392	Hinduism Upanishadic stages awareness	
HH 3772	Hinduism Vaishnavism	
HM 2336	Hinduism Vaisvanara	
HH 2782	Hinduism Vanaprasthashrama	
HM 3592	Hinduism Veils delusion	
HM 8762	Hinduism Violent rasa	
HM 7298	Hinduism Vipulamati	
HM 0883	Hinduism Visada	
HH 3772	Hinduism Vishishtadvaita	
HM 0239	Hinduism Viyukta perception	
HM 2692	Hinduism Vritti	
HM 2014	Hinduism Wakeful dream	
HM 2567	Hinduism Wakefulness	
HM 1211	Hinduism Way action	
HH 0495	Hinduism Way enlightenment	
HH 0495	Hinduism Way knowledge	
HH 0628	Hinduism Way loving faith	
HH 0353	Hinduism Wealth	
HM 7308	Hinduism Yearning	
HM 5005	Hinduism Yogic perception	
HM 5005	Hinduism Yogipratyaksa	
HM 7100	Hinduism Yukta perception	
HM 1598	Hinekurata (Japanese)	
HM 2881	Hiri (Pali)	
HM 1590	Hiri (Pali)	
HM 0352	Hiribala (Pali)	
HM 3828	Hiriyati hiriyitabbena (Pali)	
HM 0552	Hiriyati papakanam akusalanam dhammanam samapattiya (Pali)	
HM 3828	Hiriyitabbena ; Hiriyati (Pali)	
HM 1023	Hiriyitabbena ; Na kiriyati (Pali)	
HM 3611	His afflicted mother ; Jesus meets (Christianity)	
HM 4009	His cross ; Jesus falls first time (Christianity)	
HM 3166	His garments ; Jesus stripped (Christianity)	
MP 1040	Historical dimension ; Integrating	
MS 8383	Historical, legendary literary characters	
HM 2616	Historical vocation (ICA)	
KC 0725	History	
HM 3075	History ; Being (ICA)	
HH 0933	History ; Cyclic	
KC 0394	History ; Cyclical theory	
HM 2682	History ; Determining (ICA)	
HM 2588	History ; Engagement (ICA)	
HM 3245	History ; Invented (ICA)	
HM 3242	Hit awareness ; Not (Yoga)	
MS 8846	Hobbies pastimes	
HM 3690	Hod-gsal (Tibetan)	
HM 2238	Hod-sephira (Kabbalah)	
HM 3168	Hokkai ; Ji (Zen)	
HM 3352	Hokkai ; Ri (Zen)	
HM 0610	Hold meditative attitude ; Failure (Yoga)	
HM 1992	Holding paramount one's view (Buddhism)	
TP 0008	Holding together	
KD 2205	Holes ; Non comprehension	
VC 0905	Holiness	
HH 0183	Holiness	
HH 0183	Holiness code	
HM 0446	Holiness ; Human (Christianity)	
HM 1723	Holiness ; State (Christianity)	
KC 0941	Holism	
VC 0907	Holism	
HH 0649	Holistic approach	
HM 2367	Holistic consciousness	
HH 0581	Holistic health care movement	
HH 0581	Holistic medicine	
KC 0096	Holistic medicine	
KD 2250	Holistic resonance ; Dissonant harmony	
KC 0937	Holocyclation	
KC 0339	Holographic logic	
KC 0592	Holographic universe	
KC 0857	Holography	
KC 0766	Holon	
KD 2220	Holonomy ; General systems	
HM 0876	Holy (Christianity)	
HM 0673	Holy fear (Christianity)	
HM 2961	Holy Ghost ; Presence	
HM 5776	Holy indifference (Christianity)	
HH 0593	Holy places	
HM 8022	Holy resignation (Christianity)	
MS 8705	Holy scriptures	
HM 3499	Holy Spirit ; Age (Christianity)	
HH 3687	Holy war (Islam)	
VC 0909	Homage	
HH 3487	Home circle	
MS 8416	Home furnishings amenities	
HM 3153	Home ox's back awareness ; Coming (Zen)	
VD 2796	Homelessness	
KC 0618	Homeomorphism	
HH 0498	Homeopathy	
KC 0226	Homeostasis	
HH 0188	Homiletics	
TC 1683	Homogeneity ; Conference	
KC 0300	Homologous series	
KC 0326	Homology	
KC 0166	Homomorphism	
KC 0325	Homonymy	
HH 0980	Homosexuality	
VC 0913	Honesty	
HH 5304	Honesty	
VC 0915	Honour	
HH 0192	Honour	
HM 2680	Honour ; Hallowed (ICA)	
HM 3029	Honouring mystery (ICA)	
HM 1809	Hooked ; Being	
VC 0917	Hope	
HH 0199	Hope (Christianity)	
HM 4477	Hope getting nearer (Sufism)	
HM 4393	Hope God (Sufism)	
VP 5888	Hope-Hopelessness	
HM 7656	Hope ; State (Christianity)	
HM 3405	Hopelessness	
VD 2798	Hopelessness	
VP 5888	Hopelessness ; Hope	
VD 2800	Horribleness	
HM 0904	Horror	
HM 2720	Horror sin (ICA)	
MP 1121	Hospitality communication pathways	
MP 1094	Hospitable common domains	
MP 1091	Hospitable contexts perspectives transition	
MP 1139	Hospitable domain perspective nourishment	
MP 1160	Hospitable interface structures external environment	
MP 1163	Hospitable non-linear domain external structures	
MP 1149	Hospitable reception external perspectives structures	
MP 1092	Hospitable transit points	
VC 0919	Hospitality	
VP 5925	Hospitality-Inhospitality	
HH 0643	Hostile feelings	
HH 0643	Hostility	
VD 2802	Hostility	
HM 0719	Hostility (Buddhism)	
HM 2658	Hostility ; Transcending (ICA)	
HM 2859	Hosts ; Divine (ICA)	
HM 2576	Hot hells awareness ; Very (Buddhism)	
HH 0408	Hotoke ; Kami (Japanese)	
HM 1167	Hours ; Canonical (Christianity)	
HM 4321	Hours compline (Christianity)	
HM 0894	Hours lauds (Christianity)	
HM 0160	Hours matins (Christianity)	
HM 0466	Hours none (Christianity)	
HM 1904	Hours prime (Christianity)	
HM 3725	Hours sext (Christianity)	
HM 2965	Hours terce (Christianity)	
HM 1468	Hours vespers (Christianity)	
TP 0037	Household (Family)	
HH 2343	Householder stage life (Hinduism)	
HM 2100	Howling-hells awareness (Buddhism)	
HM 5122	Hpho-ba (Tibetan)	
HM 2881	Hri (Buddhism)	
HH 0589	Hsin tsung (Zen)	
HM 2163	Hsin ; Wu (Zen)	
HM 0917	Hsing ; Chen (Zen)	
HM 0910	Hsing ; Wu (Zen)	
HM 4533	Hubris (Christianity)	
TC 1306	Huddle groups	
	Human [activity...]	
MS 8500	— activity ; Symbolic	
MS 8323	— ages development	
HH 2267	— aura	
	Human [behaviour...]	
HH 0447	— behaviour development	
HH 1542	— being	
MS 8400	— beings ; Non	
MS 8305	— body	
MS 8365	— body fluids	
HH 3498	— bonding	
	Human brain consciousness	
HM 2355	— Mammalian limbic	
HM 2427	— Neomammalian neocortex	
HM 2307	— Reptilian	
	Human [capabilities...]	
HH 2806	— capabilities ; Enhancement	
HH 0937	— capital formation	
HH 3607	— centred development	
MS 8333	— congenital relationship	
HM 1105	— consciousness ; Ontogenetic model	
HM 2104	— contingency (ICA)	
	Human development	
HH 4223	— atheism	
HH 2435	— (Baha'ism)	
HH 0650	— (Buddhism)	
HH 2198	— (Christianity)	
HH 5522	— Communalist	
HH 7324	— diet	
HH 3098	— (Existentialism)	
HH 1127	— General system theory	
HH 0330	— (Hinduism)	
HH 5343	— humour	
HH 5101	— index (United Nations)	
HH 1799	— (Islam)	
HH 0622	— (Jainism)	
HH 3029	— (Judaism)	
HH 6122	— Liberal capitalist	
HH 4033	— Liberal humanist	
HH 3003	— music	
HH 7637	— Mythical control	
HH 1523	— new religious movements	
HH 1908	— panentheism	
HH 5190	— pantheism	
HH 1902	— primal religion	
HH 7637	— Psychic control	
HH 1198	— religion	
HH 6292	— (Sikhism)	
HH 3997	— State socialist	
HH 0247	— strategy (United Nations)	
HH 0436	— (Sufism)	
HH 0689	— (Taoism)	
HH 4663	— theism	
HH 3290	— (United Nations)	
HH 1003	— (Zen)	
HH 1903	— (Zoroastrianism)	
	Human [engineering...]	
KC 0707	— engineering	
HH 0276	— engineering	
HH 6115	— entitlements ; Expression	
HH 0615	— evolution ; Technological implications	
HM 3188	— example (ICA)	
MS 8325	— external organs	
	Human [face...]	
HH 4009	— face ; Structural adjustment (United Nations)	
HH 0276	— factors engineering	
MS 8315	— form face	
	Human [gender...]	
MS 8385	— gender	
HH 0357	— genetic improvement	
HH 0347	— growth development ; Healthy	
HM 0446	Human holiness (Christianity)	
MS 8345	Human internal organs	
HH 0647	Human life ; Development quality	
	Human mind	
HM 6534	— Dissolving (Taoism)	
HH 0246	— Evolution	
HM 5622	— (Taoism)	
HH 4389	Human nature ; Universal	
	Human [perfectibility...]	
HH 0212	— perfectibility	
HH 0212	— perfection	
HH 0010	— personality ; Lifelong educational transformation	
HM 5155	— plane (Leela)	
HM 0933	— plane (Leela)	

Human

Human [plurality...] cont'd
MS 8313 — plurality
HH 0461 — potential
HH 0398 — potential movement
MS 8303 — presence ; Signs
HH 0447 — psychological development
Human [relations...]
HH 0176 — relations
HH 0672 — relations groups
HH 0047 — resource training
Human resources
HH 0745 — development (United Nations)
HH 0745 — Mobilizing (United Nations)
HH 0147 — (United Nations)
MS 8363 Human roles
Human [scale...]
HH 2651 — scale development
HM 3012 — service ; Consciousness
KC 0225 — settlements ; Science
MS 8355 — skeleton
MS 8373 — spiritual phenomena
HH 0517 — synergy
KC 0887 — systems management
HM 2334 Human transformation (ICA)
VC 0921 Humaneness
HM 1099 Humaneness
HH 0674 Humanism
VC 0923 Humanism
HH 4965 Humanism ; Ecological
HH 0215 Humanism ; Enlightenment
HH 0105 Humanism ; Secular
HH 0360 Humanism ; Socialist
HH 4033 Humanist human development ; Liberal
HH 0806 Humanistic biology
HH 6209 Humanistic capitalism
HH 0793 Humanistic psychology
VC 0925 Humanitarianism
KC 0311 Humanities
VC 0927 Humanity
HH 0203 Humanity
VP 5417 Humanity-Nonhumanity
HM 3492 Humanity ; World (Buddhism)
HH 0203 Humanness
HM 3112 Humanness ; Archetypal (ICA)
HM 2092 Humanness ; Inventing (ICA)
MS 8375 Humans ; Deformed diseased
VD 2804 Humiliation
HH 2329 Humiliation exaltation ; Journeying transcendence
VC 0929 Humility
TP 0015 Humility
HH 0861 Humility (Christianity)
HM 3382 Humility ; Mansions (Christianity)
VP 5905 Humility ; Pride
VC 0931 Humour
HH 1031 Humour
HH 5343 Humour ; Human development
HM 3596 Humour ; Sense
HM 3634 Hunch
HM 2317 Hundred ; Maqamat (Sufism)
HM 0472 Hunger ; Spiritual
HM 0150 Hunger ; State (Buddhism)
HM 2112 Hungry-ghosts-hell awareness (Buddhism)
TP 0017 Hunting
VD 2806 Hurtfulness
MS 8336 Husbandry domestication ; Objects animal
MS 8225 Hybrid monsters
KD 2262 Hybrids ; Interwoven alternatives: resonance
HM 4533 Hybris (Christianity)
VC 0933 Hygiene
HH 0296 Hygiene ; Mental
HM 2337 Hyper-bliss-realm awareness (Buddhism)
HH 0208 Hyper-personalisation
HM 1479 Hyperaesthesia
HM 1273 Hypercognition
HH 1562 Hypergnostic meditation
HH 0929 Hypertrance (Psychism)
HM 3511 Hypnagogic consciousness (Psychism)
HM 3511 Hypnagogic imagery (Psychism)
HM 3688 Hypnapompic state (Psychism)
HH 0512 Hypnodrama
HM 3511 Hypnogogic state (Psychism)
HM 1971 Hypnoidal state
HM 3688 Hypnopompic consciousness (Psychism)
HM 3688 Hypnopompic imagery (Psychism)
HM 2133 Hypnosis
HM 1226 Hypnosis ; Deep state
HH 0962 Hypnosis ; Guided
HM 4595 Hypnosis ; Light state
HH 2134 Hypnosis ; Mass
HM 5856 Hypnosis ; Medium state
Hypnosis [Self...]
HH 0962 — Self
HM 5856 — Semi stage
HM 1226 — Somnambulistic
HH 0962 Hypnotherapy
HM 1400 Hypnofic ecstasy
HH 0072 Hypnotic imagogy
Hypnotic [state...]
HM 4401 — state transcendence ; Mono motivational
HM 2133 — states consciousness
HH 0962 — suggestion
VD 2808 Hypocrisy
HM 0514 Hypomania
HM 2020 Hypometabolic state ; Wakeful
HM 3010 Hysteria
HM 1400 Hysteria

HM 0574 Hysteria ; Mass
HM 7608 Hysterical amnesia

I

HM 2877 I am
HM 2937 I am this (Yoga)
HH 0004 I Ching (Taoism)
HH 0840 I ; Feeling
HH 0840 I ; Sense
HM 3225 I-Thou awareness
HM 3060 I-transcendent stage life
KC 0241 Iatrodisciplines
HM 0128 'Ibadat (Sufism)
HM 2424 Ibn Abi'l Khayr ; Stations consciousness (Sufism)
HM 2429 ICA Abject helplessness
HM 2209 ICA Abounding abasement
HM 3047 ICA Absorption nothingness
HM 2331 ICA Absurd existence
HM 2466 ICA Action irrelevant
HM 3080 ICA Actional existence
HM 3151 ICA Acute inadequacy
HM 2199 ICA Affirmation
HM 3105 ICA Agape appreciation
HM 2774 ICA Agape compassion
HM 2828 ICA Agape motivity
HM 3217 ICA Agape responsibility
HM 2889 ICA Agonizing prediction
HM 3097 ICA All-being-in-myself
HM 3088 ICA All-that-ever-was
HM 3083 ICA All-that's-yet-to-be
HM 3122 ICA Ancestral obligation
HM 3054 ICA Appropriated passion
HM 3112 ICA Archetypal humanness
HM 2316 ICA Authentic relation
HM 2073 ICA Awareness allegiance
HM 3007 ICA Awareness archaism
HM 2447 ICA Awareness depth
HM 2637 ICA Awareness externality
HM 2609 ICA Awareness futurity
HM 2286 ICA Awareness spiritual poverty
HM 2619 ICA Awful encounter
HM 2983 ICA Being
HM 3075 ICA Being history
HM 2154 ICA Being myself
HM 2379 ICA Besetting sin
HM 3049 ICA Beyond morality
HM 2941 ICA Beyond success
HM 3148 ICA Blissful seizure
HM 2872 ICA Bottomless centre
HM 2008 ICA Called accountability
HM 2273 ICA Certitude centre
HM 2757 ICA Classical story
HM 3179 ICA Committed teacher
HM 2023 ICA Common earth
HM 2338 ICA Communion saints
HM 2582 ICA Concerned judge
HM 3106 ICA Contentless transformation
HM 2373 ICA Contentless word
HM 2529 ICA Contentment centre
HM 2842 ICA Contextual world-view
HM 2456 ICA Contingent eternality
HM 2694 ICA Corporate duty
HM 2945 ICA Cosmic sanctions
HM 2894 ICA Creative existence
HM 2493 ICA Creative futility
HM 2303 ICA Cruciform exaltation
HM 2824 ICA Cryptic disclosure
HM 3167 ICA Cut-off unknownness
HM 2449 ICA Dangerous intrusion
HM 2009 ICA Daring embracement
HM 3009 ICA Decisional nothingness
HM 2591 ICA Defender deeps
HM 2796 ICA Definitive effectivity
HM 2131 ICA Definitive predestination
HM 2309 ICA Destinal accountability
HM 3027 ICA Destinal elector
HM 2682 ICA Determining history
HM 2927 ICA Diaphanous intuition
HM 2552 ICA Disengagement work
HM 2812 ICA Disinterested collegiality
HM 3190 ICA Divine captive
HM 2859 ICA Divine hosts
HM 3042 ICA Divine nothingness
HM 2143 ICA Doing mystery
HM 3099 ICA Dying death
HM 3091 ICA Dynamic selfhood
HM 3056 ICA Embodying charity
HM 2442 ICA Embodying equity
HM 2876 ICA Embodying peace
HM 3199 ICA Embodying service
HM 2437 ICA Endless life
HM 2588 ICA Engagement history
HM 3096 ICA Enveloped mystery
HM 3135 ICA Essential dubiety
HM 3095 ICA Eternal friends
HM 2541 ICA Eternal identification
HM 2517 ICA Eternal insecurity
HM 2786 ICA Eternal moment
HM 2253 ICA Eternal saviour
HM 2994 ICA Eternal void
HM 2661 ICA Ethical existence
HM 2814 ICA Ever-present brother
HM 2777 ICA Everlasting community
HM 2821 ICA Everlasting enemy

HM 2791 ICA Everlasting inescapability
HM 2437 ICA Everlastingness centre
HM 2991 ICA Every situation
HM 2951 ICA Exclusive contradiction
HM 2114 ICA Existential guidance
HM 2903 ICA Expectant descendant
HM 2471 ICA External relation
HM 3206 ICA Final accountability
HM 2958 ICA Final blessedness
HM 2674 ICA Final limits
HM 2119 ICA Final situation
HM 2316 ICA Freedom awareness
HM 2721 ICA Freedom decision
HM 2894 ICA Freedom inventiveness
HM 3206 ICA Freedom obligation
HM 3117 ICA Frightful possibility
HM 3016 ICA Futuric responsibility
HM 2432 ICA Global brotherhood
HM 2817 ICA Global guardianship
HM 2042 ICA Glorious mystery
HM 3143 ICA Good stewardship
HM 3130 ICA Guardian angel
HM 2680 ICA Hallowed honour
HM 2784 ICA Heavenly advocate
HM 2918 ICA Heavenly secret
HM 2216 ICA Heavenly sorrow
HM 2616 ICA Historical vocation
HM 3029 ICA Honouring mystery
HM 2720 ICA Horror sin
HM 2104 ICA Human contingency
HM 3188 ICA Human example
HM 2334 ICA Human transformation
HM 3140 ICA Imitation Christ
HM 3124 ICA Immutable friend
HM 2619 ICA Impacted mystery
HM 2607 ICA Impactful profundity
HM 2771 ICA Imploring succour
HM 2524 ICA Impossible possibility
HM 2480 ICA Incarnate Christ
HM 2394 ICA Incarnate living
HM 2333 ICA Inclusive collegiality
HM 2256 ICA Inclusive comprehension
HM 2223 ICA Individual fatefulness
HM 2782 ICA Individual rights
HM 3096 ICA Inescapable power
HM 2234 ICA Infinite passion
HM 2892 ICA Intentional conscience
HM 2369 ICA Intentional self-negation
HM 2851 ICA Interior discipline
HM 3245 ICA Invented history
HM 3138 ICA Inventing essence
HM 2092 ICA Inventing humanness
HM 3155 ICA Irreplaceable uniqueness
HM 2249 ICA Keeping conscience
HM 2434 ICA Land mystery
HM 2706 ICA Levitational submission
HM 3131 ICA Liberation possessions
HM 3240 ICA Liberation relationships
HM 3229 ICA Life being
HM 2506 ICA Life chastity
HM 2109 ICA Life contemplation
HM 3018 ICA Life doing
HM 2801 ICA Life knowing
HM 3234 ICA Life meditation
HM 2024 ICA Life obedience
HM 2299 ICA Life poverty
HM 2511 ICA Life prayer
HM 2808 ICA Living death
HM 3178 ICA Living word
HM 2910 ICA Loyal opposition
HM 3065 ICA Luminous change
HM 2269 ICA Manifold blessings
HM 2017 ICA Meaning creation
HM 2976 ICA Missional comradeship
HM 2343 ICA Missional engagement
HM 2721 ICA Moral ground
HM 2170 ICA Mountain care
HM 3004 ICA New religious modes
HM 2005 ICA Obedient son
HM 2598 ICA Objective awareness
HM 3105 ICA Original gratitude
HM 2773 ICA Original integrity
HM 2614 ICA Other world midst this world
HM 2463 ICA Painful acknowledgement
HM 2939 ICA Particular concern
HM 2076 ICA Passing awayness
HM 3128 ICA Passionate concern
HM 2547 ICA Passionate disinterest
HM 2484 ICA People God
HM 2717 ICA Perpetual becoming
HM 2839 ICA Perpetual revolutionary
HM 2139 ICA Persistent friend
HM 2370 ICA Personal absolution
HM 2595 ICA Personal epiphany
HM 2971 ICA Personal obligation
HM 2561 ICA Personal violation
HM 2206 ICA Presence Jesus Christ
HM 2550 ICA Primal sympathy
HM 2621 ICA Primal vocation
HM 3224 ICA Primordial ancestor
HM 2352 ICA Primordial colloquy
HM 2877 ICA Primordial sociality
HM 2186 ICA Primordial wonder
HM 2747 ICA Problemless living
HM 3015 ICA Problemlessness centre
HM 3172 ICA Promissorial offering

Immensity

HM 2846	ICA Prophetic sight
HM 2019	ICA Radiant guru
HM 2477	ICA Radical contingency
HM 2758	ICA Radical identification
HM 2273	ICA Radical illumination
HM 2587	ICA Radical incarnation
HM 3195	ICA Raw reality
HM 2441	ICA Realized vocation
HM 2386	ICA Recreated mystery
HM 3061	ICA Reforged transformation
HM 2978	ICA Relational situation
HM 2577	ICA Religious function
HM 2293	ICA Religious vocation
HM 3140	ICA Replication Christ
HM 2557	ICA Representational existence
HM 2532	ICA Representational sign
HM 2631	ICA Resurrectional existence
HM 2654	ICA Revered hero
HM 3071	ICA Revolutionary sign
HM 2993	ICA River consciousness
HM 2366	ICA Sacramental universe
HM 2445	ICA Sacramental universe
HM 2754	ICA Sacrificial friendship
HM 2641	ICA Sacrificial passion
HM 2413	ICA Saving mystery
HM 2921	ICA Scorching avatar
HM 3033	ICA Sea tranquillity
HM 3175	ICA Second birth
HM 3086	ICA Secondary integrity
HM 2234	ICA Seduced mystery
HM 3076	ICA Self determining
HM 3114	ICA Self programming
HM 2584	ICA Self transcendence
HM 2056	ICA Seminal illumination
HM 2323	ICA Serious sharing
HM 3101	ICA Sheer re-creation
HM 2111	ICA Singular adoration
HM 3217	ICA Singular mission
HM 2728	ICA Social failure
HM 2404	ICA Solitary being
HM 2904	ICA Soteriological existence
HM 2037	ICA Spiritual creativity
HM 2174	ICA Spiritual denial
HM 2165	ICA Splendid vices
HM 3025	ICA Spontaneous gratitude
HM 2101	ICA Stark givenness
HM 2718	ICA Struggle confess
HM 2258	ICA Submissive obedience
HM 3126	ICA Suffering servant
HM 2922	ICA Surrender inadequacy
HM 2346	ICA Surrender one's calling
HM 3103	ICA Symbol maker
HM 3176	ICA Symbolizing eternal context
HM 2363	ICA Temporal solidarity
HM 2460	ICA Terrifying acceptance
HM 2764	ICA Total exposure
HM 3058	ICA Transcendent guru
HM 3034	ICA Transcendent immanence
HM 2658	ICA Transcending hostility
HM 3161	ICA Transfigured man
HM 2862	ICA Transformed existence
HM 2386	ICA Transformed state
HM 2736	ICA Transparent engagement
HM 3067	ICA Transparent existence
HM 2184	ICA Transparent lucidity
HM 2828	ICA Transparent power
HM 2462	ICA Transparent presence
HM 3077	ICA Transparent selfhood
HM 2761	ICA Trust intuitions
HM 2570	ICA Ubiquitous otherness
HM 2388	ICA Ultimate awareness
HM 2030	ICA Ultimate reality
HM 3108	ICA Unexplainable thereness
HM 2356	ICA Unfailing prompter
HM 2644	ICA Universal Christ
HM 2734	ICA Universal compassion
HM 2774	ICA Universal concern
HM 2687	ICA Universal fate
HM 2418	ICA Universal father
HM 2458	ICA Universal prior
HM 3110	ICA Universal responsibility
HM 3015	ICA Unknowable peace
HM 3230	ICA Unlimited commitment
HM 3200	ICA Unmitigated death
HM 2400	ICA Unspeakable joy
HM 2529	ICA Unspeakable joy
HM 2611	ICA Unveiled being
HM 2625	ICA Utter awareness
HM 2982	ICA Vibrant powers
HM 2666	ICA Virgin birth
HM 2875	ICA Vital signs
HM 2831	ICA Wealth untold
HM 3052	ICA Wonder-filled fate
HM 3062	ICA Wonder world
HM 2236	ICA Word-bearing priest
HM 2451	ICA Worldly detachment
HH 4102	Icebox effect
MS 8705	Icon
HM 2087	Ida conscious energy (Yoga)
HM 5498	Idamatthita (Buddhism)
HH 5652	Iddhi (Pali)
HH 0893	Idealism
VC 0935	Idealism
HH 0893	Idealism ; Absolute
HH 0893	Idealism ; Ethical
HM 6554	Idealism ; Way (Esotericism)
MS 8415	Ideas
	Ideas [Delusional...]
HM 2795	— Delusional
HM 4504	— Delusional
HM 1466	— Dominant
HM 3527	Ideas ; Knowledge (Buddhism)
HM 1466	Ideas ; Over valued
HH 0540	Ideas ; Pressure
MP 1014	Identifiable context
HH 0358	Identification
HM 1430	Identification ; Active emotional
HM 2843	Identification (Brainwashing)
HH 0599	Identification ; Cosmic
HM 2541	Identification ; Eternal (ICA)
HM 3520	Identification God (Sufism)
HM 0946	Identification God (Sufism)
TP 0013	Identification others
HM 2758	Identification ; Radical (ICA)
HM 3944	Identification spiritual master (Sufism)
HM 2985	Identifying reality (Sufism)
HH 0875	Identity
	Identity [Assault...]
HM 3002	— Assault (Brainwashing)
HM 3260	— attributes their essence (Sufism)
HM 3232	— Awareness spiritual (Yoga)
HH 1929	Identity ; Cultural
VP 5013	Identity-Difference
HH 0875	Identity ; Ego
HM 2214	Identity God (Sufism)
HH 1002	Identity interchange
HH 0875	Identity ; Personal
HH 0471	Identity ; Self
HM 1907	Identity-strength
HM 4341	Identity ; Totemic
KD 2325	Identity transformation ; Game comprehension
HH 0571	Ideological commitment
HH 0757	Ideological conversion
HH 2187	Ideological development
	Ideological [re...]
HH 0996	— re-education
HM 2843	— rebirth (Brainwashing)
HH 0865	— reform (Brainwashing)
KC 0601	Ideology
HH 2187	Ideology
TC 1663	Idiosyncratics
VD 2810	Idleness
HM 5087	Idyllic love
HM 2248	Iesu
HH 9760	Ignatius ; Spiritual exercises Saint (Christianity)
VD 2812	Ignobility
VD 2814	Ignominiousness
VD 2816	Ignominy
VD 2818	Ignorance
HM 3196	Ignorance (Buddhism)
HM 3035	Ignorance dependent origination formula (Buddhism)
HM 3592	Ignorance ; Descending steps (Hinduism)
HM 4470	Ignorance ; Fetters (Buddhism)
HM 3145	Ignorance ; Flood (Buddhism)
VP 5475	Ignorance ; Knowledge
HM 0310	Ignorance (Leela)
HM 3493	Ignorance nature essential self (Buddhism)
HM 2695	Ignorance ; Non (Buddhism)
HM 3829	Ignorance ; Obsession (Buddhism)
HH 0486	Ignorance ; Recognition
HM 1013	Ignorance ; Snares (Buddhism)
HM 1375	Ignorance ; Tendency towards (Buddhism)
KD 2305	Ignorance ; Uncertainty function
HM 4663	Ikhlas (Sufism)
HM 3925	Ikia
HM 3925	Ikia healing
HM 6719	Ikram ; Dhul Jalal wal (Sufism)
HM 6719	Ikram ; Dhul Jalal wal (Sufism)
HM 2476	Ilham (Sufism)
HH 2119	Ill ; Cessation (Buddhism)
HM 1440	Ill health ; Delusions
	Ill Understanding reference
HM 6870	— truth (Buddhism)
HM 8163	— truth cessation (Buddhism)
HM 7290	— truth origin (Buddhism)
HM 7645	— truth path leading cessation (Buddhism)
HH 3873	Ill-wishing
VD 2820	Illegality
VP 5998	Illegality ; Legality
VD 2822	Illegitimate
VD 2824	Illiberalism
VD 2826	Illness
TP 0018	Illness ; Responding
HM 6001	Illness ; State
VD 2828	Illogic
HM 1359	Illogical thinking
HM 6208	Illuminated consciousness (Buddhism)
HM 6208	Illuminated mind (Buddhism)
HM 2612	Illuminating intelligence ; Path (Judaism)
HM 2035	Illuminating intelligence ; Path (Judaism)
	Illumination
HM 2291	— [Illumination]
HH 0804	— [Illumination]
VC 0937	— [Illumination]
HM 2029	— [Illumination]
HM 3263	Illumination ; Buddha (Buddhism)
HM 2215	Illumination (Buddhism)
TP 0036	Illumination ; Concealment (Darkening of the light)
HM 4337	Illumination due insight (Buddhism)
HM 0486	Illumination ; Harmonizing (Taoism)
HM 2273	Illumination ; Radical (ICA)
HM 2056	Illumination ; Seminal (ICA)
HM 6162	Illumination spirit (Sufism)
HM 3039	illuminations Jami (Sufism)
HH 6030	Illuminativa ; Via (Christianity)
HM 2973	Illumined effort (Astrology)
HH 0384	Illumined thinking
HH 0804	Illuminism
HM 2510	Illusion
HM 4664	Illusion ; Coriolis
VP 5519	Illusion-Disillusionment
HM 0351	Illusion ; Elevator
HM 6124	Illusion ; Extinction (Buddhism)
HH 1004	Illusion (Hinduism)
HM 0351	Illusion ; Oculoagravic
HM 0351	Illusion ; Oculogyral
HM 0767	Illusion ; Unwillingness let go (Korean)
	Illusions [Ambiguous...]
MM 2021	— Ambiguous visual
HM 7212	— Audiogravic
HM 7212	— Audiogyral
HM 8135	Illusions ; Pareidolic
HM 1222	Illusory body ; Yoga (Buddhism)
HM 0476	'Ilm al-basir (Sufism)
HM 0930	'Ilm al-yaqin (Sufism)
KC 0316	Image
TP 0022	Image
HM 2201	Image awareness (Tarot)
HM 3028	Image ; Betrothal initiation archetypal (Tarot)
HM 1766	Image ; Body
HM 3045	Image ; Charioteer archetypal (Tarot)
TC 1359	Image development
TC 1001	Image development
HM 2285	Image ; Female sovereignty archetypal (Tarot)
HM 2260	Image ; Hierophant archetypal (Tarot)
HM 2237	Image ; Magician archetypal (Tarot)
HH 0351	Image man
HM 2202	Image ; Old wise man archetypal (Tarot)
HM 2225	Image ; Pilgrim archetypal (Tarot)
HH 0019	Image ; Primordial
KC 0527	Image system
HM 2261	Image ; Wisdom archetypal (Tarot)
HH 0816	Imageless prayer (Christianity)
HH 0627	Imagery ; Guided
HM 3511	Imagery ; Hypnagogic (Psychism)
HM 3688	Imagery ; Hypnopompic (Psychism)
HH 0072	Imagery therapy ; Mental
TC 1323	Images
HM 0794	Images ; After
KC 0847	Images ; Artistic
HH 3020	Images ; Guiding
HH 5119	Imaginal psychology
HH 4312	Imaginary relationships
VC 0939	Imagination
	Imagination [Active...]
HM 0867	— Active
HM 1409	— Alchemical (Esotericism)
HM 2901	— (Anthroposophy)
HH 2076	Imagination ; Constructive
HM 0786	Imagination ; Creative
HH 3098	Imagination ; Existential (Existentialism)
HH 2498	Imagination ; Journeying transcendence isolation
HM 0043	Imagination ; Mythical
HM 1538	Imagination ; Poetic
HM 1538	Imagination ; Sacred
HH 4510	Imagination ; Social
HM 7154	Imagination ; Understanding way (Buddhism)
HM 3616	Imagination ; World (Sufism)
HM 2901	Imaginative consciousness (Anthroposophy)
HM 2011	Imaginative intelligence ; Path (Judaism)
VP 5535	Imaginativeness-Unimaginativeness
TC 1550	Imagineer
HH 4510	Imaging future ; Social
HH 4510	Imagining alternative futures
HH 4510	Imagining utopias
HH 0072	Imagogy ; Autogenic
HH 0072	Imagogy ; Hypnotic
VD 2830	Imbecility
	Imitation [Christ...]
HH 0810	— Christ (Christianity)
HM 3140	— Christ (ICA)
HH 0810	— (Christianity)
MP 1085	— contexts developing perspectives ; Perspective
VP 5023	Imitation ; Originality
HM 1995	Imitative musical inspiration
HM 3558	Immanence
HM 3034	Immanence ; Transcendent (ICA)
	Immaterial sphere
HM 0696	— concentration (Buddhism)
HM 2110	— First absorption (Buddhism)
HM 2051	— Fourth absorption (Buddhism)
HM 0282	— functional ; Indeterminate consciousness (Buddhism)
HM 4701	— Profitable consciousness (Buddhism)
HM 4982	— resultant ; Indeterminate consciousness (Buddhism)
HM 3043	— Second absorption (Buddhism)
HM 2027	— Third absorption (Buddhism)
HH 3198	Immaterial states meditation subjects (Buddhism)
HH 0893	Immaterialism
VP 5376	Immateriality ; Materiality
VD 2832	Immaturity
HM 0102	Immaturity ; Spiritual (Christianity)
TP 0004	Immaturity (Young shoot)
HM 2507	Immeasurables ; Four (Buddhism)
VC 0941	Immediacy
HM 0035	Immediacy ; Alert
HH 0223	Immediacy fulfilled living
HM 0035	Immediate boundary awareness
VC 0943	Immensity

Imminence

VC 0945	Imminence
VP 5151	Imminence ; Eventuation
VD 2834	Immodesty
VD 2836	Immorality
	Immortality
HH 0686	— [Immortality]
HH 0800	— [Immortality]
VC 0947	— [Immortality]
HM 5265	Immortality ; Celestial (Taoism)
VC 0949	Immunity
HH 3755	Immunity ; Super
VC 0951	Immutability
HM 3124	Immutable friend (ICA)
KC 0754	Impact analysis ; Cross
TC 1850	Impact ; Conference
VP 5283	Impact-Reaction
HM 2619	Impacted mystery (ICA)
HM 2607	Impactful profundity (ICA)
VD 2838	Impairment
VP 5691	Impairment ; Improvement
HM 9132	Impairment unity self
VC 0953	Impartiality
VD 2840	Impatience
VP 5861	Impatience ; Patience
HM 6704	Imperative hallucination
VD 2842	Imperfection
HM 9722	Imperfection insight (Buddhism)
VP 5677	Imperfection ; Perfection
HM 9021	Impermanence (Buddhism)
HH 0241	Impersonality (Buddhism)
VD 2844	Impertinence
HH 5543	Imperturbable ; Formation (Buddhism)
VD 2846	Imperviousness
VD 2848	Impetuousity
VD 2850	Impiety
VP 6028	Impiety ; Piety
VD 2852	Implausibility
VC 0955	Implementation
KP 3009	Implementation transformation process
KD 2230	Implicate order ; Wholeness
KD 2353	Implications action formulation ; Alternation:
KD 2350	Implications agreement consensus ; Alternation:
HH 0615	Implications human evolution ; Technological
HM 2771	Imploring succour (ICA)
HH 1107	Implosive therapy
VD 2854	Impoliteness
VC 0957	Importance
TP 0028	Importance
KP 3011	Importance form ; Empowerment
VD 3638	Importance ; Self
VP 5672	Importance-Unimportance
VD 2856	Impossibility
VP 5509	Impossibility ; Possibility
HM 2524	Impossible possibility (ICA)
VD 2858	Impotence
VP 5157	Impotence ; Power
VD 2860	Impoverishment
VD 2862	Impracticality
VD 2864	Imprecision
VD 2866	Impressionability
VC 0959	Impressivensss
HM 4539	Imprisonment ; Psychological
VD 2868	Improbity
VP 5974	Improbity ; Probity
VD 2870	Impropriety
VC 0961	Improvement
HH 0357	Improvement ; Human genetic
VP 5691	Improvement-Impairment
HM 8290	Improvement ; Understanding skill (Buddhism)
VD 2872	Improvidence
VD 2874	Imprudence
VD 2876	Impudence
HM 7268	Impulsion (Buddhism)
VD 2878	Impulsiveness
VD 2880	Impurity
TC 1753	In-conference computer conferencing
TC 1616	In-conference computer contact system
HM 3484	Inabat (Sufism)
VD 2882	Inability
HM 0298	Inability achieve concentration (Yoga)
HM 0427	Inability compare (Buddhism)
HM 1746	Inability demonstrate (Buddhism)
VD 2884	Inaccessibility
VD 2886	Inaccuracy
VP 5705	Inaction ; Action
TP 0005	Inaction ; Calculated (Getting wet)
TP 0052	Inaction (Mountain)
HM 2620	Inactive awareness
VD 2888	Inactivity
MP 1186	Inactivity ; Appropriate configuration perspective
MP 1069	Inactivity ; Common external context
MP 1188	Inactivity ; Occupiable sites perspective
MP 1150	Inactivity ; Provision temporary perspective
VD 2890	Inadequacy
HM 3151	Inadequacy ; Acute (ICA)
KP 3001	Inadequacy formulations
HM 2922	Inadequacy ; Surrender (ICA)
KP 3018	Inadequate transformation attempts
VD 2892	Inadvertence
VD 2894	Inadvisability
VD 2896	Inanity
VD 2898	Inapplicability
HH 0984	Inappropriate patterns development
VD 2900	Inappropriateness
VP 5665	Inappropriateness ; Appropriateness
VD 2902	Inaptitude
VD 2904	Inarticulation
VD 2906	Inattention
VP 5530	Inattention ; Attention
HM 3348	Inattentive perception (Buddhism)
VP 5448	Inaudibility ; Audibility
HM 7098	Inbisat (Sufism)
HH 1257	Incantation
VD 2908	Incapability
VD 2910	Incapacity
HM 2480	Incarnate Christ (ICA)
HM 2394	Incarnate living (ICA)
HM 2587	Incarnation ; Radical (ICA)
HM 1263	Inception energy ; Physical (Buddhism)
HH 1308	Incest
HM 4110	Incipient union ; Mansions (Christianity)
VD 2912	Incivility
VD 2914	Inclemency
HM 3432	Inclination towards view (Buddhism)
VC 0963	Inclusion
VP 5076	Inclusion-Exclusion
HM 2333	Inclusive collegiality (ICA)
HM 2256	Inclusive comprehension (ICA)
VD 2916	Incoherence
VD 2918	Incompatibility
VD 2920	Incompetence
HH 0322	Incompetence ; Myth
VD 2922	Incompleteness
VP 5054	Incompleteness ; Completeness
VD 2924	Incomprehensibility
VD 2926	Incongruity
VD 2928	Inconsequence
VD 2930	Inconsiderateness
VD 2932	Inconsistency
VD 2934	Inconstancy
VD 2936	Incontinence
VD 2938	Inconvenience
MS 8415	Incorporeal formless symbols
VD 2940	Incorrectness
VD 2942	Incorrigibility
VC 0965	Increase
TP 0042	Increase
VP 5038	Increase-Decrease
HM 8290	Increase ; Understanding skill (Buddhism)
VD 2944	Incredulity
HM 5265	Incubation spiritual embryo (Taoism)
VP 5528	Incuriosity ; Curiosity
VD 2946	Incuriosity
HM 3413	Incurring guilt ; Not (Buddhism)
VD 2948	Indecency
HM 0870	Indecision (Buddhism)
HM 3074	Indecisive wavering (Buddhism)
VD 2950	Indecisiveness
HM 3745	Indecisiveness (Buddhism)
VD 2952	Indecorum
VC 0967	Indefatigableness
VD 2954	Indefensibility
VD 2956	Indelicacy
VC 0969	Independence
HH 0250	Independence
HM 6243	Independent consciousness (Buddhism)
MP 1001	Independent domains
HM 6243	Independent mind (Buddhism)
HH 0632	Independent reflection
HM 3869	Independent vehicle ; Consciousness (Yoga)
HM 3072	Indestructible Mahayana ; Emptiness (Buddhism)
KC 0190	Indeterminacy principle
MP 1025	Indeterminacy ; Relationship
	Indeterminate consciousness
HM 4761	— fine-material sphere - functional (Buddhism)
HM 0594	— fine-material sphere - resultant (Buddhism)
HM 4761	— form-realm - inoperative (Buddhism)
HM 0594	— form-realm - resultant (Buddhism)
HM 0282	— formless realm - inoperative (Buddhism)
HM 4982	— formless realm - resultant (Buddhism)
HM 0822	— immaterial sphere - functional (Buddhism)
HM 4982	— immaterial sphere - resultant (Buddhism)
	Indeterminate consciousness sense
HM 3852	— sphere - functional (Buddhism)
HM 3852	— sphere - inoperative (Buddhism)
HM 5721	— sphere - resultant (Buddhism)
HM 5129	Indeterminate consciousness supramundane plane - resultant (Buddhism)
HM 5129	Indeterminate consciousness transcendental plane - resultant (Buddhism)
HH 5101	Index ; Human development (United Nations)
KC 0303	Indexing ; Coordinate
HM 2446	Indian approaches consciousness (Hinduism)
HH 0624	Indicators development
VD 2958	Indifference
HM 7769	Indifference (Buddhism)
	Indifference [complexes...]
HM 0424	— complexes ; Knowledge (Buddhism)
HM 5008	— Concentration accompanied (Buddhism)
HM 8021	— Concentration accompanied (Buddhism)
HM 1405	— Contemplative
HM 8566	Indifference due insight (Buddhism)
HM 3803	Indifference ; Faculty (Buddhism)
HM 5776	Indifference ; Holy (Christianity)
HM 7769	Indifferent feeling (Buddhism)
VD 2960	Indignation
VD 2962	Indirection
VD 2964	Indiscretion
VD 2966	Indiscrimination
VP 5492	Indiscrimination ; Discrimination
VD 2968	Indisposition
KC 0373	Individual
MM 2020	Individual ; Changing moods
HH 1543	Individual development
HM 2681	Individual distinctions ; Beyond (Sufism)
HM 2223	Individual fatefulness (ICA)
HM 4497	Individual ; Fund (Christianity)
HM 2195	Individual knowledge character (Buddhism)
HM 2782	Individual rights (ICA)
HM 2692	Individual tendencies consciousness (Hinduism)
HH 1162	Individualism
HH 0776	Individuality
VC 0971	Individuality
HM 0344	Individuality ; Experience (Systematics)
MP 1012	Individuality multiplicity
HM 2239	Individualized awareness (Astrology)
HH 0385	Individuation
HM 3614	Individuation essence ; First (Sufism)
HM 1665	Individuative-reflective faith
HM 2113	Indivisibility essence (Sufism)
VD 2970	Indocility
VD 2972	Indolence
HM 2075	Indriya phenomena (Pali)
VC 0973	Indubitableness
	Induced attentiveness
HM 3378	— (Buddhism)
HM 3378	— Other (Buddhism)
HM 3378	— Self (Buddhism)
	Induced experience
HM 3213	— Dance
HH 0075	— Drug
HM 3559	— Light
HM 2464	— Mantra (Yoga)
HM 0062	Induced modes perception ; Sport
HM 3221	Induced religious experience
HM 1964	Induced trance ; Self (Psychism)
KC 0014	Induction
HM 0461	Induction device ; Altered states consciousness (ASCID)
HH 0025	Induction technique
HM 2583	Inductive intelligence unity (Judaism)
VD 2974	Indulgence
VD 3640	Indulgence ; Self
KC 0529	Industrial design
KC 0426	Industrial dynamics
KC 0807	Industrial engineering
KC 0859	Industrial integration
VC 0975	Industriousness
HH 0814	Indwelling ; Divine
HM 3589	Ineffability
VD 2976	Ineffectiveness
VD 2978	Ineffectuality
VD 2980	Inefficiency
VD 2982	Inelegance
VP 5589	Inelegance ; Elegance
VD 2984	Ineptitude
VD 2986	Inequality
VP 5030	Inequality ; Equality
VD 2988	Inequity
VD 2990	Inertia
TC 1615	Inertia ; Conferencing
HM 0906	Inertia ; Mental (Yoga)
HM 3421	Inertia ; Psychic (Physical sciences)
HM 3374	Inertness ; Non (Buddhism)
HM 2791	Inescapability ; Everlasting (ICA)
HM 3096	Inescapable power (ICA)
VP 5857	Inexcitability ; Excitement
VP 5539	Inexpectation ; Expectation
VD 2992	Inexpedience
VP 5670	Inexpedience ; Expedience
VD 2994	Inexperience
TP 0004	Inexperience (Young shoot)
VC 0977	Infallibility
HH 5221	Infallibility
VD 2996	Infamy
HH 3498	Infant bonding ; Parent
HH 0630	Infants ; Psychological development
HM 1683	Infatuated ; Being (Buddhism)
HM 3250	Infatuated (Buddhism)
HM 1688	Infatuated ; Not (Buddhism)
HM 3183	Infatuated ; State not being (Buddhism)
HM 5991	Infatuation
VD 2998	Infatuation
HM 4363	Infatuation ; Feeling (Buddhism)
HM 4081	Infatuation ; State not feeling (Buddhism)
VD 3000	Infelicity
HM 5562	Inferior concentration (Buddhism)
TP 0044	Inferior ; Infiltration
VD 3002	Inferiority
VP 5036	Inferiority ; Superiority
MS 8907	Infernal beings
VD 3004	Infertility
VD 3006	Infidelity
TP 0044	Infiltration inferior
	Infinite concentration
HM 0496	— (Buddhism)
HM 5214	— infinite object (Buddhism)
HM 4653	— limited object (Buddhism)
HM 3043	Infinite consciousness ; Sphere (Buddhism)
	Infinite object
HM 5214	— Infinite concentration (Buddhism)
HM 5547	— Limited concentration (Buddhism)
HM 6681	— Understanding having (Buddhism)
HM 2234	Infinite passion (ICA)
HM 2110	Infinite space ; Sphere (Buddhism)
KC 0259	Infinity
HM 0934	Infinity ; Awareness (Judaism)
HM 1288	Inflation

Integration

VD 3008	Inflexibility	
HM 6521	Inflexible mind (Buddhism)	
VC 0979	Influence	
TP 0031	Influence	
VP 5172	Influence-Influencelessness	
HM 3089	Influence ; Path intelligence mediating (Judaism)	
VP 5172	Influencelessness ; Influence	
HM 2045	Influences ; Path intelligence (Judaism)	
MP 1111	Informal context ; Blended integration formal structure	
MP 1041	Informal context formal processes	
MP 1088	Informal local perspective interface zones	
VP 5646	Informality ; Formality	
KC 0430	Information	
HM 2262	Information ; Bio (ESP)	
TC 1750	Information seeking participants	
TC 1900	Information ; Sharing	
	Information system	
KC 0526	— [Information system]	
KC 0408	— design	
KC 0518	— Distributed	
KC 0308	— Integrated	
KC 0608	— integrity	
KC 0936	— Real time	
KC 0518	Information systems ; System	
VC 1909	Informed ; Well	
MP 1214	Infrastructure ; Integrative	
VP 5135	Infrequency ; Frequency	
HM 2476	Infused awareness ; Sufi (Sufism)	
HM 2710	Infused contemplation	
VC 0981	Ingenuity	
VC 0983	Ingenuousness	
VD 3010	Ingloriousness	
VD 3012	Ingratitude	
VP 5949	Ingratitude ; Gratitude	
HM 3743	Inhalants ; Modes awareness associated use	
VD 3014	Inharmonious	
HM 1596	Inherent existence (Buddhism)	
HM 2448	Inherent existence non things ; Emptiness (Buddhism)	
VD 3016	Inhibition	
VD 3018	Inhospitality	
VP 5925	Inhospitality ; Hospitality	
VD 3020	Inhumanity	
VD 3022	Iniquity	
TP 0003	Initial difficulty	
MP 1215	Initial level formation	
HH 0230	Initiation	
HM 3028	Initiation archetypal image ; Betrothal (Tarot)	
HM 1267	Initiation baptism (Esotericism)	
HM 1337	Initiation birth (Esotericism)	
HH 6211	Initiation (Christianity)	
HM 0210	Initiation crucifixion (Esotericism)	
HM 0322	Initiation decision (Esotericism)	
HM 0210	Initiation detachment (Esotericism)	
HH 0659	Initiation (Freemasonry)	
HM 1417	Initiation ; Group spiritual (Esotericism)	
HH 1478	Initiation ; Isolation	
HH 3432	Initiation (Magic)	
HH 4502	Initiation ; Peer group	
	Initiation [refusal...]	
HM 1020	— refusal (Esotericism)	
HM 0210	— renunciation (Esotericism)	
HM 1153	— resurrection (Esotericism)	
HM 0181	— revelation (Esotericism)	
	Initiation [secret...]	
HH 9807	— secret societies	
HH 3390	— Spiritual (Esotericism)	
HH 3465	— spiritual rebirth (Yoga)	
HM 0428	Initiation transfiguration (Esotericism)	
HM 1258	Initiation transition (Esotericism)	
HH 4502	Initiation ; Work group	
VC 0985	Initiative	
TP 0019	Initiative	
MM 2030	Initiative ; Gaining/losing	
KC 0195	Initiatives ; Open systems	
TC 1491	Initiator participant type	
VD 3024	Injury	
TP 0036	Injury (Darkening of the light)	
VD 3026	Injustice	
VP 5976	Injustice ; Justice	
TP 0061	Inner confidence	
HH 0708	Inner contradictions	
HM 3863	Inner fire ; Yoga (Buddhism)	
HM 3575	Inner peace	
HM 6287	Inner space ; Plane (Leela)	
TP 0061	Inner truth	
HM 1317	Inner voices	
HH 4066	Innermost benevolent desires soul	
VC 0987	Innocence	
TP 0025	Innocence	
HM 3412	Innocence (Christianity)	
VP 5983	Innocence-Guilt	
HM 0888	Innocent cognition	
HM 6633	Innocent mind ; Recognition (Taoism)	
KC 0423	Innovation	
	Innovation Social	
KC 0167	— [Innovation ; Social]	
HH 8963	— [Innovation ; Social]	
HH 0432	— [Innovation ; Social]	
HH 0812	Innovative learning	
VD 3028	Inobservance	
VC 0988	Inoffensiveness	
	Inoperative Indeterminate consciousness	
HM 4761	— form realm (Buddhism)	
HM 0282	— formless realm (Buddhism)	
HM 3852	— sense sphere (Buddhism)	
KC 0665	Input/Output analysis	

MP 1184	Input processing ; Appropriate configuration	
MP 1199	Input processing context external insight ; Exposure	
VD 3030	Inquietude	
HM 7434	Inquiry ; Concentration due (Buddhism)	
HH 0508	Inquiry meditation ; Free self	
VD 3032	Inquisition	
HM 2920	Insan kamil ; Al (Sufism)	
VD 3034	Insanity	
VP 5472	Insanity ; Sanity	
HM 2920	Insanu'l-kamil (Sufism)	
HM 3677	Insanu'l-kamil (Sufism)	
MS 8103	Insects	
VD 3036	Insecurity	
HM 2517	Insecurity ; Eternal (ICA)	
VD 3038	Insensibility	
VP 5422	Insensibility ; Sensation	
VD 3040	Insensitivity	
HM 5479	Insertion ; Thought (Psychism)	
HH 3215	Inside ; Journeying transcendence outside	
VC 0991	Insight	
HM 1364	Insight	
TP 0036	Insight ; Adulteration (Darkening of the light)	
HM 6144	Insight ; Attachment due (Buddhism)	
	Insight [blindspots...]	
MP 1162	— blindspots ; Organization integrative superstructure minimize	
HM 5088	— Bliss due (Buddhism)	
HM 6502	— (Buddhism)	
	Insight [capturing...]	
MP 1175	— capturing non-linear extensions structures	
HM 9722	— Corruption (Buddhism)	
HH 0680	— Cultivation experiential (Buddhism)	
	Insight [Desire...]	
HM 6144	— Desire due (Buddhism)	
MP 1161	— domain ; Structure enfolded	
MP 1252	— Domains	
	Insight [Equanimity...]	
HM 8566	— Equanimity due (Buddhism)	
HM 6544	— Establishment due (Buddhism)	
HM 5986	— Exertion due (Buddhism)	
HM 1632	— Experiential (Buddhism)	
MP 1199	— Exposure input processing context external	
	Insight [Illumination...]	
HM 4337	— Illumination due (Buddhism)	
HM 9722	— Imperfection (Buddhism)	
HM 8566	— Indifference due (Buddhism)	
HM 1344	— Intuitional (Buddhism)	
HM 5221	Insight ; Knowledge due (Buddhism)	
	Insight [Occupiable...]	
MP 1180	— Occupiable sites exposed external	
MP 1107	— Organization structures enhance receptivity external	
MP 1159	— Organization structures permit two sources external	
MP 1128	— Orientation domains receive external	
MP 1105	— Orientation structures enhance receptivity external	
	Insight [Paths...]	
HM 3194	— Paths (Buddhism)	
HM 5088	— Pleasure due (Buddhism)	
HM 2623	— preparations states ; Special (Buddhism)	
HM 6544	— Presentation due (Buddhism)	
HH 0370	— provoking tales	
	Insight [Rapturous...]	
HM 4886	— Rapturous happiness due (Buddhism)	
MP 1138	— Receptivity emerging external	
HM 1593	— Religious experience meditational	
HM 7115	— Repose due (Buddhism)	
HM 5432	— Resolution faith due (Buddhism)	
HM 2207	Insight ; States special (Buddhism)	
MP 1181	Insight structures focal point ; Maintenance source direct	
HM 7115	Insight ; Tranquillity due (Buddhism)	
HM 2084	Insight ; Transrational (Zen)	
HM 5986	Insight ; Uplift due (Buddhism)	
MP 1135	Insight varying levels exposure it ; Enhancing	
MP 1223	Insight ; Zones intermediate	
MP 1238	Insights ; Filtered	
KD 2207	Insights: geometry connectivity ; Communicable	
HM 3630	Insights ; Great (Buddhism)	
HM 3630	Insights ; Principal (Buddhism)	
VD 3042	Insignificance	
VD 3044	Insincerity	
VD 3046	Insinuation	
VD 3048	Insobriety	
VD 3050	Insolence	
VD 3052	Insouciance	
	Inspiration	
HM 3125	— [Inspiration]	
VC 0993	— [Inspiration]	
HH 2139	— [Inspiration]	
HM 6006	Inspiration ; Avataric level musical	
HM 0203	Inspiration ; Egoic musical	
HM 1995	Inspiration ; Imitative musical	
HM 4805	Inspiration ; Musical	
HM 1094	Inspiration ; Psychic (Psychism)	
HM 1094	Inspirational art (Psychism)	
HM 1094	Inspirational Speaking (Psychism)	
HM 1094	Inspirational thought (Psychism)	
HM 1094	Inspirational writing (Psychism)	
HM 1094	Inspirational writing (Psychism)	
HM 3125	Inspired consciousness	
HM 1094	Inspired medium (Psychism)	
VD 3054	Instability	
VP 5112	Instantaneousness ; Perpetuity	
HM 3517	Instinct	
HM 2559	Instinct ; Herd	
HH 0750	Instinct ; Life	
TP 0025	Instinctive goodness	

KC 0597	Institutional integration ; International	
HH 1407	Instructing dying	
HH 0522	Instruction	
TC 1733	Instrumental conferencing	
HH 6131	Instrumental enrichment	
MS 8715	Instruments ; Medical	
MS 8316	Instruments ; Musical	
VD 3056	Insubordination	
VD 3058	Insubstantiality	
VD 3060	Insufficiency	
HM 2574	Insufficiency (Buddhism)	
VP 5660	Insufficiency ; Oversufficiency	
VD 3062	Insularity	
TC 1738	Insulation ; Conference	
HH 0975	Insulin coma	
VD 3064	Insult	
VP 5425	Intangibility ; Tangibility	
HM 2152	Integral consciousness	
KC 0551	Integral life	
HH 0579	Integral meditation	
HH 7123	Integral psychoanalysis	
HH 0037	Integral yoga (Yoga)	
KC 0189	Integralism	
VC 0995	Integrality	
KC 0180	Integrated	
KC 0890	Integrated	
KC 0345	Integrated automatic systems	
KC 0600	Integrated circuit	
MP 1080	Integrated contexts perspective dynamics	
	Integrated [education...]	
KC 0976	— education	
KC 0690	— education ; Racially ethnically	
KC 0782	— electronics	
KC 0308	Integrated information system	
KC 0711	Integrated methods science teaching	
KC 0600	Integrated microcircuit	
KC 0666	Integrated planning	
	Integrated science	
KC 0811	— [Integrated science]	
KC 0961	— curricula	
KC 0711	— education	
KC 0961	— Teaching concepts	
HH 0960	Integrated type	
KC 0755	Integrated world	
MP 1144	Integrating coordinated exposure irrationality structures	
MP 1040	Integrating historical dimension	
MP 1039	Integrating new dimension	
MP 1125	Integrating points perspective common domains	
HH 0204	Integrating shadow	
MP 1133	Integrating transition pathways levels structure	
	Integration	
HH 0786	— [Integration]	
KC 0992	— [Integration]	
KC 0891	— [Integration]	
KC 0133	Integration ; Adaptive cultural	
	Integration [Communicative...]	
KC 0140	— Communicative social	
KC 0686	— Conceptual	
KC 0380	— Configurational cultural	
KC 0360	— Connective cultural	
MP 1246	— context	
HH 0808	— Cosmic	
KC 0243	— Cultural	
KC 0976	— Curriculum	
KC 0976	Integration ; Educational	
HH 0832	Integration ; Existential	
	Integration [formal...]	
MP 1111	— formal structure informal context ; Blended	
KC 0133	— Functional cultural	
KC 0620	— Functional social	
KC 0859	Integration ; Industrial	
	Integration International	
KC 0397	— [Integration ; International]	
KC 0497	— attitudinal	
KC 0792	— economic	
KC 0597	— institutional	
KC 0697	— policy	
KC 0287	— political	
KC 0577	— social	
KC 0686	Integration knowledge	
	Integration [languages...]	
KC 0778	— languages	
KC 0131	— laws	
KC 0301	— level ; Theoretical	
KC 0460	— Logical cultural	
KC 0986	Integration ; Mathematical	
MP 1118	Integration non-linearity integrative superstructure	
KC 0520	Integration ; Normative social	
	Integration personality	
HH 0561	— [Integration personality]	
HH 0157	— Final	
HH 0157	— Total	
MP 1083	Integration perspective acquisition perspective maintenance dynamics	
HH 0823	Integration ; Psychic	
	Integration [Racial...]	
KC 0992	— Racial ethnic	
KC 0587	— Regional international	
KC 0143	— Regulative cultural	
MP 1178	— rejected perspective products ; Re	
	Integration [Spiritual...]	
HH 0107	— Spiritual	
HH 0082	— Structural	
KC 0145	— Stylistic cultural	
MP 1220	— superstructure	
MP 1249	— Symbols	

Integration

KC 0380 Integration ; Thematic cultural
MP 1115 Integration unstructured internal domains ; Functional
KC 0248 Integration ; Vertical
KC 0651 Integration ; Voluntary economic
KC 0545 Integrational analysis
KC 0367 Integrative
KC 0290 Integrative education
TC 1005 Integrative failure
MP 1214 Integrative infrastructure
KC 0841 Integrative levels
TC 1062 Integrative skills
Integrative superstructure
MP 1117 — Containment
MP 1118 — Integration non linearity
MP 1162 — minimize insight blindspots ; Organization
MP 1209 — Organization
MP 1231 — Overview sites
MP 1116 — Patterning
KC 0106 Integrative thinking
KC 0634 Integrism
Integrity
KC 0211 — [Integrity]
VC 0997 — [Integrity]
TP 0025 — [Integrity]
HH 0234 — [Integrity]
VT 8065 Integrity*complex
HM 4521 Integrity consciousness ; Anomalies
KC 0608 Integrity ; Information system
MP 1227 Integrity ; Inter level
HM 2773 Integrity ; Original (ICA)
HH 0129 Integrity ; Personal
HM 3086 Integrity ; Secondary (ICA)
HM 2292 Intellect ; Appropriate (Buddhism)
KC 0206 Intellect ; Augmentation
HM 3623 Intellect ; Negative (Leela)
HM 1789 Intellect ; Positive (Leela)
KC 0979 Intellect ; Unity
HM 3455 Intellectual condition (Hinduism)
KC 0555 Intellectual cooperation
VT 8045 Intellectual faculties*complex
HH 0146 Intellectual growth development
TC 1971 Intellectual momentum
VC 0999 Intellectuality
VC 1001 Intelligence
HH 0431 Intelligence
HH 2189 Intelligence ; Artificial
HM 2984 Intelligence ; Awareness self spiritual (Yoga)
HH 0662 Intelligence ; Bodily kinesthetic
HM 2105 Intelligence conciliation ; Path (Judaism)
HH 0679 Intelligence ; Creative
HH 6131 Intelligence enhancement
HM 2045 Intelligence influences ; Path (Judaism)
Intelligence [Leadership...]
HM 5451 — Leadership
HM 2107 — light ; Path (Judaism)
HH 0843 — Linguistic
HH 1456 — Logical mathematical
HM 3089 Intelligence mediating influence ; Path (Judaism)
HH 0961 Intelligence ; Musical
Intelligence Path
HM 3243 — absolute (Judaism)
HM 2218 — active (Judaism)
HM 2095 — administrative (Judaism)
HM 3013 — admirable (Judaism)
HM 3102 — cohesive (Judaism)
HM 3046 — collective (Judaism)
HM 2117 — constituting (Judaism)
HM 3149 — corporeal (Judaism)
HM 2569 — disposing (Judaism)
HM 2593 — eternal (Judaism)
HM 2218 — exciting (Judaism)
HM 2081 — faithful (Judaism)
HM 2523 — fiery (Judaism)
HM 2444 — hidden (Judaism)
HM 3013 — hidden (Judaism)
HM 2612 — illuminating (Judaism)
HM 2035 — illuminating (Judaism)
HM 2011 — imaginative (Judaism)
HM 3102 — measuring (Judaism)
HM 2071 — natural (Judaism)
HM 2444 — occult (Judaism)
HM 3243 — perfect (Judaism)
HM 2643 — perpetual (Judaism)
HM 2943 — purified (Judaism)
HM 3181 — radical (Judaism)
HM 3102 — receiving (Judaism)
HM 2047 — renewing (Judaism)
HM 2047 — renovating (Judaism)
HM 3426 — resplendent (Judaism)
HM 3567 — sanctifying (Judaism)
HM 2533 — stable (Judaism)
HM 2107 — transparent (Judaism)
HM 2593 — triumphant (Judaism)
HM 2583 — uniting (Judaism)
HM 2788 Intelligence ; Pre conscious pure
HH 1579 Intelligence revolution
HM 2105 Intelligence reward ; Path (Judaism)
Intelligence [secret...]
HM 2021 — secret ; Path (Judaism)
HH 1579 — Social
HH 1075 — Spatial
HM 2021 — spiritual action ; Path (Judaism)
HM 2503 Intelligence temptation ; Path (Judaism)
Intelligence [Unappreciated...]
TP 0036 — Unappreciated (Darkening of the light)
VP 5467 — Unintelligence
Intelligence [unity...] cont'd
HM 2583 — unity ; Inductive (Judaism)
HM 1997 Intelligence ; Way active (Esotericism)
HM 2629 Intelligence will ; Path (Judaism)
HH 0214 Intelligences ; Multiple
HH 1274 Intelligences ; Personal
VC 1003 Intelligibility
VP 5548 Intelligibility-Unintelligibility
HM 5532 Intelligible intuition (Buddhism)
VP 5992 Intemperance ; Temperance
VD 3066 Intemperence
HM 1412 Intended ; Being (Buddhism)
HM 0762 Intense longing divine unity (Sufism)
VC 1005 Intensity
MP 1010 Intensity ; Access
MP 1059 Intensity communication pathways ; Low
HM 0543 Intensity ; First order perceptions
MP 1055 Intensity relationships ; Protected low
TC 1588 Intent ; Conference
HM 3064 Intent ; Mastery
VC 1007 Intention
HM 2589 Intention (Buddhism)
HM 0705 Intention (Buddhism)
HM 3100 Intention ; Pure (Hinduism)
HM 4901 Intention (Sufism)
HM 2892 Intentional conscience (ICA)
HM 2369 Intentional self-negation (ICA)
HM 2469 Intentionality
TC 1920 Intentions ; Conferencing
VC 1009 Intentness
TC 1866 Inter-cultural communication teams
Inter [dependency...]
HH 3321 — dependency conscious existence phenomena ; Awareness (Buddhism)
MP 1198 — domain contexts perspective adjuncts
MP 1131 — domain dynamics ; Organization
TC 1602 Inter-group therapy
Inter level connections
MP 1216 — [Inter-level connections]
MP 1213 — Distribution secondary
MP 1212 — transitions boundary orientation ; Primary
MP 1227 Inter-level integrity
MP 1226 Inter-level zone
HH 0769 Inter-personality
MP 1005 Inter relationships ; Network
HH 2434 Inter-religious dialogue
MP 1185 Interaction ; Appropriate configuration perspective
MP 1074 Interaction coherent irrational perspectives
MP 1187 Interaction complementary perspectives ; Appropriate configuration
VT 8074 Interaction*complex
MP 1148 Interaction constraints ; Perspective
MP 1151 Interaction contexts ; Small scale perspective
KP 3004 Interaction ; Developmental
MP 1061 Interaction domains ; Bounded common small scale
MP 1075 Interaction ; Extended pattern nuclear
MM 2017 Interaction ; Interpersonal
KC 0127 Interaction ; Level
MP 1072 Interaction opportunities transposed concrete level ; Competitive
KC 0127 Interaction ; Order
Interaction preferences
TC 1640 — Non verbal
TC 1845 — Participant
TC 1378 — Verbal
KD 2337 Interaction ; Self limiting patterns
HM 1210 Interactional synchrony
HH 0740 Interactive discipline
MP 1034 Interchange
MP 1032 Interchange axis ; Selective
MP 1046 Interchange environment ; Diversified
HH 1002 Interchange ; Identity
TC 1650 Interchange panel
MP 1019 Interchange ; Web selective
KC 0165 Interchangeability
MP 1108 Interconnected structures
MM 2034 Intercourse ; Sexual
TC 1798 Intercultural consultants
TC 1798 Intercultural interpreter
KC 0280 Interdisciplinarity
TC 1208 Interdisciplinarity
KC 0961 Interdisciplinary science curricula
VC 1011 Interest
TC 1593 Interest groups ; Special
VC 1569 Interest ; Self
MP 1140 Interface communication pathway ; Relative isolation structural
MP 1160 Interface structures external environment ; Hospitable
MP 1088 Interface zones ; Informal local perspective
MP 1100 Interfaces ; Arrangement structures engender fruitful
MP 1124 Interfaces ; Ensuring function common domain
MP 1244 Interfaces ; Flexible
MP 1119 Interfaces ; Partially contained
MP 1166 Interfaces structure external environment ; Overview domains
MP 1240 Interfaces ; Tolerance level
MP 1052 Interfacing vehicles communication networks unmediated relationships
VD 3068 Interference
HM 3225 Interhuman awareness
HH 1409 Interior castle Saint Teresa (Christianity)
HM 2851 Interior discipline (ICA)
HH 3008 Interior life (Christianity)
VP 5224 Interiority ; Exteriority
HH 1432 Interiorization
KC 0258 Interlock
KD 2120 Interlocking cycles ; Omnitriangulation:
HM 0101 Intermediary state ; Yoga (Buddhism)
MP 1223 Intermediate insight ; Zones
MP 1242 Intermediate position
MP 1006 Intermediate scale organization
HM 2052 Interminable-hell awareness (Buddhism)
HM 1347 Internal awareness (Psychism)
MP 1235 Internal boundaries ; Unalienating
MM 2051 Internal combustion engine
MP 1194 Internal connectedness domains
Internal domains
MP 1115 — Functional integration unstructured
MP 1193 — Partially enclosed
MP 1233 — Zoning
HM 4490 Internal external ; Understanding interpreting (Buddhism)
HH 0799 Internal marriage
MS 8345 Internal organs ; Human
MP 1130 Internal transition spaces enhancing structural entry points
HM 0956 Internal ; Understanding interpreting (Buddhism)
HM 1044 Internalization awareness
KD 2210 Internalization ; Discontinuity: Comprehension
KC 0643 International
KC 0497 International attitudinal integration
International [citizenship...]
KC 0067 — citizenship
KC 0786 — community
HM 0099 — consciousness (Yoga)
KC 0850 — cooperation
KC 0897 — cooperation ; Regional
KC 0792 International economic integration
KC 0277 International economic order
International [institutional...]
KC 0597 — institutional integration
KC 0397 — integration
KC 0587 — integration ; Regional
KC 0132 International language
International [peace...]
KC 0200 — peace
KC 0697 — policy integration
KC 0287 — political integration
KC 0037 International relations
International [social...]
KC 0577 — social integration
KC 0062 — socialism
KC 0786 — system
KC 0997 — system ; Regional
International [understanding...]
KC 0845 — understanding
HH 6451 — understanding ; Education
KC 0790 — university
KD 2140 Interparadigmatic dialogue ; Logos lemma
MM 2017 Interpersonal interaction
TC 1902 Interpersonal suspicion
HH 0434 Interpersonal therapy
TC 1852 Interpersonal trust
VP 5552 Interpretability-Misinterpretability
HH 6109 Interpretation behaviour ; Ethological
MP 1003 Interpretation complementary modes organization
TC 1866 Interpretation teams ; Conceptual
HM 5026 Interpretation ; Understanding knowledge (Buddhism)
TC 1798 Interpreter ; Intercultural
HM 1128 Interpreting external ; Understanding (Buddhism)
HM 4490 Interpreting internal external ; Understanding (Buddhism)
HM 0956 Interpreting internal ; Understanding (Buddhism)
KC 0828 Interrelatedness
MP 1076 Interrelating mature emerging perspectives ; Minimal context
MP 1063 Interrelation complementary perspective common domains ; Cyclic
MP 1143 Interrelationship contexts developing perspectives
MP 1011 Interrelationship domains ; Local
MP 1016 Interrelationships ; Web general
TC 1650 Interrogation-discussion
TC 1650 Interrogator panel
TC 1866 Interruption teams
MM 2009 Interruption thematic development
MP 1054 Intersection differently paced communications
MP 1048 Interstices ; Adaptive
TC 1478 Interstitial conferencing
HH 0886 Intersubjectivity
HM 1709 Interval (Japanese)
VD 3070 Intervention
KC 0195 Interworking
KD 2262 Interwoven alternatives: resonance hybrids
KD 2260 Interwoven alternatives: tensegrity organization
VC 1013 Intimacy
TC 1952 Intimacy ; Conference
HM 1957 Intimacy God (Sufism)
VD 3072 Intimidation
VD 3074 Intolerance
VD 3076 Intoxication
HH 4111 Intoxication
HM 0992 Intoxication (Hinduism)
VD 3078 Intractability
VD 3080 Intransigence
HM 2820 Intrapsychic forces ; Negative (Psychism)
VD 3082 Intricacy
VC 1015 Intricacy
VD 3084 Intrigue
TP 0023 Intrigue (Stripping away)
VP 5005 Intrinsicality-Extrinsicality
HH 0824 Introspection
HM 7109 Introspection (Buddhism)

HM 2063	Introspection ; Non (Buddhism)	
HM 0899	Introvertive mystical experience	
VP 5234	Intrusion ; Circumscription	
HM 2449	Intrusion ; Dangerous (ICA)	
VD 3086	Intrusiveness	
	Intuition	
HM 2829	— [Intuition]	
HM 3634	— [Intuition]	
KC 0412	— [Intuition]	
HM 1632	Intuition (Buddhism)	
HM 4875	Intuition ; Clairvoyant (Hinduism)	
	Intuition Concentration	
HM 4977	— easy progress quick (Buddhism)	
HM 1010	— easy progress sluggish (Buddhism)	
HM 7859	— painful progress quick (Buddhism)	
HM 5992	— painful progress sluggish (Buddhism)	
HM 2927	Intuition ; Diaphanous (ICA)	
HM 5532	Intuition ; Direct (Buddhism)	
HM 4875	Intuition ; Flash (Hinduism)	
HM 1256	Intuition (Hinduism)	
HM 5532	Intuition ; Intelligible (Buddhism)	
HM 3104	Intuition (Japanese)	
HM 4875	Intuition ; Prophetic (Hinduism)	
VP 5481	Intuition-Reason	
HM 9124	Intuition sages (Hinduism)	
HM 3104	Intuitional cognition (Japanese)	
HM 1344	Intuitional insight (Buddhism)	
KC 0522	Intuitionism ; Mathematical	
HM 2761	Intuitions ; Trust (ICA)	
	Intuitive [cognition...]	
HM 2829	— cognition ; Spiritual	
HM 2439	— consciousness (Astrology)	
KC 0611	— Counter	
HM 2099	Intuitive knowledge ; Direct (Hinduism)	
HM 3239	Intuitive knowledge (Yoga)	
HH 0816	Intuitive meditation ; Contemplative	
HM 2785	Intuitive, meditative thinking	
HM 2829	Intuitive perception	
HM 1425	Intuitive-projective faith	
VC 1017	Intuitiveness	
VD 3088	Invalidity	
KC 0921	Invariance	
VD 3090	Invasion	
HM 3245	Invented history (ICA)	
HM 3138	Inventing essence (ICA)	
HH 4510	Inventing future	
HM 2092	Inventing humanness (ICA)	
HH 8963	Invention ; Social	
HH 0481	Inventiveness	
VC 1019	Inventiveness	
HM 2894	Inventiveness ; Freedom (ICA)	
HM 1751	Inverted views ; Grasping (Buddhism)	
HM 7324	Investigating (Buddhism)	
HM 3345	Investigation ; Analytical (Buddhism)	
HM 1177	Investigation (Buddhism)	
HM 7434	Investigation-concentration (Buddhism)	
HM 4552	Investigation ; Full understanding (Buddhism)	
TC 1650	Investigative panel	
TC 1814	Investigator ; Resource	
VC 1021	Invincibility	
VP 5444	Invisibility ; Visibility	
HH 1876	Invocation	
HH 0683	Invocation ; Contemplative (Japanese)	
HH 1988	Involuntary functions ; Control	
VD 3092	Involution	
VC 1023	Involvement	
HH 0030	Involvement	
TC 1229	Involvement ; Alienative participant	
TC 1895	Involvement ; Calculative participant	
TC 1345	Involvement ; Moral participant	
VC 1025	Invulnerability	
HM 1817	Inward attention	
HM 3605	Inward experience real (Sufism)	
HM 2738	Inward-flowing consciousness (Yoga)	
HM 3602	Inward purity (Sufism)	
HH 1134	Inward zazen meditation (Zen)	
HM 8901	Iradat (Sufism)	
HH 1903	Iranis (Zoroastrianism)	
VD 3094	Irascibility	
HM 0757	Irascibility	
HM 1972	Iriyana (Pali)	
VD 3096	Irksomeness	
HH 0239	Iron age	
MP 1074	Irrational perspectives ; Interaction coherent	
VD 3098	Irrationality	
MP 1071	Irrationality ; Access contained	
MP 1064	Irrationality ; Access patterns active	
MP 1144	Irrationality structures ; Integrating coordinated exposure	
VD 3100	Irreconcilability	
VD 3102	Irredeemability	
VD 3104	Irreformability	
VP 5137	Irregularity ; Regularity	
VD 3106	Irrelevance	
HM 2466	Irrelevant ; Action (ICA)	
HM 0670	Irreligiosity (Leela)	
VD 3108	Irreligiousness	
VD 3110	Irremediability	
HM 3155	Irreplaceable uniqueness (ICA)	
VC 1027	Irrepressibility	
VC 1029	Irreproachability	
VC 1031	Irresistability	
VD 3112	Irresolution	
VP 5624	Irresolution ; Resolution	
VD 3114	Irresponsibility	
VD 3116	Irreverence	
HM 2651	Irreversible direction time ; Consciousness	
MM 2081	Irrigating	
VD 3118	Irritability	
HM 1706	Irritability	
HM 2472	Irritability ; Chronic	
HM 2638	Irshya (Buddhism)	
HM 5124	Isakower phenomenon	
HM 3656	'Ishq (Islam)	
HM 4054	'Ishq (Sufism)	
HM 0877	Ishtiyaq (Islam)	
HM 2913	Ishvara-consciousness (Yoga)	
HM 0453	Islam Admiration	
HM 5304	Islam Affinity	
HM 1264	Islam Al-fikr fi 'l-manzur	
HM 4712	Islam Al-mahabba fi 'Ilah	
HM 5580	Islam Al-mahabba ma'a 'Ilah	
HM 5643	Islam Almsgiving	
HM 6003	Islam Amorous regard	
HH 0053	Islam Bakarah	
HM 1882	Islam Baraka	
HM 0129	Islam Concupiscible appetite	
HM 1264	Islam Contemplation object	
HH 0741	Islam Dervish	
HH 0741	Islam Derwish	
HM 0129	Islam Desire	
HM 0877	Islam Desire	
HM 4772	Islam Desire possess	
HM 2351	Islam Dhikr	
HM 5876	Islam Enslavement	
HM 1087	Islam Fana	
HH 4216	Islam Fasting	
HM 6003	Islam Gazing	
HM 6003	Islam Glancing	
HH 1265	Islam Hajj	
HM 0129	Islam Hawa	
HH 3687	Islam Holy war	
HH 1799	Islam Human development	
HM 3656	Islam 'Ishq	
HM 0877	Islam Ishtiyaq	
HM 0453	Islam Istihsan	
HM 2850	Islam Ittihad	
HH 3687	Islam Jihad	
HM 2351	Islam Khikr	
HM 7259	Islam Khulla	
HM 2351	Islam Kikr	
HH 0738	Islam Law religion	
HM 0877	Islam Longing	
HM 8303	Islam Loss discernment	
HM 7259	Islam Love	
HH 5712	Islam Love	
HM 6523	Islam Love based reason	
HM 4712	Islam Love God	
HM 5116	Islam Love God	
HM 5580	Islam Love God	
HM 6523	Islam Mahabba	
HH 5712	Islam Mahabba	
HH 0741	Islam Meditation motion	
HM 5304	Islam Munasaba	
HM 2375	Islam Nakshbandi recollection	
HH 0536	Islam Namaz	
HM 6003	Islam Nazar	
HH 6501	Islam Nomos	
HM 1306	Islam Nubuwah	
HM 1087	Islam Passing away	
HM 0129	Islam Passion	
HM 3656	Islam Passionate love	
HM 4772	Islam Physical lust	
HH 1265	Islam Pilgrimage	
HM 2341	Islam Profession faith	
HM 2351	Islam Recollection	
HH 0536	Islam Salat	
HH 4216	Islam Saum	
HM 1882	Islam Sense blessedness	
HM 2341	Islam Shahada	
HH 0738	Islam Sharia	
HM 0877	Islam Shawq	
HM 1882	Islam Spiritual energy	
HH 0738	Islam Spiritual path	
HH 5902	Islam Spirituality	
HM 4766	Islam Striving against soul	
HH 6973	Islam Striving against soul	
HH 1799	Islam Surrender will God	
HM 5876	Islam Ta'abbud	
HM 4772	Islam Tama'	
HM 5876	Islam Tatayyum	
HM 3006	Islam Thelema	
HM 5409	Islam Wadd	
HM 8303	Islam Walah	
HM 3006	Islam Will	
HH 0536	Islam Worship	
HM 5876	Islam Worship	
HM 5643	Islam Zakat	
HM 2351	Islam Zikr	
HM 3521	Islamic contemplative state (Sufism)	
HM 2850	Islamic transformation consciousness	
HM 2055	Island awareness ; Demon (Buddhism)	
HM 2175	Island awareness ; Potala (Buddhism)	
MP 1141	Isolated context each perspective ; Relatively	
MP 1152	Isolated contexts ; Partially	
VD 3120	Isolation	
MP 1136	Isolation complementary perspectives ; Relative	
MP 1237	Isolation ; Connectedness	
HH 2498	Isolation imagination ; Journeying transcendence	
HH 1478	Isolation initiation	
HH 0543	Isolation ; Psychic	
	Isolation [structural...]	
MP 1140	— structural interface communication pathway ; Relative	
HM 2743	— subtle contemplation ; Full (Buddhism)	
HM 6762	— (Sufism)	
KC 0832	Isomorphies	
KC 0832	Isomorphism	
HH 0966	Isomorphism ; Psychosocial	
HM 0453	Istihsan (Islam)	
HH 1016	Isvara pranidhana (Hinduism)	
HH 0992	Isvarapranidhanadva	
MP 1205	It ; Congruence spaces defined framework spaces defined processes	
MP 1135	It ; Enhancing insight varying levels exposure	
HM 5000	Italian Cuor contento	
HM 5000	Italian Happy-heartedness	
HM 3520	Itasal (Sufism)	
HM 3520	Itihad (Sufism)	
HM 0050	Itmianan (Sufism)	
HM 2850	Ittihad (Islam)	

J

HM 4392	Jabarut (Sufism)	
HM 0215	Jabbar ; Al (Sufism)	
HM 0148	Jada samadhi (Yoga)	
HM 0148	Jadya (Yoga)	
HM 2141	Jagrat state waking consciousness (Yoga)	
HM 6099	Jahd (Sufism)	
HM 3697	Jainism Adha-loka	
HH 0622	Jainism Human development	
HM 3948	Jainism Kevalajnana	
HM 1357	Jainism Madhya-loka	
HM 3948	Jainism Omniscience	
HM 0569	Jainism Siddha-loka	
HH 8712	Jainism Spirituality	
HM 0131	Jainism Urdhva-loka	
HM 0172	Jala-loka (Leela)	
HM 6719	Jalal wal Ikram ; Dhul (Sufism)	
HM 6719	Jalali wal Ikram ; Dhul (Sufism)	
HM 6771	Jalil ; Al (Sufism)	
HM 0234	Jam cetasikam asatam ; Manovinnanadhatusampassa (Pali)	
HM 3520	Jam (Sufism)	
HM 2318	Jam-tafriqah (Sufism)	
HM 3384	Jamais vu	
HM 2127	Jambudvipa (Buddhism)	
HM 0009	Jame' ; Al (Sufism)	
HM 0009	Jami' ; Al (Sufism)	
HM 3039	Jami ; illuminations (Sufism)	
HM 5155	Jana-loka (Leela)	
HM 4474	Janma (Leela)	
HM 0978	Janma ; Manushya (Leela)	
HH 0931	Japa yoga (Yoga)	
HH 0931	Japam (Yoga)	
HM 4390	Japanese Acheta	
HM 3387	Japanese Aesthetic attitude	
HM 2762	Japanese Aesthetic feeling	
HM 1365	Japanese Aesthetic simplicity	
HM 3597	Japanese Aesthetic understatement	
HM 3118	Japanese Akirame	
HM 3387	Japanese Artistic refinement	
HM 3522	Japanese Bewilderment	
HM 1598	Japanese Bitterness	
HM 2917	Japanese Compliance	
HH 0683	Japanese Contemplative invocation	
HM 0450	Japanese Engaku	
HM 3554	Japanese Freedom desire	
HM 3387	Japanese Furyu	
HH 0300	Japanese Giri	
HM 4001	Japanese Group harmony	
HM 2662	Japanese Hara	
HM 4350	Japanese Harmony aesthetic contrasts	
HM 2762	Japanese Having heart	
HM 1598	Japanese Hinekurata	
HM 1709	Japanese Interval	
HM 3104	Japanese Intuition	
HM 3104	Japanese Intuitional cognition	
HM 4282	Japanese Jigoku	
HH 0408	Japanese Kami hotoke	
HM 3104	Japanese Kan	
HH 0031	Japanese Ke hare kegare cycle	
HM 2700	Japanese Kensho-godo	
HM 3554	Japanese Lack deep feeling	
HM 1709	Japanese Ma	
HH 0031	Japanese Matsuri	
HM 3833	Japanese Mono no aware	
HM 0793	Japanese Muga	
HM 3554	Japanese Mushin	
HH 0912	Japanese Natural feelings	
HH 0683	Japanese Nembutsu	
HH 0912	Japanese Ninjo	
HM 2684	Japanese Outside strength	
HM 7512	Japanese Relaxed alertness	
HM 0285	Japanese Resonance	
HM 3522	Japanese Rotoo	
HM 4350	Japanese Sabi	
HH 0408	Japanese Sacred	
HH 1109	Japanese Seishin education	
HM 3037	Japanese Seriousness mind	
HM 3597	Japanese Shibui	
HM 3597	Japanese Shibumi	
HM 3104	Japanese Sixth sense	
HM 4001	Japanese Social harmony	
HH 0300	Japanese Social obligations	

Japanese

HM 0888	Japanese Sonomama	HM 0402	Journey ; Mystical (Christianity)	HM 2107	Judaism Path transparent intelligence
HH 1109	Japanese Spiritual education	MM 2074	Journey ; Path	HM 2593	Judaism Path triumphant intelligence
HM 1709	Japanese Stillness	HM 6120	Journey ; Shamanic	HM 2583	Judaism Path uniting intelligence
HM 2917	Japanese Sunao		**Journeying transcendence**	HM 2509	Judaism Paths wisdom
HM 2684	Japanese Tariki	HH 5908	— agape philo	HM 2188	Judaism Seraphim
HM 3037	Japanese Tsutsushimi	HH 1354	— blindness sight	HM 2002	Judaism Shekinah awareness
HM 2917	Japanese Uprightness	HH 2661	— bondage freedom (Christianity)	HH 4776	Judaism Spirituality
HM 2762	Japanese Ushin	HH 3352	— call mission (Christianity)	HM 4501	Judaism Talmudic ecstasy
HM 4001	Japanese Wa	HH 6505	— (Christianity)	HH 8206	Judaism Unity personality
HM 1365	Japanese Wabi	HH 2987	— death life	HM 3038	Judaism World creation
HM 1598	Japanese Warpedness	HH 3366	— discussion decision	HM 2408	Judaism World emanation
HM 0285	Japanese Yoin	HH 4110	— extraordinary ordinary (Christianity)	HM 2360	Judaism World formation
HM 7512	Japanese Zanshin	HH 5219	— forgetfulness memory	HM 2312	Judaism World manifestation
HM 1026	Jazba ; Kedesh (Sufism)	HH 4974	— head heart	HM 2582	Judge ; Concerned (ICA)
HM 3164	Je ; Ji ri mu (Zen)	HH 2329	— humiliation exaltation	VC 1037	Judgement
VD 3122	Jealousy	HH 2498	— isolation imagination	VT 8092	Judgement∗complex
HM 2638	Jealousy (Buddhism)	HH 5856	— me you	VP 5494	Judgement-Misjudgement
HM 3098	Jealousy (Christianity)	HH 4188	— object subject	HM 2290	Judgement ; Sphere (Kabbalah)
HM 6993	Jealousy ; Delusions	HH 3215	— outside inside	HM 2315	Judgment (Tarot)
HM 0554	Jealousy (Leela)	HH 1900	— prologue	TC 1574	Judicial conferencing
HM 3012	Jen	HH 3796	— secular sacred	VC 1039	Judiciousness
HM 2737	Jerusalem ; Jesus speaks daughters (Christianity)	HH 6012	— served servant (Christianity)	MM 2069	Juggling
TC 1546	Jester	HH 1773	— son father	HM 3615	Jung Abaissement niveau mental
HM 2206	Jesus Christ ; Presence (ICA)	HH 4219	— water wine	HM 3468	Jung Ambivalence
HM 2535	Jesus condemned death (Christianity)	MS 8113	Journeys ; Symbolic quests	HH 0085	Jung Collective unconscious
HM 3719	Jesus dies cross (Christianity)	VC 1033	Joy	HM 2173	Jung Depth consciousness
	Jesus falls	HM 1172	Joy	HM 2230	Jung Ego transcendence
HM 4009	— first time His cross (Christianity)	HM 8737	Joy (Buddhism)	HM 3811	Jung Mysterium tremendum
HM 3824	— second time cross (Christianity)	HM 0747	Joy (Buddhism)	HM 3811	Jung Numinosum
HM 2289	— third time cross (Christianity)	HM 8098	Joy (Hinduism)	HM 3811	Jung Numinous experience
HH 6019	Jesus ; Following (Christianity)	TP 0058	Joy (Lake)	HM 3912	Jung Self-acceptance
HM 3344	Jesus helped carry cross (Christianity)	HM 2167	Joy-land awareness (Buddhism)	HH 1187	Junior stream-winner (Buddhism)
HM 3500	Jesus made bear cross (Christianity)	HM 4376	Joy ; Plane (Leela)	VC 1041	Justice
HM 3611	Jesus meets his afflicted mother (Christianity)	HM 2267	Joy ; Subtle contemplation withdrawal (Buddhism)	HH 2117	Justice
HM 4136	Jesus nailed cross (Christianity)		**Joy [Understanding...]**	VP 5976	Justice-Injustice
HM 2246	Jesus one's being (Sufism)	HM 6245	— Understanding accompanied (Buddhism)	HM 2178	Justice (Tarot)
HM 3536	Jesus placed sepulchre (Christianity)	HM 2400	— Unspeakable (ICA)	HH 0341	Justification (Christianity)
HH 0205	Jesus prayer (Christianity)	HM 2529	— Unspeakable (ICA)		
	Jesus [saviour...]	HM 8737	Joyful feeling (Buddhism)		**K**
HH 2877	— saviour	HM 2022	Joyful-heaven awareness (Buddhism)	HH 0921	Kabbalah
HM 2737	— speaks daughters Jerusalem (Christianity)	HM 2724	Joyful satisfaction (Sufism)	HM 2372	Kabbalah Binah sephira
HM 3166	— stripped his garments (Christianity)	VC 1035	Joyfulness	HM 2420	Kabbalah Chesed sephira
HM 3410	Jesus taken down cross (Christianity)	VD 3124	Joylessness	HM 2348	Kabbalah Chokmah sephira
HM 3401	Jesus ; Veronica wipes face (Christianity)	HM 2106	Joyous awareness	HM 0126	Kabbalah Daath-sephira
HM 2282	Jeu	HM 1608	Jubilatio (Christianity)	HM 2290	Kabbalah Geburah-din sephira
HM 3063	Jewel awareness ; City shining (Yoga)	HH 0956	Jubilee spirit (Christianity)	HM 2238	Kabbalah Hod-sephira
HM 1486	Jewel wisdom (Buddhism)	HM 0934	Judaism Awareness infinity	HM 3132	Kabbalah Kether sephira
HM 2275	Jewelled-peaks-realm awareness (Buddhism)	HM 0377	Judaism Devekuth	HM 2288	Kabbalah Malkuth sephira
MS 8446	Jewellery adornments	HH 3776	Judaism Discipline worship	HM 2362	Kabbalah Netzach sephira
HH 1097	Jewish meditation (Judaism)	HM 4501	Judaism Ecstatic experience God	HM 3031	Kabbalah Sphere beauty
HH 1232	Jewish mysticism (Judaism)	HM 0934	Judaism En-sof	HM 2410	Kabbalah Sphere foundation
HM 7193	Jhana (Buddhism)	HM 4501	Judaism God wholly	HM 2238	Kabbalah Sphere glory
	Jhana Dwelling	HM 4291	Judaism God wholly same	HM 2290	Kabbalah Sphere judgement
HM 4298	— first (Buddhism)	HH 0597	Judaism Hasidism	HM 2288	Kabbalah Sphere kingdom
HM 6553	— fivefold (Buddhism)	HM 3029	Judaism Human development	HM 0126	Kabbalah Sphere knowledge
HM 8087	— fourth (Buddhism)	HM 2583	Judaism Inductive intelligence unity	HM 2420	Kabbalah Sphere mercy greatness
HM 7121	— second (Buddhism)	HH 1097	Judaism Jewish meditation	HM 3132	Kabbalah Sphere supreme crown
HM 5643	— third (Buddhism)	HH 1232	Judaism Jewish mysticism	HM 2372	Kabbalah Sphere understanding
	Jhana five Concentration	HM 2002	Judaism Kavod	HM 2362	Kabbalah Sphere victory
HM 5143	— fifth (Buddhism)	HM 4501	Judaism Kiss God	HM 2348	Kabbalah Sphere wisdom
HM 0297	— first (Buddhism)	HM 4501	Judaism Merkabah mysticism	HM 3031	Kabbalah Tiphareth sephira
HM 4882	— fourth (Buddhism)	HM 2312	Judaism Olam-asiyah	HM 2410	Kabbalah Yesod sephira
HM 4575	— second (Buddhism)	HM 2408	Judaism Olam atziluth	HM 2800	Kabd (Sufism)
HM 8532	— third (Buddhism)	HM 3038	Judaism Olam briah	HM 4437	Kabir ; Al (Sufism)
HM 6553	Jhana fivefold system ; Dwelling second (Buddhism)	HM 2360	Judaism Olam-yetzirah	HM 2749	Kairos
	Jhana four Concentration	HM 2188	Judaism Orders angelic awareness	HM 3869	Kaivalya (Yoga)
HM 4456	— first (Buddhism)	HM 3243	Judaism Path absolute intelligence	HM 3869	Kaivalyam (Yoga)
HM 7202	— fourth (Buddhism)	HM 2218	Judaism Path active intelligence	HM 2151	Kalacakra-tantra shambhala awareness (Buddhism)
HM 4380	— second (Buddhism)	HM 2095	Judaism Path administrative intelligence	KC 0580	Kaleidometrics
HM 2284	— third (Buddhism)	HM 3013	Judaism Path admirable intelligence	HH 0239	Kali yuga
HM 4298	Jhana happiness bliss born seclusion (Buddhism)	HM 3102	Judaism Path cohesive intelligence	HH 0754	Kama (Hinduism)
	Jhana Pali	HM 3046	Judaism Path collective intelligence	HM 7800	Kama (Hinduism)
HM 2536	— [Jhana (Pali)]	HM 2117	Judaism Path constituting intelligence	HM 3627	Kama-loka (Leela)
HM 0137	— [Jhana (Pali)]	HM 3149	Judaism Path corporeal intelligence	HM 4666	Kamachanda (Hinduism)
HM 2122	— [Jhana (Pali)]	HM 2569	Judaism Path disposing intelligence	HM 2211	Kamadera-vidyadhara (Buddhism)
HM 3168	Ji hokkai (Zen)	HM 2593	Judaism Path eternal intelligence	HM 2733	Kamadhatu (Buddhism)
HM 2034	Ji-ji-mu-ge (Zen)	HM 2218	Judaism Path exciting intelligence	HM 2060	Kamaloka consciousness (Buddhism)
HM 2034	Ji mu ge ; Ji (Zen)	HM 2081	Judaism Path faithful intelligence	HH 0408	Kami hotoke (Japanese)
HM 3164	Ji-ri-mu-je (Zen)	HM 2523	Judaism Path fiery intelligence	HM 2920	Kamil ; Al insan (Sufism)
HM 4282	Jigoku (Japanese)	HM 2444	Judaism Path hidden intelligence	HM 2920	Kamil ; Insanu'l (Sufism)
HH 3687	Jihad (Islam)	HM 3013	Judaism Path hidden intelligence	HM 3677	Kamil ; Insanu'l (Sufism)
HM 0853	Jihi (Buddhism)	HM 2612	Judaism Path illuminating intelligence	HH 0567	Kamma
HM 2263	Jihvavijnana (Buddhism)	HM 2035	Judaism Path illuminating intelligence	HM 2048	Kamma (Pali)
HM 3215	Jiriki (Zen)	HM 2011	Judaism Path imaginative intelligence	HM 3716	Kammannabhava (Pali)
HM 3890	Jivan-mukti (Hinduism)	HM 2105	Judaism Path intelligence conciliation	HM 1381	Kammannata ; Vedanakkhandhassa sannakkhandhassa sankharakkhandhassa (Pali)
HM 2137	Jivana mukta state	HM 2045	Judaism Path intelligence influences		
HM 3890	Jivanmukti (Hinduism)	HM 2107	Judaism Path intelligence light	HM 0979	Kammannattam (Pali)
HM 3379	Jivita (Pali)	HM 3089	Judaism Path intelligence mediating influence	HM 3104	Kan (Japanese)
HM 3404	Jivitindriya (Pali)	HM 2105	Judaism Path intelligence reward	HH 0495	Kanda ; Jnana (Hinduism)
HM 0261	Jivitindriya (Pali)	HM 2021	Judaism Path intelligence secret	HM 4347	Kankha (Pali)
HM 5321	Jna (Buddhism)	HM 2021	Judaism Path intelligence spiritual action	HH 1187	Kankhavitarana-visuddhi-niddesa (Pali) Lesser stream-enterer (Buddhism)
HM 7549	Jnana darsana ; Yatha bhuta (Buddhism)	HM 2503	Judaism Path intelligence temptation		
HH 0495	Jnana kanda (Hinduism)	HM 2629	Judaism Path intelligence will	HM 0177	Kankhayana (Pali)
HM 3846	Jnana (Leela)	HM 3102	Judaism Path measuring intelligence	HM 3934	Kankhayitattam (Pali)
HH 0495	Jnana marga (Hinduism)	HM 2071	Judaism Path natural intelligence	HM 2596	Kappa ; Samvatta (Pali)
HM 2756	Jnana ; Prajna (Buddhism)	HM 2444	Judaism Path occult intelligence	HM 3237	Kara ; Manasi (Pali)
HM 3060	Jnana samadhi (Hinduism)	HM 3243	Judaism Path perfect intelligence	HM 8667	Kardias ; Gnosis (Sufism)
HH 0610	Jnana (Yoga)	HM 2643	Judaism Path perpetual intelligence	HM 2215	Kari ; Prabha (Buddhism)
HM 0927	Jnana yoga (Yoga)	HM 2943	Judaism Path purified intelligence	HM 5289	Karim ; Al (Sufism)
HM 1456	Jnanam ; Vivekajam (Yoga)	HM 3181	Judaism Path radical intelligence	HH 0567	Karma
HM 3514	Jobutsu ; Sokushin (Buddhism)	HM 3102	Judaism Path receiving intelligence		**Karma awareness**
HM 1169	Jogging meditation	HM 2047	Judaism Path renewing intelligence	HM 2783	— Amoghasiddhi (Buddhism)
HM 1714	John Cross ; Dark night Saint (Christianity)	HM 2047	Judaism Path renovating intelligence	HM 2048	— (Buddhism)
TC 1546	Joker	HM 3426	Judaism Path resplendent intelligence	HM 2192	— Discipleship (Buddhism)
HM 2877	Joshua	HM 3567	Judaism Path sanctifying intelligence		
HM 2900	Journey ; Awareness mystic	HM 2533	Judaism Path stable intelligence		

-888-

GENERAL INDEX TO VOLUME 2

Kshipta

HM 0948 Karma-loka (Leela)
HH 1211 Karma marga (Hinduism)
HM 2491 Karma ; Niskama (Hinduism)
HH 5487 Karma ; Niskama (Hinduism)
HM 2783 Karma-paripurana (Buddhism)
HM 0948 Karma ; Plane (Leela)
HM 0130 Karma-result consciousness (Buddhism)
HM 0717 Karma ; Third chakra: theatre (Leela)
HH 0372 Karma yoga (Yoga)
HM 0198 Karmanta ; Samyak (Buddhism)
HM 5634 Karuna (Buddhism)
HM 1452 Kashf (Sufism)
HH 3246 Kasina ; Colour (Buddhism)
HH 3246 Kasina ; Earth (Buddhism)
HH 3246 Kasina meditation (Buddhism)
HH 3246 Kasina ; Nature (Buddhism)
HH 3246 Kasina ; Sound (Buddhism)
HH 3246 Kasina ; Space (Buddhism)
HH 3246 Kasina ; Water (Buddhism)
HH 2918 Kasinas ; Meditating (Buddhism)
HM 7209 Kaukritya (Buddhism)
HM 3163 Kausidya (Buddhism)
HM 2002 Kavod (Judaism)
HM 5600 Kaya ; Abstinence misconduct (Buddhism)
HM 7636 Kaya ; Buoyancy (Pali)
HM 2397 Kaya ; Dharma (Buddhism)
HM 2340 Kaya ; Nirmana (Buddhism)
HM 1455 Kaya-pagunnata (Pali)
HM 3696 Kaya ; Pliancy (Pali)
HM 5402 Kaya ; Rectitude (Pali)
HM 4704 Kaya ; Repose (Pali)
HM 2873 Kaya ; Sambhoga (Buddhism)
HM 7969 Kaya ; Wieldiness (Pali)
HM 5600 Kayaduccaritavirati (Pali)
HM 1133 Kayaduccaritehi arati virati pativirati veramani ; Tihi (Pali)
HM 0739 Kayakammannata (Pali)
HM 4395 Kayalahuta (Pali)
HM 3370 Kayamuduta (Pali)
HM 1867 Kayapassaddhi (Pali)
HM 2562 Kayavijnana (Buddhism)
HM 1424 Kayujukata (Pali)
HM 3420 Kazantzakis ; Spiritual exercises Nikos (Christianity)
HH 0031 Ke hare kegare cycle (Japanese)
HH 1026 Kedesh-jazba (Sufism)
VC 1043 Keenness
HM 2249 Keeping conscience (ICA)
HM 3327 Keeping going (Buddhism)
TP 0052 Keeping still (Mountain)
HH 0031 Kegare cycle ; Ke hare (Japanese)
HM 2700 Kensho-godo (Japanese)
HM 4256 Kerygma (Christianity)
HM 3132 Kether sephira (Kabbalah)
HM 3948 Kevalajnana (Jainism)
KD 2285 Key crop rotation ; Patterns alternation: agricultural
TC 1566 Key-note speakers
KD 2280 Key political philosopher ; Patterns alternation: musical
TC 1932 Key resource persons
HM 4400 Khabir ; Al (Sufism)
HM 3644 Khafid ; Al (Sufism)
HM 0335 Khalifa (Sufism)
HM 0034 Khaliq ; Al (Sufism)
HH 6282 Khalq ; Alam (Sufism)
HH 1875 Khalwat (Sufism)
HM 2177 Khams-gsum (Tibetan)
HM 2098 Khanda development ; Vinnana (Pali)
HM 2108 Khanda ; Rupa (Pali)
HM 2050 Khanda ; Sankhara (Pali)
HM 4143 Khanda ; Sanna (Pali)
HM 4983 Khanda ; Vedana (Buddhism)
HM 2556 Khanda ; Vinnana (Pali)
HM 2692 Khandha (Pali)
HH 3321 Khandha (Pali)
HH 3321 Khandhas (Buddhism)
HM 4277 Khat ; Modes awareness associated use
HM 1626 Khatim ; Al (Sufism)
HM 1047 Khawf (Sufism)
HM 2424 Khayr ; Stations consciousness ibn Abi'l (Sufism)
HM 8776 Khidmat (Sufism)
HM 2351 Khikr (Islam)
HM 4987 Khilwat dar anjuman (Sufism)
HM 2134 'Khon-'dzin (Tibetan)
HM 2959 Khong-khro (Tibetan)
HM 2177 Khor-ba (Buddhism)
HM 3203 Khrel-med-pa (Tibetan)
HM 2210 Khrel-yod-pa (Tibetan)
HM 2264 Khro-ba (Tibetan)
HM 7259 Khulla (Islam)
HM 4493 Khyati ; Viveka (Yoga)
HH 0620 Ki energy
HH 0620 Kiai
HM 2599 Kicca ; Vinnana (Pali)
HM 2351 Kikr (Islam)
HM 7345 Kilesappahana (Buddhism)
VC 1045 Kindness
HH 0457 Kindness
VP 5938 Kindness-Unkindness
HM 4958 Kinds knowledge ; Understanding knowledge (Buddhism)
HH 2031 Kinerhythm meditation
HH 0392 Kinesics
HH 0379 Kinesiotherapy
HH 0662 Kinesthetic intelligence ; Bodily
HM 2759 King awareness ; Wheel turner (Buddhism)
HM 2058 King awareness ; Wheel turning (Buddhism)
HM 1606 Kingdom God
HM 0563 Kingdom heaven awareness
HM 2288 Kingdom ; Sphere (Kabbalah)
HM 2082 Kings heaven awareness ; Four great (Buddhism)
MS 8333 Kinship
HM 1023 Kiriyati hiriyitabbena ; Na (Pali)
HM 3257 Kiriyati papakanam akusalanam dhammanam samapattiya ; Na (Pali)
HH 2390 Kirtan (Hinduism)
HH 1362 Kismet
HH 3773 Kismet
HM 4501 Kiss God (Judaism)
HH 1047 Klesas (Yoga)
HH 0390 Kleshas ; Five (Yoga)
MM 2049 Knotting, nets basketry
HM 1045 Knower (Sufism)
HM 7554 Knowing ; Higher (Christianity)
HM 2801 Knowing ; Life (ICA)
HM 6997 Knowing ; Lower (Christianity)
HM 3599 Knowing ; Self
HH 0873 Knowing ; Ways (Buddhism)
HH 1616 Knowing ; Ways (Christianity)
KC 0338 Knowledge
VC 1047 Knowledge
Knowledge [Adoptive...]
HM 4637 — Adoptive (Buddhism)
HM 0476 — All pervading (Sufism)
HM 0169 — analysis (Buddhism)
HM 7298 — another's thoughts (Hinduism)
HM 7114 — appearance terror (Buddhism)
HM 2858 Knowledge ; Body
KC 0528 Knowledge ; Body
Knowledge Buddhism
HM 2655 — [Knowledge (Buddhism)]
HM 5498 — [Knowledge (Buddhism)]
HM 6502 — [Knowledge (Buddhism)]
HH 1971 — [Knowledge (Buddhism)]
Knowledge [cessation...]
HM 8163 — cessation suffering ; Understanding (Buddhism)
HM 4637 — change lineage (Buddhism)
HM 2195 — character ; Individual (Buddhism)
Knowledge Concentration
HM 5992 — difficult progress sluggish direct (Buddhism)
HM 7859 — difficult progress swift direct (Buddhism)
HM 1010 — easy progress sluggish direct (Buddhism)
HM 4977 — easy progress swift direct (Buddhism)
HM 5403 Knowledge conformity truth (Buddhism)
HM 2297 Knowledge consciousness ; Tacit
Knowledge contemplation
HM 7297 — danger (Buddhism)
HM 5364 — dispassion (Buddhism)
HM 3385 — dissolution (Buddhism)
HM 0169 — reflection (Buddhism)
HM 3723 — rise fall (Buddhism)
Knowledge [Contemplative...]
HM 2254 — Contemplative (Sufism)
HM 4788 — Continuous pervading (Sufism)
KC 0528 — Corpus
HM 5187 — Cyclic (Systematics)
HM 0766 Knowledge desire deliverance (Buddhism)
HM 0766 Knowledge desire release (Buddhism)
Knowledge Direct Buddhism
HM 4297 — [Knowledge ; Direct (Buddhism)]
HM 5232 — [Knowledge ; Direct (Buddhism)]
HM 5982 — [Knowledge ; Direct (Buddhism)]
HM 7672 — [Knowledge ; Direct (Buddhism)]
HH 5652 — [Knowledge ; Direct (Buddhism)]
HM 0748 — [Knowledge ; Direct (Buddhism)]
HM 2099 Knowledge ; Direct intuitive (Hinduism)
Knowledge discernment
HH 3550 — middle way ; Purity (Buddhism)
HH 3025 — Purity (Buddhism)
HH 4007 — right path wrong path ; Purity (Buddhism)
Knowledge [Discriminative...]
HM 1234 — Discriminative (Systematics)
HM 5982 — divine ear element (Buddhism)
HM 0748 — divine sight (Buddhism)
HM 4726 — doctrine ; Understanding (Buddhism)
HM 5221 — due insight (Buddhism)
Knowledge [Ecology...]
KC 0922 — Ecology
HM 8080 — Effectual (Systematics)
HM 0344 — Eight fold (Systematics)
HM 0065 — Eleven fold (Systematics)
HM 0424 — equanimity formations (Buddhism)
HM 4349 — Experience mystical (Sufism)
HH 1809 — Extraordinary (Buddhism)
HM 8080 Knowledge ; Five fold (Systematics)
HM 1991 Knowledge ; Four fold (Systematics)
Knowledge [Game...]
HM 2278 — Game
HH 1765 — General (Buddhism)
HM 0413 — Gnostic
HM 0930 — God ; Certain (Sufism)
HM 2254 — God (Sufism)
HH 2003 — Gradations
Knowledge Higher Buddhism
HM 4297 — [Knowledge ; Higher (Buddhism)]
HM 5232 — [Knowledge ; Higher (Buddhism)]
HM 5982 — [Knowledge ; Higher (Buddhism)]
HM 7672 — [Knowledge ; Higher (Buddhism)]
HH 5652 — [Knowledge ; Higher (Buddhism)]
HM 0748 — [Knowledge ; Higher (Buddhism)]
HH 0353 Knowledge (Hinduism)
Knowledge [ideas...]
HM 3527 — ideas (Buddhism)
VP 5475 — Ignorance
Knowledge [indifference...] cont'd
HM 0424 — indifference complexes (Buddhism)
KC 0686 — Integration
HM 5026 — interpretation ; Understanding (Buddhism)
HM 3239 — Intuitive (Yoga)
HM 4958 Knowledge kinds knowledge ; Understanding (Buddhism)
Knowledge [language...]
HM 5026 — language ; Understanding (Buddhism)
HM 4726 — law ; Understanding (Buddhism)
HM 3846 — (Leela)
HM 6558 — liberated souls (Hinduism)
Knowledge [Man...]
HH 0338 — Man
HM 1663 — meaning ; Understanding (Buddhism)
HM 6997 — Mundane (Christianity)
HM 1408 Knowledge ; Nine fold (Systematics)
HM 4002 Knowledge ; Non discriminative (Systematics)
Knowledge [Organization...]
KC 0320 — Organization
HM 7290 — origin suffering ; Understanding (Buddhism)
HM 5232 — others' thoughts (Buddhism)
HM 0748 Knowledge passing away reappearance beings (Buddhism)
Knowledge path
HM 7055 — arahantship (Buddhism)
HM 6920 — non-return (Buddhism)
HM 7563 — once-return (Buddhism)
HM 1088 — stream entry (Buddhism)
Knowledge [penetration...]
HM 3527 — penetration (Buddhism)
HM 5232 — penetration minds (Buddhism)
HM 1234 — Polar (Systematics)
HM 8080 — Potential (Systematics)
HM 7114 — presence fear (Buddhism)
HM 2966 — Pure
Knowledge [recollection...]
HM 4297 — recollection past life (Buddhism)
HM 4297 — recollection previous existence (Buddhism)
HM 0811 — Relational (Systematics)
HM 5364 — repulsion (Buddhism)
KC 0197 — Restructuring
HM 0344 — Revealed (Systematics)
HM 0567 — Right (Leela)
Knowledge [Salvational...]
HM 0413 — Salvational
MS 8963 — Search
HH 0312 — Self
HM 0765 — Self (Zen)
HM 0776 — Seven fold (Systematics)
HM 5187 — Six fold (Systematics)
KC 0304 — Sociology
HM 0126 — Sphere (Kabbalah)
HM 7554 — Spiritual (Christianity)
HM 0776 — Structural (Systematics)
HM 6870 — suffering ; Understanding (Buddhism)
HM 0107 — Superconscious
HM 7672 — supernormal powers (Buddhism)
Knowledge [Ten...]
HM 1804 — Ten fold (Systematics)
HM 0811 — Three fold (Systematics)
HM 5187 — Transcendental (Systematics)
HM 7297 — tribulation (Buddhism)
HH 5767 — True (Christianity)
HH 3527 — truths ; Two fold (Buddhism)
HM 0707 — Twelve fold (Systematics)
HM 1234 — Two fold (Systematics)
HM 4958 Knowledge ; Understanding knowledge kinds (Buddhism)
KC 0686 Knowledge ; Unification
HM 1991 Knowledge ; Value (Systematics)
Knowledge vision
HH 3025 — Purification (Buddhism)
HH 3550 — way ; Purification (Buddhism)
HH 4007 — what is path what is not path ; Purification (Buddhism)
Knowledge [Way...]
HM 1201 — Way (Esotericism)
HH 0495 — Way (Hinduism)
HM 7645 — way leading cessation suffering ; Understanding (Buddhism)
HH 0538 Knowledge (Yoga)
HH 0610 Knowledge (Yoga)
HM 3490 Known ; Comprehension (Buddhism)
HM 3490 Known ; Full understanding (Buddhism)
HH 0263 Koan (Zen)
HM 0144 Koinonia (Christianity)
HM 0767 Korean Unwillingness let go illusion
HM 0767 Korean Won
HM 4803 Koro
HM 2912 Kosa ; Manomaya (Yoga)
HM 3460 Kosa ; Vijnanamaya (Yoga)
HM 1554 Kosalla (Pali)
HM 3212 Kosha ; Anandamaya (Yoga)
HM 2300 Krishna consciousness
HM 2864 Kriya ; Action tantra: (Buddhism)
HH 4309 Kriya sakti (Hinduism)
HH 2558 Kriya-tantra awareness (Buddhism)
HH 0969 Kriya yoga (Yoga)
HM 1433 Krodh (Leela)
HM 2264 Krodha (Buddhism)
HM 8665 Krodha (Hinduism)
HM 4535 Ksana (Buddhism)
HM 2247 Kshanti (Buddhism)
HH 0813 Kshanti (Zen)
HM 4058 Kshipta (Yoga)

Ku

HM 0581	Ku (Buddhism)	KC 0569	Laws ; Conservation	HM 1726	Leela Egotism
HM 0764	Ku-sang-loka (Leela)	KC 0131	Laws ; Integration	HM 0832	Leela Eirsha
HM 3189	Kufu	VD 3138	Laxity	HM 4403	Leela Entertainment
HM 9101	Kukkucca (Pali)	VD 3140	Laxness	HM 0832	Leela Envy
HM 1912	Kummagga (Pali)	HH 0782	Laya yoga (Yoga)	HM 0933	Leela Fifth chakra: man becomes himself
HH 0320	Kun-'gro (Buddhism)	HH 2552	Laying hands	HM 4103	Leela First chakra: fundamentals being
HM 0237	Kundali yoga (Yoga)	HH 2023	Laying hands (Christianity)	HM 3291	Leela Fourth chakra: attaining balance
HM 2871	Kundalini conscious energy (Yoga)	VD 3142	Laziness	HM 0008	Leela Gandha-loka
HH 0237	Kundalini yoga (Yoga)	HM 3163	Laziness (Buddhism)	HM 4403	Leela Gandharvas
HM 2067	Kuru (Buddhism)	HM 0046	Lazy	HM 3779	Leela Ganga
HM 3560	Kusala mulam ; Adoso (Pali)	HM 2263	Lce'i rnam shes pa (Tibetan)	HM 3779	Leela Ganges
HM 3335	Kushesh (Sufism)	HM 3163	Le-lo (Tibetan)	HM 6434	Leela Gaseous plane
HH 0578	Kushula (Buddhism)		**Leadership**	HM 4474	Leela Genesis
HM 0303	Kut	VC 1057	— [Leadership]	HM 1355	Leela Good company
HM 2275	Kuta ; Ratna (Buddhism)	TP 0019	— [Leadership]	HM 5197	Leela Good tendencies
HH 3443	Kyo ; Nam myoho renge (Buddhism)	TP 0008	— [Leadership]	HM 1931	Leela Greed
		TC 1944	Leadership ; Conference	HM 3846	Leela Gyana
	L	HH 0516	Leadership development	HM 5099	Leela Happiness
		HH 0617	Leadership development group	HM 4376	Leela Harsha-loka
HM 1264	'l manzur ; Al fikr fi (Islam)	HM 5451	Leadership intelligence	HM 1276	Leela Himsa-loka
HM 2208	Labdha samadhi ; Dharmaloka (Buddhism)	HM 5451	Leadership mind	HM 5155	Leela Human plane
HH 0672	Laboratories ; Training	VP 5292	Leading-Following	HM 0933	Leela Human plane
MM 2037	Labour ; Forced	HM 5265	Leaping out world (Taoism)	HM 0310	Leela Ignorance
KC 0572	Labour ; Organization	HM 5298	Learned ; Understanding consisting what is (Buddhism)	HM 0670	Leela Irreligiosity
HH 0742	Labour training	VC 1059	Learning	HM 0172	Leela Jala-loka
HM 1435	Labuta ; Vedanakkhandhassa sannakkhandhassa sankharakkhandhassa (Pali)	HH 0180	Learning	HM 5155	Leela Jana-loka
			Learning [Accelerated...]	HM 4474	Leela Janma
VD 3126	Lack	HH 4511	— Accelerated	HM 0554	Leela Jealousy
HM 4006	Lack concentration	HH 0315	— Action	HM 3846	Leela Jnana
HM 3554	Lack deep feeling (Japanese)	HM 2280	— application awareness ; Tantra union (Buddhism)	HM 3627	Leela Kama-loka
HM 0110	Lack dispassion (Yoga)		**Learning [Conscious...]**	HM 0948	Leela Karma-loka
HM 4228	Lack real grasping (Buddhism)	HM 0168	— Conscious	HM 3846	Leela Knowledge
HM 3941	Ladder divine love ; Mystical (Christianity)	KD 2275	— cycles	HM 1433	Leela Krodh
KC 0346	Lag ; Surveillance	TC 1599	— cycles ; Repetition	HM 0764	Leela Ku-sang-loka
HM 3808	Lahuparinamata (Pali)	HH 0704	Learning ; Deutero	HM 0172	Leela Liquid plane
HM 0741	Lahut (Sufism)	KD 2192	Learning ; Encompassing system dynamics:	HM 1931	Leela Lobh
HM 3403	Lahuta ; Vinnankkhandhassa (Pali)	HM 0196	Learning ; Functional	HM 3779	Leela Lunar plane
TP 0058	Lake Delight	TC 1856	Learning group ; Peer mediated	HM 0098	Leela Mada
TP 0058	Lake Joy	HH 0812	Learning ; Innovative	HM 1351	Leela Maha-loka
TP 0058	Lake Standing straight	KD 2329	Learning loss cyclic phase	HM 1351	Leela Mahar-loka
TP 0058	Lake Vitality	KD 2090	Learning ; Opening closing: alternation discontinuous	HM 0978	Leela Manushya-janma
HM 2946	Lakshanashunyata (Buddhism)	HH 0546	Learning ; Programmed	HM 0987	Leela Matsar
HH 1673	Lamarckism	KD 2295	Learning response ; Entropic crisis	HM 0987	Leela Matsarya
VP 5873	Lamentation ; Exultation		**Learning [Secondary...]**	HM 0953	Leela Maya
HM 1846	Lamp wisdom (Buddhism)	HH 0704	— Secondary	HM 3434	Leela Mercy
HM 2167	Land awareness ; Joy (Buddhism)	HH 0417	— Self	HM 0697	Leela Moha
HM 2434	Land mystery (ICA)	KD 2270	— society ; Non comprehension structuring phenomenon	HM 1485	Leela Naga-loka
HM 4587	Land shadows (Psychism)			HM 0856	Leela Narka-loka
HM 2168	Lands awareness ; Pure (Buddhism)	KD 2215	— society ; Pattern accumulation	HM 3623	Leela Negative intellect
KC 0516	Landscape ; Epigenetic	HM 3662	— State (Buddhism)	HM 4512	Leela Nullity
KC 0694	Language	HM 1244	— Superconscious	HM 3732	Leela Omkar
HH 0244	Language	HH 0326	— systems ; Open	HM 0294	Leela Parmarth
KC 0196	Language ; Analog computer	HH 0614	Learning theory therapy	HM 3907	Leela Phenomenal plane
KC 0706	Language ; Block	HM 1217	Learning ; Virtual	HM 4103	Leela Physical plane
HH 0392	Language ; Body	HM 2417	Least action principle conscious states (Physical sciences)	HM 1024	Leela Physical plane
TC 1894	Language constraints			HM 0859	Leela Plane agnih
HH 0485	Language development	HM 2604	Leaving man alone awareness ; Ox forgotten (Zen)	HM 4412	Leela Plane austerity
HM 1538	Language divine analogy		**Leaving off abstaining**	HM 5244	Leela Plane austerity
KC 0069	Language ; Energy circuit	HM 1781	— totally abstaining refraining four deviations speech (Buddhism)	HM 3291	Leela Plane balance
KC 0093	Language ; Form			HM 1351	Leela Plane balance
KC 0132	Language ; International	HM 1133	— totally abstaining refraining three deviations body (Buddhism)	HM 6901	Leela Plane bliss
MS 8115	Language literature			HM 1141	Leela Plane cosmic consciousness
KC 0123	Language ; Meta	HM 0252	— totally abstaining refraining wrong livelihood (Buddhism)	HM 1301	Leela Plane cosmic consciousness
	Language [participants...]			HM 4343	Leela Plane cosmic good
TC 1603	— participants ; Pattern	HM 3429	Leaving undone (Buddhism)	HM 0481	Leela Plane dharma
KC 0093	— Pattern	HH 0158	Lectio divina	HM 1485	Leela Plane fantasy
KD 2340	— probabilistic vision world	TC 1383	Lecture	HM 0859	Leela Plane fire
KD 2152	Language relationships sound ; Modelling	HM 2278	Leela	HM 0008	Leela Plane fragrance
HM 1538	Language tradition ; Symbolic	HM 4326	Leela Absolute plane	HM 6287	Leela Plane inner space
	Language [Understanding...]	HM 0670	Leela Adharma	HM 4376	Leela Plane joy
HM 5026	— Understanding knowledge (Buddhism)	HM 0859	Leela Agnih-loka	HM 0948	Leela Plane karma
KC 0947	— universals	HM 1726	Leela Ahamkara	HM 4050	Leela Plane neutrality
KC 0947	— Universals	HM 6901	Leela Anand-loka	HM 3732	Leela Plane primal vibrations
KD 2154	Languages ; Complementary	HM 1433	Leela Anger	HM 5028	Leela Plane radiation
KD 2150	Languages ; Emancipation particular	HM 4512	Leela Antariksha	HM 1293	Leela Plane reality
KC 0778	Languages ; Integration	HM 0145	Leela Apana-loka	HM 5965	Leela Plane sanctity
VD 3128	Languor	HM 2981	Leela Apt religion	HM 0878	Leela Plane taste
VD 3130	Lapse	HM 3651	Leela Astral plane	HM 1293	Leela Plane truth
VC 1049	Largeness	HM 4600	Leela Astral plane	HM 1276	Leela Plane violence
HM 2171	Las-dang-po-pa-tsam-kyi-yid-byed (Tibetan)	HM 3837	Leela Atonement	HM 1789	Leela Positive intellect
VC 1051	Lasting	HM 0697	Leela Attachment	HM 3907	Leela Prakriti-loka
HM 0581	Latency ; State (Buddhism)	HM 0987	Leela Avarice	HM 4399	Leela Prana-loka
VD 3132	Lateness	HM 0310	Leela Avidya	HM 6112	Leela Prithvi
KC 0260	Lateral thinking	HM 0764	Leela Bad company	HM 0856	Leela Purgatory
HM 2320	Lateral thinking ; Vertical	HM 1141	Leela Beyond chakras: gods themselves	HM 1773	Leela Purification
HM 0308	Latif ; Al (Sufism)	HM 1475	Leela Bhakti-loka	HM 3281	Leela Rajoguna
HM 1905	Latifa-i-qalb (Sufism)	HM 1024	Leela Bhu-loka	HM 0878	Leela Rasa-loka
HH 6282	Latifas (Sufism)	HM 4600	Leela Bhuvar-loka	HM 0567	Leela Right knowledge
HH 0086	Latihan	HM 0978	Leela Birth man	HM 4343	Leela Rudra-loka
KC 0554	Lattice theory	HM 4326	Leela Brahma-loka	HM 3837	Leela Saman-paap
HM 0894	Lauds (Christianity)	HM 4403	Leela Celestial musicians	HM 4050	Leela Saraswati
HM 0894	Lauds ; Hours (Christianity)	HM 0717	Leela Celestial plane	HM 4310	Leela Satoguna
VC 1053	Laughter	HM 1052	Leela Celestial plane	HM 1293	Leela Satya-loka
HM 2271	Laukikagra-dharma (Buddhism)	HM 4387	Leela Charity	HM 3651	Leela Second chakra: realm fantasy
MM 2087	Launching spacecraft	HM 0455	Leela Clarity consciousness	HM 0904	Leela Selfless service
VD 3134	Lavishness	HM 0098	Leela Conceit	HM 3627	Leela Sensual plane
HH 4536	Law ; Fivefold moral	HM 1601	Leela Conscience	HM 0754	Leela Seventh chakra: plane reality
HH 0642	Law ; Natural	HM 4387	Leela Daan	HM 4412	Leela Sixth chakra: time penance
HH 0738	Law religion (Islam)	HM 3434	Leela Daya	HM 1279	Leela Solar plane
HM 4726	Law ; Understanding knowledge (Buddhism)	HM 0697	Leela Delusion	HM 1782	Leela Sorrow
KC 0692	Law ; World	HM 0481	Leela Desire	HM 1475	Leela Spiritual devotion
VC 1055	Lawfulness	HM 1782	Leela Dharma-loka	HM 1355	Leela Su-sang-loka
VD 3136	Lawlessness	HM 3623	Leela Dukh	HM 1789	Leela Subuddhi
VP 5739	Lawlessness ; Authority	HM 0554	Leela Durbuddhi	HM 2981	Leela Sudharma
HH 3398	Laws (Buddhism)	HM 6112	Leela Dwesh	HM 5099	Leela Sukh
			Leela Earth	HM 0567	Leela Suvidya

Littleness

HM 1052	Leela Swarga-loka
HM 0455	Leela Swatch
HM 6238	Leela Tamas
HM 5561	Leela Tamoguna
HM 1773	Leela Tapah
HM 5244	Leela Tapah-loka
HM 5028	Leela Teja-loka
HM 0717	Leela Third chakra: theatre karma
HM 6287	Leela Uranta-loka
HM 5197	Leela Uttam gati
HM 1301	Leela Vaikuntha-loka
HM 6434	Leela Vayu-loka
HM 1601	Leela Vivek
HM 6002	Leela Vyana-loka
HM 3846	Leela Wisdom
HM 5965	Leela Yaksha-loka
HM 1279	Leela Yamuna
HM 2415	Left brain consciousness
KC 0060	Legal consciousness
VC 1061	Legality
VP 5998	Legality-Illegality
MS 8383	Legendary literary characters ; Historical,
KC 0770	Legislation ; Unification
VC 1063	Legitimacy
VC 1065	Leisure
HH 0410	Leisure
KD 2140	Lemma interparadigmatic dialogue ; Logos
MP 1132	Length communication pathways ; Limiting
VC 1067	Leniency
VP 5756	Leniency-Compulsion
HM 2376	Leo-consciousness (Astrology)
TC 1739	Less-haste participant type
HH 1187	Lesser stream enterer (Buddhism)
HH 0845	Lesser vehicle Buddhism (Buddhism)
HM 0767	Let go illusion ; Unwillingness (Korean)
VD 3144	Lethargy
HM 2926	Lethargy (Buddhism)
HM 6328	Leucosis (Esotericism)
HH 0656	Level ; Aspiration
HM 3005	Level awareness ; Multi (Psychism)
MP 1072	Level ; Competitive interaction opportunities transposed concrete
	Level connections
MP 1213	— Distribution secondary inter
MP 1216	— Inter
MP 1212	— transitions boundary orientation ; Primary inter
HH 0634	Level ; Ego
MP 1215	Level formation ; Initial
MP 1211	Level generation ; Boundary expansion permitting new
MP 1219	Level generation minimum tension
	Level [integrity...]
MP 1227	— integrity ; Inter
KC 0127	— interaction
MP 1240	— interfaces ; Tolerance
HH 0768	Level living
HM 6006	Level musical inspiration ; Avataric
MP 1021	Level structural limit ; Four
KC 0301	Level ; Theoretical integration
MP 1222	Level visibility ; Ground
MP 1226	Level zone ; Inter
	Levels [accessibility...]
MP 1127	— accessibility ; Nested
MP 1228	— Appropriate superstructure contain transitions
HM 2391	— awareness
MP 1135	Levels exposure it ; Enhancing insight varying
MP 1158	Levels ; External access higher structural
MP 1195	Levels ; Framework transition structural
MP 1210	Levels ; Harmonizing space distribution
KC 0841	Levels ; Integrative
MP 1096	Levels ; Limitation number structural
MP 1169	Levels ; Maintaining distinctions contextual
KC 0841	Levels organization
MP 1133	Levels structure ; Integrating transition pathways
HH 1010	Levels understanding ; Five (Buddhism)
MM 2083	Lever
HM 8634	Levitation
HM 2706	Levitational submission (ICA)
VD 3146	Levity
HM 2623	Lhag mthong manaskara (Buddhism)
HM 2207	Lhag-mthong (Tibetan)
HH 5213	Li
VD 3148	Liability
HH 6122	Liberal capitalist human development
HH 4033	Liberal humanist human development
VC 1069	Liberality
HM 8241	Liberated consciousness (Buddhism)
HM 6558	Liberated souls ; Knowledge (Hinduism)
	Liberation
HH 0613	— [Liberation]
VC 1071	— [Liberation]
TP 0040	— [Liberation]
HH 0388	— [Liberation]
HM 5977	Liberation (Buddhism)
HH 0388	Liberation ; Collective
HM 2196	Liberation consciousness (Hinduism)
HM 3244	Liberation heart (Sufism)
HM 8125	Liberation means wisdom (Pali)
HM 0125	Liberation mind (Pali)
HM 7983	Liberation (Pali)
HM 3131	Liberation possessions (ICA)
HM 3240	Liberation relationships (ICA)
HH 0907	Liberation self
HM 2907	Liberation spirit attraction divine grace (Sufism)
HH 4552	Liberation theology (Christianity)
HM 1833	Liberation ; Transcendent (Taoism)
HM 3869	Liberation (Yoga)
HM 2635	Liberations ; Eight (Buddhism)
VC 1073	Liberty
HH 0986	Libido development
HM 2512	Libra-consciousness (Astrology)
HH 2099	Licence
VD 3150	Licentiousness
VC 1075	Life
	Life [Alternative...]
HH 0477	— Alternative ways (AWL)
KC 0115	— Artistic
HH 0754	— Artistic (Hinduism)
	Life [Beginner...]
HH 0102	— Beginner spiritual (Christianity)
HM 3229	— being (ICA)
HM 0201	— (Buddhism)
HH 1987	Life ; Celibate stage (Hinduism)
HM 2506	Life chastity (ICA)
VT 8043	Life∗complex
	Life [contemplation...]
HM 2109	— contemplation (ICA)
HH 2145	— Contemplative (Christianity)
HM 6221	— continuum (Buddhism)
HH 0021	— crisis rites
HH 0754	— Cultural (Hinduism)
	Life cycle
HH 0511	— [Life cycle]
KC 0965	— [Life cycle]
HH 0285	— Sequence psychological stages
MM 2073	Life cycles ; Animal
	Life [Death...]
VP 5407	— Death
HM 1261	— Desire formless (Buddhism)
HM 3871	— Desire (Yoga)
HH 0647	— Development quality human
HM 3018	— doing (ICA)
HH 0477	— Dominant ways (DWL)
HM 1294	— Doubt efficacy good (Buddhism)
	Life [Emotional...]
HM 2954	— Emotional sexual stage
HM 2437	— Endless (ICA)
HH 0731	— energy ; Universal
HH 0958	— Eremetical
HM 3340	— essence ; Unification (Sufism)
	Life [Faculty...]
HM 0261	— Faculty (Buddhism)
HM 3543	— Family
HH 1801	— Family (Christianity)
HM 3191	— Fifth stage
HM 3320	— First stage
HM 3137	— fluid emblems - star (Tarot)
HM 3289	— fluid emblems - temperance (Tarot)
HM 3583	— form ; Desire (Buddhism)
HM 2563	— Four stages (Hinduism)
HM 2278	— Game
HM 3518	Life ; Heart awakening stage
HM 2343	Life ; Householder stage (Hinduism)
	Life [I...]
HM 3060	— I transcendent stage
HH 0750	— instinct
KC 0551	— Integral
HM 3008	— Interior (Christianity)
HM 2987	Life ; Journeying transcendence death
HM 2801	Life knowing (ICA)
HM 4297	Life ; Knowledge recollection past (Buddhism)
	Life [Mansions...]
HM 2969	— Mansions exemplary (Christianity)
HM 3234	— meditation (ICA)
HH 2056	— Mendicant stage (Hinduism)
HM 3464	— Mental volitional stage
HH 0592	— Mode
KC 0824	— Mode
HH 0767	— Mutational phases
HM 3191	— Mystical stage
HM 2888	Life ; Near death review (NDE)
HM 2024	Life obedience (ICA)
	Life [Perfection...]
HH 3652	— Perfection spiritual (Christianity)
HM 2299	— poverty (ICA)
HM 2511	— prayer (ICA)
HM 3404	— principle (Buddhism)
HH 0921	Life ; Qabalistic tree
HM 2161	Life reading (Psychism)
HH 2782	Life ; Recluse stage (Hinduism)
	Life [Saintly...]
HM 3518	— Saintly
HM 2954	— Second stage
HH 0460	— Seven stages
HM 3060	— Sixth stage
HH 0511	— Social phases
HH 0249	— space
HH 0101	— Spiritual
HH 1366	— Stages
HH 0102	— Stages spiritual
HM 2347	— state bardo awareness (Buddhism)
KC 0204	— style
HH 0941	— style
	Life [theme...]
KC 0204	— theme
HH 0474	— themes
HM 3464	— Third stage
	Life [Unitary...]
HH 0311	— Unitary
HM 2137	— Unitive
HH 0574	— urge
HH 7231	— urge
HM 3320	Life ; Vital physical stage
VD 3152	Lifelessness
HH 0955	Lifelong education
HH 0010	Lifelong educational transformation human personality
HH 0592	Lifestyle
HH 3356	Lifestyle ; Disruption
HH 0477	Lifestyles ; Alternative
VC 1077	Light
MS 8605	Light
HM 2335	Light awareness ; Death bardo clear (Buddhism)
HH 6452	Light ; Circulation (Taoism)
VP 5335	Light-Darkness
HM 3039	Light ; Fountains (Sufism)
HM 3559	Light-induced experience
HM 0361	Light-mindedness (Yoga)
HM 3381	Light ; Near death awareness (NDE)
HM 3180	Light ; Near death entering
HM 2107	Light ; Path intelligence (Judaism)
TP 0030	Light ; Shining (Fire)
HM 4595	Light state hypnosis
HM 3690	Light ; Yoga clear (Buddhism)
VC 1079	Lightness
HM 3403	Lightness aggregate consciousness (Buddhism)
HM 1435	Lightness aggregate sensation, aggregate cognition, aggregate synergies (Buddhism)
HM 4395	Lightness body (Buddhism)
HM 6033	Lightness consciousness (Buddhism)
HM 7636	Lightness mental body (Buddhism)
HM 1830	Lightness thought (Buddhism)
VP 5352	Lightness ; Weight
KC 0316	Likeness
HM 2278	Lila
HM 2355	Limbic human brain consciousness ; Mammalian
HM 1478	Limbo
HM 2934	Limbo ; Transitional (Brainwashing)
HM 1809	Limerence
MP 1021	Limit ; Four level structural
HM 3435	Limit ; Not trespassing (Buddhism)
MP 1022	Limit ; Occupiable temporary site
TP 0060	Limitation
MP 1103	Limitation number occupiable temporary sites
MP 1096	Limitation number structural levels
VP 5507	Limitation-Unlimitedness
	Limited concentration
HM 1073	— (Buddhism)
HM 5547	— infinite object (Buddhism)
HM 1175	— limited object (Buddhism)
HM 5547	— measureless object (Buddhism)
	Limited object
HM 4653	— Infinite concentration (Buddhism)
HM 1175	— Limited concentration (Buddhism)
HM 4653	— Measureless concentration (Buddhism)
HM 2495	— Understanding having (Buddhism)
VD 3154	Limitedness
MP 1134	Limiting exposure harmonious perspectives
MP 1132	Limiting length communication pathways
KD 2337	Limiting patterns interaction ; Self
HM 3043	Limitless consciousness absorption (Buddhism)
HM 2110	Limitless space absorption (Buddhism)
VC 1081	Limitlessness
HM 2674	Limits ; Final (ICA)
HH 6129	Lin chi (Zen)
TC 1972	Line ; Conference
KC 0324	Line ; World
VC 1083	Lineage
HM 4637	Lineage ; Knowledge change (Buddhism)
TC 1123	Linear agendas ; Non
MP 1173	Linear contexts communication pathways ; Protecting non
MP 1163	Linear domain external structures ; Hospitable non
MP 1175	Linear extensions structures ; Insight capturing non
MP 1007	Linear organization ; Non
MP 1051	Linear processes ; Linear relationships enhanced non
KC 0626	Linear programming
MP 1051	Linear relationships enhanced non-linear processes
HH 2287	Linear thinking
TC 1123	Linear thinking ; Beyond
MP 1060	Linearity ; Accessible non
	Linearity [Communication...]
MP 1174	— Communication pathways enfolded non
MP 1172	— Contexts self organizing non
MP 1170	— Cultivation productive non
MP 1118	Linearity integrative superstructure ; Integration non
KC 0347	Linearity ; Non
MP 1176	Linearity ; Sites grounding perspectives non
MP 1171	Linearity structures ; Appropriate relationship non
TC 1651	Liners ; Party
HM 3212	Linga stage gunas (Yoga)
HH 0843	Linguistic intelligence
HH 4872	Linguistic programming ; Neuro (NLP)
HH 0585	Linguistics
HM 8266	Linking ; Rebirth (Buddhism)
HM 0172	Liquid plane (Leela)
MS 8606	Liquids ; Foods
HH 1183	Listening dharma (Buddhism)
HH 5124	Listening heart
TC 1913	Listening process ; Creative
TC 1913	Listening process ; Systematic
TC 1742	Listening teams
VD 3156	Listlessness
MM 2053	Literacy
HM 1525	Literal faith ; Mythic
MS 8383	Literary characters ; Historical, legendary
MS 8115	Literature ; Language
MS 8423	Literature ; Symbols sacred
VD 3158	Littleness
VP 5195	Littleness ; Bigness

Liturgical

HH 4209	Liturgical prayer (Christianity)
HM 3871	Live ; Will (Yoga)
HM 7887	Livelihood ; Abstinence wrong (Buddhism)
HM 0549	Livelihood ; Correct (Buddhism)
HM 1763	Livelihood ; Correct (Buddhism)
HM 0252	Livelihood ; Leaving off, abstaining, totally abstaining refraining wrong (Buddhism)
HM 1341	Livelihood ; Right (Buddhism)
VC 1085	Liveliness
HM 3379	Living (Buddhism)
HM 2808	Living death (ICA)
HM 3404	Living ; Faculty (Buddhism)

Living [Immediacy...]

HH 0223	— Immediacy fulfilled
HH 2434	— in-dialogue
HM 2394	— Incarnate (ICA)
HH 0768	Living ; Level
HH 2345	Living present
HM 2747	Living ; Problemless (ICA)
HH 0768	Living ; Standard
HM 3178	Living word (ICA)
HM 4712	'Ilah ; Al mahabba fi (Islam)
HM 5580	'Ilah ; Al mahabba ma'a (Islam)
HH 0873	Lo-rig (Tibetan)
VD 3160	Loathsomeness
HM 1931	Lobh (Leela)
HM 3283	Lobha (Pali)
HM 0802	Lobha (Pali)
HM 1954	Lobho akusalamulam (Pali)
MP 1045	Local action network
MP 1177	Local cultivation sources perspective nourishment
MP 1044	Local focal points
MP 1011	Local interrelationship domains
MP 1157	Local opportunities perspective activity
MP 1088	Local perspective interface zones ; Informal
MP 1049	Local relationship loops
MP 1029	Local relationships ; Stable density gradient
MP 1089	Local sources perspective nourishment
VC 1087	Location
VP 5185	Location-Dislocation
MP 1093	Location sources perspective nourishment ; Transit point
HM 2794	Loci senses ; Emptiness (Buddhism)
MP 1142	Loci structure ; Sequence viewpoint

Locomotion [Animal...]

MM 2010	— Animal
MM 2025	— Aquatic animal
MM 2039	— Arboreal animal
MM 2064	Locomotion flight ; Aerial animal
MM 2033	— Fossorial animal
MM 2059	Locomotion gliding soaring ; Aerial animal
MM 2012	Locomotion ; Terrestrial animal
KC 0605	Logic
KC 0758	Logic ; Fuzzy
KC 0339	Logic ; Holographic
KC 0446	Logic ; Mathematical
KC 0603	Logic ; Multi valued
KD 2010	Logic ; Oppositional
KC 0937	Logic ; Process
KC 0776	Logic ; Relational
KC 0776	Logic relations
KC 0446	Logic ; Symbolic
KD 2105	Logic whole ; Trialectics:
TC 1965	Logical biases ; Participant pre
KC 0460	Logical cultural integration
HM 2564	Logical dishonouring (Brainwashing)
HH 1456	Logical-mathematical intelligence
HM 2320	Logical metalogical awareness
HH 1276	Logical positivism
KC 0912	Logistics
KC 0511	Logos
HM 2248	Logos ; Angelic awareness
KD 2140	Logos lemma interparadigmatic dialogue
HH 0416	Logotherapy

Loka [Adha...]

HM 3697	— Adha (Jainism)
HM 0859	— Agnih (Leela)
HM 6901	— Anand (Leela)
HM 0145	— Apana (Leela)

Loka [Bhakti...]

HM 1475	— Bhakti (Leela)
HM 1024	— Bhu (Leela)
HM 4600	— Bhuvar (Leela)
HM 4326	— Brahma (Leela)
HM 2072	Loka consciousness (Buddhism)
HM 0481	Loka ; Dharma (Leela)
HM 0008	Loka ; Gandha (Leela)
HM 4376	Loka ; Harsha (Leela)
HM 1276	Loka ; Himsa (Leela)
HM 0172	Loka ; Jala (Leela)
HM 5155	Loka ; Jana (Leela)

Loka [Kama...]

HM 3627	— Kama (Leela)
HM 0948	— Karma (Leela)
HM 0764	— Ku sang (Leela)

Loka [Madhya...]

HM 1357	— Madhya (Jainism)
HM 1351	— Maha (Leela)
HM 1351	— Mahar (Leela)
HM 1485	Loka ; Naga (Leela)
HM 0856	Loka ; Narka (Leela)
HM 3907	Loka ; Prakriti (Leela)
HM 4399	Loka ; Prana (Leela)
HM 0878	Loka ; Rasa (Leela)
HM 4343	Loka ; Rudra (Leela)

Loka [Satya...]

HM 1293	— Satya (Leela)
HM 0569	— Siddha (Jainism)
HM 1355	— Su sang (Leela)
HM 1052	— Swarga (Leela)
HM 5244	Loka ; Tapah (Leela)
HM 5028	Loka ; Teja (Leela)
HM 6287	Loka ; Uranta (Leela)
HM 0131	Loka ; Urdhva (Jainism)

Loka [Vaikuntha...]

HM 1301	— Vaikuntha (Leela)
HM 6434	— Vayu (Leela)
HM 6002	— Vyana (Leela)
HM 5965	Loka ; Yaksha (Leela)
HM 2120	Lokuttara consciousness (Buddhism)

Loneliness

VD 3162	— [Loneliness]
HM 1260	— [Loneliness]
HM 1539	— [Loneliness]
KC 0977	Long-range goals mankind
VC 1089	Longanimity
VC 1091	Longevity
HM 0762	Longing divine unity ; Intense (Sufism)
HM 0877	Longing (Islam)
HM 0762	Longing (Sufism)
MP 1049	Loops ; Local relationship
VD 3164	Looseness
HM 2118	Lord awareness ; Great black (Buddhism)
HH 1016	Lord ; Devotion
HM 2088	Lord-of-death awareness (Buddhism)
VD 3166	Lordliness
TC 1432	Losing focus meetings
HH 1433	Loss
HM 2423	Loss attention
KD 2329	Loss cyclic phase ; Learning
HM 8303	Loss discernment (Islam)
HM 2548	Loss experience reality
HM 1336	Loss personal attribution
VP 5808	Loss ; Possession
HM 2632	Loss ; Soul
HH 0210	Loss soul
HM 7602	Loss ; Understanding skill (Buddhism)
HM 0176	Lotus ; Sixteen petalled (Yoga)
HH 3443	Lotus sutra (Buddhism)
HM 3398	Lotus ; Thousand petalled (Yoga)
HM 0176	Lotus ; Twelve petalled (Yoga)
TC 1972	Loud speaker telephone
VD 3168	Loudness
VP 5451	Loudness ; Silence

Love

KC 0181	— [Love]
VC 1093	— [Love]
HH 0258	— [Love]

Love [Absence...]

HM 7344	— Absence (Hinduism)
HM 1780	— Aesthetic emotion (Hinduism)
HM 0212	— Androgynous
HM 2401	— awareness ; Divine (Sufism)

Love [based...]

HM 6523	— based reason (Islam)
HM 8772	— Blissful (Christianity)
HM 7607	— (Buddhism)
HM 1234	— (Buddhism)

Love [Carnal...]

HM 9406	— Carnal
HM 6732	— Christian self (Christianity)
HM 0973	— Companionate
HM 2000	— Conscious
HM 2000	— consciousness
HM 2278	— Cosmic
HM 6122	— Courtly
HM 3012	Love ; Degrees (Christianity)
HM 6276	Love ; Delusions
HM 0334	Love ; Existential
HM 5087	Love ; Falling
HH 5232	Love God (Christianity)

Love God Islam

HM 4712	— [Love God (Islam)]
HM 5116	— [Love God (Islam)]
HM 5580	— [Love God (Islam)]
HM 4273	Love God (Sufism)
HM 2929	Love ; Great (Buddhism)
VP 5930	Love-Hate
HM 9845	Love (Hinduism)

Love [Idyllic...]

HM 5087	— Idyllic
HM 7259	— (Islam)
HM 5712	— (Islam)

Love [Mannered...]

HM 4905	— Mannered
HM 0973	-- Mature
HM 5087	— Mutual passionate
HM 3941	— Mystical ladder divine (Christianity)
HH 0042	Love neighbour
HM 5702	Love ; Neurotic
HM 1211	Love ; Ontological

Love [pain...]

HM 6122	— pain
HM 3656	— Passionate (Islam)
HM 0074	— Pure (Christianity)
HM 5087	Love ; Romantic
HM 1809	Love ; Romantic

Love Self

HH 1066	— [Love ; Self]
HM 0478	— [Love ; Self]
HM 4671	— aggrandizing

Love Self cont'd

HH 6509	— surrender
HH 0258	— transformation

Love [Selfless...]

HH 4477	— Selfless (Christianity)
HM 9406	— Sexual
HM 8977	— Spiritual (Christianity)
HM 0074	— State perfect (Christianity)
HM 4905	— Sympathy
HM 7045	Love ; Therapeutic
HM 7045	Love ; Transference
HM 1266	Love ; Unconditional (Sufism)
HM 4671	Love ; Vanity
HM 1367	Love whole
HM 1507	Love wisdom ; Way (Esotericism)
VD 3170	Lovelessness
HM 3028	Lovers (Tarot)
HH 0628	Loving faith ; Way (Hinduism)
HM 0976	Loving mind ; Christ consciousness (Christianity)
HM 7607	Lovingkindness (Buddhism)
MP 1101	Low density communication pathways ; Protected
MP 1059	Low-intensity communication pathways
MP 1055	Low intensity relationships ; Protected
HH 0720	Low magic
HH 3667	Low paths ; Middle grade (Taoism)
HH 4289	Low paths ; Upper grade (Taoism)
HM 6342	Lower fetters (Buddhism)
HH 6997	Lower knowing (Christianity)
HM 2590	Lower manas ahamkara (Hinduism)
HH 3296	Lower middle grade (Taoism)
HM 2268	Lower path awareness ; Great vehicle (Buddhism)
HM 3274	Lower qualities ; Freedom (Sufism)
HM 2790	Lower unconscious (Psychosynthesis)
HH 4711	Lower upper grade (Taoism)
HH 4864	Lower vehicle path (Taoism)
HM 1219	Lowered state consciousness (Psychism)
HH 3013	Lowest paths (Taoism)
VD 3172	Lowness
VP 5207	Lowness ; Height
HM 2910	Loyal opposition (ICA)
VC 1095	Loyalty
TC 1345	Loyalty
HM 2542	LSD experience ; Aesthetic
HM 3271	LSD experience ; Psychodynamic
HM 2996	Lta-ba-nyon-mongs-can (Tibetan)
HM 1442	Lubbhana (Pali)
HM 1513	Lubbhitattam (Pali)
HM 0750	Lucid awareness (Psychism)
HM 3618	Lucid dreaming
HH 4087	Lucid dreams (Buddhism)
VC 1097	Lucidity
HM 2184	Lucidity ; Transparent (ICA)
VC 1099	Luck
VD 3174	Lucklessness
HH 0633	Ludo therapy
HM 3065	Luminous change (ICA)
MS 8435	Lunar nature ; Sub
HM 3779	Lunar plane (Leela)
HM 2562	Lus-kyi-rnam-par-shes-pa (Tibetan)
VD 3176	Lust
HM 4666	Lust (Hinduism)
HM 4772	Lust ; Physical (Islam)
HM 4180	Lust ; Spiritual (Christianity)
HH 0401	Lustration
HM 1547	Lustre wisdom (Buddhism)
VC 1101	Luxury
HM 3726	Lying ; Pathological

M

HM 5580	Ma'a 'Ilah ; Al mahabba (Islam)
HM 4392	Ma'arifa (Sufism)
HM 2128	Ma-chags-pa (Tibetan)
HM 2461	Ma-dad-pa (Tibetan)
HM 1709	Ma (Japanese)
HM 2254	Ma'rifa (Sufism)
HM 2254	Ma'rifat (Sufism)
HM 3196	Ma-rig-pa (Tibetan)
KC 0081	Machine systems ; Man
MS 8102	Machines supplies
VC 1103	Machismo
KD 2070	Macro complementarity ; Observer entrapment micro
KC 0571	Macroevolution
HM 2528	Mada (Buddhism)
HM 0098	Mada (Leela)
HM 0191	Maddavata (Pali)
HM 3500	Made bear cross ; Jesus (Christianity)
MS 8955	Made beings ; Beings
MS 8600	Made forms structures ; Man
HH 0777	Made person ; Self
VC 1911	Made ; Well
HM 1357	Madhya-loka (Jainism)
HH 1038	Madhyamaka (Buddhism)
HH 1010	Madhyamaka buddhism (Buddhism)
VD 3178	Madness
HM 0796	Magga-bhavana (Buddhism)
HH 4007	Maggamagga-nanadassana-visuddhi-niddesa (Pali)
HH 0720	Magic
HM 5120	Magic Ceremonial magic
HH 6053	Magic Guided visualization
HH 0720	Magic ; High
HH 3432	Magic Initiation
HH 0720	Magic ; Low
HH 4521	Magic Mediation
TC 1567	Magic ; Meeting

Code	Entry
HH 5120	Magic Ritual pattern-making
HH 3556	Magic ; White
HM 2090	Magical consciousness
HH 0720	Magical development
HM 2679	Magical-forces awareness (Buddhism)
MS 8925	Magical garments
HH 5547	Magical meditation
HM 2639	Magical-shamanic state (Buddhism)
HM 0359	Magical thinking
HH 0720	Magical training
HM 2237	Magician archetypal image (Tarot)
HH 0497	Magico-religious powers
HH 0834	Magnanimity
VC 1105	Magnanimity
TC 1800	Magnetic board
MM 2072	Magnetic containment plasma
HH 2552	Magnetic healing
VC 1107	Magnificence
HM 2237	Magus (Tarot)
HM 1351	Maha-loka (Leela)
HH 0550	Maha paramita (Zen)
HM 4712	Mahabba fi 'Ilah ; Al (Islam)
HM 6523	Mahabba (Islam)
HH 5712	Mahabba (Islam)
HM 5580	Mahabba ma'a 'Ilah ; Al (Islam)
HM 2401	Mahabba (Sufism)
HM 4273	Mahabbat (Sufism)
HM 2118	Mahakala (Buddhism)
HM 2656	Mahamudra (Buddhism)
HH 8677	Mahamudra meditation
HM 1351	Mahar-loka (Leela)
HM 3764	Maharishi effect
HM 2995	Mahashunyata (Buddhism)
HH 0900	Mahayana Buddhism (Buddhism)
HM 2232	Mahayana climax-application awareness (Buddhism)
HM 3072	Mahayana ; Emptiness indestructible (Buddhism)
HM 2208	Mahayana heat-application awareness (Buddhism)
HM 2271	Mahayana highest-teachings-application awareness (Buddhism)
	Mahayana path Buddhism
HM 2268	— [Mahayana-path (Buddhism)]
HM 3048	— [Mahayana path (Buddhism)]
HM 2160	— [Mahayana path (Buddhism)]
HM 2247	Mahayana receptivity-application awareness (Buddhism)
MP 1110	Main entry point structure ; Distinctiveness
MP 1234	Maintainable, multi-element external boundaries
MP 1169	Maintaining distinctions contextual levels
TC 1564	Maintenance conferencing
MP 1083	Maintenance dynamics ; Integration perspective acquisition perspective
MP 1181	Maintenance source direct insight structures focal point
HH 0306	Maithuna (Buddhism)
HH 2034	Maithuna (Yoga)
VC 1109	Majesty
HM 4526	Majid ; Al (Sufism)
HM 5446	Majid ; Al (Sufism)
HM 7002	Majjhattata ; Tatra (Pali)
TC 1421	Majoral conferencing
HM 3103	Maker ; Symbol (ICA)
TC 1385	Making conferencing ; Decision
KC 0340	Making ; Decision
TC 1627	Making process ; Decision
HH 5120	Making ; Ritual pattern (Magic)
HH 0142	Making ; Soul
HM 7089	Mala'ika ; Wilayat (Sufism)
VD 3180	Maladjustment
VD 3182	Maladroit
VD 3184	Malaise
HM 4280	Malamat (Sufism)
HM 0116	Malakut (Sufism)
VD 3186	Malevolence
VD 3188	Malfeasance
VD 3190	Malformation
VD 3192	Malice
HM 1111	Maliciousness (Buddhism)
VD 3194	Malignancy
HM 4061	Malik ; Al (Sufism)
HM 5902	Malik-ul-Mulk (Sufism)
VD 3196	Malingering
HM 2288	Malkuth sephira (Kabbalah)
HM 1155	Malleability consciousness (Buddhism)
HM 3696	Malleability mental body (Buddhism)
VD 3198	Malpractice
VD 3200	Maltreatment
HM 2355	Mammalian limbic human brain consciousness
HM 2604	Man alone awareness ; Ox forgotten leaving (Zen)
HM 2202	Man archetypal image ; Old wise (Tarot)
	Man [becomes...]
HM 0933	— becomes himself ; Fifth chakra: (Leela)
HM 0978	— Birth (Leela)
HM 2492	— both out sight awareness ; Ox (Zen)
	Man [Complete...]
HH 0727	— Complete
HH 0351	— Concept
HM 2920	— consciousness ; Perfect (Sufism)
HH 0134	Man ; Elevation (Christianity)
HH 1778	Man ; Fall
HH 3567	Man ; Fall
HH 2764	Man ; Generative
HH 2764	Man ; Good
HM 2190	Man ; Hanged (Tarot)
HH 0351	Man ; Image
HH 0338	Man knowledge
	Man [machine...]
KC 0081	— machine systems
MS 8600	— made forms structures
HM 2766	— man ; Relationship (Christianity)
HM 3488	Man nature ; Relationship (Christianity)
HM 2183	Man ; Perfect (Hinduism)
HH 0338	Man power
HM 3595	Man ; Relationship God (Christianity)
HM 2766	Man ; Relationship man (Christianity)
HM 3161	Man ; Transfigured (ICA)
HH 0727	Man ; Universal
MS 8955	Man woman
HM 2686	Mana
HM 2823	Mana (Christianity)
HM 2686	Mana personalities
VC 1111	Manageability
KC 0867	Management
	Management [complexity...]
KC 0604	— complexity
HH 0421	— conflict
KC 0293	— Crisis
KC 0116	— Cybernetic
KC 0887	Management ; Human systems
KC 0540	Management science
KC 0540	Management ; Scientific
TC 1313	Managerial-political conferencing
KD 2170	Managing: embracing conflicting styles ; Modes
HM 7298	Manahparyayajnana (Hinduism)
HM 2590	Manas ahamkara ; Lower (Hinduism)
HM 2902	Manas awareness (Hinduism)
HM 0251	Manasa (Pali)
HM 6644	Manasa-pratyaksa (Buddhism)
HM 3237	Manasi-kara (Pali)
HM 3237	Manaskara (Buddhism)
HM 2623	Manaskara ; Lhag mthong (Buddhism)
HM 2171	Manaskaradhikarmika (Buddhism)
HM 1736	Manayatana (Pali)
HH 0112	Mandalas
HM 7220	Mani ; Al (Sufism)
HM 2787	Mania
HM 3594	Mania
HH 1616	Manic depression
HM 3007	Manifestation conscience (Christianity)
HM 2642	Manifestation essence veils (Sufism)
HM 3170	Manifestation ; Prerequisites (Buddhism)
HM 2312	Manifestation ; World (Judaism)
KC 0854	Manifold
HM 2269	Manifold blessings (ICA)
HM 1369	Manindriya (Pali)
HH 0096	Manipulation
HH 1020	Manipulation ; Destructive
HH 0396	Manipulation ; Self
HM 3063	Manipura (Yoga)
KC 0057	Mankind ; Collectivization
KC 0977	Mankind ; Long range goals
KC 0057	Mankind ; Planetization
HM 3501	Mankind ; Third step: (Christianity)
KC 0087	Mankind ; Unity
VC 1115	Manliness
HM 4905	Mannered love
VC 1913	Mannered ; Well
VC 1117	Manners
HM 3323	Mano ; Bhavanga (Pali)
HM 1743	Mano (Pali)
VC 1113	Manoeuvrability
HM 2912	Manomaya kosa (Yoga)
HM 2838	Manovijnana (Buddhism)
HM 0223	Manovilekha ; Cittassa (Pali)
HM 1834	Manovinnanadhatu (Pali)
	Manovinnanadhatu [samphasajam...]
HM 1921	— samphasajam cetasikam satam (Pali)
HM 0404	— samphassaja cetana (Pali)
HM 1501	— samphassaja sanna (Pali)
HM 0234	Manovinnanadhatusampassa jam cetasikam asatam (Pali)
HM 3314	Manovinnanadhatusamphassajam cetasikam neva satam nasatam (Pali)
HM 2838	Manovynana (Buddhism)
HH 0745	Manpower resource development (United Nations)
HH 0745	Manpower training (United Nations)
HM 2969	Mansions exemplary life (Christianity)
HM 2470	Mansions favours afflictions (Christianity)
HM 3382	Mansions humility (Christianity)
HM 4110	Mansions incipient union (Christianity)
HM 3218	Mansions practice prayer (Christianity)
	Mansions [soul...]
HH 1409	— soul (Christianity)
HM 3443	— spiritual consolations (Christianity)
HM 3971	— spiritual marriage (Christianity)
HH 0545	Mantic arts
HM 2464	Mantra-consciousness (Yoga)
HM 2464	Mantra-induced experience (Yoga)
HH 0931	Mantra yoga (Yoga)
HH 0386	Mantras
HM 0978	Manushya-janma (Leela)
HM 7672	Many ; One becoming (Buddhism)
HM 1264	Manzur ; Al fikr fi 'l (Islam)
KC 0726	Mapping
HH 8903	Maps consciousness
HH 8903	Maps ; Mental
HH 8903	Maps mind
TC 1275	Maps ; Relevance
HM 2424	Maqamat forty (Sufism)
HM 2317	Maqamat hundred (Sufism)
HM 2341	Maqamat seven-worlds (Sufism)
	Maqamat Sufism
HM 3371	— [Maqamat (Sufism)]
HM 3415	— [Maqamat (Sufism)]
HM 4311	— [Maqamat (Sufism)]
HM 6705	Marana (Hinduism)
HH 0817	Marathon ; Encounter
HH 0817	Marathon group
HM 3439	March ; Saviors God: (Christianity)
HH 0628	Marga ; Bhakti (Hinduism)
HH 5645	Marga ; Bodhi (Buddhism)
HH 0495	Marga ; Jnana (Hinduism)
HH 1211	Marga ; Karma (Hinduism)
	Marga Sravaka
HM 2240	— darsana (Buddhism)
HM 2716	— prayoga (Buddhism)
HM 2192	— sambhara (Buddhism)
TC 1129	Marginal participant type
TP 0056	Marginality (Wanderer)
HM 3691	Marijuana ; Use
HM 8673	Mark calm (Buddhism)
KC 0659	Market ; World
KC 0804	Market ; World socialist
KC 0677	Markov systems
HH 0799	Marriage ; Internal
	Marriage [Mansions...]
HM 3971	— Mansions spiritual (Christianity)
HM 3028	— Mystic (Tarot)
HM 6321	— Mystical
HH 0465	Marriage ; Spiritual (Christianity)
HH 0362	Marriage therapy
TP 0054	Marrying maiden Elective affinity
TP 0054	Marrying maiden Propriety
HH 0085	Martial arts
HM 0220	Martial arts training ; One pointedness
HH 0085	Martial way
HH 0838	Martyr
HM 0291	Marvellous sentiment (Hinduism)
MS 8385	Masculinity
VP 5420	Masculinity-Femininity
HH 1004	Masking reality ; Sheaths (Hinduism)
HH 0659	Masonic degrees (Freemasonry)
MS 8843	Masonic symbols
	Mass [communication...]
HH 0588	— communication
KC 0387	— Critical
KC 0609	— culture
MS 8723	Mass destruction weapons ; Nuclear
HH 1098	Mass evangelism (Christianity)
HH 2134	Mass hypnosis
HH 0574	Mass hysteria
HH 0588	Mass media
KC 0661	Mass society
HH 0851	Mass therapy
HH 0284	Massage
	Master awareness
HM 2301	— Eighth vajra (Buddhism)
HM 2763	— Fifth vajra (Buddhism)
HM 2849	— Final vajra (Buddhism)
HM 2251	— Fourth vajra (Buddhism)
HM 2325	— Ninth vajra (Buddhism)
HM 2703	— Second vajra (Buddhism)
HM 2789	— Seventh vajra (Buddhism)
HM 2287	— Sixth vajra (Buddhism)
HM 2727	— Third vajra (Buddhism)
HM 2235	Master form realm awareness ; Tantra (Buddhism)
HM 3944	Master ; Identification spiritual (Sufism)
	Master [sense...]
HM 2211	— sense realm awareness ; Tantra (Buddhism)
HH 2080	— Spiritual (Sufism)
HM 2187	— stage ; First vajra (Buddhism)
TP 0029	Mastering pitfalls (Abyss)
HM 3090	Masters ; Way vajra (Buddhism)
HM 3066	Masters wisdom ascension stages game (Buddhism)
VC 1121	Mastery
HM 3173	Mastery dreaming
HM 3064	Mastery intent
HM 2304	Mastery stalking
HM 6097	Masti (Sufism)
HM 2336	Material condition (Hinduism)
HM 4968	Material qualities ; Understanding way fixing (Buddhism)
	Material sphere
HM 4265	— concentration ; Fine (Buddhism)
HM 2536	— Fine (Buddhism)
HM 2450	— First trance fine (Buddhism)
HM 2586	— Fourth trance fine (Buddhism)
HM 4761	— functional ; Indeterminate consciousness fine (Buddhism)
HM 5338	— Profitable consciousness fine (Buddhism)
HM 0594	— resultant ; Indeterminate consciousness fine (Buddhism)
HM 2038	— Second trance fine (Buddhism)
HM 2062	— Third trance fine (Buddhism)
HH 0529	Materialism
HH 0529	Materialism ; Dialectical
HH 3376	Materialism ; Medical
HH 0828	Materialism ; Spiritual
KC 0457	Materialist synthesis ; Dialectical
HM 2108	Materiality aggregate (Buddhism)
HM 8273	Materiality ; Consciousness born (Buddhism)
HH 2718	Materiality ; Defining mentality (Buddhism)
VP 5376	Materiality-Immateriality
HM 4968	Materiality ; Understanding defining (Buddhism)
VC 1123	Materialization
HH 3663	Maternal conception social responsibility
HH 4646	Maternal thinking

Maternalism

Code	Entry
HH 4646	Maternalism
VC 1125	Maternity
KC 0010	Mathematical complex
KC 0586	Mathematical elegance
KC 0054	Mathematical group
	Mathematical [integration...]
KC 0986	— integration
HH 1456	— intelligence ; Logical
KC 0522	— intuitionism
KC 0446	Mathematical logic
KC 0719	Mathematical model
KC 0386	Mathematical space ; Abstract
KC 0421	Mathematics
KC 0462	Mathematics discontinuous change
HM 2191	Mati ; Arcis (Buddhism)
HM 2292	Mati ; Sadhu (Buddhism)
HM 5633	Matin ; Al (Sufism)
HM 0160	Matins (Christianity)
HM 0160	Matins ; Hours (Christianity)
HM 0635	Matrix ; Release (Taoism)
HM 2285	Matrona (Tarot)
HM 0987	Matsar (Leela)
HM 2964	Matsarya (Buddhism)
HM 0987	Matsarya (Leela)
HH 0031	Matsuri (Japanese)
HM 2108	Matter ; Aggregate (Buddhism)
HH 0196	Maturation
TC 1394	Maturation ; Conference
MP 1076	Mature emerging perspectives ; Minimal context interrelating
HM 0973	Mature love
HH 0971	Mature personality
MP 1154	Maturing perspectives ; Semi autonomous contexts
	Maturity
KC 0013	— [Maturity]
VC 1127	— [Maturity]
HH 0971	— [Maturity]
HH 0407	Maturity genitality
HH 5434	Maturity growth
HH 0971	Maturity ; Psychological
HH 0021	Maturity rituals
HH 1325	Maturity ; Sexual
HH 5223	Maturity ; Spiritual
HM 2110	Maulasamapatti (Buddhism)
TC 1879	Maximization conferencing
HM 3246	Maya (Buddhism)
HM 1004	Maya (Hinduism)
HM 0953	Maya (Leela)
HM 3192	Mayko
HM 3306	Mayoi (Zen)
HH 5856	Me you ; Journeying transcendence
TP 0030	Mean ; Cosmic (Fire)
KC 0249	Meaning
VT 8055	Meaning*complex
HM 2017	Meaning creation (ICA)
VP 5545	Meaning-Meaninglessness
HM 1663	Meaning ; Understanding knowledge (Buddhism)
VC 1915	Meaning ; Well
MP 1253	Meaningful symbols self-transformation
VC 1129	Meaningfulness
VP 5545	Meaninglessness ; Meaning
VD 3204	Meanness
HM 0227	Meanness (Buddhism)
HM 6601	Means ; Understanding skill (Buddhism)
HM 8125	Means wisdom ; Liberation (Pali)
TP 0014	Measure ; Possession great
	Measureless concentration
HM 0496	— (Buddhism)
HM 4653	— limited object (Buddhism)
HM 5214	— measureless object (Buddhism)
	Measureless object
HM 5547	— Limited concentration (Buddhism)
HM 5214	— Measureless concentration (Buddhism)
HM 6681	— Understanding having (Buddhism)
HM 2178	Measuring emblems (Tarot)
HM 3102	Measuring intelligence ; Path (Judaism)
MM 2092	Mechanical cycles
MS 8446	Medals, decorations trophies
HM 1750	Medha (Pali)
MM 2022	Media diets
MP 1113	Media ; Harmoniously structured entry point external communication
HH 0588	Media ; Mass
TC 1862	Medial conferencing
TC 1856	Mediated learning group ; Peer
HM 3089	Mediating influence ; Path intelligence (Judaism)
VC 1131	Mediation
HH 4521	Mediation (Magic)
TC 1775	Mediator participant type
MS 8715	Medical instruments
HH 3376	Medical materialism
MS 8406	Medical preparations
HH 0581	Medicine ; Holistic
KC 0096	Medicine ; Holistic
VD 3206	Mediocrity
HM 2918	Meditating kasinas (Buddhism)
HH 0761	Meditation
HH 4008	Meditation ; Active
HM 8634	Meditation ; Ascension
HH 0551	Meditation ; Buddhist (Buddhism)
HH 3783	Meditation ; Burmese
	Meditation [Cave...]
HH 1932	— Cave
HH 5023	— Christian (Christianity)
TC 1135	— Conferencing
HH 0816	— Contemplative intuitive
	Meditation [crystals...] cont'd
HH 4908	— crystals
	Meditation [Detached...]
HH 0229	— Detached self observation
HH 1688	— Discipline (Christianity)
HH 0197	— Discursive (Christianity)
HH 0787	— Dynamic self opening
HH 1100	Meditation ; Electronic
HM 2864	Meditation emptiness ; Tantra (Buddhism)
HH 0508	Meditation ; Free self inquiry
HH 0632	Meditation ; Free thinking
HM 6227	Meditation God (Sufism)
HH 2304	Meditation ; Guided
HM 0137	Meditation (Hinduism)
HH 1562	Meditation ; Hypergnostic
HM 0579	Meditation ; Integral
HH 1134	Meditation ; Inward zazen (Zen)
HH 1097	Meditation ; Jewish (Judaism)
HH 1169	Meditation ; Jogging
HH 3246	Meditation ; Kasina (Buddhism)
HH 2031	Meditation ; Kinerhythm
HH 3234	Meditation ; Life (ICA)
	Meditation [Magical...]
HH 5547	— Magical
HH 8677	— Mahamudra
HH 0741	— motion (Islam)
HH 5256	— Motionless
HH 0621	Meditation ; Outward zazen (Zen)
HH 2054	Meditation ; Process
HH 2213	Meditation ; Pyramid
HH 4022	Meditation ; Quaker (Christianity)
HH 0976	Meditation ; Reflective rational
HH 1134	Meditation ; Rinzai zazen (Zen)
	Meditation [Seed...]
HM 3692	— Seed
HH 0621	— Soto zazen (Zen)
HM 2311	— state bardo awareness (Buddhism)
HH 3287	— Structured
HH 3987	— Subjects (Buddhism)
	Meditation subjects
HH 3534	— Divine abidings (Buddhism)
HH 3198	— Immaterial states (Buddhism)
HH 6221	— Recollections (Buddhism)
	Meditation [Tenth...]
HM 1717	— Tenth order perceptions universal oneness
HH 0976	— Thought
HH 0682	— Transcendental (TM)
	Meditation way
HM 2769	— bodhisattvas (Buddhism)
HM 2245	— four truths (Buddhism)
HM 2161	— hearers (Buddhism)
HM 2709	— solitary realizers (Buddhism)
HM 3911	Meditation ; Witness
	Meditation [Zazen...]
HH 0882	— Zazen (Zen)
HH 0785	— (Zen)
HH 0882	— Zen (Zen)
HM 1593	Meditational insight ; Religious experience
	Meditative [absorption...]
HH 0827	— absorption ; Path (Yoga)
HH 1690	— absorption (Yoga)
HH 0610	— attitude ; Failure hold (Yoga)
HH 4089	Meditative prayer
HH 5550	Meditative reading
HM 0931	Meditative stabilization ; Right (Buddhism)
HM 2968	Meditative stabilizations (Buddhism)
	Meditative states
HM 2693	— form concentrations ; Tibetan (Buddhism)
HM 2669	— formless absorptions ; Tibetan (Buddhism)
HM 2145	— mental abidings (Buddhism)
HM 2785	Meditative thinking ; Intuitive,
HM 6089	Medium concentration (Buddhism)
HM 1094	Medium ; Inspired (Psychism)
HH 0878	Medium ; Spirit
HM 5856	Medium state hypnosis
VC 1133	Meekness
HH 0414	Meekness
HM 6226	Meekness ; Spiritual (Christianity)
TP 0044	Meeting
TC 1603	Meeting analogies
	Meeting configurations
TC 1943	— Constrained
TC 1155	— Socio structural
TC 1699	— Unconstrained
	Meeting Electronic
TC 1418	— [Meeting ; Electronic]
TC 1218	— [Meeting ; Electronic]
TC 1876	— [Meeting ; Electronic]
TC 1699	Meeting energy dissipation
	Meeting [fatigue...]
TC 1776	— fatigue
TC 1351	— focal processes
TC 1285	— focus ; Varieties
	Meeting God
HM 6204	— active desire (Christianity)
HM 5309	— both resting working accordance righteousness (Christianity)
HM 0541	— (Christianity)
HM 8772	— emptiness (Christianity)
TC 1567	Meeting magic
TC 1915	Meeting ; Prayer
	Meeting [serendipity...]
TC 1567	— serendipity
TC 1304	— space
TC 1972	— Speakerphone
TC 1567	Meeting take off
	Meetings [aesthetic...]
TC 1909	— aesthetic symbols
TC 1529	— agriculture food processing
TC 1488	— analytical frameworks
TC 1603	Meetings ; Configurative models
TC 1475	Meetings ecological natural processes
TC 1465	Meetings energy-processing configurations
TC 1152	Meetings games contests
TC 1432	Meetings ; Losing focus
	Meetings [Metaphors...]
TC 1603	— Metaphors
TC 1727	— models reality
TC 1724	— Multi
	Meetings [physical...]
TC 1689	— physical constructs
TC 1367	— physical processes
TC 1922	— psycho-physical processes
TC 1789	Meetings social activities
TC 1909	Meetings symbolic configurations
HM 3611	Meets his afflicted mother ; Jesus (Christianity)
KC 0759	Megalopolis
HH 0289	Megavitamin therapy
HM 0194	Megha samadhi ; Dharma (Yoga)
HM 1266	Mehr (Sufism)
HM 3141	Melancholia
HM 3141	Melancholy
VD 3208	Melancholy
HM 5345	Melanosis (Esotericism)
VC 1135	Mellowness
KC 0822	Melody
HM 7384	Melting (Christianity)
HM 4504	Memories ; Delusional
HM 1689	Memory ; Deep penetration (Buddhism)
HH 5219	Memory ; Journeying transcendence forgetfulness
HM 3480	Memory ; Paradisal (Psychism)
HH 0528	Memory ; Racial
HM 2127	Men ; Southern continent awareness (Buddhism)
VD 3210	Menace
VD 3212	Mendacity
VD 3214	Mendicancy
HM 2056	Mendicant stage life (Hinduism)
HM 3615	Mental ; Abaissement niveau (Jung)
HM 2277	Mental abiding equipoise (Buddhism)
	Mental abiding states
HM 2181	— Disciplining (Buddhism)
HM 2729	— Full pacification (Buddhism)
HM 2753	— One pointedness (Buddhism)
HM 2205	— Pacifying (Buddhism)
	Mental [abidings...]
HM 2145	— abidings ; Meditative states (Buddhism)
HM 2122	— absorptions ; Trances (Buddhism)
HM 2050	— activities ; Aggregate (Buddhism)
	Mental body
HM 1455	— Fitness (Buddhism)
HM 7636	— Lightness (Buddhism)
HM 3696	— Malleability (Buddhism)
HM 5402	— Rectitude (Buddhism)
HM 4704	— Tranquillity (Buddhism)
HM 7969	— Wieldiness (Buddhism)
HM 0251	Mental (Buddhism)
HM 3040	Mental condition (Hinduism)
	Mental consciousness
HM 2319	— [Mental consciousness]
HM 2838	— (Buddhism)
HM 2522	— Emptiness objects sense (Buddhism)
HM 3237	Mental engagement (Buddhism)
	Mental factors
HM 7636	— Buoyancy (Buddhism)
HH 0910	— Changeable (Buddhism)
HH 0170	— Determining (Buddhism)
HH 0320	— Ever recurring (Buddhism)
HH 0320	— Omnipresent (Buddhism)
HM 3696	— Pliancy (Buddhism)
HM 1455	— Proficiency (Buddhism)
HM 5402	— Rectitude (Buddhism)
HM 4704	— Repose (Buddhism)
HH 1348	— Secondary (Buddhism)
HH 0578	— Virtuous (Buddhism)
HM 7969	— Wieldiness (Buddhism)
HM 2050	Mental formation group conscious existence ; Awareness (Buddhism)
HM 1649	Mental gladness ; Faculty (Buddhism)
HH 0146	Mental growth development
	Mental [healing...]
HH 0458	— healing
HH 0296	— health
HH 0675	— health ; Positive
HH 0296	— hygiene
HH 0072	Mental imagery therapy
HM 0906	Mental inertia (Yoga)
HH 8903	Mental maps
	Mental [perturbation...]
HM 0223	— perturbation thought (Buddhism)
HH 8672	— prayer (Christianity)
HM 4539	— prison
HM 4529	— projection
HH 1775	— psychism
HM 6149	— purity (Buddhism)
HM 5627	Mental qualities ; Understanding way fixing (Buddhism)
HM 0709	Mental quiescence (Buddhism)
HM 0831	Mental release (Buddhism)
HM 6644	Mental sensation (Buddhism)
	Mental [tiredness...]
HM 5806	— tiredness
HH 1116	— transformation
HH 4561	— transformation (Yoga)

HM 3464	Mental-volitional stage life	HH 5207	Middle middle grade (Taoism)		Mind [Serenity...]	
HM 2655	Mental wisdom (Buddhism)	HH 2160	Middle path awareness ; Great vehicle (Buddhism)	HM 0226	— Serenity (Buddhism)	
KD 2300	Mentalistic reduction ; Alternation energetic expansion	HM 3026	Middle-path tantra awareness (Buddhism)	HM 3037	— Seriousness (Japanese)	
HH 2718	Mentality materiality ; Defining (Buddhism)	HM 2306	Middle unconscious (Psychosynthesis)	HM 2217	— Setting (Buddhism)	
HH 0265	Mentality ; Teamwork	HM 6304	Middle upper grade (Taoism)	HM 1858	— Shining (Taoism)	
HM 5627	Mentality ; Understanding defining (Buddhism)	HH 6345	Middle vehicle path (Taoism)	HM 1434	— Soundness	
HH 0865	Menticide (Brainwashing)		Middle way	HM 3312	— Subtlety (Hinduism)	
HH 3219	Mentor	HH 1010	— (Buddhism)		Mind [Tao...]	
HH 0735	Mentor relationships	HH 1038	— (Buddhism)	HM 6534	— Tao ; Activating (Taoism)	
HH 0805	Mentor ; Spiritual	HH 3550	— Purity knowledge discernment (Buddhism)	HM 1858	— Tao (Taoism)	
HH 0442	Mentor ; Spiritual	HM 6089	Middling concentration (Buddhism)	HH 2387	— Training higher (Buddhism)	
VC 1137	Mercifulness	HH 1366	Midlife transition	HM 3489	— Transcending (Zen)	
VD 3216	Mercilessness	HM 2614	Midst this world ; Other world (ICA)		Mind [Unbending...]	
HH 0844	Mercy	HM 2074	Mig-gi-rnam-par-shes-pa (Tibetan)	HM 5335	— Unbending (Buddhism)	
HM 2420	Mercy greatness ; Sphere (Kabbalah)	VC 1147	Might	HM 8621	— Unconfined (Buddhism)	
HM 3434	Mercy (Leela)	MM 2044	Migration ; Cyclic	HM 5623	— Unelated (Buddhism)	
HM 2750	Merging	VC 1149	Mildness	HM 6791	— Unfettered (Buddhism)	
HM 0486	Merging ordinary world (Taoism)	HM 0191	Mildness (Buddhism)	HM 6663	— Unification (Buddhism)	
HM 1330	Merging yin yang (Taoism)	HH 0336	Milieu therapy	HM 5908	— Unoffended (Buddhism)	
HH 0859	Merit	VD 3226	Militancy	HM 1858	— Unstirring (Taoism)	
VC 1139	Merit	HM 2659	Mimamsamanaskara (Buddhism)		Mind upheld	
HH 5122	Merit ; Formation (Buddhism)	HH 0595	Mime	HM 5588	— concentration (Buddhism)	
HH 0932	Merit others ; Rejoicing (Buddhism)	HM 1430	Mimesis	HM 4396	— energy (Buddhism)	
HH 1266	Merit ; Transferring (Buddhism)	HH 0299	Mind	HM 7902	— faith (Buddhism)	
VD 3218	Meritlessness		Mind [All...]	HM 5499	— mindfulness (Buddhism)	
HH 0446	Meritorious deeds ; Ten (Buddhism)	HM 2730	— All conserving (Buddhism)	HM 5901	— understanding (Buddhism)	
HM 4501	Merkabah mysticism (Judaism)	HM 2640	— awake body-asleep trance (Psychism)	HM 5622	Mind ; Wandering (Taoism)	
VC 1141	Merriment	HM 2556	— Awareness consciousness group conscious existence senses (Buddhism)	HM 6556	Mind ; Wieldiness (Buddhism)	
HM 1061	Merriment (Buddhism)			HM 4976	Mindedness ; Absent	
HH 1306	Messengers God		Mind [Base...]	VD 2188	Mindedness ; Bloody	
HM 2248	Messiah	HM 1736	— Base (Buddhism)	HM 5008	Mindedness ; Concentration accompanied even (Buddhism)	
VD 3220	Messiness	HM 4902	— blankness			
	Meta [answer...]	HM 7019	— blindness	HM 8021	Mindedness ; Concentration accompanied even (Buddhism)	
KD 2045	— answer ; Constraints	HH 0299	— body dualism			
KD 2006	— answer: gladiatorial arena ; Prevailing	HM 3237	— Bringing (Buddhism)	HM 3362	Mindedness ; Double	
KD 2060	— answer patterning	HM 1743	— (Buddhism)	HM 7769	Mindedness ; Even (Buddhism)	
KC 0637	Meta-games	HM 6033	— Buoyancy (Buddhism)	VC 0899	Mindedness ; High	
KC 0123	Meta-language	HM 0226	Mind ; Calming (Buddhism)	HM 0361	Mindedness ; Light (Yoga)	
KC 0205	Meta-models		Mind Christ consciousness	HH 3562	Mindedness ; Open	
KC 0537	Meta-science	HM 0778	— empty (Christianity)	HM 2430	Mindedness ; Partial	
MM 2068	Metabolic pathways cycles	HM 1728	— forgiving (Christianity)	HM 3211	Mindedness ; Single	
KC 0163	Metabolism	HM 0976	— loving (Christianity)	HM 7525	Mindedness ; Understanding accompanied even (Buddhism)	
KC 0038	Metacommunication		Mind [Close...]			
KC 0862	Metagalaxy	HM 2157	— Close setting (Buddhism)	HM 0715	Mindedness ; Unperturbed (Buddhism)	
KC 0015	Metahistory	HM 4999	— Collectedness (Buddhism)	VD 4050	Mindedness ; Weak	
KC 0925	Metalogic	HM 0311	— Collectedness (Buddhism)	HM 4543	Mindfulness	
HM 2320	Metalogical awareness ; Logical	HM 2602	— Collective mob	VC 1151	Mindfulness	
KC 0969	Metamathematics	HM 7961	— concentration (Buddhism)		Mindfulness Buddhism	
KC 0544	Metamorphosis		Mind consciousness	HM 2847	— [Mindfulness (Buddhism)]	
HM 0551	Metamorphosis (Psychism)	HM 3323	— (Buddhism)	HH 6221	— [Mindfulness (Buddhism)]	
HH 0847	Metamotivation	HM 1501	— Cognition born contact element (Buddhism)	HM 1423	— [Mindfulness (Buddhism)]	
HH 0871	Metaneeds	HM 6173	— element (Buddhism)	HM 5499	Mindfulness ; Consciousness reinforced (Buddhism)	
HH 0888	Metanoia (Christianity)	HM 1834	— Element (Buddhism)	HM 2997	Mindfulness ; Establishments (Buddhism)	
HH 0947	Metapathology	HM 3314	— Neither physical ease nor unease born contact element (Buddhism)	HM 6754	Mindfulness ; Four stations (Pali)	
KC 0756	Metaphor			HM 5499	Mindfulness ; Mind upheld (Buddhism)	
HH 3020	Metaphors ; Guiding	HM 1921	— Psychical ease born contact element (Buddhism)	HH 4006	Mindfulness ; Practice presence (Buddhism)	
TC 1603	Metaphors meetings	HM 0234	— Psychical unease born contact element (Buddhism)	HM 0704	Mindfulness ; Right (Buddhism)	
KD 2342	Metaphors ; Patterns complementary	HM 0404	— Volition born contact element (Buddhism)	VD 3228	Mindlessness	
KC 0441	Metaphysics		Mind [Continuous...]	HM 0847	Mindlessness (Buddhism)	
HH 1765	Metaphysics (Buddhism)	HM 2241	— Continuous setting (Buddhism)	HM 2329	Minds ; Extra terrestrial	
KC 0937	Metaphysics ; Process	HH 3635	— control ; Silva	HM 5232	Minds ; Knowledge penetration (Buddhism)	
HH 3209	Metaprogramming ; Self	HH 1772	— cure	KD 2145	Mindscapes ; Epistemological	
HH 0355	Metapsychiatry		Mind [Development...]	HM 6119	Mindset ; Group	
KC 0769	Metasystem	HH 5973	— Development (Buddhism)	HM 5123	Mineral consciousness (Psychism)	
HH 0419	Metatalk	HM 6633	— Discovery natural (Taoism)	HM 5123	Mineral psychism (Psychism)	
KC 0876	Metatheorem	HM 8092	— Dissociated (Buddhism)	HM 0765	Ming (Zen)	
KC 0876	Metatheory	HM 6534	— Dissolving human (Taoism)		Minimal context	
HH 0686	Metempsychosis	HH 0589	— doctrine (Zen)	MP 1077	— complementary perspectives	
KC 0055	Method	HM 8336	— door adverting (Buddhism)	MP 1076	— interrelating mature emerging perspectives	
TC 1835	Method ; Art form	HM 4006	— drift	MP 1078	— single perspective	
KD 2040	Method ; Beyond		Mind [element...]	MP 1082	Minimal distance related operational control contexts	
KC 0736	Method ; Critical path	HM 2838	— element consciousness (Buddhism)	MP 1206	Minimal structural distinctions ; Efficient enclosure spaces	
TC 1420	Method ; Seminar	HM 8241	— Emancipated (Buddhism)			
HH 5342	Method ; Three vehicles gradual (Taoism)	HH 0246	— Evolution human	KD 2110	Minimal system ; Threshold comprehensibility: fourfold	
KC 0188	Methodology	HH 0075	— expanding drugs	MP 1081	Minimally-structured perspective control operations	
KC 0709	Methodology ; Social sciences	HM 2126	— expansion	MP 1162	Minimize insight blindspots ; Organization integrative superstructure	
TC 1188	Methods ; Dynamic design	HM 1369	Mind ; Faculty (Buddhism)			
TC 1056	Methods ; Event planning	HM 0666	Mind ; Flow experience	MP 1167	Minimum proportions ; Enfolded overview domains	
TC 1927	Methods ; Participation		Mind [Grass...]	MP 1219	Minimum tension ; Level generation	
KC 0711	Methods science teaching ; Integrated	HM 4312	— Grass hopper	TC 1525	Minoral conferencing	
HH 7004	Methods Taoism ; Sidetracks auxiliary (Taoism)	HM 3321	— Group	HM 0284	Miracles ; Experience (Christianity)	
TC 1449	Methods ; Workshop	KC 0372	— groups ; World	HM 7221	Miracles ; Passion (German)	
VC 1143	Meticulousness	HM 3529	Mind ; Higher (Hinduism)	HM 2327	Miradj (Sufism)	
VC 1145	Mettlesomeness	HM 5622	Mind ; Human (Taoism)	HM 0645	Mirth ; Aesthetic emotion (Hinduism)	
HM 3306	Mi (Buddhism)		Mind [Illuminated...]	HM 1421	Mirth (Buddhism)	
VD 3222	Miasma	HM 6208	— Illuminated (Buddhism)	VD 3230	Misalliance	
HM 0252	Miccha ajiva arati virati pativirati veramani (Pali)	HM 6243	— Independent (Buddhism)	VD 3232	Misapplication	
HM 1710	Micchaditthi (Pali)	HM 6521	— Inflexible (Buddhism)	VD 3234	Misapprehension	
HM 0863	Micchapatha (Pali)	HM 5451	Mind ; Leadership	HH 0361	Misappropriation ; Abstairing	
HM 1802	Micchasamadhi (Pali)	HM 0125	Mind ; Liberation (Pali)	VP 5737	Misbehaviour ; Behaviour	
HM 3356	Micchatta (Pali)	HH 8903	Mind ; Maps	VD 3236	Misbelief	
HM 1585	Micchavayama (Pali)	HM 2163	Mind ; No (Zen)	VD 3238	Miscalculation	
HM 2198	Michael ; Angelic awareness		Mind [One...]	VD 3240	Miscarriage	
HH 0689	Michi (Taoism)	HM 7843	— One centred (Buddhism)	VD 3242	Mischief	
KD 2070	Micro macro complementarity ; Observer entrapment	HM 4312	— One track	VD 3224	Mischievousness	
KC 0600	Microcircuit ; Integrated	HM 2163	— Original (Zen)	VD 3244	Misconception	
KC 0024	Microcosm		Mind [Peace...]	VD 3246	Misconduct	
TC 1743	Microphone ; Travelling	HM 3575	— Peace	HM 5600	Misconduct ; Abstinence bodily (Buddhism)	
HM 0092	Micropsia	HM 1155	— Pliancy (Buddhism)	HM 7171	Misconduct ; Abstinence verbal (Buddhism)	
HM 7921	Middha (Buddhism)	HM 1810	— Proficiency (Buddhism)	HM 5600	Misconduct kaya ; Abstinence (Buddhism)	
HM 0264	Middha (Pali)	HH 2332	— Purity (Buddhism)	VD 3248	Misconstruction	
	Middle grade		Mind [Recognition...]	VD 3250	Misdemeanor	
HH 3667	— low paths (Taoism)	HM 6633	— Recognition innocent (Taoism)	VD 3252	Misdirection	
HH 3296	— Lower (Taoism)	HM 6633	— Recognition true (Taoism)	VP 5562	Miseducation ; Education	
HH 5207	— Middle (Taoism)	HM 9001	— Rectitude (Buddhism)	HM 2964	Miserliness (Buddhism)	
HH 2633	— Upper (Taoism)	HM 2265	— Resetting (Buddhism)	VD 3254	Misery	

Misery

HH 1047	Misery ; Causes (Yoga)		TC 1853	Monger participant type ; Value		HH 3172	Movement ; New age
VD 3256	Misfeasance		TC 1642	Monitor		HH 3124	Movement ; Pentecostal (Christianity)
VD 3258	Misfortune		HM 2590	Monkey ego consciousness ; Crazy (Hinduism)		HM 5974	Movement ; Rapid eye
VD 3260	Misguidance		HM 4401	Mono-motivational hypnotic state transcendence		HM 3213	Movement ; Ritual
VD 3262	Misinformation		HM 3833	Mono no aware (Japanese)		HH 0172	Movement therapy ; Body
VP 5552	Misinterpretability ; Interpretability		TC 1383	Monologue		HH 1523	Movements ; Human development new religious
VD 3264	Misinterpretation		KD 2001	Monopolarization ; Questionable answers:		HH 0763	MPD
VD 3266	Misjudgement		MP 1097	Monotonous perspective patterns ; Concealment necessary		HH 0307	MRA
VP 5494	Misjudgement ; Judgement					HM 2605	Mraksha (Buddhism)
VD 3268	Mismanagement		HM 2423	Monotony		HM 2623	Mthong manaskara ; Lhag (Buddhism)
VD 3270	Misrepresentation		MS 8225	Monsters ; Hybrid		HM 4052	Mu'akhkhir ; Al (Sufism)
VP 5573	Misrepresentation ; Representation		KC 0926	Monte Carlo simulation technique		HM 2034	Mu ge ; Ji ji (Zen)
VD 3272	Misrule		MS 8825	Monuments		HM 4093	Mu'id ; Al (Sufism)
HH 3798	Mission (Christianity)		HM 1748	Mood		HM 1116	Mu'izz ; Al (Sufism)
HH 3352	Mission ; Journeying transcendence call (Christianity)		TC 1334	Mood motivation assumptions		HM 3164	Mu je ; Ji ri (Zen)
HM 3217	Mission ; Singular (ICA)		MM 2020	Moods individual ; Changing		HM 1245	Mu'min ; Al (Sufism)
HM 2976	Missional comradeship (ICA)		HM 2398	Moon (Tarot)		HM 0622	Mubdi ; Al (Sufism)
HM 2343	Missional engagement (ICA)			**Moral consciousness**		VD 3292	Muddle
VD 3274	Misstatement		HM 5338	— form-realm (Buddhism)		HM 1346	Mudha (Yoga)
VD 3276	Mistake		HM 4701	— formless realm (Buddhism)		HM 4510	Mudhill ; Al (Sufism)
VD 3278	Mistrust		HM 4930	— transcendental plane (Buddhism)		HH 0286	Mudras
VD 3280	Misunderstanding			**Moral [delinquency...]**		HM 3397	Muduta ; Vedanakkhandhassa sannakkhandhassa sankharakkhandhassa (Pali)
VD 3282	Misusage		HH 0565	— delinquency			
VD 3284	Misuse		HH 0565	— development		HM 1796	Muduta ; Vinnakkhandhassa (Pali)
HH 3421	Misuse scripture		HH 1086	— discipline		HM 0793	Muga (Japanese)
VC 1153	Mitigation		HH 0565	Moral education		HM 3633	Mughni ; Al (Sufism)
HH 6321	Mixing ; Phosphenic			**Moral [ground...]**		HM 0526	Muhaymin ; Al (Sufism)
HH 1809	Mngon shes pa (Tibetan)		HM 2721	— ground (ICA)		HM 4544	Muhsi ; Al (Sufism)
HM 2277	Mnyam-par-jog-pa (Tibetan)		HH 0565	— growth		HM 1007	Muhyi ; Al (Sufism)
HM 2602	Mob mind ; Collective		HH 1321	— growth		HH 0464	Mujahada (Sufism)
VC 1155	Mobility		HH 4536	Moral law ; Fivefold		HM 5215	Mujahadat (Sufism)
KC 0108	Mobilization ; Cognitive		HM 3550	Moral morbidness		HM 0504	Mujib ; Al (Sufism)
HH 0924	Mobilization ; Educational			**Moral [participant...]**		HM 7336	Mukhalafat-i nafs (Sufism)
KC 0177	Mobilization resources		TC 1345	— participant involvement		HM 2137	Mukta state ; Jivana
HH 0745	Mobilizing human resources (United Nations)		HH 5098	— perfectibility		HM 6558	Muktajnana (Hinduism)
HM 2473	Mode awareness while unconscious ; Alternative		HH 5098	— perfection		HH 0613	Mukti
HM 4389	Mode consciousness occurrence ; Sense (Buddhism)		HH 0278	— precepts ; Observing (Buddhism)		HM 3890	Mukti ; Jivan (Hinduism)
HH 0592	Mode life		HH 0307	Moral re-armament		HM 4489	Mukti ; Videha (Yoga)
KC 0824	Mode life		HM 6663	Moral thought ; Collectedness (Buddhism)		HM 2893	Muladhara (Yoga)
KC 0640	Model		HH 0491	Moral tranquality		HM 0270	Mulaklesha (Buddhism)
TP 0030	Model elegance (Fire)		HH 0565	Moral values ; Development		HM 3560	Mulam ; Adoso kusala (Pali)
HM 1105	Model human consciousness ; Ontogenetic		VC 1163	Morality		HM 3752	Mulam ; Alobhokusala (Pali)
KC 0719	Model ; Mathematical		HM 3049	Morality ; Beyond (ICA)		HM 5902	Mulk ; Malik ul (Sufism)
KC 0042	Modelling ; World		VT 8093	Morality∗complex		HM 2780	Multi-dimensional consciousness
KD 2152	Modelling language relationships sound		HH 1865	Morality ; Purity (Buddhism)		MP 1234	Multi element external boundaries ; Maintainable,
TC 1392	Models ; Conceptual		HH 0429	Morality (Zen)		MP 1239	Multi-faceted frameworks
TC 1603	Models meetings ; Configurative		VD 3286	Morbidity		HM 3005	Multi-level awareness (Psychism)
KC 0205	Models ; Meta		HM 3550	Morbidness ; Moral		TC 1724	Multi-meetings
TC 1727	Models reality ; Meetings		HM 2553	Moriendi ; Ars		KC 0257	Multi-ordinality
VC 1157	Moderation		HH 1121	Morita therapy		KC 0603	Multi-valued logic
VP 5161	Moderation ; Energy		KD 2310	Morphic resonance		KC 0191	Multidimensional space
TC 1969	Moderator		HM 4060	Morphine ; Use		KC 0007	Multidisciplinarity
	Modes awareness associated		KC 0826	Morphogenesis		HM 2242	Multifocal thought
HM 0134	— alcohol consumption		KC 0964	Morphological analysis		TP 0029	Multiple dangers (Abyss)
HM 4546	— barbiturates		KC 0884	Morphology		HH 0214	Multiple intelligences
HM 0584	— psychoactive substances		HH 4012	Mortal sin		HH 0763	Multiple personality
HM 2313	— schizophrenia		MS 8603	Mortality		HH 0763	Multiple personality disorder
HM 0803	— use amphetamines similar drugs		HH 0780	Mortification		HH 0846	Multiple therapy
HM 1368	— use caffeine		VD 3288	Mortification		MP 1012	Multiplicity ; Individuality
HM 3691	— use cannabis		HH 0464	Mortification ; Self (Sufism)		KC 0169	Multiplicity ; Organic functional
HM 1594	— use cocaine		HM 5215	Mortification self (Sufism)		KC 0027	Multistable system
HM 0801	— use hallucinogens		HM 2719	Mos-pa-las-byung-bai-yid-byed (Tibetan)		KC 0036	Multivariate analysis
HM 3743	— use inhalants		HM 2933	Mos-pa (Tibetan)		HM 4173	Mumin (Sufism)
HM 4277	— use khat		HM 3616	Moses one's being (Sufism)		HM 3868	Mumit ; Al (Sufism)
HM 1881	— use nicotine		HM 2561	Most beautiful names (Sufism)		HM 5304	Munasaba (Islam)
HM 4060	— use opium similar drugs		HM 3611	Mother ; Jesus meets his afflicted (Christianity)		HM 0652	Mundane awareness (Buddhism)
HM 4254	— use phencyclidine similar drugs		VC 1165	Motherhood		HM 7234	Mundane concentration (Buddhism)
HM 4546	— use sedatives		HH 4646	Mothering		HM 6997	Mundane knowledge (Christianity)
KP 3007	Modes change		VC 1167	Motherliness		HM 6534	Mundane ; Restoration celestial awareness (Taoism)
KD 2170	Modes managing: embracing conflicting styles		VT 8034	Motion∗complex		HM 0130	Mundane resultant consciousness (Buddhism)
HM 3004	Modes ; New religious (ICA)		VT 8035	Motion∗complex ; Relative		HM 7628	Mundane understanding (Buddhism)
HM 6720	Modes occurrence consciousness (Buddhism)		HH 0741	Motion ; Meditation (Islam)		KC 0302	Mundi ; Anima
MP 1003	Modes organization ; Interpretation complementary		HH 0512	Motion picture ; Therapeutic		HH 5450	Muni
HM 0062	Modes perception ; Sport induced		VP 5267	Motion-Quiescence		VC 1173	Munificence
MP 1056	Modes relationship ; Special		TC 1876	Motion video teleconferencing ; Full		HM 4513	Muntaqim ; Al (Sufism)
VC 1159	Modesty		HH 5256	Motionless meditation		HM 5331	Muqaddim ; Al (Sufism)
TP 0015	Modesty		VC 1169	Motivation		HM 3171	Muqit ; Al (Sufism)
VP 5908	Modesty-Vanity		HH 0181	Motivation		HM 8001	Muqsit ; Al (Sufism)
HH 0663	Modification		HH 0456	Motivation ; Achievement		HM 3576	Muqtadir ; Al (Sufism)
HH 0771	Modification ; Behaviour		TC 1334	Motivation assumptions ; Mood		HM 2305	Muraqaba (Sufism)
TC 1370	Modification ; Programme		VT 8063	Motivation∗complex		HM 2305	Muraqabat (Sufism)
KC 0165	Modularity		VP 5648	Motivation-Dissuasion		HM 1159	Murccha (Hinduism)
HM 3035	Moha (Buddhism)		HH 0924	Motivation ; Educational		HM 3887	Murid (Sufism)
HM 0992	Moha (Hinduism)		HH 0456	Motivation ; Success		HM 5786	Musawwir ; Al (Sufism)
HM 0697	Moha (Leela)		HH 0055	Motivational development		HM 3521	Mushahada (Sufism)
HM 0184	Moha (Pali)		HM 4401	Motivational hypnotic state transcendence ; Mono		HM 3521	Mushahadat (Sufism)
HM 0918	Moha (Pali)		VC 1171	Motive		HM 3521	Mushahida (Sufism)
HM 6124	Mohakkhaya (Pali)		HH 0181	Motive		HM 3554	Mushin (Japanese)
HM 6124	Mohaksaya (Buddhism)		HM 2828	Motivity ; Agape (ICA)		HM 2163	Mushin (Zen)
HM 3513	Mohammad one's being (Sufism)		MM 2043	Motors generators ; Electric		HM 2974	Mushinjo (Zen)
HM 1439	Moho akusalamulam (Pali)		TP 0049	Moulting Change		HM 3053	Mushitasmrtita (Buddhism)
HM 2196	Moksa (Hinduism)		TP 0049	Moulting Revolution		HH 0496	Music
HM 2196	Moksha (Hinduism)		HM 2170	Mountain care (ICA)		HH 3003	Music ; Human development
HM 2603	Mokshakala (Buddhism)		TP 0052	Mountain Desisting		HH 3003	Music ; Psycho social transformation
MM 2086	Molecular resonance		TP 0052	Mountain Inaction		HH 1973	Music spheres ; Hearing
HM 2749	Moment ; Auspicious		TP 0052	Mountain Keeping still		HH 0496	Music therapy
HM 4449	Moment ; Creative (Sufism)		TP 0052	Mountain Resting		HM 4805	Musical ecstasy
HM 1655	Moment ; Creative (Sufism)		VD 3290	Mournfulness		KC 0203	Musical form
HM 2786	Moment ; Eternal (ICA)		HM 1502	Mourning			**Musical inspiration**
VC 1161	Momentousness		TP 0031	Movement		HM 4805	— [Musical inspiration]
HM 1655	Moments ; Experience original (Sufism)		MM 2084	Movement ; Alluring		HM 6006	— Avataric level
TC 1971	Momentum ; Intellectual		HM 2315	Movement ; Ecology		HM 0203	— Egoic
HM 2069	Monadic awareness (Amerindian)		MM 2070	Movement ; Geography		HM 1995	— Imitative
HH 0807	Monasticism		HM 2315	Movement ; Green		MS 8316	Musical instruments
MS 8397	Money		HH 0581	Movement ; Holistic health care		HH 0961	Musical intelligence
			HH 0398	Movement ; Human potential		KD 2280	Musical key political philosopher ; Patterns alternation:

-896-

GENERAL INDEX TO VOLUME 2
Neutrality

MM 2090 Musical variations
HM 4403 Musicians ; Celestial (Leela)
VD 3294 Mustiness
HM 8019 Muta'ali ; Al (Sufism)
VD 3296 Mutability
HM 2250 Mutability emblems (Tarot)
HM 3092 Mutable awareness (Astrology)
HM 3850 Mutakabbir ; Al (Sufism)
HH 0767 Mutational phases life
VD 3298 Muteness
VD 3300 Mutilation
HH 4233 Mutual confidence
HM 5087 Mutual passionate love
HM 2657 Mutual possession ten worlds (Buddhism)
HH 4233 Mutual trust
VC 1175 Mutuality
KP 3020 Mutually constraining forms ; Significance
HM 7981 Muwafaqat (Sufism)
HM 4510 Muzill ; Al (Sufism)
HH 3443 Myoho renge kyo ; Nam (Buddhism)
HM 0882 Myopia ; Empty field
HM 3097 Myself ; All being (ICA)
HM 2154 Myself ; Being (ICA)
HH 4032 Mysteries religion
HH 0303 Mysteries ; Sacred
HM 1452 Mysteries ; Unveiling divine (Sufism)
HM 2398 Mysterious ; Emblems (Tarot)
HM 0899 Mysterium
HM 3811 Mysterium tremendum (Jung)
HH 0303 Mystery
HM 2143 Mystery ; Doing (ICA)
HM 3096 Mystery ; Envelopped (ICA)
HM 2042 Mystery ; Glorious (ICA)
HM 3029 Mystery ; Honouring (ICA)
HM 2619 Mystery ; Impacted (ICA)
HM 2434 Mystery ; Land (ICA)
HM 2386 Mystery ; Recreated (ICA)
Mystery [Saving...]
HM 2413 — Saving (ICA)
HM 2234 — Seduced (ICA)
HM 3608 — Sense
Mystic experience
HM 2272 — [Mystic experience]
HM 4398 — Deautomatization
HM 3542 — Psychological assessment (Sufism)
HM 2900 Mystic journey ; Awareness
HM 3028 Mystic marriage (Tarot)
HM 3415 Mystic stations (Sufism)
HM 3889 Mystic union God (Christianity)
HM 3889 Mystica ; Unio (Christianity)
Mystical [Christ...]
HM 2880 — Christ-consciousness (Christianity)
HM 2272 — cognition
HM 2710 — contemplation
HM 2867 Mystical detachment (Sufism)
HM 3521 Mystical ecstasy (Sufism)
Mystical experience
HM 3445 — [Mystical experience]
HM 0899 — Introvertive
HM 0530 — (Sufism)
HM 0402 Mystical journey (Christianity)
HM 4349 Mystical knowledge ; Experience (Sufism)
HM 3941 Mystical ladder divine love (Christianity)
HM 6321 Mystical marriage
HM 3241 Mystical poverty ; Consciousness
HH 0816 Mystical prayer
Mystical [seduction...]
HM 3436 — seduction ; Surrender bliss (Sufism)
HM 0062 — sensations sport
HM 3191 — stage life
HM 3607 — states (Sufism)
HH 5217 Mystical theology (Christianity)
HH 1432 Mystical transformation
HM 2272 Mysticism
HM 1635 Mysticism ; Cosmological
Mysticism [Eastern...]
HH 9219 — Eastern
HM 0800 — Environmental
HM 2272 — Epithalamian
HH 1232 Mysticism ; Jewish (Judaism)
HM 4501 Mysticism ; Merkabah (Judaism)
HH 6199 Mysticism ; Sikh (Sikhism)
HM 2272 Mysticism ; Unitive
VD 3302 Mystification
HH 0322 Myth incompetence
HM 1525 Mythic-literal faith
HH 3405 Mythic yoga (Yoga)
Mythical [aerial...]
MS 8357 — aerial beings
MS 8300 — anthropomorphic beings
MS 8917 — aquatic beings
MS 8915 — artifacts
Mythical [celestial...]
MS 8337 — celestial beings
MS 8925 — clothing
HM 0043 — consciousness
HM 2078 — consciousness
HH 7637 — control human development
MS 8937 Mythical fire-dwelling beings
HM 0043 Mythical imagination
MS 8387 Mythical terrestrial beings
MS 8225 Mythical theriomorphs
MS 8200 Mythological reality
TC 1481 Mythology ; Conference
HH 0690 Myths
HH 3020 Myths ; Guiding

N

HM 2051 N'asannayatana ; Nevasanna (Pali)
HM 1023 Na kiriyati hiriyitabbena (Pali)
HM 3257 Na kiriyati papakanam akusalanam dhammanam samapattiya (Pali)
HM 1673 Na ottappati ottappitabbena (Pali)
HM 3353 Na ottappati papakanam akusalanam dhammanam samapattiya (Pali)
HM 0606 Nabi (Sufism)
HH 0160 Nada yoga (Yoga)
HM 4049 Nadi (Balinese)
HM 6350 Nafi' ; An (Sufism)
HM 7336 Nafs ; Mukhalafat (Sufism)
HH 0464 Nafs ; Subjugation (Sufism)
HH 5092 Nafs ; Tadhkiya (Sufism)
HM 1485 Naga-loka (Leela)
HM 2031 Naga-world awareness (Buddhism)
HM 2036 Nagual state awareness (Amerindian)
HH 1324 Naikan therapy
HM 4136 Nailed cross ; Jesus (Christianity)
MS 8365 Nails ; Hair
HM 2051 Naivasamjnanasamjnayatana (Buddhism)
VD 3304 Naïvety
HM 2375 Nakshbandi recollection (Islam)
HH 3443 Nam-myoho-renge-kyo (Buddhism)
HH 0539 Nam simran (Yoga)
HH 0536 Namaz (Islam)
HH 2561 Names Allah ; Ninety nine (Sufism)
HH 1167 Names Christ (Christianity)
HH 2561 Names ; Divine (Sufism)
HH 2561 Names God (Sufism)
HH 2561 Names ; Most beautiful (Sufism)
HH 0680 Nana-dassana (Buddhism)
HM 7549 Nana dassana ; Yatha bhuta (Pali)
HM 6502 Nanadassana (Buddhism)
HH 3550 Nanadassana niddesa ; Patipada (Pali)
HH 4007 Nanadassana visuddhi niddesa ; Maggamagga (Pali)
HH 3025 Nanadassana-visuddhi-niddesa (Pali)
HM 4637 Nanam ; Gotra bhu (Pali)
HH 0277 Naqsbandi tradition (Sufism)
HM 4356 Naqshbandiyya order ; Suluk (Sufism)
HH 1066 Narcissism
HM 2630 Narcissistic equilibrium
HH 0310 Narcotherapy
HM 0856 Narka-loka (Leela)
HM 1237 Narrowing ; Perceptual
VP 5526 Narrowmindedness ; Broadmindedness
VP 5204 Narrowness ; Breadth
HM 0972 Nasatam ; Cakkhuvinnanadhatusamphassajam cetasikam neva satam (Pali)
HM 3314 Nasatam ; Manovinnanadhatusamphassajam cetasikam neva satam (Pali)
HM 3579 Nasut (Sufism)
HH 0630 Natal psychology ; Peri
HH 0160 Natha yoga (Yoga)
KC 0980 Nation states ; Unitary
HM 1370 National consciousness (Yoga)
KC 0130 Nationalism
KC 0657 Nations Organization ; United
HH 6121 Native spirituality
KC 0001 Natural
HH 0912 Natural feelings (Japanese)
HM 3127 Natural forces emblems (Tarot)
HM 2071 Natural intelligence ; Path (Judaism)
HH 0642 Natural law
HM 6633 Natural mind ; Discovery (Taoism)
TC 1475 Natural processes ; Meetings ecological
HM 7961 Natural purity consciousness ; Concentration due (Buddhism)
MS 8435 Natural symbols ; Terrestrial
HH 0702 Naturalism
VC 1179 Naturalness
VP 5900 Naturalness-Affectation
HM 3189 Naturalness bodily action
HH 0343 Naturalness (Zen)
KC 0001 Nature
HH 0463 Nature ; Attentiveness (Zen)
Nature [Balance...]
KC 0968 — Balance
HH 0343 — being (Zen)
HM 2039 — body ; Buddha (Buddhism)
HM 0915 Nature ; Communion
HH 0580 Nature ; Contemplation
HH 0700 Nature ; Divine
HM 2246 Nature ; Divine (Sufism)
HM 3205 Nature ; Emptiness (Buddhism)
HM 3493 Nature essential self ; Ignorance (Buddhism)
HH 3246 Nature kasina (Buddhism)
HM 2899 Nature phenomena ; Emptiness (Buddhism)
HM 3488 Nature ; Relationship man (Christianity)
HM 0917 Nature ; Seeing self (Zen)
MS 8435 Nature ; Sub lunar
HH 4389 Nature ; Universal human
HM 3425 Nature ; World (Sufism)
HM 2145 Navakara chittasthiti (Buddhism)
HM 6003 Nazar (Islam)
HM 0777 NDE Core-experience
HM 3381 NDE Near death awareness light
HM 2213 NDE Near death black void
HM 3349 NDE Near death body separation
HM 3349 NDE Near death detachment
HM 3059 NDE Near death encounter higher self
HM 3294 NDE Near death entering darkness
HM 0777 NDE Near death experience
HM 3177 NDE Near death fear
HM 3165 NDE Near death hell
HM 3115 NDE Near death negative detachment
HM 2494 NDE Near death peace
HM 2888 NDE Near death review life
HM 3294 NDE Near death transition
HM 0777 NDE Out-of-body experience
Near death cont'd
HM 3381 — awareness light (NDE)
HM 2213 — black void (NDE)
HM 3349 — body separation (NDE)
HM 3349 — detachment (NDE)
HM 3059 — encounter higher self (NDE)
HM 3294 — entering darkness (NDE)
HM 3180 — entering light
HM 3422 — evil force
HM 0777 — experience (NDE)
HM 3177 — fear (NDE)
HM 3165 — hell (NDE)
HM 3115 — negative detachment (NDE)
HM 2494 — peace (NDE)
HM 2888 — review life (NDE)
HM 3294 — transition (NDE)
HM 4477 Nearer ; Hope getting (Sufism)
HM 2377 Nearness-awareness (Sufism)
VP 5200 Nearness-Distance
VC 1181 Neatness
MP 1097 Necessary monotonous perspective patterns ; Concealment
VC 1183 Necessity
VP 5637 Necessity ; Choice
TP 0048 Need ; Basic (Well)
HH 0043 Needs
HH 0913 Needs hierarchy
TC 1204 Needs ; Participant personality
HH 0383 Needs superiors ; Attending (Buddhism)
VD 3306 Negation
HM 2369 Negation ; Intentional self (ICA)
HH 1171 Negativa ; Via
HM 6133 Negative capability
HM 3115 Negative detachment ; Near death (NDE)
KC 0003 Negative entropy
HM 5119 Negative feeling towards God (Christianity)
HM 5904 Negative hallucination
HM 3623 Negative intellect (Leela)
HM 2820 Negative intrapsychic forces (Psychism)
VD 3308 Negativity
KC 0003 Negentropy
HM 2421 Negha ; Dharma (Buddhism)
VD 3310 Neglect
VP 5533 Neglect ; Carefulness
VD 3312 Negligence
HH 5124 Negotiating openness
HH 0042 Neighbour ; Love
VC 1185 Neighbourliness
HM 2051 Neither cognition nor non cognition ; Sphere (Buddhism)
HM 0581 Neither existence nor non-existence (Buddhism)
Neither [perception...]
HM 2051 — perception nor non perception ; Sphere (Buddhism)
HM 3314 — physical ease nor unease born contact element mind-consciousness (Buddhism)
HM 0972 — psychical ease nor unease born contact element eye-consciousness (Buddhism)
HM 1643 Neither suffering nor pleasure born contact psychical ; Sensation (Buddhism)
HM 1352 Neither suffering nor pleasure experienced born contact psychical (Buddhism)
HH 0683 Nembutsu (Japanese)
HM 0220 Nen ; Realization
HM 2427 Neocortex human brain consciousness ; Neomammalian
HM 2427 Neomammalian neocortex human brain consciousness
HM 3390 Nepunna (Pali)
VD 3314 Nervousness
HM 1394 Nescience (Buddhism)
MP 1127 Nested levels accessibility
MM 2049 Nets basketry ; Knotting,
KC 0472 Network
MS 8933 Network
KC 0842 Network analysis
KC 0218 Network data structures
KC 0932 Network ; Equivalent
MP 1005 Network inter-relationships
MP 1045 Network ; Local action
KC 0736 Network planning techniques
MP 1018 Network redefinitions
KC 0633 Network ; Social
KC 0933 Network synthesis
TC 1275 Network ; Thesaurus
KC 0455 Networking
TC 1343 Networking conferencing
TC 1275 Networks ; Associative
MP 1120 Networks ; Focus oriented communication
TC 1275 Networks ; Relevance
MP 1052 Networks unmediated relationships ; Interfacing vehicles communication
HM 2362 Netzach sephira (Kabbalah)
HH 4872 Neuro-linguistic programming (NLP)
HH 4332 Neuromuscular therapy
HH 1024 Neurosis
VP 5690 Neurosis ; Selfactualization
HM 5702 Neurotic love
VC 1187 Neutrality
VP 5806 Neutrality-Compromise
HM 4050 Neutrality ; Plane (Leela)
HM 7002 Neutrality ; Specific (Buddhism)

HM 0972 Neva satam nasatam ; Cakkhuvinnanadhatusamphassajam cetasikam (Pali)
HM 3314 Neva satam nasatam ; Manovinnanadhatusamphassajam cetasikam (Pali)
HM 2051 Nevasanna-n'asannayatana (Pali)
HH 3172 New age movement
MP 1039 New dimension ; Integrating
MP 1043 New dimensions ; Presentation
MP 1211 New level generation ; Boundary expansion permitting
MP 1065 New perspectives ; Context emergence
HM 3004 New religious modes (ICA)
HH 1523 New religious movements ; Human development
TP 0056 Newcomer (Wanderer)
VC 1189 Newness
VP 5122 Newness-Oldness
HM 2823 Nga-rgyal (Buddhism)
HH 0170 Nges ; Yul (Buddhism)
HM 2986 Ngo-tsha-med-pa (Tibetan)
HM 2881 Ngo-tsha-shes-pa (Tibetan)
HM 2330 Nibbana (Pali)
HH 3443 Nichiren shoshu buddhism (Buddhism)
HM 1881 Nicotine ; Modes awareness associated use
HM 2407 Nid ; Chos (Buddhism)
HH 3198 Niddesa ; Aruppa (Pali)
HH 3534 Niddesa ; Brahmavihara (Buddhism)
HH 2718 Niddesa ; Ditthi visuddhi (Pali)
HH 1187 Niddesa ; Kankhavitarana visuddhi (Pali)
HH 4007 Niddesa ; Maggamagga nanadassana visuddhi (Pali)
HH 3025 Niddesa ; Nanadassana visuddhi (Pali)
HH 3550 Niddesa ; Patipada nanadassana (Pali)
HM 2957 Nidra (Hinduism)
HM 2500 Nien ; Wu (Zen)
HM 1727 Night correction (Christianity)
HM 1714 Night ; dark (Christianity)
Night [Saint...]
HM 1714 — Saint John Cross ; Dark (Christianity)
HM 1727 — senses ; Dark (Christianity)
HM 3941 — soul ; Dark (Christianity)
HM 2085 Nightmares
HM 5345 Nigredo (Esotericism)
HM 3377 Nihayat (Sufism)
HM 2830 Nihilism
HH 4509 Nihilism
HM 1496 Nikkama (Pali)
HM 3420 Nikos Kazantzakis ; Spiritual exercises (Christianity)
VC 1191 Nimbleness
HM 8673 Nimitta ; Samatha (Buddhism)
HM 1408 Nine-fold knowledge (Systematics)
HH 2561 Ninety-nine names Allah (Sufism)
HH 0912 Ninjo (Japanese)
HM 4000 Ninth order perceptions - control systems concepts
HM 2292 Ninth scriptural bodhisattva awareness (Buddhism)
HM 2325 Ninth vajra-master awareness (Buddhism)
HM 3398 Niralambapuri chakra (Yoga)
HM 4406 Nirasih (Hinduism)
HM 2125 Nirbija samadhi (Hinduism)
HM 4331 Nirguna brahman
HM 2340 Nirmana-kaya (Buddhism)
HM 2172 Nirodha (Buddhism)
HM 3437 Nirodha parinama (Yoga)
HM 6346 Nirodhasamapatti (Buddhism)
HM 3437 Niruddha (Yoga)
HM 0135 Niruddha (Yoga)
HH 5026 Nirutti-patisambhida (Pali)
HM 2330 Nirvana (Buddhism)
HM 2294 Nirvana ; Emptiness ultimate (Buddhism)
HM 2330 Nirvana ; Pari (Buddhism)
HM 4665 Nirvana ; Pari (Buddhism)
Nirvana [rati...]
HM 2546 — rati (Buddhism)
HM 4665 — remainder (Buddhism)
HM 5023 — remainder (Pali)
HM 0809 Nirvicara samadhi (Yoga)
Nirvikalpa samadhi
HM 2226 — (Hinduism)
HM 2061 — (Hinduism)
HM 3619 — Unconditional
HM 4284 Nirvitarka samadhi (Yoga)
HM 2491 Niskama karma (Hinduism)
HH 5487 Niskama karma (Hinduism)
HM 7626 Nisti (Sufism)
HM 3615 Niveau mental ; Abaissement (Jung)
HM 3280 Niyama (Hinduism)
HM 4901 Niyyat (Sufism)
HH 4872 NLP Neuro-linguistic programming
HM 3833 No aware ; Mono (Japanese)
HM 2163 No-mind (Zen)
HH 1092 No technique ; Technique
HM 2500 No-thought (Zen)
HM 3286 Noah one's being (Sufism)
HH 0499 Nobility
VC 1193 Nobility
HM 2339 Noble eightfold path (Buddhism)
HM 2543 Noble figures ; Eastern continent awareness (Buddhism)
HM 4026 Noble paths ; Four (Buddhism)
HH 1099 Noble system (Buddhism)
HH 0523 Noble truths ; Four (Buddhism)
MP 1030 Nodes ; Activity
HH 0334 Noetic superstructure
VD 3316 Noisome
HH 3778 Nominalism
HH 6501 Nomos (Islam)
Non [abiding...]
HM 3485 — abiding (Zen)
Non [action...] cont'd
HH 1073 — action
HM 8000 — affection (Hinduism)
HM 0021 — agitation
HM 3310 — analytical cessations (Buddhism)
HM 6898 — anger (Hinduism)
HH 0868 — attachment
HM 2128 — attachment (Buddhism)
Non [cognition...]
HM 2051 — cognition ; Sphere neither cognition nor (Buddhism)
KD 2205 — comprehension holes
KD 2270 — comprehension structuring phenomenon learning society
HM 2033 — conceptual cognition emptiness ; Direct (Buddhism)
TC 1663 — conformist participant type
HM 2314 — conscientiousness (Buddhism)
HM 1902 — consideration (Buddhism)
MP 1145 — current elements those reserve ; Domains
Non [delusion...]
HM 2695 — delusion (Buddhism)
HM 9330 — desire (Christianity)
HH 2009 — deterministic psychology
HH 0095 — directive counselling
HM 4002 — discriminative knowledge (Systematics)
HM 4382 — distraction (Buddhism)
HH 3772 — duality ; Differentiated (Hinduism)
HH 0518 — duality (Hinduism)
Non [emanating...]
HM 2070 — emanating consciousness (Buddhism)
HM 3203 — embarrassment (Buddhism)
KD 2080 — equilibrium structures
HM 0581 — existence ; Neither existence nor (Buddhism)
HM 2461 Non-faith (Buddhism)
HM 2847 Non-forgetfulness (Buddhism)
HM 0451 Non-grasping (Buddhism)
HM 2128 Non-greed (Buddhism)
Non [harmfulness...]
HM 2608 — harmfulness (Buddhism)
HM 2744 — hate (Buddhism)
HM 1557 — hating (Buddhism)
Non hatred
HM 2744 — (Buddhism)
HM 1399 — (Buddhism)
HM 3560 — wholesome root (Buddhism)
MS 8400 Non-human beings
Non [ignorance...]
HM 2695 — ignorance (Buddhism)
HM 3374 — inertness (Buddhism)
HM 2063 — introspection (Buddhism)
Non linear
TC 1123 — agendas
MP 1173 — contexts communication pathways ; Protecting
MP 1163 — domain external structures ; Hospitable
MP 1175 — extensions structures ; Insight capturing
MP 1007 — organization
MP 1051 — processes ; Linear relationships enhanced
Non linearity
KC 0347 — [Non-linearity]
MP 1060 — Accessible
MP 1174 — Communication pathways enfolded
MP 1172 — Contexts self organizing
MP 1170 — Cultivation productive
MP 1118 — integrative superstructure ; Integration
MP 1176 — Sites grounding perspectives
MP 1171 — structures ; Appropriate relationship
HM 2391 Non ordinary reality ; States
Non [penetration...]
HM 1637 — penetration (Buddhism)
HM 2051 — perception ; Sphere neither perception nor (Buddhism)
HH 1268 — possessiveness
HM 2089 — products ; Emptiness (Buddhism)
HM 6920 Non return ; Knowledge path (Buddhism)
HM 1028 Non-rigidity (Buddhism)
Non [sensuous...]
HM 6644 — sensuous feeling (Buddhism)
HM 2986 — shame (Buddhism)
HM 3014 — stiffness (Buddhism)
HH 0587 Non-temporal dimension being
HM 2448 Non things ; Emptiness inherent existence (Buddhism)
TC 1640 Non-verbal interaction preferences
HH 0088 Non-violence
VD 3318 Nonacceptance
VD 3320 Nonaccomplishment
VP 5722 Nonaccomplishment ; Accomplishment
VD 3322 Nonadherence
VC 1195 Nonagressivity
VC 1197 Nonalignment
TC 1581 Nonchalent participant type
VD 3324 Nonconformity
VP 5082 Nonconformity ; Conformity
HM 5265 Nondoing ; Subtlety (Taoism)
HM 0466 None (Christianity)
HM 0466 None ; Hours (Christianity)
HM 0239 Nonecstatic perception (Hinduism)
VP 5001 Nonexistence ; Existence
VD 3326 Nonfeasance
VD 3328 Nonfulfilment
VP 5417 Nonhumanity ; Humanity
KD 2160 Nonlinear cybernetics
HM 4976 Nonmeditation
VD 3330 Nonobservance
VP 5768 Nonobservance ; Observance
VD 3332 Nonsense
VD 3334 Nonuniformity
VP 5017 Nonuniformity ; Uniformity
VC 1199 Nonviolence
HH 1781 Noogenesis
HH 4621 Noogenesis ; Planetary
KC 0157 Noosphere
VC 1201 Normality
TP 0030 Normative constraint (Fire)
TC 1858 Normative power
KC 0520 Normative social integration
HM 2067 Northern-continent awareness community (Buddhism)
HM 2364 Nose consciousness (Buddhism)
HM 1503 Nostalgia
HM 3257 Not ashamed acquisition sinful unwholesome dharmas (Buddhism)
HM 1023 Not ashamed ought be ashamed (Buddhism)
Not being
HM 0684 — covetous (Buddhism)
HM 3947 — greedy ; State (Buddhism)
HM 3183 — infatuated ; State (Buddhism)
HH 0367 Not-doing
Not feeling
HM 1825 — greed ; State (Buddhism)
HM 1597 — hatred ; State (Buddhism)
HM 4081 — infatuation ; State (Buddhism)
HM 3353 — remorse acquisition sinful unwholesome dharmas (Buddhism)
HM 1673 — remorse what one ought be remorseful (Buddhism)
HM 3242 Not hit awareness (Yoga)
HM 3413 Not incurring guilt (Buddhism)
HM 1688 Not infatuated (Buddhism)
HM 1510 Not-seeing (Buddhism)
HH 0241 Not-self (Buddhism)
HM 3435 Not trespassing limit (Buddhism)
HM 0403 Not understanding (Buddhism)
HM 4382 Not-wavering (Buddhism)
VC 1203 Notability
TC 1566 Note speakers ; Key
TC 1691 Noteboard
HM 2027 Nothing ; Sphere (Buddhism)
VD 3336 Nothingness
Nothingness [Absolutely...]
HM 5265 — Absolutely open (Taoism)
HM 2027 — absorption ; Consciousness (Buddhism)
HM 3047 — Absorption (ICA)
HM 3009 Nothingness ; Decisional (ICA)
HM 3042 Nothingness ; Divine (ICA)
HM 3612 Notion one's self entity (Buddhism)
VD 3338 Notoriety
MP 1182 Nourishment ; Appropriate conditions perspective
MP 1147 Nourishment ; Coordination perspective
MP 1139 Nourishment ; Hospitable domain perspective
MP 1177 Nourishment ; Local cultivation sources perspective
MP 1089 Nourishment ; Local sources perspective
TP 0027 Nourishment ; Providing
MP 1093 Nourishment ; Transit point location sources perspective
KC 0404 Nous
VC 1205 Novelty
HH 1306 Nubuwah (Islam)
MP 1075 Nuclear interaction ; Extended pattern
MS 8723 Nuclear mass destruction weapons
HH 0482 Nude group therapy
VD 3340 Nuisance
VD 3342 Nullity
HM 4512 Nullity (Leela)
KC 0981 Number
MS 8605 Number
HM 2702 Number ; Beyond
VT 8015 Number*complex
MP 1153 Number embodied perspectives ; Structures adaptable changing
MP 1103 Number occupiable temporary sites ; Limitation
MP 1096 Number structural levels ; Limitation
KC 0079 Number symbolism
KD 2130 Number time ; Quaternary consciousness:
VP 5086 Numbered-Unnumbered
KC 0032 Numbers ; Fibonacci
MS 8306 Numeracy
HM 5118 Numerical awareness (Sufism)
KC 0079 Numerology
HH 0340 Numerology
VP 5101 Numerousness-Fewness
HM 3811 Numinosum (Jung)
HM 3811 Numinous experience (Jung)
HM 1354 Nur ; An (Sufism)
TP 0009 Nurturance small
MM 2055 Nutrition cooking ; Food,
HM 2663 Nyamsrtsal (Buddhism)
HM 2161 Nyan thos (Tibetan)
HM 2157 Nye bar jog pa (Tibetan)
HM 2729 Nye-bar-zhi-bar-byed-pa (Tibetan)
HH 0781 Nye-nyon (Buddhism)
HM 2623 Nyer bsdogs (Tibetan)
HH 0781 Nyon ; Nye (Buddhism)
HH 0270 Nyon ; Rtsa (Buddhism)

O

VC 1207 Obedience
VP 5766 Obedience-Disobedience
HM 2024 Obedience ; Life (ICA)
HM 2258 Obedience ; Submissive (ICA)
HM 2005 Obedient son (ICA)
HM 5754 Obfuscation consciousness
HM 1264 Object ; Contemplation (Islam)
HM 5214 Object ; Infinite concentration infinite (Buddhism)

HM 4653 Object ; Infinite concentration limited (Buddhism)
Object Limited concentration
HM 5547 — infinite (Buddhism)
HM 1175 — limited (Buddhism)
HM 5547 — measureless (Buddhism)
HM 4653 Object ; Measureless concentration limited (Buddhism)
HM 5214 Object ; Measureless concentration measureless (Buddhism)
HH 4188 Object subject ; Journeying transcendence
Object Understanding having
HM 3296 — exalted (Buddhism)
HM 6681 — infinite (Buddhism)
HM 2495 — limited (Buddhism)
HM 6681 — measureless (Buddhism)
HM 3296 — sublime (Buddhism)
HH 1219 Objection ; Conscientious
KC 0757 Objective
HM 2203 Objective awareness
HM 2598 Objective awareness (ICA)
HM 2203 Objective consciousness
HM 2402 Objective fourth state consciousness
TC 1704 Objective ; Rational
HM 2203 Objective reason
TC 1920 Objectives ; Conferencing
VC 1209 Objectivity
HM 3354 Objectivity ; Cessation (Hinduism)
Objects [Abstract...]
KC 0129 — Abstract
HM 0110 — Addiction (Yoga)
MS 8336 — animal husbandry domestication
MS 8306 Objects devices ; Cultural
MS 8602 Objects ; Food related
HM 2522 Objects sense mental consciousness ; Emptiness (Buddhism)
MS 8326 Objects ; Symbolic
HH 0968 Oblation ; Self
HM 3122 Obligation ; Ancestral (ICA)
HM 3206 Obligation ; Freedom (ICA)
HM 2971 Obligation ; Personal (ICA)
HH 0300 Obligations ; Social (Japanese)
VD 3344 Oblivion
VD 3346 Obscurity
VC 1211 Observance
VP 5768 Observance-Nonobservance
HH 0229 Observation meditation ; Detached self
HM 2305 Observation one's psychic stream (Sufism)
HH 0586 Observation ; Self
TC 1742 Observation teams
HH 2276 Observation therapy ; Self
KD 2070 Observer entrapment micro-macro complementarity
TP 0020 Observing
HH 0278 Observing moral precepts (Buddhism)
HM 2809 Obsession
VD 3348 Obsession
HM 3829 Obsession ignorance (Buddhism)
HM 5329 Obsessional experience
VD 3350 Obsolescence
VD 3352 Obstacle
TP 0040 Obstacles ; Elimination
HM 3182 Obstacles soul cognition (Yoga)
VD 3354 Obstinacy
Obstruction
VD 3356 — [Obstruction]
TP 0039 — [Obstruction]
TP 0012 — [Obstruction]
TP 0009 Obstructions ; Small
HM 2444 Occult intelligence ; Path (Judaism)
HM 1285 Occult perception (Hinduism)
HH 0596 Occupational therapy
Occupiable sites
MP 1180 — exposed external insight
MP 1179 — Organization structure provide
MP 1188 — perspective inactivity
MP 1202 — Structure enfolded
MP 1022 Occupiable temporary site limit
MP 1103 Occupiable temporary sites ; Limitation number
VC 1213 Occurrence
HM 6720 Occurrence consciousness ; Modes (Buddhism)
HM 4389 Occurrence ; Sense mode consciousness (Buddhism)
HM 2750 Oceanic feeling
KD 2146 Ocular vision ; De categorization poly
HM 0351 Oculoagravic illusion
HM 0351 Oculogyral illusion
HM 0425 Odagya (Pali)
HM 0966 Odious sentiment (Hinduism)
MP 1126 Off-centering focal point common domain
TC 1567 Off ; Meeting take
VD 3358 Offence
HM 3172 Offering ; Promissorial (ICA)
MS 8913 Office ; Symbols
HH 0594 Official curse
HM 1998 Okappana (Pali)
HM 2312 Olam-asiyah (Judaism)
HM 2408 Olam atziluth (Judaism)
HM 3038 Olam briah (Judaism)
HM 2360 Olam-yetzirah (Judaism)
HH 5223 Old age
HH 1672 Old soul
HM 2202 Old wise-man archetypal image (Tarot)
VP 5122 Oldness ; Newness
HM 4559 Olfactory hallucination
HM 4477 Omid (Sufism)
VD 3360 Omission
HM 3732 Omkar (Leela)
KC 0818 Omnidirectionality
HM 2750 Omnipotence
VC 1215 Omnipotence
VC 1217 Omnipresence
HH 0320 Omnipresent mental factors (Buddhism)
VC 1219 Omniscience
HM 6558 Omniscience (Hinduism)
HM 3948 Omniscience (Jainism)
HM 2475 Omniscience (Yoga)
KC 0918 Omnitopology
KD 2120 Omnitriangulation: interlocking cycles
HM 7563 Once return ; Knowledge path (Buddhism)
HH 4066 Ondinnonk
HM 7672 One becoming many (Buddhism)
HM 7843 One-centred mind (Buddhism)
HM 6663 One directedness thought (Buddhism)
KC 0179 One Earth
HM 1425 One faith ; Stage
HM 0629 One ; Focus (Psychism)
One pointedness
HM 6663 — citta (Buddhism)
HM 0220 — martial arts training
HM 2753 — mental abiding states (Buddhism)
HM 5734 — (Yoga)
Ones being
HM 3974 — Abraham (Sufism)
HM 3425 — Adam (Sufism)
HM 4570 — David (Sufism)
HM 2246 — Jesus (Sufism)
HM 3513 — Mohammad (Sufism)
HM 3616 — Moses (Sufism)
HM 3286 — Noah (Sufism)
HM 2346 One's calling ; Surrender (ICA)
HM 2305 One's psychic stream ; Observation (Sufism)
Ones self
HM 8186 — Anomalies awareness
HM 3612 — entity ; Notion (Buddhism)
HM 2548 — environment ; Anomalies experience reality
TP 0005 One's time ; Biding (Getting wet)
HM 1992 One's view ; Holding paramount (Buddhism)
HM 4312 One-track mind
KC 0039 Oneness
HM 1717 Oneness meditation ; Tenth order perceptions universal
HM 2702 Oneness ; Sense
HM 5097 Oneroid states
HM 1993 Onlooker consciousness
HM 2171 Only-a-beginner subtle contemplation (Buddhism)
HH 0718 Onto-psychology
HH 0366 Ontogenesis
HM 1105 Ontogenetic model human consciousness
KC 0516 Ontogenetic recapitulation phylogeny
HH 0366 Ontogeny
HH 0645 Ontogeny ; Psychic
HM 1211 Ontological love
HH 0794 Ontological perfection
HM 3457 Ontological reality ; Existence (Sufism)
HM 3303 Ontological reality truth (Sufism)
KC 0464 Ontology
VD 3362 Opacity
TC 1947 Opaque projector
VP 5339 Opaqueness ; Transparency
HM 1330 Open consciousness ; Spirit (Taoism)
HH 0326 Open learning systems
HH 3562 Open mindedness
HM 5265 Open nothingness ; Absolutely (Taoism)
KC 0746 Open systems
KC 0195 Open systems initiatives
VC 1221 Openheartedness
KD 2090 Opening closing: alternation discontinuous learning
VP 5265 Opening-Closure
HH 0787 Opening meditation ; Dynamic self
VC 1223 Openmindedness
HM 2770 Openness
VC 1225 Openness
HH 5124 Openness ; Negotiating
MP 1082 Operational control contexts ; Minimal distance related
KC 0840 Operational gaming
MP 1081 Operations ; Minimally structured perspective control
KC 0540 Operations research
TC 1736 Operator type ; Detached
KC 0712 Opinion ; Public
TC 1011 Opinion , Public
HM 2054 Opinions ; Current views (Buddhism)
HM 4060 Opium similar drugs ; Modes awareness associated use
MP 1157 Opportunities perspective activity ; Local
MP 1156 Opportunities perspectives decreasing activity
MP 1072 Opportunities transposed concrete level ; Competitive interaction
VC 1227 Opportunity
VD 3364 Opposition
TP 0038 Opposition
HM 7336 Opposition carnal soul (Sufism)
KP 3002 Opposition/Disagreement
HM 2910 Opposition ; Loyal (ICA)
VP 5785 Opposition ; Support
KD 2010 Oppositional logic
VD 3366 Oppression
TP 0047 Oppression
VD 3368 Opprobrium
KC 0919 Optimal planning
VC 1229 Optimism
HH 6632 Optimism
KC 0436 Optimization
MM 2085 Orbiting space
KC 0364 Orchestration
HH 1141 Ordeal
HH 1141 Ordeal ; Trial
HH 0377 Ordeals
KC 0245 Order
VC 1231 Order
KC 0222 Order chaos
VT 8014 Order*complex
VP 5059 Order-Disorder
KD 2081 Order fluctuation: dissipative structures
KD 2272 Order ; Focus variable geometries global
KC 0127 Order interaction
KC 0277 Order ; International economic
Order [Structural...]
KC 0093 — Structural
KC 0740 — system
KC 0740 — System
KC 0463 Order theorizing ; Second
KC 0278 Order universe
KD 2230 Order ; Wholeness implicate
KC 0496 Order ; World
KC 0093 Ordering ; Spatial
VD 3370 Orderlessness
KC 0245 Orderliness
VC 1233 Orderliness
HM 2188 Orders angelic awareness (Judaism)
HM 0988 Orders perception
KC 0257 Ordinality ; Multi
VC 1235 Ordinariness
HM 2266 Ordinary awareness
HH 4110 Ordinary ; Journeying transcendence extraordinary (Christianity)
HM 2391 Ordinary reality ; States non
HM 0486 Ordinary world ; Merging (Taoism)
HH 1673 Organic evolution
KC 0721 Organic form
KC 0169 Organic functional multiplicity
KC 0535 Organicism
KC 0514 Organism
KC 0536 Organismic
KC 0510 Organisms ; Artificial
KC 0240 Organization
VC 1237 Organization
KC 0808 Organization ; Arboreal
MP 1002 Organization ; Distribution
MP 1146 Organization ; Flexible domain
Organization [integrative...]
MP 1209 — integrative superstructure
MP 1162 — integrative superstructure minimize insight blindspots
MP 1131 — inter-domain dynamics
MP 1006 — Intermediate scale
MP 1003 — Interpretation complementary modes
KD 2260 — Interwoven alternatives: tensegrity
KC 0320 Organization knowledge
KC 0572 Organization labour
KC 0841 Organization ; Levels
MP 1007 Organization ; Non linear
KC 0615 Organization production
KD 2180 Organization requirements ; Cyclic self
Organization [Self...]
KC 0510 — Self
KD 2085 — self-renewal: autopoiesis
KC 0160 — Social
KC 0114 — society ; Political
MP 1109 — structure enhance autonomy sub-structures
MP 1179 — structure provide occupiable sites
MP 1107 — structures enhance receptivity external insight
MP 1159 — structures permit two sources external insight
KC 0657 Organization ; United Nations
HM 1605 Organization ; Way (Esotericism)
HH 0617 Organizational development group
TP 0003 Organizational growth pains
HH 3456 Organizations ; Spirit transformation
KC 0336 Organized complexity
HH 0692 Organized games
MP 1172 Organizing non linearity ; Contexts self
KC 0510 Organizing systems ; Self
MS 8325 Organs ; Human external
MS 8345 Organs ; Human internal
HM 0829 Orgasm ; Sexual
HH 0327 Orgone therapy
MP 1128 Orientation domains receive external insight
MP 1212 Orientation ; Primary inter level connections transitions boundary
MM 2029 Orientation ; Spatial
MP 1105 Orientation structures enhance receptivity external insight
MP 1120 Oriented communication networks ; Focus
HH 0716 Oriented group experience ; Task
TC 1657 Oriented participant type ; Process
TC 1860 Oriented participant type ; Structure
TC 1765 Oriented type ; Action
HM 7290 Origin ill ; Understanding reference truth (Buddhism)
HM 3251 Origin ; Returning (Zen)
HM 7290 Origin suffering ; Understanding knowledge (Buddhism)
HM 3105 Original gratitude (ICA)
HM 2773 Original integrity (ICA)
HM 2163 Original mind (Zen)
HM 1655 Original moments ; Experience (Sufism)
HH 1778 Original sin
HH 3567 Original sin
VC 1239 Originality
VP 5023 Originality-Imitation
HM 3035 Origination formula ; Ignorance dependent (Buddhism)
MS 8816 Ornamentation
MS 8713 Ornithological symbols
HM 2390 Orpheus awareness
HM 2980 Orthodox sleep
VC 1241 Orthodoxy

Orthodoxy-Unorthodoxy

VP 6024	Orthodoxy-Unorthodoxy
KC 0669	Orthoepy
KC 0236	Orthogenesis
KC 0863	Orthography
HH 0289	Orthomolecular psychiatry
VP 5323	Oscillation-Agitation
VD 3372	Ostentation
MP 1100	Ot engender fruitful interfaces ; Arrangement structures
HM 3378	Other-induced attentiveness (Buddhism)
HM 2614	Other world midst this world (ICA)
HM 3225	Otherness ; Confirmation
HM 2570	Otherness ; Ubiquitous (ICA)
TP 0013	Others ; Identification
HH 0932	Others ; Rejoicing merit (Buddhism)
HM 5232	Others' thoughts ; Knowledge (Buddhism)
HM 8112	Ottappa (Pali)
HM 2538	Ottappabala (Pali)
HM 1673	Ottappati ottappitabbena ; Na (Pali)
HM 1757	Ottappati ottappitabbena (Pali)
HM 3353	Ottappati papakanam akusalanam dhammanam samapattiya ; Na (Pali)
HM 1208	Ottappati papakanam akusalanam dhammanam samapattiya (Pali)
HM 1673	Ottappitabbena ; Na ottappati (Pali)
HM 1757	Ottappitabbena ; Ottappati (Pali)

Ought be

HM 3828	— ashamed ; Being ashamed what one (Buddhism)
HM 1023	— ashamed ; Not ashamed (Buddhism)
HM 1757	— remorseful ; Feeling remorse what one (Buddhism)
HM 1673	— remorseful ; Not feeling remorse what one (Buddhism)
HM 0332	Out body experience ; Spontaneous

Out of body

HM 6120	— experience
HM 0777	— experience (NDE)
HM 5534	— experience (Psychism)
TC 1776	Out ; Participant burn
HM 1280	Outlook ; Right (Buddhism)
VD 3374	Outrage
HH 3215	Outside inside ; Journeying transcendence
HH 3667	Outside paths (Taoism)
HH 4289	Outside paths (Taoism)
HM 2684	Outside strength (Japanese)
HM 4754	Outside world ; Anomalies experience self distinct
HM 2581	Outward-flowing consciousness (Yoga)
HH 0621	Outward zazen meditation (Zen)
HM 1466	Over-valued ideas
VD 3376	Overabundance
VD 3378	Overactivity
HM 5123	Overall awareness (Psychism)

Overcoming [dissension...]

TP 0059	— dissension (Flooding)
HM 2655	— doubt (Buddhism)
HH 1187	— doubt ; Purification (Buddhism)
VD 3380	Overcompensation
VD 3382	Overconfidence
VD 3384	Overconscientiousnes
VD 3386	Overdeveloped
VD 3388	Overemphasis
VD 3390	Overestimation
VP 5497	Overestimation-Underestimation
VD 3392	Overexpansion
VD 3394	Overextension
TC 1472	Overhead projector
VD 3396	Overload
KC 0517	Overload ; System
VD 3398	Overreligiousness
VP 5313	Overrunning-Shortcoming
TP 0019	Overseeing
VD 3400	Oversensitiveness
VD 3402	Oversight
VD 3404	Oversimplification
HH 0504	Oversoul
HM 0827	Oversoul awareness (Psychism)
VD 3406	Overstrain
VP 5660	Oversufficiency-Insufficiency
VD 3408	Oversupply
VD 3410	Overtax
VD 3412	Overturn
MP 1164	Overview communication pathway structure
MP 1166	Overview domains interfaces structure external environment
MP 1167	Overview domains minimum proportions ; Enfolded
MP 1192	Overview external contexts
MP 1231	Overview sites integrative superstructure
VD 3414	Overweight
VD 3416	Overwork
VD 3418	Overzealousness
HM 1596	Own-being (Buddhism)
HM 3215	Own strength (Zen)
HM 2492	Ox-and-man-both-out-of-sight awareness (Zen)

Ox awareness

HM 2979	— Catching (Zen)
HM 2560	— Herding (Zen)
HM 3036	— Searching (Zen)
HM 2755	— Seeing (Zen)
HM 2604	Ox-forgotten-leaving-man-alone awareness (Zen)
HM 3153	Ox's back awareness ; Coming home (Zen)
HM 2690	Oxherding pictures Zen Buddhism (Zen)

P

HM 2586	Pa ; Bsam gtan bzhi (Buddhism)
HM 2038	Pa ; Bsam gtan gnyis (Buddhism)
HM 2623	Pa ; Yid byed (Buddhism)
HM 3837	Paap ; Saman (Leela)
HM 2599	Paccaya (Buddhism)
HH 1099	Paccayakara (Buddhism)
HM 3955	Paccekabodhi (Buddhism)
HM 0249	Paccupalakkhana (Pali)
MP 1054	Paced communications ; Intersection differently
HM 2729	Pacification mental abiding states ; Full (Buddhism)
VC 1243	Pacifism
HM 2205	Pacifying mental abiding states (Buddhism)
VD 3420	Paganism
HM 0881	Paggaha (Pali)
HM 0872	Pagunabhava (Pali)
HM 1741	Pagunata ; Vinnanakkhandhassa (Pali)
HM 3424	Pagunattam (Pali)
HM 1810	Pagunnata ; Citta (Pali)
HM 1455	Pagunnata ; Kaya (Pali)
HM 1061	Pahasa (Pali)
HH 0010	Paideia
HM 4031	Pain
VD 3422	Pain
HM 8010	Pain (Buddhism)
HM 3094	Pain ; Ego
HM 6122	Pain ; Love
HM 1539	Pain solitude
HM 2463	Painful acknowledgement (ICA)
HM 8010	Painful feeling (Buddhism)
HM 7859	Painful progress quick intuition ; Concentration (Buddhism)
HM 5992	Painful progress sluggish intuition ; Concentration (Buddhism)
TP 0003	Pains ; Organizational growth
VC 1245	Painstakingness
HH 3498	Pair-bonding
TC 1311	Pairs
HM 0158	Pajanana (Pali)
HM 2793	Pakkhiya dhamma ; Bodhi (Buddhism)
HM 2793	Pakshika dharma ; Bodhi (Buddhism)
HM 2508	Palace wisdom (Buddhism)
HM 1508	Palana (Pali)
HH 1765	Pali Abhidhamma
HM 3563	Pali Abhijjha
HM 3432	Pali Abhinivesa
HM 1809	Pali Abhinna
HM 1687	Pali Abhippasada
HM 7171	Pali Abstinence verbal misconduct
HM 7887	Pali Abstinence wrong livelihood
HM 1734	Pali Abyapada
HM 1510	Pali Adassana
HM 2744	Pali Adhosa
HM 1399	Pali Adosa
HM 3560	Pali Adoso kusala mulam
HM 1557	Pali Adussana
HM 1597	Pali Adussitattam
HM 1724	Pali Ahirikabala
HM 1054	Pali Ajimhata
HM 1028	Pali Akakkhalata
HM 3375	Pali Akaranam
HM 2110	Pali Akasanancayatana
HM 3014	Pali Akathinata
HM 2027	Pali Akincannaayatana
HM 3429	Pali Akiriya
HM 1008	Pali Akutilata
HM 1799	Pali Alobha
HM 2128	Pali Alobha
HM 3752	Pali Alobhokusala mulam
HM 3947	Pali Alubbhana
HM 1825	Pali Alubbhitattam
HM 4065	Pali Amodana
HM 2695	Pali Amoha
HM 0684	Pali Anabhijjha
HM 4090	Pali Anabhisamaya
HM 3413	Pali Anajjhapatti
HM 0403	Pali Ananubodha
HM 3949	Pali Anattamanata cittassa
HM 1915	Pali Anattappabala
HM 0870	Pali Anekamasaggaha
HM 4131	Pali Anikkhittacchandata
HM 1403	Pali Anikkhittadharata
HM 1394	Pali Annana
HM 0188	Pali Anottappa
HM 4665	Pali Anupadisesanibbana
HM 1930	Pali Anupekkhanata
HM 0405	Pali Anussati
HM 0649	Pali Anuttassa
HM 1982	Pali Anuvicara
HM 1746	Pali Apacakkhakamma
HM 1902	Pali Apaccavekkhana
HM 4228	Pali Apariyobahana
HM 0427	Pali Apariyogahana
HM 1689	Pali Apilapanata
HM 1617	Pali Appana
HM 2586	Pali Appana samadhi
HM 1637	Pali Appativedha
HM 2012	Pali Arupaloka consciousness
HM 1969	Pali Arupinam dhammanam ayu
HH 3198	Pali Aruppa-niddesa
HM 0451	Pali Asagahana
HM 1965	Pali Asamapekkhana
HM 1725	Pali Asambodha
HM 6504	Pali Asamkhata
HM 1705	Pali Asammusanata
HM 1725	Pali Asapajjana
HM 3394	Pali Asappana
HM 1688	Pali Asaraga
HM 3183	Pali Asarajjana
HM 4081	Pali Asarajjitattam
HM 3261	Pali Asithilaparakkamata
HM 3857	Pali Asuropa
HM 0140	Pali Attamanata cittassa
HM 1663	Pali Attha-patisambhida
HM 4037	Pali Avankata
HM 0227	Pali Avarice
HM 3035	Pali Avijja
HM 3196	Pali Avijja
HM 1013	Pali Avijjalanga
HM 1375	Pali Avijjanusaya
HM 3829	Pali Avijjapariyutthana
HM 4470	Pali Avijjayoga
HM 3145	Pali Avijjogha
HM 4572	Pali Avikkhepa
HM 0449	Pali Avisahara
HM 0715	Pali Avisahatamanasata
HM 3374	Pali Avitthanata
HM 0436	Pali Avupasama
HM 4565	Pali Avyapajja
HM 1864	Pali Balya
HM 0400	Pali Bhantattam cittassa
HM 1623	Pali Bhuri
HM 6033	Pali Buoyancy citta
HM 7636	Pali Buoyancy kaya
HM 1591	Pali Byapajjana
HM 3210	Pali Byapajjitatta
HM 1111	Pali Byapatti
HM 1271	Pali Cakkhuvinnanadhatusamphassaja sanna
HM 0972	Pali Cakkhuvinnanadhatusamphassajam cetasikam neva satam nasatam
HM 0945	Pali Candikka
HM 1781	Pali Catuhi vaciduccaritehi arati virati pativirati veramani
HM 2589	Pali Cetana
HM 0805	Pali Cetasikam dukkham
HM 1862	Pali Cetasikam sukham
HM 1263	Pali Cetasiko viriyarambho
HM 4425	Pali Cetaso vikkhepo
HM 1412	Pali Cetayitattam
HM 1643	Pali Cetosamphassaja addukkhamasukha vedana
HM 1090	Pali Cetosamphassaja asata vedana
HM 4076	Pali Cetosamphassaja dukkha vedana
HM 0309	Pali Cetosamphassaja sata vedana
HM 1352	Pali Cetosamphassaja adukkhamasukham vedayitam
HM 1031	Pali Cetosamphassajam dukkam vedayitam
HM 0901	Pali Cetosamphassajamasatam vedayitam
HM 0125	Pali Cetovimutti
HM 3966	Pali Cinta
HM 1810	Pali Citta-pagunnata
HM 1584	Pali Cittakammannata
HM 1830	Pali Cittalahuta
HM 1805	Pali Cittamuduta
HM 1490	Pali Cittapassaddhi
HM 3471	Pali Cittasa avatthiti
HM 0223	Pali Cittassa manovilekha
HM 1066	Pali Cittassa santhiti
HM 0051	Pali Cittassa thiti
HM 0975	Pali Cittassa uddhacca
HM 0253	Pali Cittujukata
HM 4394	Pali Conscienceless
HM 6652	Pali Conscious duration
HM 3707	Pali Craving
HH 0680	Pali Cultivation experiential insight
HM 4487	Pali Desire
HM 4487	Pali Desire-to-do
HM 4726	Pali Dhamma-patisambhida
HM 2551	Pali Dhammas
HM 1715	Pali Dhammavicaya
HM 1423	Pali Dharanata
HM 1974	Pali Dhiti
HM 1081	Pali Dhurasampaggaha
HM 1191	Pali Ditthi
HH 2718	Pali Ditthi-visuddhi-niddesa
HM 1613	Pali Ditthigahana
HM 2054	Pali Ditthigata
HM 4548	Pali Ditthikantara
HM 1511	Pali Ditthisannojana
HM 2722	Pali Ditthivipphandita
HM 1407	Pali Ditthivisukayika
HM 4502	Pali Dosa
HM 0607	Pali Dosa
HM 5676	Pali Dosakkhaya
HM 2574	Pali Dukkha
HH 0523	Pali Dukkha
HM 3556	Pali Dukkhindriya
HM 1857	Pali Dummajjha
HM 4323	Pali Dussana
HM 1967	Pali Dussitattam
HM 1286	Pali Dvedhapatha
HM 1561	Pali Dvelhaka
HM 2638	Pali Envy
HM 6754	Pali Four stations mindfulness
HM 0700	Pali Gaha
HM 4637	Pali Gotra-bhu-nanam
HM 1421	Pali Hasa
HM 2881	Pali Hiri
HM 1590	Pali Hiri
HM 0352	Pali Hiribala
HM 3828	Pali Hiriyati hiriyitabbena
HM 0552	Pali Hiriyati papakanam akusalanam dhammanam samapattiya
HH 5652	Pali Iddhi
HM 2075	Pali Indriya phenomena
HM 1972	Pali Iriyana
HM 2536	Pali Jhana
HM 7193	Pali Jhana
HM 0137	Pali Jhana

HM 2122	Pali Jhana	
HM 3379	Pali Jivita	
HM 3404	Pali Jivitindriya	
HM 0261	Pali Jivitindriya	
HM 2060	Pali Kamaloka consciousness	
HM 2048	Pali Kamma	
HM 3716	Pali Kammannabhava	
HM 0979	Pali Kammannattam	
HM 4347	Pali Kankha	
HM 0177	Pali Kankhayana	
HM 3934	Pali Kankhayitattam	
HM 1455	Pali Kaya-pagunnata	
HM 5600	Pali Kayaduccaritavirati	
HM 0739	Pali Kayakammannata	
HM 4395	Pali Kayalahuta	
HM 3370	Pali Kayamuduta	
HM 1867	Pali Kayapassaddhi	
HM 1424	Pali Kayujukata	
HM 2692	Pali Khandha	
HH 3321	Pali Khandha	
HM 1554	Pali Kosalla	
HM 9101	Pali Kukkucca	
HM 1912	Pali Kummagga	
HM 3808	Pali Lahuparinamata	
HM 7983	Pali Liberation	
HM 8125	Pali Liberation means wisdom	
HM 0125	Pali Liberation mind	
HM 3283	Pali Lobha	
HM 0802	Pali Lobha	
HM 1954	Pali Lobho akusalamulam	
HM 2072	Pali Loka consciousness	
HM 1442	Pali Lubbhana	
HM 1513	Pali Lubbhitattam	
HM 0191	Pali Maddavata	
HH 4007	Pali Maggamagga-nanadassana-visuddhi-niddesa	
HM 0251	Pali Manasa	
HM 3237	Pali Manasi-kara	
HM 1736	Pali Manayatana	
HM 1369	Pali Manindriya	
HM 1743	Pali Mano	
HM 1834	Pali Manovinnanadhatu	
HM 1921	Pali Manovinnanadhatu samphasajam cetasikam satam	
HM 0404	Pali Manovinnanadhatu samphassaja cetana	
HM 1501	Pali Manovinnanadhatu samphassaja sanna	
HM 0234	Pali Manovinnanadhatusampassa jam cetasikam asatam	
HM 3314	Pali Manovinnanadhatusamphassaja cetasikam neva satam nasatam	
HM 0227	Pali Meanness	
HM 1750	Pali Medha	
HM 0252	Pali Miccha ajiva arati virati pativirati veramani	
HM 1710	Pali Micchaditthi	
HM 0863	Pali Micchapatha	
HM 1802	Pali Micchasamadhi	
HM 3356	Pali Micchatta	
HM 1585	Pali Micchavayama	
HM 0264	Pali Middha	
HM 0184	Pali Moha	
HM 0918	Pali Moha	
HM 6124	Pali Mohakkhaya	
HM 1439	Pali Moho akusalamulam	
HM 1023	Pali Na kiriyati hiriyitabbena	
HM 3257	Pali Na kiriyati papakanam akusalanam dhammanam samapattiya	
HM 1673	Pali Na ottappati ottappitabbena	
HM 3353	Pali Na ottappati papakanam akusalanam dhammanam samapattiya	
HH 0680	Pali Nana-dassana	
HH 3025	Pali Nanadassana-visuddhi-niddesa	
HM 3390	Pali Nepunna	
HM 2051	Pali Nevasanna-n'asannayatana	
HM 2330	Pali Nibbana	
HM 1496	Pali Nikkama	
HM 5026	Pali Nirutti-patisambhida	
HM 5023	Pali Nirvana remainder	
HM 0425	Pali Odagya	
HM 1998	Pali Okappana	
HM 8112	Pali Ottappa	
HM 2538	Pali Ottappabala	
HM 1757	Pali Ottappati ottappitabbena	
HM 1208	Pali Ottappati papakanam akusalanam dhammanam samapattiya	
HM 0249	Pali Paccupalakkhana	
HM 0881	Pali Paggaha	
HM 0872	Pali Pagunabhava	
HM 3424	Pali Pagunattam	
HM 1061	Pali Pahasa	
HM 0158	Pali Pajanana	
HM 1508	Pali Palana	
HM 4246	Pali Pamodana	
HM 1371	Pali Pamoha	
HM 1398	Pali Pamojja	
HM 1499	Pali Pandicca	
HM 4523	Pali Panna	
HM 0082	Pali Panna	
HM 1547	Pali Pannaabhasa	
HM 1846	Pali Pannaaloka	
HM 1792	Pali Pannabala	
HM 0192	Pali Pannapajjota	
HM 2508	Pali Pannapasada	
HM 1486	Pali Pannaratana	
HM 1645	Pali Pannasattha	
HM 8125	Pali Pannavimutti	
HM 3233	Pali Pannindriya	
HM 1840	Pali Pannindriya	
HM 3469	Pali Parakkama	
HM 1992	Pali Paramasa	
HM 4358	Pali Parinayika	
HM 3745	Pali Parisappana	
HM 1761	Pali Passambhana	
HM 4958	Pali Patibhana-patisambhida	
HM 3942	Pali Patiggaha	
HH 3550	Pali Patipada-nanadassana-niddesa	
HM 1738	Pali Patipassaddhi	
HM 3411	Pali Patipassambhana	
HM 1528	Pali Patipassambhatattam	
HM 3806	Pali Patissati	
HM 0719	Pali Pativirodha	
HM 0633	Pali Patoda	
HH 0932	Pali Pattanumodana	
HM 1266	Pali Pattanumodana	
HM 1602	Pali Pavicaya	
HM 2708	Pali Phassa	
HM 4212	Pali Phassa	
HM 1959	Pali Phusana	
HM 0747	Pali Piti	
HM 2038	Pali Piti	
HM 1155	Pali Pliancy citta	
HM 3696	Pali Pliancy kaya	
HM 6093	Pali Ragakkhaya	
HM 9001	Pali Rectitude citta	
HM 5402	Pali Rectitude kaya	
HM 0226	Pali Repose citta	
HM 4704	Pali Repose kaya	
HM 3800	Pali Resolution	
HM 3800	Pali Resolve	
HM 2108	Pali Rupa-khanda	
HM 5023	Pali Sa-upadisesanibbana	
HM 5403	Pali Saccanulomikanan	
HM 3158	Pali Saddahana	
HM 2933	Pali Saddha	
HM 4404	Pali Saddhabala	
HM 0066	Pali Saddhindriya	
HM 1143	Pali Sallakkhana	
HM 1759	Pali Samadhibala	
HM 3454	Pali Samadhindriya	
HM 4514	Pali Samapajanna	
HM 1498	Pali Samatha	
HM 2147	Pali Samatha	
HM 0549	Pali Samma ajiva	
HM 1763	Pali Samma ajiva	
HM 1735	Pali Samma-samadhi	
HM 3252	Pali Sammaditthi	
HM 0439	Pali Sammakammanta	
HM 0687	Pali Sammakammanta	
HM 1451	Pali Sammasankappa	
HM 0180	Pali Sammasati	
HM 4608	Pali Sammavaca	
HM 1821	Pali Sammavaca	
HM 1657	Pali Sammavayama	
HM 1956	Pali Sammoha	
HM 1874	Pali Sampajanna	
HM 1380	Pali Samphusana	
HM 0836	Pali Samphusitattam	
HM 0808	Pali Samsaya	
HM 2596	Pali Samvatta-kappa	
HM 0705	Pali Sancetana	
HM 2050	Pali Sankhara-khanda	
HM 1389	Pali Sanna	
HM 4143	Pali Sanna-khanda	
HM 3419	Pali Sannanana	
HM 3250	Pali Saraga	
HM 4363	Pali Sarajitattam	
HM 1683	Pali Sarajjana	
HM 1484	Pali Saranata	
HM 2847	Pali Sati	
HM 1563	Pali Sati	
HM 3450	Pali Satibala	
HM 4248	Pali Satindriya	
HM 0047	Pali Satindriya	
HH 4006	Pali Satipatthana	
HM 1098	Pali Setughata	
HM 6754	Pali Smrityupasthana	
HM 1649	Pali Somanassindriya	
HM 6652	Pali Steadiness consciousness	
HM 3707	Pali Tanha	
HM 7002	Pali Tatra-majjhattata	
HM 1749	Pali Thama	
HM 0077	Pali Thambhatattam	
HM 5667	Pali Thina	
HM 3707	Pali Thirst	
HM 1418	Pali Thiti	
HM 1133	Pali Tihi kayaduccaritehi arati virati pativirati veramani	
HM 3685	Pali Titthayatana	
HM 3707	Pali Trisna (Buddhism)	
HM 7983	Pali Ubhatobhagavimutti	
HM 1469	Pali Uddhacca	
HM 4250	Pali Ujuta	
HM 4394	Pali Unconscientiousness	
HM 3345	Pali Upaparikkha	
HM 1647	Pali Upavicara	
HM 7769	Pali Upekkha	
HM 2062	Pali Upekkha	
HM 3803	Pali Upekkhindriya	
HM 4487	Pali Ussaha	
HM 3357	Pali Ussolhi	
HM 1436	Pali Uyyama	
HM 2468	Pali Vattana	
HM 0229	Pali Vayama	
HM 4488	Pali Vebhabya	
HM 0349	Pali Vedanakkhandhassa sannakkhandhassa sankharakkhandhassa	
HM 1381	Pali Vedanakkhandhassa sannakkhandhassa sankharakkhandhassa kammannata	
HM 1435	Pali Vedanakkhandhassa sannakkhandhassa sankharakkhandhassa labuta	
HM 3397	Pali Vedanakkhandhassa sannakkhandhassa sankharakkhandhassa muduta	
HM 1377	Pali Vedanakkhandhassa sannakkhandhassa sankharakkhandhassa passaddhi	
HM 0409	Pali Vedanakkhandhassa sannakkhandhassa sankharakkhandhassa ujukata	
HM 3435	Pali Vela anatikkamo	
HM 5089	Pali Vicara	
HM 3408	Pali Vicaya	
HM 0529	Pali Vicikiccha	
HM 0812	Pali Vicikiccha	
HM 4008	Pali Vimati	
HM 5977	Pali Vimokkha	
HM 3617	Pali Vinnana	
HM 2556	Pali Vinnana-khanda	
HM 2098	Pali Vinnana-khanda development	
HM 2599	Pali Vinnana-kicca	
HM 3043	Pali Vinnanacayatana	
HM 4605	Pali Vinnanakkhandhassa	
HM 3403	Pali Vinnanakkhandhassa lahuta	
HM 1796	Pali Vinnanakkhandhassa muduta	
HM 1741	Pali Vinnanakkhandhassa pagunata	
HM 1338	Pali Vinnanakkhandhassa ujukata	
HM 1751	Pali Vipariyasaggaha	
HM 1632	Pali Vipassana	
HH 0680	Pali Vipassana-bhavana	
HM 1333	Pali Viriya	
HM 1943	Pali Viriyabala	
HM 1470	Pali Viriyindriya	
HM 1829	Pali Viriyindriya	
HM 4239	Pali Virodha	
HH 3875	Pali Visuddhimagga	
HM 1177	Pali Vitakka	
HM 2450	Pali Vitakka-vicara	
HM 1545	Pali Vitti	
HM 0487	Pali Vyapana	
HM 6652	Pali Weak concentration	
HM 6556	Pali Wieldiness citta	
HM 7969	Pali Wieldiness kaya	
HM 4487	Pali Wish act	
HM 2072	Pali Worlds conscious existence	
HM 5324	Pali Wrong view	
HM 3327	Pali Yapana	
HM 1691	Pali Yapana	
HM 7549	Pali Yatha-bhuta-nana-dassana	
HM 4487	Pali Zeal	
HH 0686	Palingenesis	
HM 4246	Pamodana (Pali)	
HM 1371	Pamoha (Pali)	
HM 1398	Pamojja (Pali)	
HH 0278	Panca sila (Buddhism)	
HH 4536	Pancasila	
HM 1499	Pandicca (Pali)	
TC 1531	Panel discussion	
	Panel [Interchange...]	
TC 1650	— Interchange	
TC 1650	— Interrogator	
TC 1650	— Investigative	
TC 1650	Panel ; Responsive	
TC 1650	Panel ; Zetetic	
HH 1908	Panentheism ; Human development	
VD 3424	Panic	
HM 2083	Panic	
HM 4523	Panna (Pali)	
HM 0082	Panna (Pali)	
HM 1547	Pannaabhasa (Pali)	
HM 1846	Pannaaloka (Pali)	
HM 1792	Pannabala (Pali)	
HM 0192	Pannapajjota (Pali)	
HM 2508	Pannapasada (Pali)	
HM 1486	Pannaratana (Pali)	
HM 1645	Pannasattha (Pali)	
HM 8125	Pannavimutti (Pali)	
HM 3233	Pannindriya (Pali)	
HM 1840	Pannindriya (Pali)	
HH 3876	Panpsychism	
HH 5190	Pantheism ; Human development	
	Papakanam akusalanam dhammanam	
HM 0552	— samapattiya ; Hiriyati (Pali)	
HM 3257	— samapattiya ; Na kiriyati (Pali)	
HM 3353	— samapattiya ; Na ottappati (Pali)	
HM 1208	— samapattiya ; Ottappati (Pali)	
TC 1383	Paper	
TC 1568	Papers ; Conference	
HH 0370	Parables ; Fables	
KC 0005	Paradigm	
KC 0235	Paradigmatology	
HM 3480	Paradisal condition (Psychism)	
HM 3480	Paradisal memory (Psychism)	
HH 4572	Paradise	
KC 0159	Paradox	
HH 3335	Paradox ; Way (Christianity)	
KD 2030	Paradoxes antinomies	
HM 2980	Paradoxical sleep	
HM 0003	Paraesthesia	
HM 3469	Parakkama (Pali)	
HM 5856	Parallel awareness	
MP 1023	Parallel ; Compensating relationships	
HH 0748	Paralogical thinking ; Archaic	
VD 3426	Paralysis	
HM 1067	Paramartha (Buddhism)	
HM 2294	Paramarthashunyata (Buddhism)	

Paramasa

HM 1992 Paramasa (Pali)
HH 0103 Paramatman (Hinduism)
HH 0550 Paramita ; Maha (Zen)
HH 0550 Paramita ; Prajna (Zen)
HH 0219 Paramitas (Zen)
HM 1992 Paramount one's view ; Holding (Buddhism)
HM 2070 Paranirmita-vasavartin (Buddhism)
HM 2262 Paranormal cognition (ESP)
HH 0981 Parapsychology
HH 3452 Parasitism
VC 1247 Pardon
HM 8135 Pareidolia
HM 8135 Pareidolic illusions
HH 3498 Parent-infant bonding
VC 1249 Parenthood
HH 3543 Parenthood
HM 0003 Paresthesia
HM 0286 Parhiz (Sufism)
HM 2330 Pari-nirvana (Buddhism)
HM 4665 Pari-nirvana (Buddhism)
HM 3337 Parinama ; Ekagrata (Yoga)
HM 3437 Parinama ; Nirodha (Yoga)
HM 3207 Parinama ; Samadhi (Yoga)
HH 4561 Parinama (Yoga)
HM 4358 Parinayika (Pali)
HM 2783 Paripurana ; Karma (Buddhism)
HM 3745 Parisappana (Pali)
HM 0294 Parmarth (Leela)
VC 1251 Parsimony
HH 1903 Parsis (Zoroastrianism)
KC 0879 Part whole categories
HM 8002 Partaking diminution ; Concentration (Buddhism)
HM 7363 Partaking distinction ; Concentration (Buddhism)
HM 5930 Partaking penetration ; Concentration (Buddhism)
HM 6956 Partaking stability ; Concentration (Buddhism)
HM 6956 Partaking stagnation ; Concentration (Buddhism)
HM 8002 Partaking worsening ; Concentration (Buddhism)
HM 2430 Partial-mindedness
MP 1119 Partially contained interfaces
MP 1193 Partially enclosed internal domains
MP 1183 Partially exposed perspective context
MP 1152 Partially isolated contexts
TC 1776 Participant burn-out
TC 1828 Participant change preferences
TC 1493 Participant consensus buttons
TC 1613 Participant expectations
TC 1845 Participant interaction preferences
Participant involvement
TC 1229 — Alienative
TC 1895 — Calculative
TC 1345 — Moral
TC 1204 Participant personality needs
TC 1965 Participant pre-logical biases
TC 1828 Participant strategic preferences
Participant type
TC 1164 — Accommodator
TC 1929 — Aggressive
TC 1772 — Assimilator
TC 1133 — Converger
TC 1225 — Diverger
TC 1384 — Doctrinaire
TC 1990 — Encourager
TC 1358 — Expertist
TC 1651 — Followers
TC 1430 — Grassrooter
TC 1491 — Initiator
TC 1739 — Less haste
TC 1129 — Marginal
TC 1775 — Mediator
TC 1663 — Non conformist
TC 1581 — Nonchalent
TC 1657 — Process oriented
TC 1391 — Self reflexer
TC 1860 — Structure oriented
TC 1853 — Value monger
TC 1750 Participants ; Information seeking
TC 1603 Participants ; Pattern language
VC 1253 Participation
TC 1877 Participation assumptions
TC 1490 Participation ; Conference
TC 1928 Participation dynamics
TC 1927 Participation methods
TC 1843 Participatory research
HM 2939 Particular concern (ICA)
KD 2150 Particular languages ; Emancipation
TP 0043 Parting (Flight)
VC 1255 Partnership
MS 8133 Parts ; Animal body
MS 8402 Parts ; Vehicles
TC 1651 Party-liners
KC 0315 Party unity
HM 1377 Passaddhi ; Vedanakkhandhassa sannakkhandhassa sankharakkhandhassa (Pali)
HH 0230 Passage ; Rite
HH 0021 Passage ; Rites
HM 1761 Passambhana (Pali)
HM 1087 Passing away (Islam)
HM 0748 Passing away reappearance beings ; Knowledge (Buddhism)
Passing away Sufism
HM 3799 — [Passing away (Sufism)]
HM 5775 — [Passing away (Sufism)]
HM 1270 — [Passing away (Sufism)]
HM 2076 Passing awayness (ICA)
TP 0028 Passing greatness
VC 1257 Passion
HM 6093 Passion ; Abolition (Buddhism)
HM 3054 Passion ; Appropriated (ICA)
HM 2234 Passion ; Infinite (ICA)
HM 0129 Passion (Islam)
HM 7221 Passion miracles (German)
HM 2641 Passion ; Sacrificial (ICA)
HM 3128 Passionate concern (ICA)
HM 2547 Passionate disinterest (ICA)
HM 3656 Passionate love (Islam)
HM 5087 Passionate love ; Mutual
VD 3428 Passionlessness
HH 0759 Passive resistance
HM 8123 Passive state (Christianity)
HH 3412 Passive way faith
VD 3430 Passivity
HH 3229 Passivity
HM 3959 Passivity ; Alert
HM 4469 Passivity ; Alert
HM 7203 Passivity ; Delusions
TP 0002 Passivity (Earth)
HM 7203 Passivity experiences
HM 4297 Past life ; Knowledge recollection (Buddhism)
MP 1070 Past perspectives ; Context acknowledgement
MS 8846 Pastimes ; Hobbies
HH 0735 Paternalism
VC 1259 Paternity
HH 0424 Path
Path [absolute...]
HM 3243 — absolute intelligence (Judaism)
HM 2218 — active intelligence (Judaism)
HM 2095 — administrative intelligence (Judaism)
HM 3013 — admirable intelligence (Judaism)
HM 7055 — arahantship ; Knowledge (Buddhism)
Path awareness Great
HM 3048 — vehicle higher (Buddhism)
HM 2268 — vehicle lower (Buddhism)
HM 2160 — vehicle middle (Buddhism)
HM 2228 Path awareness ; Pratyeka Buddha vision (Buddhism)
Path [cohesive...]
HM 3102 — cohesive intelligence (Judaism)
HM 3046 — collective intelligence (Judaism)
HM 2117 — constituting intelligence (Judaism)
HM 3149 — corporeal intelligence (Judaism)
HM 2569 Path disposing intelligence (Judaism)
HM 3371 Path ; Divine (Sufism)
Path [effortlessness...]
HH 2188 — effortlessness
HH 5645 — enlightenment (Buddhism)
HM 2593 — eternal intelligence (Judaism)
HM 2218 — exciting intelligence (Judaism)
Path [faithful...]
HM 2081 — faithful intelligence (Judaism)
HM 2523 — fiery intelligence (Judaism)
HM 1690 — form (Yoga)
Path [hidden...]
HM 2444 — hidden intelligence (Judaism)
HM 3013 — hidden intelligence (Judaism)
HH 7386 — Higher vehicle (Taoism)
Path [illuminating...]
HM 2612 — illuminating intelligence (Judaism)
HM 2035 — illuminating intelligence (Judaism)
HM 2011 — imaginative intelligence (Judaism)
Path intelligence
HM 2105 — conciliation (Judaism)
HM 2045 — influences (Judaism)
HM 2107 — light (Judaism)
HM 3089 — mediating influence (Judaism)
HM 2105 — reward (Judaism)
HM 2021 — secret (Judaism)
HM 2021 — spiritual action (Judaism)
HM 2503 — temptation (Judaism)
HM 2629 — will (Judaism)
MM 2074 Path journey
HM 7645 Path leading cessation ill ; Understanding reference truth (Buddhism)
HH 4864 Path ; Lower vehicle (Taoism)
Path Mahayana Buddhism
HM 2268 — [Path ; Mahayana (Buddhism)]
HM 3048 — [Path ; Mahayana (Buddhism)]
HM 2160 — [Path ; Mahayana (Buddhism)]
Path [measuring...]
HM 3102 — measuring intelligence (Judaism)
HH 0827 — meditative absorption (Yoga)
KC 0736 — method ; Critical
HH 6345 — Middle vehicle (Taoism)
Path [natural...]
HM 2071 — natural intelligence (Judaism)
HM 2339 — Noble eightfold (Buddhism)
HM 6920 — non return ; Knowledge (Buddhism)
HM 2444 Path occult intelligence (Judaism)
HM 7563 Path once return ; Knowledge (Buddhism)
Path [perfect...]
HM 3243 — perfect intelligence (Judaism)
HH 1747 — perfection ; Sufi (Sufism)
HM 2643 — perpetual intelligence (Judaism)
HH 3875 — purification (Buddhism)
HH 4007 — Purification knowledge vision what is path what is not (Buddhism)
HM 2943 — purified intelligence (Judaism)
HH 3875 — purity (Buddhism)
HH 4007 — Purity knowledge discernment right path wrong (Buddhism)
Path [radical...]
HM 3181 — radical intelligence (Judaism)
HM 3102 — receiving intelligence (Judaism)
HM 2047 — renewing intelligence (Judaism)
Path [renovating...] cont'd
HM 2047 — renovating intelligence (Judaism)
HH 3405 — renunciation (Yoga)
HM 3426 — resplendent intelligence (Judaism)
Path [sainthood...]
HM 4356 — sainthood (Sufism)
HM 3567 — sanctifying intelligence (Judaism)
HH 1867 — Spiritual
HH 0738 — Spiritual (Islam)
HM 2533 — stable intelligence (Judaism)
HM 1088 — stream entry ; Knowledge (Buddhism)
Path [tantra...]
HM 2656 — tantra awareness ; Great (Buddhism)
HM 3026 — tantra awareness ; Middle (Buddhism)
HM 4603 — Tantric formless (Buddhism)
HM 2107 — transparent intelligence (Judaism)
HM 2593 — triumphant intelligence (Judaism)
HM 2583 Path uniting intelligence (Judaism)
Path [what...]
HH 4007 — what is not path ; Purification knowledge vision what is (Buddhism)
HM 0863 — Wrong (Buddhism)
HH 4007 — wrong path ; Purity knowledge discernment right (Buddhism)
HH 0779 Path yoga ; Eightfold (Yoga)
HM 0399 Pathetic sentiment (Hinduism)
HM 3726 Pathological lying
HH 0206 Pathology ; Etheric
HM 2962 Paths accumulation application ascension stages game ; Sutra (Buddhism)
HM 2848 Paths accumulation application ascension stages game ; Tantric (Buddhism)
HM 2987 Paths calming (Buddhism)
HM 3249 Paths effect (Tibetan)
Paths [False...]
HH 3013 — False (Taoism)
HM 1252 — Four evil (Buddhism)
HM 4026 — Four noble (Buddhism)
HM 3194 Paths insight (Buddhism)
HH 3013 Paths ; Lowest (Taoism)
HH 3667 Paths ; Middle grade low (Taoism)
HH 3667 Paths ; Outside (Taoism)
HH 4289 Paths ; Outside (Taoism)
HM 1914 Paths ; Six (Buddhism)
HM 3079 Paths special qualities (Buddhism)
HM 0923 Paths ; Three evil (Buddhism)
HM 2015 Paths ; Uninterrupted release (Buddhism)
HH 4289 Paths ; Upper grade low (Taoism)
HM 2944 Paths view Buddhism ; Emptinesses (Buddhism)
HM 2240 Paths vision cultivation (Buddhism)
HH 0102 Paths wisdom
HM 2509 Paths wisdom (Judaism)
HH 0603 Paths ; Yogic (Buddhism)
MP 1165 Pathway ; Exposure structural activities communication
MP 1140 Pathway ; Relative isolation structural interface communication
MP 1164 Pathway structure ; Overview communication
MP 1112 Pathway ; Transition domain structure communication
MP 1229 Pathways automatic communications ; Provision
MM 2068 Pathways cycles ; Metabolic
MP 1122 Pathways ; Direct relationship structures communication
MP 1174 Pathways enfolded non linearity ; Communication
MP 1121 Pathways ; Hospitability communication
Pathways [levels...]
MP 1133 — levels structure ; Integrating transition
MP 1132 — Limiting length communication
MP 1059 — Low intensity communication
MP 1101 Pathways ; Protected low density communication
MP 1173 Pathways ; Protecting non linear contexts communication
HH 8003 Pathworking
HM 4958 Patibhana-patisambhida (Pali)
HH 1099 Paticca samuppada (Buddhism)
VC 1261 Patience
HH 1304 Patience
VP 5861 Patience-Impatience
HM 3418 Patience (Sufism)
HH 0813 Patience (Zen)
HM 3418 Patient endurance (Sufism)
HM 3942 Patiggaha (Pali)
HH 3550 Patipada-nanadassana-niddesa (Pali)
HM 1738 Patipassaddhi (Pali)
HM 3411 Patipassambhana (Pali)
HM 1528 Patipassambhatattam (Pali)
HM 1663 Patisambhida ; Attha (Pali)
HM 4726 Patisambhida ; Dhamma (Pali)
HM 5026 Patisambhida ; Nirutti (Pali)
HM 4958 Patisambhida ; Patibhana (Pali)
HM 3806 Patissati (Pali)
Pativirati veramani
HM 1781 — Catuhi vaciduccaritehi arati virati (Pali)
HM 0252 — Miccha ajiva arati virati (Pali)
HM 1133 — Tihi kayaduccaritehi arati virati (Pali)
HM 0719 Pativirodha (Pali)
HM 0633 Patoda (Pali)
HH 0735 Patriarchialism
HM 0932 Pattanumodana (Pali)
HH 1266 Pattanumodana (Pali)
KD 2215 Pattern accumulation learning society
MP 1102 Pattern entry points complex structures ; Distinct
HM 1408 Pattern ; Experience (Systematics)
MP 1104 Pattern harmony ; Structural development designed counteract deficiencies
KC 0093 Pattern language
TC 1603 Pattern language participants

Perspectives

HH 5120 Pattern making ; Ritual (Magic)
MP 1075 Pattern nuclear interaction ; Extended
KC 0418 Pattern perception
KC 0636 Pattern recognition
MP 1028 Pattern relationship densities ; Coherent
TC 1910 Patterned behaviour
HH 0773 Patterning
MP 1098 Patterning complex structures
HH 3945 Patterning ; Enneagram (Sufism)
MP 1116 Patterning integrative superstructure
KD 2060 Patterning ; Meta answer
Patterns [active...]
MP 1064 — active irrationality ; Access
MM 2015 — activity ; Shifting
KD 2285 — alternation: agricultural key crop rotation
KD 2280 — alternation: musical key political philosopher
Patterns [Communication...]
TC 1943 — Communication
KD 2342 — complementary metaphors
MP 1097 — Concealment necessary monotonous perspective
HH 0984 Patterns development ; Inappropriate
KD 2337 Patterns interaction ; Self limiting
MS 8845 Patterns ; Shapes
KC 0071 Patterns ; Stereotypic
HM 1602 Pavicaya (Pali)
VC 1265 Peace
TP 0011 Peace
HM 2876 Peace ; Embodying (ICA)
HM 3575 Peace ; Inner
KC 0200 Peace ; International
HM 3575 Peace mind
HM 2494 Peace ; Near death (NDE)
KC 0975 Peace research
HM 3575 Peace ; Spiritual
HM 3015 Peace ; Unknowable (ICA)
VC 1267 Peacefulness
HM 2051 Peak cyclic existence absorption (Buddhism)
HM 1400 Peak experience ; Emotional
HM 2080 Peak experiences
HM 2275 Peaks realm awareness ; Jewelled (Buddhism)
HH 0179 Pedagogy
VD 3432 Pedantry
HH 4502 Peer group initiation
TC 1856 Peer-mediated-learning group
HM 2472 Peevishness
HH 0790 Penance
HM 4412 Penance ; Sixth chakra: time (Leela)
TP 0057 Penetrating clarity (Wind)
HM 5930 Penetration ; Concentration partaking (Buddhism)
HM 3527 Penetration ; Knowledge (Buddhism)
HM 1689 Penetration memory ; Deep (Buddhism)
HM 5232 Penetration minds ; Knowledge (Buddhism)
HM 1637 Penetration ; Non (Buddhism)
HH 4653 Penitential practices
HH 4653 Penitential routines
HM 1240 Pensé ; Déjà
HM 6704 Pensées ; Echo
HH 3124 Pentecostal movement (Christianity)
MS 8233 People ; Beautiful
HM 2484 People God (ICA)
HM 0218 Perception ; Aesthetic (Psychism)
HM 4143 Perception aggregate (Buddhism)
Perception [Bare...]
HM 4651 — Bare (Buddhism)
HM 1766 — Body
HM 2931 — (Buddhism)
HM 3617 — (Buddhism)
HM 0464 Perception ; Delusional
HM 4469 Perception ; Direct
Perception [Ecstatic...]
HM 7100 — Ecstatic (Hinduism)
HM 1077 — Erroneous (Yoga)
HM 2262 — Extrasensory (ESP)
HM 3300 Perception form (Yoga)
HM 4143 Perception group conscious existence ; Awareness (Buddhism)
HM 2531 Perception hidden reality (Sufism)
HM 3348 Perception ; Inattentive (Buddhism)
HM 2829 Perception ; Intuitive
HM 0239 Perception ; Nonecstatic (Hinduism)
HM 2051 Perception nor non perception ; Sphere neither (Buddhism)
HM 1285 Perception ; Occult (Hinduism)
HM 0988 Perception ; Orders
KC 0418 Perception ; Pattern
HM 6558 Perception ; Released soul's (Hinduism)
Perception [Seemingly...]
HM 3265 — Seemingly bare (Buddhism)
HM 3764 — senses
HM 3764 — Sensory
HM 2051 — Sphere neither perception nor non (Buddhism)
HM 0062 — Sport induced modes
HM 3321 — Subliminal
HM 2827 Perception universality (Sufism)
HM 0239 Perception ; Viyukta (Hinduism)
HM 3974 Perception ; World spiritual (Sufism)
HM 5005 Perception ; Yogic (Hinduism)
HM 7100 Perception ; Yukta (Hinduism)
HM 0312 Perceptions configurations ; Third order
Perceptions control
HM 1325 — principles ; Eighth order
HM 0103 — relationships ; Sixth order
HM 0772 — sequence ; Fifth order
HM 4000 — systems concepts ; Ninth order
HM 1516 — transitions ; Fourth order
HM 0543 Perceptions intensity ; First order

HM 1001 Perceptions programme control ; Seventh order
HM 1932 Perceptions sensation ; Second order
HM 0062 Perceptions sport ; altered
HM 1717 Perceptions universal oneness meditation ; Tenth order
HM 0824 Perceptual concentration
HH 0655 Perceptual development
HM 7007 Perceptual expansion
HM 1237 Perceptual narrowing
VC 1271 Percipience
VD 3434 Perdition
HH 0665 Perennial philosophy
HM 0922 Perfect attainment (Taoism)
HM 1342 Perfect attainment (Taoism)
HM 3243 Perfect intelligence ; Path (Judaism)
HM 0074 Perfect love ; State (Christianity)
HM 2920 Perfect man consciousness (Sufism)
HM 2183 Perfect man (Hinduism)
HM 7844 Perfect wisdom (Buddhism)
HH 0037 Perfect yoga (Yoga)
HH 0212 Perfectibility ; Human
HH 5098 Perfectibility ; Moral
VC 1273 Perfection
HM 2183 Perfection ; Bodily (Hinduism)
HM 1344 Perfection discriminating awareness (Buddhism)
HM 0550 Perfection ; Great (Zen)
HH 0212 Perfection ; Human
VP 5677 Perfection-Imperfection
HH 5098 Perfection ; Moral
HH 0794 Perfection ; Ontological
HH 0722 Perfection ; Physical
HH 0718 Perfection ; Psychology
Perfection [Self...]
HH 0519 — Self
HH 7343 — soul (Christianity)
HH 0017 — Spiritual
HH 3652 — spiritual life (Christianity)
HH 1747 — Sufi path (Sufism)
HH 1466 Perfectionism
HM 0213 Perfectionism
HH 0219 Perfections zen (Zen)
VD 3436 Perfidiousness
HM 1336 Performance ; Anomalies experience self recognized personal
HM 0303 Performance ; Spiritual dance
HM 2864 Performance tantra: charya (Buddhism)
KD 2156 Performance ; Theory
HH 0630 Peri-natal psychology
MP 1217 Perimeter continuity
VC 1275 Permanence
HM 0052 Permanence annihilation ; Emptiness what is free (Buddhism)
VP 5139 Permanence ; Change
VD 3438 Permissiveness
MP 1159 Permit two sources external insight ; Organization structures
MP 1211 Permitting new level generation ; Boundary expansion
HM 2717 Perpetual becoming (ICA)
HM 2643 Perpetual intelligence ; Path (Judaism)
HM 2839 Perpetual revolutionary (ICA)
VC 1277 Perpetuation
VP 5112 Perpetuity-Instantaneousness
HM 0529 Perplexity (Buddhism)
HM 0812 Perplexity (Buddhism)
HM 4425 Perplexity ; Psychical (Buddhism)
MS 8223 Perquisites
HM 7760 Persecution ; Delusions
Perseverance
HM 4312 — [Perseverance]
HH 0729 — [Perseverance]
VC 1279 — [Perseverance]
HM 5329 Perseveration
VC 1281 Persistence
TP 0029 Persistence (Abyss)
HM 2139 Persistent friend (ICA)
HM 2553 Person ; Psychological growth dying
HH 0777 Person ; Self made
TC 1651 Person ; Yes
HM 2370 Personal absolution (ICA)
HM 1336 Personal attribution ; Loss
Personal [career...]
HH 0117 — career development
HM 2383 — consciousness field (Psychosynthesis)
HH 0064 — consecration
HH 0955 — cultural potential ; Development
HH 0839 Personal demoralization
HM 0561 Personal devotion worship ; Religious experience
HM 2595 Personal epiphany (ICA)
HH 0161 Personal growth facilitation
Personal [identity...]
HH 0875 — identity
HH 0129 — integrity
HH 1274 — intelligences
HM 2971 Personal obligation (ICA)
HM 1336 Personal performance ; Anomalies experience self recognized
MS 8516 Personal possessions
Personal [salvation...]
HM 6335 — salvation (Christianity)
HM 3152 — self ; Delusion (Buddhism)
HH 0117 — self-development
HM 3358 — sensation ; Annihilation (Sufism)
HH 0249 — space
HM 2561 Personal violation (ICA)
HH 0208 Personalisation ; Hyper
HM 2686 Personalities ; Mana

HM 0463 Personalities ; Sub
HH 0763 Personality ; Alter
Personality [change...]
HH 0476 — change ; Constructive
HH 3998 — Complete
HH 0724 — Cultural
Personality development
HH 0281 — [Personality development]
HH 0447 — [Personality development]
HH 0285 — Stages
HH 0763 Personality disorder ; Multiple
HM 7131 Personality ; Dual
HH 0157 Personality ; Final integration
HH 0971 Personality ; Fully developed
HH 0971 Personality ; Healthy
HH 0257 Personality ; Healthy
HH 0561 Personality ; Integration
HH 0769 Personality ; Inter
HH 0010 Personality ; Lifelong educational transformation human
HH 0971 Personality ; Mature
HH 0763 Personality ; Multiple
TC 1204 Personality needs ; Participant
HH 0971 Personality ; Self actualizing
HH 0198 Personality ; Strength
HH 0157 Personality ; Total integration
HH 3945 Personality types ; Enneagram (Sufism)
HH 0136 Personality ; Unity
HH 8206 Personality ; Unity (Judaism)
MM 2008 Personification
HH 3908 Personification
HM 3944 Personification (Sufism)
HM 2378 Personified evil ; Emblems (Tarot)
TC 1932 Persons ; Key resource
VC 1283 Perspective
Perspective [acquisition...]
MP 1083 — acquisition perspective maintenance dynamics ; Integration
MP 1157 — activity ; Local opportunities
MP 1079 — adaptable contexts
Perspective adjuncts
MP 1201 — Accessible facilities
MP 1200 — Facilities
MP 1198 — Inter domain contexts
Perspective [common...]
MP 1063 — common domains ; Cyclic interrelation complementary
MP 1125 — common domains ; Integrating points
KD 2095 — container alternation ; Third
MP 1183 — context ; Partially exposed
Perspective contexts
MP 1191 — Appropriate proportions
MP 1095 — Complexification
MP 1196 — Eccentric access
MP 1190 — Variation size
MP 1081 Perspective control operations ; Minimally structured
MP 1080 Perspective dynamics ; Integrated contexts
MP 1087 Perspective ; Exchange contexts controlled single
MP 1090 Perspective exchange ; Unstructured context
MP 1085 Perspective imitation contexts developing perspectives
Perspective inactivity
MP 1186 — Appropriate configuration
MP 1188 — Occupiable sites
MP 1150 — Provision temporary
Perspective interaction
MP 1185 — Appropriate configuration
MP 1148 — constraints
MP 1151 — contexts ; Small scale
MP 1088 Perspective interface zones ; Informal local
MP 1083 Perspective maintenance dynamics ; Integration perspective acquisition
MP 1078 Perspective ; Minimal context single
Perspective nourishment
MP 1182 — Appropriate conditions
MP 1147 — Coordination
MP 1139 — Hospitable domain
MP 1177 — Local cultivation sources
MP 1089 — Local sources
MP 1093 — Transit point location sources
Perspective [patterns...]
MP 1097 — patterns ; Concealment necessary monotonous
MP 1062 — Points wider
MP 1189 — presentation ; Context forms
MP 1178 — products ; Re integration rejected
MP 1141 Perspective ; Relatively isolated context each
MP 1084 Perspective reorganization ; Transitional contexts
MP 1187 Perspectives ; Appropriate configuration interaction complementary
Perspectives [Context...]
MP 1070 — Context acknowledgement past
MP 1065 — Context emergence new
MP 1086 — Contexts care premature
MP 1073 — Contexts exploratory relationship formation challenging emerging
Perspectives [decreasing...]
MP 1156 — decreasing activity ; Opportunities
MP 1155 — decreasing activity ; Semi autonomous contexts
MP 1137 — Domain developing
MP 1203 Perspectives ; Exclusive spaces emergent
MP 1114 Perspectives favouring broadest ; Hierachy
MP 1074 Perspectives ; Interaction coherent irrational
MP 1143 Perspectives ; Interrelationship contexts developing
MP 1134 Perspectives ; Limiting exposure harmonious
MP 1077 Perspectives ; Minimal context complementary
MP 1076 Perspectives ; Minimal context interrelating mature emerging
MP 1176 Perspectives non linearity ; Sites grounding

-903-

Perspectives

MP 1085	Perspectives ; Perspective imitation contexts developing						

- MP 1085 Perspectives ; Perspective imitation contexts developing
- MP 1136 Perspectives ; Relative isolation complementary

Perspectives [Semi...]
- MP 1154 — Semi autonomous contexts maturing
- MP 1068 — Spontaneous relationship formation amongst emerging
- MP 1153 — Structures adaptable changing number embodied
- MP 1149 — structures ; Hospitable reception external
- MP 1091 Perspectives transition ; Hospitable contexts
- VC 1285 Perspicacity
- VC 1287 Perspicuity
- HM 4958 Perspicuity ; Understanding discrimination (Buddhism)
- HH 0696 Persuasion therapy
- VC 1289 Persuasiveness
- VC 1291 Pertinacity
- VD 3442 Perturbation
- HM 0223 Perturbation thought ; Mental (Buddhism)
- HM 0476 Pervading knowledge ; All (Sufism)
- HM 4788 Pervading knowledge ; Continuous (Sufism)
- VD 3444 Perversion
- VD 3446 Perversity
- VD 3448 Pessimism
- VD 3450 Pestiferousness

Petalled lotus
- HM 0176 — Sixteen (Yoga)
- HM 3398 — Thousand (Yoga)
- HM 0176 — Twelve (Yoga)
- MS 8433 Pets
- VD 3452 Pettiness
- VD 3454 Petulance
- HM 0932 Phala sacchikiriya (Buddhism)
- HM 0831 Phala samadhi (Buddhism)
- HM 3726 Phantastica ; Pseudologia
- HM 2599 Phase-conditions consciousness (Buddhism)
- KD 2329 Phase ; Learning loss cyclic
- KC 0318 Phase space
- HH 0767 Phases life ; Mutational
- HH 0511 Phases life ; Social
- MM 2071 Phasing ; Project
- HM 2708 Phassa (Pali)
- HM 4212 Phassa (Pali)
- HM 4254 Phencyclidine similar drugs ; Modes awareness associated use
- HM 2551 Phenomena awareness (Buddhism)
- HH 3321 Phenomena ; Awareness inter dependency conscious existence (Buddhism)
- HM 3000 Phenomena ; Emptiness (Buddhism)
- HM 2899 Phenomena ; Emptiness nature (Buddhism)
- MS 8373 Phenomena ; Human spiritual
- HM 2075 Phenomena ; Indriya (Pali)
- HM 3907 Phenomenal plane (Leela)
- KC 0741 Phenomenology
- HM 1169 Phenomenon ; Cave
- HM 5124 Phenomenon ; Isakower
- KD 2270 Phenomenon learning society ; Non comprehension structuring
- HM 1265 Phenomenon presence (Christianity)
- HH 1206 Philanthropy
- TC 1306 Phillips 66
- HH 5908 Philo ; Journeying transcendence agape
- KD 2280 Philosopher ; Patterns alternation: musical key political
- HM 1911 Philosophical awareness
- KC 0341 Philosophy
- VC 1293 Philosophy
- HH 0591 Philosophy enlightenment
- HH 0221 Philosophy ; Hermetic (Esotericism)
- HH 0665 Philosophy ; Perennial
- HM 1509 Phobia
- HM 2393 Phoenix awareness
- HM 0794 Phosphenes
- HH 6321 Phosphenic mixing
- HH 6321 Phosphenism
- HM 2638 Phrag-dog (Tibetan)
- HM 1959 Phusana (Pali)
- KC 0516 Phylogeny ; Ontogenetic recapitulation
- HH 0993 Physical appearance

Physical [conservation...]
- HM 2357 — conservation conscious states (Physical sciences)
- TC 1689 — constructs ; Meetings
- HH 0730 — culture
- HH 0836 Physical development
- HM 2381 Physical duality conscious states (Physical sciences)
- HM 3314 Physical ease nor unease born contact element mind consciousness ; Neither (Buddhism)
- HH 0575 Physical education
- HM 2075 Physical faculties awareness ; Psycho (Buddhism)
- HH 0836 Physical growth
- HM 1263 Physical inception energy (Buddhism)
- HM 4772 Physical lust (Islam)

Physical [perfection...]
- HH 0722 — perfection
- HM 4103 — plane (Leela)
- HM 1024 — plane (Leela)
- MM 2032 — position ; Changing
- HH 0379 — process therapy
- TC 1367 — processes ; Meetings
- TC 1922 — processes ; Meetings psycho
- HH 1775 — psychism
- HM 2322 Physical relativity conscious states (Physical sciences)
- HM 2298 Physical sciences Discrete states consciousness
- HM 2417 Physical sciences Least action principle conscious states
- HM 2357 Physical sciences Physical conservation conscious states
- HM 2381 Physical sciences Physical duality conscious states

- HM 2322 Physical sciences Physical relativity conscious states
- HM 3421 Physical sciences Psychic inertia
- HM 3421 Physical sciences Resistance change
- HM 2938 Physical sciences States consciousness
- HM 3320 Physical stage life ; Vital
- HH 0379 Physical therapy
- HM 5635 Physical tiredness
- HH 0512 Physiodrama
- HH 0348 Physiognomonic thinking
- HH 0379 Physiotherapy
- HH 0126 Physique temperament
- MS 8445 Phytomorphic symbols
- HH 0512 Picture ; Therapeutic motion
- HM 2690 Pictures Zen Buddhism ; Oxherding (Zen)
- VC 1295 Piety
- VP 6028 Piety-Impiety
- HM 5339 Piety (Sufism)
- HM 2225 Pilgrim archetypal image (Tarot)
- HH 0637 Pilgrimage
- HM 1265 Pilgrimage (Islam)
- HM 0634 Pill ; Gold (Taoism)
- HH 4880 Pill ; Refining gold (Taoism)
- HM 2890 Pingala conscious energy (Yoga)
- VC 1297 Piousness
- HH 2080 Pir (Sufism)
- MS 8823 Piscatological symbols
- HM 2856 Pisces-consciousness (Astrology)
- TP 0029 Pitfalls ; Mastering (Abyss)
- HM 8737 Piti (Buddhism)
- HM 0747 Piti (Pali)
- HM 2038 Piti (Pali)
- VD 3456 Pitilessness
- VP 5944 Pitilessness ; Compassion
- VC 1299 Pity
- HM 0513 Pity (Buddhism)
- KC 0220 Place theory ; Central
- HM 3536 Placed sepulchre ; Jesus (Christianity)
- HH 0593 Places ; Holy
- HH 0593 Places ; Sacred
- VC 1301 Plainness
- KC 0171 Plan ; General

Plane [Absolute...]
- HM 4326 — Absolute (Leela)
- HM 0822 — Adi (Psychism)
- HM 0859 — agnih (Leela)
- HM 3651 — Astral (Leela)
- HM 4600 — Astral (Leela)
- HM 4412 — austerity (Leela)
- HM 5244 — austerity (Leela)

Plane [balance...]
- HM 3291 — balance (Leela)
- HM 1351 — balance (Leela)
- HM 6901 — bliss (Leela)
- HM 0717 Plane ; Celestial (Leela)
- HM 1052 Plane ; Celestial (Leela)

Plane cosmic
- HM 1141 — consciousness (Leela)
- HM 1301 — consciousness (Leela)
- HM 4343 — good (Leela)
- HM 4128 Plane culture ; Understanding being (Buddhism)

Plane [development...]
- HM 4128 — development ; Understanding (Buddhism)
- HM 0481 — dharma (Leela)
- HM 6192 — discernment ; Understanding being (Buddhism)

Plane [fantasy...]
- HM 1485 — fantasy (Leela)
- HM 0859 — fire (Leela)
- HM 0008 — fragrance (Leela)
- HM 6434 Plane ; Gaseous (Leela)
- HM 5155 Plane ; Human (Leela)
- HM 0933 Plane ; Human (Leela)
- HM 6287 Plane inner space (Leela)
- HM 4376 Plane joy (Leela)
- HM 0948 Plane karma (Leela)
- HM 0172 Plane ; Liquid (Leela)
- HM 3779 Plane ; Lunar (Leela)
- HM 4930 Plane ; Moral consciousness transcendental (Buddhism)
- HM 4050 Plane neutrality (Leela)

Plane [Phenomenal...]
- HM 3907 — Phenomenal (Leela)
- HM 4103 — Physical (Leela)
- HM 1024 — Physical (Leela)
- HM 3732 — primal vibrations (Leela)
- HM 4930 — Profitable consciousness supramundane (Buddhism)

Plane [radiation...]
- HM 5028 — radiation (Leela)
- HM 1293 — reality (Leela)
- HM 0754 — reality ; Seventh chakra: (Leela)
- HM 5129 — resultant ; Indeterminate consciousness supramundane (Buddhism)
- HM 5129 — resultant ; Indeterminate consciousness transcendental (Buddhism)

Plane [sanctity...]
- HM 5965 — sanctity (Leela)
- HM 6192 — seeing ; Understanding (Buddhism)
- HM 2354 — Sensual (Leela)
- HM 4417 — Seventh (Psychism)
- HM 1279 — Solar (Leela)
- HM 0878 Plane taste (Leela)
- HM 1293 Plane truth (Leela)
- HM 1276 Plane violence (Leela)

Plane wisdom
- HM 2001 — Fifth (Hinduism)
- HM 3100 — First (Hinduism)
- HM 3006 — Fourth (Hinduism)

Plane wisdom cont'd
- HM 2443 — Second (Hinduism)
- HM 3354 — Sixth (Hinduism)
- HM 3312 — Third (Hinduism)
- MS 8338 Planes
- HM 3298 Planes wisdom (Hinduism)
- KC 0067 Planetary citizenship

Planetary consciousness
- HH 4621 — [Planetary consciousness]
- HM 2006 — [Planetary consciousness]
- HM 1267 — Ecological
- KC 0057 Planetary culture
- HH 4621 Planetary noogenesis
- KC 0506 Planetary society
- KC 0281 Planetary synthesis
- HM 2006 Planetization consciousness
- KC 0057 Planetization mankind
- KC 0025 Planning
- KC 0666 Planning ; Integrated
- TC 1056 Planning methods ; Event
- KC 0919 Planning ; Optimal
- KC 0515 Planning ; Programme
- KC 0566 Planning-Programming-Budgeting System (PPBS)
- KC 0172 Planning ; Social policy
- KC 0306 Planning ; Strategic
- KC 0736 Planning techniques ; Network
- MM 2045 Plants
- KC 0861 Plasma
- MM 2072 Plasma ; Magnetic containment
- HM 1732 Plateau cognition
- VC 1303 Plausibility
- HH 0705 Play
- HH 0633 Play therapy
- TC 1581 Playboy
- VC 1305 Playfulness
- HM 6722 Pleasant feeling (Buddhism)
- HM 1862 Pleasant ; Psychically (Buddhism)
- VC 1307 Pleasantness
- VP 5863 Pleasantness-Unpleasantness

Pleasure
- HM 2883 — [Pleasure]
- VC 1309 — [Pleasure]
- HH 0194 — [Pleasure]

Pleasure [born...]
- HM 1643 — born contact psychical ; Sensation neither suffering nor (Buddhism)
- HM 2866 — (Buddhism)
- HM 6722 — (Buddhism)
- VP 5865 Pleasure-Displeasure
- HM 5088 Pleasure due insight (Buddhism)
- HM 1352 Pleasure experienced born contact psychical ; Neither suffering nor (Buddhism)
- VD 3458 Pleasurelessness
- VC 1311 Plenitude
- VC 1313 Plenty
- HH 0911 Pleroma
- HM 3162 Pliancy (Buddhism)
- HM 1155 Pliancy citta (Pali)
- HM 3696 Pliancy kaya (Pali)
- HM 3696 Pliancy mental factors (Buddhism)
- HM 1155 Pliancy mind (Buddhism)
- VC 1315 Pluck
- KC 0149 Pluralism
- MS 8313 Plurality ; Human
- KC 0996 Pluridisciplinarity
- HH 1401 PNI Psychoneuroimmunology
- HM 0875 Poetic enthusiasm
- HM 1538 Poetic imagination
- KC 0392 Poetry
- HM 2799 Point basic fear ; Breaking (Brainwashing)
- MP 1126 Point common domain ; Off centering focal
- MP 1113 Point external communication media ; Harmoniously structured entry
- MP 1093 Point location sources perspective nourishment ; Transit
- MP 1181 Point ; Maintenance source direct insight structures focal
- MP 1129 Point structure ; Common domain focal
- MP 1110 Point structure ; Distinctiveness main entry
- HM 2354 Point total conflict ; Breaking (Brainwashing)
- VD 3460 Pointlessness
- MP 1102 Points complex structures ; Distinct pattern entry
- MP 1053 Points entry ; Principal
- MP 1092 Points ; Hospitable transit
- MP 1130 Points ; Internal transition spaces enhancing structural entry
- MP 1044 Points ; Local focal
- MP 1125 Points perspective common domains ; Integrating
- MP 1062 Points wider perspective
- HH 6876 Poise
- VD 3462 Poisonousness
- HM 1234 Polar knowledge (Systematics)
- KD 2015 Polarity
- HM 1234 Polarity ; Experience (Systematics)
- TC 1903 Polarization ; Conference
- MM 2083 Pole
- KC 0031 Policy
- KC 0696 Policy analysis
- KC 0696 Policy analysis ; Public
- KC 0080 Policy ; Economic
- KC 0697 Policy integration ; International
- KC 0172 Policy planning ; Social
- KC 0635 Policy ; Science
- KC 0343 Policy sciences
- KD 2320 Policy ; Toward enantiomorphic
- KC 0244 Policymaking

VC 1317	Politeness		**Power [shamelessness...] cont'd**	VC 1339	Precedence
TC 1313	Political conferencing ; Managerial	HM 1724	— shamelessness (Buddhism)	HH 0442	Preceptor ; Spiritual
KC 0287	Political integration ; International	TP 0009	— small ; Taming	HH 0278	Precepts ; Observing moral (Buddhism)
KC 0114	Political organization society	HH 0729	— Staying	MS 8516	Precious things
KD 2280	Political philosopher ; Patterns alternation: musical key	HM 2828	Power ; Transparent (ICA)	VC 1341	Precision
KC 0525	Politics ; Consensus	HM 1915	Power unremorsefulness (Buddhism)	HH 0611	Precocious talent
VD 3464	Pollution	HM 0454	Power ; Way (Esotericism)	HH 0563	Predestination (Christianity)
KD 2146	Poly ocular vision ; De categorization	HM 1792	Power wisdom (Buddhism)	HM 2131	Predestination ; Definitive (ICA)
KC 0033	Polyhistory	VD 3472	Powerlessness	HM 2889	Prediction ; Agonizing (ICA)
KC 0033	Polymathy	HM 3238	Powers ; Five (Buddhism)	TC 1640	Preferences ; Non verbal interaction
KC 0728	Polyphasic unity	HM 7672	Powers ; Knowledge supernormal (Buddhism)		**Preferences Participant**
KC 0325	Polysemy	HH 0497	Powers ; Magico religious	TC 1828	— change
HH 0297	Polytheism concept	HH 5652	Powers ; Psychic (Buddhism)	TC 1845	— interaction
VD 3466	Pomposity	HH 5652	Powers ; Supernormal (Buddhism)	TC 1828	— strategic
HM 4488	Pondering (Buddhism)	HH 3159	Powers ; Ten (Buddhism)	TC 1378	Preferences ; Verbal interaction
VD 3468	Ponderousness	HM 2918	Powers ; Tenfold (Buddhism)	VD 3474	Prejudice
HH 0190	Poor spirit	HM 2982	Powers ; Vibrant (ICA)	MP 1086	Premature perspectives ; Contexts care
HM 2260	Pope (Tarot)	KC 0566	PPBS Planning-Programming-Budgeting System	VD 3476	Prematurity
VC 1319	Popularity	HM 2215	Prabha-kari (Buddhism)	HM 3606	Premonition (Psychism)
KC 0173	Population	VC 1333	Practicability	HM 4312	Preoccupation
VC 1321	Position	HH 7902	Practical wisdom	HM 4976	Preoccupation
MM 2032	Position ; Changing physical	VC 1335	Practicality	HM 2870	Preoccupied consciousness
MP 1242	Position ; Intermediate	HH 0544	Practice	HM 3264	Preparation ; Saviors God: (Christianity)
MS 8353	Position ; Social hierarchical	KC 0269	Practice	HH 5543	Preparation stationariness (Buddhism)
MP 1241	Positions ; Attractive temporary	HH 3007	Practice confidence (Christianity)	MS 8406	Preparations ; Medical
MP 1024	Positions enabling transcendence	HM 6562	Practice dhikr (Sufism)	HM 2623	Preparations states ; Special insight (Buddhism)
HH 6222	Positiva ; Via	HM 4032	Practice ; Esoteric exoteric	HM 2963	Preparedness
KC 0427	Positive feedback		**Practice [performance...]**	VC 1343	Preparedness
HM 3502	Positive feeling towards God (Christianity)	MM 2007	— performance	TP 0051	Preparedness ; Crisis (Thunderbolts)
HM 1789	Positive intellect (Leela)	HH 3218	— prayer ; Mansions (Christianity)	VP 5720	Preparedness-Unpreparedness
HH 0675	Positive mental health	HH 0992	— presence God	TP 0028	Preponderance great
HH 1088	Positive thinking	HH 4006	— presence mindfulness (Buddhism)	TP 0062	Preponderance small
HH 1088	Positive thought	HH 0983	Practice silence	HM 3170	Prerequisites manifestation (Buddhism)
VC 1323	Positiveness	HH 4298	Practices ; Ascetic (Buddhism)	HM 3606	Prescience (Psychism)
HH 1276	Positivism	HH 4653	Practices ; Penitential	VC 1345	Presence
HH 1276	Positivism ; Logical	TC 1532	Practics ; Conference	HH 0389	Presence
HM 4772	Possess ; Desire (Islam)	HM 2639	Practitioner state awareness ; Bon (Buddhism)		**Presence [Absence...]**
VC 1325	Possession	HM 2778	Pradasha (Buddhism)	VP 5186	— Absence
VT 8075	Possession*complex	HM 2935	Pradhana (Yoga)	HM 3496	— Allah (Sufism)
TP 0014	Possession great measure	VC 1337	Praise	HH 0992	— alone
VP 5808	Possession-Loss	HM 2655	Prajna (Buddhism)		**Presence [Canine...]**
HM 3111	Possession (Psychism)	HM 4523	Prajna (Buddhism)	MS 8853	— Canine
	Possession [Self...]	HM 3455	Prajna (Hinduism)	HM 2013	— Christ (Christianity)
VC 1571	— Self	HM 2756	Prajna-jnana (Buddhism)	HH 0763	— Co
HM 3594	— Spirit	HH 0550	Prajna paramita (Zen)	HM 7114	Presence fear ; Knowledge (Buddhism)
HH 0056	— Spirit	HH 0550	Prajna (Zen)		**Presence God**
HM 2657	Possession ten worlds ; Mutual (Buddhism)	HM 0107	Prajna (Zen)	HM 2961	— [Presence God]
HM 3131	Possessions ; Liberation (ICA)	HM 1344	Prajnaparamita (Buddhism)	HH 0992	— Practice
MS 8516	Possessions ; Personal	HM 3907	Prakriti-loka (Leela)	HM 6989	— Sense (Christianity)
HH 1268	Possessiveness ; Non	HM 3205	Prakrtishunyata (Buddhism)	HM 2961	Presence Holy Ghost
HM 2911	Possessiveness (Yoga)	HM 2314	Pramada (Buddhism)	HM 2206	Presence Jesus Christ (ICA)
HM 3117	Possibility ; Frightful (ICA)	HH 4388	Pramana (Buddhism)	HH 4006	Presence mindfulness ; Practice (Buddhism)
VP 5509	Possibility-Impossibility	HM 2155	Pramudita (Buddhism)	HM 1265	Presence ; Phenomenon (Christianity)
HM 2524	Possibility ; Impossible (ICA)	HM 4399	Prana-loka (Leela)	HM 3546	Presence ; Psychic (Psychism)
VP 5170	Posterity ; Ancestry	HH 0213	Pranayama (Hinduism)	HM 1166	Presence ; Sense
HM 2175	Potala-island awareness (Buddhism)	HH 1016	Pranidhana ; Isvara (Hinduism)	MS 8303	Presence ; Signs human
VC 1327	Potency	HH 4330	Prasada (Hinduism)	HM 2462	Presence ; Transparent (ICA)
TC 1362	Potency ; Conference	HH 3162	Prasrabdhi (Buddhism)	HM 2416	Presences ; Awareness divine (Sufism)
KC 0219	Potential	HM 2450	Prathamadhyana (Buddhism)	HM 2814	Present brother ; Ever (ICA)
HH 0955	Potential ; Development personal cultural	HM 4875	Pratibhajnana (Hinduism)	HH 2345	Present-centredness
HH 0034	Potential directiveness	HM 2959	Pratigha (Buddhism)	HH 2345	Present ; Living
KP 3017	Potential form ; Relationship	TC 1776	Prattle fatigue	TC 1383	Presentation
HH 0461	Potential ; Human	HH 0829	Pratyahara (Yoga)	MP 1189	Presentation ; Context forms perspective
HM 8080	Potential knowledge (Systematics)	HM 6644	Pratyaksa ; Manasa (Buddhism)	HM 6544	Presentation due insight (Buddhism)
HH 0398	Potential movement ; Human	HM 5532	Pratyaksa ; Yogi (Buddhism)	MP 1043	Presentation new dimensions
TP 0048	Potentialities ; Fulfilled (Well)	HM 2709	Pratyckabuddha (Buddhism)	HM 3606	Presentiment (Psychism)
VC 1329	Potentiality		**Pratyeka Buddha**	VC 1347	Preservation
HM 8080	Potentiality ; Experience (Systematics)	HM 3020	— application awareness (Buddhism)	HH 0711	Preservation ; Self
HM 2665	Potentiality ; Undifferentiated (Astrology)	HM 2776	— arhat awareness (Buddhism)	VC 1573	Preservation ; Self
VD 3470	Poverty	HM 2180	— awareness (Buddhism)	HH 0540	Pressure ideas
HM 2286	Poverty ; Awareness spiritual (ICA)	HM 2252	— cultivation awareness (Buddhism)	HH 0987	Pressure therapy
HM 3241	Poverty ; Consciousness mystical	HM 2228	— vision-path awareness (Buddhism)	HH 0540	Pressure ; Thought
HM 9701	Poverty ; Delusions	HM 2743	Pravivekyamanaskara (Buddhism)	HM 0265	Pressure vertigo
HM 2299	Poverty ; Life (ICA)	HH 0185	Prayer	VC 1349	Prestige
HH 0991	Poverty ; Religious	HH 1994	Prayer ; Centering (Christianity)	VD 3478	Presumption
	Poverty [spirit...]	HH 0816	Prayer ; Contemplative	HM 4032	Presumption (Buddhism)
HH 0190	— spirit	HH 1193	Prayer ; Discipline (Christianity)	HM 2112	Pretas (Buddhism)
HH 0190	— Spiritual	HM 6989	Prayer ; First degree (Christianity)	VD 3480	Pretentiousness
HM 3427	— State (Sufism)	HH 0816	Prayer ; Imageless (Christianity)	VC 1351	Prettiness
HH 0991	Poverty ; Voluntary	HH 0205	Prayer ; Jesus (Christianity)	KD 2006	Prevailing meta-answer: gladiatorial arena
VP 5837	Poverty ; Wealth	HM 2511	Prayer ; Life (ICA)	VD 3482	Prevarication
	Power	HH 4209	Prayer ; Liturgical (Christianity)	HM 4297	Previous existence ; Knowledge recollection (Buddhism)
HH 0848	— [Power]		**Prayer [Mansions...]**		
VC 1331	— [Power]	HH 3218	— Mansions practice (Christianity)	HM 2823	Pride (Christianity)
TP 0034	— [Power]	HH 4089	— Meditative	VP 5905	Pride-Humility
TC 1587	Power ; Coercive	TC 1915	— meeting	HM 2852	Pride ; Spiritual (Buddhism)
VT 8024	Power*complex	HH 8672	— Mental (Christianity)	HM 4533	Pride ; Spiritual (Christianity)
HM 1759	Power concentration (Buddhism)	HH 0816	— Mystical	HM 2260	Priest ; High (Tarot)
HH 3117	Power ; Development psychosomatic	HM 0238	Prayer repose (Christianity)	HM 2236	Priest ; Word bearing (ICA)
HH 0084	Power ; Discursive		**Prayer [Second...]**	HM 2261	Priestess ; High (Tarot)
HM 1943	Power energy (Buddhism)	HM 0238	— Second degree (Christianity)	HH 2314	Primal healing
HM 4404	Power faith (Buddhism)	HM 0238	— silence (Christianity)	HH 1902	Primal religion ; Human development
	Power [God...]	HM 0238	— simplicity (Christianity)	HH 2087	Primal scream
HM 2760	— God consciousness	HH 0816	— spirit	HM 2550	Primal sympathy (ICA)
TP 0034	— great	HH 0357	— States (Christianity)	HH 2087	Primal therapy
TP 0026	— great ; Taming	HH 0185	Prayers ; Saying	HM 3732	Primal vibrations ; Plane (Leela)
VP 5157	Power-Impotence	HH 0185	Praying	HM 2621	Primal vocation (ICA)
HM 3096	Power ; Inescapable (ICA)	HH 1608	Praying tongues (Christianity)	HM 4604	Primary delusions
HH 0338	Power ; Man	HM 2716	Prayoga marga ; Sravaka (Buddhism)	HH 6669	Primary growth
TC 1858	Power ; Normative	HM 2683	Prayoganishthamanaskara (Buddhism)	MP 1212	Primary inter-level connections transitions boundary orientation
	Power [recollection...]	HM 2788	Pre-conscious pure intelligence		
HM 3450	— recollection (Buddhism)	HM 1084	Pre-conscious thinking	MS 8600	Primary visual elements
HM 2538	— remorse (Buddhism)	HH 1339	Pre-existence	HM 1904	Prime (Christianity)
TC 1831	— Remunerative	TC 1965	Pre logical biases ; Participant	HM 1904	Prime ; Hours (Christianity)
	Power [shame...]	HM 8672	Pre-stage faith	HH 0748	Primitive thinking
HM 0352	— shame (Buddhism)	HH 1022	Preaching dharma (Buddhism)	MS 8525	Primitive weapons

Primitivism

HH 1778	Primitivism	MS 8343	Professions ; Trades	MP 1179	Provide occupiable sites ; Organization structure
HH 1778	Primitivism ; Cultural	VC 1367	Proficiency	VC 1917	Provided ; Well
HM 3224	Primordial ancestor (ICA)	HM 1554	Proficiency (Buddhism)	TP 0027	Providing nourishment
HM 2163	Primordial awareness (Zen)	HM 1455	Proficiency mental factors (Buddhism)	MP 1229	Provision pathways automatic communications
HM 2352	Primordial colloquy (ICA)	HM 1810	Proficiency mind (Buddhism)	MP 1150	Provision temporary perspective inactivity
HH 0019	Primordial image	HM 8290	Profit ; Understanding skill (Buddhism)	VD 3504	Provocation
HM 2877	Primordial sociality (ICA)	VC 1369	Profitability	HH 0370	Provoking tales ; Insight
HH 0748	Primordial thinking		**Profitable consciousness**	VC 1385	Proximity
HM 2186	Primordial wonder (ICA)	HM 5338	— fine-material sphere (Buddhism)	HM 2377	Proximity ; God (Sufism)
HM 3630	Principal insights (Buddhism)	HM 4701	— immaterial sphere (Buddhism)	VC 1387	Prudence
MP 1053	Principal points entry	HM 4447	— sense sphere (Buddhism)	HH 7902	Prudence
VC 1353	Principle	HM 4930	— supramundane plane (Buddhism)	HM 4336	Pseudo-hallucination
HH 3663	Principle ; Acceptance female	VD 3492	Profligacy	HM 3726	Pseudologia phantastica
KC 0476	Principle allometry	VC 1371	Profundity	HH 0979	Psyche
HM 2417	Principle conscious states ; Least action (Physical sciences)	HM 2607	Profundity ; Impactful (ICA)	HH 1919	Psyche consciousness (Psychism)
KC 0795	Principle control ; Hierarchial	KC 0413	Prognosis	HM 2201	Psyche ; Emblem archetypes (Tarot)
KC 0190	Principle ; Indeterminacy	KC 0836	Program ; Computer	HH 0075	Psychedelic drugs
HM 3404	Principle ; Life (Buddhism)	HM 1001	Programme control ; Seventh order perceptions	HH 0801	Psychedelic experience
KC 0146	Principle requisite variety	KC 0736	Programme evaluation review technique	HH 0801	Psychedelics ; Use
KC 0190	Principle ; Uncertainty	TC 1370	Programme modification	HH 0289	Psychiatry ; Orthomolecular
KC 0966	Principle ; Unitary	KC 0515	Programme planning	HH 2675	Psychiatry ; Radical
KC 0328	Principle ; Whole system	HH 0546	Programmed learning		**Psychic [ability...]**
HM 1325	Principles ; Eighth order perceptions control	KC 0465	Programming	HH 2334	— ability
HM 3069	Principles (Sufism)	KC 0566	Programming Budgeting System ; Planning (PPBS)	HH 0100	— abnormality
HH 1339	Prior existence	KC 0626	Programming ; Linear	HM 0401	— atavism (Psychism)
HM 2458	Prior ; Universal (ICA)	HM 4872	Programming ; Neuro linguistic (NLP)	HM 4033	— attack ; Accidental (Psychism)
VC 1355	Priority	HM 3114	Programming ; Self (ICA)	HH 0837	— attack (Psychism)
HM 4539	Prison ; Mental		**Progress**	HM 3574	— attunement (Psychism)
HM 6112	Prithvi (Leela)	HH 0520	— [Progress]	HH 1919	Psychic consciousness (Psychism)
VC 1357	Privacy	KC 0670	— [Progress]	HH 7637	Psychic control human development
HH 3238	Private revelation	VC 1373	— [Progress]	HH 0855	Psychic development (Psychism)
VD 3484	Privation	TP 0035	— [Progress]	HH 0734	Psychic energy
VC 1359	Privilege	HH 6434	Progress (Christianity)	HH 0750	Psychic energy
KC 0677	Probabilistic system	HH 0166	Progress ; Concept	HM 0885	Psychic field consciousness (FC)
KD 2340	Probabilistic vision world ; Language	HM 2318	Progress ; Eighth stage (Sufism)	HH 3117	Psychic growth
KC 0667	Probability theory	TP 0053	Progress ; Gradual	HH 5434	Psychic growth
VC 1361	Probity	HM 4977	Progress quick intuition ; Concentration easy (Buddhism)		**Psychic [hallucinations...]**
VP 5974	Probity-Improbity	HM 7859	Progress quick intuition ; Concentration painful (Buddhism)	HM 1317	— hallucinations
VD 3486	Problem	KC 0760	Progress ; Scientific technological	HH 0458	— healer
	Problem solving	HM 2149	Progress ; Sense harmony (Brainwashing)	HH 0458	— healing
HH 0481	— ability		**Progress sluggish**	HH 0108	— health ; Spiritual aspects
TC 1684	— conferencing	HM 5992	— direct knowledge ; Concentration difficult (Buddhism)	HH 0109	— health spirituality
HM 3574	— state	HM 1010	— direct knowledge ; Concentration easy (Buddhism)	HM 2182	— helplessness
KC 0614	— theory	HM 1010	— intuition ; Concentration easy (Buddhism)		**Psychic [inertia...]**
HM 2747	Problemless living (ICA)	HM 5992	— intuition ; Concentration painful (Buddhism)	HM 3421	— inertia (Physical sciences)
HM 3015	Problemlessness centre (ICA)	HM 7859	Progress swift direct knowledge ; Concentration difficult (Buddhism)	HM 1094	— inspiration (Psychism)
KC 0750	Problems ; World	HM 4977	Progress swift direct knowledge ; Concentration easy (Buddhism)	HH 0823	— integration
KC 0312	Procedure ; Decision theory	HM 3336	Progression	HH 0543	— isolation
TC 1347	Procedure feedback ; Conference	HM 1972	Progression (Buddhism)	HH 0645	— ontogeny
KC 0645	Process	VP 5294	Progression-Regression	HM 5652	Psychic powers (Buddhism)
KC 0268	Process	MP 1208	Progressive framework definition	HM 3546	Psychic presence (Psychism)
TC 1913	Process ; Creative listening	VD 3494	Prohibition	HH 0981	Psychic research
TC 1627	Process ; Decision making	HM 2153	Project ; Atman	HM 4311	Psychic states (Sufism)
HH 3221	Process ; Dehumanizing	MM 2071	Project phasing	HM 2305	Psychic stream ; Observation one's (Sufism)
KP 3009	Process ; Implementation transformation	HM 1887	Projection ; Astral (Psychism)	HM 3571	Psychic transfer (Psychism)
KC 0937	Process logic	HM 4529	Projection ; Mental	HH 0304	Psychic transformation
HH 2054	Process meditation	TC 1824	Projection ; Slide	HH 3452	Psychic vampirism
KC 0937	Process metaphysics	HH 4312	Projections	HM 1921	Psychical ease born contact element mind-consciousness (Buddhism)
TC 1657	Process-oriented participant type	HM 1425	Projective faith ; Intuitive	HM 0972	Psychical ease nor unease born contact element eye consciousness ; Neither (Buddhism)
TC 1913	Process ; Systematic listening	KC 0621	Projective geometry	HM 1352	Psychical ; Neither suffering nor pleasure experienced born contact (Buddhism)
HH 0379	Process therapy ; Physical	TC 1947	Projector ; Opaque	HM 4425	Psychical perplexity (Buddhism)
HH 1643	Process thinking	TC 1472	Projector ; Overhead		**Psychical Sensation**
KD 2009	Processes accumulation significance ; Development	VD 3496	Proliferation	HM 0309	— ease born contact (Buddhism)
KP 3014	Processes ; Cycle development	VD 3498	Prolixity	HM 1643	— neither suffering nor pleasure born contact (Buddhism)
MP 1009	Processes ; Decentralized formal	HH 1900	Prologue ; Journeying transcendence	HM 1090	— unease born contact (Buddhism)
MS 8605	Processes ; Forces, energies	HM 2760	Prometheus awareness	HM 4076	— unpleasant born contact (Buddhism)
MP 1041	Processes ; Informal context formal	VD 3500	Promiscuity		**Psychical [unease...]**
MP 1205	Processes it ; Congruence spaces defined framework spaces defined	VC 1375	Promise	HM 0234	— unease born contact element mind-consciousness (Buddhism)
MP 1051	Processes ; Linear relationships enhanced non linear	HM 3172	Promissorial offering (ICA)	HM 0901	— Unease experienced born contact (Buddhism)
TC 1351	Processes ; Meeting focal	TC 1142	Promoter role ; Team	HM 1031	— Unpleasant experienced born contact (Buddhism)
	Processes Meetings	TP 0046	Promotion	HM 1862	Psychically pleasant (Buddhism)
TC 1475	— ecological natural	HM 2356	Prompter ; Unfailing (ICA)	HM 0805	Psychically unpleasant (Buddhism)
TC 1367	— physical	VC 1377	Promptness	HH 1775	Psychism
TC 1922	— psycho physical	KC 0969	Proof theory	HM 4033	Psychism Accidental psychic attack
MM 2052	Processes ; Stick carrot	HH 0171	Propensity become	HM 0442	Psychism Active trance
HH 0530	Processes ; Thought	VT 8041	Properties*complex ; Absolute	HM 0822	Psychism Adi plane
MP 1033	Processes ; Underdefined	VT 8042	Properties*complex ; Relative	HM 1889	Psychism Adrenergia
MP 1008	Processes ; Variety forms	HH 0559	Prophecy	HM 0218	Psychism Aesthetic perception
MP 1184	Processing ; Appropriate configuration input	HH 0559	Prophesy	HM 3416	Psychism Aesthetic state
TC 1465	Processing configurations ; Meetings energy	HM 4875	Prophetic intuition (Hinduism)	HM 3590	Psychism Affection
MP 1199	Processing context external insight ; Exposure input	HM 2846	Prophetic sight (ICA)	HH 0665	Psychism After death dream state
TC 1529	Processing ; Meetings agriculture food	HH 0267	Propitiation	HM 0573	Psychism Aia
KC 0526	Processing system ; Data	KC 0747	Proportion	HM 0123	Psychism Akshara consciousness
TC 1847	Procreation ; Conference	MP 1167	Proportions ; Enfolded overview domains minimum	HM 1615	Psychism Akuhaiamio
KC 0177	Procuring resources	MP 1191	Proportions perspective contexts ; Appropriate	HH 0019	Psychism Asomatic experience
VD 3488	Prodigality	VC 1379	Propriety	HM 4550	Psychism Assumption forms
VP 5851	Prodigality ; Economy	TP 0054	Propriety (Marrying maiden)	HM 0183	Psychism Astral consciousness
VC 1363	Prodigiousness	HH 5213	Propriety ; Sense	HM 1887	Psychism Astral flight
KD 2007	Production accumulation significance ; Answer	TP 0055	Prosperity	HM 1887	Psychism Astral projection
KC 0615	Production ; Organization	VP 5728	Prosperity-Adversity	HM 5123	Psychism Attraction consciousness
VP 5167	Production-Reproduction	HH 3662	Prosthesis ; Evolution	HM 0679	Psychism Auric clairvoyance
TC 1996	Productive conferencing	VD 3502	Prostitution	HM 0819	Psychism Bewitched
MP 1170	Productive non linearity ; Cultivation	MP 1101	Protected low-density communication pathways	HM 4185	Psychism Body glory
HH 0703	Productive thinking	MP 1055	Protected low intensity relationships	HM 1114	Psychism Bridge consciousness
VC 1365	Productiveness	MP 1173	Protecting non-linear contexts communication pathways	HM 4319	Psychism Catalepsy
VP 5165	Productiveness-Unproductiveness	MP 1245	Protecting variability enhance fixity	HM 4319	Psychism Cataleptic trance
MS 8346	Products ; Animal artifacts	VC 1383	Protection	HM 3456	Psychism Channelling
HM 2414	Products ; Emptiness (Buddhism)	MP 1057	Protection emerging foci	HM 4333	Psychism Clairaudience
HM 2089	Products ; Emptiness non (Buddhism)	MS 8223	Protocol	HM 4516	Psychism Clairempathy
MP 1178	Products ; Re integration rejected perspective	MM 2057	Protocol ; Bilingual	HM 0412	Psychism Clairgustance
MM 2027	Products ; Resources, commodities	VP 5505	Provability-Unprovability	HM 0318	Psychism Clairscent
VD 3490	Profanation				
HH 2341	Profession faith (Islam)				
MS 8102	Professional equipment				

Race

HM 3498	Psychism Clairvoyance
HM 0850	Psychism Crisis apparition
HM 0921	Psychism Dawning consciousness
HM 1966	Psychism Death consciousness
HM 1886	Psychism Deep trance
HM 2044	Psychism Devachon awareness
HM 4245	Psychism Emotional clairaudience
HM 1118	Psychism Emotional clairvoyance
HM 3590	Psychism Emotional consciousness
HM 4024	Psychism Esoteric divine consciousness
HM 4152	Psychism Extreme exaltation
HM 3590	Psychism Feeling consciousness
HM 0750	Psychism Fifth state consciousness
HM 3975	Psychism Focus fifteen
HM 0629	Psychism Focus one
HM 1136	Psychism Focus ten
HM 0267	Psychism Focus twelve
HM 3606	Psychism Foresight
HM 4071	Psychism Fourth dimensional clairvoyance
HM 3667	Psychism Fringe consciousness
HM 3770	Psychism Fringe space-time
HM 1886	Psychism Full trance
HM 2565	Psychism Heaven awareness
HM 0147	Psychism High dream
HM 0866	Psychism Higher state consciousness
HM 0929	Psychism Hypertrance
HM 3511	Psychism Hypnagogic consciousness
HM 3511	Psychism Hypnagogic imagery
HM 3688	Psychism Hypnapompic state
HM 3511	Psychism Hypnogogic state
HM 3688	Psychism Hypnopompic consciousness
HM 3688	Psychism Hypnopompic imagery
HM 1094	Psychism Inspirational art
HM 1094	Psychism Inspirational speaking
HM 1094	Psychism Inspirational thought
HM 1094	Psychism Inspired medium
HM 1347	Psychism Internal awareness
HM 4587	Psychism Land shadows
HH 2161	Psychism Life reading
HM 1219	Psychism Lowered state consciousness
HM 0750	Psychism Lucid awareness
HH 1775	Psychism ; Mental
HM 0551	Psychism Metamorphosis
HM 2640	Psychism Mind-awake body-asleep trance
HM 5123	Psychism Mineral consciousness
HM 5123	Psychism Mineral psychism
HM 5123	Psychism ; Mineral (Psychism)
HM 3005	Psychism Multi-level awareness
HM 2820	Psychism Negative intrapsychic forces
HM 5534	Psychism Out-of-body experience
HM 5123	Psychism Overall awareness
HM 0827	Psychism Oversoul awareness
HM 3480	Psychism Paradisal condition
HM 3480	Psychism Paradisal memory
HH 1775	Psychism ; Physical
HM 3111	Psychism Possession
HM 3606	Psychism Premonition
HM 3606	Psychism Prescience
HM 3606	Psychism Presentiment
HM 1919	Psychism Psyche consciousness
HM 0401	Psychism Psychic atavism
HM 0837	Psychism Psychic attack
HM 3574	Psychism Psychic attunement
HM 1919	Psychism Psychic consciousness
HH 0855	Psychism Psychic development
HM 1094	Psychism Psychic inspiration
HM 3546	Psychism Psychic presence
HM 3571	Psychism Psychic transfer
HM 1461	Psychism Pure alertness
HM 0521	Psychism Pure consciousness
HM 0820	Psychism Quickened consciousness
HM 0967	Psychism Reverie
HM 2436	Psychism Self-awareness
HM 1964	Psychism Self-induced trance
HM 0887	Psychism Semi-trance
HM 4417	Psychism Seventh plane
HM 3202	Psychism Shape shifting
HM 3050	Psychism Soul vision
HM 0819	Psychism Spellbound
HH 0855	Psychism Spiritual development
HM 5479	Psychism Thought insertion
HM 3456	Psychism Trance channelling
HM 2640	Psychism Unmani
HM 1078	Psychism Vampire
HM 0020	Psychism Xenophrenia
HH 0681	Psycho-cybernetics
HM 2075	Psycho-physical faculties awareness (Buddhism)
TC 1922	Psycho physical processes ; Meetings
HH 3003	Psycho-social transformation music
MM 2048	Psycho-symbolic cycles
HM 0584	Psychoactive substances ; Modes awareness associated
HH 0951	Psychoanalysis
HH 7123	Psychoanalysis ; Integral
HH 1236	Psychobiologic therapy
HH 0512	Psychodance
HH 0273	Psychodiagnostics
HH 0512	Psychodrama
HM 3271	Psychodynamic LSD experience
HM 1763	Psychological approach transcendence
HM 3542	Psychological assessment mystic experience (Sufism)

Psychological development

HH 0447	— Human
HH 0630	— infants
HH 1976	— women
HM 2553	Psychological growth dying person
HM 4539	Psychological imprisonment
HH 0971	Psychological maturity

Psychological [rapport...]

HH 0668	— rapport
HH 2008	— reductionism
HH 0356	— reform
HH 0285	Psychological stages life cycle ; Sequence
HH 1420	Psychology ; Analytical
HH 5119	Psychology ; Archetypal
HH 0718	Psychology ; Being
HH 4099	Psychology ; Creative
HH 0090	Psychology ; Cultural science
HH 1420	Psychology ; Depth
HH 3698	Psychology ; Developmental
HH 0718	Psychology ends
HH 0793	Psychology ; Humanistic
HH 5119	Psychology ; Imaginal
HH 2009	Psychology ; Non deterministic
HH 0718	Psychology ; Onto
HH 0718	Psychology perfection
HH 0630	Psychology ; Peri natal
HH 5022	Psychology ; Religious (Christianity)
HH 0090	Psychology ; Social science
HH 0723	Psychology ; Sports
HH 0718	Psychology ; Transcendental
HH 0916	Psychology ; Transpersonal
HH 5022	Psychology ; We (Christianity)
HH 0072	Psycholytic therapy
HH 0172	Psychomotor therapy
HH 0512	Psychomusic
HM 1401	Psychoneuroimmunology (PNI)
HH 0394	Psychopharmacology
HM 3664	Psychophysiological relaxation ; Deep (DPR)
HH 0407	Psychosexual development
HM 2887	Psychosis
HH 0966	Psychosocial isomorphism
HM 3117	Psychosomatic power ; Development
HH 4455	Psychospiritual dualism
HM 2909	Psychospiritual growth (Christianity)
HH 4201	Psychospiritual therapy (Christianity)
HH 0002	Psychosynthesis
HM 2811	Psychosynthesis Collective conscience
HM 2811	Psychosynthesis Collective unconscious
HM 2123	Psychosynthesis Conscious self
HM 2970	Psychosynthesis Higher self
HM 2057	Psychosynthesis Higher unconscious
HM 2790	Psychosynthesis Lower unconscious
HM 2306	Psychosynthesis Middle unconscious
HM 2383	Psychosynthesis Personal consciousness field
HH 4221	Psychotechnology
HH 0476	Psychotherapeutic change ; Constructive
HH 0922	Psychotherapeutic self examination
HH 0003	Psychotherapy
HH 0362	Psychotherapy ; Family
HH 0851	Psychotherapy ; Group
HM 2887	Psychotic state
HH 0075	Psychotomimetic drugs
HM 2262	Psychotronics (ESP)
KC 0712	Public opinion
TC 1011	Public opinion
KC 0696	Public policy analysis
TC 1501	Public relations
VD 3506	Puerility
HH 0111	Puja yoga (Yoga)
MM 2088	Pulsational variable stars
MM 2081	Pumping
HM 7174	Punch-drunk
VC 1389	Punctuality
VP 6010	Punishment ; Atonement
TP 0021	Punitive action (Biting through)
HM 1461	Pure alertness (Psychism)
HM 4493	Pure awareness reality (Yoga)
HM 3455	Pure consciousness (Hinduism)
HM 0521	Pure consciousness (Psychism)
HM 2788	Pure intelligence ; Pre conscious
HM 3100	Pure intention (Hinduism)
HM 2966	Pure knowledge
HM 2168	Pure-lands awareness (Buddhism)
HH 0074	Pure love (Christianity)
HM 2966	Pure vision
HH 4090	Purgativa ; Via (Christianity)
HM 0856	Purgatory (Leela)
HH 0401	Purification
HM 3461	Purification awareness (Yoga)
HH 2332	Purification consciousness (Buddhism)

Purification knowledge vision

HH 3025	— (Buddhism)
HH 3550	— way (Buddhism)
HH 4007	— what is path what is not path (Buddhism)
HM 1773	Purification (Leela)
HH 1187	Purification overcoming doubt (Buddhism)
HH 3875	Purification ; Path (Buddhism)
HH 5092	Purification self (Sufism)
HM 1727	Purification senses (Christianity)
HH 2718	Purification view (Buddhism)
HH 1865	Purification virtue (Buddhism)
HH 3875	Purifications (Buddhism)
HM 2943	Purified intelligence ; Path (Judaism)
HH 3875	Purities (Buddhism)
HH 0719	Purity
VC 1391	Purity
HM 7961	Purity consciousness ; Concentration due natural (Buddhism)
HM 1612	Purity Heart
HM 3602	Purity ; Inward (Sufism)

Purity knowledge discernment

HH 3025	— (Buddhism)
HH 3550	— middle way (Buddhism)
HH 4007	— right path wrong path (Buddhism)

Purity [Mental...]

HM 6149	— Mental (Buddhism)
HH 2332	— mind (Buddhism)
HH 1865	— morality (Buddhism)
HH 3875	Purity ; Path (Buddhism)
HM 4340	Purity ; State (Christianity)
HH 0575	Purity (Sufism)
HH 1187	Purity transcending doubt (Buddhism)
HH 2718	Purity views (Buddhism)
HH 1391	Purity (Yoga)
HH 0037	Purna yoga (Yoga)
HH 0691	Purpose
KC 0730	Purpose
VC 1393	Purposefulness
VD 3508	Purposelessness
KC 0730	Purposiveness
HH 5110	Pursuits ; Alternative
HM 2396	Purusa (Yoga)
HM 2396	Purusha (Yoga)
HM 2543	Purva-videha (Buddhism)
VD 3510	Putrefaction
HM 4008	Puzzlement (Buddhism)
HH 2213	Pyramid meditation
KC 0342	Pythagoreanism

Q

HH 0921	Qabalah
HH 0921	Qabalistic tree life
HM 2767	Qabid ; Al (Sufism)
HM 1500	Qadir ; Al (Sufism)
HM 1318	Qahhar ; Al (Sufism)
HM 7622	Qalb ; Fana (Sufism)
HM 1905	Qalb ; Latifa (Sufism)
HM 6932	Qalb ; Tasfiya (Sufism)
HM 2792	Qalbi ; Dhikr (Sufism)
HM 2792	Qalbi ; Wuquf (Sufism)
HM 7024	Qana'a (Sufism)
HM 4988	Qawi ; Al (Sufism)
HM 5006	Qayyum ; Al (Sufism)
HH 3862	Qi gong
HH 3862	Qigon therapy
HH 3862	Qigong
HH 4022	Quaker meditation (Christianity)
VC 1395	Qualification
KP 3019	Qualitative transformation
HM 3274	Qualities ; Freedom lower (Sufism)
HH 0364	Qualities godly
HM 3079	Qualities ; Paths special (Buddhism)
HH 9764	Qualities ; Superhuman (Buddhism)
HM 4968	Qualities ; Understanding way fixing material (Buddhism)
HM 5627	Qualities ; Understanding way fixing mental (Buddhism)
VC 1397	Quality
HM 1741	Quality aggregate consciousness ; Good (Buddhism)
HM 0349	Quality aggregate sensation, aggregate cognition, aggregate synergies ; Good (Buddhism)
HM 0872	Quality ; Being good (Buddhism)
TP 0061	Quality ; Essential
HM 3424	Quality ; Good (Buddhism)
HH 0647	Quality human life ; Development
VT 8013	Quantity∗complex
KC 0387	Quantity ; Critical
KC 0186	Quantization
VD 3512	Quarrelsomeness
KD 2130	Quaternary consciousness: number time
HM 0235	Quddus ; Al (Sufism)
HH 1179	Quest ; Grail
HH 4994	Quest ; Secular
HH 0897	Quest ; Vision
KD 2001	Questionable answers: monopolarization
HM 1080	Questioning ; Wrong (Yoga)
MS 8113	Quests journeys ; Symbolic
HM 4977	Quick intuition ; Concentration easy progress (Buddhism)
HM 7859	Quick intuition ; Concentration painful progress (Buddhism)
HM 0820	Quickened consciousness (Psychism)
VC 1399	Quickness
HM 3771	Quies ; State (Christianity)
VC 1401	Quiescence
HM 0709	Quiescence ; Mental (Buddhism)
VP 5267	Quiescence ; Motion
HM 4383	Quiet attentiveness
HM 0798	Quiet reflection
HH 2006	Quietism
HH 4477	Quietism (Christianity)
HH 0491	Quietness
VC 1403	Quietness
VC 1405	Quietude
HM 1498	Quietude (Buddhism)
VC 1407	Quintessence
HM 2377	Qurb (Sufism)
HM 2377	Qurbat (Sufism)

R

HM 0206	Ra'uf ; Ar (Sufism)
HM 2743	Rab-tu-dben-pai-yid-byed (Tibetan)
MS 8395	Race ethnicity
HM 2953	Race ; Second step: (Christianity)

Racial

Code	Entry
HM 1070	Racial consciousness (Yoga)
KC 0992	Racial ethnic integration
HH 0528	Racial memory
KC 0690	Racially ethnically integrated education
HM 1240	Raconté ; Déjà
VC 1409	Radiance
HM 2019	Radiant guru (ICA)
HM 5028	Radiation ; Plane (Leela)
HM 2477	Radical contingency (ICA)

Radical [identification...]
- HM 2758 — identification (ICA)
- HM 2273 — illumination (ICA)
- HM 2587 — incarnation (ICA)
- HM 3181 — intelligence ; Path (Judaism)
- HH 2675 Radical psychiatry
- HH 2675 Radical therapy
- MM 2035 Radio communication
- HM 4040 Rafi ; Al (Sufism)
- HM 3914 Raga (Hinduism)
- HM 4666 Raga (Hinduism)
- HM 2433 Raga (Yoga)
- HM 6093 Ragakkhaya (Pali)
- HM 6093 Ragaksaya (Buddhism)
- HM 4043 Rahim ; Al (Sufism)
- HM 3539 Rahman ; Ar (Sufism)
- HH 0427 Raising ; Consciousness
- HM 4393 Raja (Sufism)
- HH 0755 Raja yoga (Yoga)
- HH 0413 Rajas
- HM 3281 Rajoguna (Leela)
- HM 2055 Rakshasas (Buddhism)
- VD 3514 Randomness
- HM 2709 Rang-sang-rgyas (Tibetan)
- HM 1353 Range awareness ; Diminished
- KC 0977 Range goals mankind ; Long
- TC 1916 Rap group
- VD 3516 Rapacity
- HM 2162 Raphael ; Angelic awareness
- HM 5974 Rapid eye movement
- HH 0668 Rapport ; Psychological
- VC 1411 Rapture
- HM 0747 Rapture (Buddhism)

Rapture Concentration
- HM 4433 — accompanied (Buddhism)
- HM 5767 — (Buddhism)
- HM 0730 — (Buddhism)
- HM 0931 Rapture concentration ; Right (Buddhism)
- HM 2390 Rapture ; God consciousness
- HM 1973 Rapture ; State (Buddhism)
- HM 4886 Rapturous happiness due insight (Buddhism)
- HM 6098 Raqib ; Al (Sufism)
- VC 1413 Rareness
- HM 0205 Rasa (Hinduism)
- HM 0878 Rasa-loka (Leela)

Rasa [resoluteness...]
- HM 4085 — resoluteness (Hinduism)
- HM 0966 — revulsion (Hinduism)
- HM 1780 — Romantic (Hinduism)
- HM 0399 Rasa ; Sorrowful (Hinduism)
- HM 1644 Rasa ; Terrifying (Hinduism)
- HM 8762 Rasa ; Violent (Hinduism)
- HM 5234 Rashid ; Ar (Sufism)
- VP 5894 Rashness ; Caution
- HM 7222 Rasti (Sufism)
- HM 0724 Rasul (Sufism)
- HM 2546 Rati ; Nirvana (Buddhism)
- HM 2283 Rational, calculative thinking
- HH 0976 Rational meditation ; Reflective
- TC 1704 Rational objective
- VC 1415 Rationality
- HM 2267 Ratisamgrahakamanaskara (Buddhism)
- HM 2275 Ratna-kuta (Buddhism)
- VD 3518 Ravishment
- HM 3195 Raw reality (ICA)
- HM 2665 Rays ; Seven (Esotericism)
- HM 7021 Razzaq ; Al (Sufism)
- HM 2028 Rdo-rje (Tibetan)
- HH 0307 Re armament ; Moral
- HM 3101 Re creation ; Sheer (ICA)

Re [education...]
- HH 0996 — education ; Ideological
- HH 0686 — embodiment
- HH 0342 — evaluation counselling
- MP 1178 Re-integration rejected perspective by-products
- HH 5119 Re-souling world
- HH 0142 Re-visioning
- VP 5283 Reaction ; Impact
- TC 1866 Reaction teams ; Audience
- HM 4664 Reaction ; Vestibular coriolis
- VC 1417 Readiness
- HH 0158 Reading ; Divine (Christianity)
- HH 2161 Reading ; Life (Psychism)
- HH 5550 Reading ; Meditative
- HH 0158 Reading ; Spiritual
- HM 4228 Real grasping ; Lack (Buddhism)
- HM 3605 Real ; Inward experience (Sufism)
- KC 0936 Real-time information system
- HM 3755 Realities (Sufism)

Reality
- KC 0370 — [Reality]
- VC 1419 — [Reality]
- HH 0444 — [Reality]

Reality [Arriving...]
- HH 4880 — Arriving (Taoism)
- HM 2407 — awareness ; Death bardo heaven (Buddhism)
- HM 2032 — Awareness relative (Hinduism)

Reality [Awareness...] cont'd
- HM 1456 — Awareness ultimate
- HM 3316 Reality being ; Unchanging (Sufism)
- HM 3457 Reality ; Existence ontological (Sufism)
- HM 2985 Reality ; Identifying (Sufism)
- HM 2548 Reality ; Loss experience
- TC 1727 Reality ; Meetings models
- MS 8200 Reality ; Mythological
- HM 2548 Reality one's self environment ; Anomalies experience

Reality [Perception...]
- HM 2531 — Perception hidden (Sufism)
- HM 1293 — Plane (Leela)
- HM 4493 — Pure awareness (Yoga)
- HM 3195 Reality ; Raw (ICA)
- HM 3892 Reality ; Remembrance (Sufism)

Reality [Separate...]
- HH 0489 — Separate
- HM 0754 — Seventh chakra: plane (Leela)
- HH 1004 — Sheaths masking (Hinduism)
- HH 0714 — Social
- HM 2391 — States non ordinary
- HM 6992 Reality transfer
- HM 3303 Reality truth ; Ontological (Sufism)

Reality [Ultimate...]
- HH 0504 — Ultimate
- HM 2030 — Ultimate (ICA)
- HH 0726 — Unitary
- HM 1772 Reality ; Virtual
- VC 1421 Realization
- HM 3060 Realization ; Conditional self
- HM 4327 Realization duty ; Religious experience
- KC 0467 Realization ethic ; Self
- HM 0220 Realization nen
- HM 1142 Realization ; Right (Buddhism)

Realization [Self...]
- HH 0412 — Self
- HM 2438 — Spiritual (Yoga)
- HM 0450 — State (Buddhism)
- HH 0784 Realization therapy ; Self
- HM 2665 Realization ; Ways spiritual (Esotericism)
- HM 3869 Realization (Yoga)
- HM 2441 Realized vocation (ICA)
- HM 2709 Realizers ; Meditation way solitary (Buddhism)
- HM 3651 Realm fantasy ; Second chakra: (Leela)
- HM 3121 Reaper ; Grim (Tarot)
- HM 0748 Reappearance beings ; Knowledge passing away (Buddhism)
- VC 1423 Reason
- VP 5481 Reason ; Intuition
- HM 6523 Reason ; Love based (Islam)
- HM 2203 Reason ; Objective
- VC 1425 Reasonableness
- HM 7154 Reasoned ; Understanding consisting what is (Buddhism)
- HM 1177 Reasoning (Buddhism)
- HM 6997 Reasoning ; Discursive (Christianity)
- HH 0686 Rebirth
- HH 0304 Rebirth
- HH 0698 Rebirth ego ; Spiritual
- HM 2843 Rebirth ; Ideological (Brainwashing)
- HH 3465 Rebirth ; Initiation spiritual (Yoga)
- HM 8266 Rebirth-linking (Buddhism)
- HM 2431 Rebirth seeking awareness ; Death bardo (Buddhism)
- HM 4098 Rebirth ; Spiritual (Christianity)
- HH 1330 Rebirthing
- VD 3520 Recalcitrance
- HM 3806 Recalling (Buddhism)
- KC 0516 Recapitulation phylogeny ; Ontogenetic
- MP 1128 Receive external insight ; Orientation domains
- MS 8426 Receivers ; Receptacles
- HM 7092 Receiving (Buddhism)
- HM 3102 Receiving intelligence ; Path (Judaism)
- MS 8426 Receptacles receivers
- MP 1149 Reception external perspectives structures ; Hospitable
- HM 2247 Receptivity application awareness ; Mahayana (Buddhism)
- HM 2756 Receptivity application awareness ; Tantra (Buddhism)

Receptivity [Earth...]
- TP 0002 — (Earth)
- MP 1138 — emerging external insight
- MP 1107 — external insight ; Organization structures enhance
- MP 1105 — external insight ; Orientation structures enhance
- HM 2815 Receptivity ; Undifferentiated (Astrology)
- VP 5296 Recession ; Approach
- HM 3228 Reciprocal relation absolute conditional (Sufism)
- HH 0958 Recluse
- HH 2782 Recluse stage life (Hinduism)
- KC 0870 Recoding ; Binary
- VC 1427 Recognition
- TP 0020 Recognition
- KC 0636 Recognition ; Character
- HM 2923 Recognition guilt (Brainwashing)
- HH 0486 Recognition ignorance
- HM 6633 Recognition innocent mind (Taoism)
- KC 0636 Recognition ; Pattern
- TC 1875 Recognition seekers
- HM 6633 Recognition true mind (Taoism)
- HM 1336 Recognized personal performance ; Anomalies experience self
- VC 1429 Recollection
- HM 1563 Recollection (Buddhism)
- HM 2077 Recollection ; Christian (Christianity)
- HM 0180 Recollection ; Correct (Buddhism)
- HM 4248 Recollection ; Faculty (Buddhism)
- HM 0047 Recollection ; Faculty (Buddhism)
- HM 2351 Recollection (Islam)
- HM 2375 Recollection ; Nakshbandi (Islam)

Recollection [past...]
- HM 4297 — past life ; Knowledge (Buddhism)
- HM 3450 — Power (Buddhism)
- HM 4297 — previous existence ; Knowledge (Buddhism)
- HH 6221 Recollections meditation subjects (Buddhism)
- HM 1563 Recolleection (Buddhism)
- VC 1431 Recommendation
- HM 8266 Reconception (Buddhism)
- VC 1433 Reconciliation
- HH 0164 Reconciliation (Christianity)
- HH 0993 Reconstructive surgery
- TC 1498 Recording ; Audio tape
- TP 0024 Recovery (Turning point)
- HM 2386 Recreated mystery (ICA)
- VC 1435 Recreation
- HH 0410 Recreation
- VD 3522 Recrimination
- VC 1437 Rectitude
- HM 4250 Rectitude (Buddhism)
- HM 9001 Rectitude citta (Pali)
- HM 9001 Rectitude consciousness (Buddhism)
- HM 5402 Rectitude kaya (Pali)

Rectitude [mental...]
- HM 5402 — mental body (Buddhism)
- HM 5402 — mental factors (Buddhism)
- HM 9001 — mind (Buddhism)
- HM 7222 Rectitude (Sufism)
- HH 0686 Recurrence
- HH 0320 Recurring mental factors ; Ever (Buddhism)
- HM 3631 Reddening ; Alchemical (Esotericism)
- MP 1018 Redefinitions ; Network
- HH 0907 Redemption
- VC 1439 Redemption
- HH 0167 Redemption (Christianity)
- VT 8095 Redemption*complex
- HH 0428 Redemption ; Subjective
- KD 2300 Reduction ; Alternation energetic expansion mentalistic
- TP 0041 Reduction excesses
- KC 0911 Reductionism
- HH 2008 Reductionism ; Psychological
- HH 1981 Reductionism ; Spiritual
- KC 0437 Reference ; Frame
- KC 0818 Reference ; Spherical
- HM 3222 Refined cosmic consciousness
- VC 1441 Refinement
- HM 3387 Refinement ; Artistic (Japanese)
- HH 4880 Refining gold pill (Taoism)
- HM 4007 Refining self (Taoism)
- HH 0977 Reflection
- HH 1204 Reflection
- HM 1982 Reflection ; Application (Buddhism)
- HM 1647 Reflection ; Approaching (Buddhism)
- HM 5089 Reflection (Buddhism)
- HH 0632 Reflection ; Independent
- HH 0169 Reflection ; Knowledge contemplation (Buddhism)
- HM 0798 Reflection ; Quiet
- TC 1602 Reflective conferencing ; Self
- HM 1665 Reflective faith ; Individuative
- HH 0976 Reflective rational meditation
- HH 0266 Reflex ; Conditioned
- HH 0985 Reflex therapy ; Conditioned
- TC 1391 Reflexer participant type ; Self
- HH 0977 Reflexivity
- HH 0087 Reflexivity
- HH 0987 Reflexology
- HM 3061 Reforged transformation (ICA)
- VC 1443 Reform
- HH 0865 Reform ; Ideological (Brainwashing)
- HH 0356 Reform ; Psychological
- HH 0865 Reform ; Thought (Brainwashing)
- HM 1781 Refraining four deviations speech ; Leaving off, abstaining, totally abstaining (Buddhism)
- HM 1133 Refraining three deviations body ; Leaving off, abstaining, totally abstaining (Buddhism)
- HM 0252 Refraining wrong livelihood ; Leaving off, abstaining, totally abstaining (Buddhism)
- VP 5695 Refreshment-Relapse
- VP 5775 Refusal ; Consent
- HM 1020 Refusal ; Initiation (Esotericism)
- HM 2708 Reg-pa (Tibetan)
- HM 6003 Regard ; Amorous (Islam)
- HH 0634 Regard ; Self
- HM 2823 Regard ; Self (Christianity)
- HH 0618 Regeneration (Christianity)
- MP 1004 Regenerative resource cultivation areas

Regional international
- KC 0897 — cooperation
- KC 0587 — integration
- KC 0997 — system
- KC 0997 Regional subsystem global system
- HM 1646 Registering (Buddhism)
- HM 1646 Registration (Buddhism)

Regression
- VD 3524 — [Regression]
- HH 1300 — [Regression]
- HM 0044 — [Regression]
- VP 5294 Regression ; Progression
- HH 0866 Regressive electroshock therapy
- VC 1445 Regularity
- VP 5137 Regularity-Irregularity
- VC 1447 Regulation
- TP 0060 Regulation
- KC 0226 Regulation ; Self
- MM 2063 Regulation ; Traffic
- HH 0213 Regulation vital force (Hinduism)

Responsibility

KC 0143 Regulative cultural integration
HH 0874 Rehabilitation
VC 1449 Rehabilitation
MM 2004 Reification
HH 0731 Reiki therapy
HH 0686 Reincarnation
HM 5588 Reinforced concentration ; Consciousness (Buddhism)
HM 4396 Reinforced energy ; Consciousness (Buddhism)
HM 7902 Reinforced faith ; Consciousness (Buddhism)
HM 5499 Reinforced mindfulness ; Consciousness (Buddhism)
HM 5901 Reinforced understanding ; Consciousness (Buddhism)
HH 0736 Reinforcement
VC 1451 Reinforcement
MP 1225 Reinforcement apertures ; Boundary
MP 1031 Reinforcement ; Cycle relationship
MP 1178 Rejected perspective products ; Re integration
VD 3526 Rejection
HM 4348 Rejection ; Comprehension (Buddhism)
HH 0932 Rejoicing merit others (Buddhism)
HH 0525 Rejuvenation
TP 0050 Rejuvenation (Cauldron)
HH 5223 Rejuvenescence
VP 5695 Relapse ; Refreshment
MS 8602 Related objects ; Food
MP 1082 Related operational control contexts ; Minimal distance
KC 0828 Relatedness
HM 0467 Relatedness ; Conscious self (Yoga)
HM 0811 Relatedness ; Experience (Systematics)
VP 5009 Relatedness-Unrelatedness
KC 0860 Relation
VC 1453 Relation
HM 3228 Relation absolute conditional ; Reciprocal (Sufism)
HM 2316 Relation ; Authentic (ICA)
HM 2471 Relation ; External (ICA)
HM 0811 Relational knowledge (Systematics)
KC 0776 Relational logic
HM 2978 Relational situation (ICA)
HH 0672 Relations groups ; Human
HH 0176 Relations ; Human
KC 0037 Relations ; International
KC 0776 Relations ; Logic
TC 1501 Relations ; Public
KD 2082 Relations ; Unexpected global
VT 8012 Relationship∗complex
MP 1028 Relationship densities ; Coherent pattern
MP 1036 Relationship density ; Differentiation
MP 1050 Relationship entrainment ; Three way
MP 1068 Relationship formation amongst emerging perspectives ; Spontaneous
MP 1073 Relationship formation challenging emerging perspectives ; Contexts exploratory
HM 3595 Relationship God man (Christianity)
KP 3012 Relationship ; Harmoniously transformative controlled
MS 8333 Relationship ; Human congenital
MP 1025 Relationship indeterminacy
MP 1049 Relationship loops ; Local
HM 2766 Relationship man man (Christianity)
HM 3488 Relationship man nature (Christianity)
MP 1171 Relationship non linearity structures ; Appropriate
KP 3017 Relationship potential form
MP 1031 Relationship reinforcement ; Cycle
Relationship [sharing...]
MM 2042 — sharing
MP 1056 — Special modes
MP 1122 — structures communication pathways ; Direct
HH 0095 Relationship therapy
KC 0860 Relationships
TC 1760 Relationships ; Conference
MP 1051 Relationships enhanced non linear processes ; Linear
HH 4312 Relationships ; Imaginary
MP 1052 Relationships ; Interfacing vehicles communication networks unmediated
HM 3240 Relationships ; Liberation (ICA)
HH 0735 Relationships ; Mentor
MP 1005 Relationships ; Network inter
MP 1023 Relationships parallel ; Compensating
MP 1055 Relationships ; Protected low intensity
Relationships [Sixth...]
HM 0103 — Sixth order perceptions control
KD 2152 — sound ; Modelling language
MP 1029 — Stable density gradient local
MP 1136 Relative isolation complementary perspectives
MP 1140 Relative isolation structural interface communication pathway
VT 8035 Relative motion∗complex
VT 8042 Relative properties∗complex
HM 2032 Relative reality ; Awareness (Hinduism)
MP 1141 Relatively isolated context each perspective
KC 0562 Relativism
KC 0286 Relativity
HM 2322 Relativity conscious states ; Physical (Physical sciences)
KC 0076 Relativity ; Cultural
VC 1455 Relaxation
HM 4332 Relaxation ; Bodily
HM 2135 Relaxation ; Deep
HM 3664 Relaxation ; Deep psychophysiological (DPR)
HH 2338 Relaxation sequence ; Autogenic
HM 7512 Relaxed alertness (Japanese)
HM 1274 Relaxed thinking
TP 0040 Release
HM 0766 Release ; Knowledge desire (Buddhism)
HM 0635 Release matrix (Taoism)
HM 0831 Release ; Mental (Buddhism)
HM 2015 Release paths ; Uninterrupted (Buddhism)
HM 6558 Released soul's perception (Hinduism)

VD 3528 Relentlessness
VC 1457 Relevance
TC 1275 Relevance maps
TC 1275 Relevance networks
VC 1459 Reliability
VC 1577 Reliance ; Self
VC 1461 Relief
KC 0631 Religion
HM 2981 Religion ; Apt (Leela)
HH 1198 Religion ; Human development
HH 1902 Religion ; Human development primal
HH 0738 Religion ; Law (Islam)
HH 4032 Religion ; Mysteries
HH 3124 Religion ; Syncretic
HH 1902 Religion ; Totemic
KC 0543 Religion ; Unity
HH 0746 Religion ; Universal (Yoga)
KC 0791 Religions ; Transcendent unity
HH 1890 Religious affections
HH 0058 Religious community
HM 3008 Religious consciousness
HH 2434 Religious dialogue ; Inter
HH 0709 Religious education
HM 2146 Religious enthusiasm
Religious experience
HM 3445 — [Religious experience]
HM 3221 — Induced
HM 1593 — meditational insight
HM 0561 — personal devotion worship
HM 4327 — realization duty
HM 2624 — Spontaneous
HM 2080 Religious experiences ; Core
Religious [feeling...]
HM 4335 — feeling
MM 2046 — festivals ; Cycles
HM 2577 — function (ICA)
HH 1321 Religious growth
HM 3004 Religious modes ; New (ICA)
HH 1523 Religious movements ; Human development new
Religious [poverty...]
HH 0991 — poverty
HH 0497 — powers ; Magico
HH 5022 — psychology (Christianity)
HM 4111 Religious states (Christianity)
MS 8705 Religious symbols
HH 0762 Religious therapy
HM 3341 Religious traditions ascension stages game (Buddhism)
HM 2293 Religious vocation (ICA)
HM 5974 REM sleep
HM 5974 REM ; Stage one
HM 4665 Remainder ; Nirvana (Buddhism)
HM 5023 Remainder ; Nirvana (Pali)
TP 0064 Remaining ; Tasks
TP 0018 Remedial action
HM 1872 Remembering
HM 0405 Remembering (Buddhism)
HM 2486 Remembering ; Self
HM 1484 Remembrance (Buddhism)
VP 5537 Remembrance-Forgetfulness
HM 1905 Remembrance God ; Active (Sufism)
HM 6562 Remembrance God (Sufism)
HM 3892 Remembrance reality (Sufism)
VC 1463 Remission
HM 2977 Remorse
HM 1208 Remorse acquisition sinful unwholesome dharmas ; Feeling (Buddhism)
HM 3353 Remorse acquisition sinful unwholesome dharmas ; Not feeling (Buddhism)
HM 8112 Remorse (Buddhism)
HM 2538 Remorse ; Power (Buddhism)
HM 1757 Remorse what one ought to be remorseful ; Feeling (Buddhism)
HM 1673 Remorse what one ought to be remorseful ; Not feeling (Buddhism)
HM 1757 Remorseful ; Feeling remorse what one ought be (Buddhism)
HM 1673 Remorseful ; Not feeling remorse what one ought be (Buddhism)
VD 3530 Remorselessness
TC 1831 Remunerative power
HH 1676 Renaissance ; Spiritual
VC 1465 Renewableness
KD 2085 Renewal: autopoiesis ; Organization self
Renewal [Charismatic...]
HH 3124 — Charismatic (Christianity)
KP 3006 — Coherence
KP 3013 — Creative
HM 2315 Renewal ; Emblems (Tarot)
Renewal [Social...]
HH 0542 — Social
HH 0576 — Spiritual (Christianity)
HH 3124 — Spiritual (Christianity)
HM 2047 Renewing intelligence ; Path (Judaism)
HH 3443 Renge kyo ; Nam myoho (Buddhism)
HM 2047 Renovating intelligence ; Path (Judaism)
HM 3118 Renunciation
VC 1467 Renunciation
HM 2128 Renunciation (Buddhism)
HM 0210 Renunciation ; Initiation (Esotericism)
HM 3405 Renunciation ; Path (Yoga)
HH 1223 Renunciation ; Self (Christianity)
HM 4450 Renunciation (Sufism)
HH 4210 Renunciation (Yoga)
MP 1084 Reorganization ; Transitional contexts perspective
VC 1469 Repentance
HH 0441 Repentance

HM 4062 Repentance (Sufism)
HM 5187 Repetition ; Experience (Systematics)
TC 1599 Repetition learning cycles
HM 3140 Replication Christ (ICA)
TP 0016 Repose
HM 0226 Repose citta (Pali)
HM 7115 Repose due insight (Buddhism)
HM 8123 Repose God (Christianity)
HM 4704 Repose kaya (Pali)
HM 4704 Repose mental factors (Buddhism)
HM 0238 Repose ; Prayer (Christianity)
HM 6552 Repose ; Spiritual (Christianity)
VP 5573 Representation-Misrepresentation
HM 2557 Representational existence (ICA)
HM 2532 Representational sign (ICA)
HH 0951 Repression
VD 3532 Repression
VD 3534 Reproach
KC 0856 Reproducing systems ; Self
VP 5167 Reproduction ; Production
KC 0856 Reproduction ; Self
MS 8513 Reptile
HM 2307 Reptilian human brain consciousness
VD 3536 Repudiation
VD 3538 Repugnance
VD 3540 Repulsion
VP 5288 Repulsion ; Attraction
HM 5364 Repulsion ; Knowledge (Buddhism)
HM 3620 Repulsion (Yoga)
HM 0966 Repulsive sentiment (Hinduism)
VC 1471 Reputation
HH 0168 Reputation
VP 5914 Repute-Disrepute
KD 2180 Requirements ; Cyclic self organization
KC 0146 Requisite variety ; Principle
HM 1602 Research (Buddhism)
TC 1843 Research conferencing
VP 5485 Research-Discovery
KC 0450 Research ; Futures
KC 0540 Research ; Operations
Research [Participatory...]
TC 1843 — Participatory
KC 0975 — Peace
HH 0981 — Psychic
KC 0886 Research ; Systems
VD 3542 Resentment
HM 2134 Resentment (Buddhism)
HM 7332 Resentment (Hinduism)
MP 1145 Reserve ; Domains non current elements those
VC 1473 Reservedness
HM 2265 Resetting mind (Buddhism)
HM 3118 Resignation
HM 8022 Resignation ; Holy (Christianity)
HM 5290 Resignation (Sufism)
HH 1016 Resignation will God
VC 1475 Resilience
VC 1477 Resistance
HM 3421 Resistance change (Physical sciences)
HH 0759 Resistance ; Passive
HM 5329 Resistance shift
MP 1218 Resistant boundaries ; Distortion
HM 4085 Resoluteness ; Rasa (Hinduism)
VC 1479 Resolution
TP 0040 Resolution
HM 3800 Resolution (Buddhism)
HM 5432 Resolution faith due insight (Buddhism)
TP 0043 Resolution (Flight)
VP 5624 Resolution-Irresolution
HM 3800 Resolve (Buddhism)
HM 1142 Resolves ; Right (Buddhism)
MM 2031 Resonance accelerator ; Cyclic
KD 2327 Resonance-based consensus games
KD 2250 Resonance ; Dissonant harmony holistic
KD 2262 Resonance hybrids ; Interwoven alternatives:
HM 0285 Resonance (Japanese)
MM 2086 Resonance ; Molecular
KD 2310 Resonance ; Morphic
HM 3174 Resort awareness ; Her favourite (Yoga)
MP 1004 Resource cultivation areas ; Regenerative
HH 0745 Resource development ; Manpower (United Nations)
TC 1814 Resource investigator
TC 1932 Resource persons ; Key
MM 2038 Resource sharing
HH 0047 Resource training ; Human
VC 1481 Resourcefulness
KC 0177 Resources ; Assembling
MM 2027 Resources, commodities products
HH 0745 Resources development ; Human (United Nations)
HH 0147 Resources ; Human (United Nations)
KC 0177 Resources ; Mobilization
HH 0745 Resources ; Mobilizing human (United Nations)
KC 0177 Resources ; Procuring
VP 5964 Respect-Disrespect
Respect [Self...]
VC 1579 — Self
HH 1384 — Self
HH 0604 — superiors (Buddhism)
VC 1483 Respectability
VC 1485 Respectfulness
VC 1487 Resplendence
HM 3426 Resplendent intelligence ; Path (Judaism)
TP 0018 Responding illness
HH 0266 Response ; Conditioned
KD 2295 Response ; Entropic crisis learning
VC 1489 Responsibility
HM 3217 Responsibility ; Agape (ICA)

Responsibility

HM 3016	Responsibility ; Futuric (ICA)	
HH 3663	Responsibility ; Maternal conception social	
HH 0187	Responsibility ; Self	
HM 3110	Responsibility ; Universal (ICA)	
TC 1650	Responsive panel	
TP 0002	Responsive service (Earth)	
VC 1491	Responsiveness	
MM 2005	Rest ; Exercise	
VP 5709	Rest ; Exertion	
HM 8772	Rest ; State (Christianity)	
HM 0996	Restful alertness	
VC 1493	Restfulness	
HM 9213	Resting God (Christianity)	
TP 0052	Resting (Mountain)	
HM 8004	Resting Spirit (Christianity)	
HM 5309	Resting working accordance righteousness ; Meeting God both (Christianity)	
VD 3544	Restlessness	
HM 1015	Restlessness (Yoga)	
HM 6534	Restoration celestial awareness mundane (Taoism)	
VP 5693	Restoration-Destruction	

Restraint
- VC 1495 — [Restraint]
- TP 0060 — [Restraint]
- HH 6876 — [Restraint]
- KD 2190 Restraint ; Encompassing system dynamics: sixfold
- VP 5762 Restraint ; Freedom

Restraint [Self...]
- HH 0070 — Self
- VC 1581 — Self
- HH 4522 — Spiritual self (Christianity)
- TP 0026 — strong
- TP 0009 — Subtle
- HM 7098 — (Sufism)
- TP 0009 Restraint weak
- VD 3546 Restriction
- KC 0047 Restructuring ; Hierarchical
- KC 0197 Restructuring knowledge
- HM 0130 Result consciousness ; Karma (Buddhism)
- HM 0130 Resultant consciousness ; Mundane (Buddhism)

Resultant Indeterminate consciousness
- HM 0594 — fine material sphere (Buddhism)
- HM 0594 — form realm (Buddhism)
- HM 4982 — formless realm (Buddhism)
- HM 4982 — immaterial sphere (Buddhism)
- HM 5721 — sense sphere (Buddhism)
- HM 5129 — supramundane plane (Buddhism)
- HM 5129 — transcendental plane (Buddhism)
- HH 4553 Resurrection
- HM 1153 Resurrection ; Initiation (Esotericism)
- HM 2631 Resurrectional existence (ICA)
- VD 3548 Retaliation
- VD 3550 Retardation
- HM 2064 Retention ; Four doors (Buddhism)
- KC 0121 Reticulation
- TP 0033 Retreat
- HH 0420 Retreat
- HH 1875 Retreat ; Spiritual (Sufism)
- VD 3552 Retribution
- VT 8094 Retribution*complex
- VD 3554 Retrogression
- HH 0933 Return ; Eternal
- HM 6920 Return ; Knowledge path non (Buddhism)
- HM 7563 Return ; Knowledge path once (Buddhism)
- TP 0024 Return (Turning point)
- HM 3251 Returning origin (Zen)
- HM 0344 Revealed knowledge (Systematics)
- VC 1497 Revelation
- HH 1028 Revelation
- HM 3538 Revelation ; Experience
- HM 0181 Revelation ; Initiation (Esotericism)
- HH 3238 Revelation ; Private
- VD 3556 Revenge
- HM 2654 Revered hero (ICA)
- HM 3023 Reverence
- VC 1499 Reverence
- HM 1576 Reverie
- HM 0967 Reverie (Psychism)
- HM 2701 Reversible direction time ; Consciousness
- VD 3558 Reversion
- VP 5145 Reversion ; Conversion
- HM 2888 Review life ; Near death (NDE)
- KC 0736 Review technique ; Programme evaluation
- VD 3560 Revilement
- HH 0887 Revitalization
- HM 0889 Revitalized ; Being (Thai)
- VC 1501 Revival
- HM 2516 Reviving-hell awareness (Buddhism)
- KC 0396 Revolution
- VD 3562 Revolution
- HH 1267 Revolution ; Bio
- HH 0837 Revolution ; Cultural
- VP 5147 Revolution ; Evolution
- HH 1579 Revolution ; Intelligence
- TP 0049 Revolution (Moulting)
- KC 0296 Revolution ; Scientific
- KD 2100 Revolutionary cycles alternation
- HM 2839 Revolutionary ; Perpetual (ICA)
- HM 3071 Revolutionary sign (ICA)
- HH 0950 Revolutionary vigilance
- VD 3564 Revulsion
- HM 0966 Revulsion ; Rasa (Hinduism)
- HM 3620 Revulsion (Yoga)
- HM 2105 Reward ; Path intelligence (Judaism)
- HM 2724 Reza (Sufism)
- HM 2534 Rgod-pa (Tibetan)
- HM 2528 Rgyags-pa (Tibetan)
- HM 2823 Rgyal ; Nga (Buddhism)
- HM 2241 Rgyun-du-jog-pa (Tibetan)
- KC 0714 Rhyme
- KC 0831 Rhythm
- HM 1397 Rhythm ; Circadian
- HM 6533 Rhythmic sensory bombardment
- HH 0218 Rhythmic sensory bombardment therapy
- HH 0475 Rhythms ; Biological
- HH 3544 Ri
- HM 3352 Ri hokkai (Zen)
- HM 3164 Ri mu je ; Ji (Zen)
- HM 3335 Riaza (Sufism)
- VC 1503 Richness
- TP 0055 Richness
- HM 4190 Rida (Sufism)
- VD 3566 Ridicule

Right [action...]
- HM 0198 — action (Buddhism)
- HM 0198 — acts (Buddhism)
- HM 0198 — aims action (Buddhism)
- HM 1142 — aims (Buddhism)

Right [behaviour...]
- HM 0198 — behaviour (Buddhism)
- HH 0974 — beliefs (Buddhism)
- HM 2367 — brain consciousness
- HM 0931 Right concentration (Buddhism)
- HM 1735 Right concentration (Buddhism)
- HM 1295 Right effort (Buddhism)
- HM 1295 Right endeavour (Buddhism)
- HM 0567 Right knowledge (Leela)
- HM 1341 Right livelihood (Buddhism)
- HM 0931 Right meditative stabilization (Buddhism)
- HM 0704 Right mindfulness (Buddhism)
- HM 1280 Right outlook (Buddhism)
- HH 4007 Right path wrong path ; Purity knowledge discernment (Buddhism)

Right [rapture...]
- HM 0931 — rapture concentration (Buddhism)
- HM 1142 — realization (Buddhism)
- HM 1142 — resolves (Buddhism)
- HM 1157 Right speech (Buddhism)
- HM 1142 Right thinking (Buddhism)
- HM 1142 Right thought (Buddhism)
- HM 1280 Right understanding (Buddhism)
- HM 0931 Right unification (Buddhism)
- HH 2543 Right vibratoryhood
- HM 1280 Right view (Buddhism)
- VC 1505 Righteousness
- HM 5309 Righteousness ; Meeting God both resting working accordance (Christianity)
- VD 3642 Righteousness ; Self
- VC 1507 Rightness
- VP 5957 Rightness-Wrongness
- VC 0329 Rights ; Civil
- HM 2782 Rights ; Individual (ICA)
- HM 5329 Rigidity
- VD 3568 Rigidity
- HM 1028 Rigidity ; Non (Buddhism)
- VC 1509 Rigorousness
- HM 7298 Rijumati (Hinduism)
- HH 1134 Rinzai zazen meditation (Zen)
- HH 6129 Rinzai Zen (Zen)
- HM 3723 Rise fall ; Knowledge contemplation (Buddhism)
- TP 0046 Rising
- HM 1402 Rising ; Christ consciousness dying (Christianity)
- VD 3570 Risk
- HM 1363 Risk
- TC 1553 Risk ; Conferencing
- HH 0654 Risk-taking
- TC 1777 Risk triggering activities
- TC 1777 Risk working
- VD 3572 Riskiness
- HH 0230 Rite passage
- HH 0423 Rites
- HH 0021 Rites adolescence
- HH 0021 Rites ; Life crisis
- HH 0021 Rites passage
- HM 0401 Ritual cleansing
- HM 3213 Ritual movement
- HH 5120 Ritual pattern-making (Magic)
- HM 1605 Ritual ; Way (Esotericism)
- HH 0423 Rituals
- TC 1548 Rituals ; Conference
- HH 0021 Rituals ; Maturity
- VD 3574 Rivalry
- HM 2993 River consciousness (ICA)
- HM 6690 Riyada (Sufism)
- HM 0600 Rmi-lam (Tibetan)
- HM 2926 Rmugs-pa (Tibetan)
- HM 2169 Rna-ba'i-rnam-par-shes-pa (Tibetan)

Rnam par
- HM 3154 — g'yeng-ba (Tibetan)
- HM 2608 — mi-'tshe-ba (Tibetan)
- HM 3210 — 'tshe-ba (Tibetan)
- HM 3617 Rnam shes (Tibetan)
- MS 8933 Robot
- VD 3576 Roguery

Role Team
- TC 1526 — completer
- TC 1361 — driver
- TC 1642 — evaluator
- TC 1814 — explorer
- TC 1969 — focalizer
- TC 1550 — genius
- TC 1264 — grounder

Role Team cont'd
- TC 1142 — promoter
- TC 1528 Roles ; Affinity
- MS 8363 Roles ; Human
- MS 8385 Roles ; Sex
- TC 1528 Roles ; Team
- HH 0082 Rolfing
- HH 6291 Roman spirituality
- VC 1511 Romance
- HM 6122 Romance spiritual discipline
- HM 5087 Romantic love
- HM 1809 Romantic love
- HM 1780 Romantic rasa (Hinduism)
- HH 0270 Root afflictions (Buddhism)
- HM 2893 Root base awareness (Yoga)
- HM 1439 Root ; Delusion unwholesome (Buddhism)
- HM 3752 Root ; Disinterestedness wholesome (Buddhism)
- HM 1954 Root ; Greed unwholesome (Buddhism)
- HM 3560 Root ; Non hatred wholesome (Buddhism)
- HM 3142 Rope hell ; Black (Buddhism)
- VD 3578 Rot
- MM 2013 Rotating chairmanship
- MM 2094 Rotation ; Crop
- KD 2285 Rotation ; Patterns alternation: agricultural key crop
- HM 3522 Rotoo (Japanese)
- VD 3580 Roughness
- MM 2024 Round activities ; Daily
- HH 4653 Routines ; Penitential
- VC 1513 Royalty
- HM 1177 Rtog-pa (Tibetan)
- HH 0270 Rtsa-nyon (Buddhism)
- HM 2753 Rtse-geig-tu-byed-pa (Tibetan)
- VD 3582 Rubbish
- HM 3631 Rubedo (Esotericism)
- VD 3584 Rudeness
- HM 2603 Rudra awareness black freedom (Buddhism)
- HM 4343 Rudra-loka (Leela)
- HM 6162 Ruh ; Tajliya (Sufism)
- VD 3586 Ruin
- HM 2350 Ruin (Tarot)
- TC 1910 Rule-governed behaviour
- HM 1169 Running ; Zen
- HM 2142 Rupa-dhatu (Buddhism)
- HM 2235 Rupa-dhatu-vidyadhara (Buddhism)
- HM 2108 Rupa-khanda (Pali)
- HM 2257 Rupadhatu (Buddhism)
- HM 2536 Rupaloka consciousness (Buddhism)
- VD 3588 Ruthlessness
- HH 7324 Ryori ; Shojin

S

- HM 5023 Sa-upadisesanibbana (Pali)
- HH 0538 Sabda (Yoga)
- HM 4350 Sabi (Japanese)
- HM 2308 Sabija samadhi (Hinduism)
- VD 3590 Sabotage
- HM 3418 Sabr (Sufism)
- HM 0889 Sabsung (Thai)
- HM 4495 Sabur ; Al (Sufism)
- HM 5403 Saccanulomikanan (Pali)
- HM 0932 Sacchikiriya ; Phala (Buddhism)
- HM 0592 Saccidananda
- HM 3227 Saccidananda (Hinduism)
- HH 0230 Sacrament
- HM 0843 Sacramental sex
- HM 2366 Sacramental universe (ICA)
- HM 2445 Sacramental universe (ICA)
- HH 0423 Sacraments
- HM 1538 Sacred analogy
- HM 0876 Sacred ; Awareness (Christianity)
- MS 8123 Sacred calendar
- HH 0788 Sacred drugs
- HM 1538 Sacred imagination
- HH 0408 Sacred (Japanese)
- HH 3796 Sacred ; Journeying transcendence secular
- MS 8423 Sacred literature ; Symbols
- HH 0303 Sacred mysteries
- HH 0593 Sacred places
- HH 0957 Sacred scriptures ; Study
- KC 0772 Sacred tradition
- VC 1515 Sacredness
- HM 2302 Sacredness ; Sense
- HH 0577 Sacrifice
- VD 3592 Sacrifice

Sacrifice Self
- VC 1583 — [Sacrifice ; Self]
- HH 1241 — [Sacrifice ; Self]
- HM 2615 — (Astrology)
- HM 2754 Sacrificial friendship (ICA)
- HM 2641 Sacrificial passion (ICA)
- TP 0050 Sacrificial vessel (Cauldron)
- VD 3594 Sacrilege
- HM 7120 Sad feeling (Buddhism)
- HM 2077 Sadanga yoga (Yoga)
- HM 3158 Saddahana (Pali)
- HM 2933 Saddha (Pali)
- HM 4404 Saddhabala (Pali)
- HM 0066 Saddhindriya (Pali)
- HH 0093 Sadharana dharma
- HM 2292 Sadhu-mati (Buddhism)
- VD 3596 Sadism
- HM 2403 Sadness
- VD 3598 Sadness
- HM 3602 Safa (Sufism)

Science

VC 1517	Safety
VP 5697	Safety-Danger
VC 1519	Sagacity
HM 1750	Sagacity (Buddhism)
HM 9124	Sages ; Intuition (Hinduism)
HM 2726	Sagittarius-consciousness (Astrology)
HM 0576	Saguna brahman
HM 3619	Sahaj samadhi
HM 2043	Sahaja samadhi (Hinduism)
HM 3398	Sahasrara (Yoga)
HH 1009	Saijo Zen (Zen)
HH 9760	Saint Ignatius ; Spiritual exercises (Christianity)
HM 1714	Saint John Cross ; Dark night (Christianity)
HM 1409	Saint Teresa ; Interior castle (Christianity)
HH 0331	Sainthood
HM 0560	Sainthood (Christianity)
HM 4356	Sainthood ; Path (Sufism)
HM 3647	Sainthood (Sufism)
VC 1521	Saintliness
HH 1188	Saintliness
HM 3518	Saintly life
HM 2338	Saints ; Communion (ICA)
HM 7089	Saintship angels (Sufism)
HM 3647	Saintship (Sufism)
HM 7089	Saintship ; Superior (Sufism)
HH 0820	Saivism
HM 2961	Sakina
HH 0992	Sakina
HH 4309	Sakti ; Kriya (Hinduism)
HM 3575	Salaam
HM 1221	Salam ; As (Sufism)
HH 0536	Salat (Islam)
HM 4288	Salik (Sufism)
HM 1143	Sallakkhana (Pali)
VC 1523	Salubrity

Salvation

HH 0613	— [Salvation]
HH 0907	— [Salvation]
VC 1525	— [Salvation]
HH 0173	— [Salvation]
HH 0388	— [Salvation]
HM 6335	Salvation ; Personal (Christianity)
HH 1166	Salvation ; Way
HM 0413	Salvational knowledge
HM 2731	Samachittata
HM 4980	Samad ; As (Sufism)
HM 2277	Samadhana (Buddhism)

Samadhi [Aloka...]

HM 2232	— Aloka vriddhi (Buddhism)
HM 2586	— Appana (Pali)
HM 3041	— Asamprajnata (Hinduism)

Samadhi [Bhava...]

HM 3619	— Bhava
HH 8231	— bhavana (Buddhism)
HM 2440	— (Buddhism)
HM 6663	— (Buddhism)
HM 5587	Samadhi ; Ceto (Buddhism)
HM 0194	Samadhi ; Dharma megha (Yoga)
HM 2208	Samadhi ; Dharmaloka labdha (Buddhism)
HM 2226	Samadhi (Hinduism)
HM 0148	Samadhi ; Jada (Yoga)
HM 3060	Samadhi ; Jnana (Hinduism)

Samadhi [Nirbija...]

HM 2125	— Nirbija (Hinduism)
HM 0809	— Nirvicara (Yoga)
HM 2226	— Nirvikalpa (Hinduism)
HM 2061	— Nirvikalpa (Hinduism)
HM 4284	— Nirvitarka (Yoga)
HM 3207	Samadhi parinama (Yoga)
HM 0831	Samadhi ; Phala (Buddhism)

Samadhi [Sabija...]

HM 2308	— Sabija (Hinduism)
HM 3619	— Sahaj
HM 2043	— Sahaja (Hinduism)
HM 1735	— Samma (Pali)
HM 2896	— Samprajnata (Yoga)
HM 0931	— Samyak (Buddhism)
HM 2226	— Savikalpa (Hinduism)
HM 2650	— Savikalpa (Hinduism)
HM 3526	— Savitarka (Hinduism)
HM 3619	Samadhi ; Unconditional nirvikalpa
HM 4999	Samadhi ; Upacara (Buddhism)
HM 3880	Samadhi ; Vicara (Yoga)
HM 4451	Samadhi ; Vitarka (Yoga)
HH 4321	Samadhi yoga (Hinduism)
HM 1759	Samadhibala (Pali)
HM 3454	Samadhindriya (Pali)
HM 2699	Samaja urgyan awareness ; Guhya (Buddhism)
HM 3837	Saman-paap (Leela)
HM 2623	Samantaka (Buddhism)
HM 4514	Samapajanna (Pali)
HM 0159	Samapatti (Buddhism)
HM 0552	Samapattiya ; Hiriyati papakanam akusalanam dhammanam (Pali)
HM 3257	Samapattiya ; Na kiriyati papakanam akusalanam dhammanam (Pali)
HM 3353	Samapattiya ; Na ottapati papakanam akusalanam dhammanam (Pali)
HM 1208	Samapattiya ; Ottapati papakanam akusalanam dhammanam (Pali)
HH 0710	Samatha-bhavana
HM 0709	Samatha ; Ceto (Buddhism)
HM 8673	Samatha-nimitta (Buddhism)
HM 1498	Samatha (Pali)
HM 2147	Samatha (Pali)
HM 1586	Samatva (Yoga)
HM 2192	Sambhara marga ; Sravaka (Buddhism)
HM 2873	Sambhoga-kaya (Buddhism)
HM 3263	Sambodhi ; Anuttara samyak
HM 7195	Sambodhi (Buddhism)
HM 7091	Sambodhi ; Samma (Buddhism)
HM 4291	Same ; God wholly (Judaism)
VD 3600	Sameness
HM 0419	Sami' ; Al (Sufism)
HM 2399	Samjna (Buddhism)
HM 5438	Samjnavedayitanirodha (Buddhism)
HM 1142	Samkalpa ; Samyag (Buddhism)
HH 0155	Samkhya (Hinduism)
HM 0549	Samma ajiva (Pali)
HM 1763	Samma ajiva (Pali)
HM 1735	Samma-samadhi (Pali)
HM 7091	Samma-sambodhi (Buddhism)
HM 3252	Sammaditthi (Pali)
HM 0439	Sammakammanta (Pali)
HM 0687	Sammakammanta (Pali)
HM 1735	Sammasamadhi (Buddhism)
HM 1451	Sammasankappa (Pali)
HM 0180	Sammasati (Pali)
HM 4608	Sammavaca (Pali)
HM 1821	Sammavaca (Pali)
HM 1657	Sammavayama (Pali)
HM 1956	Sammoha (Pali)
HM 1356	Samnyasa (Hinduism)
HM 1874	Sampajanna (Pali)
HM 1921	Samphasajam cetasikam satam ; Manovinnanadhatu (Pali)
HM 0404	Samphassaja cetana ; Manovinnanadhatu (Pali)
HM 1501	Samphassaja sanna ; Manovinnanadhatu (Pali)
HM 1380	Samphusana (Pali)
HM 0836	Samphusitattam (Pali)
HM 2896	Samprajnata samadhi (Yoga)
HM 2177	Samsara (Buddhism)
HM 1006	Samsara (Hinduism)
HM 0808	Samsaya (Pali)
HH 0719	Samshuddhi ; Satya
HM 2414	Samskrtashunyata (Buddhism)
HM 2241	Samsthapara (Buddhism)
HM 6221	Samtana (Buddhism)
HM 2898	Samtosa (Yoga)
HH 1099	Samuppada ; Paticca (Buddhism)
HM 2596	Samvatta-kappa (Pali)
HM 7109	Samvedana ; Sva (Buddhism)
HM 0652	Samvriti (Buddhism)
HM 1341	Samyag-ajiva (Buddhism)
HM 1280	Samyag-drishti (Buddhism)
HM 1142	Samyag-samkalpa (Buddhism)
HM 1157	Samyag-vaca (Buddhism)
HM 1295	Samyag-vyayama (Buddhism)
HM 0198	Samyak-karmanta (Buddhism)

Samyak [samadhi...]

HM 0931	— samadhi (Buddhism)
HM 3263	— sambodhi ; Anuttara
HM 0704	— smriti (Buddhism)
HH 0374	Samyama
HH 0093	Sanatana dharma
HM 0705	Sancetana (Pali)
VC 1527	Sanctification
HH 0428	Sanctification
HH 5116	Sanctifying grace (Christianity)
HM 3567	Sanctifying intelligence ; Path (Judaism)
VD 3602	Sanctimony
HM 2945	Sanctions ; Cosmic (ICA)
VC 1529	Sanctity
HM 3024	Sanctity awareness ; Arhat (Buddhism)
HM 0560	Sanctity (Christianity)
HM 5965	Sanctity ; Plane (Leela)
HM 4041	Sanctity (Sufism)
VP 6026	Sanctity-Unsanctity
HH 0593	Sanctuary
HM 0764	Sang loka ; Ku (Leela)
HM 1355	Sang loka ; Su (Leela)
VC 1531	Sanity
VP 5472	Sanity-Insanity
HM 3160	Sanjna (Yoga)
HM 2551	Sankhara (Buddhism)
HM 2050	Sankhara-khanda (Pali)
HM 1381	Sankharakkhandhassa kammannata ; Vedanakkhandhassa sannakkhandhassa (Pali)
HM 1435	Sankharakkhandhassa labuta ; Vedanakkhandhassa sannakkhandhassa (Pali)
HM 3397	Sankharakkhandhassa muduta ; Vedanakkhandhassa sannakkhandhassa (Pali)
HM 1377	Sankharakkhandhassa passaddhi ; Vedanakkhandhassa sannakkhandhassa (Pali)
HM 0409	Sankharakkhandhassa ujukata ; Vedanakkhandhassa sannakkhandhassa (Pali)
HM 0349	Sankharakkhandhassa ; Vedanakkhandhassa sannakkhandhassa (Pali)
HH 0927	Sankya yoga (Yoga)
HM 1271	Sanna ; Cakkhuvinnanadhatusamphassajam cetasikam (Pali)
HM 4143	Sanna-khanda (Pali)
HM 1501	Sanna ; Manovinnanadhatu samphassaja (Pali)
HM 1389	Sanna (Pali)

Sannakkhandhassa sankharakkhandhassa

HM 1381	— kammannata ; Vedanakkhandhassa (Pali)
HM 1435	— labuta ; Vedanakkhandhassa (Pali)
HM 3397	— muduta ; Vedanakkhandhassa (Pali)
HM 1377	— passaddhi ; Vedanakkhandhassa (Pali)
HM 0409	— ujukata ; Vedanakkhandhassa (Pali)
HM 0349	— Vedanakkhandhassa (Pali)
HM 3419	Sannanana (Pali)
HH 4210	Sannyasa (Yoga)
HM 1066	Santhiti ; Cittassa (Pali)
HM 8116	Santosa (Hinduism)
HH 2056	Sanyasashrama (Hinduism)
HM 3250	Saraga (Pali)
HM 4363	Sarajitattam (Pali)
HM 1683	Sarajjana (Pali)
HM 1484	Saranata (Pali)
HM 4050	Saraswati (Leela)
VD 3604	Sarcasm
HM 3000	Sarvadharmashunyata (Buddhism)
HM 5334	Sarvajnata (Buddhism)
HH 0320	Sarvatraga (Buddhism)
HH 0641	Sarvodaya concept
HH 1784	Sastra ; Yoga (Yoga)
HM 0592	Sat-cit-ananda
HM 3227	Sat-cit-ananda (Hinduism)
HM 0309	Sata vedana ; Cetosamphassaja (Pali)
HM 1921	Satam ; Manovinnanadhatu samphasajam cetasikam (Pali)
HM 0972	Satam nasatam ; Cakkhuvinnanadhatusamphassajam cetasikam neva (Pali)
HM 3314	Satam nasatam ; Manovinnanadhatusamphassajam cetasikam neva (Pali)
HM 2212	Satan
HM 2580	Sated attention
HM 2847	Sati (Pali)
HM 1563	Sati (Pali)
HM 2580	Satiation
HM 3450	Satibala (Pali)
HM 4248	Satindriya (Pali)
HM 0047	Satindriya (Pali)
HH 4006	Satipatthana (Pali)
VD 3606	Satire
VC 1533	Satisfaction
HM 2409	Satisfaction awareness
HM 8116	Satisfaction (Hinduism)
HM 2724	Satisfaction ; Joyful (Sufism)

Satisfaction [Self...]

HM 3288	— Self
VD 3644	— Self
HM 4190	— (Sufism)
HM 4310	Satoguna (Leela)
HH 4019	Satori (Buddhism)
HM 2326	Satori (Zen)
HM 3329	Satsang (Hinduism)
HH 0413	Sattva
HM 2245	Satvani ; Chatvari (Buddhism)
HH 0554	Satya
HM 1293	Satya-loka (Leela)
HH 0719	Satya samshuddhi
HH 0138	Satyagraha
HH 0037	Satyeswarananda (Yoga)
HH 1391	Sauca (Yoga)
HH 4216	Saum (Islam)
VD 3608	Savagery
HM 0680	Savakabodhi (Buddhism)
HM 2226	Savikalpa samadhi (Hinduism)
HM 2650	Savikalpa samadhi (Hinduism)
HM 2413	Saving mystery (ICA)
HM 3462	Saviors God: action (Christianity)
HM 3420	Saviors God (Christianity)

Saviors God:

HM 3439	— march (Christianity)
HM 3264	— preparation (Christianity)
HM 3977	— silence (Christianity)
HM 2855	— vision (Christianity)
HM 2253	Saviour ; Eternal (ICA)
HH 2877	Saviour ; Jesus
HM 3526	Savitarka samadhi (Hinduism)
VP 5427	Savouriness-Unsavouriness
HH 0185	Saying prayers
HM 2683	Sbyor-mthai-yid-byed (Tibetan)
MP 1061	Scale interaction domains ; Bounded common small
MP 1006	Scale organization ; Intermediate
MP 1151	Scale perspective interaction contexts ; Small
VD 3610	Scandal
VD 3612	Scantiness
VD 3614	Scarcity
TP 0059	Scattering (Flooding)
HH 0393	Scepticism
KC 0944	Schedule ; Thesaurus Classification
KC 0427	Schismogenesis
HM 2313	Schizophrenia ; Modes awareness associated
HH 0939	Schizophrenic fantasy
KC 0011	Science
HM 0930	Science certainty (Sufism)

Science curricula

KC 0961	— Integrated
KC 0961	— Interdisciplinary
KC 0961	— Unified
KC 0711	Science education ; Integrated
KC 0225	Science human settlements
KC 0811	Science ; Integrated
KC 0540	Science ; Management
KC 0537	Science ; Meta

Science [policy...]

KC 0635	— policy
HH 0090	— psychology ; Cultural
HH 0090	— psychology ; Social
KC 0537	Science science
KC 0537	Science ; Science
KC 0961	Science ; Teaching concepts integrated
KC 0711	Science teaching ; Integrated methods
KC 0811	Science ; Unified
KC 0811	Science ; Unity
HM 1201	Science ; Way (Esotericism)

Sciences

Code	Entry
HM 3216	Sciences ; Four (Buddhism)
KC 0709	Sciences methodology ; Social
KC 0343	Sciences ; Policy
HM 0345	Scientific consciousness
KC 0540	Scientific management
KC 0296	Scientific revolution
MS 8933	Scientific symbols
KC 0760	Scientific technological progress
HH 0106	Scientology
HM 2428	Scientology Theta-clear state consciousness
HM 2921	Scorching avatar (ICA)
HM 2615	Scorpio-consciousness (Astrology)
HH 2087	Scream ; Primal
TC 1888	Screen development ; Conceptual
TC 1002	Screen development ; Conceptual
TC 1392	Screens ; Conceptual

Scriptural bodhisattva awareness

HM 2816	— Eighth (Buddhism)
HM 2909	— Fifth (Buddhism)
HM 2155	— First (Buddhism)
HM 2191	— Fourth (Buddhism)
HM 2292	— Ninth (Buddhism)
HM 2739	— Second (Buddhism)
HM 2361	— Seventh (Buddhism)
HM 2385	— Sixth (Buddhism)
HM 2421	— Tenth (Buddhism)
HM 2215	— Third (Buddhism)
HH 3421	Scripture ; Misuse
MS 8705	Scriptures ; Holy
HH 0957	Scriptures ; Study sacred
VC 1535	Scrupulousness
HM 4552	Scrutiny ; Comprehension (Buddhism)
HM 2722	Scuffle views (Buddhism)
HM 3033	Sea tranquillity (ICA)
HM 3487	Seance
HM 3408	Search (Buddhism)
HM 1715	Search dharma (Buddhism)
MS 8963	Search knowledge
HM 3036	Searching-for-ox awareness (Zen)
MM 2054	Seasons weather
VD 3616	Seclusion
HM 6312	Seclusion (Buddhism)
HM 4298	Seclusion ; Jhana happiness bliss born (Buddhism)
HM 3043	Second absorption immaterial sphere (Buddhism)
HM 2371	Second antarabhava consciousness (Buddhism)

Second [birth...]

HH 4098	— birth (Christianity)
HM 3175	— birth (ICA)
HH 0206	— body
HM 3651	Second chakra: realm fantasy (Leela)
HM 3174	Second chakra (Yoga)
HM 0238	Second degree prayer (Christianity)
HM 2861	Second duty (Christianity)
HM 2038	Second form-realm concentration (Buddhism)
HM 3043	Second formless attainment (Buddhism)

Second jhana

HM 7121	— Dwelling (Buddhism)
HM 4575	— five ; Concentration (Buddhism)
HM 6553	— fivefold system ; Dwelling (Buddhism)
HM 4380	— four ; Concentration (Buddhism)
HM 1932	Second order perceptions - sensation
KC 0463	Second-order theorizing
HM 2443	Second plane wisdom (Hinduism)

Second [scriptural...]

HM 2739	— scriptural bodhisattva awareness (Buddhism)
HM 2954	— stage life
HM 2953	— step: race (Christianity)
HM 3824	Second time cross ; Jesus falls (Christianity)
HM 2038	Second trance fine-material sphere (Buddhism)
HM 2703	Second vajra-master awareness (Buddhism)
HH 0781	Secondary afflictions (Buddhism)
HM 2795	Secondary delusions
HM 3086	Secondary integrity (ICA)
MP 1213	Secondary inter level connections ; Distribution
HH 0704	Secondary learning
HH 1348	Secondary mental factors (Buddhism)
HM 1241	Secondary sensation
VD 3618	Secrecy
HH 0959	Secrecy
HH 0959	Secret ; Discipline
HM 2918	Secret ; Heavenly (ICA)
HM 2021	Secret ; Path intelligence (Judaism)
HH 9807	Secret societies ; Initiation
TC 1776	Secretariat fatigue
HM 3685	Sectarian bias (Buddhism)
HH 0105	Secular humanism
HH 4994	Secular quest
HH 3796	Secular sacred ; Journeying transcendence
HM 2804	Secular transcendence
HM 0050	Securing self ; State (Sufism)
VC 1537	Security
HM 4546	Sedatives ; Modes awareness associated use
VD 3620	Sedition
HM 2234	Seduced mystery (ICA)
VD 3622	Seduction
HM 3436	Seduction ; Surrender bliss mystical (Sufism)
HM 3692	Seed meditation
HM 2159	Seed state wakefulness (Hinduism)
HM 1510	Seeing ; Not (Buddhism)
HM 0917	Seeing self-nature (Zen)
HM 2755	Seeing-the-ox awareness (Zen)
HM 3302	Seeing-the-traces awareness (Zen)
HM 6192	Seeing ; Understanding plane (Buddhism)
HM 3347	Seeing visions
TC 1875	Seekers ; Recognition
HM 2431	Seeking awareness ; Death bardo rebirth (Buddhism)
TC 1750	Seeking participants ; Information
HM 3265	Seemingly bare perception (Buddhism)
VD 3624	Segregation
HH 1109	Seishin education (Japanese)
HM 3148	Seizure ; Blissful (ICA)
HH 0357	Selection ; Genetic
HM 6687	Selective filtering experience
MP 1032	Selective interchange axis
MP 1019	Selective interchange ; Web
HH 0636	Self
HH 0706	Self

Self [abandonment...]

HH 0868	— abandonment ; Spiritual
HM 3587	— abasement
VD 3626	— absorption
HM 3912	— acceptance (Jung)
HH 0584	— accusation
VC 1539	— actualization
HH 0412	— actualization
HH 0971	— actualizing personality
VD 3628	— admiration
HH 0555	— adornment
VC 1541	— advancement
HM 4671	— aggrandizing love
HM 3030	— alienation
HH 0906	— analysis

Self Anomalies

HM 8186	— awareness one's
HM 8186	— experience
HM 9132	— experience unity

Self [assertion...]

HM 3587	— assertion
HH 0314	— assessment
VC 1543	— assurance

Self awareness

HM 2486	— [Self-awareness]
HM 1874	— (Buddhism)
HM 2436	— (Psychism)
HM 2229	Self-betrayal (Brainwashing)

Self [centredness...]

VD 3630	— centredness
HM 3604	— cognition (Sufism)
HH 0471	— concept
HM 0891	— Conditions (Hinduism)
VC 1545	— confidence
HM 1406	— conscious coding
HM 2610	— conscious consciousness
HM 2123	— Conscious (Psychosynthesis)
HM 2486	— consciousness
HM 2571	— consciousness
HM 2571	— Consciousness

Self consciousness

HM 2376	— Assertive (Astrology)
HM 2200	— (Buddhism)
HM 0467	— (Yoga)
VC 1547	Self-containedness
VD 3632	Self-contradiciton

Self control

HH 0600	— [Self control]
HH 0778	— [Self control]
VC 1549	— [Self-control]
HH 3710	Self-culture (Christianity)
HM 3152	Self ; Delusion personal (Buddhism)

Self denial

VC 1551	— [Self-denial]
HH 0964	— [Self-denial]
HM 2121	— (Christianity)
HH 4522	— Spiritual (Christianity)

Self [destruction...]

VD 3634	— destruction
VC 1553	— determination
HH 0317	— determination
HM 3076	— determining (ICA)

Self development

HH 0651	— [Self-development]
HH 3710	— (Christianity)
HH 0955	— Educational
HH 0117	— Personal
VC 1555	— Self-direction

Self discipline

HH 0877	— [Self-discipline]
VC 1557	— [Self-discipline]
HM 2121	— (Christianity)
HM 3280	— (Hinduism)
HH 1126	Self discovery
HM 4754	Self distinct outside world ; Anomalies experience

Self [education...]

HH 0417	— education
VC 1559	— effacement
HH 0907	— Emancipation
HM 2016	— emptiness (Buddhism)
HM 3612	— entity ; Notion one's (Buddhism)
HM 2548	— environment ; Anomalies experience reality one's

Self esteem

HM 2750	— [Self-esteem]
HH 0634	— [Self-esteem]
VD 3636	— [Self-esteem]

Self [evaluation...]

HH 0922	— evaluation
HH 0314	— examination
HH 0922	— examination ; Psychotherapeutic
HH 0743	— existence
HH 0791	— expression
VC 1561	— expression

Self [feeling...]

HH 1384	— feeling

Self [fulfilment...] cont'd

Code	Entry
VC 1563	— fulfilment
HH 0412	— fulfilment
VC 1565	Self-government

Self [harmony...]

Code	Entry
HH 0778	— harmony
VC 1567	— help
HM 2970	— Higher (Psychosynthesis)
HH 0103	— (Hinduism)
HH 0962	— hypnosis

Self [identity...]

Code	Entry
HH 0471	— identity
HM 3493	— Ignorance nature essential (Buddhism)
HM 9132	— Impairment unity
VD 3638	— importance
HM 3378	— induced attentiveness (Buddhism)
HM 1964	— induced trance (Psychism)
VD 3640	— indulgence
HH 0508	— inquiry meditation ; Free
VC 1569	— interest

Self [knowing...]

HM 3599	— knowing
HH 0312	— knowledge
HM 0765	— knowledge (Zen)

Self [learning...]

HH 0417	— learning
HH 0907	— Liberation
KD 2337	— limiting patterns interaction

Self love

HH 1066	— [Self-love]
HM 0478	— [Self-love]
HH 6732	— Christian (Christianity)

Self [made...]

HH 0777	— made person
HH 0396	— manipulation
HM 3209	— metaprogramming
HH 0464	— mortification (Sufism)
HM 5215	— Mortification (Sufism)

Self [nature...]

HM 0917	— nature ; Seeing (Zen)
HM 3059	— Near death encounter higher (NDE)
HM 2369	— negation ; Intentional (ICA)
HH 0241	— Not (Buddhism)
HH 0968	Self-oblation

Self observation

HH 0586	— [Self-observation]
HH 0229	— meditation ; Detached
HH 2276	— therapy

Self [opening...]

HH 0787	— opening meditation ; Dynamic
KC 0510	— organization
KD 2180	— organization requirements ; Cyclic
MP 1172	— organizing non linearity ; Contexts
KC 0510	— organizing systems

Self [perfection...]

HH 0519	— perfection
VC 1571	— possession
HH 0711	— preservation
VC 1573	— preservation
HM 3114	— programming (ICA)
HH 5092	— Purification (Sufism)

Self realization

HH 0412	— [Self-realization]
HM 3060	— Conditional
KC 0467	— ethic
HH 0784	— therapy

Self [recognized...]

HM 1336	— recognized personal performance ; Anomalies experience
HM 4007	— Refining (Taoism)
TC 1602	— reflective conferencing
TC 1391	— reflexer participant type
HH 0634	— regard
HM 2823	— regard (Christianity)
KC 0226	— regulation
HM 0467	— relatedness ; Conscious (Yoga)
VC 1577	— reliance
HM 2486	— remembering
KD 2085	— renewal: autopoiesis ; Organization
HH 1223	— renunciation (Christianity)
KC 0856	— reproducing systems
KC 0856	— reproduction
VC 1579	— respect
HH 1384	— respect
HH 0187	— responsibility

Self restraint

HH 0070	— [Self restraint]
VC 1581	— [Self-restraint]
HH 4522	— Spiritual (Christianity)
VD 3642	Self-righteousness

Self sacrifice

VC 1583	— [Self-sacrifice]
HM 1241	— [Self sacrifice]
HM 2615	— (Astrology)

Self [satisfaction...]

HM 3288	— satisfaction
VD 3644	— satisfaction
HM 3587	— sentiment
HH 0190	— simplification
HM 2984	— spiritual intelligence ; Awareness (Yoga)
HM 0050	— State securing (Sufism)
HM 3587	— subjection
VC 1585	— sufficiency

Self surrender

HH 0968	— [Self-surrender]
HH 1016	— [Self surrender]

Self surrender cont'd
- HH 6509 — love
- HM 5467 — State (Christianity)
- HM 0807 — (Sufism)
- HH 0606 Self-therapy

Self transcendence
- HM 2486 — [Self-transcendence]
- HH 0526 — [Self-transcendence]
- HM 2584 — (ICA)

Self [transcendent...]
- HH 1032 — transcendent systems
- HH 0258 — transformation love
- MP 1253 — transformation ; Meaningful symbols
- HH 0693 Self-understanding
- HH 0634 Self worth
- VP 5690 Selfactualization-Neurosis
- HM 3002 Selfhood ; Deprivation (Brainwashing)
- HM 3091 Selfhood ; Dynamic (ICA)
- HM 3077 Selfhood ; Transparent (ICA)
- VD 3646 Selfishness
- VP 5978 Selfishness ; Unselfishness
- HH 4477 Selfless love (Christianity)
- HM 0294 Selfless service (Leela)
- VC 1587 Selflessness
- HM 2016 Selflessness (Buddhism)
- KC 0723 Semantic field
- KC 0542 Semantics
- HH 0585 Semantics ; General
- KC 0330 Semantics ; General
- MP 1154 Semi-autonomous contexts maturing perspectives
- MP 1155 Semi-autonomous contexts perspectives decreasing activity
- HM 5856 Semi-stage hypnosis
- HM 0887 Semi-trance (Psychism)
- HM 2056 Seminal illumination (ICA)
- TC 1420 Seminar method
- HH 0115 Seminars training ; Erhard
- KC 0927 Semiotics
- HM 2145 Sems gnas dgu (Tibetan)
- HM 2217 Sems-jogpa (Tibetan)
- HM 2589 Sems-pa (Tibetan)
- HM 9022 Sensate experience ; Trans
- VC 1589 Sensation

Sensation aggregate cognition
- HM 1377 — aggregate synergies ; Composedness aggregate (Buddhism)
- HM 3397 — aggregate synergies ; Flexibility aggregate (Buddhism)
- HM 0349 — aggregate synergies ; Good quality aggregate (Buddhism)
- HM 1435 — aggregate synergies ; Lightness aggregate (Buddhism)
- HM 0409 — aggregate synergies ; Straightness aggregate (Buddhism)
- HM 1381 — aggregate synergies ; Workability aggregate (Buddhism)
- HM 3358 Sensation ; Annihilation personal (Sufism)
- HM 2270 Sensation (Buddhism)
- HM 0309 Sensation ease born contact psychical (Buddhism)
- VP 5422 Sensation-Insensibility
- HM 6644 Sensation ; Mental (Buddhism)
- HM 1643 Sensation neither suffering nor pleasure born contact psychical (Buddhism)
- HM 1932 Sensation ; Second order perceptions
- HM 1241 Sensation ; Secondary
- HM 1090 Sensation unease born contact psychical (Buddhism)
- HM 4076 Sensation unpleasant born contact psychical (Buddhism)
- HM 0062 Sensations sport ; Mystical
- KC 0971 Sense
- HM 1882 Sense blessedness (Islam)

Sense Common
- HH 0510 — [Sense ; Common]
- KC 0735 — [Sense ; Common]
- VC 0363 — [Sense ; Common]
- VT 8044 Sense∗complex
- HM 2664 Sense consciousness (Buddhism)
- HM 2149 Sense harmony progress (Brainwashing)
- HM 3596 Sense humour
- HH 0840 Sense I

Sense [mental...]
- HM 2522 — mental consciousness ; Emptiness objects (Buddhism)
- HM 4389 — mode consciousness occurrence (Buddhism)
- HM 3608 — mystery
- HM 2702 Sense oneness

Sense [presence...]
- HM 1166 — presence
- HM 6989 — presence God (Christianity)
- HH 5213 — propriety
- HM 2211 Sense realm awareness ; Tantra master (Buddhism)

Sense [sacredness...]
- HM 2302 — sacredness
- HM 2881 — shame (Buddhism)
- HM 3104 — Sixth (Japanese)

Sense sphere
- HM 1097 — concentration (Buddhism)
- HM 3852 — functional ; Indeterminate consciousness (Buddhism)
- HM 3852 — inoperative ; Indeterminate consciousness (Buddhism)
- HM 4447 — Profitable consciousness (Buddhism)
- HM 5721 — resultant ; Indeterminate consciousness (Buddhism)
- HM 8375 — Unprofitable consciousness (Buddhism)
- HM 0284 Sense wonder (Christianity)
- VD 3648 Senselessness
- HM 1727 Senses ; Dark night (Christianity)

Senses [Emptiness...]
- HM 2671 — Emptiness five (Buddhism)
- HM 2794 — Emptiness loci (Buddhism)
- HM 1756 — Enjoying
- HM 1756 Senses ; Flow experience
- HM 2556 Senses mind ; Awareness consciousness group conscious existence (Buddhism)
- HM 3764 Senses ; Perception
- HM 1727 Senses ; Purification (Christianity)
- HH 0829 Senses ; Withdrawal (Yoga)
- VC 1591 Sensibility
- VC 1593 Sensibleness
- TP 0031 Sensing
- VC 1595 Sensitivity
- TP 0031 Sensitivity
- KC 0567 Sensitivity analysis
- HH 0672 Sensitivity training
- HH 0195 Sensorimotor development

Sensory [amusia...]
- HM 6539 — amusia
- HM 2972 — awakening
- HM 2972 — awareness
- HM 6533 Sensory bombardment ; Rhythmic
- HH 0218 Sensory bombardment therapy ; Rhythmic
- HH 1478 Sensory deprivation
- HH 0865 Sensory deprivation (Brainwashing)
- HM 3764 Sensory perception
- HM 3754 Sensory translation
- HM 2664 Sensual awareness (Buddhism)
- HH 0978 Sensual enjoyment ; Abstention craving
- HM 3627 Sensual plane (Leela)
- VD 3650 Sensuality
- HM 6644 Sensuous feeling ; Non (Buddhism)
- HH 0211 Sentics
- HM 3497 Sentiment
- VC 1597 Sentiment
- HM 0645 Sentiment ; Comic (Hinduism)
- HM 1780 Sentiment ; Erotic (Hinduism)
- HM 8762 Sentiment ; Furious (Hinduism)
- HM 4085 Sentiment ; Heroic (Hinduism)
- HM 0291 Sentiment ; Marvellous (Hinduism)
- HM 0966 Sentiment ; Odious (Hinduism)
- HM 0399 Sentiment ; Pathetic (Hinduism)
- HM 0966 Sentiment ; Repulsive (Hinduism)
- HM 3587 Sentiment ; Self
- HM 1644 Sentiment ; Terrible (Hinduism)
- VD 3652 Sentimentality
- HH 0489 Separate reality
- VD 3654 Separateness
- MP 1106 Separating complementary structures ; Functional enhancement domains
- MP 1197 Separating domains ; Substantive distinctions
- VP 5047 Separation ; Conjunction
- HM 2318 Separation ; Gathering (Sufism)
- HM 3349 Separation ; Near death body (NDE)
- HM 4112 Separation (Sufism)
- HM 2906 Separation transient (Sufism)
- HM 2372 Sephira ; Binah (Kabbalah)
- HM 2420 Sephira ; Chesed (Kabbalah)
- HM 2348 Sephira ; Chokmah (Kabbalah)
- HM 0126 Sephira ; Daath (Kabbalah)
- HM 2290 Sephira ; Geburah din (Kabbalah)
- HM 2238 Sephira ; Hod (Kabbalah)
- HM 3132 Sephira ; Kether (Kabbalah)
- HM 2288 Sephira ; Malkuth (Kabbalah)
- HM 2362 Sephira ; Netzach (Kabbalah)
- HM 3031 Sephira ; Tiphareth (Kabbalah)
- HM 2410 Sephira ; Yesod (Kabbalah)
- HH 0921 Sephiroth ; Tree
- HH 3536 Sepulchre ; Jesus placed (Christianity)
- HM 2338 Sequence ; Autogenic relaxation
- HM 0772 Sequence ; Fifth order perceptions control
- HH 0285 Sequence psychological stages life cycle
- MP 1142 Sequence viewpoint loci structure
- HM 2964 Ser-sna (Tibetan)
- HM 2188 Seraphim (Judaism)
- KC 0737 Serendipity
- VC 1599 Serendipity
- TC 1567 Serendipity ; Meeting
- HM 1732 Serene B-cognition
- VC 1601 Serenity
- HM 2147 Serenity (Buddhism)
- HM 0226 Serenity mind (Buddhism)
- HM 3150 Serial absorptions (Buddhism)
- KC 0110 Series ; Harmonic
- KC 0300 Series ; Homologous
- HM 2323 Serious sharing (ICA)
- VC 1603 Seriousness
- HM 3037 Seriousness mind (Japanese)
- MS 8513 Serpents
- HH 6012 Servant ; Journeying transcendence served (Christianity)
- HM 3126 Servant ; Suffering (ICA)
- HH 6012 Served servant ; Journeying transcendence (Christianity)
- VC 1605 Service
- HH 0232 Service
- TC 1696 Service conferencing
- HM 3012 Service ; Consciousness human
- HH 0232 Service ; Discipline (Christianity)
- HM 3199 Service ; Embodying (ICA)
- TP 0002 Service ; Responsive (Earth)
- HM 0294 Service ; Selfless (Leela)
- HM 8776 Service (Sufism)
- VD 3656 Servility
- KC 0117 Set

- KC 0332 Sets ; Fuzzy

Setting mind
- HM 2217 — (Buddhism)
- HM 2157 — Close (Buddhism)
- HM 2241 — Continuous (Buddhism)
- HM 4007 Setting up foundation (Taoism)
- MP 1251 Settings ; Different
- VC 1607 Settlement
- TP 0063 Settlement
- KC 0225 Settlements ; Science human
- HM 1098 Setughata (Pali)
- HM 0776 Seven-fold knowledge (Systematics)
- HH 2665 Seven rays (Esotericism)
- HH 0460 Seven stages life
- HM 2341 Seven worlds ; Maqamat (Sufism)
- HM 0754 Seventh chakra: plane reality (Leela)
- HM 3398 Seventh chakra (Yoga)
- HM 1001 Seventh order perceptions - programme control
- HM 4417 Seventh plane (Psychism)

Seventh [scriptural...]
- HM 2361 — scriptural bodhisattva awareness (Buddhism)
- HM 3619 — stage
- HM 3193 — state (Yoga)
- HM 2789 Seventh vajra-master awareness (Buddhism)
- HH 1108 Seventy-five dharmas (Buddhism)
- VD 3658 Severance
- VD 3660 Severity
- VC 1609 Sex appeal
- HH 0547 Sex education
- HM 0829 Sex ; Enjoying
- HM 0829 Sex flow experience
- MS 8385 Sex roles
- HM 0843 Sex ; Sacramental
- HH 2034 Sex ; Yoga (Yoga)
- VP 5419 Sexiness-Unsexiness
- HM 3725 Sext (Christianity)
- HM 3725 Sext ; Hours (Christianity)
- HH 0045 Sexual abstinence
- HM 2374 Sexual awareness
- HH 3498 Sexual bonding
- HM 0829 Sexual ecstasy
- HM 0829 Sexual experience
- MM 2034 Sexual intercourse
- HM 9406 Sexual love
- HH 1325 Sexual maturity
- HM 0829 Sexual orgasm

Sexual [socialization...]
- HH 0881 — socialization
- HM 2954 — stage life ; Emotional
- HH 0524 — synergism
- HH 9112 Sexual yoga (Taoism)
- VD 3662 Sexuality
- HM 1222 Sgyu-lus (Tibetan)
- HM 3246 Sgyu (Tibetan)
- HH 0539 Shabd yoga (Yoga)
- HH 0538 Shabda (Yoga)
- HH 0539 Shabda yoga (Yoga)
- HH 0483 Shadan therapy
- HH 0204 Shadow ; Accepting
- HH 0204 Shadow ; Integrating
- HM 4587 Shadows ; Land (Psychism)
- HM 2341 Shahada (Islam)
- HM 0375 Shahid ; Ash (Sufism)
- HH 0629 Shakti
- HH 3912 Shakti yoga (Yoga)
- HM 1934 Shakur ; Ash (Sufism)
- VD 3664 Shallowness
- VP 5209 Shallowness ; Depth
- HM 3575 Shalom
- HM 1606 Shalom
- HH 0973 Shaman
- HM 2205 Shamana (Buddhism)
- HM 6120 Shamanic journey
- HM 2639 Shamanic state ; Magical (Buddhism)
- HH 0973 Shamanism
- HM 1189 Shamanism
- HM 1189 Shamanistic trance
- HH 0710 Shamatha
- HM 2147 Shamatha (Buddhism)
- HM 2151 Shambhala awareness ; Kalacakra tantra (Buddhism)

Shame Buddhism
- HM 2881 — [Shame (Buddhism)]
- HM 8112 — [Shame (Buddhism)]
- HM 1590 — [Shame (Buddhism)]
- HM 2986 Shame ; Non (Buddhism)
- HM 0352 Shame ; Power (Buddhism)
- HM 2881 Shame ; Sense (Buddhism)
- VD 3666 Shamelessness
- HM 2986 Shamelessness (Buddhism)
- HH 0649 Shamelessness (Buddhism)
- HM 1724 Shamelessness ; Power (Buddhism)
- HM 3575 Shanti
- HM 3202 Shape shifting (Psychism)
- VD 3668 Shapelessness
- TC 1361 Shaper ; Team
- MS 8845 Shapes patterns
- HH 0738 Sharia (Islam)
- VP 5815 Sharing-Appropriation
- TC 1900 Sharing information
- MM 2042 Sharing ; Relationship
- MM 2038 Sharing ; Resource
- HM 2323 Sharing ; Serious (ICA)
- HM 3093 Shathya (Buddhism)
- HM 0877 Shawq (Islam)
- HM 0762 Shawq (Sufism)
- HH 2080 Shaykh (Sufism)

Sheaths

HH 1004	Sheaths masking reality (Hinduism)
TC 1651	Sheep
HM 3101	Sheer re-creation (ICA)
HM 2961	Shekina
HM 2002	Shekinah awareness (Judaism)
HH 0126	Sheldon's types
MM 2077	Shell games
VC 1611	Shelter
HM 2663	Shen-siddhi (Buddhism)
HM 2063	Shes-bzhin-ma-yin-pa (Tibetan)
HM 5321	Shes pa (Tibetan)
HM 2655	Shes-rab (Tibetan)
HM 3597	Shibui (Japanese)
HM 3597	Shibumi (Japanese)
HM 1793	Shift ; Consciousness
HM 5329	Shift ; Resistance
MM 2015	Shifting patterns activity
HM 3202	Shifting ; Shape (Psychism)
MM 2011	Shifting topics conversation
VD 3670	Shiftlessness
HH 0429	Shila (Zen)
HM 3162	Shin-tu-sbyangs-pa (Tibetan)
HM 3063	Shining jewel awareness ; City (Yoga)
TP 0030	Shining light (Fire)
HM 1858	Shining mind (Taoism)
MS 8304	Ships ; Boats
HH 0343	Shizen (Zen)
VD 3672	Shock
HM 1962	Shock
	Shock [therapy...]
HH 0566	— therapy
HH 0975	— therapy
TP 0051	— (Thunderbolts)
HH 7324	Shojin-ryori
KC 0717	Short-term aims
VD 3674	Shortage
VP 5313	Shortcoming ; Overrunning
HH 3443	Shoshu buddhism ; Nichiren (Buddhism)
HM 2169	Shotravijnana (Buddhism)
HM 0762	Showq (Sufism)
HM 3209	Shraddha (Buddhism)
HM 2161	Shravaka (Buddhism)
HM 0762	Shugh (Sufism)
HM 2901	Shui ; Feng (Chinese)
HM 4154	Shukr (Sufism)
HH 3452	Shunamism
HH 3452	Shunamitism
HM 3398	Shunya chakra (Yoga)
HM 2193	Shunyata (Buddhism)
HM 2899	Shunyatashunyata (Buddhism)
HM 2053	Shushumna (Hinduism)
VD 3676	Sickness
HM 0569	Siddha-loka (Jainism)
HM 1285	Siddhadarsana (Hinduism)
HM 2663	Siddhi ; Shen (Buddhism)
HM 2679	Siddhis ; Eight (Buddhism)
HH 0380	Siddhis (Yoga)
HH 7004	Sidetracks auxiliary methods Taoism (Taoism)
HM 6631	Sidq (Sufism)
HM 3274	Sifat-i bashriyya (Sufism)
HM 0748	Sight ; Deva (Buddhism)
HH 1354	Sight ; Journeying transcendence blindness
HM 0748	Sight ; Knowledge divine (Buddhism)
HM 2846	Sight ; Prophetic (ICA)
HM 2532	Sign ; Representational (ICA)
HM 3071	Sign ; Revolutionary (ICA)
VC 1613	Significance
KD 2007	Significance ; Answer production accumulation
MM 2050	Significance ; Contrast
KD 2009	Significance ; Development processes accumulation
KP 3020	Significance mutually constraining forms
MS 8303	Signs human presence
HM 2875	Signs ; Vital (ICA)
HH 6199	Sikh mysticism (Sikhism)
HH 6292	Sikhism Human development
HH 6199	Sikhism Sikh mysticism
HH 3774	Sikhism Spirituality
HM 2387	Sikkha ; Adhicitta (Buddhism)
HH 0278	Sila (Buddhism)
HH 0278	Sila ; Panca (Buddhism)
HM 3603	Silence
HM 1664	Silence ; Aesthetic
VP 5451	Silence-Loudness
HH 0983	Silence ; Practice
HM 0238	Silence ; Prayer (Christianity)
HM 3977	Silence ; Saviors God: (Christianity)
HM 2227	Silence ; Voice
HH 5450	Silence ; Way
HH 3635	Silva mind control
KC 0316	Similarity
VC 1615	Similarity
VP 5020	Similarity-Dissimilarity
KC 0147	Similarity ; Dynamical
HM 2690	Simile ; Cow herding (Zen)
HM 2341	Simnami ; Stations consciousness (Sufism)
HH 0816	Simple contemplation
KC 0486	Simplicity
VC 1617	Simplicity
HM 1365	Simplicity ; Aesthetic (Japanese)
VP 5045	Simplicity-Complexity
HM 0238	Simplicity ; Prayer (Christianity)
HH 0990	Simplicity ; Virtue
HH 0190	Simplification ; Self
HH 0539	Simran ; Nam (Yoga)
KC 0246	Simulation
HH 2189	Simulation
TC 1538	Simulation exercises
KC 0926	Simulation technique ; Monte Carlo
HM 9176	Simultanagnosia
KC 0390	Simultaneity
HM 4012	Sin ; Admission
HM 2379	Sin ; Besetting (ICA)
HM 1160	Sin ; Conviction
HM 2720	Sin ; Horror (ICA)
HM 4012	Sin ; Mortal
HH 1778	Sin ; Original
HH 3567	Sin ; Original
VC 1619	Sincerity
TP 0061	Sincerity centre
HM 4663	Sincerity (Sufism)
	Sinful unwholesome dharmas
HM 0552	— Being ashamed acquisition (Buddhism)
HM 1208	— Feeling remorse acquisition (Buddhism)
HM 3257	— Not ashamed acquisition (Buddhism)
HM 3353	— Not feeling remorse acquisition (Buddhism)
HM 2765	Single desiredness (Sufism)
HM 0220	Single hearted concentration
HM 3085	Single-heartedness (Sufism)
HM 3211	Single mindedness
MP 1087	Single perspective ; Exchange contexts controlled
MP 1078	Single perspective ; Minimal context
KC 0039	Singleness
HM 2111	Singular adoration (ICA)
HM 3217	Singular mission (ICA)
TC 1815	Sink ; Conferencing energy
HM 5353	Sirr ; Cleansing (Sufism)
HM 5353	Sirr ; Emptying (Sufism)
HM 5353	Sirr ; Takhliya (Sufism)
MP 1022	Site limit ; Occupiable temporary
MP 1180	Sites exposed external insight ; Occupiable
MP 1176	Sites grounding perspectives non-linearity
MP 1231	Sites integrative superstructure ; Overview
MP 1103	Sites ; Limitation number occupiable temporary
MP 1179	Sites ; Organization structure provide occupiable
MP 1188	Sites perspective inactivity ; Occupiable
MP 1202	Sites ; Structure enfolded occupiable
HM 2991	Situation ; Every (ICA)
HM 2119	Situation ; Final (ICA)
HM 2978	Situation ; Relational (ICA)
HH 0492	Situational therapy
HH 0587	Siva
HM 8336	Six-door adverting (Buddhism)
HM 1465	Six faith ; Stage
HM 5187	Six-fold knowledge (Systematics)
HM 2077	Six-member yoga (Yoga)
HM 1914	Six paths (Buddhism)
KD 2190	Sixfold restraint ; Encompassing system dynamics:
HM 0176	Sixteen-petalled lotus (Yoga)
HM 2431	Sixth antarabhava consciousness (Buddhism)
HM 4412	Sixth chakra: time penance (Leela)
HM 2144	Sixth chakra (Yoga)
HM 0103	Sixth order perceptions - control relationships
HM 3354	Sixth plane wisdom (Hinduism)
	Sixth [scriptural...]
HM 2385	— scriptural bodhisattva awareness (Buddhism)
HM 3104	— sense (Japanese)
HM 3060	— stage life
HM 3222	— state (Yoga)
HM 2287	Sixth vajra-master awareness (Buddhism)
TC 1555	Size ; Conference
MP 1190	Size perspective contexts ; Variation
HM 2692	Skandha (Buddhism)
HM 3321	Skandha (Buddhism)
MS 8355	Skeleton ; Human
VP 5733	Skilfulness-Unskilfulness
VC 1621	Skill
HH 4777	Skill absorption (Buddhism)
HM 7602	Skill detriment ; Understanding (Buddhism)
HH 4777	Skill ecstasy (Buddhism)
HM 8290	Skill improvement ; Understanding (Buddhism)
HM 8290	Skill increase ; Understanding (Buddhism)
HM 7602	Skill loss ; Understanding (Buddhism)
HM 6601	Skill means ; Understanding (Buddhism)
HM 8290	Skill profit ; Understanding (Buddhism)
TC 1062	Skills ; Integrative
MS 8335	Skin appearance ; Alteration
HM 0889	Slaking emotional spiritual thirst (Thai)
HM 2265	Slan-te-jog-pa (Tibetan)
VD 3678	Slant
MM 2037	Slavery
HM 2980	Sleep
HH 1008	Sleep ; Abstention
HM 7921	Sleep (Buddhism)
HM 6307	Sleep ; Deep
HM 2957	Sleep ; Deep (Hinduism)
HH 0866	Sleep-electroshock therapy
HM 2853	Sleep (Hinduism)
HM 2980	Sleep ; Orthodox
HM 2980	Sleep ; Paradoxical
HM 5974	Sleep ; REM
HM 4806	Sleep spindles
	Sleep Stage
HM 6307	— four
HM 6231	— one
HM 0348	— three
HM 4806	— two
HM 1272	Sleep walking
HM 1600	Sleepiness ; Decadent (German)
TC 1824	Slide projection
HM 0992	Slight unconsciousness (Hinduism)
HM 3116	Sloth
HM 5667	Sloth (Buddhism)
HM 0491	Sloth ; Spiritual (Christianity)
VD 3680	Slothfulness
VD 3682	Slowness
VP 5269	Slowness ; Swiftness
TC 1218	Slowscan
HM 5992	Sluggish direct knowledge ; Concentration difficult progress (Buddhism)
HM 1010	Sluggish direct knowledge ; Concentration easy progress (Buddhism)
HM 1010	Sluggish intuition ; Concentration easy progress (Buddhism)
HM 5992	Sluggish intuition ; Concentration painful progress (Buddhism)
TP 0062	Small excesses
TP 0009	Small ; Nurturance
TP 0009	Small obstructions
TP 0062	Small ; Preponderance
MP 1061	Small scale interaction domains ; Bounded common
MP 1151	Small-scale perspective interaction contexts
TP 0009	Small ; Taming power
HH 0845	Small vehicle Buddhism (Buddhism)
VD 3684	Smallness
VP 5034	Smallness ; Greatness
HM 1881	Smoking ; Tobacco
HM 2847	Smriti (Buddhism)
HM 0704	Smriti ; Samyak (Buddhism)
HM 6754	Smrityupasthana (Pali)
HM 2364	Sna'i-rnam-par-shes-pa (Tibetan)
MS 8513	Snake
HM 1013	Snares ignorance (Buddhism)
HM 3743	Sniffing ; Glue
VD 3686	Snobbery
HM 5634	Snying re (Tibetan)
MM 2059	Soaring ; Aerial animal locomotion gliding
HH 0600	Sobriety
VC 1623	Sobriety
VC 1625	Sociability
VP 5922	Sociability-Unsociability
TC 1789	Social activities ; Meetings
HH 3498	Social bonding
HH 0251	Social character
HH 0046	Social development
HH 0437	Social development ; Economic
KC 0167	Social engineering
HH 2653	Social evolution
	Social [failure...]
HM 2728	— failure (ICA)
KC 0407	— field theory
KC 0974	— forecasting
KC 0915	Social group
HH 0801	Social group-work
HM 4001	Social harmony (Japanese)
MS 8353	Social hierarchical position
HH 4510	Social imagination
HH 4510	Social imaging future
	Social innovation
KC 0167	— [Social innovation]
HH 8963	— [Social innovation]
HH 0432	— [Social innovation]
	Social integration
KC 0140	— Communicative
KC 0620	— Functional
KC 0577	— International
KC 0502	— Normative
HH 1579	Social intelligence
HH 8963	Social invention
KC 0633	Social network
HH 0300	Social obligations (Japanese)
KC 0160	Social organization
HH 0511	Social phases life
KC 0172	Social policy planning
	Social [reality...]
HH 0714	— reality
HH 0542	— renewal
HH 3663	— responsibility ; Maternal conception
	Social [science...]
HH 0090	— science psychology
KC 0709	— sciences methodology
KC 0820	— system
KC 0907	— systems ; Complex
	Social [therapy...]
HH 0874	— therapy
HH 3003	— transformation music ; Psycho
KC 0167	— transmutation
HH 0874	— treatment
KC 0062	Socialism ; International
KC 0662	Socialist collective
HH 3997	Socialist human development ; State
HH 0360	Socialist humanism
KC 0804	Socialist market ; World
HM 2877	Sociality ; Primordial (ICA)
HH 0666	Socialization
VT 8084	Socialization*complex
HH 0881	Socialization ; Sexual
KD 2335	Societal evolution ; Ecodynamics
HH 9807	Societies ; Initiation secret
KC 0101	Society
KC 0506	Society ; Global
KC 0661	Society ; Mass
KD 2270	Society ; Non comprehension structuring phenomenon learning
	Society [Pattern...]
KD 2215	— Pattern accumulation learning
KC 0506	— Planetary
KC 0114	— Political organization
KC 0506	Society ; World

Spiritual

Code	Entry
HH 0046	Socio-economic development
KC 0162	Socio-Economic structure
TC 1155	Socio-structural meeting configurations
HH 0512	Sociodrama
KC 0304	Sociology knowledge
KC 0209	Sociometry
HH 0439	Socioneurosis
HH 0439	Sociopathology
HH 0439	Sociopsychosis
HH 0336	Sociotherapy
HM 0934	Sof ; En (Judaism)
VD 3688	Softness
VP 5356	Softness ; Hardness
HM 9392	Soka (Hinduism)
HM 3514	Sokushin jobutsu (Buddhism)
HM 1279	Solar plane (Leela)
VC 1627	Solemnity
VP 5870	Solemnity ; Cheerfulness
VC 1629	Solicitude
VC 1631	Solidarity
TP 0008	Solidarity
HM 2363	Solidarity ; Temporal (ICA)
VC 1633	Solidity
HH 0191	Solipsism
HM 2404	Solitary being (ICA)
HM 2709	Solitary realizers ; Meditation way (Buddhism)
VC 1635	Solitude
HH 0333	Solitude
HM 6312	Solitude (Buddhism)
HH 0333	Solitude ; Discipline (Christianity)
HM 4987	Solitude ; Enjoyment (Sufism)
HM 1539	Solitude ; Pain
HH 1875	Solitude ; Spiritual (Sufism)
HH 0481	Solving ability ; Problem
TC 1684	Solving conferencing ; Problem
HM 3574	Solving state ; Problem
KC 0614	Solving theory ; Problem
HH 0491	Soma
HH 0788	Soma
HM 0176	Soma chakra (Yoga)
HM 1649	Somanassindriya (Pali)
HH 0379	Somatherapy
HM 1766	Somatic experience
HM 7208	Somatic hallucination
HM 1272	Somnambulism
HM 1226	Somnambulistic hypnosis
HM 3087	Son ; Age (Christianity)
HH 1773	Son father ; Journeying transcendence
HM 2005	Son ; Obedient (ICA)
HM 0888	Sonomama (Japanese)
VD 3690	Sophistry
HM 1434	Sophrosune
HM 8006	Sopor
VD 3692	Sordidness
VD 3694	Soreness
HM 2685	Sorrow
VD 3696	Sorrow
HM 2216	Sorrow ; Heavenly (ICA)
HM 9392	Sorrow (Hinduism)
HM 1782	Sorrow (Leela)
HM 0399	Sorrowful rasa (Hinduism)
HM 2904	Soteriological existence (ICA)
HH 1166	Soteriology
HH 0621	Soto zazen meditation (Zen)
HH 0928	Soto Zen (Zen)
HH 0501	Soul
HM 0212	Soul androgyne
HH 1199	Soul care (Christianity)
HM 3182	Soul cognition ; Obstacles (Yoga)
HM 3941	Soul ; Dark night (Christianity)
	Soul [flight...]
HM 6120	— flight
HM 3321	— Folk
HH 0138	— force
HH 1199	— friendship (Christianity)
HM 3321	Soul ; Group
HH 4066	Soul ; Innermost benevolent desires
HM 2632	Soul-loss
HH 0210	Soul ; Loss
HH 0142	Soul-making
HH 1409	Soul ; Mansions (Christianity)
HH 1672	Soul ; Old
HM 7336	Soul ; Opposition carnal (Sufism)
HH 7343	Soul ; Perfection (Christianity)
HM 6522	Soul's ground ; Birth God (Christianity)
HM 6558	Soul's perception ; Released (Hinduism)
	Soul [Speaking...]
HM 6522	— Speaking Word (Christianity)
HM 4766	— Striving against (Islam)
HH 6973	— Striving against (Islam)
HM 3050	Soul vision (Psychism)
KC 0302	Soul ; World
HH 5119	Souling world ; Re
VD 3698	Soullessness
HH 1264	Souls ; Cure
HM 6558	Souls ; Knowledge liberated (Hinduism)
HH 8232	Sound consciousness
HH 2543	Sound health
HH 3246	Sound kasina (Buddhism)
KD 2152	Sound ; Modelling language relationships
VC 1637	Soundness
HM 1434	Soundness heart
HM 1434	Soundness mind
HM 3251	Source awareness ; Back (Zen)
MP 1181	Source direct insight structures focal point ; Maintenance
MP 1159	Sources external insight ; Organization structures permit two
	Sources perspective nourishment
MP 1089	— Local
MP 1177	— Local cultivation
MP 1093	— Transit point location
VD 3700	Sourness
HH 0845	Southern Buddhism (Buddhism)
HM 2127	Southern-continent awareness men (Buddhism)
VC 1639	Sovereignty
HM 2285	Sovereignty archetypal image ; Female (Tarot)
KC 0673	Space
	Space [absorption...]
HM 2110	— absorption ; Limitless (Buddhism)
KC 0386	— Abstract mathematical
KC 0307	— Attribute
HH 0249	Space ; Body
VT 8025	Space∗complex
TC 1451	Space design
MP 1210	Space distribution levels ; Harmonizing
HH 3246	Space kasina (Buddhism)
HH 0249	Space ; Life
TC 1304	Space ; Meeting
KC 0191	Space ; Multidimensional
MM 2085	Space ; Orbiting
	Space [Personal...]
HH 0249	— Personal
KC 0318	— Phase
HM 6287	— Plane inner (Leela)
HM 2110	Space ; Sphere infinite (Buddhism)
	Space [time...]
HM 3770	— time ; Fringe (Psychism)
KD 2240	— time ; Health
KC 0994	— Topological
MM 2087	Spacecraft ; Launching
MP 1205	Spaces defined framework spaces defined processes it ; Congruence
MP 1205	Spaces defined processes it ; Congruence spaces defined framework
MP 1203	Spaces emergent perspectives ; Exclusive
MP 1130	Spaces enhancing structural entry points ; Internal transition
MP 1206	Spaces minimal structural distinctions ; Efficient enclosure
HM 1404	Spanish Comoción
HM 1404	Spanish Crowd emotion
HM 2708	Sparsa (Buddhism)
HM 2708	Sparsha (Buddhism)
VD 3702	Sparsity
VD 3704	Spasmodicness
VD 3706	Spasticity
HM 2176	Spatial awareness
HH 1075	Spatial intelligence
KC 0093	Spatial ordering
MM 2029	Spatial orientation
TC 1972	Speaker telephone ; Loud
TC 1972	Speakerphone meeting
TC 1566	Speakers ; Guest
TC 1566	Speakers ; Key note
HM 7384	Speaking forth (Christianity)
HM 6108	Speaking tongues (Christianity)
HM 6522	Speaking Word soul (Christianity)
HM 2737	Speaks daughters Jerusalem ; Jesus (Christianity)
	Special [insight...]
HM 2623	— insight preparations states (Buddhism)
HM 2207	— insight ; States (Buddhism)
TC 1593	— interest groups
MP 1056	Special modes relationship
HM 3079	Special qualities ; Paths (Buddhism)
TC 1269	Special taskforce
MP 1020	Specialized communications ; User determined
HM 7002	Specific neutrality (Buddhism)
VD 3708	Speciousness
HM 2530	Spectrum ; Consciousness
HM 4608	Speech ; Correct (Buddhism)
HM 1821	Speech ; Correct (Buddhism)
HM 1781	Speech ; Leaving off, abstaining, totally abstaining refraining four deviations (Buddhism)
HM 1157	Speech ; Right (Buddhism)
MM 2061	Speech ; Voice
HM 0819	Spellbound (Psychism)
HM 3031	Sphere beauty (Kabbalah)
	Sphere concentration
HM 4265	— Fine material (Buddhism)
HM 0696	— Immaterial (Buddhism)
HM 1097	— Sense (Buddhism)
	Sphere [Fine...]
HM 2536	— Fine material (Buddhism)
HM 2110	— First absorption immaterial (Buddhism)
HM 2450	— First trance fine material (Buddhism)
HM 2410	— foundation (Kabbalah)
HM 2051	— Fourth absorption immaterial (Buddhism)
HM 2586	— Fourth trance fine material (Buddhism)
	Sphere functional Indeterminate
HM 4761	— consciousness fine material (Buddhism)
HM 0282	— consciousness immaterial (Buddhism)
HM 3852	— consciousness sense (Buddhism)
HM 2238	Sphere glory (Kabbalah)
	Sphere [infinite...]
HM 3043	— infinite consciousness (Buddhism)
HM 2110	— infinite space (Buddhism)
HM 3852	— inoperative ; Indeterminate consciousness sense (Buddhism)
HM 2290	Sphere judgement (Kabbalah)
HM 2288	Sphere kingdom (Kabbalah)
HM 0126	Sphere knowledge (Kabbalah)
HM 2420	Sphere mercy greatness (Kabbalah)
	Sphere [neither...]
HM 2051	— neither cognition nor non-cognition (Buddhism)
HM 2051	— neither perception nor non-perception (Buddhism)
HM 2027	— nothing (Buddhism)
	Sphere Profitable consciousness
HM 5338	— fine material (Buddhism)
HM 4701	— immaterial (Buddhism)
HM 4447	— sense (Buddhism)
	Sphere resultant Indeterminate
HM 0594	— consciousness fine material (Buddhism)
HM 4982	— consciousness immaterial (Buddhism)
HM 5721	— consciousness sense (Buddhism)
	Sphere [Second...]
HM 3043	— Second absorption immaterial (Buddhism)
HM 2038	— Second trance fine material (Buddhism)
HM 3132	— supreme crown (Kabbalah)
HM 2027	Sphere ; Third absorption immaterial (Buddhism)
HM 2062	Sphere ; Third trance fine material (Buddhism)
HM 2372	Sphere understanding (Kabbalah)
HM 8375	Sphere ; Unprofitable consciousness sense (Buddhism)
HM 2362	Sphere victory (Kabbalah)
HM 2348	Sphere wisdom (Kabbalah)
HH 1973	Spheres ; Hearing music
HM 3214	Spheres ; World transcending (Buddhism)
KC 0818	Spherical reference
HM 4806	Spindles ; Sleep
HH 0835	Spirals ; Growth
HH 0770	Spirit
HM 3499	Spirit ; Age Holy (Christianity)
HM 2907	Spirit attraction divine grace ; Liberation (Sufism)
TC 1749	Spirit ; Conference
HH 2997	Spirit healing
HM 6162	Spirit ; Illumination (Sufism)
HH 0956	Spirit ; Jubilee (Christianity)
HH 0878	Spirit medium
HM 1330	Spirit open consciousness (Taoism)
	Spirit [Poor...]
HH 0190	— Poor
HM 3594	— possession
HH 0056	— possession
HH 0190	— Poverty
HH 0816	— Prayer
HM 8004	Spirit ; Resting (Christianity)
HH 3456	Spirit transformation organizations
HM 2483	Spirit ; Ultimate transparency (Sufism)
HM 1330	Spirit ; Valley (Taoism)
VC 0901	Spiritedness ; High
VD 3710	Spiritlessness
	Spiritual [action...]
HM 2021	— action ; Path intelligence (Judaism)
HH 1890	— affections
HM 0936	— anger (Christianity)
HH 0108	— aspects psychic health
HM 2726	— aspiration (Astrology)
HH 7143	— attentiveness (Christianity)
HM 0642	— avarice (Christianity)
HH 0269	Spiritual being
	Spiritual [childhood...]
HM 2158	— childhood (Christianity)
HM 3443	— consolations ; Mansions (Christianity)
HH 7637	— control systems
HM 2037	— creativity (ICA)
	Spiritual [dance...]
HM 0303	— dance performance
HM 2174	— denial (ICA)
HH 0017	— development
HH 0855	— development (Psychism)
HM 1475	— devotion (Leela)
HH 0442	— direction
HH 0442	— direction ; Educative
HH 3900	— discernment (Christianity)
HH 1021	— discipline
HM 6122	— discipline ; Romance
HH 0707	— disciplines
	Spiritual [education...]
HH 1109	— education (Japanese)
HM 4762	— embryo ; Formation (Taoism)
HM 5265	— embryo ; Incubation (Taoism)
HM 0672	— emptiness
HM 1882	— energy (Islam)
HM 2029	— enlightenment
HM 3818	— envy (Christianity)
	Spiritual exercises
HH 0707	— [Spiritual exercises]
HH 3420	— Nikos Kazantzakis (Christianity)
HH 9760	— Saint Ignatius (Christianity)
HM 3542	Spiritual experiences (Sufism)
HH 1199	Spiritual friendship (Christianity)
	Spiritual [gluttony...]
HM 0507	— gluttony (Christianity)
HM 0642	— greed (Christianity)
HH 5009	— growth
	Spiritual guidance
HH 0878	— [Spiritual guidance]
HH 0442	— [Spiritual guidance]
HH 1199	— (Christianity)
HH 0442	Spiritual guide
HH 0458	Spiritual healing
HM 0472	Spiritual hunger
	Spiritual [identity...]
HM 3232	— identity ; Awareness (Yoga)
HM 0102	— immaturity (Christianity)
HH 3390	— initiation (Esotericism)
HM 1417	— initiation ; Group (Esotericism)
HH 0107	— integration

Spiritual

Spiritual [intelligence...] cont'd
HM 2984 — intelligence ; Awareness self (Yoga)
HM 2829 — intuitive cognition
HM 7554 Spiritual knowledge (Christianity)
Spiritual life
HH 0101 — [Spiritual life]
HH 0102 — Beginner (Christianity)
HH 3652 — Perfection (Christianity)
HH 0102 — Stages
HM 8977 Spiritual love (Christianity)
HM 4180 Spiritual lust (Christianity)
Spiritual [marriage...]
HH 0465 — marriage (Christianity)
HH 3971 — marriage ; Mansions (Christianity)
HH 3944 — master ; Identification (Sufism)
HH 2080 — master (Sufism)
HH 0828 — materialism
HH 5223 — maturity
HM 6226 — meekness (Christianity)
HH 0805 — mentor
HH 0442 — mentor
Spiritual [path...]
HH 1867 — path
HH 0738 — path (Islam)
HM 3575 — peace
HM 3974 — perception ; World (Sufism)
HH 0017 — perfection
MS 0373 — phenomena ; Human
HH 0190 — poverty
HM 2286 — poverty ; Awareness (ICA)
HH 0442 — preceptor
HM 2852 — pride (Buddhism)
HM 4533 — pride (Christianity)
Spiritual [reading...]
HH 0158 — reading
HH 2665 — realization ; Ways (Esotericism)
HM 2438 — realization (Yoga)
Spiritual rebirth
HH 4098 — (Christianity)
HH 0698 — ego
HH 3465 — Initiation (Yoga)
Spiritual [reductionism...]
HH 1981 — reductionism
HH 1676 — renaissance
HH 0576 — renewal (Christianity)
HH 3124 — renewal (Christianity)
HH 6552 — repose (Christianity)
HH 1875 — retreat (Sufism)
Spiritual self
HH 0868 — abandonment
HH 4522 — denial (Christianity)
HH 4522 — restraint (Christianity)
Spiritual [sloth...]
HM 0491 — sloth (Christianity)
HH 1875 — solitude (Sufism)
HM 5215 — struggle (Sufism)
HH 0449 — surrender
Spiritual [theology...]
HH 3652 — theology (Christianity)
HH 5217 — theology (Christianity)
HH 0458 — therapy
HM 0889 — thirst ; Slaking emotional (Thai)
HH 0089 Spiritual unfoldment
HH 0465 Spiritual union (Christianity)
HH 3521 Spiritual vision God (Sufism)
HM 0575 Spiritual vision (Sufism)
Spiritual [way...]
HH 1867 — way
HH 0631 — ways ; Three (Christianity)
HH 0892 — world order
HH 0479 Spiritualism
VC 1641 Spirituality
HH 5009 Spirituality
Spirituality [African...]
HH 5224 — African
HH 4746 — (Amerindian)
HH 4508 — Ancient
HH 4508 — Archaic
HH 3907 Spirituality (Buddhism)
HH 0792 Spirituality (Christianity)
HH 4301 Spirituality (Confucianism)
Spirituality [Eco...]
HM 0800 — Eco
HH 7234 — Egyptian
HH 5234 — (Esotericism)
HH 4003 Spirituality ; Feminine
HH 7403 Spirituality ; Formative
Spirituality [Global...]
HH 1676 — Global
HH 4865 — Greek
HM 0800 — Green
HH 3000 Spirituality (Hinduism)
HH 5902 Spirituality (Islam)
HH 8712 Spirituality (Jainism)
HH 4776 Spirituality (Judaism)
HH 6121 Spirituality ; Native
HH 0109 Spirituality ; Psychic health
HH 6291 Spirituality ; Roman
HH 3774 Spirituality (Sikhism)
HH 6210 Spirituality tantric yoga
HH 5117 Spirituality (Taoism)
HH 0746 Spirituality ; Universal (Yoga)
HM 0062 Spiritually evocative sports
HM 2778 Spite (Buddhism)
HM 2165 Splendid vices (ICA)
VC 1643 Splendour

HM 0192 Splendour wisdom (Buddhism)
HM 6138 Splitting ; Ego
VD 3712 Spoilage
VD 3714 Spoilation
Spontaneity
HM 2940 — [Spontaneity]
VC 1645 — [Spontaneity]
TP 0025 — [Spontaneity]
HH 0418 Spontaneity ; Controlled
HM 1865 Spontaneity (Taoism)
HH 0609 Spontaneity training
MM 2026 Spontaneous ceremony ; Traditional
HM 3025 Spontaneous gratitude (ICA)
HM 0332 Spontaneous out-of-body experience
MP 1068 Spontaneous relationship formation amongst emerging perspectives
HM 2624 Spontaneous religious experience
HM 0062 Sport ; altered perceptions
HM 0062 Sport-induced modes perception
HM 0062 Sport ; Mystical sensations
MS 8846 Sports contests
HH 0723 Sports psychology
HM 0062 Sports ; Spiritually evocative
VC 1647 Sportsmanship
VC 1649 Spotlessness
VD 3716 Spunklessness
VD 3718 Spurious
VD 3720 Squalor
HM 2240 Sravaka-darsana-marga (Buddhism)
HM 2716 Sravaka prayoga-marga (Buddhism)
HM 2192 Sravaka-sambhara-marga (Buddhism)
HM 2051 Srid-rtse (Tibetan)
HM 2431 Sridpa bardo (Tibetan)
KC 0917 Stability
VC 1651 Stability
Stability [Calm...]
HM 0634 — Calm (Buddhism)
VP 5141 — Changeableness
HM 6956 — Concentration partaking (Buddhism)
TC 1424 — Conference
HH 0217 Stability ego
HM 2440 Stabilization (Buddhism)
HM 0931 Stabilization ; Right meditative (Buddhism)
HM 1930 Stabilization thought (Buddhism)
HM 2968 Stabilizations ; Meditative (Buddhism)
MP 1029 Stable density gradient local relationships
HM 2533 Stable intelligence ; Path (Judaism)
MM 2083 Staff
Stage [faith...]
HM 8672 — faith ; Pre
HM 2187 — First vajra master (Buddhism)
HM 3359 — five faith
HM 1665 — four faith
HM 6307 — four sleep
HM 3518 — Fourth
HM 4902 — fright
Stage gunas
HM 3509 — Alinga (Yoga)
HM 3460 — Avivesa (Yoga)
HM 3212 — Linga (Yoga)
HM 2912 — Vivesa (Yoga)
HM 5856 Stage hypnosis ; Semi
Stage life
HH 1987 — Celibate (Hinduism)
HM 2954 — Emotional sexual
HM 3191 — Fifth
HM 3320 — First
HM 3518 — Heart awakening
HH 2343 — Householder (Hinduism)
HM 3060 — I transcendent
HM 2056 — Mendicant (Hinduism)
HM 3464 — Mental volitional
HM 3191 — Mystical
HM 2782 — Recluse (Hinduism)
HM 2954 — Second
HM 3060 — Sixth
HM 3464 — Third
HM 3320 — Vital physical
Stage one
HM 1425 — faith
HM 5974 — REM
HM 6231 — sleep
HM 3619 Stage ; Seventh
HM 1465 Stage six faith
Stage [three...]
HM 1814 — three faith
HM 0348 — three sleep
HM 1525 — two faith
HM 4806 — two sleep
HM 2392 Stages awareness ; Upanishadic (Hinduism)
HM 2097 Stages faith
HM 4000 Stages game ; Consciousness ascension (Buddhism)
HM 2805 Stages gunas (Yoga)
Stages life
HH 1366 — [Stages life]
HH 0285 — cycle ; Sequence psychological
HH 2563 — Four (Hinduism)
HH 0460 — Seven
HH 0285 Stages personality development
HH 0102 Stages spiritual life
VD 3722 Stagnation
TP 0012 Stagnation
HH 6956 Stagnation ; Concentration partaking (Buddhism)
VD 3724 Stain
VC 1653 Stainlessness
VD 3726 Stalemate

HM 2304 Stalking ; Mastery
VC 1655 Stamina
HM 1749 Stamina (Buddhism)
TC 1776 Stamina ; Conference
MP 1038 Standard frameworks
HH 0768 Standard living
KC 0223 Standardization
VC 1657 Standing
HM 1286 Standing crossroads (Buddhism)
TP 0058 Standing straight (Lake)
TP 0012 Standstill
HM 3137 Star ; Life fluid emblems (Tarot)
HH 0805 Staretz
HM 2101 Stark givenness (ICA)
MM 2088 Stars ; Pulsational variable
TC 1566 Stars ; Super
HM 1418 Stasis (Buddhism)
HM 0051 Stasis thought (Buddhism)
State [Aesthetic...]
HM 3416 — Aesthetic (Psychism)
HM 0665 — After death dream (Psychism)
HM 2959 — anger (Buddhism)
HM 0847 — animality (Buddhism)
State awareness
HM 2091 — Barbarian (Buddhism)
HM 2639 — Bon practitioner (Buddhism)
HM 2036 — Nagual (Amerindian)
HM 2305 — (Sufism)
HM 2066 — Tonal (Amerindian)
State bardo awareness
HM 2371 — Dream (Buddhism)
HM 2347 — Life (Buddhism)
HM 2311 — Meditation (Buddhism)
HM 1225 State bodhisattva (Buddhism)
HM 1873 State buddhahood (Buddhism)
State consciousness
HM 1537 — Alternate (ASC)
HM 2553 — Death altered
HM 0750 — Fifth (Psychism)
HM 0866 — Higher (Psychism)
HM 1219 — Lowered (Psychism)
HM 2402 — Objective fourth
HM 2428 — Theta clear (Scientology)
HM 2318 — Ultimate Sufi (Sufism)
HM 1999 — Usual (USC)
KC 0827 State-determined system
HM 2781 State dream consciousness ; Swapna (Yoga)
State [feeling...]
HM 1967 — feeling hatred (Buddhism)
HM 2405 — Fifth (Yoga)
HM 2020 — Fourth (Yoga)
HM 7131 — Fugue
HM 6445 State grace (Christianity)
State hell...
HM 4282 — hell (Buddhism)
HM 1723 — holiness (Christianity)
HM 7656 — hope (Christianity)
HM 0150 — hunger (Buddhism)
HM 3688 — Hypnapompic (Psychism)
HM 3511 — Hypnogogic (Psychism)
HM 1971 — Hypnoidal
State hypnosis
HM 1226 — Deep
HM 4595 — Light
HM 5856 — Medium
HM 6001 State illness
HM 3521 State ; Islamic contemplative (Sufism)
HM 2137 State ; Jivana mukta
HM 0581 State latency (Buddhism)
HM 3662 State learning (Buddhism)
HM 2639 State ; Magical shamanic (Buddhism)
HM 3947 State not being greedy (Buddhism)
HM 3183 State not being infatuated (Buddhism)
State not feeling
HM 1825 — greed (Buddhism)
HM 1597 — hatred (Buddhism)
HM 4081 — infatuation (Buddhism)
State [Passive...]
HM 8123 — Passive (Christianity)
HM 0074 — perfect love (Christianity)
HM 3427 — poverty (Sufism)
HM 3574 — Problem solving
HM 2887 — Psychotic
HM 4340 — purity (Christianity)
HM 3771 State quies (Christianity)
State [rapture...]
HM 1973 — rapture (Buddhism)
HM 0450 — realization (Buddhism)
HM 8772 — rest (Christianity)
State [securing...]
HM 0050 — securing self (Sufism)
HM 5467 — self-surrender (Christianity)
HM 3193 — Seventh (Yoga)
HM 3222 — Sixth (Yoga)
HH 3997 — socialist human development
KC 0523 — Steady
HM 2818 — Super contemplative (Yoga)
HM 3232 — Super contemplative (Yoga)
State [tranquillity...]
HM 3492 — tranquillity (Buddhism)
HM 4401 — transcendence ; Mono motivational hypnotic
HM 0074 — transformation (Christianity)
HM 2386 — Transformed (ICA)
State [unabated...]
HM 4131 — unabated desire (Buddhism)
HM 1403 — unabated endurance (Buddhism)

GENERAL INDEX TO VOLUME 2

State [unconsciousness...] cont'd
HM 2957 — unconsciousness ; Susupti (Hinduism)
HM 3261 — unfaltering exertion (Buddhism)
HM 2596 — universal cessation awareness (Buddhism)
State [Wakeful...]
HM 2020 — Wakeful hypometabolic
HM 2159 — wakefulness ; Seed (Hinduism)
HM 2141 — waking consciousness ; Jagrat (Yoga)
State [Yama...]
HM 2088 — Yama (Buddhism)
HM 0600 — Yoga dream (Buddhism)
HM 0101 — Yoga intermediary (Buddhism)
VC 1659 Statelessness
States [Altered...]
HM 4120 — Altered ego
HM 0216 — Archaic ego
HM 5354 — Ascetic (Buddhism)
States awareness
HM 2391 — Altered
HM 3744 — biofeedback training
HM 2115 — Hindu (Buddhism)
HH 0739 States ; Causative factor altered
HM 2032 States ; Conditional consciousness (Hinduism)
States consciousness
HM 2391 — Altered
HM 2298 — Discrete (Physical sciences)
HM 0935 — Higher
HM 2365 — Higher (Sufism)
HM 2133 — Hypnotic
HM 0461 — induction device ; Altered (ASCID)
HM 2938 — (Physical sciences)
HM 2177 States cyclic existence ; Consciousness (Buddhism)
States [Disciplining...]
HM 2181 — Disciplining mental abiding (Buddhism)
HH 0763 — Dissociative
HH 3534 — Divine (Buddhism)
States [form...]
HM 2693 — form concentrations ; Tibetan meditative (Buddhism)
HM 2669 — formless absorptions ; Tibetan meditative (Buddhism)
HM 2729 — Full pacification mental abiding (Buddhism)
HM 2417 States ; Least action principle conscious (Physical sciences)
States [meditation...]
HH 3198 — meditation subjects ; Immaterial (Buddhism)
HM 2145 — mental abidings ; Meditative (Buddhism)
HM 3607 — Mystical (Sufism)
HM 2391 States non-ordinary reality
HM 2753 States ; One pointedness mental abiding (Buddhism)
HM 5097 States ; Oneroid
HM 2205 States ; Pacifying mental abiding (Buddhism)
States Physical
HM 2357 — conservation conscious (Physical sciences)
HM 2381 — duality conscious (Physical sciences)
HM 2322 — relativity conscious (Physical sciences)
HM 0357 States prayer (Christianity)
HM 4311 States ; Psychic (Sufism)
HM 4111 States ; Religious (Christianity)
HM 2207 States special insight (Buddhism)
HM 2623 States ; Special insight preparations (Buddhism)
HM 2097 States ; Tarot arcana conscious (Tarot)
HM 2660 States understanding
KC 0980 States ; Unitary nation
HM 2465 Station fear
HH 5543 Stationariness ; Preparation (Buddhism)
HM 2880 Stations awareness ; Christian (Christianity)
Stations consciousness
HM 2317 — Ansari (Sufism)
HM 2424 — ibn-Abi'l-Khayr (Sufism)
HM 2341 — Simnani (Sufism)
HM 2880 Stations cross (Christianity)
HM 3516 Stations cross (Christianity)
HM 6754 Stations mindfulness ; Four (Pali)
HM 3415 Stations ; Mystic (Sufism)
MS 8825 Statues
Status
VC 1663 — [Status]
MS 8353 — [Status]
HH 0375 — [Status]
VC 1665 Staunchness
MM 2083 Stave
HH 0729 Staying power
VC 1667 Steadfastness
HM 3471 Steadfastness thought (Buddhism)
VC 1669 Steadiness
HM 6652 Steadiness consciousness (Buddhism)
KC 0523 Steady state
VP 5435 Stench ; Fragrance
HM 3570 Step: earth ; Fourth (Christianity)
HM 3748 Step: ego ; First (Christianity)
HM 3501 Step: mankind ; Third (Christianity)
HM 2953 Step: race ; Second (Christianity)
TP 0010 Stepping carefully (Treading)
HH 3220 Steps cosmic consciousness
HH 4019 Steps enlightenment (Buddhism)
HM 3592 Steps ignorance ; Descending (Hinduism)
KC 0071 Stereotype
KC 0071 Stereotypic patterns
VD 3728 Sterility
VC 1671 Stewardship
HH 3121 Stewardship ; Christian (Christianity)
HM 3143 Stewardship ; Good (ICA)
MM 2083 Stick
MM 2052 Stick carrot processes
HM 3942 Sticking strongly (Buddhism)
TC 1800 Sticky board

VD 3730 Stiffness
HM 5667 Stiffness (Buddhism)
HM 3014 Stiffness ; Non (Buddhism)
TP 0052 Still ; Keeping (Mountain)
VC 1673 Stillness
HM 1709 Stillness (Japanese)
VC 1675 Stimulation
HH 0995 Stimulation brain ; Electrical
VC 1677 Stintlessness
HH 0489 Stopping world
HM 2730 Stored consciousness (Buddhism)
HM 2730 Storehouse ; Consciousness (Buddhism)
MS 8235 Stories fairy-tales
HM 2757 Story ; Classical (ICA)
TP 0058 Straight ; Standing (Lake)
VC 1679 Straightforwardness
HM 1338 Straightness aggregate consciousness (Buddhism)
HM 0409 Straightness aggregate sensation, aggregate cognition, aggregate synergies (Buddhism)
HM 1424 Straightness body (Buddhism)
HM 0253 Straightness thought (Buddhism)
VD 3732 Strain
VD 3734 Strangeness
MM 2056 Strategic configurations
KC 0306 Strategic planning
TC 1828 Strategic preferences ; Participant
KC 0306 Strategy
HH 0247 Strategy ; Human development (United Nations)
TC 1955 Stratification ; Conference
HM 3017 Straws ; Grasping (Brainwashing)
HH 1187 Stream enterer ; Lesser (Buddhism)
HM 1088 Stream entry ; Knowledge path (Buddhism)
HM 2305 Stream ; Observation one's psychic (Sufism)
HH 1187 Stream winner ; Junior (Buddhism)
VC 1681 Strength
HH 0797 Strength character
HH 0217 Strength ; Ego
HM 1907 Strength ; Ego
HM 1907 Strength ; Identity
HM 2684 Strength ; Outside (Japanese)
HM 3215 Strength ; Own (Zen)
HH 0198 Strength personality
HM 3127 Strength (Tarot)
VP 5159 Strength-Weakness
HM 2585 Stress
KC 0417 Stress ; System
VC 1683 Strictness
VD 3736 Stridency
TP 0006 Strife
HM 3166 Stripped his garments ; Jesus (Christianity)
TP 0023 Stripping away Destruction
TP 0023 Stripping away Deterioration
TP 0023 Stripping away Intrigue
HM 1436 Strive (Buddhism)
HM 4766 Striving against soul (Islam)
HH 6973 Striving against soul (Islam)
TP 0001 Strong action (Heaven)
HM 1081 Strong grip burden (Buddhism)
TP 0026 Strong ; Restraint
HM 3942 Strongly ; Sticking (Buddhism)
TC 1989 Strongwilled conferencing
MP 1165 Structural activities communication pathway ; Exposure
HH 4009 Structural adjustment human face (United Nations)
HM 2178 Structural change
Structural [development...]
MP 1104 — development designed counteract deficiencies pattern harmony
TC 1547 — dimensions conferencing
MP 1206 — distinctions ; Efficient enclosure spaces minimal
MP 1130 Structural entry points ; Internal transition spaces enhancing
HH 0362 Structural family therapy
KC 0365 Structural-functional analysis
KC 0545 Structural hermeneutics
HH 0082 Structural integration
MP 1140 Structural interface communication pathway ; Relative isolation
HM 0776 Structural knowledge (Systematics)
Structural levels
MP 1158 — External access higher
MP 1195 — Framework transition
MP 1096 — Limitation number
MP 1021 Structural limit ; Four level
TC 1155 Structural meeting configurations ; Socio
KC 0093 Structural order
KC 0560 Structuralism
KC 0266 Structure
MS 8700 Structure ; Body, form
MP 1129 Structure ; Common domain focal point
MP 1112 Structure communication pathway ; Transition domain
VT 8033 Structure∗complex
TC 1717 Structure ; Conference
Structure [Data...]
KC 0118 — Data
KC 0361 — Deep
MP 1110 — Distinctiveness main entry point
Structure [enfolded...]
MP 1161 — enfolded insight domain
MP 1202 — enfolded occupiable sites
MP 1109 — enhance autonomy sub structures ; Organization
HM 0776 — Experience (Systematics)
MP 1166 — external environment ; Overview domains interfaces
MP 1099 Structure ; Focal centre complex
MP 1111 Structure informal context ; Blended integration formal
MP 1133 Structure ; Integrating transition pathways levels
TC 1860 Structure-oriented participant type

Submission

MP 1164 Structure ; Overview communication pathway
MP 1179 Structure provide occupiable sites ; Organization
Structure [Sequence...]
MP 1142 — Sequence viewpoint loci
KC 0162 — Socio Economic
KC 0507 — Surface
KC 0808 Structure ; Tree
MP 1113 Structured entry point external communication media ; Harmoniously
TC 1538 Structured experiences
HH 3287 Structured meditation
MP 1081 Structured perspective control operations ; Minimally
HH 0685 Structured therapy ; Assertion
MP 1153 Structures adaptable changing number embodied perspectives
MP 1171 Structures ; Appropriate relationship non linearity
MP 1122 Structures communication pathways ; Direct relationship
MP 1102 Structures ; Distinct pattern entry points complex
Structures [engender...]
MP 1100 — engender fruitful interfaces ; Arrangement
MS 8815 — engineering works
MP 1107 — enhance receptivity external insight ; Organization
MP 1105 — enhance receptivity external insight ; Orientation
MP 1160 — external environment ; Hospitable interface
MP 1181 Structures focal point ; Maintenance source direct insight
MP 1106 Structures ; Functional enhancement domains separating complementary
MP 1168 Structures ; Grounded
MP 1163 Structures ; Hospitable non linear domain external
MP 1149 Structures ; Hospitable reception external perspectives
Structures [Insight...]
MP 1175 — Insight capturing non linear extensions
MP 1144 — Integrating coordinated exposure irrationality
MP 1108 — Interconnected
MS 8600 Structures ; Man made forms
KC 0218 Structures ; Network data
KD 2080 Structures ; Non equilibrium
KD 2081 Structures ; Order fluctuation: dissipative
MP 1109 Structures ; Organization structure enhance autonomy sub
MP 1098 Structures ; Patterning complex
MP 1159 Structures permit two sources external insight ; Organization
MS 8825 Structures ; Symbolic
KD 2270 Structuring phenomenon learning society ; Non comprehension
HH 1281 Struggle
HM 2718 Struggle confess (ICA)
HM 5215 Struggle ; Spiritual (Sufism)
VD 3738 Stubbornness
KC 0074 Studies ; Area
KC 0647 Studies ; Comparative
HH 0873 Studies ; Tibetan (Buddhism)
KC 0411 Study ; Cross cultural
HH 1323 Study ; Discipline (Christianity)
TC 1375 Study group
HH 0957 Study sacred scriptures
HH 1239 Studying sutras
VD 3740 Stuffiness
TP 0039 Stumbling
HM 1346 Stupefaction (Yoga)
VD 3742 Stupidity
HM 1857 Stupidity (Buddhism)
VD 3744 Stupor
HM 2473 Stupor ; Unconscious
VC 1685 Sturdiness
HM 2926 Styana (Buddhism)
Style
KC 0443 — [Style]
VC 1687 — [Style]
TP 0022 — [Style]
KC 0204 Style ; Life
HH 0941 Style ; Life
KD 2170 Styles ; Modes managing: embracing conflicting
KC 0145 Stylistic cultural integration
HM 1355 Su-sang-loka (Leela)
MP 1013 Sub-domain boundary
MS 8435 Sub-lunar nature
HM 0463 Sub-personalities
MP 1109 Sub structures ; Organization structure enhance autonomy
HM 2094 Subconsciousness
HM 6221 Subconsciousness (Buddhism)
HM 4194 Subject cankers ; Understanding (Buddhism)
HH 4188 Subject ; Journeying transcendence object
VD 3746 Subjection
HM 3587 Subjection ; Self
HM 2026 Subjective domains consciousness (Yoga)
HH 0428 Subjective redemption
HH 3534 Subjects ; Divine abidings meditation (Buddhism)
HH 3198 Subjects ; Immaterial states meditation (Buddhism)
HH 3987 Subjects meditation (Buddhism)
HH 6221 Subjects ; Recollections meditation (Buddhism)
VD 3748 Subjugation
TP 0044 Subjugation
HH 0713 Subjugation nafs (Sufism)
HH 0464 Sublimation
HM 4036 Sublime concentration (Buddhism)
HM 3296 Sublime object ; Understanding having (Buddhism)
HM 3331 Subliminal perception
VC 1689 Submission
HH 0225 Submission ; Discipline (Christianity)
HM 2706 Submission ; Levitational (ICA)
TP 0057 Submission ; Willing (Wind)

-917-

Submissive

HM 2258	Submissive obedience (ICA)	
VD 3750	Subordination	
HM 2395	Subramania (Hinduism)	
HM 3508	Subsequent cognition (Buddhism)	
VC 1691	Subservience	
HM 1991	Subsistence ; Experience (Systematics)	
HM 0330	Subsistence God (Sufism)	
KC 0573	Substance	
MM 2004	Substances	
VC 1693	Substantiality	
VP 5003	Substantiality-Unsubstantiality	
HM 3438	Substantiation heart ; Trans (Sufism)	
MP 1197	Substantive distinctions separating domains	
KC 0227	Subsystem	
KC 0997	Subsystem global system ; Regional	
VD 3752	Subterfuge	

Subtle contemplation

HM 2719	— Belief arising (Buddhism)
HM 2195	— (Buddhism)
HM 2659	— Critical analytical (Buddhism)
HM 2683	— Final training (Buddhism)
HM 2743	— Full isolation (Buddhism)
HM 2171	— Only beginner (Buddhism)
HM 2267	— withdrawal joy (Buddhism)
HH 6282	Subtle faculties (Sufism)
TP 0009	Subtle restraint
VC 1695	Subtlety
HM 3312	Subtlety mind (Hinduism)
HM 5265	Subtlety nondoing (Taoism)
HH 0086	Subud
HM 1789	Subuddhi (Leela)
VD 3754	Subversion
VC 1697	Success
HH 3188	Success
HM 2941	Success ; Beyond (ICA)
HH 0456	Success motivation
HM 2771	Succour ; Imploring (ICA)
HM 0808	Succumbing hesitation (Buddhism)
HM 0888	Suchness
HM 2435	Suchness (Buddhism)
HM 2500	Suchness (Zen)
VD 3756	Suddenness
HM 2168	Suddhavasa (Buddhism)
HM 2981	Sudharma (Leela)
HM 2909	Sudurjaya (Buddhism)
VD 3758	Sufferance
HM 0471	Suffering
HM 2574	Suffering (Buddhism)
HH 0523	Suffering (Buddhism)
HH 2119	Suffering ; Cessation (Buddhism)
HH 2207	Suffering (Christianity)
HM 3556	Suffering ; Faculty (Buddhism)
HM 1643	Suffering nor pleasure born contact psychical ; Sensation neither (Buddhism)
HM 1352	Suffering nor pleasure experienced born contact psychical ; Neither (Buddhism)
HM 3126	Suffering servant (ICA)
HH 0708	Suffering ; Tragic

Suffering Understanding knowledge

HM 6870	— (Buddhism)
HM 8163	— cessation (Buddhism)
HM 7290	— origin (Buddhism)
HM 7645	— way leading cessation (Buddhism)
VC 1699	Sufficiency
VC 1585	Sufficiency ; Self
VD 3760	Suffocation
HM 2800	Sufi contracted-consciousness (Sufism)
HM 2476	Sufi infused awareness (Sufism)
HH 1747	Sufi path perfection (Sufism)
HM 2318	Sufi state consciousness ; Ultimate (Sufism)
HM 0815	Sufism 'Abid
HM 0330	Sufism Abiding
HM 3974	Sufism Abraham one's being
HM 4450	Sufism Abstinence
HM 0286	Sufism Abstinence wrongdoing
HM 4190	Sufism Acquiescence
HM 1905	Sufism Active remembrance God
HM 4542	Sufism Ad-Darr
HM 3425	Sufism Adam one's being
HM 7981	Sufism Agreement
HM 2365	Sufism Ahwal
HM 3371	Sufism Ahwal
HM 3807	Sufism Al-'Adl
HM 5543	Sufism Al-'Afuw
HM 6117	Sufism Al-Ahad
HM 7112	Sufism Al-Akhir
HM 4162	Sufism Al-'Ali
HM 4010	Sufism Al-'Alim
HM 6139	Sufism Al-Awwal
HM 5198	Sufism Al-'Azim
HM 4562	Sufism Al-'Aziz
HM 4017	Sufism Al-Ba'ith
HM 4299	Sufism Al-Badi
HM 6786	Sufism Al-Baqi
HM 3636	Sufism Al-Bari'
HM 7022	Sufism Al-Barr
HM 6512	Sufism Al-Basir
HM 0115	Sufism Al-Basit
HM 1209	Sufism Al-Batin
HM 0002	Sufism Al-Fattah
HM 1257	Sufism Al-Ghaffar
HM 6255	Sufism Al-Ghafur
HM 3822	Sufism Al-Ghani
HM 1287	Sufism Al-Hadi
HM 7099	Sufism Al-Hafiz
HM 5014	Sufism Al-Hakam
HM 4799	Sufism Al-Hakim
HM 4519	Sufism Al-Halim
HM 0392	Sufism Al-Hamid
HM 3945	Sufism Al-Haqq
HM 0741	Sufism Al-haqq
HM 1467	Sufism Al-Hasib
HM 4236	Sufism Al-Hayy
HM 2920	Sufism Al-insan al-kamil
HM 0215	Sufism Al-Jabbar
HM 6771	Sufism Al-Jalil
HM 0009	Sufism Al-Jame'
HM 0009	Sufism Al-Jami'
HM 4437	Sufism Al-Kabir
HM 5289	Sufism Al-Karim
HM 4400	Sufism Al-Khabir
HM 3644	Sufism Al-Khafid
HM 0034	Sufism Al-Khaliq
HM 1626	Sufism Al-khatim
HM 0308	Sufism Al-Latif
HM 4526	Sufism Al-Majid
HM 5446	Sufism Al-Majid
HM 4061	Sufism Al-Malik
HM 7220	Sufism Al-Mani
HM 5633	Sufism Al-Matin
HM 4052	Sufism Al-Mu'akhkhir
HM 4093	Sufism Al-Mu'id
HM 1116	Sufism Al-Mu'izz
HM 1245	Sufism Al-Mu'min
HM 0622	Sufism Al-Mubdi
HM 4510	Sufism Al-Mudhill
HM 3633	Sufism Al-Mughni
HM 0526	Sufism Al-Muhaymin
HM 4544	Sufism Al-Muhsi
HM 1007	Sufism Al-Muhyi
HM 0504	Sufism Al-Mujib
HM 3868	Sufism Al-Mumit
HM 4513	Sufism Al-Muntaqim
HM 5331	Sufism Al-Muqaddim
HM 3171	Sufism Al-Muqit
HM 8001	Sufism Al-Muqsit
HM 3576	Sufism Al-Muqtadir
HM 5786	Sufism Al-Musawwir
HM 8019	Sufism Al-Muta'ali
HM 3850	Sufism Al-Mutakabbir
HM 4510	Sufism Al-Muzill
HM 2767	Sufism Al-Qabid
HM 1500	Sufism Al-Qadir
HM 1318	Sufism Al-Qahhar
HM 4988	Sufism Al-Qawi
HM 5006	Sufism Al-Qayyum
HM 0235	Sufism Al-Quddus
HM 4040	Sufism Al-Rafi
HM 4043	Sufism Al-Rahim
HM 6098	Sufism Al-Raqib
HM 7021	Sufism Al-Razzaq
HM 4495	Sufism Al-Sabur
HM 0419	Sufism Al-Sami'
HM 4334	Sufism Al-Tawwab
HM 7712	Sufism Al-Wadud
HM 0069	Sufism Al-Wahhab
HM 5506	Sufism Al-Wahid
HM 0743	Sufism Al-Wajid
HM 1382	Sufism Al-Wakil
HM 0525	Sufism Al-Wali
HM 1633	Sufism Al-Wali
HM 0113	Sufism Al-Warith
HM 4169	Sufism Al-Wasi'
HH 6282	Sufism Alam-i-amr
HH 6282	Sufism Alam-i-khalq
HM 2792	Sufism Alertness heart
HM 0476	Sufism All-pervading knowledge
HM 4500	Sufism Allah
HM 6762	Sufism Aloneness
HM 4449	Sufism An
HM 6350	Sufism An-Nafi'
HM 1354	Sufism An-Nur
HM 7626	Sufism Annihilation
HM 1270	Sufism Annihilation
HM 7622	Sufism Annihilation heart
HM 3358	Sufism Annihilation personal sensation
HM 0206	Sufism Ar-Ra'uf
HM 3539	Sufism Ar-Rahman
HM 5234	Sufism Ar-Rashid
HM 1045	Sufism 'Arif
HM 1221	Sufism As-Salam
HM 4980	Sufism As-Samad
HM 2327	Sufism Ascension awareness
HM 5563	Sufism Ascertaining truth
HM 4450	Sufism Asceticism
HM 0375	Sufism Ash-Shahid
HM 1934	Sufism Ash-Shakur
HM 5563	Sufism Assertion truth
HM 4450	Sufism Austerity
HM 6690	Sufism Austerity
HM 4426	Sufism Awareness certitude
HM 2416	Sufism Awareness divine presences
HM 4042	Sufism Az-Zahir
HM 0330	Sufism Baqa
HM 4594	Sufism Baqi
HM 2106	Sufism Bast
HM 2681	Sufism Beyond individual distinctions
HM 4280	Sufism Blame
HM 7621	Sufism Catharsis
HM 0930	Sufism Certain knowledge God
HM 4556	Sufism Certain truth
HM 4426	Sufism Certainty
HM 3759	Sufism Character
HM 6932	Sufism Cleansing heart
HM 5353	Sufism Cleansing sirr
HM 2377	Sufism Closeness God
HM 4648	Sufism Conduct
HM 0807	Sufism Confidence
HM 7622	Sufism Conscious heart
HM 2305	Sufism Constant attention
HM 3521	Sufism Contemplation unitary experience
HM 2254	Sufism Contemplative knowledge
HM 4190	Sufism Contentment
HM 7024	Sufism Contentment
HM 4788	Sufism Continuous all-pervading knowledge
HM 3484	Sufism Conversion
HM 2916	Sufism Conviction
HM 4449	Sufism Creative moment
HM 1655	Sufism Creative moment
HM 4570	Sufism David one's being
HM 0946	Sufism Deification
HM 3427	Sufism Dervish
HM 0575	Sufism Desirelessness
HM 4450	Sufism Detachment
HM 7621	Sufism Detachment
HM 8776	Sufism Devotion God
HM 2351	Sufism Dhikr
HM 2792	Sufism Dhikr-i-Qalbi
HM 6719	Sufism Dhul-Jalal-wal-Ikram
HM 6719	Sufism Dhul-Jalali wal-Ikram
HM 8901	Sufism Discipleship
HM 1026	Sufism Divine attraction
HH 2561	Sufism Divine attributes
HM 3513	Sufism Divine essence
HM 2401	Sufism Divine-love awareness
HH 2561	Sufism Divine names
HM 2246	Sufism Divine nature
HM 3371	Sufism Divine path
HM 3451	Sufism Doors
HM 6097	Sufism Drunkenness
HM 3650	Sufism Ecstasy
HM 6099	Sufism Effort
HM 2318	Sufism Eighth stage progress
HM 5353	Sufism Emptying sirr
HM 6099	Sufism Endeavour
HM 4987	Sufism Enjoyment solitude
HH 3945	Sufism Enneagram patterning
HH 3945	Sufism Enneagram personality types
HH 0436	Sufism Erfan
HM 3457	Sufism Existence ontological reality
HM 7098	Sufism Expansion
HM 1655	Sufism Experience An
HM 4349	Sufism Experience mystical knowledge
HM 1655	Sufism Experience original moments
HM 3241	Sufism Fana
HM 3799	Sufism Fana
HM 1270	Sufism Fana
HM 7622	Sufism Fana-i-qalb
HM 4064	Sufism Fani
HM 3427	Sufism Faqir
HM 3427	Sufism Faqr
HM 0568	Sufism Fayd
HM 1047	Sufism Fear God
HM 4112	Sufism Firaq
HM 3614	Sufism First individuation essence
HM 7383	Sufism Forgetfulness God
HM 3039	Sufism Fountains light
HM 3274	Sufism Freedom lower qualities
HM 3012	Sufism Futuwwa
HM 3226	Sufism Gateway
HM 2318	Sufism Gathering-separation
HM 7383	Sufism Ghaflah
HM 2243	Sufism Ghaiba
HM 2254	Sufism Gnosis
HM 8667	Sufism Gnosis kardias
HM 1045	Sufism Gnostic
HM 2377	Sufism God-proximity
HM 4154	Sufism Gratitude
HM 2416	Sufism Hadarat
HM 2416	Sufism Hadrat
HM 2365	Sufism Hal
HM 0124	Sufism Haqiqa
HM 4556	Sufism Haqq al-yaqin
HM 2365	Sufism Higher states consciousness
HM 8667	Sufism Himma
HM 4477	Sufism Hope getting nearer
HM 4393	Sufism Hope God
HH 0436	Sufism Human development
HM 0128	Sufism 'Ibadat
HM 3520	Sufism Identification God
HM 0946	Sufism Identification God
HM 3944	Sufism Identification spiritual master
HM 2985	Sufism Identifying reality
HM 3260	Sufism Identity attributes their essence
HM 2214	Sufism Identity God
HM 4663	Sufism Ikhlas
HM 2476	Sufism Ilham
HM 6162	Sufism Illumination spirit
HM 3039	Sufism illuminations Jami
HM 0476	Sufism 'Ilm al-basir
HM 0930	Sufism 'Ilm al-yaqin
HM 3484	Sufism Inabat
HM 7098	Sufism Inbisat
HM 2113	Sufism Indivisibility essence
HM 2920	Sufism Insanu'l-kamil
HM 3677	Sufism Insanu'l-kamil
HM 0762	Sufism Intense longing divine unity
HM 4901	Sufism Intention

GENERAL INDEX TO VOLUME 2   Suppression

HM 1957	Sufism Intimacy God
HM 3605	Sufism Inward experience real
HM 3602	Sufism Inward purity
HM 8901	Sufism Iradat
HM 4054	Sufism 'Ishq
HM 3521	Sufism Islamic contemplative state
HM 6762	Sufism Isolation
HM 3520	Sufism Itasal
HM 3520	Sufism Itihad
HM 0050	Sufism Itmianan
HM 4392	Sufism Jabarut
HM 6099	Sufism Jahd
HM 3520	Sufism Jam
HM 2318	Sufism Jam-tafriqah
HM 2246	Sufism Jesus one's being
HM 2724	Sufism Joyful satisfaction
HM 2800	Sufism Kabd
HM 1452	Sufism Kashf
HM 1026	Sufism Kedesh-jazba
HM 0335	Sufism Khalifa
HH 1875	Sufism Khalwat
HM 1047	Sufism Khawf
HM 8776	Sufism Khidmat
HM 2351	Sufism Khikr
HM 4987	Sufism Khilwat dar anjuman
HM 2351	Sufism Kikr
HM 1045	Sufism Knower
HM 2254	Sufism Knowledge God
HM 3335	Sufism Kushesh
HM 0741	Sufism Lahut
HM 1905	Sufism Latifa-i-qalb
HH 6282	Sufism Latifas
HM 3244	Sufism Liberation heart
HM 2907	Sufism Liberation spirit attraction divine grace
HM 0762	Sufism Longing
HM 4273	Sufism Love God
HM 4392	Sufism Ma'arifa
HM 2254	Sufism Ma'rifa
HM 2254	Sufism Ma'rifat
HM 2401	Sufism Mahabba
HM 4273	Sufism Mahabbat
HM 4280	Sufism Malamat
HM 0116	Sufism Malaqut
HM 5902	Sufism Malik-ul-Mulk
HM 2642	Sufism Manifestation essence veils
HM 3371	Sufism Maqamat
HM 3415	Sufism Maqamat
HM 4311	Sufism Maqamat
HM 2424	Sufism Maqamat forty
HM 2317	Sufism Maqamat hundred
HM 2341	Sufism Maqamat seven-worlds
HM 6097	Sufism Masti
HM 6227	Sufism Meditation God
HM 1266	Sufism Mehr
HM 2327	Sufism Miradj
HM 3513	Sufism Mohammad one's being
HM 5215	Sufism Mortification self
HM 3616	Sufism Moses one's being
HH 2561	Sufism Most beautiful names
HH 0464	Sufism Mujahada
HM 5215	Sufism Mujahadat
HM 7336	Sufism Mukhalafat-i nafs
HM 4173	Sufism Mumin
HM 2305	Sufism Muraqaba
HM 2305	Sufism Muraqabat
HM 3887	Sufism Murid
HM 3521	Sufism Mushahada
HM 3521	Sufism Mushahadat
HM 3521	Sufism Mushahida
HM 7981	Sufism Muwafaqat
HM 3415	Sufism Mystic stations
HM 2867	Sufism Mystical detachment
HM 3521	Sufism Mystical ecstasy
HM 0530	Sufism Mystical experience
HM 3607	Sufism Mystical states
HM 0606	Sufism Nabi
HH 2561	Sufism Names God
HH 0277	Sufism Naqsbandi tradition
HM 3579	Sufism Nasut
HM 2377	Sufism Nearness-awareness
HM 3377	Sufism Nihayat
HH 2561	Sufism Ninety-nine names Allah
HM 7626	Sufism Nisti
HM 4901	Sufism Niyyat
HM 3286	Sufism Noah one's being
HM 5118	Sufism Numerical awareness
HM 2305	Sufism Observation one's psychic stream
HM 4477	Sufism Omid
HM 3303	Sufism Ontological reality truth
HM 7336	Sufism Opposition carnal soul
HM 0286	Sufism Parhiz
HM 3799	Sufism Passing away
HM 5775	Sufism Passing away
HM 1270	Sufism Passing away
HM 4356	Sufism Path sainthood
HM 3418	Sufism Patience
HM 3418	Sufism Patient endurance
HM 2531	Sufism Perception hidden reality
HM 2827	Sufism Perception universality
HM 2920	Sufism Perfect man consciousness
HM 3944	Sufism Personification
HM 5339	Sufism Piety
HH 2080	Sufism Pir
HM 6562	Sufism Practice dhikr
HM 3496	Sufism Presence Allah
HM 3069	Sufism Principles
HM 4311	Sufism Psychic states
HM 3542	Sufism Psychological assessment mystic experience
HH 5092	Sufism Purification self
HM 0575	Sufism Purity
HM 7024	Sufism Qana'a
HM 2377	Sufism Qurb
HM 2377	Sufism Qurbat
HM 4393	Sufism Raja
HM 7222	Sufism Rasti
HM 0724	Sufism Rasul
HM 3755	Sufism Realities
HM 3228	Sufism Reciprocal relation absolute conditional
HM 2351	Sufism Recollection
HM 7222	Sufism Rectitude
HM 6562	Sufism Remembrance God
HM 3892	Sufism Remembrance reality
HM 4450	Sufism Renunciation
HM 4062	Sufism Repentance
HM 5290	Sufism Resignation
HM 7098	Sufism Restraint
HM 2724	Sufism Reza
HM 3335	Sufism Riaza
HM 4190	Sufism Rida
HM 6690	Sufism Riyada
HM 3418	Sufism Sabr
HM 3602	Sufism Safa
HM 3647	Sufism Sainthood
HM 3647	Sufism Saintship
HM 7089	Sufism Saintship angels
HM 4288	Sufism Salik
HM 4041	Sufism Sanctity
HM 4190	Sufism Satisfaction
HM 0930	Sufism Science certainty
HM 3604	Sufism Self-cognition
HH 0464	Sufism Self-mortification
HM 0807	Sufism Self-surrender
HM 4112	Sufism Separation
HM 2906	Sufism Separation transient
HM 8776	Sufism Service
HM 0762	Sufism Shawq
HM 2080	Sufism Shaykh
HM 0762	Sufism Showq
HM 0762	Sufism Shugh
HM 4154	Sufism Shukr
HM 6631	Sufism Sidq
HM 3274	Sufism Sifat-i bashriyya
HM 4663	Sufism Sincerity
HM 2765	Sufism Single desiredness
HM 3085	Sufism Single-heartedness
HM 3542	Sufism Spiritual experiences
HH 2080	Sufism Spiritual master
HH 1875	Sufism Spiritual retreat
HH 1875	Sufism Spiritual solitude
HM 5215	Sufism Spiritual struggle
HM 0575	Sufism Spiritual vision
HM 3521	Sufism Spiritual vision God
HM 2305	Sufism State-awareness
HM 3427	Sufism State poverty
HM 0050	Sufism State securing self
HM 2317	Sufism Stations consciousness - Ansari
HM 2424	Sufism Stations consciousness - ibn-Abi'l-Khayr
HM 2341	Sufism Stations consciousness - Simnani
HH 0464	Sufism Subjugation nafs
HM 0330	Sufism Subsistence God
HH 6282	Sufism Subtle faculties
HM 2800	Sufism Sufi contracted-consciousness
HM 2476	Sufism Sufi infused awareness
HH 1747	Sufism Sufi path perfection
HM 0575	Sufism Sufism
HM 0575	Sufism (Sufism)
HM 8330	Sufism Sultan-al-dhikr
HM 2900	Sufism Suluk
HM 4356	Sufism Suluk Naqshbandiyya order
HM 7089	Sufism Superior saintship
HM 3377	Sufism Supreme goal
HM 5290	Sufism Surrender
HM 3436	Sufism Surrender bliss mystical seduction
HH 5092	Sufism Tadhkiya-i nafs
HM 6227	Sufism Tafakkur
HM 6762	Sufism Tatrid
HM 5563	Sufism Tahqiq
HM 6162	Sufism Tajliya-i ruh
HM 7621	Sufism Tajrid
HM 5353	Sufism Takhliya-i sirr
HM 4464	Sufism Talib
HM 3371	Sufism Tariqa
HM 4450	Sufism Tasawwuf
HM 0575	Sufism Tasawwuf
HH 0436	Sufism Tasawwuf
HM 6932	Sufism Tasfiya-i qalb
HM 4190	Sufism Taslim
HM 5290	Sufism Taslim
HM 4062	Sufism Tauba
HM 0807	Sufism Tawakkul
HM 0807	Sufism Tawakul
HM 4062	Sufism Tawba
HM 4062	Sufism Tawbat
HM 3438	Sufism Tawhid
HM 5775	Sufism Total annihilation
HM 3438	Sufism Trans-substantiation heart
HM 0807	Sufism Trust God
HM 0124	Sufism Truth
HM 4556	Sufism Truth certainty
HM 6631	Sufism Truthfulness
HM 4062	Sufism Tuba
HM 4062	Sufism Turning God
HM 2318	Sufism Ultimate Sufi state consciousness
HM 2483	Sufism Ultimate transparency spirit
HM 3316	Sufism Unchanging reality being
HM 1266	Sufism Unconditional love
HM 3864	Sufism Unification Allah
HM 3438	Sufism Unification God
HM 3864	Sufism Unification God
HM 3340	Sufism Unification life essence
HM 7119	Sufism Union God
HM 1957	Sufism Uns
HM 1452	Sufism Unveiling divine mysteries
HM 2874	Sufism Valleys
HM 6631	Sufism Veracity
HM 3650	Sufism Wajd
HM 4418	Sufism Wali
HM 5339	Sufism War'a
HM 0286	Sufism Wara
HM 0530	Sufism Warid
HM 2305	Sufism Watching
HM 3647	Sufism Wilayat
HM 7089	Sufism Wilayat-i-mala'ika
HM 7089	Sufism Wilayat-i-'ulya
HM 3335	Sufism Will behavioural transformation
HM 7119	Sufism Wisal
HM 8667	Sufism Wisdom heart
HM 2243	Sufism Withdrawal
HM 4570	Sufism World beyond form
HM 3286	Sufism World forms
HM 3616	Sufism World imagination
HM 3425	Sufism World nature
HM 3974	Sufism World spiritual perception
HM 0128	Sufism Worship God
HM 5118	Sufism Wuiquf al-'adudi
HM 2792	Sufism Wuquf-i qalbi
HM 4426	Sufism Yaqin
HM 4426	Sufism Yaquin
HM 4208	Sufism Zahid
HM 2351	Sufism Zikr
HM 4450	Sufism Zuhd
HH 0962	Suggestion ; Hypnotic
HH 0796	Suggestion therapy
HH 4511	Suggestology
HH 2378	Suggestopedia
VC 1701	Suitability
HM 5099	Sukh (Leela)
HM 2866	Sukha (Buddhism)
HM 1862	Sukham ; Cetasikam (Pali)
HM 2167	Sukhavati (Buddhism)
HM 8330	Sultan-al-dhikr (Sufism)
HM 4356	Suluk Naqshbandiyya order (Sufism)
HM 2900	Suluk (Sufism)
KC 0676	Summativity
HM 3156	Summing-up (Brainwashing)
VD 3762	Sumptuousness
HM 2422	Sun (Tarot)
HM 2917	Sunao (Japanese)
HM 5134	Sunnagare abhirati (Buddhism)
HM 2193	Sunyata (Buddhism)
HM 2818	Super-contemplative state (Yoga)
HM 3232	Super-contemplative state (Yoga)
HH 3755	Super-immunity
TC 1566	Super stars
VC 1703	Superabundance
VD 3764	Superciliousness
HM 2960	Superconscious
HM 0107	Superconscious knowledge
HM 1244	Superconscious learning
HM 0107	Superconsciousness
HH 0493	Superego
HM 4340	Superessential contemplation (Christianity)
VD 3766	Superficiality
VC 1705	Superfluity
HH 9764	Superhuman qualities (Buddhism)
VC 1707	Superhumanness
HM 6327	Superior concentration (Buddhism)
HM 0935	Superior consciousness
HM 7089	Superior saintship (Sufism)
VC 1709	Superiority
VP 5036	Superiority-Inferiority
HH 0383	Superiors ; Attending needs (Buddhism)
HH 0604	Superiors ; Respect (Buddhism)
HH 2143	Superman
HH 1306	Supernatural beings ; Communication
HH 4213	Supernaturalism
HH 5652	Supernormal powers (Buddhism)
HM 7672	Supernormal powers ; Knowledge (Buddhism)
HH 0515	Superstition
VD 3768	Superstition
MP 1228	Superstructure contain transitions levels ; Appropriate
MP 1117	Superstructure ; Containment integrative
MP 1220	Superstructure ; Integration
MP 1118	Superstructure ; Integration non linearity integrative
MP 1162	Superstructure minimize insight blindspots ; Organization integrative
HH 0334	Superstructure ; Noetic
MP 1209	Superstructure ; Organization integrative
MP 1231	Superstructure ; Overview sites integrative
MP 1116	Superstructure ; Patterning integrative
MS 8102	Supplies ; Machines
VC 1711	Support
TP 0027	Support
TP 0003	Support ; Gathering
VP 5785	Support-Opposition
MP 1230	Supportive emotion ; Unmediated
VD 3770	Suppression
HM 0135	Suppression (Yoga)

Suppressive

HH 0149	Suppressive therapy
HM 2156	Supra-consciousness
HM 1067	Supramundane awareness (Buddhism)
HM 4243	Supramundane concentration (Buddhism)
HM 4930	Supramundane plane ; Profitable consciousness (Buddhism)
HM 5129	Supramundane plane resultant ; Indeterminate consciousness (Buddhism)
HM 8838	Supramundane understanding (Buddhism)
VC 1713	Supremacy
HM 2411	Supremacy ; Emblems (Tarot)
HM 3132	Supreme crown ; Sphere (Kabbalah)
HM 3377	Supreme goal (Sufism)
HM 3193	Supreme ; God consciousness
HM 2813	Supreme-heaven awareness (Buddhism)
HM 3214	Supreme heavens ascension stages game (Buddhism)
VC 1715	Sureness
VC 1717	Surety
KC 0507	Surface structure
VD 3772	Surfeit
HH 0993	Surgery ; Aesthetic
HH 0993	Surgery ; Cosmetic
HH 0993	Surgery ; Reconstructive
VD 3774	Surliness
VD 3776	Surplus
VC 1719	Surprise
HH 0449	Surrender
HM 3436	Surrender bliss mystical seduction (Sufism)
HM 2922	Surrender inadequacy (ICA)
HH 6509	Surrender love ; Self
HM 2346	Surrender one's calling (ICA)
	Surrender Self
HH 0968	— [Surrender ; Self]
HH 1016	— [Surrender ; Self]
HM 0807	— (Sufism)
	Surrender [Spiritual...]
HH 0449	— Spiritual
HM 5467	— State self (Christianity)
HM 5290	— (Sufism)
HH 1799	Surrender will God (Islam)
VD 3778	Surreptitiousness
KC 0346	Surveillance lag
VC 1721	Survival
VD 3780	Susceptibility
HM 2957	Sushupti (Hinduism)
HM 1182	Suspended animation
VD 3782	Suspense
VD 3784	Suspension
VD 3786	Suspicion
TC 1902	Suspicion ; Interpersonal
HH 3181	Sustainable development
HM 5089	Sustained thought (Buddhism)
	Sustained thought Concentration
HM 4908	— applied thought (Buddhism)
HM 5199	— applied thought (Buddhism)
HM 7543	— applied thought (Buddhism)
VC 1723	Sustenance
TC 1564	Sustenance conferencing
HM 2957	Susupti state unconsciousness (Hinduism)
HH 3443	Sutra ; Lotus (Buddhism)
HM 2962	Sutra paths accumulation application ascension stages game (Buddhism)
HH 1239	Sutras ; Studying
HM 0567	Suvidya (Leela)
HM 7109	Sva-samvedana (Buddhism)
HM 1596	Svabhava (Buddhism)
HM 3174	Svadhishthana (Yoga)
HH 0957	Svadhyaya (Hinduism)
HM 2781	Svapna (Yoga)
HH 0889	Swadeshi
HM 2781	Swapna state dream consciousness (Yoga)
HM 1052	Swarga-loka (Leela)
HM 0455	Swatch (Leela)
VC 1725	Sweetness
HM 7859	Swift direct knowledge ; Concentration difficult progress (Buddhism)
HM 4977	Swift direct knowledge ; Concentration easy progress (Buddhism)
VC 1727	Swiftness
VP 5269	Swiftness-Slowness
HM 1159	Swoon (Hinduism)
HM 1645	Sword wisdom (Buddhism)
TC 1651	Sycophants
HH 0896	Syllabic synthesis
KC 0335	Symbiosis
KC 0208	Symbiotization
HM 3103	Symbol maker (ICA)
	Symbolic [configurations...]
TC 1909	— configurations ; Meetings
MP 1232	— connection encompassing domains
MM 2048	— cycles ; Psycho
MS 8500	Symbolic human activity
HM 1538	Symbolic language tradition
KC 0446	Symbolic logic
MS 8326	Symbolic objects
MS 8113	Symbolic quests journeys
MS 8825	Symbolic structures
KC 0527	Symbolic system
KC 0660	Symbolism
MS 8923	Symbolism ; Chromatic
KC 0079	Symbolism ; Number
HM 3176	Symbolizing eternal context (ICA)
HH 0690	Symbols
MS 8943	Symbols ; Agricultural
MS 8425	Symbols ; Atmosphere celestial
	Symbols [Chrematomorphic...]
MS 8326	— Chrematomorphic
MS 8223	— Civil
MS 8803	— civilization
TC 1644	— Conference
MS 8415	Symbols ; Incorporeal formless
MP 1249	Symbols integration
MS 8843	Symbols ; Masonic
TC 1909	Symbols ; Meetings aesthetic
MS 8913	Symbols office
MS 8713	Symbols ; Ornithological
MS 8445	Symbols ; Phytomorphic
MS 8823	Symbols ; Piscatological
MS 8705	Symbols ; Religious
	Symbols [sacred...]
MS 8423	— sacred literature
MS 8933	— Scientific
MP 1253	— self transformation ; Meaningful
MS 8435	Symbols ; Terrestrial natural
KC 1010	Symmetry
VC 1729	Symmetry
VP 5248	Symmetry-Distortion
KC 0299	Symmetry ; Unitary
VC 1731	Sympathy
HH 0145	Sympathy
HM 5224	Sympathy (Buddhism)
HM 4905	Sympathy-love
HM 2550	Sympathy ; Primal (ICA)
TC 1924	Symposium
KC 0022	Synaesthesia
HM 1241	Synaesthesia
KC 0420	Synaesthesis
HH 0552	Synastry
HM 1210	Sync ; Being
KC 0026	Synchronicity
HH 4269	Synchronicity
KC 0337	Synchronization
HM 1210	Synchrony ; Interactional
HM 1210	Syncing
HH 3214	Syncretic religion
HH 0348	Syncretic thinking
KC 0794	Syncretism
HH 3214	Syncretism
KC 0023	Synderesis
KC 0253	Syndrome
KC 0313	Syndrome ; Babel
KC 0767	Synectics
KC 0718	Synergetic analysis
KC 0718	Synergetics
HH 8487	Synergic enlightenment
HM 1377	Synergies ; Composedness aggregate sensation, aggregate cognition, aggregate (Buddhism)
HM 3397	Synergies ; Flexibility aggregate sensation, aggregate cognition, aggregate (Buddhism)
HM 0349	Synergies ; Good quality aggregate sensation, aggregate cognition, aggregate (Buddhism)
HM 1435	Synergies ; Lightness aggregate sensation, aggregate cognition, aggregate (Buddhism)
HM 0409	Synergies ; Straightness aggregate sensation, aggregate cognition, aggregate (Buddhism)
HM 1381	Synergies ; Workability aggregate sensation, aggregate cognition, aggregate (Buddhism)
HH 0524	Synergism ; Sexual
	Synergy
KC 0957	— [Synergy]
VC 1733	— [Synergy]
TC 1011	— [Synergy]
HH 0517	Synergy ; Human
HM 1241	Synesthesia
KC 0452	Synnoetics
	Synthesis
KC 0255	— [Synthesis]
HH 0896	— [Synthesis]
VC 1735	— [Synthesis]
TC 1011	— [Synthesis]
KC 0083	Synthesis ; Artistic
KC 0135	Synthesis ; Chemical
HH 0896	Synthesis ; Creative
	Synthesis [Dialectic...]
KP 3003	— Dialectic
KC 0457	— Dialectical materialist
HH 0896	— Distributive
KC 0933	Synthesis ; Network
KC 0281	Synthesis ; Planetary
HH 0896	Synthesis ; Syllabic
HM 1814	Synthetic-conventional faith
KC 0621	Synthetic geometry
HH 0321	Syntone
KC 0546	System
	System [Allopoietic...]
KC 0066	— Allopoietic
HH 0018	— Anthroposophical
KC 0527	— Appreciative
KC 0956	— Autopoietic
	System [Centralized...]
KC 0122	— Centralized
KC 0646	— Closed
KC 0783	— Cognitive
KC 0183	— complexity
KC 0607	— Control
	System [Data...]
KC 0526	— Data processing
KC 0408	— design ; Information
KC 0216	— Determinate
KC 0518	— Distributed information
HM 6553	— Dwelling second jhana fivefold (Buddhism)
	System [dynamics...] cont'd
KC 0426	— dynamics
KD 2192	— dynamics: learning ; Encompassing
KD 2190	— dynamics: sixfold restraint ; Encompassing
	System [Economic...]
KC 0530	— Economic
KC 0720	— Educational
KC 0527	— Eiconic
KC 0805	System ; Global
KC 0400	System ; Hierarchical
	System [Image...]
KC 0527	— Image
TC 1616	— In conference computer contact
KC 0526	— Information
KC 0518	— information systems
KC 0308	— Integrated information
KC 0608	— integrity ; Information
KC 0786	— International
KC 0027	System ; Multistable
HH 1099	System ; Noble (Buddhism)
	System [order...]
KC 0740	— order
KC 0740	— Order
KC 0517	— overload
	System [Planning...]
KC 0566	— Planning Programming Budgeting (PPBS)
KC 0328	— principle ; Whole
KC 0677	— Probabilistic
	System [Real...]
KC 0936	— Real time information
KC 0997	— Regional international
KC 0997	— Regional subsystem global
	System [Social...]
KC 0820	— Social
KC 0827	— State determined
KC 0417	— stress
KC 0527	— Symbolic
HH 1127	System theory human development ; General
KD 2110	System ; Threshold comprehensibility: fourfold minimal
KC 0064	System ; Ultrastable
HH 1108	System ; Vaibhasika (Buddhism)
KC 0805	System ; World
TC 1913	Systematic listening process
KC 0045	Systematics
HH 2003	Systematics
HM 0707	Systematics Autocracy
HM 5187	Systematics Cyclic knowledge
HM 1234	Systematics Discriminative knowledge
HM 8080	Systematics Effectual knowledge
HM 0344	Systematics Eight-fold knowledge
HM 0065	Systematics Eleven-fold knowledge
HM 1804	Systematics Experience creativity
HM 0065	Systematics Experience domination
HM 0344	Systematics Experience individuality
HM 1408	Systematics Experience pattern
HM 1234	Systematics Experience polarity
HM 8080	Systematics Experience potentiality
HM 0811	Systematics Experience relatedness
HM 5187	Systematics Experience repetition
HM 0776	Systematics Experience structure
HM 1991	Systematics Experience subsistence
HM 4002	Systematics Experience wholeness
HM 8080	Systematics Five-fold knowledge
HM 1991	Systematics Four-fold knowledge
HM 1408	Systematics Nine-fold knowledge
HM 4002	Systematics Non-discriminative knowledge
HM 1234	Systematics Polar knowledge
HM 8080	Systematics Potential knowledge
HM 0811	Systematics Relational knowledge
HM 0344	Systematics Revealed knowledge
HM 0776	Systematics Seven-fold knowledge
HM 5187	Systematics Six-fold knowledge
HM 0776	Systematics Structural knowledge
HM 1804	Systematics Ten-fold knowledge
HM 0811	Systematics Three-fold knowledge
HM 5187	Systematics Transcendental knowledge
HM 0707	Systematics Twelve-fold knowledge
HM 1234	Systematics Two-fold knowledge
HM 1991	Systematics Value knowledge
KD 2225	Systematization ; Cognitive
HM 2795	Systematization ; Delusional
KC 0627	Systems ; Abstract
KC 0610	Systems analysis
	Systems [Complex...]
KC 0336	— Complex
KC 0907	— Complex social
KC 0187	— Comprehensibility
HM 4000	— concepts ; Ninth order perceptions control
KC 0627	— Conceptual
	Systems [ecology...]
KC 0415	— ecology
KC 0920	— education
KC 0017	— Emergent
KC 0710	— engineering
KD 2220	Systems holonomy ; General
KC 0195	Systems initiatives ; Open
KC 0345	Systems ; Integrated automatic
	Systems [Man...]
KC 0081	— Man machine
KC 0887	— management ; Human
KC 0677	— Markov
KC 0746	Systems ; Open
HH 0326	Systems ; Open learning
KC 0886	Systems research
	Systems Self
KC 0510	— organizing

Systems Self cont'd
KC 0856 — reproducing
HH 1032 — transcendent
HH 7637 Systems ; Spiritual control
KC 0518 Systems ; System information
KC 1000 Systems theory ; General
KC 0247 Systems ; Transient

T

HH 0282 T'ai Chi Ch'uan
HH 0672 T-Groups
HM 5876 Ta'abbud (Islam)
HH 1101 Taboo
HH 1101 Tabu
HM 2297 Tacit knowledge consciousness
MM 2058 Tacking
VC 1737 Tact
TC 1637 Tactics ; Accident
HM 8079 Tactile agnosia
HM 4094 Tactile hallucination
HM 6118 Tactile hallucination
VD 3788 Tactlessness
HM 5092 Tadhkiya-i nafs (Sufism)
HM 6227 Tafakkur (Sufism)
HM 6762 Tafrid (Sufism)
HM 2318 Tafriqah ; Jam (Sufism)
HH 0719 Tahara
HM 5563 Tahqiq (Sufism)
HM 3040 Taijasa (Hinduism)
HM 2858 Taiken
HM 6162 Tajliya-i ruh (Sufism)
HM 7621 Tajrid (Sufism)
TC 1567 Take off ; Meeting
HM 3410 Taken down cross ; Jesus (Christianity)
HM 5353 Takhliya-i sirr (Sufism)
HH 0654 Taking ; Risk
MM 2028 Taking turns
HM 1700 Takuan (Chinese)
VC 1739 Talent
HH 0611 Talent ; Precocious
HH 0370 Tales ; Insight provoking
MS 8235 Tales ; Stories fairy
HM 4464 Talib (Sufism)
HM 1407 Talkin
HM 4501 Talmudic ecstasy (Judaism)
HM 4772 Tama' (Islam)
HM 0413 Tamas
HM 6238 Tamas (Leela)
VC 1741 Tameness
TP 0026 Taming power great
TP 0009 Taming power small
HM 5561 Tamoguna (Leela)
VC 1743 Tangibility
VP 5425 Tangibility-Intangibility
HM 3707 Tanha (Pali)
HM 2864 Tantra: anuttarayoga ; Highest yoga (Buddhism)
Tantra awareness
HM 2656 — Great path (Buddhism)
HM 2558 — Kriya (Buddhism)
HM 3026 — Middle path (Buddhism)
HM 2452 Tantra-beginner awareness (Buddhism)
HH 0306 Tantra (Buddhism)
HM 2864 Tantra: charya ; Performance (Buddhism)
HM 2220 Tantra climax-application awareness (Buddhism)
HM 2696 Tantra heat-application awareness (Buddhism)
HM 2864 Tantra: kriya ; Action (Buddhism)
Tantra [master...]
HM 2235 — master form-realm awareness (Buddhism)
HM 2211 — master-in-sense-realm awareness (Buddhism)
HM 2864 — meditation emptiness (Buddhism)
HM 2756 Tantra receptivity-application awareness (Buddhism)
HM 2151 Tantra shambhala awareness ; Kalacakra (Buddhism)
HM 3026 Tantra ; Ubhaya carya (Buddhism)
HM 2280 Tantra union-in-learning-application awareness (Buddhism)
HM 2864 Tantra ; Yoga (Buddhism)
HM 4603 Tantric formless path (Buddhism)
HM 2848 Tantric paths accumulation application ascension-stages-game (Buddhism)
HM 1690 Tantric visualization (Yoga)
Tantric yoga
HH 0306 — (Buddhism)
HH 6210 — Spirituality
HH 2034 — (Yoga)
KC 0310 Tao
HM 6534 Tao ; Activating mind (Taoism)
HM 1771 Tao ; Following (Taoism)
HM 1858 Tao ; Mind (Taoism)
HH 0689 Tao (Taoism)
HM 5265 Taoism Absolutely open nothingness
HM 6534 Taoism Activating mind Tao
HH 5887 Taoism Alchemy
HH 4880 Taoism Arriving reality
HM 4338 Taoism Assembling five elements
HH 4997 Taoism Balancing yin yang
HH 3664 Taoism Bathing
HH 4997 Taoism Blending celestial consciousness earthly consciousness
HH 0004 Taoism Book Changes
HM 5265 Taoism Celestial immortality
HM 1233 Taoism Centre
HH 6452 Taoism Circulation light
HH 3664 Taoism Cleaning
HH 0317 Taoism Darkness

HM 6633 Taoism Discovery natural mind
HM 6534 Taoism Dissolving human mind
HH 3013 Taoism False paths
HM 1771 Taoism Following Tao
HM 1330 Taoism Formation gold elixir
HM 4762 Taoism Formation spiritual embryo
HH 0634 Taoism Gold pill
HM 0486 Taoism Harmonizing illumination
HH 7734 Taoism Higher upper grade
HH 7386 Taoism Higher vehicle path
HH 4742 Taoism Highest vehicle
HH 0689 Taoism Human development
HH 5622 Taoism Human mind
HH 0004 Taoism I Ching
HM 5265 Taoism Incubation spiritual embryo
HM 5265 Taoism Leaping out world
HM 3296 Taoism Lower middle grade
HH 4711 Taoism Lower upper grade
HM 4864 Taoism Lower vehicle path
HM 3013 Taoism Lowest paths
HM 0486 Taoism Merging ordinary world
HM 1330 Taoism Merging yin yang
HH 0689 Taoism Michi
HH 3667 Taoism Middle-grade low paths
HM 5207 Taoism Middle middle grade
HH 6304 Taoism Middle upper grade
HM 6345 Taoism Middle vehicle path
HM 1858 Taoism Mind Tao
HH 3667 Taoism Outside paths
HM 4289 Taoism Outside paths
HM 0922 Taoism Perfect attainment
HM 1342 Taoism Perfect attainment
HM 6633 Taoism Recognition innocent mind
HM 6633 Taoism Recognition true mind
HH 4880 Taoism Refining gold pill
HM 4007 Taoism Refining self
HH 0635 Taoism Release matrix
HM 6534 Taoism Restoration celestial awareness mundane
HM 4007 Taoism Setting up foundation
HH 9112 Taoism Sexual yoga
HM 1858 Taoism Shining mind
HH 7004 Taoism Sidetracks auxiliary methods Taoism
HH 7004 Taoism ; Sidetracks auxiliary methods (Taoism)
HM 1330 Taoism Spirit open consciousness
HH 5117 Taoism Spirituality
HM 1865 Taoism Spontaneity
HM 5265 Taoism Subtlety nondoing
HH 0689 Taoism Tao
HH 5342 Taoism Three vehicles gradual method
HM 1833 Taoism Transcendent liberation
HM 5265 Taoism Transcending world
HM 0346 Taoism Tso-wang
HM 1342 Taoism Ultimate accomplishment
HM 4762 Taoism Unification energy
HM 0317 Taoism Unknowing
HM 1858 Taoism Unstirring mind
HM 4289 Taoism Upper-grade low paths
HM 2633 Taoism Upper middle grade
HM 1330 Taoism Valley spirit
HH 5622 Taoism Wandering mind
HH 3664 Taoism Washing
HH 6452 Taoism Waterwheel exercise
HM 1771 Taoism Yu
HM 3530 Tapa
HM 1773 Tapah (Leela)
HM 5244 Tapah-loka (Leela)
HH 1248 Tapas (Hinduism)
HH 0877 Tapasya (Yoga)
TC 1498 Tape recording ; Audio
TC 1897 Tape ; Video
HH 0452 Taraka yoga (Yoga)
VD 3790 Tardiness
KC 0717 Targets
HM 2684 Tariki (Japanese)
HM 3371 Tariqa (Sufism)
HM 2097 Tarot arcana conscious states (Tarot)
HM 3028 Tarot Betrothal initiation archetypal image
HH 0382 Tarot Book Thoth
HM 3045 Tarot Chariot
HM 3045 Tarot Charioteer archetypal image
HM 3121 Tarot Death emblems
HM 2278 Tarot Devil
HM 2201 Tarot Emblem archetypes psyche
HM 2350 Tarot Emblems disaster
HM 2398 Tarot Emblems mysterious
HM 2378 Tarot Emblems personified evil
HM 2315 Tarot Emblems renewal
HM 2411 Tarot Emblems supremacy
HM 2387 Tarot Emblems totality
HM 2422 Tarot Emblems well-being
HM 2411 Tarot Emperor
HM 2285 Tarot Empress
HM 2285 Tarot Female sovereignty archetypal image
HM 2225 Tarot Fool
HM 2250 Tarot Fortune-wheel
HM 3121 Tarot Grim reaper
HM 2190 Tarot Hanged man
HM 2202 Tarot Hermit
HM 3028 Tarot Hierogamy
HM 2260 Tarot Hierophant archetypal image
HM 2260 Tarot High priest
HM 2261 Tarot High priestess
HM 2201 Tarot Image awareness
HM 2315 Tarot Judgment
HM 2178 Tarot Justice
HM 3137 Tarot Life-fluid emblems - star

HM 3289 Tarot Life-fluid emblems - temperance
HM 3028 Tarot Lovers
HM 2237 Tarot Magician archetypal image
HM 2237 Tarot Magus
HM 2285 Tarot Matrona
HM 2178 Tarot Measuring emblems
HM 2398 Tarot Moon
HM 2250 Tarot Mutability emblems
HM 3028 Tarot Mystic marriage
HM 3127 Tarot Natural forces emblems
HM 2202 Tarot Old wise-man archetypal image
HM 2225 Tarot Pilgrim archetypal image
HM 2260 Tarot Pope
HM 2350 Tarot Ruin
HM 3127 Tarot Strength
HM 2422 Tarot Sun
HH 0382 Tarot (Tarot)
HH 0382 Tarot Tarot
HM 2097 Tarot Tarot arcana conscious states
HM 2350 Tarot Tower
HM 2190 Tarot Victimization emblems
HM 2261 Tarot Wisdom archetypal image
HM 2387 Tarot World
Tasawwuf Sufism
HM 4450 — [Tasawwuf (Sufism)]
HM 0575 — [Tasawwuf (Sufism)]
HH 0436 — [Tasawwuf (Sufism)]
HM 6932 Tasfiya-i qalb (Sufism)
TC 1630 Task force
HH 0716 Task-oriented group experience
TC 1485 Taskforce
TC 1269 Taskforce ; Special
TP 0064 Tasks remaining
HM 4190 Taslim (Sufism)
HM 5290 Taslim (Sufism)
HM 2263 Taste awareness (Buddhism)
HM 0878 Taste ; Plane (Leela)
VP 5896 Taste-Vulgarity
VC 1745 Tastefulness
VD 3792 Tastelessness
HM 5876 Tatayyum (Islam)
HM 2435 Tathata (Buddhism)
HM 7002 Tatra-majjhattata (Pali)
HM 8003 Tattva (Buddhism)
HM 4062 Tauba (Sufism)
HM 2815 Taurus-consciousness (Astrology)
HM 0807 Tawakkul (Sufism)
HM 0807 Tawakul (Sufism)
HM 4062 Tawba (Sufism)
HM 4062 Tawbat (Sufism)
HM 3438 Tawhid (Sufism)
HM 4334 Tawwab ; Al (Sufism)
KC 0045 Taxonomy
HM 3179 Teacher ; Committed (ICA)
KC 0961 Teaching concepts integrated science
KC 0711 Teaching ; Integrated methods science
HM 2271 Teachings application awareness ; Mahayana highest (Buddhism)
KC 0206 Team augmentation
HH 0717 Team building group
TC 1526 Team completer role
TC 1361 Team driver role
TC 1642 Team evaluator role
TC 1814 Team explorer role
TC 1526 Team finisher
TC 1969 Team focalizer role
HM 1550 Team genius role
TC 1264 Team grounder role
TC 1142 Team promoter role
TC 1528 Team roles
TC 1361 Team shaper
TC 1866 Teams ; Audience reaction
TC 1866 Teams ; Conceptual interpretation
TC 1866 Teams ; Inter cultural communication
TC 1866 Teams ; Interruption
TC 1742 Teams ; Listening
TC 1742 Teams ; Observation
KC 0503 Teamwork
HH 0265 Teamwork mentality
KC 0926 Technique ; Monte Carlo simulation
HH 1092 Technique no technique
KC 0736 Technique ; Programme evaluation review
TC 1354 Techniques ; Facilitative
KC 0736 Techniques ; Network planning
HH 3662 Technological development
KC 0687 Technological forecasting
HH 0615 Technological implications human evolution
KC 0619 Technological progress ; Scientific
KC 0619 Technology
KC 0943 Technology assessment
VD 3794 Tedium
MS 8355 Teeth
HM 5028 Teja-loka (Leela)
TC 1972 Teleconferencing ; Audio
TC 1418 Teleconferencing ; Audio graphic
TC 1666 Teleconferencing ; Computer
TC 1876 Teleconferencing ; Full motion video
TC 1218 Teleconferencing ; Video
HH 0601 Teleology
KC 0642 Teleology
HM 2129 Telepathic consciousness
HM 2262 Telepathy (ESP)
TC 1870 Telephone conference
TC 1972 Telephone ; Loud speaker
KC 0297 Teletics
HH 1098 Television evangelism (Christianity)

VD 3796	Temper
HH 0831	Temperament
HH 0126	Temperament ; Physique
HH 0600	Temperance
VC 1747	Temperance
VP 5992	Temperance-Intemperance
HM 3289	Temperance ; Life fluid emblems (Tarot)
VD 3798	Tempestuousness
HH 0587	Temporal dimension being ; Non
HM 2363	Temporal solidarity (ICA)
HM 2454	Temporary-hells awareness (Buddhism)
MP 1150	Temporary perspective inactivity ; Provision
MP 1241	Temporary positions ; Attractive
MP 1022	Temporary site limit ; Occupiable
MP 1103	Temporary sites ; Limitation number occupiable
VD 3800	Temptation
HM 2503	Temptation ; Path intelligence (Judaism)
HM 2995	Ten directions ; Emptiness (Buddhism)
HM 1136	Ten ; Focus (Psychism)
HM 1804	Ten-fold knowledge (Systematics)
HH 0446	Ten meritorious deeds (Buddhism)
HM 3159	Ten powers (Buddhism)
HM 2657	Ten worlds (Buddhism)
HM 2657	Ten worlds ; Mutual possession (Buddhism)
VC 1749	Tenacity
HM 2692	Tendencies consciousness ; Individual (Hinduism)
HM 5197	Tendencies ; Good (Leela)
HM 1375	Tendency towards ignorance (Buddhism)
HM 3231	Tendency (Yoga)
VC 1751	Tenderness
HH 2918	Tenfold powers (Buddhism)
HH 0757	Tenko
KC 0028	Tensegrity
KD 2260	Tensegrity organization ; Interwoven alternatives:
VD 3802	Tension
MP 1219	Tension ; Level generation minimum
TC 1782	Tensions ; Conference
TC 1441	Tensions ; Dynamic
HM 1717	Tenth order perceptions - universal oneness - meditation
HM 2421	Tenth scriptural bodhisattva awareness (Buddhism)
VD 3804	Tenuousness
HM 2965	Terce (Christianity)
HM 2965	Terce ; Hours (Christianity)
HH 1409	Teresa ; Interior castle Saint (Christianity)
KC 0717	Term aims ; Short
MM 2012	Terrestrial animal locomotion
MS 8387	Terrestrial beings ; Mythical
MS 8435	Terrestrial natural symbols
HM 1644	Terrible sentiment (Hinduism)
HM 2460	Terrifying acceptance (ICA)
HM 1644	Terrifying rasa (Hinduism)
VD 3806	Terror
HM 7114	Terror ; Knowledge appearance (Buddhism)
HM 0889	Thai Being revitalized
HM 0889	Thai Sabsung
HM 0889	Thai Slaking emotional spiritual thirst
HM 1749	Thama (Pali)
HH 0077	Thambhatattam (Pali)
HH 0574	Thanatos
HM 8003	Thatness (Buddhism)
HH 0582	Theatre games
HH 0717	Theatre karma ; Third chakra: (Leela)
HH 4663	Theism ; Human development
HH 3006	Thelema (Islam)
KC 0380	Thematic cultural integration
MM 2009	Thematic development ; Interruption
KC 0674	Theme
KC 0204	Theme ; Life
HH 0474	Themes ; Life
HH 3652	Theology ; Ascetical (Christianity)
HH 4552	Theology ; Liberation (Christianity)
HH 5217	Theology ; Mystical (Christianity)
HH 3652	Theology ; Spiritual (Christianity)
HH 5217	Theology ; Spiritual (Christianity)
HM 1027	Theomania
KC 0715	Theorem
KC 0444	Theorem ; Gödel's
KC 0301	Theoretical integration level
KC 0463	Theorizing ; Second order
KC 0892	Theory
	Theory [Catastrophe...]
KC 0462	— Catastrophe
KC 0220	— Central place
KC 0333	— Classical field
	Theory [Game...]
KC 0916	— Game
KC 1000	— General systems
KC 0584	— Graph
KC 0394	Theory history ; Cyclical
KC 0554	Theory ; Lattice
	Theory [performance...]
KD 2156	— performance
KC 0667	— Probability
KC 0614	— Problem solving
KC 0312	— procedure ; Decision
KC 0969	— Proof
KC 0407	Theory ; Social field
KC 0612	Theory ; Transformation
KC 0833	Theory ; Unified field
KC 0693	Theory ; Unitary
KC 0356	Theory ; Value
HH 0660	Theosophy
HH 0336	Therapeutic community
HH 6112	Therapeutic double bind
HM 7045	Therapeutic love
HH 0512	Therapeutic motion picture
HH 0678	Therapy
	Therapy [Active...]
HH 0091	— Active
HH 0492	— Activity group
HH 0325	— Addiction group
HH 0507	— Analytic group
HH 0507	— Analytical
HH 0775	— Art
HH 0685	— Assertion structured
HH 0802	— Assignment
HH 0676	— Attitude
HH 0256	— Aversion
	Therapy [Behaviour...]
HH 0795	— Behaviour
HH 0962	— Biomagnetic
HH 0172	— Body movement
HH 0752	— Breathing
	Therapy [Child...]
HH 0894	— Child guidance
HH 0095	— Client centred
HH 0007	— Colour
HH 0985	— Conditioned reflex
HH 0362	— Conjoint family
HH 0566	— Convulsive
HH 0975	— Convulsive
	Therapy [Dance...]
HH 0445	— Dance
HH 0596	— Diversional
HH 2116	— Dream
	Therapy [Electric...]
HH 0866	— Electric convulsive (ECT)
HH 0866	— Electroconvulsive
HH 0866	— Electroshock
HH 0885	— Experiential
HH 0149	— Expressive
HH 0362	Therapy ; Family
HH 0362	Therapy ; Family group
	Therapy [Gedatsu...]
HH 0972	— Gedatsu
HH 1334	— Gene
HH 0751	— Gestalt
HH 0851	— Group
HH 0607	— Group centred
	Therapy [Implosive...]
HH 1107	— Implosive
TC 1602	— Inter group
HH 0434	— Interpersonal
HH 0614	Therapy ; Learning theory
HH 0633	Therapy ; Ludo
	Therapy [Marriage...]
HH 0362	— Marriage
HH 0851	— Mass
HH 0289	— Megavitamin
HH 0072	— Mental imagery
HH 0336	— Milieu
HH 1121	— Morita
HH 0846	— Multiple
HH 0496	— Music
	Therapy [Naikan...]
HH 1324	— Naikan
HH 4332	— Neuromuscular
HH 0482	— Nude group
HH 0596	Therapy ; Occupational
HH 0327	Therapy ; Orgone
	Therapy [Persuasion...]
HH 0696	— Persuasion
HH 0379	— Physical
HH 0379	— Physical process
HH 0633	— Play
HH 0987	— Pressure
HH 2087	— Primal
HH 1236	— Psychobiologic
HH 0072	— Psycholytic
HH 0172	— Psychomotor
HH 4201	— Psychospiritual (Christianity)
HH 3862	Therapy ; Qigon
	Therapy [Radical...]
HH 2675	— Radical
HH 0866	— Regressive electroshock
HH 0731	— Reiki
HH 0095	— Relationship
HH 0762	— Religious
HH 0218	— Rhythmic sensory bombardment
	Therapy Self
HH 0606	— [Therapy ; Self]
HH 2276	— observation
HH 0784	— realization
	Therapy [Shadan...]
HH 0483	— Shadan
HH 0566	— Shock
HH 0975	— Shock
HH 0492	— Situational
HH 0866	— Sleep electroshock
HH 0874	— Social
HH 0458	— Spiritual
HH 0362	— Structural family
HH 0796	— Suggestion
HH 0149	— Suppressive
HH 0325	Therapy ; Verbal attack
HH 0883	Therapy ; Will
HH 0987	Therapy ; Zone
HH 0845	Theravada (Buddhism)
HM 3108	Thereness ; Unexplainable (ICA)
MS 8225	Theriomorphs ; Mythical
KC 0944	Thesaurus Classification schedule
TC 1275	Thesaurus network
HM 2428	Theta-clear state consciousness (Scientology)
HM 2321	Theta wave consciousness
HH 0269	Thetan
HM 1613	Thicket views (Buddhism)
HH 5667	Thina (Pali)
	Things Emptiness
HM 2989	— (Buddhism)
HM 3208	— (Buddhism)
HM 2448	— inherent existence non (Buddhism)
MS 8516	Things ; Precious
HM 6119	Think ; Group
HH 0530	Thinking
	Thinking [Abstract...]
HH 0348	— Abstract
HM 4967	— Alert
HH 0748	— Archaic paralogical
HH 0644	— Associative
TC 1123	Thinking ; Beyond linear
HM 3966	Thinking (Buddhism)
	Thinking [Categoried...]
HH 0348	— Categoried
KC 0106	— Comprehensive
HH 0348	— Concrete
KC 0406	— Configurative
HM 5665	— consciousness ; Unmanifest (Buddhism)
HH 0703	— Creative
HM 2368	Thinking ; Divergent
HM 2266	Thinking ; Everyday
HM 1212	Thinking ; Fragmentation
	Thinking [Illogical...]
HM 1359	— Illogical
HH 0384	— Illumined
KC 0106	— Integrative
HM 2785	— Intuitive, meditative
KC 0260	Thinking ; Lateral
HH 2287	Thinking ; Linear
	Thinking [Magical...]
HM 0359	— Magical
HH 4646	— Maternal
HH 0632	— meditation ; Free
	Thinking [Physiognomonic...]
HH 0348	— Physiognomonic
HH 1088	— Positive
HM 1084	— Pre conscious
HH 0748	— Primitive
HH 0748	— Primordial
HH 1643	— Process
HH 0703	— Productive
	Thinking [Rational...]
HM 2283	— Rational, calculative
HM 1274	— Relaxed
HM 1142	— Right (Buddhism)
HH 0348	Thinking ; Syncretic
KC 0059	Thinking ; Trans contextual
KC 0512	Thinking ; Vertical
HM 2320	Thinking ; Vertical lateral
HM 2179	Thinking ; Wishful
VD 3808	Thinness
HM 2027	Third absorption immaterial sphere (Buddhism)
HM 2311	Third antarabhava consciousness (Buddhism)
HH 0717	Third chakra: theatre karma (Leela)
HM 3063	Third chakra (Yoga)
HM 4279	Third duty (Christianity)
HM 2062	Third form-realm concentration (Buddhism)
HM 2027	Third formless attainment (Tibetan)
	Third jhana
HM 5643	— Dwelling (Buddhism)
HM 8532	— five ; Concentration (Buddhism)
HM 2284	— four ; Concentration (Buddhism)
HM 0312	Third order perceptions - configurations
KD 2095	Third-perspective container alternation
HM 3312	Third plane wisdom (Hinduism)
HM 2653	Third realm consciousness ; Beyond (Buddhism)
	Third [scriptural...]
HM 2215	— scriptural bodhisattva awareness (Buddhism)
HM 3464	— stage life
HM 3501	— step: mankind (Christianity)
HM 2289	Third time cross ; Jesus falls (Christianity)
HM 2062	Third trance fine-material sphere (Buddhism)
HM 2727	Third vajra-master awareness (Buddhism)
HM 7308	Thirst (Hinduism)
HM 3707	Thirst (Pali)
HM 0889	Thirst ; Slaking emotional spiritual (Thai)
HM 1029	Thirstlessness (Yoga)
HM 2606	Thirty-three-god-heaven awareness (Buddhism)
HM 2937	This ; I am (Yoga)
HM 2614	This world ; Other world midst (ICA)
HM 0051	Thiti ; Cittassa (Pali)
HM 1418	Thiti (Pali)
HM 3057	Thorough abandonings (Buddhism)
VC 1753	Thoroughness
MP 1145	Those reserve ; Domains non current elements
HH 0382	Thoth ; Book (Tarot)
HM 3225	Thou awareness ; I
HH 0530	Thought
	Thought [Alienation...]
HM 6508	— Alienation
HM 1177	— Applied (Buddhism)
HH 0748	— Archaic
HM 1317	— Audible
HM 9384	Thought broadcasting
	Thought [Ceasing...]
HM 3295	— Ceasing
HM 6663	— Collectedness moral (Buddhism)
HM 1490	— Composedness (Buddhism)

Thought Concentration applied
HM 4908 — thought sustained (Buddhism)
HM 5199 — thought sustained (Buddhism)
HM 7543 — thought sustained (Buddhism)
HM 1066 Thought ; Constancy (Buddhism)
HM 3949 Thought ; Depression (Buddhism)
HM 0666 Thought ; Enjoying
HM 0975 Thought ; Excitement (Buddhism)
Thought [Fitness...]
HM 1810 — Fitness (Buddhism)
HM 1805 — Flexibility (Buddhism)
HM 0666 — Flow experience
HM 3529 Thought (Hinduism)
HM 5479 Thought insertion (Psychism)
HM 1094 Thought ; Inspirational (Psychism)
HM 1830 Thought ; Lightness (Buddhism)
Thought [meditation...]
HH 0976 — meditation
HM 0223 — Mental perturbation (Buddhism)
HM 2242 — Multifocal
HM 2500 Thought ; No (Zen)
HM 6663 Thought ; One directedness (Buddhism)
HM 7154 Thought out ; Understanding what is (Buddhism)
Thought [Positive...]
HH 1088 — Positive
HH 0540 — pressure
HH 0530 — processes
HH 0865 Thought reform (Brainwashing)
HM 1142 Thought ; Right (Buddhism)
Thought [Stabilization...]
HM 1930 — Stabilization (Buddhism)
HM 0051 — Stasis (Buddhism)
HM 3471 — Steadfastness (Buddhism)
HM 0253 — Straightness (Buddhism)
HM 5089 — Sustained (Buddhism)
Thought sustained thought
HM 4908 — Concentration applied (Buddhism)
HM 5199 — Concentration applied (Buddhism)
HM 7543 — Concentration applied (Buddhism)
VP 5478 Thought-Thoughtlessness
HM 0400 Thought ; Turmoil (Buddhism)
Thought [Unitary...]
KC 0733 — Unitary
HM 4572 — Unperplexity (Buddhism)
HM 0449 — Unperturbedness (Buddhism)
HM 0140 — Uprisedness (Buddhism)
HM 1584 Thought ; Workability (Buddhism)
VC 1755 Thoughtfulness
VD 3810 Thoughtlessness
HM 0847 Thoughtlessness (Buddhism)
VP 5478 Thoughtlessness ; Thought
HH 4001 Thoughts ; Disruptive (Christianity)
HM 7298 Thoughts ; Knowledge another's (Hinduism)
HM 5232 Thoughts ; Knowledge others' (Buddhism)
HM 0709 Thoughts ; Tranquillity (Buddhism)
HM 3398 Thousand-petalled lotus (Yoga)
VD 3812 Threat
TP 0007 Threat ; Controlled (Army)
HM 2085 Threatening dreams
HH 0397 Three ages ; Doctrine (Christianity)
HM 0768 Three ; Consciousness
HM 1133 Three deviations body ; Leaving off, abstaining, totally abstaining refraining (Buddhism)
HM 0923 Three evil paths (Buddhism)
HM 1814 Three faith ; Stage
HM 0811 Three-fold knowledge (Systematics)
HM 2606 Three god heaven awareness ; Thirty (Buddhism)
HH 0413 Three gunas
HM 0348 Three sleep ; Stage
HH 0631 Three spiritual ways (Christianity)
HH 5342 Three vehicles gradual method (Taoism)
Three [way...]
MP 1050 — way relationship entrainment
HH 2189 — world concept
HM 2026 — worlds (Yoga)
HM 2321 Threshold awareness
KD 2110 Threshold comprehensibility: fourfold minimal system
TP 0064 Threshold ; Transformation
KC 0387 Threshold value
VC 1757 Thriftiness
VD 3814 Thriftlessness
TP 0051 Thunderbolts Arousal
TP 0051 Thunderbolts Crisis preparedness
TP 0051 Thunderbolts Shock
TP 0051 Thunderbolts Vigilance
HM 3208 Tibetan Abhavashunyata
HM 2448 Tibetan Abhavasvabhavashunyata
HM 2385 Tibetan Abhimukhi
HM 2337 Tibetan Abhirati
HM 2397 Tibetan Absolute-body awareness
HM 2816 Tibetan Acala
HM 2864 Tibetan Action tantra: kriya
HM 2794 Tibetan Adhyatmabahirdhashunyata
HM 2671 Tibetan Adhyatmashunyata
HM 2813 Tibetan Akanistha
HM 2696 Tibetan Alchemical-flask
HM 2232 Tibetan Aloka-vriddhi samadhi
HM 2783 Tibetan Amoghasiddhi-karma awareness
HM 2015 Tibetan Anantaryamarga
HM 3072 Tibetan Anavakarashunyata
HM 3032 Tibetan Anavaragrashunyata
HM 2636 Tibetan Animal-hell awareness
HM 2763 Tibetan Anujna
HM 2501 Tibetan Anupalambhashunyata
HM 2519 Tibetan Apara-godaniya
HM 2292 Tibetan Appropriate intellect
HM 2191 Tibetan Arcis-mati
HM 3024 Tibetan Arhat sanctity awareness
HM 3144 Tibetan Arupa-dhatu
HM 2089 Tibetan Asamskrtashunyata
HM 2579 Tibetan Asura-world awareness
HM 0052 Tibetan Atyantashunyata
HM 2052 Tibetan Avici-hell
HM 2314 Tibetan Bag-med-pa
HM 3220 Tibetan Bag-yod-pa
HM 2522 Tibetan Bahirdhashunyata
HM 0101 Tibetan Bar-do
HM 2091 Tibetan Barbarian-state awareness
HM 0698 Tibetan Bardo consciousness
HM 4651 Tibetan Bare perception
HM 2245 Tibetan Bden-pa-bzhi
HM 2653 Tibetan Beyond-the-third-realm consciousness
HM 2989 Tibetan Bhavashunyata
HM 2155 Tibetan Bhumi
HM 3142 Tibetan Black rope hell
HM 2220 Tibetan Bodhicitta awareness
HM 2639 Tibetan Bon-practitioner state awareness
HM 2663 Tibetan Bon-wisdom awareness
HM 3053 Tibetan Brjed-nges-pa
HM 2389 Tibetan Brtson-'grus
HM 2693 Tibetan Bsam gtan
HM 2062 Tibetan Bsam-gtan-gsum-pa
HM 7769 Tibetan Btang-snyoms
HM 2735 Tibetan Buddha-consciousness
HM 3019 Tibetan Buddha emanation body
HM 3113 Tibetan Buddha enjoyment body
HM 3019 Tibetan Buddha form body
HM 3113 Tibetan Buddha form body
HM 2039 Tibetan Buddha nature body
HM 2834 Tibetan Buddha truth body
HM 2039 Tibetan Buddha truth body
HM 2834 Tibetan Buddha wisdom body
HH 4019 Tibetan Byan-chub
HM 2769 Tibetan Byang-chub-sems-dpa
HM 2058 Tibetan Cakravartin
HM 2172 Tibetan Cessation-awareness
HM 2605 Tibetan 'Chab-pa
HM 2335 Tibetan Chikhai-bardo
HM 2407 Tibetan Chonyid bardo
HM 2040 Tibetan Cold-hells awareness
HM 2220 Tibetan Conception
HM 0713 Tibetan Conceptual cognition
HH 4000 Tibetan Consciousness ascension stages game
HM 3142 Tibetan Crushing-hells awareness
HM 3209 Tibetan Dad-pa
HM 2931 Tibetan Darsana
HM 8003 Tibetan De-kho-na
HM 2033 Tibetan definitions awareness (Buddhism)
HM 2546 Tibetan Delightful-emanation-heaven consciousness
HM 2055 Tibetan Demon-island awareness
HM 2397 Tibetan Dharma-kaya
HM 2421 Tibetan Dharma-negha
HM 2208 Tibetan Dharmaloka-labdha-samadhi
HM 2033 Tibetan Direct non-conceptual cognition emptiness
HM 2716 Tibetan Discipleship-application awareness
HM 2192 Tibetan Discipleship-karma awareness
HM 2240 Tibetan Discipleship-vision awareness
HM 2668 Tibetan Distorted cognition
HM 2007 Tibetan Divine-animal-hell awareness
HM 2110 Tibetan Dngos gzhii snyoms jug
HM 2433 Tibetan 'Dod-chags
HM 2733 Tibetan Dod-khams
HM 5089 Tibetan Dpyod-pa
HM 2659 Tibetan Dpyod-pa-yid-byed
HM 2847 Tibetan Dran-pa
HM 2371 Tibetan Dream-state bardo awareness
HM 2399 Tibetan 'Du-shes
HM 2181 Tibetan Dul-bar-byed-pa
HM 2578 Tibetan 'Dun pa
HM 2361 Tibetan Duramgama
HM 2127 Tibetan Dzam-bu-gling
HM 2543 Tibetan Eastern-continent awareness noble figures
HM 2679 Tibetan Eight siddhis
HM 2775 Tibetan Eighteen unshared attributes Buddha
HM 2816 Tibetan Eighth scriptural bodhisattva awareness
HM 2301 Tibetan Eighth vajra-master awareness
HM 3032 Tibetan Emptiness cyclic existence
HM 2946 Tibetan Emptiness definitions
HM 2671 Tibetan Emptiness five senses
HM 3072 Tibetan Emptiness indestructible Mahayana
HM 2448 Tibetan Emptiness inherent existence non-things
HM 2794 Tibetan Emptiness loci senses
HM 3205 Tibetan Emptiness nature
HM 2899 Tibetan Emptiness nature phenomena
HM 2089 Tibetan Emptiness non-products
HM 2522 Tibetan Emptiness objects sense mental consciousness
HM 3000 Tibetan Emptiness phenomena
HM 2414 Tibetan Emptiness products
HM 2995 Tibetan Emptiness ten directions
HM 2989 Tibetan Emptiness things
HM 3208 Tibetan Emptiness things
HM 2294 Tibetan Emptiness ultimate nirvana
HM 2501 Tibetan Emptiness unapprehendable
HM 0052 Tibetan Emptiness what is free permanence annihilation
HM 2944 Tibetan Emptinesses paths view Buddhism
HM 2873 Tibetan Enjoyment-body awareness
HH 0320 Tibetan Ever-recurring mental factors
HM 2909 Tibetan Fifth scriptural bodhisattva awareness
HM 2763 Tibetan Fifth vajra-master awareness
HM 2849 Tibetan Final vajra-master awareness
HM 2347 Tibetan First antarabhava consciousness
HM 2155 Tibetan First scriptural bodhisattva awareness
HM 2187 Tibetan First vajra-master stage
HM 2191 Tibetan Flaming enlightenment
HM 2142 Tibetan Form-realm awareness
HM 3144 Tibetan Formless-realm awareness
HM 3044 Tibetan Four fearlessnesses
HM 2082 Tibetan Four-great-kings-heaven awareness
HM 3216 Tibetan Four sciences
HM 2191 Tibetan Fourth scriptural bodhisattva awareness
HM 2251 Tibetan Fourth vajra-master awareness
HM 3093 Tibetan G'yo
HM 7921 Tibetan Gnyid
HM 2180 Tibetan Great awakening
HM 2118 Tibetan Great-black-lord awareness
HM 2807 Tibetan Great compassion
HM 2929 Tibetan Great love
HM 2656 Tibetan Great-path tantra awareness
HM 3048 Tibetan Great-vehicle higher-path awareness
HM 2268 Tibetan Great-vehicle lower-path awareness
HM 2160 Tibetan Great-vehicle middle-path awareness
HM 2695 Tibetan Gti-mug-med-pa
HM 3863 Tibetan Gtum-mo
HM 2699 Tibetan Guhya-samaja urgyan awareness
HM 7209 Tibetan 'Gyod-pa
HH 0910 Tibetan Gzhan-'gyur
HM 2257 Tibetan Gzugs-khams
HM 2669 Tibetan Gzugs-med-kyi-snyoms-jug
HM 2281 Tibetan Gzugs-sku
HM 2130 Tibetan Heaven-without-fighting consciousness
HM 2010 Tibetan Heavenly-highway awareness
HM 2864 Tibetan Highest yoga tantra: anuttarayoga
HM 2115 Tibetan Hindu-states awareness
HM 2723 Tibetan Hindu-wisdom-holder awareness
HM 3690 Tibetan Hod-gsal
HM 2100 Tibetan Howling-hells awareness
HM 5122 Tibetan Hpho-ba
HM 2112 Tibetan Hungry-ghosts-hell awareness
HM 2337 Tibetan Hyper-bliss-realm awareness
HM 2215 Tibetan Illumination
HM 3348 Tibetan Inattentive perception
HM 3074 Tibetan Indecisive wavering
HM 2195 Tibetan Individual knowledge character
HM 3378 Tibetan Induced attentiveness
HM 2052 Tibetan Interminable-hell awareness
HM 2275 Tibetan Jewelled-peaks-realm awareness
HM 2167 Tibetan Joy-land awareness
HM 2022 Tibetan Joyful-heaven awareness
HM 2151 Tibetan Kalacakra-tantra shambhala awareness
HM 2211 Tibetan Kamadera-vidyadhara
HM 2783 Tibetan Karma-paripurana
HM 2177 Tibetan Khams-gsum
HM 2134 Tibetan 'Khon-'dzin
HM 2959 Tibetan Khong-khro
HM 3203 Tibetan Khrel-med-pa
HM 2210 Tibetan Khrel-yod-pa
HM 2264 Tibetan Khro-ba
HM 2558 Tibetan Kriya-tantra awareness
HM 2247 Tibetan Kshanti
HH 0320 Tibetan Kun-'gro
HM 2067 Tibetan Kuru
HM 2946 Tibetan Lakshanashunyata
HM 2171 Tibetan Las-dang-po-pa-tsam-kyi-yid-byed
HM 2271 Tibetan Laukikagra-dharma
HM 2263 Tibetan Lce'i rnam shes pa
HM 3163 Tibetan Le-lo
HM 2207 Tibetan Lhag-mthong
HH 0873 Tibetan Lo-rig
HM 2088 Tibetan Lord-of-death awareness
HM 2996 Tibetan Lta-ba-nyon-mongs-can
HM 2562 Tibetan Lus-kyi-rnam-par-shes-pa
HM 2128 Tibetan Ma-chags-pa
HM 2461 Tibetan Ma-dad-pa
HM 3196 Tibetan Ma-rig-pa
HM 2679 Tibetan Magical-forces awareness
HM 2639 Tibetan Magical-shamanic state
HM 2118 Tibetan Mahakala
HM 2656 Tibetan Mahamudra
HM 2995 Tibetan Mahashunyata
HM 2232 Tibetan Mahayana climax-application awareness
HM 2208 Tibetan Mahayana heat-application awareness
HM 2271 Tibetan Mahayana highest-teachings-application awareness
HM 2268 Tibetan Mahayana-path
HM 3048 Tibetan Mahayana path
HM 2160 Tibetan Mahayana path
HM 2247 Tibetan Mahayana receptivity-application awareness
HM 3066 Tibetan Masters wisdom ascension stages game
HM 2311 Tibetan Meditation-state bardo awareness
HM 2693 Tibetan meditative states form concentrations (Buddhism)
HM 2669 Tibetan meditative states formless absorptions (Buddhism)
HM 3026 Tibetan Middle-path tantra awareness
HM 2074 Tibetan Mig-gi-rnam-par-shes-pa
HH 1809 Tibetan Mngon shes pa
HM 2277 Tibetan Mnyam-par-jog-pa
HM 2603 Tibetan Mokshakala
HM 2933 Tibetan Mos-pa
HM 2719 Tibetan Mos-pa-las-byung-bai-yid-byed
HM 2031 Tibetan Naga-world awareness
HM 2823 Tibetan Nga-rgyal
HM 2986 Tibetan Ngo-tsha-med-pa
HM 2881 Tibetan Ngo-tsha-shes-pa
HM 2292 Tibetan Ninth scriptural bodhisattva awareness
HM 2325 Tibetan Ninth vajra-master awareness

Tibetan

HM 2172 Tibetan Nirodha
HM 2546 Tibetan Nirvana-rati
HM 2070 Tibetan Non-emanating consciousness
HM 2067 Tibetan Northern-continent awareness community
HM 2663 Tibetan Nyamsrtsal
HM 2161 Tibetan Nyan thos
HM 2157 Tibetan Nye bar jog pa
HM 2729 Tibetan Nye-bar-zhi-bar-byed-pa
HH 0781 Tibetan Nye-nyon
HM 2623 Tibetan Nyer bsdogs
HH 0320 Tibetan Omnipresent mental factors
HM 3378 Tibetan Other-induced attentiveness
HM 2294 Tibetan Paramarthashunyata
HM 2070 Tibetan Paranirmita-vasavartin
HM 3249 Tibetan Paths effect
HM 2240 Tibetan Paths vision cultivation
HM 2931 Tibetan Perception
HM 2864 Tibetan Performance tantra: charya
HM 2638 Tibetan Phrag-dog
HM 2175 Tibetan Potala-island awareness
HM 2215 Tibetan Prabha-kari
HM 2756 Tibetan Prajna-jnana
HM 3205 Tibetan Prakrtishunyata
HM 2155 Tibetan Pramudita
HM 3020 Tibetan Pratyeka Buddha application awareness
HM 2776 Tibetan Pratyeka Buddha arhat awareness
HM 2180 Tibetan Pratyeka Buddha awareness
HM 2252 Tibetan Pratyeka Buddha cultivation awareness
HM 2228 Tibetan Pratyeka Buddha vision-path awareness
HM 4032 Tibetan Presumption
HM 2112 Tibetan Pretas
HM 2168 Tibetan Pure-lands awareness
HM 2543 Tibetan Purva-videha
HM 2743 Tibetan Rab-tu-dben-pai-yid-byed
HM 2055 Tibetan Rakshasas
HM 2709 Tibetan Rang-sang-rgyas
HM 2267 Tibetan Ratisamgrahakamanaskara
HM 2275 Tibetan Ratna-kuta
HM 2028 Tibetan Rdo-rje
HM 2708 Tibetan Reg-pa
HM 3341 Tibetan Religious traditions ascension stages game
HM 2516 Tibetan Reviving-hell awareness
HM 2534 Tibetan Rgod-pa
HM 2528 Tibetan Rgyags-pa
HM 2241 Tibetan Rgyun-du-jog-pa
HM 0600 Tibetan Rmi-lam
HM 2926 Tibetan Rmugs-pa
HM 2169 Tibetan Rna-ba'i-rnam-par-shes-pa
HM 3154 Tibetan Rnam-par-g'yeng-ba
HM 2608 Tibetan Rnam-par-mi-'tshe-ba
HM 3210 Tibetan Rnam-par-'tshe-ba
HM 3617 Tibetan Rnam shes
HM 1177 Tibetan Rtog-pa
HM 2753 Tibetan Rtse-geig-tu-byed-pa
HM 2603 Tibetan Rudra awareness black freedom
HM 2142 Tibetan Rupa-dhatu
HM 2235 Tibetan Rupa-dhatu-vidyadhara
HM 2292 Tibetan Sadhu-mati
HM 2873 Tibetan Sambhoga-kaya
HM 2414 Tibetan Samskrtashunyata
HM 3000 Tibetan Sarvadharmashunyata
HH 0320 Tibetan Sarvatraga
HM 2683 Tibetan Sbyor-mthai-yid-byed
HM 2371 Tibetan Second antarabhava consciousness
HM 2739 Tibetan Second scriptural bodhisattva awareness
HM 2703 Tibetan Second vajra-master awareness
HH 0781 Tibetan Secondary afflictions
HH 1348 Tibetan Secondary mental factors
HM 3265 Tibetan Seemingly bare perception
HM 3378 Tibetan Self-induced attentiveness
HM 2145 Tibetan Sems gnas dgu
HM 2217 Tibetan Sems-jogpa
HM 2589 Tibetan Sems-pa
HM 2964 Tibetan Ser-sna
HM 2361 Tibetan Seventh-scriptural bodhisattva awareness
HM 2789 Tibetan Seventh vajra-master awareness
HM 3246 Tibetan Sgyu
HM 1222 Tibetan Sgyu-lus
HM 2663 Tibetan Shen-siddhi
HM 2063 Tibetan Shes-bzhin-ma-yin-pa
HM 5321 Tibetan Shes pa
HM 2655 Tibetan Shes-rab
HM 3162 Tibetan Shin-tu-sbyangs-pa
HM 2899 Tibetan Shunyatashunyata
HM 2385 Tibetan Sixth scriptural bodhisattva awareness
HM 2287 Tibetan Sixth vajra-master awareness
HM 2265 Tibetan Slan-te-jog-pa
HM 2364 Tibetan Sna'i-rnam-par-shes-pa
HM 5634 Tibetan Snying re
HM 2240 Tibetan Sravaka-darsana-marga
HM 2716 Tibetan Sravaka prayoga-marga
HM 2192 Tibetan Sravaka-sambhara-marga
HM 2051 Tibetan Srid-rtse
HM 2431 Tibetan Sridpa bardo
HH 0873 Tibetan studies (Buddhism)
HM 3508 Tibetan Subsequent cognition
HM 2195 Tibetan Subtle contemplation
HM 2267 Tibetan Subtle contemplation withdrawal joy
HM 2168 Tibetan Suddhavasa
HM 2909 Tibetan Sudurjaya
HM 2167 Tibetan Sukhavati
HM 2813 Tibetan Supreme-heaven awareness
HM 3214 Tibetan Supreme heavens ascension stages game
HM 2962 Tibetan Sutra paths accumulation application ascension stages game
HM 2452 Tibetan Tantra-beginner awareness
HM 2220 Tibetan Tantra climax-application awareness
HM 2696 Tibetan Tantra heat-application awareness
HM 2235 Tibetan Tantra master form-realm awareness
HM 2211 Tibetan Tantra master-in-sense-realm awareness
HM 2864 Tibetan Tantra meditation emptiness
HM 2756 Tibetan Tantra receptivity-application awareness
HM 2280 Tibetan Tantra union-in-learning-application awareness
HM 2848 Tibetan Tantric paths accumulation application ascension-stages-game
HM 2454 Tibetan Temporary-hells awareness
HM 3159 Tibetan Ten powers
HM 2421 Tibetan Tenth scriptural bodhisattva awareness
HM 2311 Tibetan Third antarabhava consciousness
HM 2027 Tibetan Third formless attainment
HM 2215 Tibetan Third scriptural bodhisattva awareness
HM 2727 Tibetan Third vajra-master awareness
HM 2606 Tibetan Thirty-three-god-heaven awareness
HM 2033 Tibetan Tibetan definitions awareness
HM 2440 Tibetan Ting-nge-'dzin
HM 2606 Tibetan Trayatrimsa
HM 2778 Tibetan 'Tshig-pa
HM 2270 Tibetan Tshor-ba
HM 2340 Tibetan Tulku
HM 2022 Tibetan Tushita
HM 3026 Tibetan Ubhaya-carya-tantra
HM 3020 Tibetan Unicorn
HM 2015 Tibetan Uninterrupted release paths
HH 0781 Tibetan Upaklesha
HM 2187 Tibetan Vajracarya
HM 2576 Tibetan Very-hot-hells awareness
HM 4467 Tibetan Vichikitsa - The-tshom
HM 2723 Tibetan Vidyadhara
HM 2759 Tibetan Vidyadhara-emperor
HM 3066 Tibetan Vidyahara
HM 2739 Tibetan Vimala
HM 2015 Tibetan Vimuktimarga
HM 2185 Tibetan Way Buddhas
HM 3090 Tibetan Way vajra masters
HM 2519 Tibetan Western-continent awareness cattle
HM 2759 Tibetan Wheel-turner-king awareness
HM 2058 Tibetan Wheel-turning-king awareness
HM 3214 Tibetan World-transcending spheres
HM 2130 Tibetan Yama-devas
HM 2088 Tibetan Yama state
HM 2435 Tibetan Yan-dag-pa-ji-lta-ba-bzin-du
HM 2838 Tibetan Yid-kyi-rnam-par-shes-pa
HM 3237 Tibetan Yid-la-byed-pa
HM 2864 Tibetan Yoga tantra
HM 2849 Tibetan Yugbanaddha
HM 2744 Tibetan Zhe-sdang-med-pa
HM 2205 Tibetan Zhi-bar-byed-pa
HM 2147 Tibetan Zhi gnas
VD 3816 Tightness
HM 1133 Tihi kayaduccaritehi arati virati pativirati veramani (Pali)
KC 0761 Time
MS 8605 Time
HH 1301 Time ; Aboriginal dream (Australian)
TC 1370 Time allotments ; Exceeded
TP 0005 Time ; Biding one's (Getting wet)
Time binding
HH 0585 — [Time-binding]
HH 0815 — [Time binding]
MP 1248 — [Time binding]
VT 8021 Time*complex
HM 2601 Time consciousness
Time Consciousness
HM 2751 — cyclical
HM 2651 — irreversible direction
HM 2701 — reversible direction
TC 1556 Time design ; Conference
HM 3770 Time ; Fringe space (Psychism)
HM 0502 Time-gap experience
KD 2240 Time ; Health space
KC 0936 Time information system ; Real
HM 4412 Time penance ; Sixth chakra: (Leela)
KD 2130 Time ; Quaternary consciousness: number
HM 1456 Time ; Transcendence
HM 2750 Timelessness
VC 1759 Timelessness
VC 1761 Timeliness
VP 5129 Timeliness-Untimeliness
VD 3818 Timidity
HM 2440 Ting-nge-'dzin (Tibetan)
HM 6193 Tip tongue experience
HM 3031 Tiphareth sephira (Kabbalah)
HM 5806 Tiredness ; Mental
HM 5635 Tiredness ; Physical
VD 3820 Tiresomeness
HM 3685 Titthayatana (Pali)
HH 0682 TM Transcendental meditation
HM 1881 Tobacco smoking
TP 0045 Together ; Gathering
TP 0008 Together ; Holding
Togetherness
VC 1763 — [Togetherness]
HH 0265 — [Togetherness]
HH 0455 — [Togetherness]
VD 3822 Toilsomeness
VC 1765 Tolerance
MP 1240 Tolerance level interfaces
HM 2066 Tonal state awareness (Amerindian)
TC 1061 Tone ; Conference
HM 2263 Tongue consciousness (Buddhism)
HM 6193 Tongue experience ; Tip
HM 1608 Tongues ; Praying (Christianity)
HM 1608 Tongues ; Speaking (Christianity)
MS 8102 Tools ; Craftsman
MM 2011 Topics conversation ; Shifting
KC 0994 Topological space
KC 0946 Topology
VD 3824 Torment
HM 1483 Torpid consciousness
VD 3826 Torpor
HM 1483 Torpor
HM 0264 Torpor (Buddhism)
VD 3830 Tortuousness
VD 3828 Torture
HM 5775 Total annihilation (Sufism)
HM 2354 Total conflict ; Breaking point (Brainwashing)
HM 2764 Total exposure (ICA)
HH 0625 Total fitness
HM 2001 Total freedom attachment (Hinduism)
HH 0157 Total integration personality
HM 3139 Total union (Hinduism)
KC 0402 Totalitarianism
HH 0534 Totality
HM 2387 Totality ; Emblems (Tarot)
Totally abstaining refraining
HM 1781 — four deviations speech ; Leaving off, abstaining, (Buddhism)
HM 1133 — three deviations body ; Leaving off, abstaining, (Buddhism)
HM 0252 — wrong livelihood ; Leaving off, abstaining, (Buddhism)
HM 4341 Totemic awareness
HM 4341 Totemic being
HM 4341 Totemic identity
HH 1902 Totemic religion
HH 1902 Totemism
HM 2562 Touch awareness (Buddhism)
HM 2708 Touch (Buddhism)
HM 4212 Touch (Buddhism)
HH 2552 Touch healing
HM 1959 Touching (Buddhism)
VC 1767 Toughness
VP 5358 Toughness ; Elasticity
KD 2320 Toward enantiomorphic policy
HM 2350 Tower (Tarot)
VD 3832 Toxicity
HM 3302 Traces awareness ; Seeing (Zen)
VC 1769 Tractability
MS 8343 Trades professions
VC 1771 Tradition
HH 0440 Tradition
HH 0277 Tradition ; Naqsbandi (Sufism)
KC 0772 Tradition ; Sacred
HM 1538 Tradition ; Symbolic language
HM 5298 Tradition ; Understanding way (Buddhism)
MM 2026 Traditional spontaneous ceremony
HM 1173 Traditionalism (Christianity)
HM 3341 Traditions ascension stages game ; Religious (Buddhism)
MM 2063 Traffic regulation
HH 0708 Tragedy
HH 0708 Tragic suffering
Training [Advanced...]
HH 0521 — Advanced
HH 0024 — Arica
HH 0994 — Assertiveness
HH 0506 — Autogenic
HH 0765 Training ; Biofeedback
HH 0115 Training ; Erhard seminars
Training [Habit...]
HH 0638 — Habit
HH 2387 — higher mind (Buddhism)
HH 0047 — Human resource
HH 0672 Training laboratories
HH 0742 Training ; Labour
HH 0720 Training ; Magical
HH 0745 Training ; Manpower (United Nations)
HM 0220 Training ; One pointedness martial arts
Training [Sensitivity...]
HH 0672 — Sensitivity
HH 0609 — Spontaneity
HM 3744 — States awareness biofeedback
HM 2683 — subtle contemplation ; Final (Buddhism)
HH 0047 Training ; Vocational
VD 3834 Traitorousness
HM 3236 Trance
HM 0442 Trance ; Active (Psychism)
Trance [Cataleptic...]
HM 4319 — Cataleptic (Psychism)
HM 3456 — channelling (Psychism)
HM 2502 — consciousness ; Voodoo
HM 1886 Trance ; Deep (Psychism)
HM 2079 Trance ; Double awake body asleep
Trance fine material
HM 2450 — sphere ; First (Buddhism)
HM 2586 — sphere ; Fourth (Buddhism)
HM 2038 — sphere ; Second (Buddhism)
HM 2062 — sphere ; Third (Buddhism)
HM 1886 Trance ; Full (Psychism)
HH 0458 Trance healing
HM 2640 Trance ; Mind awake body asleep (Psychism)
Trance [Self...]
HM 1964 — Self induced (Psychism)
HM 0887 — Semi (Psychism)
HM 1189 — Shamanistic
HM 2122 Trances mental absorptions (Buddhism)
VC 1773 Tranquillity
HM 0226 Tranquillity consciousness (Buddhism)
HH 0710 Tranquillity ; Cultivation

GENERAL INDEX TO VOLUME 2

Two

HM 7115 Tranquillity due insight (Buddhism)
HM 4704 Tranquillity mental body (Buddhism)
HH 0491 Tranquillity ; Moral
HM 3033 Tranquillity ; Sea (ICA)
HM 3492 Tranquillity ; State (Buddhism)
HM 0709 Tranquillity thoughts (Buddhism)
TP 0010 Tranquillity (Treading)
HM 3411 Tranquillization ; Complete (Buddhism)
HM 1528 Tranquillized ; Completely (Buddhism)
KC 0059 Trans-contextual thinking
HM 9022 Trans-sensate experience
HM 3438 Trans-substantiation heart (Sufism)
KC 0327 Transaction analysis
HH 0487 Transactional analysis
Transcendence
HM 4104 — [Transcendence]
HH 0841 — [Transcendence]
VC 1775 — [Transcendence]
HH 5908 Transcendence agape philo ; Journeying
HH 1354 Transcendence blindness sight ; Journeying
HH 2661 Transcendence bondage freedom ; Journeying (Christianity)
HH 3352 Transcendence call mission ; Journeying (Christianity)
HM 2259 Transcendence ; Cross (Astrology)
HH 2987 Transcendence death life ; Journeying
HH 3366 Transcendence discussion decision ; Journeying
HM 2230 Transcendence ; Ego (Jung)
HH 4110 Transcendence extraordinary ordinary ; Journeying (Christianity)
HH 5219 Transcendence forgetfulness memory ; Journeying
HH 4974 Transcendence head heart ; Journeying
HH 2329 Transcendence humiliation exaltation ; Journeying
HH 2498 Transcendence isolation imagination ; Journeying
HH 6505 Transcendence ; Journeying (Christianity)
HH 5856 Transcendence me you ; Journeying
HM 4401 Transcendence ; Mono motivational hypnotic state
HH 4188 Transcendence object subject ; Journeying
HH 3215 Transcendence outside inside ; Journeying
Transcendence [Positions...]
MP 1024 — Positions enabling
HH 1900 — prologue ; Journeying
HH 1763 — Psychological approach
HM 2804 Transcendence ; Secular
HH 3796 Transcendence secular sacred ; Journeying
Transcendence Self
HM 2486 — [Transcendence ; Self]
HH 0526 — [Transcendence ; Self]
HM 2584 — (ICA)
HH 6012 Transcendence served servant ; Journeying (Christianity)
HH 1773 Transcendence son father ; Journeying
HM 1456 Transcendence time
HH 4219 Transcendence water wine ; Journeying
HH 6566 Transcendence ; Way (Christianity)
Transcendent [experience...]
HM 3445 — experience
HH 1123 — experience ; Drug use
HM 2080 — experiences
HH 0324 Transcendent function
HM 3058 Transcendent guru (ICA)
HM 3034 Transcendent immanence (ICA)
HM 1833 Transcendent liberation (Taoism)
HM 3060 Transcendent stage life ; I
HH 1032 Transcendent systems ; Self
KC 0791 Transcendent unity religions
HM 0107 Transcendent wisdom
HM 2395 Transcendental awareness (Hinduism)
Transcendental [concentration...]
HM 4243 — concentration (Buddhism)
HM 2020 — consciousness
HM 2405 — cosmic consciousness
HM 2712 Transcendental experience
HM 5187 Transcendental knowledge (Systematics)
HH 0682 Transcendental meditation (TM)
Transcendental [plane...]
HM 4930 — plane ; Moral consciousness (Buddhism)
HM 5129 — plane resultant ; Indeterminate consciousness (Buddhism)
HH 0718 — psychology
HM 8838 Transcendental understanding (Buddhism)
HM 3193 Transcendental unity consciousness
HH 1187 Transcending doubt ; Purity (Buddhism)
HM 2658 Transcending hostility (ICA)
HM 3489 Transcending mind (Zen)
HM 3214 Transcending spheres ; World (Buddhism)
HM 5265 Transcending world (Taoism)
VC 1777 Transcience
KC 0796 Transdisciplinarity
HM 3571 Transfer ; Psychic (Psychism)
HM 6992 Transfer ; Reality
HH 0668 Transference
HH 0235 Transference ; Counter
HM 7045 Transference ; Erotic
HM 7045 Transference love
HM 5122 Transference ; Yoga consciousness (Buddhism)
HH 1266 Transferring merit (Buddhism)
HM 0428 Transfiguration ; Initiation (Esotericism)
HM 3161 Transfigured man (ICA)
Transformation
KC 0812 — [Transformation]
VC 1779 — [Transformation]
HH 1039 — [Transformation]
Transformation [Affective...]
HH 0260 — Affective
KP 3018 — attempts ; Inadequate
HM 3146 — Awareness angelic
Transformation [awareness...] cont'd
HM 2850 — awareness ; Islamic
Transformation [Capacity...]
HM 3808 — Capacity easy (Buddhism)
HM 4423 — consciousness (Yoga)
HM 3106 — Contentless (ICA)
HH 3120 — Cultural
HH 5634 Transformation game
KD 2325 Transformation ; Game comprehension identity
HM 2334 Transformation ; Human (ICA)
HH 0010 Transformation human personality ; Lifelong educational
HH 0258 Transformation love ; Self
Transformation [Meaningful...]
MP 1253 — Meaningful symbols self
HH 1116 — Mental
HH 4561 — Mental (Yoga)
HH 3003 — music ; Psycho social
HH 1432 — Mystical
HH 3456 Transformation organizations ; Spirit
KP 3009 Transformation process ; Implementation
HH 0304 Transformation ; Psychic
KP 3019 Transformation ; Qualitative
HM 3061 Transformation ; Reforged (ICA)
HM 0074 Transformation ; State (Christianity)
KC 0612 Transformation theory
TP 0064 Transformation threshold
HM 3335 Transformation ; Will behavioural (Sufism)
MP 1042 Transformation zones ; Chain fundamental
KP 3012 Transformative controlled relationship ; Harmoniously
TC 1332 Transformative event
MP 1066 Transformative experience ; Context
HH 0175 Transformative games
HM 2862 Transformed existence (ICA)
HM 2386 Transformed state (ICA)
VD 3836 Transgression
VD 3838 Transience
VP 5110 Transience ; Durability
HM 2906 Transient ; Separation (Sufism)
KC 0247 Transient systems
MP 1093 Transit point location sources perspective nourishment
MP 1092 Transit points ; Hospitable
MP 1112 Transition domain structure communication pathway
MP 1091 Transition ; Hospitable contexts perspectives
HM 1258 Transition ; Initiation (Esotericism)
HH 1366 Transition ; Midlife
HM 3294 Transition ; Near death (NDE)
MP 1133 Transition pathways levels structure ; Integrating
MP 1130 Transition spaces enhancing structural entry points ; Internal
MP 1195 Transition structural levels ; Framework
MP 1084 Transitional contexts perspective reorganization
HM 2934 Transitional limbo (Brainwashing)
MP 1224 Transitions across boundaries ; Emphasizing
MP 1212 Transitions boundary orientation ; Primary inter level connections
HM 1516 Transitions ; Fourth order perceptions control
MP 1228 Transitions levels ; Appropriate superstructure contain
HM 9021 Transitoriness (Buddhism)
HM 3754 Translation ; Sensory
HH 0686 Transmigration
HM 2554 Transmutation ; Cross (Astrology)
KC 0167 Transmutation ; Social
KC 0823 Transnational
HH 0673 Transparency
VP 5339 Transparency-Opaqueness
HM 2483 Transparency spirit ; Ultimate (Sufism)
HM 2736 Transparent engagement (ICA)
HM 3067 Transparent existence (ICA)
HM 2107 Transparent intelligence ; Path (Judaism)
HM 2184 Transparent lucidity (ICA)
HM 2828 Transparent power (ICA)
HM 2462 Transparent presence (ICA)
HM 3077 Transparent selfhood (ICA)
HM 0768 Transpersonal awareness
HM 2530 Transpersonal bands
HH 0916 Transpersonal psychology
MS 8402 Transport
MP 1072 Transposed concrete level ; Competitive interaction opportunities
HM 2084 Transrational insight (Zen)
HM 3724 Trauma
TP 0056 Traveller (Wanderer)
TC 1743 Travelling microphone
VD 3840 Travesty
HM 2606 Trayatrimsa (Buddhism)
VD 3842 Treachery
TP 0010 Treading Careful conduct
TP 0010 Treading Stepping carefully
TP 0010 Treading Tranquillity
VD 3844 Treason
HH 0364 Treasures godly
HH 0874 Treatment ; Social
HH 0921 Tree life ; Qabalistic
HH 0921 Tree sephiroth
KC 0808 Tree structure
HM 3811 Tremendum ; Mysterium (Jung)
HM 7805 Tremens ; Delirium
VD 3846 Tremulousness
VD 3848 Trepidation
VD 3850 Trespass
HM 3435 Trespassing limit ; Not (Buddhism)
HM 2772 Triadic awareness
HH 1141 Trial ordeal
KD 2105 Trialectics: logic whole
HM 0411 Tribal consciousness (Yoga)
VD 3852 Tribulation
HM 7297 Tribulation ; Knowledge (Buddhism)
VD 3854 Trickery
HM 2177 Tridhatu (Buddhism)
HH 0739 Trigger factor
TC 1777 Triggering activities ; Risk
HH 7123 Trilogy ; Analytical
HM 0357 Trinity ; Visions (Christianity)
HM 0818 Trip ; Bad
HH 0631 Triple way (Christianity)
HM 2955 Triplicity air (Astrology)
HM 3235 Triplicity ; Earth (Astrology)
HM 2124 Triplicity fire (Astrology)
HM 2384 Triplicity water (Astrology)
HM 3707 Trisna (Buddhism)
HM 7308 Trisna (Hinduism)
HM 2062 Tritiyadhyana (Buddhism)
VC 1781 Triumph
HM 2593 Triumphant intelligence ; Path (Judaism)
VD 3856 Triviality
MS 8446 Trophies ; Medals, decorations
VD 3858 Trouble
TP 0039 Trouble
VD 3860 Truculence
HM 4604 True delusions
HH 5767 True knowledge (Christianity)
HM 6633 True mind ; Recognition (Taoism)
VC 1783 Trust
TC 1721 Trust building activities
HM 0807 Trust God (Sufism)
TC 1852 Trust ; Interpersonal
HM 2761 Trust intuitions (ICA)
HH 4233 Trust ; Mutual
TC 1721 Trust working
VC 1785 Trustworthiness
HH 0554 Truth
VC 1787 Truth
HM 5563 Truth ; Ascertaining (Sufism)
HM 5563 Truth ; Assertion (Sufism)
HM 2834 Truth body ; Buddha (Buddhism)
HM 2039 Truth body ; Buddha (Buddhism)
Truth [Certain...]
HM 4556 — Certain (Sufism)
HM 4556 — certainty (Sufism)
HM 8163 — cessation ill ; Understanding reference (Buddhism)
VT 8053 Truth*complex
HM 2196 Truth consciousness (Hinduism)
VP 5516 Truth-Error
HM 3006 Truth ; Establishment (Hinduism)
HM 6870 Truth ill ; Understanding reference (Buddhism)
TP 0061 Truth ; Inner
HM 5403 Truth ; Knowledge conformity (Buddhism)
HM 3303 Truth ; Ontological reality (Sufism)
HM 7290 Truth origin ill ; Understanding reference (Buddhism)
HM 7645 Truth path leading cessation ill ; Understanding reference (Buddhism)
HM 1293 Truth ; Plane (Leela)
HM 0124 Truth (Sufism)
KD 2005 Truth: uniformity versus aesthetics ; Forms
VC 1789 Truthfulness
HM 6631 Truthfulness (Sufism)
VD 3862 Truthlessness
HH 0523 Truths ; Four noble (Buddhism)
HM 2245 Truths ; Meditation way four (Buddhism)
HM 3527 Truths ; Two fold knowledge (Buddhism)
HH 0928 Ts'ao tung (Zen)
HM 2778 'Tshig-pa (Tibetan)
HM 2270 Tshor-ba (Tibetan)
HM 0346 Tso-wang (Taoism)
HH 0589 Tsung ; Hsin (Zen)
HM 3037 Tsutsushimi (Japanese)
HM 4062 Tuba (Sufism)
HM 2340 Tulku (Tibetan)
HH 0928 Tung ; Ts'ao (Zen)
HM 6687 Tuning-in
HM 1237 Tunnel vision
VD 3864 Turbidity
VD 3866 Turbulence
HM 2395 Turiya awareness (Hinduism)
HM 3139 Turiyatita (Hinduism)
HM 0400 Turmoil thought (Buddhism)
HM 2759 Turner king awareness ; Wheel (Buddhism)
HM 4062 Turning God (Sufism)
HM 2058 Turning king awareness ; Wheel (Buddhism)
TP 0024 Turning point Recovery
TP 0024 Turning point Return
MM 2028 Turns ; Taking
VD 3868 Turpitude
HM 3139 Turyatita (Hinduism)
HM 2022 Tushita (Buddhism)
HM 0267 Twelve ; Focus (Psychism)
HM 0707 Twelve-fold knowledge (Systematics)
HM 0176 Twelve-petalled lotus (Yoga)
HM 7655 Twenty-four causes consciousness arising (Buddhism)
HH 0715 Twice-born
HM 2406 Twilight consciousness
HM 1561 Two alternatives ; Being doubt (Buddhism)
HM 3336 Two ; Consciousness
Two [faith...]
HM 1525 — faith ; Stage
HM 1234 — fold knowledge (Systematics)
HM 3527 — fold knowledge truths (Buddhism)
HM 4806 Two sleep ; Stage
MP 1159 Two sources external insight ; Organization structures permit
HM 3662 Two vehicles (Buddhism)

HM 0450 Two vehicles (Buddhism)
Type [Accommodator...]
TC 1164 — Accommodator participant
TC 1765 — Action oriented
TC 1929 — Aggressive participant
TC 1772 — Assimilator participant
Type [Blind...]
TC 1358 — Blind belief
TC 1875 — Bricksigner
TC 1748 — Builder
TC 1133 Type ; Converger participant
Type [Detached...]
TC 1736 — Detached operator
TC 1225 — Diverger participant
TC 1384 — Doctrinaire participant
TC 1990 Type ; Encourager participant
TC 1358 Type ; Expertist participant
TC 1651 Type ; Followers participant
TC 1430 Type ; Grassrooter participant
TC 1491 Type ; Initiator participant
HH 0960 Type ; Integrated
TC 1739 Type ; Less haste participant
TC 1129 Type ; Marginal participant
TC 1775 Type ; Mediator participant
TC 1663 Type ; Non conformist participant
TC 1581 Type ; Nonchalent participant
TC 1657 Type ; Process oriented participant
TC 1391 Type ; Self reflexer participant
TC 1860 Type ; Structure oriented participant
TC 1853 Type ; Value monger participant
HH 3945 Types ; Enneagram personality (Sufism)
HH 0126 Types ; Sheldon's
KC 0484 Typology
VD 3870 Tyranny

U

HM 7983 Ubhatobhagavimutti (Pali)
HM 3026 Ubhaya-carya-tantra (Buddhism)
HM 2570 Ubiquitous otherness (ICA)
VC 1791 Ubiquity
HM 0975 Uddhacca ; Cittassa (Pali)
HM 1469 Uddhacca (Pali)
VD 3872 Ugliness
MS 8375 Ugliness
VP 5899 Ugliness ; Beauty
HM 0409 Ujukata ; Vedanakkhandhassa sannakkhandhassa sankharakkhandhassa (Pali)
HM 1338 Ujukata ; Vinnanakkhandhassa (Pali)
HM 4250 Ujuta (Pali)
HM 5902 Ul Mulk ; Malik (Sufism)
HM 1342 Ultimate accomplishment (Taoism)
HM 2388 Ultimate awareness (ICA)
HM 2294 Ultimate nirvana ; Emptiness (Buddhism)
Ultimate reality
HH 0504 — [Ultimate reality]
HM 1456 — Awareness
HM 2030 — (ICA)
HM 2318 Ultimate Sufi state consciousness (Sufism)
HM 2483 Ultimate transparency spirit (Sufism)
HM 2856 Ultimate union (Astrology)
HM 2156 Ultra-consciousness
KC 0064 Ultrastable system
HM 7089 'ulya ; Wilayat (Sufism)
HM 4090 Un-comprehension (Buddhism)
HM 4131 Unabated desire ; State (Buddhism)
HM 1403 Unabated endurance ; State (Buddhism)
VC 1793 Unadornment
VC 1795 Unadulteration
VD 3874 Unadvisedness
HM 3375 Unaffected (Buddhism)
MP 1235 Unalienating internal boundaries
VC 1797 Unambiguity
VC 1799 Unanimity
TP 0036 Unappreciated intelligence (Darkening of the light)
HM 2501 Unapprehendable ; Emptiness (Buddhism)
HM 8092 Unassociated consciousness (Buddhism)
VC 1801 Unassuming
VP 5920 Unastonishment ; Wonder
HM 5335 Unattracted consciousness (Buddhism)
VD 3876 Unauthenticity
HM 1362 Unawakened (Buddhism)
HM 1725 Unawareness (Buddhism)
VD 3878 Unbearableness
VP 5501 Unbelief ; Belief
HM 5335 Unbending mind (Buddhism)
VC 1803 Unbigoted
VC 1805 Unblemished
HM 3051 Uncertainty
VD 3880 Uncertainty
HM 0812 Uncertainty (Buddhism)
HM 0870 Uncertainty (Buddhism)
VP 5513 Uncertainty ; Certainty
HH 1471 Uncertainty (Christianity)
KD 2305 Uncertainty function ignorance
KC 0190 Uncertainty principle
VC 1807 Unchangeable
HM 3316 Unchanging reality being (Sufism)
VD 3882 Unchastity
VP 5987 Unchastity ; Chastity
VD 3884 Uncleanliness
VP 5681 Uncleanness ; Cleanness
VD 3886 Uncomfortableness
VD 3888 Uncommunicativeness
VP 5554 Uncommunicativeness ; Communicativeness
HM 6504 Uncompounded existence (Buddhism)
HM 1266 Unconditional love (Sufism)
HM 3619 Unconditional nirvikalpa samadhi
HM 6504 Unconditioned existence (Buddhism)
HM 8621 Unconfined mind (Buddhism)
HM 4394 Unconscientiousness (Buddhism)
VD 3890 Unconscionableness
HM 2473 Unconscious ; Alternative mode awareness while
HM 0085 Unconscious ; Collective (Jung)
HM 2811 Unconscious ; Collective (Psychosynthesis)
HM 0148 Unconscious enstasy (Yoga)
HM 2332 Unconscious ; Experience
Unconscious [Higher...]
HM 2960 — Higher
HM 2057 — Higher (Psychosynthesis)
HM 2957 — (Hinduism)
HM 2790 Unconscious ; Lower (Psychosynthesis)
HM 2306 Unconscious ; Middle (Psychosynthesis)
HM 2473 Unconscious stupor
HM 2500 Unconscious (Zen)
HM 2094 Unconsciousness
HM 2602 Unconsciousness ; Group
HM 0992 Unconsciousness ; Slight (Hinduism)
HM 2957 Unconsciousness ; Susupti state (Hinduism)
TC 1699 Unconstrained meeting configurations
VP 5642 Unconventionality ; Conventionality
VC 1809 Uncorrupted
HM 4037 Uncrookedness (Buddhism)
VD 3892 Unctuousness
TP 0004 Uncultivated growth (Young shoot)
KD 2272 Undefined common focus
HM 1054 Undeflectedness (Buddhism)
HM 6521 Undejected consciousness (Buddhism)
VD 3894 Undependability
MP 1033 Underdefined processes
VD 3896 Underdevelopment
VD 3898 Underestimation
VP 5497 Underestimation ; Overestimation
HH 0658 Understanding
VC 1811 Understanding
HM 4348 Understanding abandoning ; Full (Buddhism)
Understanding accompanied
HM 7525 — equanimity (Buddhism)
HM 7525 — even-mindedness (Buddhism)
HM 6245 — joy (Buddhism)
HM 4128 Understanding being plane culture (Buddhism)
HM 6192 Understanding being plane discernment (Buddhism)
Understanding Buddhism
HM 3617 — [Understanding (Buddhism)]
HM 4523 — [Understanding (Buddhism)]
HM 0158 — [Understanding (Buddhism)]
HM 5901 Understanding ; Consciousness reinforced (Buddhism)
Understanding consisting
HM 7826 — development (Buddhism)
HM 5298 — what is learned (Buddhism)
HM 7154 — what is reasoned (Buddhism)
HM 7826 Understanding culture (Buddhism)
Understanding [defining...]
HM 4968 — defining materiality (Buddhism)
HM 5627 — defining mentality (Buddhism)
HM 4958 — discrimination perspicuity (Buddhism)
HH 6451 Understanding ; Education international
HH 1010 Understanding ; Five levels (Buddhism)
HM 0756 Understanding free cankers (Buddhism)
Understanding having
HM 3296 — exalted object (Buddhism)
HM 6681 — infinite object (Buddhism)
HM 2495 — limited object (Buddhism)
HM 6681 — measureless object (Buddhism)
HM 3296 — sublime object (Buddhism)
HM 2099 Understanding (Hinduism)
KC 0845 Understanding ; International
Understanding interpreting
HM 1128 — external (Buddhism)
HM 0956 — internal (Buddhism)
HM 4490 — internal external (Buddhism)
HM 4552 Understanding investigation ; Full (Buddhism)
Understanding knowledge
HM 8163 — cessation suffering (Buddhism)
HM 4726 — doctrine (Buddhism)
HM 5026 — interpretation (Buddhism)
HM 4958 — kinds knowledge (Buddhism)
HM 5026 — language (Buddhism)
HM 4726 — law (Buddhism)
HM 1663 — meaning (Buddhism)
HM 7290 — origin suffering (Buddhism)
HM 6870 — suffering (Buddhism)
HM 7645 — way leading cessation suffering (Buddhism)
HM 3490 Understanding known ; Full (Buddhism)
HM 5901 Understanding ; Mind upheld (Buddhism)
HM 7628 Understanding ; Mundane (Buddhism)
HM 0403 Understanding ; Not (Buddhism)
HM 4128 Understanding plane development (Buddhism)
HM 6192 Understanding plane seeing (Buddhism)
Understanding reference truth
HM 8163 — cessation ill (Buddhism)
HM 6870 — ill (Buddhism)
HM 7290 — origin ill (Buddhism)
HM 7645 — path leading cessation ill (Buddhism)
HM 1280 Understanding ; Right (Buddhism)
HH 0693 Understanding ; Self
Understanding skill
HM 7602 — detriment (Buddhism)
HM 8290 — improvement (Buddhism)
HM 8290 — increase (Buddhism)
HM 7602 — loss (Buddhism)
Understanding skill cont'd
HM 6601 — means (Buddhism)
HM 8290 — profit (Buddhism)
Understanding [Sphere...]
HM 2372 — Sphere (Kabbalah)
HM 2660 — States
HM 4194 — subject cankers (Buddhism)
HM 8838 — Supramundane (Buddhism)
HM 8838 Understanding ; Transcendental (Buddhism)
Understanding way
HM 4968 — fixing material qualities (Buddhism)
HM 5627 — fixing mental qualities (Buddhism)
HM 7154 — imagination (Buddhism)
HM 5298 — tradition (Buddhism)
HM 7154 Understanding what is thought out (Buddhism)
HM 7628 Understanding ; Worldly (Buddhism)
HM 3597 Understatement ; Aesthetic (Japanese)
VD 3900 Undesirableness
HM 8672 Undifferentiated faith
HM 2665 Undifferentiated potentiality (Astrology)
HM 2815 Undifferentiated receptivity (Astrology)
VC 1813 Undivided
HM 3429 Undone ; Leaving (Buddhism)
VP 5960 Undueness ; Dueness
Unease born contact
HM 0972 — element eye consciousness ; Neither psychical ease nor (Buddhism)
HM 3314 — element mind consciousness ; Neither physical ease nor (Buddhism)
HM 0234 — element mind consciousness ; Psychical (Buddhism)
HM 1090 — psychical ; Sensation (Buddhism)
HM 0901 Unease experienced born contact psychical (Buddhism)
VD 3902 Uneasiness
HM 5623 Unelated consciousness (Buddhism)
HM 5623 Unelated mind (Buddhism)
VP 5600 Uneloquence ; Eloquence
VC 1815 Unequivocalness
VC 1817 Unerring
VD 3904 Unevenness
KD 2082 Unexpected global relations
HM 3108 Unexplainable thereness (ICA)
VC 1819 Unfailing
HM 2356 Unfailing prompter (ICA)
VD 3906 Unfaithfulness
HM 3261 Unfaltering exertion ; State (Buddhism)
VD 3908 Unfamiliarity
VP 5855 Unfeelinglessness ; Feeling
HM 6791 Unfettered mind (Buddhism)
HH 0089 Unfoldment ; Spiritual
HM 1705 Unforgetfulness (Buddhism)
VD 3910 Unfriendliness
VD 3912 Ungodliness
VP 6013 Ungodliness ; Godliness
VD 3914 Ungraciousness
VD 3916 Unhappiness
VP 5683 Unhealthfulness ; Healthfulness
KC 0099 Unicity
HM 3020 Unicorn (Buddhism)
HM 3864 Unification Allah (Sufism)
HM 4762 Unification energy (Taoism)
HM 3438 Unification God (Sufism)
HM 3864 Unification God (Sufism)
KC 0686 Unification knowledge
KC 0770 Unification legislation
HM 3340 Unification life essence (Sufism)
HM 6663 Unification mind (Buddhism)
HM 0931 Unification ; Right (Buddhism)
KC 0180 Unified
KC 0890 Unified
HM 7843 Unified consciousness (Buddhism)
KC 0833 Unified field theory
KC 0811 Unified science
KC 0961 Unified science curricula
HH 0892 Unified world order
HH 0555 Uniform
KC 0732 Uniformity
VC 1821 Uniformity
VP 5017 Uniformity-Nonuniformity
KD 2005 Uniformity versus aesthetics ; Forms truth:
VD 3918 Unimaginativeness
VP 5535 Unimaginativeness ; Imaginativeness
VD 3920 Unimportance
VP 5672 Unimportance ; Importance
HM 5768 Unincluded concentration (Buddhism)
VC 1823 Uninfluence
VD 3922 Unintelligence
VP 5467 Unintelligence ; Intelligence
VP 5548 Unintelligibility ; Intelligibility
HM 2015 Uninterrupted release paths (Buddhism)
HM 3889 Unio mystica (Christianity)
KC 0393 Union
VC 1825 Union
HM 3889 Union God ; Mystic (Christianity)
HM 7119 Union God (Sufism)
HM 2280 Union learning application awareness ; Tantra (Buddhism)
HM 4110 Union ; Mansions incipient (Christianity)
HH 0465 Union ; Spiritual (Christianity)
HM 3139 Union ; Total (Hinduism)
HM 2856 Union ; Ultimate (Astrology)
KC 0650 Union ; World
VC 1827 Uniqueness
HM 3155 Uniqueness ; Irreplaceable (ICA)
TC 1816 Unit ; Electronic communications
KC 0777 Unitary
HM 2702 Unitary consciousness

Variety

HM 3521	Unitary experience ; Contemplation (Sufism)
HH 0311	Unitary life
KC 0980	Unitary nation-states
KC 0966	Unitary principle
HH 0726	Unitary reality
KC 0299	Unitary symmetry
KC 0693	Unitary theory
KC 0733	Unitary thought
HH 3290	United Nations Human development
HH 5101	United Nations Human development index
HH 0247	United Nations Human development strategy
HH 0147	United Nations Human resources
HH 0745	United Nations Human resources development
HH 0745	United Nations Manpower resource development
HH 0745	United Nations Manpower training
HH 0745	United Nations Mobilizing human resources
KC 0657	United Nations Organization
HH 4009	United Nations Structural adjustment human face
HM 2583	Uniting intelligence ; Path (Judaism)
HH 4100	Unitiva ; Via (Christianity)
HM 2137	Unitive life
HM 2272	Unitive mysticism
KC 0039	Unity
VC 1829	Unity
HM 2512	Unity ; Aspiration (Astrology)
	Unity being
HM 2702	— [Unity being]
HM 6111	— Essential (Christianity)
HH 0602	— Existential
	Unity [consciousness...]
HM 3193	— consciousness
HM 3193	— consciousness ; Transcendental
KC 0422	— Cultural
	Unity [diversity...]
KC 0002	— diversity
TC 1558	— diversity
KC 0194	— Dramatic
VP 5089	— Duality
KC 0378	Unity faith
KC 0164	Unity faithful
	Unity [Inductive...]
HM 2583	— Inductive intelligence (Judaism)
KC 0979	— intellect
HM 0762	— Intense longing divine (Sufism)
KC 0087	Unity mankind
	Unity [Party...]
KC 0315	— Party
HH 0136	— personality
HH 8206	— personality (Judaism)
KC 0728	— Polyphasic
KC 0543	Unity religion
KC 0791	Unity religions ; Transcendent
	Unity [science...]
KC 0811	— science
HM 9132	— self ; Anomalies experience
HM 9132	— self ; Impairment
KC 0650	Unity ; World
KC 0742	Universal
TP 0013	Universal brotherhood
	Universal [cessation...]
HM 2596	— cessation awareness ; State (Buddhism)
HM 2644	— Christ (ICA)
HM 2734	— compassion (ICA)
HM 2774	— concern (ICA)
HM 2156	— consciousness
KC 0422	— culture
HH 0629	Universal ; Dynamic
HM 2687	Universal fate (ICA)
HM 2418	Universal father (ICA)
KC 0262	Universal grammar
HH 4389	Universal human nature
HH 0731	Universal life energy
HH 0727	Universal man
HM 1717	Universal oneness meditation ; Tenth order perceptions
HM 2458	Universal prior (ICA)
HH 0746	Universal religion (Yoga)
HM 3110	Universal responsibility (ICA)
HH 0746	Universal spirituality (Yoga)
KC 0065	Universalism
VC 1831	Universality
HM 2827	Universality ; Perception (Sufism)
HM 1465	Universalizing faith
KC 0170	Universals
KC 0947	Universals language
KC 0947	Universals ; Language
KC 0228	Universe
KC 0797	Universe discourse
KC 0592	Universe ; Holographic
KC 0278	Universe ; Order
HM 2366	Universe ; Sacramental (ICA)
HM 2445	Universe ; Sacramental (ICA)
KC 0107	University
KC 0790	University ; International
VD 3924	Unkindness
VP 5938	Unkindness ; Kindness
HM 3015	Unknowable peace (ICA)
HM 0317	Unknowing (Taoism)
HM 3167	Unknownness ; Cut off (ICA)
VD 3926	Unlawfulness
VC 1833	Unlimited
HM 3230	Unlimited commitment (ICA)
VP 5507	Unlimitedness ; Limitation
HM 3094	Unlust
HM 1734	Unmaliciousness (Buddhism)
VD 3928	Unmanageability
HM 2640	Unmani (Psychism)
HM 5665	Unmanifest thinking consciousness (Buddhism)
MP 1052	Unmediated relationships ; Interfacing vehicles communication networks
MP 1230	Unmediated supportive emotion
HM 3200	Unmitigated death (ICA)
VD 3930	Unnaturalness
VD 3932	Unneighbourliness
VP 5086	Unnumbered ; Numbered
HM 1965	Unobservant ; Being (Buddhism)
VC 1835	Unobtrusiveness
HM 5908	Unoffended mind (Buddhism)
VP 6024	Unorthodoxy ; Orthodoxy
VD 3934	Unpeacefulness
HM 4572	Unperplexity thought (Buddhism)
HM 0715	Unperturbed mindedness (Buddhism)
HM 4572	Unperturbedness (Buddhism)
HM 0449	Unperturbedness thought (Buddhism)
HM 4076	Unpleasant born contact psychical ; Sensation (Buddhism)
HM 1031	Unpleasant experienced born contact psychical (Buddhism)
HM 0805	Unpleasant ; Psychically (Buddhism)
VD 3936	Unpleasantness
VP 5863	Unpleasantness ; Pleasantness
HM 3094	Unpleasure
VD 3938	Unpredictability
VD 3940	Unpreparedness
VP 5720	Unpreparedness ; Preparedness
VC 1837	Unpretention
TP 0015	Unpretentiousness
VD 3942	Unproductiveness
VP 5165	Unproductiveness ; Productiveness
HM 8375	Unprofitable consciousness sense sphere (Buddhism)
VP 5505	Unprovability ; Provability
VD 3944	Unreality
VD 3946	Unreasonableness
VP 5009	Unrelatedness ; Relatedness
VD 3948	Unreliability
HM 0188	Unremorsefulness (Buddhism)
HM 1915	Unremorsefulness ; Power (Buddhism)
HM 5908	Unrepelled consciousness (Buddhism)
VD 3950	Unruliness
HM 1957	Uns (Sufism)
VP 6026	Unsanctity ; Sanctity
HM 2574	Unsatisfactoriness (Buddhism)
VP 5427	Unsavouriness ; Savouriness
VC 1839	Unselfishness
VP 5978	Unselfishness-Selfishness
TP 0064	Unsettlement
VP 5419	Unsexiness ; Sexiness
HM 2775	Unshared attributes Buddha ; Eighteen (Buddhism)
VP 5733	Unskilfulness ; Skilfulness
VP 5922	Unsociability ; Sociability
VD 3952	Unsophistication
HM 2400	Unspeakable joy (ICA)
HM 2529	Unspeakable joy (ICA)
VC 1841	Unspotted
HM 1858	Unstirring mind (Taoism)
MP 1067	Unstructured common domain
MP 1090	Unstructured context perspective exchange
MP 1115	Unstructured internal domains ; Functional integration
VP 5003	Unsubstantiality ; Substantiality
VD 3954	Unsuitability
VD 3956	Unthoughtfulness
VP 5129	Untimeliness ; Timeliness
HM 2831	Untold ; Wealth (ICA)
HM 6791	Untrammelled consciousness (Buddhism)
VD 3958	Untrustworthiness
HM 1008	Untwistedness (Buddhism)
HM 2611	Unveiled being (ICA)
HM 1452	Unveiling divine mysteries (Sufism)
	Unwholesome dharmas
HM 0552	— Being ashamed acquisition sinful (Buddhism)
HM 1208	— Feeling remorse acquisition sinful (Buddhism)
HM 3257	— Not ashamed acquisition sinful (Buddhism)
HM 3353	— Not feeling remorse acquisition sinful (Buddhism)
HM 1439	Unwholesome root ; Delusion (Buddhism)
HM 1954	Unwholesome root ; Greed (Buddhism)
VD 3960	Unwholesomeness
HM 0767	Unwillingness let go illusion (Korean)
VP 5621	Unwillingness ; Willingness
HM 4999	Upacara samadhi (Buddhism)
HM 5023	Upadisesanibbana ; Sa (Pali)
HH 0781	Upaklesha (Buddhism)
HM 2399	Upalakkhana (Buddhism)
HM 2134	Upanaha (Buddhism)
HM 2392	Upanishadic stages awareness (Hinduism)
HM 3345	Upaparikkha (Pali)
HM 2157	Upasthapura (Buddhism)
HM 1647	Upavicara (Pali)
HH 0908	Upbringing
HM 7769	Upekkha (Pali)
HM 2062	Upekkha (Pali)
HM 3803	Upekkhindriya (Pali)
HM 7769	Upeksha (Buddhism)
VD 3962	Upheaval
VC 1843	Uplift
HM 5986	Uplift due insight (Buddhism)
	Upper grade
HH 7734	— Higher (Taoism)
HH 4289	— low paths (Taoism)
HH 4711	— Lower (Taoism)
HH 6304	— Middle (Taoism)
HM 2633	Upper middle grade (Taoism)
VC 1845	Uprightness
HM 2917	Uprightness (Japanese)
HM 0140	Uprisedness thought (Buddhism)
VD 3964	Uproar
VD 3966	Upset
HM 6287	Uranta-loka (Leela)
KC 0426	Urban dynamics
HM 0131	Urdhva-loka (Jainism)
HH 0574	Urge ; Death
HH 0574	Urge ; Life
HH 7231	Urge ; Life
VD 3968	Urgency
HM 2699	Urgyan awareness ; Guhya samaja (Buddhism)
HM 3123	Uriel ; Angelic awareness
HM 1999	USC Usual state consciousness
VC 1847	Usefulness
VD 3970	Uselessness
MP 1020	User-determined specialized communications
HM 2762	Ushin (Japanese)
HM 4487	Ussaha (Buddhism)
HM 3357	Ussolhi (Pali)
HM 1999	Usual state consciousness (USC)
VD 3972	Usurpation
TC 1831	Utilitarian
VC 1849	Utility
HH 0721	Utopia
KC 0684	Utopia
HH 4510	Utopias ; Imagining
HM 5197	Uttam gati (Leela)
HH 9764	Uttarimanussa (Buddhism)
HM 2625	Utter awareness (ICA)
HM 1371	Utter delusion (Buddhism)
HM 1436	Uyyama (Pali)

V

HM 1157	Vaca ; Samyag (Buddhism)
HM 1781	Vaciduccaritehi arati virati pativirati veramani ; Catuhi (Pali)
VD 3974	Vacillation
HM 0077	Vacillation (Buddhism)
HM 2430	Vacuum ; Existential
VD 3976	Vagrancy
VD 3978	Vagueness
HH 1108	Vaibhasika system (Buddhism)
HM 1301	Vaikuntha-loka (Leela)
VD 3980	Vainness
HM 0903	Vairagya (Yoga)
HH 3772	Vaishnavism (Hinduism)
HM 2336	Vaisvanara (Hinduism)
HM 2028	Vajra-hell (Buddhism)
	Vajra master awareness
HM 2301	— Eighth (Buddhism)
HM 2763	— Fifth (Buddhism)
HM 2849	— Final (Buddhism)
HM 2251	— Fourth (Buddhism)
HM 2325	— Ninth (Buddhism)
HM 2703	— Second (Buddhism)
HM 2789	— Seventh (Buddhism)
HM 2287	— Sixth (Buddhism)
HM 2727	— Third (Buddhism)
HM 2187	Vajra master stage ; First (Buddhism)
HM 3090	Vajra masters ; Way (Buddhism)
HM 2187	Vajracarya (Buddhism)
HH 0309	Vajrayana Buddhism (Buddhism)
HH 4388	Validities (Buddhism)
VC 1853	Validity
HM 1330	Valley spirit (Taoism)
HM 2874	Valleys (Sufism)
VC 1851	Valour
VC 1855	Value
KC 0256	Value analysis
KC 0256	Value control; Value engineering
KC 0256	Value engineering ; Value control;
HM 1991	Value knowledge (Systematics)
TC 1853	Value-monger participant type
KC 0356	Value theory
KC 0387	Value ; Threshold
HM 1466	Valued ideas ; Over
KC 0603	Valued logic ; Multi
VD 3982	Valueless
KP 3016	Values assumptions
HH 0871	Values ; B
HH 0565	Values ; Development moral
HM 1078	Vampire (Psychism)
HH 3452	Vampirism ; Psychic
HH 2782	Vanaprasthashrama (Hinduism)
VD 3984	Vandalism
VD 3986	Vanity
HM 4671	Vanity love
VP 5908	Vanity ; Modesty
VD 3988	Vanquishment
VC 1857	Vantage
VD 3990	Vapidity
MP 1247	Variability ; Embedding fixity
MP 1245	Variability enhance fixity ; Protecting
KD 2272	Variable geometries global order ; Focus
MM 2088	Variable stars ; Pulsational
VD 3992	Variance
VD 3994	Variation
MP 1190	Variation size perspective contexts
HM 1397	Variations consciousness ; Daily
MM 2090	Variations ; Musical
KD 2202	Varieties communication ; Encompassing
KD 2200	Varieties form ; Encompassing
TC 1285	Varieties meeting focus
KC 0606	Variety

Variety

VC 1859	Variety
MP 1035	Variety cyclic elements ; Adequate
MP 1008	Variety forms processes
KC 0146	Variety ; Principle requisite
MP 1135	Varying levels exposure it ; Enhancing insight
HM 3231	Vasana (Yoga)
HM 2070	Vasavartin ; Paranirmita (Buddhism)
VC 1861	Vastness
HM 2468	Vattana (Pali)
HM 0229	Vayama (Pali)
HM 6434	Vayu-loka (Leela)
HM 4488	Vebhabya (Pali)
HM 2270	Vedana (Buddhism)

Vedana Cetosamphassaja

HM 1643	— addukkhamasukha (Pali)
HM 1090	— asata (Pali)
HM 4076	— dukkha (Pali)
HM 0309	— sata (Pali)
HM 4983	Vedana-khanda (Buddhism)

Vedanakkhandhassa sannakkhandhassa

HM 1381	— kammannata (Pali)
HM 1435	— labuta (Pali)
HM 3397	— muduta (Pali)
HM 0349	— (Pali)
HM 1377	— passaddhi (Pali)
HM 0409	— ujukata (Pali)
HH 0518	Vedanta ; Advaita (Hinduism)

Vedayitam [Cetosamphassajam...]

HM 1352	— Cetosamphassajam adukkhamasukham (Pali)
HM 1031	— Cetosamphassajam dukkham (Pali)
HM 0901	— Cetosamphassajamasatam (Pali)
HH 3324	Veganism
MS 8445	Vegetation
VD 3996	Vehemence

Vehicle Buddhism

HH 0309	— Diamond (Buddhism)
HH 0900	— Great (Buddhism)
HH 0845	— Lesser (Buddhism)
HH 0845	— Small (Buddhism)
HM 3869	Vehicle ; Consciousness independent (Yoga)
HM 3048	Vehicle higher path awareness ; Great (Buddhism)
HH 4742	Vehicle ; Highest (Taoism)
HM 2268	Vehicle lower path awareness ; Great (Buddhism)
HM 2160	Vehicle middle path awareness ; Great (Buddhism)

Vehicle path

HH 7386	— Higher (Taoism)
HH 4864	— Lower (Taoism)
HH 6345	— Middle (Taoism)
MP 1052	Vehicles communication networks unmediated relationships ; Interfacing
HH 5342	Vehicles gradual method ; Three (Taoism)
MS 8402	Vehicles parts
HM 3662	Vehicles ; Two (Buddhism)
HM 0450	Vehicles ; Two (Buddhism)
HM 3592	Veils delusion (Hinduism)
HM 2642	Veils ; Manifestation essence (Sufism)
HM 3435	Vela anatikkama (Buddhism)
HM 3435	Vela anatikkamo (Pali)
TC 1800	Velcro-board
VD 3998	Venality
VC 1863	Venerableness
VC 1865	Veneration
VP 5947	Vengeance ; Forgiveness
VD 4000	Venial
VD 4002	Venom
VC 1867	Venturesomeness
VC 1869	Veracity
HM 6631	Veracity (Sufism)
HM 1781	Veramani ; Catuhi vaciduccaritehi arati virati pativirati (Pali)
HM 0252	Veramani ; Miccha ajiva arati virati pativirati (Pali)
HM 1133	Veramani ; Tihi kayaduccaritehi arati virati pativirati (Pali)
HH 0325	Verbal attack therapy
TC 1378	Verbal interaction preferences
TC 1640	Verbal interaction preferences ; Non
HM 7171	Verbal misconduct ; Abstinence (Buddhism)
VD 4004	Verbosity
VC 1871	Veritableness
MS 8103	Vermin
HM 3401	Veronica wipes face Jesus (Christianity)
VC 1873	Versatility
KD 2005	Versus aesthetics ; Forms truth: uniformity
KC 0248	Vertical integration
HM 2320	Vertical lateral thinking
KC 0512	Vertical thinking
HM 0265	Vertigo ; Alternobaric
HM 0265	Vertigo ; Pressure
HM 2576	Very-hot-hells awareness (Buddhism)
HM 1468	Vespers (Christianity)
HM 1468	Vespers ; Hours (Christianity)
TP 0050	Vessel ; Sacrificial (Cauldron)
MS 8426	Vessels ; Container
HM 4664	Vestibular coriolis reaction
VD 4006	Vexation
HH 6030	Via illuminativa (Christianity)
HH 1171	Via negativa
HH 6222	Via positiva
HH 4090	Via purgativa (Christianity)
HH 4100	Via unitiva (Christianity)
VC 1875	Viability
HM 2982	Vibrant powers (ICA)
HH 4908	Vibrational health
HM 3732	Vibrations ; Plane primal (Leela)
HM 2543	Vibratory maintenance
HH 2543	Vibratoryhood ; Right
HM 5089	Vicara (Pali)
HM 3880	Vicara samadhi (Yoga)
HM 3460	Vicara stage (Yoga)
HM 2450	Vicara ; Vitakka (Pali)
HM 3408	Vicaya (Pali)
VP 5980	Vice ; Virtue
HM 2165	Vices ; Splendid (ICA)
HM 5089	Vichara (Buddhism)
HM 4467	Vichikitsa - The-tshom (Tibetan)
HM 0529	Vicikiccha (Pali)
HM 0812	Vicikiccha (Pali)
VD 4008	Viciousness
VD 4010	Vicissitude
VD 4012	Victimization
HM 2190	Victimization emblems (Tarot)
VC 1877	Victory
VP 5726	Victory-Defeat
HM 2362	Victory ; Sphere (Kabbalah)
HM 4489	Videha-mukti (Yoga)
HM 2543	Videha ; Purva (Buddhism)

Video [tape...]

TC 1897	— tape
TC 1218	— teleconferencing
TC 1876	— teleconferencing ; Full motion
HM 2723	Vidyadhara (Buddhism)
HM 2759	Vidyadhara-emperor (Buddhism)
HM 2211	Vidyadhara ; Kamadera (Buddhism)
HM 2235	Vidyadhara ; Rupa dhatu (Buddhism)
HM 3066	Vidyahara (Buddhism)
HM 1191	View (Buddhism)
HM 2944	View Buddhism ; Emptinesses paths (Buddhism)
HM 2842	View ; Contextual world (ICA)
HM 3252	View ; Correct (Buddhism)
HM 0700	View ; Grasping (Buddhism)
HM 1992	View ; Holding paramount one's (Buddhism)
HM 3432	View ; Inclination towards (Buddhism)
HH 2718	View ; Purification (Buddhism)
HM 1280	View ; Right (Buddhism)

View [World...]

KC 0053	— World
HM 5324	— Wrong (Buddhism)
HM 1710	— Wrong (Buddhism)
MP 1142	Viewpoint loci structure ; Sequence
HM 2996	Views ; Afflicted (Buddhism)
HM 1407	Views ; Distortion (Buddhism)
HM 1511	Views ; Fetters (Buddhism)
HM 1751	Views ; Grasping inverted (Buddhism)
HM 2054	Views opinions ; Current (Buddhism)
HH 2718	Views ; Purity (Buddhism)
HM 2722	Views ; Scuffle (Buddhism)
HM 1613	Views ; Thicket (Buddhism)
HM 4548	Views ; Wilderness (Buddhism)
VC 1879	Vigilance
HH 0950	Vigilance
HH 0950	Vigilance ; Revolutionary
TP 0051	Vigilance (Thunderbolts)
VC 1881	Vigour
TP 0034	Vigour ; Great
HH 0677	Vigour (Zen)
HM 3210	Vihimsa (Buddhism)
HM 3617	Vijna (Buddhism)
HM 3160	Vijna (Yoga)
HM 2730	Vijnana ; Alaya (Buddhism)
HM 3617	Vijnana (Buddhism)
HM 3460	Vijnanamaya kosa (Yoga)
HM 3043	Vijnananantyayatana (Buddhism)
HM 3306	Vikalpa (Buddhism)
HM 4425	Vikkhepo ; Cetaso (Pali)
HM 3154	Vikshepa (Buddhism)
VD 4014	Vileness
VD 4016	Villainy
HM 2739	Vimala (Buddhism)
HM 4008	Vimati (Pali)
HM 5977	Vimokkha (Pali)
HM 5977	Vimoksa (Buddhism)
HM 2015	Vimuktimarga (Buddhism)
HM 0831	Vimutti ; Ceto (Buddhism)
HH 1376	Vinaya (Buddhism)
HH 3398	Vinaya (Buddhism)
VP 6005	Vindication-Condemnation
VD 4018	Vindictiveness
HM 0927	Vinivaranata (Buddhism)
HH 0170	Viniyata (Buddhism)

Vinnana [khanda...]

HM 2098	— khanda development (Pali)
HM 2556	— khanda (Pali)
HM 2599	— kicca (Pali)
HM 3617	Vinnana (Pali)
HM 3043	Vinnanacayatana (Pali)
HM 3403	Vinnanakkhandhassa lahuta (Pali)
HM 1796	Vinnanakkhandhassa muduta (Pali)
HM 1741	Vinnanakkhandhassa pagunata (Pali)
HM 4605	Vinnanakkhandhassa (Pali)
HM 1338	Vinnanakkhandhassa ujukata (Pali)
HM 2098	Vinnanakkhandha (Pali)
VD 4020	Violation
HM 2561	Violation ; Personal (ICA)
VD 4022	Violence
HH 0088	Violence ; Non
HM 1276	Violence ; Plane (Leela)
HM 8762	Violent rasa (Hinduism)
HM 1751	Vipariyasaggaha (Pali)
HM 2207	Vipashyana (Pali)
HM 2623	Vipashyana (Buddhism)
HH 0680	Vipassana-bhavana (Buddhism)
HM 1632	Vipassana (Pali)
HM 7298	Vipulamati (Hinduism)
HM 0903	Viraga (Yoga)

Virati pativirati veramani

HM 1781	— Catuhi vaciduccaritehi arati (Pali)
HM 0252	— Miccha ajiva arati (Pali)
HM 1133	— Tihi kayaduccaritehi arati (Pali)
HM 2666	Virgin birth (ICA)
VC 1883	Virginity
HH 0272	Virginity
HM 2439	Virgo-consciousness (Astrology)
VC 1885	Virility
HM 1333	Viriya (Pali)
HM 1943	Viriyabala (Pali)
HM 1263	Viriyarambho ; Cetasiko (Pali)
HM 1470	Viriyindriya (Pali)
HM 1829	Viriyindriya (Pali)
HM 4239	Virodha (Pali)
HM 1217	Virtual learning
HM 1772	Virtual reality
HH 0712	Virtue
VC 1887	Virtue
HH 1865	Virtue ; Purification (Buddhism)
HH 0990	Virtue simplicity
VP 5980	Virtue-Vice
HH 0712	Virtues ; Cardinal
HH 0578	Virtuous mental factors (Buddhism)
VD 4024	Virulence
HM 2389	Virya (Buddhism)
HH 0677	Virya (Zen)
HM 0883	Visada (Hinduism)
HM 1766	Visceral experience
TC 1191	Viscidity ; Conference
HH 3772	Vishishtadvaita (Hinduism)
HM 3461	Vishuddha (Yoga)
MP 1222	Visibility ; Ground level
VP 5444	Visibility-Invisibility
HM 3347	Vision
VC 1889	Vision
KC 0847	Vision ; Artistic
HM 2240	Vision awareness ; Discipleship (Buddhism)
HM 3347	Vision ; Beatific
VP 5439	Vision-Blindness
TC 1940	Vision conferencing
HM 2240	Vision cultivation ; Paths (Buddhism)
KD 2146	Vision ; De categorization poly ocular
HM 3521	Vision God ; Spiritual (Sufism)

Vision [path...]

HM 2228	— path awareness ; Pratyeka Buddha (Buddhism)
HM 2966	— Pure
HH 3025	— Purification knowledge (Buddhism)
HH 0897	Vision quest

Vision [Saviors...]

HM 2855	— Saviors God: (Christianity)
HM 3050	— Soul (Psychism)
HM 0575	— Spiritual (Sufism)
HM 1237	— Vision ; Tunnel

Vision [way...]

HH 3550	— way ; Purification knowledge (Buddhism)
HH 4007	— what is path what is not path ; Purification knowledge (Buddhism)
KD 2340	— world ; Language probabilistic
TC 1940	Visionary conferencing
HM 9218	Visionary experience
HH 0142	Visioning ; Re
HM 3347	Visions ; Seeing
HM 0357	Visions Trinity (Christianity)
HM 7019	Visual agnosia
HM 2074	Visual awareness (Buddhism)
HM 2176	Visual consciousness
MS 8600	Visual elements ; Primary
HM 0092	Visual hallucination
MM 2021	Visual illusions ; Ambiguous
KC 0418	Visualization
HH 6053	Visualization ; Guided (Magic)
HM 1690	Visualization ; Tantric (Yoga)
HM 6149	Visuddhi ; Citta (Buddhism)

Visuddhi niddesa

HH 2718	— Ditthi (Pali)
HH 1187	— Kankhavitarana (Pali)
HH 4007	— Maggamagga nanadassana (Pali)
HH 3025	— Nanadassana (Pali)
HH 3875	— Visuddhimagga (Pali)
HM 1177	Vitakka (Pali)
HM 2450	Vitakka-vicara (Pali)
HH 0213	Vital force ; Regulation (Hinduism)
HM 3320	Vital-physical stage life
HM 2875	Vital signs (ICA)
VC 1891	Vitality
HM 3404	Vitality (Buddhism)
TP 0058	Vitality (Lake)
MM 2079	Vitamins
HM 1177	Vitarka (Buddhism)
HM 4451	Vitarka samadhi (Yoga)
HM 2912	Vitarka stage (Yoga)
VD 4026	Vitiation
HM 1029	Vitrisna (Yoga)
HM 1545	Vitti (Pali)
VD 4028	Vituperation
VC 1893	Vivacity
HM 1601	Vivek (Leela)
HM 4493	Viveka khyati (Yoga)
HM 2998	Viveka (Yoga)
HM 1456	Vivekajam jnanam (Yoga)
HM 2912	Vivesa stage gunas (Yoga)
HM 0239	Viyukta perception (Hinduism)
HH 0619	Vocation (Christianity)

Wisdom

HM 2616	Vocation ; Historical (ICA)
HM 2621	Vocation ; Primal (ICA)
HM 2441	Vocation ; Realized (ICA)
HM 2293	Vocation ; Religious (ICA)
HH 0178	Vocational guidance
HH 8005	Vocational guidance
HH 0047	Vocational training
HM 2227	Voice silence
MM 2061	Voice speech
HM 1317	Voices ; Inner
HM 2994	Void ; Eternal (ICA)
HM 2213	Void ; Near death black (NDE)
HM 2193	Voidness (Buddhism)
VD 4030	Volatility
HM 0404	Volition born contact element mind consciousness (Buddhism)
HM 2589	Volition (Buddhism)
HM 3464	Volitional stage life ; Mental
KC 0651	Voluntary economic integration
HH 0991	Voluntary poverty
VC 1895	Voluptuousness
HH 0061	Voodoo
HM 2502	Voodoo trance consciousness
VD 4032	Voracity
HM 1240	Voulu ; Déjà
HM 2232	Vriddhi samadhi ; Aloka (Buddhism)
HM 2738	Vritti ; Antar (Yoga)
HM 2581	Vritti ; Bahir (Yoga)
HM 2692	Vritti (Hinduism)
HH 2772	Vritti yoga ; Abhyantara (Yoga)
HM 1240	Vu ; Déjà
HM 3384	Vu ; Jamais
VD 4034	Vulgarity
VP 5896	Vulgarity ; Taste
VD 4036	Vulnerability
HM 1015	Vyagra (Yoga)
HM 6002	Vyana-loka (Leela)
HM 0487	Vyappana (Pali)
HM 1295	Vyayama ; Samyag (Buddhism)
HM 2729	Vyvpashamana (Buddhism)

W

HM 4001	Wa (Japanese)
HM 1365	Wabi (Japanese)
HM 5409	Wadd (Islam)
HM 7712	Wadud ; Al (Sufism)
HM 0069	Wahhab ; Al (Sufism)
HM 5506	Wahid ; Al (Sufism)
TP 0005	Waiting (Getting wet)
HM 6129	Waiting God (Christianity)
HM 3650	Wajd (Sufism)
HM 0743	Wajid ; Al (Sufism)
HM 2014	Wakeful dream (Hinduism)
HM 2020	Wakeful hypometabolic state
HM 3476	Wakefulness ; Alert
HM 3510	Wakefulness ; Dream (Hinduism)
HM 3055	Wakefulness ; Great (Hinduism)
HM 2567	Wakefulness (Hinduism)
HM 2159	Wakefulness ; Seed state (Hinduism)
HM 3021	Wakening consciousness
HM 1382	Wakil ; Al (Sufism)
HM 3021	Waking consciousness
HM 2141	Waking consciousness ; Jagrat state (Yoga)
HM 1376	Waking dreams
HM 6719	Wal Ikram ; Dhul Jalal (Sufism)
HM 6719	Wal Ikram ; Dhul Jalali (Sufism)
HM 8303	Walah (Islam)
HM 0525	Wali ; Al (Sufism)
HM 1633	Wali ; Al (Sufism)
HM 4418	Wali (Sufism)
MM 2001	Walking
HM 1272	Walking ; Sleep
TP 0056	Wanderer Marginality
TP 0056	Wanderer Newcomer
TP 0056	Wanderer Traveller
HM 5622	Wandering mind (Taoism)
HM 0346	Wang ; Tso (Taoism)
VD 4038	Wantonness
VD 4040	War
MM 2006	War
HM 5339	War'a (Sufism)
HH 3687	War ; Holy (Islam)
HM 0286	Wara (Sufism)
HM 0530	Warid (Sufism)
HM 0113	Warith ; Al (Sufism)
VD 4042	Warlike
VC 1897	Warmheartedness
VC 1899	Warmth
VD 4044	Warpedness
HM 1598	Warpedness (Japanese)
HH 3322	Warrior ; Way
HH 8219	Warrior ; Way (Amerindian)
HH 3664	Washing (Taoism)
HM 4169	Wasi' ; Al (Sufism)
VD 4046	Waste
VD 4048	Wastefulness
TP 0020	Watching
HM 2305	Watching (Sufism)
HH 1008	Watchings
MM 2081	Water ; Drawing
MM 2081	Water ; Getting
HH 3246	Water kasina (Buddhism)
HM 2384	Water ; Triplicity (Astrology)
HH 4219	Water wine ; Journeying transcendence
HH 6452	Waterwheel exercise (Taoism)
HM 2384	Watery awareness (Astrology)

Wave consciousness

HM 2345	— Alpha
HM 3476	— Beta
HM 1785	— Delta
HM 2321	— Theta
HM 3074	Wavering ; Indecisive (Buddhism)
HM 4382	Wavering ; Not (Buddhism)
HH 3129	Waves ; Brain
HH 0424	Way
HM 1211	Way action (Hinduism)
HM 1997	Way active intelligence (Esotericism)

Way [beauty...]

HM 0763	— beauty (Esotericism)
HM 2769	— bodhisattvas ; Meditation (Buddhism)
HM 2185	— Buddhas (Buddhism)
HM 3516	Way cross (Christianity)
HM 6554	Way devotion (Esotericism)

Way [Eightfold...]

HM 2339	— Eightfold (Buddhism)
HH 0495	— enlightenment (Hinduism)
HM 1912	— Erroneous (Buddhism)

Way [faith...]

HH 3412	— faith ; Passive
HM 4968	— fixing material qualities ; Understanding (Buddhism)
HM 5627	— fixing mental qualities ; Understanding (Buddhism)
HM 2245	— four truths ; Meditation (Buddhism)
HH 1909	Way goddess
HM 0763	Way harmony (Esotericism)
HM 2161	Way hearers ; Meditation (Buddhism)
HM 6554	Way idealism (Esotericism)
HM 7154	Way imagination ; Understanding (Buddhism)
HM 1201	Way knowledge (Esotericism)
HH 0495	Way knowledge (Hinduism)

Way [leading...]

HM 7645	— leading cessation suffering ; Understanding knowledge (Buddhism)
HM 1507	— love-wisdom (Esotericism)
HH 0628	— loving faith (Hinduism)

Way [Martial...]

HH 0085	— Martial
HH 1010	— Middle (Buddhism)
HH 1038	— Middle (Buddhism)
HM 1605	Way organization (Esotericism)

Way [paradox...]

HH 3335	— paradox (Christianity)
HM 0454	— power (Esotericism)
HH 3550	— Purification knowledge vision (Buddhism)
HH 3550	— Purity knowledge discernment middle (Buddhism)
MP 1050	Way relationship entrainment ; Three
HM 1605	Way ritual (Esotericism)

Way [salvation...]

HH 1166	— salvation
HM 1201	— science (Esotericism)
HM 5450	— silence
HM 2709	— solitary realizers ; Meditation (Buddhism)
HH 1867	— Spiritual

Way [tradition...]

HM 5298	— tradition ; Understanding (Buddhism)
HH 6566	— transcendence (Christianity)
HH 0631	— Triple (Christianity)
HM 3090	Way vajra masters (Buddhism)

Way [warrior...]

HH 3322	— warrior
HH 8219	— warrior (Amerindian)
HM 0454	— will (Esotericism)
HH 0873	Ways knowing (Buddhism)
HH 1616	Ways knowing (Christianity)
HH 0477	Ways life ; Alternative (AWL)
HH 0477	Ways life ; Dominant (DWL)
HH 2665	Ways spiritual realization (Esotericism)
HH 0631	Ways ; Three spiritual (Christianity)
HM 5022	We-psychology (Christianity)
HM 6652	Weak concentration (Buddhism)
VD 4050	Weak-mindedness
TP 0009	Weak ; Restraint
VD 4052	Weakness
VP 5159	Weakness ; Strength
TC 1908	Weakwilled conferencing
VC 1901	Wealth
TP 0014	Wealth
HH 0353	Wealth (Hinduism)
VP 5837	Wealth-Poverty
HM 2831	Wealth untold (ICA)
MS 8106	Weapons ; Armour
MS 8723	Weapons ; Nuclear mass destruction
MS 8525	Weapons ; Primitive
VD 4054	Wear
VD 4056	Weariness
VD 4058	Wearisomeness
MM 2054	Weather ; Seasons
MP 1016	Web general interrelationships
MP 1019	Web selective interchange
HH 1073	Wei ; Wu
VP 5352	Weight-Lightness
VC 1903	Welfare
TP 0048	Well Basic need
HM 2422	Well being ; Emblems (Tarot)
HM 2409	Well being ; Feeling
VC 1905	Well-disposed
TP 0048	Well Fulfilled potentialities
VC 1907	Well-grounded
VC 1909	Well-informed

Well [made...]

VC 1911	— made

Well [mannered...] cont'd

VC 1913	— mannered
VC 1915	— meaning
VC 1917	Well-provided
KC 0053	Weltanschauung
HM 0900	Weltschmerz
HM 2519	Western-continent awareness cattle (Buddhism)
HM 5909	Wheel becoming (Buddhism)
HM 2250	Wheel ; Fortune (Tarot)
HM 2759	Wheel-turner-king awareness (Buddhism)
HM 2058	Wheel-turning-king awareness (Buddhism)
HH 3556	White magic
TC 1487	Whiteboard
HM 6328	Whitening ; Alchemical (Esotericism)
KC 0500	Whole
KC 0879	Whole categories ; Part
HM 3070	Whole ; Harmony
HM 1367	Whole ; Love
KC 0328	Whole system principle
KD 2105	Whole ; Trialectics: logic
TP 0061	Wholehearted allegiance

Wholeness

HM 2725	— [Wholeness]
KC 0866	— [Wholeness]
VC 1919	— [Wholeness]
HM 4002	Wholeness ; Experience (Systematics)
KD 2230	Wholeness implicate order
HM 3752	Wholesome root ; Disinterestedness (Buddhism)
HM 3560	Wholesome root ; Non hatred (Buddhism)
VC 1921	Wholesomeness
KC 0941	Wholism
HM 4501	Wholly ; God (Judaism)
HM 4291	Wholly same ; God (Judaism)
HH 1909	Wicca craft
HM 3789	Wickedness
MP 1062	Wider perspective ; Points
HM 6556	Wieldiness citta (Pali)
HM 6556	Wieldiness consciousness (Buddhism)
HM 7969	Wieldiness kaya (Pali)

Wieldiness [mental...]

HM 7969	— mental body (Buddhism)
HM 7969	— mental factors (Buddhism)
HM 6556	— mind (Buddhism)
HM 7089	Wilayat-i-mala'ika (Sufism)
HM 7089	Wilayat-i-'ulya (Sufism)
HM 3647	Wilayat (Sufism)
HM 4548	Wilderness views (Buddhism)
VC 1923	Will
HH 0920	Will
HM 3335	Will behavioural transformation (Sufism)
HH 0920	Will Free will ; Good

Will [God...]

HH 1016	— God ; Resignation
HH 1799	— God ; Surrender (Islam)
HH 0920	— Good will Free
HH 3006	Will (Islam)
HM 3871	Will live (Yoga)
HM 2629	Will ; Path intelligence (Judaism)
HH 0883	Will therapy
HM 0454	Will ; Way (Esotericism)
VD 4060	Wilfullness
TP 0057	Willing submission (Wind)
VC 1925	Willingness
VP 5621	Willingness-Unwillingness
TP 0057	Wind Gentleness
TP 0057	Wind Penetrating clarity
TP 0057	Wind Willing submission
HH 4219	Wine ; Journeying transcendence water
HH 1187	Winner ; Junior stream (Buddhism)
HM 3401	Wipes face Jesus ; Veronica (Christianity)
HM 7119	Wisal (Sufism)

Wisdom

HH 0623	— [Wisdom]
VC 1927	— [Wisdom]
HM 2029	— [Wisdom]

Wisdom [All...]

HM 4320	— All accomplishing
HM 2261	— archetypal image (Tarot)
HM 3066	— ascension stages game ; Masters (Buddhism)
HM 2663	— awareness ; Bon (Buddhism)

Wisdom [body...]

HM 2834	— body ; Buddha (Buddhism)
HM 1623	— Breadth (Buddhism)
HM 4523	— (Buddhism)
HM 0082	— (Buddhism)
HM 2822	Wisdom ; Conscious (Astrology)

Wisdom [Ecological...]

HH 2315	— Ecological
HM 0107	— Essential
HM 2272	— Experiential

Wisdom [factors...]

HM 6336	— factors (Buddhism)
HM 3233	— Faculty (Buddhism)
HM 1840	— Faculty (Buddhism)
HM 2001	— Fifth plane (Hinduism)
HM 3100	— First plane (Hinduism)
HM 3006	— Fourth plane (Hinduism)
HM 0633	Wisdom goad (Buddhism)
HM 4358	Wisdom guide (Buddhism)

Wisdom [heart...]

HM 0935	— heart
HM 8667	— heart (Sufism)
HM 2723	— holder awareness ; Hindu (Buddhism)
HM 1486	Wisdom ; Jewel (Buddhism)

Wisdom [Lamp...]

HM 1846	— Lamp (Buddhism)

Wisdom

	Wisdom [Leela...] cont'd
HM 3846	— (Leela)
HM 8125	— Liberation means (Pali)
HM 1547	— Lustre (Buddhism)
HM 2655	Wisdom ; Mental (Buddhism)
	Wisdom [Palace...]
HM 2508	— Palace (Buddhism)
HH 0102	— Paths
HM 2509	— Paths (Judaism)
HM 7844	— Perfect (Buddhism)
HM 3298	— Planes (Hinduism)
HM 1792	— Power (Buddhism)
HH 7902	— Practical
	Wisdom [Second...]
HM 2443	— Second plane (Hinduism)
HM 3354	— Sixth plane (Hinduism)
HM 2348	— Sphere (Kabbalah)
HM 0192	— Splendour (Buddhism)
HM 1645	— Sword (Buddhism)
HM 3312	Wisdom ; Third plane (Hinduism)
HM 0107	Wisdom ; Transcendent
HM 1507	Wisdom ; Way love (Esotericism)
HM 4309	Wise attention (Buddhism)
TP 0027	Wise counsel
HM 2202	Wise man archetypal image ; Old (Tarot)
HM 4487	Wish act (Buddhism)
HM 2179	Wish-dominated awareness
HM 2433	Wishes
HM 0460	Wishes ; Fewness (Buddhism)
HM 2179	Wishful thinking
HH 3873	Wishing ; Ill
HH 0973	Witch doctor
HH 1909	Witchcraft
VD 4062	Withdrawal
TP 0033	Withdrawal
HM 2267	Withdrawal joy ; Subtle contemplation (Buddhism)
HH 0829	Withdrawal senses (Yoga)
HM 2243	Withdrawal (Sufism)
VD 4064	Withering
TP 0025	Without Expectation
VD 4066	Witlessness
HH 0838	Witness
HH 0110	Witness
HM 2396	Witness-consciousness (Yoga)
HM 3911	Witness meditation
VC 1929	Wittiness
VD 4068	Woe
MS 8955	Woman ; Man
VC 1931	Womanliness
HH 1976	Women ; Psychological development
HM 0767	Won (Korean)
HM 3197	Wonder
VC 1933	Wonder
HM 0291	Wonder ; Aesthetic emotion (Hinduism)
HM 3052	Wonder-filled fate (ICA)
HM 2186	Wonder ; Primordial (ICA)
HM 0284	Wonder ; Sense (Christianity)
VP 5920	Wonder-Unastonishment
HM 3062	Wonder world (ICA)
VC 1935	Wonderfulness
HM 2236	Word-bearing priest (ICA)
HM 2373	Word ; Contentless (ICA)
HM 6539	Word-deafness
HM 3178	Word ; Living (ICA)
HM 6522	Word soul ; Speaking (Christianity)
VC 1937	Work
HH 1909	Work ; Craft
HM 2552	Work ; Disengagement (ICA)
HM 4877	Work ; Enjoying
HH 0758	Work ; Fitness
HM 4877	Work ; Flow experience
HH 4502	Work group initiation
HH 0801	Work ; Social group
HM 4605	Workability aggregate consciousness (Buddhism)
HM 1381	Workability aggregate sensation, aggregate cognition, aggregate synergies (Buddhism)
HM 0739	Workability body (Buddhism)
HM 1584	Workability thought (Buddhism)
HM 3716	Workable ; Being (Buddhism)
TC 1485	Workgroup
HM 0979	Working ability (Buddhism)
HM 5309	Working accordance righteousness ; Meeting God both resting (Christianity)
HH 0409	Working-over
TC 1777	Working ; Risk
HH 0409	Working-through
TC 1721	Working ; Trust
MS 8815	Works ; Structures engineering
HH 0917	Workshop ; Creativity group
TC 1377	Workshop facilitation
TC 1449	Workshop methods
	World [Anomalies...]
HM 4754	— Anomalies experience self distinct outside
HM 2579	— awareness ; Asura (Buddhism)
HM 2031	— awareness ; Naga (Buddhism)
	World [Being...]
HM 2632	— Being
HM 4049	— Being another (Balinese)
HM 4570	— beyond form (Sufism)
KC 0357	— brain
	World [citizenship...]
KC 0067	— citizenship
HH 2189	— concept ; Three
HM 3038	— creation (Judaism)
KC 0750	— crises
KC 0422	— culture

	World [cycles...] cont'd
HH 2002	— cycles (Buddhism)
HM 1973	World desires (Buddhism)
	World [economy...]
KC 0821	— economy
KC 2408	— emanation (Judaism)
KC 0357	— encyclopaedia
	World [federal...]
KC 0153	— federal government
HH 3442	— Flight
HM 1973	— form (Buddhism)
HM 2360	— formation (Judaism)
HM 1973	— formlessness (Buddhism)
HM 3286	— forms (Sufism)
KC 0153	World government
HM 3492	World humanity (Buddhism)
HM 3616	World imagination (Sufism)
KC 0755	World ; Integrated
	World [Language...]
KD 2340	— Language probabilistic vision
KC 0692	— law
HM 5265	— Leaping out (Taoism)
KC 0324	— line
	World [manifestation...]
HM 2312	— manifestation (Judaism)
KC 0659	— market
HM 0486	— Merging ordinary (Taoism)
HM 2614	— midst this world ; Other (ICA)
KC 0372	— mind groups
KC 0042	— modelling
HM 3425	World nature (Sufism)
	World order
KC 0496	— [World order]
HH 0892	— Spiritual
HH 0892	— Unified
HM 2614	World ; Other world midst this (ICA)
KC 0750	World problems
HH 5119	World ; Re souling
	World [socialist...]
KC 0804	— socialist market
KC 0506	— society
KC 0302	— soul
HM 3974	— spiritual perception (Sufism)
HH 0489	— Stopping
KC 0805	— system
	World [Tarot...]
HM 2387	— (Tarot)
HM 3214	— transcending spheres (Buddhism)
HM 5265	— Transcending (Taoism)
KC 0650	World union
KC 0650	World unity
KC 0053	World view
HM 2842	World view ; Contextual (ICA)
HM 3062	World ; Wonder (ICA)
HM 7234	Worldly concentration (Buddhism)
HM 2451	Worldly detachment (ICA)
HM 7628	Worldly understanding (Buddhism)
HM 2072	Worlds conscious existence (Buddhism)
HM 2341	Worlds ; Maqamat seven (Sufism)
HM 2657	Worlds ; Mutual possession ten (Buddhism)
HM 2657	Worlds ; Ten (Buddhism)
HM 2026	Worlds ; Three (Yoga)
VD 4070	Worry
HM 9101	Worry (Buddhism)
HM 8002	Worsening ; Concentration partaking (Buddhism)
HH 0298	Worship
HH 2768	Worship ; Corporate (Christianity)
HH 3776	Worship ; Discipline (Christianity)
HH 0298	Worship ; Discipline (Christianity)
HM 0128	Worship God (Sufism)
HH 0653	Worship ; Hero
HH 0536	Worship (Islam)
HM 5876	Worship (Islam)
HM 0561	Worship ; Religious experience personal devotion
VC 1939	Worth
HH 0634	Worth ; Self
VC 1941	Worthiness
VD 4072	Worthlessness
VD 4074	Wound
VD 4076	Wrangle
MS 8245	Writing
HM 1094	Writing ; Inspirational (Psychism)
HM 1802	Wrong concentration (Buddhism)
HM 1585	Wrong endeavour (Buddhism)
HM 7887	Wrong livelihood ; Abstinence (Buddhism)
HM 0252	Wrong livelihood ; Leaving off, abstaining, totally abstaining refraining (Buddhism)
HM 0863	Wrong path (Buddhism)
HH 4007	Wrong path ; Purity knowledge discernment right path (Buddhism)
HM 1080	Wrong questioning (Yoga)
HM 5324	Wrong view (Buddhism)
HM 1710	Wrong view (Buddhism)
HM 0286	Wrongdoing ; Abstinence (Sufism)
VD 4078	Wrongness
HM 3356	Wrongness (Buddhism)
VP 5957	Wrongness ; Rightness
HM 3485	Wu-chu (Zen)
HM 2163	Wu-hsin (Zen)
HM 0910	Wu-hsing (Zen)
HM 2500	Wu-nien (Zen)
HH 1073	Wu-wei
HM 2326	Wu (Zen)
HH 4019	Wu (Zen)
HM 5118	Wuiquf al-'adudi (Sufism)
HM 7221	Wundersucht (German)

HM 2792	Wuquf-i qalbi (Sufism)

X

HM 3003	X ; Factor
HM 3631	Xantosis (Esotericism)
HM 1608	Xenoglossis (Christianity)
VD 4080	Xenophobia
HM 0020	Xenophrenia (Psychism)

Y

HM 3022	Yahriel
HH 2877	Yahweh
HM 5965	Yaksha-loka (Leela)
HM 2130	Yama-devas (Buddhism)
HM 2088	Yama state (Buddhism)
HH 6876	Yama (Yoga)
HM 1279	Yamuna (Leela)
HM 2435	Yan-dag-pa-ji-lta-ba-bzin-du (Tibetan)
HH 4997	Yang ; Balancing yin (Taoism)
HM 1330	Yang ; Merging yin (Taoism)
HH 1224	Yantra yoga (Yoga)
HH 0112	Yantras
HH 2662	Yantras
HM 3327	Yapana (Pali)
HM 1691	Yapana (Pali)
HM 4556	Yaqin ; Haqq (Sufism)
HM 0930	Yaqin ; 'Ilm (Sufism)
HM 4426	Yaqin (Sufism)
HM 4426	Yaquin (Sufism)
HM 7549	Yatha-bhuta-jnana-darsana (Buddhism)
HM 7549	Yatha-bhuta-nana-dassana (Pali)
HM 2435	Yathabhuta (Buddhism)
HM 7308	Yearning (Hinduism)
TC 1651	Yes person
HM 2410	Yesod sephira (Kabbalah)
HM 2360	Yetzirah ; Olam (Judaism)
HM 2623	Yid byed pa (Buddhism)
HM 2838	Yid-kyi-rnam-par-shes-pa (Tibetan)
HM 3237	Yid-la-byed-pa (Tibetan)
TP 0033	Yielding
HH 4997	Yin yang ; Balancing (Taoism)
HM 1330	Yin yang ; Merging (Taoism)
HM 3871	Yoga Abhinvesa
HH 2772	Yoga Abhyantara vritti yoga
HH 2772	Yoga ; Abhyantara vritti (Yoga)
HH 0829	Yoga Abstraction
HM 2491	Yoga Action desire
HH 5487	Yoga Action desire
HH 0372	Yoga action (Yoga)
HH 0406	Yoga Actualism
HM 0110	Yoga Addiction objects
HM 2933	Yoga Adhimoksha
HH 1047	Yoga Afflictions
HH 0406	Yoga Agni yoga
HH 0406	Yoga ; Agni (Yoga)
HM 2144	Yoga Ajna
HM 3509	Yoga Alinga stage gunas
HM 3869	Yoga Aloneness
HM 3242	Yoga Anahata
HM 3227	Yoga Ananda
HM 3212	Yoga Ananda stage
HM 3212	Yoga Anandamaya kosha
HH 7219	Yoga Animistic yoga
HH 7219	Yoga ; Animistic (Yoga)
HM 2738	Yoga Antar-vritti
HH 0669	Yoga Asana
HH 0862	Yoga Ashtanga yoga
HM 2937	Yoga Asmita
HM 3509	Yoga Asmita stage
HH 3904	Yoga Asparsa yoga
HH 3904	Yoga ; Asparsa (Yoga)
HH 0779	Yoga Astanga yoga
HM 3509	Yoga Atma
HH 0774	Yoga Atma yoga
HM 3119	Yoga Attachment
HM 4058	Yoga Autism
HM 2032	Yoga Avasthas
HM 3460	Yoga Avivesa stage gunas
HM 2144	Yoga Awareness authority
HM 2032	Yoga Awareness relative reality
HM 2984	Yoga Awareness self spiritual intelligence
HM 3232	Yoga Awareness spiritual identity
HM 2581	Yoga Bahir-vritti
HH 8007	Yoga Bandhas
HM 2933	Yoga Belief
HH 0337	Yoga Bhakti yoga
HM 3227	Yoga Bliss
HM 1134	Yoga Bodily disability
HH 0484	Yoga Buddhi yoga
HM 0361	Yoga Carelessness
HH 1047	Yoga Causes misery
HH 4561	Yoga Change
HM 3011	Yoga Chitta-dependent consciousness
HM 3063	Yoga City shining jewel awareness
HM 3690	Yoga clear light (Buddhism)
HM 2032	Yoga Conditional consciousness states
HM 0467	Yoga Conscious self-relatedness
HM 3869	Yoga Consciousness independent vehicle
HM 5122	Yoga consciousness transference (Buddhism)
HM 2898	Yoga Contentment
HM 3231	Yoga Desire
HM 3871	Yoga Desire life
HM 0194	Yoga Dharma megha samadhi

HH 0827 Yoga Dhyana yoga	HM 0107 Yoga Prajna	HM 1456 Yoga Vivekajam jnanam
HM 2998 Yoga Discernment	HH 0213 Yoga Pranayama	HM 2912 Yoga Vivesa stage gunas
HM 2998 Yoga Discrimination	HH 0829 Yoga Pratyahara	HM 1015 Yoga Vyagra
HM 0903 Yoga Dispassion	HH 0111 Yoga Puja yoga	HM 3871 Yoga Will live
HH 1080 Yoga Doubt	HM 4493 Yoga Pure awareness reality	HH 0829 Yoga Withdrawal senses
HH 4087 Yoga ; Dream (Buddhism)	HM 3461 Yoga Purification awareness	HM 2396 Yoga Witness-consciousness
HM 0600 Yoga dream state (Buddhism)	HM 1391 Yoga Purity	HM 1080 Yoga Wrong questioning
HM 3620 Yoga Dvesa	HH 0037 Yoga Purna yoga	HH 6876 Yoga Yama
HM 3477 Yoga Egoism	HH 0037 Yoga ; Purna (Yoga)	HH 1224 Yoga Yantra yoga
HH 0779 Yoga Eightfold path yoga	HM 2396 Yoga Purusa	HH 0661 Yoga (Yoga)
HM 5734 Yoga Ekagra	HM 2396 Yoga Purusha	HM 3169 Yoga Yogic experience
HM 3337 Yoga Ekagrata parinama	HM 1070 Yoga Racial consciousness	HM 5532 Yogi-pratyaksa (Buddhism)
HM 3869 Yoga Enlightenment	HM 2433 Yoga Raga	HM 3169 Yogic experience (Yoga)
HM 2226 Yoga Enstasy	HH 0755 Yoga Raja yoga	HH 4398 Yogic feats
HM 1586 Yoga Equanimity	HM 3869 Yoga Realization	HM 8634 Yogic flying
HM 1077 Yoga Erroneous perception	HH 0213 Yoga Regulation vital force	HH 0603 Yogic paths (Buddhism)
HM 0610 Yoga Failure hold meditative attitude	HH 4210 Yoga Renunciation	HM 5005 Yogic perception (Hinduism)
HM 2933 Yoga Faith	HM 3620 Yoga Repulsion	HM 5005 Yogipratyaksa (Hinduism)
HM 3461 Yoga Fifth chakra	HM 1015 Yoga Restlessness	HM 0285 Yoin (Japanese)
HM 2405 Yoga Fifth state	HM 3620 Yoga Revulsion	HH 5856 You ; Journeying transcendence me
HM 2893 Yoga First chakra	HM 2893 Yoga Root base awareness	TP 0004 Young shoot Acquiring experience
HH 0390 Yoga Five kleshas	HH 0538 Yoga Sabda	TP 0004 Young shoot Darkness
HM 3242 Yoga Fourth chakra	HM 3227 Yoga Saccidananda	TP 0004 Young shoot Immaturity
HM 2020 Yoga Fourth state	HM 2077 Yoga Sadanga yoga	TP 0004 Young shoot Inexperience
HH 0927 Yoga Gnani yoga	HM 2933 Yoga Saddha (Pali)	TP 0004 Young shoot Uncultivated growth
HH 0862 Yoga Hatha yoga	HM 3398 Yoga Sahasrara	TP 0004 Young shoot Youthful folly
HM 3174 Yoga Her favourite resort awareness	HM 2226 Yoga Samadhi	VC 1943 Youth
HM 2937 Yoga I am this	HM 3207 Yoga Samadhi parinama	VP 5124 Youth-Age
HM 2087 Yoga Ida conscious energy	HH 4321 Yoga Samadhi yoga	TP 0004 Youthful folly (Young shoot)
HM 1222 Yoga illusory body (Buddhism)	HM 1586 Yoga Samatva	VC 1945 Youthfulness
HM 0298 Yoga Inability achieve concentration	HM 2896 Yoga Samprajnata samadhi	HM 1771 Yu (Taoism)
HM 3465 Yoga Initiation spiritual rebirth	HM 2898 Yoga Samtosa	HH 0239 Yuga ; Kali
HM 3863 Yoga inner fire (Buddhism)	HM 3160 Yoga Sanjna	HM 2849 Yugbanaddha (Buddhism)
HH 0037 Yoga Integral yoga	HH 0927 Yoga Sankya yoga	HM 7100 Yukta perception (Hinduism)
HH 0101 Yoga intermediary state (Buddhism)	HH 4210 Yoga Sannyasa	HH 0170 Yul-nges (Buddhism)
HM 0099 Yoga International consciousness	HM 1784 Yoga sastra (Yoga)	
HM 3239 Yoga Intuitive knowledge	HM 3227 Yoga Sat-cit-ananda	**Z**
HM 2738 Yoga Inward-flowing consciousness	HM 3329 Yoga Satsang	
HM 2913 Yoga Ishvara-consciousness	HH 0037 Yoga Satyeswarananda	HM 3022 Zacharael ; Angelic awareness
HM 0148 Yoga Jada samadhi	HM 1391 Yoga Sauca	HM 3022 Zachriel
HM 0148 Yoga Jadya	HM 2226 Yoga Savikalpa samadhi	HM 4208 Zahid (Sufism)
HM 2141 Yoga Jagrat state waking consciousness	HM 3174 Yoga Second chakra	HM 4042 Zahir ; Az (Sufism)
HH 0931 Yoga Japa yoga	HM 0467 Yoga Self-consciousness	HH 5643 Zakat (Islam)
HH 0931 Yoga Japam	HM 3398 Yoga Seventh chakra	HM 7512 Zanshin (Japanese)
HH 0610 Yoga Jnana	HM 3193 Yoga Seventh state	**Zazen meditation**
HM 3060 Yoga Jnana samadhi	HH 2034 Yoga sex (Yoga)	HH 1134 — Inward (Zen)
HH 0927 Yoga Jnana yoga	HH 9112 Yoga ; Sexual (Taoism)	HH 0621 — Outward (Zen)
HM 3869 Yoga Kaivalya	HH 0539 Yoga Shabd yoga	HH 1134 — Rinzai (Zen)
HM 3869 Yoga Kaivalyam	HH 0538 Yoga Shabda	HH 0621 — Soto (Zen)
HH 0372 Yoga Karma yoga	HH 0539 Yoga Shabda yoga	HH 0882 — (Zen)
HH 1047 Yoga Klesas	HH 0539 Yoga ; Shabda (Yoga)	VC 1947 Zeal
HH 0538 Yoga Knowledge	HH 3912 Yoga Shakti yoga	HM 4487 Zeal (Buddhism)
HH 0610 Yoga Knowledge	HH 3398 Yoga Shunya chakra	HM 4969 Zeal ; Concentration due (Buddhism)
HH 0969 Yoga Kriya yoga	HH 0380 Yoga Siddhis	VD 4082 Zealotry
HM 4058 Yoga Kshipta	HH 2077 Yoga Six-member yoga	HM 3119 Zen Attachment
HH 0237 Yoga Kundali yoga	HH 0176 Yoga Sixteen-petalled lotus	HH 0463 Zen Attentiveness nature
HM 2871 Yoga Kundalini conscious energy	HM 2144 Yoga Sixth chakra	HM 3196 Zen Avidya
HH 0237 Yoga Kundalini yoga	HM 3222 Yoga Sixth state	HM 3196 Zen Avijja (Pali)
HH 0110 Yoga Lack dispassion	HH 0176 Yoga Soma chakra	HM 3251 Zen Back-to-the-source awareness
HH 0782 Yoga Laya yoga	HM 2438 Yoga Spiritual realization	HH 0785 Zen Bombu Zen
HM 0046 Yoga Laziness	HH 6210 Yoga ; Spirituality tantric	HH 0677 Zen Bravery
HM 3869 Yoga Liberation	HM 2805 Yoga Stages gunas	HM 2690 Zen Buddhism ; Oxherding pictures (Zen)
HM 0361 Yoga Light-mindedness	HM 1346 Yoga Stupefaction	HM 2979 Zen Catching-the-ox awareness
HM 3212 Yoga Linga stage gunas	HM 2026 Yoga Subjective domains consciousness	HH 0589 Zen Ch'an
HH 2034 Yoga Maithuna	HM 2818 Yoga Super-contemplative state	HH 1234 Zen Charity
HH 0306 Yoga Maithuna	HM 3232 Yoga Super-contemplative state	HM 0917 Zen Chen-hsing
HM 3063 Yoga Manipura	HM 0135 Yoga Suppression	HH 0350 Zen Circles enlightenment
HM 2912 Yoga Manomaya kosa	HM 3174 Yoga Svadhishtana	HM 3153 Zen Coming-home-on-the-ox's-back awareness
HM 2464 Yoga Mantra-consciousness	HM 2781 Yoga Svapna	HM 2690 Zen Cow-herding simile
HM 2464 Yoga Mantra-induced experience	HM 2781 Yoga Swapna state dream consciousness	HH 1234 Zen Dana
HH 0931 Yoga Mantra yoga	HH 0306 Yoga Tantra	HM 2957 Zen Deep sleep
HM 1690 Yoga Meditative absorption	**Yoga [tantra:....]**	HM 3196 Zen Delusion
HM 0906 Yoga Mental inertia	HM 2864 — tantra: anuttarayoga ; Highest (Buddhism)	HH 0589 Zen Dhyana
HH 4561 Yoga Mental transformation	HM 2864 — tantra (Buddhism)	HM 3196 Zen Dullness
HM 2933 Yoga Mos-pa (Tibetan)	HH 0306 — Tantric (Buddhism)	HM 3620 Zen Dvesa
HM 1346 Yoga Mudha	HM 4603 Yoga Tantric formless path	HM 3477 Zen Egoism
HM 2893 Yoga Muladhara	HM 1690 Yoga Tantric visualization	HM 2193 Zen Emptiness
HM 3405 Yoga Mythic yoga	HH 2034 Yoga Tantric yoga	HM 3068 Zen Entering-city-with-bliss-bestowing-hands awareness
HH 0160 Yoga Nada yoga	HH 0306 Yoga Tantric yoga	HH 0390 Zen Five kleshas
HH 0539 Yoga Nam simran	HH 0877 Yoga Tapasya	HM 0910 Zen Formlessness
HH 0160 Yoga Natha yoga	HH 0452 Yoga Taraka yoga	HH 1234 Zen Giving
HH 0160 Yoga ; Natha (Yoga)	HM 3231 Yoga Tendency	HH 6773 Zen Gogai
HM 1370 Yoga National consciousness	HM 3063 Yoga Third chakra	HH 0429 Zen Good conduct
HM 3398 Yoga Niralambapuri chakra	HM 1029 Yoga Thirstlessness	HH 0550 Zen Great perfection
HM 3437 Yoga Nirodha parinama	HM 3398 Yoga Thousand-petalled lotus	HH 0429 Zen Harmony
HM 3437 Yoga Niruddha	HM 2026 Yoga Three worlds	HM 2560 Zen Herding-the-ox awareness
HM 0135 Yoga Niruddha	HM 4423 Yoga Transformation consciousness	HH 0589 Zen Hsin tsung
HM 0809 Yoga Nirvicara samadhi	HM 0411 Yoga Tribal consciousness	HH 1003 Zen Human development
HM 2226 Yoga Nirvikalpa samadhi	HH 0176 Yoga Twelve-petalled lotus	HM 3196 Zen Ignorance
HM 4284 Yoga Nirvitarka samadhi	HM 0148 Yoga Unconscious enstasy	HH 1134 Zen Inward zazen meditation
HM 2491 Yoga Niskama karma	HH 0746 Yoga Universal religion	HM 3168 Zen Ji hokkai
HM 5487 Yoga Niskama karma	HH 0746 Yoga Universal spirituality	HM 2034 Zen Ji-ji-mu-ge
HM 3242 Yoga Not hit awareness	HM 0903 Yoga Vairagya	HM 3164 Zen Ji-ri-mu-je
HM 3182 Yoga Obstacles soul cognition	HM 3231 Yoga Vasana	HM 3215 Zen Jiriki
HM 2475 Yoga Omniscience	HM 3880 Yoga Vicara samadhi	HH 0263 Zen Koan
HM 5734 Yoga One-pointedness	HM 3460 Yoga Vicara stage	HH 0813 Zen Kshanti
HM 2581 Yoga Outward-flowing consciousness	HM 4489 Yoga Videha-mukti	HH 6129 Zen Lin chi
HH 4561 Yoga Parinama	HM 3160 Yoga Vijna	HH 1234 Zen Love
HM 1690 Yoga Path form	HM 3460 Yoga Vijnanamaya kosa	HM 3196 Zen Ma-rig-pa (Tibetan)
HH 0827 Yoga Path meditative absorption	HM 0903 Yoga Viraga	HH 0550 Zen Maha paramita
HM 3405 Yoga Path renunciation	HM 3461 Yoga Vishuddha	HM 3306 Zen Mayoi
HM 3300 Yoga Perception form	HM 4451 Yoga Vitarka samadhi	HH 0785 Zen Meditation
HH 0037 Yoga Perfect yoga	HM 2912 Yoga Vitarka stage	HH 0882 Zen meditation (Zen)
HM 2890 Yoga Pingala conscious energy	HM 1029 Yoga Vitrisna	HH 0589 Zen Mind doctrine
HM 2911 Yoga Possessiveness	HM 2998 Yoga Viveka	HM 0765 Zen Ming
HM 2935 Yoga Pradhana	HM 4493 Yoga Viveka khyati	

Zen

HH 0429 Zen Morality	HH 1009 Zen Saijo Zen	HM 2500 Zen Wu-nien
HM 2163 Zen Mushin	HM 2326 Zen Satori	HH 0882 Zen Zazen meditation
HM 2974 Zen Mushinjo	HM 3036 Zen Searching-for-ox awareness	VC 1949 Zest
HH 0343 Zen Naturalness	HM 0917 Zen Seeing self-nature	HM 0747 Zest (Buddhism)
HH 0343 Zen Nature being	HM 2755 Zen Seeing-the-ox awareness	TC 1650 Zetetic panel
HM 2957 Zen Nidra	HM 3302 Zen Seeing-the-traces awareness	HM 2760 Zeus awareness
HM 2163 Zen No-mind	HM 0765 Zen Self-knowledge	HM 2744 Zhe-sdang-med-pa (Tibetan)
HM 2500 Zen No-thought	HH 0429 Zen Shila	HM 2205 Zhi-bar-byed-pa (Tibetan)
HM 3485 Zen Non-abiding	HH 0343 Zen Shizen	HM 2147 Zhi gnas (Tibetan)
HM 2163 Zen Original mind	HM 2193 Zen Shunyata	HM 2351 Zikr (Islam)
HH 0621 Zen Outward zazen meditation	HH 0621 Zen Soto zazen meditation	HM 2713 Zodiacal forms awareness (Astrology)
HM 3215 Zen Own strength	HH 0928 Zen Soto Zen	HM 2282 Zoharariel ; Angelic awareness
HM 2492 Zen Ox-and-man-both-out-of-sight awareness	HM 2500 Zen Suchness	MP 1226 Zone ; Inter level
HM 2604 Zen Ox-forgotten-leaving-man-alone awareness	HM 2193 Zen Sunyata	HH 0987 Zone therapy
HM 2690 Zen Oxherding pictures Zen Buddhism	HM 2957 Zen Sushupti	MP 1042 Zones ; Chain fundamental transformation
HH 0219 Zen Paramitas	HM 2957 Zen Susupti state unconsciousness	MP 1088 Zones ; Informal local perspective interface
HH 0813 Zen Patience	HM 3489 Zen Transcending mind	MP 1223 Zones intermediate insight
HH 0219 Zen Perfections zen	HM 2084 Zen Transrational insight	MP 1233 Zoning internal domains
HH 0550 Zen Prajna	HH 0928 Zen Ts'ao tung	HM 7208 Zoopathy ; Delusional
HM 0107 Zen Prajna	HM 2500 Zen Unconscious	HH 1903 Zoroastrianism Human development
HH 0550 Zen Prajna paramita	HM 2957 Zen Unconscious	HH 1903 Zoroastrianism Iranis
HM 2163 Zen Primordial awareness	HH 0677 Zen Vigour	HH 1903 Zoroastrianism Parsis
HM 3620 Zen Repulsion	HH 0677 Zen Virya	HM 4450 Zuhd (Sufism)
HM 3251 Zen Returning origin	HM 2193 Zen Voidness	
HM 3620 Zen Revulsion	HM 2326 Zen Wu	**NUMBERS**
HM 3352 Zen Ri hokkai	HH 4019 Zen Wu	
HH 1134 Zen Rinzai zazen meditation	HM 3485 Zen Wu-chu	TC 1306 66 ; Discussion
HH 6129 Zen Rinzai Zen	HM 2163 Zen Wu-hsin	TC 1306 66 ; Phillips
HM 1169 Zen running	HM 0910 Zen Wu-hsing	

Appendices   Z

Statistics

Entries (1976 - 1990)	935
Cross-references and bibliography (1976 - 1990)	936
World problems entries and cross-references by section (1990)	937
Human development entries and cross-references by section (1990)	937
International organizations by type (1990)	938
Human values (*see Section VZ*)	

Computers

Use of computers by UIA	939
Software environment for the Encyclopedia	940
Graphics environment for an "Atlas of International Relationships Networks"	941
Mapping networks	943

Union of International Assocations: Profile   949

Appendices \ Z

Statistics

TABLE 1. World problems and human potential entries (1976 - 1990)

Sections	Section code 1986	Section code 1990	Entries 1976	Entries 1986	Entries 1990	% Increase	Pages (incl. notes) 1990	% Increase
WORLD PROBLEMS								
Descriptive entries	PP	PA-PF	2653	4700	8721	85.6 %	-	-
Indexed only	PQ	PG/PJ	4791	5533	4446	-19.6 %	-	-
TOTAL WORLD PROBLEMS			7444	10233	13167	28.7 %	865	91.6 %
HUMAN DEVELOPMENT								
Concepts	HH	HH	228	628	1292	105.7 %	-	-
Modes of awareness	HM	HM	-	968	2759	185.0 %	-	-
			228	1596	4051	153.8 %	340	190.6 %
INTEGRATIVE KNOWLEDGE								
Integrative concepts	KC	KC	421	632	632	-	-	-
Embodying discontinuity	KD	KD	-	70	70	-	-	-
Patterning disagreement	CP	KP	-	20	20	-	-	-
			421	722	722	-	104	26.5 %
METAPHORS AND PATTERNS								
Forms of presentation	CF		-	528	-	-	-	-
Metaphors	CM	MM	-	88	88	-	-	-
Patterns of concepts	CP	MP	-	253	253	-	-	-
Symbols	CS	MS	-	103	103	-	-	-
			-	972	444	-	103	12.0 %
TRANSFORMATIVE APPROACHES								
Transformative conferencing	TC	TC	-	207	240	15.9 %	-	-
Transformative policy cycles	CP	TP	-	64	64	-	-	-
Multi-polarization	TM		-	11	-	-	-	-
			-	282	304	7.8 %	62	3.3 %
VALUES AND WISDOM								
Constructive values	VC	VC	704	960	960	-	-	-
Destructive values	VD	VD	-	1040	1040	-	-	-
Value polarities	VP	VP	-	225	225	-	-	-
Value types	VT	VT	-	45	45	-	-	-
			704	2270	2270	-	103	21.2 %
STRATEGIES								
Strategic polarities	SP		-	239	-	-	-	-
Collective strategies	SQ		-	7148	-	-	-	-
Strategic roles	SR		-	224	-	-	-	-
Collective strategies	SS		-	679	-	-	-	-
Strategic type	ST		-	45	-	-	-	-
			-	8335	-	-	-	-
TOTAL HUMAN POTENTIAL			1353	14177	7791	-	712	34.5 %
TOTAL "PROBLEMS + POTENTIAL"			*8797	24410	20958	-14.1 %	1577	61.1 %

* In addition to the entries indicated, the 1976 edition contained the following seperate sections which were not been included in the 1986 or 1990 editions: International organizations (3300), Traded products and commodities (241), Intellectual disciplines and sciences (1845), Economic and industrial sectors (132), Occupations and jobs (739), Multinational corporations (606), Human diseases (775), International periodicals (1197), Multilateral treaties (931). International organizations (24,180) are descibed in the current *Yearbook of International Organizations* with multilateral treaties. Volume 3 of that series classifies organizations, treaties and world problems by subject (3000 categories)

Statistics

TABLE 2. World problems and human potential cross-references and bibliography (1976 - 1990)

Sections	Section code		Cross-references				Index entries	Bibliography
	1986	1990	1976	1986	1990	% Increase	1990	1990
WORLD PROBLEMS								
Descriptive entries	PP	PA-PF	-	17636	74399	321.9 %	-	-
Indexed only	PQ	PG/PJ	-	-	5995	-	-	-
TOTAL WORLD PROBLEMS			13574	17636	80394	355.9 %	64934	4745
HUMAN DEVELOPMENT								
Concepts	HH	HH	-	917	4356	375 %	-	-
Modes of awareness	HM	HM	-	3544	10671	201.1 %	-	-
			-	4461	15027	236.9 %	*22915	2488
INTEGRATIVE KNOWLEDGE								
Integrative concept	KC	KC	-	-	-	-	-	-
Embodying discontinuity	KD	KD	-	-	-	-	-	-
Patterning disagreement	CP	KP	-	-	-	-	-	-
			-	-	-	-	1610	2200
METAPHORS AND PATTERNS								
Forms of presentation	CF		-	-	-	-	-	-
Metaphors	CM	MM	-	-	-	-	-	-
Patterns of concepts	CP	MP	-	3491	3491	-	-	-
Symbols	CS	MS	-	636	636	-	-	-
			-	4127	4127	-	1485	299
TRANSFORMATIVE APPROACHES								
Transformative conferencing	TC	TC	-	-	-	-	-	-
Transformative policy cycles	CP	TP	-	(384)	(384)	-	-	-
Multi-polarization	TM		-	-	-	-	-	-
			-	(384)	(384)	-	1191	-
VALUES AND WISDOM								
Constructive values	VC	VC	-	3568	3568	-	-	-
Destructive values	VD	VD	-	3440	3440	-	-	-
Value polarities	VP	VP	-	7231	7231	-	-	-
Value types	VT	VT	-	224	224	-	-	-
			-	14463	14463	-	2549	398
STRATEGIES								
Strategic polarities	SP		-	7590	-	-	-	-
Collective strategies	SQ		-	-	-	-	-	-
Strategic roles	SR		-	-	-	-	-	-
Collective strategies	SS		-	132	-	-	-	-
Strategic type	ST		-	237	-	-	-	-
			-	7959	-	-	-	-
TOTAL HUMAN POTENTIAL			-	31394	34001	8.3 %	*26451	5385
TOTAL "PROBLEMS + POTENTIAL"			-	49030	114395	133.3 %	91385	**10130

* The special index for Human Development (Section HX) contains only 3,356 items. The 19,559 item keyword index to Section H is incorporated with the Volume 2 index.
** There is a 1.8 % overlap between the bibliographies

Appendices \ Z

Statistics

TABLE 3. World problems entries and cross-references by section (1990)

<table>
<tr><th colspan="2"></th><th>ABSTRACT PROBLEMS PA</th><th>BASIC PROBLEMS PB</th><th>CROSS-SECTORAL PC</th><th>DETAILED PROBLEMS PD</th><th>EMANATIONS OF OTHERS PE</th><th>EXCEPTIONAL PROBLEMS PF</th><th>VERY SPECIFIC PG</th><th>PROBLEMS CONSIDERED PJ</th><th>TOTAL</th></tr>
<tr><td rowspan="3">ENTRIES</td><td>1986</td><td>104</td><td>206</td><td>575</td><td>1052</td><td>836</td><td>1423</td><td>504</td><td>-</td><td>4700</td></tr>
<tr><td>1990</td><td>344</td><td>141</td><td>732</td><td>1928</td><td>3106</td><td>2470</td><td>2909</td><td>1537</td><td>13167</td></tr>
<tr><td>Percentage 1990</td><td>2.6 %</td><td>1.1 %</td><td>5.6 %</td><td>14.6 %</td><td>23.6 %</td><td>18.8 %</td><td>22.1 %</td><td>11.7 %</td><td>100</td></tr>
<tr><td rowspan="9">CROSS REFERENCES</td><td>PA</td><td>6479</td><td>254</td><td>402</td><td>378</td><td>205</td><td>867</td><td>75</td><td>79</td><td>8739</td></tr>
<tr><td>PB</td><td>255</td><td>380</td><td>775</td><td>774</td><td>459</td><td>683</td><td>69</td><td>78</td><td>3473</td></tr>
<tr><td>PC</td><td>402</td><td>775</td><td>2658</td><td>2722</td><td>1786</td><td>2112</td><td>264</td><td>184</td><td>10903</td></tr>
<tr><td>PD</td><td>375</td><td>776</td><td>2722</td><td>5216</td><td>3911</td><td>2757</td><td>1144</td><td>349</td><td>17250</td></tr>
<tr><td>PE</td><td>204</td><td>459</td><td>1787</td><td>3907</td><td>4889</td><td>2017</td><td>1220</td><td>298</td><td>14782</td></tr>
<tr><td>PF</td><td>845</td><td>682</td><td>2112</td><td>2753</td><td>2019</td><td>8807</td><td>1021</td><td>864</td><td>19103</td></tr>
<tr><td>PG</td><td>74</td><td>70</td><td>264</td><td>1144</td><td>1220</td><td>1022</td><td>124</td><td>42</td><td>3960</td></tr>
<tr><td>PJ</td><td>79</td><td>78</td><td>190</td><td>375</td><td>346</td><td>933</td><td>67</td><td>117</td><td>2184</td></tr>
<tr><td>TOTAL</td><td>8713</td><td>3474</td><td>10910</td><td>17269</td><td>14835</td><td>19198</td><td>3984</td><td>2011</td><td>80394</td></tr>
<tr><td rowspan="8">CROSS REFERENCES</td><td>Broader problems</td><td>299</td><td>198</td><td>1120</td><td>2996</td><td>4547</td><td>3264</td><td>2722</td><td>763</td><td>15909</td></tr>
<tr><td>Narrower problems</td><td>1231</td><td>1203</td><td>3080</td><td>3996</td><td>2105</td><td>4261</td><td>54</td><td>67</td><td>15997</td></tr>
<tr><td>Related problems</td><td>6153</td><td>219</td><td>1081</td><td>1789</td><td>1906</td><td>2035</td><td>176</td><td>244</td><td>13603</td></tr>
<tr><td>Aggravates</td><td>467</td><td>791</td><td>2532</td><td>4076</td><td>3058</td><td>4877</td><td>549</td><td>517</td><td>16867</td></tr>
<tr><td>Aggravated by</td><td>522</td><td>998</td><td>2901</td><td>4144</td><td>3038</td><td>4436</td><td>455</td><td>385</td><td>16879</td></tr>
<tr><td>Reduces</td><td>20</td><td>24</td><td>96</td><td>127</td><td>85</td><td>178</td><td>16</td><td>19</td><td>565</td></tr>
<tr><td>Reduced by</td><td>21</td><td>41</td><td>100</td><td>141</td><td>96</td><td>147</td><td>12</td><td>16</td><td>574</td></tr>
<tr><td>TOTAL</td><td>8713</td><td>3474</td><td>10910</td><td>17269</td><td>14835</td><td>19198</td><td>3984</td><td>2011</td><td>80527</td></tr>
</table>

TABLE 4. Human development entries and cross-references by section (1986-1990)

<table>
<tr><th colspan="2">ENTRIES</th><th>1986</th><th>1990</th><th>% 1990</th><th colspan="4">REFERENCES</th><th colspan="6">REFERENCES</th></tr>
<tr><td colspan="2">HUMAN DEVELOPMENT SECTION</td><td>1986</td><td>1990</td><td>% 1990</td><td>HH</td><td>HM</td><td colspan="2">TOTAL</td><td>Broader</td><td>Narrower</td><td>Related</td><td>Preceded by</td><td>Followed by</td><td>TOTAL</td></tr>
<tr><td></td><td>Concepts HH</td><td>628</td><td>1292</td><td>31.9%</td><td>3109</td><td>1247</td><td colspan="2">4356</td><td>548</td><td>992</td><td>2683</td><td>68</td><td>65</td><td>4356</td></tr>
<tr><td></td><td>Modes of awareness HM</td><td>968</td><td>2759</td><td>68.1%</td><td>1250</td><td>9421</td><td colspan="2">10671</td><td>2312</td><td>1873</td><td>3577</td><td>1452</td><td>1457</td><td>10671</td></tr>
<tr><td></td><td>TOTAL</td><td>1596</td><td>4051</td><td>100%</td><td>4359</td><td>10668</td><td colspan="2">15027</td><td>2860</td><td>2865</td><td>6440</td><td>1520</td><td>1522</td><td>15027</td></tr>
</table>

Statistics

TABLE 5. International organizations by type (described in Yearbook of International Organizations, 1990/91)

Sections	No.	Intergovernmental %Secn	%IGO	No.	Nongovernmental %Secn	% NGO	Total No.	% Total
CONVENTIONAL INTERNATIONAL BODIES								
A. Federations of international organizations	1	2.4	0.4	40	97.6	1.0%	41	1.0
B. Universal membership organizations	34	7.3	11.6	430	92.7	9.2	464	9.3
C. Intercontinental membership organizations	41	5.0	14.0	770	95.0	16.5	811	16.4
D. Regionally oriented membership organizations	217	6.0	74.0	3406	94.0	73.3	3623	73.3
TOTAL "CONVENTIONAL"	293	6.0	100.0	4646	94.0	100.0	4939	100.0
OTHER INTERNATIONAL BODIES								
E. Organizations emanating from places, persons, other bodies	796	27.2	51.0	2129	72.3	18.4	2925	22.2
F. Organizations of special form	680	26.2	43.5	1915	73.8	16.6	2595	19.8
G. Internationally-oriented national organizations	87	1.1	5.5	7518	98.9	65.0	7605	58.0
TOTAL "OTHER"	1563	12.0	100.0	11562	88.8	100.0	13125	100.0
TOTAL Section A-G	1856	-	-	16208	-	-	18064	-
SPECIAL SECTIONS								
H. Dissolved or apparently inactive organizations	251	10.2	10.2	2204	89.8	36.0	2455	28.6
J. Recently reported bodies, not yet confirmed	141	12.1	5.7	1023	89.7	16.7	1164	13.6
R. Religious orders and secular institutes	0	0	0	683	100.0	11.1	683	7.9
S. Autonomous conference series	71	14.0	2.9	437	86.0	7.1	508	5.9
T. Multilateral treaties and intergovernmental agreements	1674	100.0	67.9	0	0	0	1674	19.5
U. Currently inactive nonconventional bodies	329	15.6	13.3	1779	84.4	29.1	2108	24.5
TOTAL "SPECIAL"	2466	28.7	100.0	6126	71.3	100.0	8592	100.0
TOTAL ALL SECTIONS	4322	-	-	22334	-	-	26656	-

This table suggests different answers to the question "How many international organizations are there?"

1. Conventional intergovernmental organizations, for those who attach importance to the non-existence of international non-governmental organizations in terms of international law. (Multilateral treaties, Section T, might be added as closely related international "instruments".)
2. Conventional international bodies, both governmental and non-governmental, for those who attach importance to the existence of autonomous international bodies as a social reality.
3. Conventional bodies (Sections A to D) plus special forms (Section F), for those who recognize the importance of organizational substitutes and unconventional form. (To the latter might be added conference series, Section S, and multilateral treaties, Section T, as forms of organization substitute.)
4. Conventional bodies (Sections A to D), special forms (Section F), and religious orders (Section R), for those who accept the social reality of the latter as independent actors.
5. Conventional bodies (Sections A to D), other international bodies (Sections E to G), religious orders (Section R), and multilateral treaties (Section T), for those who are interested in the international impact of semi-autonomous and nationally-ties organizations. (Documentalists might also include inactive bodies, Section H, which figure in "authority lists" of international organizations.)

Computers: use of computers by UIA

The information in this publication is maintained in computer files. This is necessary in order to facilitate the incorporation of descriptions of entries, their interrelationships, their relevant index entries, as well as the correction and updating of such information. These tasks become very onerous when it is necessary to deal with the amounts of information processed for this publication and with the multiplicity of interrelationships between its parts.

1. System

The computer system used for these purposes is the same as that used for the Union of International Associations companion publication the 3-volume *Yearbook of International Organizations*, for which it was originally developed in 1972. The system was used from 1972 to 1976 to produce the first edition of this publication under its earlier name of *Yearbook of World Problems and Human Potential*. In 1981 it was adapted to produce the related *International Congress Calendar* series. The system was upgraded for the production of the 1986 and current editions of this Encyclopedia.

From 1972 to mid-1985 the magnetic tape files were maintained on a main-frame computer operated in Brussels by an external service bureau. Additions and amendments to these files were made via a micro-computers in the UIA Secretariat. Printouts (or microfiches) arising from this operation were furnished by the service bureau, permitting a further cycle of amendments, if necessary. When the files were considered correct, special software routines were used by the bureau to sort and match the cross-references between entries (to ensure that every reference from A to B was matched by one from B to A).

From mid-1985 the main-frame files have been permanently downloaded onto an in-house local area network.

When the files are then judged correct, copies are sent to a computer typesetting bureau in the United Kingdom. There a suite of 10 programmes (3 Assembler; 7 Cobol) is used to justify and hyphenate the text (including cross-reference formatting), to generate final indexes (up to 28 types of index entry are extracted) for the entries, and to make up pages for typesetting purposes. The magnetic tape files are converted there, via a photocomposition process (on Videocomp computers), onto film suitable for normal offset printing operations (currently carried out in the Federal Republic of Germany).

As an indication of the sophistication of the United Kingdom operations alone, the 3-volume *Yearbook of International Organizations* won the first **Printing World Award** (1986) for the most innovative application of computers in typesetting. This award was specially sponsored as a Bicentenary Award by Her Majesty's Stationery Office (UK).

2. File design

The design of the files represents a compromise between the needs for: photocomposition, editorial amendment and updating, management of indexing and cross-references, extraction of data subsets, research and communication with the international organizations concerned. The sections of this publication, of the *Yearbook of International Organizations* and of the *International Congress Calendar* tend to be in separate files which are however flexibly interrelated by cross-references and relational linkages. The files are organized as a text database permitting word-processor type updating of entries as well as very flexible manipulation of the data elements, especially for proofing, indexing and cross-referencing purposes. As currently designed, the system permits hands-on exploration of the networks of cross-references between the entries. Mailing addresses, for example, are extracted from the data base and automatically re-formatted using the language version of the organization's name most appropriate to the country in which its secretariat is located (then suitably abridged according to label size required).

3. Access to files

It is the policy to make available copies of such files to university research centres to assist them in their own work. Unfortunately no satisfactory formula has yet been worked out to safeguard files distributed in this way from various forms of abuse, such as repeated copying for the benefit of other bodies unknown to the editors. Until the costs of creating these files have been covered, the files will therefore not be made available. When they are made available, this will only be done under contract insuring against unauthorized use. The editors would welcome correspondence suggesting ways of alleviating this unfortunately rigid policy.

It is hoped to take advantage of the technical possibility of making the files available on CD-ROM disk.

4. Technical information

(a) System: The local area network is based on a Novell Netware 286 v2.12, using an MS-DOS 3.1 operating system. It is presently composed of 18 independent micro-computer workstations (PC compatibles) linked by Arcnet. Main disk capacity is currently 540 megabytes. Backups and communication with UK service bureau are via a streamer tape unit for standard industry tapes (9 track, 1600 bpi, odd parity, record length 180, blocking factor 20) and an Emerald 120 MB cartridge unit. Two Hewlett Packard laser printers are used for network hardcopy. Despite experiments with links to other data bases, especially in the form of computer conferencing, no such links are currently operational.

(b) Software: The software used for management of the relational text database is Advanced Revelation 1.16, network version. An extensive suite of programmes has been specially written for UIA to take advantage of the facilities of this software for handling entries and cross-references in a text database mode. Some use is also made of other software packages such as Wordstar. Wordperfect 5.1 is used for desktop publishing (notably the Notes in this Encyclopedia).

(c) Conversion: Because of the multilingual nature of the texts processed, especially for the *Yearbook of International Organizations*, a number of less obvious character conversion problems have had to be resolved in ensuring communication between different hardware and software systems. Thus although the local area network functions using the full ASCII set, communication with the UK service bureau is done using an extended EBCDIC character set with the upper and lower cases interchanged. The Revelation environment functions with files which can be converted into and from DOS versions as required.

(d) Graphics software: The possibilities of representing the data held as network maps is a subject under continuing investigation, as described in following notes.

(e) CD-ROM: Although used in-house on an experimental basis, it has not yet proved possible to make the data available in this mode for other users. This is also subject to continuing review.

Computers: software environment for this Encyclopedia

Since 1972 the database associated with this project has been maintained on computer. Although initially on a mainframe, it was transferred onto an in-house local area network in 1984. The software developed to facilitate the complex editorial process is as much a product of the project as the Encyclopedia itself. It is an extension of that used to manage the data for the Yearbook of International Organizations.

1. Software challenges
The process of editing a network of records, whether on world problems or on modes of awareness, calls for a somewhat unusual software approach. The specific challenges are:

(a) Variable size of the database: The number of records in any of the databases (on organizations, problems, etc) can vary from several hundred to 30,000. Those in the associated indexing files can extend up to 300,000.

(b) Variable size of records: The amount of information on any one item (organization, problem, etc) can vary from several hundred characters possibly up to 80,000.

(c) Variable size of fields within records: At any one time it is difficult to predict how much information any given field will be required to hold within a record.

(d) Relational nature of the database: Although some of the files of records have well-developed links to records in other files, of special interest are those which have extensive links between records in the same file. In the case of the 13,000 problems records, they have approximately 100,000 links between them. It is not possible to predict how this number will develop.

(e) Non-standard (accented) characters: Especially in the case of international organizations, provision must be made for extensive use of accented characters.

(f) Developing nature of the database: Any software solution can only be a provisional one, since new needs emerge and new features are required as the project increases in sophistication. The software should thus permit further development beyond the features provided initially.

(g) Ease of use: It is essential that, as much as possible, the complexities of the database be concealed from editors in order that their attention can be focused on their task. In practice this means organizing the database so that editorial work is done in word-processing mode (where any learning is focused on editorial issues), whereas checks and other manipulations are done in a database mode. This calls for a hybrid "text database" structure.

(h) Record locking: Since the work is done on a local area network with up to 10 people working on records in the same file, it is essential that a record locking procedure be incorporated to prevent conflicting use of the same records.

2. Database software
In 1984 there were relatively few database packages suitable for operation on a local area network and embodying features such as those described above. The choice made was for the Revelation database package since it permitted record locking in a Novell Netware environment using MS-DOS. This decision also permitted relatively low cost development of the number of workstations since clones could be used rather than expensive machines. At that time Revelation version F was available. Subsequently the system has been upgraded to Advanced Revelation (Version 1.15).

3. Development of application software
Although the Revelation environment provides many valuable basic features, the special requirements of "text database" processing necessitated extensive investment in a whole suite of programmes. The core programme (and associated routines) is designed to facilitate the editorial process and is common to all the files from which publications are generated. Other programmes have been developed to handle special editorial checking processes (including batch spellchecking), mailing of proofs for checking, formatting of entries for typesetting, etc. In all some 200 programmes and routines have been developed.

4. Design of editorial environment
The editorial environment was designed so that it could be used by editors of different levels of sophistication. It had to facilitate both bulk data entry and complex editorial adjustments amongst a network of interrelated entries. The features relevant to the preparation of this Encyclopedia include:

(a) Keyword indexing: Any significant words appearing in the names of problems are automatically indexed when the entry is saved. Entries can be retrieved via combinations of categories (using a Boolean logic). In the case of the organization database, this may extend to special geographic fields such as city and country of secretariat.

(b) Specific subject category indexing: Significant keywords in the index are used to identify subject categories via which groups of entries can be retrieved. Entries are thus automatically retrievable under some 3,000 categories or category combinations.

(c) General subject category indexing: Using the previous facility even more general searches can be made under major groups of subject categories or their combinations.

(d) Index manipulation: Since retrieval of entries from the database is a major concern, a range of techniques has been developed to filter and sort any non-unique result of an index search, including 'peeking' at entries. The results of a search can also be sorted by subject category, optionally with the possibility of indicating major categories for which items appear to missing.

(e) Text processing: Editors are not constrained by the presence or absence of distinct fields when entering or updating an entry. When the entry is saved, the software analyzes the paragraph labels and distributes the associated text into the appropriate fields of the database record. Any general errors in the format are then reported for correction before the entry is saved. Macros are used to enter any standard paragraph labels required. More rigorous levels of checking can be undertaken optionally when saving, including spell-checking.

(f) Relationships: Within any entry relationships are indicated by the record key of the corresponding entry. This may be embedded in the text within any appropriate paragraph. If desired, the editor can work with the entry such that the names of the entries corresponding to such record keys are also displayed, although that called information is stripped from the entry before it is saved.

(g) Hypertext operation: An editor can place the cursor on any significant word or relationship, save or abandon the entry, and then be transferred into any unique entry resulting from this effective query. If there is no unique entry, as in the case of a common keyword, the editor can refine the range of alternatives which is then presented from the index. This constitutes a form of hypertext in which any significant element can be used as a key, without having been pre-programmed as such.

(h) Recursive mode: From within the text processing of any entry, an editor can go into one, or more, levels of recursive exploration of other entries which can be updated before returning to the original entry.

(i) Paragraph manipulation: Several routines can be used whilst text processing in order to extract information from other entries, to sort items into alphabetic sequence, or to check long sequences of countries (as in the membership information of organization entries).

(j) Relationship network exploration: Using any entry as a point of departure, entries in any one of 7 kinds of relationship to it can be listed out from that entry, through 8 levels of branching networks, as a means of checking the relationship context. This is most frequently used for checking hierarchical relationships.

(k) Administrative data: Although any entry can be processed in a text processing mode, associated with that entry is a set of additional fields containing administrative data on the status of the entry, some of which are updated automatically. These are used to control work on the entry.

(l) Personalized operation: Since patterns of work, features required, levels of understanding, and personal preferences, vary greatly in any complex editorial process, much use is made of memory resident keyboard enhancement software to trigger chains of operations in addition to its use for constants.

(m) Oversized descriptions: Within the Revelation software environment, records cannot exceed a theoretical maximum of 64,000 characters. To handle those few descriptions which exceed this size, a supplementary procedure has been developed.

Computers: graphics environment for an Atlas of International Relationship Networks

1. Summary
This document is concerned with presentations of information which will be possible once a particular computer software problem has been solved. The problem can be illustrated by three examples:

(a) Traffic network mapping: If a database contained entries on 300 subway stations (or airports, or bus stops) and their direct route links to one another, what is required is a software package to construct one or more possible maps of the resulting network. The important point is to be able to optimize the comprehensibility of such maps with minimum manual intervention in the construction process.

(b) Hypercard stack mapping: With the widely acclaimed introduction of the Apple hypercard, whereby complex networks of relationships between database records can be handled, the problem remains of mapping the pattern of relationships in the resulting hypercard stack. The individual entries may be said to constitute "data", but it is the pattern of relationships between them which constitutes "knowledge" and "intelligence"

(c) Mind-mapping: This is a technique currently being strongly promoted in management training and time-management courses. It consists of manually drawing circles to represent key ideas, objectives or activities and then interlinking in a netwxork of relationships. There is a clear need for a software package to facilitate this process. This could take the form of a non-hierarchical form of the standard outline package to manage chapter headings of a report, in which the graphic element is emphasized. There are some resemblances to project scheduling software except that here the emphasis is on relating concepts.

(d) Comment: Consider a relational database with records consisting of subway stations and indications of which station was directly connected to which other stations (and possibly on what "line").

The core problem is how to obtain/adapt/develop software which would generate one or more maps of the subway station network. The principal constraint is that the map should be comprehensible. It is neither required nor desirable that the map should be constrained by some equivalent to "topographic" constraints (namely the position of the stations should not be determined by some form of geographic coordinates). Rather the requirement is that the positions should be determined topologically and mapped, at least for immediate purposes, onto a two-dimensional surface.

There are additional problems which can be treated at lower levels of priority, if at all. They include:

-- A second problem is that the database in fact contains over 10,000 nodes and ways must be found to segment the network (possibly filtering out lower levels of detail) so that maps for individual segments can be interrelated. Such maps, in hardcopy form, will be bound together in a book to form an "atlas".

-- A third problem is that it is desirable that there should be some editorial interaction with the map to improve its visual quality.

-- A fourth problem is that it is desirable that it be possible to update the data base by introducing changes interactively to the map.

-- A fifth problem is to open the way to using the map as a menu via which the database can be queried for information on the nodes.

2. Constraints and possibilities
(a) Conventional approach: The conventional approach to databases, and to the reference books produced from them, is to focus on individual entries. The user is not assisted in understanding the relationships between entries, other than by fairly crude grouping of entries into categories.

(b) Hypertext approach: With the development of interactive databases, hypertext (plus the new hypercard approach of Apple) and CD/ROM, data entries can be organized so that they cross-reference one another to a high degree and in a non-hierarchical manner.

For example, the current Yearbook of International Organizations (1990/91) covering 26,656 entries indicates 65,175 relationships between them (with the major organizations having an average of 70 each) and with a further 192,552 links to membership countries. Similarly the Encyclopedia of World Problems and Human Potantial (1991), covers 13,167 world problems with 80,394 relationships between them. Users can move from entry to entry without going via an index. In database terms this is a major step towards what is being called hypertext. Both publications are maintained on a computer network and with the possibility of CD/ROM versions.

(c) User need for "maps": Because of the overwhelming volume of data, users need "maps" of the pathway between entries, especially in complex subject areas. Such maps provide a sense of context which is lost in many hierarchical presentations of data in linear text form. It is only from such maps that users can quickly obtain an adequate overview of data in an unfamiliar area to guide their efficient use of conventional information tools. Such maps are of value precisely because they are richer than simple hierarchically structured thesauri.

(d) Editorial need for a graphic inrterface: In preparing such publications, editorial researchers need to be able to graphically represent the networks of relationships they are endeavouring to clarify. This is in part strongly related to mind-mapping. Without such a tool, editors have to produce extensive mind maps in manual form before building up or modifying the network of relationships. Ideally it should be possible to communicate such maps to key resource people to obtain insights which are not so easily indicated in normal text presentations.

Interesting examples of such graph displays, prepared manually, do exist. They include the route maps of the *ABC World Airways Guide*, the concept maps in the *Encyclopedia Universalis* and the graphics displays used in the UNESCO *SPINES Thesaurus* for science policy and management. These are all hand drawn and based on relatively limited data sets. As such they are costly and difficult to modify. They do however illustrate different responses to a need felt by information users. The same may be said of networks of corporations grouped by holding companies -- as they are occasionally, and painstakingly, presented in the financial press.

(e) Existing techniques: Computer hardware and sofware for the construction and manipulation of such networks of relationships have only been developed for specific applications such as in chemistry, architecture and engineering (CAD), or electronic circuit board design (PCB). It would be possible to develop similar software to display relationships between database entries.

A number of software pacakges have been developed, especially for Apple machines, which go some of the way towards the product required. These include MORE and INSPIRATION. The disadvantage of these products is that they have primarily been designed to wor around a core concept (a "main idea") which is the point of departure for a hierarchical structure. This does not correspond to the essentially non-hierarchical presentation required.

(f) Atlas production: Once such maps can be succesfully produced and manipulated, computer tapes can be made to drive photocomposition machines (with vector generators). These make high quality maps. Alternatively such maps could be generated by standard graph plotters into camera-ready form. A series of such maps, with facing explanatory text and/or mini-index, may then be bound together as an "atlas".

Maps would be designed to cover clusters of organizations and/or problems in a given subject or geographical area. They would have the advantage of provoking input of new organizations and/or relationships when used in the form of proofs. They also have important didactic uses. Enlargements of the maps could also be distributed as wall-charts.

3. Software "modules"

(a) Relational database: The data is currently held and maintained in an Advanced Revelation database (version 1.16) running on a Novell 286 network. The database has been specially developed as a text database with facilities to manage networks of relationships between the records. It is desirable that when the data is displayed in map form, interactive changes to the map should be carried back as updates to the database. But since the prime requirement is for publishable hardcopy maps, this requirement may be sacrificed in the short term.

(b) Map design: Several approaches may be taken to the problem of map design:

(i) Network analysis This uses specialized extensions of sociometrics to take data of the type described above and to position the elements in relation to each other on the basis of various measures of distance, with those most connected tending to be placed at the centre of a network and those least connected at the periphery. The advantage of this approach is that it endeavours to mirror the network on the basis of its internal characteristics. A number of software packages exist to perform the necessary computations. Various ways of describing a network and identifying key components result from such analysis.

The disadvantage of such software is that it has been developed for relatively small networks only (100 to 300 nodes). Few of the packages are designed to permit mapping of the resultant network. Data is output in matrix form only or as indices in relation to key elements. More seriously, such networks when mapped result in maps which, although they reflect the data, are not designed to enhance the comprehensibility of the data (other than in a purely scientific sense). Such computations can consume considerable amounts of computer time, even on fast machines.

This approach has been explored using test data from the UIA Revelation database consisting of some 5,000 nodes. The work was done on a Mac II using software developed at the University of Dartmouth by Joel Levine of the Department of Mathematical Social Sciences. This software has not been adapted to run under MS-DOS.

(ii) "Crude mapping" A simplistic approach could be taken. This would involve positioning the nodes on a grid determined by the subjects with which they are associated. Such a subject grid (with positions determined by a 4 character identifier) is in use to categorize the UIA data into some 3,000 categories. Relationships would then be plotted between the nodes.

In this case comprehensibility is achieved through the link to the matrix and not through determining the shape of the network. Use of a grid could severely undermine the memorability of the network. It would however be relatively easy to develop and quick to run. A key question would be what kind of interaction it would be possible to have with such a map and whether it would be possible to shift from a detailed focus on a specialized cell of the grid to a wider focus and back (a zoom facility).

(iii) Topological manipulation In this approach, the network of relationships between nodes would be simplified using topological constraints. For example a string of interlinked nodes would be represented by a straight line. The position of the nodes on the line might be equidistant or determined by some logarithmic function based on the distance from the centre of the line. The aim would be to introduce symmetry elements into the data so that it acquires a distinct and memorable pattern or shape. Some of the algorithms required presumably correspond to those of pattern recognition problems.

(c) Plotting: Once coordinates have been determined, software is required to plot the network, whether onto the screen or onto a graph plotter. Many packages exist for this purpose. A distinction should however be made here between adequate quality plots (for working purposes) and high-quality plots for publication in book form. The latter question is discussed later.

The problem in plotting is to be able to introduce distinguishing elements into the plot. These may include variations in line thickness (corresponding to some measure of importance or proximity), variations in node size (corresponding to the number of connections to the node) and the introduction of identifying labels for the nodes.

A key requirement is that the plot be made from the data as processed by one of the above techniques, rather than from data which is manually input. A distinction must also be made between a curve fitting approach and one which passes through the nodes as is required here. A distinction also needs to be made between plotting a graph (from left to right) and plotting a network in which there is no privileged direction. The latter form is more characteristic of CAD programs (see below).

(d) Drawing: It is desirable to move towards an interactive approach to the data. In other words, once a plot is made for a segment of the overall network, editors should be able to modify the network. Such modifications might take one of two forms. The first would consist of simply moving portions of the plot to make it more comprehensible, making room for labels and improving the aestheties. The second might also involve the capacity to add or delete features from the network. It would of course be highly desirable that the latter changes should be carried back into changes to the relational database. This can raise severe problems of compatibility between the relational database and the drawing/plotting software, whether in terms of software or of intermediate files. Such features are available in many CAD programs. It is however important to recognize that the CAD software is here used to "design" logical or topological constructs rather than buildings or mechanical parts. This is not a limitation but it may permit use of simpler (and cheaper) CAD software.

It is appropriate to note that the variant of CAD software used for interactive printed circuit board design (PCB) has many features of value to the present application, especially the "auto-router" feature which positions connections on the circuit board in the most economic manner (avoiding cross-overs, etc). Unfortunately the positioning criteria do not make for maximum comprehensibility.

(e) Interface software: In the case of Advanced Revelation there exists a software product CAD/Base which offers "complete integration of CAD drawings with a database environment", via industry standard DXF files. The drawing is viewed as a Revelation file and the drawing elements as Revelation records and fields. The drawing exists as a master file in both the Revelation and CAD environments. Changes in one environment are reflected in the other automatically without any intermediate file conversion required.

Clearly this offers interesting opportunities for using the network map as a menu through which users can select individual nodes on which they can immediately access additional text data.

(f) High-quality graphic output: One objective is the production of maps to be printed in book form. To achieve this one approach might be to produce output in a form which can be handled by PC-TeX to create files for output on a high quality laser printer.

(h) Integration of features: It is possible that CAD/Base offers an appropriate means of integrating the different features discussed above (except the last). It is also possible that such a product, which is relatively expensive, can be considered as "overkill", and that a more compact approach would be more suitable and easier to make available to others. If the emphasis is on the simpler strategy of generating hardcopy, this would certainly be the case. To the extent that interaction with the data is desirable, then more features would be required, even though only a selection of standard CAD features would be necessary.

For the user, there is obviously great merit in ease of use as an adjunct to normal text editing procedures. Ideally such a package would bear some resemblance to the more sophisticated forms of "outliner", such as MORE and INSPIRATION running on Apple machines. In these an essentially hierarchical outline of topics can be opened up into standard text processing or converted into bullet charts. What is required is an equivalent which is tied into a relational database environment. The different approaches to network "map design" noted above might then be options in the way the data was manipulated for presentation, as is the case in standard business graphics (bar charts, pie charts, etc).

Computers: mapping networks

1. Analysis of networks
The data collected together in the sections of this publication has been deliberately organized in a manner which stresses the interrelationships between the entities within a section and between those in different sections. (Each section is characterized by entities of a different type, and several types of relationship may exist between the same two entities). In effect, therefore, the entities and relationships in each section constitute a network, possibly composed of many subnetworks. Similarly, since entities in each section may be linked to those in other sections, the whole is constituted by a system of interlinked networks in which the relationships have a limited number of distinct meanings. The entities and relationships are currently held in computer files in a form which should facilitate analysis of these networks. It is hoped that the availability of data in this form will encourage the development of new types of analysis more appropriate to the structural complexity portrayed, especially since both the quantitative data and the mathematical functions representing the nature of particular relationships under different conditions (which are a precondition for the application of current methods of quantitative analysis of social systems), are absent and in most cases unavailable.

As François Lorrain notes (1) the abstract notion of a network is undoubtedly called to play a role in the social sciences comparable to the role played in physics by the concept of euclidean space and its generalizations. But the poverty of concepts and methods which can currently be applied to the study of networks stands in dramatic contrast to the immense conceptual and methodological richness available for the study of physical spaces. A whole reticular imagery remains to be developed. At this time a network is understood to contain simply nodes and links and little else. An attempt to define anything like a reticular variable results in very little. This is not surprising, since to succeed would require the establishment of a general mathematical theory of networks which as yet has been little developed. In contrast to this situation, consider the multitude of spatial variables which are available: coordinates, length, surface, volume, curves, classes of curves, classes of surfaces, parameters of curves, parameters of surfaces, and so on, and all these in a space of any number of dimensions and manifesting any type of curvature.

(a) Social networks: The types of network which occur in the social sciences are of such a diverse nature that only a purely formal definition of this notion is of sufficient generality.

A network is constituted by a certain set of points. In the social sciences these points may represent any or all of the following: individuals, groups, organizations, beliefs, roles, *etc*. In this exercise they represent: international organizations, multilateral treaties, world problems, strategies, concepts (human development, integrative, patterns), metaphors, symbols, modes of awareness, values. Such points may represent the existence of entities at the present time, or they may represent the existence of entities at some past or future time (or such points may also be used to represent intervals of time).

The points in a data set may be linked by one or more kinds of relationship. In this exercise three basic types of relationship are distinguished: (i) **Simple relationship,** namely A is related to B which implies that B is related to A; (ii) **Hierarchical relationship,** namely A is a part of B which implies that B is in contextual relationship to A; (iii) **Functional relationship,** namely A acts on B which implies that B is acted upon by A.

In the first case above a relationship is further defined by the types of entity between which it occurs, namely whether they are of the same type, or whether they are of different types. In the second and third case, a relationship is further defined by distinguishing the direction of the relationship, which is further developed in the third case by distinguishing several ways in which A can act upon B.

(b) Analysis of networks: Classical mathematics, summarizing François Lorrain's (1) remarks, is not able to handle complex structural features characteristic of social systems. Organization is best depicted as a network. The mathematical theory of networks derives largely from certain branches of topology and abstract algebra rather than from analysis, which underlies classical mathematics. The theory of graphs is often presented as a kind of general theory of networks with numerous possible applications in the social sciences. However, other than in the area of operations research, the theory of graphs has not proved itself to be very useful in sociology. The reason is probably that the theory has mainly been developed in the context of relatively limited problems in such a way that the results collected under the graph theory label, although numerous and of great interest, have little unity. In addition, the theory rarely handles networks with several distinct types of relationships each with its own configuration of links. It is precisely such networks which are of most interest in sociology. The theory also tends to exclude networks in which some of the points have links back to themselves when it is often just such networks which are important in representing social structures.

A final disadvantage of the theory of graphs is that it only offers a fairly limited number of means of global analysis of networks. It seriously neglects an important aspect of the study of any type of mathematical structure, namely the level of transformation relations between graphs. Because of its composition, a category possesses a richer structure than a simple graph, and it is therefore possible to define more rigorous and fruitful criteria of transformation (namely the concepts of function and functional reduction). In addition a set of points and a set of relations can be treated in their totality and simultaneously, in contrast to the methods of graph theory which considers individual paths between particular points in the graph. In the universe of categories (the universe of objects and relationships), transformations between categories may also be considered as relationships within a category whose objects are themselves categories, and so on. All this emerges from consideration of the global structure resulting from the manner of composition which relates the relationships themselves, thus providing a dialectic of levels of structure and a new imagery of networks. At all levels of this universe, the functional relationships between categories play a central role. They are the fundamental instruments which may be used in the exploration of structural complexity and the tools for extraction of information in global studies.

(c) Use of graph theory methods: Despite the limitations noted above, graph theory methods have been applied to the analysis of social structures although such applications are not very common (see references below).

The image of a "network or web" of problems (or organizations, *etc*) to represent a complex set of interrelationships is a fairly familiar one. This use of "network", however, is purely metaphorical and is very different from the notion of a network of concepts as a specific set of linkages among a defined set of concepts, with the additional property that the characteristics of these linkages as a whole may be used to interpret the semantic significance of the concepts involved.

(d) Some features of graphs: Using graph theory, a number of characteristics of networks can be determined. Points 1 to 3 below are concerned with the shape of the network, 4 to 8 with interactions within the network.

(i) Centrality: A measure (in topological not quantitative terms) of the extent to which a given entity (*eg* a problem) is directly or indirectly "related" via links to other entities (*ie*, the extent to which it is "distant" from another entity). One can speak of a "key" problem or of an organization being "central" to the concerns of a particular complex. It may also be considered a measure of the degree of "isolation" of the entity. A systematic analysis of the centrality of entities in a network could indicate where new entities are necessary to bridge gaps and link isolated domains.

(ii) Coherence: A measure of the degree of "interconnectedness" or "density" of a group of entities. This may be considered as the degree to which a system of problems is "complete". Differences in density would reflect the tendency for more highly coherent problem systems to appear more self-reinforcing in comparison to less

organized parts of the network. In some respects this is an indication of the degree of "development" of a problem system.

(iii) Range: Some entities are directly related to many other entities, others to very few. The range of an entity is a measure of the number of other entities to which it is directly related. Range could be considered an indication of the "vulnerability" of a problem to the extent that a high range problem would be less vulnerable to attack than a low range problem, since it has more relationships anchoring it to its problem environment and preserving it in existence. High range points are therefore either key points in resistance to problem change or else key points in terms of which orderly change can be introduced.

(iv) Content: The "content" of a relationship between entities is the nature or reason for existence of that relationship. Simple graphs have only one link between any two entities; multigraphs have two or more links, each of different content.

(v) Directedness: A relationship between two entities may have some "direction" (*ie*, A to B, or B to A). There may be several types of directedness. Two types are important for this project: A is a sub-element of B; A acts on B. In a multigraph, one link may point from A to B and the other from B to A.

(vi) Durability: A measure of the period over which a certain relationship between entities is activated and used. At one extreme, there are the links activated only on a "one-shot" basis (*eg* a single crisis), at the other there are links, and sets of links, which are considered stable over centuries (*eg* between the more permanent problems).

(vii) Intensity: A measure of the strength of the link or bond between two entities. Two problems may be said to be "strongly bound together". In some cases, the intensity is a measure of the amount of the "flow" or "transaction" between the entities. The link from A to B may be strong, and that from B to A, weak.

(viii) Frequency: A link between two entities may only be established intermittently.

(ix) Rearrangeability and blocking: A connecting network is an arrangement of entitites and relationships allowing a certain set of entities to be connected together in various possible combinations. Two suggestive properties of such networks, which are extensively analyzed in telephone communications, are: (a) rearrangeability (a network is rearrangeable, if alternative paths can be found to link any pair of entities by rearranging the links between other entitities); (b) blocking (a network is in a blocking state if some pair of entities cannot be connected).

(e) Implications of artificial intelligence research: In considering the possibility of analyzing networks of problems (organizations, concepts, *etc*), it is important to benefit as much as possible from related work on artificial intelligence, and possibly pattern recognition. Artificial intelligence projects to simulate human personality or belief systems have had to develop mathematical techniques and computer programmes which can handle and interrelate entitites such as concepts and propositions, some of which may be positively or negatively loaded to represent positive values and perceived problems (the credibility and importance of a belief in a network, and the intensity with which it is held, may also be indicated). Clearly the objective of such projects is not achieved once a simple inventory of entities can be examined, even if it is highly structured in the form of a thesaurus. Of particular interest is the work on "dialogues" with such belief systems, some of which are established over a period by extensive interviews with individuals and others which are specially constructed to simulate paranoia, for example (see references). Presumably it would be possible to conduct somewhat similar dialogues with the collective beliefs constituted by problem/value netwroks such as might be developed during the course of this project.

(f) Comment: Despite the available techniques noted above, and others which have been applied to non-social networks, much would seem to remain to be accomplished, as François Lorrain's (1) remarks indicated, in order to grasp networks in their totality.

The question is what it would be useful to know about networks at this time. What indicators would it be useful to attach to individual problems (organizations, *etc*) to indicate the characteristics of their relationship to the network(s) in which they are embedded? What similar indicators would be useful in describing the relationships between relatively dense networks and the larger network in which they themselves are embedded? What sort of concept about networks need to be embodied in a **network vocabulary** so that such matters can be discussed intelligently and unambiguously in public debate? In other words, what are the elements of an adequate vocabulary of structure and in what disciplines has the basis for such a vocabulary already been established: chemistry, crystallography, architecture, design in general, *etc*? What can be learnt from biologists about the growth and development of the many reticular structures they encounter (*eg* radiolaria)? More interesting perhaps, in which occupations do some individuals develop a special (instinctive or intuitive) sensitivity to the structural and dynamic characteristics of the networks with which, or within which, they work: airline pilots, urban bus drivers, electricity grid controllers, counter-espionage directors, factory process controllers, computer-based data network designer/controllers, telephone exchange designer/controllers, institutional fund controllers, *etc*? What do such people say, or want to say, about their networks? Why has the term "networking" suddenly sprung into common use and consequently what could "to network" mean? It is questionable whether any adequate organizational response (a **network strategy**) to the world problem complex can be elaborated until such rich experience is collected together and matched to an elaborated, mathematically-based concept structure, and an associated vocabulary. A conceptual quantum jump is required to grasp problem (and other organized) structures in their totality and be able to communicate such insights.

It is hoped that the availability of the data in this publication will help to stimulate such fresh thinking on the conceptual containment of societal networks.

2. Use of interactive graphics
(a) Description: The suggestion has been made above that the representation of the relationship between theoretical entities (concepts, organizations, problems, *etc*) could best be accomplished using methods based on graph theory, network theory and topology. The relationships registered in this project could be plotted manually as networks. However, particularly since the relationships are already coded on computer tape in a suitable format, there are three major disadvantages to this manual approach:
- graphic relationships are tiresome and time-consuming to draw, and are costly if budgeted as "art work" (for a comprehensive review of the current possibilites and limitations, see reference 28);
- Since drawn, there is a strong resistance to updating them (because of the previous point) and therefore they quickly become useless (as is frequently the case with organization charts);
- when the graph is complex, multidimensional, and carries much information, it is difficult to draw satisfactorily in two dimensions. The mass of information cannot be filtered to highlight particular features - unless yet another diagram is prepared.

These three difficulties can be overcome by making use of what is known as "interactive graphics" (29). This is basically a television-type screen attached to a computer. The user sits at a keyboard in front of the screen and has at his disposal what is known as a light-pen (or some equivalent device) which allows him to point to elements of the network of concepts displayed on the screen and instruct the computer to manipulate them in useful ways. In other words the user can **interact** with the representation of the conceptual network using the full power of the computer to take care of the drudgery of:
- drawing in neat lines;
- making amendments;
- displaying only part of the network so that the user is not over-loaded with "relevant" information.

In effect the graphics device provides the user with a window or viewport onto the network of concepts. He can instruct the computer, via the keyboard, to:

(i) **Move** the window to give him, effectively, a view onto a different part of the network - another conceptual domain;

(ii) Introduce **magnification** so that he can examine (or "zoom in" on) some detailed sections of the network;

(iii) Introduce **diminution** so that he can gain an overall view of the structure of the conceptual domain in which he is interested;

(iv) Introduce **filters** so that only certain types of relationships and entities are displayed - either he can switch between models or he can impose restrictions on the relationships displayed within a model, *ie* he has a hierarchy of filters at his disposal;

(v) **Modify** parts of the network displayed to him by inserting or deleting entities and relationships. Security codes can be arranged to that (a) he can modify the display for his own immediate use without permanently affecting the basic store of data, (b) he can permanently modify features of the model for which he is a member of the responsible body, (c) and so on;

(vi) Supply **text** labels to features of the network which are unfamiliar to him. If necessary he can split his viewport into two (or more) parts and have the parts of the network displayed in one (or more) part(s). He can then use the light pen to point to each entity or relationship on which he wants a longer text description (*eg* the justifying argument for an entity or the mathematical function, if applicable, governing a relationship, and have it displayed in an adjoining viewport);

(vii) **Track** along the relationships between one entity and the next by moving the viewport to focus on each new entity. In this way the user moves through a representation of "semantic space" with each move, changing the constellation of entities displayed and bringing new entities and relationships into view;

(viii) Move up or down levels or "ladders of **abstraction**". The user can demand that the computer track the display (see point 7) between levels of abstraction, moving from sub-system to system, at each move bringing into view the context of the system displayed;

(ix) Distinguish between entities and relationships on the basis of user- selected **characteristics**. The user can have the "relevant" (to him) entities displayed with more prominent symbols, and the relevant relationships with heavier lines;

(x) Select an alternative form of **presentation**. Some users may prefer block diagram flow charts, others may prefer a matrix display, others may prefer Venn diagrams (or "Venn spheres" in 3 dimensions) to illustrate the relationship between entities. These are all interconvertible (*eg* the Venn circles are computed taking each network node as a centre and giving a radius to include all the sub-branches of the network from that node);

(xi) **Copy** a particular display currently on the screen. A user may want to keep a personal record of parts of the network which are of interest to him. (He can either arrange for a dump onto a tape which can drive a graph plotter, a microfilm plotter, or copy onto a videocassette, or, in the future, obtain a direct photocopy);

(xii) Arrange for a simultaneous search through a coded microfilm to provide appropriate slide images or **lengthy text** (which can in its turn be photocopies);

(xiii) Simulate a **three-dimensional presentation** of the network by introducing an extra coordinate axis;

(xiv) **Rotate** a three-dimensional structure (about the X or Y axis) in order to heighten the 3-D effect and obtain a better view "around" the structure;

(xv) Simulate a **four-dimensional presentation** of the network by using various techniques for distinguishing entities and relationships (*eg* "flashing" relationships at frequencies corresponding to their importance in terms of the fourth dimension);

(xvi) Change the **speed** at which the magnification from the viewport is modified as a particular structure is rotated;

(xvii) **Simulate** the consequences of various changes introduced by the user in terms of his conditions. This is particularly useful for cybernetic displays;

(xviii) Perform various **analyses** on particular parts of the network and display the results in a secondary viewport (*eg* the user might point a light-pen at an entity and request its centrality or request an indication of the interconnectedness of a particular domain delimitted with the light pen.);

(xix) Use **colour** (when a colour screen is available) to distinguish between different concepts or networks of relationships on the same display. Several hundred colour codes are available under computer control (3);

(xx) **Experiment** with the generation of paths for the construction of hypothetical larger conceptual units (*eg* organizations) from available smaller units, as suggested by equivalent work on computer-assisted design of complex organic syntheses (2).

In every current use of interactive graphics there is some notion of geometry and space, but the geometry is always the three-dimensional conventional space. There is no reason why "non-physical spaces" should not be displayed instead - and this is the domain of topology. The argument has been developed by Dean Brown and Joan Lewis (3):

"Both geometry and topology deal with the notion of space, but geometry's preoccupation with shapes and measure is replaced in topology by more abstract, less restrictive ideas of the qualities of things...Being more abstract and less insistent on fine points such as size, topology gives a richer formalism to adapt as a tool for the contemplation of ideas....Concepts can be viewed as manifolds in the multidimensional variate space spanned by the parameters describing the situation. If a correspondence is established that represents our incomplete knowledge by altitude functions, we can seek the terrae incognitae, plateaus, enclaves of knowledge, cusps, peaks, and saddles by a conceptual photogrammetry. Exploring the face of a new concept would be comparable to exploring the topography of the back of the moon. Commonly heard remarks such as 'Now I'm beginning to get the picture' are perhaps an indication that these processes already play an unsuspected role in conceptualization....By sketching tentative three-dimensional perspectives on the screen and 'rotating them on the tips of his fingers', one internalizes ideas non-verbally and acquires a sensation of sailing through structures of concepts much as a cosmonaut sailing through constellations of stars. Such new ways of creating representations break ingrained thought patterns and force re- examination of preconceived notions. A mapping is a correspondence is an analogy. Teaching by analogy, always a fertile device, can be carried out beautifully by topological means....Topological techniques are useful at even the most advanced levels of scientific conceptualization...."

The fundamental importance of interactive graphics, in whatever form, is its ability to facilitate understanding. Progress in understanding is made through the development of mental models or symbolic notations that permit a simple representation of a mass of complexities not previously understood. There is nothing new in the use of models to represent psycho- social abstractions. Jay Forrester (4), making this same point with respect to social systems, states:

"Every person in his private life and in his community life uses models for decision making. The mental image of the world around one, carried in each individual's head, is a model. One does not have a family, a business, a city, a government, or a country in his head. He has only selected concepts and relationships which he uses to represent the real system. The human mind selects a few percpetions, which may be right or wrong, and uses them as a description of the world around us. On the basis of these assumptions a person estimates the system behaviour that he believes is implied....The human mind is excellent in its ability to observe the elementary forces and actions of which a system is composed. The human mind is effective in identifying the structure into which separate scraps of information can be fitted. But when the pieces of the system have been assembled, the mind is nearly useless for anticipating the dynamic behaviour that the system implies. Here the computer is ideal. It will trace the interactions of any specified set of relationships without doubt or error. The mental model is fuzzy. It is incomplete. It is imprecisely stated. Furthermore, even within one individual, the mental model changes with time and with the flow of conversation. The human mind assembles a few relationships to fit the context of a discussion. As the subject shifts, so does the model. Even as a single topic is being discussed, each participant in a conversation is using

a different mental model through which to interpret the subject. And it is not surprising that consensus leads to actions which produce unintended results. Fundamental assumptions differ but are never brought out into the open."

These structured models have to be applied to any serially ordered data in card files, computer printout or reference books to make sense of that data. Is there any reason why these invisible structural models should not be made visible to clarify differences and build a more comprehensive visible model? The greater the complexity, however, the more difficult it is to use mental models. For example, in discussing his examination of an electronic circuit diagram, Ivan Sutherland writes (5):

"Unfortunately, my abstract model tends to fade out when I get a circuit that is a little bit too complex. I can't remember what is happening in one place long enough to see what is going to happen somewhere else. My model evaporates. If I could somehow represent that abstract model in the computer to see a circuit in animation, my abstraction wouldn't evaporate. I could take the vague notion that "fades out at the edges" and solidify it. I could analyze bigger circuits. In all fields there are such abstractions. We haven't yet made any use of the computer's capability to "firm up" these abstractions. The scientist of today is limited by his pencil and paper and mind. He can draw abstractions, or he can think about them. If he draws them, they will be static, and if he just visualizes them they won't have very good mathematical properties and will fade out. With the computer, we could give him a great deal more. We could give him drawings that move, drawings in three or four dimensions which he can rotate, and drawings with great mathematical accuracy. We could let him work with them in a way that he has never been able to do before. I think that really big gains in the substantive scientific areas are going to come when somebody invents new abstractions which can only be represented in computer graphical form."

The availability of devices to restructure information in this way would seem to offer some hope that insights could emerge which respond more adequately to the recorded complexity of societal structure, whilst at the same time being more easily comprehensible to the uninitiated - because of the ease with which such devices can be used as educational tools to develop understanding and comprehension of the same structural data from which the research insights are being derived. Such displays of course lend themselves to videotape recording for wider distribution.

(b) Implications of computer augmentation of intellect: There are important intellectual implications emerging from work on advanced computer systems. Of particular interest is the work of Douglas Engelbart's team at the Center for Augmentation of Human Intellect (Stanford Research Institute) which is a centre for the US ARPA Data Network (which links the computers of major universities in the USA). Engelbart has worked on the means of creating an "intellectual workshop" to facilitate interaction between conceptual structures (6). He considers that:

"Concepts seem to be structurable, in that a new concept can be composed of an organization of established concepts and that a concept structure is something which we might try to develop on paper for ourselves or work with by conscious thought processes, or as something which we try to communicate to one another in serious discussion....A given structure of concepts can be represented by any of an infinite number of different symbol structures, some of which would be much better than others for enabling the human perceptual and cognitive apparatus to search out and comprehend the conceptual matter of significance and/or interest to the human. But it is not only the form of a symbol structure that is important. A problem solver is involved in a stream of conceptual activity whose course serves his mental needs of the moment. The sequence and nature of these needs are quite variable, and yet for each need he may benefit significantly from a form of symbol structuring that is uniquely efficient for that need. Therefore, besides the forms of symbol structures that can be constructed and portrayed, we are very much concerned with the speed and flexibility with which one form can be transformed into another, and with which new material can be located and portrayed. We are generally used to thinking of our symbol structures as a pattern of marks on a sheet of paper. When we want a different symbol-structure view, we think of shifting our point of attention on the sheet, or moving a new sheet into position. With a computer manipulating our symbols and generating their portrayals to us on a display, we no longer need think of our looking at the symbol structure which is stored - as we think of looking at the symbol structures stored in notebooks, memos, and books. What the computer actually stores need be none of our concern, assuming that it can portray symbol structures to us that are consistent with the form in which we think our information is structured. A given concept structure can be represented with a symbol structure that is completely compatible with the computer's internal way of handling symbols, with all sorts of characteristics and relationships given explicit identifications that the user may never directly see. In fact, this structuring has immensely greater potential for accurately mapping a complex concept structure than does a structure an individual would find it practical to construct or use on paper. The computer can transform back and forth between the two- dimensional portrayal on the screen, of some limited view of the total structure, and the aspect of the n-dimensional internal image that represents this "view". If the human adds to or modifies such a "view", the computer integrates the change into the internal-image symbol structure (in terms of the computer's favored symbols and structuring) and thereby automatically detects a certain proportion of his possible conceptual inconsistencies. Thus, inside this instrument (the computer) there is an internal-image, computer-symbol structure whose convolutions and multi- dimensionality we can learn to shape to represent to hitherto unattainable accuracy the concept structure we might be building or working with. This internal structure may have a form that is nearly incomprehensible to the direct inspection of a human (except in minute chunks)."

These insights have been incorporated into the design of an **operational** computer system which is now being developed so that it will be possible to use computer devices as a sort of "electronic vehicle with which one could drive around with extraordinary freedom through the information domain. Imagine driving a car through a landscape which, instead of buildings, roads, and trees, had groves of facts, structures of ideas, and so on, relevant to your professional interests? But this information landscape is a remarkably organized one; not only can you drive around a grove of certain arranged facts, and look at it from many aspects, you have the capability of totally reorganizing that grove almost instantaneously. You could put a road right through the center of it, under it, or over it, giving you, say, a bird's eye view of how its components might be arranged for your greater usefulness and ease of comprehension. This vehicle gives you a flexible method for separating, as it were, the woods from the trees." (7)

(c) Conclusion: Application of this kind of technology to an understanding of the world problem complex has not been attempted. As explained above, such devices offer a means of developing improved conceptual (and associated organizational) structures to contain the complexity with which humanity has to deal at this point in time. Of vital importance is the ability of these devices to portray the information in a more meaningful (or "iconic") form than emerges from conventional quantitative studies. This is particularly important in communicating with the informed public but specially so with the policy-making community, as Harold Lasswell notes (8): "Why do we put so much emphasis on audio-visual means of portraying goal, trend, condition, projection, and alternative? Partly because so many valuable participants in decision-making have dramatizing imaginations...They are not enamoured of numbers or of analytic abstractions. They are at their best in deliberations that encourage contextuality by a varied repertory of means, and where an immediate sense of time, space, and figure is retained."

3. Network maps
(a) Acceptability of network maps: It is now considered quite acceptable in many major cities to print and make available to the general public (often on notice boards or in tourist literature) various schematic maps: the subway (underground, or metro) network; the urban bus network; and the suburban railroad network. Travellers are also accustomed to exposure to documents showing the airline network. Other kinds of network are mapped for the benefit of workers in specialized sectors (eg oil pipeline networks, electricity distribution networks, telephone networks, military communication networks, goods distribution networks, etc). The most complex map of this type would seem to be that used to summarize (on a surface 100 x 132 cm) the relationships between over 1000 biochemical compounds involved in metabolism (See: Gerhard Michal. Biochemical Pathways. Mannheim, Boehringer Mannheim GmbH, 1974; also, but less complex: D E Nicholson. Metabolic Pathways. Colnbrook, England, Koch-Light Laboratories, 1974).

The point is that people are now very familiar with such maps in one form or another and use them, like road maps, to organize their thinking about the movement of themselves or items with which they are concerned between distant points embedded in a complex network. No such network maps are currently available to show the relationships between distant points representing particular features of the social system. As a result thinking about the social system and its problems is somewhat chaotic, as would be any discussion about travel in the absence of adequate maps to provide the necessary frameworks for such discussion.

(b) Reasons for the lack of societal network maps:
(i) There is much confusion concerning the kinds of entities that can be distinguished in the social system, due to overlapping systems of categories, needs, and the maze of associated terminologies.

(ii) Where clarity emerges, it is usually in relation to one particular entity (eg one holding company and its network of subsidiaries, or one government agency and its associated bodies); any maps produced then have that body as the central reference point.

(iii) Much of the required information is scattered through a variety of reference books and no research has justified its systematic organization.

(iv) Systematic sociological research in the past inverts the focus so that, for example, instead of determining how many organizations (problems, etc) there are in a sample in order to determine the number per capita, the mean number of personal relationships to such entities is determined on a per capita basis, so that there is no means of determining how many distinct entities there are to which the relationships are established.

(v) Where such information is collected it is often considered secret because of its political or economic significance. Examples are (a) the collection of data on organizations in every country by the civil or military intelligence units; and (b) the secrecy associated with the subsidiaries owned by a major (multinational) corporation at any one time and their interrelationships.

(vi) Where the data can be collected, and there is a strong case for doing so, there is often reluctance to do so because of the problems of data handling. This is best seen in the (non-societal) case of mapping ecosystem food webs in which animal species are embedded. There is a multiplicity of inter-specific "food chains", together with many branches and cross- connections among food chains making a structure of interactions called "food webs". The complexity of these food webs is such that no one has yet worked out the complete pattern of food relationships and interactions in any natural community. The relationships between 50 species in a given community results in a diagram so full of lines that it is difficult to follow and this only represents one quarter of the 210 known species in a "simple" community. (David Pimental. Complexity of ecological systems and problems in their study and management. In: K E F Webb (Ed) Systems Analysis in Ecology. New York, Academic, 1966, p.15-35).

(vii) Where the research has been done, there is a reluctance to produce maps because of the tiresome, time-consuming and often costly nature of the task of doing so, particularly when the networks are complicated.

(c) Psycho-social significance of maps: a parallel: The current ability to map the societal system may be usefully compared to that of the European geographical mapping ability during the Middle Ages and earlier. The changing psycho-social significance and status of maps, since such early times, provides many clues for understanding the present situation. Maps in that period were often closely guarded secrets, for military and economic reasons. And just as the understanding in Europe of non-European continents was very limited at that time, so today there are only a few well-known problem areas (such as: population, food, peace, etc). Each such territory (or "feudal state") is more or less poorly controlled by a few major organizations (the "cities") with a few well-established links between them (the "roads" or "rivers"). The relations between these feudal states are the limit of concern. Few people travel long distances and when they do, in the absence of readily available maps, they use "experts" to guide them from point to point. Other continents are only vaguely known (and are widely held to be populated by mythical monsters). Each group is content with artistic or impressionistic two-dimensional maps centred on its own organization (or field of concern), confidently held to be the prime mover in the social system as perceived from that point of reference. The significance of any three-dimensional representation is not recognized and a flat-earth perspective prevails.

Under such conditions, it is easy to understand the psychological and communication difficulties which make it impossible to achieve any general galvanization of political will in response to world problems. Each sector is content with its own sketchy local map (if any is held to be required) of the problem environment, and there is little concern for whether such local maps mesh together with those of neighbouring territories or into a general map of the region. Communication therefore frequently breaks down and moments of solidarity are soon forgotten. Warring between feudal territories is common. The state called "energy", clashes with that called "environment". Alliances are formed and each state has imperialistic ambitions: "development" wants to incorporate "environment"; "environment" lays claim to the territory of "development", and all are claimed by the territory called "peace". Lacking maps, assemblies of individuals and groups from different problem territories are pathetic. The people from "heavy rainfall" areas cannot understand the constant harping on water by people from "desert" areas; the people from "arctic" areas cannot relate meaningfully to those from "tropical" zones.

The history of the evolution of geographical perceptions, and the tools that have been required to move humanity towards a global perception, indicate the kinds of difficulty which have to be faced. (The much-used NASA photograph of Earth from space is only significant as a symbol because people know that they can relate its features to the map of the world in their own atlas in order to be able to locate their home town, for which they also have a detailed local map, to which they can relate their personally acquired knowledge.) Local maps are needed which mesh into global maps, so that each can see his place in any world problem strategy and so that global decision-making can relate to the tactical problems of groups as perceived in each community.

Problem maps (bound together into "atlases") are needed to help individuals see and appreciate the relationships, distances and differences between problem territories. And it should be possible to relate these to organization (and other) maps, just as any atlas has contour maps, climatic maps and political maps of the same region. Individuals, whether students, executives, researchers, or policy makers, have at least as much need for such visual devices to orient themselves in the social system as they have for road and other currently available maps.

Hopefully it will be possible to reach a stage at which such maps can be produced as standard conference documentation as a means of providing background documentation for debates, and in order to sharpening the focus of debate. Clearly the debate itself should lead to proposals for the amendment of such maps (as a result of the recognition of: new issues, relationships between problems, proposals for organizations or programmes, or new relationships between organizations, etc). New versions of such maps, or hypothetical maps (eg of organizational systems) could be fed into later sessions of the same meeting or used as one form of summary of the achievements of the meeting.

(d) Production of network maps: Once the information on societal entities is held on computer it becomes possible to overcome many of the obstacles to map production noted above. Computers are currently used to plot out electronic circuit diagrams and other types of network onto large charts. The computer programmes handle the tedious problem of designing such charts, including the use of appropriate colours to distinguish between different features of the network (or networks) on the same chart. (Artists, designers and communications psychologists can also introduce an aesthetic component to facilitate comprehension).This approach has the considerable advantage that different designs (based on the same data) may be tried or used for different purposes. Some designs may be highly simplified, others may be very complex. New maps can be easily produced if the original data is modified. The data base used may be the same as that used for interactive studies of the network so that both approaches may be integrated under the control of a researcher.

However, although the computer programmes exist for the production of two- dimensional maps, there are difficulties still to be overcome in the representation of three (or n) dimensional networks on a two-dimensional surface, if such complex representations are necessary. Some of these mathematical and associated problems (of projections) have been examined by geographers interested in producing a more accurate representation of the spherical Earth on a map. Experiments have been made with a number of alternatives which each have their advantages. The data collected together on computer for this publication should encourage and facilitate similar experiments in societal network map production.

References

1. François Lorrain. Réseaux Sociaux et Classifications Sociales; essai sur l'algèbre et la géométrie des structures sociales. Paris, Hermann, 1975.
2. E J Corey and W T Wipke. Computer-assisted design of complex organic syntheses. *Science*, 166, 10 October 1969, p.178-191.
3. Dean Brown and Joan Lewis. The Process of Conceptualization; some fundamental principles of learning useful in teaching with or without the participation of computers. Educational Policy Research Center, Stanford Research Institute, Menlo Park, California, p.16-18.
4. Jay Forrester. World Dynamics. Cambridge, Wright-Allen, 1971, p.14-15.
5. Ivan Sutherland. Computer graphics. *Datamation*, May 1966, p.22-27.
6. D C Engelbart. Augmenting Human Intellect; a conceptual framework. Menlo Park, Stanford Research Institute, 1962, p.34-37 (AFOSR-3223).
7. Nilo Lundgren. Toward the decentralized intellectual workshop. *Innovation* (New York), 1971. See also: D C Englebart. Intellectual implications of multi-access computer networks. Menlo Park, Stanford Research Institute, 1970. (Conference paper).
8. Harold Lasswell. The transition toward more sophisticated procedures. In: D B Bobrow and J L Schwartz (Ed). Computers and the Policy-making Community; applications to international relations. Englewood Cliffs, Prentice-Hall, 1968, p.307-314.
9. Robert O Anderson. A sociometric approach to the analysis of interorganizational relationships. Institute for Community Development and Services, Michigan State University, 1969.
10. George M Beal et al. System linkages among women's organizations. Department of Sociology and Anthropology, Iowa State University, 1967.
11. V E Benes. Mathematical Theory of Connecting Networks and Telephone Traffic. New York Academic, 1965, p.53
12. C Berge. The Theory of Graphs and its Applications. London, Methuen, 1962.
13. D Cartwright. The potential contributions of graph theory to organization theory. In: M Haire (Ed) Modern Organization Theory, Wiley, 1959.
14. K M Colby and D C Smith. Dialogue Between Humans and an Artificial Belief System. Stanford University, Artificial Intelligence Project, 1969. (Memo AI-97).
15. K M Colby, S Weber, and F D Hilf. Artificial Paranoia. Stanford University, Artificial Intelligence Project, 1970 (Memo AIM-125).
16. K M Colby, L Tesler, H Enea. Experiments with a Search Algorithm on the Data Base of a Human Belief Structure. Stanford Univeristy, Artificial Intelligence Project, 1969 (Memo AI-94).
17. John C Fakan. Application of Modern Network Theory to Analysis of Complex Systems. Washington, NASA, 1969 (Clearinghouse for Federal Scientific and Technical Information), 45p.
18. C Flament. Applications of Graph Theory to Group Structure. Englewood- Cliffs, Prentice-Hall, 1963.
19. Lucien A Gerardin. Topological structural systems analysis. (Paper presented at the Rome Special Futures Research Conference, 1973).
20. F Harary and R Z Norman. Graph Theory as a Mathematical Model in Social Sciences. Ann Arbor, University of Michigan, 1953.
21. F Harary, R Z Norman and D Cartwright. Structural Models: an introduction to the theory of directed graphs. New York, Wiley, 1965.
22. F Harary. Graph Theory. Reading, Addison-Wesley, 1969.
23. N Jardine and R Sibson. Mathematical Taxonomy. London, Wiley, 1971.
24. A J N Judge. From Networking to Tensegrity Organization. Brussels, Union of International Associations, 1984.
25. A Kaufman. Graphs, Dynamic Programming and Finite Games. New York, Academic, 1967.
26. E Kingsley, F F Kopstein, and R J Seidel, Graph theory as a meta- language of communicable knowledge. In: M D Rubin (Ed) Man in Systems. New York, Gordon and Breach, 1971, p.43-69.
27. Manfred Kochen. Organized systems with discrete information transfer. *General Systems*, 1957, 2, p.20-47.
28. John C Loehlin. Computer Models of Personality. New York, Random House, 1968.
29. J Clyde Mitchell (Ed). Social Networks in Urban Situations, Manchester UP, 1969.
30. *Networks; an interdisciplinary journal* New York, Wiley, 1971, quarterly.
31. Anatol Rapoport and W J Horvath. A Study of a large sociogram. *Behavioural Science*, 1961, 6, 4, p.279-291.
32. N Rashevsky. Topology and life; in search of general mathematical principles in biology and sociology. *Bulletin of Mathematical Biophysics*, 1954, 16, p.317.
33. Norman Schofield. A topological model of international relations. (Paper presented to Peace Research International meeting, London, 1971).
34. L Tesler, H Enea and K M Colby. A directed graph representation for computer simulation of belief systems. *Mathematical Biosciences*, 2, 1/2, Feb 1968, p.19-40.
35. Jacques Bertin. Semiologie graphique; les diagrammes, les reseaux, les cartes. The Hague, Mouton, 1973; and Walter Herdeg (Ed). Graphis/Diagrams; the graphic visualization of abstract data. Zurich, Graphis Press, 1974
36. "Interactive graphics": This term is used widely to cover both the more common alphascopes, which can display letters and numbers on predetermined lines, and the vector displays with light-pen facility, which can also generate lines and curves. It is the latter device which is discussed here. See, for example: Interactive graphics in date processing. *IBM Systems Journal*, 7, 3 and 4, 1968, whole double issue; R E Green and R D Parslow (Ed). Computer Graphics in Management. London, Gower Press, 1970, 240p.; and R D Parslow and R E Green (Ed). Advanced Computer Graphics; economics, techniques and applications. London, Plenum Press, 1970, 1250p. nd R E Green (Ed).

Union of International Associations: Profile

40 rue Washington B-1050 Brussels, Belgium.
T. (32 2) 640.18.08 Telex 65080 INAC B Fax (32 2) 649 32 69

1. Aims
Facilitate the evolution of the activities of the world-wide network of non-profit organizations, especially non-governmental or voluntary associations.
- Promote understanding of how international bodies represent valid interests in every field of human activity or belief, whether scientific, religious, artistic, educational, trade or labour.
- Enable these initiatives to develop and counterbalance each other, creatively in response to world problems, by collecting information on these bodies and their interrelationships.
- Make such information available to them, and to others who may benefit from this network.
- Experiment with more meaningful and action-oriented ways of presenting such information as a catalyst for the emergence of more appropriate organizations.
- Promote research on the legal, administrative and other problems common to these international associations, especially in their contacts with governmental bodies.

To these ends, contact is maintained with a wide variety of bodies in both East and West, developed and developing countries.

2. Historical background
Founded in Brussels in 1907 as the Central Office of International Associations, the UIA became a federation under the present name in 1910 at the 1st World Congress of International Associations. Activities were closely associated with the Institut international de bibliographie, which later became the International Federation for Documentation. Its work contributed to the creation of the League of Nations and the International Institute of Intellectual Cooperation (the precedessor of UNESCO). During the 1920s, the UIA created an International University, the first of its kind.

3. Current status
The Statutes were modified in 1951 to give the UIA the character of an Institute with a world focus, having individuals as full members. It is an independent, non-governmental, non-profit body which is a-political in character. Its programmes are totally oriented toward the community of international associations whose actions they are designed to facilitate, whether through special studies or through new uses of information. The UIA is registered under the Belgian law of 25th October 1919 as an international association with scientific aims.

4. Finance
The UIA is more than 95 percent self-financed, through the sale of publications which it produces and through membership subscriptions. The balance is made up from grants from a number of official and private bodies. The annual budget is approximately US$ 700,000.

5. Administration
The UIA General Assembly elects an Executive Council of 21 every 2 years. The programme, under the direction of the Secretary-General, is carried out by the Secretariat in Brussels.

6. Working languages
The main working languages of the UIA are English and French, although information is received in many languages. Most publications are produced in English, with French versions where there is a demand. The Yearbook of International Organizations is indexed in all languages used by international organizations. The periodical Transnational Associations contains articles in both English and French.

7. Full Members
Individuals, whose total number may not exceed 250, are elected by the UIA General Assembly which they constitute. Members are elected on the basis of their interest and activity in international organizations, usually demonstrated by an active role in such a body over an extended period of time. They include diplomats, international civil servants, association executives, professors of international relations and directors of foundations. Members do not pay annual dues, but as trustees it is expected that they will further the interests of the UIA in their particular sphere of activity. Members are currently located in the following countries:
Africa: Algeria, Benin, Madagascar, Mauritania, Morocco, Senegal, Togo.
America: Brazil, Canada, Chile, Peru, USA, Venezuela.
Asia/Pacific: Australia, India, Japan, Sri Lanka, Thailand.
Europe: Belgium, Bulgaria, France, Germany, Greece, Hungary, Ireland, Italy, Luxembourg, Netherlands, Norway, Portugal, Sweden, Switzerland, Turkey, UK, USSR.

8. Associate Members
Any corporate bodies or individuals interested in the aims and activities of the UIA and wishing to associate themselves with its work by payment of an annual memberhip fee. Members include a wide range of organizations, foundations, government agencies and commercial enterprises and are entitled to preferential use of UIA services. Membership subject to approval by the Executive Council.

9. Corresponding Organizations
The UIA is controlled by its individual members, although its work is almost entirely with the complete range of international organizations through publications and correspondence. For those international organizations who wish to be more closely associated with this work, without any commitment of "membership", a category of "Corresponding Organizations" is provided by the UIA Statutes.

10. Collaboration with inter-governmental organizations
The UIA has Consultative Relations with UNESCO, UN/ECOSOC, and ILO. It collaborates with FAO, the Council of Europe, UNITAR, and the Commonwealth Science Council. It is one of the research institutes in the network of the UN University. A special ECOSOC resolution establishes cooperation between the United Nations and the UIA for the preparation of the Yearbook of International Organizations, for which contact is maintained with over 1,000 intergovernmental bodies. A French edition was produced with the assistance of the Agence pour la coopration culturelle et technique.

11. Relationship with international associations
Contact is maintained with over 13,000 international non-governmental organizations eligible for inclusion in the Yearbook of International Organizations. Special links exist to UIA Corresponding Organizations, to the federations of international organizations established in Belgium (FAIB), France (UOIF), and Geneva (FIIG), to the conferences of consultative NGOs for ECOSOC and UNESCO, to bodies using its secretariat facilities, or to those with which it has co-publishing arrangements.

12. International Association Centres
To increase the effectiveness and efficiency of organizations with secretariats in a particular location, the UIA encourages the creation of federations of international bodies (eg established in France), and the contact between such bodies. An International Association Centre, promoted by the UIA, with shared facilities for Brussels-based organizations, was opened in 1983.

UIA Current publications

Yearbook of International Organizations / ed. by UAI. - München, New York, London, Paris : Saur, 1990. - 27th ed. 1990/1991. - 30 cm. - ISSN 0084-3814.

Vol.1 Organization Descriptions and Index. 1776 p. + Appendices (14). - ISBN 3-598-22205-X.

Organization descriptions (24,209 entries)
The non-profit organizations included may be intergovernmental, non-governmental, or mixed in character. They cover every field of human activity. Descriptions, varying in length from several lines to several pages, are grouped into the following section
- Federations of international organizations
- Universal membership organizations
- Inter-continental organizations
- Regional membership organizations
- Semi-autonomous bodies
- Organizations of special form
- Internationally-oriented national bodies
- Inactive and dissolved bodies
- Religious orders and fraternities

Contents of descriptions
The descriptions, based almost entirely on data by the organizations themselves, include:
- Organization names in all relevant languages
- Principal and secondary addresses
- Main activites and programmes
- Personnel and finances
- Technical and regional commissions
- History, goals, structure
- Inter-organizational links
- Languages used
- Membership by country

Multilingual index
The computer-generated index provides the most detailed available means of identifying international bodies. Access is possible via:
- Organization names in English, French, and other working languages
- Former names in various languages
- Name initials or abbreviations in various languages
- Organization subject categories in English, French, German, Rusian and Spanish
- Personal names of principal executive officers

Checklist of publication titles
Periodical and non-periodical publications of international organizations

Vol.2 International Organization Participation : Country Directory of Secretariats and Membership (Geographic Volume) . - 8th ed. 1990/91. - 1760 p. - ISBN 3-598-22206-1.

Secretariat countries (Section S) This part lists by country the international organizations which maintain headquarters or other offices in that country. Address are given in each case.
Membership countries (Section M) This part lists, for each country, the international organizations which have members in that country. For each organization listed, the international headquarters address is given, in whatever country that is located.

Vol.3 Global Action Networks : Classified Directory by Subject and Region (Subject Volume). - 8th ed. 1990/91. - 1684 p. - ISBN 3-598-22203-3.

Classified by subject (Sections W, X) These parts list over 23,000 international organizations by subject according to their principal preoccupations. Subjects are grouped into both general and detailed categories, as well as on the basis of interdisciplinary subject combinations. The classification scheme highlights functional relationships between distinct preoccupations.
Classified by region (Section Y). This part lists international organizations by subject according to the region with which they are particularly concerned.

International Congress Calendar/ ed. by UAI. - Brussels: UAI, 1991. 4 vol .- 331st ed. 1989, 30 cm. - ISSN 0538-6349.

The International Congress Calendar is intended as a convenient reference work for anyone seeking information on international events. From the 23rd edition (1983) the Calendar appears quarterly. Each of the four volumes is self-contained including indexes. Amendments and additions occuring between volumes are specially indicated so that every issue contains the most up-to-date information on international events. Again, this year events listed in the Calendar have increased considerably. All the information on these events is derived from primary sources, i.e. the organizations themselves through regular questionnaire mailings. The proven structure of the Calendar remains unchanged, ensuring convenient access to all events by means of: a geographical section, a chronological section and a subject/organizations index.

Encyclopaedia of World Problems and Human Potential / ed.by UAI. - München, New York, London, Paris : Saur, 1991. - 3rd ed., 2 vols, 2140 p., - ISBN 3-598-10842-7.

World Problems and Human Potential, now in its third edition, is a comprehensive source of information on world problems that have been been recognized, on how they are perceived to be interrelated, and on the human resources available to challenge them. Detailed sections draw attention to a variety of alternative insights into the ways in which human development and the world problematique mutually inhibit, enable, and provoke each other.

International Association Statutes Series / ed. by UAI. - München, New York, London, Paris : Saur, 1988. . - 30 cm - ISSN 0933-2588. Vol.1. 1 ed. - 600 p, 30 cm. - ISBN 3-598-21671-8 (Saur München).

The first volume includes the official texts of 393 statutes of international nongovernmental organizations described in Sections A, B, C of the Yearbook of International Organizations, namely bodies with membership in countries in at least two continents. Future volumes will include statutes of organizations from other sections, namely regional bodies and those of a less conventional structure. They may also include statutes of lesser known intergovermental bodies or those of a hybrid governmental/nongovernmental nature. In contrast to the Yearbook series, each volume of the Statute series will only include information not published in previous volumes of the series.

Who's Who in International Organizations ? / ed by UAI. - München, New York, London, Paris: Saur 1991. - ISBN 3-598-10908-3.

This new Who's Who in International Organizations ? is an indispensable reference work for all international non-governmental organizations, intergovernmental organizations, journalists, libraries, universities and research institutes. This 3 volume set contains approximatively 12,000 biographies of eminent individuals from organizations in every field of human endeavour. Intergovernmental organizations; international non-governmental, non-profit bodies; international committees, centers, institutes; information systems, conference series and informal networks; and national non-profit groups concerned with international issues will be represented. The biographies contain: full name, organization, position, nationality, profession, date and place of birth, family, detailed biography, own publications, memberships, and honours. Three indexes list entries by nationality, by field of work and by organization name. The set is scheduled for publication in Summer 1991.

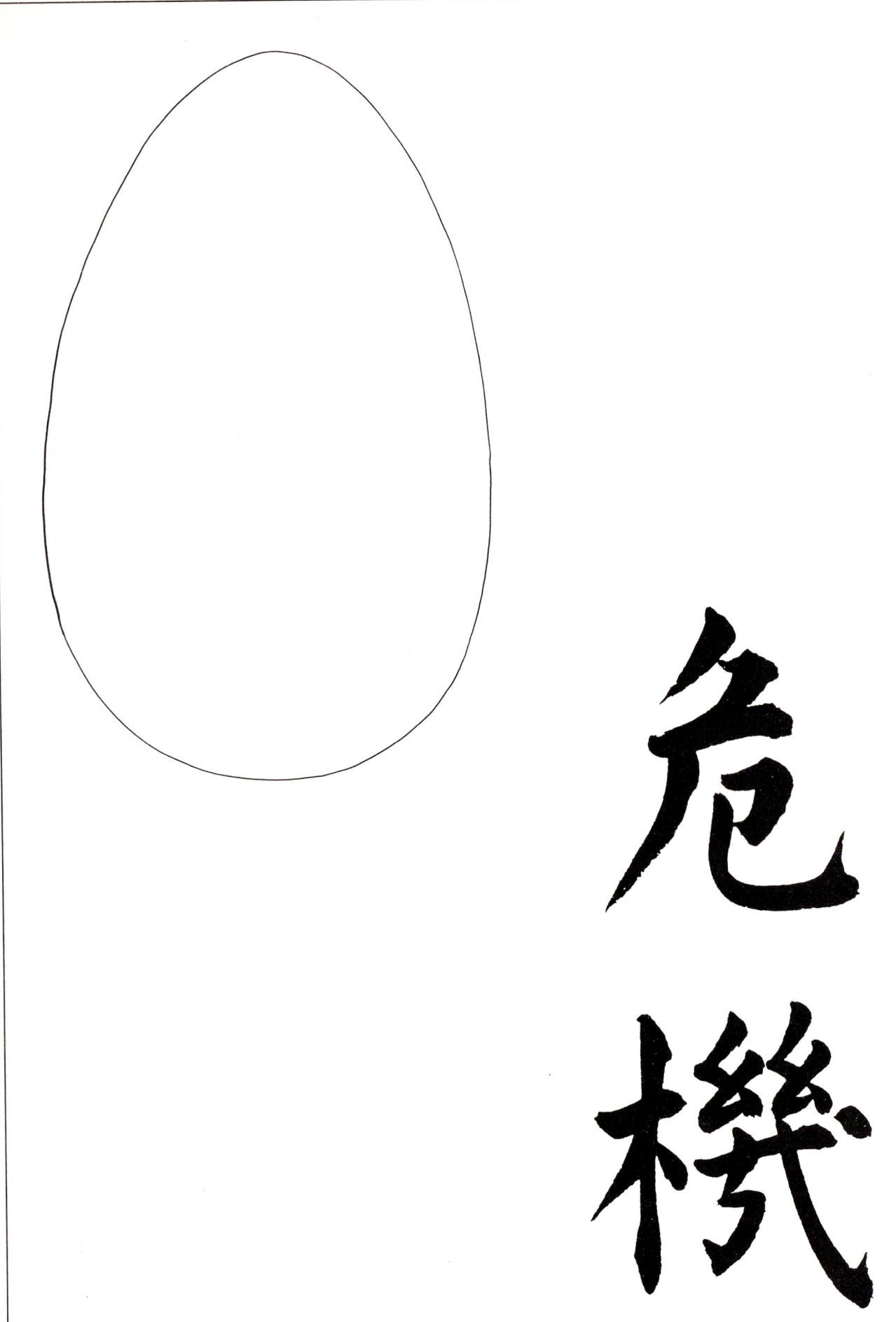

危機